U0217329

English-Chinese Dictionary of Electric Power

英汉电力词典

王邦飞　主编

·北京·

内容提要

本词典收录了电力工程国际项目招投标文件及建设与运营过程中各个环节的词汇或缩写词20余万条，涵盖火电、核电、水电、风电、太阳能发电、生物质发电、垃圾发电、地热发电、潮汐发电等各类发电项目中与热机、热控、电气、输煤、除灰、脱硫、脱硝、热电联产、空冷、循环流化床、IGCC与煤气化、土建、结构、建筑、岩土技术、勘测、设计、工程水文、总图运输、暖通与空调、供排水、水工结构、化学水处理、消防、技术经济、输电线路、变电站、电力系统、远动系统、通信与网络、仪表与控制、系统保护、计算机、施工设备、施工工具、焊接技术、煤炭、材料、HSE、招投标、合同或协议、银行与付款、信用证与保函、海运与清关、保险、外贸、财务、法律等相关的术语或缩略语。另外，编者也适当收录了其在翻译实践中所积累的常用名词或名词性短语、动词或动词性短语、形容词、副词、介词或介词短语及其他实用词组或句型等。

本词典可供从事涉外电力工程的工程技术人员、翻译人员或招投标工作者使用，也可以供与电力工程相关的大专院校学生使用。

图书在版编目（CIP）数据

英汉电力词典 / 王邦飞主编. -- 北京 : 中国水利水电出版社, 2023.11
ISBN 978-7-5170-7811-1

Ⅰ. ①英… Ⅱ. ①王… Ⅲ. ①电力工程－词典－英、汉 Ⅳ. ①TM7-61

中国版本图书馆CIP数据核字(2019)第142387号

书　　名	**英汉电力词典（附光盘）** YING - HAN DIANLI CIDIAN
作　　者	王邦飞　主编
出版发行	中国水利水电出版社 （北京市海淀区玉渊潭南路1号D座　100038） 网址：www.waterpub.com.cn E - mail：sales@mwr.gov.cn 电话：（010）68545888（营销中心）
经　　售	北京科水图书销售有限公司 电话：（010）68545874、63202643 全国各地新华书店和相关出版物销售网点
排　　版	中国水利水电出版社微机排版中心
印　　刷	涿州市星河印刷有限公司
规　　格	145mm×210mm　32开本　59.625印张　6036千字
版　　次	2023年11月第1版　2023年11月第1次印刷
印　　数	0001—1500册
定　　价	**298.00**元（附光盘1张）

凡购买我社图书，如有缺页、倒页、脱页的，本社营销中心负责调换

《英汉电力词典》编审人员名单

主　　编：王邦飞

副 主 编：刘天助

编写人员：杨静波　李　斌　李百宁　张文杰

王玉雷　赵艳丽　徐　岩　陈　颖

张静静　刘　浈　李金海　战　莉

宋　微　宋　阳　宋子恒　张　宁

任　晖　李　伟　周　辰　郜小芳

张武昌　郑天成　姜光绪　张东亮

葛惠广　张冬雪　张　垚　辛　蕾

高希伟　徐　剑　陈晓红　贾海涛

郭　丽　张博文　刘　沫　卓　维

王　林　孔丽丽　李姗姗　马　俊

王安妮　王韶华　杨　明　王帮超

王译晨　刘　冰　宋丽娜　李凤鸣

吴邓良　杨晓英　韩景文　何　昊

姜　伟　嵩　单　张永刚　李　涛

王俊凤　徐云峰　温路遥　王　楠

裴科威　乔　烨　曲　玮　张汉强

杨　威　徐　岩　董翠华　丁洋洋

王桂卿

审校人员： 王德义　张洪起　任胜军　胡志强

孟庆辉　张勤忠　段　冰　李予林

郭　峰　刘京义　韦思亮　田景奎

李焕荣　年　华　藏建敏　张铁弓

李云飞　张纯岗　孟向明　毛俊杰

杜伯承　马洪厂　李正钢　姚玉成

郭汝攀　王成林

Antony Wamukota

Brian Geoffrey Abasa Muchilwa

Ruth Kabiri Mburu

Lorena Evon Achieng Simba

前　言

本词典由编者在其所就职的中国能建集团及电力系统的若干电力专业和英语专业人士协助下，历经十余年编辑而成，旨在为广大从事涉外电力工程建设者提供“一站式”查询服务，因此本词典的特点是“全、新、实”。

在长期的国际工程招投标与国际项目的执行中，编者与他的同事们发现了一个问题：虽然某些电力词典编写得很专业，但在查取词条时往往存在一定的局限性，有时您不得不同时准备好几本词典，且仍有些新词查不到，特别是在“汉英”方向查询时尤为明显，比如在汉译英时遇到了燃机电厂方面的“一拖一（one on one）”和土建方面的“三通一平（three put - throughs and site leveling）”等。于是，编者产生了一个念头——争取中国能建集团系统一些专家的帮助、指导与审校来编辑一本“电力工程百货店式”的词汇汇编，广泛收录国际电力工程招投标及合同或协议中常用的词汇，参考、学习并吸收其他优秀词典之精华，纳入中国电力标准的规范术语，同时把编者在长期国际工程实践中所积累的新词和相关专家研究后编译出来，使这本词典的词汇更全、更新、更实用。特别是，本词典通过光盘实现了英汉-汉英的双向搜索互查，达到了 cost - effective 的目的。

本词典收录了电力工程国际项目招投标文件及建设与运营过程中各个环节的词汇或缩写词 20 余万条，涵盖火电、核电、水电、风电、太阳能发电、生物质发电、垃圾发电、地热发电、潮汐发电等各类发电项目中与热机、热控、电气、输煤、除灰、脱硫、脱硝、热电联产、空冷、循环流化床、IGCC 与煤气化、土建、结构、建筑、岩土技术、勘测、设计、工程水文、总图运输、暖通与空调、供排水、水工结构、化学水处理、消防、技术经济、输电线路、变电站、电力系统、远动系统、通信与网络、仪表与控制、系统保护、计算机、施工设备、施工工具、焊接技术、煤炭、材料、HSE、招投标、合同或协议、银行与付款、信用证与保函、海运与清关、保险、外贸、财务、法律等相关的术语或缩略语。另外，编者也适当收录了其在翻译实践中所积累的常用名词或名词性短语、动词或动词性短语、形容词、副词、介词或介词短语及其他实用词组或句型等。

感谢中国能建集团公司相关的电力专业人员的指导、协助与审核，特别感谢中国能建集团国际工程公司副总经理张洪明、中国电力工程顾问集团总经理助理杨春发、国际部副总经理赵江、华北电力设计院原董事长王毓及院长李一男、先后分管国际工程的副院长顾游、曹玉杰的鼓励，感谢华北电力设计院国际工程部的高健、侯卫东、任胜军、胡志强、李兴利、孟庆辉、湛川、彭华等历任领导及尼日利亚公司的李庆龙、赵周全、刘常青、刘京义等历任领导的支持，感谢华北电力设计院科技部在各方面给予的协调与帮助，感谢吉林送变电公司的杜伯承先生对输电线路工程的施工工具等词汇的贡献，感谢中国兵器北方化学工业公司的电厂仪控专家何金海、杨仪刚在电厂仪控与 DCS 方面提供的专业词汇，感谢美国福斯特惠勒公司主管现场调试的锅炉专家孙钦山提供的热机方面的术语，感谢徐少非等外籍华人及 Antony Wamukota 先生、Brian Geoffrey Abasa Muchilwa 先生、Ruth Kabiri Mburu 女士和 Lorena Evon Achieng Simba 女士等外国电力专家或工程师在一些新术语的编译方面给予的帮助！

由于时间仓促和能力所限，虽然编者做了巨大的努力，但本词典肯定还有若干不足或疏漏，欢迎读者批评、指正，或提供新词和意见、建议，以利于纠正或再版。编者邮箱：bangfeiwang@163. com 或 bangfeiwang@aliyun. com。

编者

2023 年 10 月

使用说明

1. 本词典附有光盘，可进行“英汉”和“汉英”双向检索，从而起到“英汉-汉英词典”的作用。

2. 无论从“英汉”方向还是“汉英”方向搜索，如果整条词没能搜索到，则可将词条拆散成几个部分进行搜索，或仅选其中的某个词进行搜索，因该词条或许被某种符号（如括号或斜杠）所阻隔，导致完整的词条无法搜索到。例如：若要搜索英文词条“corner firing burner”，而本词典中却以“corner（firing）burner”形式出现，这个时候，如果仅搜索 corner 或 firing 或 burner，最终都能找到所搜索的词条。

3. 本词典中的英文词条均按英文字母顺序排列（不考虑连字符）。

4. 当一个英文词条含有若干汉语意思时，无论是名词还是动词或形容词等，都不做词性标注，也不用分号把义群分隔，而是一律用逗号。

5. 英文复合词中圆括号（ ）内的单词表示可以省略，如：active（coal）storage 表示 active coal storage 或 active storage 均可表示“周转煤堆，周转贮量”。

6. 英文单词圆括号（ ）内带“-”者表示词尾的另一种拼写形式，如：air slide conveyer(-or) 表示 air slide conveyer 或 air slide conveyor 均可。

7. 英文单词中圆括号（ ）内不带“-”者表示圆括号内的字母可以省略，如：briquet(te) 表示 briquette 或 briquet 均可。

8. 英文单词中圆括号（ ）内的斜杠“/”表示括号内斜杠后的词与前面的词是可以相互替代的。如：abrasion（/abraded）platform 表示 abrasion platform 或 abraded platform 均可。

9. 缩写词，先以圆括号列出英文全名，再给出汉语释义。如：AAE（American Association of Engineers）美国工程师协会。

10. 英文全称只有一个单词的缩写词用等号“＝”表示，而不用圆括号。如：crs＝centers 中心距。

11. 在汉语释义中的黑括号【 】表示简要提示、说明或解释。如：beam analog（ue）梁比拟【一种计算方法】。

12. 汉语释义中圆括号（　）内的内容是可以省略的。例如：

（1）charge machine（反应堆）装料机。

（2）thermal power plant 火（力发）电厂，热（力发）电厂。

13. 汉语释义中方括号［　］内的内容表示可以替代其前面与括号内相同字数的内容。如：absolute acceptance 单纯认付［承兑］，表示“单纯认付，单纯承兑”。

14. 动词或动词词组多以动名词或其分词形式出现，也有的直接用动词原形或动词不定式表示。

15. 在所列出的汉语词义里，没有注明词性，也没用分号予以群分，而是通用逗号。

16. 在所列出的英文词条里，对名词的单复数或冠词等的用法也未做细究，而是突出名词本身及其搭配，如 maintenance expense of structure。

17. 有些重要的名词性复合词与其形状类似的动词词组是故意分出来的，如 turnover 与 turn over。

目　　录

A

AAAC(all aluminium alloy conductor) 全铝合金绞线
AAAS(American Association for the Advancement of Science) 美国科学发展协会
AAB(Accident Analysis Branch of the USNRC) 美国核管理委员会事故分析处
a=absolute 绝对的
AAC(all aluminium conductor) 全铝导线
AAC(automatic amplitude control) 自动幅度控制
AACC(American Automatic Control Council) 美国自动控制委员会
AACSR(aluminum alloy conductor steel reinforced) 钢芯铝合金绞线
AAE(American Association of Engineers) 美国工程师协会
AAh(analyzer alarm high) 分析器高值报警
alarm status 报警状态
A=ampere 安培
A=amplitude 振幅
antiseptic insulation 防腐绝缘
AAS(American Academy of Sciences) 美国科学院
AAS(Atomic Absorption Spectroscopy) 原子吸收光谱
AASW(American Association of Scientific Workers) 美国科学工作者协会
abaciscus 嵌饰
abacus 算盘，柱顶板
abac 列线图，诺谟图
abampere 电磁安培【等于10A】
abamurus 挡土墙，扶墙，支墩块
abandoned assets 废弃资产
abandoned hole 报废孔
abandoned property 废弃财产
abandoned tender 废标
abandonment clause 废弃条款，委付条款
abandonment of claim 放弃索赔
abandonment of contract 放弃合同
abandonment of right 弃权
abandonment of water rights 水权的放弃，放弃水权
abandonment 放弃，废弃，撤销，委托
abandon 委付，放弃，抛弃，废弃
abas 列线图，诺谟图
abatement claim 要求降价
abatement cost 除去成本，消除有害事物的成本
abatement of noise 噪声控制，消除噪声
abatement of pollution 消除污染
abatement of smoke 除烟，消烟，消除烟尘
abatement of water pollution 水污染的消除
abatement of wind 风力减弱
abatement of tax 减税
abatement 减少，减轻，消除，抑制，降低，扣减，失效，削弱，作废
abate 消除，减少，减轻，作废，回火，废止，缓解，抑制，削弱，降低，扣减
abatis 三角形木架透水坝
abatjour 斜片百叶窗，天窗，亮窗，遮阳
abat-vent 转向装置，折流板，固定百叶窗，通风帽，通气帽，障风装置，挡风装置
abatvoix 吸音板，反射板
abbertite 黑沥青，沥青
ABB(Asea Brown Boveri Ltd.) Group ABB集团【总部在瑞士，全球电力和自动化领域的领先企业】
abbreviated analysis 简易分析
abbreviated call letters 简缩呼号
abbreviated dialing 缩位拨号
abbreviated drawing 简图【测量】
abbreviated equation 简化方程
abbreviated method 简化法
abbreviated signal code 传输电码，缩写信码
abbreviated 简化的，缩写的，简缩的
abbreviation list 缩写表
abbreviation 缩略语，简写，缩写，缩写词，简化，略语，约分
abbreviative notation 缩写标记
abcoulomb 电磁制库仑【等于10C】
ABC 初步，入门，基本要素，基础知识
abderhalden dryer 干燥枪
abdication 活化，再生，再生作用
abed boundary condition 反照率边界条件
aberrance 反常
aberrancy 脱轨
aberrant source 异常误差来源
aberrant 异常的
aberration characteristic 相差特性
aberration curve 相差曲线
aberration 相差，偏差，光行差，失常，反常，像差
abeyance 缓办，中止（状态），停顿，暂搁，暂缓，潜态
abfarad 电磁法拉
abhenry 电磁亨利
abherent 防黏剂，防黏材料
abhesion 阻黏性，失黏性
abide by agreement 遵守协议
abide by contract 遵守合同
abide by 遵守（合同，协议），坚守（诺言等）【主语是人】
abies oil 松节油
abietic resin 松香树脂
abietin 松香脂
abietyl 松香
ability 才能，能力，性能
ability of bearing taxation 负税能力
ability test 能力测试，能力测验，才能测验
ability to bargain 议价能力
ability to compete 竞争能力
ability to function 作用能力
ability to invest 投资能力
ability to pay 偿付能力，支付能力
ability to repay in foreign exchange 外债偿付能力
abiocoen 无机生境
abiogenesis 自然发生论，无生源论
abiotic environment 非生物环境

abiotic factor 非生物因素
abiotic 非生物的，无生命的
abjoint 分隔
ablastin 抑菌素
ablation area 消融区
ablation shelter 烧蚀防护罩
ablation swamp 消融沼泽
ablation 烧蚀，切除，摘除，剥落，消融（作用），消蚀
ablative cooling material 烧蚀冷却材料
ablative insulating quality 隔热性
ablative material 烧蚀材料
ablative plastics 烧蚀性塑料
ablative polymer 烧蚀性聚合物
ablative-type protective coating 烧蚀型保护层
ablator 烧蚀挡板，烧蚀体，烧蚀剂
abluent 洗涤剂，洗净的
ablution 洗净，洗净液，清洗【透平、压气机等】，清除
abmho 电磁姆，电磁（制）欧姆，绝对欧姆
Abney level 阿布尼水准仪，手水准仪
abnormal anticlinorium 逆复背斜
abnormal bridging 非正常桥接，非正常跨接
abnormal cathode fall 反带阴极电位降
abnormal concrete 变态混凝土
abnormal condition 缺陷【运行】，异常条件，反常情况，异常工况
abnormal contact 异常接触
abnormal cost 非正常成本，异常成本
abnormal current 异常电流，事故电流，异常流
abnormal curve 非正常曲线，非正态曲线
abnormal density 反常密度
abnormal depreciation 非常折旧，特别折旧
abnormal discount 不正常折扣
abnormal distribution 非正态分布
abnormal drainage 异常水系
abnormal dump 异常转储【计算机】
abnormal end of job 作业异常终止
abnormal end 反常结束
abnormal erosion 反常冲蚀，反常侵蚀
abnormal error 不规则误差
abnormal event 异常事件
abnormal exposure 异常照射
abnormal false test 异常假测试
abnormal fan-shaped fold 逆扇状褶皱
abnormal fault 异常断层
abnormal flow 不正常流动，不规则流动
abnormal glow discharge 反常辉光放电
abnormal grain growth 晶粒反常增长
abnormal heating 异常发热
abnormal information 异常信息
abnormal input cause 异常输入原因
abnormalism 异常，变态性
abnormality 反常性，不正常性，不正常，变态，例外，不规则
abnormal load 不规则载荷
abnormal loss 非正常损失，特殊损失
abnormally 异常
abnormal network cause 异常网络原因
abnormal occurences reporting system 异常事件报告系统
abnormal operating condition 异常运行状态
abnormal operating transient 不正常运行暂态，异常运行瞬态
abnormal operation test 非正常工作试验
abnormal operation 不正常运行，非正常运行，异常运行
abnormal overload 异常过载
abnormal overvoltage 异常过电压，事故过电压
abnormal phenomena 异常现象
abnormal polarization 反常极化，反常偏振
abnormal refraction 反常折射
abnormal return 异常返回
abnormal risk 特殊风险，异常风险，异常危险
abnormal scour 异常冲刷
abnormal setting 异常凝结，异常沉降
abnormal sound 异声，异音，不正常声音，异响
abnormal state 异常状态
abnormal steel 反常钢，异常钢，非正常钢
abnormal stress 异常应力
abnormal structure steel 反常组织钢
abnormal temperature rise 不正常温升，过热
abnormal temperature 异常温度
abnormal test condition 特殊试验条件
abnormal tide 异常潮，稀遇潮
abnormal true test 异常真测试
abnormal value 异常值
abnormal voltage 异常电压
abnormal water level 非常水位
abnormal weather 反常天气
abnormal 不正常的，不规则的，异常的，反常，异常
abnormity 异常，不规则，异型
abohm 电磁欧姆，绝对欧姆
aboideau 挡潮闸，水坝，水闸，堰
abolish 废除，废止，取消
abolition 放弃，废除
abortive 无效果的，失败的
abort light 紧急停机信号，故障信号，紧急故障信号
abort sensing control unit （紧急）故障传感控制装置
abort sensing 故障测定
abort situation 故障位置，故障状态
abort 紧急停机，故障，失灵
about pitch axis balance 关于俯仰轴的平衡
above 在……之上
above-bed burner 床上方燃烧器【沸腾炉】
above-bed flame 床上方的火焰【沸腾炉】
above-bed torch 床上方点火燃烧器【沸腾炉】
above-core structure 堆芯上部构件
above-critical mass 超临界质量
above-critical state 超临界状态
above critical 超临界的，临界以上的
above earth potential 对地电位
aboveground 地面上的
aboveground cable 地面电缆
aboveground containment 地面安全壳
aboveground installation 地上敷设
aboveground power station 地面电站
above high water mark 在高水位以上

above mean sea level 平均海平面以上，平均海拔高度，平均海拔
above-mentioned 上述的，前述的
above-norm 超额定的，超标准的，超定额的，定额以上的
above requirement 上述要求
above-resonance balancing machine 超谐振平衡（试验）机
above sea level 海拔高度，海拔
above-target profit 超额利润
above-thermal neutron 超热中子
above-thermal 过热的，超热的
ABP 低压给水加热器系统【核电站系统代码】
A-bracket 人字架，A形架，推进器架
abradability 磨蚀性，磨损性，磨损度
abradable sealing strips 可研磨密封条
abradant 磨料，研磨剂，磨损的，腐蚀的，磨蚀剂
abraded coal particle 磨损炭粒
abraded fuel-element particles 燃料元件磨蚀碎屑
abraded platform 浪成台地
abrade 擦，磨，擦伤，磨损
abrading particles 打磨粉粒
abrasiometer 磨损实验计
abrasion block 磨块
abrasion by sediment 泥沙磨损
abrasion cycle 磨损周期，腐蚀周期
abrasion degree 磨耗度
abrasion hardness 耐磨硬度
abrasion index 磨损性指数，磨损指数
abrasion inspection 磨损检查
abrasion loss 磨损量，磨耗量，磨耗
abrasion machine 磨耗试验机
abrasion mark 擦痕，划痕，磨痕
abrasion material 磨料
abrasion paste 研磨膏
abrasion pattern 磨纹
abrasion performance 耐磨性
abrasion platform 浪蚀岩石阶地，浪蚀台地，磨蚀台地
abrasion powder 研磨粉
abrasion-proof component of burner 燃烧器耐磨件
abrasion-proof 耐磨的
abrasion resistance index 耐磨指数，磨耗指数
abrasion resistance material 耐磨材料
abrasion resistance 耐磨性，抗磨蚀能力，抗磨损性，耐冲刷性，磨耗阻力
abrasion resistant 耐磨的，抗磨
abrasion resistant bead welding 耐磨堆焊
abrasion resistant liner 耐磨衬板
abrasion resistant reinforcement 耐磨损加强
abrasion resistant steel 耐磨钢
abrasion running 磨耗运行
abrasion tableland 剥蚀基岩台地
abrasion terrace 浪蚀阶地
abrasion tester 磨损试验器
abrasion test machine 磨耗试验机
abrasion test 磨耗试验，耐磨试验
abrasion value 磨耗量，磨耗值
abrasion wear 磨耗，磨损
abrasion 机械磨蚀，磨损，磨耗，冲蚀，剥蚀，磨蚀，研磨，磨光，水蚀
abrasive action 磨蚀作用，磨损作用
abrasive belt 砂带，砂带式抛光机
abrasive blast cleaning 喷砂处理
abrasive blast equipment 喷砂磨光设备
abrasive blasting 磨料喷射，喷砂冲洗
abrasive capacity 剥蚀能力
abrasive cloth 砂布，擦光布
abrasive coated ball 包有磨料的球
abrasive compound 研磨材料，复合磨料
abrasive cut-off machine 砂轮切割机
abrasive cut-off wheel 切割砂轮
abrasive disc 磨盘，磨片，砂轮
abrasive disk cutter 砂轮切割机
abrasive finishing machine 抛光机
abrasive flap wheel 片状砂轮
abrasive fog 磨擦灰雾
abrasive hardness 耐磨硬度
abrasive-laden 含磨料的
abrasiveness 磨蚀性，磨损性，磨损度，耐磨性
abrasive paper 砂纸
abrasive particles 研磨粉粒，磨蚀微粒
abrasive resistance 抗磨力，磨蚀阻力，磨蚀强度，抗磨性，耐磨性，耐磨强度
abrasive surface 磨损面
abrasive tool 研磨工具
abrasive wear 磨损，磨耗，磨蚀
abrasive wheel 砂轮
abrasive 磨损的，腐蚀的，粗糙的，摩擦的，磨蚀剂，研磨剂，研磨料
abrasivity 冲蚀度，磨蚀度
abration resisting pump 耐磨泵
abrator 抛丸清理机，抛喷清理机，离心喷光机，抛丸机
abreuvoir 石块间隙缝
abridged drawing 略图
abridged general view 示意图
abridged table 简表
abridged 删减的，削减的
abridgement 缩短
abri 岩洞，洞穴，岩穴
ABRO(air bump rinse operation) 空气擦洗
abrogate original contract 废除原合同
abrogate the agreement 取消协议
abrogate 取消，废除
abrupt 陡坡的，突然的
abrupt bend 突然弯曲
abrupt curve 陡曲线，陡变曲线，急弯曲线
abrupt discharge 猝然排出
abrupt failure 突然断裂
abruption 隔断，断裂
abrupt slope 陡坡
abrupt transformation 突跃变换
abrupt wave 突变波，陡浪
ABS(Acrylonitrile Butadiene Styrene) ABS塑料，苯乙烯共聚物，丙烯腈-丁二烯-苯乙烯
ABS(American Bureau of Standard) 美国标准局
ABS(Antilock Brake System) 防抱闸系统，防抱死制动系统，防锁死刹车系统
ABS(Anti-Skid Brake System) 防滑煞车系统

abscess （金属中的）气孔，气泡，夹渣内孔，（金属中的）砂眼
abscissa axis 横坐标轴
abscissa of convergence 收敛横坐标
abscissa scale 横坐标尺度
abscissa 横坐标，横轴线，横坐标轴
abscopal radiation effect 远隔辐射效应，远位辐射效应
absence of brush 无碳刷
absence of collector 无集电环
absence of commutator 无换相器，无整流子
absence-of-ground search selector 未接地寻线器
absence of hang-up 没有大障碍
absence of offset 零静差，无静差，无偏置
absence of restriction 无约束
absence without leave 无故缺席，旷工
absence 缺乏，缺席
absentee control 无人管理
absentee 空号
absent 缺少的，缺席的，不在场的，不存在的，缺席
absolute 绝对的，绝对
absolute acceptance 单纯认付，单纯承兑，无条件认付
absolute accuracy 绝对精确度，绝对准确度
absolute activity determination 放射性活度绝对测量
absolute activity method 绝对活度测定（法）
absolute activity of ……的绝对活度
absolute addressing 绝对寻址
absolute address 绝对地址
absolute advantage 绝对优势
absolute age of groundwater 地下水绝对年龄
absolute alarm 绝对值报警
absolute alcohol 无水酒精，无水乙醇
absolute altimeter 雷达测高仪
absolute altitude 绝对标高，绝对高程，标高，绝对高度，海拔
absolute ampere 绝对安培
absolute anchor point 绝对定位
absolute annual range of temperature 温度绝对年较差
absolute assignment 无条件转让
absolute atmosphere 绝对大气压
absolute atmospheric pressure 绝对大气压
absolute black body 绝对黑体
absolute calibration 绝对刻度，绝对校准
absolute capacitivity 绝对电容率，绝对电容常数
absolute capacity 绝对容量
absolute cavity radiometer 腔体式绝对辐射表
absolute ceiling 绝对升限
absolute chronology 绝对顺序计时
absolute code 绝对代码，绝对码
absolute coefficient 绝对系数
absolute condenser pressure 凝汽器中的绝对压力
absolute constant 绝对常数
absolute construction 独立结构
absolute contract 不附带条件的合同
absolute convergence 绝对收敛
absolute coordinate 绝对坐标
absolute cost 绝对成本
absolute counter 绝对计数器
absolute counting 绝对计数
absolute cover 绝对保险额
absolute cross-section 绝对截面
absolute damping 绝对阻尼，全阻尼
absolute data 绝对数据
absolute deflection 绝对变位，绝对垂度
absolute delay 绝对延迟
absolute delivery 无条件交付
absolute density 绝对密度
absolute determination 绝对测定
absolute deviation integral 绝对偏差积分
absolute deviation 绝对偏差
absolute dielectric constant 绝对介电常数，绝对电容率
absolute differential caculus 绝对微积分学，张量分析
absolute digital position transducer 绝对数字位置转换器
absolute dimension 绝对尺寸，绝对坐标值
absolute displacement 绝对位移
absolute divergence of parameter 参数绝对误差
absolute dry condition 绝对干燥状态
absolute efficiency 绝对效率
absolute electrical unit 绝对电单位
absolute electric efficiency 绝对电效率
absolute electromagnetic system 绝对电磁单位制
absolute electromagnetic unit 绝对电磁单位
absolute electrometer 绝对静电计
absolute electrostatic system 绝对静电单位制
absolute elevation 绝对高程，绝对标高，海拔高程，海拔
absolute elongation 绝对伸长
absolute encoder 绝对编码器
absolute enthalpy 绝对焓
absolute error 绝对误差
absolute ethyl alcohol 无水乙醇，无水酒精
absolute expansion monitor of turbine 汽轮机绝对膨胀监视器
absolute expansion 绝对膨胀
absolute extreme 绝对极值
absolute farad 绝对法拉
absolute filter bank 绝对过滤（器）组件，绝对过滤器组
absolute filter package 绝对过滤（器）组件，绝对过滤器组
absolute filter 高效过滤器，高性能过滤器，绝对过滤器
absolute fission rate 绝对裂变率
absolute fission yield 绝对裂变产额
absolute flux 绝对通量
absolute frequency 绝对频率
absolute gain of an attenna 天线绝对增益
absolute galvanometer 绝对检流计
absolute gauge 绝对量计，绝对压力表
absolute heating effect 绝对加热效应
absolute heating efficiency 绝对加热效率
absolute height 绝对高度
absolute henry 绝对亨利
absolute humidity of gas 气体的绝对湿度

absolute humidity 绝对湿度
absolute inclinometer 绝对测斜仪，绝对倾斜仪
absolute instruction 绝对指令
absolute intensity 绝对强度
absolute interest 绝对权益
absolute invariance principle 绝对不变性原理
absolute language 绝对语言
absolute lethal dose 绝对致死剂量
absolute level 绝对水平，绝对标高，绝对电平，绝对级
absolute liability 绝对赔偿责任
absolute limit of error 误差绝对极限，绝对误差极限
absolute linearity 绝对线性度，绝对线性
absolute manometer 绝对压力计
absolute mass unit 绝对质量单位
absolute maximum air temperature 极端最高气温
absolute maximum fatal temperature 绝对最高致死温度
absolute maximum temperature 绝对最高温度
absolute maximum 绝对最大
absolute measurement 绝对测量
absolute method of measurement 绝对测量法
absolute minimum air temperature 极端最低气温
absolute minimum fatal temperature 绝对最低致死温度
absolute minimum temperature 绝对最低温度
absolute minimum 绝对最小
absolute modulus 绝对模量
absolute moisture content 绝对含湿量，绝对水分含量，绝对湿度
absolute moisture 绝对湿度
absolute moment 绝对矩
absolute monthly maximum temperature 月绝对最高温度
absolute motion 绝对运动
absolute net loss 绝对净损
absolute neutron flux 绝对中子通量
absolute number 绝对值
absolute ohm 绝对欧姆
absolute par 绝对平价
absolute permeability 绝对磁导率，绝对导磁率，绝对介电常数
absolute potential 绝对电势，绝对电位
absolute power 绝对功率
absolute predominance 绝对优势
absolute pressure controller 绝对压力控制器
absolute pressure gauge 绝对压力计，绝对压力表
absolute pressure head 绝对压（力水）头
absolute pressure indicator 绝对压力指示器
absolute pressure regulator 绝对压力调节器
absolute pressure sensor 绝对压力传感器
absolute pressure transducer 绝对压力转换器，绝对压力传感器
absolute pressure vacuum gauge 绝对压力真空表，绝对压力真空计
absolute pressure 绝对压力
absolute probability 绝对概率
absolute programming 绝对编程，绝对程序设计
absolute program 绝对程序
absolute radiometer 绝对辐射表
absolute reactivity 绝对反应性
absolute read out 绝对读数，绝对指示
absolute retention volume 绝对保留体积
absolute rise velocity 绝对上升速度
absolute roughness 绝对粗糙度
absolute scale of temperature 绝对温标
absolute scale 绝对标度，绝对温标
absolute scattering power 绝对散射能力
absolute self-calibrating pyrheliometer 自校准绝对直接日射表
absolute sensitivity 绝对灵敏度
absolute signal 绝对信号，纯信号
absolute similarity 完全相似性
absolute specific gravity 绝对比重
absolute spectral response 绝对光谱响应
absolute spectral sensitivity 绝对光谱灵敏度
absolute spectrum function 绝对频谱函数
absolute stability constant 绝对稳定常数
absolute stability of a linear control system 线性控制系统的绝对稳定性
absolute stability 绝对稳定度，绝对稳定性
absolute strength 绝对强度
absolute system of electrical units 绝对电单位制
absolute temperature scale of gas 气体绝对温标
absolute temperature scale 绝对温标
absolute temperature 绝对温度，热力学温度
absolute term 常数项
absolute thermodynamic scale 绝对热力学温标，绝对热力学标度
absolute thermodynamic temperature scale 绝对热力学温标
absolute thermometer 绝对温度表
absolute topography 绝对地形
absolute transmission level 绝对传输电平
absolute trust 绝对信任
absolute turbidity 绝对浊度
absolute unit of current 绝对电流单位，电磁安培
absolute unit system 绝对单位制
absolute unit 绝对单位
absolute vacuum gauge 绝对真空计
absolute vacuum 绝对真空
absolute value amplifier 绝对值放大器
absolute value computer 绝对值计算机
absolute value converter 绝对值转换器
absolute value device 绝对值装置
absolute value 绝对值
absolute velocity 绝对速度
absolute viscosity 绝对动力黏度，绝对黏度
absolute volt 绝对电压
absolute volume 绝对容积，绝对体积
absolute vorticity 绝对涡量
absolute watt 绝对瓦特
absolute wave meter 绝对波长表，绝对波频计
absolute weight 绝对重量
absolute worst case 绝对最坏情况
absolute yield 绝对产量，绝对产额
absolute zero of temperature 绝对零度
absolute zero 绝对零值，绝对零度，绝对零
absorbability 吸收性，吸收，可吸收性，吸附

力，吸收能力
absorbance index 吸收系数
absorbance 吸收率，光密度
absorbar 吸收体【反应堆】
absorbate 吸收质，（被）吸收物
absorbed dose commitment 吸收剂量负担
absorbed dose distribution 吸收剂量分布
absorbed dose equivalent rate 吸收剂量当量率
absorbed dose index rate 吸收剂量指数率
absorbed dose index 吸收剂量指数
absorbed dose rate 吸收剂量率，剂量吸收率
absorbed dose 吸收剂量
absorbed energy 吸收能
absorbed fan power 风机吸收功率
absorbed fraction 吸收份额
absorbed heat 吸热量
absorbed-in-fracture energy 冲击韧性，冲击强度，冲击功，断裂功
absorbed layer 吸收层
absorbed moisture 吸收水分
absorbed oscillation 阻尼振动，吸收振动
absorbed power 吸收能量，吸收功率
absorbed radiation dose 吸收的辐射剂量
absorbed water 吸收水分，吸收水，吸着水
absorbed 被吸收的
absorbefacient 吸收剂，吸收性的
absorbent bed 吸收床，吸收层
absorbent capacity 吸收剂能力
absorbent carbon 活性炭
absorbent charcoal 吸收性炭，活性炭
absorbent cotton 脱脂棉
absorbent factor 吸收因子
absorbent filter 吸收性过滤器，吸收过滤器
absorbent insulation 吸收绝缘，海绵状绝缘
absorbent material 吸收体，吸收剂，吸收性物质
absorbent powder 粉状吸收剂
absorbent power 吸收能力
absorbent slurry 浆状吸收剂
absorbent solution 吸收溶液
absorbent 吸收剂，吸附剂，吸湿剂，吸收体，吸收的，吸声材料，有吸收能力的
absorber area of concentrating collector 聚光型集热器的吸热体面积
absorber area of non-concentrating collector 非聚光型集热器的吸热体面积
absorber area 吸热体面积
absorber ball 吸收球
absorber bore 吸收体内孔
absorber can material 吸收体包壳材料
absorber circuit 吸收器电路
absorber cooler 吸收（器的）冷却器
absorber coupling 吸收棒接头
absorber diode 吸收二极管
absorber drum 吸收控制鼓
absorber element 吸收元件【中子】
absorber feed pump 吸收塔给料泵
absorber finger 单根吸收棒【控制组件】
absorber guide tube 吸收棒导向管
absorber headtank separator 吸收塔高位槽分离器
absorber material 吸收剂材料，吸收体材料
absorber member 吸收元件
absorber plate 板状吸收体，吸热板
absorber portion 吸收段【中子吸收体】
absorber rod bore 吸收棒内孔
absorber rod control gas 吸收棒控制气体
absorber rod coupling 吸收棒接头
absorber rod drive servicing 吸收棒驱动机构罩壳
absorber rod drive 吸收棒驱动
absorber rod guide box 吸收棒导向盒
absorber rod guide sheath 吸收棒导向套管
absorber rod guide tube 吸收棒导向管
absorber rod runaway 吸收棒失控
absorber rod structure 吸收棒结构
absorber rod system 吸收棒系统
absorber rod worth 吸收棒反应性价值，吸收棒价值
absorber rod 吸收棒，控制棒
absorber section 吸收段【中子吸收体】
absorber solution 吸收溶液
absorber tube 吸收体管
absorber type control rod 吸收型控制棒【中子】
absorber washer 吸收洗涤器
absorber 吸收塔【脱硫装置的】，缓冲器，阻尼器，减震器，减振器，吸热体，吸收电路，吸收剂，吸收体，吸收器
absorb foreign funds 吸引外资
absorb heat 吸收热量
absorbing 吸收
absorbing ability 吸收能力，减震能力
absorbing agent 吸收剂
absorbing apparatus 吸收装置
absorbing atom 吸收原子
absorbing capacity 吸收能力，吸收量
absorbing chemical 吸收剂
absorbing circuit 吸收电路
absorbing clamp 减震钳
absorbing coil 吸收线圈
absorbing column 吸收柱，吸收塔
absorbing control rod 吸收控制杆
absorbing curtain 吸收挡板，吸收屏
absorbing duct 消声器，吸声管
absorbing gas 吸收气体
absorbing load 吸收负载
absorbing material 吸收物，吸收剂
absorbing medium 吸收介质，吸收剂
absorbing power 吸收能力
absorbing rod 吸收棒
absorbing selector 吸收选择器
absorbing silencer 吸收消音器
absorbing surface 吸收表面
absorbing tower 吸收塔
absorbing well 吸水井，渗水井，渗井，污水渗井，吸收井，注水井
absorbing zone 吸收层
absorbit 微晶型活性炭，活性炭
absorb sound 隔声，吸声
absorb 吸收，减振，吞并，承受，负担
absorptance analysis 吸收率分析
absorptance for solar radiation 太阳辐射热吸收系数
absorptance 吸收系数，吸收能力，吸收率，吸

收比
absorptiometer 吸收计，液体溶气计，吸收比色计
absorptiometry 吸收测量学，吸收测量
absorption analysis 吸收分析
absorption apparatus 吸收
absorption area method 吸收面积法
absorption band 吸收谱带，吸收频带，吸收带
absorption behavior 吸收性能，吸收特性
absorption build-up factor 吸收能量积累因子
absorption by glass 玻璃的吸收
absorption capacitor 吸收电容器【消弧用】
absorption capacity 吸收能力，吸收电容【消弧用】
absorption capture 俘获吸收
absorption chamber 吸收室，填料塔
absorption characteristic 吸收特性
absorption circuit 吸收电路，吸收回路
absorption coefficient 吸收系数
absorption column 吸收柱，吸收塔
absorption constant 吸收常数
absorption-controlled reactor 吸收控制的反应堆
absorption control 用吸收中子法进行调解，吸收控制
absorption cross-section 吸收断面，吸收截面
absorption current of dielectric 介质吸收电流
absorption current （非完全介质中的）吸收电流
absorption curve 吸收曲线
absorption damper 减振器，消声器
absorption dehydration 吸收干燥
absorption delay 吸收延迟
absorption dip （通量）吸收坑，吸收曲线中的下落
absorption discontinuity 吸收曲线突变，吸收的不连续性
absorption dynamometer 吸收式功率计，吸收式测力计
absorption effect 吸收效应
absorption efficiency 吸收效率
absorption-enhancement effect 吸收增强效应
absorption equipment 吸收装置
absorption extraction 吸收萃取
absorption factor 吸收系数，吸收因子
absorption field 浸润范围，浸润区
absorption filter 吸收滤片
absorption flaw detector 吸收式探伤仪
absorption frequency meter 吸收式频谱计，吸收式频率计
absorption hardening （中子谱）吸收硬化
absorption heating 吸收加热
absorption hologram 吸收全息图
absorption hygrometer 吸收湿度计，吸收湿度表
absorption inductor 吸收电感线圈
absorption isolator 吸振器
absorption law 吸收律
absorption length 吸收长度
absorption line 吸收（谱）线
absorption liquid chiller 吸收式液体冷却器
absorption liquid chilling system 吸收式液体冷却系统
absorption loss 吸收损耗，吸收损失
absorption machine 吸收机
absorption maximum 最大吸收
absorption meter 吸收计
absorption modulation 吸收调制
absorption of gas and vapour 气体及汽体吸收
absorption of heat 吸热
absorption of moisture 吸潮，吸湿
absorption of photons 光子的吸收
absorption of radiation 辐射吸收
absorption of shocks 减震，减振，冲击波的吸收
absorption of solar radiation by pond 太阳池对太阳辐射的吸收
absorption of sound 声音吸收
absorption of the photons 光吸收
absorption of vibration 减振，振动阻尼
absorption paper 特制滤纸，吸水纸
absorption peak 吸收峰
absorption phenomenon 吸附现象
absorption pipette 吸收球管
absorption plane （无线电的）有效面
absorption power 吸收功率，吸收能力，吸收力
absorption pyrometer 吸收式光学高温计
absorption rate 吸收速度，吸水率，吸收率，吸收比，吸收系数
absorption reaction rate 吸收反应率
absorption reaction 吸收反应
absorption refrigerating machine 吸收式制冷机
absorption refrigerating plant 吸收式制冷装置
absorption refrigerating system 吸收式制冷系统
absorption refrigeration cycle 吸收式制冷循环
absorption refrigeration 吸收式制冷
absorption refrigerator system 吸收式制冷系统
absorption refrigerator 吸收式制冷机
absorption region 吸收区，吸收范围
absorption resonance 吸收共振
absorption screen 吸收滤光屏
absorption silencer 吸收式消声器
absorption spectrometer 吸收频仪，吸收光谱仪
absorption spectrum 吸收频谱，吸收光谱，吸收谱
absorption strength 吸收力，吸收强度
absorption surface 吸收面
absorption terrace 保水阶地
absorption test 吸水试验，吸收试验
absorption thickness 吸收厚度
absorption tower 吸收塔
absorption trap 吸收陷阱，陷波回路
absorption trench 吸水管道沟槽
absorption-type refrigerating machine 吸收式制冷机
absorption water 附着水，吸附水
absorption wave meter 吸收式波长计
absorption wave-trap 吸收陷阱，吸收陷波电路
absorption well 吸水井，渗水井
absorption 吸附，吸收，吸收作用，吸水作用
absorptive attenuator 吸收性衰减器
absorptive capacity 吸收率，吸收性
absorptive form lining 吸水性模板衬垫
absorptive index 吸收系数
absorptive lining 吸水衬里
absorptive material 吸附性材料

absorptive power　吸附能力，吸收能力
absorptivity　吸收能力，吸收性，吸收率，吸收系数
ABS resin　丙烯腈-丁二烯-苯乙烯树脂
abs T(absolute temperature)　绝对温度
abstain from voting　弃权
abstain　弃权
abstention　弃权
abstergent　去垢剂，有去垢性质的
abstract automata theory　抽象自动机理论
abstract code　抽象代码，理想码
abstract communication service interface　抽象通信服务接口
abstracted heat　抽出的热，放出热
abstract heat　散热
abstraction of heat　散热，排热
abstraction of water　抽取（地下）水，排水
abstraction works　引水设施
abstraction　抽出，除去，抽象（化），提取，引水工程
abstract number　不名数
abstract of particulars　细则
abstract syntax notation　抽象语法标志
abstract　抽象的，摘要，提取，提要，文摘，转移
abs visc(absolute viscosity)　绝对黏度
abt＝about　大约
abundance of offers　很多报价
abundance of water　丰水
abundance ratio　（同位素）丰度比率，丰度比
abundance sensitivity　丰度灵敏度
abundance　丰度，丰富，含量，盈余，充分
abundant precipitation　过量降水
abundant technical power　技术实力雄厚
abunits　厘米-克-秒制电磁单位
abuse failure　使用不当失效
abuse of patent　滥用专利权
abuse of right　滥用权力
abuse　滥用
abutment block of arch dam　拱坝重力墩
abutment block　人造边墩，人造坝座
abutment crane　高座起重机
abutment deformation　坝座变形【水电】
abutment joint　坝肩接缝【水电】，对接接头
abutment pier　墩式桥台，岸墩，边墩
abutment pole　中缩磁极
abutment pressure　坝座压力【水电】，两岸坝座压力，桥台压力
abutment span　（靠岸的）边跨，岸墩孔，桥台跨度
abutment stability　坝肩稳定【水电】
abutment stone　拱座石，桥台石
abutment surface　（轴承的）上推面，相接面
abutment wall　边墙，拱座墙
abutment wingwall　桥台翼墙
abutment　坝肩，坝座，坝头【水电】，桥墩，桥台，座，支承面，支座
abutted surface　邻接面
abutting building　邻接建筑物，毗连建筑物，毗连房屋
abutting ends　相邻线端，对接接头
abutting surface　端接面，对接面
abutting　毗连的，对抵的，凸出的，邻接
abut　支点，止动点，支柱
abvolt　绝对伏特，电磁伏特
ABWR(Advanced Boiling Water Reactor)　先进沸水堆
abyssal assimilation　深成同化，深同化作用
abyssal depth　深渊
abyssal fault　深断层
abyssal intrusion　深成侵入
abyssalpelagic zone　深渊带
abyssal rock　深成岩
ACA(automatic current analyzer)　自动电路分析器
AC abruptly increasing point test　交流电流激增点试验
AC(access control)　存取控制
a/c＝account　账号，计算，账单
AC＝accumulator　蓄热器，累加器
AC(address counter)　地址计数器
academic activity center　学术活动中心
academical and professional ability　学历和专业能力
academic authority　学术权威
academic certificate　学历证书
academic discussion　学术讨论
academic qualifications　学历
academy　研究院
AC(air conditioner)　空气调节器
AC(alternating current)　交流电
AC(altocumulus cloud)　高积云
AC ammeter　交流安培计，交流电流表
ACAM(Augmented Content-Addressed Memory)　可扩充的内容定址存储器
AC and DC power transmission in parallel　交直流并网输电
acanthus　叶板，叶形装饰
AC arc welder　交流弧焊机
ACAR(aluminum conductor alloy reinforced)　铝合金芯铝绞线
AC arc welding machine　交流弧电焊机
AC(automatic control)　自动控制
AC automatic recording fluxmeter　交流自动磁通记录仪
ACB(air circuit breaker)　空气断路器
AC balancer　交流电压平衡装置
AC bridge　交流电桥
ACC＝acceptance　承兑
ACC(Accepted Without Comment)　无意见接收
ACC＝accepted　已承兑
AC capacitor　交流电容器
AC cast(altocumulus castellanus)　堡状高积云
ACCA(the Association of Chartered Certified Accountants)　特许公认会计师公会，国际注册会计师协会
ACC(automatic combustion control)　自动燃烧控制
ACC(Auxiliary Core Cooling)　堆芯辅助冷却
accede to　同意，加入，就任
accelerant coatings　速燃层
accelerant　加速剂，促进剂，催速剂

accelerated aging 加速老化，人工老化，加速时效
accelerated bum-up test 加速燃耗试验
accelerated by a spring 弹簧加速的
accelerated cement 快凝水泥
accelerated clarification 加速澄清
accelerated clarifier 加速沉降器
accelerated combustion 加速燃烧
accelerated completion 加速竣工，提前完工
accelerated consolidation 加速加固
accelerated corrosion test 加速腐蚀试验
accelerated depreciation 加速折旧
accelerated distance protection 加速式距离保护
accelerated draft 加速通风
accelerated erosion 加速侵蚀
accelerated exposure testing 加速寿命试验
accelerated fatigue 加速疲劳
accelerated life testing 速效寿命试验
accelerated load test 加速载荷试验
accelerated motion 加速运动
accelerated oxidation 加速氧化
accelerated stall 过载失速，加速失速
accelerated test 加速试验
accelerated weathering 加速老化
accelerate 加速，促进
accelerated ageing 加速老化，人工老化，加速时效
accelerating 加速
accelerating anode 加速阳极
accelerating agent 加速剂，促凝剂
accelerating apparatus 加速装置
accelerating by gravity 重力加速
accelerating chamber 加速箱
accelerating coil 加速线圈
accelerating constant 加速常数
accelerating contactor 加速接触器
accelerating creep 加速蠕变
accelerating depreciation 加速折旧
accelerating electrode 加速电极
accelerating feedback 加速反馈
accelerating field 加速场
accelerating force 加速力
accelerating grade 加速坡
accelerating grid 加速栅极
accelerating load 加速负荷
accelerating nozzle 加速型喷嘴
accelerating potential 加速电位，加速电势，加速电压
accelerating power 加速功率
accelerating pressure gradient 顺压梯度
accelerating relay 加速继电器，启动继电器
accelerating signal 加速信号
accelerating thermal aging test 加速热老化试验
accelerating torque 加速扭矩
accelerating voltage 加速电压
acceleration amplitude 加速度幅值
acceleration boundary 加速度界限
acceleration by gravity 重力加速度
acceleration clause 加速条款
acceleration coefficient 加速度系数，加速系数
acceleration control unit 加速度控制器
acceleration damper 加速度阻尼器
acceleration due to gravity 重力加速度
acceleration error constant 加速度误差系数
acceleration factor 加速系数
acceleration feedback 加速反馈
acceleration governor 加速度调节器
acceleration in pitch 俯仰加速度
acceleration in roll 滚转加速度，侧倾加速度
acceleration in spectrum 加速度谱
acceleration instrument 加速表，加速传感器
acceleration in transducer 加速度传感器
acceleration in yaw 偏航加速度
acceleration lag 加速度滞后
acceleration limiter 加速限制器
acceleration losses 加速度损失
acceleration motor 加速电机
acceleration of convergence 加速收敛
acceleration of free fall 自由落体加速度
acceleration of gravity 重力加速度
acceleration of maturity 缩短到期时间
acceleration of translation 平移加速度
acceleration period 加速时间【电动机】
acceleration pulse 加速脉冲
acceleration relay 加速度继动器
acceleration response 加速度反应，加速度响应，过载反应
acceleration setter 加速度定值器，加速度给定器
acceleration spring 加速弹簧
acceleration-switching valve 快关阀
acceleration test 加速试验
acceleration time constant of unit 机组加速时间常数
acceleration time 加速时间
acceleration transducer 加速度传感器
acceleration valve 加速阀
acceleration vector 加速度矢量
acceleration vibrograph 加速型振动仪
acceleration voltage 加速电压
acceleration 加速，加速度，加速作用，再加速，加快，聚升
accelerator breeder 加速器增殖堆
accelerator driven light water reactor 加速供驱动的轻水堆
accelerator facility 加速器设备
accelerator limiter 加速限制器
accelerator plunger 加速器柱塞
accelerator valve 加速阀
accelerator 促凝剂，加速器，加速剂，速凝剂，加速电极
accelerograph 自动加速记录仪，自动加速仪
accelerometer 加速表，过载传感器，加速度表，加速度计，加速度仪，加速度传感器
accelerometric governor 加速度调速器
accelo filter 快速过滤器
accent light 强光灯
accentuator 音频强化器，频率校正电路
accept a bid 接受标书，接受投标
acceptability criteria 可接受标准
acceptability 可接受性，可承兑性，合格率
accept a bill of exchange 承兑汇票
acceptable accounting principle 可接受的会计

原理
acceptable concentration 容许浓度
acceptable condition 验收条件，合格条件
acceptable date 可接受日期
acceptable defect level 容许缺陷标准，容许缺陷程度
acceptable dose 容许剂量
acceptable emergence dose 允许事故剂量，可接受的应急剂量
acceptable environmental limit 容许环境吸入极限，容许环境极限
acceptable level 容许水平，验收标准
acceptable life 有效使用寿命
acceptable limit 可接受限值，容许极限
acceptable malfunction level 容许故障等级，容许故障水平
acceptable malfunction rate 容许故障率
acceptable material 合格材料
acceptable noise level 容许噪声级
acceptable precision 容许精度，合格精度
acceptable price 可接受价格
acceptable product 合格产品
acceptable quality level 合格品质水准，可接受的质量水平，验收质量标准
acceptable quality 合格的质量
acceptable reliability level 可接受的可靠性标准
acceptable reliability 容许可靠度
acceptable results 可接受的结果
acceptable risk level 可接受的风险度水平
acceptable risk 可接受的风险
acceptable standard 验收标准，通用标准
acceptable test 验收试验
acceptable value 容许值，可接受值
acceptable wind speed 容许风速
acceptable 合格的，容许的，可接受的，验收的
accept a claim 接受索赔
acceptance ability of the bid 中标能力
acceptance agreement 承兑协议
acceptance and delivery of cars 车辆交接
acceptance and delivery of goods 货物交接
acceptance and transfer 验收与移交
acceptance angle 接收角，接纳角
acceptance at work completion 竣工验收
acceptance bank 承付行
acceptance bill 承兑汇票
acceptance certificate 验收合格证，验收证书
acceptance charge 承兑费
acceptance check of turbine foundation 汽机基础验收
acceptance check 验收，验收检查
acceptance coefficient 合格系数
acceptance commission 承兑手续费
acceptance condition 验收（合格）条件
acceptance contract 承兑合同
acceptance credit 承兑信用，承兑信用证
acceptance criteria 验收标准，验收准则
acceptance declaration 申报承兑
acceptance draft 承兑汇票
acceptance gage 验收量规
acceptance house 期票承兑人，期票承兑行，承兑行
acceptance inspection 验收检验，验收检查
acceptance letter of credit 承兑信用证
acceptance of concrete work 混凝土工程的验收
acceptance of hidden subsurface work 隐蔽工程的验收
acceptance of offer 接受报价
acceptance of order 承接订单
acceptance of project 工程验收
acceptance of subdivisional work 分项工程验收
acceptance of tender 接受投标书
acceptance of the bid 得标，中标
acceptance of works 工程验收
acceptance on security 担保承兑
acceptance procedure 验收程序
acceptance proof test 验收证明测试，投运前的试验
acceptance quality level 可接受的质量水平
acceptance range 验收范围
acceptance report 验收报告
acceptance requirement 验收要求
acceptance specifications 验收规范
acceptance standard 验收标准
acceptance summary report 验收总结报告
acceptance survey 验收，验收测量
acceptance tested 验收试验过的【填充金属】
acceptance test of electric equipment 电气设备交接试验
acceptance test 验收试验
acceptance time 验收时间
acceptance value 合格判定值
acceptance with reservation 有保留的接受
acceptance 承兑，接受，验收，认可，同意
accept an invitation 接受邀请
accept an order 接受订货，接受命令
accept a quotation 接受报价
accept a tender 接受投标
accepted as noted 有条件接受
accepted bid 认可的投标书，已接受的投标书
accepted bill advice 承兑通知书
accepted risk 可接受的风险
accepted standards 采用的标准
accepted value 常用值，认可值
accepted 接受的，承兑的
accepter 承兑人
accept forward shipment 接受远期装运
accepting bank 承兑行
accepting charges 承兑费
accepting house 承兑行
acceptor circuit 接收器电路，带通电路
acceptor ionization energy 受主电离能
acceptor number density 受主浓度
acceptor resonance 电压谐振，串联谐振
acceptor trap 半导体受主陷阱，受主陷阱
acceptor 接收器，通波器，接受者，承兑人
accept your suggestion 接受你方建议
accept 接受，承认，同意，承兑（票据），认付【汇票】
access 附件，附属的
access adit 辅助坑道，施工支洞
access area 进入区域

access arm 磁头臂，存取臂
access authorization for 对……批准进入
access board 入口铺板
access bridge 引桥
access-control register 存取控制寄存器
access-control words 存取控制字
access control 存取控制
access corridor 入口通道，走廊
access cover 检修盖
access cycle 存取周期
access door 检查门，检修门，检修孔，人孔，便门
access duct 进入管道，进出通道
access eye 检查孔
access gallery 交通廊道
access gully 交通沟，窨井，雨水口
access hatch way 出入舱口
access hatch 检修孔，入口闸门，出入舱口，观察孔，检查孔，出入舱门
access hole 检查孔，检修孔，人孔
accessibility 可接近性，可达性，可及性，可触性，可进入性，可访问性，易维护性，可以得到，无障碍环境
accessible 可达的
accessible area 可进入区域，可接近区域
accessible canal 可通行通道
accessible cleanout 便于检修的清理口
accessible duct 通行地沟
accessible environment 可接近的环境，可进入的环境
accessible for 便于，易于
accessible hermetic compressor unit 半封闭式压缩机组
accessible-neutral 可达的中心线
accessible reactor building 可进入的反应堆厂房
accessible region 可进入的区域
accessible to sb. 为某人所能接近［到达、拿到手、看到、理解］的
accessible trench 可通行地沟，通行地沟
accession number 索取或报告号
access lock 通道门闸，出入闸门
access manhole 检查井
access mechanism 存取机构
access method 存取方法
access mode 存取方式，取数方式
access of air 空气入口，进气，空气通路，通气孔
access opening 检修孔，人孔
accessories for impact socket 风动套筒附件
accessories store 备品库
accessories 辅助设备，附件，附属件，附属设备
accessory apparatus 附属装置
accessory building 辅助建筑
accessory case 附件箱
accessory channel 附属通道
accessory constituent 副成分
accessory contract 附加合同，附加契约
accessory device 附属装置
accessory drive 附机传动
accessory ejecta 副喷出物
accessory equipment 辅助设备，附属设备
accessory gear 辅助齿轮箱
accessory hardware 附属硬件
accessory ingredient 副成分
accessory loss 附件损耗
accessory material 辅助材料，附件材料
accessory mineral 副矿物
accessory power supply 辅助电源
accessory risk 附加保险，附加险
accessory shaft 附件（传动）轴
accessory shock 副震
accessory structure 附属结构
accessory substance 副产品
accessory 附件，辅助设备，附属的，辅助的，附属设备【常用复数】
access panel 观测台，观察板，通道栅栏
access platform 通行平台
access plug 通道堵塞，通道门闸
access point 访问点
access port 入孔，检修孔，检修入孔，出入舱口
access railroad 专用铁路
access railway 铁路专用线
access ramp 进出坡道，入口坡道【指道路交叉处】
access road to the plant 入厂道路，入厂路
access road 进出厂区的道路，进出港道路，对外通道，入口道路，入厂道路，进路，通道，便道
access route 施工便道
access speed 取数速度
access system 接入系统
access time （信息）存取时间，取数时间
access to information 利用信息
access to market 进入市场
access to nacelle cabin 入机舱通道
access to sea 出入海洋，通往海洋，入海通道，出海权，入海口
access to the plant area 厂区出入口
access to tower 入塔通道
access to works site 进入工程现场
access to 有权使用
access trestle （通往码头的）栈桥
access tunnel 通行地道，人员进出地道，进出（交通）隧道
access 访问，接入，入口，通路，通道，通行，接近，存取，方法，人孔，检修孔
accidental action 偶然作用
accidental admission of air 空气偶然漏入
accidental admission of steam 蒸汽偶然漏入
accidental admission of vapour 蒸汽意外漏入
accidental air bubble 意外气泡
accidental alarm 事故报警
accidental combination for action effects 作用效应偶然组合
accidental consumption 事故消耗，意外消耗，偶然组合
accidental contamination 偶然污染
accidental cost 不可预见费用，意外费用，事故费用
accidental damage 意外损害，意外损坏

accidental depressurization 事故减压，事故泄压
accidental discharge 事故排放，事故泄放
accidental dosimetry 事故剂量学
accidental error 偶然误差，随机误差，补偿误差，偶差
accidental exposure 偶然曝光，偶然曝射，偶然辐射，事故辐照，事故照射
accidental failure 偶然失效，随机故障，偶然故障
accidental fluctuation 意外变动，意外波动，无规则变动
accidental force 不可抗力
accidental inclusion 外源包体
accidental insurance 意外保险
accidental irradiation 事故性辐照
accidental jamming 随机干扰
accidental lifting of the boom 悬臂的事故升起
accidental load 偶然负荷，偶然荷载
accidental loss 事故损失，意外损失
accidental lowering of the boom 臂架的事故降落
accidental maintenance 事故维修，事故检修
accidental operating conditions 事故运行工况
accidental overexposure 超剂量事故照射
accidental pollution 意外污染
accidental prevention 事故预防，安全措施
accidental radiation injury 事故辐射损伤
accidental radiation monitoring 事故辐射监测
accidental release 事故性排放，事故释放（装置）
accidental report 事故报告
accidental shutdown 故障停机，事故停堆
accidental spillage 偶然漏油
accidental xenolith 外源捕虏岩
accidental 偶然的，意外的，临时的，随机的，附带的
accident analysis branch of the USNRC 美国核管理委员会事故分析处
accident analysis 事故分析
accident anticipation 事故预测，事故预想
accident assumption 事故假设
accident beyond control 非人为事故，不可抗力的灾害，无法控制事故
accident brake 紧急制动器
accident compensation 意外事故赔偿
accident condition 事故工况，事故状态
accident consequence assessment 事故后果评定
accident cost 事故费用
accident damage 意外事故损坏，事故损坏
accident data recorder 事故数据记录仪
accident defect 故障缺陷，事故损坏
accident diagnosis and prognosis 事故诊断与预告
accident discharge 紧急排放口
accident dose 事故剂量
accident dosimetry 事故剂量学
accident due to poor quality 质量事故
accident error 偶然误差，事故误差
accident exposure 事故照射
accident failure 事故失效，事故损坏
accident frequency 事故频率
accident handling 事故处理
accident initiator 事故引发源
accident insurance 事故保险
accident investigation 事故调查
accident localization system 事故局部化系统
accident maintenance 事故维修
accident management 事故管理
accident mitigation 事故缓解
accident operating condition 事故运行工况，故障运行工况
accident prevention instruction 安全规程，事故预防指南
accident prevention program 事故预防计划
accident prevention 安全措施，事故预防，防止事故
accident probability 事故可能性
accident prone 易出事故的
accident protection 事故防护
accident rate 事故率
accident report 事故报告
accident risk assessment 事故风险评估，事故风险评价
accident sequence 事故序列
accident shield 事故屏蔽，事故屏障
accident signaling system 事故报警系统，事故信号系统
accident signal 事故信号
accident source term 事故源项
accident statistics 事故统计
accident survey 事故调查
accident victim 事故受害者
accident 事故，失事，意外，故障，损坏，偶然性，偶然事件
AC circuit breaker 交流断路器
AC circuit 交流电路
acclimation 风土驯化，环境适应性，气候适应，气候驯化
acclimatization 环境适应性，气候适应，气候驯化
acclivity 上斜，向上的斜坡，斜坡
accommodate 照顾，通融，接受，提供，使适应，适应，容纳，调节，和解
accommodation acceptor 贷款承兑人
accommodation allowance 膳宿津贴
accommodation bill of lading 通融提单
accommodation bill 通融票据
accommodation coefficient 适应系数，调节系数
accommodation indorsement 通融背书
accommodation ladder 扒梯，舷梯
accommodation loan 无担保融资
accommodation paper 融资票据
accommodation ring 专用环形道路
accommodation road 专用公路
accommodation vessel 生活支持船
accommodation 居住设施，住宿安排【复数】，生活设施，适应，配合，调节，供应，膳宿，住处，住宿，住房，和解，容纳，贷款
accommodator 调节器，贷款人，调解人
AC commutator motor 交流换向器式电动机
AC commutator type exciter 交流换向器式励磁机
accomodation bill 融通票据，空头票据

accompanied by shipping documents 随附装船单据
accompanied by 附有，偕同
accompanying diagram 附图
accompanying person 随行人员
accompanying sound trap 伴音陷波器
accompanying 陪伴的，附随的
accompany 陪同，伴随
accomplished bill of lading 已提货提单
AC component 交流分量
AC contactor 交流接触器
accord and satisfaction 抵偿
accordant connection 匹配连接
accordant drainage 协调水系
accordant junction 协和汇流，平齐汇流
accordant 一致的，调和的
accordingly 从而，相应地
according to the customs 根据惯例
according to the international practice 按照国际惯例
according to the normal practice 依照常规，根据惯例
according to the past practice 根据以往的惯例
according to usual practice 按照惯例
according to 根据，按照，依照，与……相一致
accordion cable 折状电缆
accordion conveyor 伸缩式输送机，折叠式输送机
accordion door 折叠式门，折叠门
accordion partition 折隔扇，折隔屏，折叠式屏风
accordion 折状插孔，折式地图，Z形插孔
accord to 给予
accord with contract 与合同符合
accord with 符合，一致
accountability system 问责制，责任追究制，衡算系统
accountability 可说明性，可计量性，责任，义务，负有责任，负有义务
accountable person 负责人
accountancy verification goal 衡算核实指标
accountant general 会计主任
accountant's handbook 会计人员手册
accountant's office 会计室
accountant 会计，会计师
account balance 账户结余，账户余额
account bill 账单
account books 账簿
account day 结账日
accountee 开证人，开证账户主
account for 40% of the total 占总量40%
accounting changes 会计变更
accounting cost control 会计成本管理
accounting cost 会计成本
accounting device 计算装置
accounting firm 会计师事务所
accounting law 会计法
accounting life 计算寿命【指核电厂】
accounting package 会计程序包
accounting policy 会计政策
accounting procedure 计量程序【后处理】，会计程序
accounting statement 会计报表
accounting system 统计报表系统
accounting unit 核算单位
accounting value 账面值
accounting year 会计年度
accounting 会计学，会计，计算
account number 账号
account of payment 支出账户
account payable ledger account 应付款分类账，应付款分类明细账账户
account payable 应付账款
account payee 收款人账户
account receivable 应收账款
accounts for settlement of claim and debt 债权债务结算账户
account system 账目系统，报表系统
account valuation 估价，估计，造价预算
account year 会计年度
account 计算，结算，算账，说明，账，账户，账目
accreditation 委任，任命，认可，鉴定为合格
accredited degree 合格程度
accredited party 授信买方【即信用证申请人】，被授权方
accredited representative 授权代理人
accredited 被认可的【组织】
accredit party 信用证受益人
accretion of levels 淤高
accretion of river bottom 河底淤高
accretion 始长，炉瘤，积冰，添加，淤高，淤积，增长作用
accrual basis accounting 以应收应付制为基础的核算
accruals 增值
accrued depreciation 增加折旧，应计折旧
accrued expenditure 应计开支
accrued expense 应计费用，增加费用
accrued income 应计收益，应计收入，应收未收收益，应付收益
accrued interest payable 应付利息，应计未付利息
accrued interest receivable 应收利息，应计未收利息
accrued interest 应计利息，应付利息，待付利息，产生的利息
accrued liability 应计负债，流动负债
accrued payroll 应付工资，应计工资
accrued profit 应计利润
accrued revenue 累计收入，应计受益
accrued tax 应计税金
accrued 增加的，权责已发生的，产生的
accrue （通过自然增长）产生、获得，（使钱款、债务）积累
accruing amount 累计金额
accumulated deficiency 累积差值
accumulated discrepancy 累积（误）差，累积不符值
accumulated dose 累积剂量，总剂量
accumulate deformation 累积变形
accumulated error 累积误差

accumulated excess 累积超量
accumulated filth 积垢
accumulated heat 蓄热
accumulated horizon 堆积层
accumulated island 堆积岛
accumulated pressure 蓄积压力
accumulated snowdrift area 吹雪堆积区
accumulated temperature 累积温度，积温
accumulated time 累积时间
accumulated value 累计值
accumulate 累积
accumulating capacity 储备能力
accumulating counter 累加计数器
accumulating diagram of water demand 累积需水量曲线（图）
accumulating register 累加寄存器
accumulating reproducer 累加复制器
accumulating tank 贮存箱
accumulating 累积，累加
accumulational platform 堆积台地
accumulational relief 堆积地形
accumulational terrace 堆积阶地
accumulation area ratio 累积面积比
accumulation area 补给区，堆积区
accumulation curve 累加曲线，累积曲线
accumulation cycle 累加循环
accumulation diagram of water demand 用水累积图
accumulation dose 堆积剂量
accumulation factor 累积因子
accumulation fund 公积金
accumulation horizon 堆积层，积聚层
accumulation line 累积线
accumulation maximum 累积最大
accumulation of cold 蓄冷
accumulation of energy 能量聚积，蓄能
accumulation of fatigue damage 累积疲劳损伤
accumulation of funds 资金积累
accumulation of gas 气体积聚，蓄气
accumulation of heat 蓄热，热量蓄积
accumulation of mud 污垢增多
accumulation of stress 应力集中
accumulation of water 积水
accumulation pattern 堆积模式
accumulation plain 堆积平原
accumulation rate 堆积速率
accumulation region 聚积区，补给区
accumulation terrace 洪积阶地
accumulation volume 堆积量
accumulation 累积，堆积，积聚，蓄压，聚集
accumulative carry 累加进位
accumulative commission 累计佣金
accumulative dose 累积剂，累积剂量
accumulative error 累积误差，累计误差
accumulative fission yield 累积裂变产额
accumulative formation 堆积层
accumulative margin state frequency 累积裕度状态频率
accumulative outage state frequency 累积停运状态频率
accumulative process 累积过程
accumulative total 累计，累加总和
accumulative 累积的，聚集的，累加的，累计的
accumulator battery 蓄电池组
accumulator blowdown 蓄热器排污
accumulator box 蓄电池容器，蓄电池箱
accumulator capacity 蓄电池容量
accumulator car 电瓶车，蓄电池汽车
accumulator cell 蓄电池
accumulator charger 蓄电池充电器
accumulator feeding 蓄电池馈电
accumulator gas cushion 蓄压箱气垫
accumulator grid 蓄电池栅板
accumulator injection 安全注射箱注入，蓄压箱注入
accumulator metal 蓄电池极板合金
accumulator plant 蓄热器室，蓄电池室，充电室
accumulator plate 蓄电板，蓄电池极板
accumulator recharging pump 蓄压箱再充压泵
accumulator rectifier 蓄电池整流器
accumulator register 累积寄存器，累加寄存器
accumulator room 蓄电池室
accumulator shift instruction 累积器移位指令
accumulator stage 累加器级
accumulator steam turbine 附蓄热器的汽轮机
accumulator switchboard 蓄电池配电盘
accumulator tank 蓄热器罐，蓄电池槽，集油罐，蓄压水箱，蓄压箱【液体】
accumulator transfer instruction 累积器转移指令，累积器传送指令
accumulator turbine 蓄能式透平【尖峰或应急用】
accumulator 安全注射箱【压水堆】，累加器，蓄压器，蓄热器，蓄电池，储压罐，蓄能器，存储器，安注箱，蓄压箱
accuracy class index 准确度等级指数
accuracy class 精度等级，精确度等级，准确度等级
accuracy control system 精度控制系统
accuracy control 精确控制
accuracy degree 准确度
accuracy for vertical control 高程控制精度
accuracy horizontal control 平面控制精度
accuracy in calibration 标定精度
accuracy in computation 计算精确度
accuracy in instrument 仪表的精确度
accuracy in measurement 测量精度
accuracy of adjustment 调整精度，调节精度
accuracy of alignment 对中精度【转子】
accuracy of calibration 标定精度
accuracy of instrument 仪表精度
accuracy of manufacture 制造精度
accuracy of measurement 测量精确度，测量准确度，测量精度
accuracy of observation 观测精度
accuracy of positioning 定位精度
accuracy of reading 读数精确度
accuracy of scale 刻度精确度
accuracy rating 额定精确度，精确度
accuracy test 精确度检查，精确度试验
accuracy to size 尺寸准确度
accuracy 精度，准确度，精确度，精密度

accurate adjustment 精密调整，精密校准，精调
accurate filter 精密过滤器
accurate grinding 精磨
accurate reading 准确读数
AC current transformer 交流电流互感器
accuse diffraction 声衍射
ACCW(alternating-current continuous wave) 交流连续波
ACD(activation information of direction protection) 方向保护启动信息
AC/DC(alternating current/direct current) 交直流
AC/DC ammeter 交直流两用电流表
AC/DC cabinet 交直流配电箱
AC/DC converter 交直流转换器
AC/DC network interconnection 交直流联网
AC/DC power transmission 交直流混合输电
AC/DC relay 交直流继电器
AC directional overcurrent relay 交流方向过流继电器
AC distribution system 交流配电系统
ACDS(acoustic crack detection system) 声裂纹检测系统
AC earth relay 交流接地继电器
ACE(automatic checkout equipment) 自动检查装置
ACE(automatic computing equipment) 自动计算装置
ACE(automatic control device) 自动控制装置
ACE(Aylesbury collaborative experiment) 埃尔兹伯里（风荷载）合作实验
AC electromagnetic relay 交流电磁继电器
AC electromagnet 交流电磁铁
AC equivalent capacitance 交流等效电容
AC equivalent inductance 交流等效电感
AC equivalent resistance 交流等效电阻
AC erasing head 交流消磁头
AC erasure 交流消除，交流消除记录
ACES(automatic checkout and evaluation system) 自动检测和估算系统
acescency 微酸味
acescent 微酸的，发酸的
acetaldehyde 乙醛
acetal plastics 乙缩醛塑料
acetal resin 缩醛树脂
acetal 乙缩醛
acetate 醋酸盐
acetic acid 醋酸，乙酸
acetified solution 再生溶液
acetifier column 再生器，再生塔
acetone oxime 丙酮肟【加到除氧器中】
acetone 丙酮
acetoxime 丙酮肟【加到除氧器中】
acetyl cellulose 醋酸纤维素，乙酰纤维素
acetylene apparatus 乙炔气焊设备
acetylene burner 乙炔灯
acetylene cutter 乙炔烧割器
acetylene cutting 乙炔切割
acetylene cylinder 乙炔瓶
acetylene generator 乙炔发生器
acetylene generating station 乙炔发生站，乙炔站
acetylene pipe 乙炔管
acetylene producer 乙炔发生器
acetylene station 乙炔站
acetylene welder 气焊机
acetylene welding 乙炔焊，乙炔焊接，气焊
acetylene 乙炔，电石气
acetylenic corrosion inhibitor 乙炔缓蚀剂
AC excitation 交流励磁
AC exciter 交流励磁机
ACF(average calibration factor) 平均校准因子
ACFG(automatic continuous function generator) 自动连续函数发生器
AC filter capacitor 交流滤波电容器
AC filter resistor 交流滤波电阻器
AC filter 交流滤波器
AC generator 交流发电机
AC grounding grid 交流地网
AC harmonic filters in converter station 换流站交流滤波装置
Acheson graphite 人造石墨【阿基逊石墨】
achievable burn-up 可达到的燃耗
achievement of quality 达到的质量
achievement 成果，成绩，功绩
achieve 取得（成就），获得，完成，到达
achromatic lens 清色差透镜
achromatic map 单色图
achromatic sheet 单色图
achromat 消色差透镜
ACS(American Chemical Society) 美国化学学会
ACIA(asynchronous communication interface adapter) 异步通信接口转接器
ACI(adjacent channel interference) 相邻信道干扰
ACI(allocated configuration index) 分配位置索引
ACI(American Certification Institute) 美国认证协会
ACI(American Concrete Institute) 美国混凝土学会
acicular bainite 针状贝氏体
acicular cast iron 针状铸铁
acicular constituent 针状组织
acicular grey cast iron 针状灰铸铁
acicular martensite 针状马氏体
acicular powder 针状粉末
acicular structure 针状结构，针状组织
acicular 针状的
acid 酸，酸性的
acid accumulator 酸性蓄电池
acid-alkali regeneration 酸碱再生
acid analyser 酸分析器
acid anhydride 酸酐
acidate 酸化
acid attack 酸蚀
acid-base balance 酸碱平衡
acid-base equilibrium 酸碱平衡
acid-base indicator 酸碱指示剂
acid-base neutralization 酸碱中和
acid-base reaction 酸碱反应
acid-base titration 酸碱滴定
acid-base treatment 酸碱处理
acid-base 酸碱
acid bath 酸洗，酸浴

acid brick lining 耐酸砖衬里
acid brick 酸性火砖，耐酸砖，酸性砖
acid brittleness 酸性脆化，氢脆【酸洗时吸氢所致】
acid carbonate 酸式磷酸盐
acid chimney 排酸筒
acid chrome black 酸性铬黑
acid clay 酸性黏土
acid cleaning 酸洗
acid cleaning for boiler propers 锅炉本体酸洗
acid cleaning system 酸洗系统
acid-cleaning water 酸洗水
acid concentration 酸浓度
acid constituent 酸性组分
acid-consumingion 耗酸离子
acid consumption 耗酸量，酸耗
acid-containing soot 含酸煤烟
acid converter steel 酸性转炉钢
acid corrosion 酸腐蚀，酸性腐蚀，酸蚀
acid covering 酸性药皮
acid cresol red 酸性甲酚红
acid decomposition 酸性分解
acid degree 酸度
acid dew point corrosion 酸露点腐蚀
acid dew point 酸露点
acid digestion 酸解
acid dipping 酸浸除锈
acid dip 酸洗，酸浸
acid dosing 加酸
acid effluent 酸性流出水，酸性排水，酸性废水
acid electric steel 酸性电炉钢
acid electrode 酸性焊条
acid embattlement 吸氢致脆
acid etching 浸蚀，酸腐蚀
acid-fast 耐酸的
acid feed controller 加酸控制器
acid feeder 加酸器
acid fog 酸雾
acid fume 酸烟
acid gas 酸气，酸烟，酸性气体
acid humus 酸性腐殖质
acid hydrolysis 酸解，加酸水解，酸性水解
acidic adjustment 酸性调节
acidic content 酸性含量
acidic group 酸性基
acidic lava 酸性熔岩
acidic material 酸性材料
acidic radical 酸根
acidic residue 酸性残渣
acidic rock 酸性岩
acidic wastewater 酸性废水
acidic water 酸性水
acidic 酸性的，酸的
acidification 酸化
acidimeter 酸比重计，酸度计，酸量计
acidimetry 酸量滴定法
acid inhibitor 酸腐蚀抑制剂
acid ion 酸离子
acidity and alkalinity test 酸碱度试验
acidity 酸度，酸性
acidity of residue 残油酸度
acid lining 酸性衬料，酸性炉衬，酸性衬里
acid medium 酸性介质
acid meta cresol purple 酸性间甲酚紫
acid mine drainage 酸性矿排水
acid mist 酸雾
acid moor 酸性沼泽
acid neutralizing unit 酸中和装置
acid number 酸值
acidometer pH计，酸液比重计，酸度计，酸定量器，酸液相对密度计
acid open hearth process 酸性平炉法
acid open hearth steel 酸性平炉钢
acid or alkali poisoning 酸或碱中毒
acid oxide 酸性氧化物
acid pickling 酸洗
acid pollution 酸污染
acid precipitation 酸雨，酸性降水，酸沉淀
acid pre-pickling 用酸预洗
acid-proof alloy 耐酸合金
acid-proof brick 耐酸砖
acid-proof cement 耐酸水泥
acid-proof ceramic pipe 耐酸陶瓷管
acid-proof ceramic tile floor 耐酸瓷砖地面
acid-proof coating 耐酸涂层
acid-proof diaphragm valve 耐酸隔膜阀
acid-proof mastic 耐酸胶泥
acid-proof material 耐酸材料
acid-proof paint 耐酸漆，防酸漆，耐酸涂料
acid-proof pump 耐酸泵
acid-proof varnish 耐酸清漆
acid-proof 耐酸的，耐酸性，抗酸性
acid-protection coating 耐酸涂层
acid pump 酸泵
acid purification system 酸净化系统
acid rain 酸雨
acid reaction 酸性反应
acid recovery 酸回收
acid refractory 酸性耐火材料
acid regression stage 酸性消退阶段
acid resistance 耐酸性
acid-resistant brick 耐酸砖
acid-resistant cement 耐酸水泥
acid-resistant ceramic tile 耐酸瓷砖
acid-resistant concrete 耐酸混凝土
acid-resistant enamel 耐酸搪瓷
acid-resistant lacquer 耐酸漆
acid-resistant material 耐酸材料
acid-resistant mortar 耐酸砂浆
acid-resistant motor 耐酸电动机
acid-resistant paint 耐酸漆
acid-resistant refractory castable 耐酸耐火浇注料
acid-resistant sheet maternal 耐酸板材
acid-resistant stainless steel 耐酸不锈钢
acid-resistant steel 耐酸钢
acid-resistant 耐酸的，耐酸，抗酸
acid resisting cement 耐酸水泥
acid resistant stainless steel 耐酸不锈钢
acid resisting stoneware 耐酸陶器
acid resisting alloy 耐酸合金
acid-resisting cement 耐酸水泥
acid-resisting concrete 耐酸混凝土

acid resisting lining 耐酸衬砌
acid-resisting pump 耐酸泵
acid resisting 耐酸的
acid rinse 酸漂洗
acid rock 酸性岩
acid salt 酸式盐
acid settler 酸沉降池
acid slag 酸性炉渣，酸性渣
acid sludge 酸性泥渣
acid smut 酸性煤尘，酸性煤烟，酸洗残渣，含硫排烟
acid soil 酸性土
acid-soluble aluminum 酸溶铝
acid solution 酸溶液
acid soot 酸性烟灰
acid spray-proof type battery 防酸式蓄电池
acid stain 酸斑，酸性沾污
acid steel 耐酸钢
acid storage battery 酸性蓄电池
acid tank 酸槽，酸洗槽
acid test 酸性试验
acid transfer vehicle 酸运输槽车
acid transformer 油已氧化的变压器，绝缘油酸化变压器
acid treatment plant 酸洗站
acid treatment 酸处理，加酸处理，酸性处理
acidulate 酸度，酸性
acid value 酸值
acid vapor 酸性蒸汽
acid wash active carbon 酸洗活性炭
acid washing liquor 酸洗液
acid washing test 酸洗试验
acid washing 酸洗
acid waste liquid 酸性废液
acid waste sludge 酸性废污泥
acid wastewater 酸性废水
acid waste 酸性废料，酸性废物，酸性废液
acid water 酸性水
acid wood 酸性木材
acierate 表面渗碳
ac impeditive spectroscopy 交流阻抗谱
AC indicator 交流指示器
AC induced polarization 交流激发极化法
acinous 细粒状的
a circular letter 通告信、通知书
ACK(acknowledge character) (消息)收到符号，肯定符号
ACK＝acknowledge 肯定，答应，告知收到
Ackers-Keller cycle 闭式燃气轮机循环
acknowledge circuit 警告电路，认可电路
acknowledge cycle 确认时间，应答周期
acknowledge interrupt 中断响应
acknowledgement of receipt 回执，收据
acknowledgement signal unit 证实信号单元
acknowledgement 回执，确认【信号，警报，收到】，回单，收条，承认，认可，收到通知，感谢
acknowledger 认收开关
acknowledge signal 确认信号，肯定信号，认可信号
acknowledge 肯定，答应，确认，承认，告知收到
acknowledging contactor 认收开关，认收接触器
acknowledging switch 认收开关
acknowledgment of receipt 接收认可，确认收到
ACL(access control level) 接入控制等级
a clear aim 明确的目标
a clear-cut stand 明确的立场
AC line fitter 交流线路滤波器
aclinic line 无倾线
AC load line 交流负载线
AC load switching 交流负载转线，交流负载切换
AC lube oil pump 交流润滑油泵
acme 最高点
AC motor 交流电动机，交流电机，交流马达
A/C name 账户名
AC network calculator 交流计算台
AC network distribution 交流网络配电
AC network 交流网络
AC noise immunity 交流抗干扰度
ACOE(automatic checkout equipment) 自动检测装置
ACOM(Automatic Coding Machine) 自动编码器，自动编码系统
a complete set of design 成套设计
a concessionary contract 特许契约
ACORE(automatic checkout and recording equipment) 自动检测记录设备，自动检测记录装置
acorn 整流罩
acoumeter 测听计
acousimeter 测听计
acoustic absorbent 吸声材料，吸音材料
acoustic absorption coefficient 吸声系数，声吸收率
acoustic absorption 声吸收，吸音
acoustic absorptivity 吸声率，吸声系数，声吸收系数
acoustic admittance 声导纳
acoustical board 吸声板，吸音板
acoustical damper 消声器
acoustical detector 声波探测器
acoustical door 隔音门
acoustical insulation 隔声
acoustical material 吸音材料
acoustical panel ceiling 吸音板吊顶
acoustical reactance 声抗
acoustical signaling device 音响信号装置
acoustical tile ceiling 吸声砖吊顶
acoustical 传音的，声(学)的，声音的，听觉的，音响的
acoustic amplitude reflection coefficient 声振幅反射系数
acoustic attenuation constant 声衰减常数
acoustic baffle 吸声板
acoustic barrier 吸声层
acoustic bearing 声向位，声相位
acoustic behavior 声学性能
acoustic board 隔音板，吸声板
acoustic burglar alarm 声防盗警报器
acoustic burner 声波燃烧器
acoustic ceiling board 吸声天花板
acoustic-celotex board 隔音板，吸声纤维板
acoustic characteristic impedance of a medium 媒

质的声特性阻抗
acoustic cloud 声波反射云
acoustic compliance 声顺，声容抗
acoustic conductance 声导
acoustic conductivity 传声性，声导率
acoustic consecution 隔声构造
acoustic coupler 声耦合器
acoustic coupling 声耦合
acoustic crack detection system 声裂纹检测系统
acoustic current meter 声学海流计
acoustic damper 消声器
acoustic damping effect 声阻尼效应
acoustic damping 声阻尼
acoustic data processor 声数据处理机
acoustic dazzle 声干扰，异常聚声
acoustic delayline storage 声延迟线存储器
acoustic delay line 声延迟线
acoustic density 声能密度
acoustic depth finder 回声探测仪，回声测深仪
acoustic depth sounding 回声测深
acoustic design 声学设计，音质设计
acoustic detector 声波检测器
acoustic determination of pressure 声学法测压
acoustic diffuser 声扩散器，声扩张器
acoustic distortion 声畸变
acoustic disturbance 声干扰
acoustic emission counter 声发射计数器
acoustic emission detection system 声发射检测系统
acoustic emission detector 声发射探测器
acoustic emission inspection 声发射检测
acoustic emission leak locator 声发射漏泄定位器
acoustic emission monitoring 声波发射监测法，声发射监测，声发射监控
acoustic emission multi-parameter analyzer 多参数声发射分析仪
acoustic emission pulser 声发射脉冲发生仪
acoustic emission rate 声发射速率
acoustic emission testing 声发射检测
acoustic emission 声发射，声发射法
acoustic environment 声环境
acoustic fatigue 由噪声导致的疲劳
acoustic feedback 声反馈
acoustic felt 吸声毡
acoustic fiber board 吸声纤维板
acoustic field 声场
acoustic filter 滤声器，声过滤器
acoustic frequency branch 声频分支
acoustic frequency 声频率
acoustic guidance 声学制导
acoustic hologram 声全息图
acoustic holography by electron-beam scanning 电子束扫描声全息术
acoustic holography 声学全息
acoustic homing guidance 声学自动寻的制导
acoustician detector 声探伤仪
acoustic image aberration 声像差
acoustic image converter 声像转换器
acoustic image 声像
acoustic imaging by Bragg diffraction 布拉格射声成像
acoustic-impact technique 音响敲击法
acoustic impedance 声阻抗
acoustic industrial stethoscope 声波工业听诊器
acoustic insertion loss 声学插入损失
acoustic insulating material 隔声材料
acoustic insulation 隔音，隔声
acoustic intensity 声强度
acoustic interferometer 声干涉仪
acoustic isolation 隔声
acoustic level 声级
acoustic logging 声波测井
acoustic mass 声质量
acoustic material 隔声材料，吸声材料，声学材料，隔音材料
acoustic maximum 最大音响
acoustic measurement 声响测量，声学量度
acoustic memory 声存储器
acoustic meter 测声计，声级计
acoustic mismatch 声失配
acoustic near field 近声场
acoustic noise measurement techniques 噪声测量技术
acoustic noise 噪声，声干扰，声学噪音
acoustic-optic scanner 声光扫描器
acoustic oscillation 声振动
acoustic plaster 吸音灰膏，吸声灰膏
acoustic pollution 噪声污染，声响污染，声污染
acoustic pressure level 声压级
acoustic pressure 声压
acoustic probe 声探头
acoustic quartz delay line 声石英延迟线
acoustic quadripole 声四极子
acoustic radar 声雷达
acoustic radiation pressure 声辐射压力
acoustic reactance ratio 声抗比
acoustic reactance 声抗
acoustic reduction factor 隔声系数，声降系数
acoustic reference wind speed 声的基准风速，声参考风速
acoustic reference 声参考
acoustic resistance 声阻
acoustic room constant 声学房间常数
acoustic screen 隔声板，声屏蔽
acoustic signal 音响信号
acoustic sounding 回声测深，声波测深法
acoustic susceptance 声呐
acoustic spectrometer 声频谱仪
acoustic speed 声速，音速
acoustic stiffness 声刚度，声劲
acoustic strain gauge 声指示【安全壳监视】，声响传感器，声波应变计，发声式应变计
acoustics 声学，音质
acoustic technology 吸声技术
acoustic testing 声学测试
acoustic thermometer 声学温度计
acoustic tile 吸声砖
acoustic transducer 声能转换器，声波传感器
acoustic transmission factor 声音传递因数
acoustic treatment 声学处理，音响处理，防声处理
acoustic type strain gage 发声式应变计

acoustic type strain gauge 发声式应变计
acoustic underwater survey equipment 水下测声设备
acoustic unit 声学单位
acoustic vault 吸声穹窿
acoustic velocity logger 声速测井仪
acoustic velocity log 声波测井
acoustic velocity 声速，音速
acoustic vibration 声波振动，声振动，声振荡
acoustic volume velocity 声体积速度
acoustic wave 声波
acoustic 声学的，音响的，有声的
acoustimeter 测声计，声强度测量器
acoustoelastic effect 声弹性效应
acoustoelectric domain 声电畴，电声学的领域
acoustoelectric effect 声电效应
acoustoelectric field 声电场
acoustoelectric 电声的
acoustoelectronics 声电子学
acoustolith title 吸音贴砖，吸声贴砖
acoustometer 测声计
acousto-optically tuned laser 声光调谐激光
acousto-optic cavity 声光角，声光腔
acousto-optic deflector 声光偏转器
acousto-optic medium 声光媒质
acousto-optic mode locker frequency doubter 声光锁模倍频器
ACO 给水加热器疏水回收系统【核电站系统代码】
ACP(asbestos cement pipe) 石棉水泥管
ACP(auxiliary control panel) 辅助控制仪表板，辅助控制仪表盘
AC pilot relaying 交流控制继电保护
AC potentiometer 交流电位计
AC power loss 交流功率损耗
AC power source 交流电源
AC power system 交流电力系统
AC power transmission system 交流输电系统
AC power transmission 交流输电
ACPR(annular core pulse reactor) 环状堆芯脉冲堆
acquaint sb. with 了解，认识
acquaint 使知道，了解，认识
acquiesce in 默认
acquired full legal force 具有法律效力
acquired right 既得权利
acquire 获得，取得
acquiring enterprise 引进方企业，购方企业
acquisition commission 资产购置手续费
acquisition cost 购置成本
acquisition mode 搜索方式
acquisition of assets 购置资产
acquisition of information 信号获得
acquisition of technology 获得技术，技术引进
acquisition probability 目标探索概率
acquisition rate 搜索率
acquisition time 采集时间
acquisition 获得，采集，购置，购置物，收购，征用
ACR(accumulator register) 累加寄存器
ACRC(additional cycle redundancy check) 附加的循环冗余码校验
ACRDS(automatic control rod drive system) 自动控制棒驱动系统
acreage 英亩数，面积
ACRE(automatic check and readiness equipment) 自动检查及待命装置
AC reclosing relay 交流重合闸继电器
AC rectifier charging 交流整流器充电
AC resistance 交流电阻
acre 英亩
acrolein resin 丙烯醛树脂
acrometer 油类比重计
acromotor 风力发动机
acronym 首字母缩略词
across bulkhead 横向隔墙
across comer 对角
across cutting 横向切割
across-effectiveness clause 交互生效条款
across flow 横向流，交叉流，错流
across grain 横向木纹
across-the-grain 横断面，横纹
across-the-line starter 全压（直接）启动器
across-the-line starting 直接启动，全压启动
across-the-line 跨接线，并行线路
across 穿过
acrotorque 最大扭矩，最大扭力
ACRS(Advisory Committee on Reactor Safeguards) 反应堆保障咨询委员会【美国】
acrylamide 丙烯酰胺
acrylic acid 丙烯酸
acrylic-based cement 丙烯基黏结剂
acrylic column 丙烯酸柱
acrylic paint 丙烯酸漆
acrylic panel 丙烯（塑料）板
acrylic plastics 丙烯酸塑料
acrylic polymers 丙烯聚合物
acrylic resin 丙烯酸树脂
acrylonitrile 丙烯腈
ACS(aluminum clad steel) 铝包钢绞线
ACS(above-core structure) 堆芯上部构件
ACS(auxiliary coolant system) 辅助冷却系统
ACSE(association conrtol service element) 关联控制服务元素
AC series motor 交流串励电动机
AC signaling 交流信令
ACSI(abstract communication service interfaces) 抽象通信服务接口
ACS-O(access opening) 检修孔，人孔
ACSR(aluminium conductor steel-reinforced) 钢芯铝绞线
ACSR/TW(aluminum conductor steel reinforced/trapezoid wire) T型钢芯铝绞线
ACSS(aluminum conductor steel supported) 钢芯软铝绞线【内含碳纤维，高温耐热】
ACSS(analog computer subsystem) 模拟计算机子系统
AC static switch 交流静态开关
ACT(auxiliary current transformer) 辅助电流互感器
ACTE(actuate) 致动，开动，使动作
AC testing equipment with triple power-frequency

三倍频试验装置
ACTG(actuating) 致动，启动，开动，驱使，激励，作用
actification 再生作用，复活作用
AC time overcurrent relay 交流时限过流继电器
acting characteristic 动作特性
acting force 作用力
acting head 作用水头，作用压头，有效水头，有效压头
acting load 作用荷载
acting manager 代经理
acting stress 作用应力
acting surface 工作面，作用面
acting time 执行时间，作用时间
acting 代理的，起作用的
actinic glass 光化玻璃，闪光玻璃，吸热玻璃
actinide 锕系元素
actinium 锕
actinography 日射计，光能测定器，辐射仪，日光强度自动记录器
actinometer （自记）曝光计，曝光表
actinometry 日射测量学
action and reaction 作用力和反作用力
action center 作用中心
action command 操作指令
action current 工作电流，作用电流
action cutout 机械断流器
action element 执行元件
action founded in contract 基于合同诉讼
action founded in tort 基于侵权的诉讼
action level 应急行动水平，应急响应的水平
action of blast 鼓风效应
action of debt 债务转让
action of gravity 重力作用
action of heat 热效应
action of law 法律诉讼
action of pile group 群桩作用
action of points 尖端效应，尖端作用
action of rust 锈蚀作用
action of surge chamber 调压井功能
action of tort 侵权行为诉讼
action-oriented negotiation 着眼于行动的谈判
action-oriented research 应用性研究
action plan 行动计划
action radius 作用半径，有效半径
action roller 活动滚轮，动辊
action routine 运行规程，操作程序
action spot 作用光点，作用点
actions 举措
action time 作用时间
action turbine 冲动式汽轮机
action wheel 主动轮
action 作用（量），影响，效应，操作，行为，行动，动作
activate button 启动按钮，起动按钮
activated absorption 活性吸附
activated agent 活化剂
activated alumina 活性氧化铝
activated alum 活性化明矾
activated carbon adsorption bed 活性炭吸附床
activated carbon filter process 活性炭过滤法
activated carbon moving bed 活性炭移动床
activated carbon pore volume 活性炭孔隙容积
activated carbon reactivation 活性炭复活作用
activated carbon 活性炭
activated charcoal absorber 活性炭吸附器
activated charcoal adsorption bed 活性炭吸附床
activated charcoal column 活性炭柱
activated charcoal delay bed system 活性炭滞留床系统
activated charcoal filler bank 活性炭过滤器组，活性炭过滤器组件
activated charcoal filter bed 活性炭过滤床
activated charcoal filter package 活性炭过滤器组，活性炭过滤器组件
activated charcoal filter 活性炭过滤器
activated charcoal glass fiber fitter 活性炭玻璃纤维过滤器
activated charcoal iodine filter 活性炭碘过滤器
activated charcoal 活性炭
activated clay 活性白土，活性黏土
activated cocoanut charcoal 椰子壳活性炭
activated corrosion product 活化腐蚀产物
activated filter film 活性化过滤膜
activated impurity 放射性杂质，活化杂质
activated magnesia 活性氧化镁
activated montmoril lonite 活性蒙脱土，活性微晶高岭土
activated nucleus 活化核
activated primary sodium 活化的一回路钠
activated process 活化过程
activated scale 活化水垢
activated sewage 活性污水
activated silica 活性硅，活性硅土，活性硅胶，活性化的二氧化硅
activated sludge aeration 活性污泥曝气作用
activated sludge bulking 活性污泥膨胀
activated sludge effluent 经活性污泥法处理后的出水，活性污泥废液
activated sludge kinetics 活性污泥动力学
activated sludge loading 活性污泥负荷
activated sludge plant 活性污泥厂，活化污泥设施，活性污泥车间
activated sludge process 活化污泥法
activated sludge settling tank 活性污泥沉淀池
activated sludge treatment 活性污泥处理
activated sludge 活性污泥
activated solids 活性固体
activated water 活化水，放射性水
activated 活性的
activate key 启动键
activate 使活化，活性化，启动，刺激，激活，触发，使活动
activated carbon method 活性碳法
activated silica 活性硅胶
activating agent 活化剂
activating radiation 活化辐射，活化辐照
activating signal 启动信号，触发信号
activation analysis 活化分析
activation calculation 活化计算
activation cross-section 活化截面
activation detector 放射性检测仪，活化检测仪

activation energy 激活能量，活化能
activation fluence 活化注量
activation heat 活化热
activation information of direction protection 方向保护启动信息
activation over potential 活化过电位
activation polarization 活化极化
activation power of wind turbines 风力机临界功率
activation power 临界功率
activation product 活化产物
activation record 活动记录，现役记录
activation rotational speed 临界转速
activation test 活化试验
activation 活动，赋活，激活，活化（作用），活性，激化，激励，启用
activator key 启动键
activator 活化剂，活化器，活性剂，激活剂，催化剂，有效面积
active acidity 有效酸度
active activation analysis 有源活化分析
active agent 活化剂
active alkali 活性碱
active analysis 活动分析
active antenna 有源天线
active appendage 主动附件
active arch 有效拱，主动拱
active area of a solar cell 单体太阳电池的有效光照面积
active area 有效面积，工作面积
active arm 通航汊道，活汊道
active assay 有源分析，活性分析
active assets 活动财产
active augmentation methods 有（外动力）源的强化换热方法
active balance return loss 有源平衡回程损耗
active balance 等效平衡
active biological surface 活性生物表面
active block 有源组件
active braking distance 实际制动距离
active braking time 实际制动时间
active campaign 热试验运行阶段【核电，后处理】
active carbon filter 活性炭过滤器
active carbon mechanical filter 活性炭机械过滤器
active carbon 活性炭
active circuit elements 主动电路元件
active circuit 有源电路
active clay 活性黏土
active (coal) storage 周转煤堆，周转储量
active coil 有效线圈
active combustion 强烈燃烧
active component of current 电流有功分量
active component 活性部分，活性部件，能动部件，有功分量，有源部件，活性组分
active condition 活性状态
active conductor 有效导体
active constituent 有功部分
active constraint 有效约束
active control technology 主动控制技术
active control 主动控制
active coolant gas 放射性气体冷却剂
active cooling surface 有效冷却表面积
active cooling 主动式冷却
active core height 堆芯活性高度，堆芯高度
active core 活性区，活性堆芯【反应堆】
active corrective maintenance time 实际矫正性维修时间
active cross-section 水流断面，有效断面
active-current compensator 有功电流补偿器
active-current 有效电流，有功电流
active DC filter in converter station 换流站有源直流滤波装置
active debt 活动债务
active decoder 有源译码器
active defect 运行故障
active deposit 活性沉积，放射性沉积，放射性沉积物，放射性沉降物
active device 有源器件
active diameter 堆芯直径
active display 主动式显示器，发光型显示器
active drag 正阻力
active drainage area 有效排水面积
active drain hold-up tank 放射性疏水储存箱
active drains pump 放射性疏水泵
active drains tank 放射性疏水箱
active dust 放射性尘埃
active earth pressure 主动土压力
active earth thrust 主动土推力
active effluent disposal 放射性排出物处置
active effluent drain pipe 放射性废液排放管道
active effluent plant area 放射性排出物处理厂区
active effluent system 放射性排放系统，放射性废液系统
active effluent 放射性排出物
active electric network 有源电气网络
active electrode 有效电极
active electronic countermeasure 积极电子对抗
active element 活性元素，有效元件，有源元件
active energy 活性能，有效能，有效能量
active equivalent network 有源等效网络
active erosion 强烈侵蚀
active face 刃面
active failure 有效损坏（故障），主动破坏，自行破坏
active fall-out 放射性落下灰，放射性沉降物
active fault 活动断层，活断层
active filler 活性填料，有源滤波器
active fission product release 放射性裂变产物的释放
active effluent hold-up tank 放射性废液暂存箱
active fluidization （沸腾床）良好流化
active force 有效力，主动力
active four terminal network 有源四端网络
active fuel bed 有效燃料床，燃烧着的燃料层
active fuel length 燃料活性长度
active furnace area 炉膛有效面积
active gas 活性气体，腐蚀性气体
active gage 电阻应变仪动作部分
active gauge 电阻应变仪动作部分
active height （堆芯）活性高度，有效高度

active in respect to 相对……呈阳性
active interlock 主动联锁装置
active jamming 有源干扰，积极干扰
active lattice 非均匀堆芯，堆芯栅格【反应堆】
active laundry 放射性洗衣房
active layer 作用层，冻融层，活动层
active leg 有源支路，放射性管段
active length 有效长度，活性长度，活性段长度
active life 有效寿命，有效寿期，工作寿命
active line 工作线路，有效线
active load 有功负荷，有源负载
active logic 常用逻辑，有效逻辑
active loop 放射性环路
active loss 有功损耗，实际损耗
actively mode-locked glass laser 主动锁模玻璃激光器
active maintenance time 实际维修时间
active mass 有效质量
active material 活性物质，放射性物质，活性材料，激活材料，放射材料
active medium 工作介质
active microorganism 活性微生物
active network 有源网络
active output 有功输出，有效输出，实际产量
active oxygen 活性氧
active partner 积极合伙人
active part of bearing 轴承活动部分
active pile 日产周转煤堆，日用煤场，周转煤堆，主动桩
active pollution 放射性污染物
active pool 活性沉淀池
active porosity 有效孔隙度
active power input 有效输入功率
active power loss 有功功率损失，有功功率损耗，有效功率损失
active power meter 有功功率计
active power relay 有功功率继电器
active power load 有功功率负载
active power 有功功率，有效功率，实际功率
active pressure 主动压力，实际压力，有效压力
active preventive maintenance time 有效预防性维修时间
active product 放射性排出物，放射性产物
active reaction 活性反应
active redundancy 工作冗余
active region 有源区，激活区，活化区，工作区，作用区，放射性区（域）
active regulation 主动监管
active remote sensing 人工源遥感
active repair time 修理实施时间，有效修理时间
active resistance 有效电阻
active return loss 有源回波损耗
active safety system 有效安全系统
active safety 能动安全，主动安全性
active salt marsh 活性盐沼
active sampling equipment 放射性取样设备
active section 活性区【反应堆】
active seismic area 地震活动区
active session 有效对话时间，工作时间
active side of coil 线圈的有效边
active site theory 活化区理论，活性中心理论
active slide area 活动性滑坡区域
active sliding surface 主动滑动面
active soil pressure 主动土压力
active solar system 主动式太阳能系统
active solvent 活性溶剂
active source 放射源
active stabilizer 活性加固剂
active stall power control 主动失速功率控制
active stall regulated wind turbine 主动失速调节风电机组
active stall 主动失速
active standard 现行标准
active state of plastic equilibrium 主动塑性平衡状态
active storage area 有效存储面积
active storage capacity 有效库容
active storage of reservoir 水库有效库容
active storage pile 周转煤堆
active storage 活动存储器，活性存储，随时取用的储存
active surface agent 表面活化剂
active surface of sliding 主动土压力的滑动面
active system 有源系统，主动系统，能动设备【如水泵、阀门等】系统
active tectonic system 活动性构造体系
active thrust of earth 主动土推力，主动土压力
active time 有效时间
active tracking system 有源跟踪系统
active trade balance 贸易顺差
active user 现时用户
active volcano 活火山
active voltage close loop control 有功电压闭环控制
active voltage 有功电压，有效电压
active volt-ampere 有功伏安
active volume 激活体积，有效容积
active waste building 固体废物贮存库，放射性废物厂房
active waste 放射性废物
active water 侵蚀性水，活性水，有腐蚀性的水
active weight 有效重量
active winding 有效绕组
active wire 有效导线
active workshop 放射性维修车间，“热”车间
active yawing 主动偏航
active yaw mechanism 主动偏航机构
active zone 活性区，活动层，有效区域，工作范围，放射性区
active 活性区的【反应堆】，放射性的，活化的，有效的，活性的，有功的，有源的，积极的，主动的
activity accident 放射性事故
activity affecting quality 影响质量的放射性
activity analysis 活性分析，放射性分析
activity build-up 放射性积累，放射性增长
activity chart 活动示意图
activity coefficient （放射性）活度系数，活性系数，功率因数，功率系数
activity concentration 放射性浓度，容积放射性

activity curve 放射性曲线，活度曲线，放射性衰变曲线
activity decay 放射性衰变
activity discharge at the stack 从烟囱排放的放射性物质
activity discharge to the environment 向环境排放的放射性物质
activity discharge 放射性排放，放射性释放
activity duration 作业持续时间，有效期限
activity escape 放射性放出，放射性逸出
activity factor of solute N 溶质 N 的活度因子
activity for defocus 散焦作用
activity index 活性指数
activity inventory 放射性总量
activity level 活性水平，活度水平，放射性能级，放射性水平，放射性浓度
activity measuring instrument （放射性）活度测量仪
activity measuring point （放射性）活度测量点
activity median aerodynamic diameter 放射性气动力中位直径，活性中值空气动力学直径
activity meter 放射性测定仪表，（放射性）活度测量仪
activity monitoring 放射性监测，放射性监控
activity of cathode 阴极激活度
activity of cement 水泥的活性
activity product 活度积
activity quotient 活度比
activity ratio method 活度比值法
activity ratio （放射性）活度比率，活度比
activity retention capability 放射性滞留能力
activity retention rabidity （燃料元件）放射性滞留能力
activity sampling 工作抽样检查
activity-sensing equipment 放射性探测设备
activity source 放射源
activity state 工作状态
activity supervisor 放射性检验员
activity surveillance of reactor coolant 反应堆冷却剂活度监测
activity test 活性试验
activity unit （放射性）强度单位，（放射性）活度单位
activity 活化率，放射性，活度，活动，活性，效力，放射强度，放射性活度
act of bankruptcy 破产法
act of God 自然灾害，不可抗力，天灾
act of infringement of regulation 违章行为
act of nature 天灾
act of tort 民事侵权行为
AC trigger 交流触发器
actual 真实的，实际的
actual acidity 实际酸度
actual address 实际地址
actual aggregate breaking strength 骨料实际断裂强度
actual air consumption 实际耗气量
actual assets 实际资产
actual breaking load 实际破坏载荷
actual breaking strength 实际断裂强度
actual breaking stress 实际破坏应力
actual budget 决算
actual capacity compressor 实际出力压气机
actual capacity 有效容量，实际容量，实际出力，实际能力
actual capital 实际资本
actual carrier 实际承运人
actual cash balance 实际现金余额
actual cash value 实际现金价值
actual clearance 有效间隙，实际间隙，实际净空
actual construction sequence 实际施工程序
actual consumption 实际消费量，实际消费
actual cooling surface 有效冷却表面积
actual cost 实际支出费用，实际成本
actual cycle 实际循环
actual damage 实际损害赔偿金
actual date 实际日期
actual debt 实际债务，实际负债
actual delivery （泵的）实际抽水量，实际交货
actual demand 实际需求
actual density 实际密度，真实密度，真密度
actual depreciation 实际折旧
actual deviation 实际偏差
actual dimension 实际尺寸，有效尺寸
actual discharge coefficient 实际流量系数
actual displacement 实际位移
actual distance 实际距离，有效距离
actual efficiency 实际效率
actual elapsed time 实际经历时间
actual elevation 实际标高，实际高程
actual energy 实际能量，实用能量
actual enthalpy drop 实际焓降
actual enthalpy rise 实际焓增
actual error 实际误差，真误差
actual evaporation 实际蒸发
actual evapotranspiration 现场蒸散发量，实际蒸散，有效蒸散
actual expenditure 实际支出
actual failure stress 实际破坏应力
actual final condition 实际最终工况，实际终态
actual flow of heat supply network 热网实际流量
actual flow 实际流量
actual flux density 有效磁感应，实际磁通密度，有效通量密度
actual gap area 磁极下气隙计算面积，实际气隙面积
actual gas 实际气体，真实气体
actual grain size 实际粒径
actual groundwater velocity 地下水实际速度
actual head 实际扬程
actual heating load 实际热负荷
actual higher measuring range value 实际测量范围上限值
actual horsepower 实有功率，实有马力
actual imbalance 实际差额
actual induction 有效感应，有效磁密
actualing element 执行机构
actual instruction 有效指令
actual internal area 实际流通截面，流通截面积，流通截面
actual length 实际长度

actual lifetime 实际寿命
actual life 实际使用寿命，实际寿命
actual lift 实际升力
actual line 实线
actual loading test 实际负荷试验，实际负载试验
actual load 有效载荷，实际负荷，实际负载
actual loss 实际损失
actual lower measuring range value 实际测量范围下限值
actual mean 有效平均值
actual measurement 实际测量，实测
actual monitor 实用监测器，有效监视
actual net heat drop 实际净热降，可用热降
actual obligations 实际债务
actual operating plant 实际操作设备
actual order 实际指令，有效指令
actual output rating 有效输出率，实际生产率
actual output 有效输出，实际输出，实际出力，实际出水量
actual parameter 实际参数
actual path 实际轨迹，真实轨线，实际路径
actual peak load 实际峰荷
actual power 实际功率，有效功率
actual pressure lapse rate 实际气力递减率
actual pressure 实际压力，有效压力
actual price 实际价格，实价
actual profile 实际断面
actual profit 实际利润
actual pump head 实际泵扬程
actual quantity 实际用量
actual raceway 实际走向
actual reserves 实际储量，实际储备金
actual resources 实际资源
actual running time 实际运行时间
actual service life 实际使用寿命
actual situation 实际情况
actual size 有效尺寸，实际尺寸
actual solid retention time 固体实际停留时间
actual source 实际（污染）源
actual specific gravity 真实比重
actual speed 实际转速
actual spent hours 实际花费时数
actual stack height 实际烟囱高度
actual state 实际状况，实际状态，现状
actual steam consumption 实际汽耗量
actual steam cycle 实际蒸汽循环
actual steam rate 实际汽耗率
actual stress 有效应力，实际应力
actual surface finish 表面光洁度【筒仓】
actual tare 皮重
actual temperature of rectum water 实际回水温度
actual temperature of supply water 实际供水温度
actual temperature 真实温度，实际温度
actual terrain height 实际地形高度
actual throat 实际厚度【焊缝】
actual thrust 实际推力
actual tide 真潮
actual time 有效时间，动作时间，实际时间，实时
actual tooth density 有效齿密度
actual torque 实际转矩
actual total evaporation 实际总蒸发
actual total loss 绝对全损，实际全损
actual value 有效值，实际值，真值
actual vegetation 现存植被
actual velocity of groundwater 地下水实际流速
actual velocity 实际速度，实际流速
actual volume 实容积
actual weight 有效重量，实际重量
actual weld-throat thickness 焊缝实际厚度，实际焊缝厚度
actual wind energy 实际风能
actual wind power 实际风能
actual wind 实际风
actual working hours 实际工时
actual working pressure 实际工作压力
actual zero point 测站基点，绝对零点，基点
actuated valve 控制阀
actuate 致动，开动，使动作，启动，驱使，激励，操纵，执行，作用
actuating apparatus 促动器，执行器，调节器
actuating arm 操作杆，驱动杆
actuating bellows 执行波纹管
actuating cable 操作电缆，控制电缆
actuating cam 主动凸轮
actuating coil 工作线圈，动作线圈
actuating current 开动电流，动作电流
actuating cylinder 驱动缸，工作缸，主动缸，动力汽缸
actuating device 执行元件，动作元件，操作装置，驱动装置
actuating element 执行元件，执行机构
actuating error signal 动作偏差信号
actuating fluid 驱动流体
actuating force 驱动力
actuating lever 驱动杆
actuating mechanism 执行机构，伺服机构，操作机构，传动机构
actuating medium 工作介质，工质
actuating member 执行元件，调节元件
actuating motor 伺服马达，伺服电动机，启动电动机，驱动电动机
actuating oil system 压力油系统，驱动油系统
actuating position 工作位置
actuating pressure 工作压力，作用压力，驱动压力，启动压力
actuating quantity 作用量【继电器】
actuating signal 启动信号，动作信号，操作信号，作用信号
actuating system 执行系统，操作系统，传动系统，传动机构
actuating time 动作时间，启动时间
actuating transfer function 有用传递函数，作用传递函数
actuating unit 传动装置，传动机构，驱动机组，动力机构，动力传递装置
actuating variable 作用变量，感应变量，影响变量
actuating 开动，启动，驱动
actuation confirming test 动作确认试验
actuation shaft 驱动轴，传动轴

actuation system　动作系统
actuation time　动作时间，启动时间，触发时间
actuation　开动，传动，启动，投运
actuator arm　传动杆
actuator cabinet　执行机构，控制柜
actuator disc　促动盘，激（动）盘，作用盘
actuator disk　促动盘，激（动）盘，作用盘
actuator generator　测速发电机
actuator governor　调速控制器
actuator level　执行级
actuator locking device　调速器锁定装置
actuator manufacturer　启动器制造商
actuator motor　驱动电动机，伺服电动机，执行电动机
actuator rating　执行器额定功率
actuator　传动装置，驱动装置，启动器，执行器，操作机构，执行元件，执行机构，油动机
act　法规，法则，规程，条例，行动，法案
ACU(add control unit)　加法控制器
ACU(arithmetic and control)　运算控制器
ACU(automatic calling unit)　自动呼叫装置
ACUPS(AC uninterruptible power supply)　交流不间断电源，交流不停电电源
acute angle blade　锐角叶片
acute angle　锐角
acute arch　尖拱，锐拱
acute discharge　急性排放
acute exposure　急性照射，强烈照射
acute irradiation　急性辐照【指短时间内的强辐照】
acute poisoning　急性中毒
acute radiation death　急性辐射死亡
acute radiation disease　急性辐射病
acute radiation sickness　急性辐射病
acute radiation syndrome　急性辐射综合症
acute toxicity　剧毒，急性毒性
acute transducer　有源换能器，有源传感器
acute　急性的，精明的
ACV(alternating current voltage)　交流电压
ACV(automatic control valve)　自动控制阀
AC voltage converter　交流电压变流器
AC voltage stabilizer　交流电压稳定器，交流稳压器
AC voltage transformer　交流电压互感器，变压器
AC voltmeter　交流电压表
ACW(anti-clockwise)　反时针方向，逆时针方向
ACW(auxiliary cooling water system)　辅助冷却水系统
AC welder　交流焊接机
AC welding machine　交流焊机
AC winding　交流绕阻
AC withstand voltage test　交流耐压试验
acyclic dynamo　单极发电机，单极电机
acyclic feeding　非周期馈电
acyclic generator　单极发电机
acyclic machine　单极电机
acyclic motor　非周期性电机，单极马达
acyclic　非周期（性）的
AC yoke magnetization　交流磁轭磁化
ADA(automatic data acquisition)　自动数据采集程序
ADAC(analog-digital-analog converter)　模拟-数字-模拟转换器
AD(aerodynamic decelerator)　气动减速器，气动减速装置，空气动力减速装置，空气动力减速器
adamantine spar　刚玉
adamant plaster　硬石灰膏
adamant　金刚石，硬石
adamellite　石英二长石
adamic earth　红黏土
AD(analog-digital)　模数混合，模拟数字
adaptability　适应性，适应，可用性，灵活性，
adaptation　适应，改进，采用，自适应，适配，适合，调配
adapter booster　后燃室，加力燃烧室，加速燃烧室
adapter bush　接头衬套
adapter connector　接头，连接器，连接器接头
adapter control　自适应控制
adapter converter　附加变频器
adapter coupling　转接，连接
adapter flange　转接法兰，配接凸缘
adapter pipe　套管，承接管
adapter plug　转接插头
adapter ring　接合环
adapter skirt　连接套
adapter sleeve　定心套筒，连接套，紧固件，紧固套，连接套管
adapter transformer　用电附加变压器
adapter　承接管，适配器，转接器，接合器，连接器，接合器，拾波器，管接头，接头，插座，控制阀
adapting control　自适应控制
adapting pipe　承接管，连接管，联结管，异径联结管
adaption binary load program　适应二进制负载的程序
adaptive beam forming　适应波束形成
adaptive capacity　适应力
adaptive communication　自适应通信
adaptive control action　自适应调节作用
adaptive control mode　自适应控制方式
adaptive control system　自适应控制系统
adaptive control welding　适应控制焊接
adaptive control　适应性控制，自适应控制
adaptive equalizer　自适应均衡器
adaptive feedback control　适应反馈控制
adaptive filter　自适应滤波器
adaptive learning network　自适应查询网络
adaptive optimal control　自适应最佳控制，自适应最优控制
adaptive optimal system　自适应最佳系统
adaptive planning　适应性规划
adaptive plate　上端适应板【燃料组件】
adaptive process　自适应过程
adaptive regulator　自适应调节器
adaptive servo system　自适应伺服系统
adaptive system of water-turbine engine　水轮机调节系统
adaptive system theory　自适应系统理论
adaptive system　自适应系统

adaptive telemetry　自适应遥测
adaptive-wall wind tunnel　自适应壁风洞
adaptive　适应的，合适的，自适应的
adaptor kit　适配件
adaptor lug　适配接线片
adaptor plate　（燃料组件上管座）配板【核电】
adaptor　适配器，适配座，转接器
adapt　使适用，使适应，适应
adarce　钙华，石灰华，泉渣
ADAT（automatic data accumulator and transfer）自动数据累加器和传送器
ADB（Asian Development Bank）　亚洲开发银行
ADCON（address constant）　地址常数
adconductor cathode　加位导体阴极
A/D converter（analog-to-digital converter）　模数转换器，模/数变换器，A/D 转换器
ADCON　模数转换器，模拟数字转换器
add　附加，加法，加
ADDAC（analog data distributor and computer）　模拟数据分配器和计算机
ADD＝addition　加法器，加法，加法指令
ad damnum clause　损害赔偿条款，索赔条款
ad damnum　主张一定数额的赔偿金，请求损害赔偿，索赔金额
add and subtract relay　增减继电器，加减继电器
ADDAR（automatic digital data acquisition and recording）自动数字数据收集和记录
ADDAS（automatic digital data assembly system）自动数字数据汇编系统
addaverter　加法转换器
add circuit　加法电路
add control unit　加法控制器
added enthalpy　焓增
added heat　附加热
added lift　附加升力
added loss　附加损失，附加损耗
added mass coefficient　附加质量系数
added mass　附加质量
added metal　附加金属
added resistance　附加电阻，扩量程电阻
added　增加的，更多的
addenda and corrigenda　补遗和勘误
addenda　补遗
addend digit　第一被加数的数字
addend partial product register　被加数部分积寄存器
addend register　被加数寄存器
addendum angle　齿顶角
addendum circle　齿顶圆，外圆
addendum line　齿顶线
addendum modification on gear　齿轮的变位
addendum of agreement　协定附录
addendum of contract　合同附录
addendum　补遗，附录，附件，附加物，补充合同，补充协议，补助金，（齿轮的）齿顶，齿顶高
addend　被加数，加数
adder accumulator　加法累加器
ADDER（automatic digital data error recorder）　数字信息误差自动记录器
adder circuit　加法器电路
adder-subtractor　加减器
adder valve　介质添加阀
adder　加法器
add gate　加法门，加法器门
adding element　加法元件
adding load　加负荷
adding machine　加法器，加法计算机
adding network　加法网络
adding stage　相加级
adding-storage register　加法存储寄存器
addition agent　添加剂，合金元素
additional air governor　补给空气量调节器，补给空气量控制器
additional air regulator　补给空气量调节器
additional air slide valve　补给空气分气阀
additional air valve　补充空气阀
additional amount　附加额
additional arbitrator　补充仲裁员
additional article　增订条款
additional bending moment　附加弯矩
additional capability　附加出力
additional cargo　加载货
additional charge　补充充电，附加费
additional clause　附加条款
additional condition　附加条件
additional contact　辅助触点
additional contract clause　合同补充条款，附则
additional control point　辅助控制点
additional cost　额外费用
additional cycle redundancy check　附加的循环冗余码校验
additional data　补充资料
additional disturbing force　附加扰动力，附加扰力
additional document　补充文件
additional drag　附加阻力
additional electromotive force　附加电动势
additional equipment　辅助设备，配套设备，附加设备
additional erection tool　附加安装工具
additional evidence　补充证据
additional expense to war risk　战争险附加费用
additional export　额外出口
additional factor for extensor door　外门附加率
additional factor for intermittent heating　间歇供热附加率
additional factor for room height　房间高度附加率
additional factor for wind force　风力附加率
additional freight　附加运费
additional harmonic field　附加谐波场
additional heat loss　附加耗热量
additional HP governing valve　辅助高压调节阀
additional insurance　附加保险
additional investment　额外投资，附加投资，追加投资
additional iron loss　附加铁损
additional level holding pump　附加液位维持泵
additional lighting　辅助照明
additional load　附加负荷，附加荷载
additional loss　附加损耗
additional memory　辅助存储器

additional order 追加订单，追加订货
additional payment 另外支出，另外支付，额外付款
additional pipe 接长管，支管
additional power source 附加电源
additional premium to reinsurance 再保险附加保险费
additional premium 附加保险费，附加费用，附加奖金，追加保费，额外保险费
additional pressure 外加压力，附加压力
additional product 附加产品，副产品
additional protocol 附加议定书，增订议定书
additional provision 附加条款
additional regulation 补充规定
additional reinforcement 附加钢筋
additional remarks 补充说明
additional resistance 附加电阻
additional risk 附加险
additional safeguard oil pressure 附加保安油压
additional service 附加服务
additional sheet 增页
additional stipulation 附加规定
additional storage 辅助存储器
additional stress 附加应力
additional tax 附加税
additional torque 附加转矩
additional transfer 追加转拨
additional transmitting contact 附加传输节点
additional unit 新增机组
additional voltage 附加电压
additional winding 附加绕组
additional work 额外工作，附加工作
additional 补充的【主要事物之外】，补给的，附加的，外加的，追加的，额外的，增加的
addition by subtraction 采用减法运算的加法
addition item 增列项目
addition of clay 掺黏土
addition of sand 掺砂
addition or deletion of variables 变量的增添或删除【焊接程度资格审定】
addition reaction 加成反应
addition safeguard oil pressure 附加保安油压
addition theorem 加法定理
addition 附加，添加，附加物，相加，加，加法，加法指令
additive agent 添加剂
additive alternative 补充比较方案
additive congruential method 相加余式法，加同余法
additive effects 叠加效应
additive error 附加误差
additive for lubricating grease 润滑脂添加剂
additive phase-sensitive rectifier 加性相敏整流器
additive polarity 加极性【变压器】
additive postulate 附加假设
additive property 可加性
additive-type oil 含添加剂油
additive 掺合料，掺合物，附加剂，添加剂，添加的，附加的
additivity law 相加性定律
additron 相加速器
add lot 零星交易
add-on plant 扩充设备，添加设备
add or subtract 加或减
add output 加法输出
add pulse 加法脉冲
add register 加法寄存器
address 地址
addressable point 可寻址的点
addressable storage 可选存储器，可寻址存储器
addressable 可寻址的，可选址的
address blank 空地址，指令地址部分的空白
address book 通讯录
address bus 地址总线
address character 地址字符
address code 地址码
address computation 地址计算
address constant 地址常数
address conversion 地址转换
address counter 地址计数器
address decoder 地址译码器
address decoding 地址译码
address display subsystem 地址显示子系统
addressed memory 地址存储器
addressed storage 编址存储器
addressed 编址的
addressee 收电人，收件人
address-enable 地址启动
addresser 发信人，署名人
address field 地址段，地址部分
address file 地址行列，地址数据存储器
address format 地址格式
address-free program 无地址程序
address generation 地址生成
address in case of need 临时地址
address indicating group 地址指示组
addressing mode 寻址模式，寻址方式
addressing system 寻址系统，编址系统
addressing 寻址
address instruction format 地址指令格式
address interleaving 地址交叉插入，地址隔行插入，地址交叉存取
address mark 地址标记
address modification 地址修改，地址变更，变址，地址改变
address part 地址部分
address-read wire 地址读出线
address reference number 地址参考数
address register 地址寄存器
address resolution protocol 地址解析协议
address search 地址检索，寻址
address selection switch 地址选择开关
address selection system 地址选择系统
address signal 地址信号
address sort routine 地址排序程序，地址分类程序
address substitution 变换地址
address track 地址磁道
address translation 地址翻译
address wire 地址线
address write tire 地址写入线

add-subtract control unit　加减控制装置
add-subtract time　加减时间
add time　加法时间
add to memory technique　向存储器加入数据字的技术
adduction　引用，内收，引证，内转，内收作用，氧化作用
ADE(automated design engineering)　自动设计工程
ADELA(Atlantic Community Development Group for Latin America)　大西洋共同体拉美开发团投资公司
AD encoder　模数编码器
ADEPT(automatic data extractor and plotting table)　自动数据提取及制表程序
adequacy　充裕性【电力系统】，足够，充足，适合
adequate allowance　适当裕度
adequate and systematic service　配套服务
adequate flow　充足的流量
adequate shielding　足够屏蔽
adequate verification　充分的核查
adequate　足够的，充分的，适当的
adequation of stress　应力均匀化
A-derrick　动臂起重机，人字起重机
ADES(automatic digital encoding system)　自动数字编码系统
ADF(African Development Fund)　非洲开发基金
adfreezing force　冻附力，冻结力
adfreezing strength　附冻强度
adfreezing　冻附，冻结，冻硬
ADG(air-driven generator)　风动发电机
ADG　给水除氧器系统【核电站系统代码】
adhered　黏附体
adherence deposit　黏性沉积物
adherence of nappe　水舌黏附
adherence surface　黏附面
adherence to agenda　遵照议程
adherence to budget　遵守预算
adherence to commitments　遵守承诺
adherence　附着（力），黏着，严格遵守，坚持，依附，黏附（力）
adherent scale　黏附性垢
adherent sludge deposit　黏渣沉积物
adhere to promise　遵守承诺
adhere to　坚持【原则等】，黏附
adhering nappe　贴附水舌
adhering slag　挂渣【切割】
adhesion agent　胶黏剂，黏附剂
adhesional wetting　黏润作用
adhesion coefficient　黏着系数
adhesion contract　服从契约
adhesion degree　黏附程度，黏着度
adhesion factor　黏附系数
adhesion strength　黏着力
adhesion tower-crane　附着式塔吊
adhesion　黏着（力），附着（力），黏附（力），黏结（力）
adhesive ability　附着能力，黏着力
adhesive agent　黏附剂，黏着剂
adhesive attraction　黏附作用，黏着力
adhesive bitumen primer　冷底子沥青
adhesive-bonded joint　黏胶接头
adhesive bonding　胶结，黏合剂
adhesive capacity　胶黏度，黏附能力
adhesive coating　黏附层
adhesive film　黏附膜
adhesive force　附着（力），黏附（力），黏着（力），内聚力
adhesive material　黏结材料，黏附材料
adhesive moisture　附黏水分
adhesiveness　黏性，黏着度，黏着性
adhesive power　黏附力，黏着力，附着力
adhesive revenue stamp　印花税票
adhesive seal　胶泥密封
adhesive strength　黏结强度，黏附强度
adhesive tape　胶带
adhesive tension　黏附力
adhesive value　黏合值
adhesive water　薄膜水，黏附水，附着水，吸附水，黏结水
adhesive wax　黏蜡，封蜡，胶黏封，胶黏蜡
adhesive wear　黏着磨损，黏附磨损
adhesive　带黏性的，黏胶，黏合剂，胶合剂，胶黏的，黏的，易黏的，黏着的，附着剂
ad hoc advisory committee of experts　特设专家咨询委员会
ad hoc arbitration　临时仲裁，特设仲裁，专家仲裁
ad hoc arrangements　特别安排
ad hoc assignment　特定任务
ad hoc committee　临时委员会，特设委员会
ad hoc expert group　特设专家小组
ad hoc group of experts　特设专家组
ad hoc group　特设小组
ad hoc meeting　特设会议，专门会议
ad hoc observer　特别观察员
ad hoc program　专门计划
ad hoc representative　临时代表
ad hoc request　特别要求
ad hoc　特定，特设，专设，特别的，特设的，临时的，专门的
adiabatic　绝热的，绝热式的
adiabatic apparatus　绝热装置
adiabatic boundary layer　绝热边界层
adiabatic calorimeter　绝热量热器，绝热式热量计
adiabatic change　绝热变化
adiabatic coefficient　绝热系数
adiabatic combustion temperature　理论燃烧温度
adiabatic combustion　绝热燃烧
adiabatic compression　绝热压缩
adiabatic condensation　绝热凝结
adiabatic condition　绝热条件，绝热工况，绝热状态
adiabatic cooling curve　绝热冷却曲线
adiabatic cooling　绝热冷却
adiabatic curve　绝热曲线，绝热线
adiabatic efficiency　绝热效率，隔热效率
adiabatic energy storage　绝热储能
adiabatic energy　绝热能
adiabatic equation　绝热方程

adiabatic equilibrium 绝热平衡
adiabatic expansion 绝热膨胀
adiabatic exponent 绝热指数，等熵指数
adiabatic flow 绝热流动，绝热流
adiabatic gradient 绝热梯度
adiabatic heat drop 绝热降
adiabatic heating 绝热升温
adiabatic horse-power 绝热马力，绝热功率
adiabatic humidification curve 绝热增温曲线
adiabatic humidification 绝热加温
adiabatic index 绝热指数
adiabatic insulation 绝热
adiabatic lapse rate 绝热递减率
adiabatic layer 绝热层
adiabatic line 绝热线
adiabatic operation 绝热操作
adiabatic potential curve 绝热势能曲线，绝热势曲线
adiabatic pressure drop 绝热压降
adiabatic process 绝热过程
adiabatic rate 绝热变化率
adiabatic region 绝热区
adiabatic relationship 绝热变化规律
adiabatic saturation curve 绝热饱和曲线
adiabatic saturation temperature 绝热饱和温度
adiabatic temperature 绝热温度
adiabatic throttling 绝热节流
adiabatic transformation 绝热变换
adiabatic wall temperature 绝热壁温
adiabatic warming 绝热增温
adiathermal 绝热的
adipic acid 肥酸，己二酸
adit collar 平洞洞口，支洞洞口
adit entrance 坑道入口
adit for draining 排水洞，排水坑道，排水坑口
adit opening 坑道入口，坑道口
adit 平洞，坑道，走道，入口，施工支洞
adjacent accommodation 厢房，附属建筑物，相邻的设备
adjacent block 相邻浇筑块
adjacent bubbles 相邻气泡
adjacent building 邻近建筑，邻近建筑物
adjacent carrier 邻信道载波
adjacent channel attenuation 相邻信号衰减，邻近信道衰减
adjacent channel interference 邻通道干扰，相邻信道干扰
adjacent channel selectivity 邻信道选择性
adjacent channel 相邻信道，相邻通道
adjacent country 相邻地区，毗邻国家
adjacent faces 相邻两个面
adjacent frequency 相邻频率
adjacent land 邻近地域，邻地
adjacent map 相邻图，邻接图
adjacent plank 门框边梃，踢脚板，门头线
adjacent sheet 邻接图幅
adjacent sound carrier 邻信道伴音载波
adjacent span 邻挡
adjacent structure 附近建筑物，附联建筑物
adjacent video carrier 邻信道图像载波
adjacent vortex 附着涡，附着旋涡，邻涡
adjacent waters 毗邻水域，相邻水域
adjacent 邻近的，毗连的，相邻的，交界的
adjoining building 邻接建筑，邻近房屋
adjoining course 结合层
adjoining plane 接合面
adjoining railway 铁路引入线，进厂线，进港线
adjoining rock 围岩
adjoining sheet 邻接图表
adjoining 邻接的，毗连的
adjoint branch 相邻支路
adjoint collision density 伴随碰撞密度
adjoint curve 伴随曲线
adjoint flux 伴随中子通量，辅助通量
adjoint function 伴随函数
adjoint matrix 伴随矩阵
adjoint of the neutron flux density 伴随中子通量密度，中子通量密度的修正
adjoint simulation 伴随法模拟
adjoint system 伴随系统
adjoint variable 伴随变量
adjourned meeting 延期会议
adjournment of the meeting 休会
adjournment 休会
adjourn 休会
adjudicate a dispute 处理争端
adjudicate 判决
adjudication of bankruptcy 裁定破产，宣告破产
adjudication of tenders 对投标书的裁定
adjudication of water right 水权判决
adjudication 裁判，宣告，审定
adjudicator 裁判员，判决者，裁定者
adjunct 添加剂，附件，附属物
adjustability 调整性，可调性
adjustable anchorage bar 可调锚杆
adjustable bearing 可调整轴承
adjustable bench level 可调台用水准仪
adjustable blade propeller pump 可调叶片式轴流泵，调节叶片螺桨泵
adjustable blade propeller runner 可调叶片式螺旋桨转子轮
adjustable blade propeller 可调（螺距）螺旋桨，可调叶片式螺旋桨，变距螺桨
adjustable blade turbine 旋桨式水轮机，转桨式水轮机，转叶式水轮机
adjustable blade 可调叶片
adjustable bolt 可调螺栓
adjustable bracket 可调托架，活动托架
adjustable brush 可调电刷
adjustable cam 可调凸轮
adjustable capacitor 可变电容器，可调电容
adjustable cistern barometer 调槽式气压表
adjustable clamp 可调夹具
adjustable clearance 可调间隙
adjustable cleat 可调夹板
adjustable compensating capacitor 可调补偿电容器
adjustable condenser 可调电容器
adjustable constant speed motor 可调恒速电动机
adjustable contact 可调接点，可调触头，可调整触点
adjustable counter balance 可调平衡器

adjustable crest 活动堰顶，可调堰顶
adjustable damper 调节挡板，可调挡板，可调减震器
adjustable die 活动扳牙，可调板牙，可调模子，活动扳手
adjustable diverter gate 可调节分液门
adjustable drive 可调传动装置
adjustable duration timer 可调持续时间继电器
adjustable fan 可调风机
adjustable feet 可调活脚
adjustable flat wrench 扁平活动扳手
adjustable flume 活动水槽，可调整水槽
adjustable frequency motor 变频电动机
adjustable fulcrum 可调节支点
adjustable gag 可调塞
adjustable grille 可调格栅
adjustable guide vane 可调导叶
adjustable heater valve 加热器调节阀
adjustable higher measuring range limit 可调测量范围上限
adjustable indicator 可调节指示器
adjustable inductance 可调电感
adjustable inductor 可变电感线圈，可调电感器
adjustable instrument mounting 可调仪表座
adjustable jet 可调喷管，可调喷口
adjustable level 活动水平尺
adjustable louver 可调百叶窗，活动百叶窗，活动百叶
adjustable lower measuring range limit 可调测量范围下限
adjustable mark 调整标记
adjustable mask 可调屏
adjustable mass balance 可调配重
adjustable micrometer calipers 活动千分尺
adjustable motor tuning 可调电机调谐
adjustable mouth 可调节口
adjustable moving-blade axial-flow fan 轴流动叶可调风机
adjustable moving blade 可调动叶片
adjustable nozzle 可调喷嘴
adjustable orifice 可调喷管，可调节孔，可调注孔
adjustable parallel 活动平垫铁，可调平垫铁
adjustable parameter 整定参数，可调参数，可调参量
adjustable pipe tong 可调管钳
adjustable-pitch propeller 变距螺旋桨
adjustable pitch 可调螺距，可调桨距，变螺距
adjustable pliers 可调手钳，可调钳
adjustable price quotation 可调报价
adjustable propeller 可调螺旋桨，变距螺桨
adjustable prop 可调支柱
adjustable rake block 活动倾斜块
adjustable range 可调范围
adjustable reamer 可调铰刀
adjustable resistor 可调电阻器，可变电阻器
adjustable rotating blade 可调动叶片
adjustable screw 调整螺钉
adjustable sectional-steel frame 可调整组合钢架
adjustable series resistance 可调节的串联电阻
adjustable shelf 活动架
adjustable shore 可调支撑
adjustable shuttering 活动支模
adjustable shutter 可调百叶窗
adjustable sieve 可调筛
adjustable slatted shutter 活动百叶窗
adjustable spanner wrench 可调开脚扳手
adjustable spanner 活扳，可调扳手，活动扳手
adjustable speed drive 可调速拖动
adjustable speed motor 调速电动机，可调速电机
adjustable speed 变速，可调速度
adjustable spring loaded piston valve 可调弹簧式柱塞阀
adjustable starter 可调启动器
adjustable starting rheostat 可调启动变阻器
adjustable static-blade axial-flow fan 轴流静叶可调风机
adjustable stroke cylinder 可变行程油缸
adjustable submerged orifice 可调潜孔
adjustable support 可调支架
adjustable surface 可调面
adjustable tap wrench 可调丝锥扳手
adjustable tax 调节税
adjustable thrust journal 可调止推轴颈
adjustable tilting angle 可调倾斜角
adjustable transformer 调节变压器，可调变压器
adjustable try square 活动矩尺
adjustable valve 调节阀，调整阀
adjustable vane turbine 转叶式水轮机
adjustable vane 可调导向叶片，可调叶片，可调导叶，可转导叶
adjustable varying speed motor 可调变速电动机
adjustable vent-glazed louver 通风玻璃百叶窗
adjustable voltage control 可调电压控制
adjustable voltage divider 分段分压器，可调分压器
adjustable voltage stabilizer 可调稳压器
adjustable wall 可调（风洞）壁
adjustable weir crest 活动堰顶
adjustable welding roller 调校滚轮架
adjustable wrench 活扳，活扳手，可调扳手，活动扳手
adjustable 可校准的，可调节的，可调整的，可校准的
adjustage 辅助设备，精调设备
adjust and control function 调控职能
adjusted angle 平差角
adjusted data 订正资料，校核资料
adjusted elevation 平差高程
adjusted position 平差位置
adjusted price 调整价
adjusted solution 调整溶液
adjusted uranium 调质铀
adjusted value 调整值，平差值
adjuster bolt 调节螺栓
adjuster for windows 撑窗杆
adjuster rod 调节棒
adjuster-type equation solver 调整器式方程求解器
adjuster 调节杆，调整器，调整装置，调节器，调节装置
adjusting bar 调整杆

adjusting block 调整块
adjusting bolt bearing 调节螺栓支承点
adjusting bolt 调整螺栓
adjusting coil 可调线圈
adjusting command 调节命令
adjusting damper 调节风门
adjusting device 调整装置，调节装置
adjusting disk 调整垫片
adjusting element 调整元件
adjusting gage 整定仪表
adjusting gear 调整机构，调整装置，调整齿轮
adjusting handle 调整手柄
adjusting instruction 调节指令
adjusting instrument 调节仪器
adjusting key 调整键
adjusting lever 调整杆
adjusting link 调节连杆
adjusting mechanism 调整机构，校准装置
adjusting needle 调整针
adjusting nut 调节螺母，调整螺母
adjusting of gap of flow path 通流间隙调整
adjusting of gap of sealing 汽封间隙调整
adjusting of generator magnetic center 调整发电机磁力中心
adjusting pad 调整垫块
adjusting pin 定位销
adjusting plate 调整板
adjusting potentiometer 调整电位计，调整电势计
adjusting range 调节范围
adjusting resistor 调整电阻器
adjusting ring set screw 调节环定位螺丝
adjusting ring 调整环
adjusting rod 调节杆，调整杆
adjusting screw 调整螺钉，调节螺丝，调整螺丝，校正螺钉
adjusting shim 调整垫片
adjusting spring 调整弹簧
adjusting thrust block 可调推力轴承
adjusting time 调整时间
adjusting tool 调整工具
adjusting valve 调节阀，调整阀，调整门
adjusting wedge 调整楔，调整楔块，调整垫铁
adjusting 调整，校准，调节
adjustment accuracy 调整精度
adjustment and calibration of individual equipment 单体调校
adjustment by direction 方向平差
adjustment computation 平差计算
adjustment controls 调整机构，调整装置
adjustment correction 平差改正
adjustment curve 缓和曲线
adjustment device 调节装置
adjustment disc 调整盘
adjustment factor 调节因素，调整因数
adjustment for altitude 高程改正
adjustment for collimation 视准改正
adjustment for definition 调整清晰度
adjustment gear 调节机构
adjustment in one cast 整体平差
adjustment mark 调整标记
adjustment method 平差法
adjustment module 调整模数，调节系数
adjustment network 调整网络
adjustment of combustion system 燃烧系统调整
adjustment of economic texture 调整经济结构
adjustment of errors 误差配赋，误差校正
adjustment of hydraulic machinery 水力机械调节
adjustment of observation 观察平差
adjustment of track gage 轨距校正
adjustment of track 轨道校正
adjustment pellet （燃烧）本高芯块，调节芯块
adjustment plate 调整板
adjustment price 调整价
adjustment stroke （吸收棒）调节行程
adjustment test of turbine by-pass system 汽机旁路系统调试
adjustment test 调整试验
adjustment 调节，调整，整定，校正，校准，平差，调节器
adjust 调整，校正，校准，调节
adjutage 放水管，排水筒，喷射管，射流管
adjust ring 调整环
ADLWR（accelerator driven light water reactor）加速器驱动的轻水堆
ADM（automated data management） 自动数据管理
admeasurements 测量，尺度，分配，度量外形尺寸
adminicular evidence 辅助证据
administered price 管制价格，控制价格
administering authorities 管理机构
administer radiation 照射
administration building 办公楼，行政楼
administration data network 数据管理网络
administration of assignment 委派管理
administration office 行政办公室
administration of justice 司法裁判
administration of power supply 供电所，供电局，供电机构
administration of water rights 水权的管理
administration responsible for payment 负责付款局
administration-welfare quarter 生产福利区
administration 管理（机构）
administrative agency 管理机构，行政机构
administrative and general expenses 行政管理费用
administrative and services area 厂前区
administrative area 行政区域
administrative assistant 行政助理
administrative authorities 行政当局，行政机关
administrative budget 管理费预算，行政预算
administrative building 行政办公楼
administrative bureau for industry and commerce 工商管理局
administrative cost 行政管理费，管理成本，行政成本
administrative data 管理数据，管理资料
administrative decision 行政判决结果，行政决定
administrative delay 管理延迟
administrative division 行政区域

administrative expense　管理费
administrative law　行政法
administrative management　行政管理
administrative manager　行政经理
administrative map　行政区划图
administrative organization of electric power industry　电业管理机构
administrative organization　管理机构，行政机构
administrative practice standard　行政管理实施标准
administrative programme　行政管理计划
administrative region　行政区域
administrative safeguard　管理性安全措施
administrative service division　行政服务部
administrative staff　行政管理人员
administrative terminal system　管理终端系统，自动管理终端系统
administrative zone　行政管理区
administrator　管理人员，行政官员
admiralty brass　海军铜
admiralty metal tube　海军铜管
admiralty metal　含锡黄铜，海军铜
admissibility　许入，准许
admissible concentration　容许浓度
admissible deviation　容许误差
admissible document　可作证据文件
admissible error　容许误差，允许偏差
admissible evidence　可接受的证据
admissible load　容许负荷，容许荷载
admissible mark　容许符号，容许数字，可容许符号
admission of alien　外国人入境许可
admissible parameter　导纳参数，容许参数
admissible pressure　容许压力
admissible stress　许用应力
admissible　容许的，允许的，可采纳的，可接受的，可容纳的
admission chest　进气室，进汽室
admission degree of stage　级的（部分）进汽度
admission end　进气端，进汽端
admission intake　进气管，进汽管，进气口，进汽口
admission line　进汽线
admission loss　入口损失，进口损失
admission mode selection　进汽方式选择
admission of goods free of duty　准许免税输入商品
admission of new member　接纳新成员
admission of steam　进汽
admission of water　进水，给水
admission passage　进水道，通流部分
admission pipe　进气管，进汽管
admission port　进气口，进汽口
admission pressure　进口压力，允许压力，进气压力，进汽压力
admission space　装填体积，容纳空间
admission steam　进汽
admission stroke　进气冲程，进汽冲程
admission valve　进汽阀，进气阀，进水阀
admission velocity　进汽速度，进气速度，入口速度
admission　进气，进汽，进汽度，接纳，准允，许可，进入
admittance bridge　导纳电桥
admittance chart　导纳图
admittance except on business　闲人免进
admittance function　导纳函数，容许函数
admittance matrix　导纳矩阵
admittance measuring instrument　导纳测量仪
admittance operator　导纳算子
admittance parameter　导纳参数
admittance relay　导纳型继电保护装置
admittance triangle　导纳三角形
admittance　进汽，允许进入，进入，进入许可，导纳
admitted operation　许可操作
admitting pipe　进汽管，进气管，输入管，进入管，进水管
admitting port　进汽口
admit　进汽，引入
admixing coefficient　混合系数
admixture　混合，掺和，混合物，混合材，混入物，掺和物，掺和料，外加剂，附加剂
admix　混合，掺和
admonition　告诫
adnate hydrophyte　固着水生植物
adobe block　土坯
adobe brick　风干砖坯
adobe construction　土坯建筑，干打垒结构，干打垒
adobe masonry　土砖圬工
adobe wall　土坯墙，土墙
adobe　风干砖，风干土坯，冲积黏土，灰质黏土，土墙，砖房，砖坯砌的建筑物
adopt a prudent policy　采取慎重态度
adopted scale　采用的比例尺
adopting pipe　承接管
adoption binary load program　适应二进制负载的程序
adoption by consensus　协商一致通过
adoption of new technology　采用新技术
adoption　采纳，采用，适合，正配，配合，匹配
adopt　采用，采纳，采取，正式接受，接受，批准
ADP(automatic data processing)　自动数据处理
ADPE(automatic data processing equipment)　自动数据处理装置
ADPS(automatic data processing system)　自动数据处理系统
ADR(American depositary receipt)　美国存托凭证
ADRE(automated diagnostics for rotating equipment)　汽轮机（旋转机械）故障诊断系统
ad referendum contract　暂定合同，草约，草签合同
ad referendum　尚待考虑，待进一步审核，有待批准
adret　阳坡，山阳，山的向阳面
adrift　漂浮的，漂移的，漂浮，漂移，漂流，失去控制
ADS(auto-depressurization system)　自动减压系统，自助卸压系统

ADS(automatic dispatch system) 自动调度系统
adsorbate 被吸附物，吸附选购
adsorbed column 吸附柱
adsorbed film 吸附膜
adsorbed gas 吸附气体
adsorbed layer 吸附层
adsorbed tracer 吸附的示踪物
adsorbed water 吸附水
adsorbent activity 吸附剂活性
adsorbent bed 吸附床
adsorbent filter 吸附过滤器
adsorbent power 吸附剂能力
adsorbent 吸附剂
adsorb 吸附
adsorption affinity 吸附力
adsorption analysis 吸附分析
adsorption bed 吸附床
adsorption capacity 吸附容量
adsorption coefficient 吸附系数
adsorption column 吸附柱
adsorption compound 吸附化合物
adsorption current 吸附电流
adsorption curve 吸附曲线
adsorption cycle 吸附周期
adsorption dip 吸收曲线中的下落，吸收引起的下落
adsorption effect 吸附效应
adsorption enthalpy 吸附焓
adsorption equilibrium 吸附平衡
adsorption equipment 吸附装置
adsorption filter 吸附式过滤器
adsorption filtration 吸附过滤
adsorption hysteretic 吸附滞后
adsorption index 吸附指数
adsorption isobar 吸附等压线
adsorption isoster 吸附等容线
adsorption isotherm 吸附等温线
adsorption layer 吸附层
adsorption method 表面吸附法
adsorption of gas and vapor 气体吸附
adsorption process 吸附过程
adsorption rate 吸附率
adsorption ratio 吸附比
adsorption site 吸附点
adsorption system 吸附系统
adsorption tank 吸附罐，吸附槽
adsorption water 吸附水，吸着水
adsorption 表面吸收，吸附，吸附作用
adsorptive capacity 吸附容量，吸附能力，吸附性
adsorptive 吸附的，被吸附物，吸附性
ADS 自动调度系统【DCS 画面】
ADT(average daily traffic) 平均每日交通量
adult education 成人教育
adulterant 掺杂物，混杂料
adulteration 掺杂
adumbration 草图，轮廓
adustion of coal 煤的可燃性
ad valorem duty 从税价
ad valorem freight 从税运价
advance angle 超前角，前进角，前置角，起前角，提前角
advance ball 滑动滚珠
advance control 超前控制，步进控制，前进控制
advanced 尖端的，领先的，先进的
advanced administrative system 先进管理系统
advanced anchor bar 超前锚杆
advanced and innovative wind system 革新型风能系统
advanced BL(advanced bill of lading) 预借提单
advanced BWR 先进沸水堆
advanced charge 预付费用
advanced collective 先进集体
advanced converter reactor 先进型转换堆
advanced culture and ideology 精神文明
advanced data management system 先进数据管理系统
advanced deposit 预付押金
advanced development 先行开发
advanced epithermal thorium reactor 先进型超热钍反应堆
advanced equipment 先进设备
advanced freight 预付运费
advanced fuel assembly 改进型核燃料组件
advanced fuel 先进核燃料
advanced gas-cooled reactor 改进型气冷反应堆，改进型气冷堆，先进气冷反应堆
advanced group 先进班组
advanced guidelines 修订过的细则
advanced individual 先进个人
advanced international level 世界先进水平
advanced level 先进水平
advanced light water reactor 改进型轻水堆，先进轻水堆
advanced linear programming system 高级线性规划系统
advanced nation 发达国家
advanced operation 先进操作
advanced payment 预付，预付款，预付款额
advanced pressure tube reactor 改进型压力管式反应堆
advanced pressurized water reactor 改进型压水堆，先进压水堆
advanced PWR 先进压水堆
advanced reactivity measuring facility 先进反应性测量装置
advanced reactor 改进型反应堆
advanced real-time instructor station 先进的实时指导员站
advanced repayment of credit 提前偿还贷款
advanced research 远景研究，前沿研究
advanced science and technology prize 科技进步奖
advanced science 尖端科学
advanced solar thermal power 先进太阳能热发电
advanced solid logic technology 先进固态逻辑技术
advanced starting valve 预启阀
advanced support processor 先进支援处理机
advanced technique 尖端技术，先进技术

advanced test loop and simulator　先进试验回路及模拟器
advanced test reactor critical　先进试验堆临界实验装置
advanced thermal reactor　先进型热动力反应堆
advanced training　高级培训
advanced transverse nodal method　先进的横向积分节快法
advanced treatment process　深度处理过程
advanced treatment　深度处理，高级处理
advanced unit　先进单位
advanced warning　事前警告
advanced waste water treatment　废水深度处理
advanced wave　移动波
advance freight　预付运费
advance grouting　超前灌浆
advance ignition　提前点火
advance in cash　预付现金
advance material request　先期材料要求
advance metal　高比阻铜镍合金，阿范斯电阻合金
advance money of contract　合同预付款，预付约金
advance money on a contract　合同预付定金，合同预付款
advance money on security　垫付保证金
advance notice　预先通知
advance of flood wave　洪水波推进
advance of sea　海侵
advance of shoreline　滨线推进
advance of the face　工作面推进
advance payment bond　预付款保函
advance payment guarantee　预付款保函
advance payment recovery　预付款的回收
advance payment security　预付款保函
advance payment　预付，预付款，预付款项
advance per round　一个循环的进度
advance pulse　推进脉冲
advance repayment　预先偿还
advancer　相位超前补偿，进相机
advance for construction　施工预付款
advance sign　前置交通标志
advance to suppliers　预付货款
advance surrender of export exchange　预缴出口外汇
advance　进度，推进，进尺，移动步距，提前量，步进量
advancing gallery　前导坑
advancing side of belt　皮带紧边
advancing slope grouting　斜坡推进灌浆
advancing slope method　斜坡推进法
advancing wave　前进波，推进波
advancing　超前，接长（皮带）
advantage factor　有利因子，峰值因数
advantage ratio　有利比
advantage　优点，利益，有利条件，优势
advection fog　平流雾
advection inversion　平流逆温
advection layer　平流层
advection region　平流区
advection scale　平流尺度
advection　平流，平流作用，平流热效，平移，对流
advective ablation　平流消融
advective cooling　平流冷却
advective region　平流区
advective term　平流项
adventitious moisture　（燃料的）外在水分，外水分
adventure　投机活动，投机买卖
adversaria　注解，注释，记事，杂记
adverse atmospheric condition　不利大气条件
adverse balance　逆差
adverse current　逆流
adverse effect　不利影响，逆效应
adverse factor　有害因素，不利因素
adverse grade　反坡
adverse impact　不利影响
adverse judgement　不利的判决
adversely　逆地，反对地
adverse physical condition　不利自然条件
adverse pressure gradient　反向压力梯度，逆压力梯度，逆压梯度
adverse report　不利报告
adverse slope　倒坡，逆坡
adverse to　不利于
adverse weather condition　不利气候条件
adverse weather　不利天气，不良天气，恶劣气候，恶劣天气
adverse wind　逆风
adverse　逆的，相反的，不利的，有害的
adversity　逆境，不幸，灾难
advertisement　广告
advertiser　信号器，信号装置
advertise　做广告
advertising contract　广告合同
advertising expenses　广告费用
advice　意见，劝告，通知，建议
advice for collection　托收通知书，托收委托书
advice note　通知单
advice of arrival　到货通知，抵港通知
advice of authority to pay　授权付款通知书
advice of charge　付款通知
advice of claim　索赔通知
advice of correction　更正通知
advice of damage　损坏通知
advice of drawing　汇票通知书，提款通知书
advice of payment　付款通知
advice of shipment　装运通知，装船通知，转船通知
advice of shortage　短缺通知
advice of transfer of L/C　信用证转让通知书
advice of transfer of letter of credit　信用证转让通知
advice on production for export　出口商品通知单
advisability　合理（性），适当
advisable　可取的，明智的，适当的
adviser　顾问，咨询，建议人
advise　咨询
advisory board　咨询委员会，顾问委员会
advisory commission　咨询委员会，顾问委员会
Advisory Committee on Reactor Safeguards　反应

堆保障咨询委员会【美国】
advisory committee　咨询委员会，顾问委员会
advisory department　咨询部
advisory engineer　咨询工程师，顾问工程师
advisory fee　咨询费
advisory group　咨询小组
advisory mission　咨询工作组
advisory opinion　咨询意见
advisory service　咨询服务，顾问服务
advisory system　顾问制
advisory　顾问的
advisor　顾问，咨询，建议人
advocacy planning　建议规划
advocate　辩护人，拥护人，辩护者，律师，主张，拥护
adz-eye hammer　小铁锤
adz　扁斧
adz plane　刮刨
A. D.　公元【习惯上自耶稣降生之年起计】
AEA(Atomic Engergy Authority)　原子能管理局【英国】
AE(absolute error)　绝对误差
AE(air entraining)　加气
AEBIOM(Association of European Biomass Energy)　欧洲生物质能协会
AEC(Atomic Energy Commission)　原子能委员会【美国】
A/E company(architectural engineering company)　建筑工程公司
AED(automated engineering design)　自动工程设计
AEEN(Agence Pour L'Energle Européenne Nucléaire)　欧洲核能局【法语缩略词】
aegirine apiite　霓细晶岩
aegirine granite　霓花岗岩
aegirine rhyolite　霓流纹岩
AEH(analog electro-hydraulic control)　模拟式电液控制系统
AEN(accepted with comment)　带意见接收
A-energy　原子能
aeolation　风化，风蚀
aeolian　风积的
aeolian accumulation　风成堆积，风积
aeolian deposit　风积物，风积土，风成沉积
aeolian erosion　风蚀，风蚀作用
aeolian excitation　风成激励
aeolian feature　风成地貌，风成特征
aeolian material　风积物
aeolian plain　风成平原
aeolian sand ripple　风积沙波纹
aeolian sand　风成砂，风积沙
aeolian sediment　风积物
aeolian soil　风积土
aeolian vibration　（导线的）风吹振动，风激振动，风成振动，微风振动
aeolotropic　各向异性的
aeolotropies　各向异性
aeolotropism crystal　各向异性晶体
aeolotropism　各向异性
aeolotropy　各向异性
aerated block wall　加气砌块墙
aerated bulk density　充气松密度
aerated concrete block　加气混凝土砌块
aerated concrete　掺气混凝土，加气混凝土，充气混凝土
aerated conduit　曝气管道
aerated contact bed　曝气接触床
aerated flow　掺气水流，充气流动
aerated hopper　冲气料斗
aerated lagoon　曝气塘
aerated nappe　掺气水舌
aerated pond　曝气塘
aerated sand column　充气沙柱
aerated sewage lagoon　曝气污水塘
aerated stabilization basin　曝气稳定池
aerated water　掺气水，碳酸水
aerated weir　掺气溢流堰，真空溢流堰
aerate　曝气，充气，通风，使透气
aerating apparatus　曝气装置
aerating system　通气系统
aerating wastewater　废水曝气
aerating　曝气
aeration air　松动风，松料空气
aeration basin loading　曝气池负荷
aeration basin　曝气池
aeration blower for ash silo　灰库气化风机
aeration blower for ESP　电除尘器气化风机
aeration blower house　气化风机房
aeration blower　气化风机
aeration coefficient　曝气系数，掺气系数
aeration cooling　鼓风冷却，通风冷却
aeration corrosion　充气腐蚀
aeration degree　掺气度，含气度
aeration drying　通风干燥
aeration factor　曝气因素
aeration jet　含气射流
aeration method　充气法
aeration mitigation cavitation　掺气减蚀
aeration nozzle　气化喷嘴
aeration of sewage　污水曝气
aeration pad　气化板
aeration period　曝气时间，曝气周期
aeration pipe　通气管
aeration pond　风干池
aeration porosity　通气孔隙度
aeration slot　掺气槽
aeration tank　曝气池
aeration tap　充气通风
aeration test　充气试验
aeration time　曝气时间
aeration tower　充气塔，曝气塔
aeration treatment　曝气处理
aeration valve　补气阀
aeration zone　饱气带，掺气层，掺气区，含气层，通气带，通气区
aeration　掺气，充气，换气，通风，曝气，（煤粉管道中）风粉混合物
aerator pipe　曝气管
aerator　充气器，充气机，曝气器，曝气机，曝气设备
AERB(Atomic Energy Regulatory Board)　原子能监管委员会，原子能管理局【印度】

aerial array　多振子天线，天线阵
aerial bundle cable　架空绝缘线
aerial cable line　架空电缆线路
aerial cableway　架空索道
aerial cable　架空电缆
aerial camera　航空摄影机
aerial capacity　天线电容
aerial chart　航空图
aerial circuit breaker　天线断路器
aerial circuit　天线电路，空架线路
aerial coil　天线线圈
aerial condenser　空气冷凝器，天线电容器
aerial conductor　架空导线，架空电线，架空线，明线
aerial construction　架空线设施
aerial contaminant　空气污染物
aerial contamination　空气污染
aerial conveyer　架空输送机
aerial conveyor　架空输送机，吊运器
aerial coupling　天线耦合
aerial crosstalk　空中串话干扰
aerial current　天线电流，气流
aerial data　航摄资料
aerial detection　空气检测，空中检测
aerial discharger　避雷器
aerial discharge　空气中放电，气体放电
aerial dump　空中卸料
aerial dust filter　空气滤尘器，空气过滤器
aerial dust　粉尘
aerial effect　天线效应
aerial efficiency　天线效率
aerial farming　航空农业
aerial feeder　架空馈线
aerial gain　天线增益
aerial growth　气生，地面生长，地上部分生长
aerial inductance　天线自感应，天线电感
aerial information　航摄资料
aerial insulator　天线绝缘子，天线隔离子
aerial ladder fire truck　云梯消防车
aerial ladder　架空消防梯，消防梯，云梯
aerial lighthouse　航空灯塔
aerial line　架空线路
aerial lug　外部连接器，外部连线
aerial map　航测图，航摄图
aerial monitoring　空中监督
aerial network　天线网，架空线路网
aerial patrol sign　空中巡视牌
aerial perspective　空中透视，立体透视
aerial photographic mapping　航空摄影制图
aerial photographic survey　空中摄影测量
aerial photograph　航摄照片，航空照片，航测
aerial photomapping　航空摄影制图
aerial photo map　空中摄影地图
aerial pollution　空气污染
aerial radiation thermometer　空气辐射温度计
aerial railway　高架铁道，高架铁路
aerial reactance　天线电抗
aerial reconnaissance　航空勘测，空中勘测
aerial resistance　天线电阻
aerial ropeway conveyer　架空索道输送机
aerial ropeway　架空索道
aerial series condenser　天线串联电容器
aerial sewer　露置下水管道
aerial surveillance　空中监测
aerial survey topographical map　航测地形图
aerial survey　航空测量，空中监测，航测
aerial switch　天线转换开关
aerial trainway　架空索道，高架电车道
aerial transmission　空中传播
aerial triangulation block　空中三角网区
aerial triangulation　空中三角测量
aerial triangulator　辐射三角仪
aerial tuning capacitor　天线调谐电容器
aerial tuning inductance　天线调谐电感
aerial tuning　天线调谐
aerial turning motor　天线旋转用马达
aerial view photo　航摄照片
aerial view　俯瞰图，俯视图，鸟瞰图
aerial wire　架空导线，架空线
aerial　架空的，航空的，大气的，生活在空气中的，空气的，高耸的，天线
aerification　掺气
aerify　充气，掺气，气化
aero　飞机的，航空的，飞行的
aero ball flux measurement　气动小球通量测量
aero ball measurement probe　气动小球测量探头
aero ball measurement room　气动小球测量室
aero ball measurement system　气动小球能量测量系统
aero ball measurement　气动小球测量
aero ball measuring probe　气动小球测量探头
aero ball measuring system　气动小球测量系统
aero ball system　气动小球测量系统
aerobe　需氧微生物
aerobic　需氧的，好气的，好氧的
aerobic bacteria　需氧细菌，好气细菌
aerobic biological fluidized bed　嗜氧生物流化床
aerobic biooxidation　需氧生物氧化作用
aerobic breakdown　需氧分解
aerobic decomposition　有氧分解作用
aerobic digestion　有氧消化
aerobic organism　需氧有机物
aerobic oxidation　需氧氧化作用
aerobic process　有氧过程
aerobic treatment　好氧处理（法），需氧处理，充氧处理，好气处理
aero chemistry　气体化学
aero chlorination　加气氯化
aeroconcrete　加气混凝土，泡沫混凝土
aero cooler　空气冷却器
aerocrete　加气混凝土
aerocurve　曲翼，曲翼飞机
aeroderivative gas turbine　航改式燃气轮机，轻型燃气轮机
aerodiscone antenna　机载盘锥天线
aerodromometer　气流流速表
aerodynamic action　气动力作用，气动力影响
aerodynamic add-on device　气动附加装置
aerodynamic admittance　气动导纳
aerodynamic airfoil　气动翼型
aerodynamically clean body　气动力流线型车身
aerodynamically control　动态控制

aerodynamically fully rough 气动充分粗糙部分
aerodynamically induced vibration 气动力致振
aerodynamically shaped 流线型的，空气动力型的
aerodynamically smooth 气动光滑的
aerodynamically stable 气动力稳定的
aerodynamically 空气动力学地
aerodynamical 空气动力学的，气体动力学的
aerodynamic analysis 空气动力分析
aerodynamic appendage 气动力附件
aerodynamic area 气动力面积
aerodynamic attachment 气动力附件
aerodynamic augmentation device 气动增强装置
aerodynamic balance 气动力平衡，气动力天平
aerodynamic bearing 气动轴承，空气动力轴承
aerodynamic behavior 气动性能
aerodynamic body 空气动力绕流体，流线型体
aerodynamic boundary layer 气动边界层
aerodynamic braking 气动制动，空气动力刹车，空气动力制动
aerodynamic calculation 空气动力计算
aerodynamic capture 气动力捕获，空气动力捕获
aerodynamic center 气动中心
aerodynamic characteristics of rotor 风轮气动特性，风轮空气动力特性
aerodynamic characteristics 空气动力特性，气动特性
aerodynamic chord of airfoil 翼型的气动弦线
aerodynamic coefficient 气动力系数，空气动力系数
aerodynamic compensation 气动补偿
aerodynamic compressor 透平压气机
aerodynamic condition 气动状态
aerodynamic configuration 气动构型
aerodynamic controls 气动控制装置
aerodynamic coupling 气动耦合，气动联轴器，气动力相互作用
aerodynamic criterion 气动判据，气动力准则
aerodynamic cross-coupling 气动交叉耦合
aerodynamic damping coefficient 气动阻尼系数
aerodynamic damping 气动阻尼
aerodynamic data 气动数据
aerodynamic decelerator 气动减速器，气动减速装置，空气动力减速器，空气动力减速装置
aerodynamic deflector 气动导流板
aerodynamic derivative coefficient 气动导数系数
aerodynamic derivative 气动导数
aerodynamic design 气动设计
aerodynamic destabilizing 气动失稳
aerodynamic device 气动装置
aerodynamic diameter 空气动力学直径
aerodynamic directional instability 气动力方向不稳定性
aerodynamic dissipation 气动力耗散
aerodynamic disturbance 气动力扰动，空气扰动
aerodynamic downwash 气动力下洗
aerodynamic drag 气动阻力
aerodynamic effectiveness 气动力效能，空气动力效能
aerodynamic effect 气动力效应
aerodynamic efficiency 空气动力准备率，气动效率
aerodynamic end effect 气动力端部效应
aerodynamic environment 气动环境
aerodynamic excitation 气动力激励
aerodynamic field 气动力场
aerodynamic force coefficient 气动力系数
aerodynamic force component 气动力分量，空气动力分量
aerodynamic force derivative 气动力导数
aerodynamic force differential 气动力差
aerodynamic force 气动力
aerodynamic form 空气动力学形状，气动形状流线型
aerodynamic handling force 气动操纵力
aerodynamic heating 空气动力加热，气动加热
aerodynamic heat 气动热
aerodynamic heat transfer 气动换热
aerodynamic hinge moment 气动铰链力矩
aerodynamic hysteresis 气动力迟滞
aerodynamic influence coefficient 气动影响系数
aerodynamic instability 气动不稳定性
aerodynamic interaction 气动力相互作用
aerodynamic interference 气动力干扰，气动力干涉
aerodynamicist 空气动力学家，气动学家
aerodynamic laboratory 空气动力实验室
aerodynamic lag 气动力滞后
aerodynamic layout 气动力外形
aerodynamic lift 气动升力
aerodynamic loading 气动力负荷，气动载荷
aerodynamic load distribution 气动力载荷分布
aerodynamic mass 气动质量
aerodynamic mean chord 平均气动弦长
aerodynamic model 空气动力模型
aerodynamic modification 气动修改，气动改型
aerodynamic moment of inertia 气动力惯性矩
aerodynamic noise 气动噪声
aerodynamic nomogram 空气动力特性图
aerodynamic nonlinearity 空气动力特性的非线性，气动特性的非线性
aerodynamic optimization 空气动力特性最优化，气动特性最优化
aerodynamic parameter 气动参数
aerodynamic performance test 气动性能试验
aerodynamic performance 气动性能
aerodynamic piston theory 气动活塞理论
aerodynamic pitch 气动桨距
aerodynamic power controller 气动功率控制器
aerodynamic power regulation method 气动功率调节方法
aerodynamic profile 气动翼型
aerodynamic property 气动特性
aerodynamic quality 气动品质
aerodynamic radius 空气动力学半径，气动半径
aerodynamic reaction 气动力反作用，气动反冲力
aerodynamic reference center 气动力参考中心
aerodynamic refinement 气动改善，气动改型
aerodynamic resistance 气动阻力
aerodynamic roughness length 气动粗糙长度

aerodynamic roughness parameter 气动粗糙参数
aerodynamic roughness 气动粗糙度
aerodynamic self-excitation 气动力自激
aerodynamic self-starter 气动自动启动装置
aerodynamic shadow 气动阴影，气动尾迹
aerodynamic shape correction 气动外形修改
aerodynamic shape optimization 气动外形最优化
aerodynamic shape 气动形状
aerodynamic shaping 空气动力造型
aerodynamic shear 气力剪切力
aerodynamic shield 气动整流罩
aerodynamic shroud 整流罩
aerodynamic similarity 气动相似性
aerodynamic simulation 空气动力模拟
aerodynamics of cooling 冷却系统的空气动力学
aerodynamics of hot system 热系统空气动力学
aerodynamic solidity 气动实度
aerodynamic sound 气动声音
aerodynamic source of sound 气动声源
aerodynamic speed control 气动速度控制
aerodynamic spoiler device 气动扰流装置
aerodynamic spoiler 气动扰流板
aerodynamic stability 气动稳定性
aerodynamic stack downwash 空气动力堆积下洗气流
aerodynamic stack height 空气动力堆积高度，气动高度
aerodynamic starting 气动力启动
aerodynamic stiffness 气动刚度
aerodynamic surface 气动力作用面
aerodynamics 空气动力学，气体动力学
aerodynamic technology 空气动力工艺，气动工艺
aerodynamic term 空气动力项，气动项
aerodynamic turbine 气动透平，空气动力涡轮
aerodynamic test 气动力试验
aerodynamic theory 空气动力学理论，气动理论
aerodynamic torque 空气动力扭矩
aerodynamic trail 气动尾迹，航迹云
aerodynamic transfer function 气动力传递函数
aerodynamic tunnel 风洞
aerodynamic twist 气动扭转
aerodynamic vane 整流片，导流片
aerodynamic wake 气动尾流
aerodynamic yaw 气动力偏航
aerodynamic 空气动力学的，空气动力的，气体动力学的，气动的
aeroelastic building 气动弹性建筑物
aeroelastic characteristic 气动弹性特性
aeroelastic deformation 气动弹性变形
aeroelastic derivative 气动弹性导数
aeroelastic effect 气动弹性效应
aeroelastic eigenfunction 气动弹性特征函数
aeroelastic eigenvalue 气动弹性特征值
aeroelastic equation of motion 气动弹性运动方程
aeroelastic feedback 气动弹性反馈
aeroelastic instability 气动弹性不稳定性
aeroelastic interaction 气动力弹性干扰
aeroelasticity 气动弹性学，气动弹性力学
aeroelastic load 气动弹性载荷
aeroelastic modeling 气动弹性建模
aeroelastic model 气动弹性模型
aeroelastic oscillations 气动弹性振动
aeroelastic phenomena 气动弹性现象
aeroelastic response 气动弹性响应，气动弹性反应
aeroelastic section model 气动弹性节段模型
aeroelastic simulation 气动弹性计算
aeroelastic stability 气动弹性稳定性
aeroelastic stiffness 气动弹性刚度
aeroelastic structure 气动弹性结构物
aeroelastics 空气弹性学
aeroelastic transfer function 气动弹性传递函数
aeroelastic vibration 气动弹性振动
aeroelastic wing theory 气动弹性翼理论
aeroelastic wing 气动弹性翼
aeroelastic 气动弹性的，气动力弹性的
aerofilter 空气过滤器
aerofoil balding 翼面叶
aerofoil camber 翼型弯度
aerofoil design 翼型设计
aerofoil double speed 轴流式双速吸风机
aerofoil fan 机翼型轴流风机，机翼型叶片风机，轴流通风机
aerofoil profile 翼型剖面，翼型
aerofoil surface finish 翼型表面光洁度
aerofoil theory 机翼理论
aerofoil thickness 翼型厚度
aero foil 翼型，翼剖面，机翼型，翼面，机翼
aero gas turbine engine 航空燃气轮机［发动机］
aero gas turbine 航空燃气轮机
aerogel 气凝胶
aerogenerator 风力发电机
aerography 大气图表，大气学，气象学
aerograph 高空气象计
aero hydraulic gun 气压水枪
aero hydrous 含有气水的，空穴水
aero indicating light 航空障碍灯
aero jet 空气喷射，喷气发动机
aerolian vibration of overhead line 架空线微风振动
aerolite 陨石，硫磺炸药
aerolith 陨石，硫磺炸药
aerological analysis 高空分析
aerological ascent 高空观测
aerological instrument 高空仪器
aerological observatory 高空观象台
aerological sounding 高空观测，高空探测
aerological station 高空观测站
aerological theodolite 高空经纬仪
aerologist 大气学家
aerology 大气学，高空气象学
aeromancy 大气预测
aeromap 航空图
aeromechanics 空气力学，气体力学
aerometeorograph 高空气象计
aerometer 风速仪，气体密度计，气体比重计，空气流量计，量气计
aerometric network 高空气象网
aerometry 气体测量
aeromobile 气垫车
aeronautical chart 航行图

aeronautical experiment 航空实验
aeronautical meteorology 航空气象学
aeronautical type wind tunnel 航空风洞
aeronautic-radio navigation station 航空无线电导航台
aeronautics 航空学
aeronomy 高层大气物理学，高空大气学
aeropause 大气上界
aero performance 气动性能
aerophoto-grammetric survey 航空摄影测量
aerophoto-grammetric map 航空摄影测量图
aerophoto-grammetric plan 航空摄影测量平面图
aerophotography 航空摄影学，航空摄影照片
aeroplane view 鸟瞰图
aeroplankton 空气浮游生物
aero-projection method 空中投影法
aero-projector multiplex 多倍投影测图仪
aero-projector 航空投影仪，立体投影器
aeropulse 脉冲式空气喷气发动机
aeropulverizer 吹气磨粉机
aero radioactivity 大气放射性，空中放射性，大气放射性强度
aero-sensitive 气动力敏感的
aeroslide 斜槽
aerosol activity 气溶胶放射性，气溶胶活度
aerosol analyzer 气溶胶分析器，气溶胶分析仪
aerosol collection 气溶胶捕集
aerosol concentration 气溶胶浓度
aerosol filter 气溶胶过滤器
aerosol generation 气溶胶生成
aerosol generator 气溶胶发生器
aerosol ionization detector 气溶胶电离探测器
aerosol mixing ratio 气溶胶混合比
aerosol monitor 气溶胶检测器【记录器】，气溶胶监测器
aerosol particle 气溶胶粒子
aerosol sampling 气溶胶取样
aerosol scope 空气微粒测算器
aerosol sedimentation 气溶胶沉淀
aerosol source 气溶胶源
aerosol spectrometer 气溶胶谱仪
aerosols 悬浮微粒
aerosol test 烟雾试验
aerosol 大气悬浮物，大气微粒，空气溶胶，悬浮微粒，悬空微尘，烟雾剂，喷雾剂，气雾剂，气溶胶，浮质
aerospheree 人工呼吸器，呼吸面具，通风面具
aerosphere 大气层，大气圈，气圈
aerostatic bearing 空气静力轴承
aerostatics 空气静力学，气体静力学
aerostatic 空气静力学的
aerostat 气球，飞艇
aerosurveying 航空勘测
aerothermodynamics duct 冲压式空气喷气发动机
aerothermodynamics 空气热力学，气动热力学，气体热力学
aerothermopressor 气动热力扩压器，气动热力增压器
aerotron 三极管
aeroturbine 航空涡轮，空气涡轮机
aerovane 风向风速计
aeroview 鸟瞰图
aerugo 氧化铜，铜绿
AES(American Electromechanical Society) 美国机电学会
AES(American Engine Society) 美国发动机学会
AESC(American Engineering Standards Committee) 美国工程标准委员会
aesthetic constraint 美学制约
aesthetic forest 风景林
aesthetic requirement 审美要求
aesthetic standard 审美标准
aesthetics 美化要求，审美，美学
aesthetic 审美的，美观的，美学的
AETR(advanced epithermal thorium reactor) 先进型超热钍反应堆
AET 主给水泵汽轮机轴封系统【核电站系统代码】
AF(as fired basis) 应用基
AF(axial flow) 轴向流动
AFBC(atmospheric fluidized bed combustion) 常压流化床燃烧
AFC(automatic following control) 自动跟踪控制
AFCEN 法国核岛设备设计和建造及在役检查规则协会
AFDB(Africa Development Bank) 非洲开发银行
a few 有些，几个
affecting factor 影响因素
affect 影响，起作用，损伤
affidavit 保证书，宣誓书
affidavit of document 文件证明书
affiliated company 附属公司，隶属公司
affiliated corporation 附属公司
affiliated enterprise 分支企业，联号企业
affiliated installations 辅助设施，隶属设施
affiliated organization 下属机构
affiliated person 关联人
affiliate 分公司，子公司，分支机构，附属企业
affiliation to 附属于
affiliation 从属关系，联系
affine deformation 均匀变形
affinity diagram 相似图
affinity law 相似定律，亲和定律
affinity 亲和力，近似，相似，相似性
affinor 反对称张量，张量
affirmation of contract 批准合同
affirmative warranty 确认保证
affirm 确认，肯定，证实，承认
affixture 粘贴，附加，加成物，附加物
affix 附录，附件，附言，添加，附标，签字
affluence 汇流
affluent level 壅高水位
affluent stream 支流
affluent 汇流的，流入的，支流，富裕的
afflux curve 汇流曲线
afflux 汇集，汇流，流入，聚集，壅水，汇入
affordable 提供得起的
affording ease of operation 便于操作的，操作方便的
afford 给予，使得到
afforestation 绿化建设，植树造林，绿化，造林
afforest 绿化，造林

AFI(Air Filter Institute)test　AFI 试验【过滤器】
a flight of stairs　一段楼梯
afloat evaporimeter　漂浮蒸发计
aforementioned　上述的，前述的，前面提到的
aforesaid　上述的，前述的
a fortnight ago　两周前
a fortnight sailing　双周班【船】
AF-PMSG(axial flux permanent magnet synchronous generator)　轴向磁通永磁同步发电机
A-frame　A 形架
A-framed derrick　人字起重机，人字吊杆
African Development Fund　非洲开发基金
a Friday sailing　周五班【船】
AFR(away-from-reactor) storage　离堆贮存，堆外贮存
AFT(atmosphere flash tank)　大气扩容器
AFT(automatic fine tuning)　自动微调
AFT(automatic frequency tuner)　自动频率调谐器
aft bearing　后轴承
aft deck　后甲板
aft draft　尾吃水
after　在……之后
after-bay dam　尾水池坝
after-bay reservoir　尾水水库，后池
after-bay　尾水池，后池，后湾
after-body　后体，后车身
after-body effect　后体效应
after-boiler corrosion　炉后腐蚀【指汽包后的汽水系统】
after-boiler system　炉后系统【指汽包后的汽水系统】
afterburner nozzle　加力燃烧室喷管，加速燃烧室喷管
afterburner　后燃室，加力燃烧室，加速燃烧室，复燃室，燃尽室
afterburning　复燃，补燃，后期燃烧，加力燃烧，加速燃烧，补充燃烧，二次燃烧
after-combustion　复燃，燃尽，后期燃烧，补充燃烧，再次燃烧
aftercondenser　后冷却器，后置冷凝器，后凝汽器
after-contraction　残余收缩，剩余变形
after-cooler　附加冷凝器，末级冷却器，后置冷却器，停堆冷却器，后冷却器
after-cooling　再冷却
after-current　余电流，剩余电流
after-drying　再次干燥，再干燥
after-effect　后效应，滞后效应，次生效应，副作用，后效
after-end　后端
after-expansion　残余膨胀，残存膨胀，余膨胀
after-exposure　后期曝光，后照射
after-filter　后过滤器，再过滤器，二次过滤器
afterflow　蠕变，残余塑性流动，残余塑性变形
after four revisions　经过四次修订
after-furnace　炉后的
afterglow duration　余辉持续时间
afterglow　余辉
after-grass　再生草
after-hardening　后期硬化
after-heat condenser　余热凝汽器
after-heater　后置加热器
after-heat output　余热功率，停堆热功率
after-heat removal chain　余热排出管系
after-heat removal　余热导出，余热排出
after-heat　剩余热量，余热，后加热，剩余释热，剩（残）余热，衰变期
after-irradiation　后辐照
aftermarket　后继市场
after-mast　后桅
aftermath removal　余热排出
after-mould stress　成形后的应力
after-power　滞后功率，剩余功率
after-precipitate　滞后沉淀物
afterprecipitation　二次沉淀，后沉淀
after-product　后产品，后产物，副产品，二次产物，副产物
after-pulse　剩余脉冲，残留脉冲
after-pulsing　主脉冲后的剩余脉冲
after-quake　余震
after-sales service　售后服务
after-service center　售后服务中心
after-service　售后服务
after-set insert　凝固后镶嵌件
after-shock activity　余震活动
after-shock　余震，后震
after-shrinkage　后期收缩
after-tax margin　税后利润
after-tax profit　税后利润
after-tow　后曳
after-treatment　后处理，二次处理
after-vibration　后振动
after-wind　余风
aft flow　残余塑性流动【金属的】，蠕变，塑性变形【停止加载后】
aft loading blade　后加载叶片
AFW(auxiliary feed-water)　辅助给水
AFWC(automatic feed-water control)　自动供水控制，自动给水控制
AFWP(auxiliary feed-water pump)　辅助给水泵
against freezing and evaporation　抗冻抗蒸发
against wind　逆风，迎风
agar-agar　琼脂
agate mortar　玛瑙研钵
AG(available generation)　可用发电量
AGCA(Associated General Contractors of America)　美国总承包商协会
AGC(automatic gain control)　自动增益控制
AGC(automatic generating control)　自动发电控制【DCS 画面】
AGC operating mode　自动发电控制运行方式
age coating of lamp　灯泡老化层
age dating　年龄测定，年代测定
aged cement　过期水泥，老化水泥
aged deterioration　时效劣化
age determination　年代测定
age diffusion approximation　年龄扩散近似
age diffusion equation　年龄扩散方程
age distribution　年龄分布
aged oil　老化油
aged uranium　衰变铀，蜕变铀

age factor 年代因数
age hardening 时效硬化，时效硬化处理
age histogram 年龄直方图
ageing behavior 时效行为，时效效应，耐老化状态
ageing characteristic 老化特性
ageing crack 时效裂纹，自然裂纹，时效断裂
ageing hardening 时效硬化
ageing of insulation 绝缘老化
ageing of ion exchange resin 离子交换树脂的老化
ageing of materials 材料老化
ageing of valve 电子管老化
ageing process 老化过程
ageing resistance 抗老化性，耐久性，时效阻力
ageing stability 老化稳定性
ageing strengthening 时效强化
ageing test 老化试验，时效试验
ageing time 老化时间
ageing treatment 时效处理
ageing unit 老机组，旧机组
ageing 时效，老化，变质
age limit 寿命，使用年限，年龄限制
agency agreement 代理协议
agency commission 代理手续费，代理佣金，代理费
agency contract 代理合同，代理契约
agency fee 代理费
agency manager 代理经理
agency marketing 代理处经销
agency project 机构援助项目
agency 代理，代理权，代办处，代理商，代理处，代理关系，机关，机构，中介，作用，介质，办事处
agenda item 议程项目
agenda of meeting 会议日程，日程，议程，日程表，议事日程
agendum 操作规程，执行规程
agent bank 代理银行
agent general 总代理
agent of fusion 熔剂
agent's remuneration 代理（人）酬金
agent 剂，作用剂，媒介物，试剂，药剂，代理人，代理商
age of big data 大数据时代
age of catalyst 催化剂寿命
age of concrete 混凝土龄期
age of diurnal inequality 日潮不等潮龄
age of diurnal tide 日潮潮龄
age of earth crust 地壳年代
age of moon 月龄
age of sludge 污泥变陈
age of specimen 试件龄期
age of stand 林龄
age peaking factor 年龄峰值因子
age pyramid 年龄统计角锥状图
age ratio 年代比，年龄比
age-strength relation of concrete 混凝土龄期-强度关系
age to indium resonance 到铟共振能量的年龄
age 老化，陈化，熟化，时效，寿命，年龄，龄期，时代，年代，使变老，使用期限
agglomerant 凝聚剂
agglomerated flux 聚结熔剂，烧结焊剂，陶质焊剂
agglomerated 结块，烧结，烧结块，凝聚，凝聚物，聚结的
agglomerate-foam concrete 烧结矿渣泡沫混凝土
agglomerate section 沉降电极【电气除尘器的】
agglomerate 成团，烧结团块，附聚物，结块，聚集，凝聚物，凝聚剂
agglomeratic ice 块冰
agglomerating coal 烧结煤
agglomerating combustor 结渣燃烧室
agglomerating index 烧结指数，黏结指数
agglomeration 结块，烧结，凝聚，成团，附聚（作用），团聚（粉末）
agglutinant 凝聚剂，胶着剂
agglutinate reaction 凝集反应
agglutinate 胶结，凝聚，附着，黏合
agglutinating 凝聚的，胶结的
agglutination reaction 凝聚反应
agglutination 烧结，凝聚，附着，凝集，成胶状，胶着作用，凝聚作用
agglutinative absorption 凝集吸收作用
aggradation permafrost 增积永冻层
aggradation plain 沉积平原，淤积平原
aggradation 淤高，沉积，淤积，加积，堆
aggraded flood plain 洪积平原
aggrading action 淤积作用
aggrading river 冲积河流
aggrading stream 冲积河流
aggregate analysis 总量分析
aggregate averaging 骨料平均粒径的测定
aggregate batcher 骨料配料机
aggregate batching bin 骨料配料仓
aggregate bin 骨料仓
aggregate blending 骨料混合
aggregate capacity 总功率，机组功率，聚集容量，总容量
aggregate cement ratio 骨料水泥比，骨灰比【混凝土配料的】
aggregate chip 石屑
aggregated duration 累计持续时间
aggregated erratum 累积误差，总误差
aggregated error 累积误差
aggregated momentum 总动量
aggregate equivalent peak interruption duration 等效峰荷累计停电持续时间
aggregate error 累积误差，总误差
aggregate feeding 骨料供给
aggregate gradation 骨料分级，骨料级配
aggregate grading 集料级配，选配集料
aggregate income 总收入
aggregate investment 投资总量，总投资
aggregate load 总负荷
aggregate material 添加材料
aggregate plant 砂石（料）厂，集料厂，骨料（加工）厂
aggregate processing plant 骨料制备厂，骨料加工厂
aggregate production system 砂石料生产系统

aggregate reactivity 骨料活性
aggregate reclaiming plant 骨料场，砂石料厂，砾石料场，骨料开采设备
aggregate recoil 集合反冲，群体反冲
aggregate roofing surface 骨料屋面
aggregate size 骨料粒径
aggregate spreader 骨料撒布机
aggregates processing 骨料的制备
aggregate state 聚集态，固粘态
aggregate storage capacity 总储存容量，总库容
aggregate storage 骨料储存场，骨料储存，骨料仓库
aggregate structure 团粒结构
aggregate tonnage 总吨位
aggregate trailer 骨料拖运车
aggregate unit 组合装置，联合结构
aggregate value method of weighting 总值加权法
aggregate washing 骨料冲洗
aggregate weighing equipment 骨料称量设备
aggregate 掺和料【水泥】，凝聚体，聚集体，成套设备，（混凝土）骨料，合计，总计，总和，总额，集合体，总的，合计的，聚合的，碎石，机组，集结，聚集
aggregating 聚集的，聚集作用
aggregation degree 团聚度
aggregation of soil particles 土粒聚集
aggregation 团粒作用【土壤】，聚集，聚合，团聚作用，集成体
aggregative method 总计法
aggregative planning 总体规划，综合规划
aggregator 组合体
aggressive agent 慢蚀剂，侵蚀剂
aggressive atmosphere 侵蚀性大气，腐蚀性气氛，污染大气
aggressive device 主动装置
aggressive gas 腐蚀性气体
aggressive medium 腐蚀介质
aggressive soil 侵蚀性土壤
aggressive water 侵害性水，腐蚀性水，侵蚀性水
aggressive 侵袭性的，侵害性的，腐蚀性的，侵蚀性的，生锈的，积极的
aggrieve 冒犯，侵害……的合法权利
aging city 老龄化城市
aging coefficient 老化系数
aging crack 时效裂纹
aging effect 老化效应
aging failure 老化破坏
aging of cathode 阴极的老化
aging phenomenon 老化现象
aging process 老化过程
aging rate 老化速度
aging resistance 抗老化性，时效电阻，抗时效性
aging retarder 防老化剂
aging test 老化试验
aging treatment 时效处理
aging voltage 老化用电压
aging 老化，时效，衰变，变质
agitated fluidized bed 搅拌流化床
agitate 搅动，搅拌
agitating lorry （混凝土）搅拌车
agitating truck 搅拌车
agitation blower 搅拌鼓风机
agitation dredging 搅动挖泥
agitation frequency 搅动频率
agitation nozzle 搅拌喷嘴
agitation 拌和，搅拌，搅动，掺混，摇动，激励
agitator arm 搅动杆
agitator truck 混凝土搅拌车
agitator 搅拌机，搅拌器，混音器
AGMA（American Gear Manufacturers Association）美国齿轮制造商协会
AGM 电动主给水泵润滑油系统【核电站系统代码】
agonic line 无偏线
AGR（advanced gas-cooled reactor） 改进型气冷反应堆，改进型气冷堆
agraff 搭钩，搭扣
agreed amount 议定数额
agreed conclusion 议定结论
agreed duty 议定关税
agreed period 商定期限
agreed port of destination 约定目的港
agreed price 议定价格
agreed sum 商定金额
agreed text 一致同意的文本，议定文本
agreement by piece 计件合同
agreement fixing price 共同定价协议
agreement form 协议书格式
agreement in written 书面协定（协议）
agreement of electric power supply and demand 供用电协议
agreement of reimbursement 偿付协议
agreement on scientific and technical cooperation 科学技术合作协定
agreement price 协议价格，议定价
agreement rate 协议运价
agreement tariff 关税协定
agreement 协议，契约，合同，协议书，协定，同意
agree to differ 求同存异
agree to disagree 同意双方存在歧见
agree to your terms 同意你方的条件
agree 同意，赞成
AGR fuel element 改进型气冷堆燃料元件
AGR fueling machine 改进型气冷堆装料机
agricultural buildings 农业建筑物
agricultural chemicals 农业化学品，农药
agricultural drainage 农业排水
agricultural drain 农用排水沟
agricultural ecosystem 农业生态系统
agricultural fire pump 农用消防泵
agricultural planting 农作物
agricultural soil 种植土
agricultural watershed 农业集水区
agricultural water 农业用水
agriculture bank 农业银行
agriculture 农业，农学
AGR nuclear power plant 改进型气冷堆核电厂
AGR 主给水泵汽轮机润滑、调节油系统【核电站系统代码】

AH(air heater) 空气预热器
AH(ampere hour) 安培小时
ahead blading 正车叶片
ahead of schedule 提前
ahead stage 正车级
ahead turbine 正车透平
ahead 提前
AH hopper(air heater hopper) 空气预热器灰斗【DCS 画面】
A-horizon 淋溶层，淋溶土层，表土层，上土层
a host country 东道国
AHP 高压给水加热器系统【核电站系统代码】
AIA(American Institute of Accountants) 美国会计师协会
AIA(American Institute of Architects) 美国建筑师协会
AIA(Association of International Accountants) 国际会计师公会
AIA(Award Intention Agreement) 议标协议
AIA contract conditions 美国建筑师协会合同体系的合同条件
AI(analog input) 模拟量输入
AI(analogue input) 模拟量输入
AI/AO(analogue input/analog output) 模拟输入、输出，模拟量
AICPI(All India Consumer Price Index) 印度全国消费价格指数
AIEE(American Institute of Electrical Engineers) 美国电气工程师协会
AID(application industry data) 应用工业资料
aid cooperation agreement 援助合作协议
aid disbursement 援助额
aided design 辅助设计
aided tracking mechanism 辅助跟踪装置
aided tracking 人工跟踪，半自动跟踪
aid equipment 援助设备
aid given gratis 无偿援助
aid giving nation 援助国
aid in kind 实物援助
aid receiving nation 受援国
aid to navigation 航路标志
aid 辅助设备，辅助装置，帮助，援助
AIEEE(American Institute of Electrical and Electronics Engineers) 美国电气和电子工程师协会
aiguille 钻孔器，钻头
AIIB(Asian Infrastructure Investment Bank) 亚投行，亚洲基础设施投资银行
AIIE(American Institute of Industrial Engineers) 美国工业工程师协会
aileron actuator 副翼促动器
aileron cable 副翼操纵索
aileron chord 副翼弦
aileron control 副翼控制
aileron hinge 副翼铰链
aileron horn 副翼杆，副翼操纵杆
aileron yaw 副翼偏航
aileron 副翼
ailment 疾病
aiming line 觇线，照准线，瞄准线
aiming point 觇点，照准点
aiming post 标杆
aiming （风轮的）对风，调向
aimless drainage 紊乱水系，无目标排水
aim of wheel 轮辐
aim 目标，目的
air accumulation 集气
air accumulator gas turbine 蓄气式燃气轮机
air accumulator 储气瓶
air acetylene welding 气焊，乙炔焊，空气-乙炔焊接
air activity messing room 空气活度测量室
air-actuated control system 气动控制系统
air admission port 进气口
air admission valve 进气阀，真空破坏阀
air admission 进气
air admitting surface 进风口面积
air after heater 空气后热器
air agitated wash 气搅动清洗
air agitation device 空气搅拌装置
air agitation 空气搅拌
air analysis 空气分析
air and bead mixture 空气微珠混合器
air and coal mixture 气粉混合物
air and fuel mixture 空气燃料混合气，空气燃料混合物
air and gas mixer 空气煤气混合器
air and gas system 烟风系统
air and steam blast 空气蒸汽鼓风
air and vacuum valve 空气真空阀
air and water system 气水系统
air-arc cutter 空气电弧切割器
air-assisted pressure burner 空气助压喷燃器
air atomization 空气雾化
air atomizer 空气雾化器
air atomizing oil burner 空气雾化油燃烧器
air atomizing 空气雾化
air at rest 静止空气
air attemperator 气冷式调温器
air away bill 空运单
air back valve 空气止回阀，止回气阀
air baffle plate 挡风板
air baffle ring 挡风环
air balance 风量平衡
air-ballast pump 气镇泵
air balloon 气球
air barrier effect 空障效应
air base 航空基地
air bath 空气浴，干燥室
air bearing 空气轴承
air bellows 风箱，空气弹簧
air bells pollution 大气污染
air bell 气流罩
air belt 气带
air bill of lading 空运提单
air binding 气封，气塞，气结，气泡堵塞
air blanketing 空气夹层
air blast atomizer 空气雾化喷油嘴，气动雾化器
air blast atomizing 空气雾化，气动雾化
air blast breaker 空气吹弧断路器
air blast circuit breaker 空气开关，空气断路器
air blaster 空气喷射器，气嘴
air blast loading 气流冲击荷载

air blast quenching 气流淬，风冷淬火
air blast refrigeration 空气喷射制冷
air blast shutdown 强制通风冷却停机
air blast transformer 风冷式变压器
air blast 鼓风，空气喷射，空气鼓风，强气流
air bleeder 放气管，排气孔，放气阀，通气小孔
air bleed set 抽气机，抽气器
air bleed valve 放气阀，泄气阀，排气阀
air bleed 空气泄放，放气，抽气，排气，通气器
air blow aeration tank 空气鼓风曝气槽
air blower room 风机室
air blower 鼓风机，吹风机，散热风扇
air blow floatation unit 鼓气浮选装置
air blowing oxidation 吹气氧化
air blowing 吹气，空气吹污，空气吹洗，空气洗涤
air blown asphalt 喷射沥青
air blown mortar 喷射砂浆
air blowout 空气灭弧
air blow 鼓风
air boiler 空气加热器，空气锅炉
air-boost compressor 增压压气机
airborne activity 大气层放射性，大气中的放射性，气载放射性，机载活动
airborne biological form 空气生物形态
airborne concentration 气载浓度
airborne contaminant 空气污染物，气鼓污染物，大气污染物，气载污染物
airborne contamination 大气污染
airborne dirt 空气污尘
airborne dryer 悬浮空气干燥器
airborne dust 大气尘埃，空气夹带的灰尘，飘尘，空气粉尘，气载尘埃
airborne electromagnetic prospecting 机载电磁勘探
airborne infection 空气传染
airborne noise 空中噪声，空气噪声
airborne particular 气载微粒，气载粒子，大气中的粒子，大气尘埃
airborne particulate control 悬浮粒子控制
airborne particulates 灰尘的，尘埃，悬浮粒子
airborne plutonium 大气中钚，大气中钚尘
airborne pollutant 气载污染物
airborne pollution 大气污染，空气污染
airborne radioactivity survey 航空放射性测量
airborne radioactivity 大气中的放射性，气载放射性，航空放射性
airborne reactor 飞机用反应堆
airborne release 空中排放，大气排放
airborne salt 空气悬浮盐类
airborne sensor system 空中传感系统
airborne waste 气载废物
airborne 空运的，飞机载的，航空的，空气所带的，空气传播的，空中的，空中浮游的，气载的
air bottle 空气瓶，气瓶，空气罐
air bound 气塞的，气结，空气阻塞，气障，气隔，空气障碍
air box 风箱，波纹管，皮老虎
air brake control valve 气动控制阀
air brake dynamometer 气闸测功器，空气制动测力计
air brake switch 气动开关
air brake system 气动系统
air brake 空气制动，空气制动器，气闸，风闸
air braking system 空气制动系统
air braking tachometer 摩擦式转速计
air braking test 空气刹车试验
air break circuit breaker 空气断路器
air breaker 空气断路器
air breakwater 气动防波堤，空气防波堤，喷气防波堤
air breather 通风装置，通气装置，通风孔，通气孔
air breathing suit 气衣
air brick 风干砖，空心砖，多孔砖
air brush 气刷
air bubble collapse 气泡破碎
air bubble pitting 气泡点蚀
air bubbler system 气泡防冻系统
air bubbler type level measurement 气泡式水平测量
air bubbler type specific gravity measurement 气泡式比重测量
air bubble viscometer 气泡黏度计
air bubble 气泡，砂眼，空气泡
air buffer 空气缓冲器，空气阻尼器，空气隔层
air bumper 空气减震器，空气缓冲器，空气阻尼器
air bump rinse operation 空气冲洗
air cannon 空气炮
air capacitor 空气电容器
air capacity 空气量，含气量，容气量
air carbon arc gouging 空气碳棒弧割
air cargo 空运货物
air casing 空气套，空气隔层，空气夹层
air cavity 气孔，气腔，空穴，空气穴
air cell 空气夹层，空气电池，气囊
air cement gun 风动水泥喷轮
air-chamber pump 气室泵
air-chamber type surge tank 气箱式调压塔
air-chamber 风室，空气室，气室，气舱
air change loss 空气交换损失
air change rate 空气更新率，换气次数，换气率，通风换气次数
air changes 换气次数
air channel 通风道，排气道，风道
air-charged accumulator 充气蓄能器
air charging system 充气装置
air charging tool 充气工具
air chimney 竖通风道，通风竖井，竖风道，排气道，通风道，通风井，通风烟囱
air chipper 风铲
air choke 空气扼流圈
air chute 风道，排气道，通风道
air circuit breaker 空气断路器
air circuit 空气管道，空气管道系统
air circulation 空气环流，空气循环
air classifier 风力分级机
air cleaner 空气过滤器，空气净化器，空气净

化机，空气滤净器，空气净化装置，滤气器
air cleaning equipment 空气净化装置
air cleaning facility 空气净化设备，空气净化装置
air cleaning system 空气净化系统
air cleaning unit 空气净化装置
air cleaning 空气净化
air cleanliness 空气洁净度
air cloth ratio 气布比
air clutch 气动离合器
air coal ratio 风煤比
air cock on the indicator 散热器放气阀
air cock 放气活塞，空气旋塞，气栓，排气旋塞，气阀，气嘴
air-coil 气冷盘管
air collector 集气罐，空气集热器
air compartment 风室，气室
air compressing station 空压站
air compressor oil 空气压缩机油
air compressor plant 空压机站
air compressor room 空气压缩机室，空压机间
air compressor station 空气压缩机房
air compressor surge 压气机喘振
air compressor 空气压缩机，压气机，空压机
air condenser 空气凝汽器，空气电容器，空气冷凝器，风冷冷凝器，风冷式冷凝器
air conditioned area 空调场所
air conditioned building 带空调的建筑
air conditioned location 空调区域
air conditioned room 空调房间
air conditioner house 空调机室
air conditioner room 空调机室
air conditioner 空气调节器，空调机，空调器，空调
air conditioning condition 空调工况
air conditioning design 空气调节设计
air conditioning equipment 空气调节设备
air conditioning installation 空调装置，空气调节装置
air conditioning labor 空调实验室
air conditioning load 空气调节负荷，空调负荷
air conditioning machine room 空调机室，空调机房，空气调节机房
air conditioning method 空气调节方法
air conditioning plant 空气调节装置
air conditioning system cooling load 空气调节系统冷负荷
air conditioning system 空气调节系统
air conditioning unit 空气调节装置
air conditioning 空调，空气调节
air conduction 空气传导
air conductivity 空气电导率，大气导电性
air conduit 风管，风道，空气管道，空气导管
air consumption 空气消耗量，耗气量，耗风量
air container 储气器
air contaminant 空气污染，大气污染物
air contamination analysis 空气污染分析
air contamination meter 空气污染测量计
air contamination monitor 大气污染监测器，空气污染监测器
air contamination 空气污染，大气污染，空气污染物
air content test 含气量试验
air content 含气量，空气量
air control 气动控制
air control equipment 气动控制装置
air controlled direction valve 气动控制换向阀
air control motor 气动控制电动机
air control system 气动控制系统
air control unit 气动控制装置
air control valve 空气调节阀，气动控制阀
air convection loss 空气对流损失
air conveying line 空气输送管道
air conveyor 气力输送机，气流式的输送机
air coolant 冷却空气，空气冷却剂
air cooled blade 气冷式叶片
air cooled cargo 冷气货
air cooled cascade 气冷叶
air cooled condenser 空冷凝结器，空冷式冷凝器，风冷式冷凝器，气冷式冷凝器
air cooled condensing unit 风冷冷凝机组
air cooled conditioner 风冷式空调器
air cooled desuperheater 空冷减温器
air cooled fin 肋片【空气侧】
air cooled generator 空冷发电机，空冷式发电机
air cooled heat exchanger 空冷换热器
air cooled heating unit 空冷供热机组
air cooled oil cooler 风冷式油冷却器
air cooled pipe penetration 空冷管贯穿件
air cooled reactor 空气冷却堆
air cooled riveting hammer 气冷式铆钉锤
air cooled rotor 气冷式转子
air cooled side shielded nozzle 气冷式侧面保护喷嘴
air cooled steam condensation system 空气冷却凝汽系统
air cooled steam turbine 空冷汽轮机
air cooled transformer 气冷式变压器，空气冷却的变压器
air cooled tube （沸腾炉）空冷管，气冷电子管
air cooled turbine unit 空冷式汽轮机组
air cooled type 风冷式
air cooled wall 空冷炉墙
air cooled 风冷的，空冷的，空气冷却的，气冷，空冷
air cooler duct 空气冷却器（空气）导管
air cooler turbine 空冷机组
air cooler 空气冷却器，风冷装置，冷风机
air cooling by evaporation 蒸发式空气冷却
air cooling condenser 空冷凝汽器
air cooling condensing installation 空冷装置
air cooling duct 空气冷却风道
air cooling equipment 空气冷却设备
air cooling fin 散热（翅）片
air cooling of hydro-generator 水轮发电机空气冷却
air cooling system 空气冷却系统，气冷系统
air cooling technology 空冷技术
air cooling tower 空气冷却塔
air cooling tubes condenser 空冷凝汽器
air cooling unit 空气冷却机组

air cooling zone 空气冷却区
air cooling 空气散热，空气冷却，空冷，气冷，风冷
air coordinates 空中坐标
air cored 空心的，空气旋涡，空气芯
air course 通风巷道，风道
air crack 干裂
aircraft crash 飞机撞击，飞机坠落
aircraft derivative 航空改装型
aircraft-derived gas turbine 航空改装型燃气轮机
aircraft gas turbine 航空燃气轮机
aircraft generator 航空发电机
aircraft-impact-resistant 抗飞机撞击的
aircraft induction motor 航空用感应电机
aircraft nuclear power facility 航空核动力装置
aircraft nuclear propulsion 航空核推进
aircraft obstruction lamp 航空障碍灯
aircraft propulsion reactor 航空推进反应堆
aircraft servomotor 航空用伺服电动机
aircraft shield test reactor 航空屏蔽试验堆
aircraft transformer 航空用变压器
aircraft turbine 航空涡轮
aircraft warning lights 飞机警告灯
aircraft warning marker （导线和地线）航空警告标志
aircraft warning sphere 航空警告球
air-cured specimen 空气养护试件
air curing 空气养护
air current 气流，空气流
air curtain method 空气幕法
air curtain 空气幕，空气屏蔽，气帘，气幕
air cushion beading 空气式轴承
air cushion belt conveyor 气垫式带式输送机
air cushioning 空气减震
air cushion shock absorber 空气弹簧隔振器，气垫减震器
air cushion surge chamber 压气式调压室，气垫式调压室
air cushion transmitter 气垫运输
air cushion 气垫
air cycle efficiency 空气循环效率
air cycle equipment 空气循环设备
air cycle 空气循环
air cylinder 压气缸，气压缸，储气筒，储气罐
air-damped 空气减震器的
air damper 风门，空气挡板，空气阻尼器，空气节制器，挡风板
air damping 空气阻尼
air dam 前扰流板
air dashpot 空气阻尼器
air defence works 防空工程
air defence 防空
air deflector 空气转向器，导风板，导流板
air delivery pipe 输气管，导气管
air delivery 排气
air demand 供气量，需气量
air density correction factor 空气密度校正因数
air density 空气密度
air dielectric 空气介质
air diffusion aerator 空气扩散式曝气器
air diffusion 空气扩散
air direction 风向
air director 空气导流器
air discharge duct 排气管道
air discharge orifice 排气孔
air discharge 排气，空中放电
air discharging acid vehicle 气卸式酸槽车
air discharging caustic vehicle 气卸式碱槽车
air disk 截气门
air displacement pump 空气排代泵，排气泵
air displacement 排气量
air distribution plate 空气分布板
air distribution system 空气分布系统
air distribution button （沸腾炉）配风风帽
air distribution pipe 配气管道
air distribution 气流分配，空气分布，气流组织
air distributor 布风板【沸腾炉】，空气分配器，配气阀，空气分布器
air dolly 气压顶铆器
air door 气门，通气门，通风孔，通风门
air dose 空气剂量
air draft 通风，气流，空气通风
air drag 空气阻力，空气阻滞作用
air drainage 排气设备
air drain 排风道，排气道，通风孔，空气泄流，通风管
air draught 气流，抽风，空气通风
air-dried basis analysis 风干基分析（法）
air-dried basis 风干基，空气干燥基
air-dried coal 风干煤，空气干燥过的煤
air-dried moisture 空气干燥基水分，风干水分
air-dried soil 风干土
air dried timber 风干木材
air-dried wood 风干木材
air-dried 风干的，空气干燥的
air-drier 空气干燥器
air drill 风钻
air-driven generator 风动发电机
air-driven grout pump 气动灌浆泵
air-driven post drill 气动柱钻
air-driven sump pump 风动小型抽水泵
air-driven 气力传动的，气动的
air driver 气动传动装置
air drop hammer 气锤
air dryer 空气干燥器
air-drying varnish 风干清漆
air-drying 空气干燥
air-dry sample 风干样品
air-dry weight 风干重
air-dry 风干的
air duct design 风道设计
air ducting 导风装置
air duct installation 风道安装
air duct system 风道系统
air duct 通气管（道），空气管道，风管，空气导管，导气管，排气道
air dump 气动倾卸
air eddy 空气涡流
air eductor 喷气器
air ejecting zone 空气抽出区
air ejecting 空气喷射，抽气
air ejector condenser 空气喷射器冷凝器，抽气

器冷凝器
air ejector fan 排气扇，抽气风扇，射气抽气器，抽风机
air ejector for starting 启动抽气器
air ejector 抽气机，空气喷射器，抽气器【真空】，喷气器
air end 气窗，通风口
air engine 空气发动机，闭循环燃气轮机
air entrained cement 加气水泥
air-entraining agent 加气剂【混凝土】
air-entraining cement 加气水泥，掺气水泥
air-entraining concrete 加气混凝土
air entrainment 吸入空气，加气处理，掺气，加气，吸入空气
air entry 进气口
air envelope 二次风，外包空气【燃料】
air equivalent ionization chamber 空气等效电离室
air equivalent material 空气等效材料，空气等颜色物质
air equivalent 空气当量，空气等效，空气当量的，空气等效的
air escape cock 空气阀，泄气阀，排气阀
air escape valve 排气阀，放气阀
air escape 放气，漏气，泄气，漏风
air evacuation valve 放气阀，排气阀
air evaporator 空气蒸发器
air exhaust blower 排气风机
air exhauster 抽风机，抽气机，排气风机，排气器，排风机，排风装置，排气装置
air exhaust fan 排风机，抽风机，排气风机
air exhaust system 排风系统
air exhaust valve 排气阀
air exhaust 排气，排风，抽风，空气抽取
air exit hole 排气孔
air express 航空快递
air extracting pump 排气泵，抽气泵
air extraction equipment 抽气设备
air extraction main 抽气总管
air extraction piping 抽空气管路，抽空气管道
air extraction system 抽空气系统，抽真空系统
air extractor 抽气器，抽气机
air face 临空面
air fan 风扇
air feeder 供气装置
air feed 风力牵引，风力推进，供气
air filled ionization chamber 充空气电离室
air-filled porosity 充气孔隙度
air film cooling 气膜冷却
air filter apparatus 空气过滤装置
Air Filter Institute 空气过滤器学会
air filter used in gas turbine 燃气轮机空气过滤器
air filter 空气滤清器，空气过滤器
air flap 气阀，断气闸
air floatation classifier 气力浮选分离器
air floatation unit 气浮单元
air float conveyer 气垫输送机
air flotation 空气浮选
air flow breakaway 气流分离
air flow characteristic 气流特性
air flow condition 空气流动条件
air flow controller 空气流量调节器
air flow control 空气流量调节，送风控制
air flow differential 空气压降，空气流动压差
air flow direction 气流方向
air flow distribution 气流分布
air flow guide 导流罩，导流板
air flow indicator 空气流量表，气流指示表
air flow in rime 覆冰气流
air flow instrument 气流仪
air flow line 流线
air flow meter 空气流量计，气流计
air flow noise 气流噪声
air flow over blade 叶片绕流
air flow pattern 流谱
air flow proving switch 空气流量开关【火焰监测系统】
air flow pulsation 气流脉动
air flow rate 气流流速，风量，气量，空气流量，空气流速，气流率
air flow regulating damper 气流调节风门
air flow resistance 气流阻力
air flow separation 气流分离
air flow stability 气流稳定性
air flow structure 气流结构
air flow test 空气流动试验
air flow velocity 气流速度
air flow wattmeter 气流瓦特计
air flow 空气（质量）流量，气流，通风量，让空气自由流动的，气流产生的
air flue 气道，风道，烟道，烟囱
air-flushing system 空气冲洗系统
air-foam fire branch 空气泡沫灭火枪
air-foam fire fighting truck 干粉消防车
air-foam fire monitor 空气泡沫灭火炮
airfoil camber 翼型弯度
airfoil center of pressure 翼型压力中心
airfoil characteristic 翼型特性
airfoil chord 翼弦
airfoil contour 翼型廓线
airfoil drag 翼型气动阻力，翼型阻力
airfoil fan 机翼型叶片风机
airfoil pressure distribution 沿翼型压力分布
airfoil profile 翼型，翼型剖面
airfoil section 翼型，翼剖面
airfoil-shaped blade 翼型叶片
airfoil-shaped vane 翼型叶片
airfoil surface 翼面
airfoil theory 翼型理论，机翼理论
airfoil thickness 翼型厚度
airfoil wind machine 翼型板风力机
airfoil with reverse camber 反弯度翼型
airfoil 风量机翼测量装置，机翼测量装置，机翼，翼型
Air Force Nuclear Engineering Test Reactor 空军核工程试验堆【美国】
air-free 无空气的，真空的
air-free concrete 密实混凝土
air freezer 空气冷冻装置
air freezing 空气冷冻（法）
airfreight 空运（货物），用运货飞机运输（货物），货物空运，货物空运费，空运的，用运

货飞机运输的
air friction loss 空气摩擦损失
air friction 空气摩阻，空气摩擦
air fuel rate 风煤比
air fuel ratio 空气燃料比
air funnel 通风斗，通气筒
air furnace 自然引风炉，热风炉，反射炉
air gage 气压计
air gap area 气隙面积
air gap asymmetry protection 气隙不对称保护装置
air gap barrier 气隙隔板
air gap breakdown 空气间隙击穿
air gap characteristic 气隙特性曲线
air gap clearance 气隙长度，气隙间隙
air gap diameter 气隙直径
air gap field 气隙磁场
air gap flux density 气隙磁通密度
air gap flux distribution 气隙磁通分布
air gap flux 气隙磁通
air gap length 气隙长度
air gap line 气隙磁化线
air gap separation 气隙间距
air gap torque 气隙转矩
air gap transformer 气隙变压器
air gap voltage 气隙电压
air gap width 气隙长度
air gap winding 气隙绕组
air gap 气隙，空气间隙
air gas 含空气可燃气体，可燃混合物，油气混合物
air gate 气门
air gauge 气压表，气压计，气压量规
air gauging 空气测量
air governor 送风调节器，空气调节器
air grate 通风花格
air gravity conveyor 空气斜槽，空气重力输送机
air grid 通风格栅，通风格
air grille 百叶风口，格栅风口
air guide ring 空气导流环
air guide 空气导流，导风板
air gun 气枪
air hammer 气锤
air handing capacity 空气调节能力
air handing equipment 空气调节设备
air handing system 空气调节系统
air handing 空气调节，空气处理
air handler 空气处理机
air handling unit 空气处理机组
air harbor 航空港
air-hardening alloy 气硬合金，自硬合金
air-hardening lime 气硬石灰
air-hardening steel 空冷淬硬钢
air-hardening unit room 空气调节机房
air hatch 通风口
air head 通风平巷，通风巷道
air header 集气管，集合管，空气总管
air heater battery 空气加热器组
air heater draft losses 空气加热流动损失
air heater element 空气预热器元件，空气加热元件
air heater fire 空气加热器再燃
air heater for slider 斜槽空气加热器
air heater hopper 空气预热器灰斗，空气加热器灰斗
air heater 空气预热器，暖风器，暖风机，空气加热器
air heat exchanger 空气热交换器
air heating collector 空气集热器
air hermetic 气密
air hoist 风动绞车，气压起重机，气动卷扬机
air hold down 空气顶压
air holder 储气罐
air hole 风罩，砂眼，空气罩，通风罩，通气孔，气罩，气孔，空穴
air hood 通风罩
air hose connection 压风软管接头
air hose 通气软管，空气软管
air humidification 空气加湿
air humidifier 空气加湿器
air humidity 空气湿度
air-hydraulic convertor 气液转换器
air impeller 空气叶轮
air impermeability 不透气性，气密性
air impingement starter 注气启动机
air inclusion 空气杂质
airiness 通风
air infiltration 空气渗透，漏风，空气掺入，空气漏入
air-inflated structure 充气结构
air inflow 进气量
airing circuit 空气干燥回路
airing 风干
air injection engine 空气喷射柴油机，压气式柴油机
air injection pump 压气喷射泵，空气喷射泵
air injection 空气喷射，喷气
air injector 气力喷射器
air inleakage 漏（入）风量，空气内漏，空气漏入，渗入空气，空气渗入
air inlet clack valve 进气瓣阀
air inlet control 进气控制，进气调节
air inlet disk valve 进气碟阀
air inlet duct 进气管，进气道
air inlet grille 空气入口格栅，空气进入格栅，进气格
air inlet housing 入口风罩，空气入口风罩
air inlet hub 进气整流罩
air inlet-outlet-housing 进出口联箱
air inlet plenum chamber 进气室
air inlet valve 进气阀
air input 进气量，进风口，进气口，通风口，空气入口，空气进口
air-in screen 进气滤网
air-insulated switchgear 空气绝缘开关柜
air insulation 空气绝缘
air intake control 空气进气调节
air intake duct 进气管，进气道
air intake efficiency 进气装置效率
air intake filtration 空气进气过滤
air intake guard mechanism 进气口防护装置
air intake loss coefficient 进气管损失系数

air intake louver 空气进气百叶窗
air intake shaft 进风道，进风竖井，通风竖井，进气竖井
air intake system 进气系统
air intake thermometer 入口空气温度计
air intake valve 进气阀
air intake 空气进口，进气口，进风口，进气箱，进气孔，吸气管，进气，进风
air interchange 空气交换
air interstice 电机铁芯里的风道
air invasion 空气侵入
air-in 供气，空气入口，供给空气
air ionization dosimeter 空气电离剂量计
air-jacketed condenser 空气套冷凝器
air jacket 气套
air jack 气压千斤顶
air jet 空气射流，空气喷射，喷气口
air lance 空气枪，空气吹灰枪
air lancing 空气吹灰
air lane 气道
air layer 空气层
air leakage coefficient 漏风系数
air leakage factor 漏风系数
air leakage flow rate 空气泄漏率
air leakage ratio 漏风率
air leakage system 漏风系统
air leakage test 漏气试验
air leakage 漏风，漏气量，空气泄漏
air leak test 漏风试验
air leak 漏气
air leg rock drill 气腿式凿岩机
air leg （风动伸缩）气腿，气动杆，风力锤，风动钻架
airless-injection engine 无压气柴油机，机力喷射柴油机
air level 气泡水准仪
air lift crusher 竖井磨
air lifter 空气升液器
air lift pump 气动泵，空气升液泵
air lift 空气吸升，空气升力，气力提升器，空气升液器
air-line assembly 空气管线组件
air-line correction 干线校正，测绳水上部分偏移校正
air-line suit 防护衣，气衣
air-line system 风管系统
air-line 空气管路，送气管路，飞行航线，架空线，航空线，进气管
air liquefier 空气液化器
air-loaded accumulator 储气器
air load testing 空气动力载荷试验
air load 气动荷载
air-locked 不透气的，密闭的
air locker 空气闸门，锁气器
airlock feeder 锁气式给料机
airlock gate 气闸闸门
airlock system 空气闸门系统
airlock valve device 气锁阀装置
airlock 锁气器，气闸，风闸，气锁闸，气锁阀，空气闸门
air loss 漏气损失
air magnetic breaker 空气磁断路器
airmail fee 航空邮件费
airmail 航空信，航空邮件
air main 主空气管道，主风道
air make-up 补给新风
air management 大气管理
air manifold 空气总管，总风箱，空气歧管，空气集管
air manometer 空气压力计
air mass analysis 空气团分析
air mass flow rate 空气质量流量
air mass flow 空气流量
air mass fog 气团雾
air mass modification 气团变性
air mass 大气质量，气团
air measurement 空气测量
air measuring and recording instrument 空气测录仪
air measuring device 供气量测定仪
air mechanics 空气力学，空气动力学
air metering 空气测量
air meter 风速计，气流计，量气计，气量计，空气流量计
air mixing chamber 空气混合箱
air mixture 空气混合物
air model study 空气模型试验，气流模型试验
air model 空气模化
air moistening 空气加湿
air moisture content 空气含湿量
air moisture （空气中）悬浮水分，空气湿度
air monitoring car 大气监测车
air monitoring instrument 大气监测仪，大气监测仪器
air monitoring network 大气监测网
air monitoring procedure 大气监测规程
air monitoring station 大气监测站
air monitoring 大气监测，空气监测
air monitor 大气监测器，空气监测器，空气监视器
air motor drill 风动钻机
air motor driven 风动的
air motor 风动发动机，航空发动机
air natural cooled 空气自然冷却的
air natural cooling 空气自然冷却（的），自然通风冷却（的）
air need for combustion 燃烧所需的空气量
air noise 空气噪声
air nozzle 空气喷嘴，风嘴
air oasis 空气净化区
air oil cooler 空气油冷却器
air oil separator 空气油分离器
airometer 气量计，气流表，气体计
air opening 风口
air-open power station 露天电站，户外电站
air-operated controller 气动调节器，气动控制器
air-operated disk cutter 气动切断机
air-operated gate 气动闸门，气动阀门
air-operated hook 气动吊钩
air-operated pilot-control valve 气动滑阀制的调节器
air-operated positioned 气动位置控制器

air-operated pump 气动泵
air-operated switch 气动开关，空气开关
air-operated tool 气动工具
air-operated valve 气动阀
air-operated 气动，气动的，风动的
air orifice 气孔
air outlet opening 排气口
air outlet valve 排气阀
air outlet with vanes 格栅式出风口，带叶片送风口
air outlet 排气口，空气出口，排气孔，放气口，出风口
air output 空气流量，送风量
air-out 空气输出口，放气，出气，排气
air-over-hydraulic brake 空气液压刹车装置
air pad 气垫
air painter 喷漆设备
air parameter 空气环境参数，空气参数
air particle monitor 空气粒子监测器
air passage 通风道，气道，空气道，排气道，空气通道
air patenting 空气淬火，风冷
air permeability apparatus 透气仪，空气渗透仪【测定水泥比表面积】
air permeability test 透气试验
air permeability 透气性，通风性，空气导磁率
airphoto duplicate 复制航摄照片
airphoto stereoscopic instrument 立体航测仪器
air photo 航空照片
air pick 风镐
air pile hammer 气动打桩机，风压打桩锤
air pilot pump 气动控制阀，气动继动阀
air pilot valve 空气导阀
air pipe 空气管道，通气管，风管道，风管
air piston 压气活塞
air pit 通风井，风井
air placed concrete 气压浇注混凝土
air placer 空气压送机
airplane generator 飞机用发电机
air plenum 空气室，风室，风箱
air pocket 气穴，气窝，（混凝土的）麻点，气囊，气孔，气袋，气阱，空中低压区
air pollutant concentration 空气污染物浓度
air pollutant emission inventory 空气污染物排放清单
air pollutant 空气污染物，大气污染物
air pollution abatement 消除空气污染
air pollution agency 大气污染管理机构
air pollution alert 空气污染警报
air pollution code 空气污染法规
air pollution complaint 空气污染诉讼
air pollution concentration 空气污染物浓度
air pollution control district 空气污染管制区，大气污染控制区
air pollution control law 空气污染防治法
air pollution control regulation 空气污染防治条例
air pollution control system 空气污染防治系统
air pollution control 空气污染控制
air pollution effect 空气污染效应
air pollution emission 空气污染排放物
air pollution episode 空气污染事件
air pollution forecasting 大气污染预报
air pollution index 空气污染指数
air pollution legislation 空气污染立法
air pollution level 空气污染水平
air pollution matter 空气污染物质
air pollution measurement program 空气污染测定规划
air pollution meteorology 空气污染气象学
air pollution modelling 空气污染模拟，空气污染建模
air pollution model 空气污染模式
air pollution monitoring system 空气污染监测系统
air pollution monitoring 空气污染监测
air pollution observation station 空气污染监测站
air pollution occurrence 空气污染（偶发）事故
air pollution potential 空气污染潜势
air pollution prediction 大气污染预报
air pollution reduction device 减轻空气污染的装置
air pollution region 空气污染地区
air pollution sensor 空气污染传感器
air pollution source 空气污染源
air pollution standard 空气污染标准
air pollution surveillance system 空气污染监测系统
air pollution surveillance 空气污染监视
air pollution survey 空气污染调查
air pollution transport 空气污染物运输
air pollution trend 空气污染趋势
air pollution zone 空气污染区
air pollution 空气污染，大气污染
airport crash fire vehicle 机场消防车
airport of destination 目的地
airport 风口，风门，航空港，航空站
air-powered pump 气力泵，气动泵
air-powered servo 气动伺服机构
air-powered 气动的
air-power silt collector 空气吸泥机
air precooled 空气预冷的
air preheated 空气预热的
air preheater ash handing system 空气预热器灰斗除灰系统
air preheater 空气预热器
air pressure amplifier 空气压力放大器
air pressure duct 压缩空气管
air pressure engine 压气发动机
air pressure gauge 气压表，气压计
air pressure gun 压缩空气枪，风枪
air pressure probe 空气压力传感器，气压探头
air pressure regulator 空气压力调节器
air pressure test of boiler 锅炉风压试验
air pressure test 风压试验
air pressure valve 空气压力阀
air pressure 空气压力，大气压，风压，空气压，气压
air proof 不透气的，气密的，密封的，气封的
air puff blower 压缩空气断续吹灰器
air pump 抽风泵，抽气泵，打气筒，压气机，气泵
air purge seal 吹气密封

air purge 空气吹扫，空气吹洗，气洗
air purification 空气净化，空气洁净
air purifier 空气净化器
air quality circulation model 大气质量循环模型
air quality control region 大气质量控制区，空气质量控制区
air quality control system 空气质量控制系统
air quality criteria 大气质量标准，大气质量判据，空气质量准则
air quality display model 大气质量显示模型
air quality impact analysis 大气质量影响分析
air quality index 大气质量指数，空气质量指数
air quality model 空气质量模式
air quality standard 空气质量标准
air quality 空气质量，大气质量
air quantity control 风量调节
air quantity 风量
air quenching 空气冷却，空气淬冷
air radiation monitor 空气辐射监测仪
air raid alarm 空袭警报
air raid shelter 防空洞
air rammer 气动夯锤
air rate 空气率，含气率，混气率，通风量
air reactor 空心扼流圈，空心电抗器
air receipt 空运收据
air receiver 储气罐，气罐，储气器，空气储存器
air recirculation fan 空气再循环风机
air recirculation system 空气再循环系统
air recirculation system for plant compartments 设备室空气再循环系统
air recirculation 空气再循环
air refrigerating machine 空气制冷机
air regenerating device 换气设备
air register vane 调风叶片
air register 调风器，空气挡板，风量调整器，可调通风口，配风器
air regulation fan 空气再循环风机
air regulation 风量调节
air regulator 空气调节器
air relay 气压继电器，气动继电器
air release valve 放气阀，放气门
air release 放气
air relief cock 放气旋塞
air relief installation 排气装置
air relief shaft 排气竖井，排风竖井，排气竖风筒，通风竖井
air relief valve 放气阀，排气阀，安全阀
air reluctance 空气磁阻
air removal 排气，除气，空气抽取
air removal jet 抽气器，空气喷射器，空气吹除喷嘴
air removal section 空抽区
air renewal rate 换气率，空气更新率，通风率
air renewal system 换气系统
air renewal 换气，通风
air requirement 空气需求量
air resistance brake 空气阻力制动器
air resistance of permeability 空气渗透阻力
air resistance 空气阻力，气动阻力，空气共振
air resource management 空气资源管理
air return grille 回风格
air return method 回风方式
air return mode 回风方式
air return system 回气系统
air return through corridor 走廊回风
air return 回风
air ring 通气环
air risk 空运险
air route map 航线图
air routing system 空气按规定路线通过系统
air saddle 气鞍
air sample analysis 空气样品分析
air sampler 空气采样器，空气取样器
air sampling device 空气取样装置
air sampling equipment 空气取样设备
air sampling history 空气取样历史
air sampling 空气取样
air sand blower 气压喷砂机
air scoop intake 进气口
air scoop 进气喇叭口，进风口，进气口，空气口
air screen 空气滤网，空气幕
air screw balance 空气螺旋桨平衡
air screw boss 空气螺旋桨毂
air screw pitch 空气螺旋桨距
air screw 空气螺旋桨，螺旋桨
air scrubber 空气洗涤器，淋水室，喷雾室，空气洗涤塔
air scrub 空气擦洗
air-sea interaction 气海作用
air-sea interface 气海界面
air seals installation 空气侧密封装置
air seal 气密，气封
air-seasoned timber 风干木材
air-seasoning 通风干燥，风干
air separator 除气器，吹气分离器，风力选矿机，气力分离器，空气分离器
air set pipe 空气冷凝管
air set pressure test 压缩空气机组压力试验【对卸压阀】
air-setting mortar 气硬性砂浆
air set 压缩空气机组，空气中凝固，常温自硬，自然硬化
air shaft 风井，排风井，通风井，通风竖管，风道，通风筒
air shed 空气污染区
air shrinkage 风干收缩
air shutter 调风门，空气阀，挡风板，风门片
air side AC sealing oil pump 空气侧交流密封油泵
air side capacity 空侧容量
air side conveyer 气流式输送器
air side DC sealing oil pump 空气侧直流密封油泵
airside road 机场禁区道路
airside trolley 禁区内的行李手推车
air side 空气侧，供风端，对空面
airside 机场空侧，登机区，机场周边活动区，机场禁区
air silencer 消声器，空气消声器
air siltometer 气挠泥沙分析仪

air sinker 风动凿岩机
air sink 气穴，气汇
air sizing 风力分级
air-slaked lime 潮解石灰，气化石灰
air slaking 风化，潮解
air slide conveyer 气力输送斜槽，气滑式输送机，空气输送斜槽
air slide disintegrating mill 气流粉碎机
air slider fan 斜槽风机
air slider 空气斜槽
air slide valve 空气分气阀
air source heat pump system 空气源热泵系统
air space cable 空气绝缘电缆
air spaced coax 空气绝缘同轴电缆
air spaced coil 空气绝缘线圈
air space for roof insulation 屋顶保温空气层
air space insulate barrier 留置空气屏障
air space insulation 空隙绝缘，空气绝缘
air space paper core cable 空气纸绝缘电缆
air space 气隙，空间，空气间层
air sparger pipe 空气喷射管
air specific gravity 空气的比重
air speed gauge 风速表
air speed head 空速管，气压感受器
air speed indicator 空速指示器
air speed meter 空气流速计，空速表
air speedometer 空气流速计，空速表
air speed static tube 空速静压管
air speed tube 空气流速管，皮托管，空速管
air speed 气流速度，风速
air spoiler 扰流板
air spray 空气喷射，空气喷雾
air spring suspension 空气弹簧悬挂，气边吊架
air spring 空气弹簧，气垫
air stagnation 大气滞止
air starter 气力启动器，气动启动机，气动启动器
air station 航空站
air-steam mixture 气汽混合物
air stone 砂滤多孔石
air stopping 挡风（墙）
air storage tank 储气罐
air storage 露天堆放，蓄气库，储气
air strainer 滤气器，空气滤清器，空气过滤器
air strata 空气层
airstream contamination 气流污染
airstream deflector 导风板
airstream quality 气流品质
airstream 气流，空气射流
air suction main 吸气总管
air suction opening with slide plate 插板式吸风口
air suction pipe 吸气管
air suction port 吸风口
air suction valve 空气吸入阀
air suction 吸入空气
air suit 气衣
air supplied mask 送气面罩
air supplied suit 气衣
air supply duct 供气道
air supply fan 鼓风机，送风机
air supply manifold 供气总管
air supply method 送风方式
air supply mode 送风方式
air supply opening with slide plate 插板式送风口
air supply pipe line 供气管，输气管
air supply pipe 供气管
air supply system 供气系统
air supply support 供气支架
air supply system 送风系统
air supply terminal 供气终端
air supply volume per unit area 单位面积送风量
air supply 供风，供气，鼓风，吹风，气源，送风，补充供气，空气流量
air-supported roof 充气屋顶，空气支撑屋顶
air-supported structure 充气结构物
air sweetening 空气氧化脱硫醇【或脱臭】
air swept pulverizer 球磨机
air swept surface 空气冲刷面
air switch-over 切换风
air switch 空气开关，空气断路器
air tack cement 封气黏胶水泥
air tamper 风动打夯机，风动夯，气动夯锤，风动捣棒
air tank 储气罐，空气筒
air temperature indicator 空气温度指示器
air temperature of generator 发电机风温
air temperature 空气温度，气温
air terminal charge 机场费
air terminal device 风道末端装置
air terminal 风道末端
air termination system 接闪器
air test 气压试验
air thermometer 气温表，气温计
air throttle 节气阀
air through tunnel 地道风
air ticket 机票
air tight access door 气密入口门，气密通道
air tight bottle 密封瓶，气密瓶
air tight chamber 气密室
air tight coefficient 严密性系数
air tight cover 气密盖，密封盖
air tight door 密封门，气密阀
air tight joint 气密连接，紧密连接，气密接合，气密接头
air tight machine 密封式机器
air tightness or water-tighthess 气密或水密封的
airtightness 不透气性，气密性，气密度，密闭度
air tight packing 气密填料，密封垫
air tight partition 气密隔墙
air tight seal 气封，密封
air tight test 气密试验，气密性检验
air tight 不漏气的，气密的，密封的，不透气的，气密，不漏气
air-to-air cooler 空气-空气冷却器
air-to-air total heat exchanger 全热换热器
air-to-air heat exchanger 空气-空气换热器
air-to-air heat recovery 空气-空气热回收
air-to-air heat transmission coefficient 空气中热传导系数
air-to-close valve 失气开启阀，通气关闭阀
air-to-close 气关式
air-to-cloth ratio 气布比

air-to-oil cooler 空气冷却式冷油器
air tool 风动工具
air-to-open on-off valve 失气关闭的开关阀
air-to-open valve 失气关闭阀，通气开启阀
air-to-open 气开式
air to push down 气压向下
air to push up 气压向上
air torrent 空气湍流
air-to-water cooler 空气-水冷却器
air-to-water heat exchanger 空气-水热交换器
air transformer 空心变压器
air transportation insurance 空运保险
air transport system 气力输送系统
air transport 空气输送
air trap 防气阀，空气捕集器，空气阱
air treatment 空气处理
air trunk 通风筒，通风道
air tube cooler 空气管冷却器
air tunnel 风洞
air turbine alternator 空气涡轮交流发电机
air turbine pump 空气透平泵，空气涡轮泵
air turbine 空气透平
air turbulence 空气湍流
air valve cage 气阀盖座
air valve camshaft 空气阀凸轮轴
air valve control lever 气阀控制杆
air valve 空气阀，控气阀，空气门，气阀
air vane guide 气流导向叶片
air vane 风标，风向标，通风机叶片，风车叶片
air-vapor mixture 空气蒸汽混合物，气汽混合物
air-vapor temperature 气汽混合物温度
air velocity at work area 作业地带空气流速
air velocity at work place 工作地点空气流速
air velocity vector 气流速度矢量
air velocity 气流速度
air ventilation window 通风窗
air ventilation 通风，换气
air vent manifold 通气多叉管
air vent pipework 辅助蒸汽管，厂用蒸汽管
air vent pipe 排气管
air vent seal 排气管座
air vent valve 放空气门，放气阀，通气阀
air vent 排气口，排风道，空气阀，出气口，放气（阀），通气孔，气孔，通风口
air vessel 气室，储气罐
air vibrator 气动振动器
airview 空瞰图
air void ratio 空气孔隙比，气隙比
air void 气孔，含气率，孔隙率
air volume displacement 空气容积排量
air volume 风量，空气体积
air-wall cavity ionization chamber 空气壁空腔电离室
air washer room 喷雾室
air washer 气洗机
air wash 空气冲洗
air-water backwashing 气水合洗
air-water emulsion 气水乳状液
air-water interaction 气水交替作用
air-water interface 空气水分界面，气水界面
air-water jet 气水混合射流
air-water mixture 气水混合体
air-water ratio 气水比
air-water surface 气水结合面
air-water system 气水系统
air-water two phase flow 气水两相流
air-water washing 气水合洗
air wave 空气波，无线电波，气波，气浪
airway beacon 航空灯塔
airway bill fee 货运单费【承运人收取此费为AWC，代理人收取此费为AWA】
airway bill of lading 空运提单
airway bill 航空货运单，空运提单，航空运单，空运单
airway delivery note 空运提货单
airway 风道，航空线，呼吸道，通气井
air wedge 气楔
air weight flow 空气重量流量
air well 通风井
air wetting 空气加湿
air winch 气动卷扬机，气动绞车
air wire 架空线，天线，架空导线
air working chamber 沉箱工作室
air zoning 分区送风，分段送风
AIS(air-insulated switch) 空气绝缘开关
AIS(alarm indication signal) 告警指示信号
AISC(American Institute of Steel Construction) 美国钢结构学会
aisle clearance 通道宽
aisle gradient 通道坡度
aisle space 过道空间【用于仓库内操作及防火】
aisle wall 过道墙
aisle width 通道宽度
aisle 通道，走廊，过道，（车间的）跨，工段
AIV(air intake valve) 进气阀
ajutage 排水筒
alabaster 蜡石
AL＝aluminum 铝
alameda 林荫散步道
A/L(analog/logic) 模拟/逻辑
ALAP(as low as possible) 尽可能地低
ALARA(as low as reasonably achievable) 尽可能低地获得，合理抑低，最优化原则
ALARA principle 合理抑低原则，最优化原则
alarm actuation value 报警动作值
alarm and protection system 报警保护系统
alarm and setback 报警及重置
alarm and status management system 报警与状态管理系统
alarm annunciation 警报
alarm annunciator 报警信号器
alarm apparatus 报警器
alarm bell 告警铃，警铃，信号铃
alarm box 报警箱
alarm buzzer 报警器
alarm check valve 警报止回阀
alarm circuit 报警电路
alarm contact 报警信号接点
alarm cut out 报警抑制
alarm dead band 报警死区
alarm delay 报警延时
alarm device 报警装置，报警设备

alarm display 报警显示，报警画面
alarm dosimeter 报警剂量计
alarmed 报警的
alarm enunciator 警报器，事故指示装置，报警信号灯
alarm float 报警浮标，浮子报警器
alarm free 警报解除
alarm gauge 警报器
alarm horn 警号
alarm indication signal 告警指示信号
alarm indication 报警信号
alarm indicator 报警指示器
alarm information 报警信息
alarming apparatus 报警装置
alarming control 告警控制
alarming relay 报警继电器
alarm lamp 警报灯，危急信号灯，危急指示灯，报警信号灯
alarm level 报警油位，报警水位
alarm light 报警灯
alarm limit 报警极限值，报警值
alarm manometer 报警压力计
alarm message 报警信息
alarm of fire 火灾警报
alarm oil pressure 报警油压
alarm pressure 极限允许压力，报警压力
alarm processing system 警报处理系统，警报信号处理系统
alarm processing 报警信号处理，报警处理
alarm relay 告警继电器
alarm setting 报警设定
alarm signal system 报警信号系统
alarm signal 报警信号
alarm summary 报警汇总表
alarm switch 报警开关
alarm system 报警系统
alarm test 报警试验
alarm unit 报警单元，报警组件
alarm water level 报警水位
alarm whistle 报警汽笛
alarm window 报警窗，报警光字牌，光字牌
alarm 报警，警报，报警器，报警信号，告警装置，警报装置，信号铃
ALARP(as low as reasonably practicable) 尽量合理地抑低，最低合理可行，要求之合理抑低，二拉平原则
ALARP principle ALARP 准则
alary 翼状的，翼形
albanite 地沥青
albedometer 反射仪，反照仪，反射系数表，反照率表
albedo neutron dosimeter 反照率中子剂量计
albedo of scattering 散射衰减率
albedo of the earth 地球反照率
albedo 反照率，反射率
albertite 黑沥青
albinism 光化作用，射线化学，光化度
albronze 铜铝合金，铝青铜
albuminoidal 硬蛋白，类蛋白
alburn （木材的）边材
alburnum 边材
alclad sheet 包结板
alclad 包铝的，铝夹板，覆铝层，衬铝，包铝，铝衣合金
alcohol blast burner 酒精喷灯，酒精灯
alcohol blow lamp 酒精喷灯
alcohol burner 酒精灯
alcohol cycle 动作周期，工作周期，工作循环
alcoholic varnish 泡立水，凡立水，酒精清漆
alcohol lamp 酒精灯
alcohol manometer 酒精压力计
alcohol resistant foam concentrate 抗溶泡沫液
alcohol thermometer 酒精温度计
alcohol 酒精，乙醇，醇
alcove （河岸的）凹壁，壁龛
aldehyde 醛
aldrin 艾氏剂
aleatory contract 投机性合同
aledo component 反照率分量
alerter 警报器
alert histogram 警报数据【柱态图方式显示】
alert history 警报历史数据
alerting signal 报警信号
alert operation 警戒运行
alertor 报警器
alert state （电力系统的）警戒状态
alert summary 警报数据全面记录
alert 警报，警报信号，发出报警
alfalfa gate 螺旋式闸门，灌水闸
alfalfa valve 螺旋式阀
alfisol 淋溶土
algaecide 除藻剂，灭藻剂
algae coal 藻煤
algae 藻类
algam 锡，铁皮
alga 水藻
algaecide 除藻剂，杀藻剂
algebraic 代数，代数的
algebraic addition 代数相加
algebraic apparent power 视在功率的代数和
algebraic average error 代数平均误差
algebraic compiler and translator 代数编译程序和翻译程序
algebraic compiler 代数编译程序
algebraic equation 代数方程
algebraic manipulation language 代数操作语言
algebraic operation 代数动算
algebraic oriented language 代数排列语言
algebraic sign 代数符号
algebraic sum 代数和
algebraic translator and compiler 代数翻译程序及编译程序
algicide 除藻剂，杀藻剂
algorithmic language 算法语言
algorithmic 算法的
algorithm 算法，验算法，计算程序，算法语言，运算法则
algorithm translation 算法翻译
algovite 辉斜岩类
aliasing signal 混淆信号
aliasing 混淆
alidade rule 照准规

alidade 照准仪，测高仪
alienate 产权转移，转让
alienation coefficient 不相关系数
alien bank 外国银行
alien corporation 外国公司
alien merchant 外国商人，外商
alien tax 外国人入境税
alien 外国人
aliform 八字墙
aligment function 补充函数
aligment test 补充试验
aligned and expanded field 齐向扩展场
aligned heat exchanger of straight tubes 排成一行的直管式热交换器
aligned tube bank 顺排管束
aligned view 旋转视图，转正视图
aligner 校准器，准直器，对准器，前轴定位器
aligning capacitor 微调电容器
aligning idler 调心托辊
aligning of piping 管子对口
aligning of turbine and generator couplings 汽轮发电机联轴器中心找正
aligning of turbine cylinders and bearing pedestals 汽轮机气缸与轴承座之间找正
aligning pin 定位销
aligning pole 定线标杆
aligning section 旋转剖视
aligning 排成直线，对中，找平
alignment chart 列线图，线算图，诺模图，计算图表，调整列线图，准线图
alignment coil 校正线圈，对准检查，微调线圈
alignment correction 定线改正，准线修正
alignment curve 定线曲线
alignment design 定线设计
alignment diagram 定线图
alignment error 定线误差
alignment hole 找正孔，准直孔，定位孔
alignment idler 调心托辊
alignment keyway 对中键槽
alignment laser 对中激光器
alignment line 列线图
alignment map 定线图
alignment mark 调整标记
alignment of a hole （轴）孔对中
alignment of bearings by taut wire 轴承拉线找正
alignment of canal 渠道定线，渠线
alignment of shafts by coupling 用联轴器找主轴中心
alignment of shafts 主轴找中心
alignment of sounding 水道测深线，测探定线
alignment of tuned circuit 调谐回路微调，调谐回路调准
alignment of tunnel 隧洞定向线
alignment pile 定位桩
alignment pin 定位销，对中销
alignment requirement 找中要求
alignment scale 对准标尺
alignment scope 校准用示波器
alignment stake 路线桩
alignment test 对中试验
alignment wire 定线钢丝
alignment 成直线，准线，对中，校直，调直，找中心，找正，中线校正，校准直线，直线排列，对准，定线，定线图
alignment clearance 调整间隙
align 对准，找中，找正，校直，定线（测量），调直
alimentation region 补给区
alinement pin 定位销
alinement 定线，准线，排列，排成直线
aliphatic 脂肪
alive 活的，有电流的，带电的，通有电流的，加有电压的
alkali 碱，强碱
alkali accumulation 碱累积
alkali aggregate reaction 骨料碱化反应
alkali chloride industry 氯碱工业
alkali cleaning 碱洗
alkali corrosion 碱性腐蚀
alkali cracking 碱性（应力腐蚀）裂纹
alkali earth meta 碱土金属
alkali earth 碱（性）土
alkali flat 碱性平地
alkalify 碱化，加碱
alkali matrix deposit 碱基沉积
alkali metal coolant 碱金属冷却剂
alkali metal 碱金属
alkalimeter 碳酸定量计
alkali metric method 碱量滴定法
alkaline accmulator 碱性蓄电池
alkaline aggregate reaction 骨料碱化反应
alkaline attack 碱性侵蚀
alkaline boilout 碱水煮炉
alkaline battery 碱性蓄电池
alkaline boil-out solution 煮炉碱液，碱煮溶液
alkaline boil-out 碱煮炉
alkaline buffer 碱性缓冲剂
alkaline cell 碱性电池
alkaline cleaning 碱液清洗，碱洗，化学除油，碱清洗
alkaline corrosion 碱腐蚀
alkaline detergent solution 碱性洗涤液
alkaline earth 碱土
alkaline earth metal 碱土金属
alkaline electrolysis 碱性电解
alkaline flushing 碱冲洗
alkaline hardness （水的）碱度
alkaline medium 碱性介质
alkaline primary cell 碱性原电池
alkaline reaction 碱性反应
alkaline reagent 碱性试剂
alkaline resistant mortar 耐碱砂浆
alkaline resistant paint 耐碱漆
alkaline resistant 耐碱的
alkaline resisting 耐碱的
alkaline rinse 碱洗，碱洗液
alkaline soil 碱性土，碱化土
alkaline storage battery 碱性蓄电池，碱性蓄电池组
alkaline treatment 碱处理
alkaline waste water 碱性废水
alkaline water conditioning 碱性水运行工况，碱

性水运行调节
alkaline water operation conditioning　碱性水运行工况
alkaline water　碱性水
alkaline　碱的，强碱的，碱性的
alkalinity　碱度，含碱量，含碱度
alkalinization　碱化（作用）
alkali-proof varnish　耐碱清漆
alkali-proof　耐碱的
alkali resistant castable refractories　耐碱耐火浇注料
alkali-resisting primer　抗碱底漆
alkali-sensitive　碱性感，碱敏，碱性感测
alkali soil　碱土
alkali treatment　碱处理
alkali water　碱水
alkalization　碱化
alkalized alumina process　碱化氧化铝法【烟气脱二氧化硫】
alkanet　链烷
alkenes　烯属烃
ALKF(airlock feeder)　气闸式给料器
alkoxide　酚盐
alkyd enamel paint　醇酸磁漆
alkyd paint　醇酸类漆，醇酸涂料
alkyd plastic　醇酸塑料
alkyd resin enamel varnish　醇酸树脂磁漆
alkyd resin　醇酸树脂，醇酸磁漆
alkyd varnish　醇酸磁漆
alkyl benzene sulfonate　烷基苯磺酸盐
alkyl sulfate　烷基硫酸酯
alkyl　烷基
all air heat recovery system　全空气热回收系统
all air system　全空气系统
all aluminium conductor　铝绞线
all aluminum gondola　全铝敞车
Allan cell　阿伦电池
all-around loading　全面加荷，四周加荷
all-around pressure　周围压力，全周压力
all-automatic developing machine　全自动显影机
all-brick building　全砖建筑
all-channel tuning　全信道调谐
all-commodity rate　通用运费率
all-concrete　全混凝土的
all consequences arising threfrom　由此产生的一切后果
all-control-rod-in condition　控制棒全插入状态
all-diffused monolithic integrated circuit　全扩散单片集成电路
allege　声称，断言
all-electric　全部电气化的，全电的
Allen key　艾伦内六角扳手
allen screw　六角固定螺丝
allen wrench　六方扳手，通用扳手
allergy　敏感症
alleviate　减轻，缓和，缓解，取消，减少
alleviation　缓和，冲积，冲积作用
alleviator　缓冲装置
all expected benefits　全部预期效益
all expected cost　全部预期费用
alley stone　矾石
alley　巷
all-function　通用，多用
allgal　藻类的
all gartering　碎裂，细裂纹化，龟裂，细裂纹
all gas turbine propulsion　全燃气轮机推进
all-geared　全齿轮的
all-glass evacuated collector tube　全玻璃真空集热管
allgovite　辉绿玢岩
all-graphite reflector　全石墨反射层
all-haydite concrete　全陶粒混凝土
allied company　联营公司
alligator crack　龟裂
alligator-hide crack　龟裂
alligatoring surface crack　表面裂痕
alligatoring　龟裂，鳄纹，皱裂，裂痕
alligator jaw　柱脚，桩靴，碎石机颚板
alligator ring　齿环
alligator wrench　管扳手，管钳
alligator　齿销【木结构用】，皱裂，裂开，齿键，鳄口式工具
all-in aggregate　毛骨料，统货集料【未经筛分的】
all-inclusive charge of transshipment　转运包干费
all-inclusive type of income statement　总括性收益表
all-in contract　交钥匙合同，整套承包合同
all in one cable　合用电缆
all in one　一体化
all-in position　全插入位置【控制棒】
all-in rate　综合利率，全程运价，全价，综合电价
all-insulation　全绝缘的
allite　铝铁岩，富铝土
allitic soil　富铝性土
all magnetic　全磁的
all-mains　可有任意电源的，有通用电源的
all-metal construction　全金属结构
all-metallic insulation system　全金属隔热系统
allobaric wind　变压风
allobar　变压区
allocable profit　可分配利润
allocated configuration index　分配位置索引
allocated price　调拨价格
allocate　分拨，分配，配给
allocation of budget　预算分配
allocation of funds　资金分配
allocation of items　项目分配
allocation of responsibility　责任分担
allocation of water resources　水资源分配，水资源配置
allocation plan　配置图，分布图，地址分配方案
allocation sheet　分配表【原料等】
allocation　拨款，调拨，分配，分摊，配置，划拨的款项
allochthonous ground water　外来地下水
allogenetic　异源的
allogenic　外源的，异源的，同种异体的
allotment advice　拨款通知
allotment drawing　调配图
allotment issue　已拨款项

allotment letter 分派认购额通知书
allotment 拨款，分配，分配数
allotriomorphic granular 它形粒状
allotropy 同素异形，同素异构体
allotter relay 分配继电器
allotter 分配器
allot 分配，配给
all-out position 全提出位置【控制棒】
allowable 容许的，许可的，允许的
allowable accuracy 许用精度
allowable amplitude 容许振幅
allowable bearing pressure 轴承许用支承压力
allowable bearing capacity of sub-grade 地耐力
allowable bearing capacity 轴承的许用负载，容许承载量，允许承载能力
allowable bearing pressure 容许承压力，容许承载力
allowable belt sag 允许的皮带垂度
allowable bending stress due to thermal compensator 许用补偿弯曲应力
allowable bending stress 许用弯曲应力
allowable bond stress 容许黏着应力
allowable buckling stress 许用扭转应力
allowable burn-up 允许燃耗
allowable combined stress due to external load 许用外载综合应力
allowable compression ratio 许用压缩比
allowable compressive stress 许用压缩应力，许用压应力
allowable concentration 容许浓度
allowable constant current 容许恒定电流
allowable crack size 许用裂纹尺寸，容许裂纹尺寸
allowable current 容许电流
allowable deflection 容许挠度
allowable deformation 容许变形量
allowable design stress 许用设计应力
allowable deviation 容许偏差
allowable elongation 容许伸长度，伸长限度
allowable environment condition 容许环境条件
allowable error 允许误差，容许误差
allowable exposure 容许照射，容许照射量
allowable factor of safety 容许安全系数
allowable flexural stress 容许弯曲应力
allowable flexural unit stress 容许弯曲单位应力
allowable flutter 容许颤动，容许颤振
allowable heat limit 容许热限度
allowable heeling 容许倾斜度
allowable inverse voltage 容许逆电压
allowable limit 容许限度，容许极限，允许限值
allowable load 许用载荷，允许负荷，容许负荷，容许荷载
allowable maximum bending deflection of pipe 管道最大允许挠度
allowable maximum circular velocity 允许最大圆周速度
allowable maximum current density 容许最大电流密度
allowable maximum distance between fixing supports 固定支座最大允许间距
allowable maximum spacing between movable supports 活动支座最大允许间距
allowable maximum velocity 最大允许流速
allowable minimal back pressure 容许最低背压
allowable misalignment 容许不同心度
allowable moisture content of steam 容许的蒸汽湿度
allowable momentary maximum pressure 容许的短时最大压力
allowable noise 容许噪声
allowable peak pressure 容许最高压力
allowable pile bearing load 单桩容许荷载
allowable pile load 桩的容许荷载
allowable pressure drop 允许压力降
allowable pressure 许用压力，容许压力
allowable range 容许界限
allowable resistance 容许阻力
allowable resultant stress 许用合成应力
allowable tolerance 容许公差
allowable settlement 容许沉降量
allowable shearing stress 容许剪切力，许用剪切应力
allowable soil pressure 容许土压力
allowable strength 许用强度，容许强度
allowable stress design 许用应力设计
allowable stress intensity 容许应力强度
allowable stress 许用应力，容许应力
allowable subsoil deformation 地基变形允许值
allowable suction vacuum high 容许吸上真空高度
allowable suction vacuum 容许吸入真空
allowable temperature drop 容许温度降
allowable temperature 许用温度，允许温度，容许温度
allowable tensional stress 许用扭应力
allowable tolerance 容许公差，允许公差
allowable twisting stress 许用扭转应力
allowable unit stress for bending 许用单位弯曲应力
allowable value 容许量
allowable variation 容许偏差
allowable velocity 容许速度
allowable voltage deviation 容许电压偏差
allowable voltage loss 容许电压损耗
allowable voltage 容许电压
allowable working pressure 许用工作压力【锅炉设计压力】，容许工作压力，允许工作压力
allowable working speed 容许工作速率，容许通信速率
allowable working stresses of steel 钢材许用应力
allowance and rebates 折让及回扣
allowance clause 短溢装条款，宽容条款
allowance erratum 容许误差
allowance error 容许误差
allowance for corrosion 腐蚀裕度
allowance for curvature 曲率容许量
allowance for depreciation 折旧备抵，备抵折旧
allowance for fabrication tolerances 容许制造公差
allowance for finish 加工余量，精加工余量，完工留量
allowance for funds used during construction 建造

期使用的资金利息备抵
allowance for machining 机械加工余量
allowance for shrinkage 收缩留量
allowance test 公差配合实验，容差试验
allowance 津贴，补贴，（价格）减让，折扣，（加工）余量，留量，公差，容差，裕量，裕度，允许量，容许量
allow clearance 容许间隙
allowed band 公差带
allowed equivalent crack length 容许当量裂纹尺寸
allowed frequency 许用频率
allowed stress 允许应力
allow for unforeseen circumstances 留余地
allow for 考虑（到），估计（到），酌情（处理），可以供……之用
alloy addition 添加合金，合金添加剂
alloyage 炼合金
alloy bipolar transistor 合金双极晶体管
alloy casting 合金铸件，合金铸造
alloy chilled iron roll 冷硬合金铸铁轧辊
alloy cost iron 合金铸铁
alloy-diffusion transistor 合金扩散晶体管
alloyed steel 合金钢
alloy elements migration 合金元素迁移
alloy film 合金薄膜
alloying element 合金元件，合金元素，掺杂元素
alloying function transistor 合金结晶体管
alloying metal 合金
alloy iron 铁合金
alloy nuclear fuel 合金核燃料
alloy phase analysis 合金相分析
alloy phase equilibrium diagram 合金相平衡图
alloy photocathode 合金光电阴极
alloy plating 合金镀层
alloy reguline chlorination oxidation process 合金块氯化氧化法
alloy rubber belt cleaner 合金橡胶清扫器
alloy rubber cleaner 合金橡胶清扫器
alloy solid solution 合金固溶体
alloy steel pipe 合金钢管
alloy steel protective plating 合金钢保护电镀
alloy steel 合金钢
alloy tool steel 合金工具钢
alloy transistor 合金晶体管
alloy-junction photocell 合金结光电池
all parties to contract 合同各方
all-pass filter 移相滤波器
all-pass structure 全通结构
all-pass 移相的，全通的
all-position electrode 全焊位焊条
all print circuit 全印刷电路
all purpose communication system 通用通信系统
all purpose computer 通用计算机
all purpose engine oil 通用机油
all purpose loader 万能装料机，万能式装载机
all purpose 多种用途，通用的
all range 全量程
all-refractory furnace 全耐火材料炉膛
all risk clause 综合险条款
all risk insurance certificate 一切险证书
all risk insurance 全损险，一切险
all risks policy 综合保险单，全损保险单
all risks 全险，总风险度，综合险
all-round advance 全面涨价
all-round pressure 全周压力，周围压力
all-round price 全包价格
all-round properties 综合性能
all-round 全方位
all speed governor 全速调速器
all-terrain vehicle 全地形车，适应各种地形的车辆，越野汽车
all-thin-film IC 全薄膜化集成电路
all-transistor computer 全晶体管计算机
all-transistor 全晶体管的
all-ups 原煤，混合煤
all-up weight 总重，最大重量，满载重量
allure 廊道
allusion 击岸波
alluvial 冲积的
alluvial apron 山麓冲积扇
alluvial channel 冲积河槽
alluvial clay 冲积黏土，冲击黏土
alluvial cone 冲积锥
alluvial dam 冲积阻塞，冲积堤
alluvial district 冲积区，冲积层，冲积土层，冲积物
alluvial fan 冲积扇
alluvial filtration 冲积过滤
alluvial fitter 冲积式过滤器
alluvial flat 冲积平地，冲积平原
alluvial formation 冲积地层
alluvial ground water 冲积层地下水
alluvial land 冲积地
alluvial layer 冲积层
alluvial meander 冲积曲折河段
alluvial piedmont plain 山麓冲积平原
alluvial plain 沉积平原，冲积平原，淤积平原
alluvial process 冲积过程
alluvial river bed 冲积河床
alluvial river 冲积河，冲积河流
alluvial sediment 洪积物
alluvial slope spring 冲积坡泉
alluvial slope 冲积坡
alluvial soil 冲积土，淤积土
alluvial stream 冲积河流
alluvial terrace 冲积阶地，冲积台地
alluvial tract 下游河段，平原河段，冲积河段
alluvial valley 冲积河谷
alluviation 冲积
alluvium grouting 冲积层灌浆
alluvium 冲积层，冲积土，洪积带，淤积层
all-veneer construction 全镶板构造
all volatile treatment 全挥发性处理【水处理，用于压水堆二次回路】
all-volatility reprocessing plant 全挥发后处理厂
all water system 全水系统
all-weather operation 全天候运行
all-weather port 不冻港
all-weather road 全天候道路
all-weather 全天候的，适应各种气候的

all-welded construction 全焊接结构
all-welded frame 全焊接框架
all-welded panel 全焊式管屏
all-welded 全焊的
all-weld-metal tension test 全焊接金属拉伸试验，全焊金属的焊缝拉力试验
all-weld-metal test specimen 焊缝金属试样
all-wheel drive 全齿轮传动
ally arm 悬臂，托架
ally 联合
ALM＝alarm 报警
ALM(asset and liability management) 资产负债管理
almen intensity 阿尔门强度
almen strip 阿尔门条带
almost periodic behaviour 准周期行为，拟周期性行为
almost-periodic function 概周期函数
almost-periodic 近周期性的
almost self-adjoint operator 准自伴算符
almost triangular matrix 准三角形矩阵
almucantar 平纬圈，等高圈，地平纬圈
alnico 铝镍钴合金，磁钢
alone PV system 独立的（太阳能）光伏系统
alone solar system 独立的太阳能系统
along-flow 顺流向的
alongshore current 顺岸流
alongshore wind 海岸风，沿岸风
alongside bill of lading 船边交货提单
alongside delivery 船边交货
along-wind 顺风向的
along-wind acceleration 顺风向加速度
along-wind buffeting 顺风向抖振
along-wind correlation 顺风向相关
along-wind cross-correlation 顺风向互相关
along-wind deflection 顺风向挠度
along-wind fluctuation 顺风向脉动
along-wind galloping 顺风向驰振
along-wind oscillation 顺风向振动
along-wind response 顺风向响应，顺风向反应
along with 随着
ALPDU(application layer protocol data unit) 应用层协议数据单元
alpenglow 高山辉
alpeth cable 聚乙烯绝缘铝芯电缆
alphabetic axis 字母轴线
alphabetic code 字母码
alphabetic shift 字母换挡
alphabetic sorter 字母分类器
alphabetic string 字母串
alphabetic type bar 字母数字打印杆
alphabetic 字母的，按字母顺序的
alpha burst detector α猝发检测器
alpha-code 字母代码
alpha contamination monitoring α污染监测
alpha decay α衰变
alpha disintegration 衰变，α蜕变
alpha disintegration energy α衰变能
alpha emission α发射
alpha emitter α发射体
alpha format 字母格式
alpha iron α铁
alphalloy 渗铬处理
alpha-neutron source α中子源
alphanumeric characters 字母数字字符
alphanumeric code 字母数字编码
alphanumeric display 字母数字显示
alphanumcric instruction 字母数字指令
alphanumeric reader 字母数字阅读机，字母数字读出器
alphanumeric readout 字母数字读出
alphanumeric representation 字母数字符号表示法
alphanumeric tube 字母数字显示管
alphanumeric 字母数字的，按字母数字顺序的
alpha-particle detector α粒子探测器，氦核探测器，α粒子检测器
alpha particle α粒子
alpha-phase producing α相生产【冶金】
alpha-photon source α光子源
alpha radiation α辐射
alpha radioactivity α放射现象
alpha ratio α比【俘获截面与裂变截面之比】
alpha ray α射线
alpha Rockwell hardness α洛氏硬度
alpha solid solution α固溶体
alphatizing 渗铬处理
alphatron α射线的
alphatron gage α射线计
alpine relief 高山地形
alpine road 山岭道路，山道，高山道路
alpine valley 高山河谷
already existing flaw 固有裂缝，固有缺陷
Alstom 法国阿尔斯通公司
alt.(altitude) 海拔，高度，高程，标高
altazimuth tracker 地平式跟踪器
altazimuth 地平经纬仪
altenrating current operation 交流操作
alter 改变，涂改
alteration of bid 投标备选方案
alteration of cross-section 截面改变
alteration of design 设计变更
alteration of port of departure 变更启运港
alteration of project 工程变更
alteration on entries in the register 变更注册事项
alterations to existing electrical installation 变更用电
alteration work 变更工作
alteration 变更，改变，改造，更改，修改，改进
alter cooling 再次冷却，再冷，停堆后冷却
alternate air inlet 备用空气入口
alternate amplifier 交流放大器
alternate angle 错角
alternate-bay construction 隔仓施工法【混凝土用的】
alternate bay 隔仓
alternate bid 备用标价
alternate cooling mode 替代冷却方式
alternate current 交流电
alternate delegate 副代表
alternate depth 交潜水深

alternate determinant 交错行列式
alternate energy sources 新能源
alternate extender angle 外错角
alternate freezing and thawing 冻融交替
alternate fuel 代用燃料
alternate immersion test （材料的）交替浸沾腐蚀试验
alternate joint 错缝接合，错列接头，错缝
alternate layout 比较布置方案，交互比较草图设计，另一方案
alternate load 交变载荷，交变负载，交替荷载，备用荷载
alternately spaced vortex 交替分布涡
alternate material 替换材料，代用材料
alternate operation 轮换操作，交替操作
alternate optima 交替最优法，交替最佳法
alternate representative 副代表，候补代表
alternate rod insertion 替代棒插入
alternate stress 交变应力，交替应力
alternate system 替换系统
alternate 代替品，替换物，交替的，交替，使交替
alternating 交替的，交变的
alternating bending moment 交替弯矩
alternating bending stress 交变弯曲应力
alternating bending test 反复弯曲试验
alternating bending 反复弯曲，交变弯曲
alternating-component 交流成分，交变分量
alternating current 交流电流，交流电
alternating current amplifier 交流放大器
alternating current analog computer 交流模拟计算机
alternating current balancer 交流均压器，交流平衡器
alternating current bridge 交流电桥
alternating current calculating board 交流计算台
alternating current characteristic 交流特性
alternating current commentator machine 交流换向器式电动机
alternating current component 交流电流分量
alternating current continuous waves 交流连续波
alternating current controller 交流控制器
alternating current converter 交流换流器
alternating current distribution 交流配电
alternating current drive 交流电驱动
alternating current dynamo 交流发电机，变流电机
alternating current easing head 交流擦除头
alternating current electromagnetic pump 交流电磁泵
alternating current erase 交流擦除
alternating current exciter 交流励磁机
alternating current feeding 交流馈电
alternating current field 交流场
alternating current generator 交流发电机
alternating current impedance 交流阻抗
alternating current machine 交流电机
alternating current mains 交流电网，交流电源，交流干线
alternating current measurement 交流测量
alternating current motor 交流电动机
alternating current pattern 交流探伤图
alternating current phase meter 相差计，相位计
alternating current rectifier 交流电流整流器
alternating current reference-voltage measuring unit 交流基准电压测量装置
alternating current relay 交流继电器
alternating current resistance 交流电阻
alternating current servomechanism 交流伺服机构
alternating current system 交流系统，交流制
alternating current transformer 交流变压器
alternating current transmission 交流输电
alternating current voltage 交流电压
alternating current watt-hour meter 交流电度表
alternating current welding machine 交流电焊机
alternating current winding 交流绕组
alternating current working voltage 交流工作电压
alternating cyclic stress 交变应力，周期应力
alternating device 交换器，分流器
alternating discharge 周期性放电
alternating electromagnet 交流电磁铁
alternating electromotive force 交变电动势
alternating energy tax law 替代能源税收法
alternating field 交变磁场
alternating flux 交变通量
alternating-gradient synchrotron 交变磁场梯度同步加速器
alternating hysteresis 交变磁滞
alternating impact bearing test 交变冲击弯曲试验
alternating impact machine 交变冲击试验机
alternating-line scanning 隔行扫描
alternating load 反复荷载，交替荷载，交变负载，交变荷载
alternating magnetic field 交变磁场
alternating magnetization 交变磁化
alternating motion 交变运行，往复运行，往复运动，交替运动
alternating plasticity 交变，塑性变形
alternating potential difference 交变电势差，交变电位差
alternating pressure 交变压力
alternating quantity 交变量，变量
alternating return trap 交替回水弯管
alternating sliding 交替滑动
alternating spark-over voltage 交变放电电压，交变闪络电压
alternating strain 交变应变
alternating stress amplitude 交变应力幅，交变应力幅度
alternating stress intensity 交变应力强度
alternating stress 交变应力
alternating tension and compression 交变拉伸和压缩
alternating track 交变磁道，更换磁道
alternating voltage 交流电压
alternating vortex shedding 交替漩涡脱落
alternation gate “或”门
alternation switch 转换开关
alternation 改变，交错，变更，更改，更替，交变，变换，交替

alternative bidder 备选投标商
alternative calculation 另算
alternative code 替代规范
alternative cost 替代费用
alternative denial gate 与非门
alternative design 比较设计
alternative energy sources 新能源，替代能源
alternative equipment 供选用设备，替代设备
alternative flow 可供选择的流程
alternative fuel 代用燃料
alternative layout 比较布置方案
alternatively 交互的
alternative material and equipment 替代材料与设备
alternative motion 往复运动
alternative power station 替代电站
alternative price 备选价格
alternative project 备选项目
alternative provision 替代条款
alternative scheme 比较方案，备用方案，替代方案
alternatives comparison 方案比较
alternative site 比较坝址
alternative solution 替代方案
alternative technology 备选技术
alternative test 替换试验
alternative 可替代的，可选择的，交换的，互换的，替换的，交替的，（二者）任取其一的选择，可能性，替换物，备选方案，替代方案，替换装置，后备机组
alternator armature winding 同步发电机电枢绕组
alternator armature 同步发电机电枢
alternator field voltage 同步发电机励磁电压
alternator overcurrent 同步发电机过电流
alternator-rectifier exciter 同步发电机-整流器式励磁装置
alternator transmitter 高频发电机式发射机，高频发电机
alternator 交流发电机，同步发电机，交替符，振荡器
alter the contract 修改合同
although 尽管
altigraph 高度记录器，测高计，气压计
altimeter lag 高度计读数滞后
altimeter setting 测高计调整
altimeter 测高器，测高仪，高程计，高度表，高度计
altimetric data 高程计资料
altimetric measurement 测高，高程测量
altimetric reading 高程表读数
altimetry 测高法，测高学，测高术，高度测量法
altitude above sea level 海拔
altitude-altitude heliostat 高度-高度定日镜
altitude angle 高度角，仰角
altitude-azimuth heliostat 高度-方位定日镜
altitude-azimuth 高度-方位
altitude circle 地平纬圈
altitude-control valve 水位控制阀，高度控制阀
altitude correction factor 高度修正系数
altitude correction 高度改正
altitude data 高程数据
altitude difference 高度差，高差
altitude gage 测高仪，测高计
altitude gauge 测高仪，测高计
altitude graph 海拔图
altitude interval 高差
altitude level 测高水准仪
altitude limit 高程界限
altitude of triangle 三角形高线
altitude parallel 地平纬圈
altitude receiver 高度接收器
altitude scale 高度比例尺
altitude slide 高度滑尺
altitude survey 高程测量
altitude table 高程表
altitude throttle 高度节流阀
altitude transmitter 高度发送机，高度数据发送装置
altitude tunnel 高空模拟风洞
altitude valve 高度节流阀
altitude wind tunnel 高空模拟风洞
altitude 标高，海拔，高度，地平纬度，海拔高度
altogether coal 原煤
alum-calcium cement 铝酸钙水泥
alum clay 明矾黏土
alum coagulation treatment 矾凝结处理
alum earth 明矾土
alumel-chromel thermocouple 镍铝-镍铬热电偶
alumel 铝镍合金
alum floc 矾花，硫酸铝絮凝物，铝矾絮凝剂
alumina aero gels 氧化铝气凝胶
alumina cement 铝酸钙水泥，矾土水泥，高铝水泥
alumina insulating pellet 氧化铝隔热片
alumina refractory 矾土耐火材料
aluminate cement 矾土水泥
aluminate 铝酸盐
alumina 矾土，氧化铝
aluminithermic welding 铝热剂焊接
aluminium alloy conductor steel-reinforced 钢芯铝合金绞线
aluminium alloy conductor 铝合金导线
aluminium alloy door in swing type 转动式铝合金门
aluminium alloy door 铝合金门
aluminium alloy herringbone ladder 铝合金人字梯
aluminium alloy sliding or side-hung casement type window 铝合金推拉窗或侧挂式平开窗
aluminium alloy 铝合金
aluminium argon arc welding machine 铝氩弧焊机
aluminium bar 铝棒
aluminium brass 铝黄铜
aluminium busbar 铝母线
aluminium cable 铝芯电缆
aluminium-cell arrester 铝避雷针
aluminium chloride 氯化铝
aluminium coating 热镀铝法
aluminium conductor steel reinforced 钢芯铝绞线
aluminium conductor 铝导线
aluminium evaporation 蒸铝

aluminium explosive 铝粉炸药
aluminium foil 铝箔
aluminium industry 铝工业
aluminium magnet wire 铝电磁线
aluminium mesh 铝板网
aluminium oxide 氧化铝
aluminium paint 银粉漆
aluminium paste 铝浆
aluminium pipe 铝管
aluminium plate 铝板
aluminium polychlorid 聚合氯化铝【絮凝剂】
aluminium sash 铝窗框
aluminium sheet roll rack 铝皮架
aluminium sheet 铝片
aluminium silicate fiber rope 硅酸铝纤维绳
aluminium solder 铝焊料
aluminium steel 渗铝钢
aluminium 铝，铝的
aluminized fluorescent screen 涂铝荧光屏
aluminized screen 涂铝荧光屏
aluminizing 镀铝，渗铝
aluminosilicate 铝硅酸盐
aluminosilicate fiber 硅酸铝纤维
aluminous cement 矾土水泥，高铝水泥，铝酸钙水泥
aluminous fire brick 铝质耐火砖，矾土耐火砖，高铝耐火砖
aluminum alloy conductor 铝合金绞线
aluminum alloy door in swing type 转动式铝合金门
aluminum alloy door 铝合金门
aluminum alloy herringbone ladder 铝合金人字梯
aluminum alloy ladder 铝合金直梯
aluminum alloy sheet 铝合金板
aluminum alloy wire grip 铝合金导线卡线器
aluminum alloy 铝合金
aluminum armature 铝线电枢
aluminum beads collector 铝珠集电极
aluminum brass tube 铝黄铜管
aluminum brass 铝黄铜
aluminum bromide 溴化铝
aluminum bronze 铝青铜
aluminum busbar 铝汇流条，铝母线
aluminum-cable steel-reinforced conductor 钢芯铝线
aluminum cable 铝芯电缆
aluminum cast iron 高铝铸铁
aluminum-cell arrester 铝管避雷器
aluminum chloride 氯化铝
aluminum clad steel conductor 铝包钢导线
aluminum clad steel 铝包钢
aluminum coil 铝线圈
aluminum-conductor-alloy-reinforced 合金芯铝导线
aluminum conductor aluminum-clad steel reinforced 铝包钢芯铝绞线
aluminum conductor fiber reinforced 碳纤维铝芯导线
aluminum conductor machine 铝线电机
aluminum conductor steel supported round wire 钢芯圆铝绞线
aluminum conductor steel supported shaped wire 钢芯型铝绞线
aluminum conductor 铝导体，铝导线，铝绞线
aluminum core plastic cable 铝芯塑料电缆
aluminum-covered steel wire 铝包钢线
aluminum cell 铝电池，铝电解槽
aluminum die-cast alloy 铝铸造合金
aluminum diffusion treatment 渗铝处理
aluminum flocculation 铝凝聚，铝絮凝
aluminum foil winding 铝箔线圈
aluminum foil 铝箔
aluminum frame door 铝框门
aluminum hydroxide precipitate 氢氧化铝沉淀物
aluminum ingot 铝锭
aluminum magnesium alloy 铝镁合金
aluminum magnet wire 铝电磁线，铝磁线
aluminum oxide 矾土，氧化铝
aluminum paint 铝粉漆，银粉漆
aluminum powder 银粉，铝粉
aluminum rectifier 铝电解整流器
aluminum retort 铝甑
aluminum sheathed cable 铝包电缆
aluminum sheath 铝包皮，铝护套
aluminum sheet roof 铝板屋面
aluminum sheet 铝板
aluminum sliding window with louver 带有百叶窗的铝推拉窗
aluminum steel conductor 钢芯铝电缆，铝电缆
aluminum sulphate 硫酸铝
aluminum tape 铝包带
aluminum tubular conductor 管型铝导体，铝导管
aluminum wedge 铝槽楔
aluminum winding foil 铝绕组箔
aluminum winding machine 铝线电机
aluminum winding transformer 铝线变压器
aluminum winding 铝绕组
aluminum wire 铝线
aluminum 铝，铝的
alumite 防蚀铝
alumize 渗铝，镀铝，铝化处理
alum shale 明矾页岩
alum stone 明矾石
alum 明矾
alunite 明矾石
alure 院廊，走廊，廊道
alustite 蓝高岭土
always safe geometry 恒安全几何条件
AMAD (activity median aerodynamic diameter) 放射性气动力中位直径，活性中值空气动力学直径
AM0(air mass zero) 大气质量为零的状态
amalgamated balance sheet 合并资产负债表
amalgamated company 合并公司
amalgamation of companies 公司合并
amalgamation 合并，混合
amalgam cell 汞齐电池
amalgam 汞合金，汞齐
AM(amplitude modulation) 调幅
amargosite 膨土岩
amateur band 业余爱护者频带
A/M(automatic/manual)output station 自动/手动输出操作站

A/M(automatic/manual) station 自动/手动操作站
A/M(automatic/manual)with bias station 带偏置自动/手动操作站
A/M(automatic/manual) 自动/手动
amber lamp 黄色信号灯
amber (指示灯)淡黄色，黄色的，琥珀
ambient 环境的，周围的，周围环境
ambient air monitoring 环境空气监测
ambient air level 周围空气水平，环境空气水平
ambient air quality standard 周围空气质量标准，环境空气质量标准
ambient air quality 环境空气质量
ambient air sampling instrument 环境空气采样仪
ambient air standard 环境空气标准
ambient air temperature 环境气温
ambient air 大气，四周空气，环境空气，周围空气
ambient atmosphere 环境大气
ambient concentration 环境浓度
ambient condition 环境状态，外界条件，外界状态，环境条件，周围条件
ambient density 周围介质密度，环境密度
ambient dose equivalent 环境剂量当量，周围剂量当量
ambient dose rate 环境剂量率
ambient environment 周围环境
ambient humidity 环境湿度，周围湿度，环境空气湿度
ambient-induced breakdown 环境感应击穿
ambient level 环境水平
ambient light illumination 环境光照明
ambient lighting 环境照明，环境光照度
ambient light 周围光线，环境光线
ambient liquid water content 环境液态含水量
ambient medium 周围介质
ambient noise level 环境噪声级
ambient noise 环境噪声，外界噪声
ambient operating temperature 环境工作温度
ambient pollution burden 环境污染负荷
ambient power density 周围风能密度
ambient pressure 环境压力，外界压力，周围压力，周压力，围压
ambient quality standard 环境质量标准
ambient stress 包围应力，周围应力，环境剪应力
ambient temperature compensation 环境温度补偿，室温补偿
ambient temperature 环境温度，周围温度，周围介质温度，室温
ambient turbulence 周围湍流，环境湍流
ambient vibration 环境振动
ambient water quality 环境水质
ambient wind angle 周围迎风角
ambient wind speed 周围风速
ambient wind stream 周围气流
ambient wind 周围风
ambiguity function 模糊度函数
ambiguity 含糊，歧义，模糊不定，非单值性，模糊性，模棱两可，意义不明确，模棱两可的话，可作两种或多种解释，多义性
ambiguous and vague 含糊不清
ambiguous case 分歧情况
ambiguous 含糊的，意义不明确的，引起歧义的
ambulance 救护车
ambulatory chattels 动产
ambulatory 回廊
Ambursea dam 安布森式坝，平板支墩坝
AMC(automatic modulation control) 自动调制控制
AMDES(automatic meter data exchange system) 自动仪表数据交换系统
AME(angle measuring equipment) 测角装置
a means to an end 达到目的的方法
amelioration 改良
amend 改变，修改，修订
amend a contract 修订合同
amended invoice 修正发货单
amended plan 修正计划，修正平面图，修正图
amended standard 修正标准
amendement 修改
amending agreement 修正协定
amending clause 修正条款
amendment advice 修改通知单
amendment charge 改单费
amendment commission 修改手续费
amendment of bidding documents 招标书修改，招标文件修改，标书修改
amendment of contract 修改合同，修正合同
amendment to authorized signature 更改印鉴，更改签字
amendment 改善，修正，修订，增订，修改，修正案，改正
amends 赔偿，赔罪，赔礼，赔偿金
amend version 修订版
amercement 罚金，罚款
American Academy of Sciences 美国科学院
American Association for the Advancement of Science 美国科学发展协会
American Association of Engineers 美国工程师协会
American Association of Scientific Workers 美国科学工作者协会
American Automatic Control Council 美国自动控制委员会
American Bureau of Standard 美国标准局
American Certification Institute 美国认证协会
American Concrete Institute 美国混凝土学会
American depositary receipt 美国存托凭证
American Electromechanical Society 美国机电学会
American Engineering Standards Committee 美国工程标准委员会
American Engine Society 美国发动机学会
American Institute of Accountants 美国会计师协会
American institute of architect 美国建筑师协会
American Institute of Industrial Engineers 美国工业工程师协会
American Institute of Steel Construction 美国钢结构协会

American International Assurance Co. Ltd. 美国友邦保险有限公司
American Investment Corporation 美国投资公司
American National Standards Institute 美国国家标准学会
American Nuclear Energy Council 美国核能理事会
American Nuclear Society 美国原子核研究学会
American Power Conference 美国电力会议
American Society for Industrial Security 美国工业安全学会
American Society for Testing and Materials 美国材料与试验协会
American Society of Civil Engineers 美国土木工程师协会
American Society of Heating, Refrigerating and Air-Conditioning Engineers 美国采暖、制冷与空调工程师学会
American Society of Mechanical Engineers 美国机械工程师学会
American Society of Nondestructive Testing 美国无损检验学会
American standard code for information interchange 美国信息交换标准代码
American standard fittings 美国标准配件
American Standard of Testing Materials 美国材料试验标准
American Standards Association 美国标准协会
American Standards Committee 美国标准委员会
American Standards of Test Manual 美国标准实验手册
American standard steel section 美国标准型钢截面
American Standard Straight Pipe Thread for Mechanical Joints 美国标准机械接头直管螺纹
American standard thread 美国标准螺纹
American standard wire gauge 美国标准线规，ASW 规
American stander pipe thread plug 美国标准管道螺纹塞
American wheel 美国水轮机
American wire gauge 美国线规
AMER STD(American Standard) 美国标准
amesdial 测微仪，千分表
amiable composition 友好调解
amianthus 石棉绒
amicable settlement 友好解决，友好结案
amicable 有好的，和睦的
amide 氨基化合物，酰胺
amidol 二氨酚显影剂
aminate 胺化，胺化产物
amination 胺基化
amine 胺
amino acetic acid 氨基乙酸
amino acid 氨基酸，胺酸
amino-alcohol 氨基醇
aminoethane 乙胺，氨基乙烷
amino group powder 氨基干粉
aminolysis 氨解
amino nitrogen 氨基氮，胺型氮
amino plastics 氨基塑料
amino resin 氨基树脂
aminosulfonic acid 氨基磺酸
amino 氨基
AMM(analog monitor model) 类比监视器模组，模拟监视器模块
ammeter shunt 电流表分流器
ammeter switch 安培计转换开关，电流表转换开关
ammeter 安培计，电流表
ammine 氨络物
ammonex 氨化运行
ammon-gelatine dynamite 铵胶炸药
ammonia 氨水，氨
ammonia absorption refrigerating machine 氨吸收式制冷机
ammonia bottle 氨瓶
ammoniacal 含氨的，氨的
ammoniac compounds 氨类化合物
ammonia chloride 氯化氨
ammonia compression refrigerator 氨压缩式制冷机
ammonia compressor 氨气压缩机
ammonia condenser 氨气冷凝器
ammonia cooler 氨气冷却器
ammonia corrosion 氨腐蚀
ammonia dynamite 氨爆炸药
ammonia maser clock 氨脉塞时钟
ammonia oil separator 氨油分离器
ammonia-potassium water regime 氨钾水工况
ammonia refrigerant 氨制冷剂
ammonia scrubbing method 氨洗涤法
ammonia solution metering pump 氨溶液计量泵
ammonia solution pump 氨溶液泵
ammonia solution storage tank 氨溶液储槽
ammonia solution transfer pump 氨溶液输送泵
ammonia spirit 氨水
ammonia synthesis 氨合成法
ammoniated mixed-bed 氨化混床
ammoniate 加氨器
ammoniation 含氨化合物，氨化（作用）
ammonia treatment 氨处理
ammonia water 氨水
ammonification 加氨，化氨，氨化
ammoniometer 氨量计
ammonite 阿芒炸药
ammonium acetate 醋酸铵
ammonium carbonate 碳酸铵
ammonium citrate 柠檬酸铵
ammonium diuranate 重铀酸铵
ammonium ferric sulfate 硫酸铁铵
ammonium ferrous sulfate 硫酸亚铁铵
ammonium fluoride 氟化铵
ammonium-hydrazine passivating solution 氨-联氨钝化液
ammonium hydrogen carbonate 碳酸氢铵
ammonium hydrogen phosphate 磷酸氢二铵
ammonium hydroxide 氨水，氢氧化铵
ammonium ion 铵离子
ammonium magnesium phosphate 磷酸铵镁
ammonium metavanadate 偏钒酸铵
ammonium molybdate 钼酸铵

ammonium nitrate explosive　硝铵炸药
ammonium nitrogen　氨氮，氨中的氮
ammonium oxalate　草酸铵
ammonium persulfate　过硫酸铵
ammonium phosphate monobasic　磷酸二氢铵
ammonium picrate　苦味酸铵
ammonium prelate　过硫酸铵
ammonium purpurate　紫脲酸铵，骨螺紫，红紫酸胺
ammonium thiocyanate　硫氰酸铵
ammonium　铵，铵盐基
ammonization　氨化
ammoxidation　氨氧化
AMMS (automated maintenance management system) 自动化维修管理系统
among other things　亦在其中，其中包括，除了……（还），其中，尤其（还）
among the rest　其中之一，也在其中
a monthly sailing　每月班【船】
amorphism　无定型，非晶形
amorphous material　无定型材料
amorphous memory array　非晶形存储器阵列，无定型存储器阵列
amorphous metal transformer　非晶合金变压器
amorphous rock　无定形岩
amorphous silica　非定晶硅，无定形氧化硅
amorphous silicon photocell　非晶硅太阳电池
amorphous silicon solar cell　非晶硅太阳电池
amorphous solid silicon solar cell　非晶硅太阳能电池
amorphous　无定形的，非晶形的，非晶的
amortisseur bar　阻尼条
amortisseur winding　阻尼绕组
amortisseur　减震器，消音器，缓冲器，阻尼器，阻尼绕组，阻尼线圈
amortization charge　分期还本付息费用，清偿费，摊销费
amortization fund　偿债基金，清偿费
amortization loan　分期偿还贷款
amortization of debt　分期偿还债务
amortization of limited-term power assets　有限期电力资产设备摊销
amortization of other electric plant　其他电力资产设备摊销
amortization of the capital　资金偿还
amortization payment　分期偿付款项
amortization period　摊销期限，清偿期，摊还期
amortization rate for repayment　预付款分期摊还比率
amortization rate　分期摊还比率，分期偿还率，摊销比率，分摊返还比例，折旧率，摊销系数
amortization　缓冲，阻尼，减震，防止，分期偿还，偿付，摊销，折旧，偿还
amortized payment rate　分期付款率
amortized payment　分期付款
amortize　缓冲，阻尼，减振，折旧
amortizing a loan　分期还贷
amort winding　制动线圈，阻尼线圈，阻尼绕组
amount at risk　风险额
amount carried over　转后期金额
amount declared　申报金额
amount due　应付金额
amount in cash　现金额
amount in figures　小写金额
amount in words　大写金额
amount of balance　账目余额
amount of camber　拱高
amount of compression　压缩量
amount of contraction　收缩量
amount of cut　挖方量
amount of daily solar radiation　太阳日辐射量
amount of dead storage　储备量，储存量，死库容，总贮量
amount of deflection　偏斜量，垂度（量），弯曲度量，挠度，变位度，偏移值，偏差值
amount of direct solar radiation　太阳直射辐射量
amount of electricity saving　节电量
amount of energy　能量
amount of evaporation　蒸发量
amount of excavation　挖方量
amount of export　出口额，输出额
amount of fill　填方数量，填方量
amount of guarantee　担保金额
amount of heat　热量
amount of inclination　倾角，倾斜量
amount of information　信息量
amount of insurance carried　实保金额
amount of insurance required　应保金额
amount of insurance　保险金额，保险价值，投保金额，保额
amount of investment　投资额
amount of live storage　周转储存量
amount of looseness　松动量，间隙值
amount of modulation　调制率，调制度百分数
amount of performance security　履约保证金额
amount of precipitation　降水量，降雨量
amount of radiation　辐射量
amount of subject　标的金额
amount-of-substance concentration of...　物质的量浓度
amount of substance　物质的量
amount of theoretical air　理论空气量
amount of traffic　交通量，运输量
amount of unbalance　不平衡量
amount of work　工作量
amount to 40% of the total　占总量40%
amount　总量，总额，数量，总计，总数，量，金额，合计
ampacity　安培容量，（以安培计的）载流量
amp=ampere　安培
amp=amplification　放大
amp=amplifier　放大器
amperage　电流强度，安培数，电流量，额定电流值
ampere　安培
ampere balance　安培秤
ampere capacity　安培容量，载流量
ampere conductor　安培导体
ampere density　电流密度
ampere gage　电流表
ampere gauge　电流表
ampere-hour capacity　安时容量

ampere-hour efficiency 安时效率
ampere-hour meter 安培小时计，安时计
ampere-hour 安培小时
amperemeter 安培表，安培计，电流表
ampere-minute 安分
ampere-second 安秒
Ampere's law 安培定律
Ampere's right-hand screw rule 安培右手螺旋定则
Ampere's rule 安培定则
Ampere's theorem 安培定理
Ampere-turns factor 安匝系数
Ampere-turns 安培匝，安匝，安培匝数，安匝数
ampere-volt 伏安
ampere-winding 安培匝数
ampere-wires 安培导体，电流导体
amperite 镇流器，限流器，镇流管，镇流电阻器
amperometric titration 电流滴定法
amperometric titrimeter 电流计式滴定计
amperometric 测量电流的，电流测定的
amperometry 电流分析法
ampersand 表示“&”符号，和
amphenol connector 电缆接线盒，电缆接头，线夹
amphibian 水陆两用的
amphibious tractor 水陆两用拖拉机
amphibious vehicle 水陆两用车
amphibole-schist 闪片岩
amphibole 闪石
amphibolite 闪岩
amphidromic centre 潮流旋转中心，潮位不定点
amphidromic region 无潮区
amphidromos 转风点，转潮点
amphoteric compound 两性化合物
amphoteric ion exchange resin 两性离子交换树脂
amphoteric ion 两性离子
amphoteric metal 两性的金属
amphoteric oxide 两性氧化物
amphoteric reaction 两性反应
amphoteric 两极的，两性的，含有阴阳电荷的
AMP-HR(ampere-hour) 安培小时
amp hr 安培-小时
amp in(amplifier input) 放大器输入端
AMPL＝amplitude 幅度，振幅
ample flow 丰水
ample rainfall 充沛降水
ample 充分的
amplidyne generator 放大发电机，微场扩流发电机，电机放大器
amplidyne regulating unit 电机放大器调节装置
amplidyne 电机放大器，交磁放大器，放大电机，磁场放大机
amplification channel 放大通道
amplification characteristic curve 放大特性曲线
amplification circuit 放大电路
amplification coefficient 放大系数
amplification constant 放大系数，加强系数，放大常数
amplification control 增益控制，放大控制
amplification degree 放大程度，（电流或电压的）放大倍数，扩大率
amplification-factor bridge 测量放大系数的电桥
amplification-factor 放大系数，放大因数，放大因子
amplification generator 放大发电机，直流扩大机
amplification matrix 放大矩阵
amplification of powerful signal 强信号放大
amplification range 放大范围
amplification ratio 放大比，放大率
amplification system 放大系统
amplification 放大，加强，放大率，增益，放大系数，放大倍数
amplified action 放大作用
amplified surge 扩大的涌浪
amplified verification 进一步核实，扩大的核查
amplifier-driver 放大器激励器，放大器驱动器
amplifier gain 放大器增益
amplifier-inverter 倒相放大器，放大器倒相器
amplifier motor 放大器电动机
amplifier noise 放大器噪声
amplifier output 放大器输出功率
amplifier panel 放大器盘
amplifier rack 放大器架
amplifier relay 放大器继电器
amplifier stage 放大级
amplifier transformer 放大器变压器，低频变压器
amplifier tube 放大管
amplifier-type meter 放大型仪表
amplifier-type voltage regulator 放大器式电压调节器
amplifier valve 放大器电子管
amplifier with time lag 带时滞的放大器
amplifier 放大器，增益器
amplifilter 放大器输出端的滤波器
amplifying generator 放大电机
amplifying lens 放大镜
amplifying power 放大率，放大倍数，放大能力
amplifying relay 继电器放大器，放大继电器
amplifying stage 放大级
amplifying transformer 放大用变压器
amplifying vibrograph 放大示振器
amplifying winding 放大绕组
amplify 放大，增强
amplirectifier 放大整流器
amplistat 内反馈或自反馈式磁放大器
amplitrans 特高频功率放大器
amplitron 特高频功率放大管
amplitude absorption-coefficient 振幅吸收系数
amplitude adder 幅度加法器
amplitude adjustment 幅度调整
amplitude characteristic 振幅特性，频率特性
amplitude code 振幅码
amplitude comparator polyphase ohm relay 振幅比较式多相欧姆继电器
amplitude comparator 振幅比较器，量值比较器
amplitude component 振幅分量
amplitude control 振幅控制，振幅调整
amplitude correction 振幅校正，幅度校正

amplitude corrector 振幅校正器
amplitude discriminator 振幅鉴别器，检波器
amplitude distortion 振幅畸变，波幅畸变，幅度畸变，幅值失真
amplitude domain 幅域
amplitude equalizer 幅度均衡器
amplitude error 幅值误差
amplitude excursion 振幅偏移
amplitude factor 振幅因数，波幅系数，幅值因数
amplitude fading 振幅衰落
amplitude filter 振幅滤波器
amplitude-frequency characteristic 振幅频率特性，幅频特性
amplitude-frequency distortion 振幅-频率畸变
amplitude-frequency response 幅频响应，幅度-频率响应
amplitude-gating circuit 振幅选通电路
amplitude growth 增幅
amplitude histogram of wind-speed 风速较差直方图
amplitude keying 振幅键控
amplitude level 振幅电平
amplitude limiter 限幅器
amplitude locus 幅值轨迹，振幅轨迹
amplitude lopper 限幅器，削波器
amplitude magnification factor 振幅放大系数
amplitude meter 振幅计
amplitude-modulated sinusoid 调幅正弦曲线
amplitude-modulated transmitter 调幅发送机，调幅发射机
amplitude-modulated 调幅的，振幅调制的
amplitude modulation double side band 双边带调幅
amplitude modulation 调幅，幅度调制，振幅调制
amplitude modulator 调幅器
amplitude of annual variation 年较差【气象】
amplitude of beat 拍幅，拍频振幅，跳动振幅
amplitude of diurnal variation 日较差【气象】
amplitude of first harmonic 基波振幅
amplitude of fluctuation 变动幅度
amplitude of oscillation 摆幅，振幅
amplitude of set-up 壅水幅度
amplitude of stress 应力幅度
amplitude of swing 摆动幅度，摆幅
amplitude of the lightning current 雷电流幅值
amplitude of the tide 潮幅
amplitude of tidal current 潮流振幅
amplitude of variation 变幅，振幅
amplitude of waves 波幅
amplitude of wind tide 风壅水幅度
amplitude permeability 振幅磁导率
amplitude ratio 振幅比，幅度比
amplitude range 振幅范围，衰减程度
amplitude resonance characteristic 振幅共振特性
amplitude resonance 振幅共振响应，振幅共振
amplitude response characteristic 振幅响应特性，振幅特性曲线
amplitude response 幅度响应
amplitude selection amplifier 振幅选择放大器
amplitude selection 幅度选择
amplitude selector 振幅选择器
amplitude-sensitive reflectometer 幅敏反射仪
amplitude separation 振幅区分
amplitude stability 振幅稳定性
amplitude step time 振幅阶跃时间
amplitude taper 振幅锥度，振幅衰减
amplitude variation 波幅变化，振幅变化
amplitude velocity 幅速
amplitude-versus-frequency curve 振幅特性曲线
amplitude versus frequency distortion 振幅对频率的畸变
amplitude versus frequency response characteristic 幅频响应特性
amplitude 振幅，幅度，幅值，距离，幅角，范围，作用半径
amp out(amplifier output) 放大器输出端
AMPTD＝amplitude 振幅，幅度
AMR(Automatic Message Recording) 自动抄表
AMS(Administration Management System) 行政事务管理系统
A/M station 自动/手动操作站
amusement hall 娱乐厅
A/M with bias station 带偏置自动/手动操作站
a. m. 午前，上午【中午 12 点之前的时间】
anabatic flow 上升气流
anabatic wind 上升风，坡风
anabatic 上升的
anabranch river fork 汊河
anabranch 交织支流，再汇流侧流
anaclinal stream 逆斜河
anacom 分析计算机
anadromous midstream 溯流回游
anaerobe 厌氧菌，厌氧生物
anaerobia 厌氧微生物
anaerobic bacteria 厌氧细菌，厌气细菌
anaerobic breakdown 厌气分解
anaerobic contact process 厌氧接触法
anaerobic corrosive bacteria 厌氧腐蚀菌
anaerobic digestion 厌氧消化，厌气消化
anaerobic fermentation 厌氧发酵
anaerobic fluidized bed 无氧流化床
anaerobic organism 厌氧有生物
anaerobic sewage treatment 污水厌氧处理
anaerobic treatment 厌氧处理，厌氧处理法
anaerobic waste treatment 废水废液厌气处理
anaerobic 没有空气而能生活的，厌氧性的，厌氧的，厌气的
anaflow 上升气流
anafront 上升锋面
anaglyphic map 补色立体图
analcime 方沸石
analcite-basalt 方沸玄武岩
anallatic lens 测距透镜
analog 模拟量，模拟的
analog adder 模拟加法器
analog back-up 模拟备用设备
analog bracket 固定托架
analog calculation module 模拟计算模块
analog calculation 模拟计算
analog calculator 模拟计算机，模拟计算器
analog channel 模拟通道，模拟电路，模拟信道

analog communication 模拟通信
analog comparator 模拟比较器
analog computation 模拟计算
analog computer subsystem 模拟计算机子系统
analog computer 模拟计算机
analog computing element 模拟计算元件
analog computing system 模拟计算系统
analog controller 模拟控制器
analog control system 模拟控制系统
analog control 模拟控制
analog correlator 模拟相关器
analog data distributor and computer 模拟数据分配器和计算机
analog data recorder 模拟数据记录器
analog data 模拟数据
analog device 模拟器件，模拟装置
analog diagram 模拟图
analog-digital-analog converter 模拟-数字-模拟转换器
analog-digital computer 模拟数字计算机
analog-digital computing system 模拟数字计算系统
analog-digital conversion equipment 模拟数字转换装置
analog-digital converter 模拟数字转换器，模数转换器
analog-digital data interconversion 模拟数字数据变换
analog-digital element 模拟数字元件
analog-digital integrating translator 模拟数字集中转换器
analog-digital 模拟数字，模拟量数字量
analog display 模拟显示
analog divider 模拟除法器
analog electro-hydraulic control 模拟式电液控制系统
analog equation solver 模拟方程解算器
analog equipment 模拟装置
analog feedback control 模拟反馈控制器
analog function generator 模拟函数发生器
analog generator 模拟装置触发发生器
analogical design 类比设计
analogical electro-hydraulic control system 模拟式电液调节系统
analogical method 模拟法
analog indicator 模拟指示器
analog information 模拟信息
analog input terminals 模拟量输入端子
analog input 模拟输入，模拟量输入
analog instrumentation 模拟测量仪表
analog integrator 模拟积分器
analog interpolation 模拟内插法
analogism 类比法，类比推理
analog/logic 模拟/逻辑
analog machine 模拟计算机，模拟机
analog master module 模拟主模件
analog measurement 模拟测量，模拟量测量
analog memory 模拟存储，模拟存储器
analog model 模拟模型
analog multiplexer 模拟多路转换器
analog multiplication 模拟乘法
analog multiplier 模拟乘法器，模拟式乘法运算器
analog nest units 模拟组合装置
analog network 模拟网络
analog operational unit 模拟运算装置
analogous column 比拟柱
analogous pole 模拟极
analogous 类似的，模拟的，类比的
analog output 模拟输出，模拟量输出
analog output module 模拟输出模件，模拟输出模块
analog recorder 模拟记录器
analog recording dynamic analyzer 模拟记录动态分析程序，模拟记录动态分析仪
analog recording 模拟记录
analog relay 模拟继电器
analog representation 模拟量表示，模拟表示法
analog sampled data filter 模拟采样数据滤波器
analog scan-converter tube 模拟扫描转换器
analog setting 模拟设定值
analog signal 类比信号，模拟信号
analog simulation 模拟式仿真，模拟仿真
analog slave module 模拟子模件，模拟子模块
analog study 模拟法研究
analog switch 模拟开关
analog telemetering 模拟遥测
analog telemeter 模拟遥测计，模拟遥测仪
analog-to-digital conversion 模拟数字变换
analog-to-digital converter 模数转换器，模拟数字转换器，A/D 转换器
analog to digital encoder 模数编码器
analog-to-digital interface 模数接口电路
analog-to-digital programmed control 模拟数字程序控制
analog-to-digital recorder 模拟数字记录器
analog-to-digital recording system 模拟数字记录系统
analog-to-digital sensing 模数传感
analog-to-digital 模拟数字
analog-to-frequency converter 模拟频率转换器
analog-to-pulse converter 模拟脉冲变换器
analog trip comparator 模拟紧急停机比较器
analogue board 模拟盘
analogue calculator 模拟计算器
analogue comparing calculating hardware 模拟（量）比较计算（控制）部件
analogue computation 模拟计算
analogue computer 模拟计算机，模拟计算器
analogue computing system 模拟装置
analogue computing unit 模拟计算单元
analogue control panel 模拟控制盘
analogue control 模拟控制
analogue-digital adaptor 模拟数字适配器
analogue-digital converter 模拟数字转换器，模数变换器，模数转换器
analogue display unit 连续显示器，模拟显示单元
analogue filter 模拟滤波器
analogue function generator 模拟函数发生器
analogue indication 模拟读数，模拟示值
analogue input 模拟量输入
analogue machine 模拟机

analogue measuring instrument with digital presentation 数字显示模拟测量仪表
analogue measuring instrument with semi-digital presentation 半数字显示的模拟测量仪表
analogue measuring instrument 模拟测量仪表
analogue method 模拟方法
analogue model 模拟模型
analogue network 模拟网络
analogue pneumatic signal 模拟气动信号
analogue readout 模拟读数，模拟示值
analogue recording 模拟记录
analogue relay 模拟继电器
analogue result 模拟结果，比拟结果，相似结果
analogue shifter 模拟移相器
analogue signal in audio-frequency band 音频模拟信号
analogue signal in carrier system 载波模拟信号
analogue signal 模拟信号
analogue simulation 相似模拟，类比模拟，仿真模拟
analogue study 模拟研究
analogue switching 模拟交换
analogue switch 模拟开关
analogue technique 模拟技术
analogue test 模拟试验
analogue-to-digital converter 模数变换器
analogue to time to digital 模拟-时间-数字转换
analogue transistor 类比晶体管
analogue translator 模拟译码器
analogue 模拟，类似物，类比，模拟设备
analog value process 模拟量处理
analog value 模拟量
analog variable 模拟变量，模拟量测量
analog voltage 模拟电压，连续变化电压
analog voltmeter 模拟伏特计
analogy model 相似模型，模拟模型，比拟模型
analogy rule 类似律
analogy switch 模拟开关
analogy year 相似年
analogy 类似，模拟，比拟，类比，模拟量，模拟装置
analphabetic notation 类字母标音【Jespersen 的标音法】
analysable assembly drawing 可分拆装配图
analyse 分析，分解
analyser 分析器，模拟装置，检偏器
analysis 分析，分析法，分析学，分解，研究
analysis by measure 容积分析
analysis by titration 滴定分析法
analysis of coal sample 分析煤样
analysis of covariance 协方差分析
analysis of elasticity 弹性分析
analysis of end point 分析终点
analysis of flexural action 弯曲作用分析
analysis in heating power 热工分析
analysis line 分析管线
analysis mode 分析式，分析型
analysis of aggregate 总体分析
analysis of business profit 商业利润分析
analysis of control system in hydropower station 水电站控制系统分析
analysis of critical factors 临界因素分析
analysis of data 分析数据，数据分析
analysis of hydrological data 水文资料分析
analysis of hydrologic data 水文资料分析
analysis of partial discharge 局部放电分析
analysis of power system subsynchronous resonance 次同步谐振分析计算
analysis of profit 利润分析法
analysis of relationship 相关性分析
analysis of result 成果分析
analysis of rigid frame 刚架分析
analysis of sewage effluent 污水分析
analysis of significant safety events terms 安全重要事件项分析
analysis of simulator of hydraulic system 液压系统的模拟机分析
analysis of strain 应变分析
analysis of technical-economic benefits 技术经济效益分析
analysis of tenders 投标分析
analysis of tube burst 爆管分析
analysis of variance 方差分析
analysis of water 水质分析
analysis of weather map 天气图分析
analysis report 分析报告
analysis system 分析系统
analysis versus design 分析与设计
analyst assistance program 分析员辅助程序
analyst programmer 程序分析人员
analyst 分析员，化验员
analytical apparatus 分析仪器
analytical balance 分析天平
analytical chemistry 分析化学
analytical computer 机械式数字计算机
analytical control 分析控制
analytical curve 分析曲线
analytical data 分析数据
analytical document 分析性文件
analytical error 分析误差
analytical function generator 分解函数发生器
analytical function 解析函数
analytical instrument 分析仪表
analytical method 解析法
analytical model 解析模型
analytical nodal method 解析节块法
analytical precision 分析精确度
analytical prediction 解析预估
analytical predictor 分析预测
analytical reagent 分析试剂，分析纯
analytical sampling 分析取样，分析采样
analytical stratigraphy 分析地层学
analytical summary 分析性摘要
analytical technique 分析技术
analytical test 分析试验
analytical 分析的，分解的，解析的
analytic balance 分析天平
analytic extrapolation 解析外推法
analytic function 解析函数，分解函数
analytic 分解的
analyzer electronic assembly 分析仪电子装置
analyzer 分析器，分析仪，模拟装置，检偏器

analyze 分析，分解
analyzing device for C/S with computer 微机碳硫分析仪
analyzing 分析
anamesite 中粒玄武岩，细玄岩
anamorphic zone 合成带
anamorphism 合成变质
an annex to a building 建筑物的扩建部分
anaphoresis 阴离子电泳，电粒升泳
anastomosing drainage 排水网
anastomosis （江河水道之间的）并接，交织水系，合流，接合，融合
anatexis 深熔作用
anchorage beam 锚梁
anchorage bearing 锚座
anchorage block 锚定块，描块，镇墩，锚墩
anchorage bolt 固定螺栓，地脚螺栓
anchorage bond stress 锚固黏结应力
anchorage cable 锚缆，锚索
anchorage deformation 锚头变形
anchorage device 锚定设备，锚具
anchorage dues 停泊税
anchorage force 锚固力
anchorage length 锚固长度，锚着长度
anchorage loss 锚固损失
anchorage pier 锚墩
anchorage shoe 锚座
anchorage space 停泊区
anchorage spud 挖泥机下的钢柱
anchorage zone 锚定区
anchorage 固定支座，固定，锚固，锚墩，锚泊，锚地，锚定，锚具，锚座，抛锚，停泊处，锚泊地
anchor and shotcrete support 锚喷支护
anchor bar anchored at head 端头锚固型锚杆
anchor bar bonded all length 全长黏结型锚杆
anchor bar 锚着钢筋，锚筋，锚固筋，锚杆，固定条
anchor beam 固定梁，锚定梁
anchor bearing 固定支座
anchor block 镇墩，锚块，锚定块，地锚
anchor bolts foundation 地脚螺栓式基础
anchor bolt 基础螺栓，地脚螺栓，锚定螺栓，固定螺栓，锚固螺栓，锚栓，锚杆
anchor boom 锚臂
anchor bracing 锚杆支撑
anchor bulkhead 锚定的挡土墙，锚定岸壁
anchor buoy 锚泊浮标，锚标
anchor chair 锚支座
anchor channel 固定槽条
anchor clamp 耐张线夹
anchor cone 固定锥，锚锥
anchor core 衔铁芯
anchor crane 锚头吊
anchor device 描定装置
anchor drilling 钻探
anchored bulkhead 锚固式驳岸，锚固式挡土墙，锚定岸壁
anchored buoy station 定泊浮标站
anchored filament 固定灯丝
anchored footing 锚桩基础
anchored plate retaining wall 锚定板挡土墙
anchored-position observation current 定点测流
anchored recording station 定泊自记测站
anchored ring girder support 支承环式支墩
anchored section on the compression-concentration type 压力集中型锚固段
anchored section on the compression-dispersion type 压力分散型锚固段
anchored section on the compression type 压力型锚固段
anchored sheet piling 锚定板桩墙，锚系板桩
anchored steel trestle 锚固式钢栈桥
anchored suspension bridge 锚定式悬索桥
anchor frame 锚架，地脚框架
anchor gap 衔铁间隙，火花隙
anchor gate 锚式闸门，锚支闸门
anchor grouting 锚固灌浆
anchor hydrographical station 定泊水文测站
anchor hydrographic station 定泊水文测站
anchor ice 底冰，锚冰
anchoring accessories 锚定件的加固钢筋，地脚钢筋
anchoring and shotcrete 锚喷
anchoring device 锚定装置
anchoring element 锚固件
anchoring fascine 锚定柴捆
anchoring point 系锚点，固定接头，锚定点，固定点，死点，支承管
anchoring strength 锚固强度
anchoring system 锚定系统
anchoring 锚，锚定，锚固，停泊，拉牢，装拉线，拉线装置
anchor insulator 拉杆绝缘子，拉桩绝缘子
anchor jack 撑柱
anchor key 固定键，止动键
anchor length 锚固长度
anchor light 停泊灯
anchor line 锚索
anchor loop 锚环
anchor nut 固定螺母
anchor pier 锚墩
anchor pile 锚桩，锚固桩，锚定桩
anchor plate （地基）锚定板，固定板，锚板
anchor point 死点，固定点
anchor post 撑柱，固定支柱，锚柱
anchor prop 锚杆支柱，固定支柱
anchor ring （压力壳的）加固环，支架锚定环，锚环，环面
anchor rod 锚筋，锚杆，拉线桩，锚桩，固定杆，拉线棒，板桩拉杆
anchor screw 基础螺栓，地脚螺栓
anchor shoe 锚床
anchor socket 锚座
anchor span 锚跨
anchor station 锚定位置，定泊站
anchor strap 锚定带，锚定索
anchor support 固定支架，锚固支架
anchor tie 地锚拉杆，锚杆，拉条
anchor tower 锚定塔，高架塔，耐张塔，拉线塔，锚塔
anchor wall 锚定墙，锚墙

anchor wire　拉线，锚线
anchor yoke　锚轭，锚索环
anchor　支座，固定，锚定，锚，地锚，锚固，抛锚，固定器，支撑点，地脚螺栓
ancillary allowance　附带津贴
ancillary attachment　特殊附件
ancillary buildings　辅助厂房
ancillary cable　辅助电缆
ancillary claim　附带要求
ancillary cooler　辅助冷却器
ancillary device　辅助装置，辅助设备
ancillary document　辅助单据
ancillary equipment　辅助设备，附属设备，外围设备，辅机
ancillary facility　附属设备，配套设施，附属设施
ancillary right　附属权利
ancillary service charge　辅助服务费
ancillary service　附加服务，辅助服务
ancillary shoring　辅助支撑
ancillary structure　附属构筑物，附属建筑物
ancillary system　附属系统，辅助系统
ancillary works　辅助工程，附属工程
ancillary　辅助的，附属的，副的，辅助设备，辅机
ancon　悬臂托梁
1 and 1/2 circuit breaker connection　1个半断路器接线
AND circuit　“与”门电路
AND-connection　“与”连接
andcsine　中长石
AND element　“与”元件，“与”门
Anderson bridge　安德森电桥【测量电感用】
andesine-andesite　中长安山岩
andesine-anorthosite　中长斜长岩
andesinite　中长岩
andesite basalt　安山玄武岩
andesite block　安山岩块体
andesite lava　安山岩熔岩
andesite porphyry　安山斑岩
andesite-tuff　安山凝灰岩
andesite　安山岩
AND-function　“与”作用，“与”功能
AND gate　“与”门，逻辑“与”门
andhi　对流性尘暴
AND NOT gate　“与非”门
AND operator　“与”算子
AND-OR-AND gate　“与或与”门
AND-OR inverter　“与或”反相器
and,or,not,complement　与、或、非、补
AND-OR-NOT gate　“与或非”门
Andrade creep　高温蠕变
AND-to-OR function　“与或”作用，“与或”功能
AND-tube　“与”门管
AND unit　“与”单元
and vice versus　反之亦然
AND　“与”【逻辑】
ANEC(American Nuclear Energy Council)　美国核能理事会
anecdotal record　简历
anechoic chamber　消声室
anechoic test　消声试验
anelasticity　滞弹性，内摩擦力
anelastic strain　塑性应变
anelectric　不能因摩擦而生电的物质，非电化体
anemobarometer　风速风压表
anemobiagraph　压管风速计，压管风速仪，风速风压计，风速风压记录器
anemochore　风播植物
anemocinemograph　电动风速计，风速风压记录仪，一种风速图
anemoclinograph　铅直风速计
anemoclinometer　铅直风速仪，垂直风速表，风斜表
anemodispersibility　风力分散率
anemo-electric generator　风力发电机
anemogram　风力自记曲线，风速记录表
anemograph　风速记录表，风力记录计，自动记风仪，测风仪，风力计，风速计
anemology　测风学
anemometer cup　风速计转杯
anemometer factor　风速计系数
anemometer mast　测风杆
anemometer tower　测风塔
anemometer with stop-watch　停表风速表
anemometer　风速仪，风速计，风速表，测风仪，流速表，风压计
anemometric　测定风力的
anemometrograph　记风仪，风向风速记录仪
anemometry　风速风向测定法，风速测量术，风力测定
anemorumbometer　风向风速表
anemoscope　风向仪，风向计，风速仪，测风仪，热线风速仪
anemovane　接触式风向风速器
anergy　无反应性，无力
aneroid barograph　空盒气压计
aneroid barometer　空盒气压表，无液气压计
aneroid battery　干电池，无液电池
aneroid manometer　无液压力计
aneroid mixture control　无液混合物控制
aneroidograph　空盒气压计，无液气压器
aneroid　无液气压表，无液气压计，真空气压盒，无液的，不利用液体的
aneutronic power　无中子动力
aneutronics　非中子学
an executory contract　尚待执行的合同
angel echo　异常回波
angle adaptor kit　角度适配器
angle-admittance　角导纳
angle bar　角钢，角铁
angle bead　角条，护角
angle beam　角钢梁，隅梁
angle beam contact unit　接触法斜探头
angle beam examination　倾斜（束）检验
angle beam method　斜射法（超声探伤）
angle beam probe　斜探头
angle beam ultrasonic examination　斜探头超声波探伤
angle-bender　钢筋弯曲机
angle bend　角形接头，角形弯管
angle block　角铁，弯板
angle bond　墙勒角外角砌合，角砌合，啮合

隅角
angle brace 角撑，角支撑，斜撑，角铁撑
angle bracket 角撑架，角形托座，角托
angle branch 肋管，肘管，弯管
angle butt weld 斜缝对接焊，斜对角焊缝
angle catch 窗插销
angle change 角变位
angle check valve 直角单向阀，直角型单向阀
angle cleat 连接角钢，角楔
angle control valve 直角调节阀
angle correct 角度校正
angle coupling 直角接头
angled concentrating roll 成角度的集中辊
angle deadend 转角终端
angle displacement 角位移
angle divider 分规
angled nozzle 斜喷管
angle dozer 侧铲推土机，万能推土机，斜角推土机
angled socket wrench 弯头套筒扳手
angle encoder 角度编码器
angle error 角度误差
angle factor 角系数，角因子
angle fillet 三角形盖板，三角焊缝
angle fishplate 角钢鱼尾板，角钢接合板
angle fitting 弯头
angle forward welding 前倾焊
angle gate valve 弯头闸阀，角闸阀
angle gage 测角仪，测角计，角规
angle gauge 测角仪，测角计，角规
angle gear drive 斜交轴伞齿轮转动
angle gear （斜交轴）伞齿轮，（传动）锥齿轮
angle grinder 角锉
angle gusset 角撑板
angle head 弯头
angle hip tile 角形坡脊瓦
angle-impedance relay 角阻抗继电器
angle increment 角增量
angle insulator 转角绝缘子
angle iron frame 角钢构架，角钢框架
angle joint 角接，角接接头，隅接，角焊缝，角钢连接
angle measuring equipment 测角装置
angle modulation 调角，角调制
angle of advance 导前角，超前角
angle of altitude 仰角
angle of application 加力角
angle of approach 渐进角【齿轮】
angle of arrival 到达角【电波】，入射角
angle of ascent 螺旋角，上升角
angle of attack 迎角，冲角，攻角
angle of attack convergence 迎角减小
angle of attack of blade 叶片攻角
angle of attack of maximum lift 最大升力攻角，临界攻角
angle of attack vane 迎角风标
angle of balance 平冲角
angle of bank 倾斜角，侧倾角
angle of bend 弯曲角，弯头角度
angle of bevel 单边坡口角度
angle of bite 压力角
angle of blade 叶片安装角
angle of brush displacement 电刷位移角
angle of brush lag 电刷后移角
angle of brush lead 电刷前移角
angle of brush shift 电刷位移角
angle of cant 喷管摆角
angle of chamfered end 斜切角
angle of chamfer 斜切角
angle of chord 弦角
angle of commutation 换向角
angle of contact 接触角
angle of coverage 象场角
angle of current 水流角
angle of curvature 曲率角，曲度角
angle of curve 曲度
angle of cut-off 遮去角，截止角
angle of declination 偏斜角
angle of deflection 偏向角，截止角，偏移角，挠度
angle of delivery 出手角度，投掷角度
angle of departure 分离角，发射角，发射离地角，偏离角
angle of depression 俯角，低降角
angle of descent 下降角，坡道角
angle of deviation 偏向角
angle of difference 差角
angle of dilatancy 膨胀角
angle of dip 磁偏角
angle of dispersion 散射角
angle of displacement 失配角，位移角
angle of distortion 歪角，扭转角，畸变角
angle of divergence 扩张角
angle of dump slope （煤）堆积坡度
angle of efflux 喷射角，流出角
angle of elevation 离地角，仰角
angle of exaggeration 偏斜角，流的偏斜角
angle of external friction 外摩擦角
angle of fall 落下角，落角
angle of field of vision 视场角
angle of firing 射角，射击角
angle of forward inclination 前倾角
angle of friction 摩擦角
angle of frustum of earth resisting uplift 土壤截头锥体抗上拔角
angle off 偏角，提前角
angle of gradient 坡度角，倾斜角，仰角
angle of hade 断层余角
angle of heel 踵角
angle of impact 命中角
angle of incidence 入射角，迎角，冲角，安装角
angle of inclination 倾斜角，倾角
angle of internal friction 内摩擦角
angle of keenness 锐利角
angle of lag 移后角，滞后角，落后角
angle of lead 导程角，导前角，超前角，前置角
angle of leeway 横漂角，偏航角
angle of move 动堆积角
angle of nip 咬入角
angle of obliquity 倾斜角，压力角，倾角
angle of optical rotation 旋光角

angle of oscillation 摆动角
angle of outlet 出口角
angle of overlap 重叠角
angle of pitch 俯仰角，桨距角，螺距角
angle of polarization 偏振角，极化角
angle of preparation 焊接坡口角度
angle of projection 投射角
angle of rack 倾角
angle of recess 渐远角【齿轮】
angle of reflection 反射角，折射角
angle of release 释放角，出手角，开角
angle of repose 安息角【堆料】，自然堆积角，静止摩擦角，休止角，静止角，自然倾角
angle of resistance 抗滑角
angle of rest 安息角，休止角
angle of retarded closing 关闭滞后角【阀门】
angle of retard 滞后角【电力整流器】
angle of ricochet 反跃角
angle of rise 上升角，螺旋角，仰角
angle of roll 滚转角，倾侧角
angle of roof 屋面倾斜角，屋面坡度
angle of rotation 旋转角，转动角
angle of rotor shaft 风轮仰角
angle of rotor 转子转角
angle of rupture 断裂角，破裂角
angle of scattering 散射角，散布角
angle of shearing resistance 抗剪角
angle of shearing strength 剪切角，剪切强度角
angle of shear 剪切角
angle of shot 投掷角
angle of sideslip 侧滑角
angle of sight 视线角，瞄准角，高低角
angle of slide 滑动角
angle of slope 倾斜角度，坡度，倾斜角
angle of spatula 刮铲角
angle of spread 展开角，扩散角，散布角
angle of stalling 失速角
angle of stream junction 合流角度
angle of surface friction 表面摩擦角
angle of sweepback 后掠角
angle of the electron gun 电子枪角度【电子束焊】
angle of torque 扭转角
angle of torsion 扭转角
angle of trail 后缘角
angle of trim 纵倾角，配平角，修角
angle of true internal friction 真内摩擦角
angle of twist 扭转角
angle of unbalance 不平衡角
angle of vee 坡口角度
angle of view 视角
angle of visibility 视界角
angle of wall friction 墙面摩阻角
angle of wave approach 波浪行径角
angle of wind approach 迎风角，来流迎角
angle of wrap 包角
angle of yaw 偏航角
angle order 角度指令
angle pattern 角型
angle pedestal bearing 斜轴承，斜托架轴承
angle pins 角销
angle pipe 弯管，弯头，曲管
angle plate 角盘
angle pole 角形支架，转角杆，剪切强度角
angle probe 斜探头
angle purlin 角檩，角钢檩条
angle relief valve 角式释放阀
angles back to back 背对背角钢
angle scale 角度标尺
90°angle scattered light method 90°角散射原理【测量浊度】
angle seated valve 角阀
angle-seat valve 角阀
angle shock 斜冲波，斜激波
angle steel 角钢
angle strain 耐张转角
angle struct 角钢支柱
angle suspension 悬垂转角
angle switching 角投励【根据功角给同步电机励磁】
angle table 角钢托座，角撑架
angle template 角规
angle thermometer 角式温度计
angle tie 角钢拉杆
angle tower 转角塔，斜塔架，倾斜支撑架
angle tracking 角跟踪，角坐标跟踪，角度跟踪
angle type axial flow pump 半贯流式轴流泵，弯管轴流泵
angle type connections 角形接线
angle type 角式
angle valve combination 角阀组件
angle valve 角阀
angle with equal legs 等边角钢，等肢角钢
angle with equal sides 等边角钢
angle with unequal legs 不等肢角钢
angle wrench 斜口扳手
angle 角，角度，角钢，角铁，夹角，倾斜
angling hole 斜炮眼
angling snubbing hole 斜掏槽眼
angstrom unit 埃【光线或辐射线波长单位】
angstrom 埃【波长单位】
angular 角的，角度的，有角的
angular acceleration 角加速度
angular accuracy 角精确度
angular adjustment 角度调整
angular aggregate 有棱角骨料
angular bearing 径向止推轴承
angular bevel gear 斜交伞齿轮
angular cleat 三角木
angular coarse aggregate 有棱角粗骨料
angular contact ball bearing 向心推力球轴承，向心止推滚珠轴承，角面接触滚珠轴承，径向止推滚珠轴承
angular contact 斜接触
angular coordinate 角坐标
angular correlation 角相关
angular coverage 扇形作用区
angular cross section 角截面
angular current density 中子流角密度
angular deflection 角偏转，角变位
angular deformation 转动变形，角变形
angular discrepancy 角度偏差

angular displacement transducer 角位移传送器
angular displacement 角位移
angular distance 角距离
angular distortion 角变形
angular distribution 角分布
angular divergence 角偏向，角散度，角误差
angular error of closure 角度闭合差
angular error 角度误差
angular field 视角范围，视角场
angular finishing grinder 角向磨光机
angular flux density 角通量密度，角能通密度
angular force 角向力，旋转力，角动力
angular freedom 角自由度
angular frequency 角频率
angular function 角函数
angular gear 锥齿轮
angular-hemispherical reflectance 角半球反射率
angular impulse turbine 斜击式水轮机
angular impulse 角冲量
angular instrument 测角仪
angularity correction 角度校正，角修正
angularity of sounding line 测深线偏斜度
angularity 弯曲度，倾斜度，倾斜角
angular kinetic energy 旋转动能
angular magnification 角放大率，角放大系数
angular meter 测角仪，测量计
angular mil 角密耳
angular misalignment 角位移，角偏差，角度失准
angular modulation 角度调制，调角
angular momentum vector 角动量矢量
angular momentum 动量矩，角动量
angular moment 转动力矩
angular motion transducer 角运动传感器
angular movement 角运动
angular oscillation 角振荡
angular perspective 斜透视
angular phase difference 角相位差，相角差
angular pipe union 弯管接头
angular position digitizer 角位置数字化装置，角位置数字化仪
angular position pickup 角位移传感器
angular position 角度位置
angular powder 角状粉末
angular rate component 角速度分量
angular rate 角速率
angular repetency 角波率
angular resolution 角度分辨能力
angular resolver 角分析器
angular rotation 角旋转，角位移，角转动
angular section 斜剖面
angular sensor 斜探头【超声波探伤试验用】
angular speed 角速度
angular spin rate 自旋角速率，角旋转速率
angular strain 角应变
angular surface 斜面
angular thread 三角螺纹
angular torsional displacement 扭转角位移
angular tower 转角铁塔，转角塔
angular variation 角度变化，角位差，同步发电机的相角位移
angular velocity vector 角速度矢量
angular velocity 角速度
angular wave number 角波数
angulometer 测角计
anharmonic curve 非调和曲线
anharmonicity 非谐性
anharmonic vibration 非谐振动
anharmonic 非谐波的，不调和的，非谐的，非调和的
anhydration 脱水，干化
anhydride 酐，酸酐
anhydrite cement 无水石膏水泥，硬石膏水泥
anhydrite surface 无水石膏饰面
anhydrite 脱水石膏，硬石膏【无水】
anhydrous acid 无水酸
anhydrous alcohol 无水酒精
anhydrous gypsum 无水石膏
anhydrous lime 无水石灰
anhydrous period 无水期，无水时期
anhydrous plaster 无水粉饰，无水石膏粉饰
anhydrous salt 无水盐
anhydrous sulfuric acid 无水硫酸
anhydrous 无水的
anhyetism 缺雨性，缺雨
a nice fat contract 一个很有利的合同
anicut 灌溉堰，水坝
aniline black 苯胺黑
aniline dye 苯胺染料
aniline-formaldehyde lacquer 苯胺甲醛漆
aniline 阿尼林，苯胺
animal carbon 骨炭
animalcule 极微动物
animal-drawn traffic 畜力车交通
animal glue 动物胶
animal husbandry 畜牧业
animal injury protection 动物危害防护
animal and plant quarantine fee 动、植检疫费
animating electrode 电离电极
animatronics 动物机器人制造技术，动物机器人学
an improved version 改进后的新式样
anion 阴离子，负离子
anion adsorption 阴离子吸附
anion bed demineralizer 阴离子交换床除盐器，阴离子床除盐装置
anion bed ion exchanger 阴离子床交换器
anion bed 阴离子交换床，阴离子床，阴床
anion exchange column 阴离子交换器，阴离子交换柱
anion exchange filter 阴离子交换滤池
anion exchange membrane 阴离子交换膜
anion exchange resin 阴离子交换树脂
anion exchanger 阴离子交换器，阴离子交换剂
anion exchange tower 阴离子交换塔
anion exchange 阴离子交换
anionic detergent 阴离子除垢剂，阴离子洗涤剂
anionic emulsifier 阴离子乳化剂
anionic fixed site 阴离子固定基
anionic polyelectrolyte 阴离子聚电解质，阴离子聚合电解质
anionic radical 阴离子根，阴离子基
anionic surface-active agent 阴离子表面活性剂

anionic surfactant　阴离子表面活性剂，阴离子表面活化剂
anion-permeable membrane　阴离子渗透膜
anion resin cleaning tank　阴树脂清洗罐
anion resin regeneration tank　阴树脂再生罐
anion resin storage tank　阴树脂储存罐
anion resin　阴离子树脂，阴离子交换树脂
anion-semipermeable membrane　阴离子半透
anion surface active agent　阴离子表面活性剂
anion vacancies　阴离子空位
aniseikon　电子探伤器，探测缺陷光电装置
anisentropic　不等熵的
anisobaric　不等压的
anisoelastic　非等弹性的
anisometric　不等轴的
anisothermal cooling　连续冷却
anisothermal heat treatment　非等温热处理，非恒温热处理
anisothermal　非等温的
anisotropically consolidated undrained test　不等压固结不排水试验
anisotropic buckling　各向异性曲率
anisotropic consolidation　各向异性固结，各向不等压固结，非等压固结，非等向固结
anisotropic crystal　各向异性结晶，异方向性结晶
anisotropic dielectric　各项异性电介质
anisotropic diffusion coefficient　各向异性扩散系数
anisotropic diffusion　各向异性扩散，各向异性变形
anisotropic ferrite　各项异性铁氧体
anisotropic fluid　各向异性流体
anisotropic hardening　各向不等硬化
anisotropic membrane　各向异性
anisotropic rock mass　各向异性岩体
anisotropic scattering　各向异性散射
anisotropic seepage　各向异性渗流
anisotropic sheet　各项异性钢片，方向性钢片
anisotropic soil　各向异性土
anisotropic turbulence scale　各向异性紊流度
anisotropic turbulence　各向异性湍流
anisotropic　各向异性的
anisotropy factor　各向异性因数
anisotropy　各向异性现象，非均质，各向异性
anlogue　模拟盘
annals　编年史
anneal crack　温裂，退火裂纹
annealed copper wire　退火铜线，韧化铜线，软铜线
annealed glass　退火玻璃
annealed zone　重结晶区
annealing furnace　退火炉
annealing glass　玻璃退火
annealing out　退火
annealing pass　退火合格
annealing temperature　退火温度
annealing twin　退火孪晶
annealing wire　软铁丝
anneal　退火【热处理】，焖火，韧化，退火处理，煨
annexation　兼并
annex building　附属厂房
annex of contract　合同附件
annex room　辅助小室，前室
annex storage　附属存储器，联想存储器
annex　（文件、合同的）附件，附录，附属品，附加，追加，合并
annihilator　灭火器，减振器，灭弧器，阻尼器，吸收器
anniversary wind　年周风，定期风
anno domini　公元【习惯上自耶稣降生之年起计】
annotate　批注
annotation　注解，注释
announced bid price　唱标价
announcement of tenders　招标通知
announcement　公报，公布，公告，通告，声明
announce　宣告
annoyance　烦恼，困扰
ann reset　报警复位
ann silence　报警消声
ann test　报警测试
annual accounts　年度决算
annual accumulation of earthquake　年地震积累数
annual accumulation of sediment　年泥沙淤积量
annual amplitude　年变幅，年振幅
annual audit　年度审计
annual available wind energy　年有效风能
annual average wind power density　年平均风功率密度
annual average　年平均
annual balance sheet　年终决算书，年终决算表，年度资产负债表
annual benefit　年效益
annual borrowing plan　年度贷款计划
annual budget estimate　年度概算
annual business expenses　年度经营费
annual capacity factor　年利用系数，年利用率
annual capital cost　年资本成本
annual charge　年支出
annual coal consumption　年耗煤量
annual coldest month　历年最冷月
annual collective dose　年集体剂量
annual compositive depreciation rate　年综合折旧率
annual consumption　年消耗量
annual contract on-grid energy　年度合同上网电量
annual contract　年度合同
annual cost　年度费用，年费用
annual crops　年收获量
annual cycle　年循环
annual depletion rate　年亏损量，年退水量
annual depreciation　年折旧（费）
annual design throughput capacity of the terminal　码头年设计通过能力
annual discharge rate　年排放率
annual discharge　年流量
annual distribution of runoff　年径流分配
annual distribution　年内分配
annual dose equivalent limit　年剂量当量限值

annual dose limit 年剂量限
annual dose 年剂量
annual efficiency 年效率
annual electric load curve 全年电负荷曲线
annual energy output 年电能输出，年发电量
annual energy production 年发电量
annual equation 年差，周年差
annual erosion 年侵蚀量
annual extreme daily mean of temperature 年最高日平均温度
annual extreme-mile wind speed 年极端里程风速
annual financial statement 年度财务报表
annual fixed charge rate 年固定费用率，年固定费用
annual fixed charge 年固定支出，年固定费用
annual flood peak series 年洪峰系列
annual flood series method 年洪水系列法
annual flood 年洪水，年径流，年流量，年洪水量，年最大流量
annual generating energy 年发电量
annual growth layer 年生长层
annual heat consumption on hot water supply 热水供应年耗热量
annual heat consumption on space-heating 供热年耗热量，采暖年耗热量
annual heat consumption 年耗热量
annual heating load curves 全年暖负荷曲线
annual hottest month 历年最热月
annual income 年收入
annual inequality 年不均衡性
annual inspection 年度检查，年度检修
annual integrated activity concentration 年度综合放射性浓度
annual interest rate 年利率
annualized capital cost 年度投资费
annualized cost estimates 分年费用估算法
annual limit of intake 年摄入量限值
annual load curve 年负荷曲线，年负载曲线
annual load diagram 年负荷曲线
annual load duration curve 年持续负荷曲线
annual load factor 年负载率，年负荷率，年负荷系数，年负荷因数
annual load forecasting 年负荷预测
annual load variation 年负荷变化
annual maintenance cost 年度维修费用，年维护费用
annual maximum flood 年最大洪水量
annual maximum load utilization hours 年最大负荷利用小时
annual maximum 年最高
annual mean temperature 年平均温度
annual mean tidal cycle 年平均潮汐周期
annual mean 年平均，年平均值
annual normal flow 年正常径流
annual objective 年度目标
annual operating charge 年运行费
annual operating cost 年经营成本，年运行费用
annual operating hours 年运行小时
annual operating time 年运行时间
annual operational cost 年运行费
annual operation and maintenance cost 年运行维修费
annual opportunity cost 年度机会成本
annual output of electric energy 年发电量
annual output 年产量，年发电量
annual overhaul 年度检修
annual payment 年支付额
annual peak load curve 年尖峰负荷曲线，年最大负荷曲线
annual peak load 年最大负荷，年峰荷
annual plan 年度计划
annual power output 年出力
annual precipitation 年降水量，年降雨量
annual progress report 年进度报告
annual rainfall 年降水量，年降雨量，年雨量，年总降雨量
annual range of temperature 温度年较差，年温度范围
annual range 年变幅，年较差
annual rate 年率
annual regulating reservoir 年调节水库
annual regulation 年调节
annual repair 全年需修量
annual report 年度报告，年报
annual revenue 年收益
annual review 年度审查
annual runoff depth 年径流深
annual runoff 年径流量，年径流
annual sales revenue 年销售收入
annual schedule 年度计划
annual sediment transport 年输沙量
annual service hours 年运行小时数
annual session 年会
annual statement 年度决算书，年度报表
annual statistic report 统计年报
annual storage plant 年调节库容电站，年调节水库电厂
annual storage 年调节库容
annual survey 年度调查
annual tax 年税额
annual temperature range 年温差，温度年较差
annual temperature variation 年温度变化
annul the contract 废除合同
annual throughput 年吞吐量【煤场通过量】，燃料年投量，年处理量，年投入量
annual tide 年周潮
annual utility factor 年利用率，年利用系数
annual utility hours 年利用小时
annual utility 年利用率
annual value of coal movements 年运煤量
annual value 历年值
annual variation 年变化
annual water loss 年损失水量
annual wind regime 年风况
annual wind speed frequency distribution 年风速频率分布
annual worth method 年值方法
annual yield 年产量
annual zone 年轮
annual 每年（的），历年的，年度的，年报，年刊，年鉴
annubar flow meter 牛巴流量计

annubar 环杆，环形棒
annuciator 声光报警系统，报警器
annuity 年金
annular air gap 环形气隙
annular area 环状面积
annular ball bearing 径向滚珠轴承，径向轴承
annular bearing 环形轴承
annular bit 环形钻头
annular burner 环形喷燃器，环形燃烧室
annular can combustion chamber 环管式燃烧室
annular cascade 环形叶栅
annular catwalk 环状小道
annular chamber 环形室
annular channel 环形流道，环形通道
annular combustion chamber 环形燃烧室
annular combustor 环形燃烧器，环式燃烧室
annular compartment 环形腔室
annular core pulse reactor 环形堆芯脉冲反应堆，环状堆芯脉冲堆
annular core 环形铁芯，环形堆芯
annular cowling 环形整流罩
annular diffuser 环形扩压器
annular drainage pattern 环状排水布置型式，环状排水系统，环状水系
annular drainage 环状排水系统，环状水系
annular duct 环形管道
annular electric scram magnet 紧急停堆环形电磁铁
annular electromagnet 环形电磁铁
annular entry burner 周环进气喷燃器
annular fir-tree blade root 杉木叶根
annular flow 环状流，环状流动
annular fluidized bed 环形流化床
annular fuel bed 环形燃料床
annular fuel compact section 环形燃料密实段
annular fuel element 环形燃料元件
annular fuel pellet 环形燃料芯块
annular gap 环隙，环状空腔
annular gasket 环形密封垫
annular gear 内齿轮
annular groove 环形槽
annular HTR 环形堆芯高温堆
annular hydraulic jump 环状水跃
annular jet 环形射流，环形喷口
annular knurl 滚花
annular ledge 环形支座
annular lighting 环形照明
annular magnet 环形磁铁
annular micrometer 圆径测微计
annular packing 垫圈，环状填料
annular passage 环形通道，环形风道
annular plate 环形板
annular recess 环形槽
annular return-pressure wind tunnel 环形回压风洞
annular-ring gypsum board nail 环纹石膏
annular rotor pole 环形转子磁极
annular seal 环状密封
annular section 环形断面
annular shroud 环形围筒
annular space 环形空间，环形通道，环形室
annular spent fuel decay storage 乏燃料衰变贮存环形室
annular transistor 环形晶体管
annular tube 套管
annular valve 环状阀
annular vault 圆锯，圆形穹顶
annular wheel 内齿轮
annular wind tunnel 环状回流式风洞
annular zone 环形区
annular 环形的，环状的
annulated bit 环形钻头
annulated column 环饰柱【有环纹的柱】，环柱
annulet 圆箍线
annulling of the contract 撤销合同
annulus area 环形空间【安全壳】，环室，全周面积，环形面积
annulus drag loss 环壁阻力损失
annulus exhaust air handling system for MCA 最大设想事故的环室排气处理系统，双层安全壳环室排气处理系统
annulus gear 内齿圈
annulus loss 环周流动损失
annulus servicing walkway 环形维修走道【安全壳】
annulus velocity 环周速度
annulus 环，圈，环带，环形空间【安全壳】，环形缝隙，环形套筒，环形物，环状面，圆环域，环形孔道，内接齿轮
annul 废除，无效，宣布无效，取消，废止
annum 年
annunciated logic cabinet 报警信号器逻辑柜
annunciate 报警
annunciation 报警信号，报警
annunciator acknowledge 报警确认
annunciator board 信号盘
annunciator light 指示灯
annunciator logic cabinet 报警信号器逻辑柜
annunciator menu 报警菜单
annunciator panel 信号盘
annunciator relay 信号继电器
annunciator system 预报信号系统
annunciator 信号灯，（报警）信号器，信号装置，回转拨码机，电铃指示装置，报警器，信号器，指示器，声光报警系统
ANO（Association of Nuclear Operators） 核电站营运者协会
anodal 正极的，阳极的
anode balancing coil 阳极平衡感应线圈
anode battery 阳极电池组，乙级电池组
anode break-down voltage 阳极击穿电压
anode casting shop 阳极体车间
anode-cathode capacitance 阳极-阴极间电容
anode characteristic 阳极特性
anode compartment 阳极室
anode conductance 阳极电导
anode control 阳极控制
anode converter 阳极电源变流器
anode copper 阳极铜
anode corrosion （金属）阳极腐蚀
anode current 阳极电流
anode dark space 阳极暗区
anode deposition 阳极沉淀

anode detector 阳极检波器
anode disk 盘型阳极
anode dissipation 阳极耗散，板极耗散
anode drop 阳极电压降
anode follower 阳极跟随器
anode forward peak voltage 阳极正向峰值电压
anode glow 阳极辉光
anode grid 阳极栅
anode half bridge 阳极半阀桥
anode heater 阳极加热器
anode inverse peak voltage 阳极反向峰值电压
anode light 阳极发光，阳极辉光
anode load 阳极负载
anode loss 阳极损耗
anode material 阳极材料
anode modulation 阳极调制
anode mud 阳极泥【电解】
anode paralleling reactor 阳极并联电抗器
anode peak current 阳极峰值电流
anode plate 阳极板
anode polarization 阳极极化
anode potential fall 阳极压降，压降电位降
anode potential 阳极电位
anode-ray current 阳极射线电流
anode reaction 阳极反应
anode reactor 阳极电抗器
anode rectification 阳极检波，阳极整流
anode resistance 阳极电阻
anode rest current 阳极静电流
anode screen control grid 阳极控制屏
anode seal 阳极密封【水银整流器】
anode slim 阳极泥
anode supply voltage 阳极电源电压
anode terminal 阳极端子
anode-to-cathode voltage 阴阳极间电压
anode transformer 阳极变压器
anode voltage drop 阳极电压降
anode voltage 阳极电压
anode 阳极，正极，极板，屏极
anodic 阳极的
anodic cleaning 阳极清洗法，阳极清洗
anodic coating 阳极镀层
anodic density 阳极电流密度
anodic deposition 阳极沉积
anodic dissolution 阳极溶解
anodic inhibitor 阳极缓蚀剂
anodic oxidation 阳极氧化
anodic passivation 阳极钝
anodic passivity 阳极钝态，阳极钝性
anodic pickling 阳极除氧化膜法，阳极酸洗
anodic protection 阳极保护
anodic zone 阳极区
anodization 阳极化处理，阳极氧化
anodized finish 阳极化抛光
anodized natural aluminium colour 阳极化处理天然铝色
anodizing 阳极处理，阳极化，阳极腐蚀，阳极氧化
anomalistic tide 异常潮
anomalous displacement current 反常移位电流，不规则移位电流
anomalous expansion 异常膨胀
anomalous field 剩余磁场，异常磁场
anomalous flocculation 反絮凝作用
anomalous high level 异常高水位
anomalous high tide level 异常高潮位
anomalous high water level 异常高水位
anomalous magnetization 反常磁化，异常磁化
anomalous propagation 反常传播，异常传播
anomalous structure 异常结构
anomalous tide level 异常潮位
anomalous 反常的，不规则的，异常的
anomaly 异常，缺陷【功能】，不规则，反常
anomorphic zone 变质带
anonymous 匿名的
anorak 防水布
anorthite 钙长石
anorthoclase 歪长石
anorthose 斜长石
anorthosite 斜长岩
anoxic fluidized bed reactor 无氧流化床反应器
anoxic 缺氧
ANPSH(available net positive suction head) 有效净正吸入压头
ANS(American National Standard) 美国国家标准
ANS(American Nuclear Society) 美国核协会
ANSI(American National Standards Institute) 美国国家标准学会
ANSI(American Nuclear Standard Institute) 美国核标准协会
anstatic agent 抗静电剂
answer 回答
answer-back code 应答代码
answer-back key 应答键，应答电键
answer-back mechanism 应答机构
answer-back signal 应答信号
answering cord 应答塞绳
answering equipment 应答设备
answering jack 应答塞孔，应答插孔
answering key 应答电键
answering lamp 应答信号灯
answering plug 应答插塞
answering time 应答时间，响应时间
answering wave 应答电波
answer signal 应答信号
antagonism 对抗（性）
antagonistic cooperation 对抗性合作
anta 壁端柱【墙砌出的一部分】
antecbamber 接待室
antecedent discharge 前期流量
antecedent moisture 前期含水量，前期水分
antecedent precipitation index 前期降水指数
antecedent precipitation 前期降水量，前期雨量
antecedent stream 先成河
antecedent valley 先成谷
antecedent wetness 前期含水量
antecedent year 上年度
antecedent 前期的，先成的
antechamber-type compression ignition engine 预燃室式压缩点火发动机
antechamber 沉沙室，前室，沉积室，前厅，

接待室，预燃室
antedate 早填日期，倒填（的）日期
antefix 瓦檐饰，装饰屋瓦，末端装饰
antehall 前厅
anteklise 陆背斜，台背斜
ante meridiem 午前，上午【中午 12 点之前的时间】
antenna 天线，触角
antenna admittance 天线导纳
antenna array 天线阵
antenna bearing 天线方向
antenna cable 天线电缆
antenna changeover 天线转换
antenna circuit breaker 天线电路断路器
antenna coil 天线线圈
antenna coincidence 天线重合
antenna condenser 天线电容器
antenna constant 天线常数
antenna counterpoise 天线衡网，天线地网
antenna coupler 天线耦合器
antenna current 天线电流
antenna curtain 天线屏幕，天线屏障
antenna directivity 天线方向性
antenna disconnect switch 天线断路开关，天线断路器
antenna downlead 天线引入线，天线下引线
antenna duplexer 天线转换开关
antenna effect 天线效应
antenna efficiency 天线效率
antenna eliminator 假天线，等效天线
antenna energy 天线辐射能量
antenna feeder 天线馈线
antenna form factor 天线形状因数
antenna for super high frequency 超短波天线，超高频天线
antenna gain 天线增益
antenna height 天线高度
antenna impedance 天线阻抗
antenna insulator 天线绝缘子
antenna lead-in 天线引入线
antenna loading coil 天线加感线圈
antenna mast 天线杆
antenna matching 天线匹配
antenna natural frequency 天线固有频率，天线自然频率
antenna oscillator 天线振子
antenna reactance 天线电抗
antenna switch 天线转换开关
antenna trailer 拖曳天线
antenna tuning 天线调谐
antenna turning motor 转动天线电动机
anteport 外门
anteroom 前厅，接待室
anthracene scintillation counter 蒽闪烁计数器
anthracene scintillation dosimeter 蒽闪烁剂量计
anthracene 蒽【一种碳氢化合物】
anthracite(coal) 无烟煤
anthracite filter 无烟煤过滤器
anthracite mine 无烟煤矿
anthracite 无烟煤
anthraconite 沥青灰岩，黑沥青灰岩
anthrafine 无烟煤屑，无烟煤末
anthropic factor 人为因素
anthropic zone 人类活动区
anthropogenic emission 人为排放
anthropogenic heat 废弃热
anti 反对，抗，逆，防，非，减，耐
anti-abrasion concrete 耐磨混凝土
anti-abrasion elbow 耐磨弯头
anti-abrasion fittings 耐磨附件
anti-abrasion liner 耐磨衬板
anti-abrasion pipe 耐磨管道
anti-abrasion 耐磨蚀
anti-abrasive 耐磨损的
anti-accident exercise 反事故演习
anti-accident measures 反事故措施
anti-acid 耐酸
anti-actinic glass 阻光化玻璃
anti-actinic 阻光化的，隔光的
anti-aircraft defence 防空工事
anti-air-pollution system 防止空气污染系统，大气污染防治系统
anti-alkaline 抗碱的
anti-attrition 减少磨损
anti-avoidance clause 反避税条款
anti-avoidance measures 反避税措施
anti-backlash 齿隙游移
anti-backlash spring 消隙弹簧
anti-bacterial agent 抗细菌剂
anti-bacterial 抗菌的
anti-baric flow 反压流
anti-biosis 抗生（作用）
anti-bird 防禽的
anti-bouncer 防跳装置
anti-breaker 防碎装置
anti-bunt action 防振作用，阻尼作用
anti-capacitance switch 抗电容开关
anti-capacitance 防电容，抗电容
anti-carbon burner 防积炭喷燃器
anti-catalyst 反催化剂，抗催化剂，逆时针方向的
anti-cathode 对阴极
anti-cavitation material 防气蚀材料
anti-chatter 抗颤震
anti-checking agent 防龟裂剂
anti-checking clamp 防裂夹具
anti-checking coating 防裂涂料
anti-checking iron 防裂钩，防裂扒钉
antichlor 脱氯剂，阴氢剂，除氯剂，去氯剂
anticipated capability 预计容量，预计功率
anticipated cost 预期费用
anticipated duty 预计负载，预期工作方式，预期负载
anticipated life 预期寿命
anticipated load 预期载荷，预期负载
anticipated operational occurrence 预期运行故障，预期运行事件
anticipated revenue 预期收入
anticipated risk 预计风险
anticipated seepage 预计渗漏量
anticipated settlement 预计沉陷
anticipated transients without scram 未能紧急停

堆的预期瞬态
anticipated transients without trip 未能紧急停堆的预期瞬态
anticipated transient 预期瞬态
anticipated 预先的，预期的
anticipate 预先考虑，预先处理，超前，期盼，盼望，使提前发生
anticipating risk 预期风险
anticipating signal 预告信号，超前信号
anticipation mode 直观式信息存储法
anticipation survey 前景调查
anticipation 期望，预料，预测
anticipatory breach of contract 预料的违约，先期违约
anticipatory control 超前控制，预先调整装置
anticipatory device 预测装置，预感装置
anticipatory gear 越前动作危急保安器
anticipatory letter of credit 提前付款信用证
anticipatory overspeed device 越前动作危急保安器
anticipatory renunciation 预先放弃
anticipatory 超前的
anticipator 预测器，预感器，超前预测器
anti-climbing device 防爬装置
anticlinal axis 背斜轴
anticlinal fault 背斜断层
anticlinal limb 背斜翼
anticlinal meander 背斜河曲
anticlinal ridge 背斜脊，背斜岭
anticlinal river 背斜河
anticlinal spring 背斜泉
anticlinal strata 背斜层
anticlinal valley 背斜谷
anticline 背斜
anticlinkering box 防焦箱
anticlinorium 复背斜
anticlockwise rotation 反时针方向旋转，逆时针旋转
anticlockwise 逆时针方向的，反时针方向的，反时针方向，逆时针方向
anticlutter 反干扰，抗近地物干扰系统
anticlutter circuit 反杂乱回波电路，抗近地干扰电路
anti-clutter gain control 抗干扰增益控制，反干扰放大调整
anti-coagulant 阻凝剂
anticoincidence circuit 防止重合电路，不重合线路
anticoincidence counter 反符合计数器
anticoincidence element 反符合元件
anticoincidence gate “异”门
anticoincidence pulse 反符合脉冲
anticoincidence unit “异”单元
anticoincidence 反符合的，反重合
anti-collision device 缓冲装置
anti collision light 防撞灯
anti-compound generator 反复激发电机，反复励发电机
anti-condensation heater 防潮加热器，防结露加热机，防凝结加热器
anti-condensation lining 防凝水内衬，防凝内衬
anti-condensation 防凝结
anti-contamination clothing 防沾污服
anti-contamination type 防污型
anti-convection device 反对流装置
anti-corona coating 防晕层
anti-corona collar 电晕保护环，防晕环
anti-corona varnish 防晕漆
anti-corona 防电晕
anticorrodent 防锈剂，防腐蚀剂
anticorrosin and antifouling paint 防蚀防污涂层
anticorrosin composition 防蚀化合物，防蚀油漆
anticorrosin insulation 防蚀层
anti-corrosion additive 防腐蚀添加剂
anti-corrosion agent 防腐剂，防蚀剂
anti-corrosion coating 防腐涂层，防腐蚀层，防蚀层
anti-corrosion insulation 防蚀绝缘，防腐蚀绝缘层，防蚀层
anti-corrosion measures 防腐蚀措施，防锈措施，防腐措施
anti-corrosion method 防蚀法
anti-corrosion protection 防腐
anti-corrosion treatment 防腐处理，防腐蚀处理
anti-corrosion venting chimney 耐腐蚀排气筒
anti-corrosion worker 防腐工
anti-corrosion 防腐，防蚀，防锈，耐蚀
anticorrosive ACSR 防腐钢芯铝绞线
anti-corrosive additive 防腐添加剂
anti-corrosive agent 防蚀剂，防腐剂
anti-corrosive coating 防腐层，防锈层，防腐涂层
anti-corrosive composition 防锈剂，防蚀油漆，防腐蚀剂
anti-corrosive，dust-proof and water-resistant lighting fixture 三防灯具
anti-corrosive insulation 防蚀绝缘，防腐绝缘层，防蚀层
anti-corrosive lacquer 防腐漆
anti-corrosive machine 防蚀电机
anti-corrosive motor 防蚀电动机
anti-corrosive painter 油漆防腐工
anti-corrosive paint 防蚀漆，防锈漆，防腐涂料，防锈油漆，防腐漆
anti-corrosive pigment 防腐涂料
anti-corrosive pump 耐腐蚀泵
anticorrosive sleeve 防腐锌套
anti-corrosive treatment 防腐处理，防蚀处理，防锈处理
anti-corrosive valve 防蚀阀门
anti-corrosive 防腐的，防蚀的，防锈的，耐腐蚀的，防蚀剂，缓蚀剂
anti-crash device 防撞装置
anti-creep device 抑制频率漂移装置
anticreeper 防爬电设备，防漏电设备，防蠕动设备
anti-creep 防爬电，防漏电，防蠕动
anticrustator 防垢剂，表面沉垢防止剂
anticyclogenesis 反气旋发生，反气旋生成
anticyclone circulation 反气旋环流
anticyclone curvature 反气旋曲率
anticyclone eddy 反气旋涡旋

anticyclone 反气流，反气旋，高气压
anticyclonic curvature 反气旋曲率
anticyclonic swirl 反气旋（型）旋涡
anticyclonic wind 反气旋风
antidamping 抗阻尼，抗衰减
anti-dated BL(anti-dated bill of lading) 倒签提单
anti-dazzle glass 防眩玻璃，遮光玻璃
anti-dazzle lighting 防眩灯光，防眩照明
anti-dazzle screen 防眩屏
anti-decaying paint 防腐漆
anti-detonating quality 抗爆性
antidetonation 防爆，抗爆
anti-detonator 防爆剂，抗爆剂
antidiffusion coating 反扩散涂层【燃料棒】
antidip river 反倾斜河
antidip stream 反倾斜河
anti-dirt 防尘
antidolorin 氯乙烷
antidote 抗电脑病毒程序
anti-drag 减阻的，反阻的
antidrip 防漏，防滴
anti-drying surface 防干裂面
anti-dumping duty 反倾销税
anti-dumping measures 反倾销措施
anti-dumping 反倾销
antidune movement 逆向沙丘运动
antidunes 逆沙丘波，逆（行）沙丘
anti-dust device 防尘装置
anti-dust 防尘，抗尘
antielastic bending 反弹性变形弯曲
anti-electron 反电子，阳电子，正电子
antiemulsification time 破乳化时间
anti-evaporant 防蒸发剂
anti-evasion measures 反逃税措施
anti-explosion 防爆
anti-explosion device 防爆设施
anti-explosion machine 防爆电机
anti-explosion of production buildings 厂房的防爆
anti-explosion panel 防爆墙板
anti-fading antenna 防衰落天线
anti-fading device 衰落补偿装置，防衰落装置
antifatigue bolt 耐疲劳螺栓
antifatigue 耐疲劳，抗疲劳剂
antiferrodistortive transition 抗铁磁失真相变
antiferroelectricity 反铁电现象
antiferroelectric 反铁电的
antiferromagnetic resonance 反铁磁共振
antiferromagnetic 反铁磁质，反铁磁体
antiferromagnitism 反铁磁现象，反铁磁性
anti-flashing equipment 防气化装置，防汽化装置
antiflex cracking 抗折裂
anti-flocculation 反絮凝作用
anti-flood facilities 防洪设施
antiflooding valve 安全阀【换热器】，满水保护阀，防洪阀
anti-fluctuator 缓冲器
antiflutter device 抗颤振装置
antifoam additive 防泡沫添加剂
antifoam agent 防沫剂，消泡剂
antifoamant 防沫剂，消沫剂
antifoamer 防沫剂，消泡剂
antifoaming additive 防沫添加剂
antifoam reagent dosing tank 防沫剂加药箱
antifoam reagent 防沫剂，消沫剂
antifog and anti-smog device 消除烟雾设施
anti-fog insulator 防雾绝缘器，抗雾绝缘子
antifog 防雾，耐雾
antifoulant additive 防污添加剂
antifoulant 防污剂
anti-fouling agent 防污剂
antifouling insulator 耐污绝缘子
antifouling motor 防污电动机
anti-fouling paint 防污涂料，防污油漆
antifouling 防垢，抗垢，防沾污，防污染，防污
antifreeze additive 防冻添加剂
antifreeze admixture 防冻附加剂
antifreeze agent 防冻剂
antifreeze insulation 防冻保温
antifreeze plate 防磁粘片，隔磁片【继电器】
anti-freeze protection 防冻
antifreezer 防冻剂
antifreeze solution 防冻液
antifreeze 防冻，抗冻，抗凝，防冻剂，抗冻剂
antifreezing agent 防冻剂，（混凝土）抗冻剂
antifreezing coat 防冻涂层
antifreezing lubricant 防冻润滑剂，抗冻润滑剂
anti-freezing mixture 防冻掺和料，防冻剂
antifreezing solution 防冻液
anti-freezing 防冻
antifriction bearing grease 减摩轴承润滑脂，轴承用润滑油脂
antifriction bearing pillow 减摩轴承垫座
anti-friction bearing 耐磨轴承，减摩轴承，抗摩轴承，滚动轴承
antifriction grease 减摩润滑脂
antifriction material 耐磨材料，润滑剂
antifriction metal 减磨轴瓦合金，减磨轴承金属，减摩金属
antifriction thrust bearing 减磨止推轴承
antifriction 减低或防止摩擦之物，润滑剂，减摩剂，减磨设备，减磨，防磨，耐磨，抗摩
anti-frost 防冻的，防霜的，防冻
antifungus insulation 防霉绝缘
antigalloping damper 抗驰振阻尼器
antiglare 遮光的，防闪光的，防眩光的
antigorite 叶蛇纹石
anti-ground 防接地的
anti-halation 防光晕作用，防光晕，防光晕的
anti-high temperature cable 耐高温电缆
antihum device 噪声消除器，静噪设备，交流声消除器
antihum 静噪器，噪声消除器
antihunt action 稳定作用，防振作用，防震作用，阻尼作用
antihunt circuit 稳定电路
antihunt filter 防振滤波器
antihunting protection 防振保护，防摆保护
antihunting transformer 镇定变压器
antihunting unit 防振装置，防摆装置
antihunt means 稳定方法，稳定器，阻尼器，防振设备

antihunt transformer 防振变压器
antihunt 阻尼，防振荡，稳定的，防震的，阻尼的，防摆动的
anti-icer 防冰器，防冰装置
anti-icing 防冰
anti-icing device 防冰设施
anti-incrustation corrosion inhibitor 阻垢缓蚀剂
anti-induction conductor 防感导线
anti-induction 防感应
anti-inductive resistance 无电感电阻
anti-interference condenser 抗干扰电容器
anti-interference equipment 抗干扰设备，抗干扰装置
anti-interference filter 抗干扰滤波器
anti-interference measures 抗干扰措施
anti-interference 抗干扰，抗干扰设备
anti-jamming stick test 防卡涩试验，抗干扰试验
anti-jamming technique 抗干扰技术
anti-jamming unit 抗干扰装置，抗干扰附加器
anti-jamming 抗干扰，反干扰
antiknock component 防爆剂
antiknock device 防爆设备
anti-knocker 防爆震剂
anti knock fuel 防爆震燃料
anti knock rating 防爆震率
antiknock 抗震，消振，抗震剂，抗爆的，防爆剂
antileak 防泄漏
antilevitation safeguard （元件）防漂浮安全装置
antilithic 去垢剂
antilogarithm 反对数
antimagnetic 抗磁性的，反磁性的，防磁的
antimagnetic effect 防磁效应
antimagnetic shield 防磁屏蔽
antimagnetic steel 反磁性钢，防磁钢
antimagnetized 消磁的
antimonsoon 反季风
antimony-beryllium neutron source 锑-铍中子源
antimony-bismuth couple 锑-铋热偶
antimony rod 锑棒
anti-motoring device 逆功率保护装置，防止倒拖装置
antinodal 波腹的，波腹，腹点
antinoise paint 吸声油漆，吸声涂料
antinoise 防噪声的，抗噪声的，防噪的
anti-nucleating agent 防泡核剂，防晶核剂
anti-oil whip 抗甩油【泵】
antiosmosis 反渗透
anti-overloading 防过载的
anti-overshoot device 防过调装置
anti-overshoot 反过冲，防过冲
anti-oxidant 抗氧化剂，防老化剂
antioxidation stability 抗氧化安定性
anti-oxygen 抗氧化剂
anti-parallax 反视差
antiparallel thyristor 反平行晶闸管，反并联晶闸管
antiparallel 反平行的，逆平行的，反并联的
antiparalyse pulse 启动脉冲
antiparasitic 防寄生振荡的
antiparticle 反粒子
antiphase dipole 反向偶极子
antiphase 反相位的，反相的
anti-pick-up factor 始动可靠系数的倒数
anti-piping compound 防缩孔剂，发热剂
antipit 防点蚀，防点蚀剂
antiplug relay 防反转继电器，防止电动机反转继电器
antipode 对极，相对极，对应体，对蹠点
antipolarizing winding 反极化线圈
antipole 反极，相反事物
antipollutant 抗污染剂
anti-pollution device 防污染装置
antipollution law 防污染法律
antipollution measure 环境污染对策
anti-pollution standard 防污染标准，抗污染标准
antipollution system 防污染系统
antipollution-type insulator 耐污绝缘子
anti-pollution 防污染，反污染
anti-popout tube 防离位管子
antipriming 防止汽水共腾
antiproton 反质子
anti-pump circuit 防跳回路
antipump device （断开运行后）重合闸闭锁装置
anti-pumping device 防跳装置
anti-pump relay 防跳继电器
antiputrefactive 防腐的
antiradar 防雷达的，反雷达的
antiratchetting groove 防松槽【燃料元件】
antirattle clip 防振线夹
antirattler 消声器，减声器，防振器
anti-reflecting 减反射的，消反射的
anti-reflection coating 减反射膜
anti-regeneration 防再生
anti reset windup 抗复位锁紧（特性）
anti-resonance 反共振，反谐振，防共振，防共鸣，并联谐振，防共鸣
anti-resonant circuit 并联谐振电路，反谐振电路
anti-resonant frequency 并联谐振频率，电流谐振频率，反共振频率
anti-resonating 反共振的，并联谐振
anti-return valve 止回阀
anti-reverse device 防倒转装置
anti-reverse rotation device （泵的）防倒转装置
anti-reverse system 不可逆系统
anti-reverse 抗反转
antiroll bar 防滚杆，防侧摆杆
anti-rolling device 消摇装置，减摇装置
anti-rolling 防滚动的
anti-roll pump coupling 减摇泵联轴器
anti-rotation device （泵的）防倒转装置，防反转装置
antirotation lug 防反转凸耳
antirotation tool 防反转工具
antirot substance 防腐材料
antirot 防腐的，防腐
antirust coat 防锈层，防锈涂层
antirust composition 防锈混合剂
antirust compound 防锈混合剂
antirust grease 防锈脂
antirusting agent 防锈剂
antirusting paint 防锈涂层，防锈漆

antirusting 防锈
antirust oil 防锈油
antirust paint 防锈漆，防锈油，防锈涂料
antirust primer 防锈底漆
antirust solution 防锈溶液
antirust varnish 防锈清漆
antirust 防锈，防锈的
antisaprobic zone 防污染带
antiscalant 防垢剂，阻垢剂，防垢
antiscale composition 防垢剂
antiscale 防垢剂，防垢
anti-scaling in ash sluicing pipe 灰管防垢处理
anti-scour trench 防冲槽
anti-scour wall 防冲墙
antiseepage 防渗（漏）
antiseep collar 防渗环
antiseep diaphragm 防渗环，防渗隔膜
anti-seismic and reinforcing design 抗震加固设计
anti-seismic apparatus 抗地震电器
anti-seismic design 防（地）震设计，抗震设计
anti-seismic joint 抗震缝
anti-seismic plate 防震板，抗震支承板
anti-seismic structure 抗震结构
anti-seismic 防震的，抗地震的
antiseize compound 防黏剂
antiseptic effect 防腐效应
antiseptic oil 防腐油
antiseptics for wood 木材防腐剂
antiseptic treatment 防腐处理
anti-septic wooden sleeper 防腐木枕
antiseptic 防腐的，抗菌的，杀菌剂，防腐剂，抗菌剂
anti-settling agent 抗沉淀剂，防沉剂
antishadowing 反荫蔽
anti-shrink 抗缩的，耐缩的
antisidetone circuit 消侧音电路
antisidetone telephone set 消侧音电话机
antisimmer pressure relief valve 延时卸压阀，防抖动卸压阀
anti-simmer 防抖动
anti-sine 反正弦
antisinging device 振鸣抑制装置
anti-single phasing scheme 防止单相运行方案
anti-skid braking device 防滑制动装置
anti-skid device 防滑装置
anti-skid factor 防滑系数
anti-skid service required 需要防滑服务，所需的防滑服务
anti-skid 防滑，抗滑
anti-slide micropile 微型抗滑桩
anti-slide pile 抗滑桩
anti-slip mat 防滑垫
antisludge 去垢
antisludging agent 阻垢剂
anti-softener 防软剂
antispark 防火花的，消火花的
antisplash guard 挡泥板
anti-squeak 消声器，防振器，减声器，减声片
anti-stall gear 防止失速装置
anti-stall 防失速，防止失速
antistatic agent 抗静电剂
antistatic coating 抗静电涂料
antistatic fluid 防静电液
antistatic paint 防静电漆
antistatic 抗静电，防电剂
antisun 防止日光照射的，抗日光的
anti-superheating system 防过热装置
anti-superheating 防止过热（装置）
antisway device 防倾斜装置
antisweat insulation 防结露保温
antiswing device 防摆动装置
anti-symmetrical load 反对称负荷，不对称荷载
anti-symmetrical matrix 反对称矩阵
anti-symmetrical mode 反对称模态，反对称振型
anti-symmetrical tensor 反对称张量
anti-symmetrical 反对称的
anti-symmetry 反对称性
anti-syphonage pipe 反虹吸管
antithetic fault 反断层，相反组断层
antithetic varieties 对偶变数
anti-thrust bearing 止推轴承
anti-thrust 止推的
anti-thunder capacity 耐雷能力
anti-torque moment 反扭力矩，抗扭力矩
anti-torque 反作用转矩，反向转矩，抗扭
anti-tracking coat 抗爬电涂层，防爬电涂层
anti-tracking 防跟踪，反跟踪，防爬电
antitrade wind 反信风
antitransmit-receive switch 反收发开关
antitransmit-receive tube 反收发管，反收发开关管
anti-trigonometric function 反三角函数
antitriptic wind 摩擦风，减速风
anti-TR switch 反收发开关
anti-TR tube 反收发管，反收发开关管
anti-twisted wire rope 防扭钢丝绳，防捻钢丝绳
antitype 模型所代表的实物，对型
anti-vandal bolt 防盗螺栓【防破坏】
anti-vaporizing equipment 防止汽化装置
antivibration bar attachment 防震条固定
anti-vibration bar 抗振杆
antivibration base 防震基础
anti-vibration clamp 防振动线夹
anti-vibration device 防振设备，减震装置
anti-vibration measures 防振措施
anti-vibration mounting 弹性安装
anti-vibration pad 防振衬垫，防振填料
anti-vibration spacer bar 防振定位条，防振定位板
anti-vibration structure 抗振结构
anti-vibration 防振，阻尼，防震，消振
antivibrator 防震器，减震器，防振器，减振器，阻尼器
anti-virus 反病毒
antivoid valve 真空安全阀
anti-vortex baffle 抗涡流挡板，抗涡流挡圈
antiwear additive 防磨剂
antiwear agent 抗磨剂
antiwear characteristic 耐磨特性
antiwear property 耐磨特性
antiwear 耐磨的，抗磨损的
antiwhip device 防冲击装置，抗甩击装置

antiwhip 稳定性
anti-wind beam 防风梁，抗风梁
antiwind grade 抗风等级
anti-windup feature 抗饱和（特）性
antomony electrode 锑电极
antonite 白云母
anvil block 砧座
anvil 电键下接点，砧，铁砧
any quantity 任意量
AOA(angle of attack) 迎角，攻角
AOA device(aerodynamic add-on device) 气动附加装置
AO(analog output) 模拟量输出
AO(assembly order) 装配指令
AO(assembly outline) 装配大纲
AOB(angle of bank) 倾斜角，侧倾角
AOB(any other business) 其他事项
AOC(air oil cooler) 空气油冷却器
AOI(and-or-inverter) “与或”反相器
AOI(and-or-invert) “与或非”
AOI gate “与或”翻转门，“与或”反相门
AOL(aircraft obstruction lamp) 航空障碍灯
AOM(analog output module) 模拟输出模件
AO/OSP(assembly order/operational sequence planning) 装配大纲/操作顺序计划
AOP(automatic operation panel) 自动操作仪表盘
AOP(auxiliary oil pump) 辅助油泵
AOQ(average outgoing quality) 检验的平均质量
AORN(assembly outline requirement notification) 装配工艺规程要求通知单
AOS(add-or-subtract) 加或减
AO-SR(assembly outline-ship record) 装配大纲-架次记录
AO-SRS(assembly outline-ship recording system) 装配大纲-架次记录系统
AOTCR(assembly outline tracing control report) 装配大纲跟踪控制记录
AOV(air operated valve) 气动门
AO water treatment(anoxic oxic water treatment) AO（水处理）工艺法，厌氧好氧水处理法
a packaged deal 整套工程，整批交易，一揽子交易
apart from 除……之外
apartment building 公寓建筑，公寓大楼，住宅
apartment 一套公寓房间，公寓，楼中单元房
AP(Ash Slurry Pump) 灰浆泵
apatite 磷灰石
APA 电动主给水泵系统【核电站系统代码】
APC(American Power Conference) 美国电力会议
APC(Atomic Power Construction Ltd.) 英国原子能建设有限公司
APC(automatic program control) 自动程序控制器
APC(Provisional Acceptance Certificate) 临时验收证书
APEC(Asia-Pacific Economic Cooperation) 亚太经合组织
aperiodic aerial 非调谐天线，非谐振天线
aperiodic circuit 非周期电路，非周期振荡电路
aperiodic component 非周期分量，非周期部分
aperiodic current 非周期电流
aperiodic curve 非周期曲线
aperiodic damped motion 非周期阻尼运动
aperiodic damping 非周期阻尼，过阻尼
aperiodic discharge 非周期放电
aperiodic disturbance 非周期性扰动
aperiodic element 非周期摆动的可动部分，非周期部件
aperiodic function 非周期函数
aperiodic galvanometer 不摆电流计，直接指示电流计，大阻尼电流计
aperiodic instrument 不摆式仪表，非周期式仪表
aperiodicity 非周期性，非周期性的，非调谐的，非调谐性
aperiodic mode of motion 非周期运动形式
aperiodic motion 非周期性运动
aperiodic oscillation 非周期性振荡
aperiodic phenomenon 非周期性现象
aperiodic recovery 非周期性恢复（至给定值）
aperiodic time constant 非周期分量时间常数，非周期时间常数
aperiodic transformer 无调谐变压器
aperiodic variation 非周期性变化
aperiodic wave 非周期波
aperiodic 非周期的，非周期性的，非调谐的
aperture angle 孔径角
aperture area of collector 集热器采光面积
aperture area of solar collector 太阳集热器采光面积，太阳灶采光面积
aperture area 采光面积【太阳能集热器】，孔径面积，开孔面积
aperture board 窗孔板
aperture card 穿孔卡，穿孔卡片
aperture effective area 有效孔径面积
aperture of window 窗口
aperture plane 采光平面
aperture ratio 开孔率
aperture seal 穿孔密封
aperture 孔，小孔，孔口，窗孔，壁孔，缝隙，开口，眼，孔径，开度，采光口
apex angle 顶角
apex climax 顶峰
apex load 顶点负荷，顶点荷载，顶点载重
apex of roof 尖屋顶
apex 分流劈，分流器，尖劈，顶点，峰，脊
APG 蒸汽发生器排污系统【核电站系统代码】
aphanite 隐晶岩
aphelion 远日点
aphotic zone 非生化带，无光带
aphrolite 泡沫岩
aphytal 深水带
API(activity performing inspection) 放射性检测
API(application programming interface) 应用程序界面
apical 顶尖的，峰顶的
APL(a programming language) 程序设计语言
aplite 细晶岩
APMS(air pollution monitoring system) 空气污染监测系统

apocynthion 远月点
apogean range 远地点潮差
apogean tidal current 远地点潮流
apogean tidal range 远地点潮差
apogean tide 远地点潮，远月潮
apogee motor 远地点控制电动机，遥控电动机
apogee 远地点
apologize 道歉
apomecometer 测角测距仪，测距仪
apophyse 岩枝
aposandstone 石英岩
a posteriori 凭经验的，归纳的，后天的，由事实推论到原理的
A-post 前立柱，A支柱
app. =apparatus 仪表，装置
apparatus constant 仪器常数
apparatus dew point 机器露点
apparatus for air conditioning 调节空气设备，空调设备
apparatus for inserting insulation 嵌绝缘机
apparatus for taping coil 线圈包带机
apparatus room 机房
apparatus 器具，仪器，仪表，器件，装置，设备，机械，机构
apparent ablation 视消融
apparent activation energy 表观活化能
apparent angle of attack 视迎角
apparent angle of friction 表观摩擦角，视摩擦角
apparent angle of internal friction 视内摩擦角，表观内摩擦角
apparent capacity 视在电容，表观电容，视在容量
apparent charge 表观电荷，视在电荷
apparent coefficient 表观系数
apparent cohesion 视凝聚力，表观凝聚力，表观黏结，假黏结
apparent component 表观分量，视在分量
apparent concentration 表观浓度
apparent crack initiation CTOD 表观启裂尖端张开位移值
apparent crack initiation toughness 表观启裂韧度
apparent density 表观密度，视在密度，视密度，堆积密度
apparent dip 视倾斜，视倾角
apparent discount rate 显现贴现率，名义贴现率
apparent distance 视距
apparent earth conductivity 视在大地导电率
apparent efficiency 视在效率，视效率，表观效率
apparent elastic modulus 表观弹性模量
apparent electromotive force 视在电动势，表观电动势
apparent emissivity 表观黑度，表观发射率
apparent emittance 视在发射率
apparent-energy meter 视在功率计，伏安时计，表观能量计
apparent energy 表观能量，视在能量
apparent error 视误差，表观误差
apparent escalation rate 显现浮动率
apparent gap area 视在气隙面积【磁极下的气隙面积】
apparent gap density 视在气隙磁通密度
apparent gravity 表观比重，视比重，视重力
apparent groundwater velocity 地下水表观流速，视地下水流速
apparent half-life 表观半衰期
apparent heat transfer coefficient 表观传热系数，近似传热系数
apparent impedance 表观阻抗，视在阻抗
apparent inductance 表观电感，视在电感
apparent interest rate 显现的利率
apparent lifetime 表观寿命
apparent life 表观寿命，近似寿命，
apparent load 视在负荷
apparent loss 表观损失，视在损失，视损失，毛估损失
apparent magnetism 视在磁性，表观磁性
apparent mass 表观质量
apparent monochromatic temperature 表观单色温度
apparent output 视在输出，视在功率，视在输出功率，表观输出，表观功率
apparent overconsolidation 表观超固结
apparent porosity 表观孔隙率，视孔隙率
apparent position 视位置
apparent power consumption 视在功率消耗
apparent power loss 视在功率损失，表观功率损失
apparent power meter 视在功率表
apparent power 表观功率，视功率，视在功率
apparent precession 视进动，自然进动
apparent preconsolidation pressure 表观先期固结压力
apparent quality factor 外观品质因数
apparent relative density 视相对密度
apparent reluctance 视在磁阻，表观磁阻
apparent resistance 表观电阻，视在电阻
apparent resistivity prospecting 视电阻率勘探法
apparent resistivity 视在电阻率，视电阻率，视在电阻系数，表观电阻率
apparent sag 视在垂度
apparent selectivity 视选择性
apparent shearing strength 表观抗剪强度
apparent size 表观尺寸，视尺寸
apparent slug 似腾涌，假腾涌
apparent solar day 视太阳日，真太阳日
apparent solar time 视太阳时，真太阳时
apparent sound power level 表观声功率级，视在声功率级
apparent specific gravity 视在比重，表观比重，视比重
apparent specific heat 表观比热
apparent specific resistance 表观电阻率，视电阻率
apparent statistical property 表观统计性质
apparent strain 视在应变，表观应变
apparent stratigraphical gap 表观层距
apparent strength of insulation 视在绝缘强度，（导线所有绝缘层的）总绝缘强度
apparent stress 表观应力，视在应力，视应力
apparent temperature difference 表观温差，视在温差

apparent throw 视纵断距
apparent tooth density 表观齿磁密，视在齿磁密
apparent unconformity 视不整合
apparent velocity 表观速度，视流速，视速度
apparent viscosity or coal water mixture 水煤浆表观黏度
apparent viscosity 表观黏度
apparent volume 表观容积，视容积，松装体积
apparent watt 视在瓦特，表观瓦特
apparent wave frequency 表观波频率
apparent wave period 表观波周期
apparent wave 表观波
apparent weight 毛重，视重量，视在重量，表观重量
apparent wind angle 表观风向角
apparent wind direction 表观风向
apparent wind velocity 表观风速
apparent wind 表观风
apparent 表面的，表观的，视在的，明显的
APP(Atomic Power Plant) 核电厂，核电站
appeal body 上诉机构
appealing 吸引人的
appeal 申诉
appearance of fracture 断口外观，断口形状
appearance potential spectroscopy 表观电势谱技术
appearance potential 表观电势，外观电势
appearance test 外形检查，外观检查
appearance 表面，出现，外表，外观，外形
appellation of origin 原产地名称，以原产地取名的产品名
appellation 称号，名称
appendage pump 备用泵，辅助泵
appendage sensor 备用传感器
appendage 附属设备，附件，附加，备用仪器，附属部分，附属物，附加物，备用仪表，备件，配件
appendant displacement 附加位移
appendices to technical specification 技术规范的附件
appendices 附录【appendix 的复数】
appendix of contract 合同附件
appendix to tender 投标书附录，招标书附录
appendix 附录，补遗，附属物，附件
appilcation protocol control information 应用协议控制信息
apple tube 苹果管【一种单束彩色显像管】
applet 小应用程序
appliance carrying fire vehicle 器材消防车
appliance circuit 仪表用电路
appliance load 生活负荷，民用负荷
appliance 器具，附件，器械，装置，设备，应用，用具，仪表，适用
applicability 适用性，可贴性，适用范围
applicable law 适用法律，准据法
applicable standard 适用标准
applicable 能应用的，可应用的，适合的，适用的，适宜的
applicant for insurance 投保人
applicant for shares 入股申请人
applicant 投保人，申请人，报名者，请求者
application and maintenance performance 应用和维护性能
application control valve 作用控制阀
application data structure 应用数据结构
application development language 应用开发语言
application drawing 操作图，应用图
application experience 应用经验
application fee 申请费
application for amendment of letter of credit 信用证修改申请书
application for credit 请求贷款，开证申请书
application for drawback 申请退还税金，申请退款
application for employment 求职申请
application for export license 出口许可证申请书
application for foreign exchange 申请外汇
application for insurance 保险申请书，投保申请
application form 申请表，申请书格式
application for negotiation of draft under L/C 出口押汇信用证申请书
application for opening letter of credit 开立信用证申请书
application for operating licence 运行许可证申请书
application for partial transfer of L/C 信用证部分转让申请书
application for patent 申请专利
application for payment 付款申请书
application for shares 认股书
application for shipment 装运申请
application for transfer of letter of credit 信用证转让申请书
application for use of funds 使用资金申请表
application for withdrawal 申请撤回，提款申请
application function 应用功能
application guide 使用导则，应用手册
application industry data 应用工业资料
application in field 现场应用
application instruction 使用说明书
application layer protocol data unit 应用层协议数据单元
application layer 应用层
application of force 施力
application of fund 资金运用
application of load 加载，施加负载，施加荷载
application of new lead seal 新铅封的作用
application-oriented language 面向应用语言
application-oriented programming language 面向应用的程序设计语言
application package 应用程序包
application point 作用点，施力点，着力点
application profile 应用协议集
application program interface 应用程序接口
application programme 应用程序
application program 应用程序
application protocol data unit 应用协议数据单元
application reference data 应用参考资料数据
application reference manual 应用参考手册
application service data unit 应用服务数据单元
application software 应用软件
application specific integrated circuit 应用型专用

集成电路
applications program　应用程序
application valve body　控制阀体
application valve cover　控制阀盖
application　使用，应用，申请（书），施加，运用
applicator　介质加热用电极，敷贴器，洒施器
applied　外加的，应用的，实施的
applied EMF　外加电动势
applied aerodynamics　实用气体动力学，应用气体动力学，工程气体动力学
applied business statistics　企业应用统计
applied climatology　应用气候学
applied dose　施予剂量
applied elasticity　应用弹性学
applied field　作用场，施加场，外加场
applied fluid dynamics　应用流体动力学
applied force　作用力，外加力
applied geology　应用地质学
applied hydraulics　应用水力学
applied hydrology　应用水文学
applied load　外加载荷，外加负荷，施加负荷
applied mechanics　应用力学
applied meteorology　应用气象学
applied moisture　应用水分
applied moment　外施力矩
applied new technology　实用新技术
applied-potential test　外加电压试验
applied power　输入功率，外加功率
applied pressure　外加压力，外加电压，施加压力
applied seismology　应用地震学
applied stratigraphy　应用地层学
applied stress　作用力，作用应力，外加应力
applied thermodynamics　应用热力学
applied thrust　外施推力
applied voltage　外加电压
apply force　施加力
apply for　申请
applying for collection　托收
applying for remittance　托付，托汇
apply oil　上油
apply work　做功
apply　使用，申请，应用，施加
appointed bank　指定银行
appointed day　指定日期
appointing authority　指派机关，认命机构
appointment system　聘任制
appointment　委派，任命，约会
appoint　委派，指定
apportioned cost　分摊成本
apportioncd depreciation　分摊折旧
apportioned tax　摊分税
apportionment of damages　损害赔偿的分担
apportionment of liability　责任分担
apportionment　分摊，拨款
apportion　分摊
apposition　并列，平排布置
appraisal company　评估公司
appraisal cost　评估费用，评价成本，鉴定成本，鉴定费用
appraisal inventory　估价财产目录
appraisal of construction project　建设项目评价
appraisal of damage　估损
appraisal of quality　质量鉴定
appraisal repot　鉴定报告，估价报告
appraisal　评价，鉴定，估价，估计
appraised price　估价，评估价格
appraisement　评价，估计，鉴定
appraiser　评价人，估价人，鉴定人
appraise　鉴定，评价
appreciable current　明显的水流，较大电流
appreciable error　显著误差
appreciable　客观的，相当大的，显著的
appreciably　客观地，相当大地
appreciate　评价，鉴定，估价，重视，感谢，感激
appreciation of currency　货币增值
appreciation of fixed assets　固定资产增值，固定资产估价
appreciation of money　货币增值
appreciation　评价，增值，正确评价，鉴别，涨价
apprehensive　有理解力的，聪明的
apprenticeship program　学徒计划
apprentice　徒工，生手，学徒，徒弟
apprise　通知，报告
apprize　通知，报告
approach bank　引堤
approach bridge　引桥
approach channel　进港航道，引航道，引槽
approach cutting　引桥挖方
approach embankment　引道路堤
approach fill　桥头填方，引道填方，引道填筑，引桥
approach flow　迎面气流
approach flume　引水槽
approach head　行近水头
approaching airstream　来流，迎面气流
approaching flow　来流，迎面气流，行近流
approaching wind　迎面风，来流
approaching　接近
approach jetty　引道突堤
approach load　引路，引道
approachment　端差，进口端温差
approach ramp　引道坡，引桥坡道，引道斜坡段
approach road　引路，引道
approach segment　临近段
approach sign　引道标志
approach span　引桥跨，引接跨，进线挡，岸跨
approach speed　行近速率
approach steam velocity　迎面汽流速度
approach to criticality phase　逼近临界状态
approach to criticality　逼近临界，接近临界
approach to equilibrium phase　接近平衡阶段，接近平衡相
approach to power　提升功率
approach trench　交通沟，窨井，引近沟，雨水口
approach trestle　引道栈桥
approach velocity　来流速度，驶近速度，行近流速，进口速度，趋进速度
approach viaduct　高架引桥
approach　近似，接近，逼近，趋近，方法，方

式，途径，通道，引道
approbation 许可，批准，认可
appropriate a fund 拨款
appropriate authorities 有关当局
appropriate body 有关机构
appropriated right 专有权
appropriate money 拨款
appropriateness of technology 技术的适用性
appropriateness 适合性，适合程度
appropriate professional 合适的专业人员
appropriate technology 适用技术
appropriate 盗用，挪用，侵吞，拨出（专款），适当的，恰当的，合适的，合理的
appropriation provisions 拨款条例
appropriation contract 拨款合同
appropriation 拨款，占用，盗用，挪用，侵吞，拨出（专款）
appropriative right 使用权
appropriative water right 专用水权
appropriator of water right 水权享用者
appropriator 拨给者，专用者
approval agreement 审批协议
approval authority for projects 项目核准权
approval for construction drawings 批准用于施工的图纸
approval notice of the contract 合同批准通知
approval number 批号
approval of import 批准进口
approval of order clause 承认订单条款
approval phase 审批阶段
approval process 审批过程
approvals to be obtained by contractor 由承包商获取的批件
approvals to be obtained by employer 由业主获取的批件
approval test 接收试验，验收试验，鉴定试验，合格性检验
approval version 批准版
approval 认可，批准，同意，批件，审核，审定，赞成，赞同，核准
approved accountant 认可的会计师
approved apparatus 合格设备，安全设备，防爆设备
approved as noted 有条件批准
approved bidders list 批准的投标人名单
approved budget 核定预算
approved by 由……认可，由……批准
approved cable 安全电缆，合格电缆
approved design 批准的设计
approved document 批件
approved investment 核定投资
approved in writing 书面同意的
approved lamp 安全灯
approved maximum capacity 认可最大容量
approved packing 认可的包装
approved product 定型产品
approved programme 核定计划，经批准的计划
approved project 批准的项目
approve the contract 审批合同
approve 批准，同意，赞同，认可，确认，授权
approx. = approximately 近似地
approximate adjustment 近似平差
approximate bill of project quantity 近似工程量清单
approximate calculation 概算，近似计算
approximate composition 大致成分
approximate configuration 大致地尺寸形状
approximate continuity 近似连续性
approximate dimension 概略尺寸
approximate error 近似误差
approximate expression 近似式，近似公式
approximate integral 近似积分
approximate location 大概位置
approximately equal 近似相等
approximately 大致地，大约
approximate method 近似方法
approximate quantities 概略工程量
approximate relief 草绘地貌
approximate representation of excitation system 励磁系统的近似（表示）法
approximate similarity 近似相似性
approximate solution 近似解
approximate theory 近似理论
approximate value 近似值，概算价值
approximate 近似的，大概的，大约
approximation theorem 逼近定理
approximation 近似，近似法，近似值，概算，逼近
appurtenance 附属物，附属部件，辅机，附件，附属设备，附属建筑，辅助工具
appurtenant structure 附属构筑物
appurtenant works 附属工程
APP 汽动主给水泵系统【核电站系统代码】
APR(automatic reactive power regulator) 自动无功功率调整器
apriori （从原因或假定）推出结果的，演绎的，推理的，先天的
apron board 裙板
apron conveyer 裙式输送器，板式输送机，链板式输送带，挡边输送机
apron crib 护坦笼框
apron extension 护坦延伸段
apron feeder 带挡板的给料机，板式给料机，带式给料机，链板式加料器，带裙边的刮板式给料机
apron flashing 遮檐板，披水板
apron piece 遮檐板，披水板
apron plate conveyer 带裙边的刮板式输送机
apron plate 裙板，闸门，点火门
apron rolls 运输机皮带滚轴，托辊
apron slab 护坦板
apron slope 护坦坡度
apron space 码头前沿区
apron stone 护脚石
apron wall 前护墙
apron wheel 履带
apron 隔板，发射板，托板，防护挡板，护底，护坦，平板，挡板，海堤，围堤，裙圈
APSA(automatic particle size analyzer) 自动粒度分析仪
apsacline 斜倾型
APPSS(Automatic Power Plant Start-Up and Shut-

down） 机组自启停控制系统
APS（auxiliary power supply） 辅助电源，自备供电设备
apse-buttress 半圆室支墩，半圆形扶垛
apse 半圆室
apsidal 半圆室的
apteral 无侧柱的
aptitude test 合格试验，鉴定试验，性能试验，智力测验
aptitude 才能，资质，天资
APU（auxiliary power unit） 辅助动力装置，辅助电源装置，辅助供电装置
APU 主给水泵汽轮机疏水系统【核电站系统代码】
APWR（advanced PWR） 先进压水堆
apyrous 不易燃的
AQ（any quantity） 任意量
AQCS（air quality control system） 空气质量控制系统
AQL（acceptance quality level） 可接受的质量水平
aqua-ammonia absorption-type refrigerating 氨-水吸收式制冷
aqua ammonia 氨水
aqua-culture 水产养殖
aquadag 导电敷层，胶体石墨，炭末润滑剂
aquafacts 水浸石
aquafluor process 水氟化法
aquage 水路
aquagraph 导电敷层
aquamarsh 水沼地
aquanaut work 潜水作业
aqua regia 王水
aquarium reactor 游泳池式反应堆
aquaseal 密封剂【电缆绝缘涂敷用】，水封，水封剂
aquastat 水温自动调节器
aqua storage tank 储水池，储水箱
aquathruster 气压扬水机
aquatic boiling slurry reactor 悬浮式沸水堆
aquatic ecosystem 水生态系统
aquatic food chain 水生食物链
aquatic growth 水生物生长物
aquatic life 水生生物
aquatic pollution 水生污染，水污染
aquatic weed 水草
aquatic 水生的
aquation 水合，水化，水合作用
aqueduct arch 输水道拱
aqueduct bridge 渡槽，渠桥
aqueduct 导水管，渡槽，输水道，水道，水管
aqueo-residual sand 水蚀残沙
aqueous 含水的
aqueous ammonia 氨水
aqueous boiling slurry reactor 悬浮式沸水堆
aqueous corrosion 水的腐蚀，水腐蚀
aqueous current ripple mark 水流波痕
aqueous deposit 水成沉积
aqueous effluent 废水
aqueous emulsion 水乳胶体，水乳状液
aqueous film forming foam concentrate 水成膜泡沫液
aqueous film 水膜
aqueous fuel reprocessing plant 水法燃料后处理厂
aqueous homogeneous reactor 水均匀反应堆
aqueous media 水介质
aqueous medium 水介质
aqueous metamorphism 水变质（作用）
aqueous phase 水相
aqueous reprocessing 水法（核燃料）后处理，湿法后处理
aqueous rock 水成岩
aqueous separation 水法分离，湿法分离
aqueous slurry reactor 悬浮式水堆
aqueous solution 水溶液
aqueous stratum 含水层，蓄水层
aqueous tension 水的张力
aqueous vapour 水汽，水蒸气，蒸汽
aquiclude 滞水层，隔水层，含水土层，半含水层
aquifer basin 含水层的集水区，含水层补给区
aquifer coefficient 含水层系数
aquifer exploration 含水层勘探
aquifer permeability 含水层渗透能力
aquifer recharge 含水层补给
aquifer storage 含水层储水量
aquifer test 含水层试验
aquifer transmissibility 含水层输水率
aquifer water-bearing stratum 含水地层
aquifer 蓄水层，含水土层，地下水位，地下含水层
aquifuge 不透水岩层，不透水层
aquitard 弱透水性岩体，弱含水层
aquo-compound 含水化合物
Arabic notation 阿拉伯数字符号
Arabic number 阿拉伯数字
arable area 可耕地区
arable land 可耕地，宜农荒地
araeometer 液体比重计
araeosystyle 对柱式的，对柱式的建筑物
aragonite 霰石，文石
araldite 环氧树脂，环氧树脂黏结剂，合成树脂黏结剂
aramid fibre 芳族聚酰胺纤维
AR（aspect ratio） 叶片展弦比
AR（augmented reality） 增强现实
arbiter 判优器，判优电路【仲裁机】，仲裁人，仲裁员
arbitrage account 套汇账户
arbitrage 套利
arbitral agreement 仲裁协议
arbitral award 仲裁裁决，仲裁判定书
arbitral court 仲裁庭
arbitral decision 裁决，仲裁
arbitral institution 仲裁机构
arbitral procedure 仲裁程序
arbitral rules of procedure 仲裁程序规则
arbitral tribunal 仲裁法庭
arbitral 仲裁人
arbitrament 裁决，仲裁，仲裁结论
arbitrarily 任意

arbitrary 随意的，任意的
arbitrary analysis 仲裁分析
arbitrary angle 任意角
arbitrary circulation distribution 任意环流分布
arbitrary constant 任意常数
arbitrary function 任意函数
arbitrary height 任意高度
arbitrary input 任意输入信号
arbitrary number 任意数
arbitrary parameter 任意参数
arbitrary phase-angle power relay 任意相角功率继电器，有功无功功率继电器
arbitrary reference value 任意基准值
arbitrary-sequence computer 任意顺序计算机
arbitrary-sequence 任意序列
arbitrary surface of revolution 任意回转面
arbitrary time behaviour 任意非稳定状态，任意时变状态
arbitrary unit 任意单位
arbitrary value 任意值
arbitration act 仲裁法
arbitration agreement 仲裁协议
arbitration analysis 仲裁分析
arbitration award 仲裁裁决
arbitration by summary procedure 简易仲裁
arbitration clause 仲裁条件
arbitration commission 仲裁委员会
arbitration convention 仲裁公约
arbitration court 仲裁法庭
arbitration expenses 仲裁费用
arbitration fee 仲裁费【总裁机构收费】
arbitration law 仲裁法
arbitration procedure 仲裁程序
arbitration protocol 仲裁协议
arbitration rules 仲裁规则
arbitration 调解，仲裁
arbitrator 仲裁员，仲裁人，调停人
arbor assembly 心轴组件
arborescent drainage pattern 树枝状排水系
arbor 柄轴，主轴，轴，心轴，刀轴，乔木
arcaded 有拱廊的，有拱顶的
arcade quay wall 连拱式驳岸墙
arcade 拱廊，拱形建筑物，封闭拱廊，有拱顶的走道，有连拱廊的街道
arc air cutting 电弧气割，压缩空气气割
arc air gouging machine 碳弧气刨机
arc air gouging power supply 碳弧气刨电源
arc air gouging 电弧气割
arc-arrester 消弧器，火花熄灭器
arc atmosphere 电弧气氛
ARC(automatic remote control) 自动遥控
arc-back 逆弧
arc blow 电弧偏吹【焊接或切割】，磁偏吹，吹弧，消弧，灭弧
arc booster （焊接）起弧稳定器
arc brazing 电弧铜焊，电弧钎焊，电弧硬钎焊
arc cathode 电弧阴极，弧光放电阴极
arc chute 电弧隔板，熄弧沟，灭弧罩，灭弧沟，灭弧
arc contact 电弧触头
arc-control breaker 有灭弧设备的断路器
arc-control device 灭弧设备，灭弧装置，灭弧室，灭弧器
arc-control 电弧控制，消除火花，控制电弧
arc converter 电弧变流器，电弧整流器，电弧负阻振荡器
arc core 电弧心
arc cosine 反余弦
arc cotangent 反余切
arc crater 弧坑，电弧焊口
arc crest 弧形堰顶
arc current 电弧电流，弧流
arc cutting machine 电弧切割机
arc cutting 电弧切割
arc-damping 电弧阻尼，熄弧的
arc deflector 电弧偏转器，熄弧隔板
arc discharger 电弧放电器
arc discharge 电弧放电，弧光放电
arc-driven supersonic laser 电弧驱动超声激光器
arc-drop loss 弧压降损耗
arc-drop 电弧压降
arc duration 电弧持续时间，燃弧时间
arc dynamo 碳弧灯用直流发电机
arc end loss 弧端损失，斥汽损失
arc end 引弧端，发弧端
arc energy 电弧能量
arc erosion 电弧侵蚀
arc extinction 灭弧，消弧
arc extinguish chamber 灭弧室，消弧室【断路器】
arc extinguisher 灭弧器，火花消除器，灭弧线圈，消弧线圈
arc extinguishing equipment 消弧设施，灭弧装置，灭弧设备
arc extinguishing medium 消弧介质，灭弧介质
arc flame 弧焰
arc force 电弧力
arc frequency 弧流频率，电弧频率
arc furnace transformer 电弧炉变压器，电弧炉
arc gap 弧隙
arc generator 电弧发生器，电弧振荡器
arc gouging 电弧刨削
arc ground 接地弧，对地电弧放电
arc guide 弧导筒，电弧波导，水银蒸汽阻隔筒
arch-abdomen dam 腹拱坝
arch abutment 拱端，拱台，拱座
arch action 拱作用
archaeological area 古文化遗址
archaeological 考古学的
arch and pier system 拱墩系统
arch apex 拱顶
arch back 拱背
arch barrel （连拱坝的）拱筒
arch bar 拱木条，拱形杆
arch block 拱面石【楔形的】，拱块
arch bond 拱砌合
archbound commutator 拱束式环形换向器
arch braced roof 拱支桁架，拱支架屋顶
arch brick 拱砖
arch buttress dam 拱式支墩坝
arch buttress 拱扶垛
arch camber 拱矢，起拱

arch cantilever bridge 拱式悬臂桥
arch centering 拱架，拱顶脚手架，拱形脚手架，拱鹰架
arch centre 拱心
arch chord 拱弦
arch covering 拱形屋架，拱盖，拱板，穹形板
arch crown 拱顶，拱冠
arch culvert 拱形涵洞，拱涵
arch dam 拱坝
arch-deck buttress dam 拱面支墩坝
arch door 拱门
arc heating 电弧加热
arched-beam bridge 拱形梁式桥
arched bridge 拱桥
arched buttress 拱扶垛，拱式扶垛，拱式支墩
arched cantilever bridge 悬臂式拱桥
arched concrete dam 混凝土拱坝
arched conduit 拱形管道
arched construction 拱形建筑物
arched cover band 拱形围带
arched crown 拱形顶
arched dome 圆顶穹隆
arched falsework 拱形脚手架
arched furnace 有拱炉膛
arched girder 拱形主梁，拱形大梁
arched gravity dam 拱形重力坝
arched mole 拱形防波堤，拱形突堤
arched plate 拱板
arched roof truss 拱形屋架
arched roof type 上拱形的
arched roof 拱形屋顶，拱形车顶
arched sheet 拱板
arched-truss bridge 拱形桁架式桥
arched 拱形的
archetype 原始型
arch face 拱面
arch fired furnace W形火焰炉膛
arch flat 平拱
arch floor 拱形底板，拱形楼板
arch gravity dam 拱形重力坝，重力拱坝
arch haunch 拱背圈
arch height 拱高
arch hinged at ends 双铰拱
archie 网络文件查询系统
Archimedean principle 阿基米德原理，阿基米德定律
Archimedean screw pump coupling 阿基米德螺旋泵联轴器
Archimedean screw pump 阿基米德螺旋泵，阿基米德螺旋抽水机
Archimedean screw 阿基米德螺线，螺旋升水泵，阿基米德螺旋泵
Archimedean spiral 阿基米德螺线
Archimedes law 阿基米德定律
Archimedes number 阿基米德数
arching effect 弯拱作用，拱作用
arching factor 弯拱因素，拱度
arching 结块【沸腾炉】，搭桥，形成拱状，起拱作用，架拱
arch in trellis work 格形拱
archipelago 列岛，群岛
architect 建筑师
architect engineering company 建筑工程公司
architect-engineer 建筑工程师
architective 关于建筑的，建筑的
architectonic 构造的，地质构造的，建筑术的
architectonic geology 构造地质学
architectonics 构造设计，建筑设计，建筑学，建筑原理
architectural aerodynamics 建筑空气动力学
architectural and structural design 建筑和结构设计
architectural appearance and elevational treatment 建筑外观和立面处理
architectural appearance 建筑外观
architectural compatibility 建筑协调性，建筑调和性
architectural complex 建筑群，建筑总体，建筑综合体
architectural composition 建筑构造，建筑构图，建筑构造方式
architectural concrete 装饰用混凝土
architectural construction drawing 建筑施工图
architectural control 建筑管理规则
architectural decoration 建筑装饰
architectural design 建筑设计
architectural details 建筑细部，建筑详图
architectural drawing 建筑图，结构图
architectural elevational treatment 建筑立面处理
architectural engineering 建筑工程
architectural groups and surrounding landscape 建筑群和四周地形
architectural groups 建筑群
architectural line weight 建筑的线载荷
architectural meteorology 建筑气象学
architectural model 建筑模型
architectural ornament 建筑装饰物
architectural perspective 建筑透视图
architectural planning 建筑规划
architectural sketch 建筑草图
architectural style 建筑风格，建筑式样，建筑形式
architectural treatment 建筑处理
architectural wind tunnel 建筑风洞
architectural 建筑上的，建筑学的
architecture design institute 建筑设计院
architecture design scheme 建筑设计方案
architecture 建筑学，建筑艺术
architrave 框缘，门头线，嵌线，下楣，柱顶过梁，（柱的）下楣
archives management 档案管理
archives 档案馆，档案室，档案库，档案集，文献，案卷，档案，存档，文件
archivolt 拱门饰
arch keystone 拱顶石
archless furnace 无拱炉膛
arch limb 拱翼，顶翼，穹翼
arch lining 拱圈衬砌
arch motor 拱式电动机
arch moulding 拱饰
arch nose 折焰角，折焰拱，鼻形拱
arch of vault 穹拱

arc horn　角形避雷器
arch pitch　拱的高跨比
arch rib　拱肋
arch ring　拱环，拱圈
arch rise　拱高
arch roof of ferro-cement　钢丝网水泥拱形屋架
arch roof truss　拱形屋架
arch roof　拱顶【炉膛，燃烧室】
arch set　拱形支架
arch sheeting　拱板
arch span　拱跨
arch springing line　起拱线
arch springing　起拱点，拱脚
arch stone　拱石
arch stress　拱应力
arch striking　拆除拱架
arch support　拱形支架，拱座
arch thrust　拱的轴向压力，拱推力
arch timbering　拱形支护
arch truss　拱形桁架
arch tube　拱形管，形成炉拱的水管
arch-type culvert　拱形涵洞
arch viaduct　拱形栈桥
arch wall　拱墙
archway　拱道
archwise　成弓形
arch with three articulations　三铰拱
archy　拱形的
arch　（使）弯成拱形，用拱连接，弓形，拱，拱门，拱形物，主要的，首要的
arc ignition　电弧点火
arcing back　回弧，逆弧
arcing brush　发弧电刷，跳火电刷
arcing butt welding　弧对焊
arcing chamber　灭弧室
arcing contact　灭弧触点，飞弧触点
arcing distance　放电距离
arcing-extinguishing contact　灭弧触头
arcing fault　闪络故障
arcing ground overvoltage　弧光接地过电压
arcing-ground suppresser　接地弧抑制器
arcing ground　电弧接地，接地弧，对地电弧放电
arcing horn gap length　弧角间隙长度
arcing horn　导弧角，消弧角，角形避雷器，招弧角
arcing-over　跳弧，弧越
arcing ring　屏蔽环，消弧环，引弧环，环形消弧器
arcing shield　电弧屏蔽环，均压环
arcing-time of a three-phase circuit-breaker　三相断路器燃弧时间
arcing-time　燃弧时间，飞弧时间，闪络时间
arcing tip　电弧接触点
arcing voltage　灭弧电压，起弧电压，跳火电压
arcing　飞弧，发弧，弧击穿，电弧作用，起弧
arc initiation　起弧
arc inverter　电弧逆变器
arc key　弧形键
arc lamp　弧光灯
arc leakage power　电弧漏过功率
arc length　弧，弧长
arc lighting　弧光照明
arc-light　弧光灯，弧光
arc line　弧线
arc loss　电弧损耗
arc melting　电弧熔化，电弧熔炼
arcograph　圆弧规
ARCO process　合金块氯化氧化法
arc oscillator　电弧振荡器
arcose　长石砂岩
arc-over voltage　电弧放电电压，飞弧电压
arc-over　电弧放电，闪络，飞弧
arc path　弧道
arc pit　弧坑，灼伤坑，弧斑
arc plasma　电弧（放电）等离子
arc potential　弧电位
arc process　电弧法
arc profile　圆弧翼型
arc-proof enamel　耐弧磁漆
arc-proof　耐电弧的
arc protection　闪络保护（绝缘子）
arc quenching　熄弧，灭弧
arc rectifier　电弧整流器
arc regulator　电弧调节器
arc resistance furnace　弧阻炉
arc resistance　耐电弧性，弧光电阻，弧阻，电弧电阻
arc-resistant brush　耐弧电刷
arc resisting property　耐电弧性
arc-resistivity　弧电阻率，耐电弧性
arc ring　电弧环，引弧环【绝缘子】
arc scale　弧度标尺
arc section　弧段
arc self-regulation　电弧自身调节
arc shooting　弧形爆炸法
arcsine　反正弦
arc source　弧光源，弧光离子源
arc spark　电弧火花
arc spectrograph　电弧摄谱仪
arc splitter　熄弧沟，分弧器，熄弧分隔片
arc spot welding　电弧点焊
arc spraying　电弧喷涂
arc stability　电弧稳定性
arc stabilizer　稳弧器，稳弧剂
arc starting　起弧
arc stiffness　电弧挺度
arc stream　弧流
arc strike　电弧闪击，弧光放电，电弧触发，电弧放电，电弧烧伤，引弧，起弧
arc-suppressing coil　消弧线圈
arc-suppressing reactor　消弧电抗器
arc-suppressing transformer　消弧变压器
arc-suppression circuit　消弧电路
arc-suppression-coil-earthed system　中性点消弧线圈接地系统
arc-suppression coil　消弧线圈，灭弧线圈
arc-suppression contact　灭弧触头
arc-suppression resistance　灭弧电阻
arc-suppression　消弧，灭弧，抑弧
arc surfacing　电弧堆焊
arc survey　弧垂观测

arc tangent 反正切
arc-through loss of control 通弧失控
arc-through 通弧，电弧通过
arc-tight 耐弧的，隔弧的
arc timer 燃弧时间测定装置
arc time totalizer 发生弧光时间计算器
arc time 燃弧时间
arc tip 弧尖，电弧触点
arc torch 电弧焊矩
arc trajectory 高弹道【飞射物】，弧形轨道
arc transmitter 电弧发射机
arcuated architecture 拱式建筑
arcuated construction 拱形建筑物，拱式构造
arcuate shuttering 弧形模板
arc valve 弧阀
arc-voltage closed loop control 弧压闭环控制
arc-voltage 弧压，电弧电压，电弧压降
arc wear 电弧侵蚀
arc welder 电弧焊机，电焊机
arc welding electrode 电弧焊条，电焊条
arc welding generator 弧焊发电机
arc welding gun 焊枪
arc welding machine 电焊机，弧焊机
arc welding rectifier 弧焊整流器
arc welding set 电弧焊机组，弧焊设备
arc welding transformer 弧焊变压器
arc welding 电弧焊，弧焊
arc 弧，弓形，电弧，圆弧，拱，弧形，弧（度），天穹，弧光，形成供状物
ARD(application reference data) 应用参考资料数据
ardometer 辐射高温计，光测高温计，辐射热测量计
arduous service 困难的运行条件
area adjustment 面积平差
area agreement 地区协议
area-altitude curve 面积高度曲线
area assessment 分区评价
area assist action 区域供电临时改进措施
area-balanced current 面积平衡电流，与直流分量对称的电流
area-capacity curve 面积容积曲线
area code 区域代码
area coefficient 面积系数
area computation 面积计算
area control error 地区控制误差
area covered by agreement 协议范围
area curve 面积曲线
area density heat loss 面积密度散热
area density 面密度
area-depth curve 面积深度曲线
area development 地区开发
area dispatcher 地区调度员
area distribution curve 面积分配曲线
area drain 地面排水沟，露天排水沟，区域排水
area effect 面积效应
area-elevation distribution curve 面积高程分配曲线
area exposed to erosion 风化区
area factor 面积因数
area flowmeter 截面流量计
area forecast 区域预报
area fraction 面积比
area generation 地区发电情况，地区发电能力
area heating device 局部加热装置，区域加热装置
area heating 区域供暖（热）
area heat release rate 面积热强度
area heat substation 小区热力分站
area law 面积定律
areal centre 分布区中心
areal coordinate 重心坐标
areal deformation 表面变形，表面面积变化
areal degradation 区域剥蚀
areal density 表面密度，面密度
areal distribution 面上分布
areal geology 区域地质学
area load-frequency characteristic 区域负荷频率特性，地区负荷频率特性
area load 面荷载
areal resistance 面电阻
areal resistivity 表面电阻率
areal standard 地区标准
areal storage density 面存储密度
areal suction effectiveness 表面吸收效能
areal variation in rainfall 雨量随面积的变化
areal velocity 掠面速度，面积速度
areal 地区的
area map 区域图
area mean pressure 表面平均压力，面积平均压力
area method 求面积法，面积法
area-moment method 力矩面积法，面矩法
area moment of inertia 面积惯性矩
area moment 力矩面积，弯矩面积，面积矩
area monitoring system 区域监测系统
area monitoring 区域监测，场所监测
area monitor 区域放射线监测器，区域监测器
area occupied buildings 建筑占地面积
area of administrative zone 厂前区面积
area of aquifer intake 含水层进水面积
area of base 地基面积
area of bearing 支承面积
area of building 建筑面积
area of contact 接触面积
area of cooling surface 冷却面积
area of coverage 影响范围，覆盖区
area of cross section of nozzle 喷嘴截面积
area of cross section 横截面积
area of dissipation 消融区
area of diversion 分水面积
area of exceeding 超标区
area of expertise 专业领域，专长领域，专长
area of fracture 断裂面积
area of greenery space 绿化面积
area of heating surface 加热面积，受热面的面积
area of heat-supply service 供热面积
area of hysteresis loop 磁滞曲线面积，磁滞环面积
area of infection 受影响区，受影响面

area of influence 影响面积
area of inlet 进口面积
area of main power building per kW 每千瓦主厂房面积
area of micro-topography 微地形区
area of minute meteorological phenomena 微气象区
area of nozzle throat 喷嘴喉部面积
area of outlet 出口面积
area of passage 通流面积，有效截面积
area of pile head 桩头面积
area of reinforcement 钢筋面积
area of roads and places 道路及广场面积
area of safe operation 安全工作区
area of section 截面积
area of site utilization 场地利用面积
area of slack water 缓流区，平潮区
area of snowmelt 融雪面积，融雪区
area of stream evaporation 河流蒸发面积
area of structural steel 型钢截面积
area of structure 建筑面积，构造物面积
area of supply 补给区
area of surface-heat transfer 表面换热面积，换热面积
area of wave generation 波发生区
area of well influence 井影响面积
area of wetted cross-section 过水横断面积
area planning 地区规划
area pollution 区域污染
area radiation monitoring system 电站辐射监测系统
area radiation monitor 区域放射线监测仪
area ratio of soil sampler 取土器面积比
area ratio 面积比
area redevelopment program 区域重新发展规划
area reduction 断面收缩率
area ruling 面积律的应用
area sampling 面积抽样法
area source 地区性空气污染源，面源，区域源
area substation 地段变电站，小区热力站
area supplementary control 区内辅助性调节，区内辅助性控制
area swept 扫掠面积
area tie line 地区联络线
area to power ratio of solar array 太阳能电池板的面积与功率比
area under crops 播种面积
area under the notch 切口面积【冲击试验】
area utilization factor 场地利用系数
area utilization of solar array 方阵的面积利用率
area velocity method 面积速度测流法，面积流速法
area-volume curve 面积容积曲线
area-volume ratio 面积体积比
areaway 空地
area-weighted averaging 面积加权平均数
area 领域，面积，区，区域，场地，范围，区域范围
areic area 无流区
areic charge 面积电荷
areic electric current 面积电流
areic heat flow rate 面积热流量
areic mass 面质量
areism 无流区
arenaceous limestone 砂质石灰岩，砂质灰岩
arenaceous rock 砂质岩
arenaceous sediment 砂质沉积物
arenaceous shale 砂质页岩
arenaceous 多砂的，砂质的
arene 风化粗砂
arenite arenaceous rock 砂质岩
arenite 砂粒碎屑岩
arenose 粗砂质，多砂的，粗砂质的，含砂砾的，红砂土
arenyte 砂粒岩，粗屑岩，砂粒碎屑岩
areometer 比重计，液体比重计
areopycnometer 稠液比重计
are uncalled for 不必要的，不适当的
AREVA 法国阿海珐集团【核能公司】
ARE 主给水流量调节系统【核电站系统代码】
ARG(Advanced Reactors Group) 先进型反应堆组【美国】
ARG (argument) 幅角
argentometer 电量计，测银比重计
argic water 中间包气带水
argillaceous cement 泥质胶结物
argillaceous formation 泥质地层
argillaceous ground 亚黏土
argillaceous limestone 泥质灰岩
argillaceous rock 泥质岩石
argillaceous sand ground 泥质沙地
argillaceous sandstone 泥质砂岩
argillaceous schist 泥质片岩
argillaceous sediment 泥质沉积
argillaceous shale 泥质页岩
argillaceous slate 泥质板岩，黏土石板
argillaceous 泥质的，含黏土的，含陶土的，黏土质的
argilla 铝氧土，高岭土
argillic horizon 黏化层
argilliferous 含泥的，泥质的
argillite 泥质岩，泥质板岩，厚层泥岩
argillization 泥化作用
argillo-calcareous 泥灰质的
argil 陶土，矾土，白土
argon activation 氩气活化
argon anti-convection device 氩气防对流装置
argon arc welding machine 氩弧焊机
argon arc welding 氩弧焊
argon bulb rectifier 氩管整流器
argon cooler 氩气冷却器
argon cover gas atmosphere 氩气覆盖气氛
argon cover gas system 氩气覆盖系统
argon cover 氩覆盖层
argon gas 氩气
argon heating system 氩气加热系统
argon hose 氩气软管
argon manifold station 氩气泄流管站
argon rectifier 氩整流器
argon shielded arc welding 氩弧焊
argon storage tank 氩气贮存罐
argon supply station 氩气供应站

argon supply system 氩气供应系统
argon system 氩气系统
argon tungsten-arc welding machine 钨极氩弧焊机
argon welder 氩弧焊机
argon 氩【元素 Ar】
argument of phasor 相量幅角
argument of vector 矢量幅角
argument principle 幅角原理
argument 幅角，幅，变元，自变数，自变量，幅度，论据，争论，论点，主题
arid area 干旱地区
arid belt 干旱（地）带
arid climate 干燥气候
arid cycle 干燥周期
arid erosion 干旱侵蚀
aridisol 干燥土
aridity index 干燥指数
aridity 干燥度，干燥，干旱
arid land 旱地
arid mist 酸雾
arid region 干旱地区
arid zone 干旱（地）带
arid 干燥的，干旱的
aright 点燃的，烧着的
ariscope 移相光电摄像管
arise from 产生于，起因于
arise out of 产生于，起因于
arise 产生，出现，产生于
arising out of or in relation to 由此产生的或与之相关的
arithmetic 运算的
arithmetic address 算术地址
arithmetical average 算术平均
arithmetical element 算术元素，运算元素，运算器
arithmetical instruction 运算指令，算术指令
arithmetical operation 算术运算
arithmetical progression 算术级数
arithmetical series 算术级数，等差级数
arithmetical unit 运算器，运算单元，运算部件
arithmetic and control unit 运算控制器
arithmetic apparent power 算术视在功率
arithmetic average 算术平均值，算数平均
arithmetic check 算术效验，运算效验
arithmetic circuitry 运算电路
arithmetic circuit 运算电路
arithmetic device 运算装置，运算器
arithmetic element 运算元件
arithmetic error 运算错误，运算误差
arithmetic logic register stack 算术逻辑寄存器栈
arithmetic logic unit 算术逻辑单元，运算逻辑单元
arithmetic mean temperature difference 算术平均温差
arithmetic mean 算术平均值，算术平均
arithmetic operation 算术运算
arithmetic operator 算术运算符
arithmetic point 小数点
arithmetic progression 算术级数，等差级数
arithmetic reactive factor 算术无功因数【多相回路】
arithmetic register 运算寄存器，运算装置寄存器
arithmetic scan 算术扫描，运算扫描
arithmetic section 运算器，运算单元，运算部件
arithmetic shift 算术移位
arithmetic symmetry 算术对称
arithmetic system 运算系统
arithmetic technique 运算装置，运算技术
arithmetic unit 运算器，运算部件，运算装置
arithmometer 四则计算机
arkose-quartze 长石石英岩
arkosic conglomerate 长石砾岩
arkosic limestone 长石灰岩
arkosic sandstone 长石砂岩
arkosite 长石石英岩
armamentarium 全套设备，全部设备
armament 武器，装备，武装力量
ARM(application reference manual) 应用参考手册
armature air gap 衔铁间隙，电枢间隙
armature assembly 电枢总成
armature backstop 衔铁后挡，衔铁复原止挡
armature band 电枢绑带
armature bar 电枢线棒，电枢导条
armature bearing pin 衔铁支座销
armature bearing 电枢轴承，电枢支座
armature bender 衔铁弯曲器，电枢弯机
armature binder 电枢绑线
armature body 电枢体
armature bore 电枢内孔，电枢孔径
armature chamber 电枢室，定子室
armature characteristic curve 电枢特性曲线
armature chatter 衔铁振动，衔铁震颤作响
armature-circuit loss 电枢绕组损耗，电枢铜耗
armature-circuit 电枢电路
armature circumference 电枢圆周
armature closed-coil 闭圈电枢
armature coil 电枢线圈，电枢绕组，衔铁线圈
armature conductor 电枢导线，电枢线圈
armature contact spring 衔铁触点弹簧
armature contact 衔铁触点，衔铁接点
armature-controlled motor 电枢电压可调的（直流）电动机
armature control 电枢电压控制
armature core disk 电枢铁芯叠片
armature core lamination 电枢铁芯叠片，电枢铁芯冲片
armature core 电枢铁芯
armature current 电枢电流
armature diameter 电枢直径
armature dropout overtravel 衔铁释放超行程，衔铁回动过调
armature drum 电枢芯子
armature duct 电枢风道
armature end-connections 电枢绕组端部接线
armature end-plate 电枢端板，电枢压板
armature end slug 衔铁端缓动铜环【继电器线圈】
armature excitation curve 电枢励磁曲线
armature-excited machine 电枢励磁电机
armature factor 电枢有效导体数

armature field 电枢磁场
armature flange 电枢凸缘，电枢端盖
armature flux 电枢磁通量
armature frame 机座【旋转磁场式电机】，电枢支架
armature frequency 电枢频率
armature gap 铁芯间隙，电枢间隙
armature head 电枢铁芯压板，电枢端板
armature hesitation 衔铁动作滞缓，衔铁动作犹豫
armature hinge 衔铁支枢，衔铁铰链
armature inductor 电枢感应线圈，电枢感应
armature insulation 电枢绝缘
armature iron loss 电枢铁芯损耗
armature iron 电枢铁芯，衔铁
armature lamination 电枢叠片
armature leakage field 电枢漏磁磁场
armature leakage flux 电枢漏磁通
armature leakage reactance 电枢漏磁电抗，电枢漏抗
armature leakage 电枢漏磁
armature length 电枢长度
armature lever 衔铁杆
armature lifter 衔铁提杆，衔铁推杆
armature lock washer 电枢锁紧垫圈
armature loop 电枢绕组元件
armature loss 电枢损耗
armature magnetic pole gap 电枢磁极间隙
armature magnetization curve 电枢磁化曲线
armature magnetization 电枢磁化
armature of cable 电缆的铠装
armature of magnet 磁铁衔铁
armature ohmic loss 电枢电阻损耗，电枢铜耗
armature paper 绝缘纸
armature periphery 电枢圆周，电枢周缘
armature pickup overthrow 衔铁吸合过度
armature pickup rebound 衔铁吸合回跳
armature plate 电枢冲片
armature play 衔铁游隙，衔铁间隙
armature pusher 衔铁推杆
armature ratio 衔铁动程比
armature reactance factor 电枢电抗系数
armature reactance 电枢电抗，电枢漏磁电抗
armature reaction-excited machine 电枢反应励磁电机，反应式电机
armature reaction harmonics 电枢反应谐波
armature reaction MMF 电枢反应磁动势
armature reaction reactance 电枢反应电抗
armature reaction 电枢反应
armature rebound 衔铁回跳
armature relay 衔铁式继电器
armature residual gap 衔铁余隙
armature resistance 电枢电阻
armature return spring 衔铁回复弹簧
armature rheostat speed control 电枢串电阻调速
armature set 电枢总成
armature shaft 电枢轴
armature side play 衔铁侧面间隙
armature sleeve 电枢轴套，电枢套筒
armature slot 电枢槽
armature spider 电枢支架，星形轮，电枢辐式机架，电枢幅臂
armature spindle 电枢心轴
armature spool 电枢线圈
armature spring 衔铁弹簧，舌簧
armature stamping 电枢冲片
armature stop 衔铁止挡，电枢止动件
armature stroke 电枢行程
armature stud 电枢螺栓
armature teeth 电枢齿
armature tester 电枢试验器，电枢测试器
armature time constant 电枢时间常数
armature travel 衔铁行程
armature type magneto 电枢式磁电机
armature type relay 衔铁式继电器
armature vent 电枢通风孔
armature voltage control 电枢电压控制
armature voltage 电枢电压
armature wedger 嵌电枢槽楔机，制造电枢槽楔机
armature wedge 电枢槽楔
armature winding cross connection 电枢绕组均压线
armature winding equalizer 电枢绕组均压线
armature winding machine 电枢绕组绕线机
armature winding 电枢绕组
armature with ventilation 通风电枢
armature 电容器板，装甲，铠装，衔铁，电枢，加强料，电枢【电机的部件】，盔甲，衔铁线圈
armchair 扶手椅
armco iron 工业纯铁
armco steel 不硬化钢
arm crane 悬臂式起重机
armlak 电枢用亮漆
arm of balance 平衡臂
arm of couple 力偶臂
arm of force 力臂
armor 铠装
armor-clad switch 金属壳开关
armor-clad 铠装的
armor clamp 电缆的铠装线夹，金具
armor coat 甲层，护层
armor course 道路护面层
armored cable 铠装电缆
armored capillary 铠装毛细管
armored concrete 钢筋混凝土
armored door 防火门
armored glass 防弹玻璃，钢化玻璃
armored hose 铠装软管，夹金属丝软管
armored lead cable 铠装铅包电缆
armored platinum thermal resistance 铠装铂热电阻
armored temperature regulator 铠装温度调节器
armored thermocouple 铠装式热电偶，铠装热电偶
armored thermometer 带套温度计，铠装温度计
armored transformer 壳装变压器
armored tubing 铠装管路
armored type 双壁壳式【指泵】
armored wall 铠装水冷壁
armored wood 包铁木材

armored　铠装的，包铁皮的
armoring steel tape　铠装钢带
armoring wire　铠装线
armor layer　护面层
armor plate　护板，装甲板
armor rod　护线条
armor tape　保护带
armour coat　保护涂层
armour course　保护层
armoured cable　铠装电缆
armoured concrete　包铁板混凝土
armoured hose　缠丝软管
armoured pump coupling　铠装泵联轴器
armoured pump　双壁壳式泵
armouring concrete block　护面混凝土块体
armourless cable　无铠装电缆
armour plate glass　钢化玻璃板
armour plate　防护板，护板
armour rock　护面石
armour rod　护线条【输电线，绝缘子】
armour stone　护面块石，防护石
armour wire　金属铠装线
armour　铠装，装甲
arm path　桥臂支路
arm rest　拾音器臂架，靠手，扶手
arm's length pricing　公平定价
arm's length transaction　正常交易
arm tie　拉板，连壁板，斜撑，横臂拉条
arm-type brush holder　悬臂式（电）刷握［柄，座，架］
arm　臂，杆，支架，指针，桨叶，手柄，柄，扶手，辐（条）
aromatic amine　芳族胺
aromatic hydrocarbon　芳（族）烃
aromatic　芳香剂
aromatization　芳构化
around-the-clock job　昼夜施工
around-the-clock　日夜不停的
around　围绕
arouse　引起，激发
ARP(address resolution protocol)　地址解析协议
arranged in cross-counterflow　交叉逆流布置
arrangement and method for construction　施工方案
arrangement and method statement　施工方案
arrangement diagram　布置图，排列图，帕累托图
arrangement drawing of power plant area　电厂厂区布置图
arrangement drawing　规划图，布置图，（设备）定位图
arrangement factor　布置系数，管子布置的修正系数
arrangement in parallel　并联布置，并列
arrangement of auxiliary electrical equipment　厂用电设备布置
arrangement of insulation　绝缘配合
arrangement of pole attachment　电杆附近的配置
arrangement of reinforcement　钢筋分布
arrangement of runs　焊道布置
arrangement of sewerage system　排水管道布置
arrangement of terminals　端子布置
arrangement of the facilities　地下设施布置
arrangement of the track on plant site　厂内铁路线布置
arrangement of tube bundles and number of passes　管束布置与管程数
arrangement of underground facilities in the power plant area　电厂厂区地下设施布置
arrangement of wires　布线
arrangement pitch　布置间距
arrangement track　编组线【铁路】
arrangement　安排，布置，布线，排列，配置，和解，装置，装配，整理，方案，协议
array curtain　天线障
array dish mirror　阵列碟形镜面
array field　方阵场
array index　数组下标
array of fuel rods in the fuel assembly　燃料组件中燃料棒排列
array parabolic mirror　阵列抛物镜面
array processor　数组处理机，阵列处理机
array reduction analysis circuit　阵列简化分析电路
array sub-field　方阵子场
array switching system　方阵联结开关系统
array　排列，栅格【堆芯】，方阵【太阳能电池】，阵列，阵，布置，列阵，大批，一系列
arrearage　欠款［债］，迟滞，拖延，延滞
arrears　欠款［债］，拖欠，（过期未付的）欠款，尚待完成的工作或任务
arrested anticline　平缓背斜
arrested crushing　拦截破碎
arrested dune　稳定沙丘
arrested failure　限制故障的扩展，制止损坏
arrested property　被扣押的财产
arrested　镇定的，自动的
arrester alternation sparkover voltage　避雷器振荡火花放电电压
arrester characteristic element　避雷器特性元件
arrester discharge capacity　避雷器放电容量
arrester discharge counter　避雷器放电计数器
arrester discharge indicator　避雷器放电指示器
arrester discharge voltage-current characteristic　避雷器放电伏安特性
arrester discharge voltage-time curve　避雷器放电电压-时间曲线
arrester disconnector　避雷器隔离开关
arrester grading ring　避雷器均压环
arrester ground terminal　避雷器接地端子
arrester ground　避雷器接地
arrester in convertor station　换流站避雷器
arrester residue voltage　避雷器残余电压
arrester　制动器，捕捉器，避雷器，捕尘器，布袋除尘器，搜集器，稳定装置，放电器
arresting device　制动装置，止挡，爪，卡子
arresting gear　制动装置，稳定装置，限动器
arresting nut　止动螺母
arresting stop　止动块，停止装置
arrestment system　制动系统
arrestment　财产扣押，阻止
arrestor　制动器，止挡器，限位器

arrest point 驻点，转变点，临界点，支承点，停止点【加热或冷却】
arrest 停止，制动，延迟，捕捉，阻止，吸引，扣留
arris of dab 板棱，板肋
arris of joint 接缝圆角
arris rail 三角轨
arrissing tool 抹角镘
arris 边棱，棱角，锐边，尖脊，隅
arrival at port 入港
arrival basis 收到基
arrival card 更改地址通知单，入境卡
arrival contract 到货合同
arrival and departure service 到发作业
arrival and departure track 到发铁路线
arrival in batches 成批到达
arrival notice 到货通知，抵港通知
arrival quality terms 到达品质条件
arrival quality 到岸品质
arrival terms 到货条件，抵港条件
arrival time 到达时间
arrival 到达
arrived weight 到达重量
arrow diagram 箭线图，箭头（线）图
arrow head 箭头
arrow height 矢高
arrow 箭，箭头，指针，测针
arroyo 旱谷
ARS(asbestos roof shingle) 石棉屋顶瓦
arsenate 砷酸盐
arsenic oxide 五氧化二砷
arsenic pentoxide 五氧化二砷
arsenic trihydride 砷化三氢
arsenic trioxide 三氧化二砷，砒霜
arsenide 砷化物
arsenite 亚砷酸盐
arsenous acid 亚砷酸
arsenous oxide 三氧化二砷，砒霜
arsine 砷化三氢
arsous acid 亚砷酸
art 技术，技巧，技艺
AR(augmented reality)technology 增强现实技术
arterial canal 总渠
arterial drainage 排水干渠，排水干管，干渠排水系统，脉状排水系统
arterial drain 排水干沟
arterial grid 干线网
arterial highway 干道，干线公路
arterial railway 干线铁路
arterial road 干线道路，主要道路
arterial street 城市干道，主要街道，干线街道
arterial traffic 干线交通
arterite 脉状混合岩
artery 干线，要道，大动脉，大路
artesian aquifer flow 自流含水层
artesian aquifer 承压含水层，有压含水层，自流含水层
artesian area 承压水地区，自流水地区
artesian basin 承压水盆地，自流水盆地
artesian capacity 自流产水率
artesian condition 承压（水）状态，自流条件
artesian discharge 自流流出，自流流量
artesian flow area 自流水区域，自流区域
artesian(flowing)well 自流井
artesian flow 承压水流，自流水流
artesian fountain 自喷喷泉
artesian ground water 自流地下水，承压地下水
artesian head 自流水头，承压水头
artesian leakage 承压渗漏
artesian pressure surface 自流水压面
artesian pressure 承压水压力，自流压力
artesian province 自流区
artesian slope 承压含水层坡降
artesian spring 自流泉
artesian waste 自流排水
artesian water circulation 自流水循环
artesian water leakage 承压水渗漏
artesian water power 自流水动力
artesian water 承压水，自流水
artesian well capacity 自流井产水率
artesian well 承压水井，自流井
artesian 自流的，承压的
article board 木屑板
articles of agreement 协定条款，协议条款，协议条文
articles of arbitration 仲裁条例
articles of association 组织章程，公司章程，组织细则，协会章程
articles of contraband 违禁物品
articles of contract 合同条款
articles of corporation 公司章程，公司条例，组织章程
articles of partnership 合伙组织章程
article 论文，文章，项目，条款，物品，物件，章程
articulation 铰接的
articulated absorber 象鼻管式吸收器
articulated arm 铰接臂
articulated bus 铰接客车
articulated camera mount 铰接的摄像机架
articulated chute 活接卸槽【浇混凝土用】
articulated concrete matting 铰接混凝土块铺面，活节混凝土块铺面
articulated conduit 分节管道，象鼻管
articulated construction 装配式结构，活节式结构
articulated coupling 铰链式联轴器，活节连接器
articulated dumper 铰接式翻斗车
articulated joint 活节接合，铰链接合，支铰
articulated manipulator 关节式机械手
articulated sleeve 活接套管
articulated tractor 铰接尾车
articulated trailer 铰接挂车
articulated unit substation 单元组合式变电站
articulating joint 铰接，关节接合
articulation index 清晰度指数
articulation test 清晰度试验
articulation 铰接，铰，活接头，清晰度，接合，咬合，关节
artifical radioactivity 人工放射性
artifical radionuclide 人工放射性核素
artificial abutment 人造支座
artificial ageing 人工老化，人工时效

artificial aggregate 人造骨料，人工骨料
artificial aging 人工时效，人工老化
artificial asphalt 人造沥青
artificial atmosphere 人造大气
artificial base 人工地基
artificial bio-membrane 人工生物膜
artificial bitumen 人造沥青
artificial brain 电脑，仿真脑
artificial bubble 人造气泡
artificial cable 仿真电缆
artificial circuit 仿真电路，模拟电路
artificial climate 人造气候
artificial commutation 人工整流，强制整流
artificial compensate implement 人工补偿器，膨胀节，伸缩器
artificial contaminant 人为污染物
artificial contamination test procedure 人工污秽试验方法
artificial control 人工控制
artificial cooling 人工冷却
artificial crack 人造裂缝
artificial crystal 人造晶体
artificial daylight 人工日光
artificial defence 人工防护
artificial delay line 人工延迟线
artificial dielectric 人造介质
artificial draft 强制通风，人工通风
artificial drainage system 人工排水系统
artificial drainage 人工排水，有组织排水
artificial draught 人工通风
artificial earth electrode 人工接地体
artificial earthing 人工接地
artificial earthquake 人工地震
artificial echo 假回波，仿真回波
artificial erosion 人为侵蚀
artificial error signal 人为误差信号
artificial fall 人造落差
artificial fiber 人造纤维
artificial filtering material 人造滤料
artificial flood wave 人为洪水波
artificial flushing 人工冲洗
artificial foundation 人工地基
artificial friction coefficient 假想摩擦系数
artificial fuel 人工燃料
artificial gas 人工瓦斯
artificial graded aggregate 人工级配骨料
artificial groundwater recharge 人工地下水补给，人工补给地下水，人工补水
artificial ground 人工地基
artificial harbor 人工港口
artificial horizon 人为地平
artificial illumination 人工照明
artificial impoundment 人工蓄水
artificial intelligence 人工智能
artificial island 人造岛，人工岛
artificial language 人工语言
artificial lighting 人工采光，人工照明，人工光照
artificial lightning 人工闪电
artificial light source 人造光源
artificial line duct 仿真线通道
artificial line 仿真线路，模拟线路
artificial load 仿真负荷，模拟负荷，虚负载
artificially excavated port 挖入式港口
artificially graded aggregate 人工级配骨料
artificially recharged groundwater 人工补给地下水
artificially replenishment 人工填充
artificial magnet 人造磁铁
artificial marble 人造大理石
artificial mica 人造云母
artificial mouth 仿真口，模拟口
artificial navigable waterway 人工可航水道
artificial network 模拟网络，仿真网络
artificial neutral point 人工中性点
artificial nourishment 人工养滩
artificial obsolescence 人为报废
artificial obstruction 人为障碍
artificial pollution tests on insulating surface 绝缘表面人工污秽试验
artificial pollution test 人工污秽试验
artificial pond 人工池
artificial precipitation stimulation 人工催雨
artificial radio aurora 人造无线电极光，模拟无线电极光
artificial radionuclide 人工放射性核素
artificial rainfall 人工降雨
artificial recharge 人工再充电，人工回灌，人工补水，人工补给
artificial reservoir 人工水库
artificial roughening 人工加糙
artificial rubber 人造橡胶，合成橡胶
artificial satellite 人造卫星
artificial shift erosion 人力侵蚀
artificial sludge 人工污泥
artificial stabilization 人为稳定
artificial stone 人造石
artificial storage 人工蓄水
artificial time history 人工时程
artificial transmission line 仿真输电线，模拟输电线
artificial ventilation 人工通风，机械通风
artificial vibration 人工振动
artificial voice 仿真语声，模拟语声
artificial watercourse 人工水道
artificial weather 人工气象
artificial 人工的，人造的，仿真的
artisan 工匠
artistic plastering sprayer 喷花枪
artistry 艺术手法
art work master 照相原图，照相底图
art work sheet 工艺单，工艺图纸
art work tape 原图信息带
art work 原图，工艺图，图模，工艺
ASA(American Standard Association) 美国标准协会
as above 如上
as advised 依照通知
as against 与……对照
as agreed by L/U(leading underwriter) 经首席承保人同意
as agreed 按照合同，依照协定，照约定

as analysed basis 分析基，分析质
ASAP(as soon as possible) 尽快
as applicable 根据具体情况，视情况而定，如果适用的话，酌情
as appropriate 根据具体情况，视情况而定，在适当的时候，酌情
as a result of 由于……的结果，由于，因为
as a result 结果，因此
as at 截止，截止日
AS(auxiliary specifications) 辅助规范
asb＝asbestos 石棉
as below 如下
asbestic title 石棉瓦
asbestic 石棉的
asbestine 滑石棉，不燃性的，石棉质的
asbestonite 石棉制绝缘材料，石棉绝热材料
asbeston 防火布，石棉棉花混纺织品，石棉防火布
asbestophalt 石棉沥青
asbestos 石棉
asbestos-bitumen thermoplastic sheet 石棉沥青热塑板
asbestos-bitumen 石棉沥青
asbestos blanket 石棉毯
asbestos board 石棉板
asbestos cement board ceiling 石棉水泥板顶棚
asbestos cement corrugated sheet 石棉水泥波形板
asbestos cement downcomer 石棉水泥落水管
asbestos cement fitting 石棉水泥配件
asbestos cement pipe 石棉水泥管
asbestos cement product 石棉水泥产品［制品］
asbestos cement sheet roof 石棉水泥瓦屋面
asbestos cement slate 石棉水泥浪板，石棉水泥板
asbestos cement sleeve 石棉水泥套管
asbestos cement 石棉水泥
asbestos cloth 石棉布
asbestos cord 石棉绳
asbestos covering 石棉覆盖
asbestos cushion 石棉垫
asbestos diatomite 石棉硅藻土
asbestos emergency seal 石棉安全密封
asbestos fabric 石棉织品，石棉布
asbestos felt 石棉毯，石棉毡
asbestos fiber filter 石棉纤维过滤器
asbestos fiber 石棉纤维
asbestos flake 石棉片
asbestos gasket 石棉垫片，石棉垫圈
asbestos insulating plate 石棉绝缘板
asbestos insulation 石棉绝缘
asbestos laminate 石棉层压板
asbestos lining 石棉衬垫，石棉衬里
asbestos mesh 石棉网
asbestos mica 石棉云母
asbestos packing 石棉垫料，石棉垫，石棉填料
asbestos pad 石棉垫
asbestos paper 石棉纸
asbestos pipe 石棉管
asbestos plaster 石棉粉饰，石棉灰泥
asbestos protection 石棉护层
asbestos ribbon 石棉带
asbestos ring 石棉箍
asbestos roofing sheet 石棉屋顶板
asbestos roof 石棉瓦屋顶
asbestos rope 石棉绳
asbestos rubber plate 石棉橡胶板
asbestos seal 石棉密封
asbestos sheet 石棉板，石棉纸，石棉垫
asbestos shingle nail 螺纹石棉瓦
asbestos shingle 石棉屋面板，石棉瓦
asbestos slab 石棉板
asbestos slate 石棉板
asbestos textolite 石棉胶布板，石棉夹心胶合板
asbestos tile 石棉瓦
asbestos twine 石棉绳
asbestos veneer plywood 石棉夹心胶合板
asbestos wallboard 石棉墙板
asbestos washer 石棉垫圈
asbestos wool 石棉绒
asbestos yarn 石棉线，石棉丝，石棉纱
asbestumen 石棉沥青
as-built conclusion 竣工结论，竣工总结
as-built condition 竣工状态
as-built dimension 竣工尺寸
as-built documents 竣工文件
as-built drawings 竣工图
as-built facility 竣工设施
as-built 建成（的），竣工（的）
ASC(American Standards Committee) 美国标准委员会
ascape of air 漏气，排气
as-cast aging 铸态时效
as-cast concrete 清水混凝土
as-cast condition 铸造状态
as-cast strip 铸钢带
as-cast 毛坯铸件，铸的
ASCE(American Society of Civil Engineers) 美国土木工程师协会
ascending air 上升空气
ascending current 上升水流
ascending eddy 上升旋涡，上升涡流
ascending fluid 上升流体
ascending grade 上坡，坡度
ascending pipe 增压管，上升管，扬水管
ascending stroke 活塞上行程，上冲程
ascending zone 上流区
ascend 上升，追溯，升高
ascensional velocity 上升流速
ascensional ventilation 上升通风
ascension pipe 上升管
ascension spring 上升泉
ascension theory 上升学说
ascent angle 上升角
ascent rate 上升速率
ascent 爬高，坡度，斜坡，上升，上进，阶梯
ascertainment error 查明的误差
ascertainment of damage 确定损失
ascertainment 确定
ascertain 确定，弄清楚，查明
aschistic 非片状的
aschistite 未分异岩

as-coated roughness 涂敷后的粗糙度
as-constructed drawing 竣工图纸
as-contracted 按照合同
ascorbic acid 抗坏血酸
as customary 按照惯例
ASD(allowable stress design) 许用应力设计
as-deposited cladding 未经处理的包壳，未经处理的对焊层
as-deposited 焊后【状态】，堆焊【状态】
asdic gear 水下探测器
as-dug gravel 原状砾石
aseismatic design 抗震设计，耐震设计
aseismatic planning 抗震规划
aseismatic structure 耐震构造，抗震架构
aseismatic 防震的，抗震的
aseismic bearing device 抗地震轴承装置
aseismic bearing pad 抗震基础，抗震垫
aseismic code 抗震规范
aseismic design 防震设计，抗震设计
aseismic frame 抗震构架
aseismic joint 防震缝
aseismic plate 抗震支撑板
aseismic region 无地震区
aseismic structure 抗震结构
aseismic technology 抗震工艺
aseismic 抗地震的，抗震
aseptic technique 防腐技术
aseptic 无菌的，防腐剂，防腐的
as expiry 照旧续保
as far as it goes 以现在的情况
as far as possible 尽可能，尽量，尽可能远地
as-fired basis 应用基，工作基，工作质
as-fired coal 入炉煤
as-fired fuel 炉前煤，入炉煤
as-fired pulverized coal sample 入炉煤粉样
as-fired sample 燃用前煤样
as-fired sampling switching 燃用前取样系统［装置］
as follows 如下
ASG 辅助给水系统【核电站系统代码】
ash agglomerating combustor 结渣燃烧器
ash air 含灰空气
ash analysis of coal 煤灰成分分析
ash analysis 灰成分分析，灰分析，灰分分析
ash and slag handling system diagram 除灰渣系统图
ash and slag yard 灰渣场
A-shaped pole A形杆
ash attack 灰渣侵蚀，灰分冲蚀
ash balance 灰平衡
ash ballast 灰渣
ash basement 出灰间
ash bed 灰床，灰渣层
ash bin 灰仓，灰坑，灰桶
ash box 灰箱，灰斗
ash bucket capacity 灰斗容积
ash bucket inlet damper 灰斗进口挡板
ash buildup 积灰，堆灰，灰的沉积
ash bunker 灰斗，灰桶
ash can 垃圾桶
ash car 灰车
ash catcher 捕灰器，除尘器，集尘器
ash cellar 灰坑
ash coal 多灰分煤，高灰煤
ash collection 集灰，捕灰
ash composition 灰分组成
ash comprehensive utilization 灰渣综合利用
ash concrete 灰渣混凝土
ash conditioner 调灰器，调湿灰机
ash cone 凝灰火山锥，灰锥
ash content 灰分，灰，灰含量
ash cutting 飞灰磨损
ash dam 灰坝
ash deformation temperature 灰变形温度
ash deposition 积灰
ash deposit 积灰，堆灰，粉灰
ash discharge gate 排灰闸门
ash discharge platform 卸灰站台
ash discharge 出灰
ash disposal area 贮灰场
ash disposal yard 贮灰场
ash disposal 灰渣处理，除灰
ash door 出灰门
ash dredger pump 灰浆泵
ash dump 灰堆，灰场
ash dust haydite 粉煤灰陶粒
ash dust 粉煤灰
ashed 灰化了的，成灰的
ash ejector 水力冲灰器
ash emission 飞灰逸出，逸灰量
ash emulsion pump 灰浆泵
ash emulsion 灰浆
ashery 堆灰场
ash exhauster 排灰器
ash extraction 除灰
ash fall 灰尘沉降，烟尘降落
ash fluidizing temperature 灰流化温度
ash fluid point 灰流动温度
ash flushing water treatment 冲灰水处理
ash-forming impurities 成灰杂质
ash fouling 堵灰
ash-free basis 无灰基
ash-free coal 无灰煤，不含灰分煤
ash-free filter paper 无灰滤纸
ash-free 无灰的
ash fusibility temperature 灰熔化温度，灰熔点
ash fusibility 灰可熔性，灰熔融性
ash fusion characteristics 灰熔化特性
ash fusion point 灰熔点，灰的熔化点
ash fusion property 灰熔化特性
ash fusion temperature 灰熔化温度，灰熔点，灰熔融温度
ash fusion 灰熔化
ash gate 灰斗闸门
ash handling pump house 除灰泵房
ash handling pump 灰浆泵，灰渣泵
ash handling system control room 除灰系统控制室
ash handling system design 除灰系统设计
ash handling system diagram 除灰系统图
ash handling system 除灰系统，灰输送系统
ash handling 除灰

ash hole 出灰口，出灰孔，灰坑
ash hopper 灰斗
ash humidifier 灰加湿器
ashing 灰化，积灰
ash intake 灰进口
ash-laden gas 含尘烟气，富尘气体
ash-lagoon 灰分处理池
ashlar brick 仿石砖，琢石砖
ashlar facing 条石面层，琢石砌面，琢石镶面
ashlaring 贴方石墙面，镶墙石板，砌琢石，砌方石墙面
ashlar masonry 条石砌体，琢石砌体，琢石圬工
ashlar 条石，琢石，方石堆，石板，方石
ash layer control 灰层控制
ash layer 灰层
ashless filter paper 无灰滤纸
ashless 无灰的
ash melting point 灰的熔化点，灰熔化温度
ash off-takes 除灰装置【沸腾炉】
a short list of five candidates 五位候选人的短名单
ash pan 灰盘
ash piping 除灰管道
ashpit damper 灰坑挡板，灰坑出灰门
ashpit losses 漏煤损失
ashpit refuse 灰渣
ash pit 灰坑
ash pocket 灰斗，灰袋
ash pond 灰沉淀池，灰池
ash pool 灰池
ash precipitation pond 沉灰池
ash pumping 水力除灰
ash pump 灰渣泵
ash pyramid 测温灰锥
ASHRAE(American Society of Heating, Refrigerating and Air-Conditioning Engineers) 美国采暖、制冷与空调工程师学会
ash recirculation 飞灰再循环
ash removal by slurry pump 水力除灰
ash removal crane 除灰吊车
ash removal system 除灰系统
ash removal 除灰
ash residue 灰渣
ash-retention efficiency 排渣率
ash-rich fuel 多灰燃料
ash room 出灰间
ash screen 滤灰网
ash separator 飞灰分离器，除尘器
ash silo aerating system 灰库气化系统
ash silo fluidizing system 灰库流化系统
ash silo unloading system 灰库卸料系统
ash silo vent system 灰库排气系统
ash silo 灰库，灰仓
ash sluice gate 炉灰闸门，出灰闸门
ash sluice way 水力冲灰沟
ash sluice 水力输灰道，水流输渣道
ash sluicing canal 灰渣沟
ash sluicing channel 灰渣沟
ash sluicing pump house 冲灰泵房
ash sluicing pump 冲灰浆（渣）泵，冲灰（水）泵，灰浆泵
ash sluicing water treatment 灰水处理系统，灰水处理
ash sluicing water 冲灰水
ash slurry pond 灰浆池
ash slurry pump 灰浆泵
ash slurry 灰浆
ash softening temperature 灰软化温度
ash specification 灰分，规格
ash storage yard 贮灰场
ash storage 灰槽
ash structure 火山灰构造
ash substance 似灰物质
ash sump 沉灰池
ash test 灰试验，灰分试验
ash transfer tank 灰转运仓
ash transportation piping lines 输灰管线
ash transport cycle 输灰循环
ash trench system 水力除灰沟道系统
ash-tuff 灰质凝灰岩
ash viscosity 灰黏度
ash water pump house 灰水泵房
ash water pump 冲灰泵
ash wetting 调湿灰
ash wharf 灰码头
ash yard dam 灰坝，贮灰场，堆灰场
ashy 灰的，含灰的
ash 灰，尘埃，炉灰，槐木，废渣，灰分
Asian Development Bank 亚洲开发银行
Asian Infrastructure Investment Bank 亚投行，亚洲基础设施投资银行
“as if” statistics “假设”统计
a single-pulley drive 单滚筒驱动
as inspected 根据验货付款
as in the case of 像（在）……一样，例如
a-Si(amorphous silicon) PV module 非晶硅光伏组件
as is 按货样，照原状
ASIS(American Society for Industrial Security) 美国工业安全学会
as is clause 现状条款
as it is customary 按照惯例
as it turns out 结果，到头来
askarel-filled transformer 充氯代联苯的变压器
askarel 氯代联苯
asked price 要价
ask for leave 请假
ask for payment 讨账，催付款，请求付款
ASL(above sea level) 海拔高度
ASLB(Atomic Safety and Licensing Boards) 核安全和审批委员会【美国】
as low as reasonably achievable principle 合理可行尽量低的原则
as low as reasonably achievable 合理可行尽量低
as maintained 维护状态
as may be paid thereon 可以赔付，可以支付
ASME(American Society of Mechanical Engineers) 美国机械工程师协会
ASME boiler and pressure vessel code 美国机械工程师协会锅炉与压力容器规程
ASME code 美国机械工程师协会标准
ASME dust test code 美国机械工程师协会除尘

效率试验规程
ASME vessel code 美国机械工程师协会容器规范
ASN(average sample number) 平均取样数
ASNT(American Society for Nondestructive Testing) 美国无损检验学会
as of the date of 自……日起【合同中正式表达法】
as of the day and year first above written 以上述书就日期【合同中正式表达法】
as ordered 订单状态
aspect angle 视界角，视线角
aspect card 标号卡片，特征卡片
aspect of slope 坡向
aspect photocell 平面型光电池
aspect raito 展弦比，长径比，长宽比，长细比，形态比，高宽比，纵横比
aspect 方面，状态，方向，方位，外观，形状
aspen 杨木
as per advice 按照通知付款
as per details 按照详图，参见详图
as per enclosed document 按照附上的文件
asperity junctions 粗糙接合
asperity 粗糙，粗糙度，不平
as per list enclosed 见附表
as per original rate 照原利率
as per sample 与样品相符
as per specifications attached 按照附上的明细单
as per 依照，根据
asphalite 沥青矿，沥青岩
asphalt-aggregate mixture 沥青骨料拌和物，地沥青集料混合料
asphalt base course 沥青底层
asphalt-based solidification 沥青固化
asphalt base oil 沥青底油，沥青冷底子油
asphalt block pavement 沥青块路面
asphalt board strip 沥青填缝条
asphalt built-up roofing 沥青组合屋面
asphalt carbon 沥青炭
asphalt cement 沥青膏，沥青胶结料
asphalt clinker 沥青熔渣
asphalt-coated pipe 沥青刷面管子
asphalt coating 刷地沥青，沥青护层，沥青涂层
asphalt-compound impregnated coil 浸胶线圈
asphalt concrete pavement 沥青混凝土路面，沥青混凝土铺面
asphalt concrete 沥青混凝土
asphalt covering 沥青面层，沥青铺面
asphalt cutback 低熔点沥青混合物
asphalt damp-proof course 沥青防潮层
asphalt dip 沥青浸渍
asphalt distributor 沥青喷洒机
asphalt emulsion 乳化沥青
asphaltene 沥青质
asphalt felt roof 油毡屋面
asphalt felt waterproof layer 油毛毡，防水层
asphalt felt 油毛毡，沥青毡，油毡，沥青油毡
asphalt filler 沥青填料，沥青填缝
asphalt flooring tile 沥青砖板
asphalt grouting 沥青灌浆
asphaltic binder 沥青结合料
asphaltic concrete batching plant 沥青混凝土配料装置
asphaltic concrete core earth-rock dam 沥青混凝土心墙土石坝
asphaltic concrete facing earth-rock dam 沥青混凝土面板土石坝
asphaltic concrete pavement 沥青混凝土路面
asphaltic concrete road 地沥青混凝土路
asphaltic concrete 地沥青混凝土
asphaltic lining 沥青衬砌
asphaltic mastic 石油沥青玛碲脂
asphaltic mulch 沥青覆盖料
asphaltic painting 沥青涂层
asphaltic scale 沥青硬壳
asphaltic sheet 沥青纸，沥青板
asphalting felt 沥青油毡
asphaltite 沥青岩
asphalt jointed pitching 沥青砌石护坡
asphalt joint filler 沥青填缝料
asphalt jute cloth 沥青黄麻布
asphalt lacquer 沥青漆
asphalt layer 沥青铺设机，沥青层
asphalt macadam 沥青碎石路
asphalt mastic 石油沥青玛碲脂，沥青胶泥，沥青砂胶
asphalt mattress revetment 沥青垫层护岸
asphalt mattress 沥青垫层
asphalt mat 沥青垫层
asphalt membrane 沥青防水膜
asphalt-mica insulation 沥青云母绝缘
asphalt mixing plant 沥青拌和厂
asphalt mixture 沥青拌和物
asphalt mortar 沥青砂浆
asphalt overlay 沥青涂层
asphalt paint 沥青油，沥青涂料
asphalt paper 油纸
asphalt pavement 沥青铺面，沥青路面，沥青路
asphalt paving block 沥青铺块
asphalt paving machine 沥青铺面机
asphalt paving plant 沥青铺面装置
asphalt powder 沥青粉
asphalt road 柏油路
asphalt rock 沥青岩
asphalt roofing 沥青屋面
asphalt sand 沥青砂
asphalt saturated felt 沥青浸润毡，沥青毡
asphalt seal 沥青止水
asphalt slab mattress 沥青块沉排
asphalt slab 沥青板
asphalt slag wool 沥青矿渣棉
asphalt soil stabilization 土的沥青固化
asphalt spreader 沥青摊铺机
asphalt surface treatment 沥青表面处理
asphalt tile 沥青砖
asphaltum 地沥青，石油沥青
asphalt varnish 沥青漆
asphalt vermiculite 沥青蛭石
asphalt well 沥青井
asphalt winding 沥青绝缘绕组，浸胶绕组
asphalt 柏油，沥青，地沥青
aspirail 通风孔

aspirated hygrometer 吸收湿度计
aspirate 吸出，吸引，吸气
aspirating air pipe 吸气管道，抽气管道
aspirating air （防燃烧烟气泄漏用）抽气或密封空气
aspirating burner 抽气燃烧器
aspirating pipe 吹气管道
aspirating stroke 进气冲程，吸气冲程
aspiration probe 抽气式探针
aspiration psychrometer 通风湿度计，抽气式干湿计
aspiration pump 吸扬式泵
aspiration 吸出，抽出，抽气，吸气，吸入
aspirator bottle 吸气瓶
aspirator combined with dust collector 吸气集尘器
aspirator 抽气器，吸气器，吸尘器，抽风机，吸尘机，吸液泵，吸气装置
aspiring pump 抽气泵，抽风泵
as-planned schedule 计划工期
ASPP(alloy-steel protective plating) 合金钢保护电镀（层）
as prescribed by law 依照法律的规定
as prescribed 依……规定
as provided in 依照……规定
ASR(asynchronous send/receive) 异步发送/接收装置
as received basis 收到基（煤），接收基，应用基
as-received coal 入厂煤
as-received sample 入厂煤采样
as regards 涉及，至于，在……方面
as requested 按照要求，照（你）要求，依照请求
as required 按照规定，根据需要
asrosol 悬浮微粒
ASS(automatic synchronizing system) 自动同步系统
ASS(auxiliary steam supply system) 辅助蒸汽供应系统
assay curve 试验曲线，试样曲线，分析曲线
assayer 化验员
assay value 分析值
assay 试验，化验，试样，分析，检验
as scheduled in the contract 按合同规定的进度【日期】
ASSEM(assembly/assemble) 组件，组装
assemblage index 图幅拼接索引，图幅接合索引
assemblage 安装，装配，集合，装置
assembled battery 电池组
assembled conductor 组装导线，组合导线
assembled monolithic concrete structure 装配整体式混凝土结构
assembled monolithic structure 装配整体式结构
assembled pieces of boiler heating surface 锅炉受热面组合件
assembled rotor 组装式转子
assembled structure 装配式结构，组装结构
assembler instruction 汇编指令
assembler language 汇编语言
assembler 汇编程序，汇编语言，装配器
assemble tower supports 杆塔组立
assemble 装配，集合，汇编，组装
assembling and disassembling 装卸
assembling bolt 装配螺栓
assembling clearance 安装间隙
assembling department 装配车间
assembling diagram 装配图
assembling jig 装配机架
assembling joint 装配式接头，装配接头
assembling loss 组合损失
assembling mark 装配记号
assembling of condenser 凝汽器组装
assembling of parts 散件组装
assembling of turbine upper cylinder 汽轮机扣大盖
assembling pin 定位销
assembling reinforcement 绑扎钢筋
assembling structure 装配式结构
assembling 装配，收集，汇编，编制程序
assembly and checkout 装配及检查
assembly and erection 装配和安装
assembly and operating corridor 组装和操作通道
assembly average 汇集平均值
assembly bay 装配场，装配间
assembly bottom fitting 组件底部配件
assembly chain 装配链
assembly channel （燃料）组件孔道
assembly chart 装配图
assembly check 组装检查
assembly clearance 组装间隙
assembly diagram 装配图
assembly dimension 组装尺寸
assembly-disassembly station （组件）装拆站
assembly dolly 移动式装配台，装配台车
assembly drawing 装配图，总装图，组装图，总装配图
assembly fixture 组装架【控制棒】
assembly for failed element localization 破损元件定位装置
assembly for installation 安装图
assembly foundation 装配式基础
assembly hall 礼堂，大会堂
assembly housing （燃料）组件盒
assembly key 装配楔块
assembly language 汇编语言
assembly line production 流水线生产
assembly line 装配线
assembly list 汇编表
assembly longitudinal section drawing 组装纵断面图
assembly mark 装配记号
assembly modulation 组件调制法
assembly on site 现场组装
assembly opening 安装开口，安装洞口
assembly parts 装配部件
assembly pitch （燃料）组件栅距
assembly procedure 装配步骤
assembly programming system 汇编程序设计系统
assembly program 汇编程序，组合程序

assembly rod 装配杆
assembly room 会议室
assembly routine 汇编程序
assembly screw 安装螺旋
assembly section 装配图剖视，装配区
assembly shop 装配车间
assembly sketch 装配草图
assembly specific enthalpy 组件比焓
assembly stress 装配应力
assembly system 汇编系统，装配系统
assembly test 组装试验
assembly-to-assembly failure propagation 组件间事故蔓延
assembly tool 组装工具
assembly transverse section drawing 组装横断面图
assembly unit 汇编程序单位，装配部件
assembly work 装配作业
assembly yard 装配厂，装配场，组装场
assembly 汇编，组合，组装，成套，装配，总成，组件，集会，大会
assertion 断定
assessable land 可征地
assess 估价，估定，评定，评价，征收
assessed failure rate 估计事故率
assessed value 评价值
assessment district 估值区
assessment of fatigue accumulation 疲劳积累的评价
assessment of fault 故障评价
assessment of loss 估损
assessment of performance 职能考核，考绩
assessment of proposal 建议书的评价
assessment of radiation protection 辐射防护评价
assessment of result 成果评定
assessment of safety significant event team 安全重要事件评价组
assessment report 评估报告
assessment 评定，估价，评价，估计数，评估，估计，确定
assessor 鉴定器，鉴定管
as set forth in 依……里规定
asset-liability ratio 资产负债率
asset quality 资产质量
assets and liability statement 资产负债表
assets cover 资产担保
asset size 资产规模
assets-liabilities ratio 资产负债比率
assets liquidity 资产流动性
assets reorganization 重组的资产，资产重组
assets revaluation 资产重估
assets valuation 资产评估
asset 资产，财产
assignable credit 可转让的信用证
assignable letter of credit 可转让的信用证
assigned frequency 规定频率
assigned risk 分摊风险
assignee 代理人，受托人
assignment and renovation 转让和更新
assignment charge 转让费
assignment check 用途核查
assignment clause 转让条款
assignment lamp 呼叫灯，联络灯
assignment of authority 授权
assignment of contract 合同转让
assignment of lease 转租
assignment of rights and delegation of duties clause 权利转让和义务让渡条款
assignment statement 赋值语句
assignment system 分配制度
assignment to third party 对第三方的转让
assignment 分配，分派，指定，任务，指定任务，用途，委派，转让（所有权）
assignor 转让人
assign trend default 指派趋势缺省值
assign 分配，指定，赋予，给定，指派，转让【财产、权利、利息等】
assimilation 同化
assimilative capacity 同化能力，同化容量
assistance 援助，协助
assistant crane 辅助吊车
assistant department 辅助部门
assistant engineer 助理工程师
assistant manager 经理助理
assistant professor translator 副译审
assistant project manager 项目经理助理
assistant shift supervisor 助理值长
assistant superintendent 助理监理员
assistant translator 助理翻译
assistant unit operator 机组助理操纵员
assistant 助手，助理
assisted circulation boiler 辅助循环锅炉【带有辅助循环泵的汽包炉】
assisted circulation 辅助循环，强制循环
assisted draft 强制通风
assisting party 援助方
assist in management 助理
associated agency 协作机构
associated assembly （燃料）相关组件
associated bank 联合银行
associated circulation boiler 辅助循环锅炉
associated company 联营公司，关联公司
associated conductor 组合导线
associated cost 投产外加费，辅助费
associated flow rule 相适应的流动法则
associated general contractors of America 美国总承包商协会
associated liquid 缔合液体
associated mineral 伴生矿物
associated phase layout 联相布置
associated structure 伴生构造
associated system 辅助系统，关联系统
associated works 相关工程，联合工程
associated 有关的
associate expert 助理专家
associate member 准会员
associate professor of translation 副译审
associate statistician 助理统计员
associate 同事，关联
association conrtol service element 关联控制服务元素
Association of Chartered Certified Accountants 国

际注册会计师协会，特许公认会计师公会
Association of International Accountants 国际会计师公会
Association of Nuclear Operators 核电站营运者协会
association 社团，联合体，联合会，结合，联合，组合，关联，协会，学会
associative array processor 相联阵列处理机
associative data processing 相联数据处理
associative key 相联关键字，相联关键码
associative memory 相联存储器，内容定址存储器
associative processor 相联处理机
associative property 结合性
associative relation 相关关系
associative storage 相联存储器，内容定址存储器
associative 结合的
associativity 结合性，缔合性
as soon as possible 尽可能快地
as soon as practical 尽快地
assorted 配套的，各种各样的，分类的
assorting process 分类过程
assortment 分级，分类，品种，种类
as stipulated in 依照……规定
ass transfer theory 质量转移理论
assumed 假设的，假定的，计算的，理论的
assume 假定，假设，采取，承担
assumed accident 假想事故
assumed architecture coordinate system 假定建筑坐标系统
assumed coordinate 假定坐标
assumed load 假定负荷，假定荷载
assumed origin 假定原点
assumed value 假定值
assume sole responsibility for profit and loss 自负盈亏
assume their respective liabilities accordingly 各自承担相应的责任
assumption diagram 结构图，理论线图
assumption 假设（值），假定，采取，承担
as-supplied state 供货时状态
assurance coefficient 安全系数，保险系数，保证系数
assurance factor of strength 强度保证率
assurance factor 安全系数，保险系数，保证系数，安全因数
assurance of compliance with 确保遵守……
assurance 保险，保证，确信，保证条件
assured system capacity 系统保证出力
assured 被保险人
assurer 保险人，保险商，保险公司，承保人
assy = assembly 装配，组装，装置，部件，组件
astable multivibrator 自激多谐振荡器，不稳定多谐振荡器
astable 不稳定的，非稳态的
AST(American Standard Thread) 美国标准螺纹
astatic action 无定位作用
astatical 非静止的，无静差的，不定向的，无定向的
astatic coil 无定向线圈
astatic control 无定向控制，无静差控制，无静差调节
astatic element 无静差元件
astatic galvanometer 无定向电流计
astatic governing 无定向调节
astatic-governor 无向调节器，不稳定调速器
astatic instrument 无定位仪表，无静差仪表，无定向式仪表
astatic magnetometer probe 无定向磁强计探头
astatic multivibrator 自激多谐振荡器
astatic regulation 无定向调节
astatic regulator 无定向调节器，无静差调节器
astatic relay 无定向继电器，无静差继电器，摆动继电器
astatic 不稳（定）的，无静差的，无定位的，无定向的
AST(automatic stop trip) 自动停机跳闸系统
astel 隧洞平巷顶支撑板
asterisk 星号
astern power 倒车功率
astern stage 倒车级
astern turbine 倒车透平
astern wheel 倒车叶轮
aster rectifier circuit 十二相（及以上）整流器回路
as the case may be 看情况，根据具体情况，视情况而定
as the case stands 事实上
asthenosphere 软流圈，岩流圈
as the usual practice 按照惯例
astigmation 像散，像差
astigmatism 像散现象，像差现象
ASTM(American Society for Testing and Materials) 美国试验与材料协会
ASTM(American Standards of Test Manual) 美国标准实验手册
A-STP(auto stop) 自动停止
astrigency 收敛性
astringent substance 收敛性物质
astrolabe 等高仪
astronomical latitude 天文纬度
astronomical longitude 天文经度
astronomical observation 天文观测
astronomical observatory 天文台
astronomical point 天文点
astronomical theodolite 天文经纬仪
astronomical tide 天文潮，特大潮汐
astronomical time 天文时
astronomical unit 天文单位
astronomical year book 天文年历
astronomic azimuth point 天文方位点
astronomic azimuth 天文方位角
astronomic geology 天文地质学
astronomic transit 中星仪
astronomy 天文学
A-STRT(auto start) 自动启动
astylar 无柱式
astyllen （坑道中的）拦水埝
ASU(automatic synchronizing unit) 自动同步装置，自动同步系统

as usual 按照惯例
as-welded 初焊状态，焊态，焊态的
as with 如同，和……一样，就……而言
asymeter 非对称计
asymmetric flow 不对称水流
asymmetric 不对称的
asymmetrical distortion 非对称畸变
asymmetrical load 不对称负荷
asymmetrical 不对称的，不平衡的，不对称，不平衡
asymmetric balance 不对称平衡
asymmetric bending 不对称弯曲
asymmetric characteristic 不对称特性
asymmetric conductivity 不对称导电性
asymmetric conductor 不对称导体，具有整流作用的导体
asymmetric configuration 不对称结构，不对称形态
asymmetric current 不对称电流
asymmetric dipole 不对称偶极子
asymmetric distribution 不对称分布
asymmetric double-chamber surge tank 不对称式双室调压井
asymmetric drainage 不对称水系
asymmetric effect 非对称效应
asymmetric fault 不对称故障，不对称短路
asymmetric fold 不对称褶皱
asymmetric hysteresis 非对称磁滞回线
asymmetric joint 不对称接合
asymmetric load 不平衡负荷，不对称负载
asymmetric motion 非对称运动，不对称运动
asymmetric network 不对称网络
asymmetric operation 不对称运行
asymmetric polyphase current 不对称多相电流
asymmetric potential 不对称电势
asymmetric short-circuit 不对称短路
asymmetric side-band transmission 不对称边带传送
asymmetric slug 不对称腾涌
asymmetric state 不对称状态
asymmetric system 不对称系统
asymmetric two-wire system 不对称两线系统
asymmetric vibrator 不对称振子
asymmetric wave 不对称波
asymmetric winding 不对称绕组
asymmetry impedance 不对称电抗
asymmetry 不对称性，不对称，不平衡度，反对称性，不对等，偏态
asymptote 渐近线，渐近
asymptotical behaviour 渐近特性，渐近行为，渐近性质
asymptotical distribution 渐近分布
asymptotical equation 渐近方程
asymptotical fouling 渐近型污垢
asymptotical neutron density 渐近中子密度
asymptotical period technique 渐近周期法
asymptotical period 渐近周期
asymptotical relaxation length 渐近弛豫长度
asymptotical stability 渐近稳定性
asymptotical value 渐近值
asymptotical voidage 渐近空隙度
asymptotical 渐近线的，渐近的
asymptotive behaviour 渐近行为
asynchronism 异步性
asynchronization 异步
asynchronized synchronous generator 异步化同步发电机
asynchronized synchronous motor 异步化同步马达
asynchronous capability 非同步能力
asynchronous circuit 异步电路
asynchronous communication interface adapter 异步通信接口转接器
asynchronous communication 异步通信
asynchronous computer 异步计算机
asynchronous condenser 异步补偿机，异步调相机，异步电容器
asynchronous excitation 异步激励
asynchronous flow 异步径流
asynchronous generator 不同步发电机，异步发电机
asynchronous impedance 异步阻抗
asynchronous input 异步输入
asynchronous linear motor 异步直线电动机
asynchronous link 异步联络线
asynchronous machine 异步电机
asynchronous motor 异步马达，非同步马达，异步电动机，异步电机
asynchronous operation of synchronous machine 同步电机的异步运行
asynchronous operation 异步运行，异步操作
asynchronous oscillation 异步振动
asynchronous paralleling 非同步并列
asynchronous phase modifier 异步整相器
asynchronous quenching 异步遏制
asynchronous reactance 异步电抗
asynchronous resistance 异步电阻
asynchronous running 异步运行
asynchronous sending/receiving 异步发送/接收（装置）
asynchronous sequential circuit 异步时序电路
asynchronous servomotor 异步伺服电动机
asynchronous slip-ring motor 滑环式感应电动机，滑环式异步电动机
asynchronous spark-gap 异步放电器，非同步火花隙
asynchronous speed 异步转速
asynchronous starting 异步启动，感应启动
asynchronous status 异步运行状态
asynchronous torque 异步转矩
asynchronous transmission 异步传输
asynchronous wind turbine generator 异步风力发电机
asynchronous working 异步工作
asynchronous 异步的，不同步的，非同步的
asynchtonous motor 异步马达
ata(absolute atmosphere) 绝对大气压单位
ATA(air turbine alternator) 空气涡轮交流发电机
A&T(acceptance and transfer) 验收与移交
AT(acceptance test) 验收试验
AT(air temperature) 空气温度
at a moment's notice 立即，马上

at an angle of θ to M 与M成θ角
at an angle of θ with M 与M成θ角
A & T(assemble and test) 装配与测试
atatchment unit interfacce 附属单元接口
at attenuation 衰减
at buyer's option(/selection) 由买方选择
ATC(acoustical tile ceiling) 吸声砖吊顶
ATC(automaitc turbine start up control) 汽轮机自动启动控制
ATC(automatic turbine control) 汽轮机自动控制
ATC(automatic turbine startup or shutdown control system) 汽轮机自启停控制系统
at consignee's risk 收货人险
at constant price 不变价
ATC system(automatic tool changing system) 自动换刀系统
at current price 按现行价格
at discretion 任意，随意
atectonic dislocation 非构造变位
at equal spacing 均匀放置
at fair price 价格公道
ATF(asphalt tile floor) 沥青砖地面
athermal transformation 无扩散型相变
athermal 冷却变态，冷却相交，无热的，非热的
athermanous 绝热的，不透热的
athletic ground 运动场地
a thousand year frequency flood 千年一遇洪水
at one's discretion 随意，任意，自行决定
at its sole discretion 自行决定
atlantes （用以支撑柱顶线盘的）男像柱
Atlantic Community Development Group for Latin 大西洋共同体拉美开发团投资公司
atlas 图册，地图册，地图集，图集
atm＝atmosphere 大气
at maturity （票据）到期
atmidometer 蒸发计，蒸发器
atmidometry 蒸发测定
atmidoscope 湿度指示器
at minimum level 最低程度
atmogenic rock 气生岩
atmoliths 气成岩
atmology 大气学，水气学
atmometer 蒸发计，蒸发表，蒸发仪，汽化计
atmometry 蒸发测量，蒸发计量学，蒸发测定
atmosphere absorption 大气吸收
atmosphere aerosol 大气气溶胶
atmosphere circulation 大气环流，大气循环
atmosphere composition 大气成分
atmosphere concentration 大气浓度
atmosphere constituent 大气成分
atmosphere contamination 大气污染
atmosphere data 大气数据
atmosphere density 大气密度
atmosphere dispersion factor 大气弥散因子
atmosphere dust 大气尘埃
atmosphere exhaust station 大气排放站
atmosphere gauge 气压表，气压计
atmosphere impurities 大气杂质
atmosphere layer 大气层
atmosphere moisture 大气湿度，大气水分
atmosphere-ocean interaction 气海相互作用
atmosphere of cooperation 合作环境
atmosphere pollutant 大气污染物
atmosphere pollution burden 大气污染负荷
atmosphere pollution 大气污染
atmosphere regenerant 大气更新剂
atmosphere relief diaphragm 防爆门，向空排气膜板
atmosphere science 大气科学
atmosphere standard 标准大气压，标准大气
atmosphere 大气，大气层，大气圈，气氛
atmospheric absorption 大气吸收
atmospheric acoustics 大气声学
atmospheric action 大气作用
atmospheric adsorption 大气吸收
atmospheric advection 大气平流
atmospheric air 大气
atmospherical composition 大气组成
atmospheric arc rectifier 大气电弧整流器
atmospheric assimilation 大气同化
atmospheric attenuation 大气衰减
atmospheric baseline observing station 大气基准观测站
atmospheric boundary layer 大气边界层
atmospheric boundary 大气边界
atmospheric carcinogen 大气致癌物
atmospheric chemistry 大气化学
atmospheric circulation 大气循环，大气环流
atmospheric composition 大气组成
atmospheric condensation 降水，降雨，雨量，大气凝结
atmospheric condenser 大气凝汽器【船用】，冲激式冷凝器，空气冷凝器
atmospheric condition 大气状况，大气条件
atmospheric constituent 大气组分
atmospheric contaminant 大气污染物
atmospheric contamination 大气污染
atmospheric convection 大气对流
atmospheric cooling tower 自然通风冷却塔，空气冷却塔
atmospheric corrosion 大气侵蚀，大气腐蚀
atmospheric damping 大气阻尼
atmospheric deaerator 大气式除氧器
atmospheric density 空气密度
atmospheric depression 气压下降，低气压
atmospheric depth 大气深度，大气厚度
atmospheric diffusion equation 大气扩散方程
atmospheric diffusion process 大气扩散过程
atmospheric diffusion test 大气扩散试验
atmospheric dilution 大气稀释，大气扩散
atmospheric discharge 大气放电
atmospheric dispersion and dilution capability 大气扩散稀释能力
atmospheric dispersion experiment 大气扩散试验
atmospheric dispersion model 大气弥散模型
atmospheric dispersion 大气弥散，大气分散
atmospheric disturbance 天电干扰，大气扰动
atmospheric drag 大气阻力
atmospheric duct 大气波导层
atmospheric dump valve （蒸汽）大气排放阀
atmospheric dust 大气尘埃
atmospheric electricity 天电，大气电，大气电学

atmospheric emission standard 大气排放标准
atmospheric emission 大气排放
atmospheric environment 大气环境
atmospheric evaporation 大气蒸发
atmospheric exhaust station 大气排放站
atmospheric exhaust valve 排空阀，向空排气阀，向空泄放阀
atmospheric exposure test 曝露试验
atmospheric extinction coefficient 大气消光系数
atmospheric extinction 大气消光，大气削弱
atmospheric fallout 大气微尘
atmospheric flow 大气流动
atmospheric fluidized bed combustion cycle 常压流化床联合循环
atmospheric gas burner 低压煤气燃烧器
atmospheric heat balance 大气热平衡
atmospheric humidity 大气湿度
atmospheric hydromechanics 大气流体力学
atmospheric impurities 大气杂质
atmospheric interaction 大气相互作用
atmospheric interference 天电干扰，大气干扰
atmospheric inversion condition 大气逆温状态
atmospheric inversion 大气逆温
atmospheric irradiance 大气（辐射）辐照度
atmospheric layer 大气层
atmospheric line 大气压力线
atmospheric mass transport 大气质量运输
atmospheric mass 大气团，大气质量
atmospheric modelling technique 大气模拟技术，大气建模技术
atmospheric model 大气模式，大气模型
atmospheric moisture capacity 大气湿度
atmospheric moisture 大气湿度
atmospheric momentum transport 大气动量运输
atmospheric monitoring system 大气监测系统
atmospheric monitoring 大气监测
atmospheric noise 大气噪声
atmospheric opacity 大气不透明度
atmospheric optical depth 大气光学厚度
atmospheric optical thickness 大气光学厚度
atmospheric over-voltage protection 大气过电压保护，防雷保护
atmospheric overvoltage 大气过电压
atmospheric oxidation 大气氧化，空气氧化
atmospheric ozone layer 大气臭氧层
atmospheric parameter 大气参数
atmospheric particulate 大气微粒
atmospheric phenomenon 大气现象
atmospheric photochemistry 大气光化学
atmospheric physics 大气物理学
atmospheric pollutant 大气污染物
atmospheric pollution 大气污染
atmospheric precipitation 大气降水，大气沉降
atmospheric pressure bed 常压床【沸腾炉】
atmospheric pressure method 真空预压法
atmospheric pressure 大气压，大气压力，气压
atmospheric process 大气过程，大气作用
atmospheric pump 抽气泵
atmospheric radiation 大气辐射
atmospheric radioactivity 大气放射性
atmospheric radon activity 大气中氡放射性
atmospheric reaction 大气反应
atmospheric refraction 大气折射
atmospheric regional station 大气地区观测站
atmospheric relief valve 大气释放阀，安全阀，排空阀，对空排气门
atmospheric rock 气成岩
atmospheric sampling 空气取样
atmospheric scale 大气尺度
atmospheric scavenging 大气净化
atmospheric science 大气科学
atmospheric shear flow 大气剪切流动
atmospheric similarity 大气相似性
atmospheric singularity 大气奇异现象
atmospheric stability class 大气稳定度类别
atmospheric stability constant 大气稳定度常数
atmospheric stability length 大气稳定度长度
atmospheric stability 大气稳定度
atmospheric stagnation pressure 大气驻压
atmospheric stagnation temperature 大气驻温
atmospheric stagnation 大气滞止，大气停滞
atmospheric static condition 标准大气条件
atmospheric steam cured concrete 低压蒸汽养护混凝土
atmospheric steam curing 低压蒸汽养护
atmospheric steam dump valve 蒸汽向大气排放阀
atmospheric steam dump 蒸汽向大气排放，蒸汽大气排放
atmospheric stratification 大气层结
atmospheric structure 大气结构
atmospheric surface layer 大气表面层
atmospherics 大气干扰，天电，自然产生的离散电磁波
atmospheric temperature 大气温度
atmospheric thermal conductivity 大气热导率
atmospheric thermodynamics 大气热力学
atmospheric tide 大气潮
atmospheric trace gas 大气痕量气体
atmospheric transmission 大气透射，大气传递
atmospheric transmissivity 大气透射率
atmospheric transmittance 大气透射度
atmospheric transparency 大气透明度
atmospheric transport model 大气运输模式
atmospheric transport 大气运输
atmospheric turbidity 大气浑浊度
atmospheric turbulence model 大气湍流模型
atmospheric turbulence spectrum 大气湍流谱
atmospheric turbulence structure 大气湍流结构
atmospheric turbulence 大气湍流
atmospheric valve 排空阀，大气释放阀，空气阀，放空阀
atmospheric vortex 大气旋涡
atmospheric water 大气水
atmospheric wave 大气波
atmospheric whirl 大气旋风
atmospheric wind tunnel 大气压风洞
atmospheric wind 大气风
atmospheric 大气的，大气压的，空气的，大气中的，大气层的
atm press 大气压，大气压力
ATMZ(atomizing) 雾化

atoll reef 环礁
atom 原子
atomic 原子的
atomic absorption method 原子吸收法
atomic absorption spectrometer 原子吸收光谱法
atomic absorption spectrophotometer 原子吸收分光光度仪
atomic absorption spectroscopy 原子吸收光谱
atomic aircraft 核动力飞行器
atomically pure graphite 原子能工业用纯石墨
atomic angular momentum 原子轨道角动量
atomic arrangement 原子的排列
atomic attenuation coefficient 原子衰减系数
atomic battery 原子能电池
atomic beam 原子束
atomic binding energy 原子结合能
atomic binding 原子键，原子键联
atomic boiler 原子锅，原子锅炉，核蒸汽发生器
atomic bond 原子键
atomic cell 原子电池
atomic charge 原子的电荷
atomic cross 原子断面
atomic disintegration 原子分裂，原子蜕变
atomic energy act 原子能法
Atomic Energy Commission 原子能委员会【美国】
atomic energy detection system 原子能探测系统
atomic energy industry 原子能工业
atomic energy installation 原子能装置
atomic energy level 原子能级
atomic energy plant 原子能电站，核电站，原子能发电厂，核电厂
atomic energy power generation 原子能发电
atomic energy power station 原子能电站
atomic energy project 原子能设计，原子能电站设计
Atomic Energy Regulatory Board 原子能管理局【印度】
atomic energy storage battery 原子能蓄电池
atomic energy 原子能
atomic excitation 原子激发
atomic field 原子场
atomic fission 原子裂变
atomic force microscopy 原子力显微镜
atomic formula 原子式，结构式
atomic frequency standard 原子频率标准
atomic fuel 核燃料，原子燃料
atomic fusion 核聚变
atomic hydrogen arc weld 氢原子电弧焊（接）
atomic hydrogen welding 原子氢焊
atomic installation 原子动力装置
atomic ionization 原子电离
atomicity 原子价，原子性
atomic layer apiary 原子层外延
atomic locomotive 核机车
atomic mass unit 原子质量单位
atomic mass 原子质量
atomic merchant ship 核商船
atomic navy 核动力海军
atomic nucleus fission 原子核裂变
atomic nucleus 原子核
atomic number 原子序数
atomic parachor 原子等张比容
atomic photoelectric effect 原子的光电效应
atomic pile 核反应堆，原子反应堆
atomic plant 核装置
atomic polarization 原子极化
Atomic Power Construction Ltd. 原子能建设有限公司【英国】
atomic power facility 原子动力设备，核动力设施
atomic power plant 原子能发电厂，核电站，核电厂，原子能电站
atomic power 核动力，原子能动力，原子能，原子动力
atomic precession angular frequency 原子进动角频率
atomic propulsion unit 核推进装置
atomic ratio 原子比率
atomic raw material 原子原料
atomic reactors in space 航天用核反应堆
atomic reactor 原子反应堆，核反应堆
Atomic Safety & Licensing Boards 核安全和审批委员会【美国】
atomic self-absorption spectrophotometer 自吸收原子吸收分光光度计
atomic spectrum 原子光谱
atomic stopping power 原子阻止本领
atomic submarine tanker 核动力潜水油轮
atomic surface burst 地面核爆炸
atomics 核工艺学，原子学，原子工艺学
atomic tritium diffusion 原子氚扩散，原子扩散
atomic waste 原子能工业废料，核废料
atomic weight 原子量，原子重量
atomistic competition 原子状竞争，单纯竞争
atomization 粉化，喷雾，雾化
atomized particle size 雾化颗粒尺寸
atomized spray injector 雾化喷嘴
atomized water 雾化水
atomized 雾状的
atomizer burner 雾化喷燃器
atomizer cone 雾化锥
atomiser 雾化器，喷嘴
atomizer tip 喷嘴，喷头
atomizer 喷油嘴，雾化器，喷雾器，喷嘴，雾化喷嘴
atomize(-se) 雾化，喷雾，粉化，喷射
atomizing angle 雾化角
atomizing cone 雾化锥
atomizing deaerator 喷雾式除氧器，雾化除气器
atomizing distribution 雾化分布
atomizing fineness 雾化细度
atomizing nozzle 雾化喷头，雾化喷嘴
atomizing pressure 雾化压力
atomizing spraying 雾化喷涂
atomizing steam 雾化蒸汽
atomizing test 雾化实验
atomizing type deaerator 喷雾式除氧器
atom probe 原子探针
at-once payment 立即付款，即时付款
at one pouring 一次浇灌

at one's own cost　由某人自己出钱，费用自理
at one's own disposal　由某人自行处置
at one's own expense　费用自理，由某人自己出钱
at one's own risk　承担风险，风险自负
at option　随意
at owner's risks and responsibilities　由业主负担一切责任
at par　按原价【票面价格】，平价
ATP(authorization to proceed)　(合同)批准执行，批准开工，开工令
ATR(advanced thermal reactor)　先进型热动力反应堆
a transcript of talks　会谈记录
ATRC(advanced test reactor critical)　先进试验堆临界实验装置
at-reactor fuel storage　堆旁燃料贮存
A-TRIP(auto trip)　自动跳闸
atrium　中庭，门廊，天井
atrmospheric pressure bed　无顶压床
A-truss　三角形屋架
ATS(automatic transform system)　厂用电源快速切换装置
ATS(automatic turbine start-up)　汽机自动启动
at seller's option　由卖方决定的，卖方可选择的
at sight　见票即付
ATS　奥地利先令
attached bow-wave　附体冲波
attached building　附联建筑物，配套建筑物
attached clause　附加条款
attached column　半柱，附柱，附墙柱，壁柱
attached drawing　附图
attached dune　附着沙丘
attached film expanded bed　固定膜膨胀床
attached great importance to　非常重视，强调了……的重要性
attached list　附表
attached metallic particles　黏结金属颗粒
attached pier　扶垛，堵墙，支墩，壁柱，扶壁
attached shock　附着激波，无脱流激波
attached sunspace　附加阳光间式【太阳光热利用】
attached superheater　布置在蒸发受热面之间的过热器，辅助过热器
attached voidage　附着空泡
attached void　附着空泡
attached vortex　附着旋涡
attached water　附着水
attach hereto　附上
attach importance to　重视
attaching organism　附着生物
attaching parts　紧固件
attaching plug　小型电源插头，电话塞子
attaching　附着
attachment effect　附壁效应
attachment flow　附着流，附壁流
attachment hardware　连接用的金属件，五金附件
attachment link　连接杆
attachment mode　固定模式
attachment plant　并置装置，辅助设备，附属建筑物
attachment plug　插头
attachment point　附着点，固定点，挂点
attachment screw　定位螺钉，装合螺钉，连接螺钉，止动螺钉
attachment steel plate　挂线板
attachment　附属装置，连接物，附件，附着，配件，夹具，固定
attach　连接，附上
attack angle　迎角，攻角，冲角
attack drill　冲抓式钻孔机
attack time　增高时间【信号电平】，上升时间，启动时间
attack　冲蚀，冲击，侵蚀，侵袭，攻击
attainable burn-up　可达到的燃耗
attainable precision　可达到的精密度，可达精确度，可达精度
attainable　可达到的
ATT=attachment　辅助设备
ATT(automatic transfer of title)　所有权自动转移
attemperating air duct　调温风管，加冷风用风管
attemperating water piping　减温水管道
attemperating water valve　减温水门
attemperating water　减温水
attemperation　减温，调温，温度调节，温度控制
attemperator　温度控制器，保温装置，保温器，减温器，恒温箱，调温装置
attemper　减温，调温，调和，调节
attempted starts　试启动
attempt　尝试
attendance bonus　加班奖金，出勤奖
attendance　看护，值班，保养，维护，入会人员，与会者，维修
attendant circumstance　有关情况
attendant of auxiliary equipment　辅机值班员
attendant of feed pump　给水值班员
attendant　值班人员，运行人员，维护人员，维修人员，伴随物，伴随的
attended substation　有值班员的变电站，有人值班的变电站
attendee　参加者，出席者，到场者，参加人员，参会人员
attention　维护，保养，注意，留心，维修
attenuater　减温器，衰减器
attenuate　衰减，稀释，减薄
attenuating cable　衰减电缆
attenuating mechanism　衰减机制
attenuation coefficient　衰减系数
attenuation compensation　衰减补偿
attenuation constant　衰变常数，衰减常数
attenuation cross-section　消减断面
attenuation curve　衰减曲线
attenuation decrement　衰减，减缩率，衰减量
attenuation distortion　衰减失真，衰减畸变
attenuation equalization　衰减均衡
attenuation equalizer　衰减补偿器，衰减均衡器
attenuation equivalent nettiness　网络传输衰耗等效值
attenuation factor　衰减因子，衰减系数
attenuation-frequency characteristic　衰减频率

特性
attenuation gauge 衰减测量计
attenuation loss 衰减损耗
attenuation measuring device 衰减测定装置
attenuation mechanism 衰减机制
attenuation of flood wave 洪水波的削弱
attenuation of seismic intensity 地震烈度衰减关系
attenuation of traveling wave 进行波衰减
attenuation of wave 波浪衰减
attenuation property 衰减性能
attenuation rate 衰减比（率），衰减率
attenuation region 衰减区，衰减范围
attenuation rule 衰减法则
attenuation 衰减，减弱，衰耗，减少，缩减，稀释，衰减因子
attenuator box 消声静压箱
attenuator 衰减器，减温器，减压器，消声器，衰耗器，阻尼器，分压器
attestation 见证，作证，认证，证词，证据
attested copy 经认证的副本
attested document 经公证或核证的文件
attest 证明
at the discretion of 由……决定，取决于，随……的意见
at the drive end 传动端的
at the kerbside 在马路牙子上
at the most favorable price 最优惠的价格
attic floor 屋顶层
attic story 屋顶层，阁楼
attic 阁楼，顶楼，屋顶室
attitude angle 姿态角，方位角
attitude control 位置控制，姿态控制
attitude of rocks 岩层产状
attitude of stratum 地层产状
attitude of the joint 焊缝特性
attitude toward work 工作态度
attitude 方位，姿态，位置
attle 废矿渣，废屑，石屑
ATTN= attention 由……收阅，收件人，联系人，注意
attorney agreement 代理协议
attorney-in-fact 代理人
attorneys fees clause 律师费条款
attorney ship 代理人身份，代理权
attorney 律师，代理人，代办人
attracted armature relay 衔铁吸合式继电器
attracted disc electrometer 吸盘式静电计
attract foreign investment 吸引外资
attracting groin 引流丁坝
attract investors 吸引投资者
attraction centre 引力中心
attraction force 吸引力，吸力
attraction 引力，吸力，吸引，吸引力
attractive design 精美设计
attractive force 吸引力，引力
attractive interaction 互相吸引
attractiveness 吸引，吸性
attractive price 具有吸引力的价格，有吸引力的价格
attractive site 优越坝址
attractive 引人注目的，有吸引力的
attract 吸引
attributable to 因为，由于，由……引起的，可归因于……的
attribute sampling 按属性抽样
attribute 特性，标志，属性
attrited 磨损的
attrition mill 碾磨机
attrition pulverizer 研磨式磨煤机
attrition rate 磨损率，损耗率
attrition resistance 抗磨耗性
attrition test 磨损试验
attrition value 磨损值
attrition 摩擦，磨损，磨耗，磨碎，损耗
attune 调谐，使调和，使一致
ATU(auxiliary test unit) 辅助测试装置
at variance 不一致，有出入，有分歧
at wharf 在码头交货
at will employment 任意聘用关系【雇主可随时解雇】
at. wt.(atomic weight) 原子量
at your convenience 根据你的方便，在你方便时
at your disposal 任意使用，任你自由支配
A type insulator A 型绝缘子
auctioneering circuit 最大值选择电路
auctioneering device 高低信号选择器
auctioneering unit 高值选择装置
auctioneering 高值选择，最大值选择
auctioneer 拍卖人，拍卖商，拍卖
auction sale 拍卖
auction 拍卖
AUD 澳元
Audiberts-Arnu dilation 奥阿膨胀度
audibility current 可听度电流
audibility factor 可听系数
audibility meter 听觉测量器，听度计，听力计
audibility threshold 听觉界限
audibility value 可听值
audibility 可听度，可闻度，清晰度
audible alarm 音响报警[警报]，音响报警装置[设备]，可闻报警信号
audible and visible alarm 声光信号警报
audible busy signal 音响占线信号
audible call 音响呼叫
audible frequency 音频，声频，可听频率
audible indication 声指示，音响指示，可闻信号
audible leak indicator 音响检漏指示器
audibleness 可听度
audible noise 声频噪声，可听声音
audible or visible warning devices 可闻或可视的警告装置
audible range 可听范围
audible reception 音响接收
audible signaling device 音频信号器，音频信号装置
audible signal 音响信号，可听信号
audible sound 可听声
audible spectrum 声谱
audible warning device 音响报警装置
audible 可听到的，听得见的，音响的
audience chamber 接见室，会见室

audience room 会见室，接见室
audifier 音响放电器，声频放大器
audio alarm 音响报警
audio amplifier 声频放大器，音频放大器
audio band 声频带，音带
audio circuit 声频电路，音频电路
audio communication 声音通信
audio demodulator 伴音信号解调器
audio equipment 声频设备，音频设备
audio fader amplifier 声频控制放大器，音量控制放大器
audio-fidelity control 音色保真控制
audioformer 声频变压器，音频变压器
audio-frequency amplification 声频放大，音频放大
audio-frequency amplifier 声频放大器
audio-frequency apparatus 声频设备，音频设备
audio-frequency circuit 声频电路，音频电路
audio-frequency current 声频电流，音频电流，通话电流
audio-frequency generator 音频产生器，声频发生器
audio-frequency millivoltmeter 声频毫伏计
audio-frequency range 声频范围
audio-frequency reception 声频接收，音频接收
audio-frequency stage 声频级，音频级
audio-frequency telegraph 声频电报机，音频电报机
audio-frequency transformer 声频变压器，音频变压器
audio-frequency 声频，音频，声响频率
audio generator 声频发生器，声频振荡器
audiogram 声波图，听力图
audiograph 声波图
audiohead 录音头，拾音头，放音头
audio indicator 声频指示器，音频指示器
audio line 声频线路，音频线路
audiolloy 铁镍透磁合金
audiometer 测听计，听力计，听度计
audiometry 听力测定法
audio monitor 监听设备，监听器
audio noise meter 噪声计
audion 三极管，三级检波管，检波管
audio oscillator 声频振荡器，音频振荡器
audio radiation indicator 音响辐射监测器
audio response unit 答话器，声音应答装置
audio signal 声频信号，音频信号
audio tape 录音磁带
audio transformer 声频变压器，音频变压器
audio-visual center 电教中心
audio-visual 视听的，直观的
audio 声频，音频的，声音的，听觉的
audit by test 抽查，抽审
audit client 审核委托方
audit completion 审核结束
audit conclusion 审核结果，审计结论
audit criteria 审核准则
auditee 受审核方
audit evidence 审核证据
audit findings 审核发现，监查结果
audit follow-up 追踪监查【质量保证】，监查跟踪
audit function 检查功能
audit-in-depth 分层检查，按层次检查
auditing clauses 审计条款，查账条款
auditing committee 审计委员会
auditing law 审计法
auditing routine 检查规程，检查程序
auditing system （数据）检查系统
auditing 审计证明书
audition 声量检查
audit notification 审计通知，监察通知书，监查通知书
audit objective 审核目标
audit opinion 审计意见
auditorium 礼堂，大会堂，音乐厅
auditor registration 审核员注册
auditory fatigue 听觉疲乏
auditory response 听觉反应
auditory sensation 听觉
auditory 听觉的
auditor 审计员，审核员，监查员，查账员，听者
audit period 审计期
audit plan 审核计划，监查计划
audit procedure 审核程序
audit program 审核程序
audit record 监查记录，审核记录，监察记录
audit report 监查报告，审核报告，审计报告
audit result 审核结果，审核结论
audit risk 审计风险
audit scope 审核范围
audit team leader 审核组长，审计组长
audit team 监查小组，审核组
audit trail 数据检查，跟踪，寻迹检查
audit 核查【裂变材料流通量】，检查，审查，监查，查账，审计
auganite 辉安岩
augend digit 第二被加数的数字，加数的数位
augend in gate 第二被加数输入门，加数输入门
augend 被加数
auger bit 螺旋钻头
auger boring 螺旋钻钻探，麻花钻钻探
auger drill 麻花钻，螺旋钻
Auger effect 俄歇效应
Auger electron spectroscopy 俄歇电子能谱
auger hole 钻孔
auger pile 螺旋桩
auger rig 螺旋式钻井机
auger sampler 钻头取样器
auger stem 螺旋钻杆
auger with valve 带阀土钻，带阀螺旋钻具【钻浅井用】
auger 麻花钻，钻，螺旋输送机，钻头
augetron 高真空电子倍增管
auget 雷管
augite-porphyrite 辉石玢岩
augite-porphyry 辉石斑岩
augite 辉石
augmentation device 增强装置
augmentation of heat transfer 强化传热
augmentation 增大，增强，增加，增长，增加量，增加物

augmented code 增信码
augmented content addressed memory 可扩充的内容定址存储器
augmented matrix 增广矩阵
augmented operator grammar 扩充算符文法
augmented reality patrol inspection system 增强现实巡检系统
augmenter 增压器
augment nozzle 加速喷嘴
augmentor condenser 抽气冷却器
augmentor 增强器
augmenter 增强器
augment 增大，增长，增加
aural carrier 伴音载波，声频载波
aural detector 声频检波器，音频检波器
aural null 无声，消声
aural signal 音响信号，音频信号
aural transmitter 录音广播发射机，伴音发射机
aural-type receiver 收音机，伴音接收机
aura 气氛，气味，电风，辉光
aureola 接触变质带，接触变质地，光轮，日晕
aureole 接触变质带，接触变质地，光轮，日晕
Ausform-annealing 奥氏体形变退火
ausforming 变形热处理，形变淬火
Ausging 奥氏体时效处理
auspices 赞助，支持
austemper case hardening 等温淬火表面硬化
Austempering 奥氏体回火，等温淬火
austemper stressing 等温淬火表面应力
Austemper 奥氏体回火
Austenite stabilization treatment 奥氏体稳定化处理
Austenite 奥氏体
Austenitic boronated steel 奥氏体硼钢
Austenitic electrode 奥氏体焊条
Austenitic heat resistant steel 奥氏体耐热钢
Austenitic nickel-chromium steel 奥氏体镍铬钢
Austenitic stainless steel pipe 奥氏体不锈钢管
Austenitic stainless steel 奥氏体不锈钢
Austenitic steel 奥氏体钢
Austenitizing temperature 奥氏体化温度
Austenization 奥氏体化
austerity program 紧缩计划
austral window 滑动窗，南窗
autecology 环境生态学
authenticate 鉴定，认证，证明……是真实的
authentication code 识别码
authentication 鉴定，认证，证明真实性，识别
authenticator 证实，证明，鉴定
authenticity 确实，确定性，可靠性，真实性
authentic text 正式文本，有效文本
authentic 可信的
authonized inspector 受权视察员
authorisation 授权，认可
authorised signature 授权的签字
authoritative body 权威机构
authoritative inquiry 权威性调查
authoritative interpretation 权威性解释
authorities concerned 有关方面，有关当局
authorities for project approval 项目批准机关
authorities 当局
authority commensurate with responsibility 权责相符
authority to negotiate 授权议付，权力委托
authority to pay 授权付款
authority to sign 授权签字
authority 管理局，管理机构，权力，权威
authorization certificate 授权书
authorization decree 建设（电厂）批文，建厂许可证
authorization to proceed with fabrication 批准开工制造
authorization to proceed 合同批准执行，批准开工，开工许可证
authorization 审定，批准，许可，授权，核准，认可，委任，授以资格或合格证书
authorized agent 全权代理人，指定的代理人
authorized bank 指定银行，授权银行
authorized capacity 核准负载
authorized capital 核准资本，法定资本
authorized company 授权公司
authorized contract for difference 授权差价合同
authorized data list 技术数据和技术情况一览表
authorized data 正式数据，正式技术资料
authorized distributor 授权经销商
authorized for testing 批准试验
authorized functions 受权的职能
authorized inspection agency 指定的检查机构
authorized inspector 授权的检查员，公认的检查员
authorized investments 经授权代办投资
authorized limit 管理限值，特许限值
authorized personnel 授权人
authorized pressure 容许压力，许用压力
authorized representative 授权代表
authorized signatory 授权签字人
authorized signature 授权签署，授权人签名
authorizer 授权人
authorize 授权
authorizing engineer 授权（给他人）的工程师
autoalarm 自动报警器，自动报警，自动警报装置
autoanalyzer 自动分析器
AUTO＝automation 自动
autobalance 自动平衡器，自动天平
auto-bias circuit 自偏压电路
auto-bias 自动偏压，自动偏置
autobiography 履历（表），自传
autobond 阻抗结合，自耦合，自动键接，自动接合
autobrake 自动制动器
autobus 机动车
auto by-pass 自动旁路
auto-capacity 本身电容，固有电容
autocartograph 自动测图仪
autocar 机动车
autocatalysis 自动催化
autocatalytic reaction 自催化反应
autochangeover 自动接通电路【备用系统】
autochrome circuit 自动色度信号电路
autochrome 彩色照片，彩色底片
autochthonous 独立的，当地的，地方性的

auto-circuit breaker 自动断路器
autoclast 自碎岩
autoclaved fly ash-lime brick 蒸压粉煤灰砖
autoclaved sand-lime brick 蒸压灰砂砖
autoclaved unit 蒸汽养护预制件
autoclave expansion 蒸压膨胀
autoclave test 蒸压试验
autoclave 高压灭菌器，高压釜，压力加热器，蒸汽养护间
autocode instruction 自编码指令
autocoder 自动编码机，自动编码语言，自动编码器
autocode 自动编码，自动代码
auto-collimating 自准值，自动对准
autocollimator 自动视准仪，自动准直仪
autocompensation 自动补偿
auto-compounded current transformer 自复绕式电流互感器
autoconnected transformer 自耦变压器
autoconnected 按自耦变压器线路接线的
autoconnection 自耦变压器的接线
autocontrol feature 自动控制特点，自调节特性
autocontrol rod 自动控制棒
autocontrol system 自动控制系统
autocontrol 自动控制，自动调节，自动调整，自调节
auto-convection 自动对流
autoconverter 自动变换器，直流-直流变换机，直流衡功率变换机
auto-correction 自动校正，自动修正
autocorrelation coefficient 自相关系数
autocorrelation function 自相关仪，自相关函数
autocorrelation 自相关
autocorrelator 自相关器
autocorrelogram computer 自相关图计算机
autocorrelogram 自相关图
autocovariance technique 自协方差技术
autocovariance 自协方差
auto crane 汽车起重机，汽车吊，汽车式起重机
autocrat 加气混凝土
auto-cut-out 自动断路器，自动截止，自动断路
AUTOCV(automatic check valve) 自动止回阀
autocycle 自动循环
autodecomposition 自动分解
auto-depressurization system 自动减压系统
autodestructive fuse 自炸引信
auto-draft 自动制图
autodumper 自卸车
autodyne oscillator 自差振荡器
autodyne receiver 自差收音机
autodyne reception 自差接收法
autodyne 自差，自拍，自差接收器，自差收音机
autoelectronic current 场致发射电流，冷发射电流
autoelectronic emission 自动电子发射
autoelectronic 场致电子发射的，自动电子发射的
autoenlarging apparatus 自动放大器，自动放大机
autoexcitation 自激，自激振荡，自励
auto-exciting 自激，自励
autoformer 自耦变压器
autofrettage 自增强，空心零件预应力处理
autogenetic topography 自成地形
autogenic 气焊的，气割的
autogenous cutting 气割
autogenous growth of concrete 混凝土自生体积增长
autogenous heating 自然发热，自热
autogenous shrinkage 自然收缩，自生收缩
autogenous soldering 气焊，熔接
autogenous volume change 自生体积变化
autogenous welding 铝焊条，自热焊，气焊，熔接，自动气焊
autogenous 气割的，自生的
autogentic drainage 自成水系
autographic apparatus 自动记录仪
autographic data acquisition system 自动数据采集系统
autographic data exchange system 自动数据交换系统
autographic data processing 自动数据处理
autographic drag spoiler 自动阻力板，自动阻力扰流板
autographic load strain recorder 负载应变自动记录器
autographic oedometer 自录固结仪
autographic pyrometer 自动记录式高温计
autographic recording apparatus 自动图示记录仪
autographometer 自动图示仪，自动地形仪
autograph reception 印码接收
autograph 笔迹，手稿，亲笔签名，签署，亲笔签署
autogyration 自转
auto heterodyne 自差，自拍，自差收音机
autohoist 汽车式起重机，汽车起重机
auto-ignition 自动点火，自燃
autoincrement addressing 自动加速寻址，自动递增寻址
autoindex 自动变址，自动索引，自动变指数，自动编索引
autoinduction 自感，自动感应，自感应
auto-inductive coupling 自耦变压器耦合，电感耦合
autoinhibition 自动抑制作用
auto-jigger 自耦变压器
autokeyer 自动键控器
autolifter 自动升降机
auto-limitation clause 自行限制条款
autoload 自动加载，自动装入
automaitc runback device 自复位装置
automaitc turbine start-up control 自动起控制系统
automanual block 半自动闭塞
automanual switch 自动-手动转换开关，自动-手动切换开关，自动-手动开关，半自动开关
automanual system 半自动系统，自动-手动方式，自动-手动系统
automanual transfer valve 自动-手动切换阀
automanual 半自动的，自动手动的，自动手动
auto-man 自动-手动
automat 自动装置，自动机，自动开关
automatable machine 可自动化机械
automata 自动机

automate　使自动（化）
automated　自动化的
automated batch mixing　自动批量混合法
automated circuit card etching layout　自动电路插件腐蚀设计
automated communication　自动通信
automated data management　自动数据管理
automated data medium　数据自动传送媒体，自动数据载体
automated design engineering　自动设计工程，自动设计技术
automated diagnostics for steam turbine　汽轮机故障诊断系统
automated direct analog computer　自动控制模拟计算机
automated emission leak locator　声发射漏泄定位仪
automated engineering design　自动工程设计
automated maintenance management system　自动化维修管理系统
automated reactor inspection system　反应堆自动检查系统
automated statistical analysis programming　自动统计分析程序设计
automated tape library　自动磁带库
automate message accounting　自动信息计算
automatic acceleration　自动加速
automatic addressing system　自动访问系统
automatic address modification　自动改变地址
automatic adjustment　自动调节，自动调整
automatic advanced breaker　自动提前断电器
automatic air brake　自动空气制动器，自动气闸
automatic air circuit breaker　自动空气断路器
automatic air eliminator　自动放气装置
automatic air valve　自动空气阀
automatic air vent valve　自动放气阀
automatic air vent　自动跑风门，自动排气口
automatic alarm receiver　自动警报接收机
automatic alarm　自动报警，自动报警设备，自动警报器，自动警报信号
automatical escape　自动退水（闸），自动排气管
automatical excitation　自动励磁作用
automatically closing fire door　自闭式防火门
automatically controlled continuous process　自动控制的连续过程
automatically controlled generator　自动控制发电机
automatically control valve　自动控制阀
automatically operated valve　自动操纵阀
automatically regulated　自动调节的
automatically reset relay　自动复归继电器
automatically synchronizer　自动同步机
automatical pumping system　自动抽水系统
automatical recording ga(u)ge　自动记录仪
automatical restart　自动再启动
automatic alternation　自动切换
automatical water-stage recording station　自记水位站
automatical　自动的
automatic amplitude control　自动幅值控制，自动振幅控制
automatic and hand operated changeover switch　自动手动转换开关
automatic antenna reel　自动天线绞车
automatic arc welding　自动电弧焊接，自动电弧焊
automatic assembly　自动装配
automatic back bias　自动反偏压
automatic background control　自动背景噪声控制
automatic balance manometer　自动平衡压力计
automatic balancing circuit　自动平衡电路，自动零点调整电路
automatic balancing　自动平衡
automatic bandwidth control　自动带宽控制
automatic barring gear　自动盘车装置
automatic batcher　自动称量器，自动配料器
automatic batching system　自动配料系统
automatic bias compensation　自动偏压补偿
automatic bias control　自动偏压控制
automatic bias　自动偏压
automatic binary computer　自动二进制计算机
automatic binary data link　自动二进制数据链接
automatic blasting　自动化喷砂
automatic block equipment　自动闭锁设备
automatic blocking　自动闭锁
automatic block signal　自动闭锁信号
automatic block system　（铁路）自动闭塞系统，自动闭锁系统
automatic blowdown system　自动排污系统
automatic blow-off valve　自动安全阀
automatic boost control　自动升压控制，自动增压调节器
automatic brake　自动闸
automatic braking　自动制动
automatic branch exchange　自动交换机
automatic breaker　自动断路器
automatic break　自动切断
automatic brightness control　自动亮度调节
automatic broadcasting control system　自动广播控制系统
automatic burette　自动滴定管
automatic by-pass valve　自动旁通阀
automatic calculation　自动计算
automatic calibration　自动校准
automatic call device　自动报警器，自动呼叫装置
automatic calling equipment　自动呼叫装置
automatic calling unit　自动呼叫装置
automatic call　自动呼叫
automatic carry　自动进位
automatic central office　自动电话总局
automatic change-over　自动切换，自动转换，自动转换开关
automatic character recognition　自动字符识别，自动字母识别，自动符号识别
automatic check and readiness equipment　自动检查及待命装置
automatic checkout and evaluation system　自动检测和估算系统
automatic checkout and recording equipment　自动检测记录设备
automatic checkout equipment　自动检测装置，

自动减负荷装置，自动检查装置
automatic checkout system 自动检测系统，自动校验系统
automatic checkout tester 自动检测仪
automatic check valve 自动逆止阀，自动止回阀
automatic check 自动检测，自动校验，自动检查
automatic chemical monitoring and control device 自动化学监控装置
automatic circuit analyzer 自动电路分析器
automatic circuit breaker 自动断路器，自动开关
automatic circuit recloser 自动重合器，自动重合闸装置
automatic circuit switching 自动电路转换，电路自动切换
automatic classification 自动分类
automatic cleaning 自动清洗
automatic clearing signal 自动终话信号，自动拆线信号
automatic clock switch 时控开关，自动时钟开关
automatic closed-loop system 自动闭环系统
automatic closing gear 自动闭合机构，自动关闭装置
automatic closing 自动闭合
automatic clutch 自动离合器
automatic coaling 自动上煤
automatic coal weigher 自动煤磅
automatic code 自动代码
automatic coding machine 自动编码机
automatic coding system 自动编码系统
automatic coding 自动编码，自动编码方法
automatic coiling of tape 自动卷带，自动绕带
automatic combustion control 自动燃烧控制
automatic come-along clamp 自动卡线钳
automatic compensation 自动补偿
automatic computer 自动计算机
automatic computing equipment 自动计算装置
automatic computing 自动计算
automatic contingency ranking 预想事故自动排列
automatic contingency selection 预想事故自动选择
automatic continuous blowdown 自动连续排污
automatic continuous function generator 自动连续函数发生器
automatic continuous process control 连续过程的自动控制
automatic control assembly 自动控制组件
automatic control block diagram 自动控制方块图，自动控制框图
automatic control chlorinator 自动控制加氯机
automatic control circuit 自动控制电路
automatic control cutout valve 自动切断阀
automatic control device 自动控制设备，自动控制装置
automatic control engineering 自动控制工程，自动控制技术
automatic control equipment 自动控制设备
automatic control error coefficient 自动控制误差系数
automatic control for start-up and shut-down 自启停控制
automatic control for the auxiliaries 公用设备自动控制
automatic controller 自动调节器，自动控制器
automatic control program 自动控制程序
automatic control relay 自动控制继电器
automatic control rod drive system 自动控制棒驱动系统
automatic control rod 自动控制棒
automatic control system 自动控制［调节］系统
automatic control theory 自动控制理论
automatic control valve 自动控制阀，自动调节阀
automatic control 自动控制，自动调整，自动调节
automatic conveying 自动传输
automatic corrective measure 自动纠正措施
automatic counter 自动计数器
automatic coupler 自动连接器，自动耦合器
automatic coupling device 自动连接器，自动耦合器，自动耦合设备
automatic crimping machine 全自动压接机
automatic current regulator 自动电路调节器
automatic cut-off 自动切断
automatic cut-out valve 自动断流阀
automatic cutout 自动切断，自动开关，自动断路器
automatic cycler 自动周期计
automatic cycle 自动循环
automatic cylinder drain valve 汽缸自动放水阀
automatic data accumulator and transfer 自动数据累加器和传送器
automatic data acquisition equipment 自动数据数据采集装置
automatic data acquisition system 数据自动采集系统
automatic data acquisition 自动数据采集程序
automatic data exchange system 自动数据交换系统
automatic data extractor and plotting table 自动数据提取及制表程序
automatic data line 自动数据传输线
automatic data-logging equipment 自动数据巡回检测装置
automatic data logging 自动数据巡回检测，自动数据记录
automatic data management information system 自动数据管理信息系统
automatic data processing center 自动数据处理中心
automatic data processing equipment and software 自动数据处理设备和软件
automatic data processing equipment 自动数据处理装置
automatic data processing program 自动数据处理程序
automatic data processing service center 自动数据处理服务中心
automatic data processing system 自动数据处理系统
automatic data processing 自动数据处理

automatic data reducer 自动数据简化器
automatic data reduction tunnel 实验数据自动处理风洞
automatic data reduction 自动数据简化
automatic data retrieval 自动数据检索
automatic data switching system 自动数据转接系统
automatic data translator 自动数据翻译程序
automatic decision-making 自动进行判断
automatic de-excitation 自动灭磁
automatic degausser 自动去磁器，自动去磁电路
automatic delivery control valve 自动输送控制阀
automatic deloading 自动卸荷，自动减载
automatic depressurization system 自动减压系统，自助卸压系统
automatic design engineering 自动设计工程
automatic desk computer 台式自动计算机
automatic detecting system 自动检测系统
automatic detecting unit 自动检测单元
automatic detector 自动检测器
automatic device 自动装置，自动仪表
automatic diagnosis of alarms 警报的自动诊断
automatic diagnosis 自动诊断
automatic diagnostic program 自动诊断程序
automatic dialer 自动拨号器
automatic digital data acquisition and recording 自动数字数据收集和记录
automatic digital data assembly system 自动数字数据汇编系统
automatic digital data error recorder 数字信息误差自动记录器
automatic digital drafting 自动数控绘图
automatic digital encoding system 自动数字编码系统
automatic digital handling system 自动数据处理系统
automatic digital input/output 自动数字输入/输出
automatic digital interchange system 自动数字交换系统
automatic digital message switching center 自动数字报文转换中心
automatic digital recording and control 自动数字记录及控制
automatic digital switch 自动数字开关
automatic digital tracking analyzer computer 自动跟踪数字分析计算机
automatic direction finder 自动测向仪，自动定向仪
automatic discharge opening 自动排气口
automatic discharge 自动卸料【反应堆】，自动排出，自动释放
automatic disconnection 自动切断，自动断开，自动脱离
automatic dispatching system 自动调度系统
automatic display and plotting system 自动显示及作图系统
automatic distillator 自动蒸馏器
automatic door 自动门
automatic double trap weir 双拼门式活动堰，自动双叠门堰
automatic drafting machine 自动制图机
automatic drain valve 自动疏水阀
automatic driver 自动传动装置
automatic drop 自动掉牌
automatic dust removal system 自动除尘系统
automatic ejection 自动抛出
automatic election 自动选择
automatic electrical arc welding 自动电弧焊
automatic electrical ignition 自动电点火
automatic electronic optical scanning 自动光电扫描
automatic elevator 自动电梯
automatic engineering design system 自动化工程设计系统
automatic equation solver 自动方程解算器
automatic error-correcting code 自动误差校正码，自动纠错码
automatic error detection 自动误差探测，自动检错
automatic exchange telephone 自动交换机
automatic excitation regulator 自动励磁调节器，自动调整励磁装置
automatic extraction turbine 自动调节抽汽式汽轮机
automatic fault detection 自动检测故障
automatic fault signaling 自动事故信号发送
automatic feeder equipment 自动馈电装置，自动送料设备
automatic feeder 自动供料机
automatic feeding 自动进料
automatic feed tray 自动给料盘
automatic feedwater by-pass equipment 给水自动旁路装置
automatic feed water by-pass system 给水自动旁路系统
automatic feed water by-pass 给水自动旁路装置
automatic feed water control 自动给水控制
automatic feed water regulator 自动给水调节器
automatic feed 自动馈给，自动进给，自动供料
automatic fidelity control 逼真度自动控制
automatic field break switch 自动磁场断路开关
automatic field forcing 强迫励磁，强行励磁
automatic field suppressing 自动灭磁
automatic field weakening 自动减弱磁场，自动减磁
automatic film processing machine 自动洗片机
automatic fine tuning 自动微调，自动细调
automatic fire alarm system 火灾自动报警系统，自动报火警系统
automatic fire-detection system 自动火警检测系统
automatic fire-extinguishing system 自动灭火系统
automatic fire pump 自动灭火泵
automatic fire signal 自动火灾信号
automatic fire sprinkler 自动喷洒灭火装置
automatic flap gate 自动活瓣闸门
automatic flap 自动襟翼
automatic flash board 自动启闭闸板
automatic flaw detection 自动探伤
automatic floating station 自动漂浮测站

automatic flow control 自动流量控制，流量自动调节，流量自动控制
automatic flushing cistern 自动冲洗水箱
automatic focusing 自动聚焦
automatic following control 自动跟踪控制
automatic following 自动跟踪
automatic fraction collector 自动分选机
automatic frequency control operation 频率自动控制运行
automatic-frequency-control synchronization 自动频率控制同步
automatic frequency control 自动频率控制
automatic frequency correction 自动频率校正
automatic frequency corrector 自动频率校正器
automatic frequency tuner 自动频率调谐器
automatic fuel control 自动燃料控制
automatic fuel drain valve 自动放油阀
automatic gain adjustment 自动增益调节
automatic gain control circuit 自动增益控制电路
automatic gain control 自动增益控制
automatic gain stabilization 自动增益稳定
automatic gas cutting 自动气割
automatic gas fired water heater 自动煤气燃烧的水加热器
automatic gate 自动闸门
automatic gauging machine 自动检验机
automatic gauging 自动检测
automatic gear gauge 自动齿轮检测计
automatic gear 自动装置
automatic generating plant 自动发电设备，自动化发电厂
automatic generating station 自动化发电站
automatic generation control 出力自动控制，自动发电控制
automatic generator 自动发电机，自动发生器，自控发电机，自控发生器
automatic gland packing 弹性汽封
automatic governing 自动调节，自动调速
automatic governor 自动调节器，自动调速器
automatic grid bias 栅压自动调节，自动栅偏压
automatic group controller 成组自动操作器
automatic grouping system 自动分组系统
automatic grouter 自动灌浆机
automatic hardness tester 自动硬度计
automatic high-voltage control 自动高压控制
automatic hoist 自动升降机，自动绞车
automatic holding device 自动保持装置
automatic hold 自保持，自动保持
automatic homing relay 自动复位继电器
automatic humidity controller 自动湿度调节器
automatic hydroelectric station 自动化水电厂
automatic hysteresis loop recorder 磁滞回线自动记录仪，磁滞回线自动记录器
automatic indexing 自动变址，自动编索引，自动索引
automatic indicator 自动指示器
automatic information processing 自动信息处理
automatic information retrieval system 自动情报检索系统
automatic input 自动输入，自动输入数据
automatic inspection 自动检查
automatic intercepting valve 自动截断阀
automatic interlocking device 自动联锁装置
automatic interlock 自动联锁，自动联锁装置
automatic internal diagnosis 自动诊断计算机内部故障，自动诊断内部故障
automatic interrupter 自动断续器
automatic interrupt 自动中断
automatic ionization chamber 自动电离室
automatic isolating valve 自动隔离阀
automaticity 自动化程度
automatic keying device 自动键控设备
automatic level control 自动液位控制，自动电平控制，液位自动控制
automatic level 自动化水平仪【测量学】，自动化水平
automatic light control 自动光量调节
automatic lighter 自动点火器
automatic line insulation tester 自动线路绝缘测试机
automatic line numbering 自动行编号
automatic line sectionalizer 自动线路分段器，自动线路分段开关，自动分段器
automatic line selection 自动选行
automatic load-frequency control 自动负载频率控制
automatic loading equipment 自动装料设备
automatic load limitation 自动减载装置，自动限载装置
automatic load regulating operation 负荷自动调整运行
automatic load regulator 自动负载调节器，自动加载调节器
automatic load-shedding control equipment 自动减负荷装置
automatic load-shedding device 自动减荷装置
automatic load-shedding equipment 自动减负荷装置
automatic local frequency control 自动本机频率控制
automatic locking device 自动闭锁装置
automatic logger 自动记录仪
automatic logging 自动记录
automatic logic testing and recording equipment 自动逻辑测试及记录装置
automatic loss-of-synchronism control equipment 失步自动控制装置
automatic loss-of-voltage tripping equipment 失压自动跳闸装置
automatic low frequency load shedding relay 自动低频减载继电器
automatic low vacuum regulator 低真空自动调节器
automatic lubrication installation 自动润滑装
automatic lubrication system 自动润滑系统
automatic lubrication 自动润滑
automatic machine 自动机
automatic main stop valve 自动主停汽门
automatic mathematical translator 自动数学方程求解计算机
automatic measurement 自动测量
automatic measuring device 自动测量器
automatic message accounting 自动信息计算

automatic message counting 自动信息计算
automatic message handling system 自动信息处理系统
automatic message recording 自动抄表
automatic message registering 自动信息记录
automatic message-switching center 自动电报交换中心
automatic message switching 自动信息转接
automatic meteorological observation station 自动气象观测站
automatic meter data exchange system 自动仪表数据交换系统
automatic meter reading technology 自动抄表技术
automatic micrometer 自动测微计
automatic mixture control 混合物成分自动控制
automatic modulation control 自动调制控制
automatic monitoring system 自动监测系统
automatic monitor 自动监测器，自动监测仪
automatic motor starter 电动机自动启动器
automatic noise canceller 自动消杂波电路
automatic noise limiter 自动噪声限制器
automatic noise suppressor 自动噪声抑制器
automatic non-return air release valve 自动止回排气阀
automatic nonreturn valve 自动逆止阀，自动止回阀
automatic nozzle 自动喷头
automatic number identification 发信号码的自动识别装置
automatic number normalization 数值自动规格化
automatic observer 自动观察记录仪
automatic oil switch 自动油开关
automatic oil temperature regulator 自动油温调节器
automatic operating and schedule program 自动操作和调度程序
automatic operating system 自动操作系统
automatic operation panel 自动操作仪表盘
automatic operation 自动操作，自动控制，自动运算
automatic Orsat 奥氏自动气体分析器
automatic oscillography 自动示波器
automatic output control 自动输出控制
automatic overload control 自动过载控制
automatic packaging 自动包装
automatic paralleling （线路的）自动并行，自动同期
automatic parallel off 自动解列
automatic particle counter 自动粒子计数器
automatic particle size analyzer 自动粒度分析仪
automatic performance 自动性能，自动操作性能，自动运行性能
automatic period control 周期自动控制
automatic phase comparison circuit 自动相位比较电路
automatic phase compensation 自动相位补偿
automatic phase control circuit 自动相位控制电路
automatic phase control 自动相位控制，相位自动控制，自动相位调整
automatic phase-locked loop 自动锁相环
automatic phase shifter 自动移相器
automatic phase synchronization 自动相位同步
automatic pH control pH 值自动控制
automatic photometer 自动光度计
automatic picture transmission system 自动图像传输系统
automatic picture transmission 自动图像传输
automatic pipet 自动吸液管，自动移液管
automatic plant coordinate control 机组自动协调控制
automatic plant 自动工厂，自动设备
automatic plotter 曲线自动描绘器，自动绘迹器
automatic plotting 自动绘图
automatic plugging meter 自动阻塞计
automatic positioning system 自动定位系统
automatic positioning 自动定位，自动对位
automatic potentiometer titrimeter 自动电势计式滴定计
automatic potentiometer 自动电位计，自动电势计
automatic power control 自动功率控制
automatic power factor regulator 功率因数自动调整器
automatic power plant 自动化发电厂
automatic power reduction 功率自动降低
automatic power station 自动化电站
automatic pressure control 自动压力控制
automatic pressure reducing valve 自动减压阀
automatic pressure suppression system 自动抑压系统，自动弛压系统
automatic pressure suppression 自动压力抑制
automatic primer 自动启动器，自动起爆装置
automatic printing 自动打印，自动印刷
automatic process cycle controller 过程周期自动控制器
automatic processing 自动处理
automatic production record system 生产自动记录系统，自动生产记录系统
automatic production 自动化生产
automatic program control 自动程序控制
automatic programmed checkout equipment 自动程序控制检查装置
automatic programming and data system 自动程序设计和数据处理系统
automatic programming and recording 自动程序设计和记录
automatic programming and test system 自动程序设计及试验系统
automatic programming tool 自动编程工具【一种自动数控编程语言】
automatic programming 自动编程序，自动程序设计
automatic program system 自动程序系统
automatic protective system 自动保护系统
automatic pulse reverse blowing device 自动脉冲反吹装置
automatic pumping system 自动抽水系统
automatic pump 自动水泵
automatic puncher 自动穿孔机
automatic punching machine 自动冲压机，自动

穿孔机
automatic pyrometer 自动指示高温计
automatic quasi-synchronization 自动准同步
automatic quick-changeover unit 自动快速切换装置
automatic ram pile driver 自动冲锤打桩机
automatic ratio control 自动比率控制，自动比例调节
automatic reactive power regulating operation 无功功率自动调整运行
automatic reading machine 自动读出器
automatic receptor 自动接收器
automatic reclosing circuit-breaker 自动重合闸断路器
automatic reclosing control equipment 自动重合闸装置
automatic reclosing equipment 自动重合设备
automatic reclosing relay 自动重合闸继电器
automatic reclosing switch 自动重合闸
automatic reclosing 自动重合，自动重合闸
automatic reclosure 自动重合闸
automatic recorder 自动记录仪，自动记录器
automatic recording device 自动记录器
automatic recording gage 自动记录仪表
automatic recording instrument 自动记录仪
automatic recording spectrophotometer 自动记录式分光光度计
automatic recording 自动记录
automatic recovery property 自动恢复特性
automatic rectifier 自动纠正仪
automatic reducing valve 自动调节阀，自动减压阀
automatic regulating device 自动调节设备，自动调节装置
automatic regulating system 自动调节系统
automatic regulating valve 自动调节阀
automatic regulation 自动调节，自动调整，自动控制
automatic regulator 自动调节器，自动调整器
automatic relay 自动中继
automatic release 自动释放
automatic relief valve 自动卸压阀
automatic remote-controlled plugging system 自动遥控堵塞系统
automatic remote-controlled plugging 自动遥控堵塞
automatic remote control 自动遥控
automatic reset relay 自动复位继电器
automatic reset 自动复位
automatic restart of HVDC power transmission system 高压直流输电系统自动再启动
automatic retrieval of incidents affecting nuclear reactors 核反应堆事故自动检索系统
automatic reverse current cutout 逆流自断器，反向电流自动断开开关
automatic reverser 自动反向器，自动转向开关
automatic reversible battery booster 蓄电池用自动可逆升压器
automatic reversing 自动倒转，自动反向
automatic rheostat 自动变阻器
automatic ringing 自动振铃，自动呼叫
automatic rod drop 自动落棒
automatic roll filter 自动卷绕式过滤器
automatic roll-type filter 自动旋滚式过滤器
automatic rotating-type filter 自动旋滚式过滤器
automatic routine 自动例行程序
automatics 自动装置，自动学
automatic safety switch 自动安全开关
automatic safety valve 自动安全阀
automatic sample changer 试样自动更换器
automatic sample handling system 样品自动处理系统
automatic sampler counter 自动取样计数器
automatic sampler 自动取样器
automatic sampling device 自动取样装置
automatic sampling system 自动取样系统
automatic sampling 自动采样
automatic scale command 自动定标命令
automatic scaler 自动定标器
automatic scale 自动秤，自动磅
automatic scan 自动扫描
automatic search circuit 自动搜索电路
automatic search 自动搜索
automatic segregator 自动分离器
automatic selection control 自动选择控制，自动局部控制
automatic selection 自动选择
automatic selectivity control 自动选择性控制
automatic self-verification 自动自检验
automatic sending 自动发送
automatic sensibility control 自动灵敏度控制
automatic sensitivity correction 自动灵敏度校正
automatic sequence computer 自动顺序计算机
automatic-sequence-controlled starting 自动顺序控制启动
automatic sequence control 自动程序控制，自动顺序控制，自动序列控制
automatic sequences 自动时序
automatic sequential operation 自动顺序操作
automatic sequential soot blower 自动顺序吹灰器
automatic sequential starting circuit 自动顺序启动电路
automatic service 自动服务
automatic servo plotter 自动伺服作图仪
automatic setting of control point 控制点的自动设置
automatic sextant 自动六分仪
automatic shim rod follow-up 补偿棒自动跟踪
automatic shutdown system 自动停机系统，自动停反应堆系统
automatic shutdown 自动停止，自动断开
automatic shut-off circuit 自动闭锁电路
automatic shut-off device 自动关闭装置
automatic shut-off valve 自动断流阀
automatic shut-off 自动断流，自动关闭
automatic shutting-down 自动关机
automatic shut valve 自动关闭阀
automatic side-dump muck car 自动侧卸出渣车
automatic signal 自动信号
automatic signaling 自动发信号
automatic signal lamp 自动信号灯

automatic siphon 自动虹吸管
automatic sizing 自动控制尺寸
automatic slag pool welding 自动电渣焊
automatic smoke alarm system 自动起烟报警系统
automatic smoke sampler 自动烟尘取样器
automatic sorting 自动分类
automatic source data 自动源数据
automatic spark timer 自动点火定时器
automatic speech recognition 自动语言识别
automatic speed adjustment 自动速度调节
automatic speed controller 自动调速器
automatic speed run up 自动升速
automatic spillway gate 自动泄水闸门
automatic spillway 自动溢洪道
automatic spiral ratchet screwdriver 自动螺旋齿杆起子
automatic sprinkler and fire protection system 自动火警检测、灭火系统
automatic sprinkler system 自动喷水管路，自动洒水系统
automatic sprinkler 自动喷灌机，自动喷水器，自动人工降雨机
automatic square rooting 自动开平方
automatic stability 自动稳定性
automatic stabilization and control system 自动稳定和控制系统
automatic stand-by control 备用设备自动控制
automatic start and stop device 自启停装置
automatic starter 自动启动器
automatic starting control 自动启动控制
automatic starting device 自动启动装置
automatic starting gear 自启动装置
automatic starting motor 自动启动马达
automatic starting relay 自动启动继电器
automatic start-up and shut-down control 自启停控制
automatic start-up and shut-down 自启停
automatic start 自启动
automatic station 自动化电站，无人值班电站，自动（水力）发电站
automatic steam-temperature control 自动蒸汽温度控制
automatic steam trap 自动疏水器
automatic step-regulator 自动分级调压器
automatic stop and check valve 自动截止和逆止阀，主汽门
automatic stop arrangement 自动停机装置
automatic stopping device 自动停止装置
automatic stop valve 自动停止阀，自动断流阀
automatic stop 自动停机装置，自动停机，自动停车
automatic storage allocation 自动存储器配置［分配］
automatic submerged arc welding machine 埋弧自动焊机
automatic submerged arc weld 自动埋弧焊
automatic substation 自动变电所
automatic supplementary air valve 自动补充空气阀
automatic surface examination 自动表面检查
automatic switchboard 自动交换机，自动装置配电盘
automatic switching control equipment 自动切换装置
automatic switching 自动交换
automatic switch-over 自动切换，自动转接
automatic switch 自动开关，自动断路器
automatic synchronization unit 自整步装置
automatic synchronization 自动同步
automatic synchronized control 自动同步控制
automatic synchronized discriminator 自动同步鉴别器
automatic synchronized system 自动同期系统，自动同期装置
automatic synchronizer 自动同步器，自动整步装置，自动同步机
automatic synchronizing indicator 自动同步指示器
automatic synchronizing relay 自动同步继电器，自同期继电器
automatic synchronizing system 自动同步系统
automatic synchronizing unit 自动同步装置
automatic system 自动系统
automatic tap-changing device 变压器分接头自动切换开关
automatic telephone room 自动电话室
automatic teleswitch 自动远动开关，自动遥控开关
automatic temperature compensator 自动温度补偿器
automatic temperature recorder 温度自动记录器
automatic temperature regulator 自动温度调节器
automatic tension balancer 自动张力平衡器
automatic tension regulator 自动电压调整器，自动张力补偿器
automatic terminal information service 自动终端信息服务
automatic terminal information 自动终端信息
automatic test button 自动测试按钮
automatic test equipment 自动测试设备，自动测试系统，自动测试装置
automatic throttle valve 自动节流阀
automatic throttle 自动节流，自动节流阀
automatic throw-over equipment 自动转换装置
automatic throw-over substation 自动切换变电所
automatic throw-over switch 自动转换开关
automatic throw-over 自动切换
automatic tide recorder 自动潮位记录计
automatic tilting gate 自动倾侧闸门，自动倾转闸门
automatic timed magneto 自动定时磁电机
automatic time element compensator 自动延时补偿器
automatic timer 自动程序装置，自动计时器，自动定时器
automatic time switch 自动定时开关
automatic timing control 自动定时控制
automatic timing corrector 自动时间校正器
automatic tipper 自动倾卸车，自动卸料车
automatic titration 自动滴定
automatic toad regulator 自动负载调节器

automatic totalling 自动总和
automatic track-follower 自动跟踪装置
automatic tracking 自动跟踪
automatic train control system 自动序列控制系统
automatic transfer equipment 自动转换装置
automatic transfer of auxiliaries （电厂）辅机自动切换
automatic transfer of title 所有权自动转移
automatic transformer switch 变压器自动开关【无负载时自动切换】
automatic transform system 厂用电源快速切换装置
automatic translation 自动翻译
automatic translator 自动翻译机，自动转换器
automatic transmitter 自动发送机
automatic triggering 自动触发，自动启动线路
automatic trimming 自动频率微调
automatic tripping 自动脱扣，自动跳闸，自动切断
automatic trouble diagnosis 自动故障诊断
automatic trouble locating arrangement 自动找寻故障点装置
automatic trouble locating 自动寻找故障
automatic tuning control 自动调谐控制
automatic tuning system 自动调谐系统
automatic turbine control 汽轮机自动控制
automatic turbine run up equipment 汽轮机自动升速设备
automatic turbine start system 汽轮机自动启动系统
automatic turbine startup and shutdown control system 汽轮机自启停控制系统
automatic turbine start-up program 汽轮机自动启动程序
automatic turbine start-up 汽机自动启动
automatic turnoff 自动断开
automatic turntable 自动转台
automatic typewriter 自动打印机，自动打字机
automatic unloading 自动去载
automatic valve 自动阀
automatic ventilation 自动通风
automatic vent 自动放气阀
automatic voltage control 自动电压控制
automatic voltage regulating operation 电压自动调整运行
automatic voltage regulator 自动电压调节器，自动稳压器，自动调压器
automatic volume controller 自动音量控制器
automatic volume control 自动音量控制
automatic warning device 自动报警装置［设备］，自动信号设备
automatic washing filter 自动冲洗滤池
automatic water level recorder 自动水位计
automatic water quality monitor system 水质自动监测系统
automatic water quality monitor 水质自动监测器
automatic water-stage recorder 自记水位计
automatic water treatment plant 自动水处理厂
automatic weather station 自动气象站
automatic weigh-batcher 自动配料秤
automatic weigher 自动磅秤，自动秤
automatic weir 自动堰
automatic welder 自动焊机
automatic welding equipment 自动焊装备
automatic welding machine 自动焊机
automatic welding 自动焊接，自动焊
automatic winch 自动绞车，自动升降机
automatic winding 自动绕线
automatic zero buret 自动复零滴定管
automatic zero set 自动调零
automatic 自动的，自动装置的，自动（化）的，自动装置
automation degree 自动化程度
automation dispatch system 自动调度系统
automation of hydroelectric power station 水力发电站自动化
automation procedure 自动化工序，自动化作业
automation shunting yard 自动化调车场
automation system at plant level 厂级自动化系统
automation 自动化，自动装置，自动学，自动器，自动操作
automatism 自动性，自动作用
automatization(-sation) 自动化
auto-metamorphism 自变质作用
automobile cable 汽车电缆
automobile crane 汽车式起重机
automobile current generator 汽车发电机
automobile insurance 汽车保险
automobile liability insurance 机动车责任险
automobile 机动车，汽车
automodulation 自调制
automonitor routine 自动监督程序
automonitor 自动监控器，自动监测器
automorphic granular 自形粒状
automotive gas turbine 汽车用燃气轮机
automotive 机动的，汽车的
automotor-generator 自电动发电机【两电枢串联直-直流变换】
autonomous channel 自组通道，独立通道
autonomous circuit 自激电路
autonomous control 自治调节，自律调节
autonomous country 自治县
autonomous municipality 自治市（区）
autonomous operation 自主运算
autonomous oscillation 自激振荡，自持振荡
autonomous power generator 自励发电机
autonomous prefecture 自治州
autonomous region 自治区
autonomous sequential circuit 自激时序电路
autonomous system 独立系统，自治系统，自主系统
autonomous 自备的，自给的，自持的，自激的
auto-operation 自动操作
autooscillation 自振
autooxidation 自动氧化
autophotoelectric effect 自生光电效应
autopilot 自动引燃器
auto-plant 自动装置，自动设备，自动化工厂
autoplotter 自动绘图机
autopneumatic circuit breaker 自动气动断路器，自动气动开关
auto-power spectral density 自功率谱密度

auto-power spectrum 自功率谱
AUTOPSY（automatic operating system） 自动操作系统
autopulse 自动脉冲
autopurification 自动净化
auto-radar 自动雷达
autoradiography 放射自显影法
autoreclose circuit-breaker 自动重合闸断路器
autoreclosing cycle 自动接通周期，自动重合周期，自动重合闸循环
auto-reclosing dead time 自动重合闸无电流时间
autoreclosing 自动重合闸
autoreclosure 自适应重合闸，自动重合闸
autorecorder 自动记录仪
autoregression 自回归
autoregressive model 自回归模型
autoregressive moving average model 自回归滑动平均模型
autoregulation induction heater 自动调节感应加热器
autoregulation 自动调节
auto-remaking 自动重接通
auto-repeater 自动重发器，自动替续增音器
autorestart 自动再启动
auto-revolving documentary letter of credit 自动旋转的跟单信用证
auto-room 自动室，自动交换机室
autorotation rate 自旋速率
autorotation 自转，自旋
autoscope 点火检查示波器
auto-selector system 主动选择系统
auto self-excitation 自动自励磁
autosevocom（automatic secure voice communication）自动安全电话通信
autosizing 自动尺寸监控
autospectral analysis 自谱分析仪
autospectral density 自谱密度
autospectrum 自谱
auto-spray 自动喷洒器
autostability 自稳定性
autostabilization of rock slide 岩石滑坡的自稳
autostabilization 自动稳定，自稳作用
autostabilizer 自动稳定器
autostarter transformer 自启动用变压器，自耦启动变压器
autostarter 自动启动机，自动启动器，自耦变压器式启动器
auto start test 自动启动试验
auto start up and supervision 自动启动和监视
auto stiletto and binding machine 自动打孔装订机
auto-stop 自动停止，自动停机，自停机装置
auto-stressing 自应力
auto-submerged arc welding machine 自动埋弧焊机
auto-switch 自动开关
autosynchronous motor 自同步电动机，同步感应电动机
autosynchronous network 自动同步网络
autosynchronous 自同步的
autosyn 自动同步的，自动同步机，交流同步器，自整角机
autotelegraph 电写，电传真机
auto tempering 自发回火
autothermic 热自动补偿的，自供热的
autotimer 自动计时器
autotrack 自动跟踪
auto-transductor 自耦磁放大器
autotransformer-coupled oscillator 自耦变压器反馈振荡器
autotransformer power supply system 自耦变压器供电系统
auto-transformer starter 自耦变压器启动器
auto-transformer 自耦变压器
autotruck 载重汽车
autotune 自动调谐
auto-type transformer 自耦式变压器
auto-unloading 自动卸荷
autovac 真空罐
autovalve arrester 自动阀型避雷器
autovalve lightning arrester 自动阀型避雷器
autovalve 自动阀
autoverify 自动检验
auto 汽车，自动（装置）
AUTO 自动【DCS 画面】
autumnal equinox 秋分
autumn equinox tide 秋分潮
autumn flood 秋季洪水，秋汛
aux. ＝auxiliary 辅助的
auxiliaries-supply circuit-breaker 辅助电源断路器，备用电源断路器
auxiliaries 辅助设备，辅助装置，辅机，附属设备，辅件
auxiliary accessory 辅助设备
auxiliary actuating device 辅助传动装置
auxiliary aerodynamic surface 辅助气动作用面
auxiliary air 辅助空气，二次空气，辅助进风
auxiliary alarm 辅助报警器，备用报警器
auxiliary alternator 厂用交流发电机，辅助交流发电机
auxiliary apparatus 辅助装置
auxiliary arc 辅助弧
auxiliary balance 辅助平衡
auxiliary bar 辅助钢筋，辅助杆
auxiliary base line 辅助基线
auxiliary battery power supply 辅助动力供应电池
auxiliary bed 副床【沸腾炉】
auxiliary board 场用配电盘，厂用配电盘
auxiliary body 附属机构
auxiliary boiler feeder system 辅助锅炉给水系统
auxiliary boiler 辅助锅炉，启动锅炉，厂用锅炉
auxiliary break-contact 辅助断路触点
auxiliary breaker 辅助断路器
auxiliary bridge 辅助吊桥【燃料处理】
auxiliary building nonradioactive ventilation system 附属［辅助］厂房非放射性通风系统
auxiliary building sump pump 辅助厂房地坑泵
auxiliary building transformer 辅助厂房变压器
auxiliary building 附属建筑物，附属建筑，辅助厂房
auxiliary burner 辅助喷燃器

auxiliary bus-bar joint 小母线端子，辅助母线端子
auxiliary bus-bar 辅助母线
auxiliary cathode 辅助阴极，副阴极
auxiliary changeover valve 辅助切换阀
auxiliary circuit 辅助电路
auxiliary coil 辅助线圈
auxiliary condition 附加条件，补充条件，辅助条件
auxiliary conductor 辅助导线
auxiliary console 辅助控制（操作）台
auxiliary contactor 辅助接触器
auxiliary contact 辅助触点，副触点，联锁触点，辅助接点
auxiliary contour 辅助等高线
auxiliary control coil 辅助控制线圈
auxiliary controller 辅助控制器
auxiliary control panel 辅助的控制仪表板，辅助控制板
auxiliary control rod 辅助控制棒
auxiliary coolant system 辅助冷却系统
auxiliary cooling water system 辅助冷却水系统
auxiliary core cooling 堆芯辅助冷却
auxiliary counter 辅助计数器
auxiliary crane 辅助起重机
auxiliary crossarm 辅助横担
auxiliary CT 辅助电流互感器
auxiliary current transformer 辅助电流互感器
auxiliary current 辅助电流
auxiliary curve 辅助曲线
auxiliary dam 副坝
auxiliary data processing system 辅助数据处理系统
auxiliary device 辅助装置［仪表］
auxiliary drive turbine 辅机驱动汽轮机
auxiliary drive 辅助驱动装置
auxiliary electric drive 辅助电力传动装置
auxiliary electric power supply 配套供电工程
auxiliary electrode 辅助电极
auxiliary elevation 辅助正视图
auxiliary energy source 辅助能源
auxiliary engine 辅助动力机
auxiliary entrance 辅助入口
auxiliary equipment control panel 辅助设备控制盘
auxiliary equipment department 辅助设备间
auxiliary equipment of convertor station 换流站辅助设施
auxiliary equipment 辅助设备，附属设备，辅机，外围设备，外部设备
auxiliary excitation 辅助励磁，备用励磁
auxiliary exciter 辅助励磁机，副励磁机
auxiliary exciting winding 辅助励磁绕组，备用励磁绕组
auxiliary fan 辅助风机
auxiliary fault 副断层
auxiliary feeder 辅助给水泵，辅助馈线
auxiliary feedwater pump room 辅助给水泵间
auxiliary feedwater pumps block 辅助给水泵房
auxiliary feedwater pump system 辅助给水泵系统
auxiliary feedwater pump 辅助给水泵，事故给水泵
auxiliary feedwater storage tank 辅助给水贮存箱
auxiliary feedwater supply 辅助给水供给【蒸汽发生器】
auxiliary feedwater system 辅助给水系统【蒸发器】
auxiliary feedwater 辅助给水，事故给水
auxiliary field 辅助场
auxiliary fire vehicle 后援消防车
auxiliary flap 辅助襟翼
auxiliary flow 附加气流，副流
auxiliary follow-up piston 辅助随动活塞
auxiliary force 辅助力
auxiliary fuse 辅助熔断器，辅助保险丝
auxiliary gearbox 辅助齿轮箱
auxiliary generator set 辅助发电机组，备用发电机组
auxiliary generator 厂用交流发电机，辅助发电机
auxiliary governor 辅助调节器
auxiliary grounding 辅助接地
auxiliary haulage 辅助运输，中间运输
auxiliary heater 辅助加热器
auxiliary heating surface 附加受热面，辅助受热面
auxiliary heat source 辅助热源
auxiliary helium circuit 辅助氦气回路
auxiliary hoist 辅助起重机，辅助提升机【燃料装卸】
auxiliary illumination 辅助照明
auxiliary impulse 辅助脉冲
auxiliary information 辅助信息
auxiliary instrument 辅助仪器
auxiliary investment 辅助投资
auxiliary jet 辅助喷口
auxiliary lead 辅助引线
auxiliary letter of credit 附属信用证
auxiliary lighting 辅助照明
auxiliary line constant 辅助线路常数
auxiliary line 辅助线
auxiliary live steam admission 辅助新汽进汽阀
auxiliary loading air 辅助充填空气
auxiliary loading force 辅助载荷力
auxiliary load 辅助负荷，厂用电
auxiliary loop 辅助回路，辅助环路
auxiliary lubrication pump 辅助油泵
auxiliary machinery 辅助机械，副机
auxiliary machine 辅机
auxiliary mass curve 副积线
auxiliary means 辅助工具
auxiliary measure 辅助措施
auxiliary memory 辅助存储器
auxiliary menus 辅助菜单
auxiliary motor-pump assembly 辅助电动泵组
auxiliary motor 辅助电动机
auxiliary oil lift pump 辅助油提升泵
auxiliary oil pump 辅助油泵
auxiliary operation 辅助操作
auxiliary operator 辅助操作员，副操纵员
auxiliary panel 辅助控制盘，辅盘
auxiliary personnel 辅助人员

auxiliary phase 辅助相，备用相
auxiliary pilot valve 辅助滑阀，辅助错油门
auxiliary piping 辅助管路
auxiliary plant 辅助厂房，辅助设备，辅助装置，辅助车间
auxiliary point 辅助点，补充点
auxiliary pole 附加极，辅助磁极，换向极，副极，辅助极
auxiliary power busbar half 厂用电半母线
auxiliary power consumption 厂用电耗，厂用电量，厂用电耗量
auxiliary power house 副厂房
auxiliary power load 厂用电负荷
auxiliary power plant 辅助电源设备，辅助电厂
auxiliary power rate of entire power plant 全电厂厂用电率
auxiliary power rate 厂用电率
auxiliary power ratio 厂用电率，自用电比例
auxiliary power source 辅助电源，厂用电源
auxiliary power station 备用发电厂，备用电源，辅助电站
auxiliary power supply system 辅助供电系统
auxiliary power supply 辅助电源，备用电源，自用电源，自备供电设备
auxiliary power system engergized from the grid 厂用电受电
auxiliary power system 辅助供电系统，厂用电系统
auxiliary power transformer 备用电力变压器，辅助变压器，厂用变压器
auxiliary power unit 辅助动力装置，辅助电源装置，辅助供电装置
auxiliary power wiring 厂用电接线
auxiliary power 厂用电
auxiliary production 辅助生产
auxiliary project 附属工程
auxiliary pump 辅助泵，辅助水泵
auxiliary reinforcement 辅助钢筋
auxiliary relay 辅助继电器，中间继电器
auxiliary report 辅助报告
auxiliary resistance 辅助电阻
auxiliary rod holdout coil 辅助棒保持线圈，辅助棒闭锁线圈【控制棒驱动机构】
auxiliary rod holdout device 辅助棒锁定装置，棒组件机构联锁装置【控制棒驱动机构】
auxiliary room 辅助厂房
auxiliary routine 辅助程序
auxiliary safety device 辅助安全设备
auxiliary section 辅助断面
auxiliary service transformer 辅助供电变压器
auxiliary service 辅助设施
auxiliary servomotor 辅助伺服马达
auxiliary set 辅机
auxiliary shaft 副井
auxiliary shop 辅助车间
auxiliary signal 辅助信号
auxiliary spark-gap 辅助火花隙
auxiliary specifications 辅助规范
auxiliary spillway 辅助溢洪道
auxiliary spray line 辅助喷淋管线
auxiliary spray valve 辅助喷淋阀
auxiliary staff gauge 辅助水尺
auxiliary stairs 便梯，辅助楼梯
auxiliary starting winding 附加启动绕组
auxiliary station 辅助电厂，备用电厂，辅助测站，副厂房
auxiliary stay 辅助拉线，辅助撑条
auxiliary steam and condensate return system 厂用蒸汽与凝结水回流系统
auxiliary steam distribution system 辅助蒸汽分配系统
auxiliary steam header 辅助蒸汽母管，辅助蒸汽联箱
auxiliary steam manifold 厂用蒸汽总管
auxiliary steam supply system 厂用蒸汽供应系统
auxiliary steam system 厂用蒸汽系统，辅助蒸汽系统
auxiliary steam 辅助蒸汽，电站自用蒸汽，厂用蒸汽
auxiliary storage allocation table 辅助存储分配表
auxiliary storage 辅助存储器，备用存储器
auxiliary structure 辅助建筑，辅助结构，辅助构筑物
auxiliary substation 备用变电所
auxiliary supply 辅助水源
auxiliary switchboard 辅助开关盘，厂用配电盘，厂用开关盘，辅助配电盘，站用配电屏
auxiliary switchgear 厂用配电装置
auxiliary switching power supply 机内辅助开关电源
auxiliary switchyard 厂用辅助配电盘，厂用辅助配电装置，厂用配电装置
auxiliary switch 辅助接点，辅助开关
auxiliary system open loop control 辅助系统开环控制（回路）
auxiliary system 辅机系统
auxiliary team system 厂用蒸汽系统
auxiliary telescope 辅助望远镜
auxiliary test unit 辅助测试装置
auxiliary thermal source 辅助热源
auxiliary traction motor 牵引辅助电机
auxiliary trailer 辅助拖车
auxiliary transformer for neutral grounded system 中性点辅助变压器接地系统
auxiliary transformer 厂用（工作）变压器，辅助变压器，站用变压器
auxiliary turbine 辅助汽轮机，辅助透平
auxiliary valve 辅助阀，辅助阀门
auxiliary view 辅助视图
auxiliary voltage 辅助电压
auxiliary water meter 辅助水表
auxiliary water pump 自用水泵
auxiliary water ratio 自用水率
auxiliary water supply 辅助供水
auxiliary winding 辅助绕组，附加绕组，辅助线圈
auxiliary windshield wiper 副刮水器，辅助雨刷
auxiliary work 辅助工程，附属工程
auxiliary 辅机，辅助设备，附属设备，辅助的，补充的，副的，附加的
auxiliary worker 辅助工

AUX＝auxiliary 辅助的【DCS画面】
AV(actual velocity) 实际速度
availability coefficient 可利用系数
availability factor 可用系数，可利用系数，可利用率，可用率，运转系数，投运率，使用系数
availability guarantee 可用性保证（值）
availability improvement 可利用率的改进
availability of a power plant 电厂可用率
availability of automatic control device 自控投入率，自控装置投入率
availability of equipment 设备利用率，设备完好率
availability of resources 可用资源，资源可用（性）
availability ratio （装置的）可利用率，有效度比
availability time ratio 时间可利用系数
availability 可用率，有效性，可用性，可利用率，可得到，可达性
available accuracy 实际精确度，有效精度
available acreage 可利用的土地面积
available berth 可利用泊位，现有码头
available capacity 可用功率，可用容量，有效容量，有效容积，有效库容，可用发电量
available channel capacity 现有［可利用］航道通过能力
available characteristics 有效特性
available coal-yard acreage 可利用的煤场面积
available cross section 有效面积
available data 可用资料，现有资料，已有资料
available deposition velocity 有效沉积速度
available device table 可用设备表
available diffusion time 有效扩散时间
available dilution 有效稀释（作用），有效稀释率，有效稀释度
available discharge 可用流量
available draft 有效通风压头
available energy output 有效输出能量，可用发电量
available energy 可用能，现有能量，有效能，可用能量，有效能量
available facility 现有设备，原有设备
available factor 可用系数，可利用系数，可用率
available for use 可加以利用的
available funds 获得的资金
available groundwater 可用地下水，有效地下水
available head 有效压头，可用压头，有效水头，可用水头
available heat drop 有效热降，可用热降
available heating value 可用热值，有效热值
available heat transfer coefficient 有效传热系数
available heat transfer 有效传热
available heat 可用热，有效热
available horsepower 可用马力，有效容量，有效功率，可用功率
available hours 可用小时，可用时数，有效小时数
available impulse 有效冲量
available information 能得到的资料，已有资料
available input power 有效输入功率
available life 可用期，有效寿命
available line 可用扫描线
available load 有效负荷，可用负荷
available machine time 计算机的正常运算时间，机器有效工作时间，机器可用时间
available moisture capacity 有效含水量
available moisture 有效水分
available net head 净可用水头
available net positive suction head 有效净正吸入压头，有效气蚀余量
available NPSH 有效气蚀裕量
available on demand 可随时索取
available output 有效输出
available oxygen 有效氧
available power curve 有效功率曲线
available power efficiency 有效功率效率，可用功率效率
available power loss 有效功率损耗，有功损耗
available power 可用功率，有效功率
available precipitation 可用降水
available pressure head in the consumer 用户预留压头
available pressure 可用压力，资用压力
available runoff 可用径流
available soil moisture 土内有效水分，有效土壤水分
available state 可用状态
available static head 有效静压头，运动压头
available storage capacity 有效库容，有效储量
available storage 可用库容，有效库容
available surface 有效面积
available technology 可用的技术
available thermal power 有效热功率
available thrust 有效推力，可用推力
available time 有效工作时间，开机时间，可用时间，有效时间
available to do 可用来做
available to M M可以采用的，可以得到的
available transfer capability 可用输电容量
available velocity 有效速度，可用速度
available volume 可利用体积，有效容积
available water capacity 有效水容量，有效含水量
available water in soil 土内有效水分
available water supply 有效供水，可利用水源
available water 有效水分，可用水
available wharves 可供用的码头【wharves为wharf的复数】
available width of navigable channel 航道有效宽度
available wind energy 有效风能，可用风能
available wind power 有效风能，可用风能
available wind 有效风
available work 可用功
available 有效的，现有的，可达到的，可利用的，可得到的
availably 有效地
avalanche alarm 崩坍警报
avalanche baffle 塌方防御设施
avalanche blast 气浪
avalanche control works 防坍工程
avalanche dam 崩坍形成的坝
avalanche defence 崩塌防护，防塌

avalanche observation station 崩塌观测站
avalanche of ionization 电离雪崩
avalanche prevention works 防坍工程
avalanche prevention 防坍
avalanche SCR 雪崩可控硅整流元件
avalanche trigger zone 崩坍触发区
avalanche zone 崩坍区
avalanche 崩坍，崩塌，雪崩
avalite 铬云母
AV(angle valve) 角阀
avant port 前港
avasite 硅铁矿
AVC(automatic voltage control) 自动电压控制
avenue 大街，大道，林荫道，林荫路
average 平均，平均的，普通的，平均值，平均数，求平均，取平均值，均分
average absolute error 平均绝对误差
average absolute value 平均绝对值
average absorbing coefficient 平均吸收系数
average absorbing power 平均吸收功率
average access time 平均选取时间，平均取数时间，平均存取时间
average active power 平均有功功率
average ambient temperature over the years 多年平均环境温度
average annual capacity factor 年平均设备利用率
average annual cost 平均年费用
average annual damage index 年平均损失指数
average annual discharge 年平均流量
average annual energy output 多年平均年发电量，年平均发电量
average annual energy 多年平均年发电量
average annual flood 平均年洪水量
average annual flow 多年平均流量
average annual loss 年平均损失
average annual output 年平均发电量，年平均产量
average annual precipitation 多年平均降水量，平均年降水量
average annual rainfall 年平均降水量，年平均降雨量
average annual runoff 平均年径流量
average annual temperature 年平均温度
average annual times of freezing-thawing cycle 年平均冻融循环次数
average annual yield 多年平均产量
average arc life 平均电弧寿命
average atmospheric pressure over the years 多年平均气压
average available discharge 平均可用流量
average available hour of equipment 设备平均利用小时
average available hour for power generation 发电设备平均利用小时
average benefit 平均效益
average blade ring diameter 叶片平均直径
average blading diameter 叶片平均直径
average bond stress 平均黏结应力
average burn-up 平均燃耗
average calculating operation 平均运算
average calibration factor 平均校准因子
average capacity factor 平均利用率
average capacity standard 平均容量标准
average capacity 平均容量
average capital plus interest 等额本息
average capital 等额本金
average channel 平均通道
average closing error 平均闭合差
average closure 平均闭合差
average cold end temperature 平均冷端温度
average computing device 平均值计算装置
average conductivity 平均电导率
average coordination number 平均配位数
average core burn-up 堆芯平均燃耗
average core exit quality 平均堆芯出口含汽率
average core transit time 平均堆芯通过时间
average corrosion rate 平均腐蚀速率
average cost 平均费用，平均成本
average current 平均电流
average daily consumption of water 平均日用水量
average daily-efficiency 平均日效率【太阳光热利用】
average daily flow 平均日流量
average daily precipitation 日平均降水量
average daily sol-air temperature 日平均综合温度
average daily traffic 平均日交通量
average demand 平均需要量
average deviation 平均偏差，平均漂移
average diameter 平均直径
average dimension 平均尺度
average discharge 平均流量
average divergence 平均散度
averaged time 平均时间
average duration curve of water level 水位平均历时曲线
average duration curve 平均历时曲线
average efficiency 平均效率，平均有效系数，平均生产率
average elevation 平均海拔高度
average energy expended per ion pair formed 形成单位离子偶消耗的平均值
average energy loss per ion pair formed 形成每对离子平均损失的能量
average energy not supplied （用户）平均停电缺供电量
average energy per ion pair formed 每形成一对离子平均能量
average enrichment 平均富集度，平均浓缩度
average equivalent availability factor 平均等效可用系数
average error 平均误差
average evaporation capacity 平均蒸发量
average evaporation over the years 多年平均蒸发量
average firm output power 平均可靠出力
average fixed cost 平均固定成本
average flow duration curve 平均流量历时曲线
average flow of domestic hot water supply 热水供应平均流量
average flow period 平水期

average flow rate 平均流量率，平均流量
average flow velocity 平均流速
average flow 平均流量
average fuel duty 燃料平均消耗量
average fuel enrichment 燃料平均富集度
average fuel temperature 燃料平均温度
average full-load power 平均全负荷功率
average gradient 平均坡降，平均比降
average grading 平均级配
average grain diameter 平均粒径
average grain size 平均粒径，平均晶粒度
average groundwater velocity 地下水平均流速
average haul 平均运程，平均运距
average head 平均水头
average heavy swell load 七级涌浪
average height of surface projections 表面凸起平均高度
average hot-water heating load 热水供应平均热负荷
average hourly steam consumption 平均小时耗汽量
average ideal efficiency 平均理想效率
average illumination 平均照明度
average infiltration capacity 平均渗入量
average infiltration 平均渗透率
average information 平均信息量
average instruction time 平均指令时间
average interruption duration 故障停电平均持续时间
average interruption hours of customer 用户平均停电时间
average interruption times of customer 用户平均停电次数
average irradiance 平均辐（射）照度
average life expectancy 平均寿命预期值
average lifetime 平均使用期限，平均寿期
average life 平均使用年限，平均寿命，平均寿期
average load curve 平均负荷曲线
average load 平均负荷，平均负载
average logarithmic energy decrement 平均对数能降，平均能量的对数衰减量
average loss of energy 平均能量损失
average low water flow 平均低水流量
average magner 平均无功功率
average maximum demand 平均最大需量
average moderate swell 平均中涌
average monthly humidity 月平均湿度
average monthly rainfall 月平均降水量，月平均雨量
average monthly wind speed 月平均风速
average navigation period 平均通航期
average noise level 平均噪声级
average number of neutrons emitted per fission 每次裂变发出的平均中子数
average number 平均数
average of the daily load curve 平均日负荷曲线
average operation time 平均运行时间
average outgoing quality limit 平均出厂质量界限[极限]，平均出厂合格率限度，平均质量抽查极限
average outgoing quality 平均抽检质量，平均出厂质量，平均逸出量
average output power 平均输出功率
average output 平均输出，平均出力，平均产量
average overall efficiency 平均总效率
average overall plant efficiency 电站平均总效率
average peak load 平均最大负荷
average phasor power 平均相量功率
average potential water power 平均水力蕴藏量
average power curve 平均出力曲线
average power demand 平均需电量
average power density 平均功率密度
average power factor 平均功率因数
average power range monitoring 平均功率区段监测
average power range monitor 平均功率区段监测器
average power 平均功率
average pressure 平均压力
average price 平均价格
average probability 平均概率
average productive capacity 平均生产量，平均生产能力
average quasi-peak value 准峰值的平均值，平均准峰值
average rainfall 平均雨量
average rate of convergence 平均收敛速度
average rate of filtering 平均滤速
average reactive factor 平均无功因数
average reactive power 平均无功功率
average reactor temperature rise 反应堆平均温升
average reactor temperature 反应堆平均温度
average relative humidity over the years 多年平均相对湿度
average resistivity 平均电阻率
average revenue 平均收入，平均收益
average risk 平均风险
averager 平均器，均衡器，中和器
average sample number 平均取样数，平均样品数
average sample 平均样品
average service conditions 一般使用条件，正常使用条件
average service life 平均使用年限
average service rate 平均使用率
average sewage 一般污水，中等程度污水
average size 平均尺寸
average-sized motor 中型电动机
average size of aggregate 骨料平均粒径
average slope 平均坡度，（河道的）平均比降
average sound absorption coefficient 平均吸声系数
average sound pressure level 平均声压级
average space-heating load during heating period 供暖期采暖平均热负荷
average specific frictional head loss 平均比摩阻
average specific heat 平均比热
average speed 平均速度，平均流速
average station efficiency 电站平均效率
average stream flow 平均河流流量，平均流量

average stress 平均应力
average sunshine time 平均日照时间
average superficial velocity of two-phase mixture 两相混合物平均表观速度
average tare 平均皮重
average temperature of the coldest month 最冷月平均温度
average temperature 平均温度
average term 一般项
average thermal conductivity （燃料）平均热导率
average thermal neutron cross-section 热中子平均截面
average thickness 平均厚度
average thunder day over the years 多年平均雷暴日数
average tidal curve 平均潮位曲线
average time between maintenance 平均维修间隔时间
average time to repair 平均修复时间
average time 平均时间
average total inspection 全数检验平均值，全部检验平均值
average transfer time 平均传送时间
average transmission rate 平均传输率
average trip mileage 平均旅程，平均里程
average unit charge 平均电价
average unit graph 平均单位过程线
average unit price 平均单位价格
average useful life 平均使用期限，平均使用寿命
average value of contamination 污染平均值
average value 平均值
average variability 平均变率
average vector power 平均矢量功率
average velocity 平均速度
average void content 平均含汽量
average voltage 平均电压
average water level 平均水位
average-weighted 加权平均的
average wind direction 平均风向
average wind velocity 平均风速
average work load 平均工作负荷
average yearly rainfall 平均年降雨量
averaging amplifier 平均放大器
averaging method 求平均值法
averaging operator 求平均算子
averaging rectifier 平均整流器
averaging spectrum 平均能谱
averaging time 风速的时距，平均时间
avertable 可避开的，可防止的
avertence 斜流
AVF(available factor) 利用系数
AVG=average 平均
AVGDIA(average diameter) 平均直径
aviation gas turbine engine 航空燃气轮机发动机
aviation industry 航空工业
aviation insurance 航空保险
aviation liability insurance 航空责任保险
aviation turbine 航空涡轮
avidity 活动性
AVL=available 可用的
Avogadro's law 阿伏伽德罗定律
Avogadro's number 阿伏伽德罗数
avoidable accident 可避免的事故
avoid a contract 废止合同
avoidance clause 宣告无效条款
avoidance of contract 宣告合同无效
avoidance of double taxation 避免双重税
avoidance of tax 逃税
avoidance 避免，取消，躲避，废止，回避，宣告无效
avoid any glare on the panels 避免盘上的任何眩光
avoid 避免，避开，取消，废止，回避，宣告无效
avometer 安伏欧计，安伏欧三用表，万用（电）表
AVR(automatic voltage regulator) 自动调压器，自动电压调整器［调节器］
AVT(all volatile treatment) 全挥发处理【用于压水堆二次回路】
avulsion cutoff 冲裂割断
avulsion 冲裂（作用），改道
AW(air waybill fee) 货运单费【承运人收取此费为 AWC，代理人收取此费为 AWA】
awaiting repair time 等待修复时间
await 期待
award 决标，裁决，授予，奖（章）
award a contract 签订合同，授予合同
award at tender opening 现场决标
award criteria 中标标准，授标标准，授标准则
award decision 授标决定
awarded contract price 中标合同价
awarded value 签约价格
award in final 最后裁决
awarding announcement 中标公告
award notification 授标通知
award of bid 授标，决标，投标裁定
award of contract 授予合同【即决定中标人】，授标，签订合同
award of tender 授标，决标，投标裁定
award version 合同版
awareness 意识
awash rock 浪刷岩
awash 与水面齐平的，齐水面的，浪刷
a. w. (atomic weight) 原子量，原子重量
away-from-reactor fuel storage 离堆燃料储存，厂外燃料储存
away-from reactor 堆外，离堆
AWB(air waybill) 航空货运单
A-weighted decibel 等效 A 声级
A-weighted sound power level A 加权的声能强度
A-weighted sound pressure level A 加权声（压）级
AWG(American Wire Gauge) 美国线规
a whopping sum of 巨大额度的
AWI(American Welding Institute) 美国焊接协会
awkward cargo 大件货物
AWN(additional work notice) 额外工作通知（单）【工程验收用语】
awning type window 篷式天窗
awning 雨篷，遮篷，遮阳篷
AWS(American Welding Society) 美国焊接学会

AWWA(American Water Works Association) 美国水工协会
axed arch 斧斩拱面
axe hammer 斧锤
axe 斧头
axial 轴的，轴向的，轴流的
axial adjustment 轴向调整
axial admission 轴向进汽
axial airflow 轴向气流
axial air-gap reluctance motor 轴向气隙式磁组电机
axial air-gap 轴向气隙
axial ampere-turn 轴向安匝
axial back-pressure turbine 轴流式背压汽轮机
axial bearing 止推轴承，轴向轴承，支撑轴承
axial blade clearance 叶片轴向间隙
axial blanket 轴向转换区，轴向再生区
axial breeder （燃料元件）轴向增殖段
axial buckling 轴向挠曲
axial bush 轴瓦
axial cam 凸轮轴
axial centrifugal 轴向离心式
axial clearance 轴向空隙，轴向间隙
axial coil 轴线圈
axial compression force 轴向压力
axial compression 轴向压力，轴向压缩，轴心受压
axial-compressor-driven wind tunnel 轴流式压气机驱动风洞
axial compressor 轴流式压缩机，轴流式压气机
axial concentration 轴向浓度
axial condenser 轴向布置凝汽器
axial conduction 轴向导热
axial deformation 轴向变形
axial diffusion coefficient 轴向扩散系数
axial dimension 轴向尺寸
axial direction 轴向
axial displacement indicator 轴向位移指示器
axial displacement limiting device 轴向位移限制器，轴向位移保护装置
axial displacement monitor 轴向位移监视器
axial displacement transducer 轴向位移传感器
axial displacement 轴向位移，轴位移
axial distance 轴向距离
axial distribution 轴向分布
axial eddy 轴向涡流
axial elongation 轴向伸长
axial entry blade 轴向插入叶片
axial entry impeller 轴向进口叶轮
axial exducer 轴向导流器
axial expansion 轴向膨胀
axial extension test 轴向拉伸试验
axial extension 轴向拉伸
axial factor 轴向因子
axial fan 轴流式通风机，轴流风机
axial field 轴向场
axial float 轴向游隙
axial flow air compressor 轴流式空压机
axial flow blading 轴流式叶片
axial flow blower 轴流风机
axial flow compressor 轴流式压缩机，轴流压气机
axial flow distribution 轴向分布
axial flow exhaust fan 轴流排气通风机
axial flow fan 轴流式风机，旋桨式风机，轴流式风扇
axial flow hydraulic 轴流式水轮机
axial flow hydroelectric unit 轴流式水轮发电机组
axial flow impeller 轴流式叶轮，轴流涡轮
axial flow impulse reaction turbine 轴流式冲动反动汽轮机
axial flow impulse turbine 轴流冲动式汽轮机
axial flow pump 轴流式泵，轴流泵
axial flow steam turbine 轴流式汽轮机，轴流透平
axial flow turbine 轴流式汽轮机
axial flow wheel 轴流式水轮
axial flow 轴向流动，轴流式，轴流
axial flux distribution 轴向通量分布
axial flux motor 轴向磁通电动机
axial flux permanent magnet synchronous generator 轴向磁通永磁同步发电机
axial flux 轴向通量，轴向磁通
axial force diagram 轴向力图
axial force 轴向力，轴心力
axial form factor 轴向形状因子
axial-gap reluctance motor 轴向间隙磁阻电动机
axial-gap 轴向间隙
axial grid spacing coefficient 轴向格架定位系统
axial heat diffusion 轴向热扩散
axial heat-flux density distribution 轴向热流密度分布
axial in phase mode 轴向同相振型
axial instability 轴向不稳定性
axial internal clearance 轴向间隙
axialiy-split casing pump 轴向拼合壳泵
axialiy-split casing 轴向拼合壳体
axialiy-split 轴向拼合
axial joint 轴节理
axial labyrinth gland 轴向迷宫式汽封，轴向曲径式汽封
axial length 轴长
axial loaded column 轴向受力柱
axial load on horizontal pipe 管道水平荷载
axial load on pipe 管道轴向荷载
axial load 轴向负荷，轴向荷载
axial location 轴向位置
axially bare 轴向无反射层的
axially magnetized stator 轴向磁化定子
axially movable motor 转子轴向可动的电动机
axially nonuniform heat flux density 轴向不均匀热流密度
axially periodic field 轴向周期场
axially space air-gap 轴向分布气隙
axially split 轴向剖分式
axially symmetrical deformation 轴对称变形
axially symmetrical jet 轴对称射流
axially symmetric bending 轴对称弯曲
axially symmetric diffusion 轴对称扩散
axially symmetric flow 轴对称流动
axially symmetric load 轴向对称荷载
axially symmetric motion 轴对称运动

axially symmetric shape 轴对称形状
axially uniform heat flux density profile 轴向均匀热流密度分布图
axially-ventilated motor 轴向通风电动机
axially 沿轴向的
axial magnetic centering force 电机定转子之间的轴向力
axial magnetic field 轴向磁场
axial magnetic flux 轴向磁通
axial magnetizing force 轴向磁化力
axial magnet 条形磁铁，磁棒
axial mean 轴向平均值
axial mixing 轴向混合，轴向交混
axial moment theory 轴向力矩理论
axial motion 轴向运动
axial movement type expansion 轴向位移型膨胀节
axial movement 轴向位移，轴向运动，轴向移动
axial neutron flux density distribution 轴向中子流密度分布
axial nuclear factor 轴向核因子
axial offset factor 轴向偏移系数，轴向偏移因子
axial offset 轴向不平衡【通量】，轴向偏移
axial of rotation 旋转轴
axial of twist 扭转轴
axial oscillation 轴向振动，轴向摆动
axial out of phase mode 轴向异相振型
axial peaking factor 轴向峰值因子
axial peaking hot spot factor 轴向峰值热点因子
axial phasor 轴向相量
axial piston pump 轴向活塞泵
axial pitch 轴向齿距，轴向节距，轴向螺距
axial plane cleavage 轴面劈理
axial plane foliation 轴面叶理
axial plane of fold 褶皱轴面
axial plane 轴面
axial play 轴向游隙
axial plunger pump 轴向柱塞泵
axial position indicator 轴向位移指示器
axial power density distribution 轴向功率密度分布
axial power distribution monitoring system 轴向功率分布监测系统
axial power distribution 轴向功率分布
axial power flattening 轴向功率展平
axial power tilt 轴向功率倾斜
axial pressure gradient 轴向压力梯度
axial pressure 轴向压力
axial pump 轴流泵，轴向泵
axial ratio 轴比
axial rearrangement of fuel 燃料轴向倒换
axial road 辐射式道路
axial runout 轴向偏差，轴向摆差
axial-scan closed circuit TV camera 轴向扫描闭路电视摄像机
axial seal 轴向密封，轴向气封
axial secondary 直线电机的轴向次级
axial section 轴向剖面
axial segment 轴向段
axial shifting machine drum 轴向转换机筒
axial shift 轴向位移
axial shuffling of fuel 轴向倒换燃料
axial split pump 轴向剖分泵
axial spotting 轴向找正
axial stagger 轴向交叉排列
axial stiffness 轴的刚度
axial stop 轴向挡块
axial strain 轴向应变
axial strength 轴向强度
axial stress 轴向应力
axial surface current 轴的轴向移位
axial symmetrical stress 轴对称应力
axial symmetry 轴对称
axial temperature distribution 轴向温度分布
axial temperature rise 轴向温升
axial temperature stress 轴向温度应力
axial tension 抗拉应力，轴向受拉，轴向拉伸，轴向拉力
axial thermal fuel elongation 燃料轴向热伸长
axial thrust balancing apparatus 轴向推力平衡装置
axial thrust bearing 轴向推力轴承
axial thrust on fixing support 固定支座轴向推力
axial thrust 轴向堆力
axial torque 轴向转矩，轴向扭矩
axial translation 轴位移，轴向位移
axial turbine 轴流式涡轮机，轴流式透平
axial type 轴流式
axial vector 轴矢量
axial velocity 轴向速度
axial ventilation 轴向通风
axial vibration amplitude 轴向振动值
axial vibration 轴向振动
axial volute 轴向蜗壳
axial vortex 轴向涡流，轴向旋涡
axiom 原理，公理，定理，原则
axiotron 一种磁控管
axis bearing 轴承
axis distance 轴距
axis of abscissa 横坐标轴，横轴
axis of bent 排架轴线
axis of column row 柱列轴线
axis of commutation 换向轴线，换向器的中性线
axis of earth 地轴
axis of equilibrium 平衡轴
axis of fold 褶皱轴
axis of inertia 惯性轴
axis of ordinate 纵坐标轴，纵轴
axis of plant site 厂址轴线
axis of principal strain 主应变轴
axis of principal stress 主应力轴
axis of reference 参考轴线，参考轴
axis of revolution 回转轴线，绕转轴
axis of rotation 旋转轴线，转动轴
axis of symmetry 对称轴线，对称轴
axis of the principal stress 主应力轴
axis of weld 焊缝轴线
axis through absorber 通过吸收器的轴
axisymmetrical shell 轴对称壳体
axisymmetrical 轴对称的
axisymmetric body 轴对称体
axisymmetric determinant 轴对称行列式

axisymmetric expansion 轴对称扩大
axisymmetric flow 轴对称流动
axisymmetric jet 轴对称射流
axisymmetric plume 轴对称羽流
axisymmetric puff 轴对称喷团
axisymmetric thermal elasto-plastic 轴对称热弹塑性
axisymmetric wind tunnel 轴对称风洞
axis 轴线【几何】，轴，坐标轴
axle arm 轴臂，驱动桥定位臂
axle base 轴距，轴座
axle basis 基轴制
axle bearing 轴承
axle box 轴箱
axle bush 轴衬
axle cap 轴罩，轴帽
axle circuit 经过车轴的电路
axle diameter 轴径
axle-driven generator 车轴发电机
axle driving motor 车轴驱动用电动机
axle elongation 轴伸长
axle fairing 轴向整流片
axle fracture 轴断裂
axle gauge 轮距
axle generator pole changer 保证车轴发电机极性不变切换开关
axle generator 轴带发电机
axle grease 轴用脂
axle housing 轴套
axle hung motor 悬轴型电动机
axle joint 轴头
axle journal 轴颈
axle load 轴荷载，轴负荷，轴载
axle lubrication pipe 轴润滑油管
axle-neck 轴颈
axle pin 轴销
axle shaft 车（后）轴，主动轴
axle sleeve 轴套
axle steel reinforcement 轴钢钢筋
axle steel 轴钢
axle suspended motor 轴悬型电动机
axle 车轴，轮轴，小轴，轴，轴杆，心棒
axonometric drawing 轴测图
axonometric perspective 不等角投影图，三向图，轴测投影图
axonometric projection 轴测投影三向图，三向投影，不等角投影
axonometry 三向正投法，轴测法
AX THR BRG(axis thrust bearing) 轴向推力，轴承
Aylesbury collaborative experiment 埃尔兹伯里（风荷载）合作实验
Aylesbury experimental building 埃尔兹伯里实验建筑物
azabache 褐煤
azeotrope 共沸混合物
azeotropic mixture refrigerant 共沸溶液制冷剂
azimuthal angle 方位角
azimuth 方位，方位角，偏振角方位
azimuthal factor 方位角因子
azimuthal instability 方位角不稳定性
azimuthal position 方位，方位位置
azimuthal projection 方位投影
azimuth angle 方位角
azimuth A 方位角 A
azimuth circle 方位圈
azimuth closure 方位角闭合差
azimuth commutator 方位角换向器，方位角转换开关
azimuth compass 方位罗盘
azimuth control 方位控制
azimuth correction 方位改正数
azimuth determination 方位角测定
azimuth dial 方位刻度盘
azimuth drive motor 方位驱动电动机
azimuth driver 方向传动器
azimuth drive 方位角驱动
azimuth-elevation 方位-高程
azimuth indicating meter 方位指示器
azimuth mark 方位标
azimuth method 方位角法
azimuth motor 方位电动机
azimuth observation 方位角观测
azimuth potentiometer 方位分压器，方位角电位计
azimuth-range 方位-距离
azimuth servomechanism 方位伺服机构
azimuth transmitting synchro 方位角传送电动机
azimuth traverse 方位角导线
azote 氮
azotic acid 硝酸
azran 方位距离
AZS(automatic zero set) 自动调零（装置）
azure spar 天蓝石
azure stone 青金石，天蓝石
azure 由横条平行线组成的，天蓝的
azygos 奇数部的
azyl 氮基
azymia 酶缺乏
A 报警【DCS 画面】

B

Babbit cushion 巴金垫
babbit metal temperature 轴承钨金温度
Babbit pad 巴金垫
babbitt metal temperature 轴承钨金温度
Babbitt metal 铜锡合金，巴氏（轴承）合金，钨金
Babbitt 巴氏合金【白合金，轴承合金】
babble 混串音，数据传输通道的感应信号，转播干扰
BA(binary add) 二进制加，二进制加法指令
BA(bottom ash) 底灰
baby compressor 小型压缩机
baby knife switch 小型刀闸开关
baby rail 小钢轨
baby track 小轨道
baby tractor 小型拖拉机
baby truck 坑道运输车，小型运输车

bachelor dormitory 单身宿舍
bachelor 学士学位
backacter 反铲挖土机
back acting shovel 反向铲，反铲
back action 反作用
back ampere-turn 反安匝数，逆向安匝
back analysis 事后分析，反分析
back-and-forth movement 往复运动
back angle 后视角
back arc profile 背弧【汽轮机】
back azimuth 反方位角
back balance 平衡重量，平衡锤
back bargain 蚀本生意，亏本买卖
backbar 支承梁
back beach 后滩，后滨
back bead 背面焊道，背面焊缝，填角焊缝，封底焊道
back bearing 反象限角，后方位角
back bias voltage 反向偏压，负偏压
back bias 反偏压，回授偏压
back blading 倒车叶片
backblowing 反吹法，倒冲洗
backboard 背后挡板，底板，后板
backbone 骨干
backbone bus 主干总线
backbone road 主干道，干道
backbone route 基干线路，基干路由
back-boundary cell 后膜光电管
back brace 后斜撑
back bracing 反斜撑
back calculation 反演计算，反算法
backcharges 决算后的各种费用
back-checking 返回校验
back chipping 清根【焊缝】，挑焊根
back coat 底层涂料
back connected instrument 接头在背面的仪表
back connected switch 背面接线的开关
back connection 背面连接，盘后连接
back contact 常闭触点，背面触点，后触点
back corona 逆电晕
back cover 下面覆盖层，（书的）封底
back current relay 反向电流继电器
back current 回流，反向电流，回电流
back cut-off power recorder 反向截止功率记录仪
back cutting 必需的超挖部分，回挖
back-data for pre-qualification 资格预审返回数据
back diffusion 反向扩散，逆扩散，反行弥散
back digger 反铲挖土机，反向铲
back discharge 反向放电，反向出料
back draft damper 反逆转风门
back drain （堤或墙的）背面排水，墙背排水管
backdraught 倒转，回程，反向通风，反向气流
back echo 底部回波，后波瓣接收的回波信号，后瓣回波
back eddy 逆涡流，出气边，后缘，下降边，后沿
backed-up auxiliary 备用辅机
backed-up busbar 备用汇流排
backed-up power supply 备用电源
backed-up supply 应急电源
backed-up 应急的，后备的，备用的
back electromotive force constant 反电动势常数
back electromotive force 反电动势
back elevation drawing 后视图，背视图
back elevation 背面立视图，背视图，后视图，背立面图
back-EMF 反电动势，逆电动势
back emission 反向发射
back-end ductwork 尾部烟道
back-end equipment 后置设备
back-end surfaces 尾部受热面
back-end 后端，尾端，后段【核燃料循环】
back energization 反送电，倒送电
backer brick 墙心砖
backer pump 备用泵
backer 背衬，衬垫，支持物，补强衬里
back-extraction 反萃取，反向萃取
back face of tower 杆塔后侧，塔身背面
backface 后表面【在阀体上与后座配合】，背面
back feed factor 反馈因数
back feed 反馈，回授，反向馈电，反向进给
backfill compaction 回填土压实
backfill compactor 回填土压实机
backfill consolidation 复土压实，回填夯实
backfill density 回填密度
backfilled earth 回填土
backfilled 反填充的，回填的
back-filler 回填机，填土机
backfill grouting 回填灌浆
back-filling in layers 分层回填
backfilling scheme 回填方案
backfilling tamper 回填夯土机
backfilling 回填，再次充气
backfill material 回填材料
backfill of the site 现场的回填
backfill soil 回填土
backfill tamper 回填夯
backfill under pressure 加压回填
backfill with mortar 灰浆回填
backfill 反向充液，回填土，回填料，（土方）回填，程序填充，复土
back filter 逆反过滤器，后置过滤器
backfin 夹层，缺陷
back fire arrester 回火防止器
back fire 逆弧，逆火，反燃，回火
backfitting 不大的改形，修合，改进，修改，稍有改变
back flap hinge 明铰链
back flap 里垂帘【里扉】
back flashover 反击雷闪络，反向闪络
backflash 反闪，回燃，逆弧，逆火，反燃，回火
backflow cell model 回流槽模型
backflow connection 倒流连接，回流管
backflow damper 回流缓冲器
backflow preventer 防回流装置，倒流防止器，防止回流设施，回流防止器
backflow-prevention device 倒流防止设备
backflow 反流，逆流，倒流，回流
backflushing 逆流洗涤，回洗，反冲洗，反向

洗涤
back focal length 后焦距
backform 顶模，后模
back freight 回程运费，退货运费
back gear 背齿轮，减速齿轮，反齿轮
back gouging of plasma arc 等离子弧清根
back gouging 背面切割，背部清根，清根，刨焊根，背刨
back grinding 背面打磨
background activity level 本底放射性水平，本底活度水平
background activity 本底放射性
background air pollution 本底空气污染
background air 本底空气
background area 本底区域
background buildup 本底积累
background color 背景颜色
background concentration 本底浓度，现状浓度
background condensation 背景凝聚，本底凝聚，(在威尔逊云室内）外来杂质凝聚
background correction 本底校正
background count 本底计数
background cross-section 本底截面
background density 本底黑度
background document 背景文件
background dose 本底剂量
background equivalent activity 本底等效活度
background excitation factor 本底激励因子
background exposure 本底照射，本底照射量
background fluorescence 荧光底色
background heater 隐闭式供暖器
background information 背景资料，依据资料
background ionization voltage 基底电离电压
background irradiation 天然辐照
background level 本底水平，自然本底值，基值电流
background limited infrared photoconductor 背景限红外光电导体
background luminance 背景亮度
background material 背景材料
background monitoring 本底监测
background neutron 本底中子
background noise criteria 本底噪声标准
background noise 背景噪声，背景杂音，基底噪声，本底噪声
background of project 工程背景
background pollution 本底污染
background processing 后台处理，背景处理
background radiation 背景辐射，本底辐射，本底放射性
background response 本底响应
background scattering 背景散射
background survey 背景调查，本底测量，本底调查
background turbulence 背景湍流
background voltage noise 基底电压噪声
background wind load 本底风载，背景风荷载
background 背景，背景噪声，本底，基础，背景，底色
back guy 后拉线，拉揽，牵索
back-hand drainage 倒转水系
backhander 贿赂
backhand welding 后倾焊，右焊法
backhaul cable 后拖缆绳，后曳索
back haul 后曳
backhoe front end loader 后挖前卸式装载机
backhoe loader 反铲装载机
backhoe tamping 反铲捣实
backhoe 反铲，反向铲
back hole 顶板炮眼
back induction （电枢的）去磁作用
backing bar 支撑条
backing bead 打底焊道
backing brick 背衬砖，墙心砖
backing coil 反接线圈，补偿线圈
backing gas 背面气体【保护】
backing groove 背面坡口
backing layer 垫层【焊接】
backing log 挡边木
backing memory 后备存储器
backing of veneer 饰面板里层
backing of wall 墙托
backing of window 窗托
backing pass 底焊焊道
backing plate 垫板【焊接】，支撑板，底板，座板
backing power 倒车功率，后退功率
backing pump 备用泵，旋转油泵，初级抽气泵，前置泵
backing ring 支承环【焊接】，衬环，垫环
backing run 背面焊缝，打底焊道，封底焊道
backing sand 填充砂
backing shoe 铜滑块
backing storage 辅助存储器，后备存储器
backing strip 衬板，垫板，支撑条
backing turbine 倒车透平
backing up 衬砌，逆行，倒车，支撑，封底焊
backing voltage 闭锁电压，阻塞电压
backing welding electrode 底层焊条
backing weld 打底焊，打底焊道，底焊，底焊焊缝，封底焊
backing wind 逆转风
backing 背衬，衬垫，底座，轴瓦，倒转，（砖衬的）内侧，消除，底板，增强，衬背，靠背，里壁，支持，支架，补强衬里
back inversion 倒转
back iron 护铁，背铁
backland 天然堤后泛滥地，内陆地区，内地，腹地，偏僻地区
backlash current 间断电流
backlash eliminator 齿隙消除装置
backlash spring 消隙弹簧
backlash unit 间隙模拟部件
backlash 齿隙游移，反冲，无效行程，偏移，后冲，回弹，回跳，游隙，间隙，空回，齿隙，后退，后冲，齿隙游移
back leg 背面支柱
back levee 背水堤，支堤
backlighted button 带灯按钮
backlighted push-button 带灯按钮
backlighted rotary pushbutton switch 带灯旋转按钮开关

backlighted rotary switch 灯光旋钮
backlighted switch 带灯开关
backlighted 带灯的
backlighting 后部照明
back lining 背衬，背衬料
backlit sign 带灯标志牌
backlit 带灯的
back loader 反铲装载机
backlog control 储备管理
backlog document 积压文件
backlog 储蓄积累，积累，储备，后备，积压代办事项，积压的订单，积压而未交货的订单，积压的工作
back magnetization 反磁化
back-mix condition 返混条件
back-mix-flow reactor 逆混流反应器，反向混流反应器
back-mix fluidized bed reactor 返混流化床反应器
back-mixing 返混
back nail 平钉
back nut 防松螺帽，锁紧螺母，限位螺母，支承螺母
back observation 向后观测
back of blade 叶背
back-of-board wiring 盘后接线
back-off angle 后让角，后角
back-off assembly 松开装置
back-off joint 倒扣短节
back-off operation 倒扣作业
back-off procedure 倒开步骤，退避过程
back off 拧松，旋出，补偿，倒扣，软化，减轻，卸下，倒转后解松，后退，避让，放弃，收回（主张、要求、承诺等）
back of levee 堤背
back order memo 延期交货通知单
backosmotic pressure 反渗压力
back out of contract 背弃合同义务，不履行合同
back out 拧松，退出，脱出，逆序操作，收回，食言，违约，不遵守（诺言、合约等）
backpass surface 尾部受热面
backpass 后烟道，尾部烟道，（焊接）封底焊道
back pitch 排距
back plan 底视图
backplastered 背面抹灰的
backplate 垫板，支承板，挡板，后板，护板，信号板，靠背板，背面板，背面电极
back play 游隙
back porch （脉冲的）后沿
back power 反向功率
back pressure boiler 背压锅炉
back-pressure condensate return system 背压凝结水回收系统，余压凝结水回水系统
back pressure extraction turbine 背压抽汽式汽轮机
back pressure fitting 背压配合
back pressure governor 背压调节器
back pressure manometer 后压测压计
back pressure of steam trap 疏水器凝结水背压力
back pressure operation 背压运行
back pressure regulating valve 背压调节阀
backpressure regulator 背压调节器
back pressure return 余压回水
back pressure steam turbine 背压式汽轮机
back pressure steam 背压蒸汽
back pressure turbine with intermediate bleed-off 抽汽背压式汽轮机
back pressure turbine 背压式汽轮机
back pressure valve 止回阀，逆止阀，背压阀，回压阀，防逆阀
back pressure 背压，回压，反压，背压力，反压力
back profile 背弧
back prop 后撑，后支柱
back pull 反拉力
back putty 打底油灰
back rack angle 纵向角
backreach （起重机）后伸距
back resistance 逆向电阻，反向电阻，背部阻力
back(/return)stroke 回程
back-roll 反绕，倒转，重算
backrun 反转，倒车，封底焊
backrush 退浪
back saw 夹背锯，脊锯，镶边手锯
backscattering correction 反散射校正
backscattering factor 反散射因子
backscattering 反向散射
backscour 反向冲刷
back sealing weld 封底焊
back seal 倒密封
back seat bushing 后座衬套
backseated valve 后座阀
back seat gasket 底座垫圈
back seat （阀门的）后座
backset bed 逆流层，涡流层
backsetting 收进【砌墙】
backset 涡流，逆流，倒退，挫折，向前赋值
backshaft 副轴，后轴
back shielding 背面屏蔽
back shore 背撑，顶撑，后滨，滨后，潮间丘，后海岸地带
back shroud 后盖板
back shutter 后快门【摄影】，百叶窗［门］，里百叶窗，内百叶窗
backsight 后视
backsiphonage 倒虹吸作用，倒虹吸能力，回吸，反吸，反虹吸
backslash 反斜号
back slope angle 反坡角度
back slope 背坡，反坡，后坡
backspace character 退格符
backspace 回退，退格，退后一格，查看上一级文件夹
back-spin timer 反转时间继电器，反旋计时器
back-spin 反转，回旋
back stairs 后楼梯
backstand 支承，构架
backstay anchor 拉索锚定
backstay cable 拉索
backstay 后拉缆，后拉索，后拉线，后支撑，

后接片，支撑，撑条，（蒸汽机车锅炉的）背撑
backstep method 分段退焊法【焊接】
backstep sequence 分段后向法，分段退焊
backstep welding 分段退焊，分段退焊法
backstop 后障，托架，支撑，防逆转，棘爪，托架止回器，棘爪，逆止器，止回器
back streaming 反流
back stroke 返回行程，回程，返回冲程
back strut 反撑，后撑
back substitution 倒转代换，回代
back suction 反吸
back surface 背面
back surface field solar cell 背电场太阳能电池
back surface field 背表面场，背面场，背场
back surface reflection and back surface field solar cell 背场背反射太阳电池
back surface reflection solar cell 背反射太阳电池
back-swamp 堤后池沼
back-swept vane 后弯叶片
back-swept 后掠，后掠角
backswing voltage 反向电压，回程电压
backswing 回程，反冲，回摆，后摆，回复，准备姿态
back tax 退税
back thrust 反推力
back titration 反滴定，回滴定
back-to-back angles T形组合角钢，背靠背组合角钢
back-to-back connection 背靠背连接
back-to-back contract 背对背合约
back-to-back converter station 背靠背换流站
back-to-back credit export first 出口方先开的对开信用证，输出为先的对开信用证
back-to-back credit import first 进口方先开的对开信用证，输入为先的对开信用证
back-to-back credit 背对背信用证，转开信用证
back-to-back HVDC transmission 背靠背高压直流输电
back-to-back impellers 背靠背叶轮
back-to-back L/C 背靠背信用证
back-to-back loan 背对背贷款
back-to-back method 反馈法，背靠背法
back-to-back starting 反旋启动
back-to-back switchboard 背靠背配电盘
back-to-back testing 背靠背试验，反馈试验，对拖试验
back-to-back 背靠背（的），紧接的
back-to-front ratio 正反比
back-to-wall switchboard 背靠墙配电盘
back-to-wall 背靠墙
backtrack cancel 返回撤销
backtrack condition summary 返回状态概要
backtracking 反程，回程，返回，回路，反馈，由原路返回，取消诺言
backtrack interval 返回间隔
backtrack menu 返回菜单
backtrack mode 返回模式
backtrack summary 返回概要
back transient 反向瞬态过程
back tunneling 反向隧道效应
backup breaker 备用断路器
backup brick 衬里砖，墙心砖
backup component 备品
backup coolant system 备用冷却剂系统
backup cycle in normal condition of the system 系统正常情况下备份周期
backup data 备查资料
backup device 辅助设备，备用装置
backup each other 相互备用
backup energy storage 后备储能
backup equipment 后备设备
backup filter 备用过滤器，辅助过滤器，后备文件
backup frequency 辅助频率
backup fuel 备用燃料，辅助燃料
backup fuse 后备熔断器
backup governor 辅助危急保安器
backup manual operator 备用手动操作器
backup mechanical brake 备用机械制动器
backup overspeed governor 备用超速保安器
backup plan 后备计划
backup plate 垫片
backup power source 备用电源
backup power supply 后备电源
backup power system 备用动力系统
backup protection 后备保护，后备保护装置
backup push button 按钮
backup reactor shutdown capability 后备停堆能力
backup relaying 后备继电保护
backup ring 支撑环【控制棒驱动机构】，垫圈，承压轴，支承辊，支撑轧辊
backup safety device 备用安全设备，备用安全装置
backup safety system 备用安全系统
backup scheme 备用方案
backup source 后备电源，备用电源
backup strip 支持板
backup system 备用系统，后备系统，备份系统
backup trap 后备阱，备用阱，辅助阱
backup vent valve 备用通风阀，备用排气阀
backup washer 支撑垫圈
backup 备份，备用，后备，支持，辅助，备件，备用的，后援的
back valve 单向阀，止回阀
back vane 后叶片
back vent 背通气管，（虹吸管的）顶背排气管
back view 背面立视图，背视图，后视图
back voltage 逆电压，反电压
backwall echo 后壁回音
backwall photovoltaic cell 后壁阻挡层光电池
backwall 背墙，后墙，后墙板
backward-acting regulator 反向作用式调制器
backward area 不发达地区，落后地区
backward bent vane 后弯式叶片
backward boost 制动，制推，降压，降压器
backward brush-lead 电刷超前后移
backward chaining 反向推理
backward channel 返回通道，反向通道
backward clearing 后向拆线
backward country 不发达国家，落后国家
backward crosstalk 后向串扰

backward current 逆向电流，反向送料
backward curved blade 后弯叶片
backward curved vane 后弯曲叶片，后弯式叶片，向后曲线的叶片
backward-difference 向后差分
backward diode 反向二极管
backward direction 逆向
backward drainage 逆向水系
backward erosion 反向冲刷，向源侵蚀
backward feeding 反向馈电，反向送料
backward-field impedance 反向磁场阻抗
backward-field torque 反向转矩，负序转矩
backward flow 逆流，反流，对流，回流，反向流
backward-forward counter 双向计数器
backward impedance 反向阻抗
backward jet 反向射流，反向喷射
backward lead 后移超前【电刷】
backward link 反向连接
backward motion 反向运动，逆动
backwardness 落后状态
backward-order reactance 负序电抗
backward phase component 反向相分量，负序分量
backward progression lap winding 左行叠绕组
backward progression winding 左行绕组
backward-rotating field 反向旋转磁场
backward-rotating wave 反向旋转波
backward running 反向运动，倒转
backward scattering 反散射
backward sequence 反向序，负序
backward shift operator 后移算子
backward shift 反移，后移
backward swept vane 反弯式叶片
backward visibility 后视度
backward vision 后视
backward-wave amplifier 返波放大管
backward-wave converter 反波变频管
backward-wave crossed-field amplifier 反波正交场放大管
backward-wave magnetron 反波式磁控管
backward-wave oscillator 回波振荡器，反波振荡器
backward-wave tube 回波管，反射波管
backward-wave 负序波，回波，反射波
backward welding 后倾焊
backward 向后，相反，向后的，相反的，背面
backwash effect 回水作用
backwash expansion ratio 反洗膨胀率
backwash expansion 反洗膨胀
backwashing filter 反洗滤池
backwashing system 反冲洗系统
backwashing 反洗，反冲，逆向冲洗，反萃，反向冲洗法，反冲洗，逆流冲洗，反向洗涤，回流冲刷，反激浪，反溅，尾流，回流，后涡流，回冲，回刷
backwash limit 反冲界线
backwash liquor 反洗液
backwash procedure 反洗程序
backwash ripple mark 回流波痕
backwash tank 反洗箱，回洗槽
backwash valve of condenser 冷凝器逆洗阀
backwash valve 反冲洗阀，逆洗阀
backwash water connection 反洗水连接装置
backwash water 反洗水，回洗水
backwater area 回水面积，回水区
backwater curve 回水曲线，壅流曲线，壅水曲线，壅水线
backwater effect 回水影响，回水作用，回水效应，壅水效应
backwater envelope curve 回水包络曲线
backwater formula 回水公式
backwater function 壅水函数，回水函数
backwater gate 回水闸，挡潮闸门，挡水闸门，逆止阀，防回门
backwater head 回水头
backwater height 壅水高度
backwater length 回水长度，壅水长度
backwater levee 壅水堤
backwater level 回水高程，壅高水位，壅水高程，壅水水位
backwater limit 回水极限
backwater pump 回水泵
backwater silting 回水区淤积
backwater slope 回水坡度
backwater storage 回水库容，壅水蓄量
backwater suppressor 回水制止设备
backwater trap 回水存水弯
backwater valve 回水阀
backwater zone 回水区
backwater 回水，潮汛壅水，壅水
back wave 反向波
back wearing ring 后耐磨环
back welded branch pipe 背面焊接的支管
back weld 底焊焊缝，底焊，打底焊道，打底焊
backwind 背后风
back-wired panel 后面布线的面板，盘后布线的配电板
back wiring 背面布线
back work ratio 回功比
back 背面［部］，后部，基座，承托，叶片背弧，衬垫，后面，后端，拖动端，背，底座，反面，支座，走向节理，后面的，背面的
bacteria bed 生化滤层，菌床
bacterial analysis 细菌分析
bacterial constituent 细菌组分
bacterial corrosion 细菌腐蚀
bacterial examination of water 水的细菌检验
bacterial examination 细菌检验
bacterial treatment 细菌处理
bacteria 细菌
bactericide 杀菌剂
bacteriological aftergrowth 细菌再生
bacteriological analysis 细菌分析
bacteriological count 细菌计数
bacteriostatic compound 抑菌化合物
bad check 空头支票
bad conductor 不良导体
bad contact 不良触点，不良接触，接触不良
bad debt 呆账，坏账，死账【无法收回的欠款】
baddeckite 赤铁黏土，含铁质白云母，斜锆石，

二氧化锆矿
bad earth 接地不良
bad geometry 不利几何条件
badger 排水管清扫器，榫接边
badge 标记，符号，表征，佩章剂量计，佩章
badigeon 嵌填灰膏，油灰，腻子
badijun 嵌填灰膏
badland 荒原，瘠地，劣地，崎岖地
badly graded sand 级配不良的砂
badly-worn idler roll 报废的托辊辊子
“bad-order” certificate “货物短缺”证明书
bad packing 包装不良
bad slip 强烈滑坡
bad weather 恶劣天气，坏天气
BAF(baffle) 隔板
BAF(bunker adjustment factor) 燃油附加费，燃油附加费系数
BAF(bunker adjustment fee) 燃油附加费
baffle aerator 挡板曝气器，阻墩式通气器
baffle assembly 围板组件【反应堆压力壳】
baffle block 消力墩，消力块
baffle board 隔音板，烟囱风帽，挡板，折流板
baffle bolting 围板栓接
baffle box 消能柜，消能箱，阻流箱
baffle brick 隔墙砖
baffled fluidized bed 接板流化床
baffled plate 挡板
baffled slot plate distributor 条状挡板分布器
baffle flame-holder 钝体稳燃器，挡板式火焰稳定器
baffle jetting 围板射流
baffle mixing chamber 隔板式混合池
baffle pier 分水墩，消力墩，砥墩
baffle plate type separator 挡板分离器，隔板式分离器
baffle plate 挡板，缓冲板，导风板，导流板，消力板，折流（挡）板，挡风板，隔板【反应堆压力壳】
baffle prong 消力锥
baffle reaction chamber 隔板式反应池
baffler 缓冲板，换向板，反击板，导流板，消力器，消声器，障壁，阻尼器，（节流）挡板，气隙隔板，折流器
baffle scrubber 挡板式洗涤器
baffle sill 门槛，消力槛
baffle structure 消力建筑物，消力设施
baffle support flange 围板支承法兰【堆芯】
baffle-type collector 隔板式除尘器
baffle-type 百叶窗式的，挡板式的
baffle wall 分水墙，隔墙，消力墙
baffle washer 折流清洗器
baffle weir 砥堰
baffle 挡板，隔板，折流板，折焰板，导流叶，障板，缓冲板，导流板，挑流物，围板，遮护物，阻力板
baffling wind 无定向风，迎面风
baffling 挡板调节，折流调节
bag 袋子
bag accumulator 囊式蓄能器
bag and spoon dredger 袋勺式挖泥船
bagasse board 蔗渣板
bagasse fiberboard 蔗渣纤维板
bagasse 甘蔗渣
bag collector 布袋除尘器
bag dam 土袋埝坝
bag dust collector 布袋除尘器
bag filter 布袋式过滤器，袋式除尘器，袋式过滤器
bagged cement 成袋水泥，袋装水泥
bagger 泥斗，装袋机
bag house dust collection 袋式除尘系统
bag house precipitator 袋式除尘器
bag house 布袋室【除尘装置】，布袋除尘器，布袋除尘装置，沉渣室，大气污染微粒吸收器，袋式除尘器
Bagnold’s threshold parameter 拜格诺阈值参数
bag-stacking machine 堆袋机
bag trap 袋形存水弯
bag type air filter 布袋滤尘器
bag type collection system 袋式集尘装置
bag type collector 袋式除尘器
bag type dust collector 袋式除尘器
bag type dust remover 布袋除尘器，袋式除尘器
bagwork 沙包，装袋工作
bahada 山麓冲积平原，山麓冲积扇
BAH(bottom ash hopper) 底灰斗
bail bond 担保书
bailee 受委托人
bailer 抽泥桶，抽泥筒，泥浆泵，水斗，委托人
Bailey bridge 活动便桥，贝雷桥
Bailey span 贝利式桥跨
Bailey truss 贝利式桁架
bailey wall with ignition belt 卫燃带水冷壁
bailey wall 水冷壁，水冷耐火壁
bailing experiment 抽水试验，扬水试验
bail lighting 球状闪电
bailor 委托人
bailsman 保释人，保证人
bail 保释，担保，保证人，保证金，保释金，吊环，横木，戽斗，夹紧箍，卡钉，水斗，水桶，委托
Bain index 贝恩指数
Bainite quenching 等温淬火，贝氏体淬火
Bainite 贝茵体，贝氏体
Bainitic 贝氏体
bajada 山麓冲积扇
baked carbon 碳精电极，碳极
baked clay 烧黏土
baked diatomite 焙烧硅藻土
baked flux 烧结焊剂
baked winding 烘干绕组
bakelite-cover porcelain-base knife switch 胶盖瓷底闸刀开关
bakelite paste 酚醛树脂胶，胶木板
bakelite powder 胶木粉，电木粉
bakelite sleeve pipe 胶木套管
bakelite tube 胶木管
bakelite wedge 胶木槽楔
bakelite 电木，人造胶木，酚醛塑料，绝缘胶木，胶木
bakelized paper 电木纸，胶木纸
bake-out furnace baker 烘箱

bake 烘烤，焙烧，烧结，烤炉
baking coal 黏结性煤
baking finish 烘漆，烤漆
baking storage 后备存储器
baking varnish 烤漆
baking 烘（焊条、焊剂），焙，烤干，脱氢处理
balance account 结算账户，结余账目，结余账户，结算，结账
balance amount 差额，差额清算，余额
balance arm 秤杆，平衡杆
balance axes system 平衡轴系
balance bar 平衡杆
balance beam 天平梁，平衡杆，平衡梁
balance bridge 平衡活动桥
balance brought forward from last year 上年余额
balance-bus 均衡母线
balance calculation of earthwork 土方的平衡计算
balance coil 平衡线圈
balance cone 平衡锥
balance crane 平衡式起重机
balance cuts and fills 挖填土方平衡，填挖平衡，平衡挖填，均衡挖填
balanced 平衡的
balanced aileron 平衡副翼
balanced amplifier 平衡放大器
balanced armature relay 平衡衔铁型继电器
balanced armature 平衡衔铁，平衡电枢
balanced attenuater 平衡衰减器
balanced-beam relay 平衡杆式继电器
balanced bridge 平衡电桥，有补偿电桥
balanced budget （收支）平衡预算
balanced cable 平衡电缆，对称电缆
balanced capacitance 平衡电容【两导线间的】
balanced check valve 平衡的单向阀，平衡式逆止阀
balanced circuit 平衡电路，平衡回路
balanced condition 平衡状态
balanced control surface 平衡的控制面
balanced controls 平衡机构，平衡控制
balanced current relay 平衡电流继电器
balanced current 平衡电流
balanced damper 平衡挡板
balanced deflection 对称偏转
balanced demodulator 平衡解调器
balanced design 平衡设计
balanced detector 平衡探测器
balanced downflow condenser 平衡进汽式凝汽器
balanced draught 平衡送风
balanced draft furnace 具有引风及鼓风的平衡通风炉膛，负压炉膛
balanced draft 负压通风，平衡通风
balanced earthworks 填挖平衡的土方工程，平衡土方工程
balanced error 平均误差，比较误差，差额
balance detector 检零器，平衡检波器
balanced excitation 平衡励磁，平衡激磁
balanced feeder 平衡馈线
balanced-filter 平衡滤波器
balanced flow 平衡流量，均衡流
balanced force relay 平衡力继电器
balanced gate 平衡门，平衡（闸）门
balanced hole 垂直炮眼
balanced inductor logical element 平衡式感应器逻辑元件
balance disc 平衡轮，平稳盘
balanced-level-type bell gauge 平衡杆式钟形压力表
balanced line system 平衡线路制，平衡线路系统
balanced line 平衡电路，平衡线路
balanced load meter 平衡负载电度表，平衡负载仪表
balanced load 平衡负载，对称负载，平衡荷载
balanced method 平衡法
balanced mixer 平衡混频器
balanced modulator 平衡调制器，对称调制器
balanced moment 平衡的力矩
balanced pair 平衡双股线，平衡线对
balanced periodic quantity 对称交变量，对称周期量
balanced phase 平衡相位，对称相位
balanced-piston governor valve 平衡式活塞调节阀
balanced-plug control valve 平衡的旋塞控制阀
balanced polyphase current 平衡多相电流
balanced polyphase load 平衡多相负荷，对称多相负荷
balanced polyphase system 平衡多相制，平衡多相系统
balanced potentiometer 平衡电位器
balanced pressure rotameter 压力补偿转子流量计
balanced pressure thermostatic trap 压力平衡式恒温疏水阀
balanced protective system 差动保护系统，平衡保护系统
balanced push-pull amplifier 平衡推挽放大器
balance draft combustion 负压燃烧
balanced reinforcement 平衡钢筋
balanced relay 差动继电器，平衡继电器
balanced ring 均衡环
balanced rotary armature 平衡旋转式衔铁
balanced rotary relay 平衡旋转式继电器
balanced running 平稳运转，无振动运转
balanced sash 平衡式窗框［窗扇］
balanced seal 平衡密封
balanced section 平衡断面
balanced signal pair 平衡信号对
balanced state 平衡状态
balanced steam admission valve 平衡式进汽阀
balanced steel ratio 平衡配筋率
balanced step 均衡踏步，平衡阶梯
balanced structure 平衡结构
balanced supply 平衡电源，对称电源
balanced surface 平衡面，补偿面
balanced system 平衡系统，平衡制
balanced three-phase current 对称三相电流，平衡三相电流
balanced three-phase load 平衡三相负载，对称三相负荷
balanced three-phase voltage 对称三相电压，平衡三相电压

balanced-to-ground 对地平衡的
balanced transformer 平衡变压器
balanced transmission line 平衡传输线，对称输电线
balance due 尚欠余额，不定额，到期余额
balanced-unbalanced transformer 平衡-不平衡变压器
balanced valve 平衡阀
balanced vane relay 平衡叶片式继电器
balanced voltage 对称电压【多相电路】
balanced weight 平衡重
balanced wicket 活叶闸门
balanced winding 均衡绕组，对称绕组
balance element 平衡元件
balance equation 平衡方程
balance error 平衡误差
balance flap 平衡襟翼
balance force 平衡力
balance gate 平衡门，平衡闸门
balance hole 平衡孔
balance indicator 平衡指示器
balance in our favour 我方受益的余额，我方结存
balance mechanism 平衡机构
balance of cut and fill 土方平衡，（土方）挖填平衡
balance of export trade 出超贸易差额
balance of feed water 给水平衡
balance of interest 利益平衡
balance of intermittent energy 间歇能量平衡
balance of international payments 国际收支差额
balance of make-up water 补给水平衡
balance of nature 自然平衡
balance of nuclear island 核岛配套设施
balance of nuclear steam supply system 核蒸汽供应系统辅助设备
balance of payment deficit 收支逆差
balance of payment 贸易支付差额，支付平衡
balance of plant （核）电厂配套设备，电站配套设施
balance of power demand and supply 电力供求平衡
balance of power plant 电厂辅助设备
balance of profit 利润结余，利润余额
balance of steam 蒸汽平衡
balance of system 系统平衡
balance of turbine island 汽轮机岛辅助设施
balance of voltage 电压平衡
balance on current account 经常账户余额
balance out 衡消，平衡掉，中和
balance phase control 相位平衡控制
balance piece 平衡块
balance pipe 平衡管
balance piston 平衡活塞
balance pit 平衡试验坑
balance pivot 杠杆支点
balance plane 平衡平面
balance point 平衡点
balance pressure-reducing valve 平衡减压阀
balance price 差价
balance process 平衡过程
balancer-booster 均压机—增压机【同轴】
balance relay 平衡继电器
balancer field rheostat 均压机励磁分流电阻
balance room 天平室
balancer set 平衡机组，均压机组
balancer-transformer 均压变压器【交流三相系统】
balancer 配重，平衡重，平衡器，均压器，补偿器，平衡机
balance sheet date 结账日，资产负债表日
balance sheet of fund 资金平衡表
balance sheet ratio 资产负债比率
balance sheet 平衡表，资产负债表，财务状况表
balance spring 游丝
balance stabilizer shaft 平衡稳定器轴
balance strut 平衡支柱
balance support 平衡支架
balance test 平衡试验
balance the account with 结清差额
balance the book 算账，结算，结账
balance the profit and loss 平衡损益
balance ticket 结算余额，决算单，余额凭单
balance to earth 对地平衡
balance to neighboring lines 邻线平衡
balance transformer 平衡变压器，平衡转换器
balance-type potentiometer 补偿式电位差计，平衡式电位计
balance unit 平衡装置
balance valve 平衡阀
balance voltage 均衡电压
balance weight 平衡锤，平衡块，配重，平衡重量，平衡器，平衡重
balance wheel 平衡轮
balance winding 平衡绕组
balance 秤，结算，平衡，天平，补偿，余额，差额，收支差额
balancing account 平衡结算，平衡账户
balancing adjustment 平衡调整
balancing and overspeed test 动平衡及超速试验
balancing antenna 除干扰天线，平衡天线
balancing apparatus 平衡仪
balancing battery 浮充电电池组，补偿电池
balancing bellows 平衡波纹管
balancing bushing 平衡衬套
balancing bus 平衡母线
balancing capacity 平衡容量
balancing cell 附加电池，反压电池
balancing chain 平衡链
balancing chamber 平衡室
balancing check 平衡检查
balancing coil 平衡线圈
balancing condenser 平衡调相机，补偿电容器
balancing connection 均压连接
balancing control 调零设备，调零装置，平衡控制
balancing current transformer 平衡变流器，平衡电流互感器
balancing cuts and fills 平衡挖填
balancing device 平衡装置
balancing disc pump 平衡盘式泵
balancing disc 平衡环，平衡盘

balancing drum　平衡锤，平衡轮鼓
balancing dynamometer　平衡测力计
balancing dynamo　平衡发电机
balancing energy supply　平衡能源供应
balancing equipment　平衡设备
balancing float　平衡浮子
balancing flux　平衡磁通
balancing heat transfer coefficient　平衡给热系数
balancing hole pump　平衡孔式泵
balancing hole　平衡孔
balancing illumination　均匀照明
balancing liquid column　平衡液柱
balancing loss　平衡损耗
balancing machine　平衡机，平衡试验机
balancing network　平衡网络
balancing of rotating machinery　旋转机械的平衡
balancing pipe　平衡管
balancing piston pump　平衡鼓式泵
balancing piston　平衡活塞
balancing pit　平衡试验坑
balancing pressure　平衡压力
balancing reactor　平衡电抗器
balancing report　平衡报告
balancing reservoir　调节水库，平衡池，平衡水库
balancing rheostat　平衡变阻器
balancing rig　平衡试验台
balancing ring　平衡环，均压环
balancing seat　平衡座
balancing sleeve　平衡套
balancing speed　均衡速度，稳定速度
balancing spring　平衡弹簧
balancing stand　平衡机，平衡架
balancing storage　调节库容，平衡库容
balancing surface　平衡面
balancing tank　调节池，调节水池，平衡水箱
balancing torque　平衡力矩
balancing valve　平衡阀
balancing weight　平衡重，平衡块，配重
balancing　对称，平差，平衡，补偿，定零装置，平衡的
balconied　有阳台的
balcony-like intake　阳台式进水口
balcony　楼厅，阳台，挑阳台，眺台
bale capacity　包装容量
bale of wire　线束
bale opener　开包机，松包机
baler　包装机，压捆机，打包机，压密机
bale tie　打包铁皮带
bale　货物【用复数】，包，捆，件
baling band　打包窄钢带
baling press　打包机，包装机
baling wire　打包钢丝
baling　打包，废物压实，脚手杆
balitron　稳定负载特性电子管
balk board　防护板，隔板，障碍板
balking　突然停止
balk ring　摩擦环，阻环
balk　粗木方，大木，故障，阻碍，障碍，（突然）卡住，枕木
ball adapter　球形连接器，电子管适配器
ball-and-disc integrator　摩擦积分器
ball-and-lever valve　杠杆球阀
ball and line float　深式浮标
ball-and-race type pulverizer mill　钢球座圈式磨煤机
ball and roller bearing　滚珠和滚柱轴承
ball and slot dovetail　球形叶根
ball and socket bearing　球窝轴承
ball and socket coupling　球窝连接，球窝连接器，球窝耦合
ball and socket joint　球窝接头，球窝活节，万向接头
ball and socket size　球窝尺寸
ball and socket type insulator　球窝形绝缘子
ballast bed　道渣路基，石渣路基
ballast cargo　压舱货物
ballast chamber　平衡重室，压块室，压载舱
ballast coil　镇流线圈，平衡线圈
ballast concrete　石渣混凝土
ballasted deck　铺渣桥面，道渣桥面
ballasted track undulation　有渣轨道起伏不平
ballasted　铺道渣的
ballast engine　运渣机车
ballast-factor　镇流器系数
ballast hammer　碎石锤
ballasting and sinking of mattress　柴排压载下沉
ballasting burner　乏气喷燃器
ballasting-up　压载调整
ballast lamp　稳流灯
ballast pump　衡重水泵，压舱泵，镇定泵
ballast resistance　镇流电阻
ballast resistor　镇流电阻器，稳流电阻器
ballast rheostat　镇定变阻器
ballast road　石渣路
ballast sand　压舱沙
ballast screen　渣筛
ballast-surfaced　麻面的
ballast tamper　捣渣机
ballast tank　衡重水箱，压载箱，气镇罐
ballast truck　砂石料车，石渣车
ballast tube　镇流管
ballast water　压舱水
ballast　渣，道渣，碎石，石渣，压舱石，压舱物，路基，石砾，压载物，镇流器，镇流电阻，平衡器，平衡块，镇重
ball bearing motor　滚珠轴承电动机
ball bearing socket　滚珠轴承套
ball bearing support　滚珠轴承支座
ball bearing torque　滚珠轴承扭矩
ball bearing washer　滚珠轴承垫片
ball bearing　滚珠轴承，球面支承，球轴承，珠承铰链
ball bond　球焊接头
ball burnishing　钢球磨光，钢球滚光
ball charge　装球量【磨煤机】
ball check valve　落球式止回阀，球形止回阀，逆止球阀
ball clack　球形阀，球瓣
ball clay　球土
ball cleaner　胶球清洗器
ball clevis　球头挂板

ball-cock assembly 浮球阀装置
ball cock device 浮子装置
ball cock 球阀，球形旋塞，浮球旋塞，浮球阀
ball collar thrust bearing 滚珠环推力轴承，滚珠止推轴承
ball collector （清洗凝汽器的）装球室，胶球收集装置，装球网
ball condenser 球形冷凝器
ball control valve 球控制阀
ball coupling 球形连接器，球形联轴节
ball cover 轴承压盖
ball diffuser 球形风口，球形散流器，球形送风口
ball distributor 填球分布器
ball electrolyte tester 色球式电解液检验器
ball-ended magnet 圆头形磁铁
ball-eye 球头挂环
ball float level controller 浮子式液面调节器
ball float level control 浮子式液位控制
ball float trap 浮球活门，浮子式疏水器
ball float valve 浮子阀，浮球阀
ball float 球状浮体，浮球，球状浮子
ball flow indicator 球水流指示器
ball hardness testing 布氏硬度测定
ball hardness 钢球硬度，布氏球测硬度，布氏球印硬度，布氏硬度
ball head governor 飞锤式调节器
ball head stay 球头撑
ball hook 球头挂钩
ball inclinometer 球形倾斜仪
balling-up of cement 水泥起球
balling-up 起球
ball isolation valve 球隔离阀
ballistic breaker 快速断路器
ballistic constant 冲击常数
ballistic electrometer 冲击静电计
ballistic galvanometer 冲击式检流计
ballistics 弹道学
ballistic work 冲击功
ball joint compensator 球形补偿器
ball joint manipulator 球承机械手
ball joint 球节，球窝接头，球形接合，球形接头
ball lift check valve 球抬升单向阀
ball lifting device 卸球装置
ball lightning 球形雷，球形电闪
ball loading facility 装球装置
ball method of testing 钢球压印试验【测定硬度】
ball mill 钢球磨煤机，球磨机，球碾磨机，钢球磨煤机
ball nozzle 球形管嘴
ballometer 雾粒电荷计
balloon construction 轻捷型构造，轻型构造
balloon densimeter 囊式密度计
balloon frame 轻型骨架
balloon observation 气球观测
balloon sounding 气球探测
balloon structure 轻型结构
balloon survey 气球测量
balloon tyre 低压轮胎
balloon 膨胀，气球
ball passing test 通球试验
ball peen hammer 球形锤头，圆头锤，圆头手锤
ball pressure relief valve 球卸压阀
ball probe 球状探头
ball pulverizer 球磨机
ball pump 球形泵
ball race mill 球式磨煤机，中速球磨机，钢球座圈式磨煤机
ball race 滚珠轴承座圈，滚珠座圈
ball relief valve 安全球阀，减压球阀，释放球阀，球形安全阀
ball resolver 球形解算器
ball ring 滚珠圈，球头挂环
ball roll pulverizer 球式磨煤机
ball saddle 滚珠支撑
ball safety valve 球形安全阀
ball screen 收球网
ball screw 圆头螺钉
ball seat 球形阀座
ball socket manipulator 球承机械手
ball spark-gap 球形火花隙
ball spline 滚珠花键
ball stop 球形止块
ball strainer 球形过滤器
ball test 球压硬度试验
ball thrust bearing 滚珠止推轴承，球形推力轴承，止推滚珠轴承，轴向止推轴承
ball tube mill 钢球滚筒式磨煤机，滚筒球磨机
ball tube pulverizer 球磨机
ball type dust collector 钢珠除尘器
ball type expansion joint 球形补偿器
ball unloading facility 卸球装置
ball up 成球
ball valve oiler 球阀油杯
ball valve 球阀，球形阀，浮球阀，漂子门
ball Y-clevis Y联球头
ball 海岸沙洲，球，滚珠，浮子，球形物
balometer 辐射热测定器
balopticon 投影放大器
balsa wood model 软木模型，轻木模型
balsa wood 轻木，软木
balsa 轻木
Baltimore truss 平行弦再分式桁架
balun(balanced-unbalanced transformer) 平衡-不平衡变压器，平衡-不平衡转换器
baluster column 栏杆柱
baluster railing 立轴栏杆，立柱栏杆
baluster 栏杆柱，栏杆
balustrade 扶手，栏杆
bamboo basket 竹笼
bamboo concrete 竹筋混凝土
bamboo crate 竹笼，竹篓
bamboo effect 竹节效应
bamboo fence 竹篱，竹篱笆
bamboo filament 竹丝
bamboo framing 竹构架
bamboo hedge 竹篱笆
bamboo mattress 竹排
bamboo-pole scaffold 竹脚手架

bamboo-reinforced concrete 竹筋混凝土
bamboo reinforcement 竹节钢筋，竹筋
bamboo ridge formation 竹节生成
bamboo scaffold 竹材脚手架，竹脚手架
bamboo wedge 竹槽楔
bamboo work 竹工
bamboo 竹，竹材
B-amplifier B类放大器，乙类放大器
ban 禁令，禁止，取缔
banana boundary zone 香蕉形外界区
banana jack 香蕉插头的插孔
banana region 香蕉状区域
banana tube 长筒形单枪阴极射线彩色显像管
band adjustment 频带调整
bandage 绑箍，绷带，铁箍
bandal 竹框架工程
band and gudgeon 长铰，大门门铰，大门合页
band approximation 谐带近似法
band-armoured cable 钢带铠装电缆
band-block filter circuit 带阻滤波器电路
band brake backstop 带式逆止器
band brake 带式制动器，带闸
band compensation 频带补偿
band conveyer 皮带运输机
band coupling 带状联轴器
band course 带层
banded architrave 带饰门头线
banded carbide 带状碳化物
banded column 带饰柱，箍柱
banded gneiss 层状片麻岩，带状片麻岩
banded granite 带纹花岗岩
banded lode 带状矿脉
banded penstock 加箍压力钢管
banded porphyry 带状斑岩
banded sandstone 层状砂岩
banded sediment 带状沉积
banded shaft 箍柱
banded steel pipe 加箍钢管
banded structure 带状构造，带状组织
banded vein 带状脉
band electrode 带极，板极
bandelet 扁带，细带
band elevator （倾斜式）皮带运输机
band elimination filter 带阻滤波器
bander 打捆机，箍工
band-exclusion filter 带阻滤波器
band filter 带式滤波器，带通滤波器
band indicator 倾斜仪，倾斜指示表
banding clearance 围带间隙
banding 打箍
banding clip 绑扎夹
banding end fixing strap 绑线端箍
banding insulation 绑带绝缘，端箍绝缘
banding ring 绑环，护环
banding wire 包扎线，绑线
band iron 扁铁，扁钢，扁铁条，带钢，窄型带钢
bandlet 细带，扁带
band lightning 带状闪电
band-limited frequency spectrum 有限带宽的频谱
band-limited spectrum 限带频谱
band limiting 频带限制
band loss （电枢）绑线损耗，端箍损耗
band microphone 带式传声器
band model 能带模型，频带模型
band molding 带状线脚，装饰线条板
band number 区域号码
band of ice 冰带，冰夹层
band of regulated voltage 调整电压范围【以百分数表示】
band-operated hoist 带式卷扬机
band origin 谱带基线
band overlap 能带重叠
band pass filter 带通滤波器
band-pass transformer 带通变压器
band-pass tuner 带通调谐器
band-pass 带通，通频带
band printer 带式打印机
band pulley 带动滑车
band-rejection filter 带阻滤波器，带除滤波器
band rope 扁钢丝绳
band-run aggregate 岸边集料【未经筛选的河边天然骨料】
bandsaw 带锯
band screen 带条筛，带形筛，环带式滤网，带式筛，转筛
band selector 波段开关，波段选择器
band-separation network 频带分离网络
band shoe 管箍座
band shroud ring 围带【叶片】
band spectra 光谱带
band spectrum 带谱，带状（光）谱
band-splitting privacy equipment 频带分段保密设备
band spreader 频带扩展，（微调）电容器
band spread receiver 带展接收机
band spread 频带展宽，波段展开，频带扩展
band stop filter 带阻滤波器
band-suppression filter 带阻滤波器
band switch 波段开关，波段转换开关
band tail localized state 带尾局域态
band tape 卷尺
band theory 能带理论
band to band transition 带间跃迁
band-tower intake 岸塔式进水口
band transformer 波段变压器
bandwidth amplifier 频带放大器，带宽放大器
bandwidth compression 频带压缩
bandwidth correlator 带宽相关器
bandwidth expansion ratio 频带展宽比
bandwidth of vortex shedding 旋涡脱落带宽
bandwidth parameter 带宽参数
bandwidth ratio 带宽比
bandwidth technique 带宽技术
bandwidth 带宽，频带宽
bandy 曲折的，带状的，曲线
band 扁钢，区，带，波段，箍，频带，范围，围带，光带，箍带，箍环
bang-bang circuit 开关式线路，乒乓电路
bang-bang control 乒乓控制，继电器控制，开关式控制，启停控制
bang-bang servo 继电伺服机构

bang 冲击，冲击声
banister brush 软毛刷
banister 栏杆支柱，楼梯扶手
banjo fixing 对接接头，对接组件
bankability 银行的认可性
bankable project 可筹措资金项目，银行肯担保项目
bank acceptance rate 银行承兑（利）率
bank acceptance 银行承兑
bank accommodation 银行贷款，银行设施
bank account 银行账户
bank act 银行条例
bank agreement 银行议定书
bank breaching 堤岸冲毁
bank buying rate 银行买入汇率
bank cable transfer 银行电汇
bank cable 多芯电缆，线弧电缆
bank caving 堤岸坍陷，坍岸，岸塌，淘岸
bank charge 银行费用，银行收费
bank check 银行支票
bank clearance 离岸净空
bank coal fire 压火
bank contact 触排接点
bank crane 岸边起重机
bank credit card 银行信用卡
bank credit transfer 银行信用过户，银行信用证转让
bank credit 银行信贷，银行信用（证）
bank cutting 河岸切割
bank debenture 金融债券
bank demand 银行即期支付【国外汇兑】
bank discount 银行贴现
bank draft 银行汇票
banked 积起的，堆积的
banked battery 蓄电池组，并联电池组
banked boiler 热备用锅炉
banked configuration 联动排列【控制棒组】
banked distribution transformers 配电变压器组
bank eddy 岸边涡流
banked earth 储存土
banked fire 闷火的，压火的
bank edge 岸边
banked period 压火期
banked radiators 分布散热器组【变压器】
banked relay 继电器组，组合继电器
banked rods 联动控制棒组
banked secondary 配电变压器组的二次网
banked-up water level 壅高水位，壅水水位
banked winding 多层连续绕组，叠层绕组，重叠绕法
banker house 煤仓间
bank erosion 岸坡侵蚀，堤岸冲刷，河岸冲刷，河岸侵蚀
banker's bill 银行汇票
banker's demand draft 银行即期汇票
banker's guarantee 银行担保
banker 堤防土工，银行家
bankette 护坡道，填土，弃土堆
banket 护脚，弃土堆
bank failure 坍岸
Bank for International Settlement 国际清算银行
bankfull discharge 满槽流量
bankfull flow 满槽水流，平岸流
bankfull stage 满槽水位，平岸水位
bankfull 平岸，齐岸
bank grading 河岸修坡，河岸修整
bank gravel 河卵石
bank guarantee bond 银行保函
bank-head 岸首，堤头，岸塌，横堤，斜井井口出车平台
bank high flow 高岸流量【即开始漫滩泛滥的流量】
bank indicator 倾斜指示器
banking agreement 银行议定书
banking angle 外缘超高角，侧倾角
banking center 金融中心
banking charge 银行费用
banking department 银行业务部
banking institution 金融机构
banking loss 机组热备用损失，封炉损失，埋火消耗量，空行时的燃料消耗
banking of distribution transformers 配电变压器组
banking process 堆积过程
banking screw 限位螺杆，止挡螺杆，止动螺钉
banking steam turbine 热备用汽轮机
banking 侧倾，外缘超高，斜度，热备用，银行业务，筑堤，停运，压火
bank interest 银行利息
Banki turbine 班克式水轮机，双击式水轮机
bank levee 岸式码头，河岸堤
bank line profile 岸线纵断面，路基边线纵断面
bank line 河岸线
bank loan 银行借款
bank material 土堤坝填筑材料
bank measure 填方量，填方数量
bank multiple cable 线弧复式电缆，触排复接电缆
bank of acceptance 承兑银行
bank of capacitors 电容器组
Bank of China 中国银行
bank of coils 线圈组
bank of commerce 商业银行
Bank of communications 交通银行【中国】
bank of condenser pipes 凝汽器管束
bank of condensers 冷凝器组，电容器组
bank of contacts 接点排，触头排，选择器线弧
bank of deposit 存款银行，开户银行
Bank of East Asia 东亚银行
bank of issue 发行银行
bank of staggered pipes 错列管束
bank of total-head tubes 全压排管
bank of transformers 变压器组
bank of tubes 管排
bank planting 堤岸栽植
bank protection works 护岸工程
bank protection 堤岸护坡，河岸保护，护岸，护坡
bank rate 贴现率，银行利率
bank reef 岸礁
bank reference 银行征信资料
bank refundment guarantee 银行偿付保证书
bank reinstatement method 复堤法

bank reinstatement 复岸
bank revetment 堤岸护坡，护岸，连续护岸
bank ruin of reservoir 水库塌岸
bank-run gravel 河岸砂砾
bank-run sand 岸沙
bankruptcy law 破产法
bankruptcy proceedings 破产诉讼
bankruptcy 倒闭，破产，破产倒闭，无偿付能力
bankrupt 破产，破产的，破产者
bank sand 岸沙，河岸沙，河砂，岸砂
bank seat 碇桩
bank selling rate 银行卖出汇率
bank settlement 堤的沉陷，河岸沉陷
bank side 岸边
bank sill 岸槛，岸坡底槛
bank sliding 河岸滑坡
bank slope 岸坡，路堤边坡，河岸斜坡
bank slough 岸坡表层脱落
bank sloughing 堤岸坍塌，坍岸
bank span 岸跨
bank's swift number 银行编码或流水号
bank stability 岸坡稳定
bank stabilization 岸坡稳定，堤岸加固，管束稳定性，河岸加固
bank statement 银行结算清单，银行对账单
bank storage discharge 河岸调蓄流量
bank storage 岸贮水，堤岸贮水，河岸蓄水，沿岸调蓄
bank strengthening 堤岸加固，河岸加固
bank subsidence 堤的沉陷
bank suction 贴岸吸力（作用）
bank telegraphic transfer 银行电汇
bank terminal 触排端子
bank the fires 压火，维持小火
bank winding 多层连续绕组，叠绕法
bank wiring 触排布线
bank-wound coil 多层连续线圈，叠层线圈
bank 岸，河岸，堤，堤岸，堆，管束，存储单元，倾斜，组合，线弧，银行，组【电容器等】
banquette 窗口凳，长条形软座，后戗，护坡道，弃土堆，戗台
bantam tube 小型管
BA ohm 英制欧姆
BAP(basic assembler program) 基本汇编语言
bar and dot generator 点划发生器
bar and key grate 横梁式链条炉排
bar and yoke method 棒轭磁性试验法
bar apparatus 杆状基线尺
bar armor 线棒保护层，线棒保护带
barb bolt 地脚螺栓，带刺螺栓，基础（棘）螺栓，棘螺栓
barbecue 烤牲炉，户外烧烤，烤肉架
barbed arrow 风矢
barbed dowel pin 带刺销钉
barbed drainage pattern 倒刺状排水系统
barbed drainage 倒钩水系
barbed nail 刺，刺钉
barbed wire border fence 带刺的铁丝网围栅
barbed wire entanglement 带刺铁丝网，有刺铁丝网
barbed wire fence 刺铁丝围墙，倒刺铁丝围栏，带刺的铁丝网围栏
barbed wire 刺钢丝，刺铁丝，带刺的铁丝网，带刺铁丝
bar bender 钢筋挠曲器，钢筋弯曲工，钢筋弯曲机，钢筋弯折机，弯筋工，弯条机
bar bending machine 钢筋弯曲机，钢筋弯折机
bar bending schedule 钢筋表
bar bending 弯钢筋
barber 大风雪，冷风暴
barbituric acid 巴比妥酸
barbotage 鼓泡，起泡的，起泡，泡式的，起泡作用
barb 毛刺，倒钩，装倒钩
bar chain 杆链
bar chair 钢筋支座
barchansdune 新月形沙丘
bar chart display 棒状图显示，条形图画面
bar chart for progress 进度条线图
bar chart 甘特图，横道图，长条图，总计划条线图
bar clamp 杆夹
bar code reader 条形码阅读器
bar code scanner 条形码扫描器
bar code system 条码系统
bar code technology 条形码技术
bar code 条形码
bar-conductor insulation 导条绝缘，线棒绝缘
bar-conductor 导条，线棒
bar copper 铜条，铜棒，母线联络开关
bar-coupling panel 母线联络控制屏，母联控制屏
bar cropper 钢筋剪断机，钢筋截断机，钢筋切断机
bar cutter 钢筋剪断机，钢筋剪切机，钢筋截断机，钢筋切断机
bar drawing test 钢筋拉拔试验
bar drill 杆钻
bard-to-get commodities 紧俏商品
bare bar copper 裸铜条
bare busbar 裸母线，裸汇流条
bare bus 裸母线
bare cable 裸电缆
bare conductor 裸线，裸导体
bare contract 无条件契约
bare copper drain wire 裸铜屏蔽线
bare copper wire 裸铜线
bare core 无反射层堆心，裸堆心【活性区】
bare-cut slope 新开挖的边坡
bared turn 裸线线匝，无绝缘线匝
bare electrode 裸电极，无药焊条，裸焊条
bare-faced tenon 裸面榫头
bare-faced tongue 裸面榫舌
bare foot 无榫骨架
bare fuel element 无包壳燃料元件，裸燃料元件【反应堆】
bare height 裸堆芯高度，裸高度
bare homogeneous reactor 均匀裸反应堆，无反射层均匀反应堆
bare karst 裸露喀斯特，裸露岩溶

bare land 白地，不毛之地，裸地
bare lattice 裸栅格，无反射层栅格
bare motor 无配件电动机
bare pipe 光管，裸管，裸露管
bare pole 裸磁极
bare reactor 裸堆，无反射层反应堆，裸反应堆
bare rock 裸露岩石
bare rod bundle 裸棒束
bare shaft 裸轴
bare solid cable 裸实心电缆
bare source 裸源，无屏蔽源
bare-strap coil 无绝缘的带绕线圈，裸扁线线圈
bare strap copper 裸铜带
bare terminal 焊条夹持端
bare thermocouple 裸露热电偶
bare-tube superheater 光管过热器
bare-tube wall 光管水冷壁
bare-tube 光管
bare weight 空重，皮重，卸去货物后的重量
bare wire 光焊丝，裸线，裸导线
bare 裸的，裸露的，空的，无设备的
bar feed lock 进给杆锁
bar-flight feeder 刮板给料机
bar formation 钢筋构成，钢筋组架
bargain money 定金
bargain renewal option 承租人续租选择权
bargain 谈判，讲价，讨价还价
bar gauge 钢筋规格，钢筋量规，钢筋行距
barge 驳船，泵船，平底船，干扰
barge bed 驳船停泊区
barge berth 驳船码头
barge board 挡风板，山墙挡风板，山墙封檐板
barge canal 驳船运河
barge couple 山墙上的椽，檐口人字木
barge course 山墙沿石板，山墙砖压顶
barge crane 浮式起重机
barge dock 驳船坞
barge dry ash unloader 驳船干灰卸载装置
barge hauling system 驳船牵引装置
barge-loading belt conveyor 驳船装料皮带运输机
barge loading dock 装货的船坞
bar generator 条形脉冲信号发生器
barge port 驳船港
barge shipment 驳船运输，驳船运量，水运
barge stone 封檐石
barge terminal 驳船码头
barge train 驳船队
barge unloader 驳船卸料机，卸船机
barge unloading suction dredge 吹泥船
bar graph 条形图
bar grate stoker 横梁式链带炉排，横梁式炉排
bar grit 条筛
bar grizzly 粗格筛，栅条筛，拦污筛，格筛
bar hammer 棒槌
bar hooked at both ends 两端带弯钩的钢筋
baric area 气压区
baric flow 压流
baric gradient 压力梯度
barie 巴列【气体压力单位】
baring wire 光焊丝，裸铜丝，裸线
bar insulation 线棒绝缘，排间绝缘
bar iron 扁铁，扁钢，钢条，条钢，型钢
barite concrete 重晶石混凝土
barite 重晶石
barium carbonate 碳酸钡
barium discharger 钡放电器
barium glass 防射线玻璃
barium hydroxide 氢氧化钡
barium-oxide-coated cathode 涂氧化钡阴极
barium plaster 钡灰浆，防射线抹灰
barium sulfate 硫酸钡
barium 钡
bark burning boiler 燃树皮锅炉
Barkhausen effect 巴克豪森效应
Barkhausen-Kurz oscillation 巴克毫森-库尔兹振荡
Barkhausen-Kurz valve 巴克豪森-库尔兹电子管
Barkhausen noise analysis 巴克豪森噪声分析
Barkhausen noise 巴克豪森噪声
Barkhausen's formula 巴克毫森公式
bar length from face of support 支座边算起的钢筋长度
barley coal 大麦级无烟煤
barley 无烟煤屑【2.5～5.0毫米】
bar list 钢筋表
bar magnet 磁棒，条形磁铁
bar manipulator 长柄操作器
bar mat reinforcement 钢筋网
bar-matrix display 正交电极杆寻址矩阵显示器
bar mat 钢筋网
bar method 贯通法磁粉探伤
bar mill 条材轧机
barn 仓
barnacles 藤壶
bar netting 钢筋网
barney 小矿车，推车器，推车铁牛，轻型牵引机，小型电动机车
barn waste water 畜圈废水
baroceptor 压力感受器，气压传感器
baroclinic atmosphere 斜压大气
baroclinic flow 斜压流动
baroclinic model 斜压模式
baroclinity 斜压性
barograph 气压计，气压记录器，气压记录仪，自计气压计
barometer altitude 气压计高度
barometer reading 气压读数
barometer 气压表，气压计
barometric coefficient 气压系数
barometric compensator 气压补偿器
barometric condenser 大气压凝汽器，气压式冷凝器
barometric depression 低气压
barometric determination 气压测定
barometric draft control 气压挡板式通风调节
barometric fluctuation 气压变动
barometric gradient 气压梯度
barometric height measurement 气压测高
barometric height 气压高度
barometric high 高气压
barometric jet condenser 喷射式大气压凝汽器

barometric leveling 气压测高
barometric low 低气压
barometric maximum 气压最高值
barometric minimum 气压最低值
barometric pointer 气压指示器
barometric pressure gradient 气压梯度
barometric pressure sensor 大气压力传感器
barometric pressure 气压，大气压，大气压力
barometric 大气的，气压的
barometrograph 气压自动记录仪
barometry gradient 气压梯度
barometry 气压测定，气压测定法
baromil 毫巴【气压单位】
baroresistor 气压电阻
barosphere 气压层
barostat 恒压器
barothermograph 气压温度记录器，气压温度记录仪，气压温度计，自动压力温度计
barothermohydrograph 气压温度湿度记录仪，气压温度湿度计
barotropic atmosphere 正压大气
barotropic flow 正压流动
barotropic fluid 正压流体
barotropic 正压的
bar pattern 线条形图形
bar placer 钢筋工
bar port 候潮港
bar pressure 大气压力
bar primary 条形一次绕组
barracks 临时工房，兵营，工棚
barrage 挡水建筑物，拦河坝，堰，拦河闸堰
barrage gate 堰顶闸门
barrage head 坝前水头，堰前水头
barrage jamming 阻塞干扰，全波干扰，抑制干扰
barrage power station 堰坝式电站
barrage receiver 双天线抗干扰接收机
barrage with stop plank 插板式拦河堰
barranca 深峡，深峪
barranco 深峡
barratron 非稳定波形磁控管
barred claim 终止时效索赔
barred door 无门板栅门
barred gate 无门板栅门
bar reinforcement 粗钢筋，钢筋棒
barrel arch 筒形拱
bar relay 棒式继电器
barrel boiler 筒形锅炉
barrel bulk 一桶【散料体积计量单位】
barrel buoy 桶式浮标
barrel cam 筒形凸轮
barrel cased turbine 桶壳式水轮机
barrel casing feed pump 筒式给水泵
barrel casing pump 筒形壳体泵
barrel casing 筒形壳体
barrel cement 桶装水泥
barrel culvert 筒形涵洞
barrel distortion 桶形畸变，桶形失真
barrel drain 筒形排水道
barrel(/drum) reclaimer 滚筒取料机
barreling 装桶
barrel outlet nozzle 吊篮出口接管
barrel pier 筒形支墩
barrel pin 圆柱，外壳
barrel printer 鼓式打印机
barrel pump 筒形泵
barrel roller bearing 鼓形滚柱轴承
barrel roller 鼓形滚柱
barrel roof 筒形薄壳屋顶，筒形屋顶
barrel-shaped generator 圆筒形发电机
barrel sheet 筒圈，外壳
barrel shell 筒壳，筒体
barrel structure 筒形结构
barrel switch 鼓形开关
barrel type casing 筒形汽缸，筒形缸体
barrel type feed pump 多级蜗壳式离心泵
barrel type pump 筒式泵
barrel type turbine 圆筒形汽缸汽轮机
barrel vault 半圆形拱顶，筒壳，筒形穹顶
barrel winding 桶形绕组，桶形绕法
barrel 桶，圆柱体部分【安全壳】，卷筒，圆筒，筒形物，泵筒
barren hill 荒山
barren inland 溶蚀盆地
barren land 不毛之地，荒地，裸地
barren upland 荒芜高地，灰岩高地，贫瘠高地
barren 荒地，荒芜的，贫瘠的，不结果实的，无价值的，不毛之地
barretter bridge 镇流电阻器电桥
barretter resistance 镇流电阻，镇流管电阻
barretter 稳流灯，稳流管，镇流电阻器
barricade 挡墙，隔板，屏蔽墙，阻塞，路障，栅栏
barrier bar 潜坝型沙洲，沿岸沙坝，沿岸沙埂，沿岸洲
barrier basin 堤堰水池
barrier beach 滨外沙埂，滨外滩，围岸浅滩
barrier chain 沙岛群
barrier dam 拦污坝
barrier device 隔离器件
barrier diffusion 穿过多孔隔板的扩散
barrier effect 势垒效应
barrier factor 势垒穿透系数
barrier film rectifier 阻挡膜整流器
barrier frequency 截止频率，阻挡频率，封闭频率
barrier function 障碍函数，闸函数
barrier grid storage tube 网垒式存储管，阻挡栅存储管
barrier grid 阻挡栅极，制动极
barrier insulation 隔板绝缘
barrier island 砂岛
barrier layer capacitance 阻挡层电容
barrier layer cell 阻挡层光电池
barrier layer condenser 阻挡层电容器，势垒电容器
barrier layer photocell 阻挡层光电管
barrier layer rectifier 阻挡层整流器
barrier layer 阻隔层
barrier pillar 安全矿柱
barrier power station 堤堰式电站
barrier reef 堡礁，堤礁

barrier resistance 阻挡层电阻
barrier spring 堤泉
barrier terminal block 隔板端子板
barrier theory 障碍学说
barrier tube 屏障管
barrier valve 隔绝阀，挡板
barrier wall 屏障，围墙
barrier 壁垒，挡板，栅栏，障碍（物），隔板，隔离（物），栏木，屏蔽栅，屏障，阻挡层
barrigged drifter 柱架式风钻
barring gear 盘车齿轮装置，盘车装置
barring motor 盘车电动机
barring speed 盘车速度，盘车转速
barring 箍紧，盘车
barrister's chambers 大律师事务所
barrister 律师
barrow truck 手推运料车，小型运货车
barrow 手推车，放线车
bar screen 棒条筛，铁栅筛，格栅，清污机，条筛，条形筛，条状栅，栅网
bar setter 钢筋工
bar shearing machine 钢筋切断机
bar spacer 钢筋定位卡，钢筋分隔器
bar spacing 钢筋间距，炉条间隔
bar stay 撑杆，撑条
bar steel 棒钢，扁钢，钢，条钢
barstock body 棒材体
bar stock 棒材，棒束，条材，条料，棒形钢材，钢筋储备
bar straightener 钢筋调直机，钢筋矫直机，钢筋校直机
bar-straightening machine 钢筋拉直机
bar support 钢筋支座
barter agreement 易货协定
barter contract 易货合同
barter market 易货贸易市场
barter trade 易货贸易
barter 易货交易
bar timbering 水平木支撑
bar-to-bar test 换向片片间试验
bar-to-bar voltage peak 片间最高电压【换向器】
bar-to-bar voltage 片间电压【换向器】
bar tracery 铁杆窗格
bar-type current transformer 单匝电流互感器，棒式电流互感器
bar-type transformer 条形铁芯变压器
bar-type trash rack 栅条式拦污栅
bar-type winding 条形绕组，棒式绕组
bar viscometer 杆式黏度计
bar winding 棒式绕组，条形绕组
bar with hooked end 带钩钢筋，带弯钩钢筋
bar-wound armature 条绕电枢，杆式绕组电枢
barycenter coordinates 重心坐标
barycenter 质心，重心
barye 巴列
barysphere 地球核心，地心圈，重圈，地心圈
barytes concrete 重晶石混凝土
baryte 重晶石
barytron 介子，重电子
bar 巴【气压单位，压强单位】，棒，条，杆，炉排片，汇流条，钢筋，型钢，栅门，撬棍
basal cleavage 底面解理
basal conglomerate 底砾岩
basal contact 底部接触
basal disc 基盘
basalt-agglomerate tuff 玄块凝灰岩
basaltic jointing 玄武岩节理
basaltic lava 玄武岩熔岩
basalt-porphyry 玄武斑岩
basal transgressive lithofacies 基本海浸岩相
basalt structure 玄武岩构造
basalt 玄武岩
bascule barrier 竖旋路栏
bascule bridge 竖旋活动桥
bascule door 吊门
bascule gate 竖旋闸门
bascule pier 竖旋桥桥墩
bascule span 竖旋孔，竖旋桥桥跨
bascule 吊桥活动桁架，竖旋桥双翼
base active power 功率因数为1时的总有效功率，额定输出功率
base address 基本地址
base adjuster 基值整流器，底座调整装置
base allowable stress intensity 应力强度，设计应力强度
base alternating voltage 基础交流电压【电力整流器】
base angle 底角
base apparent power 总的额定视在功率
base area 底面积，基础面积
baseband 基本频带
base bar 杆状基线尺，基线杆
base bench mark 水准原点
base bending moment 底部弯矩
base bias bleeder circuit 基极分压电路
base bid price 标底，基本的标价
base bid 基本报价
base block 门基石，柱脚石，柱石
baseboard 护壁板，基线板，踢脚板，底板
base brick 碱性砖
base capacity 通过能力【码头等】
base casing 下缸
base cavity 底部空穴，尾部
base charge 基极电荷，雷管药包
base coat 底漆，底涂层，底漆层
base collector 基极集电极
base concrete 基础混凝土
base-conductivity modulation 基区电导率调制
base consumption of life 基本寿命消耗，寿命的基本消耗
base contact 底部接触
base controller unit 基本控制器单元
base cost 基本投资，基本费用【指直接费和间接费之和】
base current 基极电流，基线电流，基部电流
base data 基本数据，基本资料，原始数据，原始资料
base date 基础日期【即基础价计价日期】，基准日期
base depth 基区深度
base design 基本设计
base-diffused resistor 基极扩散电阻，基区扩散

base display unit 主显示器
base document 基本文件
based on 根据，在……的基础上，以……为基础，依据
base drag 底部阻力
base earth 基极接地
base elbow 带支座弯头
base electrode 基电极
base elevation 基线标高，基底标高
base-emitter 基极发射极
base endshield 座式端盖
base exchange process 阳离子交换过程
base exchange 碱交换，阳离子交换
base excitation 基本励磁
base face 基准面
base failure of slope 坡脚坍塌
base failure 基础破坏
base field 基本场
base flashing 基层泛水
base flow depletion curve 基流退水曲线
base flow hydro-graph 地下径流过程线
base flow storage 基流储水量
base flow 基本径流，基本流量，基流
base for skidding 滑动用底板
base frame 底座，基础台架，基础构架，基架
base frequency 基本频率，固有频率
base generating set 基本发电机组
base heating load 基本热负荷
base hinge 座铰
base hum 基极交流声
base impedance 基本阻抗【基本电压除以基本电流】
base insert 基座垫片
base insulator 支柱绝缘子，托脚绝缘子
base lacquer 底漆
base language 基本语言
base-lead resistance 基极引线电阻
base leakage 管座漏电
base level of erosion 侵蚀基面
base level 基面，基准面，海平面，基数电平本，基本水位，基本坡度，基准高程
base lighting-load 基本照明负荷
base line break 基线中断
baseline correction 基线校正
baseline data 原始数据
baseline design 基准设计，基线设计
baseline study 基线研究
base line survey 基线测量
base line 底线，基线，基准线，初始的，原始的，基本的，原始资料，扫描行
base load capacity 基荷容量
base load compressor 基本负荷压缩机
base load condition 基本负荷工况
base load control 基荷控制
base-loaded 带基本负荷的
base load generating set 基本负荷机组
base load generator 基荷发动机，基底负载发电机
base load heat source 基本热源
base load hydro-plant 基荷水电站，基底负载水电站
base load operation 基本负荷运行，基荷运行，基底负载运行
base load plant 基荷发电厂，基本负荷电厂
base load power plant 基本负荷发电厂，基荷电厂
base load power station 基荷电站，基底负载发电厂
base load power 基荷动力，基荷功率
base load rated output 基本额定出力
base load regime 基荷工况，基本负荷工况
base-load station 基本负荷电站
base load supply 基荷电源
base load thermal power plant 基本负荷火力发电厂，基本负荷火电厂
base load turbine 基荷汽轮机，基本负荷汽轮机
base load unit 基本负荷机组，基荷机组
base load 低谷负荷，基本负荷，基底负载，基本荷载，基本载重，基荷
base map 底图，工作草图
base mark 基线标志，基础标志
base material test coupon 母材试样，母体金属材料试样
base material 基本材料，基底材料，母材
base mat 基础层，护面，垫板，基础底板
base measurement 基线测量
base measuring tape 带状基线尺
base measuring wire 钢丝基线尺
basement bin 地下斗，地下仓
basement complex 基底杂岩
basement floor 零米标高，底层，零米层，地下室层，地下层
basementless 无地下室的
basement wall 地下室墙
basement 底层，地下室，基础，地窖汽轮机底层，底基
base metal attack 母材金属腐蚀
base metal couple 基金属温差热电偶
base metal crack 基体金属的裂缝，母材金属的裂纹
base metal test specimen 母材试件
base metal thermocouple 基金属热电偶
base metal 基本金属，母材，母材金属，碱金属，碱基金属，底层金属，基金属母材
base moment 基底力矩
base moulding 底座线脚
base network 基线网
base notation 基本符号，基数记数法
base of dam 坝脚
base of energy source 能源基地
base of excavation 开挖基线
base of foundation 基础的底板
base of logarithm 对数底
base of road 路基
base of sea mark 潮汛基线
base of slope 坡底，坡脚，坡脚底部
base of spring 弹簧座
base package 基本组件
base pad 基底，底垫
base period 地下径流时期，基本周期
base pin 座销，管脚
base plate 底板，垫板，底座，支承板，基础，

地脚板，基板，基座
base power 基本功率【交流电机】
base pressure 底压，基础压力
base price for bid evaluation 评标基准价
base price 基础价【指未计入浮动加价的价格】，标底，基价
base quantity 基本量
base rate 基准利率，基本电价
base reference potential 基极参考电位
base regime 基本工况
base region 基极区
base register 基址寄存器
base resistance 基本电阻，基极电阻
base rock 岩石，基石，基岩，底岩
base runoff 基本径流，基（本径）流，枯水流量
base screw 底座螺钉
base self-biasing effect 基区自偏压效应
base sheet 原图
base shell 管座外壳
base slab 底板，垫板，基板，基础板，基础石板
base slag 碱性炉渣
base slope 底坡
base speed 基本转速，额定同步转速，基本速度
base spreading resistance 基区扩展电阻
base stabilization 地基稳定
base standard cost 基本标准成本
base station 基本测站，基本（水文）站，基站
base structure 基本结构，底部结构
base suction 底部吸力
base support equipment 主要辅助设备
base support 管道支架，底座支架
base tariff 基础税率
base tee 带支座三通
base thickness 底厚，基区厚度
base tilt factor 基底倾斜因数
basetone loudspeaker 低音扬声器
base transport factor 基区传输因子
base treatment 地基处理
base turbine 后置汽轮机
base unit （测量的）基本单位
base-vented hydrofoil 底面开孔水翼
base voltage 基本电压【交流电机的额定相电压】
base width modulation effect of bipolar transistor 双极晶体管基区宽度调制效应
base width modulation 基区宽度调制
base width 底宽，根开
base 地基，底，底座，基地，垫层，基本，基础，底板，碱，根据，基座，基本方式
BASE 基本【DCS 画面】
basic allowable stress intensity 基准容许应力强度
basic amplitude 基本振幅
basic apparent power 基准视在功率
basic assembler language 基本汇编语言
basic assembler program 基本汇编程序
basic automatic checkout equipment 基本自动检验装置
basic Bessemer pig iron 碱性转炉生铁
basic bid schedule 投标所要求的基本的项目执行进度表
basic calcium phosphate 碱性磷酸钙
basic capacity 基本容量，基本功率，基本通过能力，碱性
basic cave 主屏蔽室
basic charge as per installed capacity 按设备容量算的基本电费
basic circuit 基本线路，原理电路图，主接线，原理图
basic coal-flow paths 主要煤流通路
basic condition 基本条件
basic contractual responsibility 基本合同责任
basic contract 基本合同
basic control program 基本控制程序
basic control system 基本控制系统
basic counter 主计数器
basic covering 碱性药，碱性药皮
basic cycle 基本循环
basic data 基本数据，基本资料，原始资料，原始数据
basic design data 基本设计数据，设计原始资料
basic design drawing 基本设计图，基础设计图纸
basic design feature 基本设计要素
basic design information 基本设计资料
basic design load 基本设计荷载
basic design package 基础设计包，基本设计包
basic design scheme 基本设计方案，基础设计方案
basic design 初步设计，基本设计，基础设计，基准设计，原图
basic dimension 基准尺寸
basic display unit 基本显示装置
basic document 原始单据
basic earthquake intensity 地震基本烈度，基本地震烈度
basic electrode 碱性焊条
basic element 基本元件
basic encoding rules 基本编码规则
basic engineering design data 基础工程设计数据
basic engineering version 初步设计版
basic equation 基本方程
basic error 基准误差
basic event 基本事件
basic failure rate 基准故障率
basic flow diagram 基本流程图，流程图
basic flow 基元流，基本流
basic flux 碱性焊剂
basic frequency 基本频率，主频率
basic functional requirement 基本功能要求
basic function unit 基本功能元件
basic function 基本功能
basic grid strap 基本格架条带
basic grid 基本网络
basic harmonic 基波，基本谐波
basic heat loss 基本耗热量
basic hole system 基孔制
basic hydrological forecast 水文趋势预报
basic hydrologic data 基本水文数据

basic impulse insulation level 基本冲击绝缘水平，基准冲击绝缘水平
basic industry 基础产业
basic information 基本资料，基础资料，基本指令
basic instrumentation 基本检测仪表
basic insulation level 基本绝缘等级，绝缘基本冲击耐压水平
basic internal pressure 基本内压
basic ion 碱性离子
basicity index 碱度指标
basicity 碱度，碱性
basic law of heat conduction 导热基本规律
basic law 根本法，基本法
basic layout 基本布局
basic limit 基本限值，基本限度
basic line 底线
basic load heater 基本负荷加热器
basic load 基本负载，基本荷载，基本负荷，主要荷载
basic mechanical design feature 主要技术性能
basic medium-level radioactive waste 碱性中放废物
basic model 基本模型
basic network 基本网络
basic nuclear installation 基本核设施
basic open hearth furnace 碱性平炉
basic open hearth process 碱性平炉炼钢法
basic open hearth steel 碱性平炉钢
basic operating conditions 基本工作条件，基本运行参数
basic operating parameter 主要运行参数
basic operating system 基本操作系统
basic operation condition 基本运行工况，主要工作状况
basic order code 主指令码
basic oxide 碱性氧化物
basic parameter 基本参数
basic plan 底图
basic power source 基本能源，主要能源
basic power 基本容量，基本功率
basic principles of radiation protection 辐射防护基本原则
basic principles training simulator 基本原理培训仿真器
basic principle 基本原理
basic property 碱性
basic pulse 基本脉冲，基准脉冲
basic regional wind velocity 地区基本风速
basic requirement 基本要求
basic safety 基本安全
basic salt 碱式盐
basic seismic intensity 地震基本烈度
basic shaft system 基轴制
basic shaft 主轴
basic size 基本尺寸，基本规格，公称尺寸，标称尺寸
basic skill 基本技能
basic slag 碱性渣
basic solution 碱性溶液
basic specifications 主要技术规格，基本参数
basic-stage flood 基点供水
basic standard 基础标准
basic storage module 基本存储模块
basic stress 主应力
basic structure 基本结构
basic switching impulse insulation level 基本操作冲击绝缘水平
basic temperature 基础温度
basic time 第一段时限【继电保护】，基本时限
basic type 基本类型
basic unit weight 基准容重
basic value 基数，基值
basic variable 基本变量
basic wage 基本工资
basic wave current 基波电流
basic wind speed 基本风速
basic 碱的，碱性的，基本的
BASIC 一种通用的会话式计算机语言
basil 遮光屏
basin analogy 流域相似
basin configuration 流域轮廓，流域形状
basin dry dock 池式干船坞
basin entrance 泊船池入口
basin irrigation 漫灌
basin lag 流域滞时
basin management 流域管理
basin mean runoff 流域平均径流
basin mean 流域平均
basin mouth 流域出口
basin outlet 流域出口
basin perimeter 流域周长
basin range structure 断块构造
basin recession 流域退水
basin recharge 流域补给，流域再补给
basin room 洗餐具室
basin shape factor 流域形状系数
basin storage discharge 流域调蓄流量
basin storage 流域蓄水量
basin-wide program 流域整体规划
basin 盆地，槽，池，潭，水池，水槽，承盘，流域面积，漫灌塘
basis control module 基本控制模件
basis for taxation 纳税基数
basis of credit 放款根据，信用根据
basis risk 基差风险
basis triangulation network 基本三角网
basis 根据，基础，基准，基底，基数，基线
basket capital 花篮状柱头
basket coil 篮形线圈，笼形线圈
basket guard 吊篮防护器
basket handle arch 三心拱
basket purchase 整套购买
basket reactor 吊篮反应器
basket strainer 篮式过滤网
basket type centrifuge 滚筒型离心机
basket-type sampler 笼式采样器，匣式采样器
basket-type trash rack 围笼式拦污栅
basket weave armor 编织铠装层
basket winding 篮形绕组，笼形绕组
basket-wound coil 篮形线圈，笼形线圈
basket 传热组件框架【回转式空气预热器】，

铲斗，篮，筐，吊篮，笼
basque 涂层，炉衬
bass boost 低音频放大，低音增强
bass compensation 低音补偿
bass 低音，低频
bastard ashlar 粗琢石
bastard coal 劣质煤，硬煤
bastard file 粗齿锉
bastard freestone 劣质毛石
bastard jointing 粗嵌缝
bastard masonry 乱石圬工
bastard sawed board 粗锯板
bastard stucco 粗粒水泥粉刷
bastard tuck jointing 粗嵌灰缝
bastard 假的，畸形的，非标准的
basylous 碱的，碱性的，碱式的
BATAN(National Atomic Energy Agency) 国家原子能局【印尼】
BAT=battery 蓄电池，电池
batch agitator 分批搅动器，间歇式拨火板
batch bin 配料仓，配料箱
batch-bulk processing 成批处理
batch-by-batch method 分批核算法
batch charge 间断负荷，断续充电，间歇加料
batch controller 批量调节器，批量控制器
batch control sample 批量控制采样
batch counter 伴和计数器，配料计算器，选组计数器
batch data processing 成批数据处理
batch data 分批数据
batch discharge 分批卸料
batch distillation 不连续蒸馏法【间歇式蒸馏】
batched water 一次拌和用水
Batchelor constant 白切勒常数
batcher 拌和机的配料仓，计量器，计量箱，料斗，给料机，配料器，配料箱，配料楼，混凝土分批搅拌机
batch-fabricated 成批生产的
batch file 一批文件
batch filter 间歇式过滤器
batch filtration 间歇式过滤
batch firing system 间歇式燃烧系统
batch flotation 分批浮选，间歇浮选
batch-flow production 成批生产，批量生产
batch fluidized bed 间歇流化床
batch handling 分批处理，间歇处理
batch heat transfer 间歇式传热
batch hopper 配料斗
batching by volume 按体积配料，体积配料法
batching by weight 按重量配料，重量配料法
batching counter 剂量计数器
batching equipment 配料设备
batching plant （混凝土）搅拌机，拌和厂，搅拌站，分批配料装置，配料装置
batching processing 成批处理
batching pump 计量泵
batching set-up 配料装置
batching tank 配料箱，配制箱，制备箱
batching 分批，分段，分组，配料，定量，分类
batch integrator 分批积分器，选组积分器
batch ion exchange 分批离子交换，间歇离子交换
batch irradiation 分批辐照，间歇辐照
batch loading 分批加料，间断加料
batch meter 分批称量器
batch method 分批处理法，间歇处理法
batch mixer 分批搅拌机
batch mixing 分批混合
batch number 批号
batch of concrete 分批（拌和的）混凝土
batch of mortar 分批（拌和的）砂浆
batch operation 间歇操作，分批操纵，间歇操纵，间歇运行，配料作业
batch pile 分批料堆，材料储备量
batch plant 拌和厂，拌和楼，配料装置
batch pot dissolution 分批罐式溶解，间歇罐式溶解
batch processing 成批处理，间歇加工，间歇处理，批处理
batch process 间歇过程，分批法，间歇法，批量制法，批量生产
batch produce stage 批量生产阶段
batch production 分批生产，批量生产
batch reactor 间歇反应器，分批加料反应堆
batch refuelling scheme 分批换料法
batch refuse incinerator 分批垃圾焚化炉
batch settling 间歇沉降
batch size 一次换料量
batch soldering equipment 批量焊接设备
batch soldering 批量焊接
batch tank 计量箱
batch terminal 成批处理终端，批处理终端
batch treatment 分批处理，间歇处理
batch truck 分批运料车，拌和卡车，配料车
batch-type 分批式，间歇式
batch unit 间歇操作设备
batch-weighing plant 分批称重配料装置
batch weight 批重
batch-wise operation 间歇式操作
batch-wise 分批的，间歇的
batch 批处理【计算机】，批【货物】，程序组，一炉，一批，炉料，配料，批量，一次生产量，一次投料量，分批（法），份额，定量
bateau bridge 浮桥
Bate damper 贝特阻尼线
batement-light 跛窗，坡窗，楼梯斜窗
bat-handle switch 铰链式开关，手动式开关
bath composition 电解液成分
bath lubrication 油浴润滑
batholite 岩基，岩盘
bathometers string 水深仪测量绳
bathometer 测深器，水深测深器
bathothermograph 温深仪
bathroom 浴室
bath solution 电解液
bathtub capacitor 金属壳纸质电容器
bathtub curve 澡盆曲线，浴缸曲线【即产品失效率曲线】
bathtub 浴缸盆，浴盆，澡盆
bath voltage 电解槽中阴阳极间的电压
bathylith 岩盘

bathymeter 测深器，水深测量器
bathymetrical chart 等深线图，水深图
bathymetric line 等深线
bathymetric measurement 测深
bathymetric survey 水深测量
bathymetry 测深法
bathyorographical map 水底地形图，水深地形图
bathy photometer 深水光度计
bathyseism 深源地震
bathysphere 潜水球
bathythermograph 深度温度仪，水深温度自记仪
bath 浴，洗浴，槽，池，电解槽
batt 沥青质页岩，黏土质页岩
batten and button 木板接合法
batten board 板条心胶合板
batten door 板条门
battened partition 板条隔墙
battened strut 钉板条的支柱
battened wall 板条墙
batten ends 短板条
batten floor 木条地板
battening 钉板条
batten plate column 缀合柱
batten plate 缀合板
batten sheet piling 打板条桩，打条板桩
batten system 同位穿孔检查系统
batten underdrain 木板暗沟
batten wall 板壁，板条墙
batten 夹板，撑条，板条，压条，木条，顺水条
batter board 定位板，定斜度板，龙门板
batter brace 斜撑
batter chamber wall 斜面闸室墙
batter drainage 斜水沟
battered bank system 倾斜护岸工程
battered pilaster 斜面壁柱
batter gauge 斜坡样板，定斜规
batteries test loops 蓄电池试验电路
battering ram 冲击夯
battering rule 测斜器，定斜尺，坡规
battering wall 倾斜墙
batter leader pile driver 斜导架式打桩机，打斜桩机
batter leg tower 斜柱塔架
batter level 测坡仪，测斜器，倾斜仪
batter pier 斜面墩，斜面式桥墩，斜桩，斜柱
Battersea gas washing process 巴特西（电站）烟气洗涤法
battery acid 电池酸液，蓄电池用酸
battery backup module 电池备用组件
battery backup unit 电池后备单元
battery backup 电池（组）备用
battery bank 蓄电池组
battery booster 充电调压器，蓄电池用的计压器
battery capacity 蓄电池容量，蓄电池室容量
battery car 蓄电池车，电瓶车
battery cell 电瓶，原电池，（蓄电池组）电池
battery charger 蓄电池充电机，蓄电池充电器
battery charging system 蓄电池充电系统
battery charging 充电，蓄电池室充电
battery clock 电池钟
battery container 电池槽
battery cutout 电池电路自动断路器，电池断路器
battery dialing 单线拨号
battery driven motor 电池供电电动机
battery drive 电池驱动
battery eliminator 代电池，电瓶代用器，电池组代用器
battery energy forage 蓄电池贮能
battery excited generator 蓄电池励磁发电机
battery-fed motor 蓄电池供电的电动机
battery ignition 蓄电池点火，蓄电池引燃
battery input 蓄电池充电
battery in quantity 并联电池组
battery insulator 蓄电池绝缘子
battery jar 电池瓶，蓄电池容器，电瓶单元
battery limit line 装置边界线
battery limit 界区
battery locomotive 电瓶机车，蓄电池式机车
battery meter 电池安时计
battery of capacitors 电容器组
battery of wells 井组
battery-operated digital pressure monitor 电池数字压力监测仪
battery overcharge 电池过量充电
battery pack 电池包，电池组件，蓄电池组
battery pliers 蓄电池钳
battery polarity 电池端子极性
battery power drill 电池钻
battery-powered 用电池供电的
battery rack 电池架
battery receiver 电池式接收机，电池式收音机
battery-regulating switch 蓄电池调整开关
battery reverse switch 电池换向开关，电池反向开关
battery ringing 电池呼叫
battery room 蓄电池间，蓄电池室
battery-saver 电池保护元件
battery separator 蓄电池极板的隔离板
battery setting 有公用分隔墙的两台和多台锅炉装备
battery solution 电池电解液，电池溶液，电液
battery-supply coil 蓄电池供电线圈
battery-supply relay 电池供电继电器
battery terminal 蓄电池接线端子，电池接线头
battery tester 蓄电池试验器
battery timing group 电池定时组
battery traction 蓄电池牵引，蓄电池牵引机
battery truck garage 电瓶车库
battery truck 电瓶车
battery tube 直流管，电池供电管
battery-type locomotive 蓄电池式机车
battery 电池，蓄电池，组，排，电池组
batter （墙壁或路堤的）坡度，倾斜，倾斜度，斜坡，锤打
battlement 城墙牒口，防卫墙，城墙垛
batture 淤高河床
bat-wing antenna 蝙蝠翼天线
baud base system 波特基准制
Baud rate 波特率，波特速率，传输速度
Baud 波特【数据通信速度的表示单位】，波特率

baulk 粗木方，大木
Baume degree 波美度【测量液体比重用】
Baume scale 波美比重计
bauxite brick 矾土砖
bauxite cement 矾土水泥，高铝水泥
bauxite 铝矾土，铝土矿
bawke 料罐
bay bar 湾内沙坝
bay-bolt 地脚螺栓，基础螺栓
bay delta 海湾三角洲
bay harbor 港湾
bay head bar 湾头坝
bay head delta 湾头三角洲
bay head 湾头
bay joint 跨间接缝
bay level functions 间隔层功能
bay mouth bar 湾口沙坝，湾口沙洲
bay mouth 湾口
bayonet arrangement 回流管布置【热交换器】
bayonet base 卡口灯座，插口式灯座，插座
bayonet cap 卡口灯头，卡口帽，插头盖
bayonet catch 插销，插座，卡口式连接
bayonet closure 插入式封盖【改进型气冷堆立管头部】
bayonet connector （可快速拆卸的）卡口接头
bayonet coupling 插旋式接头，卡口式联结器
bayonet fixing 卡口式固定，管脚固定
bayonet gauge 插入式表计
bayonet holder 卡口灯头，卡口灯座，插座
bayonet sampler 插入式取样器
bayonet socket 卡口灯座，卡口插座
bayonet 接合销钉，卡口，接合销钉
bayou lake 长沼
bay window 八角窗，窗洞，凸窗
bay （电厂的）煤仓间【往往是露天的】，（变电站的）间隔，海湾，湾，间距，跨度，框架，开间，隔舱，柱距，底板，托架
Bazin roughness coefficient 巴森糙率系数
bazooka 导线平衡转接器，超高频转接变换器，平衡到不平衡变换装置，活动螺旋运送器
BB(ball bearing) 滚珠轴承
BBL=barrel 桶，筒，圆筒形支架
BBL(basic insulation level) 绝缘基本冲击耐压水平
BBM(break-before-make) 先断后通，先开后关
BBRG(bail baring) 滚珠轴承
BBS(behavior-based safety) 行为导向安全，行为安全
BBS(British Standard Sieve) 英国标准筛
BBS(bulletin board service) 公告板服务
BBT(ball bearing torque) 滚珠轴承扭矩
BBU(battery backup) 电池（组）备用
B/C(benefit-cost ratio) 效益成本比率，益本比
BC(between centers) 中心距，轴心距
BCC(blind carbon copy) 隐式抄送【电子邮件用语】，密件抄送，隐蔽副本，密送人
BCD(binary coded decimal) 二进制编码的十进制
B-class insulation B级绝缘
BCL(backfilling check list) 回填土检查单
BCP 炉水循环泵【DCS画面】
BCR(benefit-cost ratio) 效益成本比率，益本比
BCS(burner control system) 燃烧器控制系统
BCT(bushing current transformer) 套管式电流互感器
BCU(base controller unit) 基本控制器单元
BD(block decrease) 闭锁减
BDIG(brushless doubly-fed induction generator) 无刷双馈异步发电机
BDS(basic data sheet) 基础数据表
BDS(steam generator blowdown system) 蒸汽发生器排污系统
BDU(base display unit) 主显示器
BDV(blow-down valve) 再热旁路放汽阀，排污阀
BDV(breakdown voltage) 击穿电压
BD 排污【DCS画面】
beach barrier 海滩暗礁
beach comber 拍岸浪
beach deposit 海滨沉积
beach drifting 海滨漂流，沿滩漂移，海滩移动
beached bank 铺石护岸
beach erosion 海滩冲刷，海滩侵蚀
beach face 前滩
beach formation by waters 浪成海滨，浪成滩
beach gravel 海滨砾石
beaching 海岸堆积，砌石护坡
beach line 海滨线
beach pipe 水力输泥管
beach profile 海滩剖面
beach protection and accretion promotion 促淤保滩
beach retreat 海滩冲刷
beach sand 海砂
beach slope 海滨比降
beach 海滨
beacon antenna 指向标天线【无线电】
beacon lamp 标向灯
beacon light tower 探照灯塔
beacon light 灯标
beacon trigger generator 信标触发信号发生器
beacon 岸标，灯标，信（号）标，信号台，信号塔，信号站，灯塔（标志），航标，指示标，指路明灯
be acquainted with 了解，认识
bead blasting 喷丸
bead collecting system 微丸收集系统【喷丸硬化】
bead collector 微丸收集器
bead crack 焊道裂纹
beaded insulation 串联绝缘，垫圈绝缘
beaded screen 粒状荧光屏
beaded tube end 翻边的管端，板边的管端
bead electrode 珠状电极
bead generator 微丸发生器
beading fillet 半圆饰边，凸圆线脚
beading 焊上焊珠，玻璃熔接，压出凸缘，管口卷边，叠置焊道
bead joint 填角焊缝，圆凸勾缝
be adjudged to be... 被裁定为……
bead lightning 珠状闪电
bead machine 压片机，压锭机

bead-on plate weld　堆焊焊缝
bead resistance　珠电阻，珠状热敏电阻
beads-shaped structure　串珠状构造
bead storage　垫底库容
bead-supported line　绝缘珠支持线路
bead thermister　珠形热敏电阻器
bead thermocouple　珠形热电偶，珠形温差电偶
bead transistor　熔珠晶体管，珠状晶体管
bead welding　堆焊，珠焊
bead　叠珠焊缝，墙角圆，焊道，球，珠，卷边，磁珠，焊蚕，焊珠，压缝条
beagle　自动探测干扰台，自动搜索干扰台
beaker　量杯，烧杯
beak welding　堆焊焊道
beak　壳尖
bealock　分水岭山口
beam-addressed display　束寻址显示器
beam alignment assemble　波束校正装置
beam analogue　梁比拟【一种计算方法】
beam anchor　梁端锚栓
beam and girder construction　主次梁（式）结构
beam and girder floor　主次梁式楼板
beam and slab structure　梁板结构
beam angle of scattering　散射波束锥形角
beam antenna　定向天线
beam array　定向天线阵
beam attenuation　射线束衰减
beam balance　杠杆式天平
beam bearing block　梁垫块
beam bender　弯梁机
beam bending magnet　电子束偏转磁铁
beam bending stress　悬梁弯曲应力
beam blanker　电子束熄灭装置
beam bracket　横梁托架
beam bridge　梁式桥
beam calipers　大卡尺，卡尺，梁卡，游标卡尺
beam catcher　射线收注栅，电子束收集器
beam ceiling　露梁平顶
beam channel　槽型梁
beam clamp　抱箍，横梁夹板，梁式夹车装置，梁式夹车器
beam collimation error　波束准直误差
beam collimator　束准直仪
beam-column connection　梁柱接合，梁柱连接，梁柱结构
beam-column　梁柱，偏心受力柱
beam communication　定向通信
beam-compasses　横臂圆规，长臂规，长脚圆规
beam conductance　电子束电导
beam constant　梁常数
beam control　射束控制，电子束控制
beam current　束流，电子束电流，射束电流
beam curvature　梁的曲率
beam deflecting crystal　光束偏转晶体
beam-deflection tube　电子束偏转管
beam-deflection valve　射束偏转管
beam-deflection　射束偏转，电子束偏转，光束偏转
beam distortion　（声，光，射线）束畸变
beamed radio　定向无线电通信
be amended to read in(/as)　改为……【文字修改】
beam engine　立式蒸汽机
beam extractor　束引出装置
beamfilling　梁间墙
beam fitter　束流起伏，束流晃动
beam fixed at both ends　两端固定梁
beam fixed at one end　一端固定梁
beam flange　梁翼缘
beam focusing　束聚焦
beam-forming electrode　成束电极
beam foundation　梁式基础
beam gantry　中横担
beam hanger　梁托
beam hole　束流孔，束孔
beam homogeneity　束流均匀性
beam idler gear　惰性轮齿
beam index color picture tube　电子注引示彩色显像管
beam-indexed electrical signal storage tube　电子束字标电信号储存管
beam index　波束指数【超声】，声束入射点
beam injected crossed-field amplifier　注入式正交场放大管
beam injection magnetron amplifier　注入式磁控放大管【毕玛管】
beam intensity　光束强度，电子束强度
beam landing　射束沉陷，电子束沉陷
beam lead bonding　梁式引线连接，梁式引线连接法
beam lead integrated circuit　梁式引线集成电路
beam lead isolation　梁式引线隔离
beam lead matrix method　梁式引线矩阵法
beam length　射线长度
beam load　电子束负载
beam matching　束流匹配
beam modulation　射束调制，束流调制
beam of constant strength　等强度梁
beam of elastic support　弹性支承梁
beam-of-light transistor　光束晶体管
beam of uniform strength　等强度梁
beam of variable cross section　变截面梁
beam on elastic foundation　弹性地基梁
beam on elastic support　弹性支承梁
beam orifice　射线孔
beam pad　梁垫，梁垫板
beam penetration cathode ray tube　电子束穿透式电子束管
beam pivot　杠杆，支点
beam-positioning magnet　电子束位置调整磁铁，电子束调位磁铁
beam power tube　电子束功率管，射束功率管
beam power　波束功率，电子束功率
beam purity indicator　光束纯度显示器
beam radio communication　定向无线电通信
beam reflector aerial system　定向反射天线系统
beam relaxor　锯齿波发生器
beam relay　平衡杆式继电器，旋转衔铁式继电器
beam-restriction device　限束装置
beam retrace time　电子束回扫时间
beam rider guidance　波束制导

beams and stringers 横梁与纵梁
beam scale 杆式秤
beam separator 梁间隔材
beam slab structure 梁板结构
beam slab 梁式板
beam solar irradiance 直接日射辐照度
beam solar radiation 直接日射
beam span 梁跨度
beam spar 腹板式翼梁
beam splitter mirror 分光镜
beam splitter prism 分光棱镜
beam splitter 电子束分裂设备，分光器，分束器
beam splitting lens 分光透镜
beam splitting 光束分离
beam storage tube 射束存储管
beam storage 电子束扫描存储器
beam stud 梁托
beam supported at both ends 两端支承梁，托端梁
beam support 梁式支架
beam-switching tube 射束开关管，电子束开关管，射线开关管
beam test 梁试验
beam tetrode 射束四极管
beam theory 梁理论
beam-to-beam spacing 电子束间隔，射束间隔，梁间间隔
beam trammel 骨架
beam transmission 定向发射，定向传输，射束发射
beam transmitter 定向发射机，波束发射机
beam trap 电子束阱，射束收发器
beam tube 射束管，电子束管，束射管
beam wave 束状波
beam width 射束宽度，天线方向图宽度，束宽
beam wind 横风，侧风
beam with both ends built in 固端梁
beam with central prop 三托梁，三支点梁
beam with compression steel 双筋梁
beam with double reinforcement 双筋梁
beam with fixed ends 固端梁
beam with overhanging ends （两端）悬臂梁
beam with simply supported ends 简支梁
beam with single reinforcement 单筋梁
beam with variable cross section 变截面梁
beam 梁，次梁，横梁，光线，杆，光束，射线，束，射束，电子束
bear 常有，支持，承受，负担，负重
bearable load 可支撑荷载
bear a loss 负担损失
bearer bar 支承梁
bearer cable 承载钢索，吊索，支持钢索
bearer frame 支承架
bearer plate 座板，支承板
bearer 承木，持票人，托架，载体支承，支座，垫块，载体，支架
bear expenses 承担费用
bearing accuracy 定位精度
bearing adjustment 轴承调整
bearing alignment 方位对准
bearing alloy 轴承合金
bearing angle 方位角
bearing area 承压面，承压面积，支承面，支承面积，承载面积，承重面积
bearing baffles 轴承挡板
bearing block connection 垫板接合
bearing block 支承块，轴承座
bearing body 轴承体
bearing box 轴承箱
bearing bracket 轴承座，轴承（支）架
bearing brass 轴承合金，轴承黄铜
bearing bridge 轴承支架
bearing burnt-out 轴承损坏【烧瓦】
bearing bush 轴承衬，轴承套，轴瓦
bearing cage 轴承保持架，轴承罩
bearing camber 支承反挠度
bearing capacity factor 承载力系数［因数］【地基】，承载量系数，支承力因数，承载系数，载重因数
bearing capacity of soil 土的承载力，土壤承载力
bearing capacity 承载能力【土壤等】，承载力，承受能力，容许载荷，承载量
bearing cap 轴承盖
bearing cartridge 轴承架，轴承支架
bearing chatter 轴承颤振，轴承跳动
bearing clearance check 轴承间隙检查
bearing clearance 轴承间隙
bearing collar 轴承环，轴颈，推力环
bearing cooling water cooler 轴承冷却水冷却器
bearing cooling water head tank 轴承冷却水高位水箱
bearing cooling water pipe 轴承冷却水管
bearing cooling water pump 轴承冷却水泵
bearing cooling water system 轴承冷却水系统
bearing cooling water 轴承冷却水
bearing course 承压层，承重层，承压垫层
bearing cover 轴承盖，轴承罩
bearing deformation 承压变形
bearing deviation 方位偏差
bearing disk 止推轴承盘
bearing end cover 轴承端罩，轴承外盖
bearing end shield 轴承端盖
bearing face 支承面
bearing factor 承载因数
bearing failure 承压破坏
bearing film 轴承油膜
bearing fittings 轴承配件
bearing force 承压力，支承力
bearing frame 支承结构，轴承底座，轴承架
bearing friction loss 轴承摩擦损失
bearing friction 轴承间摩擦
bearing gear 盘车装置
bearing housing jacket 轴承箱水套
bearing gland 轴承压盖
bearing graph 承重曲线
bearing holder 轴承座
bearing housing 轴承套，轴承壳，轴承箱
bearing insert 轴承衬套
bearing inspection platform 轴承检查平台
bearing instability 轴承不稳定性

bearing insulation 轴承绝缘
bearing jacking oil 顶轴油
bearing journal 轴颈
bearing kelmet 油膜轴承合金
bearing layer 持力层
bearing length 轴承长度
bearing life 轴承寿命
bearing liner remover 拆轴瓦工具
bearing liner 轴承瓦，轴瓦，轴衬
bearing load 轴承载荷
bearing location 轴颈，轴承位置
bearing lodgement 轴承洼窝
bearing member 承重构件
bearing metal temperature 轴承钨金温度
bearing metal 轴承合金
bearing neck dipping inclination 轴颈下沉
bearing neck up-rising inclination 轴颈扬度
bearing neck 轴颈
bearing oil cooler 轴承油冷却器
bearing oil film 轴承油膜
bearing oil pressure at which pumps cut in 泵切入时的轴承油压力
bearing oil pressure trip device 轴承低油压脱扣装置
bearing oil pressure 轴承油压
bearing oil pump 轴承油泵
bearing oil seal 轴承油封，轴承油封圈
bearing oil sump 轴承油槽
bearing oil system 轴承油系统
bearing oil 轴承油
bearing pad fitting 刮瓦
bearing pad 轴瓦，轴承瓦块，乌金轴瓦
bearing partition 承重壁，承重隔墙
bearing pedestal movement indicator 轴承座位移指示器
bearing pedestal 轴承架，轴承座
bearing pile wall 承重桩墙
bearing pile 承重桩，支承桩
bearing pillow 瓦枕
bearing plate 承压板，承重板，基础底板，支承板［座］，载荷板，垫板
bearing point 方位点，支承点，控制点
bearing post 支承柱
bearing power 负载能力，承载能力
bearing pressure of foundation 地基承压力
bearing pressure 支承压力，轴承压力，承压力
bearing pressurized water system 轴承加压水系统
bearing processing equipment 轴承加工机
bearing protective device 轴承保护装置
bearing radio 承载比
bearing rating 额定承载能力
bearing ratio test 承重系数试验
bearing reaction 支点反作用，支点反作用力，支承反力，轴承反力
bearing replacer 轴承装拆工具
bearing resistance 承压强度，抗压力，轴承阻力
bearing retaining ring 轴承护圈
bearing return oil temperature 轴承回油温度，轴承加油温度
bearing ring seal 滑环密封
bearing ring 轴承套环，轴承套圈，轴承环，轴承油环，支承环
bearing running hot 轴承过热
bearing seal 轴承密封
bearing seat 轴承座
bearing shaft sleeve 轴承衬套
bearing shell 轴瓦，轴承壳体，轴承壳套
bearing shield 轴承护罩
bearing shim 轴承垫片
bearing shoe grinding 刮瓦
bearing shoe 支承底板，轴瓦
bearing sleeve 轴承衬套，轴承套
bearing span 轴承跨距
bearing spring 承重弹簧，托簧
bearing stiffener 承重加劲杆
bearing stool 轴承支架
bearing strain 承压变形，支承应变，承压应变
bearing stratum 承力地层，承重层，持力层，支承层
bearing strength 承压强度，支承强度，轴承强度
bearing stress 承压应力，挤压应力，支承应力，承载应力，轴承负荷强度
bearing structure 支承结构
bearing support tube 轴承支承管
bearing support 轴承体，轴承座，轴承支架
bearing surface of foundation 地基支承面，基础支承面
bearing surface 承压面，支撑面，支承面，轴承面，承载面
bearing temperature monitoring system 轴承温度监视系统
bearing temperature 轴承温度
bearing thrust face 支承推力面
bearing thrust 轴承推力
bearing tree 定向树
bearing-type high strength bolt 承压型高强度螺栓
bearing unit 轴承组合件
bearing value 承载能力，承载值
bearing vibration measuring instrument 轴承振动测量仪表
bearing vibration 轴承振动，轴振
bearing walled structure 承重墙结构
bearing wall 承重墙
bearing water line 轴承水管线
bearing water pressure booster pump 轴承冷却水升压泵
bearing water pump 轴承水泵
bearing water supply line 轴承水供水管
bearing water tank 轴承水箱
bearing water train 轴承水管系
bearing wear 轴承磨损
360° bearing 整圆轴承
bearing 支承【即座架】，支座，轴承，承载，支承点，方向，方位
bear legal liability 承担法律责任
bear market 熊市，空头市场
bear risk 承担风险
bear their respective responsibilities 承担各自责任
bear the responsibility for... 承担……的责任

bear the same legal status 具有同等法律效力
bear-trap dam 开合式闸坝
bear-trap gate 开合式闸门，熊阱闸门，屋顶式闸门
bear-trap log chute 熊阱（式）放筏槽
bear-trap sill 熊阱式闸门槛
bear-trap weir 熊阱（式）堰
be assigned to 被分配给……，归属于，指定，分配给
beat a bargain 成交，还价
beat cob works 夯土工程，捣土工作
beat counter 差频式计数器
beat cycle 差拍周期，拍频周期
beat down method 逐次差拍法
beat effect 拍效应，差频效应
beater mill 锤击式磨煤机，锤头形磨煤机
beater plate 风扇用锤打，打击板
beater pulverizer 锤击磨煤机
beater wheel mill 锤击式风扇磨煤机
beater wheel pulverizer 带前置击锤的风扇磨
beater 锤，打浆机，夯具
beat-frequency meter 差频计
beat-frequency oscillator 拍频振荡器
beat-frequency 差频，差拍，拍频，拍频率
beating effect 差拍现象
beating in 合拍，进入同步
beating of wave 波浪拍岸
beat interference 拍频干扰
beat note signal 拍频信号
beat note zero 零拍，拍频为零
beat oscillator 拍频振荡器
beat pattern 拍频波形图
beat phenomenon 差拍现象
beat receiver 外差式收音机，外差式接收机
beat reception 差拍接收法
beat telephone 调度电话
beat-time programming 节拍时间编程
beat voltage 跳动电压，差拍电压
beat 节拍，打，拍，跳动，脉动，迎风航行，差拍，搏动
Beaufort force 蒲福风力
Beaufort number 蒲福（级）数
Beaufort scale 蒲福风级
beautification 美化，装饰
beaver board 木纤维板，人造纤维板
beavertail 扇形雷达波速
beaver-type dam 堆木坝
beaver-type timber dam 堆木坝
beaver 干扰雷达电台
Be(billion-electron-volt) 10亿电子伏特
because of 因为，由于
be characterized by... ……的特点在于
BECHTEL （美国）柏克德工程公司【建筑巨头】
becket 环，圈，绳环
Beckman's thermometer 贝克曼温度计
beck 山涧
become compatible with internationally accepted practices 与国际惯例接轨
be compacted at optimum moisture content 以最优含水量压实
be conducive to 有助于，有利于，有益于
Becquerel 贝可勒尔【放射性活度单位】
bedabble 喷水浇地，喷水，泼溅
bed behaviour 床料活动状态
bed-building discharge 造床流量
bed-building stage 造床水位
bed capacity 交换能力，交容量【离子交换器】
bed chamber 卧室
bed charge 床底料【沸腾炉】
bed circulation 床内循环
bed collapse 床层塌落
bed component 床层组分
bed configuration 河床地形
bed course 垫层，下垫层
bed cross-sectional area 床横断面积
bed current 底层流
BEDD(basic engineering design data) 基础工程设计数据
bedded deposit 层状沉积
bedded rockfill 分层堆石
bedded rock 层状岩石
bedded salt 层状盐
bedded structure 层状构造
bedded vein 层状脉
bedded 层状的
bed density 床层密度
bed depth 床深，层高，床层高【即深度】
bed diameter 床径
bedding angle 层面方向，层面角
bedding area （轴承的）承载面
bedding cleavage 层面劈理，顺层劈理
bedding course 垫层，垫底层
bedding error 端垫误差
bedding fault 层面断层，顺层断层
bedding glide 层面冲断层
bedding in 研磨，研刮
bedding joint 层间接缝，层面节理
bedding material 垫底材料
bedding mortar 垫层砂浆，封住灰浆，砂浆垫层
bedding of brick 砖铺垫层
bedding plane slip 层面滑动
bedding plane 层理面，层面，顺面，顺层面
bedding sand 垫层砂
bedding schistosity 层面片理，顺层片理
bedding slip 顺层滑动
bedding stone 研磨石，地基砂层，垫层砂
bedding surface 层面
bedding tape 封住胶带
bedding thrust 层面冲断层，顺层冲断层
bedding 层理，成层，基础，底板，修整，研磨，研配，垫层，基床，基底，嵌入，封住
bed erosion 河床冲刷
be designated for 被指定为
bed expansion ratio 床层膨胀比
bed expansion 层床膨胀度【离子交换】
bed-forming discharge 造床流量
bed-forming water stage 造床水位
bed frame 底座
bed freezing 床层冻结
bed groin 填积堤
bed heat transfer tube 床层传热管
bed height 床高

be divorced from 脱离
bed joint 层间接缝，层面节理，平层节理
bed level 床料界面
bed load deflecting sill 导沙底槛
bed load deflection apron 挑沙护坦
bed load discharge 推移质输沙率，推移质输送量，底沙输送量
bed load equation 推移质方程
bed load formula 推移质公式
bed load movement 底沙运动，推移质运动
bed load rate 推移质输沙率
bed load sampler 底沙取样器，推移质采样器，推移质取样器
bed load sampling 底沙取样
bed load sediment 推移质沉淀物
bed load 床内装载量，推移质，底沙，底负载
bed material migration 床料迁移【沸腾炉】
bed material 床料【沸腾炉】，河床质
bed moisture 煤层水分
bed molding 底层线脚
bed of brick 砖铺垫层
bed of fluidized solid 液化固体层
bed of particles 颗粒床
bed piece 底座板，垫板
bed plane 层理面
bed plate foundation 板式基础
bed plate 座板，底板，底座，台板，炉底，地脚板
bed porosity 床层空隙度
bed profile 沙床断面
bed protection for closure 截流护底
bed regrading system 床料尺寸保证系统【沸腾炉】
bed ripple 沙波
bed rock 底岩，基岩，岩石，岩床
bedroom 卧室
beds alternation 岩层交互变化
bed separation 层状剥落
bed shear 层面剪切
bed silt 底沙
bedstead 试验台，试验装置，骨架，壳体
bed stone 座石
bed structure 床层结构
bed support surface 床层支承面
bed temperature 床层温度
bed thickness 床层厚度
bed timber 垫木，枕木
bed viscosity 床层黏度
bed voidage 床层空隙度
bed volume 床层体积
bed weight 床重
bed 床，底盘，底座，试验台，层，垫层，基础底座，机架，矿床
beech 山毛榉
beef wood 硬红木
bee hire coke 蜂巢式焦炭
bee hive gland 蜂窝式汽封
bee hive packing 蜂窝式汽封
bee hive seal 蜂窝式汽封
beehive type radiator 蜂窝式散热器
beeper 无线寻呼机，传呼机，寻呼机，电话传呼机
beep 嘟（声响）
be equally authentic 具有同等法律效力
Beer's law 比尔定律
beeswax 蜂蜡
beetle head 送桩锤
beetle 木夯
beetling cliff 悬崖
be exempted from 免税
before the wind 顺风
be forward in(/with)M 在M方面先进
BEF 比利时法郎
bega 千兆倍【10^9】
begin block 开始分程序，开始区块，起始块
beginning of breakup 解冻开始
beginning of core life 堆芯寿命初期
beginning of cycle 循环初期
beginning of fatigue failure 早期疲劳失效
beginning of freeze-up 封冻开始
beginning of information marker 信息开始标志
beginning of life 寿命初期【循环燃料】，寿命前期
beginning of month 上旬，月初
begin-of-tape marker 磁带上某一区的开始记号
begin to take shape 初具规模
begin 开始
be good at 擅长，精通
begrime 熏黑
behavior-based safety program 行为安全管理
behavior in service 运转性能，使用性能
behavior of boundary layer 边界层特性
behavior of neutron 中子行为
behaviour characteristics 性能特性
behaviour in service 运转性能，使用性能
behaviour of cross section 截面的变化曲线
behaviour of fluids dynamic 流体动力特性
behaviour 行为，性能，工况，运转，制度，状态，模式，特性，习性，机能，运行情况
beheaded river 断头河，断源河，夺流河
beheaded stream 断头河，夺流河
beheaded valley 断头谷
behead 夺流，断头
behest 命令，紧急指示
be highly principled 原则性
behind schedule 落后于预定计划，拖期
beidellite 贝得石
be in conference 正在开会
be in danger of 有……危险的
be in danger 处在危险中
be in session 正在开会
Belanger's critical flow 伯朗格临界流
Belanger's critical velocity 伯朗格临界速度
belated claim 迟索的赔款
belated 误期的
belaying pin 套索桩
belay 拴住，系牢，把缆绳拴在系索栓上
belching 汽水沸出
belch 烟柱，火焰柱
be legally binding 受法律约束
belfast roof truss 弓形屋顶桁架

belfast truss 弓形桁架，弧形桁架
be liable for 应付有责任
be liable to 应付有责任
believable 可信的
bell and hopper arrangement 钟斗装置【高炉】
bell and hopper 进料器，钟口漏斗
bell and plain end joint 平接
bell and spigot joint 承插连接［接头］，钟口式接头，(管子的) 套筒结合［接头］
bell-and-spigot 套接
bell arch 钟形拱
bell buoy 钟形浮标
bell caisson 钟形沉箱
bell center punch 钟形中心冲头
bell character 报警信号，报警符
bell cot 钟架
bell crank 曲柄，双臂曲柄
bell dolphin 钟形护墩桩
belled excavation 钟形开挖
belled-out cylindrical pile 大头圆柱桩
belled-out pile 扩底桩，扩孔桩
belled-out pit 扩大竖井
belled-out section 扩大断面
belled pier 扩底墩
belled tube-end 翻边管端，喇叭口管端
belled 套接的，钟形花的，有钟形口的，钟形口
bell end pipe 承插管，喇叭口管件
bell end 锥形管头，大小头，(管道的) 承插端，扩大端，承口，钟口，喇叭口
Belleville spring 贝莱维勒弹簧
Belleville washer 贝莱维勒垫圈
bell furnace 罩式炉
bell housing 外罩，屏蔽套，钟形罩，外壳
belling roll 胀管器扳边滚子
belling 翻成喇叭口
bell insulator 碗形绝缘子，单裙绝缘子，钟罩形绝缘子
bell jar 钟罩
bell key 振铃电键
bell mouth area 喇叭口区
bell mouth defect 喇叭口缺陷
bell-mouthed opening 漏斗口
bell-mouthed orifice 喇叭形孔口
bell-mouthed pipe 漏斗口管，承插管
bell-mouth entrance 喇叭形进口
bell mouth inlet 喇叭形入口，钟形进水口，钟形口
bell mouth intake 喇叭形进水口
bell-mouth orifice 喇叭孔口，钟形孔口
bell mouth outlet 喇叭形出水口
bell-mouth spillway 喇叭形溢洪道
bell mouth 锥形管头【大小头】，测流喷管，扩流管，胀接管，扩散口，喇叭口，锥形孔
bell-operated pressure gauge 钟形压力表
bellow expansion joint 波纹膨胀节
bellows compensator 波纹管补偿器
bellows differential flowmeter 膜盒式差动流量计
bellow sealed valve 波纹管密封阀
bellows expansion joint 波纹管膨胀接头
bellows flowmeter 波纹管式流量计，膜盒式流量计
bellows-gauge 膜盒压力计，膜盒式压力表，波纹管压力计
bellows joint 波纹管接头，波纹管连接
bellows manometer 波纹管压力计，膜盒式压力计
bellows-operated pilot valve 膜盒控制阀
bellows pipe 波纹管
bellows pressure gauge 波纹管压力表，膜盒压力计
bellows pump 膜片泵
bellows-sealed isolating valve 波纹管密封隔离阀
bellows seal valve 弹簧箱密封阀
bellows seal 波形密封，膜盒密封，波纹管密封，弹簧箱密封
bellows type expansion joint 波纹管补偿器
bellows type gas flowmeter 膜盒式气体流量计
bellows-type manometer 弹簧箱压力计
bellows type pressure gauge 波纹管压力计
bellows type regulator 波纹管式调节器
bellows type safety valve 波纹管式安全阀
bellows valve 波纹管阀，弹簧箱阀
bellows 波纹管（式），膜盒，皮老虎，风箱，(凸面) 补偿器，弹簧箱
bell pier 钟形桥墩
bell push-button 电铃按钮
bell ringing transformer 电铃变压器
bell roof 钟形屋顶
bell-shaped capital 钟形柱头
bell-shaped end 锥形管头
bell-shaped hill model 钟形山丘模型
bell-shaped insulator 裙式绝缘子
bell-shaped rotor 钟形转子
bell-shaped suspension insulator 裙式悬垂绝缘子
bell transformer 电铃变压器
bell valve 钟形阀
bell wire 电铃线
bellying in 向内凸胀
bellying out 向外凸胀
bell 承口，喇叭口，钟，铃，漏斗，圆锥体，锥形口，扩散管，钟形管
below bearing type generator 伞式发电机
below critical 低于临界状态，低于临界，次临界的
below curb 路缘石标高以下
below grade 不合格
below norm 限额以下
below proof 不合规定（的），不合格（的）
below yard reclaim hopper 煤场地下取煤斗
below zero 零下
belt advancement resistance 输送带前进阻力
belt and road initiative 一带一路倡议
belt brake 皮带制动器
belt-bucket elevator 斗带式提升机
belt carcass （皮）带芯
belt cleaner 胶带清扫器，皮带清扫器
belt cleaning equipment 皮带清扫装置
belt cleaning gear 皮带清扫器
belt cleaning scraper 皮带清扫器
belt cleaning system 皮带清扫系统
belt coal feeder 皮带给煤机，皮带式给煤机

belt conveyer floor　运输皮带层
belt conveyer scale　皮带秤
belt conveyer　带式输送机，皮带输煤机
belt conveying capacity　皮带输送量，皮带输送机的出力
belt conveyor floor　皮带层
belt conveyor stretcher　皮带输送机的伸张器
belt conveyor system　带式输送机系统
belt conveyor　传送带，带式输送机，皮带（运输）机，皮带装料机
belt course　腰线，束带层，带状层
belt-cover　皮带护罩
belt creep　皮带打滑
belt critical speed switch　皮带临界速度开关
belt differential factor　相带［绕组］系数
belt-drive mechanism　皮带传动机构
belt driven pump　皮带传动泵
belt drive　带传动，皮带传动，皮带调节
belted cable　铠装电缆
belt edge　皮带边
belted type cable　带绝缘电缆，统包型电缆
belted　束带的，装甲的
belt elevator　倾斜式皮带运输机，带式提升机，皮带提升机
belt fastener　皮带扣，皮带连接器
belt feeder　带式给料机［器］，带式给煤机［器］，皮带给料机［给煤机］，皮带送料机
belt filter　皮带过滤机
belt footing　条形基础
belt highway　带状公路，环行公路
belt hook　皮带扣
belting component　输送机部件，带芯材料
belting course　带状层
belting　（传动）带装置，皮带传动，制带的材料，带类，调带装置
belt joint　皮带接头
belt lacing　皮带接头
belt leakage flux　相带漏磁通
belt leakage　相带漏磁
beltline rail-way　环行铁道
belt line region　最大通量辐照区【压力容器】，束带区
belt loader　皮带装料机
belt magnetic separator　带式磁铁分离器
belt meander　带状河曲
belt misalignment(/deviation) switch　皮带跑偏开关
belt misalignment　皮带跑偏
belt modulus of elasticity　输送带弹性模数
belt of cementation　胶结带
belt of erosion　侵蚀（地）带
belt of fluctuation of water table　水位变动带
belt of fluctuation　波动带
belt of folded strata　褶皱带
belt of no erosion　无侵蚀带
belt of phreatic fluctuation　地下水位波动带，潜水波动带
belt of soil water　土壤水分带
belt of transition　过渡带
belt of wandering　河道蜿蜒（地）带，河流摆动带
belt of water table fluctuation　地下水位波动（地）带，潜水面波动带
belt of weathering　风化带
belt operating tension　皮带运行张力
belt planting　防护带栽植
belt press　带式压滤机
belt printer　带式打印机
belt pulley attachment　皮带传动装置
belt pulley wheel　皮带轮
belt-punch　皮带冲床
belt railroad　环行铁道
belt run-off-track switch　皮带跑偏开关
belt reactance factor　相带漏抗系数
belt reactance　相带漏抗
belt rim repair device　胶带边修补器
belt roller　皮带滚轮
belt roll　皮带卷
belt sag　输送带垂度
belt saw pipe cutter　带锯式切管机
belt saw　带锯
belt scale and calibration device　皮带秤和校验装置
belt scale　皮带秤
belt scraper　皮带清扫器，皮带刮板
belt screen　带式筛，转筛，带状拦污栅
belt side skid　皮带跑偏
belt side slip switch　带边滑动开关
belt skimmer　带式撇油器
belt slip　胶带打滑
belt span　相带宽度
belt speed　皮带速度
belt stacker　皮带装料机
belt stretcher　皮带张紧器
belt string repair device　胶带线修补器
belt sway switch　两级跑偏开关
belt system　输送带系统
belt tensile stress　输送带拉应力
belt tension　皮带张力，输送带张力
belt tension gear　皮带拉紧器
belt tensioning device　皮带拉紧装置
belt tension unit　皮带张紧装置
belt tightener　皮带张紧轮
belt travel direction　皮带运行方向
belt tripper　配煤车，皮带卸料车，皮带卸料器
belt type sampler　带式取样器
belt velocity　皮带速度
belt vulcanizer　胶带硫化器
belt weigher house　皮带秤间
belt weigher　皮带秤
belt wheel　皮带轮
belt width　皮带宽度
belt wiper　清带刷
belt　传送带，传动带，带，皮带，胶带
belvedere　平台，瞭望塔
bel　贝尔【音强单位或电平单位】
BEM(blade element momentum)　叶素动量
BEMF(back electromotive force)　反电动势
BEM theory(blade element momentum theory)　叶素动量理论
Benard cell　坝纳单体，坝纳涡胞
bench blasting　阶梯（式）爆破，台阶式爆破，

梯段爆破
bench board 操纵台，台式配电盘，控制台
bench calibration 试验台校验
bench chisel 钳工錾
bench cut 台阶式开挖
bench drill 台钻，台式钻床
benched excavation 阶梯式开挖
benched foundation 阶梯式基础，阶梯形基础，台阶式基础
benched 阶梯式的，台阶式的
bench flume 低渡槽［水槽］，台架式渡槽［水槽］，阶地水槽
bench gravel 阶地砾石
bench grinder 台式磨床
benching cut 台阶式挖土，台阶式掏槽
benching method 梯段法
benching strain 弯曲应变
benching 台阶式开挖，台阶式掏槽，台阶式挖土法，形成阶地
bench insulator 绝缘座
bench land 台地
benchmark elevation 水准点标高，水准点高程
benchmark estimate 基准估计数字，基准估量数字
benchmark list 水准点表
benchmark price 基准价格，基准价，指标价格
benchmark study 基准研究
benchmark test 基准测试，基准程序测试
benchmark yield 基准收益率
benchmark 用基准问题测试【计算机系统等】，基准，基准点，基准标志【地形测量学】，标准检查程序，基准测试程序，水准点，水准石，对比用原始样品
bench method 分层开挖法，台阶式挖土法
bench model 小型模型
bench of silt 泥沙阶地
bench reclaiming 台式取料
bench-scale experiment 实验室试验，小型试验
bench-scale facility （实验室小规模的）试验装置
bench-scale unit 实验室装置
bench-scale 小型的，实验室规模的，台秤
bench section 横断面，横剖面
bench slopes 台阶式边坡
bench terrace 梯田
bench test 工作台试验，试验台试验，台架试验，小型试验
benchtop cave 半敞开小室
benchtop 台式
bench trimmer 截锯机
bench-type 台式
bench vise 台钳
bench wall 承拱墙
bench worker 钳工
bench 长凳，工作台，试验台，台，钳工台，阶地，台架，台座，挖掘工作面的台阶
bend angle 弯度角，弯头角度
bend bar 弯曲钢筋，元宝钢筋
bend diameter 弯曲直径
bender 弯管机，弯筋工，弯筋机，弯曲工，弯曲机
bend improvement 裁弯取直，弯道改善
bending and straightening machine 弯曲矫直两用机
bending and unbending test 曲折试验
bending beam 抗弯梁
bending coefficient 弯曲系数
bending crack 弯曲断裂
bending creep 弯曲蠕变
bending critical speed 弯曲临界速度
bending deflection of pipe 管道挠度
bending deflection 挠度，挠曲变位，弯曲变位，弯度
bending die 弯曲模
bending displacement 弯曲位移
bending elasticity 弯曲弹性
bending failure 挠曲破坏，弯曲破坏
bending fatigue strength 弯曲疲劳强度
bending fatigue 挠曲疲劳，弯曲疲劳
bending force 弯曲力，挠曲力
bending formula 挠曲公式
bending frequency 弯曲频率
bending induced by thermal contraction 热收缩引起的弯曲
bending iron 弯钢筋板子，弯钢筋工具
bending line 挠曲线，弯曲线
bending load 弯曲负荷
bending machine 弯筋机，折弯机
bending mode 弯曲模态
bending modulus 抗弯模数，弯曲系数
bending moment diagram 挠矩图，弯矩图
bending moment envelope 弯矩包络线
bending moment 挠矩，弯矩，弯曲力矩
bending oscillation 弯曲振动
bending plane 弯曲平面
bending point 挠曲点
bending press 压弯机，弯管机
bending radius of cable 电缆弯曲半径
bending radius of rectangular rigid busbar 矩形硬母线弯曲半径
bending radius 挠曲半径，弯曲半径，转弯半径
bending resistance 抗弯，抗弯曲性
bending response 弯曲响应，弯曲反应
bending rigidity 抗弯刚度
bending roll 卷板机
bending schedule 钢筋规范表【钢筋混凝土用】，弯钢筋表
bending stiffness 弯曲刚度，抗弯劲度，抗弯刚度
bending strain 挠曲应变，弯曲应变
bending strength 抗挠强度，抗弯强度，弯曲强度
bending stress 挠曲应力，弯曲应力
bending table 弯曲工作台，弯铁台【管道弯曲】
bending tensile strength 抗弯拉强度
bending test 弯曲试验，屈服试验
bending test jig 弯曲试验用夹具
bending test under three-point loading 三点弯试验
bending test 挠曲试验，弯曲试验，抗弯试验，抗折试验
bending trestle 弯筋工作台

bending vibration characteristics 弯曲振动特性
bending vibration strength 弯曲震动强度
bending vibration 弯曲振动
bending 挠曲，弯曲（度），偏移，偏差，波束曲折，挠度，折弯，曲折
bend loss 弯道损失，弯管段损失，弯头（阻力）损失
bend meter 弯道流量计，弯管流量计
bend muffler 消声弯头
bend pipe 弯管
bend pulley 改向滚筒
bend radius 弯弧内径，弯曲半径
bend strength 抗弯强度
bend test of metal 金属弯曲试验
bend test of non-quench-hardening metal 金属不淬硬性弯曲试验
bend test on tubes of metals 金属管弯曲试验
bend test 抗弯试验，弯曲试验
bend waveguide 弯波导
bend 使弯曲，弯道，弯管，弯曲，弯头，变管，胀弯弯头，改向，接头
beneficial owner 受益人，受益业主
beneficial use of water 水的有效利用
beneficial wind effect 有利风效应
beneficial 有益的
beneficiary of remittance 汇款收款人
beneficiary of transferable credits 可转让信用证的受益人
beneficiary 受益人，收款人，享受保险赔偿者
beneficiated party 受益方
beneficiate 选（矿），（冶炼前）对（矿石）进行预处理
beneficiation 选矿，富集
benefit consideration 利益的考虑
benefit-cost analysis 收益成本分析
benefit-cost ratio method 收支比法，效益费用比法
benefit country 受惠国
benefited party 受益方
benefit-risk analysis 利益风险分析
benefit 利益，收益，受益，效益
be not authorized to do 无权做……
Benson boiler 本生式直流锅炉
bent bar anchorage 弯筋锚固
bent bar 弯筋，弯曲钢筋，元宝钢筋
bent frame 排架
bent gun 曲轴电子枪
benthal demand 有机沉积需氧量
benthal deposit 水底有机沉积
benthon 水底生物，底栖生物
benthos 海底生物，海洋深处，水底生物，海底，海底的动植物群
bentonite block 膨润土块
bentonite gel 膨润土凝胶体
bentonite grease 膨润土滑润脂
bentonite grouting 膨润土灌浆
bentonite powder 膨润土粉
bentonite slurry 膨润土浆
bentonite suspension 膨润土悬浮液
bentonite 斑脱土，斑脱岩，膨润土，膨土岩，皂土
bent on 下决心，热衷
bent-over jet 弯曲射流
bent-over plume 弯曲羽流
bent rung ladder 爬梯
bent steel 弯筋
bent tube boiler 弯管锅炉
bent tube 弯管
bent-up bar 弯起钢筋
bent wood 挠曲木材
bent wrench 弯头扳手
bent 弯曲的，决心的
benzene extract 苯抽出物
benzene soluble extracts 苯萃取物
benzene sulfonic acid 苯磺酸
benzene 苯
benzidine 联苯胺
benzine 挥发油
benzoate 苯甲酸盐
be out of danger 脱离危险
BEP(break even point) 盈亏平衡点，零利润点，保本点，盈亏临界点
be propitious to 有利于
BER(bit error ratio) 误码率
beresowite 碳铬铅矿
berge 冰山
bergmeal 硅藻土
BERI(business environmental risk index) 企业环境风险指数
berkelium 锫
berme ditch 傍山排水沟，边坡截水沟，护道排水沟，护路排水沟
berme for channel 渠槽戗道
berme of ditch 截水沟戗道
berme 护道，狭道，路肩，马道，戗道，戗台，阶地，小平台，滩边阶地
berm-type structure 戗台式建筑
Bernoulli 伯努利
Bernouilli force 伯努利力
Bernoulli binomial distribution 伯努利二项式分布
Bernoulli constant 伯努利常数
Bernoulli equation 伯努利方程
Bernoulli formula 伯努利公式
Bernoulli's energy equation 伯努利能量公式
Bernoulli's law 伯努利定律
Bernoulli's series 伯努利级数
Bernoulli's surface 伯努利面
Bernoulli's theorem 伯努利定理
Bernoulli test 伯努利实验
Bernoulli trail 伯努利尾迹
Bernoulli vector 伯努利矢量
be routed to 接入……道路，通向……道路
berthage 泊位，停泊费
berthing facilities 系泊设施，运输码头，停泊设施，系船场设施，码头装置
berthing fee 停泊费
berthing force 泊力
berthing impact 停泊冲击力
berthing plan 停泊地平面图
berthing space 停泊场，停泊区
berthing structure 靠船建筑物
berthing 停泊

berth terms 定期船条件
berth 泊船处，泊位，船台，停泊，停泊处，停泊地
Bertrand model 伯特兰德模型
Bertrand qualifying equation 伯特兰德验证方程式
berylliosis 铍中毒
beryllium-moderated reactor 铍慢化反应堆
beryllium moderator 铍慢化剂
beryllium oxide moderated reactor 氧化铍慢化反应堆
beryllium oxide thermoluminescent dosimeter 氧化铍热释光剂量计
beryllium reactor 铍反应堆
beryllium reflector 铍反射层
beryllium sheath 铍外壳【中子源】
beryllium-zirconium eutectic 铍锆共晶体
besel 监视窗［孔］
be short of 缺乏
Bessel's interpolation formula 贝塞尔内插公式
bessemer converter 酸性转炉
bessemer steel 转炉钢
best approximation 最佳逼近，最优逼近
best available technology 最佳可行技术
best bid 最佳标价
best decision 最佳决策
best efficiency point 最佳效率点
best engineering design 最佳工程设计
best estimate flow 最佳估算流量
best estimator 最佳估计量
best-first search 最佳优先搜索
best fit technique 最佳拟合技术
best fit line 最佳拟合线
best fitting curve 最佳拟合曲线
best fit 最佳配合，最佳匹配，最佳拟合，最佳适合
best hydraulic cross section 最佳水利断面
best in quality 品质优良
best interest 最大利益
best lift drag ratio 最佳升阻比
best load 最大效率时的负荷
best obtainable price 市场最好价
best offer 最佳报价
best performance curve 最佳特性曲线
best point load 最佳负荷【电站】，经济负荷
best quality product 优等品
best result 最佳结果
best servo 最佳伺服系统，最佳随动系统
be subjected to 遭受
be subject to 服从，以……为条件，受……的支配，须经，易遭（受），易发生
beta absorber β射线吸收器，β射线吸收体，β吸收体
beta-absorption gauge β吸收测量计
beta-active sample β放射性样品
beta background β本底
beta barrier β防护屏，β射线防护屏
beta burn （皮肤）β射线烧伤
beta contamination indicator β污染指示器
beta decay energy β衰变能量
beta decay series β衰变系
beta decay β衰变
beta disintegration energy β衰变能
beta dust and gas monitor β尘气监测计
beta-ell 贝埃尔，线路总衰减
beta-emitter β发射体，β粒子
beta function β函数
beta-gamma cell β-γ热室
beta-gamma doorway monitor 门口β-γ监测器
beta-gamma exposure doseratemeter β-γ照射剂量率计
beta-gamma monitoring β-γ监测
beta-gamma monitor β-γ监测器
beta-gamma scanning detector β-γ扫描探测器
beta-gauging technique β射线测量技术
beta measuring station β测量站
beta-minus decay 负β衰变
beta particle β粒子
beta phase β相
beta plus 稍高于第二等
beta quench β淬火
beta radiation β辐射
beta radioactivity β放射现象
beta-ray gauge β射线测量计
beta-ray monitoring β射线监测
beta-ray spectrograph β射线摄谱仪
beta-ray thickness gauge β射线测厚规
beta scintillator β射线闪烁体
betatopic 失电子的，差电子的
beta transformation β转换
beta transition β跃迁，β跃【氚衰变】
betatron 电子回转加速器，电子感应加速器
beta value β值
beta wave β波
beta 晶体管发射极短路电流放大系数，希腊字母β
bethanized wire 电镀锌钢丝
bethanizing 镀锌
bethlehem beam 宽缘工字钢梁
beton 混凝土
betray 泄露，出卖，背叛
betrunked river 断尾河
betrunk stream 断尾河
betstead 台，架
betterment works 改善工程，修缮工作
betterment 改建，改进，改善
better-off country 富裕国家
better-run enterprises 运行机制良好企业
bettle 木槌，木夯
between baffles 挡板之间
between-coil connections 线圈间的连接
between(/on) centers 中心间距，中到中
between open contacts 断口间，开路触点间
Betz coefficient 贝茨系数
Betz equation 贝茨等式，贝茨效率公式
Betz law 贝茨定律，贝茨理论
Betz limit 贝茨极限
Betz manometer 贝茨压力计
Betz 贝茨
be valued at 估价为，取值为
bevatron 高能质子同步稳向加速器
bevel angle 倒角【焊接】，坡口（面）角度，斜角，斜面角，斜削角

bevel drive 伞齿轮传动
beveled edge 倒斜板边，坡口边，斜削边
beveled end 坡口端，斜削端
beveled gear 斜面齿轮
bevel edge 斜缘，斜边，（焊管坯的）倒角边
beveled joint 斜角接头，斜接
beveled washer 楔形垫圈，斜垫圈
bevel face 斜面
bevel friction gear 斜摩擦齿轮
bevel gear differential 锥形齿轮差动装置
bevel gear drive 伞齿轮传动，斜齿轮传动，圆锥齿轮传动
bevel gear gate lifting device 斜齿轮闸门启闭机
bevel gear main drive 斜齿轮主传动
bevel gear operated valve 伞齿轮传动阀门
bevel gear 伞齿轮，伞形齿轮，斜齿轮，圆锥齿轮，锥齿轮
beveling machine 坡口机，斜边机，刨边机
beveling plane 榫槽刨
beveling tool 斜切具
bevelled brush 削角电刷
bevelled sleeper 楔形垫木
bevelled pole shoe 倒角极靴，圆边极靴
bevel pinion 小圆锥齿轮
bevel protractor 量角规，斜（量）角规
bevel sheet pile 斜角板桩
bevel siding 互搭板壁
bevel square 斜角规
bevel washer 斜垫块，斜垫圈
bevel welding 斜角焊
bevel wheel 斜齿轮
bevel 坡口，切斜角，倾斜，斜角，斜边，斜面，倒角，成斜角，斜角规，伞形轮，倾角，锥形的，倾斜的，斜的，斜面的，斜削，开坡口
be well forward with one's work 早做完了工作
beyond revoke 不能撤销的，不能作废的，不能取消的
bezel 遮光屏
BF(bandpass filter) 带通滤波器
BFBB(bubbling fluidized bed boiler) 鼓泡流化床锅炉
BF(blind flange) 盖板，闷头法兰，盲板，法兰堵头
BF(boiler follow) 锅炉跟随（方式），锅炉跟踪（方式）
BFBP(boiler feedwater booster pump) 锅炉给水前置泵，锅炉给水增压泵
BFBP current 锅炉给水前置泵电流
BFBP fail 锅炉给水前置泵故障
BFBP OTL WTR T 锅炉给水前置泵出口水温【DCS 画面】
BFBP started 锅炉给水前置泵启动
BFBP stopped 锅炉给水前置泵停
BF_3 counter tube 三氟化硼计数管
BFE(boiler front equipment) 炉前点火控制设备
BF(flat bar) 扁钢
BFI(Baltic freight index) 波罗的海运价指数
BF_3 neutron monitor 三氟化硼中子监测器
BFP(boiler feedwater pump) 锅炉给水泵
BFPM(boiler feed pump motor) 锅炉给水泵电机
BFPT(boiler feedwater pump turbine) 锅炉给水泵汽（轮）机，小汽机
BFR＝buffer 缓冲器
BFSP(boiler feedwater startup/standby pump) 锅炉给水启动/备用泵
BFV(butterfly valve) 蝶阀
BFW(boiler feedwater) 锅炉给水
BHN(Brinell hardness number) 布氏硬度数
BHO(building hand over) 厂房移交
BHP(brake horse power) 制动马力
biangular reflectance 双角反射率
biannual 一年两次的，半年一次的
bias adjustment 偏置调节
bias cell 偏压电池，偏流电池
bias check 边缘检验，偏压校验
bias circuit 偏压电路
bias combustion 偏离燃烧
bias control 偏压控制，偏压调整
bias current 偏流，偏压电流
bias data 偏离数据，偏置数据
bias detection 偏压检波，偏压探测，偏流检波
bias distortion 偏移失真，偏移畸变
biased amplifier 加偏压的放大器
biased automatic gain control 加偏压的自动增益控制
biased differential protection 偏压差动保护
biased differential protective system 偏置差动保护装置，极化差动保护装置
biased differential relaying 极化差动继电保护，偏压差动继电保护
biased differential transformer protection 变压器偏置差动保护
biased error 偏误
biased estimator 有偏估计量
biased field 偏移磁场
biased multivibrator 闭锁多谐振荡器
biased percentage differential protection 偏置式比率差动保护
biased rectifier amplifier 偏置整流放大器
biased relay 偏压继电器，极化继电器
biased transformer 偏磁变压器
biased 加偏压的，有偏见的，偏置的，闭锁的
bias electrical restrain 电气偏置制动
bias error 偏置误差，系统误差，固有误差，偏移误差
bias field 偏移磁场
bias gear 偏动装置
biasing capacitor 偏压旁路电容器
biasing circuit 偏压电路，偏移电路
biasing impedance 偏移阻抗
biasing logic 偏置逻辑
biasing magnet 偏磁磁铁
biasing transformer 偏磁变压器
biasing winding 偏置绕组，偏移绕组，辅助磁化线圈
biasing 加偏压，偏压，偏置，附加励磁
bias light 偏置光
bias meter 偏畸变器，偏流表
bias-off 偏压截止，偏置截止，加偏压使截止，截止

bias potentiometer　偏压分压器，偏压电位计
bias pulse　偏压脉冲
bias rectifier　偏压整流器
bias set circuit　偏压调节电路，偏流调节电路
bias test　偏压试验，拉偏试验，边缘试验
bias torque　偏转力矩
bias trap　偏磁陷波器
bias tube　偏压管
bias voltage control　偏压控制，偏压调整
bias voltage　偏压
bias winding　偏压绕组，辅助绕组，磁化线圈
bias　偏差，偏离，偏压，偏流，偏磁，偏置
biatomic　双原子的
biax　双轴
biax core　双轴磁芯
biaxial apparatus　双轴仪
biaxial coordinates　双轴坐标
biaxial crystal　双轴晶体
biaxial eccentricity　双向偏心距
biaxiality bending stress　双向弯曲应力
biaxiality　二轴性
biaxial stress state　二轴应力状态，双轴应力状态
biaxial stress　双向应力
biax magnetic element　双轴磁元件
biax memory　双轴磁芯存储器
bibasic　二元的，二代的
bibb　活门，小水龙头，水龙头，旋塞，龙头，弯嘴旋塞
bibcock　活塞，龙头，弯管旋塞，弯嘴龙头
bible　圣经，宝典，权威著作【书，杂志等】
bibliography　书刊目录，参考书目，书目，文献目录
BI(block increase)　闭锁增
BI(boiler island)　锅炉岛
bib valve　活塞阀，水龙头阀，弯管阀，弯嘴阀，水龙头，弯管旋塞阀
bib　活门，旋塞，活塞，水龙头，弯嘴旋塞
bicable ropeway　双缆索道
bicarbonate alkalinity　碳酸氢盐碱度，重碳酸盐碱度
bicarbonate hardness　（水的）暂时硬度
bicarbonate　重碳酸盐，碳酸氢盐，酸式碳酸盐
bicathode tube　双阴极管
bicharacteristic　双特征性的
bichromate cell　重铬酸盐电池
bichromate-treated　重铬酸盐浸渍过的
bichromate　重铬酸盐
bicirculating boiler　双工质循环锅炉，双循环锅炉
bicirculating　偶极流，双重循环的
bicirculation　偶环流
biconcave　双凹的，双面凹的
biconical antenna　双锥形天线
biconic　双锥的，二次曲线的
biconvex airfoil　双凸形翼面
biconvex　双面凸的，双凸的
bicorn　纳米【10^{-9} 米】
bicrystal　双晶体
bicubic interpolation　双三次插值
bicycle shed　自行车棚
bicycle-type multi-bladed wind machine　自行车轮型多叶片风力机
bicycle-type windmill　自行车轮型风车
bicylindrical resonator　双圆柱共振体，双圆柱谐振器
bid abstract　标价汇总表
bid addenda　投标书补遗
bid amount　投标金额
bid application document　申请文件
bid award criteria　决标准则
bid award meeting　决标会议
bid bond　投标保函，投标保证金，押标金，投标保证书
bid closing date　投标截止日期
bid conditioning meeting　评标协调会
bid condition　投标条件
bid cost　投标费用
bid currency　投标货币
bid curve　报价曲线
bid data sheet　招标数据表
bid data　投标数据
bid deadline　投标截止日期，承包截止日期
bid deposit　投标保证金，投标押金
bidder representative　投标人代表
bidder's duration of validity　投标人有效期，报价员有效期，投标有效期
bidder sheet　投标人名单
bidder's name　投标人名单，报价员名称
bidder's queries　投标人的疑问［问题、质疑］
bidder's response time　投标人响应时间
bidder　投标方，投标人，报价人
bidding advice　招标通知
bidding agency　招标代理
bidding condition　投标条件
bidding consortium　联合投标方，投标联合体
bidding contract　投标合同，投标契约
bidding data sheet　投标资料表［数据表］
bidding document　招标文件，招标书，标书
bidding form　投标书格式，投标文件格式，投标需要填写的表格
bidding group　投标集团，投标小组
bidding period　投标阶段
bidding price　投标报价
bidding procedures　招投标程序，投标手续
bidding process data　报价数据
bidding process　投标步骤
bidding proposal　投标书，投标文件，投标报价，投标方案
bidding rules　投标规则
bidding sample　投标样品
bidding sheet　标价单，投标单，投标报价书
bidding strategy　投标策略，投标决策
bidding unit　竞价机组
bidding　竞价，投标，招标
bid document compiling　投标文件编制，编制标书，编标
bid document drawing　投标文件图纸
bid document preparation　投标文件准备，标书编制
bid document production　标书文档制作
bid document　标书，投标资料，招标文件

bid drawing 投标图纸，招标图纸，标书图纸
bidet 洗身盆，坐洗器
bid evaluation board 评标委员会
bid evaluation committee 评标委员会
bid evaluation criteria 评标准则
bid evaluation factor 评标因素
bid evaluation guideline 评标细则
bid evaluation report 评标报告
bid evaluation 标书评估，评标，投标评估价
bid examination 投标核查
bid form 投标表格，投标（书）格式，投标形式，投标单
bid for the project 对工程的投标
bid guarantee 投标担保，投标保函
bidimensional flow 二维流动
bid inquiry 询标
bid invitation 招标
bid invitation contents 招标内容
bid invitation document 招标文件
bid invitation letter 招标邀请函
bid invitation specification 招标说明书
bidirectional bus 双向总线
bidirectional contract for difference 双向差价合同
bidirectional coolant flow 双向冷却剂流
bidirectional counter 双向计数器
bidirectional current 双向电流
bidirectional diode-thyristor 双向两极可控硅，双向二端闸流管
bidirectional feeder 双向给料机
bidirectional movement 双向运动
bidirectional power electronic converter 双向电力电子变流器
bidirectional power flow 双向功率流
bidirectional pulse train 双向脉冲列，双向脉冲序列
bidirectional pulse 双向脉冲
bidirectional reflectance 双向反射率
bidirectional relay 双向继电器
bidirectional switch 双向转换开关，双向开关
bidirectional thrust bearing 双向推力轴承
bidirectional transducer 双向传感器，双向变换器
bidirectional triode thyristor 双向晶闸管，双向三端可控硅
bidirectional vane 双向风标
bidirectional 双向的
bidirection changeover valve 双向切换阀
bid language 标书的语言
bid letter 投标函
bid negotiation 议标
bid opening minutes 开标纪要
bid opening price 开标价格
bid opening procedure 开标程序
bid opening record 开标记录
bid opening 开标，揭标
bid package 一揽子投标
bid phase 投标阶段
bid preparation 编标，编写标书
bid price quotation 投标报价单
bid price 标价，买价，投标报价，投标价，投标价格
bid procedure 投标程序
bid process 投标步骤
bid proposal evaluation 评标
bid proposal 投标建议书，投标报价，投标文件，投标方案，投标书
bid purchasing 招标采购
bid rate 买价，投标价，递价
bid receipt 投标书签收
bid rejection 废标
bid requirement 投标要求
bid-rigging 操纵投标
bi-drum boiler 双汽包锅炉
bid schedule of price 投标价目表，报价明细表，投标价格表，报价表格
bid schedule 报价表格，所投标项目的执行计划
bid-securing declaration 投标保证声明
bid security 投标担保（书），投标保函，投标保证（金）
bid sheet 标价单，投标单，投标人名单，投标者名单
bid specification 招标技术条款，招标说明书，标书
bid submission document 投标标书
bid substance 投标实质内容
bid sufficiency 申报充足率
bid summary 投标汇总
bid system 投标制
bid tabulation 标价汇总表，投标表册
bid technique document 招标技术文件
bid time 投标时间
bid unit price 投标单价
bid validity period 投标有效期
bid winner 投标得标人，中标人
bid-winning probability 中标概率
bid-winning rate for price cap bidding 申报最高限价时的中标率
bid 报价，投标，出价，投标书，投标文件，递盘，叫牌
biennial plant 两年生植物
bifilar bridge 双线电桥
bifilar coil 双绕线圈，双绕无感线圈
bifilar electrometer 双线静电计
bifilar inductor 双线扼流圈
bifilar oscillograph 双线示波器
bifilar suspension 双线悬置，双线悬控
bifilar transformer 双线绕制变压器，双绕变压器
bifilar winding 双线无感绕组，双线无感绕法
bifilar wire 双股线
bifilar wound transformer 双线绕制的变压器，双绕变压器
bifilar 双股的，双线的，双向的
biflow filter 双向滤池
biflow 双流
biflux 汽-汽热交换器
bifuel system 双燃料系统
bifurcated chute 分叉落煤管
bifurcated contact 分叉触点，双叉触点，双叉接插件
bifurcated line 分叉线，分叉线路
bifurcated pipe 岔管，分叉管

bifurcated tube　分叉管
bifurcated　分为两支的，分叉的，三通
bifurcate　分叉的，二分叉，分支，分路，分为两支（的）
bifurcating box　双芯线终端套管，双叉分接盒
bifurcation angle　分叉角
bifurcation gate　分水闸门
bifurcation plume　分岔型羽流
bifurcation point　分叉点
bifurcation structure　分叉结构，分水结构
bifurcation　分支，分叉，分支点，两歧状态，分路，双态（计算机），双叉管
big data storage　大数据存储
big data　大数据
big-end-down　上小下大的
big-end-up　上大下小的
BIGFET(bipolar insulated gate field effect transistor)　双极绝缘栅场效应晶体管
bight　曲线，回线，线束，弯曲，绳环，绳扣
big iron　主机电脑硬件【相对微电脑硬件而言】
bigit　二进位，位【二进制的】
big power station　大型发电厂
big repair　大修
bigrid valve　双栅管
bigrid　双栅极的
big share holder　大股东
biharmonic　双谐波的，双耦合的
bike shed　自行车棚
bilateral agreement　双边协定，双边协议
bilateral and multilateral economic cooperation　双边和多边经济合作
bilateral circuit　双向电路，对称电路，可逆电路
bilateral contract for difference　双边差价合同
bilateral contract　双边合同
bilateral cooperation　双边合作
bilateral diffusion　双向扩散
bilateral document　双边文件
bilateral force reflecting manipulator　双边力反射机械手
bilateral iterative network　双向迭代网络
bilateral levelling　对向水准测定
bilateral loan　双边贷款
bilateral matching　双向匹配
bilateral network　双向网络
bilateral servomechanism　双向式油动机，双向式伺服机构
bilateral switching　双通开关，双向转换
bilateral symmetry　两侧对称，左右对称
bilateral tolerance　双向公差
bilateral trade agreement　双边贸易协定
bilateral trading　双边交易
bilateral transducer　双通转换器，双向传感器，可逆传感器
bilateral　双通的，两边的，双向的
BILE(balanced inductor logical element)　平衡式感应器逻辑元件
bilection　凸嵌线
bilge block　舱底垫块
bilge hat　舱底水阱
bilge pump　船底排水泵，喷射器
bilge suction　舱底水吸入
bilge water　舱底污水
bilge　舱底，底舱，鼓胀，突出
bilinear floating resistor　双线性浮地电阻
bilinear form　双线性形
bilinear transformation　双线性变换
bilinear　双线的，双线性的
bill accepted　承兑的汇票
bill acceptor　票据承兑人
bill at sight　见票即付票据，即期汇票
billboard　（路边）广告牌，布告牌，招贴板
bill collector　票据收款人
billet steel　短条钢
billet　钢垫板，钢坯，金属坯段
bill for acceptance　承兑汇票
bill for clearing the project accounts　项目结算票据［单据］
bill for collection　托收汇票，托收票据
bill holder　票据持票人
billibit　十亿位，千兆位
billicapacitor　微调电容器
billicondenser　管状微调电容器
billion electron-volts　十亿电子伏特
billion　十亿【美、法】，兆【英、德】，太，万亿，10^{12}【英、德】
billisecond　毫微秒，纳秒，十亿分之一秒
billi　毫微，千兆（十亿）分之一，10^{-9}
bill of approximate estimate　概算书
bill of approximate quantities contract　估计工程量单价合同，近似工程量单价合同
bill of draft　汇票
bill of exchange　汇票
bill of freight　运单
bill of goods　货单，货品清单
bill of lading made out to order Bank of China　以中国银行指示抬头的提单
bill of lading marked freight prepaid　注明运费已付的提单
bill of lading marked freight to collect　注明运费到付的提单
bill of lading to order　指示提单
bill of lading　提单，运货证书
bill of materials　材料表，材料单，材料明细表，材料清单，用料单
bill of particulars　明细表
bill of payment　付款清单，付款单
bill of quantities contract　工程量合同，工程量单价合同
bill of quantity　工程量清单，工程量表，BQ单，工作量清单
bill of store　再入关免税单，船上用品免税单，船上用品税单，重进口免税单
bill of sufferance　免税货物运输许可证，免税装货许可证，卸货许可单，免税单，海关落货许可单
bill of three parts　三联单
bill on demand　即期汇票，见票即付
billow clouds　浪云
billow　波涛，大波，巨浪，狂飙，五级风浪，汹涌
bill payable after sight　见票即付汇票

bill payable at a fixed period after sights 见票后定期付款的汇票
bill payable by instalments 分期付款的汇票
bill payable 15 days after date 出票后十五日支付
bill payable to a specified person only 不能转让的汇票
bill payable to order 记名票据
bill quantity of works 工程量清单
bill rate 汇票贴现率，票据贴现率
bill receivable 应收票据
bill stamp 印花税票
bill sufferance 免税单
bill to order 空白抬头票据
bill without credit 非凭信用证开发的汇票，非信用证签发的汇票
bill 发票，汇票，账单，票据，清单，议案，提单
bilux bulb 双灯丝灯泡
BIMAG(bistable magnetic core) 双稳态磁芯，双磁芯
bimag 双磁芯
bimanualness 双手操作
bimbaled motor 万向架固定式电动机
BIM(beginning of information marker) 信息开始标志
BIM(building information modeling) 建筑信息模型
BIM(bus interface module) 总线接口模件
BIMCO(Baltic and International Maritime Council) 波罗的海国际海事协会
bimetal blade 双金属刀形开关
bimetal fuse 双金属熔丝
bimetal grid 双金属格架
bimetallic conductor 双金属导线，双金属线，双金属导体
bimetallic corrosion 电镀腐蚀，双金属腐蚀
bimetallic couple 双金属热电偶
bimetallic deflection thermometer 双金属偏转式温度计
bimetallic dial thermometer 双金属表盘式温度计
bimetallic element 双金属元件
bimetallic expansion steam trap 双金属膨胀式蒸汽疏水阀
bimetallic fitting 双金属金具
bimetallic joint 双金属接头
bimetallic rotor 双金属转子
bimetallic starter 双金属启动器
bimetallic strip compensation 双金属片补偿
bimetallic strip relay 双金属片继电器
bimetallic strip 双金属片，双金属带
bimetallic thermocouple 双金属热电偶
bimetallic thermograph 双金属温度记录仪
bimetallic thermometer 双金属温度计
bimetallic thermostat 双金属温度调节器，双金属恒温控制器
bimetallic type 双金属型
bimetallic wire 双金属线
bimetallic 双金属，双金属的，复合金属
bimetal relay 双金属继电器
bimetal release 双金属开关
bimetal strip 双金属条，双金属片
bimetal thermometer 双金属温度计
bimetal time-delay relay 双金属延时继电器
bimetal 双金属的，复合钢材
bimodal distribution 双峰分布
bimodal 双峰的【曲线】，双重模型
bimolecular reaction 双分子反应
bimolecular 双分子的
bimoment 复合力矩【转矩，扭矩】，双力矩，双弯矩
bimonthly 两月一次的，隔月的，双月刊
bimorph cell 双层晶体元件
bimorph crystal 双层晶体，振荡互补偿晶体
bimorph 双压电晶片，双晶
bimotored 双马达的，双电动机的，双发动机的
bina 坚硬黏土岩
bin activator 料仓促流器
bin and feeder system 中间仓储系统，中间仓储式制粉系统
binant electrometer 双限静电仪
binary add circuit 二进制加法电路
binary adder 二进制加法器
binary addition 二进制加法
binary add 二进制加，二进制加法指令
binary alloy photoconductor 二元合金光电导体，二元合金光敏电阻
binary alloy 二元合金
binary arithmetic operation 二进制算术运算
binary arithmetic 二进制算法
binary asymmetric channel 二进不对称波道，二进制不对称信道
binary automatic computer 二进制自动计算机
binary base 二进制基数
binary bit 二进制位，二进位
binary card 二进制卡件，二进制卡片
binary carry 二进制进位
binary cell 二进制单元，二进制元件
binary chain 二进制链
binary channels 二进信道，双信道
binary character 二进制符号
binary circuit 二进制电路
binary coded character 二进制编码符号
binary coded decimal counter 二-十进制计数器
binary coded decimal digit 二-十进制数
binary coded decimal notation 二-十进制表示法，二-十进制记数法
binary coded decimal representation 十进制数用二进制代码的表示法
binary coded decimal-to-decimal converter 二进制编码的十进制向十进制的转换器
binary coded decimal 二进制编码的十进制
binary coded octal system 二进制编码的八进制，二-八进制
binary coded octal 二-八进制，二进制编码的八进制
binary coder 二进制编码器
binary code 二进制代码，二进制码
binary column 二进制列
binary commutation circuit 二进制转接电路
binary computer 二进制计算机
binary control module 二位控制模件
binary control 二进制控制，两位控制

binary converter　二进制转换器，二元交流-直流变流机
binary counter circuit　二进制计数器电路
binary counter　二进制计数器
binary cycle boiler　双工质锅炉
binary cycle fluid　二元循环工质
binary cycle operation　两汽循环运行
binary cycle plant　二汽循环电厂
binary cycle system　双循环系统
binary cycle　双工质循环，双循环
binary-decade counter　二进制-十进制计数器
binary-decimal conversion　二进制-十进制变换
binary-decimal converter　二进制-十进制变换器
binary-decimal notation　二进制-十进制记数法
binary digital logic system　二进制数字逻辑系统
binary digit duration　二进制数字脉冲宽度
binary digit　二进制位，二进制数字
binary divider　二进制除法器
binary division　二进制除法
binary dump　二进制信息转储
binary electrometer　双极静电仪
binary element　双态元件
binary encoder　二进制编码器
binary encoding　二进制编码
binary eutectic　二元低共熔物，二元共晶
binary flip-flop　二进制触发器
binary fluid cycle　双工质循环
binary flutter　二维颤振，二元颤振
binary fuel　双组分燃，二元燃料
binary hologram　二元全息图
binary internal number base　内部二进制基数
binary loader　二进制装配程序
binary logic element　二进制逻辑元件
binary logic　二进制逻辑
binary magma　二元岩浆
binary message　二进制信息
binary mixture of particles　二种颗粒混合物
binary mixture　二元混合物
binary multiplication　二进制乘法
binary multiplier　二进制乘法器
binary notation　二进位符号，二进制记数法
binary number system　二进计数制
binary number　二进制数（码）
binary operation　二进制运算
binary output　二进制输出
binary parallel digital computer　二进制并行数字计算机
binary place　二进制数位
binary point　二进制小数点
binary punch　二进制穿孔
binary quantizer　二进制数字转换器
binary ring　二进制计算环，二进制计数环
binary row　二进制行
binary scaler　二进制计数器，二进制换算电路
binary search method　二分检索法
binary search　对分检索，二分检索
binary sediment　二相沉积物
binary signal　开闭信号，二进制信号，双元信号
binary state information　二元状态信息，双态信息
binary storage device　二进制存储装置
binary storage tube　二进制储存管
binary subtraction　二进制减法
binary switch　二进制开关
binary synchronous communication　二元同步通信，双同步通信，二进制同步通信
binary system　二进制，二进制码系统，二元系
binary thermodiffusion factor　双元热扩散系数
binary-to-decimal conversion　二进制-十进制变换
binary-to-decimal converter　二进制-十进制变换器，二进制-十进制转换器
binary-to-hexadecimal conversion　二进制-十六进制转换
binary-to-octal converter　二进制-八进制变换器
binary-vapor cycle　双汽循环
binary-weighted capacitor array　二进制权电容阵列
binary-weighted ramp generator　加权二进制斜波发生器
binary-weighted resistor　二进制权电阻
binary word　二进制字
binary　二进制，双的，两元的，二元的，二进制的
bin continuous level signal　煤仓连续料位信号
binder energy　结合能
binder of concrete　胶凝材料
binder resin　黏合剂树脂
binder　黏合剂，胶结物，夹子，黏结，系梁，铺路沥青
binding agent　黏合剂，黏着剂，黏结剂
binding beam　联（接）梁
binding bolt　连接螺栓，连接螺钉
binding clasp　夹紧器
binding clip　接线夹，线箍，钮夹
binding coal　黏结煤，黏结性煤
binding contract　有约束力的合同
binding course　结合层，黏结层，拉结层，联系层
binding effect　约束力
binding energy　结合能
binding face　结合面
binding fatigue　结合疲劳
binding force　结合力，束缚力，内聚力
binding head screw　圆顶宽边接头螺钉
binding material　胶结材料，黏结材料，黏合料
binding piece　线夹
binding post　接线柱，接线端子，绑杆，缚杆
binding power　结合力，约束力，黏结力
binding rivet　结合铆钉，紧固铆钉
binding screw　接线螺钉，夹紧螺钉，紧固螺钉，结合螺钉，接线螺旋
binding signature　有约束力的签字
binding strength　黏结强度
binding strip　拉筋
binding turns　绑匝
binding wire　绑扎钢丝，捆扎用钢丝，绑扎线，拉筋，绑线
binding　结合，黏合，紧固，扎带，连接，装订，捆绑，约束，有约束力的
bind up　束缚
bind　结合，连接，黏合，黏固，捆扎，约束，

束缚
BINE(Beijing Institute of Nuclear Engineering) 北京核工程研究院【核二院】
bin feeder 斗式给料机
bin flow device 料斗抖动装置
bin gate 料仓门，料斗闸门
Bingham model 宾汉模型
bing 材料堆
bin hang-up 仓内搭桥
bin high level signal 煤仓高煤位信号
binistor 四层半导体开关器件
binit 二进制位，二进制数字，二进制符号
bin level indicator 仓内料位指示器
bin low level signal 煤仓低煤位信号
bin middle level signal 煤仓中煤位信号
binocular coil 双孔线圈，双筒线圈
binocular LP line filter 双筒低压管路过滤器
binoculars sight 双筒望远镜观测
binoculars 双筒望远镜
binocular 两眼的，双目的，双筒的
binodal lateral wave 双节横波
binode 双阳极的，双阳极电子管
binomial array 双向天线阵
binomial coefficient 二项式系数
binomial data distribution 二项式数据分布
binomial distribution 二项式分布
binomial equation 二项方程，二项式方程
binomial expansion 二项式展开
binomial expression 二项式
binomial theorem 二项式定理
binomial 二项，二项式，二种名称，二项式的，二种名称的
bin outlet 料仓排料口
bin storage type pulverizing system 中贮式制粉系统
bin system 中间储仓式系统，仓储式（制粉）系统，集中储仓制，储仓系统
bin vibrator 料仓振动器
bin wall 隔仓式挡土墙
bin 仓，箱，料仓，料斗，储存仓，磁带存储器
bioaccumulation factor 生物累积因子
bio-aeration 生物曝气
bioassay 生物鉴定，生物鉴定法，生物检验，生物分析
biobattery 生物电池
biochemical action 生化作用
biochemical deposit 生物化学沉积
biochemical fuel cell 生物化学燃料电池
biochemical indicator 生化指示剂，生化指标
biochemical oxidation 生化氧化
biochemical oxygen demand 生化需氧量【指水的污染参数】
biochemical process 生化过程
biochemical rock 生物化学岩
biochemical 生物化学的，生化的
biochemistry 生物化学
biocide 杀虫剂，生物杀灭剂
bioclastics 生物碎屑岩
bioclimate 生物气候
bioclimatology 生物气候学
biocoenose 生物群落
biocolloid 生物胶体
biocommunity 生物群落
biocorrosion 生物腐蚀
biocycle 生物带
biodegradability 生物降解，生物可降解性
biodegradable 生物降解的
biodegradation 生物降解
bioelectricity 生物电流，生物电学
bioengineering 生物工程
biofilm fluidized bed reactor 生物膜流化床反应器
biofilm 生物膜
bio-filter 生物滤池
biofiltration 生物过滤
bioflocculation process 生物絮凝法
bioflocculation 生物絮凝（作用）
biofouling 生物淤积，生物污垢
biofuel 生物燃料
biogenetic deposit 生物沉积
biogenetic rock 生物岩
biogeography 生物地理学
biolith 生物岩
biological accumulation factor 生物积累因子
biological activity 生物活性
biological chain 生物链
biological clarification 生物净化
biological concentration factor 生物浓集因子
biological concentration 生物集结
biological contact oxidation 生物接触氧化
biological control 生物防治
biological corrosion 生物腐蚀
biological damage 生物损伤
biological decay constant 生物衰变常数
biological decay 生物衰变
biological decomposition 生物分解
biological depollution 生物去污染
biological deposit 生物沉积物
biological diversity 生物多样性
biological dose 生物剂量
biological dosimeter 生物剂量计
biological effectiveness 生物有效性
biological effect of radiation 辐射的生物效应
biological effect 生物效应
biological energy 生物能，生物能源
biological film 生物膜
biological filtering membrane 生物过滤膜
biological filter loading 生物滤池负荷
biological filter 生化滤层，生物滤池
biological filtration 生物过滤
biological fluidized bed 生物流化床
biological fouling 生物污染
biological fuel cell 生物燃料电池
biological geology 生物地质学
biological half-life 生物半排期，生物半衰期，生物半减期
biological hazard 生物公害
biological hole 生物洞，生物通道
biologically active floc 生物活性絮体
biologically active fluidized bed 生物活性流化床
biologically equivalent single dose 生物当量单一

剂量
biological magnification 生物密度扩大
biological method 生化法
biological oxidation 生物氧化
biological oxygen demand 生物需氧量
biological pollution 生物污染
biological process 生化过程，生物过程
biological protection 生态保护，生物防护
biological purification 生物净化
biological radiation effect 生物辐射效应
biological reserve 生物保护区
biological resource 生物资源
biological sampling 生物取样
biological shielding 生物保护，生物屏蔽
biological slime 生物黏膜，生物黏质物
biological stain 生物染色剂
biological treatment plant 生化处理厂
biological treatment 生化处理
biological tunnel 生物实验孔道【反应堆内】
biological waste treatment 生物（法）废物处理
biological weathering 生物风化
biological wind indicator 植物风力指示器
biological 生物学的
biologic effect of ultrosound 超声生物效应
biologic half life 生物半衰期
biology 生物学
biomass energy 生物质能
biomass gas power plant 生物质燃气电站
biomass Nippon strategy 日本生物质能策略
biomass power generation project 生物质发电项目
biomass power generation 生物质发电
biomass support particle 载生物质颗粒
biomass （单位面积或体积内的）生物的数量，生物质，生物量
biomechanics 生物力学
biomedical dosimetry 生物医学剂量学
biomedical radiation-counting system 生物医学辐射计数系统
biomedical radiography 生物医学射线照相
biomedical reactor 生物医学用反应堆
biometeorology 生物气象学
biomonitoring 生物侦测
bionic computer 仿生计算机
bionics 仿生学，仿生电子学
bioreactor 生物反应器
biosensor 生物传感器
biosorption method 生物吸着法
biosphere 生物层，生物圈，生命层
biospheric contamination 生物圈污染
biostatics 生物静力学
biostatistics 生物统计学
biostratigraphy 生物地层学
biota factor 生物因素
biota influence 生物区系影响
biota 生物群
biotechnology 生物技术
biotic balance 生物平衡
biotic index 生物指数
biotic influence 生物影响
biotic 生物的，生物环
biotite granite 黑云花岗岩
biotite 黑云母【矿物】
biotron 高跨导孪生管，提高互导的孪生管
Biot-Savart law 毕奥-萨伐尔定律
biparted hyperboloid 双叶双曲面
biparting doors 双扇（对开）门
bipartite cubic 双枝三次曲线
bipartite 双向的，双只的，两部分构成的
bipartition 对分，分为两部分
bi-pass 二通符，双通，双行车路
bipatch 双螺旋线的，双螺旋的，双头的
BIPCO(built-in place components) 内部元件
biphase current 两相电流
biphase equilibrium 两相平衡
biphase motor 两相电动机
biphase rectification 双相整流，全波整流
biphase rectifier 全波整流器，双相整流器
biphase 双相，两相，双相的，两相的
biphone 电话耳机
bipod 二脚架，双脚架
bipolar armature winding 双极电枢绕组
bipolar coordinates 双极坐标
bipolar dynamo 双极电机
bipolar field 双极磁场【电机】
bipolar HVDC transmission system 双极高压直流输电系统
bipolar insulated gate field effect transistor 双极绝缘栅场效应晶体管
bipolar integrated circuit 双极集成电路
bipolarity 双极性
bipolar junction transistor 双极性晶体管，双极面结型晶体管
bipolar line 双极线路
bipolar membrane 双极性膜
bipolar memory 双极型存储器
bipolar motor 两极电动机
bipolar pulse 双极脉冲
bipolar signal 双向信号
bipolar switch 双极开关
bipolar transistor 双极晶体管
bipolar tube 双极射线管
bipolar 双极的，双极性的
bipole planned outage times 双极计划停运次数
bipole unplanned outage times 双极非计划停运次数
bipotential 双电位的
bipropellant fuel 二元燃料
biquadratic equation 双二次方程，四次方程
biquadratic 双二次的，四次的
biquinary representation 二-五进制数的表示
biquinary system 二-五进制系统
biquinary 二元五进制的，二元五进位的
Biram's wind meter 翼型风速仪
birch 桦木
birdcage 鸟笼式框架【储运易裂变材料的】，运输罐，宿舍
birdcall 单边带长途通信设备
bird-dogging 被测参数缓慢偏离要求值
birdnesting deposit （管间的）结渣搭桥
birdproof 防禽的，防鸟类的
bird screen 防挥发屏，防飞禽保护网

bird's-eye drawing 鸟瞰图
bird's-eye perspective 俯视透视图
bird's-eye survey 鸟瞰测量
bird's-eye view of power plant 全厂鸟瞰图
bird's-eye view 鸟瞰图，空瞰图
bird's-eye 俯视的，鸟瞰的
birefringence 双折射，二次折射
biregular transformation 双正则变换
biregular 双正则的
Birmingham wire gauge 伯明翰线规
birotary gas turbine 双转子燃气轮机，双转子透平
birotary turbine 双转子燃气轮机，双转子透平
birotor pump 双转子泵
birotor 双转子
birth rate 形成率，出生率
bi-salt 酸式盐
BIS(Bank for International Settlement) 国际清算银行
BIS(bid invitation specification) 招标说明书
BIS(Bureau of Indian Standards) 印度标准局
bisected coil 对分分段线圈
bisection theorem 电路中分定理，二等分定理
bisection 二等分
bisector 等分线
bi-service 两用的
Bishop's simplified method of slice 毕肖普简化条分法
bi-signal zone 双信号区
bi-signal 双信号
bismuth-cooled reactor 铋冷反应堆
bismuth phosphate process 磷酸铋载体沉淀流程
bismuth spiral 铋螺线
bismuth-tin soldering 锡铋合金钎焊
bistability 双稳定性
bistable amplifier 双稳放大器
bistable circuit 双稳（态）电路
bistable contact 双稳触点
bistable device 双稳态装置
bistable element 双稳态元件
bistable flip-flop circuit 双稳态反复电路
bistable magnetic core 双稳态磁芯
bistable mode of operation 双稳态运行方式
bistable multivibrator 双稳态多谐振荡器，双稳多谐振荡器
bistable operation of double diode laser 双二极管激光器的双稳态工作
bistable operation 双稳运行
bistable optical device 双稳光器件
bistable polarization 双稳极化
bistable polarized relay 双稳极化继电器
bistable relay 双位置继电器
bistable storage tube 双稳态存储管
bistable transistor circuit 双稳态晶体管电路
bistable trigger circuit 双稳触发电路
bistable unit 双稳态部件，双稳单元
bistable 双稳态的，双稳态，双稳定的，双稳态装置
bistatic sonar 收发分置声呐
bister 砂眼
bisulfate 酸式硫酸盐，硫酸氢盐，重硫酸盐
bisulfide 二硫化物
bisymmetric 双对称的
bisynchronous motor 双倍同步速度电动机
bit adaptor 旋具头接头
bit address 位地址
BIT(built-in test) 内部测试
bit bumming 程序压入存储器
bit-by-bit memory type 逐位存储方式
bit control 按位控制
bit count appendage 位计数附件
bit cutting angle 钎子头刃角，钻头磨角
bit density 位密度
bit dressing 修整钎子，修整钻头
BITE(built-in test equipment) 机内测试设备
bite error rate 误码率
bi-telephone 双耳受话器，电话耳机
bit erasure probability 比特删除概率
bit erasure rate 比特删除率
bit error probability 比特出错概率
bit error rate 比特差错率，误码率
bit extension 旋具头延长杆
bite 最小插入值，提升极限【控制棒】
bit gauge 钻头规
bit holder 钻套
biting 咬入
bit interlacing 比特交错
bit location 位单元，位地点
bit of information 信息单位
bit oriented data transmission 面向比特数据传输
bit oriented 按位存取的
bit pattern 位组合格式
bit per second 每秒位，位/秒，每秒传送位数
bit position 位单元，位位置
bit ratchet 旋具头棘轮扳手
bit rate 比特率
bit reamer 扩孔钻
bit screwdriver 旋具头螺丝起子
bit serial 位串行
bit shank 钎子的尾端
bit sharpener 修钻头机，修钎机
bit-sliced microprocessor 位片微处理机
bit stream transmission 位流传输
bit stream （微处理机的）位片，（信息）位流
bit string 位串
bit switch 按位开关
bit synchronization 比特同步
bitter-brackish water desalination 苦咸水淡化
bittern 盐卤
Bitter pattern 比特图
Bitter type coil 比特型线圈
bit traffic 位传送，二进制信息通道
bitt 缆柱
bitulithic pavement 沥青混凝土路面
bitulith 沥青混凝土
bitumastic enamel 沥青瓷漆
bitumen 沥青
bitumen-bonded insulating board 沥青黏合绝缘板
bitumen cable 沥青电缆，沥青绝缘电缆
bitumen carton 沥青油毡
bitumen coated pile 沥青涂面桩
bitumen compound filled joint 沥青胶填充接头

bitumen embedding 沥青固化
bitumen extender 沥青延伸仪
bitumen felt 沥青油毡
bitumen heater 沥青加热炉
bitumeniferous 含沥青的
bitumen melting tank 沥青熔化罐
bitumen mica flake tape 沥青云母带
bitumen needle forcemeter 沥青针入度仪
bitumen paint 沥青漆
bitumen sand mixture 沥青砂
bitumen-sealing compound 沥青填缝料，沥青止水料
bitumen-sheathed paper cable 沥青绝缘纸电缆
bitumen sheet 油毛毯
bitumen softening point gauge 沥青软化点测定仪
bitumen solidification 沥青固化
bitumen sprayer car 沥青洒布车
bitumen sprayer 沥青撒播机
bitumen winding 沥青浸渍绕组
bituminite 烟煤
bituminization unit 沥青固化装置
bituminization 沥青化
bituminized paper 沥青纸
bituminized waste 沥青化废物
bituminizing 沥青处理，沥青浸渍
bituminous base course 沥青底层
bituminous bond 沥青黏合，沥青结合
bituminous carpet 沥青毡
bituminous cement 沥青胶结料，沥青胶体
bituminous coal 长焰煤，烟煤，肥煤，沥青煤，生煤
bituminous coating 沥青护面，沥青涂层
bituminous concrete facing 沥青混凝土面层
bituminous concrete pavement 沥青混凝土路面
bituminous concrete 沥青混凝土
bituminous distributor 沥青喷布机，沥青喷洒机
bituminous facing 沥青护面
bituminous felt 石油沥青油毡
bituminous grouting 沥青灌浆
bituminous limestone 沥青灰岩
bituminous macadam 沥青碎石路
bituminous mastic broken stone 沥青玛蹄脂碎石
bituminous mastic 沥青玛蹄脂
bituminous mat 沥青垫层
bituminous membrane 沥青薄层，沥青（防水）膜
bituminous mortar 沥青砂浆
bituminous painted surface 沥青涂刷面
bituminous paint 沥青漆，沥青涂料
bituminous pavement 沥青路，沥青面层
bituminous paver 沥青铺路机
bituminous penetration 沥青灌注
bituminous road 沥青路
bituminous seal 沥青封缝，沥青止水
bituminous shale 油页岩
bituminous sheeting 沥青薄层
bituminous shield 沥青覆盖层，沥青护面
bituminous-type ash 烟煤型灰
bituminous varnish 沥青清漆
bituminous waterproof coating 沥青防水层
bituminous 沥青的，含沥青的
bitumite 烟煤
bit 位，比特【二进位数】，钻头，凿子，旋具头
bivane 双向风向标，双向风标
bivariant function generator 二变式函数发生器
bivariant 二变式的
bivariate distribution 双变数分布
bivariate function generator 二元函数发生器
bivariate interpolation 二变量内插法
bivariate stochastic process 二元随机过程
bivariate 双变数的，双变的
bi-weekly sailing 周双班，每周两班【轮船】
BJT(bipolar junction transistor) 双极性晶体管，双极面结型晶体管
BKUP(backup) 备用
black absorber rod 黑吸收棒
black absorber 黑吸收体
black adobe soil 黑色冲积黏土
black alkali soil 黑碱土
black alum 黑矾
black-and-white work 木石构造
black annealing 初退火，黑色退火
black anodized 阳极氧化致黑
black ash 黑灰
black band test 黑带试验【电机换向】
black base 沥青基层，黑色基层
black blizzard 黑尘暴
blackbody flame temperature 绝对黑体的火焰温度
blackbody radiating temperature 绝对黑体辐射温度
blackbody radiation 黑体辐射，绝对黑体的辐射
blackbody receiver 绝对黑体吸收器
blackbody standard 黑体标准，绝对黑体标准
blackbody 黑体，绝对黑体
black bog 黑色泥沼
black boundary 黑边界
black box concept 黑盒概念
black box model 黑盒模型，黑箱模型
black box 黑匣子，黑盒子，暗箱，黑箱，未知框
black bulb thermometer 黑球温度表
black buran 黑风暴
black cable 黑色浸渍电缆
black coal 烟煤
black control element 黑体控制元件
black cotton clay 黑棉土
black culvert 铁管涵洞
blackdamp 窒息瓦斯
black detector “黑体”探测器
black diamond 黑金刚石
black discharge 无光放电
black earth 黑钙土
blackened 致黑
blackening 发黑，烧坏，发黑处理
black fog 黑雾
black frost 黑冻，黑霜
black globe temperature 黑球温度
black ice 雨凇
black ink figure nation 国际收支顺差国
black insulating tape 黑绝缘带
black japan 沥青类涂料

black lead 黑墨
black level 黑色电平
black light borescope 黑光管道镜
black light filter 黑光滤片
black lighting 黑光照明
black light meter 黑光表，黑光检测仪
black lightning 暗闪电，黑色闪电
black light 不可见光，黑光
black liquor recovery boiler 黑液锅炉
black liquor 黑液
black mica 黑云母
blackness coefficient 黑度系数
blackness 黑度
black neutron detector 黑体中子探测器
black nut 反向螺帽
black oil 重油，黑油
blackout area 盲区
blackout building 无窗房屋
blackout effect 关闭效应，放电能力瞬间损失
blackout lamp 灯火管制灯
blackout peak 黑色信号峰值
blackout pulse 消隐脉冲
blackout start （核电厂）无外电源启动
black out 熄灭，关闭，断电，停电，总电源消失，电源中断
black particle 黑磁粉
black pigment 炭黑
black pipe 黑铁管
black pitch 黑沥青
black plant start-up 电厂无外电源启动
black plate （未镀的）黑钢板，碳钢板
black poison 黑体毒物【百分之百的吸收体】
black powder 黑色火药
black rain 黑雨
black region 黑信号区
black rod 黑棒
black sand 黑砂
black schist 黑色片岩
black shale 黑色页岩
black sheet 碳钢板
blacksmith shop 锻工车间
blacksmith welding 锻工焊接，锻焊
blacksmith 锻工，炉工
black smoke 黑烟
black soil 黑土
black spot interference 黑点干扰，黑点失真
black spotter 噪声抑制器
black spot 黑斑，黑点
black start service 黑启动服务
black start-up （核电厂）全断电启动，无外电源启动
black start 黑启动，无电源启动
black storm 黑风暴
blacktop 沥青道路面层
black water 黑水，废水，污水
black wind 黑风
black 黑的，吸收全部辐射的
bladder accumulator 充气蓄能器
blade accumulator 叶片蓄能器
blade activity factor 桨叶功率因子
blade adjusting mechanism 轮叶调整机构，叶片调整装置，转桨机构
blade adjusting servomotor 轮叶调节接力器，轮叶调节伺服器，转轮叶片调节接力器
blade aerial 刀形天线
blade aerodynamics 桨叶空气动力（学）
bladeaerofoil 叶片翼型
blade anchorage 叶片定位
blade angle change 桨叶角的变化
blade angle of attack 叶片迎角
blade angle 叶片角，叶片角度，桨叶角，叶片安装角
blade antenna 刀形天线
blade arrangement 叶片配置
blade articulation 桨叶关节连接
blade aspect ratio 叶片翼弦比，叶片展弦比
blade attachment 叶片紧固，叶片装配
blade azimuth angle 叶片方位角
blade back 叶片背弧，叶背
blade balance 桨叶平衡
blade bearing 叶片轴承
blade butt 桨叶根部
blade-camber angle 叶片脊线角，叶片中弧线弯曲角
blade carrier 静叶持环，隔板套
blade cascade 叶栅
blade centrifugal bending stress 叶片偏心弯应力
blade centrifugal tensile stress 叶片离心拉应力
blade cleaning equipment 叶片清洗设备
blade clearance cavitation 叶片间隙气蚀
blade clearance 叶片间隙
blade concavity 叶片内弧
blade coning angle 叶片锥角
blade construction 叶片结构
blade convexity 叶片背弧
blade cooling by effusion 发散式叶片冷却
blade corrosion 叶片腐蚀
blade count 叶片数
blade cuff 桨叶根套
blade damper 桨叶减摆器，桨叶减震器
bladed disk vibration 叶轮振动
blade deposits 叶片积垢
blade diameter length ratio 叶片径高比
blade disc 叶轮
blade disk vibration 叶片-轮盘系统振动
blade disk 叶轮
blade dozer 刮铲推土机
blade drag 轮叶阻力，叶片阻力，叶片曳力，刀片式刮路机，铲式刮路机
bladed rotor 装有叶片的转子
blade duct 叶片流道
bladed wheel 装叶片叶轮，叶轮叶片
bladed 装有叶片的
blade edge 叶片（边）缘，桨叶缘
blade efficiency 叶片效率
blade element theory 叶素理论，桨叶剖面理论
blade element 叶素
blade end loss 叶片端部损失
blade erosion shield 叶片防蚀片
blade erosion tester 叶片侵蚀测试计
blade erosion 叶片侵蚀
blade exit 叶片出口

blade face cavitation 叶面空蚀，叶面汽蚀
blade face 叶片端面，叶面
blade failure and repair 叶片损坏及处理
blade fatigue 叶片疲劳
blade feathering test 顺桨试验
blade feathering 叶片顺桨
blade for iron saw 锯刃，铁锯片，盘刃
blade fouling 叶片结污
blade frequency test 叶片频率试验
blade geometric twist 叶片几何扭转
blade grader 刀片式平地机
blade groove 叶根槽
blade grouping 叶片分组
blade height 叶（片）高（度）
blade hub 叶片轮毂
blade incidence 叶片倾角
blade inlet angle 叶片进口角
blade inlet edge 叶片进气边，叶片前缘
blade in regulation stage 调节级叶片
blade insertion hole 叶片插入孔
blade interference 桨叶干涉
blade lattice 叶栅
blade leading edge 叶片进气边，叶片前缘
blade length ratio 叶片相对高度
blade length 叶片长度
blade lever 轮叶臂杆
blade lift coefficient 桨叶升力系数
blade lift 桨叶升力
blade loading diagram 叶片负荷图
blade load 叶片负载
blade locking device 叶片锁紧装置
blade loss factor 叶片损失系数
blade loss 叶片能量损失，叶片损失
blade magnetic domain 刀片形磁畴
blade mass factor 桨叶质量因数
blade milled from solid material 方钢铣制叶片
blade neck 叶根颈部，叶颈
blade nozzle 叶片式喷嘴
blade numbers 叶片数
blade offset stress 叶片偏置应力
blade offset 叶片偏装
blade of the steam turbine 汽轮机叶片
blade of variable cross-section 变截面叶片
blade-opening 叶道最小通流截面
blade-operating servomotor 轮叶接力器，（转）轮叶（片）伺服器，转轮叶片操作接力器
blade orbitalangle 叶片运行角
blade outlet angle 叶片出口角
blade passage 叶片汽道，叶片流道
blade passing frequency 叶片通过频率
blade passing noise 越桨噪音
blade path 叶片流道
blade pitch angle 桨距角
blade pitch change 桨距变化
blade pitching motion 桨叶俯仰运动
blade pitch 叶片螺距，叶片间距，叶片节距，叶栅栅距，叶距，桨距
blade planform taper ratio 叶片平面形状的锥度比
blade profile 叶型
blade profile characteristic 叶片轮廓特性
blade profile thickness 叶型厚度
blade regulating valve 轮叶调节阀，叶片调节阀
blade-removal opening 叶片拆装孔
blade resonant vibration 叶片共振
blade restoring mechanism 叶片复位装置
blade ring 叶片环，叶片圈，叶片持环
blade root attachment 叶根连接
blade root diameter 叶根直径
blade root pin 叶根销
blade root shoulder 叶根销钉，叶根肩部
blade root 叶根
blade rotating angle 叶片旋转角度
blade roughness 叶片粗糙度
blade row 叶列，叶栅
blade rubbing 叶片摩擦，叶片碰磨
blade seal ring 叶片密封环
blade seal 叶片汽封
blade section camber 叶片截面弯度
blade section chord 叶片截面弦，叶片剖面弦长
blade section pitch 叶片剖面桨距
blade section 叶片截面
blade segment 叶段
blade servomotor 叶片接力器，叶片伺服器
blade setting angle 叶片安装角
blade setting 装定叶片角（度），叶片位置调定
blade shank 叶柄
blade shape 叶片形状
blade shedding of ice 叶片甩冰
blade shoulder cavitation 叶肩空蚀
blade skin 叶片蒙皮
blade socket 叶根承座，叶片座
blade solidity 叶片稠度
blade spacing 轮叶间距，叶片间隔，叶片间距，栅距
blade span axis 桨叶轴线
blade span 叶展，翼展
blade spar 桨叶梁
blade stall condition 叶片气流分离状态
blade stall regime 叶片气流分离状态
blade stall 叶片失速
blade station 桨叶站位
blade stem 叶柄
blade structural testing 叶片结构试验
blade strut 叶片撑杆
blade surface 叶面
blade taper 叶片楔面，叶片锥度
blade thickness 叶片厚度
blade tilt 桨叶倾角，叶片倾角
blade tip cavitation 叶尖空蚀
blade tip eddy 叶顶旋涡，叶尖旋涡
blade tip loss 叶顶损失，叶尖损失
blade tip-root loss factor 叶端损失系数
blade tip seal 叶顶汽封
blade tip speed 叶片圆周速度，叶顶速度，叶尖速度
blade tip spoiler 叶尖扰流器
blade tip stall 叶顶失速
blade tip velocity 叶尖速度
blade tip vortex 叶顶涡流，叶梢涡旋，叶尖涡
blade tip 叶顶，叶尖
blade-to-blade analysis 跨叶片分析

blade-to-blade flow 叶间流动
blade-to-blade solution 跨叶片解
blade-to-blade vortex 叶间涡
blade-to-jet speed ratio 叶片-气流速比
blade top 叶片顶部
blade trailing edge 叶片出汽边，叶片后缘
blade tuning 叶片调频
blade twist test 叶片扭转度试验
blade twist 叶片扭转，叶片扭曲，桨叶扭转
blade type burner 叶片式喷燃器
blade type connector 刀形接头
blade type control rod 片状控制棒
blade type damper 叶片式挡板
blade vibration frequency 叶片振动频率
blade vibration 叶片振动
blade vortex 叶片涡流
blade wake 叶片尾流
blade wear 叶片磨损
blade wheel 叶轮
blade width ratio 叶片宽度比
blade width 叶宽
blade windage loss 叶片鼓风损失
blade with external fir-tree root 外包枞树形叶根叶片
blade with external inverted-T-root 外包倒T形叶根叶片
blade with external T-root 外包T形叶根叶片
blade without filler pieces 无隔金叶片
blade with pinned root 叉形叶根叶片，插入式叶根叶片
blade with reinforced root 加强叶根叶片
blade with stepped root 阶梯形叶根叶片
blade with straddle root 外包形叶根叶片
blade 叶轮，叶片，桨叶，踏板，风叶，轮叶，刀片，刀形开关，旋翼，推土铲
blading back of earth 倒向推土，回推泥土，回泥土
blading clearance 叶片间隙
blading compaction 刮平压实
blading deposits 叶片积垢
blading-efficiency curves 叶片效率曲线
blading exhaust annulus 叶片排汽环状通路
blading loss 叶片损失
blading nomenclature 叶型几何参数
blading 叶片，叶片组，叶栅，装叶片
Blagg diffraction imaging 布拉格衍射成像
Blake-type jaw crusher 布莱克型颚式破碎机
blank arcade 封闭拱廊，假拱廊，实心连拱
blank arch 假拱，轻拱，装饰拱
blank assay 空白试验，初步试验，坯件试验
blank BL(blank bill of lading) 空白提单
blank bolt 光螺栓，无螺纹栓，非切制螺栓
blank cheque 空白支票，不记名票据
blank correction 空白校正
blank credit 空额信用证，空白信用证，无具体金额的信用证
blank door 暗门，假门
blanked-off impeller 盲叶轮
blanked-off nozzle （双层箱）封闭接管
blank endorsed B/L 空白背书提单
blank endorsed 空白背书的
blank endorsement 空白背书，不记名背书
blanket acceptance 全部接受
blanket approval 全部认可，全面认可，全部同意，完全同意
blanket area 敷层面积，覆盖面积
blanket assembly 转换区燃料组件
blanket claim 全部权利要求，总括索赔
blanket contract 一揽子合同
blanket conversion ratio 转换区转换比
blanket corrosion 表面腐蚀
blanket drainage 褥垫式排水
blanketed area 遮蔽面积，气动阴影面
blanketed 包上的，覆盖的，有转换区的，有再生区的
blanket element 转换区燃料元件
blanket gas analysis 覆盖气体分析
blanket gas clean-up system 覆盖气体净化系统
blanket gas generator 覆盖惰性气体发生器
blanket gas space 覆盖气体空间
blanket gas system 覆盖气体系统
blanket gas 填充气，覆盖气体，填充气体
blanket grout hole 覆盖灌浆孔
blanket grouting 覆盖灌浆
blanketing effect 气动阴影效应
blanketing frequency 抑止频率
blanketing 核燃料再生，形成覆盖层，熄灭，强信号噪扰
blanket insulant 绝热材料【电热等】
blanket insulation 毛毡绝缘物，毡状绝缘
blanket insurance 统保，总括保险，统括保险
blanket material 铺盖材料
blanket mortgage 总括抵押，统括抵押，全部抵押，一揽子抵押
blanket of graded gravel 级配砾石铺盖
blanket pin 转换区燃料小棒
blanket policy 统保单，总括保险单
blanket power 转换区功率，再生区功率【反应堆】
blanket processing 转换区处理，转换区燃料处理
blanket purchase contract 一揽子采购合同
blanket region 转换区，再生区【反应堆的】
blanket reprocessing 转换区燃料后处理
blanket sand 冲积沙层
blanket-separation plant 转换区产物分离装置【反应堆】
blanket subassembly change 转换区分组件倒换
blanket subassembly flow rate 转换区分组件流量
blanket-to-total absorption 转换区吸收与总吸收之比
blanket-to-total fission 转换区裂变与总裂变之比
blanket unit 转换区单元
blanket vessel 转换区容器
blanket 转换区，再生区【反应堆】，铺盖，覆盖，掩盖，毯，毡，毯状物，泥渣层
blank experiment 空转试验，检验试验
blank flange 法兰盖，盲板，闷头法兰，堵板，无孔法兰，封口法兰，死法兰，盲法兰
blank form contract 空白合同
blank head 无人孔封头

blank hole　未下套管的钻孔
blanking die　冲裁模，下料模，切口冲模
blanking disc　切料圆盘
blanking gate　消隐脉冲选通电路，消隐门
blanking-off　封闭，阻塞
blanking plate　盖板，盲板
blanking pulse　消隐脉冲
blanking signal　消隐信号
blanking tube　截止管，匿影管
blanking voltage　截止电压，匿影电压
blanking wave　消隐波，匿影波
blanking　闭锁，闭塞，堵塞，熄灭，消隐，切料
blank instruction　空操作指令
blank letter of credit　不记名票据
blank map　空白地图，轮廓地图
blank medium　间隔介质，参考介质
blank off panel　封板，堵头
blank off pressure　极限低压强，极限压力
blank off　加管盖，熄灭，断开，空白，盖上
blank out　使熄灭，使无效，作废
blank page　空白页［纸］
blank panel　空白面板，空面板，备用面板
blank run　空转
blank sample　空白试样
blank sheet　空白图幅
blank signal　间隔信号，空白信号
blank table　空白表
blank tape　空白带
blank test　空白试验
blank value　空白值
blank wall　无窗墙，无门窗的墙，闷墙
blank window　暗窗，假窗
blank wire　裸线
blank　空白，空白的，盲板，毛坯，半制品，半成品，坯料，空格，熄灭脉冲
Blasius equation　布位修斯方程
blast air fan　鼓风机
blast air　鼓风，空气射流
blast area　爆破区
blast atomizer　喷射雾化器
blast burner　喷灯
blast capacity　送风量
blast charging　爆破装药
blast cleaning　喷砂清理
blast cock　放气阀，放汽阀
blast cooling　鼓风冷却，风冷
blast cover　火焰反射器【燃烧室】
blast deflector　火焰反射器
blast draft　强制通风，强迫通风，压力通风
blast draught　压力气流
blast engine　鼓风机
blaster cap　起爆雷管
blaster fuse　引线
blaster　爆破器，爆炸手，放炮工，起爆器，喷射器
blast fan　鼓风机，风扇
blast furnace　高炉，鼓风炉，炼铁炉
blast fumace gas turbine　高炉煤气燃气轮机
blast furnace cast iron　高炉生铁
blast furnace cement　高炉水泥，炉渣水泥
blast furnace cinder　高炉熔渣
blast furnace coke　高炉焦炭，鼓风炉焦炭
blast furnace gas pipe　高炉煤气管
blast furnace gas　高炉煤气
blast furnace method　鼓风炉熔炼法
blast furnace plant　炼铁厂
blast furnace slag cement　高炉矿渣水泥，炉渣水泥
blast furnace slag　高炉矿渣
blast furnace stack　高炉烟囱，高炉炉身
blast furnace tar　高炉煤焦油
blast gas cloud　爆破气体云
blast gate　风门
blast hole drill　爆孔钻
blast hole　爆破孔，炮眼
blasting agent　爆破剂
blasting air　喷气
blasting cap　火雷管，雷管，起爆雷管，爆破雷管
blasting cartridge　爆破管，爆破筒
blasting compaction method　爆炸挤密法
blasting compaction　爆破压实
blasting device　放炮用具
blasting fuse　爆炸导火索，导爆索，导火线，起爆引线，引火线，引线
blasting gear　爆破设备
blasting gelatine　爆炸胶
blasting layout　爆破点布置
blasting machine　发爆机
blasting material　炸药
blasting mat　爆破垫，爆破防护网
blasting method　爆破方法
blasting monitor　爆破监测器
blasting oil　爆炸油
blasting operation　爆破作业
blasting powder　炸药，爆破火药
blasting power　爆炸力
blasting procedure　爆破方法
blasting shock wave　冲击波
blasting technique　爆破技术
blasting vibration observation　爆破地震观测
blasting　爆破，碎裂，过载失真，（扬声器的）振声
blast loading　爆破装药
blast main　主空气管道
blast meter　风压表
blast nozzle　喷气嘴，喷砂嘴，吹风喷嘴
blastogranitic　变余花岗状
blastomylonite　变余糜棱岩
blastoporphyritic　变余斑状
blast pipe　鼓风管，排气管，放气管
blast pressure　爆炸波压力，鼓风压力
blast proof soft exhaust bellow　防爆柔性排气波纹管
blast proof　防弹的
blast protection　防爆
blast-resistant door　防爆门
blast table spreader　风力抛煤机
blast tube　送风管道
blast wall　防爆墙
blast wave　爆炸波，冲击波

blast 鼓风，爆炸，喷射，喷砂，从存储单元中消除空白
blatt 横推断层
blazer cooling 吹风冷却，鼓吹冷却
blaze 火焰，燃发，激发，刻标
B/L(bill of lading) 提单，装船提单，运货证书，提货单
BL(boundary line) 边界线
BLC(boundary layer control) 边界层控制
BLCS(boundary layer control system) 边界层控制系统
bleached hologram 漂白全息图
bleacher 漂白剂，漂白业者，露天看台
bleaching effect 漂白效果
bleaching effluent 漂白流出物
bleaching powder 漂白粉
bleach 漂白剂
bleb ingot 有泡钢锭
bled steam evaporator 抽汽加热蒸发器
bled steam feed heater 抽汽回热给水加热器
bled steam preheating 抽汽预热
bled steam tapping point 抽汽口
bled steam 抽出蒸汽，抽汽【汽轮机】
bleed air 抽出空气，抽气
bleed-and-feed system 排补系统
bleed-axial compressor 抽气式轴流压气机
bleed back 回热抽汽，回渗
bleed condenser 泄放凝汽器
bleeder chain 分压电路【串联电阻组成的降压电路】
bleeder circuit 分压电路，泄放电路
bleeder current 分压器电流，泄漏电流
bleeder heater 再生加热器
bleeder hole 出气孔，通气孔
bleeder nozzle 抽汽管（口）
bleeder piping 抽汽管线
bleeder resistance 泄放电阻，分压电阻
bleeder resistor 泄放电阻器，分泄电阻器
bleeder steam 汽轮机抽汽
bleeder tile 泄水瓦管
bleeder turbine 非调整抽汽汽轮机，抽汽式汽轮机
bleeder type condenser 溢流式大气凝汽器
bleeder valve 抽汽阀，放气阀，排泄阀，出料阀
bleeder well 渗沥井，减压井
bleeder 泄放管，分水口，分压器，分泄电阻，疏水管［阀］，泄放阀，泄出管
bleed gas turbine 抽汽式燃气轮机
bleed heating 抽汽加热
bleed hole 泄放孔
bleeding capacity 泌水能力
bleeding cement （混凝土）水泥浮浆
bleeding cock 泄放旋塞，疏水旋塞
bleeding coil 排流线圈
bleeding condition 抽（放）汽工况［条件］
bleeding line 抽汽管线
bleeding of concrete 混凝土泌浆（现象）
bleeding of waste liquor 废液的排除
bleeding point 排气口，抽汽点，抽汽口【汽轮机】
bleeding rate 泌水率
bleeding ratio 泄放比
bleeding regulation 调节抽汽
bleeding turbine 抽汽式汽轮机
bleeding 抽汽【汽机，不可调的】，疏水，排出，放气，抽出，漏泄，（混凝土的）泌浆，泌水，析水，泄放，析出，放汽
bleed line 抽汽管线，排出管线
bleed of boundary layer 边界层抽吸
bleed-off belt 抽气环形室
bleed-off passage 抽气通道
bleed-off 抽出，放水，放空，抽汽，排出，放出，排水，排泄，排流，排放，泄放
bleed orifice 放气孔
bleed out 渗出
bleed point 抽汽点
bleed position 抽汽位置
bleed steam pipework 抽汽管路
bleed steam turbine 抽汽式汽轮机
bleed steam 抽汽
bleed turbine 抽汽式汽轮机
bleed valve 抽出阀，抽汽阀，放气阀，泄放阀
bleed water 排污水，排出的水
bleed 抽出，抽汽，放气，排空，泄放，抽吸，放出
blemish 表面缺陷，缺陷，损伤，损害，损坏，瑕疵，污点
blendable 可掺和的
blend back addition 掺和添加物，返料添加掺和
blend batch 掺和配料
blend composition 组分
blended cement 混合水泥
blended coals 混煤
blended fuel 混合燃料
blended raw coal preparation 配煤入选
blended treatment 混合处理，掺和处理
blended 混合的，掺烧的
blender-reclaimer 混匀取料机
blender 混合器，掺和机，搅拌机，搅拌器
blending accuracy 混煤精度
blending bin 混煤仓
blending credit 协调信贷
blending ratio 混料比
blending station 混煤装置
blending zone 混合带，混合区
blend 混合，混匀，融合，掺和，混料，混煤，混合物，掺和物［料］，配料，混合贷款
BLF(building-life factor) 建筑寿命因子
blind alley 死胡同
blind area 无信号区，盲区
blind balustrade 实心栏杆
blind box 百叶窗匣
blind catch basin 暗截流井，盲沟式截留井
blind coal 无烟煤
blind controller system 隔离的控制器系统
blind creek 干谷
blind ditch 暗沟，盲沟
blind door 暗门，百叶门，装饰门
blind drainage 暗沟排水，盲沟排水，排水暗沟
blind drain 暗渗道，盲沟，死排水沟，排水暗沟，地下排水沟

blinder 盲区，闭锁装置，遮阳板
blind flange 法兰堵板，盲板法兰，堵塞法兰，法兰盖，盲板法兰盖，闷头法兰，盖板
blind floor 毛地板
blind gallery 死通道
blind hole 盲孔，未穿的孔
blinding bolt 连接螺栓
blinding concrete 盖面混凝土，填充混凝土，混凝土垫层
blinding material 暗沟填料
blinding tile drain 盖土瓦管
blinding 堵塞，填塞，充填，填没，盖土
blind inlet 排水暗沟进水口
blind investment 盲目投资
blind joint 无间隙接头
blind navigation 仪表导航
blindness 盲区
blind nipple 闷头短管
blind nut 闷头螺帽
blind-off cap （电缆贯穿件的）封闭盖
blind-off nozzle 封闭接管
blind-off 堵塞
blind plate 堵板，盲板
blind plug 绝缘插头，空插头
blind power 电抗功率，无功功率
blind production 盲目生产
blind rivet 实心铆钉
blind roadbed 暗道床
blind roller 暗浪
blind scale 盲区比例，盲区刻度
blind set plug 设定盲插头
blind spot 静区，盲点
blind start-up 盲区启动【反应堆】
blind subdrain 填石阴沟
blind tracery 实心窗格
blind valley 盲谷
blind wall 无窗墙，闷墙
blind well 沙底水井
blind window 百叶窗
blind zone 盲区
blind 膜片，百叶窗，窗帘，堵塞，填碎石，遮蔽，遮阳篷，挡板，盲板，堵板［头］
BLINK(backward link) 反向链接
blinker lamp 闪光信号灯
blinker 闪光警戒灯，闪光信号装置
blinking signal 闪光信号
blink 使闪光，烧化（焊接）
blip （显示屏幕上的）标志，记号，（导线的）绝缘层
blister copper 粗铜
blistered casting 多孔铸件
blister packing 起泡包装，吸塑包装
blister plastic packaging 吸塑包装
blister steel 泡钢
blister test 起泡试验
blister 隆起，砂眼，气孔，气泡，起泡，起皮，水泡，起砂眼，发泡，起气泡
blivet 接触段，接触部分
blizzard 暴雪，风暴
BLMT(build-lease-maintenance -transfer) “建设-租赁-维护-移交”模式【工程建设】
B/L No. 提单号
bloating agent 膨胀剂
bloat 起泡，隆起，鼓胀
blob 滴，小球
block access 字组存取，成组存取
blockade 封锁，阻碍
block adjustment 面积平差
blockage correction factor 阻塞修正因子
blockage correction 阻塞修正
blockage effect in cascade 叶栅的堵塞效应
blockage effect 堵塞效应
blockage ratio 阻塞比
blockage test device 堵煤检测仪
blockage test 堵塞试验
blockage 堵塞，封闭，阻塞，锁定，阻塞度，障碍（物），小方石
block and cross bond 丁砖与顺砖交叉砌合
block and tackle 滑车组
block-based design 模块化设计
block beam 拼块梁
block bearing 止推轴承
block bidding 分段竞价
block body 块体，分程序体
block bond 丁砖与顺砖隔层砌合
block brush 块状电刷，电刷块
block building 块体建筑物
block capacitor 隔直流电容器，阻塞电容器
block capital 方块式柱头
block cast 整铸
block caving 大块坍落
block chain 车链，块环链
block chart 方框图，方块图
block check character 信息组检验符
block check 信息组检验
block code 分组码，成组码
block coefficient 充满系数，填充系数，方形系数
block concreting 分块浇筑
block condenser 隔直流电容器，阻塞电容器
block constant 数字组特性常数
block construction wall 大块砌筑岸壁
block construction 大型砌块结构，大型砌块构造，部件结构
block contact 闭塞接点，阻塞接点
block control word 字组控制字
block-covered wall 有卫燃带的水冷壁，有护板炉墙
block curve 连续曲线，实线曲线
block data 数据块
block decrease 减闭锁，闭锁减
block diagram compiler 框图编译程序
block diagram 方块图，方框图，分块图，简图，结构图，区划图，主体图，草图，框图
blocked fault 块（状）断层
blocked funds 冻结资金
blocked-grid keying 栅截止键控
blocked impedance 阻挡阻抗，受挡阻抗
blocked joint 分段连接
blocked level 闭锁电平，阻挡电平
blocked record 成组记录，块式记录
blocked rotor test 转子止转试验，堵转子试验

blocked rotor 止转转子，堵转转子
blocked voltage 闭锁电压
blocked 闭锁的，封锁的
block encoding 分组编码
block end 数字组末位数代码，信息组终端
block error probability 分组差错概率
block error rate 分组差错率，信息组差错率
blockette 分程序块，分区块，子群
block-faulting 块断作用，块状断层，地块断层
block filter 阻塞滤波器
block fin 块状散热器
block flooring 分块地板
block format 程序段格式
block gap 信息组间隙，字区间隙
block gauge 量块，块规
block glide 地块滑坍，块状滑动
block harvesting 成片采伐
block head 程序块首部，分程序首部
block heating 分片供热，分片供暖
block holing 爆破大石
block house 工棚，木屋，砌块屋
block identifier 块标识符
block ignore character 信息块作废符号
block in common channel signaling 公共信道信令中的块
block-in-course 成层砌石块体，嵌入楔块层
block increase 闭锁增
blocking anticyclone 阻塞反气旋
blocking capacitor 隔直流电容器，耦合电容器
blocking circuit 间欺电路
blocking condenser 隔直流电容器，耦合电容器
blocking device 闭锁装置，闭塞装置，停机装置
blocking diode 闭锁二极管，隔离二极管，阻塞二极管
blocking effect 阻塞效应
blocking element 闭锁元件
blocking flow 阻塞流，顶压流
blocking for wind turbines 锁定风力机
blocking interference 阻塞干扰
blocking interval 闭锁期间，关断期间
blocking layer photocell 阻挡层光电池
blocking layer rectifier 阻挡层整流器
blocking layer 阻挡层，闭锁层
blocking medium 截断介质【射线照相】，遮挡介质
blocking of ice 冰塞，冰障
blocking order 闭锁指令
blocking oscillator transformer 间歇振荡器变压器
blocking oscillator 间歇振荡器，阻塞振荡器
blocking overreach distance protection system 闭锁式超范围距离保护系统
blocking period 闭锁期，阻塞周期
blocking protection 闭锁式保护
blocking signal 闭塞信号
blocking state 闭锁状态，关断状态
blocking switch 阻断开关
blocking time 阻塞时间，截止时间
blocking tube oscillator 电子管间歇振荡器
blocking valve 截止阀
blocking voltage 阻塞电压，闭塞电压
blocking wind turbine 锁定风力机
blocking 锁定【风力机】，堵塞，联锁，闭塞，阻挡，阻断，阻塞，闭合，闭锁，分块
block insulation 绝缘，块状绝缘
block interconnection diagram 互连框图
block interlacing 分组交错
block joint 预留【作接头用的】，混泥土块孔，滑块连接，预留砖孔
block journal 止推轴颈
block kit 缸体工具包
block lava 块状溶岩
block layer 阻挡层
block length 块长度，数据块长度
block level 闭塞电平，封锁电平
blocklike structure 块状构造
block lining 混凝土板块衬砌
block load increase 增负荷闭锁
block loading 程序块存入
block load 巨额负荷
block mark 块标志
block mica 块云母，云母块
block mount 组合装配，组装
block movement 地块运动，断块运动，块体移动
block multiplexer channel 数组多路通信
block multiplexing 字组多路转换的
block of code 代码块，代码组
block of decomposition 分解块
block offer 集团报价
block of information 信号字数
block of punch pins （穿孔机上）穿孔针块
block of words 字组
block-oriented 面向字组的，面向块的
block-out ore 已勘矿量，待采矿量
block out 封闭，画出略图，勾画轮廓，草拟大纲，规划，筹划，画草图，封锁，封网
block pavement 块体路面，块状路面
block placement 分区浇筑，分块浇筑【混凝土】，分区填筑
block plane 分区平面
block plan 区划图，分区图，分区规划
block power plant 单岸式电站，河床式水电站
block power station 单岸式电站
block pulley 滑车组
block relaxation 块松弛法，块张弛
block relay 闭塞继电器，闭锁继电器，联锁继电器
block representation 方块线路表示
block sample 块体试样，块状试样
block separation of system 系统分片，系统分组
block sequence welding 分段多层焊
block sequence 叠置次序【多层焊】
block-shaped fuel element 块状燃料元件
block-signal system 闭锁信号系统，阻塞信号系统
block size 块体尺寸
blockslide 块体滑坡
block sorting 分组
block stone 块石
block stream 块状岩流
block structure 分程序，分组结构

block supply 区域供电
block switch 闭塞转接器，闭塞开关
block system 闭锁系统，闭塞系统，阻塞系统，分区开采法
block terminal 分线盒，配电盒，配电箱，端子排
block thrust 块状冲断层
block tin 纯锡块
block trade 集团贸易
block transfer 块转移，字组转移，成组传送，信息组传送
block tridiagonal matrix 块三角阵
block type dock wall 砌块坞墙
block type element fuelled HTR 块状燃料元件高温气冷堆
block type fuel element 块状燃料元件
block type quay 砌块码头
block type runoff-river power station 单岸式径流式电站
block type thermal power stations 单元制热电站
block type turbine 组装式透平
block up 垫高，封闭，堵塞
block valve 插板阀，闭锁阀，截止阀，切断阀
block wall fence 砌块围墙
block wall 砌块墙（体）
block welding technique 多层焊接工艺
block welding 分段多层焊
block winding 闭锁绕组，阻塞跳闸的绕组
block work dado wall 砌块护墙板
block work 块砌体【预制混凝土】，块体结构，砖砌体
blockyard 预制（混凝土）构件厂
blocky structure 块状构造，块状结构
block 浇筑块，程序块，单元，街区，垫块，阻塞块，块，块体，团块，堵塞，段，封锁，滑轮，接线板，组件，单元机组，字组
Blondel chart 布朗德尔图表【计算机输电线弛度用】
Blondel diagram 布朗德尔图【计算机同步电机参数】
blondin 索道起重机［机］悬索道
bloom 丛生水草，晕，图像发晕，大钢坯
bloomed 无反射的，模糊的，起晕的
blooming mill 开坯机
blooming 初轧
bloom pass 初轧孔型，开坯轧槽，开坯道次
bloop 防杂音设备，杂音
BLOT(build-lease-operate-transfer) “建设-租赁-运营-移交”模式【工程建设模式】
blotch 污迹，斑点
blotting agent 吸收剂
blow 吹，吹风，鼓风，吹灭
blow away 脱火
blow back air 反吹气
blow back ring 回座压力调节环【安全阀】
blow back system 反吹系统
blow back 反喷，回座压力差【安全阀】，回火
blow count 击数，锤击计数，打击数，打桩的锤击数
blowdown adjustment 排污调整
blowdown apparatus 排污装置
blowdown cleanup system 排污净化系统
blowdown connection 排污管接头
blowdown control ring 排放控制环【可以是上部或下部调节环】
blowdown cooler 排污水冷却器【蒸汽发生器】，排放冷却器
blowdown cooling water pump 排放冷却水泵，排污冷却水泵
blowdown demineralization 排污除盐
blowdown demineralizer 排污除盐器
blowdown drum 排污罐
blowdown exchanger 排放热交换器，排污热交换器
blowdown flash tank 排污扩容器，排污扩容箱
blowdown flash vessel 排污扩容器
blowdown flow 排污量
blowdown heat transfer 喷放传热【失水事故时】
blowdown line 排放管路，排污管路，泄压管线，卸料导管，排污管线【蒸汽发生器】
blowdown losses 排污损失
blowdown nozzle 排污口【蒸汽发生器】，排污接管
blowdown percentage 排污率
blowdown phase 排放阶段
blowdown pipe 排放管，排污管
blowdown pit 泄料池，放空池
blow down port 排污孔
blowdown pressure of safety valve 安全阀的回座压力
blowdown pressure 回座压力，关闭压力【安全阀】，排污压力，排放压力
blowdown pump 排放泵，排污泵
blowdown rate 排放速率，排污率
blowdown ring 回座压力调节环【安全阀】
blowdown stack 放空烟道
blowdown system 排污系统，排放系统，放空系统，泄料系统
blowdown tank 排污罐，排污箱，排污扩容器
blowdown tap 排放口，排污口
blowdown test 风洞吹风试验
blowdown time 喷放时间
blowdown transfer pump 排污输送泵
blowdown turbine 废气透平
blowdown valve 排污阀，排污门，再热旁路放汽阀
blowdown vessel 凝汽箱
blowdown water cooler 排污水冷却器
blowdown water purification circuit 排污水净化管路
blowdown water purification system 排污水净化系统
blowdown water 排污水
blowdown 排污【锅炉】，扰动【反应堆】，放空，泄放，排放，挑顶，强制崩顶
blower assembly 鼓风机总成
blower-cooled 鼓风冷却的
blower fan 鼓风机
blower mill 风扇磨煤机
blower motor 鼓风电动机
blower room 鼓风机室
blower snow fence 导雪栏栅
blower stacking 用吹灰器消除烟道积灰

blower system 鼓风系统，送风系统
blower turbine 鼓风机透平
blower unit 鼓风机总成
blower 鼓风机，送风机，吹风机，吹风器，通风机，风箱
blow gun 喷枪
blow hole segregation 气孔偏析
blow hole 气孔砂眼【焊缝】，气孔【火床燃料层】，气眼，（铸件的）砂眼，潮吹【指海蚀洞天井】
blowing agent 发泡剂，起泡剂
blowing cooling tower 鼓风式冷却塔
blowing-current 熔断电流
blowing device 鼓风［吹风］装置
blowing dust 吹尘，高吹尘
blowing intensity 鼓风强度
blowing machine 鼓风机
blowing nitrogen on the surface 表面吹氮
blowing-out of arc 熄弧，电弧吹熄
blowing-out 熄灭，吹灭，熔断
blowing period 排污周期，吹灰周期
blowing pressure 吹灰压力
blowing promotor 发泡助剂
blowing sand 刮风沙，飞沙，流沙，吹沙，高吹沙
blowing snow 吹雪，风雪流
blowing soil 土壤吹失
blowing spray 水花，水沫
blowing time 排污时间，吹灰时间
blowing tube 排污管，吹灰管
blowing ventilation 压入式通风
blowing well 吹风井
blowing 冲洗，吹净【如吹灰】，爆发
blow-in system 压入式通风系统
blow in 吹入，送风
blow lamp 喷灯
blowland 风蚀地
blown 吹胀的，多孔的，吹出的
blown bitumen 吹制沥青
blown fuse indicator 保险丝熔断指示器
blown joint 吹接
blown-out land 风蚀区
blown-out shot 空炮
blown-out velocity 喷出速度
blown oxidized asphalt 吹制沥青
blown sand 飞沙，风成砂
blown smoke 吹烟
blown tip pile 爆端桩，爆扩桩
blow number 贯入击数
blow-off chamber 吹泄室，排泥井
blow-off circuit 排放回路
blow-off cock 排污阀，排泄龙头，排泄栓
blow-off connection 排污管接头
blow-off nozzle 排污短管
blow-off of boiler water 炉水排污
blow-off pipe 放气管，排污管，放空管，排泥管，（安全）排气管，排泄管
blow-off tank 排污罐
blow-off tee 排污三通
blow-off valve for boiler 锅炉舷外排污阀
blow-off valve 排污阀，排泄阀，排空阀，放气阀，放空阀，吹泄阀
blow-off water 排污水，吹灰水
blow-off 脱火【沸腾炉，燃烧器】，排污，排放，放水，放气，吹出
blow-out circuit 灭弧电路
blow-out coil 消弧线圈，灭弧线圈
blow-out current 熔断电流
blow-out diaphragm 防爆门薄膜，安全膜
blow-out magnet 灭弧磁铁
blow-out of spark 火花吹灭
blow-out proof stem ball valve 防脱出阀杆球阀
blow-out shaft 排气竖井
blow-out switch 放气开关
blow-out 喷出，放出，吹出，熔断，灭弧，吹灭，风蚀坑
blow pipe analysis 吹管分析
blow pipe system （阵风模拟用的）吹管系统
blow pipe test 吹管试验，吹管分析
blow pipe 吹管，直吹管，焊枪，通风管，鼓风管，送风管
blow sand 飞沙，喷砂
blow tank type pneumatic ash transportor 仓式气力输灰泵
blow tank type pump 仓泵
blow tank 充气柜【煤粉输送】
blow-test 冲击试验，戳穿试验
blow-through valve 安全阀，放泄阀，排污阀
blow-through 排污，吹洗
blow-up 爆炸，吹胀
blow valve 安全阀，通风阀，排空阀，放空阀
blow wash 吹洗，喷洗，压水冲洗
BLR＝boiler 锅炉
B/L surrender fee 电放提单费
BLT(build-lease-transfer) “建设-租赁-移交”模式【工程建设模式】
blue brick 青砖
blue brittleness 蓝脆性
blue collar employee 蓝领工人
blue copperas 胆矾，蓝矾
blue gas 水煤气
blue magnetism 磁铁南极磁性，磁铁蓝色磁性
blue mud 青泥
blueprint machine 晒图机
blue print paper 蓝图纸，晒图纸
blue print 蓝图，晒图，设计图，总体规划，计划大纲，方案
blue-red ratio 蓝红比
blue-sky 无价值的，不安全的，不健全的，不可靠的
blue stone 蓝石，青石
blue tops 坡面桩
blue-violet radiation 蓝紫辐射
blue vitriol 胆矾，蓝矾
bluff body aerodynamics 钝体空气动力学
bluff body burner 钝体燃烧器
bluff body flame holding 钝体稳定火焰法
bluff body flame 钝体稳定的火焰
bluff body 钝体，非流线型体，不良流线型物体
bluff building 钝体型建筑
bluff cylinder 钝形柱体
bluff erosion 陡壁冲刷

bluff failure 陡壁坍毁
bluffness 钝度
bluff section 钝形截面
bluff structure 钝体型结构
bluff work 边坡整平工作
bluff 悬崖，陡岸，陡峭的，钝形的
bluing 发蓝，发蓝处理
blunder away 错过
blunder 错误，故障
blunger 圆筒搅拌机
blunt angle 钝角
blunt arch 垂拱，平圆拱
blunt base 钝底，钝尾
blunt body 钝体，钝头体
blunt cowling 钝头整流罩
blunt end 钝端
blunt leading edge 钝头进气边
blunt-nosed body 钝，钝头体
blunt-nosed pier 平头墩
blunt-nosed stamp 圆形冲孔器
blunt-nosed 钝头，钝头的，不尖的
blunt object 钝头体
blunt pile 钝头桩
blunt tip 钝翼尖
blunt trailing edge 钝缘出汽边
blur circle 模糊圈
blurring effect 模糊效应
blurring 模糊
blur spot 模糊斑点
bluster 栏杆柱
BM(boiler master) 锅炉主控
BM(branch on minus) 负转移
BM(breakdown maintenance) 故障维修
BM(buffer module) 缓冲器模件
BMCR(boiler maximum continuous rating) 锅炉最大连续出力，锅炉最大连续蒸发量，锅炉最大工况
BMCR 锅炉最大连续出力【DCS 画面】
BMEP(brake mean effective pressure) 平均有效制动压力，制动平均有效压力
BMFT(boiler master fuel trip) 锅炉主燃料跳闸
BMLR(boiler minimum load rate) 锅炉最低稳燃负荷率
BMP 燃烧器管理系统【DCS 画面】
BMP 英制马力
BMS(burner management system) 燃烧器管理系统，(锅炉) 烧嘴管理系统
B/N(booking note) 托运单，订舱单
BNC(bayonet nut connector) 刺刀螺母接头
BNC terminal BNC 接线端
BNI(balance of nuclear island) 核岛配套设施
BNR(burner) 燃烧器
board and lodging 膳宿
board B/L 装运提单
board card 插件板
board chairman 董事会主席，董事长
board coal 木质煤，纤维质煤
board director 董事
board extender 插件扩充装置
board foot 板英尺【木材度量单位】
board girder 大板梁
board guide 插件导轨
boarding and lodging 食宿，膳宿，供应食宿
boarding 镶板
board meeting 董事会会议
board meter 板米
board mill 板材制板厂
board mounted instrument 盘装仪表
board ocean bill of lading 装船提单
board of administration 董事会，理事会，管理委员会
board of arbitration 仲裁委员会
board of auditors 审计委员会
board of conciliation 调解委员会
board of consultant 顾问委员会
board of directors 董事会
board of executive directors 执行董事会，常务董事会，执行理事会，执行委员会
board officer 董事会成员
Board of Governors 理事会【国际原子能机构】
board of reference 咨询委员会
board of review 检查局，评审委员会
board of supervisors 监事会
Board of Trade unit 商用电量单位【等于 1 千瓦时】
Board on Radiation Effects Research 辐射效应研究委员会【美国】
board parquetry 木板嵌镶
board rule 板尺，量木尺
board swapping 卡片交换【抢修时】
board-type insulant 毡状绝热材料
board-type insulation 板材保温
board walk 木板人行道，人行板
board 板，屏，盘，仪表盘，配电盘，控制盘，木板，委员会，董事会
boasted ashlar 粗凿石板
boaster 榫槽刨
boast 粗凿石
boat basin 锚地
boat sample 船形金属试样
boatswain's chair 吊椅【高空作业用】，高处工作台【绳系吊板】，飞车
boat tailing 钝尾，船形尾部
boat 船，艇
BO(base order) 基本指令
bobbin core 带绕磁芯，绕线管铁芯
bobbing buoy 梭式浮标
bobbing 标记的干扰性移动，摆动，浮动
bobbin height 绕线管高度
bobbin insulator 绕线管式绝缘子
bobbinite 筒管硫铵炸药
bobbin loader 线盘架
bobbin winding 圆筒式绕组【变压器】
bobbin 绕线管，点火线圈
bobble rise velocity 气泡上升速度
BO(blocking office) 隔离办
bobtail truss 截尾桁架
bobtail 截尾
bobweight 平衡重量，平衡重，平衡锤
bob 摆动，摇动，测锤，秤锤，悬锤，摇摆
BOC(beginning of cycle) 循环初期
BOD(biochemical oxygen demand) 生化耗氧量，

生化需氧量【指水的污染参数】
BOD(board) 董事会
Bode diagram 波特图，伯德图
BOD load 生物耗氧量负荷
body burden 人体承受的剂量，体内放射性物质负荷，体内积存量
body capacity effect 人体电容效应，人体电容，车身容量
body case 壳体，外壳
body-centered cubic 体心立方的
body charge 体电荷
body contact 接壳，碰壳
body contamination 人体沾污
body corporate 法人实体
body current 主体电流，人体电流，水体流动
body decontamination 人体去污
body effect dose 躯体辐射剂量
body end 物体端部
body excretions 身体排泄物
body extension 塔体的延伸【输电线路铁塔】
body feed tank 物料箱
body force 质量力，彻体力，体积力，物体力，体力
body gasket 本体接合填料
body heat loss 人体散热
body leakage 机壳漏电，管身漏电
body-mounted type solar array 壳体式方阵
body nut 阀体螺帽
body of ballast 道渣层
body of coding sheet 一叠程序纸
body of flame 焰心
body of persons 社团
body position 体位
body radiocartography 人体放射绘图法，人体放射统计法
body radiocartograph 人体放射图
body retention ratio 体内残存率
body safety current 人体安全电流
body-section radiography 分步照相，身体截面X射线照相法
body-sodium activation 体内钠活化
body stem seal 本体阀杆密封
body stress 内应力，体应力，自重应力
body swivel bearing 球形止推轴承
body-to-bonnet joint 阀体对阀盖的接合
body wave 立体波【地震】，体波
body 机体，机身，本体，轮体，人体，体，壳体
bogaz 深岩沟
bogey efficiency 假想效率
boghead coal 藻煤
boghead 烟煤
bogie bolster 转向架支承梁
bogie engine 转向机车
bogie 转向车，载重车，转向架，移车台，小车
bog land 沼泽地
bog muck 沼泽腐殖土
bog peat 沼泽泥炭
bog reclamation 沼泽改造
bog soil 沼泽土
bog 泥炭地，沼泽，酸沼
BOI(bought-out items) 外购件，外购项
boil down 蒸发，煮浓
boiled oil paint 干性油漆
boiler abnormal operation 锅炉异常运行
boiler accessory 锅炉辅助设备
boiler acid cleaning equipment 锅炉酸洗设备
boiler aerodynamic calculation 锅炉空气动力计算
boiler air flow model 锅炉冷态模化试验模型
boiler antiscaling composition 锅炉防垢剂
boiler area 锅炉区
boiler ash 锅炉灰渣
boiler assembler 锅炉安装单位
boiler auxiliaries 锅炉辅助设备
boiler availability 锅炉可用率
boiler bag separator 锅炉布袋除尘器
boiler barrel 锅筒，汽包
boiler basket zone 锅炉燃烧区域
boiler bay 锅炉房，锅炉跨度
boiler bearer 锅炉基础
boiler bed 炉床
boiler blowdown system 排污系统
boiler blowdown valve 锅炉排污阀
boiler blowdown water heat exchanger 定排排污水换热器
boiler blowdown water recovery pump 定排排污水回收水泵
boiler blowdown water 锅炉回水，锅炉冷凝水
boiler blowdown 锅炉排污
boiler blow-off piping 锅炉排污管，放汽管
boiler blow-off system 锅炉排污系统
boiler blow-off 锅炉排污
boiler brace 锅炉拉条
boiler brickwork 锅炉砖
boiler building 锅炉房
boiler capacity 锅炉出力，锅炉蒸发量，锅炉容量
boiler casing 锅炉外护板，锅炉外壳，锅炉围壁，蒸汽发生器壳体
boiler chemical cleaning 锅炉化学清洗
boiler circuit 锅炉汽水系统，锅炉循环回路
boiler circulating pump 锅炉循环水泵
boiler circumferential surface area 锅炉外表面积
boiler clinker removal system 锅炉除渣系统
boiler clinker removal 锅炉除焦
boiler code 锅炉规程
boiler combustion 锅炉燃烧
boiler component 锅炉部件
boiler compound 锅炉洗净剂，锅炉防垢剂
boiler conservation 锅炉停炉保养，锅炉保养
boiler construction code 锅炉制造规程
boiler control room 锅炉控制室
boiler control system 锅炉控制系统
boiler control 锅炉控制
boiler convection bank 锅炉对流传热管束
boiler convection tube bank 锅炉管束，锅炉对流传热管束
boiler corrosion 锅炉腐蚀
boiler covering 锅炉绝热层
boiler cradle 锅炉支座
boiler deposit 锅垢
boiler discipline 锅炉专业
boiler dome 锅炉干汽室

boiler drain and vent system 锅炉疏水及排气系统
boiler drum in position 锅炉汽包就位
boiler drum water level protection 锅炉汽包水位保护
boiler drum water level 汽包水位
boiler drum 锅炉汽包，锅筒，汽包
boiler dust 锅炉粉尘
boiler duty 锅炉出力，锅炉蒸发量
boiler dynamic response test 锅炉动态特性试验
boiler economic operation 锅炉经济运行
boiler efficiency test 锅炉（热）效率试验
boiler efficiency 锅炉热效率，锅炉效率
boiler enclosure 锅炉外壳
boiler end 锅炉封头
boiler erection 锅炉安装
boiler erector 锅炉安装单位
boiler evaporation 锅炉出力
boiler evaporative section 锅炉蒸发区
boiler external piping 锅炉外部管道，锅炉的外侧管道，锅炉范围内管道
boiler fan turbine 锅炉风机透平
boiler-feeding pump turbine 给水泵驱动用汽轮机
boiler feed main 锅炉给水母管
boiler feed pump turbine 锅炉给水泵透平，锅炉给水泵汽轮机，驱动给水泵的汽轮机，小汽机
boiler feed pump 锅炉给水泵
boiler feed system instability 锅炉给水系统不稳定性
boiler feed valve 锅炉给水阀
boiler feed water booster pump 锅炉给水前置泵
boiler feed water pipe 锅炉上水管
boiler feed water pump turbine 给水泵汽机，锅炉给水泵汽透平，驱动给水泵的汽轮机，小汽机
boiler feed water pump 锅炉给水泵
boiler feed water startup/standby pump 锅炉给水启动/备用泵，锅炉给水启备泵
boiler feed water system 锅炉给水系统
boiler feed water 锅炉给水
boiler feed 锅炉给水
boiler fire extinction 锅炉灭火
boiler-firing equipment 锅炉点火设备
boiler fittings 锅炉附件
boiler flame blackout 炉膛灭火
boiler flame failure 炉膛灭［熄］火
boiler flexibility 锅炉负荷适应性
boiler flue lining 锅炉烟道衬里
boiler flue 锅炉烟道
boiler following control system 锅炉跟踪调节系统
boiler following mode 锅炉跟踪方式
boiler follow-up control 锅炉随动控制
boiler follow 锅炉跟随
boiler foot 锅炉支座
boiler forced outage rate 锅炉事故率
boiler fouling 锅炉污垢
boiler framework 锅炉骨架
boiler front equipment 炉前点火控制设备
boiler furnace 锅炉炉膛，炉膛
boiler girder 锅炉大板梁
boiler grate 炉排
boiler hall 锅炉房
boiler header 锅炉联箱，锅炉集箱
boiler head 锅炉封头，锅炉端板
boiler heat balance calculation 锅炉热平衡计算
boiler heat balance 锅炉热平衡
boiler heating surface 锅炉受热面
boiler horsepower 锅炉马力【锅炉蒸发量单位】
boiler house bunker 锅炉房原煤仓
boiler house layout 锅炉房布置
boiler house 锅炉房
boiler hydrodynamic calculation 锅炉水力计算
boiler hydrotest 锅炉水压试验
boiler ignition system diagram 锅炉点火系统图
boiler ignition system 锅炉点火系统
boiler improvement 锅炉改进
boiler industry 锅炉制造工业
boiler in power plant 电站锅炉
boiler iron 锅炉钢
boiler island 锅炉岛
boiler lagging 锅炉保温，锅炉护板，锅炉衬套，锅炉隔热套层
boiler load curve 锅炉负荷曲线
boiler load factor 锅炉负荷因子
boiler load range 锅炉负荷调节范围
boiler maintenance 锅炉保养
boiler maker 锅炉制造厂，锅炉制造商
boiler make-up water pretreatment system 锅炉补给水预处理系统
boiler make-up water treatment 锅炉补给水处理
boiler make-up water 锅炉补给水
boiler making 锅炉制作
boiler master 锅炉主控，锅炉主控制器
boiler maximum continuous rating 锅炉最大连续蒸发量，锅炉最大工况
boiler minimum stable load without auxiliary fuel support 锅炉最低稳燃负荷
boiler mountings 锅炉装配附件，锅炉配件
boiler operating availability 锅炉可用率
boiler operation 锅炉运行
boiler output 锅炉蒸发量，锅炉出力
boiler overall efficiency 锅炉总效率
boiler package 锅炉包
boiler pass 锅炉烟道
boiler performance test 锅炉性能试验
boiler performance 锅炉性能
boiler pipe blowing 锅炉吹管
boiler plant 锅炉车间，锅炉装置，锅炉房
boiler plate 锅炉钢板，锅炉板
boiler platform 锅炉平台
boiler pod 蒸汽发生器荚式舱
boiler pressure 锅炉压力
boiler proper 锅炉本体
boiler rated capacity 锅炉额定蒸发量，锅炉额定出力
boiler rating 锅炉额定出力，锅炉额定蒸发量
boiler room header 锅炉房蒸汽母管
boiler room 生气发生器间，锅炉间，锅炉房
boiler scale 锅炉水垢，锅垢
boiler setting 锅炉炉墙，锅炉构架与护板，锅炉安装，锅炉装置
boiler shell ring 锅炉筒圈
boiler shell 锅炉筒体

boiler shutdown　停炉
boiler sizing　锅炉容量等级
boiler slag removal　锅炉排（熔）渣
boiler slag screen　防渣管
boiler slag　熔渣，炉渣，底渣
boiler sludge conditioning agent　锅炉泥渣调节剂
boiler soot-blower　锅炉吹灰器
boiler specific capacity　锅炉单位蒸发量
boiler stand still corrosion　锅炉停炉腐蚀
boiler starting　锅炉启动
boiler steam and water　锅炉汽水系统
boiler steam rate　锅炉蒸发量，锅炉出力
boiler steel frame　锅炉钢架
boiler steel structure　锅炉钢架，锅炉钢结构
boiler steel　锅炉钢
boiler storage　锅炉储备器，锅炉停炉保养
boiler strength calculation　锅炉强度计算
boiler structure　锅炉钢架，锅炉构架
boiler surge bin　锅炉缓冲仓
boiler terminal connections　锅炉范围管道的终点接头
boiler thermodynamic calculation　锅炉热力计算
boiler tube expander　锅炉胀管器
boiler tube expanding　锅炉胀管
boiler tube　锅炉管，炉管
boiler-turbine centralized control　机炉集中控制
boiler-turbine coordinated control system　机炉协调控制系统
boiler-turbine coordination control　锅炉-汽机协调控制，机炉协调控制
boiler-turbine-generator unit　单元机组
boiler unit　锅炉机组，锅炉设备，锅炉装置
boiler vibration　锅炉振动
boiler wall　炉墙
boiler water and steam quality adjustment test　锅炉汽水品质调整试验
boiler water circulating pump　炉水循环泵
boiler water circulation calculation　锅炉水循环计算
boiler water circulation　炉水循环，锅炉水循环
boiler water concentration　锅水浓度，炉水浓度，炉水含盐量
boiler water control　炉水控制
boiler water drum　锅炉水鼓
boiler water filling　锅炉上水，锅炉充水
boiler water phosphate treatment　锅炉水磷酸盐处理
boiler water separator　炉水分离器
boiler water side　锅炉水侧
boiler water supply　锅炉给水，锅炉供水
boiler water treatment　炉水处理
boiler water　炉水，锅水，锅炉水
boiler with dry-ash furnace　固态排渣锅炉
boiler with dry-bottom furnace　固态排渣锅炉
boiler with slag-tap furnace　液态排渣锅炉
boiler with stages of evaporation　分段蒸发锅炉
boiler with wet bottom furnace　液态排渣锅炉
boiler　锅炉，蒸发器，蒸煮器，蒸汽发生器
boiling bed　沸腾床
boiling burnout　沸腾烧毁
boiling channel　沸腾通道
boiling coefficient　沸腾传热系数，沸腾放热系数
boiling crises　临界沸腾
boiling crisis mechanism　沸腾危机机理
boiling crisis　沸腾危机，沸腾换热恶化
boiling-curve hysteresis　沸腾曲线滞后特性
boiling cycle reactor　沸腾循环反应堆
boiling delay　（钠）沸腾延迟
boiling detection　沸腾探测
boiling D_2O reactor　沸腾重水反应堆，沸腾重水堆
boiling down　蒸干
boiling heat transfer　沸腾传热
boiling heavy water reactor　沸腾重水反应堆
boiling height　沸腾段高度
boiling in ducts　管内沸腾
boiling light-water reactor　沸腾轻水反应堆
boiling limit　沸腾限
boiling-liquid moderated reactor　沸腾液体慢化堆
boiling-liquid reactor　沸腾液体堆
boiling molten pool　沸腾状熔池
boiling noise　沸腾噪声
boiling nuclear superheat power station　沸腾核过热反应堆，沸腾核过热电站
boiling nuclear superheat reactor　沸腾核过热（反应）堆
boiling number　沸腾数
boiling of sand　涌沙
boiling out procedure　煮炉程序
boiling out with alkaline　碱煮炉
boiling out　煮炉，蒸馏提取
boiling point test　沸点试验
boiling point　沸点，沸腾湿度
boiling range　沸腾范围，沸腾区
boiling reactor　沸水堆，沸腾反应堆
boiling regime　沸腾工况
boiling section　沸腾区，蒸发区段，沸腾段
boiling slurry reactor　沸腾浆液反应堆
boiling steel　沸腾钢
boiling sulphur reactor　沸腾硫磺反应堆
boiling-superheat fuel assembly　沸腾-过热燃料组件
boiling suppression　沸腾抑制
boiling temperature　沸点，沸腾温度
boiling transition　沸腾（工况）过渡
boiling tube　蒸发管，沸腾管
boiling up　起泡，沸腾
boiling water homogeneous reactor　均匀沸水反应堆
boiling water nuclear reactor　沸水核反应堆，沸水反应堆
boiling water reactor　沸水反应堆，沸腾循环反应堆，沸水堆
boiling water sterilization　煮沸消毒
boiling water　沸水
boiling　沸腾，煮沸
boil off　蒸煮，煮出，蒸发，汽化，煮掉
boil out　煮炉，蒸馏提取
boil-over　沸溢
boil vertical eddy　直涌旋涡
boil　冒水，沸腾，煮沸，紊腾，水浪翻花现象

BOL(beginning of life) 寿命初期【循环燃料】
BOL(boundary layer) 边界层
boldface 黑体字
bold shore 峭岸
bole 红玄武土
bollard 系船柱
Bollman truss 包尔曼桁架，多弦三角形桁架
bolometer bridge circuit 辐射热计桥式电路
bolometer 变阻测辐射热表，辐射热测定器，辐射热测量计，辐射量热计，辐射热计
bolometric instrument 测量射热计式仪表
bolson 封闭洼地，沙漠盆地，荒芜盆地，干湖地
bolster connection 垫枕接合
bolster 轨枕，枕木，垫块，垫枕，承梁板，梁托，托木，支撑杆，支承梁，支架，鞍座
bolt and nut 螺栓和螺帽
bolt blank 螺栓环件
bolt callout 螺栓名称，螺栓标注
bolt circle 螺栓分布圆
bolt clipper 断线钳
bolt cold tightening 螺栓冷紧
bolt connection 螺栓连接
bolt coupling 螺栓联轴节
bolt cross-section 螺栓截面
bolt cutter 断线钳
bolt damage sensor 螺栓损坏传感器
bolted bonnet 螺栓连接的阀盖，栓接的阀帽
bolted cap 螺栓连接的（阀）帽
bolted connection 螺栓接合，螺栓连接
bolted gland 栓接的密封压盖
bolted joint 螺栓连接，栓接
bolted link 螺栓耦合
bolted-on attachment 螺栓连接
bolted pile 螺栓接桩
bolted-pressure connector 栓接压力接头
bolted three-phase fault 金属性三相短路故障
bolted through stay 螺栓拉撑杆
bolted 螺栓连接的
bolt electric heating rod 螺栓加热棒
bolter 筛，筛选机
bolt fastening 螺栓连接
bolt head retaining slots 螺栓头固定槽
bolt head with feather 带销螺栓头
bolt head 螺栓头
bolt heater 螺栓加热器
bolt heating 螺栓加热
bolt hole 螺栓孔，螺孔
bolt hot tightening 螺栓热紧
bolt in double shear 受双剪的螺栓，双剪螺栓
bolting and shotcreting method 锚喷支护法
bolting flange 法兰环，（压力容器）螺栓连接法兰
bolting machine 筛粉机
bolting 上螺栓［螺钉，螺帽，螺杆］，用螺栓固定，栓接，用螺栓连接，栓接拧紧
bolt in single shear 单剪螺栓
bolt load at operating condition 螺栓操作负荷
bolt load at pretighten condition 螺栓预紧负荷
bolt nut 螺母
bolt-on 栓接
bolt pin 螺栓销
bolt rod 螺杆
bolts interchangeable holepatterns 孔型可互换的螺栓
bolt sleeve 螺栓套管
bolt spacing 螺孔距
bolt stress 螺栓应力
bolt wrench clearance 螺栓扳手间隙
bolt symbol 螺栓符号
bolt-threading machine 螺栓车纹机
bolt threads in bearing 支撑部分的螺栓螺纹，轴承螺栓螺纹
bolt tightener 螺栓紧固器
bolt timber 短圆木材
bolt washer 螺栓垫圈
bolt with stop 止动螺栓
Boltzman's constant 玻尔兹曼常数
bolt 用螺栓连接，插销，门闩，螺栓，栓接拧紧，吊环螺钉
bolus 胶块土
bombarding voltage 轰击电压
bombardment-induced conductivity 由电子轰击引致的电导
bombardment 撞区
bomb calorific value 弹筒发热量【煤】
bomb calorimeter 氧弹量热器
BOM(bill of material) 材料清单
bomb proof 防轰炸的
bomb praying 爆炸喷涂
bona fide applicant 有诚意的申请人
bona fide bidder 善意投标人
bona fide buyer 真诚买家
bona fide endorsee 善意被背书人
bona fide enquiries 诚意的询问
bona fide holder 合法持有人，善意持有人，善意持票人
bona fide operating data 善意（推荐）的操作数据
bona fide principle 诚实信用原则
bona fide transferee 善意受让人
bona fide 诚意，善意的，真实的，真诚的
bona transaction 公平交易，真实交易
bona 好的，善意的
bondability 结合力
bondage 束缚
bond area 黏结面积，熔合区
bond beam 结合梁
bond between concrete and steel 混凝土与钢筋间的结合力，混凝土与钢筋间的握裹力
bond breaker 防黏结材料
bond clay 黏合土
bond collateral loan 债券抵押贷款
bond course 结合层，黏结层
bonded area 保税区，海关管制区，海关管制区，保税区域
bonded cargo 保税货物
bonded deposit 黏结积灰，密实沉积
bonded element 黏结构件
bonded flux 陶质焊剂
bonded fuel element 结合型燃料元件
bonded goods 保税货物

bonded in bolts　用螺栓连接的
bonded motor　定子和端盖组合在一起的电机
bonded prestressing tendon　有黏结预应力筋
bonded roof　砌合屋顶
bonded store　保税仓库
bonded warehouse　海关保税仓库，保税仓库，保税库
bonded zone　黏结区
bonded　被结合的，黏着的，胶黏的，海关保税的，连接的
bonder wire　绑扎钢丝
bonder　顶砖，接合器，连接器，黏合器，耦合器，砌墙砖，热压焊接机，见证人
bond flux　焊剂
bonding admixture　黏合剂
bonding agent　耦合剂，焊药，黏合剂，黏着剂，黏结剂
bonding bar　等电位连接带
bonding brick　接合砖，空心墙连接砖
bonding capacity　担保额度，担保能力
bonding company　担保公司
bonding compound　黏结混合物
bonding conductor　屏蔽接地线
bonding course　结合层
bonding effect　键效应
bonding gland　填料压盖
bonding jumper　跨接
bonding machine　热压焊接机
bonding mental to mental　金属与金属的胶结
bonding pad　焊接区，焊接点
bonding point　接合点
bonding stress　黏着应力
bonding strip　电缆防电蚀的铅条
bonding wire　连接线
bonding　焊合，结合，黏合，黏接，连接，黏结，焊接
bond length　锚着长度，黏结长度，握裹长度，握固长度
bond line　(焊缝)熔合线，黏合层，黏合剂
bond master　环氧树脂类黏合剂
bond material　黏结材料，结合材料
bond-meter　胶接检验仪
bond plaster　黏结灰泥
bond quality　黏结质量
bond resistance　黏着抗力
bonds due for payment　到期应偿付的债券
bond strength　结合强度，黏着强度，黏结强度
bond stress　握裹应力，黏结应力，黏着应力
bond tester　胶结检验仪，接头电阻测试器
bond tile　搭接瓦
bond type diode　链形二极管，键合二极管
bond weld　熔合线，熔化线
bond wire　接续线【轨道电路】，接合线，焊线
bond　担保书，保证金，保函，低押，证券，债券，票据，黏合，结合，联结，黏结，砌合，黏接，通电塔接，黏合力
bone charcoal　骨炭
bone coal　煤质页岩，页岩煤，骨煤
bone deposition　骨内沉积
bone dose　骨内剂量，骨骼剂量
bone glass　乳白玻璃
bone glue　骨胶
bone marrow　骨髓
bone oil　骨油【指一种污染饮水的物质】
bone-seeker　亲骨性物质，超骨物质
bone tissue　骨组织
bone tolerance dose　骨骼容许剂量，骨骼耐受剂量
Bong-Siemens process　开路酸洗法
BONI(balance of nuclear island)　核岛配套设施
boning out　定直线
boning rod　水平尺
bonnet bearing　顶盖支承点
bonnet hip tile　屋脊弯帽状瓦
bonnet nut　阀帽螺母，阀帽螺母
bonnet plug　阀帽塞
bonnet stop　阀帽止点【防止阀帽在阀体中卡住】
bonnet　阀盖，顶盖【阀】，盖，罩，机罩，烟囱罩，发动机罩
bonus and allowances to employees　员工奖金和津贴
bonus-and-penalty clause　奖惩条款［条件］
BONUS(boiling nuclear superheat power station)　沸腾核过热反应堆
bonus clause　奖金条款，奖励条款
bonus for completion　竣工奖金
bonus fund　奖励基金
bonus income　补贴收入
bonus-penalty contract　奖罚合同
bonus-penalty provision　奖惩条件［条款］
bonus-penalty terms　奖惩条件［条款］
bonus share　红股
bonus system　奖金制度
bonus to directors　董事奖金
bonus wage　红利工资
bonus　奖金，红利，额外津贴
bon voyage　一帆风顺，一路平安【对旅行者道别所言】
bony coal　煤质页岩，页岩煤
BOO(build-own-operate)　“建造-拥有-运营”模式【工程建设模式】
boojee pump　气压灌浆泵，压浆泵
book capacitor　书形微调电容器
book condenser　书形微调电容器
booking agent　订舱代理
booking charge　订舱费
booking list　定舱清单，装货定舱表
booking note　订舱单，定舱单
booking number　订舱号
book inventory　账面存量，账面盘存，账面记载的物料量
bookkeeper　簿记员，记账人
bookkeeping machine　簿记机
bookkeeping operation　程序加工运算，内务操作
booklet　小册子
book value　账面价值，(公司或股票的)净值
book　书，手册，说明书，预定，登记，注册
Boolean algebra　布尔代数，逻辑代数
boom and bucket delivery　吊杆和庳斗输送
boom angle　吊机臂转角

boom brace 吊臂斜杆
BOOM(build-own-operate-maintain) “建造-拥有-运营-维护”模式【工程建设模式】
boom conveyer 悬臂胶带机，悬臂皮带机
boom crane 臂式起重机，桁梁起重机，吊杆起重机
boom excavator 臂式挖土机，吊杆挖土机
boom hoist 臂式吊车
boom in enterprise 企业景气
booming 闪冲砂矿法
boom length 臂长
boom luffing mechanism 悬臂仰俯机构
boom-mounted reclaimer 悬臂式取料机
boom-mounted wheel excavator 轮式吊杆挖土机
boom net 拦网，水下铁丝网
boom-out 最大伸距
boom point 伸臂末端
boom reversible belt conveyor 可逆悬臂带式输送机
boom rig 臂式起吊设备
boom slewing mechanism 悬臂回转结构
boom unit 蜂鸣报警装置
boom 吊杆，起重臂，横木，栏木，吊臂，起重机，臂架，防浪浮架，横梁，抱杆，桁，梁，悬臂，兴旺，钻臂，钻架
BOOS(build-own-operate-sell) “建设-拥有-运营-出售”模式【工程建设模式】
boost-buck response of excitation control system 励磁控制系统的增减压响应
boost-buck 升压去磁
BOOST(build-own-operate-subsidy-transfer) “建设-拥有-运营-补贴-转让”模式【工程建设模式】
boost charge 加强充电，升压充电，快速充电
boost charging voltage 大充电压
boost control 增压调节，增压控制
boosted circuit 升压电路
boosted voltage 增高电压
booster amplifier 升压放大器，辅助放大器
booster charge 补充充电，再充电
booster coil 启动线圈，升压线圈
booster compressor 增压压缩机
booster device 升压设备
booster diode 升压二极管，阻尼二极管
booster-distribution amplifier 升压器-分布放大器
booster element 增益元件
booster fan 增压风机，升压风机，增压辅助风机，辅助扇风机
booster gas turbine 加速燃气轮机
booster heater 辅助加热器
booster hose 增压器软管
booster light 增强光照
booster mill 增压提水风车
booster oil pump 升压油泵，增压油泵
booster oscillator 增压振荡器
booster pump for motor-driven feed water pump 电动给水泵前置泵
booster pump for turbine-driven boiler feed water pump 汽动给水泵前置泵
booster pump house 升压泵房
booster pumping station 升压泵站，升压抽水站，升压水泵站
booster pump 前置泵，增压泵，前置水泵，升压泵
booster reaction 增热反应
booster relay 加速继电器，升压继电器，升压器继电器
booster resistor 升压电阻器，附加电阻
booster rod 点火棒，增益棒，增强棒
booster set 增压机组
booster station 接力电台，中继电台，增压站，差转台，加压站，升压站
booster telephone circuit 电话增音电路
booster thermocouple 均衡式热电偶，均衡式温差电偶
booster transformer supply system 升压［增压，吸流］变压器供电系统
booster transformer 升压变压器，增压变压器，吸流变压器
booster transmitter 辅助发射机
booster valve 辅助阀，增压阀
booster 增压泵，升压线圈，升压电阻，升压器，升压机，前置器，传爆管，增压器
boost gauge 升压压力计，增压计
boosting battery 浮充电池，加压电池
boosting coil 升压线圈
boosting of furnace 炉膛强化，提高炉膛热强度
boosting pressure 升压，增压
boosting voltage 增压
boosting 加速，增加，升压，增压，增大
boost pressure controller 进汽压力控制器，升压调节器
boost pressure 升压，增压，进汽压力
boost pump house 升压泵房，中继泵房
boost pump 辅助泵，供油泵
boost regulator 增压调节回路
boost the economic development 促进经济发展
boost up circuit 增压电路
boost voltage 附加电压，辅助电压，增高电压，升压
boost 加强，增加，升高，升压，升举，推进，增压
boot block revision level 引导块修订级别
BOOT(build-own-operate-transfer) “建造-拥有-运营-移交”模式【工程建设模式】
booth 电话间，公用电话亭，小房子，小室
boot lid 引导盖
boot stage 孕穗期
bootstrap button 引导按钮
bootstrap circuit 自举电路
bootstrap diode 阴极负载二极管，限幅二极管
bootstrap generator 自举发生器
bootstrap instructor technique 引导指令技术
bootstrap integrator 自举电路积分器
bootstrap loader 引导装入程序
bootstrap memory 引导存储器
bootstrapping 用连续正反馈调节整机特性的方法，自举
bootstrap program 引导程序，辅助程序
bootstrap saw tooth generator 仿真线路锯齿波振荡器
bootstrap technique 引导技术
bootstrap 引导程序，引导指令，模拟线路，仿

真线，辅助程序，自举电路
boot truck 洒沥青卡车
boot up 启动
boot 引导【指电脑】，引导装入，保护罩，罩，作好使用（电脑）准备，柔性盖板，料仓，连接器，套管，拼合套筒
BOP(balance of plant) 电厂［电站］配套设施【热力车间之外的设施】，电厂辅助设施
BOP(balance of the plant) 电站配套设施
BOP boiler anciliaries 锅炉辅机工程师
BOP buildings 电厂配套设施厂房
BOPMS(balance of plant master system) 机组辅助设备主控顺序
BOP 轴承油泵【DCS 画面】
BOQ(bill of quantities) 工程量清单，工程量表【固定单价合同常用】
boral 碳化硼铝
borated concrete 含硼混凝土
borated coolant 含硼冷却剂
borated glass 硼玻璃，硼酸盐玻璃
borated graphite 含硼石墨
borated ice 含硼冰
borated paraffin 用硼酸处理过的石蜡
borated thermal shield 含硼热屏蔽
borated water header 含硼水集管
borated water storage tank 含硼水贮存箱
borated water 含硼水，加硼酸盐的水
borated 与硼砂（或硼酸）混合的，用硼酸处理过的
borate waste solution 含硼废液
borate 硼酸盐，加硼
boration 加硼
borax bead reaction 硼砂熔珠反应
borax liner 硼砂衬里，硼砂垫层
borax-paraffin collimator 硼砂-石蜡准直器
borax 硼砂
border area 边区
border check 土埂
border dike 边堤，围堤
border ditch 沿埂沟
bordereau 备忘录
border effect 边际效应，边行影响
border facies 边缘相【工程地质】
border information 图廓标注说明
bordering 缘，镶边，加边，围筑畦埂
border land 边缘地区，交界地区
borderline case 难以确定的情况
borderline mixture 饱和混合物
borderline risk 不确定风险，边境风险
border line 边界线，边线，分界线，界线，图廓线
border making implement 筑埂器
border phase 边缘相【工程地质】
border pile 边桩
border region 边区
border river 边界河
border stone 镶边石，界石
border 边界，边，缘，边沿，路缘，周界
bore a hole 镗孔
bore bit 钻头
bored cast-in-place pile 钻孔灌注桩，钻孔桩
bored cast-in-situ pile 钻孔灌注桩，钻孔桩
bored concrete pile 灌注桩，钻孔灌注桩
bore diameter 内径，定子内径
bored lock piles 钻孔咬合桩
bored pile 螺旋钻孔桩，钻孔灌注桩，钻孔桩
bored tie 钻孔枕木
bored tubular well 钻凿管井，钻凿井
bored well 螺旋钻井，钻井
borehole acoustic television logger 超声电视测井仪
borehole axial strain indicator 钻孔轴向应变指示计
borehole camera 钻孔摄影仪
borehole casing 钻孔套管
borehole core 钻孔岩心
borehole-deformation gauge 钻孔变形计
borehole depth 钻孔深度
borehole device 钻孔装置
borehole diametral strain indicator 钻孔径向应变指示计
borehole director 钻孔定向器
borehole induction log 钻孔感应电测记录
bore hole lead insulation 穿心引出线绝缘，轴心空引线绝缘
borehole loading 炮孔装药
borehole log 钻孔记录，钻孔柱状图
borehole pattern 钻孔布置型式
borehole pump 钻孔泵
borehole sample 岩芯试样，钻孔试样
borehole sealing 钻孔密封
borehole specimen 钻孔土样
borehole surveying 钻孔测量
borehole television camera 钻孔电视摄影机
borehole wall 钻孔壁
borehole water 井水原水
borehole 镗孔，钻孔
bore log 钻孔记录，钻孔柱状图
bore meal 钻孔岩粉
bore of stator 定子内径，定子内腔
bore pit 钻探坑
borer 钻孔器
borescope built into the syringe 光学孔径仪
borescope inspection 管道内孔镜检查
borescope 管道内孔镜
bore specimen 钻探岩心
bore stress 中心孔应力
bore well 钻孔井
bore 涌潮【河口】，孔眼，孔，钻孔，膛，内径，镗孔，洞，口径，钻
boric acid 硼酸
boric acid batching tank 硼酸制备箱，硼酸配制箱
boric acid batching 硼酸配制
boric acid charge tank 硼酸注入箱
boric acid circulating and metering pump 硼酸循环和计量泵
boric acid concentration 硼酸浓度
boric acid control system 硼酸控制系统
boric acid control 硼酸控制
boric acid emergency shutdown system 硼酸应急停堆系统
boric acid emergency shutdown 硼酸应急停堆

boric acid evaporator package　硼酸蒸发装置
boric acid evaporator　硼酸蒸发器
boric acid heater　硼酸加热器
boric acid injection system　硼酸注入系统
boric acid injection　硼酸注入
boric acid make-up system　硼酸补给系统
boric acid make-up tank　硼酸补给罐
boric acid make-up　硼酸补充
boric acid metering pump　硼酸计量泵
boric acid mixing station　硼酸混合站
boric acid mixing tank　硼酸混合箱
boric acid poisoning　硼酸中毒
boric acid precipitation　硼酸析出物
boric acid pump　硼酸泵
boric acid reactivity worth　硼酸反应性值
boric acid recovery system　硼酸回收系统
boric acid recovery　硼酸回收
boric acid removal　去硼
boric acid scram　硼酸应急停堆
boric acid solution　硼酸溶液
boric acid storage tank　硼酸贮存箱
boric acid surge tank　硼酸波动箱
boric acid tank　硼酸箱
boric acid transfer line　硼酸输送管线
boric acid transfer pump　硼酸输送泵
borickite　褐磷酸钙铁矿
boric oxide　硼氧化物
boride　硼化物
boring and mortising machine　钻孔眼机
boring barge　钻探驳船，钻探平底船
boring bar　钻杆
boring bench　钻台
boring casing　钻探套管
boring core　钻探岩心
boring depth　镗孔深度，钻探深度
boring equipment　钻孔设备
boring exploration　钻孔勘探
boring frame　钻井架，钻架，钻探架
boring gauge　钻孔量测计
boring heads　搪孔头
boring inclination measurement　钻孔测斜
boring lathe　镗床
boring log　钻井柱状图，钻孔地质分层图，钻孔柱状图，钻探记录，钻探柱状图
boring machine　镗床，钻机，钻探机，钻探机具
boring mill　钻机
boring motor　钻探电机
boring profile　钻探剖面图，钻探纵剖面
boring rig　钻架，钻探设备
boring rod clamp　钻杆夹
boring rod　钻杆
boring slurry test　钻孔泥浆试验
boring slurry　钻孔泥浆
boring spacing　钻孔间距
boring specimen　钻探土样
boring test　钻孔试验
boring tool　镗刀，钻探工具
boring tower　钻塔
boring work　镗孔工作
boring　钻井，镗，镗孔，钻孔，钻进，钻（探）
borocalcite　硼钙石
borocarbon resistor　硼碳电阻器
boromagnesite　硼镁石
boron　硼
boron absorption method　硼吸收法
boron absorption technique　硼吸收（测量）法
boron addition　加硼
boron and water markup system　硼与钝水补给系统
boronated stainless steel　含硼不锈钢
boronated steel　硼化钢，硼钢
boronated　含硼的
boronation　硼化
boron-bearing aggregate　含硼骨料
boron capture　硼俘获
boron carbide absorber　碳化硼吸收剂
boron carbide control rod　碳化硼控制棒
boron carbide rod　碳化硼棒
boron carbide　碳化硼
boron chamber　涂硼电离室
boron charging　加硼
boron-coated electrode　涂硼电极
boron-coated ion chamber　涂硼电离室
boron-coated proportional counter　涂硼正比计数管
boron coating chamber　涂硼电离室
boron coating counter　涂硼计数管
boron compound　硼化合物
boron concentration dial　硼浓度刻度盘
boron concentration monitoring　硼浓度监控
boron concentration　浓缩硼，硼浓度
boron-containing concrete　含硼混凝土
boron-containing gas　含硼气体
boron-containing shield　含硼屏蔽
boron content　含硼量，浓缩硼
boron control material　硼控制材料
boron control system　硼控制系统
boron control　硼控制
boron criticality search　探查硼临界点
boron dilution　硼稀释
boron equivalent　硼当量
boron-filtered neutron　硼滤过的中子
boron filter　硼过滤器
boron-free coolant　无硼冷却剂
boron glass rod　硼玻璃棒
boron heating system　硼加热系统
boron injection recirculation　硼注入再循环
boron injection surge tank　硼注入波动箱
boron injection tank　硼注入箱
boron injection　硼注入
boron isotope　硼同位素
boronizing　硼化
boron let-down　减硼
boron-lined ion chamber　涂硼电离室
boron-lined ionization chamber　衬硼电离室，涂硼电离室
boron-lined proportional counter　涂硼正比计数管
boron-lined　衬硼的
boron-loaded control rod　含硼控制棒
boron-loaded scintillation counter　涂硼闪烁计数器

boron-loaded thermocouple 载硼热电偶，涂硼热电偶
boron-loaded 含硼的，涂硼的，载硼的
boron makeup 补充硼
boron meter 硼计量仪
boron plate 硼钢板
boron poisoning 硼中毒
boron precipitation 硼结晶
boron reactivity worth 硼反应性值
boron recovery system 硼回收系统
boron recycle system 硼回收系统，硼再循环系统
boron recycle 硼回收，硼再循环
boron recycling system 硼回收系统
boron release 除硼，泄放硼液
boron removal bed 除硼床
boron removal capacity 除硼能力
boron removal 除硼，去硼
boron salt 硼盐
boron steel 硼钢
boron thermal regeneration subsystem 硼热再生辅助系统
boron thermopile 硼热电堆
boron trifluoride proportional 三氟化硼正比计数管
boron water 含硼水
boroscope 内孔表面检查仪
borosilicate glass tube 硼硅管【可燃毒物棒】
borosilicate glass 硼硅酸盐玻璃，硼硅玻璃
borosilicate tube 硼硅管【可燃毒物棒】
borosilicate 硼硅酸盐
borough 自治的市镇，自治村镇，市镇
boroxal 氧化硼铝
borrow and loan contract 借贷合同
borrow area 取土面积，取土区，采料场，取土坑
borrowed light 间接采光窗
borrower 借款人，借用者
borrowing bank 借款银行
borrowing country 借款国
borrowing plan 贷款计划
borrowing 借贷，借钱
borrow material 取用土料
borrow pit 采料场，取土坑
borrow 借，贷款，借位，借位信号，借位脉冲，采料，借入，取土
BOS(back-out system) 补偿系统
BOS(basic operating system) 基本操作系统
BOS(build-own-sell) “建设-拥有-出售”模式【工程建设模式】
Bose-Einstein statistics 玻色爱因斯坦统计法
bosom 对接连接角钢，角撑
boss bolt 轮毂螺栓
boss flange 凸法兰
boss ratio 轮毂比
boss 突起部位，轮毂，凸起部，凸台，凸缘，灰泥桶
Boston caisson 管柱沉井施工法
Bosun's chair 飞车
BOT(build-operate-transfer) “建造-运营-移交”模式【工程建设模式】
both ends thread 两端带螺纹
both normal and emergency condition 正常和事故情况
both parties 双方
both sides welding 双面焊接
bothway trunk 双向中继线
BOTI(balance of turbine island) 汽轮机岛辅助设施
botom 底部
BOTP(build-operate-transfer project) BOT 项目，“建造-运营-移交模式”的项目
botryoidal structure 葡萄状构造
bottle brush 瓶刷
bottle coal 瓦斯煤，气煤
bottle float 瓶式浮标
bottle jack 瓶式千斤顶
bottleneck problem 薄弱环节问题
bottleneck （工商发展的）瓶颈、阻碍，瓶颈段，窄路，狭窄段，（引起交通阻塞的）瓶颈路段
bottle sampler 取样瓶
bottle screw 松紧螺旋扣
bottle-shape filter 洗瓶水，瓶式过滤器
bottle silt sampler 瓶式泥沙采样器
bottle type suspension load sampler 瓶式悬移质采样器
bottle 瓶，装瓶
bottom adapter 底部插头【燃料组件】
bottom air admission 炉排底部进风
bottom and fly ash mixed handling 灰渣混除
bottom and fly ash recovery 灰渣利用
bottom and fly ash separate handling 灰渣分除
bottom articulated joint 底部铰接缝
bottom ash bin 渣仓
bottom ash cooler 冷渣器
bottom ash dewatering bin 炉渣脱水仓
bottom ash handling system 除底灰系统，除渣系统
bottom ash hopper 炉底渣斗，排渣槽，冷灰斗
bottom ash precipitation pond 沉渣池
bottom ash sluicing pump 冲渣泵
bottom ash sluicing water 底灰冲灰水
bottom ash slurry pond 渣浆池
bottom ash slurry pump 渣浆泵
bottom ash 底灰，炉渣，炉底渣，底渣，炉底灰，大渣
bottom bank 浅滩，沙滩
bottom bar 下鼠笼条，层线棒
bottom blowdown 底部排污
bottom boom 下弦杆
bottom bottle 底层浮标瓶
bottom bracing 底撑
bottom bracket 下机架
bottom bush 底衬
bottom cargo 底货载
bottom case 底座
bottom chord 下弦杆，下弦
bottom clearance 底部净距
bottom coat 底漆
bottom coil slot 定子下层线圈边
bottom contour 底部地形，底部等高线

bottom contraction 底部收缩
bottom cooling system 底部冷却系统
bottom course 底层，基层
bottom covering 底面涂层
bottom current 底层流，底流
bottom cut 底槽，下部掏槽
bottom dead centre 下死点
bottom deflecting snow fence 下导防雪棚栏
bottom die base 底模基础
bottom die 底模
bottom discharge orifice 泄水底孔
bottom discharge 下出料，底部排水
bottom door 底门，清扫孔
bottom drainage 底部排水
bottom drain valve 底部疏水阀
bottom drift method of tunneling 隧洞开挖的下导洞推进法
bottom drift 底部导洞
bottom dump bucket 底卸料斗，底卸式铲斗
bottom dump capacity 散装搬运能力
bottom dump car 底开车，底卸车
bottom dump hauler 底卸式拖运机
bottom dump hopper car 底部倾斜式漏斗车，自卸式底开车
bottom dump hopper 底卸漏斗车
bottom dump tractor-trailer 底卸拖拉机拖车
bottom dump trailer 底卸拖车
bottom dump truck 底部卸料车，底卸卡车，底卸式卡车
bottom dump type 底卸式
bottom dump unloading 车辆底卸
bottom dump wagon 底开式料车，底开车
bottom dump 底部卸，漏斗
bottom effect 底部效应
bottom elevation 底面标高
bottom end plug 底部端塞【燃料棒或阻力塞】
bottom entry control rod 底部进入的控制棒
bottom entry 底部进入
bottom feed 底部加料，底部喂料
bottom fitting 底部附件，底部接头【燃料组件】
bottom flange 底部法兰，下阀盖，下翼缘，底凸缘，(梁的)底翼线，下法兰
bottom flow 回卷流
bottom flushing of sediment 底部排沙
bottom friction 底摩阻力，底部摩擦
bottom gate 底部闸门，底孔闸门
bottom girder 底梁
bottom glade 底谷
bottom-grab 挖泥抓斗，咬合采泥器，咬合采样器
bottom-grating intake 底栏栅式取水
bottom grid assembly 下格架【燃料组件】，下格架
bottom grid 端板，堆芯支承栅格板，底部栅格板
bottom guide tube flange hole 底部导向管法兰孔
bottom guide tube flange 底部导向管法兰
bottom half-bearing 下半轴承
bottom half-shell 下半轴瓦
bottom head centre disc 底封头中心碟形盘
bottom header pipe 闸首底部放水管
bottom header 底集管箱
bottom heading 底部导洞，下导坑
bottom head instrument penetration 下封头测量贯穿件
bottom head 下封头，【压力容器】底封头
bottom hinged window 下悬窗
bottom hopper barge 底开式驳船，底卸式船
bottom hung window 下悬窗
bottom ice 底冰
bottom ignition 底层点火
bottoming tap 平底螺丝攻，平底丝锥
bottoming 石块铺底，清底
bottom inner casing 内下缸
bottom intake rack 底孔进水口拦污栅
bottom kerf 底槽
bottomland 低洼地，盆地，滩地，洼地
bottom laterals 底部横向水平支撑杆
bottom level 层底标高，底层
bottom liner plate 埋入混凝土的底部衬板【安全壳】，底部衬板
bottom loading transfer cask 底部装料用传送容器
bottom load 底沙
bottom mud 底层泥浆
bottom nozzle 下管座【燃料组件】
bottom of beam 梁底
bottom of containment extension 安全壳地下室
bottom of each riser 每根立管的底部
bottom of pipe 管底
bottom of support 支架底
bottom outer casing 外下缸
bottom outlet diversion 底孔导流
bottom outlet hole 底排水孔
bottom outlet orifice 底部放水孔口
bottom outlet 泄水底孔
bottom paint 底漆
bottom panel 底板，(风洞)底壁
bottom plenum 下腔室
bottom pour ladle 下注盛钢桶，底注式浇包
bottom pressure 底压力【隧道】，底压
bottom price 底价，最低价
bottom raft 底板【反应堆厂房】，筏基
bottom rail (门窗的)下横档，下冒头
bottom rake 后角
bottom reflector sleeve 底部反射层套管
bottom reflector 底部反射层
bottom removal system 底渣清除系统
bottom-restrained 底部受限制的立管
bottom ring 底环
bottom roller 底辊
bottom sampler 水底采样器，水底取土钻
bottom sampling device 水底取样设备
bottom sampling 底样采集器
bottom screen furnace 带炉底水帘管的炉膛
bottom section 底部截面
bottom sector gate 底部扇形闸门
bottom sediment 底部泥沙
bottom set bed 底积层
bottom slope 底坡
bottom sloping (池)底坡度
bottom slot layer 下层绕组

bottom sluice gate 底部泄水闸门
bottom sluice 底部泄水闸，泄水底孔
bottom soil 下层土，底土
bottom suction 底部吸入
bottom support casting 下部支承板【堆芯】
bottom supported boiler 自承式锅炉
bottom support plate 下部支撑板，下部支承板
bottom surface camber 下弧，下曲面
bottom surface 底部表面，底面
bottoms 残渣【蒸发器】，底渣，渣油，油脚
bottom thermal shield 底部热屏蔽
bottom topography 水底地形
bottom trash rack 底部拦污栅
bottom unit 后置机组
bottom-unloading wagon 底开车
bottom-up analysis method 自底向上分析法
bottom-up analysis 自下而上分析
bottom-up design 自下而上设计
bottom up 倒置，颠倒，干杯，自底向上
bottom-up 由下而上的，自底向上的，从细节到总体的
bottom valve 底阀
bottom velocity 底部流速
bottom view 底视图
bottom water screen 炉底（防焦）水帘管
bottom water 底层水，底水，底面水
bottom wave 水底波
bottom width 底宽
bottom wing 下翼
bottom 底板，底部，车底，舱底，底，底床，底脚
bought-in parts 外购件
bought-out 买断的，买下全部产权的
boulder base 蛮石地基，蛮石基础
boulder bed 大卵石层，卵石河床
boulder blasting 爆破大石
boulder clay 泥砾
boulderet 中砾
bouldering 铺筑巨砾
boulder setter 蛮石铺砌层
boulder stream canyon 蛮石峡谷
boulder strip 条石
boulder 大圆石，巨砾，大块石，大卵石，蛮石，漂砾，漂石
boulevard 大道
boult 筛子
bounce-back 弹回，反跳，回跳，反冲
bounce control 跳动控制
bounce-free normally open contact 防抖动常开触点
bounce-proof contact 防跳触点，防颤动触点
bounce time 接触点的调动时间
bounce 标志跳动【示波器】，蹦，跳动，反跳，弹起，弹回，弹力，脉动
bouncing of contacts 接触点的调动
bouncing 跳动
boundary beam 边梁
boundary cavitation 边界区空蚀
boundary collocation 边界配置
boundary condition for electromagnetic field 电磁场的边界条件
boundary condition 边界工况，边界条件
boundary constraint 边界约束，洞壁约束【风洞】
boundary-contraction method 边界收缩法
boundary current 边界流
boundary curve 边界曲线
boundary dimension 外形尺寸，轮廓尺寸
boundary-drag coefficient 边界阻力系数
boundary effect 洞壁效应【风洞】，界壁效应，边界效应，边界影响
boundary element method 边界单元法
boundary element 边界单元，边界元素，边缘构件
boundary flow method 边界潮流法
boundary friction 边界摩擦
boundary geometry 边界几何形状
boundary growth 边界增厚，界面增厚
boundary layer 边界层，界面层，附面层
boundary layer accumulation 边界层堆积，边界层增厚，边界层聚积
boundary layer bleed 边界层抽吸
boundary layer blowing 边界层吹除
boundary layer buildup 边界层形成，附面层的形成
boundary layer configuration 边界层形状
boundary layer control system 边界层控制系统
boundary layer control 边界层控制
boundary layer correction 边界层修正
boundary layer depth 边界层厚度，边界层深度
boundary layer development 边界层扩展
boundary layer effect 边界层效应
boundary layer energy equation 边界层能量方程
boundary layer equation of motion 边界层运动方程
boundary layer equation 边界层方程
boundary layer flow 边界层水流，边界层流动
boundary layer flux 边界层通量
boundary layer friction loss 边界层摩擦损失，附面层摩擦损失
boundary layer friction 边界层摩擦
boundary layer function 边界层函数
boundary layer geometry 边界层几何形状
boundary layer growth 边界层增长
boundary layer height 边界层高度，边界层厚度
boundary layer immersion ratio 边界层浸没比
boundary layer measurement 边界层测量
boundary layer meteorology 边界层气象
boundary layer momentum integral equation 边界层动量积分方程
boundary layer oscillation 边界层脉动，边界层振动
boundary layer phenomenon 边界层现象
boundary layer probe 边界层探针
boundary layer profile 边界层廓线
boundary layer removal 边界层的去除
boundary layer reversal 边界层回流
boundary layer separation loss 边界层分离损失，附面层脱离损失
boundary layer separation 边界层分离，边界层分离现象，边界层脱离
boundary layer skin friction 边界层表面摩擦

boundary layer structure 边界层结构
boundary layer suction 边界层抽吸
boundary layer temperature 边界层温度，边界温度
boundary layer theory 附面层理论，边界层理论
boundary layer thickness 边界层厚度
boundary layer transition strip 边界层转换绊线，边界层过渡地带
boundary layer transition 边界层过渡区，边界层转捩
boundary layer type wind tunnel 边界层型风洞
boundary layer type wind 边界层型风
boundary layer vorticity 边界层涡量
boundary limit 分界，界限
boundary line 分界线，极限线，界线，边线
boundary lubrication 面间润滑【轴承】，界面润滑，边界润滑
boundary marker beacon 边界指示标
boundary marker 边界标志
boundary node 边界节点，边界结点
boundary particle 界面颗粒
boundary perturbation problem 边界扰动问题
boundary point 边界点
boundary problem 边界问题
boundary resistance 边界阻力，边界电阻，边界层热阻，边界层阻力
boundary science 边界层科学，边缘科学
boundary separation 边界分离
boundary shear 边界剪应力
boundary similarity 边界相似
boundary spar 边缘梁，边缘翼梁
boundary state of stress 极限应力状态
boundary stress 边缘应力
boundary surface 界面，分界面，边界面，边界曲面
boundary survey 边界测量
boundary temperature 边界层温度，边界温度
boundary tension 表面张力
boundary value problem 边界值问题，边值问题
boundary value 界限值，监界品位，边值，边界值
boundary wall tube 炉墙管
boundary wall 围墙，边界墙，厂区围墙
boundary wave 边界波，界面波
boundary zone of capillarity 毛细管水边界层，毛管水边界层
boundary zone 边缘带，边界区
boundary 边界，边界的，边缘，边，界限，界面，界线，分界，分界线
bound-atom cross-section 受束缚原子的截面
bound-atom scattering cross-section 受束缚原子散射截面
bound-bound energy transition 束缚态-束缚态能级跃迁
bound charge 束缚电荷
bound circulation 附着环量
bounded above 有上限的，有上界的
bounded beds 边缘床层
bounded below 有下限的，有下界的
bounded function 有界函数
bounded variable 约束变量
bound electricity 束缚电荷
bound electron 束缚电子
bound-free energy transition 束缚态-自由态能级跃迁
bound-free 束缚态-自由态
bounding case 极端情况
bounding surface 边界表面
boundless 无界限的，无边界的
bound nuclear effect 束缚核效应
bound pair 界偶
bound scattering cross-section 束缚散射截面
bound surface 邻接面
bound vortex filament 附着涡丝
bound vortex system 附着涡系
bound vortex 附着涡，约束涡，附体涡流，约束涡流
bound vorticity 附着涡量
bound water 结合水，化合水，束缚水，吸附水，约束水
bound 约束，边界，边缘，极限，结合的，束缚，范围，限制，跳跃
bount water 附着水
bounty on export 出口津贴，出口奖励
bounty 津贴
bouquet arrangement 一揽子安排
Bourdon gauge 波登压力表，波登管式压力计，弹簧管压力表，弹簧管式压力计
Bourdon pressure element 波登管压力元件
Bourdon tube gauge 波登管压力计
Bourdon tube pressure gauge 弹簧管压力表，布尔登管式压力表
Bourdon tube 布尔登压力计管，布登管，弹簧管
bourse 证券交易所
Boussinesq approximation 布辛内斯克近似
Boussinesq coefficient 布辛内斯克系数
Boussinesq theory 布辛内斯克理论
bovey coal 褐煤
bow anchor 船首锚
bow beam 弯梁
bow blade 马刀型叶片
bow collector 集电弓，弓形滑接器
bow drill 弓形钻
bowed height 矢高
bowed rod 弓形棒
bower anchor 大锚，前锚
bower 主锚
bow girder 弓形大梁
bowing coefficient 弯曲反应性系数
bowing of loading 装料凹面
bowing under load 负载弯曲
bowing 弯曲，弓形
bowk 凿井用吊桶
bowl and shaft assembly support 球体和轴组件支承【多级泵】
bowl drain 油杯放油塞
bowling-alley test 滚球强度试验
bowl line 弓形曲线
bowl pulverizer 中速碗式磨煤机，滚磨机
bowl roll pulverizer 碗式磨煤机
bowl shaped structure 凹面浪结构

bowl 杯，滚轴，滚珠，球形体，洼地，旋转，磨碗
Bowman formula 波曼公式
bow member 弓形构件
bow pen 划线笔
bow saw and saw blades 弓锯和锯条
bow saw 弓形锯
bow shock 激波，弓形激波
bow spring 弓形弹簧
bowstring arch 弓弦拱
bowstring beam 弓弦式梁
bowstring bridge 弓弦式桥
bowstring truss 弓弦式桁架，弓形屋架
bow trolley 弓形集电器，集电弓
bow wave 脱体冲波，冲激波
bow 弓架，弓，弧，弓形（件），弯曲，弓弧，使成弓形
box abutment 箱形桥台
box beam 箱形截面大梁，箱形梁
box bed 箱式床
box breakwater 箱形防波堤
box bridge 电阻箱电桥，箱式桥
box caisson 箱形沉箱，沉箱
boxcar 厢式车，货车车厢，有盖货车
box cofferdam 箱形围堰
box column 箱形柱
box compass 罗盘仪
box conduit 箱形输水道
box coupling 套筒联轴节
box culvert 箱式涵洞
box dam 箱式围堰，箱形坝
box drain 箱形暗沟，箱形排水渠
boxear average integrator 取样平均积分器
box end wrench 套筒扳手，管钳子
box fan 箱式风扇，鸿运扇
box flume 箱形渡槽
box fold 箱状褶皱
box footing 箱型基础
box foundation 箱式基础，箱形基础
box-framed construction 箱形框架结构［建筑］
box frame motor 箱形机座电动机
box frame 箱形构架，箱形框架
box gauge 箱式潮位计
box girder bridge 箱梁桥
box girder 空心梁，箱形大梁，箱形截面大梁，箱形梁
box grab 箱形抓具
box gutter 箱形水槽，檐沟
box header boiler 整联箱式锅炉
box header 整联箱，箱式联箱
box horn 喇叭形天线
boxing type ash flusher 箱式冲灰器
box inlet 箱形进水口
box-like arm 盒形支臂
boxlike structure loader-unloader 箱形结构装卸桥
box loop 环形天线
box metal 减磨合金，轴承合金
box model 箱模式
box-out 储备，保留
box pile 箱式桩
box piston 空心活塞
box section 方形截面，箱形截面
box shear apparatus 盒式剪切仪，直剪仪
box shear test 盒式剪切试验
box socket set 棘轮套筒扳手
box spanner 套筒扳手，梅花扳手，管钳子
box spar 箱式叶梁
box staff 塔式标尺，塔尺
box steel sheet piling 箱式钢板桩，箱形钢板桩
box strut 箱形截面支柱
box switch 盒形开关，匣式开关
box type arm 盒形支臂
box type boiler 箱式锅炉
box type cofferdam 箱式围堰
box type manifold with spray nozzles 带喷雾头的箱型集管
box type manifold 箱形集管
box type rheostat 箱式变阻器
box type sampler 匣式采样器
box type spray manifold 箱形喷雾集管
box type transformer 箱式变压器
box-up fee 装箱费
box wagon 棚车，箱式车皮
boxwood folding rule 木折尺
boxwood 黄杨木
box wrench 套筒扳手，梅花扳手，管扳手
box 箱，盒，装盒，外壳，装箱，箱子
boycott 经济抵制，联合抵制
Boyden radial outward flow turbine 博伊顿式辐向外流水轮机
Boyden radial outword flow turbine 博伊顿径向出流式水轮机
BP(barometric pressure) 气压，大气压，大气压力
BP(base plate) 底版，支撑板
BP(base port) 基本港
B/P(bills payable) 应付票据
B. P. (boiling point) 沸点
BPC(back pressure control) 背压控制
BPC(blade pitch control) 叶片节距控制
BPC(bypass control system) 旁路控制系统【汽轮机】
BPI(bit per inch) 每英寸字节数
B-pillar B支柱
BPS(bit per second) 每秒比特，每秒传送位数
BPS(bypass control system) 旁路控制系统
BP sensor(barometric pressure sensor) 大气压力传感器
BPTS(International Practical Temperature Scale) 国际实用温标
BP 旁路【DCS画面】
Bq＝becquerel 贝可勒尔【放射性活度单位】
brace angle 支撑角钢
brace bit 摇钻
brace bolt and nut 拉条螺栓及螺母
braced 加支撑的
braced arch 桁架式拱
braced beam 桁架式梁
braced chain 桁链
braced core building 格架式筒体建筑
braced door 联结门

braced excavation 支撑开挖
braced frame construction 杆系框架结构
braced framing 支撑系结构，联结构架
braced girder 桁梁，联结大梁，支撑系大梁
braced pier 联结墩
braced reinforced concrete flume 桁架式钢筋混凝土渡槽
braced rib arch 桁架式肋拱
braced sheeting 支撑挡板
braced strut 联结支撑
brace nut 拉条螺母
brace rod 联结杆
bracer 撑条，拉条，撑块
brace screw 撑柱螺丝
brace summer 双重梁，支撑梁
brace system 支撑系统
brace wrench 曲柄头扳手
brace 拉条，支撑，支柱，撑住，木电杆下的垫基，十字形格架，曲柄，支架，带子
brachy-anticlinal fold 短背斜褶皱
brachy-synclinal fold 短向斜褶皱
bracing beam 联结梁，支撑梁
bracing boom 支撑杆，加劲杆
bracing frame 支撑架
bracing framing 支撑系统
bracing girder 联结梁
bracing in open cut 明挖支撑
bracing member 斜材
bracing of foundation pit 基坑围护
bracing piece 加劲杆，斜梁，加强杆，横梁，斜撑，支撑杆，支承板，拉紧板，支撑臂
bracing structure 支撑结构
bracing system 支撑体系，支撑系统，支护系统，腹杆系统
bracing wire 拉线
bracing 拉条，斜撑，撑杆，肋材，张线，支撑，加强筋，支承，支柱，背带
bracket bearing 端盖轴承，悬臂轴承
bracket bolt 托架螺栓
bracket bracing 托座链条，托座支撑
bracket connection 牛腿连接，托架结合
bracket crane 悬臂式起重机
bracket insulator 直脚绝缘子
bracket lamp 托架灯，壁灯
bracket pole 装有交叉支撑的电杆，托臂支柱
bracket post 托臂支柱，托架柱
bracket scaffold 悬臂式脚手架，挑出式脚手架
bracket stringer 挑出式楼梯斜梁
bracket support 托臂支座，牛腿
bracket type machine 座式轴承电机
bracket 支架，托座，托架，支座，括号，架，角撑架，牛腿，隅撑，支承板，支柱
bracking radiation 韧致辐射
brackish flow 微咸水流
brackish water 微咸水，半咸水，苦咸水，苦盐水，碱性水，淡盐水【盐分界于河海水之间】
brackish 微碱的
bradawl 打眼钻，小锥，锥钻
bradyseism 缓震
brad 角钉，曲头钉，无头钉
braid cable with copper conductor 铜芯编织电缆
braid conductor 金属线编织的导体
braided cable 编包电缆，屏蔽电缆，编缆，编织电缆
braided channel 辫状河道，分叉河道
braided covered cable 编包电缆，屏蔽电缆
braided flow 辫状流
braided packing 编织的填料
braided river course 分叉河道
braided river 分叉河流，辫状河
braided shielding 编织的屏蔽
braided shield 编织护套
braided stream 辫状河，叉河
braid fabric 金属丝滤网
braiding 编织，编包，(电缆的) 护层
braid 编织物，编织
brail 斜撑杆
brain drainer 外流人才
brain drain 人才外流
brain machine 自动计算机
brainpower control 智能控制
brainpower house 智能住宅
brainpower unit 智能单元
brainpower 科技人员，智能，智囊
brainstorm(ing) (决策的) 智暴法，点子会议，头脑风暴，集思广益，集体讨论
brain trust 智囊团
brain 脑，计算机，智慧
braise 焦炭屑，煤屑
braize 焦炭屑，煤屑
brake action 制动作用
brake adjuster 制动器
brake arm 制动臂
brake assisting system 制动力辅助系统
brake assistor 制动加力器
brake beam 制动梁
brake device 制动器
brake disc 刹车盘，制动盘
brake drag 制动阻力
brake dressing 制动器润滑脂
brake drum 刹车鼓，闸轮
brake dynamometer 制动测功机，制动测力计
brake efficiency 制动效率
brake electromagnet 制动电磁铁
brake equalizer 制动平衡器
brake-field tube 二次电阻抑止管，制动电场三极管
brake fluid 刹车油
brake force 制动力
brake for wind turbine 风力机制动器
brake horsepower curve 制动功率曲线
brake horsepower hour 制动马力小时
brake horsepower 制动功率，轴功率，轴马力，制动马力
brake lamp switch 刹车灯开关
brake lever 闸杆，制动杆，制动杠杆
brake lining 闸衬片，制动衬片
brake linkage 制动联动装置
brake load 制动荷载，制动负荷，刹车力
brake magnet 阻尼磁铁，制动磁铁
brake mean effective pressure 平均有效压力
brake mechanism 制动机构

brake motor 制动电动机，制动马达
brake nozzle 制动喷嘴
brake off 松开制动
brake on-off switch 制动开关
brake on 制动
brake pad thickness 制动块厚度
brake pad 闸垫
brake power 制动力，制动功率，有效功率
brake release 制动释放，松开制动器
brake resistance 制动电阻
brake ring 制动环，制动圈
brake-rod 制动杆
braker 制动器
brake setting 制动器闭合
brake shoe 制动瓦，闸瓦，制动蹄片，制动块，煞车块，制动靴
brake spring 制动弹簧
brake switch 制动开关
brake testing bench 制动试验台
brake test 制动试验
brake thermal efficiency 制动热效率
brake thrust 制动推力
brake torque 制动转矩
brake track 制动环
brake turbine 制动透平，制动涡轮
brake valve 制动阀
brake wheel 制动轮
brake 压弯成形机，制动器，刹车，闸，制动
braking absorption 阻尼吸收
braking action 制动作用
braking coefficient 制动系数
braking controller 制动控制器
braking current 制动电流
braking deceleration 制动减速率
braking device 制动装置
braking disc 制动盘
braking effect 制动作用
braking efficiency 制动效能，制动效率
braking effort 制动作用力，制动力
braking element 制动元件
braking energy 制动力，闸能，制动能
braking equipment 制动装置
braking gas 减速气体
braking jet 制动喷流，制动射流
braking magnet 制动磁铁
braking mechanism 制动机构
braking moment 制动力矩
braking path 刹车滑行距离
braking period 制动时间，制动周期
braking power 制动功率
braking releasing 制动器释放
braking rim 制动环【水轮发电机】，闸板
braking section 减速段
braking stress 制动应力
braking surface 制动面
braking system 制动系统
braking time 制动时间
braking turbine 制动涡轮
braking type 刹车形式
braking vane 刹车尾舵，刹车板
braking wheel coupling 制动轮联轴器
braking wheel 制动轮，闸轮
braking 制动的，刹车的，制动，刹车，停机制动，制动系统
Bramath's press 勃兰姆水压机
branch-admittance matrix 支路导纳矩阵
branch air duct 分支风管
branch and linkage 转移与连接
branch and store 转移和存储
branch angle 分叉角
branch bank 分行
branch box 分线盒
branch bus 分支母线
branch circuit distribution center 分支电路配电中心
branch circuit 分流电路，支路，分路，分支电路，支线
branch clamp 分线线夹
branch company 附属公司，子公司
branch conductor 支线，分支导线
branch connection 分叉连接，分接，分支接续，分支连接，支管连接
branch controller 分支控制器，转移控制器
branch current 分支电流，分路电流
branch duct 支管
branched distributor 分支配电线
branched-guide coupler 分支波导耦合器
branched pipe distributor 支管分布器
branched 分支的，分叉的
branch equation 支线方程
branch exchange 交换分机
branch heading 错车巷道，分支巷道
branch impedance matrix 支路阻抗矩阵
branch impedance 支路阻抗
branching cable 分支电缆
branching crack 分支状裂纹，分支状裂缝
branching decay 分支衰变
branching fault 分支断层
branching filter 分路滤波器
branching fraction 分支（衰变）份额，分支份额
branching operation 分路操作
branching point 分支点
branching ratio 分支比
branching switch board 并联复式交换机
branching 分路，转移，分支，分接，叉形接头，分叉
branch input signal 支路输入信号
branch instruction 转移指令
branch interval 支管间隔
branch joint 三通接头，分支连接，分支接头
branch levee 支堤
branch line heat substation 区域热力分站
branch line substation 区域热力站
branch line 分支线，支线，支路，分支管
branch office 分支机构，分局
branch-off 分流，分支，分接，分路
branch of joint 连接分支
branch on equality 相等（条件）转移
branch on minus 负转移
branch on non-zero 非零转移
branch on zero 零转移

branch order　分支指令，转移指令
branch output interrupt　转移输出中断
branch output signal　支路输出信号
branch out　分路，分出，分支，分流
branch overcurrent protection　分支过流保护
branch-path incidence matrix　支路-回路关联矩阵
branch-path　支路，分路
branch pipe joint　分支管接头
branch pipe weld　支路焊缝
branch pipe　分支管，支管
branch point　分支点，结点，转移点，分叉点
branch raceway　分支水管，分支电缆管道
branch railway　铁路支线
branch resistance　分支电阻，分流电阻
branch river　支流
branch sewer　污水支管，支沟
branch sleeve　连接套筒，管节，筒形联轴器，分支套筒
branch stiffness　电路的反电容，电路的逆电容
branch stub　分支短管
branch subroutine　转移子程序
branch switchboard　分路配电盘，电话分局总机
branch switch　分路开关，支路开关
branch terminal line　终端支线
branch terminal　分支线端
branch tube　支管
branch type switchboard　分立型配电盘
branch unconditionally　无条件转移
branch valley　支谷
branch vein　支脉
branch vent　通气支管
branch voltage　支路电压
branch wire　分支线
branchy　多支的，分支的
branch　部门，分支，分叉，支管，斜三通，分叉管，支流，分流，支路，支线
brand identification　商标识别
brand leader　占市场最大份额的品牌，名牌
brand name　标牌，铭牌，商标
brand-new　全新的，新制的，最新出品的
brandreth　铁架，铁三脚架
brand　标牌［签］，牌名，牌子，品牌，商标
brash　崩解石块，脆的，破碎片，易脆的，脆性
brass alloy　黄铜合金
brass bar　黄铜棒
brass bearing　黄铜轴承
brass board　模型的，试验性的，中间试验
brass bushing　黄铜衬套
brass lining　黄铜衬套
brass pipe　支管，黄铜管
brass plating　镀黄铜
brass pounder　黄铜鞭状天线
brass rod　黄铜条
brass solder　钎焊，黄铜焊
brass tube　黄铜管
brass tubing　黄铜管
brass wire brush　铜丝刷
brass wire　黄铜线
brassy　黄铜的
brass　黄铜，黄铜轴瓦，黄铜轴衬
brattice　临时木建筑
brat　未洗净的煤，不净煤，原煤
bray stone　多孔砂岩
brazability　钎焊性
brazed joint　焊接头，硬钎焊接
brazed part　钎焊的
brazed shield　钎焊（斯太立）防蚀片
brazed　铜焊的，硬焊的
braze interface　硬钎焊面
braze metal　硬钎缝金属
braze pressure welding of steel pipe　钢管的钎压焊
brazer　钎焊工
braze welding　钎焊，硬焊，铜焊
braze weld　硬钎焊
braze　铜焊，用黄铜镀，用黄铜制造，钎焊，硬钎缝
brazier　扁头螺钉
brazing alloy　钎焊合金
brazing blowpipe　钎炬
brazing brass　硬焊黄铜
brazing filler metal　硬钎料
brazing flux　硬钎焊钎剂，硬焊用焊剂，钎剂
brazing metal　钎焊金属
brazing paste　钎焊膏
brazing powder　粉状硬钎料，钎焊粉
brazing sheet　钎焊板
brazing(/soldering) temperature　钎焊温度
brazing torch　钎焊焊炬
brazing　钎焊，铜焊，（硬）钎焊
B/R(bills receivable)　应收票据
BR(break request)　中断请求
BRC(below-regulatory-concern)　管理当局不关心的
breached fuel element　破裂燃料元件
breach of agreement　违反协议
breach of blockade　打破封锁
breach of condition　违反条件
breach of contract　违背合同，违反合同，不履行合同，违约
breach of discipline　破坏纪律
breach of duty　失职
breach of faith　失信
breach of law　犯法，违法，违法行为，违犯法律
breach of privilege　滥用特权
breach of promise　违背诺言
breach of regulation　违反规程，违反条例，违章
breach of warranty　违背保证，违反担保
breach　破坏，破裂，违背（合同），溃决，裂口，缺口，碎浪，违反，裂缝，破口，折断，破损，不履约
breadboard design　模拟板设计
breadboarded circuit　模拟板电路
breadboard experiment　模拟板试验
breadboard　功能试验，试验性的，试验电路板
breadth coefficient　绕组系数
breadth factor　分布因子
breadth of section　截面宽度
breadth of size distribution　（煤粒）筛分尺寸范围，筛分宽度

breadth of tooth 齿宽
breadth 宽度，厚度，余隙，跨距，横幅，跨度，开度
breakable 易碎的，脆的
break a contract 违反合同
breakage bar 断条
breakage 断裂，损坏，故障，断线，损耗，破损，破裂，破碎，破损险
break an agreement 违约，破坏协定
break angular frequency 转折角频率
break arc 开断电弧
break a vacuum 破坏真空，减低真空
breakaway corrosion 腐蚀剧增
breakaway coupling 断开式联轴节，防超载联轴节
breakaway device 安全分离装置，安全脱钩装置，保险装置
breakaway force 脱离力，启动力，起步阻力，破断力
breakaway point 断裂点，脱离点
breakaway release （裂变气体的）破裂释放，破裂释放
breakaway speed 飞逸转速
breakaway starting current 初始起动电流，启动电流
breakaway torque 动力矩，起动力矩，启动转矩
breakaway 分离，脱流，气流分离，冒顶，剥裂，破裂，断开
break back contact 接触点，开路接点
break barrow 碎土机
break-before-make contact 先断后合接点，先合后开触点
break-before-make 先断后通，先断后合接点
break-break contact 断-断接点
break-break-make contact 断-断-合接点
break contact spring 断开触点簧片
break contact 常开触点，常闭触点，断开接点
break distance 断开距离
breakdown accident of heat-supply network 热网事故
breakdown cable 应急电缆，替换电缆
breakdown characteristics 击穿特性
breakdown charge 击穿电荷
breakdown current 击穿电流
breakdown field strength 击穿电场强度，绝缘强度
breakdown in cryogenic dielectric liquid 低温液体电介质击穿
breakdown in dielectric liquid caused by bridge 液体电介质小桥击穿
breakdown in electronegative gas 负电性气体击穿
breakdown lorry 抢险起重车，抢修工程车
breakdown loss 事故停机损失，故障损失
breakdown maintenance 事后维修，事故维修，故障维修
breakdown minimum 最低击穿电压
breakdown of a fuel stringer 解体燃料元件柱
breakdown of cost 成本分析，成本分解
breakdown of insulation 绝缘击穿
breakdown of long air gap 长空气间隙击穿
breakdown of oil film 油膜破坏
breakdown of oil 油的澄清
breakdown of service 服务中断，运行解体
breakdown point 破坏点，击穿点，屈服点
breakdown potential gradient 击穿电位梯度
breakdown potential 击穿电压，击穿电势
breakdown power 破坏功率，击穿功率
breakdown pressure 事故压力，停机压力
breakdown price 分项报价，分项价格
breakdown procedure 解体步骤，拆卸步骤【燃料元件柱】
breakdown region 击穿范围，击穿区
breakdown shock wave 脱体冲波，滞止冲波
breakdown signal 击穿信号，故障信号
breakdown slip 临界转差率，极限转差率
breakdown speed 损坏速度
breakdown spot 击穿点
breakdown station 解体站，拆卸站
breakdown strain 损坏应变
breakdown strength 击穿强度，破坏强度
breakdown stress 断裂应力
breakdown switch 故障开关
breakdown test 破坏试验，击穿试验，耐压试验，断裂试验，折断试验
breakdown time 破坏时间，击穿时间，破坏期
breakdown torque of the motor 电机制动力矩
breakdown torque speed 停转转速，临界转速
breakdown torque 停转力矩，极限转矩，崩溃转矩，临界转矩，损坏力矩，破坏扭矩，破坏转矩，失步转矩，停转转矩
breakdown train 抢险列车
breakdown voltage testing 击穿电压试验，耐压强度试验
breakdown voltage 击穿电压，破坏电压
breakdown 断裂，故障，分离，破裂，损坏，坍塌，击穿，打破，破坏，分成细目，分解，分类，分项，划分，事故
breaker and a half connection 一个半断路器接线
breaker and a half type 一个半开关接线
breaker back-up 备用断路器，断路后备装置
breaker bolt 安全螺栓
breaker closed 断路器合闸
breaker compartment 断路小室
breaker failure protection 断路器失灵保护
breaker failure 断路器失灵
breaker-fuse combination 断路器熔丝组合
breaker house 碎煤机房
breaker module 开关模数
breaker plate 打击板，破碎板，断路器板
breaker point cam 截断器凸轮
breaker point wrench 电流开关扳手
breaker roll 轧碎机滚筒，对轧辊
breaker strip 防断条
breaker switch 断路开关
breaker test 断路器试验
breaker time 断路器跳闸时间
breaker trip coil 断路器脱扣线圈
breaker tripping and closing test 断路器跳合闸试验
breaker trip 断路器跳闸

breaker wave 拍岸浪，破波
breaker with ultrashort breaking time 超短截断时间的断路器
breaker 断电器，断路器，缓冲衬层，破坏器，破碎机，破碎器，破浪，开关
break even analysis 盈亏临界分析，盈亏平衡分析，保本分析，损益两平分析，损益平衡分析
break even chart 盈亏平衡图
break even model 盈亏平衡模型，保本模式
break even point 盈亏平衡点，盈亏相抵点，零利润点，保本点，盈亏临界点，收支相抵点
break even revenue 保本销售额
break even value 保本值
break even 不亏不盈，盈亏平衡，盈亏相抵，收支相抵，得失相当，不赚不赔，打成平手，收支平衡
break glass units 玻璃破碎装置
break ground 破土，动土，开垦，动工
break impulse 切断脉冲
break-induced current 断路感应电流
break in electrical continuity 中断电气连续性
breaking arc 切断电弧
breaking around an obstruction 在障碍物周围断开
breaking capacity test 开断容量试验，截断容量试验
breaking capacity 截断功率，截断容量，断路容量，断流容量，开断能力，开断容量
breaking characteristic 分断特性
breaking coefficient 断裂系数
breaking coil 跳闸线圈
breaking condition 制动工况
breaking current 断路电流，分断电流，分断能力，截断电流【断路器、熔丝等】
breaking device 断路装置，断路器件，断路器
breaking-down point 破坏点，击穿点
breaking-down test 击穿试验，破坏试验，耐压实验，耐破坏试验
breaking-down 发生故障，切断，断电，打碎
breaking edge （顶板岩石的）崩落线
breaking effort 制动力
breaking elongation rate 断裂伸长率
breaking elongation 断裂伸长，破断伸长
breaking force 破断力
breaking ground 破土
breaking-in period 试验期，磨合期，溶解期
breaking-in 试运转，试车，开始使用，嵌入，插入
breaking joint 断裂节理，断裂缝
breaking limit 断裂极限，断裂限度，破坏极限
breaking link （水轮机的）脆性连杆
breaking load 断裂荷载，破坏负载，破坏荷载，破坏荷重，破损荷载
breaking moment 断裂力矩，破坏瞬间，切断瞬间，破坏力矩
breaking of contact 断接，触点断开
breaking of pigs 破碎金属块
breaking of piles up to cut-off level 破去桩头
breaking plane 破坏平面
breaking plate 破碎板
breaking point 击穿点，断裂点，破损点，强度极限
breaking step 失步，乱步
breaking strain 破坏应变，断裂应变
breaking strength 破坏强度，断裂强度，击穿强度，抗断强度，破损强度
breaking stress circle 破裂应力圆
breaking stress 致断应力，断裂应力，抗断应力，破坏应力
breaking test 断裂试验，破坏试验
breaking the pile head 破桩头
breaking time 分闸时间
breaking up 碎裂，中断
breaking water level 跌落水面
breaking wave 破浪
breaking 断开，断路，断线，破坏，阻断，破碎，轧碎
break-in keying 插话式键控
break-in oil 磨合油
break-in period 间断期，故障期，试运转期，试车期
break-in relay 插入继电器
break jack 切断塞孔
break limit 断裂极限
break line 断裂线
break location 破口位置
break-make-break contact 断-合-断触点
break-make contact 换向触点，断-合触点
break-make-make contact 断-合-合触点
break-make transfer switch 合-断转换开关
break of an earth bank 路堤滑坡
break of an engagement 毁约
break of conductor 断线
break of contract 撕毁合同
break-off diagram 断开曲线
break-off point 断裂工况点
break-off 断开，破坏
break of load 负载中断
breakout cathead 卸松螺纹猫头，卸杆猫头
breakout point 中断点，突破点，分叉点
breakout pressure 启动压力，坍塌压力
breakout 爆发，突发，冲出，排出
breakover current 转折电流，导通电流
breakover point 导通点，转折点
breakover voltage 导通电压
breakover 穿通，导通，转折
break plane 断裂面
break-point chlorination 断点加氯法，折点氯化
break-point instruction 断点指令
break-point potentiometer 断点电位计
break-point switch 断点开关，折点开关
break-point symbol 间断点符号
break-point voltage 断点电压
break point 断点，破裂点，折点，转折点，间断点
break pulse 断路脉冲
break request 中断请求
break room 休息室
break seal 拆封
break shock 断路冲击，断路时的电压冲击
break size 破口尺寸
break spark 断路火花

break step 失步，乱步
break switch 断路开关
break test 断裂试验，破坏试验，破损性试验
break the contract 毁约
break the term of agreement 违反协议条款
breakthrough capacity 漏过容量【离子交换器】，穿透容量，泄漏容量，漏过［贯流］能量
breakthrough curve 穿透曲线
breakthrough of silica 氧化硅漏过，漏硅
breakthrough point 贯通点
breakthrough 重要进展，重大的成就，炉衬裂口，突破
break time on fault 事故断电时间
break time 断开时间，切断时间，破裂时间
breakup efficiency 崩裂效率
breakup 分解，分散，分裂，破裂，蜕变，崩溃，解散
break vessel 破裂容器
breakwater capstone 防波堤压顶石
breakwater end 防波堤端
breakwater gap 防波堤缺口，堤头口门
breakwater head 防波堤堤头
breakwater pier head 防波堤堤头
breakwater pier 防波突堤
breakwater quay 防波堤堤岸
breakwater 防波堤，防浪堤
breakwind 防风林，风障，防风墙
break 断开，断裂，断路，裂缝，断线，中断，间歇，断口，溃决，破坏，破口，破裂，损坏
breast abutment 无斜翼岸墩，无斜翼桥台，胸式桥台
breast board 栏板
breast drill 胸压手摇钻，胸压钻，曲柄钻，胸钻
breast hole 中部炮眼
breasting dolphin 承冲护船桩，承冲靠船架
breast of beam 梁底
breastplate microphone 胸挂送话器，胸挂传声器
breastplate 胸前送话器
breast-shot water wheel 中射式水轮
breast summer 大木，过梁，托墙梁
breast telephone 胸挂电话机
breast transmitter 胸前送话器
breast wall 挡土墙，防浪墙，胸墙
breast 窗展，梁底
breather cap 通风帽
breather pipe 气口，通风口，排气口，通气管，气孔
breather roof 通气屋顶
breather valve 呼吸阀
breather 呼吸器，通气孔，通气管，吸潮器，通风口，呼吸阀，换气装置
breathing apparatus 排气装置，呼吸器
breathing rate 呼吸率
breathing space 呼吸区域，伸缩区
breathing well 呼吸井，透气井
breathing zone 呼吸带
breathing 呼吸，通气
bred fuel 再生核燃料
bredigite 白硅钙石
bred material 增殖材料
bred uranium 增殖铀
breeches pipe 人字形管，Y形管，叉管，叉形管
breeching 水平烟道，烟道
breechlock bonnet 热压自闭顶盖
breechlock seal 热压自闭密封
breechlock 热压自闭的
breech 烟道
breedable 可增殖的
breed doubling time 增殖加倍时间
breed draught cooling tower 压力通风冷却塔
breed element 增殖元件
breeder assembly 增殖组件
breeder blanket 增殖反应堆再生区
breeder-converter 增殖转换反应堆，再生反应堆
breeder element 增殖元件
breeder pin 转换区燃料细棒，增殖燃料细棒，增殖反应堆
breeder plant 增殖反应堆电厂，增殖反应堆电站，增殖过程
breeder processing engineering test 增殖堆燃料处理工程试验装置
breeder reactor power plant 增殖反应堆电厂
breeder reactor 增殖反应堆
breeder rod clad tube 增殖棒包壳管
breeder rod 增殖棒
breeder 增殖反应堆，扩大再生产反应堆，发起人
breed fuel 增殖燃料
breeding blanket 增殖区，转换区
breeding coefficient 增殖系数
breeding cycle 增殖循环，增殖周期
breeding factor 增殖因子
breeding gain 增殖增益
breeding pellet 增殖芯块
breeding process 增殖过程
breeding property 增殖特性【反应堆的】
breeding rate 增殖率【反应堆】
breeding ratio 再生系数，增殖系数，增殖比【核电】
breeding region 增殖区
breed particle 增殖包覆颗粒
breed 反应堆的再生，增殖，钚增殖【核电】，倍增，复制，品种，种类
breeze aggregate 炉渣骨料
breeze block 煤渣混凝土砌块，煤渣水泥砖
breeze concrete block 轻质煤渣混凝土块
breeze concrete 轻质煤渣混凝土，炉渣混凝土
breeze recirculation 煤屑再循环
breeze 焦炭屑，煤末，微风，海路风，煤渣，熔渣，微风
bremsstrahlung gauge 韧致辐射测量仪
bremsstrahlung intensity 韧致辐射强度
bremsstrahlung radiation 韧致辐射
bremsstrahlung 韧致辐射
BRER(Board on Radiation Effects Research) 辐射效应研究委员会【美国】
bressummer 托墙梁
breviate 一览表，缩写
Brewster's angle 伯里斯特角

BRG＝bearing　轴承【DCS画面】
bribery and corruption　贿赂与贪污
bribery or accept bribes　行贿或受贿
bribery　贿赂，受贿，行贿
briber　行贿人
bribes and kickbacks　贿赂和回扣
bribe　受贿人，贿赂
brick　砖
brick and a half wall　一砖半厚墙
brick and concrete composite construction　砖混结构
brick and concrete construction　砖混结构
brick and reinforced concrete construction　砖混结构
brick arch　砖拱
brick baffle　隔烟墙，砖隔墙
brick base　砖基，砖座
brick bat　砖块，砖片
brick beam　砖过梁
brick bonded arch　砖砌弧形拱
brick bonded tubular arch　砖筒拱
brick bond　砌砖法
brick-cap parapet　砖压顶女儿墙，砖顶压檐墙
brick cement　防水水泥，砌筑水泥，烧黏土水泥
brick chimney　砖烟囱
brick cladding　砖围护
brick clamp　砖夹
brick clay　制砖黏土
brick construction　砖结构
brick corbel table　砖挑檐
brick course　砖层
brick culvert pipe　砖砌涵管
brick dam　砖坝
brick dust　砖屑
bricked-in　砖衬的
brick facing　砖块面层，砖块砌面，砖砌面层
brick flat arch　砖平拱
brick flue　砖砌烟道
brick foundation　砖基础
bricking-up　用砖填塞
bricking　砖衬，砌体，砌砖，砖瓦工程
brick laid on edge　侧砌砖，竖砌砖
brick laid on flat　平砌砖
bricklayer's cleaver　瓦刀
bricklayer's hammer　瓦工锤
bricklayer's scaffold　瓦工脚手架
bricklayer　瓦工，砌筑工
bricklayer's knife　泥刀
bricklaying　砌砖
brick-lined canal　砖衬砌渠道
brick-lined shaft　砖衬竖井
brick-lined　铺砖的，砖衬的
brick lining　砖衬，铺砖，砖衬砌
brick lintel　砖过梁，砖砌过梁
brick masonry column　砖砌体柱
brick masonry structure　砖石建筑，砖石结构
brick masonry wall　砖砌体墙
brick masonry　砖砌体，砖砌圬工，砖石建筑
brickmason　砌砖工，泥水匠，瓦工，砖石匠
brick nogging building　砖填木架房屋
brick-on-edge coping　侧砖压顶
brick-on-edge paving　侧砖铺面
brick-on-edge sill　侧砌砖窗台
brick-on-edge　侧砌砖
brick-on-end soldier arch　竖砖拱
brick-on-end　竖砌砖
brick pavement　砖铺面
brick pier　大门砖墩
brick pillar　砖垛
brick press　制砖机
brick resistance　砖砌体的热阻
brick rubble　粗面砖
brick setting　砖砌，衬砖
brick set　铺砖的
brick sewer　砖砌污水沟
brick stone masonry　砖石建筑
brick tin shell　砖薄壳
brick trimmer　砖面修整，砖托梁
brick underdrain　砖砌阴沟
brick vault　砖筒拱
brick-veneered　砖镶面的
brick veneer　砖砌面层
brick wall bearing construction　砖墙承重结构
brick wall expansion joint　砖墙伸缩缝
brick wall fence　砖围墙
brick wall load bearing　砖墙承重
brick wall without plastering　清水砖墙
brick wall　砖墙
brick with groove　带槽砖
brickwork construction　砖砌结构
brickwork footing　砖基础
brickwork joint　砖缝
brickwork setting　砖砌，衬砖
brickwork　砖砌体，砖砌，砌砖工程，砌砖工作
brickwork reinforced with wire　网状配筋砌体
brickyard　砖厂，堆砖场地
bridge abutment　桥台，桥肩，桥墩
bridge across　桥接，跨接
bridge aerodynamic　桥梁空气动力学
bridge aeroelasticity　桥梁气动弹性
bridge aesthetics　桥梁建筑艺术
bridge amplifier　桥式放大器
bridge aperture　桥孔，桥洞
bridge approach　引桥
bridge approach fill　引桥路堤
bridge approach to ferry　码头引桥
bridge arch　桥拱
bridge arm ratio　电桥臂比
bridge arm　电桥臂，比例臂
bridge bar　（管子对接焊用的）管子定位夹具
bridge beam　横梁
bridge bearing　桥梁支承，桥梁支座
bridge board　梯侧板，楼梯帮，楼梯梁，斜梁
bridge boat　平底桥，浮桥
bridge box　电桥箱
bridge bracing　桥式支撑，桥梁支撑
bridge building clearance　桥梁建筑限界
bridge building　桥梁建造
bridge cable　吊桥钢缆
bridge calibration　电桥校准
bridge circuit configuration　桥形接线

bridge circuit method 电桥电路法
bridge circuit 桥路，桥接电路，桥式电路，电桥电路
bridge clamp 桥式夹
bridge clearance 桥下净空
bridge configuration 桥式接线
bridge connection 桥接，跨接
bridge connector 桥接线，桥接条
bridge construction 桥梁建造
bridge crane rail 行车轨道
bridge crane 桥吊，桥式吊车，桥式起重机，天车，行车
bridge current 桥接电流，分路电流，桥路电流
bridge-cut-off relay 断桥继电器，桥式断路继电器，分隔继电器
bridge-cut-off test 桥式断路试验，断桥试验
bridge-cut-off 断桥，桥式断路
bridge dam 桥式坝
bridged circuit 电桥电路，跨接电路
bridge deck 桥面板，桥面
bridge diagram 电桥图，桥形电路图，跨接图
bridged-T bridge 桥接T形电桥
bridged-T filter 桥接T形滤波器
bridged-T network 桥接T形网络
bridged-T tap 桥接T形抽头
bridged-T trap 桥接T形陷波器
bridge duplex system 桥接双工系统
bridge duplex 桥接双工
bridge engineering 桥梁工程
bridge feedback 桥式反馈，桥式回授
bridge financing 过渡性融资
bridge floor 桥面层，桥面
bridge flutter 桥梁抖振
bridge foot board 桥上步行板
bridge for stream-gauging 测流桥
bridge foundation 桥梁基础
bridge fuse 插接保险丝，桥接保险丝
bridge gallery 过街楼，过街桥
bridge gauge 桥形规，桥规
bridge girder 桥，电桥
bridge grab ship unloader 桥式抓斗卸船机
bridgehead 桥头（堡），桥塔，桥头
bridge joint 搭桥结合，架接
bridge limiter 桥式限制器，桥式限幅器
bridge measurement 电桥测量
bridge megger 桥式兆欧表
bridge method 电桥法，桥接法
bridge motor 桥式起重机用电机
bridge-mounted reclaimer 桥式取料机
bridge-mounted rotary-bucket wheel reclaimer 桥式旋转斗轮堆取料机
bridge network 桥形网络，桥接网络
bridge neutralizing 桥接抵消法
bridge of overhead traveling crane 天车桥（架），行车桥
bridge opening 桥洞，桥孔
bridge oscillator 桥式振荡器
bridge over 跨接，桥接
bridge piece 连接件
bridge pile 桥桩，桥洞，桥墩
bridge plate 砌工搁板
bridge railing 桥梁护栏，桥栏杆，桥式起重机轨道
bridge rectifier 桥式整流器
bridge reinforcing 桥梁加固
bridge scale 地秤
bridge scraper 桥式刮料机
bridge seat 桥座
bridge slag 渣桥
bridge sleeper 桥梁木枕
bridge span 桥跨，桥梁跨度
bridge spot welding 带接合板点焊，单面衬垫点焊，单面搭板点焊
bridge substructure 桥梁下部结构
bridge test 电桥测试
bridge thrust 拱桥推力
bridge tower 桥头堡，桥塔
bridge transformer 桥式差接变压器
bridge transition 桥式换接过程
bridge transporter 桥式装卸机，桥吊，桥式运送机，装卸桥
bridge trolley 桥式吊车滚轮
bridge truss 桥梁桁架
bridge type bucket-wheel reclaimer 桥式斗轮取料机
bridge type coal grab-unloader 桥式抓斗卸煤机
bridge type coal grab 桥式抓煤机
bridge type contact 桥式接点，桥式触头
bridge type converter 桥式换流器
bridge type crane for three-purpose 三用桥式起重机
bridge type feedback 桥式反馈
bridge type frequency meter 桥式频率计
bridge type grab crane 桥式抓斗起重机
bridge type measuring unit 桥式测量单元
bridge type movable dam 桥式活动坝
bridge type reclaimer 桥式取料机
bridge type screw unloader 桥式螺旋卸车机
bridge type unloader 桥式卸料机
bridge underpass 桥下孔道
bridge wall 挡火墙
bridge weir 桥式堰
bridge weld 接合板焊接
bridge wire 电桥标准导线
bridge works 桥梁建筑，桥梁工程
bridge 桥接线，桥，电桥，搭桥，跨接，平台，步道，桥梁，桥台，越过，中横担
bridging action 堵塞作用，架桥作用
bridging amplifier 桥式放大器
bridging bar 桥接杆
bridging beam 渡梁，横梁
bridging coil 桥接线圈
bridging condenser 桥接电容器，并联电容器，隔直流电容器
bridging contact 桥接触点，分路触点
bridging deposit 搭桥积灰
bridging finance 过渡性融资
bridging jack 桥接塞孔，并联塞孔
bridging joist 渡梁，横梁
bridging line 桥接线，跨接线
bridging loan 过渡性贷款
bridging multiple switchboard 并联复式交换机

bridging over flue 悬空烟道
bridging over 敷设，架设，跨越
bridging piece 挑板
bridging plug 过 桥插头
bridging relay contact 短接继电器接点
bridging wiper 桥接线刷，并接弧刷
bridging 形成灰桥，分路，分流，跨接，桥接，堵塞，搭桥，架桥
bridle hitch （锚索与拉索间的）鞍形连接装置
bridle iron 钢支座，箍筋
bridle road 大车道
bridle rod 拉杆
bridle rope 拉索
bridle wire 跳线，绝缘跨接线
bridle 拖绳，限动器，承接梁，束带，托梁
brief acceleration 短暂加速
brief analysis 简易分析
brief appraisal 简评
brief career 简历
briefcase company 皮包公司
brief description 简介
brief dip-wash 短暂浸洗
brief introduction 简介
brief investigation 简易调查
brief 提要
Briggs pipe thread plug 布立格管螺纹塞
bright annealing 光亮退火
bright bolt 精制螺栓，光制螺栓
bright coal 亮煤，烟煤
bright-dim lamp 全亮及半亮二用指示灯
bright emitter 白炽热发射电子管，高能热离子管
brightening pulse 照度脉冲，辉度脉冲，照明脉冲
bright heat treatment 光亮热处理
bright line 明线，亮线
brightness 光泽，亮度，辉度
brightness coefficient 亮度系数
brightness concentration 亮度聚光度
brightness-modulated cathode ray tube 辉度调制阴极射线管
brightness modulation 亮度调制
brightness of illumination 照明亮度
brightness pulse 照明脉冲，亮度脉冲
brightness-temperature pyrometer 亮度-温度高温计，光学高温计
brightness-temperature 亮度温度
brightness-voltage characteristic 亮度电压特性
bright paint 明亮的油漆
bright quenching 光亮淬火
bright spot 亮点
brights 亮煤质
briguette 标准试块
brilliance 辉度，亮度
Brillouin flux density 布里渊磁通密度
brimstone acid 硫酸
brine compartment 盐水室
brine concentration 盐水浓度
brine cooling system 盐水冷却系统
brine disposal 含盐废水处理，海水处理
brine freezing 盐水冷冻
brine gauge 盐度计
brine gauging tank 盐水计量箱
brine heater 饱和盐溶液加热器，盐水加热器
brine heat exchanger 盐水换热器
Brinell hardness number 布氏硬度值，布氏硬度（数）
Brinell hardness tester 布氏硬度计
Brinell hardness test 布氏硬度试验
Brinell hardness 布氏硬度
Brinell tester 布氏硬度计，布氏硬度试验
brine loop 盐水回路
brine pipe 盐水管
brine pumping 盐水提抽
brine regenerant 盐水再生剂
brine salt 卤盐
brine solution 盐溶液
brine system 盐水系统
brine tank 盐液箱，盐液槽
brine 盐水，海水，卤水，加盐处理
bring about 导致，引起，造成
bring a contract into effect 使合同生效
bring along 带动
bring China's economy in line with international practice 使中国经济与国际接轨
bring down 收缩，浓缩，下降
bring forward 提出
bringing into service 投入运行
bringing on stream （汽轮机）开始转动，投入生产
bringing onto load 带负载
bringing up 提升功率
bring in phase 使同相
bring into effect 贯彻，实施
bring into operation 开动，投入运行，使运转
bring into production 投产
bring into service 使工作，使运行，投入运行
bring into step 使同步，整步
bring…into 使达到……，把……拿入，使开始生效
bring in 把……拿进来，带进来
bring on 呈现……，使出现，使发展
bring-out （管道）向外引出，引出（接线）
bring to critical 使达临界，趋向临界
bring together 集合，汇集，组装
bring to power 提升功率
bring to rest 停车，停止运行
bring to （使）停驶，使用于，使达到
bring up a claim 提出索赔
bring up 提出，谈到
brinishness 含盐度
brink depth 边缘水深
briny 咸的
briquette mixture 团块混合物
briquette press 水泥试块成型机
briquette tension test 水泥试块张拉试验
briquette 煤砖，型煤，煤团，坯块，团块，煤饼，煤球，试块，水泥标准试块
briquetting 团块，型煤，压块，压制成块
briquet 水泥标准试块，煤饼
brisance 爆炸威力
brisk market 繁荣的市场，市场活跃

brisk sales 畅销
British Association ohm 英制欧姆
British horsepower 英制马力
British Standard Fine Thread 英国标准细牙螺纹
British standard sieve 英国标准筛
British Standards Institution 英国标准协会
British Standard Wire Gauge 英国标准线规
British Standard 英国标准
British thermal unit 英国热量单位
British ton 英吨
British Welding Research Association 英国焊接研究协会
brittle coating 脆性涂层，脆性涂料
brittle crack initiation CTOD 脆性启裂裂纹尖端张开位移值
brittle crack 脆裂
brittle-ductile transition 脆性延性转变
brittle failure 脆性破坏，脆断，脆性故障
brittle-fracture-oriented operating diagram 以脆性断裂为基准的运行图
brittle fracture resistance 抗脆断强度
brittle fracture strength 脆断强度
brittle fracture stress 脆性断裂应力
brittle fracture surface 脆性断裂面，脆性断口
brittle fracture test 脆性断裂试验
brittle fracture 脆性断口，脆性断裂，脆裂
brittle instability CTOD 脆性失稳裂纹尖端张开位移值
brittle material 脆性材料
brittle metal 脆性金属
brittle model 脆性模型
brittleness index 脆性指数
brittleness 脆性，脆度，脆化
brittle point 脆裂点，脆化点，脆折点
brittle rock 脆性岩石
brittle rupture 脆裂，脆性破裂
brittle substance 脆性物质
brittle temperature 脆化温度
brittle zone 脆化区
brittle 脆的，易脆的，功能不可靠的
BRK(brake) 制动器
BRKR(breaker) 断路器
BRL 锅炉额定负荷
BRM(barometer) 气压计
broach 三角锥，扩孔器，宽凿
broaching 扩孔，打眼，撬挖
broach post 桁架中竖杆
broach roof 尖塔屋顶
broad-area electrode 大面积电极
broad band amplifier 宽频带放大器
broad band excitation 宽带激励
broad band matching circuit 宽频带匹配电路
broad band response 宽带响应
broad band turbulence 宽带湍流
broad band ultrasonic pulse 宽频带超声脉冲
broad band video detector 宽带视频检波器
broad band 宽带，宽波段
broad base terrace 宽底地埂，宽广台地
broad base tower 宽基塔，分腿基础输电线路塔
broad beam 宽缘梁，宽束
broad-bottomed 平底的
broadcast command 广播命令
broadcasting frequency 广播频率
broadcasting room 广播室
broadcast relaying 广播转播，广播中继
broadcast transmission 广播传输
broadcast transmitter 广播发射机
broadcast 广播
broad cooperation 广泛合作
broad-crested measuring weir 宽顶量水堰
broad-crested weir 宽顶堰
broadened cross-section 加宽横断面
broad-flanged beam 宽翼工字钢，宽翼梁，阔翼梁
broad jet burner 宽喷口喷燃器
broadleaf evergreen 常绿阔叶树
broad pulse 宽脉冲
broad range 宽分布
broad resonance 宽共振
broad scale flow 大尺度流动
broadside array 垂射天线阵，端射天线阵
broadside directional antenna 边射天线，垂射天线，同相天线
broadside method 永久磁铁磁矩决定法
broadside slipway 侧边滑道
broadstep 楼梯踏板，平台楼梯
broad stone 石板
broad tuning 宽调频，粗调
brocatelle 彩色大理石
brochure 商品说明书，(产品)小册子
broken 断开的，断开
broken ashlar masonry 不等形琢石圬工
broken blade 断裂叶片，断叶片
broken circuit 断路，开路
broken coal 煤块
broken concrete groin 碎块混凝土丁坝
broken crest 折线堰顶
broken curve 虚线曲线
broken-down 坏了的，临时出故障的，损毁
broken face 裂面
broken ice 碎冰
broken joint 错缝结合，错缝
broken line 虚线，折线，断线，破线
broken number 分数
broken out section 破断面，切面
broken period 非标准期限
broken range masonry 断层石砌
broken shaft 断轴
broken sky 裂云天空
broken stone road 碎石路
broken stone stratification 碎石层
broken stone 碎石
broken stowage 亏舱
broken surface 断裂面
broken tap 截流龙头
broken top-chord 折现上弦，折线形上弦
broken water 浪花
broken wire condition 断线工况
broken wire 断线
brokerage fee 佣金，回扣
brokerage 中间人业务，经纪业
broker 经纪人，中介

bromate 溴酸盐
bromic acid 溴酸
bromide of lithium 溴化锂
bromide paper 放大纸
bromide streak 溴化物条斑
bromide 溴化物
bromine water 溴水
bromine 溴
bromocresol green 溴甲酚绿
bromocresol purple 溴甲酚紫
bromophenol blue 溴酚蓝
bromthymol blue indicator 溴百里酚蓝指示剂，溴麝香草酚蓝指示剂
bronchial 支气管的
brontograph 雷暴自计器，雷雨计
brontometer 雷雨表，雷暴计
bronze alloy 青铜合金
bronze bar 青铜棒
bronze brush 青铜电刷
bronze casting 青铜铸件
bronze conductor 青铜线
bronze-graphite bearing 青铜石墨轴承
bronze guide 青铜导板，青铜导承
bronze leaf brush 青铜片电刷
bronze medal 铜制奖（章）
bronze pipe 青铜管
bronze thrust collar 青销推力轴环，青铜止推环
bronze tube 青铜管
bronze welding 青铜焊
bronze weld 青铜焊焊缝
bronze wire 青铜线
bronze 青铜，青铜色，镀青铜
Brookfield viscometer 布氏黏度计
brooklet 细流
brook 溪，溪流，小溪
broom finish 扫面处理，帚处理，帚面
brooming 柱顶裂缝
broomstick 干扰抑制器
broom 自动搜索干扰振荡器
brothers 双股吊索
brought out to terminal block 引至端子板
brought out 引出，显示出
brow 边线
brow leakage 端部漏磁
Brown and Sharpe wire gauge 布朗夏普线规
brown cloud 棕色烟云
brown coal 褐煤，高水分褐煤，次褐煤
brown coal boiler 褐煤锅炉
brown coal briquette 褐煤砖
brown coal hydrogasification 褐煤气化
brown coat 二道抹灰
brown earth 棕壤
brown forest soil 棕色森林土
brown fume 棕色烟雾
brown-glazed brick 防潮砖，褐釉砖
brown haze 棕色轻雾
Brownian diffusion 布朗扩散
Brownian motion 布朗运动
Brownian movement 布朗运动
brown lime 褐石灰，棕石灰
brown madder 褐红
brownmillerite 钙铁石
brownout 电压不足，节约用电，灯火管制，减少供电
brown oxide 铀氧化物
brown spar 铁白云石，铁菱镁矿
browser 浏览器
BRS(boron recycle system) 硼回收系统，硼再循环系统
brucine 番木鳖碱
brucite 水滑石，水镁石
bruise resistance 抗机械损伤性能
bruiser 捣碎机
bruise 机械损伤
Brunt-Vaisala frequency 布伦特-维塞拉频率
brush 电刷，刷子，梢料，刷，刷除，擦光【除去多余的渣】
brush aggregate 粗糙骨料
brush and cable bank protection 梢料锚索护岸
brush and pile dike 梢料木桩堤
brush and spray discharge 刷形放电
brush angle 电刷倾角
brush arc 电刷弧度，刷弧
brush arm 刷臂
brush assembly 电刷装配
brush backward lead 电刷反向超前
brush box 刷盒
brush bracket 电刷架
brush burn 擦伤
brush carbon 电刷碳
brush-carriage 电刷托架
brush carrier ring 刷架环
brush carrier 刷握，刷架
brush changing 电刷更换
brush chattering 电刷震颤，电刷振动
brush clamp 电刷夹，电刷接线端
brush coat 刷敷涂层，刷涂
brush collector 集电刷，集流刷
brush contact loss 电刷接触损耗
brush contact resistance 电刷接触电阻
brush contact surface 电刷接触面
brush contact voltage 电刷接触电压
brush contact 电刷接触，电刷触点
brush cord 电刷软绳，电刷瓣
brush corona 刷电晕
brush covering factor 电刷覆盖系数
brush current 电刷电流
brush curve 电刷曲线
brush dam 梢料坝
brush dike 梢料堤
brush discharge 电晕放电，刷形放电
brush displacement 刷移角，电刷位移
brush drain 梢捆排水沟
brushed finish 粉制，刷面，粉刷，刷饰面
brush edge 电刷刷边
brush electrode 刷式电极，电刷
brush encoder 电刷编码器
brush end 电刷端面
brush film 电刷膜
brush finger 电刷压指，刷指
brush finish 粉刷
brush for hydrogen-cooling use 氢冷电刷

brush for use in hydrogen environment 氢环境电刷
brush forward lead 电刷正向超前
brush friction loss 电刷摩擦损失
brush friction 电刷摩擦
brush front 前刷边
brushgear housing 电刷座
brushgear lead 电刷引线
brushgear yoke 电刷架，电刷座
brushgear 电刷装置
brush glowing 电刷灼热，电刷红热
brush grade 电刷牌号
brush heel 电刷滑入边
brush-holder arm 刷臂
brush-holder plate 电刷架
brush-holder rod 刷杆，刷臂
brush-holder spring 电刷弹簧，刷握弹簧
brush-holder stud 刷握支柱，刷杆
brush-holder support 刷握支架
brush-holder yoke 刷握支架
brush holder 电刷座，刷握
brush inclination 电刷倾度，电刷倾角
brushing machine 刷洗机
brushing quality 涂刷质量
brushing test 刷损试验
brushing unit 刷洗装置
brushing 火花束放电，刷尖放电，刷光，刷洗
brush interference 电刷干扰
brush joint 电刷接头，帚状节理
brushland 丛林地，灌木地，矮丛林地
brush lead 电刷移前，电刷引线
brushless cascade alternator 无刷串级同步发电机
brushless DC electric motor 无刷直流电动机
brushless direct current motor 无刷直流电动机
brushless doubly fed induction generator 无刷双馈异步发电机
brushless excitation system 无刷励磁系统
brushless excitation 无刷励磁
brushless exciter 无刷励磁机
brushless generator 无刷发电机
brushless motor 无刷电动机
brushless power electronic circuit 无刷电力电子电路
brushless synchronous generator 无刷同步发电机
brushless thyristor excitation 无刷可控硅励磁
brushless 无电刷的，无刷的
brush life 电刷寿命
brush-lifting device 举刷装置
brush-lifting motor 举刷型电动机
brush loss 电刷损耗
brush mark 刷痕
brush mattress 梢料垫层，柴排
brush noise 电刷噪声
brush off 刷清
brush on 涂刷
brush paving 梢料护面
brush plating 刷镀
brush position 电刷位置
brush pressure 电刷压力
brush rack 碳刷架
brush resistance loss 电刷电阻损耗
brush resistance 电刷电阻
brush ring 刷环
brush-rocker ring 移动刷架环
brush-rocker 电刷摇移环，移动刷架
brush rod 刷杆，刷臂
brush sealing 刷子汽封
brush shifting control 移刷调整
brush shifting motor 移刷型电动机
brush shift 电刷调整，电刷移位
brush shoe 刷靴
brush shoulder 刷肩
brush shunt 电刷连线
brush slot 电刷槽
brush spacing 电刷间隔
brush spark 电刷火花
brush spindle 电刷柄
brush spring 电刷弹簧
brush square shaft 刷架方轴
brush station 电刷测量点，电刷站
brush stud 电刷螺杆
brush supporter 刷架
brush surface analyzer 电刷表面分析仪
brush surface discharge 刷面放电
brush toe 后刷边，刷趾
brush track 刷迹圈
brush treatment 涂刷防腐剂，电刷处理
brush-type cleanup 清扫刷
brush voltage 电刷电压，刷电压
brush wastage 碳刷磨损
brush wear 电刷磨损
brush wheel 刷轮
brush width 电刷宽度
brush wire 电刷引线，刷用钢丝
brush with tail 带刷辫电刷，带尾电刷
brush wood check dam 梢捆挡水坝
brush wood dam 梢料坝
brush wood 梢捆，梢料
brush yoke 电刷架，独立刷杆座
brute force filter 倒L形滤波器，平滑滤波器，脉冲展平滤波器
B & S(beams and stringers) 横梁与纵梁
BS(British Standard) 英国（工业）标准，英标
BS(bunker surcharge) 燃油附加费
BSD(Boiler Shut Down) 停炉
BSD(British Standard Dimension) 英国度量标准
BSF(back surface field) 背表面场，背面场，背场
BSF solar cell 背场太阳电池
BSI(British Standard Institute) 英国标准协会
B-source 板极电源，阳极电源
BST(build-subsidy-transfer) BST 模式，“建设-补贴-移交”模式【工程建设模式】
BSW(bus switch unit) 总线开关部件
BTB(back-to-back converter station) 背靠背换流站
BT(boiler tube) 炉管
BT(build-transfer) BT 模式，“建设-转让”模式【工程建设模式】
BTG(boiler-turbine-generator) 锅炉-汽轮机-发电机，电厂三大主机
BTG control panel 锅炉-汽轮机-发电机控制盘，

BTG控制盘，三大主机控制盘
BTG panel BTG盘，主机控制盘，机组控制盘
BTG 锅炉-汽机-发电机【DCS画面】
BTO(build-transfer-operate) BTO模式，“建造-移交-运营”模式【工程建设模式】
BTRC 生物质能研究中心
BTS(book of technical specification) 技术规范书
BTU(British Thermal Unit) 英国热量单位，英热单位
B type insulator B型绝缘子
bubble assemblage model 气泡汇聚模型
bubble axis length 气泡轴向长度
bubble axis 气泡轴
bubble breakdown in dielectric liquid 液体电介质气泡击穿
bubble breaking surface 气泡破裂表面，气泡破裂面
bubble breakup 气泡破裂
bubble cap distributor 泡罩分布器
bubble cap plate 泡罩板
bubble cap （沸腾炉）风帽，泡罩
bubble center 气泡中心
bubble clouding 气泡晕，气泡群集
bubble cluster 气泡群
bubble coalescence 气泡合并，气泡重合
bubble collapse 气泡破裂，气泡消失
bubble collision frequency 气泡碰撞频率
bubble column chromatography 气泡柱色谱法
bubble company 皮包公司
bubble concrete 泡沫混凝土
bubble control 气泡控制
bubble correction 水准改正
bubble crowding 气泡群
bubble delay time 气泡滞后时间
bubble density 气泡密度
bubble departure diameter 气泡跃离直径
bubble detachment frequency 气泡跃离频率
bubble detachment point 气泡跃离点
bubble detachment 气泡跃离
bubble detection 气泡探测法
bubble deviation 气泡偏差
bubble diameter 气泡直径
bubble disengagement 气泡脱离
bubble domain 泡畴，磁泡
bubble dynamics 气泡动力学
bubble eccentricity 气泡偏心率
bubble elimination circuit 消除气泡电路
bubble embryo 气泡胚
bubble eruption diameter 气泡喷出直径
bubble eruption 气泡爆裂
bubble escape 气泡逸出
bubble-filled bed 充满气泡的床层
bubble film 气泡膜
bubble flow 气泡状流动，泡状流
bubble formation 气泡形成，气泡生成
bubble former 起泡器
bubble-free operation 无气泡操作
bubble frequency 气泡频率
bubble gauge 气泡水准仪
bubble growth 气泡长大
bubble hit 气泡碰撞
bubble hole-up 气泡滞留量
bubble hydrodynamics 气泡流体力学
bubble layer 气泡层
bubble margin 气泡边沿
bubble migration 气泡迁移
bubble motion 气泡运动
bubble noise （由蓄电池电液气泡所引起的）气噪声
bubble nucleation 气泡核形成
bubble orientation 气泡取向
bubble phase 气泡相
bubble plate evaporator 泡罩板蒸发器
bubble plate 泡罩板
bubble point 始沸点，鼓泡点，泡点
bubble population 气泡群
bubble pressure method 气泡压力法
bubble profile 气泡外形
bubble quadrant 带气泡水准测角器，气泡水准测角器
bubble reading 气泡读数
bubble rise velocity 气泡上升速度
bubbler level sensor 起泡层高度传感器
bubble roof 气泡顶部
bubbler 气泡器扩散器，起泡器，鼓泡器，喷水器
bubble scrubber 鼓泡洗涤塔
bubble sextant 气泡六分仪
bubble shape 气泡形状
bubble size distribution 气泡尺寸分布
bubble slip velocity 气泡滑移速度
bubble splitting 气泡分裂
bubble stream 气泡流
bubble street 气泡串
bubble structure 气泡结构
bubble suction 空泡吸力
bubble suit 潜水服
bubble swarm 气泡群
bubble test 气泡试验
bubble track 气泡迹
bubble tube 气泡管，水准管
bubble tube pressure sensing device 吹气式压力检测装置，吹气式压力检测
bubble-type gas distributor 泡罩式气体分布板
bubble-type pneumatic gauge 气泡式气压计
bubble velocity 气泡速度
bubble viscometer 泡沫黏度计
bubble void 气泡穴
bubble volume coefficient of reactivity 气泡的反应性体积系数
bubble volume 气泡体积
bubble wake 气泡尾迹
bubble 起泡，水泡，气泡，磁泡，起沫，泡沫
bubbling bed 鼓泡床，泡床【沸腾炉】
bubbling behaviour 气泡行为【沸腾炉】
bubbling cap 风帽【沸腾炉】
bubbling fluidization 鼓泡流态化
bubbling fluidized bed boiler 鼓泡床锅炉，鼓泡流化床锅炉
bubbling frequency 鼓泡频率
bubbling point 沸点
bubbling polymerization 气泡聚合

bubbling region 鼓泡区
bubbling tower 鼓泡塔
bubbling velocity 鼓泡速度【沸腾炉】
bubbling with air 空气鼓泡
bubbling zone 鼓泡区
bubbling 起泡【焊接缺陷】，冒气【指蓄电池】，沸腾，鼓泡，冒泡
bubbly flow 泡状流【沸腾炉】
bubbly-slug flow 泡状团状流
bubbly state 鼓泡状态【沸腾炉】
BUC(buffer controller) 缓冲控制器
buchholtz peotection 巴克霍尔茨保护
buchholtz relay 巴克霍尔茨继电器
buck and boost regulator 降压升压调节器
buck and boost transformer 加减电压变压器，降压升压变压器
buck arm 转角横担
buck-boost control signal 正反控制信号，加减电压控制信号
bucker unloader 斗链卸车机
bucker 粉碎机，破碎器
bucket attachment 戽斗连接
bucket auger 勺形钻
bucket basin 戽斗式消力池
bucket-boom excavator 多斗臂式挖土机
bucket-brigade delay line 斗链式延迟线
bucket-brigade device 斗链式（电荷耦合）器件
bucket capacity 斗容
bucket carrier 斗式运料车，斗式转运机
bucket chain tightening device 斗链张紧装置
bucket chain unloader 链斗卸车机
bucket conveyer(-or) 斗式运输机，链斗式输送机
bucket cover 围带
bucket crane 吊斗起重机，抓斗起重机
bucket curve of spillway 溢洪道消力戽曲线
bucket dredger 斗式挖泥船，斗式挖泥机
bucket elevating method 斗式提升法
bucket elevator 斗式升降机，斗式提升机，链斗式提升机
bucket excavator for downward scraping 下挖式多斗挖土机
bucket excavator 斗式挖掘机，斗式挖土机
bucket grab 挖土机抓斗，抓斗
bucket height 桶高
bucket hoist conveyor 斗式提升输送机
bucketing 成组
bucket invert 戽斗反弧拱
bucket ladder dredger 多斗挖泥船
bucket ladder excavator 多斗式挖掘机
bucket ladder 斗架【多斗挖泥船】
bucket line 斗链【多斗挖掘机】
bucket lip 挑坎唇部，鼻坎
bucket loader 斗式装料机
bucket locking piece 叶片锁块
bucket loss coefficient 动叶片损失系数
bucket of water wheel 水轮戽斗
bucket pump 斗式提水机
bucket rain gauge 倾斗雨量计
bucket rig 吊罐索具，戽斗索具
bucket ring 动叶环
bucket rock grab 铲斗式抓岩机
bucket roller 戽斗水辊
bucket root 叶根
bucket rubbing 叶片摩擦
buckets pitch 斗距
bucket teeth 铲斗齿，戽斗齿
bucket tipping device （斗式提升机的）翻斗装置
bucket trenching machine 斗式挖沟机
bucket-type energy dissipater 戽斗式消能工
bucket type strainer 桶式过滤器
bucket unloader 斗式卸料机
bucket wheel excavator 戽斗转轮挖土机，斗轮挖掘机
bucket-wheel loader 斗轮装卸机
bucket wheel machine 斗轮机
bucket wheel reclaimer 斗轮取料机
bucket wheel stacker-reclaimer 斗轮（式）堆取料机
bucket wheel stacker 斗轮堆料机
bucket wheel 斗轮，戽斗转轮
bucket 吊罐【浇混凝土用】，桶，斗，铲斗，戽斗，水斗，水桶，勺斗桶，动叶，吊桶
buck frame 背有凹槽的门框
buckle 弯曲，扭歪，扣，箍，起皱
bucking circuit 补偿电路，偏移电路，抵消电路
bucking coil 反接线圈，抵消线圈，补偿线圈，反感应线圈
bucking effect 反电动势效应，抵消作用
bucking electrode 屏蔽电极
Buckingham's theorem 布金汉定理
bucking-out system 补偿系统
bucking potential 抵消电势，补偿电压，反作用电势
bucking saw 造材锯
bucking signal 抵消信号
bucking voltage 抵消电压，反作用电压
bucking winding 补偿绕组，去磁绕组，抵消绕组
buckle fold 挠曲褶皱，弯曲褶皱
buckle insulator 茶台绝缘子
buckle outward 向外翘曲
buckle plate 凹凸板，压曲板
buckle 扣住，使变弯，扣环，箍
buckling calculation 曲率计算
buckling constant 曲率常数
buckling factor 曲率因子，曲率因数
buckling failure 压曲破坏
buckling instability 曲率不稳定性
buckling iteration method 曲率迭代法
buckling length 能挠长度
buckling load 纵向弯曲荷载，压曲临界荷载，压曲荷载，屈曲荷载，屈曲负载
buckling resistance 抗屈强度
buckling stability 压曲稳定性
buckling strain （纵向）弯曲应变
buckling strength 抗压弯强度，抗挠强度，抗翘曲强度，抗屈强度，抗弯强度，抗弯曲强度
buckling stress 纵弯曲应力，压曲应力，弯曲应力，抗弯应力
buckling test 纵向弯曲试验

buckling vector 曲率矢量
buckling 屈曲，翘曲，纵弯曲，曲率，挠，弯曲，挠度，弯折，纵向挠曲，纵向弯曲
buck or boost 降压或升压，增或减
buckover （换向器的）环火闪烁，发生环火
buck scraper 弹板刮土机，横板刮土机
buckshot aggregate 大团粒
buckshot sand 圆粒砂
buckstave 前墙支柱，夹炉板
buckstay spacer 刚性拉撑垫块
buckstay 砖砌中的位条，支撑，支柱，刚性梁，刚性架，拱边支柱，加固圈
buck 冲击，撞击，消除，补偿，反向，反极性，大装配架，锯架，门边立木
budding 局部屈曲【指钢结构】
budget allocation 预算拨款
budgetary estimate adjustment 概算调整
budgetary estimate chart 概预算表
budgetary estimate of power generation project 发电工程投资概算
budgetary estimate 概算
budget audit 预算审查
budget control 预算控制
budget cost 预算成本，预算价，预算费用
budget deficit 预算赤字
budget document 预算文件，预算书
budgeteering 预算编订，编制预算，制定预算
budget engineer 预算工程师
budget estimate 概算
budget for revenues and expenditures 收支预算
budgeting 编制预算
budget item 预算项目
budget line 预算线
budget making 编制预算，预算编制
budget management 预算管理
budget method 预算法
budget of power generation project 发电工程投资预算
budget report 预算报告
budget sheet 预算表
budget year 预算年度
budget 预算，堆积，积聚，预算（表）
buffer action 阻尼作用，缓冲作用，隔离作用
buffer address register 缓冲器地址寄存器
buffer amplifier 隔离放大器，缓冲放大器
buffer and refresh memory 缓冲与重显存储器
buffer area 缓冲区
buffer bar 缓冲杆
buffer battery 浮充电池组，缓冲电池组
buffer beam 缓冲梁，缓冲杆
buffer block 缓冲块
buffer capacitor 缓冲电容器
buffer capacity 缓冲容量，缓冲能力
buffer chain 缓冲链
buffer circuit 阻尼电路，缓冲电路
buffer coil 缓冲线圈
buffer controller 缓冲控制器
buffer control unit 缓冲控制装置
buffer device 缓冲装置
buffer drum 缓冲磁鼓
buffered computer 缓冲存储的计算机
buffered I/O channel 有缓冲的输入/输出通道
buffered terminal 有缓冲的终端
buffer function 缓冲作用
buffer gap 衬垫间隙，缓冲间隙
buffer gas 缓冲气体
buffer gate “或”门
buffering capacity 缓冲容量，缓冲量
buffering substance 缓冲物质
buffering 缓冲，阻尼，减震，中间转换，缓冲记忆装置，缓冲作用
buffer interleave controller 缓冲器交错控制器
buffer inverter 隔离反相器
buffer layer 缓冲层，过渡层，中间层
buffer module 中间转换，中间转换器，缓冲器模件
buffer oil tank 缓冲油箱
buffer pool 缓冲池
buffer post 缓冲杆
buffer reagent 电机液缓冲剂
buffer register 缓冲寄存器
buffer report control block 缓存报告控制块
buffer report control class 缓存报告控制类
buffer report 缓存报告【功能约束】
buffer resistance 放电电阻，缓冲电阻
buffer section 缓冲段
buffer shooting 缓冲爆破
buffer solution 缓冲溶液，缓冲液
buffer spring 缓冲弹簧
buffer stop 轨道止挡器，车挡【铁路】，止冲器
buffer storage 缓冲存储器，缓冲存贮器
buffer store 缓冲贮存器
buffer strip 缓冲带
buffer substance 缓冲物质
buffer tank 缓冲池，缓冲箱，中间贮存箱
buffer unit 缓冲部件
buffer water tank 缓冲水箱
buffer zone 缓冲区
buffer 减振器，缓冲器，阻尼器，缓冲剂，缓冲垫，缓冲，阻尼，缓冲装置
buffeting characteristic 颤振特性
buffeting deflection 抖振烧度
buffeting excitation 抖振激励
buffeting factor 抖振因子
buffeting limit 颤振极限
buffeting response 抖振响应
buffeting 抖动，打击，颤振，抖振，冲击，颤震
buffing wheel 抛光轮
bug dust 筛屑，粉尘
buggy 小斗车
bug hole 表面气泡，表面斑眼【混凝土】，晶穴
bug key 双向报键，快键
bug 缺陷，误差，故障，损坏，干扰
buhrstone 磨石
build 建造，建型
build down 降落，衰减，降低
builder's diary 施工员日志
builder's equipment 施工设备
builder's level 施工水准仪
builder's square 施工用角尺
builder 建造者，建筑施工单位，施工人员
build-in calibrator 内部校准器，机内校准器

build-in check 内部校验
building act 建筑法规
building and health code 建筑物与卫生法规
building area 建筑面积
building berth 船台
building block design 模块化设计
building block principle 组装原理，拼装原理
building block steam turbine 积木式汽轮机
building block system 积木式，积木式系统，插入式程序系统
building block 积木，组成部件，标准组件，积木式构件，砌块，预制构件，积木式部件
building by-law 建筑条例附则
building code 建筑条例［规则、标准］，建筑规范［法规］
building coefficient 建筑系数
building complex 建筑群，综合建筑
building construction 房屋构造，建筑施工
building coordinate 建筑坐标
building cost 建筑造价
building coverage of production 厂区建筑系数
building coverage 建筑覆盖率，建筑系数
building datum line 建筑基线
building demolition 建筑拆除
building density 房屋密度，建筑密度
building depth 建筑进深，建筑深度
building design institute 建筑设计院
building drainage system 厂房疏水系统，房屋排水系统
building drain 房屋排水管，建筑排水
building envelope 围护结构
building exhaust air 厂房排气
building flow zones 建筑气流区
building foundation pit 建筑基坑
building frame 房屋构架，建筑框架
building gable 房屋山墙
building ground conductor 建筑物接地导体
building ground 建筑工地
building heating entry 热力入口
building height 建筑层高
building implements 建筑设备，建筑用具
building inspection 房屋检查，建筑检查
building investment 建筑投资
building law 建筑法规
building licence 建设许可证，建筑执照
building life factor 建筑寿命因子
building line 房基线，房屋界线，建筑线，建筑红线
building material-integrated PV module 建材型光伏组件
building material storage 基建材料库
building material 建筑材料
building model 建筑模型
building occupation factor 厂房占地因数
building office 建筑事务所
building ordinance 建筑法规，建筑条例
building-out capacitor 附加电容器
building-out circuit 补偿电路，匹配电路
building-out network 补偿网络，附加（平衡）网络
building-out 附加的，补偿的
building paper 保温纸，防潮纸，油毛毡覆面层，油纸
building permit 建设工程规划许可证，建筑施工执照，建筑许可（证），建筑执照
building pit （建筑物）基坑
building plot 房屋地区图
building reconstruction 房屋重建
building regulation 建筑规程
building research institute 房屋建筑研究所
building settlement 建筑物沉降
building sewer 室内污水排出管
building shape 建筑体型
building site 建筑地址，工地，建筑场址，施工现场
building standard 建筑标准
building stone 石材
building storm drain 房屋雨水管
building structure 建筑物结构，建筑结构
building survey 建筑测量
building tile 屋瓦
building timber 建筑木材
building-to-ground ratio on site 施工区用地建筑系数
building-type switchboard 组装式配电盘
building-up curve 增长曲线
building-up of voltage 电压建起，电压升起
building-up period 建立时间，积累周期
building-up time 增长时间，建立时间
building-up 上升，堆焊，建成，组成，安装，加强
building wake effect 建筑物尾流效应
building wake region 建筑物尾流区
building yard 建筑场地
building 建筑，建筑物，厂房
build-in inertia 固有惯性
build-in oscillation 固有振荡
build-in reliability 结构可靠性，固有可靠性，剩余反应堆，固有反应性，后备反应性
build-in strain 内应变
build-in stress 装入零件时产生的应力
build in terrain 建成地形
build-in thermocouple 埋入式热电偶
build-in 内设的，机内的，固有的，嵌入
build joint 构造缝
build-operate-transfer approach “建造-运行-移交”建设方式，BOT 模式
build-up a market 开辟市场
build-up area 建成区
build-up effect 聚集效应
build-up factor 积累因子，增长因子
build up market 拓展市场
build-up member 装配部件
build-up mica 人造云母
build-up of voltage 电压建立，电压增长
build-up sequence （焊接）熔敷顺序，形成顺序
build up the funds 筹措资金
build-up time 建立时间，延长时间，上升时间
build-up welding 堆焊
build-up 建立，建造，树立，加厚，增长，增强，结瘤，积聚，积累，集结，堆焊，形成，安装，积料，堆积，组成，组合，聚集，结

垢，构造，堵塞
build up 逐步增进
built-beam 固端梁
built-in antenna 内装天线
built-in arch 固端拱，嵌固拱
built-in bathtub 镶入式浴盆
built-in beam 固端梁，嵌固梁
built-in check 自动效验，内部效验，内部校验
built-in classifier 内置式粗粉分离器
built-in cupboard 壁橱
built-in edge 固定边，嵌固边缘，嵌入边
built-in electric field 内建电场
built-in end 嵌入端，固定端，嵌固端
built-in error correction 内部误差校正
built-in evaluation 内在评价
built-in field 内建电场
built-in fitment 固定件
built-in fittings 预埋件，埋设件，埋入件，固定件
built-in flow circuit 内装流路
built-in function 内部操作，内部功能
built-in furniture 镶壁家具
built-in inertia 固有惯性
built-in lamp 隐藏式灯，墙内灯
built-in length 埋入长度
built-in lighting 隐藏式照明
built-in meter 内装仪表
built-in motor 机内电动机，内装电动机，自附电动机
built-in oscillation 固有振荡
built-in oxide field 氧化物内建电场
built-in place components 内部元件
built-in potential 内建电势，自建电势
built-in preload 初始压力，固有的预压
built-in radiator 墙内散热器
built-in reactivity 后备反应性
built-in reference voltage 自带参考电压
built-in reliability 内在可靠性，固有可靠性，结构可靠性
built-in stress 参余应力，内部应力
built-in test 内部测试
built-in thermocouple 埋入式热电偶
built-in voltage 内建电压
built-in wood block 预埋木砖
built-in wooden brick 预埋木砖
built-in Zener protection 内建齐纳管保护
built-in 嵌入的，内置的，固有的，内装的，埋设的
built-on pump 固接式泵
built pile 组合桩
built platform 堆积台地
built-up angle 组合角钢
built-up arch 组合拱
built-up area 已建成区，回填区，建筑占地面积，已建面积
built-up backpressure 增长背压
built-up beam 组合梁
built-up channel 组合槽钢
built-up column 组合柱
built-up connection 转接
built-up diaphragm 组装式隔板
built-up die 组合模具
built-up disk rotor 轮盘组合式转子，套装叶轮转子
built-up environment 建成环境
built-up flat roofing 卷材屋面
built-up frame 组合构架
built-up gear 组合齿轮
built-up girder 组合大梁
built-up H-column 组合 H 形柱
built-up magnet 组合磁铁
built-up member 组合构件，装配件，组合杆
built-up nozzle 组装式喷嘴
built-up pile 组合桩
built-up rigid frame 组合构架
built-up roofing 组合屋面
built-up rotor 组合式转子，组装式转子，组装转子
built-up seat 组合座【硬表面】
built-up section 组合截面，组合型钢
built-up steel section 组合型钢截面
built-up truss 组合桁架
built-up weld deposit 堆焊
built-up weld 堆焊
built-up 组合的，可拆卸的，积木式的，装配的，建立的
bulb heating 灯泡加热
bulb of pressure 压力泡
bulbous portion （温度计的）测温包
bulbous root 球形叶根
bulbous zone of stress 球面应力区
bulb pile 葱头桩，护孔桩，扩底桩，球形护脚桩，球形桩
bulb resistance 灯泡电阻
bulb turbine 灯泡式水轮机
bulb type generator 灯泡式发电机
bulb type hydro generating set 灯泡式水力发电机组
bulb type hydrogenerator 灯泡式水轮发电机，灯泡型发电机
bulb unit 灯泡式机组
bulb 球，球形物件，真空管，测温计，电灯泡
bulding-out circuit 补偿电路
bulding-out 补偿值
bulge joint 隆起接缝
bulge test （焊缝的）打压试验，扩胀试验
bulge 鼓泡，膨胀，凸出，鼓包，混合防水面层
bulging failure 膨胀破坏
bulging force 膨胀力
bulging projection 凸出
bulging 膨胀
bulk-acoustic-wave delay line 体声波延迟线
bulk analysis 总分析，整体分析
bulk assembly 大部件装配
bulk avalanche thyristor 体击穿晶闸管
bulk boiling region 整体沸腾区
bulk boiling 大量沸腾，整体沸腾，容积沸腾
bulk breakdown voltage 体击穿电压
bulk-breaking 分拨费
bulk buoyancy 总体浮力
bulk capacity storage 大容量存储器

bulk cargo shipping 散货运输
bulk cargo ship 散货船
bulk cargo terminal 散装货码头，散货码头
bulk cargo transportation 散货运输
bulk cargo wharf 散装货码头
bulk cargo 散装货，堆货，散货，散装货物，散件货物
bulk carrier ship 散料运输船
bulk carrier 散装货船，散装货车，散货船
bulk cement truck 运散装水泥卡车
bulk cement 散装水泥
bulk charge transfer device 体电荷转移器件
bulk-cheap 薄利多销的
bulk circulation 主体循环
bulk compressibility 体积压缩性，体积压缩率
bulk concentration 容积浓度
bulk conductivity 体电导率
bulk container 散货集装箱
bulk coolant temperature 冷却剂整体温度
bulk coolant 主冷却剂
bulk deformation 主体变形
bulk density 毛体积密度，堆积比重，装载密度，整体密度，堆积密度，堆密度，松散密度，容重，散密度，松（堆）密度，容积密度，体积密度
bulk diffusion 容积扩散
bulk-effect device 体效应器件
bulk-effect oscillator 体效应振荡器
bulk-effect 体效应
bulk elasticity 体积弹性
bulker 舱货容量检查人
bulk excavation 大规模挖土，大开挖，大量开挖
bulk factor 松散系数，体积比
bulk fission product 总裂变产物，大量裂变产物
bulk flow flight conveyor 埋刮板输送机
bulk flow 容积流量
bulk freight 散仓水运，散运，散装运输
bulk getter pump 体积吸附泵
bulkhead connector 隔板连接物
bulkhead gate 平板闸门，检修闸门
bulkhead line 堤岸线
bulkhead of an air lock 气闸的舱盖
bulkhead quay wall 岸壁码头
bulkhead slot 堵水闸门槽
bulkhead wall 岸壁，岸壁码头，防水壁，隔墙
bulkhead wharf 堤岸式码头，堤岸码头
bulkhead 闷头【钢管】，防水壁，防风墙，岸壁，驳岸，挡水墙，堵壁，堵水闸门，堵头板，隔板，护墙，座墙
bulk heat transfer coefficient 整体传热系数
bulk heat treatment 整体热处理
bulk increase 容积增量
bulk index 体积指数
bulkiness 膨松性，庞大，笨重，膨松度
bulking curve 容积增大曲线
bulking effect 膨胀效应
bulking factor 容积增大系数，湿涨率
bulking sludge 膨胀污泥
bulking strength 压曲强度
bulking 膨胀，隆起，压屈，体胀
bulk insertion 全部插入，集体插入【控制棒】
bulk irradiation 总体辐射
bulk items 散装物项，散件
bulk loading 毛载，散装
bulkload 散货
bulk lorry 散装物载重汽车
bulk mass transfer coefficient 整体质量运输系数
bulk mass 大体积
bulk materials handling equipment 散状物料辅送设备
bulk materials handling system 散状物料输送系统
bulk material terminal 散装货码头
bulk material 散（装）（材）料，散状物料，疏松物质，松散材料
bulk memory device 大容量存储器设备
bulk memory 大容量存储器，大容量外存储器
bulk mixing 整体混合，整体交混
bulk modulus of elasticity 体积弹性模量
bulk modulus 体积模量，体积弹性模量，体积压缩性，体积弹性系数
bulk moulding compound 预制整体模塑料，块状模塑料
bulk of building 房屋体积
bulk of elasticity 体积弹性模数
bulk of the bed 床内空间【沸腾炉】
bulk-oil circuit breaker 多油（式）断路器
bulk-oil-volume breaker 多油断路器，多油开关
bulk phase 主体相
bulk potential 体电势
bulk powder 散装炸药
bulk power energy curtailment index 大电力系统电量削减指标
bulk power system 大电力系统
bulk processing facility 散料操作设施
bulk property 主体性质
bulk purchase 成批采购
bulk rate train 摘挂列车
bulk residue 干残渣
bulk resistance 体电阻
bulk Richardson number 总体理查森数
bulk sample （燃料）粗样，原始试样
bulk sampling 大量抽样
bulk separation 容积分离
bulk shielding facility 一体化屏蔽试验装置
bulk shielding 一体化屏蔽，整体屏蔽
bulk shield 立体屏蔽层，整体屏蔽层，一体化屏蔽体
bulk shipment 散装运输
bulk solids circulation 主体固体循环
bulk solid 散状物料，松散固体
bulk specific gravity 堆积比重，毛比重，容积比重，松散容重
bulk spreader 散装材料撒布机
bulk stability parameter 总体稳定度参数
bulk storage memory 大容量存贮器
bulk storage silo 散料贮存仓
bulk storage tank 大体积贮箱
bulk storage 大容量存储器，散装储存
bulk strain 体应变
bulk strength 块强度

bulk stress probe 体应力探头
bulk stress transducer 体应力传感器
bulk supply of electricity 趸售用电
bulk supply point 主要供电点，整体供电点
bulk supply 整体供电
bulk tariff 趸售电费率
bulk temperature 按体积计算的平均温度，整体温度，体积平均温度，容积温度
bulk transfer coefficient 整体运输系数
bulk transport 散货运输
bulk velocity 整体速度
bulk vessel 散货船
bulk viscosity 容积黏性，容积黏度，体积黏性系数
bulk viscous effect 主体黏滞效应
bulk voltage 体电压
bulk volume 毛体积，容积，松散体积
bulk weight 散重，总重
bulky cargo 大件，大件货物
bulky goods 笨大货物
bulky grain 大颗粒
bulky X ray tube 大功率 X 射线管
bulk 散料，容积，体积，大批，大量，散装材料，桶装材料，散装，主体，容量，大块
bull 大型的，买空，液压铆机
bulldog grip 钢丝绳夹
bulldog wrench 管子扳手
bulldozer attachment 推土机附件
bulldozer house 推煤机库
bulldozer lift valve spool 推土机升降阀阀芯
bulldozer 压弯机，推土机，粗碎机，推煤机
bullet 子弹，喷口整流机
bulletin board 布告牌，公告牌
bulletin 公报，公告
bullet loan 一次还款式贷款
bullet-nosed vibrator 球头式振捣器，圆头振捣棒
bullet-nosed welded end plug 锥形焊接端塞【套管】
bullet-proof glass 防弹玻璃
bullet-shaped end plug 锥形端塞【套管】
bullet-shaped welded end plug 锥形焊接端塞【套管】
bullet transformer 一种超高频转换装置
bullet wave 弹射波
bull float 大抹子
bull gear 大齿轮，主齿轮
bullgrader 大型平土机
bullhorn 手提式扩音器
bull market 牛市，行情看涨的市场
bullnose block 圆角块体
bullnose brick 圆头砖
bull pump 推力泵
bullrail 码头护缘
bull ring （中速磨的）磨环
bull specific gravity 毛体积比重
bull stretcher 露边侧砖
bull wheel derrick 转盘起重机
bull wheel 大齿轮，大转轮
bulwark 防浪堤，舷墙
bum-in test 老化试验
bum-off 烧化，熔化焊穿，雾消，焊穿
bump contact 块形连接，大面积接触
bumped head 冲压封头
bumper angle 缓冲角铁
bumper bar 保险杆，缓冲杆
bumper block 缓冲块
bumper 防撞器，防撞杠，缓冲器，减震器，保险杠，车挡，防撞装置，轨道止挡器，阻尼器，减振器，消声器
bumping block 缓冲块
bumping collision 弹性碰撞
bumping 碰撞
bump joint 扩口接合，隆起接头
bumpless transfer system 无冲击转移系统
bumpless transfer 无扰传输，无扰动转换，无扰切换
bumpy flow 旋涡流，涡流
bumpy running 不均衡运转
bumpy 颠簸的，气流不稳的，崎岖的
bump 碰撞，撞击，剧烈沸腾，瘤，凸缘，冲撞，隆起物，冒顶，凸起，岩石压裂
bunch 束，聚束，束线
bunch discharge 束形放电
bunched cable 集束电缆，多股电缆，束状电缆
bunched conductors 多股线，导线束
buncher resonator 输入谐振腔，聚束谐振腔
buncher voltage 聚束电压
buncher 束线机，无机变速器，合股机
bunching admittance 聚束导纳
bunching machine 束线机，合股机
bunching 成组，聚束，成群
bunch of cables 电缆束
bunch of electrons 电子束
bunding 岸堤
bundle average exit quality 棒束平均出口含汽率
bundle average mass flux density 棒束平均质量流密度
bundle conductor spacer 分裂导线，间隔棒
bundle conductor 分股导线，导线束，分裂导线
bundled bars 钢丝束，钢筋束
bundled cables 成束敷设的电缆，集束电缆
bundled conductor spacing 导线分裂间距
bundled conductor 成束导线，组合导线，导线束，分裂导线，集束电线
bundled 成束的
bundle factor 分裂因数
bundle of electrons 电子束
bundle of vortex filaments 涡丝束
bundle of wires 线束
bundle pillar 群柱
bundle power line 集束输电线
bundle spacing 分裂导线间距
bundle wrapper bunker 仓，斗，斗仓
bundle wrapper 管束围板【蒸汽发生器】
bundle 束，波束，一包，一束，一捆，线卷，线盘
bundling 捆扎的
bund 堤岸
bungalow 平房
bung 塞头
bunker and hopper 仓斗

bunker A oil　A级重油，低黏度船用油
bunker bay　煤仓间
bunker B oil　中等黏度船用油，B级重油
bunker car　仓车【运煤】，仓式矿车
bunker feeder　斗式给料机
bunker fire　煤仓着火
bunker fuel　船用燃料
bunker gate　煤斗闸板
bunkering trouble　煤仓内发生故障
bunkering　装燃料
bunker oil　重油
bunker room　煤仓间
bunker vibrator　煤仓振动器
bunker　料仓，贮仓，料斗，室，斗，仓，贮煤斗，储料仓，储藏室，把煤推进煤仓
bunkhouse　简易工棚
Bunsen burner　本生灯
Bunsen cell　本生电池
Bunsen flame　本生灯火焰
buoyance　浮动性，浮力，上升力，自然拔风力
buoyancy center　浮力中心
buoyancy chamber　浮力室，浮球
buoyancy correction　浮力修正
buoyancy deflection　浮力偏转
buoyancy displacer level transmitter　浮力沉筒式液位变送器
buoyancy effect　浮力效应，浮力作用
buoyancy factor　浮力因数
buoyancy flux　浮力通量
buoyancy force　浮力
buoyancy length scale　浮力长度尺度
buoyancy level measuring device　浮力式液位测量仪表
buoyancy lift　浮力，浮升
buoyancy moment　浮力矩
buoyancy of water　水的浮力
buoyancy pressure　浮托力
buoyancy pump　浮力泵
buoyancy term　浮力项
buoyancy　浮动性，浮力，上升力，自然拔风力，浮力作用，漂浮
buoyant acceleration　浮力加速度
buoyant convection　浮力对流
buoyant force　浮力
buoyant foundation　浮筏基础，浮基
buoyant jet　浮升射流，重力射流
buoyant lift　浮力
buoyant media filter　漂浮介质过滤器
buoyant pile　摩擦桩
buoyant plume　浮升羽流
buoyant puff　浮升喷团
buoyant rise　浮力抬升
buoyant source　浮力源
buoyant thermal　浮升热泡
buoyant unit weight　浮容重
buoyant weight　浮重
buoy light　灯浮标
buoy mooring　浮标系泊
buoy　浮子，浮标，漂浮，设置浮标
bur　毛刺，毛口
burble angle　失速角
burble　起泡，气流分离，失速，扰流，旋涡
burbling point　失速点，气流分离点，汽流分离点，骚动点
burbling zone　气流分离区
burbling　气泡分离，气流分离，流体起旋，旋涡
burden of CT　电流互感器的二次容量
burden of proof　举证责任
burden rating　额定装载量，额定负载
burden　负载，载荷，装载量，二次负载，炉料，积存量
Burdon pressure element　波登管压力元件
Burdon pressure sensor　波登管压力传感器
bureau director　局长
bureau of administration　管理局
Bureau of Indian Standards　印度标准局
bureau of water resources　水利局
bureau　局，办事处
buret clamp　滴定管夹
buret meniscus　滴定管液面
buret stand　滴定管架
buret stopcock　滴定管活塞
burette viscometer　滴定管黏度计
burette　玻璃量杯，滴定管，量管
burga　布加风
burglar alarm　防盗警报器
burgy　煤屑，细粉
burial depth　埋入深度
burial ground　地下埋藏区
burial in concrete　埋入混凝土【用封闭的导管】
burial tank　埋藏容器
burial　埋藏，掩埋
buried abutment　埋入式岸墩，埋入式桥台
buried aerial　埋地天线，地下天线
buried antenna　埋地天线，地下天线
buried cable　被埋电缆，埋设电缆，地下电缆
buried cage　鼠笼
buried channel CCD　埋沟电荷耦合器件
buried channel　地下河道，古河槽，埋藏河槽
buried concrete　地下混凝土
buried copper　槽内铜导线，槽中铜【电机】
buried crack　内埋裂纹
buried depth　埋设深度，埋置深度
buried drain　埋藏式排水，暗沟排水
buried dump　填入土
buried erosion surface　地下冲蚀面，埋藏冲刷面
buried flaw　内埋缺陷
buried grid electrode　地下的网络电极
buried ground conductor　予埋接地导体
buried ground　地下埋藏区
buried inatallation　地下设施
buried joint　暗缝
buried layer　埋层
buried line　地下线路，埋设线路
buried penstock　埋藏式压力水管，回填管，埋管，埋葬式钢管
buried pipe　暗管，地下的管道，地下管，埋设管
buried piping and cables　地下的管道和电缆
buried river　地下暗河，地下河，埋藏河
buried rod　埋设的杆

buried shelter 地下防空洞
buried sleeve 预埋套管
buried steel pipe 地埋钢管
buried structure 地下建筑（物），埋入式结构物
buried wiring 隐蔽布线，埋设线，暗线
buried 被掩埋的
burin 錾刀
burlap 粗麻布
burn 燃烧，烧伤
burnable absorber 可燃吸收体
burnable poison assembly for PWR 压水堆可燃毒物组件
burnable poison assembly storage adapter 存放可燃毒物组件适配器，可燃毒物组件存放适配器
burnable poison assembly 可燃毒物组件
burnable poison rod assembly 可燃毒物棒组件
burnable poison rod 可燃毒物棒
burnable poison 可燃毒物
burnable refuse 可燃垃圾
burnable 易燃物，可燃的
burnaway 烧完，燃尽，烧尽
burn cut 直眼掏槽，平行空炮眼掏槽
burn down 燃尽
burned lime 烧石灰
burned-up fuel 缺乏燃料
burner arch 燃烧器口
burner arm 燃烧器支架
burner band 卫燃带，燃烧带
burner bank 燃烧器组合
burner barrel 燃烧器筒身
burner basket heat release 燃烧器区域散热量［热负荷］
burner basket 燃烧器区域
burner blockage 喷油嘴堵塞
burner box 燃烧器风箱
burner can 火焰筒，管形燃烧室
burner carboning 喷油嘴结炭
burner characteristics 燃烧器特性
burner control system 燃烧器控制系统
burner cup 雾化头
burner efficiency 燃烧室效率
burner head 喷燃器头部
burner heat input 燃烧器热功率
burner housing 燃烧器壳体
burner jet 燃烧器喷嘴
burner management system （锅炉）烧嘴管理系统，燃烧器管理系统
burner nozzle 燃烧器喷嘴
burner opening （锅炉上）燃烧器安装孔
burner platform 看火台，燃烧器平台
burner port 燃烧器喷孔
burner protection 燃烧器熄火保护
burner reactor 燃烧堆【指核电】
burner spiral case 燃烧器蜗壳
burner system vortex chamber 燃烧器装置旋涡室
burner throat 燃烧器喉口
burner tilt 燃烧角度，燃烧摆动
burner tip 喷嘴，喷头，燃烧器喷嘴
burner wall 燃烧器处水冷壁
burner windbox 燃烧器风箱
burner zone wall heat release rate 燃烧器区域炉壁热负荷
burner 喷嘴，燃烧器，喷燃器，灯头，烧嘴，雾化器，燃烧室
Burnett equations 伯纳特方程
burnettizing 氯化锌防腐
Burnett stresses 伯纳特应力
burning behaviour 燃烧性能
burning capacity 炉膛或炉排热容量，燃烧率
burning coal 着火的煤
burning condition 燃烧工况
burning-down period 燃尽时间
burning equation 燃烧方程
burning gas 燃烧气体
burning in period （电子管的）老化时间
burning in 烧机测试，老化测试，烧入试验
burning life （灯泡的）使用寿命
burning loss 燃烧损失
burning mixture 可燃混合物
burning of contact 触点烧毁
burning-out zone 燃尽区
burning point 着火点，燃点
burning principles 燃烧原理，燃烧的基本定律
burning process 燃烧过程
burning rate 燃烧速率，燃烧率
burning surface area 燃烧面积
burning surface 燃烧面，燃烧表面
burning temperature 燃烧温度
burning velocity 燃烧速度
burning voltage 稳定电压，燃弧电压
burning zone 燃烧区
burning 燃烧，燃烧气割，过烧
burn-in period 烧入周期，预烧期，老化周期
burn-in screen 高温功率老化筛选
burn-in test 烧机测试，老化测试，烧入试验
burn-in 烧上，初始运行，老化，烧结，准备运行，烧焊，预烧，老炼
burnisher 打磨人，磨光器
burnishing powder 抛光粉，磨光粉
burnishing 抛光，辊光，磨光
burn-off rate 熔化率，燃尽率
burn-off 烧掉，燃尽，烧损
burn-out condition 熄火条件
burn-out device 烧毁装置
burn-out grate 燃尽炉排
burn-out heat flux 烧毁热通量，烧毁热流密度
burn-out indicator 烧断指示器
burn-out life 失效寿命，烧坏寿命
burn-out loss 燃烧损失
burn-out margin 烧毁余量，烧毁裕量
burn-out point 烧毁点
burn-out poison 可燃毒物
burn-out quality 烧毁含汽率
burn-out rate 烧断率，烧毁率
burn-out ratio 燃尽率，烧毁比
burn-out safety factor 烧毁安全因子
burn-out uncertainty factor 烧毁不确定因子
burn-out 烧毁，烧光，烧尽【燃料】
burnt brick 烧透砖
burnt clay 烧黏土
burnt diatomite 焙烧硅藻土

burnt gas 废气，燃烧过的气体
burnt gault 煅烧白垩【指混凝土的一种掺料】，烧黏土
burnt gypsum 熟石膏，烧石膏
burnt lime 烧石灰，生石灰，煅石灰
burn together 烧结
burnt shale 烧页岩
burnt through 烧穿，烧损，过热，焊穿，烧蚀
burnt-up core 烧完的堆芯
burnt-up fraction 燃耗份额，燃耗深度
burnt-up fuel 乏燃料
burn-up analysis 燃耗分析
burn-up calculation code 燃耗计算程序，燃耗计算规范
burn-up compensation 燃耗补偿
burn-up condition 燃耗状态
burn-up control assembly 燃耗控制装置
burn-up control 燃耗控制，燃耗管理
burn-up cycle 燃耗循环
burn-up distribution 燃耗分布
burn-up equation 燃耗方程
burn-up equilibrium 燃耗平衡
burn-up factor 燃耗因数，燃尽率，燃耗因子
burn-up fraction 燃耗比值，燃耗份额
burn-up level 燃耗深度
burn-up life 燃耗寿命【中子探测器】
burn-up loss 燃耗损失
burn-up measuring reactor 燃耗测量反应堆
burn-up measuring system 燃耗测量系统
burn-up monitor 燃耗监测计
burn-up rate 燃耗率
burn-up ratio 燃耗比
burn-up reduction 燃耗降低，燃耗减少
burn-up sharing fraction 燃耗分配比值
burn-up sharing 燃耗分配
burn-up stretch-out 燃耗延长
burn-up time 燃耗时间
burn-up 燃掉，燃耗，烧光，燃完
burring machine 去毛刺机，毛刺清理机，光口机
burring 清除毛边，修整，去毛刺
burrock 堤堰
burrow 穴洞，挖洞
burr 毛刺，毛口，毛边，毛面【混凝土】，焊瘤
burst acoustic emission signal 突发型声发射信号
burst amplifier 彩色同步脉冲放大器，闪光信号放大器
burst blanking pulse 彩色同步信号清隐脉冲
burst can detection system 破损包壳探测系统
burst can detection 元件包壳破损探测，包壳破损探测，破损包壳探测
burst can detector 元件包壳破损探测器
burst cartridge detection equipment 包壳破损探测设备
burst cartridge detection gear 元件包壳破损探测装置
burst cartridge detection system 元件包壳破损探测系统
burst cartridge detection 燃料元件包壳破损探测，元件包壳破损探测
burst cartridge detector 燃料元件包壳破损探测器，元件包壳破损探测器
burst detector 破损探测器
burster 起爆药
burst gate tube 内光控制管，猝发选通管
burst gate 猝发选通的，短脉冲选通的
burst generator 短脉冲群发生器
bursting behaviour 抗爆破性能
bursting diaphragm 防爆薄膜
bursting disc 防爆膜，防爆盘
bursting disk 破裂片
bursting force 爆破力
bursting of the canning material 燃料棒包壳破裂
bursting pressure 爆裂压力
bursting reinforcement 防爆钢筋
bursting strength 爆破强度
bursting stress 爆裂应力，破裂应力
bursting test 爆破试验
bursting 突然开始，爆炸，猛然打开，破裂，爆裂
burst into flame 突然着火，烧起来
burst into steam 冒出蒸汽，突然蒸发
burst mode 成组方式，脉冲串式
burst pin detecting 细棒破损探测，(燃料) 细棒包破损探测
burst pin detection system 破损燃料细棒探测系统
burst pressure 爆破压力
burst pulser 短促脉冲发生器
burst regeneration 点燃信号还原，短促信号恢复
burst rod 脉冲棒【反应堆】
burst signal 短促脉冲串信号，彩色同步信号
burst slug detection system 破损（燃料）元件探测系统
burst slug detection 破损燃料块探测，破损（燃料）元件探测，燃料棒破损探测
burst slug detector 燃料棒破损探测器
burst slug 破损（燃料）元件
burst speed 毁坏转速
burst test machine 压力试验机
burst test 爆破试验，爆裂试验
burst 爆发，炸破，爆炸，瞬爆，突发，破裂，爆裂，冲开，溃决
bury 掩埋，掩盖，埋藏，埋入
bus admittance matrix 结点导纳矩阵，母线导纳矩阵
bus allocator 总线分配器
busbar adapter 母线接线盒
busbar chamber 母线箱，母线室
busbar channel 槽形母线，母线通道
busbar clamp 母线夹
busbar connection arrangement 母线连接布置，主接线布置，主结线布置
busbar corridor 母线（检查）廊道，母线廊道
busbar coupling 母线接合，母线耦合
busbar cross-section 母线截面
busbar current transformer 母线电流互感器
busbar disconnecting switch 母线隔离开关
busbar disconnection 母线解列
busbar duct 母线沟，母线导管

busbar expansion joint 母线伸缩节
busbar fault 母线故障
busbar flashover 母线飞弧，母线闪络
busbar floor 母线层
busbar framework 母线架构
busbar frame 母线架
busbar gallery 母线廊道
busbar grounding 母线接地
busbar group 母线组
busbar insulator 母线绝缘子
busbar joint 母线接头
busbar partition 母线间隔，母线隔板
busbar potential transformer 母线电压互感器
busbar price 母线价格
busbar protection 母线保护
busbar room 母线室
busbar sectionalizing switch 母线分段隔离开关
busbar section 母线段
busbar separator 母线间隔垫
busbar setting 母线校正
busbar short-circuit 母线短路
busbar straight-through current transformer 母线用穿心式电流互感器，母线式电流互感器
busbar support 母线支撑
busbar system 母线系统
busbar transformer 母线型电流互感器，母线变压器
busbar voltage 母线电压
busbar wire 汇流条，母线，汇流排
busbar 汇流排，汇流母线，汇流条
bus compartment 母线隔间
bus-compatible 总线兼容的
bus compensator 母线伸缩接头
bus conductor 母线，汇流排
bus control unit 总线控制器
bus coupler circuit-breaker 母联断路器
bus coupler 母联开关，总线耦合器，母联
bus-coupling 母线接合，母线耦合
bus differential protection 母差保护
bus duct work 母线管道工程
bus duct 母线导管，汇流管道，母线沟，母线槽
bus gallery 母线廊道
BUSH＝bushing 衬套，套管，轴衬
bush clearance （施工项目的）清障，灌木清除
bush cutting 灌木清除
bush-faced masonry 凿面块石砌体
bush fitting 刮削轴瓦
bush gallery 冲砂廊道
bush hammer 凿石锤，气动凿毛机，汽动凿毛孔
bushing current transformer 套管式电流互感器
bushing flashover 套管闪络
bushing for arm assembly boss 臂组件轮毂用衬套
bushing insulator 绝缘套管，套管式绝缘子
bushing lead terminal 套管端子
bushing oil port 滑套油孔
bushing shell 轴套
bushing terminal 套管式端头，套管接头
bushing test tap 套管试验插头
bushing transformer 套管互感器
bushing type condenser 套管式电容器
bushing type current transformer 套管式电流互感器
bushing （绝缘）套管，轴套，轴瓦，轴衬，补心，内外螺纹接头，使不同口径的管子接合在一起的转接器，总线插座
bush 矮树林，丛生灌木，灌木丛，（金属）衬套，（绝缘）套管，轴衬，加（金属）衬套于……，轴套，轴瓦
bus impedance matrix 节点阻抗矩阵，母线阻抗矩阵
business accounting 经营核算
business card 名片
business charter 专营业务
business communications 商业通信，业务通信
business computer 商用计算机
business confabulation 商务
business district 商业区
business efficiency 经营效率
business enterprise 工商企业
business executive 企业家
business failure 企业倒闭，企业破产
business finance 企业财务
business group owned by central government 中央企业集团
business hall 营业厅
business income tax 企业所得税
business information system 企业信息系统
business insolvency 企业倒闭
business item 营业项目
business law 商业法
business licence 营业执照
business line 业务范围，行业
business liquidation 企业清算
business loss 企业亏损
business machine 商用计算机
business management 企业管理，营业管理
businessman 工商业者，商人
business method 经营方式
business negotiation 洽谈业务
business panics 经济恐慌
business planning 经营计划
business practice 商业惯例
business premise 事务所
business principle 经营原则
business profit tax 营业所得税
business quarters 事务所
business registration certificate 公司注册证书
business registration 营业执照，商业注册
business representative 商务代表
business reputation 商业信誉
business risk 经营风险
business statutes 企业章程
business strategist 商务策划师
business taxes and surcharges 营业税金及附加费
business tax 营业税
business volume 交易额
business 企业，商业，实业
bus insulator 母线绝缘器
bus interface circuit 总线接口电路

bus interface module　总线接口模件
bus line　汇流线，母线，总线
bus master　总线主控
bus-oriented　总线用的
bus protection　母线保护
bus protective relay　母线保护继电器
bus reactor　母线电抗器
bus regulator　母线电抗器，母线电压蝶器，母线电压调节装置
bus-ring　集电环
bus-rod　圆汇流条，圆条母线，母线，汇流排
bus sectional circuit-breaker　母线分段断路器
bus section breaker　母线分段断路器
bus section　母线分段，母线隔间
bussed supply　母线供电
bus separation　母线分离
bussing　高压线与母线间的连接
bus slave　总线受控
bus-structured　总线结构的
bus structure　母线结构，汇流排结构，母线桥
bus support　汇流排支撑，母线（绝缘）支撑
bus switch unit　总线开关部件
bus system　汇流排系统，母线系统，总线系统
bus terminal fault　母线终端故障
buster　铆钉铲
bus tie　母联
bus-tie breaker　母线联络断路器
bus-tie cell　母线联络间隔，汇流排房间，汇流排单元
bus-tie circuit breaker　母线开关，母线联络断路器，母联断路器
bus-tie reactor protection　母线联络电抗器保护
bus-tie reactor　母线联络电抗器
bus-tie switch　母线联络开关
bust　操作错误，错误动作，崩溃，毁坏，破产，破了产的，毁坏了的
bus voltage loss　母线电压损失
bus voltage metering section　母线电压测量隔间
bus voltage regulator　母线电压调整器，母线电压调节器
busway　配电通道，母线通道，母线导管
bus wire　汇流条，母线，汇流线
busy-back jack　占线测试塞孔，忙音塞孔
busy-back signal　占线信号，忙音信号
busy-back tone　忙音，蜂音，占线音
busy-back　忙回信号，占线信号
busy-buzz　忙蜂音，占线音
busy condition　忙状态，占线
busy drop　占线掉牌
busy-flash　占线闪光
busy hour　最忙钟点，繁忙时间
busy indicator　占用指示器，占线指示器
busy lamp　占线信号灯，忙灯
busy link　忙音接续片
busy report　占线报告
busy season　旺季
busy signal　占线信号
busy test relay　占线继电器，占线测试继电器
busy test　忙碌状态测试，占线测试，忙闲测试
busy tone trunk　忙音中继线
busy tone　忙音，占线音
busy　在操作，在工作，占线
bus　汇流排，母线，总线，汇流条，信息转移通路
butadiene-nitrile rubber　丁腈橡胶
butadiene　丁二烯
butagas　丁烷气
butane　丁烷
butanol　丁醇
butanone　丁酮
butene　丁烯
butt　平接，对接，根端
butt-and-butt joint　两头对接
butt and collar joint　套筒接合
butt-and-lap joint　互搭接头
butt and strap hinge　丁字铰链
butt contact　对接接点，对接触点，对接接触，压接式接点
butt diameter　粗端直径（木材的）
butted weld　对接焊
butte　孤山，小尖山
butt endplate　平端
butteressed wall　壁柱
butterfly bolt　双叶螺栓
butterfly check valve　蝶式止回阀
butterfly circuit　蝶形电路
butterfly control valve　蝶形控制阀
butterfly damper　蝶形挡板，蝶阀，蝶形阀
butterfly dam　蝶闸坝
butterfly gate　蝴蝶闸门，蝶式闸门
butterfly hinge　蝴蝶铰链，蝶形铰链
butterflying trowel　涂灰镘
butterfly insulator　蝴蝶式绝缘子
butterfly valve　蝴蝶阀，蝶阀，旋转挡板节流阀
butterfly　蝶形挡板，蝶形的，蝶形装置，蝶形
buttering on dissimilar metal　在不同金属上的预堆边焊
buttering welding　堆焊【为隔离作用】
buttering　辅面（表面处理），涂灰浆
butt hinge　铰接，平接铰链，珠承铰链
butting　对接
buttinski　装有拨号盘和受话器的试验器
butt joining　对头接合
butt-joint　对接（接头），端接，丁字对接，平接（接头），接口缝
butt junction　对接头，平接头，对接，平接接头
butt-muff coupling　套筒联轴节，刚性联轴器
button capacitor　纽扣式电容器，小型电容器
button cell　扣式电池，纽扣电池
button condenser　纽扣式电容器，小型电容器
button control　按钮控制
button head bolt　圆头螺栓
button head cap screw　圆头螺钉
button head rivet　圆头铆钉
button rivet　圆头铆钉
button nip machine　纽扣式咬口机
button socket　按钮灯座
button switch　按钮开关
button　按钮，风帽
butt pin　铰链销
butt resistance welder　电阻对接焊机

buttress brace 垛间支撑
buttress bracing 垛间支撑
buttress centres 坝垛中心距【水电】
buttress dam 受力型坝，扶壁式坝，支墩坝
buttressed retaining wall 扶壁式挡土墙，扶垛式挡土墙
buttressed wall 扶垛墙
buttress head 垛头
buttress spacing 坝垛间距，支墩间距【水电】
buttress strut 支墩间撑梁
buttress thread 梯形螺纹
buttress web 支墩间拉板
buttress weir 支墩堰，垛堰
buttress width 支墩宽度
buttress 肋墩，扶壁，支持物，撑墙，扶垛，支墩，坝垛【水电】
butt rivet joint 对头铆接
butt seam welding 对焊
butt splice 对头拼接
butt strap 对接板，对接搭板，平接板
butt-type core 对接式铁芯
butt weld convex flange 对焊凸法兰
butt-welded concave flange 对焊凹法兰
butt-welded end 对焊端
butt-welded joint 对焊接头，对焊连接
butt-welded seam 对焊缝
butt-welded splice 平焊接头
butt-welded tube 对口焊接钢管
butt weld end connection 对头焊端连接
butt weld ends 对头焊接端
butt welder 对焊机
butt welding machine 对焊机
butt welding 对接焊，对缝焊接，对焊，对头焊接
butt weld joint 对接焊接头
butt weld seam 对接焊缝
butt weld splice 对接焊接头
butt weld 对焊，碰焊，平式焊接，对接焊，对接焊缝，对头焊缝，对头焊接
butyl benzene 丁苯
butyl rubber 异丁烯橡胶，丁烯橡胶
butyric acid 丁酸
buy-back agreement 返销协议，回购合同
buy-back 回购贸易，补偿交易，回购，产品返销，产品返销的，补进
buyer participant 买方参与者
buyer representative 买方代表
buyer's credit 买方信贷
buyer's decision 买方决定
buyer's market 买方市场
buyer's review 买方审查
buyer 购买人，买方
buying long 买空
buying price 买价
buying rate for time bill 期票买价
buyout price 一口价，买断价
buy out 全部买下（市上产品，产权、股份等），买下……的全部产权，出钱使……放弃地位，出钱使放弃地位，购空存货，收购，买断，收购全部
Buys-Ballot's law 白贝罗定律
buy 购买
buzzer call 蜂音呼叫
buzzer generator 蜂音发生器
buzzer interrupter 蜂音断续器
buzzer oscillator 蜂音发生器，蜂音振荡器
buzzer phone 蜂音器，蜂鸣器，轻便电话机
buzzer relay 蜂鸣继电器
buzzer signal 蜂音信号
buzzer 蜂音器，蜂鸣器，汽笛，轻型掘岩机
buzz group 讨论组
buzzing 蜂音，发蜂音，嗡嗡响的，蜂鸣，嗡鸣
buzz off 搁断电话，匆匆走掉
buzz session 非正式的小组讨论会
buzz-stick method 绝缘棒探测法
buzz-stick 绝缘子测试棒
buzz track test film 蜂音统调试验片
buzz 蜂音，嗡嗡声
BV(Black & Veatch) 博莱克·威奇工程公司【总部在美国】
BVD(boiler vents and drains) 锅炉疏水放气
BW＝backwash 反洗
BW(butt weld) 对接焊，对接焊焊缝
B-wire B线，第二线
BWRA(British Welding Research Association) 英国焊接研究协会
BWR(boiling water reactor) 沸水反应堆，沸腾循环反应堆，沸水堆
BWR control rod drive 沸水堆控制棒驱动
BWR control rod 沸水堆控制棒
BWR emergency core cooling system 沸水堆应急堆芯冷却系统
BWR fuel assembly 沸水堆燃料组件
BWR nuclear power plant 沸水堆核电厂
BWR pressure vessel 沸水堆压力容器
BWSR(bucket-wheel stacker reclaimer) 斗轮堆取料机
BWV(back-water valve) 回水阀
BX cable 软电缆
by all means 尽一切办法，一定，务必
by any means 无论如何
byatt 水平木
by cashier check 支票付款
by cash 现金付款
by-channel 分流河槽，溢水槽，旁侧溢洪道，支渠
by clean bills 凭光票
by contract 按照合同，承包
by convention 按照惯例，根据惯例
by default 缺席，不到场，按缺省值，默认的
by direct remittance 直接汇款
by documentary bills 凭跟单汇票
by-effect 副作用
by-election 补选
by-estimate 据估计
by fair means or foul 不择手段地，用正当或不正当的手段
by-gravity 靠重力作用的
by installments 分期付款
by-law of corporation 公司章程
bylaw 地方法规，细则，附则，法规
byline 署名权

by means of 借助于，用，依靠
by mutual consent 双方同意
by negotiations 经过谈判［协商］
by no means 决不，并没有
by notation 注明（的）
by oneself 自行
bypass accumulator 浮充蓄电池
bypass air duct 旁通风道
bypass anode 分流阳极
bypass arm 旁通（阀桥）臂，旁路臂
bypass baffles 旁通挡板
bypass battery 补偿电池组，缓冲电池组
bypass blow off valve 旁通放气阀
bypass breaker 旁路断路器
bypass bus circuit breaker 旁路断路器
bypass bus 旁母
bypass canal 侧流渠
bypass capacitor 旁路电容器
bypass channel 并联信道，旁通渠，并联电路，旁流道，溢水渠
bypass chute 旁路落煤管
bypass clean-up system 旁路净化系统
bypass cock 旁通旋塞，转换开关
bypass coil 旁路线圈
bypass condenser 旁路电容器
bypass conductor 迂回线
bypass control system 旁路控制系统
bypass control valve 旁路控制阀
bypass control 旁通调节，旁路控制
bypass culvert 旁通涵洞
bypass damper control 旁路挡板调节
bypass damper 旁通［旁路］挡板，旁路调节闸阀，旁通阀，旁路调节
bypass device 旁路装置
bypass diode 旁路二极管
bypass filter 旁通过滤器
bypass flow rate 旁路流量
bypass flow 旁路流，分流，旁通流，旁通流量
bypass flue 旁通烟道，分支烟道
bypass gas damper 旁通烟气挡板
bypass gate valve 旁通闸阀
bypass gate 旁路闸门，旁通闸门
bypass governing 旁通调节
bypassing reflux stream 旁通回流
bypassing 走旁路，旁路，分路，旁通，分流，旁泄
bypass leakage 旁路泄漏
bypass line 旁通管，旁路管线
bypass manifold 旁通歧管，旁路集管
bypass module 旁路组件
bypass of converter 换流器旁路
bypass operation 旁通运行，旁路运行
bypass pair 旁通对，旁路对
bypass path 旁路通路【指电路】
bypass pipe 旁路管，旁通管
bypass plug 旁路插销，旁通插件
bypass pressure control 旁路压力控制
bypass purification system 旁路净化系统
bypass ratio 旁路比，旁通比，分流比
bypass rectifier 旁路整流阀，分流整流阀
bypass relief valve 旁通安全阀，旁通减压阀
bypass rotameter 旁路转子流量计
bypass route 过迴径路，支路
bypass seepage 绕渗
bypass setting 旁通阀调节
bypass stage 旁路级
bypass steam flow 旁路蒸汽流
bypass stream 旁路流，分流
bypass superheater 旁路烟道内过热器，旁通（式）过热器
bypass switch 旁路开关，旁通开关
bypass system 旁通系统，旁路系统
bypass temperature control 旁路调温
bypass to ground 旁通到地
bypass track 通过线，越行线
bypass transformer 分路变压器，连接变压器
bypass tunnel 旁通隧道
bypass valve 旁路阀，旁通阀，旁路门
bypass vent 分支排气孔
bypass weir 旁通堰
bypass 旁路，分流，旁通，支管，支流，分路，旁通道［管］，加分路，旁路的，并联的
bypath system 旁路系统，旁路制
bypath valve 旁通阀
bypath 旁路，侧管，人行道，旁通，分路，小过道
by-product coal 选煤的副产品
by-product coke 副产焦炭
by-product power 副产电力，副产功率
by-product recovery 副产品回收，副产物回收
by-product steam 副产水蒸气，副产蒸汽
by-product 副产物，副产品
BYPS＝bypass 旁路，旁通，支路，绕开，设旁路，迂回
by-reaction 副反应
byroad 小过道，支路，旁路
by rough estimate 据粗略估计
by share 按份额
bysmalith 岩柱
by some means or other 用某种方法
by stages and in groups 分期分批
by stages 分期
by sufferance 经默许，被容许
byte boundary 字节界
byte for byte addressable 字节编址
byte mode 字节式，字节方式
by tender 以投标方式
byte-oriented 按字节的
byte 字节，位组
by the wind 顺风航行
by virtue of 因为，由于，凭借……的力量，借助于
by volume 按体积【计运费等】
by-wash 河岸排水道，排水管沟，旧河床，废河道
by way of open international bid 采用国际公开招标方式
byway 间道
by-workman 临时工
by-work 副业
BZ(branch on zero) 零转移

C

cabane 翼间支架
CAB＝cabinet 小室
cab driver 出租车司机
cabinet converter 变频器柜
cabinet door 柜门
cabinet nacelle transformer 机舱变压器柜
cabinet nacelle 机舱机柜
cabinet panel 配电盘
cabinet switch 开关箱
cabinet tower 塔基机柜
cabinet 舱，橱，柜，室，间，小室，盒，框，机柜，屏，小房间，内阁
cabin 舱，小室，小间，小房，机舱，驾驶室，客舱
cable accessory 电缆附件
cable-actuated excavator 索动挖土机
cable address 电报挂号
cable anchorage 缆索锚固
cable anchor 电缆固定器，索锚
cable-and-bar winding 上层编织下层实心的绕组
cable and cable raceway system 电缆及电缆管道系统
cable and drum drive mechanism 缆鼓式传动机构
cable armor 电缆铠装
cable armouring 电缆铠装，电缆铠甲
cable basement 电缆底室，地下电缆室
cable bedding 电缆衬垫
cable bent 缆绳垂度，缆索垂度
cable bond 电缆接头，电缆连接器
cable book 电缆清册
cable box bushing 电缆盒衬套
cable box splice 电缆连接箱，电缆编接盒
cable box 电缆箱，电缆盒，分线盒
cable bracket 电缆架，电缆支架【总称】
cable branch box 电缆分支箱
cable breakdown test 电缆击穿试验
cable breakdown 电缆击穿
cable bridge 电缆桥，电缆桥架，缆桥，缆索桥，索桥
cable bundle 束，光纤束，捆，卷
cable cantilever bridge 电缆悬臂桥，斜张桥
cable capacitance 电缆电容
cable cap 电缆端帽
cable carriages 缆车
cable cart 电缆盘拖车
cable car 电缆车，缆车
cable chamber 电缆室
cable channel 电缆沟，电缆槽，电缆通道
cable charge 电报费
cable charging breaking current 电缆充电开断电流
cable chart 电缆图表
cable chase 电缆探查
cable chute 电缆槽，电缆沟
cable circuit design 电缆线路设计
cable clamp 电缆夹，电缆夹紧装置
cable cleat 电缆夹，线夹，电缆线夹，电缆夹具
cable clip 电缆夹，电缆挂钩，电缆夹子
cable coal unloader 缆索式卸煤机
cable code 水线密码，水线电码，电缆码
cable-coiling drum 电缆绕线盘
cable communication 电缆通信
cable complement 电缆对群
cable compound 电缆绝缘胶，电缆膏
cable conductor connection 电缆导体连接
cable conductor 电缆线，电缆芯线
cable conduit 电缆导管，电缆道，电缆管，电缆管道，电缆暗沟
cable confirmation 电报确认，电报确认书
cable connector 电缆连接头，电缆连接器
cable controlled scraper 索铲
cable control system 缆索控制系统
cable control unit 缆索控制机
cable conveyer 钢索输送机
cable conveyor 缆索运输机，索道输送机
cable core 电缆芯
cable-correction unit 电缆校正装置
cable corridor 电缆层，电缆夹层
cable coupler adapter 电缆套管接合器
cable coupler 电缆连接器
cable coupling capacitor 电缆耦合电容器
cable coupling 电缆耦合
cable cover 电缆护套
cable crane 缆索吊车，缆索起重机，索道起重机，索道起重设备
cable credit 电报传递信用证
cable culvert 电缆管道
cable current transformer 电缆用电流互感器
cable cutter 电缆剪，断线钳，电缆切断机
cable damp 电缆夹
cable deck ventilation system 电缆间通风系统
cable deck 电缆桥架
cable defects detector 电缆探伤仪
cable depth 电缆埋深
cable derrick crane 纤缆桅杆式起重机
cable destination 电缆目的地
cable distribution box 电缆交接箱，电缆分线箱
cable distribution equipment 电缆分线设备
cable distribution head 电缆分线头，电缆分线盒
cable distribution point 电缆分线点，电缆与架空线的汇接点
cable distributor 电缆分线盒，电缆配线架
cable dividing box 电缆分线盒
cable drag scraper 索铲
cable drill 钢丝绳冲击钻机，索钻
cable-drive handling 钢索驱动操纵装置
cable drive 钢索传动
cable dropper 电缆挂钩
cable drum carriage 电缆放线车
cable drum jack 电缆卷盘架
cable drum stand 电缆盘架
cable drum support 电缆放线架
cable drum table 卷放电缆支撑架
cable drum trailer 电缆盘拖车
cable drum 电缆盘，电缆线盘，电缆卷筒，钢

缆鼓筒
cable duct bank 电缆排管
cable duct 电缆槽，电缆沟，电缆管道，电缆道，电缆管
cable-dump truck 索式自卸卡车
cabled 铺电缆的
cable echo 电缆回波
cable end bushing 电缆端套【在机壳入口保护电缆】
cable end cap 电缆端帽
cable end connector 电缆端接盒
cable end 电缆端
cable entrance 电缆入口
cable entry 电缆入口
cable equalization 电缆特性曲线均衡化
cable excavator 索缆式挖掘器
cable exit 电缆出口
cable external damage protection 电缆线路外力破坏防护
cable failure repair 电缆故障修理
cable fault 电缆故障，电缆损伤，电缆漏电
cable fault detector 电缆故障检测器
cable fault location 电缆故障测寻
cable fault locator 电缆故障定位器，电缆故障点探测器
cable feeder pillar 电缆分支箱
cable ferry 缆索渡口，绳渡
cable filler 电缆填料
cable filling applicator 电缆填充装置
cable filling compound 电缆填充剂
cable filling equipment 电缆填充设备
cable filling machine 电缆填充机
cable filling plant 电缆填充设备
cable filling rate 电缆填充率
cable filling yarn 电缆填充麻
cable fill 电缆填充，电缆充满率，电缆占用率
cable fire protection band 电缆防火带
cable fire protection slot 电缆防火槽
cable fire protection 电缆防火
cable fittings 电缆附件，电缆配件
cable floor 电缆层
cable for communication 通信电缆
cable form 电缆模板，电缆布线板
cable frogging 电缆分叉
cable gallery 电缆廊道
cable gas feeding equipment 电缆充气维护设备
cable gear 敷电缆机
cable gland with seismic-resistant cable clamp 带抗震电缆夹紧器的电缆密封套
cable gland 电缆衬垫，电缆密封套
cable grip 电缆钳，电缆夹
cable groove 电缆槽，电缆沟
cable grounded 接地电缆
cable guide 拉线导向块，电缆引导管，缆索导向器，电缆导管
cable hanger 电缆吊架，电缆挂钩
cable haulage 钢索运输
cable head 电缆分线盒，电缆接头，电缆头，电缆终端盒
cable hoist 缆索绞车
cable holder 电缆防坠夹
cable hut 电缆放线箱，电缆配电房，电缆汇接室
cable in conduit 导管中的电缆
cable in quadruples 四芯电缆
cable installation equipment 电缆安装机具
cable installation works 电缆安装工程
cable installation 电缆安装
cable insulation 电缆绝缘
cable interlayer 电缆夹层
cable in triples 三芯电缆
cable isolator 电缆隔离开关，电缆隔离器
cable jacket 电缆套
cable joint-box 电缆套，电缆连接箱
cable joint 电缆接头，缆线接头
cable label printer 电缆标签打印机
cable ladder 电缆梯架，爬线梯
cable laid in trays 敷设在托架里的电缆
cable-laid rope 多股钢丝绳，钢丝绳，绞合绳，拧索
cable layer 电缆敷设机，电缆敷设船，布缆船
cable laying 电缆敷设
cable laying and list 电缆敷设及清册
cable laying equipment 电缆敷设设备
cable laying machine 电缆敷设机
cable laying ship 布缆船，电缆敷设船
cable laying truck 电缆敷设车
cable layout 电缆敷设图，电缆配线详图
cable length measurement instrument 电缆长度测量仪
cable length 电缆长度，链
cable-line parameter 电缆线路参数
cable line 电缆线路
cable list 电缆清单
cable loss 电缆损失
cable louding 电缆走向
cable lug 电缆接（线）头，电缆终端，电缆终端衔套
cable main distribution frame 电缆总配线架
cablemaking tools 造线机
cable manhole 电缆人孔，电缆竖井
cableman 电缆工
cable map 电缆图
cable marking tool 给电缆加标志的工具
cable messenger 电报投递员，电缆悬吊线
cable mezzanine 电缆夹层
cable net structures 索网结构
cable net wall 索网幕墙
cable network 电缆网络，电缆网，索网
cable net 索网
cable of automatic system 自动化系统电缆
cable oil pump station 电缆油泵站
cable oil 电缆油
cable paper 电缆纸
cable passage 电缆通道
cable penetration unit 电缆贯穿件
cable penetration 电缆穿墙
cable penetrator 电缆贯穿件
cable pipe 电缆管道
cable pit 电缆井，电缆竖井
cable platform 电缆平台
cable pole line 电缆架空线路

cable pothead 电缆终端套管，电缆端套
cable power factor 电缆功率因数
cable protection pipe 电缆保护套
cable puller 电缆拉出器
cable pulling eye 电缆拉环，电缆牵引端
cable pulling procedure 电缆铺设规程
cable pulling 拉电缆
cable puncture test 电缆击穿试验
cable puncture 电缆击穿
cabler 成缆机
cable raceway system 电缆通道系统
cable race 电缆走廊
cable racking 电缆装架
cable rack 电缆架，电缆支架，电缆槽架
cable reel capsule 电缆盘封壳
cable reel 电缆卷筒，电缆盘
cable refractory coating 电缆防火涂料
cable relay 电缆继电器
cable ring-system 电缆环网，环形电缆系统
cable road 索道
cable roller 电缆导轮
cable roof 索屋顶
cable room 电缆层或电缆夹层，电缆室
cable route inspection 电缆线路巡视检查
cable route selection 电缆线路路径选择
cable routing schedule 电缆路线
cable routing 电缆路由选择，电缆敷设，电缆铺设
cable run 索道线路
cable saddle 电缆鞍
cable sag 缆索垂度
cable saturant 电缆浸渍剂
cable schedule 电缆明细表
cable screening factor 电缆屏蔽系数
cable sealing box 电缆密封盒
cable sealing end 电缆密封盒
cable separator 电缆分隔物
cable shaft 电缆竖井
cable shear 电缆剪
cable sheath 电缆包皮层，电缆护套，电缆外皮，缆索鞘
cable sheaves 缆索滑轮，电缆绞轮
cable shelf 电缆支架
cable shielding 电缆屏蔽
cable shield 电缆护套，电缆屏蔽
cable ship 海底电缆敷设船
cable shoe 电缆靴，电缆终端，电缆终端套管，电缆帽
cable slack meter 电缆松紧指示器
cable sleeve 电缆套管
cable socket 电缆端头
cable specification 电缆规格，电缆规范
cable splice 电缆接头
cable splicing chamber 电缆编接室
cable splicing 电缆连接，电缆编接
cable spreading area 电缆铺设区
cable spreading room 电缆分布室，电缆铺设间
cable stayed bridge 斜拉桥
cable stayed grider bridge 斜梁桥
cable stopper 锚链制动器
cable storage yard 电缆贮存场地
cable strap 电缆固定夹板
cable subway 电缆廊道，电缆隧道
cable supported membrane roof 拉索薄膜屋顶
cable support rack 电缆引入架，电缆分线架
cable support system 电缆支吊系统
cable support 电缆托架
cable-suspended current meter 缆索流速仪
cable suspender 电缆吊架
cable tag 电缆标签
cable tank 电缆槽
cable tensioner 钢丝绳张紧轮，钢索拉伸器
cable terminal box 电缆终端盒
cable terminal station 电缆终端站
cable terminal 电缆终端，电缆终端头
cable terminating procedure 电缆收端规程
cable termination 电缆头，电缆终端，电缆封端
cable test bridge 电缆试验电桥
cable tie applicator tool 电缆箍紧工具
cable tie 电缆带，电缆箍紧卡，电缆夹
cable to cable tie 拉索扣带
cable tool 缆索钻具
cable tool drilling 钢丝绳冲击钻进法，缆索钻具钻井法
cable tower 缆索塔
cable trailing device 电缆拖拉装置
cable transfer 电汇
cable tray and conduct system 地缆槽和导线管系统
cable tray 电缆槽，电缆托架，电缆盘，电缆桥架
cable trench 电缆沟，电缆沟道
cable trolley 电缆小车
cable trough 电缆槽，电缆沟，电缆沟道，电缆走线槽，电缆暗沟渠
cable trunk 电缆管道
cable tunnel 电车隧道，电缆隧道，电缆廊道
cable twist counter 电缆扭转计数器
cable twisting 电缆扭曲
cable twist sensor 电缆扭绞传感器
cable twist 扭缆
cable tying 电缆夹紧
cable tyre 电缆卷紧装置
cable unreeler 电缆退绕器
cable untwisting 电缆解扭
cable varnish 电缆漆
cable vault 检修孔，人孔，电缆竖井，电缆室，电缆地下室，电缆夹层
cable vulcanizer 电缆硫化器
cable wax 电缆蜡
cableway bucket 索道吊罐
cableway excavator 拖铲挖土机，缆索开挖机，缆索挖掘机
cableway measurement 缆索测量
cableway tower 索道塔
cableway transporter 缆道起重运送机
cableway 电缆通道索道，缆道，索道
cable winch 电缆绞车
cable winder drum 电缆卷筒
cable with extruded insulation 挤包绝缘电缆
cable works 电缆厂
cable wrapping machine 绕缆索机

cable yard 电缆贮存场地
cable yarn 缆索股绳
cable 电缆，缆索，钢丝绳，缆绳，绳索，系缆，钢索，电报，架设电缆
cabling diagram 电缆敷设图
cabling room 电缆间
cabling 敷设电缆，电缆线路，布线，线的绞合
cabtyre cable 厚橡皮绝缘软电缆，橡皮绝缘电缆
cabtyre cord 橡皮绝缘软线
cabtyre sheathing 硬橡胶电缆包皮
cab 司机室【提升设备】，操纵室，操作室，驾驶室，小室
cacaerometer 空气污染检查器
CACD(computer aided circuit design) 计算机辅助电路设计
cache buffer memory 超高速缓冲存储器
cache 电脑高速缓冲存储器，贮存物，高速缓存
cadastral plan 地籍图
cadastral survey 地籍测量，土地测量
cadastration 地籍测量
cadastre 地籍图，水册
CAD/CAM system 计算机辅助设计与制造系统
CAD(cash against documents) 凭单付款
CAD(computer aided design) 计算机辅助设计
CADD(computer aided design and drafting) 计算机辅助设计制图
CADE(computer aided design and engineering) 计算机辅助设计与工程
cadence 步调信号
cadmium absorption 镉吸收，镉（中子）吸收
cadmium bronze 镉青铜
cadmium cell 镉电池
cadmium control rod 镉控制棒
cadmium copper wire 镉铜线
cadmium covered detector 包镉探测器
cadmium curtain 镉控制片
cadmium cut-off energy 镉截止能
cadmium cut-off 镉截止值，镉切断
cadmium-difference method 镉差法
cadmium-mercuric oxide cell 锡汞电池
cadmium-nickel accumulator 镉镍蓄电池
cadmium-oxygen battery 镉氧电池
cadmium-plated 镀镉的
cadmium ratio 镉比
cadmium regulator 镉调节棒
cadmium rod 镉棒
cadmium selenide detector 硒化镉探测器
cadmium-selenium photo-cell 镉硒光电池
cadmium shield 镉屏蔽体
cadmium standard cell 镉标准电池
cadmium strip 镉条
cadmium sulfide cell 硫化镉电池
cadmium sulfide detector 硫化镉探测器
cadmium sulfide photocell 硫化镉光电管，硫化镉光电池
cadmium sulfide photoresistor 硫化镉光敏电阻器
cadmium sulfide photovoltaic cell 硫化镉阻挡层光电池
cadmium sulfide solar cell 硫化镉太阳电池
cadmium sulfide thin film solar cell 硫化镉薄膜太阳能电池
cadmium telluride detector 碲化镉探测器
cadmium 镉
cadre 干部，骨架，核心，骨干
CADS(computer aided design system) 计算机辅助设计系统
cadweld rebar splice 钢筋机械捻接
cadweld 铜与铜或铜与钢的铝热焊，放热焊接
cadwid rebar splice 钢筋机械捻接
CAD 加元
CAE(certified as executed) 竣工状态【文件管理用语】
CAE(computer aided engineering) 计算机辅助工程
CAEP(condenser air extraction pump) 真空泵
caesium photocell 铯光电管，铯光电池
caesium 铯
CAF(currency adjustment factor) 货币附加费
CAF 冷却风机【DCS 画面】
cage antenna 笼形天线
cage assembly 升降台
cage bar 鼠笼条
cage circuit 笼形电路，网格电路
cage construction 骨架构造，骨架结构
cage control valve 笼形控制阀
caged core 笼形铁芯
caged ladder 环形梯
caged stored at bonded warehouse 进入海关监管
caged valve 顶阀，顶部传动阀
cage grid 笼形栅板，网格栅板
cage-guided valve trim 笼形导向阀内件
cage hoist 罐笼提升机
cage induction motor 鼠笼式感应电动机
cageless ball bearing 无隔圈滚珠轴承
cageless induction motor 无笼式感应电动机，非鼠笼式感应电动机
cageless plug disk globe valve 非笼形塞阀瓣球阀
cage lifter 升降机
cage-like 笼形的
cage motor 鼠笼式电动机，鼠笼式电机
cage nut 笼形螺帽
cage of reinforcement 钢筋组架，钢筋骨架，钢筋笼
cage rack 笼形拦污栅
cage ring 隔离圈，鼠笼端环
cage rotor induction motor 鼠笼式转子感应电动机
cage rotor 鼠笼式转子
cage screen 笼形拦污栅，笼筛
cage synchronous motor 带阻尼条的同步电动机，鼠笼式同步电动机
cage-type multiplier 笼形倍增器
cage-type plug disk globe valve 笼形塞阀瓣球阀
cage-type 罐笼式
cage valve trim 笼形阀内件
cage valve 套筒阀
cage walls 包墙管，包覆管
cage wheel elevator 笼轮式升运器
cage winding 鼠笼式绕组

cage 鼠笼绕组，罩，笼，盒，框，笼状物，壳体，栅，罐笼，机壳，骨架构造，笼型
caging device 限位装置
caging 制动，停止，锁定，锁停，限位
calciner plant 煅烧装置
cairn 堆石觇标
caission foundation 沉箱基础
caission method 沉箱法
caission set 沉箱套
caission structure 沉箱结构
caisson breakwater 沉箱防波堤
caisson chamber 沉箱室
caisson foundation 沉井基础，沉箱基础
caisson gate 沉箱式坞闸
caisson launching 沉箱下水
caisson monolith construction 整体沉箱结构
caisson pier 沉箱墩
caisson pile 灌注桩，沉箱桩，管柱桩
caisson quay wall 沉箱码头
caisson sinking 沉箱下沉
caisson system 沉箱系统
caisson-type pile 沉箱式桩
caisson works 沉箱工程
caisson 沉箱【基础】，船坞闸门，沉井
cake filtration 饼状过滤，滤饼过滤
cake 结块，黏结，滤饼，滤渣，块，固结
caking capacity 黏结性
caking coal 黏结性煤，黏结煤
caking index G of coal 煤黏结性G指数
caking index 黏结指数
caking power 黏结力
caking property 黏结性，烧结性
caking quality 烧结性
caking 结块，黏结，烧结，结渣
caky 成块的
cal 口径，量规，直径，卡
calamine 菱锌矿
calamity 灾害
calandria assembly 排管容器组合体
calandria shell 排管容器外壳
calandria tube 排管容器内管
calandria vessel 排管容器
calandria 排管式蒸发设备，排管容器
calathiform 杯形的
CAL(calculated average life) 平均计算寿命
cal. = calorie 卡
calcareious 石灰质的，钙质的
calcarenite 钙质岩
calcareous alluvial soil 石灰质冲积土
calcareous cementing material 石灰类黏结料
calcareous clay 石灰质黏土
calcareous concretion 钙质结核
calcareous grit 钙质粗砂岩
calcareous marl 钙质泥灰岩
calcareous rock 含钙岩
calcareous sandstone 钙质砂岩
calcareous shale 钙质页岩
calcareous sinter 石灰华
calcareous slag 石灰炉渣
calcareous slate 钙质板岩
calcareous soil 石灰质土，钙质土，含钙土
calcareous 含钙的，石灰质的
calcic horizon 钙积层
calcicole 钙生植物
calcic 石灰质，钙质
calcification 钙化
calcigranite 钙质花岗岩
calcilutyte 灰泥岩
calcimine 可赛银粉，墙粉（一种涂料），刷墙粉，石灰浆
calcimorphic soil 钙成土
calcinate 煅烧
calcination of radioactive waste 放射性废物的煅烧
calcination temperature 焙烧温度，煅烧温度
calcination 焙烧，煅烧，烧成石灰
calcinator 煅烧炉
calcine 焙烧
calcined gypsum 烧石膏
calcined magnesite 菱苦土
calcined rock 锻烧岩石
calciner 煅烧炉，焙烧炉
calcining furnace 煅烧炉
calcining 烧成石灰，焙烧
calciothermy 钙热法
calcite 方解石
calcium alkalinity 钙碱度
calcium-aluminate cement 铝酸钙水泥【即矾土水泥】
calcium-based compound 钙基化合物
calcium bicarbonate 重碳酸钙
calcium biphosphate 磷酸二氢钙
calcium carbide industry 电石工业
calcium carbide 碳化钙，电石
calcium carbonate slab 碳酸钙石棉板
calcium carbonate 碳酸钙
calcium chloride 氯化钙
calcium concentration 钙浓度
calcium fluoride dosimeter 氟化钙剂量计
calcium hardness 钙硬度，钙硬
calcium hydroxide 氢氧化钙，熟石灰
calcium hypochlorite 次氯酸钙
calcium ion 钙离子
calcium lime 未消石灰，生石灰
calcium orthophosphate 磷酸钙
calcium oxalate 草酸钙
calcium oxide 氧化钙，生石灰
calcium phosphate 磷酸钙
calcium plaster 石膏浆
calcium plumbate primer 高铅酸钙底漆
calcium pyrophosphate 焦磷酸钙
calcium quicklime 生石灰
calcium requirement 石灰需要量
calcium silicate 硅酸钙
calcium soil 钙质土
calcium sulfate dihydrate 生石膏，石膏，二水合硫酸钙，白土
calcium sulfate dosimeter 硫酸钙剂量计
calcium sulfate 硫酸钙
calcium sulphate plaster 石膏抹灰面，石膏灰浆
calcium sulphate 硫酸钙，石膏
calcium 钙【Ca】

calc-spar 方解石
calculable 可计算的
calculagraph 计时器
calculated address 合成地址，计算地址
calculated average life 平均计算寿命
calculated capacity 计算容量
calculated fuel consumption 计算燃料消耗量
calculated guess 估算
calculated height 计算高度
calculated instruction 计算说明书
calculated load 计算负荷，设计负荷
calculated main 计算主干线
calculated power of the input shaft 输入轴计算功率
calculated risk 计算危险值
calculated strength 计算强度
calculated value 计算值
calculate 计算，核算，评价，推算
calculating 计算的
calculating apparatus 计算装置
calculating area 计算面积
calculating board 计算台
calculating device 计算装置
calculating length 计算长度
calculating life 计算寿命
calculating machine 计算机
calculating outdoor relative humidity 室外计算相对湿度
calculating outdoor temperature for conditioning temperature 空调室外计算温度
calculating overturning point 计算倾覆点
calculating parameter 计算参数
calculating puncher 卡片计算机
calculation accuracy 计算精度
calculational method 计算方法
calculation capacity 计算能力，计算容量
calculation chart 计算图表
calculation documents 计算书
calculation error 计算误差
calculation list 计算项目清单
calculation mesh 计算网络
calculation method 计算方法
calculation note 计算说明，计算书，计算清单
calculation of assessment 评估计算
calculation of loading 负荷计算
calculation of present worth 兑现率计算，现值计算
calculation of short circuit current 短路电流计算
calculation point 计算点，计算位置
calculation self-weight collapse 计算自重湿陷量
calculation sheet 计算书
calculation software 计算软件
calculation time 计算时间
calculation 计算，估计
calculator 计算器，计算装置，计算者，计算机
calculous soil 砾质土
calculus of differences 差分学，差分法
calculus of residues 残数计算
calculus of variation 变分学，变分法
calculus 微积分学，算法，微积分
caldera 破火山口
calefaction 发热
calendar age 役龄
calendar date 日历日期
calendar day 日历日，历日
calendar month 日历月，历月
calendar of conference 会议日期表
calendar progress chart 日进度表
calendar year 日历年（度）
calendar 日历，月份牌，一览表
calender 碾光机，压延机，压延成型，以研光机研光
calex 钙化法
calf ice 小浮冰块
caliber rule 卡尺
caliber 量规，口径，管径，测量器，管内径
calibrated attenuator 校准的衰减器
calibrated detector 校准检波器
calibrated dial 校准度盘，分度标度盘，标准度盘
calibrated disc 刻度盘
calibrated focal length 校正焦距
calibrated instrument 校准仪表
calibrated lead 校准的引线
calibrated meter 已校准仪表
calibrated neutron flux power 已标定的中子通量功率，特定的中子通量功率
calibrated phase shifter 校准移相器
calibrated screen aperture 筛的校准孔径
calibrated sluice 校准泄水闸门
calibrated span 标定量程，标准量程
calibrate 校正，校准，调校，定刻度，标定，分度，校验，定口径，检定
calibrating arm 校准臂
calibrating cable 校准电缆
calibrating capacitor 校正电容器
calibrating circuit 校验电路，校准电路，测试电路
calibrating constant 校准常数
calibrating device 校准装置，校准器，率定装置
calibrating instrument 校准装置，校准设备，校准仪器
calibrating meter 校验仪
calibrating method 校准方法
calibrating plot 校准曲线，标定曲线
calibrating resistance 校准电阻
calibrating spring 校准弹簧
calibrating standard 率定标准
calibrating terminal 测试接线柱，测试端子
calibrating tube 标定管
calibrating 标定，率定
calibration accuracy 校准精确度，校准精度，率定精度
calibration battery 校准电池，标准电池
calibration bench 校准台
calibration block 校准试块，校准块
calibration board （仪表）校验台
calibration certificate 校验证，鉴定证（书）
calibration chart 校验图表，校验曲线，校准图表
calibration coefficient 校准系数

calibration console 效验台，校准台
calibration constant 率定常数，校正常数
calibration current transformer 校验用电流互感器
calibration current 校准电流
calibration curve of current meter 流速仪率定曲线
calibration curve 校正曲线，标定曲线，校准曲线，率定曲线，校验曲线
calibration cycle 校准周期，校验周期，校准循环
calibration data 标定数据，校准数据
calibration device 标定装置，校验装置
calibration error 刻度误差，校准误差，定标误差
calibration factor 率定系数，校准因素，校准因子
calibration flaw 校准缺陷
calibration frequency 校正频率
calibration hole 参考孔，标准孔，标定孔
calibration instrument 校验用仪表
calibration method 率定法
calibration microprocessor assembly 校正微处理器装置
calibration model test 标模试验
calibration model 检验模型，标准模型
calibration notch 校准槽
calibration of current meter 流速仪率定
calibration potential transformer 校准用电压互感器
calibration pulse 校验脉冲
calibration regulator 校验用调节器
calibration report 校验报告
calibration slope 标定斜率
calibration source 校准源，刻度源
calibration station 率定站
calibration table 校准表，校验表，校准表格
calibration tail 刻度线
calibration test 标定试验，校准试验
calibration traceability 跟踪校准
calibration trace 校正曲线
calibration tunnel 标准风洞
calibration value 标定值，校准值
calibration voltage 校验电压，校准用电压
calibration wind tunnel 校正风洞，校准风洞
calibration 标定，标度，刻度，划分，测定，校准，定口径，校验，调校，定刻度
calibrator unit 校准部件
calibrator 校准器，校验器，校验者
calibre 圆柱径，量规，口径，测量器，管径
caliche 钙积层，生硝，石灰盘
californium 锎
caliper depth ride 深度卡尺
caliper face spanner with adjustable 可调式两爪钩扳手
caliper gauge 测径规，内卡规
caliper profiler 纸张厚度计
caliper rule 卡尺
caliper 卡尺，卡钳，测径器
calked joint 嵌实缝，捻缝
calk 尖铁，铁刺
call 询问，呼叫，通话
call address 调入地址，呼叫地址，引入地址
Callan cell 卡兰电池
call announcer 呼叫示号器
call attempt 试呼
Callaud cell 卡卢得电池
call back signal 回叫信号
call back 回叫
call bell 呼叫铃
call box 公共电话间
call button 呼叫按钮
call by value 赋值
call circuit （电话）业务线
call completing rate 接通率
call counter 呼叫计数器
call delayed system 等待损失制
call display position 号码指示位置
called line 被叫线
called party 被叫用户，被叫方
called station 通信台，被叫台，被调入位置
called subscriber held alarm 被叫用户保护警报
called subscriber 被叫用户机
called 被（呼）叫的
callee 受话人，被呼叫者
caller 主叫用户，呼叫者
call finder 寻线机
call for bids 请求投标，招标
call for funds 集资
call for tender 招标，请求投标
call forth 唤起，引起，使产生，使起作用
call indicator 呼叫指示器
calling code 呼叫代码，呼叫记号
calling cord 呼叫塞绳
calling equipment 呼叫设备，呼叫装置
calling jack 呼叫塞孔
calling lamp 呼叫灯
calling line 呼叫线
calling-magneto 振铃手摇发电机，振铃磁电机
calling number 主叫号码
calling-on signal 叫通信号
calling order 发送程序，呼号指令
calling party 主叫用户，主叫方
calling pilot 呼叫灯，指示灯
calling plug 呼叫插塞
calling program 调用程序
calling relay 呼叫继电器
calling sequence 调用序列，呼叫顺序
calling signal 呼叫信号
calling station 调用位置
calling subscriber 主叫用户
calling supervisory lamp 呼叫监视灯
calling trace 调用跟踪
calling up 电台呼叫
calling 呼叫，振铃，呼号，主叫
call instruction 调用指令
call into requisition 征用
call in 收集，收回，调入（子程序）
calliper brake 线闸，卡钳制动器
calliper rule 卡尺
calliper square clasp 游标卡尺的滑尺
calliper 卡尺，卡钳，测径器，测径规

call key 呼叫键
call letters 呼叫字母，呼号
call library 调用存储库
call meter 呼叫计数器，通话计次器
call number 呼叫号码
call off 取消
call on hand 预约挂号
call on waiting list 预约挂号
call on 访问【内存储单元用】
callout 图注
callow 低沼，低沼泽（地）
call parking 呼叫暂停
call port 中途港
call processing 呼叫处理
call program 调用程序
call setting up 呼通，叫通，建立呼叫
call signal 呼叫信号
call sign 呼号【通信】，呼叫信号，识别信号
call spillover 呼叫泄漏
call statement 调用语句
call through test 接通试验
call tracing 呼叫跟踪
call-up log 召唤制表
call-up 征召，电台呼叫，召唤
call waiting 调用等待，呼叫等待
call-waiting tone 呼叫等待音
call wire 业务线，挂号线，记录线
call word 呼叫字码，引入字
calm belt 无风带
calmbreeze 静风
calmcentral eye 无风眼
calming section 平稳区
calm-inversion pollution 宁静逆温污染
calmlayer 无风层
calmnight 静夜
calm sea 无浪海面
calm-smog 宁静烟雾，无风烟雾
calm wind 静风
calm zone 稳流区，宁静区，无风带，平静区
calm 无风，零级风，平静的，平稳
calomel electrode 甘汞电极
calomel half cell 甘汞半电池
calomel reference electrode 甘汞参考电极，甘汞参比电极
calomel 甘汞
calomic 镍铬铁电热合金
caloric balance 热量平衡
caloricity 热容，热值，热量，热容量，发热量
caloric power 热值，热力，发热功率，发热量
caloric receptivity 热容量
caloric theory of heat 热质学说
caloric unit 热量单位，卡
caloric value 热值
caloric 大卡，热的，热值，热量，热量的，卡的
calorie meter 热量计
calorie 卡（路里）
calorific 热量的
calorification 发热，产生热量
calorific balance 热量平衡，热平衡
calorific capacity 热容量
calorific effect 热效应
calorific efficiency 热效率
calorific equivalent 发热当量，热当量
calorific intensity 热强度
calorific loss 热损失
calorific potential 潜热能，热量势能
calorific power 热值，发热量
calorific receptivity 热容，比热
calorific value at constant pressure 恒压发热
calorific value at constant volume 恒容发热量
calorific value determined in a bomb calorimeter 弹筒发热量，弹筒热值
calorific value in a bomb cylinder 弹筒发热量，弹筒热值
calorific value of fuels 燃料的热值
calorific value 热值，发热量，热量值，卡值
calorifier 热量计，加热器，量热器，水加热器，热风机
calorimeter bomb 测热器弹，测热弹，量热器
calorimeter 量热器，热量计，卡计，热量器
calorimetric test 测热试验，热值测定试验
calorimetric type 测热计型的
calorimetric value 发热量，热值
calorimetric wattmeter 热量计式瓦特计
calorimetry 测热学，测热法，量热法，量热学
calorisation 渗铝
calorising 表面渗铝，铝化处理
calorization 渗铝
calorized steel 渗铝钢
calorizing 铝化处理，表面渗铝，渗铝
calorstat 恒温器，温度调节器，恒温箱
calsomine 刷墙粉
Cal val(calorific value) 热值，发热量
calve layer 活化层，有效层
calx 矿灰
calyx core drill 萼状取芯钻，萼状岩芯钻
calyx drill boring 萼状钻钻探
calyx drill 萼状钻
cam 凸轮，偏心轮
CAMA(computer aided mathematics analysis) 计算机辅助数学分析
cam adapter 凸轮联轴器
cam-adjusting gear 凸轮调整装置
cam-and-counter cam 共轭凸轮
cam-and-lever mechanism 凸轮杠杆机构
cam and ratchet driver 凸轮棘轮传动装置
cam angle indicator 凸轮角指示
cam angle recorder 凸轮角记录仪
cam angle 凸轮转角
camber angle 叶型折转角
camber beam 弯曲梁，上拱梁，弓背梁，反挠梁
camber blade 弯曲叶片
camber block 拱架垫块
camber control 曲面控制
camber curvature （翼型）中弧线弯度
camber distribution 延翼展的弯度分布
cambered aerofoil 曲翼面，弯曲翼面
cambered arch 平拱，弯拱
cambered axle 弯轴
cambered beam 曲梁
cambered blade （弦向）弯曲叶片

cambered ceiling 弓形顶棚，拱形顶，平拱形顶棚，平拱形天花板
cambered plate 弓形板
cambered truss 弓形桁架
camber effect 弯度效应，前轮外倾效应
cambering 上拱度
camber line 翼型中心线，中弧线，中心线
camber of truss 桁架拱度
camber piece 砌拱垫块
camber slip 砌拱垫块
camber surface 弧面
camber 坝顶起填量，坝顶余幅【水电】，起拱，弯度，曲面，弯曲角，曲率，凸起弯度，拱形，上挠度，路拱，起拱度，反挠度
cambisol 始成土
cambowl 凸轮滚子
cambreak 凸轮制动器
cambric-insulated wire 细麻布绝缘线，黄蜡布绝缘线
cambric insulation 麻布绝缘，黄蜡布绝缘
cambric tape 黄蜡布带
cambric 细麻布，黄蜡布
cam chair 凸轮座
CAM(computer aided manufacturing) 计算机辅助制造
cam control gear 凸轮控制机构，凸轮控制装置
cam disk 凸轮盘
cam driving gear 凸轮传动装置
camel-back truss 驼背式桁架
camel's hump weir 驼峰堰
camel 打捞浮筒
camera-control monitor 摄像系统监控器
camera deflection circuit 摄像机偏转电路
camera downward 摄像机向下
camera equipment 摄像器材
camera incorporating a radial viewing head 连同径向视野头的摄像机
camera-lucida 转绘仪
camera mount 摄像机支架
camera pole 摄像机支柱
camera-read theodolite 摄影读数经纬仪
camera rotation 摄像机转动
camera stand 摄像机台架
camera tube 摄像管
camera upward 摄像机向上
camera 摄像机，小室，摄影机
camflex valve 转阀，偏心旋转阀
cam follower lever 凸轮从动杆
cam follower pin 凸轮从动销子
cam follower 凸轮随动件，凸轮跟随器
cam gear 凸轮传动装置，凸轮轴齿轮，凸轮机构，偏心轮，凸轮装置
camkometer 剑桥旁压仪
cam lever 凸轮杆
cam-operated control valve 凸轮控制阀
cam-operated counter 凸轮传动计数器
cam-operated switch 凸轮控制开关
cam-operated valve 凸轮控制阀
camp 帐篷
campaign life 使用寿命
campaign 连续运行时间，使用期限
Campbell diagram 坎贝尔图，调频倍率图，坎贝尔频谱图，叶片坎贝尔曲线
Campbell-Stokes recorder 坎贝尔-斯托克司阳光记录仪
camphor wood 樟木
cam plate 凸轮盘
camp sheeting 板桩挡土岸壁，板桩
camp site 施工生活基地，施工营地
CAMP 控制、报警、监测、保护【DCS 画面】
cam reduction gear 凸轮减速装置
cam ring chuck 三爪卡盘
cam ring 凸轮环
cam roller assembly 凸轮机构
cam roller 凸轮滚柱
cam sector gear 凸轮扇形齿轮
cam set controller 凸轮设定控制器
cam shaft thrust bearing 凸轮轴止推轴承
cam shaft 凸轮轴
cam sleeve 凸轮联轴节
cam throttle control 节流阀凸轮控制
cam-timer 凸轮式定时器
cam torque angle 凸轮转角
cam type steam distribution device 凸轮式排汽机构
Canadian General Electric Co. 加拿大通用电气公司
Canadian Society for Civil Engineers 加拿大土木工程师协会
canal appurtenances 渠道附属建筑物
canal aqueduct 输水渡槽
canal capacity 渠道过水能力
canal cover plate 沟盖板
canal head 渠首
canalization 开凿运河，运河网造管术，渠化，开挖渠道，开挖运河，梯级化【河道】
canalized river section 梯级化河段
canal lining 渠道衬砌，运河衬砌
canal lock 船闸
canal offlet 渠道斗门，渠道放水口
canal rapids 湍流渠
canal ray 极隧射线，阳极射线
canal section 渠道截面，管道断面，渠道剖面，渠道断面
canal seepage loss 渠道渗漏损失
canal theory of tide 水道潮汐论
canal 沟道，通道，波道，信道，管道，运河，渠道，槽，沟
can-annular combustion chamber 环管式燃烧室
can-annular combustor 环管式燃烧室
canard 鸭翼，前翼
canaries 高频噪声
canary lamp 充气黄色灯泡
can buoy 罐形浮标，立式圆筒浮标
can burner 管形燃烧室
can bursting temperature 包壳破裂温度【燃料元件】
can burst 包壳破裂
cancel a contract 撤销合同，取消合同
cancel circuit 消除电路
canceling clause 解约条款
cancel key 符号取消键

cancellating current　补偿电流
cancellation amplifier　对消放大器
cancellation and commitment fee　取消开庭费用
cancellation clause　（合同）解除条款，撤销条款，销约条款
cancellation network　抵消网络
cancellation notice　注销通知
cancellation of a contract　撤销合同
cancellation of bid　废标
cancellation of insurance　退保
cancellation of license　吊销执照
cancellation　作废，化为零，取消，消去，注销
cancelled structure　格框式结构，格构架，格型构架，空腹结构
canceller　补偿设备，消除器
cancelling circuit　补偿电路，消除电路
cancelling clause　解约条款
cancelling date　撤销合同日期，撤销日期
cancelling of the terms　消项，并项
cancel mark switch　符号取消开关
cancel message　撤销信息，作废信息
cancel signature　签字作废
cancel the contract　撤销合同
cancel　注销，消除，取消，抵消，减免，吊销
cancerogenic substance　致癌物
cancer　癌症
can chamber　管形燃烧室
can combustor　带火焰稳定器的燃烧室，管形燃烧室
cancrinite　钙霞石
can damage　包壳损坏，包壳缺陷
can defect　包壳破损，包壳缺陷
candela　坎德拉【发光强度单位】，烛光
candelite　焰煤
candescence　白热，热灼，炽热
candescent　白热的，炽热的
candidate firm　候选厂家
candidate for tendering　投标候选人
candidate site　候选场址
candidate species　可替代的物质
candidate　候选人
candle coal　烛煤，长焰煤
candle filter　滤烛，管型过滤器，细管型过滤棒
candle power distribution curve　配光曲线
candle power　烛光，烛光功率
cane fibre board　甘蔗纤维板，甘蔗板
cane trash　甘蔗渣
can failure detection　包壳破损探测器
can failure　包壳破损
canister　小罐，金属容器，箱式真空吸尘器，空气过滤器，集装箱
canless fuel assembly　无盒燃料组件
canless　无围板的【燃料】
can loss　外壳电气损耗
can meltdown　包壳熔化
canned data　存储信息，存储的信息
canned diameter　（元件的）包壳直径
canned motor pump　转子密封泵，屏蔽电泵，密封电动泵，全密封电泵，屏蔽电机泵
canned motor　密闭式电动机，密封电动机，全封闭电动机
canned program　特定解题程序
canned pump　密封泵，全密封泵
canned rack and pinion drive　全密封齿轮齿条传动机构
canned reactor coolant pump　屏蔽式主泵
canned rod　带包壳的棒
canned-rotor pump　全密封泵，转子密封泵
canned　装成罐的，存储的，密封的，录音的，封闭的
cannel coal　长焰煤
cannelure　环形槽，纵槽，纵沟，总槽
canning-beam current　扫描束电流
canning burst　包壳破裂
canning machine　装壳机，密封外壳装置
canning material　包壳材料
canning station　（元件盒）封装站
canning tube　包套管
canning wall thickness　包壳壁厚
canning　外壳封装，装罐，外壳密封，加保护罩壳，封装，包壳
cannon-type projector　短管式喷射器
cannon　炮
cannular burner　环管式燃烧室
cannular combustion chamber　环管燃烧室，联筒燃烧室
cannular combustor　环管式燃烧室
cannular　管状的，管式的，筒状的
cannula　套管，插管
canoe fold　舟状褶皱
canoe　独木舟
canonical conjugate variable　正则共轭变量
canonical coordinates　正则坐标
canonical correlation coefficient　正则相关系数
canonical distribution　正则分布
canonical field theory　规范场论，正则场论
canonical format identifier　规范格式标识
canonical form　标准形式，典型形式
canonical matrix　标准矩阵，典型矩阵
canonical partition function　正则配分函数
canonical path　正则轨线
canonical　正则的，标准的
canon of economy　经济规律
canon　标准，准则，教规，法规，规范
canopied　遮有天篷的
canopy flow　冠层流
canopy hatch　舱口
canopy hood　伞形罩
canopy layer　冠层，林冠覆盖
canopy sublayer　冠次层
canopy switch　盖顶开关，天棚开关
canopy top　林冠顶
canopy window　挑窗，上旋窗
canopy　雨篷，顶盖，灯罩，盖，天蓬，华盖，顶棚，雨棚，苍穹，（用天蓬）遮盖
can-pellet clearance　包壳芯块间隙
can pump　筒形泵
can rupture　包壳破裂
cantalever spot welding machine　悬臂式点焊机
cant beam　斜梁
cant column　多角柱
cant deficiency angle　欠超高角

canted blade 斜面叶片
canted disc butterfly valve 斜阀盘碟阀
canted wall 墙角，有角的墙
canteen 食堂，餐厅，小卖部，小饭馆，职工食堂
can thickness 包壳壁厚
cant hook 钩杆，平头搬钩，套环搬钩
cantilever action 悬臂作用
cantilever arch bridge 悬臂拱桥
cantilever arch truss 悬臂拱式桁架，悬臂桁架
cantilever arm 悬臂
cantilever beam 挑梁，悬臂梁
cantilever boom 悬臂架，悬臂梁
cantilever bracket 悬臂托架，悬臂托座
cantilever bridge 悬臂桥
cantilever buttress dam 悬臂式支墩坝
cantilever circuit 悬臂形电路
cantilever crane 伸臂起重机，悬臂起重机，悬臂式起重机
cantilever deck dam 悬臂平板坝
cantilevered boom 臂架
cantilevered platform 外挑平台
cantilevered 悬臂的
cantilever element 悬臂式构件
cantilever balcony 挑阳台
cantilever footing 悬臂式基础，悬臂式基脚
cantilever force 悬臂受力，悬臂力
cantilever form 悬臂式模板
cantilever foundation 悬臂基础，悬臂式基础
cantilever girder 悬臂梁
cantilevering 悬挑
cantilever lock wall 悬臂式闸墙
cantilever method of design 悬臂设计法
cantilever method of erection 悬臂架设法
cantilever moment 悬臂力矩
cantilever platform 挑台，悬臂平台
cantilever portion 悬臂部分
cantilever retaining wall 悬臂式挡土墙
cantilever rotor 悬臂风轮
cantilever sheet pile 悬臂式板桩
cantilever sheet piling 悬臂板桩，悬臂式板桩
cantilever slab 挑板
cantilever span 悬臂跨
cantilever stairs 悬挑楼梯
cantilever steps 悬挑踏步
cantilever structure 悬臂梁结构
cantilever support 悬臂架，悬臂式支架
cantilever tower 悬臂塔架，独立塔架
cantilever tray 悬臂托架
cantilever truss 悬臂桁架
cantilever-type wall 悬臂式墙
cantilever 伸臂，悬臂（梁），挑出，外伸臂，交叉撑架，悬臂支架，电缆吊线夹板
cant of curve 曲线超高
cant of rail 轨道超高
cant of sleeper 轨枕倾斜
cant saw file 三角锯锉
cant scraper 三角刮刀
cant strip 垫瓦条，披水条
can-type combustor 管形燃烧室，罐式燃烧室，筒形燃烧器
cant 弥缝嵌条，斜面，角落，术语，黑话
canvas 帆布，风帘，防水布，帐篷
canvas air conduit 帆布输气管
canvasbag 帆布垄包
canvas belt 帆布带，帆布皮带
canvas cloth 帆布
canvas connection 帆布连接
canvas conveyor 帆布带运输机
canvas dam 帆布坝
canvas hose 帆布软管
can wall thickness 包壳壁厚
canyon bench 峡谷阶地
canyon dam 峡谷坝
canyon factor 峡谷因数
canyon floor 峡谷底
canyon head 峡谷头
canyon profile 峡谷轮廓，峡谷剖面
canyon type pump 屏蔽泵
canyon wall 峡谷岸壁，峡谷壁
canyon wind 下吹风，峡谷风，山风，下降风
canyon 峡谷，街谷
can （燃料）包壳，罐，壳，单独的燃烧室，密封外壳，衬套，铁罐
CAO(cargo aircraft only) 仅限货机
CAOS(completely automatic operating system) 全自动操作系统
capability factor 能力因子，能力系数
capability for plutonium recycle 钚再循环能力
capability margin 能力裕度，系统出力裕量
capability of overload 过负荷能力
capability of switching long no-load line 分合长空载线路能力
capability of switching no-load transformer 分合空载变压器能力
capability 出力，能力，本领，容量
capable fault 能动断层
capable of gas-dynamic support （气体轴承）有气体动力支承能力的
capable 有能力的，能干的
capacitance box 电容箱
capacitance bridge 电容电桥
capacitance bushing 电容式（绝缘）套管
capacitance coefficient 电容系数
capacitance comparator 电容比较器
capacitance compensation 电容补偿
capacitance-coupled flip-flop 电容耦合触发器
capacitance-coupled 电容耦合的
capacitance coupling 电容耦合
capacitance current 电容电流
capacitance drift 电容量漂移
capacitance effect 电容效应
capacitance grading 电容等级，电容分级
capacitance law tolerance 电容量规律偏差
capacitance level meter 电容式料位计
capacitance level sensor 电容量传感器
capacitance measurement instrument 电容测量仪表
capacitance meter 电容计，法拉计
capacitance of bushing 套管电容
capacitance of lead 引线电容
capacitance oscillator 电容振荡器

capacitance plate 电容极板
capacitance potential transformer 电容电压互感器
capacitance potentiometer 电容分压器，电容电位计
capacitance pressure transmitter 电容式压力变送器
capacitance probe 电容探头，电容探针
capacitance ratio 电容比
capacitance relay 电容式继电器
capacitance-resistance coupling 电容电阻耦合，阻容耦合
capacitance-resistance filter 阻容滤波器
capacitance strain gauge 电容应变计
capacitance swing 电容量范围
capacitance switching-off overvoltage 开断电容负载过电压
capacitance synchronism tolerance 电容量同步偏差
capacitance synchronism 电容量同步
capacitance tester 电容试验器
capacitance timer 电容定时器，电容式定时装置
capacitance to earth 对地电容
capacitance tolerance 电容公差
capacitance transducer 电容传感器
capacitance tube 电容管
capacitance tuning 电容调谐
capacitance type sensor 电容式传感器
capacitance-type strain gauge 电容式应变计
capacitance 电容，容量
capacitive character 电容性
capacitive charging capacity 电容充电容量
capacitive circuit 电容电路
capacitive commutator 电容转换器
capacitive component 电容分量
capacitive coupling 电容耦合，容性耦合
capacitive current 电容电流
capacitive discharge 电容性放电
capacitive divider 电容式分压器
capacitive earth current 对地电容电流
capacitive energy storage equipment 电容能量储存装置
capacitive excitation 电容励磁，电容性自励
capacitive feedback 电容反馈，电容回授
capacitive field 电容电场
capacitive filter 电容滤波器
capacitive impedance 电容性电抗，容抗
capacitive iris 电容性膜片，电容性窗孔
capacitive leakage current 电容漏泄电流
capacitive load 容性负载
capacitive potential divider 电容分压器
capacitive potential transformer 电容式电压互感器
capacitive quiescent current 电容性静态电流
capacitive reactance 容抗，电容电抗
capacitive reactive power 容性无功功率
capacitive-resistance coupling 电容电阻耦合
capacitive sawtooth generator 电容式锯齿波发生器
capacitive-shunting effect 电容分流作用
capacitive stub 电容性短线
capacitive susceptance 电容性电纳，容纳
capacitive tuning 电容调谐
capacitive voltage divider 电容分压器
capacitive voltage 容性电压，电容电压
capacitive 电容性的，容性的，电容的
capacitivity 介质常数，电容量，电容率
capacitometer 电容测量器
capacito-plethysmograph 电容脉波计
capacitor body 电容器器身，电容器组
capacitor braking 电容器制动，电容套管
capacitor case rupture 电容器外壳爆裂
capacitor charge 电容器充电
capacitor commutated converter 电容换相换流器
capacitor depressing voltage 电容降压
capacitor divider 电容分压器
capacitor for voltage protection 保护电容器
capacitor initial voltage 电容初始电压值
capacitor installation 电容器成套装置
capacitor level detector 电容式电平探测装置
capacitor loudspeaker 电容式扬声器，静电式扬声器
capacitor microphone 电容微声器，电容话筒，电容传输器
capacitor motor 电容器式电动机
capacitor oil 电容器油
capacitor packet compressing 电容器芯子压装
capacitor packet predrying 电容器芯子预烘
capacitor packet 电容器芯子
capacitor paper 电容器纸
capacitor phase shifter 电容移相器，电容移相装置
capacitor pick-up 电容拾音器
capacitor plate 电容器极板
capacitor-resistor-diode network 电容-电阻-二极管网络，电容-电阻-二极管电路
capacitor-run motor 电容器运行式电动机
capacitor section 电容区
capacitor specific characteristics 电容器特性
capacitor split-phase motor 电容分相电动机
capacitor stack 电容器叠柱
capacitor start and run motor 电容器启动和运行电动机
capacitor starting 电容器启动
capacitor start motor 电容器启动电动机
capacitor start single-phase motor 电容启动单相电动机
capacitor start 电容器启动
capacitor storage 电容器存贮器
capacitor switching event 电容器开关事件，电容器切换事件
capacitor switch 电容器开关
capacitors with barrage layer 具有阻塞层的电容器
capacitors with thin layer 薄层电容器
capacitor tachometer 电容式转速计
capacitor tester 电容器试验器
capacitor transformer 电容变压器
capacitor trigger 电容触发器
capacitor voltage transformer 电容式电压互感器
capacitor voltage 电容器电压
capacitor 蓄电池，电容器，电瓶

capacitron 电容汞弧管
capacity antenna 电容性天线
capacity attenuator 电容衰减器
capacity balance 电容平衡
capacity balancing network 电容平衡网络
capacity benefit margin 容量效益裕度 CBM
capacity bidding strategy index 申报容量策略指标
capacity bolometer 电容式测辐射热测量计
capacity booster 电容性升压电机
capacity boost 容量增大
capacity bridge 电容电桥
capacity charge 电容充电，设备容量的建设费用
capacity coefficient 流量系数
capacity commutator 电容转换器
capacity condition 能力工况
capacity constant 设备利用常数
capacity continuous rating 能力连续工况
capacity control 功率控制
capacity coupler 电容耦合器，电容耦合元件
capacity coupling 电容耦合
capacity current 容性电流
capacity curve 库容曲线，容量曲线
capacity discharge of well 井出水率
capacity discharge 电容放电
capacity earth 电容接地
capacity effect 电容效应
capacity exceeding number 超位数
capacity expansion investment 扩充生产能力的投资
capacity factor 负荷系数，负载系数，容量系数，库容系数，容量因数，利用率，容量因子，负荷因子，设备利用率，功率，能力系数
capacity fall off 电容漏电，电容量减退
capacity for duties 承担义务的能力
capacity for heat 热容量
capacity for work 做功的能力，工作能力
capacity ground 电容接地
capacity input filter 电容输入滤波器
capacity input smoothing circuit 电容输入平滑电路
capacity lag 电容时滞
capacity loading 满载
capacity loss 容量损失
capacity meter 电容计，电容测量器
capacity multiplier 电容倍增器
capacity of arrival-departure track 到发站通过能力
capacity of competition 竞争能力
capacity of condensate purification equipment 凝结水净化装置的出力
capacity of cylinder 气缸容量
capacity of driven pile 打入桩的承载力，沉桩的承载力
capacity of grout acceptance 吸浆能力
capacity of heat conduction 热传导，导热能力
capacity of high pressure by-pass system 高压旁路容量
capacity of integral by-pass system 整体旁路容量
capacity of low pressure by-pass system 低压旁路系统容量
capacity of oil header tank 集油箱的容量
capacity of oil reservoir 油罐的容量
capacity of oil system 油系统的能力
capacity of power plant 电厂容量
capacity of pump 水泵功率，水泵容量，泵出力
capacity of reservoir 水库库容，库容
capacity of road 道路通行能力
capacity of saturation 饱和量
capacity of storage battery 蓄电池容量
capacity of stream 河流挟沙能力，河流泄水能力
capacity of synchronous condenser 调相容量
capacity of the wind 风卷挟力
capacity of track in station 车站通过能力
capacity of unit 机组容量
capacity of well 水井出水率，井流量
capacity operating rate 开工率
capacity payload 容量有效载荷
capacity pondage 蓄水量
capacity price 容量电价
capacity range 容量限度
capacity ratio 容量比
capacity reactance 容抗
capacity regulator 出力调节器，功率调节器
capacity relay 电容式继电器
capacity reserve 备用容量，备用功率
capacity standard 电容标准
capacity starting 电容式启动
capacity susceptance 容纳
capacity test 电容试验
capacity time lag 电容惯性，电容时滞
capacity to contract 缔约能力
capacity-to-load ratio of distribution system 配电网容载比
capacity to pay 支付能力
capacity unbalance 电容性失衡，电容不平衡
capacity value 功率
capacity 容量，电容，能力，出力，功率，蒸发量，容积，载重量，生产力
capadyne 电致伸缩继电器
cap-and-pin suspension insulator 帽销悬式绝缘子，球形连接盘式绝缘子
capaswitch 双电致伸缩继电器
capblock 锤垫
cap bolt 倒角螺栓，盖螺栓，紧固螺栓
CAP＝capacity 容量，功率
CAP(computer aided production) 计算机辅助生产
cap contact 灯头触点
cape chisel 扁类凿
CAPES(containment atmosphere purge exhaust system) 安全壳空气吹扫排放系统
cape sleeve （焊工）开襟短背心，蝙蝠袖，蝴蝶袖
cap/fuel clearance 包壳/燃料间隙
capillarimeter 毛管测液器，毛管检液器，毛细管仪
capillarity absorption 毛细管吸收作用
capillarity phenomenon 毛管现象
capillarity 毛细现象，毛细作用，毛管作用，毛

细管作用
capillary absorption 毛细管吸收作用
capillary action 毛管作用
capillary ascent 毛管水上升高度
capillary attraction 毛管吸力，毛细引力
capillary boiling 毛细管沸腾
capillary capacity 毛管容量
capillary channel 毛细管道
capillary condensation 毛管凝缩作用，毛管凝结
capillary condensed water 毛管凝结水
capillary conductivity 毛管传导度
capillary crack 发状裂缝，毛细裂纹
capillary depression 毛管水下降，毛细下降
capillary detector 微管检波器
capillary effect 毛细管作用
capillary electrometer 毛细管静电计
capillary elevation 毛细上升
capillary flow 毛管水流，毛细管流动
capillary force 毛管力
capillary fringe belt 毛细管水边缘带
capillary fringe 毛管作用带，毛管边缘，毛管水边缘，毛细管水上升边缘
capillary front 毛管湿锋
capillary head 毛管水头，毛细管水头
capillary height 毛细管水上升高度
capillary interstice 毛管间隙
capillary leak 毛管渗漏
capillary lift 毛管水上升高度
capillary migration 毛细管水移动，毛细管移动
capillary moisture capacity 毛管持水量
capillary movement 毛管水运动
capillary opening 毛细孔
capillary penetration 毛管渗透
capillary phenomenon 毛细管现象
capillary porosity 毛管孔隙度，毛管孔隙率
capillary porous material 多孔毛细材料
capillary potential 毛管水位能
capillary pressure 毛细（管）水压力，毛细压力
capillary pulling power 毛管张力
capillary ripple 涟波
capillary rise 毛管上升，毛管水高度，毛管水上升高度
capillary seepage 毛管渗漏
capillary siphoning 毛管虹吸作用
capillary stage 毛管水位
capillary suction head 毛管吸升高度，毛管吸水头
capillary suction 毛管吸力
capillary tension 毛管张力，毛细管张力
capillary tube guard 毛细管保护罩
capillary tube thermostat 毛细管温度自动调节器
capillary tube 毛管，毛细管
capillary tubing 毛细管，节流管束
capillary viscosimeter 毛管黏度计
capillary watering 毛管灌水
capillary water 毛管水，毛细管水，毛细水
capillary wattmeter 毛细管式功率表
capillary wave 表面张力波，涟波，毛管波
capillary wetting method 毛管湿润法
capillary zone 毛细管区
capillary 毛细作用的，毛细管的，极细的，表面张力的，毛管的，毛细管，毛细管作用
capillometer 毛细管试验仪
capister 变容二极管
capital account balance sheet 资本账户平衡表
capital adequacy 资本充足
capital and operating cost estimate 投资和运行费用估算
capital appreciation tax 资本增值税
capital assets 固定资产，资本资产，资本，资产
capital bonus 资本红利
capital charges 投资费用
capital commitment 资本承担，资本承诺，资本支出，已承担资本支出
capital construction investment 基本建设投资
capital construction loan 基建贷款
capital construction plan 基本建设计划
capital construction procedure of electric power project 电力工程基本建设程序
capital construction procedure 基本建设程序
capital construction project 基本建设项目
capital construction 基建，基本建设
capital contribution of power-supply 供电贴费
capital cost 基本建设费，投资成本，投资费，基本费用，资本成本
capital credit certificate 资本信用证明书
capital equity ratio 资本产权比率
capital expenditure 基本费用，主要费用，资本支出，基本建设费用
capital expense 基本建设费用
capital fund 基建资金，资本基金
capital gain tax 资本收益税，财产收益税，资本利得税，资本增值税
capital group 资本集团
capital income 资本收入，资本所得
capital intensive enterprise 资本密集企业
capital intensive industry 资本密集型工业
capital intensive project 资本密集型项目，资金密集项目
capital and interest price 资本和利息价格
capital investment analysis 基本建设投资分析
capital investment cost 基建投资，基建费
capital investment 投资
capitalization ratio 资本化比率
capitalized cost estimate 资本化成本［代价］估算法
capitalized cost 投资总额
capital levy 财产税，资本税
capital of individuals 个人资本金
capital of legal entities 法人资本金
capital of the state 国家资本金
capital out of budget 预算外资金
capital pay-off time 投资回收期
capital project fund 基本建设项目基金
capital project 基本投资项目
capital rating 资产评估
capital recovery factor 资金回收因子
capital recovery time 资本回收时间
capital repair 大修
capital shortage 资金短缺
capital stock certificate 股本凭证
capital stock 股本

capital structure　资本结构
capital sum　本金总额
capital surplus　资本公积金
capital transfer　资本转让
capital works　基本建设工程
capital　本金，资本，主要的，资产，资金
capitation fee　均摊的费用
caplastometer　黏度计
cap nut　盖形螺帽，罩螺母
capofthermal　热泡冠，帽热
CAPP(computer aided process planning)　计算机辅助工艺规划
capped end　（电缆）加帽端
capped pipe　封头管子
capped steel　压盖沸腾钢，封顶沸腾钢
capping beam　压顶梁
capping plate　盖板
capping run　表面焊缝
capping sand filter　顶层用无烟煤的砂滤器
capping　表面加工，封顶，盖顶石，盖面，压顶，压盖
cap plug　盖塞
caprone cable　锦纶绳
caprone　聚己内酰胺纤维，锦纶，卡普隆
cap screw　有头螺栓，内六角螺钉
capsizing moment　倾覆力矩
capstan motor　主动轮电动机
capstan　绞盘，卷扬机，电动葫芦，起锚机，主动轮
capstone　压顶石
capsular-spring gauge　蒴状弹簧式压力计，囊状弹簧式压力计
capsule building　盒式建筑
capsule-tube transmitter　容器式管道输送
capsule type manometer　膜盒式压力计
capsule　胶囊，小容器，包壳，膜盒式，膜片
captain　队长
captance　容抗
captioned　上述的，标题下的，标题所述的
caption of drawing　图标
caption　标题，标题栏，题目
captive balloon　系留气球
captive bolt　固定螺栓，锁紧螺栓
captive fluid indicator　密封磁悬液指示器
captive key　锁紧键，不脱落键
captive power plant　自备电厂，自备电站
captive screw　锁紧螺丝
captive test　静态试验，试验台试验
captive wake　可捕获的尾迹
cap top　顶盖
capture and recapture experiment　标识放流试验
capture area　捕获面积，获风面积
capture cross section　捕获截面，获风截面，俘获截面
capture efficiency　俘获率
capture efficient　捕获效率
capture gamma radiation capture　俘获γ辐射
capture gamma radiation　俘获γ辐射
capture gamma　俘获的γ粒子，俘获γ粒子
capture market　占领市场
capture of neutrons　中子俘获
capture radiation　俘获辐射
capture rate　捕获率，俘获率
capture reaction　俘获反应
capture resonance　俘获共振
capture velocity　控制风速
capture　捕获，捕捉，俘获，收缴，袭夺，赢得客户
capturing hood　外部吸气罩
capturing medium　俘获介质
cap　管帽【即封头】，帽，罩，帽状拱顶，顶盖，盖板，盖，帽子，管底，帽状拱顶
car 车辆，车皮
caracole　旋梯
car arrestor　阻车器
carbamide resin　尿醛树脂
carbamide　脲，尿素，尿素
carbide analysis　金属碳化物分析，碳化物分析
carbide bit　硬质合金刀头，硬质合金钻头
carbide blade　硬质合金刀片
carbide drill　硬质合金钻头
carbide-feed generator　乙炔发生器
carbide fuel assembly　碳化物燃料组件
carbide-fuelled breeder　碳化物燃料增殖反应堆
carbide milling cutter　硬质合金铣刀
carbide network　网状碳化物
carbide tripped reamer　硬质合金铰刀
carbide　碳化物，碳化钙，硬质合金，电石
carbocoal 半焦
car body　车身，车厢
carbolic acid　石炭酸，苯酚
carbonaceous coal　半无烟煤
carbonaceous fuel　碳质燃料
carbonaceous material　含碳物质，碳质材料
carbonaceous matter　含碳物质
carbonaceous shale　碳质页岩
carbonaceous　碳的，含碳的
carbon adsorption　碳吸附
carbon and charcoal　碳和木炭
carbon anode　碳阳极
carbon arc air gouging　碳弧气刨
carbon arc cutting　碳弧气刨，碳弧切割
carbon arc lamp　碳弧灯
carbon arc welding　碳弧焊
carbon arc　碳弧
carbon arrestor　碳质放电器，碳质避雷器
carbonatation　碳酸盐化作用
carbonate alkalinity　碳酸盐碱度
carbonate carbon dioxide　碳酸盐二氧化碳
carbonate carbon　碳酸盐碳
carbonated spring　碳酸泉
carbonate hardness　碳酸盐硬度
carbonate roasting　碳酸盐焙烧
carbonate rock　碳酸盐岩
carbonate scale　碳酸盐垢
carbonate　碳酸盐
carbonation　碳酸化，碳化，碳酸饱和
carbonatization　碳酸化作用
carbonator　碳酸化器
carbon-bearing　含碳的
carbon black　碳黑，烟黑
carbon body　电刷

carbon breaker 碳触点断路器
carbon-break switch 碳触点开关
carbon brick 碳砖
carbon brush collector 碳刷集电器
carbon brush generator 有碳刷的发电机
carbon brush holder 碳刷握，碳刷架
carbon brush yoke 碳刷架
carbon brush 碳刷
carbon built-up 积碳
carbon burn-up cell 碳燃尽床【沸腾炉】
carbon capture and storage 碳捕集与封存
carbon capture 碳捕集，碳捕捉
carbon carburizing steel 碳素渗碳钢
carbon case hardening 渗碳硬化
carbon cast steel 碳素铸钢
carbon chloroform extraction 碳氯仿萃取
carbon column 活性炭柱
carbon commutator 带碳刷的换向器
carbon constructional quality steel 优质碳素结构钢
carbon contact pickup 碳粒接触拾音器
carbon contact 碳触点
carbon content 含碳量
carbon-copper contact film 碳铜接触膜
carbon copy 复写副本
carbon core 碳心，碳化点
carbon credit 碳信用，碳信用额
carbon current collector for electric locomotive 电力机车滑块
carbon cycle 碳循环，碳素循环
carbon-14 dating 碳 14 测年法
carbon depletion 脱碳【焊缝热影响区】
carbon deposit 积碳，积焦，炭垢
carbon dioxide 二氧化碳
carbon dioxide arc welding 二氧化碳气体保护焊
carbon dioxide bottle 二氧化碳瓶
carbon dioxide capacity 二氧化碳容量
carbon dioxide cleanup system 二氧化碳净化系统
carbon dioxide combining power 二氧化碳结合力
carbon dioxide content 二氧化碳含量
carbon dioxide coolant 二氧化碳冷却剂
carbon dioxide cooled graphite moderated reactor 二氧化碳冷石墨慢化反应堆
carbon dioxide cooled natural uraniumfuelled reactor 二氧化碳冷却天然铀反应堆
carbon dioxide cooled pressure tube reactor 二氧化碳冷却压力管反应堆
carbon dioxide cooled reactor 二氧化碳冷却反应堆
carbon dioxide corrosion 二氧化碳腐蚀
carbon dioxide cycle 二氧化碳循环
carbon dioxide detector 二氧化碳探测器
carbon dioxide exhaust system 二氧化碳排出系统
carbon dioxide extinguisher 二氧化碳灭火器
carbon dioxide extinguishing agent 二氧化碳灭火剂
carbon dioxide fire extinguisher 二氧化碳灭火器
carbon dioxide fire extinguishing system 二氧化碳灭火系统
carbon dioxide fire vehicle 二氧化碳消防车
carbon dioxide fixation 固二氧化碳作用
carbon dioxide flask 二氧化碳瓶
carbon dioxide-free water 无二氧化碳水
carbon dioxide gas 二氧化碳气体
carbon dioxide in carbonate 碳酸盐二氧化碳
carbon dioxide pump truck 二氧化碳泵车
carbon dioxide purification system 二氧化碳净化系统
carbon dioxide reactivity 二氧化碳反应性
carbon dioxide release 二氧化碳释放
carbon dioxide semiauto-welder 二氧化碳半自动焊机
carbon dioxide sensor 二氧化碳传感器
carbon dioxide storage system 二氧化碳储存系统
carbon dioxide system 二氧化碳系统
carbon dioxide transpiration 二氧化碳发散冷却
carbon dioxide trap 二氧化碳冷阱
carbon dioxide treatment for alkaline waste 二氧化碳处理碱性废水
carbon dioxide universal auto-arc welder 二氧化碳万能自动弧焊机
carbon-disk microphone 碳盘传声器
carbon donor 供碳体
carbon electrode 碳电极，碳棒，石墨棒
carbon equivalent 碳当量
carbon felt blanket 含碳毡垫
carbon fiber material 碳纤维材料
carbon fiber reinforced plastic 碳素纤维增强塑料
carbon fiber 碳纤维
carbon-fibre brush 碳丝电刷，碳化纤维电刷
carbon-fibre end-ring 碳化纤维护环【汽轮发电机】
carbon-filament lamp 碳丝灯，碳丝灯泡
carbon-film potentiometer 碳膜电势计
carbon-film resistor 碳膜电阻器
carbon filter 活性炭过滤器，碳过滤器，碳滤池
carbon footprint 碳足迹
carbon-free 不含碳的
carbon glass reinforcement 碳玻璃加强件
carbon granule 碳砂
carbon-graphite antifriction material 碳石墨抗磨材料
carbon-graphite brush 碳石墨电刷
carbon-graphite cloth 碳石墨布
carbon-graphite felt 碳石墨毡
carbon-graphite fiber 碳石墨纤维电刷
carbon heating element 碳素发热元件
carbonic acid 碳酸
carbonic oxide cell 氧化碳电池
carbonic 碳的，含碳的
carboniferous system 石炭系
carboniferous 碳系的
carbonify 碳化
carbon internals 碳制内部构件
carbonisation 碳化，渗碳作用，焦化，干馏
carbonite 碳质炸药，硝酸甘油，硝酸钾
carbonitride fuel 碳氮化物燃料
carbonitriding 碳氮共渗
carbonization test 干馏试验，焦化试验，渗碳试

验
carbonization zone 碳化层
carbonization 碳化，渗碳作用，焦化，干馏
carbonized oil 碳化油
carbonized particle 碳化包覆颗粒
carbonize 碳化，焦化，炼焦
carbonizing flame 碳化焰，还原焰
carbonizing zone 碳化层，碳化区
carbon lamp 碳精弧光灯
carbon lightning arrester 碳质避雷器
carbon loss 碳损失
carbon membrane 碳膜
carbon migration 碳迁移
carbon-moly steel 钼钢
carbon monoxide canister 一氧化碳滤毒罐
carbon monoxide detector 一氯化碳检测器
carbon monoxide filter 一氧化碳过滤器
carbon monoxide index 一氧化碳指数
carbon monoxide poisoning 一氧化碳中毒
carbon monoxide pollution 一氧化碳污染
carbon monoxide recorder 一氧化碳记录仪
carbon monoxide 一氧化碳
carbonous 碳的，碳化的
carbon packing ring 碳精密封环
carbon packing 碳精汽封，碳封，炭精轴封
carbon penetration 渗碳
carbon pile automatic voltage regulator 碳堆自动电压调整器
carbon pile regulator 碳质稳压器
carbon pile resistor 碳电阻片柱
carbon pile voltage transformer 碳堆变阻器，碳堆电压变换器
carbon pile 石墨反应堆
carbon reduction 碳还原法
carbon regulator 碳调整器，碳堆变阻器
carbon resistance thermometer 碳电阻温度计
carbon resistance 碳电阻
carbon resistor 碳电阻器
carbon rheostat 碳质变阻器
carbon ring gland 碳精汽封圈
carbon rod 碳棒
carbon saturation 碳饱和
carbon sealing 石墨密封
carbon seal ring 碳精密封环
carbon segment 碳精瓦块
carbon separation 碳析出，渗碳
carbon steel pipe 碳钢管
carbon steel 碳素钢，碳钢
carbon structural steel 碳素结构钢
carbon-sulphur analyzer 碳硫分析仪
carbon test 含碳量试验
carbon tetrachloride fuse 四氯化碳熔断器
carbon tetrachloride extinguisher 四氯化碳灭火器
carbon tetrachloride 四氯化碳
carbon tool steel 碳素工具钢
carbon transducer 碳精式换能器，碳精传感器
carbon trap 碳阱
carbon-uranium reactor 铀石墨反应堆，铀石墨堆
carbonyl 羰基
carbonyl dust core 铁氯体磁芯，碳粉铁芯
carbon 碳，碳棒，碳刷，碳的

carborundum detector 金钢砂检波器，碳化硅检波器
carborundum grain 金刚砂
carboseal 捕灰用的黏附剂
carboxyl cation exchanger 羧基阳离子交换剂，羧基阳离子交换器
carboxyl group 羧酸基团
carboxylic acid group 羧酸基
carboxylic acid 羧酸
carboxylic resin 羧酸树脂
carboxylic 含羟基的
carboxyl methyl cellulose 羚甲基纤维素，羧甲基纤维素
carboxy reactivity of coal 煤对二氧化碳的反应性
carboxy reactivity 羰基化反应能力
carboy 瓮，坛
car bumper 车挡
carburate 渗碳，汽化
carburation 渗碳，汽化
carburator 汽化器，化油器
carbureter 汽化器，混合器，渗碳器
carbureting pilot 气化式引燃器，空气煤气混合式引燃器
carburetion 渗碳，汽化
carburetor bowl 汽化室的浮子室
carburetor 汽化器，渗碳器
carburet 渗碳，汽化
carburization material 渗碳剂
carburization zone 碳化层
carburization 渗碳剂，渗碳作用，渗碳，渗碳处理，碳化
carburized case depth 渗碳层深度
carburized case 渗碳层
carburized depth 渗碳层深度
carburized layer 渗碳层
carburized structure 渗碳组织
carburized zone 碳化区，渗碳区，渗碳层
carburizer 渗碳剂
carburize 渗碳，碳化
carburizing agent 渗碳剂
carburizing box 渗碳箱
carburizing by molten salts 液体渗碳
carburizing by solid matters 固体渗碳
carburizing flame 碳化焰【还原焰】
carburizing furnace 渗碳炉
carburizing gas 气体渗碳剂
carburizing liquid 渗碳液
carburizing 碳化，渗碳处理，碳化焰，渗碳
carcase-roofing 毛屋面
carcase 壳体，骨架，残骸，车架，（船、车）架子，机壳，机座，构架，定子，框架，尸体
carcase work 骨架工程，预埋（管线）工程
carcass fabric 芯体织物
carcassing 大型建筑物，预埋管线工程，骨架制作
carcass 骨架，壳体，残骸，车架，房屋骨架，框架
Carcel lamp 卡索灯
carcinogenesis 辐射致癌，致癌作用，致癌性
carcinogenic compound 致癌化合物

carcinogenic dose 致癌剂量
carcinogenic effect 致癌效应
carcinogenic industrial chemicals 致癌工业化学品
carcinogenic 致癌的
carcinogen 致癌物
carcinotron 返波管，回波管
car-clamping assembly 夹车装置
CAR(corrective action request) 纠正措施要求【质保用语】
car coupler 车钩
cardan joint 万向接头
cardan shaft 万向轴
card 插件，卡件，卡片，插件板
cardboard drain 纸板排水
cardboard 纸板
card cage 卡片箱
card checking 穿孔卡效验
card code 卡片代码
card connector 印刷板插座，卡件插座
card-controlled calculator 卡片控制计算器
card counter 卡片计数器
card counting attachment 卡片计数装置
card data converter 卡片数据转换器
carder 载体，载波
card face 卡片正面
card feed 卡片输入装置，卡片馈送机
card field 卡片范围，卡片上的段，穿孔卡片栏
card file 卡片储存器
card fluff 卡孔残留纸片
card format 卡片格式
card holder 卡片盒
card hopper 储卡机，送卡箱
cardhouse fabric 片架组构
cardhouse 势必倒塌的建筑，必然失败的计划
card image 卡片映像
cardinal change 根本性变更，重大变更
cardinal point 基点，坐标点，基本点，方位点
cardinal wind 主要风向
card input 卡片输入
cardioid diagram 心形特性曲线图
cardioid microphone 心形反向性传声器
card jam 卡片卡位，卡片阻塞
card loader 卡片装入程序
card machine 卡片机
card magazine 卡片箱，卡片盒
card pocket 卡片袋
card processor 卡片数据处理装置
card programming 卡片程序设计
card punch buffer 卡片穿孔缓冲器
card punched unit 卡片穿孔装置
card punching machine 卡片打孔机
card punching 卡片穿孔
card punch 卡片穿孔机
card random access memory 卡片随机存取存储器
card reader 卡片输入机，卡片阅读机，卡片读出器，读卡机
card recorder 卡片记录器
card recording 卡片记录
card reproducer 卡片复穿孔机，卡片复制机
card row 卡片上的一行
card sorting machine 卡片分类机
card sorting 卡片分类
card stacker 叠片机
card system 卡片系统
card test extender 功能板测试延伸器，卡件测试延伸板
card test module 功能板卡件测试模件
card-to-card transceiving 卡片-卡片发送接收
card-to-disk conversion 卡片-磁盘转换
card-to-print program 卡片-打印程序
card-to-tape conversion 卡片-磁带转换
car dumper area 翻车机作业区
car dumper building 翻车机室
car dumper house 翻车机室
car dumper pit 翻车机坑
car dumper receiving hopper 翻车机受煤斗
car dumper （翻斗车）翻斗，翻车机，汽车倾卸机，倾倒卸货车
card verifier 卡片校验机
card wreck 卡孔损坏
care and maintenance party 维护保养组
CARE(computer-aided reliability) 计算机辅助可靠性鉴定
career training 职业培训，职业训练
car ejector 推车器
careless mistake of operator 操作者的疏忽致错
carelessness 疏忽
care 关怀，照顾
car ferry 汽车渡口，汽车轮渡
car frame 车架
cargo accommodation 货载船位
cargo account settlement system 货运账目清算系统
cargo aircraft only 仅限货机
cargo availability at destination in 货物运抵目的地
cargo block 吊货滑车
cargo booking advance 国际航空货物订舱单
cargo capacity 载货容积，载货能力，载荷能力
cargo charges correction advice 货物运费更改通知
cargo damage inspection 货损检验
cargo damage prevention 货损预防
cargo handling system 货物装卸系统
cargo in bulk 散装货，散货
cargo inspection 验货
cargo insurance 货物保险，货物运输保险
cargo lift 载货电梯
cargo liner 定期货船
cargo list 货物清单
cargo marine insurance 海洋货物保险
cargo mark 货物标志
cargo net 吊货网
cargo port 货运港
cargo receipt 货运收据
cargo receiver 收货单位，收货人
cargo ship 货船
cargo steamer 货轮
cargo supplier （供）货方
cargotainer 集装箱
cargo terminal 货运终点站

cargo transfer manifest 转运舱单
cargo transhipment 货物转口
cargo transportation insurance 货物运输保险
cargo transportation 货物转运
cargo transshipment 货物转口
cargo underwriter 货物承保人
cargo volume 货量
cargo winch 起货机，起货绞车
cargo wire 吊货索
cargo 货物，负荷，荷重，重量，船货
car heater 车皮加热器
car lane 车行道
carline 短纵梁，纵梁
car load 车辆荷载
Carlson stress meter 卡尔逊应力计
carminic acid 胭脂红酸
Carnot cycle 卡诺循环
car park 停车场
CARP（computed-aided reliability program） 计算机辅助可靠性分析程序
carpenter's shop 木工车间，杠场
carpenter's square 木工直角尺
carpenter's yard 木工场
carpenter 木工
carpentry shop 木工厂
carpentry work 木工工程
carpentry 木工工作，木器
carpet herb 地皮草
carpeting 地毯状【VII级植物风力指示】
carpet strip 地毯挡条
carpet 毯
carport 汽车棚，停车场，停车库
car positioner arm 定位臂【车皮】
car positioning device 车皮定位装置
car puller 牵车器
car retarder 缓行器
carriage bolt 大车螺栓
carriage by air 航空运输，空运
carriage by land and sea 海陆运输，海陆运送
carriage by land 陆运
carriage by rail 铁路运输
carriage car case 车架
carriage contract 运输合同，运送契约
carriage drive rack 拖板传动架
carriage expenses 运费
carriage forward 运费待付，运费到付，运费由提货人支付，运费未付
carriage mounting 台车钻架
carriage paid 运费付讫，运费已付，运费已由收件人缴付
carriage return character 回车符号
carriage return contact 滑架回程接点，回车接点
carriage return 回车符号，回车，打印机的滑动架回位
carriage shed 车棚
carriage take-up 车式拉紧装置
carriage way 车道，车行道
carriage wrench 套筒扳手
carriage 车架，台架，小车，导轮架，托架，运输，货运，拖板，滑架，斜梁，运费
carrier-actuated relay 载波激励继电器，载频启动继电器
carrier air 输送空气，送粉风
carrier-amplifier 载频放大器
carrier arm 承载臂
carrier blocking 载波闭锁
carrier cable 缆车钢索，载重索
carrier capture 载流子俘获
carrier channel 载波信道，载波通路
carrier circuit 载波电路
carrier communication 载波通信
carrier compound 载体
carrier computer 载波计算机
carrier-current channel 载波通道
carrier-current communication 载波通信
carrier-current protection 高频保护装置，载波保护装置
carrier-current signaling channel 载波信道，载波信号通路
carrier-current telephony 载波电话学，载波电话
carrier-current 载波电流
carrier density 载流子密度
carrier diffusion 载流子扩散
carrier distribution 载流子分布
carrier drift transistor 载流子漂移型晶体管
carrier-free 无载流子的，无载波的无载体的
carrier frequency amplifier 载频放大器
carrier frequency cable 载频电缆
carrier frequency coupling device 载频耦合装置
carrier frequency filter coil 载频滤波器线圈
carrier frequency generator 载频发生器
carrier frequency hologram 载频全息图
carrier frequency oscillator 载频振荡器
carrier frequency repeater equipment 载频中继装置
carrier frequency synchronization 载频同步
carrier frequency terminal equipment 载频终端设备
carrier frequency 载波频率
carrier gas 载气，运载气体
carrier generator 载波振荡器，载波发生器
carrier intertripping 载波联锁跳闸
carrier ion 载体离子
carrier isolating choke coil 载波隔离扼流圈
carrier leak balancer 载波泄漏平衡器
carrier leak system 载波泄漏系统
carrier leak 载波泄漏
carrier level 载波电平
carrier life time 载流子寿命时间
carrier line 载波线路
carrier liquid 载液
carrier load control 载波负荷控制，载波功率控制
carrier loading 载波负载
carrier material 底材，衬材，载体
carrier mobility 载流子迁移率，载流子平均漂移速度，载体迁移率
carrier modulation percentage 载波调制度百分率
carrier molecule 运载粒子
carrier noise level 载波噪声电平
carrier noise ratio 载波噪声比

carrier noise 载波噪声
carrier or pilot-wire receiver relay 载波或导引线接受继电器
carrier phase 载波相位
carrier pilot relay system 载波遥控中继方式
carrier pilot system 载波导频系统，载波控制系统
carrier position 载波位置
carrier power 载波功率
carrier precipitation 载体沉淀
carrier protection equipment 载波保护装置
carrier protection 载波保护，高频保护
carrier rack for spare metal O-rings 备用金属O形环托架
carrier recovery 载波恢复
carrier reference black level 黑电平参考载波
carrier reference white level 白电平参考载波
carrier reinsertion 载波重置，载波重新插入
carrier relaying 载波继电保护，高频保护，载波中继
carrier relay protecting device 载波中继保护装置
carrier relay 载波继电器
carrier repeater 载波增音机
carrier residual modulation noise 载波残余调制噪声
carrier-ring orifice plate 环室式孔板
carrier-ring orifice 环室孔板
carrier-ring 静叶持环，隔板套
carrier screening 载流子屏蔽
carrier shift 载频漂移，载波漂移
carrier signal 载波信号
carrier's liability insurance 承运人责任保险
carrier source 载波源
carrier space charge 载流子空间电荷
carrier spectrum 载波频谱
carrier storage effect 载流子存储效应
carrier storage time 载流子存储时间
carrier storage 载流子存储
carrier suppression 载波抑制
carrier swing 载波摆值
carrier synchronization 载波同步
carrier system 载波通信系统
carrier telegraphy 载波电报，载波电报学
carrier telephone terminal 载波电话终端机
carrier telephone 载波电话
carrier telephony on power circuits 电力电路上的载波电话
carrier telephony 载波电话，载波电话学
carrier-to-noise ratio 载波噪声比
carrier transfer filter 载波转移滤波器
carrier transmission system 载波传输制
carrier transmission 载波传输
carrier vessel reactor 航空母舰反应堆
carrier voltage 载波电压
carrier wave jamming 载波拥挤
carrier wave telephony 载波电话学，载波电话
carrier wave 载波
carrier 承运人，吊具，承载器，货舱，承运单位，搬运人，搬运机，载体，载波，运载工具，输送器，运输机，托架，轮式铲运机，运料车
carry-back of loss 扭转亏损
carry chain 进位链
carry circuit 进位电路
carry clear signal 进位清除信号
carry-complete signal 进位完成信号
carry delay 进位延迟
carry detecting circuit 进位检验电路
carry detection 进位检测
carry digit 进位数，进位数字
carry failure 进位失败
carry flag 进位标记
carry flip-flop 进位触发器
carry gate signal 进位门信号
carry gate 进位门
carry gating circuit 进位门电路
carrying area 升力面
carrying bolt 支撑螺栓
carrying capacity of a pile 桩的承载能力
carrying capacity of drainage system 排水系统的过水能力
carrying capacity of pipe 管道过水能力
carrying capacity of turbine 汽轮机的通流能力
carrying capacity 承载量，装载量，载重量，挟带能力，载带能力，负载能力，输水能力
carrying current conductor 载流导线，载流导体
carrying current 极限电流
carrying idler 承载托辊，承重托辊，空释
carrying plane 支承面，承力面
carrying power 承载力
carrying rod 绑线，(发电机转子的）护环
carrying roller 拖动辊，支承辊
carrying rope 承载索
carrying side 载料侧
carrying strand 承载分支
carrying velocity 带出速度
carrying well 承重墙，承压墙
carry initiating signal 进位起始信号
carry input 进位输入
carry in 携带，带进，输入
carry logic 进位逻辑
carry line 进位线
carryload scraper 铲运机
carry out a contract 履行合同，执行合同
carry output 进位输出
carry out rate fraction 夹带率份额
carry out 实行
carry-over cinder 飞灰
carry-over effect 滞后效应
carry-over efficiency 传递效率
carry-over factor 传递系数，传递因子
carry-over flow 前期降雨径流
carry-over loss （蒸汽）带水损失，出口损失，级间余速损失【汽轮机】，飞灰损失
carry-over moment 传递力矩
carry-over rate of solids 固体带出速度
carry over storage plant 多年调节电站
carry over storage reservoir 多年调节水库
carry over storage station 多年调节电站
carry over storage 多年调节库容，多年蓄水
carry-over （蒸汽）带水，（蒸汽）带水率，携

带，汽水共沸，级间余速，继续存在
carry pulse 进位脉冲
carry reset 进位复位
carry-save adder 进位存储加法器，保留进位加法器
carry skip 跳跃进位
carry storage device 进位存储装置
carry storage register 进位存储寄存器
carry storage 进位存储器
carry time 进位时间
carry-under 水中带汽，夹带，带汽率
carry 运输，传递……到
car seal close 铅封关
car seal open 铅封开
car shake-out 翻车机
car shaker 振车器
car shed 车棚
cars in the train 列车的车皮
car spotter 车辆定位器
car starter 汽车启动器
car storage track 存车线
carst 岩溶，喀斯特，石灰岩溶洞
car switching arrangement 调车装置
car switching 调车
cart 小车
cartage fee 卡车运费
cartage 搬运费，车运，车运费，货车运费，运货费
cartburetted iron 碳化铁
carter of science and technology 科技中心
Cartesian coordinate system 笛卡儿坐标系
Cartesian coordinate 笛卡尔坐标，正坐标，直角坐标
cartesian diver 浮沉子
carte 扫气室，换气室
car thawing equipment 车辆解冻设备
cartographic feature 地物图像
cartographic grid 地图格网
cartography 绘制图表，制图学，绘图法，制图，制图法
carton 纸板箱，纸板盒
car transfer platen 迁车台
car transformer 车用变压器
car traverser 牵车机
cartridge cooling pond 燃料冷却水池
cartridge filter 管式过滤器，滤芯式过滤器，盒式过滤器，卡盘过滤器，弹筒过滤器，滤元式过滤器，心盘过滤筒，筒式过滤器，滤筒式除尘器
cartridge fuse-link 管装熔断片
cartridge fuse 保险丝管，管装熔丝，熔丝管
cartridge of blade 叶栅
cartridge skip transporter 斗式运输机
cartridge 盒式磁盘，燃料元件，卡盘，滤芯，燃料元件盒，过滤器芯子，滤筒
cartype conveyor 小车式输送机
car unloader 卸车机
car unloading device 卸车装置
carve 雕刻
carving 雕刻，雕刻物
car washer 洗车机
caryatid 女像柱
CAR 汽轮机低压缸排汽口喷淋系统【核电站系统代码】
cascade aerator 阶式曝气器，水梯通气器，梯级通气器
cascade amplifier 级联放大器
cascade battery 级联电池组
cascade blade 叶栅的叶片
cascade breakers 顺序启闭断路器组
cascade carry 逐位进位
cascade compensation 串联补偿
cascade configuration 叶栅外形，叶栅组合
cascade-connected transformer 串级变压器，串联变压器
cascade-connected 串联的，级联的
cascade connection 串联，级联，串级接线
cascade controller 串级调节器
cascade control module 串级控制模件，串级控制组件
cascade control system 级联控制系统，串级调节系统，串级控制系统
cascade control 串联调速，串级控制，级联控制，串级调节，梯级控制
cascade converter 级联变换器，串级变换器，串联变流器
cascade conveyor 串联输送机
cascade current transformer 级联电路互感器
cascade dams 梯级坝
cascaded drain 逐级疏水
cascade design 叶栅设计，叶栅计算
cascade development 河流梯级开发，梯级开发
cascade drag coefficient 叶栅气动阻力系数
cascade evaporation 串级蒸发
cascade evaporator 阶流蒸发器
cascade failure （电力系统的）事故联锁反应，联锁性故障
cascade fault conditions 逐级故障条件，逐级故障状态
cascade filter 级联滤波器
cascade flow 叶栅流动，叶栅气流，叶栅绕流
cascade generator 级联发生器，串级发电机
cascade geometry 叶栅几何参数
cascade heat exchanger 淋水式换热器
cascade hydroelectric station 梯级水电站
cascade hydropower station 梯级水电站，多级水电站
cascade impactor 多级碰撞取样器
cascade machines 级联机械，级联电机
cascade method 串级接法，级联接法
cascade motor set 串级电动机组
cascade motor 级联电动机
cascade of blades 叶栅
cascade of straightline profile 平板型叶栅
cascade oiling 油环润滑
cascade operation 级联操作，级联运行
cascade outages 串级停电事故，串级式断电
cascade performance 叶栅特性
cascade phase modulation 级联调相
cascade pitch-chord ratio 叶栅节弦比
cascade planning 梯级规划
cascade potential transformer 级联式电压互感

器，串接电压互感器
cascade power stations 梯级电站
cascade rectifier 串级整流器
cascade reservoir 梯级水库
cascade screen 级联屏蔽，级屏蔽
cascade sequence 串列顺序
cascade set 串级机组，级联式机组，串级设定，串级式机组
cascades ladder 级联的梯子
cascade solidity 叶栅稠度
cascade spacing 栅距
cascade speed control 串级调速
cascade starter 级联启动器
cascade station 串级控制站
cascade system 串级系统，逐级系统
cascade testing technique 叶栅试验法，叶栅试验技术
cascade test 叶栅试验
cascade theory 叶栅理论
cascade transformer 串级变压器，级联变压器
cascade trigger circuit 串级触发电路
cascade tripping 逐级跳闸，串级跳闸
cascade tunnel 叶栅风洞
cascade turning angle 叶栅转折角
cascade-type potential transformer 串级式电压互感器
cascade unit 级联装置
cascade voltage doubler 串级倍压器
cascade wind tunnel 叶栅风洞
cascade 串联，小瀑布，阶梯，级，级联，串级，串接，急滩，梯级，叶栅
cascading flow 梯级跌水
cascading method 级联法，串级连接法
cascading of insulators 绝缘子串级网络
cascading 串级解列，级联，串级连接，串级崩溃
CAS＝casing 缸，壳
cascode amplifier 栅阴放大器，共射共基放大器
cascode 栅地阴地放大器，共源共栅
CAS(compressed and instrument air system) 压空与仪表空气系统
CAS(computer-aided system) 计算机辅助系统
CAS(content addressable storage) 内容定址存储器
CAS(control automation system) 控制自动化系统
case bay 桁间，梁间距
case carburizing 表面渗碳
case command summary 计算机辅助仿真培训命令概要
cased bore-hole 套管钻孔
cased butt coupling 套筒式联轴器
cased column 箱形柱
case depth 表面硬化层深度，硬化层深度，渗碳层深度
cased hole 下套管的钻孔
cased impeller 闭式叶轮
cased-muff coupling 刚性联轴节[器]，套筒联轴节，刚性联轴
cased pile 套管桩
case-hardened casting 冷硬铸件，冷硬铸造
case-hardened glass 钢化玻璃
case-hardened steel 表面渗碳硬化钢，表面硬化钢
case-hardening carburizer 表面硬化渗碳剂
case-hardening steel 渗碳钢，渗碳用钢，表面硬化钢
case-hardening 渗碳处理，硬化处理，置换沉淀，表面硬化(处理)，表面淬硬
case harding 表面硬化，表面淬火
case history file 工程史料档案
case-in 下套管
caseload forecast panel 工程作业量预测小组
casement cloth 窗帘布
casement door 玻璃门
casement sash 窗扇
casement window 平开窗
casement 窗扇
case No. 箱号
case study 工程实例，实例研究
case 盒，箱，机壳，机箱，壳，情况，案例，事件，套，柜
cash 现金
cash advance 现金预付
cash against bill of lading 凭提单付款
cash against delivery 货到付款，交货付款
cash against documents 凭单付款，交单付款
cash against shipping documents 凭装运单据付款，凭船务文件付款，凭单付款
cash and delivery 付款交货，货到付款
cash before delivery 预付货款
cash bonus 现金红利
cash deficit 现金逆差
cash disbursement schedule 现金支出预算[计划]
cash discount 现金贴现，现金折扣
cash dividend 现金股利，现金红利
cash equivalent 现金等值
cash flow analysis 现金流量分析
cash flow statement 现金流量表
cash flow 现金流量，现金周转，资金流动，现金流
cash guarantee 保证金
cashier 出纳员
cash in advance 预付货款
cash in bank 银行存款
cash inflow 现金流入，现金流入量
cash items department 现金票据处【美国银行】
cash joint 平齐接缝
cash journal 现款收支簿
cash on arrival 货到付款
cash on delivery 到货付款
cash on shipment 装船付现，装运付款
cash outflow 现金流出量
cash over and short account 现金盈亏账户
cash payment 付现，付现金，现金购买，现金支付，现金结算
cash price 现金价格
cash purchase 现金购买，现购
cash register 现金出纳机，现金收银机
cash remittance 汇款单
cash settlement 现汇结算，现金结算，付现金

casing anchor point 汽缸死点，汽缸固定点
casing barrel 汽缸筒体
casing bore 汽缸洼窝
casing cover 联结蜗壳底盘，泵盖，壳盖
casing end 端盖
casing fitting 套管接头零件
casing flange and bolt heating system 汽缸法兰螺栓加热系统
casing flange 壳体法兰【拼合泵壳】
casing glass 镶色玻璃
casing half 半个壳体
casing head 螺旋管塞，套管头
casing leak 炉墙漏风量，炉墙漏风，外壳漏电
casing lifting 吊汽缸
casing liner side plate 壳衬侧板
casing liner 护板内衬
casing lug 汽缸猫爪
casing pipe installation 套管敷设
casing pipe 套管
casing procedure drilling 套管钻进
casing ring 泵壳环【即壳体上的耐磨环】，壳体密封环
casing roller 胀管器
casing shell 外壳，汽缸外罩
casing shoe 套管管靴，套管靴
casing split 汽缸中分面
casing support foot 泵壳支柱
casing tongs 套筒钳
casing top half 上汽缸
casing wall 汽缸壁
casing 外壳，护板，汽缸，蜗壳，套管，壳体，缸体，箱，盒，罩，套，壳
cask car 罐车
casket 罐，小箱，吊筒，容器，手箱
cask-flask 屏蔽容器【用于贮运放射性物质】
cask loading pit 装罐井
cask unloading pool 容器卸料池【乏燃料】
cask 盛器，容器，桶，吊斗，罐，屏蔽容器
caslox 合成树脂结合剂磁铁
Cassagrande's soil classification 卡萨格兰德土分类法
CASS(cargo account settlement system) 货运账目清算系统
Cassegrain antenna 卡塞洛林天线
casserole 勺皿
cassette-cartridge system 盒式磁带系统
cassette memory 盒式存储器
cassette tape recorder 盒式磁带录音机
cassette tape 盒式磁带
cassette （胶卷）暗盒，箱子，磁带盒，盒子，胶卷盒，盒
cassification of fire hazards 火灾危险性分类
castability 可铸性
castable concrete 耐火混凝土
castable refractory 浇灌耐火材料，可塑耐火材料，耐火浆料
castable 浇注料，可铸的，可塑的
castalloyiron 合金铸铁
cast aluminium alloy 铸造铝合金
cast aluminium rotor 铸铝转子
cast aluminium 铸铝
cast aluminum junction box 铸铝接线盒
cast basalt pipe 玄武岩铸石管
cast basalt 玄武岩铸石
cast bearing plate 铸铁支承板
cast blade 铸造叶片
cast concrete 浇筑混凝土
castellated barrier wall 城垛形挡板
castellated beam 堞型梁
castellated coupling 牙嵌式连接
castellated nut 槽顶螺母【用开口销锁定】，槽形螺母
caster 主销纵倾，铸工，小脚轮
cast ferro alloy junction box 铸铁合金接线盒
cast fitting 铸铁管件
cast-foil coil 浇铸箔式线圈
cast-in aluminum squirrel cage 铸铝鼠笼
cast-in aluminum two-cage rotor 铸铝双鼠笼转子
cast-in blade 铸入叶片
cast-in-concrete reactor 混凝土芯电抗器
cast-in diaphragm 铸造隔板
casting aluminium 铸铝
casting bed 浇制台座【混凝土预制构造件】
casting box 模板，型箱
casting cooling system 铸件冷却装置
casting copper 铸铜
casting defect analysis 铸造缺陷分析
casting die 压铸模
casting for steam turbine cylinder 汽轮机汽缸铸件
casting glass 铸玻璃
casting gray iron 铸灰口铁
casting insulation process 浇注绝缘工艺
casting malleable iron 可锻铸铁
casting other 其他铸造
casting process analysis 铸造工序分析
casting process 浇铸过程
casting solution 铸膜液，浇膜液
casting steel 铸钢
casting stress 铸造应力
casting vote 决定性投票
casting 铸造，铸件，浇铸，浇注
cast-in-place concrete pile 混凝土灌注桩，现场浇制混凝土桩
cast-in-place concrete roof 现浇混凝土屋面
cast-in-place concrete 现浇混凝土
cast-in-place pile support 灌注桩支护
cast-in-place pile 钻孔灌注桩，现场浇制桩，就地浇铸的桩，灌注桩
cast-in-place reinforced concrete pile 现场浇灌钢筋混凝土桩
cast-in-place reinforced concrete slab 现场浇注钢筋混凝土板，现浇钢筋混凝土板
cast-in-place 就地灌筑，现浇，就地浇铸，现场浇捣，现场浇注，现场浇铸
cast-in-site concrete structure 现浇混凝土结构
cast-in-site 现场浇捣，现场浇注
cast-in-situ concrete pile 现浇混凝土桩
cast-in-situ concrete 现场浇筑混凝土，现浇混凝土
cast-in-situ pile 钻孔灌注桩
cast in situ 就地浇铸，现场浇捣，现场浇注，

现浇
cost-in yield-out model 投入产出模型
cast iron ballast weight 铸铁压重
cast iron blade 铸铁叶片
cast iron clad 铸铁护衬，铸铁护甲
cast iron closet bend 铸铁马桶弯头
cast iron cover 铸铁盖
cast iron flat-gate-type sluice 铸铁平板闸门式泄水闸
cast-iron gate valve 铸铁闸阀
cast-iron gilled tube economizer 铸铁鳍片管式省煤器，铸铁省煤器
cast iron grating 铸铁箅子
cast iron pipe 铸铁管
cast iron radiator 铸铁散热器
cast iron ribbed washer 铸铁加肋垫圈
cast iron socket bend 承口铸铁弯头
cast iron socket elbow 承口铸铁弯头
cast iron socket tee 承口铸铁三通
cast iron soil pipe 排水铸铁管
cast iron strainer 铸铁箅子
cast iron tube 铸铁管
cast iron valve 铸铁阀门
cast iron washer 铸铁垫圈
cast iron water pipe 给水铸铁管
cast iron 铸铁，生铁
cast jacket 整铸套箱
cast joint 铸焊，浇铸连接
castle key 槽形键
castle manipulator 双柱窝式机械手，高架式机械手
castle nut 槽形螺母
cast low alloy steel 低合金铸钢
castor oil 蓖麻油
cast plastic basin 塑料压铸池
cast-solid 整体铸造的
cast spiral case 铸钢蜗壳，铸铁蜗壳
cast squirrel-cage winding 铸笼式绕组，铸造鼠笼式转子
cast steel 铸钢
cast steel anchor block 铸钢锚定块
cast steel body 铸钢体
cast steel case 铸钢机壳
cast steel frame 铸钢机座
cast steel gate 铸钢闸门
cast steel magnet 铸造磁钢
cast steel yoke 铸钢磁轭
cast stone 铸石
cast teeth standard 铸齿标准
cast temperature 铸造温度
cast to shape 精密铸造
cast transformer 浇注（绝缘）变压器
cast tube 铸管
cast unit 整件铸造
cast valve 铸造阀
cast welding 铸焊
cast 铸件，铸造，投掷，浇铸，浇注，投，掷
casual failure 随机故障
casual inspection 临时检查，不定期检查
casual repairs 临时维修，临时修理
casualty insurance 意外事故险
casualty 伤亡，事故
casual 偶然的，临时的，非正式的，不规则的，不定期的，随机的
cat 吊锚，猫
catabolism 分解代谢
cataclastic flow 碎裂流
cataclastic metamorphism 碎裂变质
cataclastic structure 碎裂构造
cataclysm 灾变
catadioptric apparatus 反折射器
cataleptic failure 灾难性失效
catalogue 一览表，（产品）目录，（产品）样本，清单，把……编入目录，登记分类
catalysis 催化
catalyst activity 催化剂活性
catalyst bed 催化剂床
catalyst deactivation 催化剂失活
catalyst length 催化段长度
catalyst poisoning 催化剂中毒
catalyst 催化剂
catalytic 触媒
catalytic agent 催化剂
catalytical 催化的
catalytic combustion system 催化燃烧系统
catalytic combustion 催化燃烧
catalytic cracker 催化裂化器
catalytic cracking 催化裂化
catalytic desulfurization 催化脱硫作用
catalytic exhaust purifier 废气催化净化器
catalytic filter 催化过滤器
catalytic hydrodesulfurization 催化加氢脱硫
catalytic odor treatment 臭气催化处理
catalytic oxidation 催化燃烧，催化氧化
catalytic poison 催化毒物
catalytic process 催化过程
catalytic reaction 催化反应
catalytic reactor 催化反应器
catalytic recombination 触媒催化复合
catalytic recombiner system 催化复合器系统
catalytic recombiner 催化复合器
catalytic reforming 催化重整
catalytic solid-phase contact reagent 催化固相接触剂
catalyzed conversion 催化转化
catalyzed sodium sulfite 催化亚硫酸钠
catalyze 催化
catamaran 双体船
cat and can 履带式铲运拖拉机
cataphalanx 冷锋面
cataphoresis 电泳现象，阳离子电泳，电透法，电渗，电冰
cataract action 急流作用
cataract 大瀑布，缓冲器，冲程调节器
catastrophe 大灾难，大事故，灾害，大祸
catastrophic failure 突然故障失效，严重故障，灾难性故障［失效］，突变失效，严重损坏
catastrophic flood 特大洪水
catastrophic flux jump 灾难性磁通跳跃
catastrophic hazard 灾难性危险
catastrophic oscillation 灾难性振动
catastrophic storm 灾难性暴雨

catastrophic wear 严重磨损，灾变磨损
catastrophic 大灾难的，灾难的
catastrophism 灾变说
cat. = catalyst 催化剂
CAT(centralized automatic testing) 集中式自动测试
catch-all 杂物箱
catch basin 沉淀池，集水面，截留井，雨水井，沉泥井，集水池，截水池
catch bolt 止动螺栓，制动螺栓，弹簧门锁，自动门扣
catch-drain 泄水沟，截水沟，盲沟，集水沟
catcher grid 收注栅，捕获栅
catcher 捕捉器，除尘器，稳定装置，接收器，按钮，收集器，限制器，截流器，截液器，闸门，制动装置
catch frame 格栅，截留栅
catch gear 闭锁装置
catch holder 熔断器支架，熔断器
catch hook 棘爪
catching load 制动力
catching of toothed wheels 齿轮啮合
catchment area survey 汇水面积测量
catchment area 汇水区，集水面积，流域面积，汇流面积，汇水面积
catchment basin 汇水盆地，集水区
catchment boundary 流域分界
catchment intercepting wall 挡水堤，截水墙
catchment management 流域管理
catchment of water 汇水，集水
catchment yield 流域出水量
catchment 集水，流域，汇水，集水区，集水区域
catch net 保护网
catch off guy 锚线绳
catch of used oil 废油收集器
catch pan 滴水收集器
catch pit 排水井，集水坑，截留井，集流坑
catch point 安全线道岔，安全道岔，止闭点
catch pot 收集罐，捕集阱，接受盒，收集器
catch tank 汇集箱，排水箱
catch tray 收集盘
catch-water basin 集水区，集水池
catch-water channel 集水槽
catch-water-drain 载水沟
catch-water 汽水分离器
catch 制动装置，拉手，收集，捕捉，门拉手，窗沟
CAT(clear air turbulence) 晴空湍流
CAT(component acceptance test) 元件验收测试
CAT(computer-aided teaching) 计算机辅助教学
CAT(computer-aided test) 计算机辅助测试
CAT(computerized axial X-tomography) 计算机控制的轴向X射线层析照相术
CAT(control and test) 控制和测试
CAT(controlled attenuator timer) 控制衰减计时器，被控衰减计时器
CAT(crack arrest temperature) 裂纹停止传播温度
CATE(computer-controlled automatic test equipment) 计算机控制的自动测试设备
categorized drawing 分类图纸
category of coal 燃煤品种
category of construction quality control 施工质量控制等级
category of service 服务类别
category 种类，类，类别，范畴，部门，类目
catelectrode 阴极，(电池的)负极
catenarian 悬链线，悬索
catenary action 悬链作用
catenary angle 悬垂度
catenary construction 悬索结构
catenary correction (钢尺的)垂曲改正
catenary curve 垂曲线
catenary flume 悬链形渡槽
catenary idler 吊挂托辊
catenary suspension 悬链
catenary wire 电缆吊索，悬链钢丝绳
catenary 悬链线，悬索
catenation 链接
catena 联锁，耦合，联结，键，链，链锁，链条
catenoid 悬链曲面，链状的
caterpillar budozer 履带推土机
caterpillar crane 履带(式)起重机，坦克吊
caterpillar drive 履带传动
caterpillar excavating machine 履带式挖土机
caterpillar excavator 履带式挖掘机
caterpillar gate 链轮闸门，履带式闸门
caterpillar guide 履带导承
caterpillar loader 履带式装载机
caterpillar machinery 履带式机械
caterpillar track 导轨
caterpillar traction 履带牵引
caterpillar tractor 履带式拖拉机
caterpillar 履带拖拉机，履带
catforming 催化重整
cat gold 金色云母
cathead 锚架，起吊架，套管，掉锚架，吊锚架，系锚短柱，系锚杆
cathedral glass 拼花玻璃
cathetometer 高差计，精确高差测量仪
cathetron 外控式三极汞气整流管
cathode 阴极，负极
cathode activation 阴极激活
cathode arrester 阴极放电器
cathode beam 阴极射线束，电子束
cathode-biased flip-flop 自偏压触发器
cathode bias resistor 阴极偏压电阻器
cathode bias 阴极偏压
cathode coating 阴极敷层，阴极被覆
cathode compartment 阴极室
cathode-coupled multivibrator 阴极耦合多谐振荡器
cathode coupling 阴极耦合
cathode current 阴极电流
cathode dark space 阴极暗区
cathode deposition 阴极沉淀
cathode deposit 阴极沉淀物
cathode disk 盘形阴极
cathode drop 阴极压降
cathode efficiency 阴极效率
cathode emission 阴极发射

cathode emissivity 阴极发射率
cathode emitter 阴极发射体
cathode end 阴极引出端
cathode excitation 阴极激发
cathode fail 阴极压降，阴极电位降
cathode feedback circuit 阴极反馈电路
cathode feedback 阴极反馈，阴极回授
cathode filament 阴极灯丝
cathode follower valve 阴极输出管
cathode follower 阴极输出器，阴极跟随器
cathode gate 阴极门电路
cathode glow 阴极辉光，阴极电辉
cathodegram 阴极射线示波图
cathode grid 阴极栅
cathode half bridge 阴极半桥
cathode-heater 阴极加热器
cathode heating time 阴极加热时间
cathode inductance 阴极电感
cathode injection 阴极注频
cathode interlayer impedance 阴极层间阻抗
cathode keying 阴极键控法
cathode light 阴极辉光
cathode liquor 阴极电解液
cathode load 阴极负载
cathode luminance 阴极亮度，阴极发光率
cathode luminescence 阴极辉光，电子致光，阴极场致发光
cathode modulation 阴极调制
cathode noise 阴极噪声
cathode peak current 阴极峰值电流
cathode phase inverter 阴极倒相器
cathode pickling 阴极腐蚀
cathode polarization 阴极极化
cathode-potential stabilized tube 阴极电位稳定器，阴极稳压管
cathode preheating time 阴极预热时间
cathode protection 阴极保护
cathode-ray 阴极射线
cathode-ray beam 电子射线束，阴极射线
cathode-ray direction finder 阴极射线测向仪
cathode-ray display 阴极射线管显示器
cathode-ray function generator 阴极射线管函数发生器
cathode-ray indicator 阴极射线指示器
cathode-ray magnetron 阴极射线磁控管
cathode-ray memory tube 阴极射线存储管
cathode-ray oscillograph 阴极射线示波器
cathode-ray oscilloscope tube 阴极射线示波管
cathode-ray oscilloscope 阴极射线示波计
cathode-ray pencil 阴极射线笔，阴极射线束，电子束
cathode-ray picture tube 阴极射线显像管
cathode-ray polarograph 阴极射线式极谱仪
cathode-ray recorder 阴极射线记录器，阴极射线记录仪
cathode-ray spot 阴极辉点，阴极斑点
cathode-ray storage tube 阴极射线存储管
cathode-ray tube display 阴极射线管显示器
cathode-ray tube memory 阴极射线管存储器
cathode-ray tube storage 阴极射线管存储器
cathode-ray tube 阴极射线显像管，CRT 屏幕，屏幕显示器，阴极射线管
cathode-ray tuning indicator 阴极射线调谐指示器
cathode reaction 阴极反应
cathode resistor 阴极电阻
cathode saturation current 阴极饱和电流
cathode space 阴极区
cathode spot 阴极斑点，阴极弧光，阴极炽点
cathode sputtering 阴极溅射
cathode stream 阴极电子流
cathode terminal 阴极端子
cathode trap 阴极陷波电路
cathodic corrosion inhibitor 阴极缓蚀剂，阴极型缓蚀剂
cathodic corrosion 阴极性腐蚀
cathodic evaporation 阴极蒸发
cathodic eye 电眼
cathodic inhibitor 阴极抑制剂，阴极缓蚀剂
cathodic polarization 阴极极化
cathodic process 阴极作用，阴极过程
cathodic protection equipment of screen 滤网阴极保护设备
cathodic protection equipment 阴极保护装置，防电化学腐蚀装置，防电蚀装置
cathodic protection system 阴极保护系统
cathodic protection 阴极防蚀法，阴极保护，阴极防蚀，阴极防腐
cathodic reaction 阴极反应
cathodic reduction 阴极还原
cathodic wave 阴极波
cathodic 负极的，阴极的
cathodogram 阴极射线示波图
cathodograph 端子衍射照相机，X 光相机
cathodoluminescence 阴极发光，阴极电子激发光
cathodophone 阴极送话器
catholyte 阴极电解液，阴极液
cation bed demineralizer 阳离子床除盐装置阳离子床除盐器
cation bed 阳离子床
cation exchange capacity 阳离子交换量
cation exchange filter 阳离子交换器，阳离子交换过滤器
cation exchange membrane 阳离子交换膜
cation exchange resin 阳离子交换树脂
cation exchanger 阳离子交换器，阳离子交换剂
cation exchange tower 阳离子交换塔
cation exchange 阳离子交换
cation exclusion 阳离子排斥
cationic current 正离子电流，阳离子电流
cationic exchange filter 阳离子交换过滤器
cationic fixed site 阳离子固定基
cationic layer 阳离子层
cationic polyelectrolyte 阳离子型聚合电解质
cationic surface active agent 阳离子表面活化剂
cationic surfactant 阳离子表面活化剂
cationic 阳离子的
cation interchange 阳离子交换，阳离子互换
cation interfacial active agent 阳离子界面活性剂
cationite 阳离子交换剂
cation leakage 阳离子漏泄
cation resin 阳离子树脂，阳离子交换树脂
cation 阳离子，正离子

catladder 爬倾斜屋顶用的梯子，直爬梯
cat ladder 爬梯，便梯，竖梯，猫梯
cat operator 履带式拖拉机的传动轮
catoptric apparatus 反射器【光学】
CATS(centralized automatic testing system) 集中式自动测试系统
CATS(computer-aided trouble-shooting) 计算机辅助故障查找
CATS(computer automatio test system) 计算机自动测试系统
CATV(community antenna TV system) 共用天线电视系统
catwalk （油罐顶上的）人行栈桥，小过道，高架狭窄人行道，轻便栈桥，桥形通道，狭小通道，猫步，天桥，桥，平台，人行道，步道，走道，马道，架空道，环形通道
cat whisker 触须，针电极，触针
Cauchy type distribution 柯西分布
CAUIS(computer automated ultrasonic inspection system) 计算机辅助的自动超声检测系统
cauldron subsidence 火山口沉陷
caulk 捻缝，填隙
caulked seam 嵌缝，嵌缝铆钉
caulked 嵌缝的
caulker's oakum 嵌缝的麻絮
caulking-butt 填缝对接
caulking chisel 嵌缝凿
caulking compound 嵌缝填料
caulking gun 堵缝枪，压力枪
caulking hammer 嵌缝锤
caulking iron 填缝凿
caulking material 嵌缝料
caulking metal 填隙合金，填隙合金金属材料
caulking nut 自锁螺母
caulking piece 填密件，敛缝片
caulking segment 填密片
caulking set 嵌缝工具
caulking strip 嵌缝条，塞缝条，敛缝软钢带
caulking tool 嵌缝工具，填缝凿
caulking 填……以防漏，堵缝，压紧，楔紧，填缝，嵌缝，捻缝，堵缝的，嵌缝的
caulk weld 塞焊，塞焊焊缝
causal analysis 因果分析
causal factor 起因，病原
causal model 因果模型，因果模式
cause and effect analysis 因果分析
cause and effect diagram 因果分析图，因果图
cause chart 原因图
cause-consequence analysis 因果分析
cause-consequence tree 因果树
caused this instrument to be duly executed 正式签订该文本
cause of action 案由，诉因，诉讼理由
cause of bankruptcy 破产原因
cause of controllable losses 可控损耗原因
cause of damage 致损原因
cause of formation 成因
cause of frost failure 冻胀破坏原因
cause promoter 事业的创办者
causes of controllable losses 可控损耗原因
causeway 长堤，堤道
cause 原因，理由
caustic alkalinity 氢氯根碱度
caustic alkali 苛性碱
caustical 生石灰
caustic ash 苛性苏打灰
caustic consumption 碱耗
caustic contact tower 碱接触塔
caustic corrosion 苛性腐蚀，碱腐蚀
caustic cracking 苛性裂纹，苛性裂化，苛性龟裂
caustic embrittlement 苛性脆化，碱性致脆
caustic extraction 碱抽提
caustic hydride process 苛化氢化法
causticity 腐蚀性，苛性，碱度
causticization 苛化作用
caustic lime 苛性石灰，生石灰
caustic potash 苛性钾，氢氧化钾
caustic scrubbing 碱洗
caustic sludge 碱渣
caustic soda cell 苛性钠电池
caustic soda proportioning pump 苛性钠配料泵
caustic soda tank 苛性钠贮存箱
caustic soda treatment 苛性碱处理
caustic soda 苛性苏打，氢氧化钠，烧碱，苛性钠
caustic solution cooler 碱液冷却器
caustic solution filter 碱液过滤器
caustic solution heater 碱液加热器
caustic solution 苛性碱溶液
caustic tower 碱处理塔
caustic transfer vehicle 碱运输槽车
caustic washing 碱性冲洗，碱洗
caustic wash solution 碱溶液，碱洗液
caustic 腐蚀性的，苛性的，腐蚀的，苛性药
caution against wet 切勿受潮
caution mark 小心标志，注意标志，警告标志
caution money 保证金
caution signal 注意信号，警戒信号
caution 小心
CAV＝cavitation 汽蚀
CAV＝cavity 空腔
cave animal 穴居动物
cave deposit 洞穴沉积
caved goaf 岩石坍溶带
caved material 坍落体
cave-in 沉陷，倒塌，塌方，坍塌
Cavendish experiment 卡文迪什试验
cavern containment 洞穴安全壳
cavern flow 岩溶水流，穴流，洞穴水流
cavern power house 窑洞式厂房
cavern rock 溶洞岩石
cavern water 洞穴水
cavern 大山洞，凹处，岩洞，洞，洞室，洞穴，山洞，穴，溶洞
cave 小室，洞穴，屏蔽贮藏室，洞，洞孔，山洞，穴，岩洞
caving bank 崩坍河岸，受冲岸，塌陷岸
caving ground 塌陷地
caving-in 塌陷，坍方，顶板坍落，冒顶
caving line 冒落线，坍落线
caving zone 塌陷区，坍坡区，下沉区

caving 凹槽，塌陷，淘空
cavitation bubble 空泡
cavitating flow 空腔水流
cavitational erosion 液流气泡浸蚀
cavitation bubble collapse 气泡溃灭
cavitation bubble 气蚀气泡
cavitation characteristics 气蚀特性，汽蚀特性，空蚀特性
cavitation coefficient 气蚀系数，汽蚀系数
cavitation condition 气蚀条件，汽蚀条件
cavitation control 汽蚀调节
cavitation core 空化中心，空蚀中心，气蚀中心，汽蚀中心
cavitation corrosion 空泡腐蚀，空穴腐蚀，空蚀，空化腐蚀
cavitation criteria 空化准则，气蚀标准【criteria 作为 criterion 的复数形式】
cavitation criterion 空化准则，气蚀标准
cavitation damage 汽蚀损伤，气蚀破坏，空穴损坏，空蚀，气蚀损坏
cavitation deformation 空穴变形
cavitation degree 气蚀度
cavitation effect 空化效应，汽蚀效应
cavitation erosion of hydraulic machinery 水力机械空蚀
cavitation erosion 空蚀，气蚀
cavitation flow 空穴流，气蚀水流
cavitation-free operation 无气蚀运行
cavitation-free performance 无气蚀性能
cavitation guarantee 气蚀保证
cavitation hysteresis 空蚀滞后，气蚀滞后，汽蚀滞后
cavitation inception 空化初生
cavitation index 气蚀指数，汽蚀指数
cavitation intensity 气蚀强度
cavitation level 空蚀程度，空蚀级别
cavitation meter 气蚀仪
cavitation noise 气蚀噪声
cavitation nucleus 空化核
cavitation number 空化数，气蚀数
cavitation of blade 叶片气蚀
cavitation of clearance 间隙气蚀
cavitation of water turbine 水轮机气蚀
cavitation parameter 气蚀参数，空化参数
cavitation pattern 空蚀类型
cavitation phenomenon 气蚀现象
cavitation pitting 气蚀坑，气蚀破坏，点空蚀，点气蚀
cavitation pocket 气蚀穴，气蚀坑，空蚀穴，汽蚀穴
cavitation point 气蚀点
cavitation range 气蚀区域
cavitation scale 气穴缩尺
cavitation sensor 气蚀传感器
cavitation shock 气蚀冲击
cavitation specific speed 气蚀比转速
cavitation test 气蚀试验，汽蚀试验
cavitation tunnel 空泡试验筒，空蚀试验槽，气穴风洞
cavitation zone 汽蚀区域
cavitation 汽蚀，气穴（现象），空化，空蚀，空穴，气蚀，形成空穴，空化作用
CAVIT=cavitation 汽蚀，气穴现象
cavity antenna 空腔天线
cavity block 空心块体
cavity boundary 空穴边界
cavity brick 空心砖
cavity circuit 谐振腔电路，空腔振荡电路
cavity cover 腔室盖，空穴盖
cavity drag 空穴阻力
cavity effect 空腔效应
cavity emissivity 容积辐射率，空腔辐射
cavity flow 空穴流动，空泡流，空腔水流
cavity flushing 冲洗洞壁
cavity frequency meter 谐振腔频率计，空腔频率计
cavity gap 空隙【外部燃料组件和堆芯围板之间】
cavity grouting 回填灌浆，空腔灌浆
cavity ionization chamber 空腔电离室
cavity liner 反应堆腔衬里【反应堆】
cavity magnetron 空腔谐振磁控管
cavity pocket 空腔
cavity propulsion reactor 腔式推进反应堆
cavity radiation 容积辐射
cavity radiometers 空腔辐射计
cavity ratio 室腔比
cavity reactor 腔式反应堆
cavity reactor critical experiment 腔式反应堆临界实验装置
cavity reactor-MHD generator 腔式反应堆-磁流体发电机
cavity region 空穴区
cavity resonance 空腔谐振
cavity resonator wavemeter 谐振腔波长计
cavity resonator 空腔共振器，空腔谐振器，谐振腔
cavity shield 腔室屏蔽
cavity wake 空穴尾流
cavity wall filled with insulation 夹心墙
cavity wall 空心壁，空心墙，空穴井
cavity 内腔，空穴，气室，气孔，空位，空间，洞穴，空洞，气穴，腔，洞
CAVT(constant absolute vorticity trajectory) 等绝对涡度轨迹
CBA(cargo booking advance) 国际航空货物订舱单
CBA(cost-benefit analysis) 成本一效益分析，费用-收益分析
C battery 丙电池组，C 电池组，栅偏压电池组
CBC(comparator buffer) 比较器缓冲器
CB(circuit breaker) 断路器
CB(concentration basin) 浓缩池
CB(contact breaker) 接触断路器
CB(control bus) 控制总线
CBD(cash before delivery) 预付货款
C-bias detector 栅偏压检波器
C-bias 栅偏压，C 偏压
CBP(condensate booster pump) 凝结水增压泵
CBR(California bearing ratio) 加利福尼亚承载力比，加州承载比
CCA(cargo charges correction advice) 货物运费

更改通知
C = carbon 碳
cascade effect 叶栅效应
CCB(China Construction Bank) 中国建设银行
CCB(convertible circuit breaker) 可变换断路器
CC(carbon copy) 抄送，抄送人
CCC(cargo condition certificate) 货物状态证书
C/C(center-to-center) 中心距
CC(charges collect) 运费到付
CC(closing coil) 闭式循环
C/C(code converter) 代码转换器
CC(command and control) 命令和控制
CCCW(closed circulating cooling water) 闭式循环冷却水
CCCW(closed cycle cooling water system) 闭式循环冷却水系统【DCS 画面】
CCCWP(closed circuit cooling water pump) 闭式循环冷却水泵
CCD(charge coupled device) 电荷耦合器件
CCD memory 电荷耦合存储器
CCD shift register 电荷耦合移位寄存器
CCD transversal filter 电荷耦合横向滤波器
CCFI(China container freight index) 中国出口集装箱运价指数
CCGT(combined cycle gas turbine) 联合循环燃机
CCGT power plant 燃气轮机联合循环电厂
ccircumferential prestressing tendon 环向预应力钢筋束
CCKW(counter clockwise) 逆时针方向
C clamp C形夹
C-class insulation C级绝缘
CCL(concreting check list) 灌浆检查单
complete observability 完全可观性，完全能观性
computer-aided diagnosis 计算机辅助诊断
C-core C形铁芯
core power-distribution control 堆芯功率分布控制
covered equipment removal hole 带盖板的起吊孔
CCP(centrifugal charging pump) 离心式上充泵
CCP(contractual change proposal) 商务变更建议
CCP(critical compression pressure) 临界压缩压力
CCPP(combined cycle power plant) （燃气-蒸汽）联合循环电站
CCR(capacity continuous rating) 能力连续工况
CCR(central control room) 中央控制室
CCR(counter current regeneration) 逆流再生
CCS(carbon capture and storage) 碳捕集与封存
CCS(closed loop control system) 闭环控制系统
CCS(combustion control system) 燃烧控制系统
CCS(component cooling water system) 部件冷却水系统
CCS(coordinated control system) 协调控制系统【DCS 画面】
CCS(critical compression stress) 临界压缩应力
CCSF(cross-compound single flow) 双轴单排汽汽轮机
CCSS(central converter substation) 中心换流站，中央变换所
CCT(constant current transformer) 恒流变压器
CCTF(cylindrical core test facility) 圆柱形堆芯试验装置【日本】
CCTIT(Consultation Committee of International Telegraph and Telephony) 国际电话电报咨询委员会
CCTV(Closed Circuit Television) 闭路电视
CCVO(combined certificate of value and origin) 价值产地联合证明书
CCW(closed cooling water system) 闭式冷却水系统
CCW(condenser circulating water) 凝汽器循环水
CCW(counter clockwise) 逆时针，逆时针的，逆时针方向
CCW(cycle cooling water) 循环冷却水
CCWHF(closed cooling water heater) 闭式冷却水加热器
CCWP(closed cooling water pump) 闭式冷却水泵
CCWP(condenser circulating water pump) 凝汽器循环水泵
CDCA(common data class attribute) name 公用数据类属性名
CDC(common data class) 公用数据类
CDCN Space 公用数据类名称空间
C&D(collated and delivered) 货款两清
C-display C型显像管，距离方位显示器
cdiular girder 空腹梁
CDM(clean development mechanism) 清洁发展机制
Cd-Ni battery 镉镍蓄电池
CDR(CO_2-D_2O reactor) 二氧化碳冷却重水反应堆
CD ROM 小型磁盘只读存储器
CDS((condensate system) 冷凝水系统
CDSR 凝汽器【DCS 画面】
CdTe probe 碲化镉探测器，碲化镉探头
CD test(consolidated-drained triaxial compression test) 固结排水三轴压缩试验
CDU(central display unit) 中央显示器
CEA(cost-effectiveness analysis) 费用效果分析
CEAR(construction and erection all risks) 建筑安装工程一切险
ceased and deferred project 停建缓建项目
ceased project 停建项目
cease to be effective 停止生效
cease to be in effect 失效
cease to be in force 失效
cease to be 不再是
cease to exist 不复存在，灭亡
cease to have effect 停止生效
cease 中止，间断，停止
CE(case expansion) 缸胀
CEC(Canadian Electrical Code) 加拿大电气标准
CE(common emitter) 共发射极
CEDS(control element drive system) 控制元件驱动系统
CEEC(China Energy Engineering Corporation Limited) 中国能源建设股份有限公司，中国能建
ceiling air diffuser 顶棚送风口
ceiling amount 最高限额

ceiling and floor of bidding price 投标竞价的上下限
ceiling and floor of market clearing price 市场出清（电）价的上下限
ceiling board 天花板
ceiling brandering 吊顶格栅
ceiling burner 顶置式喷燃器
ceiling capacity 最大能量，最大能力
ceiling cavity 吊顶空间
ceiling cornice 顶棚线脚
ceiling direct voltage 顶值直流电压
ceiling downlight 天花灯，顶灯，吊灯
ceiling excitation 顶值励磁，极限励磁
ceiling fan 吊扇
ceiling girder 顶梁，排梁
ceiling hanger 吊筋
ceiling insert 天花板嵌条
ceiling insulator 天棚绝缘子
ceiling joist 平顶搁栅，吊顶龙骨，天花板搁栅
ceiling lamp fixture 天棚照明配件，吊灯配件
ceiling lamp 天花灯，天棚灯，顶灯，吊灯
ceiling light 天花板照明，舱顶灯，吊灯，平顶照明
ceiling load 平顶载重
ceiling-mounted 顶置式的，吊装式
ceiling mounting 倒置安装
ceiling panel heating 顶棚辐射采暖
ceiling panel 顶棚镶板，平顶镶板
ceiling plan 顶棚平面图
ceiling plastering 顶棚抹灰
ceiling plate 顶棚嵌板
ceiling price topping ratio 最高限价到达率
ceiling price 最高价格，最高限价
ceiling rosette 天花板接线盒
ceiling stuck 天花板
ceiling superheater 顶棚过热器
ceiling suspension system 吊顶体系
ceiling switch 天棚拉线开关
ceiling void 吊顶空间
ceiling voltage 顶值电压，峰值电压
ceiling wiring 天棚布线
ceiling without trussing 无梁平顶
ceiling 顶板，炉顶，顶棚，盖板，顶篷，平顶，天花板，吊顶，（风洞的）顶壁，云幂，天棚
ceilometeris 云幕仪
celadon 灰绿色，青瓷色
celerity 波速，速度，迅速
celeste 天蓝色，天蓝色的
celestial axis 天轴
celestial coordinate system 天球坐标系
celestial equator 天球赤道，天赤道
celestial meridian 天球子午圈
celestial north pole 北天极
celestial pole 天极
celestial south pole 南天极
celestial sphere 天球
celestite 天青石
celite cylinder 硅藻土圆柱体
celite 硅藻土绝热物
cellar 地下室，地窖，窖
cell atmosphere processing system 舱内空气处理系统
cell boundary 栅元边界
cell burner 格状喷燃器
cell concrete 多孔混凝土
cell constant 电极常数
cell correction factor 栅元校正因子，栅元修正因子
cell correction 栅元校正因子
cell cover 设备室盖，热室盖
cell design 单元设计
cell distortion gauge 栅元变形规【格架】
cell fabrication 电池制造
cell feeding 电池馈电
cell-homogenized cross section 栅元均匀化截面
cell jar 电瓶
cell layout 单元设备布局
cell length 栅元长度
cell of the grid 格架栅元
cell operating face 热室操作面
cellophane 玻璃纸
cellpacking 管壳，电池外壳，元件包装物
cell parameter 栅元参数
cell pressure 四周压力，室压
cell quay wall 格式岸壁
cell size 栅元尺寸
cell structure 细胞结构
cell switch 电池开关，电池换接器
cell thickness 堆层厚度
cell-type air washer 栅元型空气洗涤机
cell-type evaporator 蜂窝状蒸发器
cellular abutment 格形桥台，空心边墩，框格式边墩
cellular beam 格型梁
cellular breakwater 格构式防波堤
cellular brick 多孔砖
cellular bulkhead 格型堤岸
cellular buttress 格型支墩，空心支墩
cellular caisson 框格式沉箱
cellular cofferdam 格形围堰
cellular concrete block 格型混凝土块体
cellular concrete raceway 分格混凝土电缆管道
cellular concrete 加气混凝土，泡沫混凝土
cellular conductor 穿管导线，管状导线
cellular construction 框格式建筑法
cellular core wall 孔格形心墙，箱格型心墙
cellular dam 格箱式坝
cellular girder 格型梁，空心梁
cellular glass 泡沫玻璃，多孔玻璃
cellular gravity dam 格箱式重力坝
cellular insulant 多空隔热材料，泡沫保温材料，多孔绝热材料
cellular insulation 蜂窝状绝缘，泡沫绝缘
cellularity 蜂窝状结构，泡沫结构
cellular lava 多孔熔岩
cellular material 多孔材料
cellular metal floor raceway 格状金属地板电缆管道
cellular metal floor 格状金属地板
cellular motion 环形运动
cellular plastics 泡沫塑料

cellular pyrite 白铁矿
cellular radiator 蜂窝式散热器
cellular rubber 泡沫橡胶
cellular structure 网格结构，蜂房（式）结构，蜂窝（状）结构，细胞状结构，箱格型结构
cellular switchgear 防火隔装开关装置
cellular texture 蜂窝状结构，网格结构
cellular type radiator 孔式散热器
cellular wave 格形波
cellular 多孔的，蜂窝状的，泡沫的，细胞的，单元的，多孔状，框格状，小细胞
celluloid paint 清漆，透明漆
celluloid 赛璐珞，电影胶片，假象牙
cellulose acetate butyrate 醋酸丁酸纤维素
cellulose acetate membrane 醋酸纤维膜
cellulose acetate 醋酸纤维，醋酸纤维素
cellulose covering 纤维素涂层
cellulose filter 纤维素过滤器
cellulose membrane filter 纤维薄膜滤布，纤维薄膜过滤器
cellulose triacetate 三醋酸纤维素
cellulose type electrode 纤维素型焊条
cellulose 纤维素
cellulosic exchanger 纤维素类离子交换剂
cell 单元（格），元件，格子，筛眼，电池，格栅，单体，（测力，测压）传感器，盒，小室，细胞
celotex board 隔热板，隔音板，纤维板
celotex 隔声材料，纤维板，隔热板，隔音板
Celsius scale of thermometer 摄氏温度计温标
Celsius scale 摄氏温标
Celsius temperature 摄氏温度
Celsius thermometer 摄氏温度计
Celsius 摄氏【温度】，摄氏的
CEMA(cement asbestos) 水泥石棉
CEM(chimney emission monitoring) 烟气排放监测（系统）
CEM(comprehensive emergency management) 综合应急管理
CEMDS(condition of equipment or material upon delivery on-site) 设备或材料在现场交货条件，设备及材料到货状况报告
cement activity 水泥活性
cement additive 水泥添加剂
cement-aggregate ratio 水泥骨料比率，水泥骨料比，灰骨比
cement-aggregate reaction 水泥骨料反应
cement and sand cushion 水泥砂浆垫层
cement asbestos board 石棉水泥板
cement asbestos pipe 石棉管
cementation box 渗碳箱
cementation of fissure 灌缝
cementation process 渗碳过程
cementation pump 混凝土泵
cementation 渗碳处理，硬化处理，表面硬化，胶结作用，水泥灌浆，水泥硬化，水泥胶结
cement bacillus 水泥针状体
cement batching bin 水泥配料仓
cement bentonite grout 膨润土掺水泥浆液
cement bin 水泥仓
cement-bound macadam 水泥碎石路
cement-bound road 水泥结碎石路
cement brand 水泥牌号
cement brick pavement 水泥砖铺面
cement brick 水泥砖
cement briquette 水泥（标准）试块
cement clinker 水泥熟料，水泥焦渣，水泥仓
cement coating 水泥盖层，水泥护面，水泥涂料
cement coat 水泥涂层
cement colours 水泥染料
cement concrete 水泥混凝土
cement consumption of power generation project per kW 发电工程每千瓦水泥消耗量
cement consumption 水泥用量
cement conveyer-or 水泥输送机
cement covering 水泥盖面，水泥护面，水泥罩面
cement deep mixing 水泥搅拌法
cement dressing 水泥粉刷，水泥饰面
cemented carbide 烧结碳化物，硬质合金
cemented soil 胶结土
cemented steel 渗碳钢，表面硬化钢
cement equivalent factor 水泥当量系数
cement facing 水泥罩面，水泥护面，水泥砂浆护面
cement factor 水泥系数，水泥厂
cement fibrillate plate 水泥纤维板
cement fineness 水泥细度
cement finish 水泥饰面，水泥罩面
cement flooring 水泥地面
cement-flyash-gravel pile 水泥粉煤灰碎石桩
cement grade 水泥标号
cement grit 粗粒水泥，粗磨水泥
cement grouted zone 水泥灌浆区
cement grouter 水泥灌浆机
cement grout filler 水泥砂浆填料
cement grouting process 水泥灌浆法
cement grouting 水泥灌浆
cement grout pump 混凝土泵
cement grout 水泥浆液，水泥浆，水泥砂浆
cement gun 水泥浆喷枪，水泥喷枪
cement hopper 水泥料斗
cement hydration 水泥水化作用
cement industry 水泥工业
cementing agent 胶结剂，接合剂
cementing power 黏结力
cementing 注水泥，胶结，表面硬化，胶合，黏合，胶接
cement injection 水泥灌浆
cement injector 水泥灌浆机
cementite 碳化体，渗碳体，胶合，硬化
cementitious material 黏结材料
cementitious sheet 石棉水泥板，水泥黏合板
cementitious 似水泥的，有黏性的
cement joggle 水泥接榫
cement joint 水泥缝
cement lime mortar 水泥石灰砂浆
cement-lined pipe 水泥衬里管，水泥衬砌管
cement manufactures 水泥制品
cement mark 水泥标号
cement mineral wool plate 水泥矿棉板
cement mixer 水泥拌和机

cement-modified soil 水泥改良土
cement mortar cushion 水泥砂装垫层
cement mortar dado 水泥台度，水泥培裙
cement mortar plaster sectioned 抹水泥砂浆分格
cement mortar screed 水泥砂浆找平层
cement mortar 水泥灰浆，水泥砂浆
cement motar finish 水泥砂浆罩面
cement needle 水泥测针
cement paint 水泥涂料
cement particle 水泥颗粒
cement paste mixture 水泥浆混合物
cement paste 水泥浆
cement pearlite 水泥珍珠岩
cement pipe 水泥管
cement plaster finish 水泥灰粉饰面，水泥抹面
cement plaster plate 石膏水泥平板
cement plaster 水泥抹面，水泥抹灰，水泥灰泥，水泥粉刷
cement product 水泥制品
cement proportioning worm conveyor 混凝土配料蜗杆传送器
cement pump 水泥泵
cement rendering 水泥粉刷，水泥抹面，水泥刷面，水泥涂料
cement retarder 水泥缓凝剂
cement-roll anchor bar 水泥卷锚杆
cement sampler 水泥取样器
cement sand cushion 水泥砂装垫层
cement sand grout 水泥砂浆
cement sand ratio 灰砂比
cement screed 水泥砂浆找平层
cement screw feeder 螺旋式水泥进料器
cement setting 水泥凝固
cement shed 水泥仓库
cement shovel 水泥铲
cement silo 水泥仓，水泥罐，水泥筒仓
cement slab revetment 水泥板护岸
cement-slag hollow brick 水泥矿渣空心砖
cement slurry 水泥浆，水泥浆液，水泥稀浆
cement-soil pipe 水泥土管
cement-soil wall 水泥土墙
cement-solidified waste 水泥固化废物
cement soundness 水泥安定性
cement-space ratio 水泥量与骨料孔隙比
cement stabilization 水泥加固
cement-stabilized soil 水泥加固土
cement standard curing box 水泥标准养护箱
cement storage silo 散装水泥存放筒仓，散装木泥存放槽
cement storage 水泥仓库
cement strength 水泥强度
cement tile pavement 水泥砖铺面
cement tile 水泥瓦
cement treated base 水泥加固基础
cement type binder 水泥型黏合剂
cement unloading equipment 水泥卸料设备
cement veneer 水泥饰面
cement vermiculite 水泥蛭石
cement wash 水泥浆刷面，刷浆
cement-water ratio 灰水比
cement 水泥，接合剂，接合，用水泥涂，巩固，黏牢，胶合，胶合剂
CEMF(counter electromotive force) 反电势
CEMR(Center for Energy and Mineral Resources) 能源和矿物资源中心【美国】
CEMS(continuous emission monitoring system) （烟气）连续排放监视系统，排放连续监测系统
CENELEC 欧洲电工标准化委员会
censor contact 监护结点
censored data 截尾数据
censored sample 截尾样本
census data 人口调查资料，人口普查资料
census 人口普查
cental 百磅
center adjustment 对中调节
center bearing 中心轴承
center bit 中心钻，转柄钻，中心钻头
center bolt 主螺栓，中心拉杆螺栓
center bored concrete piles 中心压灌桩
center bore 中心孔
center burst 中心裂纹
center calliper 测径中心卡尺，泓径规
center calm 中心无风区
center conductor 中心导线，中心线
center differential 中央差动机构
center-diffusion tube gas burner 中心管进气喷燃器
center distance between bearings 轴承中心距
center distance 中心距
center drill 中心钻
centered expansion wave 有心膨胀波
centered wave 有心波，扇形波束
center elevation 中心标高
center feed 中心馈电式，中央供电式，中间馈电
center for computing and automation 计算和自动化中心
Center for Energy and Mineral Resources 能源和矿物资源中心【美国】
center frequency 中心频率，未调制频率，对称调制的载频
center fuel melting 燃料中心熔化，中心燃料熔化
center gauge 中心规
center gear 中心轮
center gradient （输电线的）中线与相线间电位梯度
center hole 中心孔
center-hung sash 中悬窗框
center-hung swivel window 中悬窗，中旋窗
center-hung window 中悬窗
centering adjustment 对中调整，对准中心，中心调整
centering brush 调准电刷
centering check 校中检查
centering chuck 定心夹盘
centering coil 中心调整线圈，定心线圈
centering collar 定中心环
centering control 居中调节，定中调节，中心调节，中心控制
centering device 定中心装置，对心装置，找中心装置

centering error　对心误差，对中误差
centering force　集中力
centering laser　找中激光仪
centering mast　定心杆
centering module　定心模数
centering of diaphragm　隔板定中心
centering pin　中心校正杆，中心定位销
centering potentiometer　定心电位计
centering ring　（汽轮发电机护环的）中心环，定心环
centering screw　定心螺钉
centering spigot　定心槽
centering support system　滑销系统，定中系统
centering support　中分面支持
centering tool　定中心工具，定心工具
centering tube　对中管，定中心管
centering wedge　定心楔
centering winding　定心管，定心绕组，中心调整管
centering　定中心，对中，校中心，打中心孔，对中心，调准，找正
center load　中心负载
centerless grinding machine　无心磨床
center lathe　普通车床
center line average method　平均高度法
center line chord　中心弦
center line crack　纵向裂纹
center line keel beam　中央龙骨
center line of arch　拱轴线
center line of a shaft　轴的中心线
center line of bolt hole　螺孔中心线
center line of shafting　轴系中心线
center line of T/G foundation　汽轮发电机基础中心线
center line support packing head　中心键定位支承式汽封套
centerline support pump　中心支撑泵
center line　中心线，中线
center mall　道路隔离绿化带，道路分割带，路中林荫带
center matched　中心匹配的，中心配合的
center M with N　把M放在对准N的中心
center of activity　活动中心
center of attack　侵蚀中心
center of buoyancy　浮力中心
center of commerce　商业中心
center of compression　压缩中心
center of crystallization　结晶中心
center of curvature　曲率中心
center of displacement　排水中心
center of distribution　配电中心
center of disturbance　扰动中心
center of equilibrium　平衡中心
center of figure　形心
center of floatation　浮体水面质量中心，浮心
center-off polarized relay　中位断开式极化继电器
center-off position　非中心位置
center off　（极化继电器的）中间位置断开
center of gravity of section　截面重心
center of gravity path　重心运动轨迹
center of gravity　重心
center of gyration　（轴颈）回转中心，旋心
center of inertia　惯性中心
center of location　定位中心
center of lift　升力中心
center of mass system　质心系
center of mass　质量中心，质心，重心，惯性中心
center of movement　运动中心
center of oscillation　振动中心，振荡中心，摆动中心
center of pipe　管中心
center of pressure distribution　压力中心分布
center of resistance　阻力中心，压力中心
center of rotation　旋转中心，转动中心
center of shear　剪心，剪切中心
center of stiffness　刚度中心
center of stress　应力中心
center of supply　供电中心
center of symmetry　对称中心
center of thrust　压力中心
center of turn　转向中心
center of twist　扭转中心，推力中心
center of volume velocity　体心速度
center opening　中心孔
center operator　中央操作员
center phase　处于中心位置的相线【输电线】
centerpiece　十字架，十字轴
center pillar　中立柱
center pin　中心销，中心拉杆
center-pivoted window　中悬窗
center-point earth　中心点接地
center-point galvanometer　中心零位电流计
center-point　中心点
center post　中心位置【控制棒束】
center puncher　中心冲
center punch　定中冲头，中心冲孔
center rod　中心棒
center roll　中间辊子
center section　中翼段，中心截面，中心剖面
center spot　中心点
center stake　中心桩
center support base　中心支承座
center support system　中心支持系统，滑销系统，对中系统，滑键系统
center-tapped transformer　中间抽头变压器，中心抽头变压器
center-tapped winding　中心抽头的绕组
center tap　中心引线，中心抽头
center to center distance　中心间距，中心距
center to center spacing　燃料组件间的中心距离【燃料组件】，中心距
center to center　中到中（的距离），中心距，中心之间，中心至中心
center to contract　中至中距离
center to end　中心到端面，中心至端部
center to face　中心至面
center vent　中心抽气口，中部排出口
center wall　（双炉膛的）中间管壁，双面水冷壁
center zero gauge　中心零刻度压力计
center zero relay　中间零位继电器
center　放在中心，定中心，中心，中央

centesimal balance 百分天平
centesimal degree 百分度
centesimal system 百分制，百进制
centesimal 百分之一的，百进位的
centet of span 跨度中心，桥跨中点
centibar 厘巴【压力单位】
centigrade degree 摄氏温度，摄氏度，百分度数
centigrade scale 百分标度，摄氏温标，百分刻度
centigrade temperature 摄氏温度
centigrade thermal unit 磅卡
centigrade thermometer 摄氏温度计，百分温度计，摄氏寒暑表，百分温度表
centigrade 百分度，摄氏温度的，摄氏温度，百分度的，摄氏度
centile 百分位数
centimeter-gram-second system CGS 制，厘米-克-秒制
centimeter per second per second 厘米每二次方秒，每秒每秒厘米【加速度单位】
centimeter-wave device 厘米波器件
centimeter 厘米
centimetric wave 厘米波
centipois(e) 厘泊【黏度单位】
centistoke 厘斯【运动黏度单位】
centi 厘
centralab 中央实验室
central address memory 中央地址存储器
central air conditioning plant 集中空调设备
central air conditioning system 集中式空气调节系统，集中空调系统
central air conditioning 集中空调
central air cylinder actuating the 4 cams 驱动四个凸轮的中央气压缸
centralairfoil 中部翼面
central angle of arch 拱圈中心角，拱中心角
central angle 中心角
central apparatus room 中央控制室
central automatic coupler 中央自动耦合器
central automatic dispatching 集中自动调度
central axial spine tube 中心轴向支承管
central axis 中心轴线，中轴，主轴线
central bank 中央银行
central bar 江心洲，中滩
central batching plant 中心配料厂
central battery telephone 供电制电话
central battery 中央电池组
central bay 中间开间
central bearing 中间轴承
central bin system 集中仓储制
central board 中央操纵台
central bore 中心孔
central business district 中心商业区
central calm 中心宁静区，中心无风区
central channel 中心通道
central chilled water system 主冷冻水系统
central chute 中央落煤管
central city 中心城市
central clamping screw 中心固定螺旋
central collision 迎面碰撞，对心碰撞
central commercial district 商业中心区
central compression 中心受压
central computer 中央计算机，中心计算机
central concentrated load 中心负荷，集中负荷
central concrete membrane 混凝土薄板，混凝土面板，混凝土心墙
central concrete mixing plant 混凝土中心搅拌厂
central conductor method 贯通法磁粉探伤
central conductor 中间导线
central conic 有心圆锥曲线
central console 中央控制台
central contact （螺丝灯头的）中心接点
central control air conditioner 中央空调
central control and operate panel 中央控制操作盘
central control cycle 集中控制循环
central control desk 中央控制台，中心控制台
central controller 中央控制器
central control mode 集中控制模（方）式，集中控制模式
central control room for coal handling 输煤集中控制室
central control room 中央控制室，单元控制室，集中控制室
central control 集中控制，中心控制，中央控制
central converter substation 中心换流站，中央变换所
central core zone 堆芯中心区
central core 中心管，内管，中央心墙
central cross-section 中心截面
central difference formula 中心差分公式
central difference method 中心差分法
central difference 中心差分
central dispatching station 中央调度所
central display unit 中央显示器
central district 中心区
central earthquake 中心震
central electric power station 中心发电厂
central enterprise 中央企业
central equilibrium 中心平衡
central exchange 中央交换器
central-excitation system 中央励磁式，中心励磁系统
central facilities area 中心试验装置区
central fan system 集中通风系统
central-flow surface condenser 向心流动式表面凝汽器
central flux 中心通量
central force 中心力，有心力
central fringe 中心条纹
central gantry 中央门型架构
central government-owned company 中央企业
central graphite rod 中心石墨棒
central hair 中丝
central hall 中央大厅
central handling control room 中央吊装控制室
central heating supply 集中供热
central heating for region 分区集中供热
central heating plant 集中供暖站
central heating station 集中供热站
central heating system 集中供热系统，中心供热

系统
central heating 集中供热，集中加热，集中采暖
central hole nozzle 中心孔式喷嘴
central hole 中心孔
central hub substation 中心枢纽变电站
central igniter 中心点火器，中心引爆装置
central impact 对心碰撞
central incineration 集中焚烧
central input/output multiplexer 中央输入输出多路转换器
central ion 中心离子
centralization of control 集中控制
centralization of management 集中管理，统一管理
centralization 集权式，集中化
centralized absolute chronometry 集中绝对时标
centralized accounting system 集中核算制
centralized automatic message accounting 集中式自动信息计算
centralized automatic testing system 集中式自动测试系统
centralized automatic testing 集中式自动测试
centralized concrete mixing plant 混凝土中心搅拌厂
centralized control building 集中控制楼
centralized control room 集中控制室
centralized control system 集控系统
centralized control with flow varied by steps 分阶段改变流量的质调节
centralized control 集中式控制，集中控制
centralized data processing 数据集中处理
centralized dispatching 集中调度
centralized downcomer 集中下降管
centralized dust collecting system 集中集尘系统
centralized equipment remote control 集中遥控，集中遥控装置
centralized heat-supply coverage factor 集中供热普及率
centralized heat supply 集中供热
centralized inspection 集中检验
centralized instrument panel 中心仪表板
centralized management by specified department 归口管理
centralized management 集中管理
centralized monitoring system 集中监测系统，集中监控系统，集中监视系统
centralized planning 集中规划
centralized purchasing 集中采购
centralized regulation 集中调节
centralized smoked stack 集中吸烟栈
centralized telecontrol 中心遥控，远动集控
centralized wind energy system 集中式风能转化系统，集中风能系统
centralized 集中的，集中式的
centralizer 定心装置，校中心装置，中心器
centralizing force 集中力
central laboratory 中央实验室
central lab 中心实验室
central limit theorem 中心极限定理
central line 中线，中心线
central lubricating system 集中润滑供油系统
centrally administered municipality 直辖市
centrally-located 中心位置的
centrally planned economy 中央计划经济
centrally ported radial piston pump 内配流径向柱塞泵
central mast manipulator 中心杆操作器
central melting （核燃料）中心熔化
central meridian 中央子午线
central milling system 集中制粉系统
central mixer 中心拌和机
central mixing plant 中央拌和楼，中央搅拌厂
central monitoring building 中央控制楼
central monitoring control system 风电场集中监控系统
central movement 有心运动
central office 中央电话局，中心局，中心站
central plane 中心面
central plant 总厂
central power plant 中心电厂
central power station 中心发电厂
central pressure regulator 中心压力调节器
central principal inertia axis 中心主惯性轴
central principal moment of inertia 中心主惯性矩
central processing system 中央处理系统
central processing unit 中央处理机，中央处理单元，中央处理器【主机】
central processor 中心处理机，集中处理装置
central pulverized coal system 集中仓储制制粉系统
central pump house 中央水泵房
central pumping station 集中泵站
central radio office 无线电总局，无线电中心工作室
central reactivity coefficient 中心反应性系数
central reactor flux 反应堆中心通量，堆心通量，堆心中心流
central receiver system 中央吸热系统【太阳能发电】
central receiver tower 中心收集塔，中心集热塔，中央吸热塔，光热塔
central receiver 中心接收器，中心接受器
central recording station 中央记录站
central reflector 中心反射层
central register 中央寄存器
central reprocessing plant 中心核燃料后处理厂
Central Research Institute for Physics 中央物理研究中心【匈牙利】
central ring 中心环
central riser 中央升管
central section 中心片段【管口堵板】
central shaft 中心轴
central shut-down rod 中心停堆棒
central signaling installation 中央信号装置
central signaling panel 中央信号盘
central span 中间跨度，中心跨距，中间跨，中跨
central station air handling unit 组合式空气处理机组
central station control 集中式控制
central station switchboard 中心发电厂控制盘
central station system 中心站制
central station 中心发电厂，中心站，公用电

厂，中心电站，中央电站，总站，总厂
central storage unit 中央存储器
central subtropical zone 中亚热带
central system 集中系统，中心制，中心系，有心系，集中式系统
central telephone exchange 中央电话局，电话总局
central temperature （核燃料）中心温度
central tension 中心受拉
central terminal unit 中央终端装置
central terminal 中央终端
central test hole 中心试验孔道
central thermal expansion 中心热膨胀
central thimble 中央管道
central tower solar thermal power generation 塔式太阳能光热发电
central transfer point 中央转移点
central trunk terminal 中央干线终端
central tube 中心试验管道
central tubular 中心管状支承
central value 试验数据中值，中心值
central ventilation system 新风系统
central waste water treatment 集中废水处理
central zone 中心区域，中心部分
central 集中的，中心的，中央的
centre adjustment 中心调整
centre align 对中，找中心
centre angle 圆心角
centre bearing 中心轴承，中心支承
centre bit 转柄钻头，中心钻头
centre-block type joint 中心滑块式连接
centre-break disconnector 中断式隔离开关
centre by-pass 中间旁通
centre cleavage 中心裂纹
centre control board 中央控制台
centre coupler 中心联轴器
centre cut(ting) 中心掏槽
centre distance 中心距
centre drift method 中心导坑法
centre drift 中心导坑隧洞开挖
centre-fed 中心馈电的，对称供电的
centre gauge 中心规
centre line 中心线
centre mixer 中心拌和站
centre of atmospheric action 大气活动中心
centre of crest circle 坝顶曲线中心点【水电】
centre of equilibrium 平衡中心
centre offset 偏心
centre-of-gravity displacement 重心偏移
centre of parallel forces 平行力系中心
centre of population 居民中心
centre of rotation 旋转中心
centre peg 中心桩
centre pin 中心销
centre point 中点荷载
centre position 中间位置
centre rod 中心燃料棒【燃料组件】
centres locus 中心轨迹线
centre-tapped 中心抽头
centre to centre distance 中至中距离
centre to centre 中至中
centre zero instrument 零点在中心的仪表
centre 放在中心，定中心，中心，中央
centricity 同心度
centrifix purifier 旋流叶片汽水分离器
centrifugal 离心的
centrifugal analysis 离心分析
centrifugal fan 离心式风机，离心式通风机
centrifugal filtration 离心过滤
centrifugal oil cleaner 离心滤油器
centrifugal acceleration 离心加速度
centrifugal action 离心作用
centrifugal air separation 离心空气分离
centrifugal atomization burner 离心式喷嘴
centrifugal atomizer 离心式喷嘴
centrifugal-axial flow compressor 离心-轴流式压气机
centrifugal blower 离心（式）鼓风机，离心式搅拌机
centrifugal brake 离心制动装置
centrifugal breaker 离心破碎机
centrifugal breather 离心式通风器
centrifugal casing 离心浇铸
centrifugal casting process 离心式浇注法
centrifugal casting 离心式浇注法
centrifugal charging pump 离心式上充泵
centrifugal clarifier 离心澄清器
centrifugal clutch 离心式离合器
centrifugal collector 离心吸集器
centrifugal compressor 离心式压气机，离心式压缩机，离心式空气压缩机
centrifugal concrete 离心法混凝土
centrifugal contactor 离心接触器
centrifugal decantation method 离心撇清法
centrifugal dewatering of sludge 污泥离心脱水
centrifugal dewatering 离心脱水
centrifugal drier 离心干燥器
centrifugal drying 离心干燥
centrifugal dust collector 离心式除尘器
centrifugal dust separator 离心除尘器
centrifugal effort 离心力
centrifugal ejection 离心放料
centrifugal extraction analysis 离心抽提分析
centrifugal extractor 离心萃取器，离心提取器
centrifugal fan-type wet scrubber 离心扇形湿式洗涤器
centrifugal fan 离心式通风机，离心风机
centrifugal filter 离心过滤机，离心过滤器，离心式滤油器
centrifugal fluidized bed 离心流化床
centrifugal flyball governor 离心摆锤式调速机
centrifugal force 离心力，向心力
centrifugal gas cleaner 离心涤气机
centrifugal governor 离心式传感器，离心调速器
centrifugal head 离心水头
centrifugal hinge moment 离心铰链力矩
centrifugal hydroextractor 离心脱水机
centrifugal impeller web 离心叶轮体
centrifugal impeller 离心式叶轮
centrifugal inertial separator 离心惯性分离器
centrifugal loading 离心力负荷
centrifugalization 离心分离

centrifugal load 离心荷载，离心载荷
centrifugal lubricator 离心式润滑器
centrifugally cast steel 离心铸钢
centrifugally spun concrete pipe 离心式旋制混凝土管
centrifugal machine 离心机
centrifugal-mechanism weight 离心式调速器飞锤
centrifugal moisture 离心含水量【土壤】
centrifugal moment of inertia 离心惯性矩
centrifugal moment 离心力矩
centrifugal motion 离心运动
centrifugal multi-stage pump 多级离心泵
centrifugal nozzle 机械雾化喷嘴，离心式喷嘴
centrifugal oil filter 离心滤油器，离心式滤油机
centrifugal oil purifier 离心式滤油器
centrifugal oil pump 离心油泵
centrifugal oil separator 油水分离器，离心滤油机
centrifugal pendulum 离心式调速器，离心摆
centrifugal photometric analyzer 离心光度分析仪
centrifugal precipitation mechanism 离心沉淀机
centrifugal process 离心法
centrifugal pull 离心拉力
centrifugal pump of multistage type 多级离心泵
centrifugal pump of single type 单级离心泵
centrifugal pump of turbine type 水轮式离心泵
centrifugal pump with horizontal axis 卧式离心泵
centrifugal pump with vertical axis 立式离心泵
centrifugal pump 离心式水泵，离心泵
centrifugal purifier 离心净化器
centrifugal refrigerating machine 离心式制冷机
centrifugal regulator 离心调节器，离心调速器
centrifugal relay 离心式继电器
centrifugal screen 离心筛
centrifugal screw pump 离心螺旋泵
centrifugal scrubber 离心洗涤器，离心涤气器
centrifugal separation 离心分离
centrifugal separator 离心分离机，离心式分离器
centrifugal settling 离心沉降
centrifugal sewage pump 离心污水泵
centrifugal speed 离心速度
centrifugal stability 离心稳定性
centrifugal starter 离心式启动器，离心开关
centrifugal starting switch 离心式启动开关
centrifugal steam purifier 离心式汽水分离器
centrifugal steam separator 离心式汽水分离器
centrifugal stop bolt 危急保安器重锤，撞击子
centrifugal stress 离心应力
centrifugal supercharger 离心式增压器
centrifugal switching 离心切换
centrifugal switch 离心式开关
centrifugal tachometer 离心式转速计
centrifugal tar extractor 离心焦油提取器
centrifugal thrower 油环，离心式挡油环
centrifugal thrust 离心推力器
centrifugal type diatomite filter 离心式硅藻土过滤器
centrifugal unit 离心单元
centrifugal unloading 离心卸载
centrifugal water-packed gland 离心水封
centrifugal water pump 离心水泵
centrifugation 离心法，离心作用，分离作用，离心分离
centrifuge contactor 离心接触器
centrifuge effect 离心分离效应
centrifuge method 离心法
centrifuge moisture equivalent 离心含水当量
centrifuge process 离心工艺【浓缩】，离心法
centrifuger 离心机
centrifuge time 离心时间
centrifuge tube 离心管
centrifuge 离心分离的，离心机，离心，分离机
centrifuging 离心法，离心作用，分离作用
centrifugul force 离心力
centring bar 定心棒
centring 定中心，对中，打中心孔，对准中心
centripetal action 向心作用
centripetal bar 向心的
centripetal drainage 向心排水，向心水系
centripetal force 向心力
centripetal pump 向心式水泵，向心泵
centripetal turbine 辐流式水轮机，向心式水轮机，向心式透平
centripetal 向心力
centroclinal dip 向心倾斜
centroclinal fold 向心褶皱
centroidal distance 形心距离，重心距离
centroid of a plane area 平面质量中心
centroid 矩心，质心，重心，形心，质量中心
centrosphere 地心圈
centrosymmetric 中心对称的
centrosymmetry 中心对称
centrum 震源
centuple 百倍
century 世纪
cent 美分，分，百分
CEO(chief executive officer) 首席执行官
CEP(condensate extraction pump) 凝结水抽水泵，凝结水泵
cepstral analysis 倒谱分析
cepstrum method 倒谱方法
cepstrum 倒频谱，对数化倒相谱
ceramal resistance 金属陶瓷电阻
ceramal 陶瓷合金，金属陶瓷
ceramet coating 陶瓷涂层
ceramic beta source 陶瓷β源
ceramic block 陶瓷块
ceramic bond 陶瓷接合剂
ceramic brick 瓷砖
ceramic brittle coating 陶瓷脆性涂层
ceramic capacitor 陶瓷电容器
ceramic carbide 陶瓷的碳化物
ceramic cartridge 陶瓷阀芯
ceramic coating 陶瓷涂层
ceramic crucible 陶瓷纤维板
ceramic dielectric capacitor 陶瓷介质电容器
ceramic dielectric 陶瓷介质
ceramic die 陶瓷模
ceramic DIP 陶瓷双列直插封装
ceramic filter cartridge 陶瓷滤芯
ceramic filter 陶瓷过滤器

ceramic fuel 陶瓷燃料
ceramic glaze 陶瓷釉
ceramic heated tunnel 有陶瓷加热的风洞
ceramic insulated coil 陶瓷绝缘线圈
ceramic insulated wire 陶瓷绝缘线
ceramic insulation 陶瓷绝缘
ceramic insulator 陶瓷绝缘子，瓷质绝缘子，瓷绝缘子
ceramic-lined 陶瓷衬里的
ceramic magnet 陶瓷磁铁，烧结磁铁
ceramic material moderated reactor 陶瓷材料慢化反应堆
ceramic material 陶瓷材料
ceramic matrix 陶瓷基体
ceramic metal 金属陶瓷，陶瓷金属
ceramic mosaic 陶锦砖，陶瓷马赛克，马赛克，陶瓷地砖
ceramic nuclear fuel 陶瓷核燃料
ceramic permanent-magnet motor 陶瓷永磁电动机
ceramic pick-up 压电陶瓷拾音器
ceramic pipe 陶瓷管，陶管
ceramic plutonium reactor 陶瓷钚反应堆
ceramic reactor 陶瓷反应堆，金属陶瓷材料的反应堆
ceramic rotor blade 陶瓷动叶片
ceramic source 陶瓷放射源，陶瓷源
ceramics 陶瓷，陶器，陶瓷学，陶瓷制品，陶瓷工艺
ceramic tile cutter 磁砖切割机
ceramic tile flooring 陶瓷板地面
ceramic tile floor 瓷砖地板
ceramic tile 磁砖，瓷砖
ceramic tool 陶瓷刀具
ceramic UO_2 pellet 二氧化铀陶瓷芯块
ceramic uranium dioxide pellet 二氧化铀陶瓷芯块
ceramic veneer 陶瓷板
ceramic 陶瓷的，陶瓷制品，陶瓷，陶制的
ceraminator 陶瓷压电元件，伴音检波元件
ceramsite concrete wall panel 陶粒混凝土墙板
ceramsite concrete 陶粒混凝土
ceramsite 陶粒，素烧黏土
ceram 陶瓷，陶器
cerap 伴音中频陷波元件
ceraunograph 雷电计
CER(certified emission reduction) 可认证减排量，经核证的减排量
ceremony 典礼，仪式
cere 蜡，上蜡，涂蜡
cerise 鲜红色的
cerium isotope 铈同位素
cermet 金属陶瓷，陶瓷合金，烧结金属学
certainty equivalence control 确定性等效控制
certificate documentation 证明文件
certificate for completion 竣工证书
certificate for evidence （证明事故细节的）事故证书［证明］
certificate for export 出口检验，出口检验证
certificate letter 证明信
certificate of acceptance 同意证明书，验收证书
certificate of appointment 委任书
certificate of appraisal 评估证书
certificate of approval 证书，鉴定书，检验合格证
certificate of authorization 核准证书，授权书
certificate of clearance 清关证书
certificate of competency 合格证书，能力证书，资格证书
certificate of completion by stages 分期竣工证书
certificate of completion of handling over charge 交接证明书
certificate of completion of works 工程竣工证书
certificate of compliance 合格证书
certificate of conformity of welding operator 焊工合格证
certificate of conformity 证明商品符合合同中的规格的证书，商品合格证
certificate of country of origin 原产国证明书
certificate of credit standing 资信证明
certificate of damage 破损证书
certificate of delivery 交货证明书
certificate of deposit 存款单
certificate of dishonour 拒付证明
certificate of expenditure 支出证明书
certificate of export 出口签证
certificate of financial standing 金融信誉保证书
certificate of goods' origin 货物原产地证书
certificate of honor 荣誉证书
certificate of identification 身份证
certificate of import license 进口许可，进口许可证
certificate of incorporation 法人认可证，公司注册证书，公司登记执照，公司注册证
certificate of independent public surveyor 公证行证明书
certificate of insurance 保险证明书，保险证书
certificate of loss or damage 损失或损坏证明书
certificate of loss 损失证明
certificate of manufacture 厂方证明书，制造证明书
certificate of merit 奖状
certificate of occupancy 占有说明书
certificate of origin 产地证明书，原产地证书
certificate of patent 专利，专利证
certificate of payment 付款证明
certificate of provisional acceptance 临时验收证书
certificate of quality test 质量检验证书
certificate of quality 货物品质证书，质量证书，质量证明书
certificate of quantity 货物数量证明书，数量证书
certificate of receipt 收货证明书
certificate of re-export 再出口证书
certificate of registration 公司注册证，注册证书，注册执照
certificate of registry 注册证书
certificate of search 查找证明
certificate of shipment 装船许可证，出口许可证，出口证明书，装运证明
certificate of ship's survey 验船证明书，检验证

书，船舶检定书，船舶检查证书
certificate of specifications 规格证明书
certificate 合格证书，证明，证书，检验合格，鉴定书，凭证
certification body 认证机构
certification of completion 竣工合格证书
certification of compliance 合格证书
certification of fitness 合格证书
certification of inspection 检验合格证书
certification of invoice 签证发票
certification of material 材料证明书，材料证实书
certification rule 认证规则
certification scheme 认证方案
certification specification 认可规范
certification system 认证体系
certification test （资格）认证考试，职业技能鉴定考试，出厂证明试验，验证试验，验收试验
certification 检验证明书，认证，证明，证明书，鉴定书
certified accountant 注册会计师
certified check 保付支票
certified copy document 文件的正式副本
certified copy 经鉴证的副本
certified correct 证明无误
certified drawing 合格图纸
certified emission reduction 可认证减排量，经核证的减排量
certified final version 最终确认版
certified for construction 确认可以施工
certified invoice 签证发票
certified person 合格的人员
certified reference coal 标准煤样
certified reference-materials of coal 煤标准物质
certified signature 经验证的签名
certified true copy 经核证无误的副本
certified 被证明的
certifying authorities 核证机构，证明机构
certifying bank 保付银行
certifying officer 核证人
cerulean 天蓝色的
CES(condenser tube cleaning system) 冷凝器管道清理系统
CES(corporate engineering standard) 公司技术标准
cesium iodide 碘化铯
cesium separation plant 铯分离工厂
cesium vapor rectifier 铯气整流器
cessation of liability 终止责任
cessation 停工，停止，终结，终止
cesser clause 租船契约上的责任终止条款
cession of right 权利的转让
cession （产权）转让，（财产）转让，（领土）割让
cess pipe 污水管，污水竖管
cesspit 污水坑[井]，垃圾坑，粪坑
cesspool 渗水井，渗井，渗坑，污水池，污水坑，粪坑，污秽场所
CET 汽轮机轴封系统【核电站系统代码】
CEX 凝结水系统【核电站系统代码】
CFA(central facilities area) 中心试验装置区
CFA(clear for action) 可以使用，可供采用，已弄清可采用【文件状态】
CFAE(contractor furnished and equipped) 承包人供应和装备的
CFB boiler 循环流化床锅炉
CFBC(circulating fluidized bed combustion) 循环流化床燃烧
CFB(circulated fluidized bed) 循环流化床
C/F(carried forward) 转（入）下页
CFC(certified for construction) 可供施工使用【文件状态】
CFC(containment fan cooler) 安全壳通风冷却系统
CFC(cost,freight and commission) 货价、运费与佣金
CF(centrifugal force) 离心力
CFC(flip-flop circuit) 交叉膜冷子管触发电路
CFC(memory circuit) 交叉膜冷子管存储器
Cf＝confer 比较，对照，参阅，参照，与……比较
CF＝coolingfan 冷却风扇
CF(correction factor) 修正系数
C&F(cost and freight) 成本加运费价格，含运费价格，货价加运费
CFC(chloro fluoro carbon) 氯氟碳化物
CFE(criteria for evaluation) 设计准则，评价准则
CFF(critical fusion frequency) 临界停闪频率
Cfh(cubic feet per hour) 立方英尺每小时
CFI 循环水过滤系统【核电站系统代码】
CFM 凝汽器精滤系统【核电站系统代码】
CFO(chief financial officer) 首席财务官
CFR(chance failure rate) 偶发故障率
CFR(code of federal regulations) 联邦法规【美国】
CFR(cost and freight) 成本加运费（价），CFR价格，货价加运费，含运费价格
CFS charge 场站费
CFS(container freight station) 集装箱货物集散站，散货仓库
CFS(container freight station) 集装箱货运站
cfs(cubic foot per second) 立方英尺每秒
CF system(core flooding system) 堆芯淹没系统
CFT(cold function test) 冷态功能实验
CFT-RVO(cold function test-reactor vessel open) 冷试压力容器开盖
CFW(constant flux voltage variation) 恒磁通调压
cf. 参看，试比较
CG(center of gravity) 重心
CG(control grid) 控制栅极
CGCS(cover-gas cleanup system) 覆盖气体净化系统
CGE(Canadian General Electric Co.) 加拿大通用电气公司
CGGC(China Gezhouba Group Corporation) 中国葛洲坝集团股份有限公司
CGNPC(China Guangdong Nuclear Power Holding Co.,Ltd.) 中广核工程有限公司
CGR 循环水泵润滑油系统【核电站系统代码】
CGS system(Centimeter-Gram-Second system) CGS制，厘米—克—秒制
chafing fatigue 摩擦疲劳

chafing 磨损，擦痕，拂痕
chainage 测链数，链测长度，桩号
chain-and-tackle 链滑轮
chain belt 链带
chain block with steel structure 带钢结构的起重葫芦
chain block 金不落【俗语】，斤不落【俗语】，链滑车，链滑轮组，神仙葫芦，手拉链条葫芦，手扳链条葫芦
chain bond 链式搭接
chain break 通道中断
chain bridge 链索吊桥，链索桥
chain bucket car unloader 链斗卸车机
chain bucket conveyor 链斗式提升机
chain bucket elevator 链斗提升车
chain bucket excavator 链斗式挖土机
chain bucket loader 链斗式装载机
chain bucket ship unloader 链斗卸船机
chain bucket wagon unloader 链斗卸车机
chain bucket 链式挖泥船，斗式挖泥船
chain cable 链索
chain circuit system 链电路系统
chain circuit 链电路
chain code calibration device 链码校验装置
chain code check device 链码校验装置
chain code 链式码，循环码，链码
chain coil 链式线圈
chain connection 链联结，串级连接
chain conveyor 链条输送机，链式输送机，链板输送机
chain coupling 链式连接器
chain debts 三角债
chain disintegration 链式蜕变，链式衰变
chain dog 链条扳手
chain dotted line 连续点线
chain dredger 链式挖泥机
chain drive 链条传动，链拖动，链传动
chain drop 链式吊灯
chained list 链接表
chained records 链接记录
chained sequential operation 联锁顺序操作，链锁程序操作
chain elevator 链式升降机，链式提升机
chainette tower 悬索塔，悬链塔
chain-eye ball 球头长环
chain feeder 链式给料机
chain fission yield 链裂变产额
chain fission 链式裂变反应
chain gauge 链式水尺
chain grate boiler 链条锅炉，链条炉排锅炉
chain grate stoker 链带式炉排，链条炉排
chain grate 链条炉篦
chain handwheel 链条手轮
chain hoist 吊链，链式起重机，链条葫芦，手动葫芦，金不落
chaining search 链接检索
chain insulator 绝缘子串
chain intermittent fillet weld 并列断续角焊缝
chain jack 链式起重器
chain lightning 链状闪电
chain-like structure 链状结构
chain link door 钢丝网门
chain link fencing 钢丝网围栏
chain link 挂环，链节，延长环
chain making tool 造链机
chain network 链形网，链形网络
chain of egg insulators 蛋形绝缘子串
chain of particle model 颗粒群模型
chain of power plants 梯级电站
chain of relays 继电器群
chain operated valve 链条操纵的阀门
chain operated 链条操作
chain pendant 链式吊灯架
chain pin 测钎，链销
chain pipe vise 链式管钳
chain pipe wrench 链式管钳，链条管扳手
chain post 链柱
chain pulley system 链-滑轮系统
chain pulley 链轮
chain pump 链泵
chain-reacting system 链式反应系统
chain reaction 连锁反应，链式反应
chain reactor 链式反应堆
chain relay 串动继电器，联锁继电器
chain-riveting 并列铆接，链型铆接
chain screen 链式过滤器
chain sling 链式吊索，链索
chains network 三角锁网
chain sounding line 测深链
chain sprocket 链轮
chain stoker 链式加煤机
chain stopper 锚链制动器
chain survey 测链测量，链测
chain suspension bridge 链索吊桥
chain tape 测链，测链尺
chain tension device 链条张紧装置
chain tensioner 链条拉紧器
chain-tensioning device 链条张拉设备
chain termination reaction 链终止反应
chain transmission 链传动，链条传动
chain transport 链传递，链替序
chain vice 链式钳
chain wheel 链条轮，链轮，轮盘，滑轮
chain winch 链式绞车，链绞盘
chain windlass 链式卷扬机
chain wrench 链式扳手
chain 链式电路，链条，链，电路，回路，联锁，锁链
chairman of the board of directors 董事长，董事会主席
chairman of the board 董事长，董事会主席
chairman of the chamber of commerce 商会主席
chairman 董事长，主席
chair plate 座板
chair-rail 靠椅扶手
chair 座板【支承钢筋】，筋鞍子
chalcocite 辉铜矿，钢铀云母
chalcophyllite 云母铜矿
chalcopyrite 黄铜矿
chalking 粉化，灰化
chalk line 粉笔线
chalk stone 石灰岩

chalk test 白垩试验，渗透探伤试验
chalky soil 白垩土
chalk 白垩，粉笔
challenger 询问器，问答器
challenge switch 呼叫开关
challenge 挑战
challenging 复杂的，多路传输，混合的
chalybeate spring 铁盐矿泉
chamber arch 弯拱
chamber blasting 洞室爆破
chambered arch 弯拱
chambered corridor 两旁有房间的走廊
chamber filling conduit 闸室充水管
chamber floor 闸室底板
chambering 扩孔爆破
chamberlet 小室
chamber of commerce 商会
chamber of mail oil pump 主油泵环室
chamber pressure 室压
chamber retaining wall 空箱式挡土墙
Chambertiin model 张伯伦模型
chamber wall 闸室墙
chamber 室，房间，燃烧室，小室，舱，峒室，腔，箱
chambranle 门框饰
chamfer angle 坡口斜角，斜切角
chamfered edge 削角边
chamfered groove 三角形断面槽
chamfered joint 切角对接接头，斜削接头
chamfered pellet 倒角芯块
chamfer geometry 沟槽形状
chamfering machine 倒角机
chamfering tool 倒棱工具，刻槽工具
chamfer strip 倒棱板条，斜边板条
chamfer 倒角，倒棱，斜切，刻槽，切角，削角
chamotte brick 黏土质耐火砖，黏土砖，耐火砖
chamotte 熟耐火土
chanalyst 故障探寻仪
chance-constrained programming 机会约束规划
chance failure period 偶发故障期间
chance failure 意外故障，偶然故障，偶然事故
chancellery 大使馆，办事处，大臣
chance machine 概率计算机
chancery 档案馆
chance 机会，时机
chandelier 枝形灯架，集灯架
chanfered pellet 倒角芯块
changeability 互换性
change area tenting 改变营地面积
change back 倒回
change behaviour 负荷变化性能
change control procedure 变更管理
change crew 换乘（务组）
changed condition 已变更的条件
change directive 更改指令
changed memory routine 变更存储的程序
change dump 变更存储内容
change for using electricity 变更用电
change gear set 齿轮变速组
change gear train 变换齿轮系
change gear 变速齿轮，交换齿轮
change in direction 改变方向，改变高度
change in legislation 法规发生变更，律师发生变更
change in load 负荷变化
change in phase 相变
change in the scope of work 工作范围的变更
change lever 变速杆
change notice 变更通知
change of air 换气
change-of-linkage law 磁链变化定律，电磁感应定律
change of registration 注册变更
change of state announcement 状态变化通告
change of the production line 转产
change of voyage 绕航
change order 变更单，变更命令
changeout 替换，代替
change-over busbar 倒母线
change-over button 换向按钮
change-over circuit 转换
change-over cock 切换旋塞，多路阀
change-over contact 转换接点，转换触点
change-over controller 切换控制器
change-over gear 转换装置
change-over key 换向按钮
change-over mechanism 换算机构
change-over of house power source 厂用电切换
change-over pilot valve 切换滑阀
change-over plug 换路插头，转换插塞
change-over radius 过渡区半径
change-over relay 切换继电器
change-over switch controller 转换开关控制
change-over switching 换接，换向开关
change-over switch 切换开关，转换开关
change-over time 转换时间
change-over valve 切换门，转换阀
change-over 转换开关，切换，转换，改变，倒向，转接，转向，倒换，转变
change-pole motor 变极电动机
change protection to signaling 将保护改为信号
change protection to tripping 将保护改为跳闸
change rate 变化率，变率
change request form 变更申请单
change request 变更请求
change review board 修改审查委员会
changer 转换开关，变换器，换流器，换能器，转换器
changes in personnel 人员的变换，职员的变更
change-speed gear box 变速箱
change-speed gear 变带箱，变速齿轮
change-speed motor 变速电动机
change tape 更换的磁带，更换的纸带
change unit 改型装置，改装机组
change-valve 转换阀
change 变更，变化，改变，兑换，更换，修改，转变
changing deterioration rate 可变腐败率
changing load 变化荷载，交变荷裁
changing room 更衣室
changing stage 变动水位
changing switch 转换开关

changing up 升速
changing voidage model 空隙度变化模型
changing 交换
channel accretion 航道淤积，渠道淤积
channel address register 通道地址寄存器
channel address word 通道地址字，分路地址代码
channel alignment 河道整治线，水道定线
channel allocation and routing data 通道分配和路径选择数据
channel allocation 通道分配
channel assignment 信道分配
channel-associated signaling 随路信令
channel average density 通道平均密度
channel bandwidth 频道带宽
channel bank 信道处理单元，通道调制解调库
channel bar 槽钢
channel beam 槽形梁
channel bend 管道弯头
channel blockage 通道堵塞
channel border 管壁，风洞壁
channel boundary 管壁，孔道边界，河道边界
channel capacity 通道容量，通道传输能力，信道容量，河槽容量
channel change 水道改变
channel cleanout 河槽清淤，河道清淤
channel closure 管封口，管封头
channel column 槽形柱
channel command 通道命令
channel configuration 燃料元件管布置
channel controller 通道控制器
channel cross-sectional flow area 通道横截流通面积
channel current 沟道电流
channel decoder 信道译码器
channel degradation 河槽冲深
channel depletion type 沟道耗尽型
channel deposit 渠道淤积
channel design 通道设计
channeled steel plate 压型钢板
channeled upland 槽蚀高地
channeled wind 夹道风
channel effect 通道效应
channel electron multiplier 通道电子倍增器
channel encoder 信道编码器
channel end condition 通道传输结束条件
channel end shell flange 管侧壳体法兰【换热器】
channel entrance 航道入口，水道入口
channel erosion 槽蚀，槽形侵蚀，沟蚀
channel filling 河道填塞，河道淤塞
channel flow accretion 渠道流量增长
channel flow 狭管流，河道径流，河槽径流，沟流
channel for oiling 油路，油槽
channel frequency 通道频率
channel gradient 河床比降
channel head drainage 水封头排水
channel head 水侧封头，下封头【蒸汽发生器】，(一回路侧的) 封头，燃料组件顶端件
channel identification 通道识别
channel impulse 通道脉冲
channeling corrosion 沟槽腐蚀
channeling effect 夹道效应，沟道效应，孔道效应
channeling fluidized bed 沟流状流化床
channeling of liability 责任的归属
channeling pattern 沟道模式，河流类型，水系类型，河道类型
channeling 沟道作用，多路传输
channel intake 渠道进水口，取水渠
channel integral model 通道积分模型
channel interface 通道接口
channel inversion 管道内燃料块轴向倒换
channel iron 槽钢，槽铁
channel island 江心洲
channel length 通道长度
channel line 河道中泓线
channel material 槽型材料
channel monitor 通路监视器
channel of stationary end (换热器的) 固定端
channel operation availability 通道运行可用率
channel physical separation 通道实体分隔
channel pitch plate 托架间距板
channel plate 槽形板
channel precipitation 河槽降水(量)，河面降水
channel program 通道程序
channel protection 水道保护
channel pulse 通道脉冲，信道脉冲
channel radius 通道半径
channel range 通道范围
channel ratio 信道比
channel read-backward command 通道回读命令
channel rectification 河道整治
channel regime 河道状况
channel revetment 河道护岸，护岸工程，航道驳岸
channel ring 环行通道
channel roughness 河床糙率
channel runoff 河槽径流，渠道径流
channel scour 河床冲刷
channel section 槽钢，槽钢截面
channel selecting system 通道选择系统
channel selection 通道选择
channel selector 波道转换开关，信道选择器，频道转换开关
channel sense command 通道异常显示命令
channel shift switch 通道移位开关
channel slab 槽形板，槽板
channel spacing 通道间距
channel span 通道量程
channel specific-enthalpy rise 通道比焓升
channel spillway 溢流渠
channel status indicator 通道状态指示器
channel status table 通道状态表
channel status word 通道状态字
channel status 通道状态
channel steel 槽钢
channel storage capacity 槽蓄量
channel storage routing 槽蓄演算
channel stripping machine 燃料盒脱盒机【沸水堆】
channel switch 波道开关

channel synchronization 信道同步
channel synchronizer 通道同步装置
channel terminal bay 电路终端架
channel terminal decoder 信道终端解码器
channel thread 中泓线
channel throw 河床落差
channel tile 槽形瓦
channel time-slot 信道时隙
channel-to-channel adapter 通道选择器，通道—通道衔接器
channel traffic-capacity 通道通信能力
channel traffic control 通道通信量控制
channel transistor 沟道晶体管
channel transition 河道渐变段
channel trap 信道陷波电路
channel tray 槽形托架
channel trigger circuit 信道触发电路
channeltron 通道倍增器
channel type impeller 通道式叶轮
channel type power station 明渠式水电站，引水式水电站
channel type reactor 压力管式反应堆
channel type terrace 槽形阶地
channel utilization factor 信道利用系数
channel utilization index 通道利用指数
channel utilization 通道利用率
channel vocoder 谱带式声码器
channel waveguide 信道波导
channel width variation 信道频宽变化
channel width 沟宽，通道宽度
channel wiring 槽中布线
channel write command 通道写取命令
channel 槽钢，通道，河槽，航道，信道，波道，沟道，槽，海峡，狭管，涵洞，水道，沟，渠道，途径，线槽，通路
channery 碎石块
chaologist 混沌理论家，混沌学家
chaology 混沌学
chaos theoretician 混沌理论家，混沌学家
chaos theorist 混沌理论家，混沌学家
chaos theory 混沌理论
chaos 混沌，混乱，无秩序状态
chaotic attractor 混沌吸引子
chaotic dynamics 混沌动力学
chaoticist 混沌学家
chaotic vorticity 混沌涡量
chaparral jungle 丛林
chaparral 丛林，矮树林
chap. = chapter 篇，章，回
chapelet 链斗泵，链斗疏浚机，链斗提水机
chap 裂纹
character adjustment 字符调整
character and graphic display 字符图形显示器
character-at-a-time printer 每次一个字符的打印机
character boundary 字界，字符边界
character code 字符码
character cycle 字符周期
character density 字符密度
character display terminal 字符显示终端
character display tube 字码显示管，显字管
character display unit 数字字母显示器，字符显示器
character display 字符显示
character error rate 字符误码率
character fill 字符填充
character-forming tube 显字管
character generator 字符发生器，符号发生器
character index 特征指数
character-indicator tube 字符指示管
characteristic absorption peak 特征吸收峰
characteristic acoustic resistance 特征声阻
characteristic admittance 特性导纳
characteristic angle 特性角
characteristic angular phase difference （多相的）特征相角差
characteristic area 特征面积
characteristic class 特征曲线类型
characteristic combination 特征性组合，标准组合
characteristic component 特征组分
characteristic condition 特征条件
characteristic constant 特征值，特征常数
characteristic correlation length 特征相关长度
characteristic curve test 特性曲线试验
characteristic curve 特性曲线
characteristic data 特性参数
characteristic delay time 特性延迟时间，特征滞后时间
characteristic depth 特性水深
characteristic dimension 特性尺寸，特征尺寸
characteristic discharge 特性流量
characteristic dislocation 特征位错
characteristic distortion 特性失真
characteristic element 特性元件，特性元素
characteristic energy 特征能量
characteristic equation 特征方程
characteristic error 特性误差，特征误差
characteristic exponent 特征指数
characteristic family 特性曲线族
characteristic fossil 特征化石
characteristic frequency 特征频率，固有频率
characteristic function 特性函数，特征函数
characteristic harmonics 特征谐波
characteristic head 特性水头
characteristic impedance 特性阻抗，波阻抗
characteristic infrared group frequency 特征红外基团频率
characteristic length 特性长度
characteristic line 特性线，特征线，特征寿命
characteristic locus 特征轨迹
characteristic loop 特性曲线
characteristic matrix 特征矩阵
characteristic method 特性法
characteristic net 特性曲线网
characteristic odor 特性恶臭
characteristic of char residue 焦渣特征
characteristic of waste 废料的特性
characteristic oscillation 特征振动，特性振动，特征振荡
characteristic overflow 特征溢出误差条件，阶码溢出
characteristic parameter 特性参数，特征参数

characteristic peak 特征峰
characteristic point 特性点，特点
characteristic quantity 特性量，特征量
characteristic radiation 特性辐射
characteristic ratio 特性比，帕森斯数
characteristics description 特性描述
characteristics method 特征线法
characteristics of diffuse distribution 漫分布的特性
characteristics of electrical product 电工产品特性
characteristics of tar residue 焦渣特性
characteristics of the sun 太阳的特性
characteristic speed 特性转速
characteristic stage 特性水位
characteristic strength value （材料的）强度参数，强度特征值
characteristic surface 特征曲面
characteristics 特征，特性
characteristic temperature 特征温度，特性温度
characteristic test 特性试验
characteristic time interval 特征时间间隔
characteristic time response 特征时间响应，特性时间反应
characteristic time 特征时间，时间常数
characteristic type 特性型号
characteristic underflow 特征下溢误差条件，阶码下溢
characteristic value of a geometrical parameter 几何参数标准值
characteristic value of a material property 材料性能标准值
characteristic value of an action 作用标准值
characteristic value of a property of a material 材料性能标准值
characteristic value of subgrade bearing capacity 地基承载力特征值
characteristic value of CTOD 特征裂纹尖端张开位移值
characteristic value 特性值，标准值，特征值
characteristic variable 特征变量
characteristic vector 特征矢量
characteristic velocity 特征速度
characteristic vibration 特征振动
characteristic voidage 特征空隙度
characteristic water level 特性水位
characteristic wave impedance 特性波阻抗，特征波阻抗
characteristic wavelength 特征波长
characteristic wave 特征波
characteristic X-ray 标识X射线
characteristic 特性（曲线），性能，特征，特点，特有的，表示特性的，特征的，特性的
characterization factor 性能因子，特性因数，特性因素
characterization map 特性图
characterization 说明特性，表征，特性描述
characterizer 表征器
characterize 表征，描绘……的特性，具有……的特征
character keyboard 字符键盘
character light 电码灯，符号灯
character of accident 事故性质
character of boring slurry in pozzuolana sediment area 火山灰沉积地区钻孔泥浆性能【试验】
character of service 工作状态
character-oriented computer 面向字符的计算机
character outline 字符轮廓
character per second 每秒字符数，字符每秒
character pitch 字符间距
character position 符号位置，字母位置
character reader 字母读出器，字符读取装置
character recognition system 字符识别系统
character recognition 字符识别，符号识别
character repertoire 字符库，字符表
character size 字体大小
character skew 字符歪斜
character storage tube 字符储存管
character string constant 字符串常数
character string 字符串，字符序列
character stroke 字符笔画
character style 字体
character-writing tube 字体写入管
character 字母，符号，性质，特性，品质，特征，性格，品格，显字管，字符管
char adsorption 炭吸附法
charcoal adsorption process 活性炭吸附法
charcoal filter 活性炭过滤器
charcoal 炭，活性炭，木炭
char combustion 木炭燃烧
charge a battery 给蓄电池充电
chargeable downtime 应受罚的故障时间
chargeable duration of calls 通话的收费时间
chargeable failure 责任失效
chargeable heat 消耗热
chargeable-time clock 收费时间计时器
chargeable weight 计费重量
chargeable 可充电的，应负责的，可收费的
charge accumulation 电荷积聚
charge accumulator 电荷存储器，电荷积聚器
charge against gross national balance 国民总支出
charge air 增压空气
charge and discharge key 充放电开关，充放电键
charge and discharge operations 装卸操作
charge capacity 充电容量，装载量
charge carrier diffusion 载流子扩散
charge carrier 带电粒子，载流子，载荷子
charge cask 装料容器
charge characteristics 充电特性
charge coefficient 装药系数
charge conservation 电荷不灭，电荷守恒，电荷位形
charge controller 充电控制器
charge control parameter 电荷控制参数
charge control 电荷控制
charge cooler 增压空气冷却器
charge-coupled device 电荷耦合器件
charge-couple memory 电荷耦合存储器
charge current 充电电流
charged body 带电体
charged corpuscle 带电粒子，带电微粒
charge density 电荷密度

charge distribution 电荷分布
charged particle detector 带电粒子检测器
charged particle equilibrium 带电粒子平衡
charged particle 带电粒子
charged powder 带电粉体
charge drive 电荷传动，电荷激励，高频电荷激励
charged state 充电状态
charged to 计入……成本
charged 已充电的，带电的，充了电的
charge efficiency 充电效率，充气效率
charge engineer 值班工程师
charge-exchange phenomenon 电荷交换现象
charge floor 装料平台
charge for electrical supply 供电贴费
charge for telegram 电报费
charge for trouble 酬劳费，手续费
charge hole 装料孔
charge indicator 带电指示器，验电器
charge leakage 充电漏泄
charge-letdown system 上充下泄系统
charge life 装载寿期，装载寿命【核燃料】
charge machine grab 装料机抓具
charge machine isolation valve 装料机隔离阀
charge machine runway 装料机走道
charge machine vent valve 装料机放气阀
charge machine 装料机【反应堆】
charge meter 电荷计
charge number of ion 离子的电荷数
charge number 电荷数，原子序数
charge of electron 电子电荷
charge off 销账
charge of rupture 破坏荷载，破坏载荷
charge of surety 容许载荷，安全载荷
charge pattern leakage 电荷起伏漏泄，电荷分布漏电
charge pattern 充电曲线，电荷分布图
charge per unit volume 比电量
charge plug 装料塞
charge pond 装料水池
charge pump 供水泵，供油泵，辅助泵
charge ratio 装料比，充填系数，装载比
charger battery charger 充电器
charge-redistribution converter 电荷再分布型转换器
charge-resistance furnace 炉料电阻炉
charger ratio 满载系数，满载比，装填比
charger room 充电机室，装料室
charger 充电装置，装料设备，装料机，充电机，加料器，充电器
charge schedule 装料方案，装料计划
charges collect 运费到付
charge-sensitive amplifier 电荷灵敏放大器
charges for ballast 压底舱费
charges for customs clearance 清关费
charges for overtime 加班费
charges of service 劳务费
charges paid 费用已付
charge spot(/point) 充电桩
charges prepaid 预付费用，费用已预付
charges that are below a just and reasonable level 低于正当合理的收费
charge storage diode 电荷存储二极管
charge storage effect 电荷存储效应
charge storage tube 电荷存储管
charge storage 电荷存储
charge switch 充电开关
charge tank 注入箱
charge-temperature 炉料温度，着火温度
charge time constant 充电时间常数
charge time 充电时间
charge-to-mass ratio 电荷质量比
charge-transfer device 电荷传输器件
charge-transfer test 电荷转移试验
charge-transfer 电荷转移
charge transition time 电荷渡越时间
charge tube 加料管【高温气冷堆装料机】
charge unit 计费单位，电荷单位
charge up 充电，加料
charge valve 增压阀
charge voltage （电容器的）电荷电势
charge 费用【复数】，充电，负载，电荷，掌管，控告，负荷，装载，装料，加注，收费，炉料
charging accident 装料事故
charging aperture 加料孔
charging apparatus 充气设备，装料设备，充电设备，喂料装置
charging area 装燃料区，装料场
charging basket 料筐
charging board 充电盘
charging box 装料斗
charging cable 充电电缆
charging capacity 充电容量，装载量
charging car 装料车
charging characteristic 充电特性
charging choke 充电扼流圈
charging circuit 充电回路，充电电路
charging coefficient 充电系数
charging compressor 充气压气机
charging conduit 落料管，落料槽
charging cone 装料斗
charging crane 装料吊车
charging current 充电电流，电容电流
charging curve 充电曲线
charging cycle 装料循环
charging device 充电设备，充电装置，装料设备
charging diode 充电二极管
charging duration 充电时间
charging equipment 加料设备，充电设备，充电装置
charging floor 加料台，装料台
charging flow rate 硼注入率
charging flue 进气管
charging generator set 充电发电机组
charging generator 充电发电机
charging hopper 装料漏斗，供料斗，装料斗
charging impulse 充电脉冲
charging indicator 充电指示器
charging inductance 充电电感
charging interval 充电时间，充电间隔
charging limit 充电极限

charging line 上充管线【主冷却剂】，加注管路，进油管路，供电线路，补水管线，供料线，供水管
charging load 充电负载
charging loss 充电损失
charging machine 装料机
charging mechanism 装料设备
charging period 充电时间，装料周期
charging pile 充电桩，充水管，供给管，装料管
charging pump 上充泵【压水堆】，补水泵，充水泵，供料泵，上水泵，进料泵，充液泵，送料泵
charging rectifier 充电整流器
charging relay 充电继电器
charging resistor 充电电阻器
charging room 装料间，装料室
charging set 增压装置，燃气发生器，增压器，充电机组
charging source 充电电源
charging station 充电站
charging stream 上充流
charging switch 充电开关
charging system 装料系统，充电系统，补给系统
charging time constant 充电时间常数
charging tube 装料管道，充电管
charging-turbine set 透平增压装置
charging-up 加添，加注，充电，装料
charging valve 进给阀
charging voltage 充电电压
charging volt-ampere 充电伏安容量
charging wagon 装料斗车
charging wiring diagram 充电接线图
charging 增压，充气，充电，带电，装料，装炸药
chargistor 电荷管
chark 木炭，焦化，炭化
Charpy impact test 摆锤式冲击试验，夏比冲击试验
Charpy test 单梁冲击试验
Charpy transition temperature 夏氏转变温度
Charpy upper platform energy 夏比冲击能量
charred coal 焦煤
charred surface 炭化表面
char residue characteristic 焦渣特征
charring 炭化
chart board 图板
chart datum 海图基准面，图表水深基准
chart driving mechanism 记录纸驱动机构，表驱动机构
charted depth 海图水深
charterage 租船费
charter capital 注册资本
chartered accountant （有合格证书的）特许会计师，注册会计师，会计师
chartered bank 特许银行
chartered carrier transport 包机运输
chartered company 特许公司
chartered concession 特许权
chartered engineer 有执照的工程师
chartered period 租期
chartered 特许的，租的【车、船等】
charter flight 包机
charter of company 公司章程
charter of concession 特许证
charter party agreement 租约
charter 租，包租【车、船等】，宪章，特许证
chart index 图幅接合索引
charting 制图，制表
chart mechanism 制表机械
chart motor 走纸马达，走纸电动机
chart of load carrying capacity 负载能力图
chartography 制图法
chart paper width 记录纸宽度
chart-pattern 图案
chart reader 图形阅读器
chart recorder 图形记录仪，曲线记录仪，图表记录器
chart recording instrument 曲线记录仪，绘图记录仪
chart sheet 图幅，图页
chart speed shifter 图表移动速度切换装置
chart speed 走纸速度
chart 曲线图，图，计算图，地图，记录纸，略图，图表，木炭，木炭炭化，焦化，炭
chase in brick wall 竖向砖槽
chase lintel 槽楣
chase mortise 槽榫
chase sculpturing 雕刻
chase 竖沟，管槽，刻槽，雕镂，螺旋板，裂口，陷坑
chassis earth 机壳接地，底盘接地
chassis ground 底盘接地，机壳接地
chassis 起落架，底盘，底架
chatogant 闪光石
chattel 动产
chattel mortgage 动产抵押
chattering 振动
chatter of contacts 触点的跳动
chatter of switch 开关的跳动
chatter proof bolt 防震螺栓，颤振防爆螺栓
chatter vibration 颤振，振动
chatter 振动，震颤，颤震，抖动，使咔哒咔哒作声
CH(clearance charge for agency) 清关费【代理人收取此费为 CHA】
CH(coal handling) 煤的装卸，输煤
CH(crusher house) 碎煤房
cheap credit 低息贷款
cheap 廉价的，便宜的
cheat 欺诈，诈骗，骗取，欺骗
Chebyshev polynomial extrapolation 切比雪夫项式外推法
chechered grid of earthworks 土方方格网
check against 检查，核对
check analysis 成分分析，检验分析，核对分析
check and acceptance from the build to the operation 移交生产验收
check and acceptance on project completion 工程竣工验收
check and acceptance 完工验收
check and accept 验收

check and supervision 制约
check back signal 后检信号，校验返回信号
check-balance run 平衡校验运转
check bearing 校验方位
check bit 校验比特，校验位
check-board system 棋盘式街道体系
check bolt 防松螺栓
check by sampling 抽查
check chain 校验链
check character 检验符，检验字符
check coal 校核煤种，校验码
check command 检验命令
check computation 核算
check crack 收缩断裂，收缩裂纹，细裂纹
check criterion of reservoir 水库校核标准
check damper 逆风挡板，止回阀
check dam 防冲刷小坝，拦沙坝，拦水坝，节制坝，溪谷小坝
check digit 核对位，检验位，校验数字
checked operation 检验运算，检验操作
checked surface 表面龟裂，裂纹面
checkerboard pattern 棋盘式布置【堆芯，燃料】
checker board refueling 插花式换料，棋盘式换料
checker brick 格子砖
checker＝chequer 检验器，校核人
checkered floor plate 网格板
checkered iron 格纹铁
checkered plate 花纹钢板，网纹板，网纹钢板
checkered sheet 网状钢板，网纹钢板
checkered steel plate 花纹钢板
checkered 错列布置的，方格的
checker plate 网纹钢板
checker 检验员，校验器，试验装置，检验器，校核人，校核者
check experiment 核对试验，对照试验，对比试验
check fit-up （可携带的）校验台
check flood level 校核洪水位
check for leakage 泄漏试验
check gate 配水闸门，堰门，节制闸，节制闸门，斗门
checking attendance system 考勤制度
checking automatic 自动检验
checking bollard 防松式系船柱，码头带缆桩
checking certificate 检验证书，检验证明书
checking circuit 校验电路，检验电路
checking-code time 检验码时间
checking computation 检验计算，验算
checking number 检验数
checking procedure 检验步骤
checking program 检验程序
checking routine 校验程序，检查程序
checking subroutine 检验子程序
checking table 验算表，校验表
checking test 检查性试验
checking valve 止回阀，逆止阀
checking 检验，检查，校验，抑止，制止
check-in 登记入住，签到，办理入住手续，入住登记手续，验票并领取登机卡
check jump 逆止水跃，阻抑水跃
checklist 检查表，目录册，核查单，检查清单，备忘录，校审单
check man 检查人，核对人，伪造支票者
check mark 成群细裂纹
check meter 检验仪表
check nut 防松螺帽，防松螺母，锁紧螺母
check-off list 核对清单，检验单
check of foundation subsoil 验槽
check-out and automatic monitoring 检查及自动监视
check-out procedure 办理离开手续，结账手续
check-out system 检出系统
check-out time 核查时间，结账时间
check-out 调整，付账离开，校正，检查，结算，（离开旅馆前）结账，校准，（全面）清查
check pawl 止回棘爪
check payable in account 以记账支付的支票，转账支票
check payable to 支票抬头
check phase order 核对相序
check plate 止动板
check plot 对照小区
checkpoint 基准点，检测点，检查点
check position 检验点
check post start-up 启动后校核
check process 检查步骤
check program 检验程序
check rail 护轨，护轮轨，挡条，碰头挡
check receipted by the holder 持票的抬头支票
check receiver 监控接收机
check record 检验记录
check register 检验寄存器
check relay 核对继电器，检验继电器
check reset 校验复位
check ring 弹簧挡圈，挡圈
check room 衣帽间，存物室，更衣室，寄存处
check routine test 校验程序试验
check routine 检验程序
check row 检验行
check sample 对照样本
check sampling 检验取样
check sequence 检验序列，校验序列
checks for abnormalities 异常情况的校验，异常情况校验
check sluice 节制闸，逆止水闸
check source 检验用放射性源
check structure 拦沙建筑物
check sum 检验和，检查和，校验和
check surface 表面龟裂
check symbol 检验符号
checks 成群细裂纹
check table 检查表
check test 核对试验，验证试验，鉴定试验，对照试验，检查试验，校核试验
check unit 防松螺母
checkup connection 查线
checkup of seismic intensity 地震烈度复核
check-up test 检查性试验，检验性试验
check-up 检验，检查，核对，测试，校正，核查

check valve poppet　单向阀芯
check valve　单向阀，逆止阀，止回阀，关闭阀
check washer　防松垫圈
check weighing wagon　校验称量车皮
check　检查，校验，校对，支票，查验，复核，制止，校核，查对，阻止，控制，账单
cheek board　边模板
cheek　颊板，曲柄
cheese antenna　盒形天线，饼形天线
cheese-head screw　有槽凸圆柱头螺钉，圆头螺钉
cheese mold　干酪压模
chelometric titration　螯合滴定
chelate compound　螯合化合物，螯合物
chelate effect　螯合效应
chelate group　螯合基
chelate treatment of water　水的螯合处理
chelate　螯合，螯合物，螯合的，螯化
chelating agent　螯合剂
chelating effect　螯合效应
chelating polymer　螯形聚合物
chelation　螯合作用，多价螯合作用
chelatometric titration　螯合滴定
chelometry　螯合滴定法
chelon　螯合剂
CHEM(chemical)　化学
chemic　电流强度单位，化学的
chemical　化学的，化学品
chemical absorption　化学吸收，化学吸收作用
chemical action　化学作用
chemical activity　化学活性
chemical addition agent　化学添加剂
chemical addition　加化学药物
chemical additive　化学添加剂
chemical additive pump　化学添加剂供给泵
chemical add tank　化学药物添加箱
chemical adsorption　化学吸附
chemical aerosol　化学气溶胶
chemical agent　化学试剂，化学药剂
chemical analysis of water　水化学分析
chemical analysis　化学分析
chemical and allied industries　化工和有关的工业
chemical and bacteriological test　化学和细菌学试验
chemical and mechanical polishing　化学机械抛光
chemical and volume control system　化容控制系统，化学和容积控制系统
chemical and volume control　化学和容积控制
chemical attack　化学侵蚀，化学腐蚀
chemical balance　化学天平
chemical behavior　化学行为
chemical binding　化学键
chemical bolt　种植螺栓
chemical bonding agent　化学黏合剂
chemical bonding　化学结合【如芯块包壳相互作用】，化学黏合
chemical buffer　化学缓冲剂
chemical capacitor　化学电容器，电解质电容器
chemical change　化学变化，化学反应
chemical characteristics　化学特性
chemical churning pile method　旋喷桩法
chemical churning pile　旋喷桩，化学搅拌桩
chemical circulation cleaning　用化学品循环清洗
chemical cladding removal　化学脱壳，化学法去壳
chemical clarifier　化学澄清池，化学澄清剂
chemical cleaning basin　（热室内的）化学净化水池
chemical cleaning of unit　机组化学清洗
chemical cleaning　化学清洗
chemical coagulation　化学凝结，化学凝聚，化学絮凝
chemical collecting tank　化学废水收集箱
chemical combination　化合作用
chemical complex　化工总厂，化学络合物
chemical composition analysis of metal　金属化学成分分析
chemical composition analysis　化学成分分析
chemical composition of aerosol　气溶胶的化学组成
chemical composition of air　空气的化学组成
chemical composition of precipitation　降水的化学组成
chemical composition　化学成分，化学组成，化学合成
chemical compound　化合物
chemical condenser　电解质电容器，化学电容器
chemical consolidation　化学加固
chemical constant　化学常数
chemical constitution　化学结构
chemical contamination　化学污染
chemical control　化学防治，化学控制
chemical control method　化学控制法
chemical control system　化学控制系统
chemical conversion coating　化学转变护
chemical coolant　化学冷却剂
chemical corporation　化学公司
chemical corrosion　化学腐蚀，化学侵蚀
chemical creep　化学蠕变
chemical deaeration　化学除气，化学除氧
chemical decay　化学性腐烂
chemical decladding　化学去壳，化学蜕壳
chemical decomposition　化学分解
chemical decontamination system　化学去污系统
chemical decontamination　化学净化
chemical degradation of waste　废物的化学降解
chemical degradation　化学剥蚀
chemical dehydrator　化学脱水器，化学脱水剂，化学干燥器
chemical deionization system　化学除盐系统
chemical denudation　化学剥蚀
chemical deoxidization　化学除氧
chemical department　化学分场
chemical deposition　化学沉积，化学镀
chemical deposit　化学沉积
chemical derusting　化学除锈，化学脱盐法
chemical dispersant　化学分散剂
chemical dissolving box　化学溶解箱，药剂溶解箱
chemical dosage　化学剂量
chemical dose　化学剂量
chemical dosimeter　化学剂量计

chemical dosimetry 化学剂量测定法，化学剂量学
chemical dosing plant 化学加药站
chemical dosing pump 加药泵
chemical dosing 化学配料
chemical drains collecting circuit 化学排放物收集回路，化学疏水收集回路
chemical drains 化学疏水
chemical drain tank pump 化学排水箱泵，化学疏水箱泵
chemical drain tank 化学排放水箱，化学疏水箱
chemical drain 化学排水
chemical ecology 化学生态学
chemical effect 化学效应
chemical efficiency 化学效率
chemical effluent 化学排放物
chemical engineering 化学工程
chemical enrichment 化学浓缩
chemical environment 化学环境
chemical equation 化学方程式
chemical equilibrium constant 化学平衡常数
chemical equilibrium 化学平衡
chemical equivalent 化学当量
chemical erosion 化学侵蚀
chemical-etch-proof motor 防化学侵蚀电动机
chemical evolution 化学进化
chemical examination of water 水的化学检验
chemical examination 化验
chemical factor 化学因素
chemical feeder 化学品加料机，加药器
chemical feed pipe 加药管
chemical feed pump 加药泵，化学药物添加泵
chemical feed system 化学药物添加系统
chemical fertilizer industry 化肥工业
chemical fiber industry 化学纤维工业
chemical fiber 化学纤维，化学过滤器
chemical fire extinguisher 化学灭火器
chemical firm 化学公司
chemical fixation 化学固定法
chemical floc 化学絮凝剂
chemical foam 化学泡沫材料，化学泡沫
chemical formation 化学结构
chemical formula 化学式
chemical fuel 化学燃料
chemical fume 化学烟雾
chemical fungicide 化学杀菌剂
chemical gagging 化学测定
chemical gas feeder 气体加药器
chemical gauging of flow 化学测流法
chemical gauging 化学方法测定
chemical grouting 化学灌浆
chemical grout 化学浆液
chemical hazard 化学危害物，化学危害性
chemical impurity cleanup system 化学杂质净化系统
chemical impurity 化学杂质
chemical index 化学指数
chemical industry 化学工业
chemical injection pump 化学加药泵，化学注入泵
chemical injection 化学灌浆
chemical inorganic compound 化学无机化合物
chemical ionization 化学电离
chemical jacket removal 化学脱壳，化学去壳
chemical kinetics 化学动力学
chemical laboratory 化学试验室
chemical laser 化学激光
chemically coagulated sludge 化学法凝结的污泥
chemically induced artifact 化学药品引起的人为缺陷
chemically oxidizing atmosphere 化学氧化性大气
chemically-precipitated sludge 化学沉淀污泥
chemically pure water 化学纯水
chemically sedimentated sludge 化学沉淀的淤渣
chemically stabilized earth lining 化学加固的土质衬砌
chemical mechanism 化学机制
chemical method measuring discharge 化学测流法
chemical mixing tank 化学搅拌箱，化学药物混合箱
chemical monitoring 化学监测
chemical neutron poison 化学中子毒物
chemical oxygen consumption 化学耗氧量
chemical oxygen demand 化学需氧量【水的污染参数】
chemical piping 化学管涌
chemical plant 化学车间，化学工厂
chemical pollutant 化学污染物
chemical potential of M M的化学势
chemical power 化学能
chemical precipitation agent 化学沉淀剂
chemical precipitation 化学沉淀
chemical pre-processing room 化学预处理室
chemical preservation 化学防腐，化学防污
chemical prestressing 化学法预加应力
chemical processing cell 化学处理室
chemical processing loop 化学处理回路
chemical processing reactor 化学反应堆
chemical processing tank 化学处理箱
chemical product 化学产品
chemical properties of soil 土壤的化学性质
chemical property 化学性质
chemical proportioner 比例加药器
chemical proportioning pump 化学药物添加配料泵
chemical pure 化学纯，化学纯的
chemical purification plant 化学净化厂
chemical purification 化学净化
chemical radiation effect 化学辐射效应
chemical reaction control 化学反应控制
chemical reaction fire extinguisher 化学反应式灭火器
chemical reaction fouling 化学反应积垢
chemical reaction rate constant 化学反应速度常数
chemical reaction 化学反应
chemical reactivity control system 化学反应性控制系统
chemical reactivity control 化学反应性控制
chemical reagent 化学试剂
chemical recovery boiler 化学剂回收锅炉

chemical rectifier　电解整流器
chemical regeneration　化学再生
chemical reprocessing plant　化学后处理工厂
chemical resistance　耐化学性
chemicals addition system　化学药物添加系统
chemicals addition tank　化学药物添加箱
chemical sampling and dosing systems　化学品抽样与加药系统
chemical sampling system　化学取样系统
chemicals dissolving tank　溶药箱
chemicals dosing plant　加药装置
chemicals dosing tank　化学加药箱
chemical sediment　化学沉积物
chemical sewage sludge　化学污水污泥
chemicals feeder　加药器
chemicals feeding pump　加药泵
chemicals feeding room　化学加药间
chemicals feeding system　化学药物添加系统
chemicals feeding　化学加药，加药
chemical shim control of nuclear reactor　反应堆化学补偿控制
chemical shim control　化学补偿控制
chemical shimming control　化学补偿控制
chemical shimming　化学补偿反应性，化学补偿
chemical shim system　化学补偿控制系统，化学补偿系统
chemical shim　化学补偿，化学补偿剂，化学补偿物【核】
chemical shutdown system　化学停堆系统
chemical sludge　化学污泥
chemicals mixing tank mixer　化学药品混合箱搅拌器
chemicals mixing tank　化学药品混合箱
chemical smoke　化学烟气
chemical softener　化学软化剂
chemical solution spray　喷洒化学溶液
chemical solution tank　化学溶液池
chemical solution　化学溶液
chemicals proportioner　比例加药器
chemicals storage　药品仓库
chemical stability　化学稳定剂，化学稳定性
chemical stabilization　化学加固，化学杀菌
chemical storage　化学品库
chemical structure　化学结构
chemical subtraction　化学去污
chemical supply pump　化学药物输送泵
chemical synthesis　化学合成
chemicals　化学药品，化学品
chemical tank　化学槽，药剂槽，化学药物箱
chemical tendering　化学软化
chemical toxicant　化学毒素
chemical tracer　化学示踪剂
chemical transfer pump　化学药物输送泵
chemical transformation　化学变化
chemical treatment feeding　化学加药处理
chemical treatment　化学处理
chemical unsheathing　化学去套
chemical vapor deposition　化学气相沉积
chemical vapor deposition　化学汽相沉积，化学蒸发沉积，化学蒸汽沉积法
chemical vapor　化学压力灭菌法
chemical washing　化学淋洗
chemical waste drainage system　化学废液排放系统
chemical waste water　化学废水
chemical waste　化学废物
chemical water treatment building　化学水处理间
chemical water treatment plant　化学水处理装置，化学水处理室
chemical water treatment room　化学水处理间
chemical water treatment system diagram　化学水处理系统图
chemical water treatment　化学水处理
chemical weathering　化学风化
chemical weeding　化学除莠
chemical weed killing　化学除莠
chemical wetting agent　化学润湿剂
chemical wood preservation　木材化学防腐法
chemico-thermal treatment　化学表面热处理
chemihydrometry　化学测流法，化学水文测量，化学水文测验
chemiluminescence　化学发光，冷焰光
chemise　衬墙，土堤岸护面，土堤岸护墙，土堤护墙
chemisorption capacity　化学吸附容量
chemisorption　化学吸附，化学吸收作用
chemistry　化学
chemolumiosity　化学发光
chemolysis　化学溶蚀
chemometrics　化学计量学
chemonuclear reactor　化学化工用的反应堆
chemonuclear　核化学，放射化学的
chemorheology　化学流变学
chemosmosis　化学渗透
chemosphere　光化层，臭氧层
chemosterilant　化学灭菌剂
chemosynthesis　化学合成
chemosynthetic bacteria　化学合成细菌
chemotron　电化学转换器
cheque payable in account　以记账支付的支票，转账支票
cheque payable to　支票抬头
chequered plate　花纹板，拉网板
chequered　错列布置的
chequer　制成网纹
cheque　支票
Chernobyl nuclear accident　切尔诺贝利核电站事故
chernozem-like soil　黑钙土状土壤
chernozem soil　黑钙土
cherry coal　软煤，不黏结煤
cherry picker　车载起重机，车载升降台，高架车
chert aggregate　燧石骨料
cherty limestone　硅质石灰岩
cherty soil　石英质土
chert　燧石
chess board structure　棋盘格式构造
chessboard　棋盘法
chestnut soil　栗钙土
chestnut tube　栗形电子管
chestnut　一种小粒度的无烟煤

chest 柜，箱
cheveron stacking 人字形堆料法
chevron-layered pile 人字层煤堆
chevron pulley 人字沟滚筒
chevron type dryer 人字形干燥器
chevron type moisture separator 人字形汽水分离器
chevron type stacking 人字形堆料法
chevron type Y字形式
chevron 人字形，人字纹，人字形板，波形板
chews 中等大小的块煤
CHF(critical heat flux) 临界热通量
CHFR(critical heat flux ratio) 临界热流密度比
CHF 瑞士法郎，瑞郎
Chicago grip 线扣，鬼爪，芝加哥剥线钳
chief accountant 总会计师
chief appraiser 主要鉴定人
chief architect 总建筑师
chief auditor 总审计师
chief comptroller 审计长，审计主任
chief delegate 首席代表
chief designer 主要设计人，总工程师，总设计师
chief dispatcher 主调度员，调度主任
chief draftsman 主任制图员
chief economist 总经济师
chief engineering 规划，设计
chief engineer office 总工室
chief engineer responsible for project 项目主管总工程师
chief engineer's office 总工程师室
chief engineer 主任工程师，总工程师
chief executive officer 首席执行官，总裁
chief executive 最高管理者，首席执行官，董事长
chief material engineer 材料总工程师
chief of section 工段长
chief operating officer 主要执行人
chief operator 值班长，话务主任
chief quality assurance engineer 质量保证总工程师
chief reactor operator 反应堆运行班长
chief representative 首席代表，总代表
chief resident engineer 驻地首席工程师，驻工地主任工程师
chief supervisor 首席监理，总监
chief treasurer 总财务师
chief 首领，酋长，主要部分，首席的，主要的，主任的
chill 激冷，冷却，淬火
chill block 三角试块
chill casting 冷铸
chill crack 激冷裂纹，火裂
chill depth 白口深度
chill down 冷却，冷凝
chilled casting 冷硬铸造，冷硬铸件
chilled cast iron 冷硬铸铁
chilled heat exchanger 冷冻水热交换器，(硼回收系统的)冷却器
chilled iron liner 白口铁衬板
chilled iron or manganese steel liner 白口铁或锰钢衬板
chilled slag 冷硬渣
chilled steel 冷钢
chilled water circulating pump 冷冻水循环泵
chilled water head tank 冷冻水高位箱
chilled water heat exchanger 冷冻水热交换器
chilled water plant 冷冻水设备
chilled water precooler 冷冻水预冷器
chilled water pump 冷冻水泵
chilled water return tank 冷冻水回流箱
chilled water system with primary-secondary pump 一、二次泵冷水系统
chilled water system 冷冻水系
chilled water unit 冷却水装置
chilled water 冰水，冷冻水，冷却水
chilled 冷冻的
chiller plant 制冷站
chiller unit 冷冻水装置【硼热再生系统】，冷却装置
chiller 制冷装置，冷却器，冷凝器
chill hardening 冷硬化
chilling injury 冷却损伤
chilling room 冷却间
chilling water 冷水
chilling 骤冷，淬火，冷却，冷凝淬火，冷凝
chill point 冰冻点
chill-pressing 低温压制
chimney above roof 屋面烟囱
chimney capital 烟囱顶
chimney cap 烟囱风帽，烟囱帽
chimney cloud 烟云
chimney concrete shell 烟囱混凝土外筒
chimney cooler 管式冷却器
chimney-cooling tower integrated project 烟塔合一工程
chimney core 烟囱芯
chimney damper 烟道挡板，烟囱挡板
chimney draft 烟囱抽力，烟囱通风，自然拔风，烟囱拔风，(土坝的)垂直排水系统
chimney effect 自然抽风效应，吸力筒效应，烟囱效应，抽吸效应，抽吸作用
chimney effluent 烟囱排放物
chimney emission monitoring 烟气排放监测
chimney emission 烟囱排放，烟囱排烟
chimney exit diameter 烟囱出口直径
chimney exit 烟囱（出）口
chimney flue 烟囱道，烟道
chimney fume 烟囱排烟，烟囱通风
chimney gas 烟气
chimney height 烟囱高度
chimney hood 烟囱帽，烟囱遮盖
chimney kiln 熏房，烟道
chimney lining 烟囱内衬，烟囱衬壁
chimney mouth 烟囱口
chimney neck 烟道
chimney pipe 烟囱道
chimney plume 烟囱烟羽
chimney raft 烟囱基础
chimney rock 柱状石
chimney sand drain 沙质排水竖井
chimney shaft 烟囱筒体，烟囱筒身

chimney soot 烟囱烟灰，烟垢
chimney stack 丛烟囱，组合烟囱，高烟囱，总合烟囱，仅一通道的烟囱
chimney superelevation 烟囱超高
chimney ventilation 烟囱拔风，自然拔风，烟囱通风，烟囱抽风
chimney 上升筒【反应堆】，烟道，气道，烟囱
China-African Development Fund 中非发展基金
China Association of Civil Engineers 中国土木工程师协会
china clay 高岭土，瓷土
China Construction Bank 中国建设银行
China Council for the Promotion of International Trade 中国国际贸易促进委员会
China Datang Corporation 中国大唐集团公司
China Development Bank Development Bank 国家开发银行
China Electric Power Construction Association 中国电力建设企业协会
China Foreign Undertaking Contract Project Chamber Of Commerce 中国对外承包工程商会
China Gezhouba Group Corporation 中国葛洲坝集团股份有限公司
China Guodian Corporation 中国国电集团公司
China Huadian Corporation 中国华电集团公司
China Huaneng Group 中国华能集团公司
China Machinery Engineering Corporation 中国机械设备工程股份有限公司
China Merchant's Bank 招商银行
China National Electric Engineering Corporation 中国电力工程有限公司（原中国电工设备总公司），中国电工
China National Machinery Important & Export Corporation 中国机械进出口（集团）有限公司，中机公司
China National Machinery Industry Corporation Ltd. 中国机械工业集团有限公司，国机集团
China National Nuclear Corporation 中国核工业集团公司
China National Petroleum Corporation 中国石油天然气集团公司，中国石油
China National Technical Important & Export Corporation 中国技术进出口集团有限公司
China Nuclear Engineering & Construction Group Corporation Limited 中国核工业建设集团有限公司
China Petroleum & Chemical Corporation 中国石油化工集团公司，中国石化
China Power Investment Corporation 中国电力投资集团公司
China Southern Power Grid Company Limited 中国南方电网有限责任公司
chinaware 瓷器
China wood oil 桐油
china 瓷器
Chinese and foreign share holders 中外合股人
Chinese Bioenergy Association 中国生物质能协会
Chinese-character display function 汉字显示功能
Chinese-character display 汉字显示，汉字显示器
Chinese-character printer 中文打印机
Chinese cuisine 中餐
Chinese Export Commodities Fair 中国出口商品交易会
Chinese-foreign cooperative joint venture 中外合作经营
Chinese linden 椴木
Chinese locust 槐木
Chinese National Standards Code 中国国家标准代码
Chinese pump 差动式泵
Chinese scale 中图级，中国等级，中国地震烈度
Chinese seismic intensity scale 中国地震烈度表
Chinese-style 中国特色的
Chinese version shall prevail 以中文本为准
Chinese white 白色颜料
Chinese windlass 差动绞筒，辘轳
Chinese windmill 中国式竖轴风车
chinese 锌白
chink 收缩裂纹，裂缝，龟裂，漏洞，塞孔
chinley coal 块煤
Chinook wind 钦诺克风，奇努克风
Chinook 钦诺克风
chip ballast 碎石渣
chip blasting 浅孔爆破
chip-board 刨花板，碎木胶合板
chip capacitor 片状电容器
chip card 微型晶片，微型芯片
chip circuit 芯片电路
chip monolithic ceramic capacitor 片状独石瓷介电容器
chip of stone 碎石片
chipology 晶片学【集成电路】，芯片学
chip package 芯片封装
chipped stone 琢石
chipper 錾子，凿子
chipping hammer （清除焊渣用）敲渣锤，錾锤，錾石锤，凿石锤
chipping mark 凿击，雕凿痕迹
chipping pile head 剃桩头，去桩头
chipping spreader 碎石撒布机
chipping 切成小块，筛屑，碎裂，碎石片，修正，凿平，修琢，錾平，琢毛
chip removal 除去渣屑
chip select 芯片选择，选片
chip set 晶片集
chips 碎屑
chip transistor 片状晶体管
chip 薄片，芯片，切屑，缺口，片，碎片，基片，石片，石屑，碎块，削，修琢，渣屑
chirp 线性调频脉冲
chisel bit 冲击式钻头，单刀钻冠
chisel 风铲，平口凿，錾子，凿子，扁錾，砍凿
CHK VLV(check valve) 逆止阀，止回阀
chloramine 氯胺
chloration 氯化作用，加氯作用
chloric acid 盐酸
chloride accumulator 铅蓄电池
chloride content 含氯率

chloride index 氯化物指数
chloride of lime 漂白粉
chloride 氯化物
chlorimet 耐蚀合金
chlorinated chamber 氯化池
chlorinated copperas 氯化绿矾
chlorinated isocyanurates 氯化异氰尿酸
chlorinated lime 氯化石灰，漂白粉
chlorinated polyether 氯化聚醚
chlorinated polyvinyl chloride 氯化聚氯乙烯
chlorinated PVC 氯化聚氯乙烯
chlorinated resin 氯化树脂
chlorinated rubber paint 氯化橡胶漆
chlorinated rubber varnish 氯化橡胶漆
chlorinated rubber 氯化橡胶
chlorinated water 氯化水
chlorinating room 加氯室
chlorination effect 氯化效应
chlorination equipment 氯化设备
chlorination house 加氯间
chlorination of water 水的加氯消毒法
chlorination plant 加氯间，氯化厂，氯化处理设备，制氯站
chlorination process 氯化过程，氯化法
chlorination room 加氯间
chlorination 加氯，氯化作用，加氯处理，氯化处理
chlorinator 氯化器，加氯器，加氯机
chlorine-ammonia process 氯-氨法
chlorine-ammonia treatment 氯氨处理
chlorine contact chamber 氯接触池
chlorine content 含氯量
chlorine cylinder 氯瓶
chlorine demand 需氯量
chlorine detection alarm 氯气检测报警器
chlorine dioxide 二氧化氯
chlorine dosage 用氯量
chlorine dosing room 加氯间
chlorine evaporator 氯气蒸发器
chlorine gas absorption tower 氯气吸收塔
chlorine gas barrel 氯气瓶
chlorine gas poisoning 氯气中毒
chlorine hydrate 水合氯
chlorine ice 氯冰
chlorine injection equipment 加氯设备
chlorine ion 氯离子
chlorine peroxide 过氧化氯
chlorine residue 余氯量
chlorine room 加氯间
chlorine usage 氯的利用率
chlorine process 氯化-萃取过程
chlorine 氯
chlorion 氯离子
chlorite phyllite 绿泥石千枚岩
chlorite schist 绿泥片岩
chlorite shale 绿泥页岩
chlorite slate 绿泥板岩
chlorite 绿泥石
chloritization 绿泥石化
chloroform 三氯甲烷
chlorophenol 氯酚
chloroprene 氯丁橡胶
chlorosulfonated polyethylene rubber 氯磺化聚乙烯橡胶
chlorosulphonated polyethylene 氯磺酰化聚乙烯
chlorsilane 氯硅烷
CHOC(civil hand over certificate) 土建交接证书
chocking effect 堵塞效应
chocking-up degree 堵塞程度
chocking-up 塞紧，楔住
chocking 扼流（的）
chock of drawings 审图
chock pile 填塞桩
chock valve 阻气阀
chock 木楔，垫木，塞块，塞紧，楔块，楔形垫木，止挡块，卡住
chocolate bar 深褐色条标志
choice goods 精选品，上等品
choice grade 上等品，精选级
choice of dam type 坝型选择
choice of law 法律的选择
choice of setting 整定值选择
choke circuit 扼流电路，抗流电路
choke coil 扼流圈，节流圈
choke coupled amplifier 扼流圈耦合放大器，抗流圈耦合放大器
choke coupling 抗流圈耦合，扼流圈耦合
choked flow region 壅塞流区
choked flow turbine 超临界压比透平，阻流式透平
choked flow 阻流，扼流，节流，阻塞流
choked throat 壅塞喉道
choke filter 抗流圈滤波器，扼流圈滤波器
choke flange 节流孔板
choke heat air modulator 阻风门热空气调节器
choke length 节流长度
choke modulation 扼流圈调制，抗流圈调整
choke plug 闷头，塞头
choke protection 扼流圈保护装置
choker check valve 阻气单向阀，空气挡板
choker 节流挡板，节汽门，风门，捆柴排机，阻风门
choke stone 拱心石
choke-transformer coupling 抗流圈变压器耦合，扼流圈变压器耦合
choke-transformer 扼流变压器，抗流变压器
choke tube 阻气管，阻塞管，阻尼管
choke up 淤塞，堵塞
choke valve 节气阀，阻流阀，节流阀
choke voltage 抗流圈电压，扼流圈电压
choke 节流挡板，阻气门，扼流圈，阻塞，堵头，阻塞门，抑止
choking action 阻塞作用
choking cavitation 阻塞性空蚀
choking coil 扼流圈
choking effect 壅塞效应，阻塞效应
choking field 反作用场
choking flow 壅塞流，阻塞流动，阻塞水流
choking limit 阻塞极限
choking phenomenon 阻塞现象
choking point 噎塞点
choking section 阻塞截面

choking turns 扼流圈，抗流圈
choking velocity 噎塞速度
choking winding 抗流线圈，扼流线圈
choking 扼流的，阻塞的，扼流，阻塞，堵塞，节气（的），卡住，壅塞，淤塞
chokon 高频隔直流电容器
choky gas 窒息性气体
choky 窒息性的
chomiluminosity 化学发光度
chooser 选择器
choose 选择
chop and leach 切断浸取
chopass 高频隔直流电容器
chop-leach process 切断浸取过程
chopped cosine distribution 切平的余弦分布
chopped heat flux density profile 截断式热流密度分布
chopped impulse wave 斩截冲击波，斩尾冲击波
chopped pulse 削顶脉冲
chopped strand mat 短切原丝结
chopped wave impulse insulation level 截波冲击绝缘水平
chopped wave 截波，斩波
chopped 削顶的，砍
chopper circuit 斩波电路
chopper-driven DC motor 断续供电直流电动机，斩波器供电直流电动机
chopper frequency 间歇频率，遮光频率
chopper motor 斩波器供电电动机
chopper switch 刀形开关，闸刀开关
chopper synchronizing rectifier circuit 断续同步整流电路，斩波同步整流电路
chopper-type neutron velocity selector 中子转子选速器
chopper-type phase detector 斩波型相位检测器
chopper 斩波器，截波器，限止器，断路器，切碎机，变流器
chopping and dissolution 切断溶解
chopping and leaching 切断浸取
chopping bit 冲击式钻头，冲击钻头
chopping device 截波装置
chopping end leaching 切断-浸取
chopping frequency 斩波频率
chopping jump 波状水跃
chopping machine 截断机-横锯床，切断机
chopping oscillator 断续作用振荡器
chopping signal 断路信号，斩波信号
chopping 切断，截断，斩波，限幅
choppy wind 不定向风，疾风
chop stroke 削球
chop 裂缝，切断，裂口，碎块，公章，风浪的突变，短峰波
chordal addendum 弦齿高
chordal thermocouple 穿壁式热电偶
chordal thickness 弦齿厚
chord axis （叶片的）弦轴
chord deflection angle 弦线偏向角
chord deflection offset 弦线偏距
chord direction 弦向
chord force 弦向分力
chord length 弦长
chord line 弦线
chord member of truss 桁架弦杆
chord member 弦杆
chord modulus 弹性模量，弦模数
chord of arch 拱弦
chord of foil 叶弦，翼弦
chord spacing ratio 叶栅弦节距比
chord splice 桁弦接合板，弦杆结合板
chord stress 桁弦应力
chord taper ratio 翼弦锥度比
chord tunnel 弧形地槽
chord winding 弦绕组，弦绕法，短距绕组
chordwise load change 弦向载荷变化
chordwise slotted 弦向开缝的
chordwise term 弦向项
chordwise 弦向
chord 弦，翼弦，弦度，横梁，叶弦，弦杆
chore 零星工作
C-horizon 底土层
CHP（combined heat and power）热电联供，热电联产
CHP plant（combined heat and power plant） 热电（联产）厂
chromacoder 信号变换装置，彩色译码器
chromascan 一种小型飞点式彩色电视机系统
chromate-treated cadmium-plated steel 铬处理的镀镉钢
chromate 铬酸盐
chromatic aberration 色差
chromatic coherence 色相干性
chromatic dosimeter 变色剂量计
chromaticity demodulator 彩色反调制器，彩色解调器
chromaticity modulator 彩色调制器，色品信号调制器
chromatics 色彩学
chromatographia 色谱学
chromatographic adsorption 色谱吸附
chromatographic analysis 色谱分析
chromatographic column 色谱吸附柱，色谱柱
chromatographic fractionation 色谱分离
chromatographic process 色谱过程
chromatographic solution 色谱溶液
chromatographic technique 色谱技术
chromatographic 色谱的，色谱分析的
chromatography detector 色谱法探测器
chromatography mass spectrometry 色谱质谱
chromatography of gas 气象色谱
chromatography of ion 离子色谱
chromatography 色谱法，色谱分析，层析法
chromatron 彩色电视显像管
chroma 色度
chrome-base refractory 铬质耐火涂料
chrome-bearing steel 含铬钢
chromel-alumel thermocouple 铬镍-铝镍热电偶，镍铬-镍铝热电偶
chromel-alumel 镍铬-镍铝（合金）
chromel-constantan thermocouple 镍铬-康铜热电偶
chromel 镍铬耐热合金，镍铬热电偶合金
chrome magnesia brick 铬酸镁砖

chrome-molybdenum steel 铬钼钢
chrome nickel steel 铬镍钢
chrome nickel wire 铬镍线
chrome-plated steel 镀铬钢，渗铬钢
chrome 铬【Cr】
chromic 铬的
chrominance carrier 彩色载波
chrominance demodulator 色度解调器，彩色信号解调器
chrominance signal carrier 色度信号载波
chrominance 色度
chromite-containing refractory 铬质耐火材料
chromium 铬
chromium-alloy runner 铬合金转轮
chromium content 铬含量
chromium depletion 除铬
chromium diffusion treatment 渗铬处理
chromium-molybdenum steel 铬钼钢
chromium-nickel electrode 铬镍焊条
chromium-nickel steel 铬镍钢，铬镍合金钢
chromium plated blade 镀铬叶片
chromium plating 镀铬
chromium steel 铬钢
chromized steel 镀铬钢，渗铬钢
chromizing 镀铬，渗铬
chromometer 比色计
chromophotometer 比色计
chromoscope 表色管，彩色显像管，验色表
chromosome aberration 染色体畸变
chromosorb 色吸收剂
chronic discharge 慢性排放
chronic exposure 长期照射
chronic frustration 长期无效
chronic irradiation 慢性辐照
chronicle 编年史，大事记
chronic radiation effect 慢性辐射效应
chronic trouble 长期性故障
chronic 慢性的
chronograph stop watch 记时停表
chronography 时间记录法
chronograph 计时器，记时器，时间记录器
chronologically 时间顺序
chronological order 按时间次序
chronological scale 地质年代表
chronological time scale 地层年代表，地层时序表
chronology 年代学，（按时间排列的）大事记，编年表
chronometer 精密时值计，计时计，精密时计，天文钟，精密计时表
chronometric tachometer 计时式转速计
chronometry 时刻测定
chronopher 电控报时器
chronoscope 计时器，记时器
chronosequence 年龄系列【土壤】
chronotron 脉冲时间间隔测定器，延时器，瞬间计时器，摆线管
CHRS(containment heat removal system) 安全壳余热排出系统
Chubb method 交流波峰值测量法
chuck wrench 卡盘扳手
chuck 卡盘，夹盘，夹头，夹紧，卡紧，轴承座，芯轴
chugging 功率振荡，功率突变，嘎嘎声
chug 不均匀燃烧，功率突变
chunk 大块，大量，木块，石块
church window condenser 尖顶管束式凝汽器
churn drill bit 冲击钻头
churn drilling 钢丝绳冲击钻进，冲击钻进
churn drill rig 钢绳冲击式钻机
churn drill 冲击钻，旋冲钻，钻石机，石钻
churner 手摇式长钻
churn flow 乳沫状流动，环弹过渡流动
churning loss 涡动损失，搅动损失
churning 起泡【油】，搅拌，旋涡，旋涡度，激浪，旋动，搅动
churn shot drill （机）顿砂钻
churn 猛冲海岸【指波浪】，搅拌
churr 蜂音，嗡嗡作响
chute block 陡槽消力墩，分流墩
chute board 滑道板，斜板
chute discharge 斜槽出料
chute grate 倾斜式炉排
chute plug up switch 落煤管堵塞开关
chute raft 斜槽式筏道
chute spillway 斜槽式溢洪道，河岸式溢洪道，陡槽式溢洪道
chute work 加料斜槽
chute 滑道，落料管［槽］，溜槽，落煤管，斜槽，陡槽，斜管，水槽，斜通道，泻槽
chuting system 溜槽系统
chuting 溜槽运送
CIA(cash in advance) 预付货款
CI and BOP 常规岛及核电站配套设施
CIBOL(circuit board layout) 电路板布局
CI(cast iron) 铸铁
CI(convective instability) 对流不稳定性
CI(conventional island) 常规岛
C&I(cost & insurance) 货价及保险
CIE(coherent infrared energy) 相干红外能量
C&I engineer 热控工程师
CIETAC(China International Economic and Trade Arbitration Commission) 中国国际经济贸易仲裁委员会仲裁规则
CI(conventional island) excavation 常规岛开挖
CIF&C(cost, insurance, freight and commission) 货价、保险、运费及佣金
CIF(cost, insurance and freight) 到岸价格，成本、保险加运费价格【国际贸易术语】
CIF&E(cost, insurance, freight and exchange) 货价、保险、运费及汇兑费
CIF Ex ship's hold 舱底交货
CIF landed terms 到岸价格加卸货【货价、保险费、运费加卸货费】
CIF liner terms 到岸价格加班船条件【由卖方卸货】
cinch bolt 张紧螺栓，胀紧螺栓
cinch connector 小型插件【电器用】，多脚插头
CIN(component intervention notice) 设备问题处理通知
cinder aggregate 炉渣骨料，煤渣骨料
cinder bed 炉渣床，煤渣床

cinder block 煤渣砖，煤渣砌块
cinder box 渣斗，灰斗
cinder brick 煤渣砖
cinder catcher 捕灰器
cinder coal 多灰分煤
cinder concrete 矿渣混凝土，炉渣混凝土，煤渣混凝土
cinder cone 火山渣锥
cinder dump 灰场
cinder erosion 飞灰磨损
cinder fall 渣坑
cinder fill 煤渣填土
cinder inclusion 夹灰
cinder loading 含灰量
cinder loss 大渣或飞灰热损失
cinder notch 出渣口
cinder pit 灰斗，灰坑
cinder pocket 沉渣室
cinder recovery 飞灰回收
cinder reinjection 飞灰回送
cinder-return 飞灰回收
cinder-return system 飞灰复燃装置
cinder spout 落灰管
cinder trap 集灰器
cinder wool 矿渣棉
cindery coal 高灰煤
cinder 炉渣，焦渣，灰渣，煤屑，矿渣，煤渣
cineradiography 电影射线照相术，活动射线照相，活动射线照相术
cinnabar 辰砂，朱砂
CIP(carriage and insurance paid to) 运费、保险费付至目的地
cipher 零号，计数，运算，密码
cipolino 云母大理岩
CIQ brokerage fee 报检费
CIRC=circulation 循环
CIR=circuit 回路
circle bend 环形膨胀接头，圆曲管，环形弯管
circle-chain method 圆链法
circle coefficient 漏磁系数，泄漏系数
circle coordinate 圆坐标
circle illumination 度盘照明
circle left 正镜
circle level 环形水准器
circle of curvature 曲率圆
circle of reference 参考圆
circle of rupture 破裂圆
circle of stress 应力圆
circle of wall plates 承梁板凸缘
circle position 度盘位置
circle reading 度盘读数
circle reverse 圆盘回动装置
circle setting 度盘位置
circle shear 圆盘剪
circle theorem of hydrodynamics 流体动力学的圆柱绕流定理
circle vector diagram 旋转矢量图，矢量圆图
circle 度盘，领域，圈，圆，圆形
circling motion 圆周运动
circlip 环形，簧环，弹性挡圈
circuit airfoil 圆弧翼型
circuital current 回路电流，电路电流
circuital field 有旋场，涡旋场
circuital law 环流定律
circuital magnetization 环流磁化，螺线管的磁化
circuital vector field 有旋矢量场
circuital 电路的，与电路有关的，网络的
circuit analogy 电流模拟
circuit analysis 电路分析
circuit analyzer 网络分析仪，电路分析仪
circuit angle 变流器相位角，换流器相位角
circuit arrangement 电路布置
circuitary 线路的，与线路相关的，网络的
circuitation 旋转矢量，环线积分，旋转，旋度
circuit board layout 电路板布局
circuit board 电路板
circuit branch 支路
circuit breaker carriage 自动开关的滑架，断路器底架
circuit breaker cell 自动断路器开关盒
circuit breaker compartment 自动断路器开关盒
4/3 circuit breaker configuration 三分之四断路器接线
circuit breaker failure protection 断路器故障保护装置
circuit breaker failure 断路器失灵
circuit breaker oil-storage tank 断路器油箱
circuit breaker oil 开关油
circuit breaker opening 断路器分闸【手动】
circuit breaker operating mechanism 断路器操动机构
circuit breaker room 断路器室
circuit breaker tripping 断路器跳闸【自动】
circuit breaker with lock-out preventing closing 带防止闭合锁定的断路器
circuit breaker with shunt resistance 具有分路电阻的断路器
circuit breaker 断路器，保护断路器，断路开关，断流器，断路闸
circuit calculation 电路计算
circuit capacitance 电路电容
circuit changer 电路转换开关
circuit-changing switch 换路开关
circuit cheater 模拟电路
circuit closer 电路开关，闭路器，电路闭合器
circuit closing connection 闭路接法
circuit closing contact 闭路接点，通路接点
circuit component 电路元件
circuit connection 电路接线，回路接线
circuit constant 电路常数
circuit continuity 电路连续性
circuit controller 电路控制器
circuit control relay 电路控制继电器
circuit design 电路设计
circuit diagram of refrigeration 制冷原理图
circuit diagram 电路图，线路图
circuit efficiency 电路效率
circuit element 电路元件，线路元件
circuit equation 电路方程式
circuit error 环线闭合误差
circuit gap admittance 电路间隙导纳
circuit image 电路图像

circuit impedance 电路阻抗
circuiting 电路图，电路设计
circuit interlocking 电路闭锁，电路联锁
circuit interrupter 电路中断器
circuit layout 电路布线，线路布置
circuit logic 线路逻辑，电路逻辑
circuit loss 电路损失
circuit name sign 回路名称牌
circuit net loss 电路损耗，电路净损耗，线路损耗
circuit noise 电路噪声
circuit number of outgoing line 出线回路数
circuit number 电路号码
circuit opening connection 开路接法
circuit opening contact 开路接点，断路接点
circuit parameter 电路参数
circuit railroad 环形铁道
circuit recloser 电路自动重合闸，电路重合闸开关
circuit regulator 电路电压调整器
circuit resistance 电路电阻
circuit resonance curve 电路谐振曲线
circuitron 双面印刷电路
circuitry 回路，电路，电路系统，线路布置，布线
circuit switching 电路交换
circuit switch 电路开关
circuit symbol 电路符号
circuit terminal 电路接线端，电路终端
circuit tester 电路试验器
circuit theory 电路理论
circuit tracing 电路跟踪
circuit transformation 电路变换
circuit value 线路参数
circuit vent 环路通气管
circuit voltage class 电路电压等级
circuit voltage 电路电压，回路电压，线电压
circuit with distributed elements 分布元件电路
circuit with lumped elements 集中元件电路
circuit 线路，电路，回路，环路，循环，环流
circular airfoil 圆弧翼型
circular antenna 圆形天线
circular arc analysis 圆弧分析法
circular arch 圆拱
circular arc method 圆弧法
circular arc profile 圆弧叶型
circular arc section 圆弧翼型
circular arc 圆弧
circular barrel siphon 圆筒形虹吸管
circular bar 通告栏
circular beam 圆梁
circular bell mouth entrance 喇叭形进口
circular bit 圆形钻头
circular brick 圆形砖【改进型气冷堆】
circular bubble 圆水准器
circular buffering 循环缓冲
circular building 圆弧形建筑
circular burner 圆形燃烧器
circular-cap 圆形泡罩
circular-casing pump 圆形套管泵
circular catwalk 环形过道
circular cell 圆形格型结构，圆形热室
circular coal-yard 圆形煤场
circular coil 圆形线圈
circular cone 圆锥
circular conical surface 圆锥面
circular control weir 圆形控制堰
circular correlation 循环相关
circular counter 表盘式计数器
circular crab runway gilder 环形吊车梁，环形吊车道梁
circular crack 环状裂纹，环状裂缝
circular crane 旋转式吊车，环形吊车
circular culvert 圆形涵洞
circular current 环电流
circular cutting snips 圆剪
circular cut 环形掏槽
circular cylindrical coordinate 圆柱坐标
circular cylindrical shell 圆柱筒壳，圆柱形壳体
circular cylindrical void 环状圆柱形空隙
circular degree 圆度
circular disk crack 圆饼状裂纹
circular distributor ring 配电环，集电环，环形整流子
circular drainage system 环行排水系统
circular duct 圆形风管
circular electric wave 圆电波，横电波
circular electrode 滚轮电极
circular ended wrench 套筒扳手
circular equivalent of rectangular duct 矩形风管的圆形当量
circular error probability 循环误差概率
circular field 旋转磁场，圆磁场
circular file 圆锉刀
circular fin 环形的肋片
circular flow 环流，圆流，环形流动
circular flue （炉胆）圆形火筒
circular-foil type 圆箔片型的
circular footing 圆形基脚
circular foundation 圆形基础
circular frequency of vibration 振动圆频率
circular frequency 角速度，圆频率，角频率
circular Fresnel lens 圆形弗雷尔透镜
circular galvanometer 圆形检流计
circular-grained graphite 圆晶粒石墨
circular grate 环形旋转炉排
circular hollow prestressed concrete pile 环形空心预应力水泥桩
circular hoop 圆箍
circularity 圆形，环形，环状，圆板
circular jet 圆柱射流
circular kiln 环形窑
circular level 圆水准器
circular luminaire 圆形照明器
circularly polarized wave 圆偏振波，圆极化波
circular magnetic wave 圆磁波，横磁波
circular magnetization 圆形磁化
circular main 环形干管
circular metal flume 圆弧形金属渡槽
circular micrometer 圆径测微计，圆径千分尺
circular mil 圆密尔【截面积单位】
circular motion 圆周运动

circular orbit 环形轨道，圆形轨道
circular orifice 圆形孔板，圆形孔口
circular pan mixer 圆盘搅拌机
circular-pattern snips 圆形剪
circular penetration 圆形孔【核电厂燃料组件下端头】
circular pin plug 圆心插头
circular pipe 圆管
circular polariscope 圆偏振光镜
circular polarization 圆极化，圆偏振
circular polarized light 圆偏振光
circular railway 环形铁路
circular rail 环形铁路
circular rammer 圆夯
circular ring 圆环
circular rotating field 圆心旋转磁场
circular runout 圆形摆动度
circular saw 圆锯
circular scanning 环形搜索，圆周扫描
circular sealed connector 圆形密封接头
circular seam welding 环缝对接焊
circular section tunnel 圆形（断面）隧洞
circular section 圆形断面
circular sewer 圆形污水管
circular shaft 圆形竖井
circular slide damper 圆形滑动挡板
circular slide rule 计算盘
circular slide 圆弧滑动
circular slip surface 圆弧滑动面
circular sluice 圆形泄水孔
circular spirit level 圆酒精水准器
circular stairs 盘旋楼梯，螺旋梯
circular stockpile 圆形料堆
circular streamline 圆形流线
circular transmission line chart 输电线圆图
circular-type cellular cofferdam 圆格型围堰
circular velocity 圆周速度
circular vibration 圆振动
circular waveguide 圆波导，圆形波导管，圆形波导
circular wave 圆形波
circular weir 圆形堰
circular weld 环形焊缝
circular wind tunnel 圆截面风洞，圆形试验段风洞
circular working section 圆形工作段，圆形试验段，圆形截面试验段
circular 循环的，圆的，巡回的，环（形）的，通报，圆形的
circulated aeration tank 循环曝气池
circulate 流通，周转，循环
circulating 循环的，流通的，循环
circulating air cooling 循环空冷
circulating air 循环空气
circulating chamber 循环室
circulating cooling system 循环冷却系统
circulating cooling 循环冷却
circulating current bridge comparator 电桥式环流比较器
circulating current protection 差动保护装置，环流保护装置
circulating current 循环电流，环流
circulating door 旋转门
circulating ejector 循环喷射泵
circulating fan 循环风机，风扇
circulating flow 循环流
circulating fluidized bed boiler 循环流化床锅炉，循环床锅炉
circulating fluidized bed combustion 循环流化床燃烧
circulating fluidized bed 循环流化床
circulating force 循环压头
circulating fuel reactor 循环燃料反应堆
circulating fund 流动资金
circulating gas 循环气体
circulating groundwater 循环地下水
circulating head 循环水压头，循环压头
circulating load 循环荷载
circulating loss 环流损失，平衡电流损失
circulating lubrication 循环润滑
circulating memory 动态存储器，循环存储器
circulating oil filter 循环油过滤器
circulating oil lubrication 循环油润滑
circulating oil pump 循环油泵
circulating oil 循环油
circulating pipe 循环管，循环水管
circulating planetary wind 环流行星风
circulating pump 循环泵，循环水泵，主泵，控制循环泵
circulating ratio 循环倍率，冷却倍率
circulating reflux 循环回流
circulating register 循环寄存器
circulating solids 循环物料【沸腾炉】
circulating-solution reactor 循环溶液反应堆
circulating storage 循环存储器
circulating system 循环系统
circulating tube 循环管
circulating unbalance 环路不平衡
circulating water cooling station 循环水冷却站
circulating water dilution pump system 循环水稀释泵系统
circulating water filtration system 循环水过滤系统
circulating water flow 循环水量，循环水流量
circulating water intake 循环水进口
circulating water isolation system 循环水隔离系统
circulating water line 循环水管路
circulating water mains 循环水总管
circulating water makeup and drains system 循环水补充和排水系统
circulating water pipework 循环水管线
circulating water pipe 循环水管
circulating water piping 循环水管道
circulating water pit discharge pump 循环水管坑排水泵
circulating water pumping station 循环水泵站
circulating water pump lubrication system 循环水泵润滑系统
circulating water pump 循环水泵
circulating water recycle system 循环水再循环系统
circulating water screen 循环水滤网，循环水

系统
circulating water temperature 循环水温度，循环水温
circulating water treatment system 循环水处理系统
circulating water treatment 循环水处理
circulating water valve 循环水阀
circulating water velocity 循环水流速，循环水速度
circulating water 循环冷却水，活水，循环水
circulating 循环的，流通的，循环
circulation about foils 绕叶片环量
circulation around circuit 封闭环流
circulation cell 环流圈
circulation circuit 循环电路，循环回路，环路
circulation coefficient 循环倍率
circulation control airfoil 环量控制翼型
circulation controlled rotor 环量控制型风轮
circulation control 环量控制
circulation distribution 环流分布
circulation disturbance 循环干扰
circulation factor 循环倍率，循环系数
circulation fan 循环风机
circulation feed water heater 给水加热器
circulation flow of heat-supply network 热网总循环流量
circulation flow 循环流，循环水流，环流，循环流动
circulation index 环流指数
circulation layer 环流层
circulation loop 循环环路
circulation lubricating 循环润滑
circulation method 流动清洗法【化学清洗】
circulation of atmosphere 大气环流
circulation of material 物料循环
circulation of vector 矢量旋转，矢量的环流量
circulation of water vapour 水汽循环
circulation of water 水循环
circulation pattern 循环型，环流型
circulation period 循环周期
circulation principle 环流原理
circulation process 循环过程
circulation pump of heat-supply network 热网循环水泵
circulation pump 循环泵
circulation rate 循环速度，循环倍率
circulation ratio 循环倍率
circulation regime 环流状态
circulation return pipe 循环回水管
circulation section 循环段
circulation sedimentation tank 循环沉淀槽
circulation supply pipe 循环给水管
circulation supply water 循环给水
circulation test 循环试验
circulation tube 循环管
circulation velocity 循环水速，循环速度
circulation water leakage detector 循环水检漏装置
circulation 环流，循环，流通量，循环流量，流通，传播
circulator drive turbine 循环风机驱动汽轮机
circulator failure 循环风机故障
circulator outer casing 循环风机外壳
circulator seal gas 循环风机密封气体
circulatory cooling system 循环冷却系统
circulatory effect 循环效应
circulatory flow 圆流，环流，循环流动
circulatory lubrication 闭路润滑
circulatory motion 循环运动
circulatory system 循环系统
circulatory 环流的，循环的
circulator 循环泵，循环管，循环电路
circumambient 沿周冲刷的
circumcircle 外接圆
circumdenudation 环状剥蚀
circumference 圆周，周围，周界，周长
circumferential backlash 圆周侧隙
circumferential band 环箍
circumferential break 环向破裂，环形破裂，圆周断裂
circumferential butt 圆周对头焊
circumferential chip 掉边
circumferential component 圆周分量
circumferential compressive force 切向压力
circumferential crack 环形裂纹
circumferential electric field 圆周电场，周界电场
circumferential finning 周向加肋
circumferential fin 圆周肋，周向肋片
circumferential flow 环流，圆周流
circumferential flux 外围流，外围通量
circumferential force 切向力，圆周力，环向力
circumferential groove （轴承）环形槽，环形槽
circumferential highway 环形道路
circumferential joint 圆圈接缝
circumferential length of nozzle 喷嘴弧周
circumferential load 切向负荷
circumferentially-finned element 带周向肋片元件
circumferential motion 绕流，环流，圆周运动
circumferential notch 圆形缺口
circumferential piston pump 环形活塞泵
circumferential power in-pressure 环向压力，圆周压力
circumferential pressure 环向压力
circumferential prestressing 环向预加应力
circumferential prestress 环向预应力
circumferential reinforcement 环向钢筋
circumferential ridging 环脊
circumferential sealing 圆周密封，环向密封
circumferential seam 圆周接缝，环缝，环形焊缝
circumferential speed 圆周速度
circumferential strain 周向应变
circumferential street 环形道路
circumferential stress 周向应力，环向应力，箍应力
circumferential tensile stress 周向拉应力
circumferential velocity 圆周速度
circumferential weld 圆周焊缝，环焊缝，环缝
circumferential 周围的，四周的，周界的
circumferentor 地质罗盘
circumfluence 绕流，回流
circumfluent 绕流的，周流的，周流

circumpolar whirl　环极旋风
circumradius　外接圆半径
circumscribe circle　外接圆
circumscribed pipe wrench　外卡式管钳
circumscribe　划边界线
circumsolar radiation　太阳周围的辐射，环日辐射
circumstance　环境，情况，事件，事实
circumstantial evidence　间接证据，旁证
circumvolute　同轴旋转
cirecular cylinder　圆柱体
CIRIA（Construction Industry Research Information Association）建筑工业研究与情报协会【美国】
cirque　山凹
CIRR（commercial international reference rate）商业国际参考利率
cirscal meter　大转角动圈式电表
CISC（complex instruction set computer）复杂指令集计算机
CIS（consumer information system）用户信息系统
CIS（containment isolation system）安全壳隔离系统
cis-orientation　顺向定位
CISPR（International Special Committee on Radio Interference）国际无线电干扰特别委员会
cistern　容器，箱，槽，池，水槽，水箱，水窖，蓄水池
citable water capacity　汲水量
citation　引证，列举，引用，引文
CIT（corporate income tax）公司所得税
Citibank　花旗银行
citrate process　烟气的柠檬酸盐法除硫
citric acid　柠檬酸
citrus　柑橘属植物
city center　市中心
city climate　城市气候
city complex　城市建筑群
city construction　城市建设
city development　城市发展
city distribution　城市配电
city district planning　城市分区规划
city dweller　城市居民
city engineering　土木工程，城市工程
city environment　城市环境
city fog　城市烟雾
city layout　城市布局
city network　城区电力系统，城区电力网，地区网络
city noise　城市噪声
city planning and administration bureau　城市规划管理局
city planning　城市规划
city pollution　城市污染
city rehabilitation　城市改建
city scape　城市风光照片，城市风貌
city sewer　城市下水道
city's ring road　环城路
city street canyon　城市街谷
city structure　城市结构
city style and features　城市风貌
city ventilation　城市通风
city wall　城墙
city water supply　城市给水
city water tank　自来水箱
city water　城市供水，自来水
city　市（区），载重量
CIV（close interceptor valve）关中压调门
CIV（containment isolating valve）安全壳隔离阀
civil　民用的
civil architecture　民用建筑
civil building cost of power generation project per kW　发电工程每千瓦土建造价
civil building　围护建筑物，民用建筑
civil code　民事法典
civil construction　土建施工，土木建筑
civil design criteria　土建设计准则【criterion 的复数】
civil design criterion　土建设计准则
civil detail drawing　土建施工图
civil drawing　土建图纸
civil engineering construction drawing　土建施工图
civil engineering cost　土建费用
civil engineering guide drawing　土建指导图
civil engineering　土建工程学，土木工程，土木工程学
civil engineer　土木工程师
civil heat substation　民用热力分站
civil heat user　民用用热单位
civilian construction　民用建筑
civil interface drawing　土建接口图
civilized production　文明生产
civil-law-rule for water rights　水权的民法条例
civil obligation　公民义务
civil servant　公务员
civil structure　土木结构
civil substation　民用热力站
civil works cost　土建费用
civil work structure　土木工程结构
civil work　土建工程，建筑工程，土建工程学，土建，土木建筑
CKD（complete knocked down）全散件组装
CKW＝clockwise　顺时针方向
clack seat　阀座
clack valve　止回阀
clack　瓣，瓣式止回阀，阀瓣
CLA（communication line adapter）通信线路转接器
clad breach　燃料包壳破裂
clad burst　燃料包壳破裂
clad collapse　包壳坍塌
clad crecpdown　包壳向内蠕变
clad damage　包壳管损坏
cladding ballooning　包壳鼓胀
cladding balloon　包壳鼓胀
cladding breakthrough　包壳穿透
cladding bursting　包壳破裂【燃料元件】
cladding burst　包壳破裂
cladding collapse　包壳坍塌
cladding creep　包壳蠕变
cladding defect　包壳缺陷
cladding diameter　包层直径

cladding embrittlement 包壳脆化
cladding expansion detector 包壳膨胀探测器
cladding expansion 包壳膨胀
cladding glass 墙幕玻璃
cladding integrity 包壳完整性，包壳坚固性
cladding load 围护结构荷载
cladding material 包壳材料，覆盖材料，镀层，护套材料
cladding non-circularity 包层不圆度
cladding panel 骨架填充板材，外墙板，填充板材，围护墙板
cladding perforation 包壳穿透
cladding process 包壳工艺
cladding sheet 覆盖板，骨架填充板材
cladding-steam reaction 包壳蒸汽反应
cladding steel sheet 复合钢板
cladding steel 复合钢，包层钢
cladding stiffness 围护结构刚度
cladding strain 包壳应变
cladding structure 围护结构
cladding temperature variation 包壳温度变化
cladding temperature 包壳温度
cladding tube cross-sectional area 包壳管横截面积
cladding tube defect 包壳管损伤
cladding tube diameter 包壳管直径
cladding tube expansion 包壳管膨胀
cladding tube internal side 包壳管内侧
cladding tube material 包壳管材料
cladding tube rupture 包壳管破裂
cladding tube surface temperature 包壳管表面温度
cladding tube surface 包壳管表面
cladding tube temperature coefficient 包壳管温度系数
cladding tube temperature 包壳管温度
cladding tube wall thickness 包壳管壁厚
cladding tube 包壳管
cladding vault 包壳库【废燃料】
cladding wastage allowance 包壳运行允许减薄量【快堆元件】
cladding waste 废包壳，包壳废物
cladding 堆焊【为了耐热、耐蚀】，覆盖，敷层，镀层，涂覆，包层，包壳，堆覆层，堆焊层，覆层，外包层，外包金属，围护结构，饰面物，溶覆，电镀，喷镀
clad expansion 包壳膨胀
clad failure detection 燃料包壳破裂的探测
clad failure rate 包壳破裂的检测【核电厂燃料棒】，包壳破裂率
clad failure （燃料）包壳破损
clad fusion defect 覆盖层溶合缺陷【反应堆压力容器焊接】
clad integrity 包壳完整性，包壳坚固性
clad material 包壳材料，覆盖材料，覆层材料
clad meltdown 包壳熔化
clad metal 复合金属，金属保护层，包层金属板，包覆金属
clad ovality 包壳椭圆度，包壳椭圆形
clad-pellet gap 包壳与芯块中的间隙，包壳芯块间隙
clad pipe 包覆管
clad plate 装甲板，包装板，复合板
clad steel 复合钢，包层钢，包层钢板
clad temperature 包壳温度
clad thinning 包壳变薄
clad tube damage 包壳管损坏
clad tube material 包壳管材料
clad tube rupture 包壳管破裂
clad tube temperature coefficient 包壳管温度系数
clad tube wall thickness 包壳管壁厚
clad valve 蝶阀
clad 包覆的，装甲的，包壳，金属保护层
claim against carrier 向承运人索赔
claim against damage 要求赔偿损失
claim amount 赔偿金额
claimant 索赔人，原告
claim based on physical loss or damage 有形损失或损坏引起的索赔
claim based on 因……引起的索赔
claim board 索赔部门，索赔委员会
claim by subrogation 代位索赔
claim clause 索赔条款
claim commission 索赔委员会
claim compensation 索取补偿
claimee 被索赔人，索赔的债务人
claimer 取煤机，索赔的债权人，索赔人，债权人
claim expenses 清理赔偿费用
claim for compensation of damages 损坏赔偿的诉权
claim for compensation 要求补偿
claim for damages 损坏索赔，由于损坏而索赔
claim for extension of time 延长工期索赔
claim for extra cost 额外费用索赔
claim for idle plant 设备窝工索赔
claim for indemnification 要求赔偿
claim for indemnity 要求索赔，准予赔偿
claim for inferior quality 由于质量低劣而索赔
claim for loss and damage of cargo 货物损失索赔
claim for proceeds 应得价款的索款
claim for short weight 由于短重而索赔
claim for trade dispute 贸易纠纷（引起的）索赔
claim indemnity 索赔
claiming administration 索赔局
claim letter 索赔书
claim on sb. 向某人提出索赔
claim on the goods 对某（批）货索赔，对此货索赔
claim rejected 拒赔
claim report 索赔报告
claims and compensation 索赔与补偿
claims assessor 估损人
claims board 索赔委员会
claims commission 索赔委员会
claims department 索赔部门
claims document 索赔证件，索赔文件
claims for extension of time and additional costs 工期延长和费用索赔
claimsman 损失赔偿结算人
claims rejected 拒赔，理赔

claims settling agent 理赔代理人
claims settling fee 理赔代理费
claims statement 索赔清单
claims surveying agent 理赔检验代理人
claim tracer 索赔查询
claim 索赔，所有权，求偿，索取
clairecolle 打底明胶
clamp amperometer 钳形电流表
clamp beam 夹车梁
clamp bias 固定偏压，钳位偏压
clamp bolt 夹紧螺栓，紧固螺栓
clamp coupling 对开套筒夹紧联轴器
clamp device 紧固装置
clamp dog 制块
clamp earth resistance tester 钳形接地电阻测试仪
clamped edge 固定边，固定端
clamped terminal 夹子接线端
clamp electrical current adapter 钳合式电流适配器
clamper 钳位电路，接线板，压圈，气流调节器
clamping action 钳位作用
clamping apparatus 夹紧装置，夹具
clamping arrangement 夹紧装置
clamping bolt 夹紧螺栓
clamping circuit 钳位电路
clamping diode 钳位二极管
clamping disk 夹紧盘
clamping extent 衰减度
clamping filter 钳位滤波器
clamping force 锁模力，夹持力，夹紧力
clamping plate 拉紧板，夹板，压板
clamping pulse 钳位脉冲
clamping ring stop 止动夹环
clamping screw 夹紧螺栓，固定螺栓
clamping system 夹具系统，夹紧机构，夹紧装置
clamping washer 防松垫圈
clamping 夹住，夹具，夹紧，钳位，钳位电路
clamp moment 固端弯矩
clamp-on ammeter 钳形电流计
clamp pad 端子钮
clamp power factor gauge 钳形功率因数表
clamp screw 压紧螺钉，制动螺旋
clamp securing bolt 卡箍紧固螺栓
clamp terminal block 端子板
clamp timber 木夹板
clamp travel 夹紧装置行程
clamp 夹，夹紧，夹钳，钳，夹紧装置，夹线板，固定住，夹持器，线夹【输电线】，夹持装置，夹板，夹子，卡子
clamshell bucket 壳形抓斗
clamshell car 自卸吊车
clamshell coal unloader 抓斗卸煤机
clamshell crane 抓斗式起重机
clamshell dredger 抓斗式挖泥船，蛤壳式挖泥船
clamshell excavator 抓斗式挖土机
clamshell grab 合瓣式抓斗
clamshell nozzle 双活门可调喷嘴
clamshell shovel 抓斗式挖土机，抓斗铲土机
clamshell 蛤壳形抓斗，蛤壳状挖泥器，抓岩机，挖泥抓斗，蛤壳，抓斗
clam-type loader 抓斗式装载机，抓斗装载机
clam 抓斗
clapboard 护墙板，墙面板
clapotis 驻波，定波
clapper switch 铃锤式开关
clapper-type armature 拍板式衔铁
clapper-type relay 拍合式继电器
clapper 自动放水阀，锁气器
clappet valve 止回阀
clap sill 闸孔开度
clap valve 瓣阀
clarain 亮煤
clarificant 澄清剂，净化剂
clarification before bid opening 标前澄清答疑
clarification document 澄清文件
clarification filter 澄清过滤器
clarification liquor 澄清液
clarification meeting 澄清会
clarification of bidding document 招标文件澄清，标书澄清
clarification of sewage 污水的净化
clarification plant 澄清车间
clarification system （油）净化系统
clarification tank 澄清池
clarification 净化（作用），澄清（作用），纯化，说明，阐明
clarified makeup water 净化补给水，净化补充水
clarified oil 澄清油
clarified sewage 沉淀后的污水，澄清污水，净化过的污水
clarified waste water 澄清废水
clarified water pit 清净水槽
clarified water pump 清水泵
clarified water 澄清水，净水
clarifier effluent 澄清器流出液
clarifier 澄清池，澄清器，澄清剂，干扰清除器，干扰清除设备，澄清槽，净化器
clari-flocculator 澄清絮凝器
clarifying basin 澄清池，沉淀池
clarifying device 澄清装置
clarifying filtration 澄清过滤
clarifying tank 澄清池，澄清器
clarify 澄清，净化
clarion call 感人的号召
clarity 清澈度，透明度
clasp joint 搭扣接合
clasp nail 扒钉，扁钉
clasp 卡环
class A amplification 甲类放大，A类放大
class A insulation A级绝缘
class A modulation A类调制，甲类调制
class B amplifier B类放大器，乙类放大器
class B insulation B级绝缘
class B modulation B类调制，乙类调制
class boundary 分组界限，组界
class B push-pull B类推挽，乙类推挽
class C circuit C类电路，丙类电路
class C insulation C级绝缘
class D amplification D类放大，丁类放大

class 1E DC and UPS system　1E级直流电与不间断电源系统
class E insulation　E级绝缘
classer　分级机
classes of pollution　污染等级
class F insulation　F级绝缘
class F motor　F级绝缘电动机
classfy　分选，分类
class H insulation　H级绝缘
classical airfoil flutter　经典翼型颤振
classical bending torsion flutter　经典弯扭颤振
classical blockage correction　经典阻塞修正
classical bubble　典型气泡，经典气泡
classical buffeting　经典抖振
classical control system　经典控制理论
classical control　传统控制，典型控制
classical electron radius　经典电子半径
classical electron　经典电子
classical field formalism　经典场论形式
classical fluidization　经典流态化
classical fluid mechanism　经典流体力学
classical flutter　经典颤振
classical nodal diffusion theory method　经曲的节块扩散理论方法
classical setting　规则地形，经典地形布置
classical smog　经典式烟雾
classical strip theory　经典片条理论
classical system　经典系统，非量子化系统
classical theory　经典理论
classical thermal noise　经典热噪声
classical Venturi tube　标准文丘里管，古典文丘里管
classical Wagner function　经典瓦格纳函数
classical　传统的，经典的
classic statistic　古典统计
classification by region　按地区划分
classification chart　分类表
classification criterion　分类标准
classification efficiency　分级效率
classification method　分类法
classification of coal　煤的分类
classification of corrosion　腐蚀分类
classification of cost accounts　成本账目分类
classification of dynamic load　动荷载的分类
classification of fire hazards　火灾危险性分类
classification of load　载荷类别
classification of pollutant　污染物分类
classification of production　生产类别
classification of project　项目划分
classification of solar energy resource　太阳能资源等级
classification of the qualitative system　定性分类法
classification of treatment　处理分类
classification of waste　废水分类，废物分类
classification standard　分类标准
classification station　编组站
classification system　分类系统
classification table　分类表
classification test　分级试验
classification　分类，分级，等级，分等，类别
classified catalogue　分类目录
classified document　保密文件
classified income tax　分类所得税
classified worker　专职工人
classifier　分离器，分级机，分选机，筛分机，筛分器
classifying screen　分级筛
class index　等级指标
class interval　分组间隔，组距
class of damage　破坏等级
class of earthquake　地震震级
class of fit　配合等级，配合类别
class of highway　公路等级
class of impedance　阻抗等级
class of insulation　绝缘等级
class of minor damage　轻度破坏等级
class of pollution　污染等级
class-of-traffic check signal　话务等级效验信号
class rating　额定等级
classroom　教室
class two transformer　第二类变压器
class　种类，分类，等级，类别，类，级，层
clastic constituent　碎屑成分
clastic deposit　碎屑沉积
clastic rock　碎屑岩
clastic sedimentary rocks　碎屑沉积岩
clastic　碎屑的
clauses of technology contract　技术合同条款
clause　条例，条款
clave　冰裂
claw clutch　爪式离合器
claw coupling　爪形联轴器，爪形联轴节，爪形连接器
claw crane　钳式起重机
claw hammer　羊角锤，拔钉锤
claw hatchet　起钉斧
claw magnet　爪形磁铁
claw-pole motor　爪极式电动机
claw stop　止爪
claw tooth　爪形齿，梳齿
claw-type pole　爪形磁极
claw-type yaw probe　爪形方向测针
claw wrench　钩形扳手
claw　爪，爪形器具
clay and straw plaster　草泥抹面
clay band　黏土层
clay binder　黏土胶结物，黏土黏合料
clay blanket　黏土铺盖
clay brick　黏土砖
clay-cement grouting　黏土水泥灌浆
clay-cement mortar　黏土水泥砂浆
clay chunk　黏土块
clay content　黏土含量
clay core earth-rock dam　黏土心墙土石坝
clay core wall　黏土防渗墙
clay core　黏土防渗层，黏土心墙
clay cutter　黏土切削器
clay digger　挖土铲
clayed sand　黏质砂土
clay electro-gardening　黏土电化固结
c-layer　C-电离层

clayey 黏土质的
clayey gravel 黏土质砾石
clayey loam 亚黏土，黏质壤土，黏土质垆姆，黏质垆姆
clayey sand 黏质砂土
clayey silt 黏质粉沙
clayey soil 黏质土壤，黏性土，黏土类土，黏类土
clay figure modeling 泥塑
clay filled cutoff 黏土填筑的截水墙
clay film 土表黏粒薄皮
clay flow 黏土塑变，黏土流，黏土流动
clay foundation 黏土地基
clay fraction 黏土粒径组，黏土粒级，黏粒粒组，黏粒粒级，黏粒成分，黏粒部分
clay grains 黏土颗粒，黏粒
clay grouting 黏土灌浆
clay layer 黏土层，黏土夹层
clay lens 黏土透镜体
clay lining 黏土衬砌
clay marl 黏土质泥灰岩
clay mineral 黏土矿物
clay model 油泥模型
clay mortar 黏土砂浆
clay mud 黏土泥
clay overburden 黏土覆盖层
clay pan 隔水黏土层，硬黏土层，黏土盘，黏土硬层
clay particle 黏土颗粒
clay parting 黏土夹层
clay pipe 瓦管，陶管，陶土管
clay product 黏土制品
clay-puddle core 捣实黏土心墙
clay rock 黏土岩
clay sampler 黏土取样器
clay sandstone 黏土砂岩，黏土质砂岩
clay sealing 黏土密封，黏土止水
clay seam 黏土层，黏土夹层
clay sewer pipe 污水瓦管
clay shale 黏土质页岩，黏土页岩，泥页岩
clay slaking 黏土水解
clay slate 黏板岩
clay slide 黏土滑坡
clay slurrying plant 黏土浆制备厂
clay slurry 黏土浆，黏土泥浆
clay stratification 黏土层
clay suspension 黏土悬浮液
clay tile 黏土瓦，平瓦，陶瓦，黏土瓦
clay-to-day variation 日际变化
clay wall 陶土墙，土墙，土塘
clay weighed fascine 黏土压重柴捆
clay 黏土盘，陶土，黏土
CIC(compensated ionization chamber) 补偿电离室
CL(center line) 中心线
clclapotis 驻波
CL(confidence level) 置信度，可信度
CL(crane load) 吊车
CLD(cloud) 云
CLD(cold/cooled) 冷的
clean acceptance 无条件承兑
clean agent fire extinguishing system 洁净剂灭火系统
clean aggregate 清洁骨料
clean air act 空气清洁法，清洁空气条例
clean air legislation 空气洁净法规
clean air system 空气净化系统
clean air 纯洁空气，洁净大气，洁净空气，清洁空气
clean area 清洁区
clean bill of lading 清洁提单，清洁装货单
clean bill 光票
clean B/L 清洁提单
clean break 明显断裂，完全断裂，无火花断路
clean coal power-generation 洁净煤发电
clean coal technology 洁净煤技术，清洁煤技术
clean coal 洗净煤
clean cold critical reactor 干净冷态临界反应堆
clean cold reactor 新堆，冷态无毒物反应堆
clean conditions preparation room 干净条件制备室
clean conditions 清洁状态，清洁条件
clean core 净堆芯，（反应堆的）非中毒活性区，清洁堆芯
clean credit 纯信贷，清洁信用证，光票信用证，无单据信用证，无条件信用证
clean-cut indication 清晰地显示
clean development mechanism 清洁发展机制
clean dry air 干洁大气
cleaned by sand blast 用喷砂法清理
cleaned coal 精选煤，精煤
cleaned coal for coking 冶炼用炼焦精煤
clean energy resource 清洁能源
clean energy source 无污染能源，清洁能源
clean energy 清洁能源
clean equipment 清洁设备
cleaner air 净化空气
cleaner-up 清洁剂
cleaner 清洁器，净化器，除垢器，吸尘器，滤清器，夹钳刮刀，清扫器，（湿式）除尘器
cleaness 清洁度
clean flow 无旋流动
clean hardening 光亮淬火
cleaning action 净化作用，清洗作用
cleaning agent 清洁剂
cleaning cartridge 净化器
cleaning cell 净化室
cleaning compound 洗涤剂
cleaning device 净化装置，清扫器，清洗装置
cleaning door 清洗门
cleaning efficiency 净化效率
cleaning equipment 清洗装置
cleaning fee 洗仓费
cleaning fires 清灰
cleaning floor 清洗用场地
cleaning hole 清扫孔
cleaning method 清洗方法
cleaning of piping system 管道系统清洗
cleaning of reservoir zone 水库库底清理
cleaning of the heater well 加热器井的清洁处理
cleaning operation 清洗操作
cleaning system for flow path of gas turbine 燃气

轮机通流部分清洗系统
cleaning system 清洁系统
cleaning tank charges 洗舱费
cleaning vacuum plant 真空吸尘装置
cleaning waste water 清洗废水
cleaning water 洗涤水
cleaning 洗涤，填平，清理，清扫，清洗
clean lattice 净栅【反应堆】
clean letter of credit 光票信用证，清洁信用证，不跟单信用证，无跟单信用证
clean lift 清洁的电梯
cleanliness control 清洁度控制
cleanliness factor 洁净度，洁净系数，清洁系数
cleanliness of site 现场整洁状况
cleanliness of surface 表面清洁度
cleanliness 清洁（度），洁净，良好流线性，良好绕流性，洁净度，干净
cleanness factor 清洁系数
cleanness 清洁（度），洁净，良好流线性
clean on board bill of lading 已装船洁净海运提单，已装船清洁提单，清洁装运提单
clean on board inland bills of lading 清洁的已装船的陆运提单
clean opinion 无保留意见
clean-out auger 抽汲筒【一种钻探工具】，清孔钻
clean-out brush 清理刷
clean-out door 出渣，出灰门
clean-out opening 清扫孔
clean-out plug 放水龙头，塞头
clean-out 清堵，校对，检查，清洗，清除，清扫，清洁口，清洗口
clean payment credit 光票付款信用证，预支全部金额信用证，金额信誉证
clean reactivity 新堆反应性，未中毒反应性，净堆反应性
clean reactor 净反应堆
clean room 洁净室
clean sand 纯砂
cleanser 清洁剂
cleanse 净化，纯化，澄清，提纯
cleansing blower 喷气净化器，喷砂器
clean steam seal system 纯净汽封系统【汽轮机】
clean steam 纯净蒸汽，洁净蒸汽
clean-swept surface 冲刷干净受热面
clean-up bed 净化床
clean-up cell 净化室，清洗室
clean-up circuit 清洗回路，净化回路
clean-up demineralizer pump 净化除盐器泵
clean-up demineralizer 净化除盐器
clean-up dozer 清舱机，清仓机
clean-up efficiency 净化效率
clean-up flow rate 净化流量
clean-up flow 净化流
clean-up performance 清洁性能，清洁程度
clean-up plant 净化装置
clean-up pump 净化泵，洗涤泵
clean-up system heat exchanger 净化系统热交换器
clean-up system mixed-bed filter 净化系统混合床过滤器
clean-up system 净化系统
clean up the site 清理现场
clean-up work 清理工作
clean-up 净化，提纯，清洗，清扫，清除，洗涤，冲洗
clean water basin 清水池
clean water pond 清水池
clean water pump 清水泵
clean water tank 清水池
clean 清洁的，纯净的，整齐的，健全的，洁净的，清洗，洗涤
clearage 空隙，清理
clear air turbulence 晴空湍流
clearance advice 出港通知书
clearance angle 留隙角，后角
clearance area 拆迁区
clearance between guide vanes 导叶间隙
clearance cavitation 空隙气蚀
clearance channel 缝隙槽
clearance charge for agency 清关费【代理人收取此费为 CHA】
clearance checkout 间隙检查
clearance check 间隙检查
clearance depot 结关货场
clearance diagram 限界图，净空图，间隙图
clearance fee 出港费
clearance fit 动配合，间隙配合
clearance flow 间隙流
clearance for expansion 膨胀补偿间隙，膨胀间隙
clearance for pulling tube 抽管距离【凝汽器】
clearance gap 间隙
clearance gauge 量隙规，塞尺
clearance goods 货物报关，货物结关
clearance in air 气隙长度
clearance leakage 间隙泄漏，缝隙漏水，间隙漏水
clearance loss 间隙损耗，余隙损失，余隙损耗，缝隙损失
clearance meter 量隙计
clearance notice 出港通知书
clearance of faults 故障排除
clearance of goods procedures 报关手续
clearance of goods 货物清关
clearance of pollution 清除污染
clearance of site on completion 竣工时现场清理
clearance of site 现场清理
clearance order 清除指令，断路指令
clearance outward 出口结关证明
clearance paper 出港许可证
clearance permit 出港许可（证）
clearance procedure 结关手续
clearance through customs 清关
clearance to ground 对地间距，离地净空，对地距离
clearance 间隙，公隙，净空，放行证，间距，结关，净距，空隙，清理，清扫，余隙，清除，排除故障
clear and definite 明确的
clear area 有效截面，有效面积

clear away　清除
clear-back signal　后向拆线信号
clear break　明显断裂
clear buttress spacing　支墩的净间距
clear channel　专用信道，开敞信道
clearcole　打底明胶
clear-cut prohibition　明确禁止
clear-cut division of labour　明确的分工
clear-cut　明确的
clear debt　清偿债务
clear diameter　孔径
clear dimension　净尺寸
clear distance　净距，净距离
clear door　出灰口
clear-down signal　话终信号
clearence account　结算账户
clear flame　透明火焰
clear-forward signal　前向拆线信号
clear gland system　无放射性汽封系统
clear glass　透明玻璃
clear height　净高
clearing accounts　结算账户
clearing and grubbing　清除树桩
clearing balance　结算差额
clearing bank　结算银行
clearing device　清除装置
clearing energy　出清电量
clearing key　清除键
clearing lamp　话终指示灯
clearing mark　断面标志，导航标
clearing of a short circuit　清除短路
clearing of blockage　清除堵塞
clearing of fault　故障清除
clearing of the right-of-way for utility lines　线路走廊清理，管道走廊清障
clearing of the site　现场的清理，场地清理，现场清障
clearing relay　话终继电器
clearing signal　话终信号
clearing time　故障清除时间
clearing work　清除工作
clearing　清算，清洁，清除（故障），清理（现场），纯化，结算，回零，置零
clear lacquer　透明漆
clearly established ownership　产权清晰
clearness number　晴朗数
clear opening　净距
clear overflow weir　明流堰
clear space between bars　钢筋间净距
clear space　净空间
clear spacing　净间距
clear span　净跨，大跨度，净跨（度），净跨距
clear store instruction　清除存储器指令
clear store　清除存储指令
clearstory　高侧窗，通气窗，天窗
clear test section　风洞试验段
clear the circuit　切断电路，清除短路
clear trend assign　清除趋势指派
clear up accounts　（彻底）清算
clear valve diameter　阀孔径
clear varnish　透明清漆
clear water basin　清水池
clear water reservoir　清水库，清水池，给水水库
clear water　净水，清水
clear way valve　全开阀
clear well　清水池，清水库，清水井
clear width　净宽
clear window glass　透明窗玻璃
clear-write time　清除写入时间
clear　复原，回零，置零，明确的，清楚的，明显的，清澈的，清零，清除，澄清，透明的，清洗，清晰的
cleat angle　夹持角钢
cleat insulator　夹板绝缘子，绝缘夹板，瓷夹
cleat wiring　瓷夹布线
cleat　挡木，导向板，垫木，挡块，夹板，劈开，三角木，线夹，楔
cleavability　可劈性
cleavage　分层【缺陷】，劈理【岩石的】，拉轧，裂缝，裂开，分裂，劈裂，劈开
cleavage brittleness　晶间脆裂
cleavage crack　解理裂纹，劈理裂缝
cleavage fracture　劈裂，解理断口，裂碎
cleavage plane　劈理面
cleavage rupture　解理断裂，沿解理面断裂
cleavage strength　解理强度，劈裂强度
cleave chop　劈
cleaving　劈开
CLE(cycle life expenditure)　循环寿命消耗
cleft timber　顺纹劈开的木材
cleft welding　裂口焊
cleft　裂口，裂缝，裂纹，裂片，劈开的
clench　铆紧，抓紧
clerestory　天窗，长廊，通气窗
clerical cost　事务费用
clerical error　书写错误
clerical job　事务工作
clerk　办事员，职员，科员
cleuch　沟谷
clevice　夹具，V形夹，卡头，叉头
clevis drawbar　牵引环，联结钩，环卡组合式牵引装置
clevis fittings　U形夹配件
clevis insert　U形环嵌入
clevis joint　拖钩，脚架接头
clevis pin　圆柱销，U形夹销
clevis　马蹄钩，叉头，U形夹
CLF(clarifier)　净化器
CLG(cooling)　冷却
clicking　微小干扰
click stop　止动爪，制动爪
click　定位销，棘爪
client body　客户团体
client change notice　客户变更通知单，用户变更通知
client concerned　当事人
client conformance requirement　客户一致性要求
client-defined requirement　客户要求
client liaison　客户联系
client's express of interest　客户意向书
client　客户，业主，建设单位，买主，顾客，买方，委托人

cliff debris 坡积物
cliffed coast 陡岸
cliff edge 悬崖边
cliff face 悬崖壁
cliff protection 悬崖防护
cliff wall 悬崖壁
cliff 陡岸，陡壁，峭壁，悬崖，崖
climagram 气候图
climate change 气候变化
climate condition 气候条件
climate cycle 气候循环
climate data 气候资料
climate divide 气候分界，气候界限
climate effect 气候影响
climate element 气候要素
climate forecast 气候预报
climate modification 气候改善，人工影响气候
climate zones 气候带
climate 气候
climatic 气候的
climatic atlas 气候图集
climatic belt 气候带
climatic change 气候变迁
climatic chart 气候图
climatic classification 气候分类，气候分类法
climatic condition 气候条件，气候状况，气象条件
climatic control 气候控制
climatic cycle 气候循环
climatic data 气候数据，气候资料
climatic degeneration 气候恶化
climatic effect 气候效应，气候影响
climatic element 气候要素
climatic environment 气候环境
climatic factor 气候要素，气候因子
climatic fluctuation 气候变动，气候波动
climatic instability 气候不稳定性
climatic noise 气候噪声
climatic optimum 气候适宜期，气候最优期
climatic oscillation 气候变化
climatic province 气候区
climatic record 气候记录
climatic region 气候区
climatic stability 气候稳定度
climatic trend 气候趋向
climatic variation 气候变迁
climatic wind speed 气候风速
climatic wind tunnel 全天候风洞
climatic year 气候年
climatic zone 气候带
climatize 适应气候，顺应气候
climatography 气候志
climatological characteristics 气候学特性
climatological chart 气候图
climatological data 气候资料，气象资料
climatological forecast 气候预报
climatological table 气候表
climatology 风土学，气候学
climax avalanche 强烈崩坍
climax height 顶点
climax 顶峰，顶点，高峰
climber 脚扣【上杆用的】
climbing ability 爬坡能力
climbing clearance 带电作业登塔间隙
climbing-film flow 爬坡膜状流
climbing-film oil nozzle 薄膜附壁上升式油喷嘴
climbing formwork 上升模板
climbing form 滑升模板，升模
climbing iron 铁爪器【上杆用的】，钉鞋
climbing space 上升空间，上杆地位
climbing tower-crane 爬升式塔吊
climbing 功率提升
climb to full-power operation 升至满功率运行
climb 爬高
climosequence 气候系列
clinch tool 敲弯工具
cline stratum 倾斜层，倾斜地层
cline 单向渐变群
clinic 医务室，医务所
clinker aggregate 焰块骨料
clinker-bearing slag cement 矿渣硅酸盐水泥
clinker breaker 碎渣机
clinker brick 缸砖，硬砖
clinker chill 防焦箱
clinker concrete block 熔渣混凝土块
clinker concrete 熔渣混凝土
clinker crusher 碎渣机
clinker ejector 水力冲渣器
clinker elevator 灰渣提升机
clinker grinder 碎渣机
clinkering bar 清渣棍
clinkering coal 易结渣煤，熔结煤，结渣煤
clinkering property 结渣性
clinkering 烧结，结渣
clinker 熔渣块，烧结块，熔渣，煤渣，炉渣，水泥熟料
clinking 裂纹，白点
clink 钢凿
clinographic curve 坡度曲线
clinograph 孔斜计【用于测量钻孔斜度】
clinometer 测角计［仪、器］，倾斜仪，测斜计，倾斜计，测斜仪，斜度仪
clino-unconformity 斜交不整合
clintonite 脆云母
clip bolt 夹紧螺栓
clip connector 夹子接头
clip level 限幅电平，削波电平
clip-on ammeter 钳形电流表
clip-on 用夹子夹上去
clipped wave 削平波，限幅波
clipper dredger 单斗挖泥机
clipper 修剪工具，钳子，铲斗，剪刀，削波器
clipping amplifier 限幅放大器
clipping circuit 限幅电路，削波电路
clipping offset 线夹预偏
clipping of noise 噪声限制，噪声削波
clipping 切断，削剪，削波，资料剪辑
clip plate 压板
clip rail clamping 净空，空隙，间隙，限界
clip-spring switch 弹簧开关
clips sealing device （槽口）密封装置
clip 夹子，线夹，夹，夹片，夹线板，接线柱，

夹具，接夹，卡子
CLK(clock) 时钟
CLNG(cleaning) 清洁，净化
cloakroom 寄存处，衣帽间
clock circle-diagram 相量圆图，矢量圆图
clock circuit 同步脉冲电路，时钟脉冲电路
clock control system 定时控制系统，计时控制系统
clock cycle 时钟周期，同步脉冲周期
clock diagram 相量圆图，旋转矢量圆图
clock-drive mechanism 时钟传动机构
clock driver 时钟驱动器，时钟驱动机构
clock dues 码头费
clocked flip-flop 时标触发器，定时触发器
clocked logic 时钟逻辑，定时逻辑
clock-face diagram 旋转向量图，旋转矢量图
clock feedthrough 时钟馈入
clock frequency 钟频，时钟脉冲频率，时钟频率
clock generator 时钟脉冲发生器
clock-hour figure 钟时序数
clocking error 计时误差，时钟误差
clocking 计时，产生时钟信号，堵塞
clock marker track 时钟记号磁道
clock meter 钟表式计数器，时钟式电度表
clock motor 电钟用电动机，电钟马达
clock or clockwise 时钟或顺时针
clock phase diagram 直角坐标表示的矢量图
clock phase 时钟脉冲，同步脉冲，时钟功率
clock pulsed control 时钟脉冲控制
clock pulse gate 同步脉冲门
clock pulse generator 同步脉冲发生器，时钟脉冲发生器
clock pulses frequency 时钟脉冲频率
clock pulse source 时钟脉冲源
clock pulse 计时脉冲，时钟脉冲，同步脉冲，时标速率
clock relay 时钟继电器
clock repetition rate 同步脉冲重复频率，时钟脉冲重复频率
clock signal 时钟信号
clock-spring motor 钟表发条发动机
clock switch 时间开关
clock synchronizer 中心电钟，时钟同步器
clock system 同步脉冲系统，时钟脉冲系统
clock-time scheduling 时钟调度，按时调度
clock track 时钟脉冲道，时标道
clockwise direction 顺时针方向
clockwise rotation looking at the driven end 从被动端看是顺时针旋转
clockwise rotation 顺时针的轮转，顺时针方向转动，顺时针旋转
clockwise sense 顺时针方向
clockwise spin 顺旋
clockwise 顺时针方向，顺时针的，顺时针方向的，顺时针，右旋
clockwork 钟表装置
clock 计时器，时钟脉冲，同步脉冲，时钟，仪表
cloclock-pulsed control 时钟脉冲控制
clod crusher 碎土块机
cloddy structure 块状结构
clod 大块，块，泥块，土块，中砾块
clogged oil line 油路阻塞
clogged sand layer 堵塞砂层
clogged up 阻塞，堵塞
clogging capacity 容尘量
clogging method 锁定方法
clog 障碍物，阻塞物，阻塞，堵塞，堵灰，结渣
cloister arch 回廊拱
cloister 回廊
clone 克隆
close an account 结算
close breaker 合断路器
close burning coal 黏结性煤，脂煤
close button 闭合按钮
close-circuited transmission line 闭路传输线
close-circuit television 闭路电视
close clearance 紧公差，小间隙
close command 关闭命令
close-connected 直接连接的，紧密连接的
close control 密接管制，严密控制，近距离控制，仔细检查，精密控制
close-coupled collector storage system 紧凑式(太阳热水)系统
close-coupled processing and fabrication 紧凑的处理和加工
close-coupled process 一体化方法，紧凑过程
close-coupled pump 紧耦合泵
close-coupled sensor 紧密连接的传感器
close-coupled solar water heater 紧凑式太阳热水器
close-coupled switchgear 紧密连接的开关装置
close-coupled type 共轴式
close-coupled 紧凑的，一体化的，紧密连接的
close coupling 强耦合，紧密耦合
close cut 窄组分
close-cycle control 闭路控制，闭环控制
closed-air-circuit motor 密闭风路电动机
closed air circulation 空气闭合循环
closed air cooling circuit 闭式空气冷却回路
closed air cooling system 封闭式气冷系统
closed amortisseur 闭路阻尼条，闭路阻尼绕组
closed angle 锐角，夹角
closed antenna 闭路天线
closed armature winding 闭圈式电枢绕组
closed armouring 封闭式铠装，叠盖式铠装
closed-back terminal 闭合背板端子
closed-back 闭合背板
closed basin 封闭盆地
closed block 闭锁部件
closed booth 大容积密闭罩
closed centrifugal pump 封闭式离心泵
closed channel 闭式通道
closed chute 封闭溜槽
closed-circuit air cooling 闭式循环空冷
closed-circuit air turbine 闭式循环空气透平
closed-circuit battery 闭路电池组，持续作用的电池组
closed circuit cooling water expansion tank 闭式冷却水膨胀水箱

closed circuit cooling water heat exchanger 闭式循环冷却水热交换器
closed circuit cooling water pump 闭式循环冷却水泵
closed circuit cooling water system 闭路冷却水系统
closed-circuit cooling 闭路冷却，二次循环冷却
closed-circuit grinding （制粉系统的）闭路研磨
closed-circuit grouting 环流式灌浆
closed-circuit operation 闭路运行
closed-circuit pipe system 闭路管系
closed-circuit system 闭路制
closed circuit television TV monitoring 闭路电视监视
closed circuit television 闭路电视，闭路换接
closed circuit TV inspection of reactor vessel outlet nozzle 闭路电视检查堆压力壳出口管嘴
closed circuit TV inspection of stud hole 闭路电视检查螺孔
closed circuit TV system 闭路电视系统
closed circuit voltage 闭路电压
closed circuit water cooler 封闭循环式水冷却器
closed circuit wind tunnel 回路式风洞
closed circuit 闭合电路，闭合回路，闭路【电视】
closed coil armature winding 闭圈式电枢绕组
closed coil armature 闭圈电枢
closed coil winding 闭圈绕组
closed conduit flow 暗管流，封闭式管道流，压力管道流
closed conduit 封闭管道，封闭式管道
closed contact 常闭触点，闭合接触器，闭合接点，紧密接触
closed cooling 闭路冷却，密闭式循环冷却
closed cooling system 闭路冷却系统，闭式冷却系统
closed cooling tower system 闭式冷却塔系统
closed cooling water heat exchanger 闭冷水换热器
closed cooling water system elevated tank 闭路冷却水系统高架箱
closed cooling water system filter 闭路冷却水系统过滤器
closed cooling water system heat exchanger 闭路冷却水系统热交换器，闭式冷却水系统热交换器
closed cooling water system pump 闭路冷却水系统泵
closed cooling water system 闭式冷却水系统
closed core transformer 闭口铁芯变压器
closed core 闭口铁芯
closed curve 闭合曲线，闭曲线
closed-cycle control system 闭环控制系统
closed-cycle control 闭环调节，闭环控制
closed-cycle cooling system 闭环冷却系统
closed-cycle cooling water system 闭式循环冷却水系统
closed-cycle cooling 闭合循环式冷却，闭路循环冷却，闭式循环冷却
closed-cycle gas-cooled reactor 闭合循环气冷堆
closed-cycle gas turbine 闭式燃气轮机，闭式循环燃气轮机
closed-cycle magnetohydor-dynamic power generation 闭式循环磁流体发电
closed-cycle MHD generator 闭路循环磁流体发电机
closed-cycle MHD power generation 闭路循环磁流体发电
closed cycle reactor system 闭环反应堆系统，闭式循环反应堆系统
closed-cycle system 闭式循环系统
closed-cycle 闭路循环，闭环，闭式循环，闭合循环，闭合回路
closed-differential relay 近差继电器
closed drainage basin 闭合流域
closed drainage 闭合式排水，暗管排水
closed drain 阴沟，排水暗管
closed economy 封闭性经济
closed effuser 闭合式喷管
closed electrical operating area 封闭的电气操作区域
closed electric circuit 闭合电路
closed-end barrel lug 闭合筒接线片
closed-end wrench 闭口扳手
closed environment 封闭环境，密闭环境
closed expansion tank 封闭式膨胀水箱
closed expansion water tank 闭式膨胀水箱
closed fault 闭合断层
closed feed cycle 闭式回热循环
closed feed system 闭式给水系统
closed flume 封闭式渡槽
closed fold 闭合褶皱
closed force polygon 力的闭合多边形
closed full flow return 闭式满管回水
closed-gas-turbine power station 闭式循环燃气轮机电厂
closed hot water heating system 闭式热水供暖系统
closed hydraulic loop circuit 闭式液压回路
close-differential relay 近差继电器
closed impeller 有前后盘的叶轮，闭式叶轮，闭合叶轮
closed interval 闭区间
closed-iron core 闭环铁芯
closed jet 闭合流，闭合射流
closed joint 无间隙接头，密缝接头
closed linkage 强耦合，紧密耦合
closed lock 闭合船闸
closed loop control system 闭环控制系统
closed loop control 带反馈的控制系统，闭回路控制，闭环调节［控制］
closed loop cooling system 闭路冷却系统，闭式环路冷却系统，闭式冷却系统
closed loop cooling water 闭式冷却水
closed loop cooling 闭路式冷却
closed loop gain amplifier 闭环增益放大器
closed loop gain 闭环增益
closed loop mode 闭环模式
closed loop pole 闭环极点
closed loop process control 闭环过程控制［调节］
closed loop recycle 闭合环路再循环

closed loop steam cooling 闭式蒸汽冷却
closed loop system 闭环系统
closed loop test 闭环试验
closed loop transfer function 闭环传递函数
closed loop zero 闭环零点
closed loop 闭式环路，闭环，闭环回路，闭合回路
closed magnetic circuit 闭环磁路
closed magnetohydrodynamical cycle 闭式磁流体循环
closed meeting 非公开会议
closed-on-itself traverse 闭合导线
closed oscillator 闭路振荡器
close-down 关闭，退役，查封
closed path 闭合通路，闭路
closed pipe system 封闭管道系统
closed piping system 封闭容器，封闭管道系统
closed polygon of force 闭合力多边形
closed porosity 隐孔隙率
closed port 封闭港
closed position 闭合位置，停止位置
closed pressurized-water circuit 闭式加压水回路
closed programme 闭合程序
closed regional cooperation 排他性区域合作
closed return 闭式回水
closed rib floor 密肋楼板
close drift ice 流冰群
closed sandwich type panel 闭合嵌板
closed scheme of hot-water supply 闭式热水供应系统
closed-section wind tunnel 封闭试验段风洞
closed set 闭集
closed shell and tube condenser 卧式壳管式冷凝器
closed shell and tube evaporator 卧式壳管式蒸发器
closed shield 密封屏蔽
closed shipping crate （新核燃料组件）封闭运输箱
closed shop 封闭式机房，不开放式计算机站
closed-slot armature 闭口槽电枢
closed stirrups 闭合箍筋
closed storage 封闭贮存
closed subprogram 闭型子程序
closed subroutine 闭型子程序
closed surge tank 封闭式调压井
closed system of ventilation 封闭式通风系统
closed system 闭合系统，封闭系统，闭路系统
closed tank 闭式水箱
closed test section 封闭式试验段【风洞】
closed thorium cycle 闭合钍循环
closed-throat wind tunnel 闭口风洞
closed throttle 关闭节流阀
closed-top duplex lantern 闭顶双灯【一种航标灯】
closed track circuit 正常闭合的轨道电路
closed traverse 闭合导线
closed-type condensate return system 闭式凝结水回收系统
closed-type condensate tank 闭式凝结水箱
closed-type feed water heater 闭式给水加热器
closed-type heater 表面式加热器，闭式加热器
closed-type hot-water heat-supply system 闭式热水供热系统
closed-type motor 封闭型电动机
closed-type slot 闭口槽
closed-type 封闭式
closed vessel 封闭容器，封闭管道系统
closed vortex line 闭合涡线
closed water line 闭式水管路
closed water model 闭式水模
closed winding 闭合绕组
closed yoke 闭合轭
closed 封闭的，接通的，闭合的，闭路的，已合闸的
close-fallout 近区沉降物，近区落下灰
close file 关闭文件
close fit 紧密配合，紧配合
close-grained wood 密纹木材
close grain 密木纹
close ice 密集流水
close impeller 闭合式叶轮，封闭式叶轮
close-in electromagnetic field 近区电磁场
close-in fault 近区故障，近距离故障
close-in measurement 近区测量
close inspection 严格检查
close limit switch 闭合限位开关
close-linked system 环形配电系统
close-lock 关闭闭锁
closely coincide 精确符合，精确重合
closely graded soil 颗粒均匀的土
closely packed reactor 稠密栅反应堆
closely packed 密装的，密集的，稠密的
closely pitched 密排的，小间距的
closely spaced tube 密排管
closely spaced 稠密的，密集的
close magnetic trap 闭合磁阱
close nipple 螺纹接口，全螺纹短节，螺纹接套，密螺纹接套，无隙内接头
close operation 合闸操作
close out 结清，结束
close-packed hexagonal lattice 密排六方点阵
close-packed lattice 稠密栅，密积点阵
close-packed structure 密堆积结构
close pack 密集流水
close pile 密排桩
close-pitched tube 密排管
close point 闭点
close port 不开放港
close pressure of safety valve 安全阀关闭压力
close push-button 闭合按钮
close-ratio two-speed motor 近比率双速电动机
close relay 闭合继电器
close restraint 严格限制
close routine 闭型例行程序
close running fit 紧转配合
closer 塞子，闭合器，闭路器，特制闭合板桩
close shield 近屏蔽
close sizing 窄筛分
close sliding fit 转子间隙配合
close structure 闭合构造，封闭结构
close switch 合闸

closet basin 盥洗室水盆
close texture 密实结构，致密结构
close the breaker 合断路器
close the grounding knife-switch 合上接地刀闸
close the isolator 合隔离开关
close-to-critical range 近临界区，亚临界区
close together 紧密
close tolerance 高精度，紧公差
close type 封闭式
closet 壁橱，壁柜，厕所，套间
close-up fault 近距离故障
close-up inspection 直观检查，目视检查
close-up 闭路，闭合，接通，密封
close working fit 紧滑配合
close 关闭，闭合，合闸，堵塞，截止，结束，紧密的，靠近的
closing account 结算，结账
closing address 闭幕词
closing bus bar operating mode 合闸母线运行方式
closing ceremony 闭幕式
closing cock 闭锁旋塞
closing coil 闭合线圈，合闸线圈
closing contact 合闸触点，闭合触点
closing current 闭合电流，合闸电流
closing date 截止日期，结算日期，终止日期，结关日，截止申报日期
closing dike 堵口堤，合龙堤
closing down 关闭，停机，关车，停止服役
closing error in departure 横距闭合差，横向闭合差
closing error in latitude 纵距闭合差，纵向闭合差
closing force 密合力【机械密封】
closing-in-line 并入线路，并入母管
closing inventory 最终物料量
closing levee 围堤
closing line 闭合线
closing magnet 合闸磁铁
closing meeting 末次会议，闭幕会议
closing moment 闭合时刻，合闸时刻，合闸力矩
closing of circuit 电路闭合
closing of cofferdam 截流
closing of contact 触点闭合
closing organ 关闭机构
closing period 关闭时间，闭合时间，合闸周期
closing phase angle 合闸相位
closing piece 锁块，炉面板
closing pointer 合闸指针
closing point 闭合点
closing position 关闭位置
closing pressure of safety valve 安全阀的复位压力
closing pressure surge 闭合时的压力冲击
closing pressure 闭合压力，关闭压力，回座压力
closing price 收盘价格
closing pulse 合闸脉冲
closing relay 合闸继电器
closing resistor 闭合电阻器
closing ring 止动环，挡圈
closing session 闭幕会议
closing signal 闭合信号
closing solenoid 合闸螺线管
closing spark 合闸火花
closing speech 闭幕词
closing stroke 关闭行程
closing switch 闸刀开关，合闸开关
closing the account 结账，结清账户
closing the book 结账
closing time of governor 调速器关闭时间
closing time of inlet valve 进口阀关闭时间
closing time of pressure regulator 压力调节器的闭合时间
closing time of regulator 调速器关闭时间
closing time 闭合时间，关闭时间，合闸时间，截止时间
closing torque 关闭力矩
closing valve 关闭阀，闭合阀，隔离阀
closing 合闸【开关，断路器】，关闭，闭合，闭路，封闭，阻塞
closure and first filling of reservoir 封孔蓄水
closure by end dumping 立堵
closure dam 合龙坝
closure design discharge 截流设计流量
closure dike 合龙堤，截流戗堤
closure discrepancy 闭合差
closure error of traverse 导线闭合差
closure error 闭合误差
closure-gap monitoring 合龙缝监测
closure gap 龙口，合龙缝
closure gasket groove 密封环垫槽
closure grouting 封闭灌浆
closure head assembly 反应堆压力壳顶盖，封盖组合体，封头组合体
closure head cavity 顶盖舱，封头坑
closure head gasket groove cleaning equipment 顶盖密封槽清理设备
closure head insulation 顶盖保温
closure head liftoff 顶盖吊走
closure head off 顶盖打开，顶盖盖上
closure head penetration 封头顶盖贯穿件
closure head 顶盖，封头顶盖，闷头
closure insulation 顶保温层
closure maximum drop 合龙最大落差，截流最大落差
closure member 截流部件
closure nut and washer carrier rack 封头螺母和垫圈放置架
closure nut carrier rack 封头螺母放置架
closure nut 封头螺母
closure of arch 封拱
closure operation 合拢作业，截流作业
closure plate 封板，堵头，盖板，盲板
closure plug 管端封盖，管道封盖
closure stud elongation 封头螺栓伸长
closure stud sealing sleeve transport pallet 封头螺栓密封套运输托架
closure stud sealing sleeve 封头螺栓密封套
closure stud 封头螺栓，顶盖螺栓，关闭装置的销钉【压力壳】

closure test 关闭试验
closure time 关闭时间
closure upon loss of electric signal 失去电源关闭
closure upon loss of pneumatic power 失去气源关闭
closure works 合龙工程，截流工程
closure 合闸【开关，断路器】，端盖，（封盖）闭合，闭路，截流，截止，终结，关闭，合龙，封闭
cloth bag collector 布袋除尘器
cloth covering 布制蒙皮
cloth dust collector 布袋除尘器
cloth filter 布袋过滤器
clothing monitor 衣服监测仪
clothing 套，罩，外壳
cloth 织物，布
clotty 易凝结的，凝块多的
clot 凝结，凝块，泥团
cloud-burst flood 泥石流，山洪
cloud-burst 暴雨，大暴雨
cloud nuclei 云核
cloud of electron 电子云
cloud of particles 颗粒团
cloud of the bubble phase 气泡云层相
cloud point 混浊器
cloud pulse 电子云脉冲
cloud resulting from industry 工业污染云
cloud seeding 云催化
cloud shedding 气泡晕剥落
cloud street 云街
cloud test 浊点试验
cloud tower 云塔
cloud tube 云管
cloud velocity gauging 浊液测流速
cloud-wake phase 气泡晕迹相
cloudy 不透明的，混浊的
clough 山谷，水闸
clout nail 大帽钉
clout 大头钉，猛击
clover-leaf body 三叶草图形
clover-leaf crossing 三叶式交叉
clover-leaf intersection 三叶式立体交叉
clover-leaf layout 三叶式布置
CLOW(cooling water) 冷却水
CLP(China Light & Power Co. Ltd.) 中华电力有限公司【总部在香港】
CLP(container load plan) 集装箱装箱单
CLR＝cooler 冷却器
CLS＝close 关
CLSD＝closed 已关
CL test(continuous loading test) 连续加荷固结试验
club dolly 顶铆具
club-footed pile 扩底桩
club-foot electromagnet 一极上绕线圈的马蹄形磁铁
club 俱乐部
CLU＝clutch 离合器
clumped fuel 块状燃料
clumped riprap 抛石护坡，抛石体
clumped rockfill 抛石
clumping site 堆渣场，堆料场
clump of piles 集桩
clump 土块
cluster control rod 束形控制棒
cluster development 分组改进设计
clustered column 群柱
clustering 成群，成束
cluster lattice 棒束栅格
cluster of magnetization 磁化线数，磁通
cluster of piles 桩群，桩组
cluster of wind turbines 风轮机群
cluster shroud tubes 束棒套管
cluster structure 团粒结构
cluster switch 组开关
cluster type control rod 棒束型控制棒
cluster type fuel element 棒束型燃料元件
cluster weld 丛聚焊缝
cluster 组，束，团，簇，群集，组合，集结，团聚体，颗粒团，丛生
clutch bearing 离合器轴承
clutch body 离合器本体
clutch cam 离合器凸轮
clutch cone 离合器圆锥
clutch control 离合器控制
clutch disc facing 离合器盘衬片，离合器摩擦片
clutch disengagement 离合器脱开
clutch disk 离合器摩擦片
clutch drag 离合器阻力
clutch driving disc 离合器主动盘
clutch facing rivet 离合器面片铆钉
clutch facing 离合器衬片
clutch gear 离合器齿轮
clutch half 半联轴器
clutch lever 离合器杆，靠背轮杆
clutch lining 联轴器衬片
clutch magnet 啮合电磁铁
clutch mechanism 搭接机构
clutch motor 离合器电动机
clutch release bearing 离合器分离轴承
clutch release 离合器松开
clutch roller bearing 离合器滚柱轴承
clutch shaft 离合器轴
clutch throwout lever 离合器推杆
clutch throw 离合器开关
clutch thrust bearing 离合器止推轴承
clutch 离合器，联轴器，控制，紧急关头，凸轮，紧握，离合，钩爪，联动器，抓住
clutter noise 杂乱回波，杂波噪声
clutter rejection 杂波干扰抑制
clutter suppression 杂波抑制
clycure 一种提高煤粉流动性的添加剂
clydonograph 过电压摄测仪
cm＝centimeter 厘米
C/M(certificate of manufacturer) 制造商证明书
CM(construction management) 快速路径施工管理法，CM模式，边设计边施工模式【工程建设模式】
CM(construction manager) 施工经理
CM(controller module) 控制器模件
CM(corrective maintenance) 故障检修，设备维护，设备保养

CMCTL(current mode complementary transistor logic)　电流型互补晶体管逻辑
CMEC(China Machinery Engineering Corporation)　中国机械设备工程股份有限公司
C meter　电容表
CMI(committee maritime international)　国际海事委员会
CMI(computer-managed instruction)　计算机管理指令
CML(current mode logic)　电流模式逻辑
CMP(control, metering and protection)　电气二次【中国用法】
CMP(corrugated metal pipe)　金属波纹管
CMPR＝compressor　压缩机
CMR(common mode rejection ratio)　共态抑制比
CMR(continuous maximum rating)　连续最大功率，持续最大工况，连续最大蒸发量
CMS(computer monitoring system)　计算机监视系统
CMS(condenser air removal system)　冷凝器空气排放系统
CMS(condition monitoring system)　状态监测系统
CMS(couplant monitoring system)　耦合剂监控系统
CMS(cycle monitoring system)　循环监测系统
CNC bending presses　电脑数控弯折机
CNC boring machine　电脑数控镗床
CNC(computer numerical control)　计算机数字控制机床，数控机床
CNC drilling machine　电脑数控钻床
CNC EDM wire-cutting machine　电脑数控电火花线切削机
CNC electric discharge machine　电脑数控电火花机
CNC engraving machine　电脑数控雕刻机
CNC grinding machine　电脑数控磨床
CNC lathe　计算机数控车床
CNC lath　电脑数控车床
CNC machine tool fittings　电脑数控机床配件
CNC milling machine　电脑数控铣床
CN(Combustibili Nucleari)　核燃料公司【意大利】
C/N(credit note)　保值单据，信用证，贷记通知单，贷方票据
CNC tooling　电脑数控刀杆
CNC wire-cutting machine　电脑数控线切削机
CND＝conduit　管道，导线管
CNDCT＝conductivity　导电率，传导率
CNEC (China Nuclear Engineering & Construction Group Corporation Limited)　中国核工业建设集团有限公司
CNEEC(China National Electric Engineering Corporation)　中国电力工程有限公司【原中国电工设备总公司】，中国电工
CNEIC (China Nuclear Energy Industrial Company)　中国原子能工业公司
CNNC(China National Nuclear Corporation)　中国核工业总公司，中核总
CNPC(China National Petroleum Corporation)　中国石油天然气集团公司，中国石油
CNT＝counter　计数器
CNTIC (China National Technical Imp. & Exp. Corp.)　中国技术进出口集团有限公司
CNTLE＝Controller　控制器
coach screw　方头螺钉
coach wrench　活（动）扳手，可调扳手
COA(contract of affreightment)　包运租船，简称包船
coadjacent　邻接的，相互连接的
coadsorption　共吸附
COAG＝coagulant　凝结剂
coagel　凝聚胶
coagulability　凝结能力，凝结性，凝结力
coagulant aid　助凝剂
coagulant sedimentation　混凝沉淀法
coagulant　混凝剂，凝结剂，凝聚剂，絮凝剂，助凝剂
coagulated sediment　凝结沉积物
coagulate flocculating agent　混凝剂
coagulate　混凝，凝结，凝聚，凝结物
coagulating agent　凝结剂，凝聚剂，絮凝剂
coagulating basin　凝结沉淀池，凝聚池，凝结池
coagulation aid　助凝剂
coagulation chamber　凝结室
coagulation point　凝结点
coagulation tank　凝聚池（净化水用的），凝聚池，凝聚箱
coagulation threshold　凝结极限
coagulation time　凝结时间
coagulation treatment　凝结处理
coagulation value　凝结值，絮凝值
coagulation　凝结，混凝，凝固，凝集，凝聚（作用）
coagulative precipitation tank　凝结沉淀池
coagulative precipitation　凝结沉淀
coagulator　凝聚器，凝结剂，混凝剂，凝聚剂，凝聚沉淀装置
coagulum　凝块，凝结物，凝结块
coal　煤
coal air mixture　煤粉空气混合物
coal-air ratio　风煤比
coal-air temperature　煤粉空气混合物温度
coal analysis　煤分析，煤质分析
coal as fired　入炉煤
coal ash analysis　煤灰成分分析，煤灰分析
coal ash fusibility　煤灰熔融性
coal ash viscosity　煤灰黏度
coal ash　煤灰
coal as mined　原煤
coal as received　入厂煤
coal barge jetty　煤驳码头
coal barge　运煤驳船，煤驳
coal-based activated carbon　煤质活性炭，煤基活性炭
coal-based carbon material　煤制碳素材料
coal-based liquid　煤基液体燃料
coal-bearing strata　含煤地层
coal bed gas　煤层气
coal bed methane　煤层气
coal bed　煤床，煤层
coal belt conveyor floor　输煤皮带运输机层
coal belt conveyor gallery　输煤皮带栈桥，输煤

栈桥
coal bin 煤仓
coal-blended container 混煤罐
coal-blending facility 混煤设施
coal-blending for coking 炼焦配煤
coal-blending system 混煤系统
coal-blending 混煤作业，配煤
coal blockage signal 堵煤信号
coal blocking 堵煤
coal boiler 燃煤锅炉
coal breaker 碎煤机
coal Btu content 煤的发热量
coal bucket 煤斗
coal bulkhead 挡煤墙
coal bulldozer house 推煤机库
coal bulldozer 堆煤机，推煤机
coal bunker bay 煤仓间
coal bunker level relay 煤位继电器
coal bunker 煤仓，原煤斗，高位煤斗
coal burner 煤粉燃烧器，煤炭燃烧器，粉煤燃烧嘴，煤燃烧器
coal-burning gas turbine 燃煤燃气轮机
coal-burning MHD cycle 燃煤磁流体动力循环
coal-burning power plant 燃煤电厂
coal-burning unit 燃煤机组
coal caking 煤结块
coal carbonization 煤的干馏，煤炭焦化
coal carbonizing process 煤的碳化过程
coal car measuring service 煤车计量作业
coal carrier 煤轮，运煤船
coal car unloading station 煤车卸煤装置
coal car 煤车
coal-cellar 地下煤库
coal characteristic 煤质，煤质燃烧特性
coal chemical conversion 煤化工转化
coal chute 落煤管，煤渣
coal cleaning plant 选煤厂
coal cleaning 煤炭分选，选煤
coal combustion 煤燃烧，煤炭燃烧
coal complex building 输煤综合楼
coal composition analysis 煤质成分分析
coal concentrator 选煤机
coal concession 煤开采权，煤炭特许权
coal conduit 落煤槽，落煤管
coal consumption rate 耗煤率，煤耗率
coal consumption 煤耗，煤耗量，耗煤量
coal conveyer 输煤机
coal conveying belt 输煤皮带
coal conveyor 输煤机，输煤皮带
coal core sample 煤心煤样
coal-cracker 碎煤机
coal crusher house 碎煤机室
coal crusher room 碎煤机室
coal crusher 碎煤机
coal crushing-cutting machine 破碎-切碎机
coal-cutter motor 截煤用电动机
coal data 煤质资料
coal deposite 煤沉积，煤矿床
coal depot 煤栈
coal-derived liquid 煤基液体燃料
coal desulfurization 煤炭脱硫
coal direct liquefaction 煤炭直接液化
coal discharge track 卸煤线
coal discharge trench 卸煤沟
coal discharging chute 卸煤槽
coal distillation 煤的干馏
coal dock 煤码头
coal dressing 选煤
coal-drop 卸煤机
coal dry cleaning 煤炭干法分选
coal drying system 煤干燥系统
coal dumping line 卸煤场
coal dust 煤粉［屑］，煤尘
coal equivalent 煤当量，标准煤
coaler 运煤铁路［车辆］，煤商，煤船
coalescence model 汇聚模型
coalescence of waves 波浪合成
coalescence rate 汇聚速度
coalescence slamming 磁撞
coalescence 并合，聚结，汇合，聚并
coalescent debris cone 复合碎石锥
coalesce 接合，结合，聚合
coal extraction 煤矿开采
coal feeder control room 给煤机控制室
coal feeder floor 给煤机层
coal feeder 给煤机
coal feeding bridge 运煤栈桥
coal feed point 进煤口
coal field 煤田，煤矿
coal fines 煤屑
coal-fired 燃煤，燃煤的
coal-fired boiler 燃煤锅炉
coal-fired gas & steam combined cycle 燃煤燃气和蒸汽联合循环
coal-fired gas turbine 燃煤燃气轮机
coal-fired MHD generator 燃煤磁流体发电机
coal-fired plant 燃煤厂
coal-fired power plant 燃煤（发）电厂，燃煤电站，燃煤火力发电厂，燃煤火电场
coal-fired steam-electric power plant 燃煤蒸汽发电厂
coal-fired unit 燃煤机组
coal flow guard 导煤流挡板
coal flow monitor 燃煤流量监视器
coal flow signal 煤流信号
coal flow 煤流
coal for PCI 喷吹煤
coal gangue characteristic 煤矸石性能
coal gangue separation 煤矸石分选
coal gangue 煤矸石
coal gas burner 煤气燃烧器
coal gasification process 煤气化工艺
coal gasification 煤气化，煤的气化，煤炭气化
coal gasifier 煤气化器
coal gas waste 煤气废水［料、物］
coal gate 煤闸门，煤成气，煤气
coal grab 抓煤机
coal grade 煤质，煤炭品位
coal grading 煤炭分级
coal granulator 碎煤机
coal grit 煤质砂岩
coal guide trough 导煤槽

coal handling and storage system 煤炭输送贮存系统
coal handling complex building 输煤综合楼
coal handling control board 输煤系统控制盘
coal handling control building 输煤控制楼
coal handling equipment 输煤设备
coal handling facility 输煤设备［设施］
coal handling plant 输煤电厂，输煤装置，成套输煤设备
coal handling structure 输煤构筑物
coal handling system flow diagram 输煤系统流程图
coal handling system 输煤系统
coal handling trestle 输煤栈桥
coal handling 运煤，输煤，供煤，煤炭装卸，煤炭搬运
coal-hatch 上煤口
coal hauling equipment 运煤设备
coal hopper anti-freeze facility 煤斗防冻设施
coal hopper 煤斗，炉前煤斗，落煤斗
coal horizon 煤层
coal hydrogenation 煤加氢
coalification 煤化作用，煤化
coalignment 匹配装置，调整装置，校正装置
coal indirect liquefaction 煤炭间接液化
coal industry 煤炭工业
coaling cycle 上煤循环
coaling equipment 上煤设备
coaling the unit silos(/bunkers) 机组煤仓上煤
coaling 装煤，上煤
coal in pile 煤堆
coalite 焦粉
coalition government 联合政府
coal jamming 堵煤
coal jetty 煤码头
coal level meter 煤位计
coal liquefaction 煤炭液化，煤液化
coal locating level signal 煤位信号
coal locator 煤位计
coal loop track plough feeder 环式给煤机
coal mash 煤浆
coal measures plant 煤系植物
coal measures 煤层
coal meter 自动煤磅
coal mill 磨煤机
coal mine 煤矿
coal mining cost 煤炭开采成本
coal mining 煤矿开采
coal nozzle 煤粉喷嘴
coal-oil co-process 煤油共炼
coal-oil mixture 煤油混合燃料
coal particle size 煤粒度
coal pier 煤码头
coal pile height 煤堆高度
coal pile runoff 煤场雨水排放，煤堆废水
coal pile 煤堆
coal pipe 输煤管
coal pit 煤矿
coal plant control room 输煤系统控制室
coal plough 犁煤器，煤犁
coal plow 犁煤器

coal plugging 堵煤
coal preparation plant 煤的制备装置，选煤厂
coal preparation 煤粉制备，煤的制备，选煤
coal price 煤价
coal processing 煤炭加工
coal property 煤的特质，煤的性质，煤质
coal pulverizer 磨煤机，碎煤机
coal pulverizing mill 磨煤机
coal pulverizing system in type of direct injection 直吹式燃煤制粉系统
coal pulverizing system in type of storage bunker 仓储式燃煤制粉系统
coal pulverizing system 燃煤制粉系统，制粉系统
coal pump 煤泵，给煤机
coal pusher 推煤机
coal quality control 煤质管理
coal quality detection 煤质检测
coal quality management 煤质管理
coal quality statistic 煤质统计
coal quality 煤质，煤炭质量
coal rank 煤级，煤种
coal rate 煤耗率
coal receiving bunker 受煤仓，卸煤仓
coal receiving facility for truck transportation 公路运输受煤站
coal receiving facility 受煤设施
coal receiving hopper 受煤斗
coal receiving station 受煤装置
coal refuse 煤渣，煤矸石
coal region 煤炭产量
coal retaining wall 挡煤墙
coal sample 煤样
coal sample as fired 入炉煤样
coal sample as received 入厂煤样
coal sample crash 煤样破碎
coal sample division 煤样缩分
coal sample for back-check 存查（煤）样
coal sample for checking production 生产煤样
coal sample for determination of gas 用于测试气体含量的煤样
coal sample for determining the apparent specific gravity 视比重煤样
coal sample for determining total moisture 全水分煤样
coal sample for laboratory 实验室煤样
coal sample for production 生产煤样
coal sample mixing 煤样混合
coal sample preparation 煤样制备
coal sample reduction 煤样破碎，煤样缩制
coal sample sieving 煤样筛分
coal sampling device 原煤取样装置
coal sampling room 煤取样间
coal sampling tap 取煤样口
coal scale discharge 称量煤斗
coal scale 煤磅
coal screening 煤炭筛分
coal screen 筛煤机，煤筛
coal seam sample 煤层煤样
coal seam 煤层
coal separation 煤炭的分选

coal separator　选煤机
coal settlement basin　沉煤池
coal shale　煤质页岩
coal shipment　煤炭运输
coal silo bay　煤仓间
coal silo load cell instrumentation system　煤仓负荷传感器测量系统
coal silo　贮煤筒仓
coal silt　煤屑
coal size　煤颗粒度
coal-sizing analysis　煤的筛分分析，煤的筛分析（法）
coal slacking　风化煤
coal slack　末煤
coal slag　煤渣
coal slime　煤泥
coal sluice wastewater　输煤系统冲洗排水
coal slurry concentration　煤浆浓度
coal slurry pipage transport　煤浆管道输送
coal slurry pipeline　煤浆管道
coal slurry　煤浆
coal smoke pollution　煤烟污染
coal smoke　煤烟
coal sorting equipment　煤炭分选设备
coal source　煤源
coal spout　落煤管
coal stockyard　贮煤场
coal stone　无烟煤，煤石
coal storage bunker loose equipment　原煤斗疏松机
coal storage bunker　原煤斗
coal storage lifetime　煤的贮存期限
coal storage yard　储煤场，贮煤场
coal storage　贮煤，储煤，存煤，存煤量
coal store　储煤仓
coal stream　煤流
coal supply　给煤，供煤，上煤
coal-tar epoxy resin lacquer　环氧煤沥青漆
coal-tar fuel　煤焦油燃料
coal-tar gas　煤焦油煤气
coal tariff　煤价
coal-tar pitch paint　沥青漆
coal-tar pitch　柏油，煤焦油，焦油沥青
coal-tar　焦油沥青，煤焦油
coal tempering　煤加水
coal terminal　煤码头
coal testing　煤质试验
coal thawing pit　煤的解冻库
coal tipper　落煤器
co-altitude　天顶距
coal transfer house　转运站，煤转运室，输煤转运站
coal transfer system　煤炭转运设备
coal transfer tower　输煤转运站，转运塔
coal transportation distance　煤炭运距
coal transportation method　运煤方式
coal transportation mode　煤炭运输方式
coal transporting trestle　输煤栈桥
coal trimmer　堆煤机
coal tripper　配煤车
coal type　煤种
coal underground hopper　地下煤斗
coal unloader　卸煤机
coal unloading device　卸煤装置
coal unloading equipment　卸煤设备
coal unloading jetty　卸煤码头
coal unloading pier　卸煤码头
coal unloading platform　卸煤栈台
coal unloading shed　卸煤棚
coal unloading station　卸煤站
coal unloading system　卸煤装置
coal unloading wharf　煤码头，卸煤码头
coal unloading yard　卸煤场
coal washery　洗煤厂
coal washing　洗煤
coal water mixture technology　水煤浆技术
coal water mixture　水煤浆
coal-water ratio　煤水比【燃煤时】
coal-water slurry　煤水混合液，煤浆，煤水浆，水煤浆
coal weigher　煤磅
coal-weighing hopper　称量煤斗
coal weight measurement　煤量检测
coal wharf　煤码头
coal whipper　卸煤工人，卸煤机
coal winning machine　采煤机
coal yard drainage canal　煤场排水沟
coal yard equipment　煤场设备
coal yard sprinkling facility　煤场喷水设备
coal yard stockpile coefficient　煤场堆积系数
coal yard system　煤场系统
coal yard　煤场，贮煤场
coaly rashing　软页岩
coaming　舱口拦板
Coanda effect　柯恩达效应，附壁效应
coarse-adjusting rheostat　粗调变阻器
coarse adjustment　粗调节，粗调
coarse aggregate concrete　粗骨料混凝土
coarse aggregate　（混凝土的）粗骨料
coarse ash bin　粗灰库［仓］
coarse ash silo　粗灰库［仓］
coarse ash　粗灰
coarse asphaltic concrete　粗骨料沥青混凝土
coarse balance　（零点）粗调，粗平衡
coarse cherty　粗燧石的
coarse circuit　大系统，大循环
coarse clay　粗黏土
coarse coal　粗煤，大块煤
coarse compensation　粗补偿
coarse control member　粗调控制元件【反应堆】
coarse control rod　粗调控制棒
coarse control　粗控，粗调节，粗调
coarse crushing　粗破碎，粗碎，粗轧
coarse estimation　粗估
coarse filter　初级过滤器，粗反滤层，粗过滤器，粗滤池
coarse filtration　粗过滤
coarse-fine control system　粗精控制系统
coarse-fine switch　粗调细调开关
coarse fraction　粗粒含量
coarse fragment　粗碎块【直径大于 2mm】
coarse grade coefficient　粗粒径级系数

coarse grading 粗级配，粗粒径筛分
coarse-grained annealing 晶粒粗化退火
coarse-grained filter 粗料滤池
coarse-grained paving 粗粒石铺
coarse grained snow 粗粒雪
coarse-grained soil 粗颗粒土壤，粗粒土
coarse-grained wood 粗纹理木材
coarse grain film 粗粒胶片
coarse grain material 粗晶材料
coarse grain zone 粗晶区
coarse grain 粗粒
coarse granular 粗团粒
coarse gravel 粗砾石
coarse grid 粗网眼
coarse grind 粗磨
coarse ground cement 粗研水泥
coarse-group cross section 粗群截面
coarse-group 粗能群
coarse linearity control 线性粗调
coarse loam 粗壤土
coarsely ringed timber 粗纹木材，宽年轮木材
coarse mesh global reactor calculation 粗网格全堆计算
coarse mesh model 粗网模型
coarse mesh rebalancing method 粗网格再平衡法
coarse mesh 粗孔筛
coarseness 粗糙度，粒度，粗度
coarse orientation （换料机）粗调定向
coarse-particle bed 粗颗粒层
coarse particle return 粗颗粒回送，粗煤粒回送
coarse particle 粗颗粒
coarse pearlity 粗片状珠光体
coarse porosity 粗大气孔群
coarse rack 粗格，疏拦污栅
coarse reading 粗读数
coarse regulating 粗调节棒
coarse ripple 大的波动
coarse roll 粗碎辊碎机
coarse sand beach 粗砂滨
coarse sandy loam 粗砂壤土
coarse sand 粗粝，粗砂，粗沙
coarse screen 粗滤网，粗筛
coarse sieve 疏孔筛
coarse silt 粗粉砂，粗泥沙
coarse size 粗粒
coarse strainer 粗滤器，粗滤网
coarse stuff 粗填料，粗涂料
coarse texture 粗结构，粗纹
coarse thread 粗螺纹
coarse-to-fine aggregate ratio 粗细骨料比【混凝土】
coarse topography 粗切地形
coarse trash rack 粗格拦污栅
coarse vacuum 低真空
coarse 粗糙的，粗粒的，粗略的，粗的，不精确的，近似的
Coase theorem 科斯定理
coastal aids 沿岸标志，沿海航标
coastal and inland region 沿海与内陆地区
coastal area 海岸区，沿海地区
coastal belt 海岸地带
coastal bevel 上部斜面
coastal cape 海岬
coastal circulation 沿岸环流
coastal climate 沿海气候
littoral current 沿岸流，滨海流
coastal current 海岸流
coastal engineering 海岸工程
coastal environment 沿海环境
coastal facility 海岸设施
coastal feature 海岸地形，海岸特征
coastal friction 海岸摩擦
coastal geomorphology 海岸地形学
coastal groin 海岸丁坝，海岸防波堤
coastal groyne 海岸丁坝，海岸防波堤
coastal harbor 沿海港口
coastal harbour 海岸港
coastal landform 海岸地形
coastal levee 海岸堤防，海堤
coastal line 海岸线
coastal nuclear power plant 沿海核电厂
coastal nuclear power station 沿海核电站
coastal plain 海岸平原，沿岸平原，滩地
coastal pollution 沿岸污染
coastal protection 海岸防护
coastal region 滨海区，沿海地区
coastal sediment 海岸沉积物
coastal site 海边厂址，沿海厂址
coastal strip 海岸带
coastal structure 沿海结构
coastal water pollution 沿海水污染
coastal 海岸的，沿海的
coast down freely 自由惰走
coast down 减速【泵】，下降，惰走，减退
coast erosion 海岸冲刷，海岸侵蚀
coaster 单向联轴节，沿海航船，近海船
coast ice 岸冰
coasting body 惯性体
coasting operation 惰走操作
coasting regulator 惰转调整器
coasting time 惰转时间，惰走时间
coasting 惰走，惯性滑行，海岸线
coastland 沿海地区
coastline 岸线，海岸线
coast protection works 海岸防护工程
coast recession 海岸后退
coast survey 海岸测量
coast terrace 海岸阶地
coast 海岸，滑下，下坡，岸边，沿海地区
coated carbide fuel 包覆碳化物燃料
coated cathode 敷料阴极
coated collector 有涂层的集热器
coated electrode 涂料焊条，药皮焊条，包剂焊条，敷料电极，敷料焊条，涂剂焊条，有药皮的焊条
coated filament 敷料灯丝
coated magnetic tape 磁带
coated nonmetallic tape 复层非金属带
coated nuclear fuel particle 包覆核燃料颗粒
coated paper tape 复层纸带
coated particle fuel 包覆颗粒核燃料，包覆颗粒

燃料
coated particle type fuel element 包覆颗粒型燃料元件
coated particle 包覆燃料颗粒
coated tape 覆层带，涂粉磁带
coated 涂有……的，包着……的，覆层的
coating bonding 涂层黏合
coating insulation process 熔敷绝缘工艺
coating layer 包覆层
coating material 包覆材料，涂层材料，涂料，盖层
coating method 涂敷方法
coating mixture 涂料
coating system 覆盖系统
coating thickness meter 覆盖层厚度计
coating UV-cured acrylic diameter 紫外固化聚丙烯涂层直径
coating 罩面，药皮，镀层，防潮层，涂，涂料，涂层
coat of color 着色层
coat of paint 涂漆层
coat rack 衣帽架
coat room 衣帽间
coat thickness 涂层厚度
coat with varnish 涂清漆
coat 镀层，涂层，涂覆，外套，层，涂料，面层，包壳，覆盖，涂上
coax 同轴的
COAX(coaxial) 同轴的
coax connector 同轴接头
coaxial 共轴的，同轴的
coaxial antenna 同轴天线
coaxial cable power supply system 同轴电缆供电系统
coaxial cable 同轴电缆
coaxial carrier system 同轴（电缆）载波系统
coaxial coil 同轴线圈
coaxial conductor 同轴导体
coaxial configuration 同轴结构
coaxial connector 同轴（电缆）连接器
coaxial delay cable 同轴延迟电缆
coaxial directional coupler 同轴定向耦合器
coaxial drive 同轴驱动
coaxial electrical cable 同轴电缆
coaxial feeder 同轴馈电线
coaxial field 同轴电磁场
coaxial filter 同轴滤波器
coaxial flexible cable 同轴软电缆
coaxial flow reactor 同轴流反应堆
coaxial inner conductor 同轴电线芯线
coaxial jack 同轴插孔
coaxial line circulator 同轴环行器
coaxial line isolator 同轴电缆隔离器，同轴电缆去耦装置
coaxial line oscillator 同轴线振荡器
coaxial line termination 同轴线终端
coaxial line 同轴线，同轴线路
coaxial loaded waveguide 同轴加载波导
coaxial pair 同轴线对
coaxial plug 同轴插头
coaxial relay 同轴继电器
coaxial socket 同心插座，同轴电缆插座
coaxial standing-wave detector 同轴驻波检测器
coaxial stream 同轴水流
coaxial-stub filter 同轴突柱滤波器
coaxial switch 同轴开关
coaxial transformer 同轴变压器
coaxial transmission line 同轴传输线
coaxial trochotron 共轴余摆磁悬器，共轴电子开关
coaxing 预应力强化法
coax switch 同轴开关
cobalt alloy permanent magnet 钴合金永磁铁
cobalt 钴
cobbled gutter 卵石边沟
cobble pavement 鹅卵石地面
cobble soil 粗砾质土
cobblestone drain 卵石排水沟
cobblestone 鹅卵石，圆石
cobble stratification 卵石层
cobble 大块煤，鹅卵石，卵石，中砾石
cob brick 土砖
cobcoal 大圆块煤
COBOL(common business-oriented language) 一种事务数据处理用计算机语言
cob-wall 土墙
cobweb rubble masonry 蛛网形缝毛石圬工
cob 夯土建筑，草泥，黏土泥
c/o(care of) 由……转交（书信用语），转交
co(carried over) 转入
COC(carrier's own container) 船公司箱
co-channel interference 同波道干扰
co-channel 同波道，同信道
co(check out) 检查
cochran boiler 立式多横火管锅炉
cocked position 扳起位置，待发状态
cocked switch 待动开关
cocking handle 机柄
cocking 压簧杆
cockle stairs 盘旋楼梯，螺旋楼梯，螺旋爬梯
cockle 折皱，起皱，波纹，浪花翻滚形成激流
cockloft 顶层，顶楼，阁楼
cockpit 灰岩盆地，落水洞
cock tap 龙头
cock valve 旋塞阀
cock 吊车，塞门，旋塞，阀，龙头，水龙头
CO(commercial operation) 商业运行
co-compacted pellet 共挤压芯块
co-content 同容积
co-contractor 联营承包商
coconut capacitor 椰子型电容器，大型真空电容器
co-conversion 共转化
CO_2 coolant gas circulator 二氧化碳冷却剂气体循环风机
CO_2 coolant gas inlet 二氧化碳冷却剂气体入口
CO_2 coolant gas outlet 二氧化碳冷却剂气体出口
CO_2-cooled natural-uranium reactor 二氧化碳冷却天然铀反应堆
CO_2-cooled pressure-tube reactor 二氧化碳冷却压力管式反应堆

co-current contact 并流接触
co-current 平行电流，重合的，伴流
cocrystallization 共结晶，共晶体
co-current cation exchanger 顺流阳离子交换器
co-current flow regeneration 顺流再生
co-current flow 同向流，平行流，并流，平等流
co-current shallow multistage fluidized bed 顺流多层浅流化床
codan(carrier-operated device antinoise) 载频控制的干扰抑制器
codan lamp 接收指示灯
COD(cash on delivery) 货到付款，交货付现
COD(chemical oxygen demand) 化学需氧量
COD(commercial operation date) 商业运行日期
COD(crack opening displacement) 裂纹开口位移，裂纹张开位移
COD determination(chemical oxygen demand determination) 化学需氧量测定
code alphabet 码符号集
code-bar switch 码条开关
code base 编码基数
code call indicator 编码呼叫指示器
code checking time 代码检验时间
code-check 代码检验
code clination 极距，同轴磁偏角
code combination 代码组合
code command 编码指令
code control system 代码控制系统
code conversion 代码变换，译码，代码转换
code converter 译码器，代码变换器，代码转换器
coded 编成代码的
coded alternating current 已编码交流电流
coded character set 编码字符集
coded character 编码符号，编码字符
coded command 编码指令
coded data decoder 编程代码数据的译码器
coded decimal adder 编码十进制加法器
coded decimal converter 代码十进制变换器
coded decimal digit 编成十进制的数字
coded decimal notation 代码十进制记数法
coded decimal presentation 编成代码的十进制表示
coded decimal system 十进制编码系统
coded decimal 十进制代码
code detecting apparatus 代码检测装置
code device 编码装置
coded graphics 编码图形
code dialing 编码拨号
coded identification 编码表示法
coded image 编码图像
coded impulse 编码脉冲
code divider 码分器
coded-light identification 编码光识别
coded message 编码信息
coded number 编码数
coded order 编码指令
coded program 编码程序，上机程序
coded pulse signal 编码脉冲信号
coded pulse 编码脉冲
coded sequence 编码序列
coded signal 编码信号
coded stop 程序停机，编码停机
coded word 编成代码的字
code element 代码元素
code error 代码误差
code for design of building foundation 建筑地基基础设计规范
code for design of chimney 烟囱设计规范
code for design of concrete structures 混凝土结构设计规范
Code for design of masonry structures 砌体结构设计规范
code for fire protection 防火规程
code for heavy isotopes 重同位素编号
code format 代码格式，码型
code for seismic design of buildings 建筑抗震设计规范
code for soil and foundation 地基与基础规范
code for water rights 水权法规
code for water supply and sewerage 给排水规范
code hole 代码孔
code letter 代码字母
code level 代码级
code light 代码信号灯
code line 记码区
code machine 编码机
code number 代号
code of conduct 行为准则，规范
code of ethic 道德行为准则
Code of Federal Regulations 美国联邦法规
code of international law 国际法典
code of practice of IAEA 国际原子能机构的实施法规
code of practice 实施法规，实用法规，实用规范
code operator 编码器算子
code page 代码页
code pattern 码型
code point 代码点
code position 代码位置
code pulse 代码脉冲
code recognition system 识码系统
code register 代码寄存器
code relay 电码继电器
coder 编码器，编码装置
codes and standards 规程和标准
code selector 选码器
code sender 发码器
code signal converter 代码信号变换器
code signal 编码信号
code sign 代码符号
co-design 联合设计
code switch 代码开关
CO detector 一氧化碳探测器
code tolerance 规范容许偏差
code translator 译码器，代码转换器
code transmitter 电码发送机
code transparent data transmission 代码透明数据传输，透明代码数据传输
code wheel 代码轮
code word 密码字

code 规程，法规，规范，编码，代码，代号，标准
coding card 编码卡
coding contact 编码触点
coding cycle 编码循环
coding filter 编码滤波器
coding line 编码线
coding network 编码网络，编码器
coding relay 编码继电器
coding sheet 程序纸
coding system 编码系统
coding tube 编码管
coding 编码
CO_2-D_2O reactor 二氧化碳冷却重水反应堆
coefficient 系数，因素
coefficient between layers 层间系数【即管涌比】
coefficient for combination value of an action 作用组合值系数
coefficient for orifice 孔口系数
coefficient matrix 系数矩阵
coefficient of absorption 吸收系数
coefficient of accumulation of heat 蓄热系数
coefficient of active earth pressure 主动土压力系数
coefficient of adhesion 黏着系数
coefficient of adjustment 改正系数，校正系数
coefficient of admission 占积率，充满系数
coefficient of aerodynamic force 气动系数
coefficient of air flow rate 空气流量系数
coefficient of air infiltration 透风系数
coefficient of air resistance 空气阻力系数
coefficient of alienation 相疏系数
coefficient of amplification 放大系数
coefficient of area 面积系数
coefficient of atmospheric transparency 大气透明度
coefficient of attenuation 衰减系数
coefficient of autocorrelation 自相干系数，自相关系数
coefficient of basin shape 流域形状系数
coefficient of basin slope 流域坡度系数
coefficient of blackness 黑度系数
coefficient of brightness 亮度系数
coefficient of building occupation 建筑系数
coefficient of capacitance 电容系数
coefficient of capacity 电容系数
coefficient of cavitation 气蚀系数，空化系数
coefficient of charge 装载系数，占空系数，填充系数
coefficient of cohesion 黏合系数，黏聚系数
coefficient of collapsibility 湿陷系数
coefficient of combination 组合系数，系统系数
coefficient of compressibility 压缩系数
coefficient of condenser coupling 电容耦合系数
coefficient of conductivity 传导系数，电导系数，导热系数，导电率
coefficient of consolidation 压实系数，固结系数
coefficient of contraction 收缩系数
coefficient of convection 对流系数
coefficient of correction 修正系数
coefficient of correlation 相关系数
coefficient of corrosion 腐蚀系数
coefficient of coupling 耦合系数
coefficient of cubical elasticity 体积弹性系数
coefficient of cubical expansion 体积膨胀系数，容积膨胀系数
coefficient of current distribution 均流系数
coefficient of curvature 曲率系数
coefficient of damping 阻尼系数
coefficient of decay 衰减系数
coefficient of deformation 变形系数
coefficient of dependent 带眷系数
coefficient of deviation 偏差系数，离差系数
coefficient of dielectrical loss 介电系数，介质损耗系数
coefficient of diffusion 扩散系数
coefficient of dilatation 膨胀系数，体膨胀系数
coefficient of dilution 稀释系数
coefficient of discharge 排放系数【实际流量/理论流量】，流量系数，放电系数
coefficient of drag 牵引系数，阻力系数
coefficient of ductility 延性系数
coefficient of dust collection 集尘系数
coefficient of dust removal 除尘系数
coefficient of dynamic viscosity 动力黏度系数，动力黏滞系数
coefficient of earth pressure at rest 静止土压力系数
coefficient of earth pressure 土压力系数
coefficient of eddy diffusion 涡流扩散系数
coefficient of eddy viscosity 涡流黏滞系数
coefficient of effective heat emission 散热量有效系数
coefficient of effect of action 作用效应系数
coefficient of efficiency 利用系数，效率系数，有效系数
coefficient of elastic recovery 弹性恢复系数
coefficient of elastic shear 弹性剪切系数
coefficient of elastic uniform compression 弹性均匀压缩系数
coefficient of electron coupling 电子耦合系数
coefficient of electrostatic induction 静电感应系数
coefficient of elongation 拉伸系数，伸长系数
coefficient of emission 放射系数
coefficient of energy utilization 能量利用系数
coefficient of energy 能量系数
coefficient of evaporation 蒸发系数
coefficient of extension 延伸率，伸长系数，膨胀系数
coefficient of extinction 衰减系数
coefficient of fatigue 疲劳系数
coefficient of fineness 细度系数，船型系数，丰满系数
coefficient of flood recession 洪水降落系数
coefficient of flow velocity 流速系数
coefficient of flow 流量系数
coefficient of foundation ditch's rebound 基坑回弹系数
coefficient of frequency modulation 调频系数
coefficient of frictional resistance 摩擦阻力系数
coefficient of friction resistance 摩擦阻力系数

coefficient of friction　摩擦系数
coefficient of fullness　充满系数
coefficient of greening　绿化系数
coefficient of grounding　接地系数
coefficient of haze　霾系数，雾系数
coefficient of heat preservation　保温系数
coefficient of heat conductivity　导热率，导热系数
coefficient of heat expansion　热膨胀系数
coefficient of heat passage　导热系数
coefficient of heat supply　供热系数
coefficient of heat transfer　传热系数
coefficient of heat transmission　传热系数，热传导系数
coefficient of horizontal consolidation　水平固结系数
coefficient of horizontal pile reaction　水平桩反力系数
coefficient of horizontal soil reaction　水平土反力系数
coefficient of humidity　湿润系数
coefficient of hydraulic conductivity　导水系数
coefficient of hydraulic stability　水力稳定性系数
coefficient of hysteresis　磁滞损失系数
coefficient of induction　感应系数
coefficient of inductive coupling　感应耦合系数
coefficient of infiltration　下渗系数
coefficient of internal friction　内摩擦系数
coefficient of kinematic viscosity　动黏滞系数，运动黏度系数
coefficient of lateral pressure　侧压力系数
coefficient of leakage　漏泄系数，渗漏系数，泄漏系数
coefficient of lift　升力系数
coefficient of linear expansion　线胀系数，线性膨胀系数，线膨胀系数
coefficient of linear thermal expansion　线性热膨胀系数
coefficient of local heat loss　局部热损失系数
coefficient of local resistance　局部阻力系数
coefficient of losses　损耗系数
coefficient of magnetic dispersion　漏磁系数
coefficient of magnetic leakage　漏磁系数
coefficient of magnetization　磁化系数
coefficient of mean value　均值系数
coefficient of modulation　调制系数
coefficient of moisture absorption　吸湿系数，吸水系数
coefficient of moisture transition　变湿系数
coefficient of momentum transfer　动量传递系数
coefficient of multiple correlation　复相关系数
coefficient of mutual inductance　互感系数
coefficient of natural lighting　自然采光系数
coefficient of non-linear distortion　非线性失真系数
coefficient of nozzle loss　喷嘴损失系数
coefficient of opacity　不透明系数
coefficient of orifice　孔口系数
coefficient of output　出力系数
coefficient of overflow　溢流系数
coefficient of oxidation　氧化系数
coefficient of partial correlation　偏相关系数
coefficient of passive earth pressure　被动土压力系数
coefficient of performance of refrigeration　制冷的性能系数
coefficient of performance　效率【风轮】，性能系数，使用系数
coefficient of permeability　渗漏系数，渗透系数
coefficient of phase displacement　相位移系数
coefficient of phase modulation　调相系数
coefficient of pipe-line orifice　管路的孔口系数
coefficient of pore pressure　孔隙压力系数
coefficient of porosity　孔隙度系数
coefficient of potential　电势系数，电位系数
coefficient of potentiometer　系数式电位计
coefficient of pressure distribution　压力分布系数
coefficient of preventative maintenance　预防性安全系数
coefficient of proportionality　正比系数
coefficient of propulsion　推进系数
coefficient of purification　净化系数
coefficient of recovery　回收率
coefficient of reduction　折减系数
coefficient of regime　（河流的）特性系数，状态系数
coefficient of relative roughness　相对粗糙系数
coefficient of resistance　电阻系数，阻力系数，（桩的）回弹系数，弹性
coefficient of restitution　还原系数，恢复系数
coefficient of retardation　滞留系数
coefficient of rigidity　刚性系数
coefficient of rolling friction　滚动摩擦系数
coefficient of rolling resistance　滚动阻力系数
coefficient of roughness　粗糙系数
coefficient of safety　安全系数
coefficient of scattering　散射系数
coefficient of secondary compression　二次压缩系数
coefficient of secondary consolidation　次固结系数
coefficient of seepage　渗漏系数
coefficient of self-induction　自感系数
coefficient of self-weight collapsibility　自重湿陷系数
coefficient of shear　切变系数
coefficient of short tube　短管系数
coefficient of site utilization area　场地利用系数
coefficient of skewness　偏差系数，偏度系数
coefficient of soil compactness　土壤压实系数
coefficient of soil looseness　土壤松散系数
coefficient of soil reaction　土壤反力系数
coefficient of sound absorption　吸音系数
coefficient of sound energy　声能反射率
coefficient of sound transmission　传声系数
coefficient of stability　稳定系数
coefficient of stabilization　稳定系数，稳定率
coefficient of standard deviation　标准差系数
coefficient of stiffness　刚性系数
coefficient of storage　库容系数，蓄水系数
coefficient of structural change　结构变化系数
coefficient of structural influence　结构性影响系数
coefficient of structural representation　结构表达

式系数
coefficient of subgrade reaction 地基反力系数，基床反力系数
coefficient of sudden contraction 骤缩系数
coefficient of swelling 膨胀系数，湿胀系数
coefficient of thermal conductivity 导热率，导热系数，热传导系数
coefficient of thermal efficiency 热效率系数
coefficient of thermal expansion 热膨胀系数
coefficient of thermal insulation 热绝缘系数
coefficient of thermal storage 蓄热系数
coefficient of thermal transmission 传热系数
coefficient of thrust 推力系数
coefficient of torsional rigidity 扭转刚度系数
coefficient of traction 牵引系数
coefficient of transformation 转换系数
coefficient of transmission 传递系数
coefficient of transparency 透明系数
coefficient of turbulence flow 紊流系数
coefficient of turbulence 紊流系数
coefficient of turbulent transport 紊流输送系数
coefficient of underrating 做功不足系数
coefficient of uniformity 均匀系数
coefficient of utilization of installed capacity 装机容量利用系数
coefficient of utilization of maximum load 最大负荷利用系数
coefficient of utilization 利用系数，利用率
coefficient of vapor permeation 蒸汽渗透系数
coefficient of vapour permeability 蒸汽渗透系数
coefficient of variation 变异系数，变化系数，变化率
coefficient of velocity 速度系数
coefficient of vertical consolidation 竖向固结系数
coefficient of viscidity 黏滞系数
coefficient of viscosity of steam 蒸汽的黏性系数
coefficient of viscosity 动力黏度，黏滞系数，黏性系数
coefficient of voltage distribution 电压分布系数，均压系数
coefficient of volume change 体积变化系数
coefficient of volume compressibility 体积压缩系数
coefficient of volume decrease 体减系数
coefficient of volumetric expansion 体积膨胀系数
coefficient of wall friction 墙摩擦系数
coefficient of waste 废料系数
coefficient of water line 浸水系数
coefficient of water storage 储水系数
coefficient of weight 加权系数
coefficient of wet-subsidence 湿陷系数
coefficient-setting potentiometer 系数设置电势计
coefficient tape 系数带
coefficient unit 系数给定部件
co-energy density 同能量密度
co-energy 同能量
coercimeter 矫顽磁力表
coercitive force 矫顽电力
coercitive 矫顽磁力的，矫顽磁场
coercive electric field 矫顽电场
coercive field strength 矫顽磁场强度
coercive field 矫顽场，矫顽磁场
coercive force 矫顽磁力，矫顽力，矫磁力，抗磁力，保磁力
coercive intensity 矫顽磁力，矫顽磁场强度
coercivemeter 矫顽磁性测量仪
coercive stress 矫顽应力，矫顽磁力
coercive voltage 矫顽电压
coercive 矫顽磁力的，矫磁的，强迫的，强制的
coercivity 矫顽磁力，矫顽磁性，矫顽力
CO_2 exhaust system 二氧化碳排气系统
coexistence border 双相分界线
coexistence curve 共存线，分界线，饱和线
coexistence region 双相区
coexisting phase 共生相，共存相
coextrusion 共挤压，双金属挤压
cofactor 余因子
COFC(containers on flat car) 平板车装运集装箱
C. O. F.(cooling of fuel) 燃料冷却
cofferdam cell 潜水箱，金属槽，沉箱
cofferdam construction 围堰施工，修筑围堰
cofferdam dewatering 围堰排水
cofferdam piling 围堰板桩
cofferdam removal 围堰拆除
cofferdam 围堰【重要厂用水系统取水口】，沉箱，潜水箱，修筑围堰
coffered ceiling 方格天花板
coffered foundation 沉箱基础，围堰底座，箱格基础，箱形基础
coffer-wall 围墙，围堰墙
coffer work 砌片石墙
coffer 围堰
coffin 容器，屏蔽容器
coffin hoist 匣式升降机
coffinite 铀石
co-financier 财政合作部门，联合投资者，融资部门
co-financing 联合集资，合作投资，统括贷款，联合融资
cogenerating plant 热电（联产）厂
cogenerating unit 热电机组
cogeneration level 热电联产发电率
cogeneration of heat and electricity 热电联产
cogeneration plant 热电联产电厂，热电厂
cogeneration power plant 热电厂，热电站，热电联产电厂
cogeneration power station 热电联供电站
cogeneration turbine 供热（式）汽轮机，热电联产（式）汽轮机，热电联供透平
cogeneration 热电合供，热电联产，共生，双重性生产，废热发电
cogged bit 齿形钻头，冲击式凿岩器
cogged joint 榫齿接合
cogging torque 齿槽转矩，顿转扭矩，齿槽效应转矩
cogging 齿，轮齿，嵌齿，开坯，齿槽效应，雄榫装入
cognizant individual 有能力的人
cog railway 齿轨铁道
CO_2-graphite reactor 二氧化碳石墨慢化反应堆

cogwheel coupling 齿形联轴器
cogwheel gearing 齿轮传动装置
cogwheel 齿轮，嵌齿轮
cog 榫头
co-hammer milling 锤式粉碎机
COH(coefficient of haze) 霾系数，雾系数
cohered video 相关视频信号
coherence function 相干函数，凝聚函数，相干系数
coherence length 相干长度
coherence system of unit 连贯（测量）单位制
coherence vortex shedding 相干涡脱落
coherence 相干性，凝聚，黏着，相关性
coherency identification 同调识别
coherent alluvium 凝聚性冲积层
coherent boundary 共格晶界
coherent carrier system 相干载波系统
coherent collision 相干碰撞
coherent cross section 相干截面
coherent detector 相干检波器
coherent disturbance 相干扰动
coherent elastic scattering 相干弹性散射
coherent generation 相干源
coherent integral 相干积分
coherently diffracting domain 相干衍射区
coherent neutron scattering 相干中子散射
coherent noise 相干噪声
coherent optical arrangement 相干光学装置
coherent oscillator 相关振荡器，相干振荡器
coherent phase-shift keying 相干相移键控制
coherent rotation 一致转动，相关转动
coherent scattering 相干散射
coherent signal 相干信号
coherent speckle effect 相干斑纹效应
coherent system of unit 连贯（测量）单位制，一贯（测量）单位制
coherent unit （测量的）一贯单位，连贯单位
coherent wave 相干波
coherent 黏着的，黏附的，耦合的，相关的
coherer 粉末检波器，金属检波器
cohere 黏着，附着，结合，凝聚
cohesional resistance 凝聚阻力，黏聚抗力
cohesionless material 非黏性材料，松散材料，无黏性材料
cohesionless soil 无黏性土
cohesionless 无凝聚性的
cohesion 内聚力，内聚性，结合力，黏合力，内聚现象，凝聚，凝聚力
cohesive force （煤的）内聚力，黏性力，黏附力，黏合力，黏聚力
cohesive material 黏性材料
cohesiveness degree 黏聚度
cohesiveness 内聚性，黏结性，黏聚性
cohesive sediment 黏结性淤积物，黏结性淤泥
cohesive soil 黏性土，黏结性土
cohesive 凝聚性的，有黏着力的
cohobation 回流蒸馏，反复蒸馏，连续蒸馏
COHO(coherent oscillator) 相干振荡器
coigne 隅石
coign 隅，隅石
coil aerial 线圈形天线，环形天线
coil arrangement 线圈布置
coil assembling apparatus 钳线装置，下线装置
coil assembly 线圈组
coil axis 线圈轴线
coil banding 线圈绑扎
coil bar 线棒
coil bobbin 绕线管，线圈架
coil boiler 盘管锅炉
coil box 线圈箱
coil brace 线圈支架，线圈撑条
coil capacity 线圈的固有电容，线圈电容
coil clamper 线圈接线板，线圈绑环
coil comparator 线圈试验器，线圈比较器
coil condenser 盘管冷凝器，盘管式凝汽器
coil connection sequence 线圈接线顺序
coil constant 线圈常数
coil cooling pipe 蛇形冷却管，蛇型冷却管
coil core 线圈铁芯
coil crossover 线圈跨接
coil current 线圈电流
coiled-coil 双螺旋灯丝，复绕灯丝
coiled filament 卷绕式灯丝
coiled finned-tube type cooler 螺旋肋管式冷却器
coiled lamp 卷丝灯泡
coiled pipe 蛇形管
coiled radiator 盘管式散热器
coiled tube 螺圈管
coiled 卷成的，绕成的
coil encapsulation 线圈的封装
coil end bracing 线圈端部支架
coil end cover 线圈端罩
coil end leakage reactance 线圈端漏电抗
coil-end leakage 线圈端部漏磁
coil end 线圈端部
coiler motor 卷取电动机
coiler 卷轴，蛇形管
coil extension 线圈槽外直线部分
coil factor 线圈系数
coil fault 线圈故障
coil final mean temperature 线圈最终平均温度
coil final power dissipation 线圈最终功耗
coil former 线圈架，线圈成形器
coil forming 线圈成形
coil form 线圈形状，线圈管，线圈架
coil gradient 线圈温度梯度
coil grading 线圈分段
coil group 线圈组
coil head 线圈鼻端，线圈端部
coil heater 盘管加热器，线圈加热器，蛇形管加热器
coil holder 线圈支持器，线圈座
coil housing 线圈外套【控制棒驱动机构】，线圈罩，线圈壳
coil ignition 线圈点火
coil-in-box cooler 蛇形管冷却器
coil inductance 线圈电感
coiling drum 卷线盘
coiling machine 卷取机，绕线机
coiling 绕制线圈［盘管］，绕成螺旋
coil initial power dissipation 线圈起始功耗
coil inserting apparatus 嵌入设备，下线设备

coil insertion device　下线装置，嵌入机
coil in slot part　槽部线圈
coil insulation　线圈绝缘
coil in　进线，输入
coil knuckle　线圈鼻端
coil lashing　线圈端部绑扎
coil loaded cable　加感电缆
coil loaded　线圈加感的
coil loading　用线圈加感，线圈加感，线圈加载
coil load　线圈负载，线圈电感
coil loss　线圈损耗
coil mean temperature　线圈平均温度
coil movement　线圈窜动，线圈移动
coil neutralization　感应中和，线圈中和
coil nose　线圈鼻端
coil of cable　电缆卷
coil of strip　带卷
coil of wire　线卷，线圈
coil out　出线，输出
coil pack　线圈组件
coil pickup　感应传感器
coil pipe cooler　盘管冷却器
coil pipe　蛇形管，盘管
coil pitch　线圈节距
coil power dissipation　线圈功耗
coil power nomograph　线圈功率图解法，线圈功率列线图
coil power　线圈功率
coil pulling machine　线圈拉成形机
coil quality　线圈的品质因数，线圈的Q值
coil rack　线圈架
coil radiator　盘管散热器
coil resistance　线圈电阻
coil section　线圈截面，线圈元件
coil segment　线圈段
coil shape factor　线圈形状因数
coil shop　绕线车间，线圈车间
coil side separator　槽内层间绝缘，线圈边间的绝缘
coil side　线圈边
coil slot　线圈槽
coil space factor　线圈占空系数，槽满率
coil space　线圈跨距，线圈后节距
coil spacing　线圈间距
coil spreading machine　线圈拉型机
coil spread　线圈跨距，线圈节距
coil spring relay　螺簧继电器
coil spring　弹圈，螺簧，螺旋弹簧，线卷弹簧
coil stack assembly　线圈堆叠组件【控制棒驱动机构】
coil stripper　剥线圈绝缘带器，线圈剥皮器
coil support bracket　线圈支架
coil support　线圈支撑
coil tap　线圈抽头
coil temperature monitor　线圈温度监测器
coil temperature-rise　线圈温升
coil temperature　线圈温度
coil thermal equilibrium　线圈热平衡
coil throw　线圈节距
coil tie　绕线管，线圈连接线
coil time constant　线圈时间常数
coil-to-coil breakdown　线圈间击穿
coil to coil fault　线圈间故障
coil-tube boiler　盘管锅炉
coil tube header-type heater　螺旋管联箱式加热器
coil tube　绕线管，线圈绕管，盘管
coil tuning　线圈调谐
coil type heat exchanger　盘管式热交换器，蛇形管式热交换器
coil type winding　圈式绕组
coil unit　线圈组，线圈单元
coil varnish　线圈绝缘漆
coil voltage　线圈电压
coil width　线圈宽度
coil winder　绕线机
coil winding apparatus　绕线机，绕线器
coil winding machine　绕线机
coil winding　线圈绕组
coil wound rotor　线绕式转子
coil　蛇形管，线圈，盘管，绕组，绕线管，螺线管
coinbox discriminating tone　公用投币电话鉴别音
coincide　符合，一致
coincide decoding　符合译码，重合译码
coincide in phase with　与……同相
coincidence adder　重合加法器
coincidence amplifier　符合加法器
coincidence analyzer　符合分析器
coincidence-anticoincidence analyzer　符合反符合分析器
coincidence-anticoincidence gamma ray spectrometer　符合反符合射线谱仪
coincidence-anticoincidence scaler　符合-反符合（计数）定标器
coincidence-anticoincidence unit　符合反符合单元
coincidence arrangement　重合装置
coincidence array　重合电路列，符合阵列
coincidence circuit　同时计数电路，脉冲重合电路
coincidence correction　重合校正，符合校正，同频校正
coincidence counter　符合计数器
coincidence counting method　符合计数法
coincidence-counting rate　符合计数率
coincidence counting　符合计数【中子】
coincidence count probability　符合计数概率
coincidence detector　符合检测器
coincidence discriminator　符合鉴别器
coincidence factor　同时使用系数，重合系数，重合率
coincidence-flux storage system　磁通重合存储系统
coincidence frequency　相干频率，重合频率
coincidence gate　符合门，符合选通电路
coincidence loss　偶然重合事故
coincidence proportional counter　比例重合计数器
coincidence pulse　符合脉冲
coincidence register　符合寄存器
coincidence signal　符合信号
coincidence　一致，符合，重合，吻合
coincident-current core memories　符合电流磁芯

存储器
coincident-current magnetic-core storage unit 符合电流磁芯存储部件
coincident-current magnetic core storage 符合电流磁芯存储器
coincident-current memory 符合电流存储器，电流重合法存储器，
coincident-current selection 电流重合选择法，符合电流选择
coincident-current 符合电流
coincident demand power 同时需用功率
coincident demand 重合需量
coincident drive system 电流重合驱动系统
coincident-flux magnetic storage 符合磁通磁存储器
coincident temperature 同时气温
coincident 符合的
coin counting machine 自动点钞机
co-insurance clause 共同保险条款
co-insurance 共同保险
co-insured 共同投保人，联合被保险人，联合投保人
co-insurer 共保人，共同保险者
coke 焦炭【煤】，焦
cokeability 成焦性，焦化性
coke accreting fluid bed 加碳流化床
coke breeze concrete 煤渣混凝土
coke breeze 焦炭屑
coke button 焦渣，焦饼
coked resin binder 焦化树脂黏合剂
coke dross 焦煤，焦炭屑
coke-like sludge 焦状污泥
coke oven effluent 焦炉废水
coke oven gas pipe 焦炉煤气管
coke oven gas 炼焦炉煤气，焦炉煤气
coke oven tar 煤焦油
cokery 炼焦厂，炼焦炉
coke-tray aerator 焦盘曝气器
coking arch 前拱
coking capacity 成焦性，结焦能力
1/3 coking coal 1/3 焦煤
coking coal 焦性煤，成焦煤，炼焦煤
coking index 煤的结焦指数
coking plate 焦化板
coking process 焦化过程
coking property 成焦性，结焦性
coking quality 成焦性
coking stoker 焦化加煤机，焦化炉排
coking test 成焦试验
coking 炼焦，焦化，结焦【锅炉】
colasmix 沥青砂石混合物
colas 沥青乳浊液
colateral dipoles 通便偶极子排，并列偶极子
colation 过滤，渗滤
colature 粗滤产物，滤液
colclad 复合钢板
COL(construction and operating license) 建造及运行许可证
colcrete 预填骨料压浆混凝土
cold accumulation 蓄冷
cold air circulating system 冷风循环系统
cold air duct 冷风道，冷风管道
cold air machine 冷风机，冷气机
cold air mass 冷气团
cold air return 冷气回流
cold-air 气冷的，空气冷却的，冷风，冷空气
cold and hot brittleness 冷脆热脆性
cold and hot trial 冷热试车
cold application 冷灌筑，冷态喷漆
cold-applied bituminous coat 冷敷沥青层
cold area 非放射性区，冷区，寒冷区
cold belt 寒带，冷带
cold bending test 冷弯试验
cold bending 冷弯
cold bend property 冷弯性
cold box 冷盒
cold brittleness 冷脆性
cold cathode arcing 冷阴极起弧
cold cathode counter 冷阴极电子管计数器
cold cathode discharge valve 冷阴极放电管
cold cathode electronic relay 冷阴极电子继电器
cold cathode glow discharge valve 冷阴极辉光放电管
cold cathode ionization gauge 冷阴极式电离压力计
cold cathode lamp 冷阴极灯
cold cathode rectifier 冷阴极整流管
cold cathode tube 冷阴极管
cold cathode vacuum gauge 冷阴极式真空压力计
cold cathode 冷阴极电子管，冷阴极
cold caulking 冷填缝
cold change room 冷更衣房
cold chisel 凿子
cold constant 冷态常数
cold coolant accident 冷环事故
cold cracking 冷裂，冷裂纹
cold critical reactor 冷临界堆，冷临界反应堆
cold critical 冷态临界，冷态临界的
cold crushing strength 冷挤压强度
cold current 寒流
cold deformation 冷变形
cold differential test pressure 冷态试验压差
cold draft 冷风
cold-drawing seamless pipe 冷拉无缝钢管
cold-drawing 冷拔，冷拉，冷轧
cold-drawn copper 冷拉铜
cold-drawn steel wire 冷拉钢丝
cold-drawn wire 冷拉线，冷拉钢丝，冷拔钢丝
cold-drawn 不加热而拉长的，不加热力以延展的，冷拔的，冷拉的
cold driven shop rivet 冷铆车间铆
cold-drown bar 冷拔钢筋
cold drying 低温干燥
cold electrode 冷电极
cold end corrosion 低温腐蚀，冷端腐蚀
cold end element 低温元件，冷端元件
cold end temperature 冷端温度
cold endurance 耐寒性
cold end 冷端，低电位端
cold expanding 冷胀
cold extrusion 冷挤压
cold finished 冷抛光的【管子】

cold flame 冷焰，冷火焰
cold-formed thin-walled shape 冷压薄壁型钢
cold forming 冷成形【材料】
cold front rain 冷锋雨
cold front thunderstorm 冷锋雷暴，冷锋雷雨
cold front 冷锋
cold fueling 停堆加料
cold functional test 冷态功能试验
cold fusion reactor 冷聚变反应堆
cold fusion 冷核聚变
cold gagging 冷弯
cold-gas header 低温气体联箱
cold-gas measuring point 低温气体测量点
cold-gas penetration 低温气体贯穿件
cold-gas plenum 低温气体联控
cold gas plume 冷气羽流
cold-gas temperature control 低温气体温度控制
cold-gas temperature 低温气体温度
cold gas 冷间隙
cold hardening 冷作硬化，冷加工硬化
cold header 冷集流管，冷集流联箱
cold heat 冷熔炼
cold hydraulic test 常温水压试验，冷态水压试验
cold hydrostatic test 冷态水压试验，冷态静水压试验
cold hydrotest 冷态水压试验
cold insulation 保冷
cold insulator 冷绝缘材料
cold joint 建筑填料接缝【混凝土】，冷缝
cold junction box 冷端盒【热电偶】
cold junction cabinet 冷端柜
cold junction compensation 冷端温度补偿，冷端补偿
cold junction compensator 冷端补偿器
cold junction temperature 冷端温度，冷接点温度
cold junction （热电偶的）冷接点，冷连接，冷接头，冷端
cold laboratory 冷实验室【非放射性】，非放射性实验室
cold lahar 冷泥石流
cold-laid bituminous pavement 冷铺沥青路面，冷铺沥青面层
cold laundry 冷洗衣间，非放射性洗衣间，非放射性洗衣房
cold leg 冷支路，冷段【管段】
cold light lamp 冷光灯
cold lime neutralization 冷石灰中和作用
cold loading 冷态装料【停堆装料】
cold load 冷态负荷，冷态载荷
cold loop accident 冷水事故，冷环事故
cold loop 冷环路，冷回路，非放射性回路
cold metal worker 白铁
cold mill motor 冷轧电动机
cold mill waste water 低温工厂废水
cold-mix asphalt 冷拌沥青
cold-model rig 冷模设备
cold-model test 冷模试验
cold model 冷态模化
cold mud-flow 冷泥石流
cold neutron facility 冷中子装置
cold neutron source 冷中子源
cold neutron 冷中子
cold operation 冷操作，冷态运行
cold penetration bitumen 冷灌沥青
cold penetration construction 冷灌沥青路面施工
cold performance test 冷态性能试验
cold pole 寒极
cold position 冷态
cold precipitator 低温电除尘器
cold-pressed sheet steel 冷压薄钢料
cold-pressed steel plate 冷压钢板
cold-pressed steel 冷压钢
cold-pressed without the addition of binder or lubricant 不加黏结剂或润滑剂的冷压加工
cold-pressed work 冷压力加工
cold-pressed 冷冲压的，冷锻的，冷压
cold pressure welding 冷压焊
cold primer 冷底子
cold pull （管道的）冷拉
cold quenching 冷（介质）淬火，水冷淬火
cold reactivity 冷态反应性
cold reactor 冷态反应堆
cold refueling 停堆换料
cold region 寒冷区
cold reheat 低温再热
cold reheated steam pressure 再热蒸汽冷段压力
cold reheated steam temperature 再热蒸汽冷段温度，再热蒸汽冷锻温度
cold reheater 低温级再热器，再热器冷段，冷再
cold reheat header 再热器冷集管，再热器冷集箱
cold reheat line 再热冷端管道
cold reheat pipe 低温再热器进汽管
cold reheat safety valve 再热冷段安全阀
cold reheat steam conditions 冷再热蒸汽参数
cold reheat steam outlet 再热冷蒸汽出口
cold reheat steam 冷再热蒸汽
cold repair 冷态检修，冷态维修
cold reserve 冷备用，冷态备用，冷备用状态
cold resistance property 耐寒特性
cold resistance 冷态电阻，耐冻性，耐寒性
cold-resistant insulation 耐寒绝缘
cold resisting property 抗冻性
cold rolled deformed bar 冷轧变形钢筋
cold rolled grain oriented transformer 冷轧晶粒取向变压器
cold-rolled plate 冷轧钢板
cold-rolled section 冷轧型材
cold-rolled silicon steel sheet 冷轧硅钢片
cold-rolled silicon steel 冷轧硅钢
cold-rolled steel 冷轧钢
cold-rolled 冷轧的，冷轧制的
cold rolling process 冷态冲转过程【汽轮机】
cold rolling 冷轧
cold run 冷态运行
cold sector 冷区
cold service 冷态运行，冷操作
cold-setting adhesive 低温硬化胶着剂
cold-setting of coupling alignment 联轴器冷态对中心

cold settler 低温澄清池
cold shock 冷冲击
cold short iron 冷脆钢
cold short material 冷脆材料
cold shortness 冷脆性
cold short 冷脆的
cold shot cooling 冷弹冷却
cold shutdown margin 冷停堆深度
cold shutdown reactivity margin 冷停堆时的剩余反应性，冷停堆反应性裕量
cold shutdown reactivity 冷态停堆反应性
cold shutdown 冷停堆
cold shut 冷态关闭，冷疤
cold side 冷端
cold sink 冷却散热片
cold-sodium slug accident 钠凝块事故
cold source 冷源
cold spell in spring 春寒期，春寒
cold spell 寒潮
cold spring pipe 冷拉（管道）
cold spring 预拉伸，预紧，冷紧
cold stand-by state 冷备用状态
cold standby 冷储备
cold starting 冷态启动
cold start lamp 冷启动灯
cold start-up mode 冷态启动模式
cold start-up 冷启动，冷态启动
cold start 冷态启动，冷态起动
cold stepping driving experiment 冷态步跃试验
cold-storage room 冷藏间
cold-storage 冷藏
cold strain 冷应变
cold strength 冷强度
cold stretching of pipings 管道冷拉
cold strip 冷轧带材
cold subcritical condition 冷次临界条件，冷次临界
cold subcritical reactor 次临界冷堆，冷次临界反应堆
cold subcritical 冷次临界
cold temperature brittleness 低温脆性
cold temperature flexibility 低温柔性
cold temperature resistance 耐低温性
cold test 不通电试验，冷态试验，无源试验，低温试验
cold-to-hot reactivity 冷热反应性
cold tolerance 耐寒性
cold trap removal tool 冷阱拆除工具
cold trap 冷阱【真空系统】，冷捕集器
cold treatment 冷处理
cold vacuum deaeration 真空除氧
cold wall effect 冷壁效应
cold wall factor 冷壁因子
cold wardrobe 清洁区衣柜
cold waste 非放射性废物
cold-water accident 冷水事故，冷水进入事故
cold-water basin 冷水池
cold water hydrostatic test 冷态水压试验
cold water injection 冷水注入
cold water loop 冷水环路
cold water shock 冷水冲击
cold water species 冷水种
cold water transient 冷水瞬变
cold wave 寒潮
cold weather concreting 冬季浇注混凝土
cold weather construction 冬季施工
cold weather test 低温试验
cold welding 冷焊，冷压焊
cold work hardening 冷锻硬化，冷加工硬化
cold working 冷锻，冷加工，冷作
cold workshop 非放射性车间，非放射性检修间
cold zone 寒带
cold 冷态的，冷的，非放射性的，寒冷的
colidar 相干光雷达，激光雷达
colinear dipole 共线偶极子
colinear 同线的，共线的
colitic texture 鲕状结构
collaborate of association 协作合同
collaborate 合作，协作
collaboration of steel and concrete 钢筋与混凝土共同受力
collaborator 共同研究者，合作者，协作单位，协作者
collagen reverse osmosis 骨胶朊反渗透
collapsable can 坍塌型包壳
collapsable cladding 坍塌型包壳
collapsable load 坍塌载荷
collapsable radio mast 伸缩套杆式天线杆，拉杆天线
collapsable 塌陷的，可崩溃的，可拆卸的
collapse by wetting 湿陷
collapse clad fuel element 坍塌型包壳燃料元件
collapse critical deflection 临界破坏饶度
collapsed bed 塌落的床层
collapsed fuel damage 燃料坍塌损坏
collapsed 塌倒，坍塌
collapse fissure 塌陷裂缝
collapse load 破坏载荷，临界载荷
collapse overspeed 急剧坍塌
collapse pressure 失稳压力，坍塌压力
collapse settlement 湿陷量
collapse sink 塌陷落水洞，坍陷
collapse 崩溃，瓦解，断裂，破坏，故障，损坏，毁坏，（涡的）破碎，湿陷，塌陷，坍塌
collapsibility grading index 分级湿陷量
collapsibility 湿陷性
collapsible bit 活头钻头
collapsible can 坍塌型包壳
collapsible cladding 坍塌型包壳【指燃料】
collapsible flash board 自溃式闸板
collapsible form 活动模板
collapsible gate 折叠式门
collapsible load 坍塌载荷
collapsible loess 湿陷性黄土
collapsible radio mast 伸缩套杆式天线杆，拉杆天线
collapsible rubber dam 可缩坍橡胶坝
collapsible soil 湿陷性土
collapsible 塌陷的，可崩溃的，可拆卸的
collapsing cavity 溃灭空穴
collapsing force 破坏力
collapsing load 破坏载荷，失稳载荷，破坏荷载

collapsing pressure 破坏压力，失稳压力
collapsing soil 崩坍土
collapsing strength 破坏强度
collapsing stress 破坏应力
collapsing 压扁，折叠，断裂，崩溃，塌陷
collar beam roof 系梁屋架
collar beam 圈梁，系梁
collar bearing 环形止推轴承
collar cheek 磁极托板
collaring 打眼
collar journal 止推轴颈，有环轴颈
collar nut 圆缘螺母，带突缘螺母
collar step bearing 环状阶式轴承
collar stop 环形挡块
collar thrust bearing 环形推力轴承，环形止推轴承
collar thrust 止推环
collar tie beam 圈梁
collar 圈，环，轴环，垫圈，箍，凸缘，法兰盘，底梁，轭部，连接套筒，系梁，柱环
collateral agreement 附属协定
collateral contract 附属合同
collateral damages 附带损害，间接损害，间接伤害
collateral evidence 旁证
collateral loan 抵押贷款
collateral warranty 附带保证
collateral 抵押品，担保品，旁系的，附属的
collate 核对，校对，整理，分类，对照，比较
collating report 核对报告
collator 校对机，集电极，校核者
colleague 同事，同僚
collect 采集，收集
collect call 对方付费电话，接听人付费长途电话
collect compensation from performance bond 从履行保证金中扣收赔款
collected duties 征税
collected oil pump 集油泵
collected papers 论文集
collected stack 集合烟囱
collected text 校正本
collecting annulus 集水槽
collecting anode 集电极
collecting aqueduct 集水道，集水渠
collecting area 地面集水区
collecting bank 代收银行，托收银行
collecting bar 汇流排，母线
collecting basin 地面集水区，集水池
collecting belt 集煤皮带
collecting box 集箱，收集箱
collecting brush 汇流刷
collecting channel 总渠，集水槽，干管
collecting conveyor 带式输送机，集矿运输机
collecting device 集电设备，收集装置
collecting duration 通过时间
collecting electrode 集电极，集尘电极
collecting element 校正元件
collecting field 集电极场
collecting flue 总烟道
collecting fund for electricity construction 集资办电
collecting funds 集资
collecting funnel 汇集漏斗【指圆形山谷】
collecting gutter 集水沟
collecting header 集箱，联箱
collecting heat from the sun 收集太阳热能
collecting main 集管，总管，干管，汇流排，母线
collecting of drawings 图集
collecting pipe 集箱，集汽管，集管，母管，集水管，集尘电极
collecting pond 集水池【冷却塔】
collecting potential 收集电势，收集电压
collecting region 收集区
collecting ring 集电环，集水环，集汽环，集流环
collecting shoe 集电靴
collecting site 收集地点，采样地点，取样点
collecting stream-flow data 收集流量资料
collecting sump 收集坑，集水坑
collecting system 集流系统，集水系统，收集系统
collecting tank 集水池，聚集水箱
collecting trap 收集阱，捕集阱
collecting vat 收集槽，集水槽
collecting well 集水井
collecting works 集水工程，引水建筑物
collection area 集水面积
collection bill of lading 托收票据
collection chamber 集气室
collection coefficient 除尘系数
collection efficiency 收集率，捕获率，除尘效率，集尘效率
collection efficient of particulate 集尘效率
collection frequency of refuse 垃圾收集次数
collection frequency 收集频率
collection of a bill 托收
collection of buildings 建筑群
collection of data 收集数据，收集资料
collection officiency 收集效率
collection of gases 气体收集
collection of heat 蓄热
collection of payment 托收，收款
collection order 托收单
collection sump 集水坑，收集坑
collection system 收集系统
collection time 收集时间
collection 收集，集中，收集物，(银行)托收，聚集，收集品，自给能中子探测器
collective bargaining 集体谈判
collective control 集中控制
collective dose equivalent 集体剂量当量
collective dose 集体剂量
collective dosimetry 集定剂量测定
collective drawings 图集，装配图
collective ecological security 集体生态安全
collective economy 集体经济
collective effective dose equivalent commitment 集体有效剂量当量负担
collective effective dose equivalent 集体有效剂量当量

collective field　集体场
collective flow curve　汇流曲线
collective lighting　集中照明
collective maintenance　集中检修
collective-owned enterprise　集体企业
collective ownership　集体所有制
collective protection　集体保护
collective ring　集电环
collective shielded cable　共屏蔽电缆
collective shunt capacitor　集合式并联电容器
collective treatment　集中处理
collective work　集体作品
collective　共同的，集合的，集中的，收集的，集合，集体
collector array　集热器阵列
collector barrier　集电极势垒，集电极位垒
collector bow　集电弓
collector capacitance　集电极电容
collector casing　集热器外壳
collector channel　集合管，拼合槽钢
collector contact　集电环接点，集电极接点，收集电流
collector cover plate　集热器盖板
collector cover　集热器盖层
collector diffusion isolation　集极扩散隔离
collector dissipation　集电极损耗
collector efficiency equation　集热器效率方程
collector efficiency factor　集热器效率因子
collector efficiency　集热器效率
collector-electrode　集电极
collector exhaust capacity　除尘器抽风量，抓斗容积
collector film　集尘过滤器
collector flow factor　集热器流动因子
collector grid　捕获栅
collector heat removal factor　集热器热转移因子
collector impedance　集电极阻抗
collector inlet　集热器进口
collector instantaneous efficiency　集热器瞬时效率
collector junction　集电极结
collector loop　集热器回路
collector manifold　集水歧管
collector mesh　收集栅
collector outlet　集热器出口
collector overall heat loss coefficient　集热器总热损失系数
collector pan　收集盘
collector-refrigerator combination　集热器制冷机组合
collector ring brush　集电环电刷
collector ring lead　集电环引线
collector ring　汇集环，集电环，汇流环，集流环
collector screen　集电屏
collector shell　集电环罩
collector shoe gear　集电装置，集流装置，集电环，汇流环
collector shoe　集电靴
collector subsystem　集热器子系统
collector-to-base current gain　集电极基极电流增益
collector-to-emitter current gain　集电极发射极电流增益
collector unit　收集单元
collector well　集水井
collector　喇叭口【开口风洞】，集热器【太阳能】，集电器，集电极，集流器，捕集器，集电环，集箱，除尘器，集气环，集合器，集尘器，集料器，聚集器，收集器，收款人
collectron　自给能中子探测器
college entrance examination　入学考试
collet chuck　套爪夹头，套爪卡盘，弹簧夹头，筒夹
collet finger　套爪销，套爪指
collet piston　套爪活塞
collet　套爪，套筒，夹头，筒夹，弹性夹头，簧片绝缘片
collided flux　碰撞通量密度
collide with　与……相撞，冲突，抵触
collide　碰撞
colliery waste tip　煤矿废物堆
colliery waste　煤矿废物
colliery　煤矿，运煤船
collimated flux　准直的通量，准直通量
collimated lamp　准直灯
collimated light source　平行光源
collimated light　照准光线
collimate　对准，使准直，使成平行
collimating adjustment　视准改正
collimating aperture　准直孔
collimating axis　视准轴
collimating error　准直误差
collimating grid　准直栅格
collimating line collimating ray　视准线
collimating point　视准点
collimation axis　视准轴
collimation correction　视准改正
collimation error　视准差，视准误差
collimation method　视准法
collimation　对准，校准，平行校正，瞄准，视准，光轴仪，视准仪，视准管，准直仪
collinear force　共线力
collinear vector　共线向量
collinear　同线的，共线的
collineation　直射，共线
colliquable　易熔的，易溶的
collisional invariant　碰撞守恒
collision angle　撞击角
collision-broadened line　碰撞展宽线
collision bulkhead　防撞舱壁
collision cross-section　碰撞截面
collision cross　碰撞断面
collision density spectrum　碰撞密度谱
collision density　碰撞密度
collision diameter　碰撞直径
collision effect　碰撞效应
collision frequency　碰撞频率，碰撞效率
collision insurance　碰撞保险
collision integral　碰撞积分
collision ionization　碰撞电离
collision kernel　碰撞核

collisionless transport 无碰撞输运
collisionless 无碰撞的
collision liability 碰撞责任
collision loss 碰撞损失
collision mat 防撞柴排
collision post 防撞柱
collision probability 碰撞概率，碰撞效率
collision rate density 碰撞率密度
collision rate 碰撞率
collision stopping power 碰撞阻止能力
collision strut 防冲支撑
collision theory 碰撞理论
collision transition probability 碰撞转移概率
collision 冲击，碰撞，跳跃，冲突，撞击
collocation 配置，布置，排列，安排
collochemistry 胶体化学
collodion 胶棉，火胶棉
colloidal 胶态的，胶体的，胶质的
colloidal activity of clay 黏土胶态活动性
colloidal alumina 胶体氧化铝
colloidal chemistry 胶体化学
colloidal concrete 胶质混凝土
colloidal dispersion 胶态分散体，胶体分散系
colloidal electrolyte 胶态电解质
colloidal fuel 胶质燃料
colloidal gel 胶态凝胶
colloidal hydrosol 胶体水溶液
colloidal impurity 胶体杂质
colloidality 胶性，胶度
colloidal matter 胶态物质
colloidal particle 胶态粒子，胶体微粒，胶粒，胶体颗粒
colloidal silica 胶硅
colloidal solution 胶体溶液
colloidal state 胶态
colloidal suspension 胶态悬浮，胶悬体
colloid amphoion 胶态两性离子
colloid flotation 胶体浮选
colloidization 胶化，胶体化作用
colloid mask 胶体遮蔽作用
colloid mill 胶体磨
colloidor 胶体助凝器
colloid rectifier 胶质整流器
colloid silica 胶态硅石
colloid 胶体，胶质体，胶体颗粒，胶态
colloquial 口语的，通俗的
colloquium 学术讨论会
collosol 熔胶
collusion 串谋，串通
collusive tendering 串通投标
colluvial deposit 塌积物
colluvial soil 崩积土，塌积土
colluviarium 入孔
colluviation 崩积作用
colluvium 崩积层
colmatage 放淤
colmation 放淤
colo-aerogenes bacteria 大肠产气细菌
cologarithm 余对数
colonnade 柱廊
colophonium 树脂，松香
colophony 松香
colorant 显色剂，着色剂
coloration 染色，着色
color balance control 彩色平衡调整
color bar signal 色条信号，色带信号
color bleeding resistance 换色电阻，抗凝色性
color break-up 色乱，颜色分层
color burst 彩色同步信号，基准彩色副载波群
color-coated sheet 彩色涂层钢板
color-coding in the main body 主体中的颜色编码【电缆】
color-coding sleeve 颜色编码套筒
color-coding 颜色编码
color comparision tube 比色管
colored galvanized steel sheet 彩色镀锌钢板
colored lamp 有色灯泡
colored paint 彩色油漆，有色涂料
colored solid 有色固体
colored suspended matter 带色悬浮物
colored water 带色水
colorflexer 彩色电视信号编码器
color-frequency control coil 色频调节线圈
color highlight display 彩色亮度显示
colorimeter 比色计，色度计
colorimetric analysis 比色分析
colorimetric determination 比色测定
colorimetric dosimeter 比色剂量计
colorimetric method 比色测定法【测空气污染度】，比色法
colorimetric test 比色试验
colorimetry 比色法
color indexing circuit 彩色定相电路，彩色指示电路
color information 彩色信息
coloring agent 着色剂
colority 色度
colorless solid 无色固体
colorless 无色
color light signal unit 色灯信号机
colormatrix 热控液晶字母数字显示器
colormetry 色度测量
color-minus-difference voltage 色负差信号电压
color of the glaze 釉色
color phase alternation 彩色相位交变
color phase 彩色相位
color phototelegraphy 彩色传真电报学
color phrometer 颜色高温计
color pick-up tube 彩色摄像管
color purity coil 色纯度控制线圈
color removal 脱色
color signal 彩色信号
color steel plate 彩钢板
color steel tile 彩钢瓦
color strip 色带
color subcarrier 彩色副载波
color synchronizing burst 彩色同步脉冲群
color television 彩色电视
color temperature 比色温度，颜色温度，色温，示色温度
color trace tube 彩色显像管，色迹管
color transmission 彩色电视传送

colortron 荫罩式彩色电视机
color video stage 彩色视频放大级
color wall protection plate 彩色护墙板
color 色调
colotex board 隔音板
colour aerial photography 彩色航空摄影
colour band method 色带法
colour code band 色码带
colour code 色标
colour control 色度控制
coloured aluminium 彩色铝材
coloured art glass 彩色艺术玻璃
coloured cement 有色水泥，彩色水泥，着色水泥
coloured corrugated metal sheet 彩色波纹金属板
coloured drawing of pattern 彩绘
coloured filament of water 染色水丝
coloured map 彩色地图
coloured pigment 染料
coloured Portland cement 彩色硅酸盐水泥
coloured-sand experiment 染色砂试验
colour filter 滤色镜
colourimeter 色度计，比色计
colouring 涂色
colour mark 色标
colour mixture computer 混合物色度测定计算机
colour of measuring velocity method 染色测速法
colour play 色彩条纹
colour printer 彩色打印机
colour radiography 彩色射线照相术
colour unit 色度单位
colour 染色，着色，色彩［调］，颜色
Co. Ltd. 有限公司
column-adding routine 按列相加程序
column analogy method 似柱法，柱比法
column-and-panel wall 镶板式柱墙
columnar basalt 柱状玄武岩
columnar crystallization 柱状结晶
columnar crystal 柱状晶体
columnar deflection 柱变位
columnar ferrite 柱状铁素体
columnar fracture 柱状断口
columnar grain zone 柱状晶区
columnar ion exchange 塔式离子交换
columnar ionization 柱状电离
columnar joint 柱状节理
columnar pile 支承桩，柱桩
columnar placement method 柱状浇筑法
columnar section 柱状剖面，柱状图
columnar structure 柱状构造
columnar 柱状的，针状的
column aspect 柱形
column back-pressure 柱反压力
column baseplate 柱底板
column base 柱座，柱基，柱脚，柱基础
column beam 柱形梁
column binary card 竖式二进制卡片
column binary 竖式二进制数，竖式二进制码
column bracing 柱间支撑
column capacity factor 柱容量因素
column cap 柱头，柱顶，柱帽
column casing 柱筒
column chart 柱状图
column conditioning 柱老化
column configuration 柱构型
column count 行计算
column crane 塔式起重机，塔吊
column deflection 柱的纵向挠度
column diagram 直方图
column elution program 柱洗提程序
column extra 柱附加物
column filling 柱填充
column footing 柱基础，柱脚，柱基
column formula 压柱公式
column grid 柱网
column head 柱顶，柱头
columniation 列柱法，列柱式
column-jib crane 塔式旋臂起重机
column load 柱负荷，柱载重
column matrix 列矩阵
column moment 柱内力矩
column of built channels laced 槽钢缀合柱
column of pellets 芯块柱【燃料棒】
column of solid section 实心截面柱
column order 列的次序
column overload 柱过载
column parameter 柱参数
column performance 柱性能
column pile 端承桩，柱墩，柱桩，悬吊管，柱管
column plate 塔板
column printer 列打印机
column radiator 粗柱形散热器
column resistance 柱阻力
column rib 柱肋
column rising pipe 柱的立管
column scrubber 洗涤柱
column shaft 柱身
column space 柱距
column split 列的分开
column strip 柱列板带
column top 柱顶
column tray 塔板
column-type radiator 柱式暖气片
column vector 列向量
column volume 柱容积
column washer 洗涤柱
column with constant cross-section 等截面柱
column with lateral tie reinforcement 配有箍筋的柱
column with steel hooping 螺旋钢筋柱
column with variable cross-sections 变截面柱
column 柱，圆柱，柱塔，（表格的）栏，（矩阵的）列，栏目，柱状物
col 鞍形区，山坳，气压谷，垭口，山口
co-manager 联合经理
combating ageing 抗老化，防老化
combat pollution 消除污染，取缔污染
comb collector 梳形集电器
COMB(combustion) 燃烧
combed joint 马牙榫接
combe 峡谷，冲沟
comb filter 梳形滤波器，梳齿滤波器梳状滤波器

combi = combination 混合式，接合，联合，两用的
combiflow 混流
combimeter 多功能（电能）仪表
combi-motor 鼠笼-滑环组合式电动机
combinability 结合性
combination actuator 复合执行结构
combinational logic circuit 组合逻辑电路
combinational logic element 组合逻辑元件
combinational logic 组合逻辑
combinational switching network 联合开关网络
combination arch furnace 带复式拱的炉膛
combination arch 复式拱
combination automatic controller 组合自动控制器，复式自动控制器，综合自动控制器
combination baseplate 组合底座
combination beam 组合梁
combination bearing 组合轴承
combination burner 复合燃料器
combination cable 组合电缆
combination caliper 内外卡钳
combination circuit 组合电路
combination column 组合柱
combination construction 混合建筑
combination control 复合控制
combination crusher 复合式破碎机
combination cycle 联合循环
combination detector 组合探测器
combination drive 混合驱动
combination fixture 整套卫生设备
combination flow regulator 组合式流量调节阀
combination for action effects 作用效应组合
combination forbidden 禁止组合，禁用组合
combination for long-term action effects 长期效应组合
combination for short-term action effects 短期效应组合
combination frequency 复合频率，组合频率
combination furnace 组合式炉膛或燃烧室
combination holding tool 组合支架工具
combination hub 接插座
combination interlock relay 组合联锁继电器
combination lock 组合锁
combination motor 组合式电动机
combination of bids 联合投标
combination of coal on fuidized bed 煤的流化床燃烧
combination of design load 设计荷载的组合
combination of enterprise 企业合并
combination of slots 槽配合
combination of turbine cylinders 汽缸组合
combination operation 组合操作，联合运行
combination pliers 台钳，剪钳，钢丝钳，组合钳，鲤鱼钳
combination probe 组合测针
combination pump 复合泵，组合泵
combination ratchet spanner set 两用棘轮扳手套装
combination ratchet spanner 两用棘轮扳手
combination spanner extra long 加长型两用扳手
combination spanner set 两用扳手套装
combination standard gauge 万能塞规
combination starter 组合启动器【电动机接触器和熔丝】
combination structure 混合结构
combination switch 组合开关
combination swivel head wrench 两用活头扳手
combination turbine 混合式汽轮机
combination type fire detector 复合式火灾探测器
combination U and V groove UV 组合坡口
combination unit 组合单元【开关装置】
combination value of an action 作用组合值
combination value 组合值
combination valve spool 组合阀阀芯
combination waste-and-vent system 排水及通气合用系统
combination well 联合井
combination window 双侧窗，组合窗
combination woven 合成织物
combination wrench 组合扳手
combination 结合，联合，合并，化合，化合物，组合
combinatorial and scheduling problem 组合调度问题
combinatorial mathematics 组合数学
combinatory analysis 组合分析
combinatory logic 组合逻辑
combinator 配合操纵器【水轮机】，配合器，协联机构
combined action 复合作用，联合作用，综合作用，联合行动
combined agent extinguishing system 混合剂灭火系统
combined aggregate grading 混合骨料级配
combined available chlorine 化学性有效氯
combined available residual chlorine 化学性有效余氯
combined axial and radial labyrinth 轴向径向组合迷宫式汽封
combined bearing 联合轴承
combined bill 并单【提单】
combined blade 整体叶片
combined blowpipe for cutting and welding 焊割两用炬
combined boiling-superheat fuel assembly 沸腾过热燃料组合
combined bridge 两用桥【铁路，公路】，铁路公路两用桥
combined bypass 大旁路
combined centrifugal and impact separator 离心碰撞分离器
combined characteristic 综合特性，总特性
combined circle power plant 联合循环电厂
combined circulation boiler 复合循环锅炉【带有辅助循环泵的直流炉】
combined circulation 复合循环
combined city 联合城市，群集城市
combined closed-loop and once-through cooling （闭环式与直流式）混合式冷却
combined column footing 联合式柱脚
combined column 复合柱
combined compression and bending 弯曲受压

combined condensate return system 混合式凝结水再循环系统
combined conductivity 组合电导率
combined control of heat supply 供热联合调节
combined creep and fatigue 联合蠕变疲劳
combined creep 联合蠕变，综合蠕变
combined critical speed 轴系临界转速
combined crushing and screening plant 石料破碎筛分厂
combined cycle electric power generation 联合循环发电
combined cycle gas turbine power plant 燃气轮机联合循环电厂
combined cycle generating unit 联合循环发电机组
combined cycle operation 联合循环运行
combined cycle power generation 联合循环发电
combined cycle power plant 联合循环电厂，联合循环发电厂
combined cycle steam turbine 联合循环汽轮机
combined cycle 联合循环
combined cyclone and demister separator 旋风除雾分离器
combined debug 联调
combined deflection 总挠度，总变形
combined diesel and gas turbine 柴油机与燃气轮机联合装置
combined dipole 复合偶极子
combined discharge 混合排放
combined distribution frame 综合配线架
combined drainage system 合流排水系统
combined drain 综合排水
combined efficiency 综合效率，总效率，合成效率
combined emergency stop and governing valve 联合汽门
combined environmental test 组合环境试验
combined environment 综合环境，综合条件
combined error 综合误差
combined failure 复合破坏，复合故障
combined field 合成场
combined filter 结合滤波器
combined firing 混合燃烧
combined flow turbine 混流式透平
combined footing 联合式柱脚，联合式基础，联合基础
combined foundation 联合基础
combined gas and steam turbine cycle 燃气蒸汽机联合循环
combined gas-steam turbine plant 燃气蒸汽轮机联合装置
combined gas-steam turbine reheat cycle 再热燃气蒸汽联合循环
combined governing 综合调节
combined head 读写兼用头，组合磁头
combined heat and power generation 供热发电，热电联产
combined heat and power plant 热电（联产）厂
combined heat transfer coefficient 总传热系数
combined HP-IP casing 高中压合缸
combined humic acid 结合腐殖酸
combined fixed/sliding pressure operation 定压/滑压混合运行方式
combine dissipater with flaring pier 宽尾墩联合消能工
combined journal and thrust bearing 径向推力联合轴承
combined layout of piping system 管线综合布置
combined lighting 组合照明
combined load 综合荷载，混合载重
combined governing valve 联合式调节阀
combined main stop 联合式主汽
combined mechanical and electrical strength （绝缘子的）机电综合强度
combined method 组合方法
combined mica 复合云母
combined moisture 结合水
combined mooring and warping bollard 系泊牵引两用桩，码头带缆桩
combined nuclear and gas turbine power plant 核动力和燃气轮机联合发电厂
combined nuclear and gas turbine 核动力与燃气轮机联合装置
combined oil storage tank 组合油箱
combined pipe trench 综合管沟
combined pollutant index 复合污染指数
combined power 组合功率，混合功率
combined pressure and vacuum gauge 压力真空表
combined protection relay 综合保护继电器
combined punching and shearing machine 联合冲剪机
combined radial and axial bearing 径向-推力联合轴承
combined read-write head 组合读写头
combined reheat stop and intercept valve 再热联合汽阀，中压联合汽阀
combined reheat valve 再热联合汽阀，再热联合汽门
combined residual chlorination 化合性余氯氯化
combined resistance 合成电阻
combined ringing and speaking key 通话呼叫合用键
combined rotor 组合转子
combined rupture 复合断裂
combined scheme 混合式布置
combined seam locker for elbow 弯头联合咬口机
combined seam locker 联合咬口机
combined sewage system 合流污水系统，雨污水合流系统
combined sewage 混合污水，合流污水，组合污水
combined sewerage system 合流排水系统
combined sewer 合流下水道，合流污水道，雨污水合流下水道
combined shaft 组合轴
combined shipmen 混合装载
combined single-post strain tower 单柱组合耐张塔
combined sink well foundation 复合式沉井基础
combined sketch 拼接草图
combined starter 综合启动器，组合定子
combined steam and gas turbine cycle 蒸汽燃气

联合循环，蒸汽燃气轮机联合循环
combined steam and gas turbine 汽轮机与燃气轮机联合装置
combined steam-gas power plant 蒸汽燃气轮机联合循环发电厂
combined steam-power generation plant 热电联产电厂
combined steam-power generation 热电联产
combined steam valve 联合汽门
combined steel and concrete column 钢骨混凝土柱
combined steel & timber gate 钢木大门
combined stop emergency valve 快速关闭主汽门
combined strength 复合强度
combined stress 复合应力，合成应力，综合应力，组合应力
combined supercharged boiler and gas turbine cycle 增压锅炉燃气轮机联合循环
combined surge arrester 复合避雷器
combined switch 组合开关
combined system 混合系统
combined television-telephone service 电视电话联合服务
combined tension and bending 弯曲受拉
combined test 综合试验，联合试验
combined transportation bill of lading 联运提单
combined transportation 联运
combined treatment 联合处理
combined turbine 冲动反动联合式汽轮机
combined-type deaerator 复合型除氧器
combined unloading 混合卸载
combined valve 联合汽阀，联合汽门
combined voltage current transformer 仪表用变压变流器，电压电流组合互感器
combined voltage generator 交直流电压发电机
combined voltage variation 混合调压
combined wastewater overflow 混合的废水溢流
combined wastewater 混合废水
combined water treatment 综合水处理
combined water （煤中的）结合水，化合水，混合水
combined wave 合成波
combined 组合的，联合的，化合的，综合，组合
combiner 组合器
combine 抓毛【抹灰底层】，组合，联合，化合，结合，梳理
combining borer 综合成孔机
combining tube 管颈，混合管
combi stopper 双头螺旋
comb lightning arrester 梳形避雷器
comb pitot 梳状皮托管
comb pole 梳形磁极
comb-shaped transverse spreading adder 梳状横向扩展全加器
comb structure 蜂窝状结构，梳状构造
comb-tooth-type rotor 梳齿式转子
comb tooth 梳齿
comb type circuit 梳形电路
comburant 燃烧着的
combustibility index 燃烧指数
combustibility 可燃性，易燃性
combustibilty and rated fire-resisting period 可燃性和额定耐火时间
combustible analyzer 可燃物分析仪
combustible basis 可燃质，可燃基
combustible constituent 可燃物组分
combustible content 可燃物含量
combustible gas analyzer 可燃气体分析器
combustible gas detector 燃气检测器
combustible gas indicator 可燃气体指示器
combustible gas 可燃气体
combustible in refuse 残渣中可燃物
combustible loss 可燃物损失
combustible matter 可燃物
combustibleness 可燃性，易燃性
combustible rubbish 可燃垃圾，可燃废物
combustibles in fly ash 飞灰可燃物
combustible sulfur 可燃硫
combustible waste 可燃废物
combustible 可燃的，易燃的，可燃物
combustion 燃烧，氧化
combustion air 燃烧空气
combustion arch 后拱，燃烧拱
combustion block 燃烧区
combustion boat （实验室用）燃烧舟
combustion catalyst 燃烧催化剂
combustion chamber deposit 燃烧室积碳
combustion chamber draft 炉膛负压，炉膛拔风
combustion chamber hopper 冷灰斗
combustion chamber pressure loss 燃烧室压力损失
combustion chamber shape 燃烧室形状
combustion chamber soot blower 燃烧室吹灰器
combustion chamber space 燃烧室容积
combustion chamber superheater 燃烧室过热器
combustion chamber surface area 燃烧室表面积
combustion chamber 燃烧室，炉膛
combustion conditions 燃烧工况，燃烧条件
combustion control instrument 燃烧控制仪
combustion control system 燃烧控制系统
combustion control 燃烧控制，燃烧调节
combustion driven tube 燃气风洞，燃烧驱动管
combustion dust 燃烧尘埃
combustion efficiency 燃烧程度，燃尽程度，燃烧效率
combustion engineering standard reactor 燃烧工程公司标准堆
combustion engineering 燃烧工程，燃烧技术
combustion engine 内燃机
combustion equipment 燃烧设备
combustion fired MHD generator 燃烧式磁流体动力发电机
combustion flue 烟道，炉起道
combustion front 燃烧前沿
combustion furnace 燃烧炉膛
combustion gas duct 烟道
combustion gas turbine 燃气轮机
combustion gas 燃气
combustion hardware 燃烧设备
combustion header 燃气收集器，燃烧式集气管
combustion heat 燃烧热，燃烧热量，燃烧生成热量

combustion indicator 燃烧指示器
combustion intensity 燃烧热强度，燃烧强度
combustion kinetics 燃烧动力学
combustion-leach process 燃烧浸取法
combustion liner 燃烧（器）内筒
combustion load 燃烧负荷
combustion losses 燃烧损失
combustion mechanism 燃烧机理
combustion MHD generator 燃烧式磁流体发电机
combustion model 研究燃烧的模化设备
combustion monitor 燃烧监视器
combustion noise 燃烧噪声
combustion of gas and vapour 气体燃烧
combustion performance 燃烧性能，燃烧工况特性
combustion pressure 燃烧压力
combustion process 燃烧过程
combustion product gases 燃烧产物烟气
combustion products 燃烧产物
combustion property 燃烧特性，燃烧参数
combustion rate coefficient 燃烧率，燃烧速度系数
combustion rate 燃烧率，燃烧速率，燃烧速度
combustion region 燃烧区
combustion residue 燃烧残渣
combustion safeguard 燃烧（火焰）安全保护，燃烧安全设备
combustion shaft 燃烧室
combustion system design 燃烧系统设计
combustion system diagram 燃烧系统图
combustion system 燃烧系统
combustion temperature 燃烧温度
combustion train 燃烧装置
combustion turbine 燃气透平，燃气轮机
combustion velocity 燃烧速度
combustion with reduced pollutant 低污染燃烧
combustion zone 燃烧带，燃烧区
combustor efficiency 燃烧器效率
combustor 燃烧室，炉膛，燃烧器
comb 蜂窝【混凝土】，探针，梳，梳形测针【端子，插头】，电梳，螺纹梳刀，梳状物
come-along clamp for galvanized steel wire 地线卡线器
come-along clamp 卡线器，卡线钳
come-along tong 卡线器，紧线夹钳
come-along 紧线夹，伸线器，混凝土推平耙
come forward 自愿
come into effect 开始生效，开始实施，生效
come into force 付诸实施
come to an agreement 达成协议
come to terms with sb. 与某人达成协议
come to terms with sth. 终于接受，接受
come to terms 让步，妥协，达成协议
comfortable wind environment 舒适风环境
comfortable 舒适的
comfort air conditioning 舒适性空气调节，舒适性空调
comfort condition 舒适条件
comfort criteria 舒适判据
comfort factor 舒适因子
comfort index 舒适指数
comfort letter 安慰函，告慰信，信心保证书
comfort parameter 舒适参数
comfort requirement 舒适要求
comfort room 盥洗间［室］，公共厕所
comfort station 公共厕所
comfort zone 舒适区
co-milling 研磨机
coming 挡水缘围
command and communication fire vehicle 通信指挥消防车
command and control 命令和控制
command button 指令按钮
command code 操作码
command control program 命令控制程序
command control 指令控制
command decoder 指令译码器
command destruct 销毁用的命令控制，销毁命令
command direction 命令方向
command generator 指令发生器
command guidance system 指令制导系统
commanding apparatus 操纵设备
commanding impulse 指令脉冲
commanding point 调度站
command interpreter 命令解释程序
command language 命令语言
command logic 指令逻辑
command mode 命令方式
command pulse 命令脉冲
command receiver of remote control 遥控指令接收器
command register 命令寄存器，指令寄存器
command relay circuitry 指令继电控制线路
command resolution 指令分解
command retrieval system 命令检索系统
command sequence 指令序列
command signal 指令信号
command string interpreter 命令串解释程序
command 指挥，命令，指令，操作，控制，控制值，需求
commencement date of work 开工日期
commencement date 开工日（期）
commencement of commercial operation 开始商业运行，正式运行
commencement of operation 运行开始
commencement of works 工程开工，动工，开工
commencement time 开工时间
commencement 开始
commendation and penalization of employee 职工奖惩
commensurability 公度性，同量
commensurate reduction 同等减缩，按比例缩小
commensurate with 与……相应，与……相当，相称，同量
commensurate （在时间和空间上）相等的，相称的，相当的，匹配的，同等的，成比例的，同量的
commensuration 同量，相当的，通约
comments on or approval of the feasibility study report 可研的批复，可行性研究报告的批准或意见
comment 注释，评论，意见，批注，注解

commerce 商业
commercial 商业的
commercial acid 商品酸，工业用酸
commercial agent 商务代表
commercial agreement 商业协议
commercial ancillary service 有偿辅助服务
commercial article 商务条款
commercial bank 商业银行
commercial bid evaluation 商务评标
commercial bid tabulation 商务报价对比表
commercial bolt 普通螺栓
commercial call 商用电话
commercial center 商业中心
commercial city 商业城市
commercial clause 商务条款
commercial coal 商品煤
commercial computer 商业用计算机
commercial conditions 商务条件，商务条款
commercial contract 商务合同，商业合同
commercial counsellor 商务参赞
commercial credit 商业信用，商业信用证
commercial data processing 商业数据处理
commercial energy resource 商品能源
commercial event 商业活动，商务事件
commercial factor 商务因素
commercial frequency 市电频率，供电频率，工业用电频率
commercial grade coal 商品煤等级
commercial grade 商品级
commercial harbour 商港
commercial instrument 商业票据
commercial insurance 商业保险
commercial invoice 商业发票
commercialization 商品化，商业化
commercial L/C agreement 开发信用证约定书
commercial letter of credit 商业信用证
commercial load 商业负荷，企业负荷
commercial loan 商业贷款
commercially available 市场上可以买的
commercially dry sludge 商业干污泥
commercially pure 工业纯
commercial manufacture 工业生产，工业制造
commercial motor 商用电动机，工业用电动机
commercial operation data 商业运行日期
commercial operation 商业运行，投产
commercial paper 商业票据
commercial part 商务部分
commercial personnel 商务人员
commercial plant 工业设备
commercial port 通商港口
commercial power reactor 商业性动力反应堆
commercial power 市电，工业用电
commercial practice 商业惯例
commercial process 工业化生产过程
commercial proposal 商务文件，商务报价
commercial radio station 商业无线电台
commercial reactor 商用反应堆，工业反应堆
commercial representative 商务代表
commercial run 工业过程，工业方法
commercial securities 商业证券
commercial situation 商业状况
commercial size 工业规模
commercial terms and conditions 商务条款
commercial terms 商务条款
commercial test 工业试验，商用试验
commercial ties 商业关系
commercial water consumption 商业耗水量
commingler 混合器
commingle with 与……混合，掺和
commingling 混合，掺和
comminute 粉碎，磨碎
comminuted refuse 粉碎的垃圾
comminuted solids 粉碎的固体
comminuting machine 污物粉碎机
comminuting screen 粉碎筛
comminution 粉碎
comminutor 粉碎机，造粒机，粉碎器
commission agent 代理商
commission charges 手续费
commissioner-general 首席专员
commissioner 官员，专员
commission fee 手续费，代理费，经纪佣金
commissioning check list 调试检验清单
commissioning date 投产日期，运行日期
commissioning engineer 调试工程师
commissioning of fluidized bed boiler 流化床锅炉调试
commissioning of individual equipment and system 分部试运
commissioning of individual equipment 单机试运
commissioning of individual system 分系统试运
commissioning period 试运行期
commissioning program 调试程序
commissioning stage 调试阶段，试投产阶段
commissioning supervisor 调试监理
commissioning test 投产试验，投运试验，交接试验，调试，试运行
commissioning 投产运行，试运行，调试，交工试运转，投入运行，交付使用，试运转
commission of inquiry 调查委员会
commission of investigation 调查委员会
commissions or other remuneration 佣金或报酬
commission 佣金，试运行，命令，委员会，手续费，调试，交付使用，委任，委托
commissural 合缝处
commissure 结合处，合缝处，接合点
commitment and involvement 信托与介入
commitment authority 承诺（授）权
commitment charge 义务承担费，承诺费
commitment document 承诺文件
commitment fee 承诺费，承约费，承担费
commitment implementation 实行
commitment letter 委托书，承诺书，承诺函
commitment period 贷款有效期
commitments 委托，承担的义务
commitment value 约定价值，承担价值
commitment 承诺，保证，委托，承担义务，承诺款项，承担，约定，承担额，义务
committed cost 约束成本【即约束性固定成本】，已承诺费用
committed dose equivalent 待积剂量当量
committed effective dose equivalent 待积有效剂

量当量
committed fixed cost 固定承诺成本
committed load 承担负荷
committed to doing sth. 致力于做某事
committee 委员会
committee of inspection 检查委员会
committee of the whole 全体委员会
committee of ways and means 筹款委员会，财政委员会
Committee on Data for Science and Technology 科学技术数据委员会【荷兰】
committee room 会议室，委员会办公室
commit 把……托付给，保证【做某事、遵守协议或遵从安排等】，承诺，使……承担义务，犯罪
commix 混合物，混合
commode 五斗柜，洗脸台
commodities fair 商品交易会
commodity code 商品编码
commodity economy 商品经济
commodity exchange 商品交易所
commodity inspection fee 商检费
commodity inspection 商品检验
commodity supply point 商品供应点
commodity 货物，商品
common alarm contact 公用报警接点
common alarm 总告警
common area of memory 存储器公用区
common auxiliaries 公用辅助系统【双机组共用】
common bargaining 共同议价
common bar 普通炉条
common baseplate type 共座式
common base 共基极，共用基座
common battery system 共电制，中央电池制
common battery telephone set 共电式电话机
common battery 共电制电池，中央电池组
common bond 普通砌砖法
common breakdown 常见故障
common business-oriented language 面向商业的公用语言
common bus system 共汇流排制，公共母线制，共母线制
common bus 总汇流条，公共母线
common cause analysis 共同原因分析
common cause failure 共因故障
common-channel signaling terminal 公用通道信号终端
common-channel signalling 公用信道信令
common chassis 公用底盘
common circuit 公用回路
common collector 共集电极
common control system 公用控制系统
common control unit 公用控制部件，公用控制单元
common control 公共控制
common cost 公共费用
common data class attribute name 公用数据类属性名
common data class name space 公用数据类名称空间
common data class 公用数据类
common datum plane 通用基面
common declaration statement 公用说明语句
common denominator 公分母
common depth point shooting 共深点爆炸
common diagram system 集中接线系统
common difference 公差
common divisor 公约数
common earthing system 共用接地系统
common emitter 共发射极
common excavation 普通开挖
common facilities 公用设施，公共设施
common hazard 公害
common header 共用集箱，主管道，母管
common information model 公共信息模型
common ink 普通炉排片，从动炉排片
common interest 双方的利益
common joist 龙骨
common language family 公用语言类
common logarithm 常用对数
commonly 普通地，普遍地
common measure 公用量度
common mode crosstalk 共模式失真，共模串扰，共态串话
common mode failure 共模故障，共同模式故障，常见故障
common mode feedback amplifier 共模反馈放大器
common mode interference 共模干扰
common mode noise 共模噪声，共态噪声
common mode rejection ratio 共态减弱系数，共态抑制比，共模抑制比
common mode rejection 共模抑制
common mode signal 共模信号
common mode voltage 通用电压，共模电压
common mode 常规方式，共模
common natural resources 公有自然资源
common neutral 公共中性点，公共中性线
common panel 公用盘
common plenum 集箱，联箱
common point 公共点
common pole 公用电杆
common power supply 公用电源
common ram 手夯
common reactance 共用电抗
common return 共同回路，公共回线
common seal 公章
common section 共同段
common sense rule 常识性规定
common sewer 公用污水管道，公用下水道
common stack 总烟囱
common statement 公用语句
common stay 公用拉线
common storage 公用存储区，公用存储器，共用库容
common switchboard 公用配电盘
common system 常规系统，公用系统，共用系统
common-tower double-circuit line 同杆双回线路
common transformer 公用变压器，公用变
common transmission service tariff 共用网络服务

价格
common trunk　共用母线
common type expansion joint　通用型膨胀节
common version　通用形式
common voltage　常用电压，普通电压
common winding　公共线圈
common wiper　共用弧刷，共用接帚
common wire　公共导线，中性线
common　普通的，公共的，公用的
commotion　电震，扰动，动摇
communicable　可传播的，能传递的，传染性的
communicate with　与……联系，与……交往，与……相通
communicate　通信，交通，传达，表达，沟通，交流，清晰地揭示，传染，扩散
communicating canal　通航渠道
communicating pipe　连通管
communicating tube　连通管
communicating vessel　连通器
communication aerials　通信天线
communication and control　通信及控制
communication and transportation　交通运输
communication and transport　交通运输
communication apparatus　通信设备
communication area　通道面积
communication band　通信频带
communication building　通信楼
communication bus translator　通信总线转换器
communication cable　通信电缆
communication center　通信中心，通信枢纽
communication channel　通信信道
communication circuit　通信电路
communication connection　通信连接
communication control system　通信控制系统
communication countermeasures　通信的反干扰
communication cycle　通信周期
communication department　通信分场
communication effectiveness　交流效果
communication electronics　通信电子学
communication engineering　通信工程学
communication equipment　通信设备
communication failure　通信故障
communication identification　通信标识
communication ID　通信标识
communication input/output control system　通信输入/输出控制系统
communication insulation　讯路绝缘
communication interrupt rate　通信中断率
communication line adapter　通信线路转接器
communication line　通信线路，通信线
communication link　通信联络【电话机】
communication module　通信模件，通信模式
communication monitoring　通信监听
communication network　通信网络
communication net　通信网
communication planning　通信设计
communication protocol stack　通信协议堆栈
communication protocol　通信协议，通信规约
communication receiver　通信接收机
communication satellite　通信卫星
communications port　通信口
communication stack　通信协议栈，通信栈
communication subsystem　通信子系统
communication system grounding　通信系统接地
communication system in wind farm　风力发电场内通信系统
communication system　通信系统，通信设备
communication terminal module controller　通信终端模块控制器，通信终端模件控制器
communication terminal module　通信终端模件，通信终端模块
communication theory　通信理论
communication transfer　通信传输
communication tube　连通管
communication vector table　通信向量表
communication zone　通信区域
communication　交通，交通工具，联络，通信，传达，交流，通讯，消息，传递，交际
community air　周围环境
community antenna television　公用天线电视
community atmosphere　城市大气，居民区大气
community automatic exchange　区内自动电话局
community center　生活区
community development plan　社区发展规划
community facilities　社区设施
community involvement　公众参与
community noise　城市噪声
community planning　社区规划
community pollution　城市污染，居民区污染
community relation representative　社会关系代表
community　社团，界，团体，共同体，社区，公众
commutated coil　换向线圈
commutated error signal　转接误差信号
commutated potentiometer servo　带有换向电势计的伺服系统
commutate　换向，转换，转接，交换，整流
commutating brush　换向电刷
commutating capacitor　加速电容器，整流电容器，换向电容器
commutating characteristics　换向特性
commutating circuit　换向电路，整流回路
commutating coil　换向线圈
commutating condenser　整流电容器，换向电容器，加速电容器
commutating current　换向电路，整流电路
commutating device　整流装置，换向装置
commutating electromotive force　换向电动势，整流电动势
commutating field resistance　换向极磁场电阻
commutating field winding　换向极绕组
commutating field　换向场，附加极磁场
commutating flux　换向磁通
commutating group　换向组
commutating impedance　换向阻抗
commutating inductance　换向电感
commutating machine　整流子电机，换向式电机
commutating number　换向数
commutating period　换向时间，换向周期
commutating pole winding　换向极绕组
commutating pole　整流极，换向极，辅助极
commutating ratio　整流比

commutating reactance 换向电抗
commutating reactor 换向电抗器，整流电抗器，整流扼流圈
commutating resistance 换向电阻
commutating speed 换向速度，整流速度
commutating voltage 换向电压
commutating watthour meter 整流式瓦时计，整流式电度表
commutating winding 换向绕组，换向线圈，换向极绕组，整流极绕组
commutating zone 换向区
commutation angle 换向角
commutation arc 换向电弧
commutation capacity 换向能力
commutation coil 换向线圈，换向极线圈
commutation condenser 换向电容器
commutation condition 换向状况
commutation current 换向电流
commutation curve 换向曲线
commutation cycle 换向周期，整流周期
commutation element 换向元件
commutation factor 整流系数，换向系数
commutation failure 换相失败，整流破坏
commutation interference 换向干扰
commutation interval 整流间隔，换向间隔
commutation limit 换向极限
commutation loss 换向损失
commutation oscillation 整流振荡
commutation period 换向期间，换向周期
commutation phenomena 换向现象
commutation switch 换向开关，转换开关
commutation turnoff 转换开关
commutation voltage 换向电压
commutation winding 换向绕组
commutation zone 换向区，换向带
commutation 交换，转换，整流，换向，换相，配电
commutative law 对易律
commutative property 交换性
commutative ring 可换环，交换环
commutative 交换的
commutator access opening 换向器视察窗
commutator armature 带换向器的电枢
commutator bar pitch 换向片节距
commutator bar slot 换向片槽
commutator bar 换向片，整流条
commutator bore 换向器孔径
commutator brush combination 换向器电刷总线
commutator brushing 换向器套筒
commutator brush track diameter 换向器外径
commutator brush 换向器电刷
commutator change over switch 换向器，切换开关
commutator core extension 换向器结构件伸出段
commutator core 换向器固定件装配
commutator end 换向器端
commutator face 换向器轴向长度，换向器面
commutator frequency changer 整流子频率变换器，换向器式变频机
commutator generator 换向器式发电机，整流子式发电机
commutator grinder 换向器研磨机
commutator grinding 换向器研磨
commutator hood 换向器罩
commutator hub 换向器套筒
commutator induction motor 换向器式感应电动机
commutator insulating hole 换向器视察窗
commutator insulating ring 换向器绝缘环
commutator insulating segment 换向器绝缘片
commutator insulating tube 换向器绝缘筒
commutator insulation 换向器绝缘
commutatorless DC motor 无换向器式直流电动机
commutatorless excitation 无换向器励磁
commutatorless generator 无换向器式发电机，无整流子式发电机
commutatorless motor 无换向器电动机
commutator loss 换向器损耗
commutator lug 换向器接线片，换向器升高片
commutator motor meter 换向器电动机式仪表
commutator motor 换向器式电动机，整流子式电动机
commutator noise 换向器噪声
commutator phase advancer 换向器式进相机
commutator phase shifter 换向器式移相器
commutator pitch segment 换向片
commutator pitch 换向器节距
commutator pulse 定时脉冲
commutator rectifier 换向整流器
commutator ring 换向器环，换向器套筒
commutator ripple 换向器脉动电压，换向器波纹
commutator riser 换向器升高片，整流子竖片
commutator sector 换向片
commutator segment assembly 换向片装配
commutator segment 换向片
commutator shaft 换向器轴
commutator shell insulation 换向器套筒绝缘
commutator shell 换向器套筒
commutator shield 换向器护罩
commutator shrink ring 换向器压圈，换向器箍环
commutator sleeve 换向器套筒
commutator spider 换向器支架
commutator strip 换向器片
commutator surface 换向面，换向器表面
commutator switch 按序切换开关，扫描转换开关
commutator tag 换向器升高片
commutator transformer 换向器式变流器
commutator tube 转换器套筒，电子射线转换器
commutator type generator 换向器式发电机
commutator veering 换向器V形压圈，换向器V形环
commutator wear 换向器磨损
commutator 换向器，整流器，转向器，整流子，转换开关，集电环
common cause outage occurrence 共因停运事件
COMPAC(computer program for automatic control) 自动控制用的计算机程序
compact automatic retrieval device 小型自动搜寻装置

compact battery 小型电池，紧装电池
compact boiler 快装锅炉
compact conductor configuration 紧凑型导线排列
compact disc read only memory 小型磁盘只读存储器
compact disk 光碟［盘］
compact district 密集区
compacted backfill 夯实回填土，碾压填土，压实回填，压实回填土
compacted clay 夯实黏土
compacted concrete 捣实混凝土
compacted density 夯实密度，压实密度
compacted depth 夯实深度，压实深度
compacted-earth lining 夯土衬砌，压实土衬砌
compacted embankment 压实路堤
compacted fill 碾实填土，压实回填土，压实填土
compacted grains 密集颗粒
compacted graphite iron 蠕墨铸铁
compacted layer 压实层
compacted lift 压实层厚
compacted lime 夯实灰土
compacted pervious fill 压实的透水性填土
compacted rockfill 压实的堆石
compacted sand layer 压实砂层
compacted soil 夯实土，压实回填土，压实土
compacted thickness 夯实厚度，压实厚度
compacted volume 压实体积
compacted 夯实的，紧凑的，密实的，压实的
compact embankment 密实路堤
compact fold 致密褶皱
compact-grain structure 致密晶粒结构，密纹组织，细粒组织
compactibility 紧密度，可压实性，密实性
compactible waste 可压缩废物
compacting effect 压实效果，压实作用
compacting energy 压实能量
compacting equipment 压实设备
compacting factor 压实系数
compacting press 模压机，成型压制机，打包机，密实压制机
compacting 压紧，压实，打夯，夯实
compact intelligent substation 紧凑型箱式变电站
compaction ability 压实能力
compaction by driving 夯实
compaction by explosion 爆炸密实法
compaction by layers 分层填土夯实
compaction by rolling 滚碾压实，碾实，碾压
compaction by vibrating roller 振动碾压法
compaction by vibration 振动密实，振动压实
compaction by watering 注水密实
compaction coefficient 压实系数
compaction curve 击实曲线
compaction degree 压实度
compaction equipment 碾压设备，压实设备
compaction in layers 分层夯实，分层压实
compaction method 碾压方法，压实方法
compaction pile 挤密桩，密实桩
compaction rate 密实程度，压密率
compaction ratio 压缩比
compaction settlement 压密沉陷
compaction test apparatus 击实仪
compaction test of backfill soil 回填土压实试验
compaction test 夯实试验，击实试验，压密试验，压实试验
compaction 压实，压紧，压缩，堆实，夯实，压密
compactive effort 击实功，压实功能
compact layer 压脂层
compact limestone 致密灰岩
compactly structured 结构紧凑地
compact machine 轻便电机，紧凑型电机
compact material 密实材料
compactness test 压实度试验
compactness of soil 土壤密度
compactness 密集（性），紧密（性），紧凑（性），密实（性），致密（性），密实度，小巧
compactor passes 压夯遍数，碾压遍数，击实遍数【夯压机】
compactor 压实机，压土机，夯土机，夯具，压实工具，压缩机
compact power transmission line 紧凑型输电线路
compact reactor 紧凑反应堆
compactron 小型电子管
compact simulator 紧凑性仿真机
compact-stranded wire 压缩多股绞线
compact structure 密实结构
compact substation 紧凑型变电站，箱式变电站
compact system 紧凑系统
compact temperature transmitter 一体化温度变送器
compact tension specimen 小型拉伸样品
compact texture 致密结构
compact tower 紧凑型铁塔
compact transmission line 紧凑型线路
compact type reactor 紧凑反应堆
compact type valve 紧凑型阀门，小型阀门
compact 小型的，紧密的，致密的，紧凑的，压紧，压实，使紧凑，契约，合同，协定
compages 综合结构
compandor 展缩器
companion fault 副断层
companion flange 对接法兰
companion specimens 同组试样
company act 公司法，公司条例
company contract 公司合同
company director 公司董事
company dissolution 公司解散
company law 公司法
company limited by share 股份有限公司
company of limited liability 有限公司
company of unlimited liability 无限公司
company performance 公司业绩
company profile 公司概况
company promoter 公司发起人
company property 公司财产
company standard 企业标准
company strategy 公司策略
company 公司
comparable price 可比价格
comparable 可比的
comparablility 可比性

comparasion method calibration　比较法校准
comparative advantage　相对优势
comparative analysis method　比较分析法
comparative analysis　比较分析，对比分析
comparative chart　比较图
comparative cost　比价，比较造价
comparative design　比较设计
comparative efficiency　相对效率
comparative evaluation　比较评价
comparative lifetime　相对寿命
comparative measurement　比较测量
comparative observation　比较观测
comparative risk assessment　相对风险评定，相对风险评估
comparative risk　相对危险度
comparative study　对照研究，对比研究，比较研究
comparative temperature　对比温度
comparative test block　对比试块
comparator buffer　比较器缓冲器，比较缓冲器
comparator　比测器，比较仪，比较器，比较电路，场强器，比长仪
compare data　比较数据
comparer　比较装置，比较器
comparing element　比较元件
comparison amplifier　比较放大器
comparison basis　比较基础
comparison bridge　比较电桥
comparison calibration　比较检定
comparison circuit　比较电路
comparison coder　比较编码器
comparison curve　比较曲线
comparison data　对比数据，对比特性，对比参数
comparison element　比较元件
comparison lamp　比较灯
comparison measurer　比较器，比值器
comparison method　比较法
comparison of bid price　投标价比较
comparison of bids for final decision　选标比价
comparison of data　数据比较
comparison of design　设计比较
comparison of fin materials　肋材料的比较
comparison of schemes　方案比较
comparison oscillator　比较示波器
comparison oscilloscope　比较示波器
comparison post-mortem　比较检错程序
comparison potentiometer　比较电位计，比较电势计
comparison sheet　比较表
comparison table　比较表
comparison test　比较试验，比较试验法
comparison transformer　标准变压器，比较用变压器
comparison　对比性［度］，比较，对比，对照
compartment cover　封闭罩
compartmented air box　分割式风室
compartment mode　库室模式
compartment pressure　风室压力
compartment shielding　分段型屏蔽
compartment　室，间，间隔，分隔，分区，炉排间隔，分隔间，隔板，隔舱，水密舱
compass card　罗盘标度板，罗盘面
compass circle　罗盘分度圈
compass error　罗盘误差
compasses　圆规
compass needle　罗盘针
compass of proportion　比例规
compass reading　罗盘读数
compass receiver　罗盘接收机
compass roof　半圆形屋顶，跨形屋顶
compass rose　罗盘仪记录盘
compass-theodolite　罗盘经纬仪
compass timber　弯曲木材
compass traverse　罗盘仪导线
compass　指南针，罗盘（仪），圆规
compatibility condition　相容条件
compatibility margin　兼容裕量
compatibility of deformation　变形协调
compatibility technique　兼容技术
compatibility　兼容性，一致性，相容性，适应性，兼容，调和
compatible color television system　黑白彩色电视兼容制
compatible computer　兼容机
compatible event　相容的事件
compatible hardware　兼容硬件
compatible IC　兼容集成电路
compatible integrated circuit　相容集成电路
compatible reception　兼容接收
compatible single sideband　兼容单边带
compatible software　兼容软件
compatible time-sharing system　兼容分时系统
compatible with　与……兼容
compatible　相容的，协调的，可共存的，适合的，一致的
COMP(complete)　完成
COMP(compressed air)　压缩空气
compelling force　强制力
compendium　纲要
compensable accident　可给赔偿的事故
compensate　补偿，校正
compensate control　补偿控制
compensated　补偿的，被补偿（的）
compensated aneroid　补偿气压表
compensated attenuator　补偿衰减器
compensated commutator motor　带补偿的换向式电动机
compensated cross-field exciter　带补偿的横轴磁场励磁机
compensated current transformer　补偿电流互感器
compensated dynamo　补偿发电机，有补偿绕组的发电机
compensated excitation system　补偿励磁系统
compensated flow control valve　压力补偿式流量控制阀
compensated foundation　补偿式基础
compensated induction motor　补偿感应电动机，有补偿的感应电动机
compensated instrument transformer　补偿仪用互感器

compensated ion chamber lead wire 补偿电离室引线
compensated ion chamber 补偿电离室
compensated ionization chamber 补偿电离室
compensated line 补偿线路
compensated log 补偿测井
compensated loop 补偿环
compensated motor 带补偿绕组的电动机，补偿电动机
compensated pendulum 补偿摆
compensated pilot wire protection system 补偿引线保护系统
compensated pyrheliometer 补偿式直接日射表
compensated regulated 补偿调节
compensated regulator 补偿调整器
compensated relief valve 平衡式溢流阀
compensated reluctance motor 补偿的磁阻式电动机，带补偿的磁阻式电动机
compensated repulsion motor 补偿推斥式电动机
compensated semiconductor 补偿半导体，抵偿半导体
compensated series motor 补偿串激电动机
compensated voltmeter 补偿式伏特计
compensated wattmeter 补偿式瓦特表
compensate for the lack of experience 弥补经验不足
compensate sb. for sth. 因……补偿某人
compensating accumulator 补偿累加器
compensating agreement 补偿协定
compensating air valve 补偿气阀，平衡汽阀
compensating ampere-turns 补偿安匝数
compensating box 补偿器【热电偶】
compensating buffer 平衡缓冲器
compensating cable 均衡电缆，补偿电缆
compensating capacitor 补偿电容器
compensating circuit 补偿电路
compensating coil 补偿线圈
compensating computation 平差计算
compensating condenser 补偿电容器
compensating conductor 补偿导线，均衡导线
compensating current 补偿电流
compensating cylinder 补偿缸
compensating delay 补偿延迟
compensating device 补偿装置
compensating diaphragm 补偿十字线片
compensating effect 补偿作用
compensating element 补偿元件
compensating equipment 补偿设备
compensating error 补偿误差
compensating feedback 补偿反馈
compensating feedforward 补偿前馈
compensating feed stoker 连续给煤层燃炉
compensating field winding 补偿磁场绕组
compensating field 补偿磁场
compensating filter 补偿滤波器
compensating flow 补偿水流
compensating gear 补偿装置，差动齿轮装置
compensating jet 补偿喷嘴
compensating lead wire 补偿导线，补偿引线
compensating lead 补偿线
compensating line 热膨胀补偿器，膨胀伸缩弯头，膨胀节
compensating magnet 补偿磁铁
compensating network 补偿网络，补偿回路
compensating operation 调相运行
compensating pipe 调整管，补偿管，平衡管，伸缩管
compensating pole 补偿极，换向极，附加极
compensating reservoir 平衡水库，补偿调节水库
compensating resistance 补偿电阻
compensating resistor 补偿电阻器
compensating rope 平衡索
compensating runoff regulation 补偿径流调节
compensating shunt 补偿分流器
compensating sight 补偿瞄准器
compensating surface current 补偿表层流
compensating weight 平衡重块，配重
compensating winding 补偿绕组
compensating 补偿，修正
compensation adjustment 补偿调整
compensation ampere-turns 补偿安匝
compensation balance 补偿平衡
compensation claim 赔偿要求
compensation coil 补偿线圈
compensation current 补偿电流
compensation curve 补偿曲线
compensation depth 补偿深度
compensation diaphragm 调压薄膜
compensation effect 补偿作用
compensation equipment 补偿设备，补偿装置
compensation expenses 赔偿费用
compensation factor 补偿系数，补偿因子，修正系数
compensation fee 补偿费
compensation for contract violation 违约金
compensation for damages 损失补偿，损失赔偿
compensation for demolition 拆迁补偿费
compensation for injuries and damages 人员伤亡和设备损坏赔偿
compensation for removal 迁移补偿费，搬迁费
compensation for temporary land occupation 临时占地补偿
compensation grade 折减坡度
compensation joint 补强接头，调整缝
compensation method 补偿法，零反应性法
compensation of cross talk 串话补偿
compensation of distortion 畸变的补偿
compensation of excitation system 励磁系统的补偿
compensation of geomagnetic field 地磁场补偿
compensation of power factor 功率因数的补偿
compensation of thermal expansion 热补偿
compensation ratio 补偿比率
compensation reactor 补偿电抗器
compensation reservoir 补偿水库
compensation terms 补偿条款
compensation theorem 补偿定理
compensation tower 平衡塔
compensation trade agreement 补偿贸易协议
compensation trade contract 补偿贸易合同
compensation valve 平衡阀，补偿阀

compensation water 补偿水，补偿波
compensation water outlet works 补偿水出水工程
compensation winding 补偿绕组
compensation 补偿，校正，补偿作用，软反馈装置，报酬，赔偿（金）
compensative regulation 补偿调节
compensator balancer 补偿平衡器，自耦调压器
compensator control 补偿控制
compensator for pressure regulator 压力调节器用补偿器
compensator for thermal expansion 热膨胀补偿器
compensator starter 补偿启动器，自耦变压器
compensator starting 补偿启动，自耦变压器启动
compensator transformer 升压变压器，补偿变压器，自耦变压器
compensator winding 补偿器绕组，调相机绕组
compensator without stator 无定子调相机
compensatory damages 应予赔偿的损失
compensatory duty 补偿性关税
compensatory lead 补偿引线
compensatory payment 补偿报酬
compensatory tension 补偿张拉
compensatory 补偿的，补偿性的
compensator 补偿器，差动装置，膨胀接头，膨胀圈
comperssed air system 压缩空气系统
competence evaluation 竞争力评价，能力评估，胜任力评价
competence of the wind 风运能力
competence 能力，胜任
competent authorities 主管当局，主管机关，主管部门，主管机构
competent bed 强岩层
competent fold 强褶皱
competent personnel 胜任的人员
competent person 合法人，胜任的人员
competent rock 强岩
competent velocity 起动流速
competent 有能力的，有资格的，适当的，充足的，授权的，称职的，有决定权的
compete with each other 相互竞争
competing reaction 竞争反应
competition energy 竞争电量
competition 竞争
competitive advantage 竞争优势
competitive-bid contract 竞争性投标合同
competitive bidding system 竞争（性）招标制，招标制
competitive bidding 竞争性招标，竞争出价，竞标
competitive-bid 竞标的，招标的，比价的，有竞争性投标的
competitive capacity 竞争能力
competitive consciousness 竞争意识
competitive decay 竞争衰变
competitive demand 竞争需求
competitive design 竞争设计
competitive dialogue procedure 竞争性对话程序【用于欧盟国家政府采购】
competitive dialog 竞争性对话【用于欧盟国家政府采购】
competitive economy 竞争经济
competitive edge 竞争优势
competitive fee 竞争费用
competitive force 竞争实力
competitive group of enterprises 竞争性企业集团
competitive investment 竞争投资
competitive market 竞争市场
competitive mechanism 竞争机制
competitiveness 竞争力，竞争能力，竞争性
competitive price 投标价，具有竞争性的价格，具有吸引力的价格，竞争性价格，竞争价格，有竞争力的报价，公开招标价格，可竞争价格
competitive product 精品，竞争产品，拳头产品
competitive relation 竞争关系
competitive selection 竞争性选择
competitive strategy 竞争策略
competitive tender 公开投标，竞争投标
competitive 竞争的，（价格等）有竞争力的，（指人）好竞争的
competitor 竞争对手，竞争者
compfil 补偿滤波器
compilation of budget 编制预算
compilation of hydro-logical data 水文资料整编
compilation of plan 计划编制
compilation 编译，编制
compile bidding documents 编制投标文件
compile bid invitation documents 编制招标文件
compiler-level language 编译语言
compiler system 编译系统
compiler 程序编制器，编译程序，编码程序，编辑（人），编译器，汇编者
compiling program 编译程序
compiling routine 编译程序
compiling 编排
complaint record 投诉记录
complaints and claims 申诉与索赔
complaint 申诉，投诉
complanation 变成平面
complemental 互补的，补充的
complement and carry add circuit 补码和进位加法电路
complementarity between A and B A和B互补
complementarity 互补性
complementary agreement 补充协议
complementary angle 补角，余角
complementary code 补码
complementary education 辅助教育
complementary energy method 余能法
complementary energy 余能
complementary enterprise 互补企业
complementary error function 补余误差函数
complementary field 附加场，辅助场
complementary function 互补函数，余函数
complementary injunction transistor 互补单结晶体管
complementary logic switch 互补逻辑开关
complementary logic 互补逻辑
complementary metal-oxide-semiconductor 互补型金属-氧化物-半导体

complementary network 附加网络
complementary nuclear training 附加核训练
complementary product 补充产品
complementary resistor-diode-transistor logic 互补电阻-二极管-晶体管逻辑
complementary set 余集，补集
complementary shutdown system 备用停堆系统
complementary strain energy 应变余能
complementary symmetry circuit 互补对称电路
complementary trade 补偿贸易
complementary transistor logic 互补晶体管逻辑
complementary transistor-resistor logic 互补晶体管-电阻逻辑电路
complementary transistor-transistor logic 互补的晶体管-晶体管逻辑
complementary wave 补偿波，余波
complementary 互余的，互补的，附加的，补充的【在主要事物之外追加的】
complementation 补码法，补数法
complement code 补码
complement form 补码形式
complement gate 互补门
complement instruction 补码指令
complement number system 补码数制
complement pulse 补码脉冲
complement representation 补码表示
complement with respect to 10 10的补数
complement 补充【原来不足或有损失的】，补足（物），余数［角］，配套，补码，互补品，补数
complete absorption 完全吸收
complete a certificate 完成一套证明书
complete alternation 整周周期，全循环
complete analysis of water 水全分析，水质全分析
complete analysis 全分析
complete assembly 整个组件
complete attenuation 全衰减
complete audit 全部审计
complete automatic 全自动的，全自动
complete auxiliaries 配套辅机
complete blade 整叶片
complete break 完全断裂
complete bypass 完全旁通，完全旁路
complete calculation 全套计算
complete carry 完全进位
complete characteristic 全特性，综合特性
complete circuit 闭合电路，全路
complete closure 完全关闭
complete combustion 完全燃烧
complete commutation 完全换向
complete compression 完全压缩
complete consolidation 完全合并
complete contraction 完全收缩
complete controllability 完全可控性，完全能控性
complete cross-section of weld metal 焊接金属的总截面
complete cycle 全循环，完整循环，全周
complete data 完整资料
complete demineralisation plant （水的）全化学除盐装置
complete design document 完整设计文件
completed item 建成投产项目
complete diversion 完全分水
complete double-ended severance of a reactor coolant pipe 反应堆冷却剂管道双端完全断裂
completed product verification 成品验证
completed project 竣工项目
complete equipment 成套设备
complete facilities 成套设施
complete failure 完全失效，整体故障，完全破坏
complete fiber optic maintenance toolkit 成套光缆维修工具箱
complete flagging 完全旗状【Ⅳ级植物风力指示】
complete flood control 防御最大可能洪水
complete flow 全流，总流量
complete fusion 完全熔化，熔透
complete gasification 完全气化
complete induction 完全归纳法
complete investigation 全面调查
complete knocked down 全散件组装
complete line 整套线路
complete loss of coolant 冷却剂完全流失
complete lubrication 完全润滑全自动操作系统
completely circular structure 整圆结构
completely hydrated cement 完全水化水泥
completely mesh-connected circuit 完全网接电路
completely orthogonal 完全正交的
completely penetrating well 完全井
completely reducible operator 完全可约算子
completely reserved stress 周期性交变应力
completely reversed fatigue limit 全交变疲劳极限
completely reversed stress 周期性交变应力
completely self-protecting transformer 全自保护变压器
completely stabilized conductor 完全稳定导体
completely water-cooled furnace 全水冷壁炉膛【锅炉】
completely watertight 滴水不漏
complete mixing system 完全混合系统
complete mixing 全混
complete nappe 完整水舌
completeness inspection 完整性检查
completeness 完整性
complete open-ended severance of the primary pipe 一回路管道完全开口断裂
complete opening 完全开启
complete operation 完整操作，完整运算
complete outage state 完全停运状态
complete overhaul 整体检修，全面检修，大修
complete penetration butt weld 贯穿对焊
complete penetration 透焊，完全透过区域
complete period 全周期
complete piping system 完整管道系统
complete plant shutdown 全厂停堆，全厂停运，全厂停产
complete plant 成套设备
complete power package 成套发电机组

complete project 成套工程
complete quadratic combination 完全二次项平方根组合
complete radiation 全辐射
complete removal 全部拆除【核电站退役第一阶段】
complete resonance 全谐振
complete retention 完全存留【放射性废物】
complete rupture 全部断裂，完全断裂，完整程序
complete schematic 总线路图，总原理图
complete set for DC power supply 直流电源成套装置
complete set of drawings 整套图纸
complete set of equipment 成套设备
complete set 全套，全组
complete shutdown 全关
complete spare parts 全套备件
complete stall 气流完全分离，完全滞止，完全失速
complete stress relief technique 应力全部释放技术
complete system 完整系统
complete termination of contract 全部终止合同
complete tool 整套工具
complete treatment of sewage 污水完全处理
complete treatment plant （污水）完全处理装置
complete turbulence 完全紊流
complete verification 全部核实，全部核查，全面鉴定
complete water analysis 水的全分析
complete well 完整井
complete 完成，竣工，成套，全部，完全的
completion acceptance 竣工验收
completion by stages 分期竣工
completion ceremony 竣工典礼
completion certificate 完工证明书，竣工证明，竣工证书
completion date 竣工日期，完工日期
completion documents 竣工文件
completion drawing 竣工图
completion of contract 完成合同
completion of erection release 安装完工证书
completion of project 竣工
completion of works 工程竣工
completion report 竣工报告，完工报告
completion risk 完工风险
completion test 竣工试验
completion time 竣工时间
completion 完成，结束，竣工，完工
complex admittance 复数导纳，导纳复量
complex amplitude 复数振幅，复数幅值，复值幅，复振幅
complex analysis 复分析，复变函数论
complex anion 络阴离子
complex apparent permeability 复视在磁导率，复视在导磁系数
complex argument 复自变量，复自变数
complexation 络合
complex automatic control system 综合自动控制系统
complex belt-conveyor 复合皮带运输机
complex building 综合楼
complex capacitivity 复电容率，复介电常数
complex cation 络阳离子
complex column packing 复合填充柱
complex compliance 弹性变形
complex compound 络合物
complex conductivity 复电导率
complex conduit 复合水道
complex conjugate function 复共轭函数
complex conjugate matrix 复共轭矩阵
complex conjugate 复共轭
complex control system 综合控制系统
complex coupling 复耦合，感容耦合
complex cover 复合覆盖
complex curve 复合曲线
complex cycle gas turbine 复杂循环燃气轮机
complex cycle 复杂循环
complex damping 复合阻尼
complex data type 复数数据形式
complex declaration statement 复数说明语句
complex dielectric constant 复介电常数，复介电系数
complex dielectric permittivity 复介电常数
complex differentiation 复变微分
complex display 复合显示器
complex dryer 组合式干燥器
complex effect 复合效应
complex eigenvalue 复特征值，复本征值
complex elasticity 复弹性
complex equivalent impedance 复数等值阻抗，等值复阻抗
complex fault calculation 复杂故障计算
complex fault 复断层
complex feedback 复反馈
complex fold 复褶皱
complex formulation 复数表示法
complex form 复数形式
complex foundation 复杂地基
complex Fourier series 复合傅里叶级数
complex frequency plane 复频面
complex frequency response 复频反应特性，复频响应
complex frequency spectrum 复频谱
complex frequency 复频率，复合频率
complex function chip 多功能集成块
complex function 复函数，复变函数，复数值函数
complex harmonic oscillation 复杂谐波振荡
complex harmonic voltage 相量电压，复谐波电压
complex impedance 复数阻抗，复阻抗
complexing agent 复合试剂，配位剂
complexing 络合
complex instruction set computer 复杂指令集计算机
complexity analysis 成分分析
complexity 错综性，复杂性
complex modes of operation 复杂运行方式
complex multiplication 复数乘法
complex multiplier register 复数乘数寄存器

complex nitride 复合氮化物
complex nucleus 复杂核
complex number 复数
complexometric titration 配位滴定，配位滴定法
complexometry 配位滴定法
complex operation 复数运算
complex operator 复数算子
complexor admittance 复量导纳
complexor impedance 复量阻抗
complex orthogonal 复数正交
complex oscillation constant 复振荡常数
complex oscillation 复杂振荡
complex periodic 复周期的
complex permeability 复导磁率，复导磁系数
complex permittivity 复电容率，复介质常数，复介电系数
complex pipeline 复合管道
complex plane 复平面
complex pole 复极
complex potential 复电势，复电位，复位势
complex power series 复数幂级数
complex power 复数功率，相量功率
complex product 复数乘积
complex quantity 复数量，复量
complex reaction 复杂反应
complex representation 复数表示法
complex resonance 复谐振，复共振
complex root 复根
complex sample 复式采样
complex setting 复合地形布置
complex sinusoidal current 复正弦电流
complex sinusoidal quantity 复正弦量，正弦复数
complex spectrum 复谱
complex steel 合金钢，多元钢
complex strain wave 复合应变波
complex stress 复合应力，综合应力
complex supercharger 复式增压器
complex surface 复合地面
complex system 综合系统
complex terrain 复杂的地形，复杂地形，复杂地形带
complex topography 复杂地形
complex truss 复式桁架
complex utility routine 综合性服务程序
complex utilization 综合利用
complex variable 复变量，复变数
complex vector space 复矢量空间
complex vector 复矢量
complex wave 复波
complex 复合的，合成的，复杂的，复数，综合体，联合企业，复合，联合体，总厂
compliance assurance 确保遵守的措施，遵守措施
compliance certificate 合格证（书）
compliance coefficient 符合系数
compliance constant 符合常数
compliance control 合格控制
compliance fuel 可塑性燃料
compliance voltage 恒流输出电压
compliance with contract 遵守合同
compliance with requirements 符合要求
compliance 符合，顺从，依从，可塑性，听从，柔软量，柔度，柔量
complicated pole 复式杆
complication 复杂
comply with an agreement 履行合同
comply with formalities 按规定办理手续，履行手续
comply with terms of contract 按合同条款办理，符合合同条件
comply with the conventions 随俗
comply with the terms of convention 遵守公约的条款
comply with 遵从，服从，照办，依从，符合，遵照，顺应
COMPNET(compensating net) 补偿网络
compole circuit 换向极磁路，换向极电路
compole core 换向极铁心
compole winding 换向极绕组
compole 间极，换向极，整流极，附加磁极，极间极
component acceptance test 元件验收测试
component analysis 成分分析
component approach 设备承包方式，多合同承包方式
component bridge 三用电桥
component cooling filter 设备冷却水过滤器
component cooling heat exchanger 设备冷却水热交换器，设备冷却系统换热器
component cooling loop demineralized water 设备冷却水回路除盐水
component cooling loop pump 设备冷却水回路泵
component cooling loop 设备冷却水回路
component cooling pump 设备冷却水泵
component cooling room 辅助设备冷却水房
component cooling surge tank 设备冷却波动水箱
component cooling system 设备冷却系统
component cooling water heat exchanger 设备冷却水热交换器
component cooling water surge tank 设备冷却水波动水箱，设备冷却水波动箱
component cooling water system 设备冷却水系统
component cooling water 设备冷却水
component drawing 零件图
component element 元件，单元，组件，构件，环节
component event data bank 部件故障数据库
component failure 部件失效
component flow test loop 部件流动试验回路【美国】
component force 分力
component frequency 部分频率，组成频率
component generator 分量发生器
component handling and cleaning facility 设备处理及清洗装置
component important to safety 安全重要部件，对安全重要的部件
component object model 部件对象模型
component of force 分力，力分量
component of pump 泵部件
component of turbulence 湍流分量
component of vector 矢量分量

component of velocity 速度分量，分速度
component parts 附件，零件，部件，元件
component reactance 电抗分量
component specification 部件规格
component test facility 部件检验[试验]设施
component test 组件试验，部件试验
component tide 子潮
component vector 矢量的分量
component wash cell 设备冲洗室
component wire （电缆的）芯线
component 部件，元件，构件，组成部分，成分，分量，分力，组件，部分，零件，设备，装备，组成的，构成的
composed of 由……组成，包括
composertron 综合磁带录音机
compose 组成，构成，谱曲
composite action 复合作用
composite apparatus 组合电器
composite arch 复合拱
composite beam 叠合梁，组合梁
composite blade 复合材料叶片
composite board 合成板
composite boiler 组合式锅炉
composite break-water 混合式防波堤
composite brickwork 组合砌体
composite bridge 组合桥
composite bushing 复合套管
composite cable 混合多心电缆，复合电缆
composite characteristic 合成特性曲线，组合特性
composite circuit 复合电路，混合电路
composite column 混合式的柱子，组合柱
composite component 复合部件，复合元件
composite conductor 组合导线，复合导体
composite cone 复合火山锥
composite configuration 混合配置，组合布局
composite construction 复合材料结构，混合结构，组合结构
composite contact 复合触头
composite controlling voltage 复合控制电压
composite cooling 复式冷却
composite core structure 堆芯复合结构
composite core 混合堆芯，组合铁芯
composite cross section 复式横断面
composite crusher 复合式破碎机
composite cycle 混合循环
composite dam 混合式土石坝，土石混合坝
composite determinant 合成行列式
composite-dielectric capacitor 复合介质电容器
composite dike 复合岩脉，混合式堤
composite double glazing 复合双层玻璃窗
composite drainage system 复式排水系统，混合式排水系统，组合排水系统
composite drawing 综合图，复合图
composite earth dam 混合式土坝
composite electrical insulation 合成电绝缘，复合电绝缘
composite electrode 混合焊条
composite enterprise 综合性企业
composite error 综合误差，合成误差，总和误差
composite excitation 混合励磁，混合激磁
composite filter 复式反滤层，复合滤波器
composite firing 混烧
composite float 复式浮标
composite fold 复合褶皱
composite force 合力
composite forecast chart 综合预报图
composite foundation pile 复合基桩
composite foundation 复合地基
composite function 复合函数
composite generation and transmission system 发输电系统
composite geometrical section 复式几何断面
composite hologram 复合全息图
composite hydro graph 复合水文过程线
composite insulate barrier 复合保温屏障
composite insulating tape 合成绝缘带
composite insulation 复合绝缘，合成绝缘
composite insulator 合成绝缘子，复合绝缘子
composite joint 混合连接
composite laminate structure 复合夹层结构
composite layout drawing 综合布置图
composite layout 综合布置
composite lens 复合透镜
composite line 复合线路，混成线路
composite list 综合清单
composite map 综合图
composite material 组合材料，复合材料，合成材料，混合材
composite member 组合杆，组合件
composite membrane 复合膜
composite metal 复合金属
composite method 综合方法
composite of heat flow diagram 热力系统的组成
composite optical fibre ground wire 地线复合光纤
composite particles 复合颗粒
composite performance curve 综合特性曲线
composite permanent magnet 复合永磁体
composite phasor 合成相量
composite pile 组合桩
composite plume 复合羽流
composite power rate 综合电价
composite prognostic chart 综合预报图
composite rate of depreciation 混合折旧率
composite rating chart 特性汇总表，综合特性图
composite resistor 混合电阻器
composite revetment 混合护坡
composite rock 复合岩[石]
composite rotor 组合式转子
composite roughness 合成粗糙度，混合糙率
composite sampler 综合取样器
composite sample 组合试样，并合试样，混合样，混合试样
composite set 收发两用机，组合装置
composite signal 复合信号，混合信号
composite sliding surface 复合滑动面
composite slope 复式边坡
composite starter 综合启动器
composite stranded wire 组合多股绞线，复合多股绞线，复合股线

composite stress-strain relation 复合材料应力应变关系
composite structure 组合结构，复合结构
composite subgrade 复合地基
composite surface of sliding 复式滑动面
composite synchronizing signal 复合同步信号
composite system 综合系统，组合系统
composite test signal 复合测试信号
composite test 组合试验
composite text 合成文本
composite tone 合成音，复合音，混音
composite treaty 复合条约
composite truss of wood and steel 钢木组合房架
composite truss 组合桁架
composite tube 复合管
composite twisted wire 复合绞线
composite type rock-fill dam 混合式堆石坝
composite unit graph 复合单位过程线，复式单位过程线
composite unit hydrograph 复合单位（水文）过程线
composite valve at high pressure heater inlet 高压加热器入口联成阀
composite valve at high pressure heater outlet 高压加热器出口联成阀
composite vector 合成矢量
composite viewing 综合观察
composite waste sampling 混合废物采样
composite wastewater sample 混合废水水样
composite waste 混合废水，混合废物
composite wave filter 复式滤波器，集合滤波器
composite wave 复合波，合成波，综合波
composite winding 合成绕组
composite wire 复合线，双金属丝
composite working 收发混合运行，电报电话混合运行
composite 复合材料【复数】，复合物，合成物，混合物，复合的，合成的，综合的，使合成，使混合
compositional analysis 成分分析
composition backing 焊接垫板
composition deviation transmitter 成分偏差变送器
composition gradient 浓度梯度
composition material 合成材料
composition metal 合金
composition of cast concrete 铸混凝土的成分
composition of concurrent forces 汇交力系的合成
composition of conductors 导线的组成
composition of couples 力偶合成
composition of electricity consumption 电力消费构成
composition of electric power industry 电力工业构成
composition of energy for electricity generation 发电能源构成
composition of forces in plane 平面力系的合成
composition of forces 力的合成
composition of parallel forces 平行力系合成
composition of phasors 相量的合成
composition of scale 水垢成分
composition of sewage water 污水组成
composition of sewage 污水成分
composition of vectors 矢量的合成，矢量合成
composition of velocities 速度合成
composition of washed water of blast furnace 高炉废水的组成
composition of water 水的成分
composition 组分，成分，化合物，合成，组成，编制，焊剂，混合物，组合
compost 混合涂料，灰泥
compound adjustment 多级调整
compound air compressor 多缸压气机
compound alluvial fan 复合冲积扇形
compound annual interest 年复利
compound arch 复合拱
compound augmentation 复合强化法
compound bearing 组合轴承
compound bending 双向弯曲
compound bevel 复合坡口
compound boiler 复合式锅炉
compound bowed and twisted blade 复合弯扭叶片
compound breakwater 混合式防波堤
compound brush 复合电刷，金属碳混合电刷，铜碳电刷
compound bushing 复合绝缘套管，充填绝缘物套管
compound cable 混合多心电缆，分段组合电缆
compound characteristic 复合特性
compound circuit 复合电路，复激电路
compound coil 复绕线圈
compound compression 复式压缩
compound compressor 多级压气机
compound control action 复合控制作用
compound control 复合控制
compound cross-section 复合截面，复式断面
compound-curved surface 复合曲面，混合曲面
compound curve 复曲线
compound cycle 复合循环
compound cylinder 双层缸，多段组合汽缸
compound dredger 复式挖泥机
compound duty 复合关税
compound dynamo 复励电机，复励式发电机
compounded governor 组合式调速器
compounded monthly 按月计算复利
compounded oil 含添加剂的油
compounded 复励，混合的，复合
compound elastic scattering 复合弹性散射
compound excitation 复式励磁，复励，复激
compound exciter 复激式励磁机
compound fault 复合断层
compound field 复激磁场，复励磁场
compound filled bushing 绝缘膏填充套管，填充绝缘化合物套管
compound filled 浇注绝缘膏的，绝缘膏填充的
compound flap 组合襟翼
compound function 复合函数
compound gasket 组合垫
compound gauge 压力真空两用表，联成计，复合规
compound generator 复励发电机，复激发电机

compound gneiss 混合片麻岩
compound harmonic function 复谐函数，复调和函数
compound hydrograph 复合（水文）过程线
compound impulse turbine 多缸冲动式汽轮机
compound inelastic scattering 复合非弹性散射
compounding coil 复激线圈，复绕线圈
compounding curve of generator 发电机综合特性曲线
compounding effect 复激作用
compounding lightning arrester 复式避雷器
compounding 配料，配方，串联连接
compound instruction 复合指令
compound insulation 复合绝缘
compound interest amortization method 复利摊销法
compound interest factor 复利因子
compound interest 复利，复利息
compound lateral scale 复横向尺度
compound lean blade 复合弯扭叶片
compound lever 复杆
compound light 复光
compound locomotive 复式机车
compound magnet 复合磁铁
compound malfunction 多重故障
compound manometer 复式压力计
compound meter 复合水表
compound mica 复合云母
compound motion 复合运动
compound motor 复励电动机，复激电动机
compound nuclear reaction 复合核反应
compound nucleus 中间核，复核
compound oscillation 复合振动
compound parabolic concentrating collector 复合抛物面集热器
compound parabolic concentrator 复合抛物面聚光器
compound pendulum 复摆
compound pipeline 复合管道
compound pipe 复式管道【有分叉的】，复合管线
compound pressure and vacuum gauge 复合型压力真空计
compound pressure gauge 复式压力计
compound pulley 复滑车
compound pump 多缸泵，复式泵，复式水泵
compound reaction turbine 多缸反动式汽轮机
compound rectifier excitation 复合整流器式励磁
compound reinforcement 复式配筋
compound relaxation instability 组合张弛不稳定性
compound relay 复合继电器
compound resonator 复合谐振器
compound ripple 复式波痕
compound river bed 复式河槽［河床］
compound section 组合截面
compound semiconductor solar cell 化合物半导体太阳电池，化合物半导体太阳能电池
compound semiconductor 化合物半导体
compound shoreline 合成海岸
compound signal 复合信号
compound specific activity 化合物比活度
compound stage expansion 两级膨胀
compound steam turbine 复式汽轮机
compound stress 复合应力，综合应力，合成应力，组合应力
compound system doubling time 综合系统加倍时间
compound tide 复潮，复合潮，混合潮，组合潮
compound transformer 复绕式变压器
compound truss 复式桁架，合成桁架
compound tube 复合管，组合管
compound turbine 复式透平，复式涡轮机，多缸汽轮机
compound voltage test of insulation 绝缘复合电压试验
compound water meter 复式水表
compound web plate 组合式腹板
compound weir 复式断面堰
compound well 组合井
compound winding 复合绕组，复励绕组，混合绕组
compound wire 复合线
compound-wound dynamo 复激发电机，复励发电机
compound-wound generator 复激发电机，复励发电机
compound-wound motor 复绕电动机，复励电动机
compound-wound relay 复励继电器，多线圈继电器，复绕继电器
compound-wound 复励式，复绕式，复励
compound 复合物，化合物，混合物，配合，混合，合成材料，场地，（筑有围墙的）院子，复绕的，复励的，复合的，多功能的，妥协，和解，达成协议，（通过妥协）解决（债务、主权等）
compo 水泥砂浆，灰泥，混合涂料
COMPR(compression ratio) 压缩比
COMPR(compressor) 压缩机
comprehend 理解
comprehensive agreement 全面协定
comprehensive all-embracing agreement 综合总协定
comprehensive analysis 全面综合分析，综合分析
comprehensive and integrated study 全面综合研究
comprehensive audit 全面监查
comprehensive balance 综合平衡
comprehensive benefits 综合效益
comprehensive building 综合楼
comprehensive comparative analysis 综合比较分析
comprehensive data base 综合数据库
comprehensive development plan 全面发展计划
comprehensive development water resources 水资源综合开发
comprehensive development 综合开发
comprehensive economic entity 综合经济实体
comprehensive emergency management 综合应急管理
comprehensive energy consumption 综合能耗

comprehensive index 综合指数
comprehensive injury 综合危害
comprehensive insurance 综合险，综合保险
comprehensive investigation 综合调查
comprehensive liability insurance 综合责任保险
comprehensive mandatory sanction 全面的强制性制裁
comprehensive negotiating program 全面谈判方案
comprehensive planning program 全面规划方案
comprehensive planning 全面规划，综合规划
comprehensive plan 整体计划，综合计划
comprehensive price 综合电价
comprehensive program 全面计划
comprehensive proof 综合证明
comprehensive pump house 综合泵房
comprehensive review 全面审查
comprehensive river basin planning 综合流域规划
comprehensive sanction 全面制裁
comprehensive schedule 综合进度
comprehensive scientific and technical capabilities 科技综合能力
comprehensive study 综合研究
comprehensive summary 综合摘要
comprehensive survey 综合调查，综合考察
comprehensive tariff 综合电价
comprehensive test 全面试验
comprehensive treatment 综合治理
comprehensive ultrasonic examination record 超声波检查综合记录
comprehensive utilization approaches 综合利用途径
comprehensive utilization of ash and slag 灰渣综合利用
comprehensive utilization 综合利用
comprehensive water pump house 综合水泵房
comprehensive 综合的，理解的，有理解力的
compress 压缩，紧缩，碾压
compress-air-operated 气动的
compressed 被压缩的
compressed air accumulator 压缩空气储存器
compressed air bearing 压缩空气轴承
compressed air bottle 压缩空气瓶，气罐
compressed air breaker 压缩空气断路器
compressed air caisson 压气沉箱
compressed air capacitor 压缩空气电容器
compressed air circuit breaker 压缩空气断路器
compressed air drill 风钻
compressed air ejector 压气喷射器
compressed air energy storage equipment 压缩空气蓄能装置
compressed air energy storage 压缩空气储能
compressed air engine 气压发动机
compressed air hammer 压缩气锤
compressed air manometer 压缩空气式压力表
compressed air method of tunneling 压缩空气开挖隧洞法
compressed air pipe line 压缩空气管线
compressed air pipe 压缩空气管
compressed air piping 压缩空气管道
compressed air plant 空压机
compressed air reservoir 压缩空气储气罐，压缩空气储气筒
compressed air starter 压缩空气启动机
compressed air station 压缩空气站
compressed air storage power generation 空气蓄能发电
compressed air storage 压缩空气储能
compressed air system 压缩空气系统
compressed air tamper 风动夯，压气式夯具
compressed air tunnel 压缩空气风洞
compressed air winch 风动绞车
compressed air wind tunnel 增压风洞，压缩空气风洞
compressed air 压缩空气
compressed bar 受压杆
compressed gas insulated cable 压气绝缘电缆
compressed gas insulation 压缩气体绝缘
compressed iron core coil 压粉铁芯线圈
compressed iron powder core 压粉铁芯
compressed-nitrogen condenser 压缩氮气电容器
compressed-powder source 压制粉末放射源
compressed state 压缩状态
compressibility coefficient 压缩性系数，压缩系数
compressibility correction 压缩性修正
compressibility degree 可压度
compressibility effect 压缩性效应
compressibility factor 压缩系数，压缩因子
compressibility influence 压缩效应
compressibility modulus 压缩系数
compressibility of soil 土壤压缩性
compressibility stall 激波失速
compressibility 可压缩性，压缩系数，压缩性
compressible aerodynamics 可压缩空气动力学
compressible airflow 可压缩气流
compressible cascade flow 可压缩叶栅流动
compressible flow theory 可压缩流理论
compressible flow 可压缩流动，可压缩流
compressible fluid flow 可压缩流体流动
compressible fluid 可压缩的液体，可压缩流体
compressible foundation 压缩性地基
compressible jet 可压缩射流
compressible medium 可压缩介质
compressible MHD (magnetohydro-dynamical) flow 可压缩磁流体流
compressible pressure gradient 可压流压力梯度
compressible soil 可压缩土，压缩性土
compressible stratum 可压缩层
compressible turbulence 可压湍流，可压紊流
compressible 可压缩性，可压缩的
compression air pipe 压缩空气管
compression air 压缩空气
compressional-dilational wave 疏密波
compressional wave 压缩波
compression bar 受压杆
compression cable 压力充气电缆
compression chamber arrester 压缩室避雷器
compression chamber 压力室
compression chiller 压缩式冷水机组
compression chord 受压弦杆

compression coefficient 压缩系数
compression cube strength 立方抗体压强度
compression curve 压缩曲线
compression deformation 压缩变形
compression degree 压缩度
compression dehumidifier 压缩减湿装置
compression diagonal 受压斜杆
compression-dilatation wave 胀缩波
compression efficiency 压缩比，压缩系数，压缩效率
compression factor 压缩系数
compression failure 压缩破坏，受压破坏
compression fiber 受压纤维
compression fitting 压合接头，压紧配合，加压装配
compression flange 受压翼缘
compression force 压缩力，压力
compression fracture 受压破坏，受压破裂，压缩断裂
compression gauge 压力表
compression gland 填料压盖
compression ignition engine 柴油机，压燃式（柴油）发动机
compression index 压缩指数
compression intensity 压力强度
compression joint 压接
compression layer 受压层
compression limit 压缩极限
compression line 压缩线，压缩曲线
compression load 压缩负荷，压缩载荷
compression loss 压缩损耗
compression manometer 压缩式压力计，压缩式真空压强计
compression member 受压构件，压杆
compression moulding 压力成型
compression of signal 信号压缩
compression parallel to grain 顺纹压力，顺纹压缩
compression perpendicular to grain 横纹压缩
compression range 受压翼缘，受压范围［区域］
compression rate 压缩速率
compression rating （增益）压缩比
compression ratio 压缩比，压缩率
compression refrigeration 压缩式制冷
compression refrigerator cycle 压缩式制冷循环
compression reinforcement 受压钢筋
compression-release valve 减压阀，去压阀
compression relief cam 调压凸轮
compression relief cock 减压旋塞
compression ring 加压环，密封环
compression shock 压力冲击，压缩激波，压力振动
compression sleeve 承压套筒
compression splice 受压拼接，压接
compression steel 受压钢筋
compression strength 抗压强度，耐压强度
compression stress 压应力
compression stroke 压缩冲程
compression strut 受压支撑
compression system 压缩系统
compression temperature 压缩温度
compression terminal 压缩端子，压线鼻子
compression testing machine 压力试验机
compression testing （塔的）下压试验，压缩试验，抗压试验，受压试验，压力试验，抗压强度试验，压紧试验
compression tool 压力钳
compression trajectory 压应力轨迹线
compression tube fitting 压缩管装配
compression type hydrant 压紧式消防龙头，压力式消火栓
compression type refrigerating machine 压缩式水冷却器
compression type refrigeration cycle 压缩式制冷循环
compression type refrigeration 压缩式制冷
compression type terminal 压接型端子
compression type water chiller 压缩式冷水机组
compression variable capacitor 压缩型可变电容器
compression wave （土中的）压缩波，激波
compression zone 压力区，压缩区
compression 压力，压缩，压实，挤压，紧缩，浓缩
compressive creep strain 压缩蠕变应变
compressive cylinder strength 圆柱体抗压强度
compressive deformation 受压变形，压缩变形
compressive force 压力，压缩力，压力荷载，压缩载荷
compressive plastic strain 压塑性应变
compressive region 受压范围，受压区
compressive reinforcement 受压钢筋
compressive strain 受压应变，压应变，压缩应变
compressive strength at 28 day-age 28 天龄期抗压强度
compressive strength 抗压强度，压缩强度，压缩力
compressive stress 压力，压（缩）应力
compressive test 压缩试验
compressive wave 压缩波
compressive work 压缩功
compressive 压缩的
compressometer 压缩计，压缩仪
compressor 压气机，压缩机，气泵
compressor blade 压气机叶片
compressor bleed 从压气机中抽气
compressor building 空气压缩机房
compressor capacity 压气机容量
compressor car 空压机车
compressor cascade 压气机叶
compressor cleaning equipment 压缩机清洁装置
compressor delivery pressure 压气机出口压力
compressor delivery temperature 压气机出口温度
compressor disc(/disk) 压缩机轮盘
compressor discharge pressure 压气机排气压力
compressor driven tunnel computational 压气机驱动隧道计算
compressor-driving turbine 驱动压气机涡轮机
compressor efficiency 压气机效率
compressor-exhauster unit 压气机排气装置

compressor flow 压气机空气流量
compressor governor 压气机调节器
compressor house 空压机房，压气机房
compressor housing 压气机气缸
compressor impeller 压气机叶轮
compressor inlet 压气机进口
compressor input 压缩机输入功率
compressor intake anti-icing system 压缩机入口防结冰装置
compressor intake pressure 压气机进口压力
compressor map 压气机特性线图
compressor motor 压缩机用电动机
compressor plant 空气压缩机房
compressor pressure ratio 压气机增压比
compressor room 压气机室
compressor rotor 压缩机转子
compressor stage 压气机级
compressor stall 压气机失速
compressor station 空压机站，压缩机站
compressor surge control 压气机喘振控制
compressor surge limit 压气机喘振极限
compressor surge line 压气机喘振线
compressor surge 压气机喘振
compressor turbine 带压缩机透平
compressor unit 压缩机组
compressor vane 压气机叶片
compressor washing equipment 压缩机清洗装置
compressor washing 压气机清洗
compressor work 压气机功
compress the bandwidth 压缩频带宽度
comprise 包括，包含，由……组成
compromise clause 妥协条款
compromised total loss 约定全损
compromise joint 异形接头
compromise network 折中网络
compromise settlement 协商解决
compromise 折中法，综合平衡，折中办法，折中，妥协，和解，让步
Compton effect 康普顿效应
Compton electrometer 康普顿静电计
Compton electron 康普顿电子
Compton recoil 康普顿反冲
Compton scattering 康普顿散射
Compton wavelength 康普顿波长
comptroller 会计检查员，审计长，审计员
compulsion 强制
compulsory acquisition 强制获取，强制征用，强制收购
compulsory arbitration 强制仲裁，强制性仲裁
compulsory clause 强制条款
compulsory conciliation 强制和解，强制调解
compulsory course 必修课程
compulsory insurance 强制保险
compulsory land acquisition （强制性）土地征用
compulsory measures 强迫手段，强制措施
compulsory sanction 强制性制裁
compulsory settlement 强制解决
compulsory 强迫的，强制的，必须做的，规定的，义务的，强制性的，必修课
computability 可计算性
computable 可计算的，计算得出的
computational 计算的
computational aerodynamic 计算空气动力学
computational mesh 计算网格
computational method 计算方法
computational scheme 计算方案，计算线路，计算框图
computation centre 计算中心
computation chart 计算图表
computation module 计算模型，计算组件
computation of error 误差的计算
computation of HVDC power transmission system 高压直流输电系统计算
computation sheet 计算表格，计算速度
computation time 计算时间
computation 计算，运算
computative 计算的
computator 计算机，计算员，计算装置
computatron 计算机用多极电子管
computed-aided design 计算机辅助设计
computed-aided reliability program 计算机辅助可靠性分析程序
computed pulse shape 被计算的脉冲波形
computed result 计算结果
computed tomography 计算机控制层析 X 射线摄影术
computed 计算的，使用计算机的
computer 计算器，计算机
computer-aided circuit design 计算机辅助电路设计
computer-aided design and drafting 计算机辅助设计制图
computer-aided design and engineering 计算机辅助设计与工程
computer-aided design 计算机辅助设计
computer-aided instruction 计算机辅助教学
computer-aided management of instruction 指令的计算机辅助管理
computer-aided manufacturing 计算机辅助制造
computer-aided mathematics analysis 计算机辅助数学分析
computer-aided measurement and control 计算机辅助测量和控制系统
computer-aided network design 计算机辅助网络设计
computer-aided plant control system 计算机辅助的电厂控制系统
computer-aided process planning 计算机辅助工艺规划
computer-aided programming 计算机辅助程序设计
computer-aided reliability 计算机辅助可靠性鉴定
computer-aided simulation exercise 计算机辅助仿真培训
computer-aided site management 计算机辅助现场管理
computer-aided system control 计算机辅助控制系统
computer-aided system 计算机辅助系统
computer-aided test 计算机辅助试验，计算机辅助测试

computer-aided 计算机辅助的
computer analyst 计算机分析师
computer and computing hardware 计算机和计算硬件
computer and data processing system 计算机及数据处理系统
computer animation 计算机动画显示
computer application for measurement and control 测量及控制的计算机应用
computer architecture 计算机体系结构，计算机功能结构
computer-assisted learning 计算机辅助训练
computer-assisted reliability and maintainability simulation 计算机辅助可靠性和可维修性模拟
computer-assisted training 计算机辅助训练
computer-assisted 计算机辅助的
computerate 精通电脑的
computer auto-manual station 计算机自动-手动操作站
computer automated measurement and control 计算机自动测量及控制
computer automatic test system 计算机自动测试系统
computer bar code 计算机用条纹码
computer-based system control 计算机控制系统
computer-based 利用计算机的，借助计算机的
computer block diagram 计算机框图
computer board 计算机面板
computer cabinet 计算机柜
computer capacity 计算机能力，计算范围
computer carving machine 电脑雕刻机
computer center 计算机中心
computer-chronograph 计时计算机，计算机测时计
computer circuit 计算机回路，计算机电路
computer coating thickness gauge 电脑涂层测厚仪
computer code 计算机代码，计算机编码，计算机程序
computer compatibility 计算机兼容性，计算机相容性
computer control counter 计算机控制计数器
computer-controlled automatic test equipment 计算机控制的自动测试设备
computer-controlled machine tool 由计算机控制的机床
computer-controlled power station 计算机控制的电站
computer-controlled reactor 计算机控制的反应堆
computer-controlled system 计算机控制的系统
computer-controlled telephone exchange 程控电话交换机
computer-controlled traffic signal system 计算机控制的交通信号系统
computer-controlled 计算机控制的
computer control register 计算机控制寄存器
computer control state 计算机控制状态
computer control unit 计算机控制器
computer control 计算机控制
computer-dependent language 面向计算机的语言，与计算机相关的语言
computer development company 计算机开发公司
computer-directed process control technique 计算机指导的过程控制技术
computer display 计算机显示器，计算机指示器
computer down-time 计算机失效时间
computer dynamics 计算机动力学
computer efficiency 计算机效率
compute relay 计算继电器
computer engineer 计算机工程师
computer entry punch 计算机输入穿孔
computer equation 计算机方程式
computer equipment 计算机设备，计算机装置
compute-reset-hold switch 计算-复位-保持三用开关
computer file 计算机外存储器
computer-friendly 善于使用电脑的，好用电脑的，精通电脑的
computer-generated filter 计算机制作的滤波器
computer graphics 计算机制图，计算机图形学
computer hedgehog 电脑知识匮乏的人
computer instruction 计算机指令
computer integrated manufacture 计算机集成制造
computer interface station 计算机接口操作站
computer interface unit 计算机接口部件，计算机接口单元
computerism 计算机主义
computerization 计算机化，装备电子计算机的
computerized axial X-tomography 计算机控制的轴向 X 射线层析照相术
computerized control 计算机控制
computerized industrial tomography 计算机辅助的工业层析照相法
computerized NAA 计算机化中子活化分析
computerized process control 计算机化过程控制
computerized simulation 计算机化仿真
computerized ultrosonic scan system 计算机化超声扫描系统
computerize 计算机化，装电子计算机，用计算机处理
computer language translator 计算机语言翻译器
computer language 计算机语言，电脑语言
computer-limited 受计算机限制的
computer logic circuit 计算机逻辑电路
computer logic 计算机逻辑
computer-managed instruction 由计算机管理的指令
computer mechanism 计算机构
computer memory 计算机存储器
computer module 计算机模件
computer monitoring system 计算机监视系统
computer monitoring 电脑监控管理
computer network 计算机网络，计算机网
computer numerical control lathe 计算机数字控制机床，数控机床
computer numerical control 计算机数控，计算机数字控制【机床】
computer operation 计算机操作
computer operator 计算机算子，计算机操作人员
computer-oriented language 面向计算机的语言，

面向计算机语言
computer-oriented 与计算机有关的，研制计算机的
computer performance evaluation 计算机性能评价
computer photocomposition 计算机光学排版
computer picture 计算结果在显示器上的图像
computer process control 计算机过程控制，计算机程序控制
computer process interface 计算机处理接口，计算机处理分界
computer program for automatic control 自动控制用的计算机程序
computer program 计算机程序
computer programming language 计算机程序设计语言
computer programming 计算机编程，计算机程序设计
computer recording 计算机记录
computer relay tester 微机继电器测试仪
computer relay 计算机继电器，计算继电器
computer representation 计算机模拟法
computer ribbon printer 电脑色带打印机
computer room 计算机房
computer scattering coefficient 康普顿散射系数
computer science 计算机科学
computer security 计算机安全性
computer-sensitive language 计算机可用语言
computer set station 计算机设定（操作）站，计算机设定操作站
computer simulating 计算机模拟
computer simulation technique 计算机模拟技术
computer simulation 计算机模拟，计算机仿真
computer software 计算机软件
computer storage device 计算机存储器件
computer storage 计算机存储器
computer supervisory control system 计算机监控系统
computer supervisory system 计算机监控系统
computer supervisory 计算机监控
computer system 计算机系统
computer tape 计算机用磁带
computer temperature controller 电脑温控仪
computer test circuit 计算机的测试电路
computer theory 计算机理论
computer time 计算机时间
computer transformer 计算机变量器
computer transistor 计算机用晶体管
computer tube 计算管
computer utility 计算机效益，计算机效用
computer variable 计算机变量
computer virus 计算机病毒，电脑病毒
computer word 计算机字
computer workstation 计算机工作站
compute switch 计算开关
compute 计算，估算，验算，运算
computing algorithm 计算算法
computing amplifier 计算放大器
computing bill 计算单
computing capacitor 计算机用电容器
computing center 计算中心
computing circuit 计算电路
computing differential 计算微分
computing element 计算元件，运算器
computing engine 计算机
computing error 计算误差
computing hardware 计算硬件
computing impedance 计算阻抗
computing instrument 计算器，计算仪表
computing interval 计算时间
computing laboratory 计算实验室
computing machine 计算机
computing matrix 计算矩阵
computing network 计算网络
computing power 计算能力
computing relay 计算继电器
computing service 计算服务
computing sight 瞄准用计算装置
computing station 计算操作站
computing system coupler 计数系统的耦合装置，计算机系统的耦合装置
computing technique 计算技术
computing time 计算时间
computing unit 计算单元，计算装置
computing weighing scale 带有计算装置的秤
computing 计算，演算，运算，估算
computron 计算机用的多极电子管，计数管
comsat(communication satellite) 通信卫星
CONAG(combined nuclear and gas turbine) 核动力与燃气轮机联合装置
conbination spanner 两用扳手
concast 连续铸锭
concatenated frequency changer 串级变频机，级联变频机
concatenated motor 级联电动机，串级电动机
concatenate 串级，级联，串联，连在一起的
concatenation control 串级调节，级联调节
concatenation motor 串级电动机，级联电动机
concatenation 连接，结合，串联，级联，并列
concave airfoil surface 叶凹面
concave bank 凹岸，冲刷河岸
concave bit 凹头钻头，凹形凿
concave-convex bank 凹凸岸
concave-convex system 凹凸镜系统
concave-convex 凹凸的，新月形的，一面凹一面凸的
concave curvature 凹曲度
concave curves conveyor 下凹弧输送机
concave curve 凹弧段，凹曲线
concave downward 向下凹的
concave face 凹面
concave filled weld 凹形角焊缝
concave flange 凹面法兰
concave flow 凹岸水流
concave-grating spectrometer 凹面光栅分光计
concave joint 槽式接合，凹圆（接）线，凹缝
concave lens 凹面镜，凹透镜
concave mirror 凹面镜
concave perforated plate 凹形多孔板
concave profile 内弧型面
concave shore 凹岸
concave side of blade 叶片内弧侧

concave side　内弧【压力侧】
concave slope　凹坡
concave stream bank　凹河岸
concave surface　内弧面，凹面
concave upward　向上凹的
concave　凹面，凹的，内弧，凹度
concavity　（焊缝）凹度，内弧
concavo-concave　双凹形
concavo-convex　凹凸形
concavo-plane　凹底刨，平凹形
concealed air conditioner　暗装的空调器
concealed column　暗柱
concealed conductor　暗线
concealed erection　暗装
concealed gutter　暗天沟，暗管
concealed heating　隐蔽式供暖
concealed installation　暗装
concealed joint　暗缝
concealed lamp sign　间接信号，隐灯信号
concealed lamp　隐藏灯
concealed lighting　隐蔽照明
concealed nailing　藏钉
concealed pipe　暗管
concealed piping　暗管敷设
concealed radiator　暗装散热器
concealed wire　暗线
concealed wiring　布暗线，隐蔽布线
concealed　隐藏的，暗的
concealment　匿报
conceal　隐瞒
conceivable accident　可信事故
conceivable malfunction　可信误动作
concel the interlock　接触联锁
concentrate air conditioning　集中空调
concentrate cooling　集中供冷
concentrated boric acid solution　浓硼酸溶液
concentrated fall　集中落差
concentrated force　集中力
concentrated heating　集中采暖，集中供暖
concentrated impedance　集中阻抗
concentrated inductance　集中电感
concentrated investment　集中投资
concentrated liquid waste　液态废物浓缩物，浓缩的液态废物
concentrated load　集中载荷，集中荷载，集中负荷
concentrated low-activity liquid wastes　浓缩的低放液态废物
concentrated poison solution　浓缩毒物溶液
concentrated radioactive liquid waste　浓缩的放射性液态废物
concentrated refrigerating　集中制冷
concentrated smoke stack　烟突丛
concentrated solar power　聚光太阳能发电
concentrated solution　浓溶液，浓缩溶液，集中的解决方案
concentrated stream　浓水流
concentrated vortex　集中旋涡，合成旋涡
concentrated vorticity　集中涡，合成涡，集中涡量
concentrated waste storage　浓缩废物贮存
concentrated wear　局部磨损
concentrated winding　集中绕组
concentrated　集中的，浓缩的
concentrate hold-up tank　浓缩物滞留箱
concentrates holding tank　浓渣箱，浓液贮存箱，浓缩物滞留箱
concentrate storage tank　浓缩物储存箱，浓缩物贮存箱，浓渣箱，浓液贮存箱
concentrate vault　浓缩物库
concentrate　浓缩，集中，聚集，聚光，浓缩物，提浓，提浓物，浓液
concentrating coil　集中线圈
concentrating collector　聚光（型）集热器
concentrating reflector　集中反射器
concentrating roller　对中辊子【用于对中的辊子】
concentrating station　集中站
concentrating system　浓集系统，蒸发器
concentration boundary layerthickness　浓度边界层厚度
concentration by volume　体积浓度
concentration by weight　重量浓度
concentration cell　浓差电池
concentration coil　集中线圈
concentration contour　浓度分布廓线
concentration control ring　浓度控制环
concentration curve　浓度曲线
concentration diffusion　浓差扩散
concentration distribution　浓度分布
concentration exergy　浓度
concentration factor　浓集因子，浓缩系数
concentration field　浓度范围
concentration gradient　浓度梯度
concentration index of suspended dust　悬浮尘埃浓度指数
concentration index of suspended smoke and soot　悬浮烟灰和烟的浓度指数
concentration index　浓度指数
concentration intensity　浓度
concentration isopleths　等浓度线
concentration key　集合按钮，并析键
concentration limit　浓度极限
concentration method　（污水的）浓集法
concentration of acid and alkali　酸碱浓度
concentration of acid-base　酸碱浓度
concentration of boiler water　锅炉水浓度
concentration of brain power　智力密集型
concentration of coal water mixture　水煤浆浓度
concentration of droplets in gas core　气体核心中的液滴浓度
concentration of electrons　电子浓度，电子密度
concentration of enterprises　企业集中
concentration of floating dust　飘尘浓度
concentration of floating　飘尘浓度
concentration of harmful substance　有害物质浓度
concentration of pollutant　污染物浓度
concentration of population　人口集中，人口密集
concentration of stress　应力集中
concentration of tributary flow　支流汇集，支流汇水
concentration on ground level　地面浓度

concentration potential 浓差电势
concentration profile 浓度分布，浓度分布图，浓度分布廓线
concentration ratio 聚光率，浓缩比
concentration scale 浓度刻度
concentration tank 浓缩池
concentration time 汇流时间
concentration 含量，浓度，密集，浓集，集聚，聚光
concentrator solar cell 聚光太阳电池
concentrator station 集控站
concentrator system 聚光系统
concentrator 集风装置，浓缩器，聚光器，聚能器，集热器，集线器，集中器，选矿厂，浓缩机
concentrically wound coil 同心线圈
concentrical 同心的，同轴的
concentric arch 同心拱
concentric butterfly valve 同心蝶阀
concentric cable 同轴电缆
concentric chuck 同心卡盘
concentric circles 同心圆
concentric coil 同心线圈，端部连接不交叉的绕组
concentric connection 同心连接
concentric cylinder viscometer 同心圆筒式黏度计
concentric double pipe 同心双层管
concentric feeder 同轴馈电缆
concentric inlet and outlet 同心出入口【冷却剂】
concentricity adjustment 同心度调整
concentricity 同心，同心度，集中，偏心度
concentric-lay cable （心线扭绞的）同轴电缆，同心绞线
concentric-lay conductor 同心导线，同轴导线，同心绞线
concentric line 总线，公共线，同轴线
concentric orifice plate 同心孔板
concentric pipe 同心管
concentric reducer 同心渐缩管，同心异径管，同心异径管接头，同心大小头
concentric ring 同心环
concentric stranded wire 同心绞线
concentric structure 同心构造
concentric tube heat exchanger 套管式换热器
concentric tube 同心套管，同心管
concentric variable capacitor 同心型可变电容器
concentric weathering 同心风化
concentric winding 同心式绕组，同轴绕组
concentric wiring system 同心接线系统，同心绞线
concentric 同心的
concept design phase 方案设计阶段
concept design 概念设计，初步设计
concept diagram 概念图
concept system 概念体系
conceptual design 概念设计，方案设计，草图设计，初步设计
conceptual development model 初样，初步开发样机
conceptual diagram 示意图
conceptual drawing 方案图
conceptual phase 方案设计阶段，草图设计阶段，初步设计阶段，构思阶段，概念（化）阶段
conceptual planning 意向性规划
conceptual plan 初步设计，概念设计，概念规划
conceptual scheme 概念图
conceptual unit 信息单位的坐标间隔
conceptual version 概念设计版
concept 概念，观念，设想，原理，思路
CONCEPT 一种用来分析核电厂经济效益的电子计算机程序
concerning 关于，就……而论
concerted action 协同动作，一致动作
concert 和谐，一致，音乐会
concession agreement 特许权协议，特许协议
concessionaire 主办单位
concessional loan 优惠贷款
concessional price 优惠价格
concessional terms 特惠条件
concession and privatisation 特许经营与私有化
concessionary aid 优惠援助
concessionary period 特许期
concessionary 特许的，特许公司，优惠的
concession contract 特许经营合同
concession of tariff 关税减让
concessions act 特许经营法
concession tax 宽减税项
concession 让步，（政府或业主给予的）特许（权），承认，租界，特权，许可，租借地
concha 半圆穹顶
conchoidal fracture 坝壳状断口，贝壳状断口
conchoids 螺线管
conciliating current 补偿电流
conciliation commission 和解委员会，协调委员会，调解委员会
conciliation committee 调解委员会，协调委员会
conciliation procedure 调解程序
conciliation 调解，和解
conciliator 调解人
conclude a contract 订立合同，签订合同
conclude agreement 签订协定［协议］
concluded contract 签订的合同
conclude the transaction 成交
conclude 得出结论，达成协议，结束，推断出，签订
concluding provision 最后规定
conclusion letter 缔约信件
conclusion of agreement 达成协议
conclusion of business 达成交易
conclusion of contract 合同的订立，合同缔结
conclusion of fact 事实结论
conclusion 结论
conclusive comments 结论意见
conclusive evidence 确证
conclusive 最后的，结论性的，明确的
concomitant 伴生物
concordance 和谐，整合
concordant bedding 整合层面
concordant fold 整合褶皱

concordant injection 整合贯入
concordant pluton 整合深成岩体
concordant valley 整合谷
concourse 集合，中央大厅
concrete admixture 混凝土掺和剂
concrete age 混凝土龄期
concrete aggregate 混凝土集料，混凝土骨料
concrete air cooled vault 混凝土空气冷却贮存库
concrete apron 混凝土护坦
concrete arch dam 混凝土拱坝
concrete arch floor 预制板钢梁楼板
concrete backfill 混凝土回填
concrete baffle pier 混凝土消力墩
concrete bagwork 袋装混凝土护岸工程
concrete ballast 混凝土压载，混凝土压（舱）块
concrete barrow 混凝土手推车
concrete base slab 混凝土基础板
concrete base 混凝土底层
concrete batcher 混凝土配料器
concrete batching and mixing plant 混凝土配料拌和厂
concrete batching plant 混凝土搅拌站，混凝土搅拌设备
concrete beam 混凝土梁
concrete bent construction 混凝土构架结构
concrete bent 混凝土排架
concrete biological shield 混凝土生物屏蔽
concrete bleeding 混凝土泌水现象
concrete block and rock-mound breakwater 混凝土方块堆石防波堤
concrete block breakwater 混凝土块防波堤
concrete block cutter 混凝土方块切割机
concrete block protection 混凝土块护岸
concrete block revetment 混凝土块护岸
concrete block wall 混凝土砌块墙
concrete blockyard 混凝土块制造场
concrete block 混凝土块，混凝土砌块，混凝土预制块
concrete box culvert 混凝土箱涵
concrete breaker 混凝土凿碎机，凿混凝土机
concrete brick 混凝土砖
concrete bucket 混凝土吊罐
concrete buggy 混凝土手推车
concrete caisson breakwater 混凝土沉箱防波堤
concrete cart 混凝土手推车
concrete casing 混凝土外壳
concrete check dam 混凝土拦沙坝，混凝土谷坊
concrete check 混凝土配水闸
concrete chute 混凝土溜槽
concrete coating 混凝土覆盖层
concrete cofferdam 混凝土围堰
concrete composition 混凝土成分
concrete construction 混凝土建筑，混凝土施工
concrete container 混凝土容器
concrete containment 混凝土安全壳
concrete core sample 混凝土芯样
concrete core-wall 混凝土心墙
concrete cover 混凝土保护层，保护层，混凝土盖
concrete cradle 混凝土管座
concrete creep 混凝土蠕变，混凝土徐变
concrete cribbing 混凝土筐笼，混凝土箱格
concrete crib breakwater 混凝土木笼防波堤，混凝土箱格防波堤
concrete cube test 混凝土立方试块试验
concrete culvert 混凝土涵洞
concrete curing agent 混凝土养护剂
concrete curing blanket 混凝土保温覆盖
concrete curing compound 混凝土养护剂
concrete curing mat 混凝土养护盖垫，混凝土养护席
concrete curing technology 混凝土养护技术
concrete curing 混凝土养护
concrete cushion 混凝土垫层
concrete cutoff wall 混凝土截水墙
concrete cylinder sample 混凝土圆柱试样
concrete dam pouring 混凝土坝浇筑
concrete dam 混凝土坝
concrete deadman 混凝土锚桩【拉岸壁板桩的】
concrete deepwater structure 深水混凝土结构
concrete delivery pipe 混凝土输送管
concrete deposit 混凝土浇筑物
concrete design 混凝土配合比设计，混凝土设计
concrete diaphragm wall 混凝土心墙，混凝土防渗墙，地下连续墙
concrete diaphragm 混凝土隔板
concrete disintegration 混凝土崩解，混凝土风化，混凝土离析
concrete distribution mast 混凝土分配架［塔］，混凝土浇灌架
concrete distribution 混凝土的分配
concrete drain tile 混凝土排水管
concrete drill 混凝土钻
concrete drum 混凝土桶
concrete embedment system 混凝土埋入件系统
concrete encasement 外包混凝土
concrete engineering 混凝土工程
concrete equipment 混凝土设备
concrete face rockfill dam 混凝土面板堆石坝，混凝土面板坝
concrete face 混凝土面板
concrete facing dam 混凝土护面坝，混凝土面板坝
concrete fatigue 混凝土疲劳
concrete-filled caisson 混凝土充填沉箱
concrete-filled tube column 混凝土充填管柱，混凝土填塞管柱
concrete filler block 混凝土填块
concrete fillet 混凝土内补角
concrete finishing machine 混凝土整面机
concrete finishing 混凝土表面磨光，混凝土整面机
concrete fireproofing 混凝土防火性
concrete floor 混凝土底板，混凝土楼板
concrete flowability 混凝土流动性
concrete flume 混凝土渡槽
concrete footing 混凝土底座，混凝土基础
concrete forming cycle 混凝土模板周转
concrete formwork 混凝土模板作业
concrete form 混凝土模板

concrete foundation block 混凝土基础块体
concrete foundation 混凝土基础
concrete frame 混凝土排架
concrete grade 混凝土等级
concrete gravity dam 混凝土重力坝
concrete gravity overflow dam 重力式混凝土溢流坝
concrete gravity platform 混凝土重力式钻井平台
concrete grip 混凝土握固力
concrete grout 混凝土灌浆
concrete guard wall 混凝土挡墙，混凝土挂墙，混凝土护墙，混凝土导墙
concrete gun 混凝土喷枪
concrete handling 混凝土吊运
concrete hardening 混凝土硬化
concrete hauling container 混凝土运送容器
concrete haunching 混凝土外壳
concrete hollow block 混凝土空心块
concrete hopper 混凝土料斗
concrete hyperbolic cooling tower 钢筋混凝土双曲线冷却塔
concrete incorporation plant 混凝土固化装置
concrete ingredient 混凝土成分
concrete insert 混凝土预埋件，插铁，混凝土嵌入件
concrete inspection 混凝土检验
concrete jacket 混凝土套，混凝土外皮
concrete jammer 混凝土捣棒
concrete key trench 混凝土截水槽，混凝土键槽，混凝土齿槽
concrete ledge 混凝土凸台【机器支承】
concrete lift 混凝土浇筑层
concrete-lined tunnel 混凝土衬砌隧洞
concrete-lined 混凝土衬砌的
concrete lining 混凝土衬砌
concrete lintel 混凝土过梁
concrete masonry 混凝土圬工
concrete mattress revetment 混凝土沉排护岸
concrete mattress roll 混凝土排辊
concrete mattress 混凝土基础板，混凝土沉排
concrete membrane 混凝土薄层
concrete mix barge 混凝土拌和船
concrete mixer truck 混凝土搅拌车，混凝土拌和车
concrete mixer 混凝土搅拌机，混凝土拌和机
concrete mixing plant 混凝土拌和厂，混凝土搅拌站，混凝土搅拌设备
concrete mixing 混凝土搅拌
concrete mixture 混凝土拌和物
concrete mix 混凝土拌和物
concrete mobility 混凝土流动性
concrete moist room 混凝土湿养护间
concrete monolithic flooring 混凝土浇制楼面，混凝土整浇楼面
concrete monolith 混凝土大块体
concrete nail 混凝土钉
concrete objective 具体目标
concrete orifice turnout 孔口式混凝土分水闸，孔口式混凝土斗门
concrete over flow dam 混凝土溢流坝
concrete paint 刷混凝土墙用浆
concrete-pan system grid floor 钢盘模混凝土井格式楼板
concrete parameter 具体参数
concrete pavement 混凝土路面，混凝土护面
concrete paver 混凝土铺路机
concrete paving 混凝土护面
concrete pedestal 混凝土基座
concrete pier 混凝土墩
concrete pile 混凝土桩
concrete piling 混凝土桩
concrete pillar 混凝土标石，混凝土支柱，水泥标桩
concrete pipe 混凝土管
concrete piping 混凝土管道（输送）
concrete place-ability 混凝土的可浇置性
concrete placement 混凝土浇筑，混凝土浇灌
concrete placer 混凝土浇筑机，混凝土摊铺机
concrete placing boom 混凝土分配架［塔］，混凝土浇灌架
concrete placing installation 混凝土浇筑设备
concrete placing mast 混凝土分配架［塔］，混凝土浇灌架
concrete placing plant 混凝土浇筑设备
concrete placing trestle 混凝土施工栈桥
concrete placing 混凝土浇筑
concrete plant 混凝土厂
concrete platform 混凝土平台
concrete plug 混凝土塞
concrete pole 混凝土电杆，水泥杆
concrete pouring and curing 混凝土浇灌和养护，混凝土浇灌和养生
concrete pouring for bedding 垫层混凝土浇筑
concrete pouring 混凝土浇灌，混凝土浇灌作业，混凝土浇筑
concrete pour-pile 混凝土灌注桩
concrete preparation 混凝土制备
concrete pressure vessel 混凝土压力容器
concrete primer 打底混凝土浆
concrete product 混凝土制品
concrete proportioning 混凝土配合
concrete pump truck 混凝土泵车
concrete pump 混凝土输送泵，混凝土泵
concrete quality control 混凝土质量控制
concrete quay 混凝土码头
concrete rammer 混凝土捣实器
concrete reactor shield 混凝土反应堆护罩
concrete reactor vessel 混凝土反应堆容器
concrete reactor 混凝土芯电抗器
concrete reinforcement worker 钢筋工人
concrete reinforcement 混凝土配筋
concrete retarder 混凝土缓凝剂
concrete retempering 混凝土重塑
concrete revetment 混凝土护岸
concrete road paver 混凝土路面铺设机
concrete roof 混凝土顶板
concrete saddle 混凝土鞍座
concrete sample 混凝土试件
concrete scaling 混凝土剥落
concrete secondary containment 二次混凝土安全壳

concrete setting 混凝土凝固
concrete sheet-pile breakwater 混凝土板桩防波堤
concrete sheet-piling 混凝土板桩
concrete shell pile 混凝土薄壳桩
concrete shell 混凝土安全壳，混凝土薄壳
concrete shielding 混凝土屏蔽
concrete shield 混凝土屏蔽体
concrete shrinkage 混凝土干缩，混凝土收缩
concrete siphon 混凝土虹吸管
concrete skeleton 混凝土骨架
concrete slab pavement 混凝土板护面，混凝土板砌面
concrete slab revetment 混凝土板护坡
concrete slab 混凝土板，混凝土面板
concrete sleeper 混凝土轨枕
concrete sluice 混凝土节制闸
concrete small hollow block 混凝土小型空心砌块
concrete specification 混凝土规范
concrete specimen 混凝土试件
concrete spiral casing 混凝土蜗壳
concrete splitter 混凝土分离器，混凝土劈裂器
concrete-spouting plant 混凝土灌注设备
concrete spreader 混凝土平仓机，混凝土撒布机，混凝土摊铺机
concrete spreading 混凝土平仓，铺混凝土
concrete station 混凝土搅拌站
concrete structure 混凝土构筑物，混凝土结构
concrete superstructure 上层混凝土结构
concrete support 混凝土支架
concrete tamping 混凝土捣固
concrete tank 混凝土贮水池，混凝土水箱
concrete terrazzo 水磨石面混凝土，混凝土水磨石
concrete test cube 混凝土立方体试块
concrete testing laboratory 混凝土试验室
concrete tetrahedron 混凝土四面体
concrete tetra pod 混凝土四脚体
concrete-timber pile 混凝土木桩
concrete transfer car 混凝土转运车
concrete transportation system 混凝土运输系统
concrete transporting equipment 混凝土运输设备
concrete trench 混凝土沟
concrete tribar 混凝土三棱体块
concrete truck-mixer 混凝土搅拌输送车
concrete tubular pile 混凝土管桩
concrete unit 具体单位
concrete vault 混凝土穹顶
concrete vibrating machine 混凝土夯实机，混凝土振捣机
concrete vibratory tamper 混凝土振捣器
concrete vibrator 混凝土捣实器，混凝土振捣器
concrete vibro column pile 振动混凝土柱桩
concrete wall 混凝土墙
concrete waterproof 混凝土防水（性），混凝土防渗
concrete waterproof technique 混凝土防水技术
concrete water tank 混凝土贮水池
concrete workability 混凝土和易性
concrete worker 混凝土工
concrete works 混凝土工程
concrete 混凝土，浇混凝土，凝结物，使……凝固，混凝土结构，具体的
concreting in cold weather 混凝土冬季施工
concreting in freezing weather 冻期浇注混凝土，混凝土冬季施工
concreting in lifts 混凝土分层浇筑，分层灌筑混凝土
concreting lift 浇灌混凝土的提升
concreting method 混凝土浇筑方法
concreting outfit 混凝土施工设备
concreting process 混凝土浇筑过程
concreting program 混凝土浇筑程序，混凝土浇筑计划
concreting rate 混凝土浇筑速率
concreting stage 混凝土浇筑阶段
concreting tower 混凝土浇筑塔
concreting 混凝土浇筑，混凝土作业，浇灌混凝土
concretion 凝岩作用，凝块，结核
concretor 混凝土工
concurrence 同时发生，同时存在，并行，一致，同意，并发
concurrent authority 同等权利
concurrent boiler 直流锅炉
concurrent centrifuge 单向流动式离心机
concurrent computer 并行计算机
concurrent consideration 同时审议
concurrent flow 顺流，多相同流，平行流
concurrent force 共点力，汇交力
concurrent gas-liquid flow 气液混合流
concurrent input/output 并行输入/输出
concurrent liability in tort 关于侵权的共同责任
concurrent lines 共点线
concurrently 同时地
concurrent operation control 并行运算控制，并行操作控制
concurrent operation 平行操作，同时操作，并行操作，并行运算
concurrent peripheral operation 并行外围设备操作
concurrent processing 并行处理
concurrent reaction 并发反应
concurrent real-time processing 并行实时处理
concurrent 同时发生的，平行的，相合的，顺流的，平行流，顺流，一致的，并发的
concussion blasting 裸露爆破
COND＝condenser 冷凝器，凝汽器
condeep platform 水下混凝土平台
condemnation 指责
condensance current 容性电流
condensance 电容器的电抗，容抗
condensate clean-up system 凝结水净化系统
condensate automatic regulation 凝结水自动调节
condensate booster pump 凝结水升压泵
condensate circuit 凝结水系统，凝结水回路
condensate circulating pipe 凝结水循环管
condensate circulating pump 凝结水循环泵
condensate clean-up plant 凝结水净化装置
condensate clean-up 凝结水净化
condensate collection tank 凝结水集水箱

condensate control 凝结水再循环控制，凝结水控制
condensate cooler 凝结水冷却器
condensate cooling tower 凝结水冷却塔
condensate demineralization plant 凝结水除盐装置
condensate demineralization 凝结水除盐
condensate demineralizer system 凝结水除盐系统
condensate demineralizer 凝结水除盐装置
condensate depression 凝结水过冷却，过冷度
condensate drain orifice 冷凝水排水孔
condensate drain pan 凝结水盘
condensate drain pipe 凝结水疏水管
condensate drain 凝结水疏出，冷凝排水，凝结水排水
condensate extraction pump 冷凝液抽泵，凝结水抽取泵
condensate extraction system 凝结水抽出系统，凝结水抽取系统
condensate film 凝结水膜
condensate filter 凝结水过滤器
condensate flow rate 凝结水流量
condensate flow 凝结水流量，冷凝水流量，凝结水量
condensate hardness 凝结水硬度
condensate in steam pipeline 沿途凝结水
condensate line 凝结水管路
condensate make-up pump 凝结水补充水泵
condensate monitor tank 凝结水监测箱
condensate outlet 冷凝水出口
condensate pipe 凝结水管，冷凝液管，冷凝水回水管
condensate piping 凝结水管道
condensate polisher regeneration room 凝结水精处理再生间
condensate polisher system 凝结水精处理系统
condensate polishing plant building 凝结水处理厂房，凝结水净化厂房
condensate polishing plant 凝结水精处理站，凝结水净化装置
condensate polishing system 凝结水精处理系统，凝结水净化系统
condensate polishing 凝结水净化处理，凝结水精处理，凝结水净化
condensate pot 凝结水罐，冷凝容器
condensate pump 凝结水泵，冷凝水泵
condensate purification 凝结水净化
condensate recirculating line 凝结水再循环管路
condensate recovery 凝结水回收
condensate recovery percentage 凝结水回收率
condensate recovery pump 凝结水回收泵
condensate re-evaporation 凝结水再蒸发
condensate removal pump 凝结水泵
condensate return system 凝结水再循环系统
condensate return tank 凝结水集，凝结水回收箱
condensate return 凝结水再循环，冷凝水回流
condensate-scavenging installation 凝结水处理装置
condensate seal 凝结水密封
condensate sparger 凝结水喷淋器
condensate spill over pipe 凝结水溢流管
condensate storage tank 凝结水储水箱，凝结水贮存箱
condensate sump 热井
condensate system 凝结水系统
condensate tank 凝结水箱
condensate temperature 冷凝水温度，凝结水温
condensate transfer pump 凝结水回收泵，凝结水输送泵
condensate trap 凝结水除气箱，冷凝水疏水器
condensate treatment 凝结水处理【除盐】
condensate valve 凝结水阀
condensate water box 凝汽器水室
condensate water collected tank 凝结水回收水箱
condensate water discharge 凝结水排出管，疏水
condensate water 凝结水
condensate 冷凝水，凝结水，冷凝物，冷凝流，冷凝，凝结，冷凝液
condensation adiabat 湿绝热
condensational wave 凝聚波，密波
condensation by contact 表面凝结
condensation cathode 冷凝式阴极
condensation center 凝结中心，凝聚中心
condensation chamber 凝汽器的蒸汽空间
condensation cloud 凝结云
condensation coefficient 凝结系数
condensation front 凝结面
condensation heat 凝结热
condensation level 凝结面，凝结高度，凝冻高度
condensation loss 凝结损失
condensation nuclei 凝结核
condensation nucleus 凝结核
condensation number 凝结数
condensation of data 数据压缩
condensation of vapour 气体冷凝
condensation point 凝结点，凝点
condensation process 凝结过程
condensation pump 凝结水泵
condensation return pump 冷凝回水泵
condensation system 凝结水系统
condensation temperature 冷凝温度，凝结温度，缩合温度
condensation theory 凝结理论【地下水】，凝缩理论
condensation trail 航迹云，凝结尾迹，雾化尾迹
condensation trap 疏水器
condensation-type pressure suppression system 凝结水弛压系统
condensation value 凝结值
condensation water 冷凝水，凝结水
condensation wave 凝聚波，密波
condensation 冷凝，凝结，缩合，凝聚，雾化，冷凝作用，浓缩
condensator 凝汽器，冷凝器
condensed data base 简化数据库
condensed fluid 冷凝液，凝结液
condensed instruction deck 压缩指令卡片组
condensed steam 冷凝蒸汽
condensed time 压缩时间

condensed water outlet 凝结水排出
condensed water recovery system 凝结水回收系统
condensed water treatment room 凝结水处理室
condensed water 凝结水
condensed 冷凝的，凝结的，浓缩的，简明的，压缩的
condenser air pump 凝汽器抽气泵
condenser antenna 电容性开关
condenser arrester 电容避雷器
condenser auxiliaries 凝汽器辅助设备
condenser back pressure 凝汽器进口绝对压力
condenser back washing 凝汽器反向冲洗
condenser bank 电容器组
condenser block 电容器组
condenser boiler 冷凝式锅炉
condenser bushing 电容式套管
condenser capacity 凝汽器冷却能力，凝汽器出力
condenser chamber 冷凝室
condenser characteristics 凝汽器特性
condenser charge 电容器充电
condenser cleaning device with sponge rubber ball 凝汽器胶球清洗装置
condenser cleaning device 清洗装置
condenser cleaning equipment 凝汽器清洗装置
condenser connection 凝汽器接颈
condenser cooling surface 凝汽器热交换面，冷却面积，凝汽器冷却面积
condenser cooling water 凝汽器冷却水
condenser-coupled amplifier 电容耦合放大器
condenser coupling 电容耦合
condenser current 电容电流，容性电流，调相机电流
condenser deaeration 凝汽器除氧
condenser debris filter 凝汽器碎渣过滤器
condenser depression 过冷度
condenser diode storage 电容器二极管存储器
condenser discharge spot welding 电容贮能点焊
condenser discharge 电容放电
condenser divider 电容分压器
condenser duty 凝汽器热负荷，凝汽器负荷
condenser electroscope 电容式试电器
condenser equipment 凝汽设备
condenser excitation 电容器励磁
condenser expansion joint 凝汽器膨胀接头
condenser flange 凝汽器颈部法兰
condenser floor 凝汽器层，汽轮机房零米层
condenser flow 凝汽器的凝汽量
condenser fouling 凝汽器结污
condenser gasket 凝汽器垫片
condenser-heat transfer 凝汽器热交换
condenser hotwell 凝汽器热井，凝汽器热阱
condenser injection cooler 凝汽器喷射冷却器
condenser installation 凝汽器安装
condenser leakage 凝汽器泄漏，凝汽器漏水，电容器泄漏
condenser leak detection device 凝汽器检漏装置
condenser level control value 冷凝器水位调节阀，凝汽器水位调节阀
condenser load 凝汽器热负荷
condenser loss 凝汽器损失，电容器损耗，调相机损耗
condenser loudspeaker 电容式扬声器
condenser makeup and discharge system 冷凝器补排水系统
condenser make-up tank 凝结水补充水箱
condenser microphone 电容式扩音器，电容式传声器
condenser motor 电容器式电动机
condenser neck 凝汽器颈，凝汽器喉部
condenser oil 电容器油
condenser 凝汽器，冷凝器，（热管的）冷凝段，电容器
condenser paper 电容器纸
condenser performance curve 凝汽器性能曲线
condenser performance test 凝汽器性能试验
condenser pick-up 电容拾音器
condenser piping 凝汽器管道
condenser plant 凝汽设备
condenser plate 凝汽器管板
condenser pressure 凝汽器压力
condenser pump 凝结水泵
condenser reactance 容抗，电容器电抗
condenser-reboiler 凝汽式重沸器
condenser recirculating valve 凝汽器回流阀
condenser related draining system 与冷凝器相关的疏水系统
condenser relief valve 凝汽器安全阀
condenser return pipe 冷凝器回水管
condenser room 调相机室
condenser safety device 凝汽器安全装置
condenser shell 凝汽器壳体，凝汽器体壳
condenser shunt-type motor 电容分相式电动机
condenser speaker 电容式扬声器，静电扬声器
condenser steam dump valve 蒸汽向凝汽器排放阀
condenser steam dump 蒸汽向凝汽器排放
condenser steam turbine 凝汽式汽轮机
condenser support 凝汽器支架
condenser surface 凝汽表面
condenser susceptance 电容器电纳
condenser telephone 电容式受话器
condenser terminal difference 凝汽器端差
condenser tester 电容试验器
condenser throat 凝汽器喉部
condenser transmitter 电容器式送话器，静电送话器
condenser tube 凝汽器管
condenser tube ball cleaning system 凝汽器胶球清洗装置
condenser tube cleaning system 冷凝管清洁系统，凝汽器管清洗系统
condenser tube cleaning 凝汽器管清洗
condenser tube end leakage 凝汽器管端泄漏
condenser tube friction 凝汽器管道阻力
condenser tube pulling space 凝汽器抽管距离
condenser type attemperator 凝汽式恒温箱，凝汽式减温器
condenser type bushing 电容式套管【变压器】
condenser type potential transformer 电容器式电压互感器

condenser type surface attemperator 冷凝式表面减温器
condenser type terminal 电容式套管【变压器】
condenser vacuum deaeration 凝汽器真空除氧
condenser vacuum system 冷凝器真空系统
condenser voltage 电容器电压，调相机电压
condenser water box air evacuation 凝汽器水室抽真空
condenser water box vacuum pump 凝汽器水室真空泵
condenser water box 凝汽器水室
condenser water level 凝汽器水位
condenser water side vacuum pump 凝汽器水侧真空泵
condenser 凝汽器，冷凝器，电容器，聚光器，调相机
condense 冷凝，凝结，凝汽，聚集
condensing apparatus 冷凝装置
condensing bleeder turbine 凝汽式抽汽汽轮机，抽汽汽轮机
condensing chamber 冷凝室
condensing coefficient 冷凝放热系数，凝结放热系数
condensing coil 冷却器蛇形管，凝汽盘管，冷凝盘管
condensing equipment 凝汽设备
condensing extraction turbine 抽汽汽轮机
condensing medium 凝结介质
condensing nuclear power plant 凝汽式核电厂
condensing plant 凝汽设备
condensing power plant 凝汽式发电厂，复水式发电厂，凝汽式电厂
condensing power station 凝汽式发电厂
condensing pressure 冷凝压力
condensing process 凝结过程
condensing steam turbine 冷凝式汽轮机，凝汽式汽轮机
condensing system 凝汽系统
condensing temperature 凝结温度
condensing tower 凝汽式冷却塔，冷凝塔
condensing turbine generator unit 冷凝式汽轮发电机组
condensing turbine with single reheat and six steam extractions 单再热六抽汽凝汽式汽轮机
condensing turbine 冷凝式汽轮机，凝汽式汽轮机
condensing type drier 冷凝式干燥器
condensing unit 凝汽装置，压缩冷凝机组，冷凝机组
condensing water conduit 凝结水管道，冷凝水管道
condensing water flow 凝结水流量
condensing water leakage 凝结水渗漏
condensing water pump 凝结水泵
condensing water recovery unit 冷凝水回收装置
condensing water 凝结水，凝汽器冷却水，冷凝水
condensing works 冷凝装置
condensing zone 凝汽区，凝结区，冷凝区
condensing 凝结，凝汽
condensite 孔顿夕电瓷，康尔塞电木

condensive load 进相负荷，电容性负荷
condensive reactance 容抗
condensivity 介电常数
condisions of carriage 货运条件
condition adverse to quality 不利于质量的工况［条件］
conditional acceptance 有条件验收
conditional AND 条件“与”
conditional assembly 条件汇编
conditional bond 有条件保函［担保］
conditional branch instruction 条件转移指令
conditional branch 条件转移
conditional break-point instruction 条件断点指令
conditional break-point 有条件的断点
conditional coefficient 条件系数
conditional contract 有条件的合同，暂行契约
conditional control sequence interruption 有条件控制的顺序中断
conditional crack initiation CTOD 条件启裂尖端张开位移值
conditional delivery 有条件交货
conditional equilibrium 条件平衡
conditional expression 条件表达式
conditional guarantee 有条件保函/担保
conditional indorsement 附条件的背书
conditional inequality 条件不等式
conditional information 条件信息
conditional instability 条件不稳定
conditional instruction 条件指令
conditionality 制约性
conditional jump 条件转移
conditional L/G 有条件保函
conditional loan 有条件的贷款
conditionally convergent 有条件收敛
conditional mathematical expectation 条件数学期望
conditional observation 条件观测
conditional operation 条件运算
conditional order 条件指令
conditional periodicity 有条件的周期性，条件性的周期性
conditional prior distribution 条件先验分布
conditional probability density 条件概率密度
conditional probability machine 条件概率计算机
conditional probability of failure 故障条件概率
conditional probability 条件概率
conditional prohibition 有条件的禁止
conditional ratification 有条件的批准
conditional reflex system 条件反射系统【通信】
conditional sampling 条件取样
conditional short-circuit current 限制短路电流
conditional solubility product 条件（性）溶度积
conditional stability constant 条件稳定常数
conditional stability of a linear control system 线性控制系统的条件稳定性
conditional stability 条件稳定性
conditional transfer command 条件转移指令
conditional transfer instruction 条件转移指令
conditional transfer of control 控制的条件转移
conditional transfer 条件转移
conditional 有条件的，有限制的，条件性的

condition-based maintenance 状态检修，状态维修
condition code 条件码，特征码，状态码
condition curve 工况曲线，状态曲线，工况图
conditioned room 空气调节房间
conditioned stop instruction 条件停止指令
conditioned zone 空气调节区
conditioned 有条件的，限制的，经过调节的
conditioner capacity 空调器容量
conditioner 调节器，调节剂，调料槽，调湿装置，调制器
condition for employment 聘用条件
condition for validity 有效条件
conditioning burner 可调燃烧器
conditioning chamber 空气干湿调节室
conditioning condensate 空调冷凝水
conditioning signal 调整信号，调符信号
conditioning tank 调节池
conditioning treatment 预备热处理
conditioning water 处理水，调节水
conditioning 调节，（工况）调整，处理，调理，性能改善
condition line 状态线，过程曲线
condition Mark's 马克边界条件
condition monitoring system 状态监测系统
condition monitoring 状态监视，条件监视，状态监测
condition number 条件数
condition of access 加入条件，进入条件
condition of circumstances 环境条件
condition of continuity 连续（性）条件
condition of contract 承包条件
condition of convergence 收敛条件
condition of delivery 交货状态
condition of equilibrium 平衡条件
condition of equivalence 等价条件
condition of experiment 试验条件
condition of furnace 炉膛工况
condition of investment 投资条件
condition of particular application 特殊应用条件
condition of precedent 先决条件
condition of production 生产条件
condition of resonance 谐振条件
condition of rest 静止状态
condition of similarity 相似性条件
condition point （曲线图上的）状态点
condition precedent 先决条件
condition recognition 状态识别
condition requiring subsequent settlement 需后续结算的条件
condition restraint 条件约束
conditions for commencement of works 开工条件
conditions of compatibility 相容条件
conditions of constant mass flow 等流量条件
conditions of contract 合同条件
conditions of engagement 雇佣条件
conditions of grant 让与条件
conditions of hire 租用条件
conditions of labour 劳动条件
conditions of particular application 合同的特殊应用条件，专用条件
conditions of sale 销售条款
condition subsequent to the contract 合约生效后的条件
condition subsequent 后续条件，后决条件，解除条件
condition 状况，条件，工况，状态，情况，调节，环境条件
CONDTY(conductivity) 导电率
conduce to 有助于
conductance for alternating current 交流电导
conductance for direct current 直流电导
conductance loop 电导回路
conductance ratio 电导比，电导率
conductance relay 电导继电器
conductance water 电导水
conductance 导电性，传导率，电导率，传导性，传导，电导
conducted interference 传导来的干扰，自电源线来的干扰，馈电线感应干扰
conductibility 电导率，热导率，导电性，传导性
conductible 可传导的，能（被）传导的
conductimetric method 电导率测定法
conducting angle 导电角
conducting bar 导电条【鼠笼转子】
conducting bolt 导电螺栓
conducting bridge 电阻电桥，导电分路
conducting coating 导电涂层，导电漆
conducting direction 传导方向，导电方向
conducting dust 导电尘
conducting element 导电元件
conducting end-ring （鼠笼转子及阻尼绕组的）导电端环
conducting ferromagnetic material 导电铁磁材料
conducting fluid 导电流体，导电液体
conducting glass 导电玻璃
conducting interval 导电间隔
conducting layer 传导层，导电层
conducting material 导电材料
conducting medium 导电介质
conducting paint 导电涂料，导电漆
conducting particles 传导粒子，导电粒子
conducting period 导电期间，导电周期
conducting pollution layer 导电污层
conducting power 导电能力，导电本领，传导能力
conducting probe 电导探针
conducting ring 导电环，集电环
conducting screen 导电屏
conducting sleeve 导电套筒
conducting state 导通状态
conducting varnish 导电清漆
conducting wheel 导电轮
conducting wire 导线
conduction angle 导通角【可控硅】
conduction band 传导带，导带，导电区
conduction carrier 传导载波，传导载流子
conduction-cooled turbine blade 传导冷却式透平叶片
conduction cooling 传导冷却
conduction-current density 传导电流密度

conduction current of capacitor 电容器的漏导电流
conduction current 传导电流
conduction electron 传导电子，载流电子，外层电子
conduction error 传导误差
conduction heating surface 传热面，导热面积
conduction heat transfer 传导传热
conduction hole 导电空穴
conduction loss 传导损耗
conduction of heat 热传导，导热，热的传导
conduction pump 导电泵，电磁泵
conduction state 导电状态
conduction through 不熄弧，不灭弧，贯通导电
conduction time 导电时间
conduction 传导性，传导率，传导系数，传导，电导，热导，流导，导电，导热
conductive body 导电体，导体
conductive cell 传导室
conductive cement 导电水泥
conductive coating 导电涂层
conductive coupling 阻性耦合，导偶，电导耦合
conductive crystal coating 导电晶体涂层
conductive discharge 传导放电
conductive earth 导电土壤，接地
conductive film 导电薄膜
conductive fluid 导电流体
conductive glass 导电玻璃
conductive heat 导热
conductive load 电导负载
conductively-closed 屏蔽的
conductively connected 导电相连接
conductive plastic potentiometer 导电塑料电位器
conductive plastics 导电塑料
conductive synthetic fibre 导电合成纤维
conductive transfer 传导
conductive 传导的，导电的
conductivity analyzer 导电度表，电导率表
conductivity bridge 电导电桥
conductivity cell 电导池，传导单元，电导率测定用的电池
conductivity coefficient 传导系数，热导系数，电导率
conductivity connected charge-coupled device 电导联结电荷耦合器件
conductivity crystal counter 传导晶体计数器
conductivity electron 传导电子
conductivity factor 热导系数，传导因子，导热系数
conductivity for heat 导热性，热导率，导热系数
conductivity integral 积分导热率
conductivity measurement 电导率测量
conductivity measuring bridge 电导率测量电桥
conductivity meter 热导计，电导仪，电导率仪，导电率计
conductivity modulation 电导率调制
conductivity of cladding 包壳导热率
conductivity of fuel 燃料导热率
conductivity of waste water 废水电导率
conductivity recorder 电导率记录仪
conductivity test 电导率试验
conductivity transmitter 电导率变送器，电导变送器
conductivity 电导率，传导性，导电性，电导系数，传导率，热导率，电导性，导电率
conductograph 传导仪，电导计
conductometer 热导计，电导计
conductometric titrimeter 电导计式滴定计
conductometric 电导计式的
conductor ampacity 电导载流量
conductor arrangement 配线，布线，导线排列
conductor assembly 导线束
conductor bar 线棒
conductor belt 导体带
conductor bundle 分裂导线，导线束
conductor characteristic 导线特性
conductor clamp 卡线钳
conductor compressor 导线压接器
conductor configuration 配线，布线，导线排列
conductor conflict 导线间距不符合规定【两条线路之间】，导线冲突
conductor contact resistance 导体的接触电阻
conductor-cooled generator 导体直接冷却的发电机，内冷发电机
conductor corrosion prevention 导线防腐
conductor cross-section 导线截面，导体截面
conductor cutter 导线切断器
conductor dancing 导线舞动
conductor efficiency 导线效率，导线利用率
conductor element 导体元件
conductor final tension 导线最终张力
conductor fittings 导线配件【零件，附件，接头】，导线金具
conductor galloping 导线跃动，导线舞动
conductor glaze formation 导线结冰
conductor holder 夹线器
conductor ice prevention 导线防冰
conductor initial tension 导线初始张力
conductor insulation 导线绝缘，导体绝缘
conductor jointer 导线连接器
conductor jointing 导线连接
conductor joint 导线接头
conductor loading 导线荷载
conductor material 导体材料，导线材料
conductor rail 接触轨，导电轨
conductor resistance 导线电阻，导体电阻
conductor roughness factor 导线表面粗糙系数
conductor sagging 调整导线弧垂
conductor sag 导线垂度，导线弧垂
conductor sag table stress 导线应力弧垂表
conductor screen 导体屏蔽
conductor shielding 导线屏蔽
conductor sleet formation 导线结冰，导线雾凇
conductor sleeting jump 导线脱冰跳跃
conductors on quad bundled 四分裂导线
conductor spacer 间隔棒
conductor strand number 导线股数
conductor strand 多股绞线
conductor stringing works 架线工作
conductor stringing 放导线
conductor subspan oscillation 线次挡距振荡

conductor support box 导线支撑盒
conductor surface gradient 电线表面电位梯度，导线表面场强
conductor sway 电线摆动
conductor tension 电线拉力
conductor-to-conductor gap 导线与导线间隙
conductor-to-tower gap 导线与塔身间隙
conductor turn 线匝
conductor type 导线型号
conductor vibration protection 导线防振
conductor vibration 导线振动
conductor with double insulation 双层绝缘线
conductor 导电体，导体，导线，传导体，导电材料，电线
conductron 光电导摄像管
conducts and fittings 导管和配件
conduct 行为，操行，传导，引导，实施，进行，导电
conduit box 导管分线匣，导管箱
conduit connection 套管接线，套管连线
conduit coupling （电缆）管道连接
conduit entrance 管道进口
conduit entry 导管引入装置
conduit fittings 导管配件
conduit format 穿管规格
conduit jacking 顶管装置，顶管法
conduit joint 管道接头
conduit outlet 电线引出口
conduit pipe 导管，导线管，地下线管，输水管
conduit pit 管道坑
conduit regulator 管道调节闸门
conduit room 导管室
conduit run 电缆管道路线，导管走向
conduit saddle 导线鞍式夹头
conduit seal 导管密封【上部堆内构件】
conduit section 输水道断面，管道断面
conduit slope 管道坡度
conduit system 导管系统，涵管系统，输水系统
conduit tee T形导管，T形管道
conduit tube 线管，导管
conduit type hydroelectric station 引水式水电站
conduit type power plant 水道式发电厂，导管发电厂
conduit valve 水道阀门
conduit wiring 管内布线，管线
conduit 导（线）管，管道，沟渠，穿（墙）管，涵洞，输水道
conduloy 康杜洛铍镍铜合金
COND 凝结【DCS画面】
cone and plate viscometer 圆锥平面型黏度计
cone angle 锥角
cone antenna 锥形天线
cone bearing test 圆锥承重试验
cone bearing 锥形轴承
cone belt 三角皮带
cone bit 锥形钻头
cone brake 锥形制动器，锥形闸
cone centre 锥顶
cone clutch 锥形离合器
cone coating 锥形涂层
cone coupling 锥形联轴节
cone crusher 圆锥形破碎机
cone delta 锥状三角洲
cone foundation 锥形基础
cone gauge 锥度量规
cone head bolt 锥头螺栓
cone head insulator 锥形头绝缘子
cone head rivet 平锥头铆钉，锥头铆钉
cone insulator 圆锥形绝缘子，锥形绝缘子
cone karst 锥形岩溶
cone loudspeaker 锥形扬声器
CONELRAD（control of electromagnetic radiation）电磁波辐射控制
cone method 圆锥法
cone of acoustic energy 声能锥
cone of depression 下降漏斗
cone of fire 火焰炬，火焰锥
cone of ground-water depression 地下水面下陷锥，地下水下降漏斗
cone of influence 下降漏斗，影响漏斗
cone of null 盲区，无声区
cone of pressure relief 减压锥
cone of protection 保护锥【雷击】
cone of pumping depression 抽水下降漏斗
cone of water table depression 水位下降漏斗
cone penetration test 触探试验，圆锥触探试验，圆锥贯入试验
cone penetrometer 圆锥贯入仪，触探仪
cone point screw 锥头螺钉
cone pulley drive 锥轮传动
cone pulley 宝塔式滑轮
cone pyrometer 锥体高温计
cone resistance 圆锥探头阻力，探头阻力
cone roof tank 锥盖式油箱
cone seal 锥形密封
cone section 锥体截面
cone shell pile 锥形煤堆
cone test 圆锥筒试验，灰锥试验，圆锥法
cone valve 锥阀，锥形阀
cone 圆锥，风向袋，（风轮）锥角，锥体，锥形，头锥，焰芯
confabulation 交谈
confer 参看，试比较，授予
conference 会议，讨论会
conference calling equipment 会议电话设备，电话会议设备
conference call 电话会议，会议电话
conference circuit 会议电话电路，调度通信电路
conference communication system 会议通信系统
conference consultant 会议顾问
conference model 会议模式
conference of production 生产会议
conference room 会议室
conference telephone 会议电话
confidence band 可信区，可信范围
confidence coefficient 可靠系数，置信系数
confidence curve 可靠程度曲线
confidence interval 置信区间，置信域，可靠区间，置信间隔，置信节
confidence level 可信度，置信水平，置信度，可靠水平

confidence limit 置信界限，可靠界限，可靠极限，置信极限，可信度，置信限度
confidence region 置信区域
confidence test 可行性试验
confidence 置信度，可信度，可靠程度，信心，信任，秘密，信用
confidential agreement 机密协定
confidential clause 保密条款
confidential details 机密的细节，保密事项
confidential document 机密［机要］文件
confidential information 保密信息，机密情报，机密信息
confidentiality agreement 保密协议
confidentiality clause 保密条款
confidential paper 密件
confidential procedure 保密的程序
confidential 不公开的，机密的，保密的，密件
confidently 确信地
configurational asymmetry 形状不对称，结构不对称
configuration and tuning module 组态与调整模件
configuration control 构形控制【反应堆】，外形轮廓的调整，配置控制
configuration diagram 组态图
configuration factor 轮廓系数，形状因数
configuration line 轮廓线
configuration list 配置表
configuration of distribution network 配电网络结构
configuration specification 组态说明书
configuration status accounting 配置状态报表
configuration test 验证试验
configuration 组态，地形，形状，轮廓，外形，配置，布置，结构，组合，构型，排列，布局，图形，线路接法，布线
confined aquifer 承压含水层，自流含水层
confined bed 封闭层
confined compression strength test 侧限抗压强度试验
confined compression 侧限压缩
confined compressive strength 侧向限制抗压强度，侧限抗压强度
confined concrete 侧限混凝土
confined eddy 固定旋涡
confined flow 承压水流
confined groundwater 承压地下水，受压地下水，约束地下水
confined jet 有限射流，约束射流
confined plasma 受限等离子体，约束等离子体
confined space 有限空间
confined vortex wind machine 约束涡型风力机
confined vortex 约束涡
confined water well 承压水井
confined water 承压水，受压水
confined 有限制的
confinement barrier 密封屏蔽，封闭屏障
confinement 约束，限制
confine 边界，范围，限制，约束，（磁场）吸持，局限于，界限
confining bed 封闭层，隔水层，隔水底层
confining field 约束场
confining overlying bed 含水层上覆层，隔水顶层
confining pressure 侧限压力，围压，约束压力，周围压力
confining stratum 隔水层
confining underlying bed 隔水底层
confirmation in writing 书面确认
confirmation of award 确定授标
confirmation of declaration 申报确认书
confirmation of oral instruction 口头指令确认书
confirmation of order 订货确认书
confirmation test 验证试验
confirmation 确认，证实，证据，证明
confirmatory reaction 验证反应
confirmatory test 验证试验
confirmed credit 保兑信用证，保兑信用状
confirmed L/C 保兑信用证，保兑信用状
confirmed letter of credit 保兑信用证，保兑信用状
confirmed 保兑的，业经证实的
confirming bank 保兑银行
confirming charge 保兑费
confirming house 保付商行
confirm the agreement reached 确定所达成的协议
confirm 确认，批准，证实，保兑，认可
confiscate 没收，征用
conflagation 特大火灾
conflagration 燃烧，着火，快速燃烧
conflation 合并，异文合成本
conflict in use of water 用水矛盾
conflict of interests 利益冲突
conflict 冲突，抵触，矛盾，不同
confluence 汇流（点），汇合（点），合流河（处，点），汇流处，会合点，群集
confluent 合流河
confocal 共焦的，同焦点的
conformability 适应性，整合性
conformable contact 整合接触
conformable strata 整合地层
conformable stratum 整合地层
conformable 一致的，相似的
conformal making 保角映射
conformal mapping 共形映射，保形映射
conformal map 保角映象，保角变换图，正形投影地图
conformal projection 保形投影，正形投影
conformal transformation 保形变换，保角变换
conformal 共形的，准形的，保形的，保角的，正形的
conformance test 验收试验，一致性测试，符合性测试
conformance 适应性，顺应，一致，一致性
conformation 构造，形态
conformed copies with the original 与原本一致的文本
conformity assessment system 质量认证体系
conformity certificate 合格证书
conformity error 一致性误差
conformity testing 合格试验
conform to international practices 符合国际惯例

conform to 与……相符，符合，遵照，到达，一致性，一致
confriction 摩擦力
confugate foci 共轭点
confugate image 共轭像，实像
confugration 结构，排列，组合，外形，轮廓，布置
confused sea 浪涛汹涌的海面，汹涛
confuse 混乱，混淆
confusion 混淆
confute 驳斥，反驳
congealed ground 冻土
congealed 冻凝的，凝固的
congealing point 冻结点
congeal 凝结，冻凝，冻结，凝固
congelation electricities 冻结电量
congelation method 锚碇基础工程，冻结法施工
congelation point 凝固点，冻凝点
congelation zone 冻结区
congelation 冻结，凝结，冻结物
congested area 人口稠密区
congested condition 饱和状态
congestion cost 阻塞费用
congestion management 阻塞管理
congestion of shipping space 船位拥挤
congestion surplus 阻塞盈余
congestion 堆积，填充，稠密，混杂，拥挤
congestus 浓云
conglomerate company 跨业公司，综合公司，跨行业公司，集团公司
conglomerate corporation 混合公司，联合企业
conglomerate merger 混合并购，企业集团合并，多行业企业合并
conglomerate stratum 卵石层
conglomerate 集团企业，聚积的，砾岩，企业集团，综合企业，大型联合企业，聚合物，聚结，凝聚成团
conglomeration 凝集，凝聚，团块
conglutination 共凝集作用
congress Ohm 法定欧姆
congruence circuit 重合电路
congruential generator 同余数生成程序，同余数发生器
congruent segment 叠合线段
congtrol range 控制范围
conical bearing 锥形轴承
conical bolt 锥形螺栓
conical bottom outlet 锥底排料口
conical buoy 锥形浮标
conical cam 锥形凸轮
conical closure 锥形封头
conical contact 锥形接触，锥形触头
conical cowl 锥形风帽
conical cut 锥形割槽
conical draught tube 锥形尾水管
conical entrance orifice plate 圆锥形入口孔板
conical fitting 锥形端部金具
conical float 锥形浮标
conical flow 锥形流
conical fluidized bed 锥形流化床
conical head rivet 锥头铆钉
conical horn 锥形喇叭口
conical insulator 圆锥形绝缘子
conical nozzle 圆锥形喷嘴，锥形管嘴
conical pendulum governor 锥摆调速机，锥摆式调节器
conical pile 锥形煤堆，锥形桩
conical plug valve 圆锥形转塞阀，锥形塞阀
conical plug 锥形锚塞
conical projection 锥形投影
conical ring 锥形环
conical roof 圆锥形屋顶
conical rotary valve 锥形回转阀
conical sealing face 锥形密封面
conical section 锥体部分【安全壳】，锥形结构
conical sleeve 锥形套筒
conical socket 锥形套管，锥形止推座
conical steel 锥形钢筒
conical stockpile 锥形煤堆
conical support skirt 锥形支裙筒
conical surge chamber 圆锥形调压井［室］
conical tank 锥形池，锥形罐，锥形箱
conical terminal 锥形端子
conical-top bubble cap plate 锥顶泡罩板
conical tube 圆锥管，锥形管
conical turbine 圆锥形水轮机
conical ventilator 锥形风帽
conical vortex 锥形涡
conical wall nozzle 锥形喷嘴，锥壁喷管
conical washer 锥形垫圈
conical 锥形的，圆锥的
conic-cavity-target X ray tube 锥形空心靶 X 射线管
conicity 圆锥度
conicograph 二次曲线规
conifer 针叶树
coniform 圆锥形的
conimeter 空气尘量计
coning and quartering method 堆锥四分法
coning angle 锥度角
coning damping 锥旋阻尼
coning dihedral effect 圆锥二面角效应
coning hinge 锥旋铰链
coning plume 圆锥形羽流
coning 锥形，形成圆锥形，锥度，圆锥度
coniosis 粉尘病，尘埃沉着病
coniscope 计尘仪，检尘器
conjoint 相连的，共同的，结合的
conjoin 连接，结合
conjugacy 共轭性
conjugate axis 共轭轴线
conjugate beam method 共轭梁法
conjugate beam 共轭梁
conjugate bridge 共轭电桥
conjugate chord 共轭弦
conjugate circuit 共轭电路
conjugate complex number 共轭复数
conjugate current 共轭电流
conjugated chain 共轭链
conjugate depth 共轭水深
conjugated fissure 共轭裂缝
conjugate diameter 共轭直径

conjugate domains 共轭域
conjugate fault 共轭断层
conjugate foci 共轭焦点
conjugate force 驱动力
conjugate functions 共轭函数
conjugate gradient method 共轭梯度法，共轭斜量法
conjugate imaginary 共轭虚数
conjugate impedances 共轭阻抗
conjugate match 共轭匹配
conjugate matrices 共轭矩阵
conjugate pair 共轭对
conjugate parameters 共轭参数，共轭参量
conjugate phasors 共轭相量
conjugate planes 共轭平面，共轭面
conjugate points 共轭点
conjugate power law 共轭幕定律
conjugate pressure 共轭压力
conjugate roots 共轭根
conjugate stage 共轭水位
conjugate stress 共轭应力
conjugate value 共轭值
conjugate vectors 共轭矢量
conjugate 共轭的，结合的，结合，耦合，配对，使成对，使结合
conjugative effect 共轭效应
conjunction “与”，接头，逻辑乘法，逻辑乘积，会合点，连测
conjunctive search 逻辑乘法搜索
conjunctive water use 水源联合运用
conjunct polymer 混合聚合物
conk 发动机突然停止，故障，故障迹象
connate deposit 原生沉积
connate water 原生水
connect 连接，联系，接通
connected contract 连带合同
connected domains 连域
connected footing 联合基脚
connected graph 连接图
connected load 连接负荷
connected shaft 连动轴
connected to driver 连到驱动端
connected to 联结的，与……联结
connected traverse 连接导线
connected winding 连接绕组
connected yoke 连接横木，连接夹，系梁
connected 连接的
connect ground wire 安装接地线
connecting angle 连接角
connecting bar 联系钢筋，连接杆
connecting beam 连接梁
connecting bolt 连接螺栓，拉紧螺栓
connecting box 接线箱，分线箱
connecting building 连接厂房
connecting bus-bar 接续汇流条
connecting bus 连接母线
connecting cable 接线电缆，连接电缆
connecting circuit 连接电路
connecting clamp 连接线夹，接头夹
connecting device 连接装置
connecting diagram 接线图，电路图
connecting elbow 连接弯管
connecting flange 连接法兰，连接凸缘，接线插孔
connecting lever 连接杆，连杆
connecting link 连杆，连接片，连线，连接杆
connecting machine 连接机
connecting mole 连接防波堤
connecting piece 连接件，接头，连接片
connecting pipe 连接管，联通管，联络管，取源管线，导压管线，测量管线
connecting plate 连接板，接合板，接线板
connecting plug 连接插头
connecting rod 活塞杆，连接棒，连杆，联杆
connecting series 串联
connecting shaft 连接轴
connecting sleeve 连接套筒
connecting terminal 连接端子，接线端子
connecting top section 顶部连接段
connecting up 布线，装配，安装
connecting wire 连线，连接电线
connecting 连接，连接的
connect in parallel 并联
connect in series 串联
connection board 接线板，连接板
connection box 接线箱
connection cable 接线电缆，连接电缆
connection chamber 接线箱
connection cord 接线绳
connection cover cutting pliers 剥线钳
connection diagram 接线图
connection drawing 接线图
connection error 接入仪表引起的误差
connection fittings 连接配件
connection in delta 三角形连接
connection in parallel 并联
connection in series 串联
connection in star 星形连接
connection insulation 连接点绝缘，连接线绝缘
connection layout 布线，接线
connectionless 无连接
connection of instrument tubes 仪表管连接
connection on containment shell for cable penetration 安全壳壳体上的电缆贯穿件接头
connection pin 连接夹
connection pipe 连接管
connection plate 连接片
connection point 连接点
connection rod bearing 连杆轴承
connection rod 连接杆
connection schedule 连接细表
connection scheme 接线方案
connection side span 连接边节距
connection signal 接通信号
connection size 连接尺寸
connection strap 连接条，连接片
connection stud 接线双头螺栓
connection symbol 接线符号
connection to grid 联网
connection to load 投运，并负荷
connection to power system 接入系统
connection to the grid 联网，电厂并入电网，并网

connection tube 连接管【控制棒驱动机构】，连通管
connection under pressure 带压带气连接
connection 连接，接线［管］，（管）接头，接合面，连接件［法］，衔接，联接［系］，连轴节
connective 连接物，连接词，连接的，接通的
connectivity node 连接节点
connector base 接线柱，接线座
connector bend （管子）弯头，接合弯头
connector block 连接器插头块，连接件插头块
connector board 接合板，接缝板
connector clamp 接头夹
connector heat sink 连接器散热片
connector lug 连接接头，接线头，插头
connector pliers 接线器克丝钳
connector plug 插头，连接插头
connector relay 终接器继电器
connector socket 接线插座
connector switch 连接器开关，接触器
connector wire 连线
connector 接头，插头，接线柱，连接线，连接记号，接插件，连接管，接线器
connect time 连接时间
connect to earth 接地
connect to grid 并网
connect to neutral 接中性线，接零
connect to the national network by price competition 竞价上网【中国】
connode 连接节点
connotation 含义，内涵
conny 碎煤，煤末
conoid shell 圆锥壳体
conoid 圆锥体的，锥体，锥形
Conoseal gasket Conoseal 密封垫
con-out 切断，关闭，结束工作
Conpernik 康普尼克铁镍基导磁合金
conractual guarantee 合同规定的担保
conrtol 控制
conscience 良心，天良
conscientious 凭良心办事的，认真的，有责任心的
conscious erratum 已知误差
conscious error 已知误差
consciousness 意识，觉悟，神志，自觉性，知觉
conscious of 意识到，发觉，觉得
conscious that 意识到，发觉，觉得
conscious 有意识的，知觉的，故意的
consecution 连贯次序，推论
consecutive-competitive reactions 连串-竞争反应
consecutive computer 连续操作计算机
consecutive days 连续日
consecutive firing 串联爆破，连续爆破
consecutive mean 动态平均值
consecutive number 连续数，相邻数，连续号码
consecutive points 相邻点
consecutive scanning 顺序扫描
consecutive separated reaction chamber 分室连续反应室
consecutive strain section 相邻耐张段
consecutive tower 相邻塔
consecutive 顺序的，顺次的，连续的，连续操作的，连贯的
consensus 意见一致，同意
consent 万能插口，插座，塞孔，同意，赞成
consequence analysis 后果分析
consequence finding program 推论寻求程序
consequences of an accident 事故影响范围，事故后果，事故的后果
consequence 重要性，后果，结果
consequent divide 顺向分水岭
consequent drainage system 顺向水系
consequent drainage 顺向排水系
consequential amendment 相应修正案
consequential arc back 持续性逆弧，继起逆弧
consequential commutation failure 连续换相失败，两相颠覆
consequential damages 间接损害赔偿金，后果性损害，间接赔款，间接损失
consequential failure 继发故障，继发性故障，继发性事故
consequential loss 间接损失，灾后损失
consequent-poles motor 交替磁极式电动机，变极式双速电动机
consequent pole winding 中间极绕组，交替极绕组
consequent pole 交替极，屏蔽极，罩极，中间极
consequent valley 顺向谷
conservancy area 自然保护区
conservancy district 保护区
conservancy engineering 水土保持工程
conservancy system （垃圾的）存置系统
conservancy trash system 拦污系统
conservancy zone （水土）保持区，（自然）保护区
conservation and collection of water 水的保护和管理，水的保护与采集
conservation area 水土保护区，自然保护区
conservation district 自然保护区，保护区
conservation equation of energy 能量守恒方程
conservation equation 守恒方程
conservation field of force 保守力场
conservation law of energy 能量守恒定律
conservation law 守恒定律
conservation of angular momentum 角动量守恒
conservation of aquatic resource 水产源保护，水资源保护
conservation of charge 电荷守恒
conservation of circulation 环量守恒
conservation of energy theorem 能量守恒定理
conservation of energy 能量守恒，节能
conservation of mass energy 质量能量守恒
conservation of mass theorem 质量守恒定理
conservation of mass 物质守恒，质量守恒
conservation of matter theorem 物质守恒定理
conservation of momentum equation 动量方程守恒
conservation of momentum theorem 动量守恒定理
conservation of momentum 动量守恒
conservation of natural resources 自然资源保护
conservation of nature 保护自然
conservation of Reynolds stress 雷诺应力守恒
conservation of vorticity 涡量守恒
conservation parameter 保守参数
conservation plant 利用废料生产的工厂
conservation procedure 保护方法

conservation project 治理项目
conservation relation 守恒关系
conservation reservoir 蓄水库
conservation storage 蓄水库容
conservation system 保守系统
conservation 保存，保持，保护，守恒
conservative force and potential energy 保守力与势能
conservatively rate 额定值超过
conservatively 留有余地
conservative rating 余度，保守额定值
conservative sizing 余度，尺寸余量，保守尺寸
conservative system 守恒系统，保守系统
conservative value 保守数值
conservative 保守的，悲观的，守恒的
conservator transformer 带储油箱的变压器
conservator 储存器
conserver 油枕
conserve 守恒，保存，节省，防腐剂
conserving agent 防腐剂
considerable orders 大量订货（单）
considerable 相当大的，相当多的，该注意的，应考虑的
consideration of alternatives 比较方案的考虑
consideration 考虑，研究，商量
considered 预计的，深思熟虑的
considering 考虑到，就……而论
consigned to negotiating bank 以押汇银行为受货人
consignee box 收货人栏［格］
consignee code 承销人代号，收货人代号
consignee preferential treatment 收货人优惠待遇
consignee's address 收货人地址
consignee 收货单位，收货人，受托人，承销人
consignment bill of lading 托运单
consignment contract 寄售合同
consignment note 发货通知单，发货通知（书），寄售通知，托运通知
consignment place 发货地
consignment 来煤批量，托运，托运货物，委托，运送，托付物
consignor 托运人，委托人，发货人，货主，寄售人
consign 发货，托运，委托，交付，寄存
consistence regulator 稠度调节器
consistence 相容性，连续性，稠度，浓度，黏度，稠性
consistency 一致性，恒定性，稠度，浓度
consistency check 一致性效验，一致性校验
consistency equation 相容方程
consistency factor 稠度系数
consistency gauge 稠度计
consistency index 稠度指数
consistency indicator 稠度指示器
consistency limit 稠度极限，稠度界限
consistency of performance 性能的一致
consistency of reading 读数的一致性
consistency probe 稠度测定器
consistency routine 相容程序
consistency test 坚固试验，稳定试验，稠度试验
consistency transmitter 稠度变送器，浓度变送器
consistency 结特性，稳定性，稠性，坚固性，密实度
consistent echoes 一致的回波
consistent estimator 一致推算子，一致估计量
consistent grease 润滑脂
consistently ordered matrix 相容次序矩阵
consistent policy 一贯方针
consistent quality 均质性
consistent with 与……保持一致，符合
consistent 前后统一的，一致的，相容的，稠的
consist in 存在于……中，以……为主要部分
consist of 由……组成［构成］，包括
consistometer 稠度计
consist with 符合，与……一致
consoidated anisotropically undrained test 室内土工试验，各向不等压固结不排水试验
console cabinet 控制室
console control desk 落地式控制台
console debugging 控制台调整，控制台调试
console desk 控制台
console display 控制台显示器
console file adapter 控制台文件适配器
console operator 控制台操作员
console package 控制台部件
console panel 控制盘，操作盘，操纵盘，操纵板
console receiver 落地式收音机
console switch 操作开关
consolette 小型控制台，小型落地式收音机
console typewriter 控制台打字机，键盘打字机
console 控制台，操纵台，仪表盘，仪表板，操作台，托架
consolidated account 合并账户
consolidated anisotropically undrained test 各向不等压固结不排水试验
consolidated balance sheet 合并资产负债表
consolidated-drained test 固结排水试验
consolidated-drained triaxial compression lest 固结排水三轴压缩试验，三轴固结排水压缩试验
consolidated financial statement 合并财务报表
consolidated income sheet 合并收益表，合并损益表
consolidated list 综合清单
consolidated mortgage 合并抵押
consolidated nuclear system generator 一体化蒸汽发生器，一体化动力反应堆
consolidated piping material summary sheet 管道综合材料汇总表
consolidated policy 综合保单
consolidated price 综合价格
consolidated production and financial plan 生产技术财务计划
consolidated profit and loss account 综合损益账，综合损益表
consolidated quick compression test 快速固结压缩试验
consolidated quick direct shear test 固结快剪直剪试验
consolidated quick shearing resistance 固结快剪强度
consolidated quick shear test 固结快剪试验

consolidated quick shear value 固结快剪值
consolidated report 综合报告
consolidated slow compression test 慢速固结压缩试验
consolidated soil 固结土
consolidated tax 统一货物税
consolidated triaxial test 固结三轴试验
consolidated-undrained compression test 固结不排水压缩试验
consolidated-undrained shear test 固结不排水剪力试验，固结不排水剪切试验
consolidated-undrained shear 固结不排水剪切
consolidated-undrained test 固结不排水试验
consolidated-undrained triaxial compression test 固结不排水三轴压缩试验
consolidated-undrained triaxial test 固结不排水三轴试验
consolidate view 综合意见
consolidate 加固，强化，固化，合并，巩固，联合，捣固，捣实，压实
consolidating pile 强化桩【加固地基用】
consolidation apparatus 固结仪，渗压仪
consolidation by electroosmosis 电渗固结
consolidation by vacuum method 真空法固结
consolidation by vibration 振动固结
consolidation curve 固结曲线
consolidation deformation 固结变形
consolidation degree 固结度
consolidation device 固结设备
consolidation due to desiccation 干燥引起的固结
consolidation grouting 固结灌浆
consolidation line 渗压曲线
consolidation of domestic and foreign curreny accounts 本外币并账
consolidation of earth dam 土坝固结
consolidation of enterprises 企业合并
consolidation of soft foundation 软弱地基加固
consolidation pressure 固结压力
consolidation process 固结过程
consolidation ratio 固结比
consolidation sedimentation 固结沉淀
consolidation service 集箱作业
consolidation settlement 固结沉陷，固结沉降
consolidation settling 集结沉积
consolidation test 固结试验，压密试验，室内土工试验
consolidation theory 固结理论，固结原理
consolidation time 固结时间
consolidation 压密，固结（作用），（混凝土）捣固，巩固，加强，强化，统一，合并，加固
consolidometer 固结仪
consonance 谐和，共鸣
consonant 协调的，一致的，和谐的
consortium agreement 联合协议
consortium bid 组合投标
consortium leader 牵头公司
consortium of contractors 承包商联合体
consortium partner 集团合伙人
consortium 联合体，企业集团，合伙，集团，国际财团，联合企业，共同体，银行财团
conspectus 线路示意图，流程图，摘要，梗要，简介，一览表
conspicuous 明显的，显而易见的，醒目的，显著的，引人注目的
constac 自动稳压器，自动电压稳定器
constance(-cy) 恒定性，稳定性，恒定量
constance force 常力
constant absolute vorticity trajectory 等绝对涡度轨迹
constant acceleration 等加速度
constant amplitude fatigue test 等幅疲劳试验
constant amplitude fluctuation 等幅脉动
constant amplitude wave 等幅波
constantan alloy 康铜合金
constant angle arch dam 等中心角拱坝
constant angular acceleration 等角加速度
constant angular velocity 等角速度
constantan thermo-couple 康铜热电偶
constantan wire 康铜线
constantan 康铜，铜镍合金
constant area flowmeter 固定面积流量计
constant area flow 等截面流
constant area 常数区，常数存储区
constant axial offset control 轴向偏移常数的控制
constant axial velocity pattern 等轴速流型
constant backpressure 背压常数
constant cell 恒压电池
constant chord blade 等截面叶片
constant circulation flow pattern 等环量流型，自由涡流型
constant circulation 稳定循环，定常循环
constant coefficient 常系数
constant critical flow 稳恒临界水流
constant cross-section 等截面
constant current characteristic 恒流特性
constant current charge 恒流充电
constant current condition 恒流条件
constant current control 恒流控制，恒流调节
constant current discharge 恒流放电
constant current distribution 定流配电，恒流配电
constant current dynamo 恒流发电机
constant current equivalent circuit 恒流等效电路
constant current generator 恒流发电机
constant current magnetron 恒流磁控管
constant current modulation 恒流调制
constant current motor 恒流电动机
constant current power supply 恒流电源
constant current regulation 恒流调节
constant current regulator 恒流调节器
constant current source 恒流电源，定流源
constant current supply 恒流电源
constant current system 恒流制，定流制
constant current transformer 恒流变压器
constant current wire anemometer 恒电流热线风速计
constant current 恒流，直流
constant damping 恒定阻尼
constant-data-rate system 恒速数据传输系统
constant delayed neutron production rate approximation 恒定缓发中子产生率近似
constant-delivery pump 定量输送泵
constant diameter elbow 等直径肘管

constant-dimension column 等形柱
constant displacement accumulator 等容蓄压器
constant duty 恒定工作，固定工作制，不变负荷，恒定工况
constant energy removal model 恒定能量排出模型
constant-enthalpy change-over 等焓切换【直流炉】
constant-enthalpy line 等焓线
constant-enthalpy process 等焓过程，节流过程
constant entropy expansion 等熵膨胀
constant entropy process 等熵过程
constant error 恒定误差，常在误差，固定误差，系统误差
constant excitation 恒定励磁
constant extinction angle control 恒熄弧角控制
constant extinction coefficient 恒定衰减系数
constant factor 不变因子
constant failure-rate period 恒定故障率周期，恒定失效率期
constant failure-rate 恒定故障率
constant field commutator motor 恒励磁换向器式电动机
constant field 恒定场
constant flow control 恒流调节，质调节
constant flow pump 定排量泵
constant flow 稳定流，定常流，稳恒流
constant flux layer 等通量层
constant-flux-linkage theorem 磁链不变定理，磁链守恒定理
constant flux linkage 恒定磁通链
constant flux voltage variation 恒磁通调压
constant force spring 恒压弹簧
constant force 恒力
constant fraction discriminator 定比率鉴别装置
constant frequency control 恒定频率控制
constant frequency generator 恒频发电机
constant frequency motor 恒频电动机
constant frequency oscillator 恒频振荡器
constant frequency output 恒频输出
constant frequency 恒定频率，不变频率
constant-glow potential 恒定辉光放电电压
constant gradient test 等梯度固结试验
constant hanger 恒力吊架
constant head orifice turnout 常水头孔口式配水闸，常水头孔口式斗门
constant head permeability test 常水头渗透试验
constant head permeameter 常水头渗透仪
constant head 不变水头，常水头
constant heat flux 恒定热流密度
constant heating effect 持续热效应
constant horsepower drive 恒定功率驱动
constant horsepower motor 恒定功率电动机，恒定马力电动机
constant horsepower 恒定马力，恒定功率
constant humidity system 恒湿系统
constant humidity 恒湿
constant hydraulic radius channel 等水力半径渠道
constant ignition 持续点火
constant impedance circuit 定阻抗电路
constant impedance regulator 固定阻抗调整器
constant instruction 常数指令
constant ionization 恒定电离
constant-K filter 定 K 值滤波器
constant lapse rate layer 等逆减率层，不断递减率层
constant lead 固定超前
constant level balloon 定高气球
constant level chart 等高面图
constant level oiler 不变液位加油器
constant life fatigue diagram 等寿命疲劳图，恒等寿命疲劳图
constant-linkage theorem 磁链不变定理，磁链守恒定律
constant load test 常荷载下试验
constant load 固定负荷，恒定负载，定荷载
constant loss 恒定损失，不变损耗
constant matrices 常数矩阵
constant moisture line 等湿线
constant money analysis 常币值分析
constant money levelized annual cost 常币值平准化年费用
constant money 常币值
constant multiplier coefficient unit 常系数乘法部件
constant multiplier 恒定系数，标度因数，常数乘法器
constant net loss 恒定净损耗
constant of measuring instrument 测量仪表常数
constant of proportionality 比例常数
constant of the line 线路常数
constant output oscillator 恒定输出振荡器
constant-perssure operation 定压运行
constant phase shifting network 恒定移相网络
constant pitching moment point 等俯仰力矩点
constant-potential generator 定压发电机，恒压发电机
constant-potential gradient 恒定电位梯度
constant-potential modulation 恒压调制
constant-potential regulation 恒压调节
constant-potential supply 恒压电源
constant-potential transformer 恒压变压器，定压变压器
constant-potential 定（电）压的【电源的输出特性】
constant potential 固定电势，固定电位，恒电位，恒定势
constant power control 恒定功率控制
constant power factor control 恒定功率因数控制
constant power generator 恒功率发电机
constant power motor 恒功率电动机
constant power operation 恒定功率运行
constant power output region 恒定功率输出区
constant power 恒定功率
constant pressure air storage reservoir 恒压蓄气库
constant pressure burning 等压燃烧
constant pressure cell 常压力盒
constant pressure chart 气压形势图，等压面图
constant pressure combustion 等压燃烧
constant pressure compression 等压压缩
constant pressure cooling 定压冷却

constant pressure cycle 定压循环
constant-pressure deflagration 定压爆燃
constant pressure device 恒压装置
constant pressure drop 定压降，恒压降，恒压差
constant pressure expansion 等压膨胀
constant pressure flow controller 稳压流量控制器
constant pressure gas turbine 等压燃气轮机
constant pressure heating process 定压加热过程
constant pressure line 等压线
constant pressure operation 定压运行
constant pressure process 等压过程，定压过程
constant pressure pump 恒定泵
constant pressure scale 定压气体温度计标度
constant pressure specific heat 等压比热，定压比热
constant pressure surface 等压面
constant pressure tank 稳压水箱
constant pressure turbine 自由射流式水轮机，冲击式水轮机
constant pressure valve 恒压阀
constant pressure 等压，定压，恒压，等压力
constant-property 常物性
constant radius arch dam 等半径拱坝
constant-rate creep 等速蠕变率
constant-rate flow 恒定流量，常流
constant rate of loading 常速加荷
constant rate of penetration test 等贯入率试验，等速贯入试验
constant rate of up-lift test 等速上拔试验
constant rate period 恒速期，恒定变化期
constant rating 持续容量，连续运行出力
constant-reaction blade 等反动度叶片
constant-reaction design 等反动度设计
constant resistance network 恒阻网络
constant resistance 恒定电阻
constant rotating field 恒定旋转场
constant scanning 等速扫描
constant section 等截面
constant size 等尺寸
constants of equivalent T-network 等值T形网络常数
constants of four-terminal network 四端网络常数
constant specific heat 定比热
constant specific mass flow 等质量流率流，等密流
constant-speed and constant-frequency WTGS 恒速恒频风电机组
constant speed synchronous generator 恒速同步电机
constant speed control 等速控制，定速调节
constant speed drive 等速驱动，匀速转动装置
constant speed electrical generator 恒速发电机
constant speed fan 定速风机
constant speed fixed pitch blade 恒速定桨距叶片
constant speed governor 定速控制器，恒速调节器
constant speed motor 等速电动机，等速马达，恒速电动机
constant speed operation 恒速运行
constant speed pump 定速泵
constant speed regulator 等速调节器，定速调节器
constant speed rotor 恒速风轮，定转速风轮【风力发电】，恒速转子
constant speed setting 恒速调整，恒速调节，恒速设定
constant speed squirrel cage induction generator 恒速鼠笼式异步发电机
constant speed wind turbine 恒速风电机组
constant speed 定速，恒速，等速
constant strain criterion 等应变准则
constant strength 等强度
constant stress layer 等应力层
constant supply power 定时供应电力，稳定供应的电力
constant surface heat rate 常表面热流
constant temperature and constant humidity 恒温和恒湿
constant temperature and humidity machine 恒温恒湿机
constant temperature and humidity system 恒温恒湿系统
constant temperature and humidity 恒温恒湿
constant temperature bath 恒温槽，恒温浴
constant temperature compression 等温压缩
constant temperature damp test 恒温恒湿试验
constant temperature expansion 等温膨胀
constant temperature furnace 恒温炉
constant temperature-humidity unit 恒温恒湿机组
constant temperature line 等温线
constant temperature oven 恒温箱
constant temperature process 等温过程
constant temperature system 恒温系统
constant temperature wire anemometer 恒温热线风速计
constant temperature 等温，定温，恒温
constant term 常数项
constant-thickness arch 等厚（度）拱
constant time lag 定时滞后，固定延时，定时延迟
constant time range 时间常数区段，周期段，周期范围
constant time relay 定时限继电器
constant torque motor 恒力矩电动机，恒转矩电动机
constant torque 恒定转矩
constant TSR operation region 恒定TSR运行区
constant value control 定值控制，定值调节
constant velocity drive 恒速驱动
constant velocity recording 恒速记录
constant velocity scan 常速扫描
constant velocity 定速，等速
constant voltage and constant frequency 恒压及恒频
constant voltage cell 恒压电池
constant voltage charging 恒压充电
constant voltage control 恒定电压控制
constant voltage dynamo 定压发电机
constant voltage generator 定压发电机
constant voltage motor-generator 定压电动发电机
constant voltage motor 定压电动机

constant voltage power supply 恒压电源
constant voltage rectifier 定压整流器，稳压整流器
constant-voltage regulation 定压调节
constant voltage regulator 定压调节器，恒压调节器，稳压器
constant voltage source 恒压电源，恒电压制
constant voltage system 恒压制，恒定电压系统
constant voltage transformer 恒压变压器，定压变压器，恒定电压变压器
constant voltage transmission 定压输电，等电压输电
constant voltage unit 恒压装置，恒压器
constant voltage 恒压
constant volume air conditioning system 定风量空气调节系统
constant volume and variable-temperature system 定风量变温度系统
constant volume combustion 定容燃烧
constant volume curve 等容曲线
constant volume cycle 定容循环
constant volume heating 定容加热
constant volume line 定容线
constant volume process 定容过程
constant volume scale 定容气体温度计标度
constant volume specific heat 比定容热容
constant volume test 等容积试验
constant volume 等容，定容，定容积
constant 不变的，恒定的，常数，常量，不断的，系数
CONST＝constantan 康铜，铜镍合金
CONST＝ constant 恒量，常数，不变的，恒定的
CONST＝construction 构造，结构
constellation 星座
constituent company 子公司
constituent corporation 子公司
constituent element 组成元件，组成部分
constituent of capital investment 基建投资构成
constituent 构成，组成，分量，分力，组分，成分
constituted authority 合法当局
constitute 组成，制定，设立，构成
constitutional detail 结构零件
constitutional diagram 组合图，平衡图，组成图
constitutional equation 本构方程
constitutional 本质的
constitution diagram 组合图，相图，平衡图，状态图
constitution water 化合水，组织水
constitution 构成，构造，结构，章程
constitutive equation 要素方程，本构方程
constitutive law 构造定律
constitutive relation 本构关系
constrained body 约束体
constrained condition 约束条件
constrained current operation 强励运行
constrained diameter 限制粒径
constrained magnetization 强制磁化
constrained-off unit 约束下机组
constrained-on unit 约束上机组
constrained oscillation method 约束振动法
constrained oscillation 强制振荡，强迫振动，约束振荡
constrained trading schedule 有约束交易计划
constrained vibration 强迫振动
constrained vortex 约束涡
constrained 拘泥的，限定的
constraining moment 约束弯矩
constraint condition 约束条件
constraint effect 约束效应，（风洞）边界效应
constraint matrix 约束矩阵
constraint 强制，限制，约束，制约，限制因素
constrict 收缩
constricted furnace 缩腰炉膛
constricted section 收缩段，束狭段
constricting nozzle 收缩喷嘴，压缩喷嘴
constriction coefficient 收缩系数，压缩系数，收缩率
constriction meter 缩口测流计
constriction resistance 收缩电阻，集中电阻
constriction scour 束狭冲刷
constriction 压缩，收缩，收敛管道
constrictor 收缩段，缩腰段，压缩装置，收缩器
constructability 可建设性，施工可能性
construction access road 施工进场道路，施工道路
construction access 施工对外交通
construction age 建设周期
construction agreement 施工协议书
constructional column 构造柱
constructional defect 施工缺陷
constructional deficiency 施工缺陷
constructional drawing 构造图
constructional elements of electrical apparatus 电器构件
constructional features 结构特征
construction all risks 工程一切险
constructional material industry 建筑材料工业
constructional plain 堆积平原
constructional terrace 堆积阶地
constructional valley 堆积谷
construction and operating license 建造及运行许可证
construction area 施工区
construction arrangement 结构布置
construction bank 建设银行
construction branch of power industry 电力建设部门，电力工业施工单位
construction bridge 施工桥
construction brigade 工程队
construction building 基建楼
construction calendar 施工日程表，施工日历表
construction camp 施工工棚，施工临时房屋，施工营地
construction capacity 建设容量
construction change directive 施工变更指示
construction chart of floors and roofs 地面屋面构造表
construction chart of roads and yards 道路堆场构造表

construction clearance 建筑接近限界
construction company （工程）建设公司，施工单位，施工公司，工程公司
construction condition 施工条件
construction contract 施工合同
construction control 施工管理，施工控制
construction cost per kW 单位千瓦造价
construction cost 工程造价，施工费用，工程费用
construction, design, manufacture and installation stage 建造、设计、制造和安装阶段
construction design 施工设计
construction details 构造详图，建筑细部，施工详图
construction detour 施工绕行便道
construction diversion 施工导流
construction document 施工文件
construction drawing delivery schedule 施工图交付进度
construction drawing design phase 施工图设计阶段
working drawing 施工图
construction elevator 施工电梯，施工升降机
construction engineering corporation 建筑工程公司
construction engineering 建筑工程，施工工程
construction engineer 基建安装工程师，施工工程师
construction enterprise 施工企业
construction equipment 施工设备，施工装备
construction experience 施工经验
construction for nonproductive purposes 非生产性建设
construction for pole and tower foundation 杆塔基础施工
construction fund 建设资金
construction gallery 施工廊道
construction general layout 施工总平面（布置）图
construction guide 施工导则，施工准则
construction hoist 施工吊车，施工起重机
Construction Industry Research Information Association 建筑工业研究与情报协会【美国】
construction industry 建筑工业
construction inspector 施工检查员
construction and installation worker 建筑安装工
construction intensity 施工强度
construction interface 施工接口，施工界面
construction intermediate schedule 施工中间进度表
construction in the dry 陆上施工
construction item 施工项目
construction job allotting 施工作业分配
construction dispatching 施工调度
construction job sheet 施工任务单
construction joint surface 工作缝缝面
construction joint 建筑填料接缝，混凝土的接缝面，工作缝，施工缝
construction journal 施工日志
construction kicked off 施工开始
construction layout drawing 土建图
construction lead time 建造周期
construction load 施工荷载
construction loan 施工贷款
construction machinery and tools 施工机械机具
construction machinery management 施工机械管理
construction machinery 施工机具，工程机械，施工机械，施工设备
construction machine 施工机械
construction management plan 施工组织设计，施工管理方案
construction management 施工管理
construction manager 施工经理，工程经理
construction manpower 施工力量
construction map 施工场地图
construction material 建筑材料
construction method statement 施工方案，施工方法说明
construction method 施工方法
construction norm 施工定额
construction of cable 电缆结构
construction of condenser 冷凝器的构造
construction of conductor 导线结构
construction of insulation protection cover 保护层施工
construction of line 线路架设
construction of thermal insulation 保温施工
construction of winding 绕组结构
construction opening 安装开口
construction & operation company 施工和运行单位
construction operation procedure 施工程序
construction operation program 施工程序
construction operation 施工作业
construction organization design 施工组织设计
construction organization engineering 施工组织设计
construction overhead 施工杂费
construction period impounding 施工期蓄水
construction period risk 工期风险
construction period 施工工期，施工阶段，施工期
construction permit 施工许可证
construction phase 施工阶段
construction planning 施工规划
construction plant 施工辅助工厂
construction plan 建筑平面图，施工方案，施工计划，施工平面图，结构图
construction power supply 施工用电
construction preparation 施工准备
construction procedure 施工步骤，施工程序
construction program 施工程序，施工计划
construction progress report 施工进度报告
construction project evaluation programme 建造阶段电厂（安全）审评计划
construction project manager 施工项目经理
construction project 基建项目，建设项目，施工项目
construction quality 施工质量，工程质量
construction quota 施工定额
construction report 施工日报表

construction road 施工道路
construction scale 建设规模，建议规模
construction scheduler 构造调度程序，施工进度协调员
construction schedule 工地计划，施工进度，施工计划，施工进度表，施工进度计划
construction scheme 施工方案，施工计划
construction seasons 施工季节
construction sequence 施工程序，装配顺序，施工步骤
construction service 施工服务
construction shaft 施工导洞，施工竖井
construction sign 施工标志
construction site general layout 施工总布置图
construction site plan 建筑场地平面图
construction site visit 施工现场调查
construction site 施工现场，工地，施工工地，施工场地，建设工地
construction specification 施工说明（书），施工规范
construction stage 建造阶段，施工阶段，施工期，施工标桩
construction stake 施工标桩
construction steel trestle 施工用钢栈桥
construction steps 施工步骤
construction subcontract 施工分包合同
construction summary schedule 施工进度汇总表
construction superintendent 施工主任
construction supervision 施工监督，施工监理，施工检查
construction survey 施工测量
construction team 施工队
construction technique 施工技术
construction test 结构试验
construction timber for waterworks 水工建筑木材
construction tolerance 施工容许误差
construction tool 施工工具
construction unit 施工单位
construction version 施工版
construction(/working) drawing 施工图
construction work(s) 施工工程，建筑工程，基建工程，建造工程
construction wrench 大型安装扳手
construction yard 施工场地
construction 施工，建造，架设，构造，结构，建造物，建筑物
constructive comment 建设性意见
constructive interference 相长干涉
constructive negotiation 建设性谈判
constructive suggestion 建设性意见
constructive total loss 推定总损耗
constructive wave 堆积浪
constructor 施工单位，建造者，施工人员
construe 理解（为），解释
constsp(constant speed) 恒速
consttemp(constant temperature) 恒温
consular invoice 临时发票，领事签证发票
consulate 领事馆
consultancy service 咨询服务
consultancy 咨询业务
consultant agreement 咨询协议
consultant expert 咨询专家
consultant firm 咨询公司，顾问公司
consultant 咨询，顾问，咨询员，咨询人员，咨询师
consultation machinery 协商机构
consultation service 咨询服务
consultation 磋商，协商，咨询
consultative committee 顾问委员会，咨询委员会
consultative department 咨询部
consultative management 协商管理
consultative panel 咨询小组
consultative 顾问的
consulting company 咨询公司
consulting engineer 顾问工程师，咨询工程师
consulting firm 顾问公司
consulting service 咨询服务
consult 参照
consumable electrode 熔化电极，自耗电极
consumable guide electroslag welding 自耗导向电渣焊
consumable guide 消耗品手册［指南］
consumable insert ring 易熔垫圈
consumable insert 自耗嵌块【焊接，冶金】，自熔焊材，易熔垫圈，易损插件，熔化的垫板，可熔化嵌条
consumable load 消耗负荷
consumable material 消耗材料
consumable 可消耗的，消耗品
consumbuble-electrode melting 自耗电极熔炼
consumed energy 消耗能量
consumed power 消耗功率
consumed work 消耗功
consumer-classified electricity tariff 用户分类电价
consumer heat inlet 热力入口
consumer heat substation 用户热力分站
consumer information system 用户信息系统
consumer price index 消费者价格指数
consumer's installation 用户装置
consumer's kilowattmeter 用户电度表
consumer's load 用电负荷
consumer's power factor 用电功率因数
consumer's power 用电功率
consumer substation 用户热力站，热力点，用户变电站
consumer surplus 消费者剩余
consumer waste 生活垃圾
consumer 用户，消费者，消费户
consume 消费，消费额，消耗量，消耗，浪费
consummate 完善的
consumption characteristic 消耗特性
consumption curve 耗（水）量曲线，消耗曲线
consumption factor 消费因数
consumption of heat 热耗
consumption of petroleum 石油的消耗
consumption of steam 汽耗
consumption of steel, cement and timber 三材消耗量【钢，水泥，木材】
consumption peak 最大消耗量，消耗高峰，用量高峰
consumption penalty 消耗量增大

consumption quota of standard project 标准项目消耗定额
consumption rate of coal equivalent for power supply 供电煤耗
consumption rate of coal equivalent 标准煤耗
consumption rate 消耗率，耗电率
consumption ratio 厂用电率
consumption tax 消费税
consumption 消耗（量），耗量，耗损，消费
consumptive requirement 消耗需量，消耗量
consumptive use of water 耗水量，消耗用水
consumptive use 消耗性使用，消耗用（途），耗水量
consumptive water 空耗水
consutrode 自供焊条的电极，自耗电极【电焊】
contact action 接触作用，接触动作
contact actuation time 触点动作时间
contact adhesion 触点黏结
contact aerator 接触曝气池
contact aerial 接触架空线
contact agent 接触剂
contact air gap 触点间距，触点气隙
contact alignment 触点校正，触点校准
contact alloy 接触器用合金，接点合金
contact angle 接触角
contact arc 接触弧，接触弧片
contact area 触点面积，接触面积
contact arm 触点臂杆
contact assembly 触点组件
contact attemperator 混合式减温器
contact bank 触点组，触点排
contact bar 接触片，接触条
contact bed method 接触床法
contact bed 接触床，接触滤床
contact biological filter 接触生物滤池
contact blade 触头片，接触闸刀片
contact block 触点群，触点块
contact bounce time 颤动跳动时间，触头跳动时间
contact bounce 触点颤动，颤动跳动
contact breaker 接触断路器
contact bridging 触点桥接
contact brush 接触电刷
contact burn 触点烧蚀
contact capacity 触点容量
contact chatter time 触点抖动时间
contact chatter 触点抖动，触点震颤
contact clearance 触点间隙
contact clip 接触线夹，闸刀开关静触头，接线柱
contact closing capacity 触点的闭合能力
contact closure 触点闭合
contact coagulation 接触凝聚
contact combination 触点组，触点组合
contact compliance 触点顺从度，触点柔度
contact conductane 接触热导
contact conductor 接触导线，电极引线
contact contamination 触点污染
contact continuity 持续接通
contact converter 接触式换流器
contact cooling 接触冷却
contact corrosion 接触腐蚀，触点腐蚀
contact current-carrying rating 触点额定载流量
contact current 触点电流
contact desulfuration 接触脱硫
contact detector 接触检波器
contact device 接触器，接触装置
contact disk 接触盘
contact drop 接触电压降，触点电压降
contacted cladding 接触包壳，坍塌包壳
contact electricity 接触电
contact electrode 接触电极
contact electromotive force 接触电动势
contact element 接触元件
contact EMF 接触电动势
contact erosion 触头腐蚀，触点侵蚀，触头磨损，接触冲刷
contact evaporation 蒸电极
contact failure 接触失效
contact fatigue wear 接触疲劳磨损
contact fatigue 接触疲劳
contact fault 接触故障，触点损坏
contact filter 接触滤池
contact filtration 接触过滤
contact finger 接点指，接触压指
contact flocculation 接触絮凝
contact follow stiffness 触点随动劲度，触点随动强度
contact follow 触点跟踪，触点随动距离
contact force 触点压力
contact for signaling 信号接点
contact for tripping 脱扣接点
contact for vacuum electrical apparatus 真空电器触头
contact-free 无触点的
contact gap 触点间隙
contact group 联系小组
contact grouting 接触灌浆
contact head 触头
contact heating surface 触点热面积
contact heat transfer coefficient 接触传热系数
contact holding indicator 接触保持指示器
contact impact 触点冲击，接触冲击
contact impingement 触点冲击
contact indication device 触点指示装置
contact information of bidder 报价员通信信息
contacting area 接触面（积）
contacting efficiency 接触效率
contacting pattern 接触形式，接触图形，（齿轮的）接触斑点
contacting plunger 接触插棒
contacting temperature measurement method 接触式测温法
contact inspection 接触（法）检验
contact interface 接触分界面
contact interrupting capacity 触点的切断能力
contact jaw 接触夹片，接触端
contact layer 接触层
contactless pick-up 无触点传感器
contactless relay 无触点继电器
contactless switching device 无触点开关装置
contactless type automatic voltage regulator 无触

点的自动调压器
contactless-vibrating bell 无触点式振动铃
contactless 不接触的，无接触的，无触点的
contact lever 接触杆
contact line 接触线
contact load 触点负载，接触荷载
contact logic 触点逻辑
contact loss 接点损耗
contact lubricant 接触润滑剂
contact maintenance 直接维修
contact maker 接合器，断续器
contact-making clock 接触电钟
contact-making voltmeter 触点式伏特计
contact-mask read-only memory 接触掩膜只读存储器
contact melting voltage 触点融化电压
contact member 触点元件
contact metamorphic action 接触变质作用
contact metamorphism 接触变质作用
contact microphone 接触式传声器
contact miss 触点失误，接触失误
contact network 触点网络，接点网络
contact noise 接触噪声
contact nomenclature 触点命名，触点名称
contact of solar cell 太阳电池电极
contactor controller 接触器控制器
contactor control 接触器控制
contactor density 接点密度
contactor-fuse combination 接触器熔丝组合
contactor panel 接触器盘
contactor relay 中继接触器
contactor sequence 接触器接触次序
contactor servo mechanism 接触式伺服机构，继电器伺服机构
contactor starter 接触启动器
contactor 触头，接触器，触点，开关，断续器，电流接触器
contact overload capability 触点过载能力
contact overtravel 触点超行程，触点过调
contact panel 接点板
contact parting time 触点断开时间
contact person 联系人
contact pick-off 接触传感器
contact pick-up 传振器
contact piece 接触元件，接触片
contact pin 触针，触片，接触柱
contact plane 触点平面，接触平面
contact planning 密集规划
contact plate 接触片
contact pointer 接触指示器
contact point 接触点
contact potential barrier 接触势垒
contact potential 接触电势，接触电位
contact pressure fluctuation 接触压力波动
contact pressure 接触压力
contact print 晒图
contact protection 触点保护
contact rating 触点额定值，触点容量，触点规格
contact ratio 接触比
contact reactor 接触反应器
contact rectification 接触式整流
contact rectifier 接触整流器，干式整流器，金属整流器
contact relation 接触关系
contact relay 有触点继电器
contact resistance 接触电阻，触点电阻，触头接触电阻
contact ring 接触环，滑环
contact rod 接触棒
contact roughing filter 接触粗滤池
contact scanning 接触扫描
contact seal 接触密封
contact section 密实截面
contact sector 接触扇形片
contact segment 接触扇形片
contact sense module 触点输入组件
contact separation 触点间隙，触点分离
contact series 接触序列，触点序列
contact set 触点组
contact shoe 触靴，接触块
contact slider 滑动触点
contact slipper 接触滑块
contact softening voltage 触点软化电压
contact spacing 触点间距，触点间隙
contact spark 触点火花，触点跳火
contact spring pile-up 触点簧片组件
contact spring 触点弹簧，接触簧片，接触弹簧，压紧弹簧
contact stabilization process 接触稳定法
contact sticking 触点黏结
contact stress under dynamic conditions 动态接触应力
contact stress 接触应力
contact striking 接触引燃，触点燃弧
contact strip 接触滑板，接触条
contact surface of valve seat 阀座接触面
contact surface 接触面
contact switch 接触开关，触簧开关
contact tank 接触池
contact temperature 触点温度
contact tension 触点弹簧张力
contact terminal 接触端点，触头
contact therapy 接触放射治疗
contact thermal resistance 接触热阻
contact thermocouple 接触式热电偶
contact thermometer 接触式温度计，接触温度计
contact time difference 触点接触时差
contact time distribution 接触时间分布
contact time 接触时间
contact to earth 接地，触地，碰地
contact tongue 接触片
contact transfer time 触点动作时间
contact transistor 点接触式晶体管
contact travel 触点行程，接触行程
contact tube 导电嘴，焊嘴【熔化极弧焊机】
contact type cup anemometer 接触式风杯风速计
contact voltage breakdown 触点电压击穿
contact voltage drop 击穿压降，触点压降
contact voltage regulator 分级电压调节器，接触调压器

contact voltage　触点电压，接触电压，人可接触的电压
contact voltmeter　点接触式电压表
contact wear　触点磨损
contact welder　接触焊机
contact welding　接触焊
contact wipe　触点擦拭
contact wire　接触导线，接触线
contact with process fluid　与工艺流体接触
contact zone　接触带
contact　触点，接触，接点，接触器，接口，接触面，触头，联系
contaiminator　沾污物
contained fluid　容纳的流体
contained nuclear explosion　封闭式核爆炸
contained radioactivity　控制放射性，密封放射性
container cargo　货柜装货物
container clearance charge　清洁箱费
container damage charge　坏箱费用
container damage fee　坏污箱费
container detention charge　滞箱费
container dirtyness change　污箱费
container frame　容器底架
container freight station　集装箱货物集散站，散货仓库
containerization　集装箱化
containerized cargo　集装箱化的货物，集装箱（装运的）货物
containerized coal　集装箱式送煤
containerized freight　集装箱装运的货物
containerized transport　集装（箱）运输
containerize　集装化，用集装箱装，以货柜运送
container loading charge　内装箱费
containers on flat car　平板车装运集装箱
container stuffing charge　提箱费
container tipping　集装箱翻转
container truck　集装箱货车
container unloading　容器卸料
container　箱，集装箱，容器，货柜，壳体，罐，水箱，油箱，安全壳，贮存器
containing boron　含硼
containing vessel　安全壳【反应堆】
containment action　遏制措施，围堵措施
containment air cleaning system　安全壳空气清洁系统
containment air filtration system　安全壳空气过滤系统
containment and isolation of oil　油的封堵和隔离
containment annulus　环形空间【反应堆大厅】，安全壳环形空间
containment atmosphere control　安全壳内的大气监测，安全壳内的大气监视
containment atmosphere monitoring system　安全壳空气监测系统
containment atmosphere purge exhaust system　安全壳空气吹扫排放系统
containment barrier　安全壳屏蔽，安全壳屏障
containment building dome　安全壳厂房圆顶
containment building pressure test　安全壳厂房压力试验
containment building　安全壳厂房
containment by-pass accident sequence　安全壳旁路事故序列
containment by-pass sequence　安全壳旁路事故序列
containment cleanup system　安全壳内部过滤净化系统
containment cooling system　安全壳冷却系统
containment crane　安全壳内吊车，反应堆厂房吊车
containment deluge system　安全壳淹没系统
containment dike　防护堤
containment drain system　安全壳疏水系统
containment drain tank　安全壳疏水箱，安全壳泄水箱
containment emergency off-gas system　安全壳应急排气系统
containment equipment cooling water pump　安全壳设备冷却水泵
containment equipment cooling water return tank　安全壳设备冷却水回水箱
containment equipment hatch　安全壳设备口
containment facility　封隔设施，分隔设施
containment failure mode　安全壳失效模式
containment fan cooler　安全壳风机冷却器
containment flooding system　安全壳淹没系统
containment gas control system　安全壳气体控制系统
containment grouting　抑制灌浆
containment heat removal system　安全壳排热系统
containment hydrogen control system　安全壳氢气控制系统
containment hydrogen recombination system　安全壳氢复合系统，安全壳消氢系统
containment instrumentation system　安全壳仪表系统
containment instrumentation　安全壳测量仪表
containment integrated leak rate testing　安全壳整体泄漏率试验
containment integrity　安全壳的密封性，安全壳的完整性
containment iodine filtration system　安全壳碘过滤系统
containment isolating valve　安全壳隔离阀
containment isolation actuation signal　安全壳隔防触发信号
containment isolation and clean-up system　安全壳隔离净化系统
containment isolation system　安全壳贯穿件密封系统，安全壳隔离系统
containment isolation　安全壳隔离
containment leakage monitoring system　安全壳泄漏监测系统
containment leak rate test system　安全壳泄漏率测试系统
containment leak tightness　安全壳密封性
containment lockoff monitoring system　安全壳防护门泄漏监测系统
containment metal enclosure　安全壳的金属包层，安全壳金属包层

containment penetration 安全壳贯穿件
containment pressure suppression system 安全壳消压系统
containment pressure suppression 安全壳压力抑制
containment purge isolation signal 安全吹扫隔离信号
containment purge system 安全壳吹扫系统
containment recirculation cooling system 安全壳再循环冷却系统
containment shell penetration 安全壳贯穿件
containment shell （反应堆装置的）防事故外壳，包壳，安全壳壳体
containment sphere 球形安全壳，安全壳球体
containment spray 安全壳喷淋
containment spray heat exchanger 安全壳喷淋热交换器
containment spraying 安全壳喷淋
containment spray injection system 安全壳喷淋注入系统
containment spray pump 安全壳喷淋泵
containment spray recirculation system 安全壳喷淋再循环系统
containment spray system heat exchanger 安全壳喷淋系统热交换器
containment spray system 安全壳喷淋系统
containment steel liner 安全壳钢衬
containment structure 反应堆外壳结构，安全壳结构
containment subatmospheric pressure maintenance system 安全壳负压保持系统
containment sump pump 安全壳地坑泵，安全壳泵
containment sump 安全壳地坑
containment sweeping ventilation system 安全壳换气通风系统
containment system 安全壳系统，包容系统
containment ventilation and purge system 安全壳通风净化系统
containment vessel cooling water system 安全壳冷却水系统
containment vessel 安全壳壳体，保护壳，安全壳【反应堆】，密封外壳，密闭壳
containment wall 安全壳壁，安全壳墙
containment 安全壳【反应堆】，保留【裂变产物】，围堵【政策】，容量，容积，密闭度，保持，电容，遏制，包容，牵制，封隔，包含
contain 包含，含有，容纳，整除，除尽贮存，包括
contaminability 玷污性
contaminanted exhaust system 污染气体排放系统
contaminant loading 污染物负荷
contaminant release 除垢
contaminant 污染物质，污染剂，杂质，污染物，致污物，污垢
contaminated area 污染区，沾污区
contaminated clothing laundry 沾污衣服洗衣处，沾污衣服洗衣站
contaminated equipment 污染的设备
contaminated-exhaust system 污染气体排放系统，沾污气体排放系统
contaminated laundry 沾污工作服的清洗站
contaminated oil storage tank 污染油贮罐
contaminated rock 混染岩
contaminated water evaporator plant 污染水蒸发装置
contaminated water 污染的水，污染水
contaminated zone 污染区
contaminate 污染，沾污
contaminating airburst 污染性空中爆炸
contamination accident 污染事故
contamination boot 防污染鞋套
contamination control station 污染控制站
contamination control 污染控制，污染防治
contamination counter 污染计数器
contamination-decontamination cycle 污染去污循环
contamination dose 污染剂量
contamination factor 污染系数
contamination hazard 污染危险，污染危害
contamination meter 污染计量仪
contamination monitoring 污染监视
contamination monitor 污染监测器［记录器］，污染监测仪，污染监视仪
contamination rate 污染率
contamination sampling 污染采样，污染取样
contamination-suspected area 可疑污染区
contamination tester 污染程度测定器
contamination uniformity ration 污秽均匀比
contamination zone 污染带
contamination 污染，沾污，污染物，沾染，污秽
CONT(contact) 接口，接点
CONT(continuous) 连续的
CONT(control) 调节器，控制
contemplate 打算
contemporaneous erosion 同期侵蚀
contemporary records 同期记录
contender 竞争者
content-addressable memory 内容选址存储器，相联存储器
content addressable random access memory 存数寻址随机存取存储器
content addressable storage 内容定址存储器
content addressed memory 内容定址存储器
content analysis 内容分析
content gauge 液面计，水位仪，量油尺
content indicator 内容指示单元
contention 争论
content meter 含量测定器，液面计，水位仪，量油尺
content of clauses 条款目录
contents of proposal package 报价文件目录
content 含量，内容，容量，浓度，目录，容积，装载物
contestant 争辩者，争执方
contesting parties 争议各方
contest the validity of an award 对仲裁决定的有效性提出异议
context-driven line editor 上下文推动的行编辑器
context of contract 合同文本
context 上下文

contiguous sheet 邻接图幅
contiguous zone 毗连地区
continent 大陆，洲
continental air mass 大陆气团
continental air 大陆空气
continental anticyclone 大陆性反气旋
continental apron 大陆裙
continental block 大陆块
continental borderland 大陆缘边地域，大陆边缘地，大陆边界区
continental climate 大陆性气候，大陆气候
continental deposit 陆相沉积，陆地沉积，大陆沉积
continental displacement theory 大陆漂移说
continental displacement 大陆漂移，大陆移迁
continental divide 大陆分水岭
continental drift theory 大陆漂移说
continental drift 大陆漂移
continental edge 大陆边缘
continental facies 陆相
continental fringe 大陆边缘
continental hydrology 陆地水文学
continental ice sheet 大陆冰盖，大陆冰原
continental island 陆边岛
continentality index 大陆度指数
continentality 大陆性，陆性率，大陆度
continental margin 大陆外缘部，大陆边缘
continental mass 大陆块体
continental migration 大陆漂移
continental plateau 大陆高原
continental plate 大陆板块
continental platform 陆架
continental raised bog 高位沼泽
continental rise 大陆隆起
continental season wind climate 大陆性季风气候
continental sedimentation 大陆沉积，陆相沉积
continental segment 陆块
continental shelf plain 大陆架平原
continental shelf 大陆架，大陆棚，陆架，陆棚
continental shore plain 大陆架平原
continental slope 大陆坡，大陆斜坡
continental swamp soil 大陆沼泽土
continental talus 大陆架外斜面
continental terrace 大陆阶地，大陆台地
continental water balance 大陆水量平衡
continental wind 大陆风
continental 大陆的
continent-making movement 造陆运动
continent shelf 大陆架
contingence cost 意外费用
contingencies 偶然事故
contingency allowance 允许的偶然事故，不可预见费
contingency clause 偶发事故条款，意外事故条款，意外条款
contingency coefficient 列联系数
contingency cost 应急费，应急基金，不可预见费，不可预见费用
contingency evaluation 不可预见性评估
contingency fund 应急费，应急基金，意外损失基金，应急费用
contingency interrupt 偶然性中断
contingency level 意外事故级别
contingency measure 应急措施
contingency plan 临时计划，应急计划，预想事故方案
contingency reserve 应急储备金，应急准备金
contingency 应急费，意外，意外事故，可能性，偶然性，偶然误差，不可预见费，意外开支，意外事件
contingent fund 临时费
contingent on 视……而定，以……为条件
contingent survey 不定期检查
contingent 意外的，偶然的，偶然事故，意外事情
continual 不断的，连续的
continuance 持续时间，持续
continuation clause 延续条款，展现条款
continuation lead 使连通的连接线
continuation 继续，延续，扩建部分，延伸部分
continued-fraction expansion 连分式展开式
continued inspection 连续检查
continued on drawing 接续图
continued overflow 连续溢流
continue statement 连续语句
continuing appropriation 继续拨款
continuing committee 常设委员会
continuing improvement 持续改进
continuing project 持续的项目
continuity boundary condition 连续的边界条件，边界连续性条件
continuity cable bond 电缆铠甲的纵向连接器
continuity check 导通检验
continuity condition for water 水流的连续条件
continuity condition 连续条件
continuity contact 持续通电触点
continuity criteria 连续性指标
continuity equation 连续性方程，连续方程
continuity law 连续定律
continuity of command 控制的连续性
continuity of supply 供电连续性
continuity shock velocity 连续激波速度
continuity test 连续性试验，断续探查
continuity 连续性，持续性
continuous absorption 连续吸收
continuous-action servomechanism 连续作用伺服机构
continuous action 连续工作，连续作用
continuous aging 连续老化
continuous air monitor 空气连续监测器，连续大气检测器
continuous air particulate monitor 空气微粒连续监测器
continuous analyzer 连续（工作）分析器
continuous annealing 连续退火
continuous approximation 连续逼近
continuous ash-discharge grate 连续出灰炉排
continuous beam 连续梁
continuous belt weigher 连续计量的皮带秤
continuous bidding 连续报价
continuous blocks 连续块
continuous blowdown flash tank 连续排污扩容器

continuous blowdown flash vessel 连续排污扩容器
continuous blowdown piping 连续排污管
continuous blowdown 连续排污
continuous blow off equipment 连续排污装置
continuous boron analyzer 硼连续分析器
continuous breaking capability test 连续开断能力试验
continuous bucket elevator 多斗（式）提升机，链斗式提升机，斗式连续升运器
continuous bucket ladder type unloader 连续梯斗式卸煤机
continuous bucket ladder unloader 连续梯斗式卸船机
continuous canalization 梯级化【河道】
continuous capability 持续能力，稳态能力
continuous capacity 连续出力，持续功率
continuous carrying 连续动作
continuous casting and rolling technology 连铸连轧工艺
continuous casting direct rolling 连续铸造直接轧制，连铸直接热轧法
continuous casting segment 连铸的段样
continuous casting slab 连铸坯
continuous charge-discharge mechanism 连续装卸料机构
continuous charging grate 连续加煤炉排
continuous chemical feeding 连续投药
continuous circulation 连续环流
continuous cleaning 连续清洗，连续清理
continuous coal unloaded 连续卸料机
continuous coil 连续式线圈
continuous-column dissolver 柱式连续溶解器
continuous comparator 连续比较器
continuous concrete insert 连续的混凝土嵌入件
continuous concrete mixer 连续式混凝土拌和机
continuous concrete wall 连续混凝土墙，地下连续墙
continuous concreting 混凝土连续浇筑，连续浇筑混凝土
continuous contact model 连续接触型【流化床】
continuous contactor 连续接触器，持续作用接触器，连续萃取器
continuous controller 连续作用控制器，连续调节器，连续控制器
continuous control system 连续控制系统
continuous control 连续调节，持续控制，连续控制
continuous conversion kiln 连续转换窑
continuous cooling transformation curve of super-cooled austenite 过冷奥氏体连续冷却转变曲线
continuous core 连续心墙
continuous counter-current operation 连续逆流操作
continuous counterflow dissolver 连续逆流溶解器
continuous counter 连续计数器
continuous coverage 连续覆盖
continuous creep 连续蠕动
continuous current dynamo 直流发电机，直流电机
continuous current generator 直流发电机
continuous current motor 直流电动机
continuous current transformer 直流变换器，直流变流器
continuous current 恒电流，直流，恒向电流，连续流
continuous curve distance-time protection 平滑时限特性距离保护
continuous data 连续数据
continuous deformation 连续变形
continuous delivery 连续输水
continuous dew-point recorder 露点连续记录器
continuous diffusion 连续扩散
continuous discharge 持续放电
continuous disc type coil 连续盘式线圈
continuous disc winding 连续盘式绕组
continuous display 连续显示
continuous distribution 连续分布
continuous dust dislodging 连续除灰
continuous dust removal 连续除灰
continuous duty relay 连续工作继电器，持续运行继电器
continuous duty with short-time loading 短时负载持续工作制
continuous duty 持续稳定负载，恒载连续运行，连续负荷，连续工作制（度）
continuous dynamic fumigation 连续动态熏蒸
continuous echo sounding 连续回声测深
continuous electrode 连续电极
continuous electrolytic chlorination 连续电解氯化
continuous elevator 连续斗式提升机
continuous emission monitoring system 连续排放检测系统，烟气排放连续监测系统
continuous engineering education training 继续工程教育培训
continuous engineering education 继续工程教育
continuous erasure-rewrite method 连续擦除重写数据法
continuous erasure 连续消除记录
continuous extrusion 连续压挤，连续挤出，持续励磁
continuous face 连续工作面
continuous feeding 连续供给，连续供电，连续进料
continuous fillet weld 连续贴角焊
continuous filter 连续过滤器，连续滤层
continuous filtration 连续过滤
continuous fins 连续的肋片
continuous fin welder 鳍片连续焊机
continuous flip bucket 连续式挑坎
continuous flooding 持续淹没
continuous flow liquid monitor 连续液流监测器
continuous flow method 连续流动量热法
continuous flow oxygen regulator 连续供氧调节器
continuous flow pump 连续流泵，续流泵
continuous flow tank 连续流水池
continuous flow 连续水流，连续流
continuous flushing sedimentation basin 连续冲洗式沉沙池
continuous footing 连续底脚，连续基础，连续基脚

continuous foundation 连续基础
continuous frame 连续框架
continuous-frequency spectrum 连续频谱
continuous fuel replacement 连续换燃料
continuous full-load run 连续满载运行
continuous function 连续函数
continuous gap leakage monitoring 连续缝隙监漏
continuous girder 连续大梁
continuous gradation 连续级配
continuous grading 连续级配
continuous graphical recording 连续图示记录
continuous graphitization 连续石墨化
continuous guidance 连续操纵
continuous heating 连续采暖，连续供暖
continuous heavy-duty service 连续重负荷运行
continuous horizontal slit distribution 连续式水平狭缝分布器
continuous hypersonic tunnel 持续高超音速风洞
continuous information 连续信息
continuous input 连续输入
continuous insulation 连续绝缘，单质绝缘
continuous interstice 连续空隙【颗粒材料】，连续缝隙
continuous irradiation 连续辐照
continuous laminating 连续层压法
continuous layer 连续层
continuous leacher 连续浸取器
continuous level shift 连续电平移动
continuous line heat source 持续线热源
continuous line 实线
continuous liquid core 连续液芯
continuous load current 持续负载电流
continuous loaded cable 均压加感电缆
continuous loading test 连续加荷固结试验
continuous loading 持续负载，连续加载，连续加感，连续装料
continuous load 连续负荷，持续负载，连续荷载
continuous loop economizer 连续环路省煤器，盘管式省煤器
continuous loop 盘管，蛇形管圈，蛇形管，连续环路，连续循环
continuously adjustable inductor 平滑调整电感线圈，连续可变电感线圈
continuously adjustable resistor 连续可调电阻，平滑调节电阻
continuously adjustable setting range 连续可调整定值范围，连续调整范围
continuously distributed 连续分布的
continuously emission monitoring system 连续排放监测系统
continuously loaded cable 均匀加感电缆
continuously loaded consolidometer 连续加荷固结仪
continuously loaded line 均匀加感线路
continuously loading 连续装料
continuously maximum rating 持续最大功率
continuously measuring control system 连续测量的控制系统
continuously rated motor 连续额定运行电动机
continuously rated 按连续工作额定值运行，按持续负载计算的
continuously recording sensor 连续记录的传感器
continuously reinforced concrete 连续配筋混凝土
continuously-running duty 连续运行方法，连续工作方式
continuously sensing control system 连续传感控制系统，连续传感自控系统
continuously variable control 连续可变控制，连续调节控制，均匀调整
continuously variable gearbox 无级变速器，连续可变变速器
continuously variable selectivity 连续变化的选择性
continuously variable transformer 连续调节变压器
continuously variable tuning 连续可变调谐
continuously variable 连续可变的
continuous maximum rating 持续最高额定值，持续最大功率，连续最大出力，最大连续功率，持续最大工况，最大连续蒸发量
continuous measurement 连续测量
continuous medium 连续介质
continuous miner crew 连采队
continuous mixer 连续式拌和机
continuous-mixture method 连续混合（量热）法
continuous model 连续介质模型
continuous monitoring 连续监测
continuous monitor 连续监测器
continuous motion 连续运动
continuous network 连续网络
continuous neutron activation analysis 连续中子活化分析
continuous neutron spectrum 连续中子能量，连续中子能谱
continuous occupational exposure 连续职业性照射
continuous operating condition 连续运行工况
continuous operating period 连续运行期
continuous operation program 连续操作程序
continuous operation test 连续运行试验，持续操作测试
continuous operation 连续运行，连续运算，连续操作，持续运行
continuous operator 连续算子
continuous oscillation 等幅振荡，连续振荡
continuous output 连续出力，连续输出，持续出力
continuous overload limit 持续超过载荷极限
continuous overrange limit 持续超过限制范围
continuous oxygen analyzer 氧连续分析器
continuous performance test 连续作业测验，持续性操作测试
continuous permafrost 成片永冻层
continuous phase 连续相
continuous placement method 通仓浇筑法
continuous placing of concrete 连续浇筑混凝土
continuous positive disk seating force 连续正的阀瓣落座力
continuous pot dissolver 连续罐式溶解器
continuous pouring 连续浇筑
continuous power 连续功率，持续功率，连续发电

continuous process control 连续过程控制
continuous processing 连续处理，在役处理
continuous process 连续过程
continuous production 连续生产
continuous rain 连绵雨，绵雨
continuous random network model 连续无规则网络模型
continuous random network 连续随机网络
continuous random variable 连续随机变量
continuous rated machine 连续额定运行电机
continuous rating power 持续功率，连续功率
continuous rating 持续功率，连续功率，长期运转的定额值，连续出力，持续额定功率，持续额定容量，持续运转额定值，连续额定出力，连续额定值
continuous reading refractometer 连续读数式折射计
continuous recognition 连续记录
continuous recorder 连续记录器
continuous reduction 连续下降，坡式下降
continuous refuse-burial machine 连续埋垃圾机
continuous repositioning 连续位移，连续移动
continuous revetment 连续式护岸
continuous rheostat 平滑调节变阻器
continuous rigid frame 连续刚架
continuous rope drive 无级绳传动
continuous running duty-type 长期连续工作制的（设备）
continuous running duty 连续运行状态［方式］【旋转机械，马达等】
continuous running voltage 持续运行电压
continuous running 持续运转
continuous sampling device 连续取样装置
continuous sampling method 连续取样法
continuous sampling 连续采样，连续取样
continuous sensitivity control 灵敏度连续调整，灵敏度连续控制
continuous sequence 连续序列
continuous service 连续运行，连续工作
continuous servo 连续作用的随动系统，连续作用的伺服系统
continuous sheet metal 连续薄钢板
continuous sheet micrometer 连续式钢板测厚计
continuous signal 连续信号
continuous slab 连续式面板，连续板
continuous slowing-down approximation 连续慢化近似
continuous slowing-down model 连续慢化模型
continuous slowing-down theory 连续慢化理论
continuous slowing-down treatment 连续慢化处理
continuous source 连续源
continuous spectrum 连续（光）谱，连续谱
continuous speed regulation 无级调速
continuous statement 连续语句
continuous stationery reader 连续纸带阅读机
continuous stationery 连续纸带
continuous stave pipe 连续板条式水管
continuous steel truss 连续钢桁架
continuous strand mat 连续毡，连续玻璃毡，连续原丝毡
continuous stream conveyor 埋刮板输送机
continuous stress 持续应力
continuous sweep 连续扫描
continuous system 连续系统
continuous testing 持续（放电）试验，连续试验
continuous thread stud 连续螺纹杆
continuous thunder and lightning 连续雷电
continuous-time-rated motor 连续额定运行电动机
continuous time 连续时间
continuous treatment 连续处理
continuous tube 蛇形管
continuous tube element 蛇形管元件
continuous tuned AC filter in converter station 换流站连续可调交流滤波装置
continuous variable 连续变量
continuous velocity log 连续声速测井
continuous ventilation 连续通风
continuous vibration 连续振动，等幅振动
continuous-voltage-rise test 持续升压试验
continuous vorticity 连续性涡场
continuous wash basin 长条盥洗池，长条盥洗盆
continuous wave oscillator 连续波振荡器
continuous wave telegraphy 等幅波电报，连续波电报
continuous wave transmitter 等幅波发射机
continuous wave 等幅波，连续波，等幅振荡
continuous weld bead 连续焊缝头【指没有搭接，再起弧】
continuous welding 连续焊接，连续焊
continuous weld 连续焊，连续焊缝
continuous winding 连续绕组，连续式线圈
continuous window 连续框格窗
continuous wind tunnel 连续式风洞
continuous working period 持续运行时间，持续工作时间
continuous 连续的，持续的，不断的
continuum boundary 连续边界
continuum effect 连续介质效应
continuum flow 连续流，连续介质流动
continuum mechanics 连续介质力学，连续体力学
continuum model 连续介质模型
continuum regime 连续介质流动状态
continuum theory of nuclear reaction 核反应的连续构理论
continuum theory 连续介质理论
continuum 连续区域，连续流，连续介质，连续光谱，连续统一体，连续时间税
contiuous exposure 连续曝光
contiuous-field method 连续磁化法，连续磁粉探伤法
contiuous spectrum 连续谱
contiuous vibration 连续振动
contiuous-wave echo signal 连续波回波信号
contiuous X ray spectrum 连续 X 射线谱
contorl valve 调节（气）阀
contorted bed 扭曲层，褶皱层
contorted strata 扭曲地层
contortion fissure 扭曲裂缝
contour aperture area 廓采光面积
contour bank 等高堤

contour control system　外形控制系统
contour correction　（风洞）边界影响修正
contour disk　仿形阀瓣
contour ditch　等高沟
contoured-plug disk control valve　异形塞阀瓣控制阀
contoured tab　异形接头
contoured transformer　封闭型变压器
contoured unit　表面成型探头
contoured　外形的，等高线的，外围的
contour effect　轮廓效应，边缘效应
contour follower　等高线跟踪器，轮廓线跟踪器
contour furrow　等高沟
contour holography　外形全息照相术
contouring accuracy　等高线绘制精度
contouring control　轮廓控制
contouring machine　仿形机床，轮廓锯床
contouring　等高线绘制，仿型的
contour integral　围线积分
contour integration　周线积分
contour interval　等高线间距，等高距，等高线间隔
contour-length method　等高线延长法
contour line　等值线，等高线，等强线，轮廓线
contour mapping　等高线绘制
contour map　等高线地图，等高线图，等值线图
contour microclimate　地形性小气候
contour model　程序执行模型
contour of equal travel time　等流时线
contour of noise　噪声等值线
contour of profile　翼型外形
contour of valley　河谷形态
contour of water table　水位等高线
contour parameter　外形参数
contour plane　等高面
contour plan　地形图，等高线平面图
contour plate　压型板，仿形样板
contour stripping　等高开采，等高剥离
contour terrace　等高阶地
contour trench　水平沟
contour zone　稳定气流区
contour　外形，等值线，恒值线，等高线，线路，（轮廓）线，断面，周线
contraband articles　违法物品，违禁物品
contraband　禁运品，走私货，违禁物品
contraclinal valley　逆斜谷
contra-clockwise　逆时钟方向，反时针方向
contract acquired property　合同所规定的性能
contract administration　合同管理
contract agreement　合同协议书，合同修改书
contract amount　合同总额［总价］，承包额，合同金额
contract an enterprise　承包企业
contract appendixes　合同附件
contract authorization　订合同权，合同授权
contract award date　合同授予日期
contract awarded project　中标项目
contract awarded　中标合同，合同中标
contract award　授标，合同判授，合同签约，签订合同
contract bond　契约保证金，合同保证金
contract bonus system　承包奖金制
contract cancellation　取消合同
contract carrier　契约承运人
contract certificate　合同凭证
contract change notice　合同更改通知
contract change request　合同更改请求
contract clause　合同条款
contract committee　合同委员会
contract construction　发包工程，承包工程，承包施工
contract currency　合同货币
contract date　签约日期，合同日期，契约履行期限
contract debt　合同债务
contract-demand tariff　合同负荷电费率
contract document　合同文件
contract draft　合同草稿
contract drawing　施工图，发包图样
contract duty　合同义务
contracted channel　收缩通道，收缩形流道
contracted coefficient　收缩系数
contracted drawing　缩图，缩绘
contracted enterprise　承包企业
contracted jet　收缩射流
contracted load　约定负荷，合同负载
contracted notation　简化的符号
contracted opening discharge measurement　缩孔流量测定，缩孔流量测定法，缩孔测流法
contracted opening method　缩孔法
contracted project　承包工程
contracted responsibility system　承包责任制
contracted section　收缩断面，收缩截面
contracted waterway　束狭水道
contracted weir　收缩堰
contracted width　收缩宽度
contracted　简化的，合同规定的，契约的，收缩了的，狭小的
contractee　发包方
contract effective date　合同生效日期
contract energy decomposition　合同电量分解
contract engineering　工程承包
contract equipment　合同设备
contract evidenced by written form　书面证明的合同
contract expiry　合同期满
contract fee　合同费
contract files　合同卷宗
contract for difference　差价合同
contract for future delivery　期货合同，远期交货合同
contract for goods　订货合同
contract for installation work　安装工程合同
contract for labor service　劳务合同
contract for labour and materials　包工包料，包工包料合同
contract formation　合同构成
contract form　合同格式，合同模板
contract for purchase　采购合同
contract for service　劳务合同
contract for single item　单项工程合同
contract for survey and design　勘测设计合同

contract for the delivery of goods by installments 分批交货的买卖合同
contract general 合同综述
contract goods 合同货物
contractibility 收缩性
contractile strain ratio 收缩变形比
contract implementation phase 合同执行阶段
contracting agency 承包代理人，承包机构
contracting brake 抱闸，带闸
contracting current 收缩水流
contracting-expanding nozzle 缩放喷管，缩放喷嘴
contracting nozzle 收缩喷嘴
contracting parties 缔约各方，签订合同的各方
contracting party 缔约方，订约当事方
contracting project 承包工程
contracting reach 收缩段
contraction at fracture 断裂处颈缩
contraction cavity 缩孔
contraction coefficient 收缩系数，收缩率
contraction cone 风洞收缩段，收缩圆锥
contraction crack 收缩裂纹，缩裂，收缩裂缝
contraction distance 收缩段距离
contraction factor 收缩因子
contraction in area 面积收缩
contraction joint grouting 收缩缝灌浆
contraction joint 伸缩接头，收缩缝，收缩节理
contraction loss 收缩损失，缩口损失
contraction of area 断面收缩
contraction of mass concrete 大体积混凝土体积收缩
contraction percentage of area 断面收缩率
contraction pressure loss 收缩压力损失
contraction ratio （喷嘴的）收缩比，收缩系数
contraction section 风洞收缩段
contraction strain 收缩应变
contraction stress 收缩应力
contraction works 束水工程，束狭工程
contraction 定约，收缩（量），断面收缩，收敛，简略字，缩写字，内积缩并，缩短
contract item 合同项目
contractive soil 收缩性土
contract labour system 合同工制度
contract language 合同语言
contract law 合同法
contract life 合同有效期
contract lump sum 承包总额
contract management subsystem 合同管理子系统
contract management 合同管理
contract managerial responsibility system for enterprise 企业承包经营责任制
contract negotiation 合同谈判
contract notes and statements 成交单据及结单
contract note 合同证明，合同说明，买卖合同，买卖契约，成交单据，买卖单据
contract notice 招标公告
contract number 合同号
contract of adhesion 同意加入合同
contract of affreightment 海运合同
contract of arbitration 仲裁合同，联合合同
contract of carriage by sea 海运合同
contract of carriage 货运合同，运输合同
contract of employment 雇佣合同
contract of future delivery 远期交货合同
contract of guarantee 担保合同
contract of guaranty 保证合同
contract of insurance 保险合同
contract of lease 租赁合同
contract of loss and compensation 损失补偿合同
contract of payment 支付合同
contract of power supply and demand 供用电合同
contract of price 合同价格
contract of renovation 更新合同
contract of sale 销售合同
contract of supply 供货合同
contract of technology transfer 技术转让合同
contractor-financed contract 承包商带资承包合同
contractor's equipment insurance 承包商设备险
contractor's equipment mobilization plan 承包商设备进场计划
contractor's equipment 承包商的施工机械
contractor's general responsibility 承包人一般责任
contractor site manager 承包商现场经理
contractor's obligation 承包商的义务
contractor's proposal 承包商建议书，承包商报价
contractor's representative 承包商代表
contractor's responsibility 承包商的责任
contractor's risks 承包商（应承担的）风险
contractor's submittals 承包商须提交的文件
contractor's superintendence 承包商负责监理
contractor supervisor 工地监工员
contractor unit 承包单位
contractor 承包人［商，方］，合同人［商］，承包公司［单位］，收缩段［管］，订约人，接触器，包工单位
contract outline schedule 合同轮廓进度
contract parties 合同当事方
contract-path method 合同路径法
contract period 合同期限，合同有效期
contract power plant 合同电厂
contract preparation 合同准备
contract price receivable 应收合同款
contract price received in advance 预收合同款
contract price 承包价格，发包价，合同价格，合同价，合约价
contract project manager 约聘专案经理
contract project 承包工程，发包工程，合同项目
contract provisions 合同规定，合同条款
contract-rate tariff 二部电价制
contract records 合同记录
contract renewal 合同续订，续约
contract requirement 合同要求
contract review 合同评审
contract right 合同权利
contract risk 合同风险
contract sales 订约销售
contract schedule 合同进度安排，合同进度表，合同工期
contract scope change 合同范围变更

contract serial number　合同系列（序）号，合同编号
contract signing　签订合同
contract stipulated price　合同规定价格
contract stipulations　合同规定，合同条款
contract sum　承包额
contract suspension　合同暂停
contract system　包工制，承包制，合同制
contract tax　合同税
contract template　合同模板，合同范本
contract termination　合同终止
contract terms　合同条款，合同条件
contract term　合同期（限）
contract time　承包期限
contract trading　合同交易
contractual acceptance　合同验收
contractual arrangement　合同安排
contractual claims　合同规定的索赔，合同涉及的索赔，根据合同的债权
contractual commitment　按合同承担的义务，合同承诺
contractual damages　合同引起的损害
contractual delivery　合同交货
contractual dispute　合同上的争议
contractual guarantee　合同规定的担保
contractual income　合同收入
contractual investment　合同性投资
contractual-joint-venture　合作经营，契约式联合经营
contractual J. V.（contractual joint venture）　联营体【合作形式】
contractual obligation　合同规定的义务
contractual liability　合同规定的责任，合同责任
contractual limit of time　合同规定的时间限，合同规定的时间限制
contractual management responsibility system　承包经营责任制
contractual objective　合同目的
contractual obligation　合同规定的义务，合同义务，合同责任
contractual payment　合同规定的付款
contractual personnel　订约人
contractual practice　合同惯例
contractual procedure　缔约程序
contractual purchase　合同采购
contractual relationship　合同关系
contractual release　合同的解除
contractual restrictions　合同的约束【条款】
contractual scope　合同范围
contractual specifications　合同规定，合同条件，合同规范
contractual terms and conditions　合同条款和条件
contractual transfer　合同转让
contractual　合同的，契约的
contract under seal　盖章合同，正式合同
contract unit　合同单位
contract variation　合同变更
contract wages　合同工资
contract with EVN　与越南电力公司签的合同
contract with　承包
contract worker　合同工
contract works　包工工程，发包工程，承包工程
contract　合同，契约，承包，收缩，订立合同
contradiction　矛盾
contradistinction　对比，区别，截然相反，对比鉴定，对照
contraflexure　反（向）弯曲，反挠，回折，反挠曲，反弯（曲）
contraflow condenser　逆流式凝汽器
contraflow gravel washer　逆流洗砾机
contraflow heat exchanger　逆流换热器
contraflow regenerator　逆流式回热器
contraflow　反向电流，额外电流，逆流，反流，反向流动
contragradience　逆步，反步
contragradient　逆步的，反步的，负梯度
contrail　凝结尾迹，航迹云
contrajet　反射流
CONTRAN　汇编指令的计算机程序语言
contrapolarization　反极化
contrarotation　反转，反向旋转
contrary current　逆流
contrary winds　逆风
contrary　相反的
contrast aid　提高对比度的涂层
contrast amplification　（音频的）对比度放大
contrast image　反差图像
contrast law　对比定律
contra-streaming instability　反向流动不稳定性
contrast sensibility　对比灵敏度
contrast threshold value　衬度阈值
contrast　对比，对照，对比度［率，性］，反差，对照物
contravalence　共价
contravane　逆向导叶
contravariance　逆变（性），抗变（性），反变性
contravariant component　逆变分量
contravariant vector　逆变矢量，反变矢量
contravene　违背，违反
contravention　触犯，反驳，违反
contra wire　镍铜线，康特拉铜镍合金丝
contra　相反，抗，逆
contribute one's payment in capital　出资
contribute the charter capital　注入注册资本金
contribute to　有助于，起到……作用，促成，贡献，捐献，捐赠，导致，投稿
contributing area　产流面积［区域］，积水面积，水流区域
contribution clause　责任分担条款
contribution to　出资，认缴
contribution　贡献，影响，作用
contrite gear pair　圆柱齿轮端面齿轮副，端面齿轮副
contrite gear　端面齿轮
contrite wheel　端面齿轮
control ability　控制能力
control accuracy　控制精度，控制精确度，调节精度，控制准确度
control action　控制作用，调节作用，控制动作
control agent　控制剂，调节体，控制媒介
control air system　控制用压缩空气系统
control air　控制用压缩空气，控制用空气，仪用

空气，工作空气
control ampere-turns 控制安匝【磁放大器】
control amplifier 控制放大器，调节放大器
control and automation department 热工分场
control and instrumentation laboratory 热控实验室
control and instrumentation rack 控制和仪表架
control and instrumentation 仪表和控制
control and monitoring system 控制和监控系统
control and primary scavenging pump 控制和主动回油泵
control and protection strategies 控制和保护策略
control and protection system 控制和保护系统
control and read-only memory 控制和只读存储器
control and switch equipment 控制和转换设备
control and test 控制和测试
control apparatus 调节器，控制装置，控制电器
control area balancing function 控制区平衡功能
control area 控制区域，控制面积，控制区
control arm 控制杆，操作杆
control assembly 控制部件，控制组件，控制装置
control automation system 控制自动化系统
control band 调节范围，控制区域
control bank 调节组，控制棒组
control battery 操作电池
control behavior 控制状态
control bit 控制位
control block diagram 控制［调节］框图
control block 控制程序块，控制单元，控制部件，控制组合，操纵板，操纵盘，操纵台，控制板，控制盘，控制屏，控制台，控制室
control boundary 控制区边界
control box 控制箱，操作箱
control break 控制中断
control brush 控制电刷
control building block 控制结构单元，控制部件
control building in substation 变电站主控制楼
control building 控制楼，电气厂房，控制厂房
control bulb 控制管
control bus 控制母线，控制总线
control button 控制按钮
control by intermittent operation 间歇调节
control cabinet 控制箱，控制室，操作室，控制柜
control cabin 驾驶舱，控制室，操纵室
control cable gallery 控制电缆廊道
control cable routing 导线敷设
control cable tunnel 控制电缆隧洞
control cable with copper tape shield 铜带屏蔽的控制电缆
control cable 控制电缆，仪用电缆，操纵索
control cam 控制凸轮
control capacitance 控制电容
control card 控制卡片
control catchment 控制流域
control cell method 控制栅元法
control cell 控制栅元
control center 控制室，控制中心，指挥中心，调度中心
control chamber 控制室
control channel 控制通道，控制回路，调节线路，控制渠道
control characteristic of a control system 控制系统的控制特性
control characteristic 控制特性，调节特性
control character 控制字符，控制符号
control chart 控制图
control circuit transformer 控制电路变压器
control circuit 控制回路，控制电路
control cluster 控制棒束
control cock 调整旋塞，控制旋塞
control code 控制码，管理规程
control coefficient 控制系数
control coil 控制线圈
control column 操作杆，控制杆
control combination 控制组合
control command 操纵指令，控制指令，控制命令
control company 控股公司
control compartment 控制室
control complex building 控制综合楼
control complex 综合控制楼
control component 控制元件
control computer 控制用计算机，控制计算机
control console 控制台，操作台，操纵台
control criterion 控制准则
control cross-section 控制断面
control cubicle 操纵室，控制室，控制柜
control current 控制电流
control cycle 控制周期，控制循环
control damper 调节挡板，调节风门
control data 控制数据
control design 控制系统的设计
control desk 控制台，操纵台
control deviation 调节偏差
control device for condenser cleaning 凝汽器清洗控制装置
control device 控制装置，控制器件，操纵装置，调节装置
control dial 控制表盘，控制刻度盘，控制盘
control diaphragm 控制膜
control discrepancy key 不对位控制按键，不对位控制开关
control discrepancy switch 不对位控制开关
control display 控制画面
control drive 控制驱动装置，控制棒驱动
control drum span 控制鼓转幅
control drum 控制鼓
control electrode 控制电极
control element assembly 控制元件组件，控制棒组件
control element drive mechanism 控制棒驱动机构
control element drive system 控制元件驱动系统
control element worth 控制元件当量，控制元件反应性
control element 控制元件
control engineering 控制工程
control engineer 控制工程师
control equipment room 控制设备室
control equipment 控制设备，控制装置

control exciter 控制励磁机
control facility 控制设备，控制装置
control factor 调节系数，控制因数
control field 控制场，控制区，控制字段
control floating action 浮点作用控制
control float 控制浮子
control fluid cooler 抗燃油冷却器
control fluid pump 抗燃油循环泵
control fluid regenerative device 抗燃油再生装置
control fluid regenerative pump 抗燃油再生泵
control fluid tank 抗燃油箱
control fluid 调节的流体，控制流体，控制液体
control flume 测流水槽
control footing 控制合计
control force 控制力
control frequency generator 控制频率振荡器
control frequency 控制频率
control function 控制作用，控制功能
control gap 可调间隙，调整间隙，可控间隙
control gas buffer tank 控制气体缓冲箱
control gas system 控制气体系统
control gas 控制气体
control gauge 校准量规，标准试块
control gear 控制传动装置，控制设备［机构］，调整装置，操作机构［装置］，控制器
control generator 控制发电机，控制振荡器
control grid 控制栅极，控制栅
control group 控制棒组
control handle 控制旋钮，控制手柄，操纵杆，操纵手柄
control hand wheel 调节手轮
control hardware 控制硬件
control header directory 控制标题目录
control hierarchy 控制级别，控制等级，控制层次，分级控制，控制分级
control hole 标志孔，控制孔
control house 控制室，操纵室
control impulse 控制脉冲，调节冲量，调节脉冲
control index setting 控制指标设置，给定值的整定，控制值整定，被调量给定值的设置
control index 控制指数
control information 控制信息，误差校正信息
control input/output 控制输入/输出
control instruction counter 指令控制计数器
control instruction 控制指令
control instrument 控制仪表，调节仪器，控制仪器，调节仪表，操纵器械
control interaction factor 控制相关因数，控制交互作用因素
control interval 控制间隔，调节间隔
control isolate valve diaphragm 隔离阀隔膜，控制隔离阀膜片
control joint block 控制缝块
control joint grouting 控制接缝灌浆，收缩缝灌浆
control joint 控制接缝，控制缝
control key 控制键
control knob 控制旋钮，控制按钮，操纵钮
controllability 可控性，可调节性，可控制性，可操纵性
controllable-area nozzle 可调截面喷嘴
controllable check valve 可控止回阀，可调单向阀，可控检验阀
controllable convertor arm 可控换流桥臂
controllable impedance 可调阻抗
controllable liquid-crystal display 可控液晶显示装置
controllable load 可控负荷
controllable losses screen 可控损耗屏幕
controllable orifice 可控阀，可控孔
controllable pitch propeller 调距螺旋桨
controllable reactance 可控电抗
controllable reaction 可控反应
controllable reactor 可控反应堆
controllable saturated reactor 可控饱和电抗器
controllable source 可控源
controllable sweep generator 可控扫描发生器
controllable transformer 可控变压器
controllable vane 可调叶片
controllable 可控制的，可调节的，可操纵的，可管理的
control lag 调整延迟，调节延迟，调节滞后
controlled access 被监控的通道
controlled area access monitoring system 控制区出入监测系统
controlled area of increased contamination probability 可能增加污染的控制区
controlled area ventilation system 控制区通风系统
controlled area 监管区，控制区
controlled atmosphere 受控大气
controlled attenuator timer 控制衰减计时器
controlled avalanche rectifier 受控雪崩整流器
controlled bidding price 招标控制价
controlled blasting 控制爆破
controlled burning 受控燃烧
controlled carrier system 控制载波系统
controlled change room area 更衣室控制区域
controlled circulating pump 控制循环泵，强制循环泵
controlled circulation boiler 控制循环锅炉【带有控制循环泵的汽包炉】，强制循环锅炉
controlled circulation 受控循环，强制循环，辅助循环
controlled communication 受控通信
controlled company 受控公司，被控制公司，分公司
controlled condition 受控条件，受控状态，控制条件
controlled cooling 强制冷却
controlled current source 控制电流源
controlled deviation 调节偏差，控制偏差
controlled diameter 限制粒径
controlled discharge 控制泄放
controlled disposal of radioactive effluents 放射性流出物的控制处置
controlled drainage 受控排水
controlled edition 受控版本
controlled electric clock 子钟，受控电钟
controlled extraction turbine 调节抽汽式汽轮机
controlled filling of reservoir 水库控制蓄水
controlled flooding 受控漫灌

controlled fluid 被控流体
controlled fusion reactor 受控聚变反应堆
controlled gate 控制闸门
controlled humidity drier 可控湿度干燥机
controlled-leakage pump 控制泄漏泵
controlled leakage seal 检验泄漏密封
controlled leakage 控制的泄漏
controlled leakoff 可控泄漏
controlled medium 被控介质
controlled member 受控元件，受控构件，控制对象，调整对象
controlled micro-machine 受控微电机
controlled mixing history furnace 可控混合过程炉
controlled motion 可控运动，被控运动
controlled motor 可控电动机，受控电动机
controlled multicyclone 可调式多管旋风子
controlled natural ventilation 有组织自然进风
controlled overvoltage test 可控过电压试验
controlled plant 被控/控制对象，调节对象
controlled pressure 受控压力
controlled process 受控过程，被控过程，调节对象
controlled reaction 受控反应
controlled reactor 受控反应堆，可控电抗器
controlled recirculation boilingwater reactor 受控再循环沸水堆
controlled rectifier 可控整流器
controlled sender 受控发送器
controlled series compensation 可控串联补偿
controlled source 受控源，受控电源
controlled spillway 有控溢洪道
controlled starting 受控起动
controlled start-up 可控起动
controlled start 可控启动
controlled station 被控站，子站
controlled storage 受控蓄水
controlled-strain test 控制应变试验，控制应力试验
controlled switch 控制开关
controlled system 被调节系统，被控系统，受控系统
controlled temperature pressure range wind tunnel 温度压力可调式风洞
controlled thermal severity test 受控热强度试验
controlled thermonuclear reaction 受控热核反应，受控聚变反应堆
controlled value 控制值
controlled variable 受控变量，受控变数，可控变量，被控参数，被调值，被控变量
controlled voltage source 受控电压源
controlled weir 有控制溢流堰
controlled zone 控制区
controlled 受控的，被控的，被操纵的，被调节的
controller action 调节器的作用
controller case 控制器箱
controller equation 调节器方程
controller gain 控制器增益
controller lag 调节器的滞后
controller lock 控制器的锁定
controller module 控制器模件
controller of WTG 风电机组控制器
controller output air 控制器输出空气
controller output 控制器输出
controller panel 控制盘
controller pilot valve 调节器导阀
controller pilot 调节器操纵阀
controller regulator 自动控制器
controller resistance 控制器电阻，调节器电阻
controller setting 控制器整定值，调节器整定值
controller signal processing cabinet 控制器信号处理柜
controller wiring diagram 控制器布线图
controller 控制器，调节器，传感器，检验员，操纵器，操纵员，管理体系，调节装置，总监
control level of surface contamination 表面污染控制水平
control lever 操纵杆，控制杆
control limit switch 极限控制开关
control limit 控制极限，控制范围，调节范围，调节极限
control line 控制线，控制线路，操纵线
controlling board 控制台，控制板，操纵台
controlling company 控股公司，控制公司，总公司
controlling condition 控制条件
controlling deposition 控制沉积
controlling depth 控制深度
controlling device 控制设施，控制设备，调节装置
controlling electric clock 主控电钟，母钟
controlling element 控制元件
controlling factor 支配因子
controlling field 控制场
controlling machine 控制机
controlling manhole 控制入孔，控制井
controlling means 控制方法，控制机构
controlling motor 控制电动机
controlling parameter 控制参数
controlling power range 控制功率范围
controlling quantity 控制量
controlling shareholder 控股股东，持有控制权的股东
controlling station 主站
controlling system 施控系统，主控系统
controlling torque 稳定力矩
controlling unit 控制环节，控制组件
controlling value 控制值
controlling 控制的，支配性的，控股的
control logic cabinet 控制逻辑柜
control logic circuit 控制逻辑电路
control logic diagram 控制逻辑图
control logic 控制逻辑
control loop gain 控制回路增益
control loop 控制环路，控制回路，控制电路，调节回路
control magnet 调节磁铁，控制磁铁
control manager 控制经理
controlman 控制员，调节员
control margin 控制裕量，控制范围
control marker 控制标志
control mark 控制标记

control measures 控制措施，管理措施
control mechanism 控制机构，调节机构，操作机构
control medium 可控介质，控制介质
control member 控制构件，控制元件
control memory 控制存储器
control message display 控制信号显示，控制消息显示，控制信息显示
control message 控制信息
control,metering and protection 电气二次【中国用法】
control-metering point 控制记录点【电力系统】
control method 调节方式
control mode 调节方式，控制方式，控制状态，控制模式
control module 控制模块，控制模件，控制小型组件
control monitoring equipment 监控装置
control monitor 监控器
control motor actuator 马达执行机构
control motor 控制电动机，控制电机
control network 控制网
control nozzle 控制喷嘴，可调喷嘴
control observation 控制观测
control of air pollution 空气污染的控制，空气污染控制
control of floating action 浮点作用控制
control of flood 洪水控制
control of flow 流量控制
control of gas-oil ratio 油气比控制
control of generating plant 发电厂控制，发电站管理
control of HVDC power transmission system 高压直流输电功率控制，高压直流输电系统控制
control of phase-sequence 相序控制
control of river basin 流域治理
control of system 系统控制
control of thyristor controlled series compensator 晶闸管控制串联补偿装置的控制
control of walking 行走控制
control of water source 水源控制
control of wind 风害防治
control-oil pressure 脉冲油压力，控制油压力
control oil system 调节油系统，控制油系统
control on slope stability 对斜坡稳定性的控制
control operation 控制操作，控制室操纵员
control organ 控制机构
control orifice 调节孔板
control output module 调节输出组件，控制输出组件
control output 控制输出
control pair 控制线对
control panel in main control room 主控制室控制盘
control panel 控制屏，控制板，控制盘，操纵台
control parameter 调节参数，控制参数
control pedestal 控制台
control period 控制时间，调节时间
control phase 控制相位，调节相位
control piston 控制活塞
control plane 控制面，控制板
control plunger 控制活塞，调节活塞
control point adjustment 控制点调整，调节器
control point survey 控制点测量
control point synchronizer 控制点同步器
control point 控制点，调节点，检测点，检查点，测量点
control pole 控制极
control post 操纵站，控制站
control power disconnecting device 控制电源隔离装置
control power unit 控制发电机组，控制动力传动装置
control precision 控制精度
control pressure 控制压力
control procedure 管理程序
control program 控制程序
control pulse 控制脉冲
control push-button 控制按钮
control rack 控制台，控制板，调节支架
control range 控制范围，调节范围
control ratio 控制比
control read-only memory 控制只读存储器
control region 操作区，控制区
control register 控制寄存器
control regulation 管理条例
control relay circuit 继电器控制电路
control relay forward 正向控制继电器
control relay 控制继电器，监测继电器
control resistor 控制电阻器，可调电阻器
control resolution 控制分辨能力
control response 控制反应，控制响应
control ring 控制环，调速环
control rod actuation program 控制棒动作程序
control rod assembly for PWR 压水堆控制棒组件
control rod assembly 控制棒束，棒束控制组件，控制棒组件
control rod bank control loop 控制棒组的控制回路
control rod bank position control system 控制棒组位置调节系统
control rod bank 控制棒组
control rod bore 控制棒内孔
control rod bulk insertion 控制棒大量插入
control rod calibration 控制棒校准［标定］，控制棒刻度
control rod cell 控制棒单元，控制棒栅元
control rod cluster changing fixture 控制棒束装卸固定装置
control rod cluster handling station 控制棒束装卸站
control rod cluster 控制棒束，棒束控制组件
control rod configuration 控制棒排列，控制棒布置
control rod continuous withdrawal 连续提控制棒
control rod coupling socket 控制棒连接套筒
control rod density distribution 控制棒密度分布
control rod drive chamber 控制棒驱动箱
control rod drive control system 控制棒驱动控制系统
control rod drive for PWR 压水堆控制棒驱动机构

control rod drive housing　控制棒驱动罩壳
control rod drive hydraulic system　控制棒水力驱动系统
control rod drive line　控制棒驱动线
control rod drive mechanism coil　控制棒驱动机构线圈
control rod drive mechanism cooling shroud　控制棒驱动机构冷却罩
control rod drive mechanism nozzle leakage monitoring system　控制棒驱动机构管座泄漏监视系统
control rod drive mechanism protective cover　控制棒驱动机构护盖
control rod drive mechanism　控制棒驱动机构，控制棒束驱动机构
control rod drive package　控制棒驱动机构
control rod drive penetration　控制棒驱动机构贯穿管
control rod drive pump　控制棒驱动泵
control rod drive shaft handling tool　控制棒驱动轴装卸工具
control rod drive shaft latching　控制棒驱动轴锁住
control rod drive shaft short-handled unlatching tool　控制棒驱动轴短柄拆解工具
control rod drive shaft storage　控制棒驱动轴贮存位置
control rod drive shaft unlatching tool　控制棒驱动轴解扣工具
control rod drive shaft　控制棒驱动轴
control rod drive system　控制棒驱动系统
control rod drive　控制棒驱动装置，控制棒驱动
control rod effectiveness function　控制棒价值函数
control rod effectiveness　控制棒价值，控制棒效率
control rod ejection accident　弹棒事故，控制棒弹出事故
control rod ejection　控制棒弹出
control rod expulsion　抛出控制棒
control rod extension　控制棒延伸
control rod follower　控制棒动体
control-rod-free fuel assembly　无控制棒燃料组件
control rod fuel follower　控制棒燃料随动体
control rod gap　控制棒通道
control rod grab　控制棒抓手
control rod group　控制棒组
control rod guide bushing　控制棒导向筒
control rod guide plate　控制棒导向板
control rod guide structure underwater removal tool　控制棒导向结构水下拆卸工具
control rod guide structure　控制棒导向结构
control rod guide thimble　控制棒导向套管
control rod guide tube　控制棒导向管，控制棒导管
control rod guide　控制棒导向装置
control rod hang-up　控制棒卡住，卡棒
control rod insertion limit　控制棒插入限值
control rod insertion speed　控制棒插入速率
control rod insertion test　控制棒插入试验
control rod location　控制棒位置
control rod motion speed　控制棒动作速度
control rod motion　控制棒动作
control rod movement　控制棒动作，控制棒移动
control rod nozzle　控制棒管座
control rod operating range　控制棒行程
control rod oscillator　控制棒振荡器
control rod pair　控制棒对
control rod pitch　控制棒间距
control rod positioning　控制棒定位
control rod range　控制棒行程
control rod reactivity testing　控制棒反应性测量，控制棒反应性测试
control rod scram pilot valve　控制棒急停先导阀
control rod scram worth　控制棒急停值
control rod shadowing　控制棒荫蔽（作用）
control rod shaft assembly　控制棒驱动轴组件
control rod shock absorber　控制棒阻尼器，控制棒缓冲器
control rod shock load　控制棒冲击负荷
control rod shroud assembly　控制棒套管组件
control rod shroud tube　控制棒套管
control rod snubbing impact　控制棒吸收冲击，控制棒缓冲
control rod stroke　控制棒行程
control rod system function　控制棒系统功能
control rod tip　控制棒端
control rod transport container　控制棒运输容器
control rod travel　控制棒行程
control rod weighing　控制棒重量测量
control rod withdrawal accident　控制棒抽出事故
control rod withdrawal program　提控制棒程序，提棒程序
control rod withdrawal sequence　提控制棒程序
control rod withdrawal　控制棒的提升，提控制棒
control rod worth evaluation　控制棒价值计算
control rod worth minimizer　控制棒价值减少装置
control rod worth　控制棒价值
control rod　控制棒，控制杆，操纵杆，吸收棒
control room air conditioning system　控制室空调系统
control room area　控制室区
control room management system　控制室管理系统
control room operator　控制室操纵人员
control room　控制室，操纵室，调度室，操作室
control rotary diaphragm　旋转隔板【抽汽式汽轮机】
control routine　控制例行程序
control section　控制段，控制部分，控制断面
control segment　控制段，控制区
control sequence　控制程序，控制序列
control set-up　控制装置，调节装置
control sheet　控制图表
control shift register　控制移位寄存器
control signal　控制信号，控制量
control source　控制源
control spillover　控制信息漏失
control spring　控制弹簧
control stage　调节级
control statement　控制语句
control station　控制台，控制站，控制箱，操纵台
control stick　操作杆，控制杆，操纵杆
control storage　控制存储器
control structure　控制结构

control subassembly 控制组件
control supervisor 控制管理，控制管理程序
control surface 控制面
control survey 控制测量
control switchboard 控制配电盘
control switchgear 控制开关装置
control switchgroup 控制组合开关
control switch 控制开关
control symbol 控制符号
control synchro 控制同步器，控制同步机
control system analogue simulator 控制系统模拟仿真器
control system analysis 控制系统分析
control system dynamics 控制系统动力学
control system feedback 控制系统反馈
control system for wind turbine 风力机控制系统
control system type 控制系统类型
control system 控制系统，调节系统，操纵系统，调速系统
controls 控制机构，调节机构，操纵装置
control tank 控制箱
control tape 控制用纸带
control target construction 施工用地面积控制指标
control technique 控制技术
control terminal 控制终端
control test 调节试验，控制试验
control timer 控制定时器，时间控制继电器
control time 控制时间
control total 控制总数
control tower 操纵塔，控制塔
control transfer instruction 控制转移指令
control transformer 控制同步机，控制变压器
control tube rectifier 控制管整流器
control turn 控制线圈
control-type generator 控制用发电机
control unit end 控制器结束
control unit 控制单元，调节单元，调节器，控制装置，控制器，控制机组，控制部件
control valve actuator 阀控传动机构
control valve gear 调节装置
control valve plug 控制阀塞
control valve stand 调节阀阀座
control valve with disk free to rotate on stem 阀瓣对阀杆自由转动的调节阀
control valve with linear characteristics 线性特性的调节阀
control valve 调节阀，控制阀，高压阀门
control variable 控制变量
control velocity 控制速度
control verification 控制检定
control voltage 控制电压，操纵电压
control volume 控制体，控制容积
control waveform 控制波形
control wave 控制波
control winding 控制绕组
control wire 操纵索，控制线
control wiring diagram 二次回路接线图，控制接线图
control wiring 控制电路，控制线路
control with fixed set-point 定点控制，定值控制，定值调节
control with variable set-point 变定值控制
control word 控制字，控制字码
control 控制，控制段，控制装置，支配，调节，调整，检验，管理，监督
controsurge winding 防振屏蔽绕组，防浪涌的屏蔽绕组
controversial issue 有争议的事项
controversial question 有争议的问题
controversy 争论
conurbation 集合城市
CONV(converter) 转换器
CONV(conveyer) 输送机
convection bank of tubes 对流管束
convection barrier 对流栅，抗对流屏
convection boiling 对流沸腾
convection boundary condition 对流边界条件
convection cell 对流单体
convection circulation 对流性环流
convection cloud 对流云
convection coefficient 对流放热系数，对流传热系数，对流系数
convection cooler 对流冷却器
convection cooling 对流冷却
convection current 对流，对流流动，对流气流，对流电流
convection effect 对流效应
convection flow 对流流动
convection flue gas pass 尾部烟道
convection flux 对流通量
convection heater 对流加热器
convection heating surface soot blower 对流受热面吹灰器
convection heating surface 对流受热面
convection heating 对流加热，对流采暖
convection heat losses 对流热损失
convection heat transfer 对流传热
convection layer 对流层
convection loss 对流损耗
convection of air 空气对流
convection pass 对流通道，对流烟道
convection process 对流过程
convection reheater 对流再热器
convection section 对流区段
convection superheater 对流过热器
convection theory 对流学说
convection transfer rate 对流换热系数
convection tube 对流管
convection-type desuperheater 对流式减温器，面式减温器
convection 对流，迁移，传递，运流，传送
convective boundary layer 对流边界层
convective circulation 对流性环流，对流循环
convective condensation level 对流凝结高度
convective cooling 对流冷却
convective diffusion equation 对流扩散方程
convective diffusion 对流扩散
convective eddy 对流涡旋
convective element 对流元
convective equilibrium 对流平衡
convective flow 对流

convective flux 对流通量
convective heat exchange 对流热交换
convective heat loss 对流热损失
convective heat transfer 对流换热，对流传热，对流热交换
convective instability 对流不稳定（性）
convective intensity 对流强度
convective internal boundary layer 对流内边界层
convective mixed layer 对流混合层
convective mixing 对流混合
convective overturn 对流翻腾
convective parcel 对流气块
convective precipitation 对流降水
convective regime 对流流型
convective region 对流区
convective stability 对流稳定度
convective storm 对流风暴
convective transfer 对流输送
convective turbulence 对流性湍流
convective 对流的，传递性的
convector radiator 对流式放热器
convector 对流散热器，对流加热器，对流器
convegent duct 收敛形通道
convene 召集，召开
convenience receptacle 电源插座，墙插座
convenience 便利
convenient construction machine 通用结构电动机
conventional boiler 传统形式锅炉
conventional closed cooling loop heat exchanger 常规的冷却回路热交换器
conventional closed cooling loop pump 常规的封闭冷却回路泵
conventional closed cooling loop 常规的封闭冷却回路
conventional design 常规设计
conventional diagram 示意图
conventional diffused aeration tank 常规扩散曝气池
conventional duty 协定关税
conventional efficiency 通用效率，惯用效率，约定效率
conventional elastic limit 常规弹限
conventional energy 常规能源
conventional environment 一般环境条件
conventional equipment 常规设备，普通设备，惯用设备
conventional explosive 普通炸药
conventional export packing 传统出口包装
conventional flux density 常规通量密度
conventional fuel 常规燃料，化石燃料，普通燃料
conventional fusing current 约定熔断电流
conventional heat treatment 普通热处理
conventional hopper 老式漏斗车
conventional interlocking coupling 常规联锁联轴器
conventional island component cooling system 常规岛的设备冷却（水）系统
conventional island equipment 常规岛设备
conventional island erection 常规岛安装
conventional island modeling 常规岛建模
conventional island waste oil and inactive water drain system 常规岛废油及非放射性水排放系统
conventional island 常规岛【核电】
conventional letter 假设符号，代表符号
conventional loading 习用荷载，常用荷载
conventionally cooled generator 一般冷却的发电机
conventional magnet 常规磁铁，常规磁体
conventional method for insulation coordination 绝缘配合惯用法
conventional method of analysis 常规分析方法
conventional method of slices 常规条分法
conventional method 常规方法，传统方法
conventional non-fusing current 约定不熔断电流
conventional number 标志数
conventional operating current 约定动作电流，保护器件动作电流
conventional operation 常规操作
conventional power plant 常规发电厂，传统化石燃料发电厂，常规动力装置，常规电站
conventional price 协定价格
conventional programming 用标准程序语言编程，常规编程
conventional projection 惯用投影
conventional punch card machinery 通用卡片穿孔机
conventional safety factor 常用安全系数
conventional security 常规安全
conventional service 常规性服务
conventional sign 图例，习惯符号，惯用符号，通用符号
conventional solar cell 常规太阳电池
conventional style 传统风格
conventional test 常规试验
conventional thermal power plant 常规热电厂
conventional tillage 正常整地
conventional top-and-bottom guided valve trim 常规的顶和底导向阀结构
conventional top-guided valve trim 常规顶部导向阀结构
conventional training idler 普通托辊
conventional treatment 常规处理
conventional true value （量的）约定真值【放射性活度】
conventional-type sprinkler 惯用式喷洒器
conventional value of a quantity 量的约定值
conventional valve 常规阀门【核电，不按照 RCC－M 规则的】
conventional waste treatment 常规废物处理
conventional windmill 常规风车
conventional 通用的，普通的，一般的，常规的，传统的，通常的
convention operation 常规操作
convention rule 常规
convention section 对流区
convention 惯例，常规，公约，协定，传统，大会，习俗，集会
convergence angle 收缩角，收敛角，会聚角
convergence coil 收敛线圈，聚焦线圈，会聚线圈

convergence criterion 收敛性判据
convergence current 收缩流
convergence error 收敛误差
convergence field 辐合场
convergence iterative procedure 收敛迭代法
convergence rate 收敛速度
convergence speed 收敛速率
convergence transformer 收敛变压器
convergence 收缩，收敛，非周期性阻尼运动，收敛性，（风洞的）收缩段
convergency 收敛，会聚，聚集，收敛性
convergent branching system 收敛分支系统
convergent channel 收缩通道，收缩流道，收敛流道，收敛形通道，汇聚槽，辐合状水道网
convergent control system 收敛控制系统
convergent control 收敛控制
convergent current 合流，汇流，收敛流，收缩流
convergent-divergent channel 缩放管，缩放形流道，收缩扩张管，缩张通道，收缩扩散流道
convergent-divergent nozzle 缩放喷嘴，缩放喷管，拉伐尔喷管
convergent flow 收敛水流
convergent lens 辐合透镜
convergent matrix 收敛矩阵
convergent mouthpiece 收缩管嘴
convergent nozzle 收缩喷管，收缩喷嘴
convergent oscillation 减幅振动，衰减振动
convergent passage 收敛形通道
convergent pipe 收缩形水管
convergent point 辐合点，汇聚点
convergent reaction 次临界核反应，次临界反应，收敛反应
convergent reactor 次临界反应堆
convergent 次临界，亚临界【堆芯】，收敛，收敛项，收敛的，会聚的
converge 收敛，集聚，集中于
converging channel 收敛河槽
converging-diverging nozzle 缩放喷管，缩放喷嘴，拉伐尔喷管，收敛扩散喷管，超音速喷管
converging duct 收敛管道
converging flow 收束水流
converging nozzle 收缩喷嘴，收缩喷管，收束形管嘴
converging transition 收缩渐变段
converging tube 渐缩管，缩口管
converging tunnel 对称式地槽，全聚式地槽
converging beam 会聚波束，汇聚光束
conversation 会谈
conversational 对话式的，会话式的
conversational algebraic language 对话代数语言
conversational compiler 对话式编译程序
conversational mode 对话方式
conversational processing 对话处理
conversational time-sharing 对话式分时，对话式分时操作
conversation-monitoring system 对话监听系统
conversation time 通话时间
converse 谈话，逆的，反的，转换
conversion card cabinet 转换插件柜
conversion card 转换插件
conversion chart 换算图，换算表，变换图，换算图表
conversion check 反校验
conversion cluster 转换棒束元件
conversion coefficient 换算系数，变换系数，转换因子，转换系数
conversion conductance 变换电导，变频跨导
conversion constant 转换系数，转换常数，换算常数，热功当量
conversion curve 变换曲线，换算曲线
conversion cycle 转化循环
conversion device 转换装置
conversion diagram 变换特性曲线，转换图
conversion effect 变换效应，转换效应，转换系数，换算系数
conversion efficiency 转化效率，转换效率
conversion electron 转换电子
conversion equation 转换公式，换算公式
conversion equipment 转换设备，转换装置
conversion facility 转换设施，转换功能，（铀）转换工厂，（燃料）转化工厂
conversion factor 变换系数，转换系数，换算系数，折合率，转换因子，换算因子
conversion function 转换函数
conversion gain 变频增益，变换增益，转化增益
conversion level 转化深度
conversion line 转换线
conversion loss 变频损耗，变换损失，转换损耗，转换损失
conversion of currency 货币兑换
conversion of electrical energy 电能转换
conversion of energy 能量转换
conversion of heat into power 热功转换
conversion pig iron 炼钢生铁
conversion plant 转换装置，（燃料）转化厂，（铀）转换工厂
conversion price 兑换价格
conversion process 转化过程
conversion program 转换程序
conversion rate 转化率，变换率，兑换率，变换速度
conversion ratio 换算系数，再生系数，转换比
conversion routine 转换程序，转换子程序
conversion screen 转换屏
conversion table 换算表
conversion time 变换时间
conversion transconductance 变频跨导
conversion transformer 转换变压器，转电线圈
conversion unit 变换机组，变换器
conversion zone 转换区
conversion 换算，转换，变换，转化，兑换，换流
convert 转变，变换，转换，改装，改造，换算
converted geometry 变换几何
converted steel 渗碳钢，硬质钢
converted timber 成材
converter arm 换流管桥臂
converter blocking 换流器闭锁，换流器关断
converter bridge blocking 换流桥闭锁
converter bridge deblocking 换流桥解锁

converter bridge 换流器桥，换流桥
converter circuit 换流电路
converter control device 换流器控制装置
converter deblocking 换流器启用，换流器导通，换流器解锁
converter display 变换器显示装置
converter efficiency 变换效率
converter equipment 换流设备
converter-fed motor 换流器供电的电动机
converter firing phase control 换流器触发相位控制
converter module 转换器模块
converter reactor 转换堆，转换反应堆，再生反应堆
converter relay 换流器继电器
converter room 变电室
converter station 换流站
converter steel 转炉钢
converter substation 换流站，换流变电所
converter 1/4 to 3/8 四分之一转八分之三转换接头
converter transformer 换流变压器
converter unit control 换流单元控制
converter unit firing control 换流器触发控制
converter unit protection 换流器保护
converter unit sequence control 换流器顺序控制
converter unit 转换系数，转换单位，换流单元
converter valve 换流阀
converter waste heat boiler 转炉废热锅炉
converter 变换器，转换器，逆变器，换流器，变频器，转换反应堆，变流器，转化器，变压器，转炉，换能器
convertibility 互换性，可逆性，可变换性，转换性
convertible bond 可转换债券
convertible circuit breaker 可变换断路器
convertible currency 可兑换货币
convertible gas diesel engine 可转换狄塞尔煤气内燃机
convertible gas engine 可转换煤气内燃机
convertible preferred stock 可转换优先股
convertible revolving credit 可更换的循环信用证，可转换循环信用
convertible shovel 正反铲挖土机，两用铲
convertible static compensator 可转换静止补偿器
converting plant 换流变电站，整流站
converting station 整流站，变流站
converting valve 快速减温减压阀，变态阀
convert instruction 变符号指令
convert . . . into 折算
convertor 变换器，换流器，转换反应堆，转炉，变流器，转化器，转换器
convertor station 换流站
convert to binary 转换成二进制
convert to decimal 转换成十进制
convevtional diagram 习用图，示意图
convex airfoil surface 叶凸面
convex bank 凸岸
convex bend 凸弯管
convex boundary 凸面边界
convex cap 凸形帽
convex-concave 一面凸一面凹的，凸凹形
convex curves conveyor 上凸弧输送机
convex curve 凸弧段，凸曲线
convex domains 凸域
convex fillet weld 凸形角焊缝
convex flange 凸面法兰
convex flow 凸岸水流
convexity 焊缝凸度，凸度，凸状
convex joint 凸缝
convex lens 凸面镜
convexo-convex 双凸形
convexo-plane 凸平形
convex perforated plate 凸形多孔板
convex profile （叶片的）背弧型面
convex shore 凸岸
convex side of blade 叶片背弧侧
convex side 背弧侧【负压侧】，凸岸，凸侧，凸面，凸面侧
convex slope 凸坡
convex surface flange 凸面法兰
convex surface 凸面，背弧面
convex 凸的，凸面，凸面的，凸状，凸圆的，背弧
conveyance factor 输水因素
conveyance loss 输水损失，输送损失
conveyance media 输送媒体
conveyance of property 产权转让
conveyance power of water 输水能力
conveyance system of lock 船闸输水系统
conveyance 运输工具，运输，输送，搬运，传播，转让
conveyed fluid 被输流体
conveyer advancing 输送机接长
conveyer belt sorter 传输带分类器
conveyer belt 传送皮带，运输带
conveyer bucket 输送料斗
conveyer chain 输送链
conveyer charging hopper 运输机装料漏斗
conveyer distributing system 输送机分输系统
conveyer loading hopper 输送机装料漏斗
conveyer 输送机，输送器，传送装置，传送机，运输器
conveyer scale 传送带秤，皮带秤
conveyer scraper 输送机刮板
conveyers for bottom ash removal 除渣机
conveyer transfer 输送机转运器
conveyer tube process 管道输送法
conveyer tube 输送管道
conveying belt 传送带，输送带
conveying blower house 输送风机房
conveying blower 输送风机
conveying capacity test 出力试验
conveying capacity 输送能力
conveying gas blower 送气风机
conveying system 输送系统
conveying worm 输送螺旋杆，螺旋输送机
conveying 传输，传送
conveyor belt carcass 输送带带芯
conveyor belt scale 皮带秤
conveyor belt 传送带，输煤皮带，输送带，运

输皮带
conveyor car 传送机台式小车【燃料】，输送车，传送车
conveyor coasting time 输送惯性（自由）停机时间
conveyor gallery 输送机通廊，输送机栈桥
conveyor haulage 皮带机输送
conveyor head pulley 输送机头部滚筒
conveyor hood 皮带机罩
conveyor rack 传动架【即皮带机架】
conveyor trip switch 紧急开关，事故开关，传送器脱扣开关
conveyor trolley 运送小车
conveyor 输送机，运输机，传送带，传送装置，输送器，载运工具，传送机，运输器
convey 运输，输送，传达，传递，搬运，传递……到
conviction 确信，信服
convince 使确信，使信服
convoke 召集（会议）
convolutional code 卷积码
convolution integral 卷积积分
convolution of probability distribution 概率分布卷积
convolution of two functions 两个函数的卷积式
convolution theorem 卷积定理
convolution 涡流，回旋，盘旋，缠绕，匝，圈，卷积，卷绕，转数
convulsive means 强制手段
CON 连续的【DCS 画面】
COO(chief operation officer) 首席业务主管，首席运营官，运营总监
cooker 火炉，炊具
cooking stove 炉灶
cool air 冷空气
coolant activity 冷却剂放射性，冷却剂活度
coolant additive 冷却剂添加剂
coolant apparatus 冷却液装置
coolant assembly 冷却剂组件，冷却组件
coolant axial temperature rise 冷却剂轴向温升
coolant blower 冷却气体风机
coolant bypass flow 冷却剂旁路流
coolant change 冷却剂更换
coolant channel closure 冷却剂通道端塞，冷却剂通道封头
coolant channel end closure 冷却剂通道端塞
coolant channel geometry 冷却剂通道几何形状
coolant channel outlet 冷却剂通道出口
coolant channel position 冷却剂通道位置
coolant channel 冷却剂通道
coolant charging system 冷却剂补给系统，冷却剂灌注系统
coolant circuit 载热剂回路，冷却剂回路
coolant circulating system 冷却剂循环系统
coolant circulator seal gas 冷却剂循环风机密封气体
coolant circulator 冷却气体循环风机
coolant clean-up system 冷却剂净化系统
coolant coefficient 冷却剂系数，冷却剂系数
coolant concentrate 冷却液添加剂【用以降低冰点，提高沸点及防锈等】
coolant conditions 冷却剂条件
coolant density coefficient 冷却剂密度系数
coolant density reduction 冷却剂密度降低
coolant density 冷却剂密度
coolant duct 冷却剂通道
coolant ejection 冷却剂喷出
coolant evaporator plant 冷却剂蒸发装置
coolant flow path 冷却剂流动路径
coolant flow rate 冷却剂流量【堆芯】
coolant flow through the core 堆芯冷却剂流量
coolant flow 冷却剂流
coolant fluid 冷却剂流体，冷却流体，冷却液流体
coolant-fuel interaction 冷却剂燃料相互作用
coolant gas activity 气体冷却剂活度
coolant-gas-carrying auxiliary system 载带气体冷却剂的辅助系统
coolant gas channel 气体冷却剂通道
coolant gas circuit 气体冷却剂回路
coolant gas circulator 气体冷却剂循环风机
coolant gas flow rate 气体冷却剂流率
coolant gas impurities 气体冷却剂杂质
coolant gas loss make-up 气体冷却剂损耗补给
coolant gas mass flow 气体冷却剂质量流
coolant gas monitoring 气体冷却剂监测
coolant gas stagnation 气体冷却剂滞流
coolant gas stream 气体冷却剂流
coolant gas stripper 冷却剂除气器
coolant gas 气体冷却剂
coolant heat-up 冷却剂加热，冷却剂温升
coolant impurity 冷却剂杂质
coolant injection system 冷却剂注入系统
coolant inlet plenum 冷却剂入口腔【燃料组件下管座】
coolant inlet 冷却液进口
coolant leakage measuring assembly 冷却剂泄漏测量装置
coolant leakage 冷却剂泄漏
coolant leg 冷却剂管段
coolant loop activity 冷却剂环路活度
coolant loop 冷却剂环路
coolant mass flow rate 冷却剂质量流量，冷却剂质量流速
coolant mixing device 冷却剂的搅拌装置，冷却剂交混装置
coolant mixing 冷却剂交混
coolant nozzle protective cap 冷却剂接管护帽
coolant orificing 冷却剂孔板节流
coolant outlet nozzle 冷却剂出口接管
coolant outlet plenum 冷却剂出口腔【燃料组件上部座】，冷却剂出口孔【燃料组件上管座】
coolant passage 冷却剂通道
coolant poisoning 向冷却剂中添加毒物
coolant pressure control loop 冷却剂压力控制回路
coolant pressure signal 冷却剂压力信号
coolant pump 冷却剂泵
coolant purification pump 冷却剂净化泵
coolant purification system 冷却剂净化系统
coolant radiation monitor 冷却剂辐射监测仪
coolant recirculation flow 冷却剂再循环流

coolant recirculation loop 冷却剂再循环回路
coolant recirculation pump 冷却剂再循环泵
coolant recirculation system 冷却泵再循环系统
coolant recirculation 冷却剂再循环
coolant-return line 冷却剂回流管线
coolant-salt pump 盐冷却剂泵
coolant-salt 盐冷却剂
coolant specific enthalpy 冷却剂比焓
coolant storage system 冷却剂贮存系统
coolant stream 冷却剂流
coolant system leg 冷却剂系统管段
coolant system rupture 冷却剂系统破裂
coolant system 冷却剂系统
coolant temperature coefficient 冷却剂温度系数
coolant temperature control loop 冷却剂温度控制回路
coolant temperature entering reactor vessel 反应堆进口冷却剂温度
coolant temperature leaving reactor vessel 反应堆出口冷却剂温度
coolant temperature map 冷却剂温度分布图
coolant temperature 冷却剂温度，冷却液温度，冷却介质温度
coolant through-flow 冷却剂贯穿流【反应堆内】
coolant treatment and storage system 冷却剂处理与贮存系统
coolant treatment system 冷却剂处理系统
coolant tube end closure 冷却剂管道端塞
coolant velocity 冷却剂流速，冷却剂速度
coolant volume control 冷却剂窖控制，冷却剂容积控制
coolant volume 冷却剂容积
coolant 冷冻剂，冷却液，散热剂，冷却剂，冷却工质，载热剂
COOL＝coolant 冷却介质
cooldown process 冷却过程
cooldown rate 冷却率，降温速率
cooldown valve 冷却阀
cooldown 降温，衰变，冷却，冷停堆
cooled-anode transmitting valve 阳极冷却式发射管
cooled blade 冷却叶片
cooled rotor 冷却转子
cooled turbine 冷却的涡轮
cooled 稳定的【放射性物质】，被冷却的
cool end system 冷端系统
cool end 冷端
cooler 冷却器
cooler casing 冷却器外壳
cooler condenser 冷凝器
cooler performance 冷却器性能
cooler room 冷却间
cooler unit 冷却器装置
cool flame 冷焰
cooling accident 冷却事故
cooling agency 冷却液，切削液
cooling air baffle 冷却空气导流板
cooling air curtain 冷风幕
cooling air fan 冷却风机
cooling air system 冷却风系统
cooling air turbine 风冷汽轮机
cooling air 冷却风，冷却空气
cooling and cleanup system 冷却和净化系统
cooling apparatus 冷却设备，冷却装置
cooling basin 冷却水池，冷却水槽
cooling blade 散热片，冷却叶片
cooling blast 冷却气流，冷却通风
cooling block 冷却区
cooling body 散热壳体，冷却器套
cooling by circulating water 循环水冷却
cooling by circulation 循环冷却
cooling by evaporation 蒸发冷却
cooling ceiling 冷却顶板
cooling chamber 冷却室
cooling channel 冷却通道［管道］
cooling circuit 冷却回路，冷却系统
cooling coil section 冷却段
cooling coil 冷却盘管，冷盘管，冷却盘
cooling column 冷却塔
cooling crack 冷却裂纹，冷却裂缝
cooling curve 冷却曲线
cooling cycle 冷却时间，冷却循环
cooling down operation 冷却运行方式，冷却操作
cooling down period 冷却期，衰变期
cooling down rate 冷却速率
cooling down 降温，冷却下来
cooling drag 冷却系统阻力
cooling effect 冷却效果，冷却效应
cooling efficiency 冷却效率
cooling element 冷却元件
cooling equipment system 冷却设备系统
cooling equipment 冷却设备
cooling facility 冷却设备
cooling fan shroud 冷却风扇护罩
cooling fan 冷却风扇
cooling filling 淋水装置，冷却填料
cooling fin 冷却翅片，散热片，冷却片
cooling fluid 冷却液体
cooling for decay 衰变冷却
cooling installation 冷却装置
cooling jacket 冷却夹套，冷却夹层
cooling line 冷却管线
cooling load calculation 冷负荷计算
cooling load estimation 冷负荷估算
cooling load from outdoor air 新风冷负荷
cooling load from ventilation 新风冷负荷
cooling load temperature 冷负荷温度
cooling load 冷却负荷
cooling loop 冷却回路，冷却循环，冷却环路
cooling loss 冷却损失
cooling means 冷却装置，冷却方法
cooling medium consumption 冷却介质耗量
cooling medium 冷却介质，冷却剂，载热介质，冷媒
cooling of fuel 燃料冷却
cooling off 冷却
cooling of intrusives 侵入体的冷却
cooling oil 冷却油
cooling period 冷却周期，放射性衰减周期
cooling pit 冷却坑，冷却阱
cooling plant 制冷厂

cooling pond cleanup plant （乏燃料）贮存冷却水池净化装置
cooling pond （乏燃料）贮存冷却水池【消除放射性】，冷却池，冷却水池，凉水池
cooling power 冷却能力
cooling process 冷却过程
cooling pump 冷却水泵，冷却泵
cooling range 冷却温降，冷却幅度，冷却范围
cooling rate 冷却率，冷却速度，冷却倍率，冷却速率
cooling ratio 冷却系数
cooling rib 冷却翅片，散热片
cooling rod 冷却棒
cooling schedule 冷却制度
cooling section 冷却段，冷却区域
cooling shaft 冷却井
cooling shroud 通风装置，冷却罩【控制棒驱动机构】，冷却管套
cooling shutdown 冷却停机
cooling spacer tube 冷却间隔管
cooling spacer 冷却管夹
cooling space 冷却空隙
cooling speed 冷却速度
cooling spray nozzle 冷却喷嘴
cooling spray 冷却喷淋，冷却喷水系统，洒水冷却
cooling steam 冷却蒸汽
cooling strain 收缩变形，冷却变形
cooling stress 冷却应力
cooling surface area 冷却面积
cooling surface 冷却面
cooling system capacity 冷却系统容量
cooling system of gas turbine 燃气透平冷却系统
cooling system 冷却系统，降温系统，制冷系统
cooling temperature 冷却温度
cooling tower circulating water pumping station 冷却塔循环水泵站
cooling tower condenser 冷却塔式凝汽器
cooling tower drift 冷却塔漂滴
cooling tower electrical building 冷却塔电气厂房
cooling tower fan 冷却塔风机
cooling tower forced draft ventilation system 冷却水塔强制通风冷却系统
cooling tower makeup and blowdown system 冷却水塔补排水系统
cooling tower makeup pump 冷却塔补充水泵
cooling tower makeup water treatment system 冷却水塔补充水处理系统
cooling tower makeup water 冷却水塔的补给水
cooling tower packing 冷却塔填料
cooling tower plume 冷却塔羽流
cooling tower pond 冷却塔水池
cooling tower pump 冷却塔泵
cooling tower raw makeup water pumping station 冷却塔补给水泵
cooling tower raw make up water 冷却塔生水补充水
cooling tower shell 冷却塔壳体
cooling tower without packing 无填料冷却塔
cooling tower 凉水塔，冷却塔，冷却水塔
cooling trap 冷阱
cooling trough 冷却槽，泄水槽
cooling tube 冷却管
cooling unit 冷风机组
cooling vane 冷却翅片，散热片
cooling water casing 冷却水套
cooling water circulating pump 循环水泵
cooling water coil 冷却水盘管
cooling water collection box 冷却水汇集箱
cooling water connection 冷却水路连接
cooling water consumption 冷却水消耗
cooling water discharge 冷却水排放，冷却水出口
cooling water drainage pipe 冷却水排水管
cooling water flow rate 冷却水流量
cooling water flow 冷却水流量
cooling water for bottom ash system 底渣冷却水系统
cooling water head tank 高位冷却水箱
cooling water inlet temperature 冷却水进口温度
cooling water inlet 冷却水进口
cooling water intake 冷却水进口，冷却水取水口
cooling water jacket 冷却水夹层，冷却水腔，冷却水套
cooling water leakage detector 冷却水泄漏检查装置
cooling water line outlet 冷却水管线出口
cooling water outlet temperature 冷却水出口温度
cooling water outlet 冷却水出口
cooling water pipeline 冷却水管线
cooling water pipe 冷却水管
cooling water piping 冷却水管道
cooling water pond 冷却水池
cooling water pump 冷却水泵
cooling water return 轮回冷却水回水
cooling water supply system 冷却水供应系统
cooling water system 冷却水系统
cooling water tank 冷却水箱，冷却池
cooling water temperature rise 冷却水温升
cooling water temperature 冷却水温度
cooling water temp variation-power correction curve 冷却水温变化对功率的修正曲线
cooling water treatment system 冷却水处理系统
cooling water treatment 冷却水处理
cooling water trench 冷却水沟
cooling water tube 冷却水管
cooling water valve 冷却水门［阀］
cooling water with natural draught 自然通风冷却塔
cooling water 冷却水
cooling zone 凝结区
cooling 衰变，冷却，降温，冷却的，放射性衰减
cool recycle 循环冷却，再循环冷却
cool-storage mode solar photovoltaic module 蓄冷降温式太阳电池组件
cool water 冷水
cool 冷的，冷却，降低放射性活度，消除放射性
coombe 冲沟，狭谷
coom 碎煤，炭黑，煤屑，煤粉，煤烟，煤灰
cooperating plants 联合运行电站

cooperation and development 合作与开发
cooperation of concrete and steel 混凝土与钢筋的联合作用，混凝土与钢筋共同作用
co-operation 合作，协作，配合
cooperative agreement 合作协定
cooperative control 合作管理
cooperative design 合作设计
cooperative drainage 合作排水，协作排水
cooperative enterprise 合作企业
cooperative exchanges 合作交流
cooperative game 合作博弈
cooperative joint venture contract 合作经营企业合同
cooperative joint venture 合作经营企业，联营体【合作方式】
cooperative J. V.(cooperative joint venture) 合作经营企业，联营体【合作方式】
cooperative management 合作经营
cooperative partner 合作伙伴
cooperative production 合作生产
cooperative project 合作项目
cooperative research 合作研究
cooperative shop 合作社
cooperative study 合作研究
cooperative system 合作制度
cooperative venture 合作企业
cooper loss 铜耗
cooper woolfilter 紫铜毛滤清器
coordinate adjustment 坐标调整，坐标平差
coordinate axis 坐标轴
coordinate control system 协调控制系统
coordinate conversion 坐标变换
coordinated boiler-turbine control 机炉协调控制机器，机炉协调控制
coordinated control system 协调控制系统
coordinated control 协调控制
coordinated development 协调发展
co-ordinated drive 协调驱动
coordinated insulation 配合绝缘
coordinated mode 协调方式
coordinated operation 联动运行
coordinated phosphate treatment 协调的磷酸盐处理
coordinated planning 协调规划
coordinated universal time 协调世界时，协调通用时
coordinate grid 坐标方格，坐标网格
coordinate indexing 坐标检索，坐标法加标
coordinate inspection machine 坐标检查机
coordinate mesh 坐标格网
coordinate method 坐标法
coordinate neighbourhood 坐标邻域
coordinate paper 坐标纸
coordinate plane 坐标面
coordinate position 坐标位置
coordinate potentiometer 坐标式电位计
coordinates calculation 坐标计算
coordinates network 坐标网
coordinates origin 坐标起始点，坐标原点
coordinate(system) paper 坐标纸
coordinate system 坐标系，坐标系统
coordinate transformation 坐标变换，坐标换算，坐标转换
coordinate 坐标（系），坐标的，同等的，配位的，同位的
coordinating body 协调机构
coordinating committee 协调委员会
coordinating group 协调小组
coordinating piles 标桩
coordination activities 协调工作
coordination center 协调中心
coordination meeting 协调会议
coordination mode 协调方式
coordination of contract 合同的协调
coordination of fuse 熔丝配合
coordination of insulation 绝缘配合
coordination of operations 运行的协调
coordination of project 工程协调
coordination of thermal and hydroelectricity 水火电配合
coordination project 配套工程
coordination work 协调工作
coordination 坐标，调整，协作，同位，配合，匹配，同等，协调
coordinator 协调人，协调员
COO(chief operating officer) 首席运营官
copal 苯乙烯树脂
COP(coefficient of performance) 性能系数（风轮）效率
COP(continuous operation program) 连续操作程序
cope box 上砂箱，上模箱
coped joint 盖顶接头，暗缝，搭接缝
cope flask 上砂箱，上模箱
cope level 顶部高程
cope match plate 上型板
cope 顶盖
cop 圆锥形线圈
cophasal 同相的
cophase 同相
cophase array 同相天线阵
cophase component 同相分量
cophased array 同相列阵
cophased 同相的
cophase excitation 同相励磁
cophase supply 同相供电，同相电源
cophase wave 同相电波
copier 复印机，拷贝机
copies of the contract 合同副本
coping beam 压顶梁
coping stone 盖石
coping wall 拦墙
coping 顶盖，压顶，盖梁，墙压顶，墙的顶部，遮檐
copious cooling 深度冷却，深冷
coplanar force 共面力
coplanar grating 共面光栅
co-planar stabilizers 共面稳压器，共面稳定器
coplanar strain 共面应变
coplanar 共面的，共面
copolymer 共聚物
copper 铜、紫铜

copper alloy　铜合金
copper-aluminium bimetal connector　铜铝过渡连接器
copper arc welding electrode　铜焊条
copper area　铜截面积
copperas　绿矾
copper band　铜带
copper bar squirrel cage　铜条鼠笼
copper bar　紫铜棒，铜棒，铜条
copper-base alloy wire　铜基合金线
copper-base alloy　铜基合金
copper-bearing　含铜的，铜轴承
copper binding-wire　铜绑线
copper bit　焊接器
copper-braid contact pad　铜接触垫
copper brush　铜电刷
copper busbar　铜母线，铜汇流条
copper bush　铜套
copper cable jointing sleeve　同心电线连接套管，电缆铜鼻子
copper carbon brush　含铜炭刷
copper clad aluminum conductor　铜包铝线
copper clad cable　镀铜电缆
copper clad steel conductor　铜包钢线
copper clad steel wire　铜包钢线
copper clad steel　铜包钢
copper clad weld wire　铜包钢复合线材
copper clad　镀铜，铜箔，包铜的
copper coil　铜线圈
copper collector　铜集电器
copper commutator segment　铜换向片
copper conductor tube　铜导管，空心铜线
copper conductor　铜线，铜导体，铜导线
copper-constantan pair　铜-康铜对
copper-constantan thermocouple　铜-康铜热电偶，铜-康铜测温计
copper constantan　铜-康铜
copper contact　铜触点
copper core wire　铜芯线
copper-covered steel wire　铜包钢线
copper cushion　铜垫
copper damper　铜阻尼器，铜阻尼条
copper drop　铜阻压降，电阻电压降
copper earth plate　接地铜板
coppered　镀铜的【铝导体】
copper efficiency　铜导线利用率
copper electrode　铜电极
copper end ring　铜端环
copper factor　含铜率
copper filings　铜屑
copper fill factor　铜填充因数
copper flake　紫铜片
copper flux shield　铜磁通屏蔽
copper foil　铜箔
copper gasket　铜垫
copper gauze brush　铜网刷
copper industry　铜工业
coppering　镀铜
copperization　镀铜
copperizing　镀铜
copper-jacketed coil　铜屏蔽线圈，带铜罩的线圈
copper litz wire　铜编织线
copper loss　铜耗，铜损
copper magnet alloys　铜磁合金
copper mesh　铜板网
copper-nickel alloy　铜镍合金
copper-nickel tube　镍铜管，白铜管
copper-nickel　白铜
copper-oxide catalyzation　氧化铜催化
copper-oxide cell　氧化铜光电池
copper-oxide detector　氧化铜检波器
copper-oxide rectifier　氧化铜整流器
copper-oxide varistor　氧化铜压敏电阻
copper-oxide　氧化铜
copper pipe　紫铜管，铜管
copper-plated　镀铜的
copper plate　铜板
copper plating　镀铜
copper pole　铜电极
copper rod　铜棒
copper screen　铜网，铜屏蔽
copper seal　止水铜片
copper section　铜线截面
copper segment　扇形铜片
copper sheath　铜包皮，铜护套
copper sheet　铜薄板
copper shield　铜屏蔽
copper shot　铜珠，铜粒
copper-silicon welding rod　铜硅焊条
copper-silver alloy　银铜合金
copper sleeve　铜套筒，铜套管
copper slot loading　线负载
coppersmith shop　铜钳工车间
copper space factor　铜线占空系数，槽满率
copper spiral　螺旋铜带
copper spun rotor　离心铸铜鼠笼转子
copper strand　铜绞线
copper sulfate　硫酸铜
copper sulfide ceils　硫化铜电池
copper sulphate　硫酸铜
copper sulphide rectifier　硫化铜整流器
copper-surfaced　镀铜的
copper-tape winding　铜带绕组
copper-tin welding rod　铜锡焊条
copper tubing　铜管
copper voltammeter　铜电量计【沉淀，电解】
copper water stop　止水铜片
copper weld cable　铜焊电缆
copper weld pool　铜熔池
copper weld wire　铜包钢线
copper weld　包铜钢丝，铜焊
copper winding　铜线绕组
copper wire braid shield cable　铜编织线屏蔽电缆
copper wire gauze　铜丝网
copper wire screen　铜纱
copper wire　铜丝，铜线
coppery　含铜的，铜质的，似铜的
copper-zinc accumulator　铜锌蓄电池
copper-zinc alloy　锌铜合金
copper-zinc cell　铜锌电池
copper-zinc welding rod　铜锌焊条
coppice　小灌木林

copple 坩埚
coprecipitation method 共沉淀法
coprecipitation 共沉淀
co-processing 同时加工，同时处理
co-production 合作生产
co-product 副产品，副产物
copse 小灌木林
copying program 复制程序
copying 复印
copyright contract 版权合同
copyright infringement 侵犯版权
copyright law 版权法
copyright material 有版权的资料
copyright royalty 版税，稿费，版权费
copyright 著作权，版权
copy 抄本件，副本，复制图纸，复印，印刷，打印，复印件，复制品，复写，抄本，拷贝
coquina 坝壳灰岩
coral island 珊瑚岛
coralline 珊瑚石
corbel arch 突拱
corbel back slab 悬臂板，引板，翅板，岸板
corbelling 悬挑，台阶形砌体，出砖牙，挑砖
corbel piece 梁托，支臂，支架，挑出块
corbel 托座【土建】，（支承）牛腿，撑架，托臂，突出部，突出部分建筑，托肩，梁托
COR(conditional OR) 条件“或”
COR＝coordination 协调控制方式
cordage 索具
cord and plug （设备的）电源线，橡包线
cord array 引线阵列
cord circuit repeater 塞绳增音机
cord circuit 绳路，塞绳电路
cordeau 爆炸导火索，雷管线
cord fastener 塞绳接线柱
cording diagram 接线图，塞绳连接图
cordless switchboard 无塞绳式交换机，无绳交换台
cordless telephone 无绳电话
cordless 无绳的【电话】，不用电线的
cord pendant 吊灯线
cord switch 拉线开关
cordtex 爆炸导火索
corduroy road 圆木路
cordwood micromodule 积木式微机组件
cordwood steam turbine 积木式汽轮机
cordwood system 积木式（一种微型器件的组合方式）
cord 绳，电线，电源线，橡包线，软线，塞绳，索
core alteration 堆芯的修改方案
core analysis 岩芯分析
core area 铁芯面积
core array 磁芯阵列
core assemble pressure 铁芯装配压力
core assembly 铁芯装配
core auxiliary cooling system 堆芯辅助冷却系统
core average exit quality 堆芯平均出口含汽率
core average heat flux 堆芯平均热流密度
core baffle 堆芯围板
core-balance protective system 磁势平衡保护系统
core-balance transformer 磁势平衡互感器
core barrel assembly 堆芯吊篮，堆芯筒体
core barrel centering pad on internals support ledge 堆内构件支承凸缘上的堆芯吊篮对中块
core barrel emergency support pad 堆芯吊篮事故支承垫
core barrel flange 堆芯筒体法兰
core barrel shell 堆芯吊篮壳体，堆芯筒体壳
core barrel storage area 堆芯筒体储存区
core barrel support ledge 堆芯吊篮支承凸耳
core barrel support skirt 堆芯吊篮支承裙筒
core barrel 堆芯吊篮，堆芯筒体，芯管，芯筒，岩芯钻筒，钻管
core bit 岩芯钻的钻头
core-blanket reactor 双区反应堆
core-bonding oven 型芯烘干炉
core boring 岩芯取样，岩芯钻探，钻孔取样
core bottom 堆芯底板
core buckling 堆芯曲率
core burst 涡核猝发，堆芯爆裂（事故）
core cage 堆芯可移动笼，芯缆
core catcher 熔化堆芯收集器，堆芯收集器，取心钻夹具
core cavern 堆芯洞穴
core cell 堆芯栅格
core channel 堆芯通道
core charge 堆芯装料
core circuit 铁芯磁路，堆芯回路，一次回路
core clamper 铁芯压板，铁芯压圈
core clamping-plate 铁芯压板
core clamping 铁芯固定，铁芯夹紧
core clamp 堆芯夹具
core component assembly 堆芯部件组件
core components 堆芯部件
core composition 堆芯成分
core concentration 堆芯（燃料）浓度
core concentricity 芯同心度
core conditions 堆芯工况
core configuration 铁芯布置，铁芯形状，堆芯布置
core content 中心内容
core conversion ratio 堆芯转换比
core coolant flow 堆芯冷却剂流
core coolant loop 堆芯冷却环路，堆芯冷却回路
core cooling monitoring system 堆芯冷却监测系统
core cooling monitor 堆芯冷却监测器
core coupling 磁芯耦合
core cover plate 堆芯盖板
core cover plug 堆芯盖塞
core cross section 岩芯断面，钻孔断面
core cutter method （测土密度的）环刀法
core cutter 土样切割器
core cycle 堆芯循环
core damage 堆芯损坏
cored beam 空心梁
cored brick 空心砖
cored casting 型芯浇注
cored electrode 有芯焊条，管状焊条

core deluge system 堆芯淹没系统
core deluge 堆芯淹没
core design constraint 堆芯设计约束条件
core design 堆芯设计
core diameter 堆芯直径，铁芯直径，纤芯直径
core dimension 铁芯尺寸
core disassembly accident 堆芯解体事故
core discharge plug 堆芯卸料塞
core disc paper 铁芯叠片绝缘纸
core disk 铁芯叠片
core disruptive accident 堆芯碎裂事故
core drilling method 取心钻探法
core drilling tool 钻取岩芯的工具
core drilling 岩芯钻探，钻取岩芯
core drill rig 岩芯钻机
core drill 钻取岩芯，空心钻，取岩芯钻
cored tile 筒状瓦
core duct 铁芯风道，通风槽
cored winding 铁芯绕组，药芯焊
cored 有铁芯的
core edge zone 堆芯边缘区
core element 堆芯元件
core emergency cooling system 堆芯应急冷却系统
core end-plate 铁芯端板，铁芯压板
core end 铁芯端部
core exit pressure 堆芯出口压力
core exit temperature 堆芯出口温度
core exit thermocouple 堆芯出口热电偶
core exit 堆芯出口
core extraction 取岩芯
core factor 铁芯系数
core field 铁芯磁场
core flange 铁芯法兰，铁芯压圈
core flooding system 堆芯淹没系统
core flooding train 堆芯淹没管系
core flooding 堆芯灌水，堆芯淹没
core flow 中心流，岩心流动
core flow subfactor 堆芯流量分因子
core flow test loop 堆芯流量试验回路
core flux test 铁芯磁化试验
core flux 铁芯磁通量
core form transformer 内铁型变压器，铁芯式变压器
core fuel exposure 堆芯燃料辐照
core fuel temperature 燃料芯核温度，堆芯燃料温度
coregraph 岩芯图
core grid structure 堆芯栅板结构
core grid 堆芯栅板
core ground 铁芯接地
core head 堆芯盖（板），铁芯端板，铁芯压板
core heat-up accident 堆芯升温事故
core height 堆芯高度
core hold-down ring 堆芯压紧环
core hole 岩心钻孔
core honeycomb 蜂窝型材，蜂窝填料
core hydraulics 堆芯水力学
core-image library 磁芯（存储器）映像库
core inductance 铁心电感，芯线电感
core induction furnace 有芯感应电炉，铁芯式感应炉
core induction 铁芯感应，铁芯磁密
core inlet flow rate 堆芯进口流速
core inlet specific-enthalpy 堆芯进口比焓
core inlet temperature 堆芯进口温度
core inlet 堆芯进口
core instrumentation 堆芯测量仪表
core instrument 堆芯仪表设备，活性区仪表设备
core insulation 芯线绝缘，铁芯绝缘
core integrity 堆芯完整性
core internal part 堆芯部件
core internals 堆芯部件，堆芯构件
core iron 铁芯用铁，型芯铁
core lamination stack 铁芯叠片段
core lamination 铁芯叠片
core lattice 堆芯栅格
core layer 夹心层
core leg 芯柱【变压器的】
core length 铁芯长度，堆芯长度
coreless armature 无铁芯电枢，空芯衔铁，空芯电枢
coreless coil 无芯线圈
coreless induction furnace 无铁芯感应电炉
coreless 无芯的
core life 堆芯寿期，一炉燃料寿期
core lifter 岩芯提取器
core limb （变压器的）芯柱，铁芯柱
core loading 堆芯装料
core load 堆芯负荷，把程序等调入内存
core logic 磁芯逻辑
core loss conductance 铁耗电导，铁耗分量
core loss current 铁耗电流，铁损电流
core loss test 铁损试验，铁耗试验
core loss 铁耗，铁损，铁芯损耗
core mapping 堆芯通量测绘
core map 堆芯图，堆芯通量分布图，堆芯通量图
core mark 铁芯上的记号，型芯记号
core material 堆芯材料，磁芯材料
core matrix 磁芯矩阵，堆芯栅阵，堆芯矩栅阵
core maximum heat flux density 堆芯最大热流密度
core melt 堆芯熔化
core meltdown accident 堆芯熔化事故
core meltdown analysis 堆芯熔化分析
core meltdown phenomena 堆芯熔化现象
core meltdown 堆芯熔化
core-melt probability 堆芯熔化概率
core melt-through accident 堆芯全部熔化事故
core memory resident 磁芯存储器常驻区
core memory 堆芯存储器，主存储器，磁芯存储器
core mixing subfactor 堆芯交混分因子
core model 堆芯模型，堆芯形式
core module 堆芯单元组件
core monitoring instrumentation 堆芯监测仪表
core monitor 铁芯监测器
core octant 堆芯八分区
core of anticline 背斜中心
core of cable 电缆芯
core of conductor 线芯

core of jet 射流中心
core on-line surveillance monitoring and operations system 堆芯在线监督与运行系统
core outlet pressure 堆芯出口压力
core outlet specific-enthalpy 堆芯出口比焓
core outlet temperature 堆芯出口温度
core outlet 堆芯出口
core over plate 堆芯盖（板）
core packet 铁芯叠片段
core pay-off reel 中心放线盘
core peaking factor 堆芯峰值因子
core performance evaluation 堆芯性能计算
core physics 堆芯物理
core pipe 取芯管，芯管
core plane 铁芯面，磁芯面
core plate varnish 铁芯硅钢片漆
core plate 铁芯叠片，堆芯板
core power density 堆芯功率密度
core power distribution estimation 堆芯功率分布估算，堆芯功率分布预测
core power distribution 堆芯功率分布
core pressure loss 堆芯压降
core pressure vessel 堆芯压力容器
core protection system 堆芯保护系统
core protocol stack 核心协议堆栈
core punching 铁芯冲制
core quadrant 四分之一堆芯
core quarter-height 堆芯四分之一高度
core radial power distribution 堆芯功率径向分布
core radial support 堆芯径向支承
core ratio 电缆芯径比
core recharge operation 堆芯换料操作
core record 岩芯记录
core recovery ratio 岩芯采取率
core recovery 岩芯获得率
core reflector 堆芯反射层
core reflooding 堆芯再淹没
core reflooding pump 堆芯再淹没泵
core reflooding system 堆心再淹没系统
core region 堆芯区
core residence time 堆芯停留时间【燃料元件】
core restraining facility 堆芯紧固装置【高温堆】
core restraint band 堆芯箍带
core rod 芯棒
core rubber 芯胶
corer 取芯管
core sampler 取样土钻
core sample test 芯样检测
core sample 岩芯试样，岩芯样品，矿样，岩芯样本，土芯样本
core sampling method 取芯法
core-sand moulding 芯砂造型
core sandwich 构架夹芯结构
core saturation 铁芯饱和
core screen 线芯屏蔽
core secondary support 堆芯辅助支承
core section 堆芯部分
core sedimentary sample 沉积柱状样品
core shell 堆芯环段【压力容器】，压力壳堆芯环段
core shift driver 磁芯移位驱动器
core shipment type 核心装运类型
core shroud support ring 堆芯围筒支承环
core shroud 堆芯围筒
core size 铁芯尺寸
core skills 主要技能
core skirt 堆芯裙筒
core slackness 铁芯松动
core spectrum 堆芯能谱
core splitting 涡核破碎
core spray and poison injection sparger ring 堆芯喷淋和毒物注入配水环
core spray header 堆芯喷淋集管
core spray heat exchanger booster pump 堆芯喷淋热交换器升压泵
core spray heat exchanger 堆芯喷淋热交换器
core spray inlet 堆芯喷淋入口
core spray nozzle 堆芯喷淋管嘴
core spray pump 堆芯喷淋泵
core spray ring 堆芯喷淋环管
core spray sparger 堆芯喷淋器
core spray suction line 堆芯喷淋吸入管
core spray system 堆芯喷淋系统
core spray 堆芯喷淋
core stability 堆芯稳定度，堆芯稳定性
core stacking 铁芯迭装
core stack of matrices 整体填料芯子
core stack 叠片铁芯，铁芯段，磁芯体
core staff 骨干人员
core stamping 铁芯冲片
core standby cooling system 堆芯备用冷却系统
core steel 铁芯用钢
core storage element 磁芯存储元件
core storage 磁心存储器
core store cycle time 磁芯存储器读写时间，磁芯存储周期时间
core stream 堆芯束流
core structural material 堆芯结构材料
core structure lay-down location 堆内构件搁置处
core structure 铁芯结构，堆内构件
core subcooling monitor 堆芯欠热度监测器，堆芯欠热监测器
core subcooling 堆芯欠热，堆芯欠热度，堆芯欠火
core support bracket 堆芯托架，堆芯支承结构
core support column 固定支承柱【堆芯构件】，堆芯支承柱
core support floor 堆芯支承板
core support forging 堆芯支承锻件
core support grid 堆芯下栅板，堆芯支承栅板
core support ledge 堆芯支撑凸肩，堆芯支承凸肩
core support plate insert 堆芯支承板插入件
core support plate 堆芯支承板
core support spring 堆芯支承弹簧
core support structure pad 堆芯支承结构垫块
core support structure 堆芯支承结构
core support 堆芯支承，堆芯底座，铁芯支撑
core taker 取芯筒，取芯样器，取样筒
core tank 堆芯容器
core temperature monitor 铁芯温度监测器
core temperature of particle 颗粒中心温度

core temperature 铁芯温度，中心温度
core test 岩芯试验，铁芯实验，铁芯试验
core thermal design 堆芯热工设计
core thermal power 堆芯热功率
core time （弹性工作制的）基本上班时间
core trench 心墙截水槽，心墙下沟
core tube nest 堆芯管束
core tube 堆芯管，取芯管
core type induction furnace 铁芯式感应电炉
core type induction heater 磁心型感应加热器
core type reactor 内铁型变压器，（铁）芯式电抗器
core type traction transformer 心式牵引变压器
core type transformer 芯式变压器，内铁型变压器
core type 芯式，铁芯式【变压器】
core unloading 堆芯卸料
core value 核心价值观
core ventilation 铁芯通风
core vessel 堆芯容器
core vibration 铁芯振动
core void content 堆芯内空泡含量
core void distribution 堆芯内空泡分布
core void fraction 堆芯内空泡份额
core volume 铁芯体积
core wall type rockfill dam 心墙式堆石坝
core wall 心墙
core water cover 堆芯水覆盖层，堆芯水覆盖
core welding-wire 焊芯，药芯焊丝
core-wide reliability 全堆芯可靠性
core wire 焊条铁芯，焊芯，电缆心线，芯线
core-zoning 堆芯分区
core 活性区，堆芯【反应堆】，铁芯，核芯，芯子，型芯，芯线，心板，芯，岩芯
coring device 取岩芯设备
coring pellet 空心芯块
coring segregation 晶内偏析
coring 去除中心部分，地层中取出的圆形岩石样品，取岩芯，钻取岩芯，在工地钻孔取样
Coriolis parameter 科里奥利参数
corium ring 收集堆芯熔化物的混凝土漏斗形结构
corium 堆芯熔化物
cork board 软木板
cork brick 软木砖，多孔砖
cork gasket 软木垫圈
cork gauge 塞尺，塞规
corking 滞塞【风洞】
cork insulation 软木绝缘
corkscrew stair 盘梯
corkscrew 螺丝锥
cork 软木塞，塞头，软木
corner antenna 角形天线
corner armour 护角
corner batt 立模角点
corner bead 墙角护条，角焊缝，镶边角板
corner brace 转角拉条
corner bracing 转角链条，角斜撑
corner bracket 角撑
corner casing 护角铸件
corner channel 角通道
corner cooler of 发电机四角冷却器
corner effect 拐角效应
corner-fired boiler 角式布置燃烧器锅炉，四角燃烧锅炉
corner firing burner 角式喷燃器，角式燃烧器
corner firing 四角喷射燃烧，四角燃烧
corner frequency 角频率，转折频率
corner gusset 角撑板
corner head 墙角护条
cornering capability 抗横偏能力，转弯能力
cornering force 拐弯力
cornering 拐弯，横偏
corner insulator 转角绝缘子
corner joint 角接缝，角焊缝，角接接头，角接头，弯管接头
corner lamp 角灯
corner loss 转弯损失【气流】
corner pocket 死角
corner post 角柱
corner protection strip 护角条
corner radius 转角半径
corner rod 燃料边棒
corner stone 奠基石
corner strap 直角铁
corner stream 角区流
corner tapping 角接取压
corner the market 垄断市场
corner tile 角砖
corner tower transmission line 转角塔架空输电线路
corner tower 转角塔
corner tube boiler 角管式锅炉，管架式锅炉
corner valve 角阀
corner vane cascade 拐角导流片栅【风洞】
corner vane 拐角导流片【风洞】，导流叶片
corner weld 角接焊缝，角焊，角焊道
corner 角落拐角处，拐角，墙角，角
cornice lighting 壁带照明
cornice 挑檐，雨水槽，檐条，上楣，檐板，檐口板，檐口，台口线，线条
Cornish boiler 康尼许式锅炉，卧式单炉胆锅炉
cornish 上楣，檐板
corollary equipment 配套设备
corollary facilities 配套设备
corollary 推论，系定理，必然结果
corona cage 电晕试验笼
corona charging 电晕荷电［充电］
corona control ring 电晕控制环
corona counter 电晕计数管
corona current 电晕电流
corona damage 电晕损坏
corona detector 电晕探测器
corona discharge cooling 电晕放电冷却
corona discharge inception test 电晕放电起始试验
corona discharge valve 电晕放电管
corona discharge 电晕放电
corona effect 电晕放电效应，电晕效应
corona electret 电晕驻极体
corona field intensity 电晕场强度
corona field 电晕场
corona free design 无电晕设计

corona free　无电晕
corona frequency analysis　电晕频率分析
corona grading paint　防电晕漆
corona grading　防电晕
corona guard　电晕防护器，电晕防护设备
corona inception field strength　电晕起始场强
corona inception test　电晕起始试验
corona inception voltage　电晕起始电压，起晕电压
corona interference　电晕干扰
corona location　电晕定位
corona loss　电晕损耗，电晕损失
corona luminescence　电晕发光
corona luminous layer　电晕发光层
corona measurement　电晕测量
corona noise　电晕噪声
corona of hydrogenerator　水轮发电机电晕
corona onset voltage　起晕电压
corona point　电晕放电尖端
corona power loss　电晕功率损失
corona prevention　电晕防护
corona preventive screen　防晕屏蔽
corona-proof cable　防电晕电缆
corona-proof coil　防电晕线圈
corona-proof cover　防晕罩
corona-proof insulation　防电晕绝缘
corona-proof　防电晕的，耐电晕的
corona protected technique　防电晕技术
corona protection　电晕防护
corona-protective varnish　防电晕漆
corona pulse　电晕脉冲
corona radiate　辐射冠
corona resistance　耐电晕能力，耐电晕性，电晕放电电阻，耐电晕放电击穿性
corona-resistant insulation　耐电晕绝缘
corona-resistant　耐电晕，防电晕
corona ring　电晕环
corona shielding　电晕屏蔽
corona stabilizer　电晕稳定器
corona starting gradient　电晕起始梯度
corona starting voltage　起晕电压
corona suppression　消除电晕，电晕抑制
corona test　电晕试验
corona treatment　电晕处理
corona voltage　电晕电压，晕电压
corona voltmeter　电晕伏特计
corona zone　电晕区，电晕范围
corona　电晕，日冕，电晕放电，飞檐上部，挑檐滴水板
coroutine　协同程序
Corp. = corporation　公司
corporate body　法人团体，法人
corporate engineering standard　公司技术标准
corporate entity　法人实体，合作实体，公司实体
corporate income tax　公司所得税
corporate product specifications　公司产品规范
corporate tax　公司税
corporation aggregate　集体法人，社团法人，公司法人
corporation attorney　公司法律顾问
corporation by-laws　公司章程
corporation cock　切断旋塞，分水旋塞
corporation law　公司法
corporation readjustment　公司改组
corporation regulation　公司条例
corporation seal　公司印章
corporation sole　单一法人（公司），单独法人
corporation tax　公司税，法人税
corporation　团体协会，有限公司
corpuscle　粒子，微粒
corpuscular radiation　微粒辐射
corpuscular rays　微粒射线
corpuscular theory　粒子说
corpuscular　微粒的
corpuscule　微粒，粒子
corpus　（某项基金的）本金
corral　深水桩支承围栏，栅栏
CORR＝correction　修正，校正
CORR＝correspondence　对应，通信
CORR＝corrosion　腐蚀
CORR＝corrugated　波浪式的，波纹的
correct assessment　正确的评价
correct average value　精确的平均值，基准的平均值
correct direction　正确方向
corrected altitude　修正高度
corrected flow　折算流量，校正流量
corrected heat consumption　校正热耗，折合热耗
corrected impedance relay　已校正阻抗继电器
corrected oil　合格油
corrected output　折合出力，修正功率，修正出力
corrected power　修正功率，校正功率，折合出力
corrected reading　改正读数
corrected result　修正结果
corrected speed　修正速度，修正转速，折合转速
corrected text　校正文本
corrected thermal efficiency　修正热效率，折合热效率
corrected value　校正值，修正值
corrected　校正的，修正的
correct factor　校正因子，修正因子
correcting circuit　校正电路
correcting code　校正码
correcting coil　校正线圈
correcting current　校正电流
correcting element　调节机构，校正元件，执行元件
correcting equipment　校正设备
correcting feedback　校正反馈
correcting feedforward　校正前馈
correcting magnet　校正磁铁
correcting member　校正组件，调整部件，校正部件
correcting network　校正网络
correcting pulse　校正脉冲
correcting rate　校正速率
correcting signal　校正信号
correcting unit　校正装置，校正部件，执行器，校正单元
correcting weight　校正重量

correcting 校正的
correction action 校正作用
correctional screw 改正螺旋
correction and compensation device 校正补偿装置
correction block 校准块
correction channel 校正通道，校正电路
correction circuit 校正电路
correction coefficient 修正系数，校正系数，改正系数，订正系数
correction computation 校正计算
correction computing 校正计算
correction curve （测量仪表的）修正曲线，校正曲线
correction data 校正数据，修正数据
correction efficient 修正系数
correction element 校正部件
correction factor for inclination 偏荷载校正系数，倾斜校正系数
correction factor 校正因数，校正系数，修正系数，校正因子，改正系数，改正因数，修正因数
correction for blockage 阻塞修正
correction for buoyancy 浮力修正
correction for compressibility 压缩性修正
correction for direction 方向改正，方向校正
correction for earth curvature 地球曲率改正
correction for lag 滞后校正
correction for Reynolds number 雷诺数修正
correction for scale effect 缩尺效应修正，尺度效应修正
correction for temperature 温度校正
correction for tunnel wall 洞壁干扰修正
correction for 因……原因而做的修正【for 后面指修正的原因】
correction function 校正函数
correction index 校正指数
correction line 改正线
correction magnet 校正磁铁
correction map 校正图，修正图
correction mass 校正质量
correction network 校正网络
correction of defects 故障排除
correction of dynamic properties 动态性能的校正
correction of error 校正误差
correction of final design 最终设计修正
correction of frequency characteristic 频率特性校正
correction of frequency 频率校正
correction of non-linear distortion 非线性失真校正
correction of power factor 功率因数校正，功率因数补偿
correction of river 河道治理
correction plane 校正平面
correction pulse 校正脉冲
correction rate 校正速度，校正率
correction screw 校正螺钉
correction table 校正表
correction term 修正项
correction time 校正时间
correction value 修正值
correction 校正，改正，修正（值），矫正，修改，治理
corrective action notice 纠正行动通知
corrective action taken 选择的校正措施
corrective action 校正作用［动作］，纠正措施［行动］，整改措施
corrective delay 校正延迟
corrective factor 修正因数
corrective form 纠正表
corrective lag 校正滞后，校正延迟，调整延迟
corrective maintenance time 矫正性维修时间
corrective maintenance 故障检修，改善性维修，设备保养，矫正（性）维修
corrective measures 改正措施，纠正措施
corrective network 校正网络
corrective security analysis 校正安全分析，校正对策分析
corrective term 修正项
corrective value of projection 投影补正数
corrective 校正的，补偿的，补偿物，修正值，校正值，矫正物
correct M for N 对 M 作 N 方面的校正［修正］
correct M to be N 把 M 校正成 N
correctness factor 校正因数，修正因数
correct operation 正确操作，正确运行
corrector 校正器，校正电路，改正器
correct pressure 正常压力
correct time 准确时间
correct 矫正，校准，纠正，校正，正确，正确的
correlated noise 相关噪声
correlated standard 相关标准
correlate 关联
correlating function 相关函数
correlation analysis 相关分析
correlation behaviour 相关行为
correlation coefficient 相关系数
correlation control unit 相关控制器
correlation curve 相关曲线
correlation degree 相关度
correlation detection scheme 相关检测线路
correlation detection 相关检波，相关检测
correlation detector 相关检波器
correlation diagram 相关图
correlation factor 修正系数
correlation from ranks 等级相关
correlation function 相关函数
correlation integral 相关积分
correlation intensity 相关强度
correlation length 相关长度
correlation matrix 相关矩阵
correlation method 相关法
correlation model 相关性试验模型，相关模型
correlation of indices 指标相关
correlation of random noise 随机噪声的相关性
correlation procedure 相关程序
correlation properties 相关特性
correlation receiver 相关接受器
correlation reception 相关接收
correlation scale 相关尺度

correlation time 相关时间
correlation 相关性，相互关系，相关，对比性[度]
correlative investment 相关投资
correlatogram 相关曲线图
correlatograph 相关图
correlator difference 相关差分
correlator 相关器，相关仪，环形解调电路
correlogram 相关图
correspond 对应
correspondence method 信函方式
correspondence principle 对应原理
correspondence theorem 相似定理
correspondence water level 相应水位
correspondence 通信，对应，相应，相当，符合，一致，书信，来往信件，函件
correspondent agent bank 代理银行
correspondent bank 业务往来银行，代理银行，往来银行
correspondent 客户
corresponding angle 对应角，同位角
corresponding bank 代理行，往来银行，业务联系银行
corresponding corners law 对应角定律，特殊互换性定律
corresponding crest 相应最高水位
corresponding depth 对应水深
corresponding discharge 相应流量
corresponding drawing 对应图纸
corresponding pressure 相应压力
corresponding speciality 相应专业
corresponding temperature 对应温度
corresponding water level 相应水位
corresponding 相应的，相当的，对应的，符合的，通信的
corridor study 路线研究
corridor 走廊，通道，通路，过道
corrie 山凹
corrigenda 需要改正之处，勘误表，正误表，新工作项目建议，corrigendum 的复数形式
corrigendum 需要改正之处，勘误表，正误表，勘误，应改正的错误
corroborate 确证，证实
corroded area 腐蚀区，腐蚀面积
corroded crystal 熔蚀斑晶
corrodent 可腐蚀性，腐蚀剂，腐蚀性物质，腐蚀的，腐蚀介质
corrode 腐蚀，侵蚀
corrosion allowance 容许腐蚀度，容许腐蚀量，腐蚀允许量，腐蚀余度，腐蚀余量，腐蚀裕度
corrosion at a seam 缝隙腐蚀
corrosion at high temperature 高温腐蚀，低温腐蚀
corrosion attack 腐蚀，侵蚀
corrosion behaviour 腐蚀特性
corrosion by gas 气体腐蚀
corrosion cavity 腐蚀空间
corrosion cell 腐蚀电池
corrosion cladding 抗腐蚀包壳
corrosion contaminant 腐蚀污染物
corrosion contamination 腐蚀污染，腐蚀产生的污染
corrosion control equipment 防腐控制设备
corrosion control 防腐控制
corrosion cracking 腐蚀开裂
corrosion crack 应力腐蚀裂纹，腐蚀裂纹
corrosion damage 腐蚀损坏
corrosion defect 磨蚀缺陷
corrosion embrittlement 腐蚀脆化
corrosion environment 腐蚀环境，腐蚀介质
corrosion failure 腐蚀损坏
corrosion fatigue crack 腐蚀疲劳裂纹
corrosion fatigue life 腐蚀疲劳寿命
corrosion fatigue 腐蚀疲劳
corrosion fouling 腐蚀污垢，腐蚀积垢
corrosion in aqueous environment 水介质中腐蚀
corrosion-inhibiting admixture 防侵蚀外加剂
corrosion-inhibiting fluid filler 防锈蚀液态填料
corrosion inhibition addition tank 缓蚀剂添加箱
corrosion inhibitor for acid-cleaning 酸洗耐蚀剂
corrosion inhibitor 缓蚀剂【锅炉化学清洗】，防腐品，防腐剂，减蚀剂，腐蚀抑制剂，阻蚀剂
corrosion kinetics 腐蚀动力学
corrosion loop test 回路腐蚀试验
corrosion measurement 腐蚀测定
corrosion-mechanical wear 腐蚀机械磨损
corrosion nodule 腐蚀结节
corrosion of metal 金属锈蚀，金属腐蚀
corrosion peak 腐蚀尖峰
corrosion penetration 腐蚀深度
corrosion pickling 腐蚀酸洗
corrosion pitting 腐蚀麻点
corrosion potential 腐蚀电位，腐蚀潜能
corrosion preventative material 防腐材料
corrosion preventative 防锈剂，预防腐蚀的
corrosion prevention 防腐蚀，防腐
corrosion preventive additive 防锈添加剂
corrosion preventive coating 防腐涂层
corrosion preventive compound 防蚀剂，防腐化合物
corrosion preventive equipment 防腐设备
corrosion preventive oil 防腐油
corrosion preventive pigment 防锈涂剂
corrosion preventive steel plate 防锈钢板
corrosion preventive 防腐剂，防锈剂，防腐蚀，腐蚀预防剂，防腐的
corrosion product 腐蚀产品，腐蚀产物
corrosion-proof electric equipment 防腐电气设备
corrosion proof type 防腐型，抗腐蚀型
corrosion proof valve 防腐门，防腐汽门
corrosion proof 抗腐蚀的，耐蚀的，抗蚀的，抗腐蚀，防腐蚀，耐腐蚀
corrosion protection lining 防腐蚀衬里
corrosion protection 防腐保护，防腐蚀，锈蚀防护，防腐，防蚀，防锈，腐蚀防护
corrosion rate 腐蚀速度，腐蚀速率，侵蚀速度
corrosion reaction 腐蚀反应
corrosion resistance coating 耐腐蚀涂层
corrosion resistance test 耐腐试验
corrosion resistance 抗腐蚀性，耐腐蚀性，耐蚀力，抗腐蚀强度，防腐，耐蚀
corrosion resistant alloy 耐蚀合金，抗腐蚀合金

corrosion resistant chromenickel steel 抗腐蚀铬镍钢
corrosion resistant cladding 抗蚀覆层
corrosion resistant coating 防腐涂层，防腐蚀涂料
corrosion resistant film 耐腐蚀膜
corrosion resistant material 耐蚀材料，防蚀材料，防蚀物料
corrosion resistant motor 防腐蚀电动机
corrosion resistant paint 防腐蚀涂料
corrosion resistant performance 耐腐蚀性能
corrosion resistant plate 不锈钢板，耐蚀钢板
corrosion resistant pump 耐腐蚀泵
corrosion resistant steel 不锈钢
corrosion resistant type 耐腐蚀型
corrosion resistant 防腐蚀的，耐腐蚀的，抗腐蚀的，防腐蚀，耐腐蚀，防腐
corrosion resisting property 耐蚀性能
corrosion resisting steel 耐蚀钢
corrosion resistivity 耐腐蚀性
corrosion rig test 腐蚀试验台试验
corrosion scale inhibitor 缓蚀阻垢剂
corrosion spot 腐蚀点
corrosion-stable lubricant 不腐蚀的润滑剂
corrosion strength 耐蚀性
corrosion test 腐蚀试验
corrosion troubles 腐蚀故障，腐蚀破坏
corrosion velocity 腐蚀速度，腐蚀速率
corrosion voltmeter 测量腐蚀的电压表
corrosion wear 腐蚀，腐蚀耗损
corrosion 腐蚀，锈蚀，风蚀，（动力）侵蚀，溶蚀
corrosive action 腐蚀作用
corrosive atmosphere 腐蚀性气氛，污染大气
corrosive bittern brine 腐蚀性卤水
corrosive effect 腐蚀作用
corrosive environment 易腐蚀环境
corrosive gas 腐蚀性气体
corrosive medium 腐蚀介质
corrosive nature 腐蚀性能
corrosiveness 腐蚀性，腐蚀作用
corrosive pitting 点蚀
corrosive-proof cable 防蚀电缆
corrosive wear 磨蚀损耗，腐蚀磨损，侵蚀性水
corrosive 腐蚀性的，侵蚀性的，生锈的，腐蚀的，腐蚀，腐蚀剂
corrosivity of water 水的腐蚀性
corrosivity 浸蚀性，腐蚀性
corrugated asbestos board 波形石棉板
corrugated asbestos cement roofing 瓦楞石棉水泥屋面
corrugated asbestos-cement sheet roofing 石棉水波形瓦屋面
corrugated asbestos-cement sheet 波形石棉水泥瓦，石棉水波形板
corrugated asbestos sheet 波形石棉瓦
corrugated baffle 波纹挡板
corrugated bar 竹节钢筋，竹节钢
corrugated bend 折皱弯管
corrugated canal gate 波纹渠道闸门
corrugated concrete slab 波纹混凝土板
corrugated connection 波纹接头
corrugated copper plate 波纹铜片
corrugated covering 波纹蒙皮
corrugated expansion joint 波纹膨胀节，波形膨胀头，波纹伸缩器
corrugated furnace boiler 波纹火筒锅炉，波形炉胆锅炉
corrugated galvanized sheet iron 镀锌波纹铁皮
corrugated iron pipe 波纹铁管
corrugated iron roof 瓦楞铁皮屋面
corrugated iron sheet flashing 瓦楞屋面钢板
corrugated iron 波纹铁，瓦楞铁
corrugated metal culvert 波纹金属涵管
corrugated metal gasket 波纹金属垫片
corrugated metal 波纹金属
corrugated pipe 波纹管，皱纹管
corrugated plate scrubber 波形板湿式除尘器，波形板洗涤器，百叶窗分离器
corrugated plate-type heat exchangers 波纹板式换热器
corrugated plate 波形铁板，波形板
corrugated rolls 皱辊，槽纹辊，瓦楞辊，瓦楞辘
corrugated rooting 波形板屋面
corrugated screw 瓦楞钉
corrugated scrubber 百叶窗分离器
corrugated separator 波形隔板
corrugated sheeting 波纹板
corrugated sheet iron 瓦楞铁
corrugated sheet metal 波纹钢板，波纹状钢板
corrugated sheet 波形板屋面，波纹板
corrugated steel bar 竹节钢筋
corrugated steel plate 波形钢板
corrugated steel wire 刻纹钢丝
corrugated surface 波纹面
corrugated tank （变压器的）波纹油箱
corrugated tile 波纹瓦
corrugated tube 波纹管
corrugated vane 波纹翼，波纹叶片
corrugated waveguide 波形波导，波形波导管
corrugated web H beam 波纹腹板 H 型钢
corrugated 波纹状的，波形的，皱纹的，成波纹的，加工成波纹的
corrugation pitch 波峰间距
corrugation 波纹，波浪形，浅畦，褶皱
corrupt and degenerate 腐化堕落
corrupt and incompetent 腐败无能
corruption 腐败，腐化，贪污
CORR 校正，修正【DCS 画面】
corss compound turbine 双轴汽轮机
corten 低合金耐腐蚀钢
corundolite 刚玉岩
corundum 刚石，金刚砂
CO_2 safety relief system 二氧化碳安全卸压系统，二氧化碳安全阀系统
COSAG(combined steam and gas turbine) 汽轮机与燃气轮机联合装置
COS(cash on shipment) 装船付现，装运付款
COS(change-over switch) 切换开关
COS(crack opening stretch) 裂纹开口扩展
cosecant antenna 余割天线

cosecant 余割
cosec 余割
coseismic curve 同震曲线
coseismic line 同震线
coseismic zone 同震带
CO_2 semiautomatic welding machine 二氧化碳半自动焊机
CO_2 shielded arc welding 二氧化碳气体保护焊
COSH 双曲余弦
co-signatory report 联名签署的报告
cosine capacitor 余弦电容器
cosine distribution 余弦分布
cosine equalizer 余弦均衡器
cosine generator 余弦发电机，余弦发生器
cosine heat flux density profile 余弦热流（密度）分布图
cosine impulse 余弦脉冲
cosine law 余弦定律
cosine oscillations 余弦振荡
cosine potentiometer 余弦电位计，余弦分压器
cosine transform 余弦变换
cosine wave 余弦波
cosine winding 余弦绕组
cosine 余弦
cosmetic pass 磨光焊道
cosmetic welding 盖面焊
cosmicdust 宇宙尘
cosmic ray 宇宙射线
cosmic water 宇宙水
cosmopolis 国际都市
COSMOS(core on-line surveillance monitoring and operations system) 堆芯在线监督与运行系统
cospectrum 余谱，同相谱，共谱
cost accounting 成本核算，成本计算，成本会计
cost account 成本核算
cost allocation 费用分摊，经费分配，投资分配
cost analysis 经济分析，价格分析，成本分析，费用分析
cost and delivery period 费用与支付期
cost and freight invoice 货价和运费发票
cost and freight 成本加运费价格，货价加运费
cost and insurance 成本加保险，成本加保险价
cost and schedule control system 堆芯备用冷却系统费用和进度管理系统【美国】
cost benefit analysis 代价利益分析，成本效益分析，费用收益分析
cost benefit comparison 成本收益比较
cost benefit ratio 费用收益比
cost benefit 成本效益
cost calculation 成本计算
cost component 成本组成
cost concept 成本概念
cost consideration 成本的考虑
cost control manager 成本管理主管人员，费用控制经理
cost control 成本控制
cost data 成本资料
cost distribution 费用分配
cost drift 价格趋势
cost-effectiveness analysis method 技术经济分析法
cost-effectiveness analysis 费用效果分析，成本效益分析
cost effectiveness 成本效益，成本效率，经济效益，成本有效性
cost-effective 合算的，有效益的
cost-efficient 有成本效益的，划算的，合算的，成本效益比合算的，本轻利厚的
cost engineer 造价工程师
cost escalation 涨价
cost estimate of power generation project 发电工程投资估算
cost estimate sheet 成本估计单
cost estimate 造价估算，成本估计，费用概算，成本估算，估价
cost estimation 估价，成本评估，造价估算，费用概算
cost factor 造价影响因数
cost feasibility 成本可行性
cost for construction equipment's entry to and exit from the site 机械进退场费
cost for decision making 决策成本
cost free 免费（的）
cost & freight 货价及运费
cost index 费用指数
costing 成本会计，成本计算
cost, insurance and freight 货价、保险加运费价格
cost, insurance, freight, commission and interest 货价、保险、运费、佣金及利息
cost, insurance, freight and commission 货价、保险、运费及佣金，到岸价格加佣金
cost, insurance, freight and exchange 货价、保险、运费及汇兑费
cost, insurance and freight 货价、保险及运费
cost of bidding 投标费用，成本核算
cost of construction 建造费用，造价，建筑成本
cost of delay 误期费用，延误费
cost of detriment 危害代价
cost of electric energy 电能成本
cost of electricity provision 供电成本
cost of electricity supply 供电成本
cost of electricity 电价，电费
cost of electric power sold 售电成本
cost of electric power station output 发电成本[费用]
cost of electric power supply in transmission 供电成本
cost of erection 架设费，安装费，安装成本
cost of failure 故障造成的损失费
cost of goods purchased 购货成本
cost of heat-supply 供热成本
cost of interruption 停电损失
cost of inventory 盘存成本
cost-of-living index 生活费指数
cost of living 生活费
cost of main power building per kW 每千瓦主厂房造价
cost of maintenance 维护费，养护费，维护费用
cost of operation 运行费，管理费，经营成本
cost of overhaul 大修费，检修费，大修费用

cost of packaging　包装费
cost of power generation project per kW　发电工程每千瓦造价
cost of power production　电力生产成本，能量生产成本，发电成本
cost of power station　电站造价
cost of production　生产成本，防护代价，生产费用
cost of purchased electric power　购电成本
cost of repair　修理费
cost of repatriation　遣散费
cost of reproduction　再生产成本
cost of services　劳动成本
cost of set-up　开办费
cost of suspension　暂时停工费用
cost of temporary buildings　临建设施费
cost of tendering　投标费用
cost of unit start-up　机组启动费用
cost of unit working　单位工作成本
cost of upkeep　日常费用，维修费，养护费
cost of winning　开采成本
cost of works　工程造价
CO_2 storage　二氧化碳贮存
cost overrun　费用超支
cost performance index　成本指数
cost performance　成本效率，价格性能
cost per kW hour of the electricitygenerated by WTGS　风电度电成本
cost per kWh lost　每千瓦时停电损失
cost per kW　单位千瓦造价，每千瓦造价，每千瓦费用【电】
cost planning　成本计划
cost plus contract　成本加利合同
cost price index　成本价格指数
cost price　成本费，成本价格
cost-profit-volume analysis　成本-利润-产销量分析，本利量分析
cost record　成本账
cost recovery year　成本回收年限，费用回收年限
cost recovery　成本回收
cost plus fee contract　成本补偿合同
cost reimbursement contract　补偿费合同，成本补偿合同，成本偿还合同
cost reimbursement　投资偿还，补偿费
cost report　成本报告
costs are shared on a fifty-fifty basis between A and B　费用由A和B均摊
cost-saving investment　节省成本的投资
cost schedule　成本明细表
costs for maintenance　维护费用，维修费用
costs for ownership　折旧费
cost sharing formula　成本分摊公式
cost sharing　费用分配
cost sheet　成本报表，成本单
cost shifting　费用转移
costs of unloading　卸货价
cost stream　价格趋势
cost unit price　成本单价
cost variance　费用偏差
cost-yield result　投入产出结果
cost　成本，费用，价格，代价，投资，造价
cosy　舒适的
CO take off pipe　一氧化碳取气管
cotangent　余切
cotidal chart　同时潮图
cotidal hour　同时潮时
cotidal line　同时潮线，同潮线，同潮时线
cottage roof　小跨度屋盖
cotter file　开槽挫
cotter pin driver　开口销起子
cotter pin　扁销，开尾销，开口销
cotter seat　弹簧座【气门】
cotterway　销槽
cotter　开口销，楔形销，楔，制销，栓
cotton braided weather proof wire　纱包耐风雨线
cotton-covered cable　纱包电缆
cotton-covered wire　纱包线
cotton-covered　纱包的，（棉）纱绝缘的
cotton cover insulation　纱包绝缘
cotton-enamel covered wire　纱包漆包线
cotton fabrics　棉织品
cotton gloves　棉袖套
cotton goods　棉制品
cotton-insulated wire　纱包绝缘电线
cotton insulation cable　纱包绝缘电缆
cotton insulation　纱绝缘
cotton tape　布带
cotton textiles　棉制品
cotton varnished sleeve　黄蜡管
cottrell installation　静电除尘器
cottrell precipitator　静电除尘器
cottrell process　静电除尘法
cottrell　电除尘器
Couetteflow　库埃特流动
coulee　熔岩流，深冲沟，深峡谷
coulomb barrier　库仑势垒
coulomb collision　库仑碰撞
coulomb damping　库仑阻尼
coulomb energy　库仑能量
coulomb excitation　库仑激发
coulomb field　库仑场，库仑静电场
Coulomb force　库仑力，库伦力
coulomb friction　库仑摩擦
coulombian field　库仑静电场，库仑场
coulombian　库仑的
coulombic attraction　库伦吸引作用，库伦引力
coulombic efficiency　库仑效率
coulombic　库仑的，库仑定律的
coulomb meter　电量计，库仑计
coulomb repulsion　库仑斥力
Coulomb's earth pressure theory　库仑土压力理论
Coulomb's law　库仑定理，库仑定律
coulomb　库仑
coulometer　电量计，库仑计
coulometric titration　电量滴定法，库仑滴定
coulometric titrator　库仑计式滴定计
Coulter principle　库尔特原理
councilor　参赞，顾问，理事
council of director　董事会
council　理事会，委员会，协会，议会

Counihan type vortex generator　库尼汉式旋涡发生器
counselor　参赞，顾问
count capacity　计数容量
count cycle　计数循环，计数周期
count down warning　计数报警，倒计时报警
count down　倒计数，倒读数，递减计数，倒计时
counted measurand　累计式测值，计量值
counted measured　表计值
counted pulses　计数脉冲
counteracting force　反作用力
counteraction　抵消，阻碍，中和，反作用，反对的行动，抵抗，反动
counter action　对诉讼程序提出抗辩
counter-arched revetment　反拱护壁
counter arch　扶垛拱
counter balance accounts　冲账
counter balance boom　机平衡臂
counter balanced gate　平衡门
counter balanced　平衡的
counter balance jib　平衡臂
counter balance moment　平衡力矩
counter balance piece　平衡块
counter-balance valve　背压阀，反平衡阀，平衡阀
counter-balance weight　平衡块，配重
counter balance　托盘天平，平衡块，平衡物，平衡性，使平衡，抵消，平衡锤，配衡
counter bonification　反补偿
counterbored ball bearing　锁口球轴承
counterbore drill　平底扩孔钻
counterbore　沉头孔，锥口孔，扩孔，镗孔，埋头孔，斜口孔
counter box　计数器箱
counter brace　交叉撑
counterbuff　缓冲器，防撞器，反击，抵抗
counter camber　预留弯度，预变形，反变形
counter capacity　计数器容量
counter cell　反压电池
counter circuit　计数器电路
counter claim precaution　反索赔防范
counter claim　反索赔，反诉
counter-clock unit　计数器时钟单元
counter-clockwise angle　逆时针角
counter-clockwise motion　逆时针方向运动
counter-clockwise rotation　逆时针（方向）旋转
counter-clockwise　逆时针，逆时针方向（的），反时针方向（的）
counter coil　补偿线圈
counter compound-wound motor　差复激电动机
counter-controller　计数器控制器
counter control　计数器控制
counter cooling tower　逆流式冷却塔
counter coupling　计数器耦连
counter current anion exchanger　逆流阴离子交换器
counter current cation exchanger　逆流阳离子交换器
counter current centrifuge　逆流式离心机
counter current circulation　逆循环，逆流循环
counter current condenser　逆流冷凝器
counter current contact　逆流接触
counter current cooler　逆流冷却器
counter current cooling tower　逆流冷却塔
counter current flow limitation　逆向流动限制
counter current flow　逆流，反流
counter current mixing　逆流混合
counter current packed-bed filter　逆流颗粒层收尘器
counter current regeneration　逆流再生，对流再生
counter current washing　逆流洗涤
counter current　反流，反向电流，反向流，逆电流，逆流，对流，回流
counter dam　护坝
counter dead time　计数器的停歇时间
counter decade　十进制计数器
counter dial　加速器度盘
counter diffusion　逆向扩散
counter directional tractor　反向尾车
counter drain　背水面坡脚排水，辅助沟，副沟
counter drift　逆向漂流
counter dyke　月堤
counter-electrode　反电极
counter electromotive force　反电动势
counter-EMF starter　反电动势启动机
counterfeit　假冒，伪造
counter field　逆磁场，反磁场
counter flange　对接法兰
counter flashing　泛水板，泛水板防水条，屋面金属防水条，帽盖泛水
counter floor　毛地板
counterflow anion exchanger　逆流阴离子交换器
counterflow boiler　逆流式锅炉
counterflow cation exchanger　逆流阳离子交换器
counterflow combustor　逆流式燃烧器
counterflow condensation　逆流凝结
counterflow cooling tower　逆流式冷却塔
counterflow drier　逆流干燥器
counterflow heat exchanger　对流式热交换器，逆流式换热器
counterflow heating　逆流加热
counterflow preheater　逆流式空气预热器
counterflow regeneration　逆流再生
counterflow reheater　逆流再热器
counterflow steam generator　逆流式蒸汽发生器
counterflow superheater　逆流过热器
counterflow　回流，逆流，反向流动
counter force　对抗力，反作用力，反力
counterfort dam　扶垛坝
counterforted lock wall　扶垛式闸墙
counterforted wall　扶垛墙
counterfort lode wall　扶垛式闸墙
counterfort retaining wall　扶壁式挡土墙，扶垛式挡土墙
counterfort　护墙，扶壁，扶垛，拱柱，山嘴
counter-free machine　无计数器的时序机，不计数的时序机
counter frequency meter　计数式频率计
counter gate　平衡门，平衡闸门
counter gradient flux　逆梯度通量
counter-guarantee system　反担保制度

counter-guarantee　对背保证书，反担保，反赔偿
counter guarantor bank　反担保行
counter guarantor　反担保人
counter hand　计数器指针
counter impulse　计数脉冲
counter jet mill　反喷研磨机
counter jet　反向射流，反向喷射
counter knob　计算机按钮
counter magnetic field　反向磁场，反作用磁场，逆磁场
counter magnetic flux　反向磁通
counter measures　对抗措施，防范措施，反制措施，治理措施，对抗，对策
counter modulation　反调整，解调
counter monitor　计数管监测器，控制计数管
counter nozzle　反面喷嘴
counter offer　还价，还盘
counter operating voltage　计数管工作电压
counter osmosis　反渗透作用
counterpart talks　对口会谈，对口谈判
counterparty risk　对手风险，（贸易）对方风险
counterpart　对方，对应部分，对应物，对应的人，副本，配对物，极相似的人或物
counter plateau　计数管坪
counterplot　预防措施，对抗策略，对策，用反计
counter point　对点，对位，对偶
counterpoise grounding　接地带，引申体接地
counterpoise　接地装置，接地线，补偿，配重，平衡体，平衡重，平衡器，平衡锤，地网，抵消力量，使平衡，抵消
counter potential　反电位
counterpressure　背压，反压，反压力
counterproductive　起相反作用的
counter pulse　计数脉冲，计数器脉冲
counter-rail　导轨，护轨
counter range　计数管范围，计数管区段，记数范围
counter rate meter　计算速度测定计
counter reaction　逆反应
counter recovery time　计数管恢复时间
counter register　计数寄存器
counter reservoir　平衡水池
counter resetting　计数器置零，计数器复位
counter resolving time　计数管分辨时间
counter revolution　反转
counter ring　计数环
counter-rotating bladed wind machine　对转叶轮风力机
counter-rotating turbine　对转式汽轮机
counter sea　逆浪
counter shaft　副轴，中间轴
counter shear　反面剪切
counter shutter　平衡活闸板
countersign a contract　会签合同
countersignature system　会签制度
countersignature　会签
countersigning a contract　会签合同
countersign　联（名签）署，会签
countersink bolt　埋头螺栓
countersink drill　埋头钻，锥口钻
counter sinking bit　埋头钻
counter sinking　锪孔，钻孔，锥形扩孔
countersink　埋头孔，沉孔
counter spring　缓冲弹簧，平衡弹簧
counter stage　计数级
counter stream　逆流
counterstrut　抗压杆，受压杆
countersunk bolt　埋头螺栓
countersunk headed bolt　埋头螺栓
countersunk head screw　沉头螺栓，埋头螺钉
countersunk head　暗钉头，埋头，沉头
countersunk rivet　埋头铆钉
countersunk screw　埋头螺钉，埋头螺丝
countersupply　对销供货
counter switch　计数开关，计时开关
counter tank　计数存储器
counter thrust　反推力
counter tide　逆潮
counter tie　副系杆
counter function　计数功能
counter-timer　时间间隔计数测量器，计数器计时器
counter torque　反扭矩
counter trade　补偿贸易，对销外贸
counter tube lifting unit　计数管提升装置
counter tube　计数管，计数放电管
counter type adder　计数器型加法器
counter type cup anemometer　计数式风杯风速计
counter type frequency meter　计数式频率计
countervail　补偿，抵消
counter vane　导向叶片
counter voltage　反电压
counter vortex　逆涡
counter warranty for arresting ship　扣船反担保
counter weight boom　平衡重量悬臂【吊车】
counter weighted cleaner　带配重的清扫器
counter weighted unit　重锤拉紧装置
counter weight fill　镇压层填方，压重填土
counter weight hanger　重锤式吊架
counter weight lift　平衡重升降机
counter weight　砝码，配重，平衡重量，平衡锤，平衡铊，平衡块，平衡重，抵消力量，重锤
counter wheel printer　数轮打印装置
counter wheel　计数器轮
counter wind　逆风
counter　柜台，计数器，相反的，反方向的，反驳，计量器，计算员，反面的，副斜杆，配重
counting accuracy　计数精确度
counting apparatus　计数管
counting area　计数灵敏区
counting-chain　计数链
counting channel　计数通道，计数电路
counting circuit　计数电路，计算电路
counting code　计数码
counting coil　计数线圈
counting device　计数器，计数装置
counting-down circuit　分频电路，分频器
counting efficiency　计数效率
counting error　计算误差，计数误差
counting forward　顺向计数

counting grid 计数栅
counting in reverse 逆向计数
counting instrument 计数仪器
counting interval 计数间隔
counting ionization chamber 计数电离室
counting loss 计数损失
counting mechanism 计数机械结构，计数机构
counting module 计数模块
counting operation 计算
counting range 记数范围
counting rate characteristic 计数率特征曲线
counting rate computer 计数率计算器
counting-rate-difference feedback 速差反馈
counting ratemeter 计数率计
counting rate 计算速度，计数率
counting region 计数范围
counting relay 计数继电器
counting response 计数响应
counting ring 计数环
counting room 计数室
counting subroutine 计算子程序
counting system 计数系统
counting threshold 计数阈
counting time 计算时间
counting tube 计数管
counting wheel 计数轮
counting yield 计数效率
counting 计算，计数，用计数法测定的放射性强度
count probability method 计数概率法
count pulse 计数脉冲
count ratemeter 计数率计
count rate 计数率
count reverse 反向计数
country boundary 国界
country country 丘陵地带
country of delivery 发货地国家
country of destination 目的地国家，收货地国家
country of dispatch 发货国
country of embarkation 装货国
country of origin 原产国，产地
country of payment 付款国
country risk （借贷）国家风险，国别风险，外汇付款风险
country seat 别墅
countryside 农村
country under the role of law 法治国家
country 国家，乡村
counts per second 每秒钟计数
count time 计数时间
count-to-count interval distribution 计数间隔分布
county emergency operating center 地区［区域］应急控制中心
county town 县城
county 县
count 计数，算进，算入，数目，读数，次数，计算，脉冲
couplant monitor system 耦合剂监测系统
couplant 耦合剂，耦合介质【超声波探伤用】
couple arm 力偶臂
couple chock station 车钩检查站
couple corrosion 电偶腐蚀，双金属腐蚀
coupled antenna 耦合天线
coupled bending-torsion oscillation 弯扭耦合振动
coupled capacitance 耦合电容
coupled circuit 耦合电路
coupled columns 对柱
coupled degree of freedom 耦合自由度
coupled field 耦合场
coupled flutter 耦合颤振
coupled frequency 耦合频率
coupled instability 耦合不稳定性
coupled neutronic-thermohydraulic model 中子学与热工水力学耦合的模型
coupled oscillation 耦合振动
coupled oscillatory circuit 耦合振荡电路
coupled resonance 耦合共振
coupled switch 联动开关
coupled test panel 组合试片【渗透实验用】
coupled thermal-nuclear model 耦合热工-核模型
coupled to driver 向驱动端耦合的
coupled-type generator 直联型发电机
coupled vibration 耦合振动
coupled wheel 联动轮
coupled with 与……联合，结合，与……连接在一起
coupled 耦合的，连接的，成对的
couple moment 力偶矩
couple of force 力偶
coupler coupling 耦合器
coupler disconnection platform 摘钩平台
coupler yoke 耦合器卡箍
coupler 车钩，耦合器，联接器，钩钩，连接器，联轴器，联络开关，联轴节
couple unbalance 力矩不平衡
couple with 与……连接在一起，外加，加上
couple 热电偶，耦合，对
coupling and guide tube 耦合导向管
coupling axle 连接轴
coupling band 连接箍
coupling block 耦合块
coupling box 联接器箱，接线箱，连接装置，连接螺栓，接合螺栓
coupling breaker 联合开关，并列开关
coupling cable 中继电缆，耦合电缆
coupling capacitor voltage transformer 耦合电容式电压互感器
coupling capacitor 耦合电容器，结合电容
coupling capacity 耦合电容，结合电容
coupling coefficient 耦合系数
coupling coil 转电线圈，耦合线圈
coupling condenser 耦合电容器
coupling connector 水泵接合器
coupling constant 耦合常数
coupling device 连接装置
coupling drive 直接传动
coupling efficiency 耦合效率
coupling element 耦合元件
coupling end 联轴节端
coupling face 联轴器平面
coupling factor 耦合系数，耦合因数
coupling filter 耦合滤波器，结合滤波器

coupling flange 连接法兰，联结翼板，联轴器法兰
coupling frequency 耦合频率
coupling function 耦合函数
coupling guard 联轴节保护罩，联轴器护罩，耦合器罩
coupling half 半联轴器，半离合器，半个联轴节
coupling head 联接头
coupling impedance 耦合阻抗，耦合电感
coupling inductor 耦合线圈
coupling jaw 联轴器爪
coupling joint 联轴节
coupling key 耦合按钮
coupling loss 耦合损耗
coupling mechanism 耦合机构
coupling nut 联接螺母
coupling of steam turbine and generator 汽轮机、发电机对轮连接
coupling-out 耦合输出
coupling parameter 耦合参数
coupling piece 联轴节件
coupling pin 联轴器销
coupling processors 耦合的处理机
coupling puller 联轴节拆卸器
coupling reactance 耦合电抗
coupling reaction 耦合反应，耦联反应
coupling release actuator 摘钩器
coupling rigid 刚性联轴器
coupling rod 连接杆
coupling shaft 连接轴
coupling size 链接标记
coupling sleeve 联轴节套筒
coupling socket 耦合插座【控制棒】，联轴节套筒
coupling spindle 联接轴
coupling support stand 联轴器支架【泵电动机】
coupling switch 耦合开关
coupling transformer 耦合变压器，联络变压器
coupling wave 耦合电波
coupling winding 耦合绕组
coupling with resilient bolt 弹性柱销联轴器
coupling 联轴节，耦合，管接头，匹配，管箍，靠背轮，联轴器，联挂，联接
coupon test 取样试验
coupon 试样，试块，取样管，试件，息票，赠券
courier number 快递单号
courier 送信人，信使，通信员，快递员
course angle 航向角，行车方向角
coursed ashlar 成层琢石
coursed blockwork 混凝土预制块，分层铺砌块
coursed masonry 成层圬工，分层圬工
coursed pavement 成层铺面
coursed rockfill 成层堆石
coursed rubble masonry 分层块石圬工，分层毛石圬工
coursed rubble wall 分层毛石墙
coursed square rubble 分层方块堆石
course of reaction 反应过程
course of receiving 验收过程
courser 垫层
course sensibility 航向灵敏度
course 过程，经过，行程，（锅筒）筒节，环，环段，课程，路径，期间，层数，层
coursing joint 成层缝，成行缝
court dock 封闭谷
courteous service 服务周到
courtesy 礼节
courtyard house 四合院
courtyard 院子，楼群院场，天井
court 法庭，庭院
covalency effect 共价效应
covariance analysis 协方差分析
covariance function 协变函数
covariance matrix 协方差矩阵
covariance 协方差，协变性
covariant calculation 协变计算
covariant differentiation 协变微分
covariant divergence 协变散度
covariant tensor 协变张量
covariant theory 协变理论
covariant vector 协变矢量
covariant 共变，共变式，协变量
COV(cover) 外壳
COV(cut-off valve) 截止阀
cove damage 堆芯损坏
coved arch 大弧拱
cove molding 凹圆线脚
covenant of warranty 保证书
covenant 盟约，契约
coverage area 覆盖面积，覆盖区
coverage contour 等场强曲线
coverage factor of centralized heat-supply 集中供热普及率
coverage including theft, pilferage and non-delivery 包括偷窃及提不到货在内的保险
coverage 保险范围，保险项目，保险险别，保险总额，投保条款，范畴，范围，覆盖（层），涂层，覆盖物，可达范围，铺砌层
cover aggregate 盖面集料
coveralls 衣裤相连工作服，工作服，连身工作服
cover block 盖面块【混凝土】
COVER=crossover 切换管
covered arc welding 手工电弧焊
covered area 占地面积
covered car 棚车
covered conductor 被覆线，绝缘线
covered conduit 暗沟，封闭式水道，加盖水道
covered ditch 盖板明沟
covered drainage 排水暗沟
covered drain 暗沟
covered duct 覆盖的导管
covered electrode welding 药皮焊条电弧焊
covered electrode 复包电极，涂药皮焊条，有药皮的焊条
covered filter 封闭式滤池，覆盖过滤器
covered gutter 暗沟
covered knife switch 带盖闸刀开关
covered penstock 埋藏式压力水管
covered reinforced-concrete flume 封闭式钢筋混凝土渡槽

covered reservoir 地下水池，水窖，有盖水池
covered sludge drying bed 加罩污泥干燥床
covered storage 非露天贮存
covered surge tank 封闭式调压井
covered tray 有盖托架
covered way 廊道
covered wire 绝缘线，被覆线，包线
covered 被覆的，覆盖的，绝缘的，有屋顶的
cover enclosure 罩壳
cover fillet 盖条
cover flashing 披水板
cover gas cleanup system 覆盖气体净化系统
cover gas discharge line 覆盖气体排放管线
cover gas 覆盖气体
cover glass 防护玻璃罩，护眼罩
cover height 覆盖高度
covering area 覆盖面积
covering capacity 覆盖能力，遮盖能力
covering machine 包线机
covering of roadway 道路铺面
covering of the lagging 绝缘层外套
covering plate 盖板
covering reagent 保护剂
covering surface 覆盖面
covering 金属内衬，衬里，保护层，镀层，覆盖物，蒙皮，面层，罩面，罩
cover insurance 投保
cover letter 附函，面函，封面信，说明信，求职信
cover material 覆盖料，铺盖材料，铺压料
cover mould 压缝条，镶边
cover pass 盖面焊道
cover plant 覆盖植物，地被植物
cover plate splice 盖板拼接
cover plate 盖，覆板，盖板
cover slab 盖板，厚盖板
cover strip 覆盖条，防蚀镶片，压缝板，盖条，整流带，蒙皮条
cover support 盖板座
cover the shortage 弥补不足
cover tile 盖瓦
cover up with concrete 用混凝土覆盖
cover 罩壳，覆盖（物），盖，上半汽缸，（支票的）保证金，投保，盖板，覆盖层，遮蔽
cove skirting 凹圆踢脚板
cove 凹圆线，穹窿，山凹，山谷，小湾，窟窿
coving 窟窿
CO_2 volume flow rate 二氧化碳体积流速
CO_2 welding machine 二氧化碳电焊机
cowl flap control valve 整流罩通风控制阀
cowl former 整流罩框架
cowling mount 整流罩架
cowling 罩，整流罩
cowl ventilated motor 机壳风冷式电动机
cowl 套，罩，风罩，壳，外壳，伞行风帽，通风帽，挡煤板，逸气罩，整流罩
coyer 盖片
coyote blast(ing) 大量爆炸，洞室爆破，硐室爆破，药室爆破
coze film 软泥薄层
coze sucker 吸泥器
coze 软泥
CPA(certified public accountant) 注册会计师
CPA firm 注册公共会计师事务所
CPA 阴极保护系统【核电站系统代码】
CP(calorific power) 发热量
CP(candle power) 烛光
CP(case preparation) 外壳制备
CPC(clock-pulsed control) 时钟脉冲控制
CPC (compound parabolic concentrator) collector 复合抛物面集热器
CPC(computer process control) 计算机过程控制
CP=centipoise 厘泊【黏度单位】
CP(change package) 改变包装
C/P(charter party) 租船合同，租船方
CP(chemical pure) 化学纯度
CP(circulating pump) 循环水泵
CP(clock puslse) 时钟脉冲
CP(command post) 指令站
CP(condensate polisher) 除盐装置
CP(condensate pump) 凝结水泵
CP(constant potential) 恒定势，恒电位
CP(constrained procedure) 约束方法，限定程序
CP(construction permit) 建造许可证【安全法规用语】，施工许可证
CP(control panel) 控制板
CP(control point) 控制点
CP(control program) 控制程序
CP(cooling pond) 贮存水池【乏燃料】，冷却水池
CP(crack propagation) 裂纹扩展
CP(critical pressure) 临界压力
CP(current paper) 近期论文
CP(cycle permuted) 循环排列
CPDC(core power distribution control) 堆芯功率分布控制
CPDE(core power distribution estimation) 堆芯功率分布估算
CPDP(core power distribution prediction) 堆芯功率分布预测
CPE(chlorinated polyether) 氯化聚醚
CPE(chlorinated polyethylene) 氯化聚乙烯
CPE(computer performance evaluation) 计算机性能评价
CPE (construction project evaluation programme) 建造阶段电厂（安全）审证计划
CPE(core performance evaluation) 堆芯性能计算
CPF(coated particle fuel) 包覆颗粒核燃料
CPI(consumer price index) 消费者价格指数
CPI(critical path impact) 关键路径影响
C-pillar C支柱
CPIS(containment purge isolation signal) 安全吹扫隔离信号
CPL(control pannel local) 就地控制盘
CPLG=coupling 耦合器，连轴器
CPM(control pannel main) 主控盘
CPM(cost per mille) 每千人成本
CPM(critical path method) 关键路径法，统筹方法，关键路线法
CPO(concurrent peripheral operation) 并行外围设备操作

CPP(condensate polishing plant) 凝结水精处理站，凝结水净化装置
CPR(current page register) 现行页面寄存器
CPS(central processing system) 中央处理系统
CPS(character per second) 每秒字符数，字符/秒
CPS(condensate polishing system) 凝结水净化系统
CPS(corporate product specifications) 公司产品规范
CPS(cycle per second) 周/秒，赫兹
CPT(carriage paid to) 运费付至目的地
CPT(cold precritical test) 临界前冷试
CPT(cone penetration test) 静力触探试验，圆锥触探试验
CPU(central processing unit) 中央处理机，中央处理单元，中央处理器【计算机】
CPVC(chlorinated polyvinyl chloride) 氯化聚氯乙烯
CPV(concrete pressure vessel) 混凝土压力容器
CQC(complete quadratic combination) 完全二次项平方根组合
CQD(customary quick dispatch) 按港口习惯快速装卸
CQO(chief quality officer) 首席质量官
crab crane 蟹爪式起重机
crab grate 拦蟹栅
crab hole 蟹洞【基岩中的】
crab rail centers 小车轨距
crab roof 复斜屋顶
crab ship unloader 抓斗卸船机
crab 小车，绞车，卷扬机，起重机，起重绞车，吊车，起重小车，抓斗
crack and fissure 裂隙
crack arrest temperature 止裂温度，裂纹制止转变温度
crack arrest 止裂
crack border 裂纹边界
crack contamination 裂纹污染
crack control measures 防裂措施
crack control reinforcement 控制裂缝钢筋
crack control 防裂
crack density 裂纹密度
crack depth 裂纹深度【覆层以下的距离】
crack detection process 裂缝检测法
crack detection 探伤，裂纹检测，裂隙检查，裂缝检查
crack detector 裂缝检查器，裂痕探测器，裂纹测深仪，探伤仪
crack due to internal stress 内应力裂纹
crack due to settlement 沉降裂缝，沉陷裂缝
crack due to shearing force 剪力裂缝
cracked aluminum block 裂纹铝试验
cracked condition 开裂状态
cracked fuel oil 裂化重油
crack edge 裂纹边缘，断裂边缘
cracked insulator 有裂痕的绝缘子
cracked length 裂纹长度
cracked oil 裂解油，热裂油
cracked open 裂开
cracked permanent deformation 裂纹永久变形
cracked residue 裂化渣油
cracked zone 破碎带
cracked 裂化的，裂解的，有裂纹的
cracker 破碎机
crack extension force 裂缝进展力，裂纹扩展力
crack extension 裂纹扩张
crack filler 填缝料
crack filling 填缝
crack formation life 裂纹生成寿命
crack front 裂纹前缘，裂纹端部
crack-generating stress corrosion 裂纹应力腐蚀，应力裂纹腐蚀
crack growth rate 裂纹扩展速率，裂纹增长率，裂缝扩展速度
crack growth 裂缝长大，裂纹发展，裂缝的成长
crack healing 裂缝合拢
cracking catalyst 裂化催化剂
cracking chrome plated panel 镀铬裂纹试片
cracking core model 裂化核模型
cracking corrosion 裂纹腐蚀，裂化腐蚀
cracking load 破坏负荷，开裂负荷
cracking plateau 裂缝平台
cracking pressure 开启压力，裂缝压力
cracking propagation 裂纹扩展
cracking quality 裂缝性
cracking ratio 开裂系数，开裂比【混凝土】
cracking resistance 抗断裂能力
cracking-resistant material 防裂材料
cracking strength 抗裂强度
cracking test 裂纹试验
cracking threshold 破裂限值，破裂阈
cracking 裂解，裂化，裂缝，裂纹，开裂，击碎，砸碎
crack initiation site 裂纹起始位置
crack initiation 裂纹起始，裂纹形成，裂纹萌生
crack length 裂纹长度
crackle test 裂纹检验，变压器油温度检查
cracklike aperture 裂纹状的开口
crackling 碎裂，细裂纹
crack lip 断裂先端
crack measuring apparatus 裂纹测量仪表，裂纹测试仪器
crack meter 裂纹探测仪，测裂缝计
crack nucleation 裂缝的发生
crack opening angle 裂纹张开角
crack opening displacement 裂缝张开位移，裂纹张口位移，裂缝开度量测，裂缝开口变位
crack opening stretch 裂纹开口扩展
crack pattern 裂缝形状，裂纹图样
crack pouring 灌缝
crack propagation and arrest analysis 裂纹扩展和抑止的分析
crack propagation life 裂纹扩展寿命
crack propagation rate 裂纹扩展率
crack propagation 裂缝扩展，裂纹扩展，裂缝的传播
crack resistance force 裂缝抵抗力
crack resistance 抗裂性
crack sealer 封缝料
crack sensitive 易裂的

crack sensitivity 易裂性，裂缝敏感性，裂纹敏感性
crack speed 裂纹扩展速度
crack stage 裂纹阶段
crack starter technique 裂纹开裂技术
crack stopper 防裂装置
crackstress 开裂应力
crack survey 裂缝调查
cracks 裂纹，裂缝，裂痕
crack test 抗裂试验
crack tip opening displacement 裂纹尖端张开位移
crack tip plasticity 裂纹顶端塑性
crack tip 裂缝末端，裂纹末端
crack valve 瓣阀
crack width 裂缝宽度
crack 裂缝，裂纹，破裂，龟裂，缝，开裂，裂痕，裂隙，细裂缝
cradle base 支承基架
cradle guard （线路的）保护网
cradle 吊架，托架，机座，托板，托盘，支承垫块，支架，鞍座
craft labor hours 生产工时
craft paper 牛皮纸
craftsman 技术，手艺，技工，工匠
craft union 同业公会
crag and tail 鼻山尾，鼻尾丘
crag 岩石碎片
CRAM(card random access memory) 卡片随机存取存储器
cramp bar 夹紧杆
cramping apparatus 夹具，夹紧装置
cramping arrangement 紧固装置
crampon 起重吊钩，夹钳
cramp 卡，卡线板，夹紧，夹钳，扣钉
crandall 石锤
crane arm 吊杆
crane barge 起重机驳船，起重船，浮吊，浮式起重机
crane beam 吊车梁，起重机梁，行车梁，
crane boom 起重机臂，吊车臂，吊车起重扒杆，吊杆
crane brake 起重机制动闸
crane bridge 起重机桥架
crane-carrier 起重机行走架
crane car 起重机车
crane column 吊车柱
crane crab 起重机小车，起重绞车
crane dynamic weighing system 起重机动态称重系统
crane erection 起重机安装
crane-excavator 起重开挖两用机
crane frame 起重机架
crane girder 起重机大梁，吊车梁
crane hatch 起重机升降口
crane hoist type motor 起重电动机，吊车电动机
crane hook coverage 起重机吊钩工作范围
crane lifting 吊车吊装
crane load 吊车负荷，吊车荷载
crane luffing 起重机变幅
crane magnet 起重磁铁
crane manipulater 直线式机械手，高架式机械手，起重操纵器
craneman 起重工
crane maximum wheel pressure 吊车最大轮压
crane motor 吊车电动机，起重电动机
crane operator 吊车司机，起重机司机，起重工
crane output 起重能力
crane platform 起重机台
crane post 起重机柱
crane radius 起重机工作半径
crane rail 吊车轨，行车轨道，起重机轨道
crane rated capacity load 吊车额定起吊荷重
crane rating 起重机定额
crane rigger 起重工
crane runway girder 吊车梁，起重行车大梁
crane runway 吊车导轨，吊车道，起重机轨道，天车滑道
crane shovel 挖土起重两用机
crane stair 起重机楼梯
crane stalk 起重机柱
crane stanchion 吊车柱
crane support wall 吊车支撑墙，吊车支承墙
crane switch 起重开关
crane thrust 吊车水平刹车力
crane tower 起重吊塔
crane trackage system 起重机轨道系统
crane track 起重机轨道
crane travel 起重机移距
crane trolley 吊机滑车
crane truck 汽车起重机，汽车吊
crane-type loader 起重机式装载机
crane vertical impact load 吊车竖向冲击荷载
crane way 吊车导轨
crane weigher 吊车秤
crane winch 起重绞车
crane 起重机，吊车，升降起重机，龙门起重机，吊架，支架，吊机
crank-and-flywheel pump 曲轴飞轮泵
crank auger 曲柄螺旋钻
crankcase 曲柄轴箱
crank drive 曲柄传动
cranked coil 弯线圈
cranked slab stairs 板式楼梯
cranker 手摇曲柄
cranking 转动曲柄
crank pin bearing 曲柄销轴承
crank shaft engine 活塞式内燃机
crank shaft vibration damper 曲柄减振器，曲轴减振器
crank shaft 曲轴，机轴
crank ship 易倾船
crank-type power unit 曲柄执行部件
cranky 出故障的
crank 不稳定的，摇晃的，曲柄，手柄，摇动，曲轴
crash back 全负荷倒车
crash barrier 防撞栏，护栏
crasher 破碎机，粉碎机
crashing 碰撞的，撞击声，爆裂声，粉碎
crash pad 防震垫
crash program 应急计划，紧急计划，应急措施

crash stop 突然停车
crashvalue 峰值
crash voltage 峰值电压
crash 粉碎，摔毁，失败，事故，碰撞，坠落，破碎，猛撞，撞碎，坠毁
crater crack 焊口裂纹，焊口裂缝，弧坑裂纹
cratering 磨顶槽，腐蚀，侵蚀
craterlet 小火山口
crater magma 火山口岩浆
crater 喷火口，焊口，焰口，弧坑，火山口
crate （包装用的）板条箱，格框板条箱，柳条箱
crawl duct 半通行管沟
crawler bearing area 履带触地面积
crawler bearing length 履带触地长度
crawler crane 履带吊，履带（式）起重机，大型履带吊车
crawler dozer 履带式推土机
crawler drill 履带式开山机，履带式钻机
crawler dump wagon 履带式倾卸车
crawler excavator 履带式挖掘机，覆带挖土机
crawler gate 履带式闸门
crawler loader 履带式装料机［装载机］
crawler-mounted bulldozer 履带式推土机
crawler-mounted dragline 履带式索铲
crawler-mounted excavator 履带式挖土机
crawler-mounted mobile hopper 履带式活动料斗
crawler-mounted power shovel 履带式铲土机
crawler-mounted reclaimer 履带式取料机
crawler scraper 履带式铲运机
crawler shovel loader 履带式铲土装卸机
crawler shovel 履带式铲土机
crawler tractor 履带式拖拉机
crawler trailer 履带式拖车
crawler truck 履带车
crawler type grader 履带式平土机，履带式自动平地机
crawler type loader 履带式装载机
crawler vehicle 履带式车辆
crawler wagon 履带式运料车
crawler 履带式推土机，履带牵引装置，履带车，履带
crawling automatic welding machine 爬行自动焊机
crawling traction 履带牵引
crawling 爬行，蠕动，磁场谐波现象
crayon 蜡笔
craze fine crack 裂纹细牵引力
craze 裂纹，龟裂，发裂，细裂缝
crazing 破裂，细裂纹
CRBR（controlled recirculation boilingwater reactor） 受控再循环沸水堆
CR-bridge 电阻电容电桥，阻容电桥
CR（carriage return） 回车符号
CR（cathode ray） 可控整流器，阴极射线
CRC（carriage return contact） 回车接点，滑架回程接点
CRC（cyclic redundancy check） 循环冗余码校验
CRCE（cavity reactor critical experiment） 腔式反应堆临界实验装置
CR（coherent rotation） 一致转动，相关转动
CR（cold reheat） 低温再热，冷再热
CR（command register） 命令寄存器
CR（compressive ratio） 压缩比
CR（control relay） 控制继电器
CR（control room） 控制室
CR（count reverse） 反向计数
CRD（capacitor-resistor-diode network） 电容-电阻-二极管网络
CRD（control rod drive） 控制棒驱动
CRDCS（control rod drive control system） 控制棒驱动控制系统
CRDHS（control rod drive hydraulic system） 控制棒水力驱动系统
CRDM adapter 控制棒驱动机构的连接装置【即适配段/管座】
CRDM and turbine 控制棒驱动装置及透平机
CRDM（control rod drive mechanism） 控制棒驱动机构
CRDM drive shaft decontamination vat 控制棒驱动杆洗涤罐
CRDM housing 控制棒驱动机构罩壳
CRDM latch arm inspection tools 控制棒驱动机构棘爪臂检查工具
CRDM latch assembly 控制棒驱动机构棘爪组件
CRDM position indicator 控制棒驱动机构位置指示器
CRDM power supply system 控制棒驱动机构电源系统
CRDM upper canopy 控制棒驱动机构耐压壳
CRDM ventilation shroud 控制棒驱动机构通风管座
CRDM ventilation system 控制棒驱动机构通风系统
CRDTL（complementary resistor-diode-transistor logic） 互补电阻-二极管-晶体管逻辑
CRD（coil redistributed） winding 线圈重新分配绕组
cream ice 冰花
crease expansion bend 波纹膨胀（补偿）弯管
crease 变皱，老河道，折缝
creasy surface （铸件）皱纹表面
created credit 建立信用
creation of vacuum 建立真空
creation operator 生成算符
creation rate 形成率
creative initiative 首创精神
creative spirit 创造精神
credence 凭证，相信，信任
credentials （个人能力及信用的）证明书，证书，证件，凭据，信任状，文凭，国书
credibility 可靠性，可信度，可信性，可行性
credible accident 可信事故
credible 可信的
credit agreement 信贷协定，开发信用证约定书
credit association 信用社
credit bank 信贷银行
credit card 信用卡
credit conditions remain tight 信贷形势依然紧张
credit conditions 信用证条款，信贷条件，贷款条件，信用状况
credit document 信用单证，信用证券

credit fund 信贷资金
credit guarantee 贷款信用担保
crediting period 贷款期限
credit instrument 信用证凭证
credit investigation 偿债能力的调查
credit line 贷款限额，信用额度，商店等给予顾客的赊账最高额
credit market conditions 信贷市场环境
credit note 保值单据，信用证，贷记通知单，贷方账单
credit of favourable conditions 条件优惠的信贷
creditor beneficiary 债权受益人
creditor of bankruptcy 破产人的债权人
creditor 贷权人，贷方，贷项，债权人
credit period 信用期限赊欠期限
credit proceeds 信贷资本
credit right with conditions 附条件债权
credit risk 信用风险
credit side 贷方
credit terms and conditions 信用证条款
credit terms 贷款条件，贷款条款，信用条件
credit worthiness 商誉，信用度
credit 信用，贷款，贷方，信贷，银行存款，记入贷方
cred 信誉，名声，金融信用
creek 小河，潮沙沟，小湾，小溪
creep activation energy 蠕变激活能
creepage current 爬电电流，漏电电流
creepage discharge 潜流放电，漏电放电，沿面放电
creepage distance 爬电距离，爬行距离，爬距
creepage length 爬电长度，爬距
creepage path 爬行放电途径
creepage ratio 爬电比
creepage surface 爬电表面
creepage 蠕动，蠕变，蠕动转速，漏电，爬电
creep analysis 蠕变分析
creep and fatigue test 蠕变和疲劳试验
creep at sliding 滑动蠕变
creep behavior deformation 蠕变变形性能
creep behavior 蠕变性能
creep buckling 蠕变压屈
creep cavity 蠕变孔洞
creep cell 蠕变传感器，蠕变盒
creep characteristic 蠕变特性
creep collapse 蠕变破坏
creep compliance 蠕变柔量
creep curve 徐变曲线，蠕变曲线
creep damage factor 蠕变破坏因素，蠕变寿命消耗率
creep damage 蠕变损伤
creep deformation measuring ruler 蠕变测量尺
creep deformation path 蠕变变形途径
creep deformation 蠕变，蠕变变形
creep down 向内蠕变
creep ductility 蠕变延展性
creep elongation 蠕变延伸率
creep embrittlement 蠕变脆化，蠕变脆性
creep failure criterion 蠕变失效准则
creep fatigue crack 蠕变疲劳裂纹
creep fatigue interaction 蠕变疲劳交互作用
creep fatigue 蠕变疲劳，蠕变疲劳损坏
creep fracture 蠕变断裂
creeping discharge 蠕缓放电，潜流放电
creeping distance 蠕变距离，爬电距离，爬距
creeping flow 蠕状流
creeping motion 蠕动
creeping pressure 徐变压力
creeping protection 防止沿面爬电措施，沿面爬电保护
creeping speed 蠕变速度，爬行速度
creeping strength 徐变强度
creeping stress 蠕变应力
creeping 蠕变，蠕动，漂移，爬行
creep intensity 蠕变强度
creep life 蠕变寿命
creep limit 蠕变极限【金属，混凝土】，蠕动极限
creep line 蠕变线
creep mechanism 徐变机理
creep monitoring section 蠕变监视段
creep of concrete 混凝土的徐变，混凝土的蠕变
creep of snow particle 雪粒蠕动
creep out 渗出，漏出
creep point 屈服点
creep property 蠕变性能，蠕变特性
creep rate 蠕变速度，蠕变率，蠕变速率
creep ratio 蠕变比
creep recovery 蠕变回复
creep relaxation test 蠕变松弛试验
creep relaxation 蠕变松弛
creep resistance 抗蠕变强度，蠕变阻力，蠕滑阻力，蠕变强度，蠕爬极限
creep-resisting 抗蠕变的，防爬电的
creep rupture life 蠕变断裂寿命
creep rupture strength 蠕变断裂强度，持久强度，蠕变破坏强度
creep rupture stress 断裂的蠕变应力，蠕变断裂应力
creep rupture test data 蠕变断裂测试数据
creep rupture test 蠕变断裂试验，持久强度试验，蠕变破坏试验
creep rupture 蠕变断裂，蠕变破坏
creep speed 爬行速度
creep strain recovery 蠕变应变回复
creep strain 蠕变变形，蠕变应变，蠕变应力
creep strength 蠕变强度，抗蠕变强度
creep stress 蠕变应力
creep test 蠕变试验，耐蠕变试验
creep-time curve 蠕变时间曲线
creep up 渗上来，蠕升
creep 蠕变，蠕动，蠕流，漂移，爬行，打滑，徐变
cremaillere 齿轮齿条传动
cremate 焚毁，金属陶瓷，陶瓷合金
cremone bolt 长插销【门，窗】，通天插销
crenellated 锯齿形的
creosote bush 石炭酸灌木
creosoted pole 用杂酚油浸过的电杆
creosoted timber pile 油浸木桩
creosoted wood 油浸防腐木材
creosote treatment 杂酚油浸处理

creosote 杂酚油，（防腐用的）木榴油，木材防腐油
creosoting 杂酚油处理
crepe rubber 皱纹薄橡皮板
crescentdune 新月形沙丘
crescent rib 月牙肋
crescent roof truss 月牙式屋架
crescent rotary tipper 新月形转子翻车机
crescent single-car dumper 新月形单车翻车机
crescent 镰刀形的，新月形的
cresent-rib reinforced wye piece 内加强月牙肋岔管
crest ammeter 峰值安培计
crest amplitude 最大振幅
crest clearance 最大间隙
crest contraction 堰顶收缩
crest control 堰顶控制
crest curve 凸形曲线
crest depth 堰顶水深
crest discharge 过顶流量，堰顶流量
crest elevation of levee 堤顶高程
crest elevation 坝顶高程【水电】
crest factor 峰值系数，峰值因数，波峰因数
crest forward anode voltage 最大阳极正向电压
crest frequency 峰值频率
crest gate 堰顶闸门
crest indicator 峰值指示器
crest interval 波峰间隔
crest inverse anode voltage 最大阳极反向电压
crest length 坝顶长度【水电】
crest level of slag dump 灰渣堆的顶部高程
crest level 坝顶标高，堤顶高程，顶部高程，峰值水平，堰顶高程
crest line 峰线
crest load 尖峰负荷
crest meter 峰值计，幅值计
crest overflowing orifice 泄水表孔
crest overflowing 坝顶溢流
crest profile 峰顶纵剖面
crest quasi-peak value 类峰值的峰值
crest radius 坝顶曲线半径【水电】
crest reduction 洪峰消减
crest segment （过程线的）峰段，水文过程线封顶
crest spillway 坝顶溢洪道【水电】
crest stage indicator 峰顶水位指示器，洪峰水位指示器，最高水位指示器
crest stage 顶峰阶段，峰顶水位，洪峰水位，最高水位
crest tile 屋脊瓦
crest value 峰值，最大值
crest voltage measuring system 峰值电压测量系统
crest voltage 峰值电压
crest voltmeter 峰值伏特计
crest wall 坝顶防浪墙
crest 峰值，波峰，顶，峰，坝顶，脊顶，屋脊，堰顶
Cretaceous system 白垩系
cretaceous 白垩纪，白垩的，白垩纪的
crevasse crack 裂缝，龟裂
crevasse filling 裂隙填充
crevasse repair 堵口
crevasse 裂缝，破口，产生裂缝，裂隙，发生裂口
crevice corrosion 裂缝腐蚀，缝隙腐蚀
crevice drainage 裂隙排水
crevice water 裂缝水
crevice 裂纹，缝隙，裂缝，发裂，裂隙，夹层，间隙
crew leader 带班长
crew 组，工作班组，运行班组，乘务组
CRF(carryout rate fraction) 夹带率份额
CRF(clarification request form) 澄清问题申请单
CRF(control relay forward) 正向控制继电器
CRF 循环水系统【核电站系统代码】
CRGT(control rod guide tube) 控制棒导管
CRH(cold reheat) 冷再热，再热器冷段
CRH 低温再热器【DCS 画面】
crib and pile dike 木笼木桩堤
cribbing 剽窃，抄袭，金属槽，框，笼
cribble 粗筛，粗粉
crib breakwater 木笼防波堤
crib check dam 木笼谷坊，木笼拦沙坝
crib cofferdam 框格式围堰，木笼围堰
crib dam 木框坝，木笼填石坝
crib foundation 木笼基础
crib groin 木笼填石丁坝
crib groyne 木笼填石丁坝
crib pier 叠木支座，木笼桥墩
crib retaining wall 框格式挡土墙，木笼挡榫土墙
crib-type cofferdam 木笼围堰，框格式围堰
crib weir 木框堰
crib work filled with stone 石笼
crib work quay wall 木笼码头岸壁
crib work 木笼作业
crib 框箱【排渗废料用】，潜水箱，金属槽，篓筐，沉箱，叠木框，框，笼
criminal negligencewillful misconducttort 民事侵权行为
criminal responsibility 形式责任
crimp connection 压接【输电线路】
crimped lock 线夹，夹具，卡具，接线柱
crimped wire 脆性钢丝
crimper 压紧钳，卷缩机，折波钳，卷边机
crimping die 卷边模
crimping machine 折边机，弯皱机，卷边机
crimping range 压接范围
crimping tool 卷边工具，压接工具
crimp lug 曲折接线片
crimp machine 压接机
crimp test 卷曲测试
crimp 折边，卷边，卷曲，弯皱，压接，压成波浪形
crinkle 皱纹，曲折，揉皱，起皱
crinkling 起皱纹
CRIP(Central Research Institute for Physics) 中央物理研究中心【匈牙利】
cripple 搭脚架
crippling 局部失稳，断裂
crippling load 临界荷载，破坏载荷，断裂载荷，

断裂荷载
crippling stress 破坏应力，临界应力
crisis 临界，危机，紧要关头
crisping 匀边电路，匀边
criss-cross arrangement of moving stairs 自动楼梯的交叉布置
criss-cross sequence 十字交叉顺序【螺栓拧紧】
criss-cross structure 方格构造，井字形构造
criss-cross 方格形，纵横交错，十字号，纵横交错的，十字形的，方格的
cristobalite 方石英
CRIT(critical) 临界的，关键的
criteria analysis 标准分析
criteria for evaluation 评定标准
criteria for noise 噪声标准，噪声判据
criteria 标准，规范，原则，准则，criterion 的复数
criterion for noise control 噪声控制判据
criterion for stability 稳定性判据
criterion of bank stability 岸坡稳定性标准
criterion of classification 分类标准
criterion of convergence 收敛判据
criterion of degeneracy 简并性判据，简并性准则
criterion of homogeneity 均匀性准则
criterion of reservoir design 水库设计标准
criterion of resolution 分辨力判据，解像率评价
criterion of stability 稳定性判据，稳定度判据
criterion 标准，判据，准则，规范，判别式【复数 criteria】
critesister 热敏电阻
critical absorption energy 临界吸收能量
critical absorption 临界吸收
critical acceleration 临界加速度
critical accident 临界事故
critical angle of attack 临界攻角
critical angle 临界角
critical approach curve 趋近临界曲线，接近临界曲线
critical arching or bridging dimensions 结拱或搭桥的临界尺寸
critical area 危险区，关键区，关键部位
critical assembly 临界燃料组件，临界装置
critical attenuation 临界衰减
critical band 临界带宽
critical bar 险滩
critical behavior 临界行为，临界特性，临界状态
critical boron concentration 临界硼浓度
critical breakdown strength 临界击穿强度
critical buckling 临界曲率，临界弯折
critical build-up resistance 临界起建电阻
critical build-up speed 临界起建转速
critical cavitation coefficient 临界汽蚀系数
critical cavitation factor 临界汽蚀系数
critical cavitation 临界空蚀，临界汽蚀
critical charge 临界载荷
critical circle 临界圆
critical clearing time 极限切除时间
critical coefficient 临界系数
critical cold point 临界冷点
critical components 关键部件
critical compression pressure 临界压缩压力
critical compression ratio 临界压缩比
critical compressive stress 临界压应力，临界压缩应力
critical concentration 临界浓度
critical condition 临界条件，临界状态，临界情况，临界状况
critical configuration 临界布置，临界体积
critical constant 临界恒量，临界常数，临界点常数
critical cooling rate 临界冷却速度
critical corona voltage 临界起晕电压
critical coupling 临界耦合
critical crack extension force 临界裂纹扩展力
critical crack length 临界裂纹长度
critical crack size 临界裂纹尺寸
critical current 临界电流，临界流
critical curve 临界曲线
critical damping 临界衰减，临界阻尼
critical data 临界数据
critical defect size 临界缺陷尺寸
critical defect 致命缺陷
critical deformation 临界变形
critical density 临界密度
critical depth control notch 临界深度控制槽口
critical depth discharge measurement 临界水深测流法
critical depth hume 临界深度测流槽
critical depth line 临界水深线
critical depth method 临界深度法，临界水深法
critical depth meter 临界水深量测计
critical depth 临界深度
critical design case 临界设计情况
critical detection 临界探测
critical diameter 临界直径
critical dimension 临界尺寸
critical dip circle 临界滑圆
critical discharge 临界流量
critical dispersion 临界分散，临界扩散
critical divergence wind speed 临界发散风速
critical drag 临界拖力，临界阻力
critical emission rate 临界排放率
critical energy 临界能
critical equation 临界方程
critical error 关键性误差
critical excursion （功率）极限增长
critical exit-gradient 临界逸出坡降
critical experiment 临界试验，临界实验
critical exposure pathway 关键照射途径
critical extinction angle 临界熄弧角，最小熄弧角
critical facility 临界装置
critical failure 致命故障，危急故障
critical fault 严重故障
critical field resistance 临界磁场电阻
critical field 临界场，临界磁场
critical flame 临界火焰
critical flashover voltage 临界闪络电压
critical flaw size 临界缺陷尺寸
critical flicker frequency 临界闪烁频率，截止闪烁频率

critical floatation gradient 临界浮动坡降
critical flow computation 临界流态计算
critical flow criteria 临界流判别准则，临界流态判别准则
critical flow diagram 临界流计算图
critical flowmeter 临界流量计
critical flow nozzle 临界流量喷嘴【有临界截面】
critical flow rate 临界流量
critical flow regime 临界流动状况
critical flow valve 临界流量阀
critical flow 临界流量，临界流态，流量
critical fluence 临界通量
critical fluidized velocity 临界流化速度
critical flutter speed 临界颤振速率
critical flutter wind speed 临界震颤风速
critical food 关键食物，紧要食物
critical form 临界形态
critical foundation depth 地基临界深度
critical frequency 临界频率
critical fusion frequency 临界停闪频率
critical galloping wind speed 临界驰振风速
critical gas velocity 临界气速
critical gradient 临界坡度，临界梯度，击穿梯度，临界比降，临界坡降
critical grid current 临界栅极电流
critical group 关键居民组，关键人群组，关键组，鉴别组，紧要组
critical head 临界水头
critical heat flux condition 临界热流密度工况
critical heat flux correlation 临界热流密度关系式
critical heat flux density 临界热流密度
critical heat flux margin 临界热流密度裕量
critical heat flux modified cold-wall factor 临界热流密度冷壁修正因子
critical heat flux modified shape factor 临界热流密度形状修正因子
critical heat flux modified spacer factor 临界热流密度定位架修正因子
critical heat flux ratio 临界热流密度比
critical heat flux site 临界热通量位置
critical heat flux 临界热流密度，临界热通量，临界热流
critical heat 临界热，转化潜热，相态转化潜热
critical height 临界高度
critical hole 临界孔
critical horse power 临界马力
critical humidity 临界湿度
critical hydraulic gradient 临界水力坡降，临界水力梯度
critical-impulse flashover voltage 临界冲击闪络电压
critical-impulse time 临界冲击时间
critical inspection 关键检验
critical instability 临界不稳定性
critical instructor station task 关键性指导员站任务
critical inversion layer 临界逆温层
critical item list 关键项目清单
criticality accident 临界事故
criticality alarm system 临界报警系统
criticality analysis 致命度分析
criticality approach curve 接近临界曲线，趋近临界曲线
criticality equation 临界方程
criticality factor 临界系数，有效增殖系数
criticality flow 临界流
criticality incident 临界事故
criticality measurement 临界测量
criticality monitor 临界监测仪表，临界监测器
criticality number 致命度指数
criticality risk 临界危险
criticality safety 临界安全
criticality spectrum 临界谱
criticality test 临界试验
criticality 临界状态，临界，危急程度
critical length of grade 临界坡道长度
critical lift force 临界扬力
critical load distribution 临界载荷分布
critical load 临界负荷，临界负载，临界荷载，临界装载
critically damped harmonic motion 临界阻尼谐振荡
critically damped oscillation 临界阻尼振荡
critically damped 临界阻尼的，在稳定边缘的
critically safe configuration 临界安全布置
critically safe geometry 临界安全几何形状
critically safe rotary dissolver 临界安全旋转溶解器
critically safe 临界安全的，临界安全
critical Mach number 临界马赫数
critical magnetic field 临界磁场
critical magnification factor 临界放大系数
critical mass flow rate 临界质量流量
critical mass 临界物质，临界质量
critical maximum 临界最大值
critical medium 临界介质
critical moisture content 临界含水量
critical moisture point 水分临界点
critical moisture 临界水分
critical moment 临界时限，临界力矩
critical NPSH(net pressure suction head) 临界汽蚀裕量
critical nuclide 关键核素，重要核素
critical operation condition 临界工况
critical operation 临界运行
critical organ 关键器官，紧要器官【保健物理】
critical overturning wind speed 临界倾覆风速
critical parameter 临界参数，重要参数
critical part 主要机件，要害部位
critical path analysis 关键途径分析法
critical path impact 关键路径影响
critical path method 关键路径法【项目】，统筹方法
critical pathway 关键途径，紧要途径
critical path 关键路径，统筹方法
critical period 临界周期
critical pile 临界反应堆
critical point 临界点
critical porosity 临界孔隙率
critical position （控制棒的）临界棒位，临界位置
critical potential 临界电势，临界电位

critical pressure gradient 临界压力梯度
critical pressure ratio 临界压力比
critical pressure 临界压力
critical profile 临界剖面
critical radius 临界半径
critical rainfall intensity 临界降雨强度
critical range of vibration 振动的临界范围
critical range 临界范围，临界区
critical rate 临界率
critical reactor 临界反应堆
critical regeneration 临界再生
critical regime 临界状态
critical region 临界区域
critical resistance 临界电阻
critical revolution 临界转速
critical Reynolds number 临界雷诺数
critical Richardson number 临界理查森数
critical roughness 临界糙率
critical run 临界试验
critical section for moment 力矩控制截面
critical section 临界断面
critical self-excitation 临界自励磁
critical self-healing voltage 临界自复电压
critical service loss duration 临界停电持续时间
critical shoal area 临界浅滩区
critical shortage 严重短少
critical size 临界尺寸【堆芯】
critical slide angle 临界滑动角，临界倾角
critical slip circle 临界滑弧
critical slope 临界比降，临界坡度
critical sonic value 临界声速值
critical span 临界档距
critical speed of rotor 转子临界转速
critical speed of second order 二阶临界转速，第二临界转速
critical speed test 临界转速试验
critical speed 临界转速，临界速度
critical stability 临界稳定，临界稳定性
critical stall speed 临界失速速度
critical state of flow 临界流态
critical state soil mechanics 临界状态土力学
critical state 临界状态
critical steam quality 临界含汽率，临界蒸汽品质
critical steepness 临界斜度
critical stimulus 临界激发
critical stress corrosion cracking factor 临界应力腐蚀裂纹因子
critical stress for fracture 断裂临界应力
critical stress intensity factor 临界应力强度因子
critical stress 临界应力
critical suction vacuum 临界吸上真空高度
critical surface gradient 表面临界梯度
critical surface-moisture content 临界表面水分
critical surface tension 临界表面张力
critical surface 临界面
critical system 临界系统
critical table 判定表
critical temperature difference 临界温差
critical temperature 临界温度
critical test 临界试验
critical thermal load 临界热负荷
critical thickness of insulation 绝缘的临界厚度
critical tissue 关键组织，紧要组织
critical towing force 临界曳力
critical tractive force 临界曳引力
critical tractive velocity 临界起动流速
critical transfer pathway 关键转移途径
critical value 临界值
critical velocity 临界流速，临界速度
critical void ratio 临界孔隙比
critical voltage difference 临界电压差
critical voltage 临界电压
critical volume 临界体积
critical water content 临界含水量
critical water depth 临界水深
critical wave length 临界波长
critical wind speed 临界风速
critical withstand current 临界耐受电流，临界耐电流
critical withstand voltage 临界耐压
critical zone 危险地带，临界地带
critical 临界的，关键性的，紧要的，决定性的，关键的
criticism 评论，批评
criticize 批评
CR(conpacitance resistance) law 电容电阻定律
CRM(confinement of radioactive materials) 放射性物质密封
CRMS(control room management system) 控制室管理系统
CRO(cathode-ray-oscilloscope) 阴极射线示波器
crockery 陶器
crocodile clip 弹簧线夹，鳄鱼夹
crocodile squeezer 颚式压挤机
CRO(control room operator) 控制操纵员
crolite 陶瓷绝缘材料
cromansil 克拉曼希尔铬锰硅钢
croman 克拉曼铬锰钼硅钢
CROM(control read-only memory) 控制只读存储器
crony economy 裙带经济
crooked timber 弯曲木材
Crookes radiometer 克鲁克斯辐射计
Crookes tube 克鲁克斯阴极射线管
crop climate 作物气候
cropland 农田
crop out 出露地表
cropper 剪边机
cropping-out of the groundwater 地下水露头处
cropping 切料【钢锭】
crop 露头
CR oscillator 阻容振荡器
cross ampere-turns 正交安匝，交磁安匝
cross-amplitude spectrum 交叉放大谱
cross-arm brace 横担斜撑，横担撑条，角撑
crossarm chord 横担弦杆
crossarm guy 横担拉线
crossarm hanger bar 横担拉杆
crossarm hanger 横担吊架
crossarm main member 横担主材
crossarm tie 横担拉杆
crossarm 横木，横担，横臂

cross axis 水平轴
cross baffled boiler 横向冲刷锅炉
cross baffle 横隔板
cross-banding 交向排列，频率交联
crossband transponder 交叉频带发送机
crossband 交错传动，交叉频带的
crossbar addressed dot matrix 正交寻址点矩阵
crossbar connector 纵横制接线机
crossbar exchange 纵横制交换机
crossbar matrix switch 叉闩整体开关，纵横矩阵开关
crossbar selector 坐标选择器
crossbar switch 十字开关，纵横制接线器，纵横接线器
crossbar system 交叉制，纵横制，坐标制
crossbar 交叉棒，交叉杆，坐标的，横杆，横撑，横撑木，闩
cross beam 横梁，顶梁
cross-bedding 斜交成层
cross belt 交叉带
cross-bending stress 横向弯曲应力
cross-bending 横向弯曲
cross bit 十字偏铲，十字钻头，十字钻
cross-blast （油断路器的）横吹
cross-bonding 交叉互连
cross box 横联箱
cross brace 剪刀撑，人字撑，横拉条，肋板
cross bracing 叉梁，剪刀撑，交叉斜材，交叉支撑
cross-breaking strength 横向破坏强度，抗弯强度
cross burner 交叉混合式燃烧器【一、二次风】
cross cable bond 电缆铠甲的横向连接器
cross-calibration 相互校准
cross century 跨世纪
cross-channel 交叉流道
cross-check 再确认，再次复核，相互核对，交叉核对，反复核对
cross claim 相互主张权利，相互提出诉讼请求，反索赔
cross-coil defect detector 交叉线圈探伤仪
cross-coil instrument 交叉线圈仪表
cross-complaint 相互指控，交叉诉讼，反诉
cross compound cycle 交叉双轴循环
cross compound double flow 双轴双排汽汽轮机
cross compound gas turbine 双轴燃汽轮机
cross compound single flow 双轴单排汽汽轮机
cross compound steam turbine 双轴式汽轮机
cross compound turbine-generator 并列双轴式汽轮发电机，交叉复式涡轮发电机
cross compound turbine 双轴汽轮机，并联复式涡轮机，并列二段涡轮机，并列双轴式涡机
cross compound turbo-alternator 并列双轴式汽轮发电机
cross compound unit 双轴机组
cross compound 交叉轴【燃气轮机】，双轴布置的，双轴的，并列的
cross-connected generator 交磁发电机
cross-connected spray desuperheater 布置在过热器交叉管中的喷水减温器
cross-connected winding 交叉联接绕组，正交绕组
cross-connected 交叉连接的
cross-connecting cable 交叉连接电缆
cross connection area 交叉连接区
cross connection pipe 交叉接管，十字接管
cross connection 交叉连接
cross-connect valve 连通阀
cross-contamination 交叉污染，交互沾染
cross control rod 十字控制棒
cross converting globe valve 四通换向球阀
cross-correlation function 互相关函数，交叉关联函数，互函数
cross-correlation 互相关联
cross cospectrum 交叉余谱
cross-countercurrent heat exchanger 交叉逆流式热交换器
cross-counterflow 交叉逆流
cross country ability 越野性能
cross country fault 越野式接地故障【即双相同时接地式的故障】
cross country minibus 越野小客车
cross country vehicle 越野汽车
cross coupling capacitor 交叉耦合电容器
cross coupling 正交耦合，交叉耦合，相互作用
cross covariance 互协方差
cross crack 横裂纹
cross culvert 横向涵洞
cross-current extraction 错流萃取法
cross current 涡流，交叉流，横流，正交流，错流
cross cut file 交叉纹锉
cross cut 横导坑，正交，横断面，横切
cross damping 正交阻尼
cross debts 互抵债务
cross-default clause 交互废约条款
cross derivative 交叉导数
cross dike 横堤
cross ditch 横沟
cross drainage 横向排水沟，过街沟
cross-drum boiler 横汽包锅炉
crossed-coil antenna 交叉环形天线
crossed-coil measuring 交叉线圈测量
crossed controls 交叉控制机构
crossed-core type transformer 交叉铁芯式变压器
cross eddy 横向旋涡
crossed magnetic field 横向磁场
crossed-wire grid 交叉丝状格架
crossed 交叉的
cross exchange 套汇，交叉汇兑；通过第三国汇付的汇兑
cross-fall 路面横向高差，横斜度
cross fault 横断层
cross feed fuel 横向进煤，前饲式给煤
cross-feed motor 交叉馈电式电动机
cross feed 交叉供电，交叉馈，横向送进
cross-field exciter 正交磁场励磁机
cross-field generator 正交磁场发电机，交叉电场信号发生器
cross-fields multiplier 正交场乘法器
cross-field theory 正交场理论

cross-field 正交场，横向磁场
cross-finned tube 横向肋片管
cross fire radiography 交叉射线照相术
cross fire tube 交叉火管
crossfire 串扰电流
cross firing 对冲燃烧
cross flow convection 横向冲刷对流
cross-flow cooling tower 横流式冷却塔，叉流式冷却塔
cross-flow excitation 横向流干扰
cross-flow fan 贯流式通风机
cross-flow fuel element 交叉流燃料元件
cross-flow header 叉流集流管
cross-flow heat exchanger （交）叉流式热交换器，（交）叉流式换热器
cross-flow rate 交叉流流率
cross-flow ratio 错流比
cross-flow resistance coefficient 横流阻力系数
cross-flow specific enthalpy 横流比焓
cross-flow turbine 双击式水轮机
cross-flow wind turbine 贯流风机
cross-flow 横向流动，横流向的，交叉流动，横向流，交叉流，正交流动
cross-flux 正交通量，正交磁通
cross focus X ray tube 交叉聚焦X射线管
cross force 横向力，剪力
cross furrow 横沟
cross garnet butt 十字铰链
cross girder 横梁
cross grain 横纹
cross-hair cursor 十字形光标，十字标
cross-hair 十字瞄准线，十字准线，十字丝，叉丝
crosshatched area 画出交叉阴影线的区域
cross heading 横向坑道
cross headwind 侧逆风
cross head 十字头，丁字头，十字头小标题，文内小标题，子题，小标题
cross helical gear 螺旋齿轮
cross-hole method 跨孔法
cross-hole testing 跨孔实验
cross induction 横轴磁场作用
crossing angle 交叉角
crossing at right angles 正交
crossing insulator 跨越绝缘子
crossing-junction box 交叉接线盒，交叉接线箱
crossing of obstacles 交叉跨越
crossing-over 交叉，跨越
crossing point 交叉点，相交点
crossing pole 跨越杆
crossing span 跨越档距
crossing tower 跨越塔
crossing 交叉点，相交，交叉，横断，穿越，过渡段，横越，渡口，交叉跨越
cross-jet breaker 横吹灭弧断路器，横向喷油灭弧断路器
cross joint 横断节理，十字接头
cross-junction box 交叉接线箱，交叉接线盒
cross key 横销
cross lay wire-rope 交叉捻向钢丝绳
cross liability 交叉责任
cross licence contract 互换许可证合同
cross line 十字丝，正交线，十字接，交叉线
crosslinkage 交联度
cross-linked acrylic copolymer 丙烯酸交联共聚物
cross-linked insulation 交联绝缘
cross-linked polyethylene insulated cable 交联聚乙烯绝缘电缆
cross-linked polyethylene 交联聚乙烯
cross-link 交叉耦合，横向连接，交联
cross magnetization 横向磁化，正交磁化
cross-magnetizing ampere-turn 横轴磁化安匝，交轴磁势
cross-magnetizing armature reaction 横轴电枢反应
cross-magnetizing effect 正交磁化效应
cross-magnetizing force 正交磁化力
cross-magnetizing turn 正交磁化线匝
cross magnetomotive force 正交磁动势
cross mark 十字标记
cross member brace 横档撑条
cross member reinforcement 横梁加强
cross member 横梁，交叉杆，十字形构件
cross modulation factor 互串调制系数，交叉调制系数
cross modulation 交叉调制，交扰调制，互串调制
cross mounted condenser 横向布置凝汽器
cross mouthed drill 十字钻
cross-notching 对开槽
cross off 划掉
cross-operation 交叉作业
cross over bend 跨越弯管，跨越弯头
cross-over block 绝缘垫块
cross-over bridge （跨越铁路的）天桥
cross-over clamp 交叉线夹
cross-over coil 圆柱形线圈，（变压器的）饼式分层线圈
cross-over filter 分离过滤器，分相滤波器
cross-over frequency 分隔频率，交界频率，交叉频率，穿越频率，交越频率
cross-over insulator 横跨用绝缘子
crossover ladder 跨越梯
cross-over leg U形管段【堆冷却剂管】，跨越管段
cross-over network 分频网络，选频网络，交叉切换站，切换网络
crossover pipe rack 架空管道支架
crossover pipe 跨接管，跨越管，交叉混合管，联通管，架空管道
crossover platform 跨越平台
crossover pole 跨越杆
cross-over potential 临界电位
crossover region 交叉区
crossover survey 交叉跨越测量
crossover tower 跨越杆塔，跨越塔
crossover track 渡线，转线轨道
crossover tube 交叉管，跨接管，连通管，过桥管
crossover tunnel 立交隧道
crossover valve 交叉阀，立体交叉阀，切换阀

cross-over winding (变压器的)饼式分层绕组
cross over 横跨，跨越，穿越，交叉，连通，切断，渡过
cross-path current 横向电流
cross peen hammer 横头锤
cross piece 十字形构件，闩，四通，十字管，横管，四通管，过梁，横档，十字架
cross-pointer indicator 双针指示器
cross-pointer instrument 双针式测量仪表
crosspoint relay matrix 交叉点继电器矩阵
crosspoint relay 交叉点继电器
cross point 交叉点
cross-polarized voltage 横向极化电压
cross power spectrum 互功率谱
cross power 互换功率
cross product 矢量积
cross-protection 横向保护，防止碰线
cross rail 横轨
cross reference 相互核对，相互参照，对照
cross-referencing program 相互对照排错程序
cross-relation function 互相关函数
cross rib 横肋
cross-rifle tube 交叉来复线内螺纹管
cross-ringing 交扰振铃
cross-river transmission line 跨河段输电线
cross riveting 十字形铆接
cross road 十字路，十字路口，交叉路，横路
cross rod 十字形控制棒
cross rope suspension 悬索挂线
cross sea bridge 跨海大桥
cross-sectional area of flow 流体截面
cross-sectional area 横断面积，横截面积，断面（面）积，截面积
cross-sectional drawing 断面图
cross-sectional shape 横截面形状
cross-sectional velocity distribution 流速沿断面的分布
cross-sectional view 横断面图【机】，剖视图，剖面图
cross-section area of nozzle 喷嘴截面积
cross-section area of oscillating stability 波动稳定断面
cross-section area of the material on the belt 输送带上的物料横断面积
cross-section area 横截面积，截面积
cross-section at crown 拱冠断面
cross-section determination 截面测定
cross-section drawing 剖面图，横断面图，纵横断面图绘制
cross-sectioning 横断面测量
cross-section library 截面数据库
cross-section measurement 断面测量
cross-section modulus 截面模量
cross-section of air gap 气隙截面
cross-section of cable 电缆截面
cross-section of conductor 导线截面
cross-section of road 道路横断面
cross-section paper 方格纸
cross-section quality 截面含汽率
cross-section reduction 断面收缩
cross-section sketch 塔基断面图
cross-section velocity distribution 断面流速分布
cross-section view 横断面图
cross-section 横断面，横截面，断面，剖面，横切面，横剖面，剖面图
cross shaft reduction gear 交叉轴齿轮减速器
cross shafts 交叉双轴
cross shake 辐裂
cross shear 横剪切
cross sill 横槛
cross slot flux 横槽磁通
cross slotted screw 十字槽螺钉
cross slot 横向槽
cross spectral density 交叉谱密度
cross-spectrum 互相关谱
cross stability derivative 交叉稳定导数
cross staff 十字杆
cross stay 横撑
cross stratification 交错齿缝
cross stream sampling 横流取样
cross stream 相交气流
cross strut 横撑，横档
cross swell 横向海浪，横涌
cross-switch 十字开关
cross tail butt 十字铰链
crosstalk attenuation 串话衰减，串音衰减
crosstalk current 串话电流
crosstalk damping 串音阻尼
crosstalk meter 串话测试器
crosstalk suppression filter 串音抑制滤波器
crosstalk volume 串音量，串音功率
crosstalk 串话干扰，串扰，串电，串音
crosstalk 交调失真
cross terms 交叉项
cross test level 方形水平尺
cross-the-line start 跨线起动
cross tie 交叉缀筋，轨枕，枕木，横拉杆
cross travel motion 定向移动【输送】
cross travel 横向移动
cross trench 横沟
cross tripping 交叉断路法
cross tube burner 缝式燃烧器，缝隙式燃烧器
cross tube 十字管，横管
cross-under pipe 连通管，跨接管，过桥管
cross valve 三通阀，转换阀
cross variance function 互方差函数
cross ventilation 对流通风，穿堂风，前后通风，十字通风
cross wall 隔断墙，横隔墙
cross weld 交叉焊缝
cross wind 横风向的，横风
cross wind-axis wind machine 横风轴风力机
cross wind buffeting 横风抖振
cross wind diffusion 横风扩散
cross wind direction 横风向
cross wind displacement 横风位移
cross wind galloping 横风驰振
cross wind gust 横向阵风
cross winding 交叉绕组
cross wind installation 横风装置
cross wind loading 横风荷载
cross wind paddles windmachine 横风桨板式风

力机
cross wind stability 横风稳定性
cross wind test 横风试验
cross wind vibration 横风振动
cross wind wake force 横风尾流力
cross wire damper 交叉型阻尼线
cross wire 交叉线，十字线，十字丝
crosswise 斜地，成十字状地，交叉地，横斜地，十字状的
cross-wood 腕木【电杆】
cross-working 交叉作业
cross 四通，十字管，交叉，跨过，绞线，十字，十字架
crotch corner 丁形终端接续套管
crotch 交叉，叉杆，分支，T形终端连续套管，弯钩，分叉处，岔口，叉架
crotovine 填土动物穴
crowbar circuit 消弧电路，保安电路
crowbar protection circuit 短接保护电路，钳位电路，过压保护电路
crowbar protection 过压保护装置，急剧短路保护
crow bar 撬杆，撬棒，急剧短路，断裂，短接电路，钳位电路，撬杠，撬棍，铁翘
crowding gear 液压传动装置
crowfoot cracks 皱裂
crowfoot spanner 爪型扳手
crowfoot 足形电极【电解】，防滑三脚架，吊索
crown bar 顶部拉杆
crown block 定滑轮【顶部】，拱顶石，起重机顶部滑车组
crown cantilever adjustment 拱冠悬臂梁校正
crown cantilever 拱冠梁，拱冠悬臂梁
crown cornice 大屋檐
crown cover degree 树冠盖度
crown density 树冠郁闭度
crown displacement 拱顶位移
crown ditch 截水沟，天沟
crown drill 顶钻
crown elevation 顶高
crown face pulley 凸面滚筒
crown face 顶面
crown glass 上等厚玻璃
crown head 拱形封头【加热器】，上部水室
crown hinge 拱顶铰，顶铰
crowning 隆起，凸起
crown joint 拱顶接缝
crown of arch 拱顶
crown pin 顶销
crown plate 顶板，梁垫，柱顶垫板
crown post 桁架中柱
crown section 顶截面，拱顶段，拱顶截面，拱冠断面
crown sheet 顶板
crown stay 顶板拉撑
crown wall 顶墙
crown width 顶宽
crown 顶部，齿冠，凸起部，顶拱，隆起，顶部，路拱，树冠，拱顶
crow 橇杠，撬棍，乌鸦，鸡鸣
croy 护岸设施，护岸挑水栅
crozzling coal 黏结性煤
CRP(control of reactor power) 反应堆功率控制
CRP test(constant rate of penetration test) 等速贯入试验
CRS(command retrieval system) 命令检索系统
CRS(cross section) 横截面
CRSV(cold reheat safety valve) 再热器冷段安全阀
CRT alarm 阴极射线管报警
CRT(cathode ray tube) 阴极射线管，显示器，屏幕
CRTC(China Road Transportation Company) 中国汽车运输总公司
CRT＝circuit 回路
CRT display monitor 阴极射线管显示监视器
CRT display 阴极射线管显示器，荧光屏显示器，屏幕显示器
CRTM(cathode-ray tube memory) 阴极射线管存储器
CRT spatial encoder 阴极射线管编码器
CRT 显示器【DCS 画面】
crucible carrier 坩埚架座
crucible chamber 坩埚室
crucible cover 坩埚盖
crucible swelling number 坩埚膨胀序数
crucible tongs 坩埚钳
crucible type furnace 坩埚式炉膛
crucible 坩埚
cruciform absorber section 十字形吸收段
cruciform absorber 十字形吸收体，十字形控制棒
cruciform base 十字基础
cruciform control rod 十字形控制棒
cruciform core 十字形截面铁芯
cruciform cracking test 十字接头抗裂试验
cruciform groin 十字形丁坝
cruciform groyne 十字形丁坝
cruciform joint 十字接头
cruciform stepped core 十字形阶梯式铁芯，十字形铁芯
cruciform weld 十字焊缝
cruciform 十字形的
crude asphalt 生沥青
crude cost estimate 原油成本估算
crude data 原始数据
crude material 原料
crude oil 原油
crude petroleum 原油
crude regulation 粗调
crude removal 去界面污物，积垢去除
crude rubber 生胶
crude sewage 原污水
crude spirit 粗汽油
crude wastewater 粗废水
crude water 生水，原水，未经处理的水
crude 未经加工的，天然的，粗糙的
crud induced localized corrosion 积垢引起的局部腐蚀
crud 沉淀，积垢，杂质，渣滓，掺和物
cruiser dock 大型钢筋混凝土浮坞
cruising speed 巡航速度

cruising way　慢车道
cruision　巡航
crumble structure　团粒构造，团粒结构
crumble　破碎物，破裂，分裂，消失，捏碎
crumbling rock　崩解岩石
crumbling soil　团粒（结构）土壤
crumbling　粉碎，崩解，起鳞，掉皮，破碎，岩块剥落
crumb-structure　碎屑结构
crumby soil　团粒（结构）土壤
crumb　碎屑，团粒
crunode　分支，交叉点，结点，叉点
crushable structure　可压扁结构
crushable　可压扁的，可破碎的
crush breccia　压碎角砾岩
crushed aggregate　破碎的骨料，轧碎的骨料，轧碎骨料
crushed brick concrete　碎砖混凝土
crushed coal　破碎煤
crushed pebble　碎石子
crushed rock ballast　碎石道渣
crushed rock　破碎岩石，轧石
crushed sand　轧制砂
crushed slag　轧碎的熔渣
crushed stone base course　碎石底层
crushed stone concrete　碎石混凝土
crushed stone drainage layer　碎石排水层
crushed stone with asphalt grouting　碎石灌沥青
crushed stone　碎石
crushed zone　破碎带，压碎带
crusher building　碎煤房
crusher equipment foundation　碎煤（机）设备基础
crusher house　碎煤机室
crusher-run aggregate　破碎骨料
crusher-run backfill　机碎料回填
crusher-run material　机碎碎石
crusher-run rock　机碎碎石，统货碎石
crusher-run　机碎的，未筛分的
crusher surge bin　破碎机缓冲仓
crusher　碎煤机，破碎机，碎石机，轧石机，轧碎机
crushing chamber　破碎腔
crushing equipment　碎石设备，轧碎设备
crushing facility　破碎设备
crushing force　压碎力
crushing load　断裂载荷，压碎荷载，破坏荷载，断裂荷载
crushing machine　轧石机，轧碎机
crushing mechanics　断裂力学
crushing mill　破碎机，压碎机，碾碎机，碎煤机，碎石厂，轧石厂
crushing plant　轧碎厂，破碎场，碎煤装置，轧石厂
crushing rings　破碎环
crushing roll　破碎辊，辊碎机，碎石机轧辊
crushing strength parallel to the grain　纵压强度
crushing strength　抗破碎强度，抗压强度
crushing test　轧碾试验，压毁试验，压碎试验
crushing　捣碎，碾碎，磨碎，压碎，破碎，压毁，压破，咬碎，轧碎
crush resistance　耐压碎
crush　压碎，破碎，轧碎，碎石，粉碎，压榨
crustal deformation　地壳变动
crustal movement　地壳运动
crustal warping　地壳翘曲
crusting agent　结壳剂，结痂剂，黏结剂
crusting　床料表面结焦【沸腾炉】
crust　外皮，壳，沉积层，水垢，表层，地壳，表皮，外壳，硬壳
crutcher　螺旋式拌和机
crutch　吊杆，支架，支柱
CRU test(constant rate of up-lift test)　等速上拔试验
crux　难题，关键，十字，坩埚
cryocable　低温电缆
cryochemistry　低温化学
cryocoil　低温线圈
cryoconcentration　深冷凝聚
cryoconductor　超导体
cryoelectric memory　低温存储器，深冷电存储器
cryoelectronics　低温电子学
cryoengineering　低温技术
cryogen　冷冻剂
cryogenerator　低温发生器，低温发电机，冷冻机
cryogenically-cooled motor　低温冷却电动机
cryogenically-cooled parametric amplifier　低温冷却参量放大器
cryogenic cable　低温电缆，超导电缆，深冷电缆
cryogenic circulator　低温环行器
cryogenic coil　低温线圈，超导线圈
cryogenic computer　深冷计算机，低温计算机
cryogenic cooling　低温冷却
cryogenic effect　低温效应
cryogenic electromagnet　低温电磁铁
cryogenic flow　低温流
cryogenic fluid　制冷液
cryogenic insulation　低温绝缘
cryogenic liquid　低温冷却液
cryogenic magnetic field　低温磁场，超导磁场
cryogenic magnetic system　低温磁体系统
cryogenic magnet　低温磁铁，超导磁铁
cryogenic maser　低温微波激射器，冷温脉塞
cryogenic measurement　低温测量
cryogenic memory　深冷存储器
cryogenic power transmission　低温输电，超导输电
cryogenic service valve　低温用阀
cryogenic spill　低温泄露
cryogenic steel　低温钢
cryogenic superconductor　低温超导体
cryogenics　低温学，冷冻技术
cryogenic treatment　深冷处理
cryogenic valve　低温阀
cryogenic wind tunnel　低温风洞
cryogenic　冷冻的，低温的，制冷的
cryogenine　冷却剂，冷却精
cryogeny　低温学
cryoliquifier　低温液化器

cryolite 冰晶石
cryomachine 超导电机
cryomachining 冷加工
cryometer 低温温度计，低温计
cryomite 小型低温制冷器
cryomotor 超导电动机
cryonetics 低温学，低温技术
cryopedology 冻土学
cryophyllite 绿磷云母
cryophysics 低温物理学，超导物理学
cryoprobe 低温探测器
cryoprotective agent 低温防护剂
cryopumping 深冷抽吸，低温排气，冷凝排气
cryopump 低温泵
cryoresistive magnetic system 低温常导磁体系统，低温有电阻磁体系统
cryosar 低温雪崩开关，雪崩复合低温开关
cryoscope 冰点测定器，冰点测定计，冻点测定仪
cryosistor 低温晶体管
cryosixtor 冷阻管
cryosolenoid 低温电磁线圈
cryostat 制冷器，恒低温箱，低温箱
cryothermal treatment 深冷处理
cryotrapping 低温捕获，冷阱
cryotrap 低温阱
cryotron circuit 冷子管电路
cryotron computer 冷子管计算机
cryotronics 低温电子学
cryotron memory 冷子管存储器
cryotron switch 冷子管开关
cryotron 冷子管
cryoturbation 冻融搅动
cryoturbogenerator 超导汽轮发电机
cryowinding 深冷绕组
cryptical 隐蔽的，秘密的，使用密码的
cryptoclimate 室内小气候
CRYPTO(cryptography) 密码技术
cryptographic device 密码装置
cryptographic machine 密码机
cryptographic 密码的
cryptography 密码学，密码术，密码技术
cryptomere 隐晶岩
cryptorheic 隐流的
crypt 地窖，地穴
crystadyne 晶体振荡检波器
crystal amplifier 晶体放大器
crystal binding 晶体键
crystal block filter 晶体滤波器
crystal boundary corrosion 晶间腐蚀
crystal bridge 晶体检波器电桥
crystal can relay 晶体密封继电器
crystal cell 晶体光电池
crystal-checked 晶体稳定的，石英校准的
crystal clock 石英钟
crystal-controlled 晶体控制的
crystal control transmitter 晶体稳频发射机，晶控发射机
crystal control 晶体控制
crystal converter 晶体变频器
crystal counter 晶体计数器
crystal detector 晶体检波器
crystal diode 晶体二极管
crystal filter 晶体滤波器
crystal frequency indicator 晶体频率指示器
crystal frequency 晶体频率
crystal galvanometer 晶体检流计
crystal gate 晶体管门电路
crystal grain 晶粒
crystal grating 晶格
crystal growth 晶粒长大，晶粒成长，晶体生长
crystal impedance 晶体阻抗
crystal lattice 晶格，晶体点阵
crystalline crack 结晶裂纹
crystalline fracture 结晶状断口
crystalline grain 晶粒
crystalline igneous rock 晶质火成岩
crystalline lens 水晶透镜
crystalline rock 结晶岩，晶质岩
crystalline silicon PV module 晶体硅光伏组件
crystalline silicon solar cell 硅晶太阳能电池
crystalline 结晶的
crystallisation fouling 析晶污垢
crystallite 微晶，小结晶，皱晶
crystallization 结晶
crystallizer 结晶器
crystallize 结晶
crystallizing container 结晶皿
crystal-locked blocking oscillator 晶体锁相阻塞振荡器
crystalloid 晶体，晶体的
crystallometer 检晶器
crystal maser 晶体微波激射器
crystal memory 晶体存储器
crystal mixer 晶体混频器
crystal nucleus 晶核
crystal oscillator 晶体控制振荡器，晶体振荡器
crystal-piloted oscillator 晶体控制振荡器
crystal rectifier 晶体整流器，晶体检波器，晶体整流管
crystal shape 晶体形状
crystal structure 晶体结构
crystal tetrode mixer 晶体四极管混频器
crystal tetrode 半导体四极管，晶体四极管
crystal triode 晶体三极管
crystal valve 晶体管
crystal water 结晶水
crystal wavemeter 石英波长计
crystal 结晶，晶体，水晶，结晶体
crywolf 狼嚎报警
CR 柬埔寨瑞尔
CSA(carry-save adder) 保留进位加法器
CSA(core support assembly) 堆芯支承组件
CSA(crack strength analysis) 裂纹强度分析
CSAS(containment spray actuation signal) 安全壳喷淋触发信号
CSB(core support barrel) 堆芯支承吊篮
CS(carbon steel) 碳钢
CS(cascade station) 串级控制站
CS(cast steel) 铸钢
CSC(convertible static compensator) 可转换静止补偿器

CS(certification specifications) 认可规范
CS(channel status) 通道状态
CS(chip select) 芯片选择，选片
CS(closed cooling water system) 闭式冷却水系统
CS(configuration specification) 组态说明书
CS(controlled shutdown) 控制停堆
CS(control section) 控制部分
CS(control signal) 控制信号
CS(control storage) 控制存储器
CS(control switch) 控制开关
CS(control system) 控制系统
C scope C型显示器，方位角-仰角显示器
CS(core spray) 堆芯喷淋
3C-screwdriver for cruciform head pH 3C十字pH螺丝起子
3C-screwdriver for slotted head screw 3C一字螺丝起子
CSCS(core spray cooling system) 堆芯喷淋系统
CSCS(core standby cooling system) 堆芯备用冷却系统
CSCS(cost and schedule control system) 费用和进度管理系统
CS(cycle shift) 循环移位
CS(cycles) 循环数
CSD(constant-speed drive) 匀速转动装置
CSD(core shift driver) 磁芯移位驱动器
CSE(containment systems experiment) 安全壳系统实验
CSE(control and switch equipment) 控制和转换设备
CSE(core storage element) 磁芯存储元件
CSECT(control section) 控制段
CSF(core support floor) 堆芯支承底板
CSG=casing 汽缸
CSH 包覆过热器【DCS画面】
CSI(channel status indicator) 通道状态指示器
CSI(command string interpreter) 命令串解释程序
CSI(current source inverter) 电流源逆变器
CSIS(containment spray injection system) 安全壳喷淋注入系统
CSISRS(cross section information storage and retrieval system) 检索系统
CSL(current steering logic) 电流导引逻辑
CSL(current switch logic) 电流开关逻辑
CSMA(carrier sense multiple access) 载波感应多路存取［访问］
CSMA/CD carrier sense multiple access with collision detection 载波监听多路访问/冲突检测
CSP(cast steel plate) 铸钢板
CSP(concentrating solar power) 太阳能集热发电
CSPE(chlorosulfonated polyethylene) 氯磺化聚乙烯
CSP power plant 聚光太阳能发电厂
CSP technology 聚光太阳能发电技术，太阳能光热发电技术
CSR(control shift register) 控制移位寄存器
CSRI(Columbia Scientific Research Institute) 哥伦比亚科学研究院【美国】
CSRS(containment spray recirculation system) 安全壳喷淋再循环系统
CSRS(core spray recirculation system) 堆芯喷淋再循环系统
CSS(channel shift switch) 通道移位开关
CSS(communication subsystem) 通信子系统
CSS(containment spray system) 安全壳喷淋系统
cstability 铸造性，流动性
CST=centistoke 厘沱【黏度单位】
CST(channel status table) 通道状态表
CST(common-channel signaling terminal) 公用通道信号终端
CSU(central storage unit) 中央存储器
CSV(combined steam valve) 联合汽门
CSW(channel status word) 通道状态字
CTA 凝汽器管清洗系统【核电站系统代码】
CTB(convert to binary) 转换成二进制
CTC(channel traffic control) 通道通信量控制
CTC(competition transition charge) 竞争过网收费
CTC(conditional transfer of control) 控制的条件转移
CT(combustion turbine) 燃气轮机
CT(compressor turbine) 压气机透平
CT(constant temperature) 恒温，等温
CT(correct time) 准确时间
CT(critical temperature) 临界温度
CT(current transformer) 变流器，电流互感器
CT(current transmitter) 电流变送器
CTD(charge-transfer device) 电荷转移器件，电荷传输器件
CTD(convert to decimal) 转换成十进制
CTE 循环水处理系统【核电站系统代码】
CTL(complementary transistor logic) 互补晶体管逻辑
CTMC(communication terminal module controller) 通信终端模件控制器
CTM(communication terminal module) 通信终端模件
CTM(configuration and tuning module) 组态与调整模件
CTOD at maximum load 最大载荷CTOD值
CTOD(crack tip opening displacement) 裂纹尖端张开位移
CTP(central transfer point) 中央转移点
CTR(critical temperature range) 临界温度范围
CTRL(complementary transistor-resistor logic) 互补晶体管-电阻逻辑电路
CT saturation 电流互感器的饱和
CTS(compact tension specimen) 紧凑拉伸试样
CTSS(compatible time-sharing system) 兼容分时系统
CTT(central trunk terminal) 中央干线终端
CTT(containment tests) 安全壳试验
CTTL(complementary transistor-transistor logic) 互补的晶体管-晶体管逻辑
CTU(centigrade thermic unit) 摄氏温度单位
CTU(central terminal unit) 中央终端装置
C tuning 电容调谐
CTW(cooling tower) 冷水塔
C type cutter C形截样器
cubage 容积
cubature 容积，求容积法

cube crushing strength 立方体压碎强度
cube factor 风速立方因子
cube root 立方根
cube strength test 立方体强度试验
cube strength 立方体抗压强度，立方强度
cube test 立方体抗压强度试验，立方体试验
cube 立方，立方体，求体积
cubical content 容积量，立方容量
cubical dilatation 体积膨胀，体膨胀
cubical expansion 容积膨胀，体膨胀
cubical triaxial test apparatus 真三轴试验仪
cubical triaxial test 立方体三轴试验
cubical 三次方的，立方的，立方体的
cubic building 立方体建筑
cubic capacity 立方容量
cubic content 立方容量，容积
cubic deformation 体积变形
cubic elasticity 体积弹性
cubic equation 三次方程
cubic expansion coefficient 体膨胀系数
cubic expansion 立体膨胀，体积膨胀
cubic foot per second 立方英尺/秒
cubic foot 立方英尺
cubic inch 立方英寸
cubic lattice 立体网格
cubicle control 控制单元
cubicle switchgear and controlgear 箱式金属封闭开关设备和控制设备
cubicle switch 组合开关，室内开关
cubicle 开关柜，小室，配电装置间隔，柜，机壳，盒，腔，密封配电盘
cubic meters per second 立方米/秒
cubic meter 立方米
cubic parabola 三次抛物线
cubic root mean wind speed 立方根平均风速
cubic root 立方根
cubic strain 容积应变
cubic term 三次项
cubic yard 立方码
cuboid 立方体，立方形的，长方体
CU(carry-under) 带汽率
CUC(chassis usage charge) 拖车运费
CU(coefficient of utilization) 利用系数
CU (consolidated undrained triaxial compression test) 固结不排水三轴压缩试验
CU(control unit) 控制器，控制部件
cu. =cubic 立方的
CUE(control unit end) 控制器结束，控制器端
cue light 彩色指示灯
cuesta 单斜脊
cue tape 指令磁带
cue voltage 辅助电压
cue 信号，指示，提示，品质因数，滴定度
cuff 套箍，根套，封套，套头
Cuft(cubic foot) 立方英尺
cuisine spring 缓冲弹簧
cuisine steam 缓冲蒸汽
cuisine stroke 缓冲行程
cuisine valve 缓冲阀
cuisine 已烧好的食物
cull lumber 等外材
cull-tie 废枕木
cull 选择，分类，拣，剔
culm and gob banks 废煤堆
Culmann construction 库尔曼图解法【土力学】
culm bank 漏煤坑
culmiferous 含无烟煤的
culminant 子午线上的
culminate 达到顶点，达到极点
culminating point 转折点
culmination 顶点，极点
culmination 中天，顶点，高潮
culmina 山顶
culm 煤屑，无烟煤屑
cultellate 刀形
cultellation 投点，测点降低
cultipacker 碎土镇压器
cultivable land 可耕地
cultivated land 农田，耕地
cultivate 培养，陶冶，耕作
cultivation by direct seeding 直播栽培
cultivation 耕作，栽培，种植
cultrated 刀装的，剪形的
cultural activity center 文化活动中心
cultural center 文化中心
cultural exchange center 文化交流中心
cultural exchange 文化交流
culturally-induced erosion 耕作砍伐导致的冲蚀
cultural relic 文物古迹
culture symbol 地物符号
culture 文化，文明，培养
culvert box 涵箱
culvert capacity 涵洞过水能力
culvert diversion 涵洞导流
culvert drop 涵道跌水
culvert end wall 涵洞端壁
culvert-flow discharge measurement 涵管测流法
culvert gate 涵管闸门
culvert inlet 涵洞进水口
culvert of reinforced concrete 钢筋混凝土涵洞
culvert outflow 涵洞出流
culvert outlet 涵洞出水口
culvert pipe 涵管，涵筒
culvert under floor 地下涵洞
culvert valve 廊道阀
culvert wall 涵洞边墙
culvert 涵洞，暗沟，电缆管道，管线渠，（电厂的）冷却水渠，暗渠，阴沟，涵管，排水道
cumulant 累积量
cumulative ablation 累积消融
cumulative burnup 累积燃耗，总燃耗
cumulative chain yield 累计裂变产额
cumulative compound excitation 积复励
cumulative compound generator 积复励发电机
cumulative compound motor 积复励电动机
cumulative compound 积复励的，积复绕的
cumulative creep-fatigue damage 累积蠕变-疲劳损伤
cumulative damage function 累积的损伤作用
cumulative damage 累积损伤，积累性损坏
cumulative deformation 累计变形
cumulative demand meter 累积最大需量电度表，

累积最大负荷测试器
cumulative departure 累积偏差
cumulative distribution curve 累积分布曲线
cumulative distribution function 累计分布函数
cumulative distribution 积分分布(曲线)
cumulative dose 累积剂量
cumulative effect 总效果
cumulative erratum 累积误差
cumulative error 累积误差
cumulative exposure index 累积照射指标
cumulative failure frequency 累积故障频率
cumulative failure probability 累积故障概率
cumulative failure rate 累积故障率
cumulative failure ratio 累计故障比
cumulative failure 累积故障，累积损伤
cumulative fission yield 累积裂变产额，累积裂变产量
cumulative flow curve 累计流量曲线
cumulative frequency diagram 累积频率曲线图
cumulative frequency distribution 累积频率分布
cumulative frequency 累积频率，累积频数
cumulative gradation curve 累计级配曲线
cumulative heat 储存热，蓄热，积存热
cumulative hunting 自振荡(同步机的)，累积性振动
cumulative infiltration 累计渗吸量
cumulative ionization chamber 累积电离室
cumulative ionization 累积(碰撞)电离，累积的电离
cumulative letter of credit 可积累使用的信用证，可累积信用证，可累积使用的信用证
cumulatively compounded motor 积复励电动机
cumulative output 累计出力
cumulative oversize 筛余物
cumulative percent passing sieve 过筛累计百分率
cumulative pressure drop 总压降
cumulative probability 累积概率
cumulative progress 累计进度
cumulative-rainfall erosion index 累积降雨侵蚀指数
cumulative reliability 总可靠性
cumulative resistance 总电阻
cumulative runoff diagram 累计径流图
cumulative runoff 累计径流
cumulative temperature 累积温度，积温
cumulative time metering 累时计量
cumulative tripping 累积式跳闸
cumulative usage factor 累积利用系数【疲劳】
cumulative vibration 累积振动
cumulative volume curve 累计容积曲线
cumulative vorticity 累积涡量
cumulative winding 叠加绕组
cumulative 积累的，总的，渐增的，累积的
cumulose soil 腐积土壤，腐泥土，腐殖土，高腐殖质淤泥土，植积土
cumulus soil 腐积土
cuneatic arch 楔形砌拱
cunette 干壕底沟，子沟
cunico alloy 铜镍钴永磁合金
cunife alloy 铜镍铁永磁合金
CuO bed 氧化铜床
cup and ball joint 插口接合，球窝接头
cup anemometer 杯形风力计，风杯风速表，杯式风速仪，旋杯式风速计，旋转杯风速计，转杯风速表
cupboard 柜
cup contact anemometer 电接风杯风速计，计数风杯风速计
cup dolly 杯状顶铆模
cupel 灰皿
cup fracture 杯形断口
cup generator anemometer 转杯磁感风速计
cup head rivet 半球柳钉
cupholder 绝缘子螺脚，杯式绝缘子螺脚【线路用】
cup insulator 杯式绝缘子
cup joint 套接
cup lock washer 杯形锁紧垫圈
cupola 化铁炉，冲天炉，圆屋顶
cupped wind machine 杯式风力机
cupped 杯形的
cupping test 深拉试验
cupping 拉伸，杯状(缩)口
cup point screw 圆头螺钉
cupreous 含铜的，铜的
cupricchloramine 氯胺铜
cupric oxide rectifier 氧化铜整流器
cupric oxide 氧化铜
cupric sulfate 正硫酸铜
cupric sulphate 正硫酸铜
cupric 含铜的，二价铜的
cuprite 赤铜矿
cupro-nickel 白铜，镍铜
cupron 康铜，试铜灵，克普隆铜镍合金
cupro 铜，铜的
cuprum 铜
cup screen 杯形筛
cup shake 轮裂
cup-shaped armature 杯形电枢，杯形衔铁
cup-shaped bucket 杯状戽斗
cup-type anemometer 杯形风速仪，杯状风速计
cup-type current meter 旋杯式流速仪
cup valve 杯形阀
cup washer 杯形螺栓垫圈，杯状垫圈
cup weld 加搭板的缝
cup 杯，帽，座，盖，(绝缘子)外裙，杯子
curb cock 切断旋塞
curb form 路缘模板
curb guard 护角
curbing 抑制
curb level 路缘标高
curb plate 围护板，侧板
curb rafter 侧椽
curb roof 复斜屋顶
curb stone 侧石，路缘石，路边石
curb-type trench 路边电缆沟【电缆用】
curb wall 侧墙，围护墙
curb 侧石，路缘，道牙，约束，抑制，(人行道的)镶边，限制，路缘(石)，马路牙子，(建筑物)边饰，路缘石，缘饰
CUR(complex utility routine) 综合性服务程序

curdle 凝结，凝固
cured shoe 硬树脂接触块
curentt ransformedrransducer 电流互感器度送器
cure the default 纠正违约
cure 解决，消除，措施，养护，固化，硬化，处理
Curie constant 居里常数
Curie cut 居里切割，X 切割
Curie point 居里点
Curie temperature 居里温度
Curie 居里【放射性活度单位】
curing agent 熟化剂，硫化剂，养护剂，固化剂，硬化剂
curing blanket 养护铺盖，养护用毡
curing by ponding 围水养护
curing by sprinkling 喷水养护，洒水养护
curing compounds for cement concrete 水泥混凝土养护剂
curing compound 养护剂【混凝土】
curing conditions （混凝土）养护条件
curing cycle 养护周期
curing mat 养护覆盖物
curing method 养护方法
curing of concrete 混凝土养护
curing period 养护周期，养护期
curing temperature 固化温度，养护温度
curing test 养护试验
curing time 固化期，熟化期
curing 养护【混凝土】，固化
curium 锔
curl field 旋度场
curling ball 旋转球
curling of slab 板翘曲
curling stress 弯曲应力
curling 卷边，翘曲
curl up （使）卷
curl 卷边，旋涡，旋度，波纹材，涡纹
currency appreciation 货币升值
currency composition 币种构成
currency depreciation 货币贬值
currency devaluation 货币贬值，钱币降职
currency exchange rate 货币兑换率
currency inflation 通货膨胀
currency mismatch 币种搭配不当
currency of agreement 协议书规定的货币
currency of payment 支付货币
currency of settlement 结算货币
currency restriction 货币限制
currency revaluation 货币增值，货币升值
currency value 币值
currency 货币，通货，流通，行情，市价，金钱
current account deficit 经常账户逆差，经常项目赤字，收支往来账户赤字
current account surplus 经常账户顺差
current account 经常账户
current address 现地址
current amplification 电流放大
current amplifier 电流放大器
current and potential transformer 电流及电压互感器，仪用互感器
current angle 流向角，水流转角
current antinode 电流波腹
current appropriation 本期拨款
current assets 流动资产
current axis 流轴
current balance equation 电流平衡方程式
current balance relay 电流平衡式继电器，电流平衡继电器
current balance type current differential relay 电流平衡式电流差动继电器
current balance 电流平衡，电流秤
current balancing relay 电流平衡继电器
current basis 现时成本基础，现实成本基础
current-bedding 流层理
current bias 偏流
current board chairman 现任公司董事长
current breaker 电流开关
current buffer 电流缓冲器，电流阻尼器
current capacity 载流量，负荷量，载流能力
current capital 流动资金
current carrier 载流子
current carrying capability 截流能力
current carrying capacity 载流容量，载流量，负荷量
current carrying conductor 载流线，载流导体
current carrying part 载流部件
current characteristic 电流特性
current chart 水流图
current check 例行校验，及时校对
current chopping 电流切断，电流截断
current-circulation loss 环流损失
current code 现行规程
current coil 电流线圈
current coincidence system 电流一致制
current collecting device 集电装置，集流装置
current collector 受电器，集电器
current comparator 电流比较仪
current compensation 电流补偿
current-conducting 导电的
current constant 电流常数
current-controlled current source 可控流电流源
current-controlled NDR 电流控制负微分电阻率
current-controlled switch 可控电流开关
current-controlled voltage source 电流控制电压源，可控电流电压源
current control starting 电流控制启动
current control 电流调节，电流控制，水流控制
current converter 变流器，换流器
current conveyance 电流通过
current coordinate 流动坐标
current cost 时价，市价
current coupling 电流耦合
current curve 水流曲线
current cut-off 电流切断，电流切除
current damper 电流阻尼器
current data 现行数据，现行资料
current degradation 电流退化
current density distribution 电流密度分布
current density of particles 粒子流密度
current density 电流密度，扩散流密度
current derating 电流额定值的下降
current detector 电流探测器

current diagram 水流图解
current difference 潮流差，潮流差比数【海洋】
current-differencing 电流差分
current differential relaying system 电流差动继电制
current direction indicator 电流方向指示器，流向指示仪
current direction meter 流向测定仪，流向仪
current direction relay 电流方向继电器
current direction 流向
current displacement motor 深槽电动机
current displacement 电流位移
current distribution 电流分配，电流分布
current distributor 配电线，配电器
current diversion during construction stage 施工导流
current diverter 分流器
current divider 分流器
current-dividing network 分流网络
current drive 电流驱动，电流激励
current edition 现行版本
current effect 水流影响
current efficiency 电流效率
current element starter 自动变阻启动器
current element 电流元件
current ellipse 落流椭圆
current energy 水流能量
current equalizer 均流器，均流线
current expenditure 经常性开支，经费
current expense 日常费用
current failure alarm 电流中断报警，停电报警信号
current fed antenna 直流馈电天线
current feedback 电流反馈
current float 测流浮标
current floe condition 截流状态
current flow 电流，水流
current flow machine 电流磁化探伤仪
current flow method 电流方法
current flow technique 通电磁化技术
current fluctuation 电流涨落，电流波动
current fraction of the ion B 离子B的电流分数
current fund employment rate 流动资金占有率
current fund loan interest rate 流动资金贷款利率
current fund 流动资金
current gain 电流增益
current generator 发电机
current growth 当年生长
current harmonics 电流谐波
current-illumination curve 电流照度曲线
current impulse 电流脉冲
current in branch circuit 支路电流
current indicator 电流指示器，先行指示器
current induction machine 感应电流探伤仪
current-instruction register 现行指令寄存器
current instrument 电流测量仪表
current integrator 电流积分器
current intensity 电流强度
current international practice 当前国际惯例
current interruption 电流断路，断电
current in the fault 故障点电流
current in the short circuit 短路点电流
current inverter device 变流装置
current leakage 漏电
current liabilities 流动负债，经常性贷款，短期负债
current limit acceleration 电流控制加速
current limit control 限流控制
current limit device 短路电流限制器
current limiter 限制电流器，限流器
current limiting breaker 限流断路器
current limiting circuit breaker 限流断路器
current limiting fuse 限流式熔断器，限流熔丝
current limiting gap 限流间隙
current limiting inductor 限流感应线圈
current limiting reactor 限流电抗器
current limiting relay 限流继电器
current limiting resistor 限流电阻器，限流电阻
current limiting transformer 限流变压器
current limiting type automatic starter 限流式自动启动器
current limiting type 限流式【动力熔丝】
current limiting 电流限制
current limit relay 限流继电器
current linkage 电流链
current list of drawings 本期图纸清单
current loop 电流波腹，电流环路
current-luminous flux characteristic （光电器件的）光电特性
current maintenance 日常保养
current margin 电流裕度，工作电流范围
current mark 水流痕迹，流痕
current maturity 本期或一年内到期
current measurement 电流测定，电流测量
current measuring instrument 测流计，测流仪器
current meter discharge measurement 流速仪测流法
current meter method 流速仪法
current meter rating 流速仪率定
current meter 电流表，流速计，电流计，流速仪，海流计，验流计
current meter measurement 流速仪测流
current meter rating curve 流速仪率定曲线
current meter rating flume 流速仪率定槽
current meter suspension cable 流速仪吊索
current mirror 电流反射镜
current mode complementary transistor logic 电流型互补晶体管逻辑
current mode logic 电流型逻辑，电流型逻辑电路
current mode switch 电流型开关
current modulation 电流调制
current modulator 电流调制器
current moment 电流矩
current money analysis 通货分析
current money 通货
current month delivery 当月交货
current multiplication tube transistor 电流倍增型晶体管
current node 电路波节，电流节点
current noise index 电流噪声指标
current noise 电流噪声

current observation 水流观测
current of commutation 换向电流
current of heat 热流
current on contact 开关电流，合闸电流
current-operated GFCI breaker 电流操作的接地故障断路器
current order 现行指令
current overload relay 过电流继电器
current overload 电流过载，过电流
current page register 现行页面寄存器
current paper 近期论文
current path 电流路径
current pattern 流型
current penetration depth 电流透入深度
current phase-balance protection 电流相位平衡保护装置
current phase 电流相位
current phasor 电流相量
current polarity comparison relay 电流极性比较继电器
current pole 测流杆
current practice 通行做法，现行惯例
current price 时价，现行价格
current protection 电流保护装置
current pulsation 电流脉动，电流双幅值
current pulse generator 脉冲电流发生器
current pulse 电流脉冲【计数器】
current range 电流范围
current rate 电流强度，气流强度，流速，现价，成交价
current rating 额定电流，电流比，电流额定值
current ratio regulator 电流比自动调节器
current ratio relay 电流比继电器
current ratio 电流比，流动比率
current regulating relay 电流调整继电器
current regulation 电流调整
current regulator 电流调节器，稳流器
current repair 日常维修，小修，现场修理，经常性修理
current replay menu 当前重演菜单
current requirement for anodic protection 阳极保护的电源要求
current requirement for cathodic protection 阴极保护的电源要求
current resonance 电流谐振，并联谐振
current resources 现有资源
current-responsive starter 电流响应式启动器
current retard 滞留设施
current reversal 电流反向
current reverser 电流换向开关
current-reversing key 电流换向键，电流换向开关
current ripple mark 水流涟痕
current ripple 电流脉动，纹波电流，水流沙涟
current-rise time 电流上升时间
current rose 流向频度分布图，水流玫瑰图，流玫瑰图，流向图
current rush 冲激电流，电流冲动，电流骤增，水流冲击，电涌
current saturation 电流饱和
current selected switch 电流选择开关
current-sensing device 电流反应设备，电流传感器
current-sensing relay 电流继电器
current-sensing resistor 电流敏感电阻器
current sensitivity 电流灵敏度
current session 本届会议
current-sharing resistor 配流电阻器，电流分配电阻器
current shifting 水流摆度
current shunt 分流器
current source inverter 电流源逆变器
current source 电流源
current square meter 平方刻度电流表
current stabilization 电流稳定
current stabilized device 电流稳定装置
current stabilized power source 稳流电源
current stabilizer 电流稳定装置，电流镇定器，稳流器
current standard 现行标准
current steering logic 电流导引逻辑，电流控制逻辑
current strength 电流强度
current summation 电流累加，电流相加
current supply device 电源设备
current supply 电源
current surge 电流浪涌，电流冲动
current survey 水流观测
current switch logic 电流开关逻辑
current system 现行体系，现行体制
current tap 分插座，分接头
current task 现任务
current temperature coefficient 电流温度系数
current terminal 电流端子
current test point 电流测试点
current thinking 现行思路
current throttling type rotor 挤流转子，深槽转子
current-time curve 电流时间曲线
current to pneumatic converter 电气转换器
current transducer 电流转换器，电流传感器
current transductor 电流磁放大器
current transfer order 现行转移指令
current transfer ratio 电流转移比
current transformer calibrating device 电流互感器校验装置
current transformer 变流器，电流互感器，电流变换器
current transient 暂态电流
current triangle 电流三角形
current-type flowmeter 电流式流量计，速度式流量计
current-type telemeter 电流式遥测计
current value 电流值，兑现值，现行值，当前值，通用值，现值，折现值
current vector 电流矢量
current velocity measurement 流速测验
current velocity 流速
current version 现行版本
current-voltage characteristics 电流-电压特性（曲线），安-伏特性
current-voltage curve 电流-电压曲线，伏-安曲线

current voltage diagram 伏安特性曲线
current weighted factor 电流衡重系数
current width 流幅，流宽
current winding 电流绕组，电流线圈
current time product 电流时间之积
current year cumulation 本年累计数
current year precipitation 当年降水
current year runoff 当年径流
current year's actual 本年实际
current year 当年（度），本年度
current yield 电流生产力，电流效率
current-zero period 电流零区
current-zero 零值电流，电流零点
current 电流，气流，水流，液流，流，流行的，通用的，现行的
curriculum plan 课程计划
curriculum program 课程计划
cursive script 草写体
cursor 光标，指示器
curtailed inspection 抽样检查，简化抽样检查
curtailment demand 可减负荷指令
curtailment of bank facility 紧缩银根
curtailment of investment 紧缩投资
curtailment 缩短，缩减，削减
curtail 降低，减少，衰减
curtain box 窗帘盒
curtain for cutting off water 截水帷幕
curtain grout hole 帷幕灌浆孔
curtain grouting 帷幕灌浆
curtain jet 幕状射流
curtain line 帷幕线
curtain panel 管屏
curtain-pole retarder 木帘滞流装置
curtain rod 窗帘棍，门帘杆
curtain rope 窗帘绳
curtain type damper 帘式挡板
curtain wall type breakwater 幕墙式防波堤
curtain wall 屏式管墙，挡火管墙，围护墙，幕墙，护墙，填充墙
curtain 帘，幕，屏幕，隔板，屏蔽，帘子，幕布，帷幕
Curtis nozzle 柯蒂斯喷管，扩放喷管
Curtis stage 柯蒂斯级，复速级
Curtis steam turbine 冲动式汽轮机
Curtis turbine 柯蒂斯式汽轮机，复速级式汽轮机
Curtis wheel 复速式叶轮，柯蒂斯叶轮
curtis winding 无感绕组，无感绕法，无自感线圈
curvature centre 曲率中心
curvature change 曲率变化
curvature degree 弯曲度
curvature design 曲率设计
curvature factor 曲率系数，弯管损失系数，急弯系数
curvature function of airfoil 翼型弯度函数
curvature measure 曲率的测度
curvature of trajectory 轨道曲率
curvature radius 曲率半径，弯曲半径
curvature 曲度，弧度，曲率，弯度，弯曲，弯曲部分，圆弧度
curve analyzer 曲线分析器
curve bend 弧形弯管，曲线板
curve composition 曲线组合
curve correction coefficient 曲线修正系数
curved bar 曲杆
curved batter 曲斜撑，弯曲倾度
curved beam 曲梁，弯梁
curved blade 弯曲叶片
curved breakwater 弧形防波堤
curved bridge 曲线桥
curved buttress dam 弧形支墩坝
curved channel 弯曲形流道
curved-chord truss 曲弦桁架
curved crystal probe 曲面晶片探头
curved dam 弧形坝
curved grate 弯曲型炉排
curved mirror 曲面镜
curved mole 弧形防波堤，弧形海堤
curved plate 弧形板，曲面板
curved profile 曲线叶型
curved rail 弯轨
curve-drawing ammeter 曲线记录式电流表
curved reach 弯曲段
curved-rib truss 曲肋桁架，曲弦桁架
curved roof 曲面屋顶
curved runner 滑刀型开沟器
curved section 弯曲段
curved sedimentation basin 曲线形沉沙池
curved stair 弧形楼梯，曲尺式楼梯
curved stroke 曲线球
curved surface 曲面
curved tooth bevel gear 曲面齿锥齿轮
curved tooth gear coupling 弧形齿轮联轴器
curved trash rack 弧形拦污栅
curved viaduct 曲线高架桥
curved weir 弧形堰，曲线形堰
curved wire gauze screen 弯线网筛
curved 倒弧角，弯曲的，弯曲的，弧形的
curve finding 曲线拟合法
curve fitting 曲线拟合，曲线选配
curve follower function generator 曲线复线式函数发生器
curve follower 曲线跟随器，曲线输出口，曲线复示器
curve following stylus 描绘曲线笔尖
curve graduation 曲线修匀
curve of counter flexure 反向曲线
curve of deviation 偏差曲线，偏移曲线
curve of equal pressure 等压曲线
curve of equal pulsation intensity 等脉动强度曲线
curve of equal settlement 等沉陷曲线
curve of equal velocity 等速曲线，流速等值线
curve of error 误差曲线
curve of initial magnetization 起始磁化曲线
curve of load 负荷曲线
curve of magnetization 磁化曲线
curve of output loss 出力损失曲线
curve of output performance 出力性能曲线
curve of same sense 同向曲线
curve of sliding 崩滑线，滑动曲线

curve of temperature variation 温度变化曲线
curve of water consumption 用水曲线
curve of water demand 需水量曲线
curve of water supply 供水曲线
curve peak 曲线顶点
curve plotter 绘图仪
curve radius 弯曲半径
curve resistance 弯曲阻力，非线形电阻，弯道阻力
curve ruler 曲线尺
curve springing 起拱点
curve stake 曲线桩
curve superelevation 曲线超高
curve tolerance 曲线公差
curve tracing 曲线描绘
curve 曲线，曲线图，弯曲，弧线，曲线状物，绘制曲线，使弯曲，使成曲线，使成弧形，弯曲的，曲线形的
curvilinear coordinate 曲线坐标
curvilinear equation 曲线方程
curvilinear motion 曲线运动
curvilinear net 曲线网格
curvilinear square 曲线方形
curvilinear strain meter 曲线应变计
curvilinear tunnel 曲线隧道
curvilinear 曲线的
curving 弯曲的
cushion blasting 缓冲爆破
cushion block 垫块
cushion course 软垫层，垫层
cushion cylinder 缓冲气缸，缓冲油缸
cushion feeder 振动式供料器
cushion idler 缓冲托辊
cushioning 缓冲器
cushioning effect 垫层作用，缓冲作用，减震作用
cushioning fender 缓冲装置，缓冲挡板，缓冲防护器
cushioning material 缓冲垫料
cushioning pocket 缓冲腔
cushioning time 缓冲时间，时滞
cushion layer 垫层
cushion material 缓冲材料
cushion of air 空气垫
cushion of replaced soil 换土垫层
cushion pile 缓冲桩
cushion pool 水垫消力池，消力池
cushion space 缓冲空间
cushion stopper 租进器
cushion valve 缓冲阀
cushion 衬垫，装衬垫，弹性垫，缓冲垫，软垫，缓冲器，垫，垫层，坐垫，换填垫层法
cuspate bar 三角沙洲
cuspate foreland 三角岬
cusped leading edge 尖缘进汽边
cusped point 尖点
cusped trailing edge 尖缘出汽边
cusping 形成水舌，扯铃式变形，沙漏式缺陷
cusp of trailing edge 后缘尖端
cusp 尖角，交点，弯曲点，尖头，尖端
CUSS(computerized ultrasonic scan system) 计算机辅助的超声扫描系统
custodian service 保管业务
custody 监视
customary consultation 例行磋商
customary international law 国际惯例法
customary packing 习惯包装
customary 通常的，习惯的，惯例的
custom circuit interconnections 定制集成电路配制
custom circuit 用户定制的电路
custom clearance 清关，报关
custom-designed software 用户设计软件
custom-designed 定制设计的
custom design 定制，定做
customer accounts expenses 用户账务费用
customer assistance expenses 提供用户帮助费用
customer average interruption duration index 用户平均断电持续时间指标
customer dissatisfaction 顾客不满意
customer experience 客户体验
customer installation expenses 用户装置费用
customer-oriented concept 顾客至上的理念
customer-oriented 面向用户的，以客户为中心的
customer owned electric power plant 自备电厂
customer records and collection expenses 用户账册及收账费用
customer satisfaction questionnaire 顾客满意度调查表
customer satisfaction survey 顾客满意度调查
customer service and informational expenses 用户服务和信息费用
customer's load factor 用电负荷率
customers supreme 顾客至上
customer substation 用户变电站
customer survey sheet 顾客调查问卷
customer valve 用户阀
customer 用户，主顾，耗电量，消耗量，买方，客户，顾客
custom house broker 报关行
custom hybrid 定制混合电路
custom IC 用户集成电路
custom integrated circuit 用户定制的集成电路
customised wireless SCADA system 定制无线监控与数据采集系统
customised 依照客户要求而具体制造的
customization 具有特性，用户化，专用化，定制
customized definition 用户自定义
customize 按顾客的具体要求制造［改造］，顾客化，来自用户的
custom-made 定制的
custom memory 用户定制存储器
custom power 用户特定电力
custom risk 通常保险
customs agent 报关代理人
customs agreement 海关协定
customs and import duties 关税和进口税
customs authority 海关当局
customs bill of store 海关免税单
customs brokerage fee 报关费
customs certificate 海关凭证

customs check 海关检查
customs clearance fee 报关费
customs clearance 海关准行，通关，清关
customs code 海关法规
customs declaration form 报关单
customs declaration 海关申报单，报关，报关单
customs documentation 海关提货单
customs document 海关单据
customs duties 海关关税，海关税，关税
customs duty rate 关税税率
customs examination 验关
customs formalities 海关手续，海关报关手续
customs forms 海关单据，报关单
customs house 海关
customs inspection fee 海关查验费
customs inspection 海关检查，验关
customs invoice 海关发票
customs office 海关
Customs Tariff of Import of the People's Republic of China 中华人民共和国进口关税
customs 海关，关税
custom tailor 定制
custom 定制的，习惯，风俗，海关，习俗，惯例，关税
cut and cover method 明挖法
cut and cover tunnel 随挖随填的隧道
cut and cover 随挖随填
cut and fill balance （土方）挖填平衡，土方平衡
cut and fill estimate 土方计算
cut and fill slope 半挖半填的斜坡
cut and fill 半挖半填，随挖随填，挖方和填方，充填法开采
cut and joint 切割和接头
cut and over method 明挖（回填）法
cut and trial 试凑法
cut-and-try iterative method 逐次迭代渐近法
cut and try method 试算法，试凑法，渐进法，渐近法，试验法
cut-and-twisted finned tubes 带有切断扭转直肋的肋化管
cutaway drawing 剖视图，剖面图
cutaway view 剖视图，剖面图，内部接线图
cutaway 剖视图，剖面图，剖面（的）
cutback asphalt binder 轻制沥青黏合料，稀释沥青黏合料
cutback asphalt 轻制沥青，稀释沥青
cutback stream 逆向水流
cutback 截短【金属线】，减少，削减，稀释
cut bank 凹岸，切岸，挖蚀岸
cut cable 电缆端，一节电缆，电缆段
cut core 对接式铁芯
cut depth 挖方深度，挖掘深度，挖土深度
cut diameter 分割粒径
cut down expenses 节约开支
cut down first cost 压缩初期投资
cut down on the scale of capital construction 压缩基建规模
CU test（consolidated undrained triaxial compression test） 固结不排水三轴压缩试验
cut-fill balance 填挖方平衡
cut-fill section 半填半挖式断面
cut-fill 随挖随填
cut-flight conveyor screw 锯齿形螺栓
cut grass 已割的草地
cut hole 掏槽炮眼
cuticular transpiration 表皮蒸腾
cuticular 表层的
cutie pie instrument 卡蒂派仪【小型γ剂量计】
cut in relay 接入式继电器
cut in speed 投入正常运行的转速
cut in stand-by 切换备用【如泵等】
cut in wind speed 最小出功风速，切入风速
cut in 干预，插入，加塞，切入，接通，接入，插图，开始工作
cut location 切断位置
cut-off angular frequence 截止角频率
cut off a pile head 切去桩头
cutoff apron 护底截水墙，截水墙护底
cut-off attenuator 截止衰减器
cut-off basin 闭合流域，封闭盆地
cut-off bias 截止偏压
cut-off blanket 隔离层
cut-off characteristics 截止特性
cut-off collar 防渗环
cut-off computer 开关计算器
cut-off current characteristic 限流特性
cut-off current 截止电流
cut-off date 截止日期，结关日
cut-off device 断水器
cut-off energy 截止能，切割能，切断的能量
cut-off frequency 截止频率
cut-off gate 截止门，截止闸门
cut-off gear 停机机构，停汽装置
cut-off governor 停机调节器
cut-off key 切断电键
cut-off level 截止电平
cut-off meander 割断曲流
cut-off of supply 停止供电
cut-off overvoltage protection 截流过电压保护
cut-off period 不得超过的期限
cut-off piling 截水板桩
cut-off plate 节流板，挡板
cut-off point 截止点，断开点，熄火点
cut-off ratio 裁弯取直比
cut-off relay 截止继电器，断离继电器，断路继电器
cut-off signal 断开信号，停车信号，关闭发电机信号
cut-off slide valve 断流滑阀
cut-off state 截止状态
cut-off switch 切断开关，断路开关，断流开关
cut-off test 解列试验，截止试验，停车试验
cut-off time 停机时间，截关日
cut-off trench 截水沟，截水槽，齿槽
cut-off valve 截止阀，截流阀，断流阀，停汽阀，关断阀
cut-off velocity 截止速度
cut-off voltage 截止电压
cut-off wall 截水墙，齿墙，隔墙，堰板，隔心墙，防渗墙
cut-off water hammer 断流水锤
cut-off waveguide 截止波导

cut-off wave length　截止波长
cut-off wheel　切割砂轮
cut-off wind speed　最大出功风速
cut-off wind velocity　切除风速
cut-off works　裁弯工程
cut-off　截水墙，隔心墙，挡水墙，截流，切断，切口，取直，剪切掉，截止，断路（开关）
cut-out box　熔断器盒，熔断器匣
cut-out breaker point　断路器断电点
cut-out damper　隔绝挡板
cut-out diagram　开孔尺寸图
cutout drawing　断面图
cut out full travel control　切断全行程控制【以允许按下慢速行走】
cut out fuse　断流保险丝
cut-out governing　喷嘴调节，断流调节
cut-out information　开孔资料
cut-out key　切断电键，断路按钮
cutout line　切割线
cut-out relay　断路继电器
cut-out spar　切口（翼）梁
cut-out speed　截止速度，脱扣速度
cut-out switch　切断开关
cut-out valve　截止阀
cut-out wind speed　切出风速，最大出功风速，截止风速
cut-out　切断，关闭，断路，截断，断路器，断流器，熔断器，断流，断路的，关断，删除，结束工作
cut-over　切换，转换
cut payment　扣款
cut payroll to improve efficiency　减员增效
cut ridge　限幅，载幅，截峰
cut searching method　线路故障逐段寻找法
cut set matrix　割集矩阵
cut set　割集
cut sheet　开挖图
cut slide valve　切断滑阀
cut spike　大方钉
cut stone masonry　琢石砌体，琢石圬工
cut stone works　琢石工程
cut stone　琢石
cuts　挖方
cuttability　可切割性，可挖性
cutter change factor　齿轮刀具变位系数
cutter compensation　刀具补偿
cutter head dredger　铣轮式挖泥机
cutter head　切头
cutter motor　截煤机电动机
cutter suction dredger　切吸式挖泥机
cutter　切割机，切削工具，刀具，采煤器，截煤器，剪断机，切割者
cut the load　减负荷
cutting and bending of reinforcement　钢筋的断配和弯曲
cutting and bevelling　切削和坡口
cutting curb　沉井筒脚，沉箱刃脚
cutting current　切割电流
cutting edge　切削刃
cutting electrode　切割用电极
cutting face　切削面
cutting fluid　切削液
cutting gun　割枪
cutting hardness　切割硬度，切削硬度
cutting head　切头
cutting in　接通，并入线路
cutting jet　切挖射流【水力开挖】，下切水流
cutting line　切割线
cutting machine for the sleeve and connecting ring　对套筒和连接环的切断机
cutting machine　截断机，横锯床，切割机
cutting management　采伐管理
cutting nozzle　割刀，割嘴
cutting-of-flux　磁力线的切割
cutting-off of short-circuit　切断短路
cutting-off pile head　桩头截断
cutting-off　截止，断绝，截断
cutting out　切断，断开，断流
cutting oxygen　切割氧
cutting performance　切割性能
cutting range　切削范围
cutting ring　环刀
cutting round snips　切圆剪
cutting speed　切割速度
cuttings　钻屑
cutting tip　割嘴
cutting tool　切削工具，刀具
cutting torch　割炬
cutting under water　水下挖土，水下切割
cutting wheel　砂轮截管器
cutting　切割，采掘，琢磨，割，划破，切断，切削，挖掘，挖土方
cut-to-length device　切割定尺装置
cut-to-length line　定尺剪切作业线，定尺剪切机组
cut to length　按定长切割，定长裁断，定尺剪切
cut-trough　切开
cutup pool　切割池
cutwall　限流壁，步行小道，狭窄过道
cutwater　桥脚的分水角，船头破浪处，隔舌
cut-wound core　（变压器的）卷铁芯，连续带绕铁芯
cut　切割，切口，分割，切除，开挖，截断（长度），剪切，挖方
CV(calorific value)　热值
CVCF(constant voltage and constant frequency)　恒压及恒频
C-V characteristic　伏安特性（曲线）
CV(check valve)　逆止阀，止回阀
CV(common version)　通用形式
CV(constant volume)　定容，固定体积
CV(continuously variable)　连续可变的
CV(control valve)　控制阀，高压阀门，高调门
CV(converter)　转换器
CVCS(chemical and volume control system)　化学和容积控制系统
CV(curriculum vitae)　简历，履历
C-V curve(capacitance-voltage curve)　电容-电压曲线
CVD(chemical vapor deposition)　化学蒸汽沉积法，化学气相沉积
CVIS(containment ventilation isolation system)　安

全壳通风隔离系统
CVI 凝汽器真空系统【核电站系统代码】
C-V measurement 电容-电压测量
CVPETS(condensate vacuum pump effluent treatment system) 凝结水真空泵排水处理系统
CVS(chemical and volume control system) 化学与容积控制系统
CVT(capacitor voltage transformer) 电容式电压互感器
CVT(communication vector table) 通信向量表
CVU(constant voltage unit) 恒压装置，恒压器
CWBS(contract work breakdown structure) 合同工作分解图
CW(carrier wave) 载波
CW(circulating water) 循环水
CW(civil work) 土建工程
CW＝clockwise 顺时针方向的，顺时针
CW(continuous wave) operation 连续波运转
CW(continuous wave) 连续波
CW(control word) 控制字码
CW(cooling water) 冷却水
CW(cosine wave) 余弦波
CWD(control wiring diagram) 控制接线图
C-wire 丙线，C线
CWJ(catalog of welded joint) 焊接接头目录
CWO(carrier wave oscillator) 载波振荡器
CWO(continuous wave oscillator) 连续波振荡器
CWP(circulation water pump) 循环水泵
CWR(chilled water return) 冷却水回路
CWS(circulating water system) 循环水系统
C. W. system 冷却水系统
CWT(combined water treatment) 联合水处理
CW 循环水【DCS画面】
cyanaloc 氰基树脂（一种防水剂）
cyanide 氰化物
cyaniding 氰化
cybernation 自动控制
cybernetic engineering 控制工程
cybernetic machine 控制论机
cybernetic model 控制论模型，模拟控制机
cybernetic system 控制系统
cybernetics 控制论，控制学
CYCL＝cyclone 旋风分离器
cycle checking device 巡回检测装置
cycle conditions 循环参数
cycle control 循环控制
cycle counter 转数计，频率计，周波计数
cycle criterion 循环判据，循环准则，周期判据
cycle duration 周期长度
cycle economy 循济经济性
cycle efficiency 循环效率
cycle fluid （热力装置的）循环工质
cycle frequency 循环频率
cycle generator 周波发生器
cycle heat rate 循环热耗
cycle index counter 循环次数计数器
cycle index 循环次数，循环指数
cycle interrupt 循环式中断，周期性中断
cycle length 运行周期，循环时间
cycle life elapse 循环寿命消耗
cycle life expenditure 寿命损耗率，循环寿命消耗
cycle life 循环寿命
cycle loading 周期性加荷
cycle motion 周期运动
cycle of concentration 循环倍率，浓缩周期
cycle of fluctuation 波动周期
cycle of freezing and thawing 冻融循环
cycle of heating and cooling 冷热循环
cycle of initiating crack 开始断裂循环数
cycle of loads 荷载循环，荷载周期
cycle of magnetization 磁化循环，磁化周期
cycle of operation 工作周期，动作循环，充-放电循环，运行周期，操作循环
cycle of oscillation 振荡周期
cycle of stresses 应力周期
cycle of stress reversal 应力交变周期
cycle of stress 应力循环
cycle of vibration 振动周期
cycle operation 循环运行
cycle order 循环次序
cycle performance 循环性能，循环特性
cycle period （燃料的）循环周期，堆内停留时间
cycle per second 周/秒，赫兹
cycler 周期计
cycle-repeat timer 周期脉冲发送器
cycle reset 循环复位
cycle shift 循环移位
cycle of concentration 浓缩周期，循环倍率
cycle stealing 周期挪用
cycles-to-failure 疲劳损坏循环数
cycle storage 循环存储器
cycle stretch-out （燃料）周期延长，增加燃耗
cycle temperature 循环温度
cycle time of dumping 翻卸周期
cycle timer 循环计时器，触点开闭周期控制装置
cycle to crack 开裂循环数
cycle to failure 失效循环数，循环致失效
cycleway 自行车道
cycle working fluid 循环工质
cycle 循环，周期，周，周波，环，圈，轮回
cyclic action 循环作用
cyclic addressing 循环寻址
cyclic admittance 相序导纳
cyclical depletion 周期性退水消耗，周期性耗竭
cyclically alternating vortex 周期性交替涡
cyclically 循环的
cyclical pin gear drive 摆线针轮传动
cyclical pin gear planetary 行星摆线针轮减速机
cyclical process 循环过程
cyclical recovery 周期回升
cyclical storage 循环存储器，周期蓄水
cyclical stress 循环应力
cyclical variation 循环性变化
cyclical 循环的，周期性的
cyclic binary code 循环二进制码
cyclic change 循环变化
cyclic check bits 循环效验位
cyclic check 循环效验
cyclic code 循环码
cyclic control system 闭环控制系统

cyclic counter 循环计数器
cyclic creep and stress rupture 循环蠕变和应力断裂
cyclic current method 回路电流法
cyclic curve 周期曲线
cyclic daily duty 日周期负载，日周期工作方式
cyclic decimal code 循环十进码
cyclic deformation 交变变形，周期性变形，周变衰化
cyclic depletion 周期亏耗，多年亏耗，周期性退水
cyclic digital telemetering 循环数字遥测
cyclic-digit code 循环数码
cyclic duration factor 周期持续系数
cyclic duty alternator 周期性工作的同步发电机
cyclic duty capacity 变负荷能力
cyclic duty 周期性负载，周期式工作方式
cyclic failure criteria 循环失效判据
cyclic frequency 角频率
cyclic hardening and cyclic softening 循环硬化与软化
cyclic impedance 相序阻抗
cyclic integrator 循环积分器
cyclic irregularity factor 调节不均匀度
cyclic irregularity 周期不规则率，周期不规则性
cyclic iterative method 循环迭代法
cyclic life expenditure 循环（使用）寿命损耗
cyclic life test 循环寿命试验
cyclic life 循环期限，循环寿命
cyclic loading spectrum 循环载荷谱
cyclic loading 循环载荷，周期荷载，周期性加荷
cyclic load 周期性负荷，循环负荷，中间负荷
cyclic magnetic condition 循环磁状态
cyclic magnetization 周期磁化
cyclic memory 循环存储器
cyclic motion 循环运动
cyclic movement solids 固体颗粒循环运动
cyclic number 循环次数
cyclic oscillation 循环性振动，周期性振荡
cyclic permeability 周期性磁导率，正常磁导率
cyclic pitch 周变桨距
cyclic process 周期性过程
cyclic reactance 相序电抗
cyclic recovery 周期复原，周期回升
cyclic redundancy check 循环冗余检查，循环冗余码校验
cyclic redundancy check 循环冗余校验
cyclic service 周期性运行
cyclic shear strain amplitude 周期剪应变振幅
cyclic shift 循环移位
cyclic simple shear test 周变单剪试验
cyclic storage system 周期存储系统
cyclic storage 周期存储器
cyclic strain 交变应变，周期应变
cyclic strength 疲劳强度
cyclic stress limit 周期性应力极限
cyclic stress ratio 周变应力比，周期应力比
cyclic stress 交变应力，循环应力，周期应力
cyclic surge 周期性涌浪
cyclic tangential loading 周期性切向加荷
cyclic tape 循环磁带
cyclic temperature change 周期性温度变化
cyclic test 循环试验
cyclic transformation 循环变换
cyclic transmission 循环传输
cyclic triaxial apparatus 振动三轴仪
cyclic triaxial test 周期荷载三轴试验，周变三轴试验
cyclic twist stress 循环扭曲应力
cyclic twist 循环扭曲
cyclic unit 中间负荷机组
cyclic variation of stress 应力周期变化
cyclic variation 循环变化，周期性变化，周期变化
cyclic 周期性的，循环的
cycling life 循环（使用）寿命
cycling-load 周期性变负荷，循环负荷
cycling open-and-closed 循环开启和关闭的
cycling operation 周期性运行，调峰运行，周期性负荷运行
cycling service 周期性运行
cycling solenoid valve 周期电磁阀
cycling unit 循环负荷机组
cycling 循环周期工作的，循环
cyclo-converter 循环换流器，双向离子变频器
cyclodos 发送电子转换开关
cyclogenesis 气旋生成
cyclogram 周期图表
cyclography eddy-current 涡流式电磁感应检测仪
cycloidal arch 圆滚线拱
cycloidal gear reducer 摆线减速齿轮
cycloid 摆线，旋轮线
cycloinverter 双向离子变频器
cyclolysis 气旋的减弱或消失，气旋消失
cyclometer 路程计，圆弧测定器，周期计，回转计，转数计
cyclone air lock 旋风筒气封
cyclone and loop seal 分离器及回料系统
cyclone boiler 旋风锅炉
cyclone burner 旋流燃烧器
cyclone capacity 旋风筒出力
cyclone clarifier 旋液澄清器
cyclone cleaner 旋风除尘器
cyclone collar 旋风筒喉口
cyclone collector 旋风除尘器，旋风集尘器
cyclone cone 旋风筒锥形底
cyclone damage 气旋破坏
cyclone discharge line 旋流排出管道
cyclone dust collector 旋风式除尘器，旋风除尘器
cyclone dust extractor 旋风除尘器
cyclone dust separator 旋风除尘器
cyclone efficiency 旋风除尘器效率，旋风筒效率【除尘】
cyclone feeder 旋流给料机
cyclone filter 旋风过滤器
cyclone-fired boiler 旋风燃烧锅炉
cyclone firing 旋风燃烧
cyclone furnace boiler 旋风炉
cyclone furnace firing 旋风炉燃烧

cyclone furnace 旋风炉
cyclone gaswasher 旋风水膜式除尘器
cyclone muff 旋风筒筒体
cyclone opposed firing 对置旋风燃烧
cyclone path 气旋路径
cyclone performance 旋风除尘器性能
cyclone precipitator 旋风除尘器
cyclone scrubber 旋风洗涤器，旋风除尘器
cyclone separator 细粉分离器，旋风分离器，旋流分离器
cyclone steam separator 旋风汽水分离器
cyclone throat 旋风筒喉口
cyclone trace 气旋轨迹，气旋路径
cyclone type burner 旋流式燃烧器
cyclone type moisture separator 旋风式汽水分离器
cyclone wall 旋风筒壁
cyclone 旋风分离器，旋风除尘器，旋风筒，旋风器，除尘器，旋风子，旋风，气旋
cyclonic circulation 气旋式环流
cyclonic motion 气旋式运动
cyclonic rain 气旋性雨
cyclonic rotation 周期旋转
cyclonic storm 气旋（性）风暴
cyclonic swirl 气旋（性）环流
cyclonic vorticity 气旋涡量
cyclonic whirl 气旋型涡流
cyclonic wind 气旋风
cyclonic 气旋的，飓风的，旋风的
cyclopack 快装式旋风炉
cyclopaedia 百科全书，丛书
cyclopean concrete 大块石混凝土，蛮石混凝土，毛石混凝土
cyclopean masonry dam 蛮石圬工坝
cyclopean masonry 蛮石圬工
cyclopean riprap 乱石堆
cyclopean 蛮石
cyclophon 旋调管
cyclorectifier 循环整流器，单相离子变频器
cyclostrophic balance 旋衡
cyclostrophic wind 旋衡风
cyclotron angular frequency 回旋角频率
cyclotron damping 回旋加速器阻尼
cyclotron maser 回旋脉塞，回旋微波激射器
cyclotron radius of electrons 回旋加速器的电子振荡半径
cyclotron 回旋加速器
CY(container yard) 集装箱堆场
CY(cycle) 循环，周，周期
CYL(cylingder) 汽缸
cylinder actuator 液压制动器
cylinder barrel 汽缸筒，油缸筒，气缸筒
cylinder block cracking 汽缸裂纹
cylinder block 汽缸体，油缸体
cylinder body 汽缸体，缸体
cylinder bore 汽缸内孔，汽缸内径
cylinder caisson foundation 管柱基础
cylinder caisson 圆柱形沉箱
cylinder claw 汽缸猫爪
cylinder covering 汽缸扣盖
cylinder crack 汽缸裂缝
cylinder crevice 汽缸夹层
cylinder crushing strength 圆柱体抗压强度，圆柱强度
cylinder drain pipe 汽缸疏水管
cylinder expansion indicator 绝对膨胀指示器
cylinder expansion 汽缸膨胀
cylinder flange heating appliance 汽缸法兰螺栓加热装置
cylinder flange 汽缸法兰
cylinder foot 汽缸脚，汽缸猫爪
cylinder frame fit 汽缸和座架配合
cylinder gas 钢瓶气
cylinder gate intake 圆柱闸门进水口
cylinder gate 圆筒形闸门，圆筒闸门，圆柱闸门
cylinder groove 汽缸洼窝
cylinder head 汽缸盖
cylinder jacket 汽缸夹层
cylinder lens 柱面透镜
cylinder lifting 吊汽缸
cylinder locking/unlocking control 液压闭锁/解锁控制
cylinder lubricator 注油器
cylinder metal temperature 汽缸金属温度
cylinder oil 汽缸油
cylinder operated valve 气（缸操）动阀，气缸操纵的阀门
cylinder operator 活塞式执行机构，液压操作器
cylinder penetration test 柱体贯入试验
cylinder pier jetty 柱墩式突码头
cylinder pier 圆筒形码头，圆筒墩
cylinder pile foundation 管柱基础
cylinder pile 圆柱桩
cylinder ring 隔板套，持环
cylinder seal 汽缸密封
cylinder sleeve 气缸套筒
cylinder sole pin 汽缸台板［底板］
cylinder sole plate 汽缸台板
cylinder specimen 圆柱体试件，圆柱形试件
cylinder strength 圆柱体强度
cylinder test 圆柱体试验
cylinder tube 汽缸筒
cylinder type jetty 沉井式码头，圆筒式丁堤
cylinder type lock valve 筒式闸阀
cylinder valve 筒形阀，液压阀，圆筒阀，圆筒堰
cylinder wharf 筒柱式码头
cylinder wrench 圆筒扳手
cylinder 汽缸，气缸，油缸，圆筒，圆柱体，钢瓶，缸体，圆形的，圆柱
cylindrical and conical shape 圆柱和圆锥形
cylindrical antenna 圆柱形天线
cylindrical arch dam 圆筒拱坝
cylindrical arch 圆筒式拱
cylindrical barrage 圆筒形坝
cylindrical bearing 滚柱轴承，圆筒轴承，圆柱轴承
cylindrical blade 直叶片，等截面叶
cylindrical bolt anchorage 筒式螺杆锚具
cylindrical boring 镗汽缸内孔，镗圆筒孔
cylindrical boundary 圆柱形分界面

cylindrical brush 圆柱形刷子
cylindrical cavity 圆筒形腔室
cylindrical ceiling 筒形顶篷
cylindrical coil 圆筒形线圈
cylindrical compression 圆柱体压缩
cylindrical concrete shell 筒形混凝土薄壳，筒形混凝土壳体
cylindrical concrete test specimen 圆柱形混凝土试样
cylindrical controller 鼓形控制器
cylindrical coordinate notation 柱面坐标标记
cylindrical coordinates 柱面坐标
cylindrical core test facility 圆柱形堆芯试验装置【日本】
cylindrical dam 圆筒形坝
cylindrical drier 圆筒干燥机
cylindrical drill 圆筒形钻
cylindrical electrical machine 圆柱形电机，隐极转子电机
cylindrical electrode 圆柱形电极
cylindrical energy absorber 圆筒形消能器
cylindrical film memory 圆筒形胶片存储器
cylindrical flue boiler 火管锅炉
cylindrical fluidized bed 圆筒形流化床
cylindrical free vortex 圆筒状自由旋涡，柱形自由涡流
cylindrical gear differential 圆柱形齿轮差动装置
cylindrical gear drive 圆柱齿轮传动
cylindrical gear 圆柱齿轮
cylindrical geometry 圆柱体形状
cylindrical grabhead 圆筒形抓头
cylindrical grinding machine 外圆磨床
cylindrical headed type insulator 圆柱头型绝缘子
cylindrical horizontal boiler 卧式锅壳锅炉
cylindrical insulation 筒式绝缘套
cylindrical insulator 筒式绝缘套［子］
cylindrical linear induction motor 圆柱形直线感应电动机
cylindrical lock 圆柱形门锁
cylindrical missile shield 飞射物圆筒形防护罩，飞射物防护筒
cylindrical nozzle 圆柱形喷嘴
cylindrical optical mounting 圆柱形光学装置
cylindrical part 圆柱体筒身【安全壳】
cylindrical pier 圆柱形墩
cylindrical pile 筒桩
cylindrical pipe 筒形管
cylindrical polar coordinate 柱面极坐标
cylindrical press fit 柱面压力配合
cylindrical projection 圆柱投影
cylindrical puff 圆柱形喷团
cylindrical reactor 圆柱形反应堆
cylindrical reinforced-concrete missile shield 飞射物钢筋混凝土防护筒，飞射物圆筒形钢筋混凝土防护罩
cylindrical roller bearing 圆柱滚子轴承，圆柱形滚柱轴承
cylindrical roller thrust bearing 圆筒形滚柱推力轴承
cylindrical roller 滚柱
cylindrical rotor generator 隐极发电机，圆柱形转子发电机
cylindrical rotor 鼓形转子，隐极转子
cylindrical shell source 圆筒状外壳源
cylindrical shell 筒体，筒体环段，圆柱体筒身，圆筒形壳体
cylindrical sieve 圆筒形格筛
cylindrical skirt 圆柱形裙筒
cylindrical solenoid 筒形螺线管
cylindrical spiral 螺旋线
cylindrical stator 圆筒形定子
cylindrical stream surface 圆柱流面
cylindrical surface screen 圆柱面屏
cylindrical surface 圆柱面
cylindrical surge chamber 圆筒式调压室
cylindrical transformer 圆柱形变压器
cylindrical type electrolytic condenser 圆筒形电解电容器，管形电解电容器
cylindrical uranium dioxide pellet 圆柱形二氧化铀芯块
cylindrical valve 圆筒阀
cylindrical vault 半圆拱，筒拱，筒形穹隆
cylindrical ventilator 筒形风帽
cylindrical vessel 圆筒形容器
cylindrical vortex 圆筒式旋涡，柱形涡流
cylindrical wall inspection 筒壁检查，圆柱壁检查
cylindrical wall 圆柱体墙【安全壳】
cylindrical waveguide 圆柱形波导管
cylindrical wave 柱面波
cylindrical winding 筒形绕组
cylindrical 圆筒形的，圆柱形的，圆柱的，筒形的
cylindroconical ball mill 圆锥形球磨机
cylindroconical reactor 圆锥形反应器
cymmotive force 波动势
cymograph 自记波长计，自记频率计，自记波频计，波形自记器，转筒记录器
cymometer 波长计，频率计，波频计
cymoscope 检波器，振荡指示器
cypher 密码
cypress 柏木，柏树
cyrtometer 曲面测量计
cyrtometry 曲面测量法
C 凝汽器【冷凝、真空、循环水，核电站系统代码】
C 切换【DCS 画面】

D

DAA(data access arrangement) 数据存取方案
DAA(decimal adjust accumulator) 十进制调整累加器
DAA 冷、热机修理车间和仓库电梯【核电站系统代码】
dabble 溅湿，浸，润湿
DAB(display assigned bit) 显示指定位
DAB(dispute adjudication board) 争端裁决委员会，争议总裁委员会
DAB 办公楼电梯【核电站系统代码】

dab 能手
DAAC(data acquisition and control) 数据采集和控制
DAC(data acquisition computer) 数据获取计算机
DAC(data analysis center) 数据分析中心
DAC(derived air concentration) 导出空气浓度
DAC(digital arithmetic center) 数字计算中心
DAC(digital-to-analog converter) 数字-模拟转换器，数-模转换器
DAC(direct access control) 直接存取控制
DAC(distance amplitude correction) 距离幅度校正
DAC(distance amplitude curve) 距离振幅曲线
DACON(data control) 数据控制
D/A conversion 数模转换
D/A converter 数字模拟转换器
DACS(data acquisition and control system) 数据采集和控制系统
DAAC system 数据采集和控制系统
D-action(derivative action) 微分作用
D_2-action 二阶微分作用，D_2 作用，二阶导数作用
DA(data acquisition) 数据采集，数据获取
DA(decimal add) 十进制加法
DA(define area) 定义范围
DA(design automation) 设计自动化
DA(differential analyzer) 微分分析机
D-A(digital-analog) 数模，数字模拟
DA(direct action) 直接作用
DA(direct add) 直接相加
DA(discrete address) 离散地址
DA(distribution automation) 配电自动化
DA(documentary acceptance bill) 承兑汇单
DA(documents against acceptance) 承兑交单
dado 墙裙，护墙板，墩身，护壁板
DADS(data acquisition and display system) 数据获取与显示系统
DAE(data acquisition equipment) 数据采集设备
DAE timer 停激后时滞继电器
DAF(delivered at frontier) 边境交货【国际贸易术语】
d. a. f. (dry ash-free basis) 无灰干燥基
DAFT(digital analog function table) 数字模拟函数表
DAGC (delayed automatic gain control) 延迟式自动增益控制
daggle 拖曳
daily 日常的
daily allowance 每日津贴
daily amount 日产量
daily amplitude 日变幅
daily average temperature of maximum 日平均最高气温
daily average temperature of minimum 日平均最低气温
daily burn requirement 日燃煤量
daily change 日改变
daily check 每日检查
daily coal consumption 日耗量
daily compensation 日补偿
daily consumption 日耗量
daily cycling 日周期性运行，日负荷峰谷变化运行
daily demand 日需要量
daily discharge 逐日流量
daily diversity power 日参差功率
daily dosimetry 逐日剂量测定
daily duration curve 日持续曲线
daily efficiency 日效率
daily expense 每日费用
daily extremes 日极端值
daily flood peak 日洪峰
daily forecast 逐日预报
daily fuel consumption 燃料日消耗量
daily heat load 日热负荷
daily high tide 日高潮位
daily inspection 日常检验，日常检查，例行测试
daily intake 日摄入量
daily load 日负荷，昼夜负荷
daily load curve 日负荷曲线
daily load cycle 日负荷循环
daily load factor 日负荷率，日负荷系数
daily load fluctuating operation 日负荷变动运行
daily load fluctuating 日负荷波动
daily load forecasting 日负荷预测
daily load prediction 日负荷预测
daily log 日志
daily maintenance task 每日维修保养工作
daily maximum temperature 日最高温度
daily maximum 日最大量，日最大值
daily maximum rainfall 日最大降水量
daily mean temperature 日平均温度
daily mean 日平均值
daily minimum temperature 日最低温度
daily minimum 日最小值，日最小量
daily observation 日常检查，逐日观测
daily operation cost 日运营成本
daily output 日出力，日产量
daily peak load 每日峰荷，日峰荷
daily per capita consumption 每人每日耗量
daily pondage 日调节容量
daily precipitation amount 日降水量
daily precipitation 日降水量
daily progress 日进度
daily pro rate 按每天比例
daily range of air temperature 气温日较差
daily range of temperature 日温度变化范围
daily range 日变幅，日较差，逐日差程
daily record of construction 施工日志
daily regulating pond 日调节池，日调节水池
daily regulating reservoir 日调节水库
daily regulation pond 日调节池
daily regulation 日调节
daily retardation 潮迟
daily run 日运转，日运转期
daily start-stop 二班制运行，每日启停，日启停
daily statement 日报表
daily storage capacity 日调节库容
daily storage plant 日调节电站
daily storage 日蓄能，日蓄水量，日储量
daily surveillance 日常检查
daily task system of maintenance 日维修工作制度

daily temperature fluctuation 日温变化，日温波动
daily test run 每日连续测试
daily tidal cycle 每日潮汐周期
daily time report 工作时间日报表
daily-to-day management 日常管理
daily transaction reporting 每日处理报告
daily variation factor 日变化系数
daily variation graph of heat consumption in one month 每月的日热负荷图，每月热耗日变图
daily variation graph of heat consumption 热负荷图，热量消耗日变图
daily variation 日变动，日变化，逐日变化
daily water consumption 日耗水量，日用水量
daily weak report 工作日报表
daily weather chart 每日天气图
daily work report 工作日报表
daily work 每日工作
DAI 核岛厂房电梯【核电站系统代码】
dale 山谷，溪谷
dalles 急流
Dalton's law of partial pressure 道尔顿分压定律
Dalton's law 道尔顿定律
Dalton's principle 道尔顿定律
dam abutment 坝肩，坝头岸坡，坝座
dam accessories 坝体附属结构物
DAMA(demand assignment multiple access) 按需分配多址
damage accident of heat-supply network 热网故障
damage accumulation 损伤累积
damage by dragging 被风吹坏
damage by fume 烟气损害
damage by storm 风暴损失
damage by wind 风灾破坏
damage capability 破坏能力
damage claim 赔偿损害的要求，损害索赔
damage control 故障控制，损伤控制，修复损害控制，破坏性控制
damage criterion 损伤分度标准，损伤判据
damage curve 损耗曲线
damaged beyond repair 损坏到不能修理的程度
damaged cargo 残损货物
damaged conductor 受损导线
damaged fuel assembly cask 破损燃料组件罐
damaged fuel assembly 破损燃料组件
damaged insulator 受损绝缘子
damaged material 损坏的材料
damaged rate 损伤率
damaged RCC element container 损坏的棒束控制元件容器
damaged screw extraction 损坏的螺钉拔出
damaged 已损坏的，已损伤的，被损害的
damage factor 损失因子
damage for detention 迟误损失
damage-frequency curve 损害频率曲线
damage function analysis 损伤功能分析
damage index 损失指数
dam ageing 坝的老化
damage law 损伤律
damage line 破坏线，破损线
damage mechanism 损坏机理
damage oscillation 阻尼振动
damage rate 损伤率
damage ratio 破坏比，损坏率，损失比
damage repair 损坏修理
damage risk 受损危险性
damage scenario 损坏情况，事故情况
damage sequence 损坏后果，事故后果，损坏顺序
damages for breach of contract 违约赔偿金
damages for default 违约赔偿金
damage survey 损失调查
damages 损害赔偿金，损害
damage threshold 损伤极限值
damage tolerance design 破损设计
damage to works 对工程的损害
damage 损坏，损害，伤害，损伤，破坏，损失
damaging effect 损伤效应
damaging flux 损伤通量
damaging impact 破坏性冲击，危害性冲击
damaging range 损害范围
damaging stress 破坏应力
damaging 有破坏性的
dam and diversion conduit type development 混合式开发
dam appurtenance 坝体附属结构物
dam axis 坝线，坝轴线
dam back 坝的背水面
dam base width 大坝底宽
dam base 坝底，坝基
dam behavior 坝的性能，大坝特性
dam block 坝块
dam-board 挡板
dam body 坝身，坝体
dam breach 垮坝，溃坝
dam break flood 溃坝洪水
dam break 坝溃决，垮坝，溃坝
dam builder 筑坝人员
dam built by dumping soil into water 倒土筑坝
dam cement 大坝水泥
dam construction technique 坝工技术，筑坝技术
dam construction 筑坝
dam core 坝的心墙，坝心
dam crest width 坝顶宽度
dam crest 坝顶
dam damage 大坝损坏
DAM(data address memory) 数据定址存储器
DAM(data associative message) 数据相关信息
dam deflection 坝体挠度
dam deformation 坝体变位
dam deterioration 大坝损坏
DAME(data acquisition and monitoring equipment) 数据采集和监视设备
dam-embedded penstock 坝内埋管
dam face 坝面
dam failure flood wave 水坝（洪水）断裂波，溃坝洪水波
dam failure 水坝决口，水坝毁坏，坝失事，垮坝，溃坝
dam for water supply 供水用坝
dam foundation 大坝基础，坝基
dam-gap diversion 缺口导流
damgate slab 挡水闸板

dam heel 坝踵
dam heightening 坝的加高，坝体加高，堤坝加高
dam height 坝高
dam inspection 大坝检查
dam intake 坝式进水口
dam length 坝长
dam location 坝的位置
dammed water 壅水
damming lock 蓄水闸
damming shield 挡水防护体
damming technology 筑坝技术
damming with weathered rock 风化石筑坝
damming 壅高，筑坝，拦断，筑坝拦水
dam monolith 坝段
damourite 水白云母
dam out 筑坝拦水
damp air 潮湿空气，湿空气
damp condition 潮湿痕迹
damp course 防潮层
damped alternating current 减幅交流，衰减交流
damped angular frequency 衰减角频率
damped ballistic galvanometer 阻尼冲击检流计
damped cosine function 阻尼余弦函数，逐步减少的余弦函数
damped factor 阻尼系数，衰减系数
damped frequency 阻尼频率
damped harmonic motion 阻尼谐动
damped harmonic system 减幅谐波系统
damped impedance 阻尼阻抗，衰减阻抗
damped oscillation 阻尼振荡，阻尼振动，减幅振荡
damped oscillator 阻尼振荡器
damped periodic instrument 阻尼周期性仪器
damped sawtooth oscillation 阻尼锯齿波振荡
damped seiche 阻尼水面波动，阻尼假潮
damped sine function 阻尼正弦函数
damped sinusoidal quantity 衰减正弦量
damped sinusoid 阻尼正弦曲线
damped train 衰减波列
damped valve 减振阀
damped vibration 阻尼振动，衰减振动
damped wave equation 衰减波动方程
damped wave 阻尼波，减幅波
damped 阻尼的，衰减的
dampener 湿润器
dampening chamber 湿润室
dampening force 阻尼力，减震力，缓冲力
dampening pipe 消音器
dampen 阻尼，弄湿，吸声，减振
damper actuator 风门执行器，风门致动机构，挡板执行器
damper adjustment 挡板调节
damper bar 阻尼条
damper blade 挡板叶片
damper brake 制动闸
damper by friction of liquid 液体摩擦减振
damper cage starting 阻尼笼启动
damper cage 阻尼笼
damper capacity 吸振能力
damper casing 风门罩壳
damper circuit 阻尼电路
damper coefficient 阻尼系数
damper coil 阻尼线圈
damper control 挡板调节，挡板控制
damper current 阻尼电流
damper cylinder 减振筒
damper action 减振作用，阻尼作用
damper coil 阻尼线圈
damper device 减振装置
damper driven shaft 挡板轴
damper drive 阻尼传动
damper frame 挡板框架，减振框架
damper isolation 阻尼隔振
damper line fitting 防振锤
damper loss 挡板静压损失【烟气】，阻尼损失，阻尼绕组损耗，阻尼器损耗
damper of friction 摩擦式减振器
damper plate 挡板
damper position indicator 调节器位置指示器
damper regulator 风门调节器
damper ring 阻尼环
damper segment 阻尼片，阻尼环的一段
damper system 阻尼系统
damper valve 减振阀
damper winding 阻尼绕组，阻尼线圈，阻尼器绕组
damper with one shutter 带有一个百叶窗的单向阀
damper 挡板，闸板，调节风门，阻尼器，缓冲器，减震器，减振器，消震器，阻尼线圈，缓冲器，防振锤，(火炉等的）风门
damp haze 湿霾
damping action 阻尼作用
damping adjustment 阻尼调节，死区调整，衰减调节
damping apparatus 减振设备，阻尼器
damping arrangement 减振装置
damping baffle 缓冲隔板
damping bar 阻尼条
damping blanket 防振垫
damping capacity 阻尼能力，吸振能力，抗震能力，减震能力
damping channel 减振系统
damping characteristic 阻尼特性
damping coefficient 阻尼系数，衰减系数，衰减率
damping coil 阻尼线圈
damping constant 衰减常数，阻尼常数
damping controller 阻尼控制器
damping curve 阻尼曲线，衰减曲线
damping decrement 衰减率
damping degree 阻尼程度
damping device 阻尼装置，减振装置，衰减装置，制动装置，减震装置
damping diode 阻尼二极管
damping dissipation 阻尼损耗
damping driven oscillation 阻尼驱动振动，负阻尼振动
damping effect 缓冲作用，阻尼效应，阻尼作用
damping factor 阻尼系数，衰减率，衰减系数，阻尼因数，阻尼因子

damping force 阻尼力，减振力，缓冲力
damping logarithmic decrement 阻尼对数减缩量
damping loop 阻尼环
damping machine 阻尼机【自动控制用】
damping magnet 阻尼磁铁，制动磁铁
damping material 减振材料，隔声材料，隔音材料
damping matrix 阻尼矩阵
damping of instruments 仪表的阻尼
damping of system oscillation 系统振荡的衰减
damping pad 减振垫
damping parameter 阻尼参数，阻尼系数
damping power 阻尼功率
damping ratio 阻尼比，阻尼率，衰减率，阻尼系数，消能率
damping regime 阻尼状态
damping resistance 衰减阻力
damping ring 阻尼环
damping rubber 橡皮减振器
damping screen 整流网，整流，稳水栅，阻尼网
damping spring 减振弹簧
damping term 阻尼项
damping time constant 阻尼时间常数，缓冲时间常数，衰减时间常数
damping time 阻尼时间
damping torque coefficient 阻尼力矩系数
damping torque time 阻尼时间
damping torque 阻尼力矩
damping transformer 阻尼变压器
damping transition 衰减瞬变过程
damping turn 阻尼线匝
damping vibration 阻尼振动，衰减振动
damping washer 减振垫圈
damping winding 阻尼绕组
damping wire 减振拉筋，阻尼拉筋
damping 阻尼，制动，减振，衰减，抑制，减震，阻抑，减幅
dampness 湿度，含水率，潮湿，湿气，湿润
damp proof course 防潮层，防水层
dampproof foundation 防潮基础
damp proofing admixture 防潮剂
damp proofing foundation 防潮基础
damp proofing of ground 地面防潮
damp proofing 防潮
damp-proof insulation 防潮绝缘，隔潮
damp-proof machine 防潮电机
damp-proof masonry 防潮砌体
damp-proof material 防潮材料
damp proofness 耐湿性
damp-proof wall 防潮墙
damp-proof 不透水的，不透水，防潮，防潮的，耐湿的，防湿
damp steam 湿蒸汽
damp-storage closet 高温度养护室【养护混凝土用】，湿养护室
damp 阻尼，缓冲，弄湿，湿气，潮湿的，夹具，润湿，衰减
dam raising 坝体加高
dam roadway 坝上道路
dam sheeting 坝面护板
dam shell 坝壳
dam shoulder 坝肩
dam site investigation 坝址调查，坝址勘查
dam site selection 坝址选择
dam site survey 坝址测量
dam site 坝址
dam slope 坝坡
dam strengthening 坝的加固，坝体加固
dam structure 坝结构
dam thickness 坝的厚度
dam toe 坝脚，坝趾，坡趾
dam top width 坝顶宽度
dam top 坝顶
dam type development 坝式开发
dam type hydroelectric station 坝式水电站
dam type power plant 蓄水式水电站，（水）坝式发电站
dam type power station 堤坝式水电站
dam volume 坝体积
dam wall 坝墙
dam width 坝宽
dam 坝，闸，拦水坝，防洪堤，挡风板，大坝，防沙堤，拦河坝，水坝
DAM 汽轮机厂房电梯【核电站系统代码】
dan bouy 标识浮标
danger class 危险等级
danger coefficient method 危险系数法
danger coefficient 危险系数
danger line 危险线
danger of condensation 冷凝的危险
dangerous building 危险房屋
dangerous cargo 危险品
dangerous condition 危险情况
dangerous cross-section 危险截面
dangerous goods regulations 危险物品手册
dangerous goods surcharge 危险品处理费【承运人收取此费为 RAC，代理人收取此费为 RAA】
dangerous goods 危险品
dangerous hill 险坡
dangerous pressure 危险压力
dangerous section 危险截面，危险地段
dangerous structure 危险结构
dangerous voltage 危险电压
danger plate 警告牌
danger poster 危险告示
danger signal 危险信号，警告信号，危险标志
danger to M from N 因 N 而对 M 产生危险
danger warning 危险警告
danger zone 危险场地，危险地带，危险区，险区
danger 危险，障碍物
danks 黑色炭质页岩
dant 劣质煤，次煤
DAP(data acquisition and processing system) 数据获取和处理系统
DAP(data acquisition and processing) 数据获取和处理
DAP(data analysis and processing) 数据分析和处理
DAP(delivered at place) 目的地交货
DAP(direct access programming system) 直接存

取程序设计系统
dapped joint　互嵌接合
dapping　单榫，刻槽
daraf　拉法【等于1法拉】
Darboux vector　达布矢量
darby　长镘刀，混凝土刮板
Darcy flow regime　达西流态
Darcy's law　达西定律
darcy　达西【多孔介质渗透力单位】
DAR(data access register)　数据存取寄存器
DAR(Doppler Acoustic Radar)　多普勒声雷达
dark adaptation　黑暗适应
dark board　暗屏【指控制室屏上无警告信号闪光】，暗室
dark characteristic curve　暗特性曲线
dark conduction　无照电导，无照电流，暗电流
dark-conductivity　无照导电性，暗导电率
dark current　无照电流，暗流，暗电流
darkened glass spheres　深色玻璃球
darkened inspection booth　检验暗室
dark field　暗区【光弹性试验中】，暗场
dark line spectrum　暗线光谱
dark minerals　黑色矿物
darkness-triggered alarm　暗通报警器，暗触发报警器
dark patch　（射线照片上的）黑斑点
dark pressure mark　黑压痕
dark resistance　暗电阻，无照电阻
darkroom illumination　暗室照明
dark room　暗室
dark spot　黑斑
dark streak　（射线照片上的）黑条纹
dark　深色的
Darlistor　复合可控硅
darrieus generator　达里厄发动机
Darrieus machine　达里厄型风力机
Darrieus type rotor　达里厄型风轮
Darrieus type wind turbine　达里厄型风力机
DAS(data acquisition system)　数据采集系统
DASD(direct access storage device)　直接存取存储器
DAS(digital attenuator system)　数字式衰减系统
DAS(direct access storage)　直接存取存储器
DAS(distribution automation system)　配电自动化系统
DAS(diverse actuation system)　多样化驱动系统
dashboard　仪表板，控制板，操纵板，挡板，操纵盘，挡泥板，挡雨板，防波板，遮水板
dash-bond coat　砂浆涂层
dash control　缓冲控制，按钮控制
dash current　冲击电流，超值电流
dash-dot-line　点划线
dashed contour line　虚线等高线
dashed line　虚线
dasher　挡泥板，遮水板，反射板
dash line　虚线
dash-out　删去，涂掉
dash panel　仪表板，控制板
dashpot check valve　缓冲单向阀
dashpot drop time　缓冲落棒时间
dashpot governor　带缓冲器的调速器
dashpot overload relay　缓冲超载继电器
dashpot piston　减振活塞
dashpot relay　定时继电器，油壶型继电器，缓冲器继电器
dashpot rotameter　旋转式流量计，带有阻尼器的转子流量计
dashpot section　缓冲段【控制棒导管】
dashpot time delay relay　油壶型延时继电器
dashpot　减振器【控制棒】，缓冲筒，缓冲器，阻尼器
dash valve　冲击阀
dash　冲撞，冲，碰撞，挡板，控制板，操纵板，按钮
dasymeter　炉热消耗计，气体成分测定仪，气体密度测定仪
DAS　数据采集系统【DCS画面】
data　资料，数据
data access arrangement　数据存取方案
data access register　设计存取寄存器
data access　数据存取
data acquisition and control system　数据采集和控制系统
data acquisition and control　数据采集和控制
data acquisition and display system　数据获取与显示系统
data acquisition and monitoring equipment　数据采集和监视设备
data acquisition and processing system　数据获取和处理系统
data acquisition and processing　数据获取和处理
data acquisition and recording system　数据采集和记录系统
data acquisition and visual display system　数据获取与直观显示系统
data acquisition computer　数据采集计算机
data acquisition equipment　数据采集设备
data acquisition function　数据采集功能
data acquisition, process and supervisory system　数据采集、处理与监控系统
data acquisition station　数据采集站
data acquisition system　数据采集系统
data acquisition unit　数据采集设备
data acquisition　数据采集，数据收集，数据获取
data address memory　数据定址存储器
data address　数据地址
data aggregate　数据采集，数据集结
data analysis and processing　数据分析和处理
data analysis center　数据分析中心，资料分析中心
data analysis set-up　数据分析装置
data analysis　数据分析，资料分析
data array　数组，数据阵列，资料阵列
data associative message　数据相关信息
data attribute reference　数据属性引用
data attribute type　数据属性类型
data attribute　数据属性
data automation　数据处理自动化
data bank system　加工数据库系统
data bank　数据总库，资料库，数据库，资料储存系统
data base administrator　数据库管理程序

data base blockout　数据库封锁
data base definition　数据库定义
data base handling system　数据库处理系统
data base machine　数据库机器
data base management system　数据库管理系统
data base management　数据库管理
data base task group　数据库任务组
data base　基本数据，数据库，资料库
data bit　数据位
data block　数据块
data book　数据表，数据手册，数据栈
data break　数据中断
data buffer unit　数据缓冲器
data bus　数据总线
data capacity　信息容量
data carrier storage　数据载体的存储
data carrier　数据记录媒体，数据载体
data cartridge　合式数据磁盘，数据带盒
data cell drive　磁片卡机
data cell　数据单元
data center　数据中心
data chaining　数据链接
data chain　数据链
data channel　数据通道
data circuit terminating equipment　数据电路终结设备
data circuit　数据电路，数据线路
data class　数据类
data code translation　数据代码转换
data code　数据代码
data collection and analysis system　数据收集和分析系统
data collection and processing　数据收集与处理
data collection system　数据汇集系统
data collection　数据汇集，数据收集
data communication cable　数据通信电缆
data communication control unit　数据通信控制器
data communication network architecture　数据通信网络体系结构
data communication system　数据通信系统
data communication　数据通信
data comparator　数据比较器
data compilation　资料整编
data compression　数据压缩
data concentrator　数据集中分配器
data control block　数据控制块
data control　数据控制
data conversion line　数据转换线
data conversion receiver　数据转换接收器
data conversion　数据变换，数据转换
data converter　数据变换器，数据转换器
data correction　数据修正
data definition　数据定义
data demand　数据请求
data dependence graph　数据相关图
data depositing　数据存储
data descriptor　数据描述符
data design　数据设计
data determination　数据确定
data dictionary　数据字典，数据目录
data disk　数据软盘
data display　数据显示
data display module　数据显示组件
data distribution center　数据分配中心
data distribution path　数据分配通路
data distribution　数据分配
data distributor　数据分配器
data division　数据部分
data-driven execution　数据驱动执行
data dump　数据转储
data element set　数据元素集
data element　数据元
data encoding system　数据编码系统，数据译码系统
data error　数据误差
data exchange control　数据交换控制
data exchange system　数据交换系统
data exchange unit　数据交换装置
data exchange　数据交流
data extended block　数据扩充块
data fetch　取数据
data field　数据区，数据组
data file　数据文件
data flow chart　数据流程图
data flow control system　数据流控制系统
data flow detection　数据流检测
data flow diagram　数据流程图
data flow language　数据流语言
data flow machine　资料流程电脑
data flow system　数据流系统
data flow　数据流
data formatting　数据格式
data form　数据记录表
data frame　数据帧
data gathering system　数据采集系统，数据收集系统
data gathering　数据收集
data generator　数据发生器
data handling component　数据处理元件
data handling equipment　信息处理装置，数据处理装置
data handling system　数据处理系统
data handling unit　数据处理设备，数据处理装置
data handling utilities　数据处理适用程序
data handling　数据处理
data handling center　数据处理中心
data handling equipment　数据处理装置
data handling　数据处理，资料处理，资料整编
data hierarchy　数据分级结构，数据层次
data highway　数据高速公路，资讯高速公路，数据总线，数据公路，数据通道
data hold　数据保持
data independence　数据独立性
data-initiated control　数据起始控制
data input bus　数据输入总线
data input　数据输入
data inserter　数据输入器
data integrity　数据完整性
data interchange　数据交换
data interpretation　数据整理
data-in　输入数据
data item　数据项

data level 数据级
data library 数据存储中心
data line 数据传输线，数据线
data link control 数据链路控制
data link escape character 数据链路换码字
datal link layer 数据链路层
data links platform 数据链平台
data link terminal 数据链路终端
data link 数据链，数据传输器，数据中继器
data locking 数据连锁，数据定位
data logger 数据记录仪［器］，巡回检测器，数值记录表，巡回检测装置
data logging scanner 数据登陆扫描器
data logging system 巡回检测系统
data logging 数据记录，巡回检测
data management program 数据管理程序
data management system 数据管理系统
data management 数据管理
data manipulation 数据操纵，数据操作
data mapping unit 数据映射装置
data medium 数据媒体
data model 数据模型
data modem 数据调制解调器
data move instruction 数据移动指令
data multiplexer channel 数据多路转换器通道
data multiplex subsystem 数据多路转接子系统
data name space 数据名称空间
data name 数据名称
DATAN(data analysis) 数据分析
data network 数据网，数据网络
data object is mandatory or optional 数据对象必备或可选
data object reference 数据对象引用
data object 数据对象
data on installation 安装数据
data organization 数据结构，数据组织
data origination 数据初始加工，数据机读化
data-out 输出数据
data output 数据输出
data path switch 数据通路开关
dataphone 数据电话，数据通信器
data plate 铭牌，数据板
dataplotter 数据标绘器
data pool 数据库，数据源，资料库
data presentation 数据表达方式
data printer 数据打印器，数据记录器，数据打印机
data processer 数据处理机［器］
data processing centre 数据处理中心
data processing equipment 数据处理装置
data processing inventory 数据处理目录库
data processing machine 数据处理机
data processing subsystem 数据处理子系统
data processing system 数据处理系统
data processing 数据处理，数据整理，资料处理，资料整编
data processor 数据处理机，数据处理器，数据处理程序
data process subsystem 数据申报子系统
data purification 数据精化，数据精炼
data qualification 数据分类
data rate 数据传送速率
data reader 数据读出器
data reading 数据读出
data receiver 数据接收器
data reconstruction 数据重构
data recorder 数据记录器
data recording amplifier 数据记录放大器
data recording medium 数据记录介质
data record 数据记录
data reducer 数据变换器
data reduction equipment 数据处理装置
data reduction system 数据变换系统，数据整理系统，数据简化系统
data reduction 数据压缩，数据变换，数据整理，数据简化，资料折算
data reference 数据引用
data register 数据寄存器
data release update time 信息发布更新时间
data reliability 数据可靠性
data repeater 数据重发器
data repository 资料馆
data reproduction 数据复制
data retrieval 数据检索
data rules 数据规则
data sampling switch 数据采样开关
data scanner 数据扫描装置
data scanning 数据扫描
data security 数据安全，数据安全性
data selector 数据选择器
data separation 数据分离，信息分离
data set block 数据集块
data set class 数据集类
data set control block 数据集控制块
data set control 数据集控制
data set coupler 数据集耦合器
data set group 数据集组
data set label 数据集符号，数据集标号
data set migration 数据集迁移
data set modem 数据解调器
data set ready 数据集就绪
data set 数据装置，数据集，数传机
data sheet 明细表，汇总表，数据表【描述特征】
data-signaling rate 数据信号传输率
data sink 数据传输接受器
data smoothing network 数据平滑网络
data smoothing 数据平滑，数据信号平滑，数据修匀
data sorter 数据分类器，数据选择器
data source 数据源，数据发送器，资料来源
data stabilization 数据稳定化
data statement 数据语句
data station control 数据控制站
data station word 数据状态字
data station 数据站
data storage system 数据储存系统
data storage 数据储存器
data stream 数据流
data strobe 数据选通
data structure 数据结构
data switching center 数据转换中心

data switching 数据转换
data synchronization unit 数据同步器
data synchronization 数据同步，数据同步设计
data system integration 数据系统集成
data system interface 数据系统接口
data system specification 数据系统的性能指标，数据系统的要求
data system 数据系统
data tablet 数据输入板
data table 数据表
data tape 数据带
data terminal equipment 数据终端设备
data terminal ready 数据终端就绪
data terminal 数据终端
data tracks 数据道
data transceiver 数据收发机
data transcription equipment 数据转录设备
data transfer rate 数据传递率，数据传送速率
data transfer 数据传送，数据传输，数据转移
data translating system 数据变换系统
data transmission block 数据传输块
data transmission equipment 数据传输设备
data transmission network 数据传输网络
data transmission protocol 数据传输协议
data transmission system 数据传输系统
data transmission terminal equipment 数据传输终端设备
data transmission video display unit 数据传输视频显示设备
data transmission 数据传输
data transport function 数据传输功能，数据传送功能
data transport service 数据传送业务，数据传输服务
data treatment 数据处理
datatron 数据处理器【十进制计算机】
data type 数据类型
data unit 数据单位
data validation 数据正确性校验
data validity 数据真实性，数据有效性
data withdrawal 数据撤除
data word buffer 数据字缓冲器
data word 数据字
DAT(delivered at terminal) 目的地或目的港的集散站交货
DAT(dynamic address translation) 动态地址转换
date 日期
date and time of arrival 抵达时间
date commenced 开始日期
date completed 完成日期
date entry terminal 数据输入终端
date of acceptance 承兑日期
date of arrival 到达日，到港日期，抵达日期
date of balance sheet 决算日，资产负债表日期
date of certification 证明日期
date of completion 完成日期
date of contract signing 合同签字日期，合同签订日期
date of contract 签约日期
date of declaration 宣布日期
date of delivery 交货（日）期，交付日期
date of departure 启运日期
date of discharge 卸货日期
date of dispatch 发运日期
date of draft 出票日
date of enforcement 实施日期
date of entry into force 生效日期
date of establishment 成立日期
date of expiration 有效期限
date of expiry 到期，有效截止日期
date off-project 项目停止使用日期
date of grace 宽限日期
date of issue 出票日，开证日，颁发日，签发日
date of letter 发函日期
date of loading 装货日期
date of maturity 期票到期日期
date of payment 付款日期，支付日期
date of postmark 邮戳日期
date of project substantial completion 工程完成交付使用日期
date of readiness for shipment 待运日期
date of receipt 收到日期
date of registration 注册日期
date of retirement 报废日期，退役日期
date of sailing 起航日期
date of shipment 发货日期，启运日期
date of signature 签订日期
date of signing the contract 签约日期
date of substantial completion 实际完工日期
date of termination 终止日期
date of validity 限期，有效日期
date of value 起息日
date on-project 项目停止使用日期
date on which payment becomes due 付款到期日
date received 收到日期
date set for power performance measurement 测试功率特性的数据组
date set 数据组
date stamp 日期戳
dating pulse 控制脉冲，同步脉冲
dating 年代测定
datin 数据输入器
dative bond 配价键
datmation 数据自动处理
datum axis 基准轴
datum elevation 基准高度，基准高程
datum error 基准误差，数据误差
datum face 基准面
datum for reduction of sounding 测深折算基点
datum grade 零米标高，基准标高
datum level 基准面，基准水平线，基准高度，基准，水准
datum line 基线，基准线
datum mark 基准点
datum of discharge 排出特性
datum of elevation 高程基准面
datum of tidal level 潮位基准面
datum plane for construction 施工基准面
datum plane for sounding 基准水平面
datum plane of gauge 测站基面，水尺基准面
datum plane 基准面，水准基面
datum point 基准点，参考点，基点

datum speed 给定速度
datum state 基准状态
datum surface 基准面
datum temperature 初始温度，基准温度
datum water level 水准零点，基准水平面
datum 基准点，（测量）基点，数据，资料，测站零点，已知条件，已知数，基准
daubing 涂抹炉底，修补炉底，粗抹灰泥，粗抹面
daub （炉底）涂料，涂覆，草灰泥，粗灰，粗灰泥，打泥底，涂抹
daughter activity 子体放射性，子体活度
daughter board 子插件
daughter company 子公司
daughter product 子产物，子体产物【放射性】
daughter-standard 派生标准
daughter 子系，子体，子核，裂变产物
davit-mounting 吊装
davit 吊杆，扒杆
d-axis d 轴，直轴，纵轴
day-ahead market 日前市场
day-ahead trade subsystem 日前交易子系统
day-ahead trading 日前交易
day-and-night meter 昼夜费率计
day breeze 日风
day classification 黏土分类
day fuel tank 日用油箱，供油箱
day grouting 黏土灌浆
day labour construction 按日计酬施工
day labour 短工
45-day lethality 辐照后 45 天的死亡率
daylight area 采光面积
daylight illumination 自然采光，日光照明，日光仪
daylighting area 天然采光面积
daylighting design 采光设计
daylighting 采光，自然采光，日光，日光照明
daylight lamp 日光灯
daylight opening 采光口
daylight ratio 天然照明率
daylight saving time 日光节约时，夏令时
day load 日间负荷
day marking of aerial obstructions 空中障碍物的白昼标志
day mark 昼标
day mineral 地面矿
day-night effect 昼夜效应
day-night switching unit 昼夜开关
day of maximum demand 最大需水日，最大用电日
day of maximum power demand 最大用电日
day of maximum water demand 最大需水日
daypack film 日光型胶片
days away from work rate 事故缺勤率
day shale 黏土质页岩
day shift 日班
day signal 日间信号，昼标
day slip 黏土滑波
days of demurrage 延滞日数
days of grace 宽限期，优惠期
days of heating period 采暖期天数
day stone 黏土岩
day storage tank 日贮存水箱
day tank 日贮存水箱
daytime mixing layer 白昼混合
day-to-day variation 逐日变化
day-to-day 天天，日复一日
day wage 计日工资
day without frost 无霜日
daywork rates 计日工费率
daywork schedule 计日工计划
DBA(design basis accident) 设计基准事故
D-bank (Doppler bank) D 棒组，多普勒棒组
D-bank position control loop D 棒组位置控制电路
DBB(Design-Bid-Build) DBB 模式，"设计-招标-建造"模式【工程建设模式】
DB(data link) 数据链路
DBD(double-base diode) 双基极二极管
dB(decibel) 分贝
DB(design-build) 设计施工一体化，DB 模式，"设计-建造"模式【工程建设模式】
DB(disbursement fee) 垫付款手续费【承运人收取此费为 DBC，代理人收取此费为 DBA】
DB(dispute board) 争议委员会
DBE(design basis earthquake) 设计基准地震
DBE(design basis event) 设计基准事件，参考事件
DBF(design basis failure) 设计基准故障
DBF(design basis fire) 设计基准火灾
DBFL(design basis flooding level) 设计基准淹没水位
DBFO (Design-Build-Finance-operate) DBFO 模式，"设计-建设-融资-运营"模式【工程建设模式】
DBGF filter(deep-bed glass fiber filter) 深床玻璃纤维过滤器
DBHS(data base handing system) 数据库处理系统
dbk(decibels with reference to one kW) 千瓦分贝
DBLOCA (design-basis loss-of-coolant accident) 设计基准冷却剂更新丧失事故，设计根据失冷（失水）事故
DBM(data base management) 数据库管理
DBM(decibel meter) 分贝计
db meter 分贝计
DBMS(date base management system) 数据库管理系统
dbm 毫瓦分贝
DBR(descriptor base register) 描述符基本地址存寄器
DB switch (double break switch) 双断开关
DBTG(data base task group) 数据库任务组
DBTT(ductile-brittle transition temperature) 延性脆性转变温度
DBU(data buffer unit) 数据缓冲器
dbv(decibels relative to one volt) 伏特分贝
dbw(decibels relative to one watt) 瓦分贝
D cable D 形导线双芯电缆，半圆导线双芯电缆
DC-AC chopper 直流变交流斩波器
DC/AC converter 直流交流电压变换器，直流交流变换机
DC/AC inverter 直流交流逆变器

DC/AC power inversion 直流交流功率变换
DCA(drain cooler approach) 疏水冷却器通道
DCAM(discriminating content addressable memory) 可鉴别的内容定址存储器
DC ammeter 直流安培计，直流电流表
DC amplifier 直流放大器
DC analog computer 直流模拟计算机
DC and AC motor generator 直流交流电动发电机
DC arc welder 直流弧焊机
DC arrester 直流避雷器
DC automatic recording fluxmeter 直流自动磁通记录仪
DC bias 直流偏置
DC booster 直流升压机
DC bridge 直流电桥
DC bus arrester 直流母线避雷器
DC bushing 直流套管
DCC(drag chain conveyer) 刮板捞渣机
DC charge spots 直流充电桩
DC choke charging 扼流圈直流充电
DC choke rectifier charging 扼流圈整流器直流充电
DC circuit breaker 直流断路器
DC component 直流分量，直流部分
DC compound motor 直流复激电动机，直流复励电动机
DC contactor 直流接触器
DC contamination test 直流污秽试验
DC control circuit 直流操作回路，直流控制回路
DC control power supply 直流控制电源
DC corona loss 直流电晕损失
DC corona 直流电晕
DC coupled 直流耦合的
DC coupling capacitor 直流耦合电容器
DC coupling 直流耦合
DC current measuring instrument 直流电流测量装置
DC current transformer 直流变流器，直流电流互感器
DC damping circuit 直流阻尼电路
DC/DC converter 直流-直流电压变换器，直流-直流功率转换器
DC detector 直流检测器
DCD gate(diode-capacitor-diode gate) 二极管-电容-二极管逻辑门
DC dialing 直流拨号
DC(direct current) 直流电
DC directional relay 定向直流继电器
DC disconnecting switch 直流隔离开关
DC(distribution coefficient) 分布系数
DC double bridge 直流双臂电桥
DC(double contact) 双触点
DC(double cylinder) 双汽缸
DC(drag coefficient) 阻力系数
DC(drain cooler) 疏水冷却器
DC drive 直流驱动
DC dump 直流清除
DC dynamometer 直流测功器，直流测功机
DCE(data circuit terminating equipment) 数据电路终结设备
DC electronic motor 直流电动机【整流器供电】
DC erasing head 直流消磁头
DC erasure 直流消除【指记录】
DC excitation 直流励磁
DC excited gas laser 直流激发气体激光器
DC exciter 直流励磁机
DC field coil 直流磁场线圈
DC filter capacitor 直流滤波电容器
DC filter 直流滤波器
DC form factor 整流电流的波形系数
DC gate bias 整流控制极偏置
DC generator 直流发电机
DC graphitization 直流石墨化
DC grounding switch 直流接地开关
DC harmonic filters in converter station 换流站直流滤波装置
DC high potential test 直流高压试验
DC high voltage measurement 直流高电压测量
DC high voltage tester 直流高压试验器
DC high voltage transmission 高压直流输电
DC high voltage 直流高压
DC holding supply cabinet 保持直流供电箱【控制棒驱动机构】
DCI(double cycle inverse) 反相双循环
DCI(ductile cast iron) 球墨铸铁
DC injection braking 直流制动
DC insulator 直流绝缘子
DC integrator 直流积分器
DC inventer welding machine 直流逆变焊机
DCIS(distributed control and information system) 分散式控制与信息系统
DC level 直流电平
DC linear motor 直流直线电动机
DC link bus 直流母线
DC link capacitor 直流链电容器
DC link converter 直流环节变流器，直流母线变流器
DC link voltage 直流母线电压
DC link 直流链，直流母线，直流联络线
DC load 直流负载
DC lube oil pump 直流润滑油泵
DC machine 直流电机
DC magnetic field 直流磁场
DC magnetization curve 直流磁化曲线
DC mains 直流电源
DCM(daily commissioning meeting) 调试（日）例会
DC measurement 直流测量
DC meter 直流电表
DC millivoltmeter 直流毫伏表
DC motor generator 直流电动发动机
DC motor 直流电动机，直流电机，直流马达
DC multimeter 直流多用电表
DC network calculator 直流计算台
DC network 直流网络，直流电网
DC noise 直流噪声
DC no-voltage relay 直流无压继电器
D-coil D形线圈
D controller(derivative controller) 微分调节器，微分控制器
DC-operated 永直流供电的，直流电操作的

DC overcurrent relay 直流过流继电器
DC overvoltage relay 直流过压继电器
DCP (digital computer program) 数字计算机程序
DC pilot relaying 直流辅助继电保护系统
DC potential transformer 直流电压互感器
DC power flow 直流潮流
DC power source 直流电源
DC power supply 直流电源，直流供电
DC power system 直流电力系统
DC power transmission 直流输电
DC power voltage ripple 直流电源电压纹波
DC process(direct conversion process) 直流转化工艺
DCR(design change request) 设计变更请求
DC reactor 直流电抗器
DC reinsertion 直流分量恢复
DC relay 直流继电器
DC reluctance motor 直流磁阻式电动机
DC resistance 直流电阻
DC resistor 直流电阻器
DC resonance charging 直流谐振充电
DC restoration 直流分量恢复
DC reverse power relay 直流逆功率继电器
DC-RF conversion 直流射频转换
DC ringing 直流振铃【电话】
DCS application software package DCS 应用软件包
DCS configuration software package DCS 组态软件包，DCS 配置软件包
DCS(distributed control system) 分布式控制系统，分散控制系统
DC selection 直流选择
DC sensors 直流传感器
DC series generator 串励直流发电机
DC series motor 串激直流电动机
DC shunt generator 并励直流发电机，并激直流发电机
DC side （换流器或变换器的）直流侧
DC signal 直流信号
DC smoothing reactor 直流平滑滤波电抗器
DC source 直流电源
DCS recovery DCS 恢复，DCS 复原，DCS 软件恢复
DC static converter 直流静态换流器
DC supply 直流电源，直流供电
DC surge capacitor 直流冲击电容器
DC switched reluctance motor 直流开关式磁阻电动机
DC switchgear 直流配电装置，直流开关装置
DC system diagram 直流系统图
DC system panel 直流盘
DC system 直流系统
DCS 分散控制系统【DCS 画面】
DC telegraphy 直流电报
DC testing of cables 电缆直流耐压试验
DC three-wire system 直流三线制
DC to DC converter 直流直流变换器
DC torque motor 直流力矩电动机
DC transformer 直流互感器，直流变流器，直流变压器
DC transmission line 直流（输电）线路
DC transmission system 直流输电系统
DC trigger 直流触发器
DC type insulator 直流绝缘子
DC voltage level 直流电压电平
DC voltage measuring instrument 直流电压测量装置
DC voltage transformer 直流电压变换器
DC voltage 直流电压
DC welding machine 直流电焊机，直流焊机
DC withstand voltage level test 直流耐电压试验
DC withstand voltage test 支流耐压试验，直流耐受电压试验，直流耐压试验
DDC(Destination Delivery Charge) 目的港交货费
DDC(direct digital control) 直接数字控制
DDC(downhole dynamic compaction) 孔内动力压实法
DDC system 直接数字控制系统
DDD(domestic direct dialing) 国内直拨（电话）
DD(decimal divide) 十进制除法
D&D(decontamination and decommissioning) 去污和退役
DD(delay driver) 延迟驱动器
D/D(demand draft) 即期汇票【银行】
D-decomposition theory D 域分解理论
DDE(dynamic data exchange) 动态数据交换
D=density 密度，浓度
D=derivative 微分
DDF(digital distribution frame) 数字配线架
D=diameter 直径
DDP(delivered duty paid) 完税后交货【国际贸易术语】
DDRS(digital data recording system) 数字数据记录系统
DDS(Data Display and Processing System) 数据显示与处理系统
DDS(digital display scope) 数字显示器
DDT(dichlorodiphenyl-trichloroethane) 二氯二苯三氯乙烷，滴滴涕
D-D thermo-nuclear reaction D-D 热核反应
DDU(deliverded duty unpaid) 未完税交货【国际贸易术语】
DDU(display and debug unit) 显示及调试剂
DDVF(dimethyl-dichlorovinylphosphate) 敌敌畏
deaccentuator 校平器，频率校正电路，平滑器
deacidifying 脱酸，脱氧的，还原的
deacidizing 还原
deactivation method 去活化方法
deactivation 钝化，惰性化，去活作用
deactivator 减活剂，钝化剂
dead abutment 固定支座，止推轴承，隐蔽式桥台
dead account 死账
dead air insulation 静止空气绝缘
dead air pocket 滞留空气，存气，滞流区
dead air 闭塞空气，停滞空气，静止空气，（扰动气流的）停滞区，气流中死滞区，含有大量一氧化碳的空气
dead area 死区，死滞区，死水区
dead axial 不转轴，静轴，从动轴
dead band error 死区误差

dead band regulator 静区调节器，非线性调节，非线性调节器
dead band 死区，不工作区，无控制作用区，迟缓率，死频带，不灵敏区，静区
dead bar 固定炉条
dead beat algorithm 非周期算法
dead beat galvanometer 不摆电流计，速示电流计
dead beat instrument 速示仪表，不摆式仪表
dead-beat 无振荡的，非周期的，不摆的，无差拍
dead belt 无风带
dead block 固定绕线盘，固定块，缓冲板
dead burning 死角燃烧，烧到黏结，僵烧
dead center 固定中心【机械的】，静点，死点，绝对死点，零位点
dead channel 残遗河段
dead circuit 断开的电路，无效电路，非放射性电路，空路
dead coal storage 长期不动用的煤堆
dead coal 非炼焦煤，不成焦煤
dead coil 无效线圈，线圈不用部分
dead color 底色
dead contact 断开触点【不能接通和断开较大电流】，死接触，开路接点，完全接触
DEA＝deaerator 除氧器
dead earthing 完全接地，牢固接地
dead earth 完全接地，直通地，固定接地
dead-end assembly 耐张绝缘子串组
dead-end champ 耐张线夹
dead-end effect 空端效应，空匝效应
dead-end feeder 终端馈线
dead end flange 法兰管堵
dead-end guy 终端拉线
dead-end insulator 耐拉绝缘子，耐张绝缘子
dead-end loss 死元件损耗，空匝损耗
dead-end main 闷头主管，尽头管，死头管
dead-end pole 止动架，固定支架，终端杆
dead end pressure 死点压力
dead-end section 死区
dead-end steel tower 终端铁塔
dead-end switch 固定端开关，终端开关
dead-end tower 锚塔，耐张塔，终端塔，终端杆塔
dead-end tube 一端封闭的管子
dead end 空端，（管道）闭端，终点，死头非连接端，袋形走道，尽端，死胡同
deadening effect 缓冲作用
deadening 隔声的，隔声作用，隔声，隔声材料，吸声
dead-flat 完全平坦
dead freight 空舱费，亏舱费，空载运费
dead front switchboard 面板无接线的配电盘，正面无接线的配电盘，前面不带电的配电盘
dead front transformer 不露带电部分的变压器
dead front 空正面【配电盘】
dead furrow 堵头沟，死水沟
dead ground water 静地下水，死地下水
dead ground 固定接地，安全接地，无矿岩层
dead halt 安全停机
deadheading 空载运行
dead hole （爆炸后的）残眼，盲孔
dead in line 轴线对中，轴线重合，配置于一直线
dead joint 固定连接
dead knot （木材的）腐节
dead layer （导体的）死层
dead leg 支路端头，盲管段
dead level 完全水平，静态电平，绝对高程，绝对水平
dead-lever trunk 备用段干线，空层中继线
dead-light 固定天窗，木窗遮板
dead lime 失效石灰
deadline for receipt of tenders 接受投标的截止时间
deadline of bid submission 截标期限
deadline of payment for the tax payable 应缴纳税期限
deadline 闲置线路，空线，停电线路，最后期限，截止日期，截止时间，最后期限
dead-load stress 恒载应力
dead-load weight 固定负载重量
dead load 不变荷载，静负荷，固定负载，底载，静重，恒载，静荷载，静态载荷，固定负载，死荷载，自重
deadlock 闭锁，死锁，僵局，停顿
dead loss 空匝损耗，固定损失，纯损失
deadly embrace 死锁，僵局
deadly poison 剧毒
dead main 空载线，无载母线，无载母线
deadman 锚栓，锚桩，临时支撑物，拉杆锚桩
dead melted steel 镇静钢
dead money 闲置资金
dead network 去源网络，无源网络
dead oil （蒸馏石油的）残油，石油渣油
dead plate 无孔板，固定板，炉排中固定挡风板
dead pocket 死滞区，静止区
dead point 死点
dead position 极限位置
dead program 死程序
dead region 死区
dead rock 废石
dead sea depression 死海洼地
dead season 淡季，停滞季节
dead section 死元件，无用段，备用段，无电段
dead short 完全短路
dead slot （电动机的）空槽
dead-smooth file 细纹锉，油光锉
dead-soft annealing 极软退火
dead sounding 隔音层
dead space characteristic 静区特性
dead space correction 死区校正
dead space 死空间，死角，亏舱，死舱位，静区，无信号区，阴影区，不灵敏区
dead spot 死点，死区
dead steam 乏汽，废汽
dead stop process 重点停止法
dead stop 完全停止
dead storage leve 死水位
dead storage of sedimentation 长期沉积，淤积死库容
dead storage pile 长期贮存煤堆

dead storage stockpile 长期贮存煤堆
dead storage 平时不动用的贮存，死库容
dead tank circuit-breaker 接地箱壳式断路器，落地罐式断路器
dead tank oil circuit-breaker 多油断路器，落地罐式油断路器
dead tide 停潮，最低潮
dead time compensation 死区时间补偿
dead time delay 空载时延
dead time effect 死区效应
dead time for pressure regulator 调压阀滞后时间
dead time loss 空时损耗
dead time of governor 调速器不动时间
dead time of reclosing 重合闸的停歇时间
dead time 死区时间，停顿时间，滞后时间，停歇时间，失效时间，空载时间，空等时间，时滞，延迟时间
dead turn 死线匝【钢线绳】，死匝，无效线匝，空匝
dead valley 干谷，死谷
dead voltage 无效电压
dead water end 死水末端
dead water level 死水位
dead water region 死水区，死水区域
dead water space 死水区
dead water zone 死水区
dead water 死水
deadweight cargo 重货
deadweight deflection 自重挠度
deadweight gage 自重压力计
deadweight piston gauge 自重活塞式压力计
deadweight pressure relief valve 自重卸压阀
deadweight pressure tester 自重压力测试表
deadweight safety valve 重锤式安全阀，杠杆式安全阀，自重安全阀
deadweight tester 活塞压力计
deadweight tonnage freight 船舶载重吨运费
deadweight tonnage 载重公吨位，载重吨位，载重吨
deadweight 静重，固有重量，本身自重，静负载，固定负载
dead wind 逆风
dead wire 不载电导线
dead wood 呆木，死木头，枯木
dead work 非直接性生产工作
dead zone 死角区，盲区，静区，死区，死滞区，不灵敏区
dead zone effect 死区效应
dead zone lag 死区滞后
dead zone regulator 死区调节器
dead zone unit 死区单元
dead 死的，静的，无电的，固定连接的，无效的，不灵敏的，断开的，停滞的，失效的，不带电的
deaerated brick 除气砖
deaerated concrete 去气混凝土，脱气混凝土
deaerated feedwater 除氧给水
deaerated water tank 除氧水箱
deaerated water 除氧水
deaerate 除氧，除气
deaerating capacity 除氧能力
deaerating chamber 除氧室，除气室
deaerating condenser 除氧凝汽器
deaerating feed heater 除氧给水加热器，除氧器
deaerating heater 除氧加热器
deaerating hotwell 除氧热井
deaerating plant 除氧装置
deaerating type of cycle 除氧系统
deaeration concrete 除气混凝土
deaeration head 除氧头
deaeration in condenser 凝汽器除氧，真空除氧
deaeration plant 除氧器，除氧设备，除氧站
deaeration time 脱氧时间
deaeration 脱气，除氧，除气【指水中的】，排气，通气
deaerator after-cooler 除氧器附加冷却器
deaerator and coal-bunker bay 除氧煤仓间
deaerator bay 除氧间，除气间
deaerator boiler 除氧器锅炉
deaerator connecting pipe 除氧器连接管
deaerator down-take pipe 除氧器下水管
deaerator effluent 除氧给水
deaerator feed water pump 除氧器给水泵
deaerator feedwater 除氧器给水
deaerator heater 除氧加热器
deaerator level control valve 除氧器水位调节阀
deaerator level 除氧器标高，除氧器水位
deaerator nozzle 除氧器喷嘴
deaerator output 除氧器出力
deaerator overflow pipe 除氧器溢流管
deaerator platform 除氧器平台
deaerator pressure control valve 除氧器压力调节阀
deaerator rated output 除氧器额定出力
deaerator recirculation pipe 除氧器再循环管
deaerator recirculation pump 除氧器再循环泵
deaerator recirculation 除氧器再循环
deaerator shell 除氧器壳体
deaerator storage space 除氧器水箱空间
deaerator storage tank 除氧器储水箱，除氧器水箱
deaerator tank 除氧器箱
deaerator water 除氧水
deaerator 除氧器，除气器，脱氧器
DEAER(deaerator) 除氧器
deafener 消声器
deafening device 隔声装置
deafening 隔声材料，隔声装置，隔声，隔声层
deafness percent 听力损失率
deafness 听力损失
deaf ore 含矿脉壁泥
deairing tower 排气塔
deairing 排气
deal board 松木板
dealership 商品经营特权
dealer 经销商，商人
deal flooring 松木地板
dealkalization 脱碱，除碱
de-alloying 脱合金
deal 交易
deamination 脱氨基
deamplification 衰减

dean air act 空气洁净法
dean area 清洁区
dean flow 无分离流动，无旋流动
dean-up 清洗
deaquation 脱水，脱水作用
dear area 有效截面
dear-water reservoir 清水库
de-ashed coal 经脱灰处理的煤炭
de-ashed fuel 脱灰燃料
de-asher 除渣机
de-ashing 脱灰作用，除灰，清灰
deash 去灰分
death rate 死亡率
death toll 死亡人数
debacle 崩溃
debasement 贬值，降低
debate 辩论，争论
DEB(data extended block) 数据扩充块
DEB(direct energy balance) 直接能量平衡
debenture capital 信用资本，债券资本
debenture stock 借贷股份，信用债券
debenture 退税凭单，无担保债券
debit balance 借方余额
debit customer 借贷户
debit note 借方账单
debit paying ability 偿债能力
debit 借方，记人借方的款项
deblocking of funds 资金解冻
deblocking 解除封锁
debooster 限制器，减压器，限幅器
deboost 减速，制动
deborated 除硼【水堆冷却剂】
deborating demineralizer 除硼离子装置，去硼除盐器
deboration 除硼，去硼
debottleneck 加强生产中的薄弱环节，消除薄弱环节，排除故障
debouch 流出
debouchure 河流出口，山谷出口
debris avalanche 岩屑崩塌
debris basin 拦沙场，拦沙池，沉沙池
debris cone 冲积扇，冲积锥，泥石堆
debris dam 拦沙坝，拦沙谷坊，拦砂坝
debris dump 废渣堆，垃圾堆
debris fall 岩屑崩塌
debris fill 碎石填筑
debris flow 泥沙流，泥石流，岩屑流
debris from demolition 建筑拆除垃圾，建筑碎料
debris guard 拦污装置
debris-laden stream 含砂砾河流
debris retaining structure 拦沙建筑物
debris slide 岩屑滑动
debris storage capacity 沉沙容积，淤积库容
debris stream 岩屑流
debris transportation 岩屑输移
debris 碎片，有机物残余，矸石，清理场地挖出的废物，熔融而塌落的堆芯，堆积物，废弃物，废物碎片，残骸，拆除物，垃圾，碎屑，瓦砾，污物，硬杂物
debt cancellation 取消债务
debtee 债权人，债主
debt in current liabilities 流动负债
debt of honour 信用借贷，信用借款
debtor 借方，欠债人，债务人
debt outstanding and disbursed 已拨付未偿清的债务
debt paying ability 偿债能力
debt service ratio 偿债能力保障倍数
debt service 债务支付
debt 借款，欠款，欠债，债务
debugged program 调试程序
debugger 调试器
debugging on-line 联机程序的调试
debugging phase 排除故障阶段
debugging routine 调试程序，排错程序
debugging statement 调试语句
debugging 排除故障，排错，寻错，程序调整，调试，故障排除
debug on-line 联机调试，在线调试
debug time 调整时间，调试时间
debug 调试，排除（故障），调整，调谐，消除误差
deburring bevel 去刺倒棱
deburring file 去毛刺锉刀
deburring 修整，去毛刺，清除毛边，去毛刺，倒角
debur 去毛刺，修整，去毛刺，清除毛边
DEB 办公楼冷、热水系统【核电站系统代码】
decade adder 十进制加法器
decade attenuator 十进衰减器
decade box 十进电阻箱
decade bridge 十进制拨盘式电桥
decade capacitance box 十进制电容箱
decade counter tube 十进计数管，电子束开关管
decade counter 十进制计数器
decade counting tube 十进制计数管
decade counting unit 十进制计数器
decade count 十进制计数
decade divider 十进分频器
decade frequency divider 十进制分频器
decade multiplier 十进制乘法器
decadence 衰落，衰退
decadent wave 减幅波，阻尼波，衰减波
decade resistance 十进制电阻
decade ring 十进制计数环
decade scaler 十进制计数器，十进制定标器
decade selection 十进位选择
decade signal 十进制信号
decade subtracter 十进制减法器
decade switch 十进制开关
decade trochotron 十进电子转换器
decade tube 十进制计数管
decade 旬，十年，十年间
decadic 十进制的
decalage 差倾角
decanner 去壳机，脱壳机
decanning 去包壳，引出，输出，引线，提取
decantation method 撤离法
decantation 澄清【水处理】，沉淀，撤除
decanter 滗泄器，倾析器，倾注洗涤器
decanting element 澄清部件，倾析部件，溢流

部件
decanting pump 倾析泵
decanting tank 滗析槽
decanting vessel 倾析槽，溢流箱，澄清槽
decanting 事故减压法
decant 倾析
decan 去掉外壳，剥
decarbonater 除碳酸盐装置
decarbonation 除去二氧化碳，除去碳酸
decarbonator 除碳器
decarbonisation 脱碳，除二氧化碳，除碳法
decarbonized steel 低碳钢
decarbonize 脱碳
decarburization 除碳法，钢的脱碳，脱碳
decartbonated river water 除去碳酸盐的河水
decartonization 脱碳
decartonizer 脱碳剂，除碳剂
decartonizing 脱碳
decationize 除去阳离子
decatron counter 十进制管计数器
decatron scaler 十进管计数器
decatron 十进管，十进制计数管，十进制管的
decauville railway 轻便铁路，窄轨轻便铁路
decauville truck 轻轨料车，小斗车，窄轨斗车，窄轨料车
decauville wagon 轻轨料车，窄轨料车
decauville 窄轨铁路
decay chain 衰（蜕）变链
decay characteristic 余晖特性，衰变特性
decay coefficient 衰减系数
decay constant 衰变常数，蜕变常数，衰减常数
decay cooling 衰变冷却
decay curve 衰减曲线，衰变曲线
decay distance 衰减距离
decayed rock 风化岩石
decay factor 衰变系数，衰变常数，衰变因子
decay function 衰减函数
decay-heat cooler 衰变热冷却器
decay heating 衰变发热
decay heat removal circuit 衰变热排出回路
decay heat removal process 衰变热排出过程
decay heat removal system 衰变热排出系统
decay heat removal 衰变热排出，衰变热排除
decay heat system 衰变热系统
decay heat 衰变热，剩（残）余热
decaying catalyst 失活催化剂
decaying conduction current 减幅传导电流，衰减传导电流
decaying current 减幅电流，衰减电流
decaying oscillation 衰减振动
decaying wave 减幅波，衰减流
decaying 衰变，衰落，衰减，衰退，失活
decay law 衰变定律
decay length 衰变长度
decay line 滞留管线，衰变管线
decay modular 衰减模量，衰减系数
decay of power 功率降低
decay of reverberation 混响的衰减
decay of turbulence 湍流衰减
decay of wave 波的衰减，波浪衰减
decay ooze 腐泥
decay parameter 衰减参数
decay path 衰变平均自由程
decay period 衰变周期，蜕变周期
decay power 衰变功率
decay product 衰变产物
decay radiation dose 衰变辐射剂量
decay radiation 衰变辐射，蜕变辐射
decay rate 衰变率，衰减率
decay scheme 衰变图表，衰变图
decay sequence 衰变序列
decay series 衰变系，放射系
decay shutdown heat 停堆衰变热
decay storage tank 衰变贮存箱
decay storage 放射性冷却，衰变贮存
decay store cooling loop 衰变贮存冷却回路
decay store 衰变贮存
decay tank 衰变箱，滞留箱，放射性冷却箱，放射性衰减箱
decay technique 衰减技术
decay time 衰减时间，衰变时间，衰落时间
decay tube 衰变管
decay 衰落，衰减，衰变，腐烂，分解，冷却
deceit 欺诈，诡计
deceive 欺骗
decelerability 减速性能
decelerate 减速
decelerating electrode 减速电极
decelerating flow 减速气流，减速流
decelerating relay 减速继电器
decelerating voltage 减速电压
decelerating 减速性能
deceleration coefficient 加（减）速阻力
deceleration valve 减速阀
deceleration 减速，制动，熄灭，刹车
decelerator 减速器，缓冲装置，延时器，制动器
decelerometer 减速计，减速器
December bonus 年终奖
decennial outage program 十年一次停运计划
decentering 拆除拱架
decentral converter substation 地方换流站，地方变换所
decentralized control 局部控制，分散控制
decentralized data processing 分布数据处理
decentralized heat-supply 分散供热
decentralized management 分散管理
decentralized system 分散式系统
decentralized wind energysystem 分散式风能系统
dechlorination 脱氯作用，脱氯，去氯
decibel-log frequency characteristic 分贝对数频率特性曲线
decibel meter 分贝计
decibels relative to one volt 伏特分贝
decibels relative to one watt 瓦分贝
decibels with reference to one kilowatt 千瓦分贝
decibel 分贝【指音响单位】
deciduous scrub 落叶灌木
decigramme 分克
deciliter 分升
decimal accumulator 十进制累加器

decimal add circuit　十进制加法电路
decimal addition　十进制加法
decimal add　十进制加法
decimal adjust accumulator　十进制调整累加器
decimal attenuator　十进衰减器
decimal base　十进制
decimal-binary conversion　十进制-二进制转换
decimal-binary switch　十进制-二进制开关
decimal-binary　十进制-二进制
decimal carry　十进制进位
decimal classification　十进制分类（法）
decimal-coded digital information　十进制码数字信息
decimal code　十进制码，十进代码
decimal computer　十进制计算机
decimal counter　十进制计数器
decimal counting tube　十进制计数管
decimal digit　十进制数字，十进制数位
decimal divide　十进制除法
decimal floating point　浮点十进制
decimal fraction　十进制小数，十进制分数
decimal location　十进制数位
decimal multiplication　十进制乘法
decimal multiplier　十进制乘法器
decimal multiply　十进制乘法
decimal notation　十进制计数法，十进制表示法
decimal number base　十进制数基
decimal number system　十进制数制
decimal number　十进制数
decimal place　十进制数位
decimal point position　小数点位置
decimal point tamp　十进制小数点指示灯
decimal point　小数点，十进制小数点
decimal presentation　十进制表达式
decimal printing　十进制符号的印刷，十进制符号的打印
decimal processor　十进制处理器
decimal scaler　十进制计数器
decimal storage system　十进制储存系统
decimal subtract　十进制减法
decimal system　十进制
decimal to binary conversion　十进制到二进制转换器
decimal to binary　十进制到二进制
decimal to fixed binary translation　十进制到固定二进制转换
decimal　小数点的，十进制的，小数
decimeter mixer　分米波段混频器
decimeter radio communication　分米波无限通信器
decimeter range　分米波段
decimeter　分米
decimetric feeder　分米波段馈线
decimetric wave　分米波
decineper　分奈，丝米，十分之一毫米
decipherer　译码器
deciphering machine　译码器，译码机
decipher　译码
decision analysis　决策分析
decision block　判定功能块
decision box　判定框图
decision criteria　决策准则
decision element　判定元件，计算元件
decision evaluation and logic　判别鉴定及逻辑
decision feedback system　判定反馈系统
decision for bids choosing　决标
decision function　判定函数
decision instruction　判定指令，判别指令
decision integrator　判断积分器
decision logic translator　判定逻辑翻译程序
decision machine　决策计算机
decisionmaker　决策人（者）
decision-making body　决策机构
decision-making management　决策管理
decision-making process　决策程序，决策过程
decision-making system　决策系统
decision-making　决策
decision model　决策模型
decision on contract award　决标
decision problem　决策问题，判断问题
decision procedure　决策方法，决策过程
decision situation　决策情况
decision table　判定表
decision technique　决策技术
decision theory　决策论
decision tree　决策树
decision under conflict　竞争决策
decision under risk　风险型决策
decision　决策，决定，决议
deck beam　上承梁，甲板梁
deck bridge　上承桥
deck construction　地板构造，平顶结构，桥面构造
deck dam　平板坝
decked explosive　分层装炸药
deck floor　平楼板
deck form　平台式（钢）模板
deck girder steel dam　板梁式钢坝
deck girder　上承梁
deck house　舱面室
decking　桥面板，铺面，车行道，盖板
deck joint　面板接缝
deck plate girder bridge　上承板梁桥
deck plate　下隔板，盖板，底板
deck reinforcement　面板钢筋
deck risk　舱面险
deck roof　平台式屋顶，平屋顶，晒台
deck sash　顶棚窗
deck slab　桥面板，上承板
deck span　桥面跨度
deck spillway　盖板溢洪道
deck stringer　桥面纵梁
deck truss　上承桁架
deck　底板，盖板，层，层楼，盖，舱面，台面，甲板，面板，平台，控制板，平屋顶
decladding　去壳，脱壳【燃料元件】
declarant　报关人
declaration for exportation　出口报单
declaration for importation　进口报关单
declaration form　申报表，申报单
declaration inwards　进口报关单
declaration of avoidance　宣告无效

declaration of outwards 出口报关单
declaration of shipment 装船通知
declaration 申报，声明，说明，说明语句，启运通知，报单，报关
declarative macro-instruction 说明宏指令
declarative statement 说明语句
declarator 说明符
declare at the custom 报关
declare bankruptcy 宣告破产
declared capital 法定资本，设定资本
declared efficiency 标称效率
declare null and void 宣布无效
declare 申报，声明，宣告
declassification 解密，降低保密等级
declassified document 取消保密的文件
declinate 差角
declination arc 磁偏角弧【罗盘仪上的】
declination variation 倾角变化
declination 倾斜，下倾，偏差，偏角，磁偏角，赤纬，宣告
declinator 磁偏计
decline an offer 不接受报价
declined conveyor 倾斜式输送机
decline of ground-water level 地下水位的降落
decline of water level 水位下降
decline of water table 水位降低
decline phase 衰退期
decline 下倾，低落，衰退，没落，下降
declining balance depreciation 衰减平衡折旧
declining 低落，下倾
declinometer 测斜计，磁偏计，方位计
declivity 倾斜度，坡度，倾斜，斜面
declogging 排出淤积物
declutching 离合器分离
declutch 脱开
decoat 除去涂层
decocell 去污小室
decodable 可译的，可解的
decoded signal 译码信号
decoder circuit 译码电路
decoder matrix circuit 解码矩阵电路
decoder matrix 译码器矩阵
decoder 译码器
decode 译码，解码
decoding algorithm 译码算法
decoding circuit 译码电路
decoding device 译码器
decoding gate 译码门
decoding information 译码信息
decoding memory drive 译码存储驱动器
decoding network 译码网络
decoding relay 译码继电器
decoding scheme 译码线路图，译码电路
decoding theorem 译码定理
decoding with relays 继电器译码
decoding 解码，译码
decohall 去污间
decoking 脱焦炭，除焦
decolorant 脱色剂
decoloration 脱色作用
decoloriser 脱色剂
decolorization 脱色
decolorizer 脱色剂
decolorizing agent 脱色剂
decolorizing carbon 脱色碳
decolourant 漂白剂
decolourisation 漂白，脱色，脱色作用
decolour 脱色，漂白
decommissioning 退役【电站】
decommutation 反互换
decomposable 可分析的，可分解的
decomposed granite 风化花岗岩
decomposed rock 风化岩石
decompose 分解，分析
decomposition constant 分解常数
decomposition force 力的分解
decomposition heat 分解热
decomposition of erection 解体安装
decomposition product 分解产物
decomposition reaction 分解反应
decomposition voltage 分解电压，电解电压
decomposition wastewater 废水分解
decomposition 分析，分解，溶解，腐烂，风化
decompounded motor 差复励电动机
decompound 分解，再混合，差复激的
decompression chamber 减压室，降压室
decompression device 减压装置
decompression 降压，减压，分解，卸压
decompressor 减压器
decompress 降压，解压
deconcentrator 分散器，去浓器
decontaminability 可去污性
decontaminable floor covering 可去污的地面涂料
decontaminable flooring 可去污地面
decontaminable painting 可去污的涂料，可去污涂料
decontaminable paint 防污染漆，去污染漆，防污漆
decontaminable plastic coating 可去污的塑料涂层，可去污塑料涂层
decontaminant 去污剂，纯化剂
decontaminate 净化，清除污染，去污，去掉放射性
decontaminating agent 去污剂
decontaminating apparatus 去污仪器
decontaminating column 净化柱，去污柱，提纯柱
decontaminating equipment 去污设备
decontaminating reagent 去污剂
decontaminating solution 去污溶液
decontamination acid tank 去污酸罐
decontamination agent 去污剂
decontamination and decommissioning 去污和退役
decontamination and disposal 去污和处置
decontamination area 去污区，消除污染区域
decontamination cell 去污小室
decontamination centre 去污中心
decontamination chamber 净化室，去污室
decontamination cycle 净化循环
decontamination device 去污设备，去污设施
decontamination drains arising 去污疏水生成量
decontamination drains pump 去污疏水泵

decontamination drains tank 去污疏水箱
decontamination drains 去污疏水
decontamination equipment 去污设备
decontamination facility 去污设施
decontamination factor 去污因子，去污系数
decontamination fluid 去污液体，消除污染流体
decontamination hall 去污间
decontamination index 净化指数，去污指数
decontamination nozzle 去污喷嘴
decontamination pit 去污槽
decontamination plant 净化装置，去污装置
decontamination rinse tank 去污漂洗箱，去污冲洗水滞留箱
decontamination room 净化室，去污室
decontamination sheet 去污工作单
decontamination shop 去污车间
decontamination shower room 去污淋浴室
decontamination sink 去污槽
decontamination solution tank 去污溶液箱
decontamination solution 去污溶液
decontamination station 去污站
decontamination system filter 去污系统过滤器
decontamination system pump 去污系统泵
decontamination system water heater 去污系统热水器
decontamination system 去污系统
decontamination tank 去污水箱
decontamination trailer 去污拖车
decontamination unit 去污装置
decontamination 去污循环【后处理】，去污染【如除放射性】，去污，去污室，净化（作用），清除放射性污染
decontrol 解除控制
decoration standard 装饰标准
decoration 装潢，装饰
decorative architecture 装饰建筑
decorative art 装饰艺术
decorative concrete 装饰用混凝土
decorative lighting 装饰照明
decorative sky-light 饰窗
decorative wall 装饰墙
decorrelation 去相关，抗相关，解相关
decorrelator 去相关器，解相关器，解联器
decosystem 去污系统
decoupled mode （电站）孤岛运行方式
decoupled subsystem 解耦子系统
decouple 去耦，退耦，离合器脱开，断开联系
decoupling circuit 去耦电路
decoupling condenser 去耦电容器
decoupling filter 去耦滤波器
decoupling of loops 环路的去耦
decoupling rate factor 解联速度因子
decoupling resistor 去耦电阻器
deco 装饰，装饰品
DECR(decrease) 下降
decrease of the enthalpy 焓降
decrease 减去，扣除，减少，下降
decreasing amplitude 降幅，减幅
decreasing failure rate 递减故障率
decreasing vibration 减振，降振
decree 规程，法令，法规
decrement field 减量字段
decrement in reactivity 反应性减少量
decrement of damping 阻尼减少
decrement of oscillation 振荡衰减量
decrement 减量，减少，减缩，衰减量，衰减率，递减，减幅，减缩率，减少量
decremeter 减缩量计，减幅器，衰减计，减幅仪
decrepitation 破裂，烧爆，爆裂
decrosslinkage 解联
decrustation 表面净化，剥皮，除去沉积物
decryption 译码，解释编码
decrypt 译码，解码，解释
DECR 减少，降低【DCS 画面】
decussate 交叉成十字形，交叉着的
dedendum angle 齿根角
dedendum cone 圆锥齿轮啮合
dedendum line of contact 齿根接触线
dedendum line 齿根线
dedendum 齿根，齿根高
dedeuterization tank （重水堆）除氘箱
dedeuterization 除氘
dedeuterizing 除氘
dedicated 专用的
dedicated car 专用车
dedicated circuit 专线
dedicated line 专用通信线
dedicated platforms 专用平台
dedicated reactor auxiliary cooling system 专用堆辅助冷却系统
dedicated storage 专用存储器
dedicated system 专用系统
dedicated task module 专用任务模件
DE(differential expansion) 不均匀膨胀，差胀
deduce 引出，推断，演绎
deductible clause 减扣赔偿条款，免赔条款，自负责任条款
deductible excess 免赔额
deductible franchise 绝对免赔率
deductible limit 免赔额限制【保险】
deductible tax 可扣除税款，可抵扣的税金
deductible 可扣除额，可减额，可减税款，（保险的）免赔额，可扣除的
deduction 推断，推论，扣除，减少，扣除额
deduct 减去，扣除，抵扣，减掉
deduster 除尘器
dedusting pump 除尘水泵
dedusting 除尘
dedust transformer 电除尘变
dedust 除尘，除灰
deed of arrangement 财产转让协定，和解协议，债务整理证书，债务清算契据，调解契约
deed of assignment 转让契约
deed of indemnity 赔偿契约
deed of mortgage 典契
deed of partnership 合伙契约
deed 契约，契据，协议书
de-electrifying 去电
de-electronation 氧化作用，去电子作用
de-emphasis network 去加重网络
de-emphasis 去加重，削弱

de-emphasizing filter 去加重滤波器
de-emphasizing network 去加重网络
deenergization 去能，去激励，失励，断开，释放，断电
de-energized equipment 不带电设备
de-energized state 失励状态，断电状态
de-energized to operate 断电致动作
de-energized 除去电源的，不赋能的，去激励的，不带电的，不带电，失励，断电，去能，去激励，断开
de-energizing circuit 去励磁电路，去激电路
de-energizing 断电
deentrainment column 夹带物去除柱，除雾沫柱
deentrainment device 除雾沫装置，除夹带设备
deentrainment filter 除夹带过滤器，除雾沫过滤器
deentrainment 去夹带，防止夹带，除雾沫
deep affset ring slogging spanner 高颈梅花敲击扳手
deep bar effect 深导条效应，深槽效应
deep bar motor 深槽鼠笼电动机
deep bar 深鼠笼条，加强肋骨
deep basin 深池
deep bay 深水湾
deep beam weldment 焊接加强筋【堆芯内件】
deep beam 厚梁，浅梁，深梁
deep-bed filter 厚床过滤器
deep bin 深仓
deep blasting 深层爆炸
deep boring 深层钻探，深钻孔
deep-cage-bar rotor 深鼠笼转子
deep channel tray 深的槽形托架
deep circulation 深层环流
deep compaction 深层压实
deep consolidation 深层固结，深层加固
deep current 深层流
deep cut 深挖
deep densification 深层加固
deep dimension picture 深维图像
deep discharge 过放电
deep diving 深水潜水
deep dose equivalent index 深部剂量当量指数
deep-draft channel 深水航道
deep-draft harbour 深水港
deep-draft navigation 深水航运
deep-draft vessel 深水货轮
deep drawing ability 深拉效率，深拉能力
deep drawing sheet 深拉薄板
deep earth burial 深埋
deep earthquake 深发地震，深震
deep earth test 深度侵蚀试验
deeper bed 深床
deep etching 深蚀，宏观腐蚀
deep excavation 深挖方
deep fill 深填方，深填土
deep filter 深度过滤器
deep flaw 深处缺陷
deep flexural member 深受弯构
deep floor 加强肋板
deep flow 深层流
deep fluidized bed 深流化床
deep-focus earthquake 深发地震，深源地震
deep foundation pit 深基坑
deep foundation 深基础
deep freezing 深层冻结
deep geological disposal 地下深埋处置，深部地质建造处置
deep-groove ball bearing 双槽滚珠轴承
deep groove 深槽
deep ground grid 深的接地网
deep hole blasting 深孔爆破
deep hole drill 深孔钻机，深钻孔
deep holography 深层记录全息术
deep intake structure 深式取水构筑物
deep karst 深层岩溶
deep layer bed 高层，高床
deep lead-in 深入引进
deep level 深层
deep loading vehicle 平板载重车，低架车
deep-lying foundation 深埋基础
deep manhole 深井式进入孔
deep measuring stick 测深杆
deep mining 深井开采
deep mixing method 深层搅拌法【软土地基加固的】
deep mixing 深层搅拌（法）
deep opening tubular socket wrench 长开口管套筒扳手
deep penetration electrode 深溶焊条
deep penetration welding 深熔焊
deep penetration 深层渗透
deep percolation 深层渗漏，深层渗透
deep piling 深层打桩法
deep pit sewage pump 深井污水泵
deep pit 深坑
deep pool 深潭
deep potential 深电位
deep-processing 深加工
deep prospecting 深层钻探
deep pumped well 深泵井
deep ring spanner 直头梅花扳手
deep rotor-bar 转子深鼠笼条
deep sea disposal 深海废物处置
deep-sea lead 深水测深锤
deep-seated grouting 深层灌浆
deep seepage 深层渗流
deep slide 深层滑动
deep slot 深槽
deep socket wrench 长套管型套筒扳手
deep soil 深层土
deep sounder 测深仪
deep sounding apparatus 深水测深仪，深水探测器
deep-space nuclear power 太空核电源
deep stall region 深度失速区
deep stall 严重失速
deep-throat indenter 深颈压头【硬度试验】
deep trench excavation 深沟挖掘
deep trough 深槽
deep tube well 深管井
deep underground disposal 地下深埋废物处置，深地下废物处置

deep valley 深谷
deep water berth 深水泊位
deep water clapotis 深水重复波
deep water current 深层水流
deep water dock 深水码头
deep water hydrophone 深水探听器
deep water intake 深层取水
deep water pathfinder 深水海洋测量船
deep water platform 深水承台
deep water port 深水港
deep water quay 深水岸壁，深水横码头
deep water table 深地下水位
deep water terminal 深水码头
deep water wave 深水波
deep water wharf 深水码头
deep water zone 深水区
deep water 深井水
deep weathering 深层风化
deep well elevator 深井吊梯
deep well injection 深井注入【指注入废物】
deep well jet pump 深井喷射泵
deep well method 深井法
deep well plunger pump 深井柱塞泵
deep well pump house 深井泵房，深井水泵房
deep well pumping unit 深井抽水机
deep well pumping 深井抽水
deep well pump 深井（水）泵
deep well reciprocating pump 往复式深井泵
deep well turbine pump 深井渗透泵，深井涡轮水泵
deep well turbine 深井涡轮
deep well 深水井，深井
deep 深槽，深水段，深渊
deexcitation system 灭磁系统
deexcitation voltage 灭励电压，灭磁电压
deexcitation 去激，灭磁，去激作用，消励，灭励，反励，去激励，去激活，去励磁
deexciting 去激励，去激活
defacement 磨耗，磨损，磨减
defacing 表面碰伤，表面损伤
defamation 诽谤
defaulter 缺席人，违约人
default in investment 投资违约
default notice 违约通知
default of contractor 承包商违约
default of employer 业主违约
default of obligation 不履行债务
default of payment 拒绝付款
default on construction cost 拖欠工程款
default party 违约方
default rule 缺省规则
default value 缺省值
default 不履行责任，违约，缺陷，不履约，拖欠
defeated party 败诉方
defeat 挫败，宣告无效，作废，失败，废除
defecate 提净，澄清
defecation 澄清作用
defecator 澄清槽，澄清器，滤清器
defecometer 探伤仪
defect cluster 缺陷群
defect depth 缺陷深度
defect detecting test 探伤检查，缺陷检查
defect detection 探伤
defect detector 探伤仪
defect evaluation 故障评价
defect-free 无缺陷的
defect fuel element 有缺陷的燃料元件
defective channel cut-out 切除故障通道
defective contact 触点缺陷
defective coupling 耦合不良
defective fuel assembly 破损燃料组件，有缺陷燃料组件
defective fuel rod cladding tube 破损燃料棒包壳管
defective goods 次货，有毛病的货物，缺陷产品
defective index 废品率
defective insulation 不良绝缘
defective material 材料缺陷，有缺陷的材料
defective parts 有毛病的零部件，有缺陷的部件
defective thread zone 损坏的螺纹区
defective works 有缺陷的工程，有缺陷的工作
defective 有缺陷的，损坏的，次品
defect liability period 合同的缺陷责任期，故障修理责任期，缺陷责任期
defect of substance 实质性缺陷
defectogram 缺陷图
defectoscope 探伤仪
defect report 缺陷报告
defect resolution 缺陷分辨率
defects assessment 缺陷评估，缺陷评价
defects connection certificate 缺陷改正证书
defects correction period 缺陷改正期
defects in timber 木材缺陷
defects liability certificate 缺陷责任证书
defects liability period 保修期，缺陷责任期
defects liability 对缺陷的责任
defects notification period 缺陷通知期
defect treatment 缺陷处理
defect value 亏值
defect visibility 缺陷可见度
defect 缺陷，缺点，故障，不足，探伤，错误，事故，毛病，欠缺
defence-in-depth concept 纵深防御概念
defence 防护，防御
defendant 被告
defend 辩护
defense in depth 纵深防御
Defense Nuclear Agency 国防部核子局【美国】
defer payment 延期付款
deferrable outage 可延迟停运
deferred addressing 递延地址，延迟地址
deferred assets 递延资产，延期资产
deferred charges 延期费，预付款，待摊费用，递延借项
deferred construction 缓建
deferred credits 递延贷项
deferred decision concept 延缓决策概念【乏燃料后处理】
deferred delivery 迟延交货，延交合同
deferred entry 延迟输入，延迟入口
deferred income tax 递延所得税

deferred maintenance 逾期维修
deferred outage 延迟停运，第 4 类非计划停运【可延至较长时间后的停运】
deferred payment contract 递延付款合同
deferred payment credit 延期付款信用证
deferred payment 分期付款，延期付款
deferred processing 延期处理
deferred project 缓建项目
deferred restart 延迟重启动，延迟再启动
deferred 延期的，延迟的
deferrization facilities 除铁设施
deferrization 除铁，脱铁
defer 延期，延迟，拖延
defibrillator 电震发生器
deficiency curve 亏格曲线
deficiency of rain 缺雨
deficiency 不足，不足额，缺乏，缺陷，亏数，差量，不足之处，亏损，欠缺
deficient 不充分，缺乏的，不足的
deficit reactivity 反应性亏损
deficit value 亏值
deficit 赤字，亏空，亏损，逆差
defiled 污损的
defile 峡谷，隘道
define 确定，明确，定义，规定，限定
defining fixed points 定义固定点
defining standard thermometer 定义的标准温度计
definite advice 确切通知
definite design report 详细设计报告
definite integral 定积分
definitely 明确的
definite minimum time-limit relay 最小定时限继电器
definite project plan 详细工程设计
definite proportion 定比
definite purpose circuit breaker 专用断路器
definite purpose motor 专用电动机
definite purpose relay 专用继电器
definite quantity 定量
definite time delay unit 固定时间的延时单元
definite time-lag switch 定时滞开关
definite time-lag 定时滞
definite time relay 定时限继电器
definite time 定时，定时限
definite undertaking 明确的承诺
definite value 定值
definite views about 关于……明确的观点
definite 一定的，确切的，明确的，限定的
definition design 技术设计
definition domain 定义域
definition of export quotations 出口价格条例
definition phase 技术设计阶段
definition point correspondence table 定义点对应表
definition 确定，定义，清晰度，分辨率，限定
definitive schedule 最终进度
deflagrability 爆燃性
deflagrate 爆燃，突然燃烧
deflagration 爆燃，突然燃烧，爆燃过程
deflate 抽气，放气，放水
deflation basin 风蚀盆地
deflation erosion 风蚀
deflation opening 放气口
deflation plane 风蚀面
deflation valley 风蚀谷
deflation vent 放气口
deflation 放气，抽气，收缩，风蚀，吹蚀，通货紧缩
deflected air 偏流空气
deflected beam 偏转束
deflected pile 偏位桩
deflecting 使偏斜
deflecting angle 偏转角
deflecting baffle 折流板，反射板
deflecting bucket 挑流鼻坎，挑流屏斗
deflecting electrode 偏转电极，致偏电极
deflecting field 偏转场，致偏场
deflecting force 偏转力，挑流反力
deflecting gate 导向挡板
deflecting magnet 致偏磁铁，偏转磁铁
deflecting needle nozzle 偏流针形喷嘴
deflecting plate 偏转板
deflecting surface 折射面
deflecting system 致偏系统，偏转系统
deflecting voltage 偏转电压，致偏电压
deflectional instrument 偏转式仪表
deflectional stiffness 抗弯刚度
deflection anemometer 偏转风速计
deflection angle method 偏角法
deflection angle 偏转角，转折角，变位角，偏角
deflection arch 折焰角，折焰拱
deflection at break 裂断变位
deflection at rupture 破裂变位
deflection baffle 折流挡板
deflection basin 偏斜盆地
deflection coil 偏转线圈
deflection computer 偏置计算机，前置量计算机
deflection criterion 变位准则，偏移准则
deflection curve 挠度曲线，变位曲线
deflection distortion 偏转失真，致偏失真
deflection field 偏转场
deflection force of earth rotation 地球自转偏向力
deflection force 偏转力
deflection gauge 偏转计，挠度计
deflection magnitude 偏离幅度
deflection method 位移法，变位法
deflection multiplier 偏转倍增器
deflection nozzle 偏转喷嘴
deflection observation 挠度观测
deflection of sound 声波的偏转
deflection period 偏转持续时间
deflection plate 折向板，偏转板
deflection potentiometer 偏转电势计，偏转电位计
deflection rod 偏转杆
deflection scale 量角器标度，偏转标度
deflection separator 偏转分离器，折流分离器
deflection spectrum 偏转道，挠度谱
deflection test 挠曲试验，弯曲试验
deflection theory 变位理论
deflection transformer 偏转变压器

deflection tube　导向管
deflection-type cathode-ray tube storage　偏转射束阴极射线管存储器
deflection-type storage tube　射线偏转存储管
deflection　折射，折流，挠度，下垂度，挠曲，偏差，偏斜，偏转，转折角，变位，变形，下沉，偏向，偏移，折角
deflectometer　挠度计，折角
deflector apron　导向挡板
deflector baffle　折流挡板
deflector bucket　鼻坎反弧段，挑流鼻坎，消力戽
deflector hood　挑流板
deflector plate　导向板，导流板
deflector screen　翻转筛板，翻版
deflector　挡液圈，导向装置，导流板，导向板，导向器，偏向器，偏转板，转向装置，折流板，折转器，致偏板
deflectoscope　挠度计，缺陷检查仪
deflect　偏转，偏斜，挠曲，折射
deflegmate　分凝，分馏
deflexion　折射，折流，挠度，挠曲，偏差，偏斜，偏转，转折角
deflocculanting agent　抗絮凝剂，反絮凝剂，悬浮剂
deflocculantion　反絮凝，分散作用
deflocculant　反絮凝剂，分散剂，抗絮凝剂
deflocculated colloid　不凝聚胶体
deflocculating agent　反絮凝剂，悬浮剂，分散剂
deflocculation　反絮凝作用
defluidization　流态化作用停滞
defluidized cap　失流泡罩
defluidized region　失流区
defluidized　不流化的
defluorination　脱氟作用
defoamant　消泡剂，去泡剂
defoamer　去泡沫剂，消沫剂
defoaming agent　消泡剂，去泡剂
defoaming tank　去泡沫箱
defoam　去泡沫
defocussed correction　散焦修正
defogging　扫雾
deforestation　采伐森林，毁林，森林砍伐
De-Forest coil　蜂房式线圈，蜂巢线圈
deformability　可变形性，变形能力
deformable locking screw　可变形锁紧螺钉
deformable particle　可变形颗粒
deformation analysis　形变分析
deformation at failure　破坏时变形
deformation band　变形带，变形范围
deformation boundary condition　边界变形条件
deformation crack　变形裂缝
deformation curve　变形曲线
deformation drag　变形阻力，森林砍伐，流态变形拖曳力
deformation erosion　变形侵蚀
deformation fracture　变形破裂
deformation gauge　变形计
deformation gradient　变形梯度
deformation joint　变形缝
deformation limit　变形极限
deformation measurement　变形测定，变形测量
deformation meter　变形测定计
deformation modulus　形变模量，形变模数
deformation observation　变形观测
deformation of river bed　河床溃变
deformation of wave　波的变形
deformation rate　变形率，变形速度
deformation ratio　变形比
deformation stress　变形应力
deformation structure　变形构造，形变结构
deformation temperature　变形温度
deformation theory of plasticity　塑性变形理论
deformation thermometer　变形温度计
deformation under load　载荷变形
deformation　畸变，变形，失真
deformed bar　螺纹钢筋，竹节钢筋，变形钢筋
deformed metal plate　变形金属板
deformed rebar　螺纹钢筋，竹节钢筋，变形钢筋
deformed surface of turbine cylinder flange　汽缸法兰结合面变形
deformed tie-bar　变形拉杆
deformer agent　除沫剂
deformeter　变形测定计，变形计
deforming agent　除沫剂
defraud　骗取，诈骗
defray　支付，付给
defroster　防冰冻装置
defrosting system　除霜系统，融霜系统
defrost　解冻
defruiting　反干扰，异步回波滤除
defunct company　已停业公司
degased solution　除气的溶液，脱气溶液
degased steel　镇静钢
degaser　除气剂，除气器
degasification line　除气管道
degasification tank　除气箱
degasification　除气，除氧，脱气，除气作用
degasified water tank　除碳水箱
degasifier extraction pump　除气器抽吸泵
degasifier supply line　除气器供应管线
degasifier unit　脱气装置，除气装置
degasifier　除气剂，吸气剂，除气器，除氧器
degassed water　除气水
degasser　脱气器，除气器，除氧器，脱气剂
degassing enhancement　表面除气强化法
degassing equipment　脱气设备
degassing phase　脱气状态【高温气冷堆石墨】
degassing transformer　真空泵电源变压器
degassing　除气，放气，脱气，抽气
degas　除气，抽真空，除氧
degausser　去磁电路，去磁器
degaussing cable　消磁电缆
degaussing coil　退磁线圈
degaussing field　去磁场
degaussing　去磁，消磁，退磁
DEGB(double-ended guillotine break)　双端剪切破裂
DEGC(degrees celsius)　摄氏度
deg cent(degrees centigrade)　摄氏度数
DEG＝degree　度

degelatinize 脱胶
degeneracy function 衰变作用
degeneracy 变质，退化，蜕化，衰退
degenerated bankruptcy 衰退倒闭
degenerate mode 简并模式
degenerate semiconductor 衰减的半导体
degenerate specimen 简并样品
degenerate 退化，简并
degeneration feedback amplifier 负反馈放大器
degeneration 退化，衰减，简并化，负反馈，变质
degenerative feedback 负反馈
degenerative feed 负反馈，负馈入
degenerative 衰退的，退化的，负的，变质的
degeroite 硅铁石
deg F(degrees Fahrenheit) 华氏度数
deg K(degrees Kelvin) 开氏度数，绝对温度
degradation below dam 坝下河底刷深
degradation degree 冲刷程度
degradation factor 退化因数，递降系数，降格因素
degradation failure 衰变故障，退化失效
degradation loss 劣化损失
degradation of water level 水位下降
degradation reaction 降解反应
degradation regression 退化
degradation terrace 剥蚀阶地
degradation 降低【品质，等级】，退化，衰变，降解，分解，裂解，冲刷，恶化，解列
degraded core accident 堆芯性能恶化事故
degraded D_2O purification system 降质量重水净化系统
degraded D_2O system 降质重水系统
degraded flux thimble 恶化的通量测量套管
degraded heavy water 浓度下降的重水，降质的重水
degraded neutron 慢化了的中子，损失部分能量的中子，非新生中子
degraded redundancy 降低等级的冗余度，冗余度降低
degraded voltage protection system 电压下跌保护系统
degrading 刷深程度，消落
degration 降级
degreasant solution 脱脂溶液，去油溶液
degreasant 脱脂剂
degreaser 除油器，去污剂，脱脂剂，脱脂装置
degrease 脱脂，去油，除油，去垢，去污
degreasing agent 脱脂剂
degreasing compound 脱脂剂
degreasing solution 去油溶液
degreasing 脱脂，去油，除油，去垢，去污
degree Beaumy 波美度
degree Celsius 摄氏温度【℃】
degree Centigrade 摄氏温度【℃】
degree-day during heating period 采暖期度日数
degree-day factor 度日因子，度日因数，融雪度日系数
degree-day model 度日模型
degree-days of heating period 采暖期度日数
degree-day 度日【冬季采暖计算热负荷的特殊单位】，户外每日平均温度单位，温度单位，测定温度之单位，日度
degree Fahrenheit 华氏温度【℉】
degree Kelvin 开氏温度【热力学绝对温标】，绝对温度【K】
degree of accuracy 精确度，准确度
degree of adaptability 配合度
degree of admission 进汽度
degree of approximation 近似程度，逼近度
degree of asymmetry 不对称度，直流分量与对称分量峰值之比
degree of balance 平衡度
degree of blackness 黑度
degree of blame 过失程度
degree of cavitation 汽蚀度
degree of coalification 煤化程度
degree of commutation （直流电机的）换向等级
degree of compression 压缩度，压实度
degree of confidence 可信赖度，置信度
degree of consistency 均匀度
degree of consolidation 固结度
degree of contamination 污染度
degree of contraction 收缩量
degree of convergence 收敛度
degree of correlation 相关度
degree of corrosion 腐蚀程度，腐蚀度
degree of coupling 耦合度
degree of curvature 曲度，曲率，置信度
degree of demonstration 证实程度
degree of density 稠密度，密实度
degree of deviation 偏差度，偏转度
degree of differential equation 微分方程的阶次
degree of discomfort 不舒适度
degree of dispersion 分散度，弥散度
degree of dissociation 离解度，电离度
degree of distortion 失真度，畸变度
degree of disturbance 扰动程度
degree of dryness 干度
degree of dustiness 含尘度
degree of eccentricity 偏心度
degree of elastic homogeneity 弹性均匀程度
degree of electrolytic dissociation 电离度
degree of enrichment 加浓度，浓缩度，富集度
degree of excellence 优惠程度
degree of exhaustion 抽空度
degree of filtration 过滤度
degree of fineness 细度
degree of finish 表面光洁度，表面粗糙度，光洁度
degree of fouling 脏污程度
degree of freedom 自由度
degree of fullness 充满度
degree of hardness 硬度
degree of humidity 湿度
degree of hydraulic misadjustment 水力失调度
degree of illumination 照度
degree of inclination 倾斜度
degree of induration 硬化度
degree of inferiority 低劣程度
degree of ionization 离解度，电离度

degree of irregularity　不规则程度，不规则度，不均衡度
degree of long range order　长程序级
degree of metamorphism　变质程度
degree of mineralization　矿化度
degree of mixedness　混合度
degree of moderation　慢化度
degree of modulation　调制度，调制深度
degree of moisture　湿度
degree of plainness for wall surface　墙面平整度
degree of pollution　污秽等级，污染程度
degree of polynomial　多项式的阶次
degree of porosity　孔隙度
degree of protection afforded by enclosure　外壳能承受的保护程度
degree of protection　防护等级
degree of purification　净化程度
degree of reaction　反动度
degree of regulation　调整精度，调准度，调节度
degree of reliability　可靠度
degree of risk　风险度
degree of safety　安全度，安全程度，安全系数
degree of saturation　饱和度
degree of screening　屏蔽度
degree of security　安全度
degree of segregation　分离等级，分离程度
degree of selectivity　选择性
degree of sensitivity　灵敏度
degree of slope　坡度
degree of sophistication　（技术的）成熟程度
degree of sorting　分选程度
degree of stability　稳定度
degree of subcooling　过冷度，欠热度
degree of superheat　过热度
degree of supersaturation　过饱和度
degree of swelling　膨胀量
degree of treatment　处理程度
degree of turn　转角
degree of twist　偏扭度
degree of unbalance of installation　装置不平衡度
degree of unbalance　不平衡度
degree of uniformity　均匀度
degree of unsaturation　不饱和度
degree of unsolvability　不可解度
degree of vacuum　真空度
degree of wall porosity　壁开孔度
degree of ware　磨损度
degree of weathering　风化度
degree of wetness　湿度
degree of wind sensibility　风敏感度
degree scale　分度尺
degree stability　稳定度
degree　度，次，程度，等级，次数
degression　下降
DEG　核岛冷冻水系统【核电站系统代码】
DEH control system　数字电液控制［调节］系统【汽轮机】
DEH(digital electro-hydraulic)　数字电液控制［调节］【汽轮机】
DEH governing system　数字电液调节系统
DEH measurement system　数字电液测量系统
DEH system for boiler feedwater pump turbine　锅炉给水泵汽机的数字电液控制系统，小机数字电液控制系统，给水泵小汽机数字电液控制系统
dehumidification process　减湿过程
dehumidification　去湿，减湿，去潮，除湿
dehumidifier　除湿器，减湿器，干燥器，降湿机，除湿机，去湿器
dehumidifying cooling　减湿冷却
dehumidifying effect　减湿效果
dehumidifying heater　除潮加热器
dehumidifying system　除湿系统
dehumidify　除去湿气，去湿，减湿，使干燥，除湿
dehydrant　脱水剂
dehydrated alcohol　无水酒精
dehydrated lime　生石灰，氧化钙
dehydrater　除水器
dehydrate　脱水，脱水物，使脱水，除水
dehydrating breather　脱水呼吸器，吸湿器
dehydration　脱水（作用），除水，干燥，去湿
dehydrator　脱水器，除水器，烘干机，脱水机，脱水设备
dehydrogenating agent　脱氢剂
dehydrogenation　脱氢处理，脱氢，去氢
dehydrolysis　脱水
deicer　防冰器，防冰冻装置
deincrustant　除垢剂
deion air circuit breaker　去电离空气断路器
de-ionate　除离子水
deion circuit breaker　去电离断路器
deion extinction of arc　去电离灭弧
deion fuse　去电离熔断器
deionization de-exciter　去电离灭磁器
deionization plant　除离子装置
deionization　去电离（作用），去离子，除离子，消电离，除盐
deionized water　除离子水，除盐水，去离子水
deionizing potential　去电离电位
deion　消电离，去电离
de-ironing　除铁
deironization　除铁
dejacketer　脱壳装置
dejacketing cutter　脱壳切割机，去壳切割机
dejacketing solution　去壳溶液
dejacketing　脱壳，去壳
de jure or de facto authority　法律上或事实上的权威
deka ampere balance　十进安培秤
dekatron scaling unit　十进计数管标数部件
dekatron stage　用十进计数管级
dekatron　十进计数管
del　倒三角形
delaminate　分层，脱层，层离，剥离，分层剥离，裂为薄层
delamination　脱层，层离，剥离，分层剥离，裂为薄层，分层
delay-action circuit-breaker　延时开关，延时断路器
delay-action push button　延时动作按钮
delay-action relay　延时继电器，缓动继电器，

延时动作继电器
delay-action stage （炸药的）迟发期
delay-action starter 延时启动器
delay-action switch 延时动作开关
delay-action 延迟作用，延迟动作
delay adsorber 缓发吸附体
delay angle 延迟角，滞后角
delay bed assembly 延迟床装置
delay bed system 延迟床系统，（气态废物）滞留系统
delay bed train 延迟床系统，滞留装置
delay bed （废气）延迟床
delay cable 延迟线，延迟电缆
delay cap 迟爆雷管，迟发雷管
delay chamber 衰变室，延迟室
delay circuit 延迟电路
delay coil 延迟线圈
delay component 延迟元件
delay constant 延时常数
delay control 延时控制
delay correction network 延时校正网络
delay cracking 延迟裂纹
delay detonator 迟爆雷管，迟发雷管，延时雷管
delay device 延迟装置，延时装置
delay distortion 延迟畸变，时延失真
delay driver 延迟驱动器
delayed access circuit 延迟投入电路
delayed action 延迟作用，延迟动作
delayed activity 缓发（放射性）活度
delayed alarm relay 延时报警继电器
delayed alarm 延时报警
delayed alpha emission 缓发 α 发射
delayed automatic gain control 延迟自动增益控制
delayed blasting cap 缓爆雷管
delayed blast 延时爆炸
delayed boiling 延迟沸腾
delayed burning 延迟燃烧
delayed carry 延迟进位
delayed coefficient 缓发系数
delayed coincidence 延迟符合
delayed coking 延迟结焦
delayed combustion 延迟燃烧
delayed commutation 延迟换向
delayed completion 拖期竣工
delayed conduction 延迟电导
delayed crack 延迟裂缝，滞后开裂，延迟裂纹，延迟开裂
delayed criticality 缓发临界
delayed critical 缓发临界，缓发临界的
delayed deformation 延迟形变
delayed delivery contract 延交合同
delayed disintegration 缓发衰变
delayed echo 延迟回波
delayed elasticity 延迟弹性
delayed explosion 迟发爆炸
delayed factor 延迟因子
delayed feedback 延迟反馈
delayed firing 延迟着火，延迟燃烧
delayed fission neutron 缓发裂变中子
delayed gamma 缓发 γ 射线
delayed heat melting 余热熔化
delayed heat 衰变热，（放射性）余热，缓发热
delayed ignition 延迟点火，延迟着火，着火滞后
delayed junction 延缓汇合
delayed linear array 延迟线阵列
delayed neutron area 缓发中子面积
delayed neutron counting 缓发中子计数
delayed neutron damping 缓发中子衰减
delayed neutron emitter 缓发中子发射体
delayed neutron failed element monitor 缓发中子破损元件监测
delayed neutron fraction 缓发中子份额
delayed neutron group 缓发中子组
delayed neutron measurement 缓发中子测量，缓发中子刻
delayed neutron monitor 缓发中子监测器
delayed neutron period 缓发中子周期
delayed neutron precursor 缓发中子发射体【核电】
delayed neutron yield 缓发中子产额
delayed neutron 缓发中子
delayed outage 第 2 类非计划停运【可短暂延迟的停运】
delayed overcurrent trip 延迟过电流脱扣
delayed particle 缓发粒子
delayed payment 推迟付款
delayed photoneutron 缓发光中子
delayed pulse oscillator 延迟脉冲振荡器
delayed pulse 延迟脉冲
delayed quenching 预冷淬火，滞后淬火
delayed radiation effect 迟发辐射效应，缓发辐射效应
delayed radiation 缓发辐射
delayed reactivity 缓发反应性
delayed reclosure 延迟重合闸
delayed relay 延时继电器，缓动继电器
delayed runoff 延迟径流，延滞径流，滞后径流
delayed shipment 延期装运
delayed start 延迟启动
delayed strain 延迟应变
delayed subcritical reactor 缓发次临界反应堆
delayed supercritical reactor 缓发超临界反应堆
delayed sweep 延迟扫描
delayed temperature feedback 延迟的温度反馈
delayed test 被延误的试验
delayed time-base sweep 延迟时基扫描
delayed time system 延时系统
delayed trigger 延迟触发器
delayed 延迟的，第 2 类非计划停运的
delay element 延迟元件，延时元件
delay equalization 延时均衡
delay equalizer 延时均衡器
delay error 延时误差
delay factor 延迟因子
delay fuse 迟发电信管
delay generator 延迟脉冲发生器
delay in completion 竣工实施图
delay in delivery 交货逾期，延迟交付
delaying separation 延迟分离

delay in payment 付款延迟，迟付
delay in recruitment 延迟征聘
delay lens 延迟电磁透镜
delay line decoder 有延迟线的译码器
delay line fall time 延迟线降落时间
delay line memory 延迟线存储
delay line pulse stretcher 延迟线脉冲展宽器
delay line register 延迟线寄存器
delay line rise time 延迟线升起时间
delay line storage 延迟线存储器
delay line 延迟管线，延迟线，滞留管线
delay loop 延迟回路，(气态废物) 滞留系统
delay medium 延迟介质
delay multivibrator 延迟多谐振荡器
delay network 延迟网络，延时网络
delay one-shot 延时单稳触发器
delay period 延迟时期，缓发周期，滞后期，滞燃期
delay pilot valve 延滞滑阀
delay primer 迟发起爆药包
delay pulse 延迟脉冲
delay rate 迟缓率
delay recorder 延迟记录器
delay record 延时记录器
delay relay 延迟继电器，延时继电器
delays by subcontractor 分包商的误期
delay slug relay 阻尼环式延迟继电器
delay storage 延时线存储器
delay switch 延时开关
delay system 延时系统，滞留系统，延时系统
delay tank 衰变箱，滞留槽，延迟箱
delay the delivery 延期交货，延期交船
delay time relay 延时继电器
delay timer 延时计时器
delay time switch 延时时间开关
delay time 延时时间，滞后时间，时延，迟延时间，推迟时间
delay tripping 延时跳闸
delay unit 延迟装置，延迟元件，延迟部件，延时元件，延时装置
delay valve 延时阀
delay voltage 延迟电压，滞后电压
delay 时滞，延迟，推迟，滞后，拖期，误期，延缓，抑制，延期
delegate 代表
delegation of authority 权力下放，授权
delegation of duties 职责委托
delegation of power 权限的委托
delegation 代表团，使团
delete bit 删除位
delete character 删除符
delete code 删除码
deleted 已删除的
deleterious effect 有害效应
deleterious material 有害物质
deleterious 有害的，有毒的
delete where appropriate 删去不适用者
delete 删去，删除
deletion character 删掉字符
deletion operator 删除算子，注销算符
deletion record 删除记录
deletion 贫化，消耗
deliberate actions 有准备行动，预定操作
deliberate control action 及时调节，有意调节
deliberate reconnaissance 周密勘测
deliberation 考虑，审议
deliberative organ 审议机构，议事机构
delicate adjusting 精调
delicate 精细的，灵敏的，敏感的，精密的
delict 违法行为，不法行为，侵权行为
delime 脱灰
deliming agent 脱灰剂
delimitation 分界
delimiter 限制符，限定器，定义符，去限器
delineateion 轮廓
delineate 勾画外形，描述
delineator 制图者，图形，描画器
delinquent tax 滞纳税款
deliquescence 潮解，溶解
deliquescent material 易潮解物
deliquescent 潮解的，溶解的，易潮解的
deliquesce 潮解，溶解
delitescent 隐匿的，潜伏的
deliverables 可交付成果，可交付物，可交付性，产出，可交付使用的产品，交付的产品
delivered at job 工地交货
delivered ex quay 目的港码头交货价
delivered ex ship 目的港船上交货
delivered heat 放热，供热，释热
delivered horse-power 输出马力，有效马力，输出功率
delivered power 输出功率
delivered price 到货价，送货上门价格
delivered terms 交货条件
delivered 已递送的，业已交货的，已交付的
deliverer 交付者
delivering carrier 交货承运人
delivering of goods 发货
deliver over 交换，交付
deliver pump 输送泵，压送泵
delivery advice 到货通知，发货通知（书），交货通知（书）
delivery against payment 付款交货，付款通知
delivery air chamber 压缩空气室
delivery and acceptance certificate 交货验收证明书
delivery and customs agent 提货报关代理人
delivery and measuring box 配水量水箱
delivery box 放水箱，配水箱
delivery car 送货车，送料车
delivery clause 交货条款
delivery cock 排污阀
delivery conditions 交货条件
delivery conduit 输送管道
delivery contract 供货合同
delivery curve 供水曲线
delivery cycle 交货周期
delivery date 交付日期，交货日期
delivery day 交货日（期）
delivery distance 输送距离，运送距离
delivery document 交货单据
delivery end 输出端，排出端，卸料端，出口端

delivery equipment 运输设备
delivery failure 交货误期
delivery flask 分液瓶，量瓶
delivery gate 出水口，斗门，分送闸板，配水闸
delivery gauge 出口压力表
delivery head 扬程，（泵）出口压头，供水水头，输送高度
delivery in installment 分期交货
delivery inspection 交货检查
delivery interval 交货时间间隔
delivery length 交货长度
delivery lift 水头，扬程，输水高度，
delivery list 交货清单
delivery main 输水主管
delivery manifold 输送集管，排气歧管，通气多岔道
delivery mode 运输方式
delivery note 交货清单
delivery of assistance 提供援助
delivery of current 送电
delivery of energy 能量输送，供能
delivery of equipment 设备交付
delivery of goods by installments 分批交货
delivery of technical documentation 技术文件交付
delivery of the object 标的物交付
delivery on call 根据买方通知交货，通知交货
delivery on field 就地交货
delivery on payment 付款交货
delivery on term 定期交付
delivery order 提货单，小提单
delivery orifice 出水孔，输出孔口，输水孔口，送气孔
delivery period 供水期，交货期
delivery pipe line 送水管路，供水管线，输送管线，输水管道
delivery pipet 吸移管
delivery pipe 输送管
delivery port 交货港
delivery power 输出功率
delivery pressure head 供水压头
delivery pressure 输送压力，输出压力，出口压力，排气压力
delivery price 交货价格，交货价
delivery proposal 运输方案
delivery pump 输送泵，给料泵，输油泵
delivery rate of erosion 冲刷输沙率
delivery rate 供水率，输送速度
delivery receipt 交货收据，送货回单
delivery regulator 进给量调节器
delivery roll 导辊
delivery schedule 交货时间表，交货进度，交付进度，交货进度表
delivery sequence 交付顺序，交货顺序
delivery shaft 输送井
delivery side 进给端
delivery space 排气容积
delivery speed 输送速度，排气速度
delivery system 供料系统，输送系统
delivery temperature 排气温度，输送温度，出口温度
delivery term 交货期限，交货条件
delivery time 交货时间，交付时间
delivery to site 运抵现场
delivery track 供料支线，输送轨道
delivery tube 压送管，给料嘴
delivery valve 输送阀，排放阀，排气阀，输水阀
delivery van 送货车
delivery volume 给料体积
delivery wagon 送料车
delivery work 输出功，输送功
delivery 交货，发货，输出，输送，供给，传递，递送，供电，交付，运输，供电量
deload 减负荷
delta amplitude 三角振幅
delta arm 三角洲河汊
delta-connected circuit 三角形连接电路
delta connection 三角接法，三角形接法，三角形连接
delta current 三角形连接电流
delta-delta connection 双三角形接法
delta estuary regulation 三角洲河口整治
delta field 三角形场
delta function 狄拉克函数，脉冲函数
deltaic deposit 三角洲沉积
deltaic fan 扇形三角洲，三角洲冲积扇
delta impulse 三角形脉冲
delta induction 三角感应
delta-iron 三角铁
delta-matching transformer 三角形匹配变压器
delta-matching 三角匹配转换
deltamax 铁镍薄板
delta modulation 增量调制，三角调制
delta network 三角形网络
delta noise 三角噪声
delta-response 三角响应，三角响应曲线
delta-star connection 三角-星形接线，三角-星形接法
delta-star transformation 三角-星形接线变换
delta voltage 三角电压，线电压
delta winding 三角形绕组
delta wing vortex 三角翼涡
delta-wye switching 三角—Y形转换
delta-zigzag connection 三角-曲折性接法
delta 三角形，通信中用以代表D的词，三角洲
deltoid river mouth 三角洲河口
deluge spray system 雨霖喷水灭火系统
deluge system 溢流系统，喷水系统，雨淋灭火系统
deluge valve assembly 雨淋阀总成
deluge valves unit 雨淋阀组
deluge valve 雨淋阀
deluge 洪水，倾盆大雨，大洪水
delve 坑
DEL 电气厂房冷冻水系统【核电站系统代码】
demagnetisation 灭磁，去磁，退磁
demagnetiser 去磁器，去磁装置
demagnetization coefficient 去磁系数
demagnetization curve 退磁曲线，去磁曲线
demagnetization factor 去磁系数，退磁系数
demagnetization force 去磁力，退磁力
demagnetization loss 退磁损失，去磁损失
demagnetization 灭磁，去磁，退磁

demagnetized field 退磁磁场
demagnetizer 退磁装置，去磁装置，去磁器，消磁装置
demagnetize 去磁，退磁
demagnetizing ampere turns 去磁安匝，退磁安匝
demagnetizing breaker 去磁开关，灭磁开关
demagnetizing effect 去磁效应，去磁作用
demagnetizing factor tensor 退磁因子张量
demagnetizing factor 退磁系数，去磁系数
demagnetizing resistance 去磁电阻，灭磁电阻
demagnetizing switch 去磁开关，消磁开关，灭磁开关
demagnetizing turns 去磁匝数
demand analysis 需求分析
demand assignment multiple access 按需分配多址
demand bid 需量报价
demand bill 即期汇票
demand charge 按需量供电
demand curve 负荷曲线，需求曲线，需用量曲线
demand draft rate 即期汇票汇率
demand draft （银行）即期汇票
demand draught （银行）即期汇票
demand estimation 需求估计
demand factor 需用系数，需要因子，需求系数
demand function 需求函数
demand guarantees 见索即付保函
demand interval indicator 需量时限指示器
demand limiter 需求限制器，电流限制器
demand mass curve 需要量累积曲线
demand meter 占用计数计
demand note 即期票据【多指本票】
demand of a system 系统需电量
demand output 指令输出
demand paging 请求式调页，请求页面
demand price 需求价格
demand processing 请求处理，需用处理
demand reading 请求读入
demand recorder 需量记录器
demand-side management 用电需方管理，（电力）需求侧管理
demand stability 需求稳定性
demand 需用量，需要，要求，需求
demanganization 除锰
demarcated section line 施测断面
demarcate 标定
demarcation line 分界线，区界
demarcation of land 土地分界
demarcation points of contract 合同分界点
demarcation 区划，定界，分界，划界
demerit 缺点
demetallisation 除盐，去矿化作用，脱矿质作用
demetallization of condensate 凝结水出盐
demetallization plant 除盐设备
demetallization 除盐，去矿化作用，脱矿质作用，失金属化
demi-column 嵌墙柱
demidovite 青硅孔雀石
DEMIN(demineralization) 除盐
demineralization effluents neutralization system 除盐中合用化学产品储存系统
demineralization plant 除盐设备，除盐厂，除盐站
demineralization water plant （电厂）化学水装置
demineralization 除盐
demineralized water header 除盐水集管
demineralized water piping 除盐水管道
demineralized water preheater 除盐水预热器
demineralized water production system 除盐水生产系统
demineralized water pump 除盐水泵
demineralized water storage tank 除盐水箱，除盐水贮箱
demineralized water supply line 除盐水供应管线
demineralized water tank 除盐水箱
demineralized water transfer pump 除盐水输送泵
demineralized water 除盐水，软化水，去矿物质水
demineralizer building 除盐器厂房
demineralizer waste water 除盐系统废水
demineralizer 除盐器，出演设置，软化器，除盐装置
demister mat 除雾器垫
demister type separator 除雾型分离器，除雾器
demister 除雾器，去雾器，气溶胶分离器，除雾装置，除油雾器
demixing effect 反混合效应
demixing 分层，分离
demobilization 复员，遣散，撤离现场
demobilize 遣散
demoded 解码的，过时的，老式的
demo＝demonstration 展示会
demoder 解码器
demodulate demodulated signal 已调解信号
demodulate 解调，反调制
demodulating carrier frequency control 解调载频控制
demodulating function 解调函数
demodulation-remodulation transponder 解调重调转发器
demodulation 反调制，解调
demodulator amplifier 反调制放大器
demodulator band fitter 解调制带通滤波路器
demodulator 解调器，检波器
demographics 人口特征，人口统计学的
demography 人口统计学
demolding 拆膜
demolding product 脱模件【混凝土】
demolish 拆除
demolition blasting 拆除爆破
demolition expense 拆除费用
demolition layer 炸毁
demolition waste 工地废渣料
demolition 拆毁，拆除，拆迁，爆破，破坏
demonstration plant 示范厂，示范电厂，中间试验电厂
demonstration reactor 示范反应堆
demonstrated experience 已证明的经验
demonstrate 证明，证实，论证，显示，示范
demonstration area 示范区，试验区
demonstration power plant 示范电厂，示范电站
demonstration project 示范工程，示范项目

demonstration test 示范试验
demonstration 示范，演示，论证，说明，证明
demo-plant 示范厂，示范电厂
demorphismus 风化变质作用
demorphism 风化作用
demotion in grade 降级
demotion in office 降职
demotion 降低
demoulding 脱模
demountable partition 活动隔墙
demountable 可拆换的
demounting 拆卸
demoutable element bundle 可拆卸燃料元件棒束
demoutable fitter 可拆卸过滤装置
demoutable X-ray tube 可拆卸X射线管
demoutable 可拆卸的，开拆除的
demo 示范产品，爆破，示范
demulsibility 反乳化性，乳化分解性，抗乳化性
demultiplexer 多路分配器，译码器，信号分离器，分接器
demultiplex 信号分离
demurrage charges （车，船）滞期费，延迟费，疏港费
demurrage days 滞期天数
demurrage 滞留费，滞期，滞留期，过期，延迟费，滞港费，滞期费
DEM(Deutsche mark) 德国马克
denary counter 十进制计数器
denary logarithm 常用对数
denaturation 改变本性，变性作用，（核燃料）中毒
denatured alcohol 变性酒精
denatured fuel 变性燃料
denatured 变性的，（核燃料）中毒的
den＝density 密度
Den(Department of Energy) 能源部【美国】
DEN(design evolution notice) 设计修改通知
dendrite orientation 枝晶取向
dendrite 树枝状晶体
dendritical crystal 枝状晶体
dendritical shrinkage 枝状疏松
dendritic drainage 树枝状排水系统，树枝状水系
dendritic pattern 树枝状型
dendritic structure 树枝状结构，枝晶组织
Den Hartog stability criterion 德哈托稳定判据
denison motor 轴向回转柱塞液压马达
denison pump 轴向活塞泵
denitration 脱销
denitrfying agent 脱硝剂
denitriding 脱氮
denitrification 脱硝作用
denivellation 水准差，不平度
denominated in foreign currency 用外币标价
denominate 命名
denomination 名称，种类，单位
denominator 分母，共同特性
denotation 符号
denote 指示，指的是，（符号）表示
denouncing party 退约方
de-NO_x 脱硝
dense aggregates 密实骨料
dense-bed viscosity 浓相床黏度
dense binary code 紧凑二进制码
dense concrete 密实混凝土
dense condition 密实状态
dense gas spill 浓气体溢出
dense graded aggregate 密级配骨料
dense graded bituminous concrete 密级配沥青混凝土
dense graded 密级配的
dense liquid 重液
densely built up city 建筑密集城市
densely packed coil 密绕线圈，密绕绕组
densely-populated area 稠密居住区，人口稠密区
densely populated 人口稠密的
densely settled area 稠密居住区
dense material 密相物料
densener 压缩机，压紧器，内冷铁，激冷材料，浓缩器
denseness 密实度
dense particle 致密颗粒
dense phase fluidization 浓相流态化
dense phase fluidized bed 浓相流化床
dense phase transportation 密相输送
dense phase 密相，浓相
dense plume 浓羽流
dense scale 密实垢
dense slag 紧密熔渣
dense smoke 浓烟
dense state 密实状态
dense-textured filter paper 密纹滤纸
dense vapour spill 浓蒸汽溢出
dense 稠密的，浓的，密致的，密实的
densification by explosion 爆破压实
densification model 致密化模型
densification of pellets 芯块化模型
densification test 压实试验
densification 稠化，增浓，压实，致密化，密实化
densifier 密实器，增浓剂，增密剂
densimeter 密度计，比重计，显像密度计，黑度计，密度测定计
densimetric curve 密度曲线
densimetric Froude number 密度（计）弗罗德数
densimetric 密度计的，采用密度计的
densitometer 光密度计，显像密度计，比重计，密度测定计，密度计，黑度计，浓度计，测微光度计
densitometry 密度测定法，显像测密度法，浓度测量，密度测量，微光度测量
density analyzer 密度计，比重计
density anomaly 密度异常
density bottle 比重瓶，密度瓶
density control 密度控制，容重控制
density corrected strength 比强度，密度修正强度
density corrected ultimate tensile strength 比极限抗拉强度，密度修正极限抗拉强度

density current bed　异重流底层
density current　重流，密度差流动，密度流，异重流，异重水流
density difference　密度差
density diffusion　速度扩散率
density discriminator　密度鉴别器
density distribution　密度分布
density effect　密集效应，异重效应
density flow　异重水流，密度场，异重流，密度流
density fluctuation　密度脉动，密度波动
density function　密度函数
density gauge　密度计，比重计
density gradient　密度，梯度，密度梯度
density impulse　密度冲量
density index　密度指数
density level　密度级
density measurement　密度测量
density measuring probe　密度测量探头
density meter　密度计，比重计
density-moisture content curve　密度-含水量曲线
density of air　空气密度
density of canopy　林冠密度，树冠覆盖密度
density of charge　电荷密度
density of energy　能量密度
density of field　场强，场强密度
density of filling gas　充气密度
density of gas phase　气相密度
density of heat flow rate　热流密度
density of liquid phase　液相密度
density of load　荷载密集度，载荷强度
density of localized state　定域态密度
density of Ionization　电离密度
density of lump　块密度
density of momentum　动量密度
density of roughness element　粗糙元分布密度
density of soil particles　土粒密度
density of solid particles　固体颗粒密度
density of states　状态密度
density of traffic　交通密度
density perturbation　密度扰动，密度扰动
density probability plot　密度概率曲线图
density profile　密度廓线
density range　密度范围
density ratio　密度比
density reactivity coefficient　密度反应性系数
density resistance　密度阻力
density scale　密度刻度
density sensor　密度传感器
density slicing　密度分割
density spectra　密度谱
density spectrum　密度谱
density spread　密度分布
density step tablet　密度分级片
density stratification　密度层结，密度分层，密度层，密度分层作用
density temperature coefficient　密度温度系数
density testing　密度检验
density transducer　密度传感器
density tunnel　高压风洞，高密度风洞
density wave instability　密度波温度系数
density wave oscillations　密度波振荡
density wave　密度波
density　密度，比重，浓度，密实度
dental concrete　齿状混凝土，混凝土塞，挖补混凝土
dental treatment　坝基补缝填坑处理【水电】
dentated bucket lip　挑流齿槛
dentated sill　齿槛，齿形槛，消力齿
dentate　锯齿状的，有齿的
dented wagon　凹型车
denting　颈缩，凹陷【蒸汽发生器传热管】，紧缩效应
dent　凹陷，凹坑，压痕，（齿轮的）齿，凹部，坑
denudation mountain　剥蚀山
denudation plain　剥蚀平原
denudation plane　剥蚀面
denudation rate　剥蚀率
denudation slope　剥蚀坡
denudation　裸露，剥裸，剥蚀（作用），滥伐
denuded area　剥蚀地
denuded soil　剥蚀土壤
denuder　金属分解
denude　使赤裸，剥光覆盖物
denunciation clause　废除条款，废弃条款
denunciation　废除，废弃
deodorant　脱臭剂
deodorization of air　空气脱臭
deodorization　除臭，脱臭作用
deodorizer　脱臭剂
deodorize　脱臭气
deodorizing　脱臭
deoiling　脱油
deoil　除油
deoscillator　减振器，阻尼器
deoxidant　脱氧剂，还原剂
deoxidate　还原，脱氧
deoxidation　脱氧，去氧，还原
deoxidizer　还原剂，脱氧剂，脱氧器
deoxidizing agent　脱氧剂
deoxygenated water　除氧水
deoxygenation constant　脱氧系数
deoxygenation　脱氧（作用）
deoxygenization　去氧，还原，除氧作用
depair　去偶拆开对偶
deparaffination　脱蜡处理
depart from　离开
departmental standard　部门标准
department director　处长，局长
department-level review　处级评审
department manager　部门经理
Department of Energy　能源部【美国】
Department of Nuclear Safety Management of SEPA　国家环境保护总局核安全管理司【中国】
department organization　部门组织
department　部门，车间，科，室，处，局（部）
departure curve　离差曲线，偏差曲线
departure from nucleate boiling correlation　偏离核态沸腾关系式
departure from nucleate boiling flux　偏离核态沸热流密度

departure from nucleate boiling margin 偏离核态沸腾裕量
departure from nucleate boiling ratio 偏离泡核沸腾比
departure from nucleate boiling 偏离核态沸腾，偏离泡核沸腾
departure of tidal hour 潮时偏差
departure port 始发港
departure resistance 气流分离阻力
departure 离开，偏离，偏差，出发，脱离，发车，发运地
dependability 可靠性，强度，坚硬度，坚固度，可信性
dependable capacity 可靠容量，保证容量
dependable flow 保证流量
dependable power 保证功率
dependant's area 家属区
dependant 家眷，受赡养人，家庭负担
dependence 相关，相关性，关系，信任，从属，从属性，依赖，依靠
dependency of attenuation on frequency 衰减对频率的关系
dependency 相关性，关系，依赖，从属性
dependent event 非独立事件
dependent failure 相关故障
dependent on 取决于，依赖于
dependent power operation 直接动力操作
dependent time-lag relay 变时滞继电器
dependent time-lag 相倚时滞
dependent variable 因变量，函数
dependent 相倚的
deperm 消水平磁场，消除磁场
dephased current 相移电流
dephased 有相位差的，相移后的
dephase 使相位偏移，位移
dephlegmation 分凝
dephlegmator 分馏塔，分馏柱
depict 描绘
deplanation 夷平
depleted field air storage reservoir 废矿井蓄气库
depleted fraction 贫化馏分【浓缩】
depleted fuel 乏燃料，贫化核燃料，贫燃燃料
depleted material 贫化材料，用过的核燃料，某种同位素减少的材料
depleted uranium 贫铀，贫化铀
depleted water 贫化水，剥淡水
depleted 贫化的【铀】，烧过的，浓度降低的
deplete 放空，用尽，耗尽，废弃的，变质的，消耗的，贫化，减少，使衰竭
depleting-layer semiconductor 贫化层半导体
depleting-layer 贫乏层
depletion calculation 贫化计算
depletion chain 贫化链
depletion charge 燃料消耗费用
depletion curve 亏水曲线，退水曲线
depletion depth 耗尽深度
depletion factor 贫化系数
depletion hydrograph 退水过程线
depletion layer 耗尽层
depletion mode transistor 耗尽型晶体管
depletion of poison 毒物贫化，毒物的燃耗
depletion of resource 资源枯竭
depletion of snow cover 积雪消融
depletion 贫化【铀】，放空，亏损，损耗，用尽，耗尽，缺乏，降低，枯竭，退水，消蚀
deplistor 三端负阻半导体器件
deployment 部署，布置
de-poisoning 去毒物
depolariser 去极化剂，退极化剂
depolarise 去磁，去极化，去偏振
depolarization 去极化，消偏振，去磁，退磁
depolarizer 去极化剂，退极化剂
depolarize 去磁，去极化，去偏振
depose 沉淀，沉积，置放，免职
deposit accounts at bank 银行存款账目
depositary 保存国，保存人，仓库，贮藏所
deposit attack 沉积物侵袭，沉积腐蚀，进口沉积腐蚀
deposit corrosion 沉积腐蚀，垢下腐蚀，结垢腐蚀
deposited-carbon resistor 碳膜电阻器
deposited drift area 沉积漂移区
deposited metal 熔焊金属，溶敷金属
deposited seat 堆焊座【阀】
deposited weld metal 沉积的焊接金属
deposit from lease 租赁保证金
deposit gauge 沉积计，落灰计，降尘测定器
depositing concrete 浇注混凝土
depositing dock 单坞墙式浮船坞
depositing substrate 堆积场基底
depositing tank 沉淀池，沉淀箱，沉淀槽
deposit in security 押金，保证金
depositional environment 沉积环境
depositional remanent magnetization 沉积剩磁
depositional topography 沉积地形
deposition angle 熔敷角
deposition coefficient 沉积系数，沉降系数，熔敷系数
deposition efficiency 熔敷效率
deposition fabric 沉积结构
deposition flux 沉积通量
deposition mode 沉积方式
deposition of radioactive dust 放射性灰尘沉淀
deposition of suspended material 悬移质沉积
deposition plane （燃料元件包覆层）沉淀平面
deposition rate 沉淀速度，熔敷速度，沉积速率
deposition region 沉积区
deposition sedimentation 沉积，沉淀
deposition surface 沉积表面
deposition velocity 沉积速度，沉降速度
deposition welding of the cladding 覆盖层堆焊
deposition welding 堆焊
deposition 沉淀（物），结垢，积灰，附着，覆盖，沉积（作用），积存，淤积（作用）
deposit of flue dust 烟尘沉积
deposit of scale 水垢
depository 贮藏所，存放处，保管人员
depositor 沉淀器
deposits of moisture 土壤含水量
deposits of rust 生锈，锈斑
deposit 沉淀，积垢，结垢，积灰，押金，沉积（物），抵押，淤积（物）

depot 仓库（设备，器材），储藏所，储存，基地，贮藏所
depreciable life 折旧年限
depreciation base 折旧基价
depreciation charge 折旧费用
depreciation coefficient 折旧系数
depreciation cost 折旧费
depreciation factor 折旧率，减光补偿率，折旧因子
depreciation fund 折旧资金，折旧基金
depreciation rate （固定资产的）折旧率
depreciation value 折旧后价值
depreciation 贬值，折旧，减值，减振，减少，损耗
depressant 抑制剂
depressed arch 低拱
depressed area 沉陷区，萧条地区
depressed block 陷落地块
depressed flux 通量下陷
depressed nappe 受抑水舌，贴附水舌
depressed semicircle 扁半圆
depressed sewer 倒虹式下水道，下穿污水管
depressed table 沉陷水层
depression angle 俯角，俯视角
depression belt 低（气）压带
depression cone 下降漏斗（地下水面的）沉陷锥
depression curve 降落曲线，浸润曲线，水位降落线，下降浸润线
depression earthquake 陷落性地震
depression economy 不景气经济
depression factor （能量）下陷因子
depression head 降落水头
depression infrabar 低气压
depression of support 支座沉陷
depression of water table 地下水位的降落
depression region 低压区
depression shoreline 沉降岸线
depression 不景气，低气压区，递减，减少，降低，降温，降压，凹部，洼地，抑制，萧条
depressive earthquake 陷落性地震
depressor 抑制器，缓冲器，阻尼器
depressurisation 减压，卸压，弛压
depressurization accident 失压事故，卸压事故
depressurization phase （事故）卸压状态
depressurization tank 卸压箱
depressurization 减压，降压，卸压，弛压
depressurize a reactor 反应堆卸压
depressurized residual oil 减压渣油
depressurized system 卸压系统
depressurize reactor 反应堆减压
deprivation 剥夺，损失
depth 深度，厚度，进深
depth absorbed dose 深部吸收剂量
depth adjustment 调节深度
depth-area curve 面深曲线，深度面积关系曲线
depth-area formula 深度面积公式
depth-area relationship 面深关系
depth below level 料堆深度
depth chart 水深图
depth compensation 深度补偿
depth contour 等水深线
depth curve 深度曲线
depth-discharge relation 水深流量关系
depth dose distribution 深部剂量分布
depth dose equivalent index 深部剂量当量指数
depth dose 深部剂量
depth-duration-area value 时面深值
depth-duration curve 深度历时曲线
depth effect 深度效应，深度影响
depth facies 深相
depth factor 埋深系数，深度因数
depth finder 测深仪
depth float 深水浮标
depth gauge micrometer 深度千分尺
depth gauge 测深仪，测深规，深度计，深度表，深度测微计，深度尺，深度规
depth indicator 深度计
depth measurement 测深
depth micrometer 测深千分尺
depth of burying 埋深
depth of camber 弧线高度
depth of crack 裂变深度
depth of cutting 切削深度，挖方深度
depth of defect 缺陷深度
depth of embedded foundation 基础的埋置深度
depth of embedded grounding wire 接地线埋深
depth of evaporation 蒸发量，蒸发深
depth of extra dredging 挖泥超深
depth of foundation 基础深度
depth of frictional influence 摩擦影响深度
depth of frost penetration 冰冻深度
depth of fusion 熔深
depth of ground table 地下水位埋深
depth of immersion 浸入深度
depth of indentation 压痕深度
depth of insertion （控制棒的）插入深度
depth of laying 埋深
depth of modulation 调制深度
depth of navigable channel 航道水深
depth of overburden 覆盖厚度，覆盖深度
depth of overflow over weir 堰顶溢流水深
depth of penetration 贯入深度，穿透深度，贯入度，渗透深度
depth of seismic focus 震源深度
depth of slag 渣的深度【电渣焊】
depth of stratum 地层深度，岩层深度
depth of tooth 齿高
depthometer 深度计，测深仪
depth penetration 深部穿透，深部贯穿
depth pressure recorder 深度压力记录器
depth range 深度范围
depth ratio 埋深比，深度比
depth recorder 深度记录器
depth-resistance curve of pile 桩的深度抗力曲线
depth scale ratio 水深比例尺
depth scale 测深标尺
depth sounder 测深仪
depth-span ratio 高跨比
depth-to-discharge relation 水深流量关系
depth to water-table map 地下水埋深图，潜水埋深图

depth-to-width ratio　高宽比
depth tube spacing　纵向管距
depth-velocity curve　水深流速关系曲线，流速垂直分布曲线
depth-volume curve　深度体积关系曲线
depth-width ratio　高宽比
depurate　提纯，净化
depuration　纯化，净化，滤清
depute　委托，代表
deputy managing director　副总经理
deputy representative　副代表
deputy resident representative　副驻地代表
DEQ(deliverde ex quay)　目的港码头交货
derailment　出轨，脱轨
derail　脱轨器
derangement　故障，不同步，紊乱，混乱
derated power　额定功率降低，降低的功率
derated state　降额状态
de-rate　降低定额，降低出力，降低参数
derating curves of dissipation　降功耗曲线
derating factor　降容系数
derating　降低功率【设备】，降级【仪器】，定额的降低，定额的减少，仪器降级
derby　粗锭
deregenerative feedback　负反馈，非再生反馈
deregulation　放松管制
derelict land　弃料地
dereption　水下剥蚀
derivant　衍生物
derivate-action coefficient　微分作用系数
derivate-action time constant　微分作用时间常数
derivate　导数，微商，衍生物
derivation action　微分作用
derivation time　微分时间
derivation　求导，推导，导出，引出，来历
derivative action coefficient　微分系数，微分作用系数
derivative action gain　微分增益，微商作用增益
derivative action time constant　微分作用时间常数
derivative action time　微分时间，微分作用时间
derivative action　衍生的动作
derivative coefficient　诱导系数
derivative control action　微分调节（控制）作用
derivative control coefficient　微分控制系数
derivative control factor　微商调节因子
derivative controller　微分控制器，微分调节器
derivative control　微分控制，微分调节
derivative discontinue　导数不连续性
derivative equalization　利用微分环节的稳定
derivative feedback　微分反馈，微分回授
derivative measurement　导数测量
derivative of nth order　n 阶导数
derivative-proportional-integral control　微分—比例—积分控制
derivative rock　导生岩
derivative standard　衍生标准
derivative time constant　微分时间常数
derivative time　微分时间
derivative unit　微分单元
derivative　衍生物，派生物，微分
derive circuit　分支电路，分路
derived air concentration　导出空气密度
derived current　分路电流，分支电路
derived data　推算资料
derived energy　二次能源
derived estimate　导出量
derived intervention level　导出干预水平，导出调查水平
derived limit　导出限值，推定限值
derived M-type filter　M 推演式滤波器，M 导出型滤波器
derived number　导数
derived quantity　导出量
derived resistance　并联电阻
derived standard　导出标准
derived-type filter　推演式滤波器，导出型滤波器
derived unit　（测量的）导出单位
derived upper bounds　导出上界
derived working limit　导出工作限值
derive　推导，导出
derivometer　测偏仪
dermal fold　表层褶曲
dermal resistance　表面电阻，皮肤电阻
derrick barge　浮式起重机，起重船，起重机驳船，浮吊
derrick boom　起重机吊杆，人字吊杆，起重杆
derrick car　汽车吊，起重车
derrick crane　转臂起重机，扒杆吊，桅杆式起重机
derricking jib crane　转臂起重机
derrick mast　起重机桅杆
derrick pole　起重杆
derrick rope　起重机钢索
derrick stone　粗石块，大石块
derrick tower gantry　人字起重机塔式起重台架
derrick tower　起重吊塔，起重机塔
derrick　吊杆，人字起重机，悬臂扒杆，进线架，抱杆，动臂起重机，起重机，油井的井架，转臂起重机，油井的铁架塔
derusting　除锈
desalination apparatus　脱盐设备
desalination membrane　脱盐膜
desalination of sea water　海水淡化
desalination plant　脱盐车间，海水淡化厂，淡化厂，海水淡化装置
desalination　除盐，脱盐，脱盐作用，咸水淡化，海水淡化，盐水淡化
desalinization　脱盐，淡化，除盐
desalinizing reactor　海水淡化反应堆
DESALN＝desalination　除盐
desalt　除盐
desalted water　脱盐水，除盐水
desalt effect　除盐效果
desalter　脱盐器，除盐设备，淡化器
desalting　淡化，脱盐，除盐
desalting agent　脱盐剂
desalting compartment　除盐室
desalting efficiency　除盐效率
desalting environment　海水淡化环境
desalting kit　海水淡化器

desalting process 除盐过程，海水淡化过程
desalting reactor 海水淡化反应堆
desalting technology 脱盐技术
desalting-water system 脱盐水系统，（电厂）化学水系统
descale 去垢
descaling brush 除锈钢丝刷
descaling 除垢，除氧化，除水垢，去氧化皮，除锈
Descarte's coordinates 笛卡儿（直角）坐标
descendant 子体，衰变产物，后代
descending air current 下降气流
descending air 下沉空气
descending current 下降流
descending curve 下降曲线
descending grade 下降坡度，下坡
descending limb of hysteresis loop 磁滞回线的下降支
descending liquid 下行液体
descending order 递减次序
descending power 降幂
descending spring 下降泉
descending stroke 下冲程，活塞下行程，
descending velocity 下降速度
descending water 下降水
descending 下降的
descent control device 缓降器
descent （控制棒）插入，下降，降落
describe 描述，作图，做……运动，描绘
describing function 描述函数，等效频率传输函数
describing 描述的
descriminator 鉴别器，鉴相器，鉴频器
description column 摘要栏
description of article 商品名称
description of core 岩心描述
description of goods 货物说明书
description of items 分项的说明，条目的说明
description of materials 材料说明
description of operation 运行说明
description of plant 设备说明书
description of subroutine 子程序使用说明书
description of works 工程描述
description 说明书，说明，叙述，规格说明，货物名称，陈述，描述，名称
descriptor base register 描述符基本地址寄存器
descriptor 描述符，解说符，描述信息
DES(data element set) 数据元素集
DES(delivered ex ship) 目的港船上交货价
DES＝description 描述性【文件类别】
desert area 沙漠地区
desert belt 沙漠带，荒芜地带，荒漠地带，荒漠带
desert climate 沙漠气候
desert creep 沙漠蔓延
desert deposit 沙漠沉积
desert devil 沙卷风
desert encroachment 沙漠扩侵
desert erosion feature 荒漠侵蚀地形
desert grey soil 灰漠土
desertification 沙（漠）化
desert pavement 荒漠覆盖层，沙漠砾石表层
desert peneplain 沙漠侵蚀准平原
desert plant 沙漠植物
desert shelterbelt 沙漠防护林带
desert soil 荒漠土，沙漠土
desert spread 沙漠扩展
desert steppe soil 荒漠草原土
desert steppe 荒漠草原，沙漠草原
desert storm 尘暴，沙暴
desert vegetation 沙漠植被
desert zone 荒漠地带，沙漠地带
desert 荒漠，沙漠
desh coat 砂浆涂层
desheathing 脱壳，除套，脱去外壳
desiccant 干燥剂，干燥，干燥产物，干燥的，去湿的，使干燥的
desiccating agent 干燥剂
desiccation fissure 干缩裂缝
desiccation 脱水，干燥，烘干，干燥作用，晒干
desiccator 干燥器，干燥剂，吸湿剂
design accident 设计事故
design accuracy 设计精度
design activity 设计机关，设计组织
design aerodynamics 空气动力学设计
design agreement 设计协议
design aids （自动）设计工具，设计参考资料
design altitude 设计高度
design ambient condition 设计大气状态
design analysis 设计分析
design and construction rules 设计和建造规则
design and survey 勘测设计
design approach 设计方法
design approval 设计批准
design assembly 设计装配图
design assumption 设计前提［假定］
designate as 把……称为
designated bank 指定银行
designated currency 指定货币
designated port of transportation 指定中转港口
designated representative 指定的代表
designated system management group 定名系统管理组
designated type 指定型号
designate port of transshipment 指定中转港口
designate 指明，指出，指派，意味着，把……定名为，任命
designation card 标识卡
designation code 标志代号
designation number 标志数，标准指数
designation of drawing 图名
designation of terminations 引出端标志
designation strip 名牌，标条
designation 名称，指定，指示，选定，规定，符号，牌号，标识，称号，任命，委派
designator 指示符，指示器，标志符，选择器
design automation 设计自动化
design-base accident 设计基准事故
design baseline 设计基线
design basic acceleration of ground motion 设计基本地震加速度

design basis accident 设计基础事故
design basis depressurization accident 设计基准卸压事故
design basis earthquake 设计基准地震
design basis event 设计基准事件，参考事件
design basis external man-induced events 设计基准外部人为事故
design basis fire 设计基准火灾
design basis flooding level 设计基准淹没水位
design basis flood 设计基准洪水
design basis for external events 有关外部时间的设计基准
design basis loss-of-coolant accident 设计基准冷却剂丧失事故
design basis natural event 设计基准自然事件
design basis storm 设计基准风暴
design basis 设计基础，设计依据，设计基准
design bearing capacity 设计承载力
design boundary definition 设计边界定义
design boundary point 设计分界点
design brief 设计概要
design-build contract 设计建造合同
design burn-up 设计燃耗
design by deformation 按变形设计
design by modulus system 模数制设计
design calculation 设计计算
design capability 设计能力
design capacity 设计容量，设计能力
design change control 设计更改控制
design change 设计修改，技术修改，设计变更
design characteristic period of ground motion 设计特征周期
design chart 设计图表
design clarification 设计澄清
design coal 设计煤种
design code 设计规范，设计准则，设计规程
design competition 设计竞争
design concept 设计概念，设计思想，设计理念
design conditions 设计参数，设计条件，计算参数，设计工况
design considerations 设计要考虑的诸因素，设计依据，设计根据
design constraints 设计约束条件，设计制约
design-construction contract 设计施工合同
design & construction rules for mechanical components of PRW nuclear island 压水堆核岛机械设备设计和制造规则
design-construction team 设计施工小组
design contract 设计合同
design control 设计管理，设计控制，设计校核
design coordinate 设计坐标
design cost 设计成本
design crest level 设计坝顶高程
design criteria of large dams 大坝设计准则
design criterion 设计标准，设计准则
design curves 设计曲线
design data survey 设计测量
design data 设计数据，设计经验，设计资料
design department 设计部门
design dependability 设计保证率
design depth 设计水深
design description 设计说明
design details 设计细则，设计详图
design discharge 设计流量
design division 设计处，工程处
design document 设计文件，设计文献
design drawing 设计图纸，设计图
design earthquake value 设计地震值
design earthquake 设计地震，抗震设计，设定地震，设防地震
design edacity 设计能力
designed differential head 设计压差
designed elevation 设计标高，设计高程
designed for seismic intensity of X 地震烈度按X级考虑，按X级的地震烈度设防
designed level 设计标高
designed service life 设计（使用）寿命
designed to do 设计成能（做），用来……，目的是使……
designed to withstand a 350kPa explosion pressure 设计成能承受350kPa的爆炸压力
designed velocity 设计流速
designed waterline 设计水线
designed with 设计成具有［带有］……
design effect 设计效果
design efficiency 设计效率
design elevation 设计标高，设计高程
design engineering 设计工程
design entrusting 设计委托
designer 设计人，设计者，计划者，设计师，设计单位，制图者
design examination and approval 设计审批
design experience 设计经验
design failure 设计失效
design fatigue curve 设计疲劳曲线
design feature 设计特点
design fee 设计费
design flood calculation 设计洪水计算
design flood composition 设计洪水组成
design flood hydrograph 设计洪水过程线
design flood inflow 设计入库洪水流量
design flood level （水库）设计洪水位
design flood occurrence 设计洪水出现率，设计洪水重现期
design flood of construction period 施工（期）设计洪水
design flood of small catchment 小流域设计洪水
design flood 设计洪水，设计洪水量
design floor response spectrum 设计楼层反应谱，设计楼层响应谱
design floor time histories 设计楼层时程
design flow 设计流量
design flow of heat-supply network 热网设计流量
design flow rate 设计流量
design formula 计算式，（设计）计算公式
design for 打算把……做（某用途），为（某目的）而设计
design grade 设计坡度
design guide 设计指南
design head 设计水头
design heating load 设计热负荷
design height 设计高度

design high water level 设计高潮位
design horse power 设计马力，设计功率
design hourly demand of hot-water 热水供应设计小时耗水量
design humidity 设计湿度
design hydraulic regime 设计水力工况
design hydrograph 设计过程线，设计水文过程线
design ice thickness 设计冰厚
design idea 设计意图
design improvement program 设计改进程序
design improvement work 设计改进工作
design improvement 设计改进
design information 设计资料
designing institute 设计院
designing scheme 设计方案
designing 设计，有事先计划的
design input date 设计输入数据
design input document 设计输入资料
design input 设计输入
design institute 设计院，设计单位
design instruction 设计任务书，设计说明书
design intention 设计意图
design intent 设计意图
design interface control 设计接口管理
design interface 设计接口
design level 设计水平
design liaison meeting 设计联络会
design life of components 部件设计寿命
design lifetime 设计寿命
design life 设计（使用）寿命
design lift coefficient 设计升力系数
design limit 设计极限
design line of demarcation 设计分界线
design line 设计线
design load case 设计载荷工况
design load 设计荷载，设计负荷，设计载荷，设计负载，设计载重
design manual 设计手册，制图手册
design master 设计大师
design method 设计方法
design mix rate 设计配合比
design motif 设计主题
design notes 计算说明，设计说明，设计书，计算清单，计算书
design objective 设计目标
design of general layout and transportation 总图运输设计
design of hydro project 水利工程设计
design of product 产品设计
design of steel structure 钢结构设计
design of thermal power plant automation 热工自动化设计
design operating conditions 额定运行工况，设计运行工况
design optimization 设计优化
design organization 设计单位
design outline 设计提纲，设计草图
design-out maintenance 无维修设计
design output document 设计输出资料
design output 设计出力，设计输出
design overpower factor 设计超功率因子
design overpressure 设计超压
design paper 设计文件，绘图纸
design parameters of ground motion 设计地震动参数
design parameter 设计参数
design patent 设计专利
design pay load 设计有效载荷
design peaking factor 设计峰值因数
design performance 设计性能
design period 设计年限，设计期限
design personnel 设计人员
design phase 设计阶段
design philosophy 设计原理，设计理念，设计特点，设计原则
design plan 设计图
design point efficiency 设计工况点效率
design point 设计点，设计工况，计算点，计算位置
design power efficiency 设计工况效率
design power 设计能力，设计功率
design precept 设计方案
design precipitation 设计降水
design pressure coefficient 设计压力系数
design pressure 设计压力
design principle 设计原则
design problem 设计问题
design procedure 设计程序，设计步骤，设计工序
design process 设计过程
design product 设计产品
design project 设计项目
design proof 设计保证值
design proposal 设计方案
design qualification 设计资质
design quality 设计质量
design quantity 设计数量
design rainfall 设计雨量
design rating 设计额定值，设计出力，设计容量
design recurrence interval 设计重现期
design reference period 设计基准期
design reference 设计参考，设计依据，设计基准
design reliability 设计可靠性
design report 设计报告，计算书，计算说明，设计书，设计清单，设计说明
design representative on site 驻工地设计代表
design representative 设计代表
design requalification 设计复审
design requirement 设计要求
design response spectrum 设计反应谱
design responsibility 设计责任
design review committee 设计审查委员会
design review meeting 设计审查会
design review process 设计审查过程
design review 设计审查，设计评审
design right 设计权
design rotor annual mean speed 设计风轮年平均转速
design rotor overspeed 设计风轮超速

design rule 设计规则
design safety factor 设计安全系数
design safety margin 设计安全裕度
design safety standard 设计安全标准
design schedule 设计计划，计算表，设计进度，核算表
design scheme 设计方案，设计图
design section 设计断面
design seismic coefficient 设计地震系数
design service life 设计使用期限
design service 设计服务
design situation 设计工况，设计和安全参数，设计状况
design size 设计尺寸
design slope 设计坡度
design software 设计软件
design solution 设计方案，解决方案
design space-heating load 供暖设计热负荷，采暖设计热负荷
design specification summary sheet 设计规范汇总表
design specification 设计技术指标，设计规范，设计规程，设计说明书
design speed 设计速度
design stage 设计阶段
design standard of flood control 洪水设计标准
design standard of large dams 大坝设计规范
design standard 设计标准，设计规程，设计规范
design steam side pressure 设计的蒸汽侧压力
design steam temperature 设计蒸汽温度
design step 设计步骤
design storm 设计暴雨，设计风暴
design strength 设计强度
design stress intensity value 应力设计强度值
design stress intensity 允许应力强度，设计应力强度
design stress 设计应力
design subcontractor 设计分包者
design supplementary document 设计补充文件
design supplement 设计附件
design system 设计系统
design team 设计组［队］
design technical requirements 设计技术要求
design temperature of return water 设计回水温度
design temperature rise 设计温升
design temperature 设计温度
design tension 设计张拉力
design thermal efficiency 设计热效率
design tide level 设计潮位
design tip speed ratio 设计尖速比
design transition temperature 设计转变温度【压力窗口材料】
design unit 设计单位
design validation 设计确认
design value of a geometrical parameter 几何参数设计值
design value of a load 荷载设计值
design value of a material property 材料性能设计值
design value of an action 作用设计值
design value of embedded depth 嵌固深度设计
design value 设计参数，计算值，设计值
design variable 设计变量
design velocity ratio 设计速比
design verification 设计验证
design voltage 设计电压
design water level 设计水位
design wave height 设计波高
design wave 设计波，设计波浪
design wind load 设计风载
design wind pressure 设计风压
design wind speed 设计风速
design working life 设计使用年限
design 设计，图样，构思，类型，绘出，计划，图纸，计算
desilicate 除硅
desilication 脱硅作用
desiliconization 除硅
desiliconized water 除硅水
desiliconize 除硅，硅分离
desilter 沉淀池，集尘器，滤水池，沉沙池，除砂器
desilting area 沉沙面积
desilting basin 沉砂池，沉沙池
desilting strip 沉沙畦条，放淤畦条
desilting works 除沙工程【如沉沙池等】，沉砂工程
desilting 沉沙，清淤，挖除泥沙
desilt 放淤
desintegrate 分解，分裂，裂变
desintegration 机械破坏，机械分解
desirable criterion 必要标准，理想标准
desired control point 所要求的控制点
desired effect 预期效果，预期影响
desired impact 要求达到的效果
desired output 理想输出值，期待输出值
desired speed 预想转速，希望转速
desired strength 要求强度，预期强度
desired value 预期值，期待值，期望值，给定值，预定值，所需值，约定真值
desired 期待的
desk calculator 台式计算机
desk fan 台（式风）扇
desk metallographic microscope 台式金相显微镜
desk switchboard 台式配电盘
desk-top computer 台式计算机
desk type control panel 台式控制屏
desk type digital multimeter 台式数字万用表
desk 控制台，操纵台，控制板，台，桌，面板
deslag 除渣
deslagger 打渣器
desliming 除残渣，除泥渣
desludging valve 排渣阀
desludging 清除泥渣
desorption reagent 解吸剂
desorption 解吸
desoxydation 脱氧
desoxy 脱氧，脱氧的
de-SO_x 脱硫
despatch room 调度室

despiker circuit 削峰电路
despiking resistance 削峰电阻
despiking 脉冲纯化，尖峰平滑
despin device 消旋装置
despin mechanism 消旋装置
despin 降低转速，停止旋转，消自转，减慢旋转速度，反转，反自转
despite 即使，尽管
despumate 除泡沫
despun antenna 消旋天线，反旋转天线
despun motor 反自转电动机，反旋转电动机
destabilization 失稳，扰动，不安定，脱落，剥离，不稳定，剥落
destaticizer 脱静点剂，去静电器
desteamer 除汽器
destination address 目的地址
destination code 目的码
destination date 目的数据
destination delivery charges 目的地交货费
destination harbor 目的港
destination memory 目的存储器
destination port 目的港，终点港，到达口岸
destination register 目的寄存器
destination symbol 目标符号
destination time 终到时间，终点时间
destination 目的地，目的，目标，指定，预定，指定地点，终点（站）
destmctive test 破坏性实验
destoring 物料出库
destrengthening annealing 应力释放退火
destressing 消除应力，去应力
destroyer prototype reactor 驱逐舰用模式反应堆
destroy 破坏
destructional bench 侵蚀阶地
destructional valley 侵蚀谷
destruction corrosion 破坏性腐蚀
destruction limit 破坏极限（值）
destruction of insulation 绝缘损坏，绝缘破坏
destruction of soil 土壤破坏
destruction of turbulence 紊流衰减
destruction rate 损坏率，耗损率
destruction test 破坏实验，破坏性试验，断裂试验
destruction 破坏
destructive activation analysis 破坏性活化分析
destructive addition 破坏信息加法
destructive analysis 破坏性分析，损毁分析
destructive distillation 破坏性蒸馏，干馏，分解蒸馏
destructive earthquake 破坏性地震
destructive effect 破坏作用
destructive examination 破坏性检验
destructive experiment 破坏实验，破坏试验
destructive inspection 破坏性检查
destructive malfunction 破坏性故障
destructive metamorphism 破坏变质
destructive overspeed failure 破坏性超速事故【汽轮机】，（汽轮机）过速破裂
destructive oxidation 破坏性氧化
destructive physical analysis 破坏物理试验
destructive power 破坏力
destructive reading 抹去信息读数，破坏性阅读，抹掉信息读出
destructive readout memory 破坏读出存储器
destructive readout 破坏性读出，抹去信息读出
destructive read 破坏读出
destructive shock 破坏性震动
destructive test of steel tower 铁塔破坏性试验
destructive test 破坏性试验，击空试验，破坏检验，破损性试验
destructive vibration 破坏性振动
destructive wave 破坏性波浪
destructive 毁灭性的，破坏的，有害的，可抹去的，破坏性的
desulfuration 脱硫
desulfurization and denitration 脱硫脱硝
desulfurization and denitrification 脱硫脱硝
desulfurization by wet processes 湿法脱硫
desulfurization chemicals 脱硫剂
desulfurization method 脱硫法
desulfurization refinery 脱硫精炼厂
desulfurization 脱硫
desulfurized fuel 脱硫燃料
desulphidation 脱硫（作用）
desulphurisation 脱硫
desulphurization and denitrogenation 脱硫脱硝
desulphurization equipment 脱硫装置
desulphurization 脱硫
desuperheated steam 减温蒸汽
desuperheater spray 减温器喷水
desuperheater 过热减温器，过热蒸汽减温器，减压减温装置，减温器
desuperheating agent 减温介质
desuperheating and reducing device 减温减压装置
desuperheating spray control valve 过热器减温喷水控制［调节］阀
desuperheating spray 喷水减温
desuperheating station 过热蒸汽减温站
desuperheating water 减温水
desuperheating zone 减温段，过热蒸汽冷却区，过热利用区
desuperheat 过热蒸汽减温
desurfacing 除表面层，消除表层金属
desynchronizing 同步破坏，失步，去同步
desyn 直流自动同步机
DETAB(decision table) 判定表
detachability 脱渣性，可分离性，可拆卸性
detachable 可拆卸的
detachable bit 可卸钻头
detachable blade 可拆叶片
detachable coil 可拆线圈
detachable device 可拆连接
detachable pole 可拆磁极，活动磁极
detachable shoulder 可拆卸的台肩
detached breakwater 独立式防波堤，离岸堤
detached coefficient 分离系数
detached column 独立柱
detached pier 独立支墩
detached shock wave 脱体激波，离体激波
detached shock 脱体激波，离体激波

detached superstructure　独立上层建筑
detached wharf　岛式码头
detached　已拆卸的
detachment of vortices　旋涡脱落
detachment point　（气泡）脱离壁面点
detachment　分离，拆开
detach　分开，拆开，卸下，脱离，脱体
detail budget　细目预算
detail calling record　呼叫详细记录
detail chart　详细流程图
detail control network　详细控制网络
detail data　详细资料
detail design　详细设计，施工图设计，细部设计
detail drawing　细部图，详图，零件图，大样图，大样（图）纸，施工图，加工图，细部详图
detailed account　清单
detailed audit　详细审计
detailed calling record　呼叫详细记录
detailed contents　详细内容
detailed description　详细说明
detailed design　细部设计，详细设计，施工图设计
detailed diagram　详图
detailed drawing　细部图，详图
detailed engineering　施工图设计版
detailed estimate norm　预算定额
detailed estimate of construction coast　建筑造价的详细预算
detailed examination　详细检查
detailed explanation　详细说明
detailed exploration　详细勘探
detailed field performance　现场性能细节
detailed implementational procedure　具体执行规程
detailed instruction　详细说明
detailed internal procedure　详细的内部程序，内部顺序细目
detailed list　明细表，清单
detailed planning　详细规划
detailed plan　碎部图，详细计划，详图
detailed project report　详细项目报告
detailed questions　细节问题
detailed regulation　细则
detailed report　详细报告
detailed schedule　详细计划表
detailed supply list　供货明细表
detailed survey　碎部测量，细部测量，详细勘测，详细调查，详勘，详查
detailed technical specification　详细技术说明书
detail engineering design data　详细工程设计数据
detail evaluation conclusion　详评结果
detail for connection　连接点详图
detailing　大样设计，细部结构，细部设计
detail list of HVAC system　暖通空调设备明细表
detail list　明细表
detail loss　清晰度损失，清晰度降低
detail mapping　碎部测图
detail of design　设计详图
detail plotting　细部测图
detail printing　细目打印
detail quotation　详细报价
detail requirement　详细要求
detail sensitive　细节灵敏度
details of construction　构造详图
details of seismic design　抗震构造措施
detail specification　详细规范
detail status report　详细情况报告
detail structure　细部结构
detail survey　详测
detail　详图，零件图，详细，细节，细目，详细的，大样（图），细部
detainment　扣留
detain　扣留，扣押
detatching　解脱，非耦联
detectability　检测能力，检验能力，探测能力，检波能力，可检测性
detectable error　可检验误差
detected probability　探测概率
detected　被检测
detecting device　检出器，敏感元件
detecting element　探测元件，检波元件，传感元件，敏感元件，检测元件
detecting head　探头
detecting instrument　探测仪器，检波仪器
detecting system　探测系统
detecting　检测的，检波的，探测的
detection apparatus　探测仪器
detection circuit　探测电路，传感电路，检测电路
detection curve　探测曲线
detection efficiency　探测效率
detection equipment　检测装置
detection limit　探测极限，检测极限
detection of defects　探伤，缺陷检查
detection pressure　探测压力
detection probability　探测概率
detection rate　探测效率
detection reservoir　拦洪水库
detection sensitivity　探测灵敏度
detection storage　拦洪库容
detection system　检测系统
detection test　探测试验
detection threshold　检测阈值，探测阈
detection time　探测时间
detection width　探测道宽度
detection　检查，检测，探测，检波，发现，探知，侦查
detectivity　检测能力
detector diode　检波二极管
detector efficiency　探测器效率，探头效率
detector hole　探测器孔
detector noise　检测器噪声，检波器噪声，探测器噪声
detector of defects　探伤仪
detector probe　探头
detector response　探伤仪灵敏度
detector sensibility　检测器灵敏度
detector storage　探测器贮存
detector temperature setting　探测器温度设定（值）

detector 探伤仪，探测器，检波器，指示器，传感器，测量探头，检出器
detect 探测，检波，检查，检测，测定
detention basin 拦洪水库，滞洪区
detention dam 拦洪坝
detention depth 滞洪水深
detention effect 滞留效应，滞留作用
detention period 延迟时间，滞留时期
detention reservoir 滞洪水库
detention storage 滞洪蓄水
detention tank 停留池，污水滞流池
detention time 停留时间
detention volume 拦洪容量，滞洪容积，滞洪容量
detention 扣留，停滞，滞留，阻止，卡住，拖延，停留时间，误期，阻滞
detent mechanism 定位器，定位机构，闭锁机构
detent pin 定位销
detent plate 制动器板
detent plug 止动销
detent torque 定位转矩，（永磁电动机）启动转矩，无激磁保持转矩
detent 爪，棘爪，锁销，止轮器，定位的，止动装置
detergence 洗涤力，净化力，去污能力
detergent pollution 洗涤剂污染
detergent waste 洗涤废液
detergent 去污剂，洗涤剂，清洁剂，去垢剂，清洁的，使干净的，净化的
deteriorate 变坏，降低品质
deterioration of weather 天气变坏
deterioration rate of fixed assets 固定资产损耗率
deterioration rate reduction 恶化速率下降
deterioration rate 变质率，恶化率，退化率
deterioration 恶化，退化，损坏，磨损，劣化，变坏，变质，腐蚀
determinant calculation 行列式计算
determinant of matrix 矩阵行列式
determinant rank 行列式秩
determinant 决定要素，行列式
determinate hydrological model 确定性水文模型
determination data 确定数据
determination of control point 控制点测定
determination of coordinate 坐标确定
determination of deformation 变形测定
determination of frequency 频率测定
determination of output 功率确定，输出量确定，功率的测定
determination of quality 定性测定
determination of responsiveness 响应性鉴定
determination of soil parameters 土壤参数测定
determination of strength of soil 土强度测定
determination of water quality 水质测定
determination 确定，决定，测定
determine on a case-by-case basis 逐案确定，根据个别方案而定
determine 终止【合同】，决定，确定，测定，规定，求出，解决，了结，结束
determining factor 确定因子，决定性因子，决定性因素
deterministic algorithm 确定性算法
deterministic analysis 确定性分析
deterministic channel 确定的通道
deterministic criteria 确定性准则【如确定电力系统可靠性】
deterministic design method 定值设计法
deterministic excitation 非随机激振
deterministic fluctuating wind speed 主动脉风速
deterministic gust 主阵风
deterministic model 确定性模型，非随机性模型
deterministic signal 确定性信号
deterministic study 确定性研究，最终研究
deterministic variable 决定性变量
deterministic 确定性的
detersive 清净剂，洗净剂，使干净的，使清洁的
detonate tube 雷管
detonating agent 起爆剂，引爆剂
detonating charge 引爆药
detonating cord string 导火线
detonating cord 起爆软线
detonating explosives 起爆炸药
detonating fuse 导火线，引爆线
detonating gas 爆炸性气体
detonating powder 起爆火药
detonating signal 起爆讯号
detonation flame 爆燃火焰
detonation front 爆炸前沿
detonation gun coating 爆燃枪涂敷
detonation sound 爆炸声
detonation spraying 气体爆燃式喷涂
detonation velocity 爆燃速度
detonation wave 爆破波，爆炸波，燃爆波
detonation 爆炸，爆燃，引爆，爆破，起爆
detonator cap 雷管帽盖
detonator cartridge 炸药包，炸药筒
detonator 爆破雷管，雷管，起爆剂，引爆雷管，引信
detoxication 解毒
detoxification 解毒
detoxify 解毒，去沾染
detract 损伤
detrimental 有害的
detrimental effect 有害的影响，危害性，有害作用，危害影响
detrimental expansion 危害膨胀
detrimental settlement 危害沉陷，有害沉降
detrimental soil 不稳定土
detriment 损害，损伤，危害，损害物
detrital deposit 碎屑沉积
detrital rock 碎屑岩
detrital sediment 碎屑沉积
detrital slope 碎屑坡
detrital 碎散的，碎屑，碎屑的
detritiation 除氚
detrition 磨耗，磨损，损耗
detritus advance 岩屑侵入
detritus avalanche 泥石崩塌，岩崩
detritus chamber 沉沙池，沉沙室，大颗粒沉沙池，泥石沉淀室
detritus movement 岩屑移动

detritus slide 碎屑滑动
detritus stream 泥石流，岩屑流
detritus tank 沉渣池
detritus 渣，岩屑
detrusion ratio 剪切比
detrusion 外冲
detuner 解调器
detune 失谐，误调谐，解谐，解谐
detuning of resonant circuit 谐振电路失谐，谐振电路解谐
detuning 解谐，失调，失谐，解调，去谐
deuterated water 氘化水
deuterated 氘化，重水合物，氘水合物
deuteration and dedeuteration system of HWR 重水堆氘化和除氘系统
deuteric 初生变质的
deuterium-bound neutron 重氢束缚中子，氘束缚中子
deuterium critical assembly 重水临界实验装置
deuterium hydrogen oxide 释义氢氧化氘
deuterium-moderated reactor 重水慢化反应堆
deuterium oxide moderated reactor 重水慢化反应堆
deuterium oxide 氧化氘，重水
deuterium pile 重水堆【指慢化反应堆】
deuterium-sodium reactor 重水钠反应堆
deuterium-tritium reactor 氘氚反应堆
deuterium-uranium reactor 铀重水反应堆，重水铀反应堆，铀重水堆
deuterium 氘，重氢
deuterization 氘化
deuterize 氘化
deuterizing 氘化
deuterogene 后期生成，后生岩
deuterogenous 后期生成的
deuteron-binding energy 氘核结合能
deuteron-induced fission 氘核诱发裂变
deuteron 氘核，重氢核
deuteroxide 重水
Deutsche Bank 德意志银行【德国】
devaluation 贬值
de-vanning charge 拆箱费
devaporation 蒸汽凝结，止汽化
devaporizer 余汽冷却器，蒸发冷却器，清洁器
devastate 毁坏
devastating flood 毁灭性洪水
DEV＝deviation 偏差
developable surface 可展面
developed area 发达地区，展开面积
developed blade area 叶片展开面积
developed country 发达国家
developed-flow region 已发展区
developed head at runout flow 最大流量时的压头
developed head 产生的压头
developed length 构筑长度，展开长度
developed lift 已升高程
developed nations 发达国家
developed power 发出功率
developed surface flaw 发展的表面缺陷
developed turbulent flow 扩展的紊流
developed view 展开图
developed water 开发水
developed width 实际宽度
developed winding diagram 绕组展开图
developed 发达的，成熟的
developer powder 显影粉
developer 显影液，显影剂，开发者
developing bath 显影液
developing coil 嵌线，下线
developing country 发展中国家
developing economy 发展中的经济
developing fault 发展性缺陷，演变性缺陷，发展中缺陷
developing flow 未定型流动的
developing nations 发展中国家
developing time-temperature system 显影时间温度系统
developing trend 发展趋势
developing winding coil 嵌线，下线
developing winding 展开的绕组
developing 展开
development aid 开发援助
developmental length of snow drift 吹雪的展开长度，雪堆的展开长度
developmental project 开发工程
development area 开发区，展开面积
development bank 开发银行
development capital 开发资金
development corporation 开发公司
development drawing 展开图
development engineering management system 开发工程管理系统
development fund 发展基金
development group 开发小组
development in coordination of 与……协调发展
development in planned and proportion way 有计划按比例发展
development length 伸展长度
development loan 开发贷款
development of gas 放出毒气
development of heat 放热
development of land 土地开发
development of new technology 新技术开发
development of nozzle 喷嘴展开
development of reservoir zone 库区综合开发利用
development of seaside area 沿岸开发
development of technology 技术开发
development of water resources 水资源开发
development organization 项目建设单位
development plan 发展规划，发展计划
development potential 开发潜力
development program of municipal heat-supply 城市供热发展规划
development program 发展计划，开发计划
development project 发展项目，开发项目
development reactor 研究发展用反应堆
development running 研制性运行
developments and prospects 发展与展望
development stage 开发阶段，研制阶段
development status report 发展情况报告
development strategy 发展战略
development system 开发系统

development tendency 发展趋向
development test 修整试验，开发测试，发展测试，开发试验，试制时试验，试制品试验，研制试验，研究实验
development type 试验样品，试验样机，研制样品
development work 试制工作
development zone 开发区
development 进展，发展，推导，研究，开发，研制
deviated flow 偏斜水流
deviating force 偏转力，偏向力
deviating 偏离
deviation alarm 偏差报警
deviation angle 落后角，偏移角，偏角，偏向角，转角
deviation bubble 水准器气泡偏差
deviation clause 偏差条款
deviation coefficient 离差系数
deviation computer 偏差计算器
deviation equalizer 偏差补偿器，偏差平衡器
deviation factor of wave 波形畸变因子，波形畸变系数
deviation factor 偏差系数，畸变系数
deviation form 偏差表
deviation frequency 偏移频率
deviation from linearity 线性偏差
deviation from mean 离均差
deviation integral 偏差积分
deviation list 偏差表
deviation method 偏差法
deviation notice 偏差通知
deviation of synchronous time 同步时间偏差
deviation permit 偏离许可
deviation prism 偏向棱镜
deviation ratio 偏差系数，漂移率
deviation reduction factor 偏差减低因子
deviation signal 偏差信号
deviation value 偏差值
deviation 偏离，偏差，偏移，偏向，差错，跑偏，误差
deviator stress 偏应力，偏差应力，应力偏量
deviator 偏差器，致偏装置，变向装置，致偏器，偏向装置
device address 外围设备地址
device assignment command 设备赋值命令
device assignment 设备的指定逻辑号及输入输出指令
device control 设备控制
device driver 设备驱动器，设备驱动程序
device flag 设备标识
device for automatic power 自动功率调整器
device for fixing rotor 转子固定装置
device for raising assembly from containers 把燃料组件从容器中提升的装置
device independence 与外部设备无关性，设备独立性
device medium control language 设备媒体控制语言
device number 设备号
device parameter 器件参数，晶体管参数
device selector 设备选择器
device status word 设备状态字
device switching unit 设备开关设置
device under test 被测器件，被测装置
device 装置，设备，仪表，仪器，方法，器件，元件，策略
devil liquor 废液
devil 不幸，麻烦，尘暴
deviometer 偏差计
devitrification 使失去光泽，失透性
devitrify 使反玻璃化
devolatilization 脱去挥发分
devolution agreement 移交协定
devolution of power 权力下放
devolution 权力下放，转移
devolve on 移交
dew 露（水），结露
DEWA（Dubai Electricity and Water Authority） 迪拜水电局
dewar bottle 真空瓶，杜瓦瓶
dewar tube 全玻璃真空集热管
dewar vacuum flask 杜氏真空瓶
dewar vessel 杜瓦瓶，真空瓶
dewatered steam 脱水蒸气
dewaterer 脱水机
dewatering agent 脱水剂
dewatering bin 脱水装置
dewatering conduit 放水管，排水管
dewatering equipment 排水设备
dewatering hole 排水孔
dewatering machine 脱水机
dewatering method 排水方法
dewatering orifice 放水孔
dewatering outlet 放水管，放水口
dewatering pump 脱水泵，抽水泵
dewatering screen 脱水筛
dewatering system 疏干排水系统，脱水系统
dewatering 脱水，排水，失水，降地下水，去水，放水，水分分离，除水
dewaterizer 脱水机
dewater 排水，脱水，疏水，使脱水
dewaxing 脱蜡处理
dew cell 道氏电池，湿敏元件
dew-drop slot 梨形槽
dew formation 结露
dew gauge 露量器
dew point control system 露点控制系统
dew point deficit 温度露点差
dew point depression 露点降低，露点下降
dew point hygrometer 露点湿度表，露点湿度计
dew point indicator 露点指示器，露点计，露点仪
dew point measuring meter 温度露点测试仪
dew point meter 露点仪
dew point recorder 露点记录器
dew point spread 露点差
dew point temperature 露点温度
dew point 露点
dew-retardation 防露
dexterity factor 灵巧因子
dextrogyrate 右旋的

dextrogyric 右旋的
dextrorotary 右旋的
dextrorotation 右旋，顺时针反向旋转
dextrorotatory 右旋的
dextrose 右旋糖，葡萄糖
dezincification 脱锌现象，脱锌
DFA(damage function analysis) 损伤功能分析
DFC(disk file control) 磁盘文件控制器
DFCS(digital feedwater control system) 数字给水控制系统
DF(dissipation factor) 损耗因素
DFIG(double fed induction generator) 双馈感应发电机，双馈异步发电机
D flip-flop D触发器，延迟触发器
DFR(Dounreay fast reactor) 唐瑞快中子反应堆
DFSG(double fed synchronous generator) 双馈同步发电机
DFT(diagnostic function tester) 诊断功能测试程序
DFW(deaerator feedwater) 除氧器给水
DG(differential generator) 微分发生器
DGQ 联邦德国质量保证学会
DGR(dangerous goods regulations) 危险物品手册
DGS＝degasifier 脱气器，除氧器
DGS(distance-gain-size) 距离-增益-尺寸
DGS 联邦德国质量体系认证组织
DHC(data handling center) 数据处理中心
DH cogeneration unit 区域供暖热电联产机组
D＆H(dangerous and hazardous) 危险品
DH(deaerator heater) 除氧加热器
DH(district heating) 区域供暖，集中供暖
DHE(data handling equipment) 数据处理装置
DHI(diffuse horizontal irradiance) 散射辐射【在大气中散射的直接到达地面的阳光】
DHP(delivered horse power) 输出功率
DHRS(decay heat removal system) 衰变热排出系统
diabase-aplite 辉绿细晶岩
diabase lining 衬辉绿岩，衬铸石，辉绿岩衬砌
diabase 辉绿岩
diabasic plaster 辉绿岩灰泥
diabatic flow 非绝热的
diabatic two-phase flow 非绝热两相流
diabatic 非绝热的，受热
DIAC(diode alternating current switch) 双向开关二极管，双向触发二极管，二极管交流开关
diacetate 二醋酸盐
diacetyldioxime 丁二酮肟，镍试剂
diaclase 正方断裂线
diacoustic 折声的，折射散焦
diacritical current 半临界值电流
DIA＝diameter 直径
diafilter 渗滤
diagenesis 成岩作用，岩化作用
diaglyphic ornament 凹刻装饰
diaglyph 凹雕，凹刻
diagnose 诊断
diagnosis technique for residual life 寿命诊断技术
diagnosis 诊断，判断分析，识别
diagnostic and test software 诊断和测试软件
diagnostic check 诊断校验，诊断试验
diagnostic component programming 诊断程序设计
diagnostic dose 诊断计量
diagnostic error processing 诊断错误处理
diagnostic function tester 诊断功能测试处理
diagnostic parameter 诊断参数
diagnostic programming 诊断程序设计
diagnostic program 诊断程序
diagnostic radiography 放射诊断（学）
diagnostic routine 诊断程序
diagnostic software 诊断软件
diagnostics 诊断，诊断法，诊断学
diagnostic technique 诊断技术
diagnostic test 诊断实验，诊断检查，诊断试验
diagnostic 诊断的
diagnotor 诊断程序
diagometer 电导计
diagonal bar 斜钢筋
diagonal bedding 斜层理
diagonal bond 对角砌合
diagonal brace 斜撑
diagonal bracing 对角支撑，剪刀撑，斜撑，对角系杆
diagonal conducting wall generator 斜导电壁发电机
diagonal corner firing 四角燃烧
diagonal cracking 斜向开裂
diagonal-cutting pliers 斜咀钳
diagonal difference 对角差分
diagonal distance 对角距离
diagonal fault 斜断层
diagonal flow generator 斜流通风式发电机
diagonal flow turbine 对角式水轮机
diagonal flow 对角流
diagonal grain 斜纹
diagonal joint 斜节理，斜接，斜接缝
diagonal leg profile 塔基断面
diagonal matrix 对角矩阵，对角阵
diagonal member 斜杆
diagonal mirror 对角反射镜
diagonal paving 斜向铺砌
diagonal pitch 斜心距，斜节距
diagonal reinforcement 斜钢筋，弯起钢筋
diagonal rib 斜交肋
diagonal rod 斜杆
diagonal scale 斜线比例尺，斜线尺
diagonal stay 斜撑条，对角拉撑
diagonal strip 斜板条
diagonal strut 对角（支）撑，斜撑，斜支柱
diagonal tension stress 斜拉应力，主拉应力
diagonal tie 斜拉杆，斜系杆
diagonal tracking 对角裂缝
diagonal triangle 对边三角形
diagonal wall generator 斜壁式发电机
diagonal web member 斜腹杆
diagonal wrench 活扳手，瑞典式扳手
diagonal 对角线，斜杆，斜拉筋，对角线斜杆，斜构件，斜的，对角的
diagond flow 斜流（式）
diagram board 图表板

diagrammatical chart 示意图
diagrammatical 图解的，图示的
diagrammatic arrangement 原则性布置
diagrammatic curve 图解曲线
diagrammatic drawing 草图
diagrammatic layout 原则性布置，示意图，原理图
diagrammatic sectional drawing 断面草图
diagrammatic sketch 草图，示意图，简图
diagrammatic solution 图解法
diagram of curves 曲线图
diagram of external loads 外荷载图
diagram of gears 齿轮传动图
diagram of heat flow 热流图
diagram of strains 应变图
diagram of work 示功图
diagram paper 绘图纸
diagram 图表，线图，图，曲线，图解，用图表示
diagrid 堆芯栅板，堆芯容器垫板
diakoptic algorithm 网络分割算法
diakoptics 大规模电网分片求解法
dial central office 自动电话总局，自动电话中心局
dial contact 号盘触电
dial counter 表盘式计数器
DIAL(differential absorption lidar) 差异吸收光达
dial drive 号盘驱动
dial drum 刻度筒
dial extensometer 针盘式伸长计
dial gauge 针盘量规，刻度盘，刻度计，指示器，测微仪，测微表，罗盘表，指示表，千分表，圆形量表
dial illumination （仪表）刻度照明
dial impulse 号盘脉冲
dial indicator 标度盘指示器，刻度盘指示器，针盘指示表［器］，度盘式指示表，千分表
dialing impulse 拨号脉冲
dialing 拨号
dial instrument 表盘式仪表，指针式仪表，有刻度的仪表
dial in 拨号接入
dial lamp 标度灯
dial light 标度盘灯光
dial pulse incoming register 拨号脉冲入局记录器
dial pulse interpreter 拨号脉冲译码器
dial pulse signaling 号盘脉冲信号
dial pulse 标度盘脉冲
dial ratio 号盘脉冲比
dial scale 度盘刻度，带表秤
dial set computer 带拨号盘的计算机
dial snap gauge 带表卡规
dial strain gauge 针盘式应变计
dial telephone 自动电话
dial thickness gauge 针盘式测厚计
dial-type 圆盘式
dialysate 淡水流，渗出液，渗析膜，渗析器，渗析，渗析法
dialyzing compartment 渗析室
dial 标度盘，表盘，刻度盘，拨号盘，拨号，度盘，分度盘，圆形把手，转盘
diamagnetic alloy 抗磁合金，反磁合金
diamagnetic effect 反磁效应，抗磁效应
diamagnetic material 抗磁材料，反磁材料
diamagnetic polarity 抗磁极性，反磁极性
diamagnetic resonance 抗磁谐振
diamagnetic shielding 抗磁屏蔽
diamagnetic susceptibility 抗磁化率
diamagnetic 抗磁性的，反磁性的
diamagnetism 抗磁性，反磁性
diamagnet 抗磁体，反磁体
diameter inspection 直径检查
diameter-length ratio 径高比【指汽轮机】
diameter of bolt hole 螺孔直径
diameter of commutation 完全换向绕组应有的径面
diameter of cylinder 气缸直径
diameter of spiral bar 螺旋筋的直径
diameter of spiral 螺旋的直径
diameter of stator bore 定子内径
diameter of wire 线径
diameter over bark 连皮直径
diameter pitch 径节距，整节距
diameter run-out 径向跳动
diameter transducer 直径传送器
diameter 直径
diametral clearance 径向间隙
diametral compression test 劈裂抗拉试验
diametral flow 径向流动
diametral gap 直径径向间隙【燃料元件】，径向间隙，直径公差
diametral plane 切径平面
diametral voltage 对径电压，径向电压
diametral winding 整距绕组
diametrical connection 延径接线，对角接线
diametrical magnetization 径向磁化
diametrical pitch 全节距，整距，径节距
diametrical tappings 电枢绕组180°抽头
diametrical voltage 对径电压，正负对向电压
diametric arrangement 径向布置
diametric winding 整距绕组
diamine 二胺，二元胺，联氨
diammonium hydrogen phosphate 磷酸氢二铵
diammonium 联胺【加到除氧器中】
diamond antenna 菱形天线
diamond array 菱形阵
diamond bar 菱形钢筋
diamond bit 金刚钻头
diamond circuit 桥形半波整流电路，桥形半波电路
diamond crystal counter 金刚石晶体计数器
diamond cutter 金刚石刀具
diamond drill rig 钻石凿井机
diamond grinding 金刚石打磨
diamond-head buttress dam 大头坝，钻头式支墩坝
diamond mesh 菱形筛格，菱形筛孔，菱形网眼
diamond nut 菱形螺母【用来固定垫圈】
diamond penetrator hardness 维氏金刚石硬度，维氏硬度
diamond pile 菱形煤堆
diamond pulley 菱形滚筒

diamond pyramid hardness　金刚石棱锥体硬度
diamond-shaped coil　菱形线圈
diamond-shaped damper　棱形叶片调节阀
diamond spar　刚玉
diamond stacking　菱形断面堆料法
diamond-tip drill　金刚石钻头
diamond type winding　菱形绕组，兰式绕组
diamond washer　斜方形垫块【止动】
diamond weave coil　菱形编织线圈
diamond winding　菱形绕组，模绕组
diamond　金刚石，钻石，菱形，菱形的
dianhraem carrier ring　隔板套
dianodic method　双阳极法
diaphaneity　透明度
diaphanometer　透明度计
diaphragm-actuated control valve　薄膜致动控制阀
diaphragm-actuated valve　薄膜控制阀
diaphragm actuator　膜片致动器，（气动）薄膜执行机构
diaphragm and diaphragm capsule　膜片和膜盒
diaphragm-and-spring actuator　膜片和弹簧制动器
diaphragm carrier　隔板套
diaphragm chamber　膜盒
diaphragm check valve　挡板止回阀，挡板单向阀
diaphragm compressor　膜式压缩机
diaphragm control valve　（气动）薄膜调节（控制）阀，膜片控制阀
diaphragm cylinder　膜盒室
diaphragm dam　有防渗墙的坝
diaphragm deflection　隔板挠度
diaphragm element　膜片元件
diaphragm filtration unit　膜片过滤装置
diaphragm float　隔板浮子
diaphragm gauge　膜片式压力计，薄膜式压力计，膜盒式压力表
diaphragm gland ring　隔板汽封环
diaphragm gland　隔板汽封
diaphragm hydraulic accumulators　隔膜式液气蓄能器
diaphragm manometer　薄膜压力计
diaphragm motor　膜片式调压伺服机，隔膜电动机
diaphragm nozzle blade　隔板静叶
diaphragm nozzle　隔板静叶
diaphragm of the steam turbine　汽轮机隔板
diaphragm-operated control valve　膜式控制阀
diaphragm-operated valve　隔膜阀
diaphragm packing ring　隔板汽封环
diaphragm packing　隔板汽封
diaphragm plate　隔板，横隔板
diaphragm pressure gauge　膜片式压力计
diaphragm pressure　薄膜压力
diaphragm pump　隔膜泵【空气或真空，往复泵型】
diaphragm relay　膜片式继电器
diaphragm-sealed piston pump　隔膜密封式活塞泵
diaphragm sealing　隔板汽封
diaphragm seal valve　薄膜密封阀
diaphragm seal　隔板汽封，隔膜密封，薄膜密封，密封膜片，膜片密封
diaphragm sleeve　隔板套
diaphragm strength　隔板强度
diaphragm switch　薄膜开关
diaphragm-type cellular cofferdam　隔墙式格型围堰
diaphragm-type manometer　薄膜压力计
diaphragm-type pressure measuring system　薄膜式测压系统
diaphragm-type sludge pump　隔膜式污泥泵
diaphragm valve　膜片阀【大气】，膜板阀，隔膜阀，薄膜阀，给水调整阀
diaphragm wall construction　防渗墙施工
diaphragm wall　防渗墙，隔墙，隔水墙，膜式壁，阻水墙
diaphragm　孔栏【聚变堆】，隔板，薄膜，隔膜，膜片，孔板，挡板，膜板
diaphram pump　隔膜泵
diaphram valve　隔膜阀
diaschistic dyke rock　二分脉岩，分浆脉岩
diaschistite　二分岩，分浆岩
diasolid copper wire　实心铜线
diaspore　硬水铝石
diastase　淀粉糖化酶
diastimeter　测距仪
diastrophism　地壳运动
diathermancy　透热性
diathermaneity　透热性
diathermanous　透过辐射热的，透热的
diathermic coagulation　电凝法
diathermic　透热的
diatomaceous　硅藻的
diatomaceous earth　硅藻土
diatomaceous earth block　硅藻土砖
diatomaceous earth brick　硅藻土砖
diatomaceous-earth filter　硅藻土过滤器
diatomaceous ooze　硅藻软泥
diatomaceous soil　硅藻土
diatom earth　硅藻土
diatomic asbestos ash　硅藻土石棉灰
diatomic gas　双原子气体
diatomic　二氢氧基的，双原子的，硅藻土的
diatomite asbestos board　硅藻土石棉板
diatomite asbestos slab　硅藻土石棉板
diatomite brick　硅藻土绝热砖
diatomite filter　硅藻土滤层
diatomite　硅藻土
diatom ooze　硅藻软泥
diatom　硅藻
diazotype microfilm　重氮缩微胶卷
dibasic sodium phosphate　磷酸氢二钠
dibasic　二元的，二价碱的
DIB(data input bus)　数据输入总线
dicarbide　二碳化合物
dice-circuitry　小片电路
dichloride　二氯化物
dichroic beam splitter　二色分光镜
dichromate　重铬酸盐
dichromic acid　重铬酸
Dickens's formula　狄更斯公式
dick　平滑（水）面

dictating machine 指令机，录音机
dictating 指令的
dictionary code 字典代码
dictionary 字典，辞典，词典
dicyclic 双环的
DI(data input) 数据输入
DID(digital information detection) 数字信息检测
DID(digital information display) 数字信息显示
diddle 骗取，诈取
DID(drum information display) 磁鼓信息显示
DI(digital input) 数字输入，数码输入
DI(disable interrupt) 禁止中断
DI/DO(data input/data output) 数据输入/数据输出
DI/DO(digital input/digital output) 数字输入/数字输出，开关量，数字量
die away curve 衰减曲线
die away 逐渐消逝，消失，衰减
die casting squirrel cage 压铸鼠笼
die casting 压模铸件
die-cast rotor winding 压铸转子绕组
die cushion 模具缓冲器
die forging 模锻
die-formed packing 模压成形密封件
die form 模压，模锻
die handle 板牙架
die-head rivet 冲垫铆钉
die head threading machine 套丝机
die head 模头
dielectric absorption characteristic 电介质吸收特性
dielectric absorption constant 介质吸收常数
dielectric absorption current 电介质吸收电流
dielectric absorption 电介质吸收
dielectric aftereffect 介电后效
dielectrically encapsulated trench 介质包封槽
dielectrically isolated 介质绝缘的
dielectrical 不导电的
dielectric amplifier 介电放大器
dielectric antenna 介质天线
dielectric body 介电体，绝缘体
dielectric breakdown test 绝缘击穿试验，电介质击穿试验
dielectric breakdown voltage 电介质击穿电压
dielectric breakdown 电介（质）击穿，介质破坏，绝缘破坏
dielectric capacity 介电常数，电容率
dielectric coating 绝缘涂层，绝缘漆
dielectric coefficient 介电系数
dielectric conductance 介质电导
dielectric constant 介电常数，电容率，介电系数
dielectric density 电通量密度，电位移
dielectric diode 介电二极管
dielectric dispersion 介电耗散，电容率弥散
dielectric dissipation factor 介质损耗因数
dielectric dry test 介质干耐压试验
dielectric fatigue 介质老化，介质疲劳
dielectric film 电介质膜
dielectric flux density 静电通量密度，电位移
dielectric flux 电通量，电介质通量
dielectric guide feed 介质导馈电器
dielectric guide 特高频电磁介电通道，介质波导管
dielectric heating 电介质加热，介质加热
dielectric hysteresis loss 介质电滞损耗
dielectric hysteresis 介质电滞，介电滞后，电滞
dielectric isolating joint 绝缘介质接头
dielectric isolation 介质隔离
dielectric leakage 介质漏电
dielectric level 介质水平，绝缘水平
dielectric loading factor 电介质负载系数
dielectric loss angle tangent 介质损耗角正切
dielectric loss angle 介质损耗角，介电损耗角
dielectric loss characteristic 介质损耗特性曲线
dielectric loss factor 介质损耗因素
dielectric loss tangent test 介质损耗角正切试验
dielectric loss 介电损耗，介质损失，介质损耗
dielectric material 介电材料，电介质
dielectric motor 电介质电动机
dielectric multilayer filter 多层介质滤片
dielectric oil 绝缘油
dielectric paper 绝缘纸
dielectric permittivity 介质电容率
dielectric potentiometer 介质分压器，介质电位计
dielectric power factor 介质损耗系数，介质功率系数
dielectric resistance 绝缘电阻，介质电阻
dielectric stability 绝缘稳定性，介质稳定性
dielectric standoff voltage 介质隔绝电压
dielectric storage 电介质存储器
dielectric strength tester 绝缘强度测试表
dielectric strength text 绝缘强度试验，电介质强度试验
dielectric strength 绝缘（电介质）强度
dielectric stress 电介质应力
dielectric surface 介质表面
dielectric susceptibility 电介质极化率，电纳系数
dielectric suspension 介质悬浮
dielectric target 介电质屏幕
dielectric test voltage 电介质试验电压
dielectric test 绝缘试验，高电位试验，介质试验，绝缘强度试验，介质性能试验
dielectric tip 绝缘材料翼尖
dielectric withstanding voltage test 电介质耐电压试验
dielectric 电介质，绝缘材料，介电的，绝缘的
die marks 钢印，模具压印
die-pressed graphite 模压石墨
die press 模压
diesel air compressor 柴油空压机
diesel building handling equipment 柴油机房装卸运输设备
diesel buildings 柴油机房，柴油发电机厂房
Diesel cycle 狄塞尔循环
diesel dredge 柴油挖泥机
diesel-driven 柴油机驱动的
diesel-dynamo 柴油发电机
diesel-electric drive 内燃机电力驱动装置
diesel-electric plant 柴油机发电厂
diesel-electric set 柴油发电机组
diesel emergency set 应急柴油发电机组

diesel engine generator start test 柴油发电机启动试验
diesel engine generator 柴油发电机
diesel engine 柴油机，狄塞尔内燃机，内燃机
diesel exhaust 柴油机废气
diesel fork lift 柴油叉车
diesel fuels 柴油
diesel generating plant 柴油机发电厂
diesel generating power station 柴油发电厂
diesel generating set 柴油发电机组
diesel generating station 柴油机发电厂
diesel generation combustion air intake and exhaust system 柴油发电机燃烧进气与排气系统
diesel generator breaker 柴油发电机开关
diesel generator building heating and ventilation system 柴油发电机厂房供暖与通风系统
diesel generator building 柴油发电机房，柴油机房
diesel generator fuel oil storage and transfer system 柴油发电机燃料油贮存和输送系统
diesel generator room 柴油发电机室
diesel generator set 柴油发电机组
diesel generator 柴油发电机，柴油发电机组
diesel hammer 柴油打桩锤
diesel index 狄赛尔指数
diesel loader 柴油装载机
diesel locomotive 柴油机车，内燃机车
diesel oil storage tank 柴油罐
diesel oil 柴油
diesel pile-driver 柴油打桩机
diesel-powered pump 柴油泵
diesel power plant 柴油发电厂
diesel power unit 柴油动力机组
diesel propulsion 柴油机推进
diesel quarrying machine 内燃式凿岩机
diesel scavenging air 柴油机费气
diesel set 柴油发电机组
diesel shovel 柴油铲土机
diesel smoke standard 柴油机排气标准
diesel tow tractor 柴油牵引车
diesel tractor 柴油拖拉机
diesel truck 柴油机卡车
die square 大方木
diestock 螺丝攻，板牙架，板牙铰手，螺丝绞板
die welding 模焊
die 模具，冲模，硬模，钢模，凹模，板牙，消灭，模制
diferential expansion 胀差
diff debris 坡积物
difference amplifier 差分放大器
difference analog 差别类比
difference current 差动电流
difference curve 温差曲线，减差曲线
difference-differential equation 差分微分方程
difference diode 差分二极管
difference equation 差分方程
difference frequency 差频
difference gauge 极限量规，极限规
difference in elevation 高程差
difference in gradients 坡度差
difference in levels 高差
difference in opinion 分歧意见
difference in principle 原则分歧
difference method 差分法
difference of arcing time 燃弧时差
difference of elevation 高差
difference of frequency 频差
difference of level 电平差
difference of phase angle 相角差
difference of potential 电平差，电压
difference of pressure 压差，压力降
difference of water table 水位差
difference operator 差分算子
difference price 差价
difference signal 差信号，差异信号
difference transfer ratio 差信号转移系数
difference voltmeter 差值电压表
difference 差，差分，差别，差额，差异，区别
differencing and checking subroutine 求差和检验子程序
differencing 求差，差分化
different grain size fractions 不同粒级
differentia 差异
differentiable random function 可微随机函数
differentiable 可微的
differential ablation 差异消融
differential absorption cross section 差异吸收截面
differential absorption lidar 差分吸收激光雷达
differential absorption ratio 微分吸收比
differential absorption technique 差异吸收技术
differential absorption 差异吸收，选择吸附
differential accelerator 微分加速器
differential-acting steam hammer 差动汽锤，差动式汽锤
differential action 差动作用
differential advection 差异平流
differential aeration corrosion 氧差腐蚀，充气差异腐蚀
differential albedo 微分反照率
differential amplifier 微分放大器，差动放大器，推挽式放大器
differential analysis 差示分析，差异分析，差值分析
differential analyzer 微分分析器，微分分析机
differential and integral calculus 微积分
differential anode resistance 差分阳极电阻，交流电阻
differential attack 不均匀腐蚀
differential bevel gear 差动伞齿轮
differential block 差动滑车
differential booster 具有差接励磁绕阻的升压电机，差接升压机
differential bridge 差接电桥
differential calculus 微分，微分学
differential capacitor 微分电容器，差动电容器
differential cathode follower 微分阴极输出器
differential chain block 差动式链滑车
differential characteristic impedance 微变特征阻抗
differential chart 变差图
differential circuit 微分电路，差动电路

differential coefficient 微分系数
differential coherent detection 差动相干检测
differential coherent transmission system 差动相干传输系统
differential comparator 差动比较器
differential compensation 差动补偿
differential compound generator 差复励发电机
differential compound motor 差复励电动机
differential compound winding 差复激绕法，差复励绕组
differential compound wound motor 差复激电动机，差复励电动机
differential compound 差复励
differential computing potentiometer 微分计算电位器
differential concatenation 差级联，反向串联
differential condenser 差动电容器
differential cone 衍射锥
differential connection 差接
differential control rod worth 控制棒微分价值
differential control 微分控制
differential correction method 微分校正法
differential cost benefit analysis 微分代价利益分析
differential cross section 微分截面
differential current protection 差动电流保护装置
differential current transformer 差动电流互感器
differential curve 微分曲线
differential cylinder 差动油缸
differential detection 微分检波，差动探测
differential diagnosis 鉴别诊断
differential-difference equation 微分-差分方程
differential discriminator 差分鉴别器，差分鉴频器
differential displacement 差动位移
differential draft gauge 差动风压表
differential draft 通风阻力
differential drive tachometer 差动传动式转速计
differential energy fluence rate 微分能通量密度，微分能注量率
differential energy flux density 微分能通量密度
differential equation solver 微分方程解算器
differential equation 微分方程
differential erosion 不等侵蚀，差异侵蚀
differential excitation 差动激励，差动励磁，差励
differential-excited generator 差励发电机
differential expanse 胀差
differential expansion detector 差胀指示器，胀差检测器，差胀检出器
differential expansion indicator 差胀指示器，相对膨胀指示器，胀差指示剂
differential expansion monitor 相对膨胀监视器
differential expansion 差胀，相对膨胀，胀差，微分式
differential factor 微分因数，差动因数，绕组系数
differential firing 换层燃烧
differential flow reactor 差流反应器
differential flux density 微分通量密度
differential form 微分形式
differential frost heave 不均匀冻胀
differential gage 压差计
differential gain control 差动增益控制
differential gain 微风增益
differential galvanometer 差动检流计
differential gap controller 差隙控制器，差动间隙控制器，两位式调节器
differential gap control 差隙控制，隙差控制
differential gap 微差隙，不可调间隙，偏差间隙，差隙，隙差
differential gauge 差压计，微分气压计，差动压力计，压力机，
differential gear 差动齿轮，分速轮
differential generator selsyn 差动自动同步机
differential generator 微分发生器，差励发电机，差动同步机
differential hammer 差动式打桩锤
differential head 压降，压差，压力降，水头差
differential heating 差异加热
differential hoist 差动式卷扬机
differential insulation 不平衡绝缘
differential intake 差动式进水口，差动式取水口
differential ionization chamber 微分电离室
differential ionization 微分电离
differential iron tester 差分磁铁测验器
differential kinematics 微分运动学
differential leakage flux 差漏磁通，电枢齿端漏磁通
differential leakage reactance 电枢齿端漏抗，差漏抗
differential leak detector 差动式检漏仪
differential linear expansion thermometer 差示线膨胀温度计
differential load transformer 差接负载变压器
differential manometer 微分压力机，差压计，差示压力机，差示压力表
differential mass flowmeter 差压式质量流量计
differential measurement 差动测量，微分测量，微差测定法
differential measuring instrument 差动式测量仪表
differential mechanism 差动机构【电器】
differential method 微分法，差动法
differential meter 差压计
differential micro-manometer 差动式微压计
differential micrometer dial 差动测微分度盘
differential mode interference 差模干扰，微分态干扰
differential mode rejection radio 抑制比，差模抑制比
differential mode signal 差式信号
differential mode voltage 差模电压
differential motor 差绕电动机，差励电动机
differential multiplying manometer 两种液体的差示压力计
differential negative resistance 微分负阻
differential neutron cross section 中子微分截面
differential of high order 高阶微分
differential operation 差动运行，差动
differential operator 微分算子
differential parallax 视差差数
differential particle fluence rate 微分粒子注量率
differential particle flux density 粒子微分通量密

度，微分粒子通量密度
differential permeability 增量磁导率，微分磁导率
differential phase modulation 差动式相位调制
differential phase-shift keying 微分移相键控法
differential phase 微分相位
differential piston gauge 差动活塞式压力计，差动活塞式压力表
differential piston 差动活塞
differential plunger pump 差动式柱塞泵，差动柱塞泵
differential pressure alarm 差压报警
differential pressure amplifier 差压放大器
differential pressure combination valve spool 差压式组合阀阀芯
differential pressure controller 差压控制器，差压调节器
differential pressure control valve 差压控制阀
differential pressure control 差压调节
differential pressure device 差压装置
differential pressure flowmeter 差压流量计
differential pressure gauge 示差压力计，差压计，差动压力计
differential pressure indicator 差压指示器，压差指示计
differential pressure instrumentation 差动压力仪表，差压式仪表
differential pressure line 差压管线
differential pressure manometer 压差式压力计
differential pressure meter 差压计，压差计
differential pressure pickup 压差传感器，差压式传感器
differential pressure producer 差压发生器，差压激励器
differential pressure regulating valve 差压调节阀
differential pressure regulator 压差调节器，差压调节器
differential pressure switch 压差开关，差压开关
differential pressure transducer 差压传感器
differential pressure transmitter 差压传感器，差压变送器，压差变送器
differential pressure type flowmeter 差压式流量计
differential pressure valve 压差阀
differential pressure 压差，不均匀压力，差动压力，压力落差，差压
differential protection system 差动保护系统
differential protection （发电机的）差动保护，接地漏电保护
differential protective relay 差动保护继电器
differential pulley block 差动滑动
differential pulley 差动滑轮
differential pulse code modulation 微分差码调制
differential pump 差动泵
differential quotient 微商
differential reactivity 微分反应性
differential reflectivity factor 差异反射因子
differential refractometer 差作用折射计
differential regulating transformer 差接可调变压器
differential regulator 差动调节器
differential relay 差动继电器
differential resistance change 差分阻值变化
differential resistance temperature characteristic 差分电阻温度特性
differential resistance 动态电阻，微分电阻
differential scale 带有差动控制系统的天平
differential screw 差动螺旋，传动装置螺钉
differential series winding 差接串励绕组
differential set pressure 差动镇定压力
differential settlement 不均匀沉降，不规则沉降，沉降差，差异沉降，非均匀下沉
differential settling 差速沉降
differential shaft 差动轴
differential shrinkage 不均匀收缩
differential signal 微分信号，差动信号
differential sign 微分符号
differential sliding valve 差动滑阀
differentials of higher order 高阶微分
differential strain analysis 微分应变分析
differential supercharger 差动增压器
differential surge chamber 差动式调压室
differential surge tank 差动式调压塔
differential susceptibility 微分磁化率
differential synchro 差动变压器，差接同步机
differential system 差动制，差动装置，混合线圈
differential tackle 差动滑轮，神仙葫芦
differential temperature controller 温差控制器
differential temperature 差异温度，温差
differential thermal advection 差异热平流
differential thermal analysis 差热分析，差热分析法
differential thermal analyzer 差热分析仪
differential thermal expansion 不均匀热膨胀
differential thermometer 差示温度计，示差温度计
differential titration 差示滴定
differential transducer 差分传感器
differential transformer 差动变压器，插接变压器，混合线圈
differential transmitter 差动发送器，差动式传感器
differential transmitting linkage 差动传动装置
differential treatment 差别待遇
differential type 差动式
differential U-tube 差压U形管
differential valve 差动阀
differential voltage 差分电压
differential weathering 差异风化
differential weir 差动堰
differential winding 差动线圈，差动绕组，差动绕法
differential windlass 差动式卷扬机
differential worth 微分（价）值
differential wound motor 差绕电动机，差励电动机
differential 差动的，差作用的，差别的，差示的，差的，微分的，差动
differentiated dike 分异岩脉
differentiated sill 分异岩床
differentiate 求微分，区分
differentiating circuit 微分电路，差动电路
differentiating-integrating measuring system 微分

积分测量系统
differentiating network 微分网络
differentiating unit 微分部件
differentiating 微分，差动
differentiation measurement 微分测量
differentiation 微分，微分法，变异，分化，区别，分异作用
differentiator 差动电路，差动装置，微分器，微分电路
different post 不同岗位
different specialties 不同专业
difficult ash 难溶灰
difficult communication 难于通信，可听度差
difficult fault 疑难性故障
difficult fuel 难烧燃料
diffluence 分流【气】
diffluent thermal ridge 分流温度脊
diffluent thermal trough 分流温度槽
diffluent trough 分流槽
diffracted error 衍射误差
diffracted ray 绕射线
diffracted wave 绕射波
diffraction current 回折流
diffraction fringe 绕射条纹，衍射条纹
diffraction grating 绕射光，衍射光栅
diffraction meter 衍射计
diffraction mottle 衍射斑纹
diffraction mottling 衍射斑点
diffraction pattern 衍射图，绕射图
diffraction phenomenon 绕射现象
diffraction region 绕射区
diffraction spectrum 绕射谱
diffraction zone 绕射带
diffraction 绕射，衍射
diffractometer 衍射计，衍射器
diffract 绕射，衍射，分解，折射
DIFFRLY(differential relay) 差动继电器
diffusate 扩散物，渗出物
diffuse-air aeration 扩散空气曝气
diffuse band 扩散带
diffuse boundary 扩散界面
diffuse bounding surface 漫射的边界表面
diffuse-collector method 集电极扩散法
diffused air aeration 扩散掺气
diffused illumination 漫射照明
diffused light 漫射光
diffuse double layer 扩散双电层
diffused silicon pressure transmitter 扩散硅压力变送器
diffused surface water 地面扩散水
diffuse field 漫射场
diffuse front 扩散锋
diffuse illumination 漫射照明
diffuse incident intercity 漫射入射强度
diffuse intensity 漫射强度
diffuse irradiance 散射辐照度
diffuse irradiation 散射辐照量
diffuse lighting 漫射照明，软眩光照明
diffuse light 漫射光
diffuseness 扩散，漫射
diffuse radiation 扩散辐射，漫射辐射，漫辐射
diffuser air supply 散流器送风
diffuser casing 导流外壳
diffuser chamber 扩散室
diffuser cone 扩散锥
diffuser efficiency 扩散器效率
diffuse reflection 漫反射
diffuse reflector 漫反射体
diffuser grid 扩压叶栅
diffuser inlet area 扩散段入口面积
diffuser intake 扩压器进口
diffuser loss 扩散器损失，扩压器损失，扩散段损失，扩散损失
diffuser nozzle 扩散式喷嘴，喇叭形喷嘴
diffuser outlet area 扩散段出口面积
diffuser plate 导向板【反应堆压力壳内】，配流板，扩散板，导流板，扩散掺气板
diffuser priming of pump 扩散器启动泵，扩散器式水泵
diffuser pump 扩散泵，导叶泵，升压器泵，扩压泵，导叶泵
diffuser reflectance 扩散反射率
diffuser reflection 扩散反射，漫反射
diffuser section 扩压段
diffuser thermometer 扩散器温度计
diffuser throat 扩散器喉管
diffuser tube 扩散管
diffuser type centrifugal pump 扩压式离心泵，扩散式离心泵
diffuser vane 扩散器叶片，扩压器叶片
diffuser 扩压器，散流器，扩散段，雾化器，喷雾器，扩风器，扩散器，扩压段，漫射体，扩散管，漫射器
diffuse scattering 漫散射
diffuse seepage 扩散渗流
diffuse skylight 漫射天光
diffuse sky radiation 天空漫辐射
diffuse solar irradiance 散射日辐照度
diffuse solar radiation 太阳漫射，散射日射，天空辐射，太阳漫射辐射，散射太阳辐射，漫射太阳辐射
diffuse 弥散，扩散，传播，漫射，渗出，扩射
diffusibility 扩散率，扩散能力
diffusible hydrogen 扩散氢
diffusibleness 扩散性，扩散能力
diffusing action 扩压作用，扩散作用
diffusing aerator 扩散曝气器
diffusing-age approximation 扩散-年龄近似
diffusing screen 散射屏
diffusing type steam valve 挤压式汽阀
diffusion action 扩散作用
diffusion aerator 扩散通气器
diffusional resistance 扩散阻力
diffusion analogy 扩散比拟
diffusion annealing 扩散退火
diffusion area 扩散面积
diffusion barrier 扩散膜，扩散屏障
diffusion bonding 扩散结合，扩散黏结
diffusion bucket 扩散式消力戽
diffusion calculation 扩散计算
diffusion capacitance 扩散电容
diffusion capacity 扩散能量

diffusion category 扩散类型
diffusion chamber 扩散云室
diffusion cloud 扩散云
diffusion coating 扩散法涂层法，扩散涂层，渗滤护膜
diffusion coefficient for neutron density 中子密度扩散系数
diffusion coefficient for neutron flux density 中子通量密度扩散系数
diffusion coefficient for particle fluence rate 粒子注量率的扩散系数
diffusion coefficient for particle flux density 粒子通量密度扩散系数
diffusion coefficient for particle number density 粒子数密度的扩散系数
diffusion coefficient 扩散率，扩散系数，漫射系数，传播系数
diffusion combustion 扩散燃烧
diffusion constant 扩散常数
diffusion controlled step 扩散控制步骤
diffusion cooling coefficient 扩散冷却系数
diffusion cooling effect 扩散冷却效应
diffusion cooling 扩散冷却，发散冷却
diffusion diagram 扩散图
diffusion distance 扩散距离
diffusion equation 扩散方程
diffusion factory （气体）扩散工作
diffusion factor 扩散系数，扩压因子
diffusionfield 扩散场
diffusion flame burner 扩散式燃烧器
diffusion flame 扩散火焰
diffusion flow 扩散流量，扩散流
diffusion flux 扩散通量
diffusion group 扩散群
diffusion heating effect 扩散加热效应
diffusion hygrometer 扩散湿度计
diffusion layer 扩散层
diffusion length 扩散长度
diffusion level 扩散高度
diffusion meanfree path 扩散平均自由程
diffusion medium 扩散介质
diffusion model 扩散模式，扩散模型
diffusion of pollutants 污染物扩散
diffusion of smoke 烟扩散
diffusion of turbulence 紊流扩散
diffusion of vorticity 漩涡强度散布
diffusion of water vapour 水汽的扩散
diffusion parameter 扩散参数
diffusion pitting 扩散性点蚀
diffusion potential 扩散电位，扩散电势
diffusion radiation 扩散放射
diffusion rate 扩散速率
diffusion region 扩散区
diffusion steam trap 扩散汽阱
diffusion theory approximation 扩散理论近似
diffusion theory 扩散理论
diffusion-thermo effect 扩散热效应
diffusion through an arbitrary plane 通过任一平面的扩散
diffusion through a semipermeable plane 通过半透明面的扩散
diffusion time 扩散时间
diffusion transistor 扩散型晶体管，扩散晶体管
diffusion vane pump 散流叶轮泵
diffusion vane ring 导叶环
diffusion vane 导叶，扩散叶轮，扩散器叶片
diffusion velocity 扩散速度
diffusion ventilation 扩散通风
diffusion welding 扩散焊
diffusion well 地下水补给井，扩散井
diffusion 扩散，传播，漫射，散射，弥散，（气流的）滞止，渗滤
diffusiophoresis 扩散电泳
diffusiophoretic force 扩散迁移力，散泳力
diffusiophoretic velocity 散泳速度
diffusisphere 扩散层
diffusive equilibrium 扩散平衡
diffusive force 扩散力
diffusive mechanism 扩散机理
diffusive process 扩散过程
diffusivity 扩散能力，扩散性，扩散系数，扩散率
diffusor duct （鼓风机）扩散风道
diffusor 扩压器，扩散器，扩压段，漫射体
difliusion effect 扩散作用
digested sludge 消化污泥
digester coil 化污器盘管，消化盘管
digester 消化池，煮解器
digestion chamber 消化室
digestion tank 化污池，消化池，腐化池
digestion 消化
digest 蒸煮，加热浸提，消化，文摘，提要
digger plough 犁式挖沟机
digger 挖斗，挖掘机
digging depth 挖掘深度
digging ladder 卸船机的梯子形挖煤斗链
digging machine 挖掘机，挖土机
digging out 挖出
digging radius 挖掘半径
digging range 挖土机工作半径
digging reach 挖掘地段
digging spade 平锹
digging 开掘，开挖，采掘，挖掘
digimer 数字式万用表
dig-in damage 掘进损伤
dig-in 掘进
digit absorbing selector 数位吸收选择器
digit absorption 数位吸收，数字吸收
digit acknowledgment signal 数字确认信号
digital adder 数字加法器
digital-analog conversion 数字模拟转换，数模转换
digital analog converter 数字模拟转换器
digital analog function table 数字模拟函数表
digital analogue converter 数字模拟转换器，数模转换器
digital analogue data conversion 数字模拟数据转换
digital-analogy 数字模拟
digital anemometer 数字式风速仪
digital angular position 数字角位置
digital approximation 数值逼近

digital arithmetic center　数字计算中心
digital attenuator system　数字式衰减系统
digital automation　数字自动装置
digital back-up　数字（计算机）备用
digital calculation method　数字计算法
digital calculation　数字计算
digital calculator　数字计算器
digital camera　数码相机
digital channel　数字通道
digital character　数字符号
digital circuit　数字电路
digital clock　数字钟
digital code wheel　数字代码轮
digital coding　数字编码
digital communication network　数字通信网
digital communication system　数字通信系统
digital communication　数字化通信
digital comparator　数字比较器
digital complement　按位补码
digital computation　数字计算
digital computer program　数字计算机程序
digital computer simulation of transient stability　暂态稳定的数字计算机模拟
digital computer simulation　数字计算机仿真，数字计算机模拟
digital computer solution　数字计算机解法
digital computer　数字计算机
digital computing system　数字计算系统
digital controller　数字控制器
digital control station　数字控制站
digital control system　数字式控制系统，数字化控制系统，开关量控制系统
digital control　数字控制
digital conversion receiver　数字转换接受器
digital conversion　数字转换，数字变换
digital correlator　数字相关器，数字相关仪
digital counter　数字计数器，数字计算器
digital counting unit　数字计数单元
digital daily instrument　数字式显示仪表
digital data acquisition system　数字数据采集系统
digital data acquisition　数字数据采集
digital data communication message protocol　数字数据通信报文协议
digital data conversion equipment　数字数据转换装置
digital data converter　数字数据转换器，数字数据变换器
digital data logging　数字数据巡回检测，数字数据记录
digital data output conversion equipment　数字数据输出转换设备
digital data processing　数字数据处理
digital data processor　数字数据处理机
digital data receiver　数字数据接收器
digital data recorder　数字数据记录器
digital data recording system　数字数据记录系统
digital data terminal　数字数据终端
digital data transceiver　数字数据收发装置
digital data transmitter　数字数据发送器，数字数据传输器
digital data　数字数据
digital decade counter　十进制数字计数器
digital delay generator　数字时间延迟发生器
digital demultiplexer　数字分接器
digital device　数字器件，数字设备
digital differential analyzer　数字微分分析器［仪］
digital differential circuit　数字微分电路
digital differential　数字微分
digital display generator　数字显示发生器，数字显示信号发生器
digital display scope　数字显示器
digital display tube　数字显示管
digital display　数字显示
digital distribution frame　数字配线架
digital divider　数字除法器
digital dock　数字钟
digital DP system　数字数据处理系统
digital drum　数字鼓
digital electric-hydraulic control system　数字式电液控制系统
digital electric-hydraulic control　数字式电液控制
digital electric hydraulic turbine control　数字电动液压汽轮机控制
digital electrohydrauilc control system　数字式电液控制系统，数字电液调节系统
digital electrohydrauilc governing system　数字式电液调节系统
digital electrohydraulic control system of steam turbine　汽轮机数字式电液控制系统
digital electrohydraulic control system　数字式电液调节系统【汽轮机】
digital electrohydraulic governing system of steam turbine　汽轮机数字式电液调节系统
digital electrohydraulic　数字电动液压的
digital element　数字元件
digital encoder　数字编码器
digital error　数字误差
digital expansion system　数字扩展系统
digital feedwater control system　数字给水控制系统
digital filtering　数字滤波
digital filter　数字滤波，数字滤波器
digital follower　数字输出器
digital format　数字格式
digital frequency comparison　数字式频率比较
digital frequency meter　数字频率计
digital head　数字磁头
digital hologram　数字全息图
digital hybrid　数字混合集成电路
digital image processing　数字图像处理
digital incremental plotter　数字增量绘图仪
digital indicating station　数字指示站
digital indicating tube　数字指示管
digital indication　数字读数，数字示值
digital indicator　数字式指示器，数字指示器
digital information detection　数字信息检测
digital information display　数字信息显示
digital information　数字信息
digital input noise　数字输入噪声，数字输入干扰
digital input/output buffer　数字输入/输出缓冲器

digital input terminal　数字量输入端子
digital input　开关量输入，数字量输入，数字输入
digital integrating circuit　数字集成电路
digital integration　数字积分
digital integrator　数字积分器
digitalizer　数字化装置，数字变换器
digital keyboard　数字键盘
digital language　数字语言
digital line link　数字有线链路
digital line section　数字有线段
digital line system　数字有线系统
digital link　数字链路
digital logic circuit　数字逻辑电路
digital logic element　数字逻辑元件
digital logic station　数字逻辑站
digitally　用计算法，用数字计算的方法
digital magnetic tape　数字磁带
digital measured entry system　数字信息输入系统
digital measured value　数字测量值，数字值
digital measuring device　数字测量装置
digital measuring instrument　数字测量仪表
digital message entry device　数字信息输入设备
digital method　数字方法
digital microcircuit　数字微电路
digital microwave equipment　数字微波设备
digital module　数字模块
digital multimeter calibrator　数字式多用表校验仪
digital multimeter　数字式万用表，数字万用表
digital multiplex equipment　数字复用设备
digital multiplexer　数字复接器
digital multiplex hierarchy　数字复用等级系列，数字多路复用层次，数字多工分级
digital multiplex system　数字多路系统
digital multiplier unit　数字乘法器部件
digital multiplier　数字乘法器
digital network analyzer　数字网络分析器
digital network architecture　数字网络体系结构
digital network　数字通信网
digital ohmmeter　数字欧姆计
digital operational circuit　数字运算电路
digital orthogonal model　数字正交模型
digital output/input translator　数字输出/输入转换器
digital output pulse train　数字输出脉冲串
digital output terminal　数字量输出端子
digital output　数字量输出，数字输出
digital path　数字通道
digital pattern generator　数字模型发生器
digital phase shifter　数字式移相器
digital plotter　数字绘图仪，数字描绘仪
digital plot　数控绘图，数字绘图
digital point plotter　数字点描绘器
digital potentiometer　数字式电势计
digital presentation　数字表示
digital printer　数字打印机
digital process control　数字程序控制，数字式过程控制
digital processing unit　数字数据处理装置
digital process　间断过程，不连续过程，断续过程
digital protection　数字式保护
digital pulse duration modulation　数字脉冲宽带调制
digital pulse sequence　数字脉冲序列
digital pulse　数字脉冲
digital quantity　数字值，开关量
digital quantizer　数字转换器
digital radio system　数字无线系统
digital reactor protection system　反应堆数字化保护系统
digital read-out display　数字读出显示
digital read-out oscilloscope　数字读出示波器
digital read-out　数字读出，数字读数，数字示值
digital recorder　数字记录器，数字记录仪
digital recording instrument　数字记录仪表
digital recording　数字记录
digital record mark　数字记录标记
digital regenerator　数字再生器
digital relay counter　继电器数字计数器
digital relay　数字继电器
digital representation　数字表示法
digital resolver　数字式旋转变压器，数字解算器，数字分解器
digital rheostat　数字式变阻器
digital rotate speed instrument　数字转速表
digital section　数字段
digital selective communication　数字选择通信
digital sensor　数字传感器
digital servomechanism　数字伺服机，数字伺服机构
digital servosystem　数字伺服系统
digital set programmer　数字式程序器
digital shaft encoder　数字式轴编码器
digital shifter　数字移相器
digital signal analyzer　数字信号分析器
digital signal conversion　数字信号变换
digital signaling　数字信令
digital signal processing　数字信号处理
digital signal processor　数字信号处理器
digital signal　数字信号
digital simulated analog computer　数字仿真模拟计算机
digital simulation of DC power transmission system　直流输电系统数字仿真
digital simulation　数字模拟，数字仿真
digital simulator　数字仿真装置，数字模拟装置
digital slave module　数字从模件
digital smart converter substation　数字化智能换流站
digital sorting method　数字分类法
digital stepper motor　数字式步进电机
digital stepping recorder　数字步进式记录器
digital storage equipment　数字存储设备
digital storage oscilloscope　数字记忆示波器
digital storage unit　数字存储器
digital storage　数字存储器
digital subtractor　数字减法器
digital surface acoustic wave delay line　数字表面声波延迟线

digital switch delay 数字切换时延
digital switching device 数字转换装置
digital switching 数字交换机技术，数字开关，数字交换
digital synchronization 数字同步
digital tachometer 数字转速表
digital tachometry 数字式转换测定法
digital technique 数字技术
digital telemetering register 数字式遥测记录器
digital telemetering 数字远距离测量，数字遥测
digital telemetry 数字遥测术
digital telephone network 数字电话网
digital television 数字电视
digital time-interval measuring circuit 数字时间间隔测量电路
digital timer 数字定时器
digital time-slot 数字时隙
digital-to-analog conversion 数字模拟转换，数字模拟变换
digital-to-analog converter 数字模拟转换器，数模转换器
digital-to-analog ladder 数字模拟转换阶梯信号发生器
digital to analog(ue) conversion 数模转换
digital-to-synchro converter 数字同步机转换器
digital-to-video display 数字视频转换式显示
digital transducer 数字式传感器，数字转换器
digital transformer winding ratio tester 数字式变压器匝比测试仪
digital transmission system 数字传输系统
digital transmission 数字传输
digital transmitter 数字发送器
digital type protection relay 数字式继电保护
digital value 数字量，数字值，开关量
digital variable 数字变量，“通断”变量
digital video camera 数码摄（录）像机
digital video display 数字图像显示
digital voltage encoder 数字电压解码器
digital voltmeter 数字式电位计，数字电压表，数字伏特计，数字式电压表
digital volt-ohmmeter 数字电压电阻表
digital water-stage recorder 数字水位计
digitalyzer 模拟数字变换器
digital 数字，数字的，计算的，手指的，指状的
digitate drainage pattern 掌指状水系，鸟足状水系
digit capacity 数位容量
digit check 数字校验
digit-coded voice 数字编码声音
digit compression 数字压缩
digit counter 数字计数器
digit count 数位计算
digit deletion 数位删除
digit drive pulse 数字驱动脉冲
digit duration 数字脉冲宽度
digit filter 数字滤波器
digit group 数字组
digitisation rate 数字转换速度
digitization 数字化
digitized voltage 数字化电压
digitizer 数字化器，数字化装置，数字转换器
digitize （模拟值的）数字化
digitizing tablet 数字面板
digitizing 数字化
digit layout parameter 数位配置参数
digit layout 数的配置
digit order number 数字位
digit path 数码道
digit period 数字周期，连续脉冲的时间间隔
digit plane 数位面
digit position 数字位，数字位置
digit pulse 数字脉冲
digit rearrangement 数字重新排列
digit receiver 数字式接收机
digitron 数字指示管，数字读出辉光管
digits delay 数位延迟
digit selector 数字选择器
digit spacing 数位容量
digit storage relay 数字存储继电器
digit time 数字时间
digit track 数字道
digit tube 数字管
digit wave form 数位波形
digit 数字，数位，数，位
digiverter 数字模拟信息转换装置
digtizer 数字读出机
digue 防浪堤
dihedral angle 二面角，双面角
dihydrazine phosphate 磷酸双联氨
dike body 堤体
dike breaching 堤防溃决，堤坝决口
dike breach 决堤，土堤决口
dike breaking 堤防溃决
dike burst 堤防溃决，决堤
diked country 堤内地，堤内区
diked land 堤防地
dike footing 堤防底脚
dike in 围以堰堤
dike lock 堤闸
dike maintenance 堤防维护
dike management 堤防管理
dike rock 墙岩
dike system 堤防系统
dike 堤坝，沟渠，排水道，堤，岩脉，岩墙，坝
diking 筑堤
dilapidation 倒塌，失修倒塌
dilatability 膨胀性
dilatable rock and soil 膨胀岩土
dilatable soil 膨胀性土
dilatancy 膨胀性
dilatation coefficient 膨胀系数
dilatation constant 膨胀常数
dilatation fissure 膨胀裂缝
dilatation joint 膨胀缝，伸缝
dilatation modulus 膨胀模量
dilatation wave 膨胀波
dilatation 膨胀，膨胀度，扩散，扩展，剪胀，扩张
dilate 扩散，膨胀
dilation water 膨胀水
dilation 扩张
dilative soil 膨胀性土，膨胀性土壤

dilatometer measurement 膨胀计式测量
dilatometer 触头膨胀计，膨胀计
D/L(dilute/dilution) 冲淡，稀释
dilecto 电木层压材料
diligence 勤奋
diluent cooling 喷液冷却，喷流冷却【燃气轮机】
diluent plug 稀释塞
diluent zone 冷却区
diluent 稀释剂，稀释器，冲淡的
dilutability 稀释度
dilute acid wash 用稀酸洗涤
dilute acid 稀酸
dilute base 稀碱
dilute chemical decontamination process 稀释化学去污工艺
dilute compartment 淡化室
dilute concentration 稀释浓度
diluted core reactor 稀释堆芯反应堆
diluted core 稀释堆芯
diluted gasket 淡水隔板
dilute disperse and decontaminate 采用稀释、弥散和去污方法的废物处置系统
diluted reactor 稀释燃料反应堆
dilute fissile core 稀释裂变堆芯
dilute mirror 稀疏反射镜
dilute phase conveying 稀相输送
dilute phase fluidization 稀相流态化
dilute phase fluidized bed 稀相流化床
dilute phase 稀相
dilute reagent 稀释剂
dilute sewage 稀释污水
dilute solution 稀溶液
dilute suspension 稀相悬浮
dilute tank 稀释槽
dilute 稀释
diluting acid 稀酸
diluting base 稀碱
diluting 稀释
dilution air 稀释用空气
dilution analysis 稀释分析
dilution between base metal and filler metal 母材和填充金属之间的掺合
dilution coefficient 稀释系数
dilution discharge 稀释排放，稀释排放法，稀释排取
dilution effect 稀释效应
dilution factor 稀释因数，稀释系数，稀释因子，稀释率
dilution fissile core 稀释裂变堆芯
dilution gauging 溶液法测流
dilution limit 稀释极限
dilution method 稀释法
dilution of pollution 污染稀释
dilution probe 稀释探头
dilution ratio 稀释比
dilution tank drain pump 稀释箱疏水泵
dilution tank 稀释箱
dilution water flow 稀释水流
dilution water 稀释水
dilution zone 混合区域
dilution 掺合【百分之几的母材进入焊接金属】，稀释（度），冲淡
diluvial deposit 洪积土层，坡积
diluvial soil 洪积土
diluvion 大洪水
diluvium 大洪水，洪积层，洪积物
dim 模糊
dimensional accuracy 尺寸精确性，尺寸精度
dimensional allowance 尺寸容差
dimensional analysis 量纲分析，尺寸分析，尺度分析，因次分析
dimensional change 尺寸变化，几何尺寸改变
dimensional characteristic 尺寸特性
dimensional check 尺寸核对
dimensional constant 量纲常数
dimensional defect 尺寸上的缺陷
dimensional diagram 尺寸图
dimensional drawing 尺寸图，联系尺寸图
dimensional equation 量纲方程
dimensional gauging 尺寸检测
dimensional homogeneity 尺寸均一性，因次均等性，量纲齐次性
dimensional inspection 尺寸检查，尺寸检验
dimensional integrity 尺寸完整性，尺寸稳定性
dimensional interchangeability 尺寸互换性
dimensionally tolerance 尺寸公差
dimensional measuring instrument 长度测量工具
dimensional method 量纲法
dimensional output 公称输出功率
dimensional precision 尺寸精度
dimensional readings 尺寸的读数
dimensional relation 量纲关系
dimensional stability 尺寸稳定性，形稳性，几何稳定性
dimensional theory 量纲理论
dimensional tolerance 几何公差，尺寸公差
dimensional 尺寸的，维数的，维的，量纲的，因次的，空间的
dimension analysis 量纲分析
dimension chart 轮廓尺寸图
dimensioned diagram 尺寸图
dimensioned drawing 尺寸图，轮廓图
dimensioned value 有名值【即有量纲的值】
dimensioned 有因次的，有尺寸的
dimension equation 量纲方程
dimension figure 尺寸图
dimensioning 标尺寸，定尺寸，定尺度
dimensionless coefficient 无因次系数，无量纲系数
dimensionless factor 无量纲系数，无量纲因子，无因次系数
dimensionless group 无因次数群
dimensionless hydro-graph 无因次过程线
dimensionless number 不名数
dimensionless parameter 无量纲参数，无量纲量，无因次参数
dimensionless quantity 无量纲量
dimensionless recession curve 无因次退水曲线
dimensionless representation 无量纲表示法
dimensionless spectrum 无因次系列
dimensionless unit 无量纲单位
dimensionless 无量纲的，无因次的，无维数的

dimension limit indicator　极限尺寸指示器
dimension of a quantity　量纲
dimension print　外形安装尺寸图
dimension relation　量纲关系
dimension scale　尺寸比例，尺度比例
dimensions list　尺寸单
dimensions of flood-prevention dike　防洪堤设计标准
dimensions of thermal conductivity　导热系数的因次
dimension　维，尺寸，外形尺寸，量纲，因次，元，尺度，度，外包尺寸
diminish　减少，降低
diminution factor　衰减常数
diminution　减少，减缩
diminutive current meter　小型流速仪
dimmer switch　调光开关
dimmer　调节灯光的变阻器，减光器，灯罩
dimorphism　风化变质作用
dimple aluminium foil　波纹铝箔
dimpled spacer　波纹形定位件
dimple rupture　韧窝断裂
dimple　波纹弹簧【燃料组件】，波纹，微凹，陷窝
DINA(direct-noise amplifier)　直接噪声放大器
DIN(Deutsches Institut für Normung/German Standards Organization)　德国标准协会，德国标准化组织
DIN(Deutsches Institut für Normung)　德国工业标准，德国标准，德标
dineutron　双中子
dine　生态群
dinging hammer　敲击锤
dingle　封闭谷地，深溪，幽谷
dingot　直接铸锭
dining room　餐室，食堂
dinitrogen tetraoxide cooled reactor　四氧化二氮冷却反应堆
dinitrophenol　二硝基酚
dinkey locomotive　轻便机车，窄轨机车
dint　凹痕，凹坑，压痕
DIOB(digital input/output buffer)　数字输入输出缓冲器
diode　二极管
diode alternating current switch　双向触发二极管，二极管交流开关
diode AND gate　二极管"与"门
diode bridge rectifier　二极管桥整流器
diode bridge　二极管电桥
diode-capacitor memory　二极管电容存储器
diode clipper　二极管削波器
diode-driven DC motor　二极管供电的直流电动机
diode electronic function generator　二极管电子函数发生器
diode emitter follower logic　二极管射极跟随器逻辑
diode equivalent　等效二极管
diode frequency multiplier　二极管倍频器
diode function generator　二极管函数发生器
diode fuses　二极管熔丝
diode gate　二极管门电路，二极管门
diode gating network　二极管门网络
diode limiter　二极管限幅器
diode logical circuit　二极管逻辑电路
diode logic　二极管逻辑，二极管逻辑电路
diode matrix encoder　二极管矩阵编码器
diode matrix　二极管矩阵
diode mixer　二极管混波器，二极管混频器
diode modulator　二极管调制器
diode multiplexer　二极管转换开关
diode phase-sensitive detector　二极管相敏检波器
diode plug-in unit　二极管插入部件
diode-probe-type voltmeter　二极管探头式电压表
diode recovery tester　二极管再生测试器
diode rectification　二极管整流
diode rectifier　二极管整流器
diode suppresser　二极管遏抑器
diode switch system　二极管开关系统
diode switch　二极管开关
diode transformer　二极管变换器
diode-transistor logic　二极管晶体逻辑电路
diode-triode　二极三极复合管
diode valve　二极管阀
diode voltmeter　二极管电压表
dionic recorder　导电度记录仪
dionic tester　导电度仪
dionic　测量水的导电率仪表
diopside　透辉石
diopsite　透辉石
dioptase　透视石
diopter　觇孔，觇孔板，折光度，窥视孔，照准器，照准仪
diorite　闪长岩
dioritite　闪长细晶岩
dioxide　二氧化物
dip angle　倾角，俯角，磁倾角
dip brazing　浸渍硬钎焊
dip circle　侧斜仪，磁倾仪，俯角圈
dip cleat　倾斜割理
dip compass　测斜仪，倾角罗盘，倾角仪
dip counter tube　浸入式计数管
dip detector　浸入式探测器
DIP(diplomatic mail)　外交信袋
dip face　倾斜面
dip fault　倾向断层，倾斜断层
dip groin　俯头丁坝
dip groyne　俯头丁坝
diphaser　两相发电机
diphase　两相的
diphasic　两相的
diphenyl boiler　联苯锅炉
diphenylcarbazide　二苯基卡巴肼
diphosphonate　二磷酸盐
dip joint　倾向节理
dipleg　沉浸支管
diplexer　同相双工器，天线分离滤波器
diplex operation　双工运行
diplex reception　双工接收
diplex system　双工制，同方向传送二路信号制
diplex telegraph　同向双工电报技术，单向双路

电报
diplex transmission 双工传输
diplex 双工的，同相双工制，双通路的
diploid 二重的
diplomatic mail 外交信袋
dip meter 测斜仪，测倾仪，磁倾角测量仪，栅流陷落测试振荡器
dip moment 倾斜力矩
dip of load curve 负荷曲线的峡谷
dip of shore horizon 岸线俯角，岸线坡度
dip of the horizon 地平俯角
dipolar 偶极的，两极的
dipole antenna 偶极天线
dipole array 振子天线阵
dipole molecule 偶极分子
dipole moment 偶极矩
dipole oscillator 偶极子振荡器，双极振荡器，偶极振子
dipole polarization 偶极子极化
dipole-quadrupole interaction 偶极子、四极子相互作用
dipole 偶极子，二极，对称振子，双极子
dipotassium hydrogen phosphate 磷酸氢二钾
dip paint 浸漆
dipped electrode 手涂焊条
dipper capacity 铲斗容量
dipper door 铲斗活底
dipper dredger 挖泥铲斗，铲斗挖泥船，铲扬式挖泥船
dipper stick 铲斗柄
dipper teeth 铲斗齿
dipper 铲，铲斗，勺，挖土斗
dipping and heaving 升沉运动
dipping coil 浸渍线圈
dipping compass 倾度仪
dipping method of timber treatment 木材浸渍处理法
dipping refractometer 浸液折射计
dipping 浸渍，蘸
dip rod 水位指示尺
dip sampler 插入取样器，浸入取样器
dip shift 倾向移距
dip slip fault 倾向滑断层
dip slip 倾向滑距
dip slope 倾向坡
dip soldering 浸渍软钎焊，浸焊
dip stick 测杆，量油计，浸量尺，量油尺，液面测量杆，测舱柜液深用标尺
dip 浸入，浸，泡，倾斜，俯角，下垂度，沉浸，倾角，倾向，下沉，下降
direct absorber 直接吸收体
direct access control 直接存取控制
direct access library 直接存取库
direct access memory 随机存取存储器
direct access programming system 直接存取程序设计系统
direct access storage device 直接存取存储器
direct access storage 直接存取存储器
direct access 直接存取，直接访问
direct acting 直接作用的
direct acting actuator 正作用执行机构，直接作用执行机构
direct acting controller 正作用调节［控制］器，直接作用调节［控制］器
direct acting control valve 直接作用的控制阀
direct acting measuring instrument 直接作用测量仪表
direct acting pressure relief valve 直接作用卸压阀，自起动卸压阀
direct acting pump 直接驱动泵，直接联动泵
direct acting reciprocating pump 直接联动往复（式）泵
direct acting recording instrument 直接作用记录仪器
direct acting transducer 直接作用的发送器
direct acting voltage regulator 直接作用式电压调整器
direct action tunnel 直流式风洞
direct action wind tunnel 支流式风洞
direct action 正作用，直接动作，直接作用
direct addition 直接相加法
direct address processing 直接地址处理
direct address 一级地址，直接地址
direct add 直接相加【指令】
direct air conditioning system 直流式空气调节系统
direct air cooling 直接空冷
direct air-cycle reactor 直接空气循环反应堆
direct allocation 直接分配
direct analogy computer 直接模拟计算机
direct axis air-gap permeance 直轴气隙磁导
direct axis armature winding 直轴电枢绕组
direct axis characteristic 直轴特性曲线
direct axis component of transient electromotive force 瞬变电动势的直轴分量
direct axis component 直轴分量，纵轴分量
direct axis open circuit transient time constant 直轴开路瞬态时间常数
direct axis reactance 直轴电抗
direct axis short circuit subtransient time constant 直轴短路次瞬态时间常数
direct axis short circuit transient time constant 直轴短路瞬态时间常数
direct axis subtransient electromotive force 直轴次暂态电动势
direct axis subtransient impedance 直轴次暂态电抗
direct axis subtransient reactance saturated 直轴次瞬变电抗饱和值
direct axis subtransient reactance unsaturated 直轴次瞬变电抗不饱和值
direct axis subtransient voltage 直轴次暂态电压
direct axis suntransient reactance 直轴次暂态电抗
direct axis suntransient time constant 直轴次暂态时间常数
direct axis synchronous impedance 直轴同步阻抗
direct axis synchronous reactance 直轴同步电抗
direct axis time constant 直轴时间常数
direct axis transient electromotive force 直轴暂态电动势，直轴瞬变电动势
direct axis transient impedance 直轴瞬变阻抗，

直轴暂态阻抗
direct axis transient open circuit time constant 直轴瞬变开路时间常数
direct axis transient reactance unsaturated 直轴瞬变电抗不饱和值
direct axis transient reactance 直轴瞬变电抗
direct axis transient time constant 直轴暂态时间常数，直轴瞬变时间常数
direct axis transient voltage 直轴瞬变电压，直轴暂态电压
direct axis voltage 直轴电压
direct axis 直轴
direct balance method 正平衡法
direct bank protection 直接护岸
direct bearing 导（向）轴承，直接引导方位
direct benefit 直接效益
direct bilge suction 舱底水直接吸入口
direct boiling 直接蒸发，直接汽化
direct-buried cable 直埋电缆，地下电缆
direct-buried transformer 直埋式变压器
direct-buried 直接掩埋的，埋入的
direct calculation 直接计算
direct call 直接通话
direct capacitance 静电容，部分电容
direct catchment 直接集水面积
direct circulation 正环流
direct clutch 直接离合器
direct coal-fired gas turbine 直接燃煤燃气轮机
direct coal-fired MGD generator 直接燃煤式磁流体发电机
direct combustion 直接燃烧
direct communication 直接通信
direct compensation measurement 直接补偿测量
direct component 直流部分，直流分量
direct compression 单纯受压
direct condenser 回流冷凝器
direct conductor cooling 导线直接冷却，绕组内冷
direct-connected exciter 同轴励磁机，直连式励磁机
direct-connected motor 直连电动机，同轴电动机
direct-connected pump 直接泵，直接联动泵
direct-connected turbogenerator 直接传动气轮发电机
direct connection 直接连接
direct construction cost 直接工程费用
direct contact condensation 直接接触凝结，混合凝结
direct contact condenser 混合式凝汽器，直接接触凝结器
direct contact feed heater 混合式给水加热器
direct contact heater 接触式加热器，混合式加热器
direct contact heat exchanger 汽-水混合式换热器，混合式热交换器
direct contact LP heater 混合式低压加热器
direct contact spray desuperheater 接触式喷水减温器
direct contact type condenser 混合式凝汽器，接触式凝汽器
direct contact vibration pickup 直接接触拾振器
direct contact water cooler 混合式水冷却器
direct contracting 直接签订合同
direct contract stipulation 合同的直接规定
direct control 直接控制
direct conversion plant 直接转换站
direct conversion reactor 直接转换反应堆
direct cooled annular compact 直接冷却环状燃料压块
direct cooled bar 直接冷却线棒，内冷线棒
direct cooled integrated block 直接冷却整燃料块
direct cooled machine 内冷式电机，直接冷却电机
direct cooled multihole compact 直接冷却蜂窝煤状燃料压块
direct-cooling system 直接冷却系统
direct cost account 直接成本账
direct cost method 直接成本法
direct cost of project 工程直接费
direct cost 直接成本，直接费，直接费用
direct-coupled circuit turbine 直接连接式汽轮机
direct-coupled circuit 直接耦合电路
direct-coupled flip-flop 直接耦合触发电路
direct-coupled generator 直连励磁机，直连式发电机
direct-coupled leakage power 直接耦合漏泄功率
direct coupled logic 直接耦合逻辑，直接耦合逻辑电路
direct-coupled motor drive 电动机直接传动
direct-coupled pump 直接连接式泵
direct-coupled transistor logic circuit 直接耦合晶体管逻辑电路
direct-coupled transistor logic 直接耦合晶体管逻辑
direct-coupled type 直接连接式
direct-coupled unipolar transistor logic 直接耦合单极晶体管逻辑
direct-coupled 直接接合，直接耦合的
direct coupling exciter 直连励磁机，同轴励磁机
direct coupling generator 直接连接发电机
direct coupling logic circuit 直接耦合的逻辑电路
direct coupling triggering 直接耦合触发
direct coupling turbine 直接传动式汽轮机
direct coupling 直接联轴节，直接耦合
direct-current acyclic generator 直流单极发电机
direct-current amplifier 直流放大器
direct-current analog computer 直流模拟计算机
direct-current arc welder 直流电焊机
direct-current balancer 直流平衡发电机，直流平衡器，直流平衡机
direct-current biasing 直流偏压
direct-current bridge 直流电桥
direct-current cable 直流电缆
direct-current characteristic 直流特性
direct-current compensator 补偿用直流发电机，直流补偿机
direct-current component 直流分量
direct-current conductance 直流电导
direct-current converter 直流换流器，直流变换器
direct-current direct-current converter 直流换流

器，直流-直流变换器
direct-current distribution 直流配电
direct-current drive 直流电驱动
direct-current dynamo 直流电机，直流发电机
direct-current electro magnetic pump 直流电磁泵
direct-current electromagnetic relay 直流电磁继电器
direct-current excited-field generation 直流励磁发电机
direct-current generator 直流发电机
direct-current induced polarization 直流激发极化法
direct current low-voltage high-speed switch 直流低压高速开关
direct current machine 直流电机
direct-current motor 直流电机，直流马达，直流电动机
direct-current reference-voltage type measuring unit 直流基准电压测量装置
direct-current relay 直流继电器
direct-current resistance 直流电阻
direct-current reversing motor 可逆式直流电动机
direct current signaling 直流信令
direct-current stabilizer 直流稳压器
direct current surge capacitor 直流冲击电容器
direct current system 直流系统
direct-current transformer 直流直流换流器，直流变流器
direct-current transmission 直流输电
direct current voltage 直流电压
direct-current working voltage 直流工作电压
direct current 直流，直流电
direct cycle boiling-water reactor 直接循环沸水反应堆
direct cycle cooling system 直接循环冷却系统
direct cycle diphenyl reactor 直接循环联苯反应堆
direct cycle gas cooled reactor 直接循环气冷反应堆
direct cycle integral boiling reactor 直接循环燃气轮机
direct cycle plant 直接循环一体化沸水反应堆
direct cycle reactor system 直接循环反应堆系统
direct cycle reactor 直接循环电厂
direct cycle 直接循环
direct demagnetizing effect 直接循环反应堆系统
direct desulfurization 直接脱硫
direct determination 直接测定法
direct diagnosis 直接诊断
direct dialing telephone 直拨电话
direct dialing 直拨
direct digital control computer 直接数字控制机
direct digital control system 直接数字控制系统
direct digital control 直接数字控制
direct distance dialing 长途电话直接拨号
direct distribution substation 直配变电站
direct document processing 直接文件处理
direct drive cyclometer 直接传动的回转计
direct drive gear 直接传动装置
direct drive generator 直驱发电机
direct-driven exciter 直接传动的励磁机，同轴励磁机
direct drive【no gearbox】wind turbine 直驱【无齿轮箱】型风电机组
direct driven pump 直接传动的泵
direct drive permanent magnet generator 直接驱动永磁发电机
direct drive wind turbine 直接驱动风电机组
direct drive 同轴传动，直接驱动，直接传动
direct dry cooling system 直接干式冷却系统
direct dumping 直接卸料
directed branch 定向分支
directed dipole 定向偶极子
directed energy balance 直接能量平衡
directed pressure 定向压力
directed water cooling of hydro-generator 水轮发电机水内冷
directed water cooling of turbogenerator 透平发电机水内冷
direct effect 直接效果，直接影响
direct electrical analogy method 直接电拟法
direct electromotive force 直流电动势
direct energy balance control strategy 直接能量平衡控制策略
direct energy balance control 正平衡监察，直接能量平衡控制
direct energy balance method 能量正平衡法【分析热能动力装置】
direct energy balance 直接能量平衡
direct energy conversion 直接能量转换
direct energy converter 能量直接转换装置
direct energy 定向能量
direct English access and control 直接英文存取及控制
direct-entry terminal 直接输入终点
direct evaporator 直接式蒸发器
direct excitation 直接励磁
direct expansion cooling coil 直接膨胀的冷却盘管
direct expense 直接费，直接费用
direct-fed coil 直馈式线圈
direct feedback 刚性反馈
direct feeder 直接馈路
direct-feed storage 由主要料场机械直接取用的贮存
direct filtration 直接过滤
direct financing 直接筹资，直接融资
direct fired boiler 直吹式制粉系统锅炉
direct fired circulating system 直吹式制粉系统
direct fired mill 直吹式制粉系统磨煤机
direct fired pressure 直接消防压力
direct fired 直燃式
direct-firing pulverized system 直吹式制粉系统
direct-firing system 直吹式制粉系统
direct firing 直吹式燃烧，直吹式制粉系统燃烧
direct fission yield 一次裂变率，直接裂变产额
direct flooding system 直接淹没系统
direct-flow burner 直流式燃烧器
direct-flow cooling 单向流动冷却，直流冷却
direct-flow water circulation 直流供水
direct foreign investment 外国直接投资
direct function 直接函数

direct gain 直接受益式【太阳光热利用】
direct gas-cooling 气体直接冷却
direct geared 齿轮直接传动
direct glare 直接眩光
direct governing 直接调节法
direct grid connection 直接并网
direct grounded 直接接地的
direct grounding system 直接接地系统
direct grounding 直接接地（通过接地电极）
direct group coupling 群间直接耦合
direct handing 直接操作
direct-heated thermistor 直接加热的热敏电阻
direct heating system 直接式供暖系统
direct heating 直接（式）供暖，直接加热
direct illumination 直接照明
direct impulse 正向脉冲
direct incidence 直接入射
direct indication 直接示值
direct-indirect pumping 直接间接抽水
directing line 导线
directing property 定向性
directing 指导，导演
direct injection cooling system 直接注入冷却系统
direct injection engine 柴油机
direct injection type coal pulverizing system 直吹式燃煤制粉系统
direct input 直接输入
direct insert subroutine 直喷狄赛尔内燃机，柴油机
direct instruction 直接指令
direct insurance 直接保险
direct interaction type of reaction 直接相互作用型核反应
direct interception 直接截取
direct interference 直接干扰
direct international procurement 直接国际采购
direct interpolation 直接内插法
direct interruption cost 直接停电损失
direct investment abroad 国外直接投资
direction action 直接作用
directional 定向的，方向的
directional absorptance 定向吸收率
directional absorptivity 定向吸收率
directional air flow 定向气流
directional antenna 定向天线
directional attenuator 定向衰减器
directional blasting 定向爆破
directional coalescence pattern 定向聚合模型
directional coil 定向线圈
directional comparison protection system 方向比较保护系统
directional comparison system 方向比较制【继电保护】
directional control valve 定向控制阀
directional control 定向控制，方向操纵性
directional cooling 顺序冷却，定向冷却
directional correlation 方向相关
directional counter 定向动作计数器
directional coupler 定向耦合器
directional current meter 定向流速仪
directional current protection 方向电流保护
directional dependence 方向依赖，方向依赖性
directional derivation 方向导数
directional distance relay 定向距离继电器
directional dose equivalent 定向剂量当量
directional drilling 定向钻孔
directional earth-leakage relay 对地漏电方向继电器
directional echo sounding 定向回声测深
directional emissivity 定向发射率
directional emittance 定向放射，定向发射
directional explosion 定向爆破，定向爆炸
directional filter 方向滤波器，分向滤波器
directional gain 定向性增益
directional ground relay 接地方向继电器
directional gyro 航向陀螺，陀螺方向仪
directionality effect 定向效应
directional lighting 方向性照明
directional line 方向线
directional mixing effect 定向交混效应
directional overcurrent protection 定向过流保护装置，方向过电流保护
directional overcurrent relay 定向过流继电器
directional phase shifter 定向移相器，方向性移相器
directional pilot relaying 定向纵联继电保护
directional power protection 方向功率保护
directional power relay 定向功率继电器，功率方向继电器
directional projection 定向投影
directional radiation properties 定向辐射特性
directional radiation 定向辐射
directional reflectance 定向反射
directional relay 定向继电器，方向继电器
directional response 方向反应
directional shooting 定向爆破
directional source 定向源
directional spectral absorptivity 定向单色吸收率
directional spectral emissivity 光谱定向发射率
directional stability 方向稳定性
directional stop 定向限位架
directional valve 定向阀【如三通等】
directional vane 导向叶片
direction angle 方向角
direction control valve 方向控制阀
direction cosine 方向余弦
direction distribution 定向分布
direction effect 方向效应
direction erosion 定向侵蚀
direction error 方向误差
direction finder 定向器，无线电罗盘，测向器[仪]，探向器
direction finding indicator 测向指示器，探向指示器
direction finding receiver 定向接受机
direction finding station 测向台
direction finding 探向，定向
direction float 流向测验浮标
direction fluctuation 方向脉动
direction indicator 流向指示器
direct ionizing particles 直接致电离粒子

direct ionizing radiation 直接电离辐射
directionless pressure 液体静压力，无向压力，静压力
direction-listening device 声波定向器
direction method 方向观测法
direction of arrow 箭头方向
direction of current 电流方向，流向
direction of development 发展方向
direction of energy flow 能流方向，潮流方向
direction of feed 进给方向
direction of fissures 裂隙方向
direction of flow 流向，流动方向
direction of ground-water flow 地下水流向
direction of magnetization 磁化方向
direction of maximum extension 最大延伸方向【冶金】
direction of outgoing line 出线方向
direction of polarization 极化方向
direction of principal curvature 主曲率方向
direction of principal stress 主应力方向
direction of propagation 传播方向
direction of rolling 辗压方向
direction of rotation 旋转方向
direction of stranding of outer-most layer 导线最外层绞向
direction of travel around the loop track 绕环形线行进方向
direction of wave travel 波的传播方向
direction of welding 焊接方向
direction peg 导向桩
direction protection 方向保护
direction sense movement 定向性运动
direction sign 方向标志，指路标志
direction theodolite 方向经纬仪
direction 指示，指导，说明书，方向，方位
direct irradiance 直射辐照度
direct irradiation calibrating method 直接辐射标定法
direct irradiation 直射辐照量
direct irrigation area 干渠灌溉面积
directive action 定向动作，定向作用
directive effect 方向作用，方向效应
directive erosion 定向侵蚀
directive planning 指令性计划
directive rule 规程
directive support 定向支架
directive 指示的，管理的，指令的，定向的，有方向的，命令
directivity gain 指向性增益
directivity pattern 指向性图
directivity 方向性，指向性，定向性
direct leakage infiltration leakage 直接泄漏
direct leakage 直接泄露
direct leveling 直接高程测量
direct lighting stroke 直接雷击
direct lighting 直接照明
direct line 定向线
direct liquid cooling 直接液冷
direct load loss 直接负荷损失，欧姆铜耗
direct load 直接载荷
directly 直接的
directly addressable 可直接按地址存取的
directly burying 直埋敷设
directly controlled 直接被控的，直接受控的
directly controlled equipment 直控式牵引设备
directly controlled system 直接被控系统，直接受控系统
directly controlled variable 直接被控量
directly earthed neutral 直接接地中性点
directly ionizing particle 直接电离粒子
directly ionizing radiation 直接电离辐射
directly proportional to 与……成正比
directly solid grounded 直接接地
direct magnetic measurement 直接磁粉测定
direct magnetizing effect 直轴磁化作用，直轴磁化效应
direct manual control 手控，手动
direct measurement 直接测定，直接测量，直接量测
direct memory access controller 直接存储器存储控制器
direct memory access 直接存储器存取
direct memory address 直接存储器地址
direct method heat balance 正热平衡
direct method of measurement 直接测量法
direct neutron radiography 直接中子射线照相
direct-noise amplifier 直接噪声放大器
direct normal insolation 直射辐射强度
direct numerical control 直接数字控制
direct obligation 直接责任
direct observation 直接观测
direct on-line starting 直接启动，全压启动
direct on-line start 直接起动
direct on-line switching 直接启动，直接投入电网
direct on starter 直接启动器
direct operating controller 直接操作控制器，直接运算控制器，直接作用控制器
direct operational cost 运转费用，运行费用
direct operation 直接操作
director element 引向器
director general （国际原子能机构的）总干事
director liability 董事职责
director of chemical operation 化学运行主任
director of dynamics and system engineering 动力与系统工程主任
director of instrumentation and control 仪表与控制主任
director of operating engineering 运行工程主任
director of production department 生产处长
director of purchasing 采购主任
director of site management 工地主任
director of the board 董事长
director responsibility system 厂长负责制
director responsible for project 工程主管院长
direct or reverse 正面或反面的，正向或反向的
directory board 指示牌
directory service 名址服务，目录服务
directory system 索引系统
directory 目录，索引薄
director 导流器，引向器，定向器，导向器，董事，主任

direct oxidation 直接氧化作用，直接氧化
direct-path 直接波束，直接路径，直接通路
direct piezoelectric effect 正压电效应
direct printer 直接打印机
direct proportional 成正比的
direct proportion 正比例
direct pumping 直接抽水
direct,quadrature and zero axis quantities 直轴、交轴和零轴参量
direct quenching 直接淬火
direct quotation 直接报价，直接标价
direct radiation 直接照射，直接辐射
direct ratio 正比例
direct reader 直接读出器，直接读数器
direct reading dosimeter 直读式计量计，直接读数剂量计
direct-reading instrument 直读式仪器，直读式仪表
direct-reading level gauge 直接读数的水位计
direct-reading manometer 直读式压力计
direct-reading meter 直读计，直读式仪表
direct-reading tacheometer 直读视距仪，直读速测仪
direct-reading thermometer 直读式温度计
direct-reading 直接读数，直读，直接读数的，直接读出
direct-recording 直接记录的，直接记录
direct reduction 直接还原
direct refrigerating system 直接制冷系统
direct-relation telemeter 正比式遥测仪
direct release 直接释放，串联释放，一次电流释放
direct remittance 顺汇
direct resistance heating 直接电阻加热
direct return system 异程回水系统，异程式系统
direct-reverse switch 正反向开关
direct-reverse turbine 正反转式透平，倒顺车透平
directrix 准线
direct runoff 地面径流，直接径流
direct scanning 直接扫描
direct scan 直接扫描，一次波法
direct selector 简单调器
direct sewer system 直接下水管道系统
direct shear apparatus 纯剪切仪，直接剪切仪，直剪仪
direct shear test 直剪试验，直接剪切试验
direct shipment 直达运输
direct shock absorber 直接式减震器
direct shock wave 正冲波
direct short-circuit 直接短路
direct short 短路
direct simple shear test 直接单剪实验
direct solar irradiance 直（接日）射辐照度
direct solar radiation intensity 太阳直接辐射强度
direct solar radiation 太阳直接辐射，直接日射
direct sound wave 直达声波
direct spark ignition 直接火花点火
direct starting 直接启动
direct steam 直接蒸汽，新蒸汽
direct stratification 原生层理
direct stress 正应力，直接应力
direct stroke lightning overvoltage 直击雷过电压
direct stroke protection 直接雷击保护
direct subtract 直接减
direct supply reservoir 直接供水水库
direct supply 直流电源
direct surface runoff 直接地表径流
direct switching in 直接启动，直接合闸
direct system 直接系统
direct tax 直接税
direct telescope 正镜
direct tensile strength 抗正拉强度
direct tensile test 直接拉力试验
direct tide 直接潮汐
direct titration 直接滴定
direct transfer of control 控制的直接转移
direct transmission factor 直接传输系数，单向透射率
direct transmission gain 直通电路增益
direct transmission 直接输送，单向透射
direct transmitter 直接发送器
direct treatment 直接处理【炉内处理】
direct-trip circuit-breaker 直接跳闸断路器
direct-type wave 非（荧光）增感形胶片
direct ultrosonic visualization of defects 缺陷的超声波直接显示
direct vernier 顺游标
direct-viewing memory tube 直观存储管
direct-viewing receiver 直观式电视接受机
direct view storage tube 直观存储管
direct-vision spectroscope 直视分光镜
direct visual observation 直接目测
direct voltage 直流电压，正向电压
direct water hammer 直接水锤
direct wave 非反射波，直达波，直接波
direct-wire circuit 单线线路
direct-writing oscillograph 直接记录示波器
direct-writing recorder 直写式记录器，直接记录器
direct 直的，直流的，直接的，明白的，正的
dire handle 板牙架
dirt band 夹矸，污层
dirt cloud 尘云
dirt collector 吸尘器
dirt deposit 积垢
dirt inclusion 夹渣
dirtiness resistance （热交换器管壁上的）垢膜热阻，污垢热阻
dirt-proof 防尘的
dirt resistance 防尘的
dirt steel 不洁钢
dirt trap 污垢阱，污泥阱
dirty coal 高灰煤
dirty condensate tank 污凝结水器
dirty environment 多尘环境，污秽环境
dirty gas 含尘气体
dirty material 脏材料
dirty oil basin 污油池
dirty oil pump 污油泵
dirty oil tank 废油箱，污油槽
dirty oil treatment facility 污油处理设备

dirty oil 污油
dirty sodium 污染钠，沾污钠
dirty waste 楼面排污水
dirty water collecting tank 污水收集箱
dirt 污物，废屑，灰尘，污点，污垢，尘土
disability benefit 伤残补助
disability clause 伤残条例
disability insurance 伤残保险
disabled capacity 受阻容量
disabled time 不能工作时间
disabled 失去能力的
disable instruction 非法指令，不能执行的指令
disable interrupt instruction 禁止中断指令
disable interrupt 禁止中断
disable pulse 封闭脉冲
disable 终止，撤除，禁止，取消
disaccommodation 磁导率减弱，失去调节
disadjust 失谐的，失调的，失调，失协
disadvantage factor 不利因子【中子通量】
disadvantage 不利，缺点，缺陷，不利条件
disaggregate 粉碎，磨碎
disaggregation model 分散模式
disagreement 不一致，异议，意见不同
disagree 意见分歧
disalignment 不同心，不对中，定线不准，偏离轴心，偏离中心线
disappearance 崩塌【水坝】，消失，失踪，不见
disappearing filament optical pyrometer 隐丝式光学高温计，隐丝式光测高温计
disappearing filament pyrometer 隐丝式光学高温计
disappearing stream sinking 伏河
disappearing stream 地下河
disappear of echo 回波消失
disarrangement 失调，破坏，断裂，失常，紊乱，混乱
disassemble tool 拆卸工具
disassemble 拆卸，拆开，分解，拆散，卸下
disassembling equipment 拆卸设备
disassembling inspection of rotating machine 转动机械解体检查
disassembling inspection of valve 阀门解体检查
disassembling 解体
disassembly examination reassembly machine 拆卸检查组装机器
disassembly station 解体站【燃料组件】
disassembly 拆卸，拆开，解开，分解，拆迁，解体
disaster area 灾区
disaster beyond control 不可抗力的灾害
disaster box 保险盒，安全线路
disaster power 事故备用容量
disaster preparedness 灾害预防
disaster prevention 防灾
disaster radiation monitoring 事故辐射监测
disaster shutdown 事故停堆，事故停机
disaster 灾难，故障，事故，严重事故
disastrous drought 灾难性干旱
disbranch 取消支路，断开，分离，分开
disbursement fee 垫付款手续费【承运人收取此费为 DBC，代理人收取此费为 DBA】
disbursement receipt 支出收据
disbursement 支付，支出，支付款，付出额，付出款
discap 盘形电容器
discarded hull monitor 废弃燃料壳监测器
discarded material 废弃的物料
discarded product 废品
discarded soil 弃土
discarded tip 挤压尾料
discard electrode 焊条头
discard 放弃，报废，清除，抛出，抛弃，废弃物，废品，废料，废渣
disc area 桨盘面积
disc armature 盘形电枢，盘形衔铁
disc brake 盘式制动器
disc buoy （海上风车的）圆盘式浮台
disc calculator 盘式计算机
disc clutch 盘形离合器
disc coil 扁平线圈，蛛网形线圈
disc-coupled vibration 叶片轮盘整体振动，轮系振动
disc coupling 圆盘联轴器
disc crusher 圆盘破碎机
disc drive 磁盘驱动器
discern 辨别
disc filter 盘式过滤器
disc float 圆盘浮标
disc flow ratio 圆盘流动比率
disc-footed pile 扩脚桩
disc friction loss 轮盘摩擦损失
disc friction 轮盘摩擦
disc generator 盘式发电机
discharge activities to the atmosphere 向大气排放放射性物质
discharge airslide 空气卸料斜槽，气力输送机
discharge along contaminated dielectric surface 沿污染电介质表面放电
discharge along dielectric surface 沿面放电
discharge amount 排放量
discharge analyzer 放电分析器
discharge angle 出口角
discharge apparatus 卸料装置
discharge area 出口（截）面积，过水面积，过水断面面积，排料口面积，排水区
discharge baffle pipe 排放缓冲管
discharge bay 泄流段，泄水池
discharge branch 排水支管
discharge bucket 卸料斗
discharge bum-up 卸料（的）燃耗
discharge canal 排水渠道，泄水渠
discharge capacity 排放量，通流能力，放电容量，排污量，过水能力，排汽能力，排水量，排泄能力，泄流能力，泄水能力
discharge casing 排出壳，排气缸
discharge cask 卸料容器
discharge chamber 排气室
discharge channel 泄水渠，卸料管道，出口管道，排水渠道
discharge characteristic 排气特性，流量特性，放电特性曲线，放电特性
discharge charge 卸货费用

discharge check valve 排出单向阀
discharge chute 出料槽【混凝土拌和机】，排料槽，落煤槽，卸料斜槽
discharge circuit 放电电路，放电回路
discharge coefficient for orifice 节流孔流量系数
discharge coefficient of nozzle 喷嘴流量系数
discharge coefficient 放电系数，排放系数，流量系数
discharge concentration 含沙量
discharge condition 放电状态，出口状态，排放条件，泄水条件
discharge conduit 泄水道
discharge connection 出口管，放电接线
discharge contact 放电触点
discharge counter 放电记录器【避雷器】，动作记录器，放电计数器
discharge criteria 排放标准
discharge culvert 排水涵洞
discharge current capacity 通流能力，通流容量
discharge current 放电电流，泄露电流
discharge curve 放电曲线，流量曲线，水位流量曲线
discharged air 被排出的气体，排气
discharge data 流量资料
discharge device 避电器
discharged fuel 卸出的燃料【反应堆】
discharge diagram 流量曲线，流量图
discharge ditch 排水沟
discharge duct 排风管
discharge duration curve 流量历时曲线
discharge duration 放水历时，流量历时
discharge elbow 排放弯管
discharge electrode 放电极，放电电极，电晕电极
discharge-elevation relation 泄量-高程关系
discharge end pressure 排水端压力，出口端压力
discharge end 出料端
discharge energy test 放电能量测验
discharge extinction voltage 熄弧电压
discharge factor 放电因数，流量因数，排放系数，流量系数
discharge fan 排风扇，排风机，抽风扇
discharge fee 排放费用
discharge final voltage 放电末期电压
discharge flow 排出流量，泄出水流
discharge fluctuation 流量变化
discharge flue 排烟道，排气道
discharge frequency curve 流量频率曲线
discharge frequency 放电频率
discharge gap 放电间隙
discharge gas 废气，排气
discharge gate 出料口，排料口
discharge grid 排料格栅
discharge head cover 排出帽盖
discharge head relation 出口流量水头关系，出口流量压头关系
discharge head 出口压头，排出压头，（泵）扬程，出口压力，出口水头，供水压头，排放扬程
discharge hole 排出孔，出口
discharge hopper 卸料斗
discharge hydrograph 流量过程线
discharge hydrostatic head 出口静压头
discharge inception test 开始放电试验，起晕试验
discharge inception voltage 开始放电电压，起晕电压
discharge index 流量指数
discharge in insulation 绝缘放电，绝缘击穿
discharge integrator 流量积分仪
discharge intensity 流量强度
discharge interpolation 流量插补
discharge jetty 卸货突码头
discharge jet 喷射水束，射水管
discharge leg of syphon 虹吸管流出端
discharge line loss 泄水管路损失
discharge line 出口管线，卸料管，排水管线【泵】，排线管线
discharge liquid 废液
discharge loss 出口损失，放电损耗，出口损耗
discharge machine 卸料机
discharge manifold of lock chamber 闸室排水道，闸室排水总管
discharge mass curve 流量累积曲线
discharge measurement 测流，流量测定，排放测量，流量测验
discharge modulus 流量模数
discharge nappe 喷射水舌
discharge nozzle 出口管，排出管咀，排气喷管，排放管嘴
discharge observation 流量观测
discharge of capacitor 电容器放电
discharge of contract 取消合同
discharge of debt 清偿债务
discharge off 无放电，放电终止，排泄中断
discharge of hydraulic machinery 水力机械流量
discharge of oil 排泄油料
discharge of pump 泵的排水量
discharge of river 河流径流
discharge of sewage 污水排放
discharge of water 排水，排水量
discharge of well 井的出水量，井流量
discharge on 正在放电，放电期间
discharge opening 排出口，排料口，排水口，卸料口
discharge orifice 泄水孔，排放孔
discharge outfall 排水口，排出口
discharge outlet 放水口，排水出口，排水管出口，排泄口
discharge path 放电路径
discharge period 放电持续时间，排放期，放电期，泄放时期
discharge per unit width 单宽流量
discharge pipeline 排水管线
discharge pipe 排水管，排放管，排出管线，泄水管
discharge piping 出水管系，排出管
discharge plenum 排气缸，排气室
discharge point 排放点
discharge pond 卸料水池
discharge port 卸料口

discharge potential difference　放电电位差，放电电压
discharge potential　放电电位，放电电势
discharge pressure　出口压力，排出压力，排气压力，排放压力
discharge procedure　卸料程序，卸料步骤
discharge process　放电过程，排泄过程
discharge protector　击穿保险器
discharge pulley　卸料滚筒
discharge pump　排泄泵，排水［气］泵，排放泵
discharge ram　卸料推杆
discharge rate　放电率，排放率，卸料率，释放率，宣泄流量
discharge rating curve　流量率定曲线
discharge ratio　流量系数，排放系数
discharge record　流量记录，流量资料
discharge regulation　流量调节
discharge regulator　流量调节器
discharge resistance　放电电阻
discharge resistor　放电电阻器
discharge rod　放电棒
discharger　放电器，扩容器，排出管，火花间隙，排放装置，卸载器，避电器，发射器，排放管，排料装置，排料器，卸料工
discharge scale ratio　流量比尺
discharge sectional-line　测流断面
discharge section　过水面积
discharge side cover　排出侧盖
discharge site　测流地点，测流地址，测流段，测流断面
discharge-slope-depth relation　流量比降深度关系
discharge spark　放电火花
discharge standard　排放标准
discharge state　出口状态
discharge steam　排出蒸汽
discharge stroke　排气冲程
discharge structure　排水结构，排水站，排出口，排出装置，泄水建筑物，排水构筑物
discharge switch　放电开关
discharge system of pebble bed　球床卸料系统
discharge system　排水系统，排放系统，卸料系统，放电系统，（燃油）清理装置
discharge table　流量表
discharge tank　排放水箱
discharge temperature　出口温度，排气温度
discharge test　放电实验
discharge through gas　气体放电
discharge time　放电时间，排放时间
discharge to atmosphere　向大气排放
discharge trough　排水槽
discharge tube rectifier　充气管整流器，放电管整流器
discharge tube　放电管，闸流管，引出管，卸出管，泄水管
discharge tunnel　泄水隧洞
discharge uniformity　流量均匀性
discharge value　排放量
discharge valve　卸料阀，排放阀，排出阀，排水阀，排气阀，减压阀，排泄阀，放泄阀
discharge vanes　排出轮叶
discharge velocity　排气速度，排放速度，卸载速度，排水速度，泄流速度，泄水速度
discharge voltage　放电电压，闪络电压，击穿电压
discharge water level　排水高度
discharge water pump　排水泵
discharge water reservoir　排水库
discharge water　排出水
discharge wave　放电波
discharge withstand current rating　额定耐放电电流
discharge zone　排水带
discharge　排出，射出，释放，排量，放电，卸载，卸料，排放，出口，流量，输送，卸货，排料，排水，泄放
discharging airslide　压缩空气卸料槽
discharging choke　放电抗流圈，放电扼流圈
discharging cock　放出旋塞
discharging current　放电电流，泄放水流
discharging into closed system　排入闭路系统【卸压阀】
discharging line for special wagon　异型车卸车线
discharging machine　卸料机
discharging of boiler　锅炉排水
discharging of insurance　退保
discharging port　卸货港，卸货港口
discharging quay　卸货码头
discharging rod　避雷针
discharging sluice　泄水闸
discharging steam　乏汽
discharging to atmosphere　排入大气
discharging tube　排水管，出水管
discharging　出流，排出，排泄，泄放，卸料，卸载，排空，放出，排放
DISCH VLV(discharge valve)　排放阀
disc insulator　盘状绝缘子
discipline engineer　专业工程师【指某个专业的】
discipline　纪律，学科，专业
disclaimer clause　否认条款
disclaimer of responsibility　放弃责任
disclaimer　放弃（权利），放弃者，否认条款，弃权，不承诺，免责声明，免责条款，拒绝
disclaim　否认与……有关，拒绝承认，放弃权利，否认
disc load　桨盘载荷
disclose a non-conformance　发现一项不符合项
disclose　揭开
disc loss　轮盘摩擦损失
disc meter　盘式流量计，圆盘式水表，盘式仪表
disc mixer　盘式搅拌器
disc motor　圆盘形电动机
disc of conical profile　锥形轮盘
disc of constant stress　等强度轮盘
disc of constant thickness　等厚度轮盘
disc of hyperbolic section　双曲线截面叶轮
disc of variable profile　变截面轮盘
disc of variable thickness　变厚度轮盘
discoidal armature　盘形电枢，圆盘形衔铁
discoloration　脱色，变色，褪色，弄脏，污染
discolor　使变色，弄脏，脱色

discolouration　变色，脱色，褪色，污染
discoloured clay　退色黏土，变色黏土
discomfort index　不舒适指数
discomfort map　不舒适区图
discomfort parameter　不舒适参数
discomfort pattern　停止条件
discomfort threshold　不舒适阈
discomfort zone　不舒适区
discomfort　不舒适（性）
discompressor　松压器
disconformity　假整合
disconnectable busbar　无载分段母线
disconnectable coupling　活络接头
disconnectable　可拆换的
disconnect button latching finger　断路按钮闭锁销
disconnect button　断路按钮
disconnect coupling　拆开联轴节【滚子链条，法兰，连续套筒等】，联轴器脱开
disconnected cracks　斑纹状裂缝，斑纹状裂纹
disconnected position　断开位置
disconnected switch　隔离开关
disconnect fuse　切断熔丝，拉出式熔丝
disconnecting chamber well　检查井
disconnecting chamber　隔离室
disconnecting contact　开断触点
disconnecting device　切断装置
disconnecting link　隔离开关，刀闸，闸刀
disconnecting means　拆开方式，拆开工具
disconnecting switch reverser　隔离反相器
disconnecting switch　隔离开关，断路开关，刀闸
disconnecting　拆开，断开，切断，断路，拆卸，分离
disconnection fault　断电故障，断线故障
disconnection from power　与电源切断
disconnection of power　失去动力，断开电源
disconnection　拆开，断开，拆接，解列，断路，切断，开断，隔离，开路，脱落
disconnector-fuse　带熔断器的隔离开关
disconnector　隔离器，切断开关，断路器，隔离开关，断开器，刀闸
disconnect rod　拆卸杆，断开杆，离合杆
disconnect valve　截止阀，断流阀，断开阀
disconnect　断接，不连接，分离，解列，断开，拆开，分开，切断，断路，拆卸
discontinuance　中止
discontinued operation　非连续作业
discontinue　停止，中断
discontinuity between the weld and adjacent surfaces　焊接与邻近表面之间的不连续性
discontinuity condition　不连续条件
discontinuity indication　缺陷指示
discontinuity layer　不连续层，突变层
discontinuity opening　开口缺陷
discontinuity point　不连续点
discontinuity stress　突变应力，不连续应力
discontinuity surface　不连续面
discontinuity which is open to the surface　通向表面的不连续，显露缺陷
discontinuity　不连续，间断，突变，缺陷，不连续变化，不连续性
discontinuous action servomechanism　不连续作用伺服机构
discontinuous action　断续动作
discontinuous combustion　不稳定燃烧，断续燃烧
discontinuous construction　不连续结构
discontinuous control system　断续控制系统
discontinuous control type regular　断续控制型调节器
discontinuous control　不连续控制，断续控制，离散控制
discontinuous distribution　不连续分布，不连续分配
discontinuous film　不连续膜
discontinuous filter　脉冲滤波器，不连续的滤波器
discontinuous flow　断续流动，不连续流动
discontinuous function　不连续函数
discontinuous gradation　不连续级配
discontinuous heating　间歇供暖
discontinuous insulation　不连续绝缘
discontinuous interstice　不连续间隙【岩石】
discontinuous load　断续负载，断续负荷，不连接负载，突变性负载
discontinuous motion　不连续运动
discontinuous oscillation　间断振动，断续振荡
discontinuous permafrost　不连续冻土带
discontinuous phase　不连续相，间断相
discontinuous running　断续运行，断运行
discontinuous servomechanism　断续作用的伺服机构
discontinuous wave　非连续波
discontinuous　断续的，不连续的，间断的，中断的
discontinuum　密断流，间断集
discordance analysis　不一致性分析
discordance　不整合，不一致，不调和，冲突，（声音的）不和谐
discordant fold　不整合褶皱
discordant injection　不整合贯入
discordant intrusion　不整合侵入
discordant valley　不整合河谷
discount and allowances　折扣与折让［津贴］
discount bank　贴现银行
discounted cash flow rate of return　内部收益率
discounted cash flow　现金流量贴现，折现的现金流量
discounted value　贴现值，折扣值
discount government loan　政府贴息贷款，国家财政贴息贷款
discounting　打折扣
discount on a promissory note　期票贴现
discount rate　折扣率，贴现率，折算
discount　减价，贴现，折扣
discoupling　去耦，分开，拆开，关闭，停止
discover　发现
disc plane　桨盘平面
disc ratio　叶盘面积比
disc reclaimer　圆盘取料机
disc relay　盘形继电器
discrepancy and claim clause　异议和索赔条款
discrepancy in elevation　高差

discrepancy report 差异报告
discrepancy switch 差速开关，不对位开关
discrepancy 不符点【单据与信用证条件不一致】，不符规定，使用不符合，不符值，不同，不符，矛盾，差异，不一致之处，分歧
discrepant 差异的
discrete address 离散地址
discrete aggregate 间断级配骨料，松散骨料
discrete analogue 离散模拟
discrete block 松散块体
discrete bubble 孤立气泡
discrete cell 孤立单元
discrete channel 离散通道，离散信道
discrete comparator 离散比较器，数字比较器
discrete component 分立元件，分立部件，离散构件
discrete-continuous system 离散连续系统
discrete control system 离散控制系统
discrete data 不连续资料，离散数据
discrete density level 离散密度级
discrete distribution 不连续分布，离散分布
discrete eigenfunction 离散本征函数
discrete excitation 不连续激发，离散激发
discrete Fourier transformation 离散型傅立叶变换
discrete harmonic 离散协波
discretely-timed signal 离散定时信号
discrete mass 离散物质
discrete member 离散构件
discrete message state 离散信息状态
discreteness 离散
discrete ordinates method 离散坐标法，分离坐标法
discrete phase 不连续相
discrete point load 不连续点荷载
discrete programming 离散程序设计
discrete pulse 离散脉冲，不连续脉冲
discrete random variable 离散随机变量
discrete representation 离散表示法
discrete sampling 离散采样
discrete series 不连续数列
discrete source 不连续源，分立源
discrete state 离散状态，不连续状态
discrete system 离散系统
discrete-time filter 离散时间滤波器
discrete-time system 离散时间系统
discrete time 离散时间
discrete winding 不连续绕组
discrete 分离的，抽象的，离散的，不连续的，分立的，个别的，独立的
discretionary purchasing power 剩余购买力
discretion 谨慎，慎重，判断力，自行决断，处置权，斟酌决定，自行处理
discretization 离散化
discriminability 鉴别力，分辨力
discriminating breaker 方向性开关，逆流自动断路开关
discriminating content addressable memory 可鉴别的内容定址存储器
discriminating protective system 区域选择性保护系统
discriminating selector 鉴别选择器，区域选择器
discriminating threshold 甄别器
discrimination factor 甄别因子
discrimination filter 鉴别滤波器
discrimination protection 选择性保护装置
discrimination threshold 鉴别阈，鉴别阈值［限度］，鉴别灵敏度
discrimination 分辨，辨别，分解，甄别，鉴别，歧视
discriminator 鉴别器，比较装置，甄别器，鉴频器，鉴相器
disc rotor 盘式转子
disc screen 过滤片，圆筛
disc-seal triode 盘封三极管
disc-seal tube 盘封管，灯塔管
disc slide rule 计算盘
disc stroboscope 盘式频闪观测仪
disc type armature 盘型电枢
disc type flaw 圆片形缺陷
disc type insulator 悬式绝缘子
disc type relay 盘形继电器
disc type rotor 圆盘式转子
discussion 讨论，辩论，磋商，商谈
discuss separately 另行讨论
disc valve 圆盘阀
disc vibration 叶轮振动
disc water meter 转盘式水表
disc winding 饼式线圈，盘形绕组
disc 盘，圆盘，圆板，磁盘，叶轮，板，轮盘，桨盘
DIS(digital indicating station) 数字指示站
DIS(draft international standard) 国际标准草案
disease carrier 带病体
disease 疾病
diseconomy of scale 规模不经济
disembosom 泄露
disengageable coupling 能自动脱开的联轴节
disengaged line 闲线
disengagement gear 分离装置
disengagement 脱开，解脱，脱离，游离，分离，解开，卸除，非耦联，摘钩
disengage 脱扣，解扣，脱离
disengaging bar 关闭杆
disengaging gear solenoid valve 解脱机构电磁阀
disengaging surface 蒸发水面
disengaging 摘钩，解脱
disequilibrium 不平衡，失去平衡
disgregation 分散作用
disharmonic 不和谐的，不调和的
disharmony 不调和
dish crusher 圆盘式破碎机
DISH(dissolved hydrogen) 溶氢
dished bottom 碟形底，碟形底封头
dished drum end 碟形汽包封头
dished end 碟形端面【燃料芯块】，冲压封头，凹形封头
dished head 冲压封头，凹形封头，碟形端面，碟形封头，凸形头，碟形头，椭球封头
dished pellet 碟形芯块
dished perforated plate 碟状多孔板
dished section 碟形段【压力容器封头】

dished sluice gate 圆盘式泄水闸门
dished 碟形的，球面的
dish emery wheel 平碟式砂轮
dishing of surface （土壤的）表面碟状沉陷
dishing 成碟形【燃料芯块】，碟形凹陷，凹面，凹陷，表面缩穴
dishomogeneity 松散【冶金】
dishonourable conduct and practices 不名誉的行为及手法
dish pellet 碟形芯片
dish sink 洗碗池
dish solar thermal power generation 碟式光热发电，盘式光热发电
dish 盘，碟，盘形物，反射器，抛物面天线，卫星接收天线，碟形天线
disinfectant 消毒，消毒剂
disinfected wastewater 消毒废水
disinfection by chlorination 氯化法消毒
disinfection of water 水的消毒
disinfection plant 消毒设备
disinfection 消毒，杀菌，灭菌
disinfect 消毒
disinflate 排气
disinflation 通货收缩
disintegrated granite 风化花岗岩
disintegrated 碎裂的
disintegrating mill 磨煤机
disintegrating nozzle 雾化喷嘴
disintegration chain 衰变链
disintegration coefficient 衰变系数
disintegration constant 衰变常数，蜕变常数
disintegration energy 衰变能
disintegration product 衰变产物
disintegration rate 衰变率，蜕变速度，衰变速度，破碎率
disintegration time 崩解时间
disintegration 分解，分散，衰变，解体，蜕变，裂变，分裂，风化，碎裂，瓦解，剥蚀，崩解
disintegrator 粉碎机，碎渣装置
disinvestment 收回投资，减少投资，停止投资，投资缩减，负投资，蚀本
disjoint 不相交的，不连贯，拆散，分解，脱节
disjunction 断开，脱节，折断
disjunctor 分离器，断路器，开关
disk antenna 盘形天线
disk armature 盘形电枢
disk assembly 阀瓣组件
disk bit 圆盘钻
disk body 圆盘
disk-bursting test 叶轮破裂实验
disk bypass 阀瓣旁路
disk capacitor 半圈形可变电容器，盘形电容器
disk clutch 圆盘离合器
disk condenser 圆盘电容器，盘式电容器
disk controller 磁盘控制器
disk conveyor 圆盘式输送机
disk coupling 圆盘联轴节
disk crusher 圆盘破碎机，圆盘式碎石机
disk cutter 圆盆式切割机
disk discharger 盘式放电器，旋转火花间隙
disk drive 磁盘驱动器，磁盘机
disk electrode 盘形电极
diskette 磁盘，软盘，软塑料磁盘
disk feeder 圆盘式给料器
disk file control 磁盘文件控制，磁盘文件控制器
disk file 磁盘文件，盘形外存储器
disk filter element 叠片滤元，圆盘滤元
disk-flow condenser 淋盘式凝汽器
disk-footed pile 盘脚桩
disk foundation pile 盘底桩
disk free to rotate on stem 可绕阀杆自由转动的阀瓣
disk friction loss 轮盘摩擦损失
disk friction 盘面摩擦【水轮机转子】
disk generator 盘形发电机
disk grizzly 圆盘筛
disk guide 阀瓣导向
disk hardness gauge 圆盘硬度计
disk holder 阀瓣座
disk insert 阀瓣垫片
disk insulated cable 有绝缘垫圈的高频电缆
disk lightning arrester 盘式避雷器
disk memory 盘形存储器，磁盘存储器
disk nut bearing 阀瓣螺母支承
disk nut 阀瓣螺母
disk of constant stress 等强度（锥形）叶轮
disk of constant thickness 等厚度叶轮
disk of throttle valve 节流阀盘
disk operating system 磁盘操作系统
disk oriented system 面向磁盘的系统
disk pack control 磁盘组控制器
disk pack 磁盘组
disk pile 盘头桩
disk ratio 盘面比
disk recorder 磁盘记录器
disk rigidly attached to stem 阀瓣与阀杆刚性连接
disk roll mill 中速辊式磨煤机
disk roll pulverizer 平盘式中速磨煤机，辊盘式中速磨煤机
disk rotor 盘形转子
disk sander 圆盘磨光机
disk screen 圆盘筛
disk seating area 阀瓣落座面积
disk seating surface 阀瓣落座表面
disk shield 圆盘形屏蔽器
disk slag extractor 圆盘出渣机
disk slightly offset such that it closes by its own weight 阀瓣略有偏移，靠自重关闭【没流体时】
disk slightly swiveled 微旋阀瓣
disk-space cable 有绝缘垫圈的高频电缆
disk stop 阀瓣止点
disk storage 磁盘存储器
disk type reamer 盘形铰刀
disk-type steam trap 盘式疏水阀
disk-type vibration mode 节圆振型，节周式振型
disk-type watermeter 转盘式水表
disk-type winding 饼式线圈
disk unit 磁盘驱动器
disk valve 圆盘阀，碟阀

disk water meter 圆盘（式）水表
disk 盘，圆盘，圆板，磁盘，叶轮，板，阀瓣，阀盘，(蝴蝶阀的）阀叶，轮盘
dislocation basin 断层盆地
dislocation earthquake 断层地震
dislocation metamorphism 断错变质
dislocation mountain 断层山
dislocation of economy 经济（比例）失调
dislocation pile up 位错塞积
dislocation valley 断层谷
dislocation wave 位移波
dislocation 脱位，脱节，变位，错位，转换位置，线缺陷，转移位置
dislodge 取出
dismantlement 拆开，拆卸，拆除
dismantle 拆卸，拆除，拆散
dismantling and reassembling of turbine rotor disc 套装叶轮拆装
dismantling bay 拆卸间
dismantling equipment 拆卸设备
dismantling height 拆卸高度
dismantling machine （燃料元件柱）拆卸机
dismantling vertical clearance 拆卸垂直间隙
dismantling 拆卸，解体，拆开，全部拆除
dismissal agreement 解聘协议
dismissal compensation 辞退金
dismissal of legal action 驳回诉讼
dismissal 解雇
dismission of company 公司解散
dismiss 开除，免职，驳回
dismountable 可拆卸的，可更换的
dismounting 拆卸，拆除
dismount 拆卸，拆除，卸下
dismutation 歧化
DISO(dissolved oxygen) 溶氧
disodium hydrogen phosphate 磷酸氢二钠
disordered structure 无序结构
disordering of crystal lattice 晶格紊乱，晶格无序化
disordering 无序化
disorder 扰乱，骚乱，无序，失调，混乱，使失调
dispatch building 调度楼
dispatch centre 调度中心
dispatch charge 分拨费
dispatch clerk 调度员
dispatch command 调度命令
dispatching 调度，配送
dispatcher's supervision board 控制板，调度盘
dispatcher training system 调度员培训系统
dispatcher 调度员，配电器，调度程序，调度员额分配器
dispatching algorithm 调度算法
dispatching automation planning 调度自动化设计
dispatching automation 调度自动化
dispatching center of distribution system 配电网调度所
dispatching center （电厂）总调度室，发送中心，调度中心
dispatching communication system 调度通信系统
dispatching communication 调度通信
dispatching engineer 调度工程师
dispatching equipment 调度设备
dispatching graph of reservoir 水库调度图
dispatching management information system 调度管理信息系统
dispatching operation of distribution system 配电网调度运行
dispatching order 调度指令，调度命令
dispatching point 调度站
dispatching price 调度价格
dispatching priority 调度优先
dispatching room 调度室
dispatching schedule 派遣进度，派遣计划
dispatching station 调度所，调度室，发送站
dispatching system 调度系统，发送系统
dispatching telephone 调度电话，调度电话机
dispatch interval 调度时段
dispatch list 发货单
dispatch management information system 调度管理信息系统
dispatch note 邮寄清单
dispatch of personnel 人员的派遣
dispatch order 发运单
dispatch room 调度室
dispatch winch 调度绞车
dispatch winder 调度绞车
dispatch 调度【如电力负荷】，迅速处理，输送，发送，发运，发货，分派，派遣
dispensable circuit 调剂电路
dispensation 分配，配方
dispenser 分配器
dispensing apparatus 配料设备，定量设备
dispensing spoon 药匙
dispersal 分散，扩散，泄漏，散射，弥散
dispersant agent 分散剂
dispersant 分散剂
dispersed air floatation 曝气浮选
dispersed bubbly flow 离散泡状流
dispersed droplet flow 离散雾状流
dispersed flow 分散流
dispersed fluidization 散式流化【沸腾炉】
dispersed medium 分散介质
dispersed part 分散部分
dispersed phase hardening 弥散硬化
dispersed phase 分散相，弥散相
dispersed plant 分散布置核电厂
dispersed state 分散状态
dispersed suspension 分散悬浮物
disperse phase 分散相
disperser 分散剂
disperse state 分散状态，扩散状态，弥散状态
disperse system 散体
disperse 分散，弥散，色散
dispersing agent 分散剂，扩散剂
dispersing medium 分散剂
dispersion coefficient 分散系数，弥散系数，扩散系数，磁漏系数，离差系数，离散系数，色散系数
dispersion degree 弥散度
dispersion dose 散射剂量
dispersion fluidized bed 散式流化床

dispersion fuel 弥散燃料
dispersion hardening 弥散硬化
dispersion index 离散指数
dispersion loss 扩散损耗，泄露损耗，漏磁损耗
dispersion measures 分散性测度，离差的度量，散布程度，色散量，频散量度
dispersion nuclear fuel 弥散核燃料
dispersion of behavior 状态的弥散现象
dispersion of effluents from stack 废物离开烟囱后的扩散
dispersion parameter （大气）分散参数
dispersion polymerization 分散聚合作用
dispersion ratio 分散率，扩散率
dispersion reactor 弥散燃料反应堆
dispersion staining 分散着色
dispersion system 分散体系
dispersion train 分散带，分散流
dispersion type fuel element 弥散形燃料元件
dispersion 分散（作用），离散，弥散，弥散体[相]，扩散，泄漏，散射，散布，色散
dispersity 弥散度，分散度，色散度
dispersive action 分散作用
dispersive day 分散性黏土
dispersive medium 弥散介质
dispersive prism 散射棱镜
dispersive soil 分散性土
dispersive 分散的
dispersoid 弥散体，弥散相，分散胶体
displaced enhancement device 可更换的强化传热构件
displaced in phase 相位的移动
displaced phase 位相移
displacement amplitude 位移幅度
displacement angle 位移角，失配角
displacement capacity 排出量【压缩机】
displacement cliff 断崖
displacement component 位移分量
displacement compressor 容积式压气机
displacement continuity 位移连续性
displacement controller 位移控制器
displacement current 位移电流
displacement curve 位移曲线，移动曲线
displacement detector 排气量探测器
displacement diagram 变位曲线，位移图
displacement factor 偏移因子【电源供应】，位移系数，相移系数
displacement fault 平移断层
displacement field 位移场
displacement force 推力，位移力
displacement governor 位移调节器
displacement indicator 位移传感器，位移指示器
displacement line 位移线
displacement loss 部分进汽损失
displacement measurement 位移测定
displacement measuring instrument 位移测量仪表
displacement meter 容积式水表，容积式流量计，位移计
displacement method 排水法，位移法，置换法
displacement node 位移波节
displacement noise 位移噪声
displacement observation 位移观测
displacement of joint 结点位移
displacement of neutral point 中位点位移
displacement of soil mass 土体位移
displacement operator 位移算子
displacement pile 送入桩
displacement power factor 位移功率因数，相移功率因数
displacement pulse 位移脉冲
displacement pump 排量泵，容积泵，容积式泵，活塞泵
displacement recorder 位移记录器
displacements composition 位移合成
displacement seismograph 位移地震仪
displacement sensor 排气量传感器
displacement-speed control 位移速度控制
displacement spike 离位峰
displacement stress 位移应力
displacement thickness 位移厚度
displacement time curve 位移时间曲线
displacement time 排代时间
displacement tonnage 排水吨位
displacement transducer 排气量传送器，位移变换器，位移传感器
displacement transition 位移相变
displacement type deoxidating device 置换式脱氧装置
displacement type level detector 位移式液位计【可是浮子式，浮子标尺式等】
displacement type level gauge 浮筒式液面计
displacement type seismometer 位移式地震计
displacement type 位移模式，位移型
displacement ventilation 置换通风
displacement voltage 位移电压
displacement wear 错位磨损
displacement （船的）排水量，排出量，位移，移动，排量，替换，置换，变位，代替，容量
displacer 浮子液位计，置换剂，取代剂，平衡浮子，置换器，排出器，排挤器，（混凝土中的）埋石
displace the operating point 工作点的移位
displace 置换取代，移动位移
displacing tank 排挤箱
display access 显示存取
display and debug unit 显示及调试器
display assignment bit 显示指定位
display block 显示部件，显示器
display console 显示控制台
display control switch 显示控制开关
display control 显示控制器
display device 显示装置，显示器，显示设备，绘图装置，指示器
display drawing 展览图
display equipment 显示装置，显示设备
display file 显示文件
display for window 开窗口
display information processor 显示信息处理器
displaying instrument 显示仪器，显示仪表
display key 显示键
display lamp 指示灯
display of information 信息显示，情报显示
display packing 显示压缩传送

display panel 显示板
display processing unit 显示处理器
display processor 显示处理器
display resolution 显示分辨率
display room 陈列室
display screen 显示屏
display sensitivity 显示灵敏度
display station controller 显示站控制器
display storage tube 显示存储管，直观储存管
display system 显示系统
display terminal 显示终端
display time 显示时间
display tube 显示管
display unit 显示单元，显示终端，显示装置
display window 显示窗，光字信号牌
display 显示器，指示器，显示，表现，显像，示数，指示，画面，显示装置
disposability 可任意处置性
disposable cartridge 一次性通用滤筒
disposable packaging 不回收包装，可弃包装
disposable packing 一次性包装
disposable reel 易处理的卷盘
disposable shipping container 可弃装运容器
disposable wave buoy 移动式波浪浮标
disposal area 处置区【放射性废物】
disposal by dilution 稀释处置
disposal by land 陆地处置
disposal by sea 海洋处置
disposal ditch 导水沟
disposal drum 废物处置桶
disposal facilities 处理设施
disposal line 排放管线【放射性废物】
disposal of rubbish 处理垃圾
disposal of sewage 污水处理
disposal on isolated islands 隔离岛处置
disposal plant 处置装置【放射性废物】
disposal region 处置区【放射性废物】
disposal site 处置场【放射性废物】
disposal solution 废液
disposal 处置，处理，安排，消除，配置，排放
dispose train 处置线
dispositioning 处理
disposition of fixed assets 固定资产清理
disposition of heating surface 受热面布置
disposition plan drawing 配置平面图
disposition plan 布置图，配置图，平面布置图
disposition 布置，部署，排列，配置，安排，位移，意向，性格
dispositive provision 处分条款
disposure 处置
disproportionately grade 级配不良的
disproportion 不成比例，不相称，不均衡，不合比例
disputations of contract 合同纠纷
disputation 争议
dispute adjudication board 争端［争议］裁决委员会，争议调解委员会
dispute resolution 分歧的解决
dispute solution 争议处理
disputes review expert 争端审议专家
dispute 争论，争议
disputing parties 争议当事方【各方】
disqualification 取消资格，无资格
disqualified for 没有资格参加……
disqualified goods 不合格货物
disqualified product 不合格产品
disqualified 被取消资格的，丧失资格的
disqualify 使无资格，使不合格，取消……的资格，取消资格
disquisition 著名论著，专题论文，学术讲演
disrepair 失修，破费，破损
disresonance coil 非谐振线圈
disresonance 非谐振
disruption of film 氧化膜破裂
disruption of progress 扰乱进度
disruption of the chain reaction 链式反应中断
disruption plane 破碎面
disruption 破坏，破裂，分裂，爆炸，击穿，断裂，溃裂
disruptive conduction 破裂电导
disruptive critical voltage 临界击穿电压
disruptive discharge 击穿放电，火花放电
disruptive distance 火花间隙，击穿距离，火花长度
disruptive field intensity 击穿电场强度
disruptive instability 破裂不稳定性
disruptive potential gradient 击穿电位梯度
disruptive strength 介电强度，击穿强度，电气绝缘强度
disruptive test 击穿实验，耐压试验
disruptive voltage 破坏电压，击穿电压
disruptive 破裂的，分裂的
disrupt 破坏，破裂，断裂
dissatisfaction 不满意
dissected by erosion 侵蚀切割的
dissected peneplain 切割准平原
dissection 分割
dissemination of fluidizing air 流化空气分配不均
dissemination of new technology 新技术推广
dissemination of radioactive effluent 放射性排出物的散布
dissemination 分散，散步，发放，传播，普及
dissent 有不同意见，不同意，持异议，异议
dissimilarity 不相似性
dissimilar metal corrosion 相异金属腐蚀
dissimilar metal weld 不同金属的焊接，异种钢焊缝
dissimilar metal 不同金属，异种金属，双金属【焊接】
dissipance 损耗系数
dissipated heat 散失热量
dissipated power 耗损功率
dissipater 喷雾器，耗散器
dissipate 消耗，消散，分散，失散，播散，消除
dissipation area 消融区
dissipation coefficient 损耗系数
dissipation factor test 介质损耗角试验
dissipation factor 损耗因数，介质损耗角，功耗因数
dissipation function 散逸函数，损耗函数，耗损函数

dissipation heat loss 散热损失
dissipationless 能量不散逸的
dissipation of energy 能量发散，能量损耗，消能
dissipation of heat 散热，热耗
dissipation of hydration heat 水化热的消散
dissipation rating 耗散率
dissipation 消散，散逸，耗散，弥散，消耗，耗损，散热
dissipative distortion 耗散性畸变
dissipative factor 耗散因数
dissipative flow 耗能流，耗损流动
dissipative network 有耗网络
dissipative structures 耗散结构
dissipator within the tunnel 遂洞内消能工
dissociate 分离，离解，游离
dissociation constant 电离常数，离解常数
dissociation energy 离解能量
dissociation in water 水中离解
dissociation 离解，离散，分解，分离，电离
dissolubility 溶解性，溶解度，解除，可溶性
dissoluble 可溶的，能溶解的，可溶解，可解除的，可溶性的
dissolution basin 溶蚀盆地
dissolution heat 溶解热
dissolution of contact 解除合同，合同的解除，解约
dissolution process 溶解法
dissolution system monitor 溶解系统监测器
dissolution 解除，溶解，溶融
dissolvability 溶解度，溶解性，可溶性
dissolvable 可溶解的
dissolved air 溶解空气
dissolved floatation 溶气浮选
dissolved gas 溶解气体
dissolved impurities 溶解杂质
dissolved iron 溶解铁
dissolved matter 溶解物，溶解物质
dissolved oxygen analyzer 溶解氧分析仪［器］，溶解氧测定仪，溶氧表
dissolved oxygen meter 溶解氧气计
dissolved oxygen sag curve 溶解氧下垂曲线
dissolved oxygen 溶解氧
dissolved salts 溶解盐
dissolved solids 溶解物［质］【水质分析】，溶解固形物，溶解的固体（物），溶解性固体
dissolvent 溶剂
dissolver 溶解装置，溶解器
dissolve 溶解，分解，分离，取消，解除，解体
dissolving 使溶解
dissolving basket 溶解筐
dissolving tank 溶解箱
dissonance 不调和，不谐和，非谐振
dissuade 劝阻
dissymmetrical polyphase system 不对称多相制
dissymmetrical 非对称的，不对称的
dissymmetric network 不对称网络
dissymmetry 非对称现象，非对称现象，不对称性
distance amplitude characteristic curve 距离-幅度特性曲线
distance apparatus 远距离测量仪器
distance between adjacent fixing supports 固定支座间距
distance between bearings 轴承间距
distance between centers 中心点间距离，间距
distance between conductors 线间距离，导体间距离
distance between wire 焊丝间距
distance block 定距块
distance bolt 定距螺栓
distance collar 定距轴环
distance control 遥控，远距离控制，远距离操纵
distance echo correction 距离回波校正
distance freight 超程货物，增加距离运费
distance from initial point 起点距
distance from the focus 震源距
distance-gain-size 距离-增益-尺寸
distance gauge 测距器，测距计
distance impedance protection 距离阻抗保护装置
distance increment 空间增量【数学模型】
distance lag 距离延迟
distance location from entry surface 埋藏深度
distance mark 远处目标
distance measurement 测距
distance measuring meter 测距仪
distance measuring theodolite 测距经纬仪
distance measuring 远距离测量
distance meteorological station 远程气象站
distance meter 测距计
distance of distinct vision 明视距离
distance of pillar center 根开
distance out to out 外侧间距
distance piece 隔件，隔块，接颈，定距块
distance pipe 撑管
distance plate 隔板，垫片
distance protection 距离保护
distance-reading 远距离读数的，遥读的
distance recorder 遥测记录器，距离记录器
distance relay 距离继电器，测距继电器
distance ring 间隙环，定距环，垫圈
distance scale 线性比例尺
distance sleeve 定位套，隔套
distance thermometer 遥测温度计，远测温度计
distance-type 远程式，遥控式，远动式
distance-velocity lag 位移速度滞后，距离速率延迟
distance washer 间隔垫圈
distance 距离，间隔，路程，远方
distant-action instrument 遥测仪表，远距离操作仪表，远relay仪表
distant back-up protection 距离后备保护装置
distant bolt 定距螺栓
distant control 遥控，远距离控制，远距离操纵
distant exchange 远程交换
distant-indicating instrument 远距离指示［显示］仪器，远距离示值读数仪器
distant piece 定距块
distant point 远点

distant radio communication　远距离无线电通信
distant reading instrument　远距离读数仪表
distant reading tachometer　远程读数转速计
distant reading thermometer　远距离读数温度计，遥测式温度计
distant reading　远距离读数，遥测读数
distant reception　远距离接收
distant recording system　遥测记录系统
distant recording　远据记录
distant regulator　远动调节器，遥控调节器
distant signal　预告信号，远距（离）信号
distant surveillance　远距离监视
distant switching-in　远距离合闸
distant synchronization　远距离同步
distant thermometer　遥测温度计
distant tuning　远距离调谐
distant voltage regulator　远距离电压调整器
distant　远程的
DIST＝disturbance　故障，扰动
disthene　蓝晶石
distillate cooler　蒸馏水冷却器
distillate fuel　馏出燃料
distillate hold-up tank　蒸馏水滞留箱
distillate oil reference conditions　馏出油基准条件
distillate oil　分馏油
distillate pump　脱水泵，蒸馏水泵
distillate tank　蒸馏水箱
distillate　馏出物，蒸馏水［液，物］，蒸馏
distillation apparatus　蒸馏器，蒸馏装置
distillation assembly　蒸馏装置
distillation column　蒸馏柱，提浓柱
distillation flask　蒸馏瓶
distillation method　蒸馏法
distillation plant　浓缩装置，（重水）蒸馏装置
distillation process　蒸馏法
distillation residue　蒸馏残渣
distillation tower　蒸馏塔
distillation zone　蒸馏区【燃料床】
distillation　蒸馏，挥发物析出
distillatory　蒸馏器
distilled smoke　蒸馏烟
distilled water　蒸馏水
distiller　蒸馏器
distill　蒸馏，萃取
distincteness　清晰度
distinction　区别［分］，差别，特征，特性
distinctive mark　甄别符号
distinctive　甄别的
distinct　显著的，明显的，有区别的，有差别的
distinguish　辨别，区别，区分
distinguishability　分辨率
distinguishable state　可识别的状态
distinguished guest　贵宾
distomat　红外测距仪
distorted alternation current　非正弦波交流，失真交变电流
distorted element　变态单元
distorted geological section　变态地质剖面
distorted grid　扭曲格网
distorted model　变态模型
distorted reception　畸变接收
distorted scale model　比例变态模型
distorted scale　变态比例尺
distorted water　变态水
distorted wave　畸变波
distorting stress　扭转应力
distortional vane　扭曲叶片
distortion by warping　翘曲畸变
distortion curve　畸变曲线，失真曲线
distortion effect　畸变效应
distortion energy　畸变能，应变能，扭曲能
distortion factor meter　失真系数测量器，畸变系数测量器
distortion factor　畸变因子，畸变率，失真系数，畸变系数
distortion-free　无畸变的，无扭曲的
distortionless circuit　无畸变电路，无失真电路
distortionless medium　无畸变介质
distortionless　无畸变的
distortion measurement test　变形测量试验
distortion of competition　扭曲竞争行为
distortion of field　畸变场，场失真
distortion of flow pattern　流型畸变
distortion of waveform　波形畸变，波形失真
distortion of wave　波形畸变，波形失真
distortion point　扭曲点
distortion power　畸变功率
distortion ratio　变态率
distortion-resistant　抗扭曲的，抗挠曲的
distortion tester　畸变试验器
distortion transformer　失真变压器，谐振变压器
distortion　畸变，变形，挠曲，损坏，扭曲，失真
distort　畸变，失真，变形，挠曲，损坏，扭曲
distressed area　贫困地区
distress　事故
distributary canal　支渠
distributary system　分流系统
distributary　支流，分流，配水沟，河道支流，配水管
distributed capacitance　分布电容
distributed capacity　分布容量
distributed circuit　分布参数电路
distributed collector solar power plant　分布聚热太阳能电站
distributed communication architecture　分布式通信网络体系结构
distributed computer system　分布式计算机
distributed constant circuit　分布参数电路
distributed control and information system　分散式控制与信息系统
distributed control system power-on and test　分散控制系统受电和测试
distributed control system　分散控制系统，分布式控制系统
distributed control　分布式控制
distributed downcomer　分散下降管
distributed element circuit　分布元件电路
distributed emission magnetron amplifier　代码管，分布发射式磁控放大器
distributed energy resource　分布能源
distributed feedback　分布反馈

distributed flow 分布流
distributed force 分布力
distributed function 分布功能
distributed inductance 分布电感
distributed intelligence 分布式智能
distributed load 分布荷载，分布负载
distributed logic 分布逻辑，分布逻辑电路
distributed office support system 分布式办公支持系统
distributed parameter integrated circuit 分布参数集成电路
distributed parameter network 分布参数网络
distributed parameter system 分布参数系统
distributed parameter 分布参数
distributed photovoltaic power generation system 分布式光伏发电系统
distributed polar rotor 分布磁极绕组转子，隐极式转子
distributed process control system 分布过程控制系统
distributed processing system 分布处理系统
distributed processing unit 分布式处理单元
distributed steel 配筋
distributed system 配电系统，分布系统
distributed-type ground bed 分散型接地床
distributed winding 分布绕组
distributed 分布的
distribute 配汽，配水，配电，分配，分布
distributing air damper 风量挡板
distributing bars 分布钢筋
distributing board 配电板，配电盘
distributing bucket 配料斗
distributing cabinet 配电盒，分线箱
distributing cable 配线电缆，配电电缆
distributing canal 配水渠，支渠
distributing cock 分配旋塞
distributing damper 分配挡板，分流挡板
distributing ditch 配水沟
distributing groove 配水槽
distributing header 分配集箱
distributing hold 配水孔
distributing mains 分配母管，配电干线，配水干管
distributing network 配水管网，配水网，配电网
distributing pan 配水槽
distributing pilot valve 分配滑阀
distributing plate 分配板，配水盘，配水板
distributing reinforcement 分布筋
distributing reservoir 配水池
distributing ring 配水环，配气环
distributing slide valve 分配滑阀
distributing substation 配电站
distributing terminal assembly 分布终端组合
distributing valve 分配阀，配气阀，配水阀
distributing 配电的，配水的，分布的，分配的，分配
distribution and consumption 供电成本
distribution apparatus 配电装置
distribution automatic switch 自动配电开关
distribution automation system 配电自动化系统
distribution automation 配电自动化
distribution baffle 配水挡板
distribution block 接线板
distribution board room 配电室
distribution board 配电盘，配电屏
distribution box 分线盒，交接盒，配电箱，分配箱
distribution cable 配电电缆
distribution capacity 配水量
distribution center 配电中心，配水中心，分布中心
distribution channel 配线电缆管
distribution circuit 配电回路
distribution coefficient 分布系数，分配系数，分配比
distribution control box 控制配电箱
distribution conveyor 配料输送机
distribution core type transformer 分布铁芯式变压器
distribution cost 配电费用，配电成本
distribution curve 分布曲线，分配曲线
distribution cutout 配电断路装置
distribution damper 分配挡板
distribution device 分配装置
distribution diagram 散布图
distribution equipment 配电装置，配水设备
distribution error 分布误差
distribution expenses 配电费用
distribution factor 分布系数，分布因子，分配系数
distribution frame 电缆配线架，配线架
distribution function 分布函数
distribution fuse panel 配电熔丝盘
distribution gate diversion gate 分水闸门
distribution gate 分水闸，配水闸
distribution graph 分布曲线，分配曲线，分配图
distribution grid 配电网，配水管网
distribution header 分配总管，分配联箱，分配集管
distribution in altitude 按高度分布
distribution in area 按面积分配，面上分布
distribution in time 按时间分配
distribution in vertical 沿垂线分布
distribution isotherm 分布等温箱
distribution law of echo 回波分布定律
distribution law 分布律，分配律
distribution line 配电线路
distribution list 分发名单，分配单
distribution load center 配电网负荷中心
distribution load 分布荷载
distribution loss 配水损失，配电损失
distribution mains 配电干线，配电总线，配水总管，配水干管
distribution management system 配电管理系统
distribution manifold 分配总管，分配联箱，分配支管
distribution map 分布图
distribution network enhancement 电网改善
distribution network 配电网，配水网
distribution number 分配系数
distribution of blast 送风分配
distribution of electrical energy 电力分配，配电

distribution of electricity 配电
distribution of enthalpy drop 焓降分配
distribution of errors 误差分布
distribution of flow 水流的分配
distribution of grain size 颗粒级配，颗粒组成，粒度分布
distribution of heat flow 热流分布
distribution of heat flux 热流分布，热流密度分布
distribution of heat 热分布
distribution of increment 子样分布
distribution of navigable channel 航道分布
distribution of occupancies 占线的分布
distribution of pipe 管路分布
distribution of pressure 压力分布
distribution of profit and loss 损益分配
distribution of reflux 回流分布
distribution of silica 二氧化硅携带系数
distribution of strain 应变分布
distribution of stress 应力分布
distribution of temperature 温度分布
distribution of velocity 速度分布
distribution of wave energy 波能分布
distribution panel 配电盘，配电屏
distribution parameter 分布参数
distribution piller 地上接头盒
distribution pipe 分配管，配水管
distribution plate 分布板，分配板
distribution plenum 配风室，配汽室，分配联箱
distribution pressure 配水压力
distribution price 配电电价
distribution ratio of silica 二氧化硅分配系数，二氧化硅携带系数
distribution ratio 分配比，分配率，分布系数，分配系数
distribution reinforcement 分布筋，配力钢筋
distribution reservoir 配水（水）库，配水池
distribution rod 配力钢筋
distribution room 配电室
distribution shaft 分配轴，配力轴
distribution station 配电变电所，配电站
distribution steel 配力钢筋
distribution substation 配电站
distribution subsystem 输配子系统
distribution switchboard 配电盘，配电开关板
distribution system of water supply 配水管网
distribution system 配电网，配电系统［制度］，分配系统［制度］，配煤系统，配水系统
distribution transformer station 配电变电站
distribution transformer 配电变压器
distribution valve 分配阀，压力调节阀
distribution voltage transformer 配电电压变压器
distribution voltage 配电电压
distribution well 配水井
distribution winding 分布绕组
distribution 分配［布］，配煤［料］，配电［水］，配置，皮带机
distributive discharge 分布放电
distributive fault subsidiary fault 分支断层
distributive 分布的，分配的
distributor box 分线箱，配电箱，分配箱，中间寄存器
distributor breaker 配电断路器
distributor open area 分布器开孔面积
distributor plate 分布板
distributorship agreement 分销协议
distributor-transmitter 分配器发送器
distributor truck 洒沥青卡车
distributor unit 分配部件
distributor valve 分配阀，分配滑阀
distributor 配电器，分配器，配水器，分布器，布风板【沸腾炉】，分水间，分线匣
distributor 平舱机，布料机
district authority owned electric power 地方电厂
district bank 地方银行
district cable 区域电缆
district center 区中心
district commissioner 地区专员
district cooling 区域供冷
district government 区政府
district heating system heater 热网加热器
district heating atomic power plant 地区供电核电厂，核供热站
district heating plant 区域供热站
district heating power plant 区域供热火力发电厂
district heating station 区域供热站
district heating system condensate pump 热网加热系统疏水泵
district heating system 区域供热系统
district heating 地区集中供热，区域供热，地区供热，区域供暖
district heat supply 区域供热
district selector 区域选择器，第一选组器
district substation 地区变电站
district system 区域系统
district 地区，区域，街区，区
disturbance centre 扰动中心
disturbance compensation 扰动补偿
disturbance current 干扰电流
disturbance decoupling 扰动解耦
disturbance degree 扰动度
disturbance depth 扰动深度
disturbance factor 干扰系数
disturbance frequency 扰动频率
disturbance line 扰动线
disturbance of magnetic field 磁场干扰
disturbance quantity 扰动量
disturbance wave 拨动波
disturbance 扰动，干扰，妨碍，故障，骚扰，骚乱，打扰，事故
disturbed 扰动的
disturbed belt 错动带
disturbed clay 扰动黏土
disturbed flow 扰动流
disturbed land 受扰土地
disturbed motion 扰动
disturbed sample 扰动试样
disturbed sand 扰动沙
disturbed scatter flux 受扰的散射通量
disturbed soil sample 扰动土样
disturbed soil 扰动土壤
disturbing force 扰动力
disturbing signal 扰动信号

disturbing torque 失步转矩，破坏转矩
disturbing voltage 干扰电压
disturbing 扰动
disturb 扰乱
disulfide 二硫化物
disulphide 二硫化物
disused salt mine 废弃盐矿
dit 小砂眼
ditch berme 沟岸戗道
ditch check 沟挡板，明沟节制闸
ditch cleaning machine 清沟机
ditch depth 沟深
ditch drainage 明沟排水
ditch erosion 沟蚀
ditcher 挖沟机
ditch excavator 挖沟机
ditch for foundation 基槽，基坑
ditch grade 沟底坡度
ditching plough 挖沟犁
ditching 挖沟
ditch irrigation 沟灌
ditch junction 沟道交接点
ditch outlet 沟的出口
ditch oxidation 氧化沟
ditch planting 壕沟栽植
ditch plough 沟犁
ditch slope 水沟坡度
ditch 边沟，沟，沟渠，明渠，明沟
dither motor 高频振动电动机
dither 高频颤动，高频振动，高频脉动，抗滞振动，颤振
ditto 同上
diurnal age 日潮潮龄
diurnal amplitude 日变幅
diurnal cycle 日循环，昼夜循环
diurnal-fluctuation 昼夜变动
diurnal inequality 日潮不等
diurnal load diagram 昼夜负荷图
diurnal load factor 日负载因数
diurnal maximum 日最大，日最大量
diurnal range 日较差，日潮差
diurnal tidal force 周日生潮力，昼夜生潮力
diurnal tide 全日潮，日周潮
diurnal variation of atmospheric pressure 气压日变化
diurnal variation of temperature 气温日变化
diurnal variation 日变化
diurnal 每天的，白天的，日
divagation 改道
divalence 二价
divalent 二价的
divarication 分叉，分叉点
dive 下潜，潜水
dive culvert 倒涵管
diver 潜水员
divergence angle 发散角，扩张角
divergence cone angle （喷嘴的）膨胀角
divergence field 辐散场
divergence loss 发散损耗，扩散损耗，扩散损失
divergence of field 场的散度
divergence of phasor 相量的散度
divergence of series 级数发散度
divergence rate 扩张度，扩张比
divergence slit 发散狭缝
divergence 达到或超过临界【反应堆】，发散（度），差异，散度，扩散，偏差，分歧，辐散
divergent channel 扩展流道
divergent-convergent channels 渐扩渐缩通道，扩缩通道
divergent current 扩散流
divergent flow 扩散水流
divergent matrix 发散矩阵
divergent mouthpiece 扩散管嘴
divergent nozzle 扩张性喷嘴
divergent pipe 扩散管
divergent reaction 发散反应
divergent tube 发散反应
divergent unconformity 成角不整合
divergent 扩展的，渐扩的，发散的
diverge 扩散，偏离
diverging cone 扩张锥
diverging nozzle 扩展型喷嘴，扩张管，喇叭形管嘴
diverging pipe 扩散管
diverging tube 扩张管，锥形管
diversification consortium 多种经营企业
diversification 多元化经营，多种经营
diversified business 多种经营
diversified economy 多种经济
diversified industrial city 综合性工业城市
diversified investment 多样化投资
diversion aqueduct 导流渡槽，引水渡槽
diversion barrage 引水坝
diversion basin 分水池
diversion botom outlet 导流底孔
diversion box 分水箱
diversion chamber 分水室
diversion channel 导流明渠，分流，分水道，分水沟，引水明渠
diversion conduit type development 引水式开发
diversion construction 导流工程
diversion dam 分水坝，引水坝
diversion dike 导流堤，引水堤
diversion discharge frequency 导流流量标准
diversion ditch 分水沟，引水沟
diversion flume 导流水槽
diversion gate 导流闸门，分水闸门，转换（切换）闸门
diversion intake 导流进水口
diversion method 导流方法
diversion pipe 引水明管
diversion project 引水工程
diversion ratio 分水率，引水率
diversion road 分支道路
diversion sluice 分水闸
diversion structure 导流建筑物
diversion terrace 导流台地，导水防蚀阶地
diversion through powerhouse 厂房导流
diversion tunnel power station 引水隧洞式电站
diversion tunnel 导流隧洞，引水隧洞
diversion type development 引水式开发

diversion type power station 引水式电站
diversion weir 分流堰，分水堰，引水堰
diversion works 导流工程，导流建筑物，分水工程，引水工程，引水建筑物
diversion 分水，分水渠，截水渠，转换，换向，变换，导流，改道，功率增长中的反应堆，发散堆，转向，改向，转移
diversity factor variability 差异度
diversity factor 差异因子，负荷变化系数，不同时率，分散率，分散系数，差异因数
diversity principle 多样性原则
diversity 多样性，多样，参差，不同，相异性，差异，多元化，发散
diver's paralysis 潜水病
diver's suit 潜水服
divert 挪用
diverted flow rate 分流流量
diverted flow 分流量
diverter chute 分流管
diverter gate 分流门
diverter pole generator 分流极发电机
diverter relay 具有分流器的继电器
diverter switch 切换开关
diverter valve spool 分流器阀阀芯
diverter valve 分流阀，分流器阀，切换阀，双向阀
diverter 分流电阻，电阻分流器，排水沟，导向隔板，分流装置，换向器，折流器，转换器，转向器
diverting dam 分水坝
diverting flow system 导流设备
diverting gate valve 切换［转向］闸门
diverting valve 分流阀，换向阀，切换门
diverting wall 导水墙，分水墙
diverting weir 分水堰
divertor 偏滤器
divided-by-two-circuit 一比二分频电路
divided circle 分度圆
divided circuit 分流电路
divided conductor protection 绕组支路断线保护
divided conductor 分裂导线，股线，分裂导体
divided fail 分段落差
divided-flow turbine 分流式汽轮机
divided furnace 分割炉膛，双炉膛，分隔炉膛
divided gas flow 分离气体
divided inlet manifold 分散引入联箱
divided into 4 lots 分为四个标段【项目招投标】
divided iron-core 分裂铁芯
divided monitoring feedback 分支监控反馈
divided reset 分段复位
divided return duct 分支回流导管
divided steam-flow design 蒸汽分流结构，蒸汽分流设计
divided water box 对分式水室
divided winding armature 复绕组电枢，分开式绕组电枢
divided winding rotor 分开式绕组转子，复绕组转子
divided winding turbogenerator 分开式绕组转子发电机，复绕组汽轮发电机
divided winding 复绕组，分开式绕组
divided 分界，划分，隔开，除
divide into sections 分格，划分
dividend coupon 股息券
dividend distribution 股息分配
dividend payable 应付股利
dividend 被除数，股息，股利
divider resistance 分压电阻
divider 除法器，分压器，除数，分离器，分配器，划线规，分流管，分割者，分配者，分线规，两脚规，分规，间隔物
divide 分水岭，划分
dividing box 电缆头，电缆分箱
dividing circuit 除法电路
dividing crest 分水界，分水岭，分水脊
dividing frequency 分配频率，分割频率
dividing head 分度头
dividing machine 刻线机
dividing network 分频网络，选频网络，分配网络
dividing partition 配水隔板，进水花墙
dividing pier 隔墩
dividing plate 分隔板
dividing ridge 分水脊
dividing strip for terrazzo 水磨石嵌条
dividing waterwall 分隔水冷壁，双面水冷壁
diving apparatus 潜水器，潜水设备
diving bell 潜水钟
diving dress 潜水服
diving equipment 潜水设备
diving outfit 成套潜水设备
diving pump 潜水泵
diving 俯冲
divining device 探测设备
divinylbenzene crosslink 二乙烯苯交联
divinylbenzene 二乙烯（基）苯
divisibility 可分性
divisible by 可被……除尽的
divisible contract 可分割合同
divisible letter of credit 可分割的信用证
divisible 可分割的，可除尽的，可约的
divisional management 部门管理
divisional quality manual 分部的质量手册
division box 分水箱，分流箱
division contract 分项承包
division device 除法装置
division engineer 工区工程师
division manager 分部经理
division of income 收益分配
division of project items 工程项目划分
division of work 分工
division pier 隔墩
division plate 隔板
division post 分道桩
division subroutine 除法子程序
division time 除法时间
division wall 分隔墙，双面水冷壁，防火墙，分道墙
division weir 渠首堰
division 刻度，区分，分割，除法，分隔墙，除，分度，部分，段，部门，单位，分开，工区，划分，

divisor　除数，因子，分压器，分离器，分压自耦变压器
DKE　德国电工委员会【隶属于德国标准化研究所和德国电子学会】
DKK　丹麦克朗
DLC(data link control)　数据链路控制
DL(data link)　数据传输器，数据链路，数据中继器
DL(delay line)　延迟线
DLE(data link escape character)　数据链路换码字符
DLL(dynamic link library)　动态链接库
DLM(design liaison meeting)　设计联络会
DLS(digital logic station)　数据逻辑站
DMA　BOP装卸搬运设备【核电站系统代码】
DMA(direct memory access)　存储器直接存取，直接存储器存取
DMA(direct memory address)　直接存储器地址
DMB(distributed mixing burner)　分布式混合燃烧器
DM cable　双心电缆
DMC(data multiplexer channel)　数据多路转换器通道
DMC(digital microcircuit)　数字微电路
DMCL(device medium control language)　设备媒体控制语言
D. M. C. (direct mud circulation)　正向泥浆循环
DM(damage monitor)　损伤监测仪
DM(decimal multiply)　十进制乘法
DM(degraded minute)　降级分
DM(design-manage)　DM模式，"设计-管理"模式【工程建设模式】
DM(design manual)　设计手册
DMD　需求，要求【DCS画面】
DME(distance measuring equipment)　测距装置
DME　主开关站装卸搬运设备【核电站系统代码】
DMH　BOP区域内的各种起吊设备【核电站系统代码】
DMIS(design management information system)　调度管理信息系统
DMI　混凝土桶长期存放用的装卸搬运设备【核电站系统代码】
DMK　核燃料厂房装卸搬运设备【核电站系统代码】
DM(magnetic drum module)　磁鼓存储模件
DMM(digital multimeter)　数字（式）万用表
DMM　汽轮机厂房机械装卸设备【核电站系统代码】
DMNRLZR＝demineralizer　除盐装置
DMN　核辅助厂房装卸搬运设备【核电站系统代码】
DM(demineralization plant)　除盐装置
DMPR＝damper　挡板
DMP　循环水泵站装卸搬运设备【核电站系统代码】
DMR(diamagnetic resonance)　抗磁谐振
DMR　反应堆厂房装卸搬运设备【核电站系统代码】
DMS(data management system)　数据管理系统
DMS(distribution management system)　配电管理系统
DMS(document management system)　文件资料管理系统
DMS(dynamic mapping system)　动态变换系统
DMT(flat shovel side expansion instrument)　扁铲侧胀仪
DNA(Defense Nuclear Agency)　国防部核子局【美国】
DNA(Deutsche Normenausschuss)　德国标准委员会
DNA(digital network architecture)　数字网络体系结构
DNB correlation　偏离泡核沸腾的相关性
DNB F(departure from nucleate boiling flux)　偏离泡核沸腾热通量
DNB(departure from nucleate boiling)　偏离泡核沸腾
DNB heat flux　偏离泡核沸腾热流密度，DNB热流密度，烧毁热通量，偏离泡核沸腾热通量
DNB margin　DNB裕量，偏离泡核沸腾热流密度裕量
DNB ratio　烧毁比，偏离泡核沸腾比值
DNBR(departure from nucleate boiling ratio)　偏离泡核沸腾比
DNC(direct numerical control)　直接数字控制
DND(delayed neutron detector)　缓发中子探测器
DN(debit note)　借项清单
DN(decimal number system)　十进制数制
DN(domain name)　域名
DNH　正常照明系统【核电站系统代码】
DNI(direct normal irradiance)　直接辐射【阳光从太阳盘面直接照射到与光路正交的表面】
DNM(delayed neutron monitor)　缓发中子监测器
DNP(defect notice period)　缺陷通知期
DNS(department of nuclear safety)　核安全部
DOC(documentation center)　资料中心
dock　码头，船坞
dockage　入坞费
dock basin　港池
dock belt conveyor　码头带式输送机
dock bottom　坞底
dock charge　码头费，入坞费
dock drain pump　码头疏水泵
dock dues　入坞费
dock entrance　坞口
dockey engine space　辅机舱
dock gate　船坞闸门
dock harbour　闭合式港
docking accommodation　入坞设施
docking area　泊船区
docking facilities　泊船设施，入坞设施
docking　泊码头，入坞
dock quay　码头
dock railway sidings　临港铁道
dock receipt　场站收据
dock road　临港道路
dock test　系泊试车
dock trial　系泊试车
dock wall　驳岸，码头岸壁
D_2O-cooled pressure-vessel reactor　重水冷却压

力容器式反应堆
doctor 博士
doctrine-of-appropriation of water rights 水权专有准则
doctrine-of-relation in water rights 土地附有水权准则
document against acceptance 承兑交单
document against payment after sight 远期付款交单
document against payment at sight 见票付款单据
document against payment 付款单据，付款交单
document alignment 文件定位
document approval 文件批准
documentary acceptance bill 承兑汇单
documentary acceptance 跟单承兑
documentary against acceptance 承兑交单
documentary bill 跟单汇票
documentary collection 跟单托收
documentary credit 货物押汇信用证，押汇信用证，跟单信用证
documentary evidence 书面证据（证明），文字证明
documentary information 文献情报
documentary L/C 跟单信用证
documentary letter of credit 跟单信用证，押汇信用证
documentation house 文档编制室
documentation room 档案室
documentation 文件编制，文件提供，文件使用，文件，文献，资料
document attached 附单据
document authenticated by a seal 有印鉴为凭的文件
document change control 文件更改管理
document change 文件更改
document charge 文件费
document control architecture 文件控制结构
document control manager 文件控制经理
document control 文件管理
document data processing 文件数据处理
documented 文件齐备的，形成文件的
document facsimile 文件传真
document fee 换［抽］单费
document flow 文件流动
document for claim 索赔文件，索赔证件
document handling 文件处理
document identification 文件标识
document issue 文件发布
document leading edge 文件前沿
document misregistration 文件未对准，文件偏移
document name 文件名
document number 文件号
document of approval 批准文件
document of carriage 运输单据
document of contract 合同文件
document of obligation 债权证书
document of settlement 结算单，结算单据
document of shipping 装货单据
document of title 所有权凭证
document on site 存放在现场的文件
document preparation 文件编制，文件的编制
document publication 文件出版
document reader 文件读出器
document reference edge 文件基准边缘
document retention 文件保存
documents against payment at sight 即期付款交单
documents against payment 付款交单
document scanner 文件扫描程序，文件扫描器
documents not accessible to the public 公众无法接触到的文件
document sorter 文件分类器
document stamp 文件图章
document type definition 文档类型定义
document 单据，文件，文本，文献，资料
docuterm 文件项目，文件条款
DO(delivery order) 提货单，交货单，小提单
DO(demand output) 指令输出
dodge a tax 偷税
dodge 躲避
DO(digital output) 数字量输出，数字输出
DO(direct operation) 直接操作，直接控制，直接输出
DO(dissolved oxygen) 溶解氧
D_2O distillation tower 重水蒸馏塔
DO(domain object) 领域对象
D_2O dosing pump 重水配料泵，重水加料泵
D_2O drain cooler 重水疏水冷却器
D_2O drain pump 重水疏水泵
D_2O drain tank 重水疏水箱
DOE(Department of Energy) 能源部【美国】
DOF(degree of freedom) 自由度
D_2O feed pump 重水给料泵
D/O fee 抽单费，换单费
dog-bone type 八字抗拉式支架
dog clutch sleeve 爪形离合器套筒
dog clutch 爪形离合器
dog coupling 爪形联轴节，爪形联轴器
dog house enclosure 鼓形罩室
dog house 调谐箱，高频高压电源屏蔽罩，鼓形箱
dog-nail 道钉
dog point screw 凸头螺丝
dog shaft 叉轴
dog shores 水平撑柱
dog spike 道钉
dog tooth bond 犬牙式砌合
dog vane 风标
do interpretive loop 执行解释循环
DOT(digital output translator) 数字输出转换器
dolerite 粗粒玄武岩，辉绿岩
dolinae 斗淋
dolina 灰岩坑
doline 斗淋，灰岩坑，灰岩坑，落水洞
dollar acceptance 美元承付票据
dollar reactivity unit 美元反应性单位
dollar 美元，一美元纸币
dolly 小机车，圆形锻模，吊钩，辘车，垫座，垫桩，独轮台车，独轮小车，铆顶，桩垫［盘］
dolomite limestone 白云灰岩

dolomite marble　白云大理岩
dolomite refractory　白云石耐火材料
dolomite sandstone　白云砂岩
dolomite　白云石，石灰岩，大理石
dolomitic limestone　白云灰岩，白云质石灰岩
dolomitic lime　白云灰
dolomitization　白云石化作用
dolphin　系船柱，护墩桩，靠船墩，锚链擎动器
D_2O main circuit　重水主回路
domain field　领域
domain name address　域名地址
domain name server　域名服务器
domain name system　域名系统
domain name　域名
domain of convergence　收敛域
domain of dependence　相关域，依赖域
domain of function　函数域
domain of integrity　整环
domain of variation　变化区域
domain-tip memory　畴壁位移
domain　域
DOM(dryout margin)　干涸裕度
dome base　汽室垫圈
dome construction　穹形建筑
dome dam　双曲拱坝
domed arch dam　穹窿坝，穹形拱坝
domed-based　球形封头
domed salt　圆丘状盐，穹形盐
domed　圆顶的，半球形的
dome flange　汽室凸缘
dome head　(锅炉的) 干汽室，球形封头
dome illumination　天棚照明
dome lamp　顶灯，天棚灯
dome liner　汽室垫
dome manhole　汽室人孔
domestic　国产的，家庭的，家用的
domestic agent　国内代理商
domestic bidder　国内投标商
domestic bond　国内债券
domestic capital　本国资本，内资
domestic competition　国内竞争
domestic consultant　国内咨询商
domestic consumption　民用耗水量，生活用水量
domestic credit　国内信贷
domestic customer　国内客户
domestic dispute　内部争端
domestic enterprise　国内企业
domestic filter　家用过滤器
domestic fund　内资
domestic garbage　生活垃圾
domestic hot-water　生活热水
domestic installation　生活 (用电) 设施
domestic investment　国内投资
domestic legal person　国内法人
domestic load　厂用负荷，民用负荷，厂用电
domestic loan　国内贷款，内债
domestic manufactured equipment　国产设备
domestic marketing prices　国内市场价
domestic market　国内市场
domestic meter　家用电表，用户水表
domestic noise　生活噪声
domestic participation　国产化
domestic preference　国内优惠
domestic project　国内工程项目
domestic purchasing　国内采购
domestic resources　本国资源
domestic seller　国内卖方
domestic sewage　生活污水，家庭污水
domestic sewerage　生活污水，生活污水工程
domestic signaling installation　内部自用信号装置
domestic smoke　生活烟气
domestic solar water heater　家用太阳热水器
domestic subcontractor　国内分包人
domestic telegram　国内电报
domestic trade　地区内贸易，国内贸易
domestic use of water　民用水
domestic wastewater　生活废水，生活污水
domestic waste　生活废物
domestic water and fire pump house　生活消防水泵房
domestic water consumption　生活耗水量，生活用水量
domestic water meter　用户水表
domestic water supply pipe　生活供水管
domestic water supply　民用给水，生活给水
domestic water　生活用水
dome storage　穹顶仓库
dome-type indoor storage bin　穹形室内贮存仓
dome　圆顶，穹顶，拱顶，干汽室，罩，整流罩，穹地，穹窿，圆丘，圆屋顶
domical vault　半球形穹顶
domicile　住处，永久住处，期票支付场所，法定住址，住所
domiciliary account　外汇账户
dominance of strategies　策略的优势
dominant alternatives　关键方案
dominant company　主要公司
dominant discharge　控制流量
dominant factor　主导因素，主要因素
dominant formative discharge　成槽流量
dominant item　主控项目
dominant mode　重要方式
dominant resonant frequency　主共振频率
dominant strategy equilibrium　占优策略均衡
dominant strategy　占优策略
dominant vibration mode　振波基型，主振型
dominant wavelength　主波
dominant wind direction　主导风向
dominate　支配
dominating condition　控制条件
dominating influence　决定性影响，主要影响
domino failure　串级崩塌【水坝】，连珠垮坝
donation　捐款，捐赠
donga　陡岸干沟，山峡
donkey boiler　辅助锅炉
donkey engine　辅助发动机，绞车，绞盘，卷扬机
donkey pump　汽动往复泵，蒸汽往复给水泵，辅助 (水) 泵
donkey　辅助发动机，辅助泵，小型补给泵，小型发动机辅机
donor-cell method　供体网络法

donor country 捐款国
donor ionization energy 施主电离能
donor number density 施主浓度
do not crush 切勿压挤【包装标志】
do-nothing instruction 空操作指令
do not stack on top 勿放顶上
do not turn-over 不可倒置
donut 热堆快中子转换器，（加速器的）环形真空室
door 门，通道
door assembly 大门周围的设备，门的装配，整套门，门总成
door bar 门闩
door bumper 门缓冲器，门碰档，车门软垫，防撞胶粒
door butt 门铰链
door case 门框
door casing 门框，门侧装修，门框饰
door chain 门链
door check 门缓闭器，门制止器
door class 门（防火）等级
door closer-spring 门弹弓
door closer 闭门器，门缓闭器
door contact 门触点，门动触点
door curtain 门帘
door finish 门的修整
door fittings 门配件
door fixture 门用零件
door frame 门框
door handle 门拉手
door hanger 挂门钩
door head 门楣，门上梃
door hinge 门铰链
door infiltration 门缝渗入量
door jamb 门窗侧壁，门樘
door knob transformer 耦合变压器
door knob 球形门拉手，高频耦合
door knocker 门环
door leaf 门扉，门扇
door leakage 门渗漏量
door light 门上镶玻璃面积，门灯，门采光面
door lintel 门过梁
door lock 门锁
door louver 门百叶
door opening 门洞，门口
door pier 门墩
doorpost 门柱
door pull 门拉手，门横木，门扇冒头
door saddle 门槛
doors and windows 门和窗
door sill 门槛
door spindle 门轴
door stopper 门制
door-to-door service 门到门服务
door trim 门头线
doorway monitor 通道监视器
doorway 门道，门口，通道，出入口
door window lock 门窗锁具
door window 落地窗，门式窗
door with inflatable seal 带有橡皮碰垫的门
door with panic bolt set 保险门，带有保险栓的门
door with wicket 带小门的门
dopants 掺杂物
doped fabric covering 涂漆布蒙皮
doped fabric 涂油蒙布
doped fuel 加防爆剂的燃料
doped oil 加添加剂的润滑油
doped sheet 涂漆层
doper 黄油枪，注油器
dope 明胶，防爆剂，涂料，上涂料
doping coil 浸渍线圈
doping 掺杂物，加添加剂
dopolarize 退机化，消偏振
Doppler acoustic radar 多普勒声雷达
Doppler-averaged cross section 多普勒平均界面
Doppler bank 多普勒棒组
Doppler coefficient of reactivity 多普勒反应系数
Doppler coefficient 多普勒系数
Doppler constant 多普勒常数
Doppler defect 多普勒缺陷
Doppler displacement 多普勒位移
Doppler effect 多普勒效应
Doppler feedback 多普勒反馈
Doppler group of control rod 多普勒棒组，多普勒控制棒组
Doppler laser radar 多普勒激光雷达
Doppler lidar 多普勒激光雷达
Doppler radar 多普勒雷达
Doppler shift 多普勒频移，多普勒移位
D_2O pressure-tube reactor 重力压水管式反应堆
D_2O pressure-vessel reactor 重水压力容器反应堆
D_2O proportioning pump 重水加料泵
DOP test DOP试验【过滤器】
D_2O purification loop 重水净化回路
D_2O purification system 重水净化系统
D_2O reflux pump 重水回流泵
dormant area 静止区
dormant bolt 沉头螺栓
dormant equipment 闲置设备
dormant failure 潜伏故障
dormant partner 隐名合伙人
dormant screw 埋头螺钉
dormant volcano 休眠火山
dormer window 屋顶窗，老虎窗，老虎天窗，屋顶采光窗
dormitory 单身宿舍楼
dosage effect 剂量效应
dosage level 剂量水平
dosage measurement 剂量测量
dosage meter 剂量计【辐射】
dosage rate 剂量率
dosage 剂量，用量，配料，定量器，剂
DOS(disk operating system) 磁盘操作系统
DOS(dosing) 配料，加药
dose albedo 剂量反照率
dose build-up factor 剂量积累因子
dose burning coal 黏结煤
dose commitment 剂量负担
dose conversion factor 剂量转换因子

dosed channel　闭式通道
dosed circuit　闭合电路
dose detector　剂量计
dosed isobar　闭合等压线
dosed jet return flow tunnel　闭式工作段回流风洞
dosed jet wind tunnel　闭口风洞
dosed loop test　闭环试验
dosed streamline　闭合流线
dose-effect curve　剂量效应曲线
dose-effect relation　剂量效应关系
dose equivalent commitment　剂量当量负担
dose equivalent index　剂量当量指数
dose equivalent limit　剂量当量限值
dose equivalent rate　等效剂量率，剂量当量率
dose equivalent　剂量当量，等效剂量
dose estimation model　剂量估算模式
dose fractionation　剂量分配
dose history　辐射剂量史
dose limit equivalent concentration　剂量限值等效浓度
dose limit　剂量极限，剂量限值
dosemeter　剂量计［仪］，剂量测量装置，计量箱，量斗
dose modal　剂量模型
dose modifying factor　剂量修正系数
dose monitoring　剂量监测
dose of radioactivity　放射性剂量
D_2O separator　重水分离器
dose point　剂量点
dose protraction　剂量的分散，剂量迁延
doser　剂量装置
dose rate conversion factor　剂量率转换因子
dose rate dosimeter　剂量率剂量计
dose rate in　入射剂量率
dose rate meter　剂量率计
dose rate out　出射剂量率
dose rate response　剂量率响应
dose rate　剂量率
dose reciprocity theorem　剂量互易定理
dose reduction factor　剂量减低系数
dose response curve　剂量反应曲线，剂量响应曲线
dose response function　剂量响应函数
dose response model　剂量响应模型
dose response relationship　剂量效应关系
dose response relation　剂量响应关系
dose response　剂量响应，剂量反应
dose to filter aid　助滤剂的投加量
dose　剂量，加药量，加药
dosimeter　剂量设备，剂量器［计，仪］，计量箱，量斗
dosimetry　剂量学，测定法，剂量测定法
dosing apparatus　投配器，投配装置
dosing chamber　投配室
dosing of regenerating reagent　再生剂剂量
dosing pipe　加药管
dosing pump　剂量泵，加药泵，计量泵
dosing ratio　投配比
dosing room for boiler feeder　锅炉给水加药间
dosing room　化学加药间
dosing siphon　投配虹吸
dosing tank　加药箱，剂量箱，投配池，投配槽
dosing wheel　计量轮
dosing　配料，加药，计量，配量，测量
D_2O spill-over tank　重水溢流箱
dosseret　副柱头
DOS(standby diesel and auxiliary boiler fuel oil system)　应急柴油机与辅助锅炉燃油系统
D_2O storage and volume control system　重水贮存和容积控制系统
D_2O storage tank　重水贮存箱
dot and dash line　点划线
dot-bar generator　点条信号发生器
dot character printer　点阵字符打印机
dot-cycle　点循环，基本信号周期
dot-dash line　点划线
dot density　点密度
DOT(develop-operate-transfer)　DOT 模式，“开发-运营-移交”模式【工程建设模式】
dot diagram　点式图示
dot frequency　点频率
dot generator　点信号发生器，点状图案信号发生器
dot matrix display　点矩阵显示，点矩阵显示器
dot matrix printer　点阵式打印机
dot matrix　点阵，点矩阵
dot-pattern generator　点信号发生器，点状图案信号发生器
dot pitch　点距
dot printer　点式打印机
dot product　标量积
D_2O transfer pump　重水输送泵
dot raster character generator　点阵法字符发生器
dotted decimal notation　点分十进制记法
dotted line　点线，虚线
dot welding　点焊，填补焊
double absorption pipette　双吸收管
double acting brake　复式制动器
double acting compressor　双动空气压缩机
double acting cylinder　双作用汽缸
double acting damper　双动减震器
double acting door　双动门，双向门
double acting duplex pump　双联动式泵
double acting limit stop　往复定值限位架
double acting pawl　双动爪
double acting power unit　双作用执行部件
double acting pump　复动式泵【活塞或柱塞】，双向泵，双作用水泵，双动泵
double acting relay　双侧作用继动器
double acting steam hammer　双动汽锤，双击式蒸汽锤
double acting switch　双动开关
double action jack　双作用千斤顶
double action pump　双动式泵
double action reciprocating pump　双作用往复泵
double action　双重作用，双作用，复动式，双动式，往复式
double amplification circuit　来复电路
double amplitude peak　双幅度峰值，全幅值
double amplitude　倍幅，双振幅

double angle bevel 双斜面坡口
double angular ball bearing 双列向心推力球轴承
double aperture seal 双孔密封
double arc airfoil 双圆弧叶形
double-arch furnace 双拱炉膛【带前后拱】
double armature cable 双层铠装电缆
double armature DC motor 双电枢直流电动机
double armature relay 双衔铁继电器
double armature 双铠装的，双电枢，双衔铁
double-armored 双层铠装的
double-arm relay 双推杆继电器，双臂继电器
double-arm tower 双臂塔
double-axial-flow turbine 双轴流式水轮机
double back pressure steam condenser 双背压凝汽器
double-balanced relay 双平衡式继电器
double-balanced suction 双侧平衡吸入
double-bank heater 两级加热器
double-bank of trays 两组托架
double-bank switch 双触排开关
double-bank 两组
double bar and yoke method 磁棒轭两次互换测量法
double barrier containment 双层安全壳
double-base diode 双基极二极管
double-beam bridge crane 双梁桥吊
double-beam oscilloscope 双线示波器
double-beam spectrometer 双光分光计
double beat sluice 双向泄水闸
double beat valve 双开阀，双座阀
double bed anion exchanger 双层阴离子交换器
double bellows differential manometer 双波纹管差压计
double belt 双层皮带
double bend 双弯管
double bevel groove K形坡口，双斜面坡口
double-biased bistable two-position relay 双偏置双稳态二位置继电器
double block 双轮滑车
double boom level luffing grab crane 双悬臂水平调幅抓斗起重机
double-bottom tank 双层底舱
double bowstring truss 双弓弦桁架，鱼形桁架
double bracing 双支撑
double bracket 双角撑架
double branch elbow 双支管弯头
double-break circuit-breaker 复断路器，双断断路器
double-break contact assembly 双断点触点组
double-break contact 双断点触点
double-break double-make contact 双断双闭触点，桥式转换触点
double breaker configuration 双断路器接线
double-break switch 双断开关
double-break 双断
double bridge 双臂电桥
double buffered gate circuit 双缓冲门电路
double busbar 双母线
double-bus configuration 双母线接线
double-bus connection with bypass 双母线带旁路接线
double-bus connection 双母线接地，双母线接线
double-bus 双汇流排，双母线
double butt strap 双拼板
double cage lift 双笼电梯
double cage motor 双鼠笼电动机
double cancellation 双系斜杆
double-cardan joint 双万向接头
double-casement window 双扇窗
double casing machine 双机壳型电机，内外机壳型电机
double casing pump 双壳体泵
double casing turbine 双蜗壳水轮机，双缸汽轮机
double casing volute pump 双壳旋涡泵
double casing 双层外壳
double-cavity klystron 双腔调速管
double cell fluidized-bed 双室浮动流化床
double-centre theodolite 双中心经纬仪
double-centre theologize 复测经纬仪
double centrifugal pump 双吸水口离心泵
double chamber anion exchanger 双室阴离子交换器
double chamber bed 双室床
double chamber cation exchanger 双室阳离子交换器
double chamber furnace 双炉室炉膛
double chamber surge shaft 双室式调压室
double chamber surge tank 双室式调压井
double channels 双拼槽钢，双通道【计算机】
double check system 重复检测器，双重监视系统
double check valve assembly 双逆止阀装置
double check valve 双止回阀
double circuit line 双回路输电线，双回路线，双回路线路
double circuit overhead transmission line 双回路高架输电线
double circuit power supply 双回路供电
double circuit power transmission 双回路输电
double circuit suspension tower 双回直线塔
double circuit system 双回路系统
double circuit tower 双回路线塔
double circuit transmission 双回路输电
double circuit vertical configuration 双回路垂直排列
double-circuit 双回路，双电路，加倍电路
double clamp 双卡头
double-closed tube 双闭管
double coagulation 双重混凝，双重凝聚
double coated electrode 双层药皮焊条
double coaxial line 双同轴线，双同轴线路
double-coiled relay 双线圈继电器
double coil 双绕线圈
double-coincidence switch 双重合开关
double colored indicator 双色液位计
double column 双柱
double command 双命令
double commutator motor 双换向器电动机
double conductor 双股导线
double conduit 双孔水道
double cone clutch 双锥离合器
double connection 双重接法
double contact key 双触点电键

double contact wire system 双接触导线制
double contact 双触点，双触点的，双接点的
double containment 双层安全壳，双层密封，双层包装，复式包容
double contingency principle 双偶然事件原则
double converter 双换流器，反并联连接法
double cotton-covered wire 双纱包线
double coupling 双联轴节
double course pavement structure 双层式路面结构
double crack 双共线裂纹
double cranes lifting 双机抬吊
double crane 双钩起重机
double crystal probe 双晶探头
double cup insulator 双杯形绝缘子，双碗式绝缘子
double current furnace 交直流电炉
double current generator 双电流发电机，交直流发电机
double current relay 双电流式继电器，交直流继电器
double current 双流，双直流，交直流
double curvature arch dam 双曲拱坝
double curvature vane 扭曲叶片，双扭曲叶片
double-curved brick arch 双曲砖拱
double-curved flat shell 双曲扁壳
double-cut file 双纹锉，交叉纹锉
double cycle inverse 反相双循环
double cylinder 双层汽缸
double day tide 双日潮
double deck bridge 双层大桥
double decker filter 双室过滤器
double deck rotor 双鼠笼转子
double deck screen 双层筛
double deck vibrating screen 双层振动筛
double deflection grille 双层百叶风口
double deflection register 双层百叶风口
double-delta connection 双三角接线，双三角接法
double-delta winding 双三角形绕组
double density 倍密度
double-derivative control 双重微分控制
double diaphragm pressure cell 双膜式压力盒【压力器，压力元件】
double differential cross section 双微分截面
double-diffused 双扩散的
double diffusion epitaxial plane 双扩散外延平面
double diode 双二极管
double-direction angular contact thrust ball bearing 双向推力向心球轴承
double-direction contract system 双向合同制
double disc parallel seat valve 平行双闸板闸阀
double-disk inclined-seat gate valve 双阀瓣斜座闸阀
double-disk parallel-seat gate valve with spreader 带扩听器的双瓣平行座闸阀
double door 双道门，双扇门
double dose 加倍剂量
double doublet antenna 双偶极天线
double-dovetail key 双燕尾榫
double drainage 两面排水，双向排水
double-drained layer 两面排水层
double drive 双驱动
double-drum boiler 双汽包锅炉
double-drum winch 双筒卷扬机
double duct system 双风道系统
double earth-fault 两相接地故障，复接地故障
double earthing switch 双接地刀闸
double-effect absorption refrigerating machine 双效吸收式制冷机
double-elbow pipe S形管，双弯头管
double electric conductor seal 双电缆密封
double end alligator type wrench 双头鳄口扳手，双头管扳手
double-ended air cylinder 双端的气压缸
double-ended ball mill 双进双出筒式球磨机，双头球磨机，双流式球磨机
double-ended break without separation 非完全断裂
double-ended break 两端断裂【管道】
double-ended guillotine break 双端剪切断裂，双端切断破裂
double-ended locomotive 双头机车
double-ended mill 双进双出钢球磨机，双进双出球磨机
double-ended pipe break 管道双端破裂
double-ended rupture 管道双端破裂
double-ended severance 管道双端破裂，双端断裂
double-ended socket wrench 双头套筒扳手
double-ended spanner 双端扳手，双头梅花扳手（长）
double-ended substation 双终端变电所
double-ended wrench 双头扳手【开口或套筒】
double-end feed 两端馈电，两端馈线
double-end grounding 双端接地
double-end stud 双端螺杆
double-end wrench 双头扳手
double entry impeller 双入口叶轮，双吸叶轮
double entry 两侧进汽，两侧进气，双倍记录，双入口
double error correcting code 二重差检验码
double error correction 双重误差校正
double expansion 两级膨胀
double-exposure hologram 双面曝光全息图
double-extraction condense turbine 双抽汽凝汽式汽轮机
double-extraction turbine 双抽汽轮机
double extra heavy 双倍加厚的
double extra strong 双倍加强的，特强
double eye impeller 双孔叶轮
double eye 双孔
double-faced plate lighting fixture 双块板灯具
double-faced reinforced concrete cover plate 双面钢筋混凝土盖板
double fault indicator relay 双重故障指示继电器，复故障指示继电器
double fed asynchronous machine 双馈电式异步电机，定转子供电的异步电机
double fed induction generator 双馈感应发电机，双馈异步发电机
double fed synchronous generator 双馈同步发电机

double female　双阴
double fibre glass-covered wire　双玻璃丝包线
double fillet welded joint　双面焊缝
double film technique　双胶片技术
double film　双胶片【辐射照相】
double filtration　双重过滤
double flanged butterfly valve　双法兰蝶阀
double flanged　双轮缘
double flat wearing ring　双平面磨损环
double flexible coupling　双弹性联轴器，双挠性联轴节
double float　复浮子
double flow condenser　双流程凝汽器
double flow design　双流设计
double flow filter　双流（机械）过滤器，双流向滤池
double flow turbine　双流式汽轮机，双排汽汽轮机，两排汽汽轮机，双流涡轮
double flow　双流，双排气
double-flue boiler　双烟道锅炉，双炉胆锅炉
double fluid theory　双流体学说
double-framed generator　双机座型发电机，内外机座型发电机
double-frame　双框架
double frequency generator　双频发生器，双频发电机
double frequency sequence signaling system　双频顺序信号制
double frequency vibration　倍频振动
double frequency　倍频
double front　双层前墙
double furnace　双炉膛
double furnace boiler　双炉膛锅炉
double-geared drive　两级齿轮传动装置
double gib arm stocker　双悬臂式堆料机
double-glazed　双玻璃的
double glazing glasses　双层玻璃
double glazing window　双层窗
double glazing　双层玻璃，中空玻璃
double governor　双调速器，双重调速器
double grab　双抓钳
double graphite sleeve　双层石墨套筒
double gripper　双抓手
double groove　双面坡口，双槽
double ground fault　双接地故障，双线接地故障
double half-day tide　双半日潮
double-headed nail　双头圆钉
double-header boiler　双联箱横水管锅炉
double heading drilling　双导洞掘进
double head open end wrench　双头开口扳手
double head wrench　双头扳手
double height water　双重高潮
double helical gear　人字齿轮
double helical spur gear　人字齿轮
double helix gearbox　双螺旋齿轮箱
double hexagon end cap wrench　六角螺帽双头扳手
double hexagon opening double offset box socket wrench　双头弯脖梅花扳手
double hexgonal S-shaped closed double head wrench　套六角环形 两头固定扳手
double high water　双高潮
double-hinged arch　二铰拱，双绞拱
double-humped curve　双峰曲线
double-humped distribution　双峰分布
double-humped flux　双峰通量
double-humped resonance　双峰谐振
double-hump　双峰的
double hung windows　双悬窗，上下扯窗
double hung　双吊钩
double-image range-finder　双像测距仪
double-impeller impact breaker　双叶轮冲击破碎机
double impulse generator　双脉冲发生器
double impulse　双脉冲
double inclined grate　双面倾斜炉排，V形炉排
double inlet and outlet (tube) mill　双进双出钢球磨煤机，双进双出球磨机
double inlet fan　双吸风机
double inlet impeller　双入口叶轮
double insulated conductor　双层绝缘线
double insulation　双层绝缘，双重绝缘
double integral　二重积分，重积分
double-integrating gyro　双（重）积分陀螺（仪），重积分陀螺（仪）
double integration method　重积分法
double interpolation　二重插值
double jacketed gasket　双夹套垫片
double jet　双喷嘴，双射流
double-J groove　双J形坡口
double labyrinth wearing ring　双迷宫磨损环
double lacing redundant support　有辅助材的双腹杆系
double lacing　双腹杆系，双缀条
double-ladder dredger　双排多斗挖泥机
double-lane highway　双车道公路
double lath　加厚的灰板条，双面灰板条
double latticing　双缀条
double layer capacitor　双层电容器
double layer contact filter　接触式双层滤池
double layer filter　双层滤料过滤器，双层过滤滤清器，双层过滤器，双层滤池
double layer layout　双层布置
double layer lining　双层衬砌
double layer media filter　双层滤料过滤器
double layer steel with nickel cover　覆镍的双层钢板
double layer winding　双层绕组
double layer　双电层
double-leaded windmill　双叶片风车
double-leaf gate　双叶闸门
double-leaf vertical lift gate　双叶垂直向提升闸门
double lee　四通
double-length word　双倍长的字
double-length working　双字长工作，双倍长工作【计算机】
double liabilities　双重负债，双重责任
double lift　二级提升
double-linear interpolation　双线形插值，双线性内插
double line double track　双线
double line ground fault　双线接地故障

double line pen　双线笔
double line　双路，复线，重叠线
double-lobe bucket　双叶戽斗
double-log scale　重对数比尺
double loop system　双回路系统
double loop　双列管阀，蛇形管阀
double low water　双低潮，双重低潮
double-L wearing ring　双L磨损环
double machine for lifting and hanging　双机抬吊
double-main system　双干管系统，双干线系统
double-make contact　双闭合触点，双工作触点
double mass curve　双累积曲线
double mechanical seal　两道机械密封
double module damper　双模数挡板
double motion turbine　双转子汽轮机
double motor alternator　双电动发电机，同步发电机
double motor　双电动机，双电枢电动机
double nozzle　双喷嘴
double nut bolt　双帽螺栓
double nuts　双螺帽
double offset expansion U bend　双偏置U形膨胀弯管
double off set spanners　梅花扳手
double offset U bend　U形膨胀接头
double offset　双效补偿
double open ended spanner set　双开口扳手套装
double open ended spanner　双开口扳手
double open end wrench　双开口扳手
double oscillation　双振荡的
double oscillograph　双电子束示波器
double output　双输出
double packing and lantern　双密封垫和外罩
double paddle mixer　双轴搅拌机
double pancake winding　双饼式绕组
double-pass condenser　双流程凝汽器
double-pass drier　双路干燥器
double-pass-out turbine　双抽气式汽轮机
double pillar jig borer　双柱坐标镗床
double-pipe condenser　套管式冷凝器
double-pipe cooler　套管式冷却器
double-pipe heat exchanger　套管热交换器
double-pipe heat tracing　双管伴随加热法
double-pipe　套管式
double-piston　双活塞
double-pitch roller chain　双节距滚子链条
double-pitch roof　双坡屋顶
double-pitch　双面坡，双坡的
double-pivoted electrodynamometer　双枢式电功率机，双轴尖式电功率计
double pivoted gate　双枢轴闸门
double plane　二重面
double plug interlock　双竖管塞连锁
double point conrtollable status output　双点可控状态输出
double point conrtol　双点控制
double point double-throw switch　双刀双投开关
double point information　双点信息
double point sattus information　双点状态信息
double pole circuit　双极回路
double-pole double throw switch　双极双投开关，双极双掷开关，双刀双掷开关
double-pole double-throw　双极双掷【开关，继电器】
double-pole dynamo　双极发电机，两极电机
double-pole relay　双极继电器，双组触点式继电器
double-pole single-throw switch　双刀单掷开关，双极单投开关，双极单掷开关
double-pole switch　双极开关
double-pole　双极的，双刀的，双极，双杆
double ported globe valve　双座球阀
double ported valve　双口阀
double ported　双通路的
double-post type disconnecting switch　双柱式隔离开关
double power source interlocking device　双电源连锁装置
double precision number　双精度数
double precision working　双倍精度工作，双精度工作
double precision　双倍精度
double pressure condenser　双倍压凝汽器
double pressure containment　双层安全壳
double pressure suppression system　双重驰压系统，双重压力抑止系统
double probe controller　双探头控制器
double probe testing method　双探头探伤法
double profile　复式断面
double-pulse generator　双脉冲发生器
double-pulse selection　双脉冲选择
double-pulse　双脉冲
double-range recording flowmeter　双量程记录式流量计
doubler circuit　二倍器电路，倍加器电路
double reaction　双重反映，双重反馈
double reduction axle　复式减速轴
double reduction bevel-spur gear　双极圆锥-圆柱齿轮减速器
double reduction gear box　两级减速器
double reduction geared turbine　两级减速汽轮机
double reduction gear　双减速齿轮
double reduction unit　两级减速器
double regulation governor　双调整调节器
double reheat cycle　二次再热循环
double reheating　二次再热，两级再热
double reheat steam turbine　二次再热式汽轮机
double reheat turbine　二次再热汽轮机
double reheat　二次再热，二次中间再热
double reinforced beam　双筋梁
double reinforced　双筋的
double reinforcement beam　双重配筋混凝土梁，配受压筋的梁
double reinforcement　复筋，双重钢筋
double-response controller　双重作用控制器
double-return wind tunnel　二次回流风洞
double-revolving-field theory　双旋转磁场理论
double right shift　双倍右移
double-ring oil sealing　双流环形油密封
double riveted butt joint　双行柳钉对接
double riveted lap joint　双行铆钉搭接
double rod cylinder　双活塞杆油缸

double rods hanger 双拉杆吊架
double roller crusher 双辊破碎机，双滚筒碎石机
double roll grinder 双辊碎渣机
double roll out solar-generator 双卷筒式光伏发电站
double rotation turbine 双转子汽轮机
double rotor induction motor 双转子感应电动机
double rotor-stator stepper 双定子转子步进电动机
double rotor steam turbine 双转子汽轮机
double row angular contact spherical ball 双列推力向心球面球轴承
double row ball bearing 双列滚珠轴承
double row ball journal bearing 双列径向滚珠轴承，双排滚珠轴承
double row bearing 双列轴承
double row layout 双列布置
double row pile cofferdam 双排板桩围堰
double row radial ball bearing 双列向心球轴承
double row riveting 双行铆接
double row runner 双转轮
double row stage 复速级
double row tapered roller bearing 双列圆锥滚柱轴承
doubler 二倍器，倍加器，乘二装置
double safety valve 双安全阀
double salt 复盐
double sampling 复式抽样
double-scale instrument 双标度仪表
double screened coal 双重筛子筛分的煤
double-screw bolt 双头螺栓
double seal 双层密封
double seal assembly 两道密封组件
double-seated control valve 双阀座的控制阀
double-seated main stop valve 双座式主汽门
double-seated 双阀座的
double-seat valve 双座阀
double-secondary transformer 双二次绕组变压器，双负边变压器
double shaft paddle mixer 双轴式桨叶拌和机
double shaft surge tank 双井式调压井
double shaft turbine 双轴汽轮机
double shaft 双轴
double shear riveted joint 双剪铆接
double shear 双剪，双剪力
double shed porcelain insulator 双裙式瓷绝缘子
double shell casing 双层缸
double shell condenser 双壳体凝汽器
double shell cylinder 双层缸
double shell 双层壳体
double shielded winding 双屏蔽绕组
double-shift operation 两班倒运行
double-shot turbine 双喷嘴水轮机
double-sideband multiplex equipment 双边带多工制设备
double-sided band 双重边带
double-sided compressor 双侧进气压气机
double-sided impeller 双吸式的叶轮【风机或泵】
double-sided linear motor 双边直线电动机
double-side-hand modulation 双边带调制
double-side spectrum 双边谱
double-side tipping wagon 两侧倾卸车，双边倾卸车
double side-wire potentiometer 双滑线电势计，双滑线电位计
double-silk-covered wire 双层丝包线
double-skin construction 双层结构
double-skin duct 双层管道
double-skinned construction 双层蒙皮结构
double-skinned metal sheet 双面层金属板
double-skin stator 双层缸，双层静子
double sky-light 双层天窗
double sleeve valve 双套阀
double slide pump 双滑片式回转泵
double slide valve 双滑阀
double sling 双吊索
double slit 双缝
double socket 双插座
double span 双跨
double speed fan 双速风机
double speed motor 双速电动机，双速马达
double spherical roller bearings 双球面滚柱轴承，双列滚子球面轴承
double-spindle hammer crusher 双轴锤式碎石机
double split brush 双子电刷
double split large cross-section line 双分裂大截面导线
double split line 双分裂导线
double spring clip 双弹簧线夹
double square closed single head wrench 套方环形一头固定扳手
double squirrel cage motor 双鼠笼电机
double squirrel cage winding 双鼠笼式绕组
double stage compressor 二级空气压缩机，双级压气机
double stage cyclone 双级旋风除尘器
double stage pump 双级泵
double stage turbine 双级式透平
double standpipe plug interlock 双竖管塞连锁
double star connection 双星形接法，双星形连接
double star-quad cable 双星绞电缆
double star winding 双星形绕组
double star 双星形
double stepped labyrinth gland 双短齿迷宫汽封
double stop watch 双针停表
double-strength pipe 高强度管
double string 双绝缘子串，双联绝缘子串
double-stroke deep-well pump 双冲程深井泵
double stuffing box 双填料函
double-submerged arc 双面埋弧焊
double suction centrifugal pump 双吸离心泵
double suction impeller 双吸式的叶轮【风机或泵】，双吸压气机，双吸入口叶轮，双吸入式叶轮，双吸叶轮
double-suction pump 双吸口泵，双吸式泵，双吸泵
double suction 双吸
double sum 二重和
double-surface transistor 双面晶体管
doublet aerial 偶极天线，对称天线
double tank 双层容器
double-tape armour 双带铠装

double-taper wedge　双斜楔块
double tariff electricity meter　二部电价制电度计，双价电度表
double tariff meter　二部电价制电度表
double taxation relief　双重课税减免，避免双重征税，免除双重税收
double taxation　双重税收，双重征税
double tee　四通
double-test relay　双重测试继电器
double-theodolite observation　双经纬仪观测
double-throw circuit-breaker　双投断路器
double-throw switch　双掷开关
double-throw　双掷【开关，继电器】
double thrust bearing　双推力轴承，双向止推轴承
double tidal ebb　双落潮流
double tidal flood　双峰潮汛，双涨潮流
double tide　双潮
double-tier switchboard　双层配电盘
doublet impulse　对称脉冲，双向脉冲
doublet panel　偶极子鳞片
double tracer technique　双元示踪技术
double track　复线
doublet sheet　偶极子面
doublet strength　偶极子强度
double tube plates　双管板
double tube pressure gauge　套管压力计，双管压力计
double tube reactor　套管反应器
double tube sheet　双层管板
double tuned circuit　双调谐电路
double-turn　双匝的，双圈的
doublet　双重线，复制器，偶极子，对称振子
double-U groove　双U形坡口，双面U形坡口
double undulated preheater　双波纹回转式预热器
double unit alternator　定子绕组具有中抽头的同步发电机
double unit motor　双电动机机组【两台电动机同轴】
double unit traction　双机牵引
double valve gravity filter　重力式双阀滤池
double vault　双重穹隆，双重圆顶
double V butt weld　双V形对接焊
double-Vee groove　X形坡口，双V形坡口
double-Vee　双V形，X形
double vernier　双游标
double-V groove weld　双斜边坡口焊缝，双斜坡口焊
double-V groove　X形坡口
double voltage motor　双电压电动机
double voltage rating　（变压器的）双额定电压
double volute casing　双蜗外壳，双蜗形体
double volute　双蜗壳
double wall cofferdam　双排板桩围堰
double wall containment　双层安全壳
double-walled cylindrical graphite sleeve　双壁圆筒形石墨套筒【气冷堆燃料元件】
double-walled well　重壁井
double-walled　夹壁的
double wall tank　双壁容器
double wall　双壁
double water circuit　二回程水通道，双水回路
double water inner-cooling generator　双水内冷发电机
double water inner-cooling motor　双水内冷电机
double water internal cooling　双水内冷
double wave detection　全波检波
double wave　全波，双波
double-way connection　桥式接法
double-way rectifier　桥式连接整流器，双相桥式线路整流器
double wearing ring　双磨损环【蜗壳和叶轮】
double web　双肋式
double-wedge airfoil　双楔机翼
double-wedge gate valve　双楔闸阀
double-welded butt joint　双面对焊接
double-welded joint　双面焊缝【从两边焊接】
double-wind hoist　双盘卷扬机
double winding alternator　双绕组同步发电机
double winding armature　双绕组电枢
double winding synchronous generator　双绕组同步发电机
double winding　双绕线圈，双绕绕组
double window　双层窗
double wire system　双线制
double-word addressing　双字寻址
double-wound coil　双股线圈，双线绕线圈，双绕线圈
double-wound relay　双线圈继电器
double-wound rotor　双线绕式转子，双鼠笼转子
double-wound silk　双丝包的
double-wound synchronous generator　双线绕式同步发电机
double-wound transformer　双线绕线圈，双股线圈
double Y connection　双星形接线，双Y接线，双Y连接
double　成倍的，成双的，加倍，倍增，两倍的
doubling circuit　倍增电路
doubling dose　加倍剂量
doubling effect　回波效应，断开脉冲，回波
doubling of frequency　倍频
doubling plate　复板，加强板
doubling register　倍增寄存器
doubling test　弯曲试验，折叠试验
doubling time　加倍时间【增殖堆技术】，倍周期，重复时间
doubling　复制，重复，加倍
doubly-fed commutator motor　定转子供电的换向器电动机
doubly-fed motor　双馈电动机，定转子供电电动机
doubly-fed repulsion motor　双馈推斥电动机
doubly-fed series motor　双馈串激电动机
doubly-fed　双馈的
doubly fluted tubes　双侧开槽管【双凹槽管】
doubly re-entrant winding　双闭路绕组
doubly slotted electrical machine　定转子开槽的电动机
doubly　双重的，加倍的
doubtful accounts　疑账，坏账
doughnut casing　环形壳体
doughnut coil　环形线圈

doughnutting　沉淀物结环【水管内】
doughnut　环形真空室【加速器】，热堆快中子转换器
Douglas-Neumanmethod　道格拉斯—纽曼法
doulateral winding　蜂房式线圈，蜂房式绕组
Dounreay fast reactor　唐瑞快中子反应堆
D_2O upgrading plant　（重水）浓缩装置，蒸馏装置
D_2O upgrading system　重水浓缩系统
dousing system　喷淋系统
dovetail bolt　开脚螺栓
dovetail condenser　鸠尾形电容器，同轴调制电容器
dovetailed groove　燕尾榫槽，燕尾槽
dovetail joint　鸠尾连接，燕尾接合
dovetail male　燕尾榫
dovetail root　燕尾形叶根【即齿型】
dovetail shaped head type insulator　圆锥头型绝缘子
dovetail slot　燕尾槽
dovetail　燕尾榫，燕尾形
D_2O volume control system　重水容积控制系统
dowel bar　暗销杆，接头插筋，传力杆，缀缝筋
dowel crown　桩冠
dowel in concrete　混凝土中的插筋
doweling　埋插筋
dowelled joint　传力杆接缝，插筋接合，螺栓接合，榫钉接缝，销钉接合，暗销接合，木钉榫
dowelled　设传力杆的，设暗销的
dowel pin and socket　定位销和孔
dowel pin to lock washer interface　定位销对锁紧垫圈接口
dowel pin　定位销，定缝销，合销，开槽销
dowel　暗销，榫钉，夹缝钉，混凝土中的插筋，插铁，固定螺钉，两头钉，木钉，定位销，定缝销钉
down　向下，沿着，往下
downburst　下击暴流，下暴流
downcast fault　下落断层
downchute　落煤管
downcomer barrel　导流套筒［围板］【蒸汽发生器】，蒸汽发生器倒流套筒
downcomer bottom plate　环形腔下水底板
downcomer header　下降管联箱
downcomer leg　下降管支路
downcomer pipe　溢流管
downcomer spece　下水空间，环状空间
downcomer tube　下降管
downcomer type cyclone　下降管式旋流分离器
downcomer　（锅炉的）下降管，水回流管，下降段，降液管，下向引出管，泄水管
down conductor　避雷线的垂直部分，下垂导线，引下线
downcounter　扣除计数器
downcut milling　向下铣切
downcutting　下切，下切侵蚀
downdraft boiler　下部排烟锅炉，下排烟锅炉
downdraft plume　下吸式羽流
downdraft　下沉气流，倒灌风，下吸式，向下通风，炉排下进风，下降气流，下引风
downdrift　下滑漂移，向下游漂移
down extreme position　下限位置【夹车器】
downfall　（雨等的）大下特下，大雨，落下，垮台
down feed riser　下供式立管
down feed system　上分式系统，下供式系统
down firing　向下燃烧
downflow reactor　下流反应器
downflow regeneration　顺流再生
downflow weir　溢流堰
downflow　溢流管，向下流动，下降管，下行流，下沉气流
down fold　槽褶皱
downgraded heavy water　降质的重水，浓度降低的重水
downgraded　降质的，浓度降低的
downgrade　等级降低，使降级，降级，退步，下坡，下坡的，下坡地
down grade　次级品
downgradient　下阶，下游
downgrading　降低等级，（重水的）浓度下降
downhand welding　平焊，俯焊，水平焊接
downhand weld　平位焊接，平焊缝，俯焊焊缝
downhill conveyor　下（倾）坡输送机
downhill flow　下坡流
downhill pipe line　下坡管线
downhill welding　下坡焊
downhill　下坡，向下，降下，倾斜的，向下的，下山，下坡的
downhole dynamic compaction　孔内动力压实法
downhole steam generator　油田下注汽锅炉
downlead　（天线）引下线，下引线，天线馈线
down lift　负升力
downlight　顶棚里向下照射的小聚光灯，筒灯，下射式灯具，下照灯，嵌顶灯，向下发射的光
down line　下行线路
download　下载，卸负荷，卸载
down pass　下行烟道
down payment plan　押金制度
down payment　预付定金，分期付款的首次交款，分期付款的订金，预付款，定金
downpipe　水落管，落水管，下降管，雨水管
downpour　大暴雨，大雨，倾盆大雨
downpull force　下曳力
downrating　降低出力
downscale protection　下限保护
downshot burner　下射喷燃器
downshot firing　下射喷然
downshot furnace　W形火焰炉膛
downsizing for efficiency　减员增效
downslope direction　下坡方向
downslope flow　下坡气流
downslope overlapping　坠覆
downslope windstorm　下破风暴
downslope wind　下坡风
downslope　下坡，下倾，下坡的，向着坡下
downspout　落水管，落料馏槽，下部排料口，雨水管，漏斗
downstage without packer method　自上而下无塞灌浆法
downstand beam　下翻梁

down state 不可用状态
downstep check 逆止阀
downstop 向下止动装置，下部挡块
downstream approach channel 下游引航道
downstream apron 坝后防冲护底【水电】，防冲护坦，防冲铺砌，下游护坦，闸后护坦
downstream attemperator 出口减温器，后置减温器
downstream bank protection 下游护岸
downstream batter 背水坡【即下游坡】
downstream bay 下水池，下游池，下游河段
downstream benefits 下游受益
downstream cascade 下游梯级
downstream chamber 出口室
downstream conveying system 下向输送机
downstream cross section 下游断面
downstream drain 下游排水
downstream edge of stay vane 固定导叶出口边
downstream effect 下游效应
downstream erosion 下游冲刷
downstream face of dam 坝的下游面
downstream fairing 下游整流器【风机】
downstream floor 下游底板
downstream flow 顺流，下游流，（流的）下游，下游的流动，顺流（向）流动
downstream gate slot 下游闸门槽
downstream gate 下游闸门
downstream gauging section 下游水文测验断面
downstream guide vanes 下游导向叶片
downstream guide wall 下游导墙
downstream heat exchanger 顺流换热器
downstream injection system 顺流喷射系统
downstream line 下游管线
downstream lock chamber 下游闸室
downstream measuring section 下游测验断面
downstream minimum water level 下游最低水位
downstream pier nose 墩的下游端点，闸墩尾部
downstream power station 下游电站
downstream pressure controller 阀后压力调节阀
downstream pressure 阀后压力
downstream product 下游产品【制造业】
downstream profile of crown section 拱冠断面下游面轮廓
downstream protection 下游防护措施
downstream radius of crest 坝顶下游面半径【水电】
downstream reservoir 下游水库
downstream scouring 下游冲刷
downstream section 下游段，下游区段，出口段
downstream shell 下游坝壳
downstream side 下游边，下游侧
downstream slope protection 下游护坡
downstream slope 背水坡【即下游坡】，下游坡
downstream spray pattern 顺喷
downstream stress 下游应力
downstream surge chamber 下游调压井，下游调压室
downstream toe 下游坝址，下游坡脚
downstream tube 下降管
downstream valve 下游阀
downstream water level 下游水位
downstream water line 下游水线
downstream water 下游水
downstream wing wall 下游翼墙
downstream zone 下游区
downstream 下游的，顺流的，下游，下向流，下段，下行线
downsurge 下涌浪，水面下降
downs 白垩山丘
downtake chamber 下行烟道
downtake pipe 下导管，下降管
downtake tube 下降管
down the slope 下坡，沿着斜坡
down throw block 陷落地块
down throw fault 下落断层
downthrow 下降，下落
downthrust 向下推力
downtime cost 故障停机费
downtime for change-over 换班时间
downtime percentage 休风率
downtime ratio 停机时间比，（设备的）停工率
downtime 停机时间，停堆时间，不工作时间，故障时间，停用期，发生故障时间，停用时间，维修时间，（设备或人的）停工时间，不可用时间，故障停机时间，设备闲置期，（工厂等由于检修、待料等导致的）停工期，停止运行时间
down total 停机总数
downtown area 闹市区，市中心，商业区，市区
down valley windflow 下坡风，出谷风
downward communication 上情下达
downward current 下沉流
downward erosion 下切侵蚀
downward-fired furnace 下射燃烧炉膛
downward flow 向下流动，向下流，下行流，下降流动，下降流，下降的流动
downward force 向下的力
downward motion 向下运动
downward-moving agglomerate 向下运动颗粒团
downward pressure 向下压力
downward solar radiation intensity 向下短波辐射强度
downward tendency 下跌趋势
downward valve 下行阀
downward vertical position welding 垂直往下焊接【工作姿势】
downward 向下的，下降的
down warp 下翘，拗陷
downwash correction 下洗修正
downwash effect 下洗效应
downwash field 下洗场
downwash flow 下洗流，下洗气流
downwash model 下洗模式
downwash parameter 下洗参数
downwash plume 下洗型羽流
downwash region 下洗区
downwash velocity 下洗速度
downwash vortex 下洗涡
downwash 冲掏，下沉洗流，下洗，下洗气流，气流下洗，向下运动，由高处冲下的物质
downwelling 沉降流，向下涌出
downwind direction 下风向

downwind rotor 下风式风轮
downwind sector 下风向扇形区
downwind type of WECS 下风式风力机【使风先通过塔架再通过风轮的风力机】
downwind wind turbine 顺风式风电机组
downwind WTGS 下风向式风电机组
downwind 顺风，在下风，下风向
dozer blade 推土机推土板
dozer 推土机
dozing 推土
DPACS(dedicated reactor auxiliary cooling system) 专用堆辅助冷却系统
DPC(data processing center) 数据处理中心
DPC(disk pack control) 磁盘组控制器
DPCS(distributed process control system) 分部过程控制系统
DPCT(definition point correspondence table) 定义点对应表
DPC(damp-proof course) 防潮层
DP(dew point) 露点
DP(differential pressure) （流的）压差，差压，压力落差，压力降
DP(dimension print) 外形安装尺寸图
DPDM(digital pulse duration modulation) 数字脉冲宽度调制
DP(double-pole) 双极
DP(drain pipe) 泄放管，排出管
DP(drip-proof) 防滴型
DP(driving power) 驱动功率
DPDT(double-pole double-throw) 双极双掷，双刀双掷【开关，继电器】
DPDT switch 双刀双掷开关，双极双掷开关
DP(dye printing) 着色探伤
DP(dynamic programming) 动态规划
DPE(data processing equipment) 数据处理装置
DPIC(differential pressure indicating controller) 压差指示控制器
DPM(data processing machine) 数据处理机
DPO(delayed pulse oscillator) 延迟脉冲振荡器
DPP(damage protection plan) 损害修理条款
DPS(data path switch) 数据通路开关
DPS(data processing system) 数据处理系统
DPS(Daya Bay Public Security Branch) 大亚湾公安分局
DPST(double-pole single-throw) 双极单掷【开关，继电器】
DPST switch 双刀单掷开关，双极单掷开关
DPT(differential pressure transmitter) 压差变送器，差压变送器，差压转换器，差压传感器
DP transmitter(differential pressure transmitter) 差压变送器，压差变送器
DPU(display processing unit) 显示处理器
DPU(distributed processing unit) 分布式处理单元
DPU 分散处理单元【DCS 画面】
DPV(dry pipe valve) 过热蒸汽管道阀门
d/p(differential pressure) 差压
DP(documents against payment) 付款交单
DQC(dynamic quality control) 动态质量控制
draft a contract 起草合同
draft agreement 草约定
draft and telegraphic transfer 汇票及电汇
draft as-built drawings 竣工草图
draft capacity 通风能力
draft chamber 气流室
draft collection 托收汇票
draft condition 通风条件，鼓引风装置特性
draft core 抽吸核心
draft decision 决议草案
draft differential 通风阻力，抽风压差
draft equipment 通风设备
drafter 起草者
draft fan room 吸风机室
draft fan 排风机，排风扇，通风机，引风机
draft for collection 托收汇票
draft gauge 通风计，风压表，拉力计，差式压力计
draft gear 牵引装置
draft hole 通气孔
draft hood 抽气罩
draft indicator 风压表
drafting committee 起草委员会
drafting machine 绘图机，绘图仪
drafting room 绘图室
drafting sand 流沙
drafting scale 制图尺
drafting standard 制图标准，制图规程
drafting 拉拔，制图
draft loss 烟道阻力，通风阻力，通风损失
draftman 制图员
draft margin 通风裕量
draftmeter 风压表，风压计
draft of international standard 国际标准草案
draft of plan 计划草案
draft pipe 吸出管
draft-pressure drop 通风压降，空气动力阻力
draft proposal 提议草案
draft regulator 负载调节器，拉力调节器
draft report 报告草案，非正式报告
draft resistance 烟道阻力，通风阻力
draft resolution 决议草案
draftsman 描图员，起草者，制图员
draftsmanship 制图技术，制图质量
draftsperson 制图员
draft standard 标准草案
draft stop 挡风板，通风中止
draft system 通风系统
draft-tube access gallery 尾水管进入廊道
draft-tube bend 尾水管弯管段
draft-tube deck 尾水平台
draft-tube efficiency 尾水管效率
draft-tube elbow 尾水管肘管段
draft-tube entrance 尾水管进口
draft-tube exit 尾水管出口
draft-tube floor 尾水管底板
draft-tube gate 尾水闸门
draft-tube inspection passage 尾水管检查通道
draft-tube liner 尾水管里衬
draft-tube loss 尾水管损失，吸出管损失
draft-tube manhole 尾水管进人孔
draft-tube performance 尾水管性能
draft-tube roof 尾水管顶板
draft-tube storey 尾水管层

draft-tube surge 尾水管涌浪
draft-tube wall 尾水管边墙
draft-tube 通风管，提升管，尾水管，吸出管
draft 通风，引风，（炉膛及烟道）负压，草图，制图，草案，图纸，汇票，草稿，压缩量，抽风，小股气流，抽力，气流，吸出
drag acceleration 负加速度，减速，减速度
drag anemometer 曳力风速计
drag area 风阻面积
drag balance 流阻平衡
drag bar 拉杆，牵条
drag bit 十字镐，镐
drag bolt 拉紧螺栓
drag brace （翼内）阻力张线
drag break 阻力刹车板
drag by lift 阻升比
drag cable 牵引钢索
drag center 阻力中心
drag chain conveyer 牵链输送机，刮板捞渣机，链板式输送机
drag coefficient 阻力系数，牵引系数
drag convergence 阻力减小
drag conveyer 刮板（式）输送机，链板输送机
drag crisis 阻力危机
drag critical value 阻力临界值
drag cup anemometer 阻力型风杯风速计
drag cup induction machine 托杯形感应电机，空心转子电动机
drag cup rotor motor 空心转子电动机，杯形转子电动机
drag cup tachogenerator 杯形转子测速发电机，空心转子测速发电机
drag cup tachometer 杯形转子测速表，空心转子测度表
drag due to lift 升致阻力
drag effect 曳滞效应
drag equation 阻力方程
drag factor 风载体型系数
drag force 阻力，制动力，迎面阻力，拖曳力，拖拉力，牵引力
dragfriction 摩擦阻力
drag from pressure 压差阻力
dragger 牵引机
dragging feeder 槽式给煤机
dragging track 滑行轨道
dragging 拖曳，拖运
drag head （耙吸式挖泥船的）耙头，把手
drag increment 阻力增量
dragless aerial 无风压天线
drag lift ratio 阻升比
dragline bucket 索铲铲斗，吊铲抓斗
dragline cableway excavator 缆索拖铲
dragline excavator 拉索挖土机，索铲，索式挖土机，拖铲挖土机
dragline scraper 索式铲运机，带式刮土机，拉铲运载机
dragline tower excavator 塔式索铲，塔式索挖掘机
dragline 牵引绳索，拉铲挖土机，绳斗电铲
draglink conveyer scraper conveyor 刮板式输送器
draglink conveyor 链板式输送机，刮板式输送机，拉杆式输送机
draglink feeder 刮板式给煤机，刮板式给料机
dragloss 阻力损失
drag parameter 阻力系数
drag penalty 阻力增大
drag plate 拖板【链条炉排下】
drag polar 阻力极曲线
drag principle 阻力原理
drag reducing device 减阻装置
drag reduction cowling 减阻整流罩
drag reduction 减阻
drag resistance 拖曳阻力，拖阻力
dragrope 拖绳
drag saddle galloping 最小阻力驰振
drag saddle 最小阻力点
dragscraper 拉曳刮土机，拖铲，拖拉铲运机
drag seal 密封板【链条炉排风室】
drag shear 由迎面阻力产生的剪力
drag shovel 拖拉铲运机，反铲
drag suction dredge 耙吸式挖泥机
drag suction method 耙吸挖泥法
drag test 风阻试验
drag type feeder 刮板式给料机，刮板式给煤机
drag type rotor 阻力型风轮
drag type wind machine 阻力型风力机
drag variation with lift 阻力随升力的变化
drag vector 阻力矢量
drag wind load 风阻载荷
drag 拖，拉，拖拉，拖拽，阻力，制动，牵引，曳，曳力
drainability 排水能力，排水性
drainable sludge 可脱水污泥，易脱水污泥
drainable-type superheater 可疏水过热器
drainable 可疏水的，可放空的
drainage adit 排水廊道，排水通道
drainage age line 排放管，疏水管线
drainage area boundary 流域边界
drainage area precipitation 流域降水
drainage area 排水区域，排水面积，排水区
drainage arrangement 排水布置
drainage basin map 流域图
drainage basin morphology 流域形态学
drainage basin precipitation 流域降水
drainage basin shape 流域形状
drainage basin topography 流域地形
drainage basin water yield 流域产水量，流域总径流量
drainage basin yield 流域产水量，流域平均年径流
drainage basin 流域，排水区域，流域盆地
drainage blanket 排水垫层，排水铺盖
drainage by desiccation 疏干排水
drainage by electro-osmosis 电渗排水
drainage by frost action 冻结作用排水
drainage by suction 负压排水，吸力排水
drainage by surcharge 超载排水
drainage by well point 井点排水
drainage canal 排水沟
drainage channel 排水沟，下水道，排水渠
drainage coefficient 排水系数

drainage coil 排流线圈
drainage collection pond 雨水集水池
drainage collector 排水干管，污水集水管
drainage conduit in dam 坝身排水管
drainage controller 疏水调节器
drainage cooler 疏水冷却器
drainage culvert 排水涵洞，排水涵管
drainage curtain 排水幕
drainage degree 排水度，排水率
drainage density 排水密度，排水网密度，水系密度
drainage district 排水区，流域管理区
drainage ditch 排水沟，排水明沟
drainage divide 地面分水岭，分水脊，分水界，分水岭，分水线，流域分界线
drainage duct 疏水管
drainage equilibrium 排水量平衡
drainage facility 排水设备
drainage filter 排水滤层，排水滤体
drainage frequency 水系频率
drainage gallery 排水廊道
drainage gate 排水闸
drainage gauge 渗漏计
drainage header 排水集管
drainage heat exchanger 疏水热交换器
drainage hole 排水孔，疏水孔，排泄口，排泄孔
drainage intensity 排水强度
drainage layer 排水层
drainage loss of water 疏水损失
drainage map 水系图，流域图，河网图
drainage method 排水方法
drainage modulus 排水模数
drainage morphometry 水系地形测量
drainage network 排水管网，排水网
drainage of bed material 床料排放【沸腾炉】
drainage of foundation 基坑排水
drainage outlet 排水出路，排水沟出口
drainage path 排水路径
drainage pattern 排水方式，水系类型
drainage perimeter 流域周界
drainage pipeline 排水管道
drainage pipe 排泄管，排污管，排水管
drainage piping 排水管道系统，排水管路，排水管系
drainage project 排水工程，排水计划
drainage pump house 排水泵房
drainage pumping plant 排水泵站
drainage pumping station 排水泵站
drainage pump 排水泵，疏水泵，排泄泵
drainage quota 排水定额排放泵
drainage ratio 径流系数
drainage receiver 冷凝罐
drainage recovery 排水回收
drainage requirement 排水定额
drainage scheme 排水系统方案
drainage scoop 疏水槽
drainage shear test 排水剪切试验
drainage sluice 排水闸
drainage structure 排水建筑物
drainage subsystem 排放子系统
drainage sucker 排水吸收器
drainage sump 集水井，排水井
drainage system 排泄系统，排水系统，排水管系，水系
drainage tank 疏水箱，泄水槽
drainage terrace 截水地埂，排水地埂
drainage tile 排水瓦管
drainage tray 排水槽
drainage treatment facility 排水处理设施
drainage trench 排水沟
drainage triaxial test 排水三轴试验
drainage tube 排水管
drainage tunnel 排水隧洞，排水通道
drainage valve 排水阀，疏水阀
drainage water right 排水权
drainage water 排水
drainage well 排水井，竖井排水
drainage works 排水工程
drainage 疏水，排水，排水管，排泄，下水道，排水设施，排水系统，下水道水系，泄水
drain and blow-down system 疏放水系统
drain and vent system 排水与放气系统
drain and vent 排水与排气
drain approach 疏水端差
drainback system 回流系统
drain basin 排水池
drain canal 排水渠
drain channel 排水渠
drain chute 排水斜管
drain cock 疏水阀，疏水旋塞，泄放旋塞，排污阀，排水旋塞，排水栓，放水旋塞
drain collection header 疏水收集总管
drain connection 排水连接管，疏水管接头，排水管
drain cooler 排水冷却器，疏水冷却器
drain cooling zone 疏水冷却区
drain current 漏电流
drain discharge 排水量
drain district 排水区
drain ditch 排水沟
draindown system 排放系统
drained shear test 慢剪试验，排水剪切试验
drained shear 排水剪切
drained test 排水试验
drained triaxial test 排水三轴试验
drain erosion 水蚀
drainer 疏水器，排水器，放泄器，排疏口
drain flange 排油法兰
drain flash tank （疏水）扩容器，疏水扩容箱，疏水膨胀箱
drain funnel 排液漏斗
drain gate 排水闸门
drain grating 拦污，排水沟栅
drain gully 落水口
drain height of water turbine 水轮机排出高度
drain hole 排水洞，排水孔，疏水孔
draining off 排出
draining outfit 排水机组
draining sucker 排水吸除器
draining sump 泄水坑
draining transformer 抗流变压器

draining waterlogged areas　排涝
draining　排水，排泄，排空
drainless area　无流区，无排水区
drainless lake　无出流湖
drain line　排水管线
drain manifold　排水总管，疏水总管
drain masking　漏极掩蔽
drain nozzle　排水管口，疏水接管
drain off　排放，排出，流出，泄水
drain oil recovery equipment　废油回收装置
drainout　排水，排空
drain pile　排水砂桩
drain pipe　疏水管，排水管，排泄管，泄水管，排放管，排油管
drain pit　排水坑，集水坑
drain plug　疏水孔塞，泄放孔塞
drain port　疏水口，排水口，排污孔，排泄口，(沸腾炉的)排料孔
drain pot　污水井，疏水箱
drain pressure　疏水压力，泄放压力
drain pump　疏水泵，泄水泵
drain recovery pump system　疏水回收泵系统
drain recovery　疏水汇集箱，疏水集水箱
drains bedding　排水管垫层
drains collecting tank　疏水收集箱
drain separator　脱水器，气水冷却分离器
drain series resistance　漏串联电阻
drains network　排水管网
drain subcooler approach　疏水过冷段
drain subcooling zone　疏水过冷区
drain system　排放网，排放系统，疏水系统
drains　疏水，疏水设施
drain tank　疏水箱，排泄箱，排水箱，泄水箱，排水池
drain tile　排水瓦管
drain transfer pump　疏水回收泵，疏水输送泵
drain transfer tank　疏水转移箱
drain transfer　疏水汇集箱，疏水转运
drain trap　疏水罐，疏水箱，疏水器，排水器，放泄弯管，排水防气瓣，存水弯，排水井，放水弯管
drain tube　疏水管，溢流管
drain valve connections　放水装置
drain valve　排水阀，泄漏阀，疏水阀，放泄阀，放水阀，排液阀，除气装置，泄流阀
drain vessel　排水容器，排水箱
drain water circuit　泄水管路
drain water cooler　疏水冷却器
drain water flash tank　疏水扩容器
drain water piping　疏水管道
drain water pond　排水池
drain water　排水，疏水
drain well　排水井
drain wire　排水管金属线
drain　排水，排水沟，排水设施，排放，排泄，疏水，泻油
drastic loss　剧烈的亏损
draught bead　窗挡风条
draught center　阻力中心
draught control　牵引力控制，拉力控制
draught cupboard　通风柜
draught damper　气流闸
draught gauge　风压表，通风计
draught gear　牵引装置
draught head　通风压头，吸水水头
draught hole　通风孔
draught line　取水量线
draught loss　压力损失
draughtman　制图员
draught of report　报告草案
draught regulator　通风调节器，牵引力调节器
draughtsman　描图员
draught tube access shaft　尾水管进入井
draught tube gate slot　尾水闸门槽
draught tube pipe　尾水管
draught tube　吸出管
draughty　通风良好的
draught　通风，引风，(炉膛及烟道的)负压，拉力，草图，草稿，抽风，牵引，吸出
drawable currencies　可提取的货币
draw a plan　草拟计划
drawback for duties paid　退税
drawback system　退税制度
drawback　弊端，缺点，障碍，退税，退款，不利因素，缺陷
drawbar dynamometer　拉杆式测力计
drawbar horsepower　牵引功率，牵引马力
drawbar pull　挂钩拉力，拉杆牵引力，牵引杆拉力
draw bar　拉杆，牵引杆
draw bolt　牵引螺栓
draw cut　拉切，上向掏槽
draw-door weir　直升闸门堰
drawdown angle　下降角
drawdown cone of groundwater　地下水下降圆锥体
drawdown curve of groundwater　地下水位降落曲线，地下水下降漏斗曲线
drawdown curve　降深曲线【地下水位】
drawdown depth　消落深度
drawdown notice　提款通知，提款通知书
drawdown period　贷款支用期，提款期
drawdown phreatic line　下降浸润线
drawdown ratio　水位下降比
drawdown request　提款要求，提款申请
drawdown test　水面下降试验
drawdown　提取借款【贷款】，水位降低【抽水或泄水后】，减少，支出，提款，动用贷款，下降，下落
draw drum　放线盘，牵引卷筒
drawee bank　付款行
drawee　受票人，汇票付款人，票据付款人，付款人
drawer　抽屉，辅助机架，拔取机，制图员，票据出票人
draw foreign funds　吸引外资
draw forth　引起
drawf partition　矮隔断
draw gate　闸阀
draw gear　牵引装置，车钩
draw-in air　吸入风，吸入的空气
draw in bolt　拉紧螺栓

drawing apparatus　拉拔设备
drawing board　绘图板，图板，制图板
drawing change　图纸更改
drawing classification plan　卷册计划
drawing-compass　绘图用两脚规
drawing contents　图纸目录
drawing desk　绘图桌
drawing direction　控制方向【冶金】
drawing error　描画误差
drawing for order　订货图纸
drawing identification　所在图号
drawing-in box　引入箱
drawing index　图纸索引
drawing ink　绘图墨水
drawing-in wire　电缆牵引线，穿管线，穿线管【布线、拉线用】
drawing list　图纸目录
drawing machine　绘图机，拉线机
drawing name　图纸名称
drawing number　图号
drawing office　绘图室
drawing paper　绘图纸
drawing pencil　绘图铅笔
drawing pen-nib　绘图笔尖
drawing pen　绘图笔
drawing pin　图钉
drawing room　绘图室
drawing rule　绘图尺
drawings and document submittal schedule　图纸与文件提交进度表
drawing scale　图示比例尺
drawings of equipment　设备图
drawing technique　制图技术
drawing title　图名
drawing tracing　描图
drawing　牵引，吸取，拉，抽，拉拔，张拉，轧制，压延，绘图，图纸，图，制图
draw-in pit　电缆拉入坑
draw-in system　鼓风系统
draw-in winding　插入式绕组，穿入式绕组
draw-in　拉回
draw lead bushing　穿缆式套管
drawn clause　出票条款
drawn-in tandem　串列拖带
drawn-in　回缩量
drawn metal　控制金属
drawn-off pipe　卸料管
drawn tube　拉制管
drawn　拉出的，拉长的，绘好的
draw-off culvert　泄水涵洞
draw-off header　排水联箱
draw-off pipe　泄水管，卸料管
draw-off point　泄水点
draw-off pump　抽吸泵
draw-off temperature　取水温度
draw-off tower　取水塔，泄水塔
draw-off tunnel　泄水隧洞
draw-off valve　排水阀，排气阀，泄水阀
draw-off　抽吸，吸取，排出，流出，取水
draw-out breaker　抽出式断路器【可以是滚出式】
draw-out carriage　抽出式小车
draw-out metal-clad switchgear　抽出式金属封闭开关设备
draw-out relay　抽出式继电器
draw-out switchboard　抽出式配电盘
draw-out tool　抽出工具
draw-out track　牵出线
draw-out type breaker　抽出型断路器
draw-out unit　抽出单元，可抽取段
draw-out　抽出，引出
draw period　抽降期【地下水】
draw rod　拉杆，连杆
draw span　仰开跨
draw the vacuum　抽真空
draw-tongs　紧线钳
draw up a contract　草拟合同，拟订合同，起草合同
draw up an agreement　订协议
draw-well　提水井
draw　牵引，吸引，吸进，抽出，绘图，绘制，拉，提款，拖，曳引
drayage charge　拖运费
drayage　集卡运费，短驳费
DRB(disputes review board)　争端评审委员会
DR(data receiver)　数据接收器
DR(data recorder)　数据记录器
DR(data register)　数据寄存器
DRD(deep rock disposal)　（放射性废物）深岩层处置
DR(Deviation Reports)　偏差项报告
DR(digital resolver)　数字分解器
D/R(direct or reverse)　正向或反向的，正面或方面的
DR(dock's receipt)　场站收据
D reactor(deuterium-deuterium reactor)　重氢-重氢反应堆
dredged channel　疏浚过的航槽
dredged trench　挖泥槽
dredge hopper　挖泥斗
dredge level　疏浚标高
dredge machine　挖泥机
dredge pipe　吹泥管
dredge pump　排污泵，挖泥泵，吸泥泵
dredger bucket　挖泥机戽斗
dredger fill　吹填土
dredger　挖泥机，挖泥船，挖泥工
dredge spoil　疏浚弃土
dredge　清淤，疏浚，挖掘（泥土），挖泥机
dredging bit　匙形钻【用于散粒土】
dredging buoy　疏浚浮标
dredging machine　疏浚机，挖泥机
dredging plant　灰浆泵房
dredging shovel　挖泥铲，挖泥机
dredging tube　吸泥管
dredging well　挖泥井
dredging　清淤，疏浚，挖泥
dregginess　沉积物，沉渣，含渣量，沉渣沉淀物
dregs　渣滓
drencher fire extinguisher　水幕式灭火器
drencher system　水幕除尘系统，水幕系统
drench　浸湿，浸润

dressed-ashlar 细琢石
dressed brick 磨光砖
dressed masonry 敷面圬工，细琢石圬工
dressed stone 细琢料石
dressed stuff 修饰过材料
dresser 修整器
dressing agent 研磨剂
dressing by washing 洗选
dressing hammer 琢面锤
dressing of cable 电缆包扎
dressing 研配，磨合，修整，梳理，加工，选矿，修琢
dress 覆盖物，清理工作面
DRG(double reduction gear) 双减速齿轮
dribbing 零星修补
dribble pipe 泄放管
dribble 滴，滴下，滴水，滴水沟
drier efficiency 干燥器效率
drier-reheater 去湿再热器
drier sol 速干剂
drier unit 干燥器，气体回路干燥装置
drier 干燥剂，干燥器，干燥机
driest period 最干旱期，最枯水期
drift accumulation 吹雪堆积，吹沙堆积
driftage 平洞推进
drift anchor 浮锚
drift angle 滑角，偏移角，偏差角
drift area 流沙区，吹雪区
drift barrier 拦浮栅，漂浮物栏栅
drift bottle 浮标瓶，瓶式浮标
drift boulder 漂砾
drift carrier 漂移载流子
drift clay 漂积黏土
drift coefficient 偏流系数
drift compensation circuit 漂移补偿电路
drift compensation 漂移补偿
drift control 堆雪控制，流沙治理
drift correction 漂移校正
drift current 风吹流，漂移电流
drift deposition 流沙沉积，堆雪沉积
drift deposit 冲积物
drift drop plume 漂滴羽流
drifted soil 吹扬土
drift eliminator 滴水器【冷却塔】，除水器，水滴分离器
drifter drill 风钻钻机
drifter machine 凿岩机
drift error 偏流差，漂移误差
drifter 风钻工，漂流物
drift face 工作面前壁
drift field 漂移电场
drift gauge 偏差计
drift heading 导坑
drift ice 流冰
drifting dust 吹尘，低吹尘，飘尘
drifting electron 弥散电子
drifting sand filter 漂沙快滤池
drifting sand 流沙，风积沙，吹沙，低吹沙
drifting snow 吹雪，堆雪，雪暴，积雪
drifting 漂流，偏航
drift log 流放的木材，流送材
driftmeter 漂移测量器，偏差器，漂移计，偏差计
drift method 导洞掘进法
drift meter 偏流计
drift mining 导洞开采
drift of particles 颗粒漂移
drift of zero 零点漂移
drift pattern 漂沙形式，堆雪形式
drift pin 冲头，销子，心轴，锥枢
drift punch 冲头
drift rate 偏移率
drift sand 冲积沙，风积沙
drift smoke 漂移烟
drift snow transport 吹雪输运
drift stabilization 漂移稳定，漂移补偿
drift transistor 漂移晶体管
drift type photovoltaic device 漂移型光伏器件
drift vector 漂流矢量
drift velocity 漂流速度，漂移速度
drift volume 吹扬体积
drift water 挟沙水
drift wood 流放的木材，流送材
drift 漂移，偏移，偏差，偏流，偏转，漂动，吹集，吹扬，风沙流，风雪流
drillability 可钻性
drill and blast method 钻爆法
drill bar 钻杆
drill bit 钻头
drill blasting 钻孔爆破
drill-blast tunneling method 隧洞钻孔爆破法
drill carriage 钻架台车
drill chuck 钻头夹盘
drill column 钻机支柱
drill core 空心钻，岩芯钻，钻探试样，岩芯，钻孔岩芯
drill cuttings 钻井岩屑
drilled base plate 带孔底板
drilled caisson 管柱，钻孔沉井，钻孔桩
drilled drain hole 排水钻孔
drilled footage 钻进尺寸
drilled hole standard 钻孔标准
drilled pier 钻孔墩
drilled well 钻井
drilled 有孔的
driller 钻孔者，钻孔机，钻孔人员，钻床，钻机
drill extractor 取钻器
drill gauge 钻规
drill hole axis 钻孔轴
drill hole log 钻孔岩心记录
drill hole 镗孔，钻孔
drilling and coring 钻孔与取圆形岩石样品
drilling and reaming unit 钻孔和铰孔组合
drilling carriage jumbo 隧道钻车
drilling electric equipment 钻井电气设备
drilling engineering 钻探工程
drilling equipment 钻孔设备
drilling fluid 钻孔清洗液，钻探泥浆
drilling hammer 打眼锤
drilling jig 手摇钻，钻模
drilling jumbo 钻架车
drilling log 钻探记录

drilling machines bench 钻床工作台
drilling machines high speed 高速钻床
drilling machines multi-spindle 多轴钻床
drilling machines radial 摇臂钻床
drilling machines vertical 立式钻床
drilling machine 钻床
drilling and milling machine 钻铣床
drilling motor 钻探电机
drilling mud 钻孔泥浆
drilling pattern 钻孔型式
drilling plan 钻孔平面
drilling platform 钻井平台【海上】，钻架台车
drilling record 钻进记录
drilling rig 钻台，钻机，钻探设备[装置]，钻具，凿岩机，钻架
drilling rod 钻杆
drilling template 钻孔模板
drilling time log 钻进记录
drilling unit 穿孔装置
drilling 钻井，钻探，穿孔，钻进，钻孔，钻孔取样
drill log 钻孔柱状图
drill post 钻机支柱
drill record 钻探记录
drill screw 钻螺纹
drill seeding 条播
drill socket 钻头插口，钻头套筒
drill steel 钻钢
drill stem 钻杆
drill test 钻孔试验
drill tower 钻塔
drill 钻孔，钻头，锥子，钻孔机，钻床，钻，钻机
drinking fountain 饮水龙头，饮水器
drinking water standard 饮用水标准
drinking water 饮用水
drink in 吸收，吸液
drip chamber 排水室，沉淀池，沉淀室
drip cock 排污阀，泄放阀
drip collector 采酸管，液滴收集器【液滴箱，盘等】
drip cup 承油杯
drip ledge 渣栏【液态炉】
drip moulding 滴水线
drip-off 液滴
drip opening 出渣口【液态炉】
drip pan 接油滴的盘子，承滴槽，承盘
dripping cup 承油杯
dripping device 淋水装置
dripping fault 滴水断层
drip pipe 冷凝水泄出管，泄放管，排出管，滴管
drip-proof insulator 防滴式绝缘子
drip-proof machine 防滴水电机，防滴型电机
drip-proof motor 防滴型电动机
drip-proof screen-protected machine 防滴网罩防护型电动机
drip-proof type 防滴式
drip pump 排水泵，疏水泵，小型滴漏水抽水机
drip recovery 液滴回收
drip ring 排液环
drip shield 液滴挡板
dripstone 滴水石
drip-tight 防溅
drip trap 滴阀，滴水阀，冷凝水槽，滴水收集器
drip valve 集液排放阀
drip 滴，水滴，滴水器，滴水槽，浸湿，屋檐
drive assembly 传动装置[组合]
drive axle 传动轴，主动轴
drive bar gate 分流闸门
drive cable 带动裂变室电缆
drive cam 主凸轮
drive chain 主动链
drive circuit 驱动电路
drive clutch 传动离合器
drive coil 驱动线圈
drive control relay 切换继电器
drive coupling 联轴器传动
drive-down 减速
drive end 驱动端
drive factor 传动系数
drive fit 打入配合
drive flange 传动法兰盘
drive friction 驱动摩擦力
drive gear box 传动齿轮箱
drive gear 主动齿轮，传动机构，驱动齿轮
drivehead 承锤头，桩帽
drive house 驱动站
drive housing 传动箱，驱动机构外壳
drive installation 驱动装置
drive-in 打入
drive line assembly 驱动线组件【控制棒组件和驱动杆】
drive link 传动杆
drive magnet 驱动磁铁，励磁磁铁，激励磁铁
drive mechanism 驱动机构
drive motor 驱动电动机，拖动电动机，驱动马达
driven cast-in-place pile 沉管灌注桩，灌注桩
driven coupling half 被动联轴盘
driven depth 钻入深度，掘进深度
driven disk 从动轮，从动盘
driven element 激励单元，驱动子
driven flow 排送水流【沸水堆喷射泵】
driven gear 从动齿轮
driven impeller 涡轮【从动轮】
driven in-situ pile 沉管桩
driven link 从动炉排片
driven pile 打入的桩，打入桩，入土桩
driven pulley 从动滚筒
driven sender 主控振荡器，主控发生器
driven shaft 被动轴，从动轴
driven-tube cast-in-place pile 沉管灌注桩
drive nut 传动螺帽
driven welding roller 被动转台
driven well 机井
driven wheel 从动轮
driven 从动的，被激励的，受激励的，从动
drive package 驱动机构
drive pile 打桩
drive pinion 传动小齿轮，主动小齿轮

drive pin 传动销
drive pulley 驱动滚筒，传动滚筒，驱动皮带轮
drive pulse generator 驱动脉冲发生器
drive pulse 驱动脉冲
drive ratio 传动比
driver circuit 激励电路，主振荡电路
driver efficiency 传动轴效率
drive reinforced concrete piles 打钢筋混凝土桩
driver fuel 燃料活性段
driver gear 传动齿轮
driver motor 拖动电动机
drive rod assembly 驱动杆【控制棒驱动机构】，驱动轴组件
drive rod guide 传动杆导口【控制棒驱动机构】
drive rod travel housing 驱动轴行程套管
drive rod 传动杆【控制棒驱动机构】
drive rolls 传动滚筒【供焊丝】
driver pinion 主动小齿轮
driver sweep 驱动器扫描
driver tube 控制管，激励管
driver zone 活性区，驱动区【反应堆】
driver 驱动电机，传动器，驱动器，激励器，主动轮，末级前置放大器，传动轮，司机
drive sampler 击入式取土器，击入式取样器
drive screw 传动螺杆
drive seal 驱动密封
drive shaft bearing 主动轴轴承
drive shaft guide tube 驱动轴导向管【控制棒】
drive shaft handling tool 驱动轴装卸工具
drive shaft material 传动轴材质
drive shaft storage rack 驱动轴贮存架【控制棒】
drive shaft throat 传动轴颈
drive shaft unlatching tool 驱动轴拆卸工具【控制棒】
drive shaft unlatching 传动轴不闭锁
drive shaft 主动轴，驱动轴，传动轴
drive shoe 套管管靴，桩靴，打入管下端引鞋
drive side 驱动端，主动端
drive signal 驱动信号，控制信号
drive-socket wrench 传动套筒扳手
drive sprocket 主动链轮，驱动链轮，传动链轮
drive stage 激励级，驱动级
drive stud 传动螺杆
drive system 传动系统
drive-through access 通车便道
drive tool 传动工具【螺丝起子，扳手等】
drive torque 传动力矩，驱动力矩
drive train bearing 传动系统轴承
drive train 传动系统，驱动系统，牵引车
drive unit 驱动机组【钠冷快堆辅助停堆系统】，驱动装置，传动装置，驱动件
drive water 驱动水
driveway 车道
drive winding 励磁绕组
drive 驱动器，传动，驱动，激励，拖动，策动，驾车，驱动机构，拧紧，驱动装置，带动
driving axle 传动轴，主动轴，传动带，传动皮带
driving by AC motor 交流电动机拖动
driving by asynchronous motor 异步电动机拖动
driving by induction motor 感应电动机拖动
driving by linear motor 直线电动机拖动
driving by synchronous motor 同步电动机拖动
driving by universal motor 交直流两用电动机拖动
driving chain 主动链
driving cone pulley 主动塔轮，主动锥形轮
driving coupling half 主动联轴盘
driving depth 打桩深度
driving device 驱动装置
driving disk 主动轮
driving drum 驱动滚筒
driving element 激励元件，传动元件
driving energy 打桩能量
driving fit 打入配合
driving flow 喷射水流，驱动水流，驱动流
driving fluid 工作液体
driving force 推动力，动力，驱动力
driving gear 传动机构，驱动装置，传动齿轮，传动装置，主动齿轮
driving hammer 桩锤
driving head 驱动压头
driving helmet 桩帽
driving impeller 泵轮【主动轮】
driving link 主动炉排片
driving magnet 激励磁铁，驱动磁铁，传动磁铁
driving mechanism 传动机构，驱动机构，主动机构
driving motor 拖动电动机，驱动电动机
driving of sheet piling 打板桩
driving pile abutment 打桩台
driving-point admittance 驱动点导纳，策动点导纳
driving-point impedance 驱动阻抗，输入阻抗
driving potential 驱动势
driving power 励磁功率，激励功率，驱动功率，牵引能力，推进力
driving pulley 传动滚筒
driving pulse 驱动脉冲，启动脉冲
driving rain index 大雨指数
driving rain 大雨
driving resistance 打入阻力，桩的打入阻力
driving roller and control box 主动转辊及控箱
driving roll 主动辊
driving shaft 驱动轴，主动轴，传动轴
driving signal 驱动信号
driving snow 大风雪
driving speed 行车速度
driving steam source 驱动气源
driving stroke 工作冲程
driving synchro 主动同步机
driving test 打桩试验
driving torque 传动转矩
driving voltage 控制电压，激励电压，励磁电压
driving wheel 主动轮
driving winding 激励绕组，策动绕组
driving 驱动激励，激振，传动，推动，驾驶，操纵
drizzle fog 毛细雨
drizzle 毛毛雨，细雨
drizzling rain 毛毛雨
D-rod bank D 棒组，多普勒棒组

DRO(destructive read out) 破坏性读出
drogue 浮标，海流板，锥形风标
droop characteristic 下垂特性，下降特性
droop control 有差调节，有差控制
drooping 减弱【电源输出特性】
droop of speed 速度变动率
droop setter 不等率调整装置
droop setting 调差
droop snoot blade 前缘下垂式叶片
droop snoot 前缘襟翼
droop 不均匀度，不等率，下降，压降，固定偏差，静差，下垂，衰减，弯曲，转速不等率
drop accumulator 液滴聚集器
drop anchor 抛锚
drop annunciator 掉牌警报信号
drop arch 垂拱，平圆拱
drop arm 转向臂
drop-away button 脱扣按钮
drop-away current 脱扣电流【继电器】
drop-away reset 脱扣复位
drop-away time 脱扣时间
drop-away voltage 脱扣电压
drop-away 脱扣
drop bar 接地棒，短路棒
drop bottom 活络底板
drop bucket 吊斗
drop chute 跌水陡槽
drop-dead halt 完全停机，突然停机
drop-down curve 下降曲线
drop-down section 跌水段
drop energy dissipator 跌水消能工
drop forging 模锻，落锤锻
drop gate sluice 吊门（水）闸
drop gate 升降式闸门
drop hammer pile driver 落锤式打桩机
drop hammer pile driving method 落锤打桩法
drop hammer 落锤，重力打桩锤
drop hanger 传动轴吊架，吊钩
drop hardness test 坠落硬度试验
drop height 落差
drop-impact penetrometer 锤击贯入仪，落锤冲击贯入仪
drop indicator relay 跌落指示器，掉牌继电器，继电器，报警器
drop-indicator shutter 呼叫掉牌
drop inlet dam 跌水进口涵洞式土坝
drop inlet spillway 竖井跌水式溢洪道
drop inlet 跌水进口
drop-in level 高差
drop-in speed 速度下降
drop-in temperature 温度下降
drop in water surface 水面降落
drop-in winding 下进槽内的模绕绕组
drop-in 励磁【继电器，接触器】，落入，杂音信息，意外出现
droplet condensation 滴状凝结
droplet entrainment 液滴夹带
droplet evaporation 液滴蒸发
droplet mass transfer coefficient 液滴传质系数
droplet mechanics 液滴力学
droplet redeposition 液滴再沉积
droplet separator 分滴器，液滴分离器
droplet size 微滴大小
droplet 熔滴【电弧焊】，小滴，飞沫，小水滴，微滴
drop light 吊灯
drop line 深井井管
drop manhole 跌水窨井
drop-nose arch 下垂的折焰拱
drop number 跌水系数
drop-off 流出，脱离，松落，下降
drop of potential 电压降，电位降
drop of pressure 压降
drop of step 失步，不同步
drop of voltage 电压降
drop-out current 开断电流
drop-out fuse 跌落式熔断器，跳开式熔断器
drop-out line 紧急放空线，排泄线
drop-out of step 失去同步
drop-out-pick-up ratio 释放-吸合比
drop-out time 释放时间
drop-out value 下降值
drop-out voltage 灭磁电压，动励磁电压，开断电压
drop-out 脱落，下降，回动，落下，退出，信号丢失，信息由“1”变“0”
dropped ceiling 吊顶
drop penetration test 落锤贯入试验，落穿试验
dropper 滴管，使滴下的东西，点滴器，延长架
drop pile hammer 重力打桩锤
dropping bottle 滴瓶
dropping characteristic 下降特性，下降特性曲线
dropping path 下降通道【控制棒】
dropping point 滴点，初馏点
dropping resistor 减压电阻器，降压电阻器
dropping speed motor 具有下降速度特性的电动机
dropping voltage generator 降压特性发电机
dropping 落棒【控制棒】，跌落
drop pipet 滴液移液管
drop pipe 水井竖管，深井井管，下垂喷管，下落管【液体】
drop pit 集水井，整修机车底部的地坑，修车坑，跌水坑
drop relay 跌落继电器，脱扣继电器，掉牌继电器，报警器
drop shaft 跌水竖井
drop-shaped 滴状的
drop test 低压试验，冲击试验，降压试验，点滴试验
drop time 降落时间，落棒时间，落控制棒时间
drop tube furnace 垂直下降炉，滴管炉【指煤的特性研究装置】
drop type fuse link 跌落式保险丝管
drop valve 坠阀
drop weight nil ductility transition test 落重零延性转变试验
drop weight tear test 落重扯裂试验
drop weight test 落锤试验
drop well 跌水井
drop wire 用户引入线
dropwise condensation coefficient 滴状冷系数

dropwise condensation　珠状凝结，球状凝结，滴状凝结，滴状冷凝
dropwise　点滴的，滴状的
drop　滴，落下，降低，下降，落差
drosometer　露量器
dross　熔渣，浮渣，残屑，泡沫，氧化，碎屑
dross coal　下黏结性煤，劣质煤
drought duration　干旱历时
drought forecast　旱情预报
drought period　干旱期，旱期，枯水期
drought season　旱季
drought　干旱，旱季，旱灾，缺乏【长期的】，枯竭
drove　石凿
drowned coast　溺岸，淹没岸
drowned estuary　沉没河口
drowned flow　淹没流
drowned hydraulic jump　淹没水跃
drowned nappe　淹没水舌
drowned pump　潜水泵，深井泵
drowned reef　沉没礁
drowned river mouth　沉没河口，沉溺河口
drowned river valley　沉没河谷
drowned river　沉溺河
drowned shoreline　沉溺岸线
drowned stream　沉溺河
drowned valley　溺谷
drowned weir　潜堰
drown　淹没
DRR(deterioration rate reduction)　恶化速率下降
DRS(double right shift)　双倍右移
DRT(diode recovery tester)　二极管再生测试器
drum armature　鼓形电枢
drum axle　滚筒轴
drum baffle　汽包挡板
drum barrel　鼓筒，线盘，绕线架
drum blender-reclaimer　筒型混匀取料机
drum blender　滚筒式混料机
drum boiler　汽包锅炉，锅筒锅炉
drum breaker starter　鼓形断路器
drum channel output bus　磁鼓通道输出总线
drum closing　放射性废物桶封盖
drum clutch　鼓形离合器
drum coils　筒形线圈
drum controller　鼓形控制器
drum course　汽包筒节
drum digger shield　圆筒形掘进盾构，圆筒形掘进铠框
drum dump　磁鼓内容的读出或打印，磁鼓信息转储
drum end plate　鼓端板
drum end　汽包封头
drum erection　锅筒就位
drum feeder　鼓轮式给煤机，鼓轮式给料机，滚筒式给煤机
drum feed piping　汽包给水管系
drum filling station　装桶站
drum filter　鼓式过滤机，圆筒过滤机，滤鼓
drum float　鼓形浮标
drum gate　鼓形闸门，圆筒形闸门
drum handling machine　转筒【增殖堆】
drum head　汽包封头，汽鼓封头
drum heater　鼓式加热器
drum information display　磁鼓信息显示
drum internals　汽鼓内件，汽包内件，汽包内部装置，汽鼓内部装置
drum level television　汽包水位电视
drum level TV system　汽包水位电视系统
drum lid handling device　桶顶盖操作装置【废物收集器】
drum lifting date　汽包吊装日期
drum lifting　汽包的吊装
drumlin　鼓丘
drum magnetic separator　滚筒式磁铁分离器
drum manhole　汽包人孔
drum memory　磁鼓存储器
drumming area　装桶场地
drumming cell　装罐间
drumming room　装桶间
drumming station　装桶站【放射性废物】
drumming　装桶，装罐
drum mixer　鼓形混合器，筒式搅拌机
drum operating pressure　汽包工作压力
drum parity error　磁鼓奇偶误差
drum pressure　汽包压力
drum printer　鼓式打印机
drum reclaimer　滚筒取料机
drum roller　滚筒式碾路机
drum roll　传动滚筒
drum rotor　鼓形转子，鼓筒式转子
drum scanner　鼓形扫描器
drum screen　鼓形筛，鼓形筛网，圆筒筛
drum separator　鼓式分离器
drum shell　汽包筒身，锅筒筒身
drum sifter　圆筒筛
drum sluice　鼓形泄水闸门
drum starter　鼓形启动器
drum storage area　金属桶储存区【废物】
drum storage　磁鼓存储器，桶罐贮存场
drum store　桶贮存库【废物】
drum stub　汽包管接头
drum switch　鼓形开关
drum trap　鼓式存水弯，鼓式凝气阀
drum type boiler　锅筒式锅炉，汽包式锅炉
drum type concrete mixer　鼓形混凝土拌和机
drum type controller　鼓形控制器
drum type dryer　鼓式干燥机
drum type mixer　筒式搅拌机
drum type reclaimer　滚筒式取料机
drum type rotary feeder　鼓轮式给料机，鼓轮式给煤机
drum type rotor　鼓型转子
drum type surface attemperator　筒式表面减温器
drum type surface cooler　筒形面式冷却器
drum water meter　鼓式水表
drum weir　鼓形闸门堰
drum winch　筒式卷扬机
drum winding　鼓形绕组
drum-wound armature　鼓形绕组电枢
drum-wound　鼓形绕线的
drum　电缆盘，线盘，汽包，锅筒，汽鼓，圆筒，鼓，鼓形物，磁鼓，鼓形的，滚筒，卷

筒，桶
dry 干的，干燥的
dry active waste 干放射性废物
dry adiabatic equation 干绝热方程
dry adiabatic lapse rate 干绝热递减率
dry adiabatic process 干绝热过程
dry adsorption 干吸附
dry air 干空气
dry and rainless 干燥少雨
dry and wet bulb hygrometer 干湿球温度表
dry and wet bulb thermometer 干湿球温度计
dry and wet combined cooling system 干湿式联合冷却系统
dry arch 防潮基拱
dry arcing distance 干弧距离
dry arc-over 干闪络
dry ash extraction 固态排渣
dry ash free basis analysis 无水，无灰基分析（法）
dry ash free basis 干燥无灰基，无灰干燥基
dry ash furnace 固态排渣炉膛
dry-ashing 干灰化
dry ash loading chute 干灰卸料器
dry ash pipe 干灰管
dry ash removal 干式除灰
dry ash unloader 干灰卸料机，干灰卸载装置
dry ash 固态渣，干灰
dry-back boiler 干背式锅炉
dry baghouse dust collection 干式除尘器
dry-basis 干燥基，干基，干燥质
dry batched concrete 干拌和混凝土
dry battery 干电池组
dry blast 鼓风
dry bond macadam 干结碎石路
dry bottom ash hopper 冷灰斗
dry bottom boiler 固态排渣锅炉
dry bottom firing system 固态排渣燃烧方式
dry bottom furnace 固态排渣炉，固态排渣炉膛
dry bottom hopper 冷灰斗
dry bottom 冷灰斗，固态排渣炉底
dry box 干燥箱
dry breather 干燥呼吸器
dry bridge 跨线桥
dry buckling 干（栅格）曲率
dry bulb temperature 干球温度
dry bulb thermometer 干球温度表，干球温度计
dry-burning coal 不结块煤，不黏结煤，贫煤，瘦煤
dry cell 干电池
dry centrifugal collector 干式离心除尘器
dry charging 干装料
dry chemical fire extinguisher 干粉灭火器，粉剂灭火机
dry circuit contact 无开关触点，干电路触点，低电平触点
dry circuit load 干电路负载
dry circuit relay 干电路继电器
dry circuit 小功率电路，弱电流电路
dry climate 干燥气候
dry coal shed 干煤棚
dry coal 干煤
dry collection system 干式集尘装置
dry collection 布袋除尘器
dry compacted weight 密实干重，干压容重
dry concrete 干硬性混凝土
dry condenser 干式冷凝器
dry connection 干接合
dry consistency 干硬度
dry construction 干地施工
dry contact signal 干接点信号
dry contact 干状态时起弧距离，干触点，固定触点
dry conversion process 干转化工艺
dry cooling condensing installation 干式冷却凝汽装置
dry cooling condition 干工况
dry cooling system 干式冷却系统
dry cooling tower 干式（通风）冷却塔，干冷塔
dry cooling turbine 空冷式汽轮机
dry cooling 干式冷却，干式熄焦
dry core cable 空气纸绝缘电缆，干芯电缆
dry corrosion 干腐蚀，干式腐蚀，干蚀
dry criticality 干临界
dry critical 干临界的
dry curing 干养护
dry density of soil 土壤干密度
dry density 干燥密度，净质密度，干密度，干容重
dry deposit flourescent penetrant 干沉积的荧光渗透剂
dry deposition rate 干沉积率
dry deposition 干沉积
dry desulphurization 干法脱硫
dry disk clutch 干式摩擦离合器
dry distillation 干馏
dry dock lock 干坞坞闸
dry drag force 干拖曳力
dry drill method 干钻法
dry drum 干汽包【防汽水共腾】
dry dust collection system 干式除尘系统
dry dust collector 干式除尘器
dry dust separator 干式除尘器
dry electrolytic capacitor 干电解质电容器
dryer basket 干燥器组件
dryer seal skirt 干燥器密封裙
dryer 干燥剂，干燥器，干燥机，汽水分离器
dry excavation 干挖土
dry expansion evaporator 干式蒸发器
dry fallout 干放射性散落物
dry feed dosage 干加剂量
dry feeder 干加料器
dry feed 干式送料
dry filter 干燥过滤器，干式过滤器
dry flashover test of insulator 绝缘子干闪络试验
dry flashover voltage 干闪络电压
dry fog type dust suppression system 干雾型灰尘清除系统
dry friction 干摩擦
dry fuel basis 干燥基
dry gas holder 干式气柜
dry gas loss 干烟气热损失
dry gas meter 干气体计量计，液化气体计量计

dry grinder 干磨机
dry grinding 干磨【水泥】
dry haze 飞尘
dry hole 干炮眼，上斜炮眼
dry ice 干冰
drying agent 干燥剂
drying air fan 干燥风机
drying air heater 干燥空气预热器
drying bed 干燥床
drying breather 干式呼吸器【变压器】，密封式呼吸器
drying cabinet 干燥箱
drying chamber 干燥室
drying condition 干燥条件
drying crack 干裂
drying effect 干燥作用
drying filter 干滤器
drying hood 干燥罩，烘干罩
drying machinery 干燥设备
drying mark 干燥痕迹
drying medium 干燥介质，干燥剂
drying method by vacuum and eddy current 真空涡热法
drying of electric equipment 电气设备的干燥
drying oil 干燥的油，快干油，干油漆，干性油
drying out period 干燥期，烘炉期
drying out 烘炉
drying plant 干燥设备
drying principles 干燥原理
drying process 干燥过程
drying roller 干燥滚筒
drying screen 百叶窗汽水分离器，百叶窗分离器，除水网，汽水分离器
drying shrinkage 干缩
drying space isolating valve 干燥空间隔离阀
drying system 烘干系统【柴油热风循环加热/恒温控制】
drying temperature 干燥温度
drying tube 干燥管
drying-up 枯竭
drying zone 干燥区
drying 干燥，水分分离，除水，干燥作用，干化，烘干
dry instrument transformer 干式仪用互感器
dry-insulation transformer 干式绝缘变压器
dry-insulation 干式绝缘的
dry jet mixing 粉体喷搅法
dry joint process 干挂法
dry joint 干接合，干接（缝），干挂，接触不良的接头，虚焊，假焊
dry-laid masonry dam 干砌石坝
dry-laid masonry wall 干砌墙
dry-laid masonry 干砌石圬工，干砌圬工
dry-laid rubble masonry 干砌毛石圬工
dry-laid rubble 干砌毛石
dry-laid stone 干砌石
dry lattice 干栅格
dry-layer type roll filter 干层式滚筒过滤器
dry layup 干保养【蒸汽发生器】，干式贮存
dry lighting impulse withstand voltage test 干雷电冲击耐压试验
dry-lubricated bearing 干润滑轴承
dry maintenance 干维修【强放射性设备】
dry masonry dam 石谷坊，石砌坝
dry material 干料
dry matter 干渣
dry method of desulfurization 干法脱硫
dry method 干法
dry mineral-free basis 干燥无矿物质基
dry mineral-matter-free basis 干燥无矿物质基，无矿物干基
dry mixed concrete 干拌和混凝土
dry mixing 干拌，干混合
dry mix 干拌和物，干硬性拌和物
dry mortar 干砂浆
dry natural gas 干天然气
dryness factor 干燥度，干燥率
dryness fraction of steam 蒸汽干度
dryness 干度，干燥度
dryout heat flux 干涸热流密度
dryout margin 干涸裕度
dryout point 干涸点
dryout procedure 烘干处理
dryout ratio 干涸比
dryout 烧干【蒸汽发生器】，烘干，蒸干，析出，干涸
dry-packed concrete 干填混凝土
dry packing method 干填法
dry paper insulated cable 空气纸隔绝缘电缆
dry partition 隔断墙，干燥间壁
dry patch （蒸发器管束的）堵塞现象，干斑，堵塞现象
dry pellets 干粒
dry period 干涸阶段【失水事故】，干旱期，枯水期
dry pipe valve 过热蒸汽管道阀门
dry pipe 干汽管，过热蒸汽输送管
dry pitching 干砌护坡
dry pit pump 干坑泵
dry pit 枯井
dry-placed fill 干填土
dry-plate rectifier 干式整流器，干片整流器
dry powder fire extinguisher 干粉灭火器
dry precipitator 干式除尘器
dry preservation 干法保养，干式保护
dry-pressed brick 干压砖
dry pressing 干压
dry process coal cleaning 干法选煤
dry process 干处理法【燃料循环】，干法
dry purification 干法提纯
dry rectifier 干式整流器，干片整流器
dry-reed contact 干簧触点
dry-reed relay 干簧继电器
dry-reed switch 干簧开关
dry refueling 干法换燃料
dry reprocessing 干法后处理
dry residue 干残渣，蒸发残渣
dry return pipe 干式凝结水管
dry riser 气口
dry rock paving 干（石块）铺砌
dry rodded aggregates 捣实的骨料
dry route 干处理法，干转化法【燃料循环】

dry rubble masonry 干砌块石圬工，干砌毛石圬
dry rubble 干砌块石
dry run protection 无水运行保护，干燥保护
dry run 干试车【反应堆在装燃料前的试车】
dry salting 干盐化
dry saturated steam 干饱和蒸汽
dry saturated vapor 干饱和蒸汽
dry screen 干筛
dry scrubber 干式烟气脱硫装置
dry season runoff 枯季径流
dry season 干季，旱季，枯水季，枯水期
dry separator 汽水分离器，干式除尘器
dry sieving 干筛分
dry sipping facility 干啜吸装置【燃料元件破损检测】
dry sipping technique 干啜吸法
dry sipping 干啜吸，干啜吸技术
dry slag removal system 干式除渣系统
dry sludge 干燥污泥
dry snow zone 干雪地带
dry solids 干燥物质【水分析】
dry spell 干旱期，无水期
dry sprinkler system 干式喷水灭火系统
dry steam drum 干蒸汽室，干汽包
dry steam humidifier 干蒸汽加湿器
dry steam 干蒸汽，干饱和蒸汽
dry stone drain 干砌石排水沟
dry stone fill 干填石
dry stone masonry 干砌石圬工
dry stone pitching 干砌石护坡
dry stone wall 干砌墙
dry storage battery 干电池组
dry storage 干式保养【停炉】，干保护法，干法贮存
dry stove 烘箱
dry strength 干强度
dry suspended solids 干悬浮固体，干悬浮质
dry suspension reactor 干悬浮反应堆
dry-tamped concrete 干捣实混凝土
dry-tamping method 干捣（实）法
dry-tamping 干捣
dry test on the surface of the insulation 绝缘表面干试验
dry test 干试验
dry the arc room with nitrogen 用氮气干燥灭弧室
dry tower cooling system 干塔冷却系统
dry transport 干运输【乏燃料】
dry type capacitor 干式电容器
dry type containment vessel 干式安全壳，干井
dry type containment 干式安全壳
dry type cooling tower 干式冷却塔
dry type dust collector 干式除尘器
dry type filter 干式过滤器
dry type rectifier 干式整流器，固体整流器
dry type transformer 干式变压器
dry type voltage regulator 干式电压调整器
dry unit weight 干容重
dry valley 干谷
dry valve arrester 干阀避雷器，阀阻避雷器
dry valve shaft 干阀井
dry vent 干式通气管
dry volatility 干挥发度
dry-volume measurement 干容积测量法
dry wall 清水墙
dry washing 干式洗涤，干式净化
dry water meter 干式水表
dry way 干法
dry wear 干磨损
dry weight 净重，自重，干重
drywell emergency cooling system 干井应急冷却系统
drywell hatch cover 干井舱盖【安全壳】
drywell spray cooling pump 干井喷淋冷却泵【安全壳】
drywell spray cooling system 干井喷淋冷却系统
drywell spray ring 干井喷淋环管【安全壳】
drywell spray system 干井喷淋系统
drywell sump 干井污水坑，干井集水坑
drywell 干井【安全壳】
dry-wet cooling tower 干湿式冷却塔
dry panel 干盘
dry wind 干风
dry work 干地施工
dry year 干旱年，枯水年
dry zone 干旱（地）带
DSAC(digital simulated analog computer) 数字仿真模拟计算机
DSA(differential strain analysis) 微分应变分析
DSB(data set block) 数据集块
DSCB(data set control block) 数据集控制块
DSC(display station controller) 显示站控制器
DSCH(discharge) 排出
DS(data set) 数据集，数据装置
DS(decimal subtract) 十进制减法
DS(define symbol) 定义符号
DS(design specifications) 设计规范【文件编码用语】
DS (design standard) 设计标准
DS(desuperheater) 减温器，蒸汽冷却器
DS(device selector) 设备选择器
DS(direct subtract) 直接减
DSE(digital storage equipment) 数字存储设备
DSG(direct steam generator) 水工质塔吸热器【槽式太阳能发电】，直接蒸汽发生器
DSH(deactivated shutdown hours) 停用停机小时
DSH＝desuperheater 减温器
DSHE(downstream heat exchanger) 顺流热交换器
DSH 应急照明系统【核电站系统代码】
DSI 厂区保安系统【核电站系统代码】
DSL(data set label) 数据集符号
DSL(design safety limit) 设计安全限
DSM(demand side management) 需求侧管理
DSM(digital slave module) 数字从模件
DSM(dynamic stiffness modulus) 动态刚度模数
DSMG (designated system management group) 定名系统管理组
DSN(data smoothing network) 数据平滑网络
DSP(digital signal processor) 数字信号处理器
DSR(data set ready) 数据集就绪
DSR(digital stepping recorder) 数字步进式记录器
DSR(digit storage relay) 数字存储继电器
DSS(daily start-stop) 日启停

DSS(decentral converter substation) 地方换流站，地方变换所
DSS(decision support system) 决策支持系统
DSS(dynamic support system) 动态支援系统
DSTL＝distill 蒸馏
DSU(data synchronization unit) 数据同步器
DSU(device switching unit) 设备开关装置
DSU(digital storage unit) 数字存储器
DSW(data status word) 数据状态字
DSW(device status word) 设备状态字
DSW(downslope wind) 下坡风
DTB(dead tank circuit-breaker) 落地罐式断路器，罐式断路器
DT(data transmission) 数据传输
DT(destructive test) 破坏性试验
DT(digital technique) 数字技术
DTE(data terminal equipment) 数据终端设备
DTL(diode transistor logic) 二极管晶体管逻辑
DTL 闭路电视系统【核电站系统代码】
DTP(data transmission protocol) 数据传输协议
DTR(data terminal ready) 数据终端就绪
DTR(digital telemetering register) 数字式遥测记录器
D-T reactor(deuterium-tritium reactor) 氘氚热核反应堆
DTS(data transmission system) 数据传输系统
DTS(demineralized water treatment system) 除盐水处理系统
DTS(dispatcher training system) 调度员培训系统
DTV 厂区通讯系统【核电站系统代码】
dual absolute alarm card 双通道绝对值报警卡
dual-alkali-based chemicals 双碱基化学药剂
dual-axis excitation 双轴励磁
dual capacitor motor 双电容器式电动机
dual capacitor 双联电容器
dual carriageway road 复式车行道，双线车道
dual channel criterion 双通道准则
dual channel oscilloscope 双电子束示波器，双迹示波器，双通道示波器
dual channel regulator 双路调节器
dual circulation 二级循环锅炉，双循环，双回路
dual-coil motor 双线圈电动机
dual-coil relay 双线圈继电器
dual coincidence 双重符合
dual computer system 双计算机系统
dual control system 复式控制系统
dual control 双重控制，复式控制
dual conveyor system 双路皮带系统
dual cycle arrangement 双循环布置
dual cycle boiling water reactor 双循环沸水反应堆
dual cycle cooling system 双循环冷却系统
dual cycle forced-circulation-boiling-water reactor 双回路强制循环沸水堆
dual cycle plant 双循环装置
dual cycle reactor 双循环反应堆
dual cycle system 双循环系统，双回路系统
dual cycle 双循环，混合循环，双循环回路
dual-dielectric charge storage device 双介质电荷存储器件
dual-diffused MOS 双扩散金属氧化物半导体
dual discriminator 双甄别器
dual Doppler radar 双多普勒雷达
dual duct air conditioning system 双风管空气调节系统
dual duct system 双风管空气调节系统
dual electrical network 对偶电网络
dual-enrichment pellet 双浓缩芯块
dual excitation 双重励磁，双轴励磁
dual-feed 双路馈电，双端馈线
dual-field motor 复激电动机
dual film aspirating probe 双膜吸气式探头
dual-fired station 双燃料电厂
dual fission chamber 双裂变室
dual flow atomizer 双流体雾化器
dual flow burner 双流燃烧器
dual flow oil burner 双流体油燃烧器
dual flow wide range burner 双流宽范围调节燃烧器
dual flow 双流
dual-frequency dye laser 双频染料激光器
dual-frequency motor 双频电动机
dual-frequency sounder 双频测深仪
dual-fuel burner 可燃用两种燃料的燃烧器
dual-fuel engine 双燃料发动机
dual fuel system 二重燃料喷射装置
dual fuel 双燃料
dual gas-driven sample tube 两端气体推动的样品管
dual gas supply 双气源
dual glazing 双层玻璃
dual hot-wire method 双热线法
dual-in-line integrated circuit 双列直插式集成电路
dual-in-line package 双列直插式组件
dual internal water cooled generator 双水内冷发电机
dual internal water cooling 双水内冷
dualin 双硝炸药
duality theorem 对偶定理
duality theory 对偶理论
duality 对偶性，双重化，二元性，二重性
dualization of tripping circuit 双套制跳闸电路
dual leadership 双重领导
dual-loop system 双循环系统，双回路系统
dual-meter 两用表
dual-mode control system 双重方式调节［控制］系统
dual motor drives 双电机驱动
dual network 对偶网络，互易网络
dual nozzle burner 双喷嘴吹灰器
dual plate wafer type check valve 双板对夹式止回阀
dual polarity 双极性
dual power supply 双电源，双电源供电
dual pressure boiler 双压锅炉
dual pressure boiling water reactor 双压沸水堆
dual pressure condenser 双背压凝汽器
dual pressure steam turbine 双压式汽轮机
dual pressure 双压
dual-purpose medium 减速剂兼冷却剂介质，双效介质，双用途介质

dual-purpose motor　两用电机
dual-purpose nuclear power plant　核热电厂，两用核电厂
dual-purpose nuclear power station　核热电站，两用核电站
dual-purpose power plant　两用发电站
dual-purpose production　双重性生产【如供电和地区供热】
dual-purpose reactor　两用反应堆
dual-purpose　双重目的的，双重用途的
dual ramp ADC　双斜波模拟数字转换
dual-ratio transformer　双变比变压器
dual regulation　双重调节
dual reheat cycle　二次再热循环
dual reheat　二次再热
dual scaler card　双通道标度变换卡
dual scaler　双刻度定标器
dual search unit　分割式探头
dual-seat valve　双座阀
dual sensitivity voltmeter　双灵敏度伏特计
dual service　双源供电
dual speed induction generator　双速异步发电机
dual speed motor　双速电动机
dual speed squirrel cage induction generator　双速鼠笼式异步发电机
dual speed synchro system　双速同步系统
dual-temperature exchange　双温交换
dual-trace oscillograph　线示波器
dual-trace oscilloscope　线示波器
dual-trace recorder　二道记录仪
dual-track system　双轨制
dual unit　双驱动
dual use reactor　两用反应堆
dual valve　复式阀
dual vector　对偶矢量
dual vent　用通气管
dual-voltage fed　两种电压供电的
dual-voltage motor　双压电动机
dual volute　双蜗壳
dual-wall containment　双层安全壳
dual yaw servos　双摆伺服系统
dual　双重的，对偶的，二元的，双的，双路的，双数的
dubbing out　刮平
dubbin　防水油
dub hammer　锤子，榔头
duckwork　管网
duct attenuation　风道消声
duct bank　导管排，导管叠放
duct condenser　旁路电容器，耦合电容器
duct connection　连接管道
duct conveyer　管道输送机，管式输送器
ducted cooling　管道式冷却
ducted fan　涵道风扇
ducted oil cooler　管道式油冷却器
ducted wind turbine　涵道式风轮机
ducter　微阻器
duct fabrication　管道制作
duct fittings　风管接头，风管配件
duct grouting　导管灌浆
duct hopper　烟道灰斗
ductile-brittle transition temperature　延性脆性转变温度
ductile cast iron casing　球墨铸铁缸套，延性铸铁缸套
ductile cast iron　球墨铸铁
ductile crack　延性破裂
ductile failure　延韧性破坏，延性破坏
ductile fracture resistance　抗塑性破坏
ductile fracture surface　韧性断口
ductile fracture toughness　延性断裂韧度
ductile fracture　塑性断口，塑性破坏，塑性断裂，延韧性断裂，形变断裂
ductile gouging　延性腐蚀
ductile iron　球墨铸铁
ductile metal　韧性金属
ductile moment resisting space frame　抗塑性变形力矩空间框架
ductileness　延性
ductile rupture　韧性断裂，塑性破坏，塑性断裂
ductile to brittle transformation characteristics　韧脆转变特性
ductile to brittle transition behaviour　韧脆转变特性，韧性向脆性的转化特性
ductile yield　延性屈服
ductile　可锻的，可延展的，柔软的，韧性的，柔韧的
ductility limit　屈服点，延性限度，延性限
ductility machine　延度仪
ductility ratio　延伸率
ductility test　延展试验
ductility transition temperature　塑性转变温度
ductility transition　初性转变
ductility value　延展值
ductility　延展性【材料】，韧性，展性，柔韧性，延性，挠性
ductilometer　延性计，展度计
duct line　管道，电缆管道，电缆沟道，导管线
duct loss　管道阻力，管道损失
duct propagation　波导传播
duct region　叶列间隙
duct spacer　风道隔片
duct system design　风道系统设计
duct system　管道系统，风管网路，风道系统，导管系统
duct tube　元件盒
duct type　管道式，波导型
duct velocity　风道流速
duct-ventilated machine　风道通风电动机
duct wall　地沟墙
ductwork　管道系统，风管网路，管道，风管网
duct　电缆沟，管，管道，导管，风道，烟道，沟，管沟，槽，排料管，管子
dud cheque　空头支票
dud resistance　风道阻力
due authorization　正式授权
due consultation　充分协商
due course　适当的时候，适当的通道
due date for payment　付款截止日
due date　到期日，期头，支付日
due day　到期日

due diligence process　企业经营评估，尽职调查
due diligence report　评价报告【一般是委托律师所或会计师所做的】
due diligence　尽职调查【企业并购或合资前做的】，应有的注意
due payment　到期付款，定期付款，到期应付的款项
due process clause　正当程序条款
due-process principle　正当程序原则
due process　正当（法律）程序，法定诉讼程序，程序正义
due share capital and dividends　到期股金和股息
dues　费用，税，应付款，该缴款，应有的权利
due to　欠下债，由于，因为，应归于
duff　半腐层，煤粉，煤屑
dug foundation　掏挖基础
dugout　地下室，地下掩蔽部，地穴，防空洞
dull bit　钝钎头，钝钻头
dull brown coal　暗褐煤
dullish green　枯绿
dull market　萧条的市场
dull-red heat　暗红热
dull-red　暗红色的，赤热的
duly authorised representatives　正式授权代表，全权代表
duly authorized　经正式授权的
duly certified copy　正式核证的副本
duly in compliance with　完全符合
duly sign　正式签署
dumb arch　假拱
dumb waiter　小型送货升降台
dumb well　枯井，污水井，污水坑
dumet wire　铜包镍铁线
dumet　铜镍铁永磁合金
dummy antenna　假天线，仿真天线
dummy assembly　假组件【培训人员用】，模拟组件
dummy beach　消波滩
dummy bus　模拟母线
dummy cell　仿真电池，等效电池
dummy coil　死线圈，虚设线圈
dummy conductor　无载导线
dummy director　挂名董事
dummy element　假元件，模拟元件
dummy end stub　假尾支杆
dummy fuel element　假燃料元件
dummy fuse　模型熔断器
dummy gauge　补偿应变片
dummy hole　盲孔
dummy instruction　伪指令，空指令
dummy joint　假缝，伪焊缝，假结合
dummy load method　虚负载法
dummy load　假定载荷，虚负载，名义载荷，等效负载，虚荷载，假荷载
dummy-mesh current　等效回路电流
dummy model　假模型
dummy node voltage　等效节点电压
dummy node　哑节点，虚节点
dummy order　伪指令
dummy piston　平衡活塞
dummy plate　隔板
dummy print routine　伪打印程序
dummy reactor　模拟堆
dummy ring　平衡环，填密环
dummy rod　假燃料棒
dummy running　空转，惰走
dummy sealed source　假密封源
dummy shaft alignment　假轴找中
dummy shaft　假轴
dummy source　假放射源
dummy strut　假支杆
dummy stud　假螺杆
dummy support　假支架
dummy tube　假管
dummy type contraction joint　半开式收缩缝
dummy weld assembly　假的焊接组件
dummy　假的，模型，模拟的，虚构的，伪装物，虚设物，假程序，缓冲器，傀儡车
dump area　弃土场
dump barge　泥驳，自卸泥驳
dump basin　排水槽，排水池
dump body　翻斗车身
dump bottom wagon　底开车
dump bucket　翻斗，卸料斗
dump buggy　小翻斗车
dump cart　垃圾车，倾卸车，翻斗车
dump car　自动卸料车，翻斗车，自卸汽车
dump circuit　排放回路，排空回路
dump condenser　卸载旁通凝汽器，事故排放冷凝器
dump cycle　翻卸周期
dumped fill　倾倒填土
dumped rockfill　抛填堆石，抛石体
dump energy　抛弃能量，储存能量，剩余电能，剩余能量，过剩电能
dump hopper　卸料斗
dumper rotating cradle　翻车机转笼
dumper system　翻车机系统
dumper　翻车机，翻斗车，卸货车，卸料车，自卸料车
dump gate　卸载阀
dumping bar　可翻转炉排片
dumping car　自卸车
dumping crane　卸料起重机
dumping cycle　翻卸周期
dumping device　卸料器
dumping facility　卸料设备
dumping grate　翻转炉排，卸灰炉排，燃尽炉排
dumping ground　堆集场【放射性废物】，垃圾倾倒场，抛泥地，卸料场
dumping height　卸载高度
dumping hopper　翻卸斗，自动翻斗
dumping intensity　抛投强度
dumping into sea　废物投海处理
dumping line　卸车线
dumping plate　翻转板，卸灰板
dumping price　倾销价格
dumping rock-blocking still in advance　预抛栏石坝
dumping sequence　翻卸程序
dumping site　堆积场，堆料场，抛泥地，卸料场
dumping track　卸车线

dumping truck 翻斗汽车
dumping 卸灰，排放，倾卸，翻转，放射性废物的倾倒，卸料，侧倒翻卸
dump line 排气管路，排放管线
dump out 排出
dump packing 堆积填料
dump pipe 事故排放管道
dump pit 垃圾坑
dump platform 卸料台
dump power 储备功率
dump pump 回油泵
dump steam atmospheric valve 对空排气阀
dump steam condenser 卸载旁通凝汽器
dump steam 废气，排气
dump system 排放系统
dump tank 废液槽，废物槽，排水箱，汇集箱，接收槽【排料】，事故排放槽
dump test 甩负荷试验
dump to atmosphere 排至大气
dumptor 倾卸式运料车
dump tower 卸料塔
dump track 出碴线
dump trailer 自卸式拖车
dump truck 翻斗汽车，自卸货车［卡车］，翻斗车，倾卸式货车，自动倾卸（式）货车
dump tube 排放管
dump valve 泄放阀，卸载阀，事故排放阀
dump wagon 自卸料车
dumpy level 定镜水准仪
dump 转储【内存信息】，事故排放【蒸汽或重水】，堆集场，倾卸，倾倒，清除，卸载，烧毁元件存放处，垃圾场，排放，倒出
dune fixation 沙丘固定
dune flow 丘状流
dune movement 沙丘运动
dune plain 沙丘平原
dune plant 沙丘植物
dune sand 沙丘，沙丘沙
dune soil 沙丘土
dune 砂丘，丘形
dunite 纯橄榄岩
dunkometer 元件破损探测器
dunnage 填木
duocentric motor 同芯双转子电动机
duodirectional relay 双向继电器
duolateral coil 蜂房式线圈，蜂巢线圈
duolateral winding 蜂巢绕组，蜂房式绕组
duosynchronous motor 旋转电枢式同步电动机
dupe 欺骗
duplex alloy 两相合金
duplex arch 双曲拱
duplex balance 双工平衡【电路】
duplex bearing 列轴承
duplex burner 双级式喷燃器
duplex cable 双股电缆，双芯电缆，双心电缆
duplex calculating machine 两面用计算机
duplex circuit 工线路【通信】
duplex-coated fuel particle 双层包覆燃料颗粒
duplex coating 双层涂敷
duplex communication 双工通信
duplex computer 双工计算机
duplex deviation alarm card 双偏差报警卡
duplex difference alarm card 双差值报警卡
duplex engine 双发动机
duplex feedback 重反馈，并联反馈
duplex feeding 双路馈电，双端供电
duplex fuel burner 双级式喷油器
duplex-headed nail 双头圆钉
duplex heat treatment 复合热处理
duplexing assembly 双工组合，双工配置
duplex lap winding 双叠绕组
duplex operation 双工运行
duplex pellet 浓缩芯块
duplex plunger pump 双柱塞泵
duplex power feed type motor 双电源馈电式电动机
duplex pressure gauge 双针压力计
duplex pump 双缸泵，双联泵，双筒泵
duplex radio communication 双工无线通信
duplex reciprocating feeder 双向往复式给料机
duplex reciprocating pump 双缸往复式泵
duplex-RTD 双支热电阻
duplex socket 双插座，双重插座
duplex spot weld 双点点焊接头
duplex stage 双列级，复速级
duplex star connection 双星形接法，双星形连接
duplex steam pump 双缸汽动泵
duplex system 双工制，双路系统
duplex telegraph 双工电报
duplex thermocouple 双支热电偶
duplex traffic 双工传送
duplex transmission 双工传输
duplex tunnel 双试验段风洞
duplex wave winding 双波绕组
duplex winding 双重绕组
duplex wire IQI 双丝像质计
duplex 双缸【活塞或柱塞泵】，双工【通信】，加倍的，双的，双重的，双路的，两倍的，双联的
duplicate busbar system 双母线制
duplicate circuitry 重复电路，复制电路
duplicate circuit 双工电路
duplicate construction 重复建设
duplicated drawing 复制图
duplicate equipment 备用设备
duplicate feeder 第二馈路，并联馈路
duplicate invoice 发票副联，发票副本
duplicate level line 双程水准线
duplicate line 平衡线路，双线线路
duplicate part 备件
duplicate power supply 双电源
duplicate production 重复生产
duplicate pump 备用泵
duplicate relay 备用继电器【双套制】
duplicate service 双接户线
duplicate supply 双电源，双路馈线，双电源供电
duplicate tunnel 双试验段风洞
duplicate 双重的，双份的，双份，副本，加倍，复制品，抄本件，复制，重复，仿型加工
duplication check 重复校验，双重校验，双重检查

duplication of imports 重复引进
duplication 复制，重复，加倍，复制本
duplicator 复制机
du pont model 杜邦财务模型
duprene rubber 氯丁橡胶
durability index 耐久性指数
durability in operation 经久耐用（性）
durability test 耐久试验，耐久性试验
durability 耐久性，寿命，耐用期限，耐用性
durable clause 期限条款
durable goods 耐用品
durable in use 使用耐久
durable metal 耐用金属
durable years 使用年限
durable 耐久的，耐用的
dural alclad 包铝
duration-area curve 历时面积曲线
duration-area-depth 降水历时、面积、深度关系
duration clause 期限条款
duration curve 历时曲线
duration in scanning 扫描持续时间
duration in storage 贮存时间
duration limit wind wave 时间限制风浪
duration of braking 制动时间
duration of breaker contact 继电器接触时间，开关触头接触时间
duration of commission 任务期限
duration of contract 合同期限
duration of down time 停机时间
duration of fall 落潮历时
duration of fault 故障持续时间，事故持续时间
duration of filtration cycle 过滤周期
duration of fire resistance 耐火极限
duration of franchise 特许期限
duration of frost-free period 无霜期历时
duration of insurance 保险期限
duration of liability 责任期限
duration of life 寿命
duration of operation 运行持续时间
duration of oscillation 振荡时间
duration of peaking time 尖峰负荷持续时间
duration of pollution 污染持续时间
duration of possible sunshine 可照时数
duration of power-on 接通持续时间
duration of precipitation 降雨历时
duration of rainfall excess 超渗降雨历时
duration of rainfall 降雨历时
duration of regulation 整定时间
duration of release 释放持续时间
duration of rising tide 涨潮历时
duration of running 运转时间，运行期
duration of sampling 取样持续时间
duration of service 设备使用年限，使用时间，使用期限，服务期限
duration of settlement 沉降历时
duration of short-circuit 短路持续时间
duration of starting 启动持续时间
duration of storm 暴雨历时，风暴持续时间
duration of sunshine 日照时间
duration of test run 试运转持续时间，试运行持续时间，试车持续时间
duration of tidal fall 落潮历时
duration of wave-front 波前宽度
duration of working cycle 运行周期
duration on a project 工程工期
duration series 历时系列
duration 脉冲宽度，期间，持久，持续时间，历时，生存期，时段
durative 持续性
during probation 试用期间
durometer 硬度计
dust absorption 吸尘
dust accumulator 集尘器
dust air 含尘气体
dust and fume 烟尘
dust and soot 煤尘
dust arrester 除尘器，吸尘器
dust arrestment 除尘，灰尘捕集，吸尘
dust arrestor 收尘器，除尘装置，除尘器
dust arrest 除尘，吸尘，集尘
dust ash foam concrete 粉煤灰泡沫混凝土
dust baghouse 集尘室
dust bin 垃圾箱
dust borne gas 含尘气体
dust bowl 尘暴，风沙侵蚀区
dust burden 含尘量
dust canopy 防尘罩
dust capacity 粉尘量，容尘量
dust cap 防尘罩
dust cart 垃圾车，废料车
dust car 废料车，垃圾车
dust catcher 除尘器，吸尘器
dust centrifuge 灰尘离心分离器
dust chamber 除尘室，降尘室
dust chimney 排尘管
dust circulation factor 飞灰再循环率
dust cleaner 吸尘器
dust cloud 尘云
dust coal 粉煤，末煤
dust coke 焦末，焦粉
dust collecting electrode 集尘电极
dust collecting fan 除尘风机，集尘风机，吸尘器
dust collecting hood 吸尘罩
dust collecting plant 除尘装置
dust collection 集尘，除尘
dust collection efficiency 除尘效率
dust collection equipment 粉尘收集装置
dust collection point 除尘点
dust collection system 集尘装置
dust collector efficiency 集尘效率
dust collector 除尘器，捕尘器，集尘器，吸尘器
dust collect unit 除尘装置
dust concentrate 高浓度粉剂，高含量粉剂
dust concentration 含尘浓度，粉尘浓度，尘埃浓度
dust concentrator 除尘器，集尘器
dust condensing flue 沉灰烟道
dust content 含尘量，灰尘含量
dust control equipment 控制粉尘装置
dust control 尘埃控制，除尘

dust core coil 铁粉芯线圈
dust core 压粉铁芯，铁粉芯
dust counter 尘埃计，测尘计，尘度计，计尘器，尘粒计数器
dust cover 防尘罩
dust density （烟气）含尘量
dust deposit 尘土堆积，积灰，粉尘沉积
dust detector 测尘仪
dust determination 粉尘测定
dust devil 尘卷，尘旋风，尘卷风，小尘暴
dust disease 积尘病
dust emission 飞灰排出，飞灰排出量，粉尘排放，飞尘
dust entrainment 粉尘夹带
duster 除尘器，除尘机，洒粉器，清洁工，清扫器
dust exhausting fan 排尘风扇
dust exhaust 排尘
dust explosion 粉尘爆炸
dust extracting plant 吸尘设备
dust extraction equipment 除尘设备
dust extraction 除尘
dust extractor 除尘器
dustfall jar 降尘（测定）瓶
dust fall 尘降，灰尘沉降，落尘
dust filter 滤尘器，灰尘过滤器
dust filtration 滤尘
dust fineness 煤灰细度
dust-fired 粉末燃烧的
dust-firing 粉状燃料燃烧
dust flow 尘流
dust flue 灰管，尘道
dust fog 尘雾
dust-free air 无尘空气
dust-free 无灰尘的
dust-fueled reactor 粉末燃料反应堆
dust gas 含尘气体
dust generation point 生尘点
dust guard 防尘，防尘设备，防尘罩
dust-handling plant 除尘装置
dust hazard 粉尘危害
dust haze 尘霾
dust-holding capacity 容尘量
dust hood 吸尘罩
dust hopper 集尘斗
dust horizon 尘埃层顶，粉尘顶层
dust-ignition-proof motor 防尘防粉型电动机
dustiness degree 含尘度
dustiness of the air 空气污染度
dustiness 尘污，尘雾度，尘雾浓度
dust infection 粉尘传染
dusting door 手工吹灰灰孔【炉墙上】
dusting-on 涂粉
dusting 涂粉【铸造】，形成粉末，扬尘，除尘
dust iron 铁粉
dust-laden air 含尘空气，带尘空气
dust-laden exhaust gas 含尘废气
dust-laden gas 含尘烟气
dust-laden 含尘的
dust-laying measures 防尘措施
dustless unloader 湿式输灰装置
dustless 无尘的
dust loading concentration 含尘浓度
dust loading dust density 烟气含尘量
dust loading 尘埃浓度，（烟气）含尘量，积垢，阻塞【空气过滤器】
dust measurement 粉尘测量
dust meter 灰尘计【测除尘器效率用】
dust monitor 灰尘监测器
dust palliative 灭尘剂
dust particle 尘粒，灰粒
dust piece 通道片【电机铁芯】
dust plume 尘埃流
dust pocket 飞灰堆积区
dust pollution 尘土污染，粉尘污染
dust precipitator 聚尘器，除尘器
dust prevention apparatus 防尘装置
dust prevention 尘埃预防，防尘
dust preventive 防尘剂
dust proof and oil-tight construction 防尘和不透油结构［构造］
dust proof coating 防尘油漆
dust proof insulator 防尘绝缘子
dust proof machine 防尘机器，防尘型电机
dust proof measures 防尘措施
dust proof type 防尘型
dust proof 防尘的，防尘
dust protected 防尘的
dust rain 尘雨
dust ram 飞灰回收装置【柱式】
dust recirculation 灰再循环
dust recovery 飞灰回收
dust removal coefficient 除尘率
dust removal filter 除尘过滤器
dust removal 除尘
dust remover 除尘器
dust removing system 除尘系统
dust resistance 耐尘性，耐污垢性
dust respirator 除尘口罩
dust sampler 粉尘采样仪，灰尘取样器
dust sampling meter 粉尘采样仪
dust sampling 尘埃取样
dust screen 防尘
dust scrubber 洗尘器
dust section 采尘区
dust separation 除尘，灰尘分离
dust separator 飞灰分离器，除尘器
dust settler 除尘器
dust settling 尘末沉淀
dust settling chamber 灰尘沉降室
dust shaft 垃圾井筒
dust shield 防尘罩，防尘，防尘板
dust source 尘源
dust stop 防尘器
dust-storage hopper 灰斗
dust storm 尘暴
dust stratification 积灰
dust suctor 吸尘器
dust suppressing agent 抑尘剂
dust suppressing wetting spray 抑尘喷嘴
dust suppression point 抑尘点
dust suppression system 除尘装置

dust suppression 防尘飞扬，抑尘
dust supression equipment 抑尘设备
dust-tight construction 防尘建筑
dust-tight 不漏灰的，防尘的，不透尘的，防尘
dust trap 放射性灰尘收集器，除尘器，集尘器，捕尘器
dust tube 集尘管，灰尘管，测尘管
dust turbidity 粉尘混浊度
dust well 尘坑
dust whirl 尘旋风，粉尘旋涡，尘旋
dust wind 尘风
dusty air 含尘空气
dusty fuel 粉状燃料
dusty gas 含尘气体，含尘烟气
dusty 灰尘的，多尘的，粉状的
dust 灰尘，粉末，飞灰，粉尘，尘埃，空中粒子
Dutch cone penetrometer 荷兰式圆锥触探仪
Dutch cone 荷兰式圆锥仪
dutch oven 前置炉膛
Dutch windmill 荷兰风车
DUT(device under test) 被测器件
dutiable goods 应税物品
duty allowance 职务津贴
duty cycle basis 负载循环
duty cycle capacity 断续负载容量
duty cycle factor 负荷持续率
duty cycle of pulse duration 脉冲焊接电流占空比
duty cycle operation 循环使用，循环工作
duty cycle rating 断续负载定额
duty cycle time 工作时间
duty cycle 荷周，工作状态循环，负载循环，占空因数，工作循环周期，占空系数，频宽比，工作比，负载持续率，工作周期
duty factor control system 工作比控制系统
duty factor 工况系数，工作比，(脉冲的)占空因数，工作系数，负载因数
duty free certificate 免税证明
duty free goods 免税货物
duty free zone 免税区
duty free 免税
duty horse-power 标称马力，报关马力
duty of care 认真尽职
duty of water 灌溉率，需水量
duty paid certificate 完税凭证
duty paid goods 已完税货物
duty paid proof 完税凭证，完税证明
duty paid 关税已付
duty period 工作周期
duty plate 机器铭牌
duty quota 关税配额
duty ratio 平均功率与最大功率之比，脉冲平均功率对颠值功率的比，负荷比，负荷率，能效比，负载比，占空比
duty room 值班室
duty set 运行机组
duty shift room 值班室
duty standard 工作标准
duty 运行【旋转机械、马达等设备的工作状态】，工作类型，工作方式，工作状态，能力，工况，能率，税，载荷，容量，负荷，负载，功率，占空，责任
DUVD(direct ultrosonic visualization of defects) 缺陷的超声波直接显示
D value D 值【吸收剂量】
DVA 冷机修理车间和仓库通风系统【核电站系统代码】
DVC 主控室通风系统【核电站系统代码】
DV(diaphragm valve) 隔膜阀
DV(drain valve) 疏水阀
DVD 柴油机房通风系统【核电站系统代码】
DVE 电缆层通风系统【核电站系统代码】
DVF 电气厂房排烟系统【核电站系统代码】
DVG 辅助给水泵房通风系统【核电站系统代码】
DVH 上充泵房应急通风系统【核电站系统代码】
DVI 核岛设备冷却水泵房通风系统【核电站系统代码】
DVK 核燃料厂房通风系统【核电站系统代码】
DVL 电气厂房主通风系统【核电站系统代码】
DVM 汽轮机房通风系统【核电站系统代码】
DVN 核辅助厂房通风系统【核电站系统代码】
DVPS(degraded voltage protection system) 电压下跌保护系统
DVP 循环水泵站通风系统【核电站系统代码】
DVQ 核废物辅助厂房通风系统【核电站系统代码】
DVR(dynamic voltage restorer) 动态电压恢复器
DVS 安全注入和安全壳喷淋泵电机房通风系统【核电站系统代码】
DVT 除盐水车间通风系统【核电站系统代码】
DVV 辅助锅炉和空压机房通风系统【核电站系统代码】
DVW 安全壳环廊房间通风系统【核电站系统代码】
DVX 润滑油输送装置厂房通风系统【核电站系统代码】
dwarf signal 小型信号机
dwarf wall 矮墙
DWA 热修理车间和仓库通风系统【核电站系统代码】
DWB(data word buffer) 数据字缓冲器
DWB 餐厅通风系统【核电站系统代码】
DW(demineralized water) 除盐水
dwelling 寓所，住所
dwelling density 住宅密度
dwelling district 居住区
dwelling house 住所
dwelling standard 住宅标准
dwelling zone 居住区
dwell time 渗透时间【液体渗透检验】，(燃料元件在堆内的)停留时间，停延时间，(压铸)保压时间，驻留时间
dwell 无运动时间，停止，静态，停机，静止，非预定的延迟
DWE 主开关站通风系统【核电站系统代码】
DWG(drawing) 图纸，资料
DWG 其他 BOP 厂房通风系统【核电站系统代码】
Dwight chart 德维特电压调整卡【输电线用】
DWL(designed waterline) 设计水线

DWL 热洗衣房通风系统【核电站系统代码】
DWN 厂区试验室通风系统【核电站系统代码】
DWR 应急保安楼通风系统【核电站系统代码】
DWS(demineralized water transfer and storage system) 除盐水输送与存储系统
DWS(demineralized water system) 除盐水系统
DWST(demineralized water storage tank) 除盐水贮存箱
DWS 重要厂用水泵站通风系统【核电站系统代码】
DWT(dead weight tonnage) 载重公吨位，载重吨，载重吨位
3D4W(three dimensional four wire) 四根线的三度重合
3D3W(three dimensional three wire) 三根线的三度重合
DWX 油和润滑油脂贮存房通风系统（FC泵房）【核电站系统代码
DWY 制氧站通风系统【核电站系统代码】
DWZ 制氢站通风系统【核电站系统代码】
DXC(data exchange control) 数据交换控制
dyadic 并矢式，并向量，二进制的，二重的，双
DYAN(dynamic analysis) 动态分析
dye check 染色探伤，着色探伤，着色检验
dyed sand 染色沙
dyed solid 染色固体
dye etchant 着色腐蚀剂
dye injection system 染色水注入装置
dye penetrant examination 穿透性试验，液体渗透试验，着色探伤检验
dye penetrant inspection 着色探伤检查，染色渗透检查，染色观察
dye penetrant process 着色渗透探伤工艺
dye penetrant test 着色探伤【金属表面】
dye penetrant 着色渗透剂
dye penetrate test 染料渗透试验，着色探伤检验，着色渗透检验
dye penetration test 染色探伤，着色探伤
dye stream 染液流
dye tracer 染料示踪剂
dye 着色
dyke boring 堤防钻探
dyke breach 土堤决口
dyke defect detecting 堤防隐患探测
dyke failure 堤防溃决
dykes and dams 堤坝
dyke strengthening by warping 放淤固堤
dyke 堤坝，沟渠，排水道，堤，堤防，围堤，岩脉，岩墙
DYNA(dynamic analysis) 动态分析
dynaflow 流体动力传动
dynaform 同轴开关
dynamax 镍钼铁合金
dynamic accuracy 动态精度，动态精确度
dynamic action 动力作用，动态作用
dynamic address translation 动态地址转换
dynamic address translator 动态地址转换器
dynamic adjustment 动态调整
dynamic admittance 动力导纳
dynamic aerodynamics 动态空气动力学
dynamic aeroelasticity 动力气动弹性力学
dynamic air field test of boiler furnace 动力场试验
dynamical condition 动力学条件
dynamical equation 动态方程
dynamical equilibrium 动力平衡，动态平衡
dynamical friction 动摩擦
dynamical loaded structure 动力荷载结构
dynamical load 动荷载
dynamic allocation 自动存储器配置
dynamically balanced 动平衡的
dynamically similar model 动力相似模型
dynamical model 动力学模型
dynamical parameter 动力参数
dynamical pressure 动压力
dynamical property 动态特性
dynamical similarity 动力相似性
dynamical stability 动态稳定
dynamical storage 动库容
dynamical stress concentration 动应力集中
dynamical system 动力系统
dynamical water pressure 动水压力
dynamical watershed model 动态流域模型
dynamical 电动的，动力学的，动态的，动力的
dynamic amplification 动力放大
dynamic analog device 动态模拟装置
dynamic analog test 动态模拟试验
dynamic analogue for power system 电力系统动态模拟
dynamic analogue method 动力相似法
dynamic analogue 动态模拟（设备）
dynamic analogy 动态模拟
dynamic analysis and design 动力分析与设计
dynamic analysis method 动态分析法
dynamic analysis model 动态分析模型
dynamic analysis 动态分析，动力分析，动态特性分析
dynamic analyzer 动态分析程序，动态分析器
dynamic arc 动态电弧
dynamic balance equation 动态平衡方程
dynamic balancer 动平衡机，动平衡器
dynamic balance test 动平衡试验
dynamic balance 动平衡，动态平衡，动力平衡
dynamic balancing machine 动平衡机
dynamic balancing test 动平衡实验
dynamic balancing 动平衡
dynamic ball indentation test 落球硬度试验
dynamic behavior 动态性能，动力性能，动态特性曲线，动态行为
dynamic behavio(u)r of a reactor 反应堆动态性能
dynamic bidding 动态投标
dynamic bottom 动力河床
dynamic boundary condition 动力边界条件
dynamic boundary layer 动力边界层
dynamic braking resistor 制动电阻
dynamic braking 动力制动，动态制动
dynamic brittle-coating test 动态脆性涂层试验
dynamic buckling 动力失稳，动力屈曲，动态屈曲，动力压屈
dynamic buffering 动态缓冲

dynamic burnout 动态烧毁
dynamic business analysis 商情分析
dynamic calculating method 动态计算法
dynamic calibration 动态校准，动态标定
dynamic capacity 动态电容
dynamic carrying capacity 动态承载能力
dynamic characteristic of arc 电弧动特性
dynamic characteristic test 动态特性试验
dynamic characteristic 动力特性，动态特性，动态特性曲线
dynamic check 动态校验
dynamic circuit 动态电路
dynamic classifier 动态分离器
dynamic climatology 动力气候学
dynamic CMOS 动态互补型金属氧化半导体
dynamic coefficient of viscosity 动力黏性系数
dynamic coefficient subgrade reaction 地基反力动力系数
dynamic coefficient 动力系数，动态系数
dynamic cohesive degree 运动黏度程度
dynamic compaction method 强夯法
dynamic compaction 动力压实，强夯，强夯法
dynamic compensator 动态补偿器
dynamic consolidation 强夯法
dynamic constant 动态常数
dynamic contact resistance 触点动态电阻，动态接触电阻
dynamic control system 动态控制系统
dynamic control 动态控制
dynamic convection 动力对流，动力性对流
dynamic cooling 动力冷却
dynamic corrector 动态校正器
dynamic corrosion 动态腐蚀
dynamic coupling 齿啮式连接，动态耦合
dynamic crack propagation 裂纹动态传播
dynamic current 持续电流，动态电流
dynamic curve 动态曲线，动态特性曲线
dynamic cycling loading 动态周期载荷
dynamic damper 动力阻尼器，动力减震器，动力消振器
dynamic data exchange 动态数据交换
dynamic debugging routine 动态诊断程序，动态调试程序
dynamic deflection 动挠度，动载挠度，动态挠曲
dynamic delivery head 动输水水头
dynamic derivative 动导数
dynamic design 动力设计
dynamic deviation 动态偏差
dynamic device reconfiguration 动态设备重新配置
dynamic discharge head 动泄水水头
dynamic draught head 动吸出水头
dynamic dump 动态转储，动态打印
dynamic economic analysis method 动态经济分析方法
dynamic ecosystem 动态生态系
dynamic effect factor 动作用系数
dynamic effect 动力效应，动态效应，动力影响
dynamic elasticity 动弹性
dynamic elastic modulus 动弹性模量，动力弹性模量
dynamic electricity 动电，动电学
dynamic endurance test 动力耐久试验
dynamic equation of equilibrium 动态平衡方程式
dynamic equation 动力方程，动态方程
dynamic equilibrium ratio 动态平衡比
dynamic equilibrium 动平衡，动态平衡，动力平衡
dynamic error-free transmission 动态无误差传输
dynamic error 动态误差
dynamic evaluation 动态评价
dynamic excitation 动态激励
dynamic experiment 动态试验，动力学实验
dynamic factor 动力因数
dynamic field test 动力场试验
dynamic flexural stiffness 动力弯曲劲度
dynamic flip-flop 动态触发器
dynamic flowmeter 动态流量计
dynamic force 动力，动态力
dynamic formula for piling 打桩动力公式
dynamic friction factor 动摩擦因数
dynamic friction 动摩擦
dynamic gain 动态增益
dynamic geology 动力地质学
dynamic hardness 冲击硬度
dynamic head 液压头，动压头，动力水头，速度头，动力头
dynamic heating 动力增温
dynamic holography 动态全息术
dynamic horsepower 指示马力，动态马力，指示功率，动态功率
dynamic ice pressure 动冰压
dynamic imbalance 动态不平衡
dynamic impedance 动态阻抗
dynamic indentation 动力压凹
dynamic inductance 动态电感
dynamic influence 动力影响
dynamic instability 动态不稳定性，动力学不稳定性，动力不稳定性
dynamic investment cost of power generation project 发电工程动态投资
dynamic investment 动态投资
dynamicizer 动态转换器
dynamic link library 动态链接库
dynamic load characteristics 动载特性，负荷动态特性，动态负载特性
dynamic loaded structure 承受动力荷载结构
dynamic load factor 动态负荷因子
dynamic loading test 动荷试验，动态负荷试验
dynamic loading 动载荷
dynamic load test of pile 桩的动荷载试验，桩的动载试验
dynamic load test 动力载荷试验
dynamic load 动态负载，动力荷载，动载荷，动态负荷，动负荷
dynamic logic circuit 动态逻辑电路
dynamic loss 动态损失
dynamic magnification factor 动力放大系数
dynamic magnification 动力放大
dynamic mapping system 动态变换系统
dynamic material accountancy 实时材料衡算
dynamic matrix 动力矩阵

dynamic measurement of pile 桩动测法
dynamic measurement 动态测定，动态测量
dynamic membrane 动态膜
dynamic memory 动态存储器
dynamic metamorphism 动力变质作用
dynamic meteorology 动力气象学
dynamic method 动力法
dynamic mockup 动力模型
dynamic modeling 动态模拟
dynamic model 动态模化，动态模型
dynamic modulus of elasticity 动弹性模量，动弹性模数
dynamic oedometer 动力固结仪
dynamic optimization 动态最优化
dynamic overload 动力过载
dynamic parameter 动态参数，动力参数
dynamic penetration test 动力贯入试验
dynamic penetration 动力触探
dynamic performance 动态特性
dynamic photoelasticity 动力光测弹性力学
dynamic pickup 动态传感器
dynamic pile-driving formula 动力打桩公式
dynamic pile formula 动力桩公式
dynamic point resistance 动力探头阻力
dynamic power consumption 动态功耗
dynamic pressure anemometer 动压风速计
dynamic pressure computer 动压力计算器
dynamic pressure recover 动压恢复
dynamic pressure sensor 动压传感器
dynamic pressure 动态压力，动压，动压力，动压强
dynamic programming 动态规划
dynamic program relocation 动态随机存储器
dynamic property 动态特性
dynamic purchasing system 动态采购系统
dynamic quality control 动态质量控制
dynamic random-access memory 动态随机存取存储器
dynamic range of magnetic tape 磁带动态范围
dynamic range 动态范围，动力区
dynamic ratio 动态比
dynamic reactor behavior 反应堆动态性能
dynamic reference model 动态参考模型
dynamic register 动态寄存器
dynamic regulation characteristic 动态调节特性
dynamic regulation 动态调整率，动态调整
dynamic regulator 动态调节器
dynamic relocation 动态再定位，动态再分配
dynamic replacement 动力置换，强夯置换法
dynamic representation 动态显示
dynamic resilience 动回弹性，动力回弹能
dynamic resistance 动态阻力，动态电阻
dynamic resonance 动态响应
dynamic response test 动力响应试验，动态特性试验，动态响应试验，动作性试验
dynamic response 动态特性，动态响应，动态谐振，动力反应，动态反应
dynamic rig 动力试验台
dynamic rotor balance 转子动平衡
dynamic sampling 动态抽样
dynamic scale 动力缩尺
dynamic scheduling 动态调度
dynamic sealing point 动密封点
dynamic seal 动密封
dynamic sensitivity 动态灵敏度
dynamic sequential control 动态序列控制
dynamic shear modulus 动剪切模量
dynamic shift register 动态移位寄存器
dynamic signal 动态信号
dynamic similarity principle 动力相似原理
dynamic similarity 动力相似，动力相似性，动态相似性
dynamic similitude 动力相似
dynamic simulation of HVDC power transmission system 高压直流输电系统动态模拟
dynamic simulation 动态模拟
dynamic simulator 动态模拟
dynamics of streams 河流动力学
dynamic solubility 动态溶解度
dynamic sounding 动力测深，动力触探
dynamic speed control 动态转速控制
dynamic stability criterion of voltage-loop 电压环动态稳定判据
dynamic stability derivative 动态稳定导数
dynamic stability 动态稳定度，动力稳定性，动力稳定，动态稳定，动稳定
dynamic stabilization 动态稳定
dynamic stable current 动稳定电流
dynamic stall 动态失速
dynamic state operation 动态运行
dynamic state 动态
dynamic status 动态
dynamic steel lamination 电工钢片，硅钢片
dynamic stiffness 动力刚度
dynamic storage allocation 动态存储分配
dynamic storage volume of reservoir 动库容
dynamic storage 动态存储器
dynamic strain 动应变
dynamic strength 疲劳强度，动态强度，动力强度
dynamic stress 动应力，动荷应力
dynamic structural analysis 动力结构分析
dynamic sublayer 动力副层，机械乱流层
dynamic subroutine 动态子程序
dynamic suction head 动力吸水头，动力吸头，动吸入水头，动吸升水头
dynamic suction lift 动力吸升高度，动吸程，动吸升高度
dynamic support system 动态支援系统
dynamic system performance 系统动态运行特性
dynamic system 动态系统，动力系统，动力学系统
dynamics 动力学
dynamic tear test 动态撕裂试验
dynamic temperature change 动态温度变化，温度动态变化
dynamic temperature correction 动态温度修正
dynamic temperature 动力温度，动温
dynamic test 动态试验，超声波试验【混凝土】，冲击试验，动力试验，动平衡试验
dynamic theory of gas 气体动力学理论
dynamic theory of tide 潮汐动力理论

dynamic threshold 动力气动值
dynamic transfer function 动态传递函数
dynamic transfer system 动态传输系统
dynamic triaxial test 振动三轴试验
dynamic trigger 动态触发器
dynamic trim 动力调整
dynamic trough 动力槽
dynamic unbalance measurement 动态不平衡测试方法
dynamic unbalance 动态失衡
dynamic valve curve 动态阀门曲线
dynamic variable 动力变量
dynamic variation 动态变化
dynamic vibration absorber 动态减振器
dynamic vibration reducer 动态减震器
dynamic viscosimeter 动黏度
dynamic viscosity coefficient 动黏度系数
dynamic viscosity 动态黏滞性，动力黏度，动黏性（系数）
dynamic voltage restorer 动态电压恢复器
dynamic warming 动力升温
dynamic water level 动水位
dynamic water pressure 动水压力
dynamic wave 动力波
dynamic wind load 动态风载
dynamic 动力的，动态的，动态，动力
dynamite pack 炸药包
dynamite 炸药
dynamo braking 发电制动
dynamo brush 电刷
dynamo condenser 调相机
dynamoelectric machine 电动发电机，电机
dynamoelectric motor 旋转换流机
dynamoelectric relay 电动式继电器
dynamoelectric 机械变电能的，电动的，机电的
dynamofluidal 动力流动的
dynamo for electrolyzer 电解用发电机
dynamo governor 发电机调速器，发电机调节器
dynamo ignition 发电机引燃，发电机点火
dynamo magneto 永磁发电机
dynamo-metamorphic rock 动力变质岩
dynamo-metamorphism 动力变质作用
dynamometer instrument 电力表
dynamometer machine 测功机，测功电机
dynamometer test 测功试验，测功机试验
dynamometer-type multiplier 测力计式乘法器
dynamometer 功率计，测力计，动力计，测功器，测力杆，测力盒，测力环，测力仪［器］
dynamometry 测功法，测力法
dynamomotor 共定子电动发电机，旋转换流机
dynamo output 发电机输出
dynamo panel 发电机控制盘，直流发电机盘
dynamo room 发电机舱
dynamo steel sheet 电机用硅钢片，电机硅钢片，电工钢片
dynamo steel 电机钢，电机用钢
dynamo-thermal metamorphism 动热变质
dynamotor 共定子电动发电机，旋转换流机
dynamo （直流）（发）电机，电动机
dynatron 负阻管
dyne 达因【力的单位】
dynode 倍增器电极，倍增管电极，二次放射极，打拿极
dysgeogenous 不易风化的
dysprosium 镝
D 通信，装卸设备，通风，照明【核电站系统代码】

E

EAC(executive alarm control) 超限警报监控
each other 相互
eactivity balance equation 反应性平衡方程
EA(environmental assessment) 环境评价
EAF(effective attenuation factor) 有效衰减系数
EAF(equivalent availability factor) 等效可用系数，等效可利用因子
EAG(equipment advisory group) 设备咨询组
eager 潮水上涨，涛
EAN(equivalent atomic number) 当量原子序数
EAN(external access network) 外部存取网络
EAON(except as otherwise noted) 除非另有说明，除非另有记载
ear 耳，吊耳，搭耳
EAR(engineering action request) 技术工作请求书
EAR(environmental appraisal report) 环评报告
earlier measurement 前一量度
earliest breakup 最早解冻日
earliness of forecast 预见期
early commutation 超前换向
early contact 先动触点，预动触点
early delivery of goods 早交货
early effect 近期效应，早期效应
Early equivalent circuit 阿莱型等效电路
early failure period 早期故障期间，早期故障阶段，早期故障期，早期失效期
early failure 早期失效，早期故障
early fallout 早期（放射性）落下灰
early fault 早期故障
early gate 前闸门
early profit 初期利润
early radiation effect 早期辐射效应
early setting cement 快凝水泥，速凝水泥
early setting 早凝
early shutdown phase 早期停堆阶段
early-stage diversion 初期导流
early strength admixture 早强剂
early strength cement 早强水泥
early strength concrete 早强混凝土
early strength 早期强度
early termination 中止
early warning 预警，早期报警
earmarking 指定用途
earmark 指定（资金的）用途
earmuffs 耳机上的橡皮护套
earn a profit 获得利润
earned hours 挣得时数
earnest money 定金
earnest 保证金，定金
earning rate 利润率，收益率

earnings after tax 税后收益额
earnings before interest and taxes 息税前利润
earnings per share 每股赢利
earnings ratio 市盈率
earning statement 盈余表
earnings yield 收益率
earning 利润，收入，所得，所赚的钱，挣钱，获利
EAROM (electrically-alterable read only memory) 电可改写只读存储器
earphone 耳机
ear protector 护耳器
ear-radio 耳式收音机
earth-air current 地空电流
earth albedo 地球反照率
earth and rock cofferdam 土石围堰
earth and rock excavation 土石方开挖
earth and rockfill dam 土石混合坝，土石坝
earth antenna 受信用的简单架空线
earth arrester 接地火花隙避雷器
earth auger 麻花钻，土钻
earth backing 还土，复土
earth bank 路堤，土堤岸
earth bar 接地棒
earth-based 地面的，陆地的
earth beam test 土梁试验
earth bearing capacity 地耐力
earth bending test 土梁试验
earth block 土块
earth boring outfit 钻土器具，钻土设备
earth bus 接地母线
earth capacity 对地电容，大地电容
earth circuit 接地电路
earth clamp 地线夹头
earth clip 接地夹
earth cofferdam 土围堰
earth coil 测地磁的线圈，接地线圈
earth column 土柱
earth compaction factors 土料的压实参数
earth concrete 掺土混凝土
earth conductivity 大地导电率，大地电容率
earth conductor 接地线，接地导线
earth connection 接地，接地线
earth continuity conductor 接地导线
earth core 地核，地心
earth crust stress 地壳应力
earth crust 地壳
earth current meter 地电流测量器，泻地电流表
earth current 泄地电流，地电流，大地电流
earth curvature 地球曲率
earth cutting 挖土方
earth dam compaction 土坝压实
earth dam paving 土坝护面
earth dam 土坝
earth detector 接地探测器，漏电检测器
earth dike 路堤，地面坝
earth ditch 土沟
earth drill 钻土机
earthed cathode 接地阴极
earthed circuit 接地电路
earthed concentric wiring system 同心线外线接地系统
earthed conductor 接地导线，地导体
earthed input 接地输入
earthed neutral 接地中线，单线制，接地中性点
earthed output 接地输出
earthed point 接地点
earthed pole 接地极
earthed system 接地系统，单线制
earthed 接地的
earth electrode 地电极，接地电极，接地极，接地体
earth embankment 路堤，地面坝，土堤
earth end 接地端
earthen path 土路
earthen road 土路
earthen structure 土工构筑物
earthenware haydite 陶粒
earthenware pipe 陶管，陶土管
earthenware tile pavement 陶砖铺面
earthenware 陶器
earth excavation 挖土工程，土方开挖
earth fall 土崩
earth fault current 接地故障电流
earth fault protection 接地故障保护
earth fault relay 接地故障继电器
earth fault transient resistance 接地瞬态电阻
earth fault 接地故障
earth fill cofferdam 土围堰
earth fill dam 土坝
earth fill 填土方，填方，填土
earth fixed axes system 地轴系
earth flax 石棉
earth flow 土流，泥流，泥石流
earth foundation 土基
earth frame of balance 地球平衡架
earth free 不接地的，不接地
earth grid 接地栅极，抑制栅极
earth endurance 地耐力
earth ground 地，接地
earth indicator 接地指示器
earth induction 地磁感应
earth inductor 地磁感应器
earthing autotransformer 中性点接地补偿器，中性点接地自耦变压器
earthing brush 接地电刷
earthing busbar 接地汇流排，接地母线
earthing bushing 接地套管
earthing cable 接地电缆
earthing circuit 接地回路
earthing clip 接地线夹，地线夹
earthing conductor 接地线，接地导体
earthing contact 接地触点
earthing device 接地装置
earthing fault 接地不良，接地故障
earthing for work 临时接地
earthing knife-switch 接地刀闸
earthing mat 接地网
earthing network 接地网
earthing of casing 外壳接地，机壳接地
earthing of lightning arrester 避雷器接地装置
earthing of shield 屏蔽层接地

earthing pad 接地端板
earthing position 接地位置
earthing reactance 接地电抗
earthing rector 接地电抗器
earthing reference points 接地基准点
earthing resistance 接地电阻
earthing resistor 接地电阻器
earthing rod 接地棒，接地柱
earthing steel angle 接地角钢
earthing strip 接地条，接地片
earthing switch for busbar 母线接地开关
earthing switch 接地开关
earthing system 接地系统，接地网络
earthing terminal 接地端子，接地线端
earthing thimble 接地套管
earthing transformer for neutral grounded system 中性点接地变压器接地系统
earthing transformer 接地变压器
earthing 接地
earth insulation 对地绝缘，主绝缘
earth interior 地球内部
earth lead 地线【焊接接地用】，接地引线，接地地线，接地线
earth leakage breaker 接地漏电断路器
earth leakage circuit-breaker 对地漏电断路器，接地保护断路器
earth leakage current 对地泄漏电流
earth leakage fault 对地漏电故障
earth leakage meter 对地漏电测量器
earth leakage protection 接地漏电保护
earth leakage relay 漏地电流继电器，接地保护继电器
earth leakage trip 接地保护自动断路
earth leakage 接地漏电，对地泄漏
earth light 地球反照光
earth loop 接地回路
earth magnetic field 地磁场
earth magnetism 地磁
earth mantle 地幔
earth manual 土工手册
earth mass 土体
earth material 土料
earth mat 地网
earth membrane 土质防渗层
earth mortar 土灰浆
earth mound 土墩
earth mover 大型推土机，运土机械
earth moving equipment 运土设备
earth moving machine 运土机械，运土机械设备
earth moving operation 运土工序，运土工作
earth moving plant 土方机械设备
earth moving scraper 土料铲运机，运土铲土机
earth moving 运土，运土的，土方工程
earth neutral system 中性点接地系统
earth nucleus 地核
earth observation satellite 地球观测卫星
earth pile 土挤密桩，土桩
earth pillar 土柱
earth pitch 地沥青
earth pit 地窖，土坑
earth plate 接地板
earth platform 土平台
earth plug 带接地触头的插头
earth point 接地点
earth potential difference 对地电位差
earth potential working 地电位作业
earth potential 地电位，地电势
earth pressure at rest 静止土压力
earth pressure cell 土压力盒，土压力计
earth pressure coefficient 土压力系数
earth pressure gauge 土压力计
earth pressure line 土压力线
earth pressure measurement 土力测定
earth pressure wedge 土压力楔形体
earth pressure 土压，土压力
earthquake 地震
earthquake acceleration 地震加速度
earthquake action 地震作用
earthquake analysis 地震分析
earthquake belt 地震带
earthquake bracing 防震拉撑【立柱间的】
earthquake center 震源，震中
earthquake damage 地震破坏，震害
earthquake design criteria 地震设计准则
earthquake design 地震设计
earthquake disaster 地震灾害，震害
earthquake distribution 地震分布
earthquake due to collapse 崩陷地震
earthquake effect 地震效应，地震影响，地震作用
earthquake engineering 地震工程学
earthquake epicenter 地震震中
earthquake fault 地震断层
earthquake-felt area 有感地震范围
earthquake focus 震源
earthquake force 地震力
earthquake geology 地震地质学
earthquake ground motion 地震地面运动
earthquake hazard 地震危害
earthquake hydrodynamic force 地震动水作用力
earthquake hydrodynamic pressure 地震动水压力
earthquake hypocenter 震源
earthquake inertia force 地震惯性力
earthquake instrumentation system 地震仪表系统
earthquake intensity map 地震烈度图
earthquake intensity occurrence probability 地震烈度发生概率
earthquake intensity scale 地震烈度表
earthquake intensity 地震烈度，地震强度
earthquake isoseismal 地震烈度包络线
earthquake light 地震光
earthquake loading 地震荷载
earthquake magnitude 地震等级，震级，地震规模
earthquake measuring 7.3 on the Richter scale 里氏7.3级的地震
earthquake mechanism 地震机制，发震机制
earthquake model 地震模型
earthquake motion 地震动，地震运动
earthquake observing station 地震观测站
earthquake of intermediate focal depth 中深地震
earthquake origin 震源

earthquake period　地震周期
earthquake precursor　地震先兆现象
earthquake prediction　地震预报
earthquake premonitory phenomenon　地震前兆现象
earthquake-proof construction　防震建筑
earthquake-proof foundation　防震基础，抗震基础
earthquake-proof joint　防震缝
earthquake-proof site　抗震试验场
earthquake-proof type steel rack　防震型钢支架
earthquake-proof　防震
earthquake protective measure　抗震措施
earthquake record　地震记录
earthquake region　地震区，震区
earthquake resistance　抗震能力，防震
earthquake-resistant design　防震设计，抗震设计
earthquake-resistant　抗地震的，抗震的
earthquake response spectrum　地震反应谱
earthquake shock　地震冲击
earthquake source　地震震源
earthquake strength　地震强度
earthquake stress　地震应力
earthquake swarm　地震群，群发地震，震群
earthquake tremor　地震颤动
earthquake warning　地震警报
earthquake wave　地震波
earthquake zone　地震带，地震区
earth radiation　地球辐射
earth reference plane　参考接地板
earth resistance meter　接地电阻计，接地电阻表
earth resistance　接地电阻，地电阻，地面电阻，土壤阻力，土抗力
earth resistivity　大地电阻率，大地电阻系数，土壤电阻率
earth resources satellite　地球资源卫星
earth-retaining wall　挡土墙
earth-retaining　挡土的
earth return circuit　接地回流电路，单线送电大地返回电路
earth return system　地回路制，地回路输电方式
earth return　接地回线，地电回路
earth revolution　地球公转
earth ridge　土垄
earth road　土路
earth-rock cofferdam　土石围堰
earth-rock dam　混合式土石坝，土石坝
earth-rock excavation　土石方开挖
earth-rockfill dam　堆石土坝
earth's atmosphere　地球大气
earth's axis　地轴
earth's boundary layer　地球边界层
earth science　地球科学，地学
earth scoop　挖土铲斗
earth scraper　刮土机
earth screen　接地网，接地屏蔽
earth's crust　地壳
earth-shielded　接地屏蔽的
earth shield　接地屏蔽
earth-shine　地球反照光
earth side of insulator string　绝缘子串接地侧
earth signaling device　接地信号装置
earth slide　土滑
earth slip　土滑
earth slope　土坡
earth's magnetism　地磁
earth's surface　地表，地面
earth structure　土工建筑物，土工构筑物，地球构造，地球结构
earth subsidence　地面沉降
earth surface　地面
earth switch　接地开关
earth-synchronous orbit　地球同步轨道
earth system of axes　地轴系
earth system　接地系统
earth tamper　夯土机
earth temperature　地温
earth terminal　接地端子，地线接线柱
earth termination system　接地装置，接地系统
earth tester　接地测试器
earth thermometer　地温计
earth thrust　土推力
earth to be borrowed　缺土
earth volume　土方量
earth wall　土墙
earthware pipe　瓦管
earth wax　地蜡
earth wire clamp　地线夹
earth wire fittings　地线金具，地线配件
earth wire for pole　电杆地线
earth wire hanger　地线吊架
earth wire　地线，接地线，架空地线
earthwork and stonework　土石方工程
earthwork balance sheet　土方平衡表
earthwork calculation　土方计算
earthwork operation　土方工程工序，土方工程作业
earthwork quantity sheet　土方表
earthwork section　土方断面，土方截面
earthworks　土方工程，土石方，土石方量
earthwork volume　土方工程量
earthy coal　土煤
earth　大地，接地，地面，泥土，地线，地，地球，土
easel　黑板架，框
easement boundary　使用权范围【土地】
easement curve　缓和曲线
easement　缓和
EAS(Energy Audit Scheme)　能源审计计划【英国】
ease of maintenance and repair　便于维护和修理
easer　掏槽爆破孔，辅助炮眼
ease the traffic pressure　缓解交通压力
ease　缓解
easily wear part　易磨损件
easily weathered　易风化的
easily worn part　易损零件
easing gear　松动装置【安全阀的】，阀门提升装置
easing lever　检查棒，听棒，试验杆
easing valve　减载阀，溢流阀
easing wedge　易脱楔块，对垫楔块

EAS(spray eductor) 喷淋引射器，化学添加剂引射器
east by north 东偏北
east elevation 东立面图，东向立面图
easterlies 东风带
easterly wave 东风波
eastern hemisphere 东半球
eastern suburb 东郊
east longitude 东经
east northeast 东北东
east southeast 东南东
east-west gas transmission project 西气东输工程
east-west oriented 东西走向的
east-west road 东西向的路
east wind 东风
easy axis 易磁化轴
easy curve 平缓曲线
easy dismantling 便于拆卸
easy grade 平缓坡度
easy money 低息贷款
easy processing channel 易处理通道
easy push fit 滑动配合
easy running fit 轻转配合
easy running 平衡运转，平稳运转，无振动运转
easy servicing 小修
easy slide fit 轻滑配合
easy to handle 易处理的，易操纵的
easy-to-sell goods 抢手货
EAS 安全壳喷淋系统【核电站系统代码】
EAT(earnings after tax) 税后收益额
eating away 侵蚀，腐蚀
eating 腐蚀，蚀坏
eat treatment 热处理
EAU 安全壳仪表系统【核电站系统代码】
eaves board 封檐板，风檐板，屋檐板，檐板
eaves ceiling 檐头顶
eaves fascia 封檐板，檐口板
eaves flashing 檐口泛水，披水，屋檐泛水，挡水板
eaves gutter 檐沟
eaves plate 封檐板，檐口板
eaves trough 檐沟
eaves 檐口，屋檐，檐
EBA 安全壳换气通风系统【核电站系统代码】
ebb and flow 涨落潮
ebb axis 落潮流轴
ebb channel 落潮水道
ebb current 落潮流，退潮流
ebb duration 落潮历时
ebb gate 落潮闸门
ebbing and flowing spring 潮汐泉
ebbing well 涨落井
ebb interval 落潮间隙
ebb stream 退潮流
ebb surge 落潮涌浪
ebb tide current 落潮流
ebb tide gate 落潮闸门
ebb tide 落潮，退潮
ebb volume 落潮量
ebb 退潮
EBF(externally blown flap) 外吹式襟翼
EBITDA(earnings before interest, tax, depreciation and amortization) 未计利息、税项、折旧及摊销前的利润息税折旧前受益
EBIT(earnings before interest and taxes) 息税前利润
ebonite clad cell 胶木包覆的电池
ebonite reed relay 胶木簧片继电器
ebonite 胶木，硬质橡胶，硬橡胶
EBOP(emergency bearing oil pump) 事故轴承油泵
EBP(exhaust back pressure) 排汽背压
EBR(electron beam recording) 电子束记录
EBSS(electron beam scanning system) 电子束扫描系统
EBT(earnings before tax) 税前收益
ebuillometer 沸点测定计
ebullated bed 沸腾床，流化床
ebullience 沸腾
ebulliometry 沸点测定（法）
ebullioscopy 沸点升高测定
ebullition chamber 沸腾室
ebullition 沸腾，起泡，冒泡
EBW(electron beam welding) 电子束焊
e-carbide e 碳化物
ECB(event control block) 事件控制块
ECC(eccentricity) 偏心度，偏心
ECC(emergency core cooling) 紧急堆芯冷却
eccentric action 偏心作用
eccentrically compressed member 偏心受压杆件
eccentrically loaded column 偏心荷载柱
eccentrically loaded footing 偏心载荷基础
eccentrically 偏心的
eccentric angle 偏心角
eccentric bit 偏心钻
eccentric bolt emergency governor 偏心锤式危急保安器
eccentric butterfly valve 偏心阀板蝶阀，偏心蝶阀
eccentric cam 偏心凸轮
eccentric compression 偏心受压
eccentric connection 偏心联结
eccentric crasher 偏心破碎机
eccentric disc 偏心盘
eccentric distance 偏心距
eccentric force 偏心力
eccentric gear 偏心齿轮，偏心机构
eccentric impact 偏心碰撞
eccentricity angle 偏心角
eccentricity correction 偏心率改正
eccentricity detector 偏心度指示器
eccentricity foundation 偏心基础
eccentricity indicator 偏心计
eccentricity of fire 火焰偏斜
eccentricity of loading 荷载偏心距
eccentricity of rest 静态偏心率
eccentricity ratio 偏心比
eccentricity recorder 偏心记录仪
eccentricity 偏心率，偏心距，偏心度，偏心
eccentric jaw crusher 颚式偏心碎石机
eccentric load 偏心负载，偏心荷载
eccentric mass vibrator 偏心质量激振器

eccentric observation　偏心观测
eccentric orifice plate　偏心孔板
eccentric plug　偏心塞子
eccentric pole face　偏心磁极面
eccentric pump　偏心轮泵
eccentric reducer　偏心异径管接头，偏心大小头，偏心异径管，偏心锥形管
eccentric ring emergency governor　偏心环式危急保安器
eccentric ring　偏心环
eccentric rotary valve　偏心旋转阀
eccentric rotation　偏转
eccentric rotor　偏心转子
eccentric shaft　偏心轴
eccentric tension member　偏心受拉杆件
eccentric tension　偏心受拉
eccentric valve　偏心阀
eccentric vibrating screen　偏心振动筛
eccentric weight　偏心重锤
eccentric wheel　偏心轮
eccentric　偏心的，偏心
ECC(equipment condition certification)　设备状态证书
ECC(equipment configuration control)　设备外形检查
ECC(error checking and correction)　误差检验与校正
ecclesiastical building　宗教建筑
ECCS(emergency core cooling system)　堆芯应急冷却系统
ECD(energy conversion device)　能量转换设备
EC(elasticity coefficient)　弹性系数
EC(electronic computer)　电子计算机
EC(environmental control)　环境控制
ECEP(engineering custom engineering proposal)　用户技术建议
EC(error correcting)　错误纠正
EC(extended control)　扩充控制
ECG(Electricity Company of Ghana)　加纳电力公司
echelon grating　阶梯光
echo altimeter　回波测高计，回声测高仪
echo amplifier　反射信号放大器
echo attenuation　反射回波，回波衰减
echo bearing　回波定位，回波方位
echo-box　回波共振器，回波谐振腔
echo check　返回检查
echo depth sounder　回声测深仪
echo distortion　回波畸变
echoed signal　反射信号
echo elimination　回波消除
echogenicity　回声反射性
echogram　回声测声曲线，回声波图
echograph　回声测深仪，音响测深自动记录仪，回声深度记录器
echo-image　回波像
echoing　回波现像，反照现象
echo intensity　回波强度
echolation　电磁波反射法
echolocation　回声测定，回声定位法，回声定位
echo loss　回波损耗
echo matching　回波匹配
echometer　回声测深仪
echo principle　信息反馈传输
echo pulse　回波脉冲
echo ranging　回波测距法，回声测距法
echo reverberation ratio　混响比
echo rings　回声振荡
echo signal of sounder　测深仪回波信号
echo signal　回波信号，反射信号
echo sounder　回声探测器
echo sounding apparatus　回声测深仪
echo sounding device　回声探测仪
echo sounding　回声测深，回声探测
echo suppressor　回波抑制器，反射信号抑制器
echo survey　回声探测
echo technique　回波法
echo trap　回波陷波器，回波抑制设备，回波阱，功率均衡器
echo wave　回声波，回波
echo　回声，回波，仿效，反射，重复，反复，回声波
ECI(emergency coolant injection)　冷却剂应急注入
Eckert number　埃克特数
E-class insulation　E 级绝缘
ECL(emitter-coupled logic)　电流开关逻辑，发射极耦合逻辑
eclipse　掩盖，月食
eclogite　榴辉岩
ECM(electronic counter-measure)　电子干扰，电子对抗
ECM(flue gas emission continuous monitoring system)　烟气监测系统
ECN(engineering change notice)　工程改变注释
ecoclimate forecasting　生态气候预测
ecoclimate　生态气候
ecocycle　生态循环
ECO＝economizer　省煤器
ecological balance　生态平衡
ecological capacity　生态容量
ecological character　生态性状
ecological crisis　生态危机
ecological cycle　生态循环
ecological damage　生态破坏
ecological destruction　生态破坏
ecological disaster　生态灾祸
ecological distribution　生态分布
ecological environment　生态环境
ecological equilibrium　生态平衡
ecological factor　生态因素
ecological group　生态群
ecological impact　生态冲击，生态影响
ecological indicator　植物风力指示
ecological map　生态地理图
ecological pollution limit　生态污染极限
ecological process　生态过程
ecological system　生态系
ecological type　生态型
ecological　生态的
ecology　生态学
ecolyte　光解塑胶

econometric method 经济计量学方法
econometrics 计量经济学
economic accounting of group 班组经济核算
economic accounting unit 经济核算单位
economic accounting 经济核算
economic activity analysis 经济活动分析
economic activity 经济活动
economic advisability 经济合理性
economic aid 经济援助
economic ailment 经济失调
economical and technical index 经济技术指标
economical current density 经济电流密度
economical effect of heat-supply 供热经济性
economical evaluation 经济性的评价
economical load dispatcher （电力）经济负载调度装置
economical load dispatching operation 经济负荷运行
economical load dispatching 经济负荷分配
economically-developed country 经济发达国家
economically-developed region 经济发达地区
economically undeveloped region 经济不发达地区
economical operation 经济运行
economical power factor 经济功率因数
economical principle 经济原则
economical rating 经济功率，经济出力
economical ratio of reinforcement 经济配筋率
economical running 经济工况运行
economical section area 经济截面积
economical speed 经济航速
economical thickness of insulating layer 保温层经济厚度
economical 经济的，节省的，故意隐瞒的，节俭的
economic analysis 经济性分析，经济分析
Economic and Social Council 联合国经社理事会
economic and social determinant 经济和社会决定因素
economic and technical development district 经济技术开发区
economic and technological cooperation 经济技术合作
economic and technological development area 经济技术开发区
economic and technological exchange 经济技术交流
economic appraisal 经济评价
economic assessment 经济性评价
economic base 经济基础
economic benefit analysis 经济效益分析
economic benefit evaluation 经济效益评价
economic benefits of enterprises 企业经济效益
economic benefit 经济效益
economic bid evaluation 经济评标
economic blockage 经济封锁
economic boiler 经济式锅炉【如回火管锅炉】
economic boom 经济繁荣
economic category 经济范畴
economic coefficient 经济系数
economic condition 经济工况
economic construction 经济建设
economic continuous rating 运转额定值，连续经济出力，连续经济工况
economic contract with foreigner 涉外经济合同
economic contract 经济合同
economic control 经济控制
economic cooperation 经济合作
economic coordination among regions 地区间经济协调
economic criteria 经济性准则
economic current density 经济电流密度
economic depreciation 经济贬值，经济折旧
economic depression 经济不景气，经济萧条
economic design 经济设计
economic development by leaps and bunds 经济腾飞
economic development plan 经济发展计划
economic development 经济发展，经济发展过程
economic diameter 经济直径
economic dispatcher 经济调度员
economic dispatch 经济调度
economic effect evaluation 经济效果评价
economic effectiveness 经济效益
economic effect 经济效应
economic efficiency 经济效益
economic entity 经济实体
economic environment 经济环境
economic equilibrium 经济平衡
economic evaluation of project 工程经济评价
economic evaluation 经济评价，经济比较
economic expressions 经济用语
economic feasibility 经济可行性
economic feature 经济性能
economic flux density 经济磁密
economic forecasting 经济预测
economic forum 经济论坛
economic framework 经济框架
economic gain 经济效益
economic groundwater yield 地下水经济产水量，地下水经济开采量
economic groundwater 地下水的经济出水量
economic growth point 经济增长点，经济集团
economic growth rate 经济增长率
economic growth 经济增长
economic impact evaluation 经济影响估计
economic impact 经济影响
economic index 经济指标
economic indicators 经济指标
economic information 经济信息
economic interest 经济权益
economic internal rate of return 内部经济收益率
economic jurisdiction 经济司法
economic justification 经济论证
economic law and regulation 经济法规
economic law 经济法则，经济规律
economic legislation 经济法规，经济立法
economic life 经济寿命
economic load dispatching 经济负荷调度
economic load dispatch 经济负载调度
economic load 经济负荷

economic management 经济管理
economic measure 经济措施
economic merit 经济价值
economic model 经济模式
economic net present value ratio 经济净现值率
economic net present value 经济净现值
economic operating mechanism 经济运行机制
economic operating mode 经济运行方式
economic operation 经济运行
economic optimization study 经济优化研究
economic order 经济秩序
economic output 经济出力，经济功率
economic performance 经济性能
economic planning 经济规划
economic potential 经济潜力
economic power 经济实力
economic prediction 经济预测
economic principle 经济原则
economic rating 经济出力
economic recession 经济衰退
economic reckoning 经济核算
economic recovery 经济恢复
economic reform 经济改革
economic region 经济区
economic rejuvenating plan 经济复苏计划
economic rejuvenation 经济复苏
economic resistance of heat transfer 经济传热阻
economic resources 经济资源
economic responsibility system 经济责任制
economic resuscitation 经济复苏
economic retrenchment policy 经济紧缩政策
economic right 财产权
economic running 经济运行
economic stability 经济稳定
economic status 经济地位
economic storage 经济库容
economic strength 经济实力
economic structural reform 经济体制改革
economic structure 经济结构
economic superiority 经济优势
economic system 经济体制
economics 经济学
economic thickness 经济厚度
economic turbine stages 经济级组
economic value 经济价值
economic velocity 经济流速
economic viability 经济可行性，经济能力
economic water yield 经济出水量
economic withholding 经济持留
economic worth of project 工程经济价值
economic yield 经济开采量
economist 经济师，经济学家
economize on electricity 节约用电
economize on manpower 节约人力
economize on raw materials 节约原料
economizer ash handing system 省煤器除灰系统
economizer coil 省煤器盘管
economizer hopper 省煤器灰斗
economizer junction 省煤器接口
economizer relay 节电继电器
economizer tube 省煤器管
economizer 省煤器，省油器，省热器
economizing land in use 节约用电
economy energy 经济能量
economy of design 设计的经济性
economy of energy 能源经济
economy of material 节约材料
economy of scale 规模经济
economy operation 经济运行
economy power 经济功率
economy resistance 电路启动电阻
economy resistor 经济电阻器
economy 经济性
ECON 省煤器【DCS 画面】
ecophysiology 生态生理学
E core E 形铁芯，山形铁芯
ECOSOC(Economic and Social Council) 联合国经社理事会
ecosphere 生态界，生态学领域，生态圈，生物大气层，生物域
ecosystem component 生态系要素
ecosystem conservation 生态保护
ecosystem dynamics 生态系动力学
ecosystem 生态系统，生态系
ecotage 生态破坏
ecotope 生态区
ecotoxicology 环境毒理学
ecotype 生态型
E-coupler E 形钩
ECPD(Engineers' Council for Professional Development) 工程师专业发展委员会
ecphlysis 破裂
ECR(economical continuous rating) 经济连续出力
ECROS(electrically-controllable read-only storage) 电可控只读存储器
ECS(electrical control system) 电气控制系统
ECS(extended core storage) 扩充的磁芯存储器
ECTL(emitter-coupled transistor logic) 射极耦合晶体管逻辑
ECU(European Currency Unit) 欧洲货币单位
ecumene 人类经常居住区
ecumenical 世界范围的
EDAC(error detection and correction) 错误检测与校正
edacity operating rate 全容量操作率
edacity 容量
EDBS(engineering data base system) 工程数据库系统
EDC(economic dispatch control) 经济调度控制
EDC(electronic digital computer) 电子数字计算机
EDC (engineering design change) 工程设计修改
EDC (engineering design collaboration) 工程设计协作
EDC(engineering drawing change) 工程图纸修改
EDC(exceed drum capacity) 超过磁鼓容量
EDC(external device code) 外部设备代码
EDCP(engineering design change proposal) 工程设计修改提议
EDCS(engineering design change schedule) 工程设计修改日程表
EDCW(external device control word) 外部设备控

制字
EDD(engineering design data) 工程设计数据
EDD(envelope delay distortion) 包线延迟失真
eddies frequency 涡旋频数
EDDP(engineering design data package) 工程设计资料包
eddy advection 涡动平流
eddy axis 旋涡轴线
eddy band effect 边缘带效应
eddy blurry 边缘模糊
eddy card memory 涡流卡片存储器
eddy conductivity 涡动传导率，涡团传导率
eddy contour 边缘轮廓
eddy correlation 涡旋相关
eddy current brake 涡流制动
eddy current coefficient 涡流系数
eddy current confinement 涡流抑制
eddy current coupling 涡流联轴节
eddy current damper 涡流阻尼器
eddy current damping 涡流阻尼
eddy current dynamometer 涡流测功机，涡流测功计
eddy current effect 涡流效应，涡流作用
eddy current examination 涡流探伤，涡流检验
eddy current factor 涡流系数
eddy current flaw detection 涡流探伤
eddy current flaw detector 涡流探伤机
eddy current flowmeter 涡流流量计
eddy current gauge 涡流计
eddy current heating 涡流发热，涡流加热
eddy current inspection instrument 涡流探伤仪
eddy current inspection 涡流探伤检查
eddy current instrument 涡流测量仪器
eddy current loss 涡流损失，涡流损耗，涡电流损耗
eddy current method 涡流法
eddy current motor 涡流电动机
eddy current probe 电涡流传感器，涡流探头
eddy current revolution counter 涡流转速计
eddy current screen 涡流屏蔽
eddy current testing method 涡流探伤法
eddy current test 涡流探伤检验［试验］，涡流探伤，涡流探测［检测］
eddy current transducer 涡流传感器
eddy current 涡流，涡流电流，涡电流
eddy-deposited silt 旋涡沉积泥沙
eddy diffusion 涡流扩散，紊流扩散，扰动扩散，涡动扩散，旋涡扩散
eddy diffusivity for heat 涡流热扩散率
eddy diffusivity 涡流扩散度，紊流扩散率，涡流扩散率，涡团扩散率，涡动扩散率
eddy echo 边缘回声
eddy effect 涡流作用
eddy energy 涡动能，旋动能，涡动能量
eddy error 涡流误差
eddy exchange coefficient 涡动交换系数
eddy filter 流线式过滤器
eddy flow 紊流，涡流
eddy flux 边缘通量，涡度通量，涡旋通量
eddy generation 发生涡流
eddy gradient 边缘梯度
eddying effect 紊流影响，涡流效应
eddying flow 紊流流动，涡流流动，涡流，旋流
eddying motion 紊流运动，涡流运动
eddying resistance 涡流阻力
eddying turbulence 涡旋，涡流
eddying wake 尾涡流
eddying 涡流的，紊流的，旋转的，旋涡，涡流，紊流度
eddy iron 角铁
eddy kinetic energy 涡旋动能
eddy length stale 涡旋长度尺度
eddy line 涡流线
eddy loss 涡流损耗，涡流损失
eddy mill 涡流粉碎机，涡流式碾磨机，涡穴
eddy mixing 涡动混合，涡流交混
eddy momentum flux 涡动量通量
eddy motion 涡动，涡流，涡流运动
eddy peak echo method 棱边波法
eddy Prandtl number 涡动普朗特数
eddy production 旋涡形成
eddy region 涡流区
eddy resistance 涡流阻力，防止涡流
eddy scale 涡流尺度
eddy shear stress 涡动剪应力
eddy shedding 涡流散发，紊流分离
eddy-sonic test method 涡流声检验法
eddy spectrum 涡谱
eddy stress 湍流应力，涡动应力，涡流应力
eddy transfer coefficient 涡动传递系数
eddy transfer theory 涡动传递理论
eddy transfer 涡流输送
eddy velocity 涡动速度，涡流速度
eddy viscosimeter 涡流黏度，湍流黏度
eddy viscosity 涡动黏性，涡流黏度，涡流黏滞度
eddy water 涡流
eddy with horizontal axis 横轴旋涡
eddy with vertical axis 立轴旋涡
eddy zone 紊流区，涡流区
eddy 涡度，涡动，旋涡，涡流，涡团，涡旋，涡，旋转，起漩涡
ED(economic dispatching) 经济调度
ED(engineering description) 技术说明书
ED(executive director) 执行董事，副总经理【主抓某一领域】
EDF(Electricite De France) 法国电力公司【法语】
edge action 边缘作用
edge-bar reinforcement 边缘钢筋
edge beam 边梁
edge block 端部线匝间的垫块
edgeboard connector 边缘板连接器，卡片边沿连接器，印刷电路板插头，板边插头
edge bolt 边缘螺栓
edge capacitance 边缘电容
edge channel 边通道
edge coil 扁绕线圈
edge condition 边界条件
edge contour 边界形状
edge control element 周边控制元件
edge corrosion 边缘腐蚀

edge crack 边缘裂缝，边裂缝
edge current probe 涡流探头
edge current 边缘电流
edge dislocation 边缘错位
edge distance 边距，孔边距
edge disturbance 边缘干扰
edge effect 边缘效应，边际效应，边缘裂缝，边界影响，边界效应
edge etching 腐蚀边
edge fairing 边缘整流
edge filter 流线式过滤器
edge-flange joint 卷边接头
edge flux 边缘磁通，边缘通量
edge forming 棱边成形
edge frequency 截止频率
edge fuel assembly 周边燃料组件
edge illuminated 边缘照明
edge-iron 角铁
edge joint 边缘接头，端部连接，边缘接缝，端接接头
edge leakage current 边缘漏电流
edge-lighted display 边缘发光显示
edge lighting 边缘照明
edge load stress 边载应力
edge load 边缘荷载
edge loss 涡流损失
edge member 边缘构件
edge of continental shelf 大陆架边缘
edge of pipe 管壁
edge of the right-of-way 走廊边缘
edge phenomena 边缘现象
edge plate 镶边平板
edge preparation of tube welding ends 管子坡口加工
edge preparation 边缘加工，刨边，开坡口，坡口加工
edge protection 保护边
edge reinforcement 边缘钢筋
edge shedding 涡流散发
edge stress 边界应力，边缘应力
edge-thickness coefficient 阻塞系数【叶片厚度造成的】
edge to be welded 需要焊接的棱边
edge trace 边缘迹线
edge water 层边水
edge wave 边缘波
edge weld 端接焊缝
edge winding 扁绕绕组
edge wind 扁绕
edgewise bending 缘向弯曲，扁弯，扁绕
edgewise indicator 边缘读数式指示表
edgewise meter 边缘读数式仪表
edgewise welding 沿边焊接，扁绕，扁绕绕组
edgewise wound coil 扁绕线圈
edgewise 沿边，边对边地
edge-wound coil 扁绕线圈
edge-wound field coil 扁绕磁极线圈
edge 边缘，边，棱，边界，边沿，刃口，使锐利，棱边，优势
edging machine 折边机，弯边机，削边机，磨边机
edible aquatic organism 食用水生生物
EDI(electron diffraction instrument) 电子衍射仪
edifice 大型建筑
E-D inverter (enhancement-depletion inverter) 增强-耗尽型倒相器
EDIS(engineering data information system) 工程数据信息系统
E-display E 型显示
edit capability 编辑能力
edition 出版，版本，刊行，版，编排，编辑
editor program 编辑用程序
editor 编辑程序，编辑器，编者
EDM(electrical discharge machining) 放电加工，电火花加工，放电机械加工
EDM(electric dipole moment) 电偶极矩
EDNS(expected demand not supplied) 期望缺供电力
EDO(executive director for operations) 业务执行主任
EDPC(electronic data processing center) 电子数据处理中心
EDP center(electronic data-processing center) 电子数据处理中心
EDPE(electronic data processing equipment) 电子数据处理设备
EDP(effective depth of penetration) 有效穿透深度
EDP(electronic data processing) 电子数据管理，电子数据处理
EDP machine 字段数据处理机
EDPM(electronic data processing machine) 电子数据处理机
EDPS(electronic data processing system) 电子数据处理系统
EDR(engineering documentation release) 技术文件发放
EDRS(engineering data retrieval system) 工程数据检索系统
EDS(extended data set) 扩充数据集
EDST(electric diaphragm switch technique) 电膜片开关技术
educational analogue computer 教学用模拟计算机
educational background 学历
educational fund 教育经费
educational investment 教育投资
educational qualifications 学历
education and training planning 教育培训计划
education for all-round development 素质教育
educe 排出
eduction gear 排出装置
eduction pipe 乏汽管，放气管
eduction port 排出口，排气口
eduction 离析，引出，排出，提取，排泄
eductor type feeder 喷射型加药器
eductor well point 喷射井点
eductor 喷射器，引射器，排泄器，注射器
EDU(electronic display unit) 电子显示器
edulcorate （从数据文件中）剔除无关数据，使纯化
EDX(energy dispersive X-ray analysis) 能量分散

X射线分析
ED 配电【DCS画面】
EEA(Electrical Equipment Association) 电气设备协会
EEC (Electronic Equipment Committee) 电子设备委员会
EEC(end of erection certification) 安装竣工证书
EEC(European Economic Community) 欧洲经济共同体
EECS(electrical engineering and computer service) 电力工程和计算机服务
EED(Electrical Engineering Department) 电机工程部
EEDM(external event detection module) 外面事件检测模块
EEE(electrical and electronics engineering) 电气和电子工程
EEE(electronic engineering equipment) 电子工程设备
EEIC(emergency feed water initiation and control) 应急给水触发与控制
EEIC(European Electronic Intelligence Center) 欧洲电子情报中心
EEI(Energy Efficiency Index) 能源效率指数，能效指数
EEI(essential elements of information) 情报要点
EEMF to current transmitter 电动势—电流变送器
EEMUA(Engineering Equipment and Materials Users' Association) （欧洲）工程设备与材料用户协会
EEPD(effective full power days) 有效满功率天数
EEPH(effective full power hours) 有效满功率小时
EEPY(effective full power years) 有效满功率年
EER (energy efficiency ratio) 能效比，能源效率比值
EER(European Exhibition Reactor) 欧洲展览反应堆
EEROM (electrically-erasable read-only memory) 电可擦只读存储器
EES(emergency exhaust system) 应急排放系统
EESR(end of erection status report) 安装竣工报告
EFDH(equivalent forced derated hour) 等效强制降低出力小时
EFDL(emitter follower diode logic) 发射极跟随器二极管逻辑
EFDS(equipment and floor drain system) 设备和地面疏水系统
EFDTL(emitter follower diode transistor logic) 发射极跟随器发射极耦合逻辑
effectance 效量
effect and maintain the insurance of 对……投保并维持其有效
effect insurance 保险，投保
effective 有效的，起作用的，有效
effective abstractions 有效降雨损失量
effective acoustic center 有效声中心
effective active fuel length 燃料有效长度
effective actuation time 有效激励时间
effective address 有效地址
effective admittance 有效导纳
effective aerodynamic downwash 有效气动下洗
effective angle of attack 有效迎角
effective angular field 有效视场
effective aperture area 有效口径区域
effective aperture 有效孔径
effective application 有效应用
effective area of antenna 天线有效面积
effective area of concrete 混凝土有效面积
effective area of orifice 孔口有效面积
effective area of reinforcement bar 配筋的有效面积
effective area 有效面积
effective aspect ratio 有效展弦比
effective atomic number 有效原子序数
effective attenuation factor 有效衰减系数
effective barrier 有效屏蔽，有效屏障
effective buoyancy 有效浮力
effective cadmium cutoff 镉有效吸收界限
effective camber 有效弯度
effective capacitance 有效电容
effective capacity 有效能力，有效容量，有效功率，实际出力，有效出力
effective case depth 有效硬化层深度
effective chimney height 有效烟囱高度
effective chord length 有效弦长
effective circumference length 有效周长【焊接】
effective clearance 有效间隙
effective coefficient of local resistance 折算局部阻力系数
effective cohesion 有效内聚力
effective collision cross-section 有效碰撞截面，有效撞击截面
effective column height 柱的有效高度
effective column length 柱的有效长度
effective conductance 有效电导
effective conductivity 有效电导率
effective contact area 有效接触面积
effective control 有效控制
effective coverage 有效作用区域，有效范围
effective cross current 有效横流，无功电流
effective cross-section area 有效截面积
effective cross-section 有效截面，有效截面积
effective current 有效电流，有效电流值
effective cycle time 有效循环时间
effective data-transfer rate 平均数据传输速度，有效数据传输率
effective date of agreement 协议的生效日期
effective date of regulations 条例生效日期
effective date 合同生效日，生效日期，有效日期
effective day 有效日
effective decay constant 有效衰变常数
effective decontamination factor 有效去污系数
effective decontamination 有效净化
effective delayed neutron fraction 缓发中子有效份额，有效缓发中子份额
effective density 有效密度
effective deposition velocity 有效沉积速度
effective depth of section 截面有效高度
effective depth 有效深度

effective diameter of grain 有效粒径
effective diameter of pipe 有效管径
effective diameter 有效直径
effective diffusion coefficient 有效扩散系数
effective diffusivity 有效扩散率
effective directivity factor 有效方向因子
effective discharge 有效排放量，有效流量
effective dispersion coefficient 有效扩散系数
effective distance 有效距离
effective dose equivalent commitment 有效剂量当量负担
effective dose equivalent 有效剂量当量
effective dose 有效剂量
effective downwash 有效下洗
effective drainage area 有效汇水面积，有效排水面积
effective drop 有效热降，可用热降
effective due 生效日期
effective duration of delay 实际延误的工期
effective duration 有效历时
effective edition 有效版本
effective efficiency 有效效率，有效功能
effective electromotive force 有效电动势
effective elongation 有效伸长
effective energy 有效能量，有效功率
effective entrainment coefficient 有效夹带系数
effective environment 有效环境
effective evapotranspiration 有效蒸散
effective evidence 有效证明［证据］
effective field intensity 有效场强
effective field 有效场
effective filtration depth 有效过滤深度
effective flange width 有效翼缘宽
effective flow area of core 堆芯有效流通面积
effective flow parameter 有效流动参数
effective flow rate 有效流量
effective flow resistance 有效流动阻力，有效流阻，动态流阻
effective flow 有效流量
effective focal length 有效焦距
effective full power days 有效满功率天数
effective full power hours 有效满功率小时数
effective furnace temperature 有效炉温
effective gap spacing 有效气隙间距
effective grain diameter 有效粒径
effective grain size 有效粒径
effective grounded line 有效接地线
effective ground level 有效地面高度
effective guarantee 有效保证
effective half-life 有效半减期，有效半衰期
effective Hall parameter 有效霍尔参数
effective hardening depth 有效淬硬深度
effective head 有效落差，有效水头
effective heating surface 有效受热面
effective heat transfer coefficient 有效传热系数
effective heat transfer surface 有效传热表面
effective height of stack 有效烟囱高度
effective height 有效高度
effective horsepower 有效功率，有效马力，轴功率
effective impedance 有效阻抗
effective incidence 有效冲角
effective inductance 有效电感
effective inertial-mass 有效惯性质量
effective input admittance 有效输入导纳
effective instruction 有效指令，有效互作用
effective interest rate or yield 实际利率或收益率
effective kilogram 有效千克
effective length of pipe 管有效长度
effective length of tube 管子有效长度
effective length 有效长度，计算长度，折算长度
effective lens aperture 透镜有效孔径
effective life 有效寿命
effective lifetime 有效寿命，有效寿期
effective load 有效负荷，有效负载，有效荷载
effective luminous intensity 有效光强
effectively grounded 有效接地，有效接地的
effective management 有效管理
effective mass 有效质量
effective measure 有效措施
effective mixing distance 有效交混距离
effective mixing layer 有效混合层
effective modulation 有效调制
effective modulus of elasticity 有效弹性模量
effective multiplication constant 有效放大因子，有效增殖常数
effective multiplication factor 有效倍增因子，有效增殖系数，有效放大因子，（中子）有效增殖系数
effectiveness factor 角系数【炉膛计算】
effectiveness of contract 合同生效
effectiveness of heat exchanger 换热器的有效度
effectiveness 效用，效率，有效性，有效度，效果，效能，有效温比，效力，效应
effective neutron lifetime 有效中子寿期
effective neutron temperature 有效中子温度
effective normal stress 有效法向应力
effective number of fission neutrons 有效裂变中子数
effective number of turns per phase 每相有效匝数
effective of dust removal 除尘效率
effective opening 有效孔径
effective output admittance 有效输出导纳
effective output 有效出力，有效输出，有效功率，有效输出功率，有效产量
effective overburden pressure 有效覆盖压力，有效上覆压力
effective passage throat 有效喉部截面，有效喉道截面
effective penetration 有效穿透深度
effective period 有效期，有效周期，有效期限，有效使用期
effective permeability 有效磁导率
effective phreatic line 有效浸润线
effective pillar length 柱的有效长度
effective plume height 有效羽流高度
effective plume rise 有效羽流抬升
effective plume 有效羽流
effective porosity 有效孔隙度，有效孔隙率
effective power factor 有效功率因数
effective power head 有效发电水头
effective power output 有效功率输出

effective power 有效动力，有效功率
effective precipitable water 有效可降水量
effective precipitation 有效降水量
effective pressure 有效压力
effective prestress 有效预应力
effective prevention distance 有效防护距离
effective price 有效价格
effective projected radiant surface 有效投影辐射受热面
effective projection radiation area 有效投影辐射区域
effective radiated power 有效辐射功率，有效发射功率
effective radius 有效半径
effective rainfall 有效雨量
effective range of spray nozzle 水雾喷头的有效射程
effective range 使用范围，有效测量范围，有效距离，有效范围
effective reactance 有效电抗
effective reinforcement 有效钢筋
effective relaxation length 有效张弛长度
effective relay actuation time 继电器实际动作时间
effective removal cross-section 有效移出截面
effective resistance 有效电阻，交流电阻
effective resistivity 有效电阻率
effective resonance integral 有效共振积分
effective rod radius 控制棒的有效半径
effective saturation line 有效浸润线
effective screen aperture 有效筛孔
effective screening area 有效筛分面积
effective sectional area 有效截面面积
effective section 有效截面
effective shadow zone 有效声影区
effective shearing rigidity 有效抗剪刚度
effective shear strength parameter 有效抗剪强度参数
effective signal 有效信号
effective size of sand 砂的有效粒径
effective size 有效尺寸，有效粒径
effective sky temperature 天空（有效）温度
effective slab width 有效板宽
effective slit width 有效缝隙宽度
effective source area 有效源面积
effective source height 有效源高
effective source strength 有效源强
effective span 有效跨度
effective speed 有效速度
effective stack height 烟囱有效高度
effective storage for reservoir 有效库容
effective storage 有效蓄水量，有效库容，有效蓄水
effective strain 等效应变
effective stress path 有效应力路径
effective stress theory 有效应力理论
effective stress 有效应力
effective summed horsepower 总有效功率，总有效马力
effective superficial porosity 有效表面空隙率
effective superstructure 有效上层建筑
effective support length of beam end 梁端有较支承长度
effective surface 有效表面，有效表面积
effective tax rate 实际税率
effective temperature difference 送风温差
effective terrain height 有效地形高度
effective thermal conductivity 有效导热系数，有效热导率
effective thermal cross-section 有效热中子截面
effective thermal efficiency 有效热效率
effective threshold energy 有效阈能
effective throat thickness 有效焊缝厚度
effective throat 风洞工作段有效面积，焊缝有效厚度，焊缝计算厚度
effective thrust 有效推力
effective till cancelled 在未注销前有效
effective time constant 有效时间常数
effective time 有效工作时间
effective torque 有效力矩
effective tractive power 有效牵引功率，有效牵引能力
effective transmission rate 有效传输率
effective tube length 有效管长，有效管子长度
effective turbulent mixing model 有效湍流交混模型
effective turn 有效线匝，有效匝数
effective unit weight 有效容重
effective value 有效值，均方根值
effective velocity 有效流速，有效速度
effective voltage gradient 有效电压梯度
effective volume 有效容积，有效体积，有效容量
effective water-holding capacity （土壤的）有效持水量，有效持水能力
effective wind speed 有效风速
effective work 有效功
effect of action 作用效应
effect of an accident 事故的影响范围，事故的影响
effect of boundary layer 边界层效应
effect of climate 气候效应
effect of contraction 收缩影响，约束作用
effect of corrosion inhibition 腐蚀抑制作用
effect of current 流动影响
effect of delayed neutrons 缓发中子效应
effect of dilution 稀释效应
effect of holes 孔的应力集中效应
effect of irradiation 辐照效应
effect-of-load power factor 负载功率因数
effect of partial invalidity 部分无效的效力
effect of pile group 桩群作用
effect of pollution 污染影响
effect of radiation on convection effect 辐射对对流作用的影响
effect of streaming squeezing 流线密集效应
effect of surrounding 环境影响
effect of tilt 倾斜角效应
effect of variable conductivity 变导热系数的影响
effect payment 支付
effect price 有效价格
effects of an accident 事故的影响范围

effects of building downwash 建筑物下洗效应
effects of saturation 饱和效应
effect 效应，作用，效能，影响，效果
EFF＝efficiency 效率
effervescence 泡腾（现象），冒泡，起沫，发泡
effervescency 泡腾（现象），冒泡，起沫，发泡
effervescing steel 沸腾钢
efficacious 有效的
efficacy 效力，功效，有效
efficiency band 效率频带
efficiency calculation 效率计算
efficiency characteristic 效率特性
efficiency curve 效率曲线
efficiency diode 高效率二极管，阻尼二极管
efficiency estimation 效率估算，有效估计
efficiency extrapolation method 效率外推法
efficiency factor 效率因数，效率因素
efficiency formula 效率公式
efficiency hill 等效率（曲线）图
efficiency indication 效率指标
efficiency of blending 配料效率，混合效率
efficiency of compression 压缩比，压缩系数
efficiency of conversion 变换效率
efficiency of cycle 循环效率，热效率
efficiency of heat engine 热机效率
efficiency of heating surface 加热面效率
efficiency of heat transfer 传热效率
efficiency of heat utilization 热利用率
efficiency of investment 投资效率
efficiency of iron removal 除铁效率
efficiency of labor 劳动生产率，运行效率
efficiency of operation 工作效率
efficiency of plant 电站效率
efficiency of propulsion 推进效率
efficiency of pump 泵的效率
efficiency of regeneration 回热系数
efficiency of scale 规模效益
efficiency of solar array 方阵效率
efficiency of thermal cycle 热力循环效率
efficiency of transformer 变压器效率
efficiency of transmission 输电效率，传输效率
efficiency of turbine 汽轮机效率，透平效率
efficiency of utilization 热利用效率
efficiency parameter 效率参数
efficiency-power curve 效率出力曲线，效率功率曲线
efficiency ratio 效率比
efficiency reduction 效率减小
efficiency-related wage 效益工资
efficiency test 效率试验
efficiency 效率，效能，效力，实力，有效性，功能，效益
efficient circuit 有效电路
efficient combustion 高效率燃烧
efficient energy 高能
efficient manner 有效方式
efficient market 有效市场
efficient operation 有效运行
efficient plant operation 工厂有效运转
efficient radiated power 有效辐射功率
efficient 有效的
efflorescence 风化【混凝土】，风化物，凝霜，渗斑，粉化，霜斑
effloresce 风化，粉化
effluence 流出物，排出流，射出物，发出，射出，流出，排出
effluent activity meter 流出液活度测量计，流出液放射性测量计
effluent bed volume 流出液床体积
effluent characteristics 出水质量
effluent control system 污水控制系统，流出液控制系统
effluent criteria 排放准则
effluent dilution 废水稀释
effluent disposal 废液处理，废液处置
effluent emission 废气排放
effluent farm 排水区［段］，废水区
effluent flow 废气流
effluent fluid 流出流体
effluent fractioning 流出物的分馏
effluent from nuclear power plant 核电厂流出物
effluent gas 废气，烟道气，排出气体
effluent hopper 疏水斗，排水漏斗
effluent-impounded body 地下水补给水体
effluent limitation criteria 排放标准
effluent limitation guideline （液、固、气）废物限制准则
effluent monitor 排出流监测计，污水监测计
effluent oil recovery 污油回收
effluent oil treatment 污油处理
effluent per unit length 单位长度出流
effluent pipe 出水管
effluent plant 污水处理厂
effluent plume 废气羽流
effluent quality standard 排出水质标准
effluent radiation monitor 排出物的辐射监测仪
effluent release permit 排出物的释放许可
effluent seepage 坡面渗流，渗出流，渗出，污水渗漏
effluent sewage 流出污水
effluent standard 排放标准，排污标准
effluent stream 流出气流，流出液流，出流河
effluent treatment 废水处理，排出物处理
effluent trough 出水槽
effluent water 滤液，流出水
effluent weir 出流堰，出水堰
effluent 废水【复数】，流出的，溢流，排出的，流出物，流出液，出口水，支流，排出流，污水，废液，废气，流出液体，排出水，排出物，排放物
effluve 静电电机或高频发电机的电晕放电
effluvium 无声放电，放出，恶臭气
efflux angle 流出角
efflux area 出流面积
efflux coefficient 流量系数，射流系数，流出系数
efflux equation 出流方程
effluxing flow rate 排出流量
efflux of solids 固体流
efflux point 排放点
efflux pressure 排出压力
efflux shell 流出势面

efflux time 喷射时间
efflux tube 出水管
efflux velocity 排出速度，排放速度，射流速度
efflux-windspeed ratio 排放风速比
efflux 流出，射流，流出物，废气流，排出水，荷载，结束，告终，期满，届期
effuser 扩散管，喷管，加速管，扩散喷管，风洞收敛段
effusion cooling 发散冷却，多孔蒸发冷却
effusion 流出，溢出，泄流，流出物，隙透，渗出（量）
effusive rock 喷发岩
effusive 射流的，喷发的，溢出的，流出的
EFL(emitter follower logic) 发射极跟随器逻辑
EFL(error frequency limit) 错误频率极限
e-folding rate 指数倍率，e倍率
e-folding time e倍时间【即反应堆周期】
e-fold length e部（衰减）长度
EFOR(equivalent forced outage rate) 等效强迫停运率
EFPD(equivalent full power days) 等效满功率日
EFPH(equivalent full power hours) 全等效功率小时
EFRUG(European Fast Reactor Utilities Group) 欧洲快堆电力公司集团
EFTP(Ethernet File Transfer Protocol) 以太网文件传输协议
EFT/POS(electronic funds transfer at point of sale) 销售处款项［资金］电子转账
EFV(electric field vector) 电场矢量
e.g.(exempli gratia) 例如，举例，作为例子
eggbeater wind turbine 打蛋器型风轮机【即达里厄型风轮机】
egg coal 中块煤（25～50mm）
egg crate arrangement 蛋篓型排列，栅元排列【燃料组件格架】
egg-crate grid 蛋篓型格架
egg-crate spacer 蛋篓型定位架
egg-crate type 蛋篓型
egg end 球形封头
egg-insulator string 蛋形绝缘子串
egg-insulator 蛋形绝缘子
egg-shaped cross section 卵形截面
egg-shaped sewer 蛋形污水道
egress of heat 放热，热传导
egress 出口，出路，运出，溢出，流出，排出
EHC(electro-hydraulic control) 电液调节，电动液压控制，电液控制
EHCS(electro hydraulic control system) 电液调节系统
EHD(electrohydrodynamic) generator 电流体动力发电机
EH=electrohydraulic 电液
EHF(extremely-high frequency) 极高频
EHP(effective horsepower) 有效功率
EHV AC power system 超高压交流电力系统
EHV AC power transmission 超高压交流输电
EHV bushing 超高压套管
EHV DC power system 超高压直流电力系统
EHV DC power transmission 超高压直流输电
EHV(extra high voltage) 极高压，超高压
EHV power transmission line 超高压输电线路
EHV substation 超高压变电站
EHV transformer 超高压变压器
EHV transmission line 超高压输电线路
EHV transmission 超高压输电
EHWL(extreme high water level) 极高水位
EHX(external heat exchanger) 外部热交换器
EH 电液【DCS画面】
EIA(Electronic Industries Association) 美国电子工业协会
EIA(Energy Information Administration) 美国能源信息署
EIA(environmental impact assessment report) 环境影响评估报告
EIA(environment impact assessment) 环境影响评价，环境影响评估
EIA(error in address) 地址错误
EIA qualification 环境影响评价资质
eiectrodynamic relay 电动继电器，电动式继电器
EI(enable interrunt) 允许中断
Eiffel type wind tunnel 埃弗尔式开路风洞
eigen frequency of vibration 振荡本征频率
eigen frequency 固有频率，特征频率，本征频率
eigen-function 特征函数，本征函数
eigen mode 本征波形
eigen period 固有周期
eigenphasor 本征相量
eigen value problem 特征值问题
eigen value 本征值，特性值，特征值，固有值
eigen vector 本征矢量，特征矢量
eigenvibration 固有振动
eigenwert 特征值
eight-hour shift 八小时工作制，八小时一班制
eight hours rating 八小时额定量
eight-place 八地址的
EIL(error in label) 标号错误
EIO(error in operation) 操作错误
EIR(engineering information retrieval) 工程信息检索
EIR(Environmental Impact Report) 环境影响报告
EIR(equipment interchange receipt) （集装箱）设备交接单
EIRP(equivalent isotropic radiated power) 等效各向同性辐射功率
EIS(end interruption sequence) 结束中断序列
EIS(environment impact statement) 环境影响报告书【预可研阶段】
EJCTR(ejector) 抽气器，喷射器
ejecta 喷出物
ejected scoria 喷出火山渣
ejection accident 弹棒事故，控制棒弹出事故
ejection pump 喷射泵
ejection stroke 排气冲程
ejection type wind tunnel 引射式风洞
ejection velocity 弹射速度【飞射物】
ejection 喷出，排斥，排出，发射，抛出，发出，喷射，喷出物，抛出物，射出，弹射
ejector air pump 气喷射泵，打气泵，喷气式水泵，气流喷射泵

ejector condenser 喷水凝汽器，射汽抽气式凝汽器，喷射式凝汽器［冷凝器］，喷射凝汽器［冷凝器］
ejector cooler 抽气冷却器
ejector jet pump 喷射泵，注油器
ejector motive steam 抽气器工作蒸汽
ejector nozzle 喷口，喷射器，喷嘴，射流喷嘴
ejector pump 引射泵，喷射泵
ejector system 喷口送风系统，射流抽气系统
ejector 引射器，发射器，射水器，喷射器，抽气器，喷射泵，喷水器，射流器，推顶器，排出器，推力器
eject 喷出物，排出物，排出
Ekman boundary layer 埃克曼边界层
Ekman layer 埃克曼螺线层
Ekman number 埃克曼数
Ekman spiral 埃克曼螺线
Ekman turning 埃克曼偏转
EKW(electrical kilowatts) 千瓦
elaborate 详尽说明，详细制定，详尽阐述
elapsed time controller 消逝时间调节器
elapsed time indicator 经过时间指示器
elapsed time 耗用时间，消逝时间，经过时间
elapse 消逝，过去
elastance 电容的倒数，倒电容
elastic abutment 弹性拱座
elastic after effect 弹性后效
elastically yielding bearing 弹性退让轴承
elastic analogy 弹性模拟
elastic analysis scheme 弹性方案
elastic analysis 弹性分析
elastic axis 弹性轴
elastic axle 弹性轴
elastic beam 弹性梁
elastic bearing 弹性轴承
elastic body 弹性体
elastic boundary condition 弹性边界条件
elastic breakdown 弹性失效
elastic brittle material 弹性脆性材料
elastic buckling 弹性失稳，弹性屈曲，弹性弯曲
elastic buffer 弹性缓冲器
elastic center 弹性中心
elastic coefficient 弹性系数
elastic collision 弹性碰撞
elastic compliance 弹性顺度，弹性柔度
elastic compression 弹性压缩
elastic constant 弹性常数
elastic coupling 弹性联轴节，弹性联轴器，弹性连接
elastic criterion 弹性失效准则
elastic curve 弹性曲线
elastic deformation 弹性变形
elastic design 弹力计算，弹性设计
elastic displacement 弹性位移
elastic distortion 弹性畸变
elastic drift 弹性残留变形
elastic elongation 弹性伸长，弹性延伸
elastic energy 弹性能
elastic equilibrium 弹性平衡
elastic failure 弹性破坏，弹性失效，弹性疲劳
elastic feedback 弹性反馈【调节系统】
elastic fiber 弹性纤维
elastic fluid 弹性介质，可压缩流体
elastic force 弹力，弹性力
elastic foundation 弹性地基，弹性基础
elastic frame 弹性机座
elastic hardness 弹性硬度
elastic homogeneous material 弹性均质材料
elastic hysteresis 弹性滞后，弹性磁滞
elastic impact 弹性冲击
elastic instability 弹性不稳定性，弹性不稳定
elastic isotropic material 弹性各向同性材料
elasticity constant 弹性常数
elasticity coupling 弹性耦合
elasticity factor 弹性因数
elasticity limit 弹性极限
elasticity matrix 弹性矩阵
elasticity modulus 弹性模量
elasticity of demand for electricity 电力弹性系数
elasticity of demand 需求弹性
elasticity 弹性力学，弹性，弹力
elastic joint 弹性接头
elastic limit 弹性限制，弹性限度，弹性极限
elastic line method 弹性线法
elastic line 弹性线
elastic load 弹性载荷
elastic loss 弹性损失
elastic membrance 弹性膜片
elastic module 弹性模数
elastic modulus 弹性模数，弹性模量，杨氏模量
elastic packing 弹性填料，弹性密封
elastic plastic behavior 弹塑性能，弹塑性状态
elastic plastic boundary in flexure 弯曲弹塑性边界
elastic plastic deformation 弹塑性变形
elastic plastic fracture mechanics 弹塑性断裂力学
elastic plastic fracture 弹塑性断裂
elastic plastic material 弹塑性材料
elastic plastic solid 弹塑性固体
elastic plastic state 弹塑性状态
elastic plastic stress distribution 弹塑性应力分布
elastic pressure gage 弹性压力表
elastic range 弹性范围，弹性区
elastic ratio 弹性比率，弹性模数，弹性系数
elastic reaction 弹性抗力
elastic rebound theory 弹性回弹理论
elastic recovery 弹性回复，回弹
elastic region 弹性区，弹性区域
elastic resilience 弹性回弹，回弹性
elastic resistance 弹性抗力，弹性阻力
elastic response 弹性反应
elastic restoring force 弹性回复力
elastic rotor 弹性转子
elastics 弹性物，松紧带
elastic scaling 弹性缩尺
elastic scattering collision 弹性散射碰撞
elastic scattering cross section 弹性散射截面
elastic scattering resonance 弹性散射共振
elastic scattering 弹性散射
elastic seminfinite body 弹性半无限体
elastic settlement 弹性沉降量

elastic shakedown 适应性试验【应力分析】，弹性稳定试验
elastic shell equation 弹性薄壳方程
elastic shock 弹性冲击
elastic shortening of concrete 混凝土的弹性压缩
elastic side wall 弹性边墙
elastic similarity 弹性相似
elastic stability 弹性稳定性，弹性稳定
elastic stage 弹性阶段
elastic state of equilibrium 弹性平衡状态
elastic stiffness 弹性刚度
elastic strain energy 弹性应变能
elastic strain 弹性应变，弹性形变，弹性变形
elastic strength 弹性强度
elastic stress 弹性应力
elastic subgrade 弹性路基，弹性地基
elastic support 弹性支撑，弹性支座
elastic suspension 弹性支承
elastic system 弹性体系，弹性系统
elastic theory 弹性理论
elastic twist 弹性扭转
elastic vibrating element 弹性振动元件
elastic vibration 弹性振动
elastic washer 弹簧垫圈
elastic wave 弹性波
elastic 弹性的，有弹力的，有弹性的，灵活的
elastivity 倒电容系数，倒介电常数，介电常数的倒数
elastomer element 弹性元件
elastomeric gasket 弹性垫圈，合成橡胶垫圈，弹性体密封垫
elastomeric network 弹性网络
elastomer 弹性体，弹胶物，塑料混凝土掺合料
elastometer 弹性计
elastometric flexible coupling 挠性联轴器
elastoplastic analysis 弹塑性分析
elastoplastic damage 弹塑性损伤
elastoplastic deformation 弹塑性变形
elastoplastic fracture 弹塑性断裂
elastoplasticity of soil 土的弹塑性
elastoplasticity theory 弹塑性理论
elastoplasticity 弹塑性
elastoplastic model 弹塑性模型
elastoplastic performance 弹塑性
elastoplastic range 弹塑性范围
elastoplastic stability 弹塑性稳定
elastoplastic strain correction factor 弹塑性变形修正因子
elastoplastic 弹性塑料，弹塑性的
elastoplast 弹性绷带，弹性塑料
elastoresistance 弹性电阻
elasto-viscous field 弹性黏性场
elaterite 弹性地沥青
ELB＝elbow 弯管，弯头，肘管
elbow bend pipe 肘形弯管
elbow bend 管子弯头
elbow board 窗台板
elbow draft tube 肘形尾水管
elbow draught tube 肘形尾水管
elbow equivalent length 弯管当量长度
elbow jointed lever 肘节杆
elbow joint 弯管接头，肘接头
elbow piece 弯管配件
elbow pipe welder 弯头焊接机
elbow pipe 弯管，肘管
elbow tube 肘管，弯管
elbow-type combustor 角形燃烧室，弯形燃烧器
elbow union 弯头套管，弯头套环
elbow unit 弯管活接头
elbow 弯管，弯头，肘管，弯管接头
ELCB（earth leakage circuit breaker） 漏电开关
ELCO（electrolytic capacitor） 电解电容
ELD（economical load dispatcher） 经济负载分配装置
ELD（edge-lighted display） 边缘发光显示
ELD（extra-long distance） 超远程
elecric fusion welding 电熔焊
elecrto-magnetic compatibility 电磁兼容
election 选择
elective tender 选择性招标
electrc magnetoacoustic probe 电磁声探头
electromechanical device 机电装置，机电设备
electromechanical oscillograph 机电示波器
electret 驻极体，电介体，永久极化的电介质
electric 电动的，电气的，电的
electric absorption 电吸收
electric accounting machine 电动计算机
electric actuator 电力传动装置，电动执行元件，电动执行器，电动执行机构
electric air heater 电空气加热器，空气电加热器
electrical accident 电气故障
electrical adder 电加法器
electrical air compressor 电动空压机
electrical analog computer 电子模拟计算机
electrical analog method 电模拟法
electrical analog 电模拟
electrical analogy to stream-flow 河川水流的电模拟
electrical analogy 电模拟
electrical anemometer 电传风速计
electrical angle 电角度
electrical angular frequency 电角频率
electrical anti-fracture testing machine 电动抗折试验机
electrical apparatus characteristics 电器特性
electrical apparatus insulation 电器绝缘
electrical apparatus measuring device 电器试验测试设备
electrical apparatus release 电器脱扣器
electrical apparatus 电气设备，电气装置，电机，电器
electrical arc welding 电弧焊
electrical auxiliary control panel 电气辅助控制屏
electrical auxiliary power 辅助电功率
electrical back-to-back test 背对背电气试验，对组电气试验
electrical balance 电平衡
electrical bandspread 电气频带扩展
electrical bias 电偏置
electrical board 电气盘板
electrical breakdown in vacuum 真空击穿

electrical breakdown 电击穿
electrical bridge 电桥
electrical brush 电刷
electrical building fire protection system 电气厂房消防系统
electrical building smoke exhaust system 电气厂房排烟系统
electrical building 电气大楼，电气厂房，电气楼
electrical cable 电力电缆，电缆
electrical capacitance level measuring device 电容料［液］位测量仪表
electrical carbon contact material 电碳接触材料
electrical carbon material 电碳材料
electrical carbon mixing equipment 电碳混合设备
electrical carbon product 电碳制品
electrical carbon 电碳
electrical characteristics 电气特性
electrical chief engineer 电气总工
electrical clearance 电气间隙
electrical condenser 电容器
electrical conductance level measuring device 电导料［液］位测量仪表
electrical conductance 电导纳
electrical conduction determination 电导测定
electrical conduction 电导
electrical conductivity 导率，导电率，导电性，导电系数
electrical conduit 电缆，导线管
electrical connector 电连接器，电气接头，插座，接线盒
electrical contact controller 点开关控制器
electrical contact liquid level indicator 电接触式液位指示器
electrical contactor 电气接触器
electrical contact 电连接，电接点，电触点，电触头
electrical continuity 电的连续性
electrical contractor 电气承包商
electrical control and protection 电气控制和保护
electrical control rack 电气控制架
electrical control room 电气控制室
electrical control 电气控制
electrical corona 电晕
electrical coupler 电耦合器，电联轴器
electrical creepage path 爬电通路
electrical degree 电角度
electrical department 电气分场
electrical design 电气设计
electrical device 电气设备
electrical dew-point hygrometer 电子露点湿度计
electrical die-head threading machine 电动套丝机
electrical differentiation 电微分
electrical discharge machining 电火花加工，放电机械制作
electrical discharge wire 电晕导线，放电导线
electrical discharge 放电
electrical distance 电磁波两点间的传播距离
electrical distribution network 配电网
electrical disturbance 电气干扰，电气骚动
electrical double layer 双电荷层，偶极子层，双电层
electrical drawing 电气图纸
electrical dry oven 电烘箱
electrical dynamic mechanical endurance 耐电-机械动应力
electrical efficiency 电效率
electrical & electronics engineering 电气和电子工程
electrical endurance test 电寿命试验
electrical endurance 电气寿命，电气耐久性
electrical energy consumption quota of whole plan 全厂电耗定额
electrical energy generation 发电量
electrical energy supply 供电量
electrical energy transformer 电能变换器
electrical energy used by auxiliaries 厂用电量
electrical energy used 用电量
electrical energy 电能
Electrical Engineering Department 电机工程部
electrical engineering 电气工程，电机工程，电力工程，电工技术，电工学
electrical engineer 电气工程师，电机工程师
Electrical Equipment Association 电气设备协会
electrical equipment hatch 电气设备舱门
electrical equipment room 电气设备间
electrical equipment 电气设备，电力装置
electrical equivalent of calorie 热电当量
electrical equivalent 电等效
electrical erection 电气安装
electrical feedback 电反馈
electrical feedthrough 电气馈通线
electrical field 电场
electrical filter 滤波器
electrical fittings and accessories 电气配件和附件
electrical flat rail cart 电动平车
electrical footing resistance 接地电阻
electrical force 电动力，电磁力
electrical fuse 电力熔断器
electrical generating system 发电系统
electrical generator efficiency 发电机效率
electrical generator speed control 发电机转速控制
electrical generator 发电机
electrical governor 电调节器，电调速器
electrical grade cold-rolled steel 冷轧电工钢
electrical grade copper 电气等级铜
electrical grade sheet 电工钢板【含1%硅】，硅钢片
electrical ground 电气接地
electrical heater 电热器
electrical heating method 电加热法
electrical heating 电加热
electrical heat tracing system 电加热保温系统，电伴热系统
electrical heat tracing 电加热保温，电加热
electrical horsepower 电马力，电功率
electrical hygrometer 电湿度计
electrical ignition 电火花点火，电气点火
electrical independent contact 无源接点
electrical industry 电力工业
electrical inertia 电气惯性
electrical installation 电气设备

electrical & instrumental disciplines 电仪专业
electrical instrumentation 电检测仪表，电气仪表设备
electrical insulating material 电工绝缘材料
electrical insulating tape 绝缘带
electrical insulation 电绝缘，电工绝缘
electrical interconnection 电的相互联系，联络线
electrical interface 电器接口
electrical interlocking 电气联锁，电气闭锁
electrical interlock relay 电联锁继电器
electrical interlock valve 电气联锁阀
electrical interlock 电气联锁，电气闭锁
electrical island 电气岛
electrical isolation 电隔离
electrical laboratory 电气实验室，电气试验室
electrical leakage 漏电
electrical length 电长度
electrical level 电平
electrical life 电气寿命
electrical line tower 电线塔
electrical loading of stator 定子电负荷
electrical load 电力负载，电力负荷
electrical locomotive 电力机车
electrical logging 电测井
electrical loss 电气损耗
electrically-alterable read only memory 电可改写只读存储器
electrically-controllable read-only storage 电可控只读存储器
electrically-controlled operator 电气控制的操作器
electrically driven feed pump 电动给水泵
electrically driven valve 电动阀
electrically driven 电驱动的，电力牵引的
electrically erasable PROM 电可擦（除）可编程只读存储器
electrically erasable read-only memory 电可擦只读存储器
electrically heated boiler 电（加）热锅炉
electrically heated coil 电热线圈，加热线圈
electrically heated distilling apparatus 电加热蒸馏器
electrically heated pressurizer 电加热稳压器
electrically heated rod 电加热棒
electrically heating device 电加热装置
electrically-independent 电气独立的
electrically operated valve 电动阀门，电动阀
electrically operated 电动的，电牵引的，电气操作的
electrically oriented wave 定向电波
electrically programmable read only memory 电可编程（序）只读存储器
electrically programmable ROM 电可编程存储器
electrically trip-free 自动断电，自动跳闸
electrically welded tube 电焊管
electrically 电学上，用电力，用电气
electrical machine constant 电机常数
electrical machinery 电机，电力机械
electrical machine 电机
electrical mains 主电源，电源干线
electrical mass filter 电学滤质器，电质谱仪
electrical measurement 电气测量，电工测量，电测
electrical measuring instruments 电测仪表
electrical-mechanical parts counter 机电式零件计数器
electrical metallic tubing 电气金属套管
electrical metering 电气计量
electrical meter 电表，电气仪表
electrical method of prospecting 电法勘探
electrical method 电法
electrical motor power input 电动机功率输入
electrical multiplication 电乘法
electrical network analyzer 电网分析仪
electrical network 电网
electrical neutral axis 电气中性线
electrical noise 电气干扰，电噪声
electrical null 零位电压
electrical operation 电气操纵，电动
electrical output storage tube 电输出存储管
electrical output 电力输出，输出电功率【电厂】，电输出，电功率
electrical overhead travelling crane 电动桥式起重机
electrical panel 配电盘
electrical parameters 电气参数
electrical penetration assembly 电气贯穿组件
electrical penetration 电气贯穿件
electrical performance 电气性能
electrical phase angle 电相角
electrical pick-off 电传感器
electrical-pneumatic converter 电气转换器
electrical polarization 电极化
electrical pole 电杆
electrical potential barrier 截止电位
electrical potential 电势，电位
electrical power cable 电缆线，输电线
electrical power consumption 耗电量
electrical power engineering 电力工程
electrical power network 电力网，电网
electrical power system 电力系统
electrical power transmission 输电，电力传输
electrical power unit 电功率单位
electrical power 电功率
electrical precipitator 静电除尘器
electrical pressure test pump 电动试压泵
electrical primary system 电气一次系统【中国用法】
electrical prospecting 电法勘探
electrical pulley block hoister 电动葫芦吊
electrical pulse generator 脉冲发生器
electrical pump 电动泵
electrical pyrometer 电高温计
electrical quadripole 四端网络，电四极子
electrical radian 电弧度
electrical release free 电气开关释放【不自动合闸的】
electrical remote control 电遥控
electrical repair shop 电修车间
electrical repair 电气维修
electrical resistance heating element 电阻加热元件

electrical resistance heating 电阻加热
electrical resistance thermometer 电阻温度计
electrical resistance 电阻
electrical resistivity 电阻率
electrical resolver 旋转变压器
electrical response 电气响应，电反应
electrical room 电气间
electrical rotating machine 旋转电机
electrical secondary circuit 电气二次回路【中国用法】
electrical secondary system 电气二次系统【中国用法】
electrical sensing 电读出，电气传感
electrical shaping machine 电动起鼓机
electrical sheet steel 电工钢片，硅钢片
electrical shield 电屏蔽
electrical shock process 电击工艺【高温气冷堆燃料元件后处理】
electrical signal 电信号
electrical soil moisture meter 电传土壤湿度计
electrical sounding 电测深
electrical spark working 电火花加工
electrical spark 电火花
electrical speed governor 电调节器，电调速器，电气式调速器
electrical storage heating 电气蓄热
electrical storage 电的储存，电存储器
electrical strength 电气强度，抗电强度
electrical stress 静电应力，介质应力，电气强度
electrical substation 电力变电站
electrical supervisor 电气监理
electrical survey 电法探测
electrical switchgear room 电气配电室
electrical system pooling 电力系统的统筹
electrical system 电力系统，电系统，电气系统
electrical tapper 电动攻丝机
electrical teletachometer 电遥测转速计
electrical test bench 电气试验台
electrical test 电器试验
electric altimeter 电测高度计
electrical torque 电磁转矩
electrical trace heating system 电加热保温系统
electrical tracing 电伴热
electrical transmission 电力传输
electrical trench digger 电动挖沟机
electrical trip and monitoring 电气跳闸与监测系统
electrical two-speed synchro 双速同步电机
electrical varnish 电工漆，绝缘漆
electrical vibrator 电振动器
electrical viscometer 电黏度计
electrical warming 电热
electrical welding roller 电动转台
electrical winding 绕组
electrical zero adjuster 零电位调整器
electrical zero autotransformer 有电气零点的自耦变压器
electrical zero 零电位
electrical 有关电的，电气科学的，电气的，电的
electric analog model 电模拟模型
electric analogy testing 电模拟试验
electric analog 电模拟
electric angle 电角度
electric apparatus 电气设备，电机，电器
electric appliance 电器用具，电气设备
electric arc ignition 电弧点火
electric arc marking pencil 做标记用的电弧棒
electric arc resistance 电弧电阻
electric arc saw 电弧锯
electric arc spot welding 电弧点焊
electric arc welded tube 电弧焊缝管
electric arc welding 电弧焊
electric arc 电弧
electricator 电触式测微表
electric automation 电气自动化
electric auxiliary power 厂用电
electric axis 电轴（线）
electric balance 电天平，电平衡
electric battery 电池
electric bell testing set 电铃试验装置【连续试验】
electric bell 电铃
electric bevelling machine 坡口机
electric blasting machine 电爆机
electric blasting unit 电力放炮机
electric blasting 电火花引爆，电炮
electric block 触电，电击，电动葫芦
electric blower 电动鼓风机，电风扇
electric blowing oven 电热鼓风干燥箱
electric boiler 电加热锅炉，电热锅炉
electric bolt heater 螺栓电加热器
electric brake motor 电动制动马达
electric brake 电力制动器，电气制动器
electric braking 电气制动，电制动（法）
electric brazing 电热铜焊
electric breakdown 电击穿
electric brush 电刷
electric butt welding 电阻对焊
electric cable groove 电缆沟
electric cable tester 电缆测试仪
electric cable 电缆
electric capacitance altimeter 电气高程计
electric capacity moisture meter 电容湿度计
electric capacity 电力容量，电容
electric capstan 电动绞车，电动绞盘，电动起锚机
electric cap 电雷管
electric carbon 电碳
electric carburizing 通电渗碳
electric cell 光电管，电池
electric center 电气中心
electric ceramics 电瓷
electric chain block 电动葫芦
electric chain hoist block 电动链式滑车，电动吊链
electric chain hoist 电动吊链，电动葫芦，环链电动葫芦
electric charge density 电荷密度
electric charge quantity of electricity 电荷量
electric charge time constant 充电时间常数
electric charge 电荷，电费
electric circuit complexity 电路复杂程度

electric circuit 电路，路电
electric cleaner 电动清洁器
electric clock synchronizer 电钟同步器
electric clock 电钟
electric code 电工规程
electric coil winder 电动绕线机
electric comb 电梳
electric communication engineering 电通信工程
electric commutator 换向器，整流器，转换开关
electric component 电气组件，电气元件
electric conductance 电导
electric conductivity 电导率，电导性，导电率，导电性
electric conductor 导电体，导线
electric consolidation 电气固结法
electric constant 电气常数
electric contact pressure gauge 电接点压力表
electric contact thermometer 电接点温度计
electric contact 电接点，触头，电接触，电触头
electric control building 主控制楼
electric controller 电动调节器，电动控制器，电力控制器
electric control valve 电动调节阀，电动控制阀
electric control 电（气）控（制），电控
electric copying device 电模拟装置
electric corona 电晕
electric corrosion 电蚀，电解腐蚀
electric coupling 电耦合，电连接，耦合器
electric crane weighing machine 电动吊称式横器
electric crane 电动起重机
electric curing of concrete 混凝土电热养护
electric current density 电流密度
electric current intensity 电流强度
electric current loop 电流环
electric current meter 电测流速仪
electric current per unit length 线电流密度
electric current phasor 电流相量
electric current testing 电流检测法
electric current 电流
electric deflection 电偏移
electric dehydration 电脱水
electric delay fuse 延期电信管
electric delay line 电延迟线
electric density 电荷密度
electric depolarization 通电去极化
electric desalination 电气脱盐
electric detonator 电雷管
electric device take-up 电动拉紧装置
electric diagnosis 电气诊断
electric dialysis 电渗析
electric dialyzator 电渗析器
electric diaphragm switch technique 电膜片开关技术
electric die head threading machine 电动套丝机
electric dipole moment of molecule 分子电偶极矩
electric dipole moment 电偶极矩
electric dipole 电偶极子
electric dipper 电铲
electric discharge time constant 放电时间常数
electric discharge 放电
electric displacement 电位移
electric dissociation 电离
electric distance 距离的光学单位
electric domain 电畴
electric doublet 电偶极子
electric drill 电钻
electric drive control system 电力拖动控制系统
electric driven diverter gate 电动挡板三通管
electric driven one side plough unloader 电动单侧犁式卸料机［器］
electric driven trough-changeable plough unloader 电动可变槽角犁式卸料机［器］
electric driven two-side plough unloader 电动双侧犁式卸料机［器］
electric drive 电动，电驱动，电力传动，电力拖动，电气传动
electric drying 电气干燥
electric dust collector 静电除尘器
electric dynamometer 电测力计，电测功计，测功电机，电动测力计，电动测功器
electric effect of ultrasonic 超声波的电学效应
electric efficiency 电效率
electric elastivity 介电常数的倒数
electric-electric 纯电的
electric elevator 电梯，电力升降机
electric energy consumed by auxiliaries 厂用电量
electric energy consumption 耗电量
electric energy loss 电能损耗
electric energy measuring management 电能计量管理
electric energy price 电力价格
electric energy transducer 电能转换器
electric energy 电量，电能
electric engineering 电气工程
electric engineer 电气工程师
electric enterprise 电气事业
electric equilibrium 电平衡
electric equipment for highland 高原电气设备
electric equipment for tropical use 热带电气设备
electric equipment 电气设备
electric erosion 电蚀，电解腐蚀
electric etcher 电蚀器
electric excitation 励磁，激磁，电激发，电激励
electric exciter 电激振器
electric exploder 电爆器
electric eye 电光池，电眼
electric fan 电风扇，送风机
electric fathometer 电测深计
electric fault 电气故障
electric feeder 电动给料机
electric feed water pump 电动给水泵
electric fence 电篱笆，电丝网，电围栏
electric fidelity 电信号保真度
electric field distribution 电场分布
electric field effect 电场效应
electric field force 电场力
electric field intensity 电场强度
electric field strength at ground level 地面场强
electric field strength 电场强度
electric field vector 电场矢量
electric field 电场
electric filter 滤波器

electric firing　电点火，电放炮
electric fishing　电捕鱼法
electric fluctuation　电波动，电脉动
electric fluviograph　电测水位计
electric flux density　电通密度，电位移
electric flux　电通量
electric focusing　电聚焦
electric folder　电动折边机
electric force gradient　电场梯度，电势梯度
electric force　电力，电场强度
electric fork lift truck　电动叉车
electric fuel gauge　燃料消耗电测表
electric furnace steel　电炉钢
electric furnace transformer　电炉变压器
electric furnace with a base bottom and lining　有基底和内衬的电炉
electric furnace　电炉
electric fuse　熔断器
electric-fusion-welded steel-plate pipe　电熔焊钢板卷管
electric fusion weld　电熔焊
electric gas flowmeter　电气体流量计
electric gas heater　气体电加热器
electric gate valve　电动闸阀
electric geared motor　齿轮驱动式马达，齿轮驱动式电动机
electric generator　发电机
electric governing system　电调装置
electric hammer　电锤
electric hand drill　手电钻
electric hand planer　手电刨
electric hand saw　手电锯
electric harmonic analyzer　电谐波分析器
electric healing coil　电热盘管
electric heat blower　电热鼓风箱
electric heated air-blowing oven　电热鼓风干燥箱
electric heated oven　电（热）炉
electric heater case　电热箱
electric heater in duct　风管电加热器
electric heater section　电加热段
electric heater　电加热器，电炉，电热器，电热炉
electric heating element　电热元件
electric heating system　电加热系统
electric heating unit　电供暖装置
electric heating　电加热
electric heat tracing　电伴随加热，电加热保温
electric hoist　电动卷扬机，电力起重机，电动吊车，电（动）葫芦
electric hot plate　电热板
electric hydraulic controller　电动-液压控制器
electric hydraulic control　电液控制系统
electrician's gloves　电工手套
electrician's knife　电工刀
electrician's pliers　电工钳
electrician's tool　电工工具
electrician　电工，电气技术员，照明员，电气技师
electric ignition　电点火
electric image　电像，电位起伏图
electric immersion heater　浸没式电热器，浸没式电加热器
electric impulse counter　电脉冲计数器
electric impulse　电脉冲
electric induced line　电感应线
electric induction　电感应
electric inductivity　介电常数
electric inertia　电惯性，电惯量
electric installation　电气设备，电气装置
electric instrument　电工测量仪表
electric insulation oil　绝缘油
electric insulation supervision　绝缘监督
electric integration　电积分
electric intensity vector　电场强度矢量
electric intensity　电场强度
electric interlocking device　电动联锁装置
electric iron　电烙铁
Electricite De France　法国电力集团
electricity bill　电费
Electricity Company of Ghana　加纳电力公司
electricity company　电力公司
electricity consumption for irrigation and drainage　排灌用电
electricity consumption for lighting　照明用电
electricity consumption for power equipment　动力用电
electricity consumption for road lighting　路灯用电
electricity consumption for rural resident　农村居民生活用电
electricity consumption for township industry　乡工业用电
electricity consumption rate　用电单耗
electricity consumption structure　用电构成
electricity consumption　用电，用电量
electricity elasticity　电力弹性系数
electricity end-users tariff　销售电价
electricity fee　电费
electricity future price　期货电价
electricity marketing　电力销售，电力营销
electricity market model　电力市场模式
electricity market operation system　电力市场运营系统
electricity market regulation　电力市场监管
electricity market　电力市场
electricity meter　电量计，电表，电度表
electricity on-grid tariff　上网电价
electricity price structure　电价结构
electricity price　电价
electricity production reactor　发电（反应）堆
electricity rate　电价
electricity resistance probe method　电阻探头法
electricity sales　售电量
electricity saving technology of pump　泵类节电技术
electricity saving technology　节电技术
electricity saving　节约用电，节电
electricity selling price　售电价格
electricity spot price　现货电价
electricity supply　电量供应，电源，电力供应
electricity tariff　电价
electricity　电，电能，电学，电流
electric lamp bulb　电灯泡

electric lamp filament 灯丝
electric lamp 电灯，电灯泡
electric leakage 漏电
electric lift 电梯，电动升降机
electric lighting load 照明负荷
electric lighting 电气照明
electric light signal 电气照明
electric light 电灯，电光
electric line of force 电力线
electric line 电力线路
electric load curves 电负荷曲线
electric loading 电负载，电负荷
electric lock 电（气闭）锁
electric locomotive 电气机车
electric log 电测记录
electric loss 电损耗
electric luminous heater 电光发热器
electric machine designer 电机设计师
electric machine regulating system 电机调节系统
electric machine regulator 电机调节器
electric machinery 电力机械，电机
electric machine 电机
electric magnetic interference 电磁干扰
electric magnetic iron remover 电磁除铁器
electric magnetic lock 电磁闭锁
electric magnetic 电磁的
electric mains 主电线，输电干线
electric material 电气材料，电工材料
electric measurement 电气测量，电工测量，电测学
electric measuring instrument 电工测量仪表
electric mechanical sieve shaker 电动振筛机
electric metallurgical works 电（气）冶（炼厂）
electric metal 高频金属，电工金属
electric meter 电表
electric microbalance 电微量天平
electric micrometer 电测微计
electric moisture meter 电测湿度计
electric moment 电矩
electric monitor 电动监控器，电动监测器
electric monorail crane 电动单轨吊车
electric monorail hoist 单轨电动葫芦，电动单轨吊
electric motor actuator 电动执行机构
electric motor car 电动机车，电平车
electric motor operation 电动机运行
electric motor operator 电动操作器，电动机操作器
electric motor 电动机，电机
electric muffle 马弗炉，电高温炉
electric-net control room 电网控制室
electric network reciprocity theorem 电网互易定理
electric network 电力网，电网
electric nonconductors 非导电体
electric oil 绝缘油，变压器油
electric opener 电动开关器
electric operator 电动执行机构
electric oscillation 电振荡
electric oscillator 电力激振器，电振荡器
electric osmosis 电渗（透）
electric oven controller 电炉控制器
electric oven 电热炉
electric overspeed tripping device 电超速保护装置
electric panel heating 电热辐射采暖
electric penetration assembly 电导体穿入装置
electric pin-and-socket coupler 电插头插座连接器
electric pipe cutting and bevelling machine 电动切管坡口机
electric pipe threading machine 电动套丝机
electric pitching mechanism 电动调桨机构
electric polarity 电极性
electric polarizability of molecule 分子电极化率
electric polarization intensity 电极化强度
electric polarization 电极化
electric porcelain 电工瓷，绝缘瓷
electric potential difference 电势差，电位差
electric potential gradient 电势梯度
electric potential 电势，电位
electric power acceptor 受电设备
electric power bureau 电力局
electric power company 电力公司
electric power construction 电力建设
electric power controller 电力定量器
electric power control panel 电力控制面板
electric power design institute 电力设计院
electric power distribution apparatus 配电电器
electric power distribution panel 电力配电盘
electric power distribution 电力分配，配电
electric power economy 电力经济，动力经济
electric power engineering 电力工程
electric power enterprise 电力企业
electric power equipment 电力设备，发供电设备
electric power generating station 发电厂
electric power generation 发电
electric power grid 电力电网
electric power industry 电力工业
electric power network at county level 县级电网
electric power network composition 电网结构
electric power network expansion planning 电网发展规划
electric power network 电网
electric power of ocean energy from concentration gradient 海水浓度差发电
electric power output 静电功率输出
electric power panel 电力控制面板
Electric Power Planning & Engineering Institute Co. Ltd. 电力规划设计院有限公司
electric power plant 电厂，发电厂，发电站
electric power pool 联合电网
electric power project 电力工程，电力工程项目
electric power quality 电能质量，发电量
electric power receiving station 受电站
Electric Power Research Institute 电力研究院
electric power source 电源
electric power station 发电站
electric power storage 蓄电
electric power substation 变电所
electric power supply circuit 电源电路，馈电电路

electric power supply 电力供应，电源，供电
electric power system design 电力系统设计
electric power system planning 电力系统规划，电力系统设计
electric power system reliability criteria 电力系统可靠性准则
electric power system reliability 电力系统可靠性
electric power system stability 电力系统稳定性
electric power system 电力系统，电网
electric power transmission 电力输送，输电
electric power 电能，电力，电功率，电源
electric precipitator 电沉淀器，电除尘器，静电除尘器，电气除尘器
electric pressure testing pump 电动试压泵
electric probing device 电测装置
electric profiling 电测剖面，电测剖面法
electric prospecting 电法勘测
electric pulse motor 脉冲电动机
electric pulse 电脉冲
electric pump 电动泵
electric radiant heater 电辐射加热器
electric radiant heating 电热辐射采暖
electric radiator 电热器
electric rammer 电夯
electric ramming impact machine 电动夯拍机，电动打夯机
electric rate 电价，电费率
electric reactor 电抗器
electric regeneration 电再生
electric regenerative braking 电力再生制动，电力反馈制动
electric relay 继电器
electric relief valve 冲量式安全阀
electric remote-control valve 电动遥控阀
electric repair 电修车间
electric reset relay 电复位继电器
electric reset 电复位
electric resistance box 电阻箱
electric resistance butt welding 电阻对焊，碰焊机焊接
electric resistance humidifier 电阻式加湿器
electric resistance probe method 电阻探头法
electric resistance pyrometer 电阻高温计
electric resistance strain gauge 电阻应变计
electric resistance thermometer 电阻温度计
electric-resistance-welded steel pipe 电阻焊钢管
electric resistance welding 电阻焊接，电阻焊
electric resistance-wire strain gauge 电阻丝弹性应变计
electric resistance 电阻
electric resistivity method 电阻法
electric resistivity 电阻率
electric resonance 电共振，电谐振
electric revolving shovel 电动旋转式挖土机
electric riveting machine 电动铆钉机
electric robot 机器人，电动机械手，电气自动机
electric rock drill bit 电动钻岩机钻头
electric safety lamp 安全灯
electric salinity meter 电测盐度计
electric screen 电屏
electric screw driver 电动改锥，电动改刀
electric service lift 电动维护吊车
electric servomechanism 电伺服机构
electric servomotor 电动伺服电机
electric shaft 电联轴
electric shaker 电动摇筛器
electric shield 静电屏蔽，电屏蔽
electric shock absorber 电击防护器
electric shock protection 触电保护
electric shock 电击，触电，电震，电休克
electric shop 电工车间
electric shovel 电铲，电动挖土机，电力铲
electric signaling light 电信号灯
electric sign 电信号，电光标志
electric simulation 电模拟
electric siren 电警报器，电气报警器
electric slag welding 电渣焊，电熔渣焊
electric slewing crane 电动旋臂起重机
electric socket wrench 套筒电扳手
electric soldering 电焊
electric sounding machine 电气水深计
electric sounding 电测深
electric source 电源
electric space constant 真空的介电常数，绝对磁导率
electric spark 电火花
electric spectrum 电能谱，电弧光谱
electric spurious discharge 乱真放电
electric squib 电点火管，电点火器，电爆装置，电气导火管
electric staff block 电气路签闭塞
electric starter 电启动器
electric starting motor 启动电动机
electric station 电站，发电厂
electric steel 电炉钢，电工钢
electric storm 电爆
electric strength test 耐压测试，电气强度测试
electric strength 电气强度，介电强度，电绝缘强度，耐电强度
electric stress 电场强度，静电强度，电应力
electric strip heater 电热丝式加热器
electric submersible pump 电动潜水泵
electric substation 变电站
electric supply failure 电源故障
electric supply line 供电线，电源线
electric supply 电炉供应，供电，电源
electric surface density 面电荷密度
electric survey 电法勘测
electric susceptibility 电极化率
electric switching operation 电气倒闸操作，倒闸操作
electric switch oil 开关油，断路器油
electric switchyard 电力开关站
electric symbol 电工符号
electric system 电气系统
electric tachometer 电力转速计，电测速计，电动转速计，电流速计
electric tamper 电动打夯机，电动夯土机
electric tamping machine 电动打夯机
electric tape gauge 电测水位计
electric tapping machine 电动攻丝机

electric technical standard 电气技术标准
electric telemeter 电控测距仪，电遥测计，电遥测仪
electric temperature control 电气温度控制
electric tension 电压
electric terminals of a machine 电机接线端子
electric testing instrument 电气试验仪器
electric test 电气试验
electric thermal-storage 电蓄热器
electric thermostatic drying oven 电热恒温干燥箱
electric thermostat 电动恒温器，恒温开关
electric threading machine 电动套丝机
electric thruster 电动推杆
electric tool 电动工具
electric to pneumatic converter 电气转换器
electric torch 手电筒
electric towing trolley 电气牵引车
electric tracer heating system 电伴热系统
electric tracing 电伴热
electric traction equipment 牵引电气设备
electric traction line 电气化牵引线
electric traction network 电气化牵引网
electric traction 电力牵引
electric transducer 电变换器，换流器，换能器，传感器，变送器
electric transient 电气瞬变（过程）
electric transmission line 输电线
electric transmission 输电
electric traveling crane 移动式电动起重机，电动桥式起重机
electric trench 电缆沟
electric typewriter 电动打字机，电传打字机
electric unit 电气单位
electric utility industry 电气公用事业
electric utility 公用电业，公用电厂，电力公司
electric vacuum cleaner 电动真空吸尘器
electric valve operator 电动阀执行器
electric vector 电矢量，电场矢量
electric vehicle charging station 电动车充电站
electric vehicle motor 电气机车电动机，牵引电动机
electric vehicle supply equipment 电动汽车供电设备
electric ventilator 电通风机，电风扇
electric vibrating feeder 电振动给料机
electric voltage 电压
electric walking dragline 移动式电动索铲挖土机
electric water heater 电热水器
electric water press 电动水压机
electric water sounder 电测水深器
electric wave filter 滤波器
electric wave recorder 电测波浪计
electric wave 电波
electric weighing coal feeder 电子称重式给煤机
electric welder 电焊机
electric welding hammer 电焊手锤
electric welding machine 电焊机
electric welding plant 电焊厂
electric welding 电焊
electric well log 电测钻井记录，电测钻井剖面
electric winch 电动绞车，电动卷扬机
electric winding machine 电动绕线机
electric windlass 电动卷扬机
electric wind 电吹风
electric wire and cable 电线电缆
electric wire 电线
electric wiring 安装电线，电气布线
electric workshop 电气车间，电工车间
electric works 电气工程
electriferous 带电的
electrifiable 可启电的
electrification scheme 电气化计划
electrification 电气化，带电，充电，起电
electrified body 带电体
electrified-particle indication 带电粒子显示
electrified-particle inspection method 带电粒子检测法
electrified railroad 电气化铁路
electrified wire netting （防护）电网
electrified 电气化的，带电的，带电
electrify 电气化，带电，充电，起电
electrion 高压放电
electrit 电铝
electrization 带电
electrizer 起电机，起电盘
electro 电镀，电镀品
electroacoustical transducer 电声换能器
electroacoustic transducer 电声变换器，电声换能器
electroacoustic transformer 电声变量器
electroacoustic 电声的，电声学的
electroaddinity 电解电势，电亲和势
electro adsorption 电吸附
electro analysis 电解分析，电（化学）分析
electroanalyzer 电分析器
electro-arc contact machine 接触放电器
electrobalance 电动天平
electrobath 电镀浴，电解浴
electro-beam-induced conductivity 电子束感应导电率
electro-beam intensity 电子束强度
electro-beam scanner 电子束扫描器，电子束析像管
electro-beam scanning system 电子束扫描系统
electro beam 电子束
electrocaloric 电热的
electrocalorimeter 电热计，电卡计
electrocapillary phoresis 毛细电泳（现象）
electro catalysis 电接触作用，电触媒作用
electrocathode 电控阴极
electrocathodoluminescence 电控阴极射线发光
electroceramic 电工陶瓷，电瓷
electrochemical action 电化学作用
electrochemical analysis 电化学分析
electrochemical attack 电化学侵袭［腐蚀］
electrochemical breakdown 电化学击穿
electrochemical capacitor 电解质电容器
electrochemical cell 电化学电池，电化电池
electrochemical corrosion 电化学腐蚀
electrochemical couple 电化电偶
electrochemical decontamination 电化学去污

electrochemical desalting 电化学脱盐
electrochemical desulfuration 电化学脱硫作用
electrochemical diffused-collector transistor 电化学扩散集极晶体管
electrochemical effect 电化学效应，电化效应
electrochemical equivalent 电化当量
electrochemical gauging 电化学测流法
electrochemical generator 电化学发电机
electrochemical method 电化学法
electrochemical power source 化学电源
electrochemical series （元素的）电位序，电化次序
electrochemical stabilization 电化学加固
electrochemical treatment 电化学处理
electrochemical 电化学的
electrochemistry 电化学
electrochronograph 电动精密计时器
electrocircuit 电路
electroclock 电钟
electrocoagulation 电凝聚
electrocoating 电涂覆，电泳涂漆
electroconductibility 电导率，导电性
electroconductive plastics 导电塑料
electroconductive resin 导电树脂
electroconductive rubber 导电橡胶
electro-conductivity 导电性，电导率，导电率
electro-contact type surface roughness tester 电接触式表面粗度试验仪
electro-control converter 电液转换器
electrocorrosion 电腐蚀
electrocratic 电稳的
electrocution 电致死，电刑
electrodata machine 电动数据处理机
electrode admittance 电极导纳
electrode arm 电极握臂，焊条夹
electrode bias 电极偏压
electrode boiler 电热锅炉，电极锅炉
electrode cable detector 电缆故障探测仪
electrode coating 焊条药皮
electrode collar 电极环，引线环
electrode compartment 极室
electrode composition 电分解作用
electrode conductance 电极电导
electrode core 焊条芯
electrode covering 焊条药皮
electrode current 电极电流
electrode dissipation 电极耗散
electrode drum level gauge 电极（式）汽包水位计
electrode edge 电极端
electrode extension 电极伸出长度
electrode force 电极压力
electrode for vertical down position welding 立向下焊条
electrode frame 极框
electrode furnace 电极炉
electrode gap 电极间隙
electrode ground bus 接地电极母线
electrode holder 焊钳，电极夹【弧焊机】，焊把，焊饼，焊条夹钳
electrode humidifier 电极式加湿器
electrode impedance 电极阻抗
electrode insulation 电极间绝缘
electrodeionization 电去电离作用
electrodeless discharge 无电极放电
electrodeless MHD generator 无电极磁流体动力发电机
electrodeless 无电极的
electrode material 电极材料
electrode oven 焊条烘干箱，电极炉
electrode pick-up 电极粘损
electrodeposition insulation 电沉积绝缘
electrodeposition 电沉积，电镀
electrode potential 电极电位
electrode regulation 电极调整
electrode regulator 电极调节装置，电极调节器
electrode resistance 电极电阻
electrode skid 电极滑移
electrode spacing 电极间距
electrode susceptance 电极电纳
electrode thermostat tank 焊条烘干箱
electrode tip 电极头
electrode travel 电极行程【接触焊】
electrode voltage 电极电压
electrode wall 电极壁
electrode width 电极宽度
electrode wrench 电极扳手
electrode 电极，电焊条，焊条
electrodialysis cell 电渗析槽
electrodialysis equipment 电渗析设备
electrodialysis plant 电渗析设备
electrodialysis process 电渗析过程
electrodialyzer 电渗析，电渗析器，电渗透
electrodics 电极学
electro-discharge machining 放电器
electrodisintegration 电蜕变
electrodispersion 电分散作用
electrodynamical potential 电动势
electrodynamic ammeter 电动式电流表
electrodynamic balance 电导平衡，电流秤
electrodynamic bridge 电动式电桥
electrodynamic destruction 电动力破坏
electrodynamic force 电动力
electrodynamic induction 动力感应，电磁感应
electrodynamic instrument 电动式仪表，测力（计型）仪表
electrodynamic mechanism 电动机构
electrodynamic meter 电动式仪表
electrodynamic multiplier 电动（式）乘法器
electrodynamic potential 电动势
electrodynamic stabilization 电动力学稳定（方法）
electrodynamics 电动力学
electrodynamic transient 电动瞬变（过程）
electrodynamic type high speed impedance relay 电动式高速阻抗继电器
electrodynamic type relay 电动式继电器
electrodynamic type reverse current relay 电动式逆流继电器，电动式反流继电器
electrodynamic type seismometer 电动式地震检波器
electrodynamic type single phase power relay 电动式单相功率继电器

electrodynamic type three phase power relay 电动式三相功率继电器
electrodynamic type wattmeter 电动式功率表
electrodynamic 电动的，电动力（学）的
electrodynamometer 电功率计，电动测功计
electroerosion 电腐蚀
electrofarming 农业中的电力应用，农（用）电
electrofax 电子摄影，电子照相
electrofilter 电滤器，（静）电过滤器
electrofining precipitator 电气澄清除尘器
electrofluid dynamic augmented tunnel 带电流体动力加速装置的风洞
electrofluid dynamic generator 电流体动力发电机
electrofluid dynamic wind driven generator 带电流体风力发电机
electrofluid 电流体
electrogasdynamic closed cycle system 电气体动力闭合循环系统
electrogasdynamics 电气体动力学的
electro gaseous dynamic wind driven generator 带电气体风力发电机
electrogas welding 气电焊，气电立焊，电气焊
electrogenerator 发电机
electrogeniometer 相序指示器，相位指示器
electrogram 电描记图
electrographic recording 电显示法记录，电子图记录，示波记录
electrographic 电显示的，电子图的，电记录器
electrographite brush 电化石墨电刷
electrographite 电炉石墨，艾奇逊石墨
electrographitic brush 电化石墨电刷，人造石墨电刷
electrogravity 电控重力
electro-gray 电灰色
electroheat installation 电热设备
electroheat 电热
electrohydraulic actuator 电液执行机构，电动液压执行机构，电液传动，电动液压执行器
electrohydraulic cabinet actuator 电动液压整体型调速控制器
electrohydraulic control system 电（气）液（压）调节系统［控制系统］
electrohydraulic control 电气液压（式）控制，电液调节，电液控制
electrohydraulic converter 电液转换器
electrohydraulic elevator 电液升降机，电液提升机
electrohydraulic forming 水中放电形成法，电水锤形成法
electrohydraulic governing system of steam turbine 汽轮机电液调节系统
electrohydraulic governor 电（动）液（压）调节器［调速器］
electrohydraulic pulse motor 电液脉冲马达，电液脉冲电动机
electrohydraulic regulator 电（动）液（压）调节器［调速器］，电动水力调节器
electrohydraulics actuator 电动液压执行机构
electrohydraulic servo loop 电液伺服回路
electrohydraulic servo system 电动液压伺服系统
electrohydraulic servovalve 电液转换器，电（动）液（压）伺服阀
electrohydraulic speed governing system 电液调速系统
electrohydraulic stepping motor 电液步进电动机
electrohydraulics thruster brake 电（动）液（压）推杆制动器
electrohydraulic transducer 电液换能器，电液转换器
electrohydraulic valve 电液阀
electrohydraulic 电（动）液（压）的
electrohydrodynamic augmentation 电流体动力学强化法
electrohydrodynamic generator 电流体发电机
electro-hydro dynamics 电流体动力学
electroilluminating 电气照明的
electroinduction 电感应
electro-ionization 电离作用
electrokinematics 动电学
electrokinetic coupling 电动偶合
electrokinetic potential 动电势，Z电势
electrokinetics 动电学，电动力学
electrokinetic transducer 电动传感器
electrokinetic 电动的
electrokinetograph 电动测速仪，电动流速仪，动电计
electrokymography 电流记波法
electrokymograph 电动转筒记录仪
electroless 无电的
electrolier 装饰灯，集灯架，枝形电灯架
electrolight 电光
electroline 电（场）力线
electrolock 电（气闭）锁
electro-locomotive 电动机车
electroluminance 场致发光，电致发光
electroluminescence 电（致）发光，场致发光，电荧光
electroluminescent display device 场致发光显示器
electroluminescent 电荧光的，场致光的
electrolyser 电解槽，电解剂，电解器，电解装置
electrolysis bath 电解槽
electrolysis cell 电解电池，电解池，电解槽
electrolysis plant 电解装置［工厂］
electrolysis unit 电解装置
electrolysis 电解
electrolyte battery 有机电解液蓄电池
electrolyte free 不含电解质
electrolyte 电解质，电解液，电离质
electrolytic bath 电解槽，电解池
electrolytic capacitor 电解电容，电解电容器
electrolytic cell 电解电池
electrolytic charger 电解液充电器
electrolytic chlorine 电解氯
electrolytic condenser 电解质电容器
electrolytic conductance 电解（液）电导
electrolytic conductivity 电解（液）导电率
electrolytic copper 电解铜
electrolytic corrosion 电解腐蚀
electrolytic detector 电解检波器
electrolytic dissociation 电离解，电离作用

electrolytic dissolution 电溶解
electrolytic generator 电解用发电机
electrolytic hardening 电解液淬火
electrolytic hydrogen energy storage 电解储氢
electrolytic image transducer 电解图像传感器
electrolytic instrument 电解式仪表
electrolytic lightning arrester 电解式避雷器
electrolytic meter 电解式电量仪，电解式仪表
electrolytic method 电解法
electrolytic nebulizer 电解雾化器
electrolytic oxidation 电解氧化（法）
electrolytic polarization 电解极化
electrolytic recovery 电解回收
electrolytic rectifier 电解液整流器
electrolytic reduction 电解还原
electrolytic regeneration 电解再生
electrolytic resistance 电解电阻
electrolytic solution 电解液
electrolytic titration 电势滴定
electrolyzer 电解槽，电解剂，电解器，电解装置
electromagnetic 电磁，电磁的
electromagnetic amplifying lens 电磁放大透镜
electromagnetic anechoic chamber 电磁无回声室，电磁消音室
electromagnetic blow-out 电磁熄弧
electromagnetic brake 电磁制动器，电闸
electromagnetic braking system 电磁制动系统
electromagnetic braking 电磁制动
electromagnetic clutch 电磁离合器
electromagnetic coil 电磁线圈，电磁铁的线圈
electromagnetic compatibility filter 电磁兼容性滤波器，EMC滤波器
electromagnetic compatibility 电磁兼容（性），电磁设备互换性，电磁适应性
electromagnetic component 电磁部件，电磁元件
electromagnetic constant 电磁常数
electromagnetic contactless relay 无触点电磁继电器
electromagnetic contactor 电磁接触器
electromagnetic control 电磁控制
electromagnetic core 电磁磁芯
electromagnetic coupling 电磁耦合，电磁联轴器
electromagnetic crane 电磁铁起重机
electromagnetic current meter 电磁流速仪
electromagnetic damping 电磁阻尼
electromagnetic data store 电磁数据存储
electromagnetic delay line 电磁延迟线
electromagnetic design 电磁设计
electromagnetic detection 电磁检测
electromagnetic discharge 电磁放电，电感放电
electromagnetic distance meter 电磁波测距仪
electromagnetic disturbance 电磁骚扰
electromagnetic eddy current braking 电磁涡流制动
electromagnetic eddy current damper 电磁涡流阻尼器
electromagnetic emission 电磁发射
electromagnetic energy beaming 电磁聚集
electromagnetic energy density 电磁能密度
electromagnetic energy 电磁能
electromagnetic environment of line 线路的电磁环境
electromagnetic environment 电磁环境
electromagnetic feeder 电磁给料机
electromagnetic field intensity 电磁场强度
electromagnetic field 电磁场
electromagnetic filter 电磁过滤器
electromagnetic flaw detector 电磁探伤仪
electromagnetic flow meter 电磁流量计
electromagnetic flux 电磁通量
electromagnetic force 电磁力，电磁场强度，电动势
electromagnetic gas valve 电磁气阀
electromagnetic gear 电磁传动，电磁联轴节
electromagnetic generator 发电机
electromagnetic induction device 电磁感应装置
electromagnetic induction testing 电磁感应检测法
electromagnetic induction 电磁感应
electromagnetic inductive interference 电磁感应干扰
electromagnetic inertia 电磁惯性，电磁惯量
electromagnetic instrument 电磁式仪表
electromagnetic integrator 电磁积分器
electromagnetic interference 电磁干扰
electromagnetic interrupter 电磁断续器
electromagnetic iron separator 电磁除铁器
electromagnetic isotope-separation unit 电磁同位素分离器
electromagnetic jack type drive mechanism 电磁力提升式驱动机构
electromagnetic latch mechanism 电磁提升机构
electromagnetic leakage 电磁漏泄
electromagnetic linkage 电磁链
electromagnetic locking 电磁锁定
electromagnetic mass 电磁质量
electromagnetic material 电磁材料
electromagnetic measuring transformer 电磁式互感器
electromagnetic meter 电磁式仪表
electromagnetic mirror 电磁波反射面
electromagnetic moment 电磁矩
electromagnetic noise 电磁噪声，电磁干扰
electromagnetic operating mechanism 电磁操动机构
electromagnetic oscillations 电磁振荡，电磁波
electromagnetic oscillograph 电磁式示波器
electromagnetic overload release 电磁过载脱扣器
electromagnetic pick-up 电磁式拾波器，电磁传感器，电磁式拾音器
electromagnetic platen 电磁平台
electromagnetic pneumatic contactor 电磁气动接触器
electromagnetic pole-piece U形电磁铁
electromagnetic potential 电磁位
electromagnetic power 电磁功率
electromagnetic pressure gauge 电磁压力计
electromagnetic property 电磁特性，电磁性能
electromagnetic prospecting 电磁勘探
electromagnetic pump 电磁泵
electromagnetic radiation 电磁辐射

electromagnetic ratchet unit 电磁棘轮装置，电磁爪轮装置
electromagnetic recorder 电磁录音机，电磁录声机
electromagnetic relay 电磁（式）继电器
electromagnetic relief valve 电磁释放（安全）阀
electromagnetic repulsion 电磁排斥
electromagnetic rotating machine 旋转电机
electromagnetics 电磁学
electromagnetic screening effect 电磁屏蔽效应
electromagnetic screen 电磁屏蔽
electromagnetic seismograph 电磁式地震计
electromagnetic separation 电磁分离
electromagnetic separator 电磁分离器
electromagnetic shield 电磁屏蔽，电磁屏幕
electromagnetic shoe brake 电磁靴制动
electromagnetic slip coupling 电磁滑差耦合器
electromagnetic slip 电磁滑差
electromagnetic sodium circulating pump 钠循环电磁泵
electromagnetic spectrum 电磁振荡的频谱，电磁谱
electromagnetic starter 电磁启动器
electromagnetic steel plate 电感钢板
electromagnetic storage 电磁存储（器）
electromagnetic stress 电磁拉力，电磁应力
electromagnetic susceptibility 电磁敏感性
electromagnetic switch 电磁开关
electromagnetic system of unit 电磁单位制
electromagnetic system 电磁制，电磁系统
electromagnetic telecommunication 电磁波长途通信
electromagnetic theory 电磁定理，电磁理论
electromagnetic thickness indicator 电磁式厚度指示器
electromagnetic time delay relay 电磁式延迟继电器
electromagnetic torque motor 电磁式力矩电动机
electromagnetic torque 电磁扭矩，电磁转矩
electromagnetic transducer 电磁传感器，电磁转换器
electromagnetic transient calculation 电磁暂态过程计算
electromagnetic transient program 电磁暂态程序
electromagnetic transient 电磁暂态过程，电磁瞬变过程
electromagnetic transmission line storage 电磁传输线存储器
electromagnetic trip 电磁脱扣
electromagnetic type relay 电磁式继电器
electromagnetic unbalance 电磁不平衡
electromagnetic unit 电磁单位，电磁单元
electromagnetic valve 电磁阀
electromagnetic velocity meter 电磁速度计
electromagnetic vibrator 电磁振动器
electromagnetic viscometer 电磁黏度计
electromagnetic voltage regulator 电磁式电压调整器
electromagnetic wave 电磁波
electromagnetic wire 电磁线
electromagnetism 电磁，电磁学
electromagnet pull 电磁拉力
electromagnet vibrating feeder 电磁振动给煤机
electromagnet 电磁铁，电磁体，电磁石，电磁
electromatic relief valve 电磁释放（安全）阀
electromatic 电气自动的
electromechanical analog computer 机电模拟计算机
electromechanical analogy 机电模拟
electromechanical brake 机电复合制动器，电子机械制动器，机电制动器
electromechanical computer 机电式计算机
electromechanical counter 机电计数器
electromechanical coupling factor 机电耦合系数
electromechanical coupling 机电耦合
electromechanical disc brake 机电盘式制动器
electromechanical drive 电动机械传动
electromechanical energy converter 机电能量转换器
electromechanical failing load test 机电破坏荷载试验
electromechanical harmonic filter 机电滤波器
electromechanical integrator 机电积分器
electromechanical interlock 机电联锁
electromechanical oscillation 机电振荡
electromechanical plotter 机电式绘图仪
electromechanical power 机电功率
electromechanical process 机电过程
electromechanical recorder 电气机械式记录仪
electromechanical recording 电气机械式记录
electromechanical regulator 机电调节器，电动机调节器
electromechanical relay system 机电继电系统
electromechanical relay 机电继电器，机电式继电器
electromechanical sine and cosine generator 正弦及余弦发电机
electromechanical switch 电动机械开关
electromechanical tensioning system 电气机械张紧装置
electromechanical transducer 机电变换器，机电换能器，机电转换器，机电传感器
electromechanical transient 电气机械暂态过程，电气机械瞬变过程
electromechanical vibrating feeder 电机振动给料机
electromechanical 电动机械的，机电的
electromechanics 机电学
electrometal furnace 电弧熔化炉
electrometer gauge 静电计
electrometer tube 静电计（用的电子）管，静电计管
electrometer 静电计，静电电位计
electrometric method 电测法
electrometrics 测电学
electrometric titration 电位滴定，电化学滴定，电势滴定
electrometric 电工测量的
electrometry 测电术，测电学，电位测量术
electromigration 电移，电迁移
electromobile 电瓶车
electromotance 电磁场强度，电磁力，电动势

electromotion 电动，电动力
electromotive field 动电场，电动场
electromotive force measurement 电动势测量
electromotive force of motion 动电动势，旋转电动势
electromotive force of rest 静电动势，变压器电势
electromotive force 电动势，电动力
electromotive series 电动次序，电化序
electromotive 电动势的
electromotor veering phase tester 电机转向相序测试仪
electromotor 电动机，电机
electron 电子的，电子
electron acceptor 电子接受体
electron admittance 电子导纳
electron affinity 电子亲和能
electronation 增加电子，获得电子，还原作用
electron-attachment 电子附着
electron avalanche 电子雪崩
electron beam 电子束，电子流，电子粒
electron beam current 电子束电流
electron beam furnace 电子束炉
electron beam generator 电子束发生器
electron beam hardening 电子束淬火
electron beam heating 电子束加热
electron beam melting 电子束熔炼，电子光束溶解法
electron beam parametric amplifier 电子束参量放大器
electron beam recording 电子束记录
electron beam scanning system 电子束扫描系统
electron beam source 电子束源
electron beam tube 电子束射管
electron beam welding 电子束焊接，电子束焊
electron belt scale 电子皮带秤
electron bombardment 电子轰击
electron bridge 电子桥
electron capture 电子俘获，E俘获
electron charge 电子电荷
electron cloud 电子云
electron collector 电子收集器
electron collision frequency 电子碰撞频率
electron compound 电子化合物
electron concentration 电子密度，电子浓度
electron conduction 电子传导
electron-coupled circuit 电子耦合电路
electron-coupling 电子耦合
electron current 电子流
electron density 电子密度
electron detection 电子探测
electron detector 电子探测器
electron device 电子设备，电子器件
electron diffraction instrument 电子衍射仪
electron discharge machining 电火花加工
electron discharge 电子放电
electron discrete variable automatic compiler 电子离散变量自动编译程序
electron donor 电子给予体
electron drift 电子漂移
electronegative ion 阴离子
electronegative 负电性的，阴电性的
electronegativity 负电性，负电度
electron emission 电子发射
electroneutrality 电中性
electron gun 电子枪
electron hole 电子空穴
electronically commutated motor 电子整流电机
electronically-controlled automatic switching system 电子控制自动交换系统
electronically-controlled 电子控制的
electronic amplifier equipment 电子放大装置
electronic amplifier 电子放大器
electronic analog and simulation equipment 电子模拟仿真设备
electronic analog correlator 电子模拟相关器
electronic analog multiplier 电子模拟乘法器
electronic and information industries 电子信息产业
electronic anemometer 电子风速仪
electronic auction protocol 电子拍卖协议
electronic auction system 电子拍卖系统
electronic auction 电子拍卖，电子竞标
electronic automatic exchange 电子自动交换机
electronic balance 电子天平
electronic batch counter 电子选组计数器
electronic beam desulfurization technique 电子束脱硫技术
electronic beam method 电子束法
electronic belt scale 电子皮带秤
electronic binary multiplying computer 二进制乘法电子计算机
electronic book 电子书
electronic brain 电脑
electronic breakdown in dielectric liquid 液体电介质电击穿
electronic building brick 电子组件
electronic calculating punch 电子计算机穿孔
electronic catalog 电子目录
electronic character sensing 电子字符读出
electronic chronometric tachometer 电子计时式转速计
electronic circuit 电子电路
electronic commutation motor 电子整流电动机
electronic commutation 电子换向，电子整流
electronic commutator 电子换向器，电子整流器，电子转换开关
electronic computer 电子计算机
electronic computing units 电子计算单元
electronic conduction 电子传导
electronic conductivity 电子导电率
electronic"contact"operate 电子“触点”操作
electronic contactor 电子接触器
electronic contact rectifier 接触整流器
electronic controller 电子调节器，电子控制器，电动调节器
electronic controlling element for current 电流的电子主控元件
electronic control system 电子控制系统
electronic control 电子控制
electronic correlator 电子相关器
electronic counter-measure 电子干扰，电子对抗
electronic counter 电子计数器

electronic crack detector 电子探伤仪
electronic current transducer 电子式电流传感器［变送器，换流器］
electronic current transformer 电子式电流互感器
electronic current 电子流
electronic damping system 电子阻尼系统
electronic data gathering equipment 电子数据收集设备
electronic data printer 电子数据打印机
electronic data processing center 电子数据处理中心
electronic data processing equipment 电子数据处理设备
electronic data processing machine 电子数据处理机
electronic data processing system 电子数据处理系统
electronic data processing 电子数据处理
electronic data transmission 电子数据传递
electronic delay storage automatic computer 延迟存储自动电子计算机，延迟自动存储电子计算机
electronic device 电子装置
electronic digital computer 电子数字计算机
electronic direct-current motor controller 电子直流电（动）机控制器
electronic discrete sequential automatic computer 电子离散时序自动计算机
electronic discrete variable automatic computer 电子离散变量自动计算机
electronic discriminator 电子鉴别器
electronic display unit 电子显示器
electronic distance measuring device 电子测距仪［器］
electronic distance measuring instrument 电子测距仪［器］
electronic divider 电子分配器
electronic editor 电子编辑器
electronic element 电子元件
electronic engineering equipment 电子工程设备
Electronic Equipment Committee 电子设备委员会
electronic equipment room 电子设备间
electronic equipment 电子设备
electronic excitation 电子激发，电子激励，电子励磁
electronic exciter 电子励磁装置
electronic fault indicator 电子故障指示器
electronic ferro-resonance 电子式铁磁谐振
electronic Fourier synthesizer 傅里叶函数电子综合器
electronic frequency changer 电子频率变换器
electronic frequency meter 电子式频率器
electronic frequency trend relay 电子频率趋势继电器
electronic gas flowmeter 电子气体流量计
electronic generator 电子振荡器，电子管振荡器
electronic governor 电子调速器
electronic ground junction box 电子地线接线盒
electronic grounding cable 电子接地电缆
electronic guidance 电子制导
electronic hit indicator 电子碰击计数器
electronic humidistat 电子恒湿器
electronic-hydraulic control system 电液调节系统
electronic hysteresis 电子磁滞
electronic ignition device 电子点火装置
electronic image 电子图像
electronic indicating equipment 电子指示设备
electronic indicator 电子指示仪，电动指示仪
electronic industrial stethoscope 电子工业听诊器
Electronic Industries Association 电子工业协会【美国】
electronic information processing 电子信息处理
electronic instrument 电子仪表，电子式仪表
electronic integrator 电子积分器
electronic interpreter 电子翻译器
electronic inverter 电子逆变器
electronic level control 信号电平电子控制
electronic level device 电子式料位器
electronic level 电子能级
electronic load 电子负载
electronic logic 电子逻辑电路
electronic magnetometer 电子磁强计
electronic mailbox 电子信箱，电子（邮件）信箱
electronic mail 电子邮件，电子邮寄，电子邮递
electronic management system 电子管理系统
electronic measurement technique 电子测量技术
electronic measurement 电子仪器量测
electronic measuring instrument 电子测量仪表
electronic metal detector 电子金属检测器
electronic microscope 电子显微镜
electronic microswitch 电子微动开关
electronic model 电子模型
electronic motor control 电动机的电子控制
electronic motor 电子电动机
electronic multiplier 电子乘法器，电子倍增器
electronic-optical input device 电子光学输入器件
electronic oscillation 电子振荡
electronic overload detector 电子过载检测器
electronic pantograph 电子比例画图仪器
electronic peak-reading voltmeter 电子峰值电压表
electronic polarization 电子极化
electronic position transmitter 电子位置传送器
electronic potentiometer 电子电位差计，电子电势计，电子电位计
electronic power research institute 电力科学研究院【美国】
electronic pressure controller 电子压力控制器
electronic print reader 电子印刷读出器
electronic private branch exchange 专用支线电子交换机
electronic processing 电子数据处理
electronic program control 电子程序控制
electronic random numbering and indicating equipment 电子随机编号及指示装置
electronic ray 电子射线
electronic reading automation 电子阅读自动化
electronic reading machine 电子阅读机
electronic recorder 电子记录器
electronic recording potentiometer 电子电位计

electronic rectifier 电子整流器
electronic regulator 电子调节器
electronic relay potentiometer 电子继电器式电势计
electronic relay slide-wire potentiometer 电子继电器滑线式电势计
electronic relay 电子继电器，电子式继电器
electronic remote switching 电子远程转接
electronic room 电子室
electronic sampling switch 电子取样开关
electronics bench 电子设备操纵台
electronics box 电子设备箱
electronics cabinet 电子设备柜
electronic scale 电子秤，电子皮带秤
electronic scanning 电子扫描，电子扫描的
electronic security 地址保密，电子保密，电子安全，电子保安
electronic selective switching unit 电子选择开关装置
electronic self-balance instrument 电子自动平衡仪表
electronics engineering technician 电子工程技术人员
electronics engineer 电子工程师
electronic servo 电子伺服系统，电子随动系统
electronic signaling device 电子信号装置
electronic simulation 电子模拟
electronic simulator 电子模拟装置
electronic sinusoid generator 电子正弦波发生器
electronic slide rule 电子计算尺
electronic sorting system 电子分类装置
electronic sound detector 电子声探测仪
electronics rack 电子设备架
electronics room 电子设备房间，电子设备室
electronic stabilizer 电子（管）稳压器
electronic storage device 电子存储装置
electronic structure 电子结构
electronic stylus 电子笔
electronic subassembly 电子组合件
electronic surge arrestor 电子避雷器
electronic sweep generator 电子扫描发生器
electronic switching system 电子开关系统，电子转换系统
electronic switching 电子开关，电子切换
electronic switch 电子开关
electronics 电子学，电子物理学
electronic tachometer 电子转速计
electronic tag 电子标签
electronic temperature contact controller 电子温度开关控制器
electronic temperature recorder 电子温度记录器，电子温度记录仪
electronic test instrument 电子测试仪表
electronic text 电子文本
electronic theodolite 电子经纬仪
electronic thermostat 电子恒温器
electronic thickness gauge 电子厚度计
electronic time-lag relay 电子延时继电器
electronic time meter 电子测时计，电子毫秒表
electronic time relay 电子时间继电器
electronic timer 电子定时器
electronic torquemeter 电子扭矩计
electronic total station 全站型电子测距仪
electronic tracer 电子描绘器
electronic transformer 电子电路变压器，电子电路变换器
electronic transit 电子经纬仪
electronic translation 电子翻译
electronic transmission system 电子传输系统
electronic transmitter 电子变送器，电动变送器
electronic tube 电子管
electronic tuning hysteresis 电子调谐滞后
electronic tuning 电子调谐
electronic type belt scale 电子式皮带秤
electronic unit injector system 电子化单体喷油器系统
electronic vacuum dilatometer 电子真空膨胀计
electronic valve type power directional relay 电子管式功率定向继电器
electronic valve 电子管，电磁阀
electronic vibrating-contact regulator 电子式振荡触点调整器
electronic video recorder 电子录像机，电子视频记录装置
electronic voltage regulator 电子式电压调整器
electronic voltage stabilizer 电子式稳压器
electronic voltage transducer 电子式电压互感器
electronic voltage transformer or transducer 电子式电压互感器或换能器
electronic voltmeter 电子伏特计
electronic volt-ohm-ma meter 电子万用表
electronic wattmeter 电子式功率表
electronic weighing device 电子秤
electronic weighting equipment 电子称量装置
electronic work function 电子功函数，电子逸出功
electronic 电子的，电子学的，电子【复数】，电子元件【复数】
electron injection 电子注入
electron jet 电子束，电子流
electron liberation 电子释放，电子逸出
electron mass 电子质量
electron-microautoradiography 电子显微放射自显影法
electron microprobe analysis 电子微探针分析
electron microprobe 电子微探针
electron microscope autoradiography 电子显微放射自显影法
electron microscope 电子显微镜
electron-microscopical technique 电子显微技术
electron microscopic technique of metal 金属电子显微技术
electron mirror 电子镜
electron multiplication 电子倍增
electron multiplier 电子倍增器，电子乘法器，光电倍增管
electron number density 电子浓度
electron optics 电子光学
electron oscillograph 电子示波器
electron pair 电子偶，电子对
electron paramagnetic resonance 电子顺磁共振
electron partial pressure 电子分压

electron path 电子轨道，电子轨迹
electron pencil 电子锥，电子窄束，电子笔
electron plasma 电子等离子体
electron pole 电子极
electron-positron field 电子正电子场，阴阳电子场
electron probe analysis 电子探测分析
electron probe microanalyser 电子探针微量分析仪
electron radiation 电子辐射
electron radiography 电子辐射透照术
electron radius 电子半径
electron ray tube 电子射线管
electron ray 电子射线，电子注，电子束，电磁波
electron relay 电子继电器
electron release 电子释放，释放电子的
electron-rich alloy 富电子合金
electron sensitive 电子敏感的
electron spectroscopy for chemical analysis 化学分析用电子能谱学
electron spectrum 电子能谱
electron switch 电子开关
electron synchrotron 电子同步加速器
electron theory 电子理论
electron trap 电子陷阱
electron tube relay 电子管继电器
electron tube 电子管
electron valve 电子管，电磁阀
electron volt 电子伏特
electron wave 电子波
electron wind volume cover 电子风量罩
electro-optical distance measuring instrument 光电测距仪
electro-optical effect 电光效应
electro-optical hot gas pyrometer 电光气体高温计
electro-optical system 光电系统
electro-optic 电光的
electro-osmotic consolidation 电渗固结
electro-osmotic dewatering 电渗排水
electro-osmotic drainage 电渗排水
electro-osmotic stabilization of soil 土的电渗稳定
electro-osmotic transmission coefficient 电渗透射系数
electropeter 整流器，转换器
electrophoresis 电泳
electrophoretic analysis 电泳分析
electrophoretic clarification 电泳澄清
electrophoretic force 电渗力，电泳
electrophoretic method 电泳法
electrophore 起电盘
electrophotography 电子照相法
electrophotometer 电子光度计【测火焰用】，光电光度计
electrophysics 电子物理学
electroplate 电镀
electroplating dynamo 电镀用直流发电机
electroplating effluent 电镀废水
electroplating shop 电镀车间
electropneumatic actuator 电动气压执行机构，电动气压传动装置
electropneumatic contactor 电动气动接触器
electropneumatic controller 电动气动控制器
electropneumatic control 电动气动控制
electropneumatic converter 电气风动转换器
electropneumatic convertor 电气转换器
electropneumatic interlocking device 电动气动联锁装置
electropneumatic master controller 电动气动制动主控制器
electropneumatic regulator 电动气动调节器
electropneumatic switch 电动气动开关
electropneumatic transducer 电动气动变换器
electropneumatic valve 电动气动阀
electropneumatic 电动气动的，电动气动，电气动的，电气的
electropolarized relay 极化继电器
electropolarized （电）极化的
electropolar 电极性的，（电）极化的
electropositive ion 阳离子
electropositive 正电性的，金属的
electroprobe 电笔，电测针，电探头
electropsychrometer 电（测）湿度计
electropyrometer 电阻高温计，电测温度计，热偶温度计
electrorecovery 电解回收
electroreduction 电还原
electroreflectance 电反射率
electroregulator 电调节器
electro resistivity prospecting 电阻探测法
electroresponse 电响应
electroscope 验电器
electroscopic powder 验电粉
electrosemaphore 电标志，电信号机
electrosensitive recording 电火花蚀刻记录机
electro-sensitivity probe 电敏探针
electro-servo control 电气随动控制，电气伺服控制
electroshock 电击，触电，电击疗法，电休克疗法
electrosilicification 电动硅化法
electroslag melting 电渣熔炼法
electroslag pressure welding tool 电渣压力焊工具
electroslag pressure welding 电渣压力焊
electroslag process 电渣焊工艺
electroslag surfacing 电渣堆焊
electroslag welder 电渣焊机
electroslag welding machine 电渣焊机
electroslag welding 电渣焊
electro slag 电渣
electro-spark hardening 电火花强化
electrosparking 电火花加工
electrostatic accelerator 静电加速器
electrostatic accumulator 静电累加器，静电存储器
electrostatic actuator 静电激励器
electrostatic adherence 静电附着
electrostatic adhesion 静电吸附
electrostatic attraction 静电引力，静电吸引，静电顺力
electrostatic capacitance 静电电容
electrostatic charge effect 静电效应

electrostatic charge 静电荷
electrostatic circuit 静电电路
electrostatic cleaner 静电净化器
electrostatic collector failed element monitor 静电收集破损元件监测器
electrostatic component 静电分量
electrostatic coupling 静电耦合
electrostatic deflection 静电偏转
electrostatic discharge 静电放电
electrostatic dispersion 静电漂移
electrostatic displacement 静电位移
electrostatic dust precipitator 静电除尘器
electrostatic effect 静电效应
electrostatic energy 静电能量
electrostatic feedback 静电反馈
electrostatic field intensity 静电场强度
electrostatic field interference 静电场干扰
electrostatic field strength 静电场强度
electrostatic field 静电场
electrostatic filter 静电过滤器
electrostatic flux density 电感应强度，静电通量密度
electrostatic flux 电位移通量，静电通量
electrostatic force 静电净化器，静电力
electrostatic generator 静电发生器，静电加速器，静电发电机
electrostatic induction 静电感应
electrostatic instrument 静电计，静电式测试仪器，静电仪器
electrostatic interference 静电干扰
electrostatic ion trap 静电离子捕集器，静电离子阱
electrostatic machine 静电起电器
electrostatic memory 静电存储器
electrostatic motor 静电电动机
electrostatic oscillograph 静电示波器
electrostatic potential 静电势，静电位
electrostatic power generator 静电发电机
electrostatic precipitation （静）电除尘，静电沉积
electrostatic precipitator （静）电除尘器，静电过滤，静电集尘器，电除尘器
electrostatic pull 静电引力
electrostatic recorder 静电记录仪
electrostatic relay 静电（式）继电器，静电式继电器
electrostatic repulsion 静电推斥
electrostatic screen 静电屏幕，静电屏，静电屏蔽
electrostatic separation 静电分离
electrostatic separator 静电分离器，静电除尘器
electrostatic shielding 静电屏蔽
electrostatic shield 静电屏蔽
electrostatic spring shields 触点弹簧静电屏蔽
electrostatic storage deflection 静电存储偏转
electrostatic storage tube 静电存储管
electrostatic storage unit 静电存储部件
electrostatic storage 静电存储器
electrostatic system of units 静电单位制
electrostatics 静电学
electrostatic transformer 静电变压器
electrostatic unbalance 静电不平衡
electrostatic unit 静电单位
electrostatic voltmeter 静电电压表，静电伏特计
electrostatic wattmeter 静电瓦特计
electrostatic winding insulation 绕组静电绝缘
electrostatic 静电的，静电学的，静电型的
electrostriction 电致伸缩
electrostrictive effect 电致伸缩效应
electrostrictive relay 电致伸缩继电器
electrostrictive 电致伸缩的
electrosynthesis 电合成
electrosyntonic switch 远方高频控制开关
electrotape 基线电测仪
electrotechnical measurement 电工测量
electrotechnical porcelain product 电工陶瓷制品
electrotechnical porcelain 电瓷
electrotechnical steel 电工钢，硅钢
electrotechnical 电工的，电工学的，电工技术的
electrotechnology 电工技术，电工学
electrotellurograph 大地电流测定器
electrothermal baffle 半导体障板
electrothermal efficiency 电热效率
electrothermal element 电热元件
electrothermal expansion element 电热膨胀元件
electrothermal load 电热负荷
electrothermal melting furnace 电热熔炉
electrothermal recording 电热记录
electrothermal relay 电热继电器
electrothermal 电热的
electrothermic 电热的
electrothermoluminescence 电控加热发光，场控加热发光
electrothermometer 电测温度计，热（电）偶温度计
electrotimer 定时继电器，电子定时器
electro-tin plated 电镀锡
electrotitration 电滴定
electrotome 电刀，自动切断器
electrotropic 向电的，屈电的
electro-vacuum gear shift 电磁真空变速装置
electrovalence 电化价
electro valve 电动阀，电磁阀
electro-vibrating feeder 电振动给料机
electrovibration screen 电动震筛机
electrovibrator 电振动器
electroviscosity 电气吸附性，电黏度，电黏滞性
electrowelding 电焊
electrowinning 电解冶金法，电积金属法
electton impulse 电子脉冲
ELEC 电的，电气的【DCS 画面】
elegible bidder 合格的投标人
elegible countries 有投标资格的国家
EL(elastic limit) 弹性极限
EL＝elevation 标高
elemental composition 化学合成，元素成分
elemental height 基本高度，初始高度
elemental iron 元素铁
elemental network 基本网络
elemental radioiodine 放射性元素碘
elemental sulfur 单质硫，元素硫

elemental value 初值
elemental wiring diagram 原理接线图
element analysis 元素分析
element antenna 振子天线
element area 单元面积
elementary 初步的，基础的
elementary analysis 元素分析
elementary ancillary service 基本辅助服务
elementary charge 质子电荷，基本电荷
elementary composition 元素成分
elementary current 元电流
elementary derivation 初步推导
elementary diagram 原理图，简图，原理性电路图
elementary event 基本事件
elementary flow pattern 基本流谱，基本流动模式
elementary frequency 基本频率
elementary gas 单质气体
elementary geodesy 普通测量学
elementarygrain size 本质晶粒度
elementary magnet 单元磁铁
elementary mass 微元质量
elementary period 基本周期
elementary reaction 基元反应
elementary regenerator section 单元再生段
elementary repeater section 单元中继段
elementary slot 虚槽
elementary step 初步阶段
elementary stream 通量线，电流线，水流线
elementary wave 原波
element buffer 缓冲元件
element characteristics 单元特性
element conveyer 元件运输机
element conveyor （燃料）元件运输机
element design 构件设计，元件设计
element displacement 单元位移
element failure 元件损坏
element interface 单元交界面
element jacket （燃料）元件包壳
element-node incidence matrix 元件节点关联矩阵
element of fluid 流体单元
element of phase shifter 移相器元件
element spacing （燃料）元件间距
element specific activity （放射性）元素比活度
element stiffness 单元刚度，单元劲度
element storage rack （燃料）元件存放架
element stress 单元应力
element subdivision 单元剖分
element 元件，部件，单元，元素，要素，零件，构件
ELEP(expansion-line end point) 膨胀线终点【汽轮机】
elephant transformer 无套管式高压室内变压器
elephant trunk 溜管
eletric beading grooving machine 电动压筋机
eletric threading machine 电动套丝机
eletrified-particle gun 带电粒子枪
eletrified-particle testing 带电粒子检测
eletro-pneumatic positioner 电气阀门定位器
elevated approach 高架引道
elevated basin 高水池
elevated beach 岸边高地，岸边台地，上升阶地
elevated block 上升地块
elevated concrete floor 零米以上混凝土楼层
elevated crossing 高架交叉道
elevated flume 高架渡槽
elevated footway 高架人行道
elevated grounded counterpoise （天线的）架空地网
elevated highway 高架公路
elevated house 高架房屋
elevated inversion 高空逆温
elevated line bridge 高架线路桥
elevated line source 高架线源
elevated oil tank 高位油箱
elevated peneplain 上升准平原
elevated pipeline 架空管道，高架管道
elevated plume 抬升羽流
elevated railway 高架铁道，高架铁路
elevated range 范围负迁移
elevated release 高空释放
elevated reservoir 高架水柜，水塔
elevated source 高架源
elevated span 量程负迁移
elevated spur track 高架分支轨道
elevated steam conditions 高蒸汽参数
elevated storage 高位蓄水
elevated supply tank 高架供水塔
elevated tank 高架（水）箱，高架水柜
elevated temperature impact test 高温冲击试验
elevated temperature properties 高温性能，高温强度
elevated temperature tension test 高温拉伸试验
elevated temperature 高温，升高温度，提高温度
elevated track 高架轨道
elevated tramway 高架电车道
elevated water tank 高架水箱
elevated-zero range 零点提升范围
elevated 架空的，提高的，抬高的
elevate 升高，增加
elevating and depression buffers 高低机减震器
elevating belt conveyor 斜升输送机
elevating check valve 升降式止回阀
elevating dredge 提升式挖泥机
elevating fire truck 升降平台消防车，升降式消防车
elevating gear 提升凿轮，提升装置
elevating grader 升运式平土机
elevating loader 提升式装料机
elevating platform fire truck 登高平台消防车
elevating scraper 升装铲运机
elevating screw 螺旋升运器，升运螺旋
elevation above sea level 海拔标高
elevational point 高程点
elevation amplitude 仰角
elevation and depression 上提及下压
elevation angle 高度角，仰角
elevation at top of jetty 码头平面标高
elevation burner 顶棚燃烧器

elevation-capacity curve 高程库容关系曲线，高程库容曲线
elevation-capacity relation 高程容积关系
elevation control network for deformation observation 高程监测网
elevation control network 高程控制网
elevation control 高程控制
elevation correction 高程改正，海拔校正
elevation depression 标高沉降，倾角
elevation deviation 标高偏差
elevation difference 标高差，高差
elevation-distance curve 高程距离曲线
elevation drawing 立面图，正视图，前视图
elevation effect on atmospheric pressure 海拔高度对大气压力的影响
elevation frequency curve 高程频率曲线
elevation head 位势水头，静压水头，高程水头，水力静压头，升水头
elevation height 提升高度
elevation indication 高程读数
elevation indicator 高度指示器
elevation limit 仰角变化范围
elevation-net storage curve 高程净蓄量曲线【水库的】，高程净库容曲线
elevation number 高程注记
elevation oblique drawing 立面斜视图
elevation of borehole 孔口标高
elevation of building 建筑立面图
elevation of in-door grade 室内地坪标高
elevation of main building area 主厂房区标高
elevation of natural ground 自然地形标高
elevation of normal water level 正常水位高程
elevation of operating floor 运转层标高
elevation of outdoor ground 室外地面标高
elevation of road-intersection 道路交叉点标高
elevation of temperature 温度升高
elevation of water 水位
elevation of well 井口标高
elevation of wharf apron 码头前沿高程
elevation of zero of gauge 水尺零点高程
elevation pressure drop 提升压降
elevation scheme 高程图
elevation selsyn transformer 仰角自动同步机变压器
elevation shoreline 上升岸线
elevation treatment 立面处理
elevation view 正视图，立视图，立面图
elevation zero 高程零点
elevation 海拔，正视图，立视图，立面图，标高，高程，上升，提高，仰角
elevator hoistway 电梯间，电梯井，升降机井
elevator layout drawing 电梯布置图
elevator machine room 电梯机房
elevator motor 升降电动机，起重电动机
elevator periscope 升降式潜望镜
elevator pit 电梯坑
elevator pump 提升泵
elevator shaft 电梯间，电梯井，升降机（竖）井
elevators outside power block 非厂用其他电梯
elevator system 升运系统
elevator tower 升降机塔，电梯塔
elevator type loader 提升式装料机
elevator well 升降机井，电梯井
elevator 提升机，升降机，电梯，起重机
EL(exhaust loss) 排汽损失
ELF(extremely low frequency) 极低频
eligibility certificate 资格证书
eligibility criterion 合格标准
eligibility 合格，资格，条件，投标人合格性
eligible tenderer 合格投标者［商］，符合条件的投标者［商］
eligible bidder 合法投标商
eligible countries 符合条件的国家【指有资格投标或供货或提供服务】
eligible for 合格，够资格
eligible goods 合格货物
eligible source country 合格来源国
eligible 符合条件的，合格的，合适的，有资格当选的，适当的，合格者，有资格者
eliminate 排除，除去，消除，拆除，消去，除法
eliminating faults 排除故障
elimination by substitution 代入消元法
elimination 消除，除去，避免，除水器，消除器，抑制器，挡水板
elliplical arch dam 椭圆拱坝
ellipse of stress 应力椭圆
ellipse 椭圆形，椭圆
ellipsis 省略号
ellipsograph 椭圆规，椭圆仪
ellipsoidal head 椭圆封头，椭圆形封头
ellipsoidal 椭球面
ellipsoid 椭球，椭圆体
elliptical arch dam 椭圆形拱坝
elliptical arch 椭圆拱
elliptical bearing 椭圆形轴承
elliptical cam 椭圆凸轮
elliptical cap 椭圆形泡罩
elliptical collineation 椭圆直射
elliptical coordinate 椭圆坐标
elliptical cross section 椭圆截面
elliptical field 椭圆场
elliptical flow 椭圆流动
elliptical function 椭圆函数
elliptical head 椭圆形封头【压力容器】
elliptical integral 椭圆积分
elliptically polarized wave 椭圆偏振波，椭圆极化波
elliptically polarized 椭圆极化的
elliptical opening 椭圆孔口
elliptical orbit 椭圆轨道
elliptical path 椭圆轨迹
elliptical pointed arch 椭圆尖头拱
elliptical polarization 椭圆极化，椭圆偏振
elliptical projection 椭球投影
elliptical ring 椭圆环
elliptical rotating field 椭圆形旋转场，椭圆旋转场
elliptical section 椭圆截面
elliptical shape 椭圆形
elliptical slot 椭圆形槽
elliptical throttling slide damper 椭圆形节流挡板

elliptical type weight　铅鱼，椭圆形测深重锤
elliptical vibration　椭圆振动
elliptical vortex generator　椭圆涡发生器
elliptical waveguide　椭圆形波导
elliptical wind tunnel　椭圆试验段风洞
elliptical　椭圆的，椭圆形的
ellipticity　椭圆度，椭圆形，椭圆率
ell　字母 L，侧房，管子弯头，L 形短管
elmillimess　电动测微仪
elm　榆木
elongated bubble flow　拉长气泡流
elongated cavity　长气孔【焊接缺陷】
elongated gas porosity　长形气孔
elongate　拉长，延长，伸长
elongation at break　裂断伸
elongation at failure　破坏伸长
elongation at rupture　断裂延伸率，断裂伸长
elongation at yield　屈服伸长
elongation factor　延伸系数
elongation in tension　受拉伸长
elongation modulus　伸长模量
elongation of tendon　预应筋的延伸
elongation pad　延长器，外延衰减器
elongation per unit length　单位长度的延伸，伸长率
elongation ratio　伸长率，细长率
elongation strain　伸长应变
elongation test　拉伸试验，伸长试验
elongation　拉长，延长，伸张度，延伸率，拉伸延长，伸长，伸长率，延伸，相对伸长
ELPHR(experimental low-temperature process heat reactor)　低温供热试验反应堆
ELR(engineering laboratory report)　工程实验试报告
ELSEC(electronic security)　地址保密，电子保密
elude　避免，逃避
elute　洗提，冲洗
elution leakage　洗提泄漏
elution　洗提，洗脱
elutriate　洗涤，淘析，扬析
elutriation by air　空气淘析
elutriation by water　水力冲洗
elutriation method　冲洗法，水析分级法，淘析法
elutriation rate constant　扬析速率常数
elutriation rate　扬析速率
elutriation-reducing baffle　扬析阻挡板
elutriation　分选，（沸腾炉飞灰）扬析，淘洗，淘分，冲洗器，分选器，淘析器，沉淀分析法
eluvial deposit　残积
eluvial facies　残积相
eluvial horizon　淋滤带
eluvial soil　残积土，淋滤土，淋溶土
eluvial　残积
eluviation　淋滤酌，淋溶【土壤】，淋溶作用，残积作用
eluvium　残积层
EMA(electromegnetic acoustic)　电磁声的
EMA(electron microprobe analysis)　电子显微探针分析
E-mail　电子邮件，电子函件
emanate　放射
emanating power　辐射功率，发射功率
emanation　辐射，放射，射气
emanometer　测氡仪，射气仪
EMAR(experimental memory address register)　试验存储地址寄存器
embanked area　围堤岸，围堤区，围垦区
embankment dam with asphalt concrete sealing　沥青混凝土防渗土石坝
embankment dam　填筑坝，填筑式坝，土石坝
embankment failure　土堤破坏
embankment foundation　堤基
embankment material　坝体土石料【水电】，堆筑材料，填筑材料
embankment mower　堤坡割草机
embankment protection　路堤护坡
embankment quay　堤岸
embankment section　路堤截面
embankment seeding　路堤植草
embankment slope　堆筑体边坡，路堤斜坡，填土边坡
embankment top　路堤顶
embankment wall　堤墙
embankment zoning　坝体分区
embankment　堤坝，沟渠，堤，堤岸，路堤，筑堤，填土，填土堤，填筑，（土石坝）坝体
embargo　封港，禁运
embassy　大使馆
embattlement　城垛
embayment　港湾，横越地槽
embed crack　埋藏裂纹
embedded bar　预埋钢筋
embedded bolt　预埋螺栓
embedded channel　水下隧道
embedded coil　已嵌入的线圈，槽内线圈
embedded column　暗柱
embedded cost pricing　会计成本定价
embedded depth　埋入深度，埋设深度，埋深，嵌固深度
embedded generation system　嵌入式发电系统
embedded hook　吊钩
embedded insert　预埋件
embedded instrument　埋设仪器
embedded iron part　预埋铁件
embedded length of bar　钢筋埋入长度
embedded length　埋入长度，锚固长度
embedded metal　预埋钢板，预埋件
embedded part of coil　线圈有效部分，槽部线圈
embedded part　埋件，预埋件，埋入部分，埋入件
embedded penstock　埋藏式压力水管
embedded pipe cooling　埋管冷却
embedded pipe　暗装管道，埋藏管道
embedded plate　预埋板
embedded pointer　嵌入指示字，嵌入指针
embedded reinforcement　预埋钢筋
embedded resistance thermometer　埋入式电阻温度计
embedded-resistor　埋入式电阻器
embedded scraper transporter　埋刮板输送机
embedded steel　埋置钢筋
embedded stone pitchin　埋藏式砌石护坡

embedded temperature detector 埋入式温度计
embedded wood brick 预埋木砖
embedded 镶嵌的，埋设的，嵌入式的，预埋（在混凝土内）的
embedment depth 埋置深度
embedment length 埋入长度，埋置长度
embedment 预埋件，埋入件，嵌入，埋入，砌住，浇牢，预埋
embed 埋置，嵌进，放入，嵌入，灌封，埋藏，埋入
embezzlement of public funds 挪用公款，侵吞公款
embezzlement 侵吞，贪污，盗用
embezzle 侵吞（公款），贪污，盗用
emblem 图解符号，图形符号
embodiment 预埋件【指金属件】
embossed glass 浮雕玻璃
embossed 模压加工
embranchment 支流，分支机构
embrasure 喷口，口
embrittlement cracking 脆裂，脆性裂缝
embrittlement detector 脆化指示器
embrittlement 脆化，脆性，脆变，变脆，脆裂
embryo deposit 初始沉积
embryonic droplet 凝结胚，滴胚，凝结核
embryonic dune 初期沙丘
CMC（China National Machinery Imp. & Exp. Corp.）中国机械进出口（集团）有限公司，中机公司
EMC（electromagnetic compatibility） 电磁适应性，电磁兼容性
EMC（end manufacturing certificate） 制造完工合格证
EMC-filter（electromagnetic compatibility filter） 电磁兼容性滤波器，EMC 滤波器
EMCON（emergency condition） 事故状态
EMCON（emergency control） 紧急控制
EM（emergency maintenance） 事故维修
EM（emergency） 紧急
emend 修订，校正
emergancy containment system 应急安全壳系统
emergence angle 出射角
emergence coast 隆起海岸
emergence shoreline 离水海岸线，上升岸线，上升滨线
emergence 显露，出射，排出，紧急情况意外事故
emergency action 紧急行动
emergency aerial 事故用天线，应急用天线
emergency air cleaning system 应急空气净化系统
emergency air inlet 安全气孔
emergency air lock 人员应急通道，应急气闸
emergency alarm bell 报警铃
emergency allocation 应急拨款
emergency and back-up mechanisms 应急和后备措施
emergency apparatus 应急备用设备，备用设备
emergency arrangements 应急安排
emergency ash yard 事故灰场
emergency auxiliaries 应急辅助系统
emergency auxiliary power 事故备用电源
emergency back-up fuel 应急燃料
emergency bearing oil pump 事故润滑油泵，备用润滑油泵
emergency bell 事故铃
emergency belt conveyor 备用皮带输送机
emergency bilge pump 应急舱底水泵
emergency blowdown valve 紧急排污阀，危急放气阀，危急排气阀
emergency boration system 紧急硼化系统，应急加硼系统
emergency boration 应急加硼，应急硼化
emergency boron injection 应急硼注入
emergency brake 事故制动器
emergency braking system 紧急制动系统
emergency braking 紧急刹车
emergency breaker 应急断路器
emergency bridge 临时应急桥
emergency button 事故按钮，备用按钮，应急按钮
emergency cable 事故电缆，应急电缆
emergency capability 应急能力，应急力量
emergency capacity 事故备用容量
emergency case 紧急情况
emergency cell 事故备用电池
emergency circumstances 紧急情况
emergency closing device 事故关闭装置
emergency closing valve 应急阀，事故关闭阀
emergency closure 事故关闭，事故停机，事故关闭设备
emergency condensation 应急凝汽
emergency condenser drain 应急凝汽器疏水
emergency condenser system 应急凝汽系统
emergency condenser vent 应急凝汽器出气口
emergency condenser 应急凝汽器
emergency condition 事故工况，事故状态，应急工况，故障工况，紧急状态
emergency connection 事故接线，应急接线
emergency containment system 应急安全系统
emergency control center 应急控制中心
emergency control room 应急控制室
emergency control 事故调节，危急控制，紧急控制，事故控制，应急控制
emergency coolant injection 应急冷却剂注入
emergency coolant system 紧急冷却剂系统
emergency cooler 应急冷却器
emergency cooling and residual heat removal system 应急冷却与余热排出系统
emergency cooling circuit 应急冷却回路
emergency cooling heat exchanger 应急冷却热交换器
emergency cooling loop 应急冷却回路
emergency cooling period 应急冷却阶段【失水事故】
emergency cooling system 应急堆芯冷却系统
emergency cooling water 应急冷却水
emergency cooling 事故冷却，应急冷却
emergency core cooling criteria 应急堆芯冷却（设计）准则
emergency core cooling injection system 应急堆芯冷却剂注入系统
emergency core cooling system 堆芯紧急冷却系

统，紧急堆芯冷却系统
emergency core cooling 紧急堆芯冷却，应急堆芯冷却
emergency crop 短生长期植物
emergency cutout 应急切断
emergency dam 应急坝
emergency decree 安全技术规程
emergency device 危急保安装置
emergency diesel building 应急柴油发电机房
emergency diesel generator room 事故柴油发电机房
emergency diesel generator set 事故柴油发电机组，应急柴油发电机组
emergency diesel generator 应急柴油发电机
emergency diesel station 应急柴油机站
emergency door 安全门，太平门
emergency dose 应急剂量
emergency drain equipment 危急疏水装置，应急疏水装置
emergency drain system 危急疏水系统
emergency drain valve 危急疏水阀
emergency drain water 事故疏水
emergency drill 应急演习
emergency drum drain valve 汽包事故放水阀
emergency dump area 应急卸煤区
emergency dump steam 事故排汽
emergency duty 事故备用，事故时的运行状态
emergency electric supply unit 事故备用供电设备
emergency electric system 事故备用供电系统
emergency engine 紧急备用机
emergency escape ladder 救生梯
emergency evacuation 紧急撤离，应急疏散
emergency exciter set 事故备用励磁机组
emergency exercise 应急演习
emergency exit ladder 安全梯
emergency exit 事故出口，紧急出口，安全出口，太平门，备用引出端
emergency exposure 事故辐照，应急照射
emergency feather 应急顺桨
emergency feed control system 应急给水控制系统
emergency feeder 备用馈电线
emergency feed pipe 应急给水管
emergency feed pump 事故给水泵
emergency feedwater supply system 应急给水系统
emergency feedwater supply 事故给水，应急给水
emergency feedwater system 应急给水系统
emergency feedwater 事故给水，辅助给水，应急给水
emergency flooding 应急淹没
emergency fuel trip 紧急停料装置，燃料自动切断器
emergency gas turbine generator 备用燃气轮发电机
emergency gate slot 事故闸门槽
emergency gate 事故闸门
emergency generating set 事故备用发电机组，应急发电机组
emergency generating unit 应急发电机装置
emergency generator set 应急发电机组
emergency generator 紧急发电机，备急发电机，事故备用发电机，应急发电机
emergency governor gear 危急保安器，危急截断器
emergency governor hand trip 危急手动脱扣
emergency governor oil filling test 危急截断器充油试验
emergency governor pilot valve 危急截断油门，危急保安器滑阀
emergency governor test 危急截断器试验
emergency governor 应急调速器，危急保安器，危急切断器
emergency grounding 应急接地，事故接地
emergency hand-drive 事故手动装置
emergency handling 事故操作处理
emergency hand trip 危急手动脱扣
emergency heat removal system 应急排热系统
emergency hydraulic regime 事故水力工况
emergency identification light 危急标志灯
emergency illumination 事故照明
emergency insertion device 紧急插入装置
emergency interconnection 紧急联络线
emergency job cider 紧急工作命令
emergency key 应急电键，事故备用电键
emergency lagoon 应急池
emergency lamp 应急灯
emergency lighting box 事故照明箱
emergency lighting storage battery 应急照明蓄电池
emergency lighting system 事故照明系统
emergency lighting unit 应急照明装置
emergency lighting 紧急照明，事故备用照明，事故照明，应急照明
emergency light 事故信号灯，故障信号灯，事故照明灯，应急照明
emergency line 事故备用线路
emergency load dump 事故甩负荷，紧急甩负荷
emergency load shedding 紧急减负荷，事故减负荷
emergency loan 紧急贷款
emergency lock 事故闸门，安全闸门，安全闸，安全间
emergency maintenance time 排除故障时间
emergency maintenance 排除故障，紧急维护，紧急维修，应急维修
emergency material 应急物资
emergency maximum transfer capability 备用最高交换功率【电力系统】
emergency measures 应急措施，紧急措施
emergency network 事故供电网，备用电网
emergency off local 现场事故断开
emergency oil drain 事故排油管
emergency oil pump 事故油泵
emergency oil relieve valve 紧急放油阀
emergency oil tank 事故油箱
emergency operated valve 事故阀
emergency operation 事故操作，紧急操作，事故处理，事故运行，应急操作
emergency organization 应急组织

emergency outage　紧急停运，事故停机，紧急停机
emergency outlet　安全出口，安全出水口，事故出口
emergency overflow box　紧急溢流水箱
emergency overload　事故过负荷
emergency overspeed governor　危急保安器，危急截断器
emergency overspeed trip　超速紧急脱扣，超速危急截断
emergency panel　事故配电盘，备用配电盘，应急停堆盘
emergency pasture　临时放牧场
emergency personnel access　人员应急进出通道
emergency planning　应急计划制定
emergency plant cooldown system　电站紧急冷却系统，电站应急冷却系统
emergency plan　应急计划
emergency-powered busbar　应急汇流排
emergency powering　应急供电
emergency power interruption　事故停电
emergency power plant　事故备用电站
emergency power source　保安电源，应急电源
emergency power supply system　紧急供电系统，备用电源
emergency power supply　事故电源，备用电源，应急电源，紧急供电，应急供电
emergency power　紧急备用动力，备用电源，应急动力，应急电源
emergency preparedness　应急准备
emergency project　应急计划
emergency protection　事故保护
emergency pulse　呼救信号，呼救脉冲
emergency pump rating　应急泵功率
emergency pump　备用泵，应急泵，事故备用泵
emergency push-button switch　事故按钮开关，备用按钮开关
emergency push button　事故按钮
emergency receiver　应急容器
emergency reclaim facility　事故取料设备
emergency reclaiming hopper　事故取煤斗
emergency reference level　应急基准水平
emergency release　事故脱扣，事故释放
emergency relief valve　紧急安全阀
emergency relieve oil valve　紧急放油阀
emergency repair　事故修理，事故抢修，紧急修理，事故检修
emergency rescue fire vehicle　抢险救援消防车
emergency reserve capacity　事故备用容量
emergency reserve funds　紧急预备金
emergency reserve station　应急备用电站
emergency reserve　紧急备用，事故备用
emergency residual heat operation　应急余热冷却运行
emergency residual heat removal　应急余热排出（系统）
emergency response center　应急响应中心
emergency response force　应急响应工作队
emergency response plan　应急响应计划
emergency response team　应急响应队伍，应急响应工作队
emergency review　事故追忆
emergency rod　事故棒
emergency room　急救室
emergency rope pull stop switches　紧急（拉线）停机开关
emergency safety evaluation　应急安全评价
emergency service order　紧急工作单
emergency service　事故供电，紧急供电，事故运行
emergency set　事故备用机组，应急设备
emergency shower and eye washer　事故沐浴洗眼器
emergency shower unit　事故淋浴装置
emergency shower　事故淋洗器
emergency shutdown coefficient　故障停运率
emergency shutdown cooling system　紧急停堆冷却系统
emergency shutdown cooling　应急停堆冷却
emergency shutdown device　紧急停车系统，紧急停机装置，紧急停堆装置
emergency shutdown for wind turbine　风力涡轮机紧急关机
emergency shutdown limit　紧急停堆限值，应急停堆限制
emergency shutdown member　紧急停堆棒
emergency shutdown of reactor　事故停堆
emergency shutdown panel　紧急停堆盘，紧急停堆控制盘
emergency shutdown rod transport mechanism　紧急停堆棒传送机构
emergency shutdown rod　紧急停堆棒
emergency shutdown system　紧急停堆系统
emergency shutdown valve　事故停机阀
emergency shutdown　紧急停堆，落棒，紧急关机，紧急停机，紧急停运，事故停炉，事故停机，事故停堆，事故停车，应急停堆
emergency shut-off rod　紧急停堆棒
emergency shut-off valve　危急关闭阀
emergency shut-off　事故停机，紧急停机，紧急切断，紧急停堆，事故关闭
emergency signaling　事故信号装置
emergency signal transmitter　事故信号发送器，危险信号发送器
emergency signal　事故信号
emergency special session　紧急特别会议
emergency speed variation　事故速度变动率
emergency spillway　非常溢洪道，紧急溢洪道
emergency spray　紧急喷水，事故喷水
emergency stacker　备用堆
emergency staircase　太平梯，疏散楼梯，安全梯
emergency standby capacity　事故备用容量
emergency standby　事故备用
emergency state　紧急状态，事故状态
emergency station service transformer　事故厂用变压器
emergency station　事故急救站
emergency stockpile　事故煤堆
emergency stop cock　紧急制动旋塞
emergency stop device　紧急停机装置
emergency stop hand off　危急手动停机
emergency stop push button　紧急停车按钮

emergency stop switch 紧急停机开关，事故停机开关
emergency stop valve 危急切断阀，紧急止流阀，主汽门，紧急刹车阀
emergency stop 紧急停运，事故停机，紧急停止
emergency storage pile 事故煤堆，事故备用煤堆
emergency storage 应急贮存
emergency supply breaker 应急供电断路器
emergency supply sources 紧急备用电源
emergency supply 应急供应
emergency switchboard 应急（电话）总机，应急开关板，应急配电盘
emergency switching-off 事故切断
emergency switch 紧急开关，事故开关
emergency system 应急系统
emergency terminal switch 终端安全开关
emergency throw-over equipment 事故转换设备
emergency throw-over of auxiliaries 辅机事故停机【电厂】
emergency transfer capability 紧急转换能力，备用转换功率
emergency transformer 应急电源变压器
emergency trip pilot valve 危急截断器滑阀
emergency trip system 紧急切跳闸系统【汽轮机】，事故跳闸系统
emergency trip testing handle 危急截断的试验手柄
emergency trip valve 危急截断阀，紧急止流阀，主汽门
emergency trip 事故切断，保护装置，紧急脱扣，紧急跳闸，危急截断，紧急保护停堆，紧急事故跳闸
emergency turning gear 后备盘车装置，事故盘车装置，应急盘车装置
emergency unit 备用机件
emergency valve 危急截断阀，紧急止流阀，安全阀，紧急阀，事故阀
emergency ventilation system 事故通风系统
emergency ventilation 紧急备用通风，事故通风，应急通风
emergency vent 紧急排气口
emergency water spray valve 事故喷水阀
emergency weir 临时应急堰
emergency 紧急情况，应急，危险，紧急的，事故，备用，紧急，后备，应急状态
emergent declearation change 冲关费
emerging defect 显露缺陷
emerging 现出，涌出
emeritus professor 荣誉教授
emery cloth 砂布，金刚砂布
emery paper 砂纸
emery wheel 金刚砂旋转磨石，砂轮
emery 金刚砂，刚玉砂
EM(external memory) 外部存储器
EMF(electromagnetic field) 电磁场
EMF(electromotive force) 电动势
EMF to pneumatic transmitter 电动势气动变送器
EMG(emergency) 事故的
EMI(electromagnetic interference) 电磁干扰
emigration 迁移
eminence 高（地）处，高地
eminently clean 良好流线型的
eminent person 知名人士
emissary 排水道
emission band 发射频带，放射频带
emission behavior 放射性能，辐射性能，发射状态，放射状态，排放性能，发射性能
emission by high chimney 高烟囱排放
emission center 放射中心
emission concentration 排放浓度，排放物浓度
emission control equipment 排放控制设备
emission control strategy 排放控制对策
emission control 放射控制，排放控制
emission current density 放射电流密度
emission current 放射电流
emission detector 排放检测器，发射检测器
emission efficiency 排放效率，发射效率
emission factor 排放系数，排放因素，散布系数
emission height 发射高度，排放高度
emission inventory 排放源清单，日放逸量
emission level 排放水平【污染物】
emission limit 排放极限，排放限度
emission line 发射线
emission of heat 热辐射，散热
emission photometry 发射光谱法
emission point 排放点
emission power 发射功率【电子束焊接】
emission rates 污染质放出率
emission rate 发射率，辐射率，排放量，排放率
emission Reynolds number 排放雷诺数
emission source 发射源，辐射源，排放源
emission spectrography 发射光谱测定法
emission spectrum 发射（光）谱，发射谱
emission standard 排出标准，排放标准
emission steam 排汽，废汽
emission temperature 排放温度
emission 排放，放射，发射，放射物，排放物，散发物，放出，散放，辐射
emissive fission 发射裂变
emissive power 辐射能力，发射力，发射本领
emissivity at a specified wavelength 光谱发射率
emissivity emittance 黑度
emissivity factor 黑度，辐射率
emissivity 辐射率，放射率，辐射能力，发射率，辐射系数
EMIT(engineering management information technique) 工程管理信息技术
emittance coefficient 发射系数
emittance 发射度，辐射度，辐射率，辐射强度，辐射，发射，放射，发射比，发射率【ε】
emitter-coupled circuit 射极耦合电路
emitter-coupled logic operator 射极耦合逻辑运算电路
emitter-coupled logic 电流开关逻辑，发射极耦合逻辑，射极耦合逻辑电路
emitter-coupled transistor logic 射极耦合晶体管逻辑
emitter-emitter coupled logic 发射极—发射极耦合逻辑
emitter follower diode logic 射极跟随器二极管逻辑

emitter follower diode transistor logic　射极跟随器二极管晶体管逻辑
emitter follower logic　发射极跟随器逻辑
emitter follower　射极跟随器
emitter-to-base voltage　基极发射极间电压
emitter　辐射体，发射体，发射器，放射体，发射管，放射器，放射极，发射极，辐射源，放射源
emitting area　辐射面积
emitting medium　散射介质
emitting surface　辐射面
emit　辐射，放射，放出，逸出，辐射传播
EML(equipment modification list)　设备改变表
EMO(equipment manufacturing operation)　设备制造工序
emolliate　软化，使柔软
emollient　软化剂
EMOP(environmental monitoring plan)　环境监测计划
EMP(effective mean pressure)　有效平均压力
EMP(electro mechanical package)　机电（工程）安装包
EMP(electromechanical power)　机电功率
EMP(environmental management plan)　环境管理计划
emphasis　强调，重点
emphasizer　加重电路，频率校正电路
emphasize　强调
empire cloth　绝缘油布
empire paper　绝缘纸
empire tape　绝缘带，黄蜡带
empire tube　绝缘套管
empire　绝缘，绝缘漆
empirical coefficient　经验系数
empirical constant　经验常数
empirical correction　经验修正
empirical correlation　经验关联，实验关联
empirical curve　经验曲线
empirical data　经验数据，实验数据
empirical design　经验设计
empirical equation　经验方程，经验方程式
empirical evidence　经验数据
empirical exponential function　经验指数函数
empirical fit　经验拟合
empirical formula　经验公式，实验公式
empirical law　经验定律，经验规律
empirical mass formula　经验质量公式
empirical rule　经验定则
empirical shape factor　经验形状系数
empirical value　经验值
empirical　经验主义的，完全根据经验的，经验的，实验的
emplacement　就位，定位，放置，安装
emplace　安装就位
emplaster　灰膏
employability　受雇就业能力
employ all available means　千方百计，用尽所有办法
employee benefit　职工福利
employee bonus　职工奖金
employee dormitory　职工宿舍
employee insurance fund　职工保险基金
employee pensions and benefits　离退休金和福利费
employee welfare fund　职工福利基金
employee　工作人员，雇员
employer liability insurance　雇主责任险
employer's liability insurance　雇主责任保险，业主责任险
employer's requirements　业主要求
employer's taking over　业主接收（工程）
employer　业主，建设单位，雇主
employment agreement　聘约
employment contract　聘约
employment injury insurance　工伤保险
employment　使用，雇用，职业，工作
employ　雇用，使用
empower　授权
empty car ejection　推出空车，空车组
empty car　空车
empty-cell process　不完全浸注法【防腐用】
empty drum store　空桶仓库
emptying culvert　放空涵洞，泄水涵洞
emptying device　放空设备，排空设备
emptying running　空转
emptying time　放空时间
emptying tunnel　放空隧洞
emptying valve　排放阀，排空阀，泄油阀，放空阀
emptying　放空，放水，排空
empty load test　空载试验
empty name plate　空白铭牌板
empty promise　空头支票
empty running　空载运行，空转
empty-track line　空车线
empty train　空列
empty tunnel　空风洞【没有装模型】
empty wagon pusher　空车推车器
empty wagon track　空车线
empty wagon　空车
empty weight　皮重，自重，空重
empty working section　空风洞试验段，空水洞试验段
empty　空的，空载的，排空的，放空
EMR(end of manufacturing report)　制造完工报告，设备出厂报告
EMS(effluent management software)　排出流管理软件
EMS(electrical management system)　电气管理系统
EMS(electronic management system)　电子管理系统
EMS(emergency switch)　紧急开关
EMS(energy management system)　能量管理系统
EMS(express mail service)　特快专递
EMT(electromagnetic transducer)　电磁探头，电磁变换器
EMT(end of magnetic tape)　磁带结束
EMU(electromagnetic unit)　电磁单位（制）
emulate system　仿真系统
emulate　模仿，仿真，仿效
emulational language　仿真语言

emulation mode 仿真模式，仿真状态
emulation program 仿真程序，仿效程序
emulation 仿真
emulator 仿真器，仿真程序
emulsification 乳化
emulsified asphalt 乳化地沥青，乳化沥青
emulsified bitumen 乳化沥青
emulsifying agent 乳化剂
emulsion density 乳胶光密度，乳胶黑度
emulsion membrane 乳化液涂层
emulsion packet 乳化团
emulsion paint 乳液涂料，乳胶漆，乳化漆
emulsion phase 乳化相
emulsion polymer 乳浊聚合物
emulsion track 乳胶（中的）径迹
emulsion varnish 乳胶清漆
emulsion zone 密相区【沸腾炉】
emulsion 乳化液，乳胶，乳剂，乳浊液
emulsive 乳化的，乳状的
emulsoid 乳胶，乳胶体
EMUST（electromagnetic-ultrosonic testing） 电磁超声检测
enabled instruction 启动指令
enabled interruption 允许中断
enabled 被允许的，使能……的，激活的
enable fault 能动断层
enable input 允许输入
enable interrupt 允许中断
enable position 启动位置，起动位置
enable pulse 启动脉冲
enabling key 赋能开关，启动电键
enabling pulse 启动脉冲
enabling signal 启动信号，恢复操作信号
enabling switch 启动开关，赋能开关
enamel-bonded single-cotton wire 单纱漆包线
enamel-covered wire 漆包线
enamel double-cotton covered wire 双纱漆包线
enamel double-silk covered wire 双丝漆包线
enameled cast iron 搪瓷铸铁
enameled heating surface 涂搪瓷受热面
enameled pressed steel 搪瓷压制钢
enameled wire 漆包线
enamel for magnet wire 电磁线漆，漆包线漆
enamel insulated aluminium wire 漆包铝线
enamel insulated wire 漆包线
enamel insulation 漆绝缘
enamel lacquer 磁漆
enamelled and cotton-covered wire 纱包漆包线
enamelled brick 瓷砖，釉面砖，釉瓷砖
enamelled cable 漆包线，漆包电缆
enamelled copper wire 漆包铜线
enamelled leather 漆皮
enamelled magnet wire 漆包线，电磁线
enamelled resistor 珐琅电阻器
enamelled slate 涂漆瓷电盘石板
enamelled tile 瓷板
enamelled wire 漆包线
enamelling machine 漆包机
enamel manganin 漆包锰线
enamel paint 搪瓷漆，亮漆，磁漆
enamel paraffin wire 蜡浸漆包线
enamel silk-covered wire 丝包漆包线
enamel wire 漆包线
enamel 磁漆，瓷釉，珐琅，搪瓷，釉，上釉，涂漆
encapsulant 密封剂
encapsulated chip capacitor 密封薄片电容器
encapsulated circuit 封装电路
encapsulated coil 密封线圈，封闭线圈
encapsulated electric motor 密封型电动机
encapsulated fuel unit 加密封套的燃料元件
encapsulated source 密封源，包封源
encapsulated winding 浇注绝缘绕组，密封绕组
encapsulate 密闭，封装，密封，包装，用胶囊包
encapsulation in bitumen 沥青固化
encapsulation 封装，密封，灌封，密封包装，固化【废物】
encapsulator 封装器，密封器
encap 加保护帽，封装
encased conduit 埋入的导管
encased duct 封闭的导管
encased magnet 铠装磁铁
encased pile 包外壳的桩，有壳桩
encased steelwork 包有外壳的钢结构
encased turbine 有壳水轮机
encased 埋入的
encasement 包壳，包装，箱，套，膜，包，包外壳，箱子，外壳，装箱，包装物
encase 装入，插入
encasing concrete 基础混凝土保护层
encasing 砌面，饰面，护壁，模板，外壳，罩子，装入
encaustic brick 彩砖，釉面砖，琉璃砖
enchainment 抓住，匹配连接，束缚
enchancement factor 可能性系数，随机因素
enchase 嵌花，镶嵌，雕刻，雕镂，镶花
enchiridion 手册，指南
encl. = enclosure 附件
enclose a cheque to your order 附寄贵方抬头支票一张
enclosed airslide conveyor 封闭式空气输送斜槽
enclosed area 封闭区
enclosed bearing 封闭式轴承
enclosed belt gallery 封闭式栈桥
enclosed building or structure 围合式建筑
enclosed busbar 封闭母线
enclosed bus 封闭母线
enclosed coal pile 封闭煤堆
enclosed conveyer conduit 封闭式输送管道
enclosed cooling system 封闭式冷却系统
enclosed cutter 带罩切削器
enclosed electric machine 封闭式电机，封闭型电机
enclosed fuse 封闭式保险丝，管形熔断片
enclosed ground 荫蔽地区
enclosed hood 密闭罩
enclosed impeller 封闭式叶轮
enclosed instrument 封闭仪器，密封仪器
enclosed knife switch 封闭式刀闸，铁壳开关
enclosed medium 孤立介质
enclosed motor 密封马达，防水马达，封闭式

电动机，封闭型电动机
enclosed nonventilated motor 无风扇封闭式电动机
enclosed relay 封闭式继电器
enclosed-scale thermometer 内标式玻璃温度计
enclosed self-cooled machine 封闭式自冷却电机
enclosed self-dumping truck 密封自卸卡车
enclosed separately ventilation 闭路强迫通风
enclosed slot 闭口槽
enclosed space 密闭舱室
enclosed spray type cooler 密闭式喷淋冷却器
enclosed staircase 封闭楼梯间
enclosed stair well 封闭楼梯间［井］
enclosed structure 围护结构
enclosed switchboard 封闭式配电盘，封闭式开关板
enclosed switchgear panel 封闭式开关柜
enclosed system 密闭系统
enclosed therein 随信附来的
enclosed trickling filter 封闭式滴滤池
enclosed truck-type switchboard 密封车载式配电盘
enclosed type induction motor 封闭型感应电动机
enclosed type motor 封闭式电机
enclosed type turbine 封闭式水轮机
enclosed type 内封装式，闭锁式
enclosed ventilated machine 防护型电机，封闭通风式电机
enclosed ventilated motor 防护型电动机，封闭通风式电动机
enclosed wake 封闭尾流
enclosed waters 闭合水区，封闭水域
enclosed with wall 砌墙封闭
enclosed with 内附
enclosed you will find 随信附上
enclosed 封闭的，封装的，包围的，内封的，内封
enclose herewith 随信附上……，用（某物）将……围住，将……与……封入同一信封［包裹］
enclose 用……围起，密闭，封闭，包围，包装，封装，包入，封入
enclosing cover 外罩
enclosing wall 围护墙，围墙，小围墙
enclosure bus 封闭母线
enclosure method 封闭方法
enclosure wall tube 包覆管，包墙管
enclosure wall 填充墙，围护墙，围墙
enclosure 附件，外壳，围墙，围护墙，四壁，套，罩，壳，护栏，包围，包封，包壳，封闭，包套，封套，围栏，围栅，围护结构，围护，罩壳
encoded （被）编码的
encoder matrix 译码器矩阵
encoder 编码器
encode 编码
encoding circuit 编码电路
encoding strip 编码条
encoding 编码
encompass 包围，围绕
encounter frequency 遭遇频率
encounter 相遇
encouragement of investment 鼓励投资
encouragement 鼓励，激励
encroachment line 侵入线
encroachment 侵入，侵犯，侵蚀，霸占，前浸
encroach on the public interests 触犯公共利益
encrustation 结壳，垢，积垢，外皮层
encrust 结垢
encryption 编码，加密
encyclopaedia 百科全书
end adjustment 端面调整
end anchorage of bar 钢筋端部锚固
end anchorage 端部锚固，末端镇墩
end anchor block 终点锚墩
end-anchored reinforcement 端部锚固的钢筋，端锚钢筋
endangered animal 濒临灭绝动物
endangered specie 濒危物种
end around carry 循环进位
end around error 舍入误差
end battery 端电池
end bay （靠岸的）边跨，端部闸孔，端部桥孔
end beam 端梁
end bearing capacity of pile 桩端承载能力
end bearing cover 端轴承盖
end bearing layer 端持力层
end bearing pile 端轴承桩，端承桩
end bearing steel H-pile 端部承载的工字钢桩
end bearing 端轴承，末端支承
end bell 护环【用于汽轮发电机转子】，钟形端板，终端盒，端箍，绑环
end-blacket 端部支架，端盖，轴承座
end block 引线端子
end box 端盒
end bracing 端盒，端部支撑
end bracket type bearing 端部支架
end bracket 端部牛腿，端部支承，端盖，尾轴承架
end built-in 末端嵌固
end bushing 端套筒
end cap 端塞，端盖式轴承，封头，端盖
end cell switch 端电池转换开关
end cell 附加电池，端电池
end chip 掉盖
end closure 端部封闭，端塞，封头
end coil 端部线圈
end-coincidence method 端点符合法
end column 端柱
end conditions 终参数，排汽参数，边界条件，末端条件，终端条件，最终状态
end conductor 线圈端部，导体端部
end connection leakage flux 绕组端部漏磁通
end connection reactance 绕组端部漏抗
end connection 端部接线，端部连接
end connector 端接头，端接器，端部接线，终端接头
end contact method 通电磁化法
end core 端部铁芯
end correction 终端校正，端部修正
end cover 端盖，封头
end crater 端部弧坑

end-cutting pliers　端面切剪钳
end delivery date　最后交货日期
end device　终端设备
end discharge　终端放电，端部排放
end-dump car　后卸车，尾卸车
end-dump closure　立堵截流
end-dump truck　后倾自卸车，后卸式卡车
endeavor　努力
end effect phenomena　末端效应
end effect　末端效应，端部效应
end electrode　端电极
end elevation　端视图，侧视图，侧立面图
endemic　地方病
end error　终端误差
end face　端面
end feed　端部馈电，端部给风
end fitting　端板，栅格板，端部配件
end fixed　固定端
end float coupling　浮动联轴器
end float　轴向间隙
end flow　（沿集箱或汽包的）轴向流动，末端出流
end frame　端盖，端罩，端架，端框，端盘
end gas　废气，尾气
end gauge unit　端测仪表组件
end girder　端梁
end grid　端部格架
end heat loss　端部热损失
ending book inventory　期末账面存量
ending physical inventory　期末实物存量
ending　端接法，终端，终端设备，中止，末期
end instrument　终端设备，终端仪表
end interruption sequence　结束中断序列
end joint　端部接缝，端接，平接
end lap joint　端搭接
end leakage　端部漏泄，端部漏磁
endless　无端的
endless belt conveyor　循环式皮带运输机
endless belt screen　行进式滤网
endless belt　无接头皮带，循环带
endless chain trench excavator　环链式多斗挖沟机
endless grate　链条炉排
endless loop　无端回线
endless rack　环道
endless saw　带锯
endless screw　蜗杆，螺旋
endless tape　环形磁带，环形带，无端带，无端磁带
end limit switch　终端开关
end load with valve closed　阀门关闭的终端荷载
end load with valve relieving　阀门释放的终端荷载
end load　端荷载，终端荷载
end loss　端部损耗
end magnetic field　端部磁场
end manhole　端部人孔
end matched lumber　端头拼接板
end milling　最终铣削
end moment　端力矩
end node　末端结点
endochronic theory　内时理论
endodynamomorphic soil　内动力型土壤
endoenergic　吸热的
end of address　地址结束
end of block　字组结束，块结束
end of conversion pulse　变换端脉冲
end of curve　曲线终点
end of cycle　循环末期
end of data mark　数据结束标志
end of data　数据结尾
end of field mark　字段结束标志
end of file gap　文件结束间隔
end of file indicator　文件结束指示标志，文件结束指示器
end of file mark　文件结束标志
end of file routine　文件结束例程
end of file　文件结束
end of job control card　作业控制结束卡
end of job　作业结束
end of life　寿命终止，寿期末
end of manufacturing report　最终制造报告
end of medium　信息终端的，介质终端的
end of message　信息结束
end of operation report　运行结果报告
end of program　程序结束
end of project evaluation　项目期末评估
end of record word　记录结束字
end of record　记录结束
end of run routine　运行结束例程
end of run　运行终结
end of scan　扫描结束
end of stroke　行程终端
end of tape routine　磁带结束例程
end of tape　磁带结束
end of text　文本结束
end of transmission block　块传输结束，传输块结束符
end of transmission recognition　传输结束识别
end of transmission　传输结束
end of travel limit switch　行程终端开关
end of travel　（装料机）行程末端
endogenetic action　内力作用，内成作用
endogenetic force calculation　内力计算
endogenetic force　内营力
endogenetic rock　内成岩
endogenic deposit　内成矿床
endogenic force computation　内力计算
endogenic force　内营力
endogenic process　内成作用
endogenous variable　内生变量
endokinetic fissure　内成裂缝
end-on armature　端头相对衔铁
end-on coupling　端头耦合，端对接法
end-on directional serial　端射天线
end-on relay armature　端对动作式继电器衔铁
end-on　端头对准的，端头相对，端头向前的，一端向前的，正对着的
end opposite coupling　终端反向联轴器
endorheic basin　内陆盆地，内流盆地
endorheic region　无出流地区，内陆河地区，内流区

endorse clause 背书条款
endorsee 被背书人，受让人
endorsement 背书，担保，批注，汇票背书
endorse over 背书转让权，将所有权让与
endorser of a bill 票据背书人
endorse 背书【用于支票等】，支持，认同，认可，签署，赞同，在背面签名，开证明文件，核准，批准
endoscope 内视镜，内窥镜
endosmometer 内渗透计
endothermal 吸热反应的，吸热的，吸能的
endothermic disintegration 吸热转化
endothermic effect 吸热效应
endothermic endothermal reaction 吸热反应
endothermic reaction 吸热反应
endothermic 吸热反应的，吸热的，吸能的
end outline 后视轮廓
endowment 捐赠的基金，资助
end packet 端部铁芯段
end panel column 抗风柱
end panel 侧板
end piece 端部铁芯段，端件，（金属）包头，套圈
end pin 尾销，端销
end plate 压板，端板，端盖，压圈，封头
end-play 轴向游隙，端隙，径向游隙，端板串动
end plug 端塞【燃料组件导向套管】
end point boron concentration 硼浓缩极限点
end point control problem 终端控制问题
end point control 端点控制
end point indicator 终点指示剂，终点指示器
end point of effective electrical travel 有效电行程端点
end point of exhaustion cycle 运行终点
end point of theoretical electrical travel 理论电行程端点
end point of titration 滴定终点
end point voltage 终点电压
end point 终点，终端
end pole 终点杆
end post 端压杆
end pressure oil 高压润滑油
end pressure 端压力，末端压力
end printing 末端打印
end print 终端打印，终端印刷
end products storage 成品库
end product 最终产物【放射链】，最终产品，最后结果，制成品，成果
end protector 端部保护器
end reaction 最终反应，末端反应
end restraint effect 端限制效应
end restraint moment 端部嵌固弯矩
end restraint 端部约束
end ring guide rail 端环导轨
end ring in dumper 内端环翻车机
end ring out dumper 外端环翻车机
end ring resistance 端环电阻
end ring rotor 带端环的转子，带短路环的转子
end ring winding 短路环【鼠笼转子的】，端环
end ring 端环【转子】，压圈，端板，护环，压板，短路环
end-scale value 满刻度值【仪表】
end seal 端部密封
end section 端件，端截面，终端部分
end shield assembly 端罩装配，端部屏蔽装配
end shield seal 端盖密封
end shield 端盖，端罩，端部屏蔽，挡风板
end sill 尾槛
end sleeve 终端套管
end slope of groin 丁坝头部坡度
end slope of groyne 丁坝头部坡度
end span 端跨
end spigot 端塞
end stiffener angle 端部加劲角钢
end stop 止动块
end suction pump 单侧吸入泵
end suction 吸入端【单级泵】
end tab 引弧板【焊接】
end thrust bearing 推力轴承
end thrust 轴端推力，端推力
end tie-bar 末端拉杆
end tightening 端部紧固
end-tipped closure method 立堵截留法
end tipper 尾卸式自动倾卸车，后倾自卸车
end-tipping barrow 端卸式手推车
end-tipping lorry 尾卸式卡车
end to end communication 终端站间通信
end to end HVDC power transmission 端对端高压直流输电
end to end joint 端点连接
end to end signaling 端到端信令
end to end switchboard 并列配电盘
end to end 不断的，首尾衔接的，并列，端到端
end-to-reel marker 磁带卷结束标记
end tube section 压力管延伸段【重水堆冷却管道】，端管段
end turn 线圈端部，端部线匝
endurance bending strength 弯曲疲劳极限
endurance capability 耐受能力
endurance crack 疲劳裂缝，疲劳裂纹，疲劳断裂
endurance expectation 期望寿命
endurance failure diagram 疲劳强度曲线
endurance failure 疲劳破坏，疲劳断裂，疲劳破损
endurance life 疲劳寿命，耐久（性）寿命，疲劳损坏期限，持久寿命
endurance limit 耐久极限，持久限度，疲劳极限，疲劳限度，持久极限
endurance period 持续时间
endurance ratio 疲劳强度比，疲劳系数，耐久比
endurance running 持续运行
endurance strength 疲劳强度，持久强度，耐久强度
endurance testing machine 持久试验机
endurance test 持久（性）试验，耐久极限测试，耐久试验，耐久性试验
endurance 耐疲劳性，耐用度，耐久性，疲劳强度，持续时间，持久性，耐久

enduring　持续的，耐久的，永久的
end use customer　终端用户
end user recovery　终端用户恢复，使用者复原
end user support　最终用户支持
end user　末端用户，最后用户，终端用户，最终用户，直接用户
end vacuum　极限真空
end velocity　终速度
end view　侧面图，端面视图，端视图
endwall boundary layer　端壁边界层
endwall loss　端壁损失
end wall　端墙
end warning area　结束警告区
endways　末端向前，末端向前地竖着，向着两端，在末端
end wear　端部磨损
end winding region　绕组端部区域
end winding retaining ring　端箍，护环
end winding support structure　端部绕组支撑结构
end winding support　端部绕组支撑，端箍
end winding ventilation　端部绕组通风
end winding　端部绕组，绕组端部
end-window counter　钟罩形计数管，端窗计数管
end-window detector　端窗探测器
end wire insulation　端部线圈绝缘
end wire stripper　顶切剥线钳
end wire winding　端部线圈，端部绕组，绕组端部
end wise　向着两端的，末端向前的
end wrench　平扳手
end　端，封头，末端，结束，金属包头，套圈，终点，终结，终止
ENEC(European Norms Electrical Certification)　欧洲标准电气认证
ENE(east-northeast)　东北东
energetic fluid　驱动水，高能流液
energetic plasma　高能等离子体
energetic self-shielding factor　高能自屏蔽系数
energetic start-up　功率启动
energetics　动力工程，力能学，动能学，动力技术，动力学，水能学
energetic wind　强风
energization of I/O control cabinet　I/O 控制柜受电
energization overvoltage　合闸过电压
energization　激磁，激发，激励，供给能量，使带电，给予……电压，通电
energize brake　通电制动器
energized circuit　励磁回路，带电回路，激励回路
energized coefficient　放热系数，涌质系数
energized field generator　电磁激励发电机
energized for holding　保持通电
energized network　通电网络，赋能网络
energized position　激励位置
energized relay circuit　继电器磁路
energized system　励磁系统，激励系统
energized-to-operate　通电运行
energized　已激励的，已通电的，带电的，通电的，已激磁的
energizer　增能器
energize　激励，供给能量，通电，激发，励磁，使通电
energizing apparatus　励磁设备，激励设备
energizing circuit relay　电源电路继电器，激励电路继电器
energizing circuitry　激励电路，电源电路
energizing circuit　励磁电路，激励电路，电源电路
energizing coil　励磁线圈
energizing current　励磁电流，激励电流
energizing cycle　励磁循环，激励循环
energizing of house power system　厂用电系统受电
energizing quantity　激励量，激励值
energizing the auxiliary power system　厂用电受电
energizing voltage　励磁电压，激发电压
energy absorber assembly　能量吸收组件
energy absorber　能量吸收器，能量吸收体，消能器
energy absorbing capacity　能量吸收能力
energy absorbing material　吸能材料
energy absorption coefficient　能量吸收系数
energy absorption cross-section　能量吸收截面
energy absorption　能量吸收作用，吸能，能量吸收
energy advection　能量平流
energy albedo　能量反照率
energy and gases supply for construction　施工力能供应
energy and load balance　能载平衡
energy attenuation　能量衰减
Energy Audit Scheme　能源审计计划【英国】
energy availability factor　能量利用率，能量可利用因子，能量可用率
energy availability　能量可用率
energy balance climatology　能量平衡气候学
energy balance　能量平衡
energy barrier　能峰，能障
energy belt　能源带，能源地带
energy-billing system　电能量计费系统
energy breakdown　能量的品质降低
energy budget method　能量平衡法
energy budget　能量平衡，能量收支，能量预算
energy buffer　能量缓冲器
energy build-up factor　能量积累因子
energy calibration standard　能量校准标准，能量刻度标准
energy capacity　储能容量【电力】
energy change　换能，能量交换
energy charge　耗能量，用电量，电费，耗能费
energy component　有功分量，有功部分，有效部分
energy conservation equation　能量守恒方程式，能量守恒方程
energy conservation law　能量守恒定律
energy conservation program　节能计划
energy conservation　能量守恒，节能，能量节约，能源节约
energy constant　能量常数
energy consumption structure　能源消费结构

energy consumption　能耗，能量消费，能量消耗，能源消耗
energy containing eddy　含能涡旋
energy content　内能，含能量
energy control system　能量控制系统
energy conversion device　换能器，能量转换设备
energy conversion efficiency　能量转换效率
energy conversion factor　能量转换系数
energy conversion　能量变换，能量转换，能量转变
energy converter　能量转换装置，电能转换器
energy correction factor　能量校正系数
energy-cost effectiveness　能源成本效率
energy crisis　能源危机
energy current　有效电流，有功电流，能流
energy curtailment　电量削减
energy cut-off　能量截止，能量阈
energy cycle　能量循环
energy decay　能量衰减
energy deficiency　能量不足
energy deficit　能量不足，能亏，电力缺口
energy degradation　能量递降，能的递降，能量降级，能量退降
energy-delivering　输出能量的，传送能量的
energy demand analysis　能源需求分析
energy demand　能量需求，能源需求
energy density of sound　声能密度
energy density　能量密度，（蓄电池）储能密度
energy dependence　能量依赖
energy deposition event　能量沉积事件
energy deposition　能量沉积，能量吸收
energy development　能量开发，能源开发
energy diagram　能量图，能量图解，热力图
energy direction contacts　电量方向节点
energy discharge　能量释放
energy disperser　缓冲器，减震器，能量扩散器
energy-dispersive X-ray fluorescence　能量色散X射线荧光
energy dissipater　消能结构
energy dissipating bucket　消能屏
energy-dissipating dents　消力齿
energy-dissipating sill　消力槛
energy dissipation baffle　消力墩
energy dissipation below spillway　溢洪道下消能
energy dissipation device　耗能装置，消能设施
energy dissipation ratio of jump　水跃消能
energy dissipation ratio　耗能率
energy dissipation spacer　耗能隔离子【电缆线】
energy dissipation　能量损耗，能量消耗，能量耗散，消力，消能，能量消失，能量散逸，能量损失
energy dissipator　消能工，消能建筑物，消能结构
energy distribution　能谱，能源分布，能量分布
energy-economic system　能源经济系统
energy economics　能源经济学
energy economy　动能经济
energy efficiency index　能源效率指数，能效指数
energy efficiency ratio　能效比，能源效率比值
energy efficiency　能量效率，电能效率，能源效率
energy elasticity　能源弹性系数
energy equation　能量方程式，能量方程
energy equilibrium　能量平衡
energy equivalent　能量当量
energy exchange time　能量交换时间
energy exchange　能量交换
energy extraction　获能
energy flow rate　能流速率
energy flow　能流
energy fluence rate　能注量率
energy fluence　能量流，能注量
energy flux density　能（量）通量密度
energy flux　能（量）通量，能流
energy frictional heat energy　摩擦热能
energy gain of collector tubes　集热器管的能量收益
energy gain　能量增益
energy gap　能极距离，能隙
energy grade line　能量坡降线，能坡线，能头线，水能线
energy grade　能量等级
energy gradient line　水能线
energy gradient　能量梯度，水能梯度，能源坡度
energy group　能量群，能群【中子】
energy head loss　能头损失
energy head　能头【水流】，有效水头
energy imparted to matter　传递至物质的能量
energy imparted　授予能
energy import　能量输入
energy impulse　能脉冲
Energy Information Administration　美国能源信息署
energy input　能量输入
energy intensity　能源强度
energy-intensive diffusion process　能源密集扩散过程
energy-intensive product　能源密集型产品
energy interchange　能量交换
energy level diagram　能级图
energy level　能量等级，能级
energy liberation　能量释放
energy line　能线
energy load　能量负荷
energy loss by radiation　辐射引起的能量损失
energy loss distribution　能量损失分配
energy loss of radiation　辐射能量损失
energy loss time　能量损失时间
energy loss　能量损耗，能量损失，有功损耗，能耗
energy management system　能量管理系统
energy management　能源管理
energy metering point　电度表连接点【电力系统用】
energy metering system　能量计量系统
energy meter　电度表，电能表，能量计，功率计，能量表
energy method　能量法
energy mix　能源混合体

energy of activation 活化能
energy of adhesion 黏附能
energy of blow 锤击能
energy of deformation 变形能
energy of electron motion 电子运动能
energy of pile driving 打桩能量
energy of position 位能
energy of resonance 共振能
energy of volume 容能
energy operation 能源经营，能源管理
energy output 能量输出，输出能量，发电量，电能生产量
energy pattern factor 能源结构因数
energy payback period 能源投资回收期
energy payback time 能量生产还本时间，能量偿还时间
energy policy 能源政策
energy potential 能势
energy power 经济功率
energy price 电量电价
energy-producing 产生能量的
energy radiance 能量辐射度
energy range 能量范围
energy ratio 能量比
energy released per fission 每次裂变释放出的能量
energy release rate 能量释放率
energy release 能量释放
Energy Research and Development Administration 美国能源研究开发署
energy resolution 能量分辨率
energy resources 能源，能量资源，能源资源
energy return period 能量生产还本周期
energy rose 能量玫瑰图
energy saving investment 节能投资
energy saving 节能，节约能源
energy shortage 能源短缺
energy shortfall 电量短缺
energy slope 能量梯度
energy source composition 能源构成
energy source development 能源开发
energy source planning 能源规划
energy source 能源，能源资源，能源来源
energy spectrum density function 能谱密度函数
energy spectrum 能谱
energy stability 能量稳定性
energy state 能量状态
energy storage capacitor bank 蓄能电容器组
energy storage capacitor 储能电容器，蓄能电容器
energy storage device 储能设备
energy storage equipment 能量储存装置，蓄能装置
energy storage for electric power system 电力系统储能
energy storage generator set 蓄能发电机组
energy storage medium 储能介质
energy storage system of wind power 风力发电储能系统
energy storage technique 蓄能技术
energy storage 蓄能，储能，能量贮存，能量储存
energy supply 能源，电源，供能，供电
energy synthesis 能量合成
energy systems engineering 动力系统工程
energy tax credit 能量投资免税
energy thickness of boundary layer 边界层能量厚度
energy thickness 能量厚度
energy threshold 能量阈，能量极限
energy transfer as heat 以热的形式传能
energy transfer as work 以功的形式传能
energy transfer coefficient 能量转换系统，能量转移系数
energy transfer equation 能量交换方程
energy transfer function 能量转移函数
energy transfer mechanism 能量转换机理
energy transfer 能量传输，能量转移，能量传递，能量交换
energy transformation 能量转换
energy transition 能级跃迁
energy transmission test 能量透射法试验
energy transmission 能量传送
energy transport 能量输送
energy uitilzation coefficient 能量利用系数
energy unavailability 能量不可用率
energy unit 能量单元，能量单位
energy utilization 能量利用，能量利用率
energy without pollution 无污染能源，清洁能源
energy yield 能量输出，能量产额，发电量
energy 能量，能，能力，功率，能源
EN(European Norm) 欧洲标准
enforceability of rights and obligations 权利和义务的实施
enforceable contract 可强制履行的合同，有效合同
enforceable 有强制力的
enforcement measure 强制措施，执行措施
enforcement 执行，实施，贯彻
engage 啮合，填入，接通
engaged lamp 占线信号灯
engaged line 占线，忙线
engaged signal 占线信号
engaged switch 接通开关
engagement letter 聘书
engagement mesh 啮合
engagement 约定，接合，啮合，保证，结合，约会，雇用，契约
engaging and disengaging gear 离合装置
engaging dog 接合夹头
engaging gear solenoid valve 啮合机构电磁阀
engaging lever 合闸杆
engaging means 接通机构
engaging pressure 启闭压力【阀门】
engaging the clutch 啮合离合器
engender 产生，形成
EDPM(engineering design procedure manual) 工程设计程序手册
engine alternator 发动机驱动同步发电机
engine attendant 火车司机
engine backplate 发动机隔板，发动机护板
engine base 机座

engine compartment 发动机舱
engine cooling airflow 发动机冷却气流
engine cover 发动机罩
engine cowl 发动机罩，发动机整流罩
engine-driven generator 发动机驱动的发电机
engine-driven 发动机驱动的
engine efficiency 发动机效率
engineer 工程师
engineered barrier 专设屏障
engineered fill 质控回填土
engineered safeguard feature switchboard 设计有安全保障特性的配电盘
engineered safeguard feature 设计的安全（保障）特性
engineered safeguards actuation system 专设安全设施触发系统
engineered safeguard system 设计的安全保障系统，专设安全设施系统
engineered safeguard train 安全保障系列，专设安全设施序列
engineered safeguard 专设安全设施，设计的安全保障
engineered safety feature 设计的安全特点，专设安全设施，专设安全设施，特设安全装置
engineered safety system 设计的安全系统，专设安全系统
engineered storage 工程加固贮存
engineer in charge 主管工程师
engineering acceptance specification 工程验收规范
engineering action request 技术工作请求书
engineering aerodynamics 工程气体动力学
engineering agreement 工程协议
engineering analysis report 工程分析报告
engineering analysis 工程分析
engineering and technical personnel 工程技术人员
engineering and technical services 工程及技术服务
engineering approximation 工程近似（法）
engineering aspects 工程展望
engineering atmospheric pressure 工程大气压力
engineering change notice 工程变更通知
engineering change 工程改变
engineering characteristic of rock 岩石的工程性能
engineering company 工程公司
engineering compromise 工程综合考虑，工程折中方案，工程折衷方案
engineering condition 工程条件
engineering construction 工程建设，工程施工
engineering consulting contract 工程咨询合同
engineering consulting firm 工程咨询公司
engineering contractor 工程合同商
engineering contract 工程合同
engineering co-ordination meeting 工程设计协调会
engineering cost 工程造价，工程费
engineering council 工程协会
engineering custom engineering proposal 用户技术建议
engineering cybernetics 工程控制论
engineering data base system 工程数据库系统
engineering data information system 工程数据信息系统
engineering data micro-reproduction system 工程数据微型再生系统
engineering data retrieval system 工程数据检索系统
engineering data 工程数据，技术数据，工程技术数据，工程技术资料，工程资料
engineering department interface control technique 工程部接口控制技术
engineering department 工程部，工程处，技术科
engineering description 技术说明书
engineering design assignment 工程设计任务书
engineering design certificate 工程设计证书
engineering design change proposal 工程设计修改提议
engineering design change schedule 工程设计修改日程表
engineering design change 工程设计更改［修改］
engineering design collaboration 工程设计协作
engineering design data package 工程设计资料包
engineering design data 工程设计数据
engineering design department 工程设计部门
engineering design level 工程设计水平
engineering design modification 工程设计修改
engineering design plan 工程设计方案
engineering design project 工程设计项目
engineering design review 工程设计审查
engineering design standard 工程设计标准
engineering design 工程设计
engineering detail 工程细目
engineering division 工程处
engineering documentation release 技术文件发放
engineering document 工程文件
engineering drawing change 工程图纸修改［改变］
engineering drawing list 工程图纸清单
engineering drawing 工程图，工程制图
engineering economics 工程经济学
engineering economy 工程经济
engineering education 工程教育
engineering enthalpy rise factor 工程焓升因子，设计焓提升因子
engineering enthalpy rise hot channel factor 工程焓升热通道因子
Engineering Equipment and Materials Users' Association 工程设备与材料用户协会【欧洲】
engineering exploration & survey certificate 工程勘察证书
engineering factor 工程因素，工程因子，设计因子
engineering feasibility 工程可行性
engineering fluid mechanics 工程流体力学
engineering geological condition 工程地质条件，工程地质评价
engineering geological investigating for reservoir

水库工程地质勘察
engineering geological monitoring 工程地质监测
engineering geological phenomena 工程地质现象
engineering geological problem 工程地质问题
engineering geological prospecting 工程地质勘探
engineering geologic condition analysis 工程地质条件分析
engineering geologic investigation for dam site 坝址工程地质勘察
engineering geology information 工程地质资料
engineering geology 工程地质（学），工程地质
engineering guideline 工程设计导则
engineering heat flux hot channel factor 工程热流密度热通道因子，设计热通量热管因子
engineering heat flux hot spot factor 工程热流密度热点因子，设计热通量热点因子
engineering high-level laboratory 高放射性工程实验室
engineering hot channel factor 工程热通道因子
engineering hot spot factor 工程热点因子
engineering hydraulics 工程水力学
engineering hydrology 工程水文学
engineering improvement time 工程改进时间
engineering information retrieval 工程信息检索
engineering information 工程资料
engineering instruction 工程（设计）说明书，技术说明书
engineering insurance 工程保险
engineering integral management 工程综合管理
engineering item description 工程项目说明
engineering item 工程项目
engineering judgement 工程判断
engineering kick-off meeting 工程设计开工会
engineering kinematics 工程运动学
engineering laboratory report 工程实验试报告
engineering legislation 工程法规，工程法制
engineering management information technique 工程管理信息技术
engineering management work 工程管理工作
engineering management 工程管理
engineering manager 工程部经理，技术经理，设计经理
engineering manual 工程手册
engineering measure 工程措施
engineering mechanics 工程力学
engineering meteorology 工程气象学
engineering model 工程模型
Engineering News-Record 《工程新闻记录》杂志
engineering oceanography 工程海洋学
engineering performance standard 技术性能标准
engineering plastics 工程塑料
engineering preliminaries 工程准备事项【用于勘察、设计、计算等】
engineering project contract 工程项目合同
engineering projection 工程设想
engineering project 工程计划，工程项目
engineering property 工程特性，工程性质
engineering prototype 正样
engineering reactor 工程反应堆
engineering record 工程记录
engineering regulation 工程规则
engineering reliability 工程可靠性，结束可靠性
engineering report 工程技术报告，工程报告，技术报告
engineering review meeting 设计审核会
engineering sample 样机，样品，模型，试样，试件
engineering scale 工程比例尺，工程规模
engineering schedule 工程设计进度表，工程进度表
engineering science 技术科学，工程科学
engineering seismology 工程地震学
engineering service car 工程宿营车
engineering service 工程服务
engineering simulator 工程仿真机
engineering standard 工程标准，技术标准
engineering stress-strain curve 设计应力变形曲线
engineering structure 工程结构，工程构筑物
engineering supervision 工程监理，工程检查，技术监督
engineering survey 工程测量
engineering system of units 工程单位制
engineering system 工程系统
engineering technical department 工程技术部
engineering test facility 工程试验设备
engineering test requirements 工程试验要求
engineering test 工程试验，技术试验
engineering thermodynamics 工程热力学
engineering time 预检时间，维修时间
engineering tool 工程工具
engineering uncertainty 工程不确定性
engineering unit conversion 工程单位换算
engineering unit 工程单位
engineering workstation 工程师工作站
engineering 工程，工程技术，工程设计设计，工程学
engineer-in-training 见习工程师
engineer level 工程师用水准仪
engineer of specialized field 专业工程师
engineer on probation 见习工程师
Engineers Council for Professional Development 工程师专业发展委员会
engineer's representative 工程师代表
engineer's station 工程师站
engineer transit 工程经纬仪
engine exhaust gas 发动机排气
engine exhaust trail 发动机排气尾迹
engine gauge 引擎指示器
engine generator 柴油机发电机，发动机驱动发电机，引擎发电机
engine governor 引擎调速器
engine-hour indicator 发动机小时指示器
engine oil 机油
engine power 发动机功率
engine racing 发电机逸转，发电机高速空转
engine room 机房，汽轮机房，工程师室
engine running light 发动机运行指示灯
engine section 发动机部件
engine shaft 发动机轴
engine-speed indicator 发动机转速指示器
engine-type generator 发动机式发电机

engine valve 发动机阀，发动机气门，引擎汽门，引擎门
engine vibration 发动机振动
engine volumetric efficiency 发动机容积效率
engine windmill 发动机风转
engine yaw 发动机偏航
engine 发动机，机器，引擎，机车，火车头，工具
Engler viscosity 恩氏黏度
English unit 英制单位
engraving machine 刻模机
engraving process 照相（刻模）排版工艺
engraving 刻模，雕版，雕刻，刻度
engrossment 独占（市场）
enhanced effect 增强的影响
enhancement effect 增强效应
enhancement factor 增强因子
enhancement filter 增强滤片
enhancement of environment 环境改善
enhancement transistor 增强型晶体管
enhancement type 增强型
enhancement 改善，加强，增强
enhance 提高，增加，加强，增进
enlarged base （桩的）扩底
enlarged detail 大样
enlarged drawing 放大图
enlarged elevation 放大立面图
enlarged preliminary design 扩大初步设计
enlarged section 放大断面，放大断面图
enlargement factor 放大系数，放大倍数，放大因数
enlargement loss 放大损失，扩大损失
enlargement ratio 放大率
enlargement 扩大，放大，增补，扩建
enlarge test 扩管试验
enlarging paper 放大纸
enlightened persons 开明人士
enlivening the economy 搞活经济
en-masse conveyor 埋刮板输送机，链运机
enormous business opportunity 无限商机
enough 足够的
ENPEP(energy and power evaluation programme) 能源和动力评价计划
ENQ(enquiry character) 询问字符
enquire 查询，调查
enquiry character 询问字符
enquiry circuit 查询电路
enquiry document 询价文件
enquiry form 查询表格
enquiry note 询价单
enquiry 询问，查询，询价，询价单
enregistor 记录器
ENR(Engineering News-Record) 工程新闻纪录【全球工程建设权威杂志，美国】
enriched bucking 富集燃料堆芯曲率
enriched fuel reactor 富集燃料反应堆
enriched fuel 富集燃料，加浓的燃料
enriched material 浓缩物，富集材料，浓缩材料
enriched nuclear fuel 富集核燃料
enriched reactor 浓缩燃料反应堆
enriched uranium reactor 富集铀反应堆，加浓铀反应堆
enriched uranium 富集铀，浓缩铀，加浓铀
enriched water 加浓水
enriched 浓缩的，富集的，加浓的
enriching section 富集段，浓集段，浓缩区
enrichment coefficient 富集系数，浓缩系数
enrichment control 改变富集度控制
enrichment factor 富化系数，浓缩系数，富集系数，浓集因于，浓缩因子
enrichment identification 浓缩度标志
enrichment plant 富集装置【燃料】，提浓装置【重水】，浓缩厂，同位素分离厂
enrichment process 富集过程
enrichment region 富集区
enrichment set 浓缩单机
enrichment tails 浓缩尾料
enrichment throwout 浓缩次品
enrichment verification 浓缩度检验
enrichment zone 富集带，浓缩区
enrichment 浓缩，丰富，改进，肥沃，浓缩度，加浓度，富集度，富集（作用）
enrich 加浓，浓缩，富集，加料
enrockment 抛石护岸，基底填石
en route 在（运输）途中
ensemble average concentration 系统平均浓度
ensemble average 总体平均值
ensemble mean field 系统平均场
ensemble of communication 信息集合
ensemble 总体，集合，采集，集，（信号）群，系统
ensure quality as well as quantity 保质保量
ensure 确保
enter 进入，输入，记录
enter a bid 投标
enter a code 输入编码
enter a port 入港
enterclose 通道
entering edge 前缘，前沿，上升边
entering tap 进入开发
entering to the register 注册登记
entering 输入，进入，插入，记录
enter into a contract 订合同，签订合同
enter into an agreement 订约
enter into commitment 承担义务
enter into force 生效
enter into negotiation 进行谈判，开始谈判，参与谈判
enter into 签订
enter market 进入市场
enter new order 接受新订单
enterprise above designated size 规模以上企业
enterprise bonds 企业债券
enterprise directly under the central government 中央企业
enterprise entity 企业实体
enterprise environment 企业环境
enterprise executive staff 企业经营人员
enterprise group 企业集团
enterprise income tax 企业所得税
enterprise in the end 亏损企业
enterprise investment plan 企业投资计划

enterprise management 企业管理
enterprise mergence 企业兼并
enterprise operating on shareholding system 股份制企业
enterprise profit drawing 企业利润提成
enterprise profit partly reserved 企业利润留成
enterpriser 企业家
enterprise's appearance 企业形象
enterprises joint-stock system 企业股份合作制
enterprises led by science and technology 科技先导型企业
enterprise standardization 企业标准化
enterprise standard system 企业标准体系
enterprise under a specified administrative department 归口企业
enterprise with sole foreign investment 外资企业
enterprise 事业单位，事业，企业
entertainment expenses 交际费，招待费
enter the cubicle from the bottom 柜底进线
enter the cubicle from the top 柜顶进线
enthalpy control system 焓值控制系统
enthalpy difference 焓差
enthalpy drop 焓降，热降
enthalpy entropy chart 焓熵图
enthalpy-entropy coordinates 焓熵图，H-S 图，I-S 图
enthalpy-entropy diagram 焓熵图，H-S 图，I-S 图
enthalpy flux 热流量，焓流，热通量
enthalpy gradient 焓降梯度
enthalpy-humidity diagram 焓湿图
enthalpy of saturated liquid 饱和液比焓
enthalpy of subcooling 欠热焓
enthalpy potential method 焓差法
enthalpy rise engineering hot channel factor 焓升工程热管因子，焓升工程热通道因子
enthalpy rise factor 焓升因子，焓增长因子
enthalpy rise hot channel factor 焓升热通道因子
enthalpy rise hot spot factor 焓升热点因子
enthalpy rise uncertainty factor 焓升不确定因子，焓升高不确定因子
enthalpy rise 焓升
enthalpy thickness 焓厚度
enthalpy titration 温度滴定，热焓滴定
enthalpy transport by diffusion 扩散焓传递
enthalpy 焓，热含量，热函
enthrakometer 超高频功率计
enthusiasm 积极性，热情
entire agreement clause 完整合约条款，完整协议条款
entire contract 完整的合同
entire life 总寿命
entirely shut 完全闭塞
entirely ventilation 全面通风
entirely 完全的
entire model 整体模型
entire system 整个系统
entire thermal resistance 总热阻
entirety 整体
entire 全体，完全的
entitlement 授以权利
entity 实体
entive 全部
entombment decommissioning 退役埋葬
entombment 埋存【反应堆】，埋葬
entrained air 混入空气
entrained bed 载流床
entrained flow gasification 气流床气化，夹带床气化，载流床气化
entrained fluid 夹带流体
entrained gas 夹带气
entrained leakage 间接泄漏，携带泄漏
entrained mass 夹带质量
entrained particle 被夹带的颗粒
entrained steam 残留蒸汽
entrained water droplets 夹带的小水滴
entrained water 蒸汽带水，携带水，夹带水，汽夹水
entrained 夹带的，携带的
entraining gas 输送气，载气
entraining 夹带，混入
entrainment block 卷吸区【沸腾炉】
entrainment coefficient 夹带系数
entrainment constant 夹带常数
entrainment effect 夹带效应
entrainment filter 雾沫过滤器，夹带物过滤器
entrainment layer 夹带层
entrainment limit in heat pipe 热管中的携带极限
entrainment loss 夹带损失
entrainment of frequency 频率诱导
entrainment parameter 夹带参数
entrainment rate 夹带率
entrainment separator 雾沫捕集器，雾沫分离器，夹带物分离器
entrainment theory 夹带理论
entrainment velocity 夹带速度，载速
entrainment 带走，携带，卷吸，传输（液体），夹带，输送，挟带
entrain 带走，夹带，输送，传输，卷吸，携带
entrance angle of conductor 导线悬垂角
entrance angle 入口角，进口角
entrance bushing 进线套管
entrance cable 引入电缆
entrance channel 进水渠
entrance coil 引入线圈
entrance concentration 入口浓度
entrance condition 入口工况，入口条件
entrance culvert 进水涵管
entrance effect 进口效应
entrance friction 进口阻力
entrance hall 前厅，门厅，穿堂
entrance head loss 入口头损失
entrance head 进口水头，入口水头
entrance length 进口（稳定）段长度
entrance loss factor 失口损失因子
entrance loss 进口损失，入口损失
entrance of cooling water 冷却水进口
entrance of elevator 电梯入口
entrance orifice 入口节流圈
entrance point 输入点
entrance pressure drop 进口压降
entrance pressure 进口压力

entrance region 入口区域
entrance road 进厂公路
entrance section 进口段
entrance sleeve 进口接管
entrance slit 入口狭缝
entrance stagnation temperature 进口滞止温度
entrance step 台阶
entrance well 进水口，进水井
entrance zone 入口区
entrance 入口，进口，加入，开始，引入，进风口，门入口
entrant gas 进入气体
entrapment 截留，捕集
entrapped penetrant 残留的渗透剂
entrapped slag 夹渣
entrap 捕捉，截留
entrefer 铁间空隙
entrenched meander 嵌入曲流
entrenchment 防御设施，挖壕，挖壕沟，下切
entrepot 仓库，转口港，货物集散地，商业集中区，中转港，中转站，转运港口
entrepreneur 企业家
entresol 中间楼层
entropy change 熵变
entropy coefficient 熵系数指标
entropy derivative 熵的导数
entropy elasticity 熵弹性
entropy flux 熵流
entropy generation 熵产
entropy increase principle 熵增原理
entropy production 熵产，熵增
entropy-temperature curve 温熵曲线图
entropy-temperature diagram 温熵图，T-S 图
entropy 熵，平均信息量，信息熵
entrusted agent 委托代理人
entrusted electricity supply to subconsumers 委托转供电
entruster 委托人
entrustment agreement 委托协议书
entrustment 委托，托管，委托
entry block 项目表，项目块
entry certificate 入境证书
entry condition 入口条件，入口点
entry control system 厂区和办公楼出入监督系统
entry declaration 入港申报
entry device 输入设施
entry-end effect 入口效应，入端效应
entry track 进车线
entry guide vane 进口导叶
entry instruction 入口指令
entry into force 生效
entry into service 投入运行
entry loss 入口损失，进口损失
entry of contract into force 合同生效
entry of goods inward 申报进口
entry of goods outward 申报出口
entry of permit 入境许可证
entry point 进入点，转移点，子程序入口
entry pressure 进口压力
entry query control console 输入询问控制台
entry side of the dumper 翻车机进车侧
entry slot 进口缺槽
entry sorting 项目块分类
entry spin （气流的）预旋，进气预旋
entry temperature 进口温度
entry visa 入境签证
entryway 入口
entry 进入，入口，进口，记录，报关单，表列值，项目，巷道
entwine 盘绕，缠住
enumerate 计算，列举
enumeration 计数，列举，枚举，细目
envelop curve 外包线
envelop delay 包络时延
envelope curve of turn voltages 线匝电压的包络线
envelope curve 包络线
envelope delay distortion 包线延迟失真
envelope delay 包络线延迟
envelope detector 包络检波器
envelope feedback 包迹反馈
envelope function 包络函数
envelope inspection fixture 外壳检查夹具
envelope of curves 曲线的包络，外包
envelope oscilloscope 包线示波器，包迹示波器
envelope pf failure 破坏包线
envelope shape 包络线形状
envelope terminal 管壳接线端
envelope test 静态试验，充氮试验，容器试验
envelope type suspension clamp 提包式悬垂线夹
envelope wide-scope 包迹宽带示波器，视频示波器
envelope 包络线［面］，信封，封装包，外壳，围护结构，封皮［套］，包壳［迹］，包络，外包线
enveloping air 外包空气
enveloping box 外形包络箱
enveloping curve 包络曲线
enveloping surface 包络面
enveloping 包络，外包，包封
envelop velocity 包络速度
enviromental impact statement 环境评价报告书
environment aerodynamics 环境空气动力学
environmental 环境，环境的
environmental activity 环境放射性
environmental adaptation 环境适应性
environmental agency 环保机构
environmental amenity 环境舒适性
environmental analysis 环境分析
environmental appraisal 环境评价，环境鉴定
environmental area 环境区域
environmental aspect 环境状况
environmental assessment 环境评价
environmental audit 环境审计
environmental baseline study 环境基线研究
environmental behavior of toxic elements 有毒元素环境行为
environmental benefit 环保效益
environmental capacity 环境容量
environmental climate 环境气候，室内小气候
environmental coefficient 环境系数

environmental complex 环境综合体
environmental conditions 环境条件
environmental consequence 环境影响
environmental conservation 环境保护，环境研究
environmental contamination 周围介质污染，环境污染
environmental control 环境控制
environmental correlation 环境关联
environmental crisis 环境危机
environmental criteria 环境标准
environmental damage 环境性损坏，环境损害
environmental data 环境资料
environmental decay 环境破坏，环境衰退
environmental degradation 环境退化
environmental design 环境设计
environmental deterioration 环境恶化
environmental determinism 环境决定论
environmental deviation 环境差异
environmental dilemma 环境困境
environmental diseconomy 环境不经济
environmental disfunction 环境功能丧失
environmental disruption 环境破坏，环境失调
environmental disturbance 环境失调
environmental disutility 环境失效
environmental dosemeter 环境剂量计
environmental dose 环境剂量
environmental dosimeter 环境剂量计
environmental ecology 环境生态学
environmental effect 环境效应，环境影响
environmental electricity 环境电学
environmental emergency 环境紧急事故
environmental engineering 环境工程学，环境工程
environmental entity 环境本质
environmental entomology 环境昆虫学
environmental error 环境误差
environmental evaluation 环境评价
environmental exposure 环境照射，环境照射量
environmental factor 环境因素
environmental fatigue 介质疲劳
environmental field test 环境现场试验
environmental forecasting 环境预报，环境预测
environmental gap 环境间隔
environmental geology 环境地质学，环境地质
environmental guideline 环境方针
environmental hazard 环境公害，环境危害
environmental health engineering 环境卫生工程
environmental health 环境卫生
environmental hydro-biology 环境水生物学
environmental hydrology 环境水文学
environmental hygiene 环境卫生
environmental impact assessment report 环境影响评价报告
environmental impact assessment 环境影响评价
environmental impact report 环境影响报告
environmental impact statement 环境影响报告书
environmental impact study 环境影响研究
environmental impact 环境影响
environmental improvement 环境改良，环境改善
environmental index 环境指数
environmental indicator 环境指示物
environmental influence 环境影响
environmentalism 环境决定论
environmentalist 环境工作者，环境学家
environmental law 环境法
environmental legal and regulatory framework 环保法规体系
environmental legislation 环境法规
environmental limit 环境极限
environment allocation 环境区域，环境场所
environmentally-friendly design 有利于环境的设计
environmentally-friendly 环境友好的
environmentally sensitive area 环境敏感区
environmental management 环境管理
environmental medium 环境介质
environmental meteorology 环境气象学
environmental monitoring system 环境监测系统【厂区】
environmental monitoring 环保检测，环境监测，环境现场监测
environmental noise 环境噪声
environmental objective 环境目标
environmental pattern 环境模式
environmental planning 环境规划
environmental policy 环境政策
environmental pollutant 环境污染物
environmental pollution control measure 环境污染对策
environmental pollution 环境污染
environmental profile 环境概貌
environmental program 环境规划
environmental project 环境规划
environmental protection 环境保护
Environmental Protection Agency 环境保护局
Environmental Protection Bureau 环保局
environmental protection for geothermal development 地热开发环境保护
environmental protection 环境保护
environmental quality index 环境质量指数
environmental quality pattern 环境质量模式
environmental quality standard 环境质量标准
environmental quality 环境质量
environmental radiation monitoring 环境辐射监测，辐射环境监测
environmental radiation 环境辐射
environmental radioactivity 环境放射性
environmental receptivity 环境受纳能力
environmental reform 环境改造
environmental report 环境报告
environmental resistance 环境阻力
environmental resource 环境资源
environmental sampling 环境取样
environmental sanitation 环境卫生
environmental science 环境科学，环境学科
environmental sensitive area 环境敏感区
environmental simulation test 环境模拟试验
environmental simulation 环境模拟，环境仿真
environmental sound 对环境无害的，合乎环境要求的
environmental standard 环境标准
environmental statement and addendum 环境报告

及其附录
environmental statement 环境报告
environmental stress 环境重点
environmental study 环境研究
environmental surveillance 环境监测，环境监视
environmental survey satellite 环境勘测卫星
environmental survey 环境调查
environmental system engineering 环境系统工程
environmental temperature 环境温度
environmental test chamber 环境试验箱
environmental testing facility 环境试验设备
environmental test 环境试验
environmental variable 环境变数
environmental variation 环境变异
environmental wind 环境风
environment capacity 环境容量
environment condition 环境条件
environment contamination 环境污染
environment control 环境控制
environment correction factor 环境校正系数
environment element 环境因素
environment factor 环境因素
environment-friendly 有利于环保的
environment health 环境卫生
environment impact assessment 环境影响评估［评价］
environment impact statement 环境影响报告书【预可研阶段】
environment impact study 环境影响调查
environment impact 环境影响
environment management 环境管理
environment mitigation project 环境治理工程
environment pollution 环境污染
environment protection measures 环境保护措施
environment protection 环境保护
environment recording 现场记录
environment return 环境恢复
environment science 环境科学
environment specification 环境规格
environment water quality 环境水质
environment wind tunnel experiment 环境风洞实验
environment wind tunnel 环境风洞
environment 环境，外界，围绕，周围，周围环境
environs radiation monitoring 环境放射性监测，环境辐射监测
environs 郊区，近郊，城市周边地区，附近范围
environ 围绕，环绕，包围
envisage 重视，正视，设想，展望
enwind 绕线，缠绕，包
enzyme fermentation 酶发酵
enzyme-support particle 载酶颗粒
enzyme 酶
EN(European Norm) 欧标，欧洲标准
EOA(end of address) 地址结束
EOB(end of block) 字组结束，块结束
EOB(end of transmission block) 传输块结束符
EOC(Emergency Operation Center) 应急操作中心
eocene epoch 始新世
EOD(end of data) 数据结尾
EOE(errors and omissions excepted) 误差和遗漏除外，错漏备查
EO (electric operate) 电气操作
EO(executive order) 执行指令
EOF(end of file mark) 文件结束标志
EOF(end of file) 文件结束
EOF gap(end-of-file gap) 文件结束间隔
Eogene System 下第三系，古近系
EOH(Equivalent operating Hours) 等效运行小时数
EOI(expression of intent) 投标意向函
EOI(express of interest) 兴趣函，意向表达函
EOJ(end of job) 作业结束
EOL(end of life) 寿命终止
eolian soil 风积土
eolotropic 各向异性的
EOM(end of message) 信息结束
EOMM(equipment operation and maintenance manual) 设备运行维修手册
EOP(emergency oil pump) 事故油泵
EOP(end of program) 程序结束
EOP 紧急油泵【DCS 画面】
EO relief valve 电动释放阀
EOR(end of record) 记录结束
EOR(enhanced oil recovery) 油强化回收热发电
EOR(evidence of receipt) 收据
EOR gap(end-of-record gap) 记录结束间隔
EOS(electro-optical system) 光电系统
EOST(electrical output storage tube) 电输出存储管
EOT(end of test) 试验结束
EOT(end of transmission) 传输结束
EOT(extension of time) 工期延长
Eozoic era 始生代
Eozoic group 始生界
EPA(electron probe analysis) 电子探针分析
EPA(Environmental Protection Agency) 美国国家环境保护局
EPA standard 美国环保标准
EPBX(electronic private branch exchange) 专用支线电子交换机
EPC(easy processing channel) 易处理通道
EPC(electronic program control) 电子程序控制
EPC(engineering, procurement and construction) EPC 模式，“工程设计-采购-施工总承包”模式【工程建设模式】
EPCM (engineering-procurement-construction-management) EPCM 模式，“设计、采购与施工管理”模式【工程建设模式】
EPCO (engineering-procurement-construction-operate) EPCO 模式，“设计-采购-施工-运营”模式【工程建设模式】
EPCS(emergency plant cooldown system) 电站应急冷却系统
EPDCC(Electric Power Development Coordinating Council) 电力发展协调协会
EPDH(equivalent planned derating hour) 等效计划降低出力小时
EP(earth plate) 接地板
epeiric sea 陆缘海
epeirogenesis 造陆作用

epeirogenetic 造陆的
epeirogenic movement 造陆运动
epeirogeny 造陆作用
EP(electrical static precipitator) 静电除尘器
EP(emergency preparedness) 应急准备
EP(end of program) 程序结束
EP(end point) 终端
EP(extreme power) 极限功率
EP(extreme pressure) 极限压力
EPF(energy pattern factor) 能源结构因数
EPFM(elasto-plastic fracture mechanics) 弹性塑性断裂力学
EPG(error pattern generator) 错误模式生成程序
ephemeral stream 季节性河流
ephemeral 短暂的
epicadmium energy region 超镉能区
epicadmium fission 超镉（中子）裂变，超镉裂变
epicadmium flux 超镉通量
epicadmium neutron efficiency 超镉中子效率
epicadmium neutron 超镉中子
epicadmium resonance integral 超镉共振积分
EPIC (Engineering-Procurement-Installation-Construction) EPIC模式，"设计-采购-安装-施工"模式【工程建设模式】
epicenter 地震震中，震源，震中
epicentral area 震中区
epicentral distance 震中距离
epicentral region 震中区
epicentral zone 震中带
EPIC(epi-planar integrated circuit) 外延平面集成电路
epicontinental sea 陆缘海
epicontinental sedimentation 陆缘沉积
epicritical 超临界的
epicyclic gearing 行星齿轮
epicyclic 外摆线的，周转圆的
epicycloid 圆外旋轮线
epidosite 绿帘石岩
epidote 绿帘石
epifocus 震源，震中
epigene action 外力作用
epigenesist 外力变质
epigene 外成的
epi-planar integrated circuit 外延平面集成电路，表目集成
epipolic 荧光的
epirock 浅带变质岩
episcotister 节光器，截光器
episode 事件
epitaxial mesa transistor 外延台面式晶体管
epitaxial planar transistor 外延平接型晶体管
epitaxial silicon variable capacitance diode 外延硅变容二极管
epitaxy 外延，晶体取向接长
epithermal absorption 超热（中子）吸收
epithermal activation 超热中子激活
epithermal activity 超热（中子激发的）活性
epithermal atoms 超热原子
epithermal capture 超热俘获，超热区的俘获
epithermal collision 超热（中子）碰撞
epithermal energy range 超热能范围
epithermal energy 超热能
epithermal fission 超热中子裂变
epithermal leakage 超热中子漏失，超热中子泄漏
epithermal neutron absorption 超热中子吸收
epithermal neutron activation analysis 超热中子活化分析，超热中子激活分析
epithermal neutron yield 超热中子产额
epithermal neutron 超热中子
epithermal range 超热能区
epithermal reactor 超热中子堆，超热中子反应堆
epithermal region 超热能区，超热区
epithermal scattering cross-section 超热中子散射截面
epithermal 超热（能）的
epithreshold energy 超阈能量
epizone 浅成带
epoch angle 初相角
epoch 出现时间，恒定相位延迟，时代，新纪元
EPOF(equivalent planned outage factor) 等效计划停用因子
epoxide resin 环氧树脂
epoxy 环氧树脂
epoxy adhesive 环氧黏合剂
epoxy asphalt concrete 环氧沥青混凝土
epoxy asphalt 环氧沥青
epoxy bitumen 环氧沥青
epoxy bituminous lacquer 环氧沥青漆
epoxy bonded fiber-glass board 环氧玻璃布板
epoxy bushing 环氧套管
epoxy cable end box 环氧树脂电缆终端盒
epoxy coating 环氧涂层
epoxy composition 环氧树脂组成
epoxy enamel 环氧瓷漆
epoxy finished paint 环氧面漆
epoxy foam 环氧泡沫
epoxy glass laminated sheet 环氧玻璃布板
epoxy glass 玻璃环氧树脂
epoxy-grouted rock bolt 环氧灌浆岩石锚杆
epoxy insulation 环氧绝缘
epoxy mastic 环氧胶泥
epoxy membrane 环氧涂层
epoxy mica paper 环氧云母板
epoxy phenolic fibre glass plate 环氧酚醛玻璃布板
epoxy plastic 环氧塑料
epoxy powder 环氧树脂粉末
epoxy resin adhesive 环氧树脂黏合剂
epoxy resin insulated 环氧树脂绝缘
epoxy resin paint 环氧树脂漆
epoxy resin pattern 环氧树脂铸模
epoxy resin 环氧树脂
epoxy seal transistor 环氧树脂密封晶体管
epoxy-silicone rubber 环氧硅橡胶
epoxy tape of glass fibre tape 环氧玻璃布带
epoxy varnish 环氧漆
EPPE(Electric Power Planning & Engineering Institute Co. Ltd.) 电力规划设计院有限公司

EPP　安全壳泄漏监测系统【核电站系统代码】
EPRC(ethylene propylene rubber cable)　乙丙橡胶电缆
EPR(earth potential rise)　地电位升高
EPRI(Electric Power Research Institute)　美国电力研究所
EPROM(electrically programmable read only memory)　电可编程只读存储器
EPROM(erasable programmable read only memory)　可擦可编程只读存储器
EPS(earnings per share)　每股赢利
Epsom salt(magnesium sulfate)　泻盐，硫酸镁
EPT(end of performance test)　性能试验结束
EPU(electrical power unit)　电功率单位
EPUT(event per unit time)　事件单位时间
EPZ(export processing zone)　出口加工区
EP　电除尘器【DCS 画面】
EQCC(entry query control console)　输入询问控制台
EQDB(equipment qualification data bank)　设备鉴定数据库
EQ(environmental qualification)　环境鉴定
EQ(environmental quality)　环境质量
EQ＝equipment　设备
equal altitudes method　等高法
equal angle projection　等角投影
equal angle steel　等边角钢
equal angle　等边角钢，等角
equal area chart　等面积图
equal area criterion of two machine system　两机系统的等面积准则
equal area criterion　瞬态稳定中的等面积法则
equal area map　等积投影地图
equal area method　等面积法
equal area projection　等积投影
equal-arm bridge　等臂电桥
equal competition　平等竞争
equal delay angle control　等滞后角控制
equal education degree　同等学历
equal energy depth　等能量水深
equalisation　相等，相等的
equality and mutual benefit　平等互利
equality　同等，相等，等式，平等
equalization charge with timer　定时等值充电
equalization circuit　均衡电路，均压电路
equalization condenser　均衡电容器，平衡电容器
equalization of discharge　平衡出水
equalization of level　均衡电平，电平的均衡
equalization　均等，均衡，平衡，使均匀，一致，稳定，补偿
equalized delay line　均衡延迟线
equalized heat distribution　等热分配（法）
equalizer assembly　平衡装置
equalizer beam　平衡梁
equalizer block　平衡器功能块
equalizer circuit　均衡电路，均压电路，补偿电路
equalizer coil　均压线圈，补偿线圈
equalizer connection　均压连接，均压线
equalizer of the first order　甲种均压线【直流电机的】
equalizer of the second order　乙种均压线【直流电机的】
equalizer of the third order　丙种均压线【直流电机的】
equalizer pipe　平衡管
equalizer pitch　均压线节距，均压线跨距
equalizer ring　均衡环，均压环
equalizer set　均衡机组
equalizer switch　均衡开关，均压开关
equalizer valve　均压阀，平衡阀
equalizer　平衡器，补偿器，均值器，均衡器，补偿电路，均压线，平衡装置，均压器，使相等的东西
equalize　平衡，均衡，补偿，使平衡
equalizing amplifier　平衡放大器
equalizing bar　均压母线
equalizing basin　反调节池，均衡池，平衡池
equalizing bed　垫床【铺管道时用的】，均匀垫层
equalizing bus bar　均压母线
equalizing charge　均衡放电【蓄电池】
equalizing conductor　均压线，补偿导线
equalizing connection　平衡连接，均压连接
equalizing culvert　平水涵洞，平压涵洞
equalizing current circuit　平衡电路，均压电路，补偿电路
equalizing current　平衡电流，均压电流，补偿电流
equalizing gear　平衡装置
equalizing hole　平衡孔，卸负荷孔
equalizing line　平衡导管，平衡线
equalizing main　平衡管道
equalizing network　均衡网络
equalizing pipe　压力平衡管，平压管，调压管
equalizing pond　调节水池
equalizing pulse　平衡脉冲，均衡脉冲
equalizing reservoir　平衡水池，调节水池，平衡水库，调节水库，反调节水库
equalizing resistance　均衡电阻，均压电阻
equalizing ring　均压环
equalizing switch　均压开关
equalizing tank　调压水塔，调压水箱，平水塔
equalizing tube　平衡管
equalizing valve　均压阀【差压传感器】，预启阀，均衡阀，卸载阀，平衡阀
equalizing vent　汽平衡接口，平衡通风口
equalizing winding　均压绕组，补偿绕组
equal leg angle　等边角钢，等肢角钢
equal-length code　等长电码
equally authentic　具有同等效力
equally likely possibility　同概率的可能性，等可能性
equally loaded　均衡负载的，等载的
equally probable　等可几的，等概率的，同等可能的
equal mutual flux linkage　等互磁通链
equal observation　等精度观测
equal-order digit　等位数字
equal percentage flow characteristic　等百分比流量特性

equal percentage valve characteristic 等于百分率阀门特性
equal percentage 等百分比
equal-phase 等相的，同相的
equal potential working 等电位作业
equal potential 等电位的
equal-pressure method 等压法
equal ripple approximation 等波纹逼近
equal settlement 等量沉陷，均匀沉降
equal strain method 等应变法
equal strength 等强度
equal treatment 平等待遇
equal 相等的，均匀的，对等物，相等
equation of continuity 连续方程，连续性方程
equation of continuous flow 流动连续方程
equation of diffusion 扩散方程
equation of dynamics 动力方程
equation of equilibrium of stress 应力平衡方程式
equation of gaseous state 气体状态方程
equation of higher degree 高次方程
equation of ideal gas 理想气体方程
equation of line 线路方程
equation of mass conservation 质量守恒定律方程
equation of motion 运动方程
equation of network 网络方程
equation of regression 回归方程
equation of small disturbance motion 小扰动运动方程
equation of state equilibrium 静平衡方程
equation of state 状态方程式，状态方程
equation of through flow 连续性方程，流动方程
equation of time 时间方程，时差
equation set-up 排方程式
equations of motion 运动方程
equation solver 方程解算器
equations set 方程组
equation 反应式【化学】，等式，方程，方程式，公式
equatorial air 赤道空气
equatorial calm belt 赤道无风带
equatorial calms 赤道无风带
equatorial climate 赤道气候
equatorial coordinate system 赤道坐标系
equatorial coordinate 赤道坐标
equatorial day 赤道日
equatorial depression belt 赤道低压带
equatorial depression 赤道低压
equatorial front 热带锋
equatorial low 赤道低压
equatorial mount 赤道式跟踪器
equatorial parallax 赤道视差
equatorial projection 赤道投影
equatorial tracker 赤道式跟踪器，赤道仪
equatorial velocity 赤道速度，大圆速度
equatorial vortex 赤道涡旋
equator principle 赤道原则
equator 赤道，大圆
equiamplitude surface 等振幅面
equiamplitude 等幅
equiangular spiral 等角螺线
equiangulator 等高仪
equibalance 平衡，补偿，匹配
equidistance 等距
equidistant line 等距线
equidistant projection 等距投影
equidistant pulse 等距离脉冲
equidistant 等距，等间距的
equi-error contour 等误差线
equifield intensity curve 等场强曲线
equiflow rate line 等流率线
equiform 相似，相似的
equifrequent 等频率的
equigranular 等粒状
equilateral arch 等边二心拱
equilateral hyperbola 等轴双曲线
equilateral 等边的
equilibalance 平衡，补偿
equilibrant 平衡力
equilibration 平衡
equilibrated valve 预启阀，均压阀，均衡阀，减压阀
equilibrate 平衡，补偿，匹配
equilibration flask 平衡瓶
equilibrium axis 平衡轴线
equilibrium boundary layer 平衡边界层
equilibrium burn-up 平衡燃耗
equilibrium concentration distribution 平衡浓度分布
equilibrium concentration 当量浓度，平衡浓度
equilibrium condition 平衡条件
equilibrium conductivity 平衡导热系数
equilibrium constant 平衡常数
equilibrium convection 平衡对流
equilibrium core 平衡堆芯
equilibrium curve 均压曲线，平衡曲线
equilibrium cycle 平衡循环
equilibrium data 平衡数据，平衡状态时的数据
equilibrium dew point 平衡露点
equilibrium diagram 平衡图，状态图，相图
equilibrium dialysis 平衡透析
equilibrium distribution coefficient 平衡分布系数
equilibrium distribution 平衡分布，均衡分布
equilibrium enrichment 平衡富集度，平衡浓缩，平衡浓度
equilibrium equation 平衡方程，平衡方程式
equilibrium factor 平衡因子
equilibrium fuel cycle 平衡燃料循环
equilibrium layer 平衡层
equilibrium level 平衡水平
equilibrium moisture content 平衡含水量，平衡含水率，平衡水分
equilibrium neutron shape 平衡中子分布
equilibrium of force 力的平衡
equilibrium of moments 力矩平衡
equilibrium of stock 资源平衡
equilibrium of supply and demand 供求平衡
equilibrium operation 稳定运行，均衡运行
equilibrium phase diagram 平衡相图
equilibrium plasma 平衡等离子体
equilibrium point 平衡点
equilibrium poisoning 平衡中毒
equilibrium position 平衡位置

equilibrium potential 平衡电势，等势，平衡电位
equilibrium price 均衡价格
equilibrium quality 平衡含汽率，热平衡含汽量
equilibrium running 平衡运转
equilibrium sand flow 均衡沙流
equilibrium slope 平衡坡降
equilibrium state 平衡态，平衡状态
equilibrium temperature 平衡温度
equilibrium theory of tide 潮汐均衡理论
equilibrium tide 平衡潮
equilibrium value 平衡值，均衡值
equilibrium valve 平衡阀，双座阀
equilibrium vapor pressure 平衡蒸汽压
equilibrium water surface 平衡水面
equilibrium xenon poisoning 平衡氙中毒
equilibrium 平衡，均衡，补偿
equi-magnetic potential surface 等磁位面
equi-magnetic 等磁的
equi-mass diffusion 等质量扩散
equi-mass model 等质量模型
equimolal diffusion 等克分子量扩散
equimolal 重量克分子浓度相等的，当量克分子的
equimolar diffusion 等克分子量扩散
equimolar mixture 等克分子混合物
equimolar 当量克分子的，克分子当量的
equimolecular 当量分子的，等分子的
equinoctial storm 二分点风暴
equinoctial 昼夜平分的
equinotical gale 二分点风暴
equinox 二分点
equip 配备，装备
equipage 设备，装备
equipartition law 均匀分布定律
equipartition of energy 能量均布
equipartition 均衡分配，均匀分布
equiphase surface 等相位面，等相面
equiphase zone 等相位区
equiphase 等相位
equipluve 雨量等比线，等雨量线，等降水量线
equipment acceptance 设备验收
equipment advisory group 设备咨询组
equipment airlock 有空气闸的设备通道，设备运输气闸
equipment and piping drain 设备及管道疏水
equipment and pump area 设备和泵区
equipment and tool 设备和工具
equipment arrangement 设备布置
equipment augmentation 设备扩充
equipment bonding conductor 设备连接电缆
equipment breakdown analysis 设备分项分析，设备损坏分析
equipment breakdown 设备事故
equipment capacity factor 设备利用率，设备利用系数，设备负载因数
equipment capacity 设备能力，设备容量，设备功率
equipment characteristic 设备性能
equipment classification 设备评级
equipment compartment 设备舱
equipment compatibility 设备兼容性，设备互换性
equipment complex 复合设备，成套设备
equipment component list 设备元件表，设备元件明细表
equipment concrete block 设备混凝土砌块
equipment configuration control 设备外形检查
equipment cost 设备成本，设备费（用）
equipment data sheet 设备数据表
equipment data 设备数据
equipment decontamination 设备去污
equipment defect 设备缺陷
equipment deficiency 设备缺陷
equipment depredation 设备折旧
equipment design layout drawing 设备设计布置图
equipment design variable 设备设计变量
equipment diagnosis technique 设备诊断技术
equipment diagnosis 设备诊断
equipment drainage system 设备疏水系统
equipment economic life 设备经济寿命
equipment efficiency 设备效率
equipment equivalent availability 设备等效利用率
equipment erection works 设备安装工程
equipment failure information 设备故障信息
equipment failure rate 设备故障率
equipment failure 设备故障，设备事故
equipment fault diagnosis 设备缺陷诊断，设备缺诊断
equipment fault rate 设备故障率，设备事故率
equipment feet 设备地脚
equipment for QC department 检测公司机械
equipment for replacement of thermal sleeves 为更换热套筒的设备
equipment for terminal station 终端站设备
equipment foundation 设备基础
equipment gauge 设备限界
equipment ground conductor 设备接地线
equipment ground 设备接地
equipment guarantee 设备担保［保证］
equipment hatch cover 设备舱口盖板
equipment hatch 设备空气闸门，设备运输通道，设备舱口，设备运输舱
equipment ID code 设备编码
equipment important for safety 安全重要设备【国际原子能组织】
equipment in power plant 电厂设备
equipment in reserve 设备备用
equipment inspection 设备检查，设备检验
equipment installation 设备安装
equipment in the heliostat field （太阳能发电的）定日镜场设备
equipment investment plan 设备投资计划
equipment investment 设备投资
equipment item 设备物项，设备细目
equipment laydown 设备搁置
equipment leasing 设备租赁
equipment life management 设备寿命管理
equipment life 设备寿命
equipment list 设备清单
equipment load 设备荷载
equipment location drawing 设备位置图
equipment location 设备定位

equipment maintenance record system 设备维护报告系统
equipment maintenance record 设备维护记录
equipment maintenance 设备检修，设备维修
equipment management 设备管理
equipment-misuse error 设备误用错误
equipment modification list 设备改变表
equipment name 设备名称
equipment number 设备编号
equipment opening 设备孔
equipment operating availability 设备完好率
equipment operating status 设备运行状况
equipment option 设备选择
equipment performance inspection 设备性能检查
equipment prestart-up inspection 设备启动前的检验
equipment problem 设备问题
equipment procurement 设备采购
equipment protection GFCI breaker 设备保护 GFCI 断路器
equipment quality 设备质量
equipment quotation 设备报价
equipment reliability 设备可靠性
equipment replacement 设备更新
equipment requalification 设备再鉴定
equipment room 设备室，维修间，设备维修间
equipment schedule 设备一览表
equipments cost of power generation project per kW 发电工程每千瓦设备造价
equipment selection 设备选型，设备选择
equipment specification 设备规范书，设备说明书
equipment supplier 设备供应商
equipment system ID number 设备系统编码
equipment title 设备名称
equipment transfer airlock 设备运输空气闸门
equipment warranty 设备担保［保证］
equipment 设备，仪器，装备，装置，器件，附件，设备费
equipoise 配重，平衡
equipollent load 等代荷载，等力荷载
equipotential bonding conductor 等电位连接导体
equipotential bonding 等电位连接
equipotential connection 等电位连接，等位连接
equipotential contour 等势线
equipotential layer 等势层，等位层
equipotential line 等势线，等位线
equipotential live line working 等电位作业
equipotential method 等势法，等位法
equipotential plane 等势面，等位面
equipotential point 等电位点
equipotential space 等位空间，等势空间
equipotential surface 等势面，等位面，等压面
equipotential winding 均压绕组
equipotential 等势的，等电位的，等电势的，等位的
equipower line 等功率线
equipower 等功率的
equipped with 配备有
equipping 装备
equipressure cycle 等压循环
equipressure line 等压线
equipressure surface 等压面
equiprobability curve 等概率曲线
equiprobability 等概率
equisignal localizer 等信号式定位器
equisignal zone 等信号区
equisignal 等信号的
equi-spaced 等间隔的，等管距的
equitable apportionment 公平分配
equitable treatment 公平待遇
equitable 公平的，公正的，平衡法的
equity capital 股东资本，股权资本，权益资本，（企业主的）股本，股本权益，股票，自有资本
equity investment 股本投资，实际投资，直接投资
equity joint venture 合资经营企业
equity J. V.（equity joint venture） 联营体【投资入股】，股权式合营企业
equity ownership 产权所有权，业主权益
equity right 股权
equity 产权，公平，股本，权益，资产净值
equivalence principle 当量定律，等值原理
equivalence 等价，当量，等积，等值，对等物，相当
equivalency 等同性
equivalent absolute nozzle flow 喷嘴绝对等值流
equivalent absorbed dose 当量吸收剂量
equivalent absorption 等效吸收
equivalent acidity 当量酸度
equivalent activity 等效放射性，等效活度，当量活度，当量放射性
equivalent admittance 等效导纳
equivalent air opening 等量风口
equivalent altitude 等效高度
equivalent antenna 等效天线
equivalent area 当量面积，等效面积
equivalent availability factor 当量可用率，等效可用系数
equivalent available factor 等效可用系数 EAF
equivalent background input 等价后台输入
equivalent background 当量背景，当量噪声
equivalent bare core 等效裸堆
equivalent basicity 当量碱度
equivalent bed 同位层
equivalent bending moment 等效弯曲力矩，等效弯矩
equivalent binary digit 等效二进制数字
equivalent capacity 等效电容
equivalent cell 等效栅元
equivalent characteristic 等效特性
equivalent circuit diagram 等效电路图，等值电路图
equivalent circuit of solar cell 太阳电池的等效电路
equivalent circuit technique 等值电路法
equivalent circuit 等效电路，等值电路
equivalent coefficient of local resistance 当量局部阻力系数
equivalent concentration 当量浓度
equivalent condenser 等效电容器
equivalent conductance 等效电导，等值电导，当量电导率

equivalent consolidation pressure 等效固结压力
equivalent constant 等效常数
equivalent continuous duty 等价持续工作负载
equivalent continuous sound level 等效连续声级
equivalent core diameter 堆芯等效直径
equivalent cross-section 等效断面，等效截面
equivalent current source theorem 等效电流源定理
equivalent current 等效电流，等值电流
equivalent curve 等效曲线，等值曲线
equivalent damping ratio 等效阻尼比
equivalent decay constant 等效衰变常数
equivalent depth 等效深度
equivalent diameter 等效直径，当量直径
equivalent diffusion velocity 等效扩散速度
equivalent distance of the oblique exposure 斜接近段的等效距离
equivalent disturbing current 等值干扰电流
equivalent disturbing voltage 等值干扰电压
equivalent dose 剂量当量，等效剂量
equivalent effective radius 当量有效半径【控制棒】
equivalent equation 等价方程
equivalent evaporation 当量蒸发量，蒸发当量
equivalent flaw diameter 当量缺陷直径
equivalent flocculation 当量絮凝作用
equivalent focal length 等效焦距
equivalent forced derated hours 等效强迫降低出力小时
equivalent forced outage rate 当量强迫停机率
equivalent force 当量力，等效力
equivalent full power days 等效满功率日，等效满功率天数
equivalent full power hours 全等效功率小时数
equivalent generator 等效发电机，等效发生器
equivalent grain size 等效粒径，相当粒径
equivalent head wind 等效逆风
equivalent heat conductivity 当量导热系数
equivalent heat value 当量热值，等效热值
equivalent height 等效高度
equivalent homogeneous parameters 等效均匀参数
equivalent hydrostatic pressure 当量水压，等效流体静压力
equivalent impedance 等效阻抗，等值阻抗
equivalent inductance 等效电感，等值电感
equivalent interruption duration 等效停电时间
equivalent laminate film model 等效层流膜模型
equivalent leakage reactance 等效漏抗，等值漏抗
equivalent length of pipe for local heat loss 局部热损失当量长度
equivalent length 等效长度，当量长度
equivalent life loss 等效寿命损耗，等效寿命损失
equivalent line admittance 线路等值导纳
equivalent linear span 等效线性跨距
equivalent line capacitance 线路等值电容
equivalent line inductance 线路等值电感
equivalent line susceptance 线路等值电纳
equivalent load 等效负载，等值负载，等效荷载
equivalent mass 等效质量
equivalent material 等效材料，替代材料
equivalent mean diameter of aggregate 骨料等效平均粒径
equivalent mean interruption duration 等效平均停电持续时间
equivalent moisture content 等效湿度
equivalent motor 等效电动机
equivalent network 等效网络，等值网络
equivalent neutron source 标准中子源
equivalent noise input 等效噪声输入
equivalent noise method 等值噪声法
equivalent noise pressure 等效噪声压
equivalent noise resistance 等效噪声声阻
equivalent observation 等效观测
equivalent opacity 等值不透明度
equivalent operating hours 等效运行小时数
equivalent parallel resistance 等效并联电阻
equivalent parameter 等效参数，等值参数
equivalent particle diameter 当量粒径
equivalent parts per billion 微克当量每升
equivalent parts per million 毫克当量每升
equivalent peak interruption duration 等效峰荷停电持续时间
equivalent performance parameters 等值性能参数
equivalent per million 百万分之当量
equivalent Pi circuit 等值 π 电路
equivalent Pi network 等值 π 形网络
equivalent planned derating hours 等效计划降低出力小时
equivalent point 当量点，滴定终点
equivalent-potential temperature 相当位势温度
equivalent pressure line 等压线
equivalent profile 等效断面
equivalent reactance 等效电抗，等值电抗
equivalent rectangular stress distribution 等值矩形应力分布
equivalent relative nozzle flow 喷嘴相对等值流
equivalent replacement method 等量置换法
equivalent resistance 等效电阻，等值电阻
equivalent roughness coefficient 等值糙率系数
equivalent roughness 当量粗度
equivalent salt deposit density 等值附盐密度
equivalent separation distance 等值隔距
equivalent series resistance 等效串联电阻
equivalent sine wave 等效正弦波
equivalent single conductor （分裂导线的）等值单根导线
equivalent source theorem 等效电源定理
equivalent source 等效电源
equivalent span 等效跨距，等值跨距，等值翼展
equivalent sphere diffusion model 等效球扩散模型
equivalent stack height 等效烟囱高度
equivalent static force method 等值静态力方法
equivalent static pressure 等效静压
equivalent steam 标准蒸汽，当量蒸汽
equivalent stress 等效应力，当量应力
equivalent surface source 等效面源
equivalent tail wind 等效顺风

equivalent T-circuit　T形等值电路
equivalent temperature　等效温度
equivalent test　等效试验
equivalent thermal conductivity　等效热导率
equivalent time　等效时间
equivalent T-network　等值T形网络
equivalent uniform live load　等效均布荷载
equivalent uniform load　等效均布荷载
equivalent unit derated hours　机组等效降低出力小时
equivalent unplanned derated hours　等效非计划降低出力小时
equivalent uranium capacity　当量铀容量
equivalent value　当量值
equivalent valve　等效电子管
equivalent voltage flicker　等值电压闪变
equivalent voltage fluctuation　等值电压波动
equivalent voltage source theorem　等效电压源定理
equivalent voltage　等效电压，等值电压
equivalent water level　等值水位
equivalent weight　等效重量，换算重量
equivalent wind speed　等效风速
equivalent　当量，等效，等值，对等，等价，当量的，相等的，相当的，等值的，相等物，等价的，等价物
equivelocity contour　等流速线
equi-volume model　等体积模型
equi-volume sphere　等体积球体
ERAB(Energy Research Advisory Board)　能源研究咨询委员会【美国】
eradiate dosage alarm　辐射剂量报警器
eradiate dosage instrument　辐射剂量仪
eradication　根除，灭绝
ERA(electronic reading automation)　电子阅读自动化
erasability of storage　存储器的可擦性
erasability　可擦度
erasable holographic memory　可擦全息存储器
erasable memory　可擦除存储器，可擦存储，可擦存储器
erasable programmable read only memory　可擦可编程只读存储器
erasable programmable ROM　可擦除可编程只读存储器
erasable storage　可擦存储器，可擦存储
erasable　可擦除的
erase character　清除符号
erased tape　擦去记录带
erase input　清除信号输入
erase pulse　清除脉冲
eraser gate　擦去装置门
eraser　消除器，擦去器，消除，擦除，擦去
erasing head　消磁头
erasure signal　清除信号
erasure　消除，擦去
erawler-tractor-mounted bulldozer　履带式拖拉-推土机
erawler-tractor-mounted shovel　履带式拖拉-铲土机
era　时代，纪元，年代
ERDA(Energy Research and Development Administration)　美国能源研究开发署
ERDS(European Reliability Data System)　欧洲可靠性数据系统
erect image telescope　直角图像望远镜
erecting bay　装配间
erecting by overhanging　悬臂拼装
erecting by overhanging　吊挂装配
erecting crane　装配吊车
erecting engineer　安装工程师
erecting frame　脚手架，装配用构架
erecting joint　安装接头
erecting platform　装配平台
erecting shop　装配间
erecting staff　安装人员
erecting welding　装配焊接
erecting yard　装配场
erection and cable pulling supervisor　安装和拉电缆检查员
erection and dismantling bay　安装拆卸间
erection and test of circuit breaker　断路器的安装与调试
erection area　安装区
erection arm lifting　扒杆吊装
erection bay　安装间
erection bolt　安装螺栓，装配螺栓
erection bracing　安装支撑
erection bridge　安装桥架，装配桥架
erection clearance　安装净空
erection contractor　安装合同商
erection cost　安装费，安装费用
erection data sheet　安装数据表
erection design　安装设计
erection diagram　装配图，安装图
erection drawing　安装图，布线图，装配图，吊装图
erection engineering　安装工程
erection engineer　安装工程师
erection equipment　安装设备
erection error　安装误差
erection frame　装配用框架，安装用框架
erection hatch cover　安装舱口盖
erection instructions　安装说明
erection jack　安装用千斤顶
erection joint　伸缩节
erection & lifting　吊装
erection load　安装荷载
erection loop　吊环，安装环
erection of anchorage system　锚定系统安装【汽轮机】
erection of anchor bolts　地脚螺栓安装
erection of auxiliary and ancillary equipment　附属机械安装
erection of boiler with preassembled pieces　锅炉组合安装
erection of busbar　母线安装
erection of control panel, desk and cabinet　控制盘、台、箱柜安装
erection of electric panel and cabinet　电气盘柜安装
erection of I&C　热工自动化安装

erection of local measuring and control instruments 就地检测和控制仪表安装
erection of pole and tower 杆塔组立
erection of pressure measuring system 压力测量系统安装
erection of temperature measuring system 温度测量系统安装
erection of tower 组塔
erection of turbine unit 汽轮机本体安装
erection of water level measuring system of differential pressure type 差压式水位测量系统安装
erection of water turbine 水轮机安装
erection opening 安装用孔
erection plan 安装计划
erection platform 安装平台
erection procedure 安装程序
erection scaffolding 安装用支架
erection schedule 安装计划，安装进程，安装进度
erection site 施工安装现场，安装工地
erection specification 安装规范
erection stress 安装应力，装配应力
erection supervision 安装监督
erection support 安装支座
erection team 安装工程队，安装队
erection tolerance 安装公差，安装容许误差
erection tool 安装工具
erection welding 安装焊接
erection weld 安装焊缝，安装焊接
erection well 安装竖井
erection works 安装
erection 吊装，耸立，装配，安装，架设，安装工程
erector arm 起重臂
erector 安装者，装配工人，安装工，安装工人，装配工
erect 安装，建造，垂直的，架设，建立，竖立，装配
ER(environment recording) 现场记录
ER(epoxy resin) 环氧树脂
ER=error 误差
ERFDH(equivalent reserve shutdown forced derated hours) 等效强迫降低出力备用停机小时
ERF(emergency response facility) 应急响应装置
ERFS/EESR(Reservation Follow-up Sheet after EESR) EESR签字后保留项目跟踪文件
ergodic hypothesis 各态历经假说
ergodicity 各态历经性，遍历性
ergogram 示功图
ergograph 示功器
ergometer 测力计，测功计，测功器
ergonomics 人机工程学，人机工效学
ergonomic 人机工程的
erg 尔格【能量单位】
ERIS(emergency response information system) 应急响应信息系统
ERL(emergency reference level) 应急参照水平
erlenmeyer flask 三角烧瓶，锥形烧瓶
ERNIE(electronic random numbering and indicating equipment) 电子随机编号及指示装置
Ernst & Young 安永会计师事务所
eroded view 分解图
eroded 被侵蚀的
erodent 侵蚀剂，腐蚀剂
erode 侵蚀，使侵蚀
erodibility 侵蚀度，可侵蚀性，冲刷性，侵蚀性，易蚀性
erodible material 易蚀材料
erodible soil 易冲蚀土壤
eroding bank 侵蚀河岸
eroding velocity 冲刷速度
erosional basin 冲刷盆地，侵蚀盆地
erosional lake 侵蚀湖
erosional plain 侵蚀平原
erosional valley 侵蚀谷
erosion base 侵蚀基面
erosion basin 侵蚀盆地
erosion belt 侵蚀地带
erosion by current 水流冲刷
erosion by water 水蚀
erosion class 侵蚀等级
erosion control below dam 坝下冲刷防治
erosion control dam 防冲坝
erosion control measure 防冲措施
erosion control of land 陆地水土保持
erosion control 腐蚀防治，浸蚀防止，侵蚀防治，侵蚀控制，冲刷防治，防冲
erosion-corrosion 侵蚀腐蚀，冲蚀
erosion cycle 冲蚀循环，侵蚀轮回，侵蚀循环
erosion effect 侵蚀作用
erosion equation 侵蚀方程
erosion factor 侵蚀因素
erosion gulley 冲沟，侵蚀沟
erosion index 侵蚀度，侵蚀指数
erosion intensity 侵蚀强度
erosion landform 侵蚀地形
erosion level 侵蚀剥削面
erosion loss 冲蚀流失，侵蚀流失
erosion mark 侵蚀痕
erosion of levee slope 堤坡冲刷
erosion of river banks 河岸的侵蚀
erosion of thermal 热腐蚀
erosion pattern 风蚀图案
erosion picture 风蚀图案
erosion pit 侵蚀坑
erosion plain 侵蚀平原
erosion plateau 侵蚀高原
erosion process 侵蚀过程
erosion-proof motor 防蚀电动机
erosion protection shield 防蚀片
erosion protection 冲刷防护，防冲，侵蚀保护，侵蚀防护
erosion rate 侵蚀率，侵蚀速度，侵蚀速率
erosion ratio 侵蚀率
erosion-ravaged area 侵蚀破坏地区
erosion resistance 耐蚀的，耐磨的，抗侵蚀性，耐腐蚀性
erosion-resistant 抗冲蚀的，抗磨蚀的
erosion resisting insulation 耐腐蚀绝缘
erosion resisting 耐侵蚀性，耐磨性
erosion shield for blade 叶片上的防蚀片
erosion shield 防蚀焊片，司太立防蚀焊片

erosion surface 冲蚀面，风化面，侵蚀面
erosion technique 侵蚀技术
erosion terrace 侵蚀阶地
erosion thrust 侵蚀冲断层
erosion 磨蚀，浸蚀，磨损，水蚀，侵蚀，剥蚀，冲刷，风化，冲蚀（作用），溶蚀
erosive action of sand 砂蚀作用
erosive agent 侵蚀力，侵蚀剂
erosive velocity 冲刷流速
erosive wear 侵蚀
erosive 侵蚀的，磨蚀的，浸蚀的
ERP(effective radiated power) 有效辐射功率
ERP(electric reliability panel) 电力可靠性小组
ERP(error recovery procedure) 错误校正过程
ERPIB(error recovery program interface byte) 错误校正程序的接口字节
errata 勘误表
erratic block 漂砾，漂块，巨漂砾
erratic boulder 漂砾，外来岩块
erratic current 不稳定电流，涡流
erratic deposit 漂砾沉积
erratic erratum 偶然误差
erratic error 不规律误差
erratic flow 涡流
erratic form 漂积层
erratic 不稳定的，无规则的，无规律的，错误的，杂散的
erratum in measurement 测量误差
erratum 写错，错字勘误表，错误
erroneous conclusion 错误结论
erroneous 错误的
error-actuated system 误差控制系统
error alarm list 故障信号表
error allowance 容许误差
error analysis 误差分析
error and trial method 试差法
error bandwidth 误差带宽
error burst 错误段，错误群，差错突发，突发差错
error character 误差特征，错误字符
error checking and correction 误差检验与校正
error checking and recovery 错误校验和改正
error checking code 错误校验码
error code 错误代码
error coefficient 误差系数，误差率
error compensation 误差补偿
error control 误差控制
error correcting capacity 纠错能力
error correcting code 误差校正码，纠错码，错误校正码
error correcting routine 错误校正程序
error correcting system 错误校正系统
error correcting 错误纠正，误差校正
error correction circuit 误差改正电路
error correction 误差校正，误差修正
error criterion 误差准则，误差判据
error curve （测量仪表的）误差曲线
error deletion by interactive transmission 用交互传输法消除误差
error detecting code 错误检测码，检错码，误差检验码
error detecting system 检错系统
error detecting 错误检测
error detection 误差检测
error detection and correction 错误检测与校正
error detection routine 错误检测程序
error detector 误差检测器，误差指示器，检错器
error diagnostic 错误诊断
error distribution principle 误差分布原理
error distribution 误差分布，误差分配
error equation 误差方程
error estimate 误差估计
error excepted 容许误差，允许误差
error exception 出错故障，例外错误
error expressed as a percentage of the fiducial value 以引用值百分数表示的误差
error expressed as percentage 百分数表示的误差
error-free data transmission 无差错数据传输
error-free operation 正常操作，无误差操作
error-free running period 正常运行期，无误运转周期，正常工作期间
error-free 安全的，可靠的，无误差的，无差错的
error frequency limit 错误频率极限
error function integral 误差函数积分
error function table 误差函数表
error function 误差函数
error graph 误差曲线
error handling 出错处理
error in address 地址错误
error in calculation 计算误差
error indicating circuit 误差指示电路
error indicator 误差指示器
error in indication 仪表指示误差
error in label 标号错误
error in operation 操作错误
error interrupt 错误中断
error in viewing 视差
error limit 误差极限
error matrix 误差矩阵
error measuring element 温差测量元件，误差测量元件
error measuring system 误差测量系统
error message repertoire 错误信息清单
error message warning 错误信息警告
error message 错误信息，差错信息，出错信息
error method 误差法，尝试法
error of behaviour 系统品质误差
error of calibration 校验误差
error of caution 错误警告
error of centering 对心误差
error of closure in azimuth 方位角闭合误差
error of closure of angles 方位角闭合差
error of closure 闭合差
error of coordinates 坐标误差
error of fact 事实错误
error of indication of a measuring instrument 测量仪表示值误差
error of indication 示值误差
error of mean square 均方误差
error of measurement 测量误差

error of observation data 观测数据误差
error of observation 观测误差
error of reading 读数误差
error of scale 比例尺误差，刻度误差
error of sighting 瞄准误差
error of survey 测量误差
error of traverse 导线闭合差
error on safe side 偏于安全的误差
error pattern generator 错误模式生成程序
error percentage 误差百分数
error per digit 每个数位误差
error probability function 误差概率函数
error propagation 错误传播，误差传播
error pulse 误差信号脉冲
error range 误差范围
error rate 误差率，出错率，误码率，差错率
error ratio 误差比
error recovery procedure 错误校正过程
error recovery program interface byte 错误校正程序的接口字节
error reset 误差复原
error-sampled control system 误差采样控制系统
errors and omissions excepted 误差和遗漏除外
error sensing device 误差敏感元件
error sensing element 误差信号传感器
errors excepted 允许误差
error signal detection 误差信号检测
error signal 误差信号，出错信号，偏差信号，差错信号
error source 误差来源
error-spectral density 误差谱密度
error-squared criterion 误差平方准则，误差平方判据
errors table 误差表
error state 错误状态
error status condition 错误状态条件
error system 误差检测系统
error transfer function 误差传递函数
error 差错，误差，偏差，误操作，错误
ERS(earth resources satellite) 地球资源卫星
ERST(error state) 错误状态
erupting bubble 喷出气泡
eruption time 喷发年代
eruption 喷出，迸发，喷发
eruptive material 喷发物质
eruptive rock extrusive rock 喷出岩
eruptive symptom 喷发征兆
erupt （火山）爆发，喷发
Esaki current 隧道电流，江崎电流
Esaki diode 隧道二极管，江崎二极管
escalating coefficient of material consumption 耗材价格上涨系数
escalation cost 价格浮动开支
escalation multiplier 涨价系数
escalation of fault 故障升级，故障扩大，故障扩张
escalation rate 浮动率【一般指年率】
escalation 浮动（加价），（物价）滑动，逐步升级
escalator 自动楼梯，电梯，电动扶梯
escapable 免除的
escapage 泄水量
escape air lock 应急人员空气闸门，应急撤离通道
escape canal 排水沟，排水渠
escape character 换码字符
escape clause 例外条款，免责条款
escape cock 放水旋塞
escape cross-section 逃脱截面
escape door 太平门
escaped plume 泄放羽流
escape factor 逃逸因子，泄漏因子
escape gate 放水闸门
escape hatch 逃生人孔
escape hole 排气口，排水口
escape ladder 安全梯，脱险梯
escape of radioactive fission products 放射性裂变产物逸出
escape orifice 排泄口，逸出孔，逸出口
escape peak 逃逸峰
escape pipe 安全管，泄水管，溢水管
escape probability 逃脱概率，泄漏概率
escape provision 撤离办法
escape rate coefficient （碘的）逃逸率系数
escape routes 撤离路线
escape staircase 安全逃脱梯，应急撤离梯
escape stairs 太平梯
escape valve 泄放阀
escape velocity 出流速度
escape weir 泄水堰
escape works 泄水建筑物
escape 泄漏，逃逸，排出，逃脱，排泄，换码，出口，紧急排水口，太平门，退水闸
escaping neutron 逸出中子
escaping radiation 漏失辐射，辐射漏失
escarpment 陡壁，悬崖，陡坡，陡崖，峭壁
escarp 内壕，筑陡坡
ESC(error status condition) 错误状态条件
ESC(escape character) 换码字符
ESC＝escape 逃逸，超出
ESCOM(electricity supply commercial reactor) 供电用商业反应堆
escort 押运，护送，护卫队
escrow account agreement 第三方账户协议
escrow credit 托付信用证，寄托信用证
escrow letter of credit 条件交付信用证，托付信用证
escrow 暂管
escutcheon plate 铭牌板
ESDA(Emergency Services and Disaster Agency) 应急服务与救灾机构【美国】
ESD(echo-sounding device) 回声探测仪
ESD(electrostatic discharge) 静电放电
ESD(electrostatic storage deflection) 静电存储偏转
ESD(emergency shutdown device) 紧急停车系统，紧急停机装置，应急停堆装置
ESD(emergency shutdown) 事故停机，应急停堆
ESD garments 防静电连体服
ESD generator 静电放电发生器
ESDH(equivalent seasonal derated hour) 等效季节性降低出力小时

ESD protection　静电保护，静电防护
ESDS(electrostatic discharge sensitivity)　静电放电敏感性
ESE(east-southeast)　东南东
ES(electromagnetic storage)　电磁存储器
ES(environmental statement)　环境报告
ES(extraction system)　抽气系统
ESFAS(engineered safety features actuation system)　专设安全设施触发系统
ESF(engineered safety features)　专设安全设施
ESF switchboard　设计有安全保障特性的配电盘
ESG(electric slag welding)　电熔渣焊
ESG(electronic sweep generator)　电子扫描发生器
E shaped iron core　山形铁芯，E 形铁芯
esker　蛇形丘
espalier drainage　格状水系
ESPD(European Single Procurement Document)　欧洲单一采购文件
especially large part　特大件
ESP(electrostatic precipitator)　静电除尘器
ESP hoppers fluidizing system　电除尘器灰斗气化系统
ESPI(electronic speckle pattern interferometry)　电子斑纹图样干涉测量
esplanade　广场，平坦的空地，岩性阶地
ESP switchgear room　电除尘配电室
ESP　西班牙比塞塔【欧元前的货币】
essay　论文，实验
ESS(electronic switching system)　电子开关系统，电子转换系统
essence　基本，本质
ESS(engineering safety system)　专设安全系统，专设保安系统
essential auxiliaries　主要辅助设备
essential clauses　必备条款
essential control of HVDC power transmission system　高压直流输电系统基本控制
essential element of information　信息基本元素
essential hole of radiography　射线照明基本孔
essential ingredients　主要成分
essential load　基本负荷
essential parameter　基本参数
essential provisions　必备条款
essential raw cooling water　重要原冷却水，可靠原冷却水
essential service water coarse filtration and trash removal system　核岛重要厂用水粗滤和除渣系统
essential service water discharge canal　重要厂用水排水渠，重要厂用水系统排水渠
essential service water discharge channel　重要厂用水排水渠
essential service water pumping station　生水泵站，应急生水泵站，重要厂用水泵站
essential service water system　核岛应急生水系统，重要厂用水系统
essential service water　重要厂用水
essential system　基本系统，重要系统
essential variables　主要变量
essential　本质的，实质的，不可少的，基本的，概要，必要的
ESS(environmental survey satellite)　环境勘测卫星
ESSM(essential systems status monitor)　主系统状态监测器
Esson coefficient　埃松系数，比转矩系数
ESSU(electronic selective switching unit)　电子选择开关装置
establish an irrevocable letter of credit　开立不可撤销的信用证
established angle　安装角
established conventions　既定惯例
established design　定型设计
established program　既定方案
establish market　建立市场
establishment charges　开办费
establishment of L/C　开证，开立信用证
establishment of letter of credit　开立信用证，信用证的开立
establishment　产业单位，建立
establish　建立，创立，确定
estate development　房（地）产开发，地产开发
estate　不动产，财产
ester　酯
esthetical area　风景区
estimate budget　概算
estimated additional reserves　估计附加资源
estimate data　估算数据
estimated complete date　预计完工日期
estimated cost　估计费用，预算成本
estimated data　估算数据
estimated design load　估算的设计负荷，估计的设计负荷
estimated flux　估计通量
estimated horse power　估计马力
estimated income tax payable　估计应付所得税
estimated life　估算寿命
estimated performance　估算性能，计算性能
estimated price　估价
estimated reasonably assured reserves　估计合理保证储量
estimated service life　估计使用年限
estimated standard deviation　估计标准差
estimated time of arrival　估计到达时间，预计到达时间
estimated time of completion　估计完工时间
estimated time of delivery　预计交货时间
estimated time of departure　估计出发时间，预计离岗时间，预估起航时间
estimated time　估算时间
estimated typical quantities and manhours　估算的典型工作量和工时数
estimated value　估计值
estimated　计算的，估计的
estimate maximum load　最大估算负荷
estimate of cost　成本估算
estimate　预测，估计，估算，估计数，概算，估价，预算
estimation error　估测误差，估算误差
estimation list　估算表
estimation of construction　工程预算

estimation of geothermal field's reserve　地热田储量估算
estimation　估计，估算，评价，估价，判断，测定估价，概算
estimator　估算人员
estuarial model　河口模型
estuarine deposit　三角港沉积
estuarine harbour　湾港
estuary cable　过河电缆，水下电缆
estuary deposit　三角港沉积
estuary　河口，港湾
e. s. u.（electrostatic unit）　静电单位
ESVCD（epitaxial silicon variable capacitance diode）　外延硅变容二极管
ESV(electrostatic voltmeter)　静电伏特计
ESV(emergency shut-off valve)　危急关闭阀
ESV(emergency stop valve)　危急截断阀，主汽门，紧急切断阀
ESW(electro-slag welding)　电渣焊
ESW(emergency service water)　应急厂用水系统
ETA(estimated/expected time of arrival)　预计到达时间
eta factor　η因子【吸收中子后产生的中子数】
etalon optical power　标准光强
etalon　标准，规格，校准器，标准量具，标准器
ETA of vessel　货船预期到达的时间
ETB(end of transmission block)　信息块传输结束
ETB(estimated/expected time of berthing)　预计靠泊时间
et cetera　等等
ETC(excess three code)　余三代码
etchant　蚀刻剂，酸蚀剂
etch cleaning　酸洗
etched groove　侵蚀的沟槽
etching agent　侵蚀剂
etching crack　腐蚀裂纹
etching primer　（钢材的）防护涂料
etching solution　蚀刻剂，蚀刻溶液，浸蚀溶液【标记】
etching structure　侵蚀构造
etching　浸蚀，酸腐蚀，腐蚀法，蚀刻法，腐蚀酸洗
etch marking　蚀刻标记
etch-proof resist　防蚀剂
etch-proof　防腐蚀的
etch test　蚀刻试验，酸蚀试验
etch　蚀刻，酸蚀，浸蚀，侵蚀，腐蚀
etc.　等等
ETD district　经济技术开发区
ETD(estimated/expected time of departure)　预计离港时间，预定出发时间，估计出发时间
ETE（electro-thermal nondestructive examination）　热点无损检测
ET(effective temperature)　实际温度
ET(engineering test)　工程试验
eternal frost climate　永冻气候
eternal tensile load　永存张拉荷载
eternit　石棉水泥
ETF(engineering test facility)　工程试验设备
ETF(engine test facility)　发动机试验设备
ETF(equipment test facility)　设备检验装置，设备试验装置
ETF(execution tracing file)　执行跟踪档案
ethane　乙烷
ethanoic acid　乙酸，醋酸
ethanol(ethyl alcohol)　乙醇，酒精
ethanol-in-glass thermometer　酒精温度计
ethanol　乙醇
ethene　乙烯
Ethernet file transfer protocol　以太网文件传输协议
Ethernet　以太网
ether wave　以太波，电磁波
ether　醚，乙醚
ethical standard　道德标准
ethics　道德准则，伦理标准
ethnic affairs commission　民族委员会
ethnic　种族的
ethoxyline resin　环氧树脂
ethylacetate　乙酸乙酯
ethyl alcohol　乙醇，酒精
ethylamine　乙胺，氨基乙烷
ethylbenzene　乙基苯
ethylene diamine　乙二胺
ethylene dichloride　二氯乙烯
ethylene hydrogenation　乙烯加氢
ethylene perchloride paint　过氯乙烯漆
ethylene propylene rubber cable　乙丙橡胶电缆
ethylene　乙烯
ethyl ether　乙醚
ethyl　乙烷基
ethyne　乙炔
ETIM(elapsed time)　消去的时间
ETL(effective tube length)　有效管长
ETMD(extended tuned mass damper)　扩展调谐质量阻尼器
ETOS(extended tape operating system)　扩充的磁带操作系统
E transformer　误差信号交换器【自动控制系统中的】
ETR（engineering test requirements）　工程试验要求
ETS(emergency trip system)　事故跳闸系统，紧急跳闸系统【汽轮机】
ETX(end of text)　文本结束
E-type mill　E型磨，中速球磨机
ETY　安全壳内大气监测系统【核电站系统代码】
EUDH(equivalent unplanned derated hour)　等效非计划降低出力小时
EU(engineering unit)　工程单位
EUF(equivalent unavailable factor)　等效不可用系数
eugeogenous　风化岩屑的，易风化的
eugranitic　花岗岩状
euhedral-granular　自形粒状
Eulerian correlation coefficient　欧拉相关系数
Eulerian correlation function　欧拉相关函数
Eulerian covariance　欧拉协方差
Eulerian cross correlation　欧拉交叉系数
Eulerian system　欧拉系统

Eulerian time scale　欧拉时间尺度
Eulerian wind　欧拉风，测点风
Euler number　欧拉数
EUNDH(equivalent unit derated hour)　机组等效降低出力小时
Euo-currency　欧洲货币
EUOF(equivalent unplanned outage factor)　等效非计划停用因子
Eurasian Continental Bridge　欧亚大陆桥
Eurasian plate　欧亚大陆块
Eureka wire　康铜导线
Eurobond　欧洲债券
European Atomic Energy Community　欧洲原子能共同体
European Company for the Chemical Processing of Irradiated Fuels　欧洲辐照燃料化学处理公司
European Currency Unit　欧洲货币单位
European Economic Community　欧洲经济共同体
European Norm　欧洲标准
European Nuclear Energy Agency　欧洲核能机构
European Organization for Nuclear Research　欧洲核研究组织
European Reliability Data System　欧洲可靠性数据系统
European Single Procurement Document　欧洲单一采购文件
European Wind Energy Association　欧洲风能协会
Europe-Asia continental railway transportation　欧亚大陆铁路运输
Euro　欧元
euryhaline　广盐性的
euryhalinous　广盐性的
euryoecic　广栖性的
eurytope　广生境的
eutectic alloy　共晶合金，低共熔合金
eutectic point　共晶点，低共熔点
eutectic structure　共晶组织
eutectic　低共熔的，易熔的，共晶的，共熔的，共熔合金，共晶体
eutectoid steel　共析钢
eutectoid structure　共析组织
eutectoid　共析
eutrophication　富营养化
eutrophic environment　富营养环境
eutrophic lake　富营养湖
eutrophic state　富营养状态
eutrophic stream　富营养河道
eutrophic　富营养的
eutrophy　营养丰富
euxinic deposition　富有机质沉积
evacuated capsule　真空膜盒
evacuated chamber　真空室
evacuated collector tube　真空集热管
evacuated collector　真空集热器
evacuated transistor　真空晶体管
evacuated tube collector　真空管集热器
evacuated tubular collector　真空管集热器
evacuate　抽空，排空，消除，抽真空，排泄
evacuating area boundary　撤离区边界
evacuation and means of evacuation　疏散及疏散设施
evacuation chute　疏散滑槽
evacuation distance　疏散距离
evacuation planning zone　计划撤离区
evacuation plan　疏散计划，撤离计划
evacuation route pressurized　正压疏散路线
evacuation route protected　防火疏散路线
evacuation route　疏散路线
evacuation signal　疏散信号
evacuation sign　撤离信号灯
evacuation time　疏散时间
evacuation valve　排气阀
evacuation　抽真空，撤离，疏散，抽成真空，排空，抽空，排气，抽气装置，抽空装置，真空器[泵]，抽气器
evade a tax　偷税
evade foreign exchange　逃汇
evaluated bid price　评标价
evaluated error　估计误差
evaluate　估计，判断，计算，评定，评价
evaluation and qualification criteria　评标和资格预审标准
evaluation chart　评估表
evaluation conclusion　评价结果
evaluation criteria　评标标准，评估准则
evaluation factor　评估因素
evaluation file　评价档案
evaluation group　评审小组
evaluation kit　评价工具
evaluation model　评价模型
evaluation of bids　评标
evaluation of crane runway horizontal force　吊车导轨水平力的估计
evaluation of employee　职工考核
evaluation of material　材料评价
evaluation of mechanical endurance　机械老化性能评估
evaluation of new technology　新技术评价
evaluation of professional rank　职称评定
evaluation of professional title　职称评定
evaluation of resources　储量评价
evaluation of sub-contractor　分包商评价
evaluation of tender　评标
evaluation of test　试验的评价
evaluation of thermal endurance　长期耐热性能评估
evaluation of welding technology　焊接工艺评定
evaluation on site safety and stability　厂址地质灾害评价
evaluation price　评标价
evaluation process　评估过程，评价过程
evaluation report　评价报告
evaluation situation　评标情况
evaluation stage　评标阶段
evaluation system　考核制度，评价制度
evaluation test of characteristic　性能鉴定试验
evaluation test　鉴定试验，评价试验
evaluation　评定，鉴定，计算，求值，评价，估价，测定，估计，估量，定值，估算
evaluator of remote-control system　遥控信息鉴定器，遥控系统求值器
evanescent voltage　衰减电压

evaporable water 易蒸发水
evaporable 可蒸发的，易蒸发的
evaporated clean water 汽化洁净水
evaporated cooling motor 蒸发冷却电机
evaporated film 蒸发薄膜
evaporated make-up 蒸发补给水，蒸馏补给水
evaporated neutron 蒸发出的中子
evaporated waste 浓缩的废物，已蒸浓废物
evaporated water 蒸馏水
evaporate 蒸发，汽化，脱水
evaporating capacity 蒸发量
evaporating circuit 蒸发回路
evaporating coil 蒸发段管圈
evaporating dish 蒸发器，蒸发皿
evaporating field 蒸发场
evaporating heating surface 蒸发受热面
evaporating pressure 蒸发压力
evaporating region 蒸发区段
evaporating surface 蒸发面，汽化面
evaporating temperature 蒸发温度
evaporation area 蒸发面积，蒸发区
evaporation basin 蒸发池
evaporation boat 蒸发舟
evaporation bottom 蒸发底渣，蒸发器底渣
evaporation capacity 蒸发能力，蒸发量
evaporation coefficient 蒸发系数
evaporation control 蒸发调节，蒸发控制
evaporation cooling 蒸发冷却
evaporation curve 蒸发曲线
evaporation discharge 蒸发量
evaporation factor 蒸发系数，汽煤比，煤水比
evaporation-fission competition 蒸发裂变竞争
evaporation fog 蒸发雾
evaporation formula 蒸发计算公式
evaporation from land surface 陆面蒸发
evaporation from land 地面蒸发
evaporation from soil 土壤蒸发
evaporation from vegetation 植被蒸发
evaporation from water surface 水面蒸发
evaporation gauge 蒸发计
evaporation heat 蒸发热，汽化热
evaporation intensity 蒸发强度
evaporation loss 蒸发耗损，蒸发损失
evaporation measurement 蒸发测定
evaporation model 蒸发模型
evaporation nuclear model 蒸发核模型
evaporation observation 蒸发观测
evaporation of antireflective coating 蒸镀减反射膜
evaporation opportunity 蒸发机会
evaporation pan coefficient 蒸发皿校正系数
evaporation pan 蒸发器，蒸发皿，蒸发盆
evaporation pond 蒸发池
evaporation power 蒸发能力
evaporation process 蒸发过程
evaporation-rainfall ratio 雨量蒸发率，蒸发雨量比，蒸发雨量率
evaporation rate 蒸发率，蒸发速度
evaporation ratio 蒸发倍率
evaporation residue 蒸发残渣，蒸发残余物，蒸发器底部残渣
evaporation retardant 防蒸发剂
evaporation stage 蒸发段，蒸发级
evaporation surface 蒸发面，汽化面
evaporation tank 蒸发皿
evaporation temperature 蒸发温度，汽化温度
evaporation test 测定蒸发量的试验，汽化度的测定，蒸发试验
evaporation zone 蒸发区
evaporation 汽化，脱水，消散，蒸发率，沸腾，蒸发，蒸发量
evaporative capacity 蒸发量，蒸发能力，蒸发容量，蒸发率
evaporative centrifuge 蒸发式离心机
evaporative circuit 蒸发回路
evaporative concentration 蒸发浓缩
evaporative condenser 蒸发式凝汽器，蒸发式冷凝器
evaporative cooler 蒸发冷却器
evaporative cooling generator 蒸发冷却发电机
evaporative cooling 蒸发冷却
evaporative duty 蒸发量，蒸汽产量
evaporative flask 蒸发瓶
evaporative rate 蒸发率
evaporativity 蒸发度
evaporator blowdown pump 蒸发器排污泵
evaporator blowdown receiver tank 蒸发器排污接收箱
evaporator bottoms storage tank 蒸发器底部沉积物贮存箱
evaporator bottoms 蒸发器底渣，蒸发器底部浓缩物
evaporator coil 蒸发器管圈
evaporator column 蒸发器柱
evaporator concentrate pipe 蒸发器浓缩物管
evaporator concentrate processing 蒸发器浓缩物处理
evaporator concentrate tank 蒸发器浓缩物储藏箱
evaporator concentrate treatment plant 蒸发器浓缩物处理装置
evaporator concentrate 蒸发器浓缩物，蒸发器底部沉积物
evaporator condensate demineralizer 蒸发器凝结水除盐器
evaporator condensate pump 蒸发器凝结泵，蒸发器凝结水泵
evaporator condensate tank 蒸发器凝结水箱
evaporator condensate 蒸发器凝结水
evaporator condenser 蒸发器冷凝器
evaporator decontamination factor 蒸发器去污系数
evaporator effect 蒸发器的级，蒸发器的效应
evaporator element 蒸发器元件
evaporator feed and neutralizing tank 蒸发器给水与中和箱
evaporator feed filter 蒸发器给水过滤器
evaporator feed pump 蒸发器给水泵
evaporator feed tank 蒸发器给水箱
evaporator outlet pipe 蒸发器出口管
evaporator outlet temperature 蒸发器出口温度
evaporator overhead condenser 顶部冷凝器

evaporator plant （废水）蒸发装置，浓集系统
evaporator residue 蒸发底渣
evaporator section 蒸发段【蒸汽发生器】
evaporator sludge 蒸发器淤渣
evaporator slurry tank 蒸发器淤渣箱
evaporator station 蒸发站
evaporator train 蒸发器管系
evaporator vapor condensation 蒸发器蒸汽凝结
evaporator vapor 蒸发器的二次汽
evaporator 蒸发器，蒸馏设备，汽化器，汽化机
evaporimeter 蒸发仪，蒸发计，蒸发器
evaporite 蒸发岩
evaporization 蒸发
evapotranspiration gauge 蒸散量计
evapotranspiration 蒸发，蒸腾，蒸散，蒸散发，蒸散作用
evapotranspirometer tank 蒸散发器
evapotron 涡动通量仪
evase stack 渐缩管，风机出口扩展段
evase 渐扩段
evasion of taxation 逃税
evasion of tax 漏税，偷税
evasion 回避
EVC 反应堆堆坑通风系统【核电站系统代码】
eV(electron volt) 电子伏特
even 均匀的，偶数的
even check 偶校验
even coating 均匀涂层
even-controlled gate 偶数控制门
even fracture 平整断口，细粒状断面，平断口
even function 偶函数
even harmonic 偶次谐波
even illumination 均匀照明
evening tide 晚潮，汐
even load 均匀载荷，均匀负载
even location 偶数存储单元
evenly distributed inductance 均匀分布电感
evenly distributed load 均布荷载
evenly hung lines 平衡挂线
evenness of surface 切割面平面度
evenness 光滑度，平面度，光滑，平滑度，均匀性，平整度
even-numbered line 偶数行
even number 偶数
even-odd check 奇偶校验
even-odd 偶数奇数
even-order harmonic 偶次谐波
even parity check 偶数奇偶校验，偶数奇数位校验
even pitch 均匀的坡度，平坦的坡度
even running 均匀运转
even symmetry 偶对称
event control block 事件控制块
event counter 信号计数器，转换计数器
event counting 事件计数
event data 事件数据
event information 事件信息
event list 事件表
event log 日志记录
evenly distributed load 均匀分布负载
event of default 违约事件，违约事项
event of low probability 低概率事件
event of moderate frequency 中等频率事件
event per unit time 事件/单位时间
event recorder 事件记录仪
event sequential record 事件顺序记录
events pending 事件挂起
events record 事件记录
events sequence 事件的时序
event synchronization 事件同时性，事件同步
event tracer 事件跟踪程序
event tree analysis method 事件树分析法
event tree analysis 事件树分析
event tree 事件树【形图】
event trigger menu 事件触发器菜单
event trigger summary 事件触发器概要
event trigger 事件触发器
eventual failure 最后破坏
event 事件，事例，案例，情况，事故
even value of contamination 污染平均值
evergreen plant 常绿植物
evergreen species 常青植物
everlube 耐寒性润滑油
eversafe tank 几何核安全容器
evershed ducter 小电阻测量表
EVF 安全壳内空气净化系统【核电站系统代码】
EVHM(ex-vessel handling machine) 容器外燃料转运机
evidence in chief 主要证据
evidence of conformity （货物）符合要求的证明
evidence of damage 残损证明
evidence of origin 原产地证明
evidence of payment 付款凭证
evidence of title 所有权证据
evidence 证词，证据，凭证，迹象，数据
evident of payment 付款凭据
evident 明显的
EVN(Electricity of Vietnam) 越南电力集团，越南国家电力公司
evolute 展开线，渐屈线，发包线
evolutionary operation 渐近操作
evolution of heat 放出热量，放热
evolution 发展，展开，进化，演变，放出，沿革，渐开线，渐伸线
evolving fault breaking test 开断发展性故障试验
evolving fault 进展性故障
EVOP(evolutionary operation) 渐近操作
evorsion 涡流侵蚀
EVR(electronic video recorder) 电子录像机
EVR 安全壳连续通风系统【核电站系统代码】
EVSE(electric vehicle service equipment) 电动车充电设备
EVSE(electric vehicle supply equipment) 电动汽车供电设备
EVTM(ex-vessel transfer machine) 容器外燃料传送机
EWA(end warning area) 结束警告区
EWD(elemental wiring diagram) 原理接线图
Ewing curve tracer 依文磁滞曲线画线器
Ewing permeability bridge 依文导磁率测量电桥
EWS(engineer working station) 工程师工作站

EWS room 工程师站室
exacerbate 加重，使恶化，使加剧
exact analytical method 严格解析法
exact data 精确数据
exact date due after six months 六个月后到期
exact date 准确日期
exact differential equation 正合微分方程式
exact differential 正合微分，恰当微分
exactitude 正确性，精密度
exact line constant 精确的线路常数
exact location 准确位置
exact measurement 精密测量
exact quantity 准确数量
exact similarity 精确相似
exact solution of transmission circuit 输电线路的精确解
exact 精密的，精确的，正合的
ex-aerodrome 机场交货价格
exaggerated scale 扩大比例
exaggerated test 超定额试验，最不利条件下的试验，超常试验
exaggerated vertical scale 放大垂直比例
exaggerate 放大，扩大
exaltation 纯化，精炼，提升，提拔，举起
exalted carrier receiver 恢复载波接收机
exalted carrier 恢复载波
examination and verification 审核
examination,inspection and testing 检查，探伤和试验
examination of bids 投标审核
examination of waste water 废水检验
examination of water quality 水质检验
examination record 审查记录
examination table 调查表
examination 试验，检查，调查，查看，研究，审查，检验，校验，考察，考试
examine and approve 检查和批准，审批
examined copy 校正本
examine for leakage 泄漏检验
examine for wear 磨损检验
examine 查验
example 例证
excavate 开挖，挖掘，挖土
excavated material 开挖出的材料，废土石，清理场地挖出的废物
excavated section 开挖断面
excavated volume 挖方
excavate volume 土石方量，挖掘体量
excavating chute 挖土滑运槽
excavating engineering 开挖工程
excavating loader 开挖装料机
excavating machinery 挖掘机械
excavating machine 挖土机械设备，挖土机，挖掘机
excavating plant 挖土机械设备
excavating pump 挖泥泵，挖吸泵，吸泥泵
excavating with timbering 支撑挖掘
excavation and replacement 换土法【软基处理】，挖填法
excavation cavity 冲挖穴
excavation depth 挖方深度，挖掘深度
excavation in open-cut 明渠开挖
excavation of earth 挖土
excavation of foundation pit 基坑开挖
excavation of the site 现场的开挖
excavation treatment 开挖处理
excavation without timbering 无支撑开挖
excavation with timbering 有支撑开挖
excavation works 开挖工程，挖方工程，挖掘工程
excavation 土石方工程，开挖，挖掘，冲挖，洞，穴，挖，挖方
excavator base machine 挖土机主机
excavator bucket 挖土斗
excavator 挖掘机，挖土机
exceed acceptable limit 超过允许极限
exceedance statistics 超过数统计
exceed drum capacity （程序）超过磁鼓容量
exceeded 超过的【范围、数值等】，非常的，过度的
exceeding a set point 超出整定点
exceeding a threshold 超过阈值
exceeding limit 超限
exceeding rate 超标率
exceeding times 超标时次
exceed 超出，超过
excellence 优点
excellent design 优秀设计
excellent pay and conditions 优惠待遇
excellent service 优质服务
excellent 优越的
excel tester 电池测验器，蓄电池电压表
excentricity 偏心率，偏心距
excentric orifice plate 偏心孔板
except as hereinafter provided 下文另有规定者除外
except as otherwise noted 除非另有说明，除非另有记载
except as otherwise provided herein 除非本文另有规定
except as provided in 除非另有规定
excepted package 例外货包
excepted risk 例外风险
except for otherwise stipulated 除另有规定外
exception 例外
exceptional circumstances 特殊情况，特别情况，异常情况
exceptional condition 例外工况，特殊工况
exceptional discount 额外折扣
exceptional length 超长
exceptional loads 异常荷重
exceptional measure 非常措施
exceptional operating condition 特殊运行工况，异常操作条件
exceptional overload 额外超载
exceptional service 额外服务，例外服务
exceptional water level 非常水位
exceptional 独特的，优越的
exception clause 除外条款，例外条款
exception interceptor 异常拦截器
exception-item encoding 异常项编码
exception-principle system 异常原则系统

exception principle 例外原则
exception reporting 异常报告
except where the context requires otherwise 除上下文另有要求外
excergic process 放热过程
excess absorption 过剩吸收
excess activated sludge 过剩活性污泥
excess air coefficient 过量空气系数，空燃当量比
excess air control 过量空气控制
excess air quantity 过剩空气量
excess air rate 过剩空气率
excess air ratio 过量空气系数，过剩空气系数
excess air 过量空气，过剩空气，剩余空气
excess ampere-turn 多余安匝，过剩安匝
excess capacity 超额容量
excess carrier 多余载流子
excess-3 coded decimal system 余三码的十进制
excess-3 code 余三代码
excess conduction 过剩型电导
excess convexity （焊缝）过高
excess cost 额外费用
excess crude oil account 溢价原油账户
excess current 过电流，过剩电流
excess dislocation 过剩错位
excess dose 超剂量
excess draught 过量气流
excess electron 多余电子
excess employee 超编人员，冗员
excess energy 过剩能量，多余能量
excess flow test 过流量试验
excess flux shutdown 通量过高停堆
excess freight train 超限货物列车
excess heating 过量供暖
excess heat 过剩热，余热，过热
excess hydrostatic pressure 超静水压力
excessive corrosion 过度腐蚀
excessive damage 过度损坏
excessive flux density 磁通密度过大
excessive gas recovery power plant 剩余煤气火电厂【利用高炉剩余煤气发电】
excessive grade 过大坡度，过陡坡度
excessive gradient 过大坡度
excessive load stepup accident 阶跃超负荷事故，越限升荷事故，升负荷过快事故
excessive loss 过大损耗
excessive machining allowance 过大的加工余量
excessive moisture 多余水分
excessive neutron flux 过高中子通量
excessive peak electrical load 过度高峰电力负荷
excessive penetration （焊缝金属）塌陷，下塌
excessive pressure 超量压力
excessive radiation 过度照射
excessive reinforcement 格外加强
excessive reliance 过于依赖
excessive residual of chlorine 过剩的余氯
excessive settlement 过度沉陷
excessive steam pressure 汽压过高
excessive sulfur content 过高的含硫量
excessive surface penetrant 多余的表面渗透剂
excessive velocity 超速
excessive ventilation 过量通风
excessive vibration 振动过大
excessive wear 过量磨损
excessive 过度的，过分的，过多的
excess letdown flow 过剩下泄流量
excess letdown heat exchanger 过剩下泄热交换器
excess letdown 过剩下泄
excess load 过载，过负荷，超载，过荷载，超荷载，超负荷
excess metal 多余金属【摩擦焊等】
excess moisture 过多水分
excess multiplication constant 过量倍增常数，盈余放大常数
excess multiplication factor 过剩倍增系数，过量倍增因子，盈余放大因子
excess neutron flux shutdown 中子通量过高停堆
excess neutron 过剩中子
excess noise 过量噪声，高噪声
excess nutrient 营养过剩
excess oxygen 氧量过剩
excess payment 多付
excess penetrant removal 多余渗透剂的去除
excess pore pressure 超孔隙压力
excess pore water pressure 超孔隙水压力
excess power peak 剩余功率的峰值
excess power 超功率，剩余功率
excess pressure valve 过压阀
excess pressure 余压，超压，过压，剩压，超出压力
excess profit duty 超额利润税
excess profit tax 过分得利税
excess profit 超额利润
excess reactivity 剩余反应性
excess reserves 备付金
excess resonance integral 过剩共振积分
excess risk 过分危险
excess semiconductor 过剩型半导体
excess sludge 过剩污泥，剩余污泥
excess sound pressure 峰值声压
excess speed test 超速试验
excess steam 旁路蒸汽，多余蒸汽
excess surface water 地面积水
excess temperature shutdown 温度过高停堆
excess temperature 超温，过余温度
excess-3 value 余 3 的数值
excess voltage protection 过电压保护
excess voltage 过电压
excess water removal 脱（去）多余水【混凝土】
excess water 过剩水量
excess weld metal （焊缝的）余高
excess weld reinforcement 超焊加强
excess 过分，过度，过量，过剩，剩余，余数，过多，盈余，余量，超额，超过，超出量，额外的
excetion 排泄
exchangeability 交换性
exchangeable disk storage 可交换磁盘存储器
exchangeable ion 可交换的离子，可交换离子
exchangeable power 可交换功率，可转换功率

exchange adsorption 交换吸附
exchange at par 平价兑换
exchange before talks 谈判前交换意见
exchange buffering 更换缓冲，交换缓冲，交换中间寄存
exchange capacity 交换容量，交换能力
exchange center 交换中心
exchange coefficient 交换系数
exchange control 外汇管制
exchange cycle 交换周期
exchange data 交换数据
exchange distillation 交换蒸馏法
exchange energy 交换能量
exchange factor 交换系数
exchange fee for CIP 商检换单费
exchange force 交换力
exchange frequency 交换频率
exchange integral 交换积分
exchange license 外汇许可证
exchange loss during start-up period 筹建期汇兑损失
exchange magnetic moment 交换磁矩，互换磁矩
exchange message 交换信息
exchange of comments 交换意见
exchange of communication 交换函电
exchange of documents 换文
exchange of experience 经验交流
exchange office 交换局
exchange of heat 热交换
exchange of needed goods 互通有无
exchange of ratification 交换批准书
exchange of stocks 股份交换
exchange of talents 人才交流
exchange of texts 交换文本
exchange of views 交换意见
exchange payment 外币付款
exchange processes 交换过程
exchange quota 外汇配额
exchanger 交换器，交换剂
exchange rate equation 交换速率方程
exchange rate fluctuations 外汇汇率波动
exchange rate quotation 外汇牌价
exchange rate 交换率，交换速度，外汇率，汇率，汇价
exchange reaction 交换反应
exchange reserves 外汇储备
exchange resin （离子）交换树脂
exchange restriction 外汇限额
exchange room 交换机房
exchange settlement 结汇
exchange site 交换基
exchange surrender certificate system 外汇转移证制度
exchange surrender certificate 外汇转移证，外汇结账单
exchange tool 更换工具
exchange views 交换意见，商谈
exchange 交换，调换，置换，交换机，交流，交易
EXCH＝exhange 交换
excide battery 铅电池组，糊制极板蓄电池
excircle 外圆
excise duty 营业税，货物税，消费税
excise tax 消费税，执照税
excision 切除，分割
excitability 可激发性，励磁性，灵敏性
exciting current 励磁电流
excitation and voltage regulation system 励磁与电压调节系统
excitation anode 励弧阳极，激励阳极
excitation arc current 励弧电流
excitation arc reactor 励弧电抗器
excitation capacitor 励磁电容器
excitation capacity 励磁容量
excitation circuit 励磁回路
excitation coefficient 激励系数，励磁系数
excitation coil 励磁线圈
excitation control of alternator SCR system 发电机可控硅制的励磁控制
excitation control 励磁控制
excitation cubicle 励磁柜
excitation current 激励电流，励磁电流
excitation curve 励磁曲线
excitation density 励磁强度，激励强度
excitation device 励磁装置，励磁设备
excitation energy 激发能，励磁能量，激励能
excitation fault 励磁故障
excitation field 励磁磁场
excitation flux 励磁通量，励磁磁通
excitation force 激励力，激振力
excitation forcing limiter 强磁限制器
excitation forcing 强迫励磁
excitation frequency 励磁频率，激发频率
excitation function 激励函数
excitation initiating equipment 起励设备
excitation loss operation 失磁运行
excitation loss relay 失磁继电器，励磁损失继电器
excitation loss 空载损耗【变压器】，励磁损耗
excitation mechanism 励磁机构
excitation of oscillation 激振
excitation of vibration 激振
excitation power 励磁功率
excitation probability 激励概率
excitation reduction 减励磁
excitation-regulating winding （双重功用）励磁-调节绕组
excitation regulation device 励磁调节装置
excitation regulation 励磁调整
excitation regulator 励磁调节器
excitation response ratio 励磁响应系数，（励磁机的）电压上升速度
excitation response 励磁响应，电压上升速度
excitation rotor 励磁转子
excitation source 激发源，励磁电源，励磁源
excitation stabilizer 励磁稳压器
excitation suppression 减励磁，灭磁
excitation system diagram 励磁系统图
excitation system response ratio 励磁系统响应比
excitation system stability 励磁系统稳定性
excitation system stabilizer 励磁系统稳定器，励

磁系统稳定装置
excitation system voltage response ratio 励磁系统电压反应系数
excitation system voltage time response 励磁系统电压
excitation system voltage 励磁电压
excitation system 励磁系统
excitation transformer 励磁变压器，激励变压器
excitation voltage 励磁电压
excitation volt-ampere 励磁伏安，磁化功率
excitation waveform 激励波形
excitation winding pole coil 励磁绕组线圈，磁极线圈
excitation winding 激励绕组，励磁绕组
excitation 激发，激励，励磁，磁化电流，激磁，触发
excitator side 励端
excited atom 受激原子
excited field loudspeaker 励磁式扬声器
excited field 励磁场，激励场，受激场
excited in phase 同相励磁
excited mode 激励模态
excited state 受激状态，励磁状态，激励状态
excited winding 激磁绕组
excited 励磁的，激励的，激发的
excite oscillation 励磁振动
exciter alternator 同步励磁机，交流励磁机
exciter armature 励磁机电枢
exciter brush 励磁机电刷
exciter ceiling voltage 励磁机峰值电压
exciter circuit 励磁机电路
exciter control system 励磁控制系统
exciter dome 励磁机上端盖
exciter end of shaft 轴的励磁机端
exciter end 励磁机端
exciter field breaker 励磁机磁场电流断路器，励磁电流断路器
exciter field 励磁机磁场
exciter loss 励磁机损耗
exciter output 励磁机输出
exciter panel 励磁盘
exciter rectifier 激励整流器
exciter response 励磁机响应速度，励磁机电压上升速度
exciter selsyn 励磁自动同步机
exciter set 励磁机
exciter time constant 励磁机时间常数
exciter transformer 励磁变压器，激磁变压器
exciter tube 激励管，主排管
exciter turbine 励磁用透平
exciter voltage 励磁机电压
exciter winding 励磁机绕组
exciter 激励器，励磁器，激振器，激发器，励磁机，主控振荡器
excite 激发，激励，励磁，使感光，引起，刺激
exciting anode 激励阳极，辅助阳极
exciting autotransformer 励磁自耦变压器
exciting choke 激发扼流圈
exciting circuit 励磁电路，激发电路
exciting current 励磁机电路，励磁电流
exciting dynamo 励磁机
exciting electrode 励磁电极，激励电极
exciting field 励磁场，激励场
exciting flux 励磁磁通
exciting force 激振力
exciting frequency 激励频率，激振频率
exciting harmonic 激振谐波
exciting magnet 励磁磁铁
exciting nozzle 激振喷嘴
exciting oscillation 激振
exciting power 励磁功率，激励本领，励磁率
exciting rush current 励磁冲击电流
exciting steam 激振用蒸汽
exciting surge limiter 励磁涌流抑制器
exciting unit 激励单元
exciting voltage 励磁电压，激发电压
exciting watts 励磁安匝数，励磁功率
exciting winding 励磁绕组，激励线圈
exciting 激励的，励磁的，振奋人心的，励磁
excitory input 激励输入
excitory 激励
excitor 励磁机，励磁器
excitron 激弧管，激励管
excluded liability 除外责任
exclude 排除，不包括
exclusion area boundary 禁区边界，隔离区边界
exclusion area control 禁区管制
exclusion area 非居住区，禁止区，禁戒区，隔离区，禁区
exclusion clause （保险）除外责任条款，除外责任条款，免责条款
exclusion limit 排斥极限
exclusion principle 不相容原理
exclusion section 非居住区，禁区，隔离区
exclusion zone boundary 禁区边界，隔离区边界
exclusion 排除，除去，切断，除外，禁止，免除，排斥
exclusive agency 唯一代理，独家代理
exclusive agent 特约总代理
exclusive and non-exclusive rights 独占和非独占权利
exclusive circuit 闭锁电路，专用电路
exclusive clause 专一条款，专属条款
exclusive dealership 独家经销
exclusive distribution 独家经销，总经销
exclusive distributorship agreement 独家销售协议
exclusive economic zone 专属经济区
exclusive event 互斥事件
exclusive liability 独家赔偿责任，专属赔偿责任
exclusive licence contract 独家许可证合同
exclusive license 独家许可证
exclusively foreign-owned enterprise 外商独资企业
exclusively 独占
exclusive privileges 专利权，专有权
exclusive representative 全权代表
exclusive right 独家经营权，专有权
exclusive sales contract 包销合同
exclusive sales 包销
exclusive selling agency 独家经销商
exclusive service tariff 专项服务价格

exclusive spur railroad 铁路专用线
exclusive trail car 专用拖车
exclusive use 独家使用
exclusive 除外的，单独的，专用的，闭锁的
exclusivity clause 独家经营权条款
ex-contractual claims 超越合同规定的索赔，非合同规定的索赔
ex-core detector 堆芯外探测器
ex-core flux instrumentation 堆外中子通量测量仪表
ex-core instrumentation 堆外测量仪表，堆芯外测量仪表
ex-core 堆芯外，堆芯外的
excrement and urine 粪便
excretion 排泄，排出，分泌，排泄物，分泌物
exculpatory 无过错的，无责任的
excursion accident （反应堆）功率剧增事故
excursion in reactor pressure 反应堆压力激增
excursion 偏离，偏移，偏差，振幅，摆幅，变化范围，功率剧增，漂移
ex custom yard 海关交货价格
excution of works 施工
EXD(ex dock) 目的地码头交货，码头交货价（格）
EXD(external device) 外部设备
ex dock 码头交货价（格），目的地码头交货
execute a contract 执行合同
execute an order 执行订单，履行义务
executed agreement 生效协议
executed contract 生效合同
executed in triplicate 签署一式三份
executed quadruplicate 签发一式四份
execute input/output 执行输入/输出
execute 行使
executing agency 执行机构
execution 签署，完成，执行
execution cycle 完成周期，执行周期
execution of contract 履行合同，契约生效，执行合同，合同的履行
execution of production plan 执行生产计划
execution program of works 施工程序
execution stage 执行阶段
execution strategy of the project 项目执行策略
execution surveying 现地施工测量
executive body 执行机构
executive committee 执行委员会
executive communication 执行通信
executive compensation 行政补偿
executive component 执行机构，执行元件
executive control language 执行控制语言
executive control program 执行控制程序
executive cost 实施费用
executive cycle 完成周期
executive device 执行机构，执行装置
executive director of the board 执行董事
executive director 常务董事，执行董事
executive employment system 干部聘用制
executive file-control system 执行文件控制系统
executive guard mode 执行防护方式
executive instruction 执行指令，管理指令
executive logging 执行录入，执行存入
executive module 执行模块，执行元件
executive order 执行指令
executive program 管理程序，执行程序
executive routine 主程序，执行程序，检验程序
executive schedule maintenance 执行调度维护
executive session 秘密会议，执行会议，常务会议，不公开会议
executive staff 行政职员
executive summary 摘要，概要，（供高层看的）执行摘要，经营综合报告，行动纲要，内容提要，前情提要
executive supervisor 执行管理程序
executive system concurrency 操作系统并发性
executive system control 执行系统控制
executive system utilities 执行系统应用率
executive system 操作系统，执行系统
executive vice president for operation 运行执行副总裁
executive vice president of project 项目执行副总裁
executive 执行的
executory contract 待执行的合同
executor 执行人
exemplary damages 惩罚性赔偿金
exempli gratia 例如，举例，作为例子
exempt from 免征
exemption certificate 免税单
exemption clause 免税条款，免责条款
exemption from customs duty 关税减免
exemption from liability 免除责任
exemption 解除，免除，免税，豁免，免责
exempt 豁免，免除，免除的
exercisability 可操纵性【阀门】
exercise control 进行控制，进行操纵
exercise description 程序描述，训练说明
exercise end time 程序结束时间
exercise number 程序号
exercise of authority 行使职权
exercise start time 程序开始时间
exercise status 程序状态
exercise 操纵，操作，行使，训练
exergic efficiency 放能效率
exergy analysis 火用分析，有效能分极
exergy balance 火用平衡
exergy consumption 火用消耗
exergy efficiency 火用效率，效率
exergy flux density 火用流密度
exergy in substance flow 物流
exergy loss 火用损失
exergy objective function 火用目标函数
exergy of heat flow 热流有效能
exergy optimization 火用优化
exergy 放射本能，有效能，可用能
exert influence on... 对……有影响
exert 施加
EX＝exhaust 排气
EX＝expansion 膨胀
ex-factory price 出厂价，工厂交货价格
exfiltration 渗漏，漏风，渗出
exfoliated graphite 片状石墨
exfoliate 剥落，脱落

exfoliation corrosion　片状剥落腐蚀
exfoliation-type corrosion　剥落腐蚀
exfoliation　剥离，剥离作用，剥落，风化，页状剥落
ex gratia claim　道义索赔
ex gratia payment　优惠给付
exhalation coefficient　呼出系数
exhalation　呼气
exhale　呼出，发出，发散，使蒸发
exhaust adapter　排气接管，排汽接管
exhaust air decontamination system　排风去污系统
exhaust air duct　排气管道
exhaust air filter monitor　排气过滤器监测器
exhaust air filter　排气过滤器
exhaust air monitor　废气监测器，排气监测器
exhaust air plume　排气烟羽
exhaust air rate　排风量
exhaust air requirement　抽风量
exhaust air shaft　排气竖井，抽风竖井
exhaust air system　排气系统，抽风系统
exhaust air　排出的空气，废气，排气，抽气
exhaust analyzer　排气分析器
exhaust annulus area　排汽面积，排汽环形面积
exhaust area　排汽面积
exhaust back pressure　排汽背压
exhaust blading　排汽叶片
exhaust blanking plate　排汽管护板
exhaust blower　排气风机，排粉机，抽气机，吸风机
exhaust box　排汽箱【汽机】
exhaust branch　排汽支管
exhaust bypass valve　排气旁通阀
exhaust casing　排气室，排气管，排汽缸，排气缸，后汽缸，排汽壳体【汽机】
exhaust chamber　排汽室，排气室
exhaust chimney　排气烟囱，排风塔
exhaust collector　排汽总管，排汽联箱，排气集合器，排气收集器，排气总管
exhaust cone assembly　喷管
exhaust cone　排气圆锥管
exhaust configuration　排气管结构，排气管布置【如横向】
exhaust contaminant　排气污染物
exhaust curve　排气线
exhaust cylinder water-spraying valve　排汽缸喷水门
exhaust diffuser　排气扩压器，排汽扩压器
exhaust-driven gas turbine　排气驱动式燃气轮机
exhaust duct　排汽管，排气管，排风管
exhausted air　废气，抽出的空气，排出的空气
exhausted developer　耗尽的显影剂
exhausted enclosure　密闭罩
exhausted ion exchange resin　失效的离子交换树脂
exhausted liquid　废液
exhausted lye　废碱液
exhausted resin　失效树脂
exhausted solution　废液
exhausted water　废水，排放水
exhausted　用过的，废的，无用的，用尽的，剥尽的
exhaust control　排风控制
exhaust emission　废气排放
exhaust enthalpy　排汽焓
exhauster-fired-boiler　排气补燃锅炉
exhauster　排风机，排粉机，真空泵，排气机，排气器，抽风机
exhaust fan room　排风机室
exhaust fan　排气风机，排风机，抽风机，排气扇，抽气扇，排风扇
exhaust fired boiler　排气燃烧锅炉
exhaust flow　废气流，排气流，排烟，废气，排汽流【汽轮机】
exhaust flue gas　废烟气，排出烟气，排放烟气
exhaust fume　排烟，废气
exhaust gas afterburner　排气后燃烧器
exhaust gas boiler　废气锅炉，废热锅炉
exhaust gas duct　排烟管道
exhaust gas fan　排烟风机
exhaust gas filter　排气过滤器
exhaust gas purifier　排气净化器
exhaust gas temperature　排烟温度
exhaust gas　排烟，排气，废气
exhaust grille　排风格栅
exhaust header　排气联箱，排水总管
exhaust heat boiler　废热锅炉
exhaust heated cycle　废气回热循环
exhaust heat exchanger　废热热交换器
exhaust heat loss　排气热损失
exhaust heat recovery cycle　废热利用循环，废热回收循环
exhaust heat recovery equipment　排热回收装置
exhaust heat recovery system　排热回收系统
exhaust heat　废热
exhaust hood temperature　排气罩温度
exhaust hood　排汽缸，排气罩，排风罩，通风柜
exhaust inlet　吸风口
exhaustion column　失效塔，失效柱
exhaustion degree　抽空度
exhaustion phase　失效段
exhaustion run　失效周期，工作周期
exhaustion test　抽气试验
exhaustion　失效，枯竭，抽空，耗尽，抽出（气），排出（气）
exhaustively　用尽一切地
exhaustive test　抽气试验
exhaust jet　废气射流
exhaust line　耗损线，排废管路
exhaust loss　排汽损失，排气损失
exhaust main　排汽总管
exhaust manifold reactor　排气支管反应器
exhaust manifold　排汽总管，排气歧管，排气集管，排气总管
exhaust muffler　排汽消音器
exhaust noise　排气噪声
exhaust odor　排气味道
exhaust opening exhaust orifice　排气口
exhaust opening　排气孔，吸风口
exhaust outlet casing stage blade　末级叶片
exhaust outlet　排气口，排风口

exhaust passage 排气通道
exhaust pass 出口烟道
exhaust pipe 排汽管，排气管
exhaust plenum 排气缸，排气室
exhaust plume 废气羽流
exhaust port 排汽口，排气口
exhaust pressure loss 排汽压力损失
exhaust pressure ratio 排汽压比
exhaust pressure turbine 排汽压力损失
exhaust pressure 排汽压力，乏汽压力，排气压力
exhaust pump 排气泵
exhaust purification 废气净化
exhaust purifier 排气净化器
exhaust region 排汽侧
exhaust regulator 抽汽调节器，排汽调节器，排气调节器
exhaust reheat 排气重热
exhaust resistance 出口阻力，排汽阻力
exhaust room 排汽室
exhaust screen 排汽挡板，排气网
exhaust shaft 排风竖井
exhaust shroud 排气管套
exhaust side 排汽侧
exhaust silencer 排气消声器，排气消音器
exhaust smoke 排烟
exhaust stack 排气烟囱，排风塔
exhaust stage blade 末级叶片
exhaust stage 排汽级，末级
exhaust steam casing 排汽缸，排汽室
exhaust steam end 排汽端
exhaust steam heating 废汽加热，乏汽供暖
exhaust steam moisture 排汽湿度
exhaust steam pipe 排汽管
exhaust steam pressure 排汽压力
exhaust steam safety valve 排汽安全门
exhaust steam separator 排汽分离器，废汽分离器
exhaust steam turbine 乏汽汽轮机，废汽汽轮机
exhaust steam valve 排气门，排汽门
exhaust stream 排出流，抽出汽流，排汽，废汽，乏汽
exhaust stroke 排气冲程
exhaust system of ventilation 通风排气系统
exhaust system 排气系统，排风系统
exhaust temperature 排气温度
exhaust thrust 排气推力
exhaust trail 排气尾流
exhaust turbine 乏汽涡轮机
exhaust valve guide bushing 排气门导管衬套
exhaust valve guide 排气门导管
exhaust valve lifter 排气阀挺杆
exhaust valve 排汽阀，排气阀
exhaust velocity 排气速度
exhaust ventilation 抽出式通风，抽风，排气通风
exhaust vent 排汽孔
exhaust vertical pipe （排气）烟囱
exhaust volume 排气量
exhaust water pipe 排水管
exhaust water 排水
exhaust wetness 排汽湿度
exhaust 排出，抽空，耗尽，用完，废气，排汽，排汽口，排气的，排气口
EXHD(exhaust hood) 排汽罩，排汽缸，排汽室，后汽缸
EXH(exhaust) 排气，排汽
exhibition hall 展览馆
exhibition reactor 展览反应堆
exhibition structure 展示结构
exhumation 剥露（作用）
exhumed topography 剥露地形
existence information 实在信息
existence 实在的
existing administration system 现行领导体制
existing building 现有房屋，现有建筑物
existing construction 现有建筑物
existing data 现有资料
existing equipment 现有设备
existing facilities 现有设施，现有设备
existing line 已有线路
existing plant 老厂
existing price 现行价格
existing system 现行体制
existing 现存的
exit air 排气
exit angle 出口角，出射角
exit area of nozzle 喷嘴出口面积
exit concentration 出口浓度
exit condition 出口工况，出口条件，出口参数
exit direction sign 出口方向标志
exit dose 出射剂量，引出端剂量，出口剂量
exit edge 出口边，出汽边，出气边
exit facilities 疏散设施
exit flue 排烟烟道，排烟道
exit gas temperature 出口烟气温度
exit gradient 出逸坡降，逸出坡降
exit guide vane 出口导叶
exit head loss 出口水头损失
exit instruction 退出指令，引出指令，出口指令
exit lamp 太平门指示灯
exit length 出口段长度
exit lighting 太平门指示灯，出口指示灯
exit loss 出口损失，出口损耗，输出端损耗，排汽损失
exit meeting 退场会
exit nozzle 出口喷嘴
exit pipe 排汽管，出口管，出口管道
exit port 出口，出口孔
exit pressure drop 出口压降
exit pressure 出口压力
exit quality 出口含汽率
exit receipt 出境回执
exit region 出口区
exitron 激励管
exit section 出口段
exit side of the dumper 翻车机出车侧
exit sign 出口标志
exit temperature 出口温度
exit velocity 出口速度，排汽速度，出口流速，逸出流速，逸出速度
exit visa 出境签证

exit void fraction 出口空泡份额
exit 出口，排气管，引出端，排风口，太平门，退出，安全出口
ex lighter 驳船交货价
exoelectron dosimetry 外逸电子剂量学，外逸电子剂量测定
exoelectron emission 外激电子发射
exoelectron 外激电子
exoergic 放热的，放能的
exogenetic action 外成作用
exogenetic rock 外成岩
exogenic force 外力
exogenic process 外成作用
exogenic 外生的，外源的，外因的
exogenous metallic inclusion 外来金属夹杂物
exogenous variable 外生变量
exomomental 发射脉冲的
exomorphism 外变质，交接触变质作用
exonerate 免除，免罪
exoneration clause 免责条款
exoneration 免责
exonerative 免除的
exorheic region 外流区
exosmose 外渗现象
exosmosis 外渗现象
exosphere 外大气层
exothermal body 发热体
exothermal 放热的，发热的，放能的
exothermic chemical reaction 放热化学反应
exothermic disintegration 放热衰变，放热转化
exothermic reaction 放热反应
exothermic 放热的，发热的，放能的
exotherm （因释放化学能而引起的）温升
exotic block 外来岩块
exotic energy 外来能源
exotic 外来的
expand 扩展
expandability 可扩展性，可扩性，扩展性
expandable-comb tap 可膨胀的螺纹梳刀
expandable material 膨胀性材料
expandable-tip pile 扩端桩
expand and contact freely 自由膨胀和接触
expanded bed 膨胀床
expanded channel 扩展流道，扩张形流道，扩压流道，扩张式流道
expanded concrete 膨胀混凝土
expanded conductor 扩建导线
expanded connection 胀管接头
expanded diameter conductor 扩径导线
expanded foam plastic 多孔塑料
expanded height 膨胀高度
expanded metal lath 钢板网，拉展金属网，金属拉网
expanded metal 多孔金属，钢板网，网眼钢板
expanded pearlite aggregate 膨胀珍珠岩骨料
expanded pearlite 膨胀珍珠岩
expanded perlite aggregate 膨胀珍珠岩骨料
expanded phase 扩散相
expanded plastic 多孔塑料，发泡塑料
expanded plug mandrel 膨胀的塞轴
expanded scale 扩展刻度，展宽表盘，扩展标度，膨胀页岩
expanded slag 多孔矿渣，膨胀矿渣
expanded slate aggregate 膨胀板岩骨料
expanded tenon 扩大接头
expanded time base 扩展扫描
expanded tube joint 胀管接头，胀接，伸缩缝
expanded 扩展的
expander roll 胀管器滚子，扩管器滚子
expander 扩张器，胀管器，蒸发器，扩音器，膨胀透平，扩容器，扩展器，扩管器，膨胀器
expanding agent 膨胀剂
expanding auger 扩孔钻
expanding cement 膨胀水泥
expanding clutch 膨胀离合器
expanding collar 夹紧环
expanding cone 膨胀塞
expanding current 扩展水流
expanding drill 扩孔钻
expanding grade 胀管度
expanding grout 膨胀性水泥浆
expanding mortar 膨胀性水泥砂浆
expanding nozzle 扩散喷管，扩张型喷管
expanding reach 扩大段
expanding reamer 扩张式铰刀，可调铰刀
expanding section 扩大段
expanding shell anchor bar 胀壳式锚杆
expanding slot atomizer 可调隙式雾化器
expanding slot burner 扩展缝隙式燃烧器
expanding universe theory 宇宙膨胀学说
expand test 扩管试验
expand tube 胀管
expansibility 扩展性，膨胀率，膨胀性
expansion agent 膨胀剂
expansional cooling 膨胀冷却
expansion anchor 膨胀锚固
expansion and contraction joint 伸缩缝
expansion and contraction 伸缩
expansion angle variable capacitor 扩展角可变电容器
expansion angle 膨胀角
expansion apparatus 膨胀仪
expansion appliance 补偿器
expansion bearing 膨胀轴承，伸缩支座
expansion bellows 波纹膨胀节，伸缩软管，膨胀补偿波纹管，膨胀节
expansion bend 膨胀接头，膨胀弯头，弯管补偿器，膨胀弯管，胀缩弯管
expansion bolt 伸缩栓，扩开螺栓，自攻螺丝，膨胀螺栓，伸缩螺栓
expansion chamber 扩大室，膨胀室
expansion circuit breaker 气体膨胀灭弧断路器
expansion clearance 膨胀间隙
expansion cock 节流阀，节流旋塞
expansion coefficient 膨胀系数
expansion compensation 膨胀补偿
expansion compensator 膨胀补偿器
expansion control box 膨胀控制箱
expansion coupling 胀缩联轴节，伸缩接头
expansion crack 膨胀裂缝，膨胀裂纹
expansion dead point 膨胀死点
expansion deaerator 膨胀式除氧器

expansion degree　膨胀度
expansion device　膨胀装置，补偿器
expansion difference　胀差
expansion drum　膨胀箱
expansion eccentric　膨胀偏心轮
expansion efficiency　膨胀系数，膨胀率，膨胀比
expansion factor　膨胀率，膨胀系数
expansion flow　扩张水流
expansion force　膨胀力
expansion funds　企业发展基金
expansion gallery　廊道扩大段
expansion groove　膨胀槽
expansion index　回弹指数，膨胀指数
expansion indicator　膨胀指示器
expansion investment　扩建投资
expansion joint filler　伸缩缝填料
expansion joint sealant　伸缩缝止水层
expansion joint　膨胀节，膨胀补偿节，胀接，膨胀接头，伸缩接头，补偿器，伸缩缝，膨胀缝，膨胀接合，伸缩器，伸缩节
expansion line　膨胀线，膨胀过程
expansion link　伸缩杆
expansion loop　膨胀伸缩弯头，热胀补偿器，膨胀节，膨胀环，膨胀圈
expansion mandrel　膨胀心轴
expansion of investment　扩大投资
expansion pedestal　伸缩支座
expansion piece　膨胀节
expansion pile　扩孔桩
expansion pin　膨胀销
expansion pipe bend　伸缩管弯头
expansion pipe　膨胀补偿器，膨胀管
expansion pressure ratio　膨胀压力比
expansion pressure　膨胀压力
expansion process　膨胀过程
expansion-producing admixture　膨胀性掺和料
expansion project　扩建项目
expansion rate　膨胀率
expansion ratio　膨胀比，膨胀率
expansion ring　膨胀环，膨胀圈
expansion rocks and soils　膨胀岩土
expansion roller　活动支承滚柱，伸缩滚轴
expansion seam of insulating layer　保温层伸缩缝
expansion seam　伸缩缝
expansion-shell bolt　胀壳式锚杆
expansion shock　膨胀突跃，膨胀波
expansion sleeve bolt　胀壳式锚杆
expansion sleeve　伸缩套筒
expansion stay　伸缩拉撑
expansion steam trap　恒温式疏水器
expansion strain　膨胀应变
expansion strength　膨胀强度，真空波强度
expansion stress　膨胀应力，拉应力
expansion strip　伸缩缝嵌条
expansion stroke　工作冲程，膨胀冲程
expansion suspender　伸缩悬杆
expansion tank　膨胀箱，扩容箱，疏水箱，缓冲箱，膨胀水箱
expansion tap　可调丝锥
expansion test　膨胀试验
expansion thermometer　膨胀式温度计，膨胀温度计
expansion tool　膨胀工具【套筒的装配】
expansion turbine　膨胀式汽轮机，膨胀透平
expansion type steam trap　膨胀式疏水器
expansion U bend　方形伸缩器，U形伸缩器
expansion valve　膨胀阀
expansion vessel　膨胀箱，油枕，储油柜，膨胀水箱
expansion washer　伸缩垫圈
expansion wave　膨胀波
expansion work　膨胀功
expansion　扩张，膨胀，胀管，展开，扩充，扩建，扩展，展开式
expansive agent　（水泥）膨胀剂
expansive cement　膨胀水泥
expansive clay　膨胀性黏土，膨胀黏土
expansive concrete　膨胀性混凝土
expansive force　膨胀力
expansive mortar　膨胀砂浆
expansive rock and soil　膨胀岩土
expansive soil　膨胀性土，膨胀土
expansivity　膨胀性
ex parte investigation　片面的调查
expascoop　推卸式装运机
expectance　期待
expectancy　预期值，期待，预期
expectation of client　客户意愿
expectation value　期望值，期待值
expectation　预期，期待，期望，期望值
expected consumption　预期消费
expected data　预期日期
expected demand not supplied　期望缺供电力，电力不足期望值，停电功率期望值
expected energy not supplied　电量不足期望（值）
expected failure duration　故障持续时间期望
expected internal rate of return　期望的内部收益率
expected life　期望寿命，预期使用年限
expected loss of energy　电量不足期望值
expected loss of load　电力不足期望值
expected mean life　预定平均寿命
expected output　预想出力
expected result　预期结果
expected time　预定执行时间
expected value　预期值，期望值
expected wind speed　期望风速
expected yield to maturity　期望的到期收益率
expected　预期的
expedance　负阻抗
expedient measure　权宜之计
expedient　应急的手段，权宜之计的，权宜之计
expediter　催交员
expedite　催交，加快进度
expediting engineer　催货工程师
expeditiously　迅速的，敏感的
expeditious measure　应急措施
expeditious　高效的，敏捷的
expeditor　催交员
expedor phase advancer　进相感受器
expellant gas　排出的气体
expeller　排除器，向心式透平叶轮，排出器，

排气机
expel 排除，驱逐，发射，放出
expendable 消耗材料
expended energy 消耗能量
expenditure of construction 建造费用
expenditure of energy 能量消费
expenditure of idleness 窝工费用
expenditure on construction 建造费用
expenditures reduction 压缩开支
expenditure 费用，消耗，耗损，损耗，支出
expense of comprehensive commission test 联合试运转费
expense of production 生产费用
expenses for administration 行政管理费
expenses for survey and design 勘测设计费
expense 费用，开支，损耗
experience and lessons 经验教训
experienced staff 经验丰富人员
experienced worker 熟练工人
experienced 有经验的
experience list 业绩表
experience record 使用业绩
experience 经验，体验，经历
experiment 试验，实验
experimental aerodynamic 实验空气动力学
experimental analysis 实验分析
experimental arrangement 实验布置
experimental assembly 实验本体
experimental basin 实验流域
experimental bench 实验台
experimental building 实验建筑物
experimental catchment 试验流域
experimental channel 实验孔道
experimental condition 实验工况，实验条件
experimental construction 试点建设
experimental data handling equipment 试验性数据处理设备
experimental data 实验数据，试验数据
experimental design 试验性设计
experimental error 实验误差，试验误差
experimental facility 实验装置，试验设备
experimental field 实验地
experimental fit 实验拟合
experimental geology 实验地质学
experimental hole 实验孔道
experimental installation 实验装置
experimental laboratory 实验室
experimental loop 实验回路
experimental medium 实验介质
experimental memory address register 实验存储地址寄存器
experimental method 实验方法
experimental MHD generator 实验磁流体发电机
experimental model 实验模型
experimental nuclear data 实验性核数据
experimental nuclear power plant 实验性核电厂
experimental point 实验点
experimental port 实验孔
experimental power reactor 实验性动力反应堆
experimental power station 实验电站
experimental project 实验性项目
experimental prototype 实验样机
experimental provision 实验装置
experimental reactor 实验堆
experimental report 实验报告
experimental research 实验研究
experimental section 实验部分
experimental sequence 实验程序，试验程序
experimental set-up 实验装置
experimental smoke plume 实验烟羽
experimental stage 实验阶段
experimental starting 试启动
experimental station 实验站
experimental stress analysis 试验应力分析，应力的实验分析
experimental study 实验研究，试验研究
experimental tank 实验池，实验水槽，试槽，试验水池，实验罐
experimental turboalternator 试验性汽轮发电机
experimental verification 实验校验，试验验证
experimental watershed 实验流域，实验性集水区
experimental 试验性的，实验性的，实验的，试验的
experimentation 试验，实验
experiment condition 实验工况，实验条件
experiment data management system 试验性数据管理系统
experiment data 试验资料，实验数据
experiment error 试验误差，实验误差
experiment model 实验模型
experiment reactor 实验反应堆
experiment rig 试验台架
expert appraisal 专家鉴定［评价］
expert appraisement 专家鉴定［评价］
expert assessment on internal threads 内部螺纹的专家评定
expert comments 专家意见
expert committee 专家委员会
expert conclusion 鉴定结论
expert display system 专家显示系统
expert group 专家小组
expertise report 鉴定书
expertise 专门知识或技能，专家的意见，专家评价，鉴定，专长，专家，专门知识，专业知识，（业务技术）经验，专家鉴定
expert mode command syntax 专家方式命令语法
expert mode 专家方式
expert opinion 专家意见，鉴定意见
expert power 专家权威
experts and scholars 专家学者
expert's statement 专家鉴定书［评价书］
expert system 专家系统
expert 鉴定人，专家，内行，能手
exp. = expansion 膨胀
exp(exponential function) 指数函数
exp = export 输出，出口
ex pier 码头交货价格
expiration clause 到期条款
expiration date for presentation of document 交单到期日
expiration date of letter of credit 信用证有效期，

信用证有效期截止日
expiration date 截止日期，有效日期
expiration notice 到期通知书
expiration of contract 合同到期，合同期满
expiration of L/C 信用证到期
expiration of license 许可证满期
expiration of period 期满
expiration of policy 保险单到期
expiration 到期，期满，截止，终止
expiratory 呼吸的
expire 期满
expiry date of L/C 信用证有效期
expiry date 终止日期，到期日，失效日期，有效期
expiry （期限、协定等的）满期，终止，期满，逾期
EXP JT(expansion junction) 膨胀节
explain in detail 申述
explain 说明
explanation 解释
explanatory legend 说明图例
explanatory memorandum 解释性备忘录
explanatory note 解释性说明
explanatory 解释性的
ex plane 空运飞机上交货价格
explicit declaration 显式说明，显示说明
explicit definition 显定义
explicit directions 明确的指示
explicit function 显（式）函数，显函数
explicit program 显程序
explicit views 明确的意见，明确的观点
explicit 显的，显式的，明确的
explode 爆炸
exploded view 剖视图，分解图
exploder 爆炸物，爆炸装置，引信
exploding gun 起爆枪
exploding plug 自溃式安全溢洪道
exploding wire 引爆线
exploitation cost 开采成本
exploitation 捕获利用，使用，开采，开发，操作，运行
exploit 开发，利用
exploration data 勘察资料，勘探数据
exploration forecast 探索性预测
exploration of stratum 地层探测
exploration procedure 查勘程序
exploration program 勘测大纲
exploration survey 勘测
exploration 探测，勘探，探险，研究，查勘，勘察
exploratory adit 探洞
exploratory borehole 探孔
exploratory consultation 探讨性磋商
exploratory development 探索研究，探索性研制，应用研究
exploratory mission 考察团
exploratory pit 探坑
exploratory sampling 探查抽样
exploratory talks 探讨性会谈
exploratory test 探查试验
exploratory tunnel 探洞
exploring brush 测试电刷，辅助电刷
exploring coil 测试线圈，探测线圈
exploring tube 测针，探针，探管，喷嘴
explosibility 可爆炸，容易爆炸，可爆性，爆炸性
explosible 易爆的
explosion accident 爆炸事故
explosion action 爆炸作用
explosion chamber 灭弧室，消弧室
explosion-deflated area 泄爆面积，泄炸面积
explosion detection and suppression means 爆炸探测和抑制手段
explosion detector 爆炸探测器，测爆器
explosion diaphragm 防爆
explosion door 防爆门
explosion factor 爆炸度，爆炸指数
explosion flap 防爆门
explosion forming 爆炸成型
explosion fuse 爆炸式熔断器
explosion gas turbine 爆燃式燃气轮机
explosion hazard 爆炸危险
explosion index 爆炸指数
explosion limit 爆炸极限
explosion mixture limit （可燃混合物）爆炸界限
explosion pot 油断路器中起爆灭弧室
explosion pressure 爆破压力，爆炸压力
explosion prevention measures 防爆措施
explosion-proof dust-tight motor 防爆，防尘电动机
explosion-proof electric equipment 防爆电气设备
explosion-proof electric machine 防爆型电机
explosion-proof electric 防爆电气设备
explosion-proof housing 防爆机壳
explosion proofing 防爆
explosion-proof lamp 防爆灯
explosion-proof luminaire 防爆照明装置
explosion-proof machine 防爆机械
explosion proof motor 防爆型电动机，防爆电动机，防爆电机，防爆马达
explosion-proof switch 防爆开关
explosion-proof transformer 防爆型变压器
explosion-proof tube 防爆筒，安全气道，防爆型
explosion-proof vent 防爆筒，（变压器的）安全气道
explosion-proof 防爆的，防爆，防爆作用
explosion protection system 防爆系统
explosion protection 防爆
explosion relief valve 爆炸安全阀
explosion resistance 抗爆
explosion-resisting frame 防爆机座，防爆机壳
explosion-safe frame 防爆型机座
explosion-safe 防炸的
explosion sensor 爆炸传感器
explosion site 爆炸位置
explosion sound 爆炸声
explosion stroke 爆发冲程，工作冲程
explosion suppression 抑爆
explosion suppressor 抑爆器
explosion test 爆破试验，爆炸试验
explosion tube 爆管
explosion vent 防暴门，防爆门

explosion wave 爆炸冲击波，爆破波，爆炸波
explosion welding 爆炸焊，爆炸焊接
explosion 爆炸，爆发，外爆，激增，爆破
explosive agent 爆炸剂
explosive bonding 爆炸连接
explosive chamber 炸药室
explosive charge 炸药装填量
explosive charging method 装炸药方法
explosive cladding 爆炸堆焊
explosive compound 爆炸化合物
explosive effect 爆炸威力
explosive expansion 爆炸膨胀
explosive fume 爆炸雾
explosive gas detector 爆炸气体检测器
explosive gas indicator 爆炸性气体指示器
explosive gas 爆炸性气体
explosive mixture 爆炸性混合物，易爆混合物
explosiveness 爆炸性
explosive oil 爆炸油
explosive plugging 爆炸堵塞
explosive signal 爆炸信号
explosive welding 爆炸焊
explosive 爆炸性的，炸药，爆炸物
explosivility 爆炸性
EXPO(extremely problem-oriented) 极端面向问题的
exponent curve 指数曲线
exponent equation 指数方程
exponential absorption law 指数吸收定律，指数吸收律
exponential absorption 指数吸收
exponential approximation 指数近似
exponential assembly 指数实验装置，指数堆
exponential attenuation method 指数衰减法
exponential attenuation 指数衰减
exponential curve 指数曲线
exponential damping factor 指数衰减因子
exponential decay function 指数衰减函数
exponential decay law 指数衰变律
exponential decay 指数衰减，指数衰变
exponential distribution 指数分布
exponential doubling time 指数倍增时间
exponential equation 指数方程
exponential experiment 指数实验
exponential extinction coefficient 指数衰减系数，指数消光系数
exponential factor 指数因子
exponential failure law 指数故障率，故障指数率
exponential flux rise 通量指数上升
exponential function 幂函数，指数函数
exponential growth 指数增长
exponential integral function 指数积分函数
exponential lag 指数滞后，指数延迟
exponential law of attenuation 衰减指数定律
exponential law 指数律
exponential line 指数线路
exponential operator 指数算子
exponential pile 指数反应堆
exponential reactor 指数反应堆
exponential region 指数分布区
exponential series 指数级数
exponential smoothing device 指数平滑装置
exponential subroutine 指数子程序
exponential time base 指数式时基
exponential time delay 指数特性时间延迟
exponential transmission line 指数式传输线
exponential velocity distribution 指数速度分布
exponential 指数的，幂的，指数，指数曲线
exponent sign 幂指数
exponent 指数，幂阶
export acceptance 出口承兑汇票
export advance 出口预付
export agency system 出口代理制
export agent 出口代理，出口代理商
exportation 输出，出口
export availability 可出口量
export bill of lading 出口单，出口提单
export bounty 出口补贴，出口津贴，出口奖励金
export broker fee 出口经纪人佣金，出口代理费
export by turnkey contract 以统包合同方式出口
export cargo 出口货
export certification 出口证书
export commission 出口代理佣金，出口代理费
export contract 出口合同
export control 出口管制
export credit insurance 出口信贷保险
export credit 出口信贷
export declaration 出口报关，出口申报书
export duty 出口关税
export exchange 出口外汇
export inspection certificate 出口检验证明单
export L/C 出口信用证，出口结汇单
export letter of credit received 收到的出口信用证，开来的出口信用证
export letter of credit 出口结汇单，出口信用证
export license 出口许可证
export management 出口管理，出口控制
export order 出口订单
export-oriented development 外向型发展
export-oriented enterprise 外向型企业
export-oriented 外向型的（经济）
export packing 出口包装
export permit 输出许可
export price 出口价格
export processing zone 出口加工区
export quotas 出口配额
export quotation 出口估价单
export regulations 出口条例
export sales 外销
export standard packing 出口标准包装
export subsidy 出口补贴，出口津贴
export tax 出口税
export trade act 出口交易法
export trade 出口贸易
export value 出口额
export volume 出口额
export 出口，输出
exposed and concealed pipe 明管和暗管
exposed cable 外露的电缆
exposed coal pile 露天煤堆
exposed conduit 明管

exposed deck 露天甲板
exposed end of covered electrode 覆盖电极的外露端
exposed face of the storage pile 贮煤堆的外露面
exposed installation 承受大气过电压的安装方式
exposed intake 明流进水口
exposed joint 明缝，明接
exposed material 受辐照材料
exposed penstock 露天（式）压力水管，明管
exposed pipe 露天式管道，明管，明装管道
exposed rock surface 岩石出露面
exposed surface area 外露表面积
exposed surface 曝光面，暴露面，冲刷面，暴露表面，外露表面
exposed to weather 露天
exposed wiring 明线，外露布线
exposed 暴露的
expose to radiation 受照射
expose 暴露，曝光，陈列，不掩蔽，不加保护
ex post evaluation 事后评价，实施后评估
ex post investment 已实现的投资，事后投资
exposure albedo 照射反照率
exposure buildup factor 照射积极因子，照射积累因子
exposure cage 照射容器
exposure cavity 照射腔
exposure cell 照射室
exposure coefficient 暴露系数
exposure container 照射容器
exposure control 曝光控制
exposure dose rate 照射剂量率
exposure dose 辐射剂量，照射剂量
exposure draft 征求意见草案
exposure factor 地貌开敞度，曝光因子
exposure field 辐射场，照射场，照射野
exposure hazard 暴露危险，辐照危害
exposure hole 照射孔道
exposure meter 照射量计，曝光表，曝光计
exposure operations 暴露次数
exposure pathway 照射途径
exposure period 暴露周期
exposure rate constant 照射量率常数
exposure ratemeter 辐照率计，照射量率计
exposure rate 照射量率，辐照率，暴露率
exposure temperature 空晒温度
exposure time 辐照时间【射线照相】，老化时间，照射时间，暴露时间
exposure 暴露，曝光，空晒，曝光量，辐照，辐射，露头，裸露，曝露，照射（量）
expound 阐述，详细说明
express abrogation 明文废除
express analysis 快速分析
express consent 表明同意，明示同意
express delivery 快递
express fee 快递费
express goods/cargo 快件货物，快运货
express highway 高速公路
expression ability 表达能力
expressional algebraic expression 表达的代数表达式
expression of interest 表示兴趣
expression 表示，表达式，表达
express laboratory 快速实验室
express lift 快速电梯，快速吊机
express mail service 特快专递
express post 快递邮件
express provision 明文规定
express telegram 急电
express type machine 高速电机
express way road 特快车道
express way 高速公路，快车道
express 特快的，明确的，快运，快车，表示，表达
expropriation of land 征收地
expropriation 征收，征用
expulsion element 吹弧元件，灭弧室
expulsion fuse 冲出式熔丝，喷射式熔断器
expulsion gap 与冲击熔丝串接的灭弧间隙
expulsion of arc 弧的吹熄
expulsion protective gap 冲出式保护放电间隙
expulsion tube 管型避雷器
expulsion type arrester 吹弧型避雷器，吹熄型避雷器
expulsion type surge arrester 排气式避雷器，管式避雷器
expulsion type 冲出式【动力熔丝】
expulsion 排斥，喷溅，排出，推出
expulsive force 斥力
EXQ(ex-quay) 目的港码头交货（价）
ex rail 铁路边交货（价）
ex reactor experiment 堆外实验
ex seller's warehouse 卖方仓库交货价格
EXS(ex-ship) 目的港船上交货价格
ex ship terms 船上交货条件
ex ship （目的港）船上交货（价），船上交货价
exsiccator 干燥器
ex store 仓库交货（价）
ext. dia.（external diameter） 外径
external pressure loss 外部压力损失
extend beyond quarter point of beam 伸出梁的四分之一处
extend contract 延长合同
extended aeration 延时曝气
extended area service 扩大服务范围，额外服务
extended area 扩散区
extended bar rotor 深槽鼠笼转子
extended binary-coded decimal interchange code 扩充二-十进制交换码
extended bond 长键
extended capability 增加发电容量【故障运行时】
extended channel status word 可扩充的通道状态字
extended control 扩充控制
extended core storage 扩充的磁芯存储器
extended credit 展期信用证，附带预支条款的循环信用证
extended cycle 延长的周期
extended data set 扩充数据集
extended definition 扩展定义【功能约束】
extended delta connection 延伸三角形接线
extended diaphragm 加长隔膜【又叫插入式隔

膜】
extended end wall　扩建端墙
extended end　扩建端
extended furnace　前置炉膛
extended metal transistor　延伸金属晶体管
extended outage　第5类非计划停运【超过计划停运期限的延长停运】
extended precision word　扩展精度字
extended processing functions　扩展处理功能，扩展处理能力
extended range　扩展范围
extended reduced power operation　延长的降功率运行
extended root-blade　长脚叶根
extended service　长期运行
extended source　扩展源
extended surfaces　扩展表面
extended surface tube　鳍片管，肋片管，翅片管
extended surface　扩展受热面，添加受热面
extended tape operating system　扩充的磁带操作系统
extended term insurance　延长保险
extended tuned mass damper　扩展调谐质量阻尼器
extended valley　延长谷
extender　延长器，延伸板
extend flip-flop　扩充触发器
extending ladder　伸缩梯
extending control　伸长控制
extending sleeve　延长套管
extend the time limit　放宽期限
extend　伸长，延长，扩大，扩展，延伸，蔓延
extensibility　延展性，伸长率
extensible language　可扩充语言
extensible mark-up language　可扩展标记语言
extensible system　可扩充系统
extension advice　推广咨询
extensional vibration　拉伸振动，纵振动
extension arm　伸缩杆
extension bonnet　加长阀盖，延伸阀盖
extension cable　延长电缆，延伸电缆
extension circuit　增设电路，扩充电路
extension coefficient　（电机磁路中空气间隙的）增长系数，伸长系数
extension cord　延伸电线
extension end wall　扩建端墙
extension end　扩建端
extension project　扩建工程［项目］
extension fee　延期费
extension furnace　前置炉膛
extension ladder　伸缩梯，消防云梯，延长梯
extension links　延伸链【输电线，绝缘子】
extension meter　延长米
extension neck　延长颈
extension of power generation project　扩建发电工程
extension of short term record　短系列记录展延
extension of the L/C　延长信用证【指有效期】
extension of time　延期
extension of works　工程扩建
extension pipe　延伸管
extension project　扩建工程
extension ringer　备用铃流，分机振铃器
extension rod　伸缩尺
extension shaft　中间轴，延伸段【汽轮机】，中介轴
extension sliding tripod　伸缩三脚架
extension spring　拉簧
extension telephone　电话分机
extension test　伸长试验
extension tube　延长管
extension wire　延伸导线【热电偶】，补偿导线
extension works　扩建工作，扩建工程
extension　伸长，延长，外延，扩展，推广，扩建，扩大，扩充，调伸长度，电话分机，扩建部分，延长部分，伸展，延伸，发展，延期
extensive cultivation　大面积耕作
extensive grassland　大草原
extensive order　大批订货
extensive property　广延量
extensive quantity　外延量
extensive structure model　扩展的结构模型
extensive system　分支系统
extensive　粗放型，广泛的
extensometer　伸长计，延伸计，应变计
extent of agreement　协议范围
extent of contract　承包范围
extent of damage　受损程度，损坏程度
extent of examination　检验范围
extent of fuel burn-up　燃料燃耗深度
extent of investigation　调查范围
extent of power　权限
extent of reaction　反应进度，反应程度
extent　程度，限度，范围，距离
extenuation　减少，降低，衰减，低估
exterior concrete form　外部混凝土模板
exterior cooling　外部冷却
exterior drainage　外流水系，外排水
exterior finishing　室外装修
exteriority　外形，外表面，外界，外观
exterior ladder　外部梯子
exterior magnetic field　外在磁场
exterior missile　外部飞射物
exterior plaster　外粉刷
exterior sewer system　外部排水系统
exterior stairs　室外楼梯，外部楼梯
exterior structure　外部结构
exterior stucco　外墙拉毛粉刷
exterior trim　室外装修，外部装饰
exterior wall tile　外墙砖
exterior wall　外墙
exterior window　外窗
exterior　外部，外部的，外表的
external absorbent method　外吸收法
external access network　外部存取网络
external admission　外进汽
external aerodynamic　外流空气动力学
external aid　外部援助
external armature alternator　外电枢式同步发电机，旋转磁场式同步发电机
external armature circuit　电枢外电路
external assets　国外资产

external auditor 外部审核员
external audit 外部审核
external bearing 外部轴承【外侧或内侧】
external block effect 外部块效应
external breeding ratio 外增殖比
external cable 外部电缆
external calibration 外部校准
external calipers 外卡钳
external calling 外部呼叫
external cathode resistance 阴极输出器外电阻
external characteristic curve 外特性曲线
external characteristics 外部特性，外特性
external circuit 外回路，外部回路，外电路
external circulating system 外部循环系统
external cleaning 外部清洗
external combustion gas turbine 外燃式燃气轮机
external concrete vibrator 附着式混凝土振动器
external condenser 外凝汽器
external condition 外部条件
external conductive casing 导电外壳
external conductive coating 导电外涂层
external conductor 同轴布线系统中外层接地导体
external constitutive law 外构造定律
external contamination 外部沾污
external control 外部控制
external conversion ratio 外区转换比
external coolant pressure 外部冷却剂压力
external coolant recirculation 外部冷却剂再循环
external cooling circuit 外部冷却回路
external cooling 外部冷却，外冷
external coordination 对外协调
external corona 外电晕
external corrosion 外部腐蚀，烟气侧腐蚀
external crack 表面缝，外表裂纹
external current 外电路电流
external cyclone 外置式旋风筒，外置式旋风分离器
external deposit 外部沉积物，烟气侧沉积物
external device address 外部设备地址
external device code 外部设备代码
external device control word 外部设备控制字
external device control 外部设备控制
external device operands 外部设备操作数
external device response 外部设备应答
external device 外部设备
external diameter 外径
external diffusional resistance 外部扩散阻力
external dimension 外形尺寸
external discharge 外部放电
external dose 外照射剂量
external dosimetry 外照射剂量学，外照射剂量测定法
external economizer 外置式省煤器
external energy interrupter 外能灭弧室
external energy 外能
external environment 外部环境，外界
external evacuation route 室外疏散路线
external event detection module 外部事件检测模块
external event 外部事件
external examination 外部检查，表面检查
external excitation circuit 外部励磁回路
external exposure 体外照射
external factor 外界因素
external failure cost 外部损失成本
external fan 外部风扇
external fault 外部故障
external feedback magnetic amplifier 外反馈式磁放大器
external feedback 外反馈
external field influence 外部场影响
external field 外场，外部流场
external finance 外来资金
external financing 外部筹措资金
external finishes 外部面漆，外面光制，外面装修
external finish of buildings 建筑物的外部装修
external firing boiler 外燃锅炉
external fission product trap （高温气冷堆）裂变产物堆外收集阱
external floods 外界洪水
external flow 外部绕流
external flue 外烟道
external fluidized bed heat exchanger 外置流化床换热器
external forced circulation 外部强制循环
external force 外力
external friction 外摩擦
external funds 外来资金
external furnace 外置炉膛
external gas pressure cable 压气电缆
external gear 外齿轮
external gland 端部汽封
external grinder 外圆磨床
external guard vessel 外保护容器【增殖堆用】
external handling gantry 外部吊装龙门架【反应堆厂房】
external hazard 外部危害，外照射危害
external heat loss 散热损失
external heat source 外部热源
external heat 外来热
external impedance 外阻抗
external initiator 外部引发源【地震，洪水等】
external inspection 外部检查
external insulation 外绝缘
external interrupt status word 外部中断状态字
external interrupt 外部中断
external investment 国外投资
external irradiation hazard 外照射危害
external irradiation 外部照射，外辐照，外照射，外部辐照
externality 外部性
external label 外部标号
external leakage 外部泄漏
external lighting protection system 外部防雷系统
external load 外载荷，外部负载，外加负载，外部荷载，外加荷载
external loss 外损失
external lug 外部接头
externally applied load 外载
externally biased ring modulator 外偏置环形调

制器
externally blown flap 外吹式襟翼
externally caused contact chatter 外因所致接点震颤
externally fired boiler 外燃锅炉，外置炉膛锅炉
externally fired superheater 外燃式过热器
externally programmed computer 外程序控制计算机，外程序式计算机
externally reversible motor 双向启动可逆式电动机
externally stored program 外部存储的程序
externally temperature influence （仪表的）外部温度影响
externally ventilated machine 外通风电机，强迫通风电机
external magnetic circuit 外磁式磁路
external magnetic field 外磁场
external memory 外存储器，外部存储器
external mitre 外斜角接合
external mix oil burner 外混式油燃烧器
external moderated reactor 外部慢化反应堆
external moisture 外水分，外在水分
external neutron 外部中子
external observation 外部观测
external oil cooler 外部油冷却器
external oil line system diagram 外部油管路系统图
external operating ratio 外部操作时间比
external organization 外部机构
external oscillator 外部激振器
external overvoltage 外部过电压
external packing leakage 端汽封漏泄
external paint 外表油漆，外部油漆
external perpendicular 外部垂直
external phase 外连续相
external photoeffect 外光电效应
external photoelectric effect 外光电效应
external piping loop 堆外管道回路
external plaster 外粉刷
external plate impedance 板极电路阻抗，外屏极阻抗
external potential energy 外部势能
external power supply 外部动力源
external premise 外在前提
external pressure coefficient 外部压力系数
external pressure stress 外压应力
external pressure 外压力
external protection 外保护
external radiation exposure 外照射
external reactance 外电抗
external recirculation flow 外部再循环流动
external recirculation loop 堆外再循环回路
external regeneration high rate mixed bed 体外再生高速混床
external regeneration 体外再生
external remote monitoring system 外部远距离监测系统
external rendering 外墙抹灰
external reset 外部复位，外部积分
external resistance 外阻力，外电阻
external riveting 外（蒙皮）铆接
external rotor motor 外转子（式）电动机
external rotor reluctance motor 外转子式磁阻电动机
external rotor 外转子
external rubber 外摩擦橡皮
external safety feature 外部安全设施
external scale thermometer 外标式玻璃温度计
external seal 外部密封
external series gap 外串联间隙
external shield 外屏蔽
external short-term liabilities 对外短期负债
external signal 外部信号
external sleeve 外壳，外圈
external sorting 外排序，外部分类
external sources of impurities 杂质的外部来源
external source 外源，外电源
external stairs 室外楼梯
external stator 外部定子，外定子
external steam tracing 蒸汽外伴热
external storage 外部存储器，外存储器
external structure 外部结构，外部构造
external subroutine 外部子程序
external suction 外部吸力
external summing 外部加法
external superheater 外置式过热器
external superheating 外部过热，外过热
external surface 外表面
external symbol dictionary 外部符号字典
external temperature 环境温度
external terminal 外部接头，出线端子
external thermal efficiency 外热效率，相对有效效率
external thread 外螺纹
external tooth washer 外部带齿垫圈
external transportation 厂外运输
external treatment 外部处理，炉外（给水）处理
external trigger pulse 外触发脉冲
external trigger 外触发
external vane pump 外部叶水泵
external vibration 外部振动
external vibrator 表面振捣器，附着式振捣器，外部振捣器
external wall 外墙
external water pressure 外水压力
external water recirculation 外部水再循环
external wind load 外部风载
external window 外窗
external work 外功
external zone 外围地带
external 外观【复数】，外形，外部，外表，外部的，表面的
extern （汇编语言中的）伪指令
EXT(extract) 抽出
exthermic 放热的，发热的
extinction angle 消光角，消弧角
extinction coefficient 消光系数，吸光系数，衰减系数
extinction coil 消弧线圈
extinction efficiency 衰减效率
extinction factor 衰减系数

extinction of waves 波浪平息
extinction potential 消电离电位，消弧电位
extinction voltage 熄灭电压，熄弧电压
extinction 熄灭，消灭，消弧，熄火，消声，消光，衰减，灭绝，光消散
extinct volcano 死火山
extinct 绝迹
extinguisher 灭火器
extinguishier 灭火机
extinguishing agent 灭火剂
extinguishing coefficient 衰减系数
extinguishing coil 消弧线圈
extinguishing voltage 灭弧电压，熄弧电压
extinguishment 灭火
extinguish 熄灭，消除，压制
extortionate interest rates 不合理的利息率
extra 附加的，外加的，额外的
extra allowance 额外津贴
extra amount 额外量，过剩量
extra best quality 特等品
extra charges on heavy lift 起吊附加费
extra charges 额外费用
extra-coarse 超粗的
extracorporeal irradiation 离体辐照，体外辐照
extra cost 额外费用
extractable energy 可获能量
extract air shaft 抽风竖井
extracted condensate 抽取凝结水
extracted steam from turbine 汽轮机抽汽
extracted steam 抽气，抽汽
extracting oil 分离油
extract instruction 开方指令，析取指令
extraction agent 萃提剂
extraction back pressure turbine 抽汽背压式汽轮机
extraction check valve 抽汽逆止门，抽汽逆止阀
extraction condensing turbine 抽汽凝汽式汽轮机
extraction condensing 抽汽凝汽式【汽轮机】
extraction control valve 抽汽调节阀
extraction cycle 回热循环，抽汽循环，萃取循环
extraction display 录取显示器
extraction efficiency 萃取效率
extraction fan 排气风扇
extraction flow 抽汽流量
extraction gas turbine 抽汽式燃气轮机
extraction heater pipework 抽汽加热器管路系统
extraction heater 抽汽加热器，回热器，回热加热器
extraction jack 拔桩机
extraction line 抽汽管道
extraction lock 引出闸门【高温气冷堆装料机】
extraction method 萃取法
extraction non-return valve 抽汽逆止阀
extraction of pile 拔桩
extraction of root 求根
extraction opening 抽汽口
extraction point 抽汽点，抽汽口【汽机】
extraction pressure governor 抽汽压力调节装置
extraction pressure regulator 抽汽压力调节装置
extract pressure 抽汽压力
extraction press variation-power correction curve 抽汽压力变化对功率的修正曲线
extraction pump 抽汽泵，凝结水泵，抽取泵，抽气泵
extraction rate 浸取率，提取率
extraction regulating valve 抽汽调节阀
extraction separation 萃取分离
extraction shell pressure 抽汽室压力
extraction stage 抽汽级
extraction steam conditions 抽汽参数
extraction steam for processing 工业抽汽
extraction steam from turbine 汽机抽汽
extraction steam heating 抽汽加热
extraction steam pipe 抽汽管
extraction steam regulation 抽汽调节
extraction steam system 抽汽系统
extraction steam turbine 抽汽式汽轮机
extraction steam 抽汽
extraction stop and nonreturn valve 抽汽截止逆止阀
extraction stop valve 抽汽截止阀
extraction system 引出系统，引漏系统
extraction temperature 抽汽温度
extraction tool 拔出工具
extraction tube 引出管【高温气冷堆】
extraction turbine 抽汽式汽轮机
extraction valve 抽汽阀
extraction ventilation system 抽气式通风系统
extraction 抽汽，抽气，抽出，萃取，提取，排出，从数据库取出数据，取出，提炼
extractive corrosion 选择性腐蚀
extractive reaction 提取反应
extractive solvent 萃取溶剂
extractive 抽取物
extractor 提取器，取出装置，抽汽装置，分离器，拔桩机，抽出器，萃取器，脱水机
extract ventilation 抽气通风
extract 开方，萃取物，萃取，抽出，榨取
extra current 额外电流
extra design tension 超张拉力
extra dividend 额外股利，特别股息
extrados of arch 拱外圈
extrados radius 拱外弧半径，外拱圈半径
extrados springing line 拱外弧起拱线
extrados springing 拱背起拱线点
extrados 拱背线，外拱圈
extra duty 超负荷，过载，附加负载，加班，额外工作，额外税金
extra equipment 额外设备
extra extraction 额外抽汽
extra-fill 超填
extra-fine grain film 超微粒胶片
extra-fine grinding 超细研磨
extra-fine thread 特细牙螺纹
extra flexible cable 超软电缆
extra-heavy duty crane 特重型起重机
extra-heavy duty type 超重型，特重型，超重级
extra-heavy duty 超重级工作制，超重级
extra-heavy pipe 特强管
extra-heavy sheet 特厚（玻璃）板
extra-heavy 超重，加厚的

extra high-pressure mercury vapor lamp 超高压水银灯
extra-high pressure 超高压
extra-high tension unit 超高压设备
extra-high tension 超高压
extra-high voltage generator 超高压发生器，超高压振荡器
extra-high voltage transformer 超高压变压器
extra-high voltage 超高压，特高电压
extra-instruction 广义指令
extra-interpolation 超插入法
extra investment 额外投资
extra length 超长度
extra levy 额外索取
extra-light lift 特轻型升降机
extra limiter 附加限幅器
extra load bearing capacity 额外荷载量
extra load 额外荷载
extra long distance 超远程
extra long-range forecast 超长期预报
extra long stud 超长杆
extra loss 附加损失，额外损耗，附加损耗，额外损失
extra-low voltage lighting 超低压照明
extra-low voltage 特低电压
extraneous 外加的
extraneous ash 外在灰分
extraneous information 外加信息
extraneous interference 外来干扰
extraneous loading 附加荷载
extraneous noise 外来噪声，外部噪声
extraneous risk 附加险
extranet 外联网
extra neutron 额外中子
extraordinarily high temperature 超高温
extraordinary expense 临时开支，特别经费
extraordinary flood 特大洪水
extraordinary high tide 特大高潮
extraordinary meeting 非常会议
extraordinary payment 意外支出
extraordinary revenue 临时收入，非常收入，特殊收入
extraordinary scour 特大冲刷
extraordinary session 非常会议，特别会议
extraordinary storm 特大暴雨
extraordinary wave 非常波
extraordinary wind 异常风
extraordinary 临时的，非凡的，特别的
extra output order 附加输出指令
extra over price 额外加价
extra packing 特殊包装
extra pair 备用电缆对
extrapolated boundary 外推边界
extrapolated cut-off 外推截止电压
extrapolated height 外推高度
extrapolated length 外推长度
extrapolated power curve 外推功率曲线
extrapolated radius 外推半径
extrapolated range 外推射程，外推路程
extrapolated side 外推面
extrapolated value 外推值，外插值
extrapolate 外推，外插，推断
extrapolation distance 外推距离
extrapolation formula 外推公式
extrapolation ionization chamber 外推电离室
extrapolation length 外推长度
extrapolation method 外推法
extrapolation number 外推数
extrapolation of discharge curve 流量曲线外延
extrapolation of rating curve 水位流量关系曲线外延，率定曲线外延
extrapolation 外推法，外推的，外插法，推论，外插
extra price 附加价格
extra print order 附加打印指令
extra pseudo order 附加伪指令
extra pure reagent 超纯试剂
extra quality 特优质量
extra-sensitive clay 特敏黏土
extra-serious accident and failure 特大事故及破坏
extra slack running fit 松动配合，松转配合
extra-strength pipe 超强度管
extra strong 加强的，特强的，高强度的
extraterrestrial disposal 地外处置【放射性废物】，外层空间处置法
extraterrestrial irradiance 地球外辐照
extraterrestrial solar irradiance 地外［日射］辐照度，地外太阳辐射
extraterrestrial solar radiation 地外日射
extraterrestrial solar spectrum 地球外太阳光谱
extraterrestrial 地球外的，地球大气圈外的
extra thickness 超厚度【机加工】
extratropical cyclone 温带气旋
extratropical storm 温带风暴
extratropic belt 温带
extravation pipe 引出管
extra work 额外工作，加班
extremal point 极值点
extremal 极值的，极值曲线，极限值
extreme annual wind speed 年极端风速
extreme atmosphere event 极端大气现象
extreme climate 极端气候
extreme downsurge 最低下涌浪
extreme duty 极端工作状态
extreme emergency state （电力系统的）严重事故状态
extreme fibre stress 最外纤维应力，轮缘应力
extreme fibre 最外边的纤维
extreme gradient wind 极端梯度风
extreme high level 超高料位
extreme hot start 极热态启动
extreme hydrological year 非常水文年
extreme lifetime wind speed 寿命极端风速
extreme low tide 最低潮位
extremely high frequency 极高频，超高频
extremely high voltage 极高电压
extremely inverse current relays 极端反时限电流保护装置
extremely inverse relay 极端反时限继电器
extremely low frequency 极低频，超低频
extremely low-sag invar-reinforced aluminium conductor

极低弧垂钢芯铝绞线
extremely problem-oriented 极端面向问题的
extremely 极端地
extreme maximum temperature 极端最高温度
extreme maximum 极端最高
extreme mile wind speed 极端英里风速
extreme minimum temperature 极端最低温度
extreme position 极限位置
extreme pressure 极限压力
extreme return period （风速）极端重现期，极端回报期
extreme situation 极端条件
extreme surface wind 地面极端风
extreme temperature refractory 超高温耐火材料
extreme temperature 极端温度
extreme upsurge 极限上涌浪，最高上涌浪
extreme vacuum 超高真空
extreme value 极端值，极值
extreme wind speed 极大风速，极端风速
extreme wind 极大风，极端风
extreme 极端的，非常的，极限值，外项，极端，极限，极限的
extremum control system 极值控制系统
extremum principle 极值原理
extremum regulator 极值调节器
extremum 极值，最大值，最小值
EXTR＝extraction 抽取
EXTR＝extractor 抽汽器
extrinsic conductivity 外赋传导率
extrinsic contaminant 外来的夹杂，外来的杂质
extrinsic detector 非本征探测器
extrinsic evidence 附加证明，外来证明
extrinsic problem 外在问题
extrinsic semiconductor 非本征半导体，杂质半导体，含杂质半导体
extrinsic 含杂质的，有杂质的，非本征的
extrudable insulation 压挤绝缘注射式绝缘
extruded aluminium blade 挤压铝叶片
extruded graphite （各向异性）挤压石墨
extruded load 挤压载荷
extruded section 挤制叶型
extruded 压出的
extruder 压出机，挤压机
extrude 挤压，挤出，喷出
extruding insulation 挤压绝缘
extrusion flow 喷流
extrusion forming 挤压成型
extrusion stress 挤压应力
extrusion 挤压，冲出，突出，挤压成形
EXTR 抽取【DCS 画面】
EXTSN＝extension 延伸，扩展
exudation-pressure test 渗水压力试验
exudation 渗出（量），流出，析出，渗出物
exude 流出，发出，散发，渗出
exurban area 城市远郊地区
ex-venting 向外排气
ex-vessel fuel handling machine 容器外燃料装卸机
ex-warehouse 仓库交货价，仓库交货
EXW(ex works) 工厂交货，工厂交货价，出厂价
ex works price 工厂交货价格，出厂价
ex works 工厂交货（价），出厂价
eye bar 眼杆
eye base 眼基线
eye bolt and key 带插销螺栓
eye bolt 吊环螺栓，带眼螺栓，有眼螺栓，带孔螺栓，环首螺栓，螺丝圈
eye brows （燃烧器的）眉状结焦
eye-catching 引人注目的，醒目的
eye diameter 入口直径
eye estimation 目测，目估
eye fatigue 眼睛疲劳
eye hole 窥视孔，小孔
eye joint 眼圈接合
eyelet work 打孔眼
eyelet 小孔，窥视孔，孔眼
eye measurement 目测
eyenut 吊环螺母
eye observation 目测，目视观测，自测
eye of storm 风暴眼
eye of typhoon 台风眼
eyepiece collective lens 目镜聚光透镜
eyepiece screw 目镜螺丝
eyepiece 目镜
eyepoint 视点
eye protector 护目镜
eye reach 视野
eye ring 套环
eye rod 带环头拉杆
eye shield 防护眼镜
eye sight 观察孔，视野
eye sketch 目测草图
eye splice 索眼
eye visible crack 肉眼可见裂缝
eye wash unit 洗眼装置
eyewitness 目击者
E 安全壳【核电站系统代码】

F

façade wall 正面墙
fabric analysis 结构分析
fabric anisotropy 组构各向异性
fabricated bar 钢筋网，网格钢筋
fabricated building 装配式房屋
fabricated component 制造的部件
fabricated construction 装配式施工
fabricated element 预制构件
fabricated equipment 制造的设备
fabricated frame 制造的框架
fabricated language 人工语言
fabricated pipe bend 预制弯管
fabricated rotor 组合式转子
fabricated runner 组合式转轮
fabricated shaft 组合轴
fabricated structure 装配式结构
fabricated 制造的，加工制造的，装配式的
fabricate 建造，制作，装配，加工，捏造
fabricating yard 制作场，施工现场
fabrication and erection of cable tray 电缆桥架的配制与安装

fabrication cost　制造费用，制造成本，造价
fabrication drawing　制造图纸，制作图，加工图
fabrication procedure　加工工艺性
fabrication process　加工过程
fabrication sequence　制造顺序，生产顺序
fabrication technology　制造工艺
fabrication test specimen　制造试样
fabrication tolerance　制造公差，制造容差
fabrication tool　制作工具，制造模具
fabrication weldability　工艺可焊性，工艺焊接性
fabrication workshop　加工车间
fabrication　加工，制造，制作，装配，
fabricator　制造者，制造商，装配面
fabric collector　袋式除尘器，织物除尘器
fabric dust aspirator　织物集尘器
fabric filter　布袋除尘器，布袋（纤维）过滤器，织物过滤器
fabric form　织物模板
fabric roof　软屋顶
fabrics woven with cotton　棉织物
fabric　纤维，织物
facade　正面，正立面，立面，外观，建筑物正面
facade sealing　端面密封
face advance　工作面推进，工作面进度
face air velocity　工作面风流速度
face area　工作面【过滤网】，正过滤截面，过滤投影截面，迎风面积
face ashlar　表面琢石
face bend test　表面弯曲试验，正面弯曲试验
face brick　墙面砖，面砖
face-centered cubic crystal structure　面心立方晶体结构
face cleat　面割理
face coat　表面涂层
face crack　表面裂缝，表面裂纹
faced　饰面的
face end　工作面终点
FACE(field alterable control element)　现场可变的控制部件
face-fired boiler　对冲燃烧锅炉
face gear　平面齿轮
face grinding machine　平面磨床
face grinding　平面磨削
face guard　（过滤器的）个面保护网
face hammer　琢面锤
face-hardened　表面硬化的
face-hardening　表面淬火
face joint　出面接缝，表面接缝
face line of teeth　齿顶线
face mask　防毒面具，面罩
face mounted motor　端面安装式电动机
face mounting　端面安装
face of cut　切割面
face of screen　屏蔽面，荧光屏面
face of tool　切削面
face of weld　焊缝表面，焊接表面
face parallel cut　Y切割，平行面切割
face plate　（配电盘的）面板，卡盘，花盘，荧光屏
face pressure　表面压力
face reinforcement　表面加强，增强面
face runout　端面跳动
face seal type　表面密封型
face seal　端面密封
face seam　表面接缝
face shield　防护面罩，焊工面罩，（过滤器的）外面保护，电焊用手持面罩
face shovel　正铲挖土机
face slab　面板
face stone　面石
faceted collector　多反射平面集热器
face-to-face arrangement　纵向排列
face to face　面到面
facette　刻面，凸线，柱槽筋
face tube　皮托管
face-type motor　凸缘型电动机
facet　（多面体的）面，小平面，槽面，面
face value　票面价额
face velocity　投影面速度【过滤器】，罩口风速，迎风风速，迎面风速，迎风面流速
face veneer　表层装饰薄板
face ventilation　工作面通风
face width　表面宽度，齿宽
face work　抹面工作
face　表面，正面，外观，面向，端面，面，前面
FAC(file access channel)　文件存取通道
FAC(final acceptance certificate)　最终验收证书
FAC(final acceptance criteria)　最终验收标准
facies　相，岩相【工程地质中用】
facilitate　便利于
facilities layout　设施布置
facilities　设备，设施
facility cost　设备成本，设备费用
facility for automatic sorting and testing　自动分类及试验装置
facility for passage over dam　过坝设施
facility management system　设备管理系统
facility modernization　设备现代化，设施现代化
facility of repair　维修设备
facility organization　组织机构
facility power control　设备功率控制
facility power panel　设备电源板
facility reliability　设备可靠性
facility request　设备要求
facility　设备［施］，装置，机构【复数】，工具，生产或科研单位，方便，容易
facing bond　砌面
facing brick　面砖
facing concrete　面层混凝土
facing expansion joint　面层伸缩缝
facing joint　面层接缝
facing machine　镶面机
facing material　表面装饰料，护面材料
facing plate　护面板，花盘，卡盘
facing slab　镶面板
facing stone　护面石料，面石
facing tile　贴面板，贴面砖
facing-up　滑配合，对研，配刮
facing　表面加工，饰面，盖面层，朝向，面板，墙面，衬面，刮面，镶面，罩面
facsimile chart　传真图
facsimile receiver　传真接收机

facsimile signal 传真信号
facsimile system 传真机
facsimile telegram 传真电报
facsimile transmission 传真传输
facsimile transmitter 传真发射机
facsimile 传真
fact 论据
fact data base 事实数据库
FACT(factual compiler) 实在编译程序
fact finding mission 实况调查组
fact finding organ 实况调查机构
FACT(fully automatic compiler translator) 全自动编译翻译程序
FACT(fully automatic compiling technique) 全自动编译技术
factor affecting runoff 影响径流的因素
factorage 代理厂商，代理商业务，手续费，代理商佣金，佣金
factor analysis 因子分析，因素分析
factor capacity 功率，容量因子
factored load 已考虑荷载系数的荷载，系数荷载
factored resistance 设计风阻
factored wind load 设计风载
factor for overcapacity 过载系数
factorial design 析因设计
factorial experiment 析因实验
factorial 阶乘，因子的，阶乘的
factoring 因子分解，托收信贷行
factorization 因式分解，因子分解
factor of adhesion 黏附系数，黏着系数
factor of air resistance 空气阻力系数
factor of assurance of cable insulation 电缆绝缘保险系数
factor of assurance 安全系数，保险系数
factor of expansion 膨胀系数
factor of fatigue 疲劳系数
factor of foundation bearing capacity 地基承载系数
factor of inertia 惯性系数
factor of influence 影响因子
factor of merit 品质系数，优良因数，灵敏度
factor of overcapacity 超荷载因数
factor of quality 质量因数
factor of rigidity 刚性系数
factor of safety against overturning 抗倾覆安全系数
factor of safety 安全系数，保险系数，安全因子
factor of saturation 饱和度，饱和系数
factor of stress concentration 应力集中因数
factor of utilization 利用系数
factors to consider 需要考虑的因素
factory acceptance gauge 验收规，验收样板
factory acceptance test and demonstration 工厂验收试验和指示
factory acceptance test certificate 厂验证书
factory acceptance test 工厂验收测试，工厂验收试验
factory assembly 工厂装配
factory cost sheet 工厂成本单
factory-data collection 工厂数据处理，工厂数据收集
factory director 厂长
factory entrance examination 入厂考试
factory for prefabrication 预制厂
factory lighting 工厂照明
factory load 工厂负荷
factory-made components 工厂预制构件
factory noise 工厂噪声
factory-owned cogeneration power plant 工厂自备热电厂
factory planks 工厂加工的板材
factory sewage 工厂废水，工厂污水
factory special railway 工厂专用铁路线
factory steam 工业用汽
factory test 工厂试验
factory weld 厂内焊接，车间内焊接
factory 制造厂
factor 系数，倍率，率，因子，因数，因素，要素
FACTS equipment 灵活交流输电设备
FACTS(flexible AC transmission system) 灵活交流输电系统
factual basis 事实依据
factual compiler 实在编译程序
factual report 事实报告
factual survey 实情调查
facula 光斑
facultative anaerobic bacteria 兼性厌气菌
faculty 才能，全体从业人员
fade area 衰落区
fade out 渐弱，逐渐消失，淡出
fader 音量控制器
FAD (floating add) 浮点加
fading channel 衰落信道
fading control 衰落控制
fading effect 衰落效应
fading end point 褪色终点
fading frequency 衰落频率
fading peak 衰落峰值
fading region 衰落区，衰减区
fading 衰减，衰退，褪色，衰落
FA(failure alarm) 故障报警
FA(feeder automation) 馈线自动化
FA(final assembly) 最后装配
FA(fine ash) 细灰
FA(full arc) 全周，全周进汽
faggot dam 柴捆坝
faggot dike 柴堤
faggot 柴捆，柴束
Fahrenheit degree 华氏温度
Fahrenheit scale 华氏温标，华氏温度标尺
Fahrenheit temperature 华氏温度
Fahrenheit 华氏温度
FAI(fresh air inlet) 进气口
fail address 失效地址
fail as is operator 故障原状操作器
fail as is 断电时不改变原有位置，故障原状
fail category 失效类型
fail-closed operator 故障关闭操作器
fail-closed 出故障时自动关闭的，故障关闭的
fail close 出故障时自动关闭，故障时关闭

fail data 失效数据
failed element detection and location 破损元件的探测和定位
failed element detection 破损元件探测
failed element 破损燃料元件，破损元件
failed fuel detection and location 破损燃料元件探测和定位
failed fuel detection system 破损燃料探测系统
failed fuel detection 破损燃料探测，破损燃料元件探测
failed fuel detector 破损燃料探测器，破损燃料元件探测器
failed fuel element detection system 燃料元件破损探测系统
failed fuel element locator 破损燃料元件定位器，燃料元件破损探测器
failed fuel element monitor 燃料元件破损探测仪
failed fuel element 破损燃料元件
failing load 破坏荷载，破坏载荷
failing strength 破坏强度，断裂强度
fail in one's duty 失职
fail in negotiation 谈判失败
fail-last-position 故障最后位置
fail-locked operated 故障闭锁操作的
fail-locked 故障闭锁的
fail lockout 故障时闭锁
fail-opened 出故障时自动打开的，故障开启的
fail-open operator 故障开启操作器
fail-open 出故障时自动打开，故障时打开，故障时开启，故障开启
fail operational 故障后能操作的
fail-operator 故障安全操作器【故障开启或故障关闭】
fail-out 故障后果，故障结果
fail point 破坏点，失效点
fail-safe alarm 事故报警
fail-safe brake pressure 故障保护制动压力
fail-safe breakout point 保险冲破点
fail-safe condition 安全情况
fail-safe control 防止控制装置，失效保险控制
fail-safe default 故障保护默认值，安全默认值
fail-safe design 故障安全设计，安全设计
fail-safe device 故障时仍能安全运行的自动保护装置，故障自动防护装置
fail-safe disc brake 故障自动防护盘式制动器
fail-safe facility 安全装置，保险装置
fail-safe instrument 安全仪器，故障时仍能安全运行的仪器
fail-safe position 故障安全位置
fail-safe principle 故障时仍能安全运行原则
fail-safe program 故障安全程序
fail-safe structure 故障自动防护结构
fail-safe system 故障安全系统，元件有故障仍能工作的系统
fail-safe test （工作）安全性试验
fail-safe transformer 损坏后仍安全的变压器
fail-safety 故障安全性，系统可靠性，故障保险
fail-safe 故障安全的【故障开启或故障关闭的】，自动防故障装置的，可靠的，安全的，自动防故障装置，故障安全防护装置，故障安全，失效安全，失效保险，安全装置
failsafe 破损安全，失效保护，故障自动保险的
fail test 可靠性试验，故障测试
failure absorbent actuator 故障防护装置
failure accident 破坏性故障
failure against sliding 滑动破坏
failure alarming signal 事故报警信号
failure analysis report 失败分析报告，事故分析报告
failure analysis 失效分析，故障分析
failure by heaving 隆起破坏
failure by piping 管涌破坏
failure by plastic flow 塑流破坏
failure by shear 剪切破坏
failure by subsurface erosion 地下侵蚀破坏，下层土冲刷破坏
failure by tilting 倾斜破坏
failure cause data report 故障原因数据报告
failure cause 故障原因
failure checking 故障检查
failure condition 破坏状况，故障条件，破坏条件
failure correction system 故障校正系统
failure correction time 故障纠正时间
failure cost 故障成本，损失费用
failure crack 断裂纹
failure criteria 失效判据，失效标准，损坏标准，故障准则
failure criterion 故障判据，破坏判据，破坏准则
failure deformation 破坏变形
failure-density distribution 故障密度分布
failure-density 故障密度
failure detection system 破损探测系统
failure diagnosis 故障诊断
failure diagnostic time 故障诊断时间
failure due to fatigue 疲劳破坏
failure effect analysis 故障影响分析，故障后果分析
failure envelope 破断包络线
failure free operation 无故障运行，正常运行
failure free period 无故障工作期
failure-frequency distribution 故障频率分布
failure-frequency 故障频率
failure indication 故障指示
failure indicator 失灵指示器
failure in service 使用中的故障
failure in shear 剪切破坏
failure limit state 破坏极限状态
failure limit 破坏极限
failure load 破坏荷载，破坏载荷
failure logging 故障记录，失效记录
failure mechanism 故障机理，失效机理
failure mode analysis 破损方式分析
failure mode and effect analysis 故障模式与影响分析，故障模式与后果分析法
failure mode 故障模式，故障类型，事故形态，损坏形式，破坏模式，失效模式
failure monitor 故障检测仪
failure of a component 元件故障
failure of continuously required function 规定的连续功能失效
failure of contract 落空的合同

failure of earth slope　土坡滑坍，土坡坍毁
failure of oscillation　停止振荡
failure of response function　响应功能失效
failure of vital plant components　电厂重要设备损坏
failure plane　破坏面
failure prediction　故障预测，故障预报，破损预测，失事预测，失效预测
failure probability analysis　故障概率分析
failure probability density　故障概率密度
failure probability distribution　故障概率分布
failure probability　故障概率，失效概率，故障机率
failure-prone　易于发生故障的
failure propagation　事故扩大，事故蔓延
failure rate acceleration factor　故障率加速因数
failure rate test　故障率试验
failure rate　损坏率，故障率，衰减率，失效率，事故频率
failure ratio　破坏比
failure record　故障记录
failure recovery　故障排除
failure strain　破坏应变
failure strength　破坏强度
failure stress　断裂应力，破坏应力，破坏原应力
failure surface　破坏面
failure test　故障试验，可靠性试验，破坏性试验，破坏试验
failure to function　（功能）失效，失灵
failure to operate　保护装置拒动，拒动
failure to perform　未履行合同
failure to safety principle　故障安全原则
failure to trip　脱扣受阻，未能跳闸
failure tree　故障树
failure warning indicator　故障警告器
failure warning relay　故障警报继电器
failure warning　故障警告
failure wedge　破裂楔体
failure zone　破坏范围，破坏区
failure　故障，破坏，缺乏，失事，损坏，事故，失效，中断，停车，失败，缺少
faint　衰弱的，无力的，模糊的
fair and adequate compensation　公平合理的补偿
fair and reasonable price　公平合理的价格
fair and sound agreement　公平合理的协议
fair and sound relationship　公正合理的关系
fair competition　公平竞争
faircrete　纤维加气混凝土
fair dealing　公平交易
faired　流线型的，减阻的，整流片的
fairer fee　更公平的费用
fair-faced brickwork　清水混凝土勾缝的砖砌体
fair-faced concrete　光面混凝土，饰面混凝土，清水混凝土
fairfield radial stacking core　成捆皮带的卷芯
fairing cap　整流罩，导流帽
fairing spoke　流线型辐条，整流支柱
fairing　流线型罩，（风洞支架）风挡，减阻装置，整流罩，导流帽，整流片，流线型的，减阻的
fairlead　引线孔，引线管，导引片，导缆孔，导索板，导线管，导向滑轮
fairlight　门顶窗，气窗
fairness of fee structure　费用结构合理性
fairness　流线型，适当
fair use　合理使用
fairway　安全通航区，水路，通路
fair-weather corona limit　好天气时的电晕限值
fair-weather power loss　好天气时电晕功率损失
fair wind　顺风
fair　使流线型化，整流，博览会，公平，商品交易会，展览会
fait-safe operation　完全安全运行，故障时安全运行
fake and faulty product　假冒伪劣产品
fake　盘索，线圈，软焊条，云母质砂岩
FAK(freight all kinds)　均一包箱费率
falality rate　致死率
FAL(frequency allocation list)　频率分配表
fall-away　分开，散开，衰落，脱落，消失
fallback controller station　后馈控制器站
fallback mode　后馈模式
fallback state　低效运行状态
fallback　后馈，降落原地，回降原地，低效运行
fall belt　瀑布带
fall block　动索滑车
fall diameter　沉降球直径
fall-down test　跌落试验
fall down　落下
fall dust　落尘
fall flood　秋季洪水，秋汛
fall head　压头，落差，降落水头
fall increaser　增差设备
falling apron　水下堆石护坡，堆石防冲护坦，堆石下落护坡
falling ball impact test　落球冲击试验，落球碰撞试验
falling ball method　落球法
falling characteristic　下降特性
falling clearance　倒杆距离
falling cone method　沉锥法
falling current of air　气流下降
falling fouling　降落型污垢
falling gradient　坡度，下降坡度
falling head permeability test　变水头渗透试验
falling head permeameter　变水头渗透率仪，降水头渗透率仪
falling height　落差
falling into step　进入同步，同步
falling-in　滑坍，陷落
falling limb　过程线下降段，退水曲线
falling main　竖管
falling-off in speed　转速下降
falling of synchronism　失步
falling of water table　水位下降
falling out of step　失步
falling pond concept　降池原理
falling pressure　下降压
falling protection　下跌保护，坠落保护
falling segment　落洪段，退水段，消落段

falling sluice 跌落式泄水闸，自动泄水闸
falling sphere viscometer 落球式黏度计
falling tide 落潮
falling time 退水历时
falling torque 下降转矩
falling velocity 沉降速度
falling weight test 落锤试验
falling 下落，减退，凹陷，落下，降落
fall into step 进入同步，同步
fall in water level 水位降落
fall-in 进入（同步），一致，下沉，塌陷，坍方
fall line 瀑布线
fall of channel 沟道坡度
fall off on one wing 横侧失速
fall off 突然失速，降落，衰减，分离，排出，开始顺风，转向下风处
fall of pipe 管道坡度
fall of potential method 电位降法，电压降测阻法
fall of potential test 电压降测试法
fall of potential 电压降
fallout deposition 沉降物积存
fallout dose rate 散落物计量率
fallout front 沉降前锋，降水垂界
fall out of step 失去同步，失步
fall out of synchronism 失步，失去同步
fallout particulate 放射性沉降粒子
fallout pattern 降落型，沉降物分布型，落下灰分布形式
fallout sampling network 沉降物取样网
fallout shelter 降落掩蔽所
fallout wind 沉降风
fallout 非预期的收获，散落，放射性沉降（物），附带成果，放射性尘埃，沉降物，放射坠尘，散落物，（微粒）回降，（放射性）沉降
fallow board 模板
fall pipe 水落管
fall time 降落时间【脉冲波】，下降时间，衰减时间
fall trap 陷阱
fall tube 排水管
fall velocity 沉降速度
fallway 吊物竖道，楼面井
fall wind 下吹风，下降风，下坡风
fall 下降，落下，落差，落距，坡降，斜度，瀑布，坍落
false action 误动作
false air 漏风
false alarm of fire 假火警
false alarm 虚警告，错误警告，假报警，误报警
false anomaly 假异常
false arch 虚拱
false bottom bucket 活底料斗
false bottom 双层底，活动底板，假底
false ceiling 假平顶，假天花板
false contact 假接触，虚接触
false drop 假检索，误查
false echo 假回波
false ellipse arch 三心装饰拱
false equilibrium 假平衡
false exclusion 错误排除
false filter bottom 滤池假底
false firing 误触发，误燃
false floor （风洞）假地板，高架地板【上升面】
false ground 误接地
false jaw 虎钳口
false neutron 假中子，不稳中子
false ogive 整流罩，风帽
false operation probability 误动概率
false operation 错误动作
false quay 辅助码头
false settlement of a claim 假理赔
false signal 错误信号
false sort 假分类
false spar 假梁
false start 启动失误，错误启动，空载启动
false surface 辅助面
false triggering 错误触发
false tripping 误脱扣，假脱扣，误动，误启动
false wall （风洞）假洞壁
falsework 脚手架，临时支架，支撑，支柱，工作架
false zero method 虚零法
fame 声誉
family living quarters 家属宿舍
family of characteristics 特性曲线族
family of curve 曲线族
FAMOS（floating grate avalanche injection MOS） 浮动栅雪崩注入型金属氧化物半导体
famous and high quality product 名优产品
famous person 知名人士
famous product 名牌产品
famous trademark 驰名商标
fan and ring 风扇及风筒
fan antenna 扇形天线
fan assisted natural draft air cooled condenser 机力辅助自然通风（直接）空冷系统
fan beam 扇形射束
fan belt 风扇皮带
fan blade 风扇叶片，风机叶片，风叶
fan blower 扇风机，吹风机，通风机，风扇式增压器，送风机
fan casing 风机导流装置
fan cleavage 扇状劈理
fan coil air-conditioning system 风机盘管空气调节系统
fan coil unit 风机盘管机组，热风装置
fan coil 风机盘管
fan convector heater 风机对流加热器
fan convector 风机对流器
fan-cooled machine 扇冷电机，风冷电机
fan cooler 冷风机
fancy lump coal 大块煤，精选块煤
FANDACC(fan assisted natural draft air cooled condenser) 机力辅助自然通风（直接）空冷系统
fan delivery 风机送风量
fan drier 鼓风干燥机
fan drift 扇形堆积物
fan drive assembly 风机驱动部件
fan-driven generator 风力发电机
fan-driven turbine 风机透平，风扇涡轮

fan-duty resistor 具有风扇功能［特性］的电阻器
fan dynamometer 风扇式测功计
fan exhauster 排气风机
fang bolt 地脚螺栓，棘螺栓，锚栓
fanglomerate 扇砾岩
fan governor 风轮调速器，离心调速器
fan heater 风扇加热器
fan house 通风机室
fan housing 风机蜗壳
fan hub 风扇轮毂
fan inlet area 风机进口截面
fan inlet 通风机进气口，风机入口
fan-in 输入，扇入，输入端数
fanion （标识测量的）小旗，测量旗
fanless cooling system 无风扇冷却系统
fanlight catch 扇形窗插销
fanlight 窗楣，扇形窗，气窗，上亮子，腰窗
fan load 风扇特性负载
fan mill 风扇磨，风扇式磨煤机
fan motor 风扇电动机
fanning loss 鼓风损失
fanning out 扇形展开
fanning plume 成扇形烟羽
fanning strip 扇形端子板
fanning 扇形展开，扇形编组【电缆】
fan noise 风扇噪声
fan outlet area 风机出口截面
fan-out logic function 输出端数的逻辑作用，扇出逻辑函数
fan-out 扇出，输出，输出端，输出负载能力
fan performance curve 风机特性曲线
fan performance 风机特性，风机性能
fan pitch drive control 风机导叶驱动控制
fan power factor 风扇功率系数
fan room 风机室，通风机室，鼓风机室，通风机房
fan section 风机段
fan sewer system 辐射形下水道系统，扇形下水道系统
fan-shaped dock 扇形码头
fan-shaped fold 扇形褶皱
fan-shaped round 扇形炮眼
fan-shaped structure 扇状构造
fan shooting 扇形爆破
fantail burner 扇形火焰燃烧器
fantailroof 扇形屋顶
fantail 尾风轮，扇状尾，扇状
fan terrace 扇形阶地
fan-type stay cable 扇形拉索
fan-ventilated motor 风扇冷却电动机
fan ventilator 叶片式通风机，风扇通风机
fan window 通风窗
fan 扇，扇状物，风机，排风机，风扇，鼓风机，风箱
FARADA(failure rate data) 失效率数据，故障率数据
Faraday cage 法拉第笼
Faraday constant 法拉第常数
Faraday effect 法拉第效应
Faraday's law of induction 法拉第电磁感应定律
Faraday's law 法拉第定律，电磁感应定律
Faraday-type generator 法拉第型发电机
Faraday 法拉第
farad bridge 电容电桥
faradic current 感应电流，法拉第电流
faradic electricity 感应电，法拉第电
faradism 感应电流，感应电疗法，感应电治疗
faradmeter 法拉计
farad 法拉
faratsihite 铁高岭石
far back maximum thickness （翼剖面）靠后的最大厚度【位置】
fare 运费
FAR(failure analysis report) 故障分析报告，失败分析报告
FAR(Federal Acquisition Regulations) 美国联邦采购法
far field concentration 远场浓度
far field condition 远场条件
far field flow model 远场水流模型
far field flow 远场流动
far field holography 远场全息术
far field plume dispersion 远场羽流弥散
far field 远域【放射性烟云】，远端场，远场
far from mine power plant 无煤区火电厂
far future 远景的
fargite 钠沸石
far infrared radiation 远红外辐射
far infrared ray 远红外线
far infrared region 远红外区
far infrared 远红外
far leading dynamo 在线路远端的增压发电
farm building 农业建筑
farm distribution system 田间配水系统
farm electirfication 农村电气化，农业电气化
farming plume 扇形羽流
farmland 耕地，农田
farm pond 池塘
farm power plant 农村发电厂
farmstead 农场及其建筑物，农民住的房子，农庄
faroelite 杆沸石
far-reaching impact 深远影响
far-seeing plan 远景规划
FARS(Federal Acquisition Regulations System) 美国联邦采购法规体系
far wake 远尾流
fascia board 封檐板，挑口板
fascia 封檐板
fascicle 分册
fascine barge 柴捆船
fascine bundle 柴捆，梢捆
fascine contracting works 梢料束狭工程
fascine dam 梢捆坝
fascine dike 柴捆护堤
fascine dyke 柴笼堤坝
fascine foundation 柴排基础，梢料基础
fascine groin 梢捆丁坝
fascine groyne 梢捆丁坝
fascine layer 梢料填层
fascine mattress 柴排，柴排席

fascine revetment 柴排护岸，梢捆护岸
fascine roll 柴笼，梢笼
fascine weir 梢料堰
fascine whip 梢鞭，梢龙
fascine 柴捆，柴笼，梢料
FAS(Federation of Atomic Scientists) 原子科学家联合会【美国】
FAS(free alongside ship) 船边交货价格，船边交货，装运港船边交货
fashion parts 异型配件
fashion 方式，流行，时尚
fast 快速的
fast-access disk system 快速存取磁盘系统
fast-access memory 快速存取存储器
fast-access storage 快速存取存储器
fast-acting circuit 快速动作电路
fast-acting force-closure seal 速动加压封闭密封
fast-acting fuse 快速熔断，快速熔断器
fast-acting relay 快动作继电器
fast-acting spring actuator （闸阀）快动作弹簧驱动器
fast-acting 快速动作
fast address 固定地址
fast advantage factor 快中子有利因子
fast assembly 快中子组件
fast bi-directional switch 快速双向开关
fast breeder power station 快中子增殖反应堆电站
fast breeder reactor nuclear power plant 快中子增殖堆核电厂
fast breeder reactor 快中子增殖堆，快中子增殖反应堆
fast breeder 快中子增值堆
fast breeding reactor 快中子增殖反应堆
fast bubble region model 快气泡区模型
fast burst reactor 快（中子）脉冲反应堆
fast burst （燃料包壳）快速破裂，突然断裂【燃料包壳】
fast capture 快俘获，快中子浮获
fast ceramic reactor 快中子陶瓷（燃料）反应堆
fast chopper 快速断路器
fast closing 快关，速断
fast coincidence circuit 快速符合电路
fast coincidence 快符合
fast control rod insertion 控制棒快速插入
fast conversion 快（中子）转换
fast cooling 快速冷却
fast coupling 硬性联轴节，刚性联轴节，刚性联轴器
fast critical assembly 快中子临界装置
fast cut back 甩负荷保护，快速切负荷，快速切回，机组快速甩负荷
fast digital processor 快速数字处理机
fast diluted-fuel reactor 稀释燃料快中子反应堆
fast drain system 快速排放系统
fast drain （慢化剂）事故快速排放
fast electron 快电子
fastener nut 紧固螺帽
fastener plate 固定板
fastener 接线柱，线夹，钩扣，固定器，紧固件，皮带扣，扣件，连接物，夹头
fastening bolt 连接螺栓，紧固螺栓
fastening piece 连接件，紧固件
fastening pile 系定桩
fastening screw 紧固螺钉
fastening wire 绑扎用钢丝
fasten 紧固，接合，连接，卡紧，固定
faster exciter 快速励磁装置
fastest mile of wind 最大英里风速
fastest mile wind speed 最大英里风速
fastest mile wind 最大英里风
fastest mile 最大英里风速
fastest 最快速的
fast exponential experiment 快中子指数实验
FAST(facility for automatic sorting and testing) 自动分类及试验装置
fast film 快速胶片
fast fissionability 快中子致裂变性
fast fission capture 快中子裂变俘获
fast fission factor 快中子裂变系数［因子］，快中子增殖系数
fast fission ratio 快中子增殖比，快中子裂变比
fast fission reactor 快（中子）裂变反应堆
fast fission 快中子裂变
fast flange 固定法兰
fast fluidization 快速流化
fast fluidized bed 快速流化床
fast flux 快中子通量
fast flux test 快（中子）通量试验
fast Fourier transformer 快速傅里叶变换器
fast Fourier transform 快速傅里叶变换
fast fracture analysis 快速断裂分析
fast fracture 突然断裂
fast governor 快调速器
fast group 快中子群
fast hardening concrete 快硬混凝土
fastigium 屋脊，尖顶，山墙，高峰期
fast-insertion piston 快插活塞
fast ionization chamber 快速电离室
fast leakage factor 快（中子）泄漏因子
fast load control 快速负载控制
fast memory 快速存取存储器
fast motion gear 变速齿轮
fast-moving depression 快速移动的低气压
fastness 坚牢度，坚固性
fast neutron 快中子
fast neutron activation analysis 快中子活化分析
fast neutron activation method 快中子活化法
fast neutron breeding cycle 快中子增殖循环
fast neutron capture 快中子俘获
fast neutron collimator 快中子准直器
fast neutron cross section 快中子截面
fast neutron cycle 快中子循环
fast neutron diffusion length 快中子扩散长度
fast neutron dose equivalent 快中子剂量当量
fast neutron embrittlement 快中子脆化，快中子脆化效应
fast neutron fission increase rate 快中子裂变增加率
fast-neutron fission 快中子作用下的分裂，快中子裂变
fast neutron fluence 快中子注量

fast neutron flux 快中子通量
fast neutron generator 快中子发生器
fast neutron group 快中子群，快中子能组
fast neutron leakage 快中子泄漏
fast neutron non-leakage probability 快中子不泄漏概率
fast neutron range 快中子能区
fast neutron reactor 快中子反应堆，快堆，快中子堆
fast neutron region 快中子区
fast neutron response 快中子响应
fast neutron source 快中子源
fast neutron spectrometer 快中子谱仪
fast neutron spectrum 快中子能谱，快中子谱
fast-operated relay 快动作继电器
fast-operate fast-release relay 快吸快放继电器
fast-operate relay 快速动作继电器
fast-operate slow-release relay 快动缓释继电器
fast operation 快动作
fast period 短周期，快周期
fast pin hinge 紧销合页
fast plutonium reactor 快（中子）钚反应堆
fast power breeder 快中子动力增殖反应堆
fast pulley 固定滑轮，定滑轮，定轮，紧轮
fast reaction 快反应，快速反应
fast reactor core test facility 快堆堆芯试验装置
Fast Reactor Joint Committee 快堆联合委员会【英国】
fast reactor rocket 快中子反应堆火箭发动机
fast reactor 快堆
fast recovery diode 快速恢复二极管
fast recovery silicon rectifier 快速恢复硅整流器
fast-reflected 快中子反射的
fast relay 高速继电器
fast-release relay 快释放继电器
fast response instrument 快速响应仪表
fast response time 快速响应时间
fast response 快速响应
fast rinse 快速冲洗
fast rod insertion loop 控制棒快插回路
fast rod insertion （控制棒）快速插入
fast-safety channel 快速安全保护通道，快速安全保护电路
fast scan 快速扫描
fast scram 快速停堆，紧急停堆
fast-setting cement 快凝水泥
fast-setting concrete 快凝混凝土
fast-setting 快凝
fast shutdown 快速停堆，紧急停堆【反应堆】
fast start 快速启动
faststep bearing 立轴承
fast sweep 快速扫描
fast switch 快速开关
fast time constant 快速时间常数
fast-track team 催货队伍，快速清关小组
fast-track 快速通道，快速跟踪
fast-transfer pneumatic tube system 气动快传输系统
fast trouble logging system 快速故障记录系统
fast trouble recorder 快速扰动记录器
fast utilization factor 快中子利用因子
fast valve 快关阀，高速阀，快速动作阀
fast valving protection 调节（汽）阀快控保护
fast valving 调节汽门快控［快关，快动］
fat 脂肪，积余
fatal accident 人身事故，死亡事故
fatal dose 致命剂量，致死剂量
fatal error 严重错误
fatality rate 致死率
fatality 死亡事故
fat coal 高挥发分煤，肥煤，富黏土，油质煤
fat concrete 富混凝土
fate of particles 颗粒衰亡
fate 最终结果，最终去向
FAT(factory acceptance test) 工厂验收试验
fathometer sounding 测深仪测深
fatigue at high temperature 高温疲劳
fatigue auditory 听觉疲劳
fatigue behaviour 疲劳性能，疲劳特性
fatigue bending machine 弯曲疲劳试验机
fatigue breakdown 疲劳破坏
fatigue break 疲劳断裂
fatigue consumption indicator 疲劳消耗指示器
fatigue corrosion 疲劳腐蚀
fatigue crack growth rate test 疲劳裂纹扩展速率试验
fatigue crack growth rate 疲劳裂纹扩展速度，疲劳裂纹扩展速率
fatigue crack growth threshold 疲劳裂纹扩展门槛
fatigue crack growth 疲劳裂纹增
fatigue crack propagation 疲劳裂纹扩展
fatigue crack 疲劳龟裂，疲劳裂纹，疲劳裂缝
fatigue crescent 新月状疲劳痕
fatigue criterion 疲劳判据
fatigue curve 疲劳曲线
fatigue damage 疲劳损伤
fatigue data 疲劳实验数据
fatigue durability 耐疲劳性
fatigue effect 疲劳效应
fatigue endurance limit 疲劳持久极限
fatigue endurance 疲劳寿命
fatigue failure 疲劳破坏，疲劳断裂，疲劳损伤
fatigue fracture toughness 疲劳破坏韧性
fatigue fracture 疲劳断裂，疲劳断口，疲劳破裂
fatigue indicator 疲劳度指示器
fatigue life for P% survival P%存活率的疲劳寿命
fatigue life gage 疲劳寿命计
fatigue life 疲劳期限［寿命］，疲劳寿命
fatigue limit for P% survival P%存活率的疲劳极限
fatigue limit 疲劳极限，耐劳极限，持久极限，疲劳限度，疲劳寿命
fatigue loading 疲劳加载
fatigue mechanism 疲劳机理
fatigue notch factor 疲劳缺口系数
fatigue notch sensitivity 疲劳缺口敏感度
fatigue point 疲劳点，疲劳极限
fatigue precracking technique 疲劳预裂技术
fatigue range 疲劳限度
fatigue rating 疲劳额定值【材料】

fatigue ratio　耐久比，疲劳比，疲劳系数，疲劳应力比值
fatigue resistance　抗疲劳性，疲劳抗力，疲劳强度，疲劳阻力
fatigue rupture　疲劳断裂，疲劳破坏
fatigue strength at N cycles　N次循环的疲劳强度
fatigue strength for P% survival at N cycles　N次循环的P%存活率的疲劳强度
fatigue strength under oscillation stresses　振动疲劳强度
fatigue strength under reversed stresses　交变疲劳强度
fatigue strength　疲劳强度，疲乏强度
fatigue stress ratio　疲劳应力比值
fatigue stress　疲劳应力
fatigue striation　疲劳条纹
fatigue testing machine　疲劳试验机
fatigue testing　疲劳试验，疲乏试验
fatigue wear　疲劳磨损
fatigue　疲劳，使疲劳，疲劳度，疲乏
fatty acid　脂肪酸
faucet joint socket joint　套筒接合
faucet joint　套筒接头，承插式接头，管口接头，套管接头，套筒式接合
faucet　水嘴，龙头，承口，插口，开关，旋塞，水龙头，承插口【管子的】
fault alarm　故障报警
fault analysis　故障分析，事故分析
fault and events recorder　故障及事件记录装置
fault arc　故障电弧
fault basin　断层盆地
fault bench　断层阶地
fault block basin　断块盆地
fault block mountain　断块山
fault block　断裂地块，断块
fault breccia　断层角砾层
fault bus　故障母线
fault clay　断层黏土，断层泥
fault clearance time　故障消除时间
fault clearance　故障切除
fault clearing time　故障断开时间
fault clearing　故障消除，排除故障，抢修
fault cliff　断层崖
fault-closing capacity　故障关闭能力
fault coast　断层岸
fault conductivity　故障部位电导率
fault control stream　断层控制河流
fault control　事故监督
fault correction time　故障矫正时间，故障纠正时间
fault crevice　断层裂隙
fault current circuit breaker　故障电流断路器
fault current　故障电流，事故电流
fault detection on transmission line　线路故障探测
fault detection　探伤，缺陷探测，故障检测
fault detector　故障探测器，故障指示器
fault detect　故障检测
fault diagnosis　事故分析，故障诊断，事故诊断
fault dip　断层倾角，断层倾斜
fault display　故障显示
fault distance　故障距离
fault earthing　故障接地
fault earthquake　断层地震
faulted bedding plane　错动层面，错断层面
faulted condition　事故工况
faulted pole　故障极
faulted zone　断层带
fault effect　断层作用
fault finder　故障寻找器，故障探测器，探伤器，探伤仪
fault finding　故障查找
fault fissure　断层裂缝，断层裂纹，断层裂隙
fault-folding　断层褶皱
fault-free　无故障的，无缺陷的
fault gap　断层峡谷
fault ground bus　故障接地母线
fault grounding　故障接地
fault impedance　故障阻抗，短路阻抗
fault-indicating contact　故障指示电路触点
fault-indicating system　故障指示系统
fault indicator　故障探测器，探伤器，故障指示器
faultiness operation　带故障运行
faultiness　故障，缺陷
fault in feed　带故障馈电
faulting　断层错动，断层作用
fault in material　材料缺陷
fault inspection　故障巡视
fault-interrupting capacity　故障断开容量
fault isolation　故障隔离
faultless operation　无事故运行
fault level　故障率水平
fault liability　易出故障
fault line valley　断线谷
fault line　裂纹线，故障线，断层线
fault localization time　故障定位时间
fault localization　确定故障点，确定事故地点，故障定位，故障测距技术
fault localizing bridge　故障勘测电桥
fault-locating method　故障定位法
fault-locating technology　故障定位技术
fault-locating test equipment　故障位置测定装置
fault-locating test　故障定位测试
fault-locating Wheatstone bridge　寻找故障的慧斯顿电桥
fault-locating　故障定位
fault location gear　故障探测装置
fault location　故障定位，缺陷定位，故障部位，故障位置，故障位置测定
fault locator　故障定位器，故障探测器，故障定位装置，故障距离探测器，故障位置测定器
fault mountain　断层山
fault movement　断层错动，断层作用
fault operation　误操作
fault outcrop　断层露头
fault phase　故障相
fault pit　断层坑
fault plane　断层面
fault plate　带疵板
fault point　故障点
fault prevention　故障防止，防止事故，预防事故
fault progression　故障发展

fault protection device 故障防护装置
fault protection 故障保护，防止故障
fault rate 故障率，损坏率，失效率
fault rating 故障等级
fault recognition 故障识别
fault recollection and recording device 事故追忆记录装置
fault recorder 故障记录仪，故障录波器
fault record 故障记录
fault relay 故障继电器，事故继电器
fault resistance 故障电阻，故障部分电阻
fault ride through 故障穿越
fault rock 断层岩
fault scarp 断层崖
fault section 故障区段，障碍区间，障碍区段
fault shoreline 断层岸线
fault signal 故障信号
fault-slip cleavage 破劈理
faults of HVDC power transmission system 高压直流输电系统故障
fault space 断层间隔
fault state range 故障状态范围，异常状态界限
fault statistics 故障统计，事故统计
fault striation 断层擦痕
fault strike 断层走向
fault surface 断层面
fault switch 故障开关，故障模拟开关
fault system 断层系统
fault terrace 断层阶地
fault through 地堑
fault throwing switch 人工短路跳闸开关
fault throwing 人工短路跳闸
fault throw 断层落差，断距
fault time 故障时间，停机维修时间
fault tolerance performance 容错性能
fault tolerance system 容错系统
fault tolerance 故障容许水平，容错，容错性
fault-tolerant computer 容错计算机
fault-tolerant technique 容错技术
fault-tolerant 容错的，容许故障的
fault trace 故障跟踪
fault transient 故障瞬间，故障过渡过程，暂态故障
fault tree analysis 故障树分析
fault tree system 故障树系统
fault tree 故障树，故障树形图，事故树
fault trough 地堑，断层槽
fault type 故障类型
fault valley 断层谷
fault versus overload discriminator 故障过载鉴别装置
fault warning system 缺陷报警系统
fault wire 故障线
faulty circuit 故障电路
faulty coal 劣质煤
faulty component 故障件
faulty concrete 劣质混凝土
faulty fuel detection equipment 破损燃料包壳探测设备
faulty fuel detection system 破损燃料元件探测系统
faulty fuel location system 包壳破裂探测系统【燃料元件】
faulty goods 劣货
faulty indicator 探伤器
faulty insulation 有故障的绝缘，有缺陷的绝缘
faulty insulator detection 故障绝缘子探测
faulty insulator 故障绝缘子，缺陷绝缘子
faulty operation 误操作
faulty packing 有缺陷的包装
faulty phase 故障相
faulty section 故障段
faulty soldered joint 不良的焊接点
faulty state information 不定状态信息，无效故障信息
faulty switching 开关误操作
faulty transformer 出事故的变压器，发生故障的变压器
faulty unit 次品
faulty weld 有缺陷的焊缝
faulty 有错误的，不合格的，有故障的，无用的
fault zone 断层带
fault 故障，漏电，缺陷，误差，断裂，错误，断层，损伤，事故，毛病，缺点，瑕疵
fauna 动物区系
fauts group 断层群
favorable balance of trade 贸易顺差
favorable exchange 顺汇
favorable geometry 安全的几何形状，有利几何尺寸
favorable price 优惠价格
favorable terms 优惠条件，有利条件
favorable treatment 优惠待遇
favorable 优惠的
favor 有利于
favourable atmospheric condition 有利大气条件
favourable interference 有利干扰
favourable loan terms 优惠贷款条件
favourable pressure gradient 顺压梯度
favourable pressure interference 顺压差
favourable price 优惠价格
favourable terms 优惠条件
favourable treatment 优惠待遇
favourable 优惠的，有利的
favoured rate of credit 优惠贷款率
fawshmotron 微波振荡管
fax machine 传真机
fax 传真
faying surface 接合面，接触面，贴合面，密合面【实际或几乎接触】
fazotron 相位加速器
FBC(fluidized bed combustion boiler) 流化床锅炉，沸腾炉
FBC(fully buffered channel) 全缓冲通道
FB＝feedback 反馈
FB(fluidized-bed) 流化床
FBFM(fractional batch fuel management) 相对批量燃料管理
FBG(fractional breeding gain) 相对增殖增益
FBHE(fluidized-bed heat exchanger) 流化床换热器

FBH(flat bottom hole) 平底孔
FBR(fast breeder reactor) 快中子反应堆，快中子增殖堆
FBRNPP(fast breeder nuclear power plant) 快增殖堆核电厂
FBRNPS(fast breeder nuclear power station) 快增殖堆核电站
FBS data set 至少有一个截尾块的 FB 数据组
FBS(function breakdown structure) 功能分解结构，技术能力构成
FCA(free carrier) 货交承运人【指定地点】
FCAA(frequency control and analysis) 频率控制和分析
FCAN(full capacity above normal) 高于正常值的全容量
FCAW(flux cored arc welding) 药芯焊丝电弧焊
FCB(fast cut back) 甩负荷保护，快速切负荷，快速切回，快速减负荷，机组快速甩负荷
FCB(forms control buffer) 格式控制缓冲器
FCB(freight for class and basis) 基于商品等级和计算标准的包箱费率
FCBN(full capacity below normal) 低于正常值的全容量
FCB 甩负荷【DCS 画面】
FCC(fluid cracking catalyst) 流化床催化裂化剂
FCDA(function constrained data attribute) 功能约束数据属性
FCD(first concrete date) 浇筑第一罐混凝土日
FCD(function control diagram) 功能控制图
FCDR(failure cause data report) 故障原因数据报告
FCE(forced circulating evaporator) 强制循环蒸发器
FC(fail close) 故障关，故障时关闭
FC(fast closing) 快关
FCFC(full coverage film cooling) 全汽膜冷却
FC(fictitious carry) 假进位
FC(film cooling) 膜式冷却
FC(flow control) 流量控制
FC(foot-candle) 呎烛光【照明单位】
FC(for construction) 按文件作施工准备
FC(functional code) 操作码
FCL(feedback control loop) 反馈控制回路
FCL(full container load) 整箱货
FCM(fiber composite materials) 纤维复合材料
FCN(field change notice) 现场变更通知
FCO(field change order) 现场更改单
F-coupler F 形钩
FCPLG(fluid coupling) 液力耦合器
FCR(field change request) 现场修改申请
FCR(full core reserve) 一炉料储备
FCS(feedwater control system) 给水控制系统
FCS(field bus control system) 现场总线控制系统
FCS(freight for class) 基于商品等级的包箱费率
FCTA(federal capital territory administration) 尼日利亚联邦首都特区管理局
FCT(Federal Capital Territory) 联邦首都特区【尼日利亚阿布贾地区】
FCU(function conversion unit) 函数转换部件
FCV(flow control valve) 流量控制阀
FDBK＝feedback 反馈
FDC(floppy disk controller) 软盘控制器
FDE (failure detection and estimation) 故障探测与诊断
FDF (forced draft fan) 送风机
FD(floor drain) 运转层排水
FD(forced draft) 强制通风
FD(frequency doubler) 倍频器
FD(fuel delivery) 燃料交付
FDF 送风机【DCS 画面】
FDM(frequency division multiplexing) 频分多路转换
FDP(fast digital processor) 快速数字处理机
FDP(file definition processor) 文件定义处理程序
FDR＝feeder 给料机
FDR(Fortschrittlicher Druckwasser Reaktor) 改进型压水堆【德国】
FDS(fast-access disk system) 快速存取磁盘系统
FDS(floor drain system) 地面疏水系统，运转层排水系统
FDS(function-distributed system) 功能分布系统
FDT(floor drain tank) 地面疏水箱
FDV(floating divide) 浮点除（法）
FDW(feed water) 锅炉给水
FDWS(feed water system) 给水系统
feasibility analysis 可行性分析
feasibility appraisal 可行性评估［评价］
feasibility evaluation 可行性评价
feasibility investigation 可行性调查研究
feasibility proof 可行性论证
feasibility reliability check 可行可靠性检验
feasibility report 可行性报告
feasibility study report 可行性研究报告
feasibility study 可行性研究，可行性调查，可行性论证【技术经济论证】
feasibility 可行性，可以实现性
feasible 可实行的，有可能的，可行的
feather direction 顺桨方向
featheredge brick 楔形砖
feathered pitch 顺桨桨距
feathering vane runner 斜翼式转轮
feathering vane 斜翼叶片
feathering 顺桨
feather joint 羽状节理
feather key and keyway 滑键与键槽
feather key 导向键，滑键
feather position 顺桨位置
feathery bainite 羽毛状贝氏体
feathery microshrinkage 羽毛状显微疏松
feather 滑键，凸起部，铸造毛刺，羽毛
feature size 形体尺寸
feature 特征，特性，部件，装置，特点，特色，外貌，细节
FEB(function electronic block) 电子功能块
FEDAL(failed element detection and location) 破损元件的探测和定位
Federal Aviation Administration 美国联邦航空管理局
Federal Water Pollution Control Act 联邦水污染控制法【美国】
federation 联合会
fee 费用

feeble current 弱电流
feeble field 弱场
feed adjustment 调料，进料调整
feed and bleed procedure 补水与排水程序
feed and bleed 补水与排水
feed apron 裙板进料机
feed auger 螺旋加料器，螺旋式输送机
feedback adjustment 反馈调整
feedback admittance 反馈导纳
feedback amplifier 反馈放大器
feedback automation 反馈自动化
feedback bellows 反馈波纹管
feedback branch 反馈分支，反馈回路
feedback channel 反馈通道，反馈信道
feedback characteristics 反馈特性，反馈特征
feedback circuit 反馈电路，反馈回路
feedback coil 反馈线圈
feedback communication 反馈通信
feedback compensation 反馈补偿
feedback component 反馈元件
feedback condenser 反馈电容器
feedback-controlled impedance test 反馈控制阻抗试验
feedback control loop 反馈控制回路
feedback control signal 反馈控制信号
feedback control system 反馈控制系统
feedback control 反馈控制，反馈调节
feedback coupling 反馈耦合
feedback decoding 反馈译码器，反馈解码器
feedback diode 反馈二极管
feedback effect 反馈效应，反馈效用
feedback element 反馈元件
feedback encoding 反馈编码
feedback envelope 反馈包迹
feedback excitation 反馈励磁
feedback factor 反馈系数，反馈因子，反馈因素
feedback filter 反馈滤波器
feedback fraction 反馈部分
feedback gain 反馈增益
feedback impedance 反馈阻抗
feedback information 反馈信息
feedback inhibition 反馈抑制
feedback integrator 反馈积分器
feedback limiter 反馈限幅器
feedback linkage 反馈联杆
feedback loop 反馈环路，反馈回路
feedback mechanism 反馈机构
feedback method of stabilization 用反馈法稳定
feedback method 反馈法
feedback on design quality 设计质量反馈，设计回访
feedback oscillator 反馈振荡器
feedback path 反馈通路，反馈路径
feedback phase 反馈相
feedback pilot valve 反馈滑阀
feedback positive 正反馈
feedback potentiometer 反馈电势计
feedback ratio 反馈比
feedback sawtooth generator 反馈锯齿波发生器
feedback servomechanism 反馈伺服机构
feedback shift register 反馈移位寄存器
feedback signal 反馈信号
feedback suppressor 反馈抑制器
feedback system 反馈系统
feedback time 反馈时间
feedback transducer 反馈转换器，反馈传感器
feedback transfer function 反馈传递函数
feedback transfer locus 反馈转移函数轨迹图，反馈传递函数根轨迹
feedback transformer 反馈变压器
feedback type current transformer 反馈式电流互感器
feedback voltage 反馈电压
feedback winding 反馈绕组
feedback 反馈，成果，资料，回复
feed belt conveyor 皮带送料机
feed belt 皮带给料机，送料带
feed box 给水槽【汽包】，装料斗
feed breed concept 投料增殖原理
feed breed cycle 投料增殖循环
feed breed 投料增殖
feed cable 馈电电缆，电源电缆
feed check valve 给水逆止阀
feed chute 进料管，进煤管，送料斜槽，给料溜槽
feed circuit 馈电电路，给水回路
feed controller 反馈调节器，给水调节器
feed control 馈给控制
feed conveyor 给料机
feed current 板极电流的直流分量，馈电电流
feed distributor 料液分配器，给水分配器
feed element 投料元件
feeder automation 馈线自动化
feeder bay 馈线间隔
feeder box 分线箱，馈电箱，电源箱
feeder breaker 馈线断路器
feeder bus-bar 馈电母线
feeder bus 馈电母线
feeder button 馈电按钮
feeder cable 馈电电缆，馈电线
feeder chute 给煤机滑槽
feeder circuit-breaker 馈线断路器
feeder circuit 馈电电路
feeder contactor 馈线接触器
feeder conveyer 皮带给料机
feeder current 补流
feeder drop 馈路电压降，馈线电压降
feeder floor 给煤机平台
feeder hundreds dial 进给器百分度盘
feeder line 馈电线，馈线，支线，馈电线路
feeder lock 装料机闭锁，进料机闭锁
feeder main 馈电总线，馈电干线
feeder panel 馈电盘，馈电板
feeder pan 给料机槽板
feeder pillar 馈电杆，地线分支箱
feeder pipe 进口跨接管，进料管
feeder pitch 加料器间距
feeder preselection 馈电线路预选
feeder protection 馈线保护
feeder reactor 馈电线路电抗器
feeder roll 供线卷
feeder station 进料机，给料站

feeder switch 馈路开关
feeder system 供电系统，供料系统
feeder tens dial 进给器十分度盘
feeder terminal unit 馈线终端单元
feeder tube 供水管【进口集箱】
feeder tunnel 给料机地槽
feeder units dial 进给器分度盘
feeder voltage regulator 馈线电压调整器
feeder weigher 称料给料器
feeder 给料机，给煤机，馈线，进给器，馈电电路，馈电线，供电户，送料器
feedforward controller 前馈控制器
feedforward control system 前馈控制系统，前馈调节系统
feedforward control 前馈控制
feedforward signal 前馈信号
feedforward 前馈
FEED(front end engineering design) 即前端工程设计合同
feed fuel 给燃料
feed gas pipeline （供）送料气管线
feed gas 原料气
feed gear 送料装置
feedhead 进料口【铸造】
feed head 馈给压头【水、油等】，浇灌突出口，冒口，补给头，进料口，进料头
feed heater 给水加热器，进料加热器
feed heating system 给水加热系统
feed hopper 供料斗，送料斗
feeding area 加料区
feeding branch to radiator 散热器供热支管
feeding conduit 供水道
feeding current 馈电电流
feeding equipment 给料设备
feeding mechanism 进给机构
feeding point 馈电点
feeding pump 送料泵，给料泵
feeding rate 给料量
feeding reservoir 蓄水池
feeding transformer 电源变压器
feeding 给料，馈电，供电，供给，输给，送料
feed injector 给料注射器
feed-in tariff 上网电价
feed-in 送进，输入，馈入，淡入
feed kenel 燃料芯核
feed line 给水管路，给水管线，馈电线，供给线，馈线
feed main 给水母管，给水干管
feed material 加入物料，馈给料，进料
feed mechanism 进给机构，送料机构
feed motor 供给电动机
feed nozzle 进气喷嘴
feed of one size 单一粒度进料
feed of wide size distribution 宽粒度分布给料
feed pan 给料槽
feed pipe 锅炉给水管道，给水管，进料管，供水管
feed plant 核燃料生产工厂
feed point impedance 馈电点阻抗
feed point 给煤点【沸腾炉】，供电点，给料点
feed port 加料口
feed pot 加料罐
feed pump capacity 给水泵出力
feed pump turbine 给水泵汽轮机
feed pump 给水泵，供水泵
feed rate indicator 进给率指示器
feed rate 加料率
feed regulating valve 进给调节阀
feed regulation station 给水调节站
feed regulator isolator 电源调整器，隔离器
feed regulator 供水调节装置
feed screw 螺旋给料机，丝杆，进给螺杆
feed size 加入物料的大小，给料粒度，进料粒度
feed steam 进气
feedstock pipeline （供）送料管线
feedstock storage 原料库
feedstock 原料，入料，添加原料
feed system 馈电系统，供电系统，给水系统，进给系统，投料系统，送料系统
feed tank 给水箱，供料箱
feed through capacitor 旁路电容器
feed through connector 传输用的电连接器，直接插头座
feed through signal 馈通信号
feed through terminal 引线端
feed through wiring 馈通布线
feed through 馈通，连通线，馈穿，穿通，引线，连接线
feed tube 装料管
feed unit 供给装置
feed valve 供给阀【水，气】，给水阀
feed-voltage modulation 馈压调制
feed water and condensate plant 给水和凝结水设备
feed water brake 给水流量控制阀
feed water by-pass 给水旁路
feed water chemical sampling system 给水取样系统
feed water chemistry 供水化学
feed water cock 给水龙头
feed water condition 给水品质
feed water connection 给水引入，给水管接头
feed water control system 给水控制系统
feed water control valve 给水调节阀
feed water control 给水控制
feed water cycle 回热循环，给水系统，给水循环
feed water deaerator 给水除氧器
feed water distribution orifice 给水分布孔
feed water distribution ring 给水分配环，给水配水环
feed water equipment 供水设备，给水设备
feed water filter 给水过滤器
feed water flashing 给水汽化
feed water flow control system 给水流量调节系统
feed water flow control 给水流量控制
feed water flow 给水流量
feed water heaters drain recovery system 给水加热器疏水回收系统
feed water heater 给水加热器，再生加热器，

给水预热
feed water heating heat recovery combined cycle 给水加热回热联合循环
feed water heating loop 给水加热回路
feed water heating system 给水加热系统
feed water heating 给水加热
feed water individual bypass valve 给水小旁路门
feed water injector 给水喷射器
feed water inlet header 给水进口集管
feed water inlet nozzle 给水进口接管，给水进口管嘴
feed water inlet valve 给水入口阀
feed water inlet 给水进口
feed water intake 给水入口
feed water isolation valve 给水隔离阀
feed water line 给水管路，给水管线
feed water main 给水母管
feed water make-up 补给水，补水
feed water manifold 给水汇流管
feed water operating stand 给水操作台
feed water overall bypass valve 给水大旁路门
feed water penetration isolation valve 给水贯穿管隔离阀
feed water pipeline 给水管道
feed water pipe 给水管
feed water piping 给水管道
feed water plant 给水设备，给水站
feed water pressure loss 给水压力损失
feed water pump turbine drain system 主给水泵汽机疏水系统
feed water pump turbine electr-hydraulic control system 主给水泵小汽轮机电液控制系统
feed water pump turbine gland system 主给水泵汽机轴封系统
feed water pump turbine lubrication and control fluid system 主给水泵汽机润滑及调节液系统
feed water pump 给水泵
feed water quality 给水品质，给水质量
feed water recirculating system 给水再循环系统
feed water regulator 给水调节器
feed water ring manifold 给水环形汇流管
feed water ring with inverted J-tubes 倒J型管给水环
feed water ring 给水环【蒸汽发生器】，给水环管
feed water softening 给水软化
feed water sparger ring 给水分配环，给水配水环
feed water sparger 给水喷淋器
feed water storage tanks 给水贮水箱
feed water system 给水系统
feed water tank and gas stripper system 给水除气器系统
feed water tank 给水箱
feed water temperature rise 给水温升
feed water temperature 给水温度
feed water treatment 给水处理
feed water valve 给水阀
feed water 给水，供水，二回路水，进料水
feed way 给料方式
feed well 给水井，给料口，供料口
feed wheel 进给轮
feed wire 馈电线
feed worm 螺旋给料机
feed 馈，供给，馈给，供电，馈电，给料，给水，加料，进料，装入，注入，给……提供
FEE(fast exponential experiment) 快中子指数实验
fee for compensation of crops 青苗赔偿费
fee for confirmation 保兑费
fee for removal of houses 房屋迁移费
fee for the compensation of crops 青苗赔偿费
feeler blade 塞尺，测隙片
feeler gage 厚薄规，塞尺
feeler gauge 塞尺，厚薄规，测厚规，测隙规
feeler lever 触杆，探杆
feeler pin 探针
feeler 塞尺，测隙规，厚薄规，探针，探头，触头，触探针
fee resistant 难燃的
fees 检验费
feet head 英尺压头
feet per minute 英尺每分
feet sheet 情况表
Fe(ferrum) 铁
FE(field engineer) 现场工程师
FE(flow element) 流量元件
FE(format effector) 格式控制符
feigh 废渣
felder 镶嵌块
feldspar 长石
feldspathic grit 长石粗砂岩
feldspathic sandstone 长石砂岩
feldspathic shale 长石质页岩
feldspathic 长石质
feldspathization 长石化（作用）
FELIX(fusion electromagnetic induction experiment) 聚变电磁感应实验装置
Fellenius method of slices 费伦纽斯条分法
felsite 致密长石
felt and gravel roof 油毡撒绿豆砂屋顶
felt metal disk 金属毛毡片
felt plug 毡塞【控制棒】
felt 油毛毡，毡
female connector 孔插座，阴连接管，插入式连接器，母连接器，连接器插口
female face 凹面
female-female angle 阴-阴角
female-female globe 阴-阴球【阀】
female-female 阴-阴
female fitting 内螺纹
female flange 凹法兰，阴法兰
female joint 插承接合，套筒接合
female multipoint connector 多点母连接器，多点连接器插口
female rotor 凹形转子
female screw thread 内螺纹
female screw 阴螺旋，螺母
female thread 内螺纹，阴螺纹
female union 管内接头
FEM(finite element method) 有限元法
feminine 阴性
femto second 飞秒，毫微微秒

femto volt 毫微微伏
femto 飞，毫微微，微型基站
fence rail 栏杆，棚栏
fence wall 围墙
fence 挡板，栅栏，篱，篱笆，围栏，围墙
fencing wall 护墙，围墙
fender beam 护舷木
fender block 缓冲块
fender board 翼子板，挡泥板
fender log 护木
fender pier 防冲突堤，护墩，码头防冲桩，防护桩，防御桩，护桩，靠船墩
fender post 防护柱
fender skirt 翼子板裙板
fender structure 围护结构
fender support 翼子板支架
fender wall 防护墙
fender 防撞桩【码头】，防护板，挡泥板，护舷材，缓冲材，缓冲装置，排障器，翼子板
fending groin 防冲堤，防护堤
fending groyne 防护堤
fenland 干沼泽地
fen 沼地
FEP(fluorinated ethylene propylene) 氟化乙丙烯
FEP(front end processor) 前端处理机
FERC(Federal Energy Regulatory commission) 联邦能源管理委员会【美国】
FERD(fuel element rupture detection) 燃料元件破损探测
FERD(fuel element rupture detector) 燃料元件破损探测器
fermentation 发酵
Fermi age theory 费米年龄理论
Fermi age 费米年龄
Fermi energy 费米能
fermitron 微波场射管
fermi 费米【长度单位】
fern 蕨
ferractor 铁淦氧磁放大器，铁电磁振荡器
ferramic core 铁氧体磁芯
ferramic 铁氧体的，粉末状的铁磁物质
Ferrari ridge 法拉利脊，鸭尾脊
ferrate （高）铁酸盐
ferreed relay 铁簧继电器
ferreous 含铁的
ferric alloy steel pipe 铁合金钢管
ferric carbide 碳化铁
ferric cement 高铁水泥
ferric chloride 三氯化铁，氯化铁
ferric citrate 柠檬酸铁
ferric ferricyanide 铁氰化铁
ferric hydroxide 氢氧化铁
ferric induction 铁磁感应
ferric iron 三价铁，正铁，高铁离子
ferric oxide 氧化铁，三氧化二铁
ferricpozzolan cement 铁质火山灰水泥
ferric sesquioxide 三氧化二铁
ferric sulfate 硫酸铁
ferric sulfide 硫化铁
ferric sulphate 硫酸铁
ferricyanide 铁氰化物
ferric 三价铁的，正铁的，含铁的
ferriferous oxide 四氧化三铁
ferriferous substrate 铁基胶片，铁基衬底
ferriferous 正铁亚铁的，含铁的，产生铁的
ferrimagnetic limiter 铁磁限制器
ferrimagnetic 亚铁磁的，铁淦氧磁体
ferrimagnetism 亚铁磁性，铁氧体磁性
ferrimag 一种铁磁合金
ferristor 自饱和磁放大器，铁磁电抗器
ferrite aerial 铁氧体天线
ferrite bolometer power meter 铁氧体测热电阻微波功率计
ferrite cast iron 铁素体铸铁
ferrite coil 铁淦氧磁芯线圈
ferrite core matrix 铁氧体磁芯矩阵
ferrite core memory 铁氧体磁芯存储器
ferrite core sensor 铁氧体磁芯传感器
ferrite core 铁氧体磁芯
ferrite-excited machine 铁素体励磁电机
ferrite film 铁氧体薄膜
ferrite magnet field 铁磁磁场
ferrite magnet 铁淦氧磁铁
ferrite measuring instrument 铁素体测量仪
ferrite memory 铁氧体存储器
ferrite microwave detector 铁氧体微波探测器
ferrite permanent magnet 铁氧体永磁体
ferrite phase modulator 铁氧体调相器
ferrite phase shifter 铁氧体移相器
ferrite-rod antenna 铁氧体棒形天线
ferrite storage 铁氧体存储器
ferrite switch 铁氧体开关
ferrite toroid 铁氧体磁环
ferrite 铁素体，铁淦氧，铁氧体，亚铁酸盐，纯铁体
ferritic 铁素体的
ferritic cast iron 铁素体铸铁
ferritic electrode 铁素体钢焊条
ferritic heat-resistant steel 铁素体耐热钢
ferritic spheroidal graphite cast iron 铁素体球墨铸铁
ferritic stainless steel 铁素体不锈钢
ferritic steel 铁素体钢
ferrito-martensitic 贝氏体
ferroacoustic storage 铁声存储器
ferro-alloy 铁合金
ferrocart coil 纸卷铁粉芯线圈
ferrocart 纸卷铁粉芯
ferro-cement construction 钢丝网水泥结构
ferro-cement grid plate 钢丝网水泥板网板，钢丝网水泥格子板
ferro-cement panel 钢丝网水泥板
ferro-cement 钢丝网水泥
ferroceramic-magnetic element 铁陶瓷磁性元件
ferroceramic-magnetic 铁陶瓷磁的
ferrochrome 铬铁合金
ferroconcrete tower 钢筋混凝土塔架
ferroconcrete 钢筋混凝土，钢骨水泥
ferrocrete 速凝水泥，快硬水泥
ferrocyanide 亚铁氰化物，氰亚铁酸盐
ferro-dynamic relay 铁磁电动式继电器
ferrod 铁磁杆

ferroelectric ceramics 铁电陶瓷
ferroelectric condenser 铁电电容器
ferroelectric memory matrix 铁电存储矩阵
ferroelectric memory 铁电存储器，铁氧体存储器
ferroelectrics 铁电体，铁电质，铁电材料
ferroelectric transformer 铁电变压器
ferroelectric 铁电的，铁电体
ferroferric oxide 四氧化三铁
ferro-glass 钢丝玻璃
ferrograph 铁磁示波器，铁粉记录图
ferromagnetic alloy 铁磁体的，铁磁性合金
ferromagnetic armature 铁磁电枢
ferromagnetic core 铁磁芯
ferromagnetic crack detector 铁磁探伤器
ferromagnetic excitation 铁磁励磁
ferromagnetic flux return path 铁磁通量回路
ferromagnetic liquid 铁磁流体
ferromagnetic material 铁磁材料，铁磁性材料
ferromagnetic particle 铁磁性颗粒
ferromagnetic resonance 铁磁共振
ferromagnetic shield 铁磁屏蔽，电磁屏蔽
ferromagnetic 铁磁的，铁磁质，铁磁物
ferromagnetism 铁磁性，铁磁学
ferromagnetoelectric 铁磁电体
ferromagnetography 铁磁性记录法
ferro-manganese-silicon 硅锰铁
ferro-manganese 锰铁合金
ferromanganin 铁锰铜合金
ferrometer 磁滞测定器，磁滞损失计
ferro-molybdenun 钼铁
ferro-nickel accumulator 铁镍蓄电池
ferro-nickel alloy 铁镍合金
ferro-nickel 镍铁合金
ferron 试铁灵
ferro-phosphorus 磷铁合金
ferro-resonance overvoltage 铁磁谐振过电压
ferro-resonance 铁磁共振，铁磁谐振
ferro-resonant circuit 铁磁谐振电路
ferro-resonant computing 铁磁谐振计算
ferro-resonant flip-flop 铁共振触发器
ferro-resonant servomotor 铁磁谐振伺服电动机
ferrosilicon 硅铁合金
ferrous alloy 铁合金
ferrous ammonium citrate 柠檬酸亚铁铵
ferrous ammonium sulfate 硫酸亚铁铵
ferrous fail-pipe 铁落水管
ferrous hydroxide film 氢氧化亚铁膜
ferrous hydroxide 氢氧化亚铁
ferrous iron 二价铁，亚铁
ferrous metal industry 黑色金属工业
ferrous metallurgy 黑色冶金
ferrous metal 黑色金属
ferrous oxide 氧化亚铁
ferrous sulfate dosimetry 硫酸亚铁剂量测定法
ferrous sulfate 硫酸亚铁
ferrous sulfide 硫化亚铁
ferrous sulphate filming system 硫酸亚铁涂膜系统
ferrous sulphate treatment 硫酸亚铁处理
ferrous sulphate 硫酸亚铁
ferrous 含铁的，亚铁的，二价铁的
ferroxcube 立方结构铁淦氧，铁氧体软磁性材料
ferroxplana 六角晶体铁淦氧
ferroxyl test 孔隙率试剂试验
ferruginous quartz 铁石英
ferruginous 含铁的
ferrule 节流圈，箍，套圈，密套，环，金属包头，水管口密套，金属箍，金属环，铁箍
ferrum 铁
ferry berth 渡船码头
ferry boat 渡船
ferry 摆渡，渡船，渡口，浮桥
fertile assembly 增殖组件
fertile blanket 转换区
fertile element 燃料原料，再生同位素，可转换元素
fertile material 燃料增殖性物质，可转换材料，增殖性材料
fertile nucleus 可转换核
fertile nuclide 可转换核素
fertile 可转换的，能增殖的
fertility 增殖能力【燃料】
fertilization 次级核燃料的制备
fertilizer 肥料
$FeSO_4$ filming wheel 硫酸亚铁涂膜小车
festoon cable trolley 拖挂电缆小车
festoon cable 铁丝网
festooning cable 悬索电缆
festooning system 悬索系统
festoon system 拖挂电缆系统
festoon trolley 拖挂电缆的滑车
festoon tube 防渣管【弗斯顿管】
festoon 伸缩电缆卡环，垂花饰
FESW(fuel element storage well) 燃料贮存井
fetch front 风区前沿
fetch length control 取长度控制
fetch length 吹程，风区长度
fetch limited wind wave 有限风区风浪
fetch rear 风区后沿
fetch 风区，吹程，风区长度，风程，风浪区，风浪区长度，取来，取物
FET(field-effect transistor) 场效应晶体管
fettler 维修工
fettling material 炉衬材料，补炉材料
fettling 补炉，去毛边，修补，炉衬，砌面，清铲，修整，去毛刺，去除毛边
few-group analysis 少群分析
few-group constant 少群常数
few-group diffusion equation 少群扩散方程
few-group diffusion theory 少群扩散理论
few-group model 几组试样，几组型式，少群模型
few-region approximation 少区近似
few 少数的，不多的
F=Fahrenheit 华氏【温度】
FFDLC(fully funded documentary letter of credit) 全额跟单信用证
FFF(freight forwarding fee) 货代佣金
FF(flat flange) 平（焊）法兰
f=foot 英尺

f=force 力
FFT(fast Fourier transformer) 快速傅里叶变换
FGD(flue-gas desulfurization) 烟气脱硫
FGDP(flue gas desulfurization plant) 烟气脱硫装置，烟气脱硫系统
FG(flow gage) 流量计
FGS-FP(fixed generator speed fixed-pitch) 恒速定桨距
FGS-VP(fixed generator speed variable-pitch) 恒速变桨距
FHS(fuel handling and refueling system) 核燃料装卸与换料系统
fiber-bed filter 纤维层过滤器
fiberboard ceiling 纤维板顶棚
fiberboard 纤维板，木丝板
fiber composite blade 纤维复合材料叶片
fiber composite materials 纤维复合材料
fiber composite 纤维加强的复合材料
fiber dust 纤维性粉尘
fiber glass blade 玻璃钢叶片
fiber glass braided wire 玻璃丝编织线
fiber glass end-shield 玻璃纤维端盖
fiber glasse poxy blade 玻璃钢叶片，纤维玻璃环氧叶片
fiber glass epoxy (reinforced) plastic 环氧玻璃钢
fiber glass epoxy 环氧玻璃钢
fiber glass fabric 玻璃布
fiber glass filament 玻璃纤维丝
fiber glass insulation 玻璃纤维绝缘
fiber glass magnet wire 玻璃丝包电磁线
fiber glass material 玻璃纤维材料
fiber glass mat 玻璃纤维薄毡
fiber glass motor 玻璃纤维绝缘电动机
fiber glass reinforced plastic container 玻璃钢集装箱
fiber glass reinforced plastic flange 玻璃钢法兰
fiber glass reinforced plastic pipe 玻璃钢管
fiber glass reinforced plastic pump 玻璃钢泵
fiber glass reinforced plastics products 玻璃钢制品
fiber glass reinforced plastics 玻璃钢，玻璃纤维增强塑料
fiber glass-reinforced plywood container 玻璃钢集装箱
fiber glass reinforcement 玻璃纤维加强
fiber glass septic tank 玻璃钢化粪池
fiber glass 玻璃纤维，玻璃棉，纤维玻璃，玻璃丝
fiber insulation 纤维绝缘
fiber material 纤维材料
fiber metal 纤维状金属，金属丝
fiber optic bundle 光导纤维束
fiber optic-photo transfer 光电纤维变换，纤维光电变换
fiber optic probe 光导纤维探头
fiber optic sensor 光纤传感器
fiber optic 纤维光学
fiber paper 纤维纸
fiber placement 纤维铺放
fiber reinforced composite materials 纤维加强复合材料
fiber reinforced composite 玻璃纤维增强复合材料
fiber reinforced plastics 纤维增强塑料，玻璃纤维加强聚酯
fiber reinforced rubber 纤维加强橡胶
fiber scope 纤维式观测器
fiber stress 纤维应力
fiber 纤维，纤维板，硬化纸板，钢纸
FIB(free into barge) 驳船上交货价格
fibre-air-entrained concrete 纤维加气混凝土
fibre board 纤维板
fibred asbestos 石棉纤维
fibre electrometer 悬丝静电计
fibre glass reinforced plastic 玻璃纤维增强塑料
fibre glass 玻璃纤维
fibre-insulated wire 纤维绝缘线
fibre optically coupled cascaded image intensifier 光导纤维耦合级联式图像增强器
fibre-optic cable 光纤电缆
fibre-optic communication 光纤通信
fibre-optic coupler 纤维光学耦合器
fibre-optic coupling 纤维光学耦合
fibre-optic faced tube 纤维光学板显示管
fibre-optic light guide 纤维光导
fibre-optic light-transmission system 光纤光传输系统
fibre-optic memory 纤维光学存储器
fibre-optic system 光纤系统
fibre-optic 光学纤维的，光导纤维的
fibre packing 纤维衬垫
fibre reinforced material 纤维补强材料，纤维强化材料
fibre strength 纤维强度
fibre 纤维
fibril 微丝
fibrocement 石棉水泥
fibroid 纤维状的
fibrous covering 纤维绝缘层，纤维绝缘套
fibrous dust 纤维性粉尘
fibrous filter 纤维过滤器
fibrous fracture 纤维裂痕，断口
fibrous glass 玻璃纤维
fibrous insulant 纤维绝热材料
fibrous insulation 纤维绝缘
fibrous material 纤维材料
fibrous root system 须根系
fibrous soil 纤维性土
fibrous-stroked dressing stone 细纹面琢石
fibrous weld 纤维状焊缝
fibrous 纤维的，纤维质的，纤维状的
FIC(film integrated circuit) 薄膜集成电路
FIC(flow indicate controller) 流量指示控制器
fiche 卡片
Fickian diffusion 斐克扩散
fictile 陶土制的，陶制品
fictitious boundary 假想边界
fictitious carry 假进位
fictitious equation 虚拟方程
fictitious load 假负载，假想负载，虚拟负荷，假荷载，模拟荷载

fictitious magnetic current 假想磁流
fictitious power 非有功功率，无功功率，虚假功率，虚功率
fictitious 虚拟的，虚设的，虚构的
fiddle 台架
fidelity curve 逼真度曲线
fidelity 忠实，忠诚，诚实，逼真度，保真，保真度
FID（flame ionization detector） 火焰离子化检测器
FIDIC(Federation Internationale Des Ingenleus Conseils) 国际咨询工程师联合会【简称"菲迪克"】
FIDIC（international federation of consulting Engineers） 国际咨询工程师联合会
fiducial error 引用误差，基准误差，置信误差
fiducial value 置信值，基值
fiduciary level 置信电平，标准电平
fiduciary loan 信用贷款
fiduciary 置信的，标准的
field-acceleration relay 激磁加速继电器
field adjusting device 野外校正装置
field adjustment 励磁调节，磁场调节
field adviser 实地顾问，现场顾问
field alterable control element 现场可变的控制部件
field ampere-turns 励磁安匝
field amplifier 励磁电流放大器，磁场放大器
field angle 张角
field annotation 野外调绘
field apparatus for ground photogrammetry 地面摄影测量仪
field apparatus 现场设备
field application relay 激励继电器，励磁继电器
field-assembled 现场装配的，现场组装，现场组装的
field assembly yard 工地装配场
field assembly 现场组装，现场装配，工地装配
field axis 磁极轴线，直轴
field balance 现场平衡
field balancing technique 现场平衡法，现场平衡技术
field balancing 现场平衡
field board 外业测图版
field bobbin 磁极线圈架，磁场线圈座
field bolt 现场安装螺栓
field boundary conditions 场地边界条件
field breaking switch 分场开关，削磁开关，励磁分段开关
field burning 田野燃烧
field bus control system 现场总线控制系统
field bus 现场总线
field cable 被覆线，野外电缆
field calibration 野外率定，野外校准
field capacity 田间持水量
field change 现场更改，现场修改
field checking 现场检验
field circuit breaker 磁场开关，励磁断路器
field circuit loss 励磁电路损耗
field circuit 磁极电路，磁极磁路
field coenergy 电场共同能量，磁场共同能量
field coil bracer 磁极线圈支撑块
field coil flange 磁极线圈托板
field coil resistance 励磁线圈电阻
field coil 励磁线圈，场线圈，磁场线圈
field computation 外业计算
field conditions 现场条件
field connection 工地装配，现场连接，现场联结
field control 励磁控制，磁场调整，地面控制
field copper loss 磁场铜耗
field core 磁场铁心
field cost 现场费用
field cross-section 产额截面
field curing （混凝土）现场养护，工地养护
field current relay 励磁电流继电器
field current 场电流，励磁电流，磁场电流
field curve 磁场曲线
field data 现场资料，现场数据，外业资料，外业数据，野外观测资料
field decelerating relay 激磁减速继电器
field deficiency report 现场偏差报告
field demonstration model 现场演示模型
field density test 现场密度试验
field density 场密度，磁通量密度，场强
field design office 现场设计室，现场制图室
field discharge protection 磁场放电保护装置
field discharge resistance 磁场放电电阻，消磁电阻
field discharge switch 磁场放电开关，灭磁开关，消磁开关
field discharge voltage 磁场放电电压
field discharge 灭磁，消磁，磁场放电
field displacement 场位移
field disposal 野外处置，旷野处置【放射性废物】
field distortion switch 场畸变开关
field distortion 场畸变，场失真
field distribution 场分布
field dividing switch 励磁分段开关
field document 外业资料
field drafting office 现场制图室
field-drying transformer 场干燥变压器
field economizing relay 弱励磁继电器
field effect transistor 场效应管，场效应晶体管
field effect 场效应，电场效应
field emission 电场发射，场致发射，静电发射
field engineering office 现场工程办公室
field engineering 现场安装工程
field engineer 安装工程师，现场工程师，工地工程师
field equipment 外业装备
field-erected boiler 散装式锅炉，工地安装的锅炉，就地安装锅炉
field-erected 工地安装的
field erection 工地架设，现场安装，现场装配
field examination 野外检查
field excitation 励磁，激磁
field exciter 励磁机，场激励器
field experiment 现场试验，野外试验
field extinction 灭磁
field fabricated 工地制造的，现场装配的，现场

制造的
field fabrication 现场制造
field factor 产额因子
field failure relay 失磁继电器
field failure 失磁故障
field flashing equipment 起励设备
field-forcing characteristic 强励特性，强励特性曲线
field-forcing limiter 磁场强励限制器
field-forcing relay 强励继电器
field-forcing voltage 强励电压
field-forcing 磁场强励，强迫励磁
field for pulling 牵引场
field for tensioning 张力场
field frame 磁极框架
field frequency 场频
field fuse 励磁保险丝，磁场保险丝
field generator 场发生器，励磁发电机
field geological investigation 野外地质调查
field geological survey 野外地质调查
field geology 野外地质学
field gradient 场梯度
field groundwater velocity 地下水实际流速
field identification 现场鉴定
field indicator 现场指示牌
field inductance 场感应
field inspection staff 现场检查人员
field inspection 现场检查，外场检查，工地视察，现场踏勘，巡回检查，野外检查
field installation 现场安装
field instrument 就地仪表
field intensity diagram 场强图
field intensity distribution 场强分布
field intensity map 场强分布图
field intensity meter 场强计
field intensity 场强
field investigation 现场试验，运转试验，实地调查，现场调查，野外调查
fieldistor 场控晶体管，场效应晶体管
field joint 安装接头
field lab manager 现场实验室主任
field laboratory 工地实验室
field labor cost 工地安装费用，现场安装费用
field lab 现场实验室
field-lead insulation 磁极引线绝缘
field leakage coefficient 漏磁系数
field leakage factor 漏磁系数
field leakage flux 漏磁通
field leakage 场漏泄，漏磁
field length 字段长度
field levelling 野外水准测量
field line 场力线
field locking 场同步
field-loss protection 失磁保护
field-loss relay 失磁继电器
field-loss 漏磁损耗，磁场损失，失磁的
field magnetomotive force 磁场磁动势
field magnet 场磁铁
field maintenance equipment 现场维修设备
field maintenance personal 现场维修人员
field maintenance 现场维修，就地维修
field manual 外业指南
field mapping 野外测图
field measurement 实地测量，现场量测，野外测量
field method 区域法，网络法
field mix 工地搅拌，现场拌和
field modulation generator 现场调制发电机
field modulation 磁场调制
field moisture equivalent 工地含水当量，室外含水当量，现场含水当量
field monitoring instrument 现场监测仪表
field monitoring 野外监测
field-mounted 现场安装的，工地安装的
field note 现场记录
field observation 现场调查，野外观测
field of application 应用范围
field of behaviour 相位场
field of excitation 漏磁场，激励场
field office 现场办公室
field of force 力场
field of lines behaviour 相轨迹场
field of real numbers 实数域
field of self-induction 自感场
field of turbulent flow 湍流场，紊流场
field of view angle of a pyrheliomter 直接日射表视场角
field of view 视野
field of vision 视野
field operation 就地操作，外业，野外操作
field operator 现场运行人员
field painting 现场喷涂
field panel 现场镶板
field pattern 场图
field performance 现场工作性能
field permeability coefficient 现场渗透系数
field phase 场相位
field photogrammetric apparatus 地面摄影测量仪
field photogrammetry 地面摄影测量
field pitch 极距
field planetable map 外业原图
field plotter 场绘迹器
field pole 磁场线圈架，磁极
field poured 现场浇灌
field power supply 漏磁电源，现场供电
field programmable logic array 现场可编程逻辑阵列，字段可编逻辑阵列
field protective relay 激磁保护继电器，磁场保护继电器，场保护继电器
field pumping test 野外抽水试验
field quality control 现场质量控制
field questions 答复提问
field radiometer 工作辐射表
field railway 工地轻便铁道
field recompression curve 现场再压缩曲线
field reconnaissance 野外踏勘
field record 外业记录，外业资料
field rectifier 外业纠正仪
field reduction 场衰减
field regulator 场强调整器，漏磁变阻器
field relay 励磁继电器
field reliability test 现场可靠性试验

field removal relay 去磁继电器
field report 现场报告
field representative 现场代表
field resistance line 磁场电阻特性曲线
field resistance 漏磁绕组电阻，漏磁回路电阻，励磁电阻
field reversing switch 漏磁换向开关
field RHEO control 磁变阻器控制
field rheostat control 磁变阻器控制
field rheostat 漏磁变阻器，磁场变阻器，激磁变阻器
field ridge 输水沟，引水工程，引水沟，引水渠
field ring 机座环形部分
field riveting 工地铆接
field sales 现场销售
field sampling and preparation 现场取样和制备
field service expert 外勤专家，现场服务专家
field service 现场服务，现场维修
field shear box 野外剪力仪
field sheet 现场测绘图，现场航测图
field shunting 磁场分流，磁场分路
field situation 现场情况
field sketch 外业草图，现场草图
field speed 场速
field spider 转子支架，磁极支架
field splice （塔架）现场接合面，现场拼接，现场接头
field spool insulation 磁场线圈框架绝缘，（磁极）极身绝缘
field spool support 漏磁线圈支撑
field spool 漏磁线圈架，磁场线圈架，励磁线圈架
fieldstone 毛石
field strength contour map 等场强线图
field strengthening 强场
field strength map 场强图
field strength meter 场强计
field strength 电场强度，场强，磁场强度，场强度
field structure 磁极结构，场结构
field supervision 现场监督
field suppressing resistor 灭磁电阻器
field suppression equipment 灭磁装置，灭磁设备，消磁装置
field suppression switch 灭磁开关
field suppression 灭磁
field survey data 野外测量数据
field survey 现场调查，野外调查
field-swept 场扫描
field switch 磁场开关，漏磁开关
field synchronizing signal 场同步信号
field tapping 磁场分接
field terminal 磁极线圈出线端
field testing 现场试验
field test with turbine 外联机试验
field test 工业性试验，现场试验，现场检测，野外试验，野外测试，工地试验
field theory 场论
field-to-field survey 分界测量
field topographical survey team 野外地形测量队
field transistor 场效应晶体管
field transmitter 就地变送器
field trial 现场试验
field triangulation 野外三角测量
fieldtron 一种场效应器件
field tube 矢量场管
field-turn insulation 磁极线圈匝间绝缘
field-turn 磁极线匝
field vane test 现场十字板试验
field vector 场矢量
field velocity MHD generator 场速磁流体发电机
field velocity 实地流速
field voltage 激励电压，漏磁电压，磁场电压，励磁电压
field wave 漏磁波，激励波
field weakening 磁场削弱
field-welded 工地焊接的，现场焊接
field welding 工地焊接，现场焊接
field winding 磁场绕组，励磁绕组，磁场线圈，磁极绕组，漏磁绕组
field work 现场作业，工地工作，野外工作，野外作业，现场工作
field yoke 磁轭
field 场，电场，工地，（安装）现场，场地，野外，野外的，现场的，字段，领域
fierce competition 激烈竞争
fierce wind 狂风
fiery coal 易燃煤，气煤
fiery fraction 粗晶粒断口
fife extinguisher 灭火器
FI(flow indicator) 流量指示件
FIFO(first-in first-out) 先进先出（法）
FI(free in) 船方不负担装货费
fifth overtone 五次泛音，六次谐波
fifth wheel 承重旋转接头
fifty percent disruptive discharge voltage 击穿电压的50%
fifty-year-return period 50年一遇
fig. =figure 数值，形状
fighting vehicle 消防车
fightness 密封性
figuline 陶器，瓷器，陶制的
figurate stone 琢纹石
figuration outline 轮廓
figuration 形状
figure adjustment 数值调整，图形平差
figure centre 图形中心
figured glass 图案玻璃，压花玻璃
figured iron 型钢
figured plate glass 图案玻璃
figured rolled glass 压花玻璃
figure of loss 能量损耗系数
figure of merit frequency 质量系数频率
figure of merit 品质因数，灵敏值，质量指标，质量指数
figure of noise 噪声指数
figure of performance 性能指数
figure out at 总计，合计
figure out 做出，计算出，想象出，合计，弄清楚
figure-reading electronic device 电子读数装置，电子图形阅读器

figure shift 变换符号，换数字档，变数字位
figure 形态，形状，图像，图，图形，数字，轮廓，图案，图解
figurine 小雕像
fiinflation 充气
filamental flow 线流
filamentary cathode 灯丝式阴极
filamentary shrinkage 丝状疏松，纤维状疏松
filamentary spark 丝状火花
filamentary structural composite 纤维结构复合材料
filamentary transistor 线状晶体管
filamentary 线状的
filament battery 灯丝电池
filament cathode 直热式阴极
filament circuit 灯丝电路
filament control 灯丝调节
filament current 灯丝电流
filament fuse 灯丝熔断器
filament lamp 白炽灯
filament line 流线
filamentous flow 流管，线流
filament power supply 灯丝电源
filament power 灯丝功率
filament reinforced metal 纤维加强金属
filament transformer 灯丝变压器
filament winding 灯丝电源绕组，丝绕，纤维缠绕
filament 丝，线，灯丝，阴极，细线，单纤维，游丝
filbore 基础轴承
file access channel 文件存取通道
file a claim 提出索赔
file computer 编目计算机
file control system 文件控制系统
file conversion 外存储器信息变换
file copy 存档原件
file data 档案资料
filed bus 现场总线
file definition processor 文件定义处理程序
filed measurement 现场测量
file drum 存储磁鼓
file event 文件事件
file gap 文件间隔
file header 文件头，文件标题
file identification 文件识别
file layout 文件格式，文件布局
file maintenance 文件处理，文件维护
file management 档案管理，文件管理
file memory 外存储器
file name 文件名
file opening 文件打开
file organization routine 文件组织程序
file organization 文件组织
file-oriented system 面向文件系统
file printout 文件打印输出
file processing 文件处理
file processor 外存储器信息处理机
file protection ring 文件夹保护环，文件存储保护环
file protection 文件保护
file reference 文件参考
file search 文件搜索
file security 文件安全性，文件保密
file separator 文件分隔符
file storage unit 文件存储单元
file storage 外部存储器，外存储器
file transfer protocol 文件传输协议
file unit 外存储器部件
file 外存储器，文件夹，文件，档案，存档，锉刀
filfer paper 滤水纸
filigree glass 嵌丝玻璃，银丝玻璃
filing cabinet 档案柜，公文柜
filing case 档案柜
filing room 档案室
filing up effect 堆积效应
filing 立卷，（文件的）整理汇集，锉磨，磨琢，锉屑【复数】
fill and draw basin 中间容器
fill and soak method 浸泡法
fill away 顺风行驶
fill block 填块
fill compaction 填土压实
fill concrete 回填混凝土，填充混凝土，填筑混凝土
fill construction 填土工程，填土施工
fill dam 填筑坝，填筑式坝
filled and enclosed cartridge fuse-link 有填料封闭管式熔断器
filled bitumen 加填料沥青
filled composite 填充复合材料
filled earth 填土
filled out a form 填表
filled pipe column （混凝土制的）填实管柱，填充式管柱
filled spandrel arch 实肩拱
filled system thermometer 充填式温度计，压力（计）式温度计
filled system 充满系统【太阳能光热利用】
filled thermal system 充灌式感温系统
filled-up ground 填高地
filled with wall 砌墙填充
filled with water 向……注水
filler aggregate 填充骨料
filler block 衬块，填料，止水塞
filler for cement 水泥惰性掺和料，水泥填料
filler gate 充水闸门
filler layer 填料层
filler metal 熔焊金属，填充金属，焊料，焊丝，熔填金属
filler piece 隔叶块，填料，充填块，填隙片
filler plate 填隙板，密封垫，垫板，填板
filler port 加油口，加水口
filler ring 垫圈
filler rod 焊条
fillers of cable 多芯电缆填充材料
filler toe 坝趾反滤层
filler valve 阀门充满
filler wall 填充墙
filler 加油口，填充金属，填料，填板，垫片，填充物，填充料，填缝料，注入孔，充填剂

filleting 角隅填密法，嵌缝法
fillet in parallel shear 侧面焊缝
fillet joint 角焊接头
fillet radius 倒角半径，圆角半径
fillet welding in the flat position 船形焊
fillet welding 角焊，填角焊
fillet weld in normal shear 正面角焊缝
fillet weld in parallel shear 侧面角焊缝
fillet weld leg 焊脚
fillet weld size 焊脚尺寸
fillet weld 角焊缝，角焊，填角焊缝，填角焊，贴角焊
fillet 圆角，嵌条，压边条
fill factor 填充系数[因子]，装填系数，占空因数，填充率，占空系数
FILL(filling) 充水，充汽
fill gas 填充的气体
fill height 填土高度
filling and emptying system （船闸的）充水和放水系统
filling compound 填料
filling culvert 充水涵洞
filling device 充水装置
filling for groin 丁坝填筑
filling for groyne 丁坝填筑
filling grouting 回填灌浆
filling in ring 势圈
filling machine 充填机械，灌注机器
filling material 填料，电缆膏
filling medium 充填介质
filling pile 灌注桩，填补桩
filling plant 装油设备
filling plate 填密板，衬板
filling point 装油站
filling port 充水孔
filling pressure 充气压力
filling pump 充水泵，装油泵
filling run 填充焊缝
filling SF_6 gas in vacuum 真空充注六氟化硫气体
filling station 加油站
filling strainer 注液过滤器
filling system 加注系统
filling time 充水时间
filling-up piece 隔叶块，隔金
filling valve 充水阀，进料阀
fillingwell 进料井
filling 填料，填充，上水，充注，填充物，填塞，填土方，填筑，装料，充气
fill in the amount in figures and words 用大小写填写
fill-in 插入，塞入，填入
fillister head screw 有槽凸圆头螺钉
fillister joint 凹槽接合
fillister 凹刨，凹槽
fill joints with mastic 玻璃脂填缝
fill packing 填料【冷却塔】
fill section 路堤截面，填土截面
fill settlement 填土沉陷
fill slope 路堤边坡，填土边坡
fill thermal system 填充式热敏系统
fill-type dam 填筑式坝
fill-up 填补
fill valve 阀门充满，灌注阀
fill 灌注，填充，填方，填料，充满，填土方，装满
film analysis 薄膜分析
film and paper condenser 纸绝缘薄膜电容器
filmatic bearing 油膜轴承
film badge service 胶片佩章管理处
film badge 胶片剂量计，个人剂量仪，胶片佩章
film boiling heat transfer 薄膜沸腾传热
film boiling 膜态沸腾，膜状沸腾
film breakdown 薄膜击穿
film cassette 胶片暗盒
film coated magnet wire 薄膜绝缘电磁线
film coefficient 膜系数，（剂量）胶片系数
film composite insulation 绝缘薄膜复合制品
film condensation 膜式冷凝
film conductance coefficient 膜放热系数
film contrast 底片对比度
film-cooled blade 膜冷却叶片，油膜冷却叶片
film-cooled heat shield 冷却隔热屏蔽
film-cooling effectiveness 薄膜冷却有效性
film-cooling 膜状冷却，膜态冷却，薄膜冷却
film cryotron 膜片冷子管
film degasification 薄膜除气
film density 底片黑度
film dielectric 介电
film diffusion 薄膜扩散
film dosimeter 胶片剂量计
film dryer 底片烘干机
film effect 薄膜效应
film fill 薄膜式填料
film filtration 薄膜过滤
film flow 薄膜流动，水膜流，薄层流动，薄膜水流
film formation 膜的形成
film forming material 成膜材料
film forming 成膜
film gradient 底片（黑度）梯度
film heat-transfer coefficient 对流换热系数，薄膜换热系数
film illuminator 底片观察光源
filming amine 成膜胺，薄膜胺
filming system 涂膜系统
filming 成膜
film integrated circuit 薄膜集成电路
filmistor 薄膜电阻
film lighting arrester 膜片避雷器
film lubrication 液体润滑，薄膜润滑，油膜润滑
film maintenance 膜的维护
film mass-transfer coefficient 膜质量转移系数，膜传质系数
film memory 薄膜存储器，胶片存储器
film moisture 薄膜水分，薄膜水
film observation device 观片灯
film of fine grain 微粒胶片
film of oxide 氧化物膜，氧化膜
film optical scanning device for input to computer 计算机胶卷光扫描输入装置
film overlap 底片搭接【射线照相】

film packing　薄膜式填料
film parametron　薄膜参变元件
film plotting　摄影测绘
film pressure　（凹状液面的）薄膜压力，水膜压力
film processing　底片处理
film radiography　胶片射线照相术
film reader　微缩胶片阅读器
film record　缩微胶片记录
film resistor　薄膜电阻器，薄膜电阻
film-riding face seal　油膜悬浮表面密封，水膜面密封
film ring　指环式胶片剂量计
film stability　膜的稳定性，油膜稳定性
film storage　薄膜存储器
film strength　油膜强度
film stripping test　膜稳定性试验
film thickness　膜的厚度
film type condensation　膜状凝结，膜状冷凝，膜态凝结
film type deaerator　膜式除氧器
film unsharpness　底片不清晰度
film varistor　膜式压敏电阻器
film viewing equipment　底片观察设备
film water　薄膜水
filmwise condensation　膜状凝结，膜状冷凝
filmwise condensing　膜状冷凝
film　薄膜，膜，胶片，软片，涂层，涂膜
FILO(free in ,liner out)　船方不负担装货费但负担卸货费
filterability　过滤率，过滤性
filterable isotope　可过滤掉的同位素
filterable　可过滤的
filter aid layer　助滤器层
filter aid　助滤剂，助滤器
filter apparatus　过滤器
filter area　过滤面积
filter attenuation band　滤波器衰减带，滤波衰减带
filter bag　滤袋
filter bank　过滤器组
filter bed　过滤层，滤床，滤池，沉沙池，滤水层
filter blanket　反滤铺盖，滤水垫层
filter block　过滤器件，过滤元件，滤块
filter board　滤板
filter bottom　滤池底
filter box　过滤箱，过滤盒
filter brick　滤砖
filter cake layer　滤渣层，滤饼层
filter cake　滤饼，滤渣
filter candle　烛式过滤器，过滤棒
filter capacitor　滤波电容器
filter capacity　过滤能力
filter cartridge descent tube　过滤器芯下落管
filter cartridge transfer tube　过滤器芯输送管
filter cartridge　过滤器芯筒，滤芯
filter cell　过滤池
filter changing equipment　过滤器更换设备
filter changing station　过滤器更换站
filter characteristic　滤波器特性曲线
filter choke　过滤器阻塞，滤波扼流圈，滤波阻流圈
filter circuit　滤波器电路，滤波电路
filter clogging　滤层堵塞，过滤器堵塞
filter cloth　滤料，过滤布
filter coke　过滤用焦炭
filter collector　机动除尘器
filter concentrate processing　过滤器浓缩物处理
filter concentrate storage　过滤器浓缩物储存
filter concentrate tank　过滤器浓缩物箱
filter condenser　滤波电容器
filter course　过滤进程
filter crib　滤水框
filter crucible　滤埚
filter cycle　过滤周期
filter cylinder for sampling　滤筒采样管
filter dam　透水坝
filter dehydration　过滤脱水法
filter drain　滤水暗管
filtered air　过滤后的空气，滤过的空气
filtered containment vent system　安全壳滤过排气系统
filtered flue gas　滤过的烟道气
filtered hologram　已滤波全息图
filtered model　滤波模式
filtered radiation　滤过辐射
filtered silicon reference solar cell　带滤光片的标准太阳电池
filtered vent　滤过排气
filtered wastewater　过滤废水，滤过废水
filtered water　过滤水
filtered water pump　过滤水泵，清水泵
filtered water reservoir　过滤水贮水池，过滤水池
filtered water supply pump　过滤水供应泵
filtered water tank　过滤水箱，清水箱，滤水箱
filter efficiency　过滤器效率，滤池效率，过滤效率
filter element　过滤器元件，滤波器元件，过滤器流出液
filter failure　过滤器失效
filter feed pump　过滤器水泵
filter film　滤膜
filter flask　吸滤瓶
filter flooding　（生物）滤池漫灌
filter function　滤波函数
filter funnel　过滤漏斗
filter gallery　滤池廊道
filter gauze　滤网
filter glass　（焊接用）黑玻璃，过滤器窗
filter housing　过滤器壳
filtering accuracy　过滤精确度
filtering bed　滤床
filtering candle　滤烛
filtering capacity　过滤能力，过滤容量
filtering cartridge　滤筒
filtering efficiency　过滤效率
filtering element　滤元，滤芯
filtering frequency　滤波频率
filtering layer　滤层
filtering material　滤料

filtering media 过滤用的材料
filtering medium 过滤介质，滤料
filtering rate 过滤速率
filtering resistance 过滤阻力
filtering surface 滤面
filtering tank 过滤罐
filtering velocity 过滤速率
filtering 过滤，滤波，滤清
filter layer 滤层，滤床
filter leaf 滤叶
filter line 滤线
filter liquor 滤液
filter loading 滤池负荷，滤器负荷
filter mask 过滤面罩
filter material 过滤料，过滤材料，反滤料，滤料
filter medium 反滤料，过滤介质，过滤料，过滤器介质
filter membrane 滤膜
filter operating table 滤池操作台，滤池控制台
filter operation 过滤操作
filter out 滤出
filter paper drain 滤纸排水
filter paper pulp 过滤纸浆
filter paper 滤纸
filter pass-band 滤波通带
filter passer 滤过性病毒
filter photometer 过滤光度计，滤色光度计
filter plant 过滤车间，过滤设备，滤水车间，滤水池
filter plate 滤光玻璃，滤光板，滤板
filter pocket 滤袋
filter ponding 滤池成塘
filter pooling 滤池积水
filter pool 滤水池
filter pre-coated with ion exchange resin powder 离子交换树脂粉预覆盖［预涂层］过滤器
filter pre-coated with pulp 纸浆覆盖过滤器
filter-pressed sludge 滤压污泥
filter press 滤压机，压滤机
filter process 过滤过程
filter rate 滤速
filter reactor 过滤器阻流，滤波电抗器
filter regulator 过滤调节器
filter residue 滤渣
filter resistance 滤波电阻
filter run 过滤器工作周期，过滤周期
filter saturation 过滤器饱和
filter screen 过滤筛网，滤网
filter section 过滤段
filter skin 滤水表层
filter slime 滤渣
filter stand 漏斗架
filter sterilized 过滤法消毒
filter stick 滤棒
filter stop band 滤波器阻带，滤波阻带
filter strainer 滤池滤器，滤管
filter strip 滤水畦条
filter surface 过滤面积
filter tank 过滤池，滤池
filter toe 滤水坝趾，滤水坡脚
filter transformer 滤波器变压器
filter transmission band 滤波器通带
filter type mask 过滤器式面罩
filter types 过滤方式
filter underdrain system 滤池排水系统
filter underdrains 滤池底排水道
filter unit 过滤器单元，过滤器装置
filter unloading 滤池卸膜
filter velocity 过滤速度
filter washer 洗滤器
filter wash 滤池冲洗，洗滤
filter wastewater 洗滤（废）水
filter well 渗水井
filter zone 反滤带
filter 过滤器，滤波器，滤光器，滤光片，滤除，滤池，滤纸，过滤，滤波，滤层，滤网
filth 污垢，污物，垃圾
filtrability index 滤过性指数
filtrability 过滤性
filtrable agent 可滤过因素
filtrable bacteria 滤过性细菌
filtrable virus 滤过性病毒
filtratable isotope 可过滤掉的同位素
filtratable nuclide 可过滤核素
filtrated stock 过滤母液
filtrate receiver 滤液接收器
filtrate 过滤，滤清，滤出液，滤液
filtrating equipment 过滤设备
filtration aid 助滤剂
filtration area 过滤面积
filtration bed 滤水层
filtration capacity 过滤能力
filtration factor 过滤系数，过滤因子
filtration mechanism 过滤机理
filtration method 过滤法
filtration of sound 声滤
filtration plant 过滤设备，过滤厂，滤水厂，滤水池
filtration pressure 过滤压力
filtration rate 过滤速度，滤过率，滤除率，滤速
filtration screen 滤网
filtration stand 漏斗架
filtration tank 过滤箱
filtration type 滤过型
filtration velocity 过滤速度，滤速
filtration virus 过滤性病毒
filtration 滤除性，过滤，筛选，渗透，渗流
filtrator 过滤器，滤清器
FIM 芬兰马克
final acceptance certificate 最后验收证书
final acceptance report 最后验收报告
final acceptance 移交验收，最后验收，最终验收
final account for completed project 竣工结算
final accounting of revenue and expenditure 决算
final accounting 最后结算，决算
final accounts of project cost 工程投资决算
final account stage 最后结算阶段
final accounts 决算账户，年终账目
final actuation time 最后激励时间

final agreement 最后协议
final and binding 决定性并有约束力的
final and firm price 最终实价
final approval 最后校准，最后批准，最终批准，验收
final article 最终条款
final assembly 最后装配，总装配图
final blade length 末级叶片长度
final blade 末级叶片，末叶片
final blowdown pressure 最后排空压力【阀瓣回座】
final blowdown 最后排空【为避免锤击】
final budget 核定预算
final carry digit 最后进位的数字
final certificate of payment 最终支付证书
final certificate 最终证书
final check 最终检查
final clarifier 终端澄清器
final clauses 最终条款
final clearing statement 最后结算表
final closure 最终关闭，退役
final coat of paint 最后一道漆
final completion certificate 最后竣工证书
final completion 竣工，完工
final concentration system 最后浓集系统
final concentration 最终浓度
final condition 边界条件
final construction report 竣工报告
final contraction value 最终收缩度
final control element 末级控制元件，末控元件，最终控制元件，(控制系统）执行元件
final controlled condition 最后控制条件
final controlling drive 控制棒有效部分
final controlling element 末控元件，控制棒有效部分，终端执行器
final copy 最终本
final cost 最终成本
final cut 最后开挖线
final data processing 数据最终处理
final delivery pressure 排汽压力
final design stage 最终设计阶段
final design 最终设计，竣工图纸，施工设计
final dimension 最终尺寸
final disposal 最终处置
final dividend 终结股利
final drawing 终版图纸
final drive bevel box 末级驱动伞形齿轮箱
final edition 最终版
final effluent 最终水，最终流出物
final estimate 结算
final evaporation point 末级蒸发点，最终蒸发点
final evaporator 末级蒸发器
final excavation line 最终开挖线
final feedwater temperature 最终给水温度
final filter 末级过滤器，终端过滤器
final filtration 终端
final forge 终锻
final fracture 最后断裂，最终断裂
final frequency 最终频率
final gap-closing 合龙
final grind 磨光
final grouting 二次浇灌，二次灌浆
final heat treatment 最后热处理
final ignorance 最终不知
final impulse operating relay 末端脉动继电器，终端继电器
final infiltration capacity 最终渗水量，终渗能力
final inspection 最终检查
final investment 期末投资
final invoice 最终发票
finalized 出版原图
final location survey 最后定线测量
final location 最后定线，定测
finally settled accounts 最后结算的账户
final maturity 最终偿还期，借款期限
final mean winding temperature 绕组最终平均温度
final moisture content 蒸汽终湿度，末级湿度
final neutralization pit 最终中和槽
final output 终端输出，最终结果
final payment certificate 最终付款证书
final penetration 终击贯入度，最终贯入度
final plant closure 最终关闭，退役
final plan 最终方案
final plume rise height 羽流最终抬升高度
final port of destination 最后目的港
final posterior distribution 最终后验分布
final pressure 终压力，最终压力
final price 最后价格
final product 最终产品，最终乘积，最终产物
final project result 项目最终结果
final protocol 最终议定书，最终协议
final purification 最终净化
final quantity 终值
final reading 末次读数，最后读数
final refining 最终精制
final reheater 末级再热器，再热器热段
final report 总结报告，最终报告
final residuum 最终残渣
final resistance of filter 过滤器终阻力
final review 终审
final rinse 最终淋洗
final safety analysis report 最终安全分析报告
final sag 最终弧垂
final sample 最终煤样
final scheme 最终方案
final sedimentation tank 最终沉淀池，最终沉淀槽
final sedimentation 最终沉淀
final selector 终结器
final setting time 终凝时间
final settlement of account 决算，最终结算
final settlement of a claim 索赔的最终结案，索赔的清偿
final settlement price 最后结算价
final settlement 最后结算，最终沉降，最终沉降量
final settling tank 最终沉淀池，最终沉降槽
final settling 最终沉降
final set 终凝
final shop inspection 工厂最终检验
final shutdown 末次停机，最终关闭

final signal unit　最终信号单元
final soil moisture　最终土壤水分
final solution　最终溶液，成品溶液
final stage deaerator　末级除氧器
final stage reheater　末级再热器
final stage superheater　末级过热器
final statement　最终报表
final state　终态
final static level　最终静水位
final storage　（文件的）最终保存
final subsidence　最终沉陷
final sum　最终和
final survey　最终勘测，终测
final take-over certificate　最终移交证书
final temperature diference of heater　加热器终温差
final temperature difference　端差，终端温差，终温差，出口温差
final temperature　最终温度
final tightening　最后旋紧
final treatment　最终处理
final tripping　最终跳闸
final value　终值
final velocity　末速度
finance analysis　财务分析
finance company　金融公司，财务公司
finance contract　信贷合同，融资合同
finance corporation　金融公司
finance evaluation　财务评价
finance reserve funds　财政储备基金
finance section　财务科
finance statement　财务报告
finance　财政学，金融，财政，财务，筹措资金，融资，资金，财政的
financer　金融家，融资公司
financial account of a completed project　竣工决算
financial affairs　财务
financial aid　经济援助
financial analysis　财务分析，经济性分析
financial appropriation　财政拨款
financial assistance　财政援助
financial backing　财政支援
financial bid　商务标
financial bond　金融债券
financial capability　财务能力
financial capital　财政资本，金融资本
financial center　金融中心
financial chief　财务科长
financial circle　财经界，金融界
financial claim　金融债权，财务金融要求权，债权
financial clique　财团，金融集团
financial commitment　财政承担额，财政承担，财务承诺，财政保证
financial condition　财务状况
financial contribution　财政捐款
financial control　财务控制
financial cost　财务费用
financial data　财务数据
financial duty　财务关税
financial efficiency analysis　财务效益分析
financial evaluation　财务评价
financial expenses　财务费用
financial feasibility　财务可行性
financial group　财团
financial guarantee　财政担保
financial institution　金融机关
financial instrument　财务契约，财务证书
financial internal rate of return　财务内部收益率，财务回款率
financial management　财务管理
financial market　金融市场
financial modeling　财务模型建立
financial model　财务模型
financial net current value　财务净现值
financial net present value ratio　财务净现值率
financial net present value　财务净现值
financial paper　金融票据，财务报告
financial planning system　财务计划系统
financial plan　财务计划
financial power trading　电力金融交易
financial projection　财务预测，财政保护
financial proposal　财政计划，财务建议书，商务报价
financial rate of return　财务回收率
financial ratio　财务比率
financial reconciliation　财务对账
financial reporting department　财务报告部门
financial resources　财力资源
financial result　财务结果
financial revenues and expenditures　财务收支
financial revenue　财务收入，财政收入
financial risk　财务风险，金融风险
financial security　财政保障，财政安全，财政担保，金融安全，金融证券
financial settlement　财务结算
financial standing　财务状况
financial statement　财务报表，财政报告，财政决算，财力证明，决算表
financial stringency　银根紧缩，信用紧缩
financial subsidy　财政补贴
financial terms　融资条件
financial tsunami　金融海啸，金融风暴
financial year　财政年度，会计年度
financing charges　融资费
financing condition　融资条件
financing institution　贷款机构，融资机构
financing notes　融资性票据，财政票据
financing of project　工程筹资
financing　资金筹措，融资，筹款
fin area　尾翼面
fin arrangement　肋片的布置
fin-cooled machine　散热筋型电机
finder's fee　财务经纪费，中介佣金，中间人佣金
finder　定向器，测距器，探测器，瞄准装置，测距仪，中间人
finding　发现，发现物，探测，研究结果，调查结论
find out　查明
fine adjustment screw　微调螺丝
fine adjustment　细调，微调，精调（整），精密

平差
fine aggregate concrete　细石混凝土
fine aggregate　细骨料，细粒料
fine air filter　高性能空气过滤器
fine appearance　外形美观
fine ash bin　细灰库
fine ash silo　细灰库
fine ash　细灰
fine balance　精密平衡，（零点）细调，微调，精调
fine breeze　煤尘
fine cleaning　细净化
fine coal　粉煤【6 毫米】，煤粉，煤屑，细煤
fine compensation　精密补偿
fine concrete　细骨料混凝土，细石混凝土
fine control element　精密调节元件
fine control member　精调控制棒
fine control rod　精密控制棒，细调棒
fine control　精密控制，细调节，微调，精密调节
fine cooling hole　细冷却孔
fine crack　细裂纹，微小裂纹
fine crusher　细碎机
fine dispersion　细分散
fine droplet　细滴
fine-dusty coal　细煤屑，含细粉的煤屑
fin effectiveness　肋片的效率，肋效能
fin efficiency　肋效率
fine file　细锉
fine filter　细过滤器
fine finished surface　精加工表面
fine fissure　微裂隙
fine fitment stage　精装修阶段
fine fit　二级精度配合
fine for delayed payment　滞纳罚金
fine gas cleaning　气体高度净化，烟气高度净化
fine grading　细级配
fine-grained aggregate　细粒集料
fine-grained alluvial deposit　细粒冲积物
fine-grained crack　细晶裂口
fine-grained dropwise condensation　细滴状冷凝
fine-grained film　细粒胶片
fine-grained fracture　细粒状断口
fine-grained soil　细粒土
fine-grained　细粒的
fine grain　细粒度，细粒
fine-granular　细颗粒的，细粒状，细团粒
fine granule　微粒剂
fine gravel　细砾，细砾石，细砂砾
fine grinding　精磨
fine ground cement　细磨水泥
fine-group averaged cross section　精细群平均截面
fine-group structure　精细群结构
fine leak　微泄漏
fine line　细线
finely broken stone　细碎石
finely-divided coal　煤粉
finely-divided mineral admixture　细磨矿物掺合料
finely laminated clay　薄层状黏土
finely-pulverized coal　细煤粉
finely stratified clay　薄成层黏土
fine mesh filter　细滤网
fine mesh model　细网格模型
fine mesh steel fabric　细眼钢丝网
fine mesh　细格筛，细网格
fine motion control rod drive　微调控制棒传动
fine needle thermistor　微针型热敏电阻器
fineness modulus　细度模量
fineness of pulverization　雾化细度
fineness of pulverized coal　煤粉细度
fineness ratio　粒度比，细度比
fineness regulator　细度调节器
fineness test　细度试验
fineness　精度，细长比，径长比，煤粉细度，细度，纯度，光洁度
fine nozzle　细孔喷油嘴
fine particle collection　细颗粒收集
fine particle　细颗粒，细粒
fine phase shifter　精密移相器，精密调相器
fine pitch　小螺距，细牙螺距
fine pore　小孔隙
fine-powder bed　细粉层
fine pulverized coal separator　细粉分离器
fine quality　上等品，优良品质
fine rain　细雨
fine regulation　细调，精调，高分辨能力
finery　精炼炉
fine sand filter　细砂过滤器
fine sand　细砂
fine scale turbulence　微尺度湍流
fine scale turbulent flow　微湍流
fines component　微粒组分
fines concentration　细粉浓度
fine screen　细孔筛，细筛网
fine-sediment load　细悬移质
fines fraction　细颗粒组分
fine size　细度
fine soil　细土
fines recycle　细灰再循环
fine streaming body　细长流线体
fine striped memory　微带存储器
fine structure constant　精细结构常数
fine structure　精细结构
fine stuffing　细填料
fines　细屑，细料，细颗粒，细土粒
fine thread　细牙螺纹
fine-tooth flexible coupling　细牙挠性联轴器
fine trash rack　密拦污栅
fine tuning　微调
fine vacuum　高真空
fine wire　细金属线
fine　精密的，精制的，罚金，罚款，优良，粉煤，筛屑，粉尘
fin generation ratio　肋片释热比
finger badge　指环式胶片剂量计
finger baffle　导向隔板
finger-board　键盘，键板
finger control　手动控制
finger gullying　初期沟蚀，指状沟蚀
finger housing　样品盒
finger joint　指形接合，指形接头

finger-like absorber 指形吸收体
fingernailing 指甲状，半月形焊接变形，焊条药皮的不均匀熔化
finger pier 突堤码头
finger plate gauge 指板规
finger plate 指孔板，压板，齿压板
finger post 指路标
finger-ring badge 指环剂量计
finger stone 小石块
finger tight 指压致密
finger-tip control 按钮控制
finger type contact 指形触点
finger type control rod 指形控制棒
finger walker 指形爬行器，指状步行器
finger 指针，销，手动的，指状物，棘爪
fining agent 净化剂
fining 净化，澄清
finish allowance 加工余量
finish coat 面漆，表面涂层，面层漆，罩面层，终饰层，面层
finish construction survey 竣工测量
finished bolt 精制螺栓
finished bright 抛光
finished cement 水泥成品
finished concrete 完工的混凝土
finished cross-section 竣工断面
finished grade 修整后坡度
finished oil 精制油
finished product review form 成品校审单
finished products storage 成品库
finished product 成品，制成品
finished steel 成品钢
finisher 抹面机，修整机，涂层
finishing cement 水泥抹面
finishing coat mortar 抗裂砂浆
finishing coat 最后一道涂工，表面修饰层，面漆，（油漆）面层，罩面层，罩面抹灰层，装饰涂层
finishing cut 精加工
finishing drawing 竣工图
finishing layer 终饰层
finishing machine 抹面机，修整机
finishing material 装饰材料
finishing mortar 抹面灰浆
finishing paint 面漆，饰面漆
finishing pass 精轧孔型
finishing polish 装修抛光
finishing strip 补强胶条
finishing superheater 末级过热器，最终过热器
finishing temperature 最终温度，终轧温度
finishing work plans 竣工图，装修图
finishing 精磨，最后一道加工，精加工，抛光，抹面，饰面，涂料，修整，最后的
finish lamp 完成操作的信号灯，工程结束信号灯
finish lathing 细车
finish lead 线圈外接头，线圈出线端
finish machining 精加工
finish mark 加工符号
finish painting 面漆
finish to a smooth surface 压光
finish work 最后工序，装修工作，精加工
finish 精加工，抛光，磨光，完成，结束，修饰抹面，装饰，终结
finite 有限的
finite amplitude convection 有限幅度对流
finite amplitude wave 有限振幅波
finite aquifer 有限含水层
finite asymptote 有限渐近线
finite beam 有限长梁
finite blade spacing 有限叶片栅距
finite cascade 有限叶栅
finite cylindrical reactor 有限圆柱形反应堆
finite delay time 有限延迟时间
finite difference equation 有限差分方程
finite difference method 有限差分法
finite difference techniques 有限差分法
finite difference 有限差，有限差分（法）
finite disturbance 有限扰动
finite element analysis 有限元分析
finite element mesh 有限单元网格
finite element method 有限单元法，有限元法
finite element numerical modeling 有限单元数值模拟
finite element 有限元
finite gain 有限增益
finite iteration 有限迭代
finit element method 有限元法
finite-length effect 有限长效应
finite life 有限寿命
finite memory filter 有限存储滤波器
finite memory 有限存储器
finite probability 有限概率
finite reactor 有限体积反应堆
finite region 有限区域
finite rod bundle 有限棒束
finite shield build-up factor 有限屏蔽层积累因数
finite span effect 有限展长效应
finite superposition 有限叠加
finite-thickness type 有限壁厚型的
finite-time filtering 有限时间滤波
finite transit time 电子有限通过时间
finite tube bundle 有限管束
finite wave 有限辐波
finite-width effect 有限宽效应
finit life 有限寿命
fink-ring 调整环，调整圈
fink truss 芬克式屋架
finned can 带肋片包壳，带散热片的包壳
finned coil 肋片盘管，鳍片蛇形管
finned cooler 翅片冷却器
finned fuel element 带肋片燃料元件
finned heater 翅片加热器
finned heating surface 带肋片的受热面
finned motor 带散热片的电机
finned pipe 鳍片管，翅片管
finned radiator 翼形散热器，圆翼形散热器
finned sheet exchanger 板翅式换热器
finned spacer 带肋片的定位件
finned surface 鳍片受热面，翅片受热面
finned tube cooler 翅管冷却器
finned tube economizer 鳍片管省煤器

finned tube　鳍片管，（燃料元件）翅管，带肋管，翅片管，肋片管，鳍状散热管，带散热片管
finned type heating coil　翅片式加热盘管
finned　带散热片的，鳍片的，有鳍的，带肋片的
finning　加鳍片，加肋片
finny distribution pipe　鳍形配水管
finny tail　鳍状尾巴
fin of labyrinth　曲径汽封片，迷宫汽封片
fin of minimum weight　最小重量的肋片
fin pitch　肋片节距
fin-plate heat exchanger　肋板式热交换器，肋管式热交换器
fin removal number　肋片排热数
fin superheater　鳍片管过热器
fin tube bundle　翅片管束
fin tube cleaning equipment　翅片管清洗装置
fin tube heat exchanger　翅片管热交换器，鳍片管热交换器
fin tube wall　鳍片管炉墙
fin tube　翅片管，鳍片管
fin　肋片，鳍片，翅片，尾翼，飞翅，散热片，尾翅，鳍状物，飞边，叶片，风叶
FIO(free in and out)　船方不负担装卸费（用）
FIO(free in & out term)　装卸船方免责
fiord coast　峡湾岸
fiord　峡湾
FIOR(fluid iron ore reduction)　铁矿石流态化还原
FIOS(free in,out and stowed)　船方不负担装卸及平舱费
FIOST(free in,out, stowed and trimmed)　船方不负担装卸费、平舱费和堆舱费
fir-blade mounting　枞树形叶根叶片装配
fire alarm control unit　火灾报警控制器
fire alarm device　火灾警报装置
fire alarm installation　火灾报警装置
fire alarm panel　火灾报警盘
fire alarm sounder　火灾报警器
fire alarm system　消防报警系统，火警报警系统
fire alarm　火灾报警（器），火警
fire and explosion prevention　防火防爆（措施）
fire annihilator　灭火器
fire appliance　消防器材
fire area　防火区
firearmor　镍铬铁锰合金
fireball　火球
fire barrier　挡火墙，挡火层，防火屏蔽，防火间隔
fire bed　火床
fire behavior　燃烧性能
fire bond　耐火泥浆
fire box crown　火箱顶，火室顶
fire box quality steel　火箱钢板，锅炉钢板
fire box　火箱，火室
fire branches and monitors　消防枪炮
fire branch　消防枪
fire break distance of production buildings　厂房的防火距离
fire break distance　防火距离
fire break　挡火墙，挡火层，防火墙，防火间隔，防火间距，隔火设施，防火障，防火空隙地带
fire brick lining　耐火砖衬
fire brick　耐火砖
fire bridge　炉门坎，火墙
fire brigade access window　消防窗
fire brigade axe　消防手斧
fire brigade　消防队
fire bucket　消防水桶
fire bulkhead　挡火墙，防火墙
fire bund　防火堤
fire catchinge　着火
fire cement　耐火水泥
fire check door　半防火门
fire check　挡火闸
fire cistern　消防贮水池，消防水池
fire classification　火灾分类
fire clay lining　耐火泥炉衬
fire clay mortar　耐火砂浆，耐火泥浆
fire clay refractory　火泥耐火材料
fire clay　耐火泥，耐火土
firecoat　氧化皮
fire cock　消防龙头
fire compartmentation　防火分隔
fire compartment　防火分区
fire consumption　消防用水量
fire control car　消防汽车
fire control line　火灾控制线
fire control regulation　防火规程
fire control system　消防系统
fire control unit　防火装置，消防装置
fire control　燃烧调节，消防，防火
fire coupling　消防接口
fire crack　热裂纹，高温应力裂纹，干裂
fire damper　防火风门，防火挡板，消火栓，防火阀
fire-damp-proof machine　沼气式电机，防火防腐型电动机，防沼气电动机
fire damp　防火挡板，爆炸气体，防火装置，沼气
fired clay brick　烧制的黏土砖
fired common brick　烧结普通砖
fire demand rate　消防需水率
fire demand　消防需水量
fire detection and alarm　火灾探测和报警
fire detection system　火警监测系统
fire detection　火警探测
fire detector　火灾探测器［指示器］，火警指示器［探测器］
fire dike　防火堤
fire division wall　防火墙，隔火墙
fire door　防火门，挡火门，炉门
fired perforated brick　烧结多孔砖
fired pressure vessel　受火压力容器
fired steam boiler　燃烧有机燃料的蒸汽锅炉
fire endurance　耐火性，火灾持续性，耐火极限，耐火力
fire engine garage　消防车库
fire engine house　消防车库
fire engine　消防车
fire escape ladder　防火梯

fire escape staircase 防火梯
fire escape 安全出口，太平门，防火梯
fire-exit bolts 太平门闩
fire extinction equipment 灭火设备
fire extinction 灭火
fire extinguisher system 灭火系统
fire extinguisher 灭火机，灭火器
fire extinguishing agent 灭火剂
fire extinguishing diesel pump 柴油机消防泵
fire extinguishing sand 灭火砂
fire extinguishing with chemical agent 化学药剂灭火
fire extinguishing 灭火
fire extinguishment 灭火
fire-fighting access 消防通道
fire-fighting control system 消防控制系统
fire-fighting equipment 消防器材，消防设备
fire-fighting passage 消防通道
fire-fighting passway 消防通道
fire-fighting pump house 消防泵房
fire-fighting pump 消防泵，灭火泵
fire-fighting ring mains 灭火环形主管
fire-fighting system 消防系统
fire-fighting vehicle 灭火消防车
fire-fighting water standpipe 消防栓
fire-fighting water storage tank 消防水贮存箱
fire-fighting water 消防水
fire-fighting 防火的，消防的
fire flow 消防流量
fire forcible entry tool 消防破拆工具
fire gas 可燃气体
fire grate 炉排
fire hazard 火灾危害性，火灾，失火危险
fire hole 火口，炉口
fire hose and fireplug cabinet 消火栓柜
fire hose reel 灭火水龙带卷
fire hose station 消防站
fire hose 水龙带，防火软管，消防水龙管，消防软管，消防水龙带，消防箱
fire hydrant box 消火栓柜
fire hydrant chamber 消防栓箱
fire hydrant 消防栓，消防龙头，灭火龙头，防火栓
fire insulation 耐火隔热性
fire insurance 火灾保险
fire integrity 挡火性，防火完整性，耐火完整性
fire ladder 消防梯
fire layer 燃烧层
fire lift 消防电梯
fire limit 消防界限
fire line 最高火界
fire load density 热负荷密度【火灾】，火灾荷载密度
fire load 可燃材料荷载，火荷载，火灾荷载，燃烧负荷
fire main pipe 主消防水管
fire main 消防总管，消防干管，消防用的主水管
fireman general protection equipment 消防员常规防护装备
fireman's axe 消防斧
fireman special protection equipment 消防员特种防护装备
fireman 司炉，消防人员，消防员
fire material 防火材料
fire monitor 消防炮
fire-off 熄火
fire parameter 火灾参数
fire partition 隔火墙
firepath 防火通道
fireplace 炉膛空间，壁炉
fire plug 防火栓，消火栓，消防龙头
fire point of oil 燃油燃点
fire point 着火点，燃点
firepot 坩埚
fire precaution measures 防火措施
fire precaution 防火注意事项，防火装置
fire pressure 火压，灭火水压力，消防水压（力）
fire presuppression 防火预备工作
fire prevention measures 防火措施
fire prevention 防火
fire priming pump 引水消防泵
fire-proof blockage 防火堵料
fire-proof bulkhead 防火墙，防火壁，防火舱壁
fire-proof cable 防火电缆
fire-proof cement 耐火水泥
fire-proof construction 防火建筑，耐火构造
fire-proof distance 防火距离
fire-proof door 防火门
fire-proof dyke 防火堤
fire-proof engine 防火式电机
fire-proof frame 防爆机座
fire-proof glass 耐火玻璃
fire proofing of cables 电缆的防火措施，电缆的防火
fire proofing wood 防火木材
fire-proof limit 耐火极限
fire-proof machine 防爆式电机
fire-proof material 耐火材料
fire-proof oil 抗燃油
fire-proof packing 防火填料
fire-proof paint 防火漆
fire-proof panel 防火板
fire-proof partition 耐火隔板
fire-proof wood 防火木材
fire-proof 防火，耐火，耐火的，防火的，抗燃的
fire-propagation 火灾传播
fire protected cable 防火电缆
fire protection and anti-explosion 防火与防爆
fire protection code 防火规范，消防规范
fire protection control room 消防控制室
fire protection data system 火焰光度检测器
fire protection for cable 电缆防火
fire protection pillow 防火枕，阻火包
fire protection pilot valve 防火滑阀
fire protection requirement 防火要求
fire protection standard 防火标准
fire protection system of converter station 换流站防火系统
fire protection system 防火系统
fire protection wall 防火墙

fire protection　防火，消防，消防措施，防火装置，火灾防护装置
fire pump　消防泵，消防水泵
fire-rated acoustic door　防火隔音门
fire-rated door　标准防火门
fire rate　防火率
fire rating　防火程度，耐火等级，耐火极限，耐火率
fire rescue equipment　消防救援器材
fire resistance cable　防火电缆
fire resistance rating　防火程度，耐火等级，耐火额定值，耐火值
fire resistance test　防火性试验，耐火试验
fire resistance　耐火性，耐火能力
fire resistant cable　耐火电缆，阻燃电缆
fire-resistant concrete　耐热混凝土
fire-resistant door　防火门
fire-resistant fluid　耐燃液体，抗燃油，抗燃液
fire-resistant housing　防爆机壳，耐火机壳
fire-resistant hydraulic fluid　抗燃液
fire-resistant hydraulic oil　抗燃油
fire-resistant lubricant　抗燃润滑油
fire-resistant oil　抗燃油
fire-resistant wire　耐火绝缘导线
fire-resistant　耐火的，耐高温的，耐火
fire resisting cable　耐火电缆
fire resisting ceiling　耐火天花板
fire resisting column　耐火柱
fire resisting concrete　耐火混凝土
fire resisting damper　防火挡板，防火阀，防火挡板
fire resisting duct　耐火管道
fire resisting element　耐火件
fire resisting finishes　耐火饰面，耐火油漆涂层
fire resisting floor　耐火楼板
fire resisting glazing　耐火玻璃
fire resisting layer　防火层
fire resisting limit of main structural member　主结构构件的耐火限度
fire resisting material　耐火材料
fire resisting partition　耐火隔墙
fire resisting roof　耐火屋顶
fire resisting shaft　耐火竖井
fire resisting shutter　防火卷帘
fire resisting suspended ceiling　耐火吊顶
fire resisting window　防火窗
fire resistive construction　耐火建筑
fire resistive grade of buildings　建筑物的耐火等级
fire resistive grade　耐火等级
fire resistive material　耐火材料
fire resistive wall　耐火墙
fire-retardant coating　耐火涂层，阻燃层，阻燃涂料
fire-retardant conveyor belt　阻燃输送带，难燃输送带
fire-retardant design　防火设计，防火阻燃设计
fire-retardant material　阻燃材料
fire-retardant paint　防火漆，耐火漆，耐火涂料，阻燃型涂料
fire-retardant type belt　阻燃型皮带
fire-retardant　难燃，难燃的，阻燃的
fire risk　火灾危险
fire safety equipment　防火安全设备
fire safe type ball valve　耐火型球阀
fire safety switch　防火安全开关
fire safety　防火安全措施
fire screen　防火墙，防火栅
fire separation　防火分隔
fire service water　消防水
fire shutter　防火百叶窗
fireside performance test facility　炉内过程试验装置
fireside surface　向火面，烟气侧表面
fire-side　烟气侧
fire sign　火灾信号
fire-smoke detection　火烟检测
fire-spreading　火灾传播
fire spread rate　火灾蔓延速度
fire stability　燃烧稳定性，耐火稳定性
fire station　消防站
fire stone　火石，耐火石
fire stop screen　挡火屏障
fire stop　挡火墙，挡火层，防火墙，防火屏障
fire stream　消防喷射水流
fire suction hose　消防吸水管
fire suppression　灭火
fire system　消防系统
fire telephone line　火警电话线
fire telephone　火警电话
fire terrace　防火平顶
fire testing　燃烧试验，燃点测定
fire trap　无太平门的建筑物
fire truck　消防车
firetube boiler　火管锅炉
firetube smoke tube　烟管，火管
firetube-watertube boiler　水管火管合并式锅炉
firetube　火管，烟管
fire valve　防火阀，消防阀
fire vehicle pump　车用消防泵
fire wall with 5-hour fire rating　耐火5小时的防火墙
fire wall　防火墙，网络安全通路
fire warning system　火灾报警系统
fire water branch　消防水枪
fire water jockey pump　消防水增压泵
fire water monitor　消防水炮
fire water pump　消防水泵
fire water supply　消防给水
fire window　防火窗
firewire sensing element　缆式火灾探测器敏感元件
fire　燃烧，点火，生火，火，火灾，着火
firing angle　触发角，点火角，点弧角，引燃角，点弧角，点火角
firing cable　引火线
firing cell　燃烧室
firing chamber　燃烧室，燃烧箱，引爆室，炸药室
firing circuit　触发电路，点火电路
firing condition　燃烧工况
firing device　燃烧装置

firing door （炉膛）加煤门
firing downward burner 下射喷燃器
firing equipment 加热设备，点火设备
firing floor 运行层
firing key 引爆器钥匙
firing order 点火次序，点火指令
firing pattern 引爆方式
firing point 着火点
firing position 燃料进给位置
firing potential 发火电位，点火电位
firing pulse 燃烧脉动
firing range 射程，靶场，着火范围
firing rate 燃烧强度，燃烧率
firing sequence 燃烧顺序
firing shrinkage 加热收缩，烧缩
firing system 燃烧系统，燃烧设备
firing tangential circle 燃烧切圆
firing test 点火试验
firing tool 司炉工具
firing-up period 升火时间
firing voltage 点火电压，触发电压
firing 开火，烧制，点火，燃烧，引火
firm agreement 不可撤销的协议
firm bargain 确实的买卖，实盘交易
firm bid 递实盘
firm capacity 可靠容量，保证出力
firm clay 硬黏土
firm contract 不可撤销的合同，束缚性合同
firm demand 固定需量
firm discharge 固定流量
firm energy 保证电能，保证能量，稳定电能
firmer chisel 凿子，直边凿
firmer 凿子
firm load demand 固定负荷
firm load 固定负荷
firm lump sum contract 固定总价合同
firm offer 实盘，不可撤销的发盘，报实盘
firm order 确定订货
firm output 水电站保证出力，保证出力，固定出力，恒定输出
firm peak capacity 恒定最高容量，恒定最高出力
firm peak discharge 恒定最大输出
firm power dependable power 保证功率
firm power energy 恒定电能
firm power 恒定功率，可靠功率
firm price bid 固定价格投标
firm price 确定的价格，固定价格
firm sale contract 确定销售契约，确定销售合同
firm soil 坚硬土壤，硬土
firm transfer capability 保证转换能力
firm ware 固件【指软件硬件相结合】，稳固设备，固态件
firm yield 稳产
firm 坚固的，牢固的，坚定的，公司，商行，实盘
firncrash 雪冰崩落
first admission of steam into the turbine 汽轮机首次进汽，汽机第一次进汽
first aid repair 初步修缮，紧急修理，急修
first aid worker 救援工人，急救工人
first aid 急救
first assembly 初步装配
first azimuth method 第一方位角法
first beneficiary of a transferable credit 可转让信用证的第一受益人
first beneficiary 第一受益人
first bit 小钻
first boiler firing date 锅炉第一次点火日期
first charge 首次装量，始装料，首次装料
first class insulated wire 第一类绝缘线
first class pyranometer 一级（工作）总日射表
first class pyrheliometer 一级（工作）直接日射表
first class pyrradiometer 一级（工作）全辐射表
first class service 一流服务，优质服务
first class workmanship 第一流的工艺
first class world standard 世界一流水平
first class 第一流
first coal firing date 第一次燃煤日期
first coat of paint 第一道漆
first coat 底漆
first collision dose 首次碰撞剂量
first collision neutron 首次碰撞中子
first collision probability 首次碰撞机率，首次碰撞概率
first concrete 第一批浇灌的混凝土
first content 基波分量
first core inventory 首次堆芯铀装载量，首次堆芯装载量
first core loading 首次堆芯装料
first core uranium inventory 首次堆芯铀装载量
first core 初堆芯，第一炉核燃料，第一炉料，首次堆芯
first cost 初步费用，开始费用，生产成本，初期费用，原始成本，初投资费用
first criticality 初次临界
first critical speed 一阶临界转速
first-degree price discrimination 一级价格歧视
first-degree 第一度的
first demand guarantee 见索即付保函
first demand 首次要求
first demonstration 首次演示，首次验证
first derivative 一阶导数
first detector 第一检波器
first difference equation 一阶差分方程
first difference 一次差分
first divergence 首次临界试验
first D_2O inventory 初次重水装载量
first draft 初稿
first driver 初级激励器
first filling （水库等的）初次蓄水，初期蓄水
first flight collision probability 首次飞行碰撞概率
first flight correction 首次飞行校正
first flight leakage 首次飞行泄漏
first flight neutron loss 首次飞行中子损失
first flight neutron 首次飞行中子
first floor 底层
first forbidden 一次禁戒
first fuel inventory 初始燃料装载量
first fuel loading 初始燃料装量，首次装燃料，首次装料

first generation neutron　第一代中子
first generation　第一代
first grade timber　一级木材
first gust　阵风前阵
first half year　上半年
first-hand data　第一手资料，原始数据
first-hand exchange　直接交流
first-hand information　第一手资料
first-hand　第一手的，直接得来的
first hard copy print publication right　首次印刷权
first harmonic oscillation　基本振荡，基波振荡
first harmonic resonance　基波共振，基波谐振
first harmonic　一次谐波，基波
first impound period　首次蓄水期
first-in first-out algorithm　先进先出算法
first-in first-out buffer　先入先出缓存器
first-in first-out discipline　先进先出项目
first-in first-out method　先进先出法，先入先出法
first-in first-out queue　先进先出排队
first-in first-out register　先入先出寄存器
first-in first-out rule　先进先出规则
first-in first-out　先进先出（法）
first in last out　先进后出
first inspection　首次检验
first installation　第一批安装
first installment　第一期分期付款
first integral　初积分
first-in　先进
first jitter　初始颤动
first law of thermodynamics　热力学第一定律
first-level address　直接地址，一级地址
first-level code　直接代码，一级码
first-level controller　一级控制器，局部控制器
first line maintenance　小修
first loading of the reactor core　堆芯初始装料
first mode　一阶模态，基本振型
first mortgage　第一抵押，首次抵押
first motion of p-wave　p 波初动
first motion wave　初至波
first motion　初动
first of a kind　一类中的第一个，同类别首个
first-off component　首先生产的部件，首先生产的设备
first-off test　初次试验
first operating period　初期运行阶段
first order approximation　一阶近似
first order closure method　一阶封闭法
first order differential equation　一阶微分方程
first order equation　一次方程，一阶方程
first order lag　一阶滞后，一次滞后
first order lead　一阶超前
first order levelling　一等水准
first order perturbation theory　一阶微扰理论
first order reaction　一级反应
first order station　一等测站
first order subroutine　第一级子程序，直接输入的子程序
first order system　一阶系统
first order traverse　一等导线
first order triangulation　一等三角测量
first order weather station　一等测候站，一级气象站
first order　一阶，一级，第一阶
first oscillation period of power system　电力系统第一振荡周期
first out alarm and signal system　第一（原因）信号报警
first outlay　初投资
first out　首出原因
first overhaul　第一次大修
first part inspection　第一部分检查
first permanent structural concrete　第一批浇灌的永久性结构混凝土
first phase of construction　一期工程
first-pitch of coil　线圈第一节距
first-pole-of-clear factor　首相断开因数
first pour　第一期浇筑
first principal stress　第一主应力
first priority　绝对优先（权）
first prize　头等奖，一等奖
first probability distribution　第一概率分布
first quarter neap tide　上弦小潮
first radiation constant　第一辐射常量
first rank　（第）一流的
first rate　（第）一流的
first reactor core　第一炉燃料，第一个堆芯
first reactor startup　反应堆首次启动，反应堆初次启动
first requirement　首次要求
first resonance　基波共振
first responder　第一急救者
first row　第一排
first shell pressure　调节汽室压力
first shipment　第一批装船
first sort tempering brittleness　第一类回火脆性
first stage blade　第一级叶片
first stage cofferdam　第一期围堰
first stage concrete　第一期混凝土
first stage cooling　一期冷却【初期冷却】
first stage installation　（电站的）第一期装机
first stage steam bypass　一级蒸汽旁路
first startup test　首次启动试验
first synchronisation of the unit with the network　机组第一次并网
first tap　首道丝锥
first terrace　一级阶地
first vibration　第一振型
first visible red　波长最长的可见光
first voltage range　第一电压范围，第一电压级
first winding　原边绕组，初级绕组
firth　三角港，港湾
fir-tree aerial　树形天线
fir-tree blade root　枞树形叶根
fir-tree connection　枞树形连接
fir-tree fastening　树枝形固定
fir-tree straddle root blade　外包枞树形叶根叶片
fir-tree-type groove　枞树型叶根槽
fir-type groove　枞树形叶根槽
firt　圈梁
fiscal agent　财务代理银行
fiscal and economic discipline　财经纪律

fiscal deficit 财政赤字
fiscal evasion 漏税
fiscal expenditure 财政支出
fiscal policy 财政政策
fiscal quality metering 会计质量计量
fiscal stamp 印花税票
fiscal subsidies 财政补贴
fiscal year 财政年度，会计年度
fiscal 国库的，国库岁收的，财政的
FIS(fee into storage) 仓库交货价格
fish-bellied beam 鱼腹梁
fish bolt 鱼尾（板）螺栓，轨节螺栓
fishbone diagram 鱼骨图
fishery resource 水产资源
Fisher 费希尔仪表公司【美国】
fish eye 白点
fishing wire 电缆牵引线，蜿蜒导线
fishing 蜿蜒导线
fish ladder 鱼梯
fish lift 升鱼机
fish line conductor 螺旋形导线
fish lock 鱼闸
fish pass structure 过鱼建筑物，过鱼设施
fish pass 鱼道
fish place 鱼尾板
fish plate bolt 鱼尾板连接螺栓
fish plate 鱼尾板，接合板
fish-pole antenna 钓鱼竿式天线
fish-pole technique 钓竿法，钓竿技术
fish scale fracture 鱼鳞形断裂
fish stairs 鱼梯
fishtail bit 鱼尾（形）钻头
fishtail bolt 鱼尾螺栓
fishtail burner 扇形火焰燃烧器
fishtail wind 不定向风
fishtail 鱼尾状的，扇形火焰
fish tape 敷线牵引线，蜿蜒带
fish way 鱼道
fish winding 螺旋绕组
fish wire 电缆牵引线
fisser 可分裂物质，可裂变物质
fissile element 易裂变元素
fissile fuel content 易裂变燃料含量
fissile fuel inventory 裂变燃料存量
fissile fuel producer 易裂变燃料生产堆
fissile fuel 裂变燃料，易裂变燃料
fissile inventory ratio 易裂变燃料存量比
fissile investment 易裂变燃料投入
fissile material flow rate 易裂变材料流通率
fissile material flow 裂变材料流通量
fissile material measurement 易裂变材料测量
fissile material production reactor 易裂变材料生产堆
fissile materials safeguard system 易裂变材料保障体系
fissile material 易裂变材料，裂变物质，核燃料，可裂变的物质，裂变材料
fissile metal 裂变金属
fissile nucleus 裂变核，易裂变核
fissile nuclide 裂变核素，裂变产物，易裂变核素
fissile phase 裂变相
fissile region 裂变区，活性区
fissile resistor 易裂变材料电阻温度计
fissile 易裂变的，裂变的，分裂的，可裂变的
fissility 易裂性
fissionable fuel 核燃料，可裂变燃料，可分裂燃料
fissionable gas 可裂变气体
fissionable material inventory 可裂变物质存量
fissionable material 核燃料，裂变物质，可裂变材料
fissionable nucleus 可分裂的核，可裂变核
fissionable plutonium 可裂变钚
fissionable 可分裂的，可裂变的
fission action 裂变分裂，裂变作用
fission barrier 裂变屏障
fission break monitoring 裂变产物逸散监测
fission break monitor 裂变产物逸散监测器
fission burner 裂变燃烧堆【燃烧器】，裂变燃烧器
fission chain 裂变链，链式反应
fission chamber 裂变室
fission chemistry 裂变化学
fission contribution 裂变贡献
fission converter 裂变转换器
fission corrosion 裂变腐蚀
fission cross-section resonance 裂变截面共振
fission cross section 裂变截面
fission deposition 裂变沉积
fission efficiency 裂变效率
fission energy spectrum 裂变能谱
fission energy 裂变能
fissioner 可裂变物质，可分裂物质，裂变物质
fission event 裂变作用
fission fraction 裂变份额，裂变份数
fission fragment detection 裂变碎片探测
fission fragment 裂变碎片
fission-fusion hybrid reactor 裂变聚变混合堆
fission-fusion symbiote 裂变聚变共生
fission gas formation 裂变气体形成
fission gas laden waste gas 含裂变气体的废气
fission gas monitor 裂变气体监测器
fission gas plenum 裂变气体气腔【燃料棒】
fission gas pressure relieved fuel rod 已释放裂变气体压力的燃料棒
fission gas pressure 裂变气体压力
fission gas release rate 裂变气体释放率
fission gas release 裂变气体释放
fission gas 裂变气体
fission heating 裂变释热
fission hole 裂变中子辐照孔
fissioning core 裂变堆芯
fissioning distribution 裂变中子分布
fissioning nucleus 裂变核
fission-initiating neutron 裂变触发中子
fission material 可裂变材料
fission mean-free path 裂变平均自由程
fission neutron absorption 裂变中子俘获，裂变中子吸收
fission neutron flux 裂变中子通量
fission neutron source 裂变中子源

fission neutron spectrum　裂变中子谱
fission neutron yield　裂变中子产额
fission neutron　裂变中子
fission nucleus　裂变核
fission nuclide　裂变核素
fission particle　裂变碎片
fission physics　裂变物理学
fission poison　裂变毒物
fission process　裂变过程
fission product accumulation　裂变产物聚集
fission product activity　裂变产物放射性，裂变产物活度
fission product adsorption filter　裂变产物吸附过滤器
fission product adsorption　裂变产物吸附
fission product build-up　裂变产物积累
fission product carry-over　裂变产物夹带
fission product catalysis　裂变产物催化
fission product chain　裂变产物链
fission product concentration　裂变产物浓度
fission product contaminant　裂变产物污染物，裂变生沾污物
fission product contamination　被裂变产物污染
fission product debris　裂变产物碎片
fission product decay heat　裂变产物衰变热
fission product dispersion　裂变产物弥散
fission product gas release　裂变产物气体释放
fission product inventory in the core　堆芯裂变产物存量
fission product isotope　裂变产物同位数，裂变产物同位素
fission product monitoring system　裂变产物监测系统
fission product nucleus　裂变产物核
fission product poisoning　裂变产物中毒
fission product refining plant　裂变产物精制厂
fission product release and transport　裂变产物释放与输运
fission product release rate　裂变产物释放率
fission product release　裂变产物释放
fission product retaining fuel　滞留裂变产物的燃料
fission product retention capacity　裂变产物滞留容量
fission product retention property　裂变产物滞留特性
fission product retention　裂变产物滞留
fission product separation　裂变产物分离
fission product spectrum　裂变产物能谱
fission products track detection　裂变产物径迹探测法
fission product stripper　裂变产物分离装置，裂变产物剥离器
fission product trap　裂变产物捕集器，裂变产物阱
fission product waste　裂变产物废物
fission product yield　裂变产物产额
fission product　裂变产物
fission rate　裂变率
fission reaction rate　裂变反应率
fission reaction　裂变反应
fission reactor　核裂变反应堆
fission recoil damage　裂变反冲损伤
fission resonance　裂变反应共振，裂变共振
fission segment　裂变碎片
fission source term　裂变源项
fission source　裂变源
fission spectrum　裂变能谱，裂变中子谱，裂变谱
fissions per initial fissionable atoms　已裂变燃料原子数占初始装料总的易裂变原子数的份额
fissions per initial heavy metal atoms　已裂变的燃料原子数占初始装料总的重金属原子数的份额
fission spike　(辐照损伤)裂变尖峰，裂变峰值
fission-suppressed blanket　压抑裂变的转换区
fission-type reactor　裂变反应堆
fission unit　裂变单元
fission width　裂变幅度
fission yield curve　裂变产额曲线
fission yield　裂变产额
fission　裂变，分裂，裂变产物合金，裂片合金
fissuration structure　裂缝结构
fissuration　裂缝，龟裂，形成裂隙
fissured clay　裂隙黏土
fissure direction　裂隙方向
fissure drainage　裂隙排水
fissured rock　裂隙岩石
fissure eruption　裂缝喷发
fissure filling　裂缝充填物
fissures closeness　裂隙密集度
fissures system　裂隙系
fissure structure　裂缝结构
fissure vein　裂缝脉
fissure water　裂缝水
fissure　龟裂，缝隙，裂纹，裂缝，裂隙，裂痕
fist aid　急救
fit clearance　配合间隙
FITG＝fitting　连接件
fit goodness　拟合良度，拟合优度
fit in　使适合
fit key　配合键
fitness　恰当，适当，适应性
fit out　装备，装配，以……装备，供给……以必需品
fit quality　配合等级
fits and tolerances　配合与公差
fittage　装配任务
fitted bolt　配螺栓，定位螺栓
fitted capacity　装配容量，设备容量
fitted curve　根据试验点描出的曲线，拟合曲线
fitted key　导向键，装配键
fitted-on-closure stud tensioner　配装的封头螺栓拉伸机
fitted with instrumentation　用仪表装备的
fitter bolt　装配螺栓
fitter's bench　钳工平台
fitter's work　设备安装工作
fitter　装配工，钳工
fitting allowance　装配余量，配合公差
fitting assembly　装配总成
fitting constant　拟合常数
fitting criterion　拟合准则
fitting fixture　装配工具

fitting metal 配件，附件
fitting-out 装置
fitting parameter 拟合参数
fitting part 配件，零件
fitting piece 配件
fitting pipe 连接管
fittings and accessories in heating pipeline 供热管路附件
fitting shop 钳工车间
fittings storage 零件库
fitting surface 安装面
fittings 管件，配件【T 接头、弯头、龙头等】，备件，附件，零件，金具，连接件，五金件，接头
fitting tolerance 配合公差，装配公差
fitting-up team 安装队
fitting-up 装配
fitting work 整修工程，装配工作
fit tolerance 配合公差
fit-up gap 装配间隙
fit-up lug 装配吊耳
fit-up tolerance 装配公差
fit-up 装备，装配
fit 配合，装配，匹配，适合，接头，适应，镶嵌
five-address instruction 五地址指令
five digit series airfoil NACA 五位数字系列翼型
five finger fork type blade root 五叉插入式叶根，叉型叶根
five-hole punched tape 五孔穿孔带
five-legged transformer 五芯柱变压器
five limb type core 五柱式铁芯
five-plywood 五夹板
five put-throughs and site leveling 五通一平
five-yearly outage program 五年一次停运计划
five-yearly outage 五年一次停运
fixation 固定，安置，固定的，固定剂
fix-bed gasification 固定床气化
fixed 定位的，固定的，恒定的
fixed action 固定作用
fixed adsorbent bed 固定吸附剂床
fixed air gap 固定空隙
fixed anchor 固定锚，固定锚头，固定支座
fixed and oblique ladders 固定式直梯和斜梯
fixed antenna 固定天线
fixed aperture lens 固定光圈镜头
fixed arch 固定拱，固端拱，无铰拱
fixed armature 固定电枢
fixed ash 固有灰分
fixed assets accounting 固定资产核算
fixed assets depreciation period 固定资产使用年限，固定资产折旧年限
fixed assets depreciation range system 固定资产折旧法
fixed assets depreciation range 固定资产折旧幅度
fixed assets under book value 固定资产盘亏
fixed assets under operating lease 租入固定资产
fixed assets 固定资产
fixed axial flow type steam separator 固定轴流式汽水分离器
fixed-bar stoker 固定炉排
fixed beam 固定梁
fixed bearing 固定支座，固定轴承
fixed bed ion exchanger 固定床离子交换器
fixed bed model test 定床模型试验
fixed bed model 定床模型
fixed bed reactor 固定床反应器
fixed bed 定床，固定床
fixed bias 固定偏压
fixed bid price 固定投标价
fixed blade propeller turbine 定轮叶旋桨式水轮机
fixed blade propeller wheel 定轮叶旋桨式水轮
fixed blade ring 静叶环
fixed blade turbine 定轮叶式水轮机，旋桨式水轮机
fixed blade 固定轮叶，静叶片
fixed block 定滑轮，固定滑车
fixed-boom crane 定臂式起重机
fixed boundary 固定墙，固定边界
fixed bracket 固定支架
fixed brush type motor 固定电刷式电动机
fixed buoy 固定浮标
fixed cantilevered stacker 固定悬臂堆料机
fixed capacitor 固定电容器
fixed capital investment 固定资本投资
fixed carbon 固定炭，固定碳
fixed cavitation 固定空穴
fixed charged rate 固定费用率
fixed charge 固定费（用），固定开支，固定电荷
fixed coil 固定线圈
fixed constant 不变常数
fixed contact 固定触点
fixed contamination 固定污染
fixed contract 定额合同
fixed coordinate system 固定坐标系
fixed cost contract 固定价格合同
fixed cost 固定费用，固定成本，固定开支，不变成本
fixed counter weight 死配重
fixed counter 固定配重，固定式车钩
fixed coupling 固定联轴节
fixed crane 固定式起重机
fixed-crest spill-way 固定堰顶式溢洪道
fixed-crest weir 固定堰顶式堰
fixed-cycle operation 固定周期操作，固定周期运算
fixed cycle 恒定周期
fixed dam 固定坝
fixed-date delivery 定期交付
fixed datum 固定基准
fixed delivery pump 定容量泵
fixed derrick 固定式动臂起重机
fixed differential reducing valve 定差减压阀
fixed dimension 固定尺寸
fixed disk driver 硬盘驱动器
fixed displacement motor 定容量马达，定排量马达
fixed displacement pump 定排量泵
fixed edge slab 嵌固板

fixed edge 固定边
fixed electrode 固定电极
fixed end arch 定端拱
fixed end beam 固定端梁，定端梁
fixed end discharge conveyor 尾部固定的卸料皮带机
fixed end moment 固端弯矩，固端力矩
fixed end wall 固定端墙
fixed end 固定端
fixed equipment 固定设备
fixed excitation （发电机的）固定励磁
fixed expenditure 固定开支
fixed extinguishing system 固定式灭火系统
fixed facilities 固定设施
fixed field 固定场，参考场
fixed-film reactor 固定膜反应器
fixed filter unit 固定过滤装置
fixed filter 固定式过滤器
fixed fire extinguisher 固定式灭火器
fixed flange 固定法兰盘，固定法兰
fixed fluidized bed 固定流化床
fixed-focus operation 定焦点操作
fixed-focus pyrometer 定焦点高温计
fixed-focus total-radiation pyrometer 定焦点全辐射高温计
fixed form 固定模板
fixed framed arch 构架式固端拱
fixed frequence continuous wave transmission 固频连续波反射
fixed frequency 固定频率
fixed gain multi-variable excitation control 固定增益多变量励磁控制
fixed gang-condenser 固定电容器组
fixed gantry crane 固定门式起重机
fixed gas 定成分气体
fixed gate generator 固定选通脉冲发生器
fixed gate 固定选通脉冲
fixed generator speed fixed-pitch wind turbine 恒速定桨距风电机组，FGS－FP 风电机
fixed generator speed fixed-pitch 恒速定桨距
fixed generator speed variable-pitch wind turbine 恒速变桨距风电机组，FGS－VP 风电机组
fixed generator speed variable-pitch 恒速变桨距
fixed grate 固定炉排
fixed ground-board 固定地板【风洞】
fixed guide vane 固定导叶
fixed-handle breaker 有固定手柄的断路器
fixed handling boom 固定的起重杆
fixed hopper 固定式料斗
fixed hub 固定桨毂
fixed-incidence vane 固定叶片
fixed incline boom stacker 固定斜升臂堆料机
fixed index measuring instrument 固定标度测量仪表
fixed interest 固定利率
fixed internal 固定间隔
fixed interval reinforcement 定期强化
fixed interval 固定间隔
fixed investment 固定投资
fixed ion exchange site 固定的离子交换基
fixed joint 刚性连接，固接点
fixed ladder 固定扶梯
fixed-length field 定长字段，定长信息组
fixed-length record 定长记录
fixed light 固定窗，固定灯光
fixed load 固定负荷，固定荷载
fixed logic 固定逻辑
fixed loss 固定损耗
fixed louver 固定百叶窗
fixed lump sum price contract 固定总价合同
fixed lump sum price 固定总价
fixed mask 固定屏
fixed mass division 定质量缩分
fixed mast 固定式桅杆
fixed memory 固定存储器
fixed mirror system 固定反射镜系统
fixed mixer 固定式拌和机
fixed negatively charged exchange site 固定的带负电荷交换基
fixed nozzle 固定喷嘴
fixed obstacle 固定障碍物
fixed order 定期订货
fixed orifice 固定节流孔
fixed overflow dam 固定式溢流坝
fixed packing 固定填料
fixed payment rate 定额制电价
fixed phase 固定相，参考相
fixed-piping 固定管道
fixed piston sampler 固定活塞式取样器
fixed pitch blade 定桨距叶片
fixed pitch propeller 定距螺旋桨
fixed pivot 固定支点
fixed plug ring （钢冷快堆）固定盖环，固定塞环
fixed plug （钠冷快堆容器）固定盖，固定塞
fixed point arithmetic 定点运算
fixed point calculation 定点计算，定点运算
fixed point computation 定点计算
fixed point computer 定点计算机，固定小数点计算机
fixed point number 带定点的数
fixed point operation 定点运算
fixed point representation 定点表示法
fixed point system 定点系统
fixed point 定点，固定点【管道，结构】
fixed poison 固定核毒物
fixed-position addressing 固定位置寻址，固定位置访问
fixed positively charged exchange site 固定的带阳电荷交换基
fixed preassigned multiple access 固定预分配多址
fixed premium 固定保险费
fixed pressure control 定压控制
fixed pressure operation of deaerator 除氧器定压运行
fixed pressure operation 定压运行
fixed pressure reducing valve 定值减压阀
fixed pressure 定压【启动】，固定压力，恒压力，恒压
fixed price competition 固定价格竞争
fixed price contract 固定价合同

fixed price lump sum contract 固定总价合同
fixed price 定价，固定价格
fixed program computer 固定程序计算机
fixed protective rail 固定式防护栏杆
fixed pulley 定滑轮，固定滑轮
fixed quantity 定量
fixed range 固定范围
fixed rate flow 恒定流量，定速流动
fixed rate 固定利率
fixed ratio division 定比缩分
fixed ratio 固定比率
fixed reactor vessel closure head insulation 反应堆容器封头的固定保温层
fixed recording head 固定记录头
fixed residue 固定残渣
fixed resistance 固定电阻
fixed resistor 固定电阻器
fixed ring consolidometer 固定环式固结仪
fixed ring 固定环
fixed roller gate 定轮闸门
fixed roll 固定辊【辊碎机】
fixed rotating stacker 固定旋转式堆料机
fixed saddle 固定鞍座
fixed sash 固定窗扇，固端框扇
fixed screen 固定筛
fixed separation point 固定分离点
fixed service 定点通信
fixed set-point control 定值控制，定值调节
fixed shutter 固定百叶窗
fixed slewable stacker 固定式回转堆料机
fixed slit 固定狭缝
fixed source entropy 固定源熵
fixed source iteration 固定源迭代
fixed span 固定跨，固端跨
fixed spot welding machine 固定式点焊机
fixed staff 固定标尺
fixed stop 固定挡块
fixed storage 固定存储器
fixed sulfur 固定硫
fixed support 固定支座，固定支架，固定支点
fixed symmetrical arch 固定对称拱
fixed tab 固定翼片
fixed telemetering station 固定遥测站
fixed temperature level 不变的温位
fixed term contract 定期合同
fixed time humidity 定时湿度
fixed time lag 定时滞
fixed time temperature 定时温度
fixed time test 定时寿命试验
fixed-trip circuit-breaker 手动跳闸断路器，非自动跳闸断路器
fixed-trip switch 手动跳闸开关，非自动跳闸开关
fixed tube plate 固定管板
fixed tube-sheet heat exchanger 固定管板式热交换器
fixed unit price contract 固定单价合同
fixed value 定值
fixed vane turbine 定叶式水轮机
fixed vane 固定叶片
fixed vessel head insulation canning 固定的封头保温层罩壳
fixed voltage 固定电压
fixed water 固着水
fixed weir 固定（式）堰
fixed wet well pumping station 固定湿井式取水泵房
fixed wharf 固定码头
fixed wheel gate 定轮闸门
Fixed Winchester 温切斯特盘
fixed window 固定窗
fixed wing 固定翼
fixed wire wound resistor 线绕式固定电阻器
fixed word length 固定字长
fixed wrench 固定扳手
fixed yaw rotor 定向风轮
fixer 定影液，固定器
fixing agent 固定剂
fixing device 固定装置
fixing dimension 装配尺寸
fixing of instrument tube 仪表管固定
fixing screw 固定螺杆，定位螺钉
fixing session 议价会议
fixing support 固定支座
fixing the price 规定价格，定价
fixing trestle 固定支架
fixity 稳定性，硬度
fix stopper 定位销，固定销
fixtingbolt 固定螺栓
fixture drain 卫生器具排水管
fixture load 器具负荷
fixture shaft 固定轴
fixture-supply pipe 卫生器具给水管
fixture wire 电器引线，设备引线
fixture 配件，附件，固定值，固定量，定位器，固定装置，固定架，夹具，焊接夹具，卡具，定期放款，定期存款，固定在某位置的人或物，定期定点举行的活动
fix up a contract 签订合同
fix with screw 螺钉固定
fix 固定，安置，装配，调整，定位，紧固，确定，修理，使固定
fizzum 富锆裂片合金
fjord 峡湾
flag bridge 信号桥楼
flag flip-flop 标记触发器
flagging plume 旗形羽流
flagging 旗状【植物风速指示】，旗形羽流
flaggy 薄砂岩的
flagman 测旗工
flag operand 标旗操作数，标志操作，标志运算
flagpole antenna 桅杆式天线
flagpole 旗杆
flagstone 板石，薄层砂岩，石板
flag 旗子，标记，标志，特征，特征位，信号旗，石板，旗标，标旗，板层
flaked graphite 片状石墨
flake off 呈鳞片状剥落
flake-shaped particle 片状颗粒
flake texture 片状结构
flake theristor 薄片热敏电阻器
flake 薄片，火花，剥落，白点【焊接缺陷】，

薄层，雪花，成片剥落，鳞片，石片
flaking 剥落【燃料芯块】，成片剥落
flaky constituent 絮片状成分【土壤或砂粒内的】，片状成分
flaky structure 鳞片状构造
flaky 鳞片状的
flame 火焰，燃烧
flame ability 可燃性
flame arc lamp 弧光灯
flame arrester 火焰消除器，灭火器，阻火器，消焰器
flame attachment 火焰装置
flame blowoff velocity 脱火速度
flame boring 火焰穿孔
flame bridge 火墙
flame center 火焰中心
flame cleaning 火焰清除【表面清理】，火净化，焰烧净化，火焰净化
flame coal 长焰煤
flame collapse 灭火
flame combustion 热力燃烧
flame condition 火焰工况，燃烧条件
flame connector 联焰管
flame control 火焰调节
flame cutter 火焰切割器，火焰切割
flame cutting torch 火焰切割炬
flame cutting 火焰切断，火焰隔离，氧炔切割，火焰切割，气矩切割
flame damper 灭火器，消焰器，防火器
flame deflector 火焰反射器，火焰导向器，火焰偏转器
flame detector system 火焰检测系统
flame detector 火焰监测器，辉光探测器，火焰检测器，火焰探测器
flame diverter 折焰器
flame drilling 火钻钻进，火焰钻孔
flame edge 火焰边界
flame emissivity 火焰辐射能力，火焰黑度
flame envelope 火焰包络，火焰包围面，外焰
flame extinction 火焰熄灭，熄火
flame eye 火焰监视器
flame failure alarm 熄火报警
flame failure control 火焰消失控制器，熄火保护装置，火焰调节器
flame failure limit 熄火极限
flame failure protection 熄火保护
flame failure 熄火，灭火
flame front 火焰前沿
flame-generated turbulence 火焰引起的湍流
flame gouging 火焰气刨，表面火焰切割
flame gun 火焰喷枪
flame hardening torch hardening 火焰淬火
flame hardening 火焰淬火法，火焰硬化法，火焰硬化
flame heart 焰心
flame holder 火焰稳定器，保焰板
flame-holding 火焰稳定
flame ignition 点火
flame ignitor 点火装置，点火器
flame impingement 火焰冲刷
flame indicator 火焰指示器
flame instability 火焰不稳定
flame intensity 火焰强度
flame ionization detector 火焰电离探测器
flame jet 火焰喷射
flame length 火焰长度
flameless combustion 无焰燃烧
flameless refractory curing 无焰烘炉
flame monitoring system 火焰监察系统
flame monitor 火焰监视器，火焰监察器，火焰监测器
flame noise 燃烧噪声，火焰噪声
flame normalizing 火焰正火
flame-out protection device 熄火保护装置
flame-out tripping device 熄火跳闸装置，熄火停机保护装置
flame-out 脱火，灭火，熄火
flame pattern 火旋焰类型，焰型
flame photometer 火焰光度计
flame plate 折焰板
flame probe 火焰探头
flame profile 火焰轮廓
flame-proof cable 防火电缆
flame-proof enclosure 隔爆外壳，防火外壳
flame-proof engine 防火式电机
flame-proof fluorescent lighting fixture 防爆耐火荧光灯
flame-proof machine 防爆电机，防火电机
flame-proof material 耐火材料
flame-proof motor 防爆型电动机，防火型电动机
flame-proof starter 防爆型启动器，防火型启动器
flame-proof tacho-generator 隔爆型测速发电机
flame-proof terminal box 隔爆式接线盒
flame-proof transformer 防火型变压器，防爆型变压器
flame-proof wire 耐火绝缘导线
flame-proof 不易燃的，防火的，耐火的
flame propagation rate 火焰传播速度
flame propagation velocity 火焰传播速度
flame propagation 火焰传播
flame reflector 火焰反射器，火焰脉动，火焰辐射
flame resistance 耐燃性
flame-resisting 耐火
flame retardant bedding 阻燃层理
flame retardant cable 难燃电缆，阻燃电缆
flame retardant oversheath 阻燃外护套
flame retardant paint 耐燃漆
flame retardant plastic pipe 阻燃塑料管
flame retardant property 阻燃性能
flame retardant rating 阻燃等级
flame retardant test 阻燃试验
flame retardant textile 阻燃织物
flame retardant 阻燃的，阻燃剂，火焰稳定
flame safeguard 熄火保护装置
flame-safety device 火焰安全装置
flame scanner 火焰监视器，火焰检测仪，火焰扫描器
flame scanning cooling fan 火焰检测冷却风机
flame speed 火焰传播速度

flame spraying　火焰喷涂，金属喷涂
flame spread angle　火焰扩散角
flame stability　火焰稳定性
flame stabilizer　火焰稳定器，稳燃器
flamestat control　熄火监视
flamestat　温度自动控制电门
flame surface quenching　火焰表面淬火
flame-swept　火焰冲刷
flame temperature detector　火焰温度检测器
flame temperature　火焰温度
flame test　用火焰法测定不气密性，焰色试验，火焰试验
flame trap　阻焰装置，消焰器
flame tube　火筒，火管，火焰管，火焰筒，燃烧器内筒
flame welding　气焊
flame weld　气焊，火焰焊接
flame zone　燃烧区
flaming arc lamp　弧光灯
flaming　燃烧的
flammability point　燃烧点
flammability　易燃性，可燃性
flammable gas　可燃气，可燃气体
flammable mixture　可燃混合物
flammable substance　可燃物，可燃物质
flammable vapor　可燃蒸汽
flammable　可燃的，易燃的
flanch　翼缘
flange and face　法兰端面
flange area　法兰面积
flange beam　工字钢，工字梁
flange bending machine　法兰弯曲机
flange blind　法兰堵头
flange bolt hole　法兰螺栓孔
flange bolt　法兰螺栓【用于拼合壳体】，凸缘螺栓
flange bushing　法兰衬套
flange clamping plate　法兰夹固板
flange connection　法兰连接，凸缘连接
flange coupling　凸缘联轴节，法兰联轴节，法兰联结，凸缘联轴器
flange cover plate　梁翼盖板
flange cover　法兰盖
flanged angle iron　卷边角钢
flanged beam　工字梁
flanged channel　卷边槽钢
flanged connection　法兰连接
flanged coupling　法兰联轴节
flanged edge welding　卷边焊
flanged edge　法兰边
flanged end connection　法兰端接
flanged end　法兰连接端，法兰端
flange detector　轮缘检测器
flanged fitting　法兰连接件
flanged head screw　凸缘螺母
flanged head　卷边封头
flanged joint　法兰连接［接头］
flanged nozzle　法兰管接头，法兰管喷头
flanged nut　凸缘螺母
flanged pipe　法兰连接的管子，法兰管，带法兰管
flanged shaft　凸缘轴，法兰轴
flanged tee　法兰三通
flanged-type joint　法兰接头
flanged union　凸缘连接，法兰联管节，折缘管节
flanged wheel　凸缘轮
flanged　装有法兰的，翻边的
flange face　法兰表面
flange follower　填料函法兰压盖
flange gasket　法兰垫片
flange heating　法兰加热
flange joint　法兰接合，法兰接头，法兰连接，凸缘接头
flange leak-off system　法兰密封引漏系统【压力容器】
flangeless butterfly valve　无凸缘的蝶阀
flangeless check valve　无凸缘单向阀
flangeless　无凸缘的
flange-mounted motor　凸缘型电动机，法兰安装型电动机
flange of beam　梁的翼缘
flange of coupling　联轴器凸边
flange of pipe　管子法兰，管子凸缘
flange of wheel　轮缘
flange pipe　法兰管，凸缘管
flange plate　凸缘板，翼板
flange pressure toppings　法兰取压口
flange protective disc　法兰盖贴面
flange protective disk　法兰盖贴面
flange quality steel　可卷边钢板
flange ring　法兰环，法兰圈
flange scaling face　法兰密封面
flange seal　法兰密封
flange surface processing lathes　法兰端面加工机
flange tap　法兰分接头，突缘分接头
flange tee　法兰三通
flange test　卷边试验
flange-type motor　法兰型电动机
flangeway　轮缘槽
flange welding　合卷边焊
flange with screw　螺纹法兰
flange wrench　凸缘扳手
flange　凸缘，法兰，卷边，法兰盘，翼板，翼缘，装法兰，给……装凸缘
flanging machine　折边机
flanging test on tubes of metals　金属管卷边试验
flanging test　（管口）卷边试验
flank clearance　侧向间隙
flank hole　侧面钻孔
flank of tooth　齿面
flank wall　山墙，侧墙
flank　侧面，侧腹，侧翼，侧墙，齿侧面，后面，肋部，厢房
flap actuating gear　挡板操作机构
flap check valve　旋启式止回阀
flap chord　襟翼弦
flap drag　襟翼阻力
flap drive shafting　襟翼传动轴系
flap electric motor　襟翼操纵电动机
flap gasket　平垫圈
flap gate　翻转板，舌瓣形闸门，翻板闸门，舌

瓣闸门
flap gauge 襟翼操纵机构
flap hinge moment 襟翼铰矩
flap hinge 明铰链
flap-lag 挥舞摆振
flap leading edge 襟翼前缘
flapper valve controller 瓣阀控制器
flapper valve 瓣阀，锁气器
flapper 挡板，阀，锁气器
flapping angle 挥舞角
flapping force 拍打力
flapping 抖动，挥舞
flap reference plane 襟翼基准平面
flap retraction 收襟翼
flap shutter 转动挡板
flap tilting gate 倾倒式闸门
flap-top bubble cap plate 平顶泡罩板
flap trap 逆止阀
flap-type attenuator 片形衰减器
flap-type damper 转动叶片式挡板
flap valve 翻板阀，挡板阀，瓣阀，舌形阀，锁气器
flap wheel 扬水轮
flap 挡板，阀，折翼，瓣，挡水板，活门，阀瓣，襟翼
flare angle 扩张角
flareback 回火
flared casing 扩张形汽缸
flared end 台锥管形，大小头【管道】
flared fittings 喇叭口管件
flared outlet 喇叭形出水口
flared pipe 喇叭口管，喇叭管
flared tube-end 翻边管端
flared tube 喇叭口管，喇叭管
flared union coupling 啦叭口接头
flared 扩张的，喇叭口的
flare nut wrench 锥孔扳手
flare nut 喇叭口螺母
flare opening 喇叭口
flare point 燃点，着火点
flare-type burner 喇叭式燃烧器，缝隙式燃烧器，扩散燃烧器
flare-wall 斜翼墙
flare wing wall 斜翼墙
flare wrench 喇叭口扳手
flare 闪光，闪光信号，突然爆燃
flaring and deflaring tool 扩口缩口工具【控制棒驱动机构】
flaring draught pipe 喇叭形尾水管
flaring funnel 喇叭形漏斗
flaring gage 卡钳【控制棒驱动机构】
flaring inlet 喇叭形入口
flaring out 突然闪亮
flaring pier energy dissipation 宽尾墩消能
flaring pier 宽尾墩
flaring test on tubes of metals 金属管扩口试验
flaring test 扩口试验
flaring tool 扩口工具
flaring （管子）扩口，卷边
flash alarming device 闪光报警装置
flash allowance 闪光留量
flash and ground protecting relay 飞弧与接地保护继电器
flash-and-ranging station 声波测距站
flash-arc 火花弧
flashback arrester 回火保险器，回火防止器
flashback arrestor 回火防止器，回火保险器
flashback tank 水封罐【防止回火】
flashback voltage 反闪电压，逆弧电压
flashback （燃烧器）回火，反闪，逆弧，回烧，闪回，逆燃
flash-barrier 隔弧板，瞬时隔板
flash board check gate 翻板节制闸门
flash board 临时挡水闸门，挡水闸板，闸板，插板
flash boiler 快速蒸发锅炉，闪蒸锅炉
flash-boiling evaporators 扩容蒸发器
flash box 扩容箱，扩容器
flash butt welder 闪光碰焊机，闪光对接焊机
flash butt welding 闪光碰焊，电弧对（接）焊，闪光对接焊，闪光焊
flash chamber 扩容箱，蒸发器
flash coat 喷浆盖层
flash condenser 快速冷凝器
flash current 闪光电流
flash density 落雷密度
flash distillation plant 闪蒸装置
flash distillation 闪蒸，突然蒸发
flash drier 急骤干燥器，扩容干燥器
flash drum （蒸汽）膨胀箱，闪蒸罐
flash dryer 闪蒸干燥器，急骤干燥器
flash drying 快速脱水，快速干燥
flash electrically erasable programmable ROM 快速电可擦写可编程只读存储器
flasher relay 闪弧继电器，闪光继电器
flasher 闪烁器，角反射器
flash evaporation 闪蒸法
flash evaporator 扩容蒸发器，闪蒸蒸发器，闪蒸器
flash fire 急剧燃烧，暴燃
flash flood 山洪暴发
flash gas 闪发气体
flash guard 防弧装置，防弧罩，防弧板
flash heat 快速加热
flashing alarm 闪光报警
flashing beacon 闪光信号标灯，闪光标灯
flashing chamber 闪蒸室
flashing deaerator 扩容除氧器
flashing effect 闪蒸效应
flashing flame 闪焰
flashing flow 急骤流动
flashing head 扩容压头
flashing indicator 闪光指示器
flashing lamp 闪光灯
flashing light signal 闪光信号
flashing light 闪光灯
flashing oscillator 闪光振荡器
flashing pipe 扩容管
flashing rate 闪烁频率
flashing signal 闪光信号
flashing stage 闪蒸级
flashing 泛水（板），挡水板，瞬时蒸发，闪蒸，

扩容，闪光，闪弧，(电弧)放电
flash lamp 闪光灯泡
flashlight 闪光灯，手电筒
flash memory 快速擦写存储器
flash mixer 快速搅拌器
flash mix 快速混合
flash-off of steam 扩容汽化
flash-off steam 闪蒸蒸汽，扩容蒸汽
flash of light 闪光
flash of steam 闪蒸汽化
flashover across terminals 线端间闪络
flashover at brush 刷弧，电刷跳火
flashover capability 殉爆敏感度
flashover distance 闪络距离
flashover ground relay 闪络接地继电器
flashover mechanism 闪络机理
flashover potential (绝缘子)闪络电位
flashover relay 闪络继电器
flashover test (绝缘子)闪络试验，(直流电机)环火试验
flashover voltage 闪络电压，击穿电压，放电电压
flashover 闪烁，飞弧，跳火，闪络，击穿，产生飞弧
flash point closed 闭口闪点
flash point opened 开口闪点
flash point tester in closed cup 闭口闪点测定仪
flash point 闪点，引火点，燃点，着火点
flash pressure 扩容压力
flash reactor 脉冲反应堆
flash runoff 暴涨径流
flash set 瞬时凝结
flash signal 闪光信号
flash steam cycle 减压扩容蒸汽循环
flash steam generator 闪蒸水蒸气发生器，快速蒸汽发生器
flash steam 扩容蒸汽
flash suppressor 环火抑制器，防闪络装置，防环火装置
flash system 闪蒸系统
flash tank 扩容器，二次蒸发箱，卸压箱，闪蒸罐，(蒸汽)膨胀箱，泄压箱，疏水膨胀器
flash test 闪点试验，瞬间高压试验，闪络试验
flash transformer 闪络变压器
flash tube stroboscope 闪光管频闪观测器
flash-up 功率激增，功率增
flash vessel 扩容器，闪蒸器
flash welding 火花对焊，闪光对焊，闪光焊
flash X ray diffraction 瞬间X射线照相，脉冲X射线照相
flashy load 瞬间负载
flash 闪光，闪蒸，突然蒸发，溢料，毛刺，飞溅，闪燃，闪烁，泄水
flask brush 烧瓶刷
flask 烧瓶，砂箱，容器，(放射性材料)屏蔽罐，长颈瓶，细径瓶
flat arch 扁拱，平拱
flat armature 扁平衔铁，边衔铁
flat-band armor 扁带铠装
flat-band 扁带
flat bank revetment 缓坡护岸
flat bar 扁材，带材，扁钢，条钢
flat beater 平板式打夯机
flat-bed crawler truck 履带式平板货车
flat-bed lorry 平板车
flat-bed trailer 平板车
flat bend test (焊件)弯曲试验
flat black absorber 平板黑吸收器
flat blade screwdriver 平刀口螺丝起子
flat blade 平叶片
flat bog 低沼泽(地)
flat bottom ditch 平底沟
flat-bottomed flask 平底烧瓶
flat-bottomed flume 平底测流槽
flat-bottomed reflecting hole 平底反射孔
flat-bottomed valley 平底谷
flat-bottomed 平底的
flat bottom furnace 平炉底炉膛
flat bottom gondola 平底的无盖货车
flat burner 平焰烧嘴，扁平喷嘴，缝隙式燃烧器
flat bus bar 扁汇流条，扁母线
flat cable 带状电缆，扁电缆，扁平电缆
flat car 平板车
flat characteristic 平缓的特性，平特性
flat chisel 冷凿，扁凿
flat coil 扁线圈，扁绕线圈
flat compound characteristic 平复励特性
flat compound excitation 平复励磁
flat compound generator 平复励发电机
flat compound 平复绕的，平复励的
flat concentration profile 平坦浓度分布
flat-conductor cable 扁线电缆
flat copper busbar 扁铜条母线
flat copper strip 扁铜线
flat copper wire 扁铜线
flat countersunk head screw 沉头螺钉
flat country 平坦地区
flat course 平砌(砖)
flat-crested measuring weir 平顶量水堰
flat-crested spillway 平顶式溢洪道
flat curve 平缓曲线
flat-deck buttress dam 平板支墩坝
flat dilatometer 扁式旁压仪
flat face 全平面，平面
flat fee 固定收费率
flat field generator 平面场发生器
flat file 扁平锉
flat filled weld in front and back 双面贴角焊
flat fillister head screw 有槽扁头螺杆
flat flame 扁平火焰，无光焰
flat floor 平肋板，平板地板，(液态炉的)平炉底
flat flux reactor 平坦通量反应堆
flat flux 平坦通量
flat frequency control 给定频率控制
flat gasket 扁平密封垫，平垫片
flat grade 缓坡
flat gradient 平缓比降，缓坡
flat grate 平炉排
flat groove weld position 平槽焊位置
flat ground 平地

flat head rivet 扁头铆钉
flat head screw 平头螺丝，扁头螺丝
flat head 平封头
flat heat collector 平板集热器
flat hole 水平炮眼
flat idler 平形托辊，平托辊
flat ingot 扁平铸锭
flat insulation material 扁平绝缘材料
flat iron 扁钢，扁铁，平顶脊，条铁，熨斗，熨斗形山，平顶山脊
flat jack test 扁千斤顶试验
flat jack 扁千斤顶
flat joint 平缝
flat keel 平板龙骨
flat key 平键
flat land 平原，平地
flat linear induction sodium pump 平面式直线感应电磁钠泵
flat load curve 平滑负荷曲线
flat membrane 平板
flat metal gasket 平金属垫片
flat metal jacketed asbestos filled gasket 金属包石棉平垫片
flat metallic gasket 金属平垫片
flatness tolerance 平直度公差
flatness 平面性，平面度，平滑性，平直度，平滑性
flat noise 白噪声，频谱上能量平均分配的噪声
flat nose plier 扁咀钳，平口钳
flat nut 扁螺母
flat on bottom 底平
flat on top 顶平
flat oval tube steel radiator 扁椭圆管式钢制散热器
flat-package integrated circuit 扁平封装集成电路
flat paint 无光漆
flat perforated plate 扁平多孔板
flat performance curve 平坦特性曲线
flat performance 平缓的特性，平特性
flat photoelectric crystal 片状光电晶体
flat planting 平面种植
flat plate cascade 平板叶栅
flat plate collector 平板（型）集热器【太阳能】
flat plate drag 平板阻力
flat plate flow 薄层水流，平板绕流
flat plate flutter 平板颤振
flat plate foundation 平板基础
flat plate-like grain 片状颗粒
flat plate margin 边缘宽，边缘余量
flat plate module 平板式组件，平板式太阳电池组件
flat plate reactor 平板元件反应堆
flat plate shaped grain 片状颗粒
flat plate solar air heater 平板太阳能空气加热器
flat plate solar cell module 平板式太阳电池组件
flat plate solar collector 平板型（太阳能）集热器
flat plate-type grain 片状颗粒
flat plate vibrator 平板式振动器
flat plate 平面底板，平板
flat position welding 顶面平卧焊，俯焊，平焊
flat position 平焊位置，平面位置
flat pulley 平面滚筒，平皮带轮，平面皮带轮
flat push-button 平形按钮
flat rate 包价收费制，按时计费制，按时计价，包价收费，比例汇率，（给排水的）毛估流量收费，统一收费率
flat relay 扁平继电器
flat response counter 水平响应计数器
flat response 平坦响应
flat ring armature 盘形电枢
flat ring dynamo 平面环形电枢发电机
flat ring gasket 环形平垫片
flat ring spanner straight pattern 平头梅花扳手（直）
flat ring 平环
flat-roofed building 平顶建筑
flat-roofed 平顶的
flat roof 平顶屋，平屋顶，平屋面，屋顶平台
flat rope 多股绳索
flat scraper 平面刮刀
flat screen bed 平直筛面
flat screen deck 平直筛面
flat shaped building 板状建筑物
flat shovel side expansion instrument 扁铲测胀仪
flat slab buttressed dam 平板（式）支墩坝，平板坝
flat slab column construction 无梁板柱结构
flat slab dam 平板坝
flat slab deck 平板式挡水面板
flat slab floor 无梁楼盖，无梁楼板
flat slab 无梁楼板，无梁板，平板
flat slope 平坦的坡度
flat source approximation 平面源近似
flat spade 扁铲
flat spiral coil 游丝形线圈
flat steel bar 扁钢条，扁钢，平钢
flat surface 平面
flat temperature profile 均匀温度分布
flattened distribution 展平的通量分布
flattened radius 展平区半径
flattened region （通量）展平区
flattened strand rope 扁平股索
flattening and bend test on sections of metals 金属型材展平弯曲试验
flattening-out of flood wave 洪水波展平
flattening radius （通量）展平区半径
flattening test on tubes of metals 金属管压扁试验
flattening test （管子）压扁试验，平整试验
flattening 展平，整平，扁平度
flatten 弄平，压扁
flat terrain 平坦地带，平坦地形
flat-top antenna 平顶天线
flat-top junction （暗沟的）平顶接头
flat-topped culvert 平顶涵洞
flat-topped curve 平顶曲线
flat-topped pulse 方脉冲
flat-topped voltage pulse 平顶电压脉冲
flat-topped wake 平顶尾迹
flat-topped weir 平顶堰
flat topping 平顶
flat top stockpile 平顶料堆

flat top wave 平顶波
flat trail car 平板拖车
flat trailer 平板车、挂车、拖车
flat trim template 平面调整垫
flat-tube heat exchanger 光管热交换器
flat tuning 粗调
flat twin cable 双芯扁电缆
flat-type aluminum 扁铝线
flat-type cable 扁电缆
flat-type relay 扁形继电器
flat-type stator 扁平式定子
flat valley 平谷，盖阀
flat wagon 平板车
flat washer 扁平垫圈
flat wave 平顶波，平缓波
flat wearing ring 扁平磨损环
flat welding 平焊，平焊接
flat wheel roller 光轮压路机
flat wire 扁线
flatwise bending 板状弯曲，平面弯曲
flatwise 平放地
flat wrench 扁平扳手
flat 平直的，扁平的，片状的，变平，弄平，平坦的，平坦，平面，平地，扁材，带材
flavor 香味，气味
flaw detectability 缺陷可检测性
flaw detecting hook 探伤钩
flaw detection （金属）探伤，缺陷探测
flaw detector 探伤仪，裂纹探测器，探伤器
flaw echo 缺陷回波
flaw in casting 铸件裂痕
flaw indication 缺陷显示
flawless 无缺陷的，无裂纹的
flaw location scale 缺陷定位标尺
flaw size 裂纹尺寸
flaw 裂缝，缺陷，伤，捩断层，裂痕，瑕疵，生裂缝
flaxe 电缆卷，一盘电缆
flax seed coal 碎粒煤
flax seed oil 亚麻油
FLB(floating-point buffer) 浮点缓冲器
FLBIN(floating-point binary) 浮点二进制
FLC(fetch length control) 取长度控制
FLC(programmable logic controller) 可编程逻辑控制器
FLDEC(floating-point decimal) 浮点十进制
FLD(field) 磁场
flea market 跳蚤市场
flea-size motor 超小型电动机，微型电动机
FLECHT(full length emergency cooling heat transfer) 全长度应急冷却传热
FLECHTSET(full length emergency cooling heat transfer system effect test) 全长度应急冷却传热系统效应试验
fleck 斑点
fleeting fault 瞬时故障
fleeting information 瞬间信息，速变信息
fleeting 短暂的，飞逝的，疾驰的，速变的
Fletter rotor 富勒特转子
flex-body valve 曲体阀
flex cracking 挠裂，疲劳裂纹，弯裂
flex gate valve 柔性闸阀
flexibility agent 增韧剂
flexibility analysis 灵活性分析，柔性分析，挠性分析
flexibility characteristic 柔度特性
flexibility constant 挠性常数
flexibility factor 挠度系数，柔性系数
flexibility influence coefficient 挠性影响系数
flexibility in operation 运行灵活性
flexibility matrix 挠性矩阵
flexibility method 柔度法
flexibility of fuel 燃料适应性
flexibility principle 弹性原则
flexibility stress 柔性应力
flexibility test 挠度试验
flexibility value 挠度值
flexibility 柔度，挠性，弹性，易弯曲性，灵活性，机动性，适应性，挠度，可挠性，柔性，挠曲性，韧性，柔韧性，伸缩性
flexibilizer 增韧剂
flexible AC power transmission 灵活交流输电，柔性交流输电
flexible AC transmission equipment 柔性交流输电设备
flexible AC transmission system 灵活交流输电系统
flexible alternating current transmission system 灵活交流输电系统
flexible armored cable 挠性铠装电缆
flexible armoured cable 软铠装电缆
flexible ball joint 活动球状接头
flexible base structure 柔性基层
flexible bearing 挠性轴承，柔性轴承
flexible bellows joint 弹性波纹管接头
flexible bend 柔性弯头
flexible blade 柔性叶片
flexible block 柔性垫块
flexible brush 软母线
flexible building 柔性建筑物
flexible busbar 软母线
flexible cable conduit 电缆软管
flexible cable 柔性电缆，软（性）电缆，柔性钢索，软电缆
flexible casing 活动套管
flexible cladding 柔性面层
flexible conductor 软导线，软线
flexible conduit for the shot peening mixture 喷丸混合物的柔性导管
flexible conduit 软管，软性导管，柔性导管
flexible connection 挠性连接
flexible connector 软性连接管，挠性连接器，弹性接头
flexible cord 软线，塞绳
flexible core wall 柔性心墙
flexible coupling of spacer type 剖分式挠性联轴器
flexible coupling 弹性联轴器，弹性连接，挠性联结器，挠性管接头，挠性联轴节，柔性联轴节，柔性联轴器
flexible dam 橡胶坝，软材料坝，柔性坝
flexible diaphragm coupling 挠性薄膜联轴节

flexible drive shaft 软传动轴
flexible duct 柔性风管，软管
flexible earth bond 地线串引流线
flexible electrode 软焊条
flexible extension 活动延长杆
flexible facing joint 柔性护面接缝
flexible facing 柔性面层
flexible fastening 弹性固定
flexible fender cushion 柔性防冲衬垫
flexible foam blanket 柔性泡沫材料衬垫，柔性泡沫毯
flexible foundation 柔性基础
flexible frame 柔性框架
flexible fuel vehicle 机动燃料车
flexible gear 挠性传动机构，柔性齿轮
flexible generation 灵活发电
flexible gland 弹性汽封
flexible governor 弹性调速器
flexible hose 挠性软管
flexible insulation 柔软绝缘
flexible jib of crane 起重机柔性臂架
flexible joint 挠性接合，挠性接头，活动接头，柔性接头，软活接，柔性接缝，软接头
flexible lateral stop 横向弹性制动
flexible life 弯曲疲劳期限
flexible mattress 柔性梢排
flexible member 柔性构件
flexible membrane spacer 挠性膜衬垫
flexible metal conduit 柔性金属导管
flexible metallic conduit 可挠金属穿管
flexible metallic tubing 柔性金属管
flexible micanite 柔软云母板
flexibleness 挠性
flexible nonmetallic conduit 柔性非金属导管【如塑料软管等】
flexible nonmetallic tubing 柔性非金属管【如塑料软管等】
flexible pavement 柔性路面，柔性铺面
flexible pipe 软管，消振管
flexible plastic drain 柔性塑料排水管
flexible plastic reactor 活动可塑体反应堆
flexible plastic sleeve 软塑料套筒
flexible plate 柔性底板
flexible pliers 万向套筒扳手
flexible power contract 灵活电力合同
flexible rigidity 抗弯刚度，弯曲刚性
flexible rolling bearing 柔性滚动轴承
flexible roof 柔性屋顶
flexible rotor 挠性转子，柔性转子
flexible rubber-insulated wire 橡胶软线
flexible screwdriver 挠性螺丝起子
flexible seal 柔性止水
flexible-shaft coupling 软轴连接
flexible-shaft turbine 挠性轴汽轮机
flexible shaft 挠曲轴，挠性轴，软轴
flexible solid wedge valve 挠性整体楔形闸板闸阀
flexible spline 柔性塞缝片
flexible staybolt 挠性拉撑螺栓
flexible stranded wire 绞合软线
flexible structure 柔性结构物
flexible support 弹性支承
flexible suspension bridge 柔性悬索桥
flexible transmission 挠性传动
flexible transport 皮带运输，无轨运输
flexible trestle 柔性支架
flexible tube pump 软管泵，蠕动泵
flexible tube 挠性管，软管
flexible tubing 挠性管路，柔性套管
flexible walled wind tunnel 柔壁风洞
flexible wall 柔性墙
flexible waveguide 柔性波导管，可弯曲波导管，软波导
flexible wedge gate valve 柔性楔形闸阀
flexible wire 软线，软导线
flexible working hours 弹性工作时间，机动工时
flexible 灵活的，易弯曲的，挠性的，柔韧的，可弯曲的，软接头
flexiglass 透明醋酸纤维素制品
flexile 易弯曲的，柔韧的，灵活的
fleximeter 挠度计
flexing life 挠曲寿命，弯曲疲劳寿命
flexing resistance 抗弯性，抗弯强度
flexing test 挠曲试验
flexi-skirt 柔性导料板
flex lance 弯曲吹管
flexometer 挠度仪，曲率计
flexo-writer 多功能打字机，打字穿孔机，快速打印装置
flex point 拐点
flextime 弹性工作时间
flex tubing 柔性套管
flex-twist section 扭转波导段
flexual 弯曲的
flexural axis 弯曲轴
flexural buckling 受弯压屈
flexural centre 弯曲中心
flexural constant 挠曲常数
flexural damping 弯曲阻尼
flexural failure 弯曲破坏
flexural fatigue 弯曲疲劳
flexural fold 挠曲褶皱，弯曲褶皱
flexural loading test 弯曲载荷试验
flexural measurement 弯曲测量
flexural mode vibration 弯曲振动，弯曲型振动
flexural mode 挠曲方式
flexural modulus 弯曲模量
flexural moment 弯矩
flexural oscillation 挠性振动，弯曲振动
flexural rigidity 弯曲刚度，抗挠刚度，抗弯刚度
flexural slip fold 曲滑褶皱
flexural slip 层面滑动，挠曲滑动
flexural stiffness 抗弯刚度
flexural strain 挠曲应变，弯曲应变
flexural strength 抗弯强度，弯曲应力，挠曲应力
flexural stress 挠（曲）应力，弯曲应力
flexural testing 弯曲试验，屈服试验
flexural vibration 弯曲振动
flexural wave 弯曲波
flexural 弯曲的

flexure centre　弯曲中心
flexure formula　挠曲公式
flexure meter　挠曲计
flexure plane　挠曲面
flexure reinforcement　受弯钢筋
flexure strength　抗弯强度
flexure stress　挠曲应力，弯曲应力
flexure strip　弹性片
flexure test machine　挠曲试验机
flexure test　弯曲试验
flexure torsion flutter　弯扭颤振
flexure　弯曲，挠曲，挠度，弯度，曲率，屈曲
flex　挠曲，弯曲，花线，拐折，挠性金属套管
FL(fatigue limit)　疲劳极限，疲劳寿命，疲劳限界
F/L(fetch/load)　取/送
FL =floor　层
FL=fluid　流体，液体
FL(fuel loading)　装料
FLG=flange　法兰
flibe　氟、锂、铍熔盐
flicker alarm over range　超量程闪烁报警
flicker alarm　闪烁性警报
flicker board　闪光器板
flicker coefficient for continuous operation　持续运行的闪变系数
flicker effect　闪变效应
flicker frequency　闪烁频率
flickering lamp　闪光灯
flicker meter　闪变仪
flicker relay　闪烁继电器【用于闪烁性警报】，闪光继电器
flicker severity factor　闪烁劣度系数，闪烁强度系数，闪烁严重程度因子
flicker step factor　闪变阶跃系数
flicker　闪烁，闪光，摇动，闪变，电压闪变
flight conveyer　链板［链动］输送机，刮板（式）输送机
flight conveyor　刮板输送机，链板传送带，刮板运输器，梯板运送机
flight feeder　刮板给料机
flight line　航线
flight number　航班号
flight rise　阶梯步高
flight route　航空路线，航线
flight run　阶梯步距
flight sewer　跌落式排水管道
3 flight stair　三跑楼梯
2 flight stair　双跑楼梯
flight-type lowering conveyor　刮板落料输送机
flight　飞行距离，阶梯步级，链板，（输送机）刮板，飞行，梯段，楼梯的一段，航班
flint　电石，火石
flint aggregate　坚硬骨料，燧石骨料
flint brick　燧石砖
flint clay　燧土
flint glass paper　粗砂纸
flint stone　打火石，燧石
flinty steel　硬钢，硅钢
flip bucket　挑坎，挑流鼻槛，挑流鼻坎，消力戽
flip-chip transistor　翻转片式晶体管，倒装片式晶体管
flip coil　探测线圈，反应线圈
flip-flop circuit　双稳多谐终端电路，触发电路，反复电路
flip-flop counter　触发器计数器
flip-flop decade ring　十进制触发计数环
flip-flop number　触发计数
flip-flop register　触发器式寄存器
flip-flop relay　脉冲继电器
flip-flop stage　触发级
flip-flop storage　触发器存储器
flip-flop　触发电路，双稳态多谐振荡器，触发器，双稳电路，双稳态触发电路
flip-top gondola　一种带开启顶盖的车皮
flitch dam　木板坝
flitch girder　组合板大梁
flitch-trussed beam　组合桁架梁
flitch　桁板，料板
FLK inspection meter　福禄克检测仪器
FLM=flame　火焰
floatability　可浮性
float-actuated recording liquid-level instrument　浮子记录式液位计
float adjusting valve　浮球调节阀
floatafion unit　浮选装置
floatage　漂浮，浮力，木材流送
float and cable level measuring device　浮标和缆索式物位测量仪表
float-and-sink analysis　浮沉分析
float-and-tape level sensor　浮子和贴尺水位传感器
float and thermostatic type steam trap　浮球与热力综合式疏水器
float-and-valve　浮子控制阀，浮筒阀
floatation center　浮体水面形心
floatation frother　浮选起泡剂
floatation line　吃水线
floatation process　浮选法
floatation reagent　浮选剂
floatation tank　浮选槽
floatation　浮力作用，浮力，飘浮，浮动，浮选，浮游，浮标
float barograph　浮子气压计
float bell pressure gauge　浮钟式压力计
float board　筏
float chain　有时差的路径，浮动路径，浮动路线【计划评审技术】
float chamber　浮子室，浮筒室，浮球室
float charge　浮充电【电池】
float charging generator set　浮充发电机组
float coal sample　浮煤样
float coefficient　浮标系数
float-controlled drainage pump　浮子控制式排水泵
float-controlled switch　浮子控制开关
float-controlled valve　浮子阀
float control　浮标控制
float course　浮标测流路线
float current meter　漂浮式流速仪
float displacement　浮子排水量

floated coat 抹灰层
floated concrete 抹面混凝土
floated finish 抹灰饰面，抹光面
floated-type gyroscope 悬浮陀螺仪
floater 浮球，漂珠，浮子，抹灰工具，飘浮物
float finish 抹光，用镘刀修整
float flowmeter 浮子流量计
float gate 浮式闸门
float gauge 浮标水尺，浮式水标尺，浮子式液面计，浮表
float gauging 浮标测流（法）
float governor 浮子水位调节器，浮子式液位调节器
float indicator 浮漂式液位计
floating absorbent 浮动吸收剂
floating accumulator 浮点累加器
floating action type servomotor 浮动式伺服电动机
floating action 浮动作用，无静差作用，无定位作用，无定向动作，不稳作用
floating address 可变地址，浮地址
floating add 浮点加
floating agent 浮选剂
floating algae 浮游藻类
floating anchor 浮锚
floating animal 浮游动物
floating ball type valve 浮动球型阀
floating battery 浮充电池
floating beacon 浮标
floating bed 浮动床
floating bivalued resistor 浮动双值电阻
floating blanket 悬浮层
floating body 浮体
floating boom 浮栅
floating breakwater 浮式防波堤
floating bridge 浮桥
floating bushing 浮动衬套
floating caisson 浮式沉箱
floating cargo landing stage 运货浮动平台
floating car positioner arm 车辆定位机臂
floating carrier modulation 浮动载波调制
floating carrier system 浮动载波机
floating charge method 浮充法
floating charging （蓄电池）浮充，浮充电
floating coat 二道油漆，二道抹灰层，二道抹灰
floating component 无定向分量，无静差元件
floating control action 无（静）差调节，浮点控制
floating controller 无定位控制器，无静差控制器
floating control mode 无静差调节模式
floating control 无静差控制，不稳定调节，无静差调节，无定位调节
floating crane 水上起重机，浮式起重机，浮吊
floating currency 浮动货币
floating dam 浮坝
floating debris pass 漂浮物过道
floating debris 浮渣，漂浮物
floating decanter 浮子式泄水器
floating decimal accumulator 十进制浮点累加器
floating decimal set-up 浮点十进制装置
floating decimal subroutine 浮点十进制子程序
floating displacement 端面瓢偏
floating dock 浮船坞，浮码头，浮坞
floating drive 浮动驱动装置
floating dust 飘尘，浮尘
floating earth 浮动接地
floating evaporation 漂浮式蒸发器
floating evaporimeter 漂浮式蒸发计，漂浮蒸发计
floating floor 浮动地板，浮床
floating flowmeter 浮子流量计
floating foam 浮动泡沫
floating foundation 浮筏基础，浮基
floating gate avalanche injection MOS 浮动栅雪崩注入型金属氧化物半导体
floating-gate metal-oxide-semiconductor 浮动栅金属氧化物半导体
floating head heater 浮动管板式加热器
floating head heat exchanger 浮头式热交换器
floating head plenum 浮动顶盖空腔
floating head type feed water heater 浮动管板式给水加热器
floating head 浮动盖，浮动管板
floating ice 浮冰
floating input floating output 浮点输入浮点输出
floating input 浮点输入，浮空输入
floating interest rate 浮动利率
floating interpretive program 浮点解释程序
floating jetty 浮式突码头
floating level gauge 浮标水尺
floating light 浮标灯
floating machine 抹平机
floating mark 浮标，浮标志
floating matter 漂浮物质
floating method of charging 浮充法
floating multiply 浮点乘
floating neutral 不接地中性线，浮电位中性线
floating nuclear power plant （海上）浮动核电厂
floating number 浮点计位数
floating octal point 浮点八进制
floating of professionals 人才流动
floating oil 浮油
floating output 浮点输出
floating packing 浮动填料
floating-paraphase circuit 阴极绝缘倒相电路
floating pile driver 水上打桩机
floating pile foundation 浮桩基础
floating pile 摩擦桩
floating plate 浮板
floating platform 浮动平台
floating point binary 浮点二进制
floating point buffer 浮点缓冲器
floating point calculation 浮点计算
floating point computation 浮点计算
floating point computer 浮点计算机
floating point decimal 浮点十进制
floating point multiplication 浮点乘法
floating point operation stack 浮点操作栈
floating point operation 浮点运算
floating point register 浮点寄存器
floating point representation 浮点表示法
floating point routine 浮点程序

floating point system　浮点系统
floating point　浮点
floating pontoon　水上起重机，浮筒
floating-potential grid　浮动电动栅极，浮动电位栅极
floating potential　位移电位，漂移电位
floating power unit　无静差执行部件
floating rate note　浮动利率票据
floating rate　无静差作用率，浮动利率
floating reference address　可变转换地址
floating response　上浮反应，无静差特性，无静差作用
floating rim　浮动磁轭
floating-ring seal　浮环密封，转动环密封，浮动环密封
floating roof tank　活动顶盖的油箱，浮顶式油箱
floating screed strip　抹灰靠尺，抹灰准线
floating seal ring　浮动密封环
floating shaft　浮轴
floating sign　浮点符号
floating slag　浮渣
floating sleeve bearing　浮套轴承
floating sludge　浮泥
floating speed　无静差作用速度，无定位速度
floating stock　浮动股票
floating storage　流动贮存
floating strainer　浮式滤头【水泵吸水用】
floating subroutine　浮点计算子程序
floating subtract　浮点减
floating switch　浮子开关，浮动开关
floating valve　浮阀
floating voltage　空载电压，浮空电压
floating wage　浮动工资
floating wind turbine　漂浮式风电机组
floating yoke　浮动磁轭
floating　浮接状态，飘浮的，浮动的，未接地的，抹平
floatless liquid-level controller　无浮子液面控制器
floatless　无浮子的
float level control valve　浮动水位控制阀
float level indicator　浮子式液位计
float level meter　浮子液位计
float material　浮体材料
float measurement　浮标测流
float on company　创办公司
float operated switch　浮子控制开关
float product　浮选产物
float regulation　浮球调节，浮子调节
float relay　浮子继电器
float run　浮标行程【指上下测流断面的间距】，浮标测距
float sample of coal　浮煤样
float sample　浮煤样
float slime　浮渣
float switch　浮动开关，浮子开关，浮动继电器，浮球开关
float tape　浮尺【指自记水位计】
float test　浮标试验
float trap　浮子式凝汽阀
float tube　浮动管，浮筒管，浮标管
float type flow meter　浮筒式流量计，浮子式流量计
float type level gauge　浮子式水位计
float type level indicator　浮子式液位指示仪
float type level sensor　浮式水位传感器
float type pneumatic measuring chain　浮标式气动测量链
float type steam trap　浮球式疏水器
float type water stage recorder　浮标式自记水位计
float valve　浮子阀，浮球阀，液面调节阀，浮动阀，浮筒阀，浮子控制阀，浮球式疏水阀
float vibrator　表面振动器，浮式振动器
float viscosimeter　浮标黏度计
float well　验潮【水利水电】，浮心井
float work　抹灰工作
float yoke　浮动磁轭
float　浮标，浮子，漂浮，轴向间隙，进度裕量，浮动，浮筒
floc carryover　絮凝物携带
flocculability　絮凝性
flocculant aid　助絮凝剂
flocculant　凝聚剂，絮凝剂
floccular　絮凝的
flocculated clay　絮凝黏土
flocculated sediment　絮凝泥沙
flocculate　絮凝粒，絮凝物
flocculating admixture　絮凝剂
flocculating agent　絮凝剂
flocculating constituent　絮凝体
flocculating settling　絮凝沉淀
flocculating tank　絮凝池
flocculation agent　絮凝剂
flocculation aid　助凝剂
flocculation basin　絮凝池
flocculation factor　絮凝因素
flocculation limit　絮凝极限
flocculation ratio　絮凝比
flocculations agent　絮凝剂
flocculation system　（废水处理的）絮凝系统
flocculation zone　絮凝区
flocculation　絮凝，絮凝作用，絮化，絮凝作用
flocculator-settler　絮凝澄清器
flocculator with revolving arms　旋臂絮凝池，旋臂絮凝器
flocculator　污水絮凝器，絮凝器，絮凝装置
flocculence　絮凝性，絮状
flocculent deposit　凝絮状沉积，絮状沉淀
flocculent precipitate　絮凝沉淀物
flocculent sludge　絮凝性污泥
flocculent soil　絮凝土
flocculent structure　絮状结构，凝絮状结构，絮凝结构
flocculent　絮凝的，絮状的，絮凝剂
floccule　絮状物，絮凝物，絮凝粒，絮状沉淀
floccus　絮状云
floc formation　絮凝体的形成，絮凝物形成
flocky precipitate　絮状沉淀物
flock　絮状沉淀，絮凝体
floc-matrix　絮体基质
floc　凝聚体，絮凝物，絮状沉淀，蓬松物质，絮凝体，絮片，大冰块，浮冰

flood absorption capacity　洪水收容能力
flood amplitude　洪水波幅，洪水幅度
flood axis　洪水流向，洪水轴线
flood bank　防洪堤，洪水河岸
flood barrier　拦洪坝
flood bed　泛滥地
flood breadth of river　洪水河面宽度
flood bridge　洪水桥
flood calamity　水灾
flood carrying capacity　排洪能力，泄洪能力
flood catastrophe　水灾
flood channel storage　洪水河槽调蓄
flood channel　洪水河槽，溢洪河道
flood characteristic curve　洪水特性曲线
flood coefficient　洪水系数
flood control capacity　防洪库容
flood control dam　防洪堤坝
flood control level　防洪限制水位
flood control planning　防洪规划
flood control project　防洪工程，防洪计划
flood control reservoir　调洪水库，防洪水库
flood control standard　防洪标准
flood control storage　防洪库容
flood control works　防洪工程
flood control　洪水控制，防洪
flood crest stage　洪峰水位
flood crest travel　峰顶行进，洪峰行径
flood crest　洪峰
flood current　洪流
flood damage　洪水损失，洪灾，水灾，淹没损失
flood dam　防洪堤坝，防洪坝
flood detention dam　拦洪坝
flood detention efficiency　滞洪效应
flood detention pond　滞洪池
flood detention pool　滞洪水池
flood detention reservoir　滞洪水库
flood detention volume　滞洪库容
flood detention works　滞洪工程
flood dike　防洪堤
flood disaster　洪灾，水灾
flood discharge channel　排洪渠，排洪沟
flood discharge forecast　洪水流量预报
flood discharge gate　泄洪闸
flood discharge tunnel　泄洪隧洞
flood discharge　泄洪量，洪水流量
flood discharging capacity　泄洪能力
flood discharging　泄洪
flood distribution　洪水分布
flood diversion area　分洪面积，分洪区
flood diversion project　分洪工程
flood diversion sluice　泄洪闸
flood diversion works　分洪工程
flood duration　洪水历时
flooded area　泛滥区，洪水泛（滥地）区，受淹面积，淹没地区，淹没面积
flooded dissolver　溢流溶解器
flooded evaporator　满液式蒸发器
flooded gravity flow　溢流性重力流
flooded land　受淹土地，水泛地
flooded weight　充满水的质量，淹没重量
flood erosion　洪水冲刷，洪水侵蚀
flood estimation　洪水估算
flood event　洪水事件
flood flow formula　洪流计算公式
flood flow　洪流，洪水径流，洪水流量
flood forecasting　洪水预报
flood forecast service　洪水预报机构
flood forecast system　洪水预报系统
flood formula of small collective area　小汇水面积洪水公式
flood formula　洪水计算公式
flood frequency curve　洪水频率曲线
flood frequency relationship　洪水频率关系
flood frequency　洪水频率
flood gate　溢洪道，排洪闸门
flood hazard　洪水险情，水灾
flood hydrograph　洪水过程线
flood information　洪讯
flooding-ability　溢流性
flooding duration　淹没期
flooding gas-solids　气固浸渍
flooding line　溢流线
flooding pipe　溢流管
flooding point　溢流点，淹没点，泛滥点
flooding potential　泛滥潜力，淹没可能
flooding rate　淹没速率
flooding system　淹没系统
flooding valve　溢流阀
flooding　溢流，淹没，满溢，注水，泛滥，灌水
flood intensity　洪水强度
flood inundation period　洪水淹没期
floodlamp　探照灯，泛光灯
flood land　漫滩地，漫滩
flood level mark　满潮标记，洪水痕迹
flood level　洪水位
flood lighting　泛光照明，强力泛光照明，强力照明，泛光灯，探照灯
flood losses　洪水损失
flood loss rate　洪水损失率
flood lubrication　浸入润滑
flood mark　洪水标记
flood occurred once in a hundred year　百年一遇洪水
flood peak forecast　洪峰预报
flood peak profile　洪峰纵剖面
flood peak rate　洪峰流量
flood peak reduction　洪峰消减
flood peak stage　洪峰水位
flood peak　洪峰
flood period　洪水期，汛期
flood periphery　洪水边缘
flood plain deposit　漫滩沉积，泛滥平原沉积
flood plain regulation　河漫滩整理
flood plain terrace　泛滥平原阶地
flood plain　泛滥平原，洪积平原，冲积平原，河床满水，河漫滩，漫滩，漫滩地
flood plane　洪水水面
flood prediction　洪水预报
flood prevention measures　防洪措施
flood prevention　防洪
flood probability　洪水概率
flood-prone areas　容易发水的地区

flood-proof electrical equipment　防浸式电气设备
flood protection during construction　施工期度汛
flood protection embankment　防洪堤
flood protection in construction period　施工期度汛
flood protection measure　防洪措施
flood protection work　防洪工程
flood protection　防洪工作，防洪
flood recession　洪水退落，退洪
flood record　洪水记录
flood regime　洪水情况，水情，汛情
flood regulation storage capacity　调洪库容
flood regulation storage　调洪库容
flood regulation　调洪，洪水调节
flood-relief channel　排洪沟
flood retarding basin　滞洪区
flood retarding dam　滞洪坝
flood retarding project　拦河工程，滞洪工程
flood retention basin　拦洪水库
flood retention capacity　滞洪能力
flood retention　拦洪
flood rise　洪水上涨，涨洪
flood routing through reservoir　水库调洪演算
flood routing　洪水演进计算，洪水演算，洪路推测，洪水过程观测研究
flood runoff　洪水径流
flood seasonal distribution　洪水季节分布
flood section　洪水断面，汛期，洪水季
flood separation works　分洪工程
flood series　洪水系列
flood-source area　成洪（地）区
flood spillway　溢洪道
flood spreading　洪水散布
flood stage forecast　洪水水位预报
flood stage　洪水位
flood storage and reclamation works　蓄洪垦殖工程
flood storage project　蓄洪工程
flood storage works　蓄洪工程
flood storage　防洪库容，蓄洪
flood stream　洪水河流
flood strength　潮流强度
flood subsidence　洪水消退
flood surcharge　洪水超高
flood surge　洪水涌浪
flood survey　洪水调查
flood synchronization　洪水同步
flood synthesis　洪水组合
flood tide channel　涨潮道，涨潮水道
flood tide current　涨潮流
flood tide depression　涨潮低降
flood tide　涨潮
flood valve　溢流阀
flood velocity　洪流速度，洪水流速
flood volume　洪量，洪水量，洪水体积，洪水总量
flood warning service　防汛站，防汛机构，洪水报警机关
flood warning　洪水警报
flood water-detention pool　滞洪水池
flood water zone　淹没区
flood water　洪水，大水
flood wave recession　洪水波消退
flood wave subsidence　洪水波消退
flood wave transformation　洪水波变形
flood wave velocity　洪水波速度
flood wave　洪水波，洪水淹没波【坝的断裂】
floodway　排洪道
floody season-dry season price　丰枯电价
flood　溢流，淹没，泛滥，洪水，涨潮
floor acceleration　楼板加速度
floorage　使用面积
floor area of building　建筑面积
floor area　建筑面积，楼层面积
floor arrangement　平面布置
floor beam　楼板梁，楼面横梁
floor board　地板
floor brick　地面砖，地砖，铺地砖
floor construction　地板构造
floor contact　平台触点，楼面触点
floor contamination monitor　地面污染监测仪
floor covering　覆土层，地面覆盖，楼面覆面层
floor crack　底板裂隙
floor damp-proof course　地面防潮层
floor damp-proofing　地面防潮
floor drainage system　地面排放系统
floor drainage　地面排水
floor drain holdup tank　楼板冲洗水排水箱
floor drain tank　楼板冲洗水排水箱
floor drain　地漏，运转层排水，（房间）地面疏水，楼板排水，楼层排水
floor elevation　室内地坪标高，底板高程
floor flange mounting　凸出地面固定
floor flange　凸出地面
floor framework　楼面构架
floor girder　楼盖主梁，桥面板主梁
floor grating　花格地板
floor gully　地板集水沟，地面集水沟
floor heating system　地板采暖系统
floor heating　地板采暖
floor heave　底板隆起
floor height　层高
flooring with checkered pattern　花纹地板
flooring with fishbone pattern　人字形地板
flooring　地板，地板材料，楼面，平台木板，室内地面，铺地板
floor insert trolley　隐轨推车
floor insert　地面嵌入件
floor lamp　落地灯
floor level　室内地坪
floor lift　底板隆起
floor light　辅助照明
floor-mounted motor　落地安装型电动机
floor mounting　地面固定
floor of manhole　人孔底部
floor pack　底板分块
floor paint　地板漆
floor panel heating　地球辐射采暖
floor paved with thin plastic plate　地板铺以薄塑料板
floor pedestal　落地式轴承台
floor plan　楼层平面图，楼面布置图，平面布

置图
floor polish 地板蜡
floor post 楼层柱
floor pressure arch 底板压力拱
floor price 最低价，底价，最低限价
floor-relay 分层继电器，楼层继电器
floor response spectrum 楼层反应谱，楼层响应谱
floor screen 炉底水帘管
floor slab 底板，楼板
floor space 运转层面积，厂房面积，楼层面积，建筑面积
floorstand handwheel 地轴承架手轮
floorstanding air conditioner 柜式空调器
floorstand 立架，落地支架，地轴承架
floor structure 楼板结构
floor tile 热炉底，地面砖，地砖，耐火砖
floor time 停机时间，空闲时间
floor-to-floor height 楼面至楼面高度，楼层高度
floor-to-roof height 楼面至楼顶高度
floor tube 炉底管
floor type bearing 落地式轴承
floor type 落地式
floor wax 地板漆
floor 楼板，底面，地板，楼板面，楼层
flop gate 翻板闸门
flopover circuit 阴极耦合多谐振荡器
flopover 触发电路，翻转电路，双稳态多谐振荡器
floppy disk controller 软盘控制器
floppy disk drive 软磁盘驱动器，软盘驱动器
floppy disk 软盘
floppy 软的
flop valve 瓣阀
flora 植物区系，植物志
florence flask 平底烧瓶
FIOR process 铁矿石流态化还原法
floss hole 出渣口
flotation cell 浮选池
flotation coagulant 浮选凝结剂
flotation collector 浮选捕收剂
flotation concentrate 浮选精矿
flotation cost 发行成本，筹资成本
flotation frother 浮选起泡剂
flotation machine 浮选机
flotation reagent 浮选剂
flotation tailings 尾煤
flotation tank 浮选池
flotation unit 浮选装置
flotation 浮选
flotrol 一种恒电流的充电机
flotsam 浮料
flourishing 兴旺
flowability of concrete 混凝土流动性
flowability 流动性
flowable flyash 粉煤灰浆
flowable solids reactor 流动固体反应堆
flow adjustment curve 流量调节曲线
flowage damage 淹没损失
flowage prevention 防止泛滥，防止淹没
flow alarm 流量信号
flow amount 流量
flow angle 流动角度，气流角
flow angularity 气流偏角，流角，气流倾斜度
flow annulus 流动环道
flow apron 导流挡板
flow area goodness factors 通流截面优度因数
flow area 过水（断面）面积，流道面积，流动截面积，流通面积，流通截面积
flow around regime 绕流状态
flow assist device 助流装置
flow axis 水流轴线
flow back 逆流，回流
flow baffle 挡流板，流动隔板
flow behavior 流动特性
flow behaviour index 流动性指数
flow blockage 流动阻塞
flow bog 浮沉泥炭沼泽
flow boiling crisis 流动沸腾危机
flow boiling 流动沸腾
flow boundary layer thickness 流动边界层厚度
flow boundary layer 流动边界层
flow brazing 铸钎焊
flow breccia 流状角砾岩
flow calibration in working section 风洞试验段的流场校准
flow calibration 流场校测
flow capacity of control valve 调节阀流通能力
flow capacity 泄水能力，流量
flow cascade 梯级跌水
flow channel inspection fixture 流道检查固定架
flow channel shroud 流道套筒【欧米茄燃料试验】
flow channel 流道【欧米茄燃料试验】
flow characteristic of control valve 调节阀流通特性
flow characteristic of regulating valve 调节阀流量特性
flow characteristics 流动特性，水流特性，水流特征
flow chart template 流程图样板
flow chart （工艺）流程图，工艺系统图，程序框图
flow choking 流动壅塞
flow circuit 流动回路
flow cleavage 流状劈理
flow coast-down 流量下降
flow coefficient of regulating valve 调节阀流通能力
flow coefficient 流量系数
flow colorimeter 流动色度计
flow condition 流动条件，流态
flow configuration 流态，流形
flow contact length 流动接触长度
flow continuity 连续性流动，流动连续性，水流连续性
flow control circuit 流量控制回路
flow control device 流量控制装置
flow controller 流量控制［调节］器
flow controlling gate 节流
flow control regular 流量控制
flow control system 流量控制系统

flow control unit 流量控制装置
flow control valve 流量调节阀，带单向阀的流量控制阀，流量控制阀
flow control 流量调节，流量控制
flow cross-section 流道截面
flow curvature 流动弯曲
flow curve 流动曲线
flow data 水流资料
flow deficiency 流量不足
flow deflector 导流片
flow demand 需要流量
flow depth 水流深度
flow detection 流量检测
flow-deviation angle 气流偏转角
flow diagram 流量图，（工艺）流程图，程序方框图，建筑安装流程图，表计和管线图
flow-directing shroud 导流围筒，导流管套
flow direction detector 流向检测装置
flow direction measurement 流向测量
flow direction 流动方向，流向
flow discharge through dam orifice 坝身孔口泄流
flow discontinuity 流动不连续性
flow distortion 气流畸变，水流变形
flow distribution baffle 流量分配板【蒸汽发生器】
flow distribution hot-channel factor 流量分配热通道因子
flow distribution 流量分配，流动分布
flow divergence 流动发散
flow diverter 料流转向器，引流器，偏流器
flow divider 分流器，流量分配器
flow dividing valve 分流阀
flow drag 流动阻力
flow driven oscillation 流动激振
flow duct 流道
flow duration curve 流量持续曲线，流量历时曲线
flow duration diagram 流量历时曲线
flow duration 风洞工作时间，流动连续时间，流量历时
flow dynamics 流体动力学，流体力学
flowed energy 气流能量
flowed equation 流动方程
flowed fluctuation 流量波动
flowed friction 流动阻力
flowed 流动
flow elbow 流量弯管
flow equalizer 流量均衡器
flow equalizing plate 均流板
flow equation 流动方程，流体运动方程，流量方程
flower bed 花坛
flower stand flower standlawn 草坪
flower terrace 花坛
flow excursion 流量偏移
flow expansion 水流扩张，气流膨胀
flow factor 流动因素
flow failure 流动破坏
flow field calibration 流场校测
flow field quality 流场品质
flow field representation 流场表示法
flow field 流场，流线谱
flow fluctuation 流量脉动，气流脉动，水流脉动，水流起伏
flow forcing device 促流装置
flow formula 流量公式
flow form 流动形态
flow frequency 流量频率
flow friction characteristics 流动摩擦特性
flow friction 流动阻力，流动摩擦阻力，流动摩擦
flow gate right 关口金融输电权
flow gauge 流量表，流量计
flow gauging station 测流站
flow gauging weir 测流堰
flow gauging 流量测定，测流
flow geometry 流动几何特性，水流几何学
flow governor 流量调节器
flow gradient 水流坡度
flow graph 流程图，流量曲线，流线图，流向图
flow guide baffle 流量导向挡板
flow guide tube 导流管
flow guide 导流装置，导流板，引流，导流
flow head 流动水头
flow history 流量历程
flow hole 进出口孔【压力容器】，水流孔【压力壳】
flow in axial direction 轴向流
flow inclination angle 气流偏角，气流倾角
flow inclination 气流倾角，水流坡降
flow in continuum 连续流
flow in convection 对流气流
flow index 流动指数
flow indicator 流量表，流量计，流量指示器，流动指示器
flow-induced vibration 气流激振，流致振动，流激振动，流动激振，水流诱发振动
flowing avalanche 流动崩塌，流动崩坍
flowing characteristic 流动特性
flowing full 满流
flowing ground 流动性地基
flowing power 流动性
flowing property 流动性
flowing resources 流转资源
flowing sand 流沙
flowing stream 流，气流，液流，流淌着的水流
flowing tide 涨潮
flowing-through chamber 流过室，穿流室
flowing-through period 流过时间
flowing-through time 流过时间
flowing water 流水，活水，流动水
flowing well 自流井
flowing 流动的
flow initiation 水流起始，流动的产生
flow inlet angle 进汽角，流入角
flow inlet 流动入口
flow in momentum 动量变化
flow in open air 明流，无压流
flow input 水流引入
flow instability burnout 流量失稳烧毁
flow instability 流量失稳，水流不稳定性

flow instrument　流量计，流量表
flow integrating system　流量积分系统
flow in three dimensions　三维流，空间流，三维（空间）流动
flow in two dimension　二维流动
flow inversion　逆流量
flow in vortex　涡流
flow in　流进，流入
flow irregularity　流动的不均匀性，水流不规则（性），水流不均匀性
flow lag angle　汽流落后角
flow layer　流动层，流层
flow limiter　流量限制器，限流器
flow-limiting venturi　限流文丘里管
flow limit　塑性流动（极限），流动极限
flow line aqueduct　渡槽
flow line　水流线，流水线，流程线
flow loss　流动损失
flow maldistribution　不均匀分布流动，流量不良分配
flow map　流谱
flow margin　流量余量
flow mark　流痕（指射线照片上的），流纹
flow mass　质量流量，流质
flow measurement at spiral case　蜗壳测流
flow measurement　流量测定，测流，流量测量，流量测验，流量量测，水流测定
flow measuring apparatus　流体流量计
flow measuring device　流量测量装置，测流设备
flow measuring flume　流量测验槽，测流槽，测流水槽
flow measuring instrument　流量量测仪器
flow measuring structure　测流建筑物
flow measuring weir　测流堰
flow mechanics　流体动力学
flow mechanism　流动机理
flow medium　流动介质
flow metering valve　节流阀
flow metering　流量测量
flow meter manometer　流量计式压力计
flow meter primary device　流量计一次仪表，流量计发信装置
flow meter secondary device　流量计二次仪表
flow meter　流量表，流速计，流量表
flow method　流水作业法
flow-mixer　流动混合器，水流搅混器
flow mixing device　水流搅混装置（翼），混流装置
flow mixing subfactor　流量交混分因子
flow modelling technique　流动模拟技术
flow model　流动模型，研究流动工况的模化设备，水流模型
flow moisture point　流动水分点
flow momentum　水流动量
flow monitor　流动监测器，流量监查器
flow net　渗流网，流网
flown line　流纹，流线
flow noise spectra　流动噪声谱
flow noise　流动噪声
flow nonuniformity　水流不均匀性
flow nozzle meter　管嘴量水计，管嘴流量计，流量喷管水表
flow nozzle　流量计喷管，流量喷嘴，水流管嘴
flow of condensate　凝结水流量
flow of energy　能流
flow of heat　热流
flow of mass　质量流，质量流量
flow of matter　质量流量
flow of momentum　动量变化
flow of power　潮流，功率流
flow of sewage　污水流量
flow of solid matter　固体流动
flow of steam　汽流，蒸汽流量
flow of water　水流量，水流
flow orifice　流出口，流量孔板
flow oscillation　流动振荡
flow-out diagram　流出量图
flow outlet angle　出汽角，流出角
flow-over　溢出，满出
flow part　通流部分
flow passage clearance　通流部分间隙
flow passage gland　通流部分汽封
flow passage　流道，通流部分，流路
flow-passing surface　过流面
flow path　流程，流道，通流部分【蒸汽通道】，流迹，流径，渗径，水流路径
flow pattern delineation　流谱概图
flow pattern efficiency　流型效率
flow pattern map　流谱图
flow pattern transition　流型过渡
flow pattern　气流结构，流动形式，流量图，流谱，流态，流型，水流流态
flow perturbation　气流扰动
flow pipe　送水管
flow plane　水流面
flow potential　流动势
flow pressure diagram　流量压力图
flow pressure　动水压力
flow processing section　（风洞）整流段
flow process　流动过程
flow production　流水作业
flow profile　水流纵剖面曲线，水流曲线
flow promoter regulator　水流调节装置
flow properties　流动参数，水流特性
flow proportioner　燃料输送调节器
flow-pulsation damping system　脉动流量阻尼系统
flow quality　流动特性，气流参数
flow quantity　流量
flow range　流动范围
flow rate alarm　流量信号
flow rate at normal running water pressure　在额定运行水压下的流量
flow rate controller　流率控制器
flow rate control　流量控制
flow rate detector　流量表传感器
flow rate increase　流量增加
flow rate indicator　流量表，流量指示仪
flow rate measurement　流量测量
flow rate meter　流量计
flow rate of a fluid　（流经管道横截面的）流体流量

flow rate of vacuum 容积流量
flow rate perturbation （反应堆）流量扰动
flow rate reduction 流量下降
flow rate 流速，流量，流率，水流速率，总流量
flow rating curve 流量率定曲线
flow ratio control 流量比控制
flow ratio 流比
flowrator （变截面）流量计，转子流量计，浮标式流量计
flow reactor 连续反应器
flow recirculation zone 回流区
flow recorder 流量自动记录仪，流量计，流量记录器，流量表
flow redistribution 流量再分配
flow regime 流动工况，流动状态，流态，水流情况，水流态
flow region 流动区域
flow regulating valve 流量调节阀
flow regulator 流量调节器，流量调节阀，水流调节器
flow relay 流量控制继电器，流量继电器
flow research 水流研究
flow resistance characteristic coefficient of pipeline 管路阻力特性系数
flow resistance factor 水流阻力系数
flow resistance 流阻，流动阻力，水流阻力
flow restricter 限流器
flow restriction 流量限制
flow restrictor 流量限制器【孔板，水流喷嘴，文丘里管等】
flow retardation 拦滞水流
flow reversal （蒸发受热面上升管中）倒流，水流逆向，回流，逆流
flow reverse 倒流
flow Reynolds number 流动雷诺数
flow routing 流量演算
flow rule 流动规则
flow sensor 流量传感器
flow separater 气流分离器
flow separation phenomenon 气流分离现象，脱流现象
flow separation 流体的分离，分流，水流分离，脱流，流动分离
flow shadow 流影
flow sheet 程序框图，工艺流程图，流程图，流程表
flow shroud 导流套，导流筒
flow sight 流动观察孔
flow simulation 流动模拟
flow skirt 导流围板，导流围筒
flow soldering 射流焊接
flow speed 流动速度，流速
flow split technique 割流法
flow spread 水流宽度，水面宽度
flow stability theory 流动稳定性理论
flow stability 流量稳定，流动稳定性，水流稳定性
flow stagnation （蒸发受热面上升管中）停滞，流动滞止，流量严重不足
flow steadying grid 稳流栅
flow steam 蒸汽流量
flow stowing 自流充填，水力充填
flow straightener 整流器
flow streamline 水流流线，流线
flow strength 流动强度，速度头，水流能量
flow stress ratio 塑流应力比
flow stress 塑流应力
flow structure 流状构造，流动结构
flow summation curve 流量累积曲线
flow superposition method 流量叠加法
flow superposition 流动叠加法
flow surface 流动表面，流面，过流面，过水面
flow survey plane 测流面
flow survey 流线谱，流动测量
flow sweeping 流动扫掠，流向转换
flow switch 流体控制开关，流量开关
flow table test 稠度台试验，流动台试验
flow table 流程表，流动性试验台【混凝土测试】
flow tank 沉淀池
flow temperature 流动温度
flow tends to close valve 水流力图关闭阀门
flow tends to open valve 水流力图开启阀门
flow test of concrete 混凝土流动性试验
flow test 流动性试验，流量试验，水流试验
flow through centrifuge 直流离心机，无逆流离心机
flow through period 水流流动时间
flow-through 流通式
flow time 流动时间
flow to close 流闭式
flow to open 流开式
flow tracer routine 流量跟踪程序
flow transient 流量瞬变，流量暂态，流型转变
flow transmitter 流量变送器，流量发送器
flow tube 流量测量管，流量管，流管
flow turbulence 紊流流动，湍流流动，水流紊动，流动紊流
flow turndown 流量调节（比），流量关小
flow turning angle 气流折转角
flow-type ionization chamber 流气式电离室
flow uniformity 水流均匀性，流动均匀性
flow uniqueness 水流唯一性
flow upwash 流动上洗，气流上升
flow valve 流量阀，流动值，流值
flow variation 流量变化
flow velocity measurement 流速测定
flow velocity 水流速度，流速
flow visualization 流动可视化，流态显示，流动显示
flow welding 铸（浇）焊
flow work 流功
flow zone 流动带，水流区
flow 流量，流动，溢出，流下，流程，潮流，流率，流体，流
flox 液氧
FLTK(flash tank) 扩容器，扩容箱
FLTR＝filter 过滤器
FLTRG＝filtering 过滤
fluctuate 起伏，涨落，波动，脉冲，脉动
fluctuating aerodynamic force 脉动气动力

fluctuating clause 波动条款
fluctuating concentration 脉动浓度
fluctuating current 波动电流，脉动电流
fluctuating data 涨落数据，起伏数据
fluctuating deflection 脉动挠度
fluctuating external wind loading 外部脉动风载
fluctuating flame 脉动火焰
fluctuating flow pattern 脉动流谱
fluctuating flow 波动流量，脉冲流量，波动水流，脉动水流，脉冲力，脉动力
fluctuating frequency 脉动频率
fluctuating internal wind loading 内部脉动风载
fluctuating load 波动负荷，不稳定荷载，变动载荷，脉动荷载
fluctuating moment 脉动力矩
fluctuating nappe 波动水舌
fluctuating noise 起伏噪声
fluctuating of service 运行不稳定
fluctuating of speed 转速波动
fluctuating plume model 脉动羽流模式
fluctuating power 波动功率
fluctuating pressure 压力波动，脉动压力
fluctuating price contract 变动价格合同
fluctuating quantity 起伏程度
fluctuating reattachment 脉动再附
fluctuating response 脉动响应
fluctuating separation 脉动分离
fluctuating stress 变化应力，变动应力，脉动应力
fluctuating trim （船的）纵倾，纵摇
fluctuating velocity 脉动速度
fluctuating wind 脉动风
fluctuating 波动的，起伏
fluctuation amplitude 波动幅度
fluctuation belt 变动区，波动带
fluctuation correction factor 涨落校正因子
fluctuation cycle 涨落周期
fluctuation in change 浮动汇率，汇率变动
fluctuation in discharge 流量变化
fluctuation in exchange 汇率变动
fluctuation of current 电流波动
fluctuation of load 负荷波动，负载波动
fluctuation of power 功率波动
fluctuation of service 运行不稳定，操作不稳定，运行不稳定性
fluctuation of stock 储量变化
fluctuation of water quality 水质波动
fluctuation period 脉动周期
fluctuation pressure 脉动压力
fluctuation rate 波动率，变动率
fluctuation ratiov 起伏比
fluctuation ratio 变动率，波动率
fluctuations in prices 价格波动
fluctuation statistics 脉动统计（量）
fluctuation stream 脉动流，不稳定流
fluctuation velocity 脉动速度
fluctuation zone 变动区
fluctuation 起伏，涨落，波动，脉冲，变动，浮动，变化，脉动，涨落
flue ash removal 除烟灰
flue belt 炉胆圈
flue block 烟道砌块，烟道砖
flue boiler 炉胆式锅炉
flue cinder 飞灰
flue collector 主烟道
flue damper 烟道闸，烟道调节挡板，烟道挡板
flue door 烟道门
flue duct 烟道
flue dust collector （烟道）除尘器
flue dust retainer 集灰斗
flue dust 烟灰，飞灰，烟尘
flue gas analysis meter 烟气分析器
flue gas analysis 烟道气分析
flue gas automatic monitoring system 烟气自动监测系统
flue gas circulation system 烟气循环系统
flue gas contaminant 烟气污染物
flue gas damper 烟气挡板
flue gas denitrification 烟气脱氮，烟气脱硝
flue gas deNO$_x$ 烟气脱硝
flue gas desulfurization plant 烟气脱硫装置
flue gas desulfurization 烟气脱硫
flue gas dew point 烟气露点
flue gas duct 烟气管道
flue gas emission continuous monitoring system 烟气监测系统
flue gas emission standard 烟气排放标准
flue gas emission 排烟
flue gases analysis 烟气分析
flue gas handling system 烟气处理系统
flue gas loss 排烟损失
flue gas moisture content 烟气水分
flue gas NO$_x$ analyser 烟气氮化物分析仪
flue gas O$_2$ analyser 烟气氧分析仪
flue gas purification 烟气净化
flue gas purifier 烟道气净化器
flue gas recirculating fan 烟气再循环系统
flue gas scrubber 烟道气洗涤器
flue gas temperature 排烟温度
flue gas treatment 烟气处理
flue gas washing 烟气清洗
flue gas 烟，烟气，烟道气，排烟，废气
flue heating surface 焰管受热面，炉胆受热面
flue lining 烟囱内衬
fluence albedo 注量反照率
fluence range 流动等级
fluence rate （粒子）注量率，排放率
fluence spectrum 注量谱
fluence 注量，影响，流
flue shutter 烟道挡板
flueway 烟道
flue 烟道，风道，炉胆，通气道
fluid bath 液槽
fluid bearing 液体轴承
fluid-bed catalytic cracker 流化床催化裂化设备
fluid-bed coking 流化床焦化
fluid-bed depth 沸腾层高度
fluid-bed dry scrubber 流化床干燥洗涤器
fluid-bed furnace 流化炉
fluid-bed heat exchanger 流化床换热器
fluid-bed heat transfer 流化床传热

fluid-bed processing 流化床处理
fluid-bed roasting reaction 流化床焙烧反应
fluid-bed 流化床，沸腾层
fluid behavior 流动特性，流动性
fluid body 流体
fluid boundary layer 流体边界层，流动边界层
fluid boundary 流体边界
fluid carry-over 流体夹带
fluid catalyst 流化催化剂
fluid circulation 液体循环
fluid clutch 液体联轴器
fluid coking 流化床焦化
fluid column 液柱
fluid composition modeling 流体组成模型（法）
fluid computer 流体计算机
fluid conditioner 流体调节器
fluid condition 流体状态
fluid control valve 流体控制阀
fluid coolant 冷却介质，冷却液，流体冷却剂
fluid-cooled blade 液冷式叶片
fluid-cooled electrical machine 液冷电机
fluid-cooled winding 液冷绕组
fluid coupling 流体联轴节【液压动力，液压静力，液力黏滞，液压动态】，液力联轴器，液力耦合器
fluid cycle 液力循环
fluid damper 流体减震器
fluid damping 流体阻尼
fluid deflection 流体偏转
fluid density meter 液体密度计
fluid density 流体密度
fluid displacement tachometer 流体转速计
fluid driven coupling 液力联轴器，液力联轴节，液压联轴器
fluid driven 液力驱动的，液压传动的，流体传动的
fluid dynamic damping 流体动力阻尼
fluid dynamic parameter 流体力学参数
fluid dynamics of suspension 悬浮体流体动力学
fluid dynamics 流体动力学
fluid elastic stability 流体弹性稳定度
fluid elastic turbulent excitation 液体弹性（湍流）扰动
fluid element 流体单元
fluid equation 流体动力学方程，流体方程
fluid feedback 流体反馈
fluid film bearing 液膜轴承，油膜轴承
fluid film 流体膜，液膜
fluid flow analogy 水动力模拟
fluid flow control 流体流量控制
fluid flowmeter 流体流量计，液体流量计
fluid flow regulator 液体流量调节器，流体流量调节器
fluid flow with heat generation 释热流体流
fluid flow 流量，液体流，流体流动，流体流量，液流
fluid friction 流体阻力，流体摩擦
fluid-fuelled reactor 液态燃料反应堆
fluid fuel thermal breeder 流体燃料热增殖堆
fluid fuel 液体燃料，液态燃料
fluid handled 液力操作的
fluid horsepower 流体功率
fluidic device 射流装置
fluidic relay 射流继电器
fluidic stabilizer 射流稳定器
fluidic stepping motor 射流式步进电动机
fluidics 应用流体学，射流学，射流技术，流体学
fluidic 流体的，射流的
fluidimeter 黏度计，流度计
fluid inclusion 液包体
fluid induced vibration 流体诱发振动，流体致振
fluid inlet temperature 工质进口温度
fluid interchange （载热剂）流动换热，载热
fluid iron ore direct reduction 流化床铁矿石直接还原
fluid iron ore reduction 铁矿石流态化还原
fluidity of the slag 熔渣流动性
fluidity value 流度值
fluidity 流动性，流动度，流度，流体状态
fluidization column 流化柱
fluidization efficiency 流化效率
fluidization medium 流化介质
fluidization pressure 液化压力
fluidization quality 流化质量
fluidization uniformity 流化均匀性
fluidization 沸腾化，流体化
fluidized adsorber 流化吸附器
fluidized bed apparatus 流化床设备
fluidized bed boiler 沸腾床锅炉，流化床锅炉
fluidized bed coating 流化床涂敷
fluidized bed combustion boiler 沸腾燃烧锅炉，流化床锅炉
fluidized bed combustion 流化床燃烧，沸腾燃烧
fluidized bed combustor 流化床燃烧器
fluidized bed crystallizer 流化床结晶器
fluidized bed drier 流化床干燥器
fluidized bed furnace 流化床炉
fluidized bed gasification 流化床气化
fluidized bed gasifier 流化床气化炉，流化床气化器
fluidized bed heat exchanger 流化床换热器
fluidized bed heat recovery 流化床热回收
fluidized bed insulation 流化床绝缘
fluidized bed pile 大型流化床装置
fluidized bed plant 流化床设备
fluidized bed process 流化床工艺
fluidized bed reactor 流化床反应器
fluidized bed roasting 流化床焙烧
fluidized bed temperature 流化床温度
fluidized bed unit 流化床设备
fluidized bed 浮动床，流化床，沸腾床
fluidized calcination 沸腾焙烧
fluidized classifying 流态化分级
fluidized combustion 沸腾燃烧
fluidized drier 沸腾干燥器
fluidized drying 沸腾干燥
fluidized economizer 沸腾式省煤器
fluidized-fuel reactor 流化燃料反应堆
fluidized packed bed 流化填充床
fluidized paste reactor 糊状燃料反应堆
fluidized phase 流化相

fluidized process technology 流态化工艺过程
fluidized state 沸腾状态，流化状态
fluidizer 流化装置
fluidize 沸腾化，流态化，流化
fluidizing air 流化空气
fluidizing blower 流化风机
fluidizing fan 流化风机
fluidizing flowrate 流化速度
fluidizing medium 流化介质
fluidizing point 流化点
fluidizing transporter 仓式泵
fluidizing velocity 流化速度
fluid jet 流体射流，射流
fluid kinetics 流体动力学
fluid level indicator 液位指示器
fluid-like property 拟流化特性
fluid loss 流体损耗，流体损失
fluid lubrication 流体润滑，液体润滑
fluid mass 流体质量
fluid mechanics principle 流体力学原理
fluid mechanics 流体力学，液压力学
fluid medium 流体，流体介质
fluid memory effect 流体记忆效应
fluid meter 流量计，流体计量器，黏度计
fluid motion 流体运动
fluid motor 液压马达，水力发动机，液动马达
fluid mud 浮泥
fluid oil 润滑油
fluid outlet temperature 工质出口温度
fluid parcel 流体团，流体块
fluid particle interaction 流体颗粒相互作用
fluid particle 流体质点
fluid passageway 流体通道
fluid poison control 液态毒物控制
fluid poison 液体毒物
fluid power motor 液力马达
fluid power society 流体动力学会
fluid power 流体动力
fluid pressure governor 液压调速器，液压调节器
fluid pressure regulator 液压调节器
fluid pressure scales 液压秤
fluid pressure sealing system 液压密封系统
fluid pressure 流体压力，液体压力，流体静压力，液压
fluid properties 流体参数，工质参数
fluid property 流体性能
fluid pumped 抽送的液体
fluid regime 流体状态
fluid resistance 流体阻力
fluid saturation 流体浸润度
fluid sealing 液体止水
fluid slowness 流体慢度
fluid sludge 流渣
fluid speed meter 液体流速计
fluid state 流态
fluid static pressure 流体静压力，静水压力
fluid statics 流体静力学
fluid strata 流体层
fluid stream 液流
fluid summation method 流体求和法
fluid temperature （灰的）流动温度，流体温度
fluid test 流体试验
fluid-tight 液密的，不漏流体的
fluidunstable 流体不稳定的
fluid velocity 流化速度
fluid vibration 流体震动
fluid viscosity 流体黏性（系数），液体黏度
fluid wax 液体石蜡，流体石蜡
fluid whirl 流体旋涡
fluid 射流，流体，流质，能流动的，流动的
flume chute 斜槽
flume crossing 渡槽交叉口
flume experiment 水槽试验
flume waste 斜槽废水
flume with both side contractions 两侧束狭测流槽
flume with side contraction 侧向测流槽
flume 斜孔道【反应堆】，引水沟［渠］，水道，斜槽，测流槽，水槽，通道，峡沟
fluming water 排放水，输送水
fluohydric acid 氢氟酸
fluoradiography 荧光射线照相术
fluoration 氟化作用
fluorescence analysis 荧光分析
fluorescence content meter 荧光度测定表，氟量计
fluorescence yield 荧光效应
fluorescence 荧光物，荧光
fluorescent crack detection 荧光探伤
fluorescent dye 荧光颜料
fluorescent fault detector 荧光探伤仪
fluorescent filament 荧光微丝
fluorescent flaw detection 荧光探伤
fluorescent lamp stabilizer 荧光灯稳压器
fluorescent lamp 日光灯，荧光灯
fluorescent magnetic particle inspection machine 荧光磁粉探伤机
fluorescent-mercury lamp 水银荧光灯
fluorescent noise generator 荧光噪声发生器
fluorescent oil flow method 荧光油流法
fluorescent particle tracer 荧光示踪粒子
fluorescent penetrant inspection 荧光检验，荧光渗透液探伤
fluorescent penetrant test 荧光探伤测定
fluorescent penetrant 荧光渗透剂
fluorescent penetrate inspection 荧光渗透检验
fluorescent penetrate test 荧光探伤
fluorescent pigment tracer 荧光示踪颜料
fluorescent screen 荧光屏
fluorescent spectrograph 荧光光谱仪
fluorescent tracer technique 荧光示踪技术
fluorescent tracer 荧光示踪剂
fluorescent tube 荧光灯，日光灯管
fluorescent 荧光的，荧光
fluorescing dyes 荧光染料
fluorhydric acid 氢氟酸
fluoric acid 氟酸
fluoridation 氟化作用
fluoride 氟化物
fluorimeter 荧光度测定表，氟量计，荧光计
fluorimetry 荧光测定法

fluorinated rubber　氟化橡胶
fluorination　氟化作用
fluorite　荧石，氟化钙，萤石，氟石，氟
fluorochemical vapor machine　氟化蒸汽冷却电机
fluoroelastomer　氟橡胶
fluorogas insulation　氟化气体绝缘
fluorometer　荧光计，氟量计
fluorometric method　荧光法
fluoro-photometer　荧光计，荧光光度计
fluoroplastics　氟塑料
fluoroscope　荧光镜
fluoroscopic screen　荧光屏
fluoroscopy　（X射线）荧光检测，（X射线）荧光屏检查
fluor-protein foam concentrate　氟蛋白泡沫液
fluorspar　氟石，萤石
fluosolids drier　沸腾干燥器
fluo-solid system　流态化系统
flurry　雪暴
flush bolt　埋头螺栓，平头螺栓
flush bottom ash cycle　冲渣循环
flush bottom tank valve　罐底排污阀，箱底排污阀
flush box　冲洗水箱
flush cistern　厕所水箱，冲洗水箱【便器用】
flusher　冲洗器，冲洗设备
flush fire hydrant　地下消火栓
flush gate　冲砂闸门
flush-head rivet　平头铆钉
flush hole　蒙皮破孔
flushing action　冲洗作用
flushing air　起泡
flushing chamber　冲洗室
flushing D_2O　冲洗重水
flushing gate　冲泄闸门
flushing gutter　冲沙沟
flushing hole　（液态炉）渣口
flushing line　冲洗导管，清洗导管
flushing manhole　冲洗检查井
flushing nozzle　冲洗喷嘴
flushing oil　洗涤油
flushing pipe　冲洗管，冲洗管道
flushing pore　冲洗孔
flushing procedure　清洗规程
flushing pump　冲洗泵，供油泵
flushing sluice　冲沙闸
flushing test　冲水试验
flushing tunnel　冲沙隧洞
flushing valve　冲洗阀，冲泄阀
flushing water pump　清洗水泵，冲洗水泵
flushing water　冲沙水
flushing　冲洗，涌流，冲刷，吹洗，清洗
flush inlet　不前伸进口，齐平式进水口
flush joint　平头接合，平直接合
flush light　吸顶暗灯
flushmeter　冲水阀，冲洗阀
flush-mounted bracket　嵌平安装的托架
flush-mounted drawout relay　平装抽出式继电器
flush-mounted fluorescent lighting fixture　嵌入式荧光灯具
flush-mounted　暗装的，齐平式，嵌入式安装的
flush nozzle　平接式接管，齐平式接管
flush off pipe　溢流管
flush off　溢出，排出
flush out stream　吹洗流，冲洗流
flush out　清洗，吹洗，冲掉，冲走
flush pipe　冲水管
flush plate　平装开关面板
flush plug receptacle　嵌入式插座，埋装式插座，埋入式插座
flush pump　清洗泵
flush rivet　平头铆钉
flush snap-switch　埋装式弹簧开关，埋装式活动开关
flush sound retardant door　齐平式隔音门
flush tank　冲洗水箱，冲洗池
flush tube　冲洗管
flush type instrument　嵌入式仪器
flush type　嵌入式
flush valve　冲洗阀
flush water　冲洗水
flush weld　无余高的焊缝，削平焊缝，平焊缝
flush　嵌平【管子对管板焊接】，清洗，冲洗，平的，埋入的，嵌入的，奔流，倾泻，排泄，排流，排放，冲水，暴涨，齐平，使齐平
fluted covering　波纹蒙皮
fluted shaft　槽轴
fluted　有凹槽的，波形的
fluting　凿槽，柱槽
flutter coefficient　颤振系数
flutter computer　颤动计算机
flutter derivative　颤振导数
flutter effect　颤动效应
flutter fading　颤动衰减，散乱反射衰减
flutter flutter　颤振，抖动【阀瓣】，飘动，扰动，脉动干扰，振动
flutter frequency　颤振频率
fluttering　振动
flutter mode　颤振模态，颤振模型
flutter moment　颤振力矩
flutter relay　振动式继电器
flutter stability　颤振稳定性
flutter tendency　颤振趋势
flutter wind speed　颤振风速
fluvial abrasion　河流冲刷
fluvial bog　河边低地
fluvial process　成河过程
fluviograph　水位计
flux agent　稀释剂
flux age　通量寿期
flux albedo　通量反照率
flux analysis　流动分析
flux and reflux　涨落潮
flux-averaged cross-section　按通量平均的截面
flux backing　焊剂垫，焊药基底
flux buildup　建立磁通，通量增大
flux-coated electrode　涂有熔剂的焊条
flux concentrating ratio　通量聚光比
flux converter　中子转换器
flux-cored arc welding　药芯焊丝电弧焊
flux-cored electrode　管状焊条，药芯焊丝
flux-cored wire　焊药芯焊丝，药芯焊丝

flux curve 通量曲线，磁通曲线
flux-cutting law 磁通切割律
flux decay 磁通衰减
flux density distribution 通量密度分布
flux density measurement 通量密度测量
flux density meter 通量计，辐射通量测量仪
flux density spectrum 通量密度谱
flux density 磁通密度，通流密度，通量密度
flux depression correction 通量压低校正
flux depression factor 通量压低因子
flux depression 通量（密度）压低
flux depressor 通量压低体
flux detector 通量探测器
flux differential relay 磁通差动继电器
flux dip fill up 填平通量坑，通量下陷填平
flux dipping 通量井，通量下陷
flux dip 通量下陷，通量坑
flux distortion 通量畸变
flux distribution 通量分布，磁通分布
flux estimator 中子通量计算器，中子通量激增
flux flattened region 通量展平区
flux flattening material 展平通量的材料
flux flattening 通量平化，通量展平
flux flow ratio 中子通量流量比
flux fluctuation 通量涨落
flux form factor 通量峰值因子，通量形状因子
flux function 流函数，通量函数
flux-gate 磁通量闸门，磁通门
flux gradient 通量梯度
flux guide 磁导
flux hardening 通量谱硬化
fluxible 易熔的，可熔的，易烧的，可流动的
flux inclusion 助熔剂杂质
fluxing 焊药，炉衬渣侵蚀，熔化，助熔，助熔剂
flux layer 焊药层
flux leakage 漏通量，漏磁通，通量泄漏
flux leaking 通量泄漏
flux level 通量水平
flux line 通量线，磁力线
flux linkage equations 磁链方程式
flux linkage 磁链，磁通匝连数
flux linking 磁链
flux map 通量图【曲线】，液［气］流图，通量图，通量曲线
flux measurement 通量测量
flux measuring channel 通量测量通道
flux measuring instrumentation 通量测量仪表［装置］
fluxmeter 磁通计，通量计，辐射热流计，辐射通量测量计，麦克斯韦计
flux monitoring 通量监测
flux monitor （中子）通量监测器
flux of energy 能通量
flux of exhaust buoyancy 排气的浮力通量
flux of induction 感应通量
flux of interlinkage 交链磁通量
flux of momentum 动量的通量
flux of radiation 辐射通量
fluxoid quantum 磁通量子
fluxoid 全磁通
fluxon 磁通量子
flux oscillation 通量震荡
flux path 磁通轨迹，磁通路径
flux peaking factor 通量峰因子
flux peaking 局部通量剧增，通量峰值
flux peak 通量峰，通量峰值
flux period 通量周期
flux perturbation 通量微扰，流的扰动，通量扰动
flux plot hole 通量测绘孔
flux plot 磁场图，通量图，磁力线，通量曲线
flux profile 通量分布图，通量廓线
flux pulsation 磁通波动，磁通脉动
flux pulse 通量脉冲
flux rating 通量额定值
flux ratio 通量比
flux recovery 通量恢复
flux recycling machine 焊剂回收机
flux recycling unit 焊剂回收机
flux-resetting magnetic amplifier 磁通复原磁放大器
flux Richardson number 通量理查森数
flux ring 磁环【控制棒驱动机构】，磁通环
flux second 通量秒
flux-sensing element 通量敏感元件
flux separator 焊剂分离机
flux shield 磁通屏蔽
flux synthesis method 通量综合法
flux thimble 通量测量套管
flux tilting test 通量斜变法
flux tilt 通量斜变
flux-time 时间积分通量
flux trap loading 通量阱装料方式
flux trap reactor 通量阱反应堆
flux trap region 通量阱区
flux trap 磁通阱，通量阱，通量栅
flux traverse measurement 通量分布测量
flux valve 溢流阀
flux wave 通量波，磁通波
flux weakening region 磁场弱化区，弱磁区
flux weakening 弱磁，磁通降低
flux weighted integral 通量权重积分
flux 焊剂，通量，质量流率，助熔剂，磁力线，注量率，熔化，磁通，焊药
FLW＝flow 流量
flw reattachment （分离的）气流再附着
fly ash bin 飞灰库
fly ash collector 除尘器
fly ash concrete 粉煤灰混凝土
fly ash control 除尘，飘尘控制
fly ash conveying system 飞灰输送系统，输灰系统
fly ash corrosion 飞灰腐蚀
fly ash emission 飞灰排放
fly ash erosion 飞灰磨蚀，飞灰磨损
fly ash extraction 飞灰分离
fly ash filter 飞灰过滤器
fly ash handling system 飞灰处理系统
fly ash refiring 飞灰再燃
fly ash re-injection system 飞灰再循环系统
fly ash separator 除尘器
fly ash silo 灰库

fly ash sluicing water　飞灰冲灰水
fly ash　扬灰，飞尘，飞灰，粉煤灰
flyback time　回扫周期
flyback transformer　反馈变压器
flyback　反馈，倒转，逆行
flyball governor　飞锤式调速器，飞球式调速器
flyball integrator　离心球积分器
flyball tachometer　离心式转速计，飞锤式转速计，飞球式转速计
flyball　飞锤，离心球，飞球，离心锤
fly drill　手拉钻
flyer　梯级
flying angle support　悬垂转角杆塔
flying boat　飞艇
flying bridge　渡船，浮桥，驾驶台，跨线桥
flying buttress　拱式支墩，拱支墩
flying falsework　悬空脚手架
flying fox　缆索渡
flying gangway　天桥
flying head　浮动磁头
flying printer　轮式打印机
flying scaffolding　悬空脚手架
flying shore　悬空槽撑
flying signal　闪光信号
flying spot scanner　飞点扫描器
flying spot storage　飞点存储器
flying-spot tube　扫描管，飞点示波管
fly off safeguard　防飘浮
fly off　上移【燃料棒】，飞起来，飞出，飞速地跑掉
flyover crossing　立体交叉
flyover type power house　厂前挑流式厂房
flyover　立体交叉
fly rock　岩石碎块
flywheel　飞轮
flywheel action　飞轮效应，飞轮作用
flywheel circuit　惯性同步电路
flywheel diode　续流二极管
flywheel effect　飞轮效应
flywheel energy storage　飞轮储能（法）
flywheel generator　飞轮发电机
flywheel inertial storage　飞轮惯性储能
flywheel magneto　飞轮式永磁发电机
flywheel moment　飞轮力矩
flywheel motor-generator　飞轮电动发电机
flywheel type alternator　飞轮式同步发电机
FMA(free mineral acidity)　无机酸的鉴定量
FM-AM multiplier　调频调幅倍增器
FMB data set　机器控制字
FMEDA(Failure Modes，Effects and Diagnostic Analysis)　故障模式、影响及诊断分析
FMEDA report　故障模式、影响及诊断分析报告
FM exciter　调频激励器
fm＝femtometer　飞米【10～15m】
FM(frequency modulation)　调频
FM receiver　调频接收机
FMS(force measuring system)　测力系统
FM wave　频率调制波，调频波
FN(fixture note)　确认备忘录，订租确认书
FNL＝final　末级，最后
FNPA(full-arc/partial-arc)　全部进汽/部分进汽
FNP(floating nuclear power plant)　（海上）浮动核电厂
FOA(forced oil air cooled)　强油风冷
FOAK(first-of-a-kind)　一类中的第一个，同类别首个
foam analysis　泡沫分析
foam article　泡沫制品
foam breaker　泡沫破碎机
foam carryover　泡沫携带
foam catcher　泡沫捕集器
foam cement　泡沫水泥
foam chamber　泡沫室
foam concrete　泡沫混凝土
foam diatomite　泡沫硅藻土
foam dust separator　泡沫除尘器
foamed asphalt　泡沫沥青
foamed concrete　泡沫混凝土
foamed materials　泡沫材料，多孔材料
foamed mortar　泡沫砂浆
foamed plastics　泡沫塑料
foamed slag concrete　泡沫矿渣混凝土
foamed slag　泡沫溶渣，泡沫矿渣
foamed vinyl resin　泡沫乙烯树脂
foamed　泡沫状的
foam extinguishing agent　泡沫灭火剂
foam extinguishing system　泡沫灭火系统
foam fire extinguisher　泡沫灭火机，泡沫灭火器
foam fire extinguishing　泡沫灭火
foam fire fighting truck　泡沫消防车
foam fire track　泡沫消防车
foam flow　沫状流
foam generator　泡沫发生器
foam glass　泡沫玻璃
foaminess　起泡沫性
foaming adjuvant　起泡剂
foaming agent　起沫剂，起泡剂，发泡剂，泡沫剂
foaming concrete　泡沫混凝土
foaming plastics insulation　泡沫塑料绝缘
foaming substance　发泡物质
foaming　泡沫共腾，起泡，起泡沫
foam inhibitor　泡沫抑制剂，抑泡剂
foam insulation material　泡沫隔热材料
foam insulation　泡沫绝缘材料，泡沫隔热，泡沫绝缘
foam liquid　泡沫液
foam maker　发泡器
foammaking agent　发泡剂，泡沫剂
foam over　起泡溢出，泡沫携带
foam plastics　泡沫塑料
foam-powder universal fire truck　泡沫干粉联用消防车
foam rubber　泡沫橡胶，海绵橡胶
foam sandwich　泡沫夹芯结构
foam separation　泡沫分离
foam separator　泡沫除尘器，泡沫发生器
foam solution　泡沫溶液
foam-spray packaging　喷泡沫包装
foam-spray packing　喷泡沫包装
foam suppressant　消泡剂
foam tube nozzle　泡沫枪，泡沫管喷嘴

foamy 起泡沫的，泡沫的
foam 泡沫，泡沫物，起泡，起泡沫，泡沫状物
FOB airport 机场交货价格
FOB destination 目的地交货价格
FOB(free on board) 船上交货，离岸价格，船上交货价格，装运港船上交货
FOB(free on board) 离岸价格
FOB price(free-on-board price) 离岸价格
focal aperture area 焦点区
focal aperture 聚焦孔径
focal back-up protection 近后备保护
focal depth 震源深度
focal distance 焦距
focal length 焦距
focal point of stress 应力集中点
focal probe with water column coupling 水柱耦合聚焦探头
focal region 震源域
focal spot size 焦点的尺寸【射线照相】
focal spot 焦斑，聚焦点，焦点，焦距【光学等】
focred outage hours 强迫停运小时
focus cylindrical Fresnel lens for solar concentrator 聚焦菲涅耳太阳聚光透镜
focusing coil 聚焦线圈【电子束焊】
focusing error 对光误差
focusing mark 调焦标，对光标
focusing monochromator 聚焦单色器
focusing point radiometer camera 聚焦式辐射照相机
focusing probe 聚焦探头
focusing ring 调焦环，对光环
focusing screw 调焦螺旋
focusing 调焦，聚焦，聚焦，对光
focus on the dispute 谈判焦点
focus on 致力于，使聚焦于，对（某事或做某事）予以注意
focus-to-film distance 焦点胶片距离，焦距
focus 焦点，聚焦，集中，震源
foehn wind 焚风
foehn 焚风
FO(fail open) 故障时自动打开，故障开
FOF(forced outage factor) 强迫停运率，强迫停运系数
FO(free out) 船方不负担卸货费用
FO(fuel oil) 燃油
fog-cooled reactor 雾冷反应堆
fog cooling 喷雾冷却
fog cured 喷雾养护的，雾室养护的
fog curing 喷雾养护
fog density 模糊密度【射线照相】
fog dispersal 雾消散，消雾
fog drip 雾滴
fog droplet 雾滴
fog duration 雾持续时间
fog flow 雾状流，雾状流动
fog frequency 雾频
fog generator 雾发生器
fogging density 灰雾度
fogging 成雾，雾化
foggy day 雾天，有雾日
foggy 模糊的，雾深的
fog lamp 雾灯
fog nozzle 喷雾管嘴，雾化喷嘴
fog plume 雾羽
fog rain 雾雨
fog reactor 雾冷反应堆
fog room 喷雾室，雾室
fog shower 雾雨
fog spraying curing 喷雾养护
fog-type insulator 防污型绝缘子，耐雾绝缘子
fog-type unit 耐雾型绝缘子
fog 雾
FOH(forced outage hours) 强制停运小时数，强迫停运小时
foil coil 箔线圈
foilcraft 水翼艇
foil detector 金属箔探测器
foil electroscope 金箔验电器
foil insulation 箔片隔热层
foil lattice 平面叶栅，薄片晶格
foil mica capacitor 云母箔电容器
foil mica 云母箔
foil port valve 全孔道阀门
foil strain gauge 箔型应变片
foil type safety valve 箔式安全阀
foil winding 箔绕组，箔式线圈
foil 箔，金属薄片，薄片，反射板，箔片
fold axis 褶皱轴
fold closure 褶曲闭合度
fold crack 折裂
folded antenna 折叠天线
folded belt 褶曲带
folded dipole 折叠偶极子
folded mountain 褶皱山
folded-plate structure 折板结构
folded plate 折板
folded region 褶皱区
folded 弄皱的，折叠的
folder pier 防冲突堤
folder 文件夹
fold fault 褶皱断层
folding axis 褶曲轴，褶皱轴
folding chart 折叠式记录纸
folding door hardware 折叠门五金零件
folding door 折叠门
folding earthquake 褶皱地震
folding frequency 折叠频率
folding gate 折叠式门
folding machine 折弯机
folding period 褶皱期
folding rule 折尺
folding staff 折标尺，折尺
folding structure 褶曲构造
folding test 折叠试验
folding time 加倍时间
folding window 折叠窗
folding 折迭，曲折，弯曲，折叠
fold inward 向内褶曲
fold-out type solar array 折叠式方阵
fold 折，折叠，褶皱
foliage 植物叶子

foliated coal　片状褐煤，层状褐煤
foliated fracture　层状断口
foliated rock　叶片状岩
foliated structure　薄片状构造，叶片状构造，叶理构造
foliation　板理，成层
follow control system　随动控制系统
follow current　续流，持续电流，跟踪电流
follower drive　随动拖动
follower lever　从动杆
follower motor　随动电动机
follower pile　送桩
follower ring　压圈，随动圈
follower rod　（压水堆）随动棒
follower　跟随器【控制棒】，随动件，从动机构，跟踪机构，跟踪仪
following core　后继堆芯，下一炉料
following distance　尾随距离
following edge　后缘
following error　随动误差
following level meter　随动水位计
following range of synchronization　同步保持范围
following train　后行列车
following wake　尾迹流，副流
following wind　顺风
follow load　跟踪负荷
follow on current　持续电流
follow-on subassembly　随动组件
follow-on　改进型的，下一代的，继承的
follow the international codes of practice　同国际惯例接轨
follow-up action　跟踪行动，后续行动，后续措施
follow-up circuit　跟踪回路
follow-up control system　随动控制系统
follow-up control　随动调节，随动控制，跟踪控制
follow-up device　随动设备
follow-up investigation　跟踪调查
follow-up investment　后续投资
follow-up meeting　后续会议
follow-up motor　随动电机
follow-up pilot valve　跟踪滑阀
follow-up pointer　随动指针
follow-up potentiometer　反馈电势计，随动系统电势计，跟踪电动计
follow-up program　后续方案
follow-up signal　跟踪信号
follow-up survey　跟踪调查
follow-up system　随动系统，追踪系统
follow up the program proceeding　跟踪进度
follow-up valve　随动阀
follow-up　跟踪，随动，跟踪装置，随动装置，随动跟踪
fomite　污染物
font size　字体大小
font　字形，字体
food chain　食物链
food dye　食用染料
food industry　食品工业
food irradiation　食品辐照
food pathway　食物路径
foodstuff　粮食，食品
foolite　石棉和树脂制成的绝缘材料
foolproofing key　确保安全的键
foolproofing shape　确保安全的型式
foolproof keying system　确保安全的键固系统
foolproof　防止错误操作
footage marking　底座标志
footage　尺寸
foot block　线脚块，柱脚块
foot board　踏板
foot bolt　地脚螺栓，基脚螺栓
foot bridge　人行桥，人行街，步行桥
foot-candle meter　英尺-烛光测定计
foot distance　根开【线路工程】
footer　页脚
foot flanges　底脚法兰
foothill　山麓丘陵，山根小丘，山麓
footing beam　底脚梁，基础梁
footing course　底层，基层
footing dressing　墙脚处理
footing foundation　柱基，底座基础
footing marker　底座标志
footing of wall　墙基梁，墙基
footing　底座【基础】，底脚，大放脚，基础，基脚
foot irons　铁爬梯
foot mattress　护底沉排
foot mat　底垫层
foot-mounted motor　落地安装型电动机，底座安装型电动机
foot-mounting cubicle　落地型开关柜
foot note　附注，脚注
foot of a perpendicular　垂运
foot pace　梯台
footpath　步道，人行道
foot plank　跳板
foot-pound-second　英尺-磅-秒
foot screw　地脚螺钉，地脚螺栓
foot stall　柱墩
foot-step bearing　立轴止推轴承，立轴承
foot-step pillow　承力瓦块球面座
footstep　轴支架
footstock　顶座，尾座
foot valve and strainer　底阀和滤网
foot valve　底阀
footwalk　桥，平台，人行道，步道
foot wall　下盘
footway　人行道
foot　底座，底脚，支点，叶根，英尺，基座
forage about　搜查
for approval version　送审版
forbearance　偿债延期，宽容
forbidden area　禁区
forbidden region　禁区
forbidden zone　禁区
for buyer's approval　由买方批准
force and energy measurement　力能测试
force application　施力
force arm　力臂
force at rupture　断裂力

force-balance transmitter　力平衡变送器
force-balance　（静）力平衡【测量设备】，测力天平
force cell　测力传感器
force centre　力心
force coefficient　风的体形系数，力系数
force-commutated　强制换流
force component　分力
force composition　力的合成
force connection　力传递
force-cooled transformer　强迫冷却变压器
forced　被迫的，强制的
forced aeration　强迫掺气
forced air blast　强制通风
forced air change　强制换气
forced air circulation system　强迫空气循环系统
forced air circulation　强制通风，强制空气循环
forced air-cooled heat exchanger　强风冷热交换器
forced air-cooled oil immersed transformer　强制风冷油浸变压器
forced air-cooled transformer　强迫风冷式变压器
forced air cooler　强制空气冷却器
forced air cooling system　强迫风冷系统
forced air cooling　强制空冷，强迫风冷，鼓风冷却，强制风冷却
forced air refrigeration return system　强制空气循环制冷系统
forced air supply　强迫供应空气，强制供气，人工供风，压力供气
forced air ventilating system　压力通风系统
forced air ventilation　强制通风，压力通风
forced argon circulation　氩气强制循环
forced back washing　强制反洗
forced circulating system　强制循环系统
forced circulating　强制循环
forced circulation air-cooling　强制循环空气冷却
forced circulation boiler　强制循环锅炉
forced circulation boiling　强制循环沸腾
forced circulation cooling　强制循环冷却
forced circulation evaporator　强制循环蒸发器
forced circulation pressurized-water reactor　强制循环压水反应堆
forced circulation reactor　强制循环反应堆，强制循环堆
forced circulation steam generator　强制循环锅炉，强制循环蒸汽发生器
forced circulation straight-tube steam generator　强制循环直管蒸汽发生器
forced circulation system　强制循环系统
forced circulation through the core　通过堆芯的强制循环
forced circulation　强制循环，强迫循环
forced closing　强迫关闭
forced coding　强制编码
forced coefficient inclined tube　倾斜管强迫对流
forced commutated inverter　强迫换流逆变器，强制转换逆变器
forced commutation of thyristor　可控硅的强迫整流
forced condensate return system　加压凝结水回收系统
forced convection boiling　强制对流沸腾，强迫对流沸腾
forced convection cooling　强迫对流冷却
forced convection heat transfer　强制对流换热，强迫对流换热，强制对流传热
forced convection vaporization　强制对流汽化
forced convection　强迫对流，强制对流
forced coolant circulation　冷却剂强制循环
forced cooling system　强制冷却系统
forced cooling　强迫冷却，强制冷却
forced derated　强迫降低出力的
forced draft air　送风
forced draft control　送风调节
forced draft cooling tower　强制通风冷却塔，机力通风冷却塔
forced draft fan inlet silencer　送风机入口消音器
forced draft fan　强制送风机，压力送风机，鼓风机，送风机
forced draft stoker　机械通风炉排
forced draft type cooling tower　强制通风冷却塔
forced draft　正压通风，强制通风，鼓风，正压送风，强力通风
forced drainage　强制排水
forced draught air cooler　强制通风冷却器
forced draught cooling　强制通风冷却
forced draught fan　送风机
forced draught　强制通风，压力通风，鼓风
force decomposition　力的分解
force derivative　力导数
forced excitation　强行励磁
forced feed lubrication　强制润滑，压力润滑
forced feed　压力进给，强制进料，加压进料，压力推进，压油润滑
forced filtering rate　强制滤速
forced flow　强制流动，受迫流动，强制流
force diagram　作用力示意图，力图
forced interruption　强行断电
force-displacement curve　力位移曲线
forced line charging　强制送电
forced line energization　线路强送电
forced line outage　线路被迫停电
forced lubrication pump　压力润滑泵
forced lubrication system　压力润滑系统，强制润滑系统
forced lubrication　强制润滑
forced magnetization condition　强制磁化条件
forced magnetization　强制磁化
forced meandering　限制弯曲，强迫蜿蜒
forced noise　强力噪声
forced non-linear oscillation　非线性强迫振荡，非线性强制振荡
forced oil-air cooling　强制油循环吹风冷却，强制油-空气冷却
forced oil-circulated air-cooled transformer　强油循环风冷变压器
forced oil-circulated water-cooled transformer　强油循环水冷变压器
forced oil circulation　强迫油循环，强制油循环，强制油冷却
forced oil-cooled bushing　强迫油冷式套管

forced oil-cooled heat exchanger 强油冷热交换器
forced oil-cooling transformer 强迫油冷变压器
forced oil-cooling 强迫油冷
forced oil forced air cooling 强迫油冷强迫风冷式冷却
forced oil transformer 强制油循环式变压器
forced opening 强迫开启
forced oscillation 受迫振荡，强迫振荡，强制振荡，强迫振动
forced outage factor 强迫停运系数
forced outage hours 强制停运小时数，强迫停运小时
forced outage incident rate 事故件数率
forced outage rate 强迫停机率，事故停用率，非计划检修停产率，强迫停运率
forced outage ratio 强迫停机比，事故停机率
forced outage state 强迫停运状态
forced outage 事故停用，强迫停机，强制停机，强迫停运
forced oxidation process 强制氧化工艺
forced partial outage 限制出力
forced power transmission 强行输电
forced programming 强制编程
forced pump 压力泵，加压泵
forced ramming 强夯（土）
forced reactor coolant circulation 反应堆冷却剂强制循环
forced recirculation type steam generator 强迫再循环锅炉，强迫再循环式蒸汽发生器
forced recirculation 强迫再循环，强制再循环
forced release 强制释放
forced response 强迫响应，强制响应
forced rolling 强制横摇
forced service 强制运行
forced shutdown 事故停机，事故停运，强迫停机，被迫停机
forced splash lubrication 强制飞溅润滑
forced stoppage 强迫停机
forced stop 强制停机
forced surface air cooling 强迫表面空气冷却
forced surface wave 强制表面波
forced synchronization 强制同步
forced synchronizing 强制同步，强迫同期
forced tamping 强夯
forced tidal wave 强制潮汐波
force due to friction 摩擦阻力
forced unloading 事故卸料
forced-ventilated motor 强制通风式电动机
forced ventilation 强迫通风，强制通风，机械通风，压力通风
forced vibration analysis 强迫振动分析
forced vibration 强迫振动，受迫振动
forced vortex 强迫涡，强制涡流，强制涡旋
forced water air heating 机械循环热水供暖
forced water circulation 强制水循环
forced water cooling 强迫水冷，压水冷却
forced wave 强制波
forced yaw system 强迫偏航系统
force dynamometer 测力器，测力传感器
force equilibrium 力平衡
force excitation factor 强励系数
force excitation 强励，强行励磁
force excited oscillation 力激振动
force factor 力因数
force feed circulation 强制循环
force feed lubrication 强迫润滑
force field 力场
force fit 压配合，压入配合
force flux 力通量
force-free field 无力场
force friction 摩擦力
forceful arc 硬电弧
force gauge 测力计，力传感器
force line 力线
force lubrication 强制润滑
force main 压力干管
force majeure clause 不可抗力条款
force majeure 不可抗力【无法预测、无法控制、无法履约】
force measuring instrument 力测量仪表
force measuring system 测力系统
force method 力法
force moment 力矩
force motor 执行电动机
force of agreement 协定效力，协议效力
force of friction 摩擦力
force of gravity 重力
force of inertia 惯性力
force of law 法律效力
force origin 力的起点
force parallelogram 力平行四边形
force payment 强迫付款
force per unit area 面分布力
force per unit length 线分布力
force per unit volume 体分布力
force polygon 力多边形
force potential 力势
forceps 镊子，钳子
force resolution 力的分解
force sensor 力传感器
force spectrum 力谱
force system 力系
force test 测力试验
force transducer 测力传感器，力传送器
force triangle 力三角形
force unbalance 力失衡
force vector 力矢量
force ventilation 强迫通风
force 力，强制，强迫，强行，促进，力量
forcible 强迫的，强制的，有力的，猛烈的
forcing control 强行控制，强迫控制
forcing fan 送风机
forcing frequency 扰动力频率，扰动频率，受迫振动频率，强迫频率
forcing function generator 功能发生器，正弦发生器
forcing function 强迫函数，外力函数
forcing method 压入通风法
forcing pump 压力泵，加压泵
forcing screw 紧定螺栓
forcing time 强励时间

forcing　强制，强迫
forded lubrication　强制润滑，压力润滑法
ford　渡口，浅滩
fore-and-aft direction　纵向
fore-and-aft level　倾斜水准器
fore-and-aft rigged vessel　纵帆船，纵帆帆船
fore-and-aft sail　纵帆，前后帆
fore axle　前桥，前轴
forebay apron　前池护坦
forebay reservoir　前池水库
fore bay　前池【燃料元件换料水池】，压力前池，船体前部
forebeach　前滩
fore bearing　前轴承
forebody drag　前体阻力
forebody　前体
fore-break distance of warehouse　库房的防火距离
forecast amendment　预报改正
forecast area　预报区
forecast curve　预报曲线
forecast demand　预估的需求，预计负荷
forecasted periods　预测期间
forecast error　预报误差
forecast estimation　预报估计
forecast for construction　施工预报
forecast hydrograph　预报水文过程线
forecasting accuracy　预报精度
forecasting agency　预报机构
forecasting for inventory control system　库存控制系统预报
forecasting runoff　预报径流
forecasting technique　预报技术
forecasting　预报
forecast map　预报图
forecast method　预报法
forecast of cold wave　寒潮预报
forecast of electric demand　电力需求预测
forecast of wave decay　波浪衰减预报
forecast period　预报期
forecast value　预报值
forecast　预报准确性，预测，预报
fore cost　预估投资
fore date　倒填（的）日期，填早日期
foredeep　陆外渊
fore filter　前置过滤器，预过滤器
foregoing information　上述的信息
foregoing statement　以上陈述
foregoing　上述的，前述的
foreground scheduler　前台调度程序
foreground　前景
forehand welding　（气焊）左焊法，左焊法
foreign activities　外事活动
foreign affiliates　外国子公司
foreign aid　外援
foreign bank　外国银行
foreign bidder　国外投标商
foreign bill　国外汇票
foreign branch　国外分部［分行］
foreign businessman　外商
foreign capital　外资
foreign consultant　国外咨询商
foreign contaminant　外来污染物
foreign contracted project　对外承包工程
foreign corporation　外国公司
foreign currency　外汇，外币
foreign customer　国外客户
foreign debt　外债
foreign diagram　外部图
foreign exchange control　外汇管理
foreign exchange earning　创外汇
foreign exchange gain or loss　外汇损益
foreign exchange instrument　外汇票证
foreign exchange permit　外汇许可证
foreign exchange rate risk　汇率风险
foreign exchange rate　外汇汇率，外汇兑换率
foreign exchange receipt　外汇收据
foreign exchange regulations　外汇管理条例
foreign exchange repaying ability　外汇偿还能力
foreign exchange settlement　外汇结算
foreign exchange swap transaction　外汇管理年度报告
foreign exchange transaction　外汇交易
foreign exchange　外汇
foreign expert　外国专家
foreign field　外磁场
foreign firm　外国公司
foreign funds　外资
foreign incursion　外来夹杂物
foreign-invested enterprise　外资企业
foreign-invested project　外商投资项目
foreign investment　外商投资，国外投资
foreign investor　外国投资者
foreign loan　国外贷款，外债
foreign material　外来杂质，外来物，杂质，掺杂物
foreign matter　外来物，异物，杂质，掺杂物
foreign object search and retrieval　异物搜寻与取回
foreign security　外国证券
foreign seller　国外卖方
foreign side provides equipment and technology　外方提供设备和技术
foreign subcontractor　国外分包人
foreign subsidiary　外国子公司
foreign substance　异物，杂质，外来物
Foreign Trade Arbitration Commission　对外贸易仲裁委员会
foreign trade corporation　外贸公司
foreign trade enterprise　外商企业
foreign trade group　外资企业集团
foreign trade　对外贸易
foreign water　客水
foreign worker　外籍工人
foreign　外来的
foreland　前沿地
forelock　开口销，扁销
foreman　工长，领班，监工，班长
forepressure　泵出口压力，预抽压力，前级真空压力
fore pump　前级泵
forerunner　预兆

foresail 前帆
foreseeability 可预见性，预见性
foresee 预见
foreset bed 前积层
foreshadowing 预示
foreshock 前震
foreshore flat 干出滩
foreshore slope 前滨坡
foreshore 潮间岸滩，海岸，前滨
fore sight 前视
forest 森林
forest area 林区
forest canopy 林冠
forest-clad 森林覆被的
forest community 森林群落
forest conservancy area 封山育林区
forest conservation 森林保护
forest cover 森林覆被，森林覆盖
forest district 森林区
forested area 绿荫面积
forest fire smoke 森林火灾烟
forest fire 森林火灾
forest for conservation of headwaters 水源涵养林
forest for protection against soil denudation 沙土防护林，防沙林
forest influences 森林影响
forest meteorology 森林气象学
forest protection 森林保护
forest region 林区，森林带
forest reservation 封山育林
forest reserve 森林资源，森林保护区，封山育林区
forest resources 森林资源
forestry 林学，林业
forest shelter belt 防护林带
forest survey 森林调查
forest zone 森林带，林区
fore treatment 前处理，预处理
fore vacuum 预抽真空，低真空，前级真空
foreword 绪言，前言，序
for example 例如
fore 前部
forfeiture clause 没收条款
forfeiture of payment 取消付款
forfeiture of water right 水权的失效
forfeiture 没收物，失效
forfeit 罚金，罚款，没收，丧失，没收物
FOR(forced outage rate) 强迫停机率，强迫停运率，事故停用率，非计划检修停产率
FOR(free on rail) 铁路交货价，火车［货车］上交货价格
forgeability 可锻性
forgeable cast iron 可锻铸铁
forge cinder 锻渣
forged body 锻造体
forged high carbon steel 锻造高碳钢
forged hollow shaft 锻造空心轴
forged hollow steel shaft 锻造空心钢轴
forged in multiple 整体锻件
forged integral rotor 整锻转子
forged low alloy steel 低合金锻钢
forged piece 锻件
forged pipe 熟铁管
forged steel clevis 锻制U形夹
forged steel 锻钢
forged valve 锻造阀
forge furnace 锻炉
forge hammer 锻锤
forge nearing temp 锻造温度
forge scale 锻渣
forge scrap 锻造废品
forge shop 锻工车间
forge time 锻压时间
forge welding 锻焊，锻接，锻接焊
forge 锻造，伪造
forging aluminium 锻铝
forging cold 冷锻
forging copper 铜锻
forging crack 锻件裂纹
forging defect 锻件缺陷
forging dies 锻模
forging direction 锻造方向
forging stress 锻件内应力
forging welding 锻焊，锻接
forging 锻造，锻件，锻接，锻
forgiveness 宽恕
for information 供参考
for instance 例如
fork channel blockage 堵塞汊道
fork channel closure 堵塞汊道
fork connection 分岔连接，插头连接，叉式接法
forked bar 叉形杆件
forked joint 叉形接头
forked lightning 叉状闪电
forked pipe 分叉管
forked tube 分叉管
forked 叉状的，双叉的，三通
fork factor of tributary 支流分叉系数
fork joint 叉形接头
fork lift hoist 叉式起重车，叉式起重机
fork lift truck operator 叉式起重车操作员
fork lift truck 叉式运输机，叉式装卸车，叉式起重车，叉车，叉式升降装卸车，铲车
fork lift 叉式起重车，叉车，高位叉车，叉形杆，铲车，堆高机，叉式升降机，用铲车搬运
fork lug 叉形接线片
fork out 耙出，付钱
fork shaft 叉轴
forkshaped tee 裤衩三通
fork type blade root 叉形叶根，插入式叶根
fork wrench 叉形扳手
fork 叉，分岔，交叉，分支点，叉子，岔流，分岔，支架，叉架
formal acceptance 正式验收
formal address 形式地址
formal agreement 正式协议
formal clause 正式条款
formal confirmation 正式确认
formal contract 正式合同
formal control 形式控制
formaldehyde 甲醛
formal inspection 正式检查

formal invitation　正式邀请
formal invoice　形式发票，正式发票
formalin　福尔马林，甲醛水
formality of contract　合同格式
formality　手续，正式手续
formalize　使正式，形式化，拘泥形式
formal language　形式语言
formal logic　形式逻辑
formal member　正式成员
formal notice　正式通知
formal notification　正式通知
formal objection　正式的反对意见
formal receipt　正式收据
formal style form standard　标准格式
formal training　正式培训
formamine　甲醛胺，乌洛托平
form anchor　模板锚定
format control　格式控制
format effector　格式控制符
formation factor　地层因数
formation horizon stratum　地层
formation line　施工线
formation lithology analysis　地层岩性分析
formation of iron in slagging furnace　液态排渣炉的析铁
formation of sand dune　沙丘形成
formation rate of marginal generation unit　边际机组形成率
formation rate of price cap by single power plant　单个电厂形成最高限价的比率
formation siding　编组线
formation station　编组站
formation transformer　电成型用变压器
formation voltage　形成电压
formation　形成，组成，构造，构成，地层
format parameter　格式参数，形式参数
format　（磁盘）格式化，形式，格式，版式
Formazin turbidity unit　福尔马津浊度单位
form brace　模板支撑，模板拉条
form coating　模板涂油，模板涂料，模板支撑
form coefficient　成形系数
form contract　标准合同
form control buffer　格式控制缓冲器
form control image　格式控制图像
form copying　靠模法
form cycling rate　模板周转速度
form drag　型阻，型面阻力，形状阻力
formed coil　成型线圈，模绕线圈
formed head　成型封头
formed steel deck　压型钢板
formed　成形加工
former groove　模型槽
former plate　辐板
former winding　模绕绕组，模绕法
former-wound coil　模绕线圈
former　模子，量规，成型设备，堆芯围板的径向支撑板，样板，型模，模型，成形设备，成型工具，从前的
for metric purpose　用于度量目的
formette coil　成型模型线圈，成型线圈
form factor of weld　焊缝成形系数
form factor　波形因子，形状因子，形状系数，波形系数，形状因数，型数，体型系数
form factor　焊缝成形系数【焊接】
form feed　格式馈给
form-fitting　形状配合的
form-fit transformer　壳式变压器
form fixer　模板工
form for recording　记录表
form freeboard　形体干舷
form hanger　模板支撑
formic acid　蚁酸，甲酸
forming method　成形法
forming rolls　成形轧制
forming winding　模绕组，模绕法模压成形绕组
form joint　模板接缝
form lateral pressure　模板侧压力
form lining　模板衬垫，模板涂层
form loss of head　形状损失水头
form of agreement　协议的格式
form of application　申请书格式
form of contractor's completion notice　承包商竣工通知的格式
form of cooperation　合作方式
form of liens waivers　放弃留置权声明的格式
form of parent company guarantee　母公司担保书的格式
form of payment　支付方式，支付形式
form of performance security　履约保函的格式
form of policy　保单格式
form of proxy　委任方式，委任书，委托书格式
form of questionnaire　调查表
form of tendering　招标形式
form of tender　投标（书）格式，投标形式
form of treaty　合同方式，特约格式
form of variation　（对标招标书提出的）变更的格式
form oil　脱模油
form panel　模板
form ratio　水流深度比
form removable　模板拆除
form removal　拆除模板，拆除模壳，脱模，拆模
form resistance　形状阻力，型阻
form sheathing　模壳衬板
forms of corrosion　腐蚀方式
form spreader　模板撑杆
form stability　形状稳定性
form stop　格式差错停机
form stripping　模板拆除，拆模，脱模
form tie　模板拉杆
formula and statement translator　公式和语句的翻译程序
formula derivation　公式推导
formula for success　成功模式
formula language　公式语言
formula manipulation compiler　公式处理编译程序
formula of pile driving　打桩公式
formula recognition　公式识别
formulation　用公式表示，作出定义，公式化，列方程式，配方
formula translation language　公式翻译程序设计

语言
formula translator 公式转换器
formula 公式，分子式，准则，打桩公式，配方
form vibrator 附着式振动器，模板振动器，外部振捣器
form winding 模压成型绕组，模绕组，模绕法
form wire 成型线
formwork drawing 模板图
formwork planning 模板布置
formwork removal 拆除模板，拆除模壳
formwork vibrator 模板振捣器
formwork 模板，支架，模板工程，盖板，量规，脚手架
form-wound coil 成型线圈，模绕线圈
form-wound motorette 模绕线圈小型电动机
form-wound stator 成型绕组定子
form 形状，构成，表格，模板，模型，形式，形态，成型，成型加工，形成
for one's account 由……支付
for prompt shipment 立即出运
for purpose of 为了……（起见），以便，来……，对……来说
for reference only 仅供参考
for reference 供参考，仅供参考
for sb.'s exclusive use ……专用的
for statement 循环语句
forsterite 镁橄榄石
fors 急流，急滩
for the avoidance of doubt 为免生疑问，为避免疑义
for the convenience of 为了……的方便起见，为了便于……
for the reason of 由于……的原因
FORTRAN 公式翻译程序设计语言
fortran 公式译码器，公式变换
fortuitous accident 意外事故
fortuitous distortion 不规则畸变，偶发失真
fortuity 偶然事件，偶然性
fortune 财富
forum 论坛，讨论会，场所
forward-acting control 预约控制，提前控制
forward active mode 正向有源状态
forward applied voltage 正向外加电压
forward azimuth 前方位角
forward-backward counter 双向可逆计数器，正反向计数器
forward bargain 期货买卖
forward bearing 前象限角
forward bent blade 前弯叶片
forward biased 正偏的
forward bias 正向偏压
forward blocking interval 正向闭锁间隔
forward blocking state 正向关断状态，正向闭锁状态
forward blocking voltage 正向阻断电压
forward breakdown 正向击穿
forward breakover voltage 正向转折电压
forward channel 正向通道
forward clearing 前向拆线
forward component 正向分量，正序分量
forward conductance 正向电导
forward construction 正向设制，正向设置
forward contract 期货合同，远期合同
forward controlling element 正向主控元件
forward control 预约控制，提前控制
forward cost 期货价格
forward crosstalk 前向串扰
forward current 正向电流，正序电流
forward curved blade 前弯（式）叶片
forward curved vane 前弯叶片
forward curve inlet blades 前弯式进口叶片
forward delivery 远期交货
forward-difference 向前差分
forward direction 正向，前向
forward drop 正向电压降
forwarder 输送器，传送装置
forward exchange contract 期货汇兑合同，远期外汇合同
forward face 内弧面，叶片凹面
forward facing front surface 向前正面锋面
forward facing surface 向前锋面
forward-field impedance 正向磁场阻抗，正序阻抗
forward-field torque 正序转矩，正向磁场转矩
forward flow zone 射流区，顺流区
forward flow 顺流，正向流，平直水流
forward gain 正向增益
forward gate 正向门
forward impedance 正向阻抗
forward inclination idler 前倾托辊
forward inclination trough idler 前倾槽形托辊
forwarding agent 货物转口代理人，运输代理人
forwarding instructions 装运说明，发运通知
forwarding operation 代运业务
forwarding order 运输委托书
forwarding station 转运站
forwarding 发运
forward intersection 前方交会
forward-inverse electronic resolver 正向-反向电子解算器
forward lead of brush 电刷前移
forward link 正向连接
forward mounted flap 前缘襟翼
forward-moving grate 倒转炉排
forward M to N 把 M 送给［转到］N
forward nodal point 前节点
forward-order current 正序电流
forward-order impedance 正序阻抗
forward overlap 前后重叠
forward path 正向通路，正向通道
forward payment 预付款
forward planning 预先计划
forward power flow 正向潮流
forward power loss 正向功率损耗
forward price 期货价格
forward progression winding 右行绕组
forward purchasing contract 预购合同
forward reaction 正反应
forward reading 前视读数
forward resistance 正向电阻
forward/retract 前进/后退
forward-rotating wave 正向旋转波

forward-sequence component 正序分量
forward-sequence reactance 正序电抗
forward shift of brush 电刷前移
forward shift 向前移动，顺旋转方向移动，前视
forward stagnation point 前驻点
forward stagnation streamline 前驻点流线
forward stroke 正冲程，前进冲程，前进行程，切削行程，正程
forward tilt angle 前倾角
forward tipping 向前倾倒
forward titled amplitude 带式输送机（托辊的）前倾角
forward titled idler 前倾托辊
forward torque 正向转矩
forward transfer function 正向传递函数
forward voltage 正向电压
forward wave 正向波，前向波
forward welding 前倾焊
forward 运送，期货，远期的，正向的，正向，向前的，前向，转发
FOSDIC(film optical scanning device for input to computer) 计算机胶卷光扫描输入装置
FOS(free on ship) 船上交货价格
fosse moat 壕沟
fosse 壕，壕沟
foss 壕沟，急滩
fossil energy resource 矿物能源
fossil energy 化石能源
fossil fired power plant design 火电厂设计
fossil fired power plant simulator 火电厂仿真机
fossil fired power plant 化石燃料电站，常规火电站，火力发电厂，化石燃料电厂，矿物燃料电厂，火电厂
fossil fired superheater 化石燃料过热器
fossil fuel boiler 燃煤锅炉，化石燃料锅炉
fossil-fueled central station 化石燃料发电厂
fossil-fueled power generation project 火力发电工程项目
fossil-fueled power plant 常规电厂，化石燃料电站，化石燃料电厂，火力发电厂
fossil fuel 化石燃料，（非核）矿物燃料
fossilization 化石作用
fossil lake 古湖
fossil meal 硅藻土，化石粉
fossil oil 石油
fossil power plant 烧化石燃料的电站
fossil power unit 火力发电机组
fossil soil 古土壤，化石土
fossil turbine 火电汽轮机
fossil water 化石水，矿物水，原生水
fossil 化石
Foster Wheeler 福斯特惠勒电力集团【总部在美国】
foster 培养
FOT(free on truck) 车上交货，敞车交货价，车上交货价
Foucault current 傅科电流，涡流
foul air duct 污浊空气导管
foul air flue 浊气道
foul air 污浊空气
foulant 污秽物，污染物
fouled resin 污染的树脂
foul gas 秽气，有害气体，不凝性气体，臭气
fouling coefficient of heat exchanger 换热器污垢系数
fouling factor 污垢系数，沾污系数，污染系数，污垢因子，沾污指数
fouling film 污塞层
fouling index 污染指数，沾污指数
fouling inhibitor 污垢抑制剂
fouling of heating surface 受热面玷污
fouling organism 污染生物，附着生物
fouling potential 积灰倾向
fouling rate 沾污速度
fouling resistance 污垢热阻
fouling resistant 耐污染
fouling 结垢【蒸汽发生器等】，污垢，积灰，污染，弄脏
foul wind 恶风
foul 使沾污，污垢，脏的，污浊的，堵塞，犯规的，犯规，污物
found 铸造，浇铸
foundational 基础的
foundation analysis 基础分析
foundation base 基础底面
foundation beam 地基梁，基础梁
foundation bed 基础垫层，基底
foundation block 台板【汽轮发电机，基座】
foundation bolt 地脚螺栓，基础螺栓，底脚螺栓
foundation bottom 基底
foundation by pit sinking 沉井基础，挖坑沉基
foundation by timber casing with stone filling 木框石心基础
foundation condition 地基条件
foundation connection 基础连接
foundation consolidation 地基加固（处理）
foundation cross 格床基础，基础格床，基础交叉线
foundation deformation modulus 地基变形模量
foundation deformation 地基变形
foundation design for pole and tower 杆塔基础设计
foundation design 基础设计
foundation disposal 地基处理
foundation ditch 基础槽，基坑
foundation dowel 基础插筋
foundation drainage system 地基排水系统
foundation drainage 地基排水
foundation drain hole 基础排水孔
foundation drawing 地基图，基础图
foundation earth electrode 基础接地体
foundation elastic modulus 地基弹性模数
foundation engineering 地基工程，基础工程
foundation excavation 地基开挖
foundation failure 基础破坏
foundation framework 地基框架
foundation frame 机座底架，基础构架
foundation girder 基础主梁
foundation grouting 基础灌浆
foundation improvement 地基改良
foundation in undisturbed soil 原状土基础
foundation investigation 地基勘察，地基勘探，

地基查勘
foundation level 基准线
foundation load 基础负载
foundation mat 基垫层，基础底板，垫板
foundation modulus 基础模量，地基系数
foundation of fuel oil tank 油罐基础
foundation of road 路基
foundation of the main building below zero level finished 主厂房基础出零
foundation of transformer platform 变压器基础
foundation of WTG 风电机组基础
foundation on raft 浮筏基础
foundation outline 基础外形
foundation pier 基座，基墩，基桩
foundation pile 基础管桩
foundation piping 地基管涌现象
foundation pit 基坑
foundation pit excavation 基坑开挖
foundation pit supporting 基坑支护
foundation plan 基础平面图
foundation plate 底板，基础板，基础底板
foundation pressure 基底压力
foundation raft 筏基垫板
foundation restraint crack 基础约束裂缝
foundation settlement 基础沉陷，基础沉降，地基下沉
foundation settling 基础沉陷
foundation slab 基础板，底板
foundation soil replacement by sand 地基的挖土换砂
foundation soil 地基土，基础土
foundation's settlement 基础沉降量
foundation stability 地基稳定（性）
foundation stabilization 地基加固
foundation stone 奠基石，基石
foundation stripping 清基
foundation treatment 基础处理，地基处理
foundation trench 基槽，基坑
foundation wall 基础墙
foundation works 基础工程
foundation 基金，基金会，基座，地基，基础
foundery alloy 铸造合金
foundery detect 铸造缺陷
founding 翻砂，铸造
foundry casting sand 型砂
foundry defect 铸造缺陷
foundry shop 铸工车间
foundry worker 铸工，翻砂工
foundry 铸造，铸造车间，铸造厂，铸工车间，翻砂
fountain aerator 喷泉式曝气池，喷水通气器
fountain bubbler 饮水口
fountain failure 涌毁
fountain head 喷水头
fountain-pen dosimeter 剂量笔
fountain-pen type dosemeter 自来水笔式剂量仪
fountain 喷水池，喷泉，泉源
four-address instruction 四地址指令
four-channel multiplier 四通道乘法器
four-channel switch 四路转换开关
four-channel 四通道的，四路的
four corners 十字路口
four digit series airfoil NACA 四位数字系列翼型
four-dimension design 四维设计
four-element ring 四合环
four-factor formula 四因子公式
four-fold coupling 四重联轴节
four-guide 四花键导向装置
Fourier amplitude spectrum 傅里叶幅值谱
Fourier analysis 傅里叶级数展开，傅里叶分析
Fourier coefficient 傅里叶系数
Fourier component 傅里叶分量
Fourier conduction equation 傅里叶传导方程
Fourier equation 傅里叶方程
Fourier phase spectrum 傅里叶相位谱
Fourier series 傅里叶级数
Fourier's integral 傅里叶积分
Fourier's law 傅里叶定律
Fourier's number 傅里叶数
Fourier's series 傅里叶级数
Fourier transformation 傅里叶变换
Fourier 傅里叶
four-jaw chuck 四爪卡盘
four-jaw concentric chuck 四爪同心卡盘
four-jaw independent chuck 四爪单动卡盘
four-jaw plate 四爪卡盘
four-layer winding 四层绕组
four-node mode 四节点振型
four paws 四爪
four-phase modulation 四相调制
four-phase stepper motor 四相步进电机
four-phase system 四相系统
four-phase 四相的
four-pipe water system 四管制水系统
four point ball bearing 四点球轴承
four-pole double-throw switch 四极双掷开关
four-pole double-throw 四极双掷
four-pole generator 四极发电机
four-pole motor 四极电动机
four-pole network 四端网络
four-pole steam turbine generator 四极汽轮发电机
four-pole turbo-generator 四极汽轮发电机
four-pole 四端网络，四极的
four-quadrant back-to-back 四象限背靠背
four-quadrant diagram 四象限图
four-quadrant multiplier 四象限乘法器
four-quadrant solid-state sensor 四象限固态传感器
four-quadrant 四象限的
four rollers reeling machine 四轮卷取机
four-run a week 一周跑 4 趟【往返】
four-shift and eight-hour cross operation 四八交叉作业
four-stage compression 四级压缩
four-stroke cycle 四冲程循环
four-terminal attenuation 四端网络衰减器
four-terminal network 四端网络
four-terminal transmission network 四端输电网络
four-terminal 四端的
four-way junction box 交叉接线盒
four-way solenoid valve 四通电磁阀，四通螺线管阀

four-way union 四通管接头
four-way valve 四通旋塞阀
four-way 四通的
four-wheel bogie 四轮转向架
four-wheel grader 四轮平地机
four-wheel trails 四轮拖车
four-wheel wagon 四轮车
four-wiredcable 四芯电缆
four-wire repeator 四线制强音机，四线增音器
four-wire three-phase system 三相四线制
four-zone four region model 四区四域模型
fovea 凹处
FOW(free on wagon) 火车货车上交货价格
fow-rate impulse 流速脉冲，调整流量脉冲
fox bolt 开尾螺栓，端缝螺栓
FOXBORO 福克斯波罗仪表公司【美国】
fox wedge 扩裂楔
foyer 门厅，休息室
FPAD freight payable at destination 目的地付运费
FPA(free from particular average) 平安险
FP(fire permit) 动火许可证
FPM(feet per minute) 英尺每分
FPS(fire protection system) 消防系统
FPS(fluid power society) 流体动力学会
FPS(foot-pound-second) 英尺-磅-秒
FPW(full penetration weld) 全透焊，全透焊焊缝
fractal dimension 分形维数
fractile 分位点，分位数，分位值，分形
fractional admission 部分进汽，部分进气
fractional breeding gain 相对增殖增益
fractional chain yield 相对链产额
fractional channel closure 流道阻塞率
fractional cold （炉膛）水冷度
fractional combustion 不完全燃烧
fractional conversion 部分转化
fractional counting loss 计数损失率
fractional crystallization 分步结晶
fractional distillation 分馏
fractional efficiency 相对效率
fractional electric motor 小功率电动机，分马力电动机
fractional energy savings 节能率
fractional error 相对误差
fractional frequency power transmission 分频输电
fractional function 分部函数，辐射函数
fractional harmonic 分数谐波，次谐波
fractional heat 摩擦热
fractional-horsepower motor 分马力电动机
fractional ionization 离子化程度，离子化度
fractional load 部分负荷，部分负载，部分荷载，轻载
fractional loss 部分损失
fractional motor 小马达，小电机
fractional number 分数，小数
fractional percentage points 千分之几
fractional pitch winding 短距绕组，分数极距绕组
fractional pitch 分数极距，分数节距
fractional power 分数幂
fractional pressure 分压
fractional rating 分数功率
fractional reaction 分级反应
fractional scanning 分段扫描
fractional-slot winding 分数槽绕组
fractional-slot 分数槽
fractional tailed test 局部破坏试验
fractional turn coil 分数槽线圈
fractional winding 分数槽线圈
fractional yield 相对产额，局部收率
fractional 分数的，部分的，组分的，分式的
fractionary 组分的，部分的，分数的，分式的
fractionated exposure 分次照射
fractionated gain 部分增益
fractionated 分部的
fractionating column （重水）精馏塔，分馏塔
fractionating tower 精馏塔
fractionation 分馏，精馏，分级
fractionator 分馏器，气体分离器，气体分离装置
fraction composition 组分
fraction dryness 含汽率
fraction power 相对功率
fraction separation efficiency 分级除尘效率
fraction solids 固体分率
fraction surviving 存活率，生存率
fraction 馏分，分数，分式，份额，部分，粒径级配，粒径组合，粒组，小数
fractography 断口分析
fractograph 断口组织照片
fracture analysis diagram 断裂分析图
fracture analysis 断口分析，断裂分析
fracture appearance transition temperature 断裂形貌转变温度，出现裂纹过渡的温度
fracture appearance transition 脆性转变温度
fracture-arrest temperature 裂缝终止温度，裂纹终止的温度
fracture behaviour 断裂特性
fracture by delayed cracking 由延迟开裂引起的断裂，由滞后裂缝引起的断裂
fracture cleavage 破劈理
fracture control technique 断裂控制技术
fracture depth 断裂深度
fracture ductility 断裂延展性
fracture elongation 断裂延伸率
fracture initiation temperature 裂缝起始温度
fracture inspection 断口检验
fracture line 断裂线
fracture mechanics 断裂力学
fracture plane 断裂面，裂面，破裂面，断面
fracture safety 防断裂裕度
fracture speed 裂纹扩展速度，裂缝扩展速度
fracture strain 断裂应变
fracture strength 破坏强度，断裂强度
fracture stress 破坏应力，断裂应力
fracture surface 破裂面，断面
fracture test 断裂试验，断口试验
fracture theory 断裂理论
fracture toughness test 断裂韧度试验
fracture toughness 断裂韧度，断裂韧性，破坏韧性
fracture zone 断裂带，破裂带，破碎带

fracture 断口，断裂，破裂，裂缝，裂痕，折断，裂口
fracturing load 致断负载
fracturing 断裂，裂缝，使破裂，破碎
fractus nimbus 碎雨云
FRA(Framatome) 法国法马通公司
fragile material 脆性材料
fragile transition temperature 脆性转变温度
fragile 脆的，脆弱的，易碎的，易损坏的，不稳定的，不能压的
fragility 脆性，易脆性，脆弱，脆度
fragmental data 不完全资料
fragmental deposit 碎屑沉积
fragmental rock 碎屑岩
fragmental structure 碎屑构造
fragmental volcanic rock 碎屑火山岩
fragmental 不全的，零碎的，破碎的
fragmentation 碎裂，（晶粒）碎化，破碎，爆裂，分裂，破碎作用
fragmented 碎裂的
fragment poisoning （裂变）碎片中毒
fragment separator 碎片分离器
fragment 裂片，碎片，碎屑，碎块，断片，片段
Frahm frequency meter 振簧式频率计，弗拉姆频率计
frailty 弱点
frame alignment 帧同步
frame and brick veneer construction 框架嵌砖建筑，框架嵌砖结构
frame and panel construction 框架墙板结构
frame and shear wall structure 框架剪力墙结构
frame antenna 框形天线
frame arc lamp 弧光灯
frame bent 排架，框式桥台
frame bore 定子铁芯内径
frame camera 分幅照相机，分格摄影机
frame coil 定心线圈，中心调整线圈
frame construction 构架建筑，框架建筑，框架结构
frame cover 护板
frame crane 龙门式起重机，龙门吊
framed barrage 构架式拦河堰
framed bent construction 排架结构
framed bent 框架桥台，排架
framed building 构架建筑，框架建筑
framed connection 构架结合
framed dam 框架坝
framed delimiter 帧定界符
framed door 镶板门
framed floor 构架底板，构架桥面
framed foundation 构架式基础
frame diagram 框架图
frame diameter 定子铁芯外径
framed mattress 木框式沉排
framed revetment 框架式挡土墙，框架式护岸
framed steel chimney 塔架式钢烟囱
framed trestle 构架式栈桥
framed wall 构架墙
framed weir 框架堰
frame flyback 帧回描
frame frequency 帧频
frame girder 构架梁，框架横梁
frame grounding circuit 框架［机架］接地电路
frame grounding 接地框架，机座接地，壳体接地
frame house 构架房屋
frame leakage protection 机壳漏电保护
frame number 机座号
frame of axes 坐标系
frame of reference 参照系统，参照构架
frame-panel building 骨架板材式房屋
frame plate liner 泵体护套
frame resistance 线绕可变电阻
frame scan transformer 帧扫描变压器
frame shear wall structure 框架剪力墙结构
frame shielding coefficient 框架遮蔽系统，框架屏蔽系数
frame size 框架尺寸
frame structure 框架，构架，框架机构，框架结构
frame support 支承框架
frame suspended motor 底座悬挂型电机
frame time base 帧扫描
frame time 帧周期
frame transmission 帧长
frame-type circuit breaker 框架式断路器
framework articles 纲要性条款
framework and casting 制模【模板支立】与浇注
framework for incoming feeder 进线架构
framework for outgoing feeder 出线架构
framework 构架工程，结构，体制，骨架，框架，机架，架，构架
frame yoke （直流电机的）磁轭
frame 框架，画面，机座，结构，帧，边框，构架，骨架，机架，框，门框，窗框，支架
framing error 帧校验误差
framing 框架，构架，构筑框架
franchise agreement 特许协议
franchise clause 免税条款，免赔条款，特权条款
franchisee 特许经营人，特许权受让人
franchise right 特许经营权
franchiser 授予特许经营权者，特许权出让人
franchise 特许经销权，特权，相对免赔率，专利权
Francis turbine 法兰西斯水轮机
Francis vane 辐流式叶片
frangibility 脆弱，易碎，脆弱性
frangible coupling 易分离耦合，易卸接头
frangible fracture 脆性断裂，脆性破断
frangible 脆的，易碎的
frangile 易碎的
frangipani soil 脆磐土
fraudulent 欺诈的，欺骗性的
fraud 欺诈，骗子，冒牌货，假货，欺骗，舞弊
Fraunhofer CSP 对硅光伏电池，弗劳恩霍夫硅光伏中心
fray 擦断，擦伤，磨损
FRC(fiber reinforced composite materials) 纤维加强聚酯
freak 畸变，变异
free 单体的，免费（的），空闲，自由

free access area　自由通道面积
free acidity　游离酸度
free acid　游离酸
free action　自由行动，可动作用
free air data　大气数据
free air diffuser　空气进气道
free air dose　自由空气剂量
free air facility　空气动力试验设备
free air ionization chamber　自由空气电离室
free air temperature　大气温度
free air thermometer　大气温度计
free air tunnel　大气湍流，自由全气风洞
free air volume　自由空间
free air　大气，自由大气，自由空间，自由空气
free alkalinity　游离碱度
free alkali　游离碱
free alongside ship　装运港船边交货价
free alternating current　自由振荡电流
free alternation　自由振荡，自由交变
free ammonia　游离氨
free and bearing　自由端支承，简支
free and force convection combined　自由和强迫相结合的对流
free aquifer　自由含水层，自由蓄水层
free area　有效截面
free ash　外灰分，游离灰分
free atmosphere wind　自由大气风
free atmosphere　自由大气
free-atom cross section　自由原子截面
free available chlorine residue　游离性有效余氯
free available chlorine　游离性有效氯
free available residual chlorine　游离性有效余氯
free axis　虚轴
free beam　简支梁
free bearing　球形支座
free bill　运费单
free blade　自由叶片
free-blowing　自由排放
free board height　稀相区高度
freeboard storage　超高库容
freeboard　悬浮段【沸腾炉】，超高，自由空间，坝顶超高
free body diagram　隔离体图
free body　隔离体，孤立体，自由体
free boundary　自由边界
free box wrench　活套筒扳手
free breathing transformer　无呼吸器的变压器，自由呼吸式变压器
free bulkhead　活动岸壁，活动挡土墙
free burning coal　不黏结煤，易燃煤，不结焦烟煤
free carbon dioxide　游离二氧化碳
free carbonic acid　游离碳酸
free carbon　游离石墨，游离碳，回火碳
free center suspension clamp　中心回转式悬垂线夹
free-center-type clamp　线路释放线夹
free charge carrier　自由带电体
free charge　自由电荷
free chlorine　游离氯
free competition　自由竞争
free convection boiling　自然对流沸腾
free convection boundary layer　自由对流边界层
free convection cooling　自然对流冷却，自然冷却，自由对流冷却
free convection factor　自然对流系数
free convection heat transfer　自然对流换热
free convection　自由对流，自然对流
free cooling　自由冷却
free currency　自由货币
free cylindrical vortex　自由柱状旋涡
free decay　自由衰减
free deformation　自由形变
free delivery price　免费送货价
free digging rate or calculated capacity　自由挖掘出力或计算出力
free discharge　自由放电，自由泄流，自由度
freedom from bias of measurement　无偏置测量
freedom from repairs　免维修
freedom from unbalance　抗不平衡性
freedom from vibration　抗震性
freedom of motion　运动自由度
freedom　自由度，间隙，摇动
free drainage　天然排水，自通排水
free draining　自由穿流，自由排水
free drop　自由跌水
free economic zone　自由经济区
free eddy　自由旋涡
free edge　自由边缘，无支承边，自由边
free electricity　自由电荷
free electron　自由电子
free end of beam　梁的自由端
free end　扩建端，膨胀端，自由端
free energy　自由能
free enthalpy　自由焓
free entry　免税报单
free expansion　自由膨胀
free fall height of the coal　煤的自由落差
free falling body　自由落体
free falling velocity　自由沉降速度
free fall insertion　自由落棒，自由下落插入，自重落棒
free fall terminal velocity　自由沉降终端速度
free fall weir　自由溢流堰
free fall　自由降落，自由落体
free field ground motion　自由场地面运动
free field sensitivity　自由场灵敏度
free field voltage response　自由场电压响应
free fit　自由配合
free flame　自由火焰，活火头
freeflight wind tunnel　自由飞行风洞
free floating pontoon　自由浮动沉箱
free flow channel　无压隧道
free flow check valve　自由流动单向阀
free flow conduit　无压管道
free-flowing coal　松散煤
free-flowing material　自由流动料，流动性好的物料
free-flow jet　自由射流
free-flow pipe　无压水管
free-flow pressure　自由流压力
free-flow tunnel development　明流隧洞式电站

free-flow tunnel　明流隧洞，无压隧洞
free-flow turbulence　自由流紊动
free-flow weir　自由溢流堰
free flow　自由流动，自由流，无约束流，明流
free-free energy transition　自由态—自由态能级跃迁
free-free transition　自由—自由跃迁
free frequency　自然频率，固有频率
free from damage　无损坏
free from duty　免税
free from particular average　平安险
free from radial runout　无径向跳动的
free from strain　无变形的，无应变的
free from unbalance　无不平衡的
free from vibration　无振动的
free from vorticity　无涡的，自由涡
free gas　自由气体，游离气体
free groundwater　自由地下水
free gyroscope　三自由度陀螺仪，自由陀螺仪
free gyro　自由陀螺
free hand drawing　徒手画，单图，示意图
free-handle breaker　具有自由释放机构的自动断路器
free hand sketch　徒手草图，手勾草图，手绘草图
free harmonic vibration　自由谐振
free head　自由水头
free height　自由高度【没有拉伸的弹簧】
free humic acid　游离腐殖酸
free hydrogen radical　游离氢根
free hydroxide ion　游离氢氧根离子
free impedance　自由阻抗
free in and out of trucks　船方不负担装货费用，船方不负担装卸费
free in and out　船方不负担装卸货费用
free inertial flow　自由惯性流
free inflow　自由进流
freeing point　凝固点
free in and out term　装卸船方免责，装卸费均免
free in term　装货免装费
free into bunker　舱内交货价格，交货到燃油舱价格
free into wagon　车厢交货价格，F. I. W. 价格，船方不负责装车费
free in　船方不负担装货费用，包括装船费在内的运费
free jet chute　自由射流斜槽
free jet MHD induction generator　自由喷射式磁流体感应发电机
free jet wind tunnel　自由喷射风洞
free jet　自由射流
free joint　万向节
free lacing wire　自由拉筋，松拉筋
free length　自由长度【弹簧】
free-level width　自由面的宽度
free lever hexagonal fixed socket wrench　活柄六角固定套筒扳手
free lever swinging socket wrench　活柄活动套筒扳手
free lift　自由升力
free lime　游离石灰
free line　空线，空闲线
free loan　无息贷款
freely bubbling bed　自由鼓泡床
freely movable bearing　自由移动支承，活动支座
freely-rising convective stream　自由上升对流
freely supported　简支的
free medical service　免费医疗服务
free-mineral acidity　游离矿物酸酸度，无机酸的鉴定量
free moisture　自由水分，外在水分，外水分
free-molecular flow　自由分子流动，自由分子流
free molecule diffusion　（气体）自由分子扩散
free motion　自由运动
free mouth　（无拦门沙的）敞开河口
free movable bearing　活动支座
free nappe　自由水舌
free of charge　免费
free of duty　免关税
free offer　自由报盘
free of loss　无损耗
free of particular average　单独海损不赔
free of tax　免税
free oil lubricating compressor　无油润滑空压机
free on board　船上交货，离岸价格，船上交货价格，装运港船上交货
free on truck　敞车交货价，车上交货价，敞车交货价格
free oscillation　自由振动，自激振荡，固有振动，自由振荡
free outfall　自由溢流（口）
free outflow　自由出流
free out unloading　卸货船方免责
free out　船方不负担卸货费用
free overfall crest　自由溢流堰顶
free overfall jet　自由溢水射流
free overfall nappe　自由溢流水舌
free overfall stilling basin　自由溢流式静水池，自由溢流式消力池
free overfall weir　自由溢流堰
free overfall　自由溢流
free path　自由行程，自由路程，自由路线，自由路径
free piston compressor　自由活塞压气机，自由活塞压缩机
free piston gas turbine　自由活塞燃气轮机，自由活塞式燃气轮机
free piston pump　自由活塞泵
free piston　自由活塞
free play　齿隙，空隙，空转
free-point tester　自由电测试器
free port　自由港
free-pressurized fuel element　未加压型燃料元件
free price　自由价格
free progressive wave　自由前进波
free pulley　惰轮，动滑轮
free quasiparticle approximation　自由准粒子近似
free radical　游离基，自由根
free replacement　免费调换
free residual chlorination　游离性余氯化
free rolling　自由横摇
free-running frequency　固有频率，自然频率

free running high frequency interference 自由振动高频干扰
free running operation 自由震荡，自由振荡
free running slag 散流渣
free running speed 空载转速，无载转速
free running 自由振荡，空转，自激，不同步
free sedimentation 自由沉降
free service 免费服务
free service model 免费服务模式
free settling 自由沉降
free shear flow 自由剪切流
free shear layer 自由剪切层
free silica 游离二氧化硅
free silicon dioxide 游离二氧化硅
free simple beam 简支梁
free slack between couplers 耦合器之间游隙
free sodium hydroxide 游离氢氧化钠
free space wavelength 自由空间波长
free space wideband transmission 自由空间宽带传播
free space 自由空间
free-spinning 自由旋转
free stagnation point 自由驻点
free-stand blade 自由叶片
free-standing arrangement 可移动布置
free-standing blade 自由叶片
free-standing cladding 自承重的包壳【燃料】，自立型包壳
free-standing column 独立柱
free-standing fuel element 自立型包壳燃料元件，稳定型包壳元件
free-standing hollow column 独立的空心柱
free-standing stack 独立烟囱
free-standing steel containment shell 自立式钢安全壳
free-standing surge tank 独立式调压罐，自由支承调压塔，独立式调压塔
free-standing system 独立系统
free-standing 自立式，独立的，无支持的，直立式
free state 游离状态，单体状态
free steam 自由蒸汽
free stone masonry 毛石圬工，毛石砌筑
free stone 乱石，毛石
free stream boundary 自由流边界
free-stream flow 自由流，无扰动流，迎面流
free streaming 自由流，无扰动流
free streamline 自由流流线，势流流线，无旋流线，位势流线，自由流线
free stream pressure 未扰动流压力，自由流压力
free stream surface 自由流面
free stream turbulence 自由来流湍流，自由来流湍流度
free stream value 自由来流值
free stream velocity 自由流速度
free stream wind speed 自由流风速
free stream wind 非扰动气流
free stream 自由来流，未受扰动流
free stroke 自由行程
free support 活动支座，自由支承
free surface condition 自由表面条件
free surface energy 液面自由能量，表面张力，自由表面能
free surface flow 无压流，自由面流
free surface level 自由表面的标高
free surface moisture 表面水分，外水分
free surface wave 自由表面波
free surface 自由表面，自由面
free surge 自由涌浪
free swelling index 自由膨胀序数，自由膨胀指数
free thermal convection 自由对流
free tidal wave 自由潮（汐）波
free time 自由时间，可利用时间，空闲
free to expand 自由膨胀
free to slide 自由滑动
free trade agreement 自由贸易协定
free trade area 自由贸易区
free trade zone 自由贸易区，保税区
free trade 自由贸易
free trip 自动脱扣，自由跳闸
free turbine design 自由透平分轴装置方案，自由涡轮设计
free turbine 自由透平，动力透平
free turbulence boundary 自由湍流边界
free turbulence 自由紊流
free vector 自由矢量
free vibration 自由振动，自振
free volume 净容积
free vortex blading 自由涡流流型叶片，自由涡流叶片组
free vortex flow pattern 自由涡流型，等环量流型
free vortex flow 自由涡流
free vortex sheet 自由涡流面
free vortex type 自由涡流，自由旋涡型，自由涡流型
free vortex 自由涡，自由涡流，自由旋涡
free vorticity 自由涡，自由涡量
free water body 自由水体
free water level 无压水平面，自由水面，自由水位
free water removal 脱除游离水
free water surface evaporation 自由水面蒸发
free water surface profile 自由水面坡线图
free water surface 自由水面
free waters 敞开水域
free water table 自由水位
free water 非结合水，自由水，表面水分，游离水，重力水，自由水分，自由波
free way transportation 高速公路运输
free way 快速道
free wheel backstop 超越离合器
free wheel clutch 空程离合器，超越离合器
free wheel diode 飞轮二极管，稳流二极管
free wheeling 单向离合器，自由旋转，自由转轮机构，惯性滑行
free wheel rectifier 飞轮整流器，稳流整流器
free wind 顺风，自由风
free working surface 自由工作面
free yawing rotor 定向风轮，自由偏航转子
free yawing 自由偏航
freeze 冰冻，冻，冻结，结冰

freeze bulge on road　道路冻胀
freeze conditioning agent　冻结调节剂
freeze drying process　冷冻干燥法，冻干工艺
freeze-free period　不冻期
freeze plug　冰冻塞
freeze-proof　防冻性
freeze protection operating mode　防冻运行方式
freezer　结冰器，冷藏箱，冷冻器
freeze seal heater　冻结密封加热器
freeze seal　冻结密封
freeze section　冻结段
freeze-thaw action　冻融（风化）作用
freeze-thaw durability　冻融耐久性
freeze-thaw test　冻融试验
freeze-up date　冻结日期
freeze-up forecast　冻结预报，封冻预报
freeze-up　冰塞，冻结，封冻
freezing and thawing cycle　冻融循环
freezing cofferdam　冻结式围堰
freezing curve　冷冻曲线，冷冻图
freezing degree-day　冻度日
freezing drizzle　雨凇
freezing duration　冻结持续时间
freezing fog　冻雾，雾凇
freezing force　冻结力
freezing fouling　凝固污垢
freezing freeze-up　冻结
freezing index　冻结指数
freezing in stages　分期冻结
freezing level　封冻水位，冰冻线，冻结高度
freezing method of tunneling　隧道冻结施工法
freezing method　冻固法，冻结法
freezing mixture　冷却剂，冷冻混合物，冷冻剂
freezing nucleus　冻结核
freezing period　冻结期，冻季
freezing point　凝固点，冻结点，冰点
freezing protection　防冻
freezing rain　冻雨
freezing retardation　延缓冻结
freezing season　冰冻季节
freezing stress　冻结应力
freezing temperature　冻结温度，固化温度，凝固温度
freezing thawing cycle　冰冻融冻循环
freezing thawing test　冻融试验
freezing trap　冷冻捕集器，冷冻阱
freezing zone　冻结区
freezing　冰冻的，严寒的，冷冻用的，冰冻，冻结，封冻，凝固
freight all kinds　均一包箱费率
freight alongside ship　船边交货
freight bill　运费清单，装货清单
freight by weight　按重量计算的运费
freight car　货车
freight charge　运费
freight clause　运费条款
freight collect　运费到付
freight container　货物集装箱
freight depot　货站
freight elevator　货物升降机，起重电梯，货梯，载货电梯
freighter　货船，货轮
freight for class and basis　基于商品等级和计算标准的包箱费率
freight for class　基于商品等级的包箱费率
freight forwarder　承运人
freight forward　运费待付，运费待收，运费到付，运费由提货人支付
freight handling facility　货物装卸设施
freight insurance　运费保险
freight lift　货物装卸机
freight note　运费单
freight paid on shipment　装船时支付运费
freight payable at destination　到付运费，目的地付运费，运费到付，货到收运费
freight payable on delivery　货到付运费
freight prepaid B/L　运费预付提单
freight prepaid　运费预付，运费已付，运费已预付
freight rate　商品运价，运费率，海运价，海运费率
freight rebate　运费回扣
freight shed　货棚
freight ship　货船，运货船
freight station　货站
freight tariff　运价表，运费率表
freight terminal　货运站，货运码头
freight to be collected　运费待收
freight to collect　运费到付，到付运费
freight tonnage　载重吨位
freight track　货物线
freight traffic volume　货流量
freight traffic　货运交通
freight train　货运列车
freight transport　货物运输
freight truck　运货卡车
freight volume　货量
freight wharf　货运码头
freight yard　堆货场，货场，货运场
freight　运输，货运，运费，运费通知，装运货物，运输费
French casement　玻璃落地窗
French chalk　滑石粉
French drain　乱石排水沟，排水沟，盲沟，石砌排水沟，暗沟
French grey　浅灰色
French sash　落地窗
freon chiller unit　氟利昂冷冻机组
freon cooler　氟利昂冷却器
freon refrigerant loop　氟利昂制冷剂回路
freon refrigeration and brine system　氟利昂制冷及盐水系统
freon　氟利昂，氟氯烷【冷却剂】
FREQ＝frequency　频率
frequency　频率，次数，周波
frequency adjustment　频率调节
frequency allocation list　频率分配表
frequency allocation　频率分配
frequency allotment　频率分配
frequency analyser　频率分析仪
frequency analysis compaction　频率分析数据精简法

frequency analysis 频谱分析，频率分析，波谱分析
frequency analyzer 频率分析仪
frequency assignment 频率分配
frequency band range 频带范围
frequency bandwidth 频率频宽，带宽，频带宽
frequency band 波段，频带，频率带
frequency basis flood 频率计算出的洪水
frequency behaviour 频率表现，频率变化，频率特性
frequency breakdown 频率急降，频率崩溃
frequency bridge 频率电桥
frequency change-over switch 频率转换开关
frequency change rate relay 频率变化率继电器
frequency changer crystal 变频晶体
frequency changer 变频机，变频器
frequency channel 频道
frequency characteristics 频率特性
frequency code 频率码
frequency comparator 频率比较器
frequency compensation 频率补偿，频率校正
frequency content 频率含量，频谱，频率组成
frequency-controlled thyristor motor 频率控制闸流晶体管电动机
frequency controller 频率调节器，频率控制器
frequency control 频率调节【电厂】，频率控制【电网】，变频控制
frequency conversion station 变频站
frequency conversion 频率变换，变频
frequency converter 变频机，变频器，频率变换器，频率校正器
frequency converter set 变频机组
frequency converter station 变频站
frequency converter substation 变频站【电力】
frequency converter tube 变频管
frequency convertor 变频机，变频器，频率变换器，频率校正器
frequency counter 频率计
frequency counting 频率计数
frequency curve 频率曲线
frequency decay 频率衰减，频率下降
frequency deceleration 频率减慢
frequency demodulation 调频
frequency demultiplier 降频器，分频器
frequency dependence 与频率相关
frequency dependent damper 依赖于频率的阻尼器
frequency detector 频率探测器
frequency deviation 频移，频差，频率偏移
frequency device 频率敏感元件
frequency diagram 频率图
frequency dip 频率下降
frequency discrimination telegraph 鉴频式电报
frequency discrimination 鉴频
frequency discriminator 鉴频器
frequency distortion 频率失真，频率畸变
frequency distribution of the wind speed 风速频率分布
frequency distribution 次数分布，频率分布
frequency diversity 频率分隔，频率分集，频率划分
frequency divider 分频器，频分器
frequency division modulation 频率划分调制
frequency division multiplex 分频多路传输
frequency division switching 频分交换
frequency division 分频，频率划分
frequency domain analysis 频域分析
frequency domain 频率范围，频率域，频域
frequency doubler 倍频器
frequency drift compensation 频移补偿
frequency drift 频率漂移，频移
frequency drop 频率跌降，频率下降
frequency effect 频率效应
frequency entrainment 频率捕捉现象
frequency equation 频率方程式
frequency error 频率误差
frequency exchanger 变频机，变频器
frequency factor 频率因子，振动因子
frequency fall 频率下降
frequency feedback 频率反馈
frequency filter 滤频器，频率滤波器
frequency fluctuation 频率波动，频率起伏
frequency function 频率函数
frequency-gain curve 频率增益曲线
frequency generator 频率发生器
frequency group 频群
frequency-halving circuit 二分频电路
frequency harmonic 谐和频率
frequency hysteresis 频率滞后，频滞
frequency identification unit 频率识别装置
frequency-indicating device 频率指示器
frequency indication 频率指示
frequency indicator 频率计，频率表，频率指示器，示频器
frequency interlace 频率交错
frequency interleaving system 频率交错制
frequency interval 频程
frequency inversion 频率倒置
frequency jump 频率跃变
frequency keyer 频移键控器
frequency keying 频率键控
frequency locus 频率轨迹
frequency marker pip 频率标志脉冲
frequency measurement 频率测定
frequency measuring equipment 频率计
frequency measuring relay 测频继电器
frequency meter anemometer 频率表式风速表，频率表式风速计
frequency meter with double scale 双标度频率计
frequency meter 频率计，频率表
frequency mixing 混频
frequency modulated carrier current telephony 调频式载波电话
frequency modulated station 调频电台，调频发电厂
frequency modulated system 调频系统
frequency modulated 调频的，频率调制的
frequency modulation characteristic 调频特性
frequency modulation motor 调频电动机，频率调制电动机
frequency modulation reserve 调频备用机组，调频备用

frequency modulation tube 调频管
frequency modulation with feedback 反馈调频
frequency modulation 频率调制，调频
frequency modulator 调频器
frequency monitor 频率监视器
frequency multiplication 频率倍增，倍频
frequency multiplier klystron 倍频调速管
frequency multiplier 倍频器
frequency of drawdowns 提款频率，提款频繁性
frequency of flood 洪水频率
frequency of gust 阵风频率
frequency of occurrence 出现频率
frequency of optimum traffic 最优通信量频率，最佳通信使用频率，最佳通信频率
frequency of rare floods 稀遇洪水频率
frequency of reimbursement rates amortized loan 分期偿还贷款
frequency of sampling 采样频率，取样频率
frequency of turbulence 湍流频率
frequency of vortex shedding 旋涡脱落频率
frequency of wind direction 风向频率
frequency of windspeed 风速频率
frequency parameter 频率参数
frequency-phase characteristic 频率相位特性曲线，频率相位特性
frequency-phase method 频率相位法
frequency planning 频带分配
frequency property 频率特性
frequency protection 频率保护
frequency pulling 频率牵引
frequency quality 频率质量
frequency range 频率范围，频带
frequency rating 额定频率
frequency recognition 频率记录
frequency record 频率记录
frequency recovery 频率复原，频率回升
frequency reduction 频率下降
frequency regulation capability 频率调节能力
frequency regulation power plant 调频发电厂
frequency regulation station 调频电站
frequency regulation 频率调整，调频
frequency regulator 频率稳定器，频率调节器
frequency-rejection amplifier 滤波放大器
frequency relay 频率继电器【超或低频保护】，谐振继电器
frequency reproduction 频率重现
frequency resolution constant 频率分辨常数
frequency response characteristic 频率响应特性
frequency response curve of regulator 调节器频率特性曲线
frequency response curve 频率响应曲线
frequency response function 频率响应函数
frequency response of network 网络频率响应
frequency response test 频率响应试验
frequency response tracer 频率响应显示仪
frequency response 频率响应，频率特性
frequency restoration 频率复原，频率回复
frequency run 频率特性试验
frequency selection 频率选择
frequency selector 选频器
frequency sensitive relay 频率继电器
frequency sensitive rheostat 频敏变阻器
frequency sensitive starter 频敏启动器
frequency sensitivity 频率灵敏度
frequency separation 频率分离，频率分隔
frequency separator 分频器，频率分离器
frequency setting drift 频率定值漂移
frequency setting matrix 频率定值矩阵
frequency setting 频率整定
frequency sharing circuit 频率划分电路
frequency sharing scheme 频率划分制
frequency sharing 频率共用
frequency shift converter 频率变换器
frequency shift keying 频移键控法，调频器，移频键控
frequency shift modulation 频移调制，移频调制
frequency shift system 频移制，频率移位制
frequency shift 频差，频移
frequency signal 频率信号
frequency span 频带宽度
frequency spectrum analyzer 频谱分析仪
frequency spectrum 频谱
frequency splitting 频率分割
frequency stability 频率稳定性，频率稳定度
frequency stabilization 频率稳定，频率稳定性
frequency stabilized power source 稳频电源
frequency stabilizer 频率稳定器，稳频器
frequency standard 频率标准
frequency support 频率保持
frequency-sweep generator 扫频振荡器
frequency swing 频率摆动
frequency synchronism 频率同步
frequency synthesizer 频率合成器
frequency terminal 频率端子
frequency test 频率试验
frequency time clock 电频时钟
frequency time control 频率时间控制
frequency time modulation 频率时间调制
frequency tolerance 频率容差，容许频差，频率容限
frequency-to-number converter 频率数字变换器
frequency-to-voltage converter 频率电压转换器
frequency tracker 频率跟踪器
frequency transformation 频率变换
frequency transformer 变频器，频率变换器
frequency transient response 频率瞬态响应
frequency-trend relay 频率趋势继电器
frequency trimming 频率微调
frequency tripler 三倍倍频器，频率三倍器
frequency-type telemeter 频率式遥测计
frequency variation 频率变动
frequency vs. time response 频率-时间响应
frequency weighting network 频率加权网络
frequency wobbling 频率摆动
frequent combinations 频遇组合
frequent operation 频繁操作，频繁工作
frequent value of action 频遇值，作用频遇值
frequent value 频遇值
frequent wind speed 常现风速
fresh air 新鲜空气
fresh 淡水河，淡水水流
fresh air filter 新鲜空气过滤器

fresh air handling unit　新风机组
fresh air inlet　新鲜空气入口
fresh air intake　新鲜空气进口，新鲜空气入口
fresh air louvers　新风百叶窗，净气百叶窗
fresh air make up　新鲜空气补充量
fresh air proportion　新风比例
fresh air rate　新鲜空气率，换气率
fresh air requirement　新风量
fresh air supply　新风供应
fresh breeze　五级风【29～38 千米每小时】
fresh coal　新添加的煤，未经干燥的煤，新煤
fresh concrete　新浇的混凝土
fresh cycle　新的周期
fresh element storage drum　新元件贮存筒
freshen　增强，变强
freshet canal　溢洪道
freshet　山洪，淡水河流
fresh fuel　新（鲜）燃料
fresh gale　大风，八级风【62～74 千米每小时】
freshly fired coal　新添加的煤，新抛入煤
freshly mixed concrete　新拌混凝土，新浇混凝土
freshly set mortar　新凝砂浆
freshness　清洁度
fresh regenerant solution　新鲜的再生溶液
fresh shot　淡水水流
fresh steam　新蒸汽
fresh water deposit　淡水沉积
fresh water flow　淡水流量
fresh water geomagnetic electrokinetograph　淡水地磁动电测流器
fresh water origin　淡水成因
fresh water pool　淡水池
fresh water system　淡水系统，清水系统
fresh water tank　淡水池，淡水箱
fresh water　淡水，软水
Fresnel collector　菲涅尔集热器
Fresnel integral　菲涅耳积分
Fresnel lens　菲涅尔透镜
Fresnel reflection method　菲涅尔反射法
Fresnel region　菲涅尔区
fret saw　钢丝锯
frettage　摩擦腐蚀
fretting corrosion defect　（燃料元件）磨蚀缺陷
fretting corrosion effect　磨蚀效应
fretting corrosion　擦伤腐蚀，咬蚀，磨损腐蚀，微震磨损，微擦磨蚀
fretting wear　微震磨损
fretting　擦伤腐蚀，咬蚀，微动磨损，微振磨损，侵蚀，摩擦
fretwork weathering　蜂窝式风化，粒状岩石风化
fret　回纹饰，腐蚀处，磨损处
FR(failure rate)　故障率，失效率
FRF　法国法郎
FRH　末级再热器【DCS 画面】
friability　（煤的）易碎性，脆性
friable deposit　疏松积灰，疏松沉积物
friable gypsum　脆性石膏
friable material　易碎物料
friable slag　脆性渣
friable soil　酥性土
friable　易碎的，脆性的
friction adjuster　摩擦力调节器
frictional behavior　摩擦性能
frictional boundary layer　摩擦边界层
frictional brake　摩擦制动器
frictional coefficient　摩擦系数
frictional damping　摩擦阻尼
frictional drag coefficient　滑动摩擦系数，摩擦阻力系数
frictional drag　摩擦阻力
frictional drop　摩擦压降
frictional electricity　摩擦电
frictional error　摩擦误差
frictional factor　摩擦系数
frictional flow　粒性流，摩擦流
frictional force　摩擦力
frictional head loss　摩阻水头损失
frictional head　流动损失压头，摩擦阻力压头
frictional heat gain　摩擦增热
frictional heating　摩擦加热
frictional heat　摩擦热
frictional loss of disk　叶轮摩擦损失
frictional loss of head　水头摩擦损失
frictional loss　摩擦损失
frictional moment　摩擦力矩
frictional power　摩擦功率
frictional pressure loss　摩擦压力损失
frictional pressure　摩擦压力
frictional reheat　摩擦再热
frictional resistance　摩擦抗力，摩阻力
frictional strength　摩阻强度
frictional stress　摩擦应力
frictional torque　摩擦扭矩
frictional work　摩擦功
friction amplitude　摩擦角
friction anchor bar　摩擦型锚杆
friction and windage loss　摩擦及通风损耗
friction angle　摩擦角
friction bearing　滑动轴承
friction-breccia　擦碎角砾岩
friction centre　摩擦中心
friction circle　摩擦圆
friction clutch coupling　摩擦离合器
friction clutch　摩擦离合器
friction coat　耐磨涂层
friction coefficient　摩擦系数
friction corrosion effect　磨蚀效应
friction coupling　摩擦离合器，摩擦联轴器，摩擦联轴节
friction current　摩擦流
friction damping　摩擦阻尼
friction disk　摩擦片
friction drag　摩擦阻力
friction drive　摩擦传动
friction dynamometer　摩擦测力计
friction effect　摩擦效应
friction electric machine　摩擦起电器
friction energy loss　摩擦能量损失
friction factor in laminar flow　层流摩擦系数
friction factor in turbulent flow　紊流摩擦系数
friction factor　摩擦力，摩擦因子，摩擦率，摩

擦系数
friction governor　摩擦式调节器
friction head loss　摩擦水头损失
friction head　摩擦水头
friction index　摩擦指数
friction in governor　调速器阻尼
friction layer　摩擦层
frictionless flow　无摩擦流动，理想液体流，无摩擦流，无滞性流
frictionless fluid　理想流体，无黏性流体，无摩擦流体
frictionless wind　无摩擦风，理想风
frictionless　无摩擦的
friction loss in transition section　过渡段摩擦损失
friction loss　摩擦损失，沿程摩擦损失
friction measuring weir　摩阻量水堰
friction of motion　动摩擦，滑动摩擦
friction of piping　管道的水力阻力
friction of rest　静摩擦
friction of rolling　滚动摩擦
friction of steam　蒸汽摩擦
friction pad　摩擦垫，摩擦块
friction pile　摩擦桩，磨擦桩
friction pressure drop　摩擦压降
friction pressure loss　摩擦压力损失
friction ratchet　摩擦棘轮扳手
friction resistance　摩擦阻力
friction-revolution counter　摩擦式转速计
friction roller drive　摩擦滚柱传动
friction roll　摩擦滚筒
friction sheave　摩擦盘
friction slope　摩擦比降，摩擦坡角
friction surface　摩擦面
friction tachometer　摩擦式转速计
friction tape　胶布，摩擦带，绝缘胶带
friction torque　摩擦力矩
friction type high strength bolt　摩擦型高强度螺栓
friction type ratchet with ring with bi-hexagon　摩擦式梅花棘轮扳手
friction type ratchet with square ring　摩擦式四方棘轮扳手
friction type torque　摩擦式扭矩
friction value　摩擦值，摩擦系数
friction velocity　摩擦速度
friction wake　摩擦尾流，摩擦损耗
friction wear　磨损
friction welding　摩擦焊，摩擦焊接
friction wheel integrator　摩擦轮积分器
friction wheel speed counter　摩擦轮转速计
friction work　摩擦功
friction　摩擦，摩擦力，摩擦阻力，摩阻，阻力
friendly consultation　友好协商，友好合作
friendly exchanges　友好交流
friendly negotiation　友好协商
fright surcharge　海洋运输附加费，空运附加费
frigid zone　寒带
frigorie　千卡每小时【冷冻能力的单位】
frigorimeter　低温计
frigostabile　耐低温的
fringe benefit　额外福利，额外津贴
fringe cost　附加费
fringe effect　边缘效应
fringe intensity　条纹强度
fringe load　短周期变负荷，高频波动负荷，扰动负荷
fringe order　条纹级数
fringe pattern　条纹图形
fringe region of atmosphere　大气边缘区
fringe value　条纹值
fringe water　（毛细管的）边缘水
fringe　边沿，边缘，端，条纹，干扰带
fringing current　岸边流
fringing effect　边缘效应
fringing field　边缘区域，边缘场
fringing flux　边缘通量，边缘磁通
fringing reef　岸礁
fringing　边缘通量，边缘效应
frith　海湾
fritted glass　融结玻璃
frit　烧结
FRJC(Fast Reactor Joint Committee)　快堆联合委员会【英国】
frog brick　凹槽砖
frogging　互换，变换，测验用端子
frog-leg coil　蛙腿线圈
frog-leg machine　蛙腿绕组电机
frog-leg winding　波叠绕混合绕组，蛙腿绕组
frogman　潜水员
frog rammer　跳跃式打夯机，蛙式打夯机，蛙式夯
frog　辙岔
FROM(fusible read-only memory)　熔性只读存储器，可熔只读存储器
from this day (/time) forward　从此以后
from time to time　不时，偶尔，间或，时常，时而，随时地，(可)随时，不时地
from-to chart　来去图
front-accessible switchboard　可前面进入的配电盘
front-accessible　可前面进入的
frontage line　临街建筑线
frontage road　街面道路
frontage　屋前空地，正面空地
frontal action　锋的活动，锋的作用
frontal analysis　前缘分析，前端分析，前沿分析，锋面分析
frontal appearance　前部外形
frontal apron　冰（川）前堆积层
frontal area　正面面积，迎风面积，迎风面，正面
frontal cyclone　锋面气旋
frontal drag　迎面阻力
frontal dumping closure method　平堵截流法
frontal dumping　平堵
frontal inversion　锋面逆温
frontal line　锋线
frontal low　锋面低压
frontal member　（探头）保护面
frontal passage　锋面过境
frontal precipitation　锋面降水，锋面降水量
frontal projected area　迎面投影面积
frontal rain　锋面雨
frontal resistance　正面阻力，迎面阻力

frontal sill 侧溢洪道堰顶
frontal startup （热管）前端启动
frontal surface 锋面
frontal version 锋面逆温
frontal wave 锋面波
frontal weather 锋面天气
frontal zone 锋带
frontal 前沿的，前面的，正面的
front and rear accessible switchboard 前面和后面可进入的配电盘
front and rear accessible 前面和后面可进入的
front and rear gland 前后汽封
front and rear view 前后视图
front aperture elevation drawing 前视图
front apron 闸前护坦
front arch furnace 有前拱的炉膛
front arch 前拱
front area of the plant 厂前区
front axle 前轴，前桥
front base plate 前底板
front bay 前湾
front bearing housing 前轴承箱
front bearing 前轴承
front brick 正面砖
front burner 前墙燃烧器
front canopy 前探梁
front connected switch 前面接线开关
front connection 盘前接线，前面接线
front connector 前接头
front contact 前触点，动合触点
front damper 前阻尼器
front discharge stoker 炉前出渣炉排
front edge （脉冲）前沿
front elevation drawing 前视图
front elevation 前视图，前立面图，正面，正视图
front embankment 正面堤，前堤
front end bearing pedestal 前轴承座
front end cost 先期（开发）费用，收尾成本，前端费
front end equipment 前端设备
front end fee 前期费，开端费用，启用费，一次性手续费
front end loader 前端装载机
front end processor 前端通信处理机，前端处理机［器］，前置处理机
front end style 头部式样
front end 前置端，前端
front entrance 正面入口
front entry 主平巷
front face of the tower 杆塔前侧
front feed 前进给，炉前给煤
front fender 前翼子板，前叶子板
front-fired boiler 前墙燃烧锅炉
front-flame burner 前墙燃烧器
front gate 前门
front gear box 前齿轮箱
front girder 前梁
front grinding 正面研磨
front guide vane 前置导叶
frontier science 尖端科学
frontier 边境，边界，边缘，边疆
front intercept valve 前截门
front inversion 锋面逆温
frontispiece （房屋的）主要立面，插画，卷头
frontlift 前轮升力
front-line system 前沿系统
front-loading project 前期投资大的项目
front nodal point 前节点
front of blade 叶片的额线
front of motor 电动机的前端，电机的非连接端
front-of-wave impulse sparkover voltage （避雷器的）波头冲击闪络电压
front-of-wave test （绕组的）脉冲波试验，陡波试验
frontogenesis 锋面生成
frontology 锋面学
frontolysis 锋面消灭
front panel mounting 前面板的安装
front panel 前面板
front pillar A支柱，前风窗支柱
front pitch of winding 绕组的前节距
front platen superheater 前屏过热器
front plate 面板
front quarter area 厂前区
front quarter 厂前区
front shroud 前盖板
front side force 前轮侧力
front side of vane 叶片工作面
front side 正面
front slope 前坡
front span （线圈的）前节距
front spoiler 前扰流板
front standing pillar A支柱，前风窗支柱
front tank 前置箱［前置槽］【一回路排污系统】
front-to-back ratio 方向性比，前后比
front vane 前叶片
front view 前视图，正视图，正面图
front wall enclosure superheater 前包墙过热器
front wall firing 前墙燃烧
front wall 前墙，正面墙
front water wall 前墙水冷壁
front wearing ring 前面磨损环【靠近泵吸入口】
front wiring 明线布线
front 前锋，锋面，前端面，波前，前沿，前面的，前面，正面
frost belt 冻土带
frostbite 霜冻
frost-bound 封冻
frost box 防冻箱
frost cleft 冻裂隙
frost-cracking 冻裂作用
frost crack 冰冻裂缝，冻裂
frost damage 冻害
frost day 霜日
frost depth 冰冻深度，冻结深度
frosted bulb 磨砂灯泡
frosted glass window 磨砂玻璃窗
frosted glass 毛玻璃，水花玻璃，磨砂玻璃，毛糙玻璃
frosted incandescent lamp 磨砂白炽灯
frosted lamp globe 磨砂球形灯泡

frosted lamp 磨砂灯
frost fissure 冻裂缝
frost flower 冰花
frost free day 无霜日
frost free growing season 无霜生长期
frost free period 无霜期
frost free 无霜冻
frost heave capacity 冻胀量
frost heave force 冻胀力
frost heaven property 冻胀性
frost heave pressure 冻胀力
frost heave 冻胀
frost heaving amount 冻胀量
frost heaving damage 冻胀破坏
frost heaving meter 冻胀仪
frost heaving soil classification 冻胀土分类
frost heaving soil 冻胀土
frost heaving 冻胀，冻胀现象，冻拔【地质】，冰冻膨胀【公路】
frost-injury 冻害
frostless season 无霜期
frostless zone 无霜带
frostless 无霜
frost line 冰冻线
frost penetration 冰冻深度
frost period 冰冻期
frost point 霜点
frost prevention 防冻
frostproof depth 防冻深度
frostproof motor 耐寒式电动机
frost protection measure 防冻措施
frost protection 防冻，防霜
frost-resistant concrete 防冻混凝土
frost-resisting property 耐寒性
frost season 霜期
frost smoke 冻雾，冻烟
frost splitting 冻裂
frost susceptibility 冻敏性，易冻胀性
frost-susceptible soil 易冻土
frost thawing 冻融
frost upheaval 冻胀
frost valve 防冻阀
frost weathering 冰冻风化
frost zone 冰冻区
frost 结霜，冰冻，霜，霜冻，霜状表面
frother 起沫剂，泡沫发生器
froth floatation test 浮选试验
froth flotation 泡沫浮选，浮沫选矿
froth flow 沫状流，泡沫流
frothiness 起泡沫性
frothing agent 泡沫剂，起泡剂
frothing 起沫
froth level 泡沫液位
frothy texture 泡沫状结构
froth 起泡，泡沫，渣滓，废物
Froude number 弗劳德数
Froude scaling 弗劳德缩尺
Froude similarity 弗劳德相似（性）
frozen asset 冻结资产
frozen battery 不充电电池
frozen boundary layer 冻结边界层
frozen chunk 冻块
frozen coal breaker 冻煤破碎机
frozen coal cracker 冻煤破碎机
frozen coal crusher 冻煤破碎机
frozen coal 冻煤
frozen conductivity 冻结热导率
frozen crust 冻结层，冻壳
frozen depth 冰冻深度，冻结深度
frozen earth 冻土
frozen ground depth 地冻（结）深度
frozen ground 冻地，冻土
frozen gust 冻结阵风
frozen-heave factor 冻胀率
frozen-heave force 冻胀力
frozen lump of coal 冻煤块
frozen precipitation 固态降水
frozen rain 冻雨
frozen soil 冻土
frozen specific heat at constant pressure 冻结等压比热
frozen specific heat 冻结比热
frozen turbulence 冻结湍流
frozen 冻结的
FRP door and window 玻璃钢门窗
FRP(fiber-glass reinforced plastic) 玻璃钢
FRP(fuselage reference plane) 机身水平基准面
FRP(glass fiber reinforced plastics) 玻璃钢
FRP pipe 玻璃钢管道
FRP product 玻璃钢制品
FRP profiles 玻璃钢型材
FRP sandwich structure panel 玻璃钢夹层结构板
FRP septic tank 玻璃钢化粪池
FRQ 频率【DCS 画面】
FRR(fiber reinforced rubber) 纤维加强橡胶
fruitful 富有成果的
fruit orchard 果园
fruit 成果
frustrated contract 不能执行的合同
frustrated 落空的合同
frustration failure 落空
frustration of contract 合同落空，合同失效
frustration 无效，中止
frustum 平截头体，平截头锥体，柱身，锥台
FSA(fuel supply agreement) 燃料供给协议
FSAR(final safety analysis report) 最终安全分析报告
FSC(fuel surcharge) 燃油附加费
FSD(formed steel deck) 压型钢板
FS(feasibility study) 可行性研究
FS(feedback sheet) （经验）反馈单
FS(Fourier's series) 傅里叶级数
f-shaped slab f形板
FSH 末级过热器【DCS 画面】
FSK(frequency shift keying) 移频键控
FSS(fuel safety system) 燃料安全系统
FSS(full scope simulator) 全范围模拟机
FSS(furnace safety guard system) 炉膛安全系统
FSSS(furnace safety guard supervisory system) （锅炉）炉膛安全监控系统【DCS 画面】
FS(frequency shift) system 频移系统
FST=first 第一级

f-stop 光圈值
FSU(final signal unit) 最终信号单元
FTA(free trade agreement) 自由贸易协议
FT(flow totalizer) 流量累加器
FT(flow transmitter) 流量变送器
FTG(function training guide) 岗位培训导则
FTOC(final take-over certificate) 最终移交证书
FTP(file transfer protocol) 文件传输协议
FTU(feeder terminal unit) 馈线远方终端
FTU(Formazin Turbidity Unit) 福尔马津浊度单位
fuctional fidelity 功能逼真度
fuel 燃料，加燃料
fuel access tube 燃料传送管
fuel addition 加燃料，供料
fuel-air mixture 燃料空气混合物
fuel-air ratio control 燃料空气比控制
fuel-air ratio indicator 燃料空气比指示器
fuel-air ratio 燃料空气比（例），风煤比
fuel alloy 燃料合金
fuel and power base 燃料动力基地
fuel array 燃料排列，燃料布置
fuel assay reactor 燃料试验反应堆
fuel assemblies 燃料组件【核电厂】
fuel assembly arrangement 燃料组件排列
fuel assembly burn-up 燃料组件燃耗
fuel assembly channel 燃料组件工艺管
fuel assembly cladding tube 燃料组件包壳管
fuel assembly cluster 燃料组件束
fuel assembly composite structure 燃料组件组合构件
fuel assembly damage 燃料元件破损
fuel assembly dismantling 燃料组件拆卸
fuel assembly drop 燃料组件下坠
fuel assembly fabrication 燃料组件制造
fuel assembly fall 燃料组件跌落
fuel assembly gripping tool 燃料组件抓具
fuel assembly guide tube 燃料组件导管
fuel assembly heat release 燃料组件释热
fuel assembly holddown spring 燃料组件压紧弹簧
fuel assembly insertion 燃料组件插入，燃料组件倒换
fuel assembly levitation safeguard 燃料组件防漂浮安全保护装置
fuel assembly manipulator 燃料组件操作器
fuel assembly reception 燃料组件接收
fuel assembly reload 换燃料组件
fuel assembly relocation 燃料组件倒换
fuel assembly sampler 燃料组件取样器
fuel assembly shuffling schedule 燃料组件倒换计划表
fuel assembly sipping test equipment 燃料组件啜吸探漏装置
fuel assembly surface temperature 燃料组件表面温度
fuel assembly tilting device 燃料组件翻转机
fuel assembly transfer station 燃料组件运送站
fuel assembly vibration 燃料组件振动
fuel assembly （核）燃料组件【指压水堆】
fuel associated assembly 燃料相关组件
fuel balance 燃料平衡
fuel balls reprocessing 燃料球后处理
fuel banding bridge 燃料操作天车
fuel band 燃料品种
fuel basket 笼形燃料元件架
fuel batch 一批燃料
fuel battery 燃料电池
fuel-bearing graphite kernel 含有燃料的石墨芯核
fuel-bearing 含核燃料的
fuel bed clinker 燃料层炉渣，燃料层底
fuel bed control 燃料层厚度调节
fuel bed depth 燃料层厚度
fuel bed firing 层燃
fuel bed resistance 燃料床通风阻力
fuel bed 燃料层，燃料床
fuel behaviour 燃料行为
fuel body 燃料层
fuel bound NO_x 燃料耦合型氮氧化物
fuel bound sulfur 燃料硫
fuel box 燃料盒
fuel breakdown cell 燃料拆卸室
fuel breeding 燃料增殖
fuel brick 煤块，燃料块
fuel building 核燃料厂房，燃料厂房
fuel bundle adjustment plate 燃料棒束调整板
fuel bundle 燃料棒束
fuel bunker 煤斗，燃料贮槽
fuel burnout 燃料烧毁，燃料元件烧毁
fuel burn-up 燃料燃耗
fuel canning material 燃料包壳材料
fuel canning 燃料包壳，燃料封装
fuel can surface temperature 燃料包壳表面温度
fuel can 燃料包壳
fuel capacity 燃料容积
fuel-carrying moderator 含燃料的慢化剂
fuel-carrying 载有燃料的
fuel cartridge 燃料元件盒
fuel cassette 燃料盒
fuel cell battery 燃料组元电池
fuel cell energy storage equipment 燃料电池蓄能装置
fuel cell energy storage 燃料电池储能
fuel cell power generation 燃料电池发电
fuel cell power station 燃料电池电站
fuel cell 燃料电池
fuel centerline melting 燃料中心线熔化
fuel central temperature 燃料中心温度
fuel changing 换装燃料
fuel changing chamber 换燃料室
fuel changing gear 换料机构
fuel channel bore 燃料通道孔
fuel channeling 燃料元件装盒
fuel channel 燃料组件盒，燃料元件管道，燃料通道［孔道，管道］
fuel characteristic 燃料特性
fuel charger 装料机
fuel charge 燃料装载量，燃料投料量，燃料费用
fuel charging machine 装料机
fuel charging 装燃料
fuel chute 卸燃料斜槽

fuel circuit 燃料回路
fuel circulating pump 燃料循环泵
fuel circulating system （熔盐）燃料循环系统
fuel clad ballooning 核燃料装箱的变形
fuel cladding bond 燃料包壳结合层
fuel-cladding stress 燃料包壳应力
fuel cladding temperature meter 燃料包壳温度计
fuel cladding tube 燃料元件包壳管
fuel cladding 核燃料装箱，燃料包覆层
fuel-clad stress 燃料包壳应力
fuel-clad 核燃料装箱，（核反应堆中的）核燃料装箱（或套罩）的
fuel clogging 燃料输送系统阻塞
fuel clump 燃料块
fuel cluster 燃料棒束
fuel coal 燃（料）煤
fuel coating 燃料涂层
fuel coefficient 燃料消耗系数
fuel coffin 燃料运输容器【核电】，运送核燃料的容器
fuel column （高温气冷堆用）燃料柱
fuel compact （颗粒）燃料密实体，燃料压块
fuel composition 燃料组成
fuel compound 混合燃料
fuel conditioning 燃料形态调整
fuel configuration 燃料布置
fuel conservation 燃料保存
fuel consumption charge 燃料消耗费
fuel consumption rate 燃料耗率，燃料耗量
fuel consumption 燃料消耗量，燃料消耗，燃耗
fuel control system 燃料控制装置
fuel control valve 燃料控制阀
fuel control 燃料控制
fuel coolant interaction 燃料冷却剂相互作用
fuel cooled oil cooler 用燃油冷却的冷油器
fuel cooling facility 乏燃料冷却设备，燃料冷却水池【后处理厂】
fuel cooling installation 乏燃料冷却设备
fuel cooling pond （乏）燃料冷却水池【后处理厂】
fuel cooling pool （乏）燃料冷却水池【后处理厂】
fuel cooling 乏燃料元件的冷却，燃料冷却
fuel core 燃料核芯，燃料元件芯体
fuel cost 燃料成本，燃料费用
fuel cracking 燃料破裂
fuel cycle center 燃料循环中心
fuel cycle cost 燃料循环成本
fuel cycle economics 燃料循环经济学
fuel cycle period 燃料循环周期
fuel cycle plant 燃料循环厂
fuel cycle technology 燃料循环工艺学
fuel cycle 燃料循环
fuel damage 燃料损伤
fuel degradation 燃料破损
fuel demand 燃料需求
fuel densification 燃料密实，燃料致密化
fuel density coefficient of reactivity 反应性的燃料密度系数
fuel density coefficient 燃料密度系数
fuel density 燃料密度
fuel depletion 燃料贫化，燃料燃耗
fuel detergenting 燃料净化
fuel discharge 卸燃料
fuel-discharging machine 卸料机
fuel dispersion reactor 弥散燃料反应堆
fuel-doubling time 燃料倍增时间
fuel economy 燃料经济性，燃料节省
fueled prototype mockup system 防火数据库，装燃料的原模型系统
fuel efficiency 燃料效率
fuel element abrasion fines 燃料元件磨蚀碎屑
fuel element bowing 燃料元件弯曲
fuel element building 乏燃料元件贮存厂房
fuel element canning machine 燃料元件外壳封装机
fuel element carrier 燃料元件支架
fuel element channel 燃料管道，工作管道
fuel element charge tube 燃料元件装料管
fuel element cluster 燃料元件束
fuel element column 燃料元件立架
fuel element crud 燃料元件上的沉积物
fuel element deposits 燃料元件上的沉积物
fuel element drying facility 燃料元件干燥装置
fuel element fabrication 燃料元件制造
fuel element flask 燃料元件运输容器
fuel element grab 燃料元件卡爪
fuel element grapple 燃料元件抓钩
fuel element handling device 燃料元件装卸装置
fuel element jacket 释热元件外套［外壳］
fuel element loop 燃料元件回路
fuel element purge condensible trap 燃料元件的吹洗可凝结物收集器
fuel element purge helium cooler 燃料元件氦吹洗冷却器
fuel element purge stream 燃料元件吹洗气流
fuel element scrap 燃料元件碎屑
fuel element shuffling procedure 燃料元件倒换程序
fuel element stringer 燃料元件吊架，燃料元件立架
fuel element transport cask 燃料元件运输罐
fuel element transport 燃料元件运输
fuel-element warpage subfactor 燃料元件翘曲子因子
fuel element withdrawal 燃料元件提出
fuel element 释热元件，燃料元件
fuel elevator 燃料升降机
fuel engineering 燃料工艺学
fuel enrichment 燃料富集度
fuel evaporation rate 燃料产汽率
fuel examination facility 燃料检验设施
fuel exchange 燃料倒换
fuel expense 燃料费（用）
fuel exposure 燃料辐照量，燃料辐照量
fuel fabrication cost 燃料制造成本
fuel fabrication facility 燃料元件制造设备
fuel fabrication plant 燃料制备［加工］厂，燃料元件制造厂
fuel fabrication 燃料制备，燃料元件制造
fuel fabricator 燃料制造厂商
fuel failure detection system 燃料（元件）破损

探测系统
fuel failure detection 燃料元件破损探测
fuel failure monitoring system 燃料元件破损监测系统
fuel failure 燃料破损
fuel feedback 燃料再循环，燃料再利用
fuel feeder 给煤机
fuel feed tube 燃料球供料管
fuel flexibility 燃料的适应性，可以改烧各种燃料的灵活性
fuel flow characteristic 燃料流动特性
fuel flow control valve 燃料调节阀
fuel flowmeter 燃料消耗计，燃料流量计
fuel for power generation 动力燃料
fuel fragment 燃料碎片
fuel-free shell 无燃料壳
fuel gas generator 燃气发生器
fuel gas inlet pressure 烟气入口压力
fuel gas inlet temperature 烟气入口温度
fuel gas metering station 烟气计量站
fuel gas outlet pressure 烟气出口压力
fuel gas outlet temperature 烟气出口温度
fuel gas pipeline 燃料气管线
fuel gas reference conditions 燃料基准条件
fuel gas supply system 烟气供给系统
fuel gas 气体燃料
fuel gate 燃料水池门
fuel grab 燃料夹钳
fuel-grade 燃料等级，可作反应堆燃料的
fuel growth 燃料增长
fuel handling and lifting device 燃料装卸及升降装置
fuel handling and storage system 燃料装卸和贮存系统
fuel handling block 燃料装卸区
fuel handling bridge 燃料操作天车
fuel handling building crane 燃料操作厂房吊车
fuel handling building 燃料操作厂房
fuel handling control room 燃料装卸控制室
fuel handling device 燃料装卸设备
fuel handling equipment 燃料装卸设备
fuel handling machine base 燃料抓取机底座
fuel handling machine 换料机，燃料操作机，燃料抓取机
fuel handling platform 燃料输送操作台
fuel handling purge system exhaust filter 燃料装卸吹洗系统排气过滤器
fuel handling ramp 燃料装卸斜通道
fuel handling system 燃料元件操作系统，燃料装卸搬运系统，燃料元件装卸系统【输送线路】
fuel handling tool 燃料吊装工具，燃料吊装机械
fuel handling 燃料处理，燃料装卸
fuel heat utilization 燃料热利用率，热利用率
fuel hopper 煤斗
fuel impurity 燃料的杂质
fuel industry 燃料工业
fueling machine 装料机
fueling 装料【反应堆】，加油，加燃料
fuel injection nozzle 燃料喷嘴
fuel injection pressure 燃料喷射压力
fuel injection 燃料喷射
fuel injector 燃料喷嘴，燃料喷射器
fuel insert 燃料投入【球形燃料元件】
fuel-intensive coal plant 燃料密集燃煤发电厂
fuel inventory 燃料总装量，燃料总投入量，燃料装载量
fuel inversion 燃料倒换
fuel investment 燃料装载
fuel irradiation level 燃料辐照水平，燃料辐照度
fuel jacket 燃料包套
fuel kernel 燃料芯核
fuel-laden air 输送煤粉的空气
fuel lattice configuration 燃料栅格排列
fuel lattice 燃料栅格
fuel leaker 泄漏燃料元件
fuel-lean operation 燃用低值混合燃料运行
fuel level 油位
fuel lifetime 反应堆连续运行时间，燃料（元件）寿期，核燃料半衰期
fuel limiter 燃料限制器
fuel line 燃料管线
fuelling cycle 装料循环
fuelling grab head 装料抓斗
fuelling machine bridge 装料机桥架
fuelling machine lowdown 装料机排气
fuelling machine magazine 装料机料仓
fuelling machine nose unit 装料机突出部分组件
fuelling machine tubular pressure vessel 装料机管式压力容器
fuelling machine 装料机
fuelling mode 装料方式
fuelling room 装料室
fuel loading and unloading 装料和卸料
fuel loading authorization 批准装料
fuel loading cycle 装料循环
fuel loading device 装料装置，燃料装卸装置
fuel loading facilities 装料设备
fuel loading permit 允许（准许）装料，装料许可证，允许装料
fuel loading schedule 装料方案，燃料倒换方案
fuel loading scheme 装料方式
fuel loading 反应堆装料
fuel load 燃料负荷
fuel magazine 燃料仓
fuel make-up 燃料补给，燃料补充
fuel management 燃料管理
fuel management programme 装燃料方案，燃料管理程序
fuel management schedule 燃料管理计划
fuel manipulator crane 装卸料机
fuel master 主燃料
fuel material 含核燃料的材料
fuel matrix 燃料基体
fuel meat （元件的）燃料部分
fuel meltdown 燃料熔化
fuel migration 燃料迁移
fuel mispositioning accident 燃料棒错位事故
fuel mixture 可燃混合物，燃料混合物
fuel mix 燃料构成
fuel-moderator mixture 核燃料慢化剂混合体
fuel nozzle 喷燃器，燃料喷嘴

fuel nuclear auxiliary and reactor building 核岛厂房
fuel oil additive 燃料添加剂
fuel oil cost 燃油成本
fuel oil discharge station 油卸站
fuel oil discharge track 卸油线
fuel oil filter 燃油过滤器
fuel oil for ignition 点火用油
fuel oil handling equipment control room 燃油装卸设备控制室
fuel oil heater 燃油加热器
fuel oil pipeline 燃料油管线
fuel oil piping 燃油管道
fuel oil pump house 燃油泵房
fuel oil pump 燃油泵
fuel oil residue 残渣油，重油
fuel oil storage tank 燃油储罐，重油库
fuel oil supply pump 供燃油泵
fuel oil system 燃油系统
fuel oil tank 燃料油罐，燃料油箱
fuel oil 燃料油，重油，燃油
fuel park 燃料循环中心
fuel particle 燃料颗粒
fuel pellet compression spring 燃料芯块压紧弹簧
fuel pellet diameter 燃料芯块直径
fuel pellet holddown spring 燃料芯块压紧弹簧
fuel pellet stack 燃料芯块叠堆，燃料芯块柱
fuel pellet 核燃料芯块，燃料芯块
fuel pencil 燃料元件细棒
fuel pin can 燃料细棒包壳
fuel pin 燃料元件细棒，燃料细棒
fuel pipe 燃料管道，燃料管
fuel pit area 燃料水池面积
fuel pit backwash water 燃料水池回洗水
fuel pit bridge 燃料水池桥架
fuel pit compartment 燃料水池隔间
fuel pit cooling loop 燃料水池冷却回路
fuel pit cooling system 燃料水池冷却系统
fuel pit cover 燃料水池顶盖
fuel pit demineralizer 燃料水池除盐器
fuel pit filter 燃料水池过滤器
fuel pit gate shield slab 燃料水池闸门屏蔽板
fuel pit hall 燃料水池大厅
fuel pit heat exchanger 燃料水池热交换器，燃料贮存水池热交换器
fuel pit inside edge guard rail 燃料水池内侧护栏
fuel pit lining 燃料水池衬里
fuel pit pump 燃料水池泵
fuel pit sluice gate 燃料水池闸门
fuel pit sprinkling line 燃料水池喷淋管道
fuel pit support structure 燃料水池支承结构
fuel pit top shielding slab 燃料水池上部屏蔽盖板
fuel pit underwater lamp 燃料水池水下灯
fuel pit water cooling and clean-up system 燃料水池水冷却和净化系统
fuel pit water pump 燃料水池泵
fuel pit water 燃料水池水
fuel pit 乏燃料水池，乏燃料元件冷却水池
fuel planar smear density 燃料平面有效密度
fuel plate fabrication 燃料板制造
fuel plate 板状燃料元件，燃料元件板，燃料板
fuel plug unit with associated stringer 连带燃料柱的栓塞单元
fuel pond clean-up filter 燃料水池净化过滤器
fuel porosity 燃料孔隙率
fuel port tube 装料管，燃料装料管道
fuel port 燃料口，装料口
fuel post-irradiation examination 燃料辐射后检验
fuel preparation 燃料制备
fuel pressure indicator 燃油压力指示器
fuel process cell 燃料处理热室
fuel processing 燃料处理
fuel property 燃料特性
fuel pulverizing plant 制粉装置，煤粉制备装置
fuel pump bowl 燃料泵碗
fuel pump sump 燃料泵槽
fuel pump tank 燃料泵槽
fuel pump 燃料泵
fuel purging 燃料吹扫
fuel quality specification 燃料质量规格
fuel rack 燃料元件架
fuel ratcheting 燃料的棘轮效应
fuel rate controller 燃料量调节器
fuel rating 燃料比功率
fuel ratio 燃料比
fuel receiving and storage station 燃料接收贮存站
fuel reconstitution 更换泄漏燃料棒
fuel recording system 燃料记录系统
fuel recovery cell 燃料回收热室
fuel recovery 燃料回收
fuel recycle 燃料再循环
fuel refreshment 燃料补充，燃料补给
fuel regeneration （核）燃料再生
fuel region （高温气冷堆）换料组，换料单元，燃料区
fuel regulating gate 燃料调节闸门
fuel regulator 燃料量调节器
fuel rejuvenation 燃料更新
fuel reliability 燃料可靠度
fuel relocation effect （在元件芯块中）燃料再分布效应
fuel relocation 燃料倒换
fuel replacement energy 火电的代用电能
fuel reprocessing 燃料后处理
fuel reprocessing cell 燃料后处理室
fuel reprocessing facility 燃料后处理设施
fuel reprocessing loop 燃料后处理生产线
fuel reprocessing plant 燃料后处理厂
fuel reproduction factor 燃料再生因子
fuel reproduction 燃料再生
fuel reshuffling 燃料倒换
fuel residence time 燃料滞留堆芯时间
fuel resynthesis 燃料再合成
fuel rich firing 燃料充分燃烧，燃料富燃
fuel rich operation 燃用高值混合燃料运行
fuel rod bundle 燃料棒束
fuel rod cladding material 燃料棒包壳材料
fuel rod cladding tube 燃料棒包壳管
fuel rod cluster 燃料棒束
fuel rod coating 燃料棒涂层，燃料棒表面涂层
fuel rod consolidation 燃料棒密集化

fuel rod diameter 燃料棒直径
fuel rod flattening 燃料棒压扁
fuel rod insertion 装芯块【燃料制造中】
fuel rod interim spacer 燃料棒中间定位架
fuel rod lattice 燃料棒栅格
fuel rod locating plate 燃料棒定位板
fuel rod outer diameter 燃料棒外径
fuel rod shearer 燃料棒剪切机
fuel rods insertion 燃料棒插入
fuel rod surface temperature 燃料棒表面温度
fuel rod surface 燃料棒表面
fuel rod tray 燃料棒托架
fuel rod 条形燃料，燃料棒
fuel safety and burner control system 燃料安全和燃烧器控制系统
fuel safety system 燃料安全系统
fuel salt pump 燃料熔盐泵
fuel salt 熔盐燃料，燃料熔盐
fuel sampler 燃料液取样器，燃料取样器
fuels and materials examination facility 燃料与材料检查装置
fuel scanning 燃料扫描
fuel section 燃料段，燃料区
fuel segment （高温气冷堆的）燃料区段，燃料支区
fuel shattering 燃料粉碎
fuel sheath 燃料包壳，燃料外壳
fuel sheet 燃料片，片状燃料元件
fuel shortage 燃料缺乏，燃料不足
fuel shuffling device 燃料倒换装置
fuel shuffling scheme 装料方案，燃料倒换方案
fuel shuffling tube 倒换燃料管，倒料管
fuel shuffling 燃料倒换
fuel shut off valve 燃料切断阀
fuel sipping 燃料啜吸
fuel slab 板状燃料元件，燃料元件板，燃料板
fuel slug 燃料元件块，燃料元件短棒，燃料块
fuel slurry 燃料浆液
fuels materials facility 燃料材料装置
fuel-sodium reaction 燃料-钠反应
fuel solution reactor 燃料溶液反应堆
fuel solution 燃料溶液
fuel-specimen capsule 燃料样品辐照盒
fuel sphere 燃料球，球状燃料
fuel spreading 播煤，燃油雾化
fuel stack 燃料柱
fuel station 燃料处
fuel storage cell 燃料贮存室
fuel storage drum 燃料贮存筒
fuel storage pit crane 燃料贮存池吊车
fuel storage pool bridge 燃料贮存水池桥架
fuel storage pool cooling and cleaning system 燃料贮存水池冷却与净化系统
fuel storage pool cooling system 燃料水池冷却系统
fuel storage pool filter holding pump 燃料贮存水池过滤器压力保持泵
fuel storage pool filter 燃料贮存水池过滤器
fuel storage pool heat exchanger 燃料贮存水池热交换器
fuel storage pool lining 燃料贮存水池衬里
fuel storage pool pump 燃料贮存水池泵
fuel storage pool recirculating pump 燃料贮存水池再循环泵
fuel storage pool 燃料贮存池
fuel storage rack 燃料元件贮存架，燃料储存架
fuel storage yard 燃油存放场
fuel storage 燃料元件贮存处，燃料储存，燃料油库
fuel stringer assembly 燃料棒组件
fuel stringer 燃料元件柱，燃料柱，燃料吊架
fuel subassembly decay store 燃料组件衰变贮存
fuel subassembly flow alarm 燃料组件流量报警器
fuel subassembly top fitting 燃料组件顶部配件
fuel subassembly transport flask 燃料组件运输罐
fuel subassembly 燃料组件
fuel supply agreement 燃料供给协议
fuel supply system 燃料供给装置
fuel support grid 燃料支承格架
fuel surface heat flux density 燃料表面热流密度
fuel surface temperature 燃料表面温度
fuel surface 燃料表面
fuel suspension 燃料悬浮液
fuel swelling 燃料膨胀，燃料肿胀
fuel swiveling chute 摆动式加料槽
fuel tank 油箱，油槽，燃料油箱
fuel temperature coefficient 燃料温度系数
fuel temperature 燃料温度
fuel-to-cavity 燃料空腔比
fuel to fertile material ratio 燃料与可转换材料比
fuel-to-moderator ratio 燃料慢化剂比
fuel transfer canal 燃料转运管道
fuel transfer carriage 燃料转运小车
fuel transfer cask 燃料转运容器
fuel transfer chute 燃料转运斜管
fuel transfer device 燃料输送装置，燃料转运装置
fuel transfer facility 燃料输送装置，燃料转运设施
fuel transfer flask 燃料转运罐
fuel transfer machine 燃料转运机
fuel transfer pond 燃料元件转运池
fuel transfer system 燃料转运系统
fuel transfer tube drying space 燃料转运通道干燥间
fuel transfer tube 燃料转运通道
fuel transient response 燃料瞬态响应
fuel transportation and storage system 燃料储运系统
fuel transport flask 燃料元件运输罐
fuel treating equipment 燃料处理装置
fuel trip 燃料切断
fuel tube loading chamber 燃料管装载室【燃料制备】
fuel tube 管状燃料元件，燃料包壳管
fuel unit 燃料单元
fuel unloading device 卸料机构
fuel unloading machine 卸料机
fuel use charge 燃料费用，燃料使用费
fuel utilization factor 燃料利用因数
fuel utilization 燃料利用，燃料利用率

fuel value 燃烧值
fuel vent 燃料排气
fuel volumetric growth 燃料体积膨胀
fuel chip power plant 薪柴电站
fuel zone 燃料区
fuel zoning 燃料分区
FU(for use) 供使用【文件状态】
fugacity of M in a gaseous mixture 在气体混合物中M的逸度
fugacity 挥发性，逃逸性，有效压力
fugitive dust control 飞尘控制设备
fugitive dust emission 飞尘污染
fugitive dust 外逸粉尘，漂泊的粉尘
fugitive resource 短效资源
fugitive water 漏出水，漏失水
fugitive 挥发的，不固定的，短效的
fuil oil 燃料油
Fukushima nuclear accident （日本）福岛核泄漏事故
Fukushima Nuclear Power Plant （日本）福岛核电站
fulchronograph 闪电电流特性记录器
fulcrum bearing 支承，支点承座
fulcrum pin 支座枢轴
fulcrum 支点，支轴，支点轴，转轴
fulfill a contract 履行合同，执行合同
fulfill an agreement 履行协议
fulfill due diligence 履行尽职调查
fulfillment date 履行日期
fulfillment of a contract 履行合同
fulfillment of a schedule 完成进度计划
fulfillment 实现，完成
fulfill the capital contribution obligations 履行注入资本金的义务
fulfil 实现，履行，遵守，完成，满足
fulgurate （闪电般）发光
full 全部，全部的，完全的
full admission turbine 全周进汽汽轮机，整周进水式水轮机
full admission 全周进汽，全周进气
full advance 全断面掘进
full aeroelastic approach 全气动弹性法
full air admission 全周进汽
full and by 满帆，扯满帆地，满帆顺风地
full and change high water 大潮高潮间隔
full annealing 完全退火
full-arc admission starting device 全周进气启动装置，全周进汽启动装置
full-arc admission turbine 全周进汽式透平［汽轮机］，全周进气式透平［汽轮机］
full-arc admission 全周进气，全周进汽
full-arc turbine 全周进汽式透平［汽轮机］，全周进气式透平［汽轮机］
full-arc 全周进汽［气］
full authority 全权
full automated portable difratometer 全自动便携式衍射仪
full automatic processing 全自动处理
full automatic 全自动的
full automation and cybernation 完全自动控制
full automation 全自动化，完全自动化
full ball 整球
full binary adder 全二进制加法器
full-blown power plant 大型配套发电厂
full-body exposure 全身照射量，全身照射
full bore valve 全腔阀门
full bore 全孔径
full bridge converter 全桥变流器
full bridge model 全桥模型
full capacity above normal 高于正常值的全容量
full capacity tap 满载抽头，全容量抽头
full capacity valve trim 全容量阀门结构
full capacity 全容量，满载容量
full car section 重车组
full car 重车，整车
full-cell process 充细胞法【木材防腐】
full circle crane 周转式起重机
full coiled winding 整圈绕组，整圈绕法
full container load 集装箱整箱货
full coordinates 完全坐标
full core （反应堆的）一炉料，全堆芯
full coupling 全管接头，双头管箍
full coverage 全额担保
full cruciform core 十字形铁芯
full current 全电流
full cut out 全断，全闭，全停
full damping 全阻尼，全衰减
full density 理论密度
full depth gear 标准齿高齿轮
full depth involute system 全高齿渐开线制
full depth rolling 全深度胀管
full depth simulation 全厚度模拟
full depth tooth 全齿高齿
full depth 全高
full developed boundary layer 充分发展边界层
full developed flow 充分发展流动
full developed region 充分发展区
full duplex operation 同时双向操作
full duplex 全双向通信，全双工
full-efficiency 满载效率
full electrification 全盘电气化
full emitter 全辐射体
fullering tool 压槽锤，凿密工具，捻缝工具
fullering 凿密，压槽锤开槽，捻缝
full face digging 全面开挖
full face driving method 全断面掘进法
full face mining 整个工作面采煤
full face tunnel boring machine 全断面掘进机
full face tunneling 全断面掘进
full feathered position 完全顺风位置
full feathering 顺桨
full feather 完全顺桨
full field relay 满励磁继电器
full field speed 全励磁转速
full field 满载磁场，全磁场
full fillet welding 满角焊
full fired heat recovery combined cycle 排气再燃式联合循环
full flap 全襟翼
full-fledged member 正式成员
full floating bearing 全浮动式轴承
full-floating weighbridge 全浮动秤桥

full flow flowmeter 满流流量计
full flow geothermal power generation 全流式地热发电
full flowrate area 全流量面积
full flow valve 全流阀
full flow 全流量
full furnace flame detection 全炉膛火焰检测
full gale 强风
full gate opening 门孔全开度，闸门孔全开，闸门全开度
full gate operation 全开度运行
full ground reflection 地面全反射
full hairpin spring 全发针弹簧
full hardening 完全淬火，淬透
full head 全水头
full hood 全罩
full impulse wave 全波冲击
full income 全部所得
full-in position 全插入位置
full insulated winding 全电压绝缘绕组
full insurance 全保险
full integral simulation test 整体全模拟试验，整体全仿真试验
full internal water recirculation 完全内部水再循环
full length absorber rod 全长吸收棒
full length control rod assembly 长控制棒组件
full length control rod drive mechanism 长控制棒驱动机构
full length control rod 全长控制棒
full length emergency cooling heat transfer 全长度应急冷却传热
full length emergency cooling 全长度应急冷却
full length rod control system 长棒控制系统
full lift pressure 完全抬升压力
full lift 完全抬升
full line 实线
full liquid-cooling generator 全液冷发电机
full load capacity 满载容量
full load characteristic 满负荷特性
full load condition 满负荷状态，满载工况
full load current 满载电流
full loaded through-running train 满载直达列车
full load efficiency 满负荷效率
full load excitation 满载激励，满载励磁
full load field 满载磁场
full load house 装料间
full load line 重车线
full load loss 满负荷损失，满载损失
full load operation 满负荷运行，满载运行，全负荷运行
full load output 满载出力
full load period 满载阶段，全负荷阶段，满载期
full load power peak 满载功率峰值
full load power 满载功率
full load rating 额定全负荷
full load refuelling 满负荷换料
full load rejection capability 甩满负荷能力
full load rejection 甩满负荷，甩全负荷
full load running 满载运行，满载运转
full load saturation curve 满载饱和曲线
full load slip 满载滑差
full load speed 满载速度，满载转速，满负荷速度
full load starter 满载启动器
full load starting 满载启动
full load test 满负荷试验，满负载试验，满载试验
full load thermal efficiency 满负荷热效率
full load torque 满载转矩，额定负载下的转矩，全负载扭矩
full load trial operation 满负荷试运
full load voltage 满载电压
full load 满负荷，满载，全负荷
full-magnetic controller 全磁控制器
full moon spring tide 满月大潮
full moon 满月，望月
fullness coefficient 充满系数，满蓄率
full node calibration 全跨距校正
fullnut 整螺母【与锁紧螺母连用】
full-opening valve 全开阀
full opening 完全开启
full-open throttle 全开节流
full-out position 全提出位置
full output 全出力
full payment 全部付款
full penetration weld 全透焊接，全焊透焊缝，全透焊
full penetration 完全贯穿
full-pitched coil 整距线圈，全距线圈
full pitch winding 整距绕组，全距绕组
full pitch 全节距，整节距
full plant discharge 电站满载泄流量
full pool level 满库水位
full potential 全电势，全电位
full power day of operation 满功率运行日
full power torque 满功率力矩
full power 满功率
full-pressure containment 全压安全壳
full-pressure design containment 全压设计安全壳
full-pressure steel containment shell 全压钢安全壳
full pre-stressing 全预应力
full Pu core 钚燃烧堆，全钚堆芯
full radiator 全辐射体
full range fuse 全范围熔断器
full rated speed 满载额定转速
full reactor power 反应堆满功率
full-rear-arch furnace （无前拱的）全后拱炉膛
full reflected system 全反射系统
full reflector 完全反射层
full release position 全松位置，完全释放位置
full replica simulator 全复制型模拟机，全复制型仿真机
full-revolving loader 全转式装料机
full river by-pass diversion 断流围堰导流
full-rotating derrick 周转式起重机
full running speed 全速运行
full scale construction 全面施工
full scale data 全尺寸数据
full scale deflection 满标偏转
full scale equipment 工业装置

full scale experiment 全尺寸实验，工业性试验，生产性试验
full scale input 最大输入信号
full scale irradiation 全剂量辐照
full scale measurement 实尺测量，全尺寸测量
full scale modeling 足尺模拟
full scale model test 足尺模型试验
full scale model 全尺寸模型，足尺模型
full scale plant 实物尺寸试验装置
full scale prototype 全尺寸原型物
full scale range 全量程范围
full scale reactor 全尺寸反应堆，一比一模拟堆
full scale real time simulation 全工况实时仿真
full scale Reynolds number 全尺寸雷诺数
full scale simulator 全尺寸模拟机
full scale template 足尺样板
full scale test 全面试验，满标试验，实样试验，实物试验，原型试验，真型试验，真实条件试验，足尺试验
full scale tower test 真型铁塔试验
full scale tower 真型塔
full scale value 满刻度值，(仪表的) 满标值
full scale version 工业模型
full scale wind tunnel test 全尺寸风洞试验
full scale wind tunnel 全尺寸风洞
full-scale 满刻度的，全部的，实物尺寸，足尺
full scene 全景
full scope high realism simulator 全范围、高逼真度电厂仿真机
full scope safeguards 全面保障
full scope simulator 全范围仿真机，全规模模拟机，全范围模拟机
full scram 全速急停，完全停堆
full screen height 荧光屏满刻度高度
full sea 满潮
full-section pipe-flow 满管流
full set of originals 全套正本
full set 整套
full signature 签全名
full sized model 全尺寸模型
full size print 全尺寸的图纸
full size reproduction 等尺寸复制
full size 实际尺寸 (的)，足尺，最大尺寸
full span blade pitch 变桨距，全翼展叶片间距
full span bridge model 全桥模型
full span flap 全翼展襟翼
full span flutter 全跨颤振
full span galloping 全跨驰振，整个挡距舞动
full span ground board 全跨地板
full span pitch control 全跨度桨距控制
full span 整个叶片
full speed operation 全速运行
full speed saturated steam turbine 全速饱和蒸汽汽轮机
full speed turbine stages 汽轮机全速级组，全速级组
full speed turbine 全速汽轮机
full speed 全速
full spiral case 完整形蜗壳
full stabilized conductor 全稳定导体
full stall 全失速
full station meter 全站仪
full storage 全量储冰
full-stream sampling 全物流取样
full stroke 全冲程
full subsidence 完全下沉
full subtractor 全减法器
full supply level 最高供水位
full supporting velocity 最小悬浮速度
full symmetry 完全对称
full synchronizing 准同期，准同步
full thread 通长螺纹
full throttle 全节流
full tide cofferdam 完全挡潮围堰
full tide 满潮
full-time service 全日工作
full-time staff 专职人员
full-time storage plant 多年调节电站，完全调节电站
full-time worker 全日工作者
full train test 全系列试验【即泵、齿轮箱、冷却器和马达】
full travel 全行程【控制棒】
full-up position 全提出位置
full utilization 充分利用
full value insurance 全值保险
full view 全景，全视图
full voltage starter 全电压启动器
full voltage starting motor 全压启动电动机
full voltage starting 全压启动
full voltage 满电压，全电压，全压
full wagon puller 重车牵车器
full wagon pusher 重车推车器
full wagon 重车，整车
full water cooling generator 全水冷发电机
full-wave 全波，全波的
full-wave bridges 全波整流电桥
Full wave impulse insulation level 全波冲击绝缘水平
full-wave lighting control 全波照明控制
full-wave oscillation 全波振荡
full-wave phase control 全波相位控制
full-wave rectification 全波整流
full-wave rectifier 全波整流器
full-wave voltage doubler 全波倍压器
full-wave voltage impulse 全波电压电击【避雷器】
full-way valve 闸阀
full weight 满载重量，毛重
full welding 满焊
full width at half maximum amplitude 最大半振幅的全宽度
full width at half maximum 半最大幅度的阔度，半高宽
full width at half peak amplitude 半峰值振幅的全宽度
full width rising closure 平堵截流
fully aerated flow 完全掺气水流
fully attached flow 完全附着流 (动)
fully austenitic structure 全奥氏体组织
fully automatic compiler translator 全自动编译翻译程序

fully automatic compiling technique 全自动编译技术
fully automatic control 全自动控制，自动控制
fully automatic operation 全自动操作
fully buffered channel 全缓冲通道
fully charged 全充（电）
fully confined gasket 全封闭密封垫
fully controlled bridge 全控电桥
fully controlled convertor 全可控变流器
fully controlled device 全控型器件
fully developed boundary layer 充分发展边界层
fully developed flow 充分发展流动
fully developed laminar flow 充分发展层流
fully developed nucleate boiling 充分发展泡核沸腾
fully developed profile 完全发展速度廓线
fully developed turbulence 完全紊动
fully developed turbulent flow 充分发展湍流
fully developed wake 完整的尾流
fully developed 定型流动的，充分发展的
fully-drainable superheater 全疏水型过热器
fully-enclosed motor 全封闭型电动机
fully-enclosed shield 包封屏蔽
fully-enclosed staircase 全封闭楼梯间
fully-enclosed stairs 全封闭楼梯
fully energized 全激励，满励磁，全带电
fully enriched 高浓缩的，高富集的
fully-flameproof motor 全防爆型电动机
fully funded （项目）资金全部到位
fully-graded aggregate 全部级配骨料，全级配骨料
fully-guarded machine 全防护型电机
fully impervious clothing 全密封工作服
fully impregnated insulation 全浸渍绝缘
fully-killed steel 镇静钢
fully laminar 全部层流的
fully loaded 满载
fully open 全开
fully reflected reactor 全反射层反应堆
fully reinforced 实力雄厚
fully rigid model 完全刚性模型
fully-shielded 全屏蔽的，全铠装的
fully-studded 布满销钉的
fully tracking mountings 全跟踪支架
fully trapped gasket 全封闭密封垫
fully water-cooled turbogenerator 全水冷式汽轮发电机
fulminate 雷酸盐，电闪雷鸣，爆炸
fulmination 爆炸
fulvic acid 黄腐殖酸，棕黄酸，褐菌素，富维酸
fumarole （硫磺气）喷出孔，喷气孔
fume abatement 烟雾消除
fume chamber 通风柜
fume cupboard 通风柜，通风橱
fume emission 烟气排放
fume extraction 烟雾回收，烟雾提取
fume height 烟柱高度
fume hood 通风柜，排烟罩，通风罩
fumeless dissolution 无烟溶解
fume-off 排出气体，排气
fume shape 烟雾形
fume 烟气，水汽，烟雾
fumigant 熏蒸剂
fumigating plume 下熏型羽流
fumigation fee 熏蒸费
fumigation 熏烟，熏沉
fuming nitric acid 发烟硝酸
fuming sulfuric acid 发烟硫酸
fuming 发烟的，烟化
function 作用，职能，函数，功能
functional 函数的，功能的，泛涵
functional analysis 泛函分析
functional area 职能范围
functional arrangement 功能图，操作电路
functional assignment 分配职责
functional authority standard 职能标准
functional authority 职能权力
functional block diagram 原理框图，功能方框[块]图
functional block 功能块，功能器件，功能框
functional character 功能符，控制符
functional clearance 操作间隙
functional code 操作码
functional constrained data attribute 功能约束数据属性
functional constrained data 功能约束数据
functional constraint 功能约束
functional counter 函数计数器
functional department 职能部门
functional dependence 函数关联
functional description 功能介绍
functional design 功能设计
functional determinant 函数式行列式，函数行列式
functional diagram 方块图，工作原理图，方框图，功能图，施工图
functional element 函数元素，作用元件
functional equation 函数方程
functional examination 功能性检查
functional failure 功能失效
functional flow diagram 功能图，操作流程图
functional generator 函数发生器
functional group 官能团
functional guarantee 功能保证（值）
functional insulation 功能绝缘，内绝缘
functional ionizable group 能电离的官能团
functional logic diagram 功能逻辑图
functional management 职能管理
functional modularity 功能调剂性，功能积木性
functional planning 功能规划
functional plan 单项规划
functional potentiometer 函数分压器，函数电位器
functional pressure differential 操作压力差
functional relation 函数关系
functional scheme 工作原理图
functional specification 功能规范
functional standard 功能标准
functional symbol 操作符号
functional test 功能试验
functional trainer 功能培训器

functional unit　功能单元，函数装置，基本功能元件
functional zoning　功能分区
function block　功能框，功能模块
function card　功能卡（件）
function chain　功能链
function chart　功能图
function check-out　功能核查
function code　函数码，操作码
function command　功能命令
function constraint data attribute　功能约束数据属性
function constraint data　功能约束数据
function constraint　功能约束
function control diagram　功能控制图
function conversion unit　函数转换部件
function cost analysis　功能成本分析
function digit　操作位
function-distributed system　功能分布系统
function electronic block　电子功能块
function generating potentiometer　函数发生电势计
function generator　函数发生器
function group control level　功能组控制级
function group control　功能组级控制
function group　功能组，专门小组
functioning capability　使用能力，运行能力
function key　功能键
function letter　函数字母
function multiplier　函数乘法器
function of time　时间函数
function of variables　多变量函数
function of　……的函数，随……而变的（东西）
function part　操作部分，功能部分
function plotter　函数描绘器
function potentiometer　函数电势计
function relay　函数继电器
functions of complex variable　复变函数
function space　函数空间
function switch　工作转换开关，函数开关
function table　函数表
function test　功能试验
function theory　函数论
function unit　操纵部件，控制部件
function value　函数值
function words　虚词
functor　功能元件，逻辑元件，算符
fundamental active power　基波有功功率
fundamental assumption　基本假设
fundamental band　基带
fundamental breach　根本违约，根本性违约
fundamental combination　基本组合
fundamental component　基波分量，基频分量，基本分量
fundamental construction　基本建设
fundamental corporate change　公司重大变更
fundamental criteria　基本准则
fundamental current　基波电流
fundamental data　依据资料
fundamental document　依据文件
fundamental element　主要部件，基本元件
fundamental flux density　基波磁通密度
fundamental formula　基本公式
fundamental frequency magnetic modulator　基频磁调制器
fundamental frequency　基本频率，基频，固有频率
fundamental harmonic　基波，一次谐波
fundamental law　基本定律
fundamental magnetization curve　基本磁化曲线
fundamental mode rebalance　基模态再平衡
fundamental mode　基谐模，基谐型，主模态
fundamental oscillation　固有振动，基本振荡，基波
fundamental particle　基本粒子
fundamental period　基本周期
fundamental point　基本点
fundamental power　基波功率
fundamental resonance　基频谐振
fundamental survey　基本测量
fundamental tolerance　基本公差
fundamental undamped natural mode of vibration　振动基本固有模态
fundamental vibration　基本振动
fundamental wave length　基准波长，基波长
fundamental wave　基波，主波
fundamental way　根本途径
fundamental　原理，基本单位，基本的，主要的，根本的
fundament　基础，基本原则
fund balance　基金余额
fund collection　集资
fund for compensation awards　补偿金基金，定额补偿基金
fund for construction　建设资金
fund for special purpose　专用基金
fund for technical improvement　技术改造基金
funding commitment　资金承付，资金承付额
funding　筹集资金
fundless company　皮包公司
fund raising　集资，筹措资金，资金筹措
funds from parent　公司投入资金
funds of enterprises　企业基金
funds retained by enterprise　企业留用资金
funds turned over to state　上交国家的资金
funds used for specified purposes　专款专用
fund　资金，基金，经费，现款，财源
fungicide　杀菌剂
fungus-proofing motor　防霉电动机
funicular car　缆车
funicular curve　垂曲线，索状曲线，悬链线
funicular force　（索的）张力
funicular polygon　索多边形，索线多边形
funicular water　纤维水
funnel cloud　漏斗云
funnel coast　漏斗形海岸
funnel drain system　漏斗形疏水系统
funnel erosion　漏斗状侵蚀
funnelling effect　漏斗效应
funnelling plume　漏斗型羽流
funnel pipe　铁烟囱
funnel sea　漏斗海

funnel-shaped aerial　喇叭形天线
funnel-shaped bay　漏斗状海湾，漏斗湾
funnel-shaped delivery tube　漏斗形卸料管
funnel-shaped　喇叭形的
funnel stand　漏斗架
funnel　漏斗，烟囱，（液态炉）渣口，缩孔管，仓斗，通风井
funotional diagramming　功能图
furacana　飓风
furane resin　呋喃树脂
furcated tube　分叉管
furling device　停车装置【风力机】，收帆装置
furling handle　停车手柄【风力机】
furling wind speed　收帆风速，停车风速【风力机】
furling　收拢
furlong　浪【长度单位】
furnace adjustment　炉温调节
furnace arch　炉拱，折焰角
furnace atmosphere　炉膛气氛
furnace attachment　炉胆连接附件，炉膛附件
furnace bar　炉条
furnace bin　日用仓，炉前煤斗
furnace brazing　炉中钎焊
furnace burden　炉料
furnace casing　炉膛炉板
furnace cavity　炉膛容积
furnace clinker concrete　炉渣混凝土
furnace combustion black　烟炱
furnace condition　炉膛工况
furnace cooling factor　炉膛水冷程度
furnace cooling　炉冷
furnace cross-section heat release rate　炉膛截面积热负荷，炉膛截面热强度
furnace crown　炉顶
furnace depth　炉膛深度
furnace door　炉门
furnace draft　炉膛负压
furnace drying　烘炉
furnace enclosure design pressure　炉膛设计压力
furnace enclosure　炉膛外壳
furnace evaporator　炉内蒸发段，炉内蒸发受热面
furnace exhaust　炉膛排烟
furnace exit gas temperature　炉膛出口烟气温度
furnace exit NO_x concentration　炉膛排出氮氧化物浓度
furnace extension　前置炉膛，前炉
furnace flame television　炉膛火焰监视电视
furnace flame TV system　炉膛火焰电视系统
furnace floor　炉底
furnace flue　炉胆
furnace gas　炉膛烟气
furnace header　炉膛上联箱
furnace heating absorption rate　炉膛吸热率
furnace heat liberation　炉膛容积热强度，炉膛放热
furnace heat release rate　炉膛放热率，炉膛释热率
furnace hopper　冷灰斗
furnace intensity　炉膛热强度
furnace lining　炉衬
furnace load　炉膛热负荷
furnace lower header　炉膛下联箱
furnace mantle　炉壳
furnace module　分炉膛，炉室组件
furnace oil　重油
furnace outlet gas temperature　炉膛出口烟气温度
furnace outlet screen　（炉膛出口）垂帘管
furnace pipe　炉管
furnace plan area　炉膛横断面积
furnace plan heat release rate　炉膛截面热负荷
furnace pressure control　炉膛压力控制
furnace pressure　炉膛压力，炉膛正压
furnace profile　炉膛轮廓，炉膛断面
furnace puff　炉膛爆燃
furnace purge　炉膛吹扫
furnace rating　炉膛热负荷
furnace refuse　炉渣
furnace release rate　炉膛热负荷
furnace safeguard supervisory system　炉膛安全保护监控系统
furnace safeguard system　炉膛安全系统
furnace safety guard supervisory system　锅炉炉膛安全监控系统
furnace safety guard system　炉膛安全系统
furnace safety supervisory system　炉膛安全监控系统
furnace section heat release　炉膛横断面热负荷
furnace sheet　炉胆板
furnace slag screen　炉膛捕渣管屏
furnace slag　炉渣
furnace soldering　炉中钎焊
furnace staying　炉胆拉撑
furnace superheater　炉膛过热器
furnace temperature　炉膛温度
furnace throat　炉膛缩口，炉膛缩腰
furnace transformer　电炉用变压器，电炉变压器
furnace volume　炉膛容积
furnace volume heat release rate　炉膛容积热负荷
furnace volumetric heat release　炉膛有效容积热负荷
furnace wall cooling surface　炉膛水冷壁受热面
furnace wall heat flux density　炉壁热流密度
furnace wall heat release rate　炉壁热负荷
furnace wall　炉墙，水冷壁
furnace water cooling　炉膛水冷
furnace　炉，炉膛，炉胆，燃烧室，燃烧器，窑，烘箱
FURN＝furnace　炉膛
furnished component　装备（配置）部件
furnish power　发电，供电
furnish statement of account　提供费用单据
furnish　提供，供给
furniture　家具
furred ceiling　吊顶，顶棚［篷］，悬吊天花板
furred wall　混水墙
furring brick　衬垫砖，面砖，贴面砖
furring clip　龙骨卡
furring insert　固定龙骨用的预埋件，龙骨衬垫
furring tile　内衬瓷砖，内墙衬砌层，墙面瓷砖
furring　磁粉起毛，钉板条，副龙骨

furrow dam 沟堤
furrow drain 排水毛沟
furrow irrigation 沟灌
furrow 沟槽
further document 补充文件
further notice 另有通知，再次通知
further subcontractor 进一步的分包商，继续分包商，再分包商
fuse alarm 熔丝烧断报警
fuse block 熔线盒，保险丝盒，熔断器
fuse blow 熔断器熔断
fuse board 熔断器盘，熔线板
fuse box 保险丝盒
fuse carrier 熔线座，熔线架
fuse cartridge 熔线盒
fuse combination unit 熔断器组合单元
fuse coordination 熔丝配合
fuse cutout 熔丝闸刀开关，熔丝断流器
fused breaker 熔丝断路器
fused catalyst 熔融催化剂
fused contactor 熔丝接触器
fused cutout 熔丝闸刀开关，熔丝断流器
fused disconnect switch 熔断开关
fused disconnect 熔断
fused flux 熔炼焊剂，熔融焊剂
fused interrupter switch 熔丝断路开关
fuse disconnecting switch 熔丝隔离开关
fused potential transformer 带熔断器的电压互感器
fused salt electrolyte 熔盐电解质
fused salt fuel 熔融盐燃料
fused salt reactor 熔盐反应堆
fused salt 熔盐
fused short-circuit current 熔断短路电流
fused solution 熔融体
fused stag 熔渣
fused 熔化的
fuse-element 熔丝，熔体，熔片
fuse grip 熔丝夹
fuse holder 保险丝盒
fuse link 熔断片保险丝管
fuse metal 保险丝合金，易熔合金
fuse plug emergency spillway 自溃式非常溢洪道
fuse plug spillway 自溃坝
fuse plug 熔线塞，保险塞，安全塞
fuse point 熔点
fuse protection 熔断器保护，熔丝保护
fuse puller 熔丝拉出器
fuse rated voltage 保险丝额定电压
fuse rating 保险丝额定值
fuse strip 熔线，熔线片
fuse switch disconnector 熔断器式隔离负荷开关
fuse switch 熔断器式开关，熔丝开关
fuses 保险丝
fuse terminal 熔丝端钮
fuse tube 熔丝管
fuse wire 熔线，保险丝，熔丝
fuse wrench 引信扳手
fuse 熔断器，熔化，保险丝，引信，火药线，熔丝
fusibility curve 熔度图
fusibility 可熔性，熔度
fusible circuit-breaker 熔丝断路器
fusible cone 熔锥【测温】
fusible cutout 熔断器
fusible disconnecting switch 熔丝隔离开关
fusible element 可熔元件【熔断器】
fusible link valve 易熔连杆阀
fusible plug 易熔塞
fusible read-only memory 可熔只读存储器
fusible 可熔的，易熔的
fusile fuel 聚变燃料
fusing agent 熔剂
fusing characteristic 熔断特性
fusing current 熔丝熔断电流
fusing factor 熔断系数
fusing point 熔点
fusing resistor 熔断电阻器，熔阻丝
fusing 熔解，熔化，熔融
fusionable material 聚变材料
fusion accelerator （电子束）聚变加速器，熔化加速器
fusion agent 熔剂
fusion boundary 熔化边界
fusion breeder 聚变增殖堆
fusion crucible 熔化坩埚
fusion curve 熔化曲线
fusion cutting 熔化切割
fusion depth 熔化深度
fusion energy 聚变能
fusion face 坡口面
fusion fission fuel factory 聚变裂变燃料工厂
fusion fission hybrid reactor 聚变裂变混合反应堆
fusion fission reaction 聚变裂变反应
fusion fuel 聚变燃料
fusion gas weld 气熔焊
fusion heat 熔化热
fusion line （焊缝）熔合线，熔化线
fusion method 熔化法
fusion penetration 熔深，熔透
fusion point 熔化温度，熔点
fusion power plant 热核发电厂，聚变发电厂
fusion power 热核能，聚变能
fusion product 聚变产物
fusion pyrometer 熔化高温计
fusion ratio 融合比
fusion reaction 聚变反应
fusion reactor material 聚变堆材料
fusion reactor 热核反应堆，核聚变反应堆，聚变堆
fusion system 聚变系统
fusion temperature 熔化温度，聚变温度，熔点
fusion-welded wall 熔焊式水冷壁
fusion welding 溶焊【熔化焊】，熔焊
fusion zone 母材熔合区，熔化区
fusion 熔化，熔解，熔接，融合，核聚变，融合物，融化
fuss type automatic voltage regulator 振动式自动电压调整器
fust 柱身
future contract 期货合同
future delivery 远期交货

future enlargement 远景扩建
future expansion 远景扩建
future flow 远景流程
future load demand 远景负荷需要
future planning 远景规划
future system 未来系统
future trading 期货交易
future units 预留机组
future 远期，将来
futurology 未来学
fuze 熔线，熔断器，保险丝，引信，火药线
fuzzy area 模糊区域
fuzzy control 模糊控制
fuzzy earthquake intensity 模糊地震烈度
fuzzy language 模糊语言
fuzzy matching 模糊匹配
fuzzy mathematics 模糊数学
fuzzy relation 模糊关系
fuzzy response spectrum 模糊反应谱
fuzzy set 模糊集合
fuzzy 模糊的
FV(flow control valve) 流量控制阀
FW＝feedwater 给水
FW(field weld) 现场焊
FW(fillet weld) 角焊缝，填角焊焊缝
FW(finishing work) 装修工程，收尾工程
FW(fixed wing) 固定翼
FWH(feed water heater) 给水加热器
FWHM(full width at half maximum) 半最大幅度的阈度
FWPCA(Federal Water Pollution Control Act) 联邦水污染控制法【美国】
FWP(feedwater pump) 给水泵【DCS 画面】
FWR(full wave rectifier) 全波整流器
FWS(feedwater system) 给水系统
FW 给水【DCS 画面】
FY(flow relay or valve) 流量传送器
FYI(for your information) 供你参考
F 故障【DCS 画面】

G

Ga-As diode 砷化镓二极管
gabbro-norite 辉长苏长岩
gabbro-pegmatite 辉长伟晶岩
gabbro-syenite 辉长正长岩
gabbro 辉长岩
gabion boom 石笼挡栅
gabion dam 石笼坝
gabion 篓筐潜水箱，石筐，石笼，填石笼
gable board 山墙封檐板
gable end 山墙端
gable roof 双坡屋顶，双坡屋面
gable-shape rigid frame 人字形刚架
gable wall 山墙
gable 人字墙，山墙，双坡的
GAC system(gap automatic control system) 间隙自动控制系统
gadder 穿孔器，凿岩机，钻孔器
gad flue 烟道
gadget 设备，装置，零件，小配件
gadiometer 磁强梯度计
gadolinia fuel 含钆燃料，钆毒物燃料
gadolinium oxide 氧化钆
gadolinium poisoned fuel 钆毒物燃料
gadolinium poison 钆毒物
gadolinium 钆
GA drawing(general assembly drawing) 总装图
gad tongs 平口钳
gaffer 领班，工头，照明电工
GAF(goods acceptance form) 货物验收单
gage block 块规
gage length 标准长度【张力试验】，测量长度
gage mark 伸长标志【张力试验】
gage pressure 表压力
gage 厚度，直径，测量仪表，规格，计量，度量，量规，测量仪器，计量器
gagged 闭锁的，闭塞的
gagging screw 封闭螺纹
gagging 堵塞，闭塞
GAG(gross available generation) 可用发电量
gaging 计量，精确计量
gag 塞子，堵头，塞铁，关闭，流量限制器
GAH alkali wash water pump 空气预热器碱洗水泵
GAH alkali wash water tank 空气预热器碱洗水箱
GAH(gas air preheater) 空气预热器
GAH soot blower 空气预热器吹灰器
gain a larger share of market 赢得市场份额
gain amplification 增益放大
gain around a feedback 反馈回路增益
gain-band merit 增益带宽指标
gain-band width product 增益频带宽积
gain-band width 增益频带宽，增益带宽
gain changer 传动比变换装置
gain characteristic 增益特性
gain coefficient 增益系数
gain constant 放大系数，增益常数，放大率
gain controller 增益控制器
gain control 增益调整，增益控制
gain crossover frequency 增益交越频率，增益交叉频率，增益剪切频率
gain crossover 增益窜渡
gaine 套，罩，壳
gain factor 增益因数，增益因子，增益系数，放大系数
gain-frequency characteristic 增益-频率特性
gaining 收入，利益【复数】
gain in weight 重量增加
gain margin 增益余量，增益容限，增益裕量
gain measuring set 增益测量器
gain on exchange 外汇获利，外汇盈余
gain oriented 增益去向的
gain profit 得利益，获利
gain setting 增益设定
gain time control 增益时间控制
gain 增益，获得，放大系数，腰槽
galactic clusters 银河星群
galaxies 银河系

galaxy 银河
gale damage 大风损失，风灾
gale pollution 大风污染
gale 大风【7～10级】，强风，暴风
gal =gallongale 大风
gall 磨损，擦伤
gallery driving 坑道开凿
gallery system 坝内廊道系统
gallery-type 外廊式
gallery 管廊，长廊，廊道，通廊，输煤栈桥，走廊，架空过道，输煤通廊
galleting 碎石片嵌灰缝
gallet 碎石，石屑
galling 表面机械损伤，擦伤，轻度擦伤，机械性损伤，黏结，磨损，咬住
gallium arsenide cell 砷化镓电池
gallium arsenide solar cell 砷化镓太阳电池
gallon 加仑
gallop bridge 立体交叉桥
galloping criterion 驰振判据
galloping excitation 驰振激励
galloping flutter 驰振型颤振
galloping force coefficient 驰振力系数
galloping inflation 恶性通货膨胀
galloping instability 驰振不稳定性
galloping of iced conductor 导线覆冰舞动
galloping response 驰振响应
galloping torsion 驰振扭转
galloping vibration 驰振型振动
galloping write and recovery 跃步写和恢复，跃步写与恢复
galloping （导线的）舞动，驰振，（导线）舞动
gallop 飞奔，不正常运转，运转不稳定
gallows frame 龙门起重架，门式吊架
gallows 门形吊架，挂架
galvanic battery 蓄电池，原电池组，蓄电池组，原电池
galvanic ceil 原电池
galvanic corrosion 自发电流腐蚀，电化腐蚀，电蚀，电偶腐蚀，电化学腐蚀
galvanic coupling 电流耦合
galvanic current 稳定的直流电，伽伐尼电流，动电电流
galvanic electricity 流电，动电
galvanic element 原电池
galvanic etching 电腐蚀
galvanic isolation 电流隔离
galvanic pile 电堆
galvanic potential 电位，伽伐尼电位
galvanic protection system 镀锌保护系统，牺牲阳极保护系统，防腐保护系统
galvanic series 电势序，电位序，电压序列
galvanic 电流的，电镀的
galvanism 由原电池产生的电，电流，电疗法
galvanist 流电学家
galvanization 电镀，镀锌，电疗
galvanized carbon steel pipe 镀锌碳钢管
galvanized corrugated sheet 镀锌波纹铁，镀锌波纹铁皮
galvanized iron pipe 白铁管，镀锌铁管
galvanized iron plate 镀锌铁板
galvanized iron sheet 镀锌铁皮
galvanized iron wire 镀锌铁丝
galvanized iron 白铁皮，马口铁
galvanized nail 镀锌钉
galvanized plain sheet 镀锌平铁皮
galvanized sheet iron roof 白铁屋面
galvanized sheet iron 镀锌铁
galvanized sheet metal 镀锌薄钢板，镀锌钢板
galvanized sheet 镀锌薄板，镀锌铁，白铁皮，镀锌薄钢板，镀锌钢板，镀锌铁皮
galvanized steel helix 镀锌螺旋锚
galvanized steel iron 镀锌白铁皮
galvanized steel pipe 镀锌钢管
galvanized steel sheet 镀锌钢板
galvanized steel wire strand 镀锌钢绞绳，镀锌钢绞线
galvanized steel wire 镀锌钢丝，镀锌铁丝
galvanized steel 镀锌钢，镀锌钢板，镀锌钢材
galvanized wire mesh 镀锌铁丝网
galvanized wire 镀锌线，镀锌铅丝，镀锌钢丝
galvanized 电镀的，镀锌的
galvanize 电镀，镀锌，通电流
galvanizing kettle 镀锌槽
galvanograph 电流记录图
galvanoluminescence 电解发光
galvanolysis 电解
galvanomagnetic 电磁的
galvanomagnetic effect 电磁效应
galvanomagnetic signal transmitter 电磁信号发生器
galvano magnetism 电磁，电磁学
galvanometer contact potentiometer 检流计式接触电势计
galvanometer modulator 电流计式调制器
galvanometer oscillograph 检流计式示波器
galvanometer recorder 检流计记录器
galvanometer relay 检流计式继电器
galvanometer shunt 检流计用分流器
galvanometer-type relay 检流计式继电器，电流计式继电器
galvanometer 检流计，电流计，微电流计
galvanometry 电流测定法
galvanoplastics 电铸，电铸技术，电镀
galvanoplasty 电铸，电铸技术，电镀
galvanoscope 验电器
galvanotaxis 向电性，趋电性
galvanotropism 向电性，趋电性
galvano-voltammeter 伏安计
gambrel roof 斜折线形屋顶
gambrel 复斜屋顶
game equilibrium 博弈均衡
game playing 博弈
gamer 谷仓
game theory 对策论，博弈论
gaming simulation 博弈模拟，博弈仿真，对策模拟，对策仿真
gamma absorber γ射线吸收体
gamma absorption analysis γ射线吸收分析，γ射线吸收分析法
gamma absorptionmetry γ射线吸收剂量学
gamma absorption γ射线吸收

gamma activation analysis γ射线活化分析
gamma background γ射线本底，γ本底
gamma-compensated tritium monitor γ射线补偿式氚监测仪
gamma-compensate ionization chamber γ补偿电离室
gamma contamination γ放射性物质污染，γ污染
gamma-correction circuit 伽马校正电路
gamma distribution γ分布，伽马分布
gamma dose rate γ射线剂量率
gamma emitter γ发射体，γ放射源
gamma energy γ能量
gamma escape γ辐射漏泄
gamma exponent 伽马指数，传输特性等级指数
gamma flux density indicator γ通量密度指示器
gamma flux density meter γ通量密度计
gamma flux density monitor γ通量密度监测器
gamma flux meter γ通量密度计
gamma flux γ通量
gamma-free flux 无γ通量
gamma fuel scanning 燃料γ射线扫描
gamma-function distribution 伽玛函数分布
gamma function γ函数
gamma-generated pulse 伽马形成的脉冲
gammagraphy γ射线照相法
gammagraph γ照相装置，γ射线照相
gamma hazard γ辐射危害，γ辐照危害
gamma heating effect γ加热效应，γ射线加热效应
gamma heating （堆芯）γ加热
gamma-induced internal heating 伽马产生的内部热量
gamma infinity 最大反差系数
gamma instrument γ测量仪
gamma iron γ铁，相铁
gamma measuring instrument γ测量仪
gamma-neutron reaction γ中子反应
gamma noise γ噪声
gamma peak 伽马射线峰值
gamma-phase producing 产生的伽马相位【冶金】
gamma-phase 伽马相
gamma quantum γ量子
gamma quench 伽马淬火
gamma radiation beacon 伽马辐射标志
gamma radiation detector γ辐射探测器
gamma radiation emitter γ辐射发射体，γ放射源
gamma radiation 伽马辐射，γ辐射
gamma radiator γ辐射体，γ辐射源
gamma radiography 伽马射线照相术，γ射线照相术，γ射线照相法
gamma ray absorption coefficient γ射线吸收系数
gamma ray absorption γ射线吸收
gamma ray detection apparatus γ射线探伤机［仪］
gamma ray dose rate constant γ射线剂量率常数
gamma ray dosimeter γ射线剂量计，γ剂量计
gamma ray dosimetry γ射线剂量测定法
gamma ray emission γ射线发射
gamma ray energy γ射线辐射能
gamma ray flaw detector γ射线探伤机［仪］
gamma ray level measuring device 伽马射线物位测量仪表，γ射线物位测量仪表
gamma ray logging 伽马γ射线测井
gamma ray projector γ射线照相检验
gamma ray quantum γ量子
gamma ray radiography γ射线透照术
gamma ray scattering γ射线散射
gamma ray source container γ射线源容器
gamma ray source γ源，γ射线源
gamma ray spectrograph γ射线摄谱仪
gamma ray spectrometer γ射线谱仪
gamma ray spectrum γ射线谱
gamma ray teletherapy γ射线远距治疗
gamma ray 伽马射线，γ射线
gamma scan 伽马扫描，γ射线扫描
gamma shielding γ辐射屏蔽
gamma shield γ射线屏蔽体，γ射线防护屏
gamma solid solution γ固溶体
gamma spectrograph γ射线摄谱仪
gamma spectrometry 伽马射线能谱测定法
gamma spectroscopy γ谱学，γ能谱学
gamma spectrum 全范围，全程γ射线谱，γ能谱
gamma 伽马，微克
gamut 全音阶
gang boss 工长
gang condenser 同轴电容器
gang control 同轴控制，联动控制
ganged 成组的，成套的
gang forms 成套模板
ganging 同轴，成组，同轴连接
gang operated 同轴操作，联动的
gangplank 跳板
gang saw 直锅
gang switch 联动开关，同轴开关
gangue content 含矸率
gangue power plants 煤矸石电厂
gangue 矸石，脉石，煤矸石
gangway cable 主巷道电缆
gang way 出入口，过道，平台，工作走道，运输平洞
gang 组，套，接合，同轴的，浮冰
gannister 硅砖，致密硅岩
gantry beam 龙门架梁
gantry coal grab 门型抓煤机
gantry column 龙门架柱
gantry crab bucket 门型抓斗，龙门抓斗
gantry crane in turbine hall 汽机房行车
gantry crane with electric hoist 电动葫芦门式起重机
gantry crane 龙门吊（车），龙门起重机，高架（移动）起重机，门式起重机，桥式吊车
gantry grab crane 门式抓斗起重机
gantry tower 门型架，门型铁塔
gantry traveler 门式移动吊车，移动式龙门起重机
gantry type coal grab 门式抓煤机
gantry 龙门架，塔架，门架，吊机架，起重机门（架），门型架构，起重（机）台架
Gantt chart 甘特图，线条图

gap action controller 间隙调节器，间隙控制器
gap adjustment 间隙调整
gap admittance 间隙导纳
gap ampere turns 气隙安匝
gap area 气隙面积
gap arrester 气隙避雷器，气隙放电器
gap automatic control system 间隙自动控制系统
gap barrier （汽轮发电机的）气隙隔板
gap capacitance 间隙电容
gap choke 空气隙铁芯轭流圈
gap-chord ratio 节弦比，相对栅距
gap clearance 间隙，对缝间隙
gap closure （燃料棒间的）间隙填满
gap conductance 间隙热导
gap diameter 气隙直径，定子内径
gap energy 禁带宽度
gap factor 间隙因数
gap field 气隙场，缝隙场
gap flux 气隙磁通，气隙通量
gap frame C形框架
gap gage 隙规
gap-gas segregating ring （汽轮发电机的）气隙隔板
gap gauge 塞尺，塞规
gap gradation 不连续级配
gap-graded aggregate 间断级配骨料
gap heat transfer coefficient 间隙传热系数
gap heat transfer 间隙传热
gap in record 观测记录间断
gap inventory 间隙中裂变产物总量
gap length 气隙长度
gapless metaloxide type 无间隙金属氧化型
gap loss 间隙损失
gap of rolls 辊隙
gapped aggregate grading 骨料间断级配
gapped core 带气隙的铁芯
gapped tape 分间隔录入数据块的磁带
gap pickup cooling 气隙取气式冷却
gapping fissure 张开裂缝
gap probe 间隙式探头
gap purge system 间隙吹洗系统
gap reluctance 气隙磁阻
gap scanning 间隙扫描
gap seal 间隔封罩，间隙密封
gap structure 气隙结构
gap suitable for air extraction 抽风间隙【安全壳】
gap surface 间隙面
gap test 间隙检测，间隙检验，间隙探伤，空隙试验
gap thickness inventory 间隙中裂变产物总量
gap thickness 间隙厚度
gap voltage 隙缝电压
gap welding 断续焊接
gap width 间隙宽度
gap 间隙，缺口，差距，凹陷，裂缝，缝隙，峡口，山口，空隙，裂口，操作间隙
garage door 车库门，汽车库门
garage for bulldozer 推煤机库
garage 车库
garbage bin 垃圾箱
garbage can 垃圾桶，垃圾箱
garbage chute 垃圾道
garbage collection 垃圾收集
garbage disposal plant 垃圾处理场
garbage disposal 垃圾处置
garbage funnel 垃圾斗
garbage furnace 垃圾焚化炉
garbage grinder 垃圾磨碎机
garbage incineration plant 垃圾焚烧厂
garbage incineration power station 垃圾焚烧电站
garbage-in garbage-out 杂乱输入杂乱输出
garbage-in 无用输入
garbage-out 无用输出
garbage pass 垃圾道
garbage power plant 垃圾电厂
garbage time 无效时间
garbage truck 垃圾车，清洁车
garbage 无用数据，垃圾，无用信息
garboard 龙骨翼板
garden architecture 庭园建筑学，园林建筑学
gardening 园艺
garden lighting fixture 庭院灯
garden planning 园林规划
gardens 园林
garnet 石榴石，金刚砂
garnet-gneiss 石榴片麻岩
garnished wage 扣发工资
garnish 装饰
garret floor 顶楼层
garretting 填塞石缝
garret window 顶楼窗，老虎窗
garret 阁楼，屋顶层，顶楼
gas absorber 气体吸收器
gas absorption 气体吸附
gas activity meter 气体放射性活度计
gas activity reduction system 气体放射性衰减系统
gas-actuated relay 气体继电器，瓦斯继电器
gas-air mixture 燃气空气混合物
gas alarm 毒气警报
gas analysis indicator 气体分析仪
gas analysis meter 气体分析仪［器］，烟气分析仪［器］
gas analysis 气体分析，烟气分析
gas and power investment company 天然气与电力投资公司
gas and steam combined cycle installation 燃气蒸汽联合循环装置
gas and steam combined cycle power plant 燃气蒸汽联合循环电厂
gas and steam combined cycle unit 燃气蒸汽联合循环机组
gas and steam combined cycle 燃气蒸汽联合循环
gas area 燃气面积
gas attack 烟气侵蚀，气体腐蚀
gas backing 气垫轴承
gas backmixing 气体返混
gas backstreaming 气体回流
gas baffle 烟气挡板
gas bearing auxiliary blower 气体轴承辅助风机

【高温气冷堆】
gas bearing circulator 气体轴承循环风机
gas bearing 气垫轴承
gas bell 气钟
gas-bending stress 气流起弯应力
gas black 气黑，气烟末，天然气炭黑
gas blanket 气体覆盖（填充），气垫，气层
gas blast arc-quenching 气吹灭弧
gas-blast breaker 气体吹弧断路器
gas-blast circuit breaker 气吹断路器
gas-bleed 抽气，放气
gas blow-off system 气体排放系统
gas blow-off 排气
gas blowout 气体喷出
gas bomb 气体旁路，气窜，气体钢瓶
gas bubble detection 气泡探测器
gas bubble injection 气泡喷入
gas bubble 气泡
gas buffer tank 气体缓冲箱
gas burner 气体燃烧器，气燃烧器，瓦斯燃烧器
gas bypass 烟气旁路，旁通烟道
gas cable hut 充气电缆线路储气站
gas cap 气顶
gas carbonitriding 气体碳氮共渗
gas carburizing 气体渗碳
gas cartridge fire extinguisher 贮气瓶式灭火器
gas cavity 气孔，气穴【焊接缺陷】
gas cell 充气光电池，离子光电池
gas-chamber 烟箱，（燃料元件）气室
gas chromatographic analysis 气相色谱分析
gas chromatographic analyzer 气相色谱分析仪
gas chromatographic column 气相色谱柱
gas chromatographic method 气相色谱法
gas chromatography 气相层析，色谱法分析气体，气体色层分析法
gas chromatograph 气体色谱分析仪，气相色谱仪
gas circuit dryer plant 气体回路干燥装置
gas-circulating system 气体循环系统
gas circulation cloud 气泡晕
gas circulator 气体循环泵，循环风机
gas cleaner 烟气净化器，气体净化器
gas cleaning 烟气清洗［除尘］，气体净化
gas clean-up 气体净化，除气
gas cloud 空气层，烟气层
gas coal 气煤
gas collection tank 集气箱
gas collection 气体收集
gas collector 集气器
gas composition 气体组成，烟气成分，气体成分
gas compressor 压气机
gas concentration gradient 气体浓度梯度
gas concrete 加气混凝土
gas condensate tank 烟气凝结水罐
gas conditioner 烟气处理装置
gas conditioning 气体调节，气体净化处理
gas conductance 气体热导
gas conductivity 气体电导
gas conduit 烟道
gas constant 气体常数
gas consumption 气体消耗量
gas container 气体容器，储气罐
gas contamination 气体污染
gas content 含气量，气体含量
gas control assembly 烟气控制装置
gas control instrument 气体检测仪表
gas control system 气体控制系统
gas converting test 气体转换试验
gas coolant 气体冷却剂
gas-cooled 气冷的
gas-cooled breeder reactor 气冷增殖反应堆
gas-cooled breeder 气冷增殖（反应）堆
gas-cooled electrical machine 气冷电机
gas-cooled fast breeder 气冷快中子增殖反应堆，气冷快中子增殖堆
gas-cooled fast reactor 气冷快中子反应堆，气冷快堆
gas-cooled generator 气冷发电机
gas-cooled graphite-moderated high-temperature reactor 石墨慢化高温气冷反应堆
gas-cooled graphite-moderated reactor 石墨慢化气体冷却反应堆，石墨气冷堆
gas-cooled pebble-bed reactor 球床气冷反应堆
gas-cooled reactor 气冷反应堆，气冷堆
gas-cooled rotor 空冷转子，气冷转子
gas-cooled solid-moderated high-temperature reactor 固体慢化高温气冷反应堆
gas-cooled thermal reactor 气冷热反应堆
gas-cooled twin reactor station 气冷双堆电站
gas cooler 煤气冷却器，气体冷却器
gas cooling system 气体冷却系统
gas cooling 气冷，气体冷却
gas corrosion 气体腐蚀
gas-coupled turbine 分轴式透平
gas current 气流，烟气流，电流
gas cushion 气垫【防止锤击】
gas cutting 气割
gas-cycle reactor 气体循环反应堆
gas cylinder 气瓶
gas decay tank 放射性废气衰减箱，气体衰变箱
gas dehumidification 气体除湿
gas delay line 气体延迟［衰减］管线
gas delay system 气体延迟［衰减］系统
gas density balance 气体密度计
gas density meter 气体密度计
gas detector relay 气体继电器，瓦斯继电器
gas detector 气体探测器，气体检测器
gas dew point heating station 烟气露点加热站
gas dielectric breakdown 气体介质击穿
gas dielectric capacitor 气体介质电容器
gas-diesel engine 狄塞尔煤气内燃机
gas diffusion 气体扩散
gas diode 充气二极管
gas discharge current 气体放电电流
gas discharged tube 气体放电管
gas discharge plasma 气体放电等离子体
gas discharge tube 气体放电管
gas discharge 气体放电
gas dispersion 气体弥散，气体扩散
gas distribution grid 气体分布板

gas distribution 配气，配气装置
gas distributor 气体分布器，气体分配器，配气器
gas dome 储气盖【用于污泥消化池】
gas drier 气体干燥器
gas duct （锅炉外部）烟道
gas dynamic behaviour 气体动态特性，气体动力性能
gas dynamic equation 气体动力方程
gas dynamic facility 气体动力研究设备
gas dynamic function 气体动力函数
gas dynamic lubrication 气体动力润滑
gas dynamics tunnel 气体动力研究风洞
gas dynamic system 气动系统
gas dynamics 气体（动）力学
gas dynamic theory 气体动力学理论，气动理论
gas eddy 气体涡流，气涡
gas efflux speed 气体排放速度
gas efflux velocity 气体射流速度
gaseity 气态
gas ejector 抽气器
gas emergency trip valve 煤气紧急截止阀
gas engine 煤气机
gas entrainment （钠中）气体夹杂物
gaseous absorption 气态吸收，气体吸收
gaseous absorptivity 气体的吸收性
gaseous activity 气体放射性，气体放射性活度
gaseous conductor 气态导体
gaseous contaminant 气态污染物
gaseous contamination 气体污染
gaseous coolant 气态冷却剂
gaseous diffusion process （同位素分离）气体扩散法
gaseous diffusion 气体扩散
gaseous diluent 气体稀释剂
gaseous discharge 气体放电
gaseous effluent 气体排放物
gaseous emission 气体发散，气体排放，气体排放物
gaseous envelope 气膜
gaseous fission product release rate 气体裂变产物释放率
gaseous fission product 气体裂变产物
gaseous fluid 气体介质，气体
gaseous fuel 气体燃料
gaseous impurity 气态杂质
gaseous insulant 气态绝缘剂
gaseous insulation 气体绝缘
gaseous ion 气体离子
gaseous medium 气体介质
gaseous phase 气相
gaseous plume 气态羽流
gaseous pollutant 气态污染物，气体污染物
gaseous protector tube 充气保管箱
gaseous radioactivity 气体放射性
gaseous radwaste system 废气处理系统
gaseous reactant 气体反应剂
gaseous reagent 气体试剂
gaseous rectifier 充气管整流器
gaseous state 气态
gaseous waste arising 气体废物产生
gaseous waste disposal system 气体废物处置系统
gaseous waste holdup system 气体废物滞留系统
gaseous waste management system 气体废物管理系统
gaseous waste processing system 气体废物处理系统
gaseous waste processing 气体废物处理
gaseous waste treatment 废气处理，气态废物处理
gaseous waste 气体废物，气态废物，废气
gaseous 气态的，气体的，过热的
gas equation 气体方程式，气体状态方程
gas equilibrium 气体平衡
gas escape tube 排气管
gases detector 可燃气测定仪
gases mixture 气体混合物
gases poisonous detector 有毒气体测定仪
gases 气体
gas etching 气体腐蚀
gas exchange rate 气体交换速率
gas exchange 气体交换
gas exhauster 抽气机，排气机，排风机，通气器
gas expander 气体膨胀机
gas exploder 气爆雷管，气爆引信
gas fat coal 气肥煤
gas-filled cable 充气电缆
gas-filled compartment 充气隔室
gas-filled dry-type transformer 干式充气变压器
gas-filled generator 充气发电机
gas-filled lamp 充气灯泡
gas-filled phototube 充气光电管
gas-filled pipe cable 充气管道电缆
gas-filled proportional detector 充气正比探测器
gas-filled rectifier 离子整流闸，离子整流器
gas-filled relay 充气继电器，离子继电器，电子继电器
gas-filled solid cable 充气实心电缆
gas filled thermal system 充气感温系统，充气热系统
gas-filled thermometer 充气式温度计
gas-filled transformer 充气变压器
gas-filled tube arrester 充气管型避雷器
gas-filled type explosion-proof machine 充气防爆型电机
gas-filled 充气的
gas fill hole 充气孔
gas film lubrication 气膜润滑
gas film 气膜
gas filter 滤气器，气体过滤器
gas-fired boiler 燃气锅炉
gas-fired combined cycle power plant 燃气联合循环发电厂
gas fire detector 气体火灾探测器
gas-fired incinerator 燃气焚烧炉
gas-fired infrared heating 煤气红外线辐射采暖
gas-fired power plant 燃气电厂，燃气电站，燃气（火力）发电厂
gas-fired radiant heater 瓦斯辐射加热器
gas-fired waste combustible incinerator 可燃废物

燃气焚烧炉
gas-fired 燃气的
gas fission product monitor 气体裂变产物监测器
gas fission product 气体裂变产物
gas flame welding 气火焰焊接
gas flow bending stress 气流弯应力
gas flow computer 气体（烟气）流量计算器
gas flow counter 流气型计数器
gas flow detector 气流检测仪
gas flow equation 气体流量方程，气体流动方程
gas flow expansion 气流膨胀
gas flow neutron flux density measuring assembly 流气型中子通量密度测量装置
gas flow pattern 气体流型
gas flow proportional counter 气流正比计数器
gas flow rate 气体流量，燃气流量
gas flow recorder 气体流量记录器
gas flow test 烟气流动试验
gas flow 气体流量，气流
gas flue 烟道
gas fluidized bed 气体流化床
gas-forming agent 发泡剂，泡沫剂
gas fuel 气体燃料
gas gap 气隙
gas gauge 气体压力表，气量计，煤气表
gas gauging 气割槽，气刨
gas generating station 煤气发生站
gas generator 气体发生器，煤气发生器，煤气发生炉
gas graphite reactor 气冷石墨反应堆
gas-handling system 气体处理系统
gas heater 煤气加热器，烟气加热器，气体加热器
gas holder 煤气罐，气体集存器，气柜
gas hold-up 气体滞留
gas hole 气孔，气泡
gas hood 气罩
gas house tar 煤蒸馏副产品
gas house 煤气站
gas hydrate 气体水合物
gash 裂纹，深痕
gasifiable 可气化的
gasification degree 气化程度
gasification reaction 气化反应
gasification 气化
gasifier 煤气发生器，气化器，自由活塞发动机，气体发生器
gasifying blower 气化风机
gasifying medium 气化介质
gas ignitor 气体点火器
gas infrared rays thawing equipment 燃气红外线解冻装置
gas injection 气体回注，气体喷射，天然气回注，注气【油气田】
gas-insulated bushing 气体绝缘套管
gas-insulated cable 气体绝缘电缆
gas-insulated circuit breaker 气体绝缘断路器
gas-insulated metal-enclosed substation 气体绝缘金属封闭的变电站
gas-insulated metal-enclosed switchgear 气体绝缘开关设备，气体绝缘金属封闭开关设备
gas-insulated substation 气体绝缘变电站
gas-insulated switchgear 气体绝缘（封闭）开关设备，气体绝缘开关
gas-insulated transformer 气体绝缘变压器
gas insulation cable 气体绝缘电缆
gas insulation 气体绝缘
gas integral process 全汽化过程
gas interchange 气体交换
gas ionization 气体电离
gas jet 瓦斯喷燃器，气体射流，燃气喷嘴，气焊枪
gasket dam 篮形填石坝
gasketed 加填料的
gasket material 填料
gasket micanite 衬垫云母板
gasket removal tool 密封垫拆除工具
gasket sealed relay 垫片密封继电器
gasket seat 阀座
gasket 填料，垫密片，垫片，垫圈，接合垫，衬垫，密封垫，密封垫圈
gas kinetics 气体动力学
gas law 气体定律，气体状态方程
gas layer produced surface shear 气流附面层引起的表面剪切力
gas leakage return line 泄漏气体回流管
gas leakage 漏气
gas leak detector 漏气检验器，漏气检查器
gas leak indicator 漏气指示器
gas leak test 气密性试验，气体检漏试验
gas length 气隙长度
gas-lift flow area 气升出流区
gas-lift pump 气升泵，气体提升泵
gas lighter 气体点火器
gas light 煤气灯，密封的
gas line 气体管道
gas-liquid chromatoghraphy 气液色谱法
gas-liquid reaction 气液反应
gas lock 气塞，气栓，气闸，气锁
gas-lubricated bearing 气体润滑轴承
gas mask 防毒面具
gas mass flow 气体质量流量，燃气质量流量
gas metal arc welding 气体保护电弧焊，气体金属电弧焊，金属极惰性气体保护电弧焊
gas meter 气体流量计，煤气表，气量表
gas mixture 气体混合物
gas model 气模
gas moisture analyzer 气体湿度分析仪
gas motor 煤气发动机
gas-moving device 送气装置
gas multiplication factor 气体放大系数，气体放大因子
gas multiplication 气体倍增，气体放大
gas nitriding 气体渗氮
gas nozzle 伴喷嘴
gaso＝gasoline 汽油
gas oil 汽油，粗柴油
gasolene 汽油
gasoline blow lamp 汽油喷灯
gasoline engine generator 汽油发电机
gasoline engine 汽油发动机，汽油机
gasoline locomotive 汽油机车

gasoline shovel 汽油机动力铲
gasoline 汽油
gasol 气态烃类，液化石油气
gasometer 煤气表，气量计，气量表
gas-operated device 气动操纵设备
gas orifice 喷气孔
gas outlet header 气体出口总管
gas outlet port 气体出口孔
gas outlet temperature 气体出口温度
gas outlet 烟气出口，排气管
gas pass （锅炉内部）烟道，烟气流程
gas path 烟气流程，烟道
gas penetration port 气体贯穿孔
gas penetration slot 气体贯穿缝
gas permeability 气相渗透率，透气性，气体渗透性
gas phase combustion 气相燃烧，均相燃烧
gas phase compressibility 气相可压缩性
gas phase mass flux density 气相质量流密度
gas phase oxidation 气相氧化，气相腐蚀
gas phase volume flow rate 气相体积流量
gas phase 气相
gas pick-up 吸气
gas pilot 煤气引燃器
gas pipeline 煤气管道，气体管路，天然气管道，输气管道
gas pipe 煤气管，瓦斯管
gas plenum 烟气室，气腔
gas plume 气态羽流，烟流，烟缕
gas pocket 气袋，气孔，气泡，气囊
gas poisoning 煤气中毒
gas pool 气池
gas pore 气孔【焊接缺陷】
gas power engine 煤气发动机
gas power plant 煤气发电厂，燃气发电厂
gas pressure loss across HRSG 穿过余热锅炉的烟气压力损失
gas pressure reducing equipment 煤气减压装置
gas pressure relay 瓦斯压力继电器，气压继电器
gas pressure welding 加压气焊
gas pressure 气体压力
gas pressurization system 气体加压系统
gas-press welding 气压焊
gas probe 烟气试样，烟气取样管
gas producer 煤气发生炉
gas proof machine 气密电动机，防瓦斯型电机
gas proof 气密的，防毒气的，不透气的，防气的
gas proportioning damper 烟气比例调节挡板
gas pulsation 气流脉动
gas pump 气体泵
gas purger 不凝性气体分离器
gas purge system 气体吹洗系统
gas purge 气体清洗，气体吹洗
gas purification loop 气体净化回路
gas purification plant 气体净化装置
gas purification system reflux 气体净化系统回流
gas purification 煤气净化，气体净化
gas-purifier 气体净化器
gas purifying system 气体净化系统
gas purity 气体纯度
gas quench system 气体冷却装置
gas radiation monitor 气体辐射监测仪
gas radiation 气体辐射
gas reactor 气冷反应堆
gas recirculation fan 烟气再循环风机
gas recirculation 烟气再循环
gas recombination 气体复合
gas recombiner 气体复合器
gas refrigerating machine 气体制冷机
gas regulator 减压器
gas reheating 气体再热
gas relay （变压器的）瓦斯继电器，气体继电器，瓦斯继电器，闸流管继电器
gas release 排气，放气
gas relief line 减压管道，排气管线
gas removal chamber 除气室
gas retention system 气体滞留系统
gas reversal chamber 烟气转向室
gas-ring burner 环形管进气喷燃器
gas rise 气体抬升
gas sampler 气体取样器
gas sample 气体样品，气体取样
gas sampling probe 烟气取样管
gas sampling system 气体采样系统，气体采样系统堆
gas sampling tap 气体取样分接头
gas sampling tube 气体采样管
gas scale 气体温标
gas scrubber 湿式洗涤器，除尘器，湿式烟气脱硫，气体洗涤器，湿式除尘器
gas seal mechanical sodium pump 气体密封式机械钠泵
gas seal 气封
gas segregating baffle （汽轮发电机的）气隙隔板
gas sensitive resistor 气敏电阻器
gas sensory semiconductor 气敏半导体
gas separation circuit 气体分离回路
gas separator 气体分离器
gas shielded arc welding 气体保护电弧焊
gas shielded magnetic flux arc welding 磁性气体保护焊
gas shield welding 气体保护弧焊
gas shock 气流激波，气流冲击
gas shuttle pipe 气体传送器
gas side corrosion 烟气侧腐蚀
gas side fouling 气侧积垢
gas side 烟气侧
gassing 放气，充气，吹气
gas-slab 气体导板
gas slug 气栓，气节
gas-solid chromatography 气固色层法
gas-solid feed ratio 气固比
gas-solid phase fluidized bed process 气固两相流化床工艺
gas-solid two-phase flow 气固两相流动
gas space （燃料元件中）气腔，气体空间
gas sphere 气界，气圈
gas state 气态
gas-steam combined cycle cogeneration power plant

燃气蒸汽联合循环热电联产电厂
gas-steam combined cycle "one-on-one"unit 燃气蒸汽联合循环一拖一机组
gas-steam combined cycle unit 燃气蒸汽联合循环机组
gas-steam combined cycle 燃气蒸汽联合循环
gas storage tank 气体贮存箱
gas storage 储气，气体贮存
gas stratification 气体停滞
gas stream 除气器，气体洗提器，气流
gas stripper column 除气柱
gas stripper condenser 除气器冷凝器
gas stripper extraction pump 除气器吸入泵
gas stripper feed pump 除气器给水泵
gas-stripper gas cooler 除气器气体冷却器
gas stripper preheater 除气器预热器
gas stripper reflux condenser 除气器回流冷凝器
gas stripper 除气器，脱气塔
gas stripping 除气，脱气
gas substitution test 气体置换试验
gas suction nozzle 气体吸入接管
gas-surface interaction 气体与表面之间的相互作用
gas swept surface 烟气冲刷的受热面
gas system 气体系统
gassy 气态的，气体的
gas table 气体热力性质表
gas tank 储气罐，煤气罐，气柜
gas tar 煤焦油
gas temperature control 燃气温度控制
gas temperature probe 烟气温度测枪
gas tempering 烟气温度调节
gas thermometer 气体温度计
gas throughput 气体通过量
gas tight casing 气密砖墙，气密式外壳
gas tight coating 气密包覆
gas tight door 防毒门
gas tight double isolation 双层气密隔离层
gas tight machine 封闭式电机
gas tightness test 气密性试验，气体检漏试验
gas tightness 气体密封，气密性，不透气性
gas tight shielded enclosure 气密屏蔽壳，气密屏蔽包层
gas tight withdrawal tube 气密抽出管
gastight 不漏气的，不透气的，气密的
gas-to-gas heat exchanger 气气换热器
gas-to-particle heat transfer 气体颗粒之间的换热
gas-touched perimeter 气体冲刷周边，气体接触周界
gas tracing experiment 气体示踪试验
gas transfer 气体传递，气体输运
gas transportation 输气
gas trap 凝气阱，气体收集器
gas treating system 气体净化系统，气体处理系统
gas tube boiler 火管锅炉
gas tube counter 充气管计数器
gas tube 充气管，煤气管
gas tungsten arc welding-pulsed arc 钨极气体保护焊-脉冲焊【钨极气体保护焊的变形，焊接采用脉冲电流】，钨极脉冲氩弧焊
gas tungsten arc welding 钨极惰性气体保护焊，钨极气体保护电弧焊
gas turbine and ancillary equipment 燃机及附属设备
gas turbine blade 燃气轮机叶片，燃气透平叶片
gas turbine burning low heating value gas 燃用低热值煤气燃气轮机
gas turbine compressor 燃气轮机压缩机
gas turbine cycle 燃气轮机循环
gas turbine driven generator 燃气轮机驱动的发电机
gas turbine generating set 燃气轮机发电机组
gas turbine generator package 燃气轮机发电机组
gas turbine generator start test 燃气轮机发电机启动试验
gas turbine generator 燃气轮发电机
gas turbine locomotive 燃气轮机车
gas turbine module 燃气轮机组件
gas turbine operating parameter 燃气轮机的运行参数
gas turbine packaged set 燃气轮机发电机组
gas turbine power generation 燃气轮机发电
gas turbine power plant 燃气轮机动力装置，燃气轮机电厂，燃气轮机电站
gas turbine power test code 燃气轮机动力试验标准
gas turbine power unit 燃气轮机动力装置
gas turbine project 燃机项目
gas turbine unit 燃气轮机组［装置］
gas turbine 燃气透平，燃气轮机
gas turbo-alternator 燃气轮发电机
gas turbo-compressor 燃气轮机压气机
gas ultracentrifuge process 气体超离心法
gas ultracentrifuge 气体超离心机
gas-vapor cycle 燃气蒸汽循环
gas-vapor 燃气蒸汽，气汽的
gas velocity at stack exit 在烟囱出口的烟气速度
gas velocity monitor 烟气流速监视仪表
gas velocity 气流速度
gas venting 放气
gas vent 气体出口，通气口，排气孔
gas volume 气体容积
gas wall 气壁【预应力混凝土容器】
gas warfare 毒气战
gas washer 湿式除尘器，气体洗涤器，气体净化器
gas washing bottle 洗气瓶
gas wash tower 洗气塔
gas waste processing system 气体废物处理系统
gas-water ratio 气水比
gas-water separating drum 汽水分离筒
gas-water separator 气水分离器
gas-water surface 气水交界面
gas-wave analogy 气浪模拟法
gas weight flow 燃气重量流量
gas welder's goggles 气焊眼镜
gas welder 气焊机
gas welding unit 气焊工具
gas welding 气焊，气焊接

gasworks　煤气厂
gas　气，气体，煤气，汽油，烟气
gatage　门叶开度
gate actuating rod　导叶操作杆
gateage indicator　导叶开度指示计，闸门开度指示器
gateage　导叶开口面积
gate amplifier　选通脉冲放大器
gate arch　门拱
gate area　导叶面积，闸门面积
gate arm　支臂架
gate bay　闸门室，闸门段
gate bias　控制极偏置
gate blade　锁紧叶片
gate chamber　传达室，闸门井，闸门室，闸室
gate charge　选通电极充电，门控充电，门电荷
gate circuit　门电路，选通电路
gate closure　闸门关闭
gate contact　门触点，栅极接点
gate control house　闸门控制室
gate controlled switch　门脉冲控制的开关
gate current　选通电流，门电流
gate dam　闸门式坝，闸坝
gate distributing valve　导叶分配阀
gate diversion works　闸门分水设备
gate dogging device　导叶锁定装置
gated outlet　有闸门的泄水孔，有闸泄水口
gate driver　门驱动器
gate drive　可控硅转换开关【控制棒用】
gated spillway　有闸门控制溢洪道
gated weir　装有闸门的堰
gate feed hopper　带挡板的装料斗
gate flap　门叶，舌瓣
gate flow indicator　活门流量指示器
gate generator　门脉冲发生器，时钟脉冲发生器
gate groove　门槽
gate gudgeon　闸门活页
gate guide vane journal　导叶轴颈
gate guide　闸门导轨
gate height　导叶高度
gate hoist　启门机
gate hook　耳轴
gatehouse　警卫站，传达室，警卫室，门房，启门机室，闸门控制室，门卫房，保安室
gate housing　闸门小室
gate inverter　门脉冲放大逆变器
gatekeeper's lodge　传达室，门房
gate latching device　闸门锁定设备
gate leaf　门叶，整流叶栅
gate leakage test　闸门漏水试验
gate leakage　导叶漏水量，闸门漏水量
gate level simulation　门级仿真，门级模拟
gate lifting device　闸门启闭装置，启门机，闸门提升装置
gate linkage　导叶传动装置
gate link pin　导叶连杆销
gate lock　导叶锁定装置
gate lowering mechanism　闸门降落机构
gate off　断开
gate on　导通，门通
gate opening　闸门孔，闸门开度，闸孔开度，导叶开度
gate operating mechanism　导叶操作机构，闸门操作机构，闸门操作机械
gate operating ring　导叶操作环，调速环
gate pier　门墩
gate pivot　闸门枢轴
gate pole　门极
gate position limiter　导叶位置限制器
gate position　导叶开度
gate post　门柱
gate pulse　门冲，选通脉冲
gate rail　门轨
gate recess　门槽，闸门凹座
gate seal　闸门止水
gate seat　门槛，门座
gate shaft　闸门井
gate signal　门信号，选通信号
gate sill　门槛，闸门槛
gate slot　门槽，闸门槽
gate terminal　门极端子
gate throttle　节流门
gate tower　启门塔
gate trigger circuit　控制门触发电路
gate trigger　控制门触发器
gate tube　选通管，闸门管，门电路管
gate turn-off switch　控制极关断开关
gate turn-off thyristor　门极可关断晶闸管
gate turn-off　闸门电路断开，矩形脉冲断开
gate-type crane dynamic weighing　门座起重机动态称重
gate-type hydrant　阀门式给水栓，阀门式消防龙头
gate unit　选通部分
gate valve with fast-acting spring actuator　弹簧驱动快关闸阀
gate valve　闸阀，平板阀，滑阀，闸板阀，插板阀，滑门阀，闸门，门阀
gate vault　闸门室
gate vibration　闸门振动
gate voltage　栅压，触发电压
gateway processor　网间连接处理机
gateway　门路，网间连接器，网关，门电路
gate wedge　门楔
gate well　闸门坑，闸板井
gate width　选通脉冲宽度，门选通脉冲宽度
gate winding　选通线圈
gate with flap　带舌瓣闸门
gate　门，闸门，门电路，选通电路，闸板，阀门，出入口
gathering ground　集水区，流域
gathering system　采集系统
gather money　集资
gather　采集，搜集，收集
gating impulse　门脉冲，选通脉冲
gating matrix　控制矩阵
gating pulse　门选通脉冲，门控制脉冲
gating signal　选通脉冲，选通信号
gating　闸，选通，开启，控制，选择作用
GATT(general agreement on tariffs and trade)　关税与贸易总协定，关贸总协定
gaugeable　可计量的

gauge block 块规
gauge board 仪表盘，测量仪表盘，标准尺，样板，模板规准尺
gauge bonding agent 应变片黏合剂
gauge bush 测量用轴瓦
gauge cock 量水旋塞，试水位旋塞，表计旋塞
gauged arch 规准拱【指用规准砖建筑成的拱】
gauge datum 测站基面，水位基点，水位零点
gauged brick-work 清水砖工
gauged brick 规准砖
gauged drainage basin 施测流域
gauged water shed 施测流域
gauge factor 仪器灵敏度，应变系数
gauge feeler 量规
gauge float 水位指示浮标
gauge glass 量水玻璃管，玻璃示位表，水位表玻璃，水位计玻璃，液位指示玻璃管，雨量杯
gauge lath 挂瓦条
gauge length 标距长度，标距，标点距离
gauge line pillar 桩式水尺
gauge line 压力表管路，计量管，规线
gauge mortar 速凝砂浆
gauge of spiral 螺旋的螺距
gauge piping 表计管系
gauge plug 塞规
gauge pressure gauge 标准压力计
gauge pressure 表压，计示压力，表压力，指示压力
gauge reading 水尺读数
gauger 计量器，测规
gauge staff 石膏面层
gauge stuff 速凝砂浆
gauge tap 仪表阀
gauge terminal 应变片接头
gauge tube 表计管
gauge vale 仪表阀
gauge water 定量水
gauge well 测井
gauge zero 测站零点，水尺零点
gauge 标准尺寸，量器，量计，轨距，标准尺，量规，量表，测量，计器，仪表，表
gauging cable car 测流缆车
gauging foot-bridge 水文测验便桥
gauging hole 测孔
gauging section 测流断面
gauging site 测流地址
gauging station density 水文站网密度
gauging station equipment 水文站设备
gauging station 测流站，水位站
gauging system 度量系统
gauging tank 量水箱
gauging weir 量水堰
gauging 喉结比【喉部】，校准，测量，计量，测量检验，测定
gault clay 泥灰质黏土
gauntlet （电焊工的）金属手套，长手套，防护手套
gauntree 桥式吊车
gauntry 台架，构台
gausistor 磁阻放大器
gaussage 高斯数
Gauss curve 高斯曲线
Gauss distribution 高斯分布
Gauss elimination 高斯消去法
Gaussian diffusion 高斯扩散
Gaussian distribution 高斯分布，正态分布
Gaussian plume distribution 高斯烟羽分布
Gaussian plume equation 高斯羽流方程
Gaussian plume model 高斯羽流模式
Gaussian plume parameter 高斯羽流参数
Gaussian plume 高斯羽流
Gaussian white noise 高斯白噪声
Gauss theorem 高斯定理
gauss 高斯
gauze boundary 网状边界
gauze brush 铜丝布电刷
gauze screen 金属丝网，（风洞）阻尼网
gauze strainer 网式过滤器，丝网粗滤器
gauze 网，纱布，金属丝网，线网，薄纱，金属网
gavel 槌，锤子
gazebo 眺望台，露台，塔楼，信号台
GBP(gain-bandwidth product) 增益频带宽积
GBP(Great Britain Pound) 英镑
GBX (gear box) 齿轮箱
GB （中国）国家标准
GCA 汽轮机和给水停运期间的保养系统【核电站系统代码】
GCC(general conditions of contract) 通用条款，一般条款
GCF(gross capacity factor) 毛容量系数
GCFR(gas cooling fast reactor) 气冷快堆
GC(gas compressor) 气体压缩机
GC(generator cooling) 发电机冷却
GCR(gas cooled reactor) 气冷反应堆
GCR(general cargo rate) 普通货物运价
GCR(grid control room) 网控室
GCR(group code recording) 成组编码记录
GC/S(gigacycles per second) 千兆周每秒
GCT 汽轮机旁路系统【核电站系统代码】
GC 高压调门控制【DCS 画面】
GDC(general design criteria) 总设计准则
GDC(graphic display control) 图形显示控制
GDC(gross dependable capacity) 毛保证容量
GDCS(gravity driven cooling system) 重力驱动冷却系统
GDDS(gamma dose detector system) 伽马剂量探测系统
GD(gate driver) 门驱动器
GDG(generation data group) 分级数据组
GD(grown-diffused) 扩散的
GDMS(generalized data management system) 广义数据管理系统
GDO(grid-dip-oscillator) 栅陷振荡器
GDP(gross domestic product) 国内生产总值
GDS(generalized data structure) 广义数据结构
gear backlash 齿轮啮合背隙，齿侧隙
gear bearing not driving end 齿轮轴承非驱动端
gearbox bearing 齿轮箱轴承
gearbox casing 齿轮箱体，变速箱体
gearbox lubricant oil 齿轮箱润滑油
gearbox manufacturer 齿轮箱制造厂商

gearbox nominal power 齿轮箱额定功率
gearbox oil volume 齿轮箱润滑油体积
gearbox oil 齿轮箱油
gearbox ratio 齿轮箱速比，传动比，齿轮箱减速比，齿轮箱变比
gearbox shock absorber 齿轮箱减震器
gearbox 变速箱，齿轮箱，减速器
gearcase 齿轮箱
gear chamfering 齿轮倒角
gear chattel 轴承颤动
gear clutch 齿轮离合器
gear composite 齿轮组合
gear coupling 齿轮联轴节，齿轮联轴器
gear cutting machine 齿轮切削机
gear cutting operation 切齿操作
gear decreaser 减速齿轮，降速齿轮，降速器
gear down 齿轮减速，换低挡，起落架放下
gear-driven exciter 齿轮传动励磁机
gear-driven pump 齿轮传动泵
gear driving 齿轮传动
geared bar-bender 齿轮式钢筋弯折机
geared brake motor 齿轮传动制动马达
geared-down 用齿轮减速的
geared exciter 齿轮传动励磁机
geared motor 带减速齿轮的电动机，齿轮传动电动机
geared non-condensing turbine 齿轮传动背压式汽轮机
geared ratio 速比，传动比
geared ring 齿圈【大牙轮】
geared rotary switch 齿轮旋转开关
geared turbine 齿轮传动式汽轮机，齿轮接汽轮机
geared turbogenerator 齿轮传动汽轮发电机
geared wind turbine 齿轮驱动风电机组
geared 用齿轮传动的
gear efficiency 传动装置效率
gear fan 齿轮箱冷却风扇
gear grease 齿轮润滑脂
gear head 齿轮头【电机】
gear hobbing machine 滚齿机
gear hoist 齿轮式起重机
gear housing 齿轮箱
gear hub 齿轮毂
gear increaser 齿轮增速器
gearing down 减速传动
gearing in 啮合
gearing oil 齿轮油
gearing up 增速传动，齿轮增速，增速装置
gearless motor 直接传动电动机，无齿轮电动机
gearless WTGS 无齿轮箱式风电机
gear lever 变速杆
gear lubricant 齿轮润滑油
gear machining 齿轮加工
gear motor 齿轮油马达，齿轮传动电动机，齿轮马达
gear oil pump high-speed 齿轮油泵高速
gear oil pump low-speed 齿轮油泵低速
gear oil pump 齿轮油泵
gear oil 齿轮油
gear pair with parallel axes 平行轴齿轮副
gear pair 齿轮副
gear pump 齿轮泵【旋转型泵】
gear rack 齿条
gear ratio order 设置传动比指令
gear ratio 齿轮速比，传动比，齿轮比，齿数比，齿轮传动比
gear reducer 减速齿轮，减速装置，齿轮减速器
gear-reduction drive 齿轮变速装置
gear-reduction unit 齿轮减速器
gear ring 齿圈
gear set 齿轮组
gearshaft 齿轮轴
gear shift cover 变速器盖
gear shifter 变速器
gear shift housing 变速箱
gear shifting technology 变速技术
gear shift 换挡，变速，变速机构
gear speed reducer 齿轮减速器
gears with addendum modification 变位齿轮
gear-tooth vernier 齿距卡规
gear train 齿轮系，轮系
gear transmission 齿轮传动
gear type coupling 齿轮联轴器
gear type motor 齿轮式马达
gear type pump 齿轮泵
gear up 加速传动，齿轮增速，增速传动
gear water pump 齿轮水泵
gear wheel 齿轮，大齿轮
gear 齿轮，传动装置，装上齿轮，用齿轮连接，装置，传动机构，机构
GECA(GEC Alsthom) 通用电气阿尔斯通公司
gecalloy 铁粉磁芯用镍铁合金
GEC(General Electric Corporation) 通用电气公司
gedendum circle 齿根圆
Ge detector 锗探测器
gehlenite 钙黄长石
Geiger counter 盖格计数器
gel adsorber 凝胶吸附剂
gelatination 凝胶，凝胶化
gelatin dynamite 明胶炸药
gelatinization 凝胶化
gelatinous precipitate 凝胶状沉淀物
gelatinous structure 凝胶结构
gelatinous 胶凝的
gelatin 明胶，动物胶
gelation 凝胶，凝胶化
gel coated 胶衣
gel coat 表面涂漆
gel filtration 凝胶过滤
gel layer 凝胶层
gel-sphere-pac process 凝胶球法
gel type resin 凝胶型树脂
gel type strong basic anion exchange resin 凝胶型强碱性阴离子交换树脂
gelular resin 凝胶型树脂
gel 冻胶，胶质体，凝胶
gemel arch 对拱
gemmho 微姆欧
gender （语言的）性，性别
generic substation event 通用变电站事件

general acceptance 普通承兑
general administration 管理总局
general administrative expense 一般管理费
general administrative expensive 总行政管理费
general affairs engineering analysis 事务工程分析
general affairs office 总务室
general affairs section 总务科
general agency 一般代理，总代理人
general agent 总代理人
general agreement on commodity 商品总协定
general agreement on tariffs and trade 关税与贸易总协定
general air change 全面通风
general analysis coal sample 分析煤样
general analysis test sample of coal 一般分析试验煤样
general analysis 全面分析
general and selective calling 群呼和选呼
general arrangement drawing 总布置图，总平面图
general arrangement of pipeline 管道总布置
general arrangement of reactor room 堆舱总体布置
general arrangement of works 工程总体布置
general arrangement plan 总平面布置图，总布置图
general arrangement 总（平面）布置（图），通用装置，总体布局，总体布置
general articles 一般性条款，总则，一般条款
general assembly drawing 总装配图
general assembly program 通用汇编程序
general assembly 总装配
general assets 资产总额
general astronomy 普通天文学
general auxiliary building elevators 公用辅助厂房电梯
general auxiliary building 通用辅助厂房
general average counter guarantee （保险）共同海损反担保
general average 共同海损
general base level of erosion 总侵蚀基准面
general boundary condition 广义边界条件
general budget 总预算
general carbon steel 普通碳素钢
general cargo rate 普通货物运价
general cargo terminal 杂货码头
general cargo 普通货物，杂货
general catalogue 公司目录【标书】，总目录
general characteristic 一般特性
general chart of coast 海岸总图
general check-up on the financial work 财务大检查
general circulation 环流，大气环流
general claim agent 索赔总代理人
general clause 通用条款，普通条款
general compiler 通用编译程序
general components 一般金属部件
general comprehensive operating system 通用综合操作系统
general conditions of construction 施工总则，施工概况
general conditions of contract 合同的通用条件，合同的一般条件
general conditions of delivery 一般交货条件
general conditions 通用条款，通用条件，一般条件，一般条款
general conference 大会
general connection diagram 全部设备接线图，总接线图
general construction organization design 施工组织总设计
general contents 总目录
general contract for project 工程总合同
general contractor 总承包人，总承包者
general contract 总承包合同，总承包，总合同
general control panel 总控制盘
general control system 总控制系统
general corporate 一般法人
general corrosion 一般腐蚀，均匀腐蚀
general council 普通理事会
general criterion and rule 一般标准和规则
general damages 一般损害赔偿
general decision 一般决策
general description of construction 构造说明，施工总说明，施工说明（书）
general description 概述，总论，总说明
general design 总体设计
general detuning 一般调解，一般失调
general diagram 总图
general dimension 主要尺寸，总尺寸
general director 总监
general drawing 总图，概图
general-duty 通用的
General Electric Corporation 通用电气公司
general employee training in radiological protection （核电厂的）一般雇员辐射防护培训
general environment 总环境，一般环境
general equilibrium theory 一般（供求）均衡理论
general escape 广义换码
general estimate 概算，总概算
general evacuation 全部撤离，总体排放
general exhaust ventilation 全面通风
general expense 总务费用，综合费用，一般费用，日常开支
general extension 总伸长，全面扩建
general factory 总厂
general feedback servomechanism 一般的带有反馈的伺服机构
general filtering problem 一般过滤问题
general financial inspection 财务大检查
general flowchart 综合流程图
general forecasting 一般预报
general foreman 总工长
general gas law 通用气体定律
general geology 普通地质学
general heliostat 一般定日镜
general income tax 一般所得税
general information processing system 通用信息处理系统
general information system 通用信息处理系统

general inspection　一般检查，总清单
general instruction to tenders　投标须知
general interconnecting network　总联络网
general interrogation command　总查询命令
general interrogation　总查询，总召唤
general investigation　普查
generalist　多面手
generality　一般性，普遍性，概括性
generalization　归纳
generalized admittance　广义导纳
generalized aerodynamic force　广义气动力
generalized aerodynamic moment　广义气动力矩
generalized constant　通用常数，标准化常数
generalized coordinates　广义坐标，普通坐标
generalized corrosion　广义的腐蚀
generalized data management system　广义数据管理系统
generalized data structure　广义数据结构
generalized eigenvalue problem　广义本征值问题
generalized fault table　通用故障表
generalized fluidization　广义流态化
generalized force　广义力
generalized Fourier analysis　广义傅里叶分析
generalized heavy gas mode　广义重气体模型
generalized impedance　广义阻抗
generalized inverse matrix　广义逆矩阵
generalized laminar flow　广义层流
generalized load　广义荷载
generalized newtonian model　广义牛顿模型
generalized Ohm's law　普遍化的欧姆定律
generalized phasor　一般化的相图
generalized procedure of slices　广义条分法
generalized programming language　通用程序设计语言
generalized programming　通用程序设计
generalized reaction modulus　综合反应模数
generalized relation　综合关系
generalized routine　广义子程序
generalized simulation language　通用模拟语言
generalized sinusoidal quantity　广义正弦量
generalized sort　通用分类
generalized stochastic matrix　广义随机矩阵
generalized system of preferences　普遍优惠制
generalized thermal force　广义热力
generalized two-phase model　广义两相模型
generalized　广义的，普遍的
general design criteria　总设计准则
general layout and transportation　总图运输
general layout design　总平面设计
general layout drawing　总平面图，总图，装配图，总平面布置图
general layout of construction site　施工总（平面）布置（图），施工总平面图
general layout of pipeline　管道总平面布
general layout of power plant area　电厂厂区总布置
general layout of power plant　厂区总布置，全厂总平面布置图
general layout of power station　电厂总平面布置
general layout of project　工程总体布置
general layout plan of construction organization design　施工组织设计总平面布置图
general layout plan　平面布置总图，总平面布置图
general layout　总布置（图），总平面布置图，总平面图，总图，总体布置
general leveling　连续式平整
general license　一般许可，通用许可证，普通许可证
general light and power distribution　一般照明及配电
general list　总清单
general locking　强制同步
general machining centers　通用加工中心
general maintenance　大修
general management department　总管理部门
general management　综合管理
general manager　总经理
general modeling approach　通用模拟方法
general mortgage　一般抵押，总抵押
general nitrogen storage system　氮气储存系统
general notes to annex tables　附表总说明
General Nuclear Services Inc.　通用核服务公司
general objective　总目标
general object-oriented substation events　通用面向对象的变电站事件
general obligation bonds　普通责任债券
general obligation　基本义务
general offer　一般报价，总报价
general office　总办公室
general operational requirement　普通操作要求
general order notice　一般订货通知
general oscillating control servomechanism　广义振荡控制伺服机构
general overhaul　大修，（设备的）全面检查，全面检修，总翻修
general peripheral control　通用外围控制
general personality right　（法人）一般人格权
general planning　总规划，总体布局，总体规划
general plan　总体规划，总布置图，总图
general plot plan　总图，总平面图，总平面布置图
general port　综合性港口，通用港
general power of attorney　全权委托书
general precaution　一般注意事项
general primary membrane stress　一次综合薄膜应力，初期综合膜应力
general principles　通用原则，一般原则，总则，通则，总纲
general probability　总概论，一般概率
general procedures of bidding　招标基本程序
general procurement　总采购
general programme　总纲，大纲，通用程序
general proportion　总体比例
general protocol　总议定书
general provisions　一般条款，通则，一般规定，总纲，总则
general purpose analog computer　通用模拟计算机
general purpose computer　通用计算机
general purpose digital computer　通用数字计算机

general purpose interface bus 通用接口总线
general purpose keyboard and display control 通用键盘和显示控制器
general purpose language 通用语言
general purpose machine 通用机械
general purpose macrogenerator 通用宏生成程序，通用宏处理程序
general purpose manipulator 通用机械手，万能机械手
general purpose memory 通用存储器
general purpose motor 通用电动机
general purpose pliers 通用钳
general purpose processor 通用处理机
general purpose reactor 通用反应堆
general purpose register 通用寄存器
general purpose relay 通用继电器
general purpose simulation system 通用模拟系统
general purpose timer 通用定时器
general purpose tipper 通用自动倾卸车
general purpose tractor 通用拖拉机，万能拖拉机
general purpose transformer 通用变压器
general purpose valve 通用阀
general purpose 通用目的
general register 通用寄存器
general regulation 普遍监管
general representative 总代表
general requirement 总的要求
general responsible personnel 总负责人
general risk analysis 一般风险分析
general rolling country 一般丘陵地区，缓坡丘陵
general routine 标准程序，通用程序
general rules and regulations 总规定
general rules of construction 施工总则
general rules on loans 贷款通则
general rules 通则，总则
general schedule 总进度
general scour 普遍冲刷，一般冲刷
general service building 总服务楼，附属系统厂房
general service desk 总操作台，通用控制台
general service system 附属系统，公用系统
general single point 普通单分道岔
general site switchboard 厂区配电盘
general sketch 总略图
general slope of basin 流域总坡降
general solution 通解
general specification 通用规范，总规范
general stability criterion 一般稳定判据，一般稳定准则
general state of strain 总的应变状态
general station 总厂，总站
general storage 总仓库，综合库
general stream 主流，自由流
general structure low-alloy steel 普通低合金结构钢
general suitability 总体适合性
general summary of accounts 总决算
general superintendent 总裁，总管，总监
general supply 总电源，普查
general table 综合表
general terms of fire 一般性消防术语
general terms 一般术语
general theory 一般理论
general use 通用
general utility balance 普通天平
general utility function 通用辅助操作
general ventilation 全面通风
general view 综观，总图，大纲，全景，概貌，全视图
general vocabulary 一般术语
general wage level 一般工资水平
general works（factory） 总厂
general yield strength 总屈服强度
general 一般的，普通，通常，总的，总论，综述
generated address 合成地址，形成地址
generated code 形成码，派生码
generated electromotive force 内生电势
generated energy 发电量
generated matrix 生成矩阵
generated output 发电容量，发电出力，发电机输出功率
generated power 发电功率
generated quantity 输出量，被产生的量
generated voltage 感应电压，感应电势
generated 输出的，被产生的
generate 发生，发电，产生，引起
generating auxiliary power 发电厂厂用电功率，厂用电
generating capacity 发电量，发电容量
generating cost 电价，发电成本
generating efficiency 发电效率
generating equipment 发电设备
generating function 母函数，生成函数
generating line 母线
generating operation 发电运行
generating plant 电站，发电厂，发电设备
generating routine 生成程序
generating set shutdown 停机
generating set 发电设备，发电机组
generating station auxiliaries 发电厂辅助设备
generating station efficiency 发电站效率
generating station 发电站，电厂
generating surface 蒸发受热面
generating system 发电系统
generating tube 蒸发管
generating unit 发电机组
generating voltage 发电电压
generation capability 生产能力，发电能力
generation competition 发电竞争模式
generation cost 发电成本［费用］
generation data group 分级数据组
generation deficiency 发电不足
generation deficit 发电功率亏数，发电不足容量
generation engineering 发电工程
generation equipment 发电设备
generation forecast management 发电预测管理
generation load relationship 发电-负载关系
generation load 发电负荷
generation loss 发电损耗
generation market 发电市场
generation of electricity 发电，电力生产

generation of turbulence 紊流生成
generation outage 发电事故，停止发电
generation output 净发电容量，发电出力
generation program of graphic display 画面生成程序
generation rate 产生率，生成率
generation re-scheduling 发电再计划
generation right transfer trading 发电权转让交易
generation schedule 发电计划
generation set 发电机组［设备］
generation station 发电站
generation system 发电系统
generation time 产生时间，每代时间【中子】
generation 代，发生，产生，发电，改进阶段，生成
generator active power 发电机有效功率
generator adjusting arm 发电机调节臂
generator air control system 发电机气体控制系统
generator alignment 发电机找正
generator armature 发电机电枢
generator bearing 发电机轴承
generator braking equipment 发电机制动设备
generator breaker 发电机断路器，发电机负荷开关
generator brush 发电机电刷
generator bus-bar 发电机母线
generator bus 发电机母线
generator capability curve 发电机容量曲线，发电机输出效能曲线
generator capability 发电机能力，发电机容量，发电机出力
generator capacity 发电机容量
generator casing 发电机外壳，发电机机壳
generator choke coil 发电机轭流圈
generator circuit breaker 发电机电路断路器
generator circuit 发电机电路
generator coil 发电机线圈
generator commutation 发电机整流，发电机换向
generator connecting in parallel with system 发电机并网
generator control panel 发电机控制屏
generator control 发电机控制
generator cooling system 发电机冷却系统
generator cooling water device 发电机冷却水装置
generator coupling 发电机联轴节
generator cover 发电机罩
generator current limiter 发电机限流器
generator current 发电机电流
generator cut-out relay 发电机断流继电器，发电机断路继电器
generator cut-out 发电机断流器
generator drying out operation 发电机干燥运转
generator duct 磁流体发电机通道
generator efficiency 发电机效率
generator end 发电机端
generator excitation and voltage regulation system 发电机励磁及电压调节系统
generator excitation circuit 发电机励磁回路
generator exciter 发电机励磁机
generator exciting winding 发电机励磁绕组
generator fan external 发电机外部风扇
generator fan internal 发电机内部风扇
generator field accelerating relay 磁场调节增速继电器
generator field control 发电机磁场控制
generator field decelerating relay 磁场调节减速继电器
generator field 发电机磁场
generator floor 发电机层
generator for pendulum motor 飞摆电动机用发电机
generator frame 发电机机座
generator gas 发生炉煤气
generator gate 门脉冲发生器
generator group 发电机组，发电机群
generator housing 发电机机座
generator hydrogen cooling system 发电机氢供给系统
generator hydrogen leakage rate 发电机漏氢量
generator hydrogen supply system 发电机氢气冷却系统
generator inner cooling water pump 发电机内冷水泵
generator inner cooling water system 发电机内冷水系统
generator inner cooling water tank 发电机内冷水箱
generator input 发电机输入功率
generator load-break switch 发电机负荷中断开关
generator load curve 发电机负荷曲线
generator load factor 发电机负载因数
generator load 发电机负荷，发电机负载
generator loss 发电机损失
generator main leads gallery 发电机主出线廊道
generator main protection 发电机主保护
generator matrix 生成矩阵
generator-motor for pumped storage power station 抽水蓄能电站发电电动机
generator-motor 发电电动机，发电电动两用机
generator neutral grounding equipment 发电机中性接地设备
generator neutral point 发电机中性点
generator no load test 发电机空载试验
generator operating condition 发电机运行状况
generator outage 发电机停机
generator outgoing terminal cubicle 发电机出线小室，发电机出线小屋
generator output 发电机出力，发电机输出功率
generator overload 发电机过负荷
generator panel 发电机盘，发电机屏
generator parallel operation 发电机并列运行
generator pedestal 发电机基座
generator performance test 发电机性能试验
generator pit 发电机机坑，发电机机座
generator power angle 发电机功角
generator power 发电机功率
generator protection for negative sequence current 发电机负序电流保护

generator protection system 发电机保护系统
generator protection 发电机保护
generator protective equipment 发电机保护设备
generator raft foundation 发电机筏形基础
generator rating 发电机额定值
generator reactive power 发电机无功功率
generator reactor 发电机电抗器
generator regulation 发电机调节
generator regulator 发电机调节器
generator room 发电机室
generator rotor shaft 发电机转子轴
generator rotor 发电机转子
generator running in synchronization 发电机与系列并列运行，发电机并网运行
generator running out of synchronization 发电机与系列解列
generator's disconnection with the grid 发电机解列
generator sealing oil system 发电机密封油系统
generator seal oil unit 发电机密封油装置
generator set 发电机组
generator shaft 发电机轴
generator shielding 发电机屏蔽，发电机罩
generator speed range 发电机转速范围
generator speed 发电机转速
generator's synchronization with the power grid 发电机并网
generator star point earthing system 发电机星点接地系统
generator star point 发电机星形接法的中性点
generator stator core 发电机定子铁芯
generator stator terminal lead 发电机定子引出线
generator stator winding 发电机定子绕组
generator stator 发电机定子
generator step-up transformer 发电机升压变压器
generator storey 发电机层
generator support 发电机支架，发电机支座
generator system 发电机系统
generator temperature 发电机温度
generator terminal lead 发电机引出线
generator terminal 发电机端子
generator termination room 发电机出线小室
generator three-phase short circuit test 发电机三相短路试验
generator-transformer block 发电机变压器组
generator-transformer extension unit 发电机变压器扩大单元，发电机变压器组扩大单元
generator-transformer unit protection 发电机变压器组保护
generator-transformer unit 发变组，发电机变压器单元，发电机变压器组
generator-transformer 主变压器，发电机出口端变压器，发电机变压器
generator triode 振荡三极管
generator type test 发电机型式试验
generator unit 发电机组
generator voltage regulator 发电机电压调节器，发电机电压调整器
generator voltage 发电机电压
generator water cooling system 发电机水冷却系统
generator winding 发电机绕组
generator winding thermistor protection 发电机绕组热敏电阻保护装置
generator with disk armature 盘形电枢发电机
generator withstand voltage test 法电机耐压试验
generator withstand 发电机耐电压
generator 发电机，发生器，生成器，振荡器，生产者
generatrix 基体，母线，发生器，发电机
generette 大型发电机线圈寿命试验装置
generic analysis 普遍性分析，广义分析
generic design 通用设计
generic object models for substation and feeder equipment 变电站和馈线设备通用对象模型
generic object oriented substation event 面向通用对象的变电站事件
generic power plant 通用电厂
generic relation 属种关系
generic reliability parameter data bank 通用可靠性参数数据库
generic safety report 通用安全报告
generic seismic response spectrum 通用地震反应谱
generic simulator 通用型仿真机
generic substation event model 通用变电站事件模型
generic substation state event 通用变电站状态事件
generic term 通用术语
genescope 频率特性观测仪
genesis 成因，发生，起源
genetically significant dose 遗传有效剂量，有效遗传剂量
genetic analysis 成因分析
genetic classification 成因分类
genetic dose equivalent 遗传剂量当量
genetic effect of radiation 辐射的遗传效应
genetic effect 遗传效应
genetic radiation effect 辐射遗传效应
genetic sexing mechanism 遗传选性机理
genetics 遗传学
genetic 遗传学的
GEN Ⅲ technology 三代技术
geniometer 晶体量角器
genlock 集中同步系统，强制同步系统，同步耦合器，同步锁相
gentle breeze 轻风，三级风
gentleman's agreement 君子协定
gentle slope 缓和坡度，缓坡，平缓边坡，平缓坡度
gentle wind 和风
gentle 温和的
gently rolling country 缓丘地区
gently sloping surface 缓坡面，缓斜面
GEN 发电机【DCS 画面】
geoanticline 地背斜
geobotany 地植物学
geocentre 地心
geochemical cycle 地球化学循环
geochemistry 地球化学
geochronic geology 地史学

geochronology 地质年代学
geodata 地理数据
geodesic method 短程线法，测地线法
geodesic satellite 大地测量卫星
geodesic 测地线，短程线，测大地线，大地测量学的
geodesy 测地学，大地测量，大地测量学
geodetical control 大地测量控制
geodetical datum 大地基准点
geodetic astronomy 大地天文学
geodetic azimuth 大地方位角
geodetic base line 大地测量基线
geodetic book 大地测量手簿
geodetic chain 大地测量控制锁
geodetic circle 大地圆
geodetic construction 轻型受拉杆系结构
geodetic control 大地测量控制
geodetic coordinates 大地坐标
geodetic curve 大地测量曲线
geodetic datum 大地基准点
geodetic latitude 大地纬度
geodetic leveling 大地水准测量
geodetic level 大地水准仪
geodetic longitude 大地经度
geodetic network 大地控制网
geodetic point 大地点
geodetic position 大地位置
geodetic satellite 大地测量卫星
geodetic station 大地点
geodetic surveying 大地测量（学），测量统计
geodetic survey station 大地测量点
geodetic survey 大地测量
geodetic theodolite 大地经纬仪
geodetic triangulation 大地三角测量
geodetic 大地测量学的，测地学的
geodimeter 光电测距计，光电测距仪
geodynamics 地球动力学
geo-electric measurement 地电测量
geognosy 地球构造学
geographer 地理学家
geographical condition 地理条件
geographical coordinates 地理位置坐标
geographical distribution 地理分布
geographical drilling at location of WTG 机位钻探
geographical element 地理要素
geographical horizon 地平线
geographical information system 地理信息系统
geographical isolation 地理隔离
geographical latitude 地理纬度
geographical location 地理位置
geographical longitude 地理经度
geographically load-shedding 分区减载，按地区减负荷
geographical map 地图
geographical position 地理位置
geographical system connection diagram 电力系统地理接线图
geographical 地理学的，地理的，地区的
geographic and geomorphic conditions 地形地貌条件
geographic coordinates 地理坐标
geographic distribution 地理分布
geographic location 地理位置
geographic meridian 地理子午圈
geographic region 地理区
geographological map 地貌图
geography 地理学，地貌
geogrid 土工格栅
geoguide 岩土指南
geohydrological condition 水文地质条件
geohydrological environment 水文地质环境
geohydrology 地下水水文学，水文地质学
geoidal rise 大地水准面隆起
geoidal surface 大地水准面
geoid 大地水准面，地球形体
geoisotherm 等地温线
geokinetics 地球运动学
geological action 地质作用
geological age 地质年代
geological anomaly 地质变态
geological body 地质体
geological chronology 地质年代学
geological climate 地质气候
geological column 地质柱状图
geological compass 地质罗盘
geological condition 地质条件
geological cycle 地质循环
geological disposal 地质处置
geological distribution 地质分布
geological drilling and sampling 地质钻探与取样
geological drilling 地质钻探
geological engineer 地质工程师
geological erosion 地质侵蚀
geological examination 地质勘测［勘查，勘察］
geological exploration 地质勘测［勘探，勘查，勘察］
geological factor 地质因素
geological formation 地质建造，地质构造
geological hammer 地质锤
geological horizon 地质层位
geological investigation 地质调查
geological log of drill-hole 钻孔地质剖面
geological log 地质记录，地质柱状图
geological map 地质图
geological period 地质时期
geological profile 地质剖面
geological prospecting 地质勘探
geological report 地质报告
geological section 地质剖面
geological setting 地质背景，地质结构
geological store （核废物）地质贮存
geological structure 地质构造
geological succession 地质顺序
geological survey map 地质调查图
geological survey report 地质勘测报告
geological survey 地质测绘，地质调查
geological time scale 地质时标
geological time 地质时期
geologic body 地质体
geologic disposal concept 地质处置概念，地质处置方法
geologic disposal 地质处置

geologic drawing 地质图
geologic fracture 地质断裂
geologic log 地质测井记录
geologic medium 地质介质
geologic norm 地质标准矿物分类
geologic rift 地质断裂
geologic section 地质剖面
geologic time 地质年代
geologist 地质学家
geology 地质，地质学，地质情况
geomagnetic electrokinetograph 地磁动电测流计
geomagnetic field 地磁场
geomagnetic survey 地磁测量
geomagnetic variation 地磁变化
geomagnetism 地磁，地磁学
geomechanics 地质力学
geomembrane 土工（薄）膜，土工隔膜
geometric 几何学的，几何体
geometric accuracy 几何准确度
geometric airfoil 几何翼型
geometrical boundary 几何边界
geometrical buckling factor 几何曲率因子
geometrical buckling 几何曲率，几何形状翘曲
geometrical capacity 几何容量
geometrical concentration ratio 几何聚光率
geometrical configuration 几何布置，几何构形，几何形状
geometrical control 几何控制
geometrical edge 几何边界
geometrical error 几何误差
geometrical factor 几何形状因数，几何系数
geometrical figure 几何图形
geometrically safe system 几何安全系统
geometrically safe 几何形状安全的
geometrically similar model 几何相似模型
geometrical mean 几何平均值
geometrical model 几何模型
geometrical neutral line 几何中性线
geometrical position 几何位置
geometrical progression 几何等比级数，几何级数
geometrical property 几何特性
geometrical relationship 几何关系
geometrical safe container 几何安全容器
geometrical safe evaporator 几何安全蒸发器
geometrical safe 几何形状安全的
geometrical significance 几何意义
geometrical similarity 几何相似
geometrical stability 几何稳定性
geometrical 几何的
geometric angle of attack 几何迎角
geometric attenuation 几何衰减
geometric beam length 几何射线长度
geometric blurring effect 几何模糊效应
geometric boundary condition 几何边界条件
geometric buckling 几何形状翘曲
geometric center 几何中心
geometric centroid 几何中心，几何形心
geometric chord of airfoil 几何弦长
geometric concentrating ratio 几何聚光比
geometric cross-section 几何截面
geometric data of cascade 叶栅的几何参数
geometric dimension 几何尺寸
geometric discontinuity 几何不连续性
geometric distortion in hydraulic model 水工模型几何畸变
geometric drawing 几何图形
geometric factor 几何因子
geometric height 几何高度
geometric incidence 几何迎角
geometric law of reflection 几何反射定律
geometric leading edge 几何前缘
geometric mean distance 几何均距
geometric mean size 平均几何尺度
geometric mean 几何等比中项，等比中项，几何平均值，几何平均数
geometric power diagram 相量功率图，复功率图，矢量功率图
geometric progression 几何级数，等比级数
geometric proportion 等比（例），几何比（例）
geometric scale factor 几何缩尺因子
geometric scale model 几何缩尺模型
geometric series 几何级数，等比级数，几何形状
geometric similarity 几何相似，几何相似性
geometric stairs 螺旋形楼梯
geometric stiffness 几何刚度
geometric twist 几何扭转
geometric unsharpness 模糊的几何形状【射线照相】，几何不清晰度
geometric variation 几何变动
geometry control 几何控制
geometry factor 几何因子
geometry of interface 界面几何形状
geometry 几何形状，几何尺寸，几何学，几何外形
geomorphic accident 地貌突变
geomorphic blow 地貌渐变
geomorphic element 地貌要素
geomorphic feature 地貌，地理特征
geomorphic geology 地貌学
geomorphic occurrence 地貌突变
geomorphic profile 地貌剖面图
geomorphography 地貌叙述学
geomorphology 地貌，地形学，地貌学，地球形态（学）
geomorphy 地貌
geopause 地球同步卫星
geophone 地震测波器，地震检波器
geophysical exploration 地球物理勘探，地球物理探测
geophysical investigation 地球物理勘探
geophysical logging 地球物理测井
geophysical method 地球物理方法
geophysical log 地球物理测井记录
geophysical prospecting 地球物理勘探，物探
geophysical seismic exploration 地球物理地震探测
geophysical vortex 地球物理旋涡
geophysical 地球物理学的
geophysicist 地球物理学家
geophysics 地球物理学，地域物理学，地质

力学
geopolymer 土工聚合物
geopotential height 位势高度
geopotential 地球重力势
geopressed geothermal electricity generation 地压地热发电
geoscience 地球科学
geosphere 地圈，陆界，岩石圈
geostationary operational environmental satellite 地球同步环境卫星
geostationary orbit （通信卫星的）静止轨道
geostationary satellite 地球静止卫星，地球同步卫星，通信卫星
geostenogram 地理速测图，草测图
geostenography 地理速测
geostrophic acceleration 地转加速度
geostrophic advection 地转平流
geostrophic balance 地转平衡
geostrophic current 地转风气流，地转偏向流
geostrophic departure 地转偏差
geostrophic drag coefficient 地转阻力系数
geostrophic equilibrium 地转平衡
geostrophic flow 地转风气流
geostrophic force 地转力
geostrophic Richardson number 地转理查森数
geostrophic shear 地转切变
geostrophic transport 地转运输
geostrophic wind field 地转风场
geostrophic wind height 地转风高度
geostrophic wind vector 地转风矢量
geostrophic wind velocity 地转风速
geostrophic wind 地转风
geosurvey 大地测量
geosynchronous meteorological satellite 地球同步气象卫星
geosynclinal axis 大向斜轴，地槽轴
geosynclinal subsidence 地槽沉陷
geosynclinal synclinorium 地槽内复向斜
geosynclinal system 地槽系
geosynclinal 地槽的，地向斜的
geosyncline chain 大向斜山脉，地槽山链
geosyncline 地槽，地向斜
geotechnical data 岩土数据
geotechnical engineering report 岩土工程勘察报告
geotechnical engineering 土工，土力学工程，岩土工程
geotechnical engineer 土工工程师，岩土工程师，土力工程师
geotechnical erosion 岩土侵蚀
geotechnical hazards 岩土危害
geotechnical investigation and analysis 岩土勘察分析
geotechnical investigation report 岩土工程勘察报告，地质调查报告
geotechnical investigation 岩土工程勘察，地质勘测查［察］
geotechnical laboratory 岩土实验室
geotechnical layering 岩土分层
geotechnical map 岩土工程图
geotechnical nature 岩土性质
geotechnical process 土工处理方法
geotechnical record 岩土记录
geotechnical report 岩土工程勘察报告
geotechnical slope 岩土斜坡
geotechnical stratified 岩土分层
geotechnical test 土工试验
geotechnical tool 土工器具
geotechnical 岩土工程技术的，地质技术的
geotechnics 土工技术，土工学，地质技术学，土力学，岩土工程，地质技术
geotechnique 土力学，土工学，地质技术学，土工技术，岩土工程，地工技术
geotechnology 岩土工程学，地下资源开发工程学，岩土技术，土工技术，土工学，工艺地质学
geotectonic geology 大地构造地质学
geotectonic map 地质构造图
geotectonics 大地构造，大地构造学，大地构造地质学
geotectonic valley 大地构造谷
geotectonic 大地构造的，地壳构造的
geotextile fabrics 土工织物
geotextile 土工织物
geothermal area 地热地带，地热区
geothermal electric plant 地热发电厂［站］
geothermal electric power station 地热发电站
geothermal energy conversion 地热能转换
geothermal energy 地热，地热能
geothermal fluid 地热流体，地热田
geothermal generating equipment 地热发电设备
geothermal gradient 地内增热率，地热梯度，地热增温率
geothermal metamorphism 地热变质作用
geothermal mixture condenser 地热混合凝汽器
geothermal power generation by dry-heat rock 高温热岩地热发电
geothermal power generation with flash cycle system 扩容法地热发电
geothermal power generation with total flow system 全流式地热发电
geothermal power generation 地热发电
geothermal power plant 地热发电厂［站］，地热电站
geothermal power station 地热发电厂［站］，地热电站
geothermal power system using binary cycle 双循环地热发电系统
geothermal power using steam flashed from hot brine 闪蒸地热发电【热卤水】
geothermal prospecting 地热勘探
geothermal reservoir 地热水库，地热资源
geothermal steam plant 地热蒸汽发电厂
geothermal steam turbine 地热汽轮机
geothermal steam 地热蒸汽
geothermal survey 地热调查
geothermal temperature increment ratio 地热增温率
geothermal vapor 地热蒸汽
geothermal water anticorrosion 地热水防腐
geothermal water reinjection 地热水回灌
geothermal water scale prevention 地热水防垢

处理
geothermal water 地热水，地下热水
geothermal well drilling technique 地热井钻探技术
geothermal well 地热井
geothermal 地热的
geothermic gradient 地温梯度
geothermics 地球热学，地热学
geothermo-electric plant 地热电站
geothermometer 地热计，地温计
geothermy 地球热学，地热学
GEPB(Guangdong Environmental Protection Bureau) 广东省环保局
germanium diode 锗二极管
germanium film 锗薄膜
germanium mesa transistor 锗台面型晶体管
germanium photocell 锗光电管
germanium photoconductor 锗光电导体
germanium rectifier 锗整流器
germanium semiconductor triode 锗半导体三极管
germanium transistor 锗晶体管
germanium triode 锗三极管
germanium 锗
germproof paint 抗菌漆
gesso 石膏粉
get a contract 得到合同
get adrift 脱节
get consent of 取得……的同意
get rid of 清除
get rusty 生锈
getter action 除气作用
gettered kernel 加吸氧剂的芯核
getter pump 抽气泵
get word from string 从行取字
GEV 输电系统【核电站系统代码】
GEW 主开关站超高压母线（400/500kV）配电装置【核电站系统代码】
GEX 发电机励磁和电压调节系统【核电站系统代码】
geyser 热水锅炉，热水器，水加热器
g-factor g因数，朗德因子
g-factor of atom or electron 原子或电子的g因数
g-factor of nucleus or nuclear particle 原子核或核子的g因数
GFCI(ground fault circuit interrupter) 接地故障电路中断，接地故障电流漏电保护器
GF(glass fiber) 玻璃纤维
GFR(glass fiber reinforced) 用玻璃纤维加强的
GFRP(glass fiber reinforced plastics) 玻璃钢，玻璃纤维增强塑料
GFR 汽轮机调节油系统【核电站系统代码】
gate operating platform 闸门操作平台
gearing 传动装置，齿轮装置，啮合，（齿轮）传动
GGPC(Guangdong General Power Company) 广东省电力总公司
g=gram 克
GGR 汽轮机润滑、顶轴、盘车系统【核电站系统代码】
Ghana National Fire Service 加纳国家消防局
GHC 加纳塞地
GHE 发电机密封油系统【核电站系统代码】
GHG(greenhouse gas) 温室气体
GHI(global horizontal irradiance) 总（水平）辐射
ghost echo 虚假回波
ghost effect 寄生效应，幻想效应
ghost line 重影，叠影【射线底片】
ghost point 起泡点
ghost pulse 虚假脉冲，寄生脉冲
GHR(gross heat rate) 总热耗率，毛热耗率
GHV(gross heating value) 总发热量
GHz=gigahertz 千兆赫兹
giant brain 大型电脑
giant nuclei 巨核
giant particle 巨大粒子
giant pulse 窄尖大脉冲，巨脉冲
giant resonance 巨共振
giant size 特大号
giant source 巨源
giant transistor 巨型晶体管
gib arm of crane 吊车起重扒杆，吊车臂
gibbet 撑架，吊机臂，起重杆
gibbsite 水铝
Gibbs 吉布斯【吸收单位】
gib crane 悬臂起重机，挺杆起重机，扒杆起重机
gib hoist 悬臂式起重机
gibrous glass 玻璃丝
gib 扁栓
GIC(gas insulated cable) 压气绝缘电缆
Gieseler fluidity 吉氏流动度
gigabit ethernet 千兆位以太网
gigabyte ethernet 千兆位以太网
gigacycle 千兆周，千兆赫
giga-electron-volt 吉电子伏，十亿电子伏
gigahertz 千兆赫
GI(gas insulation) cable 气体绝缘电缆
gigawatt hour 十亿瓦时
gigawatt 千兆瓦，十亿瓦，吉瓦，兆千瓦
giga 千兆，十亿
GIGO(garbage-in garbage-out) 杂乱输入杂乱输出
gig 吊桶
gilbert 吉伯【磁势单位】
gilding 镀金，装金
gill cooling 散热片冷却
gilled pipe 肋片管
gilled radiator 翅片式散热器
gilled ring type economizer 圆环肋片管省煤器
gilled tube 肋片管，鳍片管
gilled 肋片的
gill 加强筋，鳃，加肋片，肋片，鳃状物，散热（翅）片
gimbal lock 框架自锁
gimbals 平衡环，平衡架
gimbal type suspension （提升臂）万向接头悬挂装置
gimbal 万向支架，平衡环，万向接头，平衡架，（罗盘的）常平环
gimlet 手钻，螺丝锥，手锥
gin-block 单轮滑车

gin pole derrick 桅杆式起重机
gin pole 安装拔杆，起重拔杆，三脚起重机，安装用起重架，储罐支柱
gin wheel 起重滑轮
gin 三脚起重机
girder and beam connection 主次梁连接
girder and connection 主次梁连接
girder beam 梁式桁架
girder bearing plate 梁垫板
girder bent 梁式排架
girder bridge 梁式桥，桥梁桥
girder flange 大板梁翼缘
girder frame 横梁，桁架梁
girder framing into column 梁与柱的连接
girder grillage 钢梁格床，梁式格排
girder head 枪头
girderless floor 无梁楼板
girder main beam 主梁
girder pole 桁架杆，桁架杆柱
girder stay 梁撑
girder structure 大梁式结构，梁式结构
girder truss 桁架梁，梁构桁架
girder 桁架，大梁，纵梁，梁，钢桁的支架，槽钢，中横担，主梁
girdle 环带
gird 圈梁，横梁，保安带，护环，围梁
girth baffle plate 环形导流板
girth joint 环焊接头，环向接头，环形接缝
girth rail 圈栏
girth seam 环焊焊缝，环缝，周向焊缝
girth superheater 包箍过热器【卧式回火管锅炉】
girth welding 环形焊，环缝焊接
girth 圈梁，围绕
girt 箍梁，围梁
GIS(gas-insulated metal-enclosed switchgear) 气体绝缘金属封闭开关设备，气体绝缘开关设备，封闭式组合电气柜
GIS(gas insulated switchgear) 气体绝缘开关设备，气体绝缘金属封闭开关设备，气体绝缘全封闭组合电器【由断路器、隔离开关、接地开关、互感器、避雷器、母线、连接件和出线终端等组成】
GIS(general information system) 通用信息处理系统
GIS(geographic information system) 地理资讯系统，地理信息系统
GIS(Global Information System) 全球情报系统
give an impact to 对……发生影响
give an order 发布命令
give a week's notice 在一星期前通知
give bond 担保
give... influence on... 对……有影响
given dose 施予剂量
given period 特定期
given size 规定尺寸
given that 假定，设，已知，给定，在……条件下
given value 给定值，已知值
given...,it follows that... 已知……，则……【可以得出……】
given 指定的，假设的，假定的，考虑到【表示原因】，倘若，假定
give priority to 优先考虑
give rise to 引起，使发生
give serious consideration to 给予认真考虑
give tacit consent to 默认
give up a business 停业
give up his position 放弃他自己的立场，让位
give-up 放弃，中止
glacial abrasion 冰川磨蚀
glacial acetic acid 冰醋酸
glacial action 冰川作用
glacial age 冰期
glacial debries 冰碛物
glacial drift 冰碛物
glacial epoch stage of freezing 冰期
glacial erosion 冰川侵蚀
glacial karst 冰川喀斯特
glacial period 冰期
glacial river 冰河
glacial snout 冰川末端
glacial stream 冰河
glacial tongue 冰舌
glacial 冰川的，冰河的
glaciated 冰覆盖的
glaciation 冻结成冰，冰川作用，冰蚀
glacier bands 冰川带
glacier burst 冰坝溃决
glacier deposit 冰川沉积
glacier stream 冰河
glacier terminus 冰舌
glacier 冰川
glacis 缓斜坡，斜岸，斜堤
glance coal 亮煤，无烟煤，镜煤
gland air extractor 轴封排风机
gland air fan 轴封风机
gland bonnet 密封盖，轴端密封盖
gland box 填料箱，填料函
gland bush 汽封套
gland casing 汽封体，压盖箱
gland chamber 汽封腔室
gland condenser 汽封冷凝器
gland cooler 轴封抽汽冷却器
gland cooling water 盘根冷却水
gland cover 填料压盖，密封套，密封压盖
gland ejector 汽封抽气器
gland end cover 盘根端盖，密封端盖
gland exhauster system 汽封排汽系统
gland flange 密封法兰
gland follower 密封套，密封套衬圈
gland heater 轴封加热器
gland housing 汽封体，压盖箱
gland lead-off system 填料函引漏系统
gland leakage losses 汽封漏汽损失
gland leak-off steam condenser 汽封冷凝器
gland leak-off 填料函引漏，密封套引漏，汽封漏汽
glandless circulating pump 无填料函循环泵
glandless pump 无填料泵，无轴封泵
glandless 无密封垫的
gland nut 压紧螺母，填料函螺母

gland over shroud 围带汽封
gland packing leakage 汽封漏汽，轴封漏汽
gland packing 压盖填料，汽封装置，压盖密封，轴封装置，盘根，轴封填料
gland piping 汽封管路
gland plate 密封盖板
gland pocket 填料函槽，汽封腔室
gland pump 填料函泵，水封泵
gland retainer plate 轴封盖板，轴密封盖
gland ring 汽封环
gland seal condenser 填料密封冷凝器，轴封蒸汽冷凝器
gland seal cooling water 轴封冷却水
gland sealing ring 汽封环
gland sealing steam 轴封蒸汽
gland sealing valve 轴封供汽阀
gland sealing 汽封
gland seal steam condenser 压盖密封蒸汽冷凝器
gland seal 迷宫密封，汽封，轴封
gland seat system 压盖密封系统，轴封系统
gland segment 汽封弧段
gland sleeve 汽封套，汽封套筒
gland spring 汽封弹簧
gland stealing system 轴封系统【汽轮机】
gland steam collecting pipe 汽封蒸汽集气管
gland steam condenser exhauster 轴封冷却器抽汽机
gland steam condenser extraction fan 轴封冷却器抽风机
gland steam condenser fan 轴封冷却器风机
gland steam condenser 汽封冷却器，汽封蒸汽冷凝器，轴封冷却器，汽封蒸汽凝汽器，轴封凝汽器
gland steam connection 汽封蒸汽连接管
gland steam control system 汽封蒸汽调节系统
gland steam control valve 汽封蒸汽调节阀
gland steam cooler drains pipe 轴封冷却器疏水管道
gland steam cooler vents pipe 轴封冷却器排汽管道
gland steam cooler 汽封冷却器
gland steam desuperheater 汽封蒸汽减温器
gland steam exhauster 汽封蒸汽排汽管，汽封抽汽器，轴封抽汽器
gland steam exhaust 汽封排汽
gland steam heater 轴封蒸气加热器，汽封蒸汽加热器，汽封加热器
gland steam pocket 汽封室
gland steam regulator 汽封压力调节器
gland steam supply 汽封供汽
gland steam system 轴封蒸汽系统，压盖汽封系统
gland steam 汽封蒸汽，轴封蒸汽，密封蒸汽
gland strip 汽封片［齿］
gland system 汽封系统
gland top cover 盘根顶盖，盘根压盖
gland-type joint 压盖型连接
gland 填料盖，密封垫，填料箱，密封压盖，密封装置，汽封装置，汽封
glare-free 无眩光
glare ice 薄冰
glare on the panels 表盘上的眩光
glare 眩光
glaring glittering 闪光
glass 玻璃，观察窗
glass-backed mica tape 玻璃云母带
glass ball 玻璃球
glass-banded commutator 玻璃丝绑带换向器
glass bead 玻璃珠
glass block panel 玻璃砖壁
glass block （固化）玻璃块
glass break 玻璃状断口
glass brick 玻璃砖
glass-bulb rectifier 玻璃壳整流器
glass carbon 玻璃碳
glass ceramic 玻璃陶瓷
glass cloth insulation 玻璃布绝缘
glass cloth tape 玻璃布带
glass cloth 玻璃丝布，玻璃布
glass-coated wire 玻璃丝包线
glass collectors 玻璃集热器
glass condenser 玻璃介质电容器
glass container 玻璃容器
glass cotton 玻璃棉
glass curtain 玻璃幕墙
glass cutter 玻璃刀
glass dosimeter 玻璃计量计，光电计量笔
glassed flume 玻璃引水槽
glass electrode 玻璃电极
glass epoxy 玻璃钢板，环氧树脂玻璃
glass fabric 玻璃布
glass fiber board 玻璃纤维板
glass fiber cloth 玻璃纤维布
glass fiber coat 玻璃纤维敷层
glass fiber epoxy laminate 玻璃纤维环氧层压板
glass fiber felt 玻璃纤维毡
glass fiber filter 玻璃纤维过滤器
glass fiber insulation 玻璃纤维绝缘
glass fiber mat wool 玻璃纤维棉垫
glass fiber reinforced composite materials 玻璃丝加强复合材料
glass fiber reinforced epoxy resin 玻璃纤维增强环氧树脂
glass fiber reinforced plastic 玻璃纤维增强塑料，玻璃钢
glass fiber reinforced polyester products 玻璃钢制品
glass fiber reinforced polyester resin 玻璃纤维增强聚酯树脂
glass fiber reinforced resin 玻璃钢
glass fiber reinforced thermoplastic materials 玻璃纤维加强的热塑材料
glass fiber reinforced 用玻璃纤维加强的
glass fiber reinforcement 玻璃纤维加固
glass fiber thermal insulation material 玻璃纤维保温材料
glass fiber 玻璃丝，玻璃纤维，玻璃棉
glass fibre cable 玻璃纤维光缆
glass fibre laser 玻璃纤维激光器
glass fibre reinforced plastic pipe 玻璃钢管道
glass fibre thermal insulation material 玻璃纤维保

温材料
glass filter mat 玻璃滤网
glass filter 玻璃滤器，玻璃工
glass flume 玻璃水槽
glass former 玻璃成形剂
glass foundry 玻璃厂
glass funnel 玻璃漏斗
glass gage 玻璃管水位计
glasshouse effect 暖房效应
glasshouse 暖房
glassification 玻璃固化【核废物】，玻璃化
glassine 薄玻璃纸
glass insulation 玻璃绝缘
glass insulator shatter 玻璃绝缘子自爆
glass insulator 玻璃绝缘子
glass isolator 玻璃绝缘体，玻璃绝缘子
glass jalousie 波形板屋面
glass level gauge 玻璃液位计
glass-lined 搪玻璃
glass micanite 玻璃云母板
glass mica-paper tape 玻璃云母纸带
glass mica tape 玻璃云母带
glass microsphere 玻璃微珠
glass-panelled tilting flume 活动玻璃水槽【水工试验用】
glass paper 玻璃纸，砂纸
glass partition 玻璃隔断
glass phosphate dosimeter 玻璃面砖
glass-plate capacitor 玻璃电容器
glass putty 窗用油灰
glass reinforced plastics 玻璃增强塑料，玻璃钢
glass retort 玻璃甑，玻璃曲颈瓶
glass rod 玻璃棒，玻璃杆
glass-seal terminal 玻璃封口端子
glass-Si heterojunction solar cell 硅玻璃异变结太阳能电池
glass slot-liner 玻璃纤维槽衬
glass-stopped bottle 玻璃塞瓶
glass strip 玻璃条
glass textolite 层压玻璃布板，层形树脂浸制玻璃布
glass thermistor 玻璃热敏电阻器
glass tiff 方解石
glass tile 玻璃瓦
glass-to-metal seal 玻璃金属封接
glass-to-metal vacuum seal 玻璃金属真空封接
glass transmissivity and absorptivity 玻璃透过率和吸收率
glass trimmer （管式）玻璃微调电容器
glass tube 玻璃管
glass tube fuse 玻璃管熔断器
glass tube manometer 玻璃管式压力计
glass tube pressure gauge 玻璃管式压力计
glass-U-tube manometer U形玻璃管压力计
glassware 玻璃器皿
glass water gauge 玻璃水位表，水位计
glass window 玻璃窗
glass wool filter 玻璃棉过滤器
glass wool sight flow 玻璃流量计
glass wool 玻璃棉
glass works 玻璃厂
glass woven fabric felt 玻璃布油毡
glass woven fabric 玻璃布
glassy carbon electrode 玻璃碳电极
glassy scale 玻璃状垢
glassy surface 光泽面，镜面
Glauert-Den Hartog criterion 葛劳渥-德哈托判据
glazed brick 瓷砖，釉面砖，釉面瓷砖
glazed ceramic tile 上釉瓷砖
glazed ceramic wall tile 釉面瓷砖
glazed collector 带透明盖板集热器
glazed door 玻璃门
glazed earthenware pipe 釉面陶管
glazed facing tile 釉面瓦，玻璃瓦
glazed frost 雨凇，冻雨
glazed ornament 玻璃花饰
glazed partition 装玻璃的隔板，玻璃隔墙，玻璃隔断
glazed pottery 上釉的陶器，釉陶
glazed rain 雨凇
glazed tile 瓷板，釉面瓦，釉面砖
glazed wall tile 釉面砖
glazed window 玻璃窗
glazed 装上玻璃的
glaze ice 雨冰
glaze 眩光，釉（料），雨凇
glazier's putty 镶玻璃用油灰，镶玻璃油灰
glazing bar 窗芯条
glazing panel 玻璃幕墙
glazing window 玻璃窗
glazing 上光，磨光，瓷釉，上釉，涂釉，镶玻璃
GLD＝gland 密封
gleam 闪光，闪烁，闪亮
gleithretter 滑片
gleization 灰粘作用
glen 平底谷，深谷
glicerine 甘油
glider 滑翔机，滑行艇，滑翔运动员
glide 滑动，滑移
gliding fracture 韧性断裂
gliding ratio 滑翔比
gliding spark discharge 滑闪放电
gliding 滑翔滑行的，流畅的，滑顺的
glime 半透明冰，雨雾凇
glim lamp 辉光放电管，阴极放电管
glimmerite 云母岩
glimmer ton 伊利石
glimmer 闪烁
glint 回波起伏
glist 云母，闪耀
glitch 短时脉冲波形干扰，低频干扰
GLND＝gland 密封装置，压盖
GLND ST(gland steam) 密封蒸汽，轴封蒸汽
global air pollution 全球大气污染
global asymptotic stability 总体渐近稳定，整体渐进稳定
global beam antenna 全球覆盖天线
global calibrating method 总辐射标定法
global circulation pattern 全球环流型
global circulation 全球环流

global climate　全球气候
global communication coverage　全球通信覆盖，全球通信范围
global communications system　全球通信系统
global contaminant　全球性污染物
global coordinates　整体坐标
global criticality　总体临界
global dispersion model　全球弥散模式
global dispersion　全球性弥散
global environment　全球环境
global error　全局误差
global irradiance　总辐照度
global irradiation　总辐照量
globally asymptotically stable system　总体渐近稳定系统
global migration　地表迁移
global network of research station　全球研究观测站网
global noise　总体噪声
global optimization　全局优化
global parameter　总体参数
global positioning system　全球（卫星）定位系统
global radiation　总辐射
global reactivity　静态反应性，总反应性
global reactor calculation　反应堆全堆计算
global roughness　宏观粗糙度
global rule　全球尺度，普遍原则
global scale　全球尺度
global search and replace　全局搜索和替换
global semaphore　公用信号
global solar irradiance　总日射辐照度，总日辐照度【Eg】
global solar radiation　总日射，总日辐射
global stability　总体稳定性
Global System for Mobile Communications　全球移动通信系统
global tide　全球潮汐
global warming　全球增温
global weather recon-naissance　全球天气侦察
global　整体的，球状的，全球的，全局的
globe bearing　球面轴承
globe cased turbine　球壳式水轮机
globecom　全球通信系统
globe diaphragm valve　截止式隔膜阀
globe insulator　球形绝缘子
globe isolation valve　球隔离阀
globe lift check valve　球体抬升单向阀
globe lightning　球形闪电
globe mill　钢球磨煤机，球磨机
globe stop valve　球形截流阀，球形截止阀
globe-tee　球形三通
globe valve with disk free to rotate on stem　阀瓣对阀杆自由转动的球阀
globe valve with disk rigidly attached to stem　阀瓣对阀杆不自由转动的球阀
globe valve　（球形）截止阀，球心阀，球形阀，球阀
globe　地球，球，球形，球形物，球形的
globular arc　球状弧
globular discharge　球形放电，球状电闪
globular pearlite　粒状珠光体
globular transfer　粗滴过渡，颗粒过渡
globule　小球，小珠，水珠
glory hole spillway　漏斗式溢洪道，竖井式溢洪道
glory hole　中心辐照孔
glossary of terms　术语集
glossary　词汇表，小词典，术语集
glossy coal　辉煤，发亮烟煤，无烟煤
glossy　有光泽的，光滑的，似是而非的
gloss　抛光
glove box　手套箱【热室】
glove bucket　抓斗，挖斗
glove port　手套箱操作孔
glove talcing unit　防护手套穿脱装置
glove　手套
glowboy　核电厂检修
glow corona　电晕，辉光电晕
glow current　辉光电流
glow discharge cathode　辉光放电阴极
glow discharge current　辉光放电电流
glow discharge display panel　辉光放电显示板
glow discharge rectifier　辉光放电整流器
glow discharge spectrometry　辉光放电光谱测定法
glow discharge stabilizer　辉光放电稳定器
glow discharge　辉光放电方法，辉光放电
glower　白炽灯丝，白炽体
glowing combustion　灼热燃烧
glowing heat　白热，炽热，白炽
glow lamp　辉光灯，辉光放电管
glow plug　火花塞，热线引火塞
glow potential　辉光电位，辉光放电电位
glow starter　辉光灯启动器
glow tube　辉光放电管，辉光管
glow-type negative corona　辉光型负电晕
glow　辉光，灼热，发光
glucose　葡萄糖
glued board　胶合板
glued construction　胶合结构
glued-laminated construction　胶合层板结构
glue-laminated lumber　胶合层木材
glue　胶，胶合，粘牢，胶水，黏结剂
glutamic acid　谷氨酸
glutaric acid　戊二酸
glycerin　甘油，丙三醇
glycerol　甘油，丙三醇
glycine　甘氨酸，氨基乙酸
glycolic acid　羟基乙酸，甘醇酸，乙醇酸
glycol water　乙二醇溶液
glycol　乙二醇
glyptal paint　甘酞类漆
GMAW(gas metal arc welding)　气体保护金属极电弧焊
Gm-meter　电子管电导测量仪
GMP(guaranteed maximum price)　保证最大工程费用【用于 CM 模式】
GMT(generator main transformer)　发电机主变压器
GMT(Greenwich mean time)　世界标准时，格林尼治标准时间，格林尼治标准时
GND＝grounding　接地
GNE(gross national expenditure)　国民总支出
gneissic　片麻状

gneissoid 片麻状
gneissose granite 片麻状花岗岩
gneissose 片麻状
gneiss 片麻岩
GNFS(Ghana National Fire Service) 加纳国家消防局
GN(general notes) 说明书【文件编码用语】
GNIC(Guangdong Nuclear Power Investment Co. Ltd) 广东核电投资有限公司
GNI(gross national income) 国民总收入
gnomonic projection 球心投影
GNP(gross national product) 国民生产总值
GNPGSC(Guangdong Nuclear Power General Services Co.) 广东核电服务总公司
GNPJVC(Guandong Nuclear Power Joint Venture Company) 广东核电合营有限公司【中国】
GNRB(General Nuclear Review Board) 核安全评审委员会
GNSI(General Nuclear Services Inc.) 通用核服务公司
goaf 采空区，废矿，空岩
go against 不利于……
go-ahead signal 放行信号，向前信号
goalpost 龙门架，门柱
goal programming 目标规划
goal-seeking behaviour 寻找目标性能
goal 目的，终点，球门，目标
go-and-return line 来回线
go-and-return resistance 环线电阻
go back on one's words 反悔
go bankrupt 破产
gob bleeder 采空区排气孔
gob caving 采空区落顶
go-between 中间网络，连杆，连接环
go beyond 超出
gobi 戈壁，戈壁滩，荒漠，沙漠
go by the book 按规矩行事，照章办事，照书本行事，墨守成规地行事
gob 采空区，矿内废石，空岩，杂石
GOCB(Government-Owned Commercial Bank) 国有商业银行
go cheap 廉价出售
go critical 达到临界
go-devil 堵塞检查器
go down 下降，下跌
godown charge 仓储费
godown 仓库，货栈
goethite 铁黄，针铁矿
gauffer 皱褶，作绉褶，压制波纹
gofer 起皱，皱褶
go-gauge 过端量规，通过规
goggles 护目镜，墨镜
going aground 搁浅
going critical 趋向临界
go into business 经营业务，下海
go into compensation 达成补偿贸易协议
go into force 生效
go into operation 投产
gold bullion trading 黄金买卖
gold certificate 金券
gold clause agreement 黄金条款协定
gold fixing 议定金价
gold-leaf electroscope 金箔验电器
gold medal 金质奖（章）
gold-plated contact 镀金触点
gold-plated floating contact 镀金浮动触点
gold-plated 镀金的
gole 水沟，溪谷
goliath crane 移动式大型起重机，巨型起重机
goliath 大型起重机
go line 投入运行
GOMSFE(generic object models for substation and feeder equipment) 变电站和馈线设备通用对象模型
gonad dose equivalent 生殖腺剂量当量
gonad 性腺，生殖腺
gondola car 无盖货车
gondola 货车，无盖货车，敞篷货车，平底船，长平底船，漏斗状卡车【运输混凝土用】
goniometer 测向器，角度计，测角器，测角计，测角仪
go-no-go gauge 过端不过端量规，通过或不通过规，极限规，通过不通过验规，进退塞尺，厚薄规
go-no-go logic 通过不通过逻辑
go-no-go 事情的最后决定，通过不通过，决定继续进行或需要停止的
Gon 哥恩【角度单位，等于直角的百分之一】
good delivery 合格交货，妥善交付
good engineering practice 成功的工程经验，工程范例
good investment 有利的投资
good management 高水平管理
good merchantable quality 上好可销品质
goodness factor 品质因数
goodness of circuit 电路的品质因数
good performance 良好的性能
good practice 好的实践
good quality 优良品质
good river 畅流河道
goods and materials 物资
goods conforming with the contract 符合合同的货物
goods consigned 托运货物
goods declaration 货物报关单
goods elevator 货物升降级，起重电梯，货梯
good service 良好（周到）的服务
goods exchange and payments agreement 换货和付款协定
goods in bond 保兑货物
goods in bulk 散装货物
goods lift 货梯
goods of the contract description 符合合同规定的货物
goods on order 订购货物
goods reception department 收货部
goods reception 收货
goods rejected 退回货物，退货
goods shed 货棚
goods tax 货物税
goods track 货物线
good stream shape 良好的流线型

goods under customs supervision　海关监管货物
goods unpacked　未包装的货物
goods unpaid for　未付款的货物
goods yard　堆货场，货场，货物堆场
goods　货物，商品，物品
goodwill mission　友好代表团
goodwill visit　友好访问
goodwill　商誉，信誉，企业信誉
goose conrtol block　通用面向对象的变电站事件对象控制块
goose conrtol　通用面向对象的变电站事件控制
GOOSE（General object oriented substation event）　面向通用对象的变电站事件
gooseneck connection　鹅颈式接合，乙字形连接管，鹅颈管
gooseneck crane　鹅颈式起重机
gooseneck faucet　鹅颈龙头
gooseneck jib tower crane　折臂式塔式起重机
gooseneck ventilator　鹅颈通风筒
gooseneck　鹅颈弯，S形弯
go out of limits　越界，超过极限值，超过定值
gopher protected cable　防鼠咬电缆
gorge type reservoir　峡谷型水库
gorge　峡谷，河谷，急流
Gortler instability　戈特勒不稳定性
Gothic arch　哥德式拱
Gottingen type wind tunnel　哥丁根型风洞【单回路】，开口回流风洞
gouge　半圆凿，擦伤，凿槽，断层泥，弧口凿
gouging abrasion　碰撞磨损
gouging wear　碰撞磨损
gouging　碰伤，拉毛，凿槽，皮带砸伤，硬伤
goundless　毫无依据的
Gounrot model　古诺模型
go up　上升，上涨
gouy　戈尤【一种电学单位】
go-valve　启动阀
governed engine speed　发动机限速
governing and protection system　调节保安系统
governing board　董事会，理事会，管理委员会
governing body　决策机构
governing by cutting out nozzle　喷嘴调节，断流调节
governing capacity　执政能力，治理能力
governing category　管辖域，管制范畴
governing characteristics　调节特性
governing device　调节装置，调节设备
governing difference equation　控制差分方程
governing equation　主导方程，控制方程
governing error　调整误差，控制误差，调节误差
governing factor　决定因素
governing gear　调节装置
governing intercept valve　调节截止阀
governing language　支配语言
governing law clause　适用法律条款
governing law　管辖法律，准据法，适用法律
governing loop　调节环节，调速环节
governing mechanism　调节机构
governing oil piping　调速油管
governing oil pressure　市政速油压
governing oil　调速油
governing parameter　控制参数
governing response　调节响应，调节冲量
governing speed　调节速度
governing stage　调节级【复速级】
governing system dynamic performance test　调速系统动态特性试验
governing system dynamic performance　调节系统动态特性
governing system pre-operation test　调速系统静止试验
governing system static performance test　调速系统静态特性试验
governing system static performance　调节系统静态特性
governing system　调节系统，调速系统，控制装置
governing valve　调速阀，调节阀，调节汽门，调节器，调速汽门
governing　调速，调节，控制，操纵
government agency　政府机构
government bond　公债，政府债券
government building　政府大楼
government commitment　政府的承诺
government credit　政府信贷
government department　政府部门
government electronic auction systems　政府电子竞价系统
government enterprise　国营企业
government expenditure　政府开支
government grant fund for special purpose　政府专用拨款
government guarantee　政府保证，政府担保
government intervention　政府干预
government investment　政府投资
government license　政府许可
government office　国家机关
government organ　政府机构
government overall balance　政府总差额
government post　公职
government procurement agreement　政府采购协议
government procurement directives　政府采购指令
government procurement law　政府采购法
government procurement　政府采购
government regulations　政府法令，政府规定
government's bailout of banks　救市
government subsidy　政府补贴
government　政府
governor actuator　调速器的传动装置，调速器执行机构
governor block　调节系统图，调节器组件
governor cabinet　调速器柜
governor characteristic　调速器特性曲线，调速器特性
governor-controlled motor　有离心调速器的电动机
governor control safety valve　调速器控制安全阀
governor dead time　调速器死区
governor deflection　调速器偏转

governor droop 调速器下降特性
governor gallery 调速器廊道
governor gear with oil relay 带油继电器调节装置
governor gear without link 无连杆调节装置
governor gear 调节装置，调速装置
governor generator 调速器发电机
governor head 调速器头
governor impeller 脉冲泵，调速泵，旋转阻尼，调速叶轮泵
governor linkage 调速器连杆
governor mechanism 调节机构
governor motor 调速马达，调速器电动机
governor oil pressure 调速油压，调速器油压
governor oil system 调速器油系统
governor oil 调速油
governor-operated 自动调速的，自动调节的
governor pump 调速器泵
governor regulation 调速器不等率
governor sleeve 调速器滑套
governor slide valve 调速器滑阀，调节器滑阀
governor speed changer 调速器速度变换器
governor stop 调速器挡块
governor switch 调速器开关
governor test 调速器试验
governor valve position indicator 调节阀行程指示器，调节阀开度计
governor valve position recorder 调节阀行程记录仪
governor valve 调速阀，调节阀
governor weight 调速器重锤
governor 调速器【汽轮机】，调节器，控制器
govern 支配，居支配地位，控制，主导
GPAC(general purpose analogue computer) 通用模拟计算机
GPA countries 签订政府采购协议的国家
GPA(Government Procurement Agreement) 政府采购协议
GPA 发电机和输电保护系统【核电站系统代码】
GPC(general peripheral control) 通用外围控制
GPC(general purpose computer) 通用计算机
GPDC(general purpose digital computer) 通用数字计算机
GPIB(general purpose interface bus) 通用接口总线
GPIC(Gas and Power Investment Company) 天然气与电力投资公司
GPKD(general purpose keyboard and display control) 通用键盘和显示控制器
GPL(generalized programming language) 通用程序设计语言
GPL(general purpose language) 通用语言
GPL(graphic programming library) 图形程序设计库
GPM(general purpose macrogenerator) 通用宏生成程序
GPM(general purpose macroprocessor) 通用宏处理程序
GPP(general purpose processor) 通用处理机
GPR(general purpose register) 通用寄存器
GPS(global position system) 全球定位系统
GPVC(chlorinated polyvinyl chlorite) 氯化聚氯乙烯
GPV 汽轮机蒸汽和疏水系统【核电站系统代码】
GQS(general quality specifications) 通用质量规范
grabbing groove 抓取槽
grabbing （操作机）抓取，抓入
grab bucket capacity 抓斗容积
grab bucket conveyer 抓斗式运送机
grab bucket crane 抓斗起重机
grab bucket dredger 抓斗式挖泥船
grab bucket excavator 蛤壳式戽斗挖掘机
grab bucket gantry 抓斗龙门吊
grab bucket loader 抓斗装载机
grab bucket type crane 抓斗门式起重机
grab bucket 挖土机抓斗，抓岩机抓斗，抓戽，抓斗，挖泥抓斗
grab clamshell bucket type unloader 抓斗式卸煤机
grab crane 抓斗起重机，抓岩机吊车，抓斗吊车，抓斗式起重机
graben fault 地堑断层
graben 地沟，地堑，裂谷
grab excavator 抓斗挖掘机
grab hoist 抓具升降机
grab hook 起重抓钩
grab linkage 抓具连杆
grab sample 定时取集的样品，定时取样样品，定时试样
grab ship unloader 抓斗卸船机
grab test 抓斗测试，抓样法
grab wrench 抓具扳手
grab 抓斗【起重机】，抓住，抓取，抓具，抓爪，抓取机，抓取装置，抓起
graceful performance degradation 实现得体的性能下降
grace of payment 支付宽限
grace period （政府特许、债务等的）宽限期，免息期，优惠期，宽缓期，优惠期
grad 百分度【角】
gradation curve 大小颗粒分布曲线，级配曲线
gradation of aggregates 集料的级配
gradation test 颗粒级配筛分试验
gradation （从一事物到另一事物的）渐变，（事物划分的）阶段，等级，刻度，粒度测定，粒度测定术，分级，分类，级配，级数
grad bucket 抓斗
grade beam 地基梁，基础梁，合格梁，斜坡梁
grade block 合格混凝土块
grade change 粒径变化
grade contour 等坡线
grade crossing 平交道，平面交叉
graded airgap 阶梯形气隙
graded bedding 粒级层，级配基床
graded blockage 分层变阻塞，用于风速廓线风洞模拟的分层变阻塞
graded cable 分层绝缘电缆
graded coal 分级煤，筛选煤
graded coil 分段线圈
graded filter 分级过滤器

graded insulated winding 分段绝缘绕组
graded insulation 分层绝缘，分段绝缘，分阶段辐照，分阶段照射
graded potentiometer 非线性电势计，非线性电位计
graded sandstone 粒级砂岩
graded slot 梯形槽
graded standard sand 标准级配砂
graded texture 粒级结构
graded time-lag relay 分段延时继电器，可调延时继电器
graded-time step-voltage test 按时升压试验
grade efficiency 分级除尘效率
grade elevation of plant site 场平标高
grade elevation 地坪标高，路面标高，坡度线高程
grade elimination 减缓坡度，高架桥，质量评定
grade force 不同压力
grade-insulated cable 分阶绝缘绝缘电缆
grade level emergency reclaim hopper 事故地面取料斗
grade level 零标高【地面】，地坪标高
grade number 标号
grade of balance 平衡度
grade of brick 砖标号
grade of concentration 浓度，含量
grade of concrete 混凝土标号
grade of filtration 过滤粒度
grade of hydraulic structure 水工建筑物级别［分级］
grade of insulation 绝缘等级
grade of lubricating oil 润滑油的等级
grade of rated voltage 额定电压等级
grade of slope 坡度
grade of steel 钢号
grade of surface preparation 表面预处理等级
grade of timber 木材等级
grade product quality 产品创新创优
grade ratio 坡度比
grader blade 平路机铲刀
grade resistance 坡道阻力，坡度摩阻
grader 推土机，平地机，平土机，平路机，分选机
grade scale 分级标准
grade separation structure 立体交叉结构
grade separation 等级分类，分级配，立体交叉
grade stake 坡度桩
grade tunnel 缓坡隧洞
grade 度【角】，等级，程度，坡度，梯度，斜度，地面标高，定坡度，零标高，分级，速度，（钢的）牌号，级，品位，粒径，（水泥的）标号，斜坡
GRAD=gradient 梯度
gradient angle 坡度角
gradient board 测斜板
gradient current 梯度流
gradienter 水准仪，倾斜度测定仪，斜度仪
gradient layout 平坡式布置
gradient measurement 斜度测量
gradient meter 测坡仪
gradient of gravity 重力梯度
gradient of groundwater table 地下水比降，地下水位坡降
gradient of pipe 管道坡度
gradient of potential 电位梯度，磁位梯度
gradient of slope 斜坡度
gradient of temperature 温度梯度
gradient of water table 地下水位坡度，地下水位坡降，水位坡降
gradient Richardson number 梯度理查德森数
gradient series factor 等差系列换算因子
gradient theory 梯度理论
gradient transfer theory 梯度运输理论，梯度转移理论
gradient wind height 梯度风高度
gradient wind speed 梯度风速
gradient wind 梯度风
gradient 坡度，倾斜度，梯度，比降，坡降，斜率，陡
grading analysis 筛分分析，粒度分析
grading current 等级电流，阶梯电流
grading curve （颗粒）级配曲线，全波曲线
grading diagram 级配图，分级图
grading elevation 路基高程
grading fraction 粒径分级
grading instrument 测坡水准仪
grading of aggregates 骨料级配
grading of river bank 河岸整坡
grading range 粒度测定范围
grading resistance 分段电阻
grading ring 屏蔽环，均压环
grading shield 分段屏蔽
grading standard 分级标准
grading 分等级，校准，分段，等级，分级，筛分，土地平整
gradiometer 陡度计，倾斜计，测坡仪，重力梯度仪，坡度测定仪
gradual contraction 逐渐收缩
gradual declining 逐渐衰退
gradual expansion 逐渐放大
gradual failure 劣化故障，渐发性故障，渐变失效
gradual load reduction 逐渐减负荷
gradually applied load 渐加荷载
gradually varied flow 缓变流，渐变流
graduated arc 分度弧
graduated bottle 刻度瓶
graduated cable 测绳，分度缆
graduated circle 刻度盘
graduated cylinder 量筒
graduated disk 分度盘
graduated float tape 分度浮尺
graduated horizontal circle 水平刻度盘
graduated income tax 累进所得税
graduated plate 分度盘
graduated ring 分度圈
graduated rod 分度标杆
graduated scale 分度尺
graduated 刻度的，分度的
graduate engineer 有学位的工程师
graduate 分度，刻度，校准，量筒，量杯
graduating machine 刻线机

graduation line 刻度线，分度线
graduation 定标，分度，刻度
graduator 分度器
grafting tool 平锹
graft 贪污
grail 砂砾
grain boundary attack 晶界侵蚀，晶间侵蚀，晶界腐蚀
grain boundary brittleness 晶间脆化
grain boundary corrosion 晶间腐蚀，晶界腐蚀
grain boundary cracks 晶间疏松，晶界裂纹
grain boundary decohesion 晶界分离
grain boundary diffusion 晶界扩散
grain boundary strengthening 晶界强化
grain boundary 晶粒间界，晶粒边界，晶界
grain composition 颗粒级配，颗粒组成
grain counting 晶粒计数
grain diameter 粒径，粒径大小
grained catalyst 粒状催化剂
grained coal 粒煤
grained iron 粒铁
grained stone facing 粗粒石面
grain growth 晶粒长大，颗粒放大
graininess 粒度
graining 漆画木纹【木纹漆面】
grain model 团粒模型
grain of crystallization 晶核
grain separation 土粒分离，颗粒离析，颗粒分离
grain size accumulation curve 粒径累积曲线
grain size analysis with sieve method 筛法粒度分析
grain size analysis 粒度分析，筛分析，粒径分析
grain size and pore morphology 颗粒大小和气孔形态
grain size classification 粒度分级，粒径分级
grain size composition 粒度组成
grain size consist 颗粒分布
grain size curve 粒径曲线
grain size distribution curve 粒径分布曲线
grain size distribution 粒度组成，晶粒度组成，粒度分布，粒径分布
grain size fraction 粒径分级
grain size frequency curve 颗粒频率曲线，粒径频率曲线
grain size grading 颗粒级配
grain size number 粒度号数，晶粒度级别数
grain size 晶粒度，粒度，粒径，粒径大小
grain skeleton 颗粒骨架
grain structure change 颗粒结构变化
grain structure 晶状组织，晶粒结构，粒状构造
grain-to-grain stress 粒间应力
grain 粒度，颗粒，晶粒，纹理，岩脉
gram equivalent 克当量
grammol 克分子
gram radium equivalent 克镭当量
gram-rad 拉德-克
grand canyon 大峡谷
grand partition function 巨正则配分函数
grand summary （各子项的）大汇总，总计
grand total 总数，总和，总计，总数之和
grand 总的
graniphyric 花斑状
granite block 花岗石块，花岗岩块
granite facing 花岗石面，花岗石铺面
granite-gneiss 花岗片麻岩
granite-greisen 花岗云英岩
granitelle 二元花岗岩，辉石花岗岩
granite pavement 花岗石铺面
granite-porphyry 花岗斑岩
granite wash 花岗岩冲积物
granite 花岗石，花岗岩
granitic formation 花岗岩地层
granitic plaster 人造花岗石面，水刷石
granitite 黑云花岗岩
granitization 花岗岩化
granitoid 花岗岩状的，人造花岗石
granitotrachytic 花岗粗面状
granoblastic 花岗变晶状
granodiorite 花岗闪长岩
granogabbro 花岗辉长岩
granolithic finish 人造石铺面
granolith 碎花岗岩混凝土铺面，人造铺地石
grant an allowance of n% 同意付给百分之n的贴补
grant a waiver 授权废除，解除义务
grant-back provision 回授规定，回授条款
grant certificate 签发证书
granted patent 批准的专利
granted 颁发的
granting a licence to 颁发营业执照
granting fee coefficient （土地）出让金系数
granting fee 出让金
grant of a patent 授予专利权
grant of safety licence 颁发安全许可证
grantor 授予者
grant 授予，同意
granular activated carbon 粒状活性炭
granular ash 粒状灰
granular bainite 粒状贝氏体
granular bed filter 颗粒层过滤器
granular calcium chloride 粒状氯化钙
granular coating 颗粒覆盖层
granular deposit 颗粒沉积物
granular disintegration 粒状崩解
granular-filled fuse unit 粉末灭弧熔丝
granular filter 粒料过滤器
granular flux 粉状焊剂
granular fracture 粒状断口，粒状断裂，结晶状断口
granular fuel 颗粒燃料
granular-graphite-cooled reactor 颗粒状石墨冷却反应堆
granular graphite 石墨粒
granularity 颗粒性，粒度，粒性，粒形
granular limestone 粒状灰岩
granular magnesium oxide barrier 颗粒氧化镁阻挡层
granular material 粒状物质，粒状材料
granular measurement curve 颗粒级配曲线
granular metric analysis 粒径分析

granular pesticide 粒状农药
granular snow 粒雪，雪珠
granular soil 粒状土
granular structure 团粒状结构，粒状结构
granular 颗粒的，成颗粒状的，粒面的，粒状的，有细粒的
granulated basalt 颗粒玄武岩
granulated blast-furnace slag 粒状高炉渣
granulated slag 粒化渣，粒状熔渣
granulated soil 团粒（结构）土壤
granulate 粒化，使成粒状
granulating apparatus 成粒装置
granulating grading 粒度组成，颗粒分布
granulating hammer 麻面锤
granulating screen 水帘屏
granulation 成粒，粒化作用，制粒
granulator 制粒机，成粒器，破碎机，成粒机
granule 颗粒料，颗粒，团粒，细粒，粒，粒砂，土团
granuliform 细粒状的，粒状构造的
granulitic texture 等粒结构
granulometer 粒度计，颗粒测量仪
granulometric composition 粒径组成
granulometric distribution 粒径分布
granulometric range 颗粒范围
granulometry 粒度测定，粒度测定术，颗粒分析，粒度测量术，颗粒测量法
grapery 葡萄园
grapevine drainage 葡萄藤状水系，格状水系
graph data 图解数据
graphechon 阴极射线存储管
graphechon storage tube 双电子光学系统的存储管
grapher 记录仪器，自动记录仪
graph follower 图形复示器，图形跟踪器，图形变示器
graphical accuracy 图解精度
graphical adjustment 图解平差，图解平差法
graphical alphanumeric display 图形字母数字显示器
graphical analysis procedure for system simulation 模拟系统的图形分析过程
graphical analysis 图解
graphical calculation 图算，图解法
graphical chart 曲线图，图解，图表
graphical computation 图解计算
graphical data processing 图解数据处理
graphical data 图表资料
graphical derivation 图解推求法
graphical determination 图表确定法
graphical differentiation 图解微分法
graphical-extrapolation method 图解外推法
graphical input for network analysis 网络分析的图形输入
graphical integral 图解积分
graphical integration 图解积分法
graphical interpolation 图解内插法
graphical measurement 图解量测
graphical method 图解法，图示法
graphical plot 图示法，图表
graphical recording unit 图表记录装置
graphical recording 图示记录
graphical representation 图示
graphical solution 图解法
graphical symbol 图解符号，图例
graphical visual display device 图形显示器件
graphical water-hammer 水击图解法
graphical 图的，图解的，图示的
graphic chart 曲线图，图标
graphic diagram 模拟图，曲线图
graphic display control 图形显示控制
graphic display unit 图形显示装置
graphic display 图形显示器，图形显示，画面
graphic dynamic information update 画面动态数据更新
graphic element configurator 图形元件组态器，图素组态器
graphic expression 图示，图解
graphic granite 文象花岗岩
graphic information 图像信息
graphic instrument 图示器，图示仪，自动记录仪
graphic interpolation 作图内插法
graphic jet printer 图形喷墨打印机
graphic manual 图表手册
graphic mechanics 图解力学
graphic meter 自动记录仪，自动记录器
graphic-mode display 图形方式显示
graphic panel 图解式面板，自动记录盘，图形板，图表板，模拟盘
graphic programming library 图形程序设计库
graphic scale 图解比例尺，图示比例尺
graphic solution 图解
graphic statics 图解静力学
graphics 制图学，画面，图像
graphic terminal 图形终端
graphic triangulation 图解三角测量
graphic wall 图解墙板
graphic 形象的，图表的
graph instrument 自动记录仪，图示仪
graphite 石墨
graphite abrasion 石墨磨蚀
graphite anode 石墨阳极
graphite block 石墨块
graphite brick work 石墨砌体，石墨砖结构
graphite bridge 石墨桥
graphite brush 石墨电刷，石墨炭刷，石墨套管，石墨套，石墨衬套
graphite burning 石墨燃烧
graphite casting die 石墨铸模
graphite cast iron 灰口铸铁，石墨铸铁
graphite centre rod 石墨芯棒
graphite-clad fuel element 石墨包壳燃料元件
graphite-coated cladding 石墨包覆包壳
graphite-coated uranium particle 石墨包覆的铀颗粒
graphite coating 石墨覆盖层
graphite column 石墨热柱
graphite core matrix 堆芯石墨栅阵
graphite core 石墨活性区，石墨堆芯
graphite corrosion 石墨腐蚀
graphited silver brush 银石墨电刷

graphite dust 石墨粉尘
graphite electrode 石墨电极
graphite fiber composite blade 石墨纤维复合材料叶片
graphite fiber 石墨纤维
graphite fill 石墨充填物
graphite-gas reactor 气冷石墨堆
graphite guide tube 石墨导管
graphite-like structure 石墨状结构，类石墨结构
graphite liner 石墨衬套
graphite lubricant 石墨润滑剂
graphite matrix 石墨基体
graphite-moderated gas-cooled reactor nuclear power plant 石墨气冷堆核电厂
graphite-moderated helium-cooled high temperature reactor 石墨慢化氦气冷却高温反应堆
graphite-moderated lattice 石墨慢化栅格
graphite-moderated reactor 石墨反应堆，石墨慢化反应堆，石墨慢化堆
graphite-moderated water-cooled reactor nuclear power plant 石墨水冷堆核电厂
graphite-moderated water-cooled reactor 石墨水冷堆
graphite moderator matrix 石墨慢化剂基体
graphite moderator stringer 石墨慢化棒
graphite moderator 石墨慢化剂
graphite oil 石墨油
graphite oxidation 石墨氧化
graphite oxide membrane 氧化石墨膜
graphite packing 石墨衬垫
graphite paint 石墨涂料
graphite pebble 石墨球
graphite permeability 石墨渗透性
graphite phenolic plastics 石墨酚醛塑料
graphite pig iron 灰口铸铁，石墨生铁
graphite potential energy 石墨潜能
graphite powder 石墨粉
graphite reactor 石墨反应堆
graphite reflector 石墨反射层
graphite seal 石墨密封件
graphite septum 石墨隔板
graphite sheath 石墨套管
graphite shield 石墨屏蔽体
graphite shrinkage 石墨收缩
graphite sleeve disposal void 石墨套管处置库
graphite sleeve 石墨套管
graphite slip ring 石墨滑环
graphite-sodium reactor 石墨—钠反应堆
graphite spine 石墨细棒
graphite stack 石墨砌体
graphite sublimation point 石墨升华点
graphite support sleeve 石墨支承套管
graphite thermal column 石墨热柱
graphite to silicon-carbide couple 石墨和碳化硅温差热偶
graphite uranium lattice 石墨铀栅格
graphite uranium mass ratio 石墨铀质量比
graphite-water reactor 石墨水反应堆
graphite whisker 石墨晶须
graphitization under pressure 加压石墨化
graphitization （燃料元件）石墨化，包覆石墨
graphitized coke 石墨化焦炭
graphitizing treatment 石墨化退火
graph of reservoir operation 水库调度图
graphometer 半圆仪
graph paper 方格纸
graph plotter 图表复制机，绘图仪
graph 图，图表，图形，表格，记录纸，曲线图，图解，过程线
grapnel 四爪锚
grappler 抓具，抓爪
grapple 抓机
grappling fixture 抓头
grappling 抓取
grard ring 防护环
grass barrier 草障
grass cutting 杂波抑制，草状波抑制
grassed slope 草皮护坡
grassed spillway 草皮溢水道
grassed waterway 草地泄水道，草皮泄水道
grasshopper rig 轻型钻机
grassland 草场，草地，草原，牧场
grassot flux meter 动圈式磁通计
grassroots unit 基层单位
grass shelter belt 防风草带
grate area 炉排面积
grate bar 炉栅，栅条
grate burning rate 炉排热强度
grated inlet 格栅式进水口
grate-fired furnace 层燃炉膛
grate firing 层燃，火床燃烧
grate frame 炉排框架
grate funnel 炉排风箱
grate furnace 炉排炉，层燃炉
grate heat release rate 炉排面积热负荷，炉排热负荷
grate overheating 炉排过热
grate ring 炉排座圈
grater 粗齿木锉
grate surface 炉排面
grate type throttling surge tank 栅格节流式调压室
grate type throttling 栅格式节流
grate type wet scrubber 湿式洗涤栅除尘器
grate 炉箅，炉排，格栅，护栅，护栏，铁箅子
graticulation 方格缩放法
graticule line 方格线
graticule 方格图，分度线，网格
grating inlet 帘格式进水口
grating plate 网格板
grating structure 格形构造
grating texture 格状结构
grating with louvered damper 格栅连动百叶风口
grating 格子板【地面】，格栅，晶格，格子，栅，栅栏，光栅，花纹地板
gratis 免费的，无偿的，免费
gravel 砾石，砾砂，卵石
gravel aggregate 砾石骨料
gravel bank 砾石岸
gravel bar 砾石滩，砾石洲
gravel beach 砾滨，砂砾滩
gravel bed filter 颗粒层除尘器

gravel box 砾石笼
gravel concrete 砾石混凝土，卵石混凝土
gravel core fascine roll 石心埽枕，石心梢辘
gravel core fascine 石心梢捆
gravel covering 砾石盖面
gravel dam 砾石坝
gravel deposit 砾石沉积
gravel desert 砾质沙漠
gravel drain 砾石排水沟
gravel-filled drain trench 砾石排水沟，填卵石排水沟
gravel-filled trench 填砾石排水沟
gravel filter well 砾石滤水井
gravel filter 砾石过滤器，砾石滤层
gravel fraction 砾石成分
gravel gabion 砾石笼
gravelling of road 砾石铺路
gravelling 铺砾石
gravelly loam 砾质壤土
gravelly soil 砾质土
gravel packing 砾石衬垫
gravel pile 碎石桩
gravel pit 采砾坑，砾石料坑
gravel plant 砾石筛选厂
gravel pump 砾石泵
gravel riffle 卵石滩
gravel road 砾石路
gravel roll 砾石帚枕，砾石梢龙
gravels 碎石子
gravel sand cushion 砂砾垫层
gravel sand stratification 砂砾层
gravel sand 砾砂
gravel scoop 砾石铲斗
gravel screen 砾石筛
gravel soil 砾土
gravel sorter 砾石分选机
gravel spreader 砾石撒布机
gravel stone 小圆石
gravel stratum 砾石层，砂砾层
gravel trap 砾石拦截坑
gravel washer 洗砾机
gravel washing screen 洗砾筛
gravel well 砾壁井
graveyard （放射性废物）埋葬场
gravimeter measurement 容重法测流
gravimeter 比重计，重力计，重力仪
gravimetric altimeter 重力测高计
gravimetric analysis 重力分析，重量分析
gravimetric feeder 重力式给料机
gravimetric method 重量分析法，重量法
gravimetric survey 重力测量
gravimetric 重量的，重量分析的，比重测定的
gravimetry 重力测量，重量分析，重量分析法
gravitate downwards 重力作用下向下移动
gravitate 重力沉降
gravitating bed 下降层
gravitational acceleration 重力加速度
gravitational bed 重力层
gravitational collapse 引力坍缩
gravitational constant 万有引力常数，引力常数
gravitational convection 自然对流
gravitational coupling 重力耦合，重力耦合器
gravitational drop 自重下降，重力跌落
gravitational energy 重力能，重力势
gravitational exploration 重力勘探
gravitational field 万有引力场，重力场
gravitational flow drainage 重力排水
gravitational flow 重力流
gravitational force 万有引力，重力，地球引力
gravitational method of exploration 重力探测法
gravitational method 重力法
gravitational potential energy 重力势能
gravitational pressure drop 重力压降
gravitational prospecting 重力勘探
gravitational separation 重力分离
gravitational separator 重力式分离器
gravitational settling 重力沉降
gravitational slip 重力滑坡
gravitational system of units 重力单位制，工程制
gravitational tide 引力潮
gravitational unit system 工程制
gravitational ventilation 重力通风，自然通风
gravitational water 重力水
gravitational wave 重力波
gravitational 引力的，万有引力的，重力的
gravitation energy 位能
gravitation inlet pipe 重力式引水管
gravitation 地心吸力，万有引力，重力，引力
gravitative differentiation 重力分异
gravitional force 引力
gravitometer 比重计
gravity abutment 重力式岸墩，重力式拱座
gravity action 重力作用
gravity and roof drain collection system 自重泄油管路与浮顶油罐排水系统
gravity anomaly 重力反常，重力异常
gravity arch dam 重力拱坝
gravity axis 重心轴
gravity bottle 比重瓶
gravity breakwater 重力式防波堤
gravity bucket conveyor 斗式重力输送机
gravity bulkhead 重力式挡土墙，重力式岸壁
gravity centre 重心
gravity chute 斜槽，重力溜槽
gravity circulation 自然循环，重力循环，重力环流
gravity clamp 重力夹，重力式夹车器
gravity coefficient of coating 药皮重量系数
gravity concentrate 重选精矿
gravity condensate return system 重力凝结水回收系统，重力自流凝结水回收系统
gravity core sampler 重锤式岩心取样器
gravity correction 重力订正，重力修正
gravity culvert 重力排水涵洞，重力流
gravity curve 抛物线
gravity dam of triangular section 三角形断面重力坝
gravity damper 重力调节风门
gravity dam 重力坝
gravity density 重力密度
gravity distribution 重力分布

gravity dock 重力式坞
gravity drainage system 自流排放系统
gravity drainage 重力排水，自然排水，自由排水
gravity draining 重力排水，自流排水
gravity driven cooling system 重力驱动冷却系统
gravity drop 自重下降，重力下降
gravity dump body 重力卸料车身
gravity effect 重力效应
gravity electrode 重力焊条
gravity extrapolation 重力外推
gravity fault 重力断层
gravity feeder 重力加药器
gravity feeding 重力送料
gravity feed oiler 重力注油器
gravity feed stoker 重力加煤机
gravity feed welding 重力焊
gravity feed 重力注入，重力进给，重力送料，自流供水
gravity filter 重力过滤池，重力式过滤器，重力式滤池
gravity flow pipe line 自流管路
gravity flow 重力流动，自流，重力流
gravity force 重力
gravity fractionation 重力分级
gravity gradient 重力梯度
gravity groundwater 地下重力水，重力地下水
gravity hammer 重力锤
gravity haulage 滑动运输，重力运输
gravity head 水头，重力压头
gravity-induced flow 重力诱导流
gravity insertion 自由落棒，自由下落插入
gravity line 重力线
gravity loaded accumulator 重力蓄力器
gravity lockwall 重力式闸墙
gravity main 重力管路
gravity measurement 重力量测
gravity meter 比重计，重力仪
gravity mill 重力式磨煤机
gravity oil filter 重力式滤油器
gravity packing 重力填充法
gravity pendulum 重力摆
gravity percolation 重力式渗滤
gravity power 重力能
gravity regulator 重力调速器
gravity reinjection nozzle 重力回燃喷管
gravity retaining wall 重力式挡土墙
gravity return 重力回水
gravity roller conveyor 重力式滚轴运输机
gravity rope way 重力式缆索道
gravity sag 重力下垂
gravity sand filter 重力砂滤器
gravity scram 重力落棒快速停堆
gravity separator 沉降室，重力式分离器
gravity settler 重力澄清器
gravity settling 重力沉降澄清，重力沉降
gravity slide 重力滑坡
gravity solution 重液
gravity spillway dam 重力式溢流坝
gravity spiral conveyer 螺旋下落式输送机
gravity spread 重力散布
gravity stowing 重力充填
gravity survey 重力探查
gravity suspended water 重力悬着水
gravity system water cooling 热虹吸水冷，重力水冷
gravity system 自流给料系统，重力系统，重力制
gravity tank 重力供油箱
gravity tension gear 重锤拉紧装置
gravity tide 引力潮
gravity type bank indicator 重力型倾斜指示器
gravity type coil 重力式盘管
gravity type diversion weir 重力式引水堰
gravity unit 重锤单位
gravity unloading 重力卸载
gravity wall 重力式墙
gravity water supply 重力供水，重力式给水，自流供水
gravity water wheel 重力式水轮
gravity water 重力水
gravity wave 重力波
gravity wharf 重力式码头
gravity wind 重力风
gravity 重力，燃料重量指数，重量
gray absorber rod 灰吸收棒
gray absorber 灰吸收体
gray area 灰色区
gray body 灰体
gray brick 青砖
gray-brown podzolic soil 灰棕色灰化土
gray cast iron 灰口铁，灰口铸铁
gray desert soil 灰钙土，灰色沙漠土
gray earth 灰钙土
gray emitter 灰辐射体
Gray-King assay 格金干馏试验
Graylock seal 格雷洛克密封
graystone 玄武岩
gray 灰体，灰体的，灰的，灰色（的）
grazing angle 掠射角
grazing land 牧场
grazing 牧草
GRC＝gearcase 齿轮箱
GRC（glass fiber reinforced compositematerials）玻璃丝加强复合材料
GRD(general rate decrease) 运价下调
GRDR＝grinder 碎渣机
grease capacity 注脂量
grease cup 油脂杯，油杯
grease fitting 加脂配件
grease gun 黄油枪，润滑脂枪
grease lubricant 润滑脂，润滑油脂
grease lubrication 油脂润滑
grease lubricator 油脂润滑器
grease melting plant 油脂熔化炉
grease nipple 滑脂嘴
grease pump 润滑油泵，润滑脂枪，黄油枪
grease packing 润滑填料
grease pump 油脂泵
grease relief fitting 油脂安全阀配件
grease removal tank 油脂清除池，除油池
greaser 润滑工，润脂杯，润滑脂注入器

grease seal 油脂密封
grease separator 除油器，分油器
grease trap 隔油池
grease 润滑脂，牛油，油脂，浓雾
greasing 涂油脂，润滑
greasy filth 油泥
greasy 油脂的，润滑的
great calorie 大卡
greater coasting area 近海区域
greatest diurnal tidal range 最大日潮差
great potential 巨大潜力
great prize 特奖
green 绿色（的）
green area 绿化面积，绿化带，绿地，绿化区
green ash 绿灰
green belt 绿化地带，绿化带
green clause letter of credit 绿色条款信用证，部分预支绿条款信用证
green coal 未经处理的煤，新添加的煤，新抛入煤
green compact 生坯，生压块，未烧结的坯块
green concrete 未凝固混凝土，新拌混凝土，新混凝土，新浇混凝土
green coverage ratio 绿化覆盖率
green density 未烧结坯块密度，生还密度
greenery factor 绿化系数
green fence 树篱笆，绿篱
green-field condition 绿色田野状况
green fuel 新添加的燃料
green hand 生手，没有经验的人
green horn 外行
green house effect 温室效应
greenhouse gas 温室气体
green house 暖房，温室，花房
greening coefficient of power plant area 厂区绿化系数
greening design 绿化设计
greening factor 绿化系数
greening of plant area 厂区绿化
greening 绿化
green lumber 生材，湿材
green mud 绿污泥
green pellet 未烧结的芯块，生芯块，生坯，芯块坯
green run 试车，试运行
green salt 绿盐，四氟化铀
green sand 新采砂，绿砂
green schist 绿色片岩
green space 绿化面积
Green's strain tenser 格林应变张量
greensward 草皮，植物
green test 试车
green timber 湿材
green vitriol 绿矾
green water 新水，鲜水
Greenwich mean time 世界标准时，格林尼治标准时间，格林尼治标准时，格林尼治平时【GMT】
Greenwich time 格林尼治时间
gregale 格雷大风
GRE(glass fiber reinforced epoxy resin) 玻璃纤维增强环氧树脂
Gregorian calendar 格里历
grenz rays 跨界射线
grey absorber rod 灰吸收棒
grey absorber 灰吸收体
grey body 灰体
grey brick 青砖
grey casting 灰口铸铁件
grey cast iron shielding 灰铸铁屏蔽
grey cast iron shield 灰铸铁屏蔽体
grey cast iron 灰铸铁，灰口铸铁
greyhound 快速船，远洋快船
greying 石墨化
grey iron 灰口铁
grey level 灰度值
grey neutron detector 灰中子探测器
grey pig iron 灰口生铁
grey radiator 灰体辐射体
grey receiver 灰体吸收体
grey rod 灰棒
grey tile 青瓦
greywacke 硬砂岩，杂砂岩
grey 灰体【反应堆技术】，灰体的，灰的，灰色（的）
GRE 汽轮机调速系统【核电站系统代码】
grgrey scale 灰度等级
GRH 发电机氢气冷却系统【核电站系统代码】
grid access tariff 接入价
grid accumulator 栅条蓄电池
grid amplitude 格网幅度【东西方向的】
grid assembly 格架【燃料组件】，格架组件
grid attachment 贴栅
grid bar 带栅格的炉条，栅条
grid bias 栅偏压
grid bucking strength 格架挠曲强度
grid ceiling 井式天棚
grid cell 格架栅元
grid chart 方格图，格网图
grid circuit 栅极电路
grid code 电网导则，电网规程
grid compatibility 电网兼容性
grid computation 格网计算
grid connected wind turbine 并网风电机组
grid connected 并网运行
grid-connection 并网
grid control building 电网（系统）控制楼，网络控制楼
grid control center 电网控制中心
grid control room 网络控制室
grid coordinates 网格坐标
grid coupling 电网连接器，栅极耦合
grid-dip-oscillator 栅陷振荡器
grid dropout 掉网
grid failure 电网故障
grid fault 电网事故
grid feeding hopper 带格栅的装料斗
grid floor 格栅楼面，格子楼板
grid following 电网跟踪
grid foundation 格栅式基础，交叉梁基础，网格基础
grid frame 格栅架

grid frequency 电网频率
grid gate 栅极选通脉冲
grid hold time 联网时间
grid injection 栅极注频
grid interval 格网间距
grid ionization chamber 屏栅电离室
gridiron arrangement of pillars 柱网布置
gridiron system 网格式系统
gridiron 高压输电网，格，环状管网，框格
grid jet 分布板射流
grid junction 格网连接线
grid levelling 方格水准测量，面水准测量
grid line 格网线，坐标格网线，网路线【超声波检查】
grid management charge 电网管理费
grid mesh 网眼，栅网
grid method 格网法
grid mismatch 格架失配
grid model 格网模型
grid partition 网状隔板
grid plate structure 栅格板构件
grid plate （沸腾炉）布风板，堆芯栅板，栅格板
grid plug 栅板节流塞
grid point 网格点
grid power line 输电线网，输电线
grid reading 网格读数
grid roller 方格压印滚筒，网格碾
grid security check 网络安全校核
grid security constraint 网络安全约束
grid side converter 电网侧变流器
grid spacer 定位格架
grid spacing 格架间距
grid spring force 栅格弹簧力，棒夹持力
grid spring protrusion 格架弹簧突出
grid station control building 网控楼
grid status 电网状态
grid strap 格架条带【燃料组件】
grid structural material 格架结构材料
grid subtransmission 网络状二次输电，次输电网
grid support plate 栅格支承板
grid synchronization and connection system 并网与接线系统
grid synchronization 电网同步
grid system 电网系统，格构体系，方格系统
grid-to-grid section 格架到格架的区段，跨距
grid type extraction valve 栅型抽汽阀
grid type network 格式网络
grid unit 格网单元
grid valve control 旋转挡板调节
grid valve 栅形阀，栅型阀
grid voltage 线电压，厂外供电电压，栅极电压，电网电压
grid zone 分布板区
grid 格子，格栅，栅格板，（高压输）电网，方眼网，网络，网路，系统，栅栏
GRI(general rate increase) 运价上调
Griggs-Putnam index 格里戈-普特南指数【植物风力指示】
grillage column base 格排柱基
grillage floor 格构式楼板
grillage foundation 格床基础，格排基础，格形基础
grillage 格架结构，网格结构，格床，格架，格排，格栅
grilled pipe 肋片管
grille member 花格构件
grille wall 花格墙
grille 格栅，花格子，格子窗，华格墙
grill 铁箅子，铁丝格子
grill with articulated bars 铰接的栏杆
grindability 可磨度，可磨性
grindability index 可磨性指数，可磨度系数，可磨系数，可磨性系数
grinder bench 磨床工作台
grinder cylinder 碎木机压力缸
grinder sludge 磨屑
grinder 研磨机，粉碎机，砂轮，磨床，磨机，磨碎机，砂轮机，碎石机，破碎机
grinding agent 研磨剂，水泥颗粒变细剂
grinding all over 磨光【焊缝】
grinding ball 研磨球，磨煤钢球
grinding capacity 磨煤机出力
grinding crack 磨削裂纹
grinding disk 摩擦盘，砂轮片
grinding flush 磨光【如焊缝】
grinding goggles 研磨护目镜
grinding machine 研磨机械，磨床
grinding mark 研磨痕迹
grinding mill 磨粉机，磨煤机，研磨机，磨碎机，辊轧机，破碎机，压碎机
grinding-off 磨去，磨削
grinding oil 润磨油
grinding ring 磨环【中速磨】
grinding roll 磨辊【中速磨】
grinding table 磨盘【磨煤机】
grinding tool 磨削工具
grinding wheel cutting machine 砂轮切割机
grinding wheel mill 砂轮机
grinding wheel 磨轮，砂轮，砂轮片
grinding 磨光，研磨，磨，抛光，粉磨，碾碎，磨碎，磨削，粉化，粉碎，磨细
grind off 磨掉
grindstone 磨石，砂轮
griotte 大理石，大理岩
grip between concrete and steel 钢筋混凝土（之间的）握裹力，钢筋与混凝土间的握固力
grip end （机械手）握物端，抓柄
grip gauge 夹紧装置
grip length 锚固长度，握固长度
grip member （机械手）抓手
grip nut 防松螺母，夹紧螺母，固定螺母
grip of concrete 混凝土握固力
gripper assembly 棘爪组件
gripper cable 抓手缆绳
gripper cam 抓手凸轮
gripper carriage 抓手托架
gripper coil 夹持线圈【控制棒驱动机构】
gripper expander 抓手扩张器
gripper head 牙板夹头
gripper jaw 夹爪
gripper latch 抓具闩

gripper long handing tool 棘爪长柄工具
gripper mechanism 机械手，伸缩机构【在役检验设备】，抓手机构，抓取机构
gripper sensing device 抓手敏感装置
gripper trolley 门式抓斗滑架
gripper tube mast 伸缩杆抓头【燃料装卸】，抓取机伸缩杆
gripper 抓爪，夹头，抓具，夹紧器，夹具，夹子，抓器
gripping 抓取，抓住，夹紧，啮合
gripping area 抓卡部位
gripping device 抓取器，固定器，抓取机，夹具
grips 夹钳
grip wrench 管子钳
grip 柄，手柄，夹，夹紧，夹具，夹子，把手，啮合，钳，洗涤槽
grit-arresting 除尘
grit arrestor 除尘器
grit blasting 喷砂【表面处理】，喷砂清理
grit carry-over 粗粒携带
grit catcher 沉沙池
grit chamber 沉沙池，沉渣池
grit collector 集沙器，除沙器
grit compartment 沉渣池，管井沉淀管
grit emission 尘粒排出量，尘粒排除物，尘粒排放
grit entrapment 砂粒截留
grit blasting 吹砂处理，喷砂处理
grit recirculation 飞灰再循环
grit reservoir 沉砂池，沉砂砾池，沉渣池
gritter 铺砂机
grittiness 砂砾性
gritty 含砂砾的
grit 粗砂岩，飞灰，砂砾，砂子，粗沙［砂］，砾石，砂粒，金属屑
grizzle 高硫煤，含硫铁矿煤
grizzly bar screen 格栅筛
grizzly bar 筛条
grizzly 铁栅，铁栅筛，劣质煤，格筛，篦子
grocery 杂货店
grog brick 耐火砖
grog firebrick 熟料制的耐火砖
grog （耐火材料）熟料，耐火黏土
groin basin 丁坝间水区
groin blanket 灌装铺盖
groined slab 带拱肋的板，井字形梁板
groin head 丁坝坝头
groining 筑丁坝
groin spacing 丁坝间距
groin works 丁坝工程
groin 防砂堤，交叉拱，折流坝，丁坝
grommet 垫圈，密封垫，衬垫，索环，绝缘环，垫环，孔眼，绝缘孔圈，麻丝垫环
groove and tongue joint 槽舌接合
groove and tongue 企口，槽舌
groove angle 坡口角度【焊接】，坡口角度
grooved and tongued flooring 企口地板，槽舌地板
grooved and tongued joint 企口接合，槽舌合
grooved armature 有槽电枢
grooved cylindrical pin 开槽的圆柱销，凹槽销
grooved drum 绳沟鼓筒
groove design 坡口设计，接头设计
grooved metal gasket 槽形金属垫片
grooved pile 企口板桩
grooved pulley 三角皮带轮，槽轮
grooved roll 槽辊
grooved rotor 有槽转子
grooved straight pin 开槽的圆柱销，凹槽销
grooved tube-seat 带槽管座
grooved tube 内螺纹管
grooved upland 槽蚀高地，沟切高原
groove edge 坡口边
groove face 坡口面，槽面
groove joint pliers 套接钳，活动钳
groove joint 槽式接合，凹缝
groove of stator 定子槽
groove pin 开槽的圆柱销，凹槽销
groover 挖槽机
groove shape 坡口形状，坡口设计
groove to be provided 预留槽
groove welding 槽焊，坡口焊，开坡口焊接
groove weld 坡口焊，坡口焊缝，开坡口焊接
groove 凹槽【下管座】，槽，沟坡口，凹口，开槽，凹槽，导向槽，坡口
grooving and tonguing 企口榫接
grooving machine 刻槽机
grooving rotor 有槽转子
grooving 倒棱，倒角，企口，企口接合，企口连接
gross 总的，总（重），毛（重）
gross activity 总放射性，总放射性活度
gross actual power generation 实际总发电量
gross aggregate 累计总额
gross amortization charges 总摊销费
gross amount 毛计，毛额，总量，总额，总计额，总计
gross analysis 全量分析
gross aperture area 总采光面积【聚光集热器】
gross area 毛面积，总面积
gross available capacity 毛可用容量
gross available generation 毛可用发电量
gross average 平均总额，总平均值
gross benefit 毛效益
gross beta spectrum β总能谱
gross bubbling bed 全鼓泡床
gross building area 总建筑占地面积
gross caloric value 高位发热量，高热值，总热值
gross calorific power 高位热值
gross calorific value at constant volume 恒容高位发热量
gross calorific value 高位热值，高位发热量
gross capability 总出力
gross capacity factor 总容量因子
gross capacity 总装机容量，总容量，总功率
gross capital formation 资本形成总值，总的资本形成
gross carryover 大量携带【指蒸汽带水】
gross cash flow 现金总流量
gross coal sample 总样

gross collector area 集热器总面积
gross collector array area 集热器阵列总面积
gross commanded area 可灌总面积
gross contamination 总污染
gross control 粗调，总量控制
gross decontamination factor 总去污系数，总净化系数
gross demonstrated capacity 总实证容量
gross discharge 毛流量，总流量
gross domestic product 国内生产总值
gross duty of water 毛灌水定额
gross effect 总效应
gross efficiency 总效率
gross electrical capacity 总电功率，总装机容量
gross electrical output 总输出（电）功率【电厂】，总装机容量
gross energy requirement 能量总需求量
gross erosion 总侵蚀
gross evaporation 毛蒸发量，总蒸发量
gross export value 出口总值
gross failed fuel detector 燃料元件总破损探测器
gross fission product 总裂变产物
gross floors area 总建筑面积
gross flux distribution 宏观通量分布
gross flux variation 粗通量变化，宏观通量变化
gross freeboard 总超高
gross freight 毛运费
gross gas turbine generator rating 总燃气轮发电机额定功率
gross generation 总发电量
gross head 毛水头，总水头
gross heating value 总发热量
gross heat input 总供热量
gross heat rate 毛热耗率，总热耗率，总热耗损
gross horsepower 总功率，总马力
gross import value 进口总值
gross income tax 总所得税
gross industrial production 工业生产总值
gross instability 整体不稳定性
gross installed capacity 总装机容量
gross investment 总投资，投资总额
gross load 毛重，总重，全负荷，总负荷，总负载
gross loss 总损耗，毛损耗，毛损
gross margin 毛利，毛利率
gross maximum capacity 毛最大容量，最大总容量
gross maximum generation 毛最大发电量
gross national expenditure 国民总支出
gross national income 国民总收入
gross national product 国民生产总值
gross operating head 总运行水头
gross output 总输出功率，总出力，毛出力
gross payroll 工资总额
gross plant efficiency 电站总效率，电站毛效率
gross porosity 总孔隙度
gross power 总功率
gross product 总产量
gross profit 毛利，总利润
gross properties 总体性质
gross rated capacity 总额定容量
gross receipts tax 总收入税
gross receipt 总收入，总收益，收入总额，总收入款
gross reserve generation 备用总发电量
gross reservoir capacity 毛库容
gross residential area 总居住区面积
gross Richardson number 总体理查森数
gross salary 工资总额
gross sample 总样品，毛样，总样
gross section 毛断面
gross site area 总建筑基地面积
gross solids circulation 总体颗粒循环
gross stage efficiency 总级效率
gross standard coal consumption rate 发电标准耗煤率
gross station heat rate 全厂热耗
gross steam turbine generator rating 总汽轮发电机额定功率
gross storage capacity 总库容
gross storage 毛库容，总蓄水量
gross stress concentration factor 总应力集中系数
gross stress 毛应力
gross structure design 总体结构设计
gross structure 粗视构造，宏观构造
gross thermal efficiency 发电端热效率，总热效率
gross thickness 毛厚度
gross thrust 总推力
gross tonnage 总吨位
gross ton 长吨，总吨
gross tractive power 总牵引能力
gross turbine heat rate 透平总热耗，汽轮机毛热耗
gross unit unavailable generation 机组毛不可用发电量
gross vehicle load 车辆总重量
gross volume 毛体积，总容积
gross weight 毛重，总重量
gross working capital 总周转资金
grotto 洞室，石窟，岩洞
ground acceleration 地面加速度
ground adhesion 地面附着力
groundage 停泊费
ground anchor 地锚
ground and floor 地面和楼面
ground antenna 地面天线
ground arc 接地电弧
ground area 占地面积
ground avalanche 大坍方
ground axes 地轴
ground bar 接地棒
ground base 地基
ground-batching plant 泥浆拌和机
ground beam 地基梁，地梁，地面梁，基础梁
ground bearing capacity 地基承载力
ground bearing pressure 地基承压力
groundbed 接地床
ground blizzard 地面雪暴
ground board （风洞）地板
ground bolt 地脚螺栓，锚栓，板底栓

ground breaking ceremony 动工仪式，开工典礼
ground breaking 破土
ground bus 接地母线
ground cable 地下电缆
ground camera 地面摄影机
ground capacitance 对地电容
ground circuit 接地电路
ground clearance line 离地间隙线，地面弧垂线
ground clearance 地面清理，离地净高，离地距离，离地高度，对地距离，最小离地间隙
ground clutter 地面杂乱电波，地物反射波
ground coal bin 落地煤仓
ground coal 底层涂料
ground coil 接地线圈
ground compensation 接地补偿，地面浓度
ground condition 基本状态，基本条件
ground conductivity 大地导电率
ground conductor 接地导线
ground connection 接地装置，接地
ground constant 大地常数
ground contour 地形
ground control survey 地面控制测量
ground control 地面控制
ground cover 地被物，地面被覆，地面覆盖
ground current 大地电流
ground detector relay 接地检测继电器
ground detector 接地检测器，接地探测器
ground directional relay 接地定向继电器
ground-displacement 地电位位移，地面位移
ground disposal area （放射性废物）地下处置区
ground disposal facility 地下处置设施
ground disposal of effluent （放射性）排出物的地下处置
ground disposal （放射性废料的）埋地处理
ground distance relay 接地距离继电保护装置
ground distance 地面水平距离
ground drag 地面阻力
ground dust 土尘
ground echo 地面反射波，地面回波
grounded bivalued resistor 接地双值电阻器
grounded capacitance 被接地的电容
grounded circuit 接地电路
grounded concentric wiring system 接地同轴电缆制
grounded conductor 接地的导体，地线
grounded counterpoise 接地平衡网路，接地地网
grounded input 接地输入
grounded neutral system 中线接地制，中性点接地系统
grounded neutral 接地中性点
grounded output 接地输出
grounded 3-phase 4-wire system 接地的三相四线系统
grounded plate amplifier 阳极接地放大器，屏极接地放大器
grounded-resistance tester 接地电阻测定器
grounded shield 接地屏蔽
grounded short circuit 接地短路
grounded system 接地系统
grounded through resistor 通过电阻接地
grounded winding 接地绕组
grounded 接地的
ground effect 地面效应
ground electrode 接地电报，接地电极，接地极，接地体
ground elevation 室外地面标高，地面标高，地面高程，地平面标高
ground engineering 地基工程
ground experiment 地面实验
ground fall 地表崩坍
ground fault circuit interrupter 接地故障电路断流器，接地故障电路漏电保护器
ground fault circuit interrupt 接地故障电路断开，接地漏电断路
ground fault current 接地故障电流
ground fault detector 接地故障探测器，接地探测器
ground fault neutralizer grounded 接地故障中和器接地
ground fault neutralizer 接地故障中和器
ground fault production 接地故障差动保护
ground fault protection 接地故障保护
ground fault relay 接地故障继电器
ground fault 接地不良，接地故障
ground flash density 地面落雷密度，地闪密度
ground flatness 地面平整度
ground floor construction 底层地面构造，底层构造
ground floor elevation 地坪标高
ground floor of boiler house 锅炉房底层
ground floor oil piping 底层油管
ground floor passageway 底层车道，底层通道
ground floor plan of main power building 主厂房底层平面布置图
ground floor plan 底层平面图，地面层平面图
ground floor 底层
ground fog 低雾，地面雾
ground freezing 地面冻结
ground frequency 基本频率
ground gate 底孔闸门
ground glass stoppered flask 带磨口塞的玻璃瓶
ground glass stopper 磨口玻璃塞
ground glass 毛玻璃，磨砂玻璃
ground grid 接地网
ground-hog 牵引料车，土拨鼠
ground hopper 卸煤沟
ground ice 底冰
ground levelling 地面水准测量，地面整平
ground inclination 地面倾斜
ground indication 接地指示
grounding apparatus 接地设备
grounding bar 接地棒
grounding bushing 接地套管
grounding cable 接地电缆
grounding circuit 接地电路，接地系统
grounding conductor 接地导体，接地导线
grounding contact 接地触点
grounding continuity 接地连续性
grounding current compensation 接地电流补偿
grounding current 接地电流
grounding design 接地设计
grounding device 接地器

grounding distance protection 接地距离保护
grounding electrode 接地电极，接地体
grounding fault 接地故障
grounding for lightening protection 防雷接地
grounding for work 临时接地
grounding impedance 接地阻抗
grounding insulation 对地绝缘
grounding jumper 接地跳线
grounding kit 接地卡
grounding lead 接地引线，接地线
grounding main line 接地干线
grounding megger 接地电阻测试仪
grounding network 接地网
grounding net 接地网
grounding of high-voltage laboratory 高电压实验室接地
grounding of pole 电杆接地，磁极接地
grounding out 落地【荷重】
grounding pad 接地板，接地垫，接地端子
grounding plate 接地板
grounding protection 接地保护
grounding reactor 接地电抗器
grounding relay 接地继电器
grounding resistance 大地电阻，接地电阻
grounding resistor 接地电阻器
grounding screw 接地螺钉
grounding short circuit current 接地短路电流
grounding short circuit relay 接地短路继电器
grounding short circuit 接地短路
grounding signal circuit 接地信号回路
grounding surge 接地过电压
grounding switch 接地开关
grounding system 接地系统【地下】，接地网路
grounding terminal 接地端子
grounding transformer 接地变压器
grounding voltage 接地电压
grounding wire 接地线
grounding 接地，接地装置，地线，搁浅
ground insulation tester 接地绝缘测试器，绝缘测试器
ground insulation 主绝缘，对地绝缘
ground interference 地面干扰
ground inversion 地面逆温
ground lamp 接地指示灯
ground layer 近地层
ground lead 地线，接地引线，接地线
ground leakage protection 接地故障漏电保护
ground leakage resistance 接地漏电电阻
ground level concentration 地面浓度
ground level hopper 地面漏斗
ground leveling 场地平整，平整场地
ground level inversion 地面逆温
ground level source 地面源
ground level wind environment 地面风环境
ground level 地面高度，地面高程，地面标高，地平面，地坪标高，室外地坪
ground line 地面线，地平线
ground location 接地探测
ground loop 接地环路
ground lug 接地片，接地端钮
ground mat 接地网
ground moistening 土壤地下水浸润，地下湿润
ground motion 地动
ground mounted 落地式
ground movement 地层运动，地动，地面运动
ground net of electrical railway 电气化铁路接地网
ground net system 地线网络制，地线网络系统
ground net 地线网，接地网
ground noise 大地噪声，背景噪声
ground object 地物
ground ohmer 接地电阻表，接地欧姆表，精密测地仪
groundometer 接地电阻表，接地欧姆表，精密测地仪
ground over-current relay 接地过流继电器
ground pad 接地座
ground path error 接地电阻误差
ground peg 量地木栓，小桩子
ground penetrating radar 探地雷达
ground photogrammetric survey 地面摄影测量
ground photogrammetry 地面摄影测量学
ground pipe 地下管道
ground piping installed lower than freeze depth of the ground 土壤冻结深度以下的地下管道
ground plane 地平面，（风洞）地板，平面布置图
ground plate 基础板，接地板
ground plot 地形平面图
ground plug 接地插头，接地穿针
ground point 地面点，接地点
ground potential rise 地电位升
ground potential 大地电位
ground preference relay 对地保护继电器
ground pressure 地压力
ground prop 地下支撑
ground protection 接地保护
ground quartz 石英粉
ground radiation 地面辐射
ground receiving station 地面接收站
ground reception bunker 地下煤斗
ground reflection factor 地面反射因子
ground relaying 接地继电保护
ground relay 接地继电器
ground resistance decreasing agent 接地降阻剂
ground resistance test 接地电阻试验
ground resistance 接地电阻，大地电阻
ground resonance 地面共振
ground response spectrum 地面反应谱，地面响应谱
ground return circuit 大地返回电路，接地回路
ground return current 大地返回电流
ground return system 大地回路制
ground return 接地回路
ground rod 接地棒
ground roller 底部旋滚【水流】
ground roughness height 地面粗糙度值
grounds 底材
ground screen 接地屏蔽
ground setting of sub-grade 地基下沉
ground setting 地面沉降，地基下沉
ground signaling 接地信号

ground sill 底槛，地槛
ground sluicing 水力开采，地面冲采
ground stab 接地插头，接地穿针
ground state 基本状态，基本条件
ground station 地面站
ground stereophotogrammetric survey 地面立体摄影测量
ground-stirring transmission line 大地返回的送电线
ground storage area 地面贮存面积
ground storage （放射性废物）地下埋存，地面贮存
ground strap 接地母线
ground strip 接地片
ground subsidence 地层塌落，地层塌陷，地基下沉，地面陷落
ground support equipment 地面辅助设备，地面支撑设备
ground surface 地表，地面
ground survey control 地面测量控制
ground swell 岸浪，长涌浪
ground switch 接地开关
ground system diagram 接地系统图
ground system 接地系统【零电位】，接地系统
ground tackle 锚泊索具
ground temperature 地温
ground terminal 接地端子
ground test reactor 地面试验反应堆
ground treatment 地基处理
ground-type socket 接地式插座，中点接地的插座
ground vegetation 地面植被
ground vegetative cover 地面植被
groundwater 地下水，潜水
groundwater artery 地下水干道
groundwater balance 地下水量平衡，地下水平衡
groundwater basin 地下水流域
groundwater budget 地下水平衡
groundwater cascade 地下水跌差，地下水梯级
groundwater cone of exhaustion 地下水下降漏斗
groundwater contour 地下水等水位线，地下水（位）等高线，地下水（位）等深线
groundwater controlling 地下水控制
groundwater dam 地下水坝
groundwater decrement 地下水减退
groundwater deflector 地下水导流装置
groundwater depletion curve 地下水耗竭曲线
groundwater depletion 地下径流衰竭，地下水耗（损），地下水枯竭，地下水匮乏，地下水消耗
groundwater depth 地下水深度
groundwater discharge area 地下水出流区，地下水流出面积，地下水排泄区，地下水排泄面积
groundwater discharge 地下水出流量，地下水流量，地下水溢出量
groundwater divide 地下水分水界，地下水分水岭，地下水分水线
groundwater drain 地下水排水管
groundwater drawdown cone 地下水下降漏斗
groundwater drawdown 地下水面降落，地下水水位泄降，地下水水位下降
groundwater equation 地下水方程
groundwater exploitation 地下水开采［开发，勘探］
groundwater feed 地下补给，地下水补给
groundwater flow into foundation pit 基坑涌水
groundwater flow 地下水流量，地下水流（量）
groundwater fluctuation 地下水面波动，地下水位升降变化
groundwater forecast 地下水情预报
groundwater gradient 地下水面坡度
groundwater hydrograph 地下径流过程线
groundwater hydrology 地下水水文学
groundwater increment 地下水补给量，地下水增补
groundwater index 地下水指数
groundwater ingression 地下水侵入
groundwater intake facility 取地下水设施
groundwater intake 地下水进水口
groundwater intrusion 地下水侵入
groundwater inventory 地下水平衡计算表
groundwater isobath 地下水位等深线
groundwater level fluctuation 地下水位变动，地下水位波动
groundwater level 地下水水面，地下水水位，地下水位，地下水面，潜水位
groundwater lowering 地下水降低，地下水面降落，降低地下水位
groundwater mining 地下水开采［开发，勘探］，地下水抽取
groundwater mound 地下水面弯起，地下水壅高
groundwater movement 地下水运动
groundwater net 地下水网
groundwater observation well 地下水观测井
groundwater origin 地下水成因
groundwater outflow 地下水出流
groundwater overdraft 地下水超抽，地下水超采，地下水过度抽取
groundwater piracy 地下水夺流，地下水截夺
groundwater plane 地下水面
groundwater pollution 地下水污染
groundwater power plant 地下水发电厂
groundwater pressure head 地下水压力水头
groundwater profile 地下水位剖面图
groundwater protection 地下水保护
groundwater province 地下水区
groundwater recession curve 地下水位下降曲线
groundwater recession 地下水亏损，地下水位下降
groundwater recharge 地下水补给，地下水回灌，地下水重蓄，地下水再补充
groundwater regime 地下水状态，地下水分布，地下水动态
groundwater regression 地下水消落，潜水位下降
groundwater replenishment 地下水补给，地下水再补给，地下水重蓄
groundwater reserve 地下水储量
groundwater reservoir 地下水库
groundwater resources 地下水资源
groundwater ridge 地下水脊，地下分水界，地

下分水线
groundwater runoff 地下径流，地下水流
groundwater seepage 地下水渗漏
groundwater seep 地下水出口
groundwater solution 地下水溶蚀
groundwater storage curve 地下水储量曲线
groundwater storage 地下储水量，地下水储量，地下蓄水量
groundwater supply 地下给水，地下供水，地下水供应
groundwater surface 地下（水）水面
groundwater table gradient 地下水位坡降
groundwater table map 地下水位图
groundwater table profile 地下水位剖面图
groundwater table rise 地下水上升
groundwater table 地下水水位，地下水位，潜水位
groundwater tapping 地下水流出口
groundwater tracer 地下水示踪剂
groundwater transmissibility 地下水导水系数
groundwater transport of waste 地下水转移废物
groundwater trapping 地下水截流
groundwater treatment 地下水处理
groundwater trench 地下水位槽陷
groundwater turbulent flow 地下水紊流
groundwater velocity 地下水流速
groundwater washing system 地面冲洗水系统
groundwater wave 地下水波
groundwater yield 地下水出水量
ground wave reflection 地面电波反射
ground wave 地面波
ground wire gradient 架空线的电位梯度
ground wire peak 地线顶架
ground wire 地线，避雷线，接地线
groundwork 路基，基础
ground 大地，土壤，地面，地，接地，地线，地板，场地
group after group at different times 分期分批
groupage container services 散货拼箱运输
group alarm 成组报警
group ambient temperature 电缆群环境温度
group amplifier 群放大器，组合放大器
group analysis 成组分析
group-averaged cross section 群平均截面
group-averaged velocity 群平均速度
group-averaged 按组平均的，按群平均的
group battery 电池组
group bonus 集体奖金
group by group 分批
group carrier frequency 群载波频率，组载波频率
group classification 群体分类，品种分类【起重机，机构等】
group code recording 成组编码记录
group code 群码，组码
group command 成组命令
group company 集团公司
group compensation 成组补偿
group condensation 并群，群归并
group constant evaluation 群常数求取
group constant 群常数
group control 群控，组控，分组控制，群控制
group corporation 集团公司
group cross section 群截面
group delay correction 群延时校正
group delay （波的）群时延
group demodulation 群解调器
group diffusion equation 群扩散方程
group diffusion kernel 分群扩散核
group diffusion method 分群扩散方法
group discussion 集体讨论
group display 分组显示【DCS组态】，成组显示
group drive 成组拖动
grouped alarm 成群报警，分组报警
group energy interval 群能间隔
group energy 群能
group equation 群方程
group flux 群通量
group frequency 群频率
group heating 集中供热，联片供热
group Ⅱ-Ⅵ solar cell Ⅱ—Ⅵ族太阳电池
grouping error 分群误差，分组误差
grouping method 分组法
grouping theory 群论
grouping 分组，一组
group method of data handling 成组数据处理法
group method 分群法，分组法
group model 群模型
group modulation 群调制
group nozzle governing 喷嘴调节法
group of boundary layer 边界层组
group of building 建筑群
group of coils 线圈组
group of conductors 导体组
group of electrons 电子群
group of lines 线束，群桩，桩束
group of transformations 变换群
group of waves 波群
group of well 井群
group operation 成组操纵
group-overlap 群重叠
group parameters 成组参数
group pile action 桩群作用
group printing 成组打印
group record 成组记录
group relaxation 成组松弛，组松弛法
group removal cross section 群分出截面
group sampling 分组抽样
group selector 选组器
group separator 组分隔符
group shower 组合淋浴器
group switch 成组开关
group system 组合系统，组合制
group terminal block 端子箱，端子排
group theory 群论
group total advance 集团总放款
group total deposit 集团总存款
group towers effect 群塔效应
group transfer scattering cross section 群传递散射截面
group velocity （波的）群速度，波群速度，群速

group yield 群份额
group 集团，小类，组，基，族，群，团
groutability ratio 可灌（浆）比（值）
groutability 易灌注性
grout acceptance 吸浆量
grout consumption 耗浆量
grout curtain 灌浆帷幕
grouted-aggregate concrete 骨料灌浆混凝土
grouted anchor bar 砂浆锚杆
grouted concrete 压浆混凝土
grouted joint 浆锚接头
grouted micropile 注浆微型桩
grouted roof-bolt 洞顶灌浆锚杆
grouted rubble masonry 毛石混凝土建筑
grouted scarf joint 砂浆嵌缝接头
grouter 灌浆机
grout filler 灌缝砂浆，灌缝砂装
grout for bonded tendon 粘着预应力钢筋的灌浆
grout for concrete small hollow block 混凝土砌块灌孔混凝土
grout hole 浇注孔，灌注孔
grouting agent 胶结剂，灌浆材料
grouting gallery 灌浆廊道
grouting material 灌装材料
grouting of equipment 二次灌浆
grouting pressure 灌装压力
grouting pump 灌浆泵
grouting radius 灌装半径
grouting socket 灌浆管口，（水泥浆）注入接口
grouting test 灌浆试验
grouting 灌浆，水泥砂浆填平
grout injection apparatus 注浆设备
grout in 灌入
grout mixer and placer 砂浆搅拌喷射器
grout mixer 拌浆机
grout mixing plant 灰（水泥）搅拌机
grout mix 混合浆液
grout outlet 出浆口，回浆口
grout stop 止浆片
grout 薄泥浆，水泥浆，灌入，灌浆，灰浆，砂浆，浆液
grove cinder 炉渣
grower washer 弹簧垫圈
growing degree-day 生长度日，生长期有效积温
growing particle 颗粒生长
growing season 生长季节
growler 电机转子试验装置
grown-diffused 生长扩散的
growth constant 增长常数
growth curve 增长曲线
growth factor 增长因子
growth form 生长型
growth of boundary layer 边界层增长
growth of concrete 混凝土的膨胀
growth rate 增长速率，生长速率，生长率，增长率
growth shake 木材心裂
growth 生长，增长，增长率，发育
groyne basin 丁坝间水区
groyne head 丁坝坝头
groyne spacing 丁坝间距
groyne works 丁坝工程，折流坝工程
groyne 丁坝【土建】，折流坝
GRP blade 玻璃钢叶片
GRP(glass-fiber reinforced plastic) 玻璃钢
GRP(glass fiber reinforced polyester resin) 玻璃纤维增强聚酯树脂
GRT(gross registered tonnage) 总登记吨，总吨
GRTP(glass fiber reinforced thermoplasticmaterials) 玻璃纤维加强热塑材料
grubbing 树根掘除，挖土，掘出
grub screw 木螺丝，平头螺丝
grummet 麻丝垫环
grush 花岗岩碎砾，粒状岩石
gruss 花岗岩碎砾，似花岗砂岩，粒状岩石
grus 风化花岗质砂岩，粒状岩石
GRV 发电机氢气供应系统【核电站系统代码】
Gr. Wt(gross weight) 毛重
GSCB(generic substaion status event) 通用变电站状态事件控制块
GSCE(gland steam condenser exhauster) 轴封抽气机
GSC(gland steam condenser) 轴封加热器
GSE(general substation events) 通用变电站事件
GSEM(general substation events model) 通用变电站事件模型
GSE 汽轮机保护系统【核电站系统代码】
GS(gland steam) 轴封蒸汽
GSH(gland steam heater) 汽封蒸汽加热器，轴封蒸气加热器
GSM(Global System for Mobile Communications) 全球移动通信系统
GSSE conrtol 通用变电站状态事件控制【功能约束】
GSSE control block 通用变电站状态事件控制块
GSSE(general substation state event) 通用变电站状态事件
GSS(gland seal system) 轴封系统
GSS 汽轮水分离再热器系统【核电站系统代码】
GST 发电机定子冷却水系统【核电站系统代码】
GSY 同步并网系统【核电站系统代码】
GTAW(gas tungsten arc welding) 钨极惰性气体保护焊，钨极气体保护电弧焊
GTC(general terms and conditions) 合同通用条款
GTCS(gas turbine control system) 燃机控制系统
GT(gas turbine) 燃气轮机
GTG(gas turbine generator) 燃气轮机发电机
GTH 汽轮机轮润滑油处理系统【核电站系统代码】
GTO(gate turn off thyristor) 门极可关断晶闸管
GTR 汽轮发电机遥控系统【核电站系统代码】
GT(generator transformer) unit 发变组
GTV(gate valve) 闸门
Guandong Nuclear Power Joint Venture Company 广东核电合营有限公司
guarantee agreement 担保协定，担保协议
guarantee and security 保证和担保
guarantee bond 保证书，担保书
guarantee company 担保公司
guarantee contract 担保契约
guaranteed capacity 保证出力，保证容量
guaranteed efficiency 保证效率

guarantee deposit 押金，保证金
guaranteed export paper 保付出口票据
guaranteed heat consumption 保证热耗
guaranteed horsepower 保证马力
guaranteed index 保证指数
guaranteed letter of credit 保兑信用证，凭保证开立的信用证，凭保单开出的信用证，凭保证开出的信用证
guaranteed minimum output 保证最低功率
guaranteed output 保证输出功率【电厂】，保证出力，保证输出功率
guaranteed performance data 性能保证数据
guaranteed performance 保证性能
guaranteed period 保用期，保证期，保修期
guaranteed price 保证价格
guaranteed qualities 保证品质
guaranteed speed regulation 保证速度调节范围
guaranteed turbine efficiency 水轮机保证效率
guaranteed value 保证值
guaranteed yield stress 保证屈服应力
guarantee flow 保证流量
guarantee for contract 合同的担保
guarantee fuel 保证燃料，备用燃料
guarantee fund 保证金
guarantee head 保证压头
guarantee incident to contract 合同附带的保证
guaranteeing association 担保协会
guaranteeing delivery 保证交货
guarantee law 担保法
guarantee letter 保函
guarantee money paid on contract 按合同付保证金
guarantee of acceptance 承兑保证
guarantee of export credit 出口信贷担保
guarantee of insurance 保险担保书
guarantee of payment 付款保证
guarantee of performance 履约保证，履约保函，履行义务保证
guarantee of quality 质量保证书
guarantee on first demand 保证债权人提出时立刻履行
guarantee payment 保证支付
guarantee period 保证期，保质期，保修期，担保期限
guarantee point 保证点
guarantee pressure 保证压力【由制造商保证的】
guarantee reagent 保证试剂
guarantee speed 保证转速
guarantee the reliability of the buyer 担保买方的信用
guarantee to measure fully up to the contract specifications 保证完全符合合同条款
guarantee to pay compensations 包赔
guarantee 保证，保证书，担保，保函，保证期，确保
guarantor 保证人，担保人
guaranty funds 保证金
guaranty period 保证期
guard against damp 防潮
guard band 保护带，防护频带
guard bit 保护位
guard board 挡板，护板
guard boom 防护栅
guard cable 安全索，防护索
guard circuit 保护电路，安全保护通道
guard dam 防护坝
guarded experimental subassembly 带保护装置的试验燃料组件
guarded multi-pipe 套装管
guarded oil pipe 套装油管
guarded switch 保险开关
guarded type motor 防护型电动机
guard electrode 保护电极
guarder and janitor room 警卫和传达室
guard gate 安全闸门，防护闸门
guard grating 安全栅，保护
guard house 警卫室
guardian 监护人
guarding counter 屏蔽计数器
guard lamp 保护灯
guard lock 防护闸
guard net 保护网
guard of circuit 电路防护
guard pipe 护管【贯穿件】
guard plate 护板
guard post 防护柱
guard rail 防护栏，护轨，护栏，保护栏杆
guard relay 保持继电器
guard-ring 保护环，电晕保护环
guard room 警卫室
guard signal 保护信号
guard space 保障区间
guards-rails 围栏护轨
guard stake 防护桩
guard's van 守车
guard valve 速动阀，事故阀，防护阀
guard vessel system 泄漏拦截系统，保护容器系统
guard vessel 保护容器
guard wall 护墙
guard wire 保护线
guard 防护，保护，挡板，防护器，护罩，防护装置，警卫，看守
gudgeon pin 耳轴销
gudgeon 舵枢，耳轴，轴柱
guesswork 推测
guest house 宾馆
guest room 客房
guest worker 外来工人
guhr 硅藻土
guidance 制导，指导，导航
guidance note 指导性说明
guidance system 导航系统
guidance to the direction 引导投资方向
guide angle 喷嘴叶栅的出口角
guide apparatus 导向器，导向装置
guide assembly 导向筒组件
guide baffle 导流叶片
guide bars 导向棍，导杆，导向杆
guide bearing bracket 导向轴承支架
guide bearing of hydro-generator 水轮发电机导轴承

guide bearing of water turbine　水轮机导轴承
guide bearing　导向轴泵【立式泵】，导向轴承
guide blade carrier　导叶环
guide blade loss　导叶损失
guide blade ring　导叶环
guide blade row　导叶列
guide blade support ring　导叶支撑环
guide blade　导流片，导流叶片，导向叶片，导叶
guide book　指导手册，导向块
guide bracket　导架
guide bush　导向套筒
guide cable　引示电缆
guide cradle　导向吊架
guide crest　导流堰顶
guide curve for flood control　防洪限制线
guide curve　导向曲线
guide dam　导水坝
guide dike　导流堤
guide dimple　导向陷窝
guide disk　导向盘
guided propagation　波导传播
guide drawing　指导图
guided wave　被导波，循迹波
guided wideband transmission　制导宽带传输
guide fixture　导向夹具
guide for navigation　导航
guide fossil　标准化石
guide frame　导向框架
guide groove　导槽
guide hole　超前炮眼，导孔
guide housing　导向框，导向罩
guide key　导键
guide levee　导堤
guideline for procurement　采购导则
guide lines　导则，守则，指南
guideline　导向器，导向图，指南，导线，导则，准则，标线，控制线
guide link　导环
guide pad　导向基座，导向底板，导向垫块
guide passage　导流段
guide pile　定位桩，导桩
guide pillar for casing　汽缸导柱
guide pillar　导杆，导销，导柱
guide pin　导向销，定位销
guide plate　导向板，支撑板，导流板，导板
guide post　导柱【堆芯辅助支撑】，导杆，路标
guide pulley　导向轮，压带轮，导向滑轮，导向滑车
guide rail　导轨
guide ray　定向射线
guide rib　导向肋片
guide ring　（水轮机的）调速环，导环，控制环，导向绳，导向环
guide rod type diesel pile hammer　导杆式柴油打桩锤，导杆式柴油桩锤
guide rod　导杆，导向棒，导轮，导向杆
guide roller　导辊，导轮
guide rule　导则，准则
guide seam　标准层，标准矿层
guide sheath　导向护套【一般为非圆形管】
guide shoe　导向底板，导向基座
guide slot　导槽，导向槽
guide specification　指导性技术要求
guide spindle bearing　导轴轴承
guide spring　定位弹簧
guide stem　导柱，导杆
guide structure　导航建筑物
guide stud　导向杆【压力壳盖】，导向螺栓
guide support　导向支架
guide system　导向系统
guide thimble　导向套管【燃料组件】
guide tool　导向工具
guide tube insertion　导向管插入
guide tube　导向筒【控制棒】，导管，导向管
guide valve　导向阀
guide vance　导向器叶片
guide vane arm　导叶拐臂
guide vane bearing　导叶轴承
guide vane control valve　导叶控制阀
guide vane lever　导叶臂杆
guide vane opening　导叶开度
guide vane packing　导叶填料
guide vane ring　导叶环
guide vane stem　导叶叶柄
guide vane trunnion　导叶轴（颈）
guide vane　泄水台【排水】，引流，导叶，导向叶片，导叶片，导流板
guide wall　导流墙，导水墙
guide wavelength　波导管波长
guide way　导向槽，导向套筒
guide wheel bucket　导向叶片
guide wheel　导叶环，导轮
guide wire　尺度定距索，准绳
guide work　导航建筑
guide　入门，导向，指南，导轨，准则，导则，导引器，导架，导槽，手册，导杆
guiding device　导向装置
guiding document　指导文件
guiding drawing　（设计）司令图
guiding field　导向场，控制场
guiding groove　导槽
guiding hole　中导孔
guiding idler　导向托辊
guiding jetty　导航堤
guiding plan　指导性计划
guiding price　指导价格
guiding principle　指导方针
guiding shaft　导向轴
guiding surface　导向表面
guiding trestle　导向支架
guild　行会
guillotine　截流器，剪断机
guillotine break　剪切断裂，切断破裂
guillotine door　煤闸门
guillotine rupture　切断破裂
guillotine shear　剪切机
guillotining　截断
gulch　冲沟，沟谷，峡谷
gulder　（双）低潮，最低水位
gulf circulation　旋涡循环
gulf stream　湾流

gulf 海湾，深坑，湾
gullet 海峡
gulley control planting 沟蚀防治种植
gulley control 冲沟控制，沟蚀防治
gulley correction 沟壑整治
gulley cutting 沟蚀冲切
gulley erosion form 沟蚀形态
gulley erosion 冲沟侵蚀，沟形侵蚀，沟状侵蚀，沟蚀
gulley head erosion 沟头冲蚀，沟头冲刷，沟头侵蚀
gulley stabilization structure 沟壑稳定建筑物
gulley trap 水沟排污井，集水沟隔气弯管
gulley 冲沟，冲刷沟，沟壑，路沟窨井，排水沟，屋檐槽，落煤管斜槽，集水口，地漏，雨水口
gullied area 沟蚀地区
gullied surface 沟壑地表
gull wing sail 海鸥翼式帆
gully control 冲沟控制，沟蚀
gully grating 集水口箅子
gullying 沟蚀，沟蚀作用
gully pit 集水坑
gully-stabilization works 沟壑稳定工程
gully water （楼面）泄水，（楼面）排水
gully 落煤管斜槽，屋檐槽，集水口，地漏，雨水口
gulp 字节组
gumbo bank 泥滩
gummy 树胶状的
gum rubber 天然橡胶
gum 树胶，胶接，树脂，橡胶
gunboat 自翻斗车，料斗
gun finish 喷枪修整【混凝土表面的】
gun-injected anchoring 枪射铆固
gunite coat 喷浆面层
gunite covering 喷浆面层
gunited concrete 喷浆混凝土
gunite layer 喷浆层
gunite lining 水泥喷射衬砌【灌浆】，喷浆衬砌
gunite material 喷浆材料
gunite work 喷浆
gunite （耐火）喷浆，喷射灌浆
guniting 喷浆
gun-mix bond 喷浆黏合料
gunned castable 喷浆
gunning 喷浆，喷射
gunny felt 麻布油毡
gunny sack 粗麻布袋，麻袋
gun oil burner 枪式油喷燃器
gunpowder 火药
gun-type coring machine 枪式取土器
gunwale 舷边，舷缘
gun 喷射器，枪，喷雾枪，拉索
gurgitation 起伏
Gurney flap 格尼襟翼
gusher 自喷油井
gushing bed 喷射床
gushing 喷出，涌出
guss asphalt 流态地沥青
guss concrete 流态混凝土
gusset connection 结点板连接
gusset plate 角撑板，加固板，节点板，联结板，梁腋，结点板
gusset stay 角撑条
gusset 补强板，角板，角撑板，节点板，结点板，连接板，联接板
gust 阵风
gust alleviation factor 阵风衰减因子
gust amplitude 阵风变幅
gust and lulls 阵风阵息，风阵风歇
gust anemometer 阵风风速计
gust averaging time 阵风平均时间
gust component 阵风分量
gust decay time 阵风衰减时间
gust downwash 阵风下洗
gust duration 阵风延时，阵风持续时间
gust effect factor 阵风响应因子
gust effect 阵风效应，阵风因子
gust energy factor 阵风能量因子
gust environment 阵风环境
gust excitation 阵风激励
gust factor approach 阵风因子法
gust factor 阵风因子
gust formation time 阵风形成时间
gust frequency 阵风频数
gust front 阵风锋面，阵风锋
gust generator 阵风发生器
gustiness effect 阵风效应
gustiness factor 阵风系数，阵风因子
gustiness wind 阵风
gustiness 阵发性
gust influence 阵风影响
gust intensity 阵风强度
gust lapse rate 阵风递减率
gust lapse time 阵风递减时间
gust loading 阵风荷载
gust load 突风荷载，阵风荷载
gust measuring anemometer 阵风风速计，阵风测量风速计
gust of rain 阵雨
gust peak speed 阵风最大风速
gust recorder 阵风记录
gust response factor 阵风响应因子，阵风影响系数
gust scale 阵风尺度
gustsize 阵风尺寸
gust spectrum 阵风谱
gust velocity variation 阵风速度变化
gust volume 阵风容积
gust wind speed 阵风风速
gust wind tunnel 阵风风洞
gusty wind 阵风
gutta-percha 杜仲胶，古塔胶
gutter board 挑口板
gutter drainage 明沟排水，檐沟排水
gutter hanger 檐沟托
gutter hook 檐沟托，檐口
guttering 深剥沟槽
gutter inlet 雨水口
gutter spout funnel 落水斗
gutter 沟，槽，排水口，集水口，沟槽，雨水

槽，排水沟，狭沟，小冲沟，檐沟
gut 狭窄水道
guy anchor 拉条地锚
guy cable anchor 拉索地锚
guy cable 拉索，拉缆，缆风
guy clamp 拉线夹
guy clip 拉线卡子，线卡子
guy derrick 拉索式桅杆起重机，牵索起重机
guyed building 拉索建筑
guyed cantilever 拉索悬臂
guyed delta tower 拉锚塔
guyed-mast 拉线式电杆
guyed pole 拉线杆
guyed portal steel tower 带拉索的门式铁塔
guyed steel chimney 拉索式钢烟囱
guyed steel stack 拉线式钢烟囱
guyed structure 拉索结构
guyed tower 带拉线塔，拉索塔架，拉线塔
guyed vee tower 牵拉V字形塔，V形拉线塔
guyed V tower V形拉线塔，牵拉V字形塔
guyed Y tower Y形拉线塔
guy insulator 拉线绝缘子
guy line 电线杆的斜拉线
guy rope 牵索，拉线，钢缆，拉索，张索，拉绳
guy strain insulator 拉线耐拉绝缘子
guy tension 拉索张力
guy wire 拉线，牵索，张索，拉索
guy 拉线，拉杆，牵索，拉索，拉绳，拉住，缆索，用支索撑住
G-value G值
GV(governor valve) 高压调门，主汽调节门
GV （高压）调节汽门【DCS画面】
GW(gigawatt) 千兆瓦，十亿瓦，百万千瓦
GWh(giga watt hour) 十亿瓦时
gymbal 万向接头
gymnasium structure 体育馆结构
gymnasium 体育馆
gypseous 石膏的
gypsite 土石膏
gypsum block soil moisture meter 石膏土壤湿度计
gypsum block 石膏块材
gypsum board lath 石膏板条
gypsum board sheathing 石膏衬板
gypsum board 石膏板
gypsum concrete 石膏混凝土
gypsum insulation 石膏保温
gypsum model 石膏模型
gypsum panel 石膏墙板
gypsum plasterboard 石膏墙板
gypsum plaster 石膏粉饰
gypsum-retarded cement 石膏缓凝水泥
gypsum sheathing 石膏盖板
gypsum tile 石膏瓦
gypsum trowel finish 石膏粉刷饰面
gypsum wall board 石膏墙板
gypsum wood-fibered plaster 石膏木丝灰泥
gypsum 用石膏处理，石膏
gyradisc crusher 转盘式粉碎机
gyral 环流，涡流
gyrate 回转
gyrating current 旋流
gyrating mass （飞轮）旋转质量
gyration centre 回转中心
gyration moment 转动力矩
gyration radius 回转半径
gyration 回转，环动，旋转
gyratory breaker 回转式碎石机，回转轧碎机
gyratory compactor 回转式夯具，回转式压实机
gyratory crusher 回转式碎石机，旋摆式碎石机
gyratory-lifting tower crane 自升式塔式起重机
gyratory screen 旋转筛
gyrator 方向性移相器，回转仪，旋转子
gyre force 环动力
gyre 涡流，旋转的，回旋的，大旋涡
gyro 陀螺，陀螺仪的
gyro axis 回转轴
gyroclinometer 回转式倾斜仪
gyro-compass 回转罗盘
gyroelectric medium 旋电介质
gyroelectric 回转电路的，旋电的
gyro frequency 旋转频率
gyrolevel 陀螺倾斜仪
gyrolite 白钙沸石
gyromagnetic coefficient 磁旋系数
gyromagnetic effect 旋磁效应
gyromagnetic medium 旋磁介质
gyromagnetic ratio 磁旋比
gyromagnetic resonance 旋磁共振
gyromagnetic 回转磁的，旋磁的
gyromixer 回转拌和机
gyro rotor end play micrometer 陀螺转子端隙测量计
gyroscope sextant 陀螺六分仪
gyroscope 陀螺仪，回旋装置，回转仪，纵舵调整器，回转器
gyroscopic 陀螺的
gyroscopically balanced flowmeter 陀螺作用的流量计
gyroscopically controlled 用陀螺操纵的
gyroscopic effect 陀螺效应
gyroscopic force 陀螺力，回转力
gyroscopic integrator 陀螺积分器
gyroscopic load 回转负荷
gyroscopic moment 回转力矩
gyroscopic motion 回转运动
gyroscopic pendulum 回转摆
gyroscopic precession 陀螺仪进动，陀螺进动
gyroscopic stabilizer 陀螺稳定器
gyroscopic torque 回转力矩
gyroscopic total station 陀螺全站仪
gyroscopic turn meter 陀螺回转指示仪
gyrosextant 回转式六分仪，陀螺六分仪
gyro sight 陀螺瞄准器
gyrostat 回转轮，回转仪
gyrotron maser 陀螺振子微波激射器
gyrotron 振动陀螺仪
gyrounit 陀螺环节
gyttja 湖底软泥，腐泥
G 汽轮发电机【核电站系统代码】

H

habeas corpus act　人身保护法
habeas corpus　人身保护权，人身保护法
habitable room　住所
habitat diversification　生境多样化
habitation　居住，住宅
habitat　栖息地
habit　习惯
hackberry　朴树
hack hammer　劈石斧
hacking off　打毛
hacking　凿毛
hack joining surface　凿毛接缝面
hackly surface　粗糙不平表面
hackly　锯齿状的，粗糙的
hacksaw blade　钢锯条
hacksaw frame　钢锯架
hacksaw　可锯金属的弓形锯，钢锯，弓锯，平背手锅
hack　凹口，切槽
hade angle　伸角，伸向
hade-slip fault　倾向滑断层
hade　断层伸角，断层伸向，断层余角，伸角，伸向
haeremai　欢迎辞
haft cycle　半周期
haft-duplex　半双工，半双工的，半双向的
haft-excited core　半激励磁芯
haft　手柄，旋钮
haggle　争论，论价，讨价还价
Hague Rules　海牙规则
hail day　雹日
hail-destroying rocket　防雹火箭
hail shooting　降雹
hail stage　成雹阶段
hailstone　雹块
hailstorm days　冰雹日
hailstorm　雹暴，雹暴般的降临，风雹，大冰雹
hail　雹，冰雹
hair brush　毛刷
hair checking　发状辐裂，发裂，细裂纹，发丝裂缝，发状裂缝，细缝
hair felt　毛毯，毛毡
hair fibered plaster　麻刀灰泥
hair hygrometer　毛发湿度计
hair line crack　发裂，细裂纹［缝］，毛发状裂缝，毛细裂纹［裂缝］，收缩裂纹，发纹
hairline　极细微的【裂缝、缝隙等】，瞄准线，发际线，发丝线
hair mortar　麻刀灰泥
hair of sight vane　视准丝
hairpin bend　发夹形弯头
hairpin coil　发夹式线圈
hairpin tube bundle　发夹形管束
hairpin tube　发夹形管
hairpin vortex　发卡涡
hairpin　夹发针，发夹
hair spring　发丝簧，细弹簧，游丝，细测量线
hair　发丝，金属细丝，毛发，麻刀，麻丝
halation　晕光作用，成晕现象
Halden effect　哈顿效应
half adder　半加法器，半加器
half add　半加
half-adjust　舍入
half and half joint　对拼接头
half angle　半角
half-bearing　半轴承
half-breadth　（船的）半宽
half-brightness effect　半亮度效应
half bubble　半泡，半磁泡
half capacity fan　50%总风量的风机
half capacity operation　半负荷运行
half-carry　半进位
half-casing　半缸
half cell potential　半电池电势
half cell　半电池
half-coiled winding　半圈式绕组，半圈绕法
half coil　半节线圈，线圈边
half-compression relief cam　降压调节凸轮
half coupling　联轴节半部，半个联轴节，半联轴器，半管接头，半联结器，半边联轴器
half cross belt　半交叉皮带，直角挂轮皮带
half cruciform core　T形铁芯
half-crystal can relay　半晶体密封继电器
half cycle　半循环，半周期
half-degree reaction　50%反动度
half-desert　半荒漠，半沙漠
half-dug foundation　半掏挖基础
half-duplex operation　半双工作业，半双向操作
half-duplex traffic　半双工传送
half-duplex transmission　半双工传输
half-duplex　半双工，半双向的，半双工的，半双工
half energized　半激励的，半激励状态的
half-finished product　半成品
half-formed winding　半成型绕组，半成型线圈
half frequency　半频
half gap　半间隙
half hairpin spring　半个发夹弹簧
half hard copper　半硬铜
half-intensity width　半强度宽度
half-interval contour　半距等高线
half irradiation　半剂量辐照
half-lapping　半叠包，（线圈的）半叠绕，楼梯平台
half lap scarf joint　半叠拼接，半搭接
half-lethal dose　半数致死剂量，半致死剂量
half-life measurement　半衰期测量
half-life period　半衰期
half-life　半衰期，半摊出期
half load　半负荷，半负载
half log　对开木
half mean spring range　大潮平均半潮差
half-metal　半金属
half-mitre joint　半斜接
half model test　半模试验
half-moon ring spanner　半月形梅花扳手

half nut 对开螺母
half-opened impeller 半开式叶轮
half open interval 半开区间
half-pace 上楼梯转弯处平台，梯台
half peak width 半峰宽度
half-period average-current 半周平均电流
half-period 半周期，半周
half-phase 半相，半相的
half-picture storage tube 半像存储管
half-power frequency 半功率频率
half-power point 半功率点
half-power 半功率
half-principal （伸不到屋脊的）半椽条
half reading pulse 半读出脉冲
half retractable type soot blower 半伸缩式吹灰器
half ring 半环
half round file 半圆锉
half round iron 半圆钢
half round pointing 半圆缝
half round timber 半圆木
half-saturated 半饱和的
half-section network 半节网络
half-section 半截面
half-selected core 半选磁芯
half-selected element 半选单元
half-shift register 半移位寄存器
half sine-wave 半正弦波
half-sinusoid 半正弦曲线
half-space 半空间
half-span roof 单坡屋面，单坡屋顶
half-span winding 半节距绕组
half speed generator 半速发电机
half speed saturated steam turbine 半速饱和蒸汽汽轮机
half speed steam turbine 半速汽轮机
half speed synchronization 半速同步
half speed turbine 半速汽轮机
half speed 半速
half story 顶楼，屋顶层
half stroke 半冲程
half subtracter 半减器
half-thickness 半厚，半值厚度
half-timbered construction 砖木混合结构，砖木结构
half-time 半衰期
half-track tractor 半履带式拖拉机
half-turn coil winding 半匝绕组
half-turn coil 半匝线圈
half-turn 半匝，半圈
half-value depth 半值层深度，半值深度
half-value layer 半值层
half-value period 半衰期，半值期
half-value thickness 半值厚度，半值层厚度
half-value width 半值宽度，半值层宽度
half vortex blading 半涡流流型叶片
half warmed turbine 半热态汽轮机
half-wave antenna 半波天线
half-wave diode rectifier 半波二极管整流器
half-wave dipole antenna 半波偶极天线
half-wave doubler 半波倍压器
half-wave line 半波线
half-wave mode 半波模态
half-wave potential 半波电位
half-wave power transmission 半波长输电
half-wave rectification 半波整流
half-wave rectifier 半波整流器
half-wave symmetry 半波对称
half-wave thyristor rectifier 半波可控硅整流器
half-wave transformer 半波整流变压器
half-wave transmission 半波传输
half-wave 半波
half way of the mountain 山腰
half winding 半圈绕组
half-write pulse 半写脉冲
halide leak detector test 卤化物检漏试验
halide 卤化物
halite 岩盐，石盐类，天然的氯化钠，石盐
Hall displacement transducer 霍尔位移传感器
Hall effect relay 霍尔效应继电器
halloysite 多水高岭土，埃洛石
hall pressure transmitter 霍尔压力变送器
hallway 门厅
Hall motor 霍尔电机
haloalkylated resin 卤化烷基化树脂
halocline 盐度突变层，盐跃层
halo effect 晕圈效应
halogen leak detector 卤素检漏器
halogen leak testing 卤素检漏试验
halogen sniffer test 卤素嗅漏试验
halogen test 卤素试验，卤素探漏试验
halogen 卤素，卤
halomorphic soil 盐成土
halon extinguishing agent 卤代烷灭火剂
halon extinguishing system 卤代烷灭火系统
halon fire extinguisher 卤代烷灭火器
halon fire extinguishing system 卤代烷灭火系统
halon suppressant 卤代烷抑爆剂
halo phase 晕相
halophilic bacteria 嗜盐细菌
halo region 晕区
halotrichite 铁明矾
halt instruction 停机指令
halt 暂停，停止，停机
halved joint 对搭接式接头
halving 平分
halving circuit 半分电路
halving register 平分寄存器
hamada 石质沙漠
hammada stone desert 石质沙漠
hammer apparatus 机械锤
hammerblow action 锤击动作
hammerblow effect 锤击效应
hammerblow handwheel 锤击手轮
hammerblow tamper 锤击夯
hammerblow 锤击，水击，锤打，撞击，打击
hammer breaker 锤碎机
hammer carrier 锤盘
hammer crusher 锤击式破碎机，锤式破碎机，锤式碎石机，捣碎机
hammer-dressed quarry stone 锤琢块石
hammer dressing 锤击修整

hammer drill 风钻，冲击式钻机，凿岩机，锤钻，电锤，冲击钻机
hammer finish 锤纹漆
hammer grab 冲击式抓斗
hammer head crane 塔式起重机，锤头式起重机，塔式悬臂吊车
hammer head groin 锤型防波堤
hammer head groyne 锤型防波堤
hammer head key 锤头形键槽
hammer head 锤头
hammering device 锤击装置
hammering test 敲打检查
hammering 锤打，锤击，锻打
hammer in 打入
hammer man 打桩工
hammer mill crusher 锤击破碎机
hammer mill 竖井式磨煤机，锤击式磨煤机，锤式破碎机，锤磨机
hammer mine 锤式破碎机
hammer pulse 锤击脉动
hammer ram 夯锤
hammer roll 锤碎机
hammer stroke 桩锤冲程
hammer type coal crusher 锤式碎煤机
hammer type vibrator 锤式振动器
hammer welding 锻接
hammer 锤，榔头，锤击
Hamming data window 数据海明窗
Hamming distance 海明距离
hance arch 平圆拱，三心拱
hance 拱腋，梁腋
hand adjustment 手调节，人工调整，手工调整
hand and foot monitor 手足监测器，手足放射性监测器
hand anemometer 手持风速计
hand back 退还
hand-barring 手动盘车
hand bending machine 手工弯筋机
handbook 手册
hand boring 手摇钻探
hand brake wheel 制动手轮
hand brake 手动制动器，手动闸
hand buggy 手拉车
hand by-pass valve 手动旁通阀
hand calculator 手摇计算机
handcart 手车，手推车，单轨小车
hand centrifuge 手摇离心机
hand chain block 手动滑轮（组）
hand-cleaned rack 人工清除格栅
hand clearing 手工清扫
hand clutch cyclometer 带有离合器的手控回转计
hand-coloured map 手工着色地图
hand-compacted concrete 人工捣实混凝土
hand contamination monitor 人工污染监测器，手污染监测器，手污染检测器
hand-controlled 手控的，手动控制的
hand control valve 手动阀
hand control 手动控制，人工控制，手工控制
hand counter 手控计数器
hand crane 手摇起重机
hand crimping tools 手工压接工具
hand-drawn original 清绘原图
hand drill 手摇钻，手钻，摇臂钻
hand driven generator 手摇发电机
hand drive 手操作驱动，手驱动
hand dug well 手工挖掘井
hand electrical drill 手电钻
hand emergency tripping level 手动紧急脱扣杆
hand feed punch 人工馈送穿孔机
hand feed 手工加料，手动进给
hand finisher 手工整修机
hand fired boiler 手烧锅炉
hand fired furnace 手烧炉膛
hand fired grate 手烧炉排
hand-firing 手烧，人工燃烧
hand float 手镘，（手）镘板
hand folder 手动折边机
hand-foot and clothing monitor 手足及衣服检测器
hand gear 手动装置
hand generator 手摇发电机
hand-goniometer 手持测角仪
hand grinder 手摇砂轮
hand guard 扶手
hand held computer 手提电脑，掌上电脑
hand held console 手旋支柱
hand held GPS map 掌上 GPS 定位地图
hand held terminal 手旋端钮
hand held wind anemometer 手持式风速计
hand hoist 手动绞车
handhold 扶手，栏杆，握住，线索，把柄，手柄，手把
hand hole cover 手孔盖板
hand hole 手孔，检查孔
handiness 操纵性轻便，灵巧，敏捷
hand insertion 手工嵌线，手工下线
hand in 提交，递交
hand jack 手动起重器，手摇千斤顶
hand lamp transformer 手提灯变压器
hand lamp 手提灯，手电筒，行灯
hand lance 手提式喷枪，手动吹灰枪
hand lancing 手工吹灰
handleability 操纵性，可运用性，操作能力，可处理性，可输送性
hand lead 测深锤
handle bar 把手
handle contractual disputes 处理合同争议
handle exclusively 独家经营
handle magneto 手摇永磁发电机
handle-operated breaker 长柄操作的断路器
handle procedures 办理手续
handler 近距离操纵机械手，处理机，处理器
hand level 手水准，手柄，手持水平仪
handle with care 小心轻放
handle with great care 特别当心搬动
handle 把手，柄，手柄，搬运，处理，拉手，提手，手轮，手把，管理
hand lift 手动启门机，手摇起重机
handling ability 装卸能力，输送能力
handling agent 操作代理
handling airlock 操作气闸
handling and slinging 搬运与吊装

handling apparatus 起吊装置
handling area 装卸区，装卸大厅
handling axis 装卸轴线
handling bay 装卸大厅
handling button 装卸按钮
handling capacity 转运能力，加工能力，操纵能力，输送能力，装卸能力，处理容量，处理能力
handling cask 贮运容器，装卸容器
handling charge 操作费用，管理成本
handling control centre 操作控制中心
handling costs 操作费用
handling crane 装卸吊车，桥式装卸吊
handling device 操作设备，装卸设备，起吊装置
handling dolly 燃料传送车
handling equipment 装卸设备，操作装置，起重装卸机械，装卸运输设备，输送设备
handling facility 装卸设备
handling fees 操作费用，管理成本
handling gantry 吊装龙门架
handling hall 操作大厅，装卸大厅
handling hole 装卸孔
handling in field 现场搬运
handling knob 操作球头，承抓球头
handling lug （元件盒）操作头
handling machinery 搬运机械，装卸机械
handling machine 操作机，装卸机，起吊机械
handling mast 起吊杆
handling of materials 搬运材料
handling of nonconformances 不符合项的处理【质保期】
handling of radioactive waste 放射性废物处理
handling opening 吊装孔
handling operation 装卸操作
handling pump 运料泵
handling quality 操纵性能
handling radius of crane 起重机臂工作半径
handling room 操纵室
handling skid 搬运滑道
handling station 装卸站
handling stress 起吊应力
handling system 搬运系统，装卸系统
handling time 处理时间，装卸时间
handling tongs 操作钳
handling tool 装卸工具，操作工具
handling 搬运，输送，处理，操纵，吊运，装卸，装卸运输，装运，管理
handling 处理，加
hand-manipulated 用手操作的
hand mapping camera 手提测图摄影机
hand microtelephone 手持微型电话机
hand mixing 人工拌和
hand money 订金，保证金
hand off 手动切断
hand of rotation 旋转方向
hand opener 手动开关器
hand-operated auger 手摇螺旋钻，手摇麻花钻
hand-operated auxiliary bridge 手动辅助桥架
hand-operated impact driver 手动操作冲击手柄
hand-operated monorail hoist 手动单轨吊
hand-operated pump 手摇泵
hand-operated regulator 手动调节器
hand-operated starter 手动启动器，手动起动器
hand-operated switch 手动开关
hand-operated valve 手动门，手动阀
hand-operated 手操作，手工操作的，手动的，人工操作的，手动控制的
hand-operating mechanism 手动操作机构
hand-operation 手动控制，人工操作
hand operator 手操作器
hand over against payment of the price 凭付款凭证交付
hand over procedure 移交程序
hand over 移交
hand packing 人工充填
hand-paint 手工着色
hand pallet truck 手摇平板车
hand pile driver 人工打桩机
hand pipe bender 手弯管机
hand placed riprap 人工抛石护坡
hand plane 手刨
hand powder bulb 手工喷粉器
hand pulley block 手动滑轮（组）
hand pump 手动泵，手摇泵
hand punched card 人工穿孔卡
hand punch 手动穿孔机，手动穿孔器
hand pyrometer 手提式高温计
handrail mounted 扶手安装
handrail of staircase 楼梯扶手
handrail 扶手，栏杆
hand rake 手耙
hand rammer 人工夯
hand ram 人工夯
hand ratchet without handle 手用棘轮头
hand regulation 手动调整，人工调整
hand reseting 手动复归，手动复位
hand reset relay 手动复位继电器
hand reset 手动复归，人工复位
hand riveting tool 手松铆接工具
hand riveting 手铆
hand-rotate 手旋
hand rule 手定则
hand sampling 手工取样
handset 听筒，受话器
hand shake cycle 交换过程处理周期，信号交换周期
hand shoveling of coal 人工投煤
hands off 手动断路，请勿动手，请勿用手摸
handsome reward 重奖
hands queeze stripping tool 手压剥线工具
hand starter 手动启动器
hand stirring 手动搅拌
handstone 小石子
hand synchronizer 手摇同步器
hand tally 计数器
hand telephone 电话手机
hand tight 手压致密
hand tool 手工工具
hand tracking 手动跟踪
hand trip control 手动跳闸控制
hand trip gear 手动脱扣器

hand tripping 手动跳闸，手动打闸停机，手动脱扣
hand trolley 手摇车
handtruck 手推车
hand tuning 手动调谐
hand valve 手动阀
hand welding machine 手工焊机
hand welding 手工焊
hand wheel-brake 手轮制动
handwheel switch 手轮开关
handwheel 手轮，操纵盘，手动转轮
hand winch 手动绞盘，手动卷扬机，手摇绞车，手动绞车起重机
hand winding 手绕法，人工嵌线
hand windlass 手摇卷扬机
handwork hood 手工操作通风柜
handwork 手工，手工作业
handwriting 笔迹
handwritten signature 亲笔签字
handy conveyer 轻便传送装置
handy conveyor 轻便传送装置
handy rule of thumb 经验定律
handy 方便的，轻便的
hand 手，手柄，指针，手动的，人工的，工人，劳动力
hanged rod 吊杆
hanger adjustment 吊架调整
hanger bearing 悬挂支承，吊轴承
hanger connection 吊架接头
hanger frame 吊架
hanger hook 吊钩
hanger kit 悬挂卡
hanger location 吊架定位
hanger rod 吊杆，拉杆
hanger tube 悬吊管
hanger 悬杆，吊架，吊钩，吊卡，挂钩，支吊架
hanging bridge 悬桥，吊桥
hanging buttress 悬扶垛
hanging cable 吊索
hanging clinometer 悬式测斜器
hanging crane 悬挂式起重机
hanging gutter 悬吊槽
hanging mercury drop electrode 悬汞滴电极
hanging mode 悬挂方式
hanging railway 悬索铁道
hanging roof 悬挂屋盖
hanging scaffolding 悬空脚手架
hanging stairs 悬空楼梯，悬挑楼梯
hanging steps 悬挑踏步
hanging structure 悬挂结构
hanging-up 堵塞，搭桥，吊挂，障碍，意外停机
hanging valley 悬谷
hanging valve 翻板阀
hanging wall 断层上盘，顶壁
hanging water 悬着水
hang-over 释放延迟，残余物
hang-type breather 悬挂式吸湿器，悬挂式呼吸器
hang up 吊具，挂起，中止，暂停，障碍，堵塞，挂，意外停机
hang 吊起，下垂物，悬吊，挂，悬挂
hank （电缆）盘
happenstance 偶然事件
harbor basin 港池
harbor charge 港建港杂费
harbor dues 停泊费
harbor power plant 港口电厂
harbor reach 港区河段
harbor 港口
harbourage 停泊处
harbour basin 港池，港域
harbour breakwater 港湾防洪堤
harbour construction plant 筑港设备
harbour construction 港口建筑
harbour district 港区
harbour dues 港税
harbour entrance jetty 港口进口导堤
harbour entrance 港湾进口，港口
harbour light 港口灯
harbour limit 港界
harbour line 港区界线，港区线
harbour model 港口模型
harbour mole 海港防波堤
harbour plan 港口图
harbour reach 港区
harbour terminal facilities 港口码头设施
harbour volume 港域容积
harbour 港，港湾，港口，码头
hard 坚硬的，硬，硬的
hard aggregate 硬质骨料
hard alarm 硬报警
hard alloy 硬质合金
hard and free expansion sheet making plant 硬板材及自由发泡板加工装置
hard arc 强电弧
hard asphalt 硬沥青
hard axis 难磁化轴
hard bargain 苛刻的交易
hardboard 硬纸板，硬质纤维板
hard-bonded furnace deposit 炉内硬块沉积
hard brazing 硬钎焊
hard burned brick 过火砖
hard busbar 应母线，硬母线
hard carbide 硬质合金
hard cash 现金，硬通货
hard casting 白口铸铁
hard chromium plating 硬铬镀层
hard clay 硬黏土
hard coal 无烟煤，硬煤
hard component 硬部件【射线】
hard concrete 干硬性混凝土
hard contact 金属触点，硬触点
hard control 硬控制
hard copy 硬副本，硬拷贝
hardcore 石填料【路基】
hard currency 硬币，硬通货，硬货币
hard cushioning 硬式减震
hard cycling 严格的循环变化
hard damping 强阻尼
hard disk capacity 硬盘容量

hard disk drive 硬盘驱动器
hard disk 硬盘
hard drawing 冷拔，冷拉
hard drawn copper wire 硬拉铜线
hard drawn wire 冷拉钢丝
hard drawn 冷拔的，冷拉的
hard electron 硬电子
hardenability band 淬透性带
hardenability characteristic 淬透性
hardenability limit 可淬透性极限
hardenability test 淬透性试验
hardenability value 硬化指数
hardenability 淬透性，可淬性，硬化程度，硬化度，淬火性，淬硬性
hardened and tempered steel 调质钢，淬火回火钢
hardened case 硬化表面
hardened concrete 已硬化混凝土，硬化的混凝土
hardened plate 硬钢板，淬硬钢板
hardened resin 硬树脂
hardened right out 使完全硬化
hardened safe storage 加固安全贮存埋葬
hardened steel 淬火钢，硬化钢
hardened voice channel 有线声音通道
hardened voice circuit 有线声音线路
hardened zone 硬化区
hardener for concrete floor 混凝土楼板硬化剂
hardener 固化剂，硬化剂
hardening-accelerating admixture 硬化加速剂
hardening agent 硬化剂，淬硬剂，速凝剂
hardening and tempering 调质
hardening at subcritical temperature 低温淬火
hardening bath 淬火浴
hardening break 淬火开裂
hardening by isothermal heat treatment 等温处理硬化
hardening capacity 淬硬性，硬化度
hardening crack 淬火裂缝，淬裂，硬化裂
hardening depth 硬化深度
hardening dispersion 弥散硬化
hardening distortion 硬化畸变
hardening flaw 淬火裂纹
hardening heat 硬化热
hardening induration 硬化作用
hardening modulus 硬化模量
hardening of concrete 混凝土硬化
hardening process 硬化过程
hardening rule 硬化规则
hardening strain 淬火应变
hardening temperature 淬火温度
hardening （金属）硬化法【增加硬度、淬火、渗碳、锻炼等】，金属淬火处理，变硬，淬火，硬化
hard-facing alloy 表面硬化用合金
hardfacing electrode 耐磨堆焊焊条
hard-facing 表面硬化【淬火】，耐磨堆焊
hard finished plastering 压光磨面，压光饰面
hard flame 硬火焰
hard flutter 强颤振
hard formation 硬岩层
hard frost 黑冻，黑霜
hard-gas circuit-breaker 自产气断路器
hard glass 硬玻璃
hard goods 耐用品
hard grained 粗粒状的
hard ground 岩底
Hardgrove grindability index 哈氏可磨性指数，哈氏可磨指数
Hardgrove grindability 哈氏可磨系数，哈氏可磨性，哈氏可磨性指数
Hardgrove index 哈氏指数，哈氏可磨度
hard hat 安全帽
hard igelite tuberigid polyvinyl chloride tube 硬聚氯乙烯管
hard igelite tube 硬聚氯乙烯塑料管
hard lead 硬铅
hard limiting 硬极限
hard loan 硬贷款
hard logic 硬逻辑
hard manual operation 费力的人工操作，硬手操，硬手动操作
hard manual 硬手操（作），硬手动
hard metal product 硬质合金制品
hard metal sheathed cable 硬金属铠装电缆
hard metal 硬性金属
hard mortar 硬性灰浆
hardness 硬度，硬性
hardness coefficient of rock 岩石硬度系数
hardness curve 硬度曲线
hardness gauge 硬度计
hardness leakage （给水）残留硬度
hardness level 硬度等级
hardness meter 硬度计
hardness method 硬度法
hardness number 硬度值，硬度指数
hardness of water 水的硬度
hardness penetration 淬透性
hardness profile 硬度分布
hardness scale 硬度标，硬度等级，硬度计
hardness standard block 硬度标准块
hardness tester 硬度试验器，硬度试验机
hardness testing 硬度测试，硬度实验，硬度试验
hardness test 硬度试验
hard oil finish 干性油饰面
hardometer 硬度计
hardpan 硬质土，底价
hard polyurethane foam plastics 聚氨酯硬质泡沫塑料
hard quench 淬硬
hard radiation 硬辐射
hard rain 暴雨
hard rime 雾凇，霜凇，硬雾凇，混合凇
hard rolled 冷轧的
hard roller unit 硬轧辊单元
hard roller 硬压辊
hard rolling at the bottom 在底部冷滚压
hard rolling at the top 在顶部冷滚压
hard rolling 冷滚压，冷轧
hard rubber roller 硬橡皮滚轮
hard rubber 硬橡皮，硬质胶，硬橡胶
hard self excitation 硬自励，硬自激

hard service 超负荷工作状态，不良使用，困难工作条件，重超荷载工作
hard site 硬质场地，坚固发射场
hard slag mass 硬渣团
hard soldering 硬焊，硬焊接
hard solder 硬焊料，钎焊，硬钎料
hard spot 硬斑，死区
hard stock 硬砖
hard stopping 硬质填塞料
hard surfacing 表面淬火，表面硬化，硬质面层
hard terms 苛刻条件
hard tube pulser 高真空电子管脉冲发生器
hard tube 硬性管
hard vacuum 高真空
hard wall plaster 墙的水泥抹面，水泥粉饰
hard wall 石膏抹底墙
hardware and fittings for power transmission line 输电线路金具
hardware and fittings 金具与配件，五金连接件，五金件【输电线路】
hardware compatibility 硬件兼容性
hardware configuration 硬件配置，硬件组态
hardware control 硬件控制
hardware diagnostic 硬件诊断
hardware fabrication drawing 金属构件加工图
hardware interrupt system 硬件中断系统
hardware logic simulation 硬件逻辑模拟
hardware monitor 硬件监督程序，硬件监视器
hardware multiplexing 硬件复用
hardware partitioning 硬件配置
hardware platform 硬件平台
hardware storage 五金库
hardware 五金，硬件，硬设备，元件，小五金，金具【指线路工程】，五金器具，五金件，金属构件
hard water 硬水
hard winding 硬绕组
hard wired 硬接线的，困难布线的
hard wood flooring 硬木地面
hard wood parquet flooring 硬木拼花地面
hard wood rough floor boards 硬木毛地面
hard wood 硬木
harm 伤害，损害，危害
Harman air cooling system 哈蒙式空冷系统
H armature H形电枢，H形衔铁
harmful agent 有害物剂
harmful bacteria 有害细菌
harmful effect 有害后果，有害影响
harmful effluents of power station 电厂的有害废物
harmful element in coal 煤中有害元素
harmful gas and vapor 有害气体
harmful gas 有害气体
harmful impurity 有害杂质
harmful ingredient 有害成分
harmful refuse 有害垃圾
harmful substance 有害物质
harmful waste 有害废物
harmful 有害的影响，危害性，有害作用，有害的
harmless flaw detector 无损探伤仪
harmless odour 无害气体
harmless test 无损检验
harmonic amplifier 谐波放大器
harmonic analysis 调和分析，谐波分析
harmonic analyzer 谐波分析器
harmonic balancer 谐波平衡器
harmonic bias 谐波偏流，偏置谐波电流
harmonic blocking relay 谐波闭锁继电器
harmonic bunching 谐波聚束，谐波束
harmonic cancellation 谐波消除
harmonic circulating current 谐波环流
harmonic compensation 谐波补偿
harmonic component 谐波分量
harmonic constant of tide 潮汐调和常数
harmonic constant 谐波常数
harmonic content of AC power supply 交流电源的谐波含量
harmonic content 谐波含量
harmonic converter 谐波变换器
harmonic correction 谐波校正
harmonic current 谐波电流
harmonic curve 谐波曲线，正弦曲线，调和曲线
harmonic cycling 谐波振荡
harmonic displacement 谐波位移
harmonic distortion 谐波失真，谐波畸变
harmonic disturbance 谐波扰动
harmonic effect 谐波效应，谐波影响
harmonic eliminator 谐波消除器
harmonic excitation 简谐激振，谐波励磁
harmonic excited generator 谐波励磁发电机
harmonic excluder 谐波抑制器，谐波滤波器
harmonic expansion 谐波展开，谐波级数展开
harmonic field 谐波磁场
harmonic filters for static var compensator 静止无功补偿滤波装置
harmonic filter 谐波滤波器
harmonic flux density 谐波磁通密度
harmonic flux 谐波磁通
harmonic force 简谐力
harmonic frequency response 谐频响应
harmonic frequency 谐波频率
harmonic function 谐函数
harmonic generation 谐波产生，谐波振荡
harmonic generator 谐波振荡器，谐波发生器
harmonic leakage reactance 谐波漏电抗
harmonic linearization 谐波线性化
harmonic loss 谐波损耗
harmonic mean particle diameter 调和平均粒径
harmonic mean 调和平均值，谐波平均值，调和中项，调和中间值
harmonic motion 谐振动，谐波运动，正弦运动
harmonic number 谐波数
harmonic oscillation 谐振，谐波振荡，谐和振荡
harmonic oscillator 谐波振荡器，正弦波发生器，谐振子
harmonic power flow calculation 谐波潮流计算
harmonic power 谐波功率
harmonic progression 调和级数
harmonic quantity 周期量，谐波量
harmonic relaying 谐波继电保护
harmonic response characteristic 谐波响应特性
harmonic response diagram 谐波响应图

harmonic response　谐波响应
harmonic current　谐波电流
harmonic series　调和级数，谐级数，谐波系
harmonics of flux　通量谐波
harmonic suppression　谐波抑制
harmonic voltage　谐波电压
harmonic synthesis　谐波合成
harmonics　谐波，调和函数，谐函数
harmonic test　谐波试验
harmonic torque　谐波转矩
harmonic tremor　谐和颤动
harmonic vibration　谐振，谐波振动，谐和振动
harmonic voltage　谐波电压
harmonic wave analyzer　调和波分析器
harmonic wave　调和波，谐波
harmonic winding　谐波绕组
harmonic　谐波，谐波的，谐音，调和，调和的，和谐的
harmonization of customs procedure　海关手续的一致化
harmonizing　调谐
harmonograph　谐振记录仪
harmony　谐波，谐振，调谐，协调，和谐，一致
harnessable power　可用风能，有效风能【风力发电】
harnessing wind　风能利用，利用风能
harnessing　导线束，导线分编，治理
harness of river　河道治理
harness　导线，导线系统，装具
harp antenna　扇形天线
harp type cable　竖琴形拉索
harrower　耙料机
harrow　耙路机，耙
harsh blowing flame　尖头火焰
harsh concrete　干硬性混凝土
harsh environment　恶劣环境，艰苦环境，苛刻环境，严峻环境
HART（highway addressable receiver transmitter）　高速公路可寻址的接收发送器
HART(highway addressable remote transducer)　可寻址远程传感器高速公路
Harvard miniature compaction test　哈佛小型击实试验
harvesting energy　获能
hash coding　散列码
hashing　散列法
hash time　闪光时间
hash total　无用数位总和
hash　杂乱脉冲，无用信息
hasp and staple　搭扣和锁环
hasp　搭扣
hassock　草丛，草垫
Hastelloy steel　耐蚀耐热镍基合金
Hastelloy　哈斯特洛伊合金【美国牌号】
hat-and-coat hook　衣帽挂钩
hatchet plan meter　斧状求积仪
hatchet　短柄小斧，短柄斧
hatch grafting　升降口花格，人孔格栅
hatch housing　销爪罩，销闩壳
hatching line　斜线，阴影线
hatch mast　舱口起重杆
hatchway　升降口，舱口，地板，天花板出人口，进出通道
hatch　舱，进出通道，舱口，舱门，气闸，门，闸门，车底排料门，人孔
hatfield time yield　短期徐变试验准则
hat orifice　帽形孔板
hat　随机编码
haugh　滩地，河岸平台
haulage appliance　拖运工具
haulage cost　运行费用，运输费用
haulage drive　牵引驱动装置
haulage motor　电机车
haulage plate　拖运板
haulage route　运输路线
haulage tunnel　运输隧洞
haulage way　运输线
haulage winch　拖运绞车
haulage　搬运，托运，牵行，牵引量，托，牵引力，运输，运费，拖曳，拖运费
haul distance　运距
hauler　运输工，推车工，机车司机，拖运机，拖车，牵引车，牵引机
hauling　牵引，拖运
hauling cable　拉绳
hauling distance　运程
hauling gear　卷扬机
hauling hope　曳引绳
hauling machine　牵引机
hauling truck　运输卡车
hauling unit　运输工具，运输设备
hauling wire　牵引绳
haul yardage　运土方数，运土量
haul　改变航向，拉，拖，拖运，牵运，运输
haunch board　梁腋侧板
haunched beam　变截面梁，加腋梁，起拱梁，井道内牛腿，托臂梁
haunched member　突起（托梁）构件
haunch of beam　梁腋
haunch　腰腿，拱石段，托肩，梁腋，拱腋
have a sense of principle　原则性
have influence on　对……有影响
haven　泊船处，港
have precedence over　优先于
have priority of　优先
have recourse to law　诉诸法律
have recourse to　依靠，求助于，求援于
have the corner on　垄断
have the lowest price　最低报价
have the same legal effect　具有同等法律效力
havoc　大破坏，严重灾害，灾害
HAWB(house air waybill)　航空分运单
hawkbill　焊钳
Hawksley's formula　霍克斯累公式【用于计算水库浪高】
hawk　托灰板
hawser　粗缆，钢缆，钢丝索，缆索
hawse　锚链孔
HAWT(horizontal axis wind turbine)　水平轴风力机
haydite concrete wall panel　陶粒混凝土墙板
haydite concrete　陶粒混凝土

haydite 陶粒
Haynes satellite 哈氏合金（铅-锡-锑）
hay rope 草绳
haywire power supply 临时线供电
haywire wiring 临时布线
haywire 临时电线
hazard analysis 危害分析
hazard and operability 危险性与可操作性分析
hazard assessment 危害评价，灾害评估
hazard beacon lamp 危险警告信号灯，危害航标灯
hazard evaluation 危害评价
hazard-free circuit 无危险电路
hazard identification 危害识别，危险标志
hazard index 危险指数，危害指数
hazard of crop 农作物受害
hazard of forest 森林受害
hazard of organism 生物受害
hazard of plant 植物受害
hazardous air pollutant 有害的空气污染物
hazardous area 危险场所，危险范围
hazardous article 危险品
hazardous chemicals 危险的化学药品
hazardous classified location （分类的）危险场所
hazardous compound 有害化合物
hazardous condition 危险情况
hazardous element 有害元素
hazardous live part 危险带电部件
hazardous location 危险场地
hazardous mechanical part 危险机械部件
hazardous operation 危险性作业
hazardous particle 有害微粒
hazardous protection device 安全保护装置
hazardous substance 危险物品，有害物质
hazardous waste 危险废物，有害废物
hazardous 危险的
hazard probability 受害概率
hazard rate 故障率
hazard rating 危害等级，危险等级
hazard reducing device （防触电）减危装置
hazard report 危害报告书
hazards in mine 矿山危险源
hazard to environment 对环境的危害
hazard to personnel 对人员的危害
hazard to the environment 对环境的危险
hazard to the power station environment 对电站周围环境的危险性
hazard 危害，公害，危险，冒险，风险，灾害，易燃易爆品【复数】
haze coefficient 霾度【一种能见度单位】
haze dome 汽室
haze horizon 霾层顶
haze layer 霾层
haze line 霾线
haze 轻雾，霾
HAZID（hazard identification） 危害识别，危险标志
HAZOP（hazard and operability） 危险性与可操作性分析
hazy atmosphere 被烟雾污染的大气
H-bar （宽型）工字钢
HB（Brinell hardness） 布氏硬度
H-beam 工字梁
HB（heat balance） 热平衡
HBL（house bill of lading） 无船承运人提单
HBr（Brinell hardness） 布氏硬度
H-cable 屏蔽电缆
HCFCs 氯氟烃，氟氯烃化合物
H class insulation H级绝缘
HCL（hydrochloric acid） 盐酸
H-column 工字柱
HCS（generator hydrogen and CO_2 systems） 发电机氢气与二氧化碳系统
HCS（hierarchical control system） 分级控制系统
HD（heater drain） 加热器疏水
HDPE（high density polyethylene） 高密度聚乙烯
HDR＝header 联箱，母管，集管
hdroelectric power station 水力发电站
HDR 联箱，集箱【DCS画面】
HDS（heater drain system） 加热器疏水系统
head amplifier 前置放大器，前级放大器
head at zero capacity （泵的）关死扬程
head bay 蓄水前池，上游池，上闸首
head bent adjustable wrench 弯头活动扳手
head block 垫块
headboard 斜帆首板
head-capacity characteristic 压头流量特性
head-capacity curve 水头容量曲线，压头容量曲线
head characteristic 压头特性
head core 磁头心
head cover 顶盖，端盖
head difference 水头差
head-discharge relation 水头流量关系
head down 顺风航行，朝向，向下
head drop 水头降落，水头下降
head drum 头部滚筒
head duration curve 水头历时曲线
headed sphere anemometer 球状头部风速计
head end drive 头部驱动
head end facility 首端处理设施
head end monitor 首端处理监测器
head end plant 首端处理车间
head end process （核燃料后处理）首端过程
head end reprocessing 头端部再加工
head end treatment 首端处理
head end 首端
header and lateral system 母管支管制
header bond 丁砖砌合法
header card 标题卡片
header course 露头层，丁砌层
header enclosure 集箱罩壳
header joist 横托梁
header pipe 集管
header sheet 管板
header space insulation 集管空间保温层
header space shielding 集管空间屏蔽
header support 集箱支座，集箱支架
header system 母管制，母管制系统
header type boiler 分联箱式锅炉
header type feed heater 联箱式给水加热器

header type heater　联箱式加热器
header valve　总管阀
header　丁砌，丁砖砌合，标头，过梁，集水管，集流管，集管，页眉，标题，联管箱，集箱，联箱，头部，集水器，母管，汇流总管
head-flow characteristic　压头流量特性
head-flow curve　压头流量曲线
head flowmeter　压头（式）流量计【用孔板，文丘里管，毕托管等】
head flume　头部水槽
head frame of shaft　竖井顶架
headframe　井架
head-gate duty of water　毛灌水定额，毛灌水率
head house　井口建筑物
heading and bench method　台阶掘进法
heading and cut method　导洞掘进法
heading block　起始字组
heading face　工作面前壁，掌子面
heading printing　标题印刷，标题打印
heading stope　回采工作面
heading　标题，信头，导坑，方向
head insulation　封头绝热层，封头热绝缘
head joint　端部接缝
head keyway　封头键槽
head lamp　头灯，大灯
headlap　搭接重叠部分
headless nail　无头钉
headless plain steel hanger screw　无头碳钢吊钩螺丝
headless threaded fastener　无头螺纹固定器
headlight relay　头灯继电器
headlight switch　头灯开关
headlight　大灯
headline　头条新闻，新闻提要，标题
head loop　（重水精馏柱）头部回路
head loss due to contraction　收缩段水头损失
head loss due to enlargement　扩大段水头损失
head loss in bend　弯管水头损失
head loss of pipeline　管路水头损失
head loss　水头损失，压头损失，压力损失，失压
headman　工长，工头
head mast　前栓
head metal　金属切头
head molding　（孔口的）顶部装饰线条
head movable device　头部可移动装置
head of division　部门负责人
head office　总部，总局，总公司，总行，总社
head of ingot　锭头
head of liquid　液柱压力
head of mast　杆塔顶部，电杆顶部
head of water over weir　堰顶水头
head of water　水头，水柱高度
head oil tank　高位油箱
head-on collision　对头碰撞
head-on rotor　上风型风轮
head on the spillway　溢流堰顶水头
head on wind machine　上风型风力机，迎风风力机
head on wind　迎面风，逆风迎面风
head oscillation curve　水头摆动曲线
headphone　头戴式受话机，耳机
head piece　（换料机）头部构件
head plate　封头，端板
head pond　前池
head pressure　排出压力，出口压力
head pulley　头部滚筒
headquarters　总公司，总部，本部
headquarter　总公司，总部，指挥部
headrace bay　进口河湾，前池
headrace channel　引水沟，引水率
headrace conduit　上游输水道
headrace surge chamber　上游调压室
headrace surge tank　上游调压井
headrace tunnel　上游隧洞，引水隧洞
head range　水头范围
head recovery　压头恢复
head reservoir　上水库
head room of flight　梯段净高
head room under a bridge　桥梁出水净高，桥下净高
head room　净空（高度），顶房，顶部空间，峰值储备，上游闸室
head safety injection pump　高压安全注入泵
head screw　主轴螺杆
head sea　顶浪，顶头浪，逆浪
head sensor　温度传感器，传感器
headset receiver　头戴式接收器，头戴耳机
headset　头戴式耳机
head shaft　端轴
head shifter　磁头移行器
head sink　散热片，压头沉落
headspace analysis　液面上部气体分析
headspace　顶部空间，顶端空间，预留空间
head spray line　封头喷淋管道
head spray system　封头喷淋系统
headstock gear　启闭机
headstock　头架，车床头，主轴箱
head stone　奠基石，墙基石
head storage　前部储存
headstream　源头
head surge basin　上游调压池
head surge chamber　上游调压室
head switch　读头开关
head tank　高位水池，稳压水箱，压力罐，高位水箱
head telephone　头戴受话器
head tide　逆潮流
head tower　顶塔，端塔，首塔
head up　迎风航行，艏向上，船首线向上
head vent　封头排气
head-vs-capacity characteristic　压头容量特性（曲线）
head wall　端墙
headward erosion　溯源侵蚀，向源侵蚀
headwater area　上游地区
headwater channel　引水渠
headwater control　水源控制，水头控制，上游水位控制
headwater pond　蓄水前池
headwater region　河源区
headwater reservoir　上游水库

headwater survey　水源调查
headwater tributary　上游支流
headwater　上游源头，河源头
head wave　顶头波
headwind　顶头风，顶风，逆风
head　顶部，端盖，封头，端盖，压头，头，磁头，水头，落差损失，（门窗的）上框，头部，顶，水压，扬程
healing agent　修补剂
healing surface area　换热面积
healing trench　暖气管沟
healing　合龙，封口
health care measure　保健措施
health care　保健
health certificate　健康证书，检疫证书
heal the breach　调停
health effect　健康效应，健康影响
health examination　健康检查，体格检查
health hazard　对健康危害
health insurance　事故及医疗险，健康保险
health monitoring　保健监测
health physicist　保健物理员
health physics and hot machine shop HVAC system　保健物理与热机械车间高压交流系统
health physics assistant　保健物理助理员
health physics equipment　辐射防护设备
health physics information system　保健物理信息系统
health physics laboratory　保健物理实验室
health physics monitoring　保健物理检测，保健物理监测
health physics research reactor　保健物理研究反应堆，保健物理研究用反应堆
health physics technician　保健物理技术人员
health physics　保健物理（学），有害辐射防护学
health protection　卫生防护
health safety and environment　健康、安全与环境
health service building　医务室
health standard　卫生标准
health surveillance　保健监护，保健监督
healthy stream　无污染河流，清洁河流
healthy　健康的
health　卫生，健康
heaped capacity　装载容量
heaped load　堆积荷载
hearing acuity　听敏度
hearing aid　助听器
hearing loss　听力损失
hearing test　听力检查
hearing threshold shift　听觉阈变化，听觉最低值变化
hearing threshold　听觉阈，听觉最低值，听力阈值
hearing　听觉，听闻，听证，听力
heart cam　心形凸轮
heart check　心裂
hearth area　炉底面积
hearth　火炉，烟囱，炉底，炉床
heart wall　心墙
heat absorber　吸热器
heat-absorbing reaction　吸热反应
heat-absorbing surface　吸热面，受热面
heat absorption area　受热面积，吸热面积
heat absorption capacity　吸热能力，吸热性能
heat absorption distribution　吸热量分配
heat absorption efficiency　吸热效率
heat absorption of heating surface　传热面热负荷
heat absorption rate profile　吸热分布图形
heat absorption rate　（受热面）吸热强度
heat absorption surface　吸热面
heat absorption　吸热，热吸收，热量吸取，散热，排热
heat abstractor　吸热设备，散热装置，散热器
heat account　热力计算，热平衡计算
heat-accumulating type electric boiler　蓄热式电锅炉
heat accumulation　蓄热
heat accumulator　蓄热器
heat action zone crack　热影响区裂纹
heat-affected zone crack　热影响区裂纹
heat-affected zone　热影响区，温度影响区域
heat ageing　热时效，热老化
heat air thawing equipment　热风解冻装置
heat alarm　过热报警信号，过热警报
heat analysis　热分析，抽样分析【炉前】，熔炼分析
heat and mass balance　热质平衡
heat and mass transfer　热质传递
heat and moisture transfer　热湿传递，热湿交换
heat and power cogeneration　热电联产
heat application　供热
heat area　发热区
heat availability factor　热量可用率，热利用系数
heat availability parameter　热量有效系数，热可用系数
heat balance calculation　热平衡计算
heat balance diagram　热平衡图
heat balance equation　热平衡方程
heat balance　热平衡，热收支，热量平衡，回热系统
heat barrier　热障
heat battery　蓄热器
heat bomb　测热弹
heat booster　加热器，预热器
heat box test　热跑试验
heat box　热跑箱，加热室
heat budget method　热量平衡法
heat budget　热量平衡
heat budge　热量收支，热量平衡
heat capacity curve　（泵或风机）扬程特性曲线，压头流量曲线
heat capacity rate ratio　热容率比
heat capacity　热容，热容量
heat carrier　载热体，载热介质，热载体
heat-carrying agent　载热剂，载热质
heat checking　热裂
heat choke　热壅塞
heat circumference　热周长
heat coefficient　热系数
heat coil　热线圈，发热线圈
heat collection storage wall　集热（蓄热）墙式

【太阳光热利用】
heat collector　集热器
heat conductance　导热性，导热率
heat-conducting fluid　导热流体
heat conduction analogy　热传导模拟
heat conduction coefficient　导热系数
heat conduction equation　热传导方程
heat conduction　导热，热传导
heat-conductive material　导热材料
heat conductivity factor　导热系数
heat conductivity pressure gauge　热传导式压力计
heat conductivity　导热性，导热率，导热系数
heat conductor　导热体
heat consuming installation　热用户
heat consumption quota　耗热定额，热耗定额
heat consumption rate　热耗率
heat consumption test　热耗试验
heat consumption　热耗，热耗量
heat content entropy chart　焓熵图
heat content table　焓表，热焓表
heat content volume chart　焓容图
heat content　焓，热含量，发热量，含热量
heat convection　对流换热
heat conversion　热能转换，热转换，热对流
heat converter　热变换器
heat cracking　热裂，热裂解
heat crack　热裂缝，热龟裂
heat credit　外来热量
heat-cured insulation　热固性绝缘
heat current　热流
heat cycle test　热循环试验
heat cycle　热循环
heat cycling　周期性加热
heat deaerator　热力除氧器，加热式除氧器
heat decomposition　热分解
heat degradation　热降解
heat-delivery surface　受热面，传热面，散热面
heat detector　热量探测器
heat deviation　热偏差
heat diffusion　热扩散，热散射
heat discoloration　（金属的）受热变色，退火颜色
heat-dispersing surface　散热面
heat dissipating　散热
heat dissipation circuit　散热回路
heat dissipation ratio　散热比例
heat dissipation through convection　对流散热
heat dissipation　热量耗散，热消散，散热，热扩散
heat distortion　热变形
heat distributing network　热网
heat distribution　热流分布，热分布
heat-driven oscillation　热致振动，热振荡
heat drop　热降
heat drying　加热干燥
heat due to friction　摩擦热
heat dumping　排热，散热
heat-dump system　排热系统【汽轮机】
heat duty control　热负荷控制
heat duty　热负荷
heat economizer　节热器
heat-economy figure　热经济性指标
heat-economy　热经济性
heated air return flow　热空气回流
heated and(/or) cooled enclosed location　升温和（或）降温封闭场所［空间］
heated container　加热的容器
heated effluent　热废水
heated fluidized bed　加热流化床
heated scrub　加热洗涤，热洗涤
heated thermometer anemometer　热风速仪
heated waste water　热废水
heated winding　发热绕组
heated wire's air speed anemometer　热线风速仪
heated wire's type anemometer　热线风速仪
heat efficiency　热效率
heat egress　排热
heat-eliminating medium　冷却介质
heat elimination　放热，散热，去热
heat emission　放热，热辐射，散热，热量发射
heat emitter　散热器
heat endurance　耐热性，热稳定性
heat energy dynamics　热能动力学
heat energy power　热能动力
heat energy station　热力站
heat energy　热能
heat engineering　热力工程，热工学
heat engine　热机，热力发动机
heat-entropy diagram　焓-熵图
heat equation　热力方程
heat equator　热赤道
heat equivalent　热当量
heater air vent system　加热器空气排放系统
heater assembly　加热组件
heater bank　加热器组
heater by-pass　加热器旁路
heater cathode　旁热式阴极
heater condensate　加热器疏水
heater condensing zone　加热器凝汽区
heater control panel　加热器控制盘
heater desuperheating zone　加热器过热蒸汽冷却区
heater drain and vent　加热器疏水及放气
heater drain cooling zone　加热器疏水冷却区
heater drain pump　加热器疏水回收泵，加热器疏水泵
heater drain system　加热器疏水系统
heater drain tank　加热器疏水箱
heater element　加热元件，加热器
heater for heating network　热网加热器
heater friction　加热器摩擦阻力损失
heaterless tube　直热式电子管
heater level control valve　给水加热器水位调节阀，加热器水位调节阀
heater lower cover　加热器下盖
heater operation rate　加热器投运率
heat erosion　热侵蚀，高温表面破坏
heater performance characteristic　加热器特性
heater power　加热器功率
heater pump　加热器泵
heater replacement tool　加热器更换工具
heater rod bundle　加热器棒束
heater rod　加热器棒

heater steam coil 加热器蒸汽盘管
heater storage space 加热器水空间
heater storage 加热器容积
heater support plate 控制棒支撑板【稳压器】
heater temperature controller 加热器温度控制器
heater temperature decrease rate 加热器温降率
heater temperature rise rate 加热器温升率
heater transformer 灯丝变压器
heater tube-blocking rate 加热器堵管率
heater tube 旁热式电子管
heater upper cover 加热器上盖
heater valve cover 加热器阀盖
heater well 稳压器加热棒的套
heater 加热器，加热炉，发热体，热源，发热器，暖气装置，加热元件
heat evolution 放热
heat exchange area 散热面，换热面积，热交换面积
heat exchange capacity 热交换额定功率，热交换容量，热交换器的额定功率
heat exchange coefficient 热交换系数，传热系数
heat exchange cycle 回热循环，热交换循环
heat exchange dynamics 热交换动力学
heat exchange effectiveness 热交换率
heat exchange facility 热交换器，换热装置
heat exchange fluid 载热剂
heat exchanger circuit 热交换器回路
heat exchanger coil 热交换器蛇形管，热交换器盘管
heat exchanger effectiveness 换热器有效度，换热器效能，换热器效率
heat exchanger fluid 载热体，液体载热剂
heat exchanger for district heating 热网加热器
heat exchanger for district 热网加热器
heat exchanger lag 热交换器滞后
heat exchanger leak detection probe 热交换器探漏器
heat exchanger leaktightness restoration 热交换密封性的恢复
heat exchanger loop 热交换器回路
heat exchanger plate 换热板
heat exchanger rig 换热装置
heat exchanger tube 热交换器管，换热管
heat exchanger type regenerator 换热式回热器
heat exchanger 热交换器，换热器
heat exchange surface 受热面，换热面，散热面
heat exchange tower 换热塔
heat exchange tube 换热管
heat exchange 热交换，换热
heat exchanging process 热交换工作过程
heat expansion coefficient 热膨胀系数
heat expansion 热膨胀
heat extraction system 抽热系统
heat extraction 除热，抽出热量，热量提取
heat extractor 热交换器
heat fatigue 热疲劳
heat fire detector 感温火灾探测器
heat flow diagram 热流图，热力系统图
heat flow gauge 热量计，热流计
heat flow meter 热流计，热量计
heat flow probe 热流测针，热流探针
heat flow rate 热流量
heat flow vector field 热流向量场
heat flow 热流量，热流
heat flush 热冲洗
heat flux density distribution 热流密度分布
heat flux density shape factor 热流密度形状因子
heat flux density 热流密度
heat flux dissipation 热流散耗
heat flux gauge 热流密度测量仪
heat flux hot channel factor 热流密度通道因子，热流热通道因子，热通量热点因子
heat flux hot spot factor 热流密度峰值因子，热流热点因子，热通量热点因子
heat flux meter 热量计，热流计
heat flux peaking filter 热流峰值因子
heat flux shape factor 热流形状因子
heat flux 热通量，热流量，热流密度，热流
heat function 焓
heat fusion 熔解热
heat gain from appliance and equipment 设备散热量
heat gain from lighting 照明散热量
heat gain from occupant 人体散热量
heat gain 增热，热量增加，热增量
heat generation reaction 放热反应
heat generation 发热，热源
heat guard 绝热体
heat gun 热风器，煤气喷枪
heat-hardening 热凝固
heating agent 载热剂，加热介质
heating alternator 加热同步发电机
heating and cooling load 供热和供冷负荷
heating and cooling rate 加热和冷却速率
heating and cooling 采暖和制冷
heating condition 加热调节【加化学添加剂】
heating and ventilating discipline 暖通专业
heating and ventilation design 暖通设计
heating apparatus 供热装置
heating appliance 采暖设备，电热器
heating area 加热面积
heating body 发热体，散热器
heating boiler 供暖锅炉
heating box 加热室
heating by exhaust gases 排气加热
heating cabinet 加热柜
heating cable 加热电缆
heating calculation 热力计算，热平衡计算
heating capacity 供热能力，热容量，热值
heating channel 供暖地沟
heating characteristic of building 建筑物的采暖特性
heating circuit 加热回路，加热电路
heating coil section 加热段
heating coil 加热线圈，热盘管，加热螺管
heating collector 集热器
heating controller 加热调节器，加热控制器，加热计算器
heating control 加热控制
heating curve 加热曲线
heating cycle 加热循环

heating degree-day 加热度日，采暖度日，热度日，采暖用日平均温度
heating design 供暖设计
heating duct 加热管道
heating effect of current 电流热效应
heating effect 热效应
heating efficiency 加热效率，热出力，加热程度，热生产率
heating element 供热元件，加热元件，加热器
heating engineering 供热工程
heating equipment 采暖设备
heating furnace 加热炉
heating fuse 热熔断器，热熔丝
heating instability 加热不稳定性
heating installation 供暖装置
heating in winding 绕组发热
heating jacket 加热套
heating load data for load estimation 热负荷估算数据
heating load data 热负荷指标
heating load diagram 热负荷图
heating load duration graph 热负荷延续时间图
heating load 供暖负荷，热负荷
heating main 供暖干管
heating management 热工管理
heating medium parameter 热媒参数
heating medium 载热体，载热介质
heating network hot water heater 热网加热器
heating network hot water pump 热网水泵
heating network system 热网系统
heating network water piping 热网水管道
heating of turbine 汽轮机暖机
heating period 供暖期
heating pipeline 暖汽管道，热力管道
heating pipe trench 供热管沟
heating pipe 加热管，暖气管
heating power 发热量，加热功率，热力
heating process 暖机过程，加热过程
heating radiator 供暖散热片，供暖散热器，暖气散热片
heating rate curve 加热速率曲线
heating rate 加热速度，升温速率，热耗率
heating rating 加热额定值
heating register 加热记录器
heating resistor 加热电阻
heating riser 供暖立管
heating section 加热段，受热段
heating sensitive cable 热敏电阻
heating sludge-digestion tank 加热污泥消化池
heating station 加热站
heating steam 加热蒸汽，供暖蒸汽
heating surface area 换热面积，加热面积，传热面积，受热面积
heating surface bank 传热管束
heating surface tube 传热管
heating surface 加热面，受热面
heating system 供热系统，采暖系统，加热系统，供暖系统
heating temperature 供暖温度
heating the turbine 汽轮机暖机，暖机
heating treatment furnaces 熔热处理炉
heating trench 供暖地沟
heating tube bundle 加热管束
heating tube 加热管
heating unit 供暖机组，供热机组
heating-up period 暖机阶段，加热阶段
heating value of the as fired coal 入炉煤发热量
heating value 热值，发热量
heating-ventilating assembly (/unit) 供暖通风（两用）机组
heating, ventilation and air conditioning 采暖、通风与空调，暖通系统
heating water field 热水田
heating wire 电热丝
heating 发热，加热，采暖，供暖，供热，增温
heat injection noise 热喷射噪声
heat input method 热输入方法
heat input 热量输入，供热量，供热，热输入，输入热量
heat instability 加热不稳定性
heat-insulated 被绝热的，隔热的
heat-insulating and noncombustible coating 绝热和非燃性涂层，绝热和非燃性涂料
heat-insulating layer 保温层，隔热层
heat-insulating material 隔热材料，绝热材料，保温材料
heat-insulating plate 隔热板，绝热板
heat-insulating properties 绝热性质
heat-insulating window 保温窗
heat-insulating 绝热的
heat insulation course 隔热保温层，隔热层
heat insulation plate 隔热板，绝热板，保温板
heat insulation 绝热，热绝缘，保温层，隔热，保温
heat insulator 热绝缘材料，热绝缘体，保温材料
heat interchanger 热交换器
heat interchange 热交换，换热
heat island center 热岛中心
heat island circulation 热岛环流
heat island effect 热岛效应
heat island 热岛
heat lag 热延迟时间
heat leakage 热渗透量，散热损失
heat leak 漏热，热漏失
heatless dryer 无热干燥器
heat level 温度，热度
heat liberation per unit furnace volume 炉膛单位容积热负荷
heat liberation per unit grate area 炉排单位面积热负荷
heat liberation per unit heating surface 单位受热面热负荷
heat liberation rate 热负荷
heat liberation 释热，放热
heat load demand 热负荷需求
heat load 热负荷，热载荷
heat loss by infiltration 渗入热损失
heat loss calculation 热损失计算
heat loss due to chemical incomplete combustion 化学未完全燃烧热损失
heat loss due to exhaust 排烟热损失

heat loss due to flue gas 排烟热损失
heat loss due to mechanical incomplete combustion 机械未完全燃烧热损失
heat loss due to radiation 散热损失
heat loss due to sensible heat in refuse 灰渣热损失
heat loss due to sensible heat in slag 灰渣物理热损失，灰渣显热损失
heat loss due to unburned carbon in refuse 固体未完全燃烧热损失，机械未完全燃烧热损失
heat loss due to unburned carbon 灰渣含碳热损失
heat loss due to unburned gases 气体未完全燃烧热损失，化学未完全燃烧热损失
heat loss factor 热损失系数
heat loss of protection structure 围护结构热损失
heat loss recovery 热损失回收
heat loss 热耗，热损耗，热损失，耗热量
heat management 热管理
heat medium 热媒
heat melting 热熔化
heat meter 量热计
heat mixing coefficient 热交混系数
heat motion 热运动
heat network 热网
heat-of-absorption 吸收热
heat of combination 化合热
heat of combustion 燃烧热
heat of compression 压缩热
heat of condensation 凝结热，冷凝热
heat of decomposition 分解热
heat of dissipation 耗散热
heat of dissociation 离解热
heat of evaporation 蒸发热，汽化热
heat of formation 生成热
heat of freezing 凝固热
heat of friction 摩擦热
heat of fusion 熔化热，聚变热
heat of hardening 凝固热【水泥或混凝土】
heat of hydration 水合热
heat of liquefaction 溶解热，液化热
heat of liquid 液体热，液体热焓
heat of oxidation 氧化热
heat of phase change 相变热
heat of radiation 辐射热
heat of reaction 反应热，溶解热
heat obsorption detector 吸热检测器
heat of sublimation 升华热
heat of superheating 过热热量，过热热
heat of transformation 相态转化热，相变热
heat of vaporization 汽化热
heat only reactor 只供热反应堆
heat-operated refrigerating system 热力制冷系统
heat-operated refrigeration 热力制冷
heat output method 热输出方法
heat output 锅炉有效利用热量，热出力
heat passage 传热通路，传热，热传递
heat passive homing guidance 热辐射被动寻的制导
heat pickup 吸热
heat pipe air heater 热管空气预热器
heat pipe evacuated tube collector 热管式真空管集热器
heat pipe evacuated tube 热管是真空管
heat pipe heat exchanger 热管式换热器
heat pipe motor 热管电动机
heat pipe solar water heater 热管太阳热水器
heat pipe 热管
heat point source 点热源
heat pole method 热线法
heat pollution 热污染
heat potential 热势
heat power 热功率
heat power engineering automation 热工自动化
heat power engineering control 热工控制
heat power engineering 热动力工程，热力工程
heat power piping system 热动力管线系统
heat power plant 火电站，火力发电厂
heat preservation 保温
heat preserving furnace 保温炉
heat-producing capability 热出力，产热能力
heat-producing reactor 热生产反应堆
heat-producing 产热量，产生热量的
heat production reactor 供热反应堆
heat production 产生热量，发热量
heat productivity 发热量
heat proof insulation 耐热绝缘
heat proof quality 热稳定性，耐热性
heat proof 耐热的，防热的，耐热，防热
heat pump air conditioner 热泵式空气调节器
heat pump air conditioning 热泵式空调
heat pump heating 热泵式供暖
heat pump 热力泵，热泵
heat quantity 热量
heat-radiating 热辐射的
heat radiation 热辐射，辐射换热，散热
heat ran test 热跑试验
heat rate guarantee 保证热耗
heat rate of heating surface 换热面热负荷
heat rate test 热耗试验
heat rate 热耗，热耗率，耗热率，热效率，发热量，热容量，热价
heat rating 热功率
heat-receiving surface 受热面，吸热面
heat-reclaiming device 热回收装置
heat reclaim unit 热量回收装置
heat reclaim 热量回收，废热回收
heat-recovering 余热回收，热回收
heat recovery 余热利用，热量回收，废热回收
heat recovery area 尾部受热面
heat recovery boiler 余热锅炉，废热锅炉
heat recovery combined cycle 回热联合循环
heat recovery equipment 热回收设备
heat recovery power generation 余热发电
heat recovery steam generator for combined cycle 联合循环余热锅炉
heat recovery steam generator 余热锅炉
heat recovery surface 尾部受热面
heat-reflecting surface 热反射面
heat regeneratives system 回热系统
heat rejection circuit 排热回路
heat rejection ratio 放热系数

heat rejection 排热，热量放出，散热，热损耗，热消耗
heat release rate 热释放速率，放热率，释热率，炉膛容积热强度
heat release to the atmosphere 热量释放到大气
heat release 放热，释热，散热量
heat-releasing fluid 载热剂
heat removal after shutdown 停堆后排热
heat removal capacity 排热能力
heat removal equipment 排热设备
heat removal factor 排热因数
heat removal fluid 排热流体
heat removal load 排热载荷
heat removal loop 排热回路
heat removal rate 排热率
heat removal system 除热系统，排热系统
heat removal 除热，排热，散热
heat-removing 排热的，散热的
heat reservoir 储热器，蓄热器
heat resistance paint 耐热漆
heat resistance test 耐热测定
heat resistance 耐热性，热阻
heat-resistant alloy 耐热合金
heat-resistant cement plate 耐热水泥板
heat-resistant current meter 耐热流速计
heat-resistant enameled wire 耐热漆包线
heat-resistant enamel paint 耐热瓷漆
heat-resistant glass 耐热玻璃
heat-resistant insulation 耐热绝缘
heat-resistant material 耐热材料
heat-resistant metal 耐热金属
heat-resistant motor 耐热电动机
heat-resistant paint 耐热涂料，耐热漆
heat-resistant quality 耐热（特）性，隔热性能，热稳定性
heat-resistant rubber 耐热橡胶
heat-resistant steel welding rod 珠光体耐热钢焊条
heat-resistant steel 耐热钢
heat-resistant varnish 耐热绝缘漆
heat-resistant 耐热的
heat-resisting cable 耐热绝缘电缆
heat-resisting cast iron 耐热铸铁
heat-resisting glass 耐热玻璃
heat-resisting material 耐热材料，耐高温材料
heat-resisting property 耐热性能
heat-resisting quality 热稳定性，耐热性，耐高温性
heat-resisting steel 耐热钢
heat-resisting test 耐热试验
heat-resisting 耐热的，耐火的，高熔点的
heat retainer 保热器，蓄热体
heat-retaining capacity 保热容量，蓄热能力
heat-retaining mass 蓄热体
heat-retaining 蓄热的
heat return 热回收
heat rise in mass concrete 大体积混凝土内热量升高，大块混凝土水化热
heat running test 热跑试验
heat-rupture test 热破坏性试验，高温强度试验
heat screen 隔热屏
heat sensing device 灵敏装置，感温器，热敏
heat-sensitive component 热敏成分，热敏元件
heat-sensitive element 热敏元件
heat-sensitive material 热敏材料
heat-sensitive paint 热变涂料，示温漆，热敏涂料
heat-sensitive paper 热敏纸
heat-sensitive sensor 热敏传感器
heat-sensitive 热敏的
heat sensitivity 热敏性
heat sensitization 热敏化
heat sensor 热传感器，热敏元件
heat shield 挡热板，热屏蔽，隔热屏，热屏，避热罩
heat shock 热冲击，热震
heat-shrinkable tubing 热缩套管
heat-shrinkable 热缩的
heat-shrink plastic tubing joint 热缩塑性套管接头
heat-shrink tubing 热缩套管【采用套筒】
heat-shrink 热缩式
heat similarity 热相似
heat sink shield 吸热式屏蔽
heat sink 热散片【电子回路】，散热器，热汇，散热片，吸热部件，冷源，热沉，热井，吸热装置，排热系统
heat-soaked turbine 热态汽轮机
heat soak period 暖机阶段
heat soak time 暖机时间
heat soak 吸热，暖机
heat source density 热源密度
heat source of heat-supply system 供热系统热源
heat source 热源，过热点
heat stability test 热稳定试验
heat stabilization test 热稳定性试验
heat-stable 热稳定的
heat storage boiler 蓄热锅炉
heat storage capacity 蓄热能力，蓄热容
heat storage device 蓄热装置
heat storage media 储热介质
heat storage 热量储存，蓄热
heat-storing device 蓄热器，蓄热装置
heat stress 热应力
heat summation 总热量，热量总和
heat supply cost 供热成本
heat supply engineering 供热工程
heat supply load factor 平均热负荷系数
heat supply network 供热网，热力网
heat supply pipeline 供热管线
heat supply pipe 供热管
heat supply piping network 热力管网
heat supply reactor 供热反应堆
heat supply system based upon heating plant 区域锅炉房供热系统
heat supply system based upon industrial waste heat 工业余热供热系统
heat supply turbine 供热式汽轮机
heat supply 供暖（热），供热
heat switch 热开关
heat tape 发热带
heat test 温升试验

heat thunderstorm 热雷暴
heat-tight machine 隔热电机，耐热电机
heat time 加热时间
heat tinting 加热着色法，热变色
heat-to-work conversion 热功转换
heat trace piping 伴热管道
heat tracing cable 伴热电缆，加热用电缆，热量跟踪电缆
heat tracing philosophy for molten salts system 熔盐系统的伴热原理
heat tracing pipe 伴热管线
heat tracing system 伴热系统
heat tracing zone （管道）加热保温区
heat tracing （管道）加热保温，伴热
heat transfer agent 传热介质，载热剂
heat transfer apparatus 换热器，热交换器
heat transfer area 换热面，传热面积，散热面
heat transfer behavior 传热工况
heat transfer by convection 对流换热
heat transfer by direct conduction 直接传导导热
heat transfer by radiation 辐射传热
heat transfer characteristic 传热特性
heat transfer coefficient in subcooled boiling 欠热沸腾传热系数
heat transfer coefficient 热交换系数，传热系数
heat transfer conductance 热导
heat transfer curve 确定传热系数值的线图，表示热传递的曲线，传热曲线
heat transfer cycle 传热循环
heat transfer equipment 传热设备
heat transfer fluid loop 传热流体回路
heat transfer fluid 传热工质，传热介质，导热流体【太阳能发电】，载热体，传热流体
heat transfer loop 传热回路
heat transfer mechanism 传热机理
heat transfer media 传热工质，传热介质【media是medium的复数】
heat transfer medium 传热介质，载热介质
heat transfer performance 传热性能
heat transfer process 传热过程
heat transfer property 传热性能
heat transfer rate distribution 传热速度分布
heat transfer rate 传热系数，传热率
heat transfer reactor experiment 传热研究用实验反应堆
heat transfer resistance 热阻
heat transfer rig 传热试验台架
heat transfer ring surface 传热面
heat transfer system effect-test 传热系统效应试验
heat transfer system 传热系统
heat transfer theory 传热理论
heat transfer tube material 传热管材料
heat transfer with vapor condensation 有蒸汽冷凝的传热
heat transfer 传热，热传递
heat transmission coefficient 传热系数
heat transmission factor 传热系数
heat transmission 传热，热传导，热传递，热传送，热交换
heat transport medium 热传输介质
heat transport property 热传输性质
heat transport system 热传输系统
heat transport 热量输运，传热，热交换，热传输
heat trap 吸热，吸热器，热阱，吸热气
heat treat 热处理
heat treat crack 热处理裂纹
heat treated forged steel 热处理锻压钢
heat-treating machine 热处理机
heat-treating waste 热处理废物
heat treatment 热处理
heat treatment after welding 焊后热处理
heat treatment for mechanical properties 提高机械性能的处理
heat-treatment furnace 热处理炉
heat treatment in protective gases 保护气氛热处理
heat-treatment operator 热处理工
heat treatment report 热处理报告
heat treatment scale 热处理鳞皮
heat treatment temperature control cabinet 热处理温控箱
heat unbalance vibration 热不平衡振动
heat unbalance 热不平衡
heat-up rate 加热速率
heat-up 加热
heat utilization rate 热能利用率
heat utilization 热能利用率
heat value 发热量，热值
heat-variable resistor 热敏电阻，热变电阻
heat wave 热浪
heat withdrawal 排热
heat work ratio 热功比
heave fault 横推断层
heavenly tunnel 高空大气湍流
heave rate 隆起速率
heave ratio 冻胀比
heaver 杠杆，叉簧
heave stake 隆起标桩
heave 波浪，鼓胀，平错
heaviest piece 最重件
heavily compound-wound motor 过复激电动机
heavily iced power transmission line 重冰输电线路
heavily polluted area 严重污染区域
heavily reinforced 超配钢筋的
heavily stressed 高应力的
heavily 猛烈地，大量地，缓慢地
heaving bottom 冻胀土基
heaving motion 起伏运动
heaving oscillation 起伏振荡，升沉振荡
heaving pitching motion 升沉纵摇运动
heaving sand 胀砂
heavy aggregate concrete 重骨料混凝土
heavy aggregate shield 重混凝土屏蔽体
heavy air 压缩空气
heavy armature relay 重力式继电器，重衔铁继电器
heavy-bodied oil 高黏度油
heavy boiling water reactor 沸腾重水反应堆
heavy caking coal 强黏结性煤

heavy-cargo transportation 大件运输
heavy casting 大型铸件
heavy clay 重黏土
heavy coated electrode 厚药皮焊条
heavy compaction instrument 重型击实仪
heavy component 大件，重件
heavy concrete 重混凝土
heavy construction 大型工程，重型结构
heavy corrosion 严重腐蚀
heavy covered electrode 厚药皮焊条
heavy cover 厚层覆盖岩层
heavy crane 重型吊车
heavy crawler dozer 重型履带式推土机
heavy crude 重质原油
heavy current circuit breaker 大电流断路器
heavy current control 强电控制
heavy current disconnecting switch 大电流隔离开关
heavy current engineering 强电工程
heavy current low-voltage electrical apparatus 大电流低压电器
heavy current relay 强电继电器
heavy current slow-speed generator 大电流低转速发电机
heavy current 大电流，强电流
heavy cut 深挖
heavy damping 强阻尼
heavy diesel oil 重柴油
heavy drifter 重型风钻
heavy duty bogie wagon 载重平板车
heavy duty contact 重负载触点
heavy duty crane 大型吊车，重型吊车，重型起重机
heavy duty design 重级工作制
heavy duty frame 重型构架
heavy duty gas turbine 大功率燃气轮机
heavy duty generator 大容量发电机，大型电机
heavy duty idler 重型托辊
heavy duty loader 重型装载机
heavy duty manipulator 重型机械手
heavy duty motor 重型电动机，大功率电机
heavy duty power electronic 高功率的电力电子器件，重型电力电子
heavy duty rectifier 大功率整流器
heavy duty relay 重负载继电器，大功率继电器
heavy duty scaffold 重载脚手架
heavy duty service 困难的运行条件，高负荷运行
heavy duty socket 重型插座
heavy duty truck 载重汽车
heavy duty tyre 重载轮胎
heavy duty 大容量，大功率，高功率，重型，重载，重级工作制，重负荷的，重型的
heavy earthwork 大量土方工程
heavy edge reinforcement 边缘加强钢筋
heavy excavation 大开挖
heavy fog 浓雾
heavy fuel oil 重油，重质燃料油
heavy fuel 重燃料油，重油
heavy gas turbine 重型燃气轮机
heavy gauge plate 特厚钢板
heavy gauge wire 粗号线
heavy-gauge 大口径的，大尺寸的
heavy goods wharf 大件运输码头
heavy grade 陡坡
heavy gust 强阵风
heavy ice condition 重冰工况
heavy ice 厚堆积冰，重冰【固态重水】
heavy icing area 重冰区
heavy impulse current 冲击大电流
heavy industrial district 重工业区
heavy industrial 重工业的
heavy industry 重工业
heavy intermittent test 大负载断续试验
heavy-ion beam fusion 重离子束聚变
heavy-ion fusion system 重离子聚变系统
heavy-ion fusion 重离子束聚变，重离子聚变
heavy item 重件
heavy joist 重型搁栅
heavy jute mattress 重麻席
heavy lift charges 起吊费
heavy lift equipment 重型起重设备
heavy lift 重件吊装
heavy line 粗实线
heavy liquid separation 重液体分离
heavy load relay 重负载继电器
heavy load truck 重型卡车
heavy load 重载，重负荷
heavy loam 重亚黏土，重壤土
heavy maintenance 大修
heavy media separation 重介质分离
heavy metal pollution 重金属污染
heavy metal 重金属
heavy mineral soil 重矿物质土壤
heavy money 大笔款
heavy mortar 稠灰浆
heavy nucleus 重核
heavy oil fired boiler 燃重油锅炉
heavy oil gasification 重油气化
heavy oil heater 重油加热器
heavy oil pump 重油泵
heavy oil tank 重油罐
heavy oil 重油
heavy oscillation 剧烈振荡
heavy overhaul 大修
heavy package 重件密封包箱，超重货物
heavy part 重型部件，重件
heavy peak 重峰，重裂变产物峰
heavy penetration 重型触探
heavy plate 厚钢板
heavy pollutant 重质污染物
heavy power 大功率的
heavy rail 重型轨
heavy rainfall 大雨，暴雨，强降雨
heavy rain 大雨
heavy reinforcement 大量配筋
heavy relay 大电流继电器
heavy repair 大修
heavy residue 重渣油
heavy ring 承力环
heavy rutile covering 重钛型药皮
heavy sand loam 重砂壤土

heavy section casting 厚壁铸件
heavy section steel 厚壁钢材，厚壁钢制件
heavy separation 严重分离
heavy sheet glass 厚玻璃
heavy shower 大阵雨
heavy silty sand-soil 重粉质砂壤土
heavy silty soil 重粉质土
heavy snow 大雪
heavy soil 重黏性土壤
heavy statics 强烈天电干扰
heavy storm 大暴雨
heavy strap hinge 重型带式铰链
heavy surf 拍岸大浪
heavy tamping 重锤夯实
heavy test 满载试验，重载试验
heavy timber construction 重型木结构
heavy vehicle 重型车辆
heavy-walled pipe 厚壁管
heavy-walled tube 厚壁管
heavy-walled 厚壁的，重型炉墙的
heavy water boiling reactor 沸腾重水反应堆，沸腾重水堆
heavy water catch tank 重水收集箱
heavy water components test reactor 重水部件试验反应堆
heavy water coolant 重水冷却剂
heavy water cooled reactor 重水冷却反应堆，重水冷却堆
heavy water homogeneous reactor 重水型均匀堆
heavy water inventory 重水总量
heavy water moderated boiling-light-water-cooled reactor 重水慢化沸腾轻水冷却反应堆
heavy water moderated gas-cooled reactor 重水慢化气冷（反应）堆
heavy water moderated natural-uranium-fuelled reactor 重水慢化天然铀燃料反应堆
heavy water moderated reactor 重水减速反应堆，重水慢化反应堆，重水慢化堆
heavy water moderated water cooled reactor 重水慢化水冷堆
heavy water moderated 重水慢化的
heavy water moderator 重水慢化剂，重水减速剂
heavy water organic cooled reactor 重水慢化有机物冷却反应堆
heavy water plant 重水工厂，重水厂
heavy water pressure tube reactor 重水压力管式反应堆
heavy water pressure vessel reactor 重水压力容器反应堆
heavy water process facility 重水处理设备
heavy water reactor power station 重水堆核电站
heavy water reactor 重水反应堆，重水堆
heavy water reflector 重水反射层
heavy water suspension breeder reactor 悬浮重水增殖反应堆
heavy water suspension reactor 悬浮重水反应堆
heavy water upgrading plant 重水加浓厂
heavy water vapour 重水蒸气
heavy water 重水
heavy wear 严重磨损
heavyweight concrete 特重混凝土
heavyweight 特重物件
heavy work 重型作业
heavy 重型的，大功率的，繁重的
hectare 公顷
hectogram 百克
hectoliter 百升
hectometer 百米
hectorite 锂蒙脱石
hectostere 百立方米
hedge clause 套头交易条款，避责条款
hedgehog spine anchorage 刺猬式锚固【压力钢管】
hedgerow 树篱，绿篱，矮树篱
hedge 篱笆，障碍物，篱，树篱
hedging 套期保值
heel air gap 尾部气隙，根部气隙
heel-end slug （继电器的）根端缓动铜环
Heeler cooling system 赫勒冷却系统
heel gap 尾部气隙，根部气隙
heeling 横倾
heel of achilles 唯一的弱点
heel of brush 刷根，电刷倾斜
heel of tooth 齿根面
heel piece 跟片，铁芯底座
heel yoke 跟片，尾轭，铁芯底座
heel 跟部，边，残渣，尾部
height above datum 基面以上高度
height above ground level 地面以上高度
height above lowest foundation of dam 坝基最底部以上高度
height above sea level 拔海高度，海拔高度，海拔
height adjustment 高度调整
height-area curve 高度面积曲线
height computation 高程计算
height difference 高差
height equivalent to a theoretical plate 理论塔板（等效）高度
height gain factor 高度增益系数
height gain 高度增量
height gauge 高度尺，高度计，测高器
height-index circuit 标高电路
height indicator 测高仪，高度计
height mark 标高，高度标记
height measurement 测高法，高程测量
height of arc 弧高，拱高
height of bed expansion 床层膨胀高度
height of bed 床高
height of building 房屋高度
height of capillary rise 毛管水上升高度
height of capillary water 毛管水高度
height of chimney 烟囱高度
height of conductor above ground 导线对地高
height of cutting 挖土高度
height of damming 筑坝高度
height of dam 坝高
height of filling 填土高度
height of free fall 自由落程
height of gate opening 闸门开启高度
height of high tide 高潮高，满潮高

height of instrument 仪表高度
height of lift 提升高度
height of liquid column 液柱高度
height of low water 枯水位高程，低潮高
height of release 排放高度
height of section 截面高度
height of sight line 视线高程
height of smoke outlet 排烟口高度
height of source 源高
height of tide 潮高
height of transfer unit 传递单元高度
height of water column 水柱高度
height point 高程点
height scale 高度比例尺
height system 高程系统
height-to-span ratio 高跨比
height-volume curve 高度体积曲线
height 高度，顶点，高
heilophilous plant 喜阳植物
heir 继承人
held instrument 携带式仪表，野外仪器
Hele-Shaw cell 多孔性材质
heliarc 氦弧
helical bevel gear 斜齿锥齿轮
helical coil arrangement （热交换器的）螺旋管式布置
helical coil-operated pressure gauge 螺旋管式压力表
helical coil spring 螺旋形盘簧
helical coil steam generator 螺旋管式蒸汽发生器
helical coil type heat exchanger 螺旋管型热交换器
helical coil 螺旋状管，螺旋形线圈
helical conveyer 螺旋式输送机
helical current meter 螺旋式流速仪
helical delay cable 螺旋延迟电缆
helical dislocation 螺旋状错位
helical duct 螺旋形孔道，螺管
helical electric power feeder 螺旋电动进料器
helical end winding 螺旋形端部绕组
helical finning 螺旋形肋片加工
helical fin 螺旋形肋片，螺旋形散热片
helical flow turbine 回流式汽轮机
helical flow 螺线流
helical fluid 旋流
helical gas-holder 螺旋式气柜
helical gear pump 斜齿轮泵
helical gear shaft 螺旋齿轮轴
helical gear 斜齿轮，柱齿轮，螺旋齿轮，斜齿式齿轮
helical groove 螺旋槽
helical instability 螺旋不稳定性
helical line 螺旋线
helical lobe 螺旋叶片
helically coiled plain tube 螺旋形平滑盘管
helically coiled tube element 螺旋盘管元件
helically-wound 螺旋绕法的，螺旋绕组的
helical-orbit beta-ray spectrometer 螺旋形轨道β射线谱仪
helical or herringbone type 斜齿或人字齿型
helical pressure tube 螺旋形压力弹簧管
helical protuberance 螺旋突条
helical pump 螺旋（电磁）泵
helical ribbon mixer 螺旋叶片式混合器
helical scan 螺旋形扫描
helical screw compressor 螺杆压缩机
helical spring 螺旋弹簧
helical spur gear 斜齿正齿轮
helical stairs 盘梯
helical-steam generator 螺旋管式蒸汽发生器
helical strake 螺旋箍条
helical-tube type heat exchanger 螺旋管型热交换器
helical-tubing steam generator 螺旋管式蒸汽发生器，盘管式蒸汽发生器
helical-type staircase 螺旋式楼梯
helical vortex sheet 螺旋涡面
helical weld 螺旋形焊缝
helical wheel 斜齿轮
helical winding 螺旋线圈，螺旋形绕组
helical xenon-arc lamp 螺旋形氙弧灯
helical 螺旋状的，螺旋线的，螺线的，螺旋的，螺旋纹
helicline 盘旋斜坡道
helicoflex gasket 螺旋形密封垫
helicoflex seal 螺旋形密封环
helicoidal anemometer 螺旋桨式风速计
helicoidal flow 螺旋状水流，螺旋流
helicoidal vortex sheet 螺旋面形涡面
helicoid screw 丝杠，蜗杆
helicoid 螺圈，螺旋面，螺旋体，螺旋状的
helicoil insert 螺旋垫圈
helicopter rotor 直升机旋翼
helicopter 直升机
helioelectric 日光生电的
heliogram 日光反射信号，回光信号
heliograph 日光反射信号器，反光通信，日光仪，日照计
heliohydroelectric power generation 太阳能水力发电
heliophobous plant 嫌阳植物
heliostat array field 定日镜（阵列）场
heliostat cluster 定日镜簇，定日镜组
heliostat field system 定日镜（场）系统
heliostat field 定日镜场［区域］
heliostat 定日镜，日光反射镜，日光反射装置
heliothermometer 日温量测计【测太阳温度】
heliotrope 日光回照器
heliotropic wind 日成风，日转风
helio 日光反射信号，日光反射信号器，赫利奥，海立尔
heliport 直升机场，直升机停机坪
helipot 螺旋线圈电位计，螺旋线圈分压器
helium backfill 回充氦气
helium balance line 氦平衡管
helium blanket pressure 氦覆盖层压力
helium blanket 氦覆盖层
helium bleed line 氦泄放管路
helium blower 氦气增压器
helium bombing 氦气轰炸【探漏】
helium bonding 氦结合

helium bottle 氦气瓶
helium circuit 氦气回路
helium circulator 氦气循环风机
helium closed loop 氦封闭回路
helium compressor auxiliary seal oil pump 氦气压缩机辅助密封油泵
helium compressor 氦气压缩机
helium control valve 氦控制阀
helium-coolant gas temperature 氦冷却剂温度
helium coolant 氦冷却剂
helium-cooled breeder (reactor) plant 氦气冷却增殖反应堆电厂
helium-cooled breeder reactor 氦冷增殖反应堆
helium-cooled fast breeder reactor 氦冷快增殖堆
helium-cooled reactor 氦冷反应堆
helium cooler 氦气冷却器
helium cooling 氦冷却
helium cover gas system 氦气覆盖系统
helium cycle 氦循环
helium damage 氦损伤
helium dehydrator 氦干燥器
helium detection examination 氦气探测检验
helium-displacement method 氦置换法
helium drier 氦气干燥器
helium dump tank 氦气排放箱
helium dump valve 氦气排放阀
helium embrittlement 氦脆化
helium family element 氦族元素
helium gas blanket 氦气覆盖层
helium gas circulator 氦气风机
helium gas turbine 氦气轮机
helium handling and storage system 氦气处理及贮存系统
helium heat exchanger 氦气换热器
helium ionization detector 氦电离检测器
helium layer 氦气层
helium leak-check 氦气检漏
helium leak detector 氦检漏仪，氦检漏器
helium leak testing 氦渗泄试验【新燃料】，氦气泄漏测试，氦检漏
helium mass flow 氦气质量流
helium mass spectrometer vacuum testing by dynamic method 氦质谱仪真空动态测试法
helium nuclei 氦核
helium outlet temperature 氦气出口温度
helium permeation testing 漏氦试验
helium permeation test 漏氦试验【新燃料】，氦渗透试验
helium purification system 氦气净化系统
helium purification 氦净化
helium shielded arc welding 氦弧焊，氦气保护弧焊
helium sniffer test 氦吸入探漏试验，嗅漏试验
helium summation test （设备内）充氦探漏试验，充氦试验
helium supply system 氦气供应系统
helium system 氦气系统
helium to boiling water heat exchanger 氦气沸水换热器
helium transfer compressor 氦气输送压缩机
helium turbo-generator set 氦气轮发电机组
helium vacuum test 真空喷氦探漏试验
helium-water heat exchanger 氦水换热器
helium 氦，氦气
helix slot blade-type burner （旋风炉用）螺旋叶片式喷燃器，螺旋叶片式喷燃器
helix type steam generator 螺旋管式蒸汽发生器，盘管式蒸汽发生器
helix volute 螺旋蜗壳
helix waveguide 螺旋线波导管
helix 螺旋，螺旋线，螺旋形，螺旋管，螺线，螺旋箍条，螺旋弹簧
heliodon 日影仪
Heller air cooling system 海勒式空冷系统
Heller system 海勒系统
helm angle 舵角
helmet shield 焊工面罩，护目头罩，盔式护罩
helmet 机罩，箍，环，头盔，安全帽，盔，帽，面罩
helm gear 舵机装置
Helmholtz resonance frequency 亥姆霍兹共振频率
helm wind 舵轮风
helm 舵轮，舵
helper steam turbine 辅助汽轮机
help forward 促进
helpmate 助手，合作人员
HELP screen “帮助”屏幕
hematite 低磷生铁，赤铁矿，三氧化二铁锈层
hemic material 半分解纤维质
hemi-crystalline 半晶质
hemicycle 半循环，半轮回
hemi-ellipsoidal 半椭球的
hemihydrate plaster 熟石膏灰泥
hemihydrate 半水化物
hemimorphite 异极矿
hemisphercal emissivity 半球黑度，半球发射率
hemisphere 半球
hemispherical absorptance 半球吸收率
hemispherical absorptivity 半球吸收率
hemispherical angular reflectance 半球-角反射率
hemispherical directional reflectance 半球定向反射率
hemispherical dome 半球顶【安全壳】
hemispherical earth electrode 半球形接地电极
hemispherical emittance 半球黑度，半球发射率
hemispherical equipotential surface 半球等势面
hemispherical head inspection 半球形顶盖检查
hemispherical head （压力容器）半球封头，半球形封头
hemispherical radiation properties 半球辐射特性
hemispherical radiation 半球（向）辐射
hemispherical reflectance 半球反射率，半球向反射比
hemispherical rotor pole 半圆形转子磁极
hemispherical solar irradiance 半球向日射辐照度【Eh】
hemispherical solar radiation 半球向辐射，半球向日射
hemispherical spectral absorptivity 半球单色吸收率
hemispherical temperature （灰的）半球化温度，

半球温度
hemispherical　半球形的，半球的
hemitropic armature winding　半圈式电枢绕组
hemitropic winding　半圈绕组，半圈绕法
Hemlock Semiconductor Corporation　美国 HSC 公司
hemp bag　麻袋
hemp core　麻心【钢缆的】
hemp cut lime mortar　麻刀灰浆
hemp cuts and lime as base　麻刀灰打底
hemp cut　麻刀，麻筋，麻丝
hemp packing　麻垫料，麻绳填料，蔴填料
hemp rope　麻绳
hem reinforced branched pipe　贴边岔管
henrymeter　亨利计，电感表
henry　亨利【电感单位】
H=enthalpy　焓
HEPA filter　高效过滤器
heptad　七价物
heptagon copper wire　七边形铜线
heptavalence　七价
heptode convertor　七极变频器
heptode mixer　七极混频管
heptoxide　七氧化物
herbaceous cover　草皮覆被，草皮护面
herbaceous soil covering　草皮覆盖层
herbage cover　草皮覆被
herbicide　除莠剂
herbosa　草本群落，草本植被
herb　草本植物
hereafter　今后，从此以后
hereby　特此，由此，兹
hereinabove　在上文，以上
hereinafter referred to as　以下简称为……
hereinafter　以下，在下，在下文，以下
hereinbelow　在下文
herein　在本文中，在本合同中，此中，于此
hereof　关于此点，在本文件中
hereto　本文件的，对于这个，关于这个，到此为止
hereunder　在下（文），在此之下，本协议下的，本文件规定
herewith　与此，附此
Herfindahl-Hirschman index　赫希曼指数，市场集中度指数
hermaphrodite calipers　单边卡钳
hermetically sealed cable　密封电缆
hermetically sealed casing　密封机罩
hermetically sealed motor　密封电机
hermetically sealed relay　密封式继电器，气密式继电器
hermetically sealed transformer　密闭型变压器
hermetically sealed　密封的，气封的，密闭的
hermetical　气密的，密封的
hermetic enclosure header　密封体引线座，密封体引线头
hermetic integrating gyroscope　密封式积分陀螺仪
hermetic machine　密封型电机
hermetic motor　密封型电动机
hermetic seal leakage　密封泄漏
hermetic seal　气封，气密，密封，密封止水
hermetic　气密封
hermetization　密封，封闭
hermeticity test　密封性试验
hermeticity　密封性
heron　苍鹭
herringbone bracing　斜十字撑
herringbone bridging　人字撑
herringbone fins　人字形肋
herringbone gear　人字齿轮，人字形齿轮
herringbone parquetry　人字拼木地板，席纹地面
herringbone pattern rubber lagging　人字形包胶
herringbone system　人字形系统
herringbone tooth　人字齿
herringbone wheel　人字齿轮
herringbone winding　人字形绕组
herringbone　人字形，人字形的，人字形条纹，鱼刺形的
hertz-frequency　赫兹频率
hertz　赫兹【频率单位】
hesitating relay　缓动继电器
hesitation switch　暂停开关
hesitation　暂停，暂时停机，临时停机
heterocoagulation　异质凝聚
heterodyne action　外差作用
heterodyne converter　外差变频器
heterodyne detector　外差检波器
heterodyne frequency　外差频率
heterodyne interference　外差干扰
heterodyne oscillator　外差振荡器，差频振荡器
heterodyne reception　外差接收法，外差接收
heterodyne wavemeter　外差式波长计
heterodyne　外差振荡器，外差法
heterogeneity of rock　岩石的非均质性
heterogeneity test　非均匀性试验
heterogeneity　多相性，不纯性，不均匀性，不统一性，异构性，非均匀性
heterogeneous atmosphere reaction　多相大气反应，非均相大气反应
heterogeneous azeotrope　非均相共沸混合物
heterogeneous body　不均匀体，非均质体
heterogeneous catalysis　多相催化
heterogeneous catalyst　多相催化剂
heterogeneous circuit　不均匀电路
heterogeneous composition　多相组合
heterogeneous decomposition　多相分解
heterogeneous equilibrium　多相平衡
heterogeneous fluidization　多相流态化，非均匀流态化
heterogeneous fluid　非均质流体
heterogeneous flux reconstruction　非均匀通量重建
heterogeneous foundation　非均匀地基
heterogeneous fuel configuration　燃料非均匀排列
heterogeneous lattice　非均匀栅格
heterogeneous materials　非均匀性物质
heterogeneous material　多相材料，非均质材料
heterogeneous medium　非均匀介质
heterogeneous mixture　非均匀混合物
heterogeneous molten-salt reactor　非均匀熔盐反应堆

heterogeneous natural uranium reactor　非均匀天然铀反应堆
heterogeneous reaction　多相反应
heterogeneous reactor　非均匀反应堆
heterogeneous region　多相区
heterogeneous rock mass　非均质岩体
heterogeneous shield　非均匀屏蔽
heterogeneous soil　不均质土，非均质土
heterogeneous state　非均态
heterogeneous strain　非均匀应变
heterogeneous system　多相体系，非均匀体系，非均匀系
heterogeneous　多相的，不均匀的，异型的，非均匀的，非均质的
heterogenerty　不均匀度
heterojunction solar cell　异质结太阳电池
heteromorphism　复形性，同质异象
heteropolar dynamo　异极电机
heteropolar generator　异极发电机
heteropolar inductor generator　异极感应子发电机
heteropolar　异极的，多极的
heuristic algorithm　探试算法
heuristic program　探试程序，启发式程序
heuristic routine　探试程序，启发式程序
hew　切割，中断，砍，劈
hexadecimal multiplication　十六进制乘法
hexadecimal number　十六进制数
hexadecimal　十六进位，十六进制的，十六进制的
hexadecylamine　十六烷（基）胺
hexafluoride　六氟化物
hexafluoroaluminum ion　六氟化铝离子
hexagonal bar iron　六角钢条
hexagonal brick　六角砖
hexagonal can　六角形外壳
hexagonal closed double bead wrench　六角环形双头扳手
hexagonal closed-packed structure　六角形密集结构
hexagonal column　六角柱
hexagonal configuration　六角形布置
hexagonal duct tube　六角形导向管
hexagonal fuel assembly　六角形燃料组件
hexagonal head bolt　六角（头）螺栓
hexagonal inserted handle box wrench　六角插柄套筒扳手
hexagonal lattice　六角形栅格
hexagonal nut　六角螺母
hexagonal pattern　六角形栅格【反应堆】
hexagonal prism　六角形棱柱体
hexagonal socket wrench　六角套筒扳手
hexagonal steel bar　六角钢
hexagonal　六边的，六角的
hexagon bolt　六角头螺栓
hexagon head bolt　六角头螺栓
hexagon head screw　六角头螺丝
hexagon long pattern socket　加长六角套筒
hexagon screw die　六角螺丝板牙
hexagon socket　标准六角套筒
hexagon spanner　六方（扳手），六角扳手
hexagon voltage　对称六相系统的线电压
hexagon　六边形，六角形，六角体
hexahedron　六面体
hexametaphosphate　六偏磷酸盐
hexamine　六胺，乌洛托品
hexaphase turbogenerator　六相汽轮发电机
hexaphase　六相的
hex bolt　六角螺栓
hex head bolt　六角头螺栓
hex head screw　六角头螺钉
hex nut　六角螺母
hexode　六极管
hex socket head cap screw　内六角螺钉
hex washer　六角垫圈
HFCs　氢氟碳化物
HFO(heavy fuel oil)　重燃料油，重质燃料油，重燃油，重油
HF process　氢氟酸法
HFT(hot functional test)　热态功能试验
H-girder　宽缘工字梁
HC(heat conservation)　保温
heat indication test　热耗试验【指汽机】
HHF(hyper-high frequency)　超高频
HH(high-high)　高高【极高】，高值的高限【DCS画面】
HH nut(hexagon-headed nut)　六角螺母
HHV(high heating value)　高位发热量，高位热值，高热值
hhydrological element　水文要素
hibernation　冬眠
Hi-cap(high capacity)　高容量
hickey　螺纹接合器
hick joint　平缝
hidden cost　额外费用
hidden heat　潜热
hidden ore body　隐藏的矿体
hidden periodicity　潜周期性
hidden rock　暗礁
hidden　潜在的，隐蔽的
hide out corrosion　隐藏性腐蚀
hide out　暂时消失，隐藏
hierarchical computer control system　分级计算机控制系统
hierarchical control system　分级控制系统
hierarchical control　分层（级）控制
hierarchical distributed processing system　分级分布处理系统
hierarchical file　分级文件
hierarchical organizational diagram　组织结构图
hierarchical structure　分级结构
hierarchy control system　分层控制系统
hierarchy of skill　技术层次
hierarchy structure of HVDC power transmission control system　高压直流输电控制系统分层结构
hierarchy system　分级的多计算机控制系统
hierarchy　体系等级制度，层次，等级制度，分级体系，体系
HIFI(high-fidelity)　高置信度，高保真
higgle　争论，论价，讨价还价
high-accuracy survey　高精度测量

high-active waste 高放射性废物
high activity cell 强放射性热室，高放射性热室，高放热室
high activity waste container 高放废物容器
high activity waste 强放射性废物，高放射性废物，高放废物
high alarm 高位报警
high-alkali cement 高碱水泥
high alloy steel 高含金钢
high-altitude machine 高海拔电机，高原电机
high-altitude operation 高空作业
high-altitude test 高海拔试验
high-altitude transformer 高海拔变压器
high alumina cement 高矾土水泥，高铝水泥，矾土水泥
high alumina clay 高铝黏土
high alumina refractory 高铝耐火材料
high amplitude detector 强信号检波器
high amplitude vibrating screen 高幅筛，高辐振动筛
high amplitude wave 强大振幅波
high and low matching method 高低匹配法
high and low voltage alarm 高低电压报警器
high angle fault 陡角断层
high angle of attack 大冲角
high antiknock fuel 高抗爆燃料
high-arrangement of outdoor switchgear 高型屋外配电装置
high-ash coal 高灰煤
high-ash-fusion coal 高灰熔点煤
high-ash 高灰分，高灰分的
high aspect ratio building 细长建筑
high auctioneer module 高脉冲选择模块
high auctioneer unit 高脉冲选择模块
high background contamination 高本底污染
high bainite 上贝菌体，上贝氏体
high bank 高堤
high barometric maximum 高气压
high-boiling oil 高沸点油
high-boiling point 高沸点
high-boiling 高沸点的
high bond reinforcing bar 高握裹力钢筋
high breaking capacity fuse 高截断功率熔丝
high breaking capacity 高截断公率
high burn fuel rod 深燃耗燃料棒
high burn-up 深燃耗，高燃耗
high-calcium lime 高纯度钙石灰，高钙石灰
high capacity boiler 大容量锅炉
high capacity breeder 高功率增殖反应堆，高功率增殖堆
high capacity communication 大容量通信
high capacity generator 大容量发电机，大型发电机
high capacity motor 大型电动机，大功率电动机
high capacity pump 大流量泵
high capacity resin 高交换容量树脂
high capacity turbine 大容量透平，大功率透平
high capacity water power station 大容量水电站，大型水电站
high capacity water turbine 大容量水轮机
high capacity 高容量
high-carbon steel 高碳钢
high-carbon 高碳
high cellulose potassium electrode 高纤维钾型焊条
high cellulose sodium electrode 高纤维钠型焊条
high chemical churning pile 高压旋喷桩
high chimney 高烟囱
high-coercive 高矫顽磁性的
high commissioner 高级商务代表，高级专员
high concentrate dust 高浓度粉剂
high concentration formulation 高浓度制剂
high concentration 高浓度
high concessional terms 高度优惠条件
high conductivity 高导电率
high conversion light water reactor 高转换轻水反应堆
high current arc welding 大电流弧焊
high current load 大电流负荷
high current supply 大电流电源
high current 高强度电流，强电流的，大电流的
high cut-off frequency 高截频率
high-cycle fatigue 高周波疲劳，高循环疲劳，高周疲劳
high-cycle 高频的
high cylindrical valve 高圆筒阀
high dam 高坝
high definition radiography 高清晰度射线照相术
high density and high-strength graphite 高密高强石墨
high density concrete 高密度混凝土，致密混凝土，重混凝土
high density digital recording 高密度数字记录
high density digital tape recorder 高密度数字带记录器
high density digital tape 高密度数字记录带
high density electronic packaging 高密度电子组装
high density fuel storage rack 燃料密集贮存架，高密度贮存燃料
high density fuel storage 核燃料密集储存
high density layer 致密层
high density logic 高密度逻辑
high density polyethylene 高密度聚乙烯
high density recording 高密度记录
high density resistivity measurement system 高密度电阻率测量系统
high density resistivity prospecting 高密度电阻率探测，高密度电阻率勘探法
high density solids leg 高密度固体料腿，高密度粒子管
high density storage （燃料）密集贮存
high density 高密度
high dip angle 陡倾角
high direct voltage 高直流电压，直流高压
high-dispersion nozzle 高雾化细度喷嘴
high dose corridor 高剂量走廊
high dose rate irradiation 高剂量率辐照
high drag turbulent-flow regime 大阻力紊流状态
high-ductility 高韧性
high duty boiler 大容量锅炉，高出力锅炉
high duty seal 高压密封

high duty 高负荷的，重载的
high early strength cement 快硬水泥，高早强水泥，高标号早强水泥，高标号早强水泥
high earthquake-intensity area 高地震烈度区
high efficiency aerosol filter 高效气溶胶过滤器
high efficiency concentrator 高效浓缩机
high efficiency fiber filter 高效纤维过滤器
high efficiency filter 高效过滤器
high efficiency flap 高效襟翼
high efficiency operation 高效运行
high efficiency particle filter 高效微粒过滤器
high efficiency particulate air filter 高效粒子空气过滤器，高效微粒过滤器
high efficiency particulate filter 高效粒子过滤器
high efficiency 高效，高效率
high efficient fibre filter 高效纤维过滤器
high efflux velocity tuyere 高射流速度喷嘴
high-elastic limit steel 高弹性极限钢
high energy advisory committee 高能咨询委员会
high energy arc ignition 电火花高能点火
high energy arc ignitor 高能电弧点火器，电火花高能点火器
high energy battery 高能量电池
high energy collision 高能碰撞
high energy cross-section 高能截面
high energy electron 高能电子
high energy fission 高能粒子引起的分裂，高能裂变
high energy fluid system 高能流体系统
high energy group 高能群
high energy heat treatment 高能束热处理
high energy ignition device 高能点火装置
high energy injection 高能注射
high energy line break 高温高压管线破裂
high energy magnet 高能磁体
high energy neutron reaction experiment 高能中子反应实验装置
high energy neutron 高能中子
high energy radiation 高能辐射，穿透辐射
high energy spark ignitor 高能点火器
high energy total absorption detector 高能全吸收探测器
high energy 高能
high enriched uranium 高浓缩铀，高浓铀，高富集铀
high enrichment fuel 高富集燃料
high enrichment leacher 高富集燃料浸取器
high enrichment reactor 高富集燃料反应堆
high enrichment 高浓缩度，高富集度
high enthalpy drop profile 大焓降叶型
high entry porous stone 高进气透水石
higher chain product 超铀核素
higher critical speed 高临界转速
higher critical velocity 高临界流速，高临界速度
higher geodesy 高等大地测量学
higher harmonic voltage 高次谐波电压
higher harmonic 高次谐波
higher high tide 高高潮
higher-level language 高级语言
higher low tide 高低潮
higher management 高级主管
higher mathematics 高等数学
higher measuring range value （测量范围）上限值
higher mode excitation 高次谐波励磁
higher model coupling 高次谐波激励，高次谐波耦合
higher mode vibration 高频振动
higher mode 高阶模态，高阶振型，高次模
higher operating range control 上限法调节
higher order boundary-layer theory 高阶边界层理论
higher order closure 高阶封闭
higher order harmonic 高次谐波
higher order lag 高次滞后
higher order language 高级语言
higher order wave 高次谐波
higher-pressure deaerator 高压除氧器
higher specific speed 高比转速
higher surveying 高等测量学
highest altitude 最高海拔
highest astronomical tide 最高天文潮位
highest bidder 出高价的投标人
highest bid 最高报价，最高标，最高价投标
highest common divisor 最高公约式
highest detector 高值检测装置，高值指示装置
highest-enriched fuel assembly 最高富集燃料组件
highest ever known discharge 历史最大流量
highest flood control water level 防洪高水位
highest flood level 最高洪水位
highest high-water 最高高水位
highest historical discharge 历史最大流量
highest output stage 高值输出级
highest point 最高点
highest possible frequency 最高可能频率
highest possible suction head （泵）最大可能吸入高度
highest price 最高价格
highest quotations 顶盘，最高行市
highest record stage 最高记录水位
highest runoff 最大径流
highest stage 最高水位
highest system voltage 系统最高电压
highest temperature 最高温度
highest up surge level 最高涌波水位
highest upsurge 最高上涌浪
highest voltage of a system 系统最高电压
highest water level 最高水位
highest wave 最大波浪
high expansion foam extinguishing system 高倍数泡沫灭火系统
high expansion foam fire truck 高倍泡沫消防车
high expansion mortar 膨胀砂浆
high explosive 高级炸药，烈性炸药
high-fidelity amplifier 高保真放大器
high-fidelity 高保真度，高置信度，高保真
highfield superconductor 高磁场超导体
highfield type booster 强场式升压机
highfield 强磁场，强电场，高磁场
high fill 高填土
high flash oil 高闪点油

high flash tank 高位水箱
high flood level 高洪水位
high flow period 丰水期
high flow rate mixed bed 高速混床
high flow rate profile 大流量叶型
high flow rate 高流速率
high flow year 多水年，丰水年
high flow 高流量
high flush tank 高架冲洗水箱，高水箱
high flux beam reactor 高通量束反应堆，高通量中子束堆
high flux bombardment 高通量轰击
high flux core 高通量堆心，高磁密铁芯
high flux heater 高热负荷加热器
high flux reactor 高通量堆，高通量材料试验堆，高中子通量反应堆
high flux research reactor 高通量研究反应堆
high flux 高密度流，高强度流
high-flying highway 高架公路
high frequency 高频，高频的
high frequency alternator 高频发电机，高频发生器
high frequency amplification 高频放大
high frequency buffeting 高频率颤振
high frequency cable 高频电缆
high frequency choke 高频抗流圈，高频扼流圈
high frequency circuit 高频电路
high frequency coil 高频线圈
high frequency communication 高频通信
high frequency control 高频控制
high frequency core 高频铁芯
high frequency current 高频电流
high frequency electric field heating 高频电场加热
high frequency end 高频端
high frequency filter 高频滤波器
high frequency furnace 高频电炉
high frequency generator 高频发电机，高频发生器
high frequency harmonic 高次谐波
high frequency heating 高频加热
high frequency impulse 高频脉冲
high frequency induction furnace 高频感应电炉，高频感应炉
high frequency induction welder 高频感应焊机
high frequency instability 高频不稳定性
high frequency insulation 高频绝缘
high frequency insulator 高频绝缘子
high frequency interference 高频干扰
high frequency iron core 高频铁芯
high frequency modulation 高频调制
high frequency motor 高频电机
high frequency noise 高频噪声
high frequency oscillation 高频振荡
high frequency oscillator 高频振荡器
high frequency oscillography 高频示波术
high frequency protection 高跨保护
high frequency relay 高频继电器
high frequency resistance welding 高频电阻焊
high frequency response 高频响应
high frequency switching 高频开关
high frequency transformer 高频变压器
high frequency upset welding 高频电阻焊
high frequency vibration 高频振动
high frequency voltage 高频电压
high frequency wave 高频波
high frequency welding 高频电焊
high frequent value 最频值
high-fusion-ash coal 高熔点煤
high gain cryotron 高增益冷子管
high gain linear circuit 高增益线性电路
high gain 高增益
high-gloss enamel 高光瓷漆
high-gloss 高光洁度
high grade cement 高等级水泥，优质水泥
high grade coal 优质煤
high grade concrete 高等级混凝土
high grade energy welding 高能焊
high grade energy 高势能，高级能
high grade fuel 高级燃料
high grade heat 高位热，高势热
high grade insulation 高级绝缘
high grade mica 高级云母
high grade nuclear fuel 高浓核燃料
high grade product 优质产品
high grade steel 优质钢
high grade 优质产品
high gradient magnetic separation 高梯度磁场分离法
high harmonic 高次谐波
high-head dam 高水头坝
high-head hydraulic turbine 高水头水轮机
high-head hydroelectric power station 高水头水电站
high-head hydro-power station 高水头水力发电站
high-head injection pump 高压注入泵
high-head injection 高压注入
high-head power plant 高水头电站
high-head power station 高水头电站
high-head pump 高压泵，大流量泵
high-head safety injection pump 高压安全注入泵
high-head safety injection 高压安全注入
high-head scheme 高水头电站方案
high-head turbine 高水头水轮机
high-head water turbine 高水头水轮机
high-head 高压头
high heating value of design coal 设计煤的高热值
high heating value 高热值，高位发热量，高（位）热值
high-heat waste 高释热废物
high-heat zone 高热区
high-heat 难熔的，耐热的
high-high pressure switch 高-高压开关
high-high 极高信号【设定值、报警、水位等】，高值的高限
high humidity curing 高湿度养护
high humidity treatment 高湿度处理
high impedance grounded system 中性点高阻抗接地系统
high impedance module 高阻抗模块
high impedance relay 高阻抗继电器
high impedance rotor 高阻抗转子

high impedance 高阻抗
high incidence 大冲角，正冲角，大迎角
high initial response exciter 高起始反应速度励磁机
high integrity container 坚固容器
high intensity AOL 高光强航空障碍灯
high intensity current 高强度电流
high intensity discharge lamp 高强气体放电灯
high intensity oscillation 高强度振荡，高强度振动
high intensity rainfall 高强度降雨
high intensity 高强度
high intermediate pressure cylinder 高中压缸
high internal resistance voltmeter 高内阻电压表
high inversion fog 高逆温雾
high iron oxide type electrode 氧化铁型粗焊条，氧化铁型焊条
high iron portland cement 高铁硅酸盐水泥
high-lag thermometer 高滞效温度表
highland motor 高原电机
highland 高地
high-leakage core 高泄漏堆芯
high-leakage transformer 高漏磁变压器
high LET radiation 高线性能量转换照射
high level acid storage tank 高位酸贮存槽
high level alarm 高位报警，高料位［液面］报警
high level anticyclone 高空反气旋
high level caustic storage tank 高位碱贮存槽
high level cave 高放射性水平热室
high level cell 高放射性热室
high level data link control 高级数据链控制
high level detector 高料位探测器
high level dosimeter 高位剂量计，高水平辐射剂量计，高剂量率剂量计
high level instrumentation 高功率测量仪器，强放射性测量仪器
high level intake 浅孔式进水口
high level inversion 高空逆温
high level irradiator 高强度辐照器，高放射性辐照器
high level language 高级语言
high level modulation 高电平调制，高功率调制
high level radiation source 高放射性辐射源，高强度辐射源
high level radiation 高放射性辐射，高强度辐射
high level solid waste 高放射性固体废物，高放固体废物
high level talks 高层谈判
high level tank 高架水箱，高位箱
high level transistor logic 高电平晶体管逻辑
high level trip 高功率停堆
high level waste store 高放（射性）废物贮存，高放废物贮存库
high level waste 高放废物，高放射性废物
high level 高水位，高电平，高能级，高放射性水平
high lift airfoil 高升力翼型
high lift concrete construction method 混凝土高块浇筑法
high lift construction 高空作业，高（浇注）块施工
high lift flap 高升力襟翼
high lift pumping station 高扬程抽水站
high lift pump 高扬程水泵，高扬程泵
highlighter 荧光笔，标志笔
high ligthning incidence 多雷
highlight 加亮，使显著，醒目，特殊效果，亮点，强光，最亮点，重点，要点，强调，使突出
high limit adjustment 上限调整
high limit control 上限值控制
high limit 上限值，高限
highline shuttle conveyor 高架往复移动输送机
highline 高架线，高架索，高压线，高架的，显着标题
high-low action 高低值作用，高-低值作用
high-low bias check 高低偏压校验
high-low generator 高低频发电机
high-low lamp 变光灯泡
high-low level control 高低液位控制，双位控制
high-low limit 高低限值，上下限值
high-low-range switch 高低量程转换开关
high-low selector 高低值选择器
high-low voltage relay 高低压继电器
high-low voltmeter （电源的）高低压警报电压表，高低压电压表
high-low water alarm 高低水位报警
highly acidic 强酸性的
highly active waste 高放射性废物
highly cambered blade 大曲度叶片
highly collimated beam 高度准直束
highly conducting plasma 高导等离子体
highly cross-linked resin 高交联树脂
highly damped instrument 高阻尼仪表
highly effective particle of atmosphere filter 高效粒子大气过滤器，高效粒子空气过滤器
highly enriched fuel cycle 高富集燃料循环
highly enriched nuclear fuel 高富集核燃料
highly enriched reactor 高富集燃料反应堆
highly enriched uranium 高富集铀，高浓缩铀【80%～100%】
highly pervious soil 高透水性土壤
highly plastic clay 高塑黏土
highly radioactive 高放射性的
highly rated reactor 高功率密度反应堆
highly sensitive 高灵敏度的
highly-stressed rotor 高应力转子，高强度转子
highly-stressed 高应力的
highly viscous oil 高黏度油
high-magnification seismograph 高倍率地震计
high mechanization 高度机械化
high megohmmeter 超高阻表，超绝缘测试计
high-melting alloy 高熔点合金
high-melting 高熔点的，难熔的
high modulus inclusion 高模量插入体，高模量包体
high moisture coal 高水分煤
high molecular coagulant 高分子凝结剂
high molecular 高分子的
highmoor 高沼地
high mountain station 高山测站

high neutron flux trip 中子高通量紧急停堆
high neutron flux 高中子通量
high nickel-iron base alloy 高镍铁基合金
high noise immunity logic 高抗扰度逻辑电路
high observing tower 高测标
high oil content circuit breaker 多油式断路器
high oil temperature trip device 油温过 高断路装置
high-order add circuit 高阶加法电路
high-order digit 高位数字
high-order elasticity 高阶弹性力学
high-order fission product 高次裂变产物
high-order harmonic 高次谐波
high order merge 高阶合并
high or highest 高或最高
high output 高出力，大功率，大容量，高产量
high pass filter 高通滤波器
high pass 高通
high peak current 峰值电流
high penetration bitumen 高渗透性沥青
high performance fuel 高性能燃料
high performance material 高性能材料
high performance 高性能的，高性能，高效率
high-permeability alloy 高磁导率合金
high-permeability material 高磁导率材料
high pitch cone roof 大坡度圆锥形顶盖
high pitch roof 陡坡屋顶
high plain 高平原
high plastic limit 高塑限
high point 高水位点，凹凸不平，高点
high polymer electrolyte 高聚合物电解质
high polymer emulsion 高聚合物乳剂
high polymer 高聚合物
high position 高位
high potential test 高压试验，耐高压试验
high potential 高电位，高电压
high-power channel type reactor 大功率管式反应堆
high-power-density reactor 高功率密度反应堆
high-power electric capstan 大功率电动绞盘
high-power electron beam 高能电子束
high-power factor transformer 高功率因数变压器
high-power generator 大容量发电机，大型发电机
high-power laboratory 大功率实验站
high-power modulation 高电平调制，高功率调制
high-power plant 大功率电厂
high-power range monitor 高功率区段监测器
high-power range trip 高功率区段紧急停堆，高功率水平紧急停堆
high-power reactor 大功率反应堆，高功率反应堆
high-power station 大容量发电厂
high-power testing 大功率试验
high-power transformer 大容量变压器，大型变压器
high-power tube 大功率电子管
high-power water boiler 高功率水管锅炉
high-power 大功率，高功率的，大容量
high precision levelling 高精度水准测量
high precision regulator 高精确度调整器
high precision 高精度，高精密度
high pressure 高压，高气压
high pressure air 压缩空气
high pressure area 高压区
high pressure automatic close gear 高压自动断路器
high pressure automatic closing gear 高压自动关闭器
high pressure blowpipe 等压式焊（割）炬
high pressure body 高压壳体
high pressure boiler-feed pump 高压锅炉给水泵
high pressure boiler 高压锅炉
high pressure bypass system 高压旁路系统
high pressure bypass 高压旁路，高压再热旁路
high pressure cable 高压电缆
high pressure casing （汽轮机）高压缸，高压缸
high pressure cement grouting 高压水泥灌浆
high pressure charging pump 高压上充泵，高增压充气泵
high pressure charging 高增压
high pressure cleaning machine 高压清洗机
high pressure compressor 高压压气机，高压压缩机
high pressure coolant injection pump 高压冷却剂注入泵
high pressure coolant injection system 高压冷却剂注入系统
high pressure coolant injection 高压冷却剂注入
high pressure core spray system 高压堆芯喷淋系统
high pressure cylinder expansion 高压缸膨胀
high pressure cylinder 高压缸
high pressure deaerator 高压式除氧器
high pressure draft 高压抽吸，高压气流
high pressure draught 高压气流
high pressure drilling 高压钻进
high pressure duct system 高压管路系统
high pressure electrolysis 高压电解
high pressure element 高压缸【汽轮机】
high pressure emergency steam valve 高压主汽门，高压危急截断阀
high pressure exhaust check valve 高压排汽逆止阀
high pressure extraction valve 高压抽汽阀
high pressure extraction 高压抽汽
high pressure feedwater heater system 高压给水加热器系统
high pressure feedwater heater 高压给水加热器
high pressure flange 高压法兰
high pressure flushing pump 高压冲洗泵
high pressure gate 高压闸门
high pressure gauge 高压压力计
high pressure governor valve 高压调速汽门
high pressure grouting 高压灌浆
high pressure heater availability rate 高压加热器投入率
high pressure heater available rate 高压加热器投入率
high pressure heater bypass valve 高压加热器旁路门

high pressure heater 高压加热器
high pressure heating system 高压供暖系统
high pressure heating 高压加热
high pressure heat transfer rig 高压传热设备，高压传热试验台架
high pressure hose 高压软管
high pressure hot water heating system 高压热水供暖系统
high pressure hot water heating 高压热水供热
high pressure hydraulic actuator 高压油动机，高压执行机构
high pressure hydraulic testing pump 高压管道试压泵
high pressure hydrogen 高压氢气
high pressure impregnation 高压浸渍
high pressure injection grouting 高压喷射灌浆
high pressure injection system 高压注入系统
high pressure injection 高压注射，高压注入
high pressure intake 高压进水口
high pressure ion exchange column 高压离子交换柱
high pressure ion exchange 高压离子交换
high pressure jet washing machine 高压冲洗机
high pressure leak detector 高压系统测漏器
high pressure main stop valve 高压主汽阀，高压主汽门
high pressure main 高压总管，高压母管
high pressure nozzle box 高压喷嘴室
high pressure oil jacking equipment 高压油顶轴装置
high pressure packing leakage 高压汽封漏汽
high pressure painting machine 高压喷漆机
high pressure penetration 高压贯穿件
high pressure pump 高压泵
high pressure reactor 高压反应堆
high pressure recirculation system 高压再循环系统
high pressure rotor 高压转子
high pressure safety injection pump 高压安全注入泵
high pressure safety injection system 高压安全注入系统
high pressure safety injection 高压安全注入
high pressure seal 高压密封
high pressure section 高压段，（汽轮机）高压缸
high pressure service water system 厂用高压水系统
high pressure servo-motor 高压油动机，高压伺服马达
high pressure shaft seal 高压轴密封
high pressure shaft 高压轴
high pressure sluice valve 高压闸阀
high pressure sodium lamp 高压钠灯
high pressure sodium vapour lamp 高压钠灯
high pressure sprinkler 高压喷灌机
high pressure stage 高压段，高压级
high pressure steam 高压蒸汽
high pressure steam heating system 高压蒸汽供暖系统
high pressure steam heating 高压蒸汽采暖
high pressure steam power plant 高压火力发电厂
high pressure steam turbine 高压汽轮机
high pressure tester 高压试验机
high pressure tunnel 高压隧洞
high pressure turbine stop valve 高压汽轮机截止阀
high pressure turbine 高压汽轮机，高压透平
high pressure valve 高压阀
high pressure water derusting 高压水除锈
high pressure water jet 高压喷水器，高压水枪
high pressure water pump 高压水泵
high pressure water 高压水
high pressure welder 高压焊工
high pressure wind tunnel 增压风洞，高压风洞
high pressurizer water level trip 稳压器高水位紧急停堆
high-priced service 高价劳务
high price winning ratio 高价中标率
high priority investment project 最优先投资项目
high priority project 重要优先项目
high priority 最优先【项目等】
high-profile layout 高型布置
high purity graphite 高纯石墨
high purity oxygen aeration system 高纯氧换气系统
high purity steam 高纯度蒸汽
high purity water 高纯水
high purity 高纯度
high quality and new product 优质新产品
high quality carbon steel 优质碳素钢
high quality coal 优质煤
high quality concrete 优质混凝土
high quality product 优质产品
high quality region 高含汽区
high quality reuse water 优质再生水
high quality service 高质量的服务，优质服务
high quality steel 优质钢
high quality 高质量，优良品质，（蒸汽）高含汽率
high radiation flux 高辐射通量
high radio activity 高水平放射性，强放射性
high random access 高速随机存取
high range dosimeter 高量程剂量计
high range water reducing agent 高效减水剂
high-rank coal 优质煤
high rate activated sludge process 高速活性污泥法，高负荷活性污泥处理法
high rate aeration basin 高负荷换气池
high rate aeration settling 高负荷曝气沉降法
high rate aerobic treatment 高负荷需氧处理
high rate battery 高速放电蓄电池，高速电池
high rate filter 高效率滤池，高负荷过滤池
high rate filtration 高负荷过滤，高速过滤
high rate rapid filter 高速快滤池
high rate settling tank 高负荷沉淀池
high rate trickling filter plant 高负荷滴滤池装置
high rate trickling filter 高负荷滴滤池
high rate water wash 高强度冲洗
high-rating generator 大功率发电机，大型发电机
high-ratio transformer 大变比变压器
high-reactance rotor 高电抗转子

high reactor start-up rate trip　启动过速事故保护停堆
high recovery thermocouple　高恢复系数热电偶
high recovery　高回收
high reliability power line　高可靠性电力线路
high reliability price　高可靠性电价
high reliability relay　高可靠继电器
high reliability　高可靠性
high-reluctance commutating pole　高磁阻换向极
high-remanence　高顽磁性的
high reputation　卓著的信誉
high residue level　高残留度
high-resistance alloy　高电阻合金
high-resistance material　高电阻材料
high-resistance resistor　高阻电阻器
high-resistance spacer　高电阻衬垫
high resolution　高分辨率，高分辨能力，高电阻的，高阻
high Reynolds number water tunnel　高雷诺数水洞
high Reynolds number wind tunnel　高雷诺数风洞
high rise block　高层大楼，多层建筑
high rise building power supply　高层建筑物供电
high rise building　高层建筑物，高层建筑
high rise disk　高抬升阀瓣
high rise structure　高层结构
high rupture capacity fuse　高断路容量熔丝，高截断容量熔丝
high salinity　高含盐量
high salt content liquid waste　高含盐量废液
high salt content　高含盐量
high salt waste　高含盐量废水
high seas　公海，远洋，外海，狂浪
high security zone　高度安保区
high selector　高选择器
high sensitive relay　高灵敏度继电器
high shoulder　超高路肩
high-sided open-top car　高帮敞车
high side gondola car　高帮无门敞车
high side gondola　高帮敞车
high-signal selector　高值信号选择器，高信号选择器
high silicon cast iron　高硅铸铁
high silicon sheet iron　高硅钢片
high-sill spillway　高槛溢洪道
high slip induction generator　高滑差异步发电机
high slip motor　高转差电动机，高滑差率电动机
high slope excavation　高边坡开挖
high slope　高边坡
high solidity cascade　大稠度叶栅，稠叶栅
high solidity rotor　高实度风轮
high sound insulation　高度隔声
high specific speed pump　高比速泵，高特有速度泵
high speed　高速的，高速
high speed A/D converter　高速模数转换器
high speed aerodynamics　高速气体动力学
high speed airfoil　高速翼型
high speed alternator　高速交流发电机
high speed analog computer　高速模拟计算机
high speed balancing　高速动平衡
high speed brake　高速制动器
high speed buffer register　高速缓冲寄存器
high speed buffer　高速缓冲器
high speed calculator　快速计算机
high speed camera　高速摄影机
high speed channel　高速通道
high speed cinematography　高速摄影术
high speed circuit breaker　高速断路器
high speed coagulative precipitation unit　高速凝结沉淀装置
high-speed-content demineralizer system　凝结水高速除盐系统
high speed controller　高速控制器
high speed correlator　高速相关器
high speed counter　高速计数器
high speed counting circuit　高速计数电路
high speed data acquisition　高速数据采集
high speed data channel　高速数据通道
high speed differential protection　高速差动保护
high speed digital computer　高速数字计算机
high speed disintegrator　高速粉碎机
high speed earthing switch　快速接地开关
high speed excitation　高速励磁，强行励磁
high speed exciter　高速励磁机
high speed flash photography　高速闪光照相术
high speed Francis turbine　高转速辐向轴流式水轮机，高转速混流式水轮机
high speed fuse　高速熔丝，高速熔断器
high speed gas centrifuge　高速气体离心机
high speed generator　高速发电机
high speed governor　高速调节器，高速调速器
high speed grounding switch　高速接地开关
high speed impedance relay　高速阻抗继电器
high speed indicator　高速指示器
high speed jet　高速射流
high speed low torque shaft　高速低扭矩轴
high speed memory　高速存储器，快速存储器
high speed mill　高速磨煤机，高速磨
high speed mixer　高速拌和机
high speed motor　高速电动机，高速电机
high speed neutron　高速中子
high speed nozzle　快速割嘴
high speed numerical counter　高速数字计数器
high speed oscillograph　高速示波器
high speed oscilloscope　高速示波器，快速扫描示波器
high speed photographic technique　高速摄影技术
high speed photography　高速摄影
high speed plotter　高速描绘器，高速绘图仪
high speed printer　高速打印机，快速印刷装置
high speed print　高速打印
high speed pulverizer　高速磨煤机
high speed pump　高速泵
high speed rail line　高速铁路
high speed reader　高速阅读器
high speed reclosure　快速重合闸
high speed recorder　快速记录仪
high speed rectifier　高速整流器
high speed regulator　高速调节器
high speed relay　速动继电器，高速继电器
high speed repetitive operation　高速重复操作
high speed rotor　高速风轮

high speed runner　高速转轮
high speed shaft system　高速轴系统
high speed shaft　高速轴
high speed starting　快速启动，高速启动
high speed steel　高速钢
high speed stop　高速停机，高速
high speed storage　高速存储器
high speed switching　高速切换
high speed switch　速断开关，快速开关
high speed tap changer　高速抽头切换开关，高速分接开关
high speed tripping　快速脱扣机构
high speed tunnelling　隧洞快速掘进法
high speed turbine　高速透平，高速涡轮
high speed WECS　高速风力机，额定叶尖速率比不小于 3 的风力机
high speed winding　高速绕组
high speed wind tunnel test　高速风洞试验
high speed wind tunnel　高速风洞
high stability insulation　高稳定绝缘，高可靠性绝缘
high-stage bleed　高压级抽汽
high standard of accuracy　高精度
high-static-pressure differential pressure gauge　高静压差压计
high steam flow　高蒸汽流量
high steam generator water level trip　蒸汽发生器高水位紧急停堆
high strain dynamic pile testing　高应变动力试桩
high strain dynamic testing method　高应变动测法
high strain dynamic testing pile　高应变动力试桩
high strain dynamic testing　高应变（动力测试）法【桩基】
high strain integrity testing　高应变法，高应变完整性测试
high strain pile foundation nondestructive tester　高应变桩基无损测试仪
high strength alloy bolting　高强合金螺栓
high strength alloy steel　高强度合金钢
high strength belt　高强度皮带
high strength bolting steel　高强螺栓钢
high strength bolting　螺栓高强度固定
high strength bolt　高强度螺栓，高强螺栓
high strength cement　高标号水泥，高强水泥
high strength concrete　高标号混凝土，高强度混凝土，高强混凝土
high strength conveyor belt　高强度输送带
high strength insulator　高强度绝缘子
high strength low alloy steel　低合金高强度钢
high strength tension bolt　高抗拉强度螺栓，高强度拉紧螺栓
high strength toughened heat-treated steel　高强度韧化热处理钢
high strength　高强度
high-sulfur　高硫
high-sulfur coal　高硫煤
high-sulfur combustor　高硫燃烧器
high-sulfur content fuel　高硫燃料
high-sulfur crude oil　高硫原油
high-sulfur fossil fuel　高硫化石燃料
high sulphur coal　高硫煤
high-tank　高位水箱，高位槽
high-tech enterprises　高科技企业
high-tech industrial development zone　高科技产业开发区
high technology　高技术
high-tech product　高科技产品
high-tech research and development plan　高技术研究发展规划
high-tech　高技术，高精尖技术，高科技
high temperature aerodynamics　高温气体动力学
high temperature alloy　耐热合金
high temperature and low-sag conductor　高温低驰度导线
high temperature behaviour　高温性能
high temperature brazing　高温钎焊
high temperature cable　高温电缆
high temperature cemented carbide　耐高温硬质合金，高温硬质合金
high temperature cement　耐高温水泥，耐火水泥
high temperature condition　高温条件
high temperature corrosion　高温腐蚀
high temperature cracking　高温裂纹
high temperature creep　高温蠕变
high temperature curing　高温养护法
high temperature ductility　热延展性
high temperature embrittlement　高温脆化【材料】
high temperature experimental subassembly　高温试验燃料组件
high temperature fatigue　高温疲劳
high temperature field　高温场
high temperature fuel cell　高温燃料电池
high temperature gas-cooled pebble-bed reactor　球床高温气冷反应堆
high temperature gas-cooled reactor nuclear power plant　高温气冷堆核电厂
high temperature gas-cooled reactor with helium turbine　带氦气轮机的高温气冷反应堆
high temperature gas-cooled reactor　高温气冷反应堆，高温气冷堆
high temperature helium cooled reactor　高温氦冷反应堆
high temperature and high pressure rack　高温高压架
high temperature hot water heat-supply system　高温水供热系统
high temperature hydrofluorination　高温氢氟化
high temperature hydrolysis　高温水解
high temperature insulated wire　高温绝缘线
high temperature insulation　高温绝缘，耐高温绝缘
high temperature level　（热力装置循环中的）高温热源
high temperature material　高温材料，耐高温材料，耐热材料
high temperature molten salt reactor　高温熔盐反应堆
high temperature motor　高温电动机，高温电机
high temperature neutron diffraction　高温中子衍射
high temperature nuclear power station　高温核

电站
high temperature operation 高温作业
high temperature oxidization 高温氧化
high temperature pebble-bed reactor 高温球床反应堆
high temperature plasma 高温等离子体
high temperature processing 高温处理
high temperature property 高温性能，高温特性，高温强度
high temperature PRT standard 高温铂电阻温度计
high temperature radiant emitter 高温辐射体
high temperature radiation embrittlement 高温辐射脆化
high temperature reactor 高温反应堆，高温气冷反应堆
high temperature region 高温区
high temperature reheat pipe 高温再热蒸汽管
high temperature resistant relay 耐高温继电器
high temperature resistant steel 耐热钢
high temperature resistant 耐高温的
high temperature stability 高温稳定性
high temperature steam curing 高温蒸汽养护
high temperature steel 耐热钢，高温钢
high temperature strain gauge 高温应变计，高温应变仪
high temperature strength 抗高温性能，抗高温强度，高温强度
high temperature superconductivity 高温超导性
high temperature tempering 高温回火
high temperature tension test 高温拉伸试验
high temperature thermal source 高温热源，高势热源
high temperature thermistor 高温热敏电阻器
high temperature thermomechanical treatment 高温形变热处理
high temperature thermometer 高温温度计
high temperature thorium reactor 高温钍反应堆
high temperature turbine 高温透平，高温涡轮
high temperature visible column thermometer 视柱式高温计
high temperature waste water 高温废水
high temperature water heating 高温热水采暖
high temperature wire 高温电线
high temperature 高温
high tensile bolt 高张拉螺栓
high tensile steel bar 高强度钢筋
high tensile steel electrode 高强度钢焊条
high tensile steel 高强钢
hightensile 高强度
high tension ammeter 高压电流计，高压安培计
high tension battery 高压电池
high tension bolt 高强螺栓
high tension busbar 高压母线
high tension bushing 高压套管
high tension cable 高压电缆
high tension circuit 高压电路
high tension coil 高压线圈
high tension conductor 高压导电棒
high tension current transformer 高压电流互感器
high tension current 高压电流
high tension damper 高压阻尼器
high tension direct current 高压直流
high tension distribution center 高压配电中心
high tension distribution equipment 高压配电设备
high tension distribution system 高压配电系统
high tension distributor 高压配电设备
high tension fuse 高压熔断器
high tension generator 高压发电机
high tension ignition conductor 高压点火导体
high tension insulation 高压绝缘
high tension insulator 高压绝缘子
high tension laboratory 高压实验室
high tension lead 高压引线
high tension lightning arrester 高压避雷器
high tension lightning-rod 高压避雷针
high tension line 高压电线，高压线路
high tension low sag conductor 高压低驰度导线
high tension machine 高压电机
high tension magneto 高压永磁发电机
high tension measurement 高压测量
high tension motor 高压电动机
high tension network 高压电力网
high tension pole 输电线路电杆，高压电杆
high tension porcelain insulator 高压电瓷绝缘子
high tension porcelain 高压陶瓷
high tension power line 高压输电线
high tension regulating transformer 高压调节变压器
high tension side 高压侧
high tension supply 高压电源
high tension switchgear 高压开关装置
high tension switch 高压开关
high tension testing apparatus 高压试验设备
high tension test 高压试验
high tension transformer 高压变压器
high tension transmission line 高压输电线（路）
high tension transmission 高电压输电
high tension voltmeter 高压伏特计
high tension winding 高压绕组
high tension wire 高压线
high tension 高张力，高电压，高强度，高压的，高拉力，高压
high threshold logic 高阈逻辑
high thrust motor 高推力电动机
high thrust reactor 高推力反应堆
high tick level 高潮水位
high tide alarm 高潮报警器
high tide forecasting 高潮预报
high tide interval 高潮间隔
high tide level 高潮面，高潮（水）位
high tide line 高潮线
high tide period 高潮期
high tide shoreline 高潮滨线
high tide slack water 高潮憩流
high tide 满潮，高潮，高潮线，涨潮
high titania potassium electrode 高钛钾型焊条
high titania sodium electrode 高钛钠型焊条
high trestle 高支架
high usage line 利用率高的线路
high vacuum diffusion pump 高真空扩散泵
high vacuum discharge 高真空放电

high vacuum gauge 高真空压力计
high vacuum pump 高真空泵
high vacuum technique 高真空技术
high vacuum 高真空
high value cargo 高价货
high velocity conduit 高速水道
high velocity duct 高速管道
high velocity flow 高速水流
high velocity injection nozzle 高速喷口
high velocity jet flow 高速射流
high velocity jet 高速喷射，高速射流
high velocity neutron 高速中子，快中子
high velocity system 高速风道系统，高速流
high velocity thermocouple 小惯性热电偶
high velocity 高速度，高速
high-viscosity fluid 高黏性流体
high viscous oil 高黏度油
high volatile bituminous coal 高挥发分烟煤
high volatile coal 高挥发分煤
high voltage 高电压，高压
high voltage arc 高压电弧
high voltage bushing 高压套管
high voltage cable 高压电缆
high voltage capacitor 高压电容器
high voltage cathode-ray oscilloscope 高压电子示波器
high voltage ceramic capacitor 高压陶瓷电容器
high voltage circuit 高压电路
high voltage coil 高压线圈
high voltage connection 高压接线
high voltage contact 高压触头
high voltage corona 高压电晕
high voltage current limiting fuse and vacuum contactor 高压限流熔断器及真空接触器
high voltage current shunt 高电压分流器
high voltage DC generator 高电压直流发生器
high voltage direct current transmission system 高压直流输电系统
high voltage direct current transmission 高压直流输电
high voltage direct current 高压直流
high voltage distribution installation 高压配电装置
high voltage distribution line 高压配电线路
high voltage distribution network 高压配电网
high voltage distribution 高压配电
high voltage divider 高电压分压器
high voltage earth discharge 高压接地放电
high voltage electrical apparatus 高压电器
high voltage electroscope 高压验电器
high voltage electrostatic field 高压静电场
high voltage fuse 高压熔断器
high voltage generator 高电压发电机，高压发电机
high voltage house service operating transformer 高压厂用工作变压器
high voltage instrument transformer 高电压互感器
high voltage insulation test 高电压绝缘试验
high voltage insulation 高压绝缘
high voltage insulator 高压绝缘子
high voltage isolator 高压隔离开关
high voltage laboratory 高（电）压试验室
high voltage line 高压线路
high voltage measurement 高电压测量
high voltage megohmmeter 高压兆欧表
high voltage motor 高压电动机
high voltage oil-filled cable 高压充油电缆
High voltage power MOSFET gate driver 高压功率MOSFET门极驱动电路【MOSFEI指金属氧化物半导体场效应晶体管】
high voltage power supply 高压电源，高压供电
high voltage power transmission line 高压输电线路
high voltage pulser 高压脉冲发生器
high voltage rectifier 高压整流器
high voltage relay 高压继电器，过电压继电器
high voltage resistor 高压电阻器
high voltage section 高电压段
high voltage silicon rectifier stack 高压硅堆
high voltage standby transformer 高备变
high voltage substation 高压变电所，高压变电站
high voltage switchgear interlocking device 高压开关设备联锁装置
high voltage switch gear 高压开关设备
high voltage switch 高压开关
high voltage technique 高电压技术
high voltage testing equipment 高电压实验设备
high voltage testing laboratory 高压试验室
high voltage test pencil 高压试电笔
high voltage test 高电压测试，高压试验
high voltage transformer 高压变压器，高压互感器
high voltage transmission line 高压输电线，高压线路
high voltage unit 高电压设备
high voltage winding 高压绕组
high voltage wire 高压电线，高压线
high volume filter 高容量过滤器
highwall-drilling machine 立式钻机
high water alarm 高水位报警器
high water condition year 丰水年
high water control 高水位控制
high water elevation 高水位
high water full and change 朔望高潮
high water head 高水头
high water interval 高水位间隔
high water level alarm 高水位报警
high water level of spring tide 大潮高潮位
high water level 高水位
high water line 高潮水位线，高潮线，高水位线
high water mark 高潮标记，高潮线，高水位线，洪水标记，洪水痕迹
high water of ordinary spring tide 一般大潮高潮位
high water period 丰水期，洪水期
high water revetment 高水位护岸
high water river bed 洪水河床
high water season 洪水季
high water slack 高潮憩流，高水平潮
high water spring 大潮高潮
high water stand 高潮水位，高潮停潮

high water table 高水位
high water untidily interval 高潮月潮间隙
high water velocity tunnel 高流速隧洞
high water 大水，洪水
high-watt 高瓦的，大功率的
highway addressable receiver transmitter 高速公路可寻址的接收发送器
highway addressable remote transducer 可寻址远程传感器高速公路
highway administration 公路局
highway bridge 公路桥
highway bureau 公路局
highway construction 公路建设
highway crossing 公路交叉
highway engineering 公路工程
highway erosion control 公路防冲措施
highway grade crossing 公路平面交叉
highway grade separation 公路立体交叉
highway location 公路定线
highway marking 公路标志，公路路标
highway network 公路网
highway over crossing 公路立体交叉
highway subgrade 公路路基
highway switching 多路交换
highway traffic 公路交通
highway transportation 公路运输
highway 总线，信息通道，大道，公路，公路型道路
high weir 高堰
high-width ratio 高宽比
high wind fumigation 强风下熏蒸
high wind 大风，疾风
high worth control rod 高价值控制棒
high yield stress steel 高屈服强度钢
hike 在高空检修电线，飞起，升起
Hiley's formula 黑莱公式
hill 丘陵
hill brow 山顶，山眉
hillcreep 山坡蠕变
hill crest 山峰
hill duster 群山，丘陵群
hillock plain 低丘平原
hillock 小丘
hillside connection 山坡联接
hillside dam 山麓小坝
hillside swamp 山腰沼泽
hillside 山坡
hillslope erosion 山坡冲刷
hillslope 山坡
hilltop reservoir 高山水库
hillwash 坡地坍滑
hilly 多丘陵的
hilly area 丘陵
hilly cross country 丘陵原野
hilly ground 丘陵地带，丘陵地
hilly land terrain 丘陵（地）
hilly land 丘陵，山地
hilly region 山区
hilly terrain 丘陵地，丘陵地带
hi-lo signal alarm 高低位报警
hilt 手柄
H_2 impurity 氢杂质，氢沾污
hindcasting 追算，倒推法【借鉴往事预测未来】，后报
hindered sedimentation 受阻沉降
hindered settling 受阻沉降
hinder land 背地
hinder 紧固件
hindrance factor 干扰因子
hindrance 延滞，延迟，阻碍
hinge armature 枢轴衔铁
hinge bolt 铰链栓
hinge clip 铰接线夹
hinged arch 铰接拱，有铰拱
3-hinged arch 三铰拱
hinged armature relay 旋转衔铁式继电器
hinged arm 枢杆
hinged bearing 铰支承
hinged blade 铰接叶片
hinged door 铰链门
hinged end 铰接端
hinged expansion joint 带铰链膨胀节
hinged frame 铰接框架
hinged gate 回转式闸门，旋开式闸门
hinged immovable support 铰接固定支座
hinged joint 铰接，铰接点
hinged moment balance 铰链力矩天平
hinged moment 铰接力矩
hinged movable support 铰接活动支座
hinged pressure relief panel 铰接卸压盘
hinged shoe 铰支座
hinged support 铰接支座
hinged truss 铰接桁架
hinged tube furnace 转动管道窑
hinged-type portal 铰接式门座
hinged-type trestle 铰接支架
hinged 有铰链的，铰接
hinge jaw 铰接夹头
hingeless arch 无铰拱
hinge-out 铰链，枢轴
hinge-pin-type fundament 铰（链）销子式皮带扣
hinge pin 铰链销
hinge post 铰链柱
hinge 铰链，关键，中枢，转折点，折叶，铰接，枢纽，合页
H_2 injection （冷却剂）气体喷射，氢喷射
hinterland 内陆，腹地，内地
hip and staple 搭扣
hi-pot test 高电位试验
hipot 高压绝缘试验
hipped-plate structure 折板结构
hipped roof 四坡屋顶，四坡屋面
hip slope 塔腿坡度
hip 坡屋顶屋脊，斜脊，屋脊
HIRAC(high random access) 高速随机存取
hired equipment 租用设备
hire paid in advance 预付租金
hire purchase accounting 租购会计
hire purchase agreement 分期付款协议，租购合约，分期付款购买契约
hire purchase credit insurance 租赁信用保险

hire purchase　分期付款购货，租购，分期付款购买法
hire　租赁，雇佣，租用
hiring cost　租赁成本
H-iron process　氢铁法
H-iron　宽缘工字钢
Hirox　希罗克思电阻合金
hissing arc　啸声电弧
histogram　直方图，频率分布图，柱状图，柱形图解
histograph　等流时线
historical culture city　历史文化名城
historical data memory　历史数据存储
historical data server　历史数据服务器
historical data storage and retrieval　历史数据存储和检索，历史数据检索和存储
historical data trending curve　历史数据趋势曲线
historical data　历史数据，历史资料
historical flood damage　历史洪水灾害
historical flood investigation　历史洪水调查
historical flood　历史洪水
historical geology　历史地质学，地史学
historical highest water level　历史最高水位
historical information　历史资料
historically　从历史角度，在历史上，根据历史事实
historical record　历史记载
historical site　历史遗迹
historical trend display　历史趋势画面
historical value　历史价值
historic city　历史名城
historic　有历史性的
history of operation　运行史
history of price　价格的规律
history　历史，过程
HITASS(hitichi turbine automatic start-up system)　日立汽轮机自启动系统
hitch iron　牵引板，牵引杆，牵引器梁
hitch　索眼，套，系扣，牵引装置
hit indicator　碰击指示器
hitless switching　无损伤切换
hit　碰击
Hi-VHF(high-very-high frequency)　上限甚高频
Hi-volt(high voltage)　高压
HKD　港币
HKSAR(Hong Kong Special Administrative Region)　香港特别行政区
HL(heat loss)　热损失
H/L monitor　高位/低位监视器
HMDY＝humidity　湿度
HMI(human machine interface)　人机接口，人机界面
H_2O adsorber　轻水吸附剂
H_2-O_2 analyzer　氢-氧分析器
hoarding　广告牌，临时围篱，囤积，贮存，围篱
hoar frost icing　霜成冰
hoar frost　白霜，冰霜
hob-type magnetron　柑橘形磁控管
hock clamp　钩式夹车器
hock sign　安全标志
hock spanner with lug　钩头扳手
hoc zone　放射性区（域）
hodograph transformation　速度矢端曲线变换
hodograph　速矢端线，矢端曲线，速端曲线，速度图，速度矢端图，高空风速分析图
hodometer　路程计
hodoscope　描迹仪
H_2O drainage pump　轻水疏水泵
H_2O drainage tank　轻水疏水箱
hod　灰浆桶，化灰池，砂浆桶
hoeing　挖掘
hoe　锄，耙
hog-frame　弓背构架
hogged fuel　废木屑燃料，薪材
hogger　木材切碎机，多报进尺的司钻，快速启动空气喷射器，大流量喷射器，钻工
hogging ejector　快速启动空气喷射点
hogging moment　负弯矩
hogging rotor shaft　拱曲的转子轴，弯曲的大轴
hogging　挠曲，挠度，弯曲，翘曲
hoggin　夹砂砾石，含沙碎石
hoghorm　平滑匹配装置
hog　弯曲，扭曲，弯头，软管，弯拱机，使弯曲，拱曲
HO(head office)　总公司，（公司）总部
HOH(holiday outage hours)　节日检修停运小时
hohlraum　空腔
hoist block　起重滑车
hoist bridge　升降桥
hoist bucket　吊斗，吊罐
hoist cable　启门索
hoist chain　启门链
hoist chamber　启门机室
hoist drive mechanism　起重机驱动机构
hoist drum　起重机鼓轮
hoist engine　卷扬机
hoist equipment　起吊设备
hoister　卷扬机
hoist eye　吊耳
hoist frame　启门机架，起重机架
hoist hook　提升钩，起重钩，吊钩
hoisting and conveying　起重运输
hoisting bar for casing　吊汽缸工具
hoisting bar for diaphragm　吊隔板工具
hoisting bar for rotor　吊转子工具
hoisting box　起重机操纵小室
hoisting cable　起重索
hoisting capacity and elevation　起吊能力和高度
hoisting capacity　起重能力，起重量
hoisting chain　起重链
hoisting controller　起重控制器
hoisting crane　起重机，提升机，吊车
hoisting device　启闭设备，起吊装置，起重设备，升降装置，提升设备
hoisting electric motor　提升电动机
hoisting equipment　卷扬机，启闭设备，起重设备
hoisting facilities　起吊设施，起吊装置
hoisting gear　提升机构，起重机，起重装置，

提升绞车
hoisting grip 提升夹
hoisting guider 起重
hoisting height 提升高度
hoisting hook 起重吊钩，起重钩，提升钩
hoisting jack 千斤顶
hoisting limit 起重极限
hoisting line 起重索
hoisting machine 起重机，卷扬机
hoisting motor 卷扬电机
hoisting ring 吊环
hoisting speed 提升速度
hoisting tackle 起吊滑轮组，起重滑车
hoisting tool 吊具，起重工具
hoisting winch 卷扬机，起重绞车
hoisting 起重，提升，起吊
hoist lifter 起重机
hoist linkage 起吊装置，起重装置
hoist motor 起重机电动机，吊车电动机
hoist-mounted shaker 跨式振车器
hoist over-wind device 吊车过头防止装置
hoist tower 起重塔，塔式卷扬机，提升塔，吊机塔
hoist unit 升降装置，起重装置
hoistway 电梯间，电梯井，吊运井，提升间
hoist 升起，吊装，升降机，起重吊车，卷扬机，起重机，起重葫芦，提升机
holard 土壤总水量
hold-all 工具包，工具袋，工具箱
hold area 待检区
holdback agent 抑制剂
holdback 抑制，钳制，滞留量，逆止器
hold circuit 自保持电路，保持电路
hold-closed mechanism 保持合闸机构
hold condition 持恒状态
hold control 同步调节，同步控制
hold-down assembly 夹持装置，压紧组件【控制棒】
hold-down barrel 压紧筒体
hold-down bolt 压紧螺栓，底脚螺栓
hold-down column （堆芯）压紧柱，固定柱
hold-down cylinder 固定圆筒，压紧圆筒
hold-down device 压紧装置
holddown force 压紧力
hold-down grid 压紧栅板
hold-down mechanism 压紧机构
hold-down ring 压紧环，固定环
hold-down spring 压紧弹簧（环）【燃料棒，反应堆压力外壳】，固定弹簧
hold down 抑制，压制，压紧，夹紧，保持，压具
holden permeability bridge 导磁率测量电桥
holder of a bill of lading 提单持有人
holder of a document 单据持有人
holder-up 铆钉托
holder 把，柄，托，座，支架，持有人，扶手，夹具，托架，占有人，支持物
holdfast 支架，夹钳
hold fire 压火
hold harmless clause 免责条款
holding action 保持作用
holding armature 吸持磁铁
holding arm 夹持臂
holding bin 贮存仓
holding boom 拦污埂
holding bracket 托架
holding capacitor 保持电容
holding capacity 保持能力，存储量，贮仓量，储备量，可蓄量
holding circuit 吸持电路，保持电路
holding coil 吸持线圈，保持线圈
holding company 股权公司，控股公司，母公司
holding contact 吸持触点
holding control 同步调整
holding current 吸持电流，保持电流
holding device 夹持装置，吸持装置
holding-down bolt 地脚螺栓
holding end turn 线圈端部绑扎［固定］
holding force 保持力，支持力，吸持力，握力
holding frequency 固定频率运行方式，保持频率
holding load 恒定负载
holding magnet 吸持磁铁
holding plate 夹板
holding pond 存储槽，收集池，存贮池
holding position 吸持位置，自保持位置
holding power 握力，吸持功率，保持电源
holding relay 吸持继电器，保持继电器
holding ring 定位环，调整环
holding strip 压板
holding system 保持系统
holding system 夹具［支持］系统
holding tank transfer pump 保存水箱输送泵
holding tank 存料罐，贮槽
holding temperature 维持温度，均热温度【热处理】，保温温度
holding time 中转时间，转移时间【热处理】，（继电器）吸持时间，存留时间，保温时间
holding torque 保持力矩
holding-up hammer 铆钉撑锤
holding voltage 吸持电压
holding winding 吸持绕组，吸持线圈
holding wire 测试线，信号线
holding 保持，支持，固定，调整，同步，拥有，承受，存贮，止挡
hold in pledge 抵押
hold instruction 保持指令
hold-in 抑制，阻止，压住，保持同步
hold magnet 吸持磁铁，保持磁铁
hold mode 保持状态
hold-off circuit 释抑电路
hold-off diode 闭锁二极管
hold-off rectifier 偏压电流整流器
hold-off 脱出同步，失步，延迟
hold on 坚持，继续，拉住，等一等，别挂电话，抓紧不放
holdout coil 保持线圈
holdout device 锁定装置
hold-over command 保持命令
hold over 延期，保存
hold point 控制点，停止点，停工待检点，质检点
hold price 保持价格不变

hold queue range 同步范围
hold range 牵引范围，同步范围
hold relay 保持继电器
hold starting 热态启动
hold station 保存站
hold summary sheet 待定事项汇总表
hold tank 贮留槽，贮料槽，收集槽，接收器
hold the market 垄断市场
hold time with creep 蠕变保持时间
hold time with relaxation 松弛保持时间
hold time 保持时间，维持时间
hold-up capacity 滞留容量，滞流容量
hold-up line 滞留管线，衰变管线
hold-up pipe 滞留管
hold-up system 滞流系统，滞留系统
hold-up tank 收集箱，滞留箱，暂存箱，贮存箱，中间水箱
hold-up time 滞留时间，截留时间，中转时间，转移时间，占用时间，保持时间
hold-up train 滞留系列
hold-up vessel （废料）储存罐，中间贮槽，滞留容器
hold-up 保留，保持，保存，提出，举起，拦住，滞留，停顿，耽误，抢劫
hold 抓，把握，支持，保持，停住，同步
hole and slot resonator 槽孔型谐振器
hole armature 带通风槽的电枢
hole basis 基孔制
hole collar 炮眼口
hole conduction 空穴传导
hole current 空穴电流
hole-depth gauge 孔深计
hole digger 钻孔器
hole dilating drill 扩孔钻头
hole-electron pair 电子空穴对
hole for hoist 吊装孔
hole gauge 测孔规，内径规
hole injection 空穴注入
hole irrigation 点浇，穴灌
hole lifetime 半导体的空穴寿命
hole loading 炮眼装药
hole man shooter 炮工
hole man 爆炸手，放炮工
hole number density 空穴浓度
holes pitch 钻井倾角，钻孔间距
hole to be provided 预留洞
hole top 孔口
hole-type image quality indicator 孔型景象质量指示器，孔型穿透计，孔型透度计
hole-type penetrameter 孔型透度计，孔型穿透计，孔型透光计，空洞型穿透计，空洞型透光计
H_2O level 水位，含水量
hole 洞，洞孔，孔板，注孔，开口，穴，孔，眼，槽，孔道
holiday outage hours 节日检修停运小时
holiday repair outage hours 节日检修停运小时
holidays and festivals 节假日
holiday 假期，假日
hollow arch dam 空腹拱坝
hollow-backed flooring 底空楼板
hollow blade 空心叶片
hollow block 空心板，空心块体，空心砌块
hollow brick partition 空心砖隔墙
hollow brick wall 空心砖墙
hollow brick 空心砖
hollow buttress 空腹支墩
hollow cable 空心电缆
hollow casting 空心铸件
hollow cathode discharge tube 空心阴极放电管，空心阴极灯
hollow cathode lamp 空心阴极灯
hollow chamfer 凹斜面
hollow clay brick 多孔黏土砖
hollow clay tile 空心陶土砖
hollow coil 空心线圈
hollow concrete block 空心混凝土砌块，空心混凝土块
hollow conductor cable 空心导线电缆
hollow conductor 空心导体，空心导线
hollow cone 空心圆锥体
hollow copper conductor 空心铜线，空心铜导体
hollow copper tubing 空心铜管
hollow copper wire 空心铜线
hollow core coaxial Ge/Li detector 空心同轴锗/锂探测器
hollow core conductor 空心导体
hollow core door 空心门
hollow core rock bolt 空心岩石锚杆，空心岩栓
hollow core valve 空心阀
hollow core 空心铁芯
hollow cylinder source 空心圆柱体源
hollow cylinder 空心圆筒
hollow cylindrical fuel element 空心圆柱状燃料元件
hollow dam 空腹坝，空心坝
hollow drill steel 空心钻杆
hollow drill 空心钻
hollow drum rotor 空心鼓形转子
hollow fiber module 中空纤维组件
hollow fiber 空心纤维
hollow fuel rod 空心燃料棒，中空燃料棒
hollow gravity dam 空腹重力坝，空心重力坝
hollow helical-wrap drive cable 空心螺旋型控制电缆
hollow inclusion 空心包体
hollow insulator 空心绝缘子
hollow jet valve 空注阀
hollow load cell 空心压力盒
hollow out 掏空
hollow pellet 中空芯块
hollow pile 空心桩
hollow pushrod 空心推杆
hollow quoin 空心墙角基石
hollow rod fuel pin 中空燃料细棒
hollow rotor motor 空心转子电动机
hollow rotor 空心转子
hollow section strand 空心股线
hollow section 空心型材
hollow shaft motor drive 空心轴电动机驱动
hollow shaft motor 空心轴电动机

hollow shaft　空心轴
hollow slab　空心板
hollow slug lattice　中空棒栅格，中空短棒栅格
hollow slug　中空元件块，中空元件短棒
hollow socket head screw　空心管座头部螺钉【导向套管螺纹管嘴连接】
hollow sphere　空心球体，空心轴
hollow square　空心四方块体
hollow tile　空心砖
hollow unit masonry　空心砌块圬工
hollow walling　空心墙
hollow wall　空心墙
hollow wedge　空心槽楔
hollow weir　空心堰
hollow winding　空心绕组
hollow wire　空心线
hollow wood construction　空心木结构
hollow　空心的，凹陷的，穴，孔，凹坑，凹槽，坑穴，空的，山谷
holocamera　全息照相机
holocellulose　全纤维素
holofilm　全息胶片
hologram recording material　全息照相记录材料
hologram　全息照片，全息图
holographic memory device　全息照相存储装置
holographic memory　全息存储器，全息照相存储器
holographic nondestructive testing　全息无损检测
holograph recording　全息记录
holography principle　全息摄影原理
holography　全息摄影，全息术
holograph　亲笔文件，亲爱的
holohedry　全对称
holomagnetization　全磁化
holoseismic method　全息地震法
holosteric　无水的，全部固体的
home address　标识地址
home-load supply of substation　变电站自用电
home-made equipment　国产设备
home-made　自制的，国产的
home motor　家用电动机，国产电动机
home office switch　安全开关
home office　总部，总公司
homeostasis　相对稳定平衡，动态静止，动态平衡，同态
homeostatic mechanism　适应性机能
homeostatic　体内平衡的，稳态的
homeostat　同态调节器
homeowner's insurance　房主保险
home page　主页
home registration certificate　本国注册证明
homestead　屋基
homing action　导归作用，（选择器）还原动作
homing device　寻的装置
homing on　瞄准
homing relay　复位继电器
homing　复原，归位，引导，导航
homocentric　同心的，同轴的
homocharge　纯号电荷
homoclime　相同气候
homoclinal valley　单斜谷
homodyne　零差，零拍
homoentropic flow　均熵流
homogeneity defect　不紧密【焊接】
homogeneity fluid　均质流体
homogeneity of insulation　绝缘均匀性
homogeneity range　均匀区
homogeneity rule　同一性，相似律
homogeneity substance　等质体
homogeneity　同类，同性，均匀性，均质性，同次性，同一性，同质性，均匀化
homogeneous aqueous power reactor　水均相动力反应堆
homogeneous aqueous reactor　水均相反应堆
homogeneous assembly zero energy level reactor　均匀组件零功率反应堆
homogeneous beam　均匀束
homogeneous catalyst　均相催化剂
homogeneous chemical equilibrium　单相化学平衡
homogeneous combustion　均匀燃烧
homogeneous deformation　均匀变形
homogeneous distribution　均匀分布，单一分布
homogeneous earth dam　均质土坝
homogeneous equation　齐次方程
homogeneous equilibrium critical flow model　均匀平衡态临界流动模型
homogeneous equilibrium　均相平衡
homogeneous expansion　均匀膨胀
homogeneous field　均匀场
homogeneous flow model　均质流模式
homogeneous flow　均相流，均匀流
homogeneous fluidization　均匀流态化
homogeneous fluid　均匀流体
homogeneous graphite reactor　石墨均匀反应堆
homogeneous insulation　均匀绝缘
homogeneous ionization chamber　均质电离室
homogeneous isotropic turbulence　均匀各向同性湍流
homogeneous light　单色光
homogeneous liquid membrane　均质液体膜
homogeneous mass　均质体
homogeneous membrane　均匀膜，均相膜
homogeneous mixture　均匀混合物
homogeneous model　均相模型
homogeneous phase　单相，均匀相
homogeneous poison　均匀核毒物
homogeneous precipitation　均匀沉淀
homogeneous quantity　同类量
homogeneous radiation　均匀辐射
homogeneous ray　单色射线
homogeneous reaction　均相反应，单相反应
homogeneous reactor　均匀反应堆，均匀反应堆
homogeneous ring compound　碳环化合物
homogeneous section　均匀段
homogeneous shield　均匀屏蔽
homogeneous soil　均质土
homogeneous solution-type reactor　均匀溶液反应堆
homogeneous straining　均匀应变
homogeneous stress　均匀应力
homogeneous system　均匀系统
homogeneous terrain　均布地形

homogeneous thorium reactor 均匀钍反应堆
homogeneous tunnel 均匀紊流度
homogeneous turbulence 均匀湍流，均匀紊流
homogeneous two-phase flow 均匀两相流
homogeneous water moderated 均匀水慢化的
homogeneous 均匀的，同类的，相似的，齐次的
homogenization 同质性，均匀性，均质性，均匀化，均化
homogenized 均匀化的，均质的
homogenize 使均匀化，变均匀，使均匀
homogenizing section 均化段
homogenous rock 匀质岩石
homo-ionic solution 同离子溶液
homo-ion 同离子
homoiothermic 恒温的，温血的
homoiothermism 温度调节
homojunction solar cell 同质结太阳电池
homologous compound 同系化合物
homologous series 同系列
homologous turbine 同系列的水轮机，同系列水轮机
homologous turbulence 均匀湍流
homolographic projection 等面积投影
homology 同调，透射
homolysis 均裂
homophase 同相
homopolar alternator 单极同步发电机
homopolar component 单极性分量，零序分量
homopolar dynamo 单直流发电机
homopolar field magnet 同极场磁铁
homopolar generator 单极发电机
homopolar HVDC system 同极高压直流输电系统
homopolar inductor alternator 单极感应子同步发电机
homopolar inductor motor 单极感应子电动机
homopolar machine 单极电机
homopolar motor 单极电动机，单极电机
homopolar operation 同极运行
homopolar power 零序功率
homopolar tachogenerator 单极测速发电机
homopolar 同极的，无极的，单极的
homostasis 稳态
homostrobe 零差频选通，零拍门，单闸门
homothermal condition 同温条件
hondrometer 粒度计，微粒特性测定计
honesty 诚实，正直
honeycomb and blasted surface 蜂窝麻面
honeycomb board 蜂窝夹心胶合板
honeycomb coil 蜂房式线圈
honeycomb collector 蜂窝集热器
honeycomb corrosion 蜂窝状腐蚀
honeycomb cracks 网状裂缝
honeycombed wall 蜂窝式墙
honeycomb laminate 蜂窝夹层板，蜂窝状叠层布
honeycomb memory 蜂房形存储器
honeycomb plate 蜂巢板
honeycomb sandwich construction 蜂窝夹芯结构
honeycomb sandwich structure 蜂窝夹层结构
honeycomb sandwich 多孔层结构，蜂窝夹层结构
honeycomb seal 蜂房密封
honeycomb-section beam 蜂窝形钢梁
honeycomb shape 蜂窝形状
honeycomb structure 蜂窝结构，蜂窝状结构，蜂窝式构造
honeycomb texture 蜂窝结构
honeycomb tubular filter element 蜂房式管状滤芯
honeycomb wall 地龙墙
honeycomb weathering 蜂窝式风化
honeycomb winding 蜂房式绕组
honeycomb 蜂房形的，蜂窝结构，蜂巢
Honeywell 霍尼韦尔公司【总部在美国】
hone 金属表面磨损，油石，细磨刀石
honing machine 搪磨机
honing 搪磨，金属表面磨损，桁磨
honor a bill 承兑票据
honorary certificate 荣誉证书
honorary president 名誉主席
honor 兑付，荣誉，信用
honourable prize 荣誉奖
honourable professor 荣誉教授
honour agreement 履行协定，履行协议
honour an application 批准申请
honour claim 承付赔偿要求
honour one's liability 承担赔偿责任
honour the contract 履行合同，信守合同，重合同
honour 承兑，信守，尊敬
hood beam 圈梁
hood for fume 烟罩，集气罩
hood loss 排汽口损失，排汽罩损失
hood-type steam washer 罩式洗汽器
hood 防护罩，外罩，烟囱帽，排汽口，（压力容器）顶盖，发动机罩，烟囱风帽，帽盖，帽，套，机罩，通风帽，外壳
hook and eye hinge 钩扣铰链
hook block 吊杆滑轮组，带钩滑车
hook bolt 吊耳，钩头螺栓，钩形螺栓，钩螺栓
hook clearance 吊钩下的空间高度
hook crack 钩裂纹
hooked bar 带弯钩钢筋，弯钩钢筋，弯起钢筋
hooked-groin contraction works 钩形丁坝束水工程
hooked nail 钩头钉
Hooke's law of elasticity 胡克弹性定律
Hooke'slaw 胡克定律
hook gauge 钩尺【量液体高度的】，钩形测针
hook groin 钩形丁坝
hook groyne 钩形丁坝
hook joint 钩环接头，钩接
hook lug 钩形接线片
hook operation （线路）中继操作，挂钩操作
hook spanner 钩，弯脚扳手
hookstick-operated breaker 带钩杆操作的断路器
hook stick （电气）操作棒，操作杆，带钩杆
hook-type current transformer 钩形电流互感器
hook-up drawing 连接图
hook-up wire 布线用电线，架空电线，（仪表）连接线，架空线
hook-up 挂钩，悬挂装置，连接线，试验线路，电路耦合，接线图

hook 钩住，弯钩，吊钩，箍圈，线路中继，转播，锁钩，挂钩，钩
hooped column 箍柱
hooped concrete 配箍筋混凝土
hooped pile 配箍筋桩
hooped reinforcement 箍筋
hooping stirrup 箍筋
hooping 配箍筋
hoop iron 带钢，铁箍
hoop ladder 环形梯
hoop prestressing 环向预应力
hoop reinforcement 环箍加强【预制桩】，箍筋
hoop stress 切向应力，周向应力，环状应力，环向应力，箍应力
hoop tension 箍拉力，环箍张力，环向拉力
hoop 箍圈【圆形轴的加强】，圈，环，箍，集电弓，集电环，加箍，铁环
hooter 声响信号器，警报器
hopper ash pan 灰斗
hopper bottom car 漏斗底卸车
hopper bottom freight car 底卸式货车
hopper bottom furnace 有冷灰斗的炉膛，固态排渣炉膛
hopper bottom 冷灰斗
hopper capacity 料斗容量
hopper car 自动卸料车，漏斗车
hopper chute 漏斗式斜槽
hopper-cooled 连续水冷的
hopper cooling jacket 料斗冷却套
hopper dredger 漏斗式挖泥机
hopper dryer 料斗干燥器
hopper freight car 底卸式货车
hopper gate 料斗门
hopper gritter 斗式铺砂机
hopper grizzly 漏斗灰
hopper knuckle 灰斗
hopper level detector 料斗位探测器
hopper loader 斗式装料机
hopper mouth 斗口
hopper outlet opening 料斗排料口开度
hopper scale 料斗秤
hopper shaker 漏斗振动器
hopper-shaped bottom 冷灰斗
hopper slope 灰斗斜面
hopper spreader 斗式撒布机
hopper throat opening 料斗缩口大小
hopper valley angle 灰斗斜面角度
hopper wagon 漏斗车，底开门车，底卸车，漏斗料车
hopper window 倒开窗，下悬窗
hopper 斗轮，料斗，煤斗，漏斗，仓斗，斗，翻斗小车，贮料斗，装料斗
hopping dike 月堤
horary angle 时角
HORIZ＝horizontal 水平的，卧式的
horizon closure 水平闭合差
horizon dial 水平指示盘
horizon glass 地平镜
horizon line 地平线
horizon plane 地平面
horizon refraction 水平折射
horizontal acceleration （地震）水平加速度
horizontal accuracy 平面位置精度
horizontal agreement 横向协议，水平校正
horizontal alignment 水平定线
horizontal angle 水平角
horizontal arch element 水平拱单元，拱环
horizontal axial loads of fixing support 固定支座水平轴向荷载
horizontal axis current meter 横轴流速仪
horizontal axis mixer 横轴拌和机
horizontal axis-rotor WECS 水平轴风轮式风能转换系统，水平轴风力机【风轮轴线与水平面夹角不大于15°】
horizontal axis rotor 水平轴风轮
horizontal axis wind machine 水平轴风力机
horizontal axis wind turbine 水平轴风力机
horizontal axis 横轴，平面轴，水平轴，水平轴线
horizontal axle 水平轴
horizontal baffle 水平挡板
horizontal beam 水平射束
horizontal belt conveyor 水平皮带机
horizontal blanket drainage 褥垫排水
horizontal boiler 卧式锅炉
horizontal boring machine 水平钻孔机
horizontal bottom furnace 平炉底炉膛
horizontal bracing 水平支撑
horizontal branch 水平支管
horizontal break switch 水平断路开关
horizontal buoyancy correction 水平浮力修正
horizontal bus 水平母线
horizontal centering control 水平居中调整
horizontal center line 水平中心线
horizontal centrifugal pump 卧式离心泵
horizontal check valve 水平逆止阀
horizontal circle 水平度盘，地平圈
horizontal clearance 横向净宽
horizontal coherence 水平相干性
horizontal combination 横向合并
horizontal comparator 水平比测器
horizontal component seismograph 水平动地震计，水平向地震仪
horizontal component 水平分量
horizontal configuration 水平排列
horizontal control station 平面控制点
horizontal control survey network 平面控制测量网
horizontal control 平面控制
horizontal coordinate system 地平坐标系
horizontal coordinate 横坐标，水平坐标
horizontal coordination 横向协调
horizontal correction 水平校正
horizontal crack 水平裂缝
horizontal crest 平顶波峰，平缓峰顶
horizontal cross brace 水平剪刀撑
horizontal curve 水平曲线
horizontal deflecting voltage 水平偏压
horizontal diffuser 水平扩散器
horizontal diffusion 水平扩散
horizontal disk crusher 卧式圆盘破碎机
horizontal dislocation offset 水平断错

horizontal displacement 水平位移
horizontal distance 水平距离，平距
horizontal distribution 水平分布
horizontal ditch 水平沟
horizontal drainage blanket 水平排水铺盖
horizontal drum mixer 卧筒式拌和机
horizontal earth pressure 水平土压力
horizontal earthquake load 地震水平载荷，水平地震荷载
horizontal elbow 水平弯头
horizontal evaporator 卧式蒸发器
horizontal expansion 水平伸缩
horizontal experimental hole 水平试验孔道
horizontal fault 水平断层
horizontal fillet weld position 水平角焊装置
horizontal filter 水平反滤层，水平滤层，卧式过滤器
horizontal fire-box boiler 卧式火筒锅炉
horizontal firing 水平燃烧
horizontal flange 水平中分面，水平中分面法兰
horizontal flow basin 平流池
horizontal flow 水平流
horizontal fluid layer 水平流层，水平流体层
horizontal flyback 水平回扫
horizontal force 水平力
horizontal gear-driven exciter 卧式齿轮传动激振器
horizontal girder gate 横梁式闸门
horizontal gradient of gravity 重力水平梯度
horizontal grid 水平栅格
horizontal grizzly 水平格筛
horizontal ground acceleration 水平地面加速度
horizontal guy 水平拉线
horizontal hair 横丝
horizontal heater 卧式加热器
horizontal hinged door 水平铰链门
horizontal hold control 行同步调整
horizontal hole 水平孔道
horizontal induction motor 卧式感应电动机
horizontal input to vertical axis output wind turbine 水平输入垂直轴输出式风电组
horizontal joint flange 水平中分面法兰
horizontal joint 水平中分面，水平接缝，水平节理，横接合，水平接头，水平接合
horizontal lateral loads of fixing support 固定支座侧向水平荷载
horizontal layer 水平层
horizontal lens angle 透镜水平角
horizontal line 水平线
horizontal load 水平荷载
horizontally baffled boiler 烟道水平分隔的锅炉
horizontally mounted 水平装配的
horizontally sliding window 水平推拉窗
horizontally-split casing pump 水平拼合泵壳
horizontally-split casing 水平拼合壳体，水平中分式缸体
horizontally-split 水平中分的，水平剖分的
horizontally stratified sandstone 水平成层砂岩
horizontal machine 卧式电机，卧式机器
horizontal machining center 卧式加工中心
horizontal mill 卧式磨煤机
horizontal motion 水平运动
horizontal motor 卧式电动机
horizontal movement 水平移动
horizontal multijunction solar cell 水平多结太阳电池
horizontal overhead position welding 仰角焊
horizontal overhead position 仰角焊位置
horizontal pendulum tiltmeter 水平摆动倾斜计
horizontal phasing control 行相位调整
horizontal pitch 水平间距
horizontal plane 水平面，地平面
horizontal polarization 水平极化
horizontal position welding 横焊，平焊
horizontal position weld 水平位置焊缝
horizontal position 横焊位置，水平位置
horizontal pressure tube reactor 卧式压力管反应堆
horizontal projection 水平投影
horizontal puled gate 横拉闸门
horizontal pull 水平拉力
horizontal pump 卧式泵
horizontal raceway 水平电缆管道
horizontal range 水平距离
horizontal reactor 卧式反应堆
horizontal refraction 地平折射
horizontal resultant 水平合力
horizontal return tubular boiler 卧式火管锅炉，卧式外燃回火管锅炉
horizontal rod 横杆
horizontal sampler 水平式采样器
horizontal scale 水平比例尺，水平尺度
horizontal scanning circuit 水平扫描电路
horizontal screen baffle 水平筛网板
horizontal screen 水平挡网，水平筛
horizontal screw pump 卧式螺旋泵
horizontal seam 水平焊缝
horizontal section 水平断错，水平截面，水平剖面
horizontal sedimentation tank 平流沉淀池
horizontal seismic force 水平地震力
horizontal shaft arrangement 横轴式布置
horizontal shaft current meter 横轴流速仪，水平轴流速仪，旋桨流速仪
horizontal shaft double turbine 横轴式双水轮机
horizontal shaft generator 卧式发电机
horizontal shaft hydraulic turbine 卧式水轮机
horizontal shaft hydrogenerator 卧式水轮发电机
horizontal shaft mixer 卧式拌和机
horizontal shaft 横轴，水平轴
horizontal shaper 牛头刨床
horizontal shear machine 水平剪力仪
horizontal shear 水平剪力
horizontal sheave 卧式滑轮
horizontal shoring 水平支撑
horizontal skew 水平偏斜
horizontal sleeve bearing 水平套筒轴承
horizontal sliding door 水平推拉门
horizontal sliding window 左右推拉窗
horizontal slope 平坡
horizontal spacing 水平间距
horizontal split joint 水平中分面，水平接缝，水

平节理，横接合，水平接头，水平接合
horizontal split pump 水平中分泵
horizontal split 水平中分面，水平分割，水平分裂，水平剖分
horizontal stabilizer 水平稳定器
horizontal steam generator 卧式蒸汽发生器
horizontal stiffener 水平加劲杆
horizontal stratification 水平层理
horizontal superheater 卧式过热器
horizontal system water heater 卧式热网水加热器
horizontal tabulation character 横向列表字符
horizontal tank 平流水槽，卧式储罐
horizontal thrust fault 平冲断层
horizontal thrust on fixing support 固定支座水平推力
horizontal thrust 水平推力
horizontal tipping method 平堵
horizontal transfer track 水平输送轨道
horizontal transfer 横向转移，水平传递
horizontal transport 水平输送
horizontal trim 水平调整
horizontal tube boiler 卧管式锅炉
horizontal tube coil 水平蛇形管，水平管圈
horizontal tube evaporator 卧式管形蒸发器，平管蒸发器
horizontal turbine 卧式水轮机，卧式透平
horizontal turbulent diffusion 水平湍流扩散
horizontal type deaerator 卧式除氧器
horizontal type evaporator 卧式蒸发器
horizontal type feed water heater 卧式给水加热器
horizontal type generator 卧式发电机
horizontal type motor 卧式电动机
horizontal type wind turbine 水平轴风力机
horizontal valve 水平阀
horizontal velocity distribution 水平流速分布
horizontal velocity gradient 水平速度梯度
horizontal velocity 水平流速
horizontal vertical alternate layout 平立交替布置
horizontal vertical machining centers 卧式及立式加工中心
horizontal vertical position welding 平角焊
horizontal vertical position 水平垂直位置，平角焊位置
horizontal vibration 水平振动
horizontal vibration amplitude 水平振动值，水平振幅
horizontal visibility 水平能见度
horizontal water-collecting layout 水平集水布置
horizontal water-film cyclone 卧式旋风水膜除尘器
horizontal water tube boiler 卧式水管锅炉
horizontal wave 水平波
horizontal welding 水平焊接
horizontal well 水平井
horizontal wind field 水平风场
horizontal wind shear 水平风切变
horizontal wind vector 风的水平矢量
horizontal wind 水平风
horizontal yoke heliostat 水平架定日镜
horizontal 横的，水平的，卧式的，水平分力
horizon 层位，地平，地平面，地平线，水平，水平线
horn antenna 喇叭形天线
horn-break switch 有灭弧角的开关，锥形开关
horn fuse 角式保险器
horn gap lightning arrester 角隙避雷器
horn gap switch 角形火花隙开关，角隙开关，有灭弧的开关
horn gap 角隙，角放电器火花源
horning 喇叭形
horn mouth 喇叭口
horn tube 喇叭管
horn 角，喇叭，扬声器，报警器，蜂鸣器，角状物，角楔
horse latitude high pressure 副热带高压，马纬度高压
horse latitudes 副热带无风带，回归线无风带，马纬度
horsepower characteristic 功率特性
horsepower curve 功率曲线
horsepower hour 马力小时
horsepower input 马力输入，功率输入，输入功率
horsepower loading 动力负荷，马力负荷，功率负荷
horsepower nominal 标称马力
horsepower of equipment 设备马力，设备功率
horsepower of transmission 传动马力
horsepower output 输出马力，功率输出
horsepower rating 额定功率，计算功率
horsepower 马力，功率
horseshoe arch 马蹄形拱
horseshoe bend 马蹄形弯道
horseshoe conduit 马蹄形水管，马蹄形管道
horseshoe electromagnet 马蹄形电磁铁
horseshoe magnet 马蹄形磁铁
horseshoe riveter 马蹄形铆钉机
horseshoe sewer 马蹄形排水管，马蹄形阴沟
horseshoe-shaped sewer 马蹄形排水管
horseshoe-shaped 马蹄形的
horseshoe tunnel 马蹄形隧洞
horseshoe type boiler 带马蹄型炉膛的锅炉
horseshoe type furnace 马蹄型炉膛
horseshoe vortex system 马蹄涡系
horseshoe vortex 马蹄涡
horst fault 地垒断层
horst mountain 地垒山
horst 地垒
horticulture planning 园艺规划
horticulture 园艺
hose clip 管夹
hose connection 软管接头
hose coupling 软管连接器，软管接头
hose levelling instrument 软管水准器
hose line 软管线，水龙带
hose pipe 软管
hose-proof enclosure 防水型外壳
hose-proof machine 防水型电机
hose reel 水龙带
hose station 软管站，消防带站

hose type pump　软管式泵
hose valve　软管阀
hose wrench　软管扳手
hose　软管，胶皮管，蛇管，水龙带
hospital bus bar　（备用）旁路母线
hospital relay group　故障线切换继电器群
hospital switch　事故自动切换开关
hospital　医院
host computer　主计算机，上位机
host country　东道国
hostel　招待所
hostile environment　腐蚀环境，有害环境
hostile water　腐蚀性水
hostile　腐蚀性的，侵蚀性的，生锈的
host interface module　主机接口模件
host language data-base management　用主机语言的数据库管理系统
host machine　主机
host processor　主处理机
host rock　主岩
host-satellite system　主机卫星系统
host system　主系统
host　基质，主机
hot air blast hole　热风鼓风口
hot air blower　热风鼓风机
hot air chamber　热风室
hot air damper　热空气挡板，热风挡板
hot air distribution　热风分布
hot air dry　热空气干燥，热风干燥
hot air duct　热风道
hot air heater　热空气加热器
hot air heating system　热风采暖系统
hot air heating　热风供热，热风采暖
hot air jacket　热空气夹套
hot air producer　热风发生器
hot air radiant heating system　热风辐射供暖系统
hot air seasoning　热风干燥
hot air temperature　热风温度
hot air turbine combined cycle　高温空气透平联合循环
hot air　热风
hot alignment　热态对中
hot and cold steeping　冷热浸渍
hot and cold workshop　冷热机修厂
hot application　热态喷涂
hot-applied bituminous coat　热敷的沥青层
hot area　加热面积，受热面积，高放射性区，热区
hot asphalt　热沥青
hot atom chemistry　热原子化学
hot attack　热腐蚀
hot bank　压火，热备用
hot bending test　热弯曲试验，热弯试验
hot bending　热弯曲，热压弯头
hot blast furnace　热风炉
hot blast heater　热风加热器
hot blast technique　热风技术
hot blast　热风，鼓热风
hot box dome　（改进型气冷堆）热箱顶盖
hot breaking-in　热磨合，热试车，热试验
hot brittleness　热脆性，热脆
hot bulb engine　热球发动机
hot bulb　燃烧室，点火室
hot buttering　热预堆边焊
hot canyon　热（地下）设备室
hot cathode discharge　热阴极放电
hot cathode lamp　热阴极灯
hot cathode rectifier　热阴极整流器
hot cathode tube　直热式电子管，热阴极管
hot cathode　热阴极
hot cell manipulator　热室机械手
hot cell　热室，高放射性物质工作屏蔽室，热栅元
hot cement　热水泥
hot change room　热更衣间
hot channel engineering factor　工程热通道因子
hot channel factor　热管因子
hot channel subfactor　热通道分因子
hot channel　热管，热管道，热通道
hot charging　带负荷加料，热装料
hot clean criticality　热态净临界
hot clean loading　热态干净装载（量）
hot clean reactor　热净反应堆
hot coil　热盘管
hot-compacting　热压，热压实
hot conductor　热导体
hot constant　热状态下常数【反应堆】，热态常数
hot cooling　沸腾冷却
hot corrosion　热腐蚀，高温腐蚀
hot cracking　高温裂缝，热裂纹，热裂
hot critical experiment　热临界实验
hot critical reactor　热临界反应堆
hot critical　热态临界
hot cross-section　热中子截面
hot cure　热校正
hot deformation　热变形，高温变形
hot-dip galvanized　热浸镀锌的，热浸渍镀锌的
hot-dip galvanizing　热镀锌，热浸镀锌
hot dipping　热浸，热浸渍，热浸渍的，热浸镀
hot drain　放射性排放，热排放
hot drawing　热拔
hot-drawn tubing　热拔管
hot-drawn　热拉的，热拔的
hot-driven rivet　热铆铆钉
hot dry rock　干热岩，热干岩
hot ductility　热态可塑性，热态可锻性
hot dump　渣场
hot-electron transistor　热电子晶体三极管
hotel waste pump　公用废水泵
hot end tier　受热面的热段【再生式空气预热器】
hot end　热接点，（热电偶的）热端
hot equipment　热设备
hot equivalent diameter　热当量直径
hot exchanger　热交换器
hot extruded tubing　热挤压管
hot extrusion of billets produced from ingots　从钢锭生产的钢坯的热挤压
hot extrusion　热挤压，热压
hot face insulation　耐火绝缘
hot face　受热面
hot film anemometer　热膜风速计
hot forging　热锻

hot forming 热加工，热成型
hot fuel examination facility 热燃料检查设备
hot fuelling 带负荷加料
hot functional test 高温功能试验，热态试验，热态功能试验
hot function test 热态功能试验
hot fusion 热核聚变
hot galvanizing 热镀锌法
hot gas blower 高温气体鼓风机
hot gas boiler 高温燃气锅炉
hot gas channel 高温气体管道
hot gas duct 热气管
hot gas generator 不冷却炉膛锅炉，高温烟气发生器，高温燃气发生器
hot gas header 高温气体联箱
hot gas measuring point 高温气体测量点
hot gas-path inspection 高温烟气通道检测
hot gas penetration 高温气体贯穿件
hot gas plenum 高温气体压力室
hot gas purification 热气净化
hot gas supply duct 高温气体供气管道
hot gas temperature control 高温气体温度控制
hot gas temperature 高温气体温度
hot gas weld 热风焊接
hot gas 热烟气，高温燃气
hot header 热联箱，热集管
hot inspection 热检验
hot insulation 保温
hot-jacket 热套
hot junction temperature （热电偶）热接点温度
hot junction （热电偶）热端，热接点
hot laboratory 热室，放射性实验室，热实验室
hot lab 热实验室
hot-laid asphaltic concrete 热铺沥青混凝土
hot-laid mixture 热铺混合物
hot laundry and decontamination system 放射性洗涤间和去污染系统
hot laundry ventilation 热洗衣房通风
hot laundry 热洗衣房，放射性洗衣房，热洗衣间，放射性洗衣间
hot layup （蒸汽发生器）热停用
hot leg 热支路，热段，热管段
hot length 热区长度
hot lime-soda process 热石灰苏打法
hot line clamp 带电线夹
hot line job 带电操作，带电作业
hot line maintenance 带电作业，带电维护
hot line tool 带电作业工具
hot line washing 带电清扫
hot line working 带电作业
hot line 热线，带电线路
hot loading 带负荷加料
hot load test 热负荷试验，热荷载试验
hot loop 热回路，堆内试验回路，放射性环路，热环路
hot machine shop 放射性机加工车间
hot maintenance 放射性维修，热维修
hot material 热物质，热材料
hot-melt adhesive 热熔黏结剂
hot metallurgical laboratory 热冶金实验室，放射性冶金实验室
hot mill motor 热轧电动机
hot-mixed asphalt concrete 热拌沥青混凝土
hot mix plant 热拌厂
hot mix 热拌和
hot model 热态模化
hot money 热钱，游资
hot-moulding mica 热塑云母
hotness 热度
hot oil circle 热油循环
hot operation 热态运行【反应堆】，高温运行，热操作，放射性操作
hot particle 热颗粒，放射性颗粒
hot parts 高温零件
hot penetration bitumen 热灌沥青
hot penetration method 热灌法
hot plate 电热板，烘板
hot point factor 热点因子
hot point 热点
hot position 热态
hot precipitator 高温静电除尘器，电气除尘器
hot-pressed elbow 热压弯头
hot-pressed work 热压力加工
hot pressing 热压，热挤压
hot pressure welding 热压焊
hot quenching 高温淬火，热淬火
hot reaction moderator 热反应慢化剂
hot reactivity 热态反应性
hot reactor 热态反应堆
hot refuelling 带负荷换料
hot reheated steam pressure 再热蒸汽热段压力
hot reheated steam temperature 再热蒸汽热段温度
hot reheater 再热器热段
hot reheat line 再热热端管道
hot reheat pipe 高温再热蒸汽管
hot reheat steam outlet 再热热蒸汽出口
hot reheat steam 热再热蒸汽
hot reheat system 高温再热系统
hot repair 热态维修
hot reserve 热备用，热备用状态，热态备用
hot resistance 热电阻
hot restart 热态再启动
hot riveting 热铆
hot rock 热岩
hot-rolled deformed bar 热轧变形钢筋
hot-rolled seamless pipe 热轧无缝钢管
hot-rolled silicon steel sheet 热轧硅钢片
hot-rolled steel member 热轧型钢构件
hot-rolled steel 热轧钢
hot-rolled workblank 热轧毛坯
hot-rolled 热轧的，热辊压榨的
hot rolling 热轧
hot roll welding 热轧焊
hot running-in 热试车，热试验
hot running test 热跑试验【指汽机】
hot running 热运行，热跑
hot section inspection 高温部检测
hot set 热固
hot shop 热车间，放射性车间
hot shortness 热脆性，热脆
hot shrunk fit 热套配合

hot shutdown condition　热停堆状态
hot shutdown reactivity margin　热停堆时的剩余反应性，热停堆反应性裕量
hot shutdown system　热停堆系统
hot shutdown　热停堆
hot side　高压侧，高电位侧，热端，热的一侧
hot-slug dejacketer　热元件脱壳装置
hot source　热源
hot spike　热峰，热钉，热尖
hot spot corrosion　热点腐蚀
hot spot engineering factor　工程热点因子
hot spot factor　热点因子，热点安全因子
hot spot map　热点分布图
hot spot model　热点模型
hot spot monitor　热点监测仪
hot spot safety factor　热点安全因子
hot spot subfactor　热点分因子
hot-spotting　热点，局部加热
hot spot　热点
hot spraying　热喷射，热喷涂
hot spring　热紧
hot standby condition　热停堆状态【反应堆】
hot standby duty　热备用状态
hot standby state　热备用状态
hot standby　热备用，热备份
hot start-up mode　热态启动模式
hot start-up procedure　热启动程序
hot start-up time　热态启动时间
hot start-up　热启动，热态起动
hot start　热态启动，热态起动
hot-state strength test of heat-supply network　热网热态强度试验
hot streaking　热条纹
hot strength　热强度
hot-strip ammeter　热片式安培计
hot subcritical condition　热次临界状态，热亚临界状态
hot subcritical reactor　热态亚临界反应堆
hot tarring　热灌柏油
hot tear　热裂
hot-temperature zone　高温区
hot tension test　高温拉力试验
hot test loop　热试验回路【反应堆】，放射性试验回路
hot test run　热试车
hot test　热态试验，热试验
hot trapping　热捕集
hot trap　热阱
hot upset test　热顶锻试验，热镦粗试验
hot vulcanization　热硫化
hot wall effect　热壁效应【下降流道】
hot wardrobe　放射性区衣柜
hot wastewater　高温废水
hot waste　放射性废物，热废物
hot water bath　热水浴
hot water boiler　热水锅炉
hot water circulation pipe　热水循环管
hot water cure　热水治疗，热水硫化
hot water heated superheater　热水加热的过热器
hot water heater　热水加热器
hot water heating load　热水供暖负荷
hot water heating system　热水供暖系统
hot water heating temperature　热水供暖温度
hot water heating　水暖，热水供暖
hot water heat-supply system　热水供热系统
hot water line　热水管路
hot water pipe　暖水管，热水管
hot water return　热水回路
hot water storage tank　热水供应贮水箱
hot water supply system　热水供给系统
hot water tank　热水箱
hot water tracing　热水伴热
hot water treatment　热水处理，热水处理法
hotwell depression　热井内凝结水过冷
hotwell dump valve　热井排放阀
hotwell makeup valve　热井补充阀
hotwell pump　热井泵，凝结水泵
hotwell sampling pump　热井取样泵
hotwell　热井，（凝汽器）热水井
hot wet layup　热的湿保养【蒸汽发生器】
hot wind tunnel　热风洞
hot wind　热风
hot wire ammeter　热线式电流表
hot wire anemometer　热线风速表［计］，热线室风速仪
hot wire coil　热线圈，通电线圈
hot wire direction meter　热线风向计
hot wire flow measuring device　热丝流量测量仪表
hot wire galvanometer　热线检流计
hot wire gauge　热线压力计
hot wire manometer　热线压力计
hot wire meter　热线式仪表
hot wire method　电阻法，热线法
hot wire oscillograph　热线示波器
hot wire pressure gauge　热线压力计
hot wire probe　热线测针，热线探针
hot wire relay　热线式继电器
hot wire technique　热线技术
hot wire voltmeter　热线式伏特计
hot wire　热线，热电阻线，带电电线
hot working　热加工
hot workshop　热车间，放射性车间，放射性维修间，热机修间
hot work　热加工
hot zone　热区，放射性区
hot　热的，热烈的，强放射性的，带电的
H_2O/U molecular ratio　水/铀分子比
hour-ahead market　一小时前市场
hour-ahead trading　时前交易
hour angle　时角
hour circle　时圈
hourglassing　沙漏式缺陷
hourglass　沙漏
hourly capacity　每小时出力
hourly coal consumption　小时耗煤量
hourly cooling load　逐时冷却负荷
hourly earnings　每小时工资
hourly heat consumption on hot-water supply　热水供应小时用热量
hourly height　逐时潮高
hourly load　时负荷

hourly log 时报
hourly mean wind speed 每小时平均风速
hourly observation 逐时观测
hourly precipitation 每小时降水量
hourly ratio 小时出力
hourly sol-air temperature 逐时综合温度
hourly tidal height 逐时潮高
hourly variation coefficient 时变化系数
hourly variation factor of heating load 热负荷小时变化系数
hourly variation factor 时变系数
hourly variation graph of heat consumption in one day 每日小时热负荷图，日耗热量变化图
hourly variation 每小时变化
hourly water consumption 每小时用水量
hourly wind speed 每小时风速
hour meter 计时器，小时计
hour norm 工时定额
hour of labour 工时
hours of wind 刮风小时数
hours waiting 停机待修时数
house 建筑物，室，车间，站
house air waybill 航空分运单
house cable 室内电缆，用户电缆
house connecting box 用户分线箱
house consumption 厂用电
housed joint 封装接头，套入接合
housed sting 嵌入式楼梯斜梁
house generating set 厂用发电机组
house generator 厂用发电机
house heating 住宅供暖
household 家用的，家庭，户
household appliance 家用电器
household demand 生活用电量
household electrical appliance 家用电器
household fuel gas 民用煤气，家用煤气
household fuel 民用燃料，家用燃料
household garbage grinder 生活垃圾粉碎机
household garbage 家庭垃圾，生活垃圾
household pesticide 家庭用农药
household refuse 家庭垃圾
household sewage 家庭污水
household wastewater 生活污水
household waste 家庭用过的废物
household water filter 民用滤水器，家庭用水过滤器
household water meter 家用水表
household wind turbine 家用风电机组
house keeping instruction 管理指令，辅助指令
house keeping operation 辅助操作，内部管理操作
house keeping 内务处理，（程序的）内务操作，保管，辅助工作
house lead-in 进户线
house load operation threshold 带厂用电运行的定值，带厂用电运行定值
house load operation 带厂用电运行
house load 厂用电，厂用负荷，厂用电负荷
house main （配电）干线
house microclimate 室内小气候
house power 厂用电
house service boiler 电厂自用锅炉
house service bus 厂用电母线
house service circuit 厂用电路
house service consumption rate 厂用电率
house service consumption 厂用电，厂用动力消耗
house service equipment 厂用电设备
house service generator 厂用发电机
house service network 厂用电力网
house service switchgear 厂用开关装置
house service switch room 厂用配电间
house service transformer 自用变压器，厂用变压器
house service turbo-generator 厂用汽轮发电机组
house service wire 进户线
house service 厂用
house substation 厂用变电所，专用变电所
house supply 厂用电源
house telephone system 内部电话交换系统
house telephone 内部电话
house transformer 自用变压器，厂用变压器
house turbine 厂用汽轮机
house wastewater 生活污水
house wiring 室内布线
housing and land tax 房地产税
housing area 住房区域，居住区，住宅区
housing case 外壳，机壳
housing density 房屋密度
housing element 外壳
housing estate 住宅区，居民点
housing load operation threshold 带厂用电运行定值
housing plate 壳套板
housing project 住房建造规划
housing tank （变压器的）油箱
housing 房屋，壳，罩，套，外壳，箱，机壳，汽缸，包壳，外套，护盖，壳体，机舱，住房，外罩
hovel 杂物间
hovercraft 气垫船
Howe truss 豪威式桁架，豪氏桁架
howler 蜂鸣器
HPBFP（high-pressure boiler-feed pump） 高压锅炉给水泵
HPBP（high pressure by-pass） 高压再热旁路
HP bypass loop 高压旁路回路
HP bypass pressure control valve 高旁压力控制阀
HP bypass temperature control valve 高旁温度控制阀
HP casing （汽轮机）高压缸
HP charging pump 高压上充泵
HPCI pump 高压冷却剂注入泵
HPCI system 高压冷却剂注入系统
HP control fluid 高压控制油
HP cooler 高压冷却器
hp cylinder 高压缸【汽轮机】
HP-ESV 高压主汽门，高压危急截断阀
HP evaporator 高压蒸发器
HP-EV（high pressure extraction valve） 高压抽汽阀

HP heater 高压给水加热器，高加
HPH(high pressure heater) 高压加热器
HP(high pressure) 高压
HP(horsepower) 马力，功率
HP/IP/ LP rotor 高/中/低压转子
HP/IP/ LP turbine 高/中/低压缸
HP-LP integrated rotor 高低压一体化转子
H point(hold point) 停工待检点
HP outlet header 高压出口总管
HPR＝hopper 斗，漏斗
HP sampling cooler 高压取样冷却器
HPSI(high pressure safety injection) pump 高压安全注入泵
HP steam superheater 高压蒸汽过热器
HPSV(high pressure stop valve) 高压主汽门
HPT(high pressure turbine) 高压汽轮机
HPT (hot precritical test) 临界前热试验
HPV(high pressure valve) 高压阀
HPW(hot pressure welding) 热压焊
HRB （德国）高温堆制造有限公司
HR(harmonic ratio) 谐波含有率
HR(heat bate) 热耗率
HRH(hot reheater) 再热器热段，再热热段
hr＝hour 小时
HR(humidity relative) 相对湿度
HRSG auxiliary systems 余热锅炉的辅助系统
HRSG(heat recovery steam generator) 余热锅炉，余热回收蒸汽锅炉，余热回收蒸汽发生器
HRT(heat rate) 热耗率
HRU(heat recovery unit) 回热式机组
HRV(hydraulic relief valve) 液压安全阀
HSBC(Hong Kong and Shanghai Banking Corporation) 汇丰银行
H. S.(high-strength)bolt 高强螺栓
H-section 宽缘工字钢，工字形断面，H形截面
HSE(Health,Safety and Environment) 健康、安全与环境
HS(hand switch) 手动开关
H shape cutting machine H形钢切割机
H shape work shed H形钢工棚
HS(high speed) 高速
HSR(historical data storage and retrieval) 历史数据检索和存储
HSS(hydrogen seal oil system) 氢气密封油系统
H steel pile H形钢桩
H steel 宽缘工字钢，H形钢
HSV(hot reheat safety valve) 再热器热段安全阀
HS(shore hardness) 硬度，肖氏硬度
HTDC testing transformer 高压直流试验变压器
HTG＝heating 加热
HTGR nuclear power plant 高温气冷堆核电厂
HTGR technology 高温气冷堆技术
HTLS conductor(high tension low sag conductor) 高压低驰度导线【内含碳纤维，高温耐热】
HTLS(high torque low speed) 高扭转力低速
HT(high tension) neon tester 高压试电笔
HTR(high temperature reactor) absorber rod 高温气冷堆吸收棒
HTR cogeneration 高温堆热电联产
HT rectifying tube 高压整流管
HTR＝heater 加热器
HTR(homogeneous thorium reactor) 均匀钍反应堆
HTR with a gas turbine 使用燃气轮机的高温气冷反应堆
HTR 高温气冷堆【核电站】
HT switch-board 高压配电盘
HT testing transformer 高压试验变压器
H type cable 屏蔽电缆
Huaneng group 华能集团
Huaneng international power development corporation 华能国际电力开发公司
Huaneng Power Generation Corporation 华能发电公司
hub and spigot joint 中心向联轴节，套塞接头
hub assembly 毂组件
hub bore 轮毂孔
hubcap 轮毂罩
hub controller 轮毂控制器
hub height 轮毂高度
hub insulation 套筒绝缘
hub material 轮毂材料
hub plate 后盖板，（泵或风机叶轮的）毂衬，毂衬
hub precone 桨毂预锥角
hub ratio 轮毂比
hub rigidity 轮毂刚度
hub-shroud ratio 轮毂比
hub spider assembly 桨毂辐状装配
hub spider 辐射形桨毂
hub-tip ratio 轮毂比
hub type 轮毂类型
hub wrench 轮毂螺母扳手
hub 标桩，毂，中心，柄，集线器，轮毂，衬套，枢纽
huddling chamber 混合室
huff-duff 高频无线电测向仪
huge project 大型工程
huge surge bin 大型缓冲仓
HU＝humidifier 加湿器
hull armour plate 船身（装）甲板
hull 壳体【燃料包壳-后处理】，外壳，船体
hum 灰岩残丘
human activity 人类活动
human body impedance 人体阻抗
human capital development 人力资本开发，智力开发
human capital investment 智力投资
human comfort 人体舒适性
human counter 全身计数器
human discomfort 人体不舒适性
human element accident 责任事故
human engineering 人体工程学，运行工程学，人类工程学，人机工程学
human environment 人类环境
human error 人为差错，人为误差
human factors engineering 人机系统工程学
human factors review group 人为因素评审组
human factor 人为因素，主观因素
human failure 人为故障，人为事故
human health 人体健康

human initiated failure 人为故障
human investment 人力投资
human longevity 人类寿命
human machine interface 人机接口，人机界面
human machine system 人机系统
human nutrition 人体营养
human perceptible 人可觉察的
human performance evaluation system 人力绩效评价体系
human pollution burden 人体污染负荷
human pollution 人体污染
human power evaluation system 人力资源评价体系
human relationship 人际关系
human resources 人力资源
human responsiveness 人体反应
human sensitivity 人体敏感性
human sensory system 人的感觉系统
human tissue 人体组织
human tolerance to wind 人体耐风性
human tolerance 人体耐力
humate 腐殖酸盐
hum balance resistor 哼声抵消电阻
hum-bucking coil （交流）哼声抑制线圈
humectant 润湿剂
humectation process 增湿
humectation 湿润
hum eliminator 哼声抑制器
hume 水滑道
hum filter 哼声滤波器
humic 腐殖的
humic acid 腐殖酸
humic coal 腐殖煤
humic matter 腐殖质
humic mulch 腐殖覆盖物
humic substance 腐殖物质，腐殖质
humics 腐殖质
humid air 湿空气
humid analysis 湿法分析
humid area 湿润区
humid climate 潮湿气候，湿润气候
humidification process 增湿过程
humidification 加湿作用，增湿，加湿，增湿作用
humidified combustion 增湿燃烧
humidified screw conveyer 湿式螺旋输灰机
humidifier section 加湿段
humidifier 增湿器，润湿器，加湿器，空气增湿器
humidifying and air conditioning equipment 空调湿调设备
humidifying process 加湿过程
humidify 加湿，调湿（空气）
humidiometer 湿度计
humidistat 恒湿器，湿度调节器
humidity controller 湿度调节器
humidity control 湿度控制
humidity correction factor 湿度校正系数
humidity corrosion 湿腐蚀
humidity dampness 潮湿
humidity deficit 湿度差
humidity-density meter 湿度密度计
humidity effect 湿度效应
humidity factor 湿度因子，湿润系数
humidity gradient 湿度梯度
humidity indicator 湿度指示器
humidity measurement 湿度测量
humidity meter 湿度计
humidity ratio 湿度比
humidity recorder 湿度记录器
humidity-sensitive capacitor 湿敏电容器
humidity-sensitive element 湿敏元件
humidity sensor 湿度传感器
humidity test 湿度测定
humidity 湿度，水分含量，潮湿，湿气，湿润
humid mesothermal climate 湿润中温气候，湿温气候，温湿气候
humidostat 湿度调节器，湿度调节仪
humid region 湿润区
humid room 保湿室，雾室
humid temperate climate 湿润温和气候
humid 潮湿的，湿的，湿润的
humify 腐殖化
hum in output 输出的交流声
hum interference 哼声干扰
humiture 温湿度
hum level 哼声电平
humming elimination 交流哼声消除
hummock 波状地
hum modulation factor 哼声调制系数
hum modulation 哼声调制
hump joint 弯部接头
hump-type stilling basin 驼峰式静水池，驼峰式消力池
hump voltage 驼峰电压，峰值电压
hump （曲线的）顶点，峰，峰值，隆起，山岗，驼峰
humus layer 腐殖层
humus sludge 腐殖污泥
humus soil 腐殖土
humus stratum 腐殖层
humus tank 腐殖质沉积池
humus 腐殖质
hundreds dial 百分度盘
hundred-year return period 百年一遇
hung ceiling 吊顶天花板，吊顶
hung fender 悬式护舷木
hung lines 挂线
hung start 升速失败的启动
hung wood fender 悬木防护器
hunk of cable 电缆盘
hunt effect 摆动效应
hunting angle 振荡角，摆动角
hunting contact 寻线触点
hunting frequency 振荡频率，摆动频率
hunting loss 搜索损耗
hunting motion 不规则振荡
hunting of frequency 频率波动，频率振荡
hunting of generator 发电机振荡
hunting of governor 调速器振荡
hunting of load 负荷波动，负荷振荡
hunting period 振荡周期，摆动周期
hunting power 振荡功率
hunting range 振荡范围，摆动范围

hunting slot　空槽
hunting speed　追逐速率，振荡速度，寻线速度
hunting system　自振系统
hunting time　振荡时间，振荡持续时间，寻线时间
hunting tooth limit switch　摆齿限位开关
hunting vibration　摆动
hunting zone　搜索范围
hunting　摆动，振荡，寻找故障，寻线，寻找平衡，震荡
hurdle dike　栅栏透水堤
hurdle dyke　栅栏透水堤
hurdle groin　栅栏透水丁坝
hurdle groyne　栅栏透水丁坝
hurdle　篱笆，栅栏，障碍
hurlbarrow　双轮手推车
hurricane boundary layer　飓风边界层
hurricane core　飓风核心
hurricane eye　飓风眼
hurricane lamp　防风灯
hurricane sampler　旋风式取样器
hurricane wind　飓风【12级以上】
hurricane　飓风【十二级以上】，龙卷风，狂飙，十二级风
hurst　树林，丛林，小丘
hurter　防护短柱
hurt　损害
husbandry　农业，资源管理，饲养
husk　壳，支架
hut　工棚，棚屋
hutch　棚屋
Hutter wind turbine　赫特风轮机
HVAC equipment room　采暖通风空调设备室
HVAC equipment　高压交流设备
HVAC(heating,ventilation and air conditioning)　供暖、通风和空调，暖通与空调
HVAC(heating ventilation and air conditioning)　暖通空调
HVAC(high voltage alternating current)　高压交流电
HVAC power transmission line　高压交流输电线路
HVAC system design　暖通设计
HVAC system　暖通空调系统
HV auxiliary transformer　高厂变
HVAWT(horizontal input to vertical axis output wind turbine)　水平输入垂直轴输出式风电机组
HVDC coupling station　高压直流联结站
HVDC coupling system　高压直流联结系统，高压直流耦合系统
HVDC equipment　高压直流设备
HVDC(high voltage direct current transmission)　高压直流输电
HVDC(high voltage direct current)　高压直流
HVDC light power transmission　轻型直流输电
HVDC link　高压直流联络线路
HVDC master control　高压直流输电主控制
HVDC model　高压直流输电模型
HVDC power transmission cable line　直流输电电缆线路
HVDC power transmission earthing electrode line　高压直流输电接地极（引）线
HVDC power transmission earthing electrode　高压直流输电接地电极
HVDC power transmission line　高压直流输电线（路）
HVDC power transmission overhead line　高压直流输电架空线路
HVDC power transmission simulator　高压直流输电模拟装置
HVDC power transmission system operation and maintenance　高压直流输电系统运行维护
HVDC power transmission system simulator　高压直流输电系统仿真装置
HVDC power transmission system　高压直流输电系统
HVDC power transmission　高压直流输电
HVDC substation control　高压直流变电所控制
HVDC system control　高压直流系统控制
HVDC system pole　高压直流极
HVDC system　高压直流电力系统
HVDC transmission control　高压直流输电控制
HVDC transmission line　高压直流输电线路
HVDC transmission substation　高压直流输电变电所
HVDC underground transmission　高压直流地下输电
HV electric equipments and cables for generation, transformation, transmission, distribution and power-consumption device　电气一次设备【中国术语】，耗电设备
HV(hand control valve)　手动控制器
HV(high voltage)　高压
HVPS(high voltage power supply)　高压电源
HV transmission line　高压输电线路
H wave　水力波，H形波
HWB(house air waybill)　航空分运单
H-wedge gate valve　H楔闸阀
HW＝hotwell　热井
HWR and fuel channel assembly　重水堆燃料通道组件
HWR heat transport system　重水堆热传输系统
HWR moderator cover gas system　重水堆慢化剂覆盖气体系统
HWR moderator system　重水堆慢化剂系统
HWR nuclear power plant　重水堆核电厂
HWR refueling machine　重水堆装卸料机
HWR refueling system　重水堆燃料装卸系统
HWR safety system　重水地安全系统
HWR shutdown cooling system　重水堆停堆冷却系统
HWR(heavy water reactor)　重水堆
HXCC(China Nuclear Industry Huaxing Construction Co., Ltd.)　中国核工业华兴建设有限公司
HX(heat exchange)　热交换器
hybrid analysis　混合分析
hybrid chip　混合芯片
hybrid circuit　混合电路
hybrid coil transformer　混合线圈变压器
hybrid coil　混合线圈，差动线圈
hybrid comparator　混合型比较器
hybrid computer simulation　混合计算机仿真
hybrid computer　混合计算机，混合式计算机

hybrid computing system 混合型计算系统
hybrid control computer 混合控制机
hybrid control 复合调度，混合计算机控制
hybrid digital analog computer 数字模拟混合计算机
hybrid digital analog pulse time 数字模拟混合脉冲时间
hybrid digital analogue circuit 数字模拟混合电路
hybrid distributed processing system 混合分布处理系统
hybrid electromagnetic mode 电磁混合模式
hybrid electromagnetic wave 混合电磁波，复合电磁波
hybrid electronic layout program 混合电子布线程序
hybrid-element method 混合单元法
hybrid fusion-fission breeder 聚变裂变混合增值反应堆，混合式聚变裂变反应堆
hybrid fusion-fission reactor 聚变裂变混合反应堆，混合式聚变裂变反应堆
hybrid grid 混合材料格架
hybrid input/output 混合输入/输出
hybrid integrated circuit 混合（式）集成电路
hybrid integrator 混合型积分器
hybrid interface 混合接口
hybridization 杂化
hybrid junction 混接，混合连接
hybrid library 混合程序库
hybrid LSI 混合式大规模集成电路
hybrid mode 复合模式
hybrid operation 复合运行，混合式运行
hybrid parameter 混合参数
hybrid programming 混合编程
hybrid reactor 混合反应堆
hybrid reinforced plastics 混合纤维增强塑料
hybrid relay 混合式继电器，混合继电器
hybrid rock 混染岩
hybrid simulation 混合模拟
hybrid simulator 混合仿真装置
hybrid sliding-pressure operation 复合滑压运行
hybrid start-up of high-medium pressure cylinder couplet 高中压缸联合启动
hybrid system checkout 混合系统校验
hybrid system 混合系统
hybrid tee T形波导
hybrid time delay relay 混合延时继电器
hybrid tower 混合式塔架
hybrid transformer 混合变压器
hybrid transistor-diode logic 混合型晶体管二极管逻辑回路
hybrid winding 混合绕组
hybrid 混合物，混合电路，混合，混合动力，混合型，混合的
hydatogenesis 水成作用
hydaulic pressure fluctuation 水压脉动
HYD＝hydraulic 液力的
hydraguide 油压转向装置
hydralime 熟石灰
hydramix dust conditioner 湿式灰调料机
hydrant barrel 消防龙头套筒
hydrant cabinet 消火栓柜
hydrant hose 水龙带
hydrant with double outlet 双头消火栓
hydrant with single outlet 单头消火栓
hydrant 消防栓，取水管，消火栓
hydratability 水合性
hydrate alkalinity 氢氧根碱度，水合物碱度
hydrated cation 水合阳离子
hydrated electron dosimeter 水合电子剂量计
hydrated electron dosimetry 水合电子剂量测定法
hydrated ferric oxide film 水合氧化铁膜
hydrated hydrogen ion 水合氢离子
hydrated ionic radius 水合离子半径
hydrated ion 水合离子
hydrated lime 消石灰，熟石灰，氢氧化钙，水化石灰
hydrated matter 水合物
hydrated oxide 氢氧化物
hydrated proton 水合质子
hydrated water 水合水，结合水，结晶水
hydrate of aluminium 氢氧化铝
hydrate of barium 氢氧化钡
hydrate 水合物，使成氢氧化物，使水合，水化物，水化作用
hydrating coal gasification 煤的水合气化
hydration energy 水化能
hydration heat 水化热
hydration product 水化物
hydration water 结合水，化合水
hydration 水合，水化作用，水合作用，水化
hydraucone 喇叭形尾水管
hydraulic 水力学的，液力的
hydraulic accelerated clarifier 水力加速澄清器
hydraulic accumulator charging pump 液压蓄压箱充注泵
hydraulic accumulator 液压蓄压箱【应急堆芯冷却系统】，液压蓄能器
hydraulic action 水力作用
hydraulic-actuated excavator 液压操纵挖掘机
hydraulic actuator 液压执行机构，液压执行器，液压制动器，液压传动
hydraulic admixture 水硬性掺料
hydraulic air filter 液压空气滤清器
hydraulic air pump 射水抽气器
hydraulical analysis of heat-supply network 热网水力计算
hydraulically expandable antivibration bar 用液压可膨胀的防振条
hydraulically-induced noise 由液压引起的噪声
hydraulically operated controller 液动调节器，液压控制器
hydraulically operated gate lifting device 水力启门机
hydraulically operated isolation valve 液动隔离阀
hydraulically operated rear dump truck 液压操纵后卸卡车
hydraulically operated valve 液动阀
hydraulically operated 液压操作的，水力操纵的
hydraulically smooth regime 水力光滑流态
hydraulically smooth 水力光滑
hydraulical mean radius 平均水力半径

hydraulical propeller 喷水推进机
hydraulical retaining structure 挡水构筑物
hydraulical 水力的，水压的，液压的，液力的，水力学的
hydraulic analogy 流体动力相似，流体动力模拟，水力模拟
hydraulic ash conveyer 水力除灰装置
hydraulic ash handling system 水力除灰系统
hydraulic ash handling 水力除灰
hydraulic ash removal system · 水力除灰
hydraulic ash removal 水力除灰
hydraulic ash sluicing 水力冲灰，水力除灰
hydraulic ash transmission system 水力输灰系统
hydraulic backhoe 液压反向铲
hydraulic back-pressure valve 液压止回阀，水压逆止阀
hydraulic balance device 液压平衡装置
hydraulic balancing device 液力平衡装置
hydraulic balancing 水力平衡【与平衡盘或平衡锤】
hydraulic ball control system 水力滚珠式控制系统
hydraulic barrier 折流挡板
hydraulic bender 液压弯管机，液压弯曲机
hydraulic bending machine 液压弯管机
hydraulic binding agent 液力胶结剂
hydraulic blast 水力清沙
hydraulic block 液压块
hydraulic bolt stretcher 液压螺栓拉伸机
hydraulic bolt tightener 液压螺栓上紧器
hydraulic borehole gauge 液压钻孔应变计
hydraulic borehole stressmeter 液压钻孔应力计
hydraulic bottom ash removal 水力排渣
hydraulic boundary condition 边界水力条件，水力边界条件
hydraulic brake 液压刹车，液压制动器，水力制动器
hydraulic braking system 液压制动系统
hydraulic buffer 水力缓冲器，水力消能器，液压缓冲器
hydraulic bulge forming 液压膨胀成形
hydraulic calculation 水力计算，水利计算
hydraulic capstan 水力绞盘，水力起锚机
hydraulic capsule 液压测力计，液压传感器
hydraulic car clamp 液压夹车器
hydraulic cartridge 液体炸药筒
hydraulic cell 液压传感器
hydraulic cementing agent 水硬性胶结料
hydraulic cement 水硬性水泥
hydraulic characteristic curve 水力特性曲线
hydraulic characteristics 水力特性，水力特性曲线
hydraulic check cylinder 带单向止回阀油缸
hydraulic check nut 水力防松螺帽
hydraulic check valve 液压逆止门
hydraulic classification 水力筛分，水力分级
hydraulic cleaning 水力清洗
hydraulic closure stud tensioner 液压封头螺栓拉伸机
hydraulic clutch 液压离合器
hydraulic complex 水利枢纽
hydraulic component 液压元件
hydraulic compressor 水力压缩机
hydraulic computation 水力计算
hydraulic concrete 水工混凝土
hydraulic condition 水力条件
hydraulic conductivity 水力传导系数，水力传导性，导水性，透水性
hydraulic controlled bulldozer 液压操纵推土机
hydraulic controlled scraper 液压式铲运机
hydraulic controller 液压调节器
hydraulic control rod drive system 液压式控制棒驱动系统
hydraulic control 液动调节，液压调节，液动控制，液压控制，水力控制
hydraulic core design 堆芯水力学设计
hydraulic coupler 液力偶合器
hydraulic coupling driven type 液力联轴器传动方式
hydraulic coupling 液力联轴节，液压联轴器，液力联轴器，液力耦合器，液压联轴节
hydraulic crane 液压起重机
hydraulic crimper 液压卷边机，液压折边机
hydraulic crimping pliers for cable terminals 液压压线钳
hydraulic crimping tools 液压压接工具
hydraulic current 落差流
hydraulic cushion 液压缓冲器
hydraulic cyclone 水力旋流器
hydraulic cylinder hoist 液压圆筒式启门机
hydraulic cylinder 液压千斤顶，液压唧筒，液压缸
hydraulic design criteria 水力学设计准则
hydraulic design of water engine 水轮机水力设计，水轮水力设计
hydraulic design 水力计算，液力设计，水力设计
hydraulic diameter 水力学直径，水力直径
hydraulic disorder 水力失调
hydraulic drag 水力牵引，水阻力
hydraulic dredger 水力疏浚机，水力挖泥机，水力吸泥机
hydraulic drilling 水力钻探
hydraulic drive 水力传动，液压传动
hydraulic drop 跌水
hydraulic dust removal 水力除灰
hydraulic dynamometer 水力测功器，液压测力计
hydraulic efficiency 水力效率
hydraulic ejector 水力排泥管，水力喷射器
hydraulic electrogenerating 水力发电
hydraulic element 液压元件，水力因素
hydraulic elevating platform 液压升降平台
hydraulic elevator 水力提升机，液压升降机
hydraulic energy 水利能源，水能
hydraulic engineering 水工学，水利工程
hydraulic engineer 水力工程师
hydraulic engine 水力发动机
hydraulic equipment 液压设备
hydraulic equivalent diameter 水力当量直径
hydraulic equivalent 水力当量
hydraulic evapotran-spirometer 水力式蒸腾计

hydraulic excavation and filling 水力冲填
hydraulic excavation 水力冲挖，水力掘进，水力开挖，水力挖掘
hydraulic excavator 水力探掘机
hydraulic expansion 液力膨胀
hydraulic exponent 水力指数
hydraulic extraction 水力采矿，水力提取
hydraulic factor 水力因素
hydraulic feedback 水力反馈
hydraulic feed 水力传送，液压进料
hydraulic fill dam 水力冲积坝，水力冲填坝
hydraulic fill method 水力冲填法
hydraulic fill process 水力冲填法
hydraulic fill 水冲法填土，水力冲填
hydraulic filter 液压过滤器
hydraulic flange bender 液压弯排机
hydraulic flap 液压操纵襟翼
hydraulic flattening 水力展平
hydraulic flow net 水力流网
hydraulic flow 水流，液流，湍流
hydraulic fluid pump 液压泵
hydraulic fluid 工作液体，水力液体，液压用流体，液压油
hydraulic flume 水力试验槽
hydraulic flume transport 水槽运输
hydraulic flushing sedimentation basin 水力冲洗式沉沙池
hydraulic forging 水压锻造，液压锻造
hydraulic form 流线型
hydraulic fracture test 水力劈裂试验，液压断裂试验
hydraulic fracturing 水力劈裂，水力破碎法
hydraulic friction 水力摩阻，水力阻力
hydraulic front-end loader 液压掘进装载机
hydraulic gate unit 液压装置
hydraulic gate 液压闸门
hydraulic gauge 水压计
hydraulic gear 液压传动装置
hydraulic generator 水力发电机，水轮发电机
hydraulic geometry 水力几何学
hydraulic gland 水封
hydraulic governing system of steam turbine 汽轮机液压调节系统
hydraulic governing 液压调节
hydraulic governor 液压调速器
hydraulic grade line 水力坡度线，水力坡降线，水压线，压力水头线，测压管水头线
hydraulic grade 水力坡降
hydraulic gradient line 水力比降线，压力水头线
hydraulic gradient slope 水力坡降
hydraulic gradient 落差梯度，水力比降，水力梯度
hydraulic gun 水力冲射器，水枪
hydraulic hammer 液压锤，水锤
hydraulic hardening 水硬性作用
hydraulic head 水头，压头，水位差，水力提升机，液压启门机
hydraulic hold-down system 液压压紧系统
hydraulic impact 水冲击力
hydraulic inclination 水力坡降
hydraulic indicator 水压计
hydraulic inserted strength measurement instrument 液压附着力测量仪
hydraulic instability 水力不稳定性
hydraulic intensifier 液压增强器
hydraulicity 水硬性
hydraulic jack 液压千斤顶，液压唧筒，水力千斤顶
hydraulic jam 水锤扬水机
hydraulic jet pump 水力喷射泵
hydraulic jetting 水力钻探
hydraulic jump 水跃
hydraulicking 水力开采
hydraulic laboratory 水工试验室，水力实验室
hydraulic lifter 液压提升机
hydraulic lifting device 液压提升装置
hydraulic lift 水力提升，水力提升机，液压升降机
hydraulic limestone 水硬性石灰石，水硬性灰岩
hydraulic lime 水硬性石灰
hydraulic load cell 液压测力仪，液压传感器
hydraulic loading 水力装料
hydraulic load 水力载荷
hydraulic lock 液压闭锁装置，液压锁定
hydraulic loss 水力损失，水压损失
hydraulic luffing 液压变幅
hydraulic machinery power 水力机械功率
hydraulic machinery 水力机械
hydraulic mean depth 平均水力深度
hydraulic medium 液压介质
hydraulic mining 水力采矿
hydraulic misadjustment 水力失调
hydraulic model testing 水力模型试验，水力学模型试验
hydraulic model 水工模型
hydraulic motor 水力发动机，水力马达，液压马达
hydraulic navy 水力挖掘机
hydraulic non-chain traction 液压无链牵引
hydraulic nozzle-tapping machine 液压开口机
hydraulic oil 高压油，液压油
hydraulic opener 液压开孔器
hydraulic operated gate lifting device 液压启门机
hydraulic operated valve 液压阀
hydraulic operated 液压操作的
hydraulic operating gear 液压千斤顶
hydraulic packing 液压填密，液压密封
hydraulic parameter 水力参数
hydraulic percussion method 水力冲击法
hydraulic permeability 水力渗透性，透水性
hydraulic pile driving hammer 液压打桩锤
hydraulic pile driving 液压打桩
hydraulic pipe bender 液压弯管机
hydraulic pipe bending machine 液压弯管机
hydraulic pipe fitting machine 液压对口器
hydraulic pipeline 水力管道
hydraulic piston 液压活塞
hydraulic pitching lever 液压调桨杠杆
hydraulic pitching mechanism 液压调桨机构
hydraulic pitching rod 液压调桨杆
hydraulic pitching system 液压调桨系统
hydraulic pitch lever mechanism 液压调桨杠杆

机构
hydraulic potential 水力势能，水力位能，水力位势
hydraulic power pack 液压动力装置
hydraulic power plant 水电厂，水电站，水力发电机
hydraulic power pool 水电电网
hydraulic power regulator 液压功率调节器
hydraulic power scheme 水力发电开发方案，水力发电枢纽
hydraulic power supply 水电供电
hydraulic power system 液压动力系统
hydraulic power tools 液压工具
hydraulic power transmission 液压传动
hydraulic power units 液压动力元件
hydraulic power 水力，水能
hydraulic press 水压机，液压机
hydraulic press-extracting machine 液压式压拔桩机
hydraulic pressing-jointing machine 液力压接机
hydraulic pressure 水压力，液压
hydraulic pressure control system 液压控制系统
hydraulic pressure gauge 液压计
hydraulic pressure head 水压头，水力压头
hydraulic pressure regulator 液压调节器
hydraulic pressure snubber 液压缓冲器，水压缓冲器
hydraulic pressure test pump 水压试验泵
hydraulic pressure test 水压试验
hydraulic prime mover 水力原动机
hydraulic profile 水力断面
hydraulic propeller 喷水推进机，水力推进机，水力推进器
hydraulic property 水力学特性
hydraulic propulsion system 液压推进系统
hydraulic protection 水工保护
hydraulic puller 液压拉拔器
hydraulic pumping circuit 液压泵送回路
hydraulic pump 液压泵，水泵
hydraulic punching machine 水力冲孔机，液压冲孔机
hydraulic quarrying machine 液压式凿岩机
hydraulic radius 水力学半径，水力半径
hydraulic ram pump 水锤泵
hydraulic ram 水力夯锤，压力吸扬器，液压油缸
hydraulic regime 水力工况
hydraulic regulator 水力调节器，液压调节器
hydraulic relay valve 液压继动阀
hydraulic relay 液压替续器
hydraulic relief valve 液压安全阀，液压减压阀
hydraulic research station 水力学研究站
hydraulic resistance balance 流动阻力平衡，液力平衡
hydraulic resistance 流动阻力，水力阻力，水阻，流体阻力，水力摩阻，水力阻抗
hydraulic retaining structure 挡水水工建筑物
hydraulic riveter 液压铆钉机
hydraulic rock cutting 水力凿岩
hydraulic roller 水辊，水滚
hydraulic rolling 液力滚压
hydraulic rotary cylinder 液压回转缸
hydraulic rotor lifting pump 液压顶轴油泵
hydraulic sand sizer 水力分砂器
hydraulic scale model 水工缩尺模型
hydraulic scouring 水力冲刷
hydraulic seal 水封，液封，液压密封，止水阀
hydraulic section 水力部分，水力断面
hydraulic selector 液压选择阀
hydraulic self-resetting buffer 液压自动缓冲器
hydraulic sensing pad 液压传力垫块
hydraulic separation 水力分离
hydraulic separator 水力分离器，液压分离器
hydraulic servo actuator 液压伺服执行机构
hydraulic servo motor 液压伺服电动机，液压伺服马达，液压伺服机构，油动机
hydraulic servo 液压伺服机构，液压伺服系统
hydraulic set 液压装置，水凝
hydraulic shaft jacking device 液压顶轴装置
hydraulic shield 液压掩护式支架
hydraulic shock absorber 水力减振器，液压除振器
hydraulic shock eliminator 水力减振器，液压除振器
hydraulic shock 水击，水冲击，水力冲击，水力震动
hydraulic shutoff valve 水力关闭阀，液压关闭阀
hydraulic similarity 水力相似性，水力相似
hydraulic slave system 液压随动系统
hydraulic slip coupling 液压补偿联轴节，液压滑动联轴节
hydraulic slope 水力比降，水力坡度
hydraulic sluicing 水力冲沙法
hydraulic sorting 水力分选
hydraulic speed governor 液压式调速器
hydraulic speed transmitter 液压转速传感器
hydraulic stability 液压稳定性，水力稳定性
hydraulic static pile pressing machine 液压静力压桩机
hydraulic stepping motor 液压步进电机
hydraulic stratification 水力分层
hydraulic strength 水硬强度
hydraulic stripping 水力剥夺，水力剥离，水力清除
hydraulic structure 水工构筑物，水工结构，水工建筑（物）
hydraulic stud tensioner shield ring 液压螺栓拉伸机屏蔽环
hydraulic stud tension 液压式螺栓张拉器【堆顶盖螺栓】
hydraulic subsiding value 水力沉降值
hydraulic suspension 水力悬浮
hydraulic sway brace 液压式防摇支柱
hydraulic system 液压系统
hydraulics 水力学
hydraulic tachometer 液压式转速表
hydraulic take-up 液压自动拉紧装置
hydraulic tapping machine 液压开孔机，液动攻牙机，油压攻丝机，液动攻丝机
hydraulic telemotor 液压遥控马达，液压遥控传动装置
hydraulic tensioner 液压紧线器，液压张紧装置，

液压拉伸机
hydraulic tensioning 水力张拉
hydraulic test facility 液力试验装置
hydraulic testing 水压试验
hydraulic test of boiler 锅炉整体水压试验
hydraulic test pump 水压试验泵
hydraulic test 水工试验，水力试验，液压试验，水压试验
hydraulic thruster 液压推杆
hydraulic thrust trip device 液压轴向推力脱扣装置
hydraulic tightening 液压夹紧
hydraulic torque converter 液力变扭器，液力耦合器
hydraulic torque 水力转矩
hydraulic traction system 液压驱动系统，液压牵引系统
hydraulic traction 水力牵引，液压牵引，水力曳引
hydraulic trailer 液压拖车
hydraulic transient 水力瞬变过程
hydraulic transmission gear 水力传动装置，液压传动装置
hydraulic transmission 液压传动
hydraulic transportation 水力输送
hydraulic transport 水力输送
hydraulic tripping device 液压跳闸装置，液压脱钩装置
hydraulic trouble 液压系统故障
hydraulic tube 液压管
hydraulic tunnel 水工隧洞，水工隧道
hydraulic turbine blade 水轮机叶片
hydraulic turbine generator 水轮发电机
hydraulic turbine 水轮机
hydraulic turning gear 液压盘车
hydraulic unit 水力单元，水力设备，液压装置
hydraulic universal tester 液压万能试验机
hydraulic uplift pressure 水力扬压力
hydraulic vacuum producer 水力真空抽气器
hydraulic valve 液压阀，水力阀
hydraulic vane pump 水力叶轮泵
hydraulic variable-speed coupling 液压变速联轴器
hydraulic variable-speed gear 液压变速齿轮
hydraulic vibration 水力振动
hydraulic winch 水力绞车
hydraulic wire cutter 液压压线钳
hydrauservo system 液压伺服系统
hydrazine analyzer 联氨表，联氨分析表，联氨分析仪
hydrazine chloride 盐酸联氨
hydrazine hydrate 水合联氨，水合肼
hydrazine injection 联氨注入
hydrazine sulfate 硫酸联氨
hydrazine 肼，联氨
HYDR＝hydraulic 液力的
hydride crack 氢脆，氢化裂纹
hydride defect 氢化损伤
hydride moderated reactor 氢化物慢化反应堆，氢化慢化反应堆
hydride moderator 氢化物慢化剂
hydride orientation factor 氢化物取向因子
hydride orientation 氢化倾向性
hydride separation 氢化物分离
hydride 氢化，氢化物
hydriding failure 氢化破损
hydriding 氢化
hydro-alternator 水轮发电机
hydro ash flusher 水力冲灰器
hydrobarometer 测深仪
hydrobin 脱水仓
hydroborate 硼氢化物
hydrocarbonaceous 含碳氢化合物的，含烃的
hydrocarbon synthesis 烃类合成
hydrocarbon 烃，碳氢化合物
hydrochart 水文图
hydrochemical map 水化学图
hydrochemical survey 水文化学调查，水化学调查
hydrochemical 水化学的
hydrochemistry 水化学
hydrochloric acid cleaning 盐酸清洗
hydrochloric acid 氢氯酸，盐酸
hydroclastic 水成碎屑的
hydrocleaning 水力清洗
hydroclimate 水文气候
hydroclimatic factor 水文气候因素，水候因素
hydroclimatology 水文气候学
hydroclone separator 水力旋流分离器
hydroclone 水力旋流分离器
hydrocol process 铁催化流化床合成过程
hydrocomplex 水利枢纽
hydroconsolidation 水固结
hydrocooler 水冷却器
hydrocracking 加氢裂化，氢化裂解
hydrocrane 水力起重机，液压起重机
hydrocyclone 水力旋风子，水力旋流器
hydro-cylinder 油缸，液压缸
hydro-development 水力开发
hydrodynamical equation of motion 流体动力学运动方程
hydrodynamical instability 动力不稳定性
hydrodynamically rough surface 水动力粗糙面
hydrodynamically smooth surface 水动力光滑面
hydrodynamical 水压的，水动力的，流体动力学的，流体的
hydrodynamic analogy 流体动力比拟，水动力模拟
hydrodynamic approximation 流体动力近似
hydrodynamic bearing 流体动力轴承，动压轴承
hydrodynamic characteristics 流体动力特性，流体动力学特性
hydrodynamic computation 流体动力计算
hydrodynamic constant torque converter 液力恒转矩变矩器
hydrodynamic coupling 液力联轴器
hydrodynamic crisis 流体动力危机
hydrodynamic differential equation 流体动力微分方程
hydrodynamic drag 水动力阻力，动水拖曳力
hydrodynamic drift instability 流体动力漂移不稳定性

hydrodynamic equation 流体动力方程
hydrodynamic form 流线型，绕流体
hydrodynamic gauge 动水压力计
hydrodynamic governor 液压调节器
hydrodynamic head 动压头，动水头，流体动压头，动水压头
hydrodynamic instability 流体动力不稳定性
hydrodynamic lag 水力延滞
hydrodynamic load 动水荷载
hydrodynamic lubrication 流体动力润滑
hydrodynamic mechanical seal 动压机械密封
hydrodynamic model 流体动力学模型
hydrodynamic power drive 动液压传动
hydrodynamic pressure 流体动压力，动水压力
hydrodynamic research 流体动力研究
hydrodynamic seal 动压密封
hydrodynamic shock 流体动力冲击
hydrodynamics of core disruptive accident 堆芯破裂事故流体力学
hydrodynamic stability 动水稳定性
hydrodynamics 流体动力学，液体动力学，水动力学
hydrodynamic test 流体动力试验
hydrodynamic torque converter 液力变矩器
hydrodynamic wave 水力波【即H波】
hydrodynamic 水力的，液压动态的，动水压的，流体动力的
hydrodynamometer 流速计
hydroejector 水力喷射器，水抽子
hydroelasticity 流体弹性理论
hydroelectric development 水电开发
hydroelectric generating set 水力发电机组
hydroelectric generating station 水力发电厂，水电站
hydroelectric generation plant 水力发电厂
hydroelectric generation set 水力发电机组
hydroelectric generation 水力发电
hydroelectric generator 水力发电机
hydroelectricity 水电
hydroelectric plant 水电厂
hydroelectric potentiality 水电蕴藏量，水力蕴藏量
hydroelectric power development 水力发电开发
hydroelectric power engineering 水力发电工程，水电工程
hydroelectric power facility and related works 水力发电设备和有关工程
hydroelectric power generation 水力发电
hydroelectric power planning 水力发电规划
hydroelectric power plant 水电站，水电厂，水力发电，水力发电厂
hydroelectric power station after dam 坝后水电站
hydroelectric power station in dam 坝内水电站
hydroelectric power station 水力发电站
hydroelectric power 水电，水力发电
hydroelectric project 水电工程
hydroelectric resources 水电资源，水力资源
hydroelectric scheme 水电开发计划
hydroelectric station at dam-toe 坝后式水电站
hydroelectric station 水电厂，水电站，水力发电站
hydroelectric synchronous machine 水轮发电机
hydroelectric system 水电系统
hydroelectric use of water 水力发电用水
hydroelectric wave generator 波浪水力发电机
hydroelectric 水电的，水力发电的
hydroelectrodynamic analogy 水电比拟
hydroelevator 水力提升器
hydroenergetics 水能学
hydro energy gradient 水能梯度
hydro energy 水能
hydro-extraction 水力提取，水力采矿
hydroextractor 脱水器
hydrofluidic 液体射流
hydrofluoric acid 氢氟酸
hydrofluorination 氢氟化作用，氢氟化
hydrofoil cascade 水力翼栅
hydrofoil 水翼
hydroform 加氢重整，液压形成
hydrofracture emplacement 水力压裂安置
hydrofracturing 水力劈裂
hydro-gasification 水汽化，加氢煤气化
hydrogel 水凝胶
hydrogen 氢气，氢
hydrogen absorption 吸氢，氢吸收
hydrogen analyser 氢分析器，氢表
hydrogenate effluent 含氢排出流
hydrogenate 氢化，氢化物
hydrogen attack 氢脆，氢蚀
hydrogen battery 氢电池
hydrogen blanket 氢覆盖层
hydrogen brittleness 氢脆性，氢脆
hydrogen bubble method 氢气泡法
hydrogen charging 充氢
hydrogen chloride 氯化氢，盐酸
hydrogen content 含氢量
hydrogen controlled covering 低氢型药皮
hydrogen-cooled generator 氢冷发电机
hydrogen-cooled reactor 氢冷堆，氢冷反应堆
hydrogen-cooled synchronous compensator 氢冷同步调相机
hydrogen-cooled turbo-alternator 氢冷汽轮发电机
hydrogen-cooled 氢冷的，氢冷却的
hydrogen cooler 氢冷却器，氢气冷却器
hydrogen cooling system 氢冷却系统
hydrogen cooling 氢冷却，氢冷
hydrogen corrosion 氢腐蚀
hydrogen cracking 氢裂
hydrogen cycle 氢循环
hydrogen damage 氢脆，氢破坏，氢损坏
hydrogen dehumidifier 氢气除湿机
hydrogen detection 氢探测，氢检测
hydrogen detector 氢探测器
hydrogen detonation 氢气爆炸，氢氧爆炸
hydrogen discharging 排氢
hydrogen draining water seal 氢气排水水封
hydrogen electrode 氢电极
hydrogen embrittlement rupture 氢脆断裂
hydrogen embrittlement 氢脆性，氢脆变，氢脆
hydrogen energy storage 制氢贮能
hydrogen engine 氢气发动机
hydrogeneration 水力发电

hydrogenerator set 水轮发电机组
hydrogenerator shaft 水轮发电机轴
hydrogenerator unit 水轮发电机组
hydrogenerator 水力发电机，水轮发电机
hydrogen exchange 氢交换
hydrogen-filled cloud chamber 充氢云室
hydrogen filling 充氢
hydrogen fluoride 氟化氢
hydrogen fuel cell 氢燃料电池
hydrogen fuel 氢燃料
hydrogen gas barrel 氢气瓶
hydrogen gas bottle 氢气瓶
hydrogen gas purity control 氢气纯度调整
hydrogen gas system 氢气系统，制氢系统
hydrogen generating station 氢气发生站，制氢站
hydrogen generation assembly control panel 制氢装置控制柜
hydrogen generation plant 制氢站
hydrogen generation 制氢
hydrogen generation station control system 制氢气站控制系统，制氢站控制系统
hydrogen generation system 制氢系统
hydrogen generator station 制氢站
hydrogen generator 氢气发生器，氢发生器
hydrogenic rock 水成岩
hydrogen-induced cold cracking 氢致冷裂纹
hydrogen inner cooled 氢内冷
hydrogen inner cooling generator 直接氢冷发电机，氢内冷发电机
hydrogen inner cooling system 直接氢冷系统，氢内冷系统
hydrogen inner cooling 直接氢冷的，氢内冷的
hydrogen ion activity 氢离子活度
hydrogen ion concentration 氢离子浓度
hydrogen ion exponent 氢离子指数
hydrogen ion index 氢离子指数
hydrogen ion indicator 氢离子指示剂
hydrogenization 氢化，加氢
hydrogenize 氢化，还原，与氢化合
hydrogen leakage detector 漏氢测定仪
hydrogen leakage 漏氢
hydrogen manifold 氢分流管，氢汇流管
hydrogen manufacturing 制氢
hydrogen meter 测氢计
hydrogen monitor 氢监测仪
hydrogen nuclei 氢核
hydrogen number 氢值
hydrogenous coal 褐煤，高水分煤
hydrogenous moderator 含氢慢化剂
hydrogen outer cooling turbogenerator 氢外冷汽轮发电机
hydrogen outer cooling 氢外冷的
hydrogen oxygen cell 氢氧电池
hydrogen oxygen fuel cell 氢氧燃料电池
hydrogen oxygen recombination 氢氧复合
hydrogen peroxide conditioning 过氧化氢调节
hydrogen peroxide 过氧化氢，过氧化氢
hydrogen pickup 吸氢
hydrogen pipe 氢气管
hydrogen piping 氢气管道
hydrogen plant 氢气站，制氢装置，制氢站
hydrogen power generation 氢能发电
hydrogen producer 制氢设备
hydrogen production plant 制氢站
hydrogen production 制氢
hydrogen purification 氢净化
hydrogen recombination 氢气复合
hydrogen recombiner 氢气复合器
hydrogen relief annealing 预防白点退火
hydrogen-rich recycle gas 富氢循环气
hydrogen scale 氢温标
hydrogen seal oil unit 氢密封油装置
hydrogen separating cleaner 氢分离洗漆器
hydrogen separator 氢分离器
hydrogen sulfide 硫化氢
hydrogen sulphide 硫化氢
hydrogen supply manifold 氢气供应主管
hydrogen supply valve 补氢阀，补氢门
hydrogen system 氢气系统
hydrogen vent valve 排氢阀，排氢门
hydrogen zeolite 氢沸石
hydrogeochemistry 水文地球化学，水文地质化学
hydrogeological chart 水文地质图
hydrogeological condition 水文地质条件
hydrogeological data 水文地质资料，水文地质数据
hydrogeological exploration 水文地质勘探
hydrogeological map 水文地质图
hydrogeological nature 水文地质特性
hydrogeological observation 水文地质观测
hydrogeological system 水文地质系统
hydrogeological zoning 水文地质分区
hydrogeological 水文地质的
hydrogeologist 水文地质工作者，水文地质学家
hydrogeology 水文地质，水文地质学
hydrogovernor 水轮机调速器，液压调速器
hydrograph ascending limb 水文过程线上升段，过程线上升段
hydrographer 水道测量学家，水文地理学家
hydrographical chart 水道图，水文图，水系图
hydrographical feature 水文地理特征
hydrographical map 水道图
hydrographical station 海洋观察站，海洋水文站
hydrographical survey 水道测量，水文测量
hydrographical table 水道测量表
hydrographical 水文地理的
hydrographic chart 水路图
hydrographic data 水道测量资料
hydrographic datum 水深基面，水深基准点，水深基准面，水文测量基准面
hydrographic feature 水文特征
hydrographic map 海洋测量图，水文地理分区图，水道图
hydrographic measurement 水文测验
hydrographic parameter 水文地质参数
hydrographic separation method 水文过程线分割法
hydrographic station 定点观测站
hydrographic survey 水道测量，水文测量
hydrographic vessel 水文测量船

hydrograph of surface runoff 地面径流过程线
hydrograph recession limb 过程线下降段
hydrograph separation method 水文过程线分割法
hydrograph separation 水文过程线分割，过程线分割
hydrograph synthesis 过程线合成
hydrography 水道测量学，水文地理学
hydrograph 水文过程线，水文图，自记水位计
hydrohematite 水赤铁矿
HYDRO＝hydrostatic 静水的
hydro-isobaric line 等水压线
hydroisobath 等水深线，地下水等水深线
hydroisohypse 等水深线，地下水等高线
hydroisopleth map 水文等值线图
hydrojet nozzle 水力喷嘴
hydrojet 水力喷射器，液力喷射器
hydrokinematics 流体运动学
hydrokinetic power transmission 流体动力传送
hydrokinetic symmetry 流体动力对称
hydrokinetics 流体动力学
hydrokinetic 液体动力学的，液压动力的
hydrokinetor 循环助动器
hydrological analogy 水文比拟，水文模拟
hydrological analysis 水文分析
hydrological and meteorologic survey 水文与气象测验
hydrological apparatus 水文仪器
hydrological atlas 水文图集
hydrological balance 水文平衡
hydrological basin 水域
hydrological bench-mark 水文用水准点
hydrological budget 水文平衡
hydrological characteristics 水文特性
hydrological computation 水文计算
hydrological condition 水文条件，水文状况
hydrological cycle 水文循环
hydrological data 水文资料
hydrological design 水文设计
hydrological divide 水文分水界
hydrological engineering 水文工程
hydrological equation 水文学方程
hydrological experiment 水文实验
hydrological exploration 水文查勘
hydrological factor 水文因素
hydrological forecasting 水文预报
hydrological frequency analysis 水文频率分析
hydrological handbook 水文手册
hydrological investigation 水文调查，水文查勘，水文研究
hydrological isolation 水文隔绝
hydrological manual 水文手册
hydrological materials 水文资料
hydrological model 水文模型
hydrological network 水文站网
hydrological observation 水文观测
hydrological prediction 水文预测
hydrological process 水文过程
hydrological property 水文特性
hydrological regime 水文情势
hydrological region 水文区域
hydrological research 水文研究
hydrological section 水文剖面
hydrological sensor 水文传感器
hydrological specification 水文规范
hydrological station 水文测站，水文观测站，水文站
hydrological survey of glacier 冰川水文调查
hydrological survey of lake 湖泊水文调查
hydrological survey of river basin 流域水文调查
hydrological survey of watershed 流域水文调查
hydrological survey 水文调查
hydrological yearbook 水文年鉴
hydrological year 水文年
hydrological 水文学的
hydrologic analogy 水文模拟
hydrologic balance 水分平衡
hydrologic benchmark 水文用的水准点，水文参证点
hydrologic computation 水文计算
hydrologic cycle 水分循环
hydrologic data 水文资料，水文数据
hydrologic effect 水文效应
hydrologic exploration 水文勘查
hydrologic forecasting centre 水文预报中心
hydrologic forecasting 水文预报
hydrologic investigation 水文查勘
hydrologic map 水文图
hydrologic model 水文模型
hydrologic observation 水文观测
hydrologic process 水文过程
hydrologic property 水文特性
hydrologic record 水文记录
hydrologic regime 水文状况
hydrologic region 水文区域
hydrologic routing 水文演算
hydrologic sensor 水文等值线图，水文探测设备
hydrologic service 水文服务机构
hydrologic storm 水文暴雨
hydrologic survey 水文查勘
hydrologist 水文工作者，水文学家
hydrology of ground water 地下水水文学
hydrology 水文学，水文地理学
hydrolysate 水解产物
hydrolysis constant 水解常数
hydrolysis current 水解电流
hydrolysis effect 水解效应
hydrolysis phenomena 水解现象
hydrolysis product 水解产物
hydrolysis 水解，水解作用
hydrolyte 水解质
hydrolytic action 水解作用
hydrolytic bacteria 水解细菌
hydrolytic constant 水解常数
hydrolytic stability 液压稳定性，水解稳定性
hydrolytic tank 水解槽，水解箱
hydrolytic treatment 水解处理
hydrolytic 水解的
hydrolyze 水解
hydromagnetic DC converter 磁流体直流变流器
hydromagnetic oscillation 磁流体振荡
hydromagnetics 电磁流体动力学

hydromagnetic 磁流体的，水磁的
hydromanometer 流体压力计
hydroman 液压操作器，液压机械手
hydromechanical actuator 液压机械执行机构
hydromechanical control system 液压调节系统，液压机械式控制系统
hydromechanical 液压机械的，流体力学的，水力机械的
hydromechanics 流体力学，水力学，液压流体力学，流体力学
hydrometallurgy 湿法冶金学，水冶
hydrometeological forecast 水文气象预报
hydrometeorological relationship 水文气象关系
hydrometeorology information 水文气象资料
hydrometeorology 水文气象学，水文气象
hydrometeor 水汽凝结物，水汽凝结体
hydrometer analysis 液体比重计分析法
hydrometeorological data 水文气象资料，水文气象数据
hydrometeorological 水文气象学的
hydrometer 比重计，浮秤，流速计，液体比重计，(液体）比重计
hydrometric aerial ferry 水文测验架空缆道，水文架空缆道
hydrometrical station 测流站
hydrometric basic station 水文基本测站
hydrometric boat 水文测艇
hydrometric cableway 水文测验架空缆道
hydrometric dingey 水文测艇
hydrometric float 水文测验浮标
hydrometric forecast 水文预报
hydrometric measurement 水文测验
hydrometric section line 水文测验断面线
hydrometric section 水文测验断面
hydrometric station control 水文观测站控制，水文测站控制
hydrometric station 水文测量站，水文站，水文测站
hydrometric survey 水文测量
hydrometrist 水文测验工作者
hydrometry 水文测验，水文测验学，液体比重测定法
hydromix dust conditioner 加水卸灰机，湿式搅拌器
hydromodulus 流量模数
hydromorphic soil 水成土
hydromorphous process 水成过程，水渍过程
hydromotor 液压马达
hydro peening 高压水喷丸处理，高压水去氧化皮
hydroperiod 水文周期，土壤地区积水的时期
hydrophile 亲水物
hydrophilic aggregate 亲水集料
hydrophilic colloid 亲水胶体
hydrophilic dust 亲水性粉尘
hydrophilicity 亲水性
hydrophilic polymer 亲水聚合物
hydrophilic substance 亲水物质
hydrophilic 亲水的，亲水性的
hydrophilous nature 亲水性质
hydrophilous plant 喜水植物
hydrophobe 疏水物，疏水的
hydrophobic admixture 憎水性外加剂
hydrophobic aggregate 憎水集料
hydrophobic colloid 疏水胶体，憎水胶体
hydrophobic compound 疏水化合物
hydrophobic dust 疏水性粉尘
hydrophobic hydration 疏水水合作用
hydrophobicity 憎水性，疏水性
hydrophobic nature 疏水性
hydrophobic portland cement 憎水水泥
hydrophobic substance 疏水物质
hydrophobic 疏水的
hydrophone 水下测听器，漏水检查器
hydrophonic detector 水声探测器
hydrophotometer 水下光度计
hydrophysical law 水文物理定律
hydrophysical process 水文物理过程
hydrophysics 水文物理学
hydro plant with dam 有坝电站
hydropneumatic riveter 液压气动铆钉机
hydropneumatic system 液压气动系统
hydropneumatics 液压气动学
hydropower base 水电基地
hydropower complex or hydroelectric project 水电枢纽工程
hydropower development 水电开发
hydropower economy 水能经济
hydropower generation 水力发电
hydropower plant 水电厂，水电站，水力发电厂，水力发电站
hydropower project 水力发电枢纽，水力发电项目，水电工程，水电枢纽，水力发电工程
hydropower station 水电厂，水电站，水力发电厂，水力发电站
hydropower tunnel 水力发电隧洞
hydropower utilization 水能利用
hydropower 水电，水力，水力发电，水能
hydro project 水利工程，水利枢纽
hydropsis 海洋预报
hydroquinone 氢醌，对苯二酚
hydroscience 水科学
hydroscopic coefficient 吸水系数
hydroscopic water coefficient 吸湿水系数
hydroscopic water 吸湿水，吸着水
hydroscopic 吸湿的，收湿的
hydroseparator 水力分离器
hydrosequence 水文序列
hydrosol 脱水溶胶，水溶胶
hydrosphere 水界，水圈，水层
hydrostatic accelerometer 流体静力加速仪
hydrostatical balance 液压比重计
hydrostatical overpressure test 超静水压试验
hydrostatical pressure ratio 静水压力比，静水压力系数
hydrostatical uplift 静水上托力，静水扬压力
hydrostatical 静水力学的，流体静力学的，水静力的，静液压的
hydrostatic bearing at impeller wearing ring 在叶轮磨损坏处的流体静力轴承
hydrostatic bearing 流体静力轴承，静压轴承
hydrostatic burst testing 水压爆破试验

hydrostatic cold leak test 冷态水力检漏试验
hydrostatic deformation test （验证性）水压变形试验
hydrostatic deformity 验证性水压试验
hydrostatic drive 静液压传动
hydrostatic end force 流体静力端的力
hydrostatic equation 流体静力学方程
hydrostatic equilibrium 流体静力平衡
hydrostatic excess pressure 超流体静压力
hydrostatic force 流体静压力
hydrostatic fuel-level gauge 静液压式燃料量计
hydrostatic gauge 流体静压压力计
hydrostatic head 静（水）压头，静压头，（流体）静压头，静压力水头
hydrostatic joint 水压插承接头
hydrostatic non-contact bearing 静压非接触轴承
hydrostatic non-contact seal 静压非接触密封
hydrostatic pressure ratio 流体静压率，静水压力系数
hydrostatic pressure test on tubes of metals 金属管液压试验
hydrostatic pressure test 静水压试验，液压试验，流体静压试验，水压试验
hydrostatic pressure 水压，液压，（流体）静压力，静水压力
hydrostatic press 水压机
hydrostatic radial bearing 静压径向轴承
hydrostatic seal 静压密封
hydrostatic sodium bearing 钠润滑静压轴承
hydrostatic stability 流体静稳定性
hydrostatic state of stress 静水应力状态
hydrostatics test 流体静力试验
hydrostatic structure 水工结构
hydrostatics 流体静力学
hydrostatic test 静水压试验，流体静力学试验，水压试验
hydrostatic transmission 静液压传动，液压传递，液压传动
hydrostatic 水静压，流体静力学的，流体静力的，静水力学的，液压静力的
hydrostat 恒湿仪
hydrostorage 水能储蓄
hydro-structure 水工结构
hydrotasimeter 电测水位指示器，电测水位计
hydrotechnics 水工学，水利技术
hydrotechnological protection 水工保护
hydrotest application 水压试验申请
hydrotest pressure 试验压力，水压试验压力
hydrotest pump 液压试验泵，水压试验泵
hydrotest 水压试验
hydrothermal balance 水热平衡
hydrothermal condition 水热条件
hydrothermal metamorphism 热液变质
hydrothermal method of curing 湿热养护
hydrothermal process 热液作用，水热过程
hydrotimeter 水硬度计
hydrotransport 水力输送
hydrotreatment 氢化处理
hydroturbine 水轮机
hydrous mica 含水云母
hydrous water 水合水，结合水，结晶水
hydrous 含结晶水的，水合的，水化的，水状的，含氢的
hydrovalve 水阀，水龙头
hydroviscous assistant braking 液体黏性辅助制动
hydroviscous coupling 流体黏性联轴器
hydroviscous drive soft starter 液黏软启动
hydroviscous driving 液黏传动
hydroviscous dynamometer 液黏测功机
hydroviscous speed-regulating clutch 液体黏性调速离合器
hydroviscous variable speed clutch 液黏调速离合器
hydroviscous variable speed drive 液黏变速传动，液黏调速
hydroviscous 液黏的，液力黏滞的
hydroxide alkalinity 氢氧根碱度
hydroxide anion 氢氧离子，羟离子
hydroxide ion 氢氧离子，羟离子
hydroxide moderator 氢氧化物慢化剂
hydroxide radical 羟基，氢氧根
hydroxide-regenerated form 氢氧根再生型
hydroxide 氢氧化物
hydroxyacetic acid 羟基乙酸，甘醇酸
hydroxyacetic-formic acid 羟基醋甲酸
hydroxylamine hydrochloride 盐酸羟胺
hydroxylamine sulfate 硫酸羟胺
hydroxylamine 羟胺，胲
hydroxyl-carboxylic acid 羟基羧酸
hydroxyl ion 氢氧离子，羟离子
hydroxyl 羟基
hyetal coefficient 雨量系数
hyetograph 雨量记录表，雨量计，雨量图
hygiene standard 卫生标准
hygiene 卫生学，保健学
hygienical evaluation 卫生学评价
hygienic basis 卫生基础
hygienic characteristic 卫生特性
hygienic chemical 卫生化学的
hygienic evaluation 卫生学评价
hygienic feature 卫生特征
hygienic standardization 卫生标准化
hygienic standard 卫生标准
hygienics 卫生学
hygroautometer 自记湿度计
hygrogram 湿度自记曲线，湿度图
hygrograph 自动湿度记录计，自动湿度计，湿度记录仪，湿度计，湿度仪
hygrology 湿度学
hygrometer 湿度计，湿度表，测湿法，湿度测定法
hygronom 湿度仪
hygrophyte 湿生植物
hygroscope 湿度计，湿度仪，湿度测量器，湿度表，测湿器
hygroscopic absorption 吸湿作用
hygroscopic coefficient 吸湿系数
hygroscopic effect 吸湿作用
hygroscopicity 吸湿性，水湿性，吸湿度，吸湿率，吸水性，吸潮性，润湿度
hygroscopic material 吸湿性材料

hygroscopic moisture 吸湿含水量，吸湿水
hygroscopic nature 吸湿性，收湿性
hygroscopic nucleus 吸湿性核
hygroscopic salt 吸湿盐，潮解盐，吸湿性盐
hygroscopic soil water 吸湿土壤水
hygroscopic 吸湿的，吸水的，湿度计的
hygro-stability 湿稳性
hygrostat 湿度恒定器，恒湿器
hygrothermograph 温湿计，温湿自记仪
hygrothermostat 恒温恒湿器
hyjector 高压水力除灰器，高压水力除渣器
hymatomalenic acid 棕腐殖酸
hymograph 示波器
hypabyssal rock 半深成岩，浅成岩
HYPCL＝hypochlorite 氯化物
hyperacoustic 超声波的，超声的
hyperbola 双曲线
hyperbolical coordinates 双曲坐标
hyperbolic cooling tower 双曲冷却塔
hyperbolic coordinates 双曲坐标
hyperbolic cosine 双曲余弦
hyperbolic curve 双曲线
hyperbolic decay 双曲线衰减
hyperbolic disc 双曲线叶轮
hyperbolic flow equation 双曲流方程
hyperbolic form 双曲线型
hyperbolic function 双曲线函数
hyperbolic motion 双曲线运动
hyperbolic paraboloid shell 双曲抛物面薄壳
hyperbolic paraboloid 双曲抛物面
hyperbolic shell 双曲线薄壳
hyperbolic type cooling tower 双曲线形冷却塔
hyperbolic 双曲线的
hyperboloid of revolution 回转双曲面
hyperboloid 双曲面
hyper-concentration 超浓缩
hypercritical flow regime 过临界流态，高超临速流态
hypercritical flow 超临界水流
hyperdisk 管理磁盘
hyperelasticity 超弹性
hyperelastic law 超弹性规律
hyperelliptic 超椭圆的
hyper-eutectic 过共晶的，过共晶
hyper-eutectoid steel 过共析钢
hyper-eutectoid 过共析
hyperfiltrate 超滤
hyperfiltration membrane 超滤膜
hyperfiltration 超滤，超滤法
hyperfine structure quantum number 超精细结构量子数
hyperflow lift system 密相床层中催化剂输送系统，密相气升系统
hyper-fluidized bed 超流化床
hyperforming 超重整
hyperfrequency wave 微波，超高频波
hyperfrequency 超高频
hypergeometric function 超几何函数，超比函数
hypergeometric series 超比级数
hypergeometric 超几何的，超比的
hyper-high frequency 超高频
hyperinflation 极度通货膨胀
hypermedia 超媒体
hyperoxide 过氧化物
hyper-plane 超平面
hyperplasy 增生，增殖
hypersensor 超敏感元件，过流或过电压保护敏感元件
hypersil 一种磁性合金
hypersonic flow 高超音速流动，超声速流
hypersonic frequency 特超声频率
hypersonic speed 超音速
hypersonic tube 高超音速风洞
hypersonic 超声的，高超音速的
hypersorption 超吸附法
hyperspace 多维空间
hyperstatical structure 超静定结构
hyperstatical 超静定（的）
hyperstatic structure 超静定结构
hyperstatic system 超静定体系
hypersynchronous motor 超同步电动机，有补偿的异步电动机
hypersynchronous 超同步的
hypertape control unit 快速磁带控制部件
hypertape drive 快速磁带机，快速磁带驱动器
hypertext transfer protocol 超文本传送协议
hypertext 超级存储系统，超文本
hyperthermic temperature regime 高热温度状况
hyperthermic （土温的）超热状况
hyperthermometer 超高温温度计
hypertonia 高渗压，压力过高
hypertonic solution 高渗溶液
hypervelocity free flow 超高速自由流
hypervisor 管理程序
hyphen 连字符
hypidiomorphic granular 半自形粒状
hypidiomorphic texture 半自形结构
hypo acid 次酸
hypobatholitic zone 深成岩基带
hypoborate 连二硼酸盐
hypocenter （地震）震源
hypocentral region 震源域
hypochloric acid 次氯酸
hypochlorite 次氯酸盐
hypochlorous acid 次氯酸
hypocycloid 内摆线，圆内旋轮线
hypoeutectic 亚共晶的
hypo-eutectoid steel 亚共析钢
hypofiltration 深成渗透作用
hypogee 岩洞建筑
hypogene action 内力作用
hypogene rock 深成岩
hypogene water 深成水
hypogene 上升的，深成
hypogenic 上升生成的，深成的
hypogeum 岩洞建筑
hypoid bevel gear 准双曲面斜齿轮
hypolimnile 湖下层
hypolimnion 湖下层
hyponeuston 次飘浮生物
hypophosphate 连二磷酸盐
hypophosphite 次磷酸盐

hyposulfate 连二硫酸盐
hyposynchronous resonance 次同步谐振
hyposynchronous 低于同步的，次同步的
hypothecated assets 抵押资产
hypothermal 低温的，降温的
hypothermophilous 喜低温的，适低温的
hypothesis of limiting fragmentation 极限碎裂假设
hypothesis 前提，假说，假设
hypothetical accident 假想事故
hypothetical core disruptive accident 假想堆芯破裂事故
hypothetical event 假想事件
hypothetical fluid velocity 假想流体速度
hypothetical hydrograph 设计水文过程线
hypothetical 假想的，假定的，假说的，假设的
hypsogram 电平图
hypsographical map 等深线图
hypsometer 沸点测高计，沸点气压计，气压测高计
hypsometric formula 测高公式
hypsometric method （地图的）分层设色法
hypsometry 沸点测高法，测高术
hysteresigraph 磁滞回线记录仪
hysteresimeter 磁滞测定仪，磁滞计
hysteresis active current 磁滞有功电流
hysteresis advance 磁滞超前
hysteresis band 磁滞带，滞环
hysteresis characteristic 磁滞特性，磁滞特性曲线
hysteresis clutch 磁滞离合器
hysteresis coefficient 磁滞系数，滞后系数
hysteresis constant 磁滞常数，滞后常数
hysteresis cope 磁滞回线仪
hysteresis cycle 磁滞回线，磁滞循环，磁滞周期，滞后曲线，磁滞曲线
hysteresis damping 迟滞阻尼，滞后阻尼
hysteresis distortion 磁滞失真
hysteresis effect 迟滞效应，磁滞效应
hysteresis error 磁环误差，磁滞误差，回程误差，回差，滞后误差，滞环误差
hysteresis form factor 磁滞形状系数
hysteresis free 灭磁滞，无磁滞
hysteresis graph 磁滞曲线图
hysteresis loop 滞后回线，磁滞回线，磁滞环，滞后环，迟滞回线，滞回环
hysteresis loss coefficient 磁滞损耗系数，滞后系数
hysteresis loss 磁滞损失，磁滞损耗
hysteresis material 磁滞材料
hysteresis meter 磁滞计，磁滞测量仪
hysteresis modulus 滞后模量
hysteresis motor 磁滞电动机
hysteresis of phase transformation 相变滞后
hysteresis power loss 磁滞功率损耗
hysteresis ring 磁滞回环
hysteresis synchronous motor 磁滞同步电机
hysteresis unit 滞后模拟部件
hysteresis voltage 滞后电压
hysteresis 延滞性，迟滞，滞后作用［现象］，磁滞（现象），滞变，滞环
hysteretic angle 磁滞角
hysteretic error 磁滞误差，滞环误差
hysteretic lag angle 磁滞角
hysteretic phase transition 相变滞后现象
hysteretic 磁滞的，滞后的
hysterset 功率电感调整
hythergraph 温湿图
HYUNDAI E&C（Hyundai Engineering & Construction Co.，Ltd）（韩国）现代工程建设公司
Hz＝hertz 赫兹

I

I-action(integral action) 积分作用
IAEA inspection 国际原子能机构监督
IAEA(International Atomic Energy Agency) 国际原子能机构
IA(implementation agreement) 执行协议
IALA(International Association of Lighthouse Authorities) 国际航标协会
IAMAP (International Association of Meteorology and Atmospheric Physics) 国际气象与大气物理协会
IAM(International Association of Machinists) 国际机械师协会
IATA(International Air Transport Association) 国际航空运输协会
IAWE(International Association for Wind Engineering) 国际风工程学会
I-bar 工字钢
I-beam bar 工字钢
I-beam steel 工字钢
I-beam 工字梁，工字钢
IBL(internal boundary layer) 内边界层
IBRD(International Bank of Reconstruction & Development) 国际复兴与开发银行，世界银行
ICAO(international civil aviation organization) 国际民用航空组织
I/C attendant room 热工值班室
ICBC(Industrial and Commercial Bank of China) 中国工商银行
ICB(international competitive bidding) 国际竞争性招标
ICBV(interceptor control bypass valve) 中调门旁路门
ICC(International Chamber of Commerce) 国际商会
ice 结冰，冰
ice abrasion 冰磨蚀，覆冰，积冰
ice adiabat 冰绝热线
ice admixture 冰掺和料【预冷混凝土用】
ice age 冰期
ice apron 冰挡，冰覆面，挡冰板
ice atlas 冰图
ice auger 冰钻
ice avalanche 冰崩
ice bank 冰块组
ice barrier 冰障
ice bath 冰点槽，冰浴，冰槽

iceberg 冰山
ice boating 冰帆运动
ice boom 防冰栅
icebox effect 冰箱效应
icebox 冰箱，严寒地带
ice-breaking isolator 破冰式隔离开关
ice breakup 解冻，破冰，开冻，开河
ice-cap climate 冰盖气候
ice-cap 冰覆盖的，冰盖
ice chute 泄水槽
ice-coated power cable 裹冰电缆
ice-coated transmission line 裹冰输电线
ice-coated 冰封着的，覆冰的
ice coating 覆冰
ice-cold water 冰冷水，零度水
ice condensating containment 冰冷凝式安全壳
ice condenser containment system 冰冷凝器安全壳系统
ice condenser containment 冰冷凝安全壳
ice condenser module 冰冷凝舱
ice condenser reactor containment 冰冷凝器反应堆安全壳
ice condenser system 冰冷凝器系统
ice condenser 凝冰器，冰冷凝器
ice condition 冰情
ice cone 积冰
ice control measure 防冰措施
ice-covered 冰封着的，冰覆盖的
ice cover 冰盖
ice crack 冰裂
ice-crusted ground 积冰地面
ice crystal 冰晶
ice damage 冰灾
iced containment 置冰安全壳
ice deposit 积冰
ice disposal 冰层处理
ice-dome 冰穹，冰丘
ice drop off on adjacent span 邻挡脱冰
ice drop 脱冰
ice evaporation level 冰汽转相高度，高空冰汽转相高度
icefall 冰崩
ice floe 大浮冰，凌汛
ice flow control 冰流控制
ice flower 冰花
ice fog 冰雾
ice formation 结冰
ice-free harbor 不冻港
ice-free harbour 不冻港
ice-free port 不冻港
ice hail 小雹
ice harbour 冻港
icejam stage 冰坝水位
ice jam 冰塞
ice load 覆冰负载，覆冰负荷，冰荷载，冰载
ice logging 冰塞
ice melting current （导线）溶冰电流
ice melting point 冰溶点
ice mixing 加冰拌和
ice movement 冰川运动
ice nucleus 冰核
ice pellet 小冰雹
ice plug 冰塞
ice point 冰点
ice pressure 冰压力
ice prevention 防凌
ice prism 冰针，飘降冰晶
ice quake 冰震，冰崩
ice rain 冻雨
ice-removal salt 防冻盐
ice runoff 融冰径流
ice saturation 冰饱和
ice sensor 结冰传感器
ice shedding 脱冰
ice sheet 冰盖
ice situation 冰情
ice spar 透长石
ice storm 冰暴
ice stream 冰川，冰河
ICE(The Institution of Civil Engineers) 土木工程师协会【英国】
ice thickness 覆冰厚度
ice tongue 冰舌
ice trap 冰阱，冷阱
ice unloading 脱冰
ice wind tunnel 结冰风洞
ice yachting 冰帆运动
I channel I信道
ichnography 平面图法
icicle 毛刺，焊接时管子接头内的上部金属突出物，冰柱
icing index 积冰指数
icing intensity 积冰强度
icing level 积冰高度
icing 冰丘
I&C(instrument and control) 仪表与控制，热控
IC(integrated chip) 集成芯片
IC(integrated circuit) chip 集成电路芯片
I&C island(instrumentation and control island) 仪表与控制岛，I&C岛
I/C laboratory 热工试验室
IC menu 初始工况菜单
I-column 工字柱
iconolog 光电读像仪
iconometer 测距镜
iconoscope 光电显像管，光电摄像管
I controller(integral controller) 积分控制器，积分调节器
ICONTT(international conference on nuclear technology transfer) 国际核技术转让会议
icon 图像，(电脑)图标，图示影像
ICP(international comparison project) 国际比较项目
ICRC(International Committee of the Red Cross) 国际红十字会
ICRP(International Commission on Radiological Protection) 国际放射性防护委员会，国际辐射防护委员会
ICS(International Chamber Shipping) 国际航运公会
ICV(intermediate control valve) 中压控制阀
ICV(interceptor control valve) 中调门

ICWE（International Conference on Wind Engineering） 国际风工程会议
IC 中压调门控制【DCS 画面】
IDA(International Development Association) 国际开发署，国际开发协会
IDB(Inter-American Development Bank) 泛美开发银行，美洲开发银行【总部设在华盛顿】
IDC(interest during construction) 建设期利息
IDD(International Direct Dialing) 国际直拨【电话】
ideal air 理想空气
ideal atmosphere 理想大气
ideal black body 绝对黑体，理想黑体
ideal body 理想体
ideal boundary 理想边界
ideal burning 理想燃烧
ideal circulation 理想环流
ideal climate 理想气候
ideal code 理想码
ideal coding 理想编码
ideal combustion 理想燃烧
ideal condition 理想条件，理想状态，标准条件
ideal constraint 理想约束
ideal cycle 理论循环，理想循环
ideal DC machine 理想直流电机
ideal dielectric 理想介质
ideal drag 理想阻力
ideal efficiency 理想效率，理论效率
ideal energy grade line 理想能坡线
ideal enthalpy drop 理想焓降，等熵焓降
ideal exhaust velocity 理想排气速度
ideal flow 理想流动
ideal fluid theory 理想流体理论
ideal fluid 理想流体
ideal focusing field 理想聚焦场
ideal formula 理想公式，标准公式
ideal frequency domain filter 理想频域滤波器
ideal function 理想函数
ideal gas constant 理想气体常数
ideal gas equation of state 理想气体状态方程
ideal gas law 理想气体定律
ideal gas state equation 理想气体状态方程
ideal gas temperature scale 理想气体温标
ideal gas 理想气体
ideal hinge 理想铰
ideal investment condition 理想的投资条件
idealized characteristic 理想化特性
idealized elastic continual mass 理想弹性连续体
idealized fluidization 理想流态化
idealized fluid 理想流体
idealized fragmental mass 理想碎屑体
idealized stratified bed 理想分层床
idealized system 理想系统
ideal jet velocity 理想射流速度，理想速度
ideal line 理想线路，无损耗线路
ideal liquid 理想液体
ideal load curve 理想负荷曲线
ideal load 理想负荷，理想负载
ideally-elastic member 理想弹性构件
ideal mixing stage model 理想混合级模型
ideal model 理论模型
ideal network 无损耗网络
ideal paralleling 理想并车，（发电机）准同步并车
ideal plasma 理想等离子体
ideal power output 理想功率输出
ideal power 理想功率
ideal reactor 理想反应器
ideal receiver 理想接收机
ideal rectifier 理想整流器
ideal refrigeration cycle 理想制冷循环
ideal refrigerator 理想制冷机
ideal regenerative cycle 理想回热循环
ideal sea level 理想海水面
ideal separation factor 理想分离系数
ideal source 理想电源
ideal superconductor 理想超导体
ideal synchronizing 理想同步
ideal synchronous machine 理想同步电机
ideal thermal cycle efficiency 理想热循环效率
ideal thermodynamic efficiency 理想热力学效率
ideal thrust coefficient 理想推力系数，理论推力系数
ideal thrust equation 理论推力方程
ideal time domain filter 理想时域滤波器
ideal transformer 理想变压器
ideal value 理想值
ideal Venturi 理想文丘里管
ideal viscous fluid 理想黏性流体
ideal voltage amplifier 理想电压放大器
ideal working substance 理想工质
ideal 理想的，完美的，想法
I-demodulator I 信号解调器
idem 同上，同前
identically priced 价格相同的
identical order 同序
identical product 同一批产品
identical 相同的，相等的，恒等的
identifiable assets 可识别资产
identifiable 可辨认的
identification and processing report of situations adverse to quality 对质量有害的情况的发现和处理报告
identification card 身份证，标识卡
identification code 识别码，编码
identification division 识别部分，标识部分
identification equipment 识别设备，识别装置
identification light 标志灯
identification marker 识别标志【射线照相】，识别符号
identification number 标识号，识别号，身份证号码，成套设备编号，编号
identification of phase 物相鉴定，相位鉴定
identification of position 位置识别
identification point 识别点，标识点
identification requirement 辨别要求
identification section 识别部分，标识部分
identification stamp 专用戳记
identification switch 识别开关
identification system 识别系统
identification test 鉴定试验，鉴别试验
identification threshold 识别阈

identification 辨识，识别，鉴别，鉴定，证实，符号，标记，辨别，标识，标志，身份证明，认同
identified leakage 识别泄漏，查明的泄漏
identified standard 确定的标准
identifier circuit 识别电路
identifier 标识符，标识号，识别符号，识别器，鉴定人
identify code 识别码
identifying brand 厂家检验标证
identify 辨别，辨认，鉴定，使等同
identity card 身份证
identity certificate 身份证
identity document 身份证件
identity matrix 单位矩阵
identity of views 观点一致
identity tag 识别标志
identity （质量的）均匀性，同一性，本体，恒等式，恒等
id est 即，也就是，换言之
ID fan 引风机
IDF(induced draft fan) 引风机【DCS画面】
id. ＝idem 同前，同上
ID＝ identification 标志
ID(induced draft) 抽风，引风
idioelectric 非导体的，能摩擦生电的，非导体
idiostatic method 同位差连接法，同势差连接法
idiostatic 等位差的，同电位差的
I-display I型显示器
idle bar 死导条，启动笼条，无载导条
idle battery 无载电池，闲置电池
idle boiler 停用锅炉
idle capacity 空转功率，备用功率，空载耗汽量，备用容量，空载功率，闲置生产能力
idle circuit 空载电路，空闲电路
idle coil 空置线圈，死线圈
idle component 无功分量
idle condition 空载工况，空载条件
idle contact 空触点
idle control 惰转调节
idle cost 窝工费用
idle current 空转电流，空载电流，无功电流，无效电流
idle equipment 闲置设备
idle frequency 中心频率，未调制频率
idle fund 闲散资金，闲置资金，游资
idle gear 惰轮，空转轮，中间齿轮
idle hour 停机时间，惰走时间
idle interval （换向器）闭锁期间
idle line 空线，空转
idle load 空载
idle loss 空转损失
idle money 闲散资金
idle motion 空转，空载，空载运转
idleness 空闲时间，空闲率
idle operator lamp 空位表示灯
idle period 停运时间，无功周期
idle position 空转位置，空载位置
idler brace 空转柄
idler bracket 托辊架
idler car 空车
idle revolution 空转转速
idler load rating 托辊承载能力
idler pulley 空转轮，惰轮
idler revolution 空转轮转速
idler spacing 托辊间距
idle running time 空转时间
idle running 惰行，空负荷运行，空转
idler valve 空载阀，止回阀
idler wheel 空转轮，惰轮
idler 空转轮，惰轮，托辊
idle signal unit 闲置信号单元
idle-speed adjustment 怠速调整，空转调整
idle state 闲置状态，空载状态，空载转速
idle stroke 空行程，空冲程
idle time 空等时间，停机时间，惰转时间，空载时间，惰走时间，故障时间，中断运转时间
idle trunk 空中继线
idle unit 空转机组，备用机组，闲置机组，停运装置
idle wheel 惰轮
idle wire 空线
idle work 虚功
idle 空转的，空载的，闲置的，备用的，无效的，空转，窝工，闲置
idling adjustment 空转调整
idling condition 空载工况，空载状态，空转工况
idling current 空载电流
idling curve 惰走曲线
idling cut-off 空载切断
idling frequency 无效频率，空载频率
idling loss 空载损失，低速损失，空转损失
idling running 空载运行
idling speed 惰转速度，空转速度，空转转速
idling time 停机时间，空转时间，惰走时间
idling torque 空转扭矩
idling 空转，空载，惰转，闲置，无效，窝工，无功
IDMT(inverse definite minimum time) 最短动作时间的反时限特性
idometer 测量仪表，探测仪
IDR 印尼盾
ID 标志，标识【DCS画面】
IEA(International Energy Agency) 国际能源机构【属OECD】
I/E bonded charge 转境费/过境费
IEC Conformity Assessment for Electrotechnical Equipment and Components 国际电工委员会电工产品合格测试与认证组织
IECEE(IEC Conformity Assessment for Electrotechnical Equipment and Components) 国际电工委员会电工产品合格测试与认证组织
IECEX（International Electrotechnical Commission System For Certification To Standards Relating To Equipment For Use In Explosive Atmospheres) 国际电工委员会防爆电气产品认证体系
IEC(International Electric Company) 国际电气公司【美国】
IEC（International Electrotechnical Commission） 国际电工（技术）委员会
IECQ(IEC Quality Assessment System for Electronic Components) 电子元器件质量评定体系

IECS（International Electrotechnical Commission Standard） 国际电工技术委员会标准
IEC Standard 国际电工技术委员会标准
IEC system for conformity testing and certification of wind turbines 风电机组的合格试验和认证的IEC系统
IED configuration description 智能电子设备配置描述
IED（intelligent electronic device） 智能电子设备
IED-parameter set 智能电子设备参数集
IEEE（Institute of Electrical and Electronics Engineers） 电气和电子工程师协会【美国】
IEE（Institute of Environmental Engineers） 环境保护工程师协会
IEE（Institution of Electrical Engineers） （英国）电气工程师学会
IEF（internal experience feedback） 内部经验反馈
i. e. =idest 即，就是，也就是，换言之
IEP 爱尔兰镑
IETF（Internet Engineering Task Force） 因特网工程特别工作组
IF adaptive slope equalizer 中频自适应斜率均衡器
IFAN 国际标准用户联合会
IFB（invitation for bid） 招标，招标书，招标邀请函
IFCC（incremental fuel cycle cost） 增量燃料循环成本
IFC（International Finance Corporation） 国际金融公司
IFDH（in-service forced derated hour） 强迫降低出力运行小时
IF（intermediate frequency） 中频
IFIs（International Financing Institutions） 国际金融机构
IFT（ignition fuel trip） 点火燃料跳闸
IF transformer 中频变压器
IGBT（insulated gate bipolar transistor） 绝缘栅双极型晶体管
IGCC generator unit 燃气蒸汽联合循环发电机组
IGCC（integrated gasification combined cycle） 整体煤气化联合循环（技术）
IGCC plant 整体煤气化联合循环电厂
IGCC power generation 整体煤气化联合循环发电
IGCT（integrated gate commutated thyristor） 集成门极换流晶闸管，集成门极换向晶闸管
igneous intrusion 岩浆侵入
igneous magma 岩浆
igneous rock 火成岩
igneous 火成的
ignifluid boiler 烧可燃性流体的锅炉
ignifluid process 烧可燃性流体法
IGN＝ignition 点火装置
ignitability 着火性，可燃性
ignitable 可燃的
ignited residue 烧余残渣
igniter 点火器，点火极，触发器
ignite 点火，点燃，引燃
ignitibility 可燃性
igniting circuit 点火电路
igniting fuse 传爆信管
igniting point 着火温度，着火点
igniting torch 点火炬，点火喷燃器
ignition accumulator 点火蓄电池
ignition advance 提前点火
ignition-aid burner 点火喷燃器
ignition anode 点火阳极，引燃阳极
ignition arch 点火拱
ignition belt 燃烧带，引燃带，卫燃带
ignition charge 导火炸药，点火器充电
ignition circuit 点火电路，引燃电路
ignition coil 点火线圈，发火线圈
ignition current 点火电流
ignition device 点火装置
ignition electric nozzle 点火电嘴
ignition energy 点火能量
ignition equipment 点火装置
ignition explosive 导火炸药
ignition-failure tripping device 点火故障跳闸装置
ignition flame off 点火器火焰失去
ignition flame on 点火器火焰存在
ignition fuel trip 点火燃料跳闸
ignition hazard 着火危险
ignition heat 着火热，燃烧热
ignition inhibitor 延爆剂，阻爆剂
ignition lag 延期引爆
ignition lock 点火开关
ignition loss 灼失量，灼烧损失，点火损失
ignition magneto 点火用磁电机
ignition muffle 点火马弗炉
ignition pin 点火电极，发火针
ignition plug 火花塞
ignition point 着火温度，着火点，燃点
ignition powder 点火药
ignition preparation 点火准备
ignition range 着火范围
ignition residue 烧余残渣
ignition scope 点火检查示波器
ignition spark detector 点火火花探测器
ignition spark 点火火花，引燃火花
ignition speed 点火速度
ignition switch 点火开关
ignition system of converter valve 换流阀触发系统
ignition system 发火系统，点火装置，点火系统
ignition temperature 着火温度，燃点，点火温度，着火点
ignition test 点火试验
ignition torch 点火炬，点火棒
ignition voltage 点火电压，引燃电压
ignition wire 点火线
ignition 着火，点火，引燃，发火装置，起爆
ignitor control 引燃极控制
ignitor discharge 点火极放电
ignitor electrode 引燃极
ignitor flame off 点火器火焰失去
ignitor flame on 点火器火焰存在
ignitor fuel trip 点火器燃料跳闸
ignitor pad 点火药包
ignitor 点火器，点火装置，点火极，引火器

ignitron　点火管，引燃管，放电管，水银半波整流器
ignorance factor　不知因子
ignorance　不知，无知
ignore　忽视，不理睬
IGO(inter-government organization)　政府间国际组织
IGSCC(intergranular stress corrosion crack)　晶界应力腐蚀裂纹
IGSC(intergranular stress corrosion)　晶间应力腐蚀
IIB(International Investment Bank)　国际投资银行
II(independent inspection)　独立检查
impulsion　冲击，冲动，推动，推力，脉冲，冲量
induction voltage regulator　感应式电压调整器，感应调压器
IIS(incore instrumentation system)　堆芯仪表系统
II-type support　II 形管架
IIW(International Institute of Welding)　国际焊接学会
ILC(International Law Commission)　国际法委员会
ill-conditioned equation　病态方程
ill-effect　恶果
illegal action　非法行为
illegal contract　非法合同
illegal drug　非法药物
illegal nature　非法性质
illegal operation　非法操作
illegal payment　非法支付
illegal profit　非法利润
illegal speculation　非法投机行为
illegal trade　非法贸易
illegal　非法的
illicit　禁止的，非法的，违法的
illite　伊利石，伊利水云母
illogical deduction　不合理的推论
illuminance　光照度，照明度，照度
illuminant　发光体，发光物，照明装置，发光的，光体
illuminated circuit diagram　照明电路图，照明线路图
illuminated dial(ammeter)　照明度盘式安培计
illuminated key　带灯按键
illuminated pushbutton　带灯按钮
illuminated rotary switch　灯光旋钮，带灯开关
illuminated　带灯的，照明的
illuminate　照明，说明，装饰，点亮
illuminating apparatus　照明器
illuminating line　照明线路
illuminating mark　光点测标
illuminating power　照明能力
illumination curve　照度曲线
illumination distance　光照距离
illumination distribution board　照明配电盘
illumination engineering　照明工程，照明技术
illumination factor　照明系数
illumination intensity　照度，光照强度，辐照强度，照明强度
illumination level　光照水平，照度级
illumination load　照明负荷
illumination measurement　照明测量
illumination mirror　照明镜
illumination photometer　照度计
illumination photometry　照度测定
illumination standard　照明标准
illumination time　光照时间
illumination　灯饰【用复数】，照明，光照度，亮度，照度，说明，阐明
illuminator　发光器，照明装置，反光镜，照明器
illuminometer　光照度计，照度计
illustrated catalogue　附图产品目录，附图说明
illustrate　用图说明，举例说明
illustration diagram　图解
illustration　插图，注解，说明，实例，图解
illustrative　直观的，说明性的，特征性的
illuvial horizon　淀积层
illuviation　淀积作用
illuvium　淤积层
ilmenite electrode　钛铁矿型焊条
ilmenite type electrode　钛铁矿型焊条
image　镜像，映像，图像
image analysis　图像分析
image brightness　图像亮度
image channel　图像通道
image converter tube　光电图像变换管
image converter　光电图像变换器
image data　图像数据，图像数据
image deburring　消除图像模糊
image detect　图像缺陷
image diaphragm　图像屏
image display　图像显示
image distance　像距
image distortion　图像畸变
image enhancement　图像增强
image filtering　图像滤波
image force model　像力模型
image formation　成像
image frequency interference　镜频干扰，像频干扰
image frequency　像频，视频
image hologram　图像全息图
image impedance　影像阻抗，镜像阻抗
image intensifier gamma-ray camera　影像增强 γ 射线照相机
image intensifier　图像增强管
image interference　图像干扰，镜像干扰
image location　图像位置
image method　镜像法
image minification　图像缩小率
image modulation　像频调制
image motion compensation　图像运动补偿
image of system　系统的映射
image orthicon　超正摄像管，移像正摄像管
image output transformer　图像信号输出变压器
image parameter　图像参数
image processing　图像处理
image quality indicator　像质计，图像质量指示器［显示器］，影像质量指示器【焊缝射线照

相】
image quality 图像质量【探伤用语】
image reactor 镜像反应堆，虚反应堆
image recognition 图像识别
image reconstruction 图像再现
image restoration 图像恢复
image segmentation 图像分割
image sensing 图像传感
image sharpness 图像清晰度，图像清晰度
image shield 镜像屏蔽
image source 镜像源
image storage array 影像存储阵列
image storage device 图像存储设备
image storage translation and reproduction 图像存储变换和再生
image storage tube 图像存储管
image system 镜像系统
image theory 镜像原理
image well 映射井
imaginary accumulator 虚数累加器
imaginary axis 虚数轴，虚轴
imaginary circle 假想切圆
imaginary component 无功部分，无功分量，虚数部分
imaginary emission point 镜像排放点
imaginary hinge 虚铰
imaginary loading 无功负载，虚负载，假想荷载，虚载，虚荷载
imaginary number 虚数
imaginary part 虚部，虚数部分
imaginary quantity 虚量
imaginary unit 虚数单位
imaginary 虚数的，虚的，假想的，虚数
imagination 想象力
imbalance 不平衡，失衡
imbedded temperature-detector 埋入式测温器，埋入式温度计
imbedded winding 嵌入的绕组，下线后的绕组
imbedded 嵌镶的
imbedding method 嵌入法
imbed 埋置，埋入，嵌入
imbibition water 吸入水，吸浸水，渗吸水，吸涨水
imbibition 吸入，吸收，吸液，渗入，吸涨作用
imbricated winding 链形绕组
imbricate structure 叠瓦构造
IMDG Code（international maritime dangerous goods code） 国际海运危险货物规则
IMDH（in-service maintenance derated hours） 维护降低出力运行小时
IMF（International Monetary Fund） 国际货币基金组织
I&M（inspection & maintenance） 检查与维修
imitate 模仿，模拟，仿制
imitation leather 人造革
imitation marble 人造大理石，仿云石
imitation stone 人造石
imitation 仿制器，仿造品，模仿
imitator 模拟器，仿造者
immaterial assets 无形资产
immature accumulation loess 新近堆积黄土
immature concrete 未凝结混凝土
immature residual soil 新残积土
immature slug flow 未发展的涌节流
immature 未成熟的
immeasurable 不可计量的，不能测量的
immediacy 暂时性，瞬时性
immediate access circuit 快速投入电路
immediate access memory 快速存取存储器，快速存储器
immediate access 立即访问，快速存取
immediate address 立即地址，零级地址
immediate bank 紧靠水边的堤岸
immediate cause 直接原因
immediate compensation 立即赔偿
immediate compression 瞬时压缩
immediate consequence 直接后果
immediate delivery 即刻交货，立即交货
immediate destination 直接目的地
immediate environment 直接环境
immediately dangerous to life or health 对生命与健康有即刻危险
immediate operand 立即操作数
immediate operation use 立即操作用
immediate outage 第1类非计划停运【立即停运】
immediate oxygen demand 直接需氧量
immediate packing 直接包装
immediate participation guarantee contract 立即参与保证合同
immediate passivation 快速钝化
immediate payment 即时付款
immediate processing 立即处理，快速加工
immediate runoff 地表径流
immediate settlement 瞬时沉降量，初始沉陷，瞬时沉降
immediate shipment 立即装运
immediate supervisor 顶头上司
immediate 立即的，直接的，最接近的
immersed density 潜容重
immersed heat transfer surface 沉浸受热面
immersed method 水浸法
immersed superheater 沸腾层过热器【沸腾炉】
immersed tube pouring pile 沉管桩
immersed tube 沉浸管
immersed tunnel 沉埋隧道
immersed 浸没的
immerse 浸入，浸渍，浸没，浸
immersible concrete vibrator 插入式混凝土振捣器
immersible motor 浸入型电动机，潜水电动机
immersible switchgear 潜水开关装置，浸入式开关装置
immersion coating 化学涂层，电化学涂层
immersion cooler 浸入式冷却器
immersion counter 浸入式计数管
immersion depth 浸没深度
immersion dose 浸没剂量
immersion heater 浸没式电热器，浸入式加热器
immersion heat exchanger 浸入式热交换器
immersion length 浸没长度
immersion method 油浸法

immersion plate 浸入板
immersion plating 化学浸镀
immersion probe 液浸探头
immersion pyrometer 埋入式高温计
immersion straight pipe type level indicator 浸入直管式液位指示器
immersion technique 浸没法
immersion test 浸没试验，浸没探漏试验，浸水试验
immersion thermocouple 埋入式热电偶，埋浸式热电偶
immersion ultrasonic testing 浸入式超声检查
immersion vibrator 振捣棒，插入式振捣器
immersion 水浸，浸没，浸渍，埋入式的，潜入，浸泡
imminent 即将来临的，迫切的，急迫的
immiscibility 难混溶性，不混合性，不溶混性，不混溶性
immiscible fluid 不混合流体
immiscible solvent 不混溶溶剂
immittance 导抗，阻纳
immix 混合，掺和
immobile ion exchange site 不移动的离子交换基
immobile 固定的，不动的，静止的
immobilization 固定，降低流动性，阻塞
immobilized enzyme 固定化酶
immobilize waste 固化废物
immoderate 过度的，不适中的
immovable bed model test 定床模型试验
immovable bed model 定床模型
immovable property 不定性
immovable 不动的，固定的，不动产的，静止的
immune 免疫的，不敏感的
immunity against radiated fields 抗电磁辐射干扰度
immunity to interference 抗干扰，不受干扰，抗扰性
immunity 不敏感性，抗扰性，免除性，豁免，免除，免疫，免疫性
IMO(International Maritime Organization) 国际海事组织
impact absorber 缓冲器，减振器
impact absorbing energy 冲击吸收功
impact absorbing idler 缓冲托辊
impact acceleration 碰撞加速
impact action 撞击作用
impact air pressure 气流冲击压力
impact allowance 冲击留量，冲击容许量，容许冲击荷载
impact bending test 冲击弯曲试验，冲弯试验
impact breaker 冲击式碎石机，锤击式破碎机，反击式破碎机
impact brittleness 冲击脆性
impact burner 冲击式喷燃器
impact coefficient 冲击系数
impact compactor 冲击式夯具
impact counter system 冲击计数系统
impact crusher 冲击式碎石机，反击式破碎机，锤击式破碎机
impact current 冲击电流
impact damper 冲击阻尼器，缓冲器，减震器
impact ductility test 冲击韧性试验
impact ductility 冲击韧性
impact dust collector 冲击式除尘器
impacted SG tube 碰撞的蒸汽发生器管子
impact elasticity 冲击韧性，冲化性，冲击弹性
impact endurance test 耐冲击试验，耐冲试验
impact excitation 冲击励磁，冲击激励
impact exciter 冲击励磁机，强行励磁机
impact extrusion 冲挤
impact factor 冲击系数，碰撞系数，影响因素
impact failing load 打击破坏荷载
impact failure 冲击损坏
impact fatigue 冲击疲劳
impact fluorescence 撞击荧光
impact force 冲击力，撞击力
impact forging property 冲压性
impact fracture 冲击断口，冲击断裂
impact fretting wear 碰撞微震磨损
impact generator 冲击发电机，冲击电压发生器
impact hammer crusher 反击锤式破碎机
impact hammer mill 反击式锤式磨机
impact idler 缓冲托辊
impact induced vibration 冲击振动
impact ionization 碰撞电离
impaction loss 碰撞损失
impaction range 冲击量程
impaction 冲击，压紧
impact loading 冲击荷载，冲击负荷，突加负载
impact loss 冲击水头损失
impact meter 冲击计
impact mill 竖井式磨煤机，锤击式磨煤机
impact mixer 冲击式混合机
impact noise 碰撞噪声
impactometer 碰撞式空气取样器，冲击仪
impactor 冲击器，冲击机
impact pile driving 锤击打桩
impact plate 撞击板，反射板，反击板
impact potential 冲击电势，冲击电位
impact pressure tube 冲击压力管
impact pressure velocity meter 冲压式流速仪
impact pressure 冲击压力，动压力
impact printer 冲击式打印机
impact recorder 冲击记录器
impact resilience 冲击回弹性
impact resistance 冲击抗力，冲击阻力，抗冲击
impact resistant battery 抗震电池
impact screen 冲击筛，振动筛
impact shock 冲击震动
impact socket 风动套筒，长冲击套筒，压配套筒
impact spot welding 冲击点焊法
impact strength 冲击强度，冲击韧性，抗冲强度
impact stress 撞击力，冲击应力
impact temperature 滞止温度
impact tester 冲击试验机，冲击试验器
impact testing apparatus 冲击试验机
impact testing machine 冲击试验机
impact test specimen 冲击试件
impact test 冲击试验，碰撞试验

impact tool　冲击工具
impact torque　冲击扭矩
impact toughness　冲击韧度，冲击韧性
impact transition temperature　（金属）从延性断裂到脆性断裂的转变温度
impact tube　全压管，冲压管，皮托管
impact-type energy dissipater　冲击式消能工
impact value　冲击值
impact velocity　冲击速度
impact wave　冲击波
impact wrench　拧紧扳手，套筒扳手，冲头，机动扳手，套管扳手
impact zone　冲击区域，冲击区
impact　碰撞，冲击，压紧，影响，冲力，撞击
impairment　损害，削弱，损伤，毁损，减损
impair　削弱，损伤，损害
impale　钉住
imparity　不同，不等
impartiality　公正性
impartial judgement　公正判断
impartial　公正的
impart M to N　把 M 给［赋予、传给、通知］N
impassable trench　不通行沟
impassable　不通行的
impedance analysis　阻抗分析
impedance angle　阻抗相角，阻抗角
impedance bond　阻抗搭接
impedance bridge　阻抗电桥
impedance-capacity coupled amplifier　阻抗电容耦合放大器
impedance characteristic　阻抗特性
impedance chart　阻抗圆图
impedance circle　阻抗圆
impedance coil　扼流圈，电抗线圈
impedance compensator　阻抗补偿器
impedance conversion　阻抗变换
impedance converter　阻抗变换器
impedance coupling　阻抗耦合
impedance diagram　阻抗图
impedance drop test　阻抗压降试验，短路试验
impedance drop　阻抗压降
impedance earthed neutral system　中性点阻抗接地系统
impedance factor　阻抗系数
impedance function　阻抗函数
impedance grounded　阻抗接地的
impedance grounding　阻抗接地
impedance kilovolt-amperes　阻抗千伏安
impedance locus　阻抗轨迹
impedance loss　阻抗损耗，负载损耗
impedance matching load box　阻抗匹配负载箱
impedance matching transformer　阻抗匹配变压器
impedance matching　阻抗匹配
impedance matrix　阻抗矩阵
impedance meter　阻抗计，阻抗仪
impedance method　阻抗法
impedance muffler　阻抗复合消声器
impedance noise　阻抗噪声
impedance of slot　槽阻抗
impedance operator　阻抗算子
impedance protection　阻抗保护
impedance relay　阻抗继电器
impedance spectrum　阻抗谱
impedance starter　阻抗启动器
impedance-time relay　延时阻抗继电器
impedance transformer　阻抗变换器
impedance triangle　阻抗三角形
impedance unbalance　阻抗失衡，阻抗不平衡
impedance vector analysis system　阻抗矢量分析法
impedance void meter　阻抗空泡计
impedance voltage　短路电压，阻抗电压【变压器】
impedance voltage　阻抗电压
impedance　电阻抗，阻抗，全电阻
impede　阻碍，阻止，妨碍
impediment　阻碍，障碍，障碍物
impedometer　阻抗测量仪，阻抗计
impedor　二端阻抗元件，阻抗器
impellent　推动力，推进器，推进的
impeller anemoscope　叶轮风速仪
impeller blade　叶轮叶片，动叶片
impeller coal feeder　叶轮给料机
impeller cutting　叶轮切割
impeller diameter reduction　叶轮直径缩小
impeller eye　叶轮入口，叶轮孔
impeller hub ratio　叶轮轮毂比
impeller hub　叶轮轮毂
impeller impact breaker　叶轮冲击式破碎机
impeller inlet area　叶轮入口面积
impeller inlet guide vane　叶轮进口导向叶片
impellerless burner　直流式喷燃器
impeller nut　叶轮螺帽
impeller outlet area　叶轮出口面积
impeller passage　叶轮通道
impeller pump　叶轮泵
impeller ring　叶轮环【叶轮上的抗磨环】
impeller shaft　叶轮轴
impeller suction　叶轮吸入口
impeller type turbine　叶轮式水轮机
impeller vane　叶轮叶片，叶轮，叶片
impeller wheel　叶轮
impeller　叶轮，推进器，转子，叶轮，转轮
impelling power　推进力
impend　逼近，即将发生
impenetrability　不可贯透性，不能贯穿，不可贯入性
imperative　势在必行的，急需的
imperfect combustion　不完全燃烧，不适当竞争
imperfect competitive market　不完全竞争市场
imperfect contact　不良接触
imperfect dielectric　非理想介质
imperfect earth　接地不良
imperfect fluid　不完全流体
imperfect frame　不稳定框架
imperfect gas　非理想气体，实际气体
imperfection　不完美，缺点，瑕疵，残次品，不完整性，缺陷，不足，欠缺
imperfect mixing　不完全混合，非理想混合
imperfect shape　有缺陷的外形
imperfect tape　缺陷磁带，缺陷带
imperfect　不完全的，未完成的，不完善的，有

缺陷的，不适当的
impermanent 非永久的，暂时的
impermeability factor 不透水系数
impermeability 不渗透性，不透水性，防水性，隔水性
impermeable barrier 不透水隔层，不透水层
impermeable bed 不渗透层
impermeable break water 不透浪防波堤
impermeable confining bed 不透水隔层
impermeable groin 不透水折流坝
impermeable groyne 不透水折流坝，不透水防波堤
impermeable layer 不透水层
impermeable material 防渗材料，不透水材料
impermeable rock stratum 不透水岩层
impermeable rock 不透水岩
impermeable seam 不透水层
impermeable spur dike 不透水折流坝
impermeable stratum 不透水层
impermeable wall 不透水墙
impermeable 不渗透的，不可渗透的，防水的，不透水的，抗渗的
impermissible 不许可的
impersonal entity 非个人实体
impersonal account 不记名账号
impersonal 非个人的
impertinent 不恰当的，不适合的，无关的
impervious area runoff 不透水区径流
impervious barrier 防渗层
impervious bed 不透水河床，不透水层
impervious blanket 不透水覆盖，不透水铺盖，防渗铺盖
impervious boundary 不透水边界
impervious core 不透水心墙，防渗心墙
impervious curtain 防渗帷幕
impervious day blanket 不透水黏土铺盖
impervious element 防渗设施
impervious facing 不透水盖面，防渗面层
impervious foundation 不透水地基
impervious layer 不透水层，防渗层
impervious machine 密封型电机
impervious material 不透水材料，防渗材料，隔水层
imperviousness coefficient 不透水系数
imperviousness 不透过性，不透水性，不透水岩石
impervious rolled fill 不透水的碾压填土
impervious sandy clay core 砂质黏土防渗心墙
impervious soil 不透性土
impervious stratum 不透水层，不透水地层
impervious to moisture 防潮的
impervious wall 防渗墙
impervious zone 不透水带，不透水区
impervious 不可渗透的，不透水的，不透水岩层，抗渗的
impetus 动力，冲击，冲量
impingement angle 冲击角，入射角
impingement attack 磨损，侵蚀，冲蚀
impingement cooling 冲击冷却
impingement corrosion 冲击腐蚀
impingement height （羽流）拍撞高度
impingement plate scrubber 板式冲击除尘器
impingement ring 挡油环
impingement separator 撞击式分离器
impingement 拍撞，冲击，影响，侵犯，碰撞
impinger 尘埃测定器，冲击式采样器
impinge 冲击，碰撞
impinging neutron 撞击中子
impinging plume 拍撞羽流
impleader claim 引入诉讼
implement a contract 执行合同
implementary provisions 施行细则
implementation agreement 执行协议
implementation cost 实施费用
implementation department 执行部门
implementation of contract 合同的履行，履行合同
implementation period of new technique 新技术的采用期
implementation procedure 实施程序
implementation schedule 执行进度表，工程程序表，执行工作计划，执行进度
implementation 工具，器具，执行程序，实施，施行，实现
implement 工具，仪器，履行，执行，实现
implication 蕴含，隐含，含蓄，牵连，蕴涵
implicit computation 隐函数法计算
implicit definition 隐定义
implicit differentiation 隐函数微分
implicit function generation 隐函数发生
implicit function theorem 隐函数定理
implicit function 隐函数
implicit solution 隐函数解
implicit synchronizing signal 内隐同步信号
implicit 隐含的
implied consent 默示同意
implied contract 默认合同
implied terms 默示条文
implosion test （后汽缸）刚性试验
implosion 内爆，内燃
IMPLR＝impeller 叶轮
imporosity 无孔性，不透气性
import 输入，进口，入口，引进
import advisory committee 进口咨询委员会
import agent 进口代理商
importance function （中子）价值函数
importance 重要性，重大，显著，价值
import and export company 进出口公司
import and export corporation 进出口公司
import and export customs duty 进出口关税
import and export duty and tax 进出口关税和其他税项
import and export merchant 进出口商
import and export restrictions 进出口限制
import and export 进出口
important technical equipment 重大技术装备
important 重要的
importation forbidden 不准进口
import authorization 进口许可
import bill of lading 进口单
import bill 进口汇票
import charges 进口费用

import commission house 进口代理商
import commitment 进口承诺
import commodity 进口商品
import contract 进口合同
import control 进口管制
import declaration 进口声明书，进口报关单
import deposit 进口保证金
import duties 进口税
import duty and taxes 进口税
import duty memo of customs 海关进口税缴税单
import duty memo 海关进口税缴纳证
imported backfill 换土回填
imported equipment 进口设备
import entry 报关
importer's agent 进口代理人
importers' association 进口商协会
importer 进口商
import exchange 进口外汇
import letter of credit 进口货信用证，进口信用证
import licence 进口许可证
import permit 进口许可
import regulation 进口条例
import restraints 进口限制
import scope 进口范围
import surcharge 进口附加费，进口附加税
import surtax 进口附加税
imports 进口货物
import taxes and duties 进口税
import trade 进口贸易
import value 进口值
imposed deformation 外加变形
imposed load 作用荷载，强加荷载，外加荷载
impossibility 不可能性
impost springer 拱脚石，起拱石
impost 税，关税，拱端托，拱墩，拱脚
impotable water 非饮用水
impounded area 蓄水面积
impounded body 静止水体，蓄水池
impounded surface water 地面积水
impounded water 拦蓄水
impoundment 蓄水
impounding dam 蓄水坝
impounding reservoir 蓄水池，水库
impounding scheme 蓄水方案
impounding 拦蓄
impoundment 蓄水，积水，蓄水池，拦蓄，蓄水量
impound 没收，蓄水
impoverished county 贫困县
impracticability 无法实行
imprecise terms 不明确的条文
impregnant 浸渍剂
impregnated cable 浸渍电缆
impregnated carbon 浸渍碳棒
impregnated coil 浸渍线圈
impregnated glass cloth 浸渍玻璃布
impregnated graphite 浸渍石墨
impregnated paper insulated cable 油浸纸绝缘电缆
impregnated paper insulated 绝缘的浸渍纸
impregnated paper 浸渍纸
impregnated winding 浸渍绕组
impregnated wood 浸渍木材
impregnated 浸透的，饱和的，嵌装的，浸渍的
impregnate with bitumen 浸沥青，浸胶
impregnate with varnish 浸漆
impregnate 渗透，浸透，浸渍
impregnating bath 浸渍池，浸渍槽
impregnating compound 浸渍胶，复合胶
impregnating equipment 浸渍设备
impregnating insulation process 浸渍绝缘工艺
impregnating machine 浸渍机
impregnating mechanism 浸渍设备
impregnating paper 浸渍纸
impregnating resin 浸渍树脂
impregnating varnish 浸渍漆
impregnating vessel 浸渍槽，浸胶槽
impregnation crack detector 浸透式探伤器，浸透探伤器
impregnation liquid 浸渍液
impregnation of insulation 绝缘浸渍
impregnation technology 浸渍工艺
impregnation 浸渍，浸透，浸染，饱和
impregnator 浸渍机，浸渍设备
impressed current cathodic protection 外加电流阴极保护
impressed current groundbed 泄漏电流接地床
impressed current 外加电流
impressed electromotive force 外加电动势
impressed pressure 外加压力
impressed torque 外加转矩
impressed voltage 外加电压
impression of legal seal 盖印鉴
impression 印痕，凹槽，压痕，痕迹
impress 刻记号
imprest current funds 定额流动资金
imprimitive 非本原的，非原始的
imprinter 印刷器，刻印器
imprint 盖印
improper 不适当的，不合理的
improper integral 反常积分
improper operation 误操作
improper packing 包装不良
improper shipment 装运不当
improper ventilation 通风不良
improved data interchange 改进的数据交换
improved gas-cooled reactor nuclear power plant 改进型气冷堆核电厂
improved nuclear material 改进的核材料
improved Venturi flume 改进文丘里槽
improvement action 改进措施
improvement cycle 改进环
improvement factor 改进系数，改进因子
improvement patent 改进专利
improvement project 改建工程
improvement research 革新研究
improvement 改进，完善化，改良，改善
improve 改善，改进，提高
improvised house 临时住房
imprudent expenditure 不适宜的支出
impulsator 脉冲发生器，脉冲传感器

impulse action 冲动作用
impulse amplitude 脉冲振幅
impulse analyzer 脉冲分析器
impulse attenuation 脉冲衰减
impulse blade 冲动式叶片
impulse breakdown 冲击击穿
impulse breaker 冲击开关
impulse cascade 冲动式叶
impulse circuit-breaker 冲击断路器
impulse circuit 脉冲电路
impulse code modulation 脉冲编码调制
impulse code 脉冲码
impulse coding 脉冲编码
impulse-conservation equation 冲量守恒方程
impulse control 脉冲控制
impulse counter 脉冲计数器
impulse current generator 冲击电流发生器
impulse current relay 脉冲电流继电器
impulse current shunt 冲击分流器
impulse current 脉冲电流，冲击电流
impulse discharge 冲击放电
impulse duration system 脉冲时间系统
impulse duration 脉冲宽度，脉冲持续时间
impulse earth impedance 冲击接地阻抗
impulse equation 冲量方程
impulse excitation 脉冲激发，脉动激励
impulse exciter 冲击激励器
impulse factor 冲击因数
impulse flashover voltage 冲击闪络电压
impulse flashover 冲击闪络
impulse-forced response 脉冲强迫响应，脉冲响应，强制脉冲响应
impulse force 冲击力，冲力
impulse frequency telemetering 脉冲频率遥测法
impulse frequency 脉冲频率
impulse front 脉冲前沿
impulse function 脉冲函数
impulse fuse 冲击熔丝
impulse generator 冲击电压发生器，脉冲发生器
impulse hunting 脉冲摆动，脉冲振荡
impulse inertia 冲击惯性
impulse insulation level 冲击电压绝缘水平
impulse insulation strength 冲击绝缘强度
impulse law 冲量守恒定律
impulse level 冲击水平
impulse line 脉冲管路，脉冲管线
impulse load tests 冲击动载荷试验
impulse load 冲击负荷
impulse machine 脉冲发生器
impulse measurement 冲击测量
impulse mechanical strength 冲击机械强度
impulse meter 脉冲计数器，脉冲计算器
impulse method 脉冲法
impulse-momentum equation 冲量动量方程
impulse noise limiter 脉冲噪声限幅器
impulse noise 脉冲噪声，冲击噪声，冲击干扰
impulse of electron current 电子流脉冲
impulse oil pressure 脉动油压
impulse oscillograph 脉冲示波器
impulse overvoltage 冲击过电压
impulse phase-locked loop 脉冲锁相环
impulse power 脉冲功率
impulse protection level （避雷器的）冲击保护水平
impulse pump 冲击式泵
impulse ratio 脉冲比
impulse reactance 冲击电抗
impulse-reaction turbine 脉冲反动式涡轮机，冲动反动式汽轮机，冲击反击式水轮机
impulse recorder 脉冲记录器，脉冲记录仪
impulse regenerator 脉冲再生器
impulse register 脉冲寄存器，脉冲计数器
impulse regulator 脉冲调节器
impulse relay 脉冲继电器
impulse repeater 脉冲重发器
impulse response function 脉冲响应函数
impulse response 脉冲响应
impulser 脉冲发生器，脉冲传感器
impulse scaler 脉冲计数器
impulse section blade 冲动式叶片
impulse sender 脉冲发送器
impulse sequence 脉冲序列
impulse set 冲动式机组
impulse shaper 脉冲形成器
impulse shape 脉冲波形
impulse signal 脉冲信号
impulse source 脉冲源
impulse sparkover characteristics 冲击闪络特性
impulse sparkover voltage time curve 脉冲击穿电压时间曲线
impulse sparkover voltage 冲击闪络电压
impulse sparkover 脉冲闪络，冲击闪络
impulse speed 脉冲速度
impulse stage 脉冲级【汽轮机】
impulse starting 脉冲启动
impulse steam trap 脉冲式蒸汽疏水阀，脉冲式蒸汽疏水器，脉冲式疏水器
impulse steam turbine 冲动式汽轮机
impulse stepping motor 脉冲步进电动机
impulse strength 脉冲强度，冲击强度
impulse stress 冲击应力
impulse summation 脉冲相加
impulse switch 冲击式断路器，脉冲开关
impulse system 脉冲系统
impulse test 冲击试验，脉冲试验
impulse time margin 脉冲时间裕度
impulse timer 脉冲时间继电器，脉冲定时器
impulse transformer 脉冲变压器
impulse transmission 脉冲传输
impulse transmitting relay 脉冲发送继电器
impulse trap 脉冲式疏水器
impulse turbine 冲动式汽轮机，冲击式水轮机
impulse type relay 脉冲式继电器
impulse type stage 冲动级
impulse type steam trap 脉冲式疏水器
impulse type steam turbine 冲动式汽轮机
impulse type telemeter 脉冲式遥测计
impulse type turbine 冲动式汽轮机
impulse type voltage regulator 脉冲式电压调整器
impulse unit 冲击式机组

impulse voltage divider 冲击分压器
impulse voltage generator 冲击电压发生器，脉冲电压发生器
impulse voltage oscilloscope 脉冲电压示波器
impulse voltage response 冲击电压响应
impulse voltage test 冲击电压试验
impulse voltage 冲击电压，脉冲电压
impulse water turbine 冲击式水轮机
impulse water wheel 冲击式水轮
impulse wave test 冲击波试验
impulse wave 冲击波，脉冲波
impulse welding 脉冲焊接
impulse wheel 冲动式叶轮，冲斗式水轮
impulse withstand level 耐冲击水平
impulse withstand test 耐冲击试验
impulse withstand voltage test 冲击耐压试验
impulse withstand voltage 冲击波耐压，冲击耐受电压，冲击耐压
impulse 冲击，冲动，冲力，撞击，脉冲，冲量，推动，激发，推力，推动
impulsing power source 脉冲电源
impulsing 发送脉冲，脉冲激励
impulsive discharge 脉冲放电
impulsive excitation 脉动冲
impulsive load 脉冲荷载
impulsive noise 冲击噪声，脉冲干扰
impulsive rotatory motion 冲击旋转运动
impulsive sound 冲击声
impure steam 污染蒸汽
impure 不纯的，不洁的，有杂质的
impurity concentration 杂质浓度
impurity conduction 杂质导电
impurity damage 污染损伤，杂质损性
impurity in ionized state 离子态杂质
impurity level 杂质水平，杂质度
impurity semiconductor 杂质半导体
impurity 杂质，夹杂物，污染，有杂质的，不纯洁的，不纯性，混杂度，污物，杂物
imputable to 可归罪于……的，可归因于……的
imputed cost 应付成本
imputed income 应计收入，估算收入
IMP 冲动式（级）【DCS 画面】
IMS(information management system) 信息管理系统
inability 无能力
inaccessible area 不可接近区域，不可进入区域，不可进入区
inaccessible reactor building 不可进入的反应堆厂房
inaccessible 不通达的
in accordance with 根据，与……一致，按照，依照
inaccuracy dimension 尺寸不准
inaccuracy 不精确性，不准确度，误差，偏差，不精密，不准确
inaction period 无作用期间，钝化周期
inaction 故障，无行动，无作用，不活泼，不活动
inactivate 钝化
inactivation cross-section 钝化截面，非活化截面
inactivation 钝化
inactive alkali 惰性碱
inactive area 非放射性区，非活性区
inactive coal storage 长期不动用的煤堆
inactive component 非能动部件，无外源设备
inactive fault 不活动断层
inactive gas 惰性气体
inactive laboratory 非放射性实验室，冷实验室
inactive member 无效杆
inactive money 呆滞资金
inactive pile 非自流煤堆
inactive state 不活泼状态，待用状态，关闭状态，停用状态
inactive stockpile 非自流煤堆，长期不动煤堆
inactive storage 无效库容
inactive substance 非活性物质
inactive test 非放射性试验，冷试验
inactive time 无效时间，不可用时间
inactive waste 非放射性废物
inactive 不活泼的，钝性的，非放射性的，非活性的
inactivity 钝性，失去活性，不活动
inadapted river 不相称河
in addition to 除了……外
in addition 另外，此外，而且
inadequacy of data 数据不足
inadequacy 不适当，不合适，不足，不够
inadequate design 不合理设计
inadequate trip 误动作
inadvertent maloperation 偶然误动作，意外操作
inadvertent protective action 偶然保护作用
inadvertent rod withdrawal 误提棒
inadvertent safety action 偶然安全作用
inadvertent trip 误动作，乱真跳闸
inadvertent 偶然的，误动作，计划外的
in-air test 在空气中试验
in all the event 无论如何
in all 总共，共计
in an inert atmosphere 在惰性气体中
in any case 无论如何
in any event 不管发生什么事，在任何情况下，无论如何
inapplicable 不能应用的，不适用的
inappreciable error 不显著误差
inappreciable 微不足道的，不足取的
in appreciation of 鉴于，酬谢
inapprehensible 难以理解的，难了解的
inappropriate leakages of steam, water, pulverized coal, oil, air, heat and ash in power plant 电厂的七漏
inappropriate 不适当的，不相称的
inapt 不适当的，不合适的，不熟练的
in a reverse direction 相反的方向
inarmoured cable 未铠装电缆
inarmoured 非铠装的
in arrears 拖欠
in assessing terms 评定条款
in a true scale 按真实的比例
inaudible vibration 无声振动
inauguration ceremony 就职典礼，开幕典礼，落成典礼
inauguration 就职，就职典礼，落成典礼，开幕

式，开始
in a word 综上所述
in axial alignment 轴线对齐
in bad condition 处于不良状态
in-band frequency assignment 带内频率分配
in-band signaling 带内信令
in bankruptcy proceedings 在破产程序中
in batches 分批的，批式的，分批
in-bed tube 沉浸管，(沸腾炉）床内埋管
in-between position 中间位置
inblock 整体，整块
inboard bearing 内置轴承，内侧轴承
inboard end 内侧端
inboard rotor 两端支承转子
inbond brick 丁砖
in-boxes 箱装，手套箱内
inbreak angle 陷落角
in bridge 旁路，并联，跨接
in bulk 散装，大批
incaging 释放，松开
incandescence 白炽，白热，炽热
incandescent arc lamp 白炽弧光灯
incandescent filament lamp 白炽灯
incandescent globe 球形白炽灯
incandescent lamp 白炽灯
incandescent light 白炽灯光
incandescent spot 炽热点
incandescent state 白炽状态，白热状态
incandescent 白炽的，炽热的，白热的
in-can melter 罐内熔融器
incapability of meeting obligations 无法履行义务的情况
incapability 不胜任，无资格
in capacity of 作为……，以……资格
in cascade 串级排列
in case of fire 如遇火灾
in case of 万一，以防，如果发生，假使，如果
in cases 箱装
in case that 万一，如果
INCC (International Nuclear Cooperation Center) 国际核合作中心【日本】
in-cell piping 热室管线，设备室管线
in-cell valve 设备室内阀门
in-cell 热室内，设备室内
incendivity 可燃性，引火性能
incentive bonus 激励性奖金
incentive method 奖励办法
incentive plan 奖励计划
incentive standard 奖励标准
incentive system 奖励制度
incentive 动机，刺激，诱因
inception of oscillation 起振，起始振荡
inception voltage 起始电压【局部放电】
inceptisol 始成土
inceptor 拦截设施
in-channel moderator coefficient 通道内慢化剂反应性系数
in-channel water 通道内水
inches per year of penetration 年侵蚀英寸
inching button 点动按钮，微动按钮
inching clutch 微调离合器
inching control valve 微动控制阀
inching control 微动控制
inching relay 微动继电器
inching speed （调整阶段间歇启动时的）爬行速度
inching starter （电动机的）低速启动器
inching switch 微调开关，微动开关
inching valve 微调阀，微动阀
inching 缓动，寸进，微动，极低速度转动，（发动机）低速转动，瞬时断续接电
inch permeance 每一英寸槽的磁导率
inch 英寸，少量
incidence angle 入射角，迎角，安装角
incidence indicator 倾角指示器
incidence loss coefficient 碰撞损失系数，入射损失系数
incidence loss 冲角损失
incidence matrix 关联矩阵
incidence measuring gear 冲角测量装置
incidence of stabilizer 稳定器倾斜角
incidence plane 入射面
incidence point 入射点
incidence 入射角，冲角，倾角，迎角，入射，安装角
incidental damages 附带损害赔偿金
incidental expense 偶然费用，零星支出，杂费
incidental frequency modulation 临时频率调制
incidental to 易发生的，难免的，附带的，属于……的
incidental 偶然发生的，伴随的，偶然的，偶然（发生）的
incident angle modifier 入射角修正系数
incident angle 入射角
incident beam 入射光束，入射流
incident detection 事故探测
incident electron 入射电子，轰击电子
incident field intensity 入射场强
incident flow turbulence 来流湍流（度）
incident flow 入射流，来流，迎面气流
incident flux 入射通量，事故通量
incident galloping 初始驰振
incident gust 来流阵风
incident instability 初始不稳定性
incident investigation team 事故调查组
incidentmotion 初始运动
incident neutron energy 入射中子能量
incident of moderate frequency 中等频率故障
incident radiation 入射辐射，投射辐射
incident reporting system 事故报告系统
incident shock front 入射激震前沿
incident shock 入射突变，入射激波
incident solar ray 入射太阳射线
incident turbulence 来流湍流（度）
incident wave 入射波
incident wind 来流风
incident 易发生的，附带的，入射的，事故，事件，故障
incinerate 焚烧
incinerating solution tank 焚烧液箱
incineration firing 垃圾燃烧
incineration 烧尽，焚烧，焚化

incinerator residue　焚化炉残渣
incinerator stoker　垃圾焚烧炉
incinerator　垃圾焚烧炉，（放射性发料）焚烧炉，煅烧炉，焚化炉
incipient accident　起始事故，早期事故
incipient boiling　初始沸腾
incipient break　初始断裂
incipient bubble　初始气泡
incipient cavitation number　初生空穴数
incipient cavitation　初始空蚀，初始汽蚀
incipient combustion　起始燃烧
incipient crack　初裂纹，初始裂缝，初始裂纹
incipient erosion　早期冲蚀
incipient failure　初期故障
incipient fault　早期故障，初始故障
incipient flame　初期火焰
incipient fluidization　起始流化
incipient fluidizing velocity　初始流化速度
incipient gullying　初期沟蚀
incipient leakage　初期泄漏
incipient period　开始阶段，初期
incipient piping　初始管涌
incipient pressure regulation　初压控制
incipient sediment motion　初始泥沙运动
incipient slugging　初始腾涌
incipient spoilage　初期腐败
incipient stability　初始稳定
incipient stage　初始阶段
incipient turbulence　初始紊流
incipient　开始的，初期的，刚出现的，早期的
in-circuit emulation　内部电路仿真
in-circuit　线路中的，内部电路
incised meander　嵌入曲流，峡谷曲流
incise　切刻
incision　下切作用，深切
inciting　雕刻
inclement weather　恶劣天气
inclination angle　倾斜角
inclination compass　倾角罗盘
inclination correction　倾斜改正
inclination error　倾斜误差
inclination method　倾度法
inclination of ground　地面坡度
inclination of pipe　管道坡度
inclination of pole　电杆斜度
inclination of roof　屋顶斜度
inclination of wave front　波前倾角
inclination of weld axis　焊缝倾角
inclination　倾斜，斜度，倾角，偏角，坡度，倾斜度
incline conveyor　斜升输送机
inclined　倾斜的
inclined antenna　倾斜天线
inclined apron　斜护坦
inclined arch barrel　斜拱筒
inclined arch　斜拱
inclined axis mixer　斜轴拌和机
inclined axis rotor WECS　斜轴风力机【风轮轴线与水平面夹角在 15°～90°之间】
inclined baffle　倾斜挡板
inclined bedding　倾斜层理
inclined belt conveyor　斜升皮带输送机，斜皮带机
inclined cableway　斜缆道
inclined channel　斜孔道
inclined chord of truss　桁架斜弦杆
inclined core earth dam　斜墙土坝
inclined core　斜墙
inclined damper　斜插板阀
inclined distributor　倾斜布风板
inclined drop　斜槽跌水
inclined fault　倾斜断层
inclined floor　倾斜底板
inclined flow turbine　斜流式水轮机
inclined flow　偏斜水流
inclined fold　倾斜褶皱
inclined force　斜向力
inclined fuel assembly transfer tube　燃料组件斜传动管
inclined gauge　倾斜计，斜坡水尺
inclined grate　倾斜炉排
inclined guide apparatus　斜向式导水机构，锥形导水机构
inclined hole　斜孔
inclined intake　斜坡式进水口
inclined jet turbine　斜击式水轮机
inclined joint　斜缝
inclined multitube manometer　倾斜多管压力计
inclined pan conveyer　倾斜盘式输送机
inclined pile　斜桩
inclined pipe inlet　卧管式进水口
inclined pipe　倾斜管
inclined plane　倾斜面，斜面
inclined plate　斜板
inclined position welding　倾斜焊
inclined pump　斜式泵
inclined reciprocating grate　倾斜式往复炉排，往复炉排
inclined ropeway　斜缆道
inclined-seat gate valve　斜座闸阀
inclined shaft excavation　斜井开挖
inclined shaft tubular turbine　斜轴贯流式水轮机
inclined shaft　斜井
inclined shock wave　斜冲波，斜激波
inclined span　斜挡距，不等高挡
inclined staff gauge　斜水尺
inclined steel　斜钢筋
inclined-stem butterfly valve　斜阀杆蝶阀
inclined stirrup　斜向箍筋
inclined stratum　倾斜地层
inclined surface　斜面
inclined thermal　倾斜热泡，倾斜热
inclined transfer tube　斜输送管
inclined trough chute　斜槽
inclined tube gauge　倾斜式微压计，斜管风压计
inclined tube manometer　倾斜式微压计，斜管压力计，斜管式测压器
inclined tube settler　斜管沉降器
inclined weir　斜背堰
inclined wharf　斜码头
incline　斜坡，斜面，倾斜，坡度，倾斜面
inclinometer　倾斜仪，倾角计，斜度仪，磁倾

计，测斜仪
inclosed body　箱式车身
inclosed　密闭的，封闭式的
inclosure wall　围墙
inclosure　包壳，包入物，附件
in-cloud icing　冻雾覆冰
included angle of arch　拱的包角【指中心角】
included angle of crest　堰顶包角
included angle　夹角【焊接坡口】，开角
include into the state plan　纳入国家计划
including but not limited to　包括但不限于
including　包括
inclusion complex　包合配合物
inclusion relation　包含关系
inclusion texture　包体结构
inclusion　夹杂物，杂质，包含，夹杂，包括，包体，（个人对团体的）融合度
inclusive of　把……包括在内，包括在内的，也算入的
incoagulable　不可凝结的
incoalation　煤化作用
incoherence　不相干性，不连贯性
incoherent alluvium　无黏性冲积层
incoherent approximation　不相干近似
incoherent detection　不相干检波
incoherent integral　不相干积分
incoherent interaction　不相干相互作用
incoherent interference　不相干干涉
incoherent-light holography　非相干光全息术
incoherent optical information processing　非相干光学信息处理
incoherent radiation　不相干辐射
incoherent rock　不胶结岩石
incoherent scattering cross section　不相干散射截面
incoherent scattering　不相干散射，非相干散射
incoherent　不连贯的，不相干的，松散的
in collaboration with　与……合作
Incoloy　因科洛依合金，高温镍铬铁合金
incombustibility　不燃性
incombustible building material　阻燃建筑材料
incombustible construction type　防火的结构型
incombustible construction　不燃烧结构
incombustible material　不燃烧材料，非燃性材料，不可燃物质
incombustible matter　不燃物
incombustible mixture　不可燃混合物
incombustible substance　不可燃物
incombustible　不可燃的，不可燃物，不燃物，不能燃烧的
income beneficiary　收入受益人
income distribution　收入分配，所得分配
income statement　损益表
income tax allocation　所得税分配，所得税扣除
income taxes expense accrued　发生的所得税
income tax payable　应缴所得税
income tax return　所得税申报表
income tax　所得税
income　收入，所得
incoming beam　入射束
incoming breaker　进线开关
incoming call　吸入，来话呼叫
incoming carrier　进入载波
incoming charge　进料，进料费用
incoming circuit　输入电路
incoming coal non-uniformity coefficient　来煤不均衡系数
incoming coal non-uniformity factor　来煤不均衡系数
incoming condensing water　（凝汽器）进口凝结水
incoming current　输入电流
incoming data　输入数据
incoming direction　引入方向
incoming feeder　电源进线【开关，配电装置】，进馈线，输入馈路，进线，进线馈线
incoming flow　来流，迎面流，来水流量，入流
incoming gas　进气
incoming generator　准备并车的发电机
incoming group selector　入局选组器
incoming highway　进厂公路
incoming level　输入电平
incoming line　电源进线，进线，引入线，输入线路
incoming neutron flux　入射中子通量
incoming panel　受电盘
incoming particle　入射粒子
incoming partner　新合伙人
incoming pulse　输入脉冲
incoming road　进厂道路
incoming section or cubicle　进线隔舱或小室
incoming signal　输入信号
incoming solar radiation　射入太阳辐射
incoming stock　进料，进入原料
incoming traffic　输入通信量，入局通信量
incoming trunk　收话专用中继线，入局中继线
incoming turbulence　来流湍流（度）
incoming wave　来波
incoming wind　来流，迎面风
incoming　引入的，入射的，输入的，进来的，进入的
in commission　投入使用，服役
in common　共同的，公共的
in comparison with　与……比较
incompatibility　不相容性，不协调性
incompatible events　不相容事件
incompatible　不相容的，不能共存的，不一致的
incompetent attendance　不合格的维护
incompetent bed　软岩层
incompetent witness　无资格的证人
incomplete annealing　不完全退火
incomplete burning　不完全燃烧
incomplete circuit　开路，不闭合电路
incomplete combustion　不完全燃烧
incomplete commutation　不完全换向，不完全固结
incomplete contraction　不完全收缩
incomplete data　不完全资料
incomplete diffusion　不完全漫射，不完全扩散
incomplete effective collective dose　不完全有效集合剂量
incomplete expansion　不完全膨胀
incomplete fusion　不完全熔焊［焊接］，未焊透，

未完全熔合，未溶合，未熔合
incomplete joint penetration 未焊透
incompletely contracted orifice 不完全收缩孔
incompletely filled groove （焊缝）未焊满，未填满的坡口【焊接】
incompletely saturated 未完全饱和的
incompletely thermal reactor 非完全热力反应堆
incomplete penetration 未全部焊透，夹生焊，未焊透
incomplete phase operation 非全相运行
incomplete rarefaction wave 不完全稀疏波
incomplete reaction 不完全反应
incomplete root penetration 根部未焊透
incomplete well 非完整井
incomplete 不完全的，不完善的，未完成的
in compliance with drawings 与图纸相符
in compliance with standards 达标准
in compliance with 遵照，和……一致，按照，顺从，依照，根据
incompressibility modulus 不可压缩性系数
incompressibility 不可压缩性
incompressible boundary layer 不可压缩边界层
incompressible energy equation 不可压缩能量方程
incompressible flow 不可压流动，不可压缩流，不可压缩水流
incompressible fluid 不可压流体，不可压缩流体
incompressible wake 非压缩流尾流
incompressible wind 来流风，迎面风，不可压缩风
incompressible 不可压缩的
in conclusion 综上所述
incondensable gas 不冷凝的气体，不凝结气体
Inconel alloy 因科内尔镍铬铁耐热耐蚀合金，因科镍合金，英高镍镍铬合金
Inconel bush 因康镍合金衬套
Inconel reflector 因康镍合金反射层
Inconel spring clip 因康镍合金弹簧夹
Inconel 因康镍合金，因科内尔铬镍铁合金，铬镍铁合金
in confirmation of 为确认……起见
in conformity with the specifications in the contract 符合合同的技术要求
in conformity with 和……一致，和……相适应，遵照，本着，符合，吻合，接轨，步调一致地
in-connector （流线）内接符【计算机的】
inconsequent stream 非顺向河
in consideration of 考虑到，由于，鉴于，以……为约因，考虑到，作为……的报酬
inconsistency 不协调性，不一致性，不一致，不相容性，前后矛盾
inconsistent 不一致的，不相容的
in constant dollar value 按不变价值美元计算
inconstant 不稳定的
in-containment fuel storage area 安全壳内的燃料贮存区，安全壳内燃料储存区
in-containment fuel storage 安全壳内的燃料贮存，安全壳燃料储存
in-containment heat transport system 安全壳内热传输系统
incontestability clause 不可争议条款，不可异议条款
incontrollable 难以控制的，不能控制的
inconvertibility 不可逆性
inconvertible 不可逆的，不能交换的
in-core analysis system 堆内分析系统
in-core cable 堆芯电缆
in-core control structure 堆内控制构件
in-core detector 堆内探测器，堆芯探测器
in-core experiment 堆内实验
in-core fission product inventory 堆内裂变产物总量
in-core flux detector 堆内通量探测器
in-core flux instrumentation 堆芯通量测量仪表
in-core flux monitor 堆芯通量监测器
in-core fuel management 堆内燃料管理
in-core instrumentation assembly 堆芯测量组件
in-core instrumentation detector 堆芯测量探测器
in-core instrumentation drive system 堆内测量驱动系统
in-core instrumentation flux thimble exercising test 堆内测量通量套管运用试验
in-core instrumentation lead 堆内测量导线
in-core instrumentation room 堆芯仪表室
in-core instrumentation support structure 堆芯（测量）仪表支承结构
in-core instrumentation system 堆芯测量仪表系统，堆芯仪表系统
in-core instrumentation 堆芯测量仪表，堆芯测量装置
in-core instrument position 堆芯测量仪表位置
in-core instrument test facility 堆芯仪表试验装置
in-core ionization chamber 堆芯电离室
in-core measurement system 堆芯测量系统
in-core monitoring system 堆芯监测系统
in-core monitoring 堆芯监测
in-core movable flux detector 堆内可移动式中子通量探测器
in-core neutron detector 堆芯中子探测器
in-core neutron flux instrumentation 堆芯中子通量测量仪表
in-core neutron flux measurement 堆芯中子通量测量仪表，堆芯中子注量率测量
in-core neutron flux monitoring system 堆芯中子通量监测系统
in-core temperature measurement 堆内温度测量
in-core thermionic reactor 堆内热离子反应堆
in-core thimbles retraction 堆内套管缩回
in-core 芯内的，铁芯内，堆芯内部
incorporated company 股份有限公司
incorporated radionuclide （体内）结合的放射性核素
incorporated scanning 插入扫描
incorporate in bitumen 掺入沥青，（放射性废物）沥青固化
incorporate integrative 一体化
incorporate 组成公司，把……合并，合为一体的，结合，编入，插入，包含，使混合
incorporation in concrete （放射性废物）水泥固化
incorporation into bitumen （放射性废物）沥青固化

incorporation of foreign matter （晶格间）杂质掺入
incorporation 结合，合并，掺和，插入，公司，固化
incorporator 公司创办人
incorrect connection 错误接线
incorrect in size 尺寸不准
incorrect manipulation 误操作
incorrect operation 误动作，不正确动作，误操作
incorrect switching 开关误操作
incorrect trip 误跳闸
incorrect 不正确的，错误的，不恰当的
incorrodible 不腐蚀的，抗腐蚀的
Incoterms(International rules for the Interpretation of Trade Terms) 国际贸易术语解释通则【国际商会】
INCOT(in-core instrument test facility) 堆芯仪表试验装置
in course of 正在……，在……过程中
increase active power 增加有功功率
increase by a factor of 5 提高到5倍，提高了4倍
increase by a factor of 1/5 增加20%，增加到(1+1/5)倍
increase by M times 增加到M倍，乘以M
increased output 提高出力，超发
increased safety electrical instrument 增安型电动仪表
increased-safety motor 增安型电动机
increase frequency 提高频率
increase in direct ratio with 随……成正比地增加
increase in potential 电势增长，电位增
increase in 在……方面增加
increase M by N 使M增加N
increase M to N 把M增加到N
increase of charge 电荷增长
increase of frequency 频率升高，提高频率
increase of power 功率增加，提高功率
increase of pressure 升压
increase of reactor power 提升反应堆功率
increase of service life 延长使用寿命，延长使用期限
increase of speed 升速，增速，转速上升
increase of voltage 升压，电压升高
increase rate 增长率
increase reactive power 增加无功功率
increaser 扩径水管，异径接头
increase the accuracy of measurement 提高测量精度
increase with years 逐年增加
increase 增长，增加
increasing amount 增加量
increasing forward wave 前进波，增大向前波
increasing function 递增函数
increasing income 增加收入
increasing sequence 递增系列
increasing velocity 递增速度
incredible event 盛事，难以置信的事件，不可信事件，不可思议的事件
incredible 不可思议的，惊人的，难以置信的
incremental 增量的
incremental coder 增量编码器
incremental compiler 增量编译程序
incremental computer 增量计算机
incremental cost 边际成本，增加费用
incremental delivered power 输出功率增量，供电增量
incremental digital computer 增量数字计算机
incremental digital recorder 增量数字记录器
incremental dose 剂量增量
incremental dump tape 增量转储磁带
incremental encoder 增量编码器
incremental energy 能量增值
incremental equivalent 等效增量
incremental feed ram 渐增式加料推杆
incremental flow changes 增量流量变化
incremental generating cost 发电成本增量
incremental hysteresis loop 增量磁滞回线
incremental hysteresis loss 磁滞损失增量，增量磁滞损耗
incremental induction 增量电感，增量磁感应
incremental item 增量信息
incremental load deviation 增量负荷偏差
incremental loading test 逐级加荷试验
incremental loading 分级加荷，负荷增量，装料增量，递增加荷
incremental magnetizing force 增量磁化力，增量磁强
incremental maintenance cost 维护费用增量
incremental manner 递增方式
incremental method of sampling 增量取样法
incremental momentary speed droop 增量暂态转速下降
incremental permanent speed variation 稳态转速变化增量
incremental permeability 微分磁导率，磁导率增加
incremental plotter 增量绘图器
incremental representation 增量表示，增量表示法
incremental resistance 增量电阻
incremental response time 增量反应时间
incremental speed governing droop 局部转速不等率
incremental speed regulation 增盘转速调节
incremental speed 转速增量
incremental step loading test 逐级加力法
incremental stiffness 增量刚度
incremental system 增量系统
incremental transmission loss 输电损耗增量
incremental type 递增型
increment borer 生长锥
increment command 增量命令
increment list 增量表
increment luffing distance 逐次调幅距离
increment of coordinate 坐标增量
increment of plastic strain 塑性应变增量
increment rate 增量率，增率
increment reduction and division 逐次缩分
increment travel distance 逐次走行距离
increment 增加，增大，增长，增量，步进量，

微增，增额，定期的加薪
incrustate 结水垢，长硬皮
incrustation scale 水垢
incrustation 水垢，积垢，结壳，结痂，结硬皮，水锈，紧密沉积物，硬壳，镶嵌物，痂
incrusting solid 表面变硬固体
INCR 提高，增加【DCS 画面】
incubator thermometer 恒温箱温度计
incumbency 现任职，义务
incumbent chairman 现任公司董事长
incumbent 负有责任的，负有义务的
incur debt 借债
incur expense 支出费用
incur obligation 承担债务
incurred cost 已发生成本，应计成本
incurred losses 蒙受的损失
in current price 按时价计算，按现价计算
incur 招致，引起，遭受，承担，惹起
indebtedness 负债
in default 不履行责任
indefinite integral 不定积分
indefinitely small 无穷小
indefinite scale 任意比例尺
indefinite shipment 不定期装运
indefinite 不定的，无穷的
indeformable 不变形的
indelible ink 消不去的墨水，不褪色的墨水
indelible 不易擦掉的
indemnification clause 保证赔偿豁免条款
indemnification 赔偿，补偿
indemnify for damage 赔偿损失
indemnify for 赔偿……
indemnify from 保护……
indemnifying obligation 赔偿义务
indemnify sb. against sth. 保障【保护】某人不受…
indemnify sb. for the loss incurred 赔偿造成的损失
indemnify sb. from damage 保护某人不受伤害
indemnify sb. from sth. 保护某人不受……
indemnify 保障，保护，使免于受罚，补偿，赔偿
indemnity agreement 补偿协议
indemnity for damage 损坏赔偿
indemnity for risk 风险赔偿
indemnity recoverable 可获得的赔偿
indemnity refused 拒绝赔偿
indemnity 保证（书），保障，赔偿，赔款，赔偿金，免罚
indentation hardness 刻痕硬度，压痕硬度
indentation method 凹痕法，刻痕法
indentation 刻痕，压痕，凹陷，凹入，低凹，凹痕
indented arch 三角拱
indented bar 刻痕钢筋，螺纹钢筋
indented bolt 带纹螺栓
indented coast line 曲折岸线
indented joint 齿合接缝
indented wire 齿痕钢丝，刻痕钢丝
indenter 压头【硬度试验】
identification number 标识号
indenting punch 压头，冲头
indenting tool 压凹工具
indenting 压凹，压痕
indenture of lease 租赁合同
indenture worker 合同工
indenture 双联合同，正式凭单，契约，合同
independence criterion 不相关准则
independence of random variables 随机变量的独立性
independence of the letter of credit 信用证的独立性
independence test 独立性检验
independence 独立性，（功能）独立性
independent accounting unit 独立核算单位
independent accounting 独立核算
independent agent 独立代理人
independent arch method 纯拱法
independent circulating component 独立回路部件
independent condition 独立条件
independent conformity 独立一致性
independent consultant 独立咨询人员
independent contact 独立触点，单独触点
Independent contractor 独立承包商，独立缔约人
independent control 独立控制，自治控制
independent development trust 独立发展信托（公司）
independent digit 独立数字
independent driven exciter system 独立励磁制
independent electric supply 独立电源
independent enterprise 独立自主的企业，独立自主企业
independent equipment 独立设备
independent event 独立事件
independent excitation 单独励磁，他励
independent failure 独立故障，单独故障
independent fission yield 独立裂变产额
independent footing 独立底脚，独立基础
independent foundation 独立基础
independent ground electrode 单独接地极
independent legal corporate 独立公司法人地位
independent linearity 独立线性，无关线性，独立线性度
independent manual operation 手动操作，独立人工操作
independent observation 独立观测
independent package 独立包
independent party 独立一方
independent pole breaker 独立极断路器
independent power producer 独立发电项目，独立发电站，独立电力生产者，独立发电商
independent professional firm 独立的专业公司
independent review of a document 独立对文件的审查
independent spread footing 独立扩展式基础
independent stairs 不靠墙楼梯
independent subtree equation 独立子树方程
independent subtree 独立子树
independent synchronizing system 独立同步制
independent time control 独立的时间控制
independent time-lag relay 定时限继电器
independent time-lag 独立时滞

independent variable unit　自变量部分
independent variable　自变量，独立变量
independent yield　独立份额，直接份额
independent　独立的，自主的，自备的
in-depth filtration　深层过滤
in-depth　详细的，深入的
indeterminacy condition　不确定性条件
indeterminacy degree　不定度
indeterminate coefficient　不定系数，未定系数
indeterminate equation　不定方程
indeterminate error　不定误差
indeterminateness　不定性，不确定
indeterminate principle　测不准原理
indeterminate structure　超静定结构
indeterminate-term liability　不定期负债
indeterminate　不定的，不明确的，模糊的
index-area method　指标面积法
index arithmetic unit　变址运算单元
index array　变址阵列
index bed　标准层
index bit　变址位
index card　索引卡片
index chart　指标图
index contour　示量等高线
index correction　仪表误差校正
index diagram　索引图
index dial　表盘，刻度盘，指示盘
indexed access　变址存取，变址访问
indexed addressing　变址寻址
indexed address　变址，结果地址
indexed number　索引号
indexed sequential data set　加下标顺序数据组
indexed sequential organization　加下标顺序结构
index error　指示误差，分度误差
indexer　调车机，分度器
index file　索引文件
index gauge　分度规
index glass　指标镜
index hand　指针
indexing plate　标度盘
indexing revolving station　刻度转动盘
indexing stroke　对位行程
indexing system　定位系统
indexing　检索，索引，加索引，指向，分度法，变址，标刻度
index intensity　标记亮度
index marker　索引标记
index mark　指示器，指标
index method　指数法
index mineral　标志矿物，指示矿物
index motion　刻度移动
index movement　（财务）指数变化
index number　指数
index of atmospheric purity　大气纯度指数
index of biotic integrity　生物完整性指数
index of cooperation　合作指数
index of coupling　耦合指数
index of discharge　流量指数，排量指数，泄流指数
index of fuzziness　模糊性指数
index of harm　危害指数
index of land occupation　土地占用指标
index of leading indicator　主要指标的指数
index of performance　性能指数，性能指标，效率指数
index of plasticity　塑性指数
index of pollution　污染指数
index of precision　精确度指数
index of price　物价指数
index of quality　质量指标，质量指数
index of refraction　折射指数，折射率，反射率
index of reliability　可靠性指数
index of resistance　阻力指数，风阻指数
index of stability　稳定性指数
index of stabilization　稳定指数
index of technical economics in general layout　总平面布置的技术经济指标
index of technical economics　技术经济指标
index of terms of trade　贸易条件指数，进出口商品比价指数
index of thermal inertia　热惯性指数
index of turbulence　湍流指数
index plate　分度盘，标度盘，刻度盘
index properties　指标特性
index property　特征性质
index register　变址寄存器，指数寄存器
index ring　刻度环
index storage　变址存储器
index stress　特征应力
index strip　（元件盒的）标号条
index table　索引表
index tag　标牌
index value　给定值，要求值，指标
index variable　下标变量
index word　变址字，下标字，索引字
index　指数，指标，系数，索引，变址，标牌，标签，检索，标志，对位
india-rubber cable　橡皮绝缘电缆
india rubber　橡胶，橡皮
indicated claim　规定的索赔
indicated efficiency　指示效率
indicated horsepower　指示马力
indicated mean effective pressure　平均有效指示压力
indicated cargo　指装货
indicated output　指示出力
indicated power　指示功率
indicated pressure　指示压力
indicated resources　推定资源，可控资源
indicated thermal efficiency　指示热效率
indicated thrust　指示推力
indicated value　指示值
indicated wind speed　指示风速
indicated work　指示功
indicate　指示，表明，表示
indicating bell　指示铃
indicating calipers　指示卡规
indicating controller　指示控制器
indicating device　指示器件，指示设备，指示装置
indicating dial　指示盘
indicating error　示值误差，显示误差，指示

误差
indicating fuse 指示熔断器
indicating gauge 指示计
indicating head 仪表刻度盘
indicating hole gauge 指示测孔规
indicating instrument 指示仪表，指示仪器
indicating lamp for escape 疏散指示灯
indicating lamp 指示灯，信号灯
indicating light 指示灯，光指示
indicating meter 指示仪表，指示器，指示剂
indicating needle 指针
indicating oscillograph 指示式示波器
indicating pointer 指针
indicating potentiometer 指示电位器计
indicating range 指示范围
indicating relay 指示继电器，信号继电器
indicating thermometer 指示温度计
indicating transmitter 指示式传送器
indicating value 示值
indicating 指示的，显示的
indication device housing 指示装置套筒
indication error 示值误差，指示误差，读数误差
indication lag 指示延迟
indication of a measuring instrument 测量仪表的示值
indication of manufacture 制造标记
indication of price 标明价格
indication of source 产地标记
indication 指示，表示，示数，读数，标记，示值，征兆
indicative mark 指示性标志
indicative of 指示的，预示的，可表示的
indicator board 指示板，指示屏
indicator diagram 指示图，示功图
indicator dial 指示器标度盘，指示盘
indicator display 计数器的读数
indicator drum 指示筒
indicator electrode 指示电极
indicator element 示踪元素，指示元素
indicator gear 指示器的传动装置
indicator lamp 指示灯
indicator light 指示灯
indicator of pollution 污染指示物
indicator panel 指示屏，信号显示板
indicator paper 试纸
indicator pencil 指示器的记录头
indicator post 指示器支杆
indicator rod 标志杆，指示杆
indicator test 指示器试验
indicator travel 指示仪表行程
indicator valve 带指示器的阀
indicator 显示器，指示器，指示仪表，指示剂，指针，（经济）指标，指压器，指示仪
indicatrix of diffusion 漫射指示量
indicatrix 指标，指示量，指示线
indices correlation 指标相关
indicial admittance 单位阶跃导纳
indicial motion 示性运动
indicial response 单位阶跃响应，指数响应
indicial 指示的，指数的，单位阶跃的
indifference curve 无差异曲线
indifferent equilibrium 随遇平衡，中性平衡
indifferent gas 惰性气体
indifferent 冷淡的，中性的，惰性的，无关紧要的
indiffusible ion 不扩散离子
in-diffusion 向内扩散
indigenous fuel 当地燃料
indigenous people 当地居民
indigenous raw materials 当地原料
indigenous soil 原生土
indigenous 本地的，土产的，当地的，土生土长的，国产的，本土的
indigo carmine 靛蓝胭脂红，靛蓝，胭脂靛，酸性靛蓝，靛胭脂红
indigo 靛蓝
indirect 间接的
indirect absorber 间接吸收体
indirect acting measuring instrument 间接作用式测量仪表，间接作用式测仪表
indirect acting 间接动作的，间接作用的
indirect action 间接作用
indirect A/D converter 间接模数转换器
indirect address 间接地址
indirect addressing 间接寻址，间接编址
indirect air cooling 间接空冷，空气外冷
indirect analog 间接模拟，函数模拟，非直接模拟
indirect analysis 间接分析
indirect-arc furnace 间接电弧炉
indirect balance method 反平衡法
indirect benefit 间接效益
indirect braking method 间接制动法
indirect business 间接贸易
indirect carrier 间接载体
indirect catchment 间接集水面积
indirect charge 间接费用
indirect circulation 逆环流
indirect component 间接分量
indirect connection 间接连接
indirect contact desuperheater 表面式减温器
indirect contract 间接合同
indirect control 间接控制
indirect-cooled desuperheater 间接冷却减温器
indirect cooling generator 外冷发电机，间接冷却发电机
indirect cooling water 间接冷却水
indirect cooling 间接冷却
indirect cost of project 工程间接费
indirect cost 间接成本，间接费，间接费用
indirect cycle boiling-water reactor 间接循环沸水反应堆
indirect cycle reactor 间接循环反应堆
indirect cycle 间接循环
indirect damage 间接损害，间接损失
indirect data address 间接数据地址
indirect desulfurization 间接脱硫
indirect determination 间接测定
indirect draft 间接通风
indirect drive 间接传动
indirect dry cooling system of surface condenser

表面冷凝器间接干冷却系统
indirect dry cooling system 间接干式冷却系统
indirect dry cooling tower 间接空冷塔
indirect effect 间接效果
indirect electromagnetic wave 间接电磁波
indirect excitation 间接激励，间接励磁
indirect expense 间接费用
indirect-fired （用中间载热介质）间接加热的
indirect firing system 仓储式制粉系统
indirect firing 非直接受火，仓储式系统燃烧，间接加热
indirect flood damage 洪水间接损失，间接洪水损害
indirect flow control 间接流量控制
indirect forms of solar energy 太阳能的间接形式
indirect frequency modulation 间接调频
indirect heat exchanger 表面式换热器
indirect heating 间接加热
indirect illumination 反射照明，间接照明
indirect influence 间接影响
indirect input 间接输入
indirect interruption cost 间接停电损失
indirect ionization 间接电离
indirect liability 间接责任
indirect lighting 间接照明
indirect lightning strike 感应雷击
indirect loss 间接损失
indirectly coal fired combined cycle 外燃式燃煤联合循环
indirectly controlled system 间接受控系统，间接被控系统
indirectly controlled variable 间接被控变量，间接受控变量
indirectly cooled by hydrogen 间接氢冷，氢外冷
indirectly heated cathode 间接加热阴极
indirectly heated thermistor 间接加热的热敏电阻
indirectly ionizing particles 间接致电离粒子，间接电离粒子
indirectly ionizing radiation 间接电离辐射
indirectly 间接，间接地
indirect maintenance 间接维修，远距离维修
indirect measurement of efficiency 间接测量效率法
indirect measurement 间接测量
indirect-method heat balance 反平衡，反平衡法
indirect method of measurement 间接测量法
indirect neutron radiography 间接中子射线照相
indirect observation 间接观测
indirect or consequential damages 间接或连锁损害
indirect private investment 私人间接投资
indirect refrigerating system 间接制冷系统
indirect self-excitation 间接自励
indirect simple cycle 简单间接循环
indirect steam 二次蒸汽
indirect stroke 间接雷，感应雷
indirect system 间接系统
indirect titration 间接滴定
indirect trade 间接贸易
indirect transmitter 间接发送器
indirect treatment 间接处理，炉外处理
indirect vapor cycle 间接蒸汽循环
indirect wave 空间波，间接波
indirect welding 单面焊接，间接点焊
indirect working 间接作业【绝缘操作杆作业】
indispensable 不可缺少的
indissolubility 不溶解性
indissolvableness 不溶解性
indissolvable 不溶解的，难溶解的
indistinct 模糊
indisturbed flow 未扰流动
inditron 指示管，示数管
indium resonance flux 铟共振中子通量
indium resonance neutron 铟共振中子
indium steel level 铟钢水准尺
indium 铟
individual axis drive 单轴驱动，独立驱动
individual batcher 分批计量器
individual bucket volume 单斗容积
individual burner flame detection 单燃烧器火焰检测
individual bypass 小旁路
individual calling 个别呼叫
individual car shipments 单车载重量
individual channel flow control 单个通道流量控制
individual compensation 单独电抗补偿
individual construction 个体建筑
individual consumer 单独用户，个体用户
individual control level 执行控制级
individual control system 分级控制系统，分级控制
individual control 单独控制，个别控制，独立控制
individual defect 单个缺陷
individual dose 个人剂量
individual drive motor 单独传动电动机
individual drive 单独传动，单个驱动
individual excitation system 单独励磁制，他励制
individual footing 独立基础
individual foundation 单独基础
individual gust 单阵风
individual income tax 个人所得税
individual income 个人所得
individual inspection 单项检查，单独检查，逐个检验，单个检查
individual investment 个人投资
individual lead 单独引入线
individual leg extension 专用的塔腿延长段
individual lighting 独立供电照明
individual line 用户线路，专用线路
individual load 单一负荷
individual loss 单项损耗
individual measurement 单独测量
individual monitoring 单独监测
individual operation 单独运行
individual-owned enterprise 个体企业
individual ownership proprietorship 独资企业
individual package delivery 分别包装交货
individual price 单价
individual project 单项工程
individual proprietorship 独资，独资经营
individual responsibility 个人责任

individual sewage disposal system 专用污水处理系统
individual shield 分屏，单个屏蔽，单屏，独立屏蔽【电缆】
individual shot 单孔爆破
individual synchronizing system 单独同步方式
individual system operation 单个系统操作
individual system 单独系统
individual test 单件试验，个别试验
individual transmission 单独传动，单独传输
individual unit 单个单元
individual vent 单独通气管
individual weight 自重，净重
individual well yield 单井开采量
individual wire （导地线的）单丝
individual 单独的，个别的，独特的，个人的，不可分的
indoor air design conditions 室内空气计算参数
indoor air pollution 室内空气污染
indoor air velocity 室内空气流速
indoor and outdoor decoration of buildings 建筑物的室内外装饰
indoor apparatus 户内设备
indoor arrangement 室内布置
indoor boiler 室内锅炉
indoor coal storage yard 干煤棚，干煤场
indoor controlgear 户内控制设备
indoor design temperature 室内设计温度
indoor environment 室内环境
indoor equipment 室内设备
indoor fire-fighting water distribution system 建筑物内部消防水分配系统
indoor fire hydrant 室内消火栓
indoor floor tile 室内地砖
indoor gust 室内阵风
indoor humidity 室内湿度
indoor hydrant station 室内消火栓站
indoor hydroelectric station 室内式水电站
indoor illumination 室内照明
indoor insulator 户内绝缘子
indoor lighting 室内照明
indoor reference for air relative humidity 室内相对湿度基数
indoor reference for air temperature 室内温度基数
indoor substation 户内变电站，室内变电站，屋内变电所
indoor switchgear 屋内配电装置
indoor switching station 屋内开关站
indoor temperature 室内温度，室温
indoor thermal comfort 室内热舒适性
indoor thermal power plant 屋内式火力发电厂
indoor transformer 户内型变压器
indoor turbine 室内汽轮机
indoor type pot-head 户内式端套
indoor type switchgear 户内型配电装置
indoor type 户内式
indoor voltage regulator 户内电压调整器
indoor wall bushing 户内穿墙套管
indoor wiring 室内布线
indoor work 内业，室内工作
indorsed bill 已背书票据
indorsee 被背书人
indorsement （票据等后面的）署名、背书，认可
indorser of a bill 票据背书人
indraft 引入，吸入，流入
indraught 流入
in-drum mixing 桶内混合
induce 感应，诱导，诱发，引起
induced 感应的，诱导的，诱发的
induced activity 感应放射性
induced air 吸气，引入空气，引风，吸入空气
induced angle of attack 诱导攻角，诱导迎角
induced charge 感应电荷
induced circuit 感应电路
induced circulation 强制循环
induced cleavage 诱生劈理
induced cooling tower 机力通风冷却塔
induced current 感应电流
induced downwash 诱导下洗
induced draft control 引风调节
induced draft cooling tower 吸风冷却塔
induced draft fan 引风机
induced draft tower 引风式机力通风塔
induced draft 引风，负压通风，负压送风
induced drag coefficient 诱导阻力系数
induced drag of lift 升力诱导阻力
induced drag 引风阻力，诱导阻力
induced-draught fan 吸风风扇
induced earthquake 诱发地震
induced effect 感应效应
induced electric current 感生电流，感应电流
induced electromotive force 感应电动势
induced EMF 感应电动势
induced emission 诱导发射
induced enzyme 诱导酶
induced fan 引风机
induced fission 诱发裂变
induced flow effect 诱导气流影响
induced flow heater 强制流动加热器
induced flow 诱导流动
induced flux 感应通量
induced insulation test 感应绝缘试验
induced lightning overvoltage 感应雷过电压
induced loss 感应损耗
induced magnetism 感应磁，感应磁性
induced noise 感应噪声，感生噪声
induced overvoltage 感应过电压
induced polarization method 激发极化法
induced potential test 感应电压试验
induced potential 感应电势
induced power 诱导功率
induced precipitation 诱导沉淀
induced pressure gradient 诱导压力梯度
induced pressure 诱导压力
induced radiation 感应辐射
induced radioactivity 辐照感生放射性，感生放射性
induced reverse osmosis 诱导反渗透
induced test 感应试验
induced velocity 扰动速度，诱导速度
induced vibration 诱导振动

induced voltage by rotating field 旋转磁场感应电压
induced voltage test 感应电压试验
induced voltage 感应电压
Inducement 引导
inducer casing 叶轮壳体
inducer 电感器，导流轮，叶轮，诱导轮，诱因，诱导体，进口段
induce vertical flow 诱导垂直流动
inducing charge 加感电荷
inducing coil 加感线圈
inducing current 加感电流
inducing field 加感场
inducing winding 加感绕组，励磁绕组
inducing 加感的
induct 引入，引导，感应
inductance-bridge flowmeter 感应电桥式流量计
inductance-capacitance coupling 感容耦合
inductance-capacitance tuning 感容调谐
inductance coil 电感线圈
inductance commutation 感性换向
inductance coupling 电感耦合
inductance feedback 电感反馈
inductance figure 电感系数
inductance meter 亨利计，电感表
inductance of air gap armature coil 气隙电枢电感，气隙电枢电感
inductance reactance 感抗
inductance switching-off overvoltage 开断电感负载过电压
inductance switch 电感转接开关
inductance tuning 电感调谐
inductance type transducer 电感式传感器
inductance 电感，感应系数，感应现象，感应
in-duct installation 管沟敷设
induction accelerator 感应加速器
induction air conditioning system 诱导式空气调节系统
induction alternator 感应式同步发电机
induction and hysteresis motor 感应磁滞电动机
induction balance 感应平衡电路
induction blazing 感应钎焊
induction bridge 电感电桥
induction by current 电流感应
induction chamber 吸气室
induction coefficient 诱导系数，电感应系数，感应系数
induction coil 电感线圈，感应线圈
induction compass 感应罗盘
induction control device 感应控制设备
induction cup relay 感应杯式继电器
induction disc relay 感应圆盘式继电器
induction disc type directional power relay 感应圆盘式方向功率继电器
induction drive motor 感应电动机
induction dynamometer 感应式测力计，感应测功计
induction effect 感应作用，诱导作用
inductioner 电感调谐设备
induction factor 电感因数
induction field calculation 感应场计算法
induction field 感应场
induction flux 感应通量
induction frequency converter 感应式变频机，异步变频机
induction frequency meter 感应频率计
induction furnace 感应电炉，感应炉
induction generator 感应发电机，异步发电机
induction hardening 感应硬化，感应淬火
induction harden 感应硬化
induction heater 感应加热器
induction heating coil 感应加热线圈
induction heating equipment 感应加热设备
induction heating method 感应加热法
induction heating stress improvement 感应加热应力改善
induction heating 感应加热
induction heat treatment 感应热处理
induction instrument 感应式仪表
induction interrupter 感应断续器
inductionless conductor 无感导体，无感导线
inductionless 无感应的
induction machine 感应电机，异步电机
induction manifold 入口联箱
induction meter 感应式仪表，感应式电表
induction motor-generator 感应电动机拖动的发电机
induction motor meter 感应式电度表
induction motor type synchronous motor 感应式同步电动机
induction motor 感应电动机，异步电动机，感应马达，异步马达
induction period 诱导期
induction phase shifter 感应移相器
induction phenomena 感应现象
induction phenomenon 感应现象
induction pipe 吸入管，进入管
induction pump 感应泵，交流电磁泵
induction regulator 感应调压器，感应调整器
induction relay 感应继电器
induction servomotor 感应伺服电动机
induction sheath 磁感应屏蔽层
induction shielding coil 感应屏蔽线圈
induction shield 感应屏蔽
induction soldering 感应钎焊
induction starter 感应启动器，启动用自耦变压器
induction stroke 吸入冲程，进气冲程
induction tachogenerator 感应测速发电机
induction time （污垢）诱导期
induction torque 异步转矩，感应转矩
induction type alternator 感应式交流发电机
induction type automatic voltage regulator 感应式自动电压调整器
induction type over-voltage relay 感应式过电压继电器
induction type relay 感应式继电器
induction unit 诱导器
induction valve 送气阀，送水阀
induction velocity 诱导速度
induction winding 感应线圈，感应绕组，感应熔焊

induction 感应，感应现象，感应密度，归纳法，诱导，引入，归纳
inductive acceleration 感应加速度计，感应加速
inductive action 感应作用
inductive branch 电感性支路
inductive character 电感性
inductive choke 电感线圈
inductive circuit 电感电路，有感电路，感应电路
inductive coil 有感线圈，电感线圈
inductive component 电感分量，感应分量，电感性分量，感性（无功）分量
inductive control 感应控制，感应操纵
inductive coordination （输电线的）电感配合
inductive coupling 电感耦合，电磁耦合，感性耦合
inductive current 电感电流，感应电流
inductive drop 感应电压降，电感电压降
inductive effect index 诱导效应指数
inductive effect 感应效应，感应作用，诱导效应
inductive electromotive 感应电动热
inductive EMF 感应电动势
inductive energy storage 电感储能
inductive feedback 电感反馈
inductive heating 感应加热
inductive leak detector unit 感应式泄漏探测装置
inductive level detector 电感式电平探测装置
inductive loading 加感
inductive load 电感负载，感性负载
inductively coupled plasma 感应耦合等离子体
inductive pick-off 感应传感器
inductive potential divider 感应分压器
inductive reactance 感性电抗，感抗
inductive reactive power 感性无功功率
inductive reactor 感抗线圈
inductive reluctance 感应磁阻
inductive rise 电感性电压升
inductive shunt 感应分流器
inductive strain gauge 感应式应变仪
inductive surge 感应冲击电压，感应性电涌
inductive susceptance 电感电纳
inductive switching 电感开关，电感转接
inductive ventilation 诱导通风
inductive voltage 感应电压
inductive winding 电感线圈，感性线圈
inductive zone 感应区
inductive 电感的，感应的，吸入的，电感性的，诱导的
inductivity 感应性，介电常数，感应率
inductometer 电感计
inductor-alternator excitation system 感应发电机励磁系统
inductor coil 感应线圈，电感线圈
inductor form 线圈架，线圈管
inductor generator 感应子发电机，旋转磁铁式发电机
inductorium 感应线圈
inductor machine 感应子电机
inductor motor 感应子式电动机
inductor type generator 感应子式发电机
inductor type synchronous alternator 感应子式同步发电机
inductor 感应体，感应器，感应线圈，电感器，电感线圈
inductosyn 感应式传感器，感应同步器，感应整步机
inductuner 感应调谐装置
in due time 及时地，适时地，在适当的时候
indulgence 付款延期
in duplicate 一式两份
industrial aerodynamics wind tunnel 空气动力学风洞
industrial aerodynamics 工业空气动力学
industrial air conditioning 工业用空气调节
industrial air pollution 工业空气污染
Industrial and Commercial Bank of China 中国工商银行
industrial and commercial income tax 工商所得税
industrial architecture 工业建筑
industrial area 工业区
industrial automation instrument 工业自动化仪表
industrial automation 工业自动化
industrial boiler 工业锅炉
industrial brush 工业用电刷
industrial building 工业建筑物，厂房
industrial chemistry 工业化学
industrial chimney 工厂烟囱
industrial city 工业城市
industrial cleaning 工业清洗
industrial climatology 工业气候学
industrial complex 工厂群
industrial computation 工业计算
industrial computer 工业用计算机，工控机
industrial consumer 工业用户
industrial consumption 工业耗水量
industrial contaminant 工业污染物
industrial control 工业控制，工业控制装置
industrial cooling water system 工业冷却水系统
industrial data processing 工业数据处理
industrial density 工厂密度
industrial design 工业设计
industrial development area 工业发展区
industrial device 工业装置
industrial distribution equipment 工业配电装置
industrial district 工业区
industrial dust 工业粉尘
industrial dynamics 工业动态
industrial effluent 工业废物，工业外排物，工业流出物，工业废水，工业污水
industrial electric heating 工业电热
industrial electronics 工业电子学
industrial engineering 工业管理，工业管理学，工业工程学，企业管理学，生产管理技术
industrial engineer 企业工程师
industrial enterprise above designated size 规模以上工业企业
industrial environment 工业环境
industrial estate 工业用地
industrial ethernet 工业以太网
industrial exhaust heat 工业余热
industrial frequency 工业频率，工频

industrial furnace 工业用电炉
industrial gammagraphy 工业γ射线照相法
industrial gas turbine 工业用燃气轮机，工业燃气轮机
industrial grade micro-computer 工业微机
industrial graphic display system 工业图形显示系统
industrial haze 工业烟雾
industrial heat reactor 工业用热反应堆
industrial heat substation 工业热力分站
industrial heat user 工业热用户
industrial inspection 工业检查，工业检验
industrial instrument 工业仪表
industrial irradiator 工业辐照设备，工业辐照器
industrialist 实业家，工厂主
industrialization 工业化
industrialized aircraft gas turbine 飞机改装型燃气轮机
industrialized building system 工业化建筑系统
industrial lighting 工业照明，工厂照明
industrial load 工业负荷
industrial management program 工业管理程序
industrial mathematics 工业数学
industrial meteorology 工业气象学
industrial packing 工业包装
industrial park 工业园区
industrial PC 工业用 PC 机，工业用个人电脑
industrial plume 工业烟羽
industrial pollution 工业污染
industrial port 工业港
industrial power distribution 工业配电
industrial power plant 企业自备电厂，工业电站
industrial power supply 工业供电
industrial process control system 生产过程控制系统
industrial process control 生产过程控制
industrial process measurement and control instrument 工业过程检测控制仪表，工业自动化仪表
industrial process supervisory system 生产过程监督系统
industrial process 工业生产方法，工业过程，生产过程
industrial production 工业生产
industrial property right 工业产权
industrial property 工业产权，工业财产，工业特性
industrial pure 工业纯
industrial radiography 工业射线照相法
industrial radiology 工业辐射学，工业放射学
industrial reactor 工业用反应堆
industrial refrigeration 工业制冷
industrial refuse 工业垃圾
industrial relay 工业继电器
industrial robot 工业机器人
industrial scale 工业规模
industrial screening 工业筛分
industrial self-contained power station 工业自备电站
industrial service plant 工业自备电厂
industrial sewage 工业污水
industrial sewerage system 工业污物排放系统
industrial siding 工业企业专用线
industrial smoke 工业烟雾
industrial stack 工厂烟囱
industrial standard 行业标准
industrial station 工业站
industrial steam plant 工业蒸汽动力厂，工业热电厂
industrial steam turbine 工业汽轮机
industrial substation 工业变电所
industrial telemetering system 工业遥测系统
industrial telemetry 工业遥测
industrial television 工业电视
industrial test 工业性试验
industrial thermometer 工业用温度计
industrial town 工业城镇
industrial truck 搬运车辆
industrial turbine 工业透平
industrial TV 工业电视
industrial urbanization 工业城市化
industrial use of water 工业用水
industrial ventilation 工业通风
industrial waste 工业废水，工业废物
industrial waste disposal 工业废水处理，工业废物处置
industrial waste drainage system 工业废水排放系统
industrial waste gas 工业废气
industrial waste heat 工业余热
industrial waste liquid 工业废液
industrial waste water treatment 工业废水处理
industrial waste water 工业废水，工业污水
industrial water law 工业用水法
industrial water piping 工业水管道
industrial water pollution 工业水污染
industrial water pump 工业水泵
industrial water quality 工业水质量
industrial water service 工业给水，工业供水
industrial water supply 工业给水，工业供水
industrial water system diagram 工业水处理系统
industrial water system 工业水系统
industrial water works 工业水厂
industrial water 工业用水
industrial worker 产业工人
industrial X-ray crack detector 工业 X 探伤机
industrial 工业的，实业的，商业的，产业的，从事工业的，供工业用的
industry and commerce administration 工商行政管理，工商行政管理局
industry and commerce 实业，工商，工商业
industry balance 工业天平
industry boiler 工业锅炉
industry closed circuit television 工业闭路电视
industry commerce united enterprise 工商合营企业
industry consolidation 产业联合，行业性重组，产业整合，产业集群
industry degraded core rule-making 工业退役堆芯规则制定
industry endoscope 内窥镜
industry gas turbine 工业燃气轮机

industry hygiene 工业卫生
industry issues 核工业界的公告
industry-owned thermal power plant （企业）自备火力发电厂
industry standard 工业标准
industry television 工业电视
industry waste water 工业废水
industry waste 工业废物
industry water collected tank 工业水回收水箱
industry water diagram 工业水系统图
industry water pump 工业水泵
industry water system 工业水系统
industry water tank 工业水箱
industry water 工业水
industry 行业，工业
in economic 在经济领域
ineffective call 无效呼叫
ineffective time 无效时间
ineffective 不起作用的，无效的，效率低的，不适当的
inefficiency 低效率，无效率，效率低
inefficient 低效的，不经济的
inelastic collision 非弹性碰撞
inelastic cross section 非弹性截面
inelastic deformation 非弹性变形
inelastic event 非弹性碰撞事件
inelasticity coefficient 非弹性系数
inelasticity 刚性，非弹性
inelastic lateral 非弹性侧向压屈
inelastic moderation 非弹性慢化
inelastic neutron 非弹性散射中子，非弹性中子
inelastic range 非弹性范围
inelastic removal cross section 非弹性移出截面
inelastic removal 非弹性移去
inelastic scattering cross section 非弹性散射截面
inelastic scattering excitation 非弹性散射激发
inelastic scattering 非弹性散射
inelastic stop 刚性限制器
inelastic wave 非弹性波
inelastic 非弹性的
inequable 不相等的，不均匀的
inequality 不等式，不等，不等性，不均衡性
inert 惰性的，不活泼的
inert aggregate 惰性骨料
inert atmosphere glove box 惰性气体手套箱
inert atmosphere processing cell 惰性气体处理热室
inert bed material 惰性床料【沸腾炉】
inert carrier 惰性载体
inerted containment 充惰性气体安全壳
inert element 惰性元素
inert filler 惰性掺合料，惰性填料
inert floating resin 惰性浮动树脂
inert fluidized bed 惰性流化床
inert gas arc welding 惰性气体保护焊
inert gas blanketing 惰性气体覆盖
inert gas blanket 惰性气体覆盖层
inert gas cooler 惰性气体冷却器
inert gas cushion 惰性气垫
inert gaseous constituents （燃料中）惰性气体成分
inert gas fission product 惰性气体裂变产物
inert gas metal-arc welding 惰性气体保护金属（熔化极）电弧焊
inert gas-shielded arc welding 惰性气体保护电弧焊，惰性气体保护焊
inert gas supply and blanketing system 惰性气体供应和覆盖系统
inert gas system 惰性气体（覆盖）系统
inert gas welding 惰性气体保护焊
inert gas 惰性气体
inertia 惯性，惰性，惯量
inertia balance 动平衡，惯性平衡，惯性秤
inertia coefficient 惯性系数
inertia compensation 惯性补偿
inertia constant of generating set 发电机组的惯性常数
inertia couple 惯性力偶
inertia current 惯性流
inertia damper 惯性阻尼器
inertia damping 惯性阻尼
inertia effect 惯性效应，惯性作用
inertia force 惰力，惯性力
inertia governor 惯性调节器，惯性调速器
inertia grade （公路的）惯性坡度
inertia head 惯性水头
inertial 惯性的
inertial axes system 惯性轴系
inertial confinement fusion 惯性约束聚变
inertial coordinate system 惯性坐标系
inertial deposition 惯性沉积
inertial dust separator 惯性除尘器
inertial effect 惯性效应，惯性作用
inertial energy storage 惯性贮能
inertial flow rate 惯性流量
inertial force 惯性力
inertial fusion power plant 惯性聚变动力厂
inertial impaction 惯性碰撞，惯性撞击
inertial instability 惯性不稳定性
inertially damped servomotor 惯性阻尼伺服电动机
inertial mass 惯性质量
inertial separator 惯性分离器
inertial settling 惯性沉降
inertial subrange law 惯性子区定律
inertial subrange 惯性子区，惯性次区
inertial switch 惯性开关
inertial transfer of energy 惯性能量输送
inertial wave 惯性波
inertia mass 惯性质量
inertia moment 惯性矩，转动惯量
inertia oscillation 惯性振动
inertia relay 惯性继电器
inertia resistance 惯性阻力
inertia revolution counter 惯性摆转速计
inertia shear 惯性剪力
inertia speed counter 惯性摆转速计
inertia system 惯性系统
inertia tachometer 惯性转速计
inertia test 惯性试验
inertia time constant 惯性时间常数
inertia type starter 惯性启动器

inertia vibrating screen 惯性振动筛
inertia wave 惯性波
inerting blanketing system 惰性气体覆盖系统
inerting system 惰化系统
inerting 惰化
inert packing 惰性填料
inert particle 惰性颗粒
inert resin 惰性树脂
inert zone unit 非灵敏区模拟装置
inert zone 不灵敏区，非灵敏区
inestimable 极贵重的，(大得) 无法估计的
inevitable accident 不可避免的意外事故，不可避免的事故
inevitable outcome 必然结果，定局
inevitable 必然发生的，不可避免的
in exchange for 以作为……的交换，交换
inexhaustible 无穷无尽的，永不枯竭的
inexpensive 价廉的
inexperience 缺乏经验，外行
inextricable 无法摆脱的，解不开的，无法摆脱地
in favor of 有利于，以……为受益人
infeasible 不可行的
infected water 含菌水
infeed 馈电，横向送进
inference engine 推理机
inference 推断，推理，推论
inferential 推理上的，推理的
inferface with sb. 与某人联络，接触
inferior arc 劣弧
inferior coal 劣质煤
inferior fuel 劣质燃料
inferior goods 低档货
inferiority of quality 质量低劣
inferiority 劣
inferior limit 下限
inferior material 劣质材料
inferior product 次品
inferior quality feedwater 低品质给水
inferior 低级的，次的，低质的，次品，劣
infernal surge 内涌浪
inferred resources 推断资源，推测资源，推定资源，隐含资源
inferred zero instrument 无零点仪器
infertile 贫瘠的
infestation 蔓延，侵扰
infidel 失真的，不精确的，不正确的
infield 安装地点，运用处
in figures 小写
infilling 填充物
infiltrant 浸渍剂
infiltrated air 渗漏空气，空气渗入
infiltrate 渗入，渗透，渗漏，渗滤
infiltration area 地下水补给区，入渗面积，渗水面积
infiltration capacity curve 渗透量曲线
infiltration capacity 入渗量，渗量，渗入量，渗入能力，渗透能力
infiltration coefficient 入渗系数，渗入系数，渗透系数
infiltration curve 入渗曲线
infiltration ditch 渗水沟
infiltration diversion 渗水管引水
infiltration experiment 渗透试验
infiltration flow 渗入流量
infiltration gallery 渗流集水廊道
infiltration heat loss 渗入热损失
infiltration index 渗入指数
infiltration intensity 入渗强度，渗入强度
infiltration leakage 直接泄漏
infiltration loss 渗漏损失
infiltration method 渗透，渗入，浸润法
infiltration pipe 渗水管
infiltration rate 地表入渗率，地表渗入率
infiltration rate 入渗率，渗入率，渗入速率，渗透率，渗透速率，下渗率
infiltration slit 渗水缝
infiltration theory 渗入理论
infiltration tunnel 地下渗水道，渗水隧洞
infiltration velocity 渗入速度，渗透速度
infiltration volume 渗透水量
infiltration water 渗入水
infiltration well 入渗井
infiltration zone 渗漏带
infiltration （地表）渗入，渗透，渗滤，渗漏，漏风，渗透作用
infiltrometer 渗入测定计，渗透计
infinite anisotropic source 无限各向异性源
infinite aquifer 无限含水层
infinite busbar （功率）无穷大母线
infinite bus 无穷大母线
infinite cascade 无限叶栅
infinite cloud model 无限烟云模式
infinite cylinder 无穷大圆柱体
infinite dilution 无限稀释
infinite freeboard 无限自由空域
infinite integral 无穷积分
infinite isotropic source 无限各向同性源
infinite lattice （堆芯）无限栅格
infinite lifetime 无限寿期
infinite line source 无限线源
infinite long line 无限长线路
infinitely long 无限长
infinitely positionable rod drive 平滑控制棒传动
infinitely reflected reactor 无限反射层反应堆
infinitely reflected 无限反射层的
infinitely safe geometry 无限安全几何条件
infinitely variable speed 无级调速
infinitely variable 无级调速的，无限可调的
infinitely 无限地
infinite medium lifetime 无限介质寿期
infinite medium multiplication constant 无限介质放大常数，无限介质增殖系数
infinite medium multiplication factor 无限介质增殖因数
infinite medium reactivity transfer function 无限介质的反应性传递函数
infinite medium 无限介质
infinite-memory filter 无限存储滤波器
infinite multiplication constant 无限介质增殖系数，无限放大常数
infinite multiplication factor 无限放大常数，无限

增殖系数
infinite multiplication （中子）无限增殖
infinite plane source 无限平面源
infinite reactor 无限反应堆
infinite rejection filter 无限衰减滤波器
infinite rod bundle 无限棒束
infinite series 无穷级数
infinite set 无穷集，无限集
infinitesimal analysis 微积分
infinitesimal bed height 无限小床高
infinitesimal calculus 微积分学
infinitesimal deformation 无穷小形变
infinitesimal displacement 无穷小位移
infinitesimal disturbance 无限小扰动
infinitesimal error probability 无限小误差概率
infinitesimal geometry 微分几何
infinitesimal 无穷小的，无限小的，无穷小
infinite slab reactor 无限平板反应堆
infinite slab source 无限平面源
infinite slab 无穷大平板
infinite solid 无限大固体
infinite span 无限翼展，无限展长
infinite tube bundle 无限管束
infinite voltage gain 无穷大电压增益
infinite vortex street 无限涧街
infinite wake 无限尾流
infinite 无穷的，无限的，无穷大，不定的
infinitude 无穷
infinity 无穷，无穷大
infix 嵌入，插入，穿入，中缀
inflame 点火，燃烧，发炎，激动
inflammability limiting concentration 可燃极限浓度
inflammability limit 易燃限度，可燃极限，着火极限
inflammability test 可燃性试验
inflammability 可燃性，易燃性
inflammable material storage 易燃物库
inflammable material 易燃材料，易燃物
inflammable mixture 可燃混合物
inflammable substance 易燃物
inflammable 可燃的，易燃的，易燃物，可燃物
inflammation 着火，燃烧
inflatable dam 可吹胀的堵塞物，充气坝，橡胶坝
inflatable packer 充气止浆塞，充气垫
inflatable seal 充气密封，肿胀密封
inflated form 充气模
inflated lava 膨胀熔岩
inflated plastic stills 可膨胀塑料蒸馏器
inflate the nitrogen 充氮
inflation rate 通货膨胀率
inflation 膨胀，通货膨胀，充气，肿胀，鼓起
inflecting point 拐点，转折点，反弯点，回折点
inflection 挠曲，弯曲，拐折，回折，反弯（曲）
inflexibility 非柔性，刚性
inflexion 挠曲，弯曲，拐折，回折，反弯（曲）
infloat switch 浮子开关
inflow angle 入流角，迎流角，流入角，进气角
inflow curve 来水曲线，入流曲线
inflow discharge curve 来水流量曲线
inflow forecast 来水预报
inflow hydrograph 来水过程线，入流过程线
inflowing air 进入的空气
inflowing sediment 流入泥沙，入库泥沙
inflow into reach 河段入流
inflow Mach number 进口马赫数
inflow rate 入流量，入流率
inflow reach 入流段
inflow-storage-discharge curve 入流-蓄水-出流曲线
inflow-storage-outflow curve 入流-储量出-流曲线
inflow 流入，进气，流入量，进水量，来流，来水，来水量，入流，入流量
influence area 影响区，影响面积，影响范围
influence basin 影响范围
influence by the tower shadow 塔影响效应【塔架造成的气流涡区对风力机产生的影响】
influence by the wind shear 风切变影响
influence chart 影响图，感应图
influence coefficient method 影响系数法
influence coefficient 影响系数
influence diagram 影响线图
influence electricity 感应（静）电
influence factor 影响因素
influence line 影响线
influence machine 静电发电机，感应起电机
influence of foreign field 外磁场影响
influence quantity 影响量
influence radius 影响半径
influence range 影响区，影响范围
influence sphere 影响范围
influence value 感应值
influence zone 影响区
influence 感应，影响，效应，作用，影响量，环境影响
influencing factor 影响因素
influencing variable 作用变量，感应变量，影响变量
influent action 渗水作用
influent header 入口联箱
influent hopper 进水漏斗
influential components 有影响部件
influent impounded body 补给地下水的水体，拦蓄的水量
influent river 地下水补给河
influent seepage 渗漏，渗入
influent stream 补给地下水的河流
influent water 渗漏水
influent 流入的，进水的，流入液体，渗流，汇合处
influx 汇集，流入量
inform 通知
informal agreement 草约，非正式合约
informality 非正式性
informal record 非正式记录
informatics 信息学，信息科学，资料学
information acquisition 信息获取，信息采集
informational 信息的，介绍情况的
information and instructional advertising expenses

信息和指导性广告费用
information asymmetry 信息不对称
information bank 资料库
information-based 信息化
information bit 信息位
information bulletin 情况通报
information capacity 信息容量
information center 信息中心，情报中心，资料中心
information channel 信道，信息通道，信息通路
information circuit 信息传送线路
information code 信息码
information collection and processing 信息收集与处理
information collector 信息收集器
information conference 信息发布会
information content 信息内容，信息量
information decoding 信息译码
information density 信息密度
information desk 查询台
information-destroying process 信息破坏过程
information disclosure 信息披露
information display rate 信息显示速度
information display 信息显示
information distributor 信息分配器
information document 资料性文件
information encoding 信息编码
information engineering 信息工程，信息工程学
information exchange list 信息交换表
information exchange 情报交流，信息交流，资料交换
information feedback 信息反馈，资料反馈
information flow 信息流
information gain 信息增益
Information highway 信息通路
informationization 信息化
information management system 信息管理系统
information materials 情报资料
information meeting 情况介绍会
information model 信息模型
information network 信息网络，信息网
information note 情况通知
information overload testing apparatus 信息过载测试设备
information paper 资料性文件
information parameter 信息参数
information processing center 信息处理中心
information processing language 信息处理语言
information processing machine 信息处理机
information processing system 信息处理系统
information processing 信息处理
information processor 信息处理机
information publishing subsystem 信息发布子系统
information pulse 信息脉冲
information rate 信息发送率，信息传输速度
information read wire 信息读出线
information read 信息读出
information redundancy 信息冗余
information register 信息寄存器
information requested 要求（的）信息
information requirements 信息要求
information retrieval system 信息检索系统
information retrieval 信息恢复，情报检索
information selection system 信息选择系统
information separator 信息分隔符
information sequence 信息序列
information sharing 信息共享
information source 信息源
information storage and retrieval 信息存储和检索
information storage means 信息存储方法
information storage 信息存储器
information-storing device 信息存储装置
information stream 信息流
information system access lines 信息系统存取总线
information system 信息系统
information theory 信息论
information transfer efficiency 信息传输效率，信息传送效率
information transfer rate 信息传输速度，信息传送速率
information transfer 信息传输
information unit 信息单位，信息单元
information wire 信息线
information word 信息字
information write wire 信息写入线
information 情报，信息，消息，资料，报导，报告，通知
informative 情报的，供给知识的
informatory 情报的，供给知识的
informer 举报人
infra-acoustic frequency 亚音频
infra-acoustic telegraphy 亚音频电报
infra-audible 亚声频的，亚音频的
infra-black synchronizing signal 黑外同步信号
infract 侵害，违犯
infradyne reception 低外差接收
infradyne 低外差法
infra-littoral deposit 远岸沉积
infranics 红外线电子学
infrared absorber 红外吸收剂
infrared absorption coefficient 红外线吸收系数
infrared absorption method 红外线吸收分析法
infrared absorption spectroscopy 红外线吸收光谱法
infrared analysis 红外分析
infrared analyzer 红外分析器
infrared camera 红外摄像机
infrared communication system 红外通信系统
infrared copy paper 红外复制纸
infrared detection system 红外检测系统
infrared detection 红外检测，红外线探测，红外检测仪
infrared diagnosis 红外诊断
infrared distancer 红外测距仪
infrared drying 红外干燥法
infrared felescope 红外线望远镜
infrared gas analyzer 红外线气体分析仪
infrared heater 红外线辐射加热器，红外线加热器
infrared homing 红外线自动寻的，红外加热，

红外线加热
infrared humidifier 红外线加湿器
infrared inspection 红外线检测，红外线检验，红外线探伤
infrared lamp 红外线灯
infrared leak detector 红外线检漏器
infrared light emitting diode 红外发光二极管
infrared microscope 红外显微镜
infrared nondestructive testing 红外无损检验
infrared oven 红外线炉
infrared-photoelastic effect 红外光弹效应
infrared photography 红外线摄影，红外照相术
infrared photograph 红外线摄影相片
infrared pyrometer 红外高温计
infrared radiant heater 红外线辐射加热器
infrared radiation pyrometer 红外辐射高温计
infrared radiation thermometer 红外辐射温度计
infrared radiation 红外辐射
infrared radiometer 红外辐射计
infrared rainbow 红外虹
infrared rays heating 红外线供暖
infrared ray 红外线
infrared reflection absorption spectroscopy 红外反射吸收光谱法
infrared scanner 红外线扫描仪
infrared sensing camera 红外传感照相机
infrared sensitive film 红外感光胶片
infrared spectrographic method 红外分光分析法
infrared technique 红外线技术
Infrared thermistor 红外热阻器，红外热敏电阻
infra-red thermometer 红外线温度计，红外温度计
infrared tracking 红外线跟踪
infrared transducer 红外传感器
infrared type burner 红外线式喷燃器
infrared vulcanization 红外线硫化
infrared 红外线，红外区，红外线的，红外的
infrasonic frequency 亚声频，亚音频
infrasonic 次声的，次声频的，亚声的
infrastructure construction 底层施工，基础建设，基础设施建设
infrastructure investment 基础设施投资
infrastructure project 基础设施项目
infrastructure rehabilitation 基础设施更新，基础设施改造
infrastructure substruction 底层结构
infrastructure 基础设施，公共建设，基础构架，下部结构，下层构造，基础，基底
infrequent drainage 非经常性排水
infrequent incident 少见的事故，稀有故障
infrequent wind 异常风
infringement of contract 违反合同
infringement of patent 侵犯专利权，专利权侵权
infringement 侵权，违反，违约，侵犯
infringe upon one's right 侵犯权利
infringe 侵害，违反，破坏，侵犯
infringing upon 侵害
in full compliance with 完全遵照……，完全按照……
in full 全部地，详细地
infuse 灌注，注入，浸渍
infusibility 不熔性，难熔性，扩散性
infusible hazard 吸入危害
infusible precipitate 不熔性沉淀物
infusible 不熔的，难熔的
infusion 灌注，注入，灌输，夹杂物
infusorial earth 硅藻土
in-gate 输入门，入口孔
in general terms 概括地说，一般地（说）
Ingersoll-Rand air compressor 英格索兰空压机
ingestion dose 摄入剂量
ingestion hazard 食入危害
ingestion 吸入量，吸入，吸收，摄取
ingle line diagram 单线图
ingoing checker 吸入阀，进入阀
ingoing power 进入功率，入射功率
ingoing 进来的，进入的，入射的
in good condition 状态良好
ingot blank 锭坯
ingot copper 铜锭
ingot iron 铁锭
ingot 锭，铸块筑块
ingredient 组成部分，成分，组分，要素，配料，拼料
ingress of heat 输入热量
ingress of light water 轻水漏入【重水回路】
ingress of water into the core 水侵入堆芯，水漏进堆芯
ingress pipe 导入管
ingress protection 入侵防护，防护等级，外壳防护等级，低压电器外壳防护等级【IEC 对电气设备外壳对异物侵入的防护等级。如 IP54，第一数字表示接触保护和外来物保护等级，第二数字表示防水等级】
ingress 进入，进口，入口
ingrown meander 增幅深切曲流，内生曲流
inhabitant 居民
inhabited environment 居住环境
inhabit 居住，栖息
inhalation dose 吸入剂量
inhalation hazard 吸入危害
inhalation 吸入剂，吸入，吸入量
inhaler 滤气器，吸入器
inhale 吸入
inharmony 不谐和，不协调，不和睦，冲突
in harmony 匹配，合拍，和谐无间
inhaul cable 头绳，拉索
inhaul 拖铲索，卷帆索，引索
inhaust 渗入
inherent ash 固有灰分，内在灰分
inherent capacitance 固有电容
inherent characteristic of a system 系统固有特性
inherent characteristic 固有特性
inherent closing time 固有闭合时间，固有合闸时间
inherent delay angle 固有引燃角，固有滞后角
inherent delay 固有滞后时间
inherent erodibility 天然易蚀性
inherent error 固有误差
inherent feedback 固有反馈
inherent filtration 固有过滤，固有过滤作用，固有滤波，自过滤

inherent flow characteristic　固有流量特性
inherent frequency　固有频率
inherent gamma flux　固有γ射线通量
inherent instability　固有不稳定性
inherent loss　固有损耗，固有损失
inherently safe reactor　内在安全性反应堆，固有安全性反应堆
inherently safe　固有安全，内在安全，自身安全的
inherently stable reactor　固有稳定性反应堆
inherently stable　固有稳定
inherent moisture content　固有水分含量
inherent moisture　固有水分，内在水分
inherent neutron source　固有中子源
inherent noise　固有噪声
inherent opening time　固有断开时间，固有分闸时间
inherent overheating protection　内部过热保护
inherent reactivity　固有反应性
inherent reactor stability　反应堆固有稳定性
inherent regulation of controlled plant　调节对象自平衡，受控装置自调节
inherent regulation　内部自动调整，自平衡调节，固有调整，自行调节，固有变动率
inherent reliability　固有可靠性，内在可靠性
inherent safety features　固有安全特性
inherent safety　固有安全性，内在安全性
inherent serf-excitation　固有自励磁
inherent settlement　原有沉陷
inherent stability　固有稳定性
inherent storage　固有存储器
inherent strain　内在应变
inherent stress　内在应力
inherent vice　内在缺陷，固有缺陷
inherent voltage change　固有电压变化
inherent voltage regulation　固有电压调整率
inherent　内在的，固有的，本来的，先天的，本质的，固有
inhering　记录
inherited drainage　遗留水系
inherited error　固有误差，遗留误差
inherited meander　嵌入曲流
inherited　累积的，固有的，继承，遗传
inhibit circuit　禁止电路，抑制电路
inhibit current pulse　阻塞电流脉冲
inhibited acid　加缓蚀剂的酸
inhibited oil　带抗氧化剂的油，抗氧化油
inhibited oxidation process　抑制氧化工艺
inhibited oxidation　抑制限制氧化
inhibited pulse　被禁止脉冲
inhibiter　缓蚀剂，抑制剂，阻止剂
inhibit gate　禁止门
inhibit halt flip-flop　禁止停机触发器
inhibiting effect　缓蚀效应，抑制作用，抑制效应，抑制剂，缓凝剂，抑制效果，抑制因素
inhibiting input　禁止输入
inhibiting interpreter　抑制的翻译程序
inhibiting signal　禁止信号，抑制信号
inhibition　缓蚀，抑制，制止，遏制，抑制作用
inhibitor desulfurization　抑制剂脱硫
inhibitor effectiveness　缓蚀剂效率，抑制剂效率
inhibitor film　缓蚀剂膜
inhibitory action　抑制作用，迟滞作用
inhibitory coating　保护层，防护层
inhibitory-gate　禁止门，与非门
inhibitory input　禁止输入
inhibitory　抑制的，禁止的，迟滞的
inhibitor　防腐蚀剂，缓蚀剂，抑制剂，阻化剂，阻垢剂，抑制器，减缓剂，防锈蚀剂
inhibit pulse　禁止脉冲
inhibit reclosing　禁止重合闸
inhibit signal　禁止信号
inhibit winding　保持线圈，封闭绕组，禁止绕组
inhibit wire　禁止线
inhibit　禁止，抑制，制止
inhomogeneity of structure　组织不均匀
inhomogeneity　多相性，不均匀，不均匀性，不同类，不同质，不同族
inhomogeneous field　非均匀场
inhomogeneous flow　非均匀流动
inhomogeneous medium　非均匀介质
inhomogeneous soil　非均质土
inhomogeneous turbulence　不均匀乱流
inhomogeneous　非同质性的，不均匀的，非齐次的，多相的，不均匀性
inhour equation　倒时方程
inhour formula　倒时公式
inhour unit　倒时单位【反应性单位】
inhour　核反应单位【每周期（小时）】，倒时数，反时
in-house standard　内部标准，厂内标准
in-house　由本机构内部产生的［地］，机构内部的［地］，自身的，内部的，固有的
in increments of　以……的增量
initial abstraction retention　（降水的）初始截留
initial abstraction　初渗，初始损失，初损
initial a contract　草签合同
initial actuation time　起始激励时间，初始动作时间
initial adjustment　初调节
initial admission of steam into the turbine　汽轮机首次进汽，汽机初始进汽
initial amplitude　初振幅，初（振）幅
initial appearance　初始状态
initial appropriation　初步拨款，首次拨款
initial approximation　初始近似值
initial ash softening point　灰初始软化温度，灰初始软化点
initial audit　初步审计
initial blasting　初次爆破
initial blowdown pressure　初始排放压力【阀瓣不回座】
initial blowdown　初始排放
initial boiling point　初始沸点
initial budget estimate　初步概算
initial calibration level　起始校正电平
initial calibration　初始校准，初始标定
initial capacity　初始容量
initial capital investment　创办资本投资
initial capital　初始资金，启动资金，创办资本
initial cavitation　初始汽蚀
initial charge oil capacity　初充油量

initial charge 首炉料，原始电荷，初充电，首次燃料装量
initial charging voltage 起始充电电压
initial coal pile 一次煤堆
initial code vector 初始码矢
initial collapse pressure 湿陷起始压力
initial collision 初始碰撞
initial compaction 初次碾压
initial compression 初始压缩
initial concentration of dust 初始粉尘浓度
initial concentration 初始浓度，起始浓度
initial condensation 初始凝结，初凝
initial condensing 初始凝结
initial condition 初始条件，原始数据，初始工况
initial conductor tension 线路导线起始张力
initial control 初步控制
initial conversion rate 初始转换率
initial conversion ratio 初始转换比
initial cooling 初始冷却，初冷却
initial core charge 首炉料，首次燃料装载
initial core 初始堆芯
initial corrosion 初始腐蚀
initial cost 初始费用，初期费用，基本建设费，原始成本，初成本
initial creep 初期蠕变，初始蠕变
initial criticality test 初次临界试验
initial criticality 首次临界
initial current 起始电流，初电流
initial data 初步资料，起始数据，原始数据，原始资料
initial deflection 初始挠度
initial deformation point 初始变形温度，初始变形点
initial deformation temperature 初始变形温度
initial deformation 初始变形
initial density 初始密度
initial design stage 初设阶段
initial detention 初始滞流
initial disturbance 初始扰动
initial dose 初始剂量
initial drying shrinkage 初始干缩
Initial energization 初始受电，首次受电，首次倒送电
initial enrichment （燃料）初始富集度，初始浓缩度，初始富集度
initial error 初期误差，起始误差
initial estimate 初步测算
initial excess reactivity 初始过剩反应性
initial excitation response 起始励磁响应，起始励磁电压上升速度
initial expense 筹建费，开办费，初次费用
initial failure 初期故障
initial fee 入门费
initial filling run 初始充填焊缝【用低热量输入】
initial filling （水库等的）初次蓄水，初期蓄水
initial firing of boiler 锅炉第一次点火
initial firing 首次点火
initial fission cross section 初始裂变截面
initial fluidizing velocity 初始流化速度
initial form 原始地形
initial fuel charge 首炉料，首次燃料装置
initial fuel-conversion ratio 初始燃料转换比
initial furnace temperature 初始炉温
initial gamma radiation 初始γ射线辐射
initial gradient 初始坡降
initial guess 初始猜测值
initial heat of hydration 初始水化热
initial impound period 初蓄期
initial impulse 起始脉冲
initial infiltration capacity 初始下渗量
initial infiltration rate 初期浸透强度，初渗（量）
initial injection 起始喷射
initial input program 起始输入程序
initial input routine 起始输入程序
initial installation 初期装机
initial internal pressure 初始内压
initial inverse voltage 起始反向电压
initial investment 初投资，创办投资
initial ionization 初始电离
initialization 初始化，置初值
initialize 起始，设定，初值，发送
initializing data 起始数据
initial learning period 试运行期
initial level 零米标高，基准面标高
initial loading 初始装载
initial load 初负荷，初始负荷，初始负载
initial loss 初始损失
initial lump sum 首付总额
initial magnetization curve 起始磁化曲线
initial margin 初始保证金
initial measurement 初始量测
initial microprogram loading 初始微程序调入
initial moisture content 初始含水量
initial moisture deficiency 初始水分不足量
initial motion 初动
initial nuclear radiation 初期核辐射，早期核辐射
initial numerical data 起始的数字数据
initial operation 初始运行
initial order 起始指令，首批订货
initial outlay 初始投资，初投资
initial output test 起始出力试验
initial parameter 初参数，最初参数
initial particle velocity 初始颗粒速度
initial payment 初付款，首期付款，入门费，初付款
initial performance 最初性能
initial period 初始周期
initial permeability 起始磁导率，初磁导率
initial phase angle 初相角
initial phase 起始相位，初相
initial pit dewatering 初期排水
initial placement condition 初始填筑条件
initial plan 初步设计，初步方案
initial plume dimension 初始羽流尺度
initial plume rise 初始羽流抬升
initial point 始点，起始点，起点
initial pore water pressure 初始孔隙压力，初始孔隙水压力
initial position 起始位置，初始位置
initial power receiving 首次受电
initial power transmission 首次送电

initial power 启动功率
initial precipitation 初期降水
initial pressure rate limiter 初压速率限制器
initial pressure regulation 新汽压力调节器
initial pressure 初始压力，初压力
initial press variation -power correction curve 初压变化对功率的修正曲线
initial prestress 初预应力
initial price 开价，基础价格
initial processing 初步处理
initial radiation 初始辐射，初辐射
initial reactivity 初始反应性
initial reading 初次读数，初始读数，起始读数
initial receiving point 初始接收点
initial release turbulence 初始释放湍流度
initial residual heat removal phase 余热排除初始阶段
initial resistance of filter 过滤器初阻力
initial resistance 起始电阻，初电阻，初阻力，起始阻力【空气过滤器】
initial response 初始响应
initial rolling 初次碾压
initial sag 初始弧垂
initial section 起始段
initial setting time 初凝期，初凝时间
initial setting 初整定值，初调，初整定，初调值，初凝
initial settlement 初始沉降，初始沉陷
initial set 初凝
initial shear stress ratio 初始剪应力比
initial shoreline 原始海岸线
initial signal unit 初级信号单元
initial slackness 原始轴承间隙
initial soil moisture 初始土壤水分
initial soil storage 初期土壤储水量
initial soil 原始土壤
initial speed 初速
initial stage 开始阶段，初期，初始阶段
initial starting current 初始起动电流
initial starting 起步，启动
initial start-up cost 首次启动费用
initial start-up test 交付运行试验，首次启动试验
initial start-up 初次启动，第一次启动
initial start 初次启动
initial state 原始状态，初始状态，起始状态
initial steady discharge 初始稳定流量
initial steam admission into the turbine 汽轮机首次进汽
initial steam admission 首次进汽【汽轮机】，首次送汽，新汽进入
initial steam conditions 蒸汽初参数
initial steam flow rate 主蒸汽流量
initial steam parameter 初始蒸汽参数
initial steam pressure regulator 新汽压力调节器
initial steam pressure 新蒸汽压力，新汽压力，进汽压力
initial steam temperature 新汽温度，进汽温度
initial steam 初蒸汽，新蒸汽
initial storage 初始库容
initial stream 初始料流
initial strength 初始强度
initial stress method 初始应力法
initial stress 初始应力，初应力
initial suppression 初始抑制，初始压制
initial survey 初测
initial susceptibility 起始磁化率
initial synchronization 初并网，首次并网
initial tangential modulus 初始切线模量
initial temperature difference of heater 加热器初温差
initial temperature difference 始端温差，初始温差
initial temperature 初始温度
initial temp variation-power correction curve 初温变化对功率的修正曲线
initial tension 初张力
initial tightening 预紧
initial torque 初始转矩
initial transient recovery voltage 起始瞬态恢复电压
initial treatment 前处理，初处理
initial tunnel turbulence 风洞初始湍流度
initial unbalance 初始不平衡
initial update 首次更新
initial value 起始值，初值，始值
initial vector display point 初始向量显示点
initial velocity 起始速度，初速，初速度
initial void ratio 初始孔隙比
initial voltage response 初始电压上升速度，初始电压响应
initial voltage 初始电压
initial water deficiency 初始水分不足量
initial water level 初始水位
initial wave 初生波，原波
initial wind speed 初始风速
initial （姓名或组织名称等的）首字母【initial 的名词复数】，姓名中的大写字母，缩写，（印刷品的章节或段落等开始的）特大的大写字母，用姓名的首字母作标记（或签名），草签，最初的，初始的
initiate a notice 发出通知
initiate key 启动键
initiate 发动，起始，起动，启动，开始，起爆
initiating element 启动元件
initiating event 触发事件，初始事件，初因事件
initiating pulse 启动脉冲，触发脉冲
initiating relay 始动继电器，启动继电器
initiating signal 触发信号，启动信号
initiating 启动的，开动的，起爆的，引发
initiation area discriminator 初始区域鉴别器
initiation control device 启动控制设备
initiation event 起始事件
initiation fee 入会费
initiation of fluidization 起始流态化
initiation switch 启动开关
initiation 启动，发起，开始，起始，起爆，励磁
initiatives for ……的举措
initiative 倡议，积极性，主动权，主动性，首创精神，创始，举措，主动的，自发的，起始的

initiator of conformance test　一致性测试委托者
initiator　激励器，点火器，引燃器，首创者，起爆剂，起爆器，引发器
injected bubble　注入气泡
injection air　喷射空气，雾化空气
injection and leak-off system　（汽轮机）轴封防泄漏系统
injection angle　喷射角，注入角
injection apparatus　喷射装置，注入器
injection burner　引射式喷燃器
injection coal　喷吹煤
injection cock　喷嘴，喷管，喷射旋塞
injection coefficient　喷射系数
injection commutator　浇注换向器
injection condenser　喷射式凝汽器
injection cooler　喷射式冷却器
injection current　注入电流
injection drop　喷嘴压差
injection flow rate　喷水量，喷注量
injection grouting　压力灌浆
injection heating　注入加热
injection hose　喷射软管
injection line　注入管线
injection molding　喷射造型法，喷射模塑法，注压法，注模
injection neutral beam　注入中性束
injection noise　喷射噪声
injection nozzle　喷射管嘴，喷嘴
injection oil engine　喷油发动机
injection orifice　喷油孔，喷嘴，射水孔
injection period　喷射时间，喷射周期
injection phase　（失水事故）注入阶段
injection pipe　喷射管
injection point　注射点，加药点
injection port　注射口，进样口
injection pressure　灌注压力，喷射压力
injection process　注液法
injection pump　喷射泵，注射泵，注入泵
injection ratio　注射系数，注射比
injection station　注液站
injection syringe　注射管，注射器
injection system　喷浆系统，喷射系统，注入系统
injection tank　注入箱
injection technique　注射技术
injection temperature　进口温度，喷入温度
injection transformer　浇注绝缘变压器
injection tube　注射管，喷射管
injection-type engine　喷油式发动机
injection-type　喷射式
injection valve　喷射阀
injection water cooling system　注水冷却系统
injection water　注入水，注射水
injection well　（废液）注入井，注水井
injection　喷射，灌注，注射，射入，注入
injector head　雾化头，喷嘴头
injector overflow　喷射器，溢流孔
injector setting　喷射器调整
injector set　喷嘴组
injector spray tip　喷嘴头
injector　喷射器，注射器，喷嘴，注水器，牵车器，喷油器
INJ＝injection　喷射
injunction　禁止令，强制令
injunctive relief　禁令性救济，强制救济
injured party　受害方，受损害方
injured powder　变质炸药
injure　伤害
injuries combined with radiation damage　辐射复合损伤
injurious consequence　损害性后果
injurious defect　有害的缺陷
injurious imperfection　有害的缺陷
injurious surface mark　表面伤痕
injurious　有害的
injury　损伤，损害，伤害，杀伤
injustice　不公正
ink box　墨盒
ink drawing　墨线图
ink fog printer　墨水雾式打印机，墨水雾式印刷机
in-kind payment　以实物支付，实物支付
inking-in　上墨
ink jet printer　喷墨式打印机
inkless recorder　无墨水记录仪
ink writer　油墨印码器，有色记录器，印字机
inlaid flooring　镶嵌地面
inland area　内陆地区
inland basin　内陆盆地
inland bill of lading　陆运提单
inland city　内陆城市
inland desert　内陆荒漠
inland earthquake　内陆地震
inland enterprise　内陆企业
inland fossil fuel power plant　内地火电厂
inland freight　内陆运费
inland inundation　内涝
inland lake　内陆湖
inland plain　内陆平原
inland river　内河
inland surcharge　内陆运输附加费
inland transportation　内陆运输
inland waterway hydraulic dredging grab　内河液压疏浚抓斗
inland waterway ports　内河港口
inland　内陆的，内地，内陆
inlaying　镶嵌
inlay　嵌入，插入，插入物
inlead　引入线，引入
inleakage of air　空气漏入，空气渗入，漏风
inleakage of cooling water　漏入冷却水
inleakage of radioactivity　放射性内漏
inleakage of wind　漏风
inleakage　吸入，漏入，渗入，贯穿内部
inlet air　进风，进气
inlet angle　进口角
inlet annulus　进汽环室，进气环室
inlet attack　进口侵蚀，进口攻角，进口冲蚀
inlet blade angle　叶片进口角
inlet box　进口风箱，入口室，进线箱，进线盒
inlet branch　进汽支管，进气支管
inlet bucket　导风轮叶片【高心式压缩机】，进

口导叶
inlet casing 进气缸，进气室，前汽缸
inlet chamber 进气室，进口水室
inlet channel 进水槽，冷却水进水渠，进水渠
inlet check valve 入口单向阀
inlet circuit 输入端电路
inlet condition 新汽参数，进口参数，进口条件
inlet control system 进口调节系统
inlet culvert 进水涵洞
inlet device 入口装置
inlet diffuser 进口扩压管
inlet duct 进风道
inlet-end attack 入口端侵蚀
inlet filter 进口过滤器，入口过滤层
inlet flow angle 进气角，进口气流角
inlet flow blockage 进口流阻塞
inlet flow maldistribution 进口流量分配不均匀
inlet for storm water 雨水口
inlet geometric angle 进口几何角
inlet guide blade 进口导叶
inlet guide vane adjustment 进口导向翼调整
inlet guide vane 进口导叶，前置导叶
inlet header 进口联箱，进口总管
inlet head 进口压头
inlet hole 进水孔
inlet hose 吸入软管
inlet import 进口阀
inlet jumper 进口跨接管，连接管
inlet length 进口长度
inlet loss 进口损失
inlet louver 进口百叶窗，入口百叶窗
inlet Mach number 进口马赫数
inlet manifold 进气总管
inlet nozzle 进口接管，进口管嘴，入口管嘴
inlet of sewer 下水道入口
inlet opening 入口，进气孔
inlet part 埋入部分，(水轮机的) 埋入件
inlet pipe 进口管，入流管，引入管
inlet plenum 进气室，进口腔室
inlet port 入口，进气口
inlet pressure control 入口压力控制
inlet pressure 进口压力，入口压力
inlet region 进口区
inlet scoop 吸气口，进气口，进风斗【汽轮发电机】
inlet scroll 进口蜗壳
inlet side room 进水室
inlet silence 入口消声器
inlet size 入料粒度，入口尺寸
inlet sleeve 进汽套管，进气套
inlet sluice 进口水闸，渠道闸
inlet steam 入口蒸汽
inlet strainer 入口过滤器，入口滤网
inlet stroke 吸入冲程，进气冲程
inlet temperature 进口温度，入口温度
inlet tube 进口接管
inlet valve of penstock 压力钢管进口阀
inlet valve 进汽阀【汽轮机】，进水阀，进口阀，进气阀，入口阀，入口门
inlet vane control (风机的) 入口叶片调节
inlet vane 进口叶片，入口叶片
inlet velocity diagram 进口速度图
inlet velocity 进口速度
inlet volute 进口蜗壳
inlet well 排水窨井
inlet whirl 进口旋涡，进口预旋
inlet 进口，进入，引入，入口，输入端，入口段
inlier 内窗层，内围层
in lieu of 代替，替代，代替
in-line alpha monitor 在线α监测器，在线α射线监测器
in-line analysis 线上分析，在线分析
in-line arrangement 串联式布置，顺列布置
in line array 行阵
inline bank 顺列布置管束
inline booster 轴向加速器，序列式增压器
inline check valve 直通单向阀
in-line closure 管线截流【管线盲板，管孔盲板等】
in-line coding 成组编码
inline data process(ing) 在线数据处理
in-line dilution 管内稀释
in line displacement 顺风向位移
in-line engine 单列式发动机
in-line gamma absorptiometer 在线γ射线吸收计
in-line gamma monitor 在线γ监测器
in-line glove box 在线手套箱
inline heater 串联在管道上的加热器
in-line neutron monitor 在线中子监测器
in-line position 顺列布置
in-line procedure 直接插入的程序，联机程序
in-line processing 在线处理
in-line pump 管道泵
in-line radiation monitor 在线辐射监测器
in-line radiation 在线辐射监测器
in line response 顺风向响应
in-line rotary car dumper 串联式翻车机
in-line sensor 管线传感器
in-line splicing sleeve 管线连接套筒
in-line splicing 管线连接
in-line spoon cutter 直插式截样器，料勺
in-line subroutine 直接插入子程序
in-line suction 管线吸入
in-line system 联机系统，成簇数据处理系统
in line with international practices 与国际惯例接轨
in line with 跟……一致，符合，本着
in-line 串联的，管线内的，并网的，联机的，顺列的，排成行的，在一直线上的，在线的
in lots 分批，分堆
INLT＝Inlet 入口
in mass flux density 进口质量通量密度
inmost layer 最内层
inmost 最深的
in-motion radiography 动态射线照相术
in-motion track scale 动态轨道衡
innate 天生的，特有的，固有的，
in nature 实质上，事实上
innavigable 不通航的
inner 内部的
inner air circuit 内部空气回路

inner anchoring section　内锚固段
inner back end　内尾端
inner bank　内岸
inner bar　内滩
inner basin of levee　堤内集水区
inner bearing face　内支承面
inner bearing ring　轴承内座圈
inner bearing　内轴承
inner bellows seal　波纹管内密封
inner blanket　内转换区
inner breakwater　内防波堤
inner bremsstrahlung　内轫致辐射
inner cage　内机座，内笼，内鼠笼
inner capillary water　内毛管水
inner cap　内盖
inner-cased　带内护板的
inner casing　（水冷壁背火侧）内护板，内缸，内壳，内汽缸
inner coil　下层线圈，内层线圈
inner conductor　内层导体，内层导线
inner cone　焰心【气体焊火焰】
inner containment　内安全壳
inner continental shelf　大陆棚内半部
inner contradiction　内在矛盾
inner conversion　内变换
inner-cooled machine　内冷电机
inner-cooled stator coil　内冷定子线圈
inner-cooling　内冷
inner cylinder　内缸
inner dead point　内死点
inner diameter　内径
inner electron　层电子
inner energy　内能
inner enrichment region　内加浓区
inner envelope　内信封
inner enveloping profile　内包络断面
inner force　内力
inner formwork　内模板
inner frame　（电机）内机座
inner fuelled zone　内加料区
inner gland　内侧汽封，内汽封
inner group flux　群内通量
inner harbor　内港
inner inspection　内部检查
inner insulation　内绝缘
inner level memory　内级存储器
inner linearization　内线性化
inner liner　衬里
inner oil seal　内油封
inner oil sump　内油槽
inner orientation　内方位
inner packing　内包装
inner photoeffect　内光电效应
inner-pole type alternator　内极式同步发电机
inner race　（滚动轴承）内座圈，内环，（轴承的）内滚道，内套，星型套
inner reflected reactor　内反射式反应堆
inner reflector　内部反射层，内反射层
inner ring and outer ring　内外定位环
inner ring of diaphragm　隔板内环
inner ring　内定位环，内环
inner rotor　（超导电机的）内转子
inners　内部
inner sealed box　内部密封箱，内部密封盒
inner sheath　内护套，内屏蔽套
inner shelf　内侧大陆棚，内陆架
inner shell electron　内壳层电子
inner shell　内缸，内壳层，内电子层
inner shield　内屏蔽
inner shroud　堆芯围板，堆芯围筒，内壳
inner sleeve　内套筒，内套管
inner stator　内定子
inner strap　内部条带【燃料组件】
inner stress　内应力
inner stroke　内行程
inner surface inspection　内表面检查
inner surface　内表面
inner tower　（觇标的）内架
inner tube　内管，内筒
inner wall　内壁，内墙
inner waterbreak　内部防波堤
inner water tube boiler　内水管锅炉
inner work　内功
in no case　决不【无论怎样都不行】
innocuous effluent　无害废水
innocuous　无毒的，无害的
in no event　决不【无论怎样都不行】
innovation ability　创新能力
innovation in science and technology　科技创新
innovation solution　创新的解决方案
innovation　新设施，改进，革新
innovative decision　革新决策
innovativeness　创新精神
innovative potential　革新潜力
innovative program　创新方案
innovative spirit　革新精神
innovative wind energy conversion system　创新型风能转换系统
innovative wind system　革新性风能转换系统
innovative　革新的，创新的
innoxious substance　无害物质
innoxious　无害的，无毒的
inodorous　无气味的
in one's favor　以……为受益人
inoperability　停堆，停机，不可用
inoperable　不在役，不能运行的
inoperating contact　停用触点，静止触点
in operation　正在实施，正在运行
inoperative period　停机期，非运行期
inoperative　不能使用的，不工作的，不运行的，无效的，不起作用的
inorganic acid　无机酸
inorganic analysis　无机分析
inorganic clay　无机黏土
inorganic coagulant　无机凝结剂
inorganic insulation　无机绝缘
inorganic ion exchange　无机离子交换
inorganic material　无机材料
inorganic phosphorus　无机磷
inorganic pollutant　无机污染物
inorganic salt　无机盐
inorganic sand　无机砂土

inorganic sediment　无机沉淀，无机沉淀物
inorganic silt　矿质泥沙
inorganic soil　无机土
inorganic solidification agent　无机固化剂
inorganic sulfur　无机硫
inorganic toxic material　无机毒物
inorganic waste water　无机废水
inorganic zinc-rich paint　无机富锌漆
inorganic　无机的
in other words　换言之，换句话说
in-out box　输入输出盒
inoxidizability　不可氧化性
inoxidizable　不可氧化的，抗氧化的
inoxidizing coating　抗氧化涂层
in pace with　随着
in parallel and series　混联，串并联
in parallel with　与……平行，与……同时
in parallel　并行的，并联的，平行的，平行，并联，并机，并行
in particular　释义尤其，特别
in-phase component　同相分量
in-phase operation　同相运行
in phase opposition　反相的
in-phase signal　同相信号
in-phase state　同相状态，等相状态
in-phase vibration　同相振动
in-phase voltage　同相电压
in-phase　同相，同相位
in-pile corrosion　堆内腐蚀
in-pile creep　（核燃料）堆内蠕变
in-pile densification　（核燃料）堆内密实
in-pile experiment　堆内实验
in-pile gamma scanning　堆内γ射线扫描
in-pile instrument　堆芯仪表
in-pile irradiation facilities　堆内辐照装置
in-pile loop　堆内回路
in-pile test loop　堆内试验回路
in-pile test　堆内试验
in-pile time　堆内停留时间
in-pile　反应堆内部，堆内
in-place density　原状密度
in-place entombment　就地埋葬，临时埋葬，现场测定
in place of　代替
in-place regeneration　体内再生，就地再生
in-place test　实地试验
in-place　原地，就地，现场，原状，在位
in place　在对的位置，在工作，准备就绪，适当（的），在适当的地方，在恰当的位置
in plane bending　面内弯曲
in-plant boiler silo　厂内锅炉筒仓
in-plant coal bunker　厂内锅炉筒仓
in-plant coal handling system　厂内输煤系统
in-plant communication system　厂内通信系统
in-plant control　厂内控制
in-plant exposure　厂内辐照量
in-plant handling　输送机械设备
in-plant noise　厂内噪声
in-plant overall economic calculation　厂内全面经济核算
in-plant system　室内系统，近距离（控制）系统
in-plant test　现场试验
in-plant training　厂内培训
in-plant transport system　厂内运输系统
INPO(Institute of Nuclear Power Operations)　核电运行研究院
inpolar　内极点
in position　就位
in principle　原则上，基本上
in-process control　生产过程控制
in-process inspection　加工过程的检查，工序检测
in-process installation tests and checks　安装过程的测试和校核
in-process inventory　过程中的存料量
in process of　正在……
in-process　加工过程中的，生产过程中的，处理过程中的
in progress　正在进行中，程序进行中
in proportion as　按……的比例，按……比例
in proportion of　按……比例
in proportion to　与……成（正）比例，与……相比，在……中所占的比例
in proportion　按比例
in purpose　故意地
in pursuance of　依，按，为执行［实行］……
input accountability vessel　进料计量容器
input admittance　输入导纳
input amplifier　输入端放大器，输入放大器
input amplitude　输入振幅
input and output with isolated common point　有公共隔离点的输入和输出
input area　输入存储区
input axis　输入轴
input bandpass filter　输入带通滤波器，输入端带通滤波器
input block　输入存储区，输入信息块，输入部件，输入信息组
input buffer　输入缓冲器
input capacitance　输入电容
input capacity　输入功率
input channel　输入通道
input checking equipment　输入检验设备
input choke　输入轭流圈
input circuit　输入电路
input computer　输入计算机
input concentration　输入物浓度，进料浓度
input conductance　输入电导
input control valve　进气调节阀
input crystal　输入晶体
input current　输入电流
input curve　输入量曲线
input data assembler　输入数据汇编程序
input data　输入数据
input device　输入装置，输入设备，输入器
input digit　输入数字，输入数位
input element　输入元件
input end　输入端
input energy　输入能
input equipment　输入设备，输入装置
input filter　输入滤波器
input flyback　输入回描

input function 输入函数
input generator 输入信号发生器
input impedance 输入抗阻，输入阻抗
input impulse 输入脉冲
input information 输入信息
input interface unit 输入接口装置
input inversion 输入信号反向
input job stream 输入作业流
input level 输入电平
input limiter 输入限幅器
input limit 输入极限
input linear group 输入线性部分
input loading factor 输入负载因数
input loop 输入回路
input loss 输入损耗
input magazine 输入卡片箱
input materials 输入材料，投料
input mechanism 输入机构
input medium 输入设备，输入装置，输入媒介
input method 投入分析法
input module 输入模块
input noise 输入噪声
input nominal speed 额定输入转速
input offset voltage 输入补偿电压
input order 输入指令
input-output adapter 输入输出转接器
input-output allocation 输入输出分配
input-output analysis 投入产出分析
input-output balance 进料出料衡算，输入输出衡算
input-output bound 受输入输出限制的
input-output buffer 输入输出缓冲器
input-output channel 输入输出通道
input-output characteristic 输入输出特性
input-output control center 输入输出控制中心
input-output control command 输入输出控制命令
input-output controller 输入输出控制器
input-output control module 输入输出控制模块
input-output control program 输入输出控制程序
input-output control system 输入输出控制系统
input-output control 输入输出控制
input-output device 输入输出设备，输入输出装置
input-output equipment 输入输出设备，输入输出装置
input-output instruction 输入输出指令
input-output interface 输入输出接口
input-output interrupt identification 输入输出中断识别
input-output interrupt indicator 输入输出中断指示器
input-output library 输入输出程序库
input-output limited system 受输入输出限制的系统，输入输出受限制的系统
input-output model 投入产出模型
input-output multiplexer 输入输出多路转换器
input-output order 输入输出次序，输入输出指令
input-output processor 输入输出处理机
input-output reference 输入输出访问
input-output register 输入输出寄存器
input-output request 输入输出请求
input-output routine 输入输出程序
input-output supervisor 输入输出管理程序
input-output switching 输入输出转接
input-output table 投入产出表，输入输出表
input-output traffic control 输入输出流量控制
input-output transfer 输入输出传送
input-output unit 输入输出部件，输入输出装置
input-output 输入输出，输入输出装置
input parameter 输入参数
input pass-band 输入通带
input plane 输入平面
input potentiometer 输入电势计
input power 输入功率
input pressure 进口压力
input program tape 程序输入带
input program 输入程序
input pulse 输入脉冲
input reactivity spectrum 输入反应性谱
input rearrangement 输入重新排列
input reference axis 输入基准轴
input register 输入寄存器
input rejection test 切断电源试验，断电试验
input resistance 输入电阻
input resonator 输入谐振器，输入共振器
input routine 输入程序
input section （数据）输入程序段
input shaft 输入轴
input side 输入侧
input signal 输入信号
input speed 输入速度
input stage 输入级
input station 输入站，输入终端
input storage 输入存储器［区］
input subroutine 输入子程序
input system 输入系统
input tax 进项税
input temperature 进口温度
input terminal 进线端子，输入端子
input time constant 输入时间常数
input transformer 输入变压器
input translator 输入翻译器
input tube 输入管
input unit 输入器，输入设备
input value 输入值
input variable 输入变量，输入量
input wattage 输入功率，输入瓦特数
input white noise 输入白噪声
input winding 输入绕组
input wiring diagram 输入端接线图
input 输入功率，输入，输入量，输入电压，输入信号
in quadrature 正交
in quadruplicate 一式四份
in question 上述提到的，正在谈论的，考虑中的
in quintuplicate 一式五分
inquire 追究，询问，打听，调查
inquiry about 查询
inquiry and communication system 询问和通信系统

inquiry and subscriber display 查询和订购显示器
inquiry circuit 询问电路
inquiry display terminal 询问显示终端
inquiry station 询问站，询问台
inquiry system 询问系统
inquiry unit 查询装置
inquiry 打听，查询，调查，询价，询价书
in-reactor experiment 堆内实验
in-reactor loop 堆内回路
in-reactor 反应堆内
in readiness for 为……准备妥当
in real terms 实质上
in reference to 关于，就……而论
in relation to 关于，涉及
INREQ(information requested) 要求信息
in respect of 涉及，至于，在……方面，就……而论，关于
in respect thereof 就此事，在此，就其而言，关于上文已提及的
in respect to 就……而论，关于
in return for 以换取，作为……的交换，作为……的回报
INRTNG=inerting 惰性的
in running nips(/pinch) points 咬入点
in running order 工作状态
inrush current 涌入电流，涌流，冲击电流，突入电流，瞬间起峰电流，合闸电流
inrush kVA 冲击千伏安，冲击负荷
inrush starting current 启动冲击电流
inrush transient current 瞬时冲击电流
inrush 侵入，流入，启动功率，启动冲量，涌入，浸入
INSAG(International Nuclear Safety Advisory Group) 国际核安全咨询组
insanitary 不卫生的
inscattering term 内散射项
inscriber 记录器
inscription 注册，登记，题词，碑文，编入名单
insecticide 杀虫剂
insectivorous bird 食虫鸟
insect pest 虫害
insect 昆虫
in security for 作为……的保证［担保］
inselberg 岛山
insensitive interval 不灵敏时间
insensitive 不灵敏的，迟钝的
insensitivity zone 不灵敏区
insensitivity 不灵敏性
insequent drainage 斜向水系
insequent valley 斜向谷
in series connection 串联安装
in series 串联的，按顺序排列，成批地，串联
insertable evaporator 可插入式汽化器
insert a thimble 插入套管
insert bolt 顶螺销，暗插销
insert cover 插入式围带
inserted blade 插入式叶片
inserted bolt 预埋螺栓
inserted casing 插入套管
inserted plate 预埋件，插入板
inserted shim 嵌入垫片
inserted sleeve 预埋套管
inserted tooth 镶齿
inserted wedge 嵌入楔块
inserter 插入物，插件，隔板，隔金，隔叶块，嵌入片
insert gauge 塞规
inserting coil 下线，嵌线
inserting tool 下线工具，嵌线工具
inserting winding 下线，嵌线
insertion depth （控制棒的）插入深度，淹没深度
insertion force 插入力
insertion gain 插入增益，介入增益
insertion in the reactor 装入反应堆
insertion loss 插入损耗，介入损耗
insertion machine 嵌线机，下线器，插入机
insertion of reactivity 反应性的输入，反应性引入
insertion of withdrawal rate 插入或抽出速率【控制棒】，移动速率
insertion piece 插件
insertion pole 插入杆
insertion power function 介入功率函数
insertion ring 填料圈
insertion sequence 插入程序
insertion shaft 插入轴【加套】
insertion switch 引入开关
insertion tool 插入工具
insertion 接入，插入，介入，嵌入，钳装物，插入物
insert panel 插件板
insert radiator 嵌入式散热器
insert reactivity 引入反应性
insert stressmeter 插入式应力计
inserts 埋入件
insert valve 嵌入式阀
insert 嵌线，嵌入，插入，接入，插入物，嵌入物，埋入件，插件
in-service behavior 运行特性，在役性能
in-service condition 工作状态
in-service examination 运行中检查
in-service factor 实际运行率
in-service forced derated hours 强迫降低出力运行小时
in-service inspection equipment 在役检查设备
in-service inspection system 在役检查系统
in-service inspection 运行中检查，在役检验，在役检查
in-service life 使用寿命
in-service maintenance derated hours 维护降低出力运行小时数
in-service maintenance 运行中维护
in-service planned derated hours 计划降低出力运行小时
in-service silo(/bunker) 使用中的煤仓
in-service staff 在职人员
in-service state 运行状态
in-service surveillance 运行监督，在役监督，在役监视
in-service training 在职培训

in-service unit derated hours 机组降低出力运行小时
in-service unplanned derated hours 非计划降低出力运行小时
in service 检修中的，运行中的
in settlement of 解决
inset 插图，插页，插入物，嵌入，插入，镶嵌
inshore engineering 沿岸工程
inshore wind 向岸风
inside admission 内进汽
inside bank slope 河岸内坡
inside cable 户内电缆
inside caliper gauge 内径千分尺
inside calipers 内卡钳，内卡规
inside casing 内框，内壳体
inside clearance ratio 内间隙比
inside conductor 内导线
inside corner weld 内角焊接
inside cylinder 内缸
inside diameter 内径
inside dimension 内尺寸
inside face of wall 墙的内侧面
inside face 内表面
inside lining 内衬砌，内衬
inside-mixing burner 内混式喷燃器
inside-out filter 外流式过滤器
inside-out motor 转枢式同步电动机
inside pitch line length 齿根高度
inside screw yoke 内螺纹轭
inside screw 内螺纹
inside trimming 内部整修
inside waters 内海，内陆水域
inside width 内部宽度
inside 内的，内部的，内侧的，内部，内面，在内部
in-situ activation analysis 就地活化分析，现场活化分析
in-situ analysis 就地分析
in-situ balancing 就地平衡
in-situ CBR test 原位加州承载比试验
in-situ concrete tower 现浇混凝土塔架
in-situ concrete 就地浇灌混凝土，现浇混凝土
in-situ conduit 原地导管，原有导管
in-situ deposit 原地沉积
in-situ inspection 就地检查
in-situ measurement 实地测量，现场测定
in-situ neutron activation 现场中子活化，就地中子活化
in-situ pile 灌注桩
in-situ recovery 现场回收，原地回收，就地回收
in-situ regeneration 原位再生，体内再生
in-situ repair 就地修理
in-situ soil test 现场土壤试验
in-situ soil 原地土
in-situ strength 现场强度，原地强度
in-situ test 实地试验，原位试验，现场试验，就地试验，原地试验
in-situ type 现场使用型
in-situ 原地的，原位的，现场的，原位，原地，在原处，在原位，在原地，现场，就地，在施工现场，在原来位置
in-slot signaling 隙内信令
insofar as ... is concerned 就……而论
insofar as 在……情况下，在……限度内，在……的范围内，到……这样的程度，既然，因为
insolation 暴晒，日照，日射，太阳辐射，曝晒【尤指为脱色】
insolation breakdown voltage 绝缘击穿电压
insolation class 绝缘等级
insolation duration 日照时间
insolation energy 太阳能
insolation parameter 日照参数
insolation resistance 绝缘电阻
insolation supervision meter 绝缘监察表
insolation weathering 日晒风化
insolubility 不溶性，不可解性，不可溶性
insoluble anode 不溶性阳极
insoluble impurities 不溶性杂质
insoluble material 不可溶物质，不可溶物，非溶解质
insoluble matter 不溶解物，不溶物质
insoluble problem 不能解决的问题
insoluble residue 不溶（性）残渣
insoluble silica 不溶性氧化硅
insoluble substance 不溶性物质
insoluble 不溶物，不溶物质，不溶解的，不可解的，不可溶的
insolvency proceedings 破产诉讼
insolvency 无力偿付债务，破产
insolvent clause 破产条例
insolvent 无力还债者，破产者，无清偿力的，破产的
in some sense 在某种意义上
inspect 视察
inspected vehicle 检查的车辆
inspecting date 检验日期
inspecting standard 检查标准
inspecting steps 检验步骤
inspection agency 探伤室，检验室，检验机构
inspection and acceptance 验收
inspection and claim clause 检验与索赔条款
inspection and maintenance 检查与维修
inspection and survey of overhead line 线路巡视检测
inspection and testing certificate 商品检验证书
inspection and testing requirements 检验与测试要求
inspection and test plan 检验与测试计划
inspection area 检查区域
inspection at delivery point 交货监督，交货检查，交货点检查
inspection authority 检验机构，检验当局，验收机构
inspection before delivery 交货前检验
inspection body 检验机构
inspection by attributes 特征检验
inspection by variables 变量检验，可调参数检验
inspection card 检验卡
inspection cell 观察室

inspection certificate of origin　产地检验证书
inspection certificate of quality　品质检验证书
inspection certificate of quantity　数量检验证书
inspection certificate of value　价值检验证书
inspection certificate of weight　重量检验证书
inspection certificate on damaged cargo　残损货物检验证书
inspection certificate　检验证（明）书
inspection chamber　检修人孔
inspection clause　检验条款
inspection code　故障等检验规程
inspection cover　视察窗盖，检查孔盖，人孔盖，检查用进出盖板
inspection device　窥测器
inspection door　观察孔，检查门，检查孔，检修门
inspection during manufacture　加工检验，技术检验
inspection earthing　检修接地
inspection expenses　检验费
inspection eye　检查孔
inspection feature　检查要点
inspection fee　检查费，检验费
inspection field　检验现场
inspection gallery　检查廊道，巡视走廊
inspection hatch　检查用进出舱口，检查用舱口
inspection hole cover　检查孔盖板
inspection hole　观察孔，检查孔，人孔，窥视孔
inspection instruction　验收说明
inspection instrument　检查仪器
inspection kit　成套检验器材
inspection lamp　检查灯
inspection lot　检验批
inspection manhole　检查人孔
inspection manipulator　检查用操纵器，检查用机械手
inspection mission　检查团，视察团
inspection nozzle　检查管嘴，检查接管
inspection of dam　大坝检查
inspection of document　文件查阅，文件审查
inspection of foundation subsoil　验槽，地基验槽
inspection of origin　产地检验
inspection of relay protection apparatus　继电保护装置检验
inspection of transformer　变压器器身检查
inspection of work　工作的检查
inspection on tightness of turbine vacuum system　汽轮机真空系统严密性检查
inspection opening　检查孔，视察孔，人孔
inspection outage　检查停堆
inspection panel　检查盘
inspection pit　检查井
inspection plan　检查计划
inspection plate　检查板
inspection plug　检查塞
inspection port　观察窗，检查孔，检修人孔
inspection procedure　检验程序
inspection progress notification　检查进度报告书
inspection record　检验记录
inspection report　检查报告，检验报告，检验报告单
inspection responsibility　检验责任
inspection service fee　查验费
inspection shaft　检查井，检查竖井
inspection shutdown　检查停堆，停堆检查
inspection specification　检查规范书
inspection spider　检查用的星形爬行器
inspection standard　检查标准
inspection steps　检查步骤
inspection supervisor　检查员，监工员
inspection surface　观测面
inspection technique　检验技术
inspection test　出厂试验
inspection train　列检
inspection trolley　检查车
inspection well manhole　检查井人孔
inspection well　检查井
inspection window　检查窗，视察窗，观察窗
inspection　检查，检验，视察，调查，探伤，监督，校验，巡视
inspector　检查员，检验员，鉴定人
inspiration　进气，吸汽，吸入，蒸浓法，吸气
inspirator burner　引射式喷燃器
inspirator mixer　引射式混合器
inspirator　水抽子，吸入器，呼吸器，喷注器，注射器
inspissation　浓缩，变浓，蒸浓
inspissator　蒸浓器，蒸发器
in spite of　即使，尽管
instability constant　不稳定性常数
instability of flow　流量不稳定
instability Reynolds number　不稳定雷诺数
instability　不安定性，不稳定性，不稳定度
instable combustion　不稳定燃烧
instable　不稳定的，不安定的
installation accuracy　安装精度
installation all risks insurance　安装一切险
installation amount　安装量
installation and commissioning stage　安装与调试阶段
installation and maintenance instruction　安装维护说明
installation area　安装区
installation axis　安装中轴线
installation bar　安装挡杆
installation capacity　安装能力
installation component　安装部件
installation contractor　安装合同商，安装承包商
installation cost　安装费用，设备费用，安装费
installation detail　安装详图
installation diagram　安装图，装配图，装置图
installation dimension　安装尺寸
installation drawing　安装图，装配图，装置图
installation elevation of water turbine　水轮机安装高程
installation engineer　安装工程师
installation error　安装误差，装置误差
installation expense　安装费用
installation hole of pipe duct　管沟安装孔
installation instruction　安装说明
installation manual　安装手册
installation material　安装材料

installation mode 安装方式
installation of conventional island 常规岛安装
installation of gauging station 水文站装置，测站装置
installation of heating pipeline 供热管道敷设
installation of overhead conductor and ground wire 架空线和地线的施工
installation plan 安装计划
installation position 安装位置
installation procedure 安装程序
installation site 安装地点，安装现场，安装位置
installation size 安装尺寸
installation specification 安装规程，安装说明书，安装检修规程
installation standard 安装标准
installation test 装配试验
installation tool 安装工具
installation zone 安装区
installation 安装，装置，设备，装备，装配，组装，设备，设施
installed capacity of power station 电站装机容量
installed capacity 装机容量，装机规模
installed cost per kW 每千瓦安装费用
installed cost 安装费用
installed generating capacity 发电装机容量
installed gross capacity 总装机容量
installed load 安装荷载，安装荷重
installed plant capacity 电站装机容量
installed power 装机功率
installed reserve margin 装机备用余量
installed thermal capacity 装置热容量
installed vacuum cleaner 固定式真空吸尘器
installer cost 安装者工资
installer 安装工
installing of condenser tubes and tube end expansion on terminal tube plates 凝汽器冷却管臌胀
installment basis 分期付款基础
installment contract 分期付款合同，分期合同
installment credit 分期付款信贷
installment delivery 分期交货
installment loan 分期付款贷款，分期支付贷款
installment payment 分期付款
installment plan 分期付款计划
installment shipment 分批装船
installment terms 分期付款条件
installment 分期付款
install shade and railings 安装遮蔽与栏杆
install 装配，安装，装入，安置，建立
instalment basis 分期付款标准，分期付款制
instalment delivery 分期分批交货
instalment payment clause 分期支付条款
instalment payment 分期付款
instalment shipment 分批装运
instalment 分期付款，每期付款额，装设
instance name 实例名
instance 例子，实例，情况
instantaneity 瞬时性
instantaneous acceleration 瞬时加速度
instantaneous acting relay 瞬动继电器
instantaneous AGC 瞬时自动增益控制
instantaneous amplitude 瞬态振幅，瞬时幅度
instantaneous area source 瞬时面源【风力发电】
instantaneous assembly 瞬时装置
instantaneous availability 瞬时可用度
instantaneous AVC 瞬时自动电压控制，瞬时自动音量控制
instantaneous blasting cap 瞬时雷管
instantaneous blasting 瞬发爆破
instantaneous break 瞬时切断
instantaneous capacity 瞬时功率，瞬时容量
instantaneous center 瞬时中心
instantaneous collapse 瞬时坍塌
instantaneous communication 瞬时通信
instantaneous complete load rejection 瞬时全部弃荷
instantaneous current 瞬时电流
instantaneous damper current 瞬时阻尼电流
instantaneous deformation 瞬时变形
instantaneous demand 瞬时需要
instantaneous detonator 瞬发雷管
instantaneous deviation control 瞬时偏移控制
instantaneous discharge 瞬时放电，瞬时流量
instantaneous elasticity 瞬时弹性
instantaneous electric power 瞬时电功率
instantaneous exciter 瞬时激励器，瞬时励磁机
instantaneous expander 瞬时扩展器
instantaneous failure rate 瞬时故障率，瞬时失效率
instantaneous firing 瞬时引爆
instantaneous frequency 瞬时频率
instantaneous heat exchanger 快速换热器
instantaneous incidence 瞬时迎角
instantaneous linear velocity 瞬时线速度
instantaneous load 瞬时负载，瞬时负荷，瞬时荷载
instantaneously-operating apparatus 瞬时运行电器
instantaneous maximum speed 瞬时最高转速
instantaneous measured 瞬时测值
instantaneous measurement 瞬时测量
instantaneous mechanical power 瞬时机械功率
instantaneous modulus of elasticity 瞬时弹性模量
instantaneous neutron pulse 瞬发中子脉冲
instantaneous neutron 瞬发中子
instantaneous output 瞬时功率，瞬时出力，瞬时输出功率
instantaneous over current or rate-of-rise relay 瞬时过流或增长率继电器
instantaneous over current protection 速断超电流保护
instantaneous over current relay 瞬时过电流继电器
instantaneous over current 瞬时过电流
instantaneous overload 瞬时过载
instantaneous overturning moment 瞬时倾覆力矩
instantaneous overvoltage 瞬时过电压
instantaneous peak demand 瞬时高峰需求量，瞬时尖峰负荷
instantaneous peak load 瞬时高峰负荷，瞬时尖峰负荷
instantaneous peak power 瞬时峰值功率
instantaneous peak value 瞬时峰值

instantaneous phase current 瞬时相电流
instantaneous phase indicator 瞬时相位指示器
instantaneous point disturbance 瞬时点扰动
instantaneous point source 瞬时电源，瞬时点源
instantaneous pore pressure 瞬时孔隙压力
instantaneous power output 瞬时功率输出
instantaneous power 瞬时功率
instantaneous pressure 瞬时压力
instantaneous reaction 瞬时反应
instantaneous reactivity 瞬时反应性
instantaneous reading 瞬时读数
instantaneous reclosing 瞬时重合闸
instantaneous recovery time 快速恢复时间，瞬时恢复时间
instantaneous relay 瞬时继电器，瞬动继电器
instantaneous release 瞬时释放，瞬时脱扣
instantaneous reoperate time 瞬时再动作时间
instantaneous repair rate 瞬时修复率
instantaneous response 瞬时反应
instantaneous return time 瞬时复原时间
Instantaneous sampling 瞬时采样
instantaneous short circuit test 瞬时短路试验
instantaneous short circuit 瞬时短路
instantaneous sound particle acceleration 瞬时声质点加速度
instantaneous sound particle displacement 瞬时声质点位移
instantaneous sound particle velocity 瞬时声质点速度
instantaneous sound pressure 瞬时声压
instantaneous source 瞬时作用源
instantaneous specific heat 瞬时比热，真实比热
instantaneous speed change rate 瞬时变速率
instantaneous speed 瞬时速度
instantaneous stand-by 瞬时备用，短时备用
instantaneous state 瞬时状态
instantaneous surge 瞬时涌浪
instantaneous torque 瞬时转矩
instantaneous total closure 瞬时全关闭，瞬时全关
instantaneous transmission rate 瞬时传输率
instantaneous trip 瞬间切断，瞬间跳闸
instantaneous turbulence energy 瞬时湍流能量
instantaneous unavailability 瞬时不可用度
instantaneous unit hydrograph 瞬时单位线
instantaneous value 瞬时值
instantaneous velocity 瞬时流速，瞬时速度
instantaneous vibration acceleration 瞬时振动加速度
instantaneous vibration displacement 瞬时振动位移
instantaneous vibration velocity 瞬时振动速度
instantaneous voltage 瞬时电压
instantaneous volume flow rate 瞬时体积流量
instantaneous volume source 瞬时体源
instantaneous wind speed 瞬时风速
instantaneous 瞬时的，同时的，即时的，瞬态的
instant automatic gain control 瞬时自动增益控制
instant complete load rejection 瞬时全甩负荷
instant development 快速发展
instantiation 实例化
instant of failure 故障瞬间
instant of time 时刻
instant-on switch 瞬时动作开关，瞬动开关
instant 瞬时的，立刻的，紧急的，迫切的，即刻，瞬间，瞬时，片刻
instauration 恢复，修复
instead of 代替，而不是
in-step condition 相位一致条件，同步状态
in-step 同相的，同步的，同级的
instinct 本能
INST＝instrument 仪表
Institute for Urban and Rural Planning 城乡规划设计院
institute of design 设计院
Institute of Electrical and Electronic Engineers 电气和电子工程师协会【美国】
Institute of Electrical Engineers 电气工程师协会
Institute of Heat Engineering Instruments and Meters 热工仪表研究所
Institute of Nuclear Power Operations 核电运行研究院
Institute of Process Automation Instrumentation 工业自动化仪表研究院
Institute of Sewage Purification 污水净化学会【英国】
institute review 院级评审
institute 学会，学院，研究所，学术会议，协会，研究院
institutional constraint 制度约束，制度限制
institutional facility 机构设施
institutional factors 制度因素
institutional investment 公共投资，机构投资
institutional investor 机构投资者
institutional reform 机构改革
institution 机构
instruction address register 指令地址寄存器
instruction address 指令地址
instructional text 指示文本
instruction area 指令存储区
instruction authority 指导机构
instruction book 说明书
instruction catalogue 指令表
instruction character 指令字符，控制符，操作符
instruction-check indicator 指令校验指示器
instruction classification 指令分类
instruction code 指令码
instruction command 调度命令，指令命令
instruction constant 伪指令，无用指令，指令常数
instruction control unit 指令控制器
instruction counter 指令计数器，指令地址寄存器
instruction cycle 指令周期
instruction deck 指令卡片组
instruction decoder 指令译码器
instruction element 指令元件
instruction fetch 取指令，指令取出
instruction format 指令格式
instruction for use 使用说明

instruction length code 指令长度码
instruction length 指令长度
instruction list 指令表
instruction manual 使用说明书，说明书
instruction modification 指令修改
instruction pointer 指令描述符，指令指针
instruction processor 指令处理机
instruction pulse 指令脉冲
instruction register 指令寄存器
instruction repertoire 指令系统，指令表，指令程序
instruction sequence 指令序列，控制序列
instruction set 指令集
instructions for variations 变更说明书
instruction sheet 指令单，通知单
instruction time 指令执行时间，指令时间
instruction to attendants 操作规程
instruction to bidders 投标人须知，投标须知
Instruction to negotiating bank 开证行对议付行的指示
instruction to tenderers 投标（人）须知
instruction transfer 指令转移
instruction type 指令类型
instruction unit 指令部件
instruction word 指令字
instruction 规程，指示，用法说明，说明书，指令，引言，指导方针，指令
instructor action log 指导员操作记录
instructor console 教员控制台
instructor station software 教练员站软件
instructor station 教练员（控制）台，指导员（控制）台
instructor 指导人员
instruct 通知，教育，讲授，指示，指导，命令
instrument adjustment 仪器的调整
instrument air system 仪表用空气系统
instrument air tubing 仪表空气管
instrument air 仪表气源，仪表用气，仪表用压缩空气，仪用空气
instrumental analysis 仪器分析
instrumental error 仪表误差，仪器误差
instrumental height 仪器高
instrumentality 工具，手段，方法，媒介
instrumental neutron activation analysis 仪器中子活化分析
instrumental observation 仪器观测
instrumental operation 自动运行
instrument amplifier 测量放大器
instrument and control system 仪表和控制系统
instrument and control 仪表与控制，热控
instrument anti-radiation 仪表抗辐照
instrumentation and control equipment 仪表和控制设备
instrumentation and control island 仪表与控制岛，I&C 岛，仪控岛
instrumentation and control system 测量控制系统
instrumentation and control 仪表和控制
instrumentation averaging time 仪表平均时间
instrumentation buoy 带仪器浮标
instrumentation cable 仪用电缆，控制电缆，测量电缆
instrumentation channel 测量通道，测量回路，测量线路
instrumentation contractor 仪表承包商
instrumentation control and power system 仪表控制和供电系统
instrumentation distribution frame 测量线路分配装置，仪表装配架
instrumentation engineering 仪器仪表工程
instrumentation engineer 仪表工程师
instrumentation manufacturer 仪表制造商
instrumentation noise 仪表噪声
instrumentation nozzle 仪表接管，仪表管口
instrumentation panel 仪表盘
instrumentation penetration 仪表穿墙，仪表穿墙管件
instrumentation plate 仪表板
instrumentation port column 测量柱【堆芯上部构件】，测量孔竖管
instrumentation port 测量孔【压力壳】
instrumentation recorder 模拟记录器
instrumentation sheath 测量封套
instrumentation tap 仪表管口，仪表接口
instrumentation technician 仪表技师［技工］
instrumentation thimble guide 测量导向套管，测量导管【堆芯】
instrumentation thimble 仪表套管
instrumentation tube 仪表管
instrumentation 仪表化，测量仪表，检测仪表，配置仪表，使用仪表，仪表装置
instrument autotransformer （仪）表用自耦变压器
instrument availability 仪表完好率
instrument background 仪器本底
instrument bay 仪表架
instrument board 仪表盘，仪表板
instrument cable 测量电缆，仪表电缆
instrument case 仪表箱
instrument centering 仪器对中
instrument chart 仪表记录图纸
instrument compressed air distribution system 仪表用压缩空气分配系统
instrument compressed air system 仪表用压缩空气系统
instrument compressed air 调节用压缩空气，仪用压缩空气
instrument constant 仪器仪表常数，仪表常数
instrument control and power system 仪表控制和供电系统
instrument & control 仪表与控制
instrument correction 仪器改正
instrument cubicle 仪表室
instrument data display 仪表数据显示
instrument deficient effect 仪表失效效应
instrument desk 仪表台
instrument dial 仪表盘
instrument earthing 仪表接地
instrumented core support column 测量的堆芯支撑柱
instrumented experimental subassembly 带测量装置的试验（燃料）组件
instrumented fuel assembly 装有测量仪表的燃料

组件
instrumented pile　装有测量仪器的桩
instrumented　用仪表装备的，装仪器的，装仪表的
instrument electrical diagram　仪表电气图
instrument enclosed cabinet　封闭式仪表保护柜
instrument engineering diagram　仪表工程图
instrument error　仪表误差
instrument fault rate　仪表故障率
instrument flange　仪表法兰，仪表安装边缘
instrument for water quality analysis　水质分析仪器
instrument front　仪表正面
instrument gas　仪表用气体
instrument incubator　仪表保温箱
instrument industry　仪表制造业
instrument inertia　仪表惯量
instrumenting　用检测仪表装备的，检测仪表装置
instrument lag　仪表惰性，仪表滞后
instrument lead　仪表导线
instrument light　仪表灯，仪表盘照明指示灯
instrument line　仪表（测量）管线，测量线
instrument location　仪表位置
instrument loop drawing　仪表回路图
instrument maintenance room　仪表检修间
instrument motor　仪表电动机
instrument nitrogen aftercooler　仪表用氮气后置冷却器
instrument nitrogen compressor　仪表用氮气压缩机
instrument nitrogen dryer　仪表用氮气干燥器
instrument nitrogen supply system　仪表用氮气供应系统
instrument nitrogen tank　仪表用氮气用罐
instrument nitrogen　仪表用氮气
instrument noise　仪表噪声
instrument of flow velocity measurement　流速量测仪器
instrument panel　仪表板，仪表盘，仪表屏
instrument pen　仪表记录笔
instrument port adapter　热电偶套管适配段【反应堆压力壳】
instrument precision　仪器精度
instrument rack　计测器支架，仪表架
instrument radiation effect　仪表辐射效应
instrument range　仪表量程
instrument reading　仪表读数，仪器读数
instrument relay　仪表继电器
instrument repair shop　仪表修理车间
instrument repair　仪器检修
instrument room　仪表室，机键室
instruments and control equipment　仪表和控制设备
instruments and control system　仪表和控制系统
instrument schematic diagram　仪表示意图
instrument shed　工具棚
instrument shelter　百叶箱
Instrument Society of America　美国仪器学会
instrument stand　仪表架
instrument stock room　仪表储藏室
instrument supervisor　仪表监理
instrument system　测量系统
instrument table　仪表桌，试验台
instrument tap　测量分接头，仪表接口
instrument technology　仪表工艺学
instrument transformer　仪（表）用互感器，仪（表）用变压器，仪表互感器［变压器］
instrument tube routing　仪表管路敷设
instrument valve　仪表阀
instrument well　仪表井
instrument with suppressed zero　无零位刻度仪表，无零点仪表
instrument　（法定）文件，文据，手段，证书，契约，工具，仪表，仪器，方法，器具
insuccation　浸渍
in succession　相继
insufficient　不足的，不够的，不充分
insulance　绝缘电阻
insulant　绝热材料，绝缘材料，绝缘物质，绝缘物
insular shelf　岛基台，岛架
insulated aluminium bar　绝缘铝棒
insulated aluminum wire　绝缘铝线
insulated armature　绝缘电枢
insulated bearing pedestal　绝缘轴承座
insulated bearing　绝缘轴承
insulated body　被绝缘体
insulated boundary　绝热边界
insulated brush holder　绝缘刷握
insulated bushing　绝缘套管
insulated bus　绝缘母线
insulated cable　被绝缘的电缆，绝缘电缆
insulated cambric pipe　麻布绝缘管
insulated coil　绝缘线圈
insulated conductor strand　绝缘股线
insulated conductor　绝缘导体，绝缘导线
insulated conduit　绝缘导管
insulated connector　绝缘连接器，绝缘接头
insulated copper bar　绝缘铜棒
insulated copper wire　绝缘铜线
insulated coupling　绝缘联轴器
insulated door　隔热的门
insulated dowel　绝缘定位销
insulated electric conductor　绝缘电线，绝缘导体
insulated evaporation pan　挡罩式蒸发皿
insulated fitting for distribution line　配电线路绝缘金具
insulated flange　绝缘法兰
insulated flexible pipe　绝缘软管
insulated gate bipolar transistor　绝缘栅双极型晶体管
insulated gate field-effect power transistor　绝缘栅场效应功率［电力］晶体管
insulated gate　绝缘栅
insulated hose　绝缘软管
insulated lug　被绝缘的接线片
insulated magnet wire　绝缘电磁线
insulated materials　绝缘材料
insulated metal roofing　隔热金属屋顶
insulated metal wall pane　带保温金属墙板
insulated neutral　不接地中性点

insulated panel 隔热板，绝缘纸
insulated part 绝缘部件，绝缘部分
insulated plastic coated corrugated steel deck siding 保温型彩色压型墙板
insulated return system 绝缘回流制
insulated ring 隔电环，绝缘圈
insulated rotor 绝缘转子
insulated slot 绝缘槽
insulated static wire 绝缘架空地线
insulated system 不接地制，不接地系统
insulated terminal 被绝缘的端子
insulated tube 绝缘管
insulated winding 绝缘绕组
insulated wire 绝缘线
insulated 绝缘的，隔热的，被绝缘的
insulate 使绝缘，隔离，绝缘，绝热
insulating 隔离的，绝缘的，绝热的，隔热的
insulating ability 绝热能力，绝缘本领，绝缘能力，隔热能力
insulating adhesive tape 绝缘胶带
insulating asbestos paper 石棉绝缘纸
insulating asbestos tape 石棉绝缘带
insulating backing 绝缘衬垫
insulating barrier 绝缘障，绝缘挡板，绝缘层，绝缘隔板
insulating bar 绝缘杆，绝缘棒
insulating base 绝缘底板
insulating bead 绝缘小珠
insulating blanket 绝缘镀层，绝缘垫层
insulating board 隔热板，绝缘板
insulating boot 绝缘靴，绝缘罩
insulating braiding 绝缘包扎，绝缘编织物
insulating brick 绝热砖，保温砖
insulating buildup 绝缘堆覆物【电缆终端】
insulating bushing 绝缘套管，绝缘衬套
insulating cardboard 绝缘纸板
insulating cell 绝缘隔板，绝缘衬垫，绝缘室
insulating cement 绝缘胶
insulating ceramic 绝缘陶瓷
insulating clamp 绝缘线夹
insulating coating 绝缘外套，绝缘涂层，绝缘漆
insulating compound 绝缘膏，绝缘胶，绝缘化合物
insulating coolant 绝缘冷却剂
insulating coupling 绝缘联轴节，绝缘联轴器
insulating cover 隔离罩，绝缘盖
insulating cylinder 绝缘套筒
insulating enamel varnish 绝缘瓷漆
insulating fiber board 绝缘纤维板
insulating fiber material 绝缘纤维材料
insulating film 绝缘薄膜
insulating foil 保温箔
insulating glaze 绝缘釉
insulating glove 绝缘手套
insulating joint 绝缘接头
insulating lagging 保温套层，绝热层
insulating laminated product 绝缘层压制品，层压塑料，积层塑料
insulating layer 绝缘层，保温层，隔绝层，隔离层
insulating lining 绝缘衬里
insulating liquid 绝缘液体
insulating material 隔热材料，绝缘物质，绝缘材料，保温材料
insulating medium 绝缘介质
insulating mica product 绝缘云母制品
insulating molded plastics 粉压塑料，绝缘模压塑料
insulating molded product 绝缘模压制品
insulating molding powder 胶木粉，电木粉
insulating oil 绝缘油
insulating oxide 绝缘氧化物
insulating packing 绝热填料
insulating paint 绝缘漆
insulating panel 绝热板，绝缘板
insulating paper plate 绝缘纸板
insulating pellet 隔热芯块
insulating piece 绝缘块
insulating plastic 绝缘塑料，粉压塑料
insulating plate 绝缘垫片，绝热板，绝缘板
insulating polymer 绝缘聚合物
insulating press board 绝缘纸板，厚纸板
insulating property 绝缘性能，绝缘性质
insulating puncture tester 绝缘击穿试验器
insulating refractory 绝热耐火材料
insulating resistance 绝缘电阻
insulating rod 绝缘棒
insulating rolled product 绝缘卷制品
insulating sheet paper 绝缘薄纸
insulating sleeve 绝缘套筒
insulating spacer 绝缘垫块，绝缘隔片，绝缘隔块
insulating space 隔离空间
insulating stand 绝缘台
insulating strength 绝缘强度
insulating substrate 绝缘衬底
insulating tape 绝缘带
insulating test 绝缘试验
insulating transformer 隔离变压器
insulating vacuum 隔离真空
insulating varnished cloth 绝缘漆布，黄蜡布
insulating varnished glass cloth 绝缘玻璃漆布
insulating varnished silk 绝缘漆绸，黄蜡绸
insulating varnished sleeving 绝缘漆管，黄蜡管
insulating varnish 绝缘漆
insulating wire 绝缘电线
insulation against ground 对地绝缘
insulation aging rate 绝缘老化速度
insulation aging 绝缘老化
insulation allowance 绝缘容差，绝缘公差
insulation and cladding of the boiler 锅炉的保温与包层［护皮］
insulation angle 绝缘角
insulation between layers 层间绝缘
insulation between phase 相间绝缘
insulation between strands 线股间绝缘
insulation between turns 匝间绝缘
insulation blanket 绝热层，绝热衬垫
insulation breakdown voltage 绝缘击穿电压，绝缘强度
insulation breakdown 绝缘击穿
insulation cladding 隔热套，保温壳

insulation classification 绝缘分类，绝缘等级
insulation class voltage 绝缘等级电压
insulation class 绝缘种类，绝缘等级【定子，转子等】
insulation coating 绝缘漆，绝缘涂层
insulation coefficient 绝缘系数，绝热系数
insulation construction 保温结构
insulation control 绝缘监督，绝缘控制
insulation coordination for HVDC transmission system 高压直流输电系统绝缘配合
insulation coordination 绝缘配合
insulation course 隔热层，保温层
insulation covering 绝缘套
insulation defect 绝缘损伤
insulation deterioration 绝缘损伤，绝缘老化
insulation distance 间隔距离，绝缘距离
insulation efficiency of heat-supply piping 供热管道保温效率
insulation engineering 绝缘工程
insulation factor 隔热因数
insulation failure 绝缘事故，绝缘损坏
insulation fault 绝缘损坏，绝缘故障
insulation for strut 支柱的绝缘
insulation grade 绝缘等级
insulation indicator 绝缘指示器
insulation in guy 拉线绝缘
insulation jacketing 隔热套，保温壳
insulation layer 隔热层，绝缘层，保温层
insulation leakage 绝缘漏电
insulation level of electrical equipment 电气设备绝缘水平
insulation level 绝缘等级，绝缘水平，绝缘强度
insulation life 绝缘寿命
insulation loss angle 绝缘损耗角
insulation material life test 绝缘材料寿命试验
insulation material 保温材料
insulation measurement after power outage 停电后测绝缘
insulation measurement 绝缘测量
insulation monitoring 绝缘监测
insulation mortar 绝热灰浆
insulation of equipment 设备保温
insulation of hydro-generator 水轮发电机绝缘
insulation of pipe 管道保温
insulation oil tank 绝缘油箱
insulation oil test 绝缘油试验
insulation oil 绝缘油
insulation partial discharge test 绝缘局部放电试验
insulation paste 绝缘胶
insulation plate 绝缘板
insulation power factor 绝缘功率因数
insulation process 绝缘工艺
insulation puncture test 绝缘击穿试验
insulation puncture 绝缘击穿
insulation quality 绝缘质量
insulation ratio 绝缘比
insulation removal 除去绝缘，除去隔热层
insulation resistance tester 绝缘电阻测量器
insulation resistance test 绝缘电阻试验
insulation resistance 绝缘电阻
insulation resistivity 绝缘电阻率，绝缘电阻系数
insulation sheath 绝缘外壳，绝缘套
insulation sleeve 绝缘套管
insulation space-factor 绝缘占空系数
insulation spacer 绝缘隔块，绝缘垫片，绝缘垫块
insulation strength 绝缘强度
insulation stud 绝缘螺杆
insulation system （悬吊）隔振系统
insulation tape 绝缘带
insulation tester 绝缘试验器，绝缘试验仪，绝缘测试表
insulation test of electrical equipment 电气设备绝缘试验
insulation test voltage 电介质试验电压
insulation test with alternating voltage 绝缘交流电压试验
insulation test with DC voltage 绝缘直流电压试验
insulation test with impulse voltage 绝缘冲击电压试验
insulation test 绝缘试验
insulation thermal life 绝缘热寿命
insulation thermal resistivity 绝缘热阻率
insulation thickness 保温层厚度，绝缘厚度
insulation varnished cloth 绝缘胶布
insulation wall 绝缘层，绝缘壁
insulation wedge 绝缘楔
insulation withstanding test 绝缘耐受试验
insulation worker 保温工
insulation 隔离，绝缘，绝热，绝缘体，绝缘材料，隔热层，隔热
insulativity 绝缘性，绝缘度，体积电阻率
insulator arcing horn 绝缘子灭弧角
insulator arcing shield 绝缘子电弧屏蔽
insulator arc-over 绝缘子放电，绝缘子闪络
insulator bracket 绝缘子托架
insulator breakdown 绝缘击穿，绝缘子击穿，绝缘子故障
insulator cap 绝缘子帽
insulator chain 绝缘子串
insulator connecting 绝缘子串组装
insulator corona 绝缘子电晕
insulator crossarm 绝缘横担
insulator fittings 绝缘子金具，绝缘子配件
insulator hardware 绝缘子金具
insulator of HVDC transmission line 高压直流绝缘子
insulator pellet 隔热块
insulator pin 绝缘子连接插销，绝缘子球头，绝缘子直脚，绝缘体销
insulator set 整套绝缘子，绝缘子串组
insulator shed inclination 绝缘子伞裙倾斜角
insulator spindle 绝缘子心轴
insulator string 绝缘子串
insulator type current transformer 绝缘子式电流互感器
insulator 隔热体【太阳能发电】，绝缘子，绝缘体，绝热体
insulband 绝缘绑带
insulcrete 绝缘板
insulex 一种绝缘漆

insullac 绝缘漆
insulosity 岛率【即岛屿面积与总水面之比】，岛屿度
insult 损伤，损害
in summary 综上所述
insurance against strike 罢工保险
insurance against total loss only 保全损险
insurance against war risk 保战争险
insurance agent 保险代理人
insurance amount 保险金额
insurance and freight 到岸价
insurance benefit 保险赔偿费
insurance certificate 保险凭证，保险证书
insurance claim 保险索赔
insurance clause 保险索赔，保险条款
insurance company 保险公司
insurance compensation 保险赔偿金
insurance contract 保险合同，保险契约
insurance cost 保险费用
insurance coverage 保险范围，保险责任范围，保险类别，保险总额
insurance cover 保险范围，保险责任范围，保险类别，保险总额
insurance expense 保险费，保费，保险费用
insurance fee 保险费，保费
insurance for land transportation 陆运保险
insurance for life 人寿保险
insurance fund 保险基金
insurance indemnity 保险赔偿，保险赔偿金
insurance law 保险法
insurance object 保险标的
insurance of export-import goods 进出口货物保险
insurance of works 工程保险
insurance policy 保险单，保单
insurance premium 保险费，保险金，保费
insurance requirements 保险要求，关于保险的要求
insurance 保险，保险费，安全保障，保险金额
insurant 被保险人，受保人，投保人，保险契约者
insure against fire 保火险
insured amount 保险金额，投保金额
insured cargo 保价货物
insured goods 保险货物
insured sum 保险金额
insured value 保险价值，投保价值
insured 被保险的，保过险的，被保险人
insurer's interest 保险人权益，保险人利益
insurer's security 保险人的保证
insurer 保险人，保险公司，承保人，保险商，承保商
insure with particular average and against war risk 保水渍险及战争险
insure with particular average 保水渍险，保单独海损赔偿险
insure 保险，投保
insurgent group 叛乱团体
insurge 向内扩流【液体】
insuring party 承保方
insuring stage 保证水位
insusceptible to ageing （材料）不受时效影响的，非时效的
in synchronism 同步
intact clay 未扰动黏土，原状黏土
intact fuel 完好燃料
intact ground 未扰动地基
intact rock 完整岩石
intact specimen 原状试件
intact 完整的，无损伤的，原封不动的，未触动的
intaglio 凹雕
intake air line 进风管线
intake air 进风，进气，进风流
intake angle 入口角
intake area of aquifer 含水层来水区，蓄水层的受水面积
intake area 受水区
intake canal 取水渠道［管道］，进水渠
intake capacity 吸入能力，吸入性能
intake chamber 冷却水进水槽，进水室，进汽室，进气室
intake channel 取水渠道［管道］，供水管道
intake coarse filtration and trash removal system 取水粗滤和除渣系统
intake condition 进口参数
intake crane 进水栓
intake culvert 供水管道，进水渠
intake dam 坝内式水电站厂房，大坝进水口，引水坝
intake deck 进水口平台
intake disturbance 进气扰动
intake duct 进气道
intake equipment 取水设备
intake guide vane 进口导叶
intake header 进口联箱，进口总管
intake heading 取水首部
intake loss 进口损失
intake louver 进气百叶窗
intake manifold 入口联箱
intake of water works 水厂进水口，水厂取水口
intake opening 进风口
intake-outlet-separated arrangement 进水口泄水口分列式布置，分列式布置
intake-outside-outlet arrangement 差位式布置，进水口泄水口重叠式布置
intake pipe 进气管，进水管
intake rate 入流率
intake screen 进口拦河栅，进口滤网
intake silencer 进口消声器
intake sluice 进水闸
intake stack 进风道
intake structure 进水建筑物，取水构筑物，引水构筑物，取水建筑物，取水结构
intake system 引水系统
intake temperature 进口温度
intake tower 取水塔
intake tunnel 取水隧洞
intake valve 进气阀，进口阀，进汽阀
intake velocity 进水速度
intake with automatic flushing 自动冲沙进水口
intake works 取水工程

intake 取水口，吸入物，入口，进口，进水口，吸入量，（放射性核素）摄入量，取水结构，进气口，进风量，取水
intangible assets 无形资产
intangible benefit 无形效益
intangible depreciation 无形损耗
intangible legacy 无形资产
intangible 无形的
in-tank-solidification 罐内固化，槽中固化
integer harmonics 整数倍谐波
integer-slot winding 整数槽绕组
integer status controllable 整数状态可控
integer status information 整数状态信息
integer 整数，总体，整体
integral absorbed dose 积分吸收剂量
integral action coefficient 积分作用系数
integral action limiter 积分作用限制器
integral action rate 积分速率
integral action time constant 积分作用时间常数
integral action time 积分作用时间，复位时间
integral action 积分作用
integral age 累计寿命
integral automation 全厂综合自动化
integral basis 整体基础
integral blade 带围带叶片，整体叶片，带叶冠叶片，带冠叶片
integral blower burner 带鼓风机的燃烧器
integral blower 内置式鼓风机
integral body 整体式车身
integral boiling-water nuclear superheating reactor 一体化沸水核过热反应堆
integral bypass system 整体旁路系统
integral bypass 整体旁路
integral cable 整根电缆
integral calculus 积分学，积分
integral cam shaft 整体凸轮轴
integral casing 整体气缸
integral collector storage solar water-heater 整体式太阳热水器
integral collector storage system 整体式（太阳热水）系统，闷晒式（太阳）热水器
integral computer 整数计算机
integral condenser 整体式凝汽器
integral control action 积分控制作用
integral control coefficient 积分控制系数
integral control factor 积分控制因子
integral controller 积分控制器，积分调节器
integral control 积分控制，积分调节，整体控制
integral convolution 积分卷积式，积分褶合式，积分卷积法，积分卷积
integral cover 整体围带，整体罩盖，整体覆盖，整体盖片
integral damping winding 全阻尼绕组
integral diode solar cell 整体二极管太阳电池
integral-disk rotor 整锻转子
integral dose 累积剂量，积分剂量
integral drain cooler 内置式疏水冷却器
integral dry route 一体化干法
integral economizer （水管锅炉）组合式省煤器
integral electrical heating 整体电加热
integral enclosure 整体密闭罩
integral equalization 积分稳定
integral equalizer 积分稳定环节
integral equation 积分方程
integral experiment 积分实验
integral fast reactor 一体化快堆
integral fit method 积分符合法
integral flange 整法兰
integral flow orifice assembly 积分流量孔板组件
integral flow orifice differential pressure transmitter 积分流量孔板差压变送器
integral forged coupling 整体锻制的联轴节，整体联轴节
integral form of the transport equation 输运方程的积分形式
integral fuel storage pool 一体化的燃料贮存水池
integral furnace boiler 立式全水冷炉膛锅炉
integral gasified combined-cycle power plant 煤气化联合循环电厂
integral governing 积分调节
integral horsepower motor 功率大于1马力的电动机，整数马力电动机
integral insulation construction 整体保温结构
integral insulation 整体绝缘
integral leak rate test 整体泄漏率试验
integral length hypothesis 积分长度假说
integral length scale of turbulence 湍流积分长度尺度
integral low-finned tube 整体低翅片管
integral lug 整体支耳
integral method of waterproofing 总体防水法
integral method 积分法
integral multiple 整倍数
integral part 整数部分，本体部分，不可拆部分，整体部件
integral pipe flange 整体管法兰
integral pitch winding 整距绕组
integral plan 整体规划
integral plow 全悬挂犁
integral pulse-height distribution 脉冲振幅积分分布
integral rate action 积分作用
integral reactivity 积分反应性
integral reactor 一体化反应堆
integral reinforcing 整体补强，整体加固
integral roadbed 整体道床
integral rotor 整锻转子，整体转子
integral seat 整阀座
integral sensitivity 积分灵敏度
integral separator 内置式分离器
integral shroud blade 整体围带叶轮，整体围带叶片
integral shroud 叶冠，整体围带
integral-slot winding 整数槽绕组
integral solar water heater 闷晒式太阳能热水器
integral spent fuel pit 一体化堆乏燃料贮存水池
integral-square-error criterion 误差平方积分准则
integral-square-error method 误差平方积分法
integral square root extractor 均方根积算器
integral superheat boiling water reactor 一体化过热沸水反应堆
integral superheater 整体过热器

integral support lug 整体支耳
integral temperature 平均温度，积分温度
integral testing 积分检验，整体试验
integral time constant 积分时间常数
integral time 积分时间
integral topping 合成顶端
integral train 固定成组列车，整列，整体列车
integral transformation 积分变换
integral transport equation 积分输运方程，积分运输方程
integral trial operation of unit plant 单元机组联合试运行
integral unit substation 组装式变电站
integral unit 成套机组
integral wheel 整体叶轮
integral with 与……联结，联到……
integral 整数的，积分的，整体的，完整的，主要的，集成的，完全的，一体化［式］的，积分，部分，完整
integrand 被积函数
integraph 积分仪
integrated analog-to-digital converter 整体模拟数字转换器
integrated approach 综合方案，一揽子方案
integrated a-Si solar cell 集成型非晶硅太阳电池
integrated band absorption 积分谐带吸收
integrated boundary-layer equation 边界层积分方程
integrated burnable poison absorber 一体化可燃毒物吸收器
integrated chimney and cooling tower 烟塔合一
integrated chip 集成芯片
integrated chopper 集成削波器，集成斩波器
integrated circuit communication data processor 集成电路通信数据处理机
integrated circuit micro-motor 集成电路电机
integrated circuit package 集成电路组件
integrated circuit 集成电路，积分电路
integrated coal gasification combined cycle 整体煤气化联合循环
integrated coil 整体式线圈
integrated company 联营公司
integrated component circuit 集成元件电路
integrated concept 一体化反应堆方案
integrated contractor 综合承包商
integrated control and display 综合控制和显示
integrated control and protection system 集成控制与保护系统
integrated control system 成套控制系统
integrated control technique 综合控制技术
integrated control 综合防治，集中控制，综合控制，全面控制
integrated coolant gas circulator 一体化的气体冷却剂循环风机
integrated cooling system 一体化冷却系统
integrated coupling 固定连接
integrated curve 总和曲线
integrated data file 综合数据文件
integrated data processing 综合数据处理，统一数据处理，集中数据处理
integrated data retrieval system 集中数据检索系统
integrated demand meter 累积需量计
integrated demand 累积需量
integrated deposition density 综合沉积密度
integrated design reactor 一体化设计反应堆
integrated design 一体化设计，综合［总体］设计
integrated development 全面发展，综合发展
integrated device 集成半导体器件，集成器件
integrated dose 积分剂量，总剂量
integrated dry route 一体化干法
integrated electronics 集成电子学
integrated energy curve 累积能量曲线
integrated exposure 积分照射量
integrated flux 积分通量
integrated form 积分形式
integrated fuel absorber 一体化的燃料吸收体
integrated gasification combined cycle power plant 整体煤气化联合循环电厂
integrated gasification combined cycle 整体煤气化燃气蒸汽联合循环，整体煤气化发电技术，整体煤气化联合循环
integrated gas-steam combined cycle power plant 整体（式）燃气蒸汽联合循环电厂
integrated gas-steam combined cycle 整体（式）燃气蒸汽联合循环
integrated gate commutated thyristor 集成门极换流晶闸管，集成门极换向晶闸管
integrated heat conductivity 积分热导率，积分导热系数
integrated industrial power plant 综合工业电站
integrated information system 综合信息系统
integrated injection logic 集成注入逻辑
integrated into 统一，归并，被结合到……中
integrated I/O adapter 整体输入输出衔接器
integrated joint venture 紧密型联营体
integrated layout 一体化布置
integrated load curve 荷载累积曲线
integrated management department 综合管理部
integrated management information system 综合管理信息系统
integrated management 综合管理
integrated mica 再生云母，黏合云母，云母塑料
integrated mill 综合型工厂
integrated model approach 整体模型方法，整体模型研究
integrated mode 整体方式
integrated motor 内装电动机，机内电动机
integrated multipurpose approach 整体多目标法，综合多目标法
integrated neutron flux 中子积分通量
integrated operation 综合运行，联合运行，整体运行
integrated plan 综合规划
integrated pollution control 污染综合治理
integrated power system 统一电网，综合电力系统
integrated pre-service cleaning 投运前整体清洗
integrated pressurized water reactor 一体化压水反应堆，一体化压水堆
integrated project 综合项目

integrated puff model 积分喷团模式，集成烟团模式
integrated quality 整体素质
integrated rate expression 总速率表达式
integrated research 综合研究
integrated safety assessment report 综合安全评价报告
integrated service digital network 综合服务数字网络
integrated service network 综合业务通信网
integrated services digital network 综合服务数字网
integrated service 整套服务
integrated signal 积分信号，综合信号
integrated square error 误差平方积分
integrated square 平方积分
integrated storage 积累存储器
integrated survey 综合考察
integrated switching and multiplexing 集中开关和多路转换
integrated system 整体系统
integrated test-run 联合试运行，整体试运行
integrated transmission system 联合输电系统
integrated treatment 综合处理
integrated type of primary system construction 一体化结构，一回路系统的
integrated type reactor 一体化反应堆
integrated value 累计值
integrated wastewater treatment plant 污水综合处理厂
integrated water resources planning 水资源综合利用规划，水资源综合规划
integrated 集成的，积分的，累积的，一体化的，完整的，综合的
Integrate M into N 使M并于N，
Integrate M over N 把M对N积分
integrate M with N 把M和N结合起来
integratety 完整性，诚实，正直
integrate 综合，积分，使……完整，使……成整体，求……的积分，表示……的总和，整合的，完全的，一体化，集成体
integrating 累计的，积分的，积算的
integrating bellows 积分膜盒，积分波纹管
integrating capacitor 积分电容器，存储电容器
integrating circuit 积分电路
integrating comparator 积分比较器
integrating counting circuit 积分计算电路
integrating depth recorder 积分式深度记录仪
integrating device 积分器，累计装置
integrating element 积分元件
integrating flow meter 累积式流量计，积分流量计
integrating frequency meter 总频率计，积分频率计
integrating galvanometer 积分电流计
integrating gyroscope 积分陀螺仪
integrating gyrounit 积分陀螺部件，陀螺积分环节
integrating instrument 积分器，积算仪器，积算仪，积算仪表
integrating measuring instrument 积分式测量仪表
integrating mechanism 积分机构
integrating meter 积分计，积分计算仪，积算仪表
integrating microbalance 积分微量秤
integrating motor 积分电动机
integrating multi-ramp converter 多级斜度转换器，积分式多级变换器
integrating network 积分网络
integrating period 积分周期
integrating photometer 积分光度计
integrating relay 积分型继电器，积算继电器
integrating tachometer 积算转速表
integrating unit 积分装置
integrating wheel 积分轮
integration circuit 积分电路
integration clause 整合条款，整体性条款，完整合约条款
integration comb 汇集排管
integration constant 积分常数
integration design 一体化设计
integration ionization chamber 积分电离室
integration method of velocity measurement 流速积测法
integration of specialist or customized software 专家集成或用户软件
integration plan 综合方案
integration step 积分步长，积分距
integration time constant 积分时间常数
integration 集合，综合，积分，集成，积累，累计，整合，总合
integrative management approach 综合管理方法
integrative management system 综合管理系统
integrative organization 综合组织
integrative suppression 综合校正
integrative 综合的，一体化的
integrator amplifier 积分放大器
integrator wheel 积分器轮
integrator 积分电路，积分器，积分仪，积累器，积算器
integrity in operation 运行完整性
integrity lifetime 完整性寿命【燃料组件】
integrity of a bulk power system 大电力系统的整体性
integrity 完整性，密封性，整体，整体性，正直
integrodifferential equation 积分微分方
integrodifferential form of the transport equation 输运方程的积分微分形式
integrodifferential 积分微分的
integrometer 惯性矩面积仪
intellectronics 人工智能电子学
intellectual property right 知识产权
intelligence data 情报资料
intelligence grid 智能电网
intelligencer 情报员
intelligence 信息，信号，智力，理解力，情报
intelligent control 智能控制器
intelligent electronic device 智能电子设备
intelligent instrument 智能仪表
intelligentized total station 智能化全站仪
intelligentized 智能化的
intelligent low-voltage circuit breaker 智能化低压

断路器
intelligent management 智能管理
intelligent sensor 智能传感器
intelligent terminal 智能终端
intelligent transformer calibrator 智能型互感校验仪
intelligent transmitter 智能变送器
intelligent workstation 智能工作站
intelligent 智能的
intelligibility 清晰度，可懂性，可理解性
intendant 管理人员，监督人员，监督人
intended for a specific use 预期的特殊用途
intended product 预期产品
intended target 指定目标
intend 打算
Intenrational Electrotechnical Commission 国际电工委员会
Intenrational Standard Organization 国际标准化组织
intenret protocol 网际协议
intense current 强电流，大电流
intense erosion 强烈侵蚀
intense gamma radiation 强γ辐射
intense neutron generator 强中子发生器
intense rain 暴雨
intensification 加强，增强，强化
intensified charging 增强充电
intensifier electrode 增光电极
intensifier stage 放大级
intensifier 增辉器，放大器，增强器，增强剂
intensifying effect 增感作用，强化效应
intensifying screen 增感屏【射线照相】
intension agreement 意向协定
intension 紧张，加剧，内涵，强度
intensitometer 曝光表，强度计
intensity attenuation 烈度衰减，强度衰减
intensity calculation of hydraulic machinery 水力机械强度计算
intensity curve 强度曲线
intensity determination 强度测定，强度计算
intensity discrimination 强度鉴别
intensity distribution 强度分布
intensity-duration curve 强度历时关系曲线
intensity-duration formula 强度历时公式
intensity factor of stress 应力强度因子
intensity level 亮度级，强度级，声强级
intensity modulation 亮度调制，强度调制
intensity of activation 活化强度
intensity of compression 压缩强度
intensity of current 电流强度
intensity of electric field 电场强度
intensity of field 场强
intensity of illumination 照度
intensity of light 光强，光度，照度
intensity of magnetic field 磁场强度
intensity of magnetic flux 磁通密度
intensity of magnetization 磁化强度
intensity of pressure 压强
intensity of radiation 射线强度，辐射强度，选择强度
intensity of sound 声强
intensity of stress 应力强度
intensity of turbulence 紊流强度，湍流强度，湍流度
intensity of vibration 振动强度
intensity of washing 冲洗强度
intensity of wash water （快滤沙池的）冲水强度，冲洗强度
intensity of wave pressure 波压强度
intensity spectrum 强度谱
intensity 亮度，强度，烈度，密集度，强烈，紧张，强烈程度
intensive arc lamp 强弧灯
intensive cooling 强冷却
intensive exploitation 集约开采
intensive fluidization condition 强烈流化状态
intensive investment 集约投资
intensive management 集约经营
intensive property 强度量
intensive 集约化型
intention agreement 意向协定
intentional pollution 有意污染
intention 意向
interacting redundancy 相联冗余度
interacting safety system 相联安全系统
interaction design 交互设计
interaction effect 干扰效应，交互作用，相互影响，相互作用效应
interaction parameter 相互作用参数
interaction 相互作用，相互影响，互相感应，交互作用，交互（型），干扰
interactive computer 交互式计算机
interactive computing （人机联系用）交互计算
interactive debug 人机对话故障排除
interactive information system 交互信息系统
interactive processing 交互处理
interactive process 交互过程，相互作用过程
interactive system 相互作用系统，交互系统
interactive terminal 交互终端
interactive 人机对话的，相互作用的，交互的，交互式，交互式的
interact 互相感应，相互作用，相互影响，交互作用，影响
inter-agency agreement 机构间协定
inter-air space 气隙空间
Inter-American Development Bank 美洲开发银行，泛美开发银行【总部设在华盛顿】
interannual variation 年际变化
inter-area transfer 地区间传输
Interatomenergo 经互会国际原子能公司
interatomic force 原子间力
Interatominstrument 经互会国际原子能仪表公司
inter-attraction 相互吸引
interbaluster 栏杆空档
interbanded coal 夹矸煤
interbank space 管束间空间
interbank superheater 管束间过热器
interbank 管束间的
interbasin area 区间面积
interbasin diversion 流域间分洪
interbasin transfer of water 流域间水迁移
interbasin water diversion project 跨流域引水

工程
interbasin water diversion 跨流域引水
interbedded strata 互层，夹层
interbedded stratum 夹层
interbedding 互层
interblend 混合，相互掺混
interblock gap 组间间隔，组间间隙
interblock space （信息组的）组间间隔
inter-boiler process 锅内过程
interbuilding coffin 厂内屏蔽运输容器
interbus 联络母线，旁路母线
intercalated bed 夹层，间层
intercalated tape 夹层绝缘带，叠式绝缘带
intercalation 夹层
intercarrier receiving system 内载波接收系统
intercell liquid transfer 热室间液体输送
inter-cell 注液电池
intercept control valve 中间截止阀，中联门
intercepted water 拦蓄水
intercept form 截距式
intercepting channel 截水沟，天沟
intercepting ditch 截水沟
intercepting drain 截流排水沟
intercepting filter 截取过滤器
intercepting sewer system 截流下水道系统
intercepting subdrain 地下截水沟，暗沟
intercepting valve 再热调节阀，再热截流阀，截流阀，截止阀
interception loss 截断损失量
interception 折射，截听，相交，截断，截取，拦截，定方位
intercept lightning discharge 截取雷电流
intercept method 截止方法
intercept neutron 截获中子
interceptor drain tube 截流排水管道
interceptor governor valve 截流调速器阀
interceptor manhole 阻截沙井
interceptor sewer 截流污水管，截流污水渠
interceptor valve 截断阀，阻止阀
interceptor 隔断器，截流井，冷凝罐
intercept valve 截止阀，节流阀【汽轮机】，截断阀，再热调节阀，再热截流阀
intercept 截线
interchangeability 互换性，可换性
interchangeable centering device 可互换的定心装置
interchangeable program tape 可互换程序带
interchangeable 可互换的
interchange center 交换中心
interchange method 替换法【桁架分析用】
interchange of air 空气交换
interchange of heat 热交换
interchange power 互换功率，交换功率
interchange price 互供电价
interchange terminal 联运站
interchange 交换，互换，转接，交替，交换器，交换装置
interchannel interference 通道内部干扰
interchannel 通道间的，孔道间的，信道间的，层道间的
intercity network 城市电网
inter-coagulation 互凝聚
intercoil connection 线圈间的连接，过桥线
intercoil damage 线圈内损坏
intercoil insulation 线圈间绝缘，层间绝缘
inter-combination 相互组合
intercom = intercommunication 对讲电话，内部通信，互通，互相往来
intercommunication flip-flop 内部通信触发器
intercommunication system 内部通信系统
intercommunication telephone 对讲电话，内部电话
intercommunication 对讲电话通话，双向通信，相互联系，相互沟通，互相联络，内部通信，互通，互相往来
intercomputer buffer 计算机间缓冲器
intercomputer communication 计算机通信
intercom 对讲电话装置
intercondenser 中间凝汽器，中间电容器
interconnected cage motor 鼠笼式电动机
interconnected control 互联控制机构
interconnected delta connection 互连三角形连接
interconnected electric power system 互联电力系统
interconnected generating station 互联发电站
interconnected network 互联电网
interconnected operation 中间连接操作，互连操作，互联运行
interconnected power system 联合电网，联合电力系统，互联电力系统
interconnected star connection Z形连接，曲折连接，曲折接法
interconnected star winding Z形连接绕组，曲折连接绕组
interconnected synchronous generator 并网的同步发电机，互连的同步发电机
interconnected systems 互联诸系统
interconnected void 相互连通的孔隙
interconnected 联结的，内连的，互连的，互相连接的，相互连接的
interconnecting cable 中继电缆，互连电缆，连接电缆
interconnecting device 接插元件
interconnecting diagram 接线图
interconnecting digital-analog converter 转接数字模拟转换器
interconnecting ductwork 炉内连接烟道
interconnecting feeder 联络馈线
interconnecting linkage 连接杆
interconnecting main 联络干线
interconnecting pipe in heat-supply network 热网连接管
interconnecting pipe 连通管，互连管
interconnecting steam main 互联蒸汽总管
interconnecting system 网内联络系统
interconnecting transformer （电力网）联络变压器
interconnecting wiring diagram 接线图，配线图
interconnecting 互连的，内连的
interconnection cable 互连电缆
interconnection capacitance 互连电容
interconnection diagram 接线图

interconnection drawing 接线图
interconnection network 互联网，网际网，网间网
interconnection of network 电力网联络线
interconnection of power system 电力系统的互联
interconnection piping 互连管道，连接管
interconnection substation 联络变电所
interconnection switch 联络开关
inter-connection tariff 联网价
interconnection tie 联络线，联络馈路
interconnection transformer 联络变压器
interconnection valve 联络门
interconnection 内连，互连，互联，相互连接，联络，联络线，联接管，互相连接
interconnector 中继馈（电）线，内部连线，内连线，联络线路，内部连接线
interconnect 互连，内连，相互连接
interconversion 互换换算，互换，互相转换
intercooled cycle 中间冷却循环
intercooler 中间［级间］冷却器【压缩机】
intercooling 中间冷却
intercoordination 相互关系
intercostal 脉间的
inter-country project 多国项目
intercoupling 寄生耦合，相互连接，相互耦合
intercross 交叉
intercrystalline brittleness 晶间脆性
intercrystalline corrosion 晶间腐蚀
intercrystalline cracking 晶间龟裂，晶界破坏，晶间裂纹
intercrystalline fracture 晶间断裂，沿晶界断裂
intercrystalline rupture 晶间断裂
intercrystalline 晶（粒）间，晶间的
intercycle cooler 循环中间冷却器
intercycle 中间周期的，中间循环的
interdeck space 管层间空间
interdeck superheater 管层间过热器
interdendritic corrosion 显微腐蚀，枝晶间腐蚀
interdendritic segregation 枝晶间偏析，显微偏析
interdendritic shrinkage porosity 枝晶间缩松，显微缩松
interdendritic 枝晶间的
interdependent flow 横连流动
interdict 闭锁，禁止，制止，阻断
interdiffusional convection 互扩散对流
interdiffusion 相互扩散
interdigital circuit 交叉指形回路
interdigited capacitor 叉指状电容器，片状分层电容器
interdisciplinary study 跨学科研究，多学科研究
interdisciplinary talent 复合型人才
interdisciplinary 边缘学科的，多种学科的，各学科间的
interdiscipline 跨学科
interdistributary area 支流间面积
interelectrode capacitance 极间电容
interelectrode leakage 极间泄漏
interelectrode transadmittance 极间互导纳
interelectrode 电极间的
interelectron repulsion 电子间斥力
interest 利息，利益，兴趣
interest component 利息部分
interest concession 让利
interest cost 利息费用，利息成本
interest coverage 利息偿还总额
interest during construction 建设期间的利息，建造期利息
interested government 有关政府
interested party 相关方，有意者
interested part 相关部分
interest expense 利息费用
interest free advance 免息预付款
interest free credit 无息信贷
interest free loan 无息贷款
interest on grace period 宽限期利息
interest on long-term debt 长期债务利息
interest per annum 年息
interest per day 日息
interest per diem 日息
interest period 计息期，利息期
interest per mensem 月息
interest per month 月息
interest rate differential 利率差异
interest rate during construction 建设期贷款利率
interest rate swap 利率互换，利率掉期，利率调期
interest rate 利率
interest risk 利率风险
interest-sensitive rate 利率敏感率
interest table 利息表
interest tax 利息税
interest upon loans 放款利息
interface adapter 接口适配器
interface boundary condition 交界面边界条件
interface card 接口插件
interface circuit 接口电路
interface collector 界面集水器，中间排液装置
interface condition 界面条件，接口条件
interface controller 界面控制器
interface control mode 接口控制方式
interface control 接触面控制，界面控制，接口控制
interface data 接口数据
interface design 接口设计
interface device 接口装置
interface distributor 界面分配器
interface drawing 接口图
interface echo 界面回波
interface information processor 接口信息处理机
interface layer 中间层，界面层
interface logic 接口逻辑
interface management 接口管理
interface message processor 接口信息处理系统
interface M with N 使 M 与 N 面接［联系］
interface of works 工作的衔接
interface pipe 界面管
interface point 分界点，接口点
interface pressure 接触面压力
interface rate 接口速率
interface region 分界面区
interface related station level functions 与接口有

关的站层功能
interface specification 接口规范
interface tension 界面张力
interface transfer 界面传递
interface unit 连接器件
interface 相互作用，接口，（交）界面，分界面，相互关系，接触面，（货物）交接，分界
interfacial agent 界面活性剂
interfacial area 界面面积
interfacial crud 界面污物
interfacial friction coefficient 界面的摩擦系数
interfacial friction factor 界面摩擦因子
interfacial heat transfer coefficient 界面传热系数
interfacial oscillation （气流）分界面振荡
interfacial polarization 界面极化
interfacial shear stress 界面切应力
interfacial sublayer 交界面副层
interfacial tension 面际张力，界面张力
interfacial turbulence 界面湍动
interfacial wave 界面波
interfacial 接触面的，分界面的，边界的
interfacing device 分界接合装置
interfenestration 窗间墙宽度，窗距布置
interference absorber 干扰吸收器
interference amplitude 干扰振幅
interference band 干扰频带
interference correction 干扰修正
interference current 干扰电流
interference diagram 叶片的频谱图
interference drag 干扰阻力
interference effect 干扰效应，干扰影响
interference elimination 干扰消除
interference eliminator 干扰抑制器，干扰消除器
interference factor 干扰因子
interference fading 干扰衰落
interference far field pattern 远场干涉图样
interference figure 干扰图形
interference filter 干扰滤光片，干扰滤波器
interference fit 干涉配合，过盈配合，静配合，压配合
interference-free model 无干扰模型
interference-free 无干扰的，抗干扰的
interference frequency 干扰频率
interference fringe pattern 干涉条纹图形
interference fringe 干涉条纹，干涉带
interference galloping 干扰驰振
interference grating 干扰光栅
interference inverter 噪声限制器
interference level 干涉电平，干扰电平
interference locator 干扰定位器，干扰探测器
interference measurement 干扰测量
interference microscope 干涉显微镜
interference noise 干扰噪声
interference of electromagnetic wave 电磁波的干扰
interference of light 光干涉
interference of sound 声干涉
interference pattern 干涉特性，干涉图
interference photocathode 干扰型光电阴极
interference power 噪声功率，干扰功率
interference prevention 防止干扰
interference-producing unit 干扰发生器
interference rejection unit 抗干扰装置
interference rejection 抗干扰能力
interference search gear 干扰探测器，干扰定位器
interference settlement 干扰性沉陷
interference signal 干扰信号
interference source 干扰源
interference spectrometer 干涉分光计
interference spectrum 干扰频谱
interference-suppression device 干扰抑制设备
interference suppressor 干扰抑制器
interference susceptibility 干扰敏感性
interference test 干扰试验
interference transmitter 干扰发射机
interference voltage 干扰电压
interference wave 干扰波，干涉波
interference with reception 接收干扰
interference with transmission 传输干扰
interference zone 干扰区，干扰范围
interference 妨碍，干涉，干扰，相互影响，相互作用
interfere 妨碍，干涉，干扰，抵触，冲突
interfering flow field 干扰流场
interfering impulse 干扰脉冲
interfering ion 干扰离子
interfering noise 干扰杂声，干扰噪声
interfering signal 干扰信号
interfering singing 干扰啸声
interfering transmitter 干扰发射机
interferogram 干扰图，干涉图
interferometer micrometer 干涉仪式测微计
interferometer microscope 干涉仪显微镜
interferometer 干涉仪，干扰仪
interferometry 干涉量度法，干涉测量法
interferric space 铁芯间隙
interflectance 空间利用系数
interfloor traffic 楼层间交通
interfloor travel 楼层间交通
interflow 合流，混流，地下径流，层间流，汇合流
interfluent flow 层间水流【不同密度的】
interfluent 汇合的
interfuse 混合，熔合
inter-government 政府间的
intergrade 中间形式，中间级配
intergranular attack 晶间腐蚀，晶间侵蚀
intergranular corrosion test 晶间腐蚀试验
intergranular corrosion 晶间腐蚀
intergranular fracture 晶间断裂，沿晶断裂
intergranular pressure 粒间压力
intergranular stress corrosion crack 晶间应力腐蚀裂纹
intergranular stress corrosion 晶间应力腐蚀
intergranular 晶间的，颗粒间的，粒间的
intergrater 积算器，积分器
interground addition 磨细掺加剂
intergrown knot 连生节
interharmonic 间谐波
interim acceptance criteria 临时验收准则
interim agreement 临时协定，临时协议，中间

阶段协议
interim audit 期中审计，中期审计
interim certificate 临时证书
interim criterion 暂行标准
interim determination 临时决定
interim dividend 期中股息
interim duty rate on export goods 出口商品暂定税率表
interim measure 临时措施，暂定措施
interim payment certificate 期中付款证书
interim payment 临时付款，其中付款，期间支付
interim procedure 暂行办法
interim provision 暂行规定
interim regulation 暂行条例
interim reliability evaluation program 中期可靠性评定计划
interim report 阶段报告，中间报告
interim results 中期业绩
interim statement 期中决算表，中期报表
interim storage 中间贮存，暂时贮存
interim trial 临时试验
interim 间歇的，过渡性的，临时的，中间，临时
interior angle 内角
interior basin 内港池
interior breakwater 内防波堤
interior circuit 屋内线路，屋内电路，内部电路
interior conduit 室内管路，室内管道
interior decoration 内部装饰
interior defect 内部缺陷
interior differential needle valve 内差式针形阀
interior drainage 内部排水，内排水
interior face 内表面
interior finish 内部整修，内装修
interior fire protection 内部防火
interior illumination 室内照明
interior lighting 内部照明
interior lining panel 内部护板
interior magnetic field 内在磁场
interior mesh 内环
interior of concrete sump 混凝土坑的内侧
interior orientation 内方位
interior point intermodal 内陆公共点多式联运
interior port 内港
interior pressure 内压
interior slope 内斜面
interior span 内跨，中间跨
interior structure 内部结构
interior surface 内表面
interior temperature 内部温度
interior trimming 内部整修
interior trim 内部装修
interior view 内视图
interior wall tile 内墙砖
interior wall 内墙
interior wire 室内线，内线
interior wiring 室内布线
interior zone 内区
interior 内部的
interisochrone area 等瞬曲线间面积，等时线间面积
interlaboratory standard 实验室标准
interlaboratory 实验室间的
interlaced block codes 交错分组码
interlaced convolutional codes 交错卷积码
interlaced network 多回线电力网
interlaced scanning 隔行扫描
interlace operation 交错操作
interlace 隔行扫描，交错，交织，交错存储排号，使交织，使交错，夹层
interlacing 隔行扫描，隔行，间行，交错操作
interlaminar insulation 铁芯叠片间绝缘，叠片漆，层间绝缘
interlaminar 层间的
interlamination resistance 层间电阻
interlamination 层间
interlayer bonding 层间连接
interlayer water 层间水
interlayer 界层，隔层，夹层
interleaved code 隔行扫描，交错码
interleaved coil 交错式绕组，纠结式线圈
interleaved winding 纠结式绕组，（变压器的）交错式绕组，纠结式线圈
interleave type core （变压器的）交叠式铁芯
interleave 隔行扫描，隔行，交错，交织
interleaving access 交叉存取
interline agreement 承运人之间互运协议
interlinear 行间的
interlinkage flux 磁链，交链磁通
interlinkage 互连，连接，交链
interlinked leakage 交链漏磁，漏磁链
interlinked voltage 互相连接的电压，相间电压
interlink 连接，结合，互连，链接，把……互相连接
interlobe space 叶间空隙
interlock circuit 联锁电路，联锁保护电路
interlock control 联锁控制
interlock device 连锁装置，联锁装置
interlocked circuit breaker 联锁断路器
interlocked control system 联锁控制系统
interlocked equipment shut 停机联锁设备
interlocked overcurrent protection 联锁过电流保护
interlocked switch 联锁开关
interlocker 联锁装置
interlocking armored cable 嵌合铠装电缆
interlocking armor 嵌合铠装
interlocking circuit 联锁电路
interlocking device 联锁装置，联锁机构，连锁控制，连锁装置
interlocking effect 联锁作用
interlocking electromagnet 联锁电磁铁
interlocking joint 联锁接合
interlocking latch 联锁销
interlocking line 联锁线路
interlocking mechanism 联锁机构
interlocking motor 自动同步机
interlocking relay 联锁继电器
interlocking steel sheet piles 联锁钢板桩
interlocking switch group 联锁组合开关，联锁开关组

interlocking system 联锁制，联锁系统，连锁系统
interlocking test 联锁试验
interlocking tile 咬接瓦
interlocking 联锁，联锁装置，咬合作用，联动，连锁，闭锁作用，联动装置
interlock limit 联锁限值
interlock protection 连锁保护，联锁保护
interlock relay 互锁继电器
interlock system 联锁系统
interlock test 连锁试验，联锁试验
interlock veto selector 否定联锁选择器
interlock 互锁，互连，连接，联锁，联锁装置，连锁，联锁器，锁口
interlude 插算，中间程序，预备程序，初始程序，间隔时间，间歇
intermag conference 国际磁学会议
intermediary language 中间语言
intermediary paper 隔层纸
intermediary 中间人
intermediate adjustable trough angles 可调槽角托辊
intermediate annealing 中间退火
intermediate approximation 中间近似
intermediate ash tank 中间灰仓
intermediate-base resin 中等碱性树脂
intermediate bay 中间开间
intermediate bearing 中继轴承
intermediate belt core 卷皮带的芯子
intermediate belt 过渡层，中间带
intermediate bleeding chamber 中间抽汽室
intermediate cable 配线电缆，中继电缆
intermediate capacitor 中间电容器
intermediate circuit 中间回路
intermediate clamp 中间定位装置
intermediate clarifier 中间澄清器
intermediate collector 中间集水装置
intermediate computer 中间计算机
intermediate condenser station （输电线的）中间电容器站
intermediate condenser 中间凝汽器
intermediate contact 中间触点
intermediate containment 中间安全壳
intermediate coolant circuit 中间冷却剂回路
intermediate coolant 中间冷却剂
intermediate cooler 中间冷却器
intermediate cooling loop 中间冷却回路
intermediate cooling medium 中间冷却介质
intermediate cooling water pump 中间冷却水泵
intermediate cooping loop pump 中间冷却回路泵
intermediate coupling 中轴耦合器
intermediate crusher 中间破碎机
intermediate CT 中间变流器
intermediate cycling load 中间循环负荷
intermediate distribution frame 中间配线架
intermediate event 中间事件
intermediate flange 中间法兰
intermediate floor 中间楼层
intermediate flow mixer grid 中间流动交混格架
intermediate frequency preamplifier 中频前置放大器
intermediate frequency transformer 中频变压器
intermediate frequency 中间频率，中频
intermediate gear box 中间减速器，中间齿轮箱
intermediate grid 中间格架，中间栅格
intermediate groin 中间丁坝
intermediate groyne 中间丁坝
intermediate header 中间集箱，中间加热器
intermediate heat exchanger 中间热交换器
intermediate heat transport system 中间热传输系统
intermediate impregnation 中间浸渍
intermediate improvement cutting 改进间伐
intermediate inflow 中间来水
intermediate language 中间语言
intermediate layer 过渡层，中层，中间层
intermediate leak-off 中间引漏
intermediate-level cave 中等放射性工作室
intermediate-level cell 中放热室，中等放射性工作室
intermediate-level radiation 中等强度辐射
intermediate-level waste 中放废物，中等放射性废物，中等放射性水平的废物
intermediate lightning arrester 中间避雷器
intermediate load power plant 中间负荷发电厂
intermediate load unit 中间负荷机组
intermediate loop 中间回路，中间环路，中间负荷
intermediate maintenance 中修
intermediate medium method 中间介质法
intermediate memory 中间存储器
intermediate metal conduit 中间金属导管
intermediate moderated 中间慢化的
intermediate neutron 中能中子
intermediate observation 中间观测
intermediate phase 中间相，居间相
intermediate pier 中墩
intermediate pod （燃料元件储存水池）中间存放台
intermediate point 中间点
intermediate pole 换向极，间极，附加极
intermediate port 中转港
intermediate pressure casing 中压缸
intermediate pressure compressor 中压压缩机
intermediate pressure cylinder expansion 中压缸膨胀
intermediate pressure cylinder 中压缸
intermediate pressure main stop valve 中压主汽门
intermediate pressure section 中压部分，中压段
intermediate pressure steam turbine 中压汽轮机
intermediate pressure turbine 中压汽轮机
intermediate pressure water electrolyzer 中压水电解槽
intermediate principal plane 中主平面
intermediate principal strain 中主应变
intermediate principal stress 第二主应力，中间主应力
intermediate product 部分乘积，中间产物，半成品，中间产品
intermediate pulverized coal banker system 仓储式制粉系统

intermediate pumping station 中间泵站
intermediate pump 中真空泵，增压泵
intermediate quantity 中间量
intermediate radius 中间半径
intermediate range channel 中间量程通道，中间区段通道
intermediate range high neutron flux trip 中间区段高中子通量紧急停堆
intermediate range monitor 中间功率区段监测器
intermediate range 中间量程【堆外中子测量】，中间区段
intermediate reactor 中能反应堆
intermediate readout 中间读出
intermediate region 中能量区，中间区域
intermediate reheater 中间再热器
intermediate relay 中间继电器
intermediate repair 中间修理，架修
intermediate repeater 中间帮电机，中间增音机
intermediate result 中间成果
intermediate rock 中性岩
intermediate rod 中间棒
intermediate rotation plug 中间旋塞
intermediate safety analysis report 中期安全分析报告
intermediate scale turbulence 中尺度端流
intermediate sedimentation tank 中间沉降槽
intermediate sedimentation 中间沉降
intermediate shaft 中间轴，中轴
intermediate shelterbelt 过渡防护林带
intermediate shutdown 中间停堆
intermediate sodium heat transfer system 中间钠传热系统
intermediate span 中间跨，中跨
intermediate spectrum reactor 中能中子反应堆，中能堆
intermediate stage bleed-off valve 中间抽头门
intermediate stage 中间级
intermediate state information 中间状态信息
intermediate station 中间站
intermediate storage period 中间贮存期
intermediate storage rack 中间贮存架
intermediate storage 中间存储器，中期贮存
intermediate store crane 中间仓库吊车
intermediate sum digit 中间和数的数字
intermediate sum 中间和
intermediate superheater 中间过热器
intermediate support 直线杆塔，中间支撑，中间支座，中间支承
intermediate switching station （输电线的）中间开关站
intermediate term 中期
intermediate tier 受热面中间段
intermediate time scale storage 中期贮存
intermediate tower platform 中间塔架平台
intermediate transformer 中间变压器
intermediate treatment 中间处理
intermediate value theorem 介值定理
intermediate valve 中间滑阀
intermediate variable 中间变量
intermediate-voltage winding （变压器的）中压线圈
intermediate wake 中间尾流，中波
intermediate weld 断续焊缝
intermediate wind machine 中型风力机
intermediate zone 中间带，中心区
intermediate 中级的，中间的，居间的，中等的，中间体，媒介物，中间物
intermedium variable 中间变量
intermedium 中间物，媒介物
intermeshing wearing ring 互相咬合的磨损环
intermetallic compound 金属间化合物
intermetallic 金属间（化合）的
intermingle 掺杂，混合
intermission 停机，中止时间，中止，中断
intermittant absorption 间断式吸附
intermittence 中断，间断，间歇
intermittency factor （湍流）间歇因子
intermittency 间歇现象，间歇性
intermittent ash-discharge grate 间歇式出灰炉排
intermittent blowdown flash tank 定期排污扩容器
intermittent blowdown pit water discharge pump 定排水池排水泵
intermittent blowdown 间歇排污，定期排污
intermittent burning 脉动燃烧，间歇燃烧
intermittent chlorination 间歇氯化，间歇加氯
intermittent contact 断续接触
intermittent controller 断续作用控制器
intermittent control 断续控制
intermittent current 间歇电流，断续电流
intermittent discharge 断续放电，间歇放电，间歇排放
intermittent disconnection 间歇断线，间歇切断
intermittent dust removal 定期除灰
intermittent duty coil 断续通电线圈
intermittent duty motor 断续工作电动机
intermittent duty 间歇负载，间歇工作方式，间断负荷，断续工作制，间断式运行【旋转机械，马达】
intermittent earth 间歇接地，断续接地
intermittent fault 断续短路，断续故障，间歇故障
intermittent feed 间歇投药
intermittent fillet weld 断续角焊缝，间断贴角焊
intermittent filter 间歇过滤器
intermittent filtration 间歇过滤
intermittent firing 间歇燃烧，脉动燃烧
intermittent flow 间歇状流，间歇流动
intermittent flushing sedimentation basin 间断冲洗式沉沙池
intermittent guidance 间歇制导
intermittent heating 间歇供暖，间歇加热
intermittent input 断续输入，脉动输入
intermittent inspection 间歇检查
intermittent jet 间歇射流
intermittent lake 间歇湖
intermittent loading 断续加载
intermittent load 断续短时负载，断续负荷，间断负荷，间歇性负载，间歇负荷
intermittently running 间歇运行的
intermittently turbulent region 间歇紊流区
intermittent manual blowdown 手动间断排污
intermittent mode operation 间歇运行

intermittent nature of wind 风的间歇性
intermittent operation control 间歇运行控制
intermittent operation 断续工作状态，间歇工况，间歇运行，断续运行，定期运行
intermittent periodic duty 断续周期工作方式
intermittent pollution 间歇污染
intermittent power source 中继动力源，间歇性电源
intermittent process 间歇过程
intermittent-rated motor 断续定额电动机
intermittent rating 间断出力，断续负载额定出力
intermittent reaction 间歇反应
intermittent rod withdrawal 间歇提出控制棒
intermittent sampling 间歇采样，间歇取样
intermittent sand filter 间歇沙滤器
intermittent sand filtration 间歇沙滤，间歇沉降
intermittent test 断续试验，间歇试验
intermittent time between layers 层间间隔时间
intermittent welding 断续焊接，断续焊
intermittent weld 间歇焊缝，间断焊，断续焊缝
intermittent wind tunnel 间歇式风洞
intermittent working 间歇工作
intermittent 断断续续的，间断的，间歇性的
intermit 使中断，中止，间断，暂停，间歇
intermixing 混杂，混合，搅拌
intermixture 混合，混合物
intermodal transportation 联合运输
intermodal 联运的
intermodulation frequency 互调差频，交叉调制差频
intermodulation 交调，相互调制
intermolecular force 分子力
intermolecular 分子间的
intermontane basin 山间盆地，山间平原，山间贫地
inter-mountainous plain 山间平原
in terms of percentage 按百分比计算
in terms of 依据，根据，按照，在……方面，用……的话，用……字眼，以……的措辞，从……角度来讲，关于……，以……为单位，就……而言，换算，折合
INTERM 定期，间断【DCS 画面】
internal absorptance 内吸收比
internal admission 内进汽
internal aerodynamic 内流空气动力学
internal and external interface 内外部接口
internal angle of friction 内摩擦角
internal angle 阴角，内角
internal arithmetic 内部运算
internal ascent 内部爬梯
internal auditor 内审员
internal audit system 内部审计制度
internal audit 内部检查，内部审核
internal axial-flow pump type recirculation system 轴流泵式内部再循环系统
internal axial-flow pump 堆内轴流泵
internal balance 内平衡，内部均衡
internal block effect 内部阻塞效应，内部块效应
internal boiler corrosion 锅内腐蚀
internal boundary layer 内边界层
internal breeder 内增殖反应堆，内增殖堆
internal bremsstrahlung 内韧致辐射
internal calipers 内卡钳
internal calling 内部呼叫
internal capacitance 内部电容
internal cause 内因
internal characteristic 内特性曲线
internal chemical cleaning 内部化学清洗
internal chill 内冷铁
internal circuit 内部回路，内部电路
internal cleaning 内部清理
internal clearance 内部间隙
internal-combustion engine driven generator 内燃机驱动的发电机
internal combustion engine power generation 内燃机发电
internal combustion engine power plant 内燃机发电厂
internal combustion engine 内燃机
internal combustion gas turbine 内燃式燃气轮机
internal components 内部构件
internal concrete vibrator 混凝土内部振捣器
internal condition 内部条件
internal conductor reactance 导线内电抗
internal connection 内部连接
internal consumption 本体消耗，内部消耗，自身消耗
internal contamination 体内污染，内污染，内部污染
internal control 内部控制
internal convection 内部对流
internal conversion coefficient 内转换系数
internal conversion factor 内转换因数
internal conversion gain 内转换增益
internal conversion neutron detector 内转换中子探测器
internal conversion ratio 内转换比
internal conversion 内转换
internal coolant recirculation 内部冷却剂再循环
internal cooling generator 直接冷却发电机，内冷发电机
internal cooling reactor 内部冷却反应堆
internal cooling 内冷，内部冷却，内冷却
internal core catcher 堆芯内收集器
internal corrosion 内部腐蚀，汽水侧腐蚀
internal coupling 内部耦合
internal crack 内部裂缝，内部裂纹
internal customs duties 国内关税
internal damping coefficient 内阻尼系数
internal damping 内阻尼
internal debt 内债
internal deformation 内部变形
internal degree of freedom 内自由度
internal deposition 体内沉积，体内附着，内部沉积
internal deposit 内部沉积物，汽水侧沉积物
internal diameter 内径
internal dielectric loss 内部介电损耗
internal diffusion 内扩散
internal discharge 内部放电
internal distributing device 内部配水装置，内部

分配装置
internal distributor 内部配水装置，内部分配装置
internal dose 内部照射剂量，内照射剂量
internal drag bracing 翼内阻力张线
internal drag 内阻
internal drainage system 内部排水系统
internal drain 内部排水
internal drainage 内排水
internal draught tube 内导管
internal drop 内压降
internal efficiency 内效率
internal energy 内能
internal environment 内部环境
internal equipment 内部设备
internal erosion 层内侵蚀
internal exchange site 内部交换基
internal exergy dissipation 内部火用损
internal exposure 体内照射量，体内照射，内照射，内辐照
internal failure cost 内部损失成本
internal fault 内部故障
internal feedback 内（部）反馈
internal feed pipe in drum 汽包中的内给水管
internal fertile element 内部可转换元件
internal field alternator 旋转磁场式同步发电机，内极式同步发电机
internal financing 内部筹资
internal finish of buildings 建筑物的内部装修
internal fires （喷燃器的）内部着火
internal fission gas pressure 裂变气体内压
internal fission product trap 元件内裂变产物收集器
internal fissure 内部裂纹
internal fittings 内部装置
internal flame 内焰
internal flaw 内部缺陷
internal flow limiting device 内限流装置
internal flow 内流
internal fluid friction 流体内摩擦
internal fluid mechanics 内流流体力学
internal force 内力
internal friction angle 内摩擦角
internal friction coefficiency 内摩擦系数
internal friction heat 内摩擦热
internal friction of soil 土的内摩擦力
internal friction 内部摩擦，内摩擦
internal fuel cycle 堆内燃料循环，（堆）内燃料循环
internal fuel-element fission product trap 燃料元件内裂变产物捕集器
internal function register 状态字寄存器，内部操作寄存器
internal furnace 内置炉膛
internal gas pressure （燃料元件）内气压
internal gauge 内径规
internal gear pair 内齿轮副
internal gear pump 内啮合齿轮泵【旋转泵的类型】
internal gear 内齿轮
internal gland 内汽封
internal grinder 内圆磨床
internal grinding 内圆磨削
internal heat 内热
internal hydrogen cooling 氢内冷，直接氢冷
internal idle time 内空闲时间
internal impedance 内阻抗
internal impeller pump 堆内轴流泵
internal indicator 内指示剂
internal inductance 内电感
internal inspection 内部检查
internal insulation 内绝缘
internal irradiation hazard 内照射危害
internal irradiation 内部照射，内辐射，内辐照
internal jet pump 内喷射泵
internal keyboard 内键盘
internal kinetic energy 内在动能
internal latent heat 内部潜热，内潜热
internal leakage 内部漏泄，内部泄漏
internal leak 内部泄漏
internal lightning protection facility 内部防雷设施
internal lightning protection system 内部防雷系统
internal light 内部照明灯具
internal load 内部负荷，内部负载
internal losses 内部损失
internal loss torque 内损转矩
internal loss 内部损耗，内部损失
internally cooled reactor 内部冷却反应堆
internally finned tube 内肋管，内鳍片管，内翅片管
internally fired boiler 内燃锅炉
internally geared motor 内装减速器的电动机
internally programmed computer 内程序计算机
internally ribbed tube 内鳍片管，内螺纹管
internally specified index 内部规定指标
internally stored program 内存储程序
internally ventilated motor 内通风式电动机
internal management 内部管理
internal memory 内存储器
internal micrometer 内径千分尺
internal-mix oil burner 内混式油喷燃器
internal modulation 内调制
internal moisture 固有水分
internal moment 内力矩
internal noise 内部噪声
internal number system 内部计数制
internal observation 内部观测
internal overvoltage measurement 内部过电压测量
internal overvoltage 内部过电压
internal oxidation 内氧化
internal packing （级间）内部汽封
internal phase angle 内相位角
internal potential energy 内势能
internal potential vibration （变压器的）内部电位振动
internal power limit of alternator 发电机内功率极限
internal power 电磁功率，内功率，内部功率
internal pressure coefficient 内压系数
internal pressure current 内压流
internal pressure loss 内压损失

internal pressure stress 内压应力
internal pressure test 内部压力试验，耐压试验
internal pressure 内部压力，内压
internal pressurization 内加压，初始内压【燃料棒】
internal protection 内保护
internal purification system （燃料元件）内部净化系统
internal quality audit 内部质量审核
internal radiation 内辐射
internal rate of return method 内部收益率法
internal rate of return 回款率，内部收益率，内部盈利率，内部回报率
internal reactance 内电抗
internal reactor characteristics 反应堆内特性，堆内特性
internal reactor vessel ledge （支承堆芯用）压力容器内凸缘
internal recirculation boiling reactor 内部再循环沸水反应堆
internal recirculation flow 内部再循环流动
internal recirculation pump 内部再循环泵
internal recirculation 内部再循环，体内再生
internal relation 内在联系
internal remote monitoring system 内部远距监测系统
internal report 内部报告
internal resistance 内电阻，内阻力，内阻
internal ribbed tube 内鳍片管，内螺纹管
internal rupture 内部断裂
internals alignment pin 堆内构件对中销
internals 堆内构件，内部构件
internal scatter 内散射
internal scouring 内部冲刷
internal scour 内部冲刷，潜蚀
internal screw gauge 内螺纹规
internal seal 内部密封
internal separation 内分离
internal settlement measurement 内部沉陷量测
internal shake 内部环裂
internal shield 内屏蔽
internal shrinkage 内部收缩
internal shroud 内盖，内罩，内屏蔽
internal sleeve 内套管，内挡油圈
internals lifting device 堆内部件起吊装置
internals lifting rig 堆内构件起吊装置，堆内部件起吊架
internal slope impedance 内梯度阻抗，动态内阻抗
internal slope resistance 动态内电阻，内梯度电阻
internal sources of impurities 内杂质源
internal sparger pipe 内喷淋管
internals support ledge 内部构件支撑横挡，内部构件支承凸缘
internals support stand 堆内构件存放台
internal stack flow 烟囱内流
internal stagnant water 内滞水
internal stator 内部转子，内转子，内定子
internal steam separation （堆）内部汽水分离
internal steam separator 堆内汽水分离器，（堆）内部汽水分离器
internal steam tracing 蒸汽内伴热
internal steam-water baffle 锅内汽水挡板
internal storage capacity 内存容量
internal storage 内存储器
internal strain gauge balance 内式应变式天平
internal strain 内应变，内应力
internal structure 内部构件，内部结构
internal superheating 内部过热
internal support column 内支承柱
internal support 内部支撑
internal surface 内表面
internal thermal conductance 内热导
internal thermal resistance 内热阻
internal thermodynamic equilibrium 内部热力平衡
internal thread 内螺纹
internal tidal current 内潮流
internal tide 内潮
internal tooth washer 内齿垫圈
internal torque 电磁转矩，内转矩
internal transmittance 内透射比
internal transportation 国内运输
internal treatment 锅内水处理，锅内处理
internal turbulence 内紊流
internal twisted tape 扰流子
internal-vane pump 内叶泵
internal vessel ledge 压力容器内凸缘，压力容器内缘
internal vibration 内部振捣
internal vibrator 插入式振捣器，内部振捣器，内插式振捣器
internal viscosity 内黏滞性
internal voltage （发电机的）电动势，（电动机的）反电动势，内电压
internal volute 内蜗壳
internal wall 内墙
internal water cooling generator 水内冷发电机
internal water cooling 水内冷，直接水冷
internal water jet pump 内喷射水泵
internal water pressure 内水压力
internal water 内水
internal wave 内波
internal weld protrusion 内部焊缝凸起
internal wiring diagram 内部接线图，部接线图
internal work 内功
internal 内部的，固有的，内部零件
international 国际的
international accounting standards 国际会计标准
international agency 国际机构
international agreement 国际协定
international air transport association 国际航空运输协会
international ampere 国际安培
International Annealed Copper Standard 国际退火铜标准
International Association for Wind Engineering 国际风工程学会
International Association of Lighthouse Authorities 国际航标协会
International Association of Machinists 国际机械师协会

International Association of Meteorology and Atmospheric Physics 国际气象与大气物理协会
International Atomic Energy Agency 国际原子能组织，国际原子能机构
International Bank for Economic Cooperation 国际经济合作银行
International Bank of Reconstruction & Development 国际复兴与开发银行，世界银行
international barter 国际实物交易
International Bidding Procedures 国际投标程序
international bidding 国际招（投）标
international bond 国际债券
international borrowing 国际借贷
international broadcasting 国际广播
international business practice 国际商业惯例，国际贸易实务
international call letters 国际呼号
international call 国际通话
International Center for Wholesale Trade 国际批发贸易中心
International Chamber of Commerce Terms 国际贸易术语解释通则
International Chamber of Commerce 国际商会
International Civil Aviation Organization 国际民用航空组织
international clearing 国际清算
international commercial custom 国际商业惯例
International Commission on Radiological Protection 国际辐射防护委员会
International Commodity Agreement 国际商品协定
international common practice 国际惯例
international community 国际社会
international comparison project 国际比较项目
international competition 国际竞争
international competitive bidding 国际竞争性招标
international conference on nuclear technology transfer 国际核技术转让会议
International Conference on the Peaceful Uses of Atomic Energy 和平利用原子能国际会议
International Conference on Wind Engineering 国际风工程会议
international conference 国际会议
international contract project 国际承包工程
international convention 国际惯例
international cooperation 国际合作
international corporation 跨国公司
international coulomb 国际库仑
international currency 国际通行货币
international custom 国际惯例
international development agency 国际开发机构
International Development Association 国际开发协会，国际开发署
International Economic Association 国际经济联合公司
International Electrotechnical Commission 国际电工委员会，国际电工技术委员会
International Energy Agency 国际能源机构【属OECD】
international environmental conventions 国际环境惯例［公约，协定］
international exchange 国际汇兑，国际交流
international fair 国际博览会
International Federation of Consulting Engineers 国际咨询工程师联合会【简称“菲迪克”】
International Finance Corporation 国际金融公司
International Financing Institutions 国际金融机构
international fusion superconducting magnet test facility 国际聚变超导磁铁试验设施
international harbour 国际港
international henry 国际亨利
international horsepower 国际马力
International Institute of Refrigeration 国际制冷学会
International Institute of Welding 国际焊接学会
international intellectual property protection 国际知识产权保护
International Investment Bank 国际投资银行
international joule 国际焦耳
international kilocalorie 国际千卡
International Law Commission 国际法委员会
international law 国际法
International Legal Regime 国际法制
international lending institution 国际贷款机构
international loan 国际贷款
international marine cargo insurance 国际海运货物保险，国际海运货物保险
international maritime custom 海事国际惯例
international market price 国际市场价格
international market 国际市场
international Maxwell 国际麦克斯韦
International Monetary Fund 国际货币基金组织
international monetary policy 国际货币政策
international Morse code 国际莫尔斯电码
international multilateral loans 国际多边贷款
International Nuclear Cooperation Center 国际核合作中心【日本】
International Nuclear Data Committee 国际核数据委员会
International Nuclear Fuel Cycle Evaluation 国际核燃料循环评价
International Nuclear Industries Fair and Technical Meetings 国际核工业展览与技术会议
International Nuclear Information System 国际核情报系统【属IAEA】
international nuclear safety advisory group 国际核安全咨询组
international ohm 国际欧姆
International Organization for Standardization 国际标准化组织
international organization 国际组织
international patent system 国际专利制度
international payments deficit 国际收支逆差
international plutonium store 国际钚储存
international practical Celsius temperature 国际实用摄氏温度
international practical Kelvin temperature 国际实用开尔文温度
international practical system of units 国际实用单位制
international practical temperature scale 国际实用温标

international practical temperature 国际实用温标
international practices 国际惯例，国际实践
international preferential duty 国际特惠关税
international protection code 国际防护代码
international purchasing 国际采购
international pyrheliometric scale 国际直接日射计标度
international quotation of price 国际报价
international reactor safety evaluation 国际反应堆安全评价
international reference atmosphere 国际标准大气压
international registration of marks 商标的国际注册
international river 国际河流
international rules for the interpretation of trade terms 国际贸易条件解释通则
international sanctions 国际制裁
International Science Organization 国际科学组织
international screw pitch gauge 国际螺距规
international settlement 国际结算
international shopping 国际采购
International Special Committee on Radio Interference 国际无线电干扰特别委员会
international standard atmosphere 国际标准大气压
International Standardization Organization 国际标准化组织
international standard 国际标准
international standing 国际地位
international steam table 国际蒸汽表
international subcontracting 国际分包
international symbol 国际标志
international syndicated loan 国际银团贷款
international system of electrical units 国际电气单位制
international system of unit 国际单位制
international telecommunication 国际电信
international telegram 国际电报
international temperature scale 国际温标
international temperature 按国际温标的温度，国际标准温度
international tendering company 国际招（投）标公司
international trade center 国际贸易中心
international trade contract 国际贸易合同
international trade custom 国际贸易惯例
international trade law 国际贸易法
International Trade Organization 国际贸易组织
international trade policy 国际贸易政策
international transactions 国际间收支往来
international treaty 国际条约
international turnkey project 国际承包工程
international unit 国际单位
international usage 国际惯例
international value 国际价值
international volt 国际伏特
international watt 国际瓦特
International Wholesale and Foreign Trade Center 国际批发和对外贸易中心
internet 国际互联网，互联网，因特网
internet engineering task force 因特网工程特别工作组
internet protocol 互联网协议，网际协议
internet service provider 因特网服务提供者
internetwork 互联网
internode 波腹，节间
interoffice communication 办公室间通信，局间通信
interoffice 局间的
interoperability 互操作性
interparticle attraction 粒间吸力
interparticle bonding 颗粒间黏结
interparticle conduction 颗粒间导热
interparticle force 粒子间作用力
interpass temperature （多层焊的）层间温度，道间温度
interpenetration 互相渗透
interphase boundary 相界面
interphase connecting rod 相间连杆
interphase connection 相间连接线
interphase dosimetry 界面剂量学
interphase-gas-convective component 相间气体对流分量
interphase gas exchange rate 界面气体交换速率
interphase gas 相间气体
interphase heat transfer 相际传热
interphase mass transfer 相间传质
interphase power controller 相间功率控制器
interphase reactor 相间电抗器，中间抽头平滑轭流圈
interphase transfer 相间传递
interphase transformer 吸收电抗器，相间变压器，相位平衡器
interphase 中间相，相间的，相界面
interphone 内部通信装置，内部电话，内线自动电话，对讲机，互通机
interpile sheeting 桩间水平支撑
interplay 相互影响，相互作用
interpluvial 间雨期
interpolar axis 极间轴线，横轴
interpolar gap 极间空隙
interpolar space 极间空间
interpolar voltage 极间电压
interpolar 极间的
interpolated contour 内插等高线
interpolated section 插补断面
interpolated value 内插值
interpolater 插入器，内插器
interpolating multipliers 插值乘法器
interpolating potentiometer 内插式电势计
interpolating 内插
interpolation by central difference 中差插值法
interpolation error 内插误差
interpolation formula 内插公式
interpolation function 插值函数，内插函数
interpolation method 内插法，插值法，插入法
interpolation 插入，内插法，插值法，内推法，插补，内插，插值
interpolator 插入器，内插器，分数计算器
interpole coil 换向极线圈，中间极线圈
interpole core 换向极铁芯，中间极铁芯

interpole flux　换向极磁通，中间极磁通
interpole generator　带中间极的发电机
interpole motor　带中间极的电动机
interpole shoe　换向极极靴，中间极极靴
interpole space　中间极间空间
interpole spool　换向极线圈，中间极线圈
interpole winding　中间极绕组，换向极绕组
interpole　（直流电机的）换向极，辅助极，中间极
interpolymer membrane　共聚物
interpolymer　共聚物，互聚物
interposed thermocouple　插入式热电偶
interpose　调解，放入，插入，干预
interposing relay　干预继电器，断路继电器，中间继电器
interpretable area of the film　底片的可解释区域【射线照相】
interpretable area　可解释的区域【射线照相】
interpretation of aerial photograph　航片判译
interpretation of contract　合同的解释，解释合同
interpretation of logs　测井曲线解释
interpretation　解释，说明，翻译，译码，口译
interpretative language　解释语言
interpretative programme　解释程序，翻译程序
interpretative routine　解释程序
interpretative subroutine　解释子程序
interpretative trace program　解释性追踪程序
interpretative version　解释方案
interpreted photograph　判读相片
interpreter code　翻译代码
interpreter　解释程序，解释器，翻译机，翻译程序
interpretive language　解释语言
interpretive program　翻译程序，解释程序
interpretive programming　解释程序设计
interpretive routine　解释程序
interpretive trace program　解释性追踪程序
interpret parity error　解释奇偶校验误差
interpret　解释，说明，翻译，译码
interprovincial highway　省级公路
interreaction　相互反应，相互作用
inter-record gap　记录间隔
inter-regional cooptation　区域间合作
interregional project　区域间项目
interregional seminar　区域间专题讨论会
interrelated and interdependent　相互联系相互依存的
interrelated layout drawing　相关布置图
interrelated logic accumulating scanner　相关逻辑累加扫描器
interrelationship　相互关系
interrelation　相互关系，相互联系
interrod spacing　棒间距
interrogation command　查询命令
interrogation　询问，触发响应的信号
interrogative telecontrol system　问答式远动系统，查询远动系统
interrogator-responder　询问应答机
interrogator　询问器，问答机，询问者，质问者
interruptable load　可断负荷，可中断负荷
interruptable power　可切功率
interruptable　可断开的，可中断的
interrupt control routine　中断控制程序
interrupt control　中断控制
interrupt-drive system　中断驱动系统
interrupted aging　间断时效，断续时效
interrupted continuous wave　间断等幅波，断续等幅波
interrupted oscillation　间歇振荡，断续振荡
interrupted quenching　分段淬火，时歇淬火，双液淬火
interrupted raceway　中断的电缆管道
interrupted ringing　信号振铃，断续振铃
interrupted wave　断续波
interrupter contact spring　断续器触点弹簧
interrupter contact　断续器触点
interrupter disk　断续器圆盘
interrupter duty　断续工作方式，断续工作
interrupter switch　断续器开关
interrupter　断续器，断路器，障碍物
interrupt handling　中断处理
interruptible forward contract　可中断远期合同
interruptible load reserve　可停电负荷备用
interruptible load　可停电负荷
interruptible power supply　可停电电源
interruptible price　可中断电价
interruptible service　可中断服务
interrupting capability　断开能力
interrupting capacity　截断功率，截断容量，断开容量
interrupting current　断开电流，开断电流
interrupting device　截断装置
interrupting means　截断手段
interrupting pulse　断续脉冲
interrupting ratings　电流切断能力，（断路器）启断额定值
interrupting speed　断路速度，跳闸速度
interrupting time　断路时间，跳闸时间
interruption arc　断弧
interruption cable　应急电缆，替换电缆
interruption duration　断电持续时间
interruption frequency　断续频率
interruption in coal supply　供煤中断
interruption of circuit　断路
interruption of power supply　供电中断，断电
interruption of service　停电，运行障碍，运行事故
interruption status　中断状态
interruption　障碍物，截断物，停止，断开，截断，中断，中断期，断路，停电
interrupt level status word　中断级状态字
interrupt mask　中断屏蔽
interrupt mode　中断方式
interruptor　断续器，断路器，障碍物
interrupt procedure　中断过程
interrupt processor　中断处理机
interrupt sensing　中断识别
interrupt signal　中断信号
interrupt system　中断系统
interrupt trap　中断捕获，中断陷阱
interrupt vector　中断矢量
interrupt　截断，断开，断电，中断，间断，妨

碍，阻止，停止
interscendental curve 半超越曲线
intersected country 起伏地形
intersected 分割的
intersecting members 交叉构件
intersecting parallels “井”字形
intersecting 相交的
intersection angle 交叉角，夹角
intersection curve 相交曲线
intersection data 相交数据
intersection legs 交叉口的交叉路段
intersection of shock waves 冲击波交会
intersection point 交叉点，交点
intersection 交点，交线，相交，交叉，道路交叉口，横断，横切，交叉点，十字路口，交集
intersect 相交，交叉，横断，横切
intersolubility 互溶性
interspace loss 间隙损失
interspace 中间，空间，间隙，留空隙
intersperse 散布，分散，交替，点缀
interstage amplifier 中间放大级
interstage annealing 中间退火
interstage attemperator （过热器）级间减温器
interstage bushing 中间衬套
interstage casing 间级泵壳【多级泵】
interstage cooler 中间［级间］冷却器【压缩机】
interstage coupling 级间耦合
interstage crossover 级间导流管，级间分隔
interstage cyclone 层间除尘器
interstage desuperheater 级间减温器
interstage diaphragm 级间薄膜，级间隔板
interstage gland 级间汽封
interstage network 级间网络
interstage shielding 级间屏蔽
interstage transformer 级间变压器
interstage 级间的，间级
interstar winding Z形绕组，曲折绕组
interstep regulating transformer 级间调节变压器
interstep 级间的
interstice 间隙，裂缝，空隙，第二气隙，缝隙，孔隙
interstitial brick 填隙砖
interstitial channel 管间流道
interstitial compound 间隙化合物
interstitial flow 裂隙水流
interstitial gas flow 隙间气流，管间气流
interstitial gas velocity 管间烟速，隙间烟速
interstitial hydraulic pressure 空隙（水）压力
interstitial phase 间隙相
interstitial site （晶体）间隙位置
interstitial water 缝隙水，空隙水，孔隙水
interstitial 间隙的，空隙的，缝隙的，隙间的，成裂缝的
interstratified bed 间层，互层
interstream area 河间地区
inter-switch link 级间链路
intersystem communication 系统间通信【计算机】
intertidal marsh 潮间沼泽地
intertidal region 潮水浸淹带，潮间带
intertidal zone 潮间带
intertidal 潮区内的，潮间的
intertie 交叉拉杆
intertrade 相互往来，互相交换，国际贸易
intertripping 联锁跳闸
intertrip 联动跳闸，联跳，连锁跳闸
intertropical convergence zone 热带辐合带
intertropical front 热带锋
intertropic convergence zone 热带辐合区
intertube burner 缝隙式燃烧器
intertube economizer 布置在对流管束间的省煤器
intertube gas velocity 管间烟气速度
intertube superheater 管间过热器
intertube 管间的
intertubular space 管间空间
interturn fault detection 匝间故障探测
interturn fault 匝间故障
interturn insulation 匝间绝缘
interturn protection 匝间保护
interturn short circuit test 匝间短路试验
interturn short circuit 匝间短路
interturn tester 匝间试验器
interturn test 匝间试验
interturn 匝间的
interunit wiring 部件间的接线
interunit 部件间的
interurban heat-supply 城际供热
interval arithmetic 区间算术运算，区间运算
interval at commutator 在换向器上的节距，换向片节距
interval between inspections 检查间隔
interval between refuelings 换料时间间隔
interval contraction 区间收缩
interval error 间隔误差
interval function 区间函数
interval in winding 绕组节距
interval linear programming 区间线性规划
interval of service 换班时间，休息时间
intervalometer 时间间隔计，定时器，间隔式读出器，定时曝光控制器
interval sampling 间隔抽样
interval signal 间隔信号，周期信号
interval timer 精确测时计，断续定时器，断续测时计，间隔时钟，间隔计时器，限时器
interval time 间隔时间，时间间隔
intervalve transformer 电子管间耦合变压器
intervalve 电子管间的
interval 空隙，间隔，间歇，区间，距离
intervane burner 旋流叶片式喷燃器，内叶式喷油嘴
intervane 旋流叶片
intervene burner 旋流叶片式喷燃器
intervene 插入，介入，干预，干涉
intervening area inflow 区间来水
intervening area 区间面积，支流间面积
intervening boring jack 插孔（口）
intervening boring 插孔，中间钻孔
intervening cooling 中间冷却
intervening spacing 交错分布干预间距
intervening transformer 中间变压器
intervention button 紧急保险按钮，应急按钮
intervention level of dose 剂量干预水平

intervention level 干预水平
intervention switch 应急保险开关
intervention 介入，插入，干涉，干预
interview appraisal 通过面试进行评价
intervisibility 通视
interweave 组合，交织
interwinding capacity 绕组间电容
interwork 互相配合
in testimony whereof 以此为证，特立此证
in the aggregate 总计，总额，总共
in the capacity of 作为，以……资格
in the case of 就……来说，关于
in the clear 内径，内尺寸，内宽
in the doldrums 萧条，消沉，不景气
in the event of (/that) 万一，即使，倘若，如果，如果发生……，在……情况下
in the event 如果，万一
in the light of 按照，依照，本着，比照，由于
in the limit 在极限情况下
in the lump 总共，全数
in the presence of 在……面前，在有……的情况下
in the proportion of 按……的比例
in the public domain 在公共领域
in the result 作为结果，到后来，结果，后果
in the south of 在……南部【指在……领域之内的南部】
in the view of 在……看来，按照……的观点，随着
intial power transmission 首次送电
intial synchronization （机组）首次并网
intimate contact 密切接触
INTLK＝interlock 连锁
INTMD＝intermediate 中间的
intolerable contamination 超过容许值的污染
intolerable dose 不可耐受剂量，非耐受剂量
intolerance 无法忍受的
into synchronism 进入同步
into the wind 逆风，迎风，顶风
in toto 全，全然，完全，整个地，完整地，完全地
intra-atomic force 原子内力
intrabundle spacing 分裂间距，分裂导线的单根导线间距
intrabundle 分裂导线单根导线间的，绝缘股线间的
intracell flux 栅元内通量
intracell 栅元内的
in traces 痕量
intraconnection 内连，互连，内引线
intracrystalline failure 晶内断裂，穿晶断裂
intractable 难处理的，难控制的，难加工的
intrados of arch 拱内圈
intrados radius 拱内弧半径，内拱圈半径
intrados springing line 拱腹起拱线，拱内弧起拱线
intrados 拱底面，拱腹线，拱内侧面，拱内圈
intraformational bed 层内夹层
intraformational breccia 层内角砾岩
intraformational conglomerate 层内砾岩
intraformational fold 层内褶皱
intragrain diffusion 粒内扩散
intragranular corrosion 穿晶腐蚀
intragranular cracking 穿晶龟裂
intragroup cross-section 群内截面
intranet 内联网，内连网
in transit 在（运输）途中，中转
intraphase conductors （每相分裂导线的）分导线
intrasonic 超低频
intratelluric water 原生水
intra-urban transportation 市内运输
intravane-type pump 内叶片泵
intregrated circuit 集成电路
intrenched meander 嵌入曲流，峡谷曲流
intricate 复杂的，交错的，交叉的
intrinsic acidity 固有酸度
intrinsic activity 固有活性
intrinsic admittance 固有导纳，内在导纳
intrinsically safe electrical instrument and wiring 本质安全型电动仪表和接线
intrinsically safe equipment and wiring 本质安全设备和接线
intrinsically safe system 安全火花型防爆系统，固有安全系统
intrinsically 本质的
intrinsic basicity 固有碱度
intrinsic breakdown 本征击穿
intrinsic chemical reaction rate 本征化学反应速率
intrinsic coercive force 内禀矫顽力
intrinsic constant 固有常数
intrinsic conversion efficiency 本征转换效率【本征效率】
intrinsic curve 包络线，禀性曲线
intrinsic damping 固有阻尼
intrinsic efficiency 固有效率，本征效率
intrinsic energy 固有能，本征能，内能
intrinsic error 内蕴误差，基本固有误差，基本误差，固有误差
intrinsic factor 内部因素
intrinsic fill factor 本征填充因子【理论填充因子】
intrinsic frequency 固有频率
intrinsic heat 固有热，内热
intrinsic impedance 固有阻抗，内在阻抗
intrinsic induction 铁磁感应，固有电感，内在感应，固有磁感应
intrinsic internal angle of friction 固有内摩擦角
intrinsic level 本征能级，内在电平
intrinsic permeability 固有导磁率，本征导磁率，固有透水性，内在透水性
intrinsic pressure 内压力，固有压力，内在压力
intrinsic property 内在特性
intrinsic reaction rate constant 本征反应速率常数
intrinsic safety barrier 本质安全防爆栅
intrinsic safety 固有安全性
intrinsic semiconductor 纯半导体，本征半导体
intrinsic separation factor 本征分离因子
intrinsic shear strength curve 固有抗剪强度包络线
intrinsic shrinkage 内在收缩

intrinsic solubility 固有溶解度
intrinsic stability 固有稳定性，内在稳定性
intrinsic viscosity 特性黏度，固有黏度，本征黏度
intrinsic wavelength 固有波长
intrinsic weight 固有权
intrinsic 本征的，本质的，内在的，内部的，固有的，基本的
in triplicate 一式三份
introduce advanced technology 引进先进技术
introduce foreign capital 引进外资
introduce foreign funds and technology 引进外国资金和技术
introduce foreign investment 引进国外投资
introduce foreign technology 引进外国技术
introduce 引进，引导，引入，掺入，介绍
introduction of the design details prior to the construction 设计交底
introduction valve 进样阀
introduction 引言，引进，采用，介绍，前言，概述
introductory article 引言条款
introductory provision 引言条款
introductory statement 介绍性说明
introflexion 向内弯曲
introscope 内窥镜，内腔检视仪，内壁检验仪，内孔检视仪
intruder detection and access control system 出入探测与门禁系统
intrusion agent 添加剂，灌入剂，混合料
intrusion pipe 灌注管【水泥浆的】
intrusion 侵入，注入，打扰，侵扰，侵入作用
intrusive breccia 侵入角砾岩
intrusive contact 侵入接触
intrusive dike 侵入岩脉，侵入岩墙
intrusive dyke 侵入岩脉，侵入岩墙
intrusive mass 侵入体
intrusive rock 贯入岩，侵入岩
intrusive sheet 侵入岩床
intrusive sill 侵入岩床
intrusive vein 侵入脉，侵入岩脉
in-tube working medium 管内工质
intumescent paint 膨胀漆
inundated area 泛滥区，受淹面积，淹没地区，淹没面积
inundated cultivated land 淹没农田
inundated district 泛滥地区
inundated land 泛滥地区，洪水泛滥区，受淹土地，淹没地，淹没地区
inundated plain 泛滥平原
inundate 淹没
inundation area 受淹面积，淹没面积，淹区
inundation damage 淹没损失
inundation map 淹没图
inundation 泛滥，洪水泛滥，淹没
invade 灌水
invaded zone 侵入带
invading air 侵入空气
invading water 侵入水
invalid clause 失效条款
invalid contract 无效合同
invalid data 无效数据
invalidity of contract 合同失效
invalidity 无效
invalid 无效的
invariability in price 价格不变性
invariable cost 不变成本
invariable linear system 不变线性系统
invariable system 不变体系
invariable 不变的
invariance 不变性
invariant factor 不变因子
invariant imbedding 不变嵌入法
invariant of stress 应力不变量
invariant relation 不变关系
invariant scalar 不变标量，不变标度，不变比例尺
invariant 恒定的，不变量，不变式，不变的
invar levelling staff 殷钢水准尺
invar plotting scale 殷钢卷
invar tape 殷钢绘图尺
invar wire 殷钢线
invasion 侵害，侵入，侵入作用
invention 发明
inventory change 存量变化
inventory control 库存管理，库存量控制
inventory cost 存货成本
inventory management fee 仓储管理费
inventory management program and control technique 库存管理程序与控制技术
inventory management 库存管理
inventory of fission-product radioactivity 裂变产物放射性总量
inventory of water resources 水资源册
inventory rating 投料量，比投料量
inventory sheet 盘存表，盘点单
inventory statistics 库存统计表
inventory survey 现况调查
inventory system 存储制度，盘存制度
inventory table 盘存表
inventory tag 盘存标签
inventory ticket 盘存票
inventory 装载量，货单，库存清单，投料量，库存量，盘存，财产清单，存货，设备清单
invent 发明人
inverse annular flow 反环状流
inverse assembler 反汇编程序
inverse back coupling 负反馈
inverse bremsstrahlung 逆韧致辐射
inverse cosine 反余弦
inverse current 逆电流，反向电流，反相电流，逆流
inverse curve 反曲线
inversed circuit 反演电路
inverse definite minimum time 最短动作时间的反时限特性
inverse definite time relay 反定时限继电器
inverse derivative controller 反微分控制器，反导数调节器
inverse direction 逆向
inverse discrete Fourier transform 离散傅里叶逆变换

inversed repulsion motor 反推斥电动机
inversed V-connection 反 V 形接法
inverse electrode current 反向电极电流
inverse electron capture 电子逆俘获
inverse energy transmission 反向能量传输
inverse estuary 逆向河口
inverse fast Fourier transform 快速傅里叶逆变化
inverse feedback 负反馈
inverse field 逆序场，反向场
inverse flow 逆流，回流，反向流动，倒流
inverse flux trap concept 反通量阱概念
inverse Fourier transformation 傅里叶逆变换
inverse function 反函数
inverse hour 倒时数，核反应的单位
inverse hyperbolic function 反双曲线函数，反双曲函数
inverse impedance 逆阻抗，反阻抗
inverse index 逆指标，倒指标
inverse inductance matrix 逆电感矩阵
inverse integrator 逆积分器
inverse interpolation 反插值（法），反内插法
inverse kinetics measurement 逆动力学测量
inversely proportional to 与……成反比
inverse matrix 矩阵求逆，矩阵反演，逆矩阵
inverse network 倒置网络，反演网络
inverse operation 反运算
inverse operator 反算子
inverse peak voltage 最大反向电压，反向峰压
inverse period 反演周期，逆电势周期
inverse photoelectric effect 逆光电效应，反压电效应
inverse plasma betatron 逆等离子体电子感应加速器
inverse position computation 后方交会计算
inverse power relay 逆功率继电器
inverse power transmission 倒送电，反送电
inverse probability 逆概率
inverse proportional 成反比的
inverse proportion 反比例
inverse ratio 反比
inverse reactor period 反应堆周期倒数
inverse relation telemeter 反比式遥测计
inverse relation 反比关系
inverse sample censusing 颠倒采样调查
inverse signal 返回信号，回答信号
inverse sine 反正弦
inverse-speed motor 串激特性电动机，反速电动机
inverse square law 平方反比定律
inverse system 逆向系统
inverse time 反时，逆时
inverse time characteristic 反时特性
inverse time current protection 反时限过流保护
inverse time definite-time limit relay 逆时定时限继电器，定时限反时继电器
inverse time delay unit 逆延时单元
inverse time induction relay 感应式反时继电器
inverse time-lag switch 反时延开关
inverse time-lag 反时限，反时延
inverse time limit 反比时限
inverse time relay 反时限继电器
inverse transfer function 反传递函数
inverse transfer locus 反传递轨迹
inverse transformation 反变换，逆变换
inverse trigonometric function 反三角函数
inverse voltage of rectifier 整流器反向电压
inverse voltage 反电压
inverse Z-transform 逆 Z 变换
inverse 倒置的，相反的，逆向的，反相的，逆的，反面，反量，倒数，倒置，倒
inversion base height 逆温层底高度
inversion base 逆温层底
inversion break up fumigation 逆温消散型熏烟
inversion centre 反演中心
inversion condition 逆变条件
inversion constant 反演常数，反演率
inversion dissipation 逆温耗散
inversion formula 反转公式
inversion fumigation plume 逆温下熏型羽流
inversion heat flux 逆温热通量
inversion height 逆温层高度
inversion layer 逆温层
inversion lid 逆温层顶
inversion mechanism 转换机构
inversion of energy 能量转化
inversion operation of converter station 换流站逆变运行
inversion penetration 逆温渗透，逆渗透
inversion point 逆转点
inversion temperature 转换温度
inversion 逆变流，倒置，颠倒，倒转，反向，逆变，反演，反转，反相，逆增
inversive tower erecting 倒装组塔
invertebrate animal 无脊椎动物
inverted alternator 旋转电枢式同步发电机
inverted angle 倒角
inverted arch floor 倒拱底板
inverted arch 倒拱
inverted argon arc welding machine 逆变氩弧焊机
inverted beam 上翻梁
inverted bell manometer 倒钟式压力计
inverted bell 倒钟
inverted bow and chain girder 倒置拱式大梁，鱼腹式大梁
inverted bucket trap 倒吊桶式疏水阀，翻斗活门
inverted bucket type steam trap 倒吊桶式疏水器，倒浮子式疏水器
inverted buck trap 倒吊桶式疏水器
inverted converter 逆换流器，反向变流机，反向变换机
inverted cusp （旋涡的）底谷，反尖顶
inverted dip 倒倾角
inverted-dome closure （蒸汽发生器夹舱）倒置的汽包盖
inverted draft 反向气流，逆向气流
inverted drainage well 反渗排水井
inverted draught 反向气流
inverted echo sounder 反回声测深仪
inverted filter 反滤层
inverted fir-tree blade root 外包式枞树形叶根
inverted hour 逆时针的，倒时数

inverted image 倒像
inverted input 反相输入
inverted J-tube 倒钩形出水管【蒸汽发生器】
inverted L network 倒L形网络
inverted-loop tube 倒U形管
inverted magnetron 反磁控管
inverted motor 旋转电枢式电动机
inverted numbering 相反数码
inverted order 反序，逆序
inverted output 反相输出
inverted penetration 倒铺路面法
inverted position of telescope 倒镜
inverted position 倒立位置【翻车机】，倒转层位
inverted relief 倒向地形
inverted repulsion motor 反推斥式电动机
inverted roof 凹屋面，倒置屋顶
inverted rotary converter （直流变交流）旋转式逆变机
inverted sequence 反序，逆序
inverted siphon culvert 倒虹吸涵洞
inverted siphon pipe 倒虹吸管
inverted siphon 倒虹吸，倒虹吸管
inverted strata 倒转岩层
inverted stream 倒流河
inverted T-beam 倒T形梁
inverted T blade root 倒T形叶根
inverted tension set 倒挂耐张串【输电线路】
inverted tide 逆潮
inverted top hat （下部堆内结构）倒置帽【核电】
inverted triangular truss 倒三角形桁架
inverted triode 倒用三极管
inverted T root blade 倒T形叶根叶片
inverted T-section floor 倒T形楼板
inverted T-section precast floor 倒T形预制楼板
inverted T-slot 倒T形槽
inverted T-type blade root 倒T形叶根
inverted T-type retaining wall 倒T形挡土墙
inverted tube 倒相管，倒用管
inverted turn transposition 线圈端部180度换位
inverted turn 扭转180度线匝
inverted valve 逆止阀
inverted welding machine 逆变电焊机
inverted welding power source 逆变焊接电源
inverted welding 仰焊接
inverted weld 凹面焊，凹面焊缝
inverted well 倒流井，反渗井，逆向井
inverted 反的，逆的，倒转的，定转子功能倒换的
invert elevation 底板高程
inverter amplifier 倒相放大器
inverter DC welding machine 逆变直流焊机
inverter efficiency 逆变效率
inverter electric welding machine 逆变式电焊机
inverter-induction motor 逆变器供电的感应电动机
inverter stage 可控大功率汞弧整流器，倒相级
inverter station 逆变站
inverter transformer 反流变压器，逆变器用变压器
inverter 逆换流器，逆变器，反流器，倒相器，逆变装置，变换器，电流换向器
invertibility 可逆性
invertible 可逆的
inverting amplifier 反相放大器，倒相放大器
inverting element 反演器，反向元件，反向器
inverting eyepiece 倒像目镜
invert level 倒拱底高程
invert masonry block 倒拱砌块
invertor operation 逆变器运行
invertor 逆换流器，逆变器，反流器，倒相器
invert 倒置的，反转，颠倒，倒置，转换，转化，相反，倒拱
in-vessel axial-flow pump （沸水堆）压力容器内轴流泵
in-vessel filter 容器内过滤器
in-vessel jet pump 内喷射泵，容器内喷射泵
in-vessel machine 容器内（燃料）装卸机
in-vessel transfer device 容器内燃料传送装置
investigation and study 调查研究
investigation level 调查水平
investigation method 调查方法
investigation of accident and failure 事故对策
investigation phase 调查研究阶段
investigation 调查，调查研究，勘探，考察
investment abroad 对国外的投资
investment adviser 投资顾问
investment analysis 投资分析
investment appraisal 投资估价
investment association 投资协会
investment bank 投资银行
investment base 投资基准
investment broker 投资经纪人
investment center 投资中心
investment charges 投资费用
Investment climate 投资环境
investment company 投资公司
investment consultant corporation 投资咨询公司
investment consultant 投资顾问
Investment contract 投资合同
investment control 投资控制
investment corporation 投资公司
investment cost allocation 投资分摊，投资费用［成本］分摊
investment cost of power transmission project 输电工程投资费用［成本］
investment cost 投资费用，投资成本，投资费
investment counsellor 投资顾问
investment decision 投资决策
investment demand 投资需求
investment direction 投资导向，投资方向
investment effectiveness 投资效益
Investment environment 投资环境
investment estimation 投资估算
investment evaluation 投资评价
Investment funds 投资资金，投资基金
investment guarantee 投资保证
investment in capital construction 基本建设投资
investment in enterprise 企业投资
investment in fixed assets 固定资产投资
investment in human capital 人力资本投资

Investment institution　投资机构
investment intention　投资意向
investment law　投资条例
investment loan　投资贷款
investment management　投资管理
investment objective　投资目标
investment operating cost　投资运行成本
investment orientation regulatory tax　投资方向调节税
investment orientation　投资方向
investment oriented project　以投资为目标的项目
investment outside the plan　计划外投资
investment phase of project　项目投资期
investment plan　投资规划，投资计划
investment potential　投资潜力
investment priority　优先投资项目，投资重点
investment program　投资规划
investment propensity　投资倾向
investment proposal　投资方案
Investment recovery period　投资回收年限
investment regulation　投资监管
investment return　投资回收
investment risk　投资风险
investment scale　投资规模
Investment structure　投资结构
investment trust company　投资信托公司
investment trust　投资信托公司，投资信托
Investment turnover　投资周转率
investment within the plan　计划内投资
investment　投资
investor owned utility　投资者拥有的电力公司
investor　投资者
invest　投资
in view of　鉴于，由于，由……看来，考虑到，在看得见的地方，基于
in view　看见，在考虑中，在观察中，作为目的
invigilator　监视器
invigorated river　多水河流
invigorate　鼓舞
invigorating economy　搞活经济
in virtue of　由于
inviscid flow　理想液体流，无黏性流，非黏性流，无黏流
inviscid fluid　非黏性流体，理想流体，无黏流体
inviscid motion　无黏流动
inviscid shear flow　无黏剪切流
inviscid　无黏性的，无韧性的，不能展延的，非黏性的
invisible assets　无形资产，账外资产
invisible capital　无形资产，无形资本
invisible radiation　不可见辐射
invitation　请帖
invitation and submission of tender　招标和提交标书
invitation card　请帖
invitation for and submission of tender　招投标
invitation for bid　招标，招标书，招标通告，招标邀请函
invitation for offer　邀请发盘，招标
invitation of tender　招标，招标书，招标邀请函
invitation telex　邀请电报
invitation to bid　招标，投标邀请书，邀标书
invitation to prequalify　资格预审邀请
invitation to tender　招标，投标邀请书，邀标书
invitation to treat　邀请投标人做交易
invite bids for　对……招标
invite bids　招标
invited bidder　特邀投标者
invited delegate　应邀代表
invited firms　被邀请的厂商
invited lecture　特邀报告
invited participant　应邀的参与者
invite tenders　招标
invite to tender　招标
invite　邀请
inviting businessmen to invest　招商
in vitro activation analysis　体外活化分析
in vitro radio-assay　体外放射性分析法
in vivo activation analysis　体内活化分析
in vivo study　体内研究
in vivo test　体内试验
in vivo　体内
invoice amount　发票额
invoice for sales　销售发票
invoice price　发票价格
invoice　发货票，发票，开发票，发货单，结算清单，装货清单
invoke　请求，调用，引用，引起，产生，寻检器
involuntary insolvency　强制破产
involute coil　渐开型线圈
involute core　渐开线铁芯
involute gauge　渐开线齿轮
involute gear　渐开线量规
involute lamination　渐开线叠片
involute profile　渐开线齿形
involute　内旋转的，渐伸线，渐开线，切展线，内卷的
involution　对合
involutory matrix　对合矩阵
involved with　涉及
involve in　包括在……中，与……有关
involvement with project　介入项目
involvement　参与
involve　包含，包括，涉及，包围，使参入，牵涉
involving enclosure　外罩
INVR＝inverter　逆变器
inwall　内壁，内衬
inward charges　入港费
inward-flow turbine　辐流式水轮机，内流式水轮机，向心式水轮机
inward-flow　向心式【水轮机】
inward fuel transfer　向心式燃料传运装置，燃料向心传送
inward heat transmission　向内加热
inward pressure　内向压力
inward radial shuffling　径向向内倒料
inward stroke　向内冲程
inward transfer procedure　向内输送程序
inward transfer room　向内运输间
in-wash　岸边淤积，冰川边缘沉积，巨厚冲积层，冲积层
in-water test　水下试验

in wind response 顺风（向）响应
in witness whereof 兹证明，以此为证，以昭信守，以资证明，作为其证据，特此证明，特立此证，“空口无凭，特立此约为证”【合同用语】
in words 大写【金额】
IOA(input/output adapter) 输入输出转接器
IOB(input/output buffer) 输入输出缓冲器
I/O cabinets 输入输出箱
I/O card 输入输出卡件
IOCC(input/output control center) 输入输出控制中心
IOCC(input/output control command) 输入输出控制命令
IOC(input/output channel) 输入输出通道
IOC(input/output controller) 输入输出控制器
IOC(input/output control) 输入输出控制
IOCP(input/output control program) 输入输出控制程序
IOCS(input/output control system) 输入输出控制系统
iodate 碘酸盐
iodic acid 碘酸
iodide light 碘灯
iodide 碘化物
iodimetry 碘滴定法
IOD(immediate oxygen demand) 直接需氧量
iodination 用碘处理，碘化（作用）
iodine adsorber 碘过滤器，碘吸收器
iodine air monitor 空气中碘监测器
iodine characterization sampler 碘特性取样器
iodine filter 碘过滤器，碘吸收器
iodine lamp 碘灯
iodine reactor 碘反应器
iodine scavenger 碘吸收器
iodine spiking 碘脉冲
iodine trap 碘吸收器
iodine well 碘吸收器，碘阱
iodine 碘
iodometric titration 碘滴定
iodometry 碘滴定法
I/O(input/output) 输入输出
I/O interface equipment 输入输出接口装置
IOM(input/output multiplexer) 输入输出多路转换器
ion 离子，离子雾
ion association 离子缔合（作用）
ion atmosphere 离子
ion balance 离子平衡
ion beam coating method 离子束涂敷法
ion beam instability 离子束不稳定性
ion beam monitor 离子束监测器
ion chamber room 电离室房间
ion chamber unit 电离室单元
ion chamber 电离室
ion-chromatographic analyzer 离子色谱仪
ion chromatography 离子色层法，离子色谱法
ion cloud 离子云
ion collector 离子收集器，离子收集极
ion concentration 离子浓度
ion current density 离子电流密度
ion current 离子电流
ion density 离子密度
ion dose rate 离子剂量率
ion dose 离子剂量
ion exchange absorbent 离子交换吸附剂
ion exchange bed 离子交换床
ion exchange breakthrough 离子交换穿透
ion exchange capacity 离子交换容量
ion exchange cellulose 离子交换纤维素
ion exchange chromatography 离子交换色谱法
ion exchange column 离子交换床，离子交换柱
ion exchange electrodialysis 离子交换电渗析
ion exchange exhaustion circuit 离子交换提取系统
ion exchange filter 离子交换过滤器
ion exchange liquid 离子交换液
ion exchange material 离子交换材料
ion exchange membrane 离子交换膜
ion exchange process 离子交换过程，离子交换法
ion exchange product 离子交换产物
ion exchanger dedeuterization process （重水堆）离子交换器除氘过程
ion exchange resin bed 离子交换树脂床
ion exchange resin fines 离子交换器树脂屑
ion exchange resin 离子交换树脂
ion exchanger 离子交换器，离子交换剂
ion exchange softener 离子交换软化器
ion exchange system 离子交换系统
ion exchange tower 离子交换塔
ion exchange water treatment 离子交换水处理
ion exchange 离子交换
ion exclusion chromatography 离子排斥色谱法
ion exclusion 离子排斥
ion flow 离子流
ion gauge 离子压力计
ionic acidity 离子酸度
ionic activity 离子活度
ionic charge 离子电荷
ionic concentration 离子浓度
ionic conduction 离子导电
ionic conductivity 离子导电性，离子导电率，电离导电率
ionic discharge 离子放电
ionic impurity 离子不纯物，离子杂质
ionic interaction 离子相互作用
ionic link 离子键
ionic mobility 离子迁移率
ionic polymerization 离子聚合
ionic reaction 离子反应
ionic relay 离子继电器
ionic semiconductor 离子半导体
ionic strength 离子强度
ionic yield 电离产额
ionic 离子的
ion implantation 离子注入，离子种入，离子移植技术
ion injection method 离子注入法
ionised particle 电离粒子
ionitriding 离子氮化法
ionizable contaminant 离子污染
ionization chamber dosimeter 电离室剂量计

ionization chamber exposure ratemeter 电离室照射率计
ionization chamber unit 电离室单元
ionization chamber 电离室
ionization coefficient 电离系数
ionization constant 电离常数
ionization current 电离电流
ionization degree 电离度
ionization density 离子化浓度
ionization detector 电离检测器
ionization dosage 电离剂量
ionization energy 电离能
ionization equilibrium 电离平衡
ionization extinction voltage 消电离电压
ionization gauge 电离压力计
ionization manometer 电离压力计
ionization potential 电离电位，电离电势，电离势能
ionization power 电离能力
ionization process 电离过程
ionization pulse 电离脉冲
ionization puncture 电离击穿
ionization rate 电离率
ionization smoke alarm 离子感烟报警器
ionization smoke and heat detector 电离烟尘和热量探测器
ionization smoke detector 电离感烟探测器，离子感烟探测器
ionization tendency 电离倾向
ionization track 电离轨迹
ionization type radiation monitor 电离式辐射监视器
ionization 电离，离子化，电离作用
ionized atom 电离原子
ionized gas anemometer 电离式风速仪
ionized gas 电离气体，离子化气体
ionized layer 电离层
ionized stratum 电离层
ionizer wire 电晕极
ionizer 离化剂
ionize 电离，离子化，使电子化
ionizing energy 电离能
ionizing event 电离事件
ionizing field 电离场
ionizing irradiation 电离辐照
ionizing particle 致电离粒子
ionizing potential 电离电位
ionizing radiation 电离辐射，致电离辐射
ionizing voltage 电离电压
ion membrane 离子膜
ion migration 离子迁移
ion number density 离子数密度
ionogenous impurities 可离子化杂质
ionogram 电离图
ionometer 离子计
ionophore 离子载体
ionosonde 电离层探测装置
ionosphere 电离层，离子层
ionospheric interference 电离层干扰
ionospheric plasma 电离层等离子体
ionotron 静电消除器
ion pair 离子对
ion plating 电离镀层
ion product 离子积
ion selective electrode 离子选择电极
ion selective measuring system 离子选择测量系统
ion sensitive electrode 离子敏感电极
ion sputtering pump 离子溅射泵
ion strength 离子强度
ion transfer 离子迁移
IOU(investor owned utility) 投资者拥有的电力公司
IOU(I owe you) 借条
I/O unit 输入输出装置
IP address IP 地址
IPB(isolated phase bus duct) 离相封闭母线
IPB(isolated phase bus) 离相母线
IPBP(intermediate pressure by-pass) 二级减温减压装置，中压旁路
IP bypass loop 中压旁路回路
IPC(industrial process control) 生产过程控制
IPC(information processing center) 信息处理中心
IPC(integrated pollution control) 污染综合治理
IP code(international protection code) 国际防护代码
I/P converter(electric to pneumatic converter) 电/气转换器
IPCV(intermediate pressure control valve) 中压调节阀
IPDH(in-service planned derated hours) 计划降低出力运行小时
IPE(individual plant evaluation) 单个电厂评价
IPE(interpret parity error) 解释奇偶校验误差
IPI(interior point intermodal) 内陆公共点多式联运
IP(ingress protection) 防护等级，入侵防护，外壳防护等级，低压电器外壳防护等级【IEC 对电气设备外壳对异物侵入的防护等级。如 IP54，第一数字表示接触保护和外来物保护等级，第二数字表示防水等级】
IP(intellectual property) 知识产权
IP(intermediate pressure) 中压
IP(internet protocol) 互联网协议，网际协议
IP(item processing) 项目处理
IPM(inches per minute) 英寸/分
IPMT(integrated project management team) 联合工程管理模式
IPN(inspection progress notification) 检查进度报告书
IPNS(intense pulsed neutron source) 强脉冲中子源
IPO(initial public offering) 首次公开发行股票
Iporka 艾波卡低温绝缘材料
IPP(independent power producer) 独立发电厂，独立电力生产者，独立发电商，独立发电项目，独立发电站
IPPV(intermittent positive pressure ventilation) 间歇正压通气
IPR(initial pressure) 初压
IPR(intellectual property right) 知识产权
IPRM(intermediate power range monitor) 中等功

率区段监测器
IPSAR(integrated plant safety assessment report) 综合电厂安全评价报告
IPS(inches per second) 英寸/秒
IPS(information processing system) 信息处理系统
IPS(international pipe standard) 国际管材标准
IPS(international plutonium storage) 国际钚贮存
IPS(international pyrheliometric scale) 国际直接日射计标度
ipso facto avoidance 当然失效
ipso facto clause 破产约定条款，自动终止条款
ipso facto party 当然当事方
ipso facto vacate office 事实离职
ipso facto 依照事实，根据事实本身，事实上
IPSW(intermediate press service water) 中压厂用水
IPSWP(intermediate press service water pump) 中压厂用水泵
IPWR(integrated pressurized water reactor) 一体化压水反应堆，一体化压水堆
IQI(image quality indicator) 像质计【焊缝射线照相】，图像质量指示器
IQI sensitivity 像质计灵敏度
I-rail 工字钢轨道
random inspection 随机抽查
iraser 红外激光器
iridium 铱【Ir】
IRI(International Roughness Index) 国际平整度指数
IR(internal report) 内部报告
iris action 阻隔作用
iris 膜片，隔膜，可变光阑
iron accumulator 碱铁蓄电池
iron-air cell 铁空气电池
iron alloy industry 铁合金工业
iron alloy 铁合金
iron bacteria 铁细菌
iron bark 硬木
iron bar 铁条
iron-bearing protective film 含铁保护膜
iron-bearing water 含铁水
iron binding wire 铁扎线，绑扎铁丝
iron-carbon equilibrium diagram 铁碳平衡图
iron chloride 氯化铁
iron choke coil 铁芯扼流圈
iron circuit 铁芯磁路
iron-clad coil 铁壳线圈
iron-clad cutout 铁壳断流器
iron-clad distribution equipment 铠装配电设备，铁壳配电装置
iron-clad galvanometer 铁壳电流计，铁壳检流计
iron-clad switchgear 铁壳开关装置，铠装开关装置
iron-clad switch 铁壳开关
iron-clad transformer 铁壳变压器
iron-clad 铁壳的，铠装的，包铁的
iron-cobalt-nickel alloy 铁钴镍合金
iron concretion 铁质结核
iron-constantan thermocouple 铁康铜热电偶
iron-container rectifier 铁壳整流器
iron content 含铁量
iron core choke 铁芯抗流圈，铁芯扼流圈
iron core coil 铁芯线圈
iron-cored type instrument 铁芯式测量仪表
iron-cored 有铁芯的
iron core inductor 铁芯感应线圈
iron core loss 铁损
iron core transformer 铁芯变量器，铁芯变压器
iron core 铁芯
iron covering 铁盖
iron cushion 垫铁
iron damage 铁芯烧毁
iron dichloride 氯化亚铁
iron door 铁门
iron dust coil 铁粉线圈
iron dust core coil 铁粉芯线圈
iron dust core 铁粉芯
iron dust 铁粉，铁屑
iron fence 铁栅
iron filing mortar 铁屑砂浆
iron filing 铁屑
iron foundry 铸铁车间，铸铁厂
iron gauze 铁纱
iron grating 铁箅子
iron grill 铁栅
iron hoop 铁箍
ironic citrate 柠檬酸铁
ironic hydroxide 氢氧化铁
ironic oxide 三氧化二铁
ironing machine for stator coil 定子线圈热压机
iron ion 铁离子
iron length 铁芯长度
ironless armature 无铁电枢，空心电枢
ironless transformer 无铁芯变压器
ironless 无铁芯的，无铁的
iron loss factor 铁损因子，铁耗因子
iron loss per unit weight 单位铁损
iron loss 铁损，铁耗
iron magnetic property 铁磁性
ironmongery 建筑五金，五金铁件，小五金
iron nail 铁钉
iron-nickel accumulator 铁镍蓄电池
iron-oilite 多孔铁
iron ore cement 矿渣水泥，铁矿水泥，铁矿渣水泥
iron ore 铁矿，铁矿石
iron oxide 氧化铁，铁丹，红丹
iron pipe 铁管
iron plated concrete 包铁混凝土
iron plate 铁板
iron portland cement 含铁硅酸盐水泥
iron powder coated electrode 铁粉焊条
iron powder electrode 铁粉焊条
iron powder titania calcium electrode 铁粉钛钙型焊条
iron-pyrite 黄铁矿，二硫化铁
iron red alkide resin paint 铁红醇酸树脂漆
iron resistance 铁电阻
iron rivet 铁铆钉
iron rust 铁锈

iron sand 铁砂
iron scaffold 金属脚手架
iron scale 铁氧化皮
iron scrap 废铁
iron shaving 铁屑
iron sheet 铁皮
iron shield 铁屏蔽
iron-silicon alloy 硅铁合金，硅钢
iron soldering 络铁钎焊
iron stain 铁锈
iron strap 铁皮条
iron sulfate 硫酸铁
iron tower 铁塔
iron-vane instrument 动铁式仪表，铁叶式仪表
iron vitriol 铁矾
iron waste 碎铁
iron wire mesh 铁丝网
iron wire net 铁丝网
iron wire 铁线，铁丝
ironwood 硬木
iron worker's shop 金属结构车间
iron worker 安装钢支架工人
iron yoke 铁轭
iron 用铁包铁，铁芯，熨斗，熨平，烙铁，铁
irradiance 辐（射）照度【E】，辐射通量密度
irradiated fuel assembly 辐照过的燃料组件，乏燃料组件
irradiated fuel examination 辐照过的燃料检验
irradiated fuel reprocessing 辐照过的燃料后处理，乏燃料后处理
irradiated fuel shipping cask 辐照过的燃料运输罐，乏燃料运输罐
irradiated fuel store 乏燃料贮存，辐照过的燃料贮存
irradiated fuel transport cask 辐照过的燃料运输罐
irradiated fuel 辐照过的燃料，乏燃料
irradiated nuclear fuel 辐照过的核燃料
irradiated nuclear fuel element 辐照过的核燃料元件
irradiated nuclear fuel storage 辐照过的核燃料储存
irradiated thimble takeup reel 辐照过的套管缠绕卷筒
irradiated 辐照过的【核燃料】，受过辐照的，照射过的
irradiate 照射，辐射，辐照
irradiation behaviour 辐照性能
irradiation breakdown 辐照杀伤，辐照损伤
irradiation capsule 辐照盒，辐照样品
irradiation channel 辐照孔道
irradiation coefficient 辐照系数
irradiation coupon 辐照样品，辐照试样
irradiation creep 辐照蠕变
irradiation damage 辐照损伤
irradiation dose 照射剂量，辐照剂量
irradiation effect of reactor material 反应堆材料辐照效应
irradiation effect 辐照效应
irradiation efficiency 辐照效率
irradiation embrittlement 辐照脆化
irradiation enhanced creep 蠕变辐照增长
irradiation equipment 辐照装置
irradiation facility 辐照装置
irradiation growth 辐照生长
irradiation hardening 辐照硬化
irradiation hazard 辐照危害
irradiation heat 辐照热
irradiation hole 辐照孔道
irradiation-induced creep 辐照感生蠕变
irradiation-induced degradation 辐照引起的降解
irradiation-induced swelling 辐照引起的肿胀
irradiation injury 辐照伤害，辐照损伤
irradiation level 燃耗深度，辐照水平
irradiation limit 辐照极限，辐照限值
irradiation loop 辐照回路
irradiation performance of fuel assembly 燃料组件辐照性能
irradiation plant 辐照工厂
irradiation plug 辐照管，辐照套
irradiation port 辐照孔
irradiation range 辐照范围
irradiation reactor 辐照用反应堆
irradiation rig 辐照台架，辐照用具
irradiation room 辐照室
irradiation sample container 装辐照试样的容器
irradiation sample handling tool 辐照试件的吊装工具
irradiation specimen handling tool 辐照试件的吊装工具
irradiation sample 辐照样品，辐照试样
irradiation service 辐照业务
irradiation specimen access plug 辐照样品进入塞
irradiation specimen basket 装辐照试样的容器
irradiation specimen capsule 辐照样品盒
irradiation specimen 辐照样品
irradiation stability 辐照稳定性
irradiation swelling 辐照肿胀
irradiation test 辐照试验
irradiation time 辐照时间，照射时间
irradiation tube 辐照管
irradiation tunnel 辐照孔道
irradiation unit 辐照装置
irradiation vessel 辐照容器
irradiation well 辐照井
irradiation 放射，照射，遭受大量辐射，辐射，辐照，光渗，辐照量【H】
irradiator 照射器，辐射器，辐照器，辐射体，辐射源
irrationality 不合理
irrational rules and regulations 不合理的规章制度
irrational 不合理的，无理性的，荒谬的，无理数
irrecoverable strain 不可恢复应变
irredeemable currency 不能兑换的货币
irreducible 不可约的
irregular bottom 不平整河底
irregular car 异型车皮
irregular-coursed rubble 乱砌毛石
irregular course 乱砌层
irregular cross section 不规则横断面

irregular erratum 偶然误差
irregular error 偶然误差
irregular fluctuation 不规则脉动
irregular grading 不规则的分级
irregularity coefficient of coal delivery 来煤不均匀系数
irregularity coefficient 不平整系数
irregularity degree 不平整度
irregularity factor （导线）表面状态因数
irregularity of wave form 波形畸变率
irregularity 不规则性，不均匀性，不均匀度，非规范性，紊乱
irregularly activated network 不规则激励网络
irregularly distributed load 不均匀分布载荷
irregularly load 不规则分布荷载
irregularly shaped 形状不规则的
irregular material 不规则物料
irregular meteoro-logical disturbance 不规则气象干扰
irregular motion 不规则运动
irregular natural stream 不规则天然河流
irregular operation 不稳定运行
irregular oscillation 不规则振荡，不规则振动
irregular pattern type cracking 不规则形开裂
irregular pulsation 不规则脉动
irregular report 不定期报告
irregular settlement 不规则沉陷
irregular steam demand 不均匀蒸汽需求量
irregular tax 杂税
irregular terrain 不平地形
irregular wave 不规则波
irregular weir 不规则堰
irregular winding 不对称线圈，不规则线圈
irregular wind 不规则风，多变风
irregular 不规则的，不对称的，不定期的，不规则物
irretrievable disposal 不可回取的处置
irreversibility 不可逆性
irreversible absorption current 减幅传导电流，不可逆吸收电流
irreversible adsorption 不可逆吸附
irreversible conversion 不可逆变化
irreversible cycle 不可逆循环
irreversible deformation 不可回复变形
irreversible element 不可逆元件
irreversible fan 不逆转风机
irreversible motion 不可逆运动
irreversible operation 不可逆运算
irreversible permeability 不可逆磁导率
irreversible process 不可逆过程
irreversible reaction 不可逆反应，不可逆旋转
irreversible transmission 不可逆传输，不可逆传动
irreversible 不可逆的，单向的
irrevocable contract 可撤销的合同
irrevocable credit 不可撤销信用证
irrevocable documentary letter of credit 不可撤销跟单信用证
irrevocable letter of credit 不可撤销的信用证
irrevocable letter of guarantee 不可撤销的保函
irrevocable 不可撤销的
irrevocably 不可撤回地，不可取消地，不可改变地
irrigated area 灌区，灌溉面积
irrigation and drainage pumping station 排灌站
irrigation area 灌溉面积，灌区
irrigation by electric power 电力灌溉
irrigation density 淋水密度
irrigation district 灌区
irrigation pumping 灌溉提水
irrigation tunnel 灌溉隧洞
irrigation 灌溉
IRR(internal rate of return) 内部收益率
irritant gas 刺激性气体
irritating compound smog 刺激性化合物烟雾
irritating pollutant 刺激性污染物
irritating smog 刺激性烟雾
irritation 刺激性
IRR method 内部回收率法
irrotational deformation 无旋形变
irrotational field 无旋场
irrotational flow 无旋流
irrotationality 无旋性
irrotational motion 无旋运动
irrotational vector field 无旋矢量场，无旋向量场
irrotational vector 无旋矢量
irrotational vortex 无旋涡流
irrotational wave 无涡波，无旋波
irrotational 无旋的，不旋转的
irruptive rock 侵入岩
IRSE(international reactor safety evaluation) 国际反应堆安全评价
irving type balance 欧文式补偿，翼内空气动力补偿
isabellin 锰系电阻材料
isabnormal line 等异常线
ISA(Industry Standard Architecture) 工业标准结构
ISA(Instrument Society of America) 美国仪器学会
ISA(international standard atmosphere) 国际标准大气
isallobaric chart 等变压图
isallobaric low 负变压中心
isallobaricwind 等变压风
isallobar 等气压图
isallohypse 等变高线
isallotherm 等变温线
isanemone 等风速线
isanomaly 等距平线
isanomal 等距常线，等地平
ISAR(integrated safety assessment report) 综合安全评价报告
ISAR(intermediate safety analysis report) 中期安全分析报告
isasteric 等容的
ISBL(inside battery limit) 界区内
ISCSO(integer state controllable state output) 整数状态可控状态输出
ISDN(integrated services digital network) 综合服务数字网
I-section 工形截面，工字钢，工字形剖面

isenthalpic temperature-pressure coefficient 绝热节流系数，绝热温压系数
isenthalpic 等焓的，等焓线
isentropic change 等熵变化
isentropic chart 等熵图
isentropic compressibility 等熵压缩率
isentropic compression work 等熵压缩功
isentropic compression 等熵压缩
isentropic curve 等熵线
isentropic efficiency 等熵效率
isentropic enthalpy drop 等熵焓降，理想焓降
isentropic expansion 等熵膨胀
isentropic exponent 绝热指数，等熵指数
isentropic flow 等熵流动，等熵流
isentropic heat drop 等熵热降
isentropic line 等熵线
isentropic nozzle flow 喷嘴等熵流动
isentropic process 等熵过程
isentropic stagnation heating 等熵滞止加热
isentropic surface 等熵面
isentropic wave 等熵波
isentropic working expansion 等熵膨胀工作过程
isentropic work 等熵功
isentropic 等熵的，等熵线，等熵线的
I-shape construction 工字形结构
I-shaped 工字形
Ishikawa's diagram 石川图【因果图】
ISI(Indian Standards Institution) 印度标准学会
ISI(industry standard item) 工业标准项目
ISI(in-service inspection) 在役检查
ISI(integer status information) 整数状态信息
ISI(internally specified index) 内部规定指标
ISI(Iron and Steel Institute) 钢铁学会
isinglass 鱼胶，云母，白云母薄片
ISK 冰岛克朗
island barrier 拱门岛，屏障岛
island breakwater 独立式防波堤，岛堤
island chain 岛链，列岛
island country 岛国
island harbour 岛港，岛式港口
island mole 岛屿防波堤
island type construction 岛式建筑，独立式建筑
island type wharf 岛式码头
island 岛，岛状物，岛屿
islet 小岛，屿
isle 小岛
isacoustic curve 等响线
isoatmic line 等蒸发线
isobaric change of gas 气体的等压变化
isobaric chart 等压线图
isobaric cooling 等压冷却
isobaric expansion 等压膨胀
isobaric line 等气压线，等压线
isobaric process 等压过程
isobaric spin 同位旋
isobaric surface 等压面
isobaric 同量异位的，同质异位的，等压的，等压线的，等气压的，等压线，等压线图
isobar 同量异位素，等压线，同质异位素，等气压线
isobase 等基线
isobathytherm 等温深度面，等温深度线
isobath 等深线
isobront 等雷（日）线，雷暴等时线
isocatabase 等下沉线
isocenter 等角点
isoceraunicline 等雷频线，等雷雨强度线
isoceraunic 等频雷暴的，等雷雨的
isocheim 冬季等温线，等冬温线
isochlor （地下水的）等含氯量线
isochore 等体积线，等容线
isochoric 等层厚的，等体积的，等容的
isochromatic fringe 等色条纹
isochromatic line 等色线，等水色线
isochromatic pattern 等色线图
isochromatic photograph 等色图像
isochronal spiral regulator 等时螺线调节器
isochronal 等时的
isochrone map 等流线图
isochrone method 等时线法
isochrones chart 等流时线图
isochrone 同时线，等时线，瞬压曲线，等流时线
isochronism oscillation 等时振荡
isochronism speed governor 同步调速器
isochronism 等时振荡，等时性，同步，同步性，同时性
isochronization 使等时
isochronous control 无差调节，无差控制
isochronous creep curve 等时蠕变曲线
isochronous distortion 同步畸变
isochronous governor 同步调节器，同步调速器
isochronous modulation 同步调制
isochronous oscillation 等时振荡
isochronous stress-strain curve 等时应力应变曲线
isochronous vibration 等时振动
isochronous 等时的，同步的
isoclinal fault 等斜断层
isoclinal fold 等斜褶皱
isoclinal method 等倾线法
isoclinal valley 等斜谷
isoclinal 等倾线的，等伏角的
isocline 等倾线，等斜线
isoclinic line 等倾线，等斜线
isoconcentration 等浓度线
isocorrelation 等相关线
isocurlus 等旋涡强度线
iso-deflection 等挠度
isodose chart 等剂量图
isodose curve flattening 等剂量曲线平整
isodose curve 等剂量曲线
isodose 等剂量，等剂量线，等剂量面
isodrome governor 等速调速器
isodromic 恒值的，等速的
isodynamic curve 等磁力曲线
isodynamic line 等磁力线
isodynamic 等磁力线，等磁力的，等能的
isodynam 等磁力线，等风力线
isodyne 等力线
isoefficiency curve 等效率曲线，等效率线
isoelectric points 等电位点
isoelectric 零电位差的，等电位的

isoelectronical 等电子的
isoenergetic 等能的
isoentnopic compression 等熵压缩
isoentnopic efficiency 等熵效率
isoentnopic process 等熵过程
isoentnopic 等熵线
isoentropic change 等熵变化
isoentropic expansion 等熵膨胀
isoentropic flow 等熵流动
isoentropic process 等熵过程
isoentropic 等熵的
ISO 9000 family ISO 9000 族【质量管理】
isoflux 等通量
isofrigid 等寒的
isogam 等重（力）线
isogeotherm 等地温线
isogonality 保角变换
isogonic line 等磁偏线，等偏角线
isogon 正多边形，同风向线，等磁偏线，等角多边形
isograde 等坡的
isogradient 等梯度线
isogram 等值线图
isogrid 地磁等变线
isohaline 等盐度线
isohedral 等面的
isohel 等日照线
isohume 等湿线，等水分线，等湿度线
isohyetal line 等雨量线
isohyetal map 雨量分布图
isohyet 等降水量线，等沉淀线，等雨量线
isohygrotherm 等水温线
isohyperthermic 等过热的
isohypse 等高线
ISO(independent system operator) 独立系统调度机构
ISO(International Science Organization) 国际科学组织
ISO(International Standardization Organization) 国际标准化组织
isokeraunic chart 年平均雷电日预测
isokeraunic level 年平均雷电数
isokinetic Pitot 等速皮托管
isokinetic probe 等动探头
isokinetic sampling 同流态取样，等速取样，等动力取样
isokinetic 等动能的
isolantite 一种陶瓷高频绝缘材料
isolate bus 绝缘汇流排，绝缘母线
isolated aerofoil 孤立翼型
isolated analogue input 隔离模拟量输入
isolated building 孤立建筑
isolated chimney 孤立烟囱
isolated construction 孤立建筑
isolated cooling tower 孤立冷却塔
isolated corrosion 局部腐蚀
isolated cycle 孤立循环
isolated fault 孤立故障，隔离故障
isolated footing 单独基础，独立底脚，独立基础
isolated foundation 单独基础，独立基础
isolated gate field effect transistor 绝缘栅场效应晶体管
isolated gate MOS field effect transistor 绝缘栅金属氧化物半导体晶体管
isolated generating plant 单独运转发电厂，孤立电厂
isolated hill 孤山
isolated interstice 单独裂隙
isolated loop 隔离回路，孤立回路，隔离环路
isolated machine 孤立机组
isolated network 独立网络
isolated neutral system 中性点绝缘制，中性点不接地系统，不接地中性点系统
isolated neutral 中性点绝缘，不接地中性点
isolated operation 不并列运行，单独运行，单机运行，孤立运行
isolated-phase bus 分相封闭式母线，隔离的相母线，分相母线，离相母线
isolated plant 单独运转电站，孤立电厂
isolated power plant 孤立电厂
isolated power system 独立电力系统
isolated process 孤立过程，绝热过程
isolated slug velocity 单个气团上升速度
isolated source 孤立源
isolated station 孤立电厂
isolated system 单独系统，孤立系统
isolated 隔离的，隔绝的
isolate map 等岩图
isolate valve 隔绝阀
isolate 绝缘，隔离，孤立，等岩
isolating air gap 绝缘气隙
isolating capacitor 隔流电容器
isolating damper 隔震调节器，缓冲阻尼器，隔离风门
isolating distance 隔离距离
isolating dyke 分隔堤
isolating layer 隔离层
isolating link 隔离开关
isolating platen 分隔屏
isolating switch 隔离开关
isolating transformer 安全变压器，隔离变压器，绝缘变压器
isolating valve 隔离阀
isolating 隔离的
isolation amplifier 隔离放大器
isolation belt 隔离带
isolation bladder （管道）隔离软外壳
isolation butterfly valve 隔离蝶阀
isolation can test 隔离容器试验
isolation capacity 分断能力
isolation condenser 隔离式冷凝器，隔离凝汽器
isolation containment 隔离安全壳
isolation converter cabinet 隔离变流器柜【反应堆保护系统】
isolation cooling system 隔离冷却系统
isolation damper closing mechanism 隔离阻尼器闭合机构
isolation damper 缓冲阻尼器，隔离减震器，隔离阀，隔振调节器，隔离风门
isolation in integrated circuit 集成电路的隔离
isolation joint 隔离接缝，隔离接头

isolation masking 隔离掩蔽，隔离屏蔽
isolation method 分离法，隔离法
isolation module 隔离模块
isolation mounting 隔振装置
isolation network 隔离网络
isolation of noise 噪声的隔离，隔噪声
isolation oil 绝缘油
isolation period 限制期，隔离期
isolation shutdown 隔绝停堆
isolation strength 绝缘强度
isolation strip 隔离带
isolation system 隔离系统
isolation technique 分离技术，隔离技术
isolation transformer 隔离变压器
isolation trench 隔离槽
isolation unit 隔离装置（电）
isolation valve 隔离阀
isolation voltage 隔离电压，绝缘电压
isolation ward 隔离室
isolation 隔绝，隔离，绝缘，孤立，隔振，分离，隔离作用
isolator disc 绝缘子盘
isolator string 绝缘子串
isolator 隔声装置，绝缘体，隔离器，隔振体，隔离开关，隔振器，隔离物，绝缘子，绝热体，刀闸
isoline map 等值线图
isoline method 等值线法
isoline 等值线，等位线
isolit 绝缘胶纸板
isologous 相同的
isolux line 等照度线
isolux 等照度的
isomagnetic 等磁的
isomagnetic chart 等磁力线图
isomagnetic line 等磁力线，等磁线
isomenal 月平均等值线
isomeric state 同质异能态
isomer 同核异能素，同分异构体，同质异能素，等比值线，等降水线
isometrical 等角的，等容的，等量的
isometric diagram 轴测图，等距图，单线图
isometric drawing 等角（投影）图，等轴图，轴测图，等距画法，等距离图，立体示意图
isometric perspective 等角透视
isometric piping diagram 管道透视图
isometric process 等容过程
isometric projection 等角投影
isometric rainfall map 等降水率图
isometric surface coordinates 等距曲面坐标
isometric system 立方晶系
isometric view 等角图，等距图，等轴图，轴测图，等角投影视图
isometric 等轴的，等比例的，等距的
isometry 等高，等距，等容
isomorphic 同型的，同晶型的
isopachyte 等厚线
isopach 等厚线
isopag 等冻期线
isoparametric element 等参（数）单元
isoparametric type 等参数型
isoparametric 等参数的
isopentane 异戊烷
isoperm 恒导磁率铁镍钴合金
isophase 等相的，等相线
isophonic contour 等音感曲线，等声强曲线
isophot curve 等照度曲线
isophot diagram 等照度图
isopic 相同的，同相的
isopiestic line 等势线，等（水）压线
isopiestic point 等压点
isopiestic surface 等压面
isopiestic 等压的，等压线
isoplanar integrated injection logic 等平面集成注入逻辑
isopleth 等值线，等浓度线，等成分面，等值线图
isopluvial line 等雨量线
isoporous anion exchange resin 均孔型阴离子交换树脂
isoporous cation exchange resin 均孔型阳离子交换树脂
isoporous ion exchange resin 均孔型离子交换树脂
isoporous type resin 均孔型树脂
isopotential line 等位线，等势线
isopotential surface 等电位面
isopotential 等势线，等电位
isopulse 恒定脉冲
isopycnic surface 等密度面
isopycnic 等密度的，等密度线，等密度面
isorad 等辐照量线，等放射量线，等拉德线
ISO rating 感光度
isoreactivity line 等反应性线
isoreactivity 等反应性
isosaline 等盐度线
isoscope 同位素探伤仪
isoseismal line 等震线
isoseismal 等震的
isoseism 等震线，等震
isosmotic pressure 等渗压
isostath 等密度线
isostatic curve 等压曲线，等压线
isostatic equilibrium 各向等压平衡
isostatic pressing 等静压
isostatics 等压线
isostere 同电子排列体
isosteric surface 等比容面
isosteric 等比体积线，等比体积的，电子等配的
isosterism 电子等配性
isostrain creep curve 等应变蠕变曲线
isostrain diagram 等应变图
isosurface 等值面
isotachophoresis 等速电泳
isotach 等风速线，等速线
isotactic polymer 等规聚合物
isotherm 恒温线，等温线
isothermal absorption 等温吸收
isothermal analysis 等温分析
isothermal annealing 等温退火
isothermal atmosphere 等温大气

isothermal change 等温变化
isothermal compression 等温压缩
isothermal condition 等温情况，等温条件
isothermal curve 等温线，等温曲线
isothermal efficiency 等温效率
isothermal energy storage 等温贮能
isothermal equilibrium 等温平衡
isothermal evaporation 等温蒸发
isothermal expansion 等温膨胀
isothermal flow 等温流动，等温流
isothermal hardening 等温淬火
isothermal heating 等温加热
isothermal humidification 等温加湿
isothermal jet 等温射流
isothermal latitude 等温纬度
isothermal layer 等温层
isothermal line 等温线
isothermal normalizing 等温正火
isothermal pressure drop 等温压降
isothermal process 等温过程
isothermal quenching 等温淬火
isothermal section 等温截面
isothermal strain 等温应变
isothermal surface 等温面
isothermal temperature coefficient 等温温度系数
isothermal tempering 等温回火
isothermal transformation curve of super-cooled austenite 过冷奥氏体等温转变曲线
isothermal transformation of super-cooled austenite 过冷奥氏体等温转变
isothermal transformation 等温转变
isothermal 等温的，等温线的
isothermic enthalpy pressure coefficient 等温节流系数，等温焓压系数
isothermic line 等温线，恒温线
isothermic 等温的
isothyme 等蒸发量线
isotime line 等时线
isotonic 等渗压的
isotope abundance 同位素丰度，同位素分布量
isotope age determination 同位素年代测定
isotope analysis 同位素分析
isotope balance 同位素平衡
isotope capsule 同位素密封盒
isotope carrier 同位素运输容器，同位素载体
isotope dilution analysis 同位素稀释分析
isotope dilution method 同位素稀释法
isotope electric power source 同位素电源
isotope exhaust cabinet 同位素抽气小室
isotope separation plant 同位素分离厂，浓缩厂，同位素分离工厂
isotope separation 同位素分离
isotope specific activity 同位素比活度
isotope-velocity method 同位素测流法
isotope water level gauge 同位素水位计
isotope 同位素
isotopic abundance by weight 同位素重量丰度
isotopic abundance 同位素丰度
isotopic activation cross section 同位素激活截面
isotopic composition 同位素组分，同位素成分，同位素组成
isotopic concentration 同位素浓缩度
isotopic contamination 同位素沾污，同位素污染
isotopic correlation safeguards technique 同位素相关保障技术
isotopic dilution analysis 同位素稀释分析
isotopic effect 同位素效应
isotopic element 同位元素
isotopic equilibrium 同位素平衡
isotopic exchange distillation 同位素置换蒸馏法
isotopic heat 同位素衰变热
isotopic inventory 同位素平衡
isotopic rate of exchange 同位素转换速度
isotopic separation plant 同位素分离厂［浓缩厂］
isotopic specific activity 同位素比活度
isotopic tracer 同位素示踪剂，同位素示踪仪
isotopic weight abundance 同位素重量丰度
isotrope 各向同性，均质，各向同性晶体
isotropically coated particle 各向同性包覆颗粒
isotropic body 各向同性体
isotropic coating 各向同性覆盖层
isotropic compression 各向均匀压缩
isotropic crystal 各向同性晶体
isotropic deep water 各向同性的深层水
isotropic dielectric 各向同性电介质
isotropic dispersion 各向同性弥散
isotropic distributed source 各向同性分布源
isotropic distribution 各向同性分布
isotropic flux 各向同性通量
isotropic hardening 各向同性硬化
isotropic line 各向同性线
isotropic material 各向同性材料，各向同性物质
isotropic medium 各向同性介质
isotropic plasma 各向同性等离子体
isotropic point source 各向同性点源
isotropic solid 各向同性固体
isotropic term 各向同性项
isotropic turbulence scale 各向同性紊流度
isotropic turbulence 各向同性湍流
isotropic 各向同性的，各向同性体，同位素的
isotropism 各向同性
isotropy 各向同性，均质性
iso-type transformer 隔离变压器
isovelocity 等风速线，等速线
isovel 等速线，等速度曲线，等流速线，等速曲线
isovols 等容线，等体积线
isovolumetric curve 等容线
isovolumetric 等容的
isowaiping 等挠曲的
ISP(Institute of Sewage Purification) 污水净化学会【英国】
ISP(internet service provider) 因特网服务提供者
ISR(information storage and retrieval) 信息存储和检索
ISS(industry standard specifications) 工业标准规范
issuance of a note 签发票据
issuance of the import licence 发给进口许可证
issuance record 发货记录
issuance 发行，颁布，发给

issue a notice　发出通知
issue a separate notice　另行通知
issue currency　发行货币
issue of letter of credit　签发信用证
issue of loan　发行公债
issuer　发行人
issue　流出，发布，发行，问题，颁布，颁发，（正式）发给，签发
issuing a credit　开立信用证，信贷发行
issuing approval or rejection or approval with comments　作出批复
issuing bank　（信用证或保函）开立行，开证行
issuing velocity　射出速度
I-steel　工字钢
isthmus　地峡
ISU(initial signal unit)　初级信号单元
ISV(intermediate pressure turbine steam valve)　中压缸进汽阀
italicized word　斜体字
italics　斜体字
ITB(instructions to bidders)　投标（人）须知
ITB(invitation to bid)　招标，招标书，邀标书
ITC(instructions to consultant)　咨询须知【指参与投标的咨询公司】
ITC(international tendering company)　国际招（投）标公司
ITC(International Trade Center)　国际贸易中心
ITD(initial temperature difference)　初始温差
item counter　操作次数计数器
item description　项目说明
item design　项目设计，项目组成
item facility　件料操作设施
item for popularizing the results　成果推广项目
item identification　物项标识
itemized account　明细账
itemized appropriation　分项拨款
itemized equipment list　设备分项表
itemized list　分项一览表
itemized price　单项价格，分项报价，详细报价
itemized project　分项工程
itemized quotation　分项报价
itemized record　明细记录
itemized schedule　项目一览表
itemized specification　分项技求规范
itemize　分项列举，逐条列记
item No.　项号，项次码
item number　物项编号，编号，件号，项号，项次码，项目号
item of payment　支付项目
item processing　项目处理
items important to safety　安全重要物项
items of equipment　设备项目
items sent for collection　托收款项
item transfer　项目传送
item　零件，项，项目，条款，物品，物项
iterated fission expectation　迭代裂变期待值
iterated fission probability　迭代裂变概率
iterated fission　反复裂变
iterated integral　迭代积分
iterated network　累接网络
iterate　迭代，累接，重复
iteration method for buckling search　曲率搜索迭代法
iteration method　迭代法
iteration process　迭代法
iteration solution　迭代解法
iteration　迭接，迭代法，迭代，反复，重复，累接
iterative addition　迭代相加
iterative analog computer　迭代模拟计算机
iterative analysis　迭代分析，反复分析
iterative attenuation　累接衰减
iterative bidding　迭代投标
iterative circuit　累接电路
iterative computing method　迭代计算法
iterative earthing　重复接地
iterative formula　迭代公式
iterative impedance　累接阻抗，迭接阻抗
iterative loop　迭代循环
iterative method　迭代法
iterative network　累接网络，迭接网络
iterative problem　迭代问题
iterative process　迭代过程
iterative sequential circuit　累接时序电路
iterative solution　迭代求解，重复求解
iterative structure　叠合结构
iterative　反复的，迭代的，累接的，重复的
IT facility　信息技术设备
itinerary map　路线图
itinerary pillar　路标
itinerary　行程，旅程，路线，旅行日程，旅行指南，旅行路线
IT(information technology)　信息技术
it is anticipated that　可以预料……
it is envisaged that　可以预见，可以看到
ITL　意大利里拉
ITP(inspection and test plan)　检验与测试计划
I-T product　电流有效值【I】与电话干扰系数【TIF】之积
IT technology solutions　IT 技术解决方案
ITT(invitation to tender)　招标书
I-type arm　工字形支臂
IUDH(in-service unplanned derated hours)　非计划降低出力运行小时
IU(international unit)　国际单位
IUNDH(in-service unit derated hours)　机组降低出力运行小时
I-V characteristic curve of solar cell　太阳电池的伏安特性曲线
I-V characteristic　伏安特性
ivernite　二长斑岩
IV(interceptor valve)　中压调节汽门【DCS 画面】
I will undertake that　我保证
izod impact test　悬臂梁式冲击试验
izod notch　V 形缺口

J

jack arch　等厚（度）拱，单砖拱
jack bolt　起重螺栓，千斤顶螺栓

jack box 转换开关盒
jacked pile 压入式桩，顶压桩
jackengine 辅助发动机，小型蒸汽机
jacket cooler 套管式冷却器
jacket cooling 水套式冷却，护套冷却
jacket cover 水套盖
jacketed cylinder 有套汽缸
jacketed evaporator 带夹套的蒸发器
jacketed insulation （管道）有包套的隔热层
jacketed intake pipe 有水套的进气管
jacketed liner 夹套衬
jacketed valve 夹套阀
jacketed wall 双层壁
jacketed 包套的，有包壳的，带壳的，有套的，备有夹套的，保温的
jacket heater 套式加热器
jacket heating system 套加热系统
jacket heating 夹套加热，套式加热
jacketing tubing 外套管
jacket insertion depth 套管插入深度
jacket layer 套层
jacket rail carrier 导管架拖车
jacket steam 夹层蒸汽
jacket tube 套管
jacket 套管，汽套，水套，夹套，套筒，包壳，外壳，护套，外套，衬套，罩，盖，盒，套，给……装护套，给……包上护封
jack fastener 插口线夹
jack hammer drill 手持式凿岩机，手提锤钻
jack hammer 手持式风钻，手提钻，风镐，气锤，锤击式凿岩机，手持式凿岩机，凿岩锤
jack-head pump 随动泵
jacking block 千斤顶垫块
jacking device 千斤顶装置，顶托设备，顶升设备，顶起设备，顶进设备
jacking dice 千斤顶垫块
jacking force 顶推力，顶托力
jacking oil pump 顶轴油泵
jacking oil system 顶轴油系统
jacking oil 顶轴油
jacking platen 千斤顶压板
jacking plate 千斤顶垫板
jacking screw 螺旋千斤顶
jacking stress 顶托应力
jacking system 提升机构装置，提升机构，顶起系统
jacking test 千斤顶试验
jacking unit 起重装置
jacking-up oil pump 顶轴油泵
jacking up 顶起，顶高
jacking 支撑，顶起，提升，顶托，顶进法，顶轴
jack-in type 插入式
jack ladder 索梯
jack lagging 承重木构件
jack leg 起重顶杆，溢流管
jack lever 顶重杠杆
jack-loading method 千斤顶加载法
jack out 用千斤顶顶出
jack pad 千斤顶承座
jack panel 塞孔接线盘，插孔接线板，插孔盘
jack plane 粗刨，大刨，粗木刨
jack post 撑杆
jack rafter 顶木，支撑椽
jack screw plate 千斤顶螺旋板
jack screw 顶起螺杆，螺旋千斤顶，起重螺杆
jack shaft （变速箱）传动轴，中间轴，起重轴，曲柄轴
jack shore 套管支柱
jackstay 支索，撑杆
jack strip 插孔簧片，塞孔簧片
jack switch 插接开关
jack system 顶起装置，转子顶起系统
jack timber 顶木
jack truss （四面坡屋顶的）半桁架，小尺寸桁架
jack 弹簧开关，千斤顶，起重器，插孔，插座，顶重器，支柱
jaff 复式干扰
jagged edges 不平坦边缘
jalousie window 波形板屋面
jalousie 百叶窗，玻璃百叶窗
jamb casing 门（窗）侧板
jamb lining 门口镶板
jamb shaft 门窗立柱
jamb 门窗侧壁，边框
jam-free 无干扰的
jammer 人为干扰，干扰发射机，电气干扰
jamming of fuel elements 燃料元件卡住
jamming of shutdown rod 停堆棒卡住
jamming 抑制，干扰，咬死，卡住，阻塞，干扰噪声，咬住
jam nut 锁紧螺母，防松螺母，压紧螺母
jamproof 防干扰的，抗干扰的
jam riveter 窄处铆机
jam-to-signal 干扰信号比，噪声信号比，信噪比
jam up 堵塞筛眼
jam weld 对头焊接
jam 卡住，压紧，挤塞，干扰，阻塞
Janney motor 轴向回转柱塞液压马达
Japan Advanced Thermal Reactor 日本先进热堆
Japan Electric Power Information Center 日本电力信息中心
Japan Industrial Standard 日本工业标准
Japan International Cooperation Agency 日本国际协力机构，日本国际协力事业团
japanning 上漆，涂漆
Japan Nuclear Fuel Industries 日本核燃料工业公司
Japan Nuclear Fuel, Ltd. 日本核燃料有限公司
Japan Nuclear Fuel Services 日本核燃料服务公司
jargon 术语，行话
jar-proof 防震的
jarring machine 振动机，震动机
jarring motion 振动，颤动
jar test 瓶试验，烧杯试验
jar 电瓶，瓶，振动，冲击，噪声，容器，加尔
JATR(Japan Advanced Thermal Reactor) 日本先进热堆
Java language Java 语言
javelin-shaped fuel rod 标枪状燃料元件棒
jaw clutch 爪式离合器，颚式离合器
jaw coupling 爪盘联轴节
jaw crusher 颚式破碎机，颚式碎石机

jaw crushing 颚板破碎，粗碎
jaw cylinder 卡盘液压缸
jaw nut 防松螺母
jaw plate （碎石机上的）颚板
jaw 卡盘【张力试验机】，夹紧装置，夹爪，颚，颚板，夹钳，虎钳，虎钳口，齿板
jayrator 移相段
JBIC(Japan Bank for International Cooperation) 日本国际协力银行
JB(junction box) 接线盒，联轴器
JCAE(Joint Committee on Atomic Energy) 原子能联合委员会【美国】
J-carrier system J电话载波系统，J电话载波系统
JCB(job control block) 作业控制分程序
JCCCNRS(Joint Co-ordination Committee for Civilian Nuclear Reactor Safety) 联合民用核反应堆安全协调委员会【美俄】
JCT contract system 一种合同体系，JCT合同体系【英国主要的合同体系之一】
JCT(Joint Contract Tribunal) 英国联合合同委员会，英国工程承包合同审定联合会
JCT＝junction 连接
JDT 火警探测系统【核电站系统代码】
jellification 胶凝，冻结，凝结
Jellif 杰利夫镍铬电阻合金
jelly-filled capacitor 充糊电容器
jelly 凝结，胶体，胶质，胶状物，成胶状
jel 凝胶，冻胶
jemmy 短撬棍，铁撬棍
Jena glass 耶拿光学玻璃，难熔玻璃
JENER(Joint Establishment for Nuclear Energy Research) 核能研究联合机构【挪威】
jenny scaffold 活动脚手架
jenny 移动式起重机，卷扬机
jeopardize 危及
JEPIC(Japan Electric Power Information Center) 日本电力信息中心
jerry builder 偷工减料的建筑商
jerry building 粗糙的建筑物
jerry-built project 豆腐渣工程
Jerusalem virus 耶路撒冷病毒
jet agitator 喷射混合器，喷射搅拌器
jet air ejector 空气喷射器
jet air exhauster 喷射式抽气器
jet air pump 空气喷射泵，喷气泵，喷射空气泵
jet air stream 急流，喷流，射流
jet angle 射出角
jet area contraction coefficient 喷口收缩系数
jet area 喷口面积，射流截面
jet atomizer 喷射雾化器
jet axis 射流轴
jet black 烟黑，乌黑
jet blade 喷气式叶片
jet blower 喷气鼓风机，喷射式鼓风机
jet boundary 射流边界
jet-bubbling deaerator 有蒸汽鼓泡装置的淋水盘式除氧器
jet burner 喷射燃烧器
jet centrifugal-pump 喷射式离心泵
jet chamber 喷雾室
jet coal 长焰煤
jet condenser 喷射式凝汽器，喷水冷凝器，喷水凝汽器
jet contraction 射流收缩
jet control 喷流控制
jet cooling 喷射冷却
jet current 射流
jet cutting 射流切割，水力开凿
jet deepwell pump 喷射式深井泵
jet deflector 喷射导流板，射流转向器，折流器
jet device 射流器
jet diameter 射流直径，喷嘴直径
jet diffuser 射流扩散器
jet dispersion 射流分散
jet divergence angle 射流扩散角
jet drilling 喷焰钻孔【岩石内的】
jet drill 喷射式凿井机
jet edge 射流边界
jet eductor 喷射器
jet effect wind 急流效应风，喷射效应风，喷流效应风
jet efflux 喷流，射流
jet engine 空气喷气发动机，喷气发动机
jet entrance point 射流进入点
jet exhauster 喷射真空泵，喷射式抽出器，喷射抽气机
jet flow gate 射流式闸门
jet flow 喷流，喷气流，射流
jet fluid 射流
jet force 射流作用力
jet fuel 喷气发动机燃料
jet grouted cutoff wall 高喷防渗墙
jet grouting 高压喷射注浆法，高压喷射灌浆，喷射灌浆
jet height 喷水高度，射流高度
jet hose 射流软管
jet humidifier 喷水湿润器
jet-impact area 射流冲击范围
jet impingement 喷射冲击
jet impulse 射流冲击
jet in a confined space 受限射流
jet injection pipe 喷射器喷管
jet injector 抽射器，喷射器，射流泵
jet length 射流长度
jet-like-flow pattern 射流状流型
jet lubrication 射流润滑
jet mixing flow 喷射混合流，混射流
jet model 射流模型
jet molding 喷（射）模（塑）法
jet noise 喷气噪声
jet nozzle amplifier 射流喷嘴放大器
jet nozzle process 喷嘴分离法
jet nozzle 喷嘴，喷射管，喷管，射流喷嘴
jet of pressure water 压力水射流
jet orifice 喷口，喷嘴，喷射口
jet pair 射流对
jet path 射流轨迹，射流路径
jet penetration length 射流穿透长度
jet pipe heat exchanger 喷管式热交换器，喷管式换热器
jet pipe 喷射管，喷管
jet pressure 喷射压力

jet primer 喷射注油器
jet priming 射流起动
jet propeller 喷气推进器，喷水推进器，射流推进器
jet propulsion 喷气推进
jet pulsion pump 脉冲喷射泵
jet pump 喷射泵，射流泵
jet range 射流射程，水股射程
jet ratio 射流比
jet reaction 喷射反作用（力）
jet rise 射流抬升
jetsam 下沉颗粒，沉料
jet separation 射流分离，射流离散
jet separator 射流分离器
jet size 喷口尺寸
jet speed 射流速度
jet splitter 射流分水器
jet spread 喷射分散
jet switching 射流转换，射流开关现象
jetted pile 射水沉桩，水冲桩
jet temperature indicator 喷流温度指示器
jetter 喷洗器
jet theory 射流理论
jetting method 水力钻探法，喷射法
jetting piling 射水打桩法
jetting process 水冲法
jetting velocity 喷射速度
jetting 喷注，射出，射水
jet trajectory height 挑射高度
jet trajectory length 射流挑出长度
jet trajectory 水股轨迹
jet trench digger 喷射挖沟机
jet-turbine engine 喷气涡轮机
jetty dock 堤式船坞
jetty harbor 突堤港
jetty head 导流堤堤头，突堤堤头，坝头【水电站】
jet-type carburetor 射流式汽化器
jet-type deaerating heater 喷雾式除氧加热器
jet-type deaerator 喷雾式除氧器，淋水盘式除氧器
jet-type humidifier 喷射式加湿器
jet-type impulse turbine 水斗式水轮机
jet-type 喷射式
jetty-type wharf 突堤码头，指形码头
jetty 防波堤，码头，导堤，建筑物的突出部分，突（出）码头，栈桥，突堤，专用码头
jet velocity 喷气速度，射流速度，喷射速度
jet viscometer 喷射黏度计
jet washing machine 喷射清洗机
jet 喷管，喷射，喷流，射流，喷嘴，喷射器，喷口，喷气式飞机
JFET(junction-type field effect transistor) 结式场效应晶体管
jib boom 起重臂，起重杆
jib crane and grab 悬臂吊车及抓取器
jib crane 挺杆起重机，摇臂起重机，悬臂（式）吊车，悬臂（式）起重机，旋臂起重机
jib head 吊机臂上端
jib loader 旋臂装料机
jib type crane 桅杆式起重机，动臂起重机，悬臂起重机，臂架型起重机
jib-type stationary hoist 转臂式固定起重机
jib 挺杆【吊车】，起重机臂，旋臂
JICA(Japan International Cooperation Agency) 日本国际协力机构，日本国际协力事业团
jig-adjusted 粗调的
jigger coupling 电感耦合
jigger 盘车，镳铲，可变耦合变压器，减幅振荡变压器，滑轮组
jigging compaction 振动致密【混凝土】
jigging conveyer 振动输送机
jigging grate 摆动炉排
jigging screen 振动筛
jig saw 往复锯
jig 装配架，夹具，成套夹具，钻模，架，衰减波群，颠簸，钻模
jim crow 人工弯轨机
jimmy 短撬棍，料车，铁撬棍
JIS(Japanese Industrial Standards) 日本工业标准，日标
jitterbug 图像跳动，图像不稳定
jitters of contact 触点的颤动[抖动]，接点抖动[颤动]
jitter 不稳定性，失稳，失步，跳动，振动，晃动，抖动，速度偏差，信号不稳定
J=joule 焦耳
JK flip flop JK 触发器
JNFI(Japan Nuclear Fuel Industries) 日本核燃料工业公司
JNFL(Japan Nuclear Fuel, Ltd.) 日本核燃料有限公司
JNFS(Japan Nuclear Fuel Services) 日本核燃料服务公司
JN(jam nut) 锁紧螺母
JNL=journal 轴径
j-number 虚数
job analysis 作业分析
job application form 职业申请表
job application specification 工作规范
job assignment notice 工作任务单
job assignment sheet 工程任务单
job awareness preparation （使工作人员）充分了解任务的准备
job batching 分批加工
jobber 临时工，批发商，杂工
job captain 工地负责人
job card 工作单
job category 职务分类，职务类别
job classification 职务分类
job class 作业分类，题目分类
job contract system 劳动合同制
job control block 作业控制分程序
job control language 作业控制语言
job control statement 作业控制语句
job control table 作业控制表
job description card 作业说明卡
job description 任务说明，作业说明
job duplication 兼职
job entry system 作业调入系统
job evaluation 工作评价
job file 业务档案

job flexibility 工作适应性，工作灵活性
job flow control 作业流控制
job initiation 作业起始
job laboratory 工地实验室
job library 作业库
job logging 工作记录
job lot method 分批法【指成本计算等】，分批成本计算法
job management 作业管理
job mixed concrete 现场拌制混凝土
job mixed paint 现场调制涂料
job mix 现场拌和，工地拌和
job number 工号
job operation manual 操作手册
job-oriented terminal 面向作业的终端
job out 分包出去
job pack area 作业装配区
job-placed 现场灌筑的，现场铺设的
job processing 作业处理，作业加工
job processing control 作业处理控制
job program 作业程序，加工程序
job responsibility system 岗位责任制
job scheduler 作业调度程序
job schedule 工程进度
job sequence 加工程序
jobsite transportation 施工场内交通
jobsite 工地，建设工地，施工现场，现场
job specification 工作任务详细说明，任务说明书
job stacking 作业堆积
job step 加工步骤，工作步骤
job subsidy 岗位津贴
job superintendent 项目负责人
job & task analysis 岗位和职责分析
job title 职别
job 工作，职业，作业，职责，施工任务
JOC(joint operation center) 联合操作中心
jockey pump 管道补压泵，稳压泵，操控泵，辅助泵
jockey valve 辅助阀，先开阀
jockey 导轮，操纵，薄膜，振动膜
jogging 频繁反复启动，缓步，重复短暂操作，颠簸，微动
joggled beam 拼接梁
joggled stones 企口石块
joggled 榫接的
joggle piece 榫接部件
joggle 啮合榫，榫接，折曲
jog switch 微动开关
joiner's glue 木工胶
joinery works 细木工程
joinery 木工装修，细木工
joiner 细木工
joining beam 系梁
joining bolt 连接螺栓
joining by rabbets 槽舌接合
joining flange 接合法兰
joining pipe 连接管
joining rivet 连接铆钉，接合铆钉
joining-up differentially 差接
joining-up in parallel 并接
joining-up in series 串接
joining with swelled tenon 扩榫接合
join on skew 斜向接合
join pole scatter diagram 节理极点分散图
join project 合营项目
joint acceptance function 联合接纳函数，结合受纳函数
joint action 接合作用
joint adventure 合资经营
joint and several guarantee 连带保证责任
joint and several liability 个别并连带责任，连带责任，共同责任
joint and several obligation 连带之责
joint approval meeting 联合审批会议
joint arrangement drawing 节点布置图，结点布置图
joint audit 联合审计
joint bar 连接板，鱼尾板
joint bid 联合报价
joint block 接头凸抓，接头凸爪，凿石块，接头块
joint bolt 连接螺栓
joint box compound 接线盒填充剂，电缆套管填充剂
joint box cover 电缆接线盒盖
joint box 电缆接线箱，连接套筒，中间持续盒
joint capital 合资
joint cap 密封盖
joint cash account 联名现金账户
joint chair 接轨座板
joint checkup of the blue prints 图纸会审
joint checkup on detail drawings 施工详图会审
joint checkup 会审
joint clearance 钎缝间隙
Joint Committee on Atomic Energy 原子能联合委员会【美国】
joint committee 联合委员会
joint compound 接头绝缘膏，填缝混合物
joint conference 联席会议
joint connector 接头连接器
Joint Contract Tribunal 英国合同审定联合会
Joint Co-ordination Committee for Civilian Nuclear Reactor Safety 联合民用核反应堆安全协调委员会【美俄】
joint cost 联合费用
joint coupling of cable 电缆接头套管
joint coupling 连接器，活节连接器，电缆接头套管，管接头，万向接头
joint current 总电流
joint denial gate “或非”门
joint design meeting 联合设计见面会
joint design 接头设计，联合设计
joint diagram 节理图
joint dimension 接头尺寸
joint disposal （工业废水与生活污水的）综合排放，（废水、污水的）共同处理，联合排放
joint elbow 弯头
joint enterprise 联合企业
joint entropy 相关平均信息量，相关熵
jointer 结合器，连接器，接合器，电缆焊接工，管子工人
Joint Establishment for Nuclear Energy Research

核能研究联合机构【挪威】
joint face 接合面，连接面
joint fastener 接合片
joint filing 堵缝
joint filler 接缝填料，填缝料
joint fissure 节理裂缝
joint flange 连接法兰
joint frequency 联合频率
joint gate 分型面浇口，“或”门
joint gauge 测缝计
joint Gaussian distribution 联合高斯分布
joint grouting 接缝灌浆
joint hinge 接合铰链
joint impedance 总阻抗，节点阻抗
jointing clamp 接线夹
jointing compound 密封剂
jointing element 连接构件
jointing material 接合密封材料，勾缝料
jointing paste 灌缝浆
jointing sleeve 连接用的套管
jointing washer 接头垫圈
jointing 填料，接合，填缝，（鱼尾板）连接
joint inspection 联合检查，联合检验
joint invitation to tender 联合招标
jointless floor 无缝地板，无缝地面层
jointless iron core 无接缝铁芯
jointless structure 无缝结构
jointless 无接头的，无接缝的
joint liability 连带责任，共同责任
joint liaison meeting 联合联络会
joint loan 联合贷款
jointly and severally 个别并连带负责
jointly ergodic random processes 联合遍历随机过程
jointly stationary random processes 联合平稳随机过程
joint management 联合管理
joint measurement 测缝，缝的量测
joint meeting 联席会议
joint meter 测缝计
joint observation 联合观测
joint of lap splice 搭接接头
joint operating device 联合运行装置，并车装置
joint operation center 联合操作中心
joint operation of heat-supply networks 热网联网运行
joint operation 合作经营，联合运行，联营
joint ownership 共同所有权
joint packing 接合填密，垫圈
joint part 连接部分，连接件
joint pin 连接销
joint plane 节理面
joint pole 组合电杆，同架电杆
joint probability distribution 联合概率分布
joint probability 联合概率
joint production 联合生产，合作生产
joint project 合办项目，合资项目
joint random variable 联合随机变数
joint responsibility 连带责任
joint ring 连接垫圈，连接环，接合密封环
joints cast-in-situ 现场浇灌的接头
joint sealing material 填缝料
joint session 联席会议
joint sheet 填密垫片，接合垫片
joint shield 接缝盖面
joint signature 合签，会签，联合签署
joint slack 联轴节
joint state-private enterprises 公私合营企业
joint stock company 联合股份公司，合股公司，股份公司
joint stock corporation 股份有限公司
joint stock enterprise 股份制企业
joint stock limited partnership 股份有限公司
joint stock partnership 股份合作制
joint stock system 股份制
joint stock 合股，合资
joint strip 压缝条，密封胶条
joint stub 管接头
joint surface 结合面
joint system 节理系
joint tendering 合作投标
joint tender 联合投标
joint thread 连接螺纹
joint use 一线多用，同杆架设
joint venture agreement 联营体协议，合营协议
joint venture bank 合资银行
joint venture company 合营公司，合资公司
joint venture investment 合资经营投资，合营投资，联合投资
joint venture of government and private citizen 公私合营
joint venture partnership 合营
joint venture partner 合资经营伙伴
joint venture 合资，合营，合资经营，联营企业，合资企业，合资公司
joint work 合作作品
joint 共同的，联合的，合办的，结合，接缝，接头，接合面，缝，接点，结点，连接
join up 连接起来，接入（电路）
join 加入，参加，联结，连接，联合
joist ceiling 搁栅平顶
joist floor 搁栅楼盖
joist steel 梁钢，工字钢
joist 工字梁，搁栅，小梁，托梁，工字钢，桁条，龙骨
Jolly balance 测密实度天平秤，比重天平，乔利秤
jolt ramming 振动夯击
jolt 震击，摇动，振动，颠簸
JOP(jacking oil pump) 顶轴油泵
Jordan sunshine recorder 暗筒日照计，乔丹日照计，约旦日照计
Joukowski airfoil 儒科夫斯基翼型
Joule coefficient 焦耳系数
Joule cycle 焦耳循环
Joule dissipation 焦耳耗散，功率耗散
Joule effect 焦耳效应
Joule energy 焦耳能量
Joule-heated ceramic melter 焦耳加热陶瓷熔融炉
Joule heating 焦耳电热效应加热
Joule heat 焦耳热
Joule-Kelvin effect 焦耳-开尔文效应

Joule-Lenz effect　焦耳-楞次效应
Joule-Lenz's law　焦耳-楞次定律
Joulemeter　焦耳计
Joule's equivalent　热功当量，焦耳当量
Joule's law　焦耳定律
Joule-Thomson coefficient　绝热节流系数，焦耳-汤姆逊系数
Joule-Thomson cooling　焦耳-汤姆逊冷却
Joule-Thomson effect　焦耳-汤姆逊效应
Joule　焦耳
journal bearing wedge　轴颈轴承楔
journal bearing　支持轴承，轴颈轴承，径向轴承
journal box　轴颈箱
journal bracket　轴颈套筒
journal brass　轴颈铜衬
journal neck　轴颈
journal radial bearing　径向支持轴承
journal rest　轴颈支承
journal speed　轴颈速度
journal with collar　有环轴颈
journal　杂志，定期刊物，轴颈，日志
journey　路程
joystick transformer　调压变压器
joystick　控制杆，控制手柄，操纵杆
JPD　消防水分配系统【核电站系统代码】
JPH　汽轮机油箱消防系统【核电站系统代码】
JPI　核岛消防系统【核电站系统代码】
JP(jet pump)　喷射器
JPL　电气厂房消防系统【核电站系统代码】
JPP　消防水生产系统【核电站系统代码】
JPS　移动式和便携式消防系统【核电站系统代码】
JPT　变压器消防系统【核电站系统代码】
JPU　厂区消防水分配系统【核电站系统代码】
JPV　柴油发电机消防系统【核电站系统代码】
JPY　日元
repeated impact test　冲击疲劳试验
JSC(joint-stock company)　股份公司
J & T analysis(job & task analysis)　岗位和职责分析
jube　隔栏，屏障
jubilee truck　小型货车
judder　强烈振动，冲击，位移，不稳定
judgement sampling　鉴定性抽样
judgement　判断
judge of soil　土壤鉴定
judge　断定，判断，评价，鉴定，下结论
judging panel　评委
judgment　意见
judicature　司法
judiciary　司法的，法院的
juice　浆汁，电流，液体燃料
jukebox storage　盘式存储器
jukebox　（电脑的）自动光盘只读存储器，光盘自动交换存储器
jumbo barge　大型驳船
jumbo boom　隧道钻车臂
jumbo brick　大型砖
jumbo group modem　巨群调制器
jumbo group　巨群（频率）
jumbo size　特大号
jumbo windmill　巨型风车
jump address　转移地址
jump condition　转移条件，跳跃条件，跃变条件
jump drilling　冲击钻井，撞钻
jumper bit　冲击锤，冲击钻头，钎子
jumper cable　跨接电缆，分号电缆
jumper clamp　跳线线夹
jumper lead　搭接片
jumper loop　跳线
jumper plug　插座
jumper support insulator　跳线绝缘子
jumper terminal　跳线端子，跳线接头
jumper tube　跨接管
jumper wire　跨接线，跳线
jumper　跨接管，跨接器，跨接片，跨接线，桥形接线，跳线，长钻，穿孔凿
jump function　阶跃函数
jump if not　条件转移，若非则转移
jump in brightness　亮度落差，亮度跃变
jumping change in frequency　频率跃变
jumping cushion　救生垫
jumping relay　跳动继电器
jumping　突变现象，跃变，跳动
jump instruction　转移指令，跳转指令
jump in temperature　温度突变
jump joint　对头接合
jump phenomenon　跃变现象
jump resonance　跳跃共振
jump rope shaped blade　跳绳形叶片
jump-spark coil　跳火感应线圈，点火线圈
jump spark　跳跃火花，跳火
jump to if not above　无符号不大于则跳转
jump to if not below　无符号不小于则跳转
jump to if not equal　不等则跳转
jump to if not greater　有符号不大于则跳转
jump to if not lower　有符号不小于则跳转
jump valve　回跳阀
jump weld　T形焊接
jump　跃变，跳跃，突变，突跃，跨接，跳线，跳变，跳步，（程序）转移
junction battery　结型电池
junction bench mark　交叉水准点
junction board　连接台，接线台
junction box　电缆套，端子箱，接线盒，套管，联轴器，分线箱，连接箱
junction cable　中继电缆
junction circuit　连接电路，中继电路
junction compensator　冷端补偿器【热电偶】
junction current　结电流
junction diode　面结型二极管
junction house　转运站
junction line　中继线
junction manhole　管路交汇处检查井，管路交汇处人孔
junction of channels　河道汇合点
junction pipe　连接管
junction plate　连接板，接合板
junction point　连接点，接点，接合点，会合点，接触点，联络点
junction pole　接线杆，分线杆
junction port　转口港
junction potential method　接点电位法，接点电

势法
junction tower 转运站
junction transistor integrator 结式晶体管积分器
junction transistor 面结型晶体管
junction transposition 接线换位
junction-type field effect transistor 结型场效应晶体管
junction 连接点，接点，接合，连接，节点，接合处，中继线，接口，接合体，链接
junctor 联络线，连接机
juncture 接合点，接合，连接，焊接，接头，接点，接缝，节点
jungle 丛林，密林
junior beam 次梁
junior cave 小屏蔽室
junior scram 紧急降功率，非完全紧急停堆
junk mail 垃圾邮件
junk ring 填料函压盖，密封圈，密封环，填料圈，压环，衬圈
junk 绳屑，(填缝用的）麻丝，废料，小块废铁
juridical organ 司法机关
juridical person 法人
jurisdiction 司法权，管辖权，管辖范围，权限，司法管辖权，管辖区域，司法
jury pump 备用泵，辅助泵，应急泵
just assessment 公正的评价
just for reference 仅供参考
justice of exchange 交易公平
justification of a practice 实践的正当性
justification 辩护，合理（性)，正当性判断，证论
justified claim 合理索赔
justified price 合理价格
justified 被证明是合理的，正当的
justify 证明正确，(数据排列位置的）调整
just-noticeable difference 最小可辨差异
just open 微开
justowriter 带穿孔打字系统
just price 公平价格
jute insulated cable 黄麻绝缘电缆
jute 黄麻，电缆黄麻包皮
jutter 摇动，振动
jut window 突出窗
jut 凸出，伸出，突出部
juvenile water 初生水，岩浆水
juxtapose 并置，并列
juxtaposition metamorphosis 接触变质
juxtaposition 邻近，接近，并置，并列，斜接
JVCA(joint venture, consortium or association) 合资企业、财团或联合体
JVC(Joint venture Co. ,Ltd.) 合营公司
JV＝joint venture 合资企业
J 消防（探测、火警)【核电站系统代码】

K

KAF(keydate achievement form) 关键日工作进度表【格式】
kaleidophone 示振器
kalimeter 碳酸定量器
kalk 石灰
kalsomine 刷墙粉，石灰浆
kaltleiter 正温度系数半导体元件
Kamm back 卡姆背
Kamm tail 卡姆尾
K-amplifier K增益放大器
kampometer 热辐射计
kankar 灰质核
kanthal wire 铁铬铝电阻丝
kanthal 铁铬铝耐热材料，铬铝钴耐热钢
kaoline 高岭土
kaolinite 高岭石，高岭土
kaolinization 高岭土化
kaolin 高岭石，高岭土
Kaplan runner 转桨式转轮
kaplan turbine 卡普兰水轮机，转桨式水轮机
Kaplan wheel 转桨式水轮
karaburan 风沙尘，黑风暴
Karat 克拉【宝石的重量单位】，开【量金单位】
Karman constant 卡门常数
Karman spectrum function 卡门谱函数
Karman's vortex street 卡曼涡街
Karman turbulence spectrum 卡门湍流谱
karroo 干燥台地
karst base level 岩溶基准
karst basin 岩溶盆地
karst cave 水蚀石灰洞
karst erosion 岩溶侵蚀
karst funnel 岩溶漏斗
karstic channel 岩溶槽
karstic feature 喀斯特地形，岩溶地形
karstic formation 岩溶地层
karstic hydrology 岩溶水文学
karstification 岩溶作用
karst peneplain 岩溶准平原
karst phenomenon 岩溶现象，喀斯特现象
karst pit 水蚀穴，岩溶井
karst region 岩溶区，喀斯特区
karst topography 喀斯特地形，岩溶地形
karst tower 灰岩残丘
karst water 岩溶水
karst 岩溶，可溶性岩石，喀斯特，石灰岩溶洞
karton 厚纸
karts 岩溶
kar 凹地
katabatic flow 下降流动
katabatic wind 下降风，下吹风
katabatic 下降的
katafront 下滑锋
katalysis 催化，触媒
katamorphic zone 碎裂变质带
katathermometer 低温温度计，冷却温度计
kathode 阴极，负极
kation 阳离子，正离子
katogene 破坏作用
Kay-Ray Inc. 伽瑞公司【Rosemount 的一个分部】
K-band K波段
kbar 千巴
KBPS(kilobits per second) 千位/秒
K bracing K腹杆系
KBS 热偶冷端盒系统【核电站系统代码】

kcal＝kilocalorie　千卡，大卡
K-carrier system　载波系统
kC＝kilocycle　千周
KCO　常规岛共用控制系统【核电站系统代码】
kcs(kilocharacters per second)　千字符/秒
kc/s(kilocycle per second)　千周/秒
KD(knocked down)　散件组装
KDO　试验数据采集系统【核电站系统代码】
kedge　小锚
keel-and-bilge block　垫船木块
keel beam　龙骨梁
keel block　龙骨墩
keel bracing　龙骨撑杆
keel　船脊骨，龙骨
keen draft　穿堂风，强空气流
Keene's cement　干固水泥
keep accounts　记账
keep a contract　遵守合同
keep-alive circuit　保弧电路，维弧电路
keep-alive current　保活电流，电离电流
keep-alive voltage　保弧电压，维弧电压
keep-alive　点火电极，保弧，维弧
keep a promise　履行诺言
keep a reactor subcritical　使反应处于次临界
keep a record of　记录
keep at a slightly negative pressure　保持稍低于大气压，使稍有负压
keep away from moisture　谨防潮湿
keep(/be) abreast of(/with)　保持与……并列，与……保持一致
keep bolt　止动螺栓
keep comfortable condition　保持舒适条件
keep cool　放置冷处，保持凉爽
keep dry　保持干燥，切勿受潮，勿受潮湿
keeper plate　护板
keeper ring　支持环
keeper　架，夹子，卡箍，保管员，锁紧螺母
keep flat　平放
keep in a cool place　在阴冷处保管【指包装外表标志】
keep in a dark place　怕光【指包装外表标志】
keep in a dry place　在干燥处保管【指包装外表标志】
keep in confidential　保密
keep in dark place　避光保存
keeping　贮藏
keep in hold　装入舱内【指包装外表标志】
keep in touch with　保持联系
keep out of the sunlight　避开阳光，避光
keep out of the sun　避免阳光
keep pace with world market level　跟上世界市场水平
keep pin　固定销
keep relay　保护继电器，止动继电器
keep strictly to the terms of the contract　严格遵守合同条款
keep subcritical　使处于次临界
keep traceability of　保持……的跟踪能力
keep upright　不可倒置，切勿倒置，竖放，勿倒置【包装标志】
keep　保持，维持，支持零件，保存，遵守
K electron　K层电子，K电子
Kelly ball　开氏球【测定新拌混凝土稠度用】
Kelly bar　开氏方形钻杆
Kelvin ampere balance　开尔文电流天秤
Kelvin bridge ohmmeter　开尔文桥式欧姆计
Kelvin bridge　双电桥，开尔文电桥，汤姆森电桥，双臂电桥
Kelvin double bridge　开尔文双电桥
Kelvin effect　趋肤效应，开尔文效应
Kelvin scale　开氏温标，绝对温标，开式温标
Kelvin's law　开尔文定律
Kelvin temperature scale　开尔文温标，开氏温标
Kelvin temperature　开氏温度，绝对温度
Kelvin　英国商务部能量单位【等于千瓦小时】，开氏温度，开尔文
kenetron　大型热阴极二极电子管，二极整流管，高压整流二极管
kentledge　压载铁，压重料
KEPCO(Korean Electric Power Company)　韩国电力公司
Keplerian motion　开普勒运动
Keple's law　开普勒定律
keraunic level　雷电水平，雷电活动水平
kerbside　街道边，马路边
kerb　路缘石，马路牙子
kerf angle　切口角
kerf width　切口宽度
kerf　切口，劈痕
kerma factor　比释动能因子
kerma(kinetic energy released in material)　比释动能
kerma rate　比释动能率
kern area　核心面积
kern distance　核心距
kern edge　核心边界
kernel fuel　燃料芯核
kernel function　核函数
kernel　零磁场强度线，零位线，核，核心，原子核，模芯，芯核，要点，把……包在核内
kerosene vapor visualization　煤油烟显示法
kerosene　火油，煤油
ketone　酮
kettle depression　锅形陷落
kettle fault　锅形断层
kettle hole　锅穴，壶穴
kettle　锅，小汽锅，锅穴，漏斗，洼地
kevatron　千电子伏级加速器
Kew-pattern magnetometer　一种地磁变化记录仪
key aggregate　嵌缝集料
key bar　键棒，定位键，定位筋
key base　键座
key bed　键槽，键座
key block　拱顶石
keyboard checking circuit　键盘检测电路
keyboard common contact　键盘公共触点
keyboard computer printer　计算机键盘打印机
keyboard data recorder　键盘数据记录器
keyboard display system　键盘显示系统
keyboard encoder　键盘编码器
keyboard entry　键盘输入
keyboard inquiry　键盘询问

keyboardless typesetter 无键盘排字机
keyboard locking 键盘锁定
keyboard lockout 键盘闭锁
keyboard perforator 键盘穿孔机
keyboard printer controller 键盘打印机控制器
keyboard printer 键盘打印机
keyboard sender(/receiver) 键盘发送（/接收机）
keyboard send(/receive) 键盘发送（/接收）
keyboard switch 键盘开关
keyboard tape punch 纸带键盘穿孔机
keyboard teleprinter 键盘电传打印机
keyboard transmitter 键盘发送器，键盘发报机
keyboard 电键板，开关板，键盘，电键盘
key-bolt 螺杆销，键螺栓
key card puncher 卡片穿孔机
key city 中心城市
key click 电键干扰声，电键“咔嗒”声
key colour 基本色
key currency 主要货币，主要通货，基本货币
key data 主要数据
key diagram 索引图，原理图，总图
key direct access 键直接存取
key-disk machine 键盘式磁盘机
key drawings of the project 设计司令图
key drawing 索引图
key-drive 键传动，键控
keyed access method 键取数法
keyed brick 槽形砖
keyed end 锁定端，锁定点
keyed joint 键接，嵌缝
keyed original trace 分色清绘原图
keyed pulse 键控脉冲
keyed wave （等幅）电报波
keyed 用键锁固的，锁着的
key emphasis in work 工作重点
key engineering project 重点工程建设项目，关键工程项目，主要工程项目
key enterprise group 骨干企业集团
key enterprise 骨干企业
keyer 计时器，键控器，调制器，定位计，定时器
key factor 关键因素，主要因素
keyframe 键架
key groove 键槽
keyhole saw 栓孔锯，键孔锯，栓孔锤
keyhole slot 钥匙孔槽
keyhole welding 穿透型焊接法
key industry 关键工业
keying by 键控旁通
keying chirps （电键使频率变化形成的）啾啾的声音
keying circuit 键控电路
keying device 键控设备
keying wave generator 键控信号发生器
keying wave 键控信号波
key instruction 引导指令
key interlocking 钥匙联锁，连锁开关，锁定键
key investment project 重点投资项目，优先投资项目
key-in 键盘输入
key item 重点项目
key joint 键形接头，键连接
key kiloelectron-volt 千电子伏
keyless ringing 无键振铃
key letter 编码键
key light 基本灯光，主灯光
key-locked pushbutton 带钥匙的按钮
key-locked switch 带钥匙的开关
key locking 连锁开关，锁定键
key measurement point 关键测量点
key of cock 旋塞键
key off 切断
key one-line diagram 主接线图
key on 接通
key operate 键盘操作
key out 切断，断开
keypad 键盘，电键，键座，键台
key panel 键盘
key parameter 关键参数
key personal attributes 主要个人特质
key personnel 主要工作人员，关键员工，骨干人员
key person 关键人员
keyphasor or transducer 键相传感器
key pile 主桩
key pin 键销
key plan 平面索引图，索引图，位置图
key plot plan 索引（导读）图
key point 重点，关键点
key pollutant 主要污染物，主要污染
key position 关键位置，枢纽位置
key problem 关键问题
key project 枢纽工程，重点工程
key pulse 键控脉冲
keypuncher 穿孔员
keypunch 键控穿孔，键控穿孔机，盘式穿孔机，盘式穿孔
key pushbutton 电键按钮
key relay 键控继电器
key removable in both positions 电键在两者位置中移动
key removable in one position only 电键只在一个位置中移动
key rendering 抹灰打底
keyseat cutter 键槽铣刀
keyseating milling machine 键槽铣床
key seating 键槽
keysender 键控发射机
keyset 配电板，转接板
key shelf 键架，键盘，电键盘
key shim 销形填隙片
key slot milling machine 键槽铣床
key slot 键槽
key station 基本测站，中心站，主台，主控台，主台站，主要台站
keystoning 梯形失真
key substation 枢纽变电站
key suppliers list 主供货商清单
key support equipment 主要的支持设备
key switch 钥匙开关，按键开关，电键开关
key-to-card 磁卡编码器
key-to-disk 键盘磁盘结合输入器

key-to-tape　磁带编码器
key trench　坝基截水墙槽【水电】，键槽
key-type switch　键式开关
key wall　齿墙，刺墙，键入墙
key washer　键垫圈
keyway　键槽，导规，销座
keyway and feather system　键槽和滑键系统
key wiring diagram　原理接线图
key word in context index　上下文内关键字索引
key word index　关键字索引
key word out of context index　上下文外关键字索引
key word　关键词，关键字
key　销，键，按钮，开关，电键，钥匙，键控，关键码，线索，纲要，用键固定，按键，调制，自动开关，主要的，关键的，键入，锁上，调节……的音调，提供线索，使用钥匙
K factor　倍增因子，中子增殖系数，基线与高度比，导热系数，导热率
K-frame structure　K形架构
kg＝kilogram　千克
KGV(knife gate valve)　刀形闸阀
khamsin　喀新风，喀新热浪【每年从撒哈拉沙漠吹向埃及的一种干热南风】
KHIC(Korean Heavy Industries Corporation)　韩国重工业公司
kHz＝kilohertz　千赫
kibbler　粉碎机
kickback transformer　回描变压器
kickback　返程，反冲
kick-down limit switch　自动跳合限位开关
kicker baffle　冲击隔板，导向隔板
kicker light　强聚光
kicker　推料机，顶出器，投掷机，快脉冲磁铁，堆料机，抛煤机，喷射器
kicking piece　垫木
kick off meeting　开工会议，启动会议，项目启动会
kick off　切断，分离，开始
kick on　跳出，不归位式寻线机跳接
kick-plate　踢脚板
kickpoint　转折点，曲折点
kicksorter　脉冲幅度分析器，脉冲振幅分析器
kicksort　脉冲幅度分析
kick transformer　急冲变压器，脉冲变压器
kick-up　翻车器，翻缸笼
kick　突跳【曲线】，跳动，抖动，冲击，反冲，急冲，启动
kidney shaped open pile　肾形露天煤堆
kidney shaped pile　肾形煤堆
kidney　肾脏
kieselguhr　硅藻土
kilkenny coal　无烟煤
killed lime　失效石灰
killed line　断线
killed steel　镇静钢，全脱氧钢
killed　已断电的，饱和了的，镇静的，截止的
killer circuit　抑制电路，熄灭电路
killer winding　（电机、电器的）灭磁绕组
killer　断路器，限制器，抑制器，吸收器
killing frost　严霜
killing pickle　酸洗废液
killing smog　致命烟雾
killing temperature　致命温度
kill the voltage　电网短路时使发电机停止发电
kill　镇静【钢】，断电，去激励，衰减，抑制，杀死，使终止，抵消
kiln gas　窑气
kiln　炉，窑，干燥器
kiloampere　千安
kilobits per second　千位/秒
kilobit　千位
kilocalorie　千卡，大卡
kilocharacters per second　千字符/秒
kilocoulomb　千库仑
kilocurie　千居里
kilocycle per second　千周/秒，千赫
kilocycle　千周
kiloelectron-volt　千电子伏
kilogauss　千高斯
kilograin　千英厘，千格令
kilogram calorie　千卡，大卡
kilogram mete　千克米
kilohertz　千赫
kilojoule　千焦耳
kiloline　千磁力线
kiloliter　千升
kilolumen-hour　千流明时
kilolumen　千流明
kilolux　千勒克司
kilomegabit　吉位，千兆位
kilomega　千兆，十亿，吉
kilometer　千米
kilometric wave　千米级波
kilo-oersted　千奥斯特
kilo-ohm　千欧姆
kilo operations per second　每秒千次运算，千次运算/秒
kiloton　千吨当量
kilovar-hour meter　千乏时计，无功电度表
kilovar-hour　千乏时，无功千伏安时
kilovar instrument　无功功率表，千乏表，无功千伏安表
kilovar　千乏，无功千伏安
kilovoltage　千伏电压，千伏特数
kilovolt-ampere-hour meter　千伏安时计
kilovolt-ampere rating　额定千伏安，千伏安额定容量
kilovolt-ampere　千伏安
kilovolt　千伏
kilowatt hour meter　电度表，千瓦小时计
kilowatt hour　度，千瓦时
kilowatt loss　千瓦损耗，功率损耗
kilowatt meter　千瓦计，功率表
kilowatt output　输出千瓦，输出功率
kilowatt rating load　比千瓦负荷
kilowatt rating　比千瓦负荷
kilowatt-year　千瓦年
kilowatt　千瓦
kind　种类，种
kindchen　黄土结核
kindergarten　幼儿园

kindle firing 点火
kindle 燃烧，着火，点火，照亮
kindling point 着火点，燃点
kindling temperature 着火温度，燃点
kindling 升炉
kind of professions 专业类别
kindred effect 邻近效应
kinegraphic control panel 远距离控制板
kine-klydonograph 雷击电流时间特性曲线记录仪
kinematical viscosimeter 运动黏度
kinematical 运动的，运动学的
kinematic analogy 运动相似
kinematic boundary condition 运动边界条件
kinematic coefficient of viscosity 运动黏性系数
kinematic coefficient 运动系数
kinematic configuration 动态配置，运动的形状［形式］
kinematic eddy viscosity 动涡流黏滞度
kinematic elasticity 动弹性
kinematic energy 动能
kinematic equation 运动方程
kinematic model 运动学模型
kinematic similarity 运动相似性，运动相似
kinematic similitude 运动相似
kinematics 运动学
kinematic viscosimeter 运动黏度计
kinematic viscosity coefficient 动黏滞系数
kinematic viscosity 运动黏性系数【流体的动力黏度与密度的比值】，运动黏度，动黏度，动黏滞度，运动黏滞性
kinematic wave 运动波
kinematic 运动学上的
kinematograph 电影片，电影摄影机，电影制片技术
kinemometer 感应式转速表，流速计，灵敏转速计
kinescope 电子显像管
kinetic analysis 动力学分析
kinetic characteristic curve 运动特性曲线，动态特性曲线
kinetic characteristic 动力学特性，动态特性
kinetic control system 动态控制系统
kinetic control 动态控制
kinetic current 动力电流
kinetic effect 动态效应
kinetic energy absorption 动能吸收
kinetic energy flow 动能流
kinetic energy loss 动能损失
kinetic energy of turbulence 紊流动能
kinetic energy rejection 排出动能
kinetic energy storage system 动能贮能系统
kinetic energy 动能
kinetic equation 动力学方程
kinetic equilibrium 动力平衡，动态平衡，动平衡
kinetic flow factor 动流因数，运动水流因数
kinetic force 动力
kinetic friction coefficient 动摩擦系数
kinetic friction factor 动摩擦因数
kinetic friction torque 动摩擦力矩
kinetic friction 动摩擦
kinetic head 动压头，速度头，动力水头，流速水头
kinetic heating 动力加热
kinetic hypothesis 分子运动假说
kinetic intense neutron generator 动态强中子发生器
kinetic metamorphism 动力变质作用，动力变质
kinetic model 动力学模型
kinetic molecular theory 分子运动论
kinetic momentum 动量
kinetic moment 动力矩
kinetic parameter 动力学参数，动态参数
kinetic potential 运动势
kinetic power 动态功率
kinetic pressure 动压力，动压
kinetic principle 动力学原理
kinetic property 动力学性质
kinetic reactivity measurement 动态反应性测量
kinetic reactivity unit 动态反应性单位
kinetic simulator 动态特性模拟器
kinetics of combustion 燃烧动力学
kinetics of flame 火焰动力学
kinetics of gases 气体动力学
kinetics of ion formation 离子形成动力学
kinetics of reaction 反应动力学
kinetic system 动力学系统，动态系统
kinetics 动力学
kinetic tank 移动式油箱
kinetic temperature 运动温度
kinetic theory of fluid 流体动力学理论，流体分子运动论
kinetic theory of gases 气体分子运动理论
kinetic theory 分子运动论
kinetic viscosity 动力黏度，动力学黏度，动黏度
kinetic 动力的，运动的，能动的，动力学的
kine 显像管
king and queen post truss 立字桁架
king bolt truss 钢吊杆桁架
king bolt 主螺栓，大螺栓，主销，中心销，中心立轴，旋转主轴
king journal 主轴颈
king pile 主桩
kingpin bush 主销衬套
kingpin 主销，中心销，中心立轴
king post antenna 主轴式天线
king-posted beam 单柱托梁
king post girder 单立柱下撑式大梁
king post truss 单立柱倒三角桁架，单柱桁架
king post 桁架中柱，主杆，主梁柱，吊杆柱，桁架中柱
king rod 桁架中吊杆
Kingsbury thrust bearing 一回路主泵双止推轴承，金斯贝利止推轴承
Kingsbury type bearing 金斯贝利型推力轴承
Kingsbury type thrust bearing 金斯贝利型推力轴承
king truss 单柱桁架
king valve 主阀，总阀
kinking 铰接，扭结

kink in surge line 喘振线转折点
kink mark 折痕
kink 扭结，弯折，曲折，铰接，转折点，结点，拐点
kiosk 电话亭，亭，小室，变压器亭，电话间
kipp-pulse 选通脉冲
kipp relay 基普继电器，冲息多谐振荡器
KIPS＝kilopounds 千磅
Kirchhoff flow 基尔霍夫流
KIR 松动部件和振动监测系统【核电站系统代码】
kiss pressure 接触压力
kiss rolling 吻合滚压，轻触滚压
KIS 地震仪表系统【核电站系统代码】
kitchen sink 厨房水池
kitchen 厨房
kite anemometer 风筝风速计
kite ascents 风筝探测
kite balloon 风筝气球
kite 风筝
kit 成套工具，用具包，工具箱，成套仪器，一组仪表，一套工具，工具包
KIT 集中数据处理系统【核电站系统代码】
kJ＝kilojoule 千焦
K(Kelvin temperature) 绝对温度
K＝Kelvin 开尔文
KKK 厂区和办公楼出入监视系统【核电站系统代码】
KKO 电度表和故障滤波器系统【核电站系统代码】
KKS(kraftwerk kennzeichen system) 电厂标识系统，电厂设备标识系统
klaxon 电喇叭，警笛
k level k 能级
klirr-attenuation 失真衰减量
klirrfactor 非线性畸变因子，波形失真因子
klirr 波形失真，非线性失真
Klockner stud tensioning machine 克劳克耐尔螺杆拉伸机
klydonogram 脉冲电压显示照片，过电压摄测显示图
klydonograph 脉冲电压记录器，浪涌电压记录器，过电压摄测仪
klystron oscillator 速调管振荡器
klystron 速度调制电子管，调速管
kmaite 绿云母
k meson k 介子
KME 试验仪表系统【核电站系统代码】
km(kilo-mega) 千兆
km＝kilometer 千米
K-modelK 模式
k multiplication factor 增殖因子，倍增因子
knapper 碎石锤
kneading action 揉搓作用
knead 揉合，搓和
knee bend 直角弯头，弯管，弯头
knee brace 角撑，斜撑，斜角撑杆，隅撑
knee iron 斜角铁，隅铁
knee of curve 曲线的弯曲处，曲线拐点
knee pipe 弯管，弯头管，直角弯管
knee point voltage （曲线）拐点电压，膝点电压
knee point 曲线弯曲点，拐点
knee-rail 中间横栏杆
knee timber 多节材
knee 弯曲处，弯头，弯管，（铣床的）升降台，扶手弯头
KNFC(Korean Nuclear Fuel Company) 韩国核燃料公司
knick point 变坡点，反弯点，拐点，裂点，坡折点
knifeblade fuse 片形熔丝
knifeblade switch 刀闸开关
knifeblade 刀片的，刀片
knife contact 刀形触头
knife edge armature 刃口式衔铁
knife edge relay 刀口继电器，刃式继电器
knife edge switch 闸刀开关，刀形开关
knife edge weir 锐缘堰
knife edge 刀刃，刃形支承，刃状物，刀口【天平的支点】
knife file 刀挫
knife gate valve 刀型闸阀
knife stone 磨刀石
knife switch terminal block 刀闸开关接线板
knife switch 刀闸，闸刀开关，刀形开关
knife 刀
knitted fabric 针织布
knit 编，织，接合
knob and kettle topography 凸凹地形
knob-and-tube wiring 瓷珠瓷管布线，穿墙布线，穿墙布电线
knob dial 鼓形刻度盘
knob down 按下按钮
knob handle 球形手柄
knob insulator 鼓形绝缘子，瓷柱
knob lock 执手锁
knob-operated control 按钮控制，旋钮控制
knob up 拔出按钮
knob wiring 瓷柱布线
knob 球形柄，按钮，旋钮，圆形把手，鼓形绝缘子，把手，调节器，球形柄
knock-down export 成套零件输出
knockdown tower 拆卸式战标
knocked-down a machine 拆卸机器
knocked-down 拆装的
knocker 敲打工具，门环
knock hole 定位销孔
knocking combustion 爆燃，爆震燃烧
knocking explosion 爆炸
knocking test 爆震试验
knocking 爆震
knock inhibiting essence 抗爆剂
knock-in 打入，敲入
knock-off 敲掉，中止，停止
knock out drum 缓冲罐
knock-out pin 顶针，顶杆，脱模销，推出销，脱模杆
knockout 分液器，脱模装置，拆卸工具，敲除，引人注目的人或物
knock pin 定位销
knock product 拳头产品
knock property 爆震性

knock rating 防爆率，爆震率，抗暴性，爆震评定，爆震定额，辛烷值
knock-sedative dope 防爆添加剂
knock-sedative 抗震的
knock test engine 燃料爆震性能试验发动机
knock test 抗震性试验，爆震性试验
knock 敲，敲打，碰撞，震动
Knoop hardness number 努氏硬度值
Knoop hardness test 努氏硬度试验
knot 波节
know all men by these presents 根据本文件，特此宣布【通常用于前言部分】
know-how agreement 专有技术协议
know-how contract 专有技术合同
know-how education 专门技术教育
know-how licence contract 专有技术许可证合同
know-how market 专有技术市场
know-how 专门技能，技术诀窍，使用知识，专门技术，小窍门，诀窍
knowledge economy 知识经济
knowledge engineering 知识工程学
knowledge intensive 智力密集型
known damage （在交货前或交货时）已知损坏
known loss （在交货前或交货时）已知损失
known point 已知点
known sample 已知试样
known solution 已知溶液
know-what 事实知识，技术知识，专有技术
know-why 原理知识，专有技术，技术知识，知其所以然
knuckle-and-socket joint 链球形连接
knuckle joint 万向接头，指骨关节，铰链接合，肘形接头
knuckle radius 转角半径，过渡半径
knuckle type coupler 钩舌式车钩
knuckle 关节，万向接头，铰链，接合，钩爪
Knudsen diffusion 努森扩散
knurled nut 凸螺母，滚花螺母，滚花螺帽
knurling cylinder 滚花圆柱
knurling tool 滚花刀具
knurling 滚花，压花纹，滚花刀，压花刀
knurl wheel 滚花轮
Kolmogorov's hypothesis 科尔莫格罗夫假说
Kolmogorov's similarity theory 科尔莫格罗夫相似理论
konimeter 测尘器，尘度计，计尘器，空气尘量计
konimetry 空气浮尘计量法，空气浮尘计量学，微尘学
koniscope 计尘仪，检尘器
konisphere 尘层
konitest 计尘试验
konstantan 康铜，铜镍合金
KOPS(kilo operations per second) 每秒千次运算
Korean Electric Power Company 韩国电力公司
Korean Heavy Industries Corporation 韩国重工业公司
Korean Nuclear Fuel Company 韩国核燃料公司
k out of n code n中取k码
KPC(keyboard printer controller) 键盘打印机控制器
KP(key pulse) 键控脉冲
KPMG 毕马威会计事务所
K-profile K廓线
KPR 应急停堆盘系统【核电站系统代码】
KPS 安全监督盘系统【核电站系统代码】
kraft capacitor paper 电容器纸
kraft paper 牛皮纸
kremastic water 含气带水
KRG 总控制模拟系统【核电站系统代码】
KRS 厂区辐射气象监测系统【核电站系统代码】
KRT 电厂辐射监测系统【核电站系统代码】
Krupp austenite steel 奥氏体铬镍合金钢
KRW 韩国元
kryptomere 隐晶岩
krypton exposure technique 氪灯曝光技术
krypton oxide 氧化氪
krypton tracer 氪示踪剂
krytron 弧光放电充气管，氪【Kr】
KSA 警报处理系统【核电站系统代码】
KSC 主控室系统【核电站系统代码】
K shell K层，K电子层，K壳层
KSN 核辅助厂房就地控制屏和控制盘系统【核电站系统代码】
KSR(keyboard send/receiver) 键盘发送接收机
KSU 应急保安楼控制台系统【核电站系统代码】
K-theory K扩散理论
KT(keyboard transmitter) 键盘发送器
KUCA(Kyoto University Critical Assembly) 日本京都大学临界装置
kurtosis 峰态，峭度，峰度
Kussner function 库斯纳函数
Kutta condition 库塔条件
Kutta-Joukowski condition 库塔—儒科夫斯基条件
Kutta-Joukowski theorem 库塔—儒科夫斯基定理
kVA(kilovolt-ampere) 千伏安
kvar-hour meter 无功电能表
kvar=kilovar 千乏
kV=kilovolt 千伏
kV-T product 电压有效值【kV】与电话干扰系数【T】之积
kWH(kilowatt-hour) 千瓦时，度
kW-hour meter 有功电能表
kW=kilowatt 千瓦
KWU(Kraftwerk Union AG) （德国）电站联盟公司
kyanite 蓝晶石
kybernetics 控制论
kymatology 波浪学
kymogram 记录图，记波图
kymograph 波形自记器，转筒记录器，波形记录器
Kyoto Protocol 京都议定书
Kyoto University Critical Assembly （日本）京都大学临界装置
KZC 控制区出入监测系统【核电站系统代码】
K 仪表和控制【核电站系统代码】

L

lab coat　实验室工作服
label clause　标签条款
label coding　标记编码，标签编码
label constant　标号常数
label data register　标签数据寄存器
label data　标号数据
labeled atom　示踪原子，标记原子
labeled particle　示踪粒子
labeled price　标明价目
label holder　标签夹
labeling and marking fee　贴标签费
labeling reader　标记卡阅读器
labeling scheme　代码电路，标号方案
labelled atom　标记原子
labelled molecule　标记分子
labelled　标记的
labelling regulation　标签条例
label record　标号记录
label　标牌，标签，签条，标明，标号，标记，加标签
lab-gown　实验室工作服
labile equilibrium　不稳定平衡
labile flow　不稳定流
labile region　不稳定区
labile state　不稳定状态
labile　易变的，易错的，不稳定的，不安定的
lability　不稳定性，易变性
labor agreement　劳动协议，劳务协议
laboratory accreditation　实验室鉴定
laboratory apparatus　实验室仪器，实验室用仪器
laboratory application data　实验室应用数据
laboratory bench　实验室工作台，实验工作台
laboratory block　实验楼
laboratory building　实验楼
laboratory clothing　实验室工作服
laboratory coal sample　实验室煤样
laboratory coat　实验室用工作服
laboratory data　实验室数据，实验资料
laboratory drains pump　实验室疏水泵
laboratory drains tank　实验室疏水箱
laboratory drain　实验室疏水
laboratory equipment and tools　化验室设备及工具
laboratory examination　实验室检验
laboratory findings　实验结果
laboratory flume　实验槽，实验水槽
laboratory frame of axes　实验室坐标系
laboratory hood　实验室排风柜
laboratory instrument　实验室仪表，实验室用仪器
laboratory investigation　实验室研究
laboratory manual　实验手册
laboratory monitor　实验室监测器
laboratory procedure　实验程序，实验室研究方法
laboratory reactor　实验室反应堆
laboratory reagent　实验室试剂
laboratory reference standard　实验室基准标准器
laboratory reliability test　实验室可靠性试验
laboratory report　实验室报告，实验报告
laboratory result　实验室结果
laboratory sample of coal　实验室煤样
laboratory services　实验室辅助设备［管线］
laboratory sifter　实验室筛分机
laboratory simulation　实验室模拟
laboratory sink　实验室洗涤槽
laboratory soil test　实验室内土工试验
laboratory system　实验室系统
laboratory technique　实验室技术，实验技术
laboratory test　实验室检验，实验室试验
laboratory type tube　实验型风洞
laboratory waste water　实验室废水
laboratory waste　实验室废物
laboratory working standard　实验室常用标准器
laboratory　研究室，实验室，试验室，研究所，试验厅
labor contracting system　包工制
labor cost　人工费，工资
labor discipline　劳动纪律
labor efficiency　劳动效率
laborer　工人，劳动力，劳动者
labor force　劳动力
labor intensity　劳动强度
labor-intensive project　劳动密集型项目
labor market　劳务市场
labor norm　劳动定额
labor resources　人力资源
labor-saving investment　节省劳力投资
labor service company　劳务公司
labor service contract　劳务合同
labor service cooperation　劳务合作
labor union　工会
labor　劳动
labour and capital　劳资
labour arbitration　劳资纠纷仲裁
labour bureau　劳动局
labour contract　劳动契约
labour cost　劳务费用，人工费
labour division　分工制
labour fee or handling charge　操作劳务费
labour force　劳动力
labour fund　劳动基金
labour hour　工时
labour importing country　劳力进口国，劳工输入国
labour insurance expenditure　劳保支出
labour insurance fee　劳动保险费
labour insurance regulations　劳保条例
labour insurance　劳动保险
labour intensity　劳动强度
labour intensive enterprise　劳动密集型企业
labour intensive industry　劳动密集型工业
labour intensive project　劳动密集型项目
labour law　劳动法
labour only subcontractor　只承包合同劳动的分包商
labour power　人力

labour productivity 劳动生产率
labour protection articles 劳保用品
labour quota 劳动定额
labour reward 劳动报酬
labour sheet 计工表
labour suppliers 劳务供应商
labour union outlay 工会经费
labour wages 劳动工资
labradorite 富拉玄武岩，拉长石
lab-scale 实验室规模
labyrinth baffle 迷宫式障板
labyrinth bearing 迷宫轴承
labyrinth box 迷宫密封室
labyrinth bushing 迷宫轴套，迷宫式衬套，曲折密封轴套
labyrinth casing 汽封体
labyrinth clearance 汽封间隙
labyrinth collar 迷宫式汽封圈
labyrinth disc 迷宫盘
labyrinth fin 曲径汽封片，迷宫汽封梳齿
labyrinth gland casing 迷宫汽封体
labyrinth gland housing 迷宫汽封体，迷宫密封箱体
labyrinth gland loss 迷宫汽封损失
labyrinth gland packing 迷宫汽封
labyrinth gland 迷宫（式）密封［汽封］，曲径汽封，迷宫式压盖，拉别林汽封
labyrinth packing 曲径汽封，迷宫式密封，迷宫式汽封，迷宫式充填
labyrinth passage 迷宫式通道
labyrinth piston 迷宫式活塞，卸荷活塞
labyrinth ring 迷宫式汽封片，汽封环，迷宫环
labyrinth screw pump coupling 迷宫螺旋泵联轴器
labyrinth sealing 迷宫式封闭，迷宫式汽封，迷宫式密封，迷宫止水，曲径汽封
labyrinth seal ring 汽封环，迷宫汽封片，迷宫止水环
labyrinth shaft seal 迷宫式轴封
labyrinth sill 迷宫型堰顶，曲径堰顶
labyrinth strip 迷宫汽封片
labyrinth trap 迷宫式疏水器
labyrinth type shaft seal 迷宫式轴封
labyrinth viewing device 迷宫式窥视装置
labyrinth wearing ring 迷宫磨损环
labyrinth 曲折密封，迷宫，曲径，迷路，难解的事物，曲折，困惑，错综复杂，费解的事
lab 试验室
laccolite 岩盖
laccolith 岩盖
laced beam 空腹梁，缀合梁
laced card 全穿孔卡片，多余孔卡片
laced column 缀合柱
laceration 裂口
lace 拉筋，花边，饰带，束带，捆扎，绑扎，束紧，系带子
lacing bar 缀条
lacing hole 拉筋孔
lacing lug 拉筋凸台，拉筋突块
lacing system 腹杆系
lacing tape 编织带
lacing tie wire 拉筋
lacing wire 绑扎用铁丝，系束穿束线，拉筋
lack experience 缺乏经验
lack of bonding 未接合，未结合
lack of bond 结合不良
lack of contrast 反差不足
lack of defination 清晰度不佳
lack of evidence 缺乏证据
lack of fusion 未熔合
lack of homogeneity 非均质性，均匀性差
lack of materials 缺料
lack of penetration 未焊透【焊接】
lack of resolution 清晰度欠佳
lack of sharpness 清晰度不佳
lack of side fusion 侧面未焊合
lack of solder 脱焊，虚焊
lack of water 缺水
lack voltage alarm 欠压报警
lack 缺少，缺乏
lacquer-coated steel sheet 涂漆薄钢板
lacquer coat 漆涂层
lacquered wire 漆包线
lacquer enamel 珐琅，瓷漆，搪瓷器
lacquer film capacitor 喷漆薄膜电容器
lacquering 上漆
lacquer resin 虫胶树脂，线形酚醛树脂
lacquer solvent 溶漆剂
lacquer varnish 亮漆，凡立水
lacquer 漆，天然漆，清漆，亮漆，漆器，使表面光泽，涂漆，光漆，挥发性漆，喷漆
lac resin 虫胶树脂
lac 虫胶，虫漆
ladder beam 梯梁
ladder bucket dredger 多斗挖泥船
ladder cable tray 梯级电缆托架
ladder circuit 梯形电路
ladder diagram 梯形图
ladder ditcher 链斗式挖沟机
ladder drilling 梯级式钻进法
ladder effect 梯形效应
ladder escape 安全梯
ladder excavator 链斗式挖土机
ladder for repairing 修理梯
ladder language 阶梯图语言
ladder network 梯形网络
ladder rung 梯级
ladder scaffold 梯式脚手架，梯台架
ladder scraper 皮带装料铲运机
ladder step 扶梯梯级
ladder string 梯帮
ladder track 梯形线
ladder tray 梯级托架
ladder trencher 链斗式挖沟机
ladder truck 带梯卡车
ladder-type filter 梯式滤波器
ladder-type network 梯形网络
ladder type trencher 梯式挖沟机
ladder winding insulation 阶梯式绕组绝缘
ladder winding 阶梯式绕组
ladder with platform 带平台的梯子
ladder 梯子，扶梯，阶梯，爬梯，梯，梯架

laden in bulk　散装的
LAD(laboratory application data)　实验室应用资料
ladle analysis　热分析，桶样分析【炉前】
lag bolt　方头螺栓
lag coefficient　滞后系数
lag coil　叠绕线圈
lag error　迟滞误差，延迟误差
lagged-demand meter　时滞需量计
lagged incidence　滞后迎角
laggedpile　套桩
lagged pulley　覆面滚筒
lagged　绝热的，滞后的，覆盖绝热层的，延迟
lagging capacity　滞后容量，滞后电容
lagging casing　绝热外壳，保温外壳
lagging circuit　滞后电路
lagging coil　滞后线圈
lagging commutation　延迟换向
lagging current　滞后电流
lagging device　滞后装置，滞相装置，延迟装置
lagging edge　后缘，后沿，下降沿，出汽边
lagging enclosure　化桩板
lagging feedback　延迟反馈
lagging filter　延迟滤波器
lagging indicator　滞后经济指标，落后指标
lagging jacket　汽缸保温套，气缸保护套
lagging jack　拱鹰架
lagging load　滞后负荷，电感性负载，电流滞后的负载
lagging material　绝热材料，保温材料
lagging motion　迟滞运动
lagging network　延迟网络
lagging phase angle　滞后相角
lagging phase　滞后相，滞后相位
lagging power-factor　滞后功率因数
lagging reactive power　感性无功功率
lagging voltage　滞后电压
lagging water　缓流水流
lagging　护板，背板，绝热层，保温，保温材料，保温层［套］，护面，防护套，罩壳，绝缘层材料，滞后，延迟，落后的，滞后的
lag gravel　残积砾石【沙漠地区】
lag in phase　相位滞后
lag module　滞后模件
lag of controlled plant　调节对象滞后
lagoon　泻湖，废水池，礁湖
lag phase　停滞期
Lagrangian equation　拉格朗日方程
Lagrangian integral time scale　拉格朗日积分时间长度
Lagrangian similarity theory　拉格朗日相似理论
Lagrangian spectral function　拉格朗日谱函数
Lagrangian strain　拉格朗日应变
lag screw　方头螺丝，拉紧螺钉
lag storage　暂时过量贮存
lag time　延迟时间，滞后时间
lag　惰性，滞后，延迟，惯性，防护套，罩，覆盖保温层
LAH　230V直流电系统【核电站系统代码】
laid down in the contract　在合同中列明
laid-off worker　下岗工人
laid on edge course　侧砌砖层
laid up in port　停泊在港
laid up returns　停泊退费
laid up tonnage　闲置吨位
laissez-passer　通行证，护照
laitance coating　浮浆层【混凝土表面的】
laitance layer　（混凝土表面的）浮浆层，（水泥浆的）翻沫层
laitance　水泥浆【混凝土】，（混凝土表面）翻沫，浮浆，浮浆皮
laitancy　水泥浆【混凝土】
lake beach　湖滩
lake breeze　湖风
lake effect　湖泊效应
lake overturn　湖水对流
lake pollution　湖泊污染
lake sediment　湖泊沉积物
LA(lead angle)　越前角，超前角
LA(length average)　平均长度
LA(level alarm)　液位报警
LA(lightning arrester)　避雷器
lambda swelling　晶间肿胀，λ肿胀
lambskin　劣质无烟煤
lamellar field　无旋场，非旋场，层流场
lamellar magnet　叠片式磁铁，多层薄片式磁铁
lamellar roof　叠层式屋顶
lamellar single crystal　层状单晶
lamellar structure　叠层结构，层状结构，片层状组织
lamellar tearing　层状图像撕裂，层状撕裂
lamellar vector　片式矢量
lamellar　多片的，层状的，片状的，多层的，成薄层的
lamellate　分层的，层状的，片状的，薄片形的，有薄层的，把……切成薄片，用薄片覆盖
lamella　薄片，薄板，薄层
lamelliform　薄片形的
lamel　薄片，薄板
lamina boards　薄片木心夹板
laminac　聚酯树脂，泡沫塑料
lamina explosion proof machine　窄隙防爆式电机
laminal　分层的，层流的，薄片状的
laminar air flow　空气层流，空气片流
laminar area　层流区
laminar boundary layer　层流边界层
laminar cellular convection　层流环型对流
laminar coating　分层包覆
laminar composite　层状复合材料
laminar condition　层流状态，层流条件，界面条件
laminar convection　层流对流
laminar current　层流
laminar damping　层流衰减
laminar diffusion　层流扩散
laminar domain　层状畴
laminar drag　层流阻力
laminar film condensation　层流膜态凝结
laminar film　层流薄膜
laminar flame　层流火焰
laminar flow film　层流层
laminar flow layer　层流层

laminar flow　片流，层流
laminar fluidized bed　层流流化床
laminar fluid　滞流流体
laminar friction factor　层流摩擦因子
laminar heat transfer　层流传热
laminar insulation　多层绝缘
laminarity　层流性，层流态
laminar layer　层流层
laminar model　层流模型
laminar motion　层流运动，层流
laminar region　层流区
laminar separation　层流分离
laminar shear　层流剪切
laminar skin friction　层流表面摩擦
laminar structure　层状结构，层压结构
laminar sublayer　层流亚层，层流底层，层流次层
laminar surface flow　表面层流
laminar-turbulent transition　层流紊流过渡段，层流-湍流过渡
laminar vector　片式矢量
laminar velocity　层流速度
laminary　由薄片组成的，薄层的，薄片状的
laminar　层流的，层式的，层流式，分层的，薄片状的
laminate and cure technology　层压固化工艺
laminated arch　叠层拱
laminated armature　叠片电枢
laminated beam　叠层梁，叠合梁
laminated board　叠合板
laminated brush　叠片电刷，分层电刷
laminated cell　叠层电池
laminated clay　成层黏土，纹泥
laminated cloth rod　层压布棒
laminated coal　层状煤，片状煤
laminated commutating pole　叠片式换向极
laminated construction　叠层结构
laminated contact　分层片触点
laminated core　叠片铁芯
laminated dielectric　薄片介质
laminated-fabric plate　层压纤维板
laminated ferrite memory plane　叠片铁氧体存储板
laminated fiber board　层压纤维板
laminated fiber wall-board　层压纤维墙板
laminated frame　叠片机座
laminated glass　夹层玻璃
laminated glued timber arch　胶合叠板拱
laminated insulation　多层绝缘，分层绝缘，叠片绝缘
laminated insulator　分层绝缘物
laminated iron core　叠片铁芯
laminated joint　马牙榫接
laminated layer method　叠片法
laminated limestone　叠层石灰岩
laminated magnetic circuit　叠片磁路
laminated magnet　叠片磁铁
laminated micarta　米卡他绝缘板，层状胶合云母纸板
laminated mica　云母片
laminated paper rod　层压纸棒
laminated plank　层夹板
laminated plastics　层状塑料，塑料层板，层压塑料，塑料贴面板
laminated plate　叠层板
laminated pole machine　叠片磁极电机
laminated pole piece　叠片极片
laminated-pole rotor　叠片磁极转子
laminated pole　叠片式磁极
laminated pressure plate　叠片式压板
laminated rotor core　叠片式转子铁芯
laminated rotor　叠合式转子，叠片式转子
laminated rubber　层压橡胶
laminated shield　分层屏蔽
laminated shingle　叠层瓦
laminated soil　层状土
laminated spring　叠板弹簧，叠片弹簧
laminated stator packing　叠片式定子铁芯
laminated stator　叠片定子
laminated steel sheet　复合钢板
laminated structure　层状结构，片状结构，胶合板结构，纹层构造
laminated synthetic insulation　层压合成绝缘
laminated synthetic plastics　层压合成塑料
laminated timber　积层木，叠层木材
laminated transducer　叠片式换能器
laminated transformer　叠片式变压器
laminated wood product　胶合板制品，叠层木板制品
laminated wood　层压木板，层压板
laminated yoke　叠片磁轭
laminated　分层的，叠片的，薄片的
laminate technology　层压工艺
laminate　薄片状的，层压制品，分层，成层，层板，叠层，层压板（塑料板的），夹层
lamination coupling　圆盘联轴节
lamination crack　夹层裂缝
lamination diagram　叠积图
lamination enamel　叠片漆
lamination factor　叠压因子，叠片因子，分层因子，叠层因子
lamination insulation　冲片绝缘，叠片绝缘
lamination of pole　磁极冲片，磁极叠片
lamination quality　叠片质量
lamination sheet　叠片
lamination slot　叠片槽
lamination stacking machine　叠片机
lamination stacking　叠片组，铁芯段
lamination steel　叠片钢
lamination thickness　叠片厚度
lamination　叠片，分层，夹层，层压，薄片，层叠，层理，成层，迭层，叠层，纹理
laminator　层压机
lamina　层状体，片流，层流，叠层，层，层次
laming　薄层，薄板
laminiferous　薄板的，由薄片组成的
laminwood　层压木板
lamp arrangement　灯位布置，照明器布置
lamp base　灯座，管座，灯头
lamp black　灯黑，烟炱
lamp body　灯壳
lamp bracket　壁灯架

lamp bridge 灯桥
lamp bulb 灯泡
lamp call 灯光呼叫
lamp cap 灯头，管帽
lamp characteristic 灯泡特性，电子管特性
lamp cord 灯线，灯绳
lamp display panel 灯光显示牌
lamp flasher 灯光闪烁器
lamp fuse 灯用保险丝
lamp globe 灯泡，球形灯罩
lamp holder 灯座，管座，灯头
lamphole 灯孔【检查下水道用】
lamp house 灯罩，光源
lamp indicator 灯光指示器
lamp jack 灯座，灯插口，管座
lampkin circuit 小型电子管电路
lamp light display 灯光显示
lamplight 灯光，灯火
lamp load 电灯负荷，电光负荷
lamp lumen depreciation 光通量衰减系数
lamp manufacturing works 灯泡厂
lamp panel 灯板
lamp plug 电灯插头
lamp post 灯柱，灯杆，路灯杆，路灯柱，照明柱
lamp puller 灯泡（管）拉出器
lamp receiver 电子管收音机，电子管接收机
lamp receptacle 灯座，管座
lamp reflector 灯管反射器，白炽灯反射罩
lamp regulator 电灯电路自动电压调整器
lamp resistance 灯泡电阻
lamp resistor 灯泡电阻器，变阻灯
lamp shade 灯罩
lamp signal 灯光信号
lamp socket 灯座，管座
lamp support 灯架
lamp switch 灯开关
lamp synchronizer 灯泡式整步器，灯泡式同步器
lamps 灯具
lamp temperature 灯温
lamp test 灯试法【连续性试验】，灯泡试验
lamp voltage regulator 灯电压调整器
lamp voltage 灯电压
lamp wire 灯线
lamp 灯管，灯，灯泡，电子管，真空管，指示灯，信号灯，光指示
lance servomechanism 射流伺服机构
lancet arch 复合拱
lance-type burner 枪式喷燃器
lance-type sootblower 枪式吹灰器
lance 用风枪吹除，用金属杆清扫，喷枪，吹管，吹灰
lancing bridge 登岸桥
lancing door 清洁孔，吹灰门，打渣孔，清渣孔，清扫孔
lancing 用水枪吹洗，用风枪吹洗
land a big contract 签了一笔大合同
land accretion 土地围垦
land a contract 得到合同，签下合同
land acquisition price 征地价格
land acquisition 征收地，征地
land agent 地产商
land allocation 占地指标
land allotment 土地核配
land and water coordinated 水陆联运
land and water-saving 节约用水用地
land area 场地面积
land-based interceptor sewer 地面污水截流管
land-based nuclear power plant 陆上核电厂
land-based pollution source 陆上污染源
land-based pollution 陆源污染
land-based prototype reactor 陆上原型反应堆
land-based prototype 陆上原型
land-based reactor 陆上反应堆
land-based sources of ocean pollution 陆地来源的海洋污染
land betterment 土地改良
land bill of lading 陆运提单
land boundary 地界
land breeze 陆地风
land bridge 陆桥
land burial 地下埋藏
land cable 地面电缆
land capability class 地力分级
land capability map 地力分类规划图
land capability 土地生产力，地力
land carriage 陆上运输，陆运
land carrying capacity 土地承载能力，土地负荷能力
land classification 土地分类
land clearing 地面清理，土地清理
land compensation 土地沉陷
land configuration-topography 地形
land configuration 陆地形状，地形
land contract 土地契约，地契
land control measure 土地管制措施
land cost 土地费用，地价
land desertification 土壤沙漠化
land deterioration recovery fee 土地损失补偿费
land development 土地开垦，土地开发
land disposal 埋入地下，地下处置，土地处置
land drainage network 土地排水网
land drainage 地面排水，土地排水
land drain 地面排水沟
landed estate 地产
landed price 卸岸价格
landed property 地产
landed quality 到岸品质
landed terms 岸上交货条件
land erosion 土地侵蚀
land evaluation 土地估价，土地评价
land evaporation 陆面蒸发
land facies 陆相
landfall mark 近岸标
land fall 地崩，山崩
land feature 地貌
landfill site 填埋场
landfill 垃圾填埋地，垃圾垫土，回填，填土，土地填埋，废渣埋填，土地填筑，掩埋
land filtration 土地浸润
land fog 地面雾
land for future extension 扩建用地，预留场地
land form 地貌，地形

land for public use 公共用地
land for temporary works 临时工程用地
land freight 陆运费
land gas turbine 陆用燃气轮机
land grader 除荆机
land grading 平整土地，土地平整
land granting fee 土地使用权出让金
land hemisphere 陆半球
landholder 土地占有者
landholding 占用
land hydrology 陆地水文学
land improvement 土地改良
landing card 入境卡
landing certificate 上岸证明书
landing charge 起货费
landing of stairs 楼梯平台
landing platform 楼梯平台
landing stage 浮码头，趸船
landing step 平台踏步
landing 上岸，着陆，起货，卸货处，登陆
land leasing 租地
land-leveling operation 平整土地
land leveller 平地机，平土机
land levelling 平整土地，土地平整
land-line charge 陆线费
land-line facilities 陆线设施
landline phone 固定电话
landline 陆上线路，陆线，陆上运输线，陆上通信线
landlocked body of water 闭合水区，闭塞水体，环抱水区
landlocked harbour 陆封港
land locked province 内陆省
landlord 土地占有者，房东
land management 土地管理
landmark 基准标志【地形测量学】，地物，（土地）界标，陆标，界桩，岸标，地面方位标，里程碑，纪念碑，划时代的事件，标志性建筑，有重大意义或影响的
land mobile service 陆上移动通信
land mobile station 陆上移动电台
land occupation act 土地占有条例
land occupation of the plant area 厂区占地
land occupation of the track on plant site 厂内铁路线站地面积
land owner 土地占用者
land pier 岸墩
land planning 土地规划
land pollution 土地污染
land preparation 整地
land price 土地征购价，地价
land productivity 土地生产力
landquake 震地
land reclamation 土壤改良，开垦荒地，土地填筑，土地改良，土地开垦
land rent 地租
land requisition examination and approval procedures 征地审批手续
land resource region 土地资源区域
land resources bureau 国土资源局
land resources 土地资源
land restoration 土地恢复
land route 陆路
land runoff 陆地径流
landsat sensor 陆地卫星传感器
landsat 陆地卫星
landscape architecture 园林建筑学
landscape conservation 环境绿化保护，园林保护
landscape engineering 环境绿化工程，环境美化，景观工程
landscape reservation 风景保护，自然保护区
landscape 风景，景观
landscaping 造景，绿化，园林化设计，环境美化设计，环境美化
land-sea interface 陆海分界面
land shaping 土地整形
land side slope 背水坡，内坡
land side 背水侧
landslide along predetermined surface 沿已有结构面的滑坡
landslide analysis 滑坡分析
landslide area 滑坡区，塌方区
landslide classification 滑坡分类
landslide dam 崩坍形成的坝
landslide monitoring 滑坡监视
landslide surveillance 滑坡监视
landslide treatment 滑坡处理
land slide 地面滑移，山崩，滑坡，塌方
land smoothing 土地细整
land spout 陆龙卷风
land subsidence 地面沉降，地面下沉，地面陷落，土地沉陷，土地下沉
land surface 地面，陆面
land surveying 大地测量
land surveyor 测量员，土地测量员
land survey 地形图，地形测绘，地形调查
land tax 土地税
land tenure 土地使用期，土地使用权
land-tied island 陆连岛
land tie 着地拉杆
land transit insurance 陆运保险
land transportation insurance 陆运保险
land transportation 陆上运输，陆运
land treatment 土地处理，土地改良
land-type boiler 陆用锅炉
land upheaval 地面隆起
land uplift 地面隆起
land usage 土地利用率
land use capability 土地利用率
land use certificate 土地使用证书
land used for construction activities 施工作业用地
land use map 土地利用图
land use planning 土地利用规划
land use ration 土地利用率
land use zoning 土地利用分区
land use 土地利用，土地使用
land utilization 土地利用
land wall 岸壁
land water 陆地水
land 地面，大地，陆地，土地，小路，巷，乡村小道，小巷，通道

Langelier Index 朗尼里尔指数
language converter 语言转换器
language of agreement 协议书语言
language of arbitration 仲裁语言
language processor 语言处理机
language translation 语言翻译
language translator 语言翻译者，语言翻译程序
language 语言，代码
LAN(local area network) 区域通信网，本地区电网，局域网（络），本地网，本地局域网
LANPC 岭澳核电有限公司
lantern gland 笼形填料盖
lantern light 灯笼式天窗
lantern ring 灯笼式环，笼形环
lantern tower 灯塔
lantern type spacer 灯笼形定位件
lantern 街灯，手灯，灯，灯笼，灯笼式屋顶，信号台
lanthanum intensity distribution 镧强度分布
lanthanum 镧【La】
lap cement 搭接胶合料
lap coil winding 叠绕组，叠绕法
lap connection winding 叠绕组
lap connection 叠式连接
lap dovetail 搭接鸠尾榫
lapel microphone 佩戴式小型话筒
lap fillet welding 搭角焊
lapillistone 火山砾岩
lapilli 火山石，火山砾
lapillus 火山砾
lap joint flange 松套搭接法兰
lap joint 重接，搭接，叠接，搭接接头，互搭接头
Laplace transform 拉普拉斯变换
lap length 搭接长度
lapless 无重叠的，无搭接的
lap of splice 搭接部分，搭接长度
lap over seam 搭接缝
lap over 搭接，重叠
lapped butt 端搭接
lapped comer joint 转角搭接
lapped face 研磨面
lapped flange 重叠的法兰
lapped insulation 绕包绝缘
lapped joint flange 松套法兰
lapped joint 松套连接，搭接缝
lapped pipe end 搭接管端
lapped seam 搭接缝
lapped splice 搭接接头
lapped tenons 搭接榫
lapped type 重叠式【指混凝土沉排型式】
lapped 研磨，互搭的，被包住的，围起来的，重叠的
lapping compound 研磨剂
lapping machine 研磨机
lapping pass 研磨合格
lapping tool 研磨具
lapping 重叠，搭接，叠包，叠绕，研磨，抛光
lap riveting 搭叠铆接
lap scarf 互搭嵌接，互搭榫接
lap seam weld 搭接焊缝
lap seam 搭接缝
lapse rate 温度垂直梯度，温度直减率
lapse 压降，经过，下降，消失，（权利等的）失效，终止
lap splice 互搭接头
lap taping （线圈的）叠包，叠绕
laptop computer 笔记本电脑，手提电脑
laptop software 手提电脑软件
lap welding 搭焊，搭接焊
lap weld 搭焊，搭接焊缝
lap winding 叠绕组，叠绕法
lap wound armature 叠绕电枢
lap wrapping 搭盖绕包【绝缘带】
lap 研磨，搭接，重叠，余面，互搭，折叠
larboard 左舷
lareactor 激光聚变堆
large achievement 巨大成就
large air gap 长空气间隙
large amplitude 高振幅
large and medium sized enterprises 大中型企业
large and medium sized project 大中型项目
large aperture antenna 大孔径天线
large area counter 大面积计数器，大面积接触
large area electronic display panel 大面积电子显示板
large area flow counter 大面积流通式计数器
large artificial nerve network 大型人工神经网络
large-batch 大批量的
large borehole 大孔径钻孔
large bore 大口径
large break LOCA 大破口失水事故，大破口冷却剂丧失事故
large break 大破口
large capacity boiler 大型锅炉，大容量锅炉
large capacity core storage 大容量磁芯存储器
large capacity generator 大容量发电机
large capacity hydroelectric plant 大容量水电站
large capacity memory 大容量存储器
large capacity point 大流量点
large capacity power plant 大型电厂
large capacity storage 大容量存储器
large capacity transmission systems 大容量输电系统
large capacity unit 大容量机组
large capacity water turbine 大容量水轮机
large capacity 大容量
large chemical complex 大型化学联合企业
large coal 大块煤
large component test loop 大部件试验回路
large core memory 大容量磁芯存储器
large core storage 大容量磁芯存储器
large crossing 大跨越
large customer 大用户
large data base system 大型数据库系统
large data base 大型数据库
large-denomination negotiable CDs 大额可转让存单
large-diameter borehole 大孔径钻孔
large-diameter evaporation pan 大口径蒸发皿
large-diameter sample 大直径试样
large-diameter trunk 大直径管

large-diameter 大直径的，大号的
large displacement matrix 大位移矩阵
large earthquake 大地震
large end bell 大头承口
large equipment 大型设备
large-headed nail 大头钉
large high-speed machine 大型高速电机
large hydraulic turbine 大型水轮机
large hydroelectric power station 大型水电站
large long-term project 大型长期项目
large lump coal 大块煤
large matrix store 大容量矩阵存储器
large memory 大容量存储器
large open well pump house 大口井泵房
large open well 大口井
large order 大批订货
large panel construction 大型格板结构，大型预制板建筑
large panel structure 大型板材建筑物
large part 大件
large path limit 大行程极限
large platform arrangement 大平台布置
large platform elevator 大平台式升降机
large pole 大圆材
large power consumption product 高电耗产品
large pressurized water reactor 大型压水堆
large project 大型工程
large radioisotope heat source capsule 大型放射性同位素热源盒
larger erectional constituent of tide 大出差分潮
large reservoir 大型水库
large riprap 大块乱石护面，大块抛石护坡
large rotating plug （增殖堆的）大旋塞
large sample 大样品
large-scale cartography 大比例尺制图
large-scale chromatography 大型色谱法
large-scale circulation 大尺度环流
large-scale compound integration 大规模混合集成电路
large-scale consumer outage 大面积停电
large-scale convection 大规模对流
large-scale detail plan 大比例尺详图
large-scale detail 大比例详图
large-scale digital computing system 大型数字计算系统
large-scale drawing 大比例图
large-scale eddy 大尺寸涡流
large-scale enterprise 大型企业
large-scale experiment 放大尺寸的模型试验，大规模试验
large-scale facilities 大型设施
large-scale hybrid integrated circuit 大规模混合集成电路
large-scale hydraulic model 大比例尺水工模型
large-scale integrated circuit 大规模集成电路
large-scale integrated memory 大规模集成存储器
large-scale integration 大规模集成
large-scale integrated circuit 大规模集成电路
large-scale map 大比例尺地图
large-scale model 大比例模型
large-scale photograph 大比例尺照片
large-scale plan 大比例尺平面图
large-scale pollution 大规模污染，大范围污染
large-scale power system 大型电力系统
large-scale production 大规模生产
large-scale project 大型工程项目，大型规划，大型项目
large-scale structure （湍流）大尺度结构
large-scale survey 大比例尺测量
large-scale system 大系统
large-scale test 大型试验
large-scale turbine model 大比例尺水轮机模型
large-scale turbulence 大尺度湍流
large-scale use 大规模利用
large-scale utilization 大规模利用
large-scale weather situation 大范围天气形势
large-scale wind energy conversion system 大型风能转换系统
large-scale wind turbine 大型风电机组
large-scale 大比例尺的，大规模的，大型的，大的，大规模，规模化，大比例尺
large-screen display 大屏幕显示（器）
large-screen picture 大屏幕图像
large-section in-situ concrete pile 大直径灌注桩
large-signal analysis 大信号分析
large-signal 大信号
large-sized cable 大型电缆，大容量电缆
large-sized coal 大块煤【粒度 50mm】
large-sized enterprise 大型企业
large-sized wind machine 大型风力机
large-sized 大尺寸的，大块的
large-size indicator 大型指示器
large-size machine 大容量电机，大型电机
large source 强放射源
large space enclosure 大容积密闭罩
large-storage memory 大容量存储器
largest peak discharge 最大洪峰流量
largest vessel to be unloaded 要卸货的最大船型
large synchronous compensator 大型同步调相机
large-systems control theory 大系统控制理论
large temporary facilities 大型临时设施
large tooth 大齿
large turbo-type generator 大容量汽轮发电机，大型汽轮发电机
large type hatch opening 大型起吊孔
large wall pane 大型墙板
large 大的，粗的，大容量的，大规模的
Larimer column 工字钢组合柱【十字形的】
LAR(land acquisition and resettlement) 土地征用和移民
LAR(live animal regulation) 活动物规则
larrying 薄浆砌筑法
larry 小车，斗底车，手推车，称量车
laser alignment measuring system 激光准直测量系统
laser alignment 激光准直，激光找中，激光定向，激光校直
laser altimeter 激光测高计
laser anemometer 激光风速计
laser anemometry 激光测速法
laser auto-setting level 激光自动调整水准仪

laser beam cutting 激光切割
laser beam welding 激光焊，激光束焊接
laser centering device 激光对中仪
laser collimator 激光准直仪
laser cutting for steel plate 激光钢板切割机
laser cutting 激光切割
laser detector 激光探测器
laser diode 激光二极管
laser distance measuring instrument 激光测距仪
laser Doppler anemometer 激光多普勒测速计
laser Doppler anemometry 多普勒激光测风法
laser Doppler velocimetry 多普勒激光测速法
laser Doppler velometer 激光多普勒测速计
laser-EDP setup 激光电子数据处理装置
laser fusion breeder 激光聚变增殖堆
laser fusion 激光聚变
laser hardening 激光淬火
laser heating 激光加热
laser holography 激光全息摄影
laser interferometry 激光干涉测量
laser isotope separation 激光同位素分离
laser memory 激光存储器
laser motor 激光电动机，激光电机
laser patterning method 激光切割法
laser power 激光功率
laser pulse 激光脉冲
laser pumping energy 激光抽运能
laser Raman spectroscopy 拉曼激光光谱学
laser range finder 激光测距仪
laser ranging 激光测距
laser receiver 激光接收设备
laser scintillometer 激光闪烁计数器
laser speckle method 激光斑纹法
laser speckle photography method 激光散斑法
laser telemetry system 激光遥测系统
laser transmitter 激光通信发送设备
laser-triggered fusion 激光引发的聚变，激光引发聚变
laser trimming 激光微调
laser tube cavity 激光管共振腔
laser velocimeter 激光测速仪
laser welding 激光焊接
laser 激光，激光器
lashing wire 捆绑线，拉筋
lash 冲击，打击，空隙，游隙，捆扎，绑紧
lasing protective eyeglass 激光防护眼镜
last accounting date 最后结算日期
last address register 最后地址寄存器
last blade 末叶片
last but one stage blade 次末级叶片
last date for receipt of bids 投标的最后截止日期
last ice date 终冰日期
last in and first out 后进先出
last-in first-out list 后进先出表
last-in first-out queue 后进先出队列
last-in first-out 后进先出
lasting quality 持久性，耐久性
lasting 延长，持久，稳定
last moment emergency shutdown system 应急停堆系统
last-period forecast 近期预测
last port 最后停泊港
last settlement 结算
last stage and penultimate stage blade 末级和次末级动叶片
last stage blade of steam turbine 汽轮机末级叶片
last stage blade of turbine 汽机末级叶片
last stage blade stress 末级叶片强度
last stage blade 末级叶片
last stage of transmitter 发射机末级，发送机末级
last stage 末级
last trunk capacity 终端中继线容量
last 最近的，最后的，延续
latch armature 锁闩衔铁
latch arm notch 挂臂槽【控制棒驱动机构】，锁闩柄凹槽
latch arm tip 锁闩柄顶端【控制棒驱动机构】
latch arm 棘爪，抓手【控制棒驱动机构】
latch assembly 升降棘齿装置【控制棒驱动机构】，棘齿插入密封总成，销爪组件，锁闩组件
latch bar 门闩
latch circuit 锁住电路
latched relay 闩锁继电器
latched shaft 被锁住的轴
latch finger 锁闩机构弹簧销
latch housing 封闭外壳，密封罩【控制棒驱动机构】
latching contactor 连锁接触器
latching current 闭锁电流，阻塞电流
latching device 连锁装置，闭锁装置，锁定装置
latching finger 连锁销
latching full adder 闩锁全加器
latching 锁闩（单元），锁住，封锁，闭锁，阻塞，连锁作用，闭锁作用，锁紧作用
latch-in relay 自保持继电器，自锁继电器
latch link 联锁机构【控制棒驱动机构】
latch lock ring 锁紧环【控制棒驱动机构】，固定环，制动环
latch mechanism 抓爪机构【控制棒驱动机构】
latch nut 防松螺母
latch-on 挂闸
latch pin 锁闩销【控制棒驱动机构】
latch relay 自锁继电器，闩锁继电器
latch-type relay 闩锁键式继电器
latch up 封闭，闭锁，锁定，计算器闭锁
latch 闩锁【控制棒驱动机构上的闭锁机构】，插销，挂钩，锁存器，锁住，止动销，弹簧锁，门闩，锁紧，挂闸，复位
late acceptance 延迟接受
late bid 迟到的标书
late commutation 延迟换向
late contact 后动接点
late delivery penalty 迟交货罚款
late delivery 迟交（货），延迟交付
late effect 远期效应
late frost 晚霜
late-in-life failure 晚期故障
late-model 新型的
latency time 等待时间
latency 等待时间，潜在，潜在因数，执行时间

latent defect 潜在事故，隐蔽缺陷，隐蔽事故，隐蔽故障，潜在缺陷
latent energy 潜能
latent fault 潜在故障
latent heat load 潜热负荷
latent heat of evaporation 蒸发潜热
latent heat of vaporization 汽化潜热
latent heat 潜热
latent instability 潜在不稳定（性）
latent lesion 潜在损伤
latent neutron 潜中子
latent period 潜伏期
latent power demand 潜在电力需量
latent productive capacity 生产潜力
latent root 本征根
latent tissue injury 潜伏组织损伤
latent vector 特征矢量，本征矢量
latent 潜在的，隐蔽的
late payment 支付延误
late radiation effect 远期辐射效应，延迟辐射效应
lateral abrasion 侧蚀，旁蚀
lateral air cylinder for actuating the sleeve 驱动轴套的侧面气压缸
lateral attitude 倾斜姿态，坡度
lateral axis 横轴，横轴线
lateral bending 侧向弯曲，横向弯曲
lateral bracing system 侧向支撑系统
lateral buckling 侧向屈曲，弯扭屈曲，横向压屈，侧向加撑，横向支撑，侧向压曲
lateral bulging 侧向膨胀，侧向凸出
lateral canal 旁支运河，旁支渠道
lateral clearance 横向间隙，侧向净空
lateral cofferdam （平行于河道的）侧向围堰
lateral combination 横向合并
lateral communication 横向交流
lateral compression 侧向压缩
lateral condenser 侧面冷凝器
lateral conductor 斜拉线，斜导线
lateral confinement 侧向限制，侧限
lateral conical beam 侧向锥形射束
lateral contraction 侧向收缩
lateral correlation function 横向相关函数
lateral corrosion 侧向磨蚀
lateral crevasse 侧向裂缝
lateral critical speed 横向临界速度
lateral cross-flow rate per unit length 线横流量
lateral cylinder surface 汽缸侧面
lateral damping 横向阻尼
lateral deflection 侧向挠度，横向偏转，旁向偏转，横向挠曲
lateral deformation coefficient （材料）横向变形系数
lateral deformation 横向变形，侧向变形
lateral deviation 横向偏移
lateral diffusion 横向扩散
lateral dimension 侧向尺寸
lateral discharge 侧面放电
lateral dispersion 横向扩散
lateral displacement 侧向变位，侧向位移
lateral distribution 横向分布
lateral ditch 边沟，配水沟
lateral drainage 侧向排水
lateral earth pressure 侧向土压力
lateral economic ties 横向经济联系
lateral erosion 侧向侵蚀，旁蚀
lateral error 横向误差
lateral escape 侧向出口
lateral exchange 横向交流
lateral exhaust at the edge of a bath 槽边排风罩
lateral expansion value 侧面膨胀值
lateral expansion 侧向膨胀，横向膨胀
lateral extensometer 横向伸长计
lateral face 侧面
lateral flapping 侧向挥舞
lateral flexure 横向弯曲，旁弯度
lateral-flow spillway 侧流式溢洪道
lateral flow 横向流动，横流，侧流
lateral-force design 横向力设计
lateral force 横向力，侧向力，剪力，侧力，水平分力
lateral galloping 侧向驰振
lateral gas flow velocity 横向气流速度
lateral gradient 横向坡降
lateral growth of plume 羽流横向增长
lateral guide rail 仪表侧面的导向槽轨
lateral gust 横向阵风
lateral hood 侧吸风罩
lateral impact 侧向冲击
lateral incaving 侧向淘刷
lateral inclinometer 横向倾斜仪
lateral inertia 横向惯量
lateral inflow 侧向来水
lateral insulator 极侧绝缘子
lateral intersection 侧方交会
lateral leaching 侧向淋滤
lateral length scale 横向长度尺度
lateral load test of pile 水平荷载
lateral load 横向载荷，横向负载，横向荷载，侧向荷载
lateral magnification 横向放大率，横向放大
lateral mixing 横向混合
lateral momentum balance 横向动量平衡
lateral motion 横向运动
lateral movement 横向移动，横向运动，侧向运动
lateral neutron shielding assembly 中子横向屏蔽支架，中子横向屏蔽装置
lateral neutron shielding 中子横向屏蔽
lateral opening 边孔
lateral oscillation 横向振动，横向振荡
lateral overlap 侧向重叠
lateral pile load test 桩的侧向荷载试验
lateral pipe 支管
lateral play 横向游隙
lateral plume spread 羽流横向展宽
lateral pressure coefficient 侧压力系数
lateral pressure difference 横向压差
lateral pressure 旁压力，横压力，侧向压力，侧压
lateral radiation 横向辐射
lateral refraction 侧向折射

lateral reinforcement 侧向钢筋，横向钢筋，箍筋
lateral resistance 横向阻力
lateral resolution 横向分辨力
lateral restraint 侧向约束
lateral scale 横向尺度
lateral scouring 侧向冲刷
lateral section 旁边片段【管嘴堵板】，横截面，断面
lateral seismic force 横向地震力
lateral sensitivity 横向灵敏度
lateral sewer 侧向暗沟
lateral shear 侧向剪切，横向剪切
lateral shrinkage 横向收缩率
lateral slide 侧向滑坡
lateral solid mixing 横向固体混合
lateral stability 侧向稳定，横向稳定性，横向稳定，侧向稳定性
lateral stay 侧拉线
lateral stiffness 横向刚度，横向刚性，侧向刚度
lateral storage 沿岸调蓄，支流蓄水
lateral strain 横向应变，侧向应变
lateral stream 侧向水流
lateral strength 横向强度
lateral stress 侧向应力，横向应力
lateral strut 横向支撑
lateral support system 横向支撑系统
lateral support 横向支承，侧面支承，横向支撑，侧向支撑
lateral surface 侧面
lateral system of braces 横支撑系统
lateral tee 斜侧三通，斜三通
lateral terrace 侧向阶地
lateral thrust 侧向推力，横向推力
lateral ties of column （钢筋混凝土）柱的横筋
lateral tie 横向箍筋，横向拉杆，横向拉条
lateral tilt 侧向倾斜
lateral truss 横向桁架，抗风桁架
lateral velocity 横向速度
lateral vibration 横向振动
lateral view 侧面图，侧视图
lateral wedge 侧向楔体
lateral weld 侧焊，侧面焊缝
lateral yielding 横向塑性变形
lateral 侧向的，侧面的，横的，横向的，旁边的，水平的
laterite soil 红土
laterite 红土，红土矿
laterization 红土化
laterolog 侧向测井，横向测井
later strength 后期强度
late shipment 延迟装运
latest achievement 最新成就
late-stage diversion 后期导流
latest allowable event occurrence time 事件容许最迟发生时间
latest breakup 最晚解冻日
latest delivery of bid 截标
latest edition 最新版本
latest entry 最新记录
latest finish date 最后完工日期
latest research 最新研究
latest revision 最新版本
latest starting time 最晚开始的时间
latest 最近的，最后的，最新的
late tenders 迟到的标书
latex paint 乳胶漆
latex 胶乳，乳液，橡浆
lath and plaster ceiling 板条抹灰顶棚
lath closet 沐浴更衣室
lathe bench 车床工作台
lathe cutter 切板机
lathed ceiling 板条顶棚，板条顶
lathe operator 车工，旋工
lather 起泡沫，泡沫，涂泡沫
lathes automatic 自动车床
lathes heavy-duty 重型车床
lathes high-speed 高速车床
lathes turret 六角车床
lathes vertical 立式车床
lathe tool 车床工具，车刀
lathe turner 旋工，车工
lathe 车床，车削，机床
lathing 车，钉板条
lath 板条
Latin American Free Trade Association 拉丁美洲自由贸易协会
latitude-arc 纬度弧
latitude circle 纬圈
latitude determination 纬度测定
latitude difference 纬差
latitude line 纬线
latitude 纬度，活动余地，纬距
LAT(lowest astronomical tide) 最低天文潮位
latrolet 斜接支管台
lattice aerial 网状天线
lattice anisotropy 栅格不均匀性【堆芯】，栅格各向异性
lattice array 点阵列，栅格排列
lattice bar 格条
lattice beam 格子梁，格构梁
lattice brick 花砖
lattice calculation 栅元计算
lattice-cell fine structure 栅元精细结构
lattice-cell parameter 栅元参数
lattice cell 反应堆栅元，栅元，晶胞
lattice characteristic 栅格特性
lattice coil 蜂房式线圈，多层线圈，梳形线圈
lattice column 格构柱
lattice constant 晶格常数【晶体】，格子常数，晶体常数，网络常数，点阵常数
latticed column 格构柱
lattice defect 晶格缺陷
lattice deformation 晶格变形
lattice design 栅格设计
latticed fence 格构围栏
lattice diffusion length 栅格扩散长度
lattice dimensions 栅格尺寸
lattice distribution 栅格分布
lattice drainage 格状排水系统
lattice drum winding 斜格式鼓形绕组
latticed structure loader-unloader 桁架结构装卸桥

latticed structure 格构结构
latticed strut 格构支撑
lattice-effect coefficient 方格影响系数
lattice eigenvalue 栅格特征值
lattice filter 格形滤波器，X形滤波器
lattice flow 叶栅流动
lattice frame 格构构架，格构框架
lattice function 栅格函数
lattice geometry 栅格几何形状
lattice girder 格构大梁，花格大梁，格（构）梁，格构式大梁，格子梁
lattice imperfection 晶格缺陷
lattice mast 格钩式桅杆
lattice-modified value 栅格改正值
lattice network 网络形线路，X形网络，桥形网络
lattice parameter 晶格参数，栅格参数，点阵参数
lattice pile 非均匀反应堆，栅格反应堆
lattice pitch 栅格间距
lattice plane spacing 晶面间距，点阵平面间距
lattice plane 点阵平面，格子平面
lattice plate 格板
lattice point 阵点，格点，网点，格网点，点阵
lattice pole 格子形电杆，X形电杆
lattice position 栅格位置
lattice reactor 栅格反应堆，非均匀反应堆
lattice reinforcement 格构配筋
lattice roof 格构屋顶
lattice screen 网格式栅
lattice spacing 栅距
lattice span 硬跨接
lattice square 格子方
lattice steel support 格子形钢支架
lattice steel tower 格子形钢杆塔
lattice structure 点阵结构，晶格结构，栅格结构，网格结构，格状构造
lattice strut 格构支撑
lattice-testing reactor 栅格试验反应堆
lattice theory 点阵理论，晶格理论，网络理论
lattice tower foundation 桁架式塔架基础
lattice tower 格子形杆塔，格构杆塔，格构式铁塔，桁架式塔架
lattice truss 格构桁架
lattice-type filter 桥式滤波器，X形滤波器，格子形滤波器
lattice-type network 叉形网络，格子形网络
lattice-type portal frame 格构式龙门架，格构式框架
lattice vacancy 晶体空位，点阵空位
lattice vibration 晶格振动，点阵振动
lattice wall 花格墙
lattice winding 栅格型绕组，篮式绕组，斜格式绕组
lattice window 斜条格构窗，花格窗，格子窗
lattice work 格构结构，栅格结构
lattice-wound coil 斜格式线圈，蜂房式线圈
lattice 网格，格构，晶格，格子，格架，点阵，格构式，格栅，框格，叶栅，栅栏，网络
latus 侧部
launcher 启动装置
launching apron 堆石防冲护坦，堆石下落护坡，水下堆石护坡
launching device 启动装置
launch 创办，开始，启动，发射，激励，激振，下水
laundering facility 洗衣设施
launder 槽，流槽，洗涤，洗濯槽
laundry and hot shower drain filter 洗衣和热水淋浴排水过滤器
laundry and hot shower drains filtration pump 洗衣和热水淋浴排水过滤泵
laundry and hot shower drains filtration tank 洗衣和热水淋浴排水过滤箱
laundry and hot shower drains tank 洗衣和热水淋浴排水箱
laundry and hot shower drain 洗衣和热水淋浴污水
laundry and hot shower effluent train 洗衣和热水淋浴污水系统
laundry and hot shower effluent 洗衣和热水淋浴污水
laundry and hot shower tank 洗衣和热水淋浴水箱
laundry and shower drains tank 洗衣和淋浴排水箱
laundry contamination monitor 洗衣房污染监测器
laundry drain 洗衣房排水
laundry effluent 洗衣房废水
laundry room 洗衣间
laundry sink 洗衣池
laundry waste monitoring tank 洗衣污水监测箱
laundry waste tank 洗衣废水箱
laundry wastewater 洗衣废水
laundry 洗衣间
lauric acid 月桂酸，十二酸
lava ash 熔岩灰，火山灰
lava bed 熔岩层
lava flow 熔岩流
Laval nozzle 拉瓦尔喷管
lava plain 熔岩平原
lava slag 熔岩渣
lava soil 熔岩土壤
lavatory basin 洗手池
lavatory waste pipe 盥洗排水管
lavatory 盥洗间，厕所，洗脸盆，盥洗室，洗手间
lava 熔岩
laves phase 莱夫斯相
law enforcement agency 执法机构
law enforcement official 执法人员
law firm 律师事务所
lawful claim 合法索赔
lawful operation 合法经营
lawful right and interest 合法权益
lawful trade 合法贸易
law governing the contract 合同的管辖法律
lawn belt 草地带，草坪带
lawn mower 草坪割草机，割草机，刈草机
lawn roller 草坪碾压器
lawn sweeper 草坪清理机
lawn 草场，草坪
law of accounts 会计法

law of audit 审计法
law of averages 大数定律，均值定律
law of basin area 流域面积定律
law of company 公司法
law of conservation of energy 能量守恒定律
law of conservation of mass energy 质能守恒定律
law of conservation of mass 质量守恒定律
law of conservation of matter 物质不灭定律
law of conservation of momentum 动量守恒定律
law of conservation 守恒定律
law of constant angular momentum 等角动量定律
law of constant proportion 定比律
law of continuity 连续性定律
law of contract 合同法
law of corresponding states 相对状态定律，对应态定律，对应态原理
law of decay of turbulence 紊流衰减定律
law of diminishing marginal 边际效用递减规律
law of economic contract 经济合同法
law of electric network 网络定律【指基尔霍夫定律】
law of electromagnetic induction 电磁感应定律【指法拉第定律】
law of electrostatic attraction 静电吸引定律【指库仑定律】
law of energy conservation 能量守恒定律
law of equal ampereturns 等安匝定律
law of error propagation 误差传播定律
law office 律师事务所
law of gas diffusion 气体扩散定律
law of gravitation 万有引力定律
law of great numbers 大数定律
law of hydraulic similitude 水力相似律
law of increasing costs 成本费递增律
law of large numbers 大数定律
law of magnetic circuit 磁路定律
law of mass action 质量作用定律
law of mass conservation 质量守恒定律
law of minimum 最小值定理
law of mobile equilibrium 流动平衡定律
law of momentum conservation 动量守恒定律
law of moment 力矩定律
law of motion 运动定律
law of partial pressure 分压定律
law of perdurability of matter 物质守恒定律，物质不灭定律
law of photochemical absorption 光化吸收定律
law of plastic flow 塑流定律
law of reciprocity 互惠法
law of similarity 相似律
law of small numbers 小数律
law of solid friction 固体摩擦定律
law of superposition 叠加定律
law of supply and demand 供求规律
law of tax 税法
law of tolerance 容限定律
law of total current 全电流定律
law of wake 尾流定律
law of wall 壁面定律，界壁定律
law on certified public accountants 注册会计师法
law on personal income tax 个人所得税法
laws and regulations 法规
laws of flux conservation 磁通守恒定律
laws of thermodynamics 热力学定律
Lawson criterion 劳森判据
lawsuit 诉讼
lawyer 律师
law 法令，法律，定律
lax discipline 纪律松弛
lay an obligation upon 使负责任
layaway （退役第一阶段）封存【核电】
laydown area 存放区，搁置区，堆放区，下管区
lay-down location 放置处，搁置处
laydown machine （摊铺沥青混凝土用的）摊铺机，铺设机
lay-down pod 搁置架
lay-down position for fuel shipping cask 燃料元件运输罐搁置处
lay-down position 搁置位置
laydown yard 存放场地，放置场
laydown 沉淀作用，搁置
layer-built battery 层叠电池
layer by layer growth 分层生
layer by layer winding 逐层绕法，分层绕组，多层绕组
layered bed 双层床
layered clay 分层黏土
layered compacted coal dead pile 分层压实的煤堆
layered construction 叠层结构
layered foundation 成层地基
layered insulating 分层绝缘
layered map （表示不同高程的）分层着色地形图
layered pavement 成层路面
layered pile 分层堆起的料堆
layered strata 层状地层
layer firing 喷燃器分层布置燃烧
layer-for-layer winding 逐层绕法，分层绕组，多层绕组
layering 分层
layer in slot 槽内绕组层
layer insulation test 层间绝缘试验
layer insulation 层间绝缘
layer molding 分层压制模制
layer of a distributed winding 分布绕组的层
layer of blinding concrete 基础混凝土垫层
layer of cloth 布绝缘层
layer of coal 煤层
layer of no motion 无流层
layer of oxide 氧化层
layer of reinforcement 加强层
layer of riprap 防冲乱石层
layer of surface detention 地表滞流层
layer resistance 层间电阻
layer short circuit 层间短路
layer splice 层编接
layer structure 层状结构
layer-to-layer 层间
layer-type winding 层式线圈

layer water　间层水
layer winding　分层绕组，分层绕法，层式线圈
layerwise summation method　分层总和法
layer wound solenoid　多层圆筒形线圈，多层绕制的螺线管
layer　涂层，衬垫，焊层，层，垫层，封闭层，料层，铺设机
lay-flat tube irrigation　平铺管喷灌
laying-and-finishing machine　铺放整修机
laying depth　埋设深度，铺筑厚度
laying of cable　敷设电缆
laying of thermal insulation material　保温层砌筑施工
laying-out　放样，敷设
laying pipe　敷设管道
laying point　照准点
laying trowel　砌砖镘刀
laying-up　停运保养，长期停用
laying work　敷设工程，布线
laying　瞄准，敷设，布置，底层，铺设
lay land　生荒地
lay length of twist　扭绞节距
lay light　间接采光
layman　外行
lay off　下岗
lay of land　地形【地面起伏的形状】，地貌，地形走向
lay of wire　（电缆的）心线铰距
layout character　格式字符，打印格式符号
layout chart　施工流程图，布置图
layout constant　位数分配常数
layout design　图纸设计，草图设计，电路图设计
layout drawing　配线图，布置图，定位图，规划图，设计图，立面布置图
layout map　布置图
layout of blast holes　炮眼布置
layout of column grid　柱网布置
layout of construction site　施工场地布置
layout of control network　控制网布设方案
layout of equipment　设备布置图
layout of hydroproject　水电枢纽布置
layout of main building　主厂房布置
layout of power station　电站布置
layout of roads　道路布置
layout of the coal handling system　运煤系统布置图
layout plan　平面图，平面布置图，总图设计，总平面图
layout sheet　总布置图
layout survey　施工放样测量，定线测量
lay out　布局设计【加工制造的一部分】，布置，设计，安排，花钱，为……划样，提议
layout　布置（图），规划，排列，布局，方案，定位，略图
lay plate　分线板
lay ratio　扭绞系数
lay shaft　中间轴，副轴
lay up procedure　（增强塑料）敷层方法
lay-up　绞合，扭绞，成层，接头，预载料坯
lay　敷设，安排，布置，安置，绞，捻，绳索扭捻方向
lazy thermometer　惰性温度表
lazy tongs　同步机构
L-bank position control loop　负荷组位置调节回路
L-bar　角铁，不等边角钢
LBG(low BTU gas)　低热值气体
LBH　125V直流电系统【核电站系统代码】
LBM(load buffer memory)　寄存缓冲存储器
LBP(lumped burnable poison)　集中的可燃毒物
LBP　大口径管
capillary opening　毛细管孔
L/C at sight　即期信用证
LCC(local communication console)　局部通信控制器，局部通信控制台
LCCS(large capacity core storage)　大容量磁芯存储器
LCCW(loss of component cooling water)　设备冷却水丧失
LCDTL(load-compensated diode-transistor logic)　负载补偿二极管晶体管逻辑
LCFI(low cycle fatigue index)　低周疲劳指数
lC filter(inductance-capacitance filter)　感容滤波器
LCH　48V直流电系统【核电站系统代码】
lC(inductance-capacitance) circuit　感容电路
lC(inductance-capacitance) ratio　电感电容比
lC(inductance-capacitance)　电感电容
L/C issued ledger　开出信用证分类账
LC(late commitment)　晚期交付
L/C(letter of credit)　信用证
LC(level controller)　水位调节器，水位控制器
LC(level control)　电平控制，水平控制，位面控制
LC(line connector)　接线器
LC(link circuit)　链式电路
LC(link controller)　链路控制器
LCL(less container load)　拼箱货
LCL＝local　就地（的）
LC(load cell)　加载测试装置【铁塔试验】，负载单元，称重传感器，测力传感器，测压元件
LC(load center)　负荷中心
LC(logic cell)　逻辑单元
LC(logic comparison)　逻辑比较
LC(loop check)　循环检查
LCL service charge　拼箱服务费
LCL transport　散货拼箱运输
LCM(large capacity memory)　大容量存储器
LCM(large core memory)　大容量磁芯存储器
LCM(link control module)　链路控制模块
LCO(light crude oil)　轻原油
LCP(link control program)　链路控制程序
LCP(local control panel)　就地控制柜
LCR(low compression ratio)　低压缩比
LCR(low cost reactor)　低成本反应堆
LCS(large capacity storage)　大容量存储器
LCS(large core storage)　大容量磁芯存储器
LCSS(lower core support structures)　堆芯下部支承构件
LCTL(large component test loop)　大部件试验回路

LCU(load control unit) 负荷控制装置
LCV(logic control variable) 逻辑控制变量
LDA 30V 直流电系统【核电站系统代码】
LDBS(large data base system) 大型数据库系统
LDC(less developed country) 欠发达国家
LDC(line drop compensation) 线路压降补偿
LDC(London Dumping Convention) 伦敦倾废公约
LDC(low-speed data channel) 低速数据通道
LDCR(line drop compensation resistance) 线路压降补偿电阻
LDCX(line drop compensation reactance) 线路压降补偿电抗
LDCZ(line drop compensation impedance) 线路压降补偿阻抗
LDC 负荷指令计算机【DCS 画面】
LDD(luminaire dirt depreciation) 照明器污秽减光系数
LDDS(low density data system) 低密度数据系统
LDE(linear differential equation) 线性微分方程
LDF(laser demonstration facility) 激光器示范装置
LD(leak detector) 测漏器
L/D(length/diameter ratio) 长径比
LD(linear decision) 线性判定
LDL(lower destruct limit) 破坏极限下限值
LD=loading 加负荷
LD=load 负荷
LD(logic driver) 逻辑驱动器
LD(long delay) 长时延迟
LD(long distance) 远程，长途
LD(low density) 低密度
LDP(load double precision) 寄存双精确度，输入双精确度
LDP(local data package) 局部数据包
LDRI(low data rate input) 低速数据输入
LDR(light-dependent resistor) 光敏电阻
LDR(low data rate) 低速数据传输
LDS(line disconnecting switch) 线路隔离开关，线路断电器
LDS(local distribution system) 局部分布系统
LD50 time 50%致死时间，半数致死时间，50%辐射平均致死时间
LD trunk 长途中继线
LD50 半数致死剂量
LD0 逻辑设备 0
leachability 可浸出性
leachable 可浸出的
leachate treatment 渗滤液处理
leachate 沥出物，沥出液，沥滤液，浸出液，淋洗液，滤出物
leached horizon 淋溶层
leached hull 浸取过的燃料包壳
leached layer 淋溶层
leached soil 淋溶土
leached zone 淋滤带
leaching cesspool 沥滤污水坑
leaching efficiency 浸出效率
leaching factor 淋洗因数
leaching field 沥滤场
leaching liquor 沥滤液
leaching loss 沥滤损失
leaching method 淋洗法
leaching operation 淋洗作业
leaching out 浸出，淋失
leaching rate 浸出率
leaching requirement （土壤的）冲洗需水量
leaching solution 沥滤液
leaching test 淋溶试验，淋洗试验
leaching well 沥滤水井，渗水井
leaching 浸析物，浸取，沥滤，过滤，水漂，浸出，浸滤，淋滤，淋溶，淋溶作用，淋洗，溶滤，渗溶作用
leach rate 沥析率
leach residue 浸出残渣
lead accumulator 铅蓄电池
lead acetate 醋酸铅，乙酸铅
lead acid battery energy storage 铅酸蓄电池储能
lead acid battery 铅酸电池组，铅酸电池，铅蓄电池
lead acid cell 铅蓄电池
lead acid storage battery 铅酸蓄电池
lead-add battery 加铅电池
lead and lag motion 超前滞后运动
lead angle 导前角，超前角，前置角，导角
lead assembly 超前组件【最大燃耗】
lead auditor 监察负责人，监察长，审核组长，审计组长，主任监察员
lead-baffled collimator 铅挡板准直器
lead battery 铅蓄电池
lead block 导向滑轮
lead box 出线盒，出线箱，引线盒
lead brick 铅砖
lead bushing 引线导管，引线孔板，铅焊
lead cable 铅包电缆，引出线电缆
lead capacitance 导线电容，引线电容
lead car positioning 第一节车定位
lead cartridge 铅芯子【套管更换】
lead cask 铅罐，铅屏蔽罐
lead chloride accumulator 氯化铅蓄电池
lead clamp 引线夹，出线夹
lead-coated 镀铅的
lead collar 出线套
lead collimator 铅准直器
lead connector 引线接头
lead contractor 主要承包商，牵头承包商
lead control 超前控制，导向调节
lead covered cable 铅包电缆，铅皮电缆
lead covered insulated cable 铅包绝缘电缆
lead covered wire 铅包线，铅皮线
lead covering 铅包
lead damp course 铅皮防水层
lead discipline engineer 专业主要设计人
lead dress （集成电路）引线上的覆盖膜，引线上保护
lead-equivalent thickness 铅等效厚度
lead equivalent 铅的等效厚度，铅等效，铅当量
leader cable 引线电缆
leader group 领导班子
leader head 水落斗，雨水斗
leader label 磁带头标记
leader pipe 水落管
leadership 领导

leader strap　水落管卡子，雨水管卡子
leader stroke　先导闪击
leader　引线，首项，领导者，导杆，导管，负责人
lead factor　超前因子，领先因子
lead fatigue test　引线疲劳试验
lead-filled epoxy resin　充铅环氧树脂
lead-filled rubber apron　铅橡皮围裙
lead flashing　铅皮泛水
lead glass Cerenkov gamma spectrometer　铅玻璃切伦科夫γ谱仪
lead glass window　铅玻璃窗
lead glass　铅玻璃
lead-in brush　引入电刷
lead-in bushing　引入套管
lead-in chamfer　入口斜面
lead-in clamp　引入线夹，引线夹，输入接线柱
leading beacon　定向标
leading brush edge　前刷边
leading brush　超前电刷
leading company　牵头公司
leading current　超前电流，导前电流
leading dimension　主要尺寸，轮廓尺寸
leading edge discontinue　前缘转折点
leading edge droop　前缘下垂
leading edge flap　前缘襟翼
leading edge locus　前缘位置
leading edge radius　前缘半径
leading edge separation　前缘分离，前缘驻点
leading edge suction　前缘吸力
leading edge vortex　前缘涡
leading edge vortices　前缘（旋）涡系
leading edge　前沿，前缘，进汽边，上升边的
leading effect　超前效应
leading-end resistance　迎面阻力，正面阻力
leading end　前端
leading expert　权威专家
leading face of blade　叶片工作面
leading features　主要特征，主要特性
leading firm　主要商行
leading fossil　主导化石
leading full service provider　领先的全方位服务提供者
leading-in box　进线盒
leading-in bracket　进线架
leading-in cable　引线电缆
leading inlet edge　进汽边
leading-in phase　超前相位
leading-in pole　引入杆
leading-in wire　引入线
leading-in　引入的，引入线
leading light　导航标灯
leading load　电流超前负载，电容性负载
leading marks method　导标法
leading mark　导标，导航标
leading-out end　引出端
leading-out terminal　输出端子，出线端
leading-out　引出端，引出线，引出，导出
leading partner　牵头方，牵头公司
leading phase angle　超前相位角，相位超前角
leading phase operation　进相运行
leading phase　超前相，超前相位
leading pile　导桩
leading pole horn　前极尖
leading pole-tip　前极边，磁极前端，前极尖
leading power factor operation　相位超前运行
leading power factor　超前功率因数
leading side of blade　叶片工作面
leading voltage　超前电压
leading wave　前缘波，头波
leading wind　顺风
leading wire conductor　导线
leading wire　导线，引线
leading　引导，指导，超前，导前，主要的，领先的
lead-in inductance　引线电感
lead-in insulator　穿墙绝缘子，引入绝缘子
lead-in nipple　引入线短接管
lead-in porcelain tube　引入线瓷管
lead-in wire　引入线
lead-in　引入线，引入，输入，输入端，进线
lead joint　填铅接合
lead/lag circuit　超前/滞后电路
lead/lag module　超前/滞后模件
lead/lag　超前/滞后
lead-lead dioxide reserve battery　铅-二氧化铅储备电池
leadless　无导线的，无引线的，无铅的
lead-lined　用铅镶护的，衬铅的，用铅覆面的
lead line　测铅绳，测深锤绳，测深绳，测深索，测深线，铅垂线
lead lining　铅衬里，铅衬
lead-loaded apron　铅围裙
lead-loaded silicone foam　加铅硅树脂泡沫
lead load　超前负载
lead loss　铅损
lead mat　铅垫
lead metal frame　引线金属座，引线金属框架
lead network　（相位）超前网络
lead of brush　电刷超前
lead oil　铅油
lead-out　引出线，引出端，引出，输出
lead paper cable　纸绝缘铅包电缆
lead pipe　铅管
lead plate　铅板
lead-plating　镀铅
lead plug　铅塞
lead protection　铅防护
lead resistance　引线电阻
lead riser　引线头
lead rod　先导棒，引导棒
lead room　铅房
lead rubber gloves　铅橡胶手套
lead rubber　含铅橡胶
lead screw　丝杠，导螺杆，进给螺杆
lead seal　引线密封，铅封
lead security seal　铅的安全密封
lead-sheathed cable　铅包电缆
lead sheet　铅皮，铅板
lead sinker　铅锤
lead sleeve　铅套筒，铅压接管
leadsman　测深员

lead spring 前导弹簧
lead storage battery 铅蓄电池
lead terminal 引线端子
lead-through filter 引线用滤波器
lead-through 引入，输入
lead time 超前时间，提前期，订货至交货的时间，研制周期，交付［交货］周期，产品设计与实际生产间相隔的时间，前置期，交货期，完成某项活动所需的时间，更换模具的时间
lead-tin soldering 铅锡焊
lead to 导致，引起
lead tube 铅管，连接管
lead unit 引出单元
lead weight 吊锤，铅锤
lead wire compensation 引线补偿
lead wire 引线【热电偶】，铅丝，导线
lead-zinc accumulator 铅锌蓄电池
lead 提前，引线，铅，引导，导致，超前，导前，带动，导线，导向柱，引出线
leaf actuator 刀形断路器，闸刀开关
leaf chain 薄板链
leaf electrometer 箔式静电计，簧片式电位计
leaf electroscope 箔片验电器
leaf filter 板式过滤机，叶片式过滤机
leaflet 散页印刷品，活页
leaf of bascule bridge 仰开桥桥翼
leaf pack 钢板弹簧组
leaf sealing 薄叶汽封
leaf spring 板弹簧，叶片弹簧，片簧，簧片，弹簧片，钢板弹簧，翼簧
leaf valve 止回阀，瓣阀，簧片阀，舌阀
leaf 片簧，薄片，门扉，门扇，叶瓣，叶片，箔
leakage at the packing box 填密函处泄漏
leakage at the seat 阀座泄漏
leakage check 密封检查，检查漏气
leakage clearance 泄漏间隙
leakage coefficient 泄漏系数，漏磁系数，漏损系数
leakage conductivity 泄漏电导率
leakage control 泄漏控制
leakage current path 漏电途径
leakage current test 泄漏电流试验
leakage current 漏电流，泄漏电流
leakage detecting 检漏
leakage detection drain pipe 泄漏探测管
leakage detection pipe 泄漏探测管
leakage detection system 检漏系统
leakage detection test 泄漏检测试验
leakage detector 泄漏监测仪，漏电指示器，检漏仪，接地指示器，漏水探测仪
leakage discharge 泄漏放电
leakage distance per unit withstand voltage 泄漏比距
leakage distance 泄漏距离
leakage efficiency 泄漏系数，密封效率
leakage end 漏磁端
leakage error 泄漏误差
leakage extraction 引漏
leakage factor 泄漏因数，漏磁因子，泄漏因子，渗漏系数，漏水系数
leakage field ring 漏磁场环
leakage field 漏磁场，泄漏场，漏电场
leakage finding apparatus 检漏器
leakage flow 漏流
leakage flux coefficient 漏磁系数
leakage flux density 漏磁密度，漏磁通量密度
leakage flux transformer 漏磁通变压器
leakage flux 漏磁通，泄漏通量
leakage gap 泄漏缝，泄漏间隙
leakage gas flow 泄漏气体流量
leakage gas return line 泄漏气体回流管
leakage hardening 泄漏谱硬化
leakage hunting 检漏
leakage impedance 漏泄阻抗，泄漏阻抗
leakage indicator 泄漏指示器，泄漏检测计，检漏计
leakage inductance of a transformer 变压器漏电感
leakage inductance 漏磁电感，漏电感
leakage interception system 泄漏阻止系统
leakage interception vessel 泄漏收集器
leakage loss 漏损，漏汽损失，泄漏损失，漏电损失，漏水损失
leakage magnetic flux 漏磁通
leakage measuring instrumentation 泄漏测量仪表
leakage method 漏磁法
leakage monitoring system 泄漏监测系统
leakage monitoring 泄漏监测
leakage neutron 泄漏中子
leakage of current 电流泄漏
leakage path 漏磁路径，漏电路径，泄漏通路，漏水路线
leakage permeance 漏磁导，漏导磁率
leakage power 耗散功率，泄漏功率，耗衰功率
leakage protection 泄漏保护
leakage radiation 泄漏辐射
leakage rate 泄漏速度，泄漏率，漏水率
leakage reactance transformer 漏电抗变压器
leakage reactance voltage 漏电抗电压
leakage reactance 漏磁电抗，漏抗
leakage recovery system 泄漏回收系统
leakage recycling 泄漏再循环
leakage relay 漏电继电器，接地继电器
leakage reluctance 泄漏磁阻
leakage resistance 漏电阻
leakage sodium tank 钠泄漏箱
leakage spectrum 泄漏光谱，泄漏声谱，泄漏频谱
leakage steam discharge valve 漏汽排放阀
leakage steam 泄漏蒸汽，漏汽
leakage stopping 闭气
leakage surface 泄漏面
leakage susceptibility 泄漏敏感性
leakage test of heat-supply network 供热管道系统严密性试验
leakage test of instrument tube 仪表管路严密性试验
leakage test tool 泄漏检测工具
leakage test 漏电试验，漏磁试验，泄漏试验，密封性试验
leakage to length ratio 爬距高度比
leakage transformer 漏磁变压器

leakage water heat exchanger　泄漏水热交换器
leakage water system　泄漏水系统
leakage water　渗漏计
leakage winding　漏磁补偿绕组
leakage　漏出量，漏电，漏磁，漏气，漏水，漏损，泄漏，漏，漏出物，漏失量，渗漏，漏卸（货物）
leakance per unit length　单位长度的泄漏电导
leakance test　漏电试验
leakance　泄漏，漏电，漏磁，泄漏电导，泄漏系数
leak-before-break criterion　先漏后破准则，破断前泄漏准则
leak checking　探漏，检漏
leak coefficient　泄漏系数
leak coil　泄放线圈
leak conductance　泄漏电导
leak cross-section　泄漏截面
leak detecting instrument　检漏仪
leak detection system　泄漏探测系统
leak detection　检漏，漏气检查，漏水检查，泄漏检测
leak detector　检漏器，渗漏指示器，测漏器，漏水探测仪，探漏器，接地指示器，泄漏探测器
leaked-in air　空气漏入
leaker can　破损燃料元件贮存罐
leaker　水压试验时的“出汗”，泄漏元件，有漏孔的燃料元件
leak finder　泄漏探测仪
leak-free　密封的，不漏的
leak hunter　检漏器
leak hunting　检漏
leak impedance　漏电阻抗
leak indicator　泄漏指示器，接地指示器
leakiness　不严密性
leaking coefficient　漏损系数
leaking dose　泄漏剂量
leaking-out　漏失，漏出
leaking tube　泄漏管
leaking well　渗漏井
leak-in　漏入
leak jacket　防漏套
leakless　不漏的，密封的
leak localizer　检漏器
leak locater　测漏器
leak monitoring tube　泄漏监测管，泄漏探测管
leak-off connection　泄漏连接【填密函】
leak-off line　引漏管线，泄漏回收管线
leak-off pipe　溢流管，引漏管
leak-off pocket　抽汽室，泄漏孔
leak-off recovery pump　泄漏回收泵
leak-off recovery　泄漏回收
leak-off steam　轴封排汽，漏出蒸汽
leak-off system　引漏系统
leak-off valve　放泄阀
leak-off　泄漏，引漏
leak oil tank　漏油箱
leak out　漏出，泄漏
leakpan　泄漏舱，泄漏盘
leak pressure　泄漏压力
leakproof device　防漏装置，密封装置
leakproof material　密封材料
leakproof motor pump　密封电动泵
leakproofness　密封性，气密性
leakproof structure　密封结构
leakproof　不漏的，防漏的，密封的，防泄漏的，不漏气的，不漏电的，不漏磁的
leak propagation　漏洞扩展，泄漏扩散
leak rate determination test　泄漏率测定试验
leak rate test　泄漏率试验
leak rate　泄漏率
leak recovery system　泄漏回收系统
leak resistance　密封性【安全壳】，防漏，防漏性能，漏电阻
leak resisting　气密的
leak steam　漏气
leak stoppage provision　堵漏条款
leak susceptibility　泄漏敏感性
leak tank　泄漏箱
leak test by filling water　充水检漏，水封试验
leak test　漏电试验，漏磁试验，泄漏试验，密封（性）试验，检漏，试漏
leaktight enclosure　不漏的壳体，不漏的容器
leaktight housing　密封壳套
leaktight membrane　密封膜
leaktightness　密封，密封性，密封度
leaktight suit　气密工作服
leaktight system　气密系统，密封系统
leaktight　气密的，水密的，密封的，紧密的，不漏的
leak transformer　漏磁变压器
leaky aquifer　漏水含水层，渗漏蓄水层
leaky foundation　漏水地基
leaky pipe　渗漏管
leaky　泄漏的，漏电的，漏磁的，不严密的，未密封的
leak　漏磁，漏水，漏损，漏气，泄漏，漏洞，漏，漏电，渗漏
leam　黏质砂土，亚黏土
lean clay　贫黏土，瘦黏土
lean coal　贫煤，瘦煤
lean concrete　少灰混凝土，贫混凝土，素混凝土
lean gas-solid mixture　稀相气固混合物
lean gas　贫气
leaning post　靠架
leaning wheel grader　车轮可倾式平地机
lean lime　贫石灰
lean mixture　贫燃分混合物
lean mortar　贫灰浆
lean-phase　稀相
lean soil　瘦土
lean to mansard roof　单面折线屋顶
lean to roof　单坡屋顶，单坡屋面，一面坡屋顶
lean year　歉收年
lean　倚靠，贫的，瘦的，倾斜
leapfrog method　跳点法，跳步法
leapfrog test　跳步检验
leapfrog　动力夯，机动夯，蛙式夯
leaping of divide　分水界移动
leaping weir　溢流堰，溢流装置，下水道溢流堰

learning curve　知识曲线
learning machine　学习机，学习用机器
lease agreement　租赁合同，租赁协议，租约
lease charges　租用费
leased car　租用车
leased-line　专用线路，租用线路
leasee　承租人，租借人
lease financing　租赁筹资
leaseholder　承租人，租赁人
leasehold　租赁期，租赁权，租赁物，租赁，租赁的
lease out　出租，租赁
lease term　租赁期
lease　出租，租契，租借权，租赁
leasing trade　租赁贸易
least absolute value　最小绝对值
least action　最小作用，最小作用量
least cost estimating and scheduling system　最低成本估算与调度系统
least cost method　最小费用法
least cost operation　最低费用运转
least count　最小计数
least developed country　最不发达国家
least drag body　最小阻力体
least effort principle　最省力原则
least-energy principle　最小能（量）原理
least erratum　最小误差
least error　最小误差
least life　最低寿命
least limit　最小极限
least mean square fit　最小二乘法拟合
least radius of gyration　最小回转半径
least radius　最小半径
least-recently-used　最近最少使用的
least resistance body　最小阻力体
least significant bit　最低有效位
least significant character　最低有效字符
least significant digit　最低有效位，最低有效数字，最低位有效数字
least-spent assembly　最小燃耗组件
least square adjustment　最小二乘法平差
least square analysis　最小二乘法分析，最小二乘方分析法
least square error approximation　最小二乘方逼近
least square method　最小二乘法，最小平方法
least squares fit　最小二乘法符合，最小二乘法拟合
least square　最小二乘方
least time principle　最短时间原理
least upper board　最小上界
least voltage coincidence detection　最小符合检波电压
least waterholding capacity　最小持水量
least work　最少功率，最小功
least　最小的
leather belt　皮带
leather diaphragm　皮膜片
leather gasket　皮密封垫片，皮衬
leathermachine belting　（机用）皮带
leather seal　皮密封垫圈
leather washer　皮密封垫片，皮衬
leather　皮的
leave allowances　休假期
leave no means untried　用尽一切方法
leave out　省去，遗漏，不考虑，删除
leave over　推迟，延期
leave some leeway　留余地
leave　假期，离开
leaving area　排汽面积
leaving energy　出口能量
leaving loss　余速损失
leaving velocity loss　余速损失
leaving velocity　余速
leaving whirl velocity　叶轮出口周向分速
lecture　报告讲稿，演讲
ledeburite　莱氏体
ledeburitic steel　莱氏体钢
ledged door　实拼门，直板门，直拼撑门
ledge joint　搭接接合
ledgement　横线条
ledger board　栅栏顶板
ledge rock　真底岩
ledge support　搭耳支承
ledge　岸边礁，壁架，含矿岩层，横档，横拉杆，石梁，凸出，凸出部分，突出部分，凸耳，中冒头
LED(light emitting diode)　发光二极管
LED(physical layout editor)　物理布线编辑器
Leeb hardness number　里氏硬度值
lee depression　背风坡低压
lee eddy　背风涡
lee face　背风面
Leeh hardness test　里氏硬度试验
leek　石状黏土
lee side vortex　背风面旋涡
lee side　背风面，下风侧，下风弦，背弧
lee slope　背风坡
lees　沉积物，废物，残渣
lee tide　顺风潮
lee trough　背风槽
leeward face　背风面，下风面
leeward side　背风侧，背风面，下风侧
leeward slope　背风坡
leeward tidal current　顺风潮流
leeward tide　顺风潮
leeward wall　背风壁
leeward yacht　下风船
leeward　顺风的，背风处，背风（的），背风面，下风（的）
lee wave　下风波，背风波
leeway　风压差，风压角，余地，时间损失，允许误差，(强风所致）偏航
lee　背风面，下风面，背风
left bank　左岸
left elevation　左视图
left-hand adder　左侧数加法器
left-hand component　左侧数
left-hand derivative　左导数
left-hand digit　左侧数位，高位数位
left-hand door　左手门
left-handed direction　左手方向

left-handed rotation　左旋，左向旋转
left-handed screw bar　左旋螺杆
left-handed screw nut　左螺帽
left-handed screw　左转螺旋，反扣螺丝
left-handed thread　左旋螺纹，倒牙
left-handed　左旋的，左侧的，反时针方向的
left-hand engine　左转发动机
left-hand loose joint hinge　左边活节合页
left-hand rotation　左旋
left-hand rotating fan　左旋风机
left-hand rule　左手定则
left-hand screw　左旋螺丝
left-hand side　左侧，左边
left-hand thread　左螺纹
left-hand winding　左行绕组，倒退绕组
left-hand zero　左边零
left justify　左边对齐，左侧调整
left marginal bank　左边岸
leftmost bit　最左位
leftover　剩余物
left-running characteristic　左伸特征线
left-running Mach wave　左伸马赫波
left shift　向左移位
left side elevation　左侧立面图
left side wall　左侧墙
legal address　合法的地址，法定地址
legal advice　法律咨询
legal adviser　法律顾问
legal advisor　法律顾问
legal agent　法定代理人，法律代理人
legal ampere　法定安培
legal arbitration　按法律进行仲裁
legal assets　法定资产，合法资产
legal assignment　合法转让
legal authority for air pollution　空气污染法律根据
legal authority　法定机构，法定的权利
legal body　法人
legal bond　法定债券，合法债券
legal capacity　法律行为能力，法定身份
legal capital　法定股本，法定资本
legal consideration　合法报酬，法律思考，合法的代价，对价
legal constraint　法律制约
legal contract　合法的合同，合法的契约
legal duty　法律职责
legal earned surplus reserve　法定盈余公积金
legal entity shareholder　法人（实体）股东
legal entity　法定单位，法人，法律实体，法人实体，企业法人
legal evidence　合法证据
legal guardian　法定监护人
legal heir　法定继承人
legal inspection　法定检验
legal instrument　法律文件，法律文书
legal interest　合法权益
legal investment　合法投资
legal knowledge　法律知识
legal liability　法律责任
legal minimum wage　法定最低工资
legal mortgage　合法抵押
legal ohm　法定欧姆
legal owner　法定所有人
legal personality　法人资格
legal person's position　法人地位
legal person　法人
legal price　法定价格
legal procedure　法律程序
legal proceeding　法律程序，法律诉讼
legal program　合法程序
legal qualification　法人资格
legal regulation　法律条款
legal remedy　法律制裁
legal representative　法定代理人，法定代表
legal reserve fund　法定公积金
legal reserve　法定储备金
legal rights and interests　合法权益
legal right　法定权利
legal sanction　法律制裁
legal security　法律保证，合法保证
legal successors and assignees　法定继任者和受让人
legal successors in title　（权益）法定继承人
legal successor　法定继承人，合法继承人
legal term　法定期限
legal unit of measurement　法定计量单位
legal unit　法定单位
legal weight　法定重量
legal　法定的，法律的，合法的
leg centerline distance　杆塔结构根开
legend plate　说明标牌
legend　符号，图例，说明书，符号表，图标
leg extension　塔腿（的）延伸【输电铁塔】
legibility　明视度，清晰度
legible　清晰（的），易读的，易辨认的
legislation　法律，立法，法规，法制
legislative pressure　立法压力
legitimate claim　合理索赔
legitimate income　合法收入
legitimate right　合法权利
legitimate　合法的，合理的，正规的，真实的，合法性
leg length　焊脚长度
leg-mounted pneumatic rock drill　风动支架凿岩机
Lego baseboard　莱戈块粗糙元板，鱼鳞板
Lego block roughness　莱戈方块粗糙元
Lego block　莱戈方块
leg of angle　角钢肢
leg of bridge　电桥臂
leg of circuit　电路的一臂，电路的一支路，相线
leg of frame　框架支柱
leg pad　下端头【燃料棒】
leg slope　塔腿斜度
4-legs portal with hydraulic compensation　带液压补偿的4腿门座
leg support　支座（板）
leg　支线，支承管，支杆，撑脚，变压器铁芯柱，腿，支柱，柱脚
Leica total station　徕卡全站仪
leisure area　休养区
leisure　娱乐
LE(leading edge)　（叶片）进汽边，前缘，（脉

冲）上升边
LE(less than or equal to) 小于或等于
lemon spot 白点【指钢材缺陷】
lender 贷方，出借人，贷款人，债权人
lending bank 贷款银行
lending rate 贷款利率
lending 借贷
lend 出租，借（给、出），贷（款）
length adjustment 长度调节
length change 长度变化
length coefficient 长度系数
length control rod 长控制棒
length correction 长度改正
lengthened pulse 加宽脉冲，展长脉冲
lengthened 延长的，展长的
lengthening coil 加长线圈
lengthening of piles 桩的接长
lengthen 加长，延长
length equation 长度方程
length metrology 法制计量学
length of air gap 气隙长度
length of arc 弧
length of a scale division 分格间距
length of blade 叶片长度
length of break 总开距
length of cantilever 悬臂长度
length of conductor 导线长度
length of dam 坝长
length of delay 延迟值
length of embedment 埋入长度
length of engagement 啮合长度
length of heat 加热持续时间
length of intervals 间隔时间
length of last stage blade 末级叶片长度
length of lay 绞距长度
length-of-life test 寿命试验
length of line 线路长度，寿命
length of magnetic path 磁路长度
length of mean turn 线匝平均长度
length of operation 操作时间
length of path 程长
length of penetration 贯入度
length of pipeline 管线长度
length of pipe section 管段长度
length of pole arc 极弧长度
length of restraint 砌入长度【入混凝土】，嵌固长度，约束层深度
length of scale division 分格间距
length of shipping space 船位长度
length of slug 涌节长度
length of stroke 冲程，行程长度
length of the track at plant site 厂内铁路线长度
length of time 持续时间，期间
length of travel 行程
length of tubes between the tube plates 管板之间的管长
length of working cycle 工作周期，工作循环期
length overall 全长，总
lengthways gradient 纵向坡度
lengthways reinforce rib 纵向加劲肋
lengthways 纵向的，纵向地，南北向的，沿长度方向的
lengthwise oscillation 纵向振荡
lengthwise 沿长度方向的，纵向的
lengthy shut down period 长期停机时间
length 长度，记录长度，持续时间，字长，距离
lens gasket 透镜式垫片
lens opacity of eye 眼晶体混浊
lens power 透镜焦度
lens pyrometer 透镜高温计
lens ring gasket 透镜式金属环垫
lens turret 透镜旋转台
lens 镜头，透镜，物镜
lenticle 扁豆状体
lenticular arch 双叶拱
lenticular beam 扁豆形组合梁，鱼腹式梁
lenticular truss 鱼腹式桁架
lenticular 两面凸的，透镜状的
lentiform beam 扁豆形组合梁，鱼腹式梁
Lenz's law 楞次定律
Lenz's rule 楞次定则
lepeth cable 聚乙烯铅皮电缆
lepidocrocite 纤铁矿，碱式氧化铁
leptokurtic 尖峰的
Lerner index 勒纳指数
lerrite 黄绿云母
lesion 损坏，故障，伤害，疾患
less container load 散货拼箱
less-developed country 不发达国家
lessee 承租人，租户，租赁人
less-enriched uranium 欠富集铀
lessen 减少，衰减，缩小
lessor 出租人
lesspollution 无污染
less thunderstorm region 少雷区
letdown coolant 下泄冷却剂
letdown flow path 下泄流路线
letdown flow rate 下泄流量，排出率
letdown flow 下泄流量
letdown heat exchanger 下泄换热器，下泄热交换器
letdown line isolation valve 下泄管线隔离阀
letdown line 下泄管路【化学和容积控制系统】
letdown orifice 下泄节流孔板，下泄孔板
letdown pipe 下泄管
letdown stream 下行气流，下泄流
letdown valve 排放阀，下泄阀，减压阀
letdown vessel 下泄箱
letdown 下泄，排出，放下
let fly 突然展开，发射，攻击
lethal action 致死作用
lethal agent 致死剂
lethal concentration 致死浓度
lethal dose-50 time 50%剂量致死亡时间，半数致死时间
lethal dose 致命剂量，致死剂量
lethal effect 致死作用
lethal exposure 致死性照射
lethal irradiation 致死辐照
lethality 死亡率
lethal period 致死期

lethal radiation dose 致死辐射剂量
lethal 致命的，致死的
lethargy 衰减系数，对数能降，不活泼
let-in brace 嵌入式斜撑
LETS(linear energy transfer system) 能量线性传递系统
letter character 字母字符
letter code 字母代码
letterhead 公司或机构的信头，信笺上方的印刷文字，印有抬头的信笺，图廓注记
letter of abandonment 委付书
letter of acceptance 中标通知，中标函，承兑函
letter of acknowledgement 回函，确认函
letter of advice 通知书
letter of application 申请书
letter of appointment 聘书，委任书
letter of assignation 转让书
letter of assignment 转让书
letter of attorney 授权书，授权委托书，委托书
letter of authority 授权书
letter of authorization 授权委托书，授权函
letter of award 授标函
letter of bid invitation 投标邀请函
letter of bid 投标函
letter of cancellation 解约书
letter of clarification 澄清函
letter of commitment 承诺书
letter of confirmation 确认书
letter of counter guarantee 反担保书
letter of credit agreement 开发信用证约定书
letter of credit amount 信用证金额
letter of credit at sight 即期信用证
letter of credit expiration date 信用证有效期限
letter of credit term 信用证条款
letter of credit 信用证
letter of delegation 代理收款委托书
letter of deposit 抵押证书
letter of discount 折扣函
letter of entrustment 委托书
letter of guarantee for bid 投标保证书
letter of guarantee 保证书，担保函，保函
letter of hypothecation 抵押证书，质押书
letter of identity 身份证明信
letter of indemnity 赔偿保证书，认赔函，担保书，保函，认赔书，赔偿担保信，保险证书
letter of independence guarantee 独立保函
letter of inquiry 询问函，询价函，询价函件
letter of instruction （国内银行发给国外银行的）信用证通知书
letter of intention 意向书
letter of intent 合同的草约，意向书，议向书
letter of introduction 介绍信
letter of invitation to tender 投标邀请函
letter of invitation 邀请信
letter of licence 延期索偿同意书，延期索债函
letter of lien 留置函，扣押权书，留置权证书
letter of notice 通知单，通知书
letter of patent 专利证
letter of proxy （对代理人的）委托书
letter of ratification 批准书
letter of recommendation 介绍信，推荐信，推荐书信
letter of reference from correspondence bank 开户行资信
letter of reference 保荐书
letter of subrogation 权益转让书，权益委托书
letter of tender 投标函
letter of understanding 谅解书
letter of undertaking 担保书，承诺书
letter patent 专利特许证
letter punch 印记冲模，钢字码
letter shift 变换字母，换字母档
letter signal 字母符号，下段符号
letter telegram 书信电报
letter transfer 信汇
letter type code 字母型代码
letter with notice of delivery 收据，回执
letter 信，函件，字母，证书，许可证，书信，出租人
let-through current characteristic 允通电流特性
let-through current 允通电流，故障时通过电流
letting rate 出租率
LET value 传能线密度值，线能量转移值
let 出租，让，允许
leucine 亮氨酸
leucite 白榴石
leucocratic dyke 淡色岩脉
LEU(less enriched uranium) 欠富集铀
LEU(low enriched uranium) 低浓缩铀，低富集铀，低加浓铀
levee back 堤防背面，堤背
levee base 堤底
levee body 堤体
levee breach 土堤决口
levee building machine 筑堤机
levee crest 堤顶
levee crown 堤顶
leveed bank 堤岸
leveed pond 堤成池
levee footing 堤防底脚
levee foundation 堤基
levee front 堤防临河面
levee gate 堤坝闸门
levee grade 堤顶纵坡
levee maintenance 堤防维护
levee management 堤防管理
levee protection 堤防保护
levee raising 堤坝加高
levee ramp 堤上坡道
levee revetment 堤防护岸
levee slide 堤身滑坡
levee slope 堤坡
levee sloughing 堤身崩坍
levee sluice 堤上泄水闸
levee spacing 堤间距
levee system 堤防系统
levee toe 堤脚
levee undermining 堤底淘刷
levee widening 堤坝培厚，堤坝加宽
levee 堤，堤坝，堤防，防洪堤，码头
level adjustment 电平调节
level alarm-high 高位警报

level alarm-low 低位警报
level alarm 液面报警
level and straight track 平直线路
level bar 水平杆，水平尺
level book 水准手簿
level bubble 水准气泡
level capacity 满平容积
level change 电平变换
level check 校验水平，水准检测
level class gate lifting device 杠杆类闸门启闭设备
level compensator 电平补偿器
level-compound excitation 平复激，平复励
level-compound excited motor 平复激电动机
level constant 水准仪常数
level control device 液位控制装置
level controller 液位控制器，液位调节器，位面控制器
level control valve 水位调节阀
level control 水位控制，电平控制，位面控制，水位调节，水准控制
level converter 电平转换器
level cross country 平原交切地区
level crossing 平交道（口），平面交叉，水平交叉
level detection 水平检测，电平测试
level detector 液位探测器，电平探测器，电平指示器，水位探测器
level deviation 水准器气泡偏差
level device 料位计
level diagram 电平图
level electrode 液位电极
level equalization 位面稳定，电平稳定
leveler 水平仪
level fall 水位下降
level fluctuation 电平变化，电平波动
level-full capacity 总容积
level gauge 料位计，液面计，水准仪，水位计，水平规，水位指示器，水平仪，液位计
level gauging 物位测量，液面测量，料位测量，级位检验
level governor 电平调节器，水平调节器
level ground 平地
level holding pipe 液位保持管
level holding system 液位保持系统
level holding tank 液位保持箱
level holding 液位保持
level indicator 液位指示器，液位计，液面计，电平指示器，料位计，水位表，水位指示器
level indicator 液位指示器，液位指示
leveling and dressing of pile location 打桩处平整与修整
leveling base 基准面
leveling coat 找平层
leveling course 找平层
leveling-culvert outlet 水面平衡涵洞出水口
leveling 水准测量，测平，平整，校平，找平
level instrumentation nozzle 液位计接管
level instrument 位面计，水平仪，水准仪
levelization 归一化，平准化
levelized annual cost 均衡年成本
levelized cost 平准化成本
levelized discounted electricity generation cost 平准化贴现发电成本
levelized energy cost 平准化电能成本
levelizing factor 平准化因子，均衡因子
levelled plant area 整平的厂区
levelled up 整平
leveller 水准测量人员，水平仪，校平仪，平土器，平地机，整平器，矫直机
level line 水平线，水准线
levelling adjustment 水准测量平差
levelling base 水准基点
levelling blanket 找平层
levelling block 校准垫片，校准垫块，校准平台
levelling course 找平层
levelling culvert 水面平衡涵洞
levelling instrument 水准器，水准仪，水平仪
levelling machine 平土机，矫直机，矫平机
levelling of the site 现场的整平
levelling plate 校准垫块，校准垫片
levelling point 水准点
levelling screw 水平调准螺丝，校准螺钉，校平螺钉，准平螺钉
levelling 调平，平土，平整，水准测量，校水平，校平，平整场地
levelling staff 水准尺，水准标杆
level luffing crane 平臂起重机，水平变幅起重机，平伸式起重机
level luffing 水平变幅
level-man 水准测量人员，水准手
level measurement nozzle 水位测量管嘴
level measurement tap 测液面管口
level-measuring set 水平仪，水位计，电平表
level meter 水位计，液面计，水平仪，电平表，电平指示器
level moving plateform 平移台
levelness 水平度，平整度
level network 水准网
level net 水准网，水准线网
level of activity 活度水平
level of addressing 定址级数
level of a field quantity 场量级
level of air quality 空气品质级
level of a power quantity 功率量级
level of cavitation 汽蚀程度
level of dense bed 密相床层料面
level of efficiency 效率高低，效率水平，有效度
level of electricity rate 电价水平
level off 整平，矫直，校水平，趋于平衡
level of illumination 照明水平
level of innovation 创新水平
level of living 生活水平
level of loudness 响度级
level of materials 料位
level of mechanization 机械化水平
level of medium 料位
level of noise 噪声级
level of pollution 污染水平
level of radio activity 放射性水平
level of residue 渣量
level of sensitivity 感觉程度

level of significance 有效级，显著性水准，有效水平
level of stability 稳定水平
level of subsoil water 地下水位
level of technical expertise 技术水平
level of upper pond 上池水位，上游水库水位
level of vibration 振动级，振动水平
level of wave-base 波基面，浪基面，浪蚀基面
level of zero wind （有效）零风面
level overloading 电平过载
level plane 水准面
level position 水平位置
level pressure control 基准压力调节
level probe 料位探头，液位探头
level recession 水位下降
level recorder 电平记录器，水位记录器，水位记录仪，自记水位计
level reduction 水准折算
level regulator 电平调节器，液位调节器，料位调节器
level retrogression 水位消落
level rise 水位上升
level 3 schedule 三级（网络）计划
level 1 schedule 一级网络计划
level security factor 耐压水平因子
level sensing device 液面传感器
level sensitivity 气泡灵敏度
level sensor 电平传感器，水位传感器
level shoe 水准尺垫
level span 等高档
level staff 水准标尺
level surface 水平面，水准面
level survey 水准测量
level switch （信号）电平开关，液位开关，水位开关，料位开关
level terrace 水平阶地
level terrain 平坦地形
level testing instrument 水准检定器
level theodolite 水准经纬仪
level-tipped closure method 平堵截流法
level transmitter 水位传感器，液位变送器，液位传感器，物位变送器
level trier 水准检定器
level tube axis 水准管轴
level tube 水准气泡管，水准管
level-up 找水平，使平整，平衡
level vial 水准仪器泡
level width 能级宽度
level 0 零标高【地面】
level 能级，程度，水平面，液位，电平，水平，级，水准，标高，标准，级别，水位，水准器，水准仪，料位，整平，定坡度，层次
leverage 力臂比，杠杆机构，杠杆作用，杠杆臂长比，扭转力矩，利用，手段，影响力
lever and weight type valve 杠杆重锤式阀
lever arm recording flowmeter 杠杆记录式流量计
lever arm 杠杆臂
lever assembly 杠杆组件
lever balance 杠杆天平
lever block 手扳葫芦
lever boards 活动气窗
lever box 联动柄箱
lever control 操纵杆控制，杠杆操纵
lever fulcrum 杠杆支点
lever gear door 升降门
lever jack 杠杆（式）千斤顶
lever lift 杠杆起重机，杠杆提升机
lever lock 杠杆锁
lever-operated bending machine 杠杆操作的弯曲机
lever-operated knife switch 杠杆操纵闸刀开关
lever-operated shear 杠杆操作的剪刀
lever-operated valve 杠杆操纵阀
lever-operated 杠杆操作的
lever switch 杠杆开关
lever type gate lifting device 杠杆式启门设备
lever type starter 杠杆式启动器
lever valve 杠杆阀
lever 杆，杠杆，手柄，控制杆
levigate 磨光，研磨，澄清，粉碎
levin 电闪，雷凌
levitability 漂浮性
levitated fluidization 漂浮流态化
levitation of the graphite spheres 石墨球漂浮
levitation safeguard 漂浮安全装置，防浮
levitation 飘浮，悬浮，漂浮
levorotatory 左旋的
levy a fine 罚款，征收罚金
Levy's criteria 利维准则
levy tax 征税
levy 征收，征收额，征税
lewis anchors 起重爪，吊楔
lewis bar 吊楔杆
lewisson 地脚螺栓，吊楔
lewis 地脚螺栓，吊楔
lex causae 准据法
LFBR(liquid fluidized bed reactor) 流化床反应堆
LFC(laminar flow control) 层流控制
LFC(load frequency control) 负荷频率调节
LFC(low frequency correction) 低频校正
LFC(low frequency current) 低频电流
LFD(low frequency disturbance) 低频干扰
LFF(low frequency filter) 低频滤波器
LFR(low flux reactor) 低通量反应堆
LFS(logic file system) 逻辑文件系统
LFV(low frequency vibration) 低频振动
LGH 6.6kV 配电系统【核电站系统代码】
L/G(letter of guarantee) 保证书，保函
LG(level gauge) 液位计
LG(line generator) 向量产生器
LGL(logical left shift) 逻辑左移
LG(low gear) 低速齿轮
LGR(logical right shift) 逻辑右移
LHD(load-haul-dump unit) 装运卸机，装运卸联合设备
LHGR(linear heat generation rate) 线功率，线（生）热功率
LHH 6.6kV 应急配电系统【核电站系统代码】
LH(latent heat) 潜热
LH(left-handed) 左侧的，左旋的
LHSI(low head safety injection) 低压安全注射，

低压安注
LHSI pump 低压安全注射泵
LHS(left-hand side) 左侧，左边
LHSP(low head safety injection pump) 低压安全注射泵
LHST(laundry and hot shower tank) 洗衣和热淋浴水箱
LHV(low heat value) 低热值，低位发热量
LHZ 380V交流发电机组【核电站系统代码】
liabilities and capital 负债与资本
liabilities dividend 负债股息
liabilities 负债
liability and responsibility 负债与责任
liability between partners 合伙人之间的责任
liability certificate 负债证明书
liability clause 责任条款
liability company 有限公司
liability for acceptance 承兑责任
liability for breach of contract 违约责任
liability for compensation 赔偿责任
liability for damage 损失赔偿责任
liability for delay 误期责任，延付责任
liability for fault 过失责任
liability for loss 损失责任
liability for nonperformance of an obligation 不履行义务引起的责任
liability insurance premium 责任保险费
liability insurance 责任保险，责任险
liability of acceptance 承兑责任
liability period 责任期
liability reserve funds 责任准备金
liability risk 负责风险
liability to cracking 易裂性
liability to frost damage 易发生冻裂损害
liability to weathering 易风化性
liability 责任，义务，债务，职责
liable for damage 对损坏应负责任的
liable for the charges 对这些费用负责
liable scope 责任范围
liable to 易受……
liaison group 联络小组
liaison meeting 联络会议
liaison office 联络处
liaison person 联络人，联系人
liaison 联络，联系
liberal profession 自由职业
liberal wages and benefits 优惠待遇
liberal 自由的，足够的
liberate 放出，释放，游离，析出，使脱离，使游离，使逸出
liberation of gas 气体释放，放气
liberation 放出，释放，脱离，游离，逸出，解放，释出
LIBOR(London inter-bank offered rate) 伦敦（银行）同业拆借利率，伦敦银行间拆放款利率
librarian 图书管理员，程序库管理员，库管理程序
library addition and maintenance point 程序库补充及维护点
library of cross section 截面数据库
library of subroutine 子程序库
library routine 库存程序，库程序
library software 库存软件
library subroutine 库子程序
library tape （程序）库带
library 库，图书馆，程序库，文件集，文库
libration 天平动，振动，摆动，平衡，平均
licence agreement 许可证协议
licencee event report 持许可证者的事件报告
licencee 执照持有者，许可证持有者
licence 许可证，许可，特许，执照，批准
licencing system 许可证制度
licensability 取得许可证的可能性
license agreement 许可证协议
license authority 审批部门，管理机关
license contract 许可证合同
licensed architect 注册建筑师
licensed contractor 注册承包商
licensed engineer 注册工程师
licensed material 领有许可证的材料
licensed pressure 容许压力
licensed technology 许可转让的技术
licensee event report 持许可证者的事件报告
license expense 发给许可证费用
license fee 执照费
license-issuing authority 发证机关
license 执照，许可证，许可，特许，批准
license on a case-by-case basis 逐项核发的许可证
license plate 牌照
licenser 出让方
license system for electric power business 电力业务许可证制度
license tax 执照税
licensing agreement 专利权使用协议，许可证贸易协议，许可证协议，特许权协议
licensing authonty 审批部门，管理机关
licensing document 许可证申请文件
licensing operations 许可证业务
licensing procedure 许可证审批程序，许可证批准程序
licensing system 许可证制度
licensing 许可证审批，执照审批，许可，特许，审批
licensor 发执照者，发许可证者，许可方，出让方
lichen 青苔
lick 冲洗，吞没
LIDAR(light detection and ranging) 光波探测与测距
lidar 激光雷达
lid tank 遮盖水箱
lid 帽，盖，罩，制止，取缔，盖子
lien letter 留置证书
lien （货物）抵押，留置权
lierne vaulting 扇形肋穹顶
lie 位于，躺，位置，状态
life assessment 寿命评价
life belt 安全带，保险带
life-boost cathode 耐久阴极
life cage 电梯箱
life-characteristic 寿命特性

life consumption 寿命损耗，寿命消耗，寿命折损
life curve 寿命曲线
life cycle cost 寿命周期成本（费用）
life cycle test 交变载荷耐久试验
life cycle time 生命周期时间
life cycle 产品使用寿命，寿命周期，生存周期，产品寿命
life cycling test （循环加载的）疲劳寿命试验，耐久性试验
life distribution 寿命分布
life duration 耐用寿命，寿命期限
life evaluation 寿命评价
life expectance 概率寿命，预期寿命
life expectancy 概率寿命，预期寿命，预期使用期限
life expenditure 寿命消耗，寿命损耗
life exponent 寿命指数，寿命指标
life extension 寿命延长
life insurance 人寿险
lifeline earthquake engineering 生命线地震工程
lifelong education 终身教育
life management 寿命管理
life net 救生网
life of equipment 设备寿命
life of loan 借款期限
life of product 产品寿命
life of winding 绕组寿命
life pack 救生袋
life performance 寿命特性
life period 寿命期，存在时间，存在时期
life pile 周转煤堆
life prediction 寿命预测
life protection GFCI breaker 生命保护接地故障断路器
life rope gun 射绳枪
life size 和实物一样大小，原尺寸
life sliding pole 救生滑竿
life-span determination 寿命测定
life-span study 寿命研究
life-span 使用期限，寿命，存在时间
life test （使用）寿命试验
life time body burden 终身体内积存剂量
life time dilatation 寿命延长
life time dilation 寿命延长
life time dose 终身剂量
life time expenditure 寿命消耗
life time load factor 寿期负荷因子，累计负荷因子
life time of reactor core 反应堆堆芯寿期
life time service 长期使用
life time 使用期限，寿命，试验时间
life zone 生物带
life 生活，实物，使用寿命，寿命，使用期限，生命
LIFO(last in, first out) 后进先出
LIFO(liner in, free out) 船方不负担卸货费但负担装货费
lift air 抬升用空气
lift-and-carry mechanism 提升移送机构
lift associated admittance 升力导纳，升力相关导纳
lift bolt 提升螺栓
lift bridge 升降桥
lift car annunciator 电梯位置指示器
lift center 升力作用点，升力中心
lift check valve 升启式逆止阀，升降式止回阀，安全排放阀
lift coefficient 升力系数
lift coil 提升线圈【控制棒驱动机构】
lift component 升力分量
lift controller 电梯控制器
lift convergence 升力减小
lift counter 往返行程计数器
lift curve slope 升力曲线斜率
lift curve 升力曲线
lift dependent drag 诱导阻力
lift diameter ratio 升力直径比
lift direction 升力方向
lift dissymmetry 升力不对称
lift distribution 升力分布
lift-drag ratio 升力阻力比，升阻比
lift effect 升力效应
lift efficiency 升力特性
lift electromagnet 起重电磁铁
lift equation 升力方程
lifter bracket 吊环螺栓
lifter of plain gate 平面闸门启闭机
lifter of radial gate 弧形闸门启闭机
lifter 提升机，起重机，升降机，电磁铁的衔铁，启闭机，启门机，升降器
lift expectancy 预期使用年限
lift force 提升力，升力
lift gate 升降式闸门
lift head 扬程
lift hook 吊钩
lift increment 升力增量
lift induced drag 升力诱导阻力
lifting action of blade 叶片的升力作用
lifting and erecting engineering 吊装工程
lifting and erection 吊装
lifting and placing of generator rotor 发电机穿转子
lifting and suspending 吊把【指线圈的下线工艺】
lifting appliance 提升装置，起重装置
lifting assembly 提升装置
lifting bail 吊耳，吊环【燃料组件】
lifting beam 起吊杆，启门梁，起重梁
lifting block 起重滑轮
lifting body 升力体
lifting capacity 起吊能力，起重容量，升力，提升能力，提升量，起重能力
lifting chain 起重链索，提升铰链
lifting collar 吊环
lifting column 升降柱
lifting curve （凸轮）提升曲线
lifting device 起吊设备，吊具，启门设备，起吊装置，提升设备
lifting drawing 起吊图
lifting electromagnet 起重电磁铁
lifting equipment 起吊设备
lifting eye 吊装孔，吊耳，吊眼，吊环

lifting facilities 起吊设施，起重设备
lifting force 上升力，提升力
lifting frame 起吊架，起吊框架，翻倒装置，吊架
lifting gantry （龙门）起重机架
lifting gate feeder 提板式加料机
lifting gate 提升式闸门
lifting gear 升降起重联动装置，升降装置，起重装置，启门机构，起吊机构，提升铰链
lifting height 提升高度，起吊高度
lifting hole 起吊孔
lifting hook 起吊钩，吊钩，提升钩
lifting jack 起重器，千斤顶
lifting jet 喷气提升机
lifting large parts 大件吊装
lifting lever 提升杆
lifting line theory 升力线理论
lifting line 升力线
lifting load 上举荷载
lifting loop 吊环
lifting lug 提升套（吊耳），起吊耳，吊环，吊耳
lifting magnet 起重磁铁
lifting mark 起吊标志
lifting moment 升力矩
lifting motor 起重电动机
lifting movement 上升运动
lifting of reservations 解除保存【关于质保】
lifting oil pressure 顶起油压
lifting pipe 引上线用管，上升分线管
lifting plane theory 升力面理论
lifting platform 升降平台
lifting power 启门力，提升力
lifting pressure oil pump 升压油泵
lifting program 吊装程序，起重程序
lifting pump 升液泵，提升泵，顶起泵
lifting reentry 升力重返
lifting rigging 吊装索具
lifting rope 吊绳，吊索
lifting shelf 托升板
lifting speed 提升速度
lifting surface theory 升力面理论
lifting surface 升力面
lifting tackle 起重滑车
lifting test 提升试验
lifting tool 起吊工具
lifting trolley 起吊滑架，起吊小车
lifting trunnion for erection 安装用起吊耳轴
lifting trunnion 起吊耳轴
lifting turbine cover 上汽缸起吊
lifting turbine rotor （汽轮机）转子起吊
lifting up 抬吊
lifting valve 提升式阀
lifting vortex 升力（旋）涡
lifting weight 吊装重量
lifting well 起吊孔
lifting winch 提升绞车，卷扬机
lifting with three cranes 三机抬吊
lifting worker 起重工
lifting （燃料元件的）提升，举起的，提升的，起吊，举起
lift irrigation area 提灌面积
lift line 上升管，立管
lift loading 升力分布
lift lug 吊耳，提升套
lift motor 电梯用电动机，起重电动机
lift-off speed 松开转速，启动转速
lift-off voltage 起动电压
lift-off 顶离
lift of pump 泵扬程，泵压头，泵的扬程
lift of valve 阀门升程
lift opening for assembled pieces 锅炉组件吊装口
lift oscillator model 升力振子模型
lift penthouse 电梯顶楼
lift pipe 提升管
lift point 升力点
lift pole 提升杆【控制棒驱动机构】，提升磁极
lift pot 提升罐
lift principle 升力原理
lift pump 升液泵，提升泵，顶起泵
lift range 升力变化范围
lift ratio 升力系数
lift ring 升力圈
lift shaft 电梯间，电梯井，升降机井，电梯竖井，电梯井道，升降机槽
lift slab construction 升板法施工
lift station 扬水站
lift thickness 铺土厚度
lift truck 起重汽车，起重车
lift tube 升水管
lift type device 升力型装置
lift type rotor 升力型风轮，升力型转子
lift valve 提升阀
lift vector 升力矢量
lift velocity 上升速度
lift well 电梯井，电梯间
lift wire 跳线
lift 起吊，提升，电梯，升降机，扬程，吸入高度，吊装，起重机，上升，上升高度，升高，升力，抬升，提高，提升力，提升装置
ligament stress 带状应力，孔桥带应力
ligament 管孔带，细线，韧带，孔带
light-absorption method 光吸收法
light accessible transistor matrix 光存取晶体管矩阵
light-activated element 光敏元件
light-activated SCR（semiconductor controlled rectifier） 光控半导体可控整流器
light-activated switch 光敏开关
light-activated thyristor 光触发晶闸管
light-actuated SCR（silicon-controlled rectifier） 光控可控硅整流器，光控晶闸管整流器
light-addressed light valve 光寻址光阀
light aggregate 轻质骨料，轻骨料，轻集料
light air 高空大气，软风，一级风
light alloy casting 轻合金铸件
light and acoustic signal 灯光音响信号
light application time 光照时间
light armoring 轻铠装
light ash 疏松灰
light beacon 灯标，灯光信号，灯塔
light beam oscillograph 光线示波器

light beam recording oscillograph 回线示波器，光束记录示波器
light beam 光束
light breeze 二级风，轻风
light bulb and torus type containment 带环状冷凝室的灯泡型安全壳
light bulb reactor 灯泡型反应堆
light bulb torus containment 灯泡环型安全壳
light buoy 发光浮标，浮标灯
light clay 轻黏土
light coal 气煤，轻煤，瓦斯煤
light-coated electrode 薄涂层电极，药皮焊条
light colour 浅色
light compaction instrument 轻型击实仪
light concentrating collector 聚光集热器
light concentration limit 聚光极限
light concentration ratio 聚光比
light concentrator 聚光器
light concrete structure 轻质混凝土结构
light condition 空载状态
light construction 轻型建筑，轻型结构
light-coupled semiconductor switch 光耦半导体开关
light crude oil reference condition 轻原油基准条件
light crude oil 轻原油
light current engineering 弱电工程
light current relay 弱电继电器
light current system 弱电控制系统，弱电系统
light current 弱电流，光电流
light dependent resistor 光敏电阻器
light detection and ranging 光波探测与测距
light diffuser 灯光扩散器，发散光，散光器
light duration 光照延续时间
light-duty building 轻型建筑
light-duty derrick 轻型吊杆装置
light-duty design 轻型化设计，轻小型化设计
light-duty electrical cable 轻载电缆
light-duty lathe 轻型车床
light-duty live center 轻型回转顶尖
light-duty power manipulator 轻型动力机械手
light-duty rack 轻型货架
light-duty scaffold 轻型的脚手架
light-duty tension string （进出线挡用的）轻荷耐张串
light-duty truck 轻型载重卡车
light-duty 小功率工作状态，小功率的，轻级，温和条件下的，轻负荷，轻型的，轻负载的
lighted push button 灯光按钮
lighted rotary pushbutton switch 带灯旋转按钮开关
lighted rotary switch 带灯旋转开关
lighted switch 带灯开关
light effect 光效应
light electric bulb 白炽灯
light-emitting diode display panel 发光二极管显示牌
light-emitting diode 电致闪光二极管，发光二极管
lightening conductor 避雷针
lightening impulse voltage test 冲击电压试验
lightening rod 避雷针，电极棒
lighten 缓解
lighterage clause 驳运条款
lighterage pier 车船间货物驳运码头
lighterage 驳船费，驳运费，驳运
lighter fee 驳船费
lighter insurance 驳船险
light erratum 微小误差
lighter's wharf 驳船码头
lighter 打火机，点火器，引燃器，照明器，驳船，更轻，发光器
light excitation 光激励
light exposure 小剂量照射
light extinction method 消光法
light field 亮区，明场
light filter material 轻质滤料
light filter 滤波器，滤色镜，滤光板，滤光镜
light fission fragments 轻裂变碎片
light fixture 灯具
light flicker 闪变
light float 灯标船
light flux 光通量
light fog 轻雾
light fragment 轻裂变碎片
light fuel oil 轻质燃油
light fuel 易挥发燃料，轻质燃料
light fuse 灯用保险丝
light gauge cold-formed steel shape 冷弯薄壁型钢
light gauge cold formed steel structural member 低温成形轻量型钢构件
light gauge plate 薄钢板
light gauge railway 轻轨铁道
light gauge sheet 薄钢板
light gauge wire 细号线
light guide 光导向设备，光控制，光制导，光波导
light gun 光笔，光（电子）枪
light hole 水冲穴
lighthouse due 灯塔税
lighthouse 灯塔，灯室，灯台
light hydrogen 轻氢
light icing area 轻冰区
light illuminant 光照明器，光源
light impulse 光脉冲
light indicator 灯光指示器
light industrial district 轻工业区
light industry 轻工业
lighting battery 照明蓄电池
lighting box 照明箱
lighting cable 照明电缆
lighting circuit 照明电路
lighting conductor 避雷器，避雷针
lighting consumer 照明用户，电光用户
lighting consumption 照明耗量
lighting control relay 照明控制继电器
lighting control 照明控制
lighting current 照明电流
lighting demand 照明需量
lighting device 照明装置
lighting diagram 照明图
lighting distribution board 照明配电盘

lighting door　点火孔
lighting efficiency　照明效率
lighting engineering　照明工程
lighting equipment　照明设备
lighting facilities　照明设施
lighting feeder　电光馈路，照明馈路
lighting fixture　照明器材，灯具
lighting fuse　灯用保险丝
lighting gap　避雷器放电间隙
lighting generator　照明用发电机
lighting installation　照明装置
lighting lamp　照明灯
lighting line　照明线路
lighting load　照明负荷，照明负载
lighting mains　照明网络，照明干线
lighting mechanism　照明机具
lighting meter　电光电度表，照明电度表
lighting network　照明网络
lighting-off torch　点火棒，点火炬，点火喷燃器
lighting-off　点火
lighting board　照明配电盘
lighting panel　照明控制板
lighting peak　照明峰荷
lighting protection　雷击防护
lighting standard　照度标准
lighting storm　雷暴
lighting switchgear　照明开关装置
lighting switch　照明开关
lighting system　照明方式，照明制度，照明系统
lighting tariff　照明费率，照明电价表
lighting time　有效光照时间
lighting tower fire vehicle　照明消防车
lighting tower of coal-yard　煤场照明塔
lighting transformer　照明变压器，照明变
lighting-up burner by pulverized coal　煤粉点火燃烧器
lighting-up burner　点火喷燃器，点火燃烧器
lighting-up rate　点火速度，点燃速度
lighting-up　点燃，开灯
lighting wiring　照明布线
lighting　照明的，照明，点灯，采光，照明设备，点亮，灯光
light intensity fluctuation　光强波动，光强起伏
light intensity meter　照度计，光强计
light intensity　光强度，光强，光度，亮度
light level　亮度级
light list　灯标位置表
light load adjustment　轻载调节
light load compensating device　轻载补偿装置
light load compensation　轻载补偿
light loading district　轻载区
light loading　轻负载
light load period　轻负荷期间，轻载期间
light load range　（核电站）低负荷区段
light load test　轻载试验
light load　轻负荷，轻载
lightly coated electrode　薄涂焊条
light marker　灯标
light material enclosure　轻型材料围护结构
light metal spirit level　光金属水平仪
light meter　照度计，光度计，曝光表，轻金属
light mild clay　轻亚黏土
light modulation detector　光调制解调器
light modulator　光调制器
lightning arrester characteristic　避雷器特性
lightning arrester gap　避雷器间隙
lightning arrester terminal box　避雷器接线盒
lightning arrester testing bridge　避雷器试验用电桥
lightning arrestor　避雷针，避雷装置，避雷器
lightning belt　避雷带
lightning center　雷电中心
lightning conductor　避雷装置，避雷针
lightning current meter　雷电电流计
lightning current　雷电波，雷电流
lightning detecting　雷电探测
lightning detector　雷电探测仪
lightning discharge　闪电，雷闪放电
lightning disturbance　雷害，雷电事故
lightning-diverting cable　分雷电缆
lightning eliminator　消雷器
lightning fault　雷电故障
lightning flash-over　雷电闪络
lightning gap　（避雷器）放电间隙，放电空隙
lightning generator　脉冲振荡器，人工闪电发生器
lightning guard　避雷器
lightning impulse voltage test　雷电冲击电压试验
lightning impulse voltage　雷电冲击电压
lightning impulse withstand voltage　雷电冲击耐受电压
lightning impulse　闪电脉冲
lightning location system　雷电定位系统
lightning overvoltage　雷电过电压，闪电过电压
lightning parameter　雷电参数
lightning performance　雷电特性，雷电性能
lightning protection engineering　避雷技术，避雷工程学
lightning protection grounding net　防雷地线网
lightning protection ground　避雷保护接地
lightning protection module of isolation transformer　隔离变压器防雷模块
lightning-protection net　避雷网
lightning protection of distribution network　配电网防雷
lightning protection of WTG　风电机组防雷
lightning protection rod　避雷针，防雷杆
lightning protection system　防雷系统
lightning protection transformer　避雷变压器
lightning protection zone　防雷区
lightning protection　防雷接地，防雷保护，避雷保护，防雷，雷电保护
lightning-protective cable　避雷电缆
lightning protector plate　避雷板
lightning protector　避雷装置，防雷保安器，避雷器，放电器
lightning recorder　雷电记录器，雷击记录器
lightning rod　避雷针
lightning shielding　避雷
lightning spike　避雷针
lightning storm　雷暴
lightning strike　雷击，闪电攻击

lightning stroke current 雷击电流
lightning stroke recorder 雷电记录器，雷击记录器
lightning stroke 雷击
lightning surge wave 雷电冲击波，雷电波
lightning surge 雷涌，雷电过电压，雷电冲击
lightning survey station 雷电观测站
lightning survey 雷电观测
lightning switch 避雷开关
lightning trip-out rate 雷电跳闸率
lightning trouble 雷电事故
lightning wave front shape 雷电波头形状
lightning wire 避雷线
lightning withstanding level 耐雷水平
lightning 闪电般的，闪电，电光雷电，快速的，雷电
light-off mode 点火方式
light-off temperature 起燃温度，熄灯温度，起火温度
light-off time 起燃时间，启动时间
light-off 点火，启动，点燃
light oil store 轻油库
light oil 轻油
light-operated switch 光控开关
light or temperatrature resources 光温资源
light output 光输出，发光效率
light overhaul 小翻修
light partition wall 轻质隔墙
light pen 光笔
light phase 轻相
light pipe 光导管，光导向装置
light pneumatic drilling machine 轻型风动式钻机
light pollution 光污染
light porous clay brick 轻质多孔黏土砖
light power motor 小型电动机，小功率电动机
light-press fit 轻压配合
light-proof 不透光的，遮光的
light quantum maser 光量子放大器
light railway van 轻轨车
light railway 轻便铁路
light rail 轻便轨
light rain 小雨
light rating 轻载运行，低功率
light ray 光线
light-reflectance apparatus 反光仪
light-regulator 灯光调节器，照明调节器
light relay 光继电器
light requirement 需光
light resources 光资源
light resource utilization 光资源利用
light rig 轻型钻机
light running 轻载运行，无载运转，单机运行
light run 空转
light saturation 光饱和
light-scattering method 光散射法
light-scattering photometer 光散射光度计
light sensation 感光，光敏
light sensitive cell 光敏电池
light sensitive detector 光敏探测器
light sensitive diode 光敏二极管
light sensitive relay 光敏继电器
light sensitive resistor 光敏电阻
light sensitivity 光敏度
light shaft 采光井
light ship 灯标船
light shower 小阵雨
light signal 灯光信号
light silty loam 轻粉质壤土
light silty sand-silt 轻粉质砂壤土
light socket 灯口
light sounding test blow count 钎探击数，轻便触探试验的锤击数
light sounding test 轻便触探试验【原位测试】
light source colour 光源光色
light source 光源
light spectrum 光谱
light splitter 分光镜
light-spot galvanometer 光点电流计
light start 轻载启动
light-textured soil 轻质土
light-tight box 暗箱
light transmission method 光透法
light transmission technique 光传输技术
light transmittance 光透射比
light-triggered alarm 光触发报警器，光通报警器
light truck 轻便货车，轻型货车，轻型卡车
light up plug 点火塞，火花塞
light up 点亮，点火，点灯，开灯，照亮，着火
light vessel 灯标船，灯船
light volatile impurities 容易挥发的杂质
light water breeder reactor 轻水增殖反应堆，轻水增殖堆
light water coolant 轻水冷却剂
light-water-cooled graphite moderated reactor 石墨慢化轻水冷却反应堆
light-water-cooled reactor 轻水冷却反应堆
light-water-cooled 轻水冷却的
light water graphite reactor 轻水石墨反应堆，轻水石墨堆
light water hybrid reactor 轻水混合反应堆，轻水混合堆
light water infiltration 轻水渗入（重水回路），轻水漏入（重水回路）
light water inleakage 轻水漏入（重水）
light-water-moderated reactor 轻水慢化反应堆，轻水慢化堆
light water PWR 轻水压水堆
light water reactor power plant 轻水反应堆核动力厂，轻水堆核电站
light water reactor 轻水堆
light water 轻水，普通水，光波
lightweight aggregate concrete 轻骨料混凝土
lightweight aggregate 轻质骨料
lightweight cladding 轻型维护结构，轻量级包层
lightweight concrete block 轻质混凝土块
lightweight concrete partition slab 轻型隔墙板
lightweight concrete 轻混凝土，轻质混凝土
lightweight diesel engine 轻型柴油机
lightweight gas turbine 轻型燃气轮机
lightweight generator 轻型发电机
lightweight grader 轻型平地机

lightweight partition board 轻质隔墙板
lightweight partition wall 轻质隔墙
lightweight partition 轻质隔墙
lightweight power unit 轻便动力设备
lightweight refractory 轻质耐火材料
lightweight shield 轻屏蔽体
lightweight steel construction 轻型钢结构
lightweight structure 轻型结构
light weld 凹形焊，浅填焊接
light well 采光井
light wind 轻风【二级风】，小风
light work 轻作业
light-zone depth 光亮带深度
light 点亮，光，光线，照明，灯，信号灯，指示灯，光指示，淡色，发光的，轻的
ligitation history 诉讼历史
ligneous coal 褐煤，木质煤
lignify 木质化
lignin 木质素，木质
lignite breeze 褐煤屑
lignite coal 褐煤
lignite pitch 褐煤沥青
lignite wax 褐煤蜡
lignite 褐煤
lignitic 褐煤的
lignitiferous 褐煤化的
lignitous coal 褐煤
lignocellulosic board 木质纤维板
lignosa 木本植被
ligroin 粗汽油，石油英
like electricity 同号电，同性电
likelihood function 似然函数，逼真函数
likelihood 可能，似然，逼真，可能发生的事物
like pole 同性极，同名极
Lil detector 碘化锂探测器
limacon 蚶线，蜗牛形曲线
limberoller conveyor 柔性滚轴输送机
limb of electromagnet 电磁铁芯
limb 插脚，零件，部件，分度盘，铁芯柱，(树的)大枝，翼
lime and cement mortar 石灰水泥砂浆
lime-ash flooring 灰渣石灰铺面
lime-base grease 石灰基油脂，钙基润滑脂
lime-base process 石灰法
lime-bearing waste 含石灰废物
lime bin 石灰库，石灰仓
lime biological treatment 石灰生物处理
lime blow-in process 石灰吹入工艺
lime brick 石灰砖
lime brush 石灰浆刷
lime burner 石灰窑
lime burning kiln 石灰窑
lime burning 石灰煅烧
lime-calcination reactor 石灰煅烧器
lime carbonate 石灰石
lime cement mortar 水泥石灰砂浆
lime cement ridge tile 石灰水泥脊瓦
lime cement 石灰胶结料
lime cinder 石灰焦碴
lime classifier 石灰消化分离器
lime coagulation 石灰凝结
lime column 石灰柱
lime concrete 石灰混凝土，三合土
lime concretion 石灰结核，石灰核
lime content 石灰含量
lime cream 石灰乳液，石灰乳
lime deposit 石灰质沉积物
lime dry process 干石灰法
lime dust 石灰粉尘，石灰粉
lime feldspar 钙长石
lime gypsum process 石灰石膏工艺
lime hard pan 石灰硬盘
lime hydrate 消石灰，熟石灰
lime hydration 石灰的消化作用
lime kiln 石灰窑
lime marl 石灰泥灰岩
lime milk 石灰乳
lime mixer 石灰拌和机
lime mortar plaster 石灰砂浆粉刷
lime mortar undercoat 白灰砂浆打底
lime mortar 石灰砂浆
limen 声差阈，色差阈，阈
lime pile 石灰桩
lime pit 石灰池
lime plaster 石灰粉饰，石灰粉刷，石灰涂层
lime precipitation 石灰沉淀法
lime purification 石灰净化
lime putty 石灰膏
lime recalcining 石灰再烧
lime recycling system 石灰循环系统
lime requirement 石灰需要量
lime rock 石灰岩
lime sand brick 石灰砂砖
lime sand 灰砂
lime saturation degree 石灰饱和系数
lime slaked in the air 气化石灰
lime slaker 石灰熟化器
lime slaking 石灰消化，石灰熟化
lime sludge 石灰渣，石灰污泥
lime slurry treatment for acid waste 废酸的石灰处理
lime slurry 石灰浆
lime-soda ash softening 石灰苏打软化
lime-soda feldspar 钙钠长石
lime-soda process 石灰-苏打法
lime-soda softening process 石灰-苏打软化水法
lime-soda water softening 石灰-苏打软水法
lime softening plant 石灰软化装置
lime softening tank 石灰软化箱
lime-soil compaction pile 灰土挤密桩
lime-soil cushion 灰土垫层
lime soil pile 灰土桩，灰土挤密桩
lime soil 钙质土，灰土，石灰土
lime stabilization 石灰加固，石灰稳定法
lime-stabilized soil 石灰加固土
limestone addition 石灰石掺合料
limestone cavern 石灰岩溶洞
limestone coarse aggregate concrete 石灰石粗集料混凝土
limestone consumption rate NO_x emission rate 氮化物排放率
limestone handling system 石灰石处理系统，石

灰石输送系统
limestone kiln 石灰窑，石灰石磨
limestone pebble 石灰石粒
limestone quarry 石灰石采石场
limestone scrubbing method 石灰石洗涤法
limestone silo 石灰石筒仓
limestone sink hole 石灰岩落水洞
limestone soil 石灰土，石灰岩土
limestone treatment for acid waste 废酸水石灰处理
limestone wet scrubber 石灰石湿式洗涤器
limestone 石灰石，石灰岩
limes zero 无毒界量
lime-titania type electrode 钛钙型焊条
lime treatment 石灰处理
limetree 菩提树
lime type covered electrode 碱性焊条
limewash 石灰水，粉刷石灰，粉刷墙壁用的稀石灰粉，用稀石灰粉刷，涂白，刷白，刷浆
lime water 石灰水，硬水
lime white 石灰浆，石灰刷浆，熟石灰
lime whiting 石灰粉刷，石灰刷白
lime 石灰，氧化钙
liming 石灰处理，加石灰
limit alarm 限值报警
limit analysis 极限分析
limitation capacity 极限容量
limitation diagram 界限图
limitation of liability 责任限度
limitation 限制，限度，极限，制约，限幅
limitator 限制器
limit bridge 窄量程电桥，测速电桥
limit capacity 极限容量
limit case 极限情况
limitcator 电触式极限传感器
limit condition 极限值，极限工况，极限条件，极限状态
limit conductance 极限电导
limit control 极限控制
limit curve 限制曲线，界限曲线
limit cycle 极限周期，极限环
limit design 极限设计
limit distribution 极限分布
limit dose 剂量限值，最大剂量
limited access area 受限制的进入区，监控区
limited amplitude response 限幅响应
limited applicability 有限可应用度
limited approval 有限批准
limited availability 有限利用度
limited channel logout 限定通道记录输出
limited company 有限公司
limited critical concentration 极限临界浓度
limited current circuit 限流电路
limited-end-float coupling 限制轴向窜动联轴器
limited energy competition 有限电量竞争模式
limited error 有界误差，极限误差
limited exclusion area 受限禁区
limited head loss 极限水头损失
limited international bidding 有限国际招标
limited international tendering 有限国际招标
limited in 在……方面受限制
limited-leakage pump 有限泄漏泵
limited liability company 有限责任公司
limited liability partnership 有限责任合伙
limited liability 有限责任
limited mandate 有限授权
limited mixing fumigation 有限混合型下熏
limited operation 限制运行
limited order 有限订单
limited output 限定输出功率，极限输出功率
limited partnership 有限合伙公司
limited profit 限制利润
limited-purpose computer 专用计算机
limited recourse 有限追索权
limited safe 有限安全
limited stability 有限稳定度
limited stay area 停留限制区
limited tender 限制性投标
limited to 局限于，被限制在
limited work authorization 有限施工授权书
limited 有限的，被限制的
limit equilibrium analysis 极限平衡分析
limit equilibrium 极限平衡
limiter characteristic 限制器特性
limiter circuit 限幅电路，限幅器电路
limiter diode 限幅二极管
limit error 极限误差
limiter stage 限制级
limiter tube 限幅管
limiter 限制器，限幅器
limit gap gauge 间隙极限验规，极限卡规
limit gauge 极限量规，极限规
limit indicator 极限指示器，限幅指示器
limit inferior 下极限，下限
limiting ambient temperature 极限周围温度，极限环境温度
limiting amplifier 限幅放大器
limiting apparent power 极限视在功率
limiting bed depth 极限床深
limiting circuit 限制电路，限幅电路
limiting concentration 极限浓度
limiting conditions for operation 运行的极限条件
limiting condition 极限条件，极限状态，限制条件
limiting control of variation rate 变化率限值控制
limiting control 限值控制，限制控制
limiting current 限制电流，极限电流
limiting depth 限制水深
limiting design value 设计限值
limiting dimension 极限尺寸，净空
limiting error 极限误差
limiting factor 极限因素，限制因素
limiting fault 极限事故，假想事故
limiting frequency 极限频率
limiting fuel assembly 极限燃料组件
limiting function 极限函数
limiting fuse 限流熔断器
limiting grade 限制坡度
limiting gradient 限制坡度，限制斜度
limiting heat-release rate 极限释热率
limiting high temperature 最高允许温度，温度上限

limiting insulation temperature　极限绝缘温度
limiting intensity　极限强度
limiting level during flood season　防洪限制水位
limiting line　极限线
limiting maximum stress　最大限制应力，最小限制应力
limiting moving contact current　动接触点极限电流
limiting noise emission　噪声发射极限
limiting of resolution　分辨能力限制
limiting orbit　约束轨道
limiting point　极限点
limiting polarization　极限极化
limiting position　极限位置
limiting potential　极限电位
limiting pressure　极限压力
limiting quality　极限质量
limiting quantity　极限量，影响量
limiting range of stress　应力极限范围
limiting resistor　限流电阻器，限流电阻，限流电抗器
limiting resolution　极限分辨力
limiting speed　极限速度
limiting strength　极限强度
limiting stress　极限应力
limiting surface　界面
limiting temperature　极限温度
limiting vacuum　极限真空
limiting value　极限值
limiting valve　限制阀
limiting velocity　极限流速，极限速度
limiting viscosimeter number　特性黏度，极限黏度计值
limiting viscosity　特性黏度
limiting　极限的，限制，界限，约束，限幅
limit-in-mean　平均极限
limit input　极限输入，限制输入
limitless　无限的
limit load factor　限制负荷因数，最大使用负荷因数
limit load　极限载荷，极限负荷，极限荷载，限制载荷，临界载荷
limit loop　极限环
limit of acceptance　验收界限
limit of accuracy of chronometry　时间分辨率
limit of accuracy　精确度极限，精度极限
limit of age　老化期限，极限使用期
limit of audibility　听力极限，能听极限
limit of backwater　回水极限
limit of bearing capacity　承载能力极限
limit of compensation　补偿范围，补偿极限
limit of creep　蠕变极限
limit of detection　检测极限
limit of driving speed　行驶速限
limit of ductility　延性限度
limit of elastic　弹性极限
limit of equilibrium　平衡极限
limit of error　误差极限，误差范围，误差限值
limit of exchange　汇兑极限
limit of explosion　爆炸极限，爆破极限
limit of fatigue　疲劳极限
limit of fire-resistance　耐火极限
limit of free haul　土方的最大免费运距
limit of integration　积分范围
limit of interference　干扰允许值，干扰极限
limit of intrinsic error　基本误差限
limit of liability　责任范围，责任限制，责任限额
limit of operating condition　极限工作条件
limit of plastic flow　塑流极限
limit of power range　功率极限，极限功率
limit of power　功率极限，极限功率，负荷极限，功率范围
limit of pressure　压力范围，压力极限
limit of proportionality　比例极限
limit of rupture　破坏极限
limit of self-extinguishing current　自熄弧电流的极限值
limit of stability　稳定极限
limit of temperature rise　温升极限，温升范围
limit of temperature　温度极限，温度范围
limit of tidal current　潮流界限
limit of time　时间限制
limit of travel　行程极限
limit of yielding　屈服极限，沉陷限度
limit on　对……限制
limit order　限价委托，限价订单
limit output　极限出力，极限功率，极限输出
limit plug gauge　极限量规塞子，极限塞规
limit point　极限点
limit power　极限功率
limit price　限价
limit range　极限范围
limit risk　危险模式
limit rod　限制杆
limit setting　终端位置调定
limit signal　限值信号
limit snap gauge　极限卡规
limits of error　误差极限
limits of one's functions and powers　职权范围，权限
limit speed switch　限速开关
limit speed　极限速度
limit state analysis　极限状态分析
limit state design　极限条件计算，极限状态计算，极限状态设计
limit state equation　极限状态方程
limit states design method　极限状态设计方法
limit state　极限状态，极限条件
limit stop　极限制动装置，限位开关，限位挡块，定值限位架
limit strength　极限强度
limit superior　上极限，上限
limit support　限位支架
limit switch hardware　限位开关硬件
limit switch　行程开关，限位开关，极限开关，终端开关，终点开关
limit temperature　极限温度，最高温度
limit value monitor　限值监控器
limit value signal　限值信号
limit value　限值，极限值
limit valve　限位阀，行程阀
limit voltage　极限电压

limit water surface　限制水面
limit　限制【DCS画面】，极限，极限值，限定，界限，极点，限度，范畴，范围，限值
LIM=limiter　限制器
limnic basin　淡水盆地，陆相盆地，湖盆
limno-geotic　淡水环境
limnograph　自记水位计，湖水水位测量计
limnology　湖沼学
limousine　大型高级轿车
limous　混浊的
limp-diaphragm gauge　挠性膜测量计
limpidity　清澈度，透明度
limpkin　涉禽的一种
limy　含石灰的，石灰的
Lincoln automatic welding machine　林肯自动焊机
Lincoln surface tension transfer welding　林肯张力焊
Lincoln welding machine　林肯电焊机
linden wood　椴木
linden　椴木，菩提树
line absorption　定能分散吸收，线吸收，一定能级的离散吸收
line addressable RAM　行访问随机存取存储器，行寻址随机存取存储器
line admittance　线路导纳
line aggregate concrete　细骨料混凝土
lineal density heat loss　线密度散热
lineal expansion coefficient　线性膨胀系数
lineal shrinkage　线性收缩
lineal　线的，线性的，直系的，正统的，线形的
line amplifier　线路放大器，行信号放大器
line angle　线路转角
linear absorption coefficient　线性吸收系数
linear acceleration　线性加速度
linear accelerator　线性加速器，直线加速器
linear activity　线性活度
linear actuator　线性执行机构
linear amplification　线性放大
linear amplifier　线性放大器
linear analysis　线性分析
linear-and-angular-movement pickup　线位移和角位移传感器
linear anisotropic　线性各向异性
linear attenuation coefficient　线性衰减系数
linear ball bearing　沿轴线滚珠轴承
linear block code　线性分组码
linear burnup　线性燃耗
linear channel　线性通道
linear characteristic　线性特性
linear circuit　线性电路
linear code　线性编码
linear combination　线性组合
linear commutation　直线换向
linear compression　线性压缩
linear computing element　线性计算元件
linear conductor　线性导体
linear control process　线性控制过程
linear control system theory　线性控制系统理论
linear control system　线性控制系统
linear control　线性控制
linear coordinates　线坐标
linear correlation　线性相关
linear coupler　线性耦合装置
linear cumulative damage hypothesis　线性累积损伤假设
linear cumulative damage　线性累积损伤
linear current density　线性电流密度
linear current　线性电流
linear damping　线性阻尼
linear DC instrument lane　线性直流仪表管
linear decision　线性判定
linear decrement　线性衰减量
linear deformation　线性变形
linear density　线密度
linear dependence　线性相关
linear detection　线性检测，线性检波，线性检波器
linear difference equation　线性差分方程
linear differential equation　线性微分方程
linear dilatation　线性膨胀
linear dimension　线性尺寸
linear dispersion　线性扩散
linear displacement detector　线性位移探测器
linear displacement　线位移
linear distortion　线性畸变，线性失真
linear drive　线性驱动
linear dynamic response　线性动态响应
linear eccentricity　偏心距
linear elastic body　线性弹性体
linear elastic fracture mechanics　线弹性断裂力学
linear elasticity　线弹性，线性弹性
linear elastic theory　线弹性理论
linear elastodynamics　线性弹性动力学
linear electric constant　线路电气参数，线路电气常数，线性电力常数
linear electric current density　电流线密度
linear electric motor　直线电动机
linear electron accelerator　线性电子加速器，电子直线加速器
linear element　线性单元，线性元件
linear energy transfer system　能量线性传递系统
linear energy transfer　传能线密度，能量线性传递，线能量转移，线性能量传递
linear energy　线能
linear equation solver　线性方程解算器
linear equation　线性方程，一次方程
linear error of closure　长度闭合差
linear expansibility　线膨胀系数
linear expansion coefficient　线性膨胀系数，线膨胀系数
linear expansion relay　线膨胀型继电器
linear expansion　线膨胀，线性膨胀
linear extinction coefficient　线性消光系数，线性衰减系数
linear extrapolation distance　线性外推距离
linear extrapolation length　线性外推长度
linear extrapolation　线性外推法
linear feedback control　线性反馈控制
linear feedback system　线性反馈系统
linear filter　线性滤波器
linear first-order differential equation　一阶线性微

分方程
linear float guide 直线浮筒导轨
linear flow characteristic 线性流量特性
linear flow 层流，线流
linear fouling 线性污垢
linear fracture mechanics 线性破坏力学
linear function 一次函数，线性函数
linear graduation 线性刻度
linear head loss 线性压头损失，沿程损失
linear heat generation rate 线功率密度
linear heating power 线功率
linear heat rating 线功率
linear independence 线性无关性
linear integrated circuit 线性集成电路
linear intercept method 线性截取法
linear interpolation 线性内插法，线性内插
linear ionization by a particle 粒子线电离
linear ionization 线性电离
linear irrotational flow 线性无旋流
linearised equation 线性化方程
lineariser 线化器
linearise 直线化，线性化
linearity circuit 线性电路
linearity control 线性控制
linearity error 线性误差，线形误差，线性度误差
linearity programming 线性程序设计
linearity 线性度，线性，直线性，直线度
linearization compensation 线性化补偿
linearization 线性化，直线化
linearized aerodynamic 线（性）化空气动力学
linearized Boltzmann equation 线性化玻耳兹曼方程
linearized equation 线（性）化方程
linearized flow 线性流，线形流动
linearized hot wire anemometer 线（性）化热线风速计
linearized model 线性化模型
linearized network 线性化网络
linearized potential field 线（性）化势流场
linearized radiation 线性辐射
linearized theory 线（性）化理论
linearized 线性化的
linearizer 线性化电路
linearizing resistance 线性化电阻
linear lag 线性滞后
linear lead 线性超前
linear lightning 线状闪电
linear load 单位长度负荷，线负荷，单位长度负载，线性荷载
linear location 一维定位，线定性
linearly dependent vector 线性相关的矢量
linearly dependent 线性相关的
linearly polarized wave 线性极化波
linear mapping 线性变换，线性映射
linear measurement 长度测量，线性测量
linear measure 长度单位，直线量度
linear measuring assembly 线性测量装置
linear memory 线性存储器
linear meter 延米，线性仪表
linear model 线性模型
linear modulation 线性调制
linear motion 直线运动
linear motor 线性电动机
linear multivadable system 线性多变量系统
linear multivariable sampled-data control system 线性多变量数据采样控制系统
linear network 线性网络
linear operator 线性运算器
linear optimal stochastic system 线性最优随机系统，线性最佳随机系统
linear optimizing 线性最优化
linear oscillator 线性振荡器，线性振子
linear perturbation 线性扰动理论，线扰理论
linear phase filter 线性相位滤波器
lincar polarization 线极化，线偏振
linear porosity 线性气孔率【焊接缺陷】
linear potentiometer 线性电势计，线性电位计
linear power amplifier 线性功率放大器
linear power controller 线性功率控制器
linear power density 线功率密度
linear power rating 额定线功率
linear power 线功率
linear pressure loss 沿程压力损失
linear process 线性过程
linear programming 线性规划
linear program 线性规划
linear propagation 直线传播
linear pulse amplifier 线性脉冲放大器
linear quantizer 线性数量转换器，线性数字转换器
linear range 线性范围，粒子群线性射程
linear ratemeter 线性速率计，线性率表
linear rating 线功率
linear reactance 线性电抗
linear rectification 线性整流，线形检波
linear rectifier 线性检波器
linear regression 线性回归
linear regulator 线性调节器
linear relationship 线性关系
linear relation 线性关系
linear reluctance motor 直线式磁阻电动机
linear repeater 线性转发器，线性中继器，线性增音机
linear resistance flowmeter 线性电阻流量计
linear resistance 线性电阻，沿程阻力
linear resolution 线形分辨率
linear resolver 线性解算器
linear resonance overvoltage 线性谐振过电压
linear responsibility chart 条形责任图
linear rod power 线功率
linear rotational flow 线性旋转流
linear scale 线性标度（尺），线形分度，线性刻度
linear scanning 线性扫描
linear select memory 线选存储器
linear sequential network 线性时序网络
linear servomechanism 线性伺服机构
linear shrinkage 线性收缩
linear simultaneous equation 线性方程组，线性联立方程
linear source 线源，线状源
linear specific power 线功率

linear speed 线速度
linear stator machine 直线定子电动机
linear stator 直线定子
linear step motor 直线步进电动机
linear stopping power 线性制动能力，线性阻止本领
linear strain 线应变
linear stratification 线性层结，线性层理
linear stress 线性应力，线应力
linear superposition 线性叠加
linear sweep generator 线性扫描发生器
linear sweep 直线扫描
linear switching 线性开关，线性切换
linear symmetrical 直线对称
linear synchronous machine 直线同步电动机
linear system of equations 线性方程组
linear system 线性系统
linear theory 线性理论
linear thermal output 线功率
linear thermistor 线性热敏电阻器
linear time base test 线性时基检验
linear time base 直线扫描
linear time-invariant control system 线性定常控制系统
linear time-invariant system 线性时不变系统，常参数线性系统
linear time-quantized control system 线性时间整量化控制系统
linear time-varying control system 线性时变控制系统
linear time-varying network 线性变参数网络，线性时变网络
linear-to-log converter 线性对数变换器
linear transducer 线性传感器，线性换能器，线形换能器
linear transformation 线性变换
linear transformer 线性变压器
linear transient analysis 线性暂态分析
linear transmission 直线发送
linear-type unloader 门架式卸煤机
linear unit 线性部件
linear valve characteristic 线性阀门特性
linear variable differential transducer 线性可变差分传感器
linear variable differential transformer 线性变换差动转换器，线性阀位传感器
linear vector equation 线性矢量方程
linear vector function 线性矢量函数
linear velocity 线速度，线速
linear vernier 线性游标
linear vibration 线性振动，线性共振
linear viscous fluid damper 线性黏滞阻尼器，线性黏性减振器
linear voltage 线性电压
linear wave 线性波
linear winch 线性绞车
linear zone 线性区
linear 线性，线性的，直线的，线形的，线状的
line-at-a-time printer 行式印刷机，宽行打印机
lineation 划线，线理
line attendant 巡线员
line attenuation coefficient 线衰减系数
line backing 轴承衬支座
line balance 线路平衡
line bank 接线排
line battery 线路电池组
line bay extension 线路间隔扩建
line bay 线路间隔【变电站】
line bearing 线轴承
line blind valve 管道盲板阀
line blind 管线盲板，盲管
line blow 强风，直线性强风
line boss 生产指挥人员
line breaker 线路断路器【开关站的主断路器】，线路开关
line break relay 断线继电器
line break 管道破裂
line broadening effect 线加宽效应
line buffer 线路缓冲器
line busy 占线
line-by-line analysis 按行分析
line-by-line calculation method 逐线计算法
line-by-line scanning 逐行扫描
line capacitance 线路电容
line capacity 线路容量
line characteristic 线路特性
line charge model 线电荷模型，线磁荷模型
line charge voltage 线路充电电压
line charging capacity 线路充电容量
line charging current 线路充电电流，线路无载电流
line chart 线路图，单线图
line check 小检修
line choking coil 线路抗流圈，线路扼流圈
line circuit-breaker 线路断路器
line circuit 用户电路
line clamp 线夹
line commutated inverter 线换流逆变器，线换向逆变器
line concentrator 用户集线网，线路集中器，用户集线器
line conductance 线路电导
line conductor insulators and fittings 导线绝缘子及配件
line conductor survey 导线测量
line conductor 线路导线，接线器
line constant （输电）线路常数
line construction 线路架设，线路施工
line contactor 线路开关，线路接触器
line contact 线接触
line corona 线路电晕
line corridor 线路走廊
line coupler 线路耦合器
line crew 线路工程队
line current tester 线路电流测试仪
line current 线路电流
line damper 线路减振锤
lined borehole 下套管的钻孔，衬壁钻孔
line defect 线缺陷
line design 线路设计
line detection 线路检测
line diagram 单线图，线路图

line diffuser　条缝风口，条缝散流器
line disconnection　切断线路，断线
line distortion　行畸变，线路畸变，线路失真，行失真
line drawing　单线图
line drilling　撬挖
line driver　线驱动器
line drop compensation impedance　线路压降补偿阻抗
line drop compensation reactance　线路压降补偿电抗
line drop compensation resistance　线路压降补偿电阻
line drop compensation　线路压降补偿
line drop compensator　线路电压降补偿器，线路电压降补偿装置
line drop voltmeter compensator　线电压降电压表补偿器
line drop　线路电压降
lined rubber　衬胶
line drum　线盘
lined tunnel　有衬隧洞
lined up with　对准，排成一行，对齐，定向
lined with diabase　衬辉绿岩
lined with lead　衬铅
lined with rubber　用橡胶衬里，衬胶
lined　衬里的，内衬的
line editor　行编辑
line electric parameter　线路电气参数
line end　出线端
line engineer　线路工程师
line equalizer　线路均衡器
line equipment　线路设备
line fault anticipator　线路故障预防装置
line fault　线路故障
line-fed motor　直接馈电电动机
line feed character　换行字符
line feeder　线路馈电线，馈电导线，电路馈电线
line feed　移行，换行，线路馈电
line filter　线路滤波器
line finder　寻线器
line fittings　线路配件
line flashover　线路闪络
line flow equation　线路潮流方程式
line flow　线路流通量，线路潮流
line flyback　行回描
line focus collector　线聚焦（型）集热器
line focus transducer　线聚焦传送器
line focus　线焦点，线聚焦
line frequency allocation　线路频率分配
line frequency　线路频率，电网频率，电源频率，市电频率，行频，行频率
line galloping prevention　线路防舞动
line gap　线路避雷器，线隙
line gauge　线规，倍数尺
line generation　行向量生成，行产生
line generator　直线发生器，矢量发生器，向量产生器
line grip　紧线夹，紧线钳
line grounding design　线路接地设计
line guy　线路拉线
line half-width　谱线半宽
line hardware　（架空）线路金具
line hook　系链钩
line impedance　线路阻抗
line inductance　线路电感
line inlet　线路入口
line inspection　巡线，线路检查，线路巡视
line inspector　巡线员
line insulation design　线路绝缘设计
line insulation　线路绝缘
line insulator　线路绝缘子
line integral　线积分
line interface　线路接口
lineless connection drawing　没有线的连接图【只表示出端子】
line lightning arrestor　线路避雷装置，线路避雷针
line lightning performance　线路雷电特性
line lightning-protection design　线路防雷设计
line like building　细长建筑
line like structure　细长结构
line load control　线路负荷控制
line load　线性荷载，线荷载
line location　线路选位，线路踏勘
line locator　管线位置探测仪
line locking　行同步
line loss computation　线损计算
line loss of voltage　线路电压损失
line loss rate　线损率
line loss　线损，线路损耗，线路损失
line manager　专业经理
lineman's detector　携带式检电器
lineman's pliers　线路工人用钳
lineman　线路工人，巡线工，测线员
line mixer　管道混合器
line modulator　线调制器
line monitor　线路监视器
line-mounted sensor　管线安装的传感器
line mutual-capacitance　线路互电容
line mutual-inductance　线路互感
line network　线路网
line noise　线路干扰，线路噪声
linen tape　布卷尺
line number　行数
linen　麻布，亚麻布
line of action　（轴承）作用线
line of admission　进汽特性曲线
line of application　作用线
line of balance　平衡线
line of bearing　轴承方位线，方位线
line of business　行业，营业范围，业务范围
line of center　中心线
line of cloud　云线
line of code　代码行
line of communication　通信线路
line of connection　连接线
line of constant entropy　等熵线
line of corresponding stages　水位相关曲线
line of creep　蠕动线
line of cut　切割线
line of demarcation　分界线
line of dielectric　介电线

line of dislocation 位错线
line of duct 管路，管线
line of electric force 电力线
line of equal inclination 等倾线
line of equal pressure 等压线
line of equal principal stress difference 等主应力差线
line of equal shear 等剪力线
line of fastest flow 水流中心线，中泓线，最大流速线
line of feedback coupling 反馈耦合线
line of flow 流量线，流线
line of flux 通量线
line of force 力线，磁力线
line of fusion 熔合线
line of induction 感应线
line of influence 影响线，感应线
line of land requisition 土地征用线
line of least resistance 最小抵抗线
line of magnetic field 磁力线数，磁力线
line of magnetic force 磁力线
line of magnetic induction 磁感应线
line of magnetization 磁化线，磁力线
line of maximum velocity 最大流速线，中泓线
line of maximum water depth 最大水深线
line of position 位置线
line of pressure 压力线
line of principal stress 主应力迹线，主应力线
line of production 流水生产线
line of reference 参考线，基准线
line of resident relocation 居民迁移线
line of resistance 抗力线，阻力线
line of segregation （铸件）偏析区
line of separation 分界线
line of shafting 轴系，轴系中线
line-of-sight communication 直视通信线
line-of-sight microwave radio relay system 视距微波中继通信系统
line of sight 视线
line of sliding 滑动线
line of tidal wave 潮波线
line of vector 矢量线
line of zero moment 零矩线
lineograph 描线规
line operation 线路运行
line original 线划原图
line outage calculator 线路停电率计算器
line outage 线路停电
line outlet 线路出口
line output transformer 行输出变压器
line overload 线路过负荷
line pair 线对
line parameter 线路参数
line performance （输电）线路运行特性
line personnel 线路人员，电厂运行人员
line planar graph 线路平面图
line plot survey 沿线分段测绘
line polarization 线极化
line pole 线路电杆
line positioning survey 定线测量
line post insulator 线杆绝缘子
line power flow 线路潮流
line pressure 管线压力，管道压力
line printer 行式打印装置，宽行打印机，行式打印机，行式印刷装置
line printing 行向打印，行式印刷，行向印刷
line program 线性规划
line protection 输电线保护，线路保护，线路防护
line pull 绳索拉力
line radio 有线载波通信
liner backing 轴承座套
liner bushing 衬套
liner cooling pipe system 衬里冷却管系统
liner cooling system 衬里冷却系统
liner disk 衬盘
line reactance 线路电抗
line record 线路记录
line regulation 电源电压调整率，线路调节，线路电压调整
line relaxation method 逐行松弛法
line relay 线路继电器
line release 线路断开，线路脱扣装置
line repair 线路检修
line residual current 大地回流，线路残余电流
line resistance compensation 线电阻补偿
line resistance 线路电阻
line-reversal method 谱线回转法
liner material 内衬材料，衬料
line route selection 线路路径选择【指架空线路】
line route 路径，线路路径，线路走向
liner plate 衬板，垫板，衬砌板
liner port 定期船港
liner ring 轴封环
liner segment 衬砌段【隧道】
liner spacer 衬套隔环
Liner Terms 班轮条款
liner tube 衬管，衬套管，衬里管
liner 衬板，衬垫，班轮，套筒，衬层，衬，衬里，衬砌，内衬，轴瓦，衬套，衬圈
line scan transformer 行扫描变压器
line sectional drawing 线路断面图
line sectionalizing 线路分段
line section 线段，线务段
line segment function generator 线段函数发生器
line segment 线段
line selector 选线器，线路选择器，寻线器
line self-capacitance 线路自电容
line self-inductance 线路自感
line service 线路保养
line shafting 传动轴系，总轴系
line shaft 总轴，中间轴，动力轴
line side breaker 线路侧断路器
line side of insulator string 绝缘子串导线侧
line side 线路侧
line signaling 线路信令
line simulator 线路模拟器
linesman 巡线工人，养路工人，线路工人，架线工人
lines of field intensity 场强线
line source equation 线源方程

line source 线光源，线发射源，线污染源，线源
line spacing 管道间距
line span 管道跨距，线路挡距
line spectrum 谱线，线谱，线状谱
lines per inch 行每英寸
lines per minute 行每分
lines per second 行每秒
line squall 线暴风
line staff 线路人员，电厂运行人员
line start motor 直接启动电动机
line start 直接启动，全压启动
line status word 行状态字
line survey 线路测量，线路勘测
line susceptance 线路电纳
line switching voltage 线路交换电压
line switch 线路开关，寻线机
line tap 线路分支接头，线路分接头
line telephony 有线电话
line terminal 主引线端子，线路终端，线路端，与火线连接端子
line time base 横向扫描装置
line-to-earth fault 线路接地事故
line-to-earth voltage 线对地电压，线路对地电压，相电压
line-to-ground fault 线路接地故障，线对地故障
line-to-ground short circuit 线路对地短路
line-to-ground voltage 线对地电压
line-to-ground 线路对地的，线路接地
line-to-line short circuit 线间短路
line-to-line voltage 线间电压，线电压
line-to-line 相间的，两线间的，线间短路
line-to-neutral voltage 线与中性点间电压，相电压
line-to-neutral 线与中性点间的
line transfer-point 转线
line transformer 线路变压器
line transmission 线路传输
line transmitter 中继发报机
line trap 线路陷波器，限波器，线列陷波电路，导线插头，管道阱，线路阻波器
line-type fire detector 线型火灾探测器
line up device 垫整[调节]装置
line up test 校准试验，校直试验
line up with 与……对齐，定位
line up 垫整，调成一直线，弄直，排成列，铺砌耐火材料，对准中心，找中心，校正，调整
line voltage drop 线路电压降
line voltage regulator 线电压调整器
line voltage up rating 线路升压
line voltage 线电压，厂外供电电压，电源电压，线路电压
line vortex 线涡
line welding 线焊
line width gauge 线径测力计，线规
line width 行距，线幅，线宽
line winding （换流变压器的）交流网侧绕组
line wiper 线路弧刷
line wire 线路导线
line with dissipation 有损耗线
line 管路，管道，直线，行，列，线，系统，衬砌，轨道，路线，管线
linguistic ability 语言表达能力
lining alloy 轴承合金
lining arch 衬砌拱
lining board 衬板
lining brick 炉内衬砖
lining carrier 轴承衬，轴承套
lining inspection 衬套检查
lining material 衬垫材料
lining metal pipe 衬里金属管
lining metal 轴承合金
lining of bearing 轴承衬，轴承方位线
lining of canal 渠道护面
lining of shaft 轴衬，轴垫
lining of slope 坡面铺砌
lining of tunnel 隧道衬砌
lining pipe 衬里管
lining plate 衬板
lining pole 花杆
lining rubber 衬里橡皮
lining sheet 衬板
lining stone 衬砌石块
lining tube 衬砌管
lining wear 衬里磨损
lining 衬料，轴瓦，衬套，内衬，衬里，砖砌，炉衬，镀层，衬，衬垫，衬砌，电镀，喷镀，镶衬，涂底
link address 连接地址
linkage assembly 连杆机构
linkage coefficient 磁链系数
linkage computer 模化计算设备，连续动作计算机
linkage fault 连接故障
linkage gear 连接机构
linkage grab 连接抓取器，链合抓具
linkage group 链组
linkage guide tube 连接导管，链合导管
linkage head 连接头
linkage map 连锁图
linkage section 连接段，连接节
linkage tube 连接管
linkage value 连锁值
linkage voltage test （电机的）连续加压试验
linkage 交链，连接，连锁，联系，联动装置，磁链，链合，键合，连杆，连杆机构，连接方法，连杆装置，联动，联杆
link allotter 链路分配器
link block 连接滑块，导块
link board 连接板
link bolt 铰链螺栓，链环插销
link butterfly valve 连杆式蝶阀
link-by-link signaling 逐段转发信令
link chain 扁节链
link circuit 链式电路
link connected switching stage 链路连接的交换级
link controller 链路控制器
link control module 链路控制模块
link control program 链路控制程序
linked subroutine 连接子程序
linked switch 联系开关，耦合开关

linked　耦合的，联系的，交链的，连接的
Linke turbidity factor　林克浑浊因子
link fuse　链熔片，熔线片，熔丝刀闸
link gearing　联杆传动
link gear　连杆装置，联动机构，摇拐机构
linking member　联系杆件
linking rod　连杆，联杆
linking route　连接路线
linking　结合，耦合，咬合，联锁，联动，连接
link insulator　串式绝缘子，链式绝缘子，绝缘子串
link in system　系统联络线
link layer service data unit　链路层服务数据单元
link layer　链路层
link line　联络线
link lockout　链路锁定
link motion　联杆运动
link order　耦合指令
link pack area　连接装配区
link pin　联节销
link press　连杆式压力机
link protocol　链路协议
link receiver　中继接收机
link service access point　连接服务访问点
link strain insulators　耐张绝缘子串
link suspension insulators　悬垂绝缘子串
link switch　联动开关
link system　联动系统，链路系统
links　海岸沙滩，链接【计算机】
link transmitter　接力发射机，中继发射机，强发射束发射机
link-up　联络，连接，联系，连接件
linkwork　铰接机构，联动装置，链系，链接机构，链系联杆装置
link　环节，连杆，链节，铰链，连接线，网络节，键节，连接，磁链，链环，关联，机械连杆，环，联杆，链路
linn　瀑布，溪谷，瀑布下的水潭，绝壁
linoleum flooring　油毡地面
linoleum　油毡，漆布，油布
linotape　黄蜡带，漆布带
linseed oil varnish　亚麻清油，亚麻仁油漆，亚麻油清漆
linseed oil　亚麻仁油，亚麻子油，亚麻籽油
lintel beam　水平横楣梁
lintel　横楣，过梁，横梁，炉墙开孔处的水平结构件，楣石，楣
lint-free cloth　不起毛的布
lint-free　不起毛的
lint　纤维屑
liparite　流纹岩
lip contamination　唇部污染
lip loss　进口边缘损失
lipophlic　亲油性
lipper　飞溅浪花，碎波
lip seal　端头密封
lip-synchronous　与语言同步的
lip　支架，口子，突出部分，凸缘，(斗）唇
LIQ＝liquid　液体
liquation crack　液化裂纹
liquation　熔融，熔解，液化，熔析
liquefaction point　液化点
liquefaction potential　液化势
liquefaction　溶化，液化，熔化
liquefiable　可熔化的，可液化的
liquefied gas　液化气，液化气体
liquefied natural gas power plant　燃液化天然气电站
liquefied natural gas　液化天然气
liquefied petroleum-gas station　液化石油气站
liquefied petroleum gas　液化石油气
liquefier　液化剂，液化器，稀释剂
liquefy　液化，冲淡，稀释
liquid acoustical lens　液体声透镜
liquid active waste　放射性液态废物
liquid activity　液体活度
liquid air tank　液态空气罐
liquid air　液态空气
liquid ash removal　液态排渣
liquid asphalt　液态沥青
liquid assets　流动资产，速动资产
liquidated damages for delay　延期（完工）的罚款
liquidated damages　约定的损失赔偿，违约罚金，预定的违约金，清偿损失额，违约赔偿金
liquidate　清理，清除，液化，流化，偿还，取消，清算（公司）
liquidation　液流化，液化，清理，（公司的）清算
liquid atomization　液体雾化
liquid balance　流动资金平衡，现金结余
liquid bath furnace　液态排渣炉膛
liquid bath　液浴，液槽
liquid bitumen　液态沥青
liquid blade cooling　叶片液冷
liquid brush rectifier　液刷整流器
liquid capacity　液体流量
liquid capital　流动资本
liquid chlorine　液态氯
liquid collector　液体集热器
liquid column gauge tester　液柱式量计
liquid column pressure gauge　液柱压力计
liquid column　液柱
liquid condenser　液体电容器
liquid control　液态控制
liquid coolant　液冷却液，液态冷却剂，冷却液
liquid cooled gas-turbine blade　液冷燃气轮机叶片
liquid cooled reactor　液冷电抗器，液冷反应堆，液体冷却反应堆
liquid cooled rotor　液冷转子
liquid cooled stator winding　液冷定子绕组
liquid cooled transformer　液冷变压器
liquid cooler　液体冷却器
liquid cooling generator　液冷发电机
liquid cooling medium　液冷介质
liquid cooling　液体冷却
liquid crystal digital display　液晶数字显示
liquid crystal display　液晶显示屏，液晶显示
liquid crystal　液晶
liquid cutting of wood　木材水力采伐
liquid damper　液体阻尼器

liquid deficient regime 缺液工况
liquid deficient region 缺液区
liquid deficient 缺水的
liquid dielectric breakdown 液体电介质击穿
liquid dielectric 液冷电介质
liquid drier 液体干燥剂
liquid droplet additives 液滴添加剂，液滴添加物
liquid droplet 液滴
liquid drop 液滴
liquid effluent 液体排出物
liquid end bearing 液体端轴承
liquid end 液体端
liquid entrainment rate 液体夹带率
liquid entrainment 液体夹带
liquid explosive 液态炸药
liquid feed pump 液体进给泵，馈液泵
liquid-filled thermal system 充液式感温系统
liquid-filled thermometer 充液式温度计
liquid-filled transformer 液冷变压器，充液变压器
liquid-filled window 充液窗
liquid film controlled diffusion 液膜控制扩散
liquid film 液膜
liquid filter 液体滤光器，液体过滤器
liquid-fired 燃液点火的，液体燃料的
liquid flooding 液体溢流，液泛
liquid flow equation 液体流量方程，液体流动方程
liquid flow measurement 液体流量测定
liquid flowmeter 液体流量表
liquid flow 液流
liquid fluidized bed 液固流化床
liquid fluidized system 液固流化系统
liquid fluid 液体，流体
liquid friction 液体摩擦
liquid fuel reactor 液态燃料反应堆
liquid fuel 液态核燃料，液体燃料
liquid fuse 液体熔丝
liquid-gas equilibrium 液气平衡
liquid globule 液珠，液粒
liquid head 液压头
liquid heating collector 液体集热器
liquid helium cooling 液氦冷却
liquid-immersed reactor 液油浸电抗器，液浸电抗器
liquid-immersed regulator 液油浸式电压调整器
liquid immersion method 液体浸渍法
liquid-immiscibility region 液体不溶混区
liquid-in-glass thermometer 玻璃壳液体温度计，玻璃管液体温度计
liquid insulated bushing 液体绝缘套管
liquid insulating material 液体绝缘材料
liquid insulation 液体绝缘
liquid insulator 绝缘液体
liquidity index 流性指数，液化指数，液流动性指数
liquidity ratio 清偿能力比率，流动比率【会计】
liquidity 流动性
liquidize 液化
liquid jet 液体射流
liquid level connection 液位计接管
liquid level controller 液位控制器
liquid level control 液位控制
liquid level detection 液面检测
liquid level gauge 液位计
liquid level indicator 液面仪，液位计，液位指示器，液面指示器，液位表
liquid level instrument 液位仪表
liquid level meter 液位计，液面计
liquid level probe 液位探头
liquid level recorder 液面记录计，液位记录仪，液面记录器
liquid level relay 液面浮动继电器
liquid level switch 液浮动开关，液位控制开关
liquid level 液位，液面
liquid-like phase 液类液相
liquid limit apparatus 流限仪
liquid limit of soil 土壤液限
liquid limit 流限
liquid-liquid extraction 液液萃取法
liquid manometer 液压表，液柱压力计，液体压力表
liquid mass stability 液体质量稳定
liquid medium sonic delay line 液体介质声延迟线
liquid membrane 液膜
liquid metal breeder reactor 液态金属增殖反应堆
liquid metal brush 液态金属电刷
liquid metal coolant circuit 液态金属冷却剂回路
liquid metal coolant 液态金属冷却剂
liquid metal cooled fast breeder reactor 液态金属冷却快中子增殖堆
liquid metal cooled reactor 液态金属冷却堆，液态金属冷却反应堆
liquid metal extraction process 液态金属萃取法
liquid metal fuel reactor 液态金属燃料反应堆
liquid metal fuel suspension reactor 液态金属燃料悬浮反应堆
liquid metal fuel 液态金属燃料，液金燃料
liquid metal heat exchanger 液态金属热交换器
liquid metal magnetohydro-dynamic power generation 液态金属磁流体发电
liquid metal MHD generator 液态金属磁流体发电机
liquid metal MHD pump 液态金属磁流体泵
liquid metal reactor 液态金属反应堆
liquid metal rotary seal 液体金属转动密封
liquid metal slip-ring 液态金属集电环
liquid metal solvent 液态金属溶剂
liquid metal 液态金属
liquid metering vessel 液体计量容器
liquid meter 液体计量器
liquid-moderator reactor 液体慢化剂反应堆
liquid nitriding 液体渗氮
liquid nitrogen cooled generator 液氮超导发电机，液氮冷却发电机
liquid nitrogen cooled trap system 液氮冷阱系统
liquid nitrogen 液态氮，液体氮，液氮
liquidometer 液位计，液面计，液面测量计
liquid-operated 液动的，液力操纵的
liquid oscillation 液面振荡

liquid outflow 液体出口
liquid overflow 液溢流
liquid oxygen explosive 液氧炸药
liquid oxygen 液态氧
liquid packing 液力密封
liquid pack 水封
liquid paraffine 液石蜡油，液态石蜡
liquid penetrant examination of the heater-well bevel 加热器井斜口的液体渗透检验
liquid penetrant examination 液体渗透（性）检验【金属表面】，液体渗透探伤
liquid penetrant inspection 液体渗透检验，着色探伤【金属表面】
liquid penetrant testing 液体着色探伤试验，液体渗透性检验【金属表面】
liquid penetrant 液体渗透剂，渗透液
liquid penetrating test 渗液探伤
liquid phase mass flux density 液相质量流密度
liquid phase pressured drop 液相压降
liquid phase 液相
liquid piezometer 液体测压计
liquid poison control 液体毒物控制
liquid poison injection pump 液体毒物注入泵
liquid poison system 液体毒物系统
liquid poison tank 液体毒物箱
liquid poison 液体毒物
liquid pressure gauge 液体压力计
liquid-pressure transducer 液压转换器
liquid-proof 不透液体的
liquid proportioner 液体比例混合器
liquid purification column 液体净化柱
liquid quenching 液体淬火处理
liquid radioactive waste concentrate 放射性废液浓缩物
liquid radioactive waste 液态放射性废物
liquid radwaste system 放射性液态废物系统，废渣处理系统
liquid radwaste treatment system 放射性废液处理系统
liquid receiver 贮液器，溶液器
liquid rectifier 电解整流器，液体整流器
liquid regulating resistor 液体调节电阻器
liquid residues 液态残渣
liquid resistor 液体电阻器
liquid rewetting front 液体再湿前沿
liquid rheostat 液体变阻器
liquid ring vacuum pump 水环式真空泵
liquid salt 熔盐
liquid-saturation curve 液体饱和线
liquid scintillation 液体闪烁法
liquid scintillator activity meter 液体闪烁活度测量仪
liquid seal 水密封，液封，水封
liquid securities 流动证券
liquid sensor 液体传感器
liquid shutdown system 液体停堆系统
liquid slag 液态渣
liquid slip-regulator 液体转差调节器
liquid sludge 液状污泥
liquid sodium cooled reactor 钠冷反应堆
liquid sodium 液态钠
liquid-soilds handling pump 杂质泵
liquid solar array 液体太阳能电池阵列
liquid-solid discharge 液固排放
liquid-solid separation 液固分离
liquid specific heat 液体比热，液态比热
liquid starter 液压启动器，液体启动器
liquid state 液态
liquid steam froth 液汽泡沫
liquid steel 钢水
liquid superheat 液体过热度
liquid surface profile 液面纵断面
liquid-suspension reactor 液态悬浮燃料反应堆
liquid-tight cell 液体密封室
liquid-tight flexible metal conduit 水密的柔性金属导管
liquid-tight flexible tubing 水密的柔性套管
liquid-tight 液密封的，不透液的，不泄液的
liquid totalizer 液体累加器
liquid-type reactor 液态反应堆
liquidus curve 液相线
liquidus 液相线，液态的
liquid-vapor coexistence curve 液汽共存曲线，饱和线
liquid-vapor coexistence 液汽共存
liquid-vapor equilibrium 液汽平衡
liquid-vapor mixture 液汽混合物
liquid-vapor surface 液汽分界面
liquid-vapour interface 液汽界面
liquid vapour pressure 液体气化压力
liquid-volume measurement 液体体积计量
liquid wall 液垒，液障
liquid waste activity discharge 放射性废液排放
liquid waste and concentrate processing system 废液及浓缩物处理系统
liquid waste area sump pump 废液区污水泵
liquid waste area sump 废液区集水坑
liquid waste arising 废液产生的
liquid waste concentration plant 废水浓缩装置
liquid waste concentration 废液浓度
liquid waste controlled discharge pump 废液控制排放泵
liquid waste cooler 废液冷却器
liquid waste discharge pump 废液排出泵
liquid waste discharge system building 废液排放处理厂房，液体废物排放系统厂房
liquid waste disposal system 废液处理系统
liquid waste disposal technique 废液处理技术
liquid waste disposal 液状废物处置
liquid waste evaporator 废液蒸发器
liquid waste filter 废液过滤器
liquid waste hold-up system 废液滞留暂存系统
liquid waste hold-up tank 废液滞留（暂存）箱
liquid waste incineration 废液焚烧
liquid waste incinerator 废液焚烧炉
liquid waste monitoring and storage tank 废液监测及贮存箱
liquid waste monitoring tank 废液监测箱
liquid waste neutralizer tank 废液中和箱
liquid waste processing system 废液处理系统
liquid waste pump 废液泵
liquid waste receiver tank 废液接受箱

liquid waste residues　废液残渣
liquid waste storage system　废液贮存系统
liquid waste storage　废液贮存库
liquid waste system　废液系统
liquid waste treatment system　废液处理系统
liquid waste treatment　废液处理
liquid waste　废水，污水，废液，液体废物
liquid　液态的，液体，流体
liquifation　液化
liquification　液化
liquified natural gas　液化天然气
liquified petroleum gas　液化石油气
liquor condensate　冷凝液
liquor　流体，水溶液，液，液体，溶液
lisimeter　测渗计
lisoloid　凝胶
list current file　列表当前文件
list data　表格数据
list directed transmission　表式传输
list directory menu　列表目录菜单
list directory　列表目录
listed company　上市公司
listening apparatus　听声器
listening device　探听器
listening method　听声法【一种检查地下管道漏水的方法】
lister furrow　犁沟
list events　列表事件
list of articles　商品目录，物品清单
list of awards　授标名单，决标单
list of balances　余额表，差额表
list of bid awards　决标单
list of bolts　螺栓表
list of buildings and structures　建构筑物表
list of charts　统计图
list of detailed drawings　施工图纸目录
list of drawings　图纸目录
list of equipments　设备明细表
list of exchange rate quotation　外汇牌价表
list of expenses　费用清单
list of freight delivered and received　货物交接单
list of installation and inspection operation　设备安装及检验顺序表
list of main equipment　主要设备表
list of major suppliers and sub-contractors　主要供货商和分包商清单
list of materials　材料表
list of outstanding items　未完成的项目一览表
list of price　价目单，牌价表
list of references　业绩表
list of reinforcement　钢筋表
list of sailings　船期表
list of shareholders　股东名单
list of symbols　符号表
list of tenderers　投标人名单
list processing language　表处理语言
list processing　表处理，表格处理
list processor　表处理程序
list structure　表结构
list　列表，清单，表，一览表，表册，列举，目录
literal coefficient　用字母表示的系数
literal contract　成文合同，书面合同
literal operand　文字操作数
literal proof　书面证据［证明］
literal title　文字标题
literal　文字的
literature search　文献检索
literature　文献
lithanode　过氧化铅
litharge　密陀僧
lithiated　加氢氧化锂的
lithic contact　母质层【即土壤岩石接触层】
lithic tuff　石质凝灰岩
lithification　岩化，石化作用
lithium battery　锂电池
lithium blanket　锂转换区
lithium borate solution　硼酸锂溶液
lithium bromide absorption-type refrigerating machine　溴化锂吸收式制冷机
lithium bromide　溴化锂
lithium chloride resistance hygrometer　氯化锂电阻湿度计
lithium concentration　锂浓度
lithium cooled reactor　锂冷却反应堆
lithium doped cell　掺锂电池
lithium doped solar cell　掺锂太阳电池
lithium-hydroxide-inhibited　氢氧化锂抑制的
lithium hydroxide　氢氧化锂
lithium-sulphur battery　锂硫电池
lithium　锂
lithofacies　岩相
lithographic limestone　石印灰岩
lithological log　岩性测井记录
lithological map　岩性图
lithologic character　岩性特征，岩性
lithologic profile　岩性剖面
lithologic similarity　岩性相似性
lithology　岩石学，岩性学
lithometeor　大气尘粒
lithorelics　土状风化岩
lithosol　石质土
lithosphere　岩石层，岩石圈，陆界
litigation　诉讼
litmus paper　石蕊试纸
litmus　石蕊
little-fuse bolometer　熔丝式测辐射热计
littoral area　滨海区，潮汐区，海岸区
littoral deposit　海岸沉积物，沿岸沉积
littoral drift　沿岸漂流，沿岸漂移
littoral facies　沿岸相
littoral landform　海岸地形
littoral plain　海滩平原，平原潮滩，沿岸平原
littoral region　滨海区，潮汐区
littoral terrace　海岸阶地
littoral tracer　沿岸流示踪剂
littoral transport　沿岸输沙
littoral zone　沿岸带
litzen wire　编织线，绞合线，李兹线
litz wire　绞合线，编织线，李兹线
live animal regulation　活动物规则
live axle　动轴，驱动轴，传动轴，主动轴

live cable test cap　带电电缆试验盖头
live circuit　放射性回路，带电回路
live coal pile　自流煤堆
live coal storage　周转煤堆
live conductor　带电导体，火线
live crack　活裂缝
live end　带电端，加电压端，有电端，有效端
live frame of balance　天平动框
live-front switchboard　活动面板式配电盘，盘面接线式配电盘，板面接线式配电板
live gas　新鲜气体
live guy　活动牵索
live input　实际输入，恒定激励输入
live line clamp　带电作业线夹
live line connection　带电接线
live line cutter　带电线剪刀
live line disconnection　带电拆线
live line insulator detecting　带电检测绝缘子
live line insulator tester　带电线路绝缘子探测器
live line insulator washing　带电冲洗绝缘子
live line maintenance　带电作业，带电维护
live line measurement　带电测量
live line tool　带电作业工具
live line working　带电作业
live line work in substation　变电站带电作业
live line　带电线路
live load　动负载，动荷载，活（动）荷载，有效负载，工作负载，变动负荷，交变载荷
live metal　活动金属，带电金属
live part　转动部分，带电部分
live pile　周转煤堆
live plate　通风孔式板状炉排
live recording　现场记录，工作记录
live-roller conveyer　辊道输送机
live spindle　旋转主轴
live steam by-pass　新汽旁通，主汽旁路
live steam condition　新汽工况
live steam heated reheater　新蒸汽加热的再热器
live steam pipe　主蒸汽管道，新汽管道
live steam pressure　新蒸汽压力，新汽压力，主蒸汽压力
live steam reheater　新蒸汽再热器
live steam valve　新汽阀，主汽门
live steam　新蒸汽，主蒸汽
livestock farm　畜牧场，牧场
live stockpile　周转煤堆
livestock reservoir　牲畜用水蓄水池，畜用水塘
livestock　家畜
live storage of reservoir　水库有效库容
live storage pile　日产周转煤堆，日耗煤堆
live storage stockpile　常用煤堆
live storage　活库容，短期存储，日用煤场，兴利库容，调节库容，有效库容，有效储水
live stress　活载应力
live tank circuit-breaker　瓷柱式断路器，柱式断路器
live time　使用期限，寿命，使用时间，实况转播时间
live volume　有效容积
live wire entanglement　有电铁丝网
live wire maintenance　带电线维修
live wire　火线，有电导线，带电的电线，生龙活虎的人
live wood　潮湿木材
live working　带电作业
live zero　非零最小输出
live zone　活区
live　活动的，有电的，带电的，通电的，正极接地的
living allowance　生活津贴
living and welfare facility zone area　生活福利区面积
living and welfare facility zone　生活福利区
living area　生活区
living environmental condition　生物生存环境条件
living filter　生物滤池
living level　生活水平
living organism　活有机体
living place　住所，住宅
living quarters for staff　职工宿舍
living quarters　生活区，宿舍，住宅区
living resources　生物资源
living room　起居室
living standard　生活水平
Lizheng Geotechnical investigation　里正岩土勘察
LKG＝leakage(/leaking)　泄漏
LKH　380V 交流电系统【核电站系统代码】
latent heat　潜热
LCD(liquid crystal display)　液晶显示屏，液晶显示，液晶显示器
LLC(liquid level controller)　液位控制器
L＝length　长度
LLFM(low level flux monitor)　低通量监测器
LLH　380V 应急交流电系统【核电站系统代码】
light water moderator　轻水慢化剂，轻水减速剂
LLI(liquid level indicator)　液位指示器
line-to-line fault　线对线故障，线间故障
L＝liter　升
LLL circuit　低电平逻辑电路
LLL(low level logic)　低电平逻辑
LL＝low-low　低低，极低【DCS 画面】
LLLP(low leakage loading pattern)　低泄漏装料模式
loan financing　贷款筹措
loss report　损失报告书
low head injection　低压注射
Lloyd-Fisher square　劳埃德—费希尔磁损仪【测量铁芯损耗】
LLR(load limiting resistor)　负载限制电阻
LLRT(local leakage rate test)　局部泄漏率试验
LLS　水压试验泵发电机组系统【核电站系统代码】
LLW(low-level waste)　低放射性废物
LMERGE(low order merge)　低阶合并
LMFBR(liquid metal fast breeder reactor)　液态金属快增殖反应堆，液态金属冷却快中子增殖堆
LMF(liquid-metal fuel)　液态金属燃料
L/MF(low and medium frequency)　低中频率
LMFR(liquid metal fuel reactor)　液态金属燃料反应堆
LMH　220V 交流电配电系统【核电站系统代码】
LMLR(load memory lockout register)　寄存存储锁

定寄存器
LMM(logic master module) 逻辑主模件
LMP(load micro-program) 装入微程序
LMR(liquid metal reactor) 液态金属反应堆
LMTC(low-speed magnetic tape controller) 低速磁带控制器
LMTD(logarithm mean temperature difference) 对数平均温差，对数平均温压
LMTU(low-speed magnetic tape unit) 低速磁带装置
LNA 220V交流重要负荷电源系统【核电站系统代码】
LNC(logical-node class) 逻辑节点类
LN Data 逻辑节点数据
L network L型网络
LNF 220V交流不间断电源系统【核电站系统代码】
LNG(liquefied natural gas) 液化天然气
LNG(logical-node group) 逻辑节点组
LNG power plant 燃液化天然气电站
LNG tanker 液化天然气运输船
LN(logic node) 逻辑节点
LN name 逻辑节点名
LNTP(Limited Notice To Proceed) 受限的开工通知
loadability 承载能力，载荷能力
load acceleration constant 加速负载常数
load accumulator 寄存累加器
load adjusting device 负荷调节装置
load adjustment 负载调节，负荷调整，负荷调节
load admittance 负载导纳，载荷接纳
loadage 装载量
load air-flow ratio 负荷风量之比
load allocation 负荷配置，荷载分配
loadamatic control 负荷变化自动控制
loadamatic 随负载变化自动作用的
load analyzer 负载分析器
load-and-go （程序）装入立即执行
load and resistance factor design 荷载和抗力系数设计
load and unload apparatus 上下料装置
load and unload 装卸
load angle characteristic 功角特性，负载角度特性
load angle 负载角，负荷角，功率角，功角
load anticipator 负荷预测器
load application 施加荷载
load area 负载区域，负载面积
load ascension 升负荷
load at failure 破坏荷载
load at first crack 初裂荷载，初现裂缝的荷载
load backfeed 负荷反馈，反馈
load-back 反馈
load balancing reactor 匹配电抗器
load balancing 负载平衡
load-band of regulated voltage 特定负载区的电压调整范围
load bank 负荷调节组
load bar 吊梁
load bearing ability 带载能力
load bearing capacity 载荷容量，承载能力
load bearing characteristic 荷载特征，负荷特性
load bearing disc 载荷圆盘
load bearing frame 承重框架
load bearing member 承重构件，承重结构构件
load bearing plate 承压板
load bearing skeleton 承重骨架
load bearing strength 承载强度
load bearing structure 承重结构
load bearing system 承重系统
load bearing wall panel 承重墙板
load bearing wall 承重墙
load bearing 承重，承载，载荷，承受负荷
load block 负载划区，负荷划区，负荷闭锁
load brake 超载制动器，超重制动器
load-break cutout 负荷电流切断器
load-break fuse cutout 负荷切断熔丝
load-break rating 截断容量
load-break switch 负荷开关
load buffer memory 寄存缓冲存储器
load bus 负荷母线
load button 加载按钮，输入按钮，负荷按钮
load calibrating device 负载检验器
load capability 负载能力
load capacitance 负载电容
load capacity 承载能力，负载能力，负荷容量，载重能力，负荷量，载重量
load carrier 负荷载波，负荷载体
load carrying ability 负载能力，起重能力，载重能力，起重量，载重量
load carrying capacity 负荷能力，容许载荷，起重能力，承载能力
load carrying capacity per bolt 单个螺栓的承载能力
load carrying duty 负载工况，承载方式，负荷率，承载率
load carrying member 承重构件
load carrying 承载的，带负荷（运行）的
load car 空中吊运车，整车货物，重车
load case 荷载工况，载荷状况，负荷情况
load cell assembly 负载元件总成
load cell simulator 载荷传感器模拟装置
load-cell weighing system 负荷元件称重系统
load cell 测力传感器，测压元件，加载测试装置【铁塔试验】，称重传感器，载荷传感器，测力盒
load center substation 枢纽变电站
load center unit substation 枢纽变电站，负荷中心配电变电站
load center 负载中心，负荷中心
load changer 负荷调节器，负载调节器
load change test 变负荷试验，负荷变化试验
load change 负荷变化，负载变化
load changing rate 负荷变化率
load characteristic curve 负载特性曲线，负荷特性曲线
load characteristics 荷载特征，负荷特性，负载特性
load characteristic test 负荷特性试验
load chart 负荷变动图，负荷图
load choke 负荷扼流圈
load circuit 负荷电路

load classification number 荷载分类指数
load code for the design of building structures 建筑结构荷载规范
load coefficient 负荷系数，荷载系数
load coil 感应加热线圈，加感线圈
load combination 荷载组合，荷重组合
load-compensated diode-transistor logic 负载补偿二极管晶体管逻辑
load compensation 负荷补偿，负载补偿
load component 负荷分量，负载分量
load compression diagram 负载压缩图
load condition 负荷条件，载荷条件，负荷状态，负载状态，载重状态，负载情况
load-consolidation curve 荷载固结曲线
load-controlled consumer 负荷受控用户
load-controlled stress 机械荷载应力
load control of ball mill 钢球磨煤机负荷控制
load control system 电力负荷控制装置
load control unit 负荷控制装置
load control 负荷调节，负载控制，荷载控制，按负荷调节，负荷调整
load criterion 负荷标准
load cross line 荷载交叉线
load current meter 负荷电流表
load current supervision 负荷电流监视
load current tap changer 有载分接开关，有载分接头切换装置
load current tap 有载抽头，有载分接头
load current 负载电流，负荷电流
load curve 负载曲线，负荷曲线，荷载曲线
load cycle operation 负荷循环运行
load cycle 负荷循环，负载循环
load cycling behaviour 负荷循环性状
load cycling performance 负荷循环性能
load cycling 负荷周期，周期性负荷
load decrease block 减负荷闭锁
load decrease 负荷降低
load decrement 负荷减量
load-deflection curve 荷载挠度曲线
load-deflection relation 荷载挠度关系
load deformation curve 应力应变曲线
load deformation path 载荷形变过程
load demand signal 负载需量信号
load demand 负荷指令，负荷需求，负荷需要量
load density of distribution system 配电网负荷密度
load density 负载密度，负荷密度，负载强度，荷载密集度，装载密度
load-dependent turbine control 根据负荷对汽轮机进行调节
load-dependent 取决于负荷的，按负荷而定的，与负荷有关的
load device 载荷装置，加载装置
load diagram 负荷曲线，负载图，荷载图，负荷图
load diffusion 负荷分布，负荷分散，负荷分配
load-disconnected switch 负荷隔离开关
load dispatcher 负荷调度员，负载调度员
load dispatching center 负荷调度中心
load dispatching expenses 调度费
load dispatching instruction 供电调度说明
load-dispatching office 负荷调度所
load dispatching 负荷分配，负载调度，供电调度
load distribution line 荷载分布线
load distribution 配电，负荷分配，载荷分布，负荷分布
load diversity power 负荷分散功率
load diversity 负荷不同率，负荷不同时率，负荷分散率
load division 负荷分配
load double precision 寄存双精度
load down 降负荷
load draught 满载吃水
load dropping test 甩负荷试验，减负荷试验
load dropping 甩负载，甩负荷，减负荷，负荷降低
load due to wind pressure 风压荷载
load dump test 甩负荷试验
load dump 甩负荷
load duration curve 荷载历时曲线，负荷持续时间曲线，负荷持续曲线
load duration 负载持序时间
load dynamic characteristic 负载动态特性
loaded antenna 加载天线，加感天线
loaded area 承载面积，荷载面积
loaded cable 负荷电缆，加感电缆
loaded car track 重车线
loaded circuit 加感电路，加载电路
loaded concrete shield 重混凝土屏蔽
loaded concrete 混凝土保护
loaded core 已装料的堆芯
loaded filter 承载反滤层
loaded governor 重锤式离心调速器
loaded impedance 负荷电抗
loaded lattice 已装料的栅格
loaded length 承载长度
loaded line 负载线，加载线路
loaded radius （堆芯）装载半径
loaded resin 饱和树脂
loaded rubber 填料橡胶
loaded solvent 加入的溶剂
loaded test specimen 加载的试样
loaded track line 重车线
loaded train 重列
loaded wagon track 重车线
loaded 加载的，有负载的，负载的，装载的，装料的
load effect combination 荷载效应组合
load effect 荷载效应
load equalization 负荷平衡，负载平衡
load equalizer 负荷平衡装置
loader-dozer 装载推土两用机
load error 负荷误差，负载误差
loader truck 起重汽车
loader-unloader 装卸机，装卸桥
loader 装载机，装料机，输入器，装料器，装货机，加载器，装入程序
load estimate 负荷估算
load estimation 负荷估算，负荷估计，负荷计算
load-extension curve 荷载-伸长曲线
load-extension diagram 负荷曲线

load face 负荷面
load factor design 荷载系数法设计
load factor method 荷载系数法
load factor 负荷率，负荷系数，荷载系数，负荷因子，利用系数，设备利用率
load feeder 负荷馈线，负载馈线
load flow calculation 潮流计算，负荷潮流计算
load flow control 电力潮流控制
load flow 电力潮流
load fluctuation 负载变化，负荷变化，荷载变化，负荷波动
load follower reactor 负荷跟踪反应堆
load follower 负荷跟踪
load following behaviour 负荷跟踪性状
load following capacity 负荷跟踪能力
load following condition 负荷跟踪工况
load following operation 负荷跟踪运行
load following regime 负荷跟踪工况
load following 负荷跟踪
load follow operation 跟踪负荷运行，连续负荷运行
load follow pattern 负荷跟踪曲线
load follow 负荷跟踪
load force 载荷力
load forecasting 负荷预测
load-frequency controller 负荷频率控制器
load-frequency control 负载频率控制，负荷频率控制
load gauge 测力计，负荷计
load-generation imbalance 负载发电失衡
load governing 负荷控制，负荷调节
load gradient 负荷梯度
load graph 负荷图，负荷曲线
load growth 负荷增长
load histogram 负荷直方图
load hoist 装卸料吊车
load hopper 装料漏斗
load hour factor 负荷时间率
load impact allowance 容许冲击荷载
load impact 负荷冲击
load impedance 负载阻抗，终端阻抗，负荷阻抗
load inclination 荷载倾斜角
load increase block 增负荷闭锁
load increase 负荷增大，负荷增量
load increasing 加负荷
load increment 负荷增量，载重增加，载荷增量
load independent 与负荷无关的
load index register instruction 输入变址寄存器指令
load-indicating bolt 荷载指示螺栓
load inducing mechanical yawing system 装载机械偏航系统，负荷诱导机械偏航系统
load information 荷载资料
loading accident 装料事故
loading airlock 装载用气闸室，装料用气闸门
loading and unloading bridge 卸煤桥
loading and unloading risk 装卸货险
loading and unloading well 装卸井
loading and unloading 装卸，装卸工作
loading area 荷重区，装料区，装载区，装载面积
loading assumption 荷载假定
loading back method 反馈法
loading bay 吊装间，装料间，进料台，装货间
loading bench 装料台
loading berm 反压平台，反压护道
loading berth 散货停留地
loading boom 吊货杆，装载臂杆
loading bridge 桥式起重机，桥式装载机
loading broker 装卸代理人
loading capability 载荷能力，负载能力，装载能力
loading capacity design 按承载能力设计
loading capacity 承载能力，承压能力，载重量，装载能力，负荷容量，载荷能力
loading car positioning 重车定位
loading cask 装料罐，装料容器
loading charges 装船费，装货费
loading chute 供料滑槽，装料溜槽
loading coefficient 装载系数，加感系数，负荷系数，荷载系数，加载系数
loading coil spacing 加感线圈间距，加感系数
loading coil 加感线圈，加载线圈
loading combination 负载组合，综合负荷
loading conditions 载荷条件，负载条件
loading conveyor 装载运输机
loading crane 装料吊车
loading curve 负载曲线，吸附曲线
loading device 装料设备，装载设备
loading diagram 荷载图
loading distribution of turbine cylinder 汽缸负荷分配
loading duration 加载时间，加载期间，受载历时
loading effect 负荷的影响，负载效应
loading error 加载误差
loading face 装料面【反应堆】
loading facility 装载设备
loading factor 充填因子【发动机】，加载因子，负荷因素
loading frame 加荷载架，载荷架
loading funnel 装料漏斗
loading hire 装货费
loading hopper 加料斗，装料斗
loading in bulk 散装
loading inductance 加感
loading list 装货单
loading machine 上料机，装料机，带载机组
loading material 加感材料
loading method 装料方法
loading of source 装源
loading of streams 河流负荷
loading oil pressure 工作油压
loading parameter 负荷参数
loading pattern 装料方式，装料图，装填模式，装载模式，加载方式
loading pit 装料井
loading plan 装载计划
loading plate 承载板
loading platform 货运站台，装货站台，装煤站台，装卸平台
loading plug 装料塞

loading point 加感点，负荷点，负载作用点，加载点，装载点
loading port and destination 装卸货物口岸
loading port 装运口岸，装货口岸，装货港口，装货港
loading procedure 装料程序
loading program 装入程序
loading range 负载范围
loading rate limiter 升负荷速率限制器，荷载率限制器
loading rate 负荷上升率，升负荷速率
loading ratio （轴承）负载系数，充填比
loading regime 负荷状态，带载方式
loading resistor 负荷电阻器
loading routine 输入程序，装入（例行）程序
loading scheme 装料方案
loading section 加感部分，加感段
loading sensor 负荷传感器
loading-settlement curve 荷载减载曲线
loading shift 载荷重新分布
loading shovel 铲装车，装料铲
loading skirt 导料槽
loading speed 加载速度，负载速度，装货速度
loading temperature 装料温度
loading test 加载试验，载重试验，荷载试验，负荷试验
loading/unloading bridge 装卸桥
loading/unloading charge 装卸费
loading,unloading,transfer point 装、卸、转运站
loading-up 升负荷
loading winch 起重绞车
loading 加负荷，载重，加感，加载，加荷，加重，装药，装料，装货，装载，装炸药
load in system 系统负荷
load intensity 荷载强度
load interrupted 停电负荷
load leveling relay 负荷平衡继电器
load leveling 负荷调整，负载均衡，负荷管理，负荷平衡，调匀负荷
load level 负荷等级
load limitation 负荷限制，负荷极限范围
load-limit changer 负荷极限变换器
load limiter circuit 负荷限制电路，负载限制电路
load limiter 负载限制器，负荷限制器，功率限制器
load limiting device 负荷限制器
load limiting operation 限负荷运行
load limiting resistor 负载限制电阻，负荷限制电阻器
load limit motor 负荷限制电动机
load limit operation 限负荷运行
load limit 负荷限度，负荷限制
load line mark 满载吃水标线
load line 负载线，载重吃水线，载重线，加感线路
load loss-factor 负荷损失因数
load loss 负载损耗，负荷损失，（变压器的）铜耗
load-magnitude 负载值，负载大小
load management system 电力负荷管理装置
load management 负荷管理，用电负荷管理
load matching 负载匹配，负荷匹配
load matrix 荷载矩阵
load memory lockout register 寄存存储封锁寄存器
load micro-program 装入微程序
load mismatch 负载失配，负荷失配
load mode 装载方式
load module library 输入模块库
load module 装入程序模块，输入模块
load moment 负载力矩
load not supplied 缺供负荷
load of accelerating 加速负荷，加速接载
load off 甩负荷
load of installation 安装荷载
load of normal running 正常运行负荷
load of starting 启动负荷，启动负载
loadometer 测压仪，测力仪，测功器
load on axle 轴心加荷
load on call 调用装入
load on top 顶端加油法
load on 加负荷，加载
load order 带负荷顺序
load out conveyor 卸载传输装置
load packer 打包机，捆包机
load pattern 负载曲线图，负荷曲线图，负荷特性曲线
load peak 负荷尖峰，尖峰负荷，负载尖峰，峰荷
load period 负荷周期，载荷期间
load per unit length 单位长度荷载
load phase 负载相，带载相，负荷相
load pick-up 升负荷，负载传感器
load plan 装载计划
load point 开始读数点，荷载作用点，负荷点，加感点
load power factor 负荷功率因数
load power 负载功率，负荷功率
load prediction curve 负荷预测曲线
load prediction 负载预计，负荷估计，负荷预测，负荷预计
load profile 负荷曲线
load program status word 装入程序状态字
load pulley block 载重滑车组
load ramp 负荷斜变，负荷斜增
load range at constant temperature （保持）恒定汽温下的负荷范围
load range of boiler 锅炉负荷调节范围
load range 负荷（调节）范围，负载范围，荷载幅度
load rate prepayment meter 根据负荷率预付电费电表
load rate 负荷率，荷载率，单位荷载
load rating 额定负载，负荷定额，额定负荷
load ratio control 带载电压调节
load ratio voltage regulator 带载电压调整器，有载调压器
load ratio voltage transformer 有载调压式变压器
load ratio 负荷比，载荷比，有效载荷
load reactance 负载电抗，负荷电抗
load reapplication 重新加荷

load recovery　负荷恢复
load redistribution　荷载重分布
load reducing　减负荷
load reduction　负载减少，负荷减少，负荷降低，减载，减负荷
load reference　参考负荷
load regulation performance of unit　机组调峰性能
load regulation　负载调节率，负荷调节，负载调节
load regulator　负载调节器，负荷调节器
load rejection overvoltage　甩负荷过电压
load rejection test　甩负荷试验
load rejection to auxiliary load　甩外负荷【只带厂用电】
load rejection　抛负荷，甩负荷【汽轮机事故】，甩负载，弃负荷，负荷中断，供电暂停，卸载
load-related force　与负载有关的力
load relay　负荷继电器，负载继动器
load relief capability　甩负荷能力，甩负载能力
load relief capacity　甩负荷能力
load relief　卸负载，减负荷，甩负荷，减负载，甩负载
load removal age　卸荷龄期
load removal　负荷卸载，卸负荷，卸载
load repeat counter　负载重复计
load representation　负载表示法，负荷模拟法
load reserve capacity　负荷备用容量
load resistance　负荷电阻，负载电阻
load response　负荷反应
load restoration　重新加负载，负荷恢复，负载复原，恢复负荷
load reversal　负载反向，负荷反向，荷载反向，反向加载
load rheostat　负荷变阻器
load saturation curve　负载饱和曲线
load selector switch　负载选择开关
load selector　负载选择器
load sensing　负荷检测
load sensor　负荷传感器
load setting gear　负荷给定装置
load settlement curve　荷载-沉降曲线
load set　一套载荷力【三向力加三向扭】
load sharing matrix switch　均分负载矩阵开关
load sharing　负载分配，负荷分担，均分负载
load shedder　自动甩负荷装置
load shedding capability　甩负荷能力，甩负载能力
load shedding equipment　减载装置，甩负荷装置
load shedding　甩负载，甩负荷，卸负荷，限电性停电，卸载
load-shifting resistor　负荷转移用电阻器，耦合电阻器
load shifting　负荷转移
load shift register　送入移位寄存器
load shock　负载冲击，负荷冲击
load shutdown test　甩负荷试验
load-side breaker　负荷侧断路器
load signal　负荷信号
load slip　负载转差率，负荷转差率，负载滑差
load spectrum　载荷谱
load speed　负载转速，负载速度
load spreading property　荷载分布特性
load stability　负荷稳定性
load stabilization　负载稳定，负载镇定，负载稳定化
load stage　荷载阶段
load-start motor　有载启动电动机，负载启动电动机
loadstone　天然磁石，磁铁矿，磁性氧化物，磁性氧化铁，吸力物，吸引人的东西
load-strain curve　荷载应变曲线，载荷应变曲线
load-strain diagram　荷载应变图，荷载应变曲线
load stress　负荷应力，负载应力，载荷应力
load structure　负载构成
load-supporting　承载的
load suppression gear　负荷限制器
load surge　负载陡增，负荷冲击，负荷大幅度摆动，负载冲击
load-swelling curve　荷载-膨胀曲线
load swing　负荷波动，负荷摆动
load switch　负载开关，负荷开关
load tap changer　有载分接头转接器，有载分接开关，有载接头转换器
load-tap-changing transformer　带有载分接开关的变压器
load test on pile　桩上荷载试验，桩载试验
load test　加载试验，带负荷试验，负载试验，负荷试验，加负载试验
load throw-off test　甩负荷试验
load throw-off　负荷卸除，甩负荷，抛负荷
load throw-on　负荷带上，快速接带负荷
load time constant　负载时间常数
load time deflection curve　荷载时间挠度曲线
load time　加载时间，负载时间，载荷时间
load torque　负荷扭矩，负载扭矩，负载转矩
load tracking capacity　负荷跟踪能力
load transducer　荷载传感器
load transfer mechanism　荷载传递机理
load transfer method　荷载传递方法
load transfer switch　负荷切换开关
load transfer　负载转移，荷载传递，负荷转移
load transformer　负载变压器
load transient　载荷瞬变值，负荷暂态，载荷瞬变
load trial　负载试验，负荷试验，试加载
load unload point　装卸料点，加料出料点
load-up condition　负荷上升状态
load-up rate　负荷上升率
load-up test　升负荷试验
load-up variation　负荷上升率
load valley　负荷谷值，负载曲线低谷
load-variant　随负荷变化的
load variation rate　负荷变化率
load variation test of analog control system　模拟量控制系统负荷变动试验
load variation　负载波动，负载变动，负荷变化，负荷波动
load vector　荷载向量
load voltage　负载电压，负荷电压
load weighing devices　称量装置
load winding　负荷绕组，负载线圈
load zero　寄存零

load 载荷，荷载，负荷，负载，装载，加载，装填，寄存，输入，装入，取数
LOA(length over all) 船舶总长度
LOA(letter of award) 授标函
loamy clay 壤（质）黏土，垆姆质黏土
loamy gravel 壤质砾石
loamy sand 垆姆质砂土，壤质砂土，砂壤土，亚砂土
loamy soil 垆姆土，壤土
loamy texture 壤质构造
loamy 壤质的，肥沃的
loam 肥土，垆姆，壤土，亚黏土
loan account 贷款账户
loan agency 贷款机构
loan agent 贷款代理人
loan agreement 贷款协定，贷款协议，放款合同
loan amount 贷款额
loan and advance 借款与预付
loan bank 贷款银行
loan bond 债券
loan capital 借贷资本
loan collateral security 有押贷款
loan commitment 承诺的贷款
loan condition 贷款条件
loan contract 贷款合约，贷款契约
loan covenant 贷款合约，贷款契约
loan delinquency 贷款拖欠
loaned items 借出的物项
loanee 借入者，债务人
loaner 债权人，借出者
loan for a short time 短期贷款
loan for consumption 消费贷款
loan for conversion of exchequer bond 国库债券
loan for exclusive use 专用贷款
loan for export 外销贷款
loan for store 储备贷款
loan for use 供使用的贷款
loan from foreign powers 外债
loan funds 借贷资金，贷放基金
loan guarantee programme 贷款担保计划
loan guarantee 贷款担保
loan interest rate 贷款利率
loan interest 贷款利息，债利，借贷利息
loan loss reserve funds 贷款损失准备基金
loan loss 贷款损失
loan money 借款
loan note 借据
loan number 贷款号
loan of development project 开发项目贷款
loan of project 项目贷款
loan on actual estate 不动产抵押贷款
loan on collateral security 抵押贷款
loan on favourable terms 优惠贷款
loan on goods 用商品抵押的贷款
loan on guarantee 保证贷款
loan on passbook as collateral 以存折抵押的贷款
loan on security 抵押贷款
loan rate 贷款利率
loan regulation 借贷条例
loan repayment schedule 归还贷款进度
loan repayments from abroad 国外还债
loan restriction 借（贷）款限制条件
loan secured by credit 以信用担保的贷款
loan secured by things 以物担保的贷款
loan secured 抵押贷款
loan services 贷款业务
loans on favorable terms 优惠贷款
loan with discounted interest 贷款贴息，贴息贷款
loan with no or low interest 无息或低息贷款
loan without security 无担保贷款
loan worthy 有偿付能力
loan 贷款，借贷，信贷，借款
lobby ventilation 前室通风
lobby 穿堂，大厅，门厅，门廊，前厅，休息室
lobe switching method 等信号法，波瓣晃动法，波瓣转换法
lobe 波瓣【正弦波的半周】，突齿，瓣，叶片，天线辐射图，凸起部
LOB(line of balance) 平衡线
local acceleration 当地加速度
local action 局部作用，局部作用量
local adjustment 局部平差
local advantage 地方优势
local agent 当地代理人
local air conditioning 局部区域空气调节
local air supply system 局部送风系统
local amplifier 本机放大器
local angle of attack 当地迎角
local angle of latitude 当地方位角
local apparent time 当地视时
local application extinguishing system 局部应用灭火系统
local area network 部局地区电网，局域网
local asymptotic stability 局部渐近稳定性
local attack 局部浸蚀
local authorities 地方当局
local automatic message accounting 局部信息自动计算
local backup 局部后备
local base level 当地基准面
local base vector 局部基矢量
local battery switch board 磁石式电话交换机
local battery telephone 磁石式电话
local battery 本机电池，自给电池
local blade chord 当地叶片弦（长）
local board 就地仪表盘
local boiling void 局部沸腾空泡
local boiling 局部沸腾
local breeze 局部风
local buckling （局部）弯曲，局部压屈
local budding 局部压曲
local building code 地方建筑规范
local building material industries 地方建筑材料工业
local burn-up 局部燃耗，局部堆积
local busy 局部电话线不空，局部占线
local carburization 局部渗碳
local call 本地通话，市内呼叫
local circulation 地方性环流，局部环流，局部

循环
local circumstance 局部情况
local civil time 地方民用时间
local climate condition 局部气候条件
local climate 地方气候，局部气候，小气候
local code and regulation 当地法规
local code 地方法规
local communication console 局部通信控制台，局部通信控制器
local company 当地公司
local competitive bidding 国内竞争性招标
local concentration 当地浓度
local conditions hypothesis 局部状态假说
local connection 局内通话，本地通话，局部连接
local consultant 当地咨询人员
local consumer-protection 当地用户的保护
local content 当地成分，当地参与制造
local contractor 当地承包商
local control board 就地控制盘
local controller 局部控制器
local control panel 就地控制柜，就地控制盘
local control room 就地控制室
local control station 就地控制站
local control unit 就地控制设备
local control 就地控制，局部控制，现场控制
local coolant channel blockage 冷却剂管道局部堵塞
local core accident 局部堆芯事故
local corrosion 局部腐蚀
local costs 当地费用
local creep stress 局部蠕变应力
local criticality 局部临界
local currency 当地货币
local current 局部电流
local damage 局部损坏
local data package 局部数据包
local deformation 局部畸变，局部变形
local dehomogenization 局部非均匀化
local destabilization 局部扰动
local diffusivity 当地扩散率
local dip （中子通量）局部坑
local distribution network 市内配电网，地区配电网
local distribution system 局部配电系统，局部分布系统
local distribution 局部分布
local disturbance 局部扰动
local dose rate measurement 局部剂量率测量
local dose rate 局部剂量率
local dose 局部剂量
local drag 当地阻力
local duties 地方税
local eddy 局地涡流，局部涡流
local effect 局部效应
local efficiency 局部效率
local elastic instability 局部弹性失稳
local electric supply 局部电源，本地电源
local elongation 局部伸长
local emergency operating centre 地区应急控制中心
local equilibrium 局部平衡
local equipment 就地设备
local exhaust system 局部排风系统
local exhaust ventilation 局部排风，局部排气通风
local exhaust 局部排风
locale 场所，地点
local fabrication 就地制造
local factor （功率分布）局部因子
local failure 局部破坏
local feedback 局部反馈
local flow condition 局部流动工况
local flow 局部气流，当地流动
local fluctuating pressure 局部脉动压力
local flux increase 局部通量增加
local flux modulation method 局部通量调制法
local friction drag 局部摩擦阻力
local fuel 当地燃料
local govemment 当地政府
local ground water 区域地下水
local gust 局部阵风
local head loss 局部压头损失
local heat flux density 局部热流密度
local heat flux 局部热流密度
local heating 局部发热，局部加热，局部采暖
local heat loss 局部热损失
local heat transfer coefficient 局部传热系数
local heat treatment 局部热处理
local highway 地方公路
local hot-cell factor 局部热栅元因子
local hydriding 局部氢化
local hysteresis loop 局部磁滞环
local illumination transformer 局部照明变压器
local income tax 地方所得税
local infection 局部传染
local inflow 地方来水，区间来水
local influence 地方影响
local input/output cabinet 就地输入输出柜
local instability 局部不稳定性，局部失稳
local installed equipment 就地安装设备
local inversion 局部逆温
local I/O cabinet 就地输入输出柜
local isotropy 局部各向同性，局部均向性
localization of boundary stress 边缘应力的局部性
localization of fault 故障点定位，确定故障点
localization 定位，位置，单元地址，定域，局限，局部化，本土化，国产化，当地化
localizator 定位发射机，定位器，定位探测器
localized air supply for air-heating 集中送风采暖
localized control 就地控制
localized corrosion 局部腐蚀
localized damage 局部损失，地区性损失，局部损伤
localized erratum 局部误差
localized favored stream 局部被促动的煤流
localized hot spots 局部热源
localized length scale 当地积分长度尺度，局部尺度
localized network 局部网络
localized pit 局部腐蚀点

localized print suppression　局部打印封锁
localized regulation　局部调节
localized salt concentration　局部盐浓缩
localized scour　局部冲刷，局部淘刷
localized vibration　局部振动
localized　定了位的，局部化的
localizer　定位发射机，定位器，定位探测器
localize　定位
local labour　本地劳力，当地劳力
local leakage test　局部泄漏试验
local letter of credit　本地信用证
local lift　当地升力
local lighting　局部照明
local loop　就地回路
local loss　局部损失
locally active　局部含源
locally-administered state enterprise　地方国营企业
locally controlled power station　部分受控电站，当地控制电站
locally-derived control source　就地分路控制电源
locally mounted　就地安装
locally-oxidized CMOS　定位氧化的互补金属氧半导体晶体管
locally passive　局部无源
locally　局部地，在本地，当地
local Mach number　当地马赫数
local manual operation　就地人工操作
local market clearing price　当地市场结算价
local mass eccentricity　局部质量偏心率
local material　当地材料，地方材料
local mean time　地方标准时，地方平均时
local measurement　当地测量，就地测量，就地测定
local microclimate　局部微气候
local momentary concentration　局部瞬间浓度
local monitoring instrument　就地监视仪表
local-mounted controller　基地式调节仪表
local natural resources　地方自然资源
local network　本地电信网
local noise　局部噪声
local operating station　就地操作站
local oscillator　本机振荡器
LOCA(loss of coolant accident)　失水事故【轻水堆】，一回路失水事故，冷却剂丧失事故
local overheating　局部过烧，局部过热
local oxidation of silicon　硅的局部氧化
local panel　现场表盘，就地仪表盘，现场配电盘
local participation　当地参与（制造），国内分包
local peaking factor　局部峰值因子
local peaking　局部峰值
local peak of pressure　当地峰压
local people　当地居民
local perturbation　局部扰动
local per unit surface　电位面积负荷
local planning　地方规划
local pneumatic controller　基地式调节仪
local pneumatic process　局部气压法
local pollution　局部污染
local porosity　局部孔隙度，局部空隙率
local port　当地港湾，地方港湾
local power network　地方电网
local power oscillation　局部功率振荡
local power peaking　局部功率峰值
local power range monitor　局部功率区段监测器
local power　局部功率
local precipitation　地方性降水，局部降水
local pressure pulsation　局部压力脉动
local pressure　当地压力，局部压力
local primary membrane stress　初期局部膜应力，一次局部薄膜应力
local processor　本地处理机
local project　地方项目
local publicly owned electric utility　当地公有电力公司
local pushbutton station　就地按钮站
local radiation effect　局部辐射效应
local railway　地方铁路，专用铁路
local rate　当地单价
local reaction　局部反应
local reception　本地接收
local rectangular pressure pulse　局部矩形压力脉冲
local register　局部寄存器
local relief system　局部送风系统
local relief　局部送风
local-remote relay　本地遥控转换继电器
local-remote switch　就地遥控开关
local resistance factor　局部阻力系数
local resistance　局部阻力
local resonance　局部谐振，局部共振
local resources　当地资源
local response technique　局部响应技术
local rod power　局部线功率
local rounding error　局部舍入误差
local runoff　当地径流，涝水，内水
local sampling rack　就地取样盘
local section　局部剖面
local seller　当地卖方
local sender　本地发射机
local separation　局部分离
local set　就地设定
local shear failure　局部剪切破坏
local shielding　局部屏蔽
local sidereal time　地方恒星时
local signal　本地信号，本局信号
local similarity　局部相似，局部相似率
local solar time　地方太阳时
local solenoid valve box　就地电磁阀箱
local stability　局部稳定性
local storage　局部存储器
local strain　局部应变
local stress relieving　局部应力消除，局部消除应力
local stress　局部应力
local subchannel blockage　局部子通道堵塞
local subcooling　局部欠热
local subscriber　本地用户
local supplier　当地供应商
local surrounding　局部环境
local tax　地方税

local temperature conditions 当地温度条件
local thermal power plant 区域火力发电厂
local thermodynamic equilibrium 局部热力平衡
local time 地方时间
local topography 局部地形
local training 就地培训
local transmission network 城市供电网
local transmitter 地方用发射机，本地发射机
local transportalion 当地运输
local trip control 本机跳闸控制，就地跳闸控制
local trip 本地跳闸
local turbulence 局部湍流（度）
local unconformity 局部不整合
local velocity fluctuation （流量计的）局部速度起伏，局部速度变动
local velocity of sound 局部声速，当地声速
local velocity 局部速度，局部流速
local ventilation 局部通风
local vent 就地排气，局部排气，局部通风
local void distribution 局部空泡分布
local water level gauge 就地水位表
local wholesaler 本地批发商
local wind characteristics 局部风特性
local wind profile 局部风廓线
local wind regime 局部风况
local wind resource 当地风力资源
local wind 局部风
local yield of the material 材料局部屈服
local 局部的，本地的，本机的，轨迹的，就地的，当地的
locate a fault 确定故障位置
located object 定线目标
locate in position 定位，固定
locate mode 定位方式
locate statement 定位语句
locate 确定，定位，探测，判明，设置，固定，位于
locating axis 定位轴线
locating bar 定位棒，定位块
locating bearing 定位轴承，止推轴承，推力轴承
locating collar 轴承定位环，定位凸缘套，定位环
locating dowel 定位销
locating elevation of tube support plate by eddy current testing 用涡流测试管子支撑板的高度
locating flange （轴承）定位凸缘，有止口的法兰，定位法兰
locating information 位置信息
locating journal 定位轴颈
locating of tower 杆塔定位
locating pin 定位销，中心校正杆
locating point 定位点
locating ring 定位环
locating screw 定位螺丝
locating slot 定位槽
locating 定位
location centre 定位中心
location constant 单元常数
location counter （存储）单元计数器，地址计数器，指令计数器
location drawing 位置图【工程平面】
location free procedure 浮动过程
location hole 定位孔
location index 图幅接合表
location map （地理）位置图，定位图
location of dam 坝址
location of fault 确定故障位置，故障定位，故障位置
location of industry 行业所在地
location of manufacturing 工业位置，出产地
location of mistakes 寻找错误，错误勘定
location of railway connection 铁路接轨地点
location of the project 工程地址
location of the water table 地下水位
location of WTG 机位
location plan 位置图
location point 方位点
location stack register 位置组号寄存器，单元栈寄存器
location survey 定线，勘址，勘线，定测，定位测量，定线测量
location 地点，场所，安置，布置，定位，方位，位置，地址，存储单元，部位
locator key 定位销
locator 定位器，定位装置，探测器
loci of root 根轨迹
lockage water 闸内水量，闸内蓄水
lockage 闸程
lock-and-dam method 闸坝法
lock approach 船闸引航道
lock bar steel pipe 箍锚钢管
lock bar 锁紧杆
lock bay 闸池
lock block 装锁木块
lock bolt 锁紧螺栓，锁紧螺钉，防松螺栓，锁定螺栓
lock button 锁紧按钮
lock cap 锁紧帽
lock chamber wall 闸室墙
lock chamber 船闸室，闸室
lock check gate 防逆闸门
lock closed 锁闭
lock-down circuit 闭锁电路，单路锁定电路
lock-down relay operation 继电器锁定操作
locked cable 锁紧缆
locked-closed 锁定在关闭状态下
locked-coil rope 箍环缆索
locked in oscillator 同步振荡器
locked in resonant oscillation 锁定共振
locked in stall 稳定失速
locked in stress 内应力
locked in vortex-induced response 锁涡响应
locked-open 锁定在开启状态下
locked out 闭锁的，切断的
locked position 闭锁的位置
locked push button 有闩锁的按钮
locked rotor condition 转子止转工况，转子制动工况
locked rotor current 止转转子电流，堵转转子电流，转子堵转电流
locked rotor temperature-rise 转子止转温升，堵转温升
locked rotor test 堵转试验，堵转转子试验，止

转转子试验
locked rotor torque　锁定转子转矩，转子锁定转矩
locked rotor voltage　止转转子电压，堵转电压
locked rotor　锁定转子，止转转子，堵转
locked spacer　止动垫圈，止动垫片止动隔件
locked to pump shaft　（用键）把泵轴锁住
locked-up stress　残余应力
locked　锁定的，闭锁的，同步的
locker room　更衣室
locker　衣帽柜，橱，（有锁的）柜，锁扣装置
lock gate hatch　排洪闸
lock gate　船闸闸门，闸门
lock-grip clamp　台钳夹板
lock-grip pliers　台钳，老虎钳
lock head　闸首
lock hopper　自锁料斗
lock-in analyzer　同步分析器
lock-in band　锁定带
lock-in circuit　强制同步电路
lock-in effect　锁定效应
lock-in excitation　锁定激励
locking apparatus　锁定装置
locking assembly　锁紧器
locking bar　锁紧杆
locking blade　锁紧叶片，锁口叶片
locking bolt　锁紧螺栓，锁定螺栓，防松螺栓
locking button　闭锁按钮【控制棒驱动机构】，锁定按钮，止动按钮
locking circuit　闭锁电路，吸持电路，强制同步电路
locking clamp gear　带有夹钳的制动装置
locking coil　（继电器）吸持线圈，闭锁线圈
locking control voltage　同步调整电压
locking cup　止动杯，制动杯
locking device　联锁装置，闭锁装置，锁定装置，锁定设施，锁紧销
locking gear　锁定装置，止动装置
locking key　止动键，锁定开关
locking link　锁定联杆
locking magnet　吸持磁铁，锁定磁铁
locking mechanism　锁定装置，止动装置
locking nut　自锁螺母，锁紧螺母，防松螺母，锁紧螺帽
locking out　堵塞
locking piece　锁定片，锁口件
locking pin　定位销，锁销
locking piston drive system　锁紧活塞驱动系统
locking piston drive　锁紧活塞传动器
locking plate　锁定板，定位板，止动板
locking press button　自锁按钮，止动按钮，锁定按钮
locking pushbutton　闭锁按钮，可定位按钮
locking range　牵引范围，同步范围
locking relay　锁定继电器
locking ring　锁环，锁紧环，紧箍
locking rotor　堵转转子，止转转子
locking screw　锁紧螺丝
locking segment　锁定块
locking signal　锁定信号，同步信号
locking sleeve　闭锁套【控制棒传动机构】
locking spring　锁紧弹簧，锁簧
locking steel wire　锁紧钢丝，止动钢丝
locking strip　止动带，锁条，锁定带
locking transfer contact　锁定切换触点
locking washer　止动垫圈，锁紧垫圈
locking wire　锁紧钢丝
locking　闭锁，锁定，止转，制动，连锁作用，闭锁作用，锁紧作用
lock in of vortex　锁涡
lock input　同步输入
lock-in range　锁定范围，同步范围
lock-in relay　闭塞继电器，锁定继电器
lock-in synchronism　牵入同步，进入同步，锁定同步
lock into step　进入同步
lock-in　锁定，关进，同步跟踪
lock joint　锁底接头
lock location　闸址
lock loop　锁定环，锁环
lock mechanism　锁定装置，止动装置
lock nut　止动螺母，锁紧螺母，锁定螺母，防松螺帽，固定螺母
lock on circuit　强制同步电路，自保持电路，自保护电路
lock on relay　同步继电器
lock on　锁定，锁住，自动跟踪，捕捉
lock open　非通电时为常开状态，连锁打开
lock out circuit　闭塞电路，保持电路，同步电路
lock out device　闭锁装置
lockout mechanism　锁定装置，锁定机构，闭塞装置
lock-out pulse　整步脉冲，同步脉冲
lockout relay reset　闭锁继电器复位
lockout relay　保持继电器，闭锁继电器，闭锁出口继电器
lockout switch　联锁开关
lockout　锁定，封锁，闭锁，停止，切断，断路，分离，寄存
lock pillar　闸座，防松栓
lock pin　锁销
lockplate　锁定板，止动板
lock plug return spring　锁塞返回弹簧
lock plug　锁塞
lock range　同步范围
lock ring　锁环
lock screw　锁紧螺丝，止动螺钉
lock seam joint　锁口缝
locks flight　多级船闸
lock sill　闸槛
lock sleeve　锁套
locksmith　钳工，锁匠
lock tube　锁紧管
lock unit　同步装置，同步器，同步单元
lockup circuit　吸持电路，闭锁电路
lockup mechanism　锁定机构
lockup relay　自保持继电器，闩锁继电器
lockup valve　锁定阀
lockup　锁定，顶托，锁定阀，封孔蓄水
lock wall culvert　闸墙内输水涵洞
lock wall　闸墙
lock washer　锁紧垫圈，止动垫圈，防松垫圈，

弹簧垫圈
lock weld 锁焊【螺帽】
lock 加锁，锁紧，锁定，自动跟踪，气闸，闸，水闸，锁，闭锁，锁紧装置，联动同步，止挡定位器，制轮楔
locomobile 锅驼机，自动机车，自动推进的
locomotive barn 机车库
locomotive boiler 机车锅炉
locomotive crane 机车吊机，机车起重机
locomotive engine 火车头，机车
locomotive gas turbine 机车燃气轮机
locomotiveness 位置变换性能，变换位置方法
locomotive running track 机车行走线
locomotive traction motor 牵引电动机
locomotive transformer 电机车变压器，电力机车变压器
locomotive 机车，火车头
locus diagram 轨迹图
locus of principal stress 主应力轨迹
locus 位置，场所，轨迹，轨线，地方
lodestone 磁石
lodestone 天然磁铁，磁石，吸引人的东西
lode 矿脉，水沟
lodge a claim 提出索赔
lodge and accommodation 膳宿
lodge claims against 向某人索赔
lodgement 住宿，寄宿处，沉积
lodge money in a bank 在银行存款
lodge one's claim 提出索赔
lodge 门房
lodging house 寄居宿舍
lodox 微粉末磁铁
loessal soil 黄土质土
loess child 黄土结核
loess concretion 姜结石
loess conservation 黄土保持
loess deposit 黄土沉积
loess-doll 黄土结核
loess erosion 黄土侵蚀
loess formation 黄土层
loessial soil 黄土性土（壤）
loessification 黄土化
loess nodule 黄土结核
loess plain 黄土平原
loess plateau 黄土高原
loess soil 黄土，大孔性土，大孔隙土，黄土性土（壤）
loess 大孔隙土，黄土，黄土结核
LOFA(loss of flow accident) 一回路流量失水事故
LOFF(leakoff) 泄漏的，已漏的
loft drier 干燥箱，箱式干燥器
lofting plume 屋脊型羽流，高耸的烟羽，高升空中的烟羽，屋脊型烟羽
LOFT(loss-of-fluid test) 流体损失试验
loft 阁楼，顶楼，放样，放线
log console 记录控制台
log amplifier 对数放大器
logarithmic amplifier 对数放大器
logarithmic attenuator 对数衰减器
logarithmic base 对数底
logarithmic calculator 对数计算尺，对数计算机
logarithmic channel 对数通道
logarithmic computing instrument 对数计算器
logarithmic coordinate paper 对数坐标纸
logarithmic coordinates 对数坐标
logarithmic count rate meter 对数计数率表
logarithmic criteria 对数判据，对数判定
logarithmic DC instrument lane 对数直流仪表管路
logarithmic decrement 对数衰减，对数衰减率，对数衰减量
logarithmic demodulation 对数解调器，对数反调制器
logarithmic differentiation 对数微分
logarithmic distribution law 对数分布律
logarithmic distribution 对数分布
logarithmic energy change 能量对数变化
logarithmic energy decrement 能量对数减缩
logarithmic energy loss 能量对数损失
logarithmic equation 对数方程
logarithmic frequency spectrum 对数频谱
logarithmic function 对数函数
logarithmic gain 对数增益
logarithmic instrumentation lane 对数仪表管线路
logarithmic integral 对数积分
logarithmic ionization chamber 对数电离室
logarithmic law 对数（变化）律
logarithmic mean temperature difference 对数平均温差
logarithmic multiplier 对数乘法器
logarithmic paper 对数纸，对数坐标纸
logarithmic protection channel 对数保护线路
logarithmic ratemeter 对数率表，对数速率计
logarithmic scale 对数标度，对数分度，对数刻度，对数尺度
logarithmic spiral arch dam 对数螺线形拱坝
logarithmic spiral 对数螺线
logarithmic table 对数表格
logarithmic-to-linear converter 对数函数线性函数变换器
logarithmic velocity profile 对数速度廓线
logarithmic voltage quantizer 对数电压转换器
logarithmic wind profile 对数风（速）廓线
logarithmic wind shear law 对数风切变律
logarithmic 对数的
logarithm of time fitting method 时间对数拟合法
logarithmoid 广对数螺线
logarithm winding motor 对数绕组电动机
logarithm 对数
logbook 运行日记，日志，记事簿，值班簿
log channel 对数通道
log chute 放木斜槽，滑木道
log control block 日志控制块
log control class 日志控制类
log conveyer 木料输送机
log-count-rate recorder 对数计数率记录器
log-crib revetment 木笼护岸
log curve 对数曲线
log deflector 木筏转向设备
log file 日志文件
logger 记录器，记录仪，自动记录器，测井仪，

仪表读数
logging cable 电测用电缆
logging desk 值班工作台，运行记录桌
logging instrument 测井仪
logging method 测井方法
logging 日志，记录，存入，联机，测井，报表打印，伐木
log grapple 木材抓钩
log haul 木材拖运
logic address 逻辑地址
logical action 逻辑作用，逻辑操作
logical analysis device 逻辑分析设备
logical"and"component 逻辑“与”元件
logical binary counter 逻辑二进制计数器
logical block 逻辑单元
logical circuit 逻辑电路
logical compare accumulator with storage 累加器和存储器逻辑比较
logical comparison 逻辑比较
logical component 逻辑元件
logical computer 逻辑计算机
logical connection 逻辑连接
logical constant 逻辑常数
logical construction 逻辑结构
logical conversion matrix 逻辑变换矩阵
logical conversion 逻辑变换
logical data independence 逻辑数据无关
logical data type 逻辑数据形式
logical decision 逻辑判定
logical design 逻辑设计
logical device class 逻辑设备类
logical device name space 逻辑设备名称空间
logical device object 逻辑设备对象
logical device table 逻辑器件表
logical device 逻辑设备
logical diagram 逻辑图
logical diode circuit 二极管逻辑电路
logical element 逻辑单元，逻辑元件
logical equipment table 逻辑装置表
logical error 逻辑误差
logical expression 逻辑表达式
logical file 逻辑文件
logical flow chart 逻辑流程图
logical function 逻辑作用，逻辑函数
logical gate 逻辑门
logic algebra 逻辑代数
logical input/output control system 逻辑输入/输出控制系统
logical instruction 逻辑指令
logical interface 逻辑接口
logical language 逻辑语言
logical left shift 逻辑左移
logical link control 逻辑链路控制
logical module 逻辑模块
logical network 逻辑网络
logical net 逻辑网络
logical node class 逻辑节点类
logical node data 逻辑节点数据
logical node group 逻辑节点组
logical node name space 逻辑节点名称空间
logical node name 逻辑节点名
logical node object 逻辑节点对象
logical node physical device 逻辑节点物理装置
logical node zero 逻辑节点 0
logical node 逻辑节点
logical "not" circuit 逻辑“非”电路
logical "not" component 逻辑“非”元件
logical number 逻辑数
logical operation 逻辑运算，逻辑操作
logical operator 逻辑算子
logical "or-and-or" pyramid 逻辑“或-与-或”角锥
logical "or" circuit 逻辑“或”电路
logical "or" component 逻辑“或”元件
logical order 逻辑指令
logical "or" 逻辑“或”
logical pattern 逻辑图
logical product 逻辑乘积
logical program 逻辑程序
logical reading system 逻辑读数系统
logical record 逻辑记录
logical right shift 逻辑右移
logical shift left 逻辑左移
logical shift right 逻辑右移
logical shift 逻辑移位
logical sum 逻辑和
logical symbol 逻辑符号
logical system 逻辑系统
logical unit 逻辑元件，逻辑装置，逻辑部件
logical 逻辑的，逻辑学的
logic analysis 逻辑分析
logic analyzer 逻辑分析器
logic AND circuit 逻辑“与”电路
logic-arithmetic unit 运算器
logic block diagram 逻辑框图
logic cabinet 逻辑柜
logic channel 逻辑电路，逻辑通道
logic circuitry 逻辑电路图
logic circuit 逻辑电路
logic command 逻辑指令
logic comparation 逻辑比较
logic control system 逻辑控制系统
logic control variable 逻辑控制变量
logic control 逻辑控制
logic core 逻辑磁芯
logic design 逻辑设计
logic diagram 逻辑图，逻辑框图
logic driver 逻辑驱动器
logic element 逻辑要素，逻辑元件
logic file system 逻辑文件系统
logic flow chart 逻辑流程图
logic flow diagram 逻辑方块图
logic function 逻辑函数，逻辑功能
logic gate circuit 逻辑门电路
logic gate 逻辑门
logic-in memory array 逻辑存储器阵列
logic-in memory 逻辑存储器
logic machine 逻辑机
logic master module 逻辑主模件
logic matrix 逻辑矩阵
logic module 逻辑组件
logic operation 逻辑操作，逻辑运算

logic operator 逻辑运算装置，逻辑电路，逻辑算子
logic oscilloscope 逻辑示波器
logic power supply 逻辑回路电源
logic processor 逻辑处理机
logic protection system 逻辑保护系统
logic relay 逻辑继电器
logic section 逻辑部件
logic signal converter 逻辑信号转换器
logic signal 逻辑信号
logic simulation 逻辑模拟
logic simulator 逻辑仿真器
logic switching 逻辑转换，逻辑切换
logic unit 逻辑单元，逻辑部件
logic word 逻辑字
logic 逻辑，逻辑学
log-in 注册，录入
logistical support 后勤支援
logistic coordination 后勤协调
logistic delay 后勤延迟
logistics 后勤，逻辑学，后勤学，物流，物流学，后勤运输
logistic 符号逻辑的，数理逻辑，后勤学，逻辑的，后勤学的
logitron 磁性逻辑元件
log jam 木块堵塞
log level-period meter 功率与周期对数表
log linear distribution 对数线性分布
log linear wind profile 对数线性风廓线
log=logarithm 对数
log-log paper 复对数坐标纸，双对数坐标纸
log mean temperature difference 对数平均温差
log normal distribution 对数正态分布
log off 注销
logometer 比率表，电流比计，对数计算尺
log-on 注册，构成信息的一个单位
log-out 注销，运行记录，记录事件
logo 标识
log paper 对数坐标纸
log pass structure 过木建筑物
log pass 筏道，放木道
log pile 圆木桩
log printer 报表打印机
log printout 运行日志打印输出
log-probability paper 对数概率纸
log raft 木筏
log sheet 对数纸，记录表，运行记录单
logway structure 过木建筑物
logway 筏道，放木道
log 日志，记录，原木，圆木，大木料，对数，运行记录，测井记录，报表记录，测程器，岩心记录
LOI(letter of indemnity) 赔偿保证书，认赔函，担保书，保函
LOI(letter of intent) 意向书
LOI(letter of invitation) 邀请函
lolly 薄冰
LO(lube oil) 润滑油
London inter-bank offered rate 伦敦银行同业拆借利率
London type smog 伦敦型烟雾
lone signal unit 单一信号单元
long acceleration 持续加速度
long-access memory 慢速存储器
long air gap 长空气隙
long and wedge shaped pile 长条楔形煤堆
long auger bored concrete pile 长螺旋钻孔压灌素混凝土桩
long blade flutter 长叶片颤振
long blade 长叶片
long-boom dragline 长臂索铲
long-boom wharf crane 码头用长臂起重机
long burn-up 深燃烧
long-chain copolymer 长链共聚物
long-chord winding 长距绕组
long column 长柱
long-continued exposure 长期连续照射
long contract 长期合同，多头合同
long-core machine 长铁芯电机
long-crested weir 长顶堰
long cut wood 长木材
long cylindrical shell 长圆柱形外壳
longdated 长期的
long-decayed 长半衰期的，长寿命的
long-delay blasting cap 长期延时引爆雷管
long-delivery item 交货期长的物项
long distance belt conveyor 长距离带式输送机
long distance cable 长途通信电缆
long distance call 长途呼叫，长途电话
long distance communication 长途通信
long distance controlling multifunctional standard power supply 可程控多功能标准电源
long distance control 远程控制
long distance indicator 远距离指示器
long distance line 长途线
long distance power transmission 远距离输电
long distance recorder 远传自记仪器，远程记录仪器，远距记录仪
long distance stage transmitter 远传水位计
long distance telephone communication 长途电话
long distance telephone 长途电话
long distance traffic 长途通信，长途通信业务
long distance transmission line 长距离输电线路
long distance transmission system 长距离输电系统
long distance transmission 长距离输电
long distance transportation 长途运输
long distance water level recorder 遥测水位计，远距水位记录仪
long distance water-stage recorder 远距水位计
long distance 远程的，远距离的，长途的，远程，长途
long drought 久旱
long duration creep rupture test 长期蠕变破坏试验
long-duration static test 长期静力试验
long-duration test 耐久试验，连续负载试验
long epoch environmental effect 远期环境效应
long-established station 长期观测站
longevity 长寿，耐久性，寿命，使用寿命
long exposure 长期照射，长期曝光
long extended pile 延长的煤堆

long flame burner　长焰燃烧器
long flame coal　长焰煤
long flame of opposed firing　长焰对冲燃烧
long format instruction　长形指令
long forma　长格式标志
long-handled tool　长柄操作工具
long haulage　长距离运输
long haul　长途的【托运，牵运，运输】，长运距的
long heavy swell　八级涌浪
long-hole method　深孔爆破法
long-hopper furnace　长灰斗炉膛
long-hour motor　持续运行电动机
long internal wave　长内波
long-irradiated　长时间辐照的
longitude　经度
longitudinal and cross beam　纵梁和横梁
longitudinal arrangement　纵向布置
longitudinal axis　纵轴
longitudinal baffle　纵向折流板
longitudinal band　纵向地带
longitudinal bar　纵向钢筋
longitudinal-bead bend test　焊道纵向弯曲试验
longitudinal-bead test　纵向珠焊试验
longitudinal beam　纵梁，纵向梁
longitudinal bent　纵向构架
longitudinal bracing　纵向支撑
longitudinal break　纵向破裂，纵向断裂
longitudinal choke　纵向扼流圈
longitudinal chromatic aberration　纵向色像差
longitudinal circuit　纵向电路
longitudinal cofferdam　纵向围堰
longitudinal compensation　纵向补偿
longitudinal component　纵向部分，轴向分量
longitudinal concentration distribution　纵向浓度分布
longitudinal crack　纵向裂缝，纵向裂纹，纵向龟裂，纵裂
longitudinal crevasse　纵向裂缝
longitudinal cross section　纵剖面，纵向截面
longitudinal current　纵向电流，轴向分量
longitudinal damping　纵向衰减，纵向阻尼
longitudinal dam　顺坝
longitudinal data　纵向数据
longitudinal deformation　纵向变形
longitudinal differential protection　纵向差动保护，纵联差动保护，纵差保护
longitudinal differential relay　纵联差动继电器
longitudinal differential　纵向差动
longitudinal diffusion　纵向扩散
longitudinal diffusivity　纵向扩散率
longitudinal dike　纵堤
longitudinal direction　纵向，轴向
longitudinal dispersion　纵弥散，纵向弥散
longitudinal ditch　纵沟
longitudinal divide　纵分水界
longitudinal drainage　纵向排放，纵向排水
longitudinal drum boiler　纵置汽包锅炉
longitudinal dune　纵向沙丘，沙垄，纵沙丘，纵向沙垄，线性沙丘
longitudinal dyke　纵堤
longitudinal electromechanical coupling factor　纵向电机耦合因子
longitudinal expansion　纵向膨胀
longitudinal fault　纵向断层
longitudinal fiber　纵向纤维
longitudinal field　纵向场
longitudinal fin　纵向肋片
longitudinal flapping　纵向挥舞，纵向拍打
longitudinal force　纵向力
longitudinal girder　纵向大梁
longitudinal grade　纵向坡度
longitudinal heat conduction　纵向热传导
longitudinal hopper　纵门漏斗车
longitudinal impedance　纵向阻抗
longitudinal inclinometer　纵向倾斜仪
longitudinal interference　纵向干扰
longitudinal joint　纵缝，纵节理，纵向接缝
longitudinal keys　纵销
longitudinal load　纵向荷载
longitudinally welded cladding tube　纵向焊接的包壳管
longitudinal magnetic field probe　纵向磁场探头
longitudinal magnetic field　纵向磁场
longitudinal magnetic flux　纵向磁通量
longitudinal magnetization　纵向磁化，纵磁化
longitudinal magnetomotive force　纵向磁动势
longitudinal magnification　轴向放大，纵向放大，纵向放大率
longitudinal member　纵向杆件
longitudinal metacenter　纵稳心
longitudinal mixing　纵向混合
longitudinal motion　纵向运动
longitudinal oscillation　纵向振荡
longitudinal passage on both sides of turbine operating floor　汽轮机运转层两侧的纵向通道
longitudinal pile　矩形煤堆
longitudinal pin　纵销
longitudinal pitch　纵向节距
longitudinal pressure gradient　纵向压力梯度
longitudinal pressure　纵向压力
longitudinal profile　纵剖面图，纵断面，纵剖面
longitudinal recording　纵向记录
longitudinal redundancy check　纵向冗余码校验
longitudinal reinforcement　纵向钢筋
longitudinal resolution　纵向分辨率
longitudinal response　纵向响应
longitudinal reversible belt conveyor　纵向可逆皮带机
longitudinal rigidity　纵向刚度
longitudinal rod　长栏杆
longitudinal scan of the weld surface　焊缝表面纵向扫描
longitudinal seam　纵向焊缝
longitudinal section　纵切面，纵剖面，纵（向）断面
longitudinal seiche　纵向假潮
longitudinal shrinkage　纵向收缩
longitudinal sill　纵槛
longitudinal slope　纵向坡度，纵坡
longitudinal slot　纵向槽，纵向缝
longitudinal stability　纵向稳定，纵向稳定性

longitudinal stay　纵向拖拉绳，纵拉线
longitudinal strain　纵向变形，纵向应变
longitudinal stress　纵向应力
longitudinal tendon　纵向钢筋束
longitudinal tie rod　架立筋，纵向拉筋
longitudinal tie　纵向联杆
longitudinal turbulence component　纵向湍流分量
longitudinal turbulence spectrum　纵向湍流谱
longitudinal vertical bracing　纵向垂直支撑
longitudinal vibration　轴向振动，纵向振动
longitudinal view　纵向视图
longitudinal vortex　纵向涡
longitudinal wall　纵剖面图，纵墙
longitudinal wave probe　纵波探头
longitudinal wave　纵波，纵向波，地震纵波
longitudinal welding　纵向焊接
longitudinal weld　纵焊，纵焊缝，纵向焊缝
longitudinal wind　纵向风，径向风
longitudinal　纵向的，轴向的，竖的，经度的
longleaf pine　长叶松
long leakage distance insulator　大爬距绝缘子
long-lever armature　长杆衔铁
long-life　长寿命的，经久耐用的
long line effect　长线效应
long line prestressed concrete process　长线预应力混凝土张拉法
long line　长线路
long list　长名单，(投标或职位申请者等的）初选名单
long-lived activity　长效放射性，长寿命放射性
long-lived delayed neutron　长寿命缓发中子
long-lived fission product　长寿命裂变产物
long-lived isotope　长寿命同位素
long-lived neutral kaon　长寿命中性 K 介子
long-lived radiation　长寿命（同位素）辐射
long-lived radioactivity　长寿命放射性
long-lived　长寿的，耐久的，经久耐用的
long longitudinal　纵向的
long low swell　二级涌浪
long moderate swell　五级涌浪
long narrow pile　长窄形煤堆
long-neck flask　开式烧瓶，长颈烧瓶
long-nose pliers　长嘴钳
long number　多位数，长数目
long-period average flow　长期平均流量
long-period delayed neutron　长半衰期缓发中子
long-period fluctuation　长周期波动
long-period forecast　长期预报
long-period oscillation　长周期振动
long-period runoff relation　长期径流关系
long-period seismograph　长周期地震计
long-period storage reservoir　长期蓄水水库
long-period surge　长周期波
long-period variation　长期变化
long-period wave　长周波，长周期波
long-period　长寿命的，长期的
long-pitch winding　长距绕组
long polymeric chain　长聚合链
long pot life　长适用期，长储存期
long progressive wave　长前进浪
long radius elbow　长半径弯头
long radius nozzle　长径喷嘴
long radius return U-elbow　长半径 U 形弯头
long-range control　大范围调节
long-range Coulomb interaction　远距离库仑效应
long-range data　长距离数据
long-range forecast　大范围预报
long-range input monitor　远程输入监控器
long-range investment　长期投资
long-range measure　长期措施
long-range meteorological forecasting　长期气象预报
long-range order parameter　长程序参量
long-range planning　远景规划，长期规划
long-range prediction　长期预报，长期预测
long-range radiation　强贯穿性辐射，强穿透性辐射
long-range transmission　远程输送，远距离输电，远距离传送
long-range transport model　远程运输模式
long-range water soot blower　远射程水吹灰器
long-range weather forecasting　长期天气预报
long-range　长距离，远程，宽量程，远程的，长距离的
long rod insulator　棒形悬式绝缘子，长棒形绝缘子
long rolling sea　长波
long-run average water flow　长期平均水流量
long-run marginal cost pricing　长期边际成本定价
long-run test　寿命试验，长时运转试验，长期运行试验，连续运行试验
long-run　长期运行的，长期的
long separation bubble　长分离气泡，长泡分离
long sequence control scheme　长顺序控制系统
long service life　长使用寿命
long shank blade root　长颈叶根
long shift　长移位，长偏移
longshore bar　海岸沙洲，沿岸沙坝
longshore current　顺岸流，沿岸流
longshore drift　沿岸漂流
longshoreman　码头装卸工人
longshore transport　沿岸输送
longshore water current　沿岸水流
longshore wind　海岸风，沿岸风
longshore　沿岸的
long shunt compound machine　长并激复电机
long shunt compound winding　外并联复励绕组，长分路复绕组，长分流复励绕组
long shunt winding　长并激绕组，长并联绕组，长分路绕组
long shunt　长分路，长并激
long slag　长渣
long-slot burner　长缝式燃烧器
long span cable　大跨度缆线
long span of overhead transmission line　架空输电线路大跨越
long span structure　大跨度结构
long span suspension bridge　大跨度悬索桥
long span　长跨度，大跨越，长挡距
long-standing　可长期存在的，长时间的，经久的
long stem insulator　长形绝缘子，长杆绝缘子
long-stemmed nozzle　长柄喷嘴，长柄水帽

long storage location 长字存储单元，大数存储单元
long strip coal pile 长条形煤堆
long strip footing 长条形基础
long-stroke engine 长行程发动机
long-stroke piston 长行程活塞
long surf 长激浪
long suspension insulator 长悬式绝缘子
long sweep ell 大弯管半径弯头
long-tailed pair 长尾对
long-term ageing 长期老化
long-term agreement 长期合同
long-term assets 长期资产
long-term average discharge 长期平均流量
long-term behaviorur 长期性能
long-term benefit 长期受益，长期效益
long-term bill 长期汇票
long term biological hazard 长期生物危害
long-term bond 长期债券
long-term budget 长期预算
long-term change 长期变化
long-term climatic cycle 长期气候循环
long-term climatic trend 长期气候趋势
long-term coal supply contract 长期供煤合同
long-term construction contract 长期建设合同，长期承包工程合同
long-term contract 长期合同
long-term cost 长期费用
long-term creditor 长期债权人
long-term creep 长期蠕变
long-term deformation process 长期变形过程
long-term dynamics 长期动力学
long-term effects of chemicals 化学品长期影响
long-term effect 长期效应，长时作用
long-term erosion 长期侵蚀
long-term exposure 长期暴露，长期照射
long-term financing 长期资金融通
long-term flow prediction 长期预测流量
long-term forecasting 长期预测
long-term global trend 全球长期趋势
long-term goal 远期目标
long-term insurance 长期保险
long-term interest-free loan 长期无息贷款
long-term investment project 长期投资项目
long-term investment 长期投资
long-term irradiation 长期辐照
long-term lease 长期租赁，长期租约
long-term liability 长期负债
long-term loan 长期贷款
long-term mean wind speed 长期平均风速
long-term memory 长期存储器，永久存储器
long-term observation system 长期观测系统
long-term observation 长期观测
long-term planning arrangement drawing of main power building 主厂房远景规划布置图
long-term planning 长期规划，远景规划
long-term poisoning 长期中毒
long-term power development plan 电力发展远景规划
long-term prediction 长期预测
long-term price 长期价格
long-term radiation 长期辐照
long-term rate 长期利率
long-term reactivity change 长期反应性变化
long-term reactor power control 长时间反应堆功率控制
long-term record 长期（观测）记录
long-term requirement 长期要求
long-term stability analysis 长过程稳定计算
long-term stability 长期稳定性
long-term station 长期观测站
long-term storage 长期储存，长期（调节）库容，长期蓄水，长期贮存
long-term strength curve 长期强度曲线
long-term strength index 长期强度指标
long-term strength 长期强度，后期强度
long-term stress rupture 长期应力作用下的损坏，持久强度
long-term test 长期（运行）试验，寿命试验
long-term trade subsystem 长期交易子系统
long-term transaction schedule data 长期交易计划数据
long-term treatment 长期处理
long-term usage 长期使用，连续使用
long-term wind data 长期风资料
long-term wind speed average 长期平均风速
long-term 长期，远期，长期的，远景的，长远的，远期的
long test section wind tunnel 长试验段（边界层型）风洞
long-time annual maximum atmosphere pressure 多年最高气压
long-time annual maximum daily range of air temperature 多年最大气温日较差
long-time annual maximum frozen earth depth 多年最大冻土深度
long-time annual maximum 多年最大值
long-time annual minimum atmosphere pressure 多年最低气压
long-time annual minimum 多年最小
long-time average annual value 多年平均值
long-time average annual water level 多年平均水位
long-time behaviour 长期运行性能
long-time creep test 长期蠕变试验，持久蠕变试验
long-time deflection 持久挠度
long-time delay 长延时
long-time diffusion factor 长期扩散因子
long-time fatigue strength 持久疲劳强度
long-time high temperature tensile strength 持久高温抗张强度，持久高温抗拉强度
long-time maximum annual water level 多年最高水位
long-time memory 长期存储器，长时间存储器
long-time minimum annual water level 多年最低水位
long-time observation 长期观测
long-time overcurrent relay 长时过电流继电器
long-time running 长期运行
long-time rupture elongation 持久断裂延伸率
long-time stability 长期稳定性

long-time strength　持久强度
long-time washout factor　长期冲洗因子
long-time　长期的，持久的
long toll call　长途呼叫，长途通话
long ton　长吨
long track tornado　长迹龙卷风
long transmission line　长传输线，长输电线路
long triangular pile　长条三角形煤堆
long trunk call　长途呼叫，长途通话
longwall mining　长臂开采法【矿业】
long warm-up fuel cost　长时间暖机燃料费用
long wave coil　长波线圈
long wave directional antenna　长波定向天线
long wave direction finder　长波测向器，长波探向器
long wave length radiation　长波长辐射
long wave radiation　长波辐射
long wave transmitter　长波发射机
long wave　长波
long　长的
look aside memory　后备存储器
look-at-me function　中断功能
look-at-me　中断信号
look forward to　期待
look forward　向前看，考虑将来，展望
looking glass　观察窗，窥视窗
look-up table　查表法
look up　检查，查找，查阅，仰望
loop activation　回路活化
loop activity　环路活度
loop admittance matrix　回路导纳矩阵
loop antenna　环形天线，框形天线
loop armature　环形电枢
loop bar　套杆
loop body　循环体，循环本体
loop box　变址寄存器
loop capacitance　回路电容，耦合环电容
loop cell　回路小室
loop checking system　回路校验系统
loop check instrument　回路校验仪
loop-circuit subtransmission　环状二次输电
loop circuit　环形电路，环路
loop coil　环状线圈，环形线圈
loop component　环路设备，环路部件
loop configuration　管道式布置，回路布置
loop connection　环形结线，回路连接
loop control　环路控制，穿孔带指令控制
loop coupling　环耦合
loop current　回路电流
loop curve　环形曲线
loop design　回路设计
loop diagram for control system　控制系统框图
loop diagram　回路图
loop drain line　回路疏水管
looped distribution network　环式网络
looped-in supply　环形供电，环路供电
looped network　环网
loop effect　环路效应
loop error　回路误差，回路偏差
loop expansion bend　回路膨胀弯管
loop expansion joint　环形伸缩器
loop feeder　环形馈线，环路馈电线
loop feeding　环形供电
loop fill line　回路充水管
loop flushing　回路吹洗，回路冲洗
loop gain characteristic　回路增益特性
loop gain　回路增益，环路增益
loop galvanometer　回线检流计，回线电流计
loop impedance matrix　回路阻抗矩阵
loop impedance　回线阻抗，环线阻抗
loop inductance　环线电感，回线电感，回路电感
looping-in　形成回路，形成环路
looping-off　解环
looping pipe　环管
looping plume　波形羽流，波浪形烟
looping　环接【形成接地环路】，成圈，构成环形，成波浪状排烟
loop interface module　回路接口模件
loop label　回路标牌，回路标号
loop leakage　回路泄漏，环路泄漏
loop line　环形线路，环形线，回线
loop-linked　环路连接的
loop-locked　闭环的
loop loss　环路损耗
loop main piping　主环网管道
loop measurement　环路测量
loop method　回路电流法
loop monitoring system　回路监测系统
loop number　回路数
loop of end coil　线圈鼻端
loop oscillograph　回线示波器
loop phase angle　回路相角
loop phase characteristic　回路相位特性
loop pipe　高压缸环形进汽管，环状管
loop plant　环路核电厂，分散布置核电厂
loop purging　环路吹洗，回路吹洗
loop rating curve　水位流量环形关系曲线
loop reactance　环线电抗，回线电抗
loop resistance tester　回路电阻测试仪
loop resistance　回线电阻，环线电阻
loop scavenged cylinder　回流换汽汽缸
loop seal system　返料系统【循环流化床】
loop seal　返料机构【循环流化床】，环封，环形管水封，流动密封阀，回料控制阀
loop service　环路供电
loop stop　循环停机，循环停止
loop system transmission line　环形输电线
loop system　闭环系统，环形线路制，回路系统，环路系统
loop test　环线试验，回路试验，循环试验，环路电流法
loop transfer function　回路传递函数
loop transmission　回线输电
loop-type economizer　蛇形管省煤器
loop-type fast breeder reactor　回路式快中子增殖反应堆
loop-type reactor structure　回路式堆结构
loop-type reactor　回路式反应堆，回路式堆
loop-type system　环形配电系统
loop winding　环形绕组
loop wire　环线
loop　环，环路，环形管路，盘形管，环形天

线，循环，圆圈，回路，回线，匝，管圈
loose aggregate　松散集料，疏松骨料
loose alluvium　松散冲积层
loose apron　松散护坦
loose ash　飞灰，疏松灰
loose brush mattress　松梢褥
loose cement　散装水泥，松散水泥
loose coal　碎煤
loose connection　不良连接，连接松动
loose contact　松动触点，不良触点，接触不良
loose core　松散夹心
loose-coupled type pipe　松接式管道
loose coupling　松耦合，松动接合，弱耦合，耦合不良
loose deposit　疏松沉积物，疏松积灰，疏松结垢
loose earth　松土
loose fiber pollution　松散纤维污染
loose fill insulation　松散料保温
loose fill type insulant　松散绝热材料
loose fit　松配合，不紧配合
loose flange　松动法兰，松套法兰，松套凸缘
loose float trap　自由浮球式疏水阀
loose foundation　松散地基
loose ground　松软土地，松散地
loose insulation　松散绝缘
loose items　散件
loose-jointed　关节活络的，可拆开的
loose joint　松接，松套
loose lacing wire　松拉筋
loosely bound　松接的
loosely coupled multi-processor system　松耦合多处理机系统
loosely filled insulation　填充式保温
loosely spaced lattice　宽间隔栅格【堆芯】，宽距栅格，大栅距栅格，稀疏栅格
loose masonry　干堆圬工
loose material　松散材料
loose measure　粗测量
loosened zone　松碎带
looseness　松度，松弛，松动
loosening blasting　松动爆破
loosening of stud nuts　螺杆螺帽的松开
loosening pressure　松动压力
loosening water　（反洗）疏松用水
loosening　松动，松螺丝
loosen up water　（反洗）疏松用水
loosen　松脱
loose oxidation product　不稳定氧化产物
loose packed　散装的
loose particles　松散颗粒
loose parts and vibration monitoring　松动部件和振动监测
loose parts detection　松动部件测量
loose-parts monitor　零件松动监测器
loose part　松动部件
loose penstock　松接式压力钢管
loose pimple　疏松丘包
loose pin hinge　松动销铰链，松销合页
loose pin　定位滑销，导向键，松动销
loose plate flange　松套板式法兰
loose pulley　游滑皮带轮，空转轮，惰轮
loose rock dam　碎石坝
loose rock fill　抛石填充
loose rock　松散岩石
loose running fit　松转配合
loose rust　疏松锈屑
loose salt　松散盐
loose sand　散砂，松砂
loose scale　疏松水垢
loose shroud　松围带，松覆环
loose soil liable to slip　滑坡体
loose soil　疏松土，松软土，松散土，松土
loose state　松散状态
loose-stone　干砌石
loose stuff　疏松岩层，松散岩层
loose tie wire　松拉筋
loose tube cable　松管光缆
loose volume　松散体积
loose wheel　游滑轮
loose wire gripper　松线夹
loose　松开的，松动的，松的，疏松的，松散的，未紧固的松开
LOP(lube oil pump)　润滑油泵，顶轴油泵【DCS画面】
lopper　斩波器，削波器
lopsided　偏重的，倾斜的，不平衡的
loran　远程导航系统
lorry loading　卡车装车
lorry　卡车，货车，载重汽车，平台四轮车
lose bid　未中标
lose money in a business　亏本，贴本，蚀本
lose synchronism　失步
losing party　败诉方
losing step　失步
losing synchronism　失步
losing　亏损
LOS(loss of signal)　信号损耗
loss advice　损失通知书，蚀本通知书
loss allocation　网损分摊
loss and gain of exchange　汇兑损益
loss and gain on retirement of fixed assets　固定资产拆除损益
loss angle　损失角，损耗角，衰减角
loss apportionment　损失分摊
loss by percolation　渗漏损失
loss by radiation　辐射损失
loss by solution　溶解损失量
loss call　未接通的呼叫
loss clause　损失条款
loss coefficient of cascade　叶栅损失系数
loss coefficient　损耗系数，(热）损失系数
loss compensator　损耗补偿器
loss conductance　损耗电导
loss contours　等损失线
loss conversion　网损折算
loss current　损耗电流
loss-delay system　等待延滞系统，混合系统
loss density　损耗密度
loss due to dressing　选矿损失
loss due to friction　磨损，摩擦损失
loss due to heat　热损失

loss due to leakage　漏磁损耗，漏电损失，漏失，漏耗
loss due to sudden contraction　突然收缩损失，突缩损失
loss due to sudden enlargement　突扩损失
loss due to valve　节流损失，阀阻损失
losses in transit　转运损耗
losses on arising from investments　投资损失
losses on scrapping of fixed assets　固定资产报废损失
loss factor　损耗因子，损失因子，损失系数，网损系数，损失率
loss-free condition　无损失工况，无损失条件
loss-free dielectric　无损耗电介质
loss-free　无损失，无损耗
loss head　损失压头
loss heat　热耗，热损失
loss in exchange　汇兑损失
loss in head　压头损失
loss in weight test　失重试验
loss in weight　失重，重量不足
lossless code　无损失码
lossless line　无损失线
lossless network　无耗网络
lossless　无损失的
loss load　负载损失，载荷损失
loss-making enterprise　亏损企业
loss measurement test　损失测量试验
loss measurement　损耗测量
loss meter　损耗表
loss of accuracy　准确度降低
loss of all flame　全炉膛火焰丧失
loss of all power　失去所有电源
loss of auxiliary power　失去辅助电源，失去厂用电源
loss of availability　利用率的损失，利用率耗损
loss of cash　现金损失
loss of charge method　电荷漏减法，放电测高阻法
loss of charge　电荷损失，充电损失
loss of circulator driving power　循环风机断电，循环器失去动力
loss of component cooling water　设备冷却水丧失
loss of coolant accident　冷却剂丧失事故，一回路失水事故，(轻水堆）失水事故
loss of coolant experiment　失水实验，冷却剂丧失实验
loss of coolant test　失水试验，冷却剂丧失试验
loss of coolant　冷却剂损失量
loss of core auxiliary cooling system accident　堆芯辅助冷却系统丧失事故
loss of ductility　失去延性
loss of electrical energy　电能损耗
loss of electrical load accident　失负荷事故
loss of electrical load　甩去电负荷
loss of energy expectation　预期缺电量，缺电量期望值
loss of energy probability　缺电时间概率，缺电概率
loss of energy　能量损失，能耗
loss of excitation of generator　发电机无励磁运行
loss of excitation　励磁损耗，失磁
loss of feedwater accident　给水丧失事故，失水事故
loss of field　磁场损耗，失磁
loss of field protection　失磁保护
loss of filter material　滤料的损耗
loss of flame to a corner　角火焰消失
loss of flame　火焰丧失
loss of flow accident　一回路流量失水事故，冷却剂流量丧失【核电】，失流事故，断流事故
loss of flow test　失流试验
loss of flow　断流，失去流量，流量损失，流量下降
loss of fluid test　流体丧失试验
loss of friction　摩擦损失
loss of head at orifice　缩孔压头损失
loss of head due fo bend　弯管压头损失
loss of head due to contraction　断面收缩产生的水头损失，收缩段压头损失
loss of head due to enlargement　扩张段压头损失，断面扩大生产的水头损失
loss of head due to entrance　管进口段压头损失
loss of head due to exit　出口段压头损失
loss of head due to friction of the circulating water　循环水摩擦所致压力损失
loss of head due to obstruction　障碍段水头损失
loss of head due to pipe fitting　管道配件压头损失
loss of head due to pipe friction　管道摩擦压头损失
loss of head entrance　进口水头损失
loss of head gauge　压头损失计
loss of head in bends　管道转弯产生的水头损失
loss of head　水头损失
loss of heat sink　热阱丧失（事故）
loss of heat　热损失，热耗，散热
loss of ignition　点火失效，灼烧减量，熄火
loss of information walk-down　信息损失，信息丢失
loss of information　信息损失
loss of life　使用寿命降低，寿命损耗，使用期缩短
loss of load accident　甩负荷事故
loss of load duration　（电力系统）缺电持续时间
loss of load expectation　（电力系统）缺电时间期望
loss of load frequency　（电力系统）缺电频率
loss of load probability　（电力系统）缺电概率，电力不足概率，失去负荷概率
loss of load transient　甩负荷瞬变
loss of load　失载，失荷，甩负荷，失负荷，负载损耗
loss of magnetic reversals　反复磁化损耗
loss of momentum　动量损失
loss of neutrons by escape　中子散逸损失，中子泄漏损失，中子逃逸损失
loss of offsite power　失去正常电源，失去厂外电源，电网输入端解联，电网断电事故
loss of operating power　丧失运行功率
loss of package　整件缺损
loss of phase　断相，失相
loss of piping integrity　失去管系完整性
loss of power accident　断电事故

loss of power network 电网损耗
loss of power 动力损失，电源消失，功率损失，功率损耗，失电，失压，断电
loss of pressure 压损，压力损失，压力下降，压降
loss of prestress 预应力损失
loss of prime 放水【泵】，进水口失水
loss of pump power accident 泵断电事故，泵失去动力事故
loss of quantity 短量
loss of reactive voltage 无功电压损失
loss of reactivity 活性下降，反应性损失
loss of revenues and profits 收益损失
loss of sight to 对……不知
loss of signal strength 信号衰减
loss of signal 信号损耗
loss of significant figures 有效数字损失
loss of sodium accident 失钠事故
loss of start-up transformer 起备变损耗
loss of suction-side flow （泵）吸入流中断
loss of synchronism protection 失步保护
loss of synchronism 失去同期，失步
loss of time 失去时间，时间损失
loss of use 失效，使用性损失
loss of vacuum 失去真空
loss of voltage 电压损失
loss of volume 容积损失
loss of weight 失重，减轻
loss of working time 工时损失
loss on defective product 废品损失
loss on ignition 灼烧减量，烧失重
loss or gain on exchange 外汇损益
loss payee clause 赔偿受益条款
loss ratio 损耗率，损失率，损耗比
loss reserve entry 赔款准备金转入
loss reserve withdrawal 赔款准备金转出
loss resistance 损耗电阻
loss standard 损耗标准
loss-summation method 损耗相加法
loss tangent test 介质损耗角试验，损耗角正切试验
loss tangent 损耗角正切，损耗因数，损耗因素，损失角正切
loss test 损耗试验
loss time 损耗时间，时间损耗，空载时间
loss to fission products 裂变碎片俘获中子损失
loss total 总损耗
lossy dielectric 有损耗电介质
lossy network 有耗网络
lossy 有损耗的，引起衰减的，耗散能量的
loss 丧失，损失，损耗，衰减，亏损
lost count 略去的读数，漏失计数
lost energy 消耗能量
lost formwork 损耗模板
lost heat 热耗，热损失
lost motion 无效运动，无效运转，空转，不做功冲程，空动
lost sodium reactivity 失钠引起的反应性
lost work 无效功，损失功
lost 失去的，损失的，磨损的，丢失的
lot area 区域面积
lotic water 活水，流水
LOT（lease-operate-transfer） LOT模式，“租赁-运营-移交”模式【工程建设模式】
lot No. 批号
lot number 批号，批数，签号，标段号
lot production 成批生产，批量生产
lot 标段，批，块，划分，场地，份额，套，分配，部件
loudness level 响度级
loudness 响度
loudspeaker 扩音器，扬声器，喇叭
loudspeaking telephone 扬声电话机
lounge hall 休息室
lounge 休息室
louver all ceiling 大面积、发光天棚
louver board 百叶窗板，活动百叶窗
louver damper 百叶窗（式）挡板
louver door 百叶门
louvered air intake 百叶式进气口，百叶式进风口
louvered air outlet 百叶式出气口
louvered fins 百叶窗式肋片
louvered rain guard 百叶窗式防雨罩
louvered tray 通风百叶窗托架
louver fins 百叶窗式肋片
louver(-re) 放气窗，放气孔，通风窗，百叶窗，调风装置，窗鱼鳞板，气窗
louver separator 百叶窗分离器，百叶式出风口
louver shutter 活动百叶窗，活百叶窗
louver stoker 鳞片式炉排
louver-type separator 百叶窗式汽水分离器
louver vane 百叶窗式导流栅
louver window 百叶窗
louvre damper 百叶窗式风门，百叶窗形阻尼器，帘式挡板
louvre stoker 鳞片式炉排
louvre ventilation 百叶窗通风，防护式通风
love wave 乐浦波
low access time 短取数时间，慢速存取时间
low active waste 低放射性废物
low activity liquid waste concentrate 低放射性废液浓缩物
low activity waste tank 低放射性废物箱
low alarm 低值信号，低位报警
low alkali cement 低碱水泥
low alloy carbon steel 低合金碳钢
low alloy steel covered arc welding electrode 低合金钢焊条
low alloy steel 低合金钢
low altitude aerial photograph 低空航摄照片
low altitude jet flow 低空急流
low altitude jet stream 低空急流
low-and-high-pass filter 高低通滤波器，带阻滤波器
low and intermediate level radioactive waste 低中放废物
low and medium frequency 中低频率
low-angle dip 缓倾角，缓倾斜
low-angle fault 缓倾角断层
low area storm 龙卷风
low-arrangement of outdoor switchgear 低型屋外

配电装置
low-ash coal 低灰煤
low-ash-fusion coal 低灰熔点煤
low-ash 低灰分，少灰的
low aspect ratio （叶片或机翼）小展弦比
low auctioneer module 低脉冲选择模块
low auctioneer unit 低脉冲选择模块
low back pressure 低背压
low battery voltage 电池欠压，低电池电压
low battery 电池电量低，电池电量不足，电池电量过低，电量警告
low bearing oil pressure governor 轴承油压低截断装置
low bearing oil pressure trip device 润滑油压过低保护装置
low bearing oil pressure trip test 轴承油压降低跳闸试验
low bed trailer 低架拖车
low-bed 低平板车
low-blast furnace 低压鼓风炉
low boiling point's substance working cycle 低沸点工质循环
low-boiling 低沸点的
low brass 下半轴瓦
low-capacitance cable 低分布电容电缆
low capacity boiler 小容量锅炉
low capacity cable 小容量电缆
low capacity plant 小容量电站
low capacity pump 小流量泵
low capacity 低容量
low-carbon steel 低碳钢
low-carbon 低碳
low cement content 低水泥用量
low circulation-ratio boiler 低循环倍率锅炉
low circulation 低倍循环
low cloud 低云
low coil power relay 灵敏继电器，低线圈功率继电器
low conductivity water 低电导率的水，纯水
low content alloy 低合金
low-cost housing 低价住房
low-cost road 简易道路，低造价道路
low current earthing system 小电流接地系统
low current toggle switch 低电流拨动开关
low current 低强度电流
low-cycle fatigue strength 低频疲劳强度，低循环疲劳强度，低周疲劳强度
low-cycle fatigue test(ing) 低循环疲劳试验
low-cycle fatigue 低周疲劳，低循环疲劳
low-cycle high-temperature fatigue 低周高温疲劳
low-cycle operation 低周运行，低频运行
low-cycle thermal fatigue 低周热疲劳
low-cycle 低周波（的），低频的，低周循环（的）
low dam 低坝
low data rate input 低速数据输入
low data rate 低速数据传输
low density binder 低浓度黏合剂，低密度黏结剂
low density data system 低密度数据系统
low density gravity flow 低密度重力流
low density wind tunnel 低密度风洞
low density 低密度
low discharge 低水流量
low-dosage sample 低剂量样品
low dose irradiation 小剂量辐照
low draft switch 低压保护开关【机械通风系统】
low drag airfoil 低阻翼型
low drag configuration 低阻构型
low drag cowl 低风阻外罩，减阻罩
low drag profile 低阻外形
low-duty-cycle switch 短时工作开关
low-duty 小容量的，小功率的，轻型的
low early strength cement 早期低强水泥
low earth barrier 低土围堤
low efficiency 低效率
low-end 低等的，低级的，下端
low energy consumption 低能耗
low energy content wind 低能量风
low energy electron 低能电子
low energy environment 低能环境
low energy fluid system 低能流体系统
low energy fuel 低能燃料
low energy gamma monitor 低能 γ 射线监测器
low energy gamma source 低能 γ 射线源
low energy neutron 低能中子
low energy path 低能源消耗途径
low energy radiation 低能辐射，软辐射
low energy reactor 小功率反应堆
low energy relay 灵敏继电器，低能耗继电器
low energy secondary radiation 低能次级辐射
low energy 低能的
low-enriched oxide fuel 低富集（铀）氧化物（核）燃料
low-enriched uranium 低富集铀，低加浓铀
low-enriched 低富集的，低浓缩的
low enrichment ordinary water reactor 低浓燃料轻水反应堆
low enrichment reactor 低富集铀反应堆
low enrichment uranium 低浓缩铀
low enrichment 低富集度，低浓缩度
lower adjusting ring 下部调节环
lower balnite 下镇流器
lower bar 下层线棒，下鼠笼条
lower bearing spider 下轴承架
lower bearing 下轴承
lower boom 下弦杆
lower border 下边线
lower bound 下界，下限
lower cage 下鼠笼
lower-capacity point 小流量点
lower casing 下汽缸，下缸，下壳
lower chamber 下闸室
lower chord 下弦
lower coil support 磁极线圈下垫板
lower control limit 下限界限
lower core barrel 堆芯下部围筒，下部堆芯筒体
lower core plate support ring 下部堆芯板支撑环
lower core plate 堆芯底板，堆芯下板
lower core structures 堆芯下部构件
lower core support barrel 堆芯下部围筒
lower core support pad 堆芯下部支承块
lower core support plate 堆芯下部支承板
lower core support structures 堆芯下部支承构

件，堆芯底部支撑结构
lower course　下游
lower cover　下端盖
lower critical magnetic flux density　下临界磁通密度
lower critical temperature　下临界温度
lower critical velocity　次临界速度，低临界速度
lower culmination　下中天
lower-cut-off frequency　下限截止频率
lower cycle fluid　（二元循环的）下级工质
lower cylinder half　下缸，下汽缸
lower destruct limit　破坏极限下限值
lower drum　下锅筒，下汽包
lower echelon automatic switchboard　低级梯阵自动配电盘
lower end cover　下端盖
lower end plug　下端塞
lower end shield　下半端盖
lower expansion chamber　下扩大室
lower explosive limit　爆炸下限
lower extreme　下限
lower fibre　下缘纤维
lower flange of girder　梁的下翼缘
lower flange　下部法兰
lower flow period　枯水期
lower foundation raft　下部基础筏基，下部底板，下部支承板
lower frequency noise　低频干扰，低频噪声
lower frequency　低频，低频率
lower-grade coal　劣质煤
lower-grade heat　低位热，低势热
lower grid plate　下栅板
lower grid　下栅板
lower guard sill　下游防护槛
lower guard wall　下游护墙
lower guide bearing bracket　下导轴承支架
lower guide bearing　下部导向轴承
lower guide tube assembly　下部导向管组件
lower guide-vane ring　下导叶环
lower half bearing　下半轴承
lower half casing　下汽缸，下缸，下壳
lower half shell　下半泵壳，下半轴瓦
lower header　下联箱，底封头，下封头
lower heating value　低位热值，低热值
lower housing tube　下部外套管
lowering motion　下降动作
lowering of groundwater level　地下水位降低
lowering of groundwater table　地下水位下降，地下水位降低
lowering of water level　水位降低
lowering the spouts　降低给料斜槽，降低喷水口
lowering well　落煤井
lowering　降低，下降，减少
lower inner casing　内下缸
lower in quotation　报价较低
lower instrumentation guide column with extension　带延长段的下部测量导向柱
lower instrumentation guide column　下部测量导向柱
lower interlock relay　下联锁继电器
lower internals assembly　下部构件组件
lower internals storage stand　（堆芯）下部构件存放台
lower internals　（堆芯）下部构件
lower lateral bracing　下弦横向水平支撑
lower layer　下层
lower level system　下级系统，下层系统
lower level　低位
lower limb　下翼
lower limit of explosion　爆炸下限
lower limit of ultimate strength　屈服强度下限
lower limit speed regulator　调速器调速下限
lower limit switch　下部限位开关
lower limit　下限，低限
lower lip　出口反弧段，出口戽斗
lower loop　下端环
lower main shielding　下部主屏蔽
lower measuring range value　（测量范围）下限值
lower mode vibration　低频振型振动，低声调振动
lower nappe profile　水舌下缘线
lower operating range control　下限控制
lower order harmonics elimination method　低阶谐波消除法
lower order harmonics　低阶谐波，低次谐波
lower Ordovician period　下奥陶纪
lower outer casing　外下缸，外下汽缸
lower plane　底面
lower plastic limit　塑性下限
lower plate　下盘
lower plenum subfactor　下腔室分因子
lower plenum　下空腔，下腔室
lower plug insertion　下部塞插头
lower pool　尾水池，下游水池
lower position detector　下部位置探测器
lower power station　下游电站
lower priority　次重要优先项目
lower radial bearing　下径向轴承
lower raft　下部筏基，下部支承板，下部底板
lower range-limit　范围下限
lower range-value　低范围值，下限值
lower reach　下游河段，下游段，下游
lower reactor internals　堆芯下部构件
lower reservoir　下游水库，下水库，下池
lower ring beam　下部环梁
lower river basin　下游流域
lower river　下游
lower roll　下托辊
lower shaft　下轴
lower shell　下部环段，下筒体
lower shield　（汽轮发电机组的）下半端盖，下半端罩
lower side-band　下边频带
lower speed ring　下速环
lower stay column　下固定支柱
lower steam inlet port　下进汽口
lower steam passage　下部汽道
lower storage reservoir　下池
lower strand　下分支
lower support column　下部支承柱
lower support ring　底部支承环，下支承环
lower surge basin　下调压池

lower surge chamber 下调压室
lower surge tank 下调压室
lower switching value 低切换值
lower-temperature carbonization for coal 低温干馏
lower thrust wall 下游承推墙
lower tie plate 下部固定板
lower-tier 较低层次的
lower to the horizontal position 在水平位置以下
lower wall 下部水冷壁，下部墙，底壁，断层下盘，下盘
lower water box 低位水箱
lower water indicator 低水位指示器
lower yield point 下屈服点【冶金】
lower yield strength 下屈服强度
lower yield stress 下屈服点，下屈服应力
lower yoke 下磁轭
lower 较低的，下部的
lowest assured natural streamflow 最小保证天然径流
lowest astronomical tide 最低天文潮位
lowest atmospheric layer 最低大气层
lowest bidder 最低报价者，出价最低的递盘者
lowest bid price 最低标价
lowest bid 最低报价，最低价递盘，最低价中标
lowest discharge 最小流量
lowest down surge level 最低涌波水位
lowest downsurge 最低下涌浪
lowest effective power 最低有效功率
lowest ever known discharge 历史最小流量
lowest high water 最低高潮位，最低高水位
lowest level 最低水位
lowest low water 最低低潮位，最低枯水位
lowest navigable stage 最低通航水深
lowest normal tide 最低正常潮位
lowest offer 最低价的报盘
lowest order digit 最低位数字
lowest possible price 最低价
lowest price limit 最低限价
lowest quotations 低盘，最低价格
lowest recorded stage 最低记录水位，最低实测水位
lowest required radiated power 最低要求的辐射功率
lowest safe water line 最低安全水位
lowest standard 最低标准
lowest tender 最低标（价）
lowest upper pool level 上池最低水位，最低上池水位
lowest usable frequency 最低可用频率
lowest voltage of a system 系统最低电压
lowest water level 最低枯水位，最低水位
low excess air operation 低过量空气运行，低氧量运行，低过剩空气运行
low excitation 低励磁
low exhaust 低（污染）排气
low expansion foam extinguishing system 低倍数泡沫灭火系统，低膨胀泡沫灭火系统
low explosive 低效炸药
low extreme 下限
low fire start 燃烧器点火时低火焰运行
low fire 弱火，微火燃烧
low-flow circulation line 低流量循环管线
low-flow forecast 低水预报
low-flow frequency curve 低流量频率曲线，枯水流量频率曲线
low-flow rate 低流量的
low-flow regulation 枯水调节
low-flow water level 低水水位
low-flow year 枯水年
low-flow 低流量
low flush tank 低位水箱，低水箱
low flux reactor 低通量反应堆
low-freezing explosive 低温炸药
low-freezing 低凝固点的
low frequency alternate action 低周反复作用
low frequency amplification 低频放大
low frequency amplifier 低频放大器
low frequency and surge test 低频率冲击试验
low frequency band 低频带
low frequency buffeting 低频颤震
low frequency cable line 低频电缆线
low frequency choke 低频扼流线圈，低频抗流圈
low frequency current 低频电流
low frequency dry-flashover voltage 低频干闪络电压
low frequency end 低频端
low frequency filter 低频滤波器
low frequency flashover voltage 低频闪络电压
low frequency generator 低频发电机
low frequency high-voltage test 低频高压试验
low frequency impedance 低频阻抗
low frequency induction furnace 低频感应电炉，低频高压电炉
low frequency interference 低频干扰
low frequency motor 低频电机
low frequency noise 低频噪声
low frequency oscillator 低频振荡器
low frequency protection 低频保护
low frequency rebroadcasting 低频转播
low frequency regulation 低频调节，低频调整
low frequency relay 低频继电器
low frequency resonance 低频谐振
low frequency ringer 低频振铃器
low frequency signal 低频信号
low frequency stage 低频段
low frequency start 低频启动
low frequency stress 低频应力
low frequency transformer 低频变压器
low frequency tube 低频管
low frequency vibration 低频振动
low frequency wet-flashover voltage 低频湿闪络电压
low frequency 低频
low fume and toxic electrode 低尘低毒焊条
low-fusing ash 低熔点灰
low-gain cryotron 低增益冷子管
low gauge 低速齿轮
low gear 低速齿轮
low-grade cement mortar 低等级水泥砂浆

low-grade cement 低等级水泥
low-grade coal fired power plant 燃烧劣质煤火电厂
low-grade coal 低品级煤，劣质煤
low-grade concrete 低标号混凝土
low-grade energy 低势能，低级能
low-grade fuel 低级燃料
low-grade heat 低品位热
low-grade mineral 低品位矿石
low-grade 劣质的，低级的
low-hardness water 低硬度水
low head centrifugal pump 低水头离心泵
low head development 低水头开发
low head hydraulic turbine 低水头水轮机
low head hydroelectric plant 低水头电站
low head hydroelectric power station 低水头水利发电厂，低水头水电站
low head hydropower station 低水头电站
low head injection pump 低压注射泵
low head injection system 低压安注系统
low head pump 低扬程泵，低压头泵
low head safety injection pump 低压头安全注入泵
low head safety injection system 低压安全注射系统
low head water turbine 低水头水轮机
low head 低压头，低扬程
low-heat cement 低热水泥
low-heat-generating waste 低释热废物
low heating value 低（位）热值，低位发热量，低发热值，低位发热量
low-heat Portland cement 低热硅酸盐水泥
low hold-up 低持留量，低滞留量，（废物）低暂存量
low hydrogen covering 低氢药皮
low hydrogen electrode 低氢焊条
low hydrogen potassium electrode 低氢钾型焊条
low hydrogen sodium electrode 低氢钠型焊条
low hydrogen type electrode 低氢型电焊条
low impedance busbar protection 低阻抗母线保护
low-impedance path 低阻抗路径
low-impedance 低阻抗的
low incidence 小攻角，小迎角，小冲角，负冲角
low incinerator 平型炉
low-inductance synchro 低电感同步机
low inertia DC motor 低惯性直流电动机
low inertia electron beam 微惰性电子束
low inertia motor 低惯量电动机
low inertia pump motor 低惯量泵电机
low in price 低价
low-intensity AOL 低光强航空障碍灯
low-intensity 低强度的
low interaction fuel 低相互作用燃料
low interest credit 低息贷款
low interest loan 低息贷款
low land forest 低湿林
low land meadow 低地草原
low land moor 低地沼泽
low land river 低地河流
low land 低洼地，低地
low latitude 低纬度
low leakage fuel management 低泄漏燃料管理
low leakage loading pattern 低泄漏装料方式
low level alarm 低料位报警，低液位报警，低位报警
low level bog 低沼
low level cave 低放射性工作室
low level cell 弱放射性工作室，低放射性工作室
low level chassis 低底盘
low level circuit 低电平电路
low level contact 低电平触点
low level cracking 轻度裂化，轻度开裂
low level detector 低料位探测器
low level dosimeter 低辐射水平剂量计，低剂量率剂量计，低水平剂量计
low level exposure 低剂量照射
low level economizer 低温省煤器
low level intake 深式进水口
low level jet condenser 低位喷射式冷凝器，低位注水凝汽器，低位喷水凝汽器
low level jet 低空急流
low level laboratory 低放射性实验室，低放实验室
low level language 低级语言
low level logic circuit 低电平逻辑电路
low level logic 低电平逻辑
low level makeup （极限）低水位补给
low level modulation 低功率调制，低电平调制
low level multiplexing 低电平放大
low level nocturnal jet 夜间低空急流
low level outlet 深式泄水孔，泄水底孔
low level pulse 低电平脉冲
low level recovery 低级热能回收
low level relay 低电平继电器
low level sodium tank 低位储钠箱
low level solid waste 低放射性固体废物，低放固体废物
low level storage tank 低位贮存槽
low level switching 低电平切换，低电平转换
low level technologies 低水平技术，低级技艺
low level test 低电平试验
low level threshold 低水位定值，低水位阈值
low level warning switch 低位报警开关
low level waste 低放射性废物，低放废物
low level wind 底层风
low level 低电平的，低能级的，低水平的，低放射性水平，低水位
low lift centrifugal pump 低扬程离心泵
low lift construction for mass concrete 大体积混凝土薄层施工
low lift lock 低水头船闸，低水位差船闸
low lift pump 低压头泵，低扬程（水）泵
low lift 低扬程，低水位差，低压的
low lightning incidence 少雷
low limb 下翼
low-limed cement 低石灰水泥，低氧化钙水泥
low limit adjustment 下限调整
low limiting control 下限值控制
low limit of speed regulator 调速器下限
low limit register 下限寄存器

low limit 下限，低限，下限值
low load adjustment 低载调整，低负荷调整
low load circulation 低负荷循环
low load period 低负荷期，低负荷周期
low load range 低负荷区段，低负荷范围
low load Vickers hardness test 小负荷维氏硬度试验
low load 低负荷
low loss cable 低损耗电缆
low loss material 低损耗材料
low loss 低损耗的，低损耗
low-low alarm 低低值报警
low-low level 低位水平，最低水位，"低-低"水位
low-low 低-低【设定值，报警，水位等】
low-lying area 低地
low-lying land 低洼地
lowly priced 价格定得低的
low-magnification seismograph 低倍率地震计
low-mark cement 低标号水泥
low-melting point alloy 低熔点合金，易熔合金
low-modulus resin 低弹模树脂
low-molecular polymer 低聚合物
low moor 低位沼泽
low neutron leakage fuel management 低中子泄漏的燃料管理
low noise and low vibration motor 低噪声低振动电机
low noise motor 低噪声电动机
low noise relay 低噪声继电器
low noise transformer 低噪声变压器
low noise 低噪声
low NO_x burner 低氮氧化物燃烧器
low NO_x combustion 低氮氧化物燃烧
low of the wake 尾流规律
low oil alarm 低油［液］位报警，低位报警
low-oil-content breaker 贫油断路器
low-oil-content circuit breaker 少油断路器
low-oil-content 少油的
low oil pressure protection test 低油压保护试验
low optical power receiving mode 低光功率接收模式
low order add circuit 低阶加法电路
low order harmonic 低次谐波
low order merge 低阶合并
low parapet 矮女儿墙
low partition 矮隔断
low pass filter 低通滤波器
low pass 低通的，低通
low-penetration electrode 浅熔深电焊条，小熔深焊条
low permeability graphite 低渗透性石墨
low permeability 低透水性
low-persistence screen 短暂余辉屏
low perviousness 低透水性
low pitched roof 缓坡屋顶
low plain 低平原
low plastic brittle crack 低塑性脆化裂纹
low point 低水位点，低点
low pollution emission 低污染排放
low pollution energy source 低污染能源
low pollution fuel 低污染燃料
low pollution technique 低污染技术
low polymer 低聚合物
low population area 低人口密度区，人口稀少地区
low population density 低人口密度
low-porosity concrete 低孔率混凝土
low positive pressure 低正压
low potential energy 低势能，低级能
low potential 低电位，低电势
low powered 小功率的，低功率的，装有小型发动机的
low power factor operation 低功率因数运转
low power factor transformer 高电抗变压器，低功率因数变压器
low power factor wattmeter 低功率因数瓦特表
low power feedwater control system 低功率给水控制系统
low power licence 低功率运行许可证
low power logic circuit 低功率逻辑电路
low power logic 低功耗逻辑
low power machine 小功率电机
low power motor 小功率电动机
low power optical receiving 低光功率接收
low power range trip 低功率区段紧急停堆
low power range 低功率区段
low power reactor 低功率反应堆
Low Power Research Reactor （美国）低功率研究堆
low power run 低功率运行
low power water boiler 低功率沸腾式反应堆，低功率水锅炉
low pressure acetylene generator 低压乙炔发生器
low pressure area 低压区
low pressure bell-type recording flowmeter 低压钟形记录式流量计
low pressure blowpipe 射吸式焊（割）炬
low pressure boiler 低压锅炉
low pressure by-pass system 低压旁路系统
low pressure bypass 低压旁路
low pressure cable 低压电缆
low pressure casing spray 低压缸喷水装置
low pressure casing 低压缸
low pressure centre 低压中心
low pressure chemical vapor deposition 低压化学气相沉积
low pressure combustion chamber 低压燃烧室
low pressure compressor 低压压缩机，低压压气机
low pressure condensate water 低压凝结水
low pressure condensing turbine 低压凝汽式汽轮机
low pressure containment 低压安全壳
low pressure coolant injection system 低压冷却剂注入系统，低压冷却剂注射系统
low pressure coolant injection 低压冷却剂注射
low pressure coolant recirculation system 低压冷却剂再循环系统
low pressure core spray system 低压堆芯喷淋系统

low pressure core spray 低压堆芯喷淋
low pressure cylinder exhaust steam flow 低压缸排汽流量
low pressure cylinder expansion 低压缸膨胀
low pressure cylinder 低压缸，低压汽缸
low pressure emergency cooling system 低压应急冷却系统
low pressure end 低压端
low pressure extraction valve 低压抽汽调节阀
low pressure feedwater heater system 给水低压加热器系统
low pressure feed water heater 低压给水加热器，低压加热器，低加
low pressure grout hole 低压灌浆孔
low pressure grouting 低压灌浆
low pressure header 低压集管，低压联箱
low pressure heater 低压加热器
low pressure heating 低压加热
low pressure injection system 低压注射系统
low pressure injection 低压注入，低压注射
low pressure intake 低压进水口
low pressure letdown valve 低压下泄阀
low pressure mercury lamp 低压汞灯
low pressure molding 低压制模法
low pressure plenum 低压联腔
low pressure pneumatic dry ash handling system 低压气力干除灰系统
low pressure pump 低压泵
low pressure reactor 低压反应堆
low pressure recirculation cooling 低压再循环冷却
low pressure recirculation system 低压再循环系统
low pressure rotor 低压转子
low pressure safety injection system 低压安全注射系统
low pressure safety injection 低压安全注射
low pressure scram 低压紧急停堆
low pressure seal 低压密封
low pressure section 低压段，低压部分，低压缸
low pressure shaft 低压轴
low pressure shell 低压缸壳体
low pressure side of blade 叶片排汽侧，叶片背面
low pressure side 低压侧
low pressure slide valve 低压滑阀
low pressure steam chest 低压蒸汽室，低压蒸汽联箱
low pressure steam curing 低压蒸汽养护
low pressure steam generator 低压蒸汽锅炉，低压蒸汽发生器
low pressure steam heating 低压蒸汽采暖，低压蒸汽供暖
low pressure steam turbine 低压汽轮机
low pressure steam 低压蒸汽
low pressure tank 低压水箱
low pressure trip 低压事故保护停堆
low pressure tunnel 低压隧洞
low pressure turbine 低压汽轮机
low pressure vacuum pump 高真空泵
low pressure water injection 低压注水
low pressure water pump 低压水泵
low pressure zone 低压区
low pressure 低压
low pressurizer pressure trip 稳压器低压停堆
low pressurizer pressure 稳压器低压，稳压器低压力（事故）
low pressurizer water level trip 稳压器低水位事故保护停堆
low price 廉价
low priority 次要优先项目
low probability events 低概率事件
low-profile layout 低型布置
low-profile 矮型的
low-purity 低纯度
low-quality coal 劣质煤
low-quality farm land 贫瘠土壤
low-quality 低品质度，（蒸汽）低含汽率
low radio frequency 无线电低频
low range 低倍率
low rank anthracite 低级无烟煤
low rank bitumite 低级烟煤
low rank coal 劣质煤
low rate code 低信息率码
low rate filter 慢滤池
low RCS pressure 反应堆冷却剂系统低压力，稳压器低压力事故
low reactance arrangement 低电抗布置，反相并列布置
low reactor coolant flow trip 反应堆冷却剂低流量紧急停堆
low recovery 低回收
low-relief terrain 低山区
low-resistance damping winding 低电阻阻尼绕组
low-resistance material 低阻材料
low-resistance varnish 低电阻漆
low-resistance 低电阻，低阻力，低阻
low Reynolds number 低雷诺数
low-ripple generator 低脉动发电机
low-ripple 低纹波
low rise building 低层建筑（物），层数较少大厦，矮楼宇
low rise structure 低层结构，低矮结构，层数较少的构筑物
low river basin 下游流域
low runaway speed 低飞逸转速
low runoff 低水径流，枯水径流，枯水流量
low-sag conductor 低弛度导线
low sag up-rating conductor 低弛度增容导线
low-salt-content liquid waste 低盐废液
low selector 低选择器
low-shrinkage concrete 低缩性混凝土
low sided ship 低弦船
low side 低侧，低端
low-signal selector 低值信号选择器
low-silicon steel 低硅钢
low-slag cement 低矿渣水泥，低熔渣水泥
low-slag concrete 低坍落度混凝土
low-slump concrete 低坍落度混凝土，小坍落度混凝土
low-solids water 低盐水
low specific-speed 低比速，低比转速
low speed aerodynamic 低速空气动力学

low speed aerofoil 低速翼型
low speed balancing 低速动平衡
low speed characteristics 低速特性
low speed data channel 低速数据通道
low speed flow 低速流
low speed francis turbine 低速混流式水轮机
low speed high torque shaft 低速高扭矩轴
low speed indicator 低速指示器
low speed logic 低速逻辑
low speed machine 低速电机，低速机
low speed magnetic tape controller 低速磁带控制器
low speed magnetic tape unit 低速磁带装置
low speed mill 低速磨煤机
low speed motor 低速电动机
low speed needle valve 低速针型阀
low speed operation 低速运行
low speed pitot-static head 低流速皮托管静压头，低速毕托管静水头
low speed pointer 低速指示器，低速打印机
low speed pump 低速泵
low speed reclosure 慢速重合闸，低速重合闸
low speed register 低速寄存器
low speed regulation 低速调节
low speed sand filtration 低速砂滤
low speed shaft 低速轴
low speed stability 低速稳定性
low speed switch 盘车开关
low speed threshold 低速定值，低速阈
low speed turbine 低速涡轮机，低速汽轮机
low speed warming 低速暖机
low speed WECS 低速风力机【额定叶尖速率比小于3的风力机】
low speed winding 低速绕组
low speed wind tunnel test 低速风洞试验
low speed wind tunnel 低速风洞
low speed 低速
low static pressure differential pressure gauge 低静压差压计
low steam pressure unloading gear 低汽压卸负荷装置
low steam 低压蒸汽
low steel proportion 低含钢率
low-storied building 低层建筑
low strain integrity method 低应变法
low strain integrity testing method 低应变整体性测试法，低应变法
low strain method 低应变（测试）法
low strength cement 低强度水泥
low strength concrete 低强混凝土
low strength 低强度
low stress brittle fracture 低应力脆断
low stress fracture 低应力断裂
low sulfur coal 低硫煤
low sulfur oil 低硫石油
low sulphur coal 低硫煤
low sulphur crude petroleum 低硫原油
low sulphur fuel 低硫燃料
low sulphur(/sulfer) coal 低硫份的煤
low supersaturation 低度过饱和
low swampy land 低沼泽（地）
lows 低频
low temperature adsorber 低温吸附器
low temperature brittleness 冷脆性，低温脆性
low temperature cable 低温电缆
low temperature characteristic 低温特性
low temperature cooling 低温冷却
low temperature corrosion on the fire side 低温烟气腐蚀
low temperature corrosion 低温腐蚀
low temperature delay bed 低温延迟床
low temperature distillation 低温蒸馏
low temperature embrittleness 低温脆裂
low temperature filter 低温过滤器
low temperature flexibility 低温韧性
low temperature heater 低温加热器
low temperature hot water heat-supply system 低温热水供热系统
low temperature hot water 低温水
low temperature impact 低温冲击
low temperature level 低温下限，（热力循环中的）冷源，低温源
low temperature liquid-level indicator 低温液面计
low temperature motor 低温电动机
low temperature nuclear heating reactor 低温核能供热堆
low temperature physics 低温物理学
low temperature reactor 低温反应堆，低温堆
low temperature receiver （热力循环中的）冷源
low temperature reheat pipe 再热冷段管道，冷再热汽管
low temperature resistance 耐低温性，低温阻力
low temperature setting 低温固化
low temperature steel electrode 低温钢焊条
low temperature stress relieving 低温消除应力，温差拉伸消除应力
low temperature system 低温系统
low temperature tempering 低温回火
low temperature test 低温试验
low temperature thermocouple 低温热电偶
low temperature thermometer 低温温度计
low temperature treatment 低温热处理
low temperature 低温
low tension bushing 低压套管
low tension cable 低压电缆
low tension coil 低压线圈
low tension distribution board 低压配电盘
low tension fuse 低压熔丝，低压熔断器
low tension large current generator 低压大电流发生器
low tension line 低压线路
low tension magneto 低压永磁发电机
low tension motor 低压电动机
low tension network 低压电力网，低压电网
low tension switchgear 低压开关装置
low tension transformer 低压变压器
low tension winding 低压线圈，低压绕组
low tension wire 低压电线
low tension 低压的，低电压的，低张力
low thermal conductivity 低热传导率
low tide level 低潮面
low tide 低潮

low to high 低到高
low torque 低转矩
low trestle 低支架
low truss bridge 下承式低架桥
low turbulence stream 弱紊动水流
low turbulence wind tunnel 低湍流度风洞
low vacuum trip test 真空降低跳闸试验
low vacuum trip 低真空跳闸，低真空脱扣
low vacuum unloading gear 低真空卸负荷装置
low vacuum 低真空
low-value consumables 低值易耗品
low velocity waveguide 低速波导管
low velocity 低速
low viscous flow 低黏性流
low visibility 低能见度
low volatile bituminous coal 低挥发份烟煤
low volatile coal 低挥发分煤
low voltage accident 低压事故
low voltage apparatus 低压电器
low voltage automatic air switch 低压自动空气开关
low voltage auxiliary transformer 低压厂用变压器
low voltage busbar 低压汇流排，低压母线
low voltage bushing 低压套管
low voltage cable line 低压电缆线路
low voltage cable 低压电缆
low voltage circuit breaker 低压断路器
low voltage common transformer 低压公用变压器
low voltage contact 低压接点，低压触头
low voltage current 低压电流
low voltage distribution line 低压配电线路
low voltage distribution network 低压配电网
low voltage distribution room 低压配电室
low voltage distribution system 低压配电系统
low voltage distribution 低压配电
low voltage electrical apparatus 低压电器
low voltage equipment 低压设备
low voltage generator 低压发电机
low voltage house service operating transformer 低压厂用工作变压器
low voltage impulse response test 低压脉冲响应测试
low voltage impulse 低电压冲击
low voltage insulated overhead line 低压架空绝缘线路
low voltage insulation 低压绝缘
low voltage lamp 低压灯泡
low voltage line 低压线，低压线路
low voltage motor 低压电动机
low voltage network 低压电力网
low voltage overhead distribution line 低压架空配电线路
low voltage overhead line 低压架空线
low voltage power consumption 低压用电
low voltage power source 低压电源
low voltage power supply 低压电源
low voltage protection 低压保护
low voltage relay 低压继电器
low voltage release relay 低压释放继电器
low voltage release 低压释放
low voltage ride through 低电压穿越
low voltage section 低电压段
low voltage switchgear cubicle 低压开关柜
low voltage switchgear 低压配电装置，低压开关柜
low voltage system 低压系统，低电压制式
low voltage tripping 低压跳闸
low voltage vacuum electrical apparatus 真空低压电器
low voltage winding 低压线圈，低压绕组
low voltage 低电压的，低电压，低压
low volt transformer 低压变压器
low volume air sampler 低容量空气采样器
low wall 矮墙
low water-bearing formation 弱含水层
low water control 低水控制
low water culvert 低水位涵洞
low water cutoff 锅炉低水位时燃烧器停运开关
low water datum 低水位基准面，低水准面，水深零点
low water degradation 低水位消落
low water discharge 低水位流量，枯水流量，低水流量
low water forecast 低水预报，枯水预报
low water gauge 低水位尺
low water head 低水头
low water investigation 枯水调查
low water level alarm 危险水位报警器，低水位报警器
low water level 低水位，枯水位
low water line 低水位线
low water loss cement 低析水性水泥
low water mark 低水位标志，低水位线
low water neap 小潮低潮面
low water observation 低水位观测
low water of ordinary spring tide 一般大潮低潮
low water period 枯水期
low water quay 低水位岸壁式码头
low water rating curve 低水位流量关系曲线
low water regulation storage 低水位调节蓄水量
low water regulation 低水治理，枯水整治
low water runoff 枯季径流，枯水径流
low water season 枯水季
low water spring 大潮低潮面
low water stand 低潮停潮
low water survey 低水位调查，枯水调查
low water(s) 低水位，低潮，低水位的，低潮的
low water training wall 低水位导流堤
low water training 低水治导
low water year 枯水年，少水年
low water 低潮，低水位，枯水
low-weight refractory castable 轻质耐火浇注料
low weir 低堰
low 低（气）压中心，低气压，低的
LP bottom head 低压底封头
LPC(low pressure cylinder) 低压汽缸，低压缸
LP cylinder （汽轮机）低压缸
LPG(liquefied petroleum gas) 液化石油气
LP heater 低压给水加热器，低加
LPH(low pressure heater) 低压加热器

LPI(lines per inch) 行每英寸
LP(low point) 低位
LP(low pressure) 低压【DCS 画面】
LPM(lines per minute) 行每分
LPRR(Low Power Research Reactor) (美国)低功率研究堆
LP-safety injection pump 低压安全注射泵
LP sampling cooler 低压取样冷却器
LPS(lightning protection system) 防雷系统
LPSD(lines per second) 行每秒
LPTF(low power test facility) (美国)低功率试验装置
LPT(liquid penetration test) 液体渗透试验
LPZ(lightning protection zone) 防雷区
LRD(long-range data) 长距离数据
LRFD(load and resistance factor design) 荷载和抗力系数设计
L-ring L 环
LR(load ratio) 载荷比，有效载荷
LSA 试验回路系统【核电站系统代码】
LSB communication 下边带【通信】
LSB(last-stage blade) 末级叶片
LSB(least significant bit) 最低有效位
LSB(lower sideband) 下边带通信
LSD system(liquid shutdown system) 液体停堆系统
L-shaped flame boiler L 形火焰锅炉
LSHI(large scale hybrid integrated circuit) 大规模混合集成电路
LSH(local switch hand) 就地开关
LSH 低温过热器【DCS 画面】
LSIC(large scale integrated circuit) 大规模集成电路
LSI(large scale integration) 大规模集成【电路】
LSI 厂区照明系统【核电站系统代码】
LS(liquid sensor) 液体传感器
LSL(logical shift left) 逻辑左移
LSL(low speed logic) 低速逻辑
LSM(linear select memory) 线选存储器
LSN(linear sequential network) 线性时序网络
LSPB(large scale prototype breeder) 大型原型增殖反应堆
LSP(low speed pointer) 低速指示器
LSP(low speed printer) 低速打印机
L-square 直角尺
LSR(load shift register) 送入移位寄存器
LSR(location stack register) 单元栈寄存器
LSR(logical shift right) 逻辑右移
LST(left store) 左存储
LSU(lone signal unit) 单一信号单元
LSW(line status word) 行状态字
LTA(lead test assembly) 领先试验组件
LTB(live tank circuit-breaker) 瓷柱式断路器，柱式断路器
Ltd=limited 有限的
LT(leakage/leak test) 泄漏试验，紧密性试验
LT(leak testing) 泄漏检验，检漏
LT(low torque) 低转矩
LTM(long term memory) 长期存储器，永久存储器
LTMS(long term maintenance of gas turbine services) 燃气轮机的长期维护
LTP(loss of total pressure) 总压力损失
LTRS(letter shift) 换字母档
LTR 接地系统【核电站系统代码】
LTSF(lid tank shielding facility) 遮盖水箱屏蔽装置
L-type controller L 形控制器
L-type support L 形管架
lubarometer 测大气压仪
lube oil aerosol 润滑油气悬体
lube oil conditioner 润滑油净化装置
lube oil cooler 滑油冷却器
lube oil piping 润滑油管
lube oil pump 润滑油泵
lube oil purifier 润滑油净化装置
lube oil reservoir 润滑油箱
lube oil supply piping 润滑油供油管道
lube oil system 润滑油系统
lube oil transfer pump 润滑油输送泵
lube oil 润滑油
lube 润滑油，润滑剂
LUB(least upper board) 最小上界
LUB=lube/lubrication 润滑
LUB(lubricating oil) 润滑油
lubricant additive 润滑油添加剂
lubricant film 润滑膜，润滑油
lubricant grease 滑脂
lubricant groove 润滑油槽
lubricant 润滑剂，润滑油，润滑的
lubricated for life 永久性润滑
lubricate 加润滑油，润滑，上油
lubricating cup 油杯
lubricating device 润滑装置，注油器
lubricating gauge 润滑装置
lubricating gear 齿轮式滑油泵，润滑装置，
lubricating grease 润滑油脂
lubricating gun 润滑油枪
lubricating medium 润滑剂，润滑介质
lubricating moisture 润滑性水分
lubricating oil circulation 润滑油循环
lubricating oil cooler 滑油冷却器
lubricating oil gun 润滑油枪
lubricating oil pipe 润滑油管
lubricating oil pressure 润滑油压
lubricating oil pump 润滑油泵
lubricating oil storage tank 汽机检修贮油箱
lubricating oil strainer 滑油过滤器，滑油滤网
lubricating oil supply system 润滑油供应系统
lubricating oil system 润滑油系统
lubricating oil tank gas exhaust fan 油箱排烟风机
lubricating oil temperature 润滑油温
lubricating oil transfer plant building 润滑油转运厂房
lubricating oil viscosity 润滑油黏度
lubricating oil 润滑油，滑油
lubricating ring 油环，润滑环
lubricating screw 黄油枪，润滑螺旋，螺旋润滑器
lubricating system 润滑系统
lubrication factor 润滑因子
lubrication failure 润滑故障

lubrication film 润滑油膜
lubrication gap 润滑油隙，润滑缝
lubrication groove 润滑油槽
lubrication line 润滑管路
lubrication nipple 润滑喷嘴
lubrication oil 润滑油
lubrication opening 加油孔
lubrication pressure regulation 润滑油压力调节
lubrication property 润滑特性
lubrication pump 滑油泵
lubrication system 润滑系统
lubrication wedge 润滑油楔
lubrication 注油，润滑（作用），润滑方式
lubricator oil strainer 滑油滤清器
lubricator 润滑器，润滑剂，注油机，注油器，加油器，油壶
lubricious 光滑的，不稳定的
lubricity 润滑性
lubric pump 润滑泵
lubric 光滑的
lubro-pump 油泵
lub system 润滑系统
lub 润滑（物质）【缩写】
LUB 润滑油【DCS 画面】
lucarne 老虎窗，屋顶窗
lucite pipe 透明塑料管
lucite 有机玻璃，人造荧光树脂
lucrative investment 有利的投资
LU decomposition （矩阵的）LU 分解
LUD(lamp lumen depreciation) 光通量衰减系数
luff crane type coal unloader 可变伸距卸煤机
luffing and slewing stacker 升降及旋转堆料机
luffing boom stacker 升降臂堆料机
luffing cableway 塔架可倾式缆索道
luffing crane coal unloader 可变伸缩卸煤机
luffing crane 鹅头式伸臂起重机，水平起重机，俯仰式起重机，动臂起重机
luffing derrick 俯仰式吊杆，俯仰式吊货杆
luffing gear 吊杆倾角调节器，摆动装置
luffing jib crane 俯仰摇臂式起重机，俯仰旋臂起重机
luffing jib tower crane 动臂塔式起重机，动臂变幅塔式起重机
luffing jib 鹅头伸臂【伸臂头可垂直升降保持载重在一水平面上】，动臂，变幅副臂
luffing mechanism 变幅机构
luffing motor 吊杆俯仰电动机
luffing part 变幅部分
luffing speed 变幅速度
luffing the helm 迎风使舵
luffing time 变幅时间
luffing winch 变幅绞车，俯仰式起重机
luffing 起重机臂的转动，升降运动，上下摆动，使起重机吊臂起落，转向迎风，抢风
LUF 卢森堡法郎
lug connector 接线片连接
luggage 行李
lug type butterfly valve 凸耳式蝶阀
lug 吊耳，猫爪，接线片，凸耳，接线柱，柄，手柄，把手，突起，凸缘，（钢筋的）突纹
lull 风暴间歇，息静，暂时平静
lumber gauge 木材厚度规
lumber grillage 木格框
lumber port 木材港
lumber room 杂物间
lumber storage yard 储木场
lumber surfacing 木料的表面处理
lumber yard 木材堆置场
lumber 成材，木料，木材，锯木
lumen 流明【光通量单位】
lumeter 照度计
lumiling lamp 管形灯
luminaire dirt depreciation 照明器污秽减光系数
luminaire efficiency 照明器效率
luminaire 光源，照明器，发光设备
luminance 发光率，亮度
luminary 照明器，发光体，杰出人物
luminescence diode 发光二极管
luminescence 发光
luminescent digital indicator 发光数字指示器
luminescent inspection 发光检查
luminescent paint 发光涂料
luminescent pigment 荧光颜料，发光颜料
luminescent screen 荧光屏
luminescent 发光的
luminosity 光度，发光度，亮度
luminous beacon 灯标，灯塔
luminous buoy 发光浮标
luminous dial 闪光标度盘
luminous diode 发光二极管
luminous discharge tube 气体放电管，气体发光管
luminous efficiency at a specified wavelength 光谱光视效率
luminous efficiency 照明效率，发光效率
luminous emission 光发射
luminous exitance 光出射度
luminous flame radiation 发光火焰辐射，发光焰照射
luminous flame 发光火焰，发光焰
luminous flux 光通量
luminous gas 发光气体
luminous indicator 发光指示器
luminous intensity 发光强度
luminous pigment 发光颜料
luminous point 亮点，发光点
luminous power 发光强度
luminous radiation 发光体辐射，发光辐射，发光气体辐射
luminous radiographic detection 发光放射自显影探测
luminous resonator 辉光共振器，辉光谐振器
luminous sensitivity 光照敏度，光灵敏度
luminous signboard 电光广告牌，发光信号牌
luminous sign 发光标志
luminous standard 标准测光器
luminous tube transformer 霓虹灯变压器
luminous 明亮的，发光的，闪光的
lumoautoradiographic detection 发光放射自显影探测
lump coal 块煤【经简单筛选后剩下的大块有烟煤】

lump-constant network 集中参数网络，集总参数网络
lumped capacitance 集中电容，集总电容
lumped capacity technique 集总容量法
lumped capacity 集总容量，集中容量
lumped characteristic 综合特性，复合特性
lumped circuit 集中参数电路
lumped constant circuit 集中常数电路
lumped-constant filter 集中常数滤波器
lumped constant 集中常数，集总常数
lumped element isolator 集总元件隔离器
lumped fuel 燃料块
lumped inductance 集中电感，集总电感
lumped load 集总负载，集中载荷，集中加感
lumped mass aeroelastic model 集总质量气动弹性模型
lumped mass system 集总质量系统
lumped model 集总模型
lumped parameter dynamic equations 集总参数动态方程
lumped parameter line 集中参数线路
lumped parameter method 集总参数法
lumped parameter model 集总参数模型
lumped parameter network 集中参数网络
lumped parameter system 集中参数系统
lumped parameter technique 集总参数法
lumped parameter 集总参数，集中参数
lumped process control system 集总过程控制系统
lumped resistance 集总电阻，集中电阻
lumped statistics 集总统计学
lumped uranium lattice 铀块栅格
lumped uranium 块状铀
lumped voltage 控制电压，集中电压
lumped winding 集中绕组，集中线圈
lumped 集总的，集中的，集中参数的
lumpiness 块度
lump lime 生石灰，块石灰
lump-loaded circuit 集总负载电路
lump material 块料
lump network 集总参数网络
lump of coal 煤块
lump offer 整体报价，综合报盘
lump parameter 集总参数，集中参数
lump penetrability factor 块穿透因子
lump price 总价值
lump sum agreement 总额承包协定
lump sum bid 总额承包投标
lump sum charge 包干费
lump sum contract 总包合同，总价合同，总额合同，总额承包包干合同，包工契约
lump sum fixed-price contract 固定总包价格合同
lump sum in advance 预付总额
lump sum item 包干项，总价项目
lump sum payment policy 一费制政策
lump sum payment 整笔支付，一次总付，总价付款
lump sum price 总承包价格，总包价格
lump sum 总额，总价
lump work 包干工作，包工
lumpy material 块状物料
lumpy soil 碎块土
lumpy 多块的，粗颗粒的
lump 块，块团，成块，块度
lunar calendar 阴历
lunar day 太阴日
lunar diurnal component 太阴全日分潮
lunar eclipse 月食
lunar fortnightly tide 太阴二周潮
lunar hour 太阴时
lunar month 朔望月
lunar semidiurnal component 太阴半日潮
lunar semidiurnal tide 太阴半日潮
lunar tidal component 太阴潮分量
lunar tide 太阴潮（汐）
lunar time 太阴时
lunar year 太阴年
lunch counter 便餐柜台
luncheon party 便宴，午餐会
lunch room 小食堂
lunitidal interval 月潮间隙
luster 光泽，闪光，发光，光栅
lustrous coal 发亮烟煤，光亮煤，辉煤
lute in 封入，嵌入，塑入
lute 水泥封涂，腻子
lux gauge 照度计
luxmeter 照度计
luxury uptake of phosphorous 磷的奢量吸收
lux 勒克司【光照度单位】
LV AC emergency power supply 低压应急交流电源
LVCD(least voltage coincidence detection) 最小符合检波电压
LVDT(linear variable differential transducer) 线性可变差分传感器
LVDT(linear variable differential transformer) 线性阀位传感器，线性可变差动变压器
LV electrical devices and cables for monitoring, controling, adjusting, protecting and signaling 电气二次设备【中国术语】
LVE(linear vector equation) 线性矢量方程
LVF(linear vector function) 线性矢量函数
LV(latent heat of vaporization) 汽化潜热
LV(linear velocity) 线速度
LVL＝level 水位，液位
LV(low voltage) 低电压
LVPS(low voltage power supply) 低压电源
LVRT(low voltage ride through) 低电压穿越
LWA(limited work authorization) 有限施工授权书
LWBR(light water breeder reactor) 轻水增殖反应堆
LWC(liquid water content) 液体水含量
LWD(larger word) 大写
LWECS (large scale wind energy conversion system) 大型风能转换系统
LWGR(light-water graphite reactor) 轻水石墨（反应）堆
LWHCR(light water high converter reactor) 轻水高转换反应堆
LWHR(light water hybrid reactor) 轻水混合反应堆
LWL(low water level) 低水位

LWOR(light water organic reactor) 轻水有机反应堆
LWPBR(light water pre-breeder reactor) 轻水预增殖反应堆
LWR(light-water reactor) 轻水反应堆
LX=lux 勒克斯【光学计量单位】
lye 碱液
lying fold 伏褶皱，平卧褶皱
lying light 天窗
Lyman tube 赖曼放电管
lyolysis 液解
lyophilic dust 亲水性物质
lyophilic solvent-loving colloid 亲液胶体
lyophobic dust 疏水性物质
lyophobic solvent-hating colloid 憎液胶体
lysimeter pit 渗透仪坑
lysimeter 测渗计，溶度计，渗漏计，渗水计，渗漏测定计
lysimetry 渗漏测定
LYS 蓄电池试验回路系统【核电站系统代码】
lzod impact test 艾氏冲击试验
L 电气系统【核电站系统代码】

M

macadam base 碎石基层
macadamization 碎石路修筑法
macadam pavement 碎石路面
macadam road 碎石路
macadam surface 碎石路面
macadam 铺路用的碎石料，柏油碎石路，碎石
macasphalt type pavement 碎石柏油路面
macasphalt 碎石沥青混合料
mace bit （钻探用的）冲击回旋钻
macerate 浸渍
maceration 浸渍，浸解
machinability 可切削性
machine address 机器地址
machine-advance distance 整机前进距离
machine-aided cognition 机器辅助识别
machine allowance 加工余量
machine and tools 机具
machine attendant 司机
machine available time 计算机的利用时间，计算机的正常运算时间
machine-banded wood stave pipe 机箍木板条水管
machine bolt 半精制螺栓，机螺栓
machine-centering tool alignment 机械定心工具对准
machine check indicator 机器检查指示器，机器检查指示仪
machine check 自动检验，自动校验，计算机的程序检验
machine code 机器代码，机器码
machine cognition 机器识别，机器条件，机器状态
machine control rack 机器控制架
machine cycle 计算机的工作周期，机器工作周期
machined cast iron 机加工铸铁
machined gate seat 机加工闸门支座
machine drill 机钻
machine-driven rivet 机铆铆钉
machined surface 机加工面
machined washer 精制垫圈
machine dynamic load 机器动力荷载
machine efficiency 机器效率，机械效率
machine enclosure 机壳
machine equation 机器方程，计算机运算方程式，计算机方程式
machine erection 机器安装
machine error 机器故障，机器失误
machine fault 机器故障，机械故障
machine floor elevation 主机运行层标高
machine for forming winding 线圈成形机
machine for inserting winding 下线机，嵌线机
machine for tube cutting 切管机
machine foundation 机器基础
machine hall （发电站）主机室，机房，机器房
machine-hours 机器工时
machine house 机房，机器房
machine-independent language 与机器无关的语言，独立于机器的语言
machine instruction statement 机器指令语句
machine instruction 机器指令
machine interruption 机器中断
machine language code 计算机语言代码
machine language program 计算机语言程序
machine language 计算机机器语言，计算机语言，机器语言
machine learning 机器学习
machine life 机器寿命
machine-limited system 机器限制系统，极限限制系统
machine logic 机器逻辑
machine-made brick 机制砖
machine-made nail 机制钉
machine-mixed concrete 机械拌制混凝土，机拌混凝土
machine mixing 机器拌和
machine oil 机油
machine operator 机器操作员，机器算子
machine-oriented language 面向机器的语言
machine-oriented programming system 面向机器的编程系统
machine or transformer thermal relay 机器或变压器热继电器
machine output 机器出力
machine potential transformer 机端电压互感器
machine processible form 机器可处理形式
machine programming 计算机的程序设计
machine rammer 夯击机
machine rating 机器额定能力
machine readable 机器可读的
machine-recognizable 机器可识别的
machine repair shop 机修车间
machine rock drill 机动岩石钻机
machine room 汽轮机房，汽轮机间
machinery apparatus 机械
machinery noise 机器噪声

machinery processing industry　机械工业
machinery resources　机械资源
machinery shift　机械台班
machinery trailer　工具拖车
machinery　机器，机械
machines and tools　机具
machine-sensible information　机器可读信息
machine-sensible　机器可识别的，机器可读的
machine set　机组
machine shop equipment　机修间设备及工具
machine shop　机修车间，修理车间，金工车间，机械加工车间
machine-spoiled time　机器故障时间
machine spreading　机器撒布
machine starting　机器启动
machine status word　机器状态字
machine switching system　机器切换系统
machine time　计算机时间
machine tool control transformer　装有熔断器或其他保护装置的电源变压器，机床控制变压器
machine tool　机床，工作母机
machine translation　机器翻译
machine tunnel boring　隧洞全断面掘进机
machine tunnelling　机械开挖隧洞
machine unit　机器单元，计算机单位，机械装置
machine variable　机器变量，计算机变量
machine winding　电机绕组
machine with inherent self-excitation　固有自励电机，内在自励磁电机
machine with natural cooling　自然冷却式电机
machine word　计算机信息元，计算机字
machine wrench　机器扳手
machine　机器，机械，机构，机，机床，机械加工，切削加工
machining accuracy　加工准确度
machining allowance of casting　铸件机械加工余量
machining allowance　加工余量，机械公差，机械加工容限
machining blank　正在加工的毛坯，半制品，半成品
machining head　加工机头
machining pass　机械加工合格
machining precision　加工精（确）度
machining repair　机加工修理
machining reproducibility　加工重复性，加工再现性
machining stress　机加工残余应力
machining tolerance　机械公差，机械加工容限
machining unit for sleeve weld preparation　为套管焊接准备的机具
machining unit　切削加工单元
machining　机械加工，机加工
machinist's scraper　钳工刮刀
machinist　机工
MACH＝machine　机器
Mach number effect　马赫数效应
Mach number　马赫数
MAC(maximum allowable concentration)　最大容许浓度
MAC(mean aerodynamic chord)　平均气动力弦
MAC(medium access control)　介质访问控制，通信介质通路控制
macroacrylic acid　巨丙烯酸
macroanalysis　总量分析，宏观分析，常量分析
macroassembler　宏汇编程序
macroassembly program　宏汇编程序
macrocircuit　宏电路
macro-circulation　宏观循环
macro-clastic rock　粗屑岩
macroclimate　大气候
macroclimatology　大气候学
macrocode　宏代码
macrocoding　宏编码，宏代码
macro control mechanism　宏观管理机制
macrocorrosion　大量腐蚀，宏观腐蚀
macrocrack　宏观裂缝，宏观裂纹
macro-creep　宏观蠕变
macrocrystalline　粗晶，粗晶质，大结晶的，粗晶的
macrodefinition　宏指令定义
macro economic management　宏观经济管理
macro economics　宏观经济，宏观经济学
macro economy　宏观经济
macroeddy current　大涡流
macroeffect　宏观效应
macroetching　深蚀，宏观腐蚀
macro flow chart　宏流程图
macro flow　宏观流动
macrofluid　宏观流体
macroflux　宏观通量
macrofold　大褶曲，大褶皱
macrofractography　断口宏观检验，断口低倍检验
macro fracture mechanics　宏观断裂力学
macrogeneration　宏功能生成（程序）
macrogenerator　宏功能处理器，宏功能生成器
macrograin　粗晶粒，宏观晶粒
macrographic examination　宏观检验，宏观检查
macrography　宏观，宏观检查，肉眼检查
macrograph　低倍照片，宏观照片，宏观组织照片
macrogroup　宏观群
macroheterogeneity　宏观不均匀性
macroinstruction operand　宏指令操作数
macroinstruction　宏指令
macrolanguage　宏语言
macrolibrary　宏库，宏程序库
macro management planning　宏观管理计划
macro management　宏观管理
macromeritic　粗晶粒状
macrometeorology　大尺度气象学，大气象学
macromodular steam generator　大模块式蒸汽发生器
macromolecular compound　大分子化合物
macromolecular　大分子的
macro-molecule　大分子
macromteorology　大气象学
macrooscillograph　常用示波器，标准示波图
macropore permeability　粗孔孔隙渗透率
macropo-reticular resin　大孔树脂
macropore　大孔，大孔隙
macroporosity　大孔隙度，宏观孔隙度

macroporous soil 大孔土，大孔隙土，大孔性土
macroporous structure 大孔结构
macroporous 大孔，多孔，多孔的，大孔的
macroprocessor 宏处理程序，宏加工程序，宏功能处理器
macroprogramming 宏程序设计
macroprogram 宏程序
macrorelief 大起伏，大地形，宏观地形
macroreticular ion exchange resin 大网络离子交换树脂，大孔型离子交换树脂
macroreticular resin 大孔（型）树脂，巨网树脂
macro routine 大程序，宏观程序
macroscale 大尺度，大规模，宏观尺度
macroscopical analysis 宏观分析
macroscopical 宏观的
macroscopic cavitation 大空穴，宏观汽蚀
macroscopic constant 宏观常数
macroscopic convection 宏观对流
macroscopic cross section 宏观截面
macroscopic diffusion flux 宏观扩散通量
macroscopic examination 宏观检验，宏观检查，肉眼检查［检验］
macroscopic fission cross section 宏观裂变截面
macroscopic flux distribution 宏观通量分布
macroscopic flux variation 宏观通量变化
macroscopic fraction 宏观的粒级
macroscopic instability 宏观不稳定性
macroscopic magnetization 宏观磁化
macroscopic method 宏观法
macroscopic neutron flux density distribution 宏观中子流密度分布
macroscopic noise 宏观噪声
macroscopic pile theory 宏观反应堆理论
macroscopic reactor parameter 反应堆宏观参数
macroscopic removal cross section 宏观移出截面
macroscopic scale 宏观规模
macroscopic streak flaw test 断面缺陷宏观检查
macroscopic stress 宏观应力
macroscopic structure inspection 宏观检验
macroscopic system 宏观系统
macroscopic target 宏观目标
macroscopic total cross-section 宏观总截面
macroscopic 低倍放大的，宏观的，大范围的
macroscopy 宏观
macrosegregation 宏观偏析
macroshrinkage 宏观缩孔，宏观收缩
macroskeleton 宏程序纲要
macro strategy decision 宏观战略决策
macrostress 宏观应力
macrostructure 宏观结构，宏观组织，大型构造
macrosystem 宏系统
macroturbulence 大尺度湍流，大尺度紊动，宏观紊流
macro-uniformity 宏观均匀性
macroviscosity 宏观黏性
macrovoid ratio 大孔隙比
macrovoid 大孔隙，大空穴
macro 宏，宏功能，宏指令，宏观的，大的，大量的
Madaras rotor 马达拉斯转子
made circuit 闭合电路，终端连接
made ground 填土地
made out to order of shipper 以发货人为指示抬头的
made out to order of 以……为指示抬头的
made out to order 空白抬头的【提单】
made out to our order 以开证行为指示抬头的【our 即指开证行的】
made over 转让
made-up 人工的，预制的，制成的
madistor 晶体磁控管，磁控型半导体等离子器件
MAD(maximum allowable dose) 最大容许剂量
maelstrom 大旋流，大旋涡
MAE(mean absolute error) 平均绝对误差
maestro 冷燥大西北风
MAF(manifest amendment fee) 舱单改单费
magamp(magnetic amplifier) 磁放大器
magazine boiler 自动加煤小锅炉
magazine feed 仓库送料
magazine flooding and sprinkling 自动溢流喷水
magazine 期刊，杂志，箱，盒，卡片箱，仓库
mag-dynamo （充电用）直流发电机，磁电机
maggie 杂煤，不干净的煤，不纯煤
magic eye 电眼，猫眼，电子射线管，光调谐指示器
magic hand 机械手
magistrate 地方行政官
MAG＝magnetic 磁性
magmatic assimilation 岩浆同化作用
magmatic corrosion 岩浆溶蚀
magmatic eruption 岩浆喷溢
magmatic explosion 岩浆爆发
magmatic intrusion 岩浆侵入
magmatic origin 岩浆成因，岩浆起源
magmatic rock 岩浆岩
magmatic water 岩浆水
MAG(maximum available gain) 最大可用增益
magma 沉渣，软块，稠液，岩浆
magmeter 直读式频率计
magnacard 磁性凿孔卡装置
magnadure 铁钡永磁合金
magnaflux inspection 磁粉探伤，磁粉检验
magnaflux test 磁力线探伤，磁力探伤，磁力线试验，磁粉探伤
magnaflux 磁通量，磁粉探伤法，磁粉探伤机，磁粉探伤器
magnaglo 磁粉，磁粉探伤用粉末
magnalite 铝基铜镍镁合金，磁性粉末
magnechuck 电磁卡盘
magnecium-carbonate asbestos slab 碳酸镁石棉板
magneform 电磁成型
magner 无功功率，无效功率
magnesia brick 镁砖
magnesia cement 菱镁土水泥
magnesia mica 镁云母
magnesian chalk 镁质白垩
magnesian limestone 白云石，镁质石灰岩
magnesia refractory 氧化镁耐火材料
magnesia 氧化镁
magnesil 用作磁放大器芯子的磁性合金

magnesite brick 镁砖
magnesite cement 菱镁土水泥
magnesite refractory 菱镁矿耐火材料
magnesite 菱镁矿，菱苦土
magnesium alloy 镁合金
magnesium calcium carbonate 碳酸镁钙
magnesium carbonate 碳酸镁
magnesium cell 镁电池
magnesium dry cell 镁干电池
magnesium hardness 镁硬度
magnesium hydrate 氢氧化镁
magnesium hydroxide 氢氧化镁
magnesium hydroxyphosphate 羟基磷酸镁
magnesium limestone 碳酸钙镁
magnesium orthophosphate （正）磷酸镁
magnesium oxide radiation barrier 颗粒氧化镁阻挡层
magnesium sulfate 硫酸镁
magnesium sulfite 亚硫酸镁
magnesium 镁
magnestat 磁调节器
magnesyn 磁电式自动同步机
magnet anisotrophy 磁各向异性
magnet arc 磁弧
magnet armature 衔铁，磁衔铁
magnet assembly 磁系统
magnet balance 磁天平
magnet coil 励磁线圈，电磁线圈
magnet conductivity 磁导率
magnet contactor 磁铁开关
magnet core array 磁芯体
magnet core 磁芯，铁芯
magnet crack detection 磁力探伤
magnet current meter 磁力流速仪
magnet damping 磁性阻尼
magnet density 磁密度
magnet dipole moment 磁偶极矩
magnet dispersion 磁漏
magnet exciting coil 励磁线圈
magnet fatigue 磁疲劳
magnet field equalizer 磁力均衡器
magnet field probe 磁场探头
magnet flaw detection ink （磁粉探伤）磁悬液
magnet flux 磁通量
magnet frame 磁芯机座，导磁机座
magnet gap 磁隙
magnet hydro dynamic process 磁流体动力过程
magnet hysteresis 磁滞（现象）
magnetic-acoustic 磁声的
magnetic after effect 剩磁效应，磁后效应
magnetic ageing 磁性陈化，磁性老化
magnetic agitator 电磁搅拌器
magnetic air gap 磁气隙
magnetic alloy 磁性合金
magnetically aged 磁性老化的
magnetically biased transistor 带磁偏压的晶体管
magnetically blown spark gap 磁吹放电间隙
magnetically coupled circuit 磁耦合电路
magnetically focussed 磁聚焦的
magnetically hard material 硬磁材料
magnetically hard steel 硬磁钢
magnetically hard 硬磁的
magnetically polarized relay 磁性极化继电器
magnetically soft material 软磁材料
magnetically soft 软磁的
magnetically stable fluidized 磁稳流化床
magnetically 磁的
magnetic amplifier regulator 磁性放大器调节器
magnetic amplifier two-speed servosystem 带有磁放大器的双速伺服系统
magnetic amplifier 磁性放大器
magnetic amplitude 磁化曲线的幅值
magnetic analysis 磁分析法
magnetic anisotropy 磁各向异性
magnetic annular shock tube （风动研究用的）磁性环形激波管
magnetic anomaly 磁异常
magnetic arc stabilizer 电磁弧稳定器
magnetic assist 磁性助推装置，助磁
magnetic attraction force 磁引力，磁吸力，磁拉力
magnetic axis 磁场轴线，磁轴
magnetic balance protection 磁平衡保护
magnetic bath 磁浴【磁粉探伤】
magnetic bearing 磁悬浮轴承，磁方位，磁象限角
magnetic belt 磁带
magnetic bias 磁偏置，偏磁
magnetic blast arc quenching 磁吹灭弧
magnetic blast circuit breaker 磁吹式断路器
magnetic blow-out circuit-breaker 磁吹断路器，磁吹灭弧断路器，磁性熄弧断路器
magnetic blow-out coil 磁吹线圈
magnetic blow-out lightning arrester 磁吹避雷器
magnetic blow-out load switch 磁吹负荷开关
magnetic blow-out 磁吹，磁性灭弧
magnetic blow surge arrester 磁吹避雷器
magnetic bobbin core 线圈铁芯
magnetic body 磁体
magnetic brake 电磁制动器，磁力制动器
magnetic bremsstrahlung 磁韧致辐射
magnetic bridge 测磁电桥，测量导磁率的电桥
magnetic bubble storage 磁泡存储器
magnetic bubble 磁泡
magnetic card file 磁卡文件
magnetic card memory 磁卡片存储器
magnetic card 磁卡件
magnetic cell 磁单元，磁元件
magnetic character reader 磁性字符阅读器
magnetic character recognition 磁性符号识别
magnetic character sorter 磁性字符分类器
magnetic chart 磁性图
magnetic circuit 磁路
magnetic clearance 磁隙
magnetic clutch 电磁离合器
magnetic coil 电磁线圈
magnetic compass 罗盘仪
magnetic compensator 磁性补偿器
magnetic compression 磁压缩
magnetic conductance 磁导
magnetic conductivity 磁导率，导磁性
magnetic confinement fusion 磁约束聚变

magnetic confinement 磁约束
magnetic contactor 电磁开关，电磁接触器
magnetic contact relay 磁触点式继电器
magnetic controller 磁控制器
magnetic control meter 磁控表
magnetic control relay 磁控继电器
magnetic control rod drive mechanism 磁力提升式控制棒驱动机构
magnetic cooling 磁致冷却
magnetic core access switch 磁芯数据存取开关，磁芯数据存储开关
magnetic core coil 励磁线圈
magnetic core gate 磁芯门电路
magnetic core matrix 磁芯矩阵
magnetic core memory 磁芯存储器
magnetic core multiplexer 磁芯多路转换器
magnetic core plane 磁芯存储板，磁芯面
magnetic core register 磁芯寄存器
magnetic core storage unit 磁芯存储部件
magnetic core storage 磁芯存储器
magnetic core switch 磁铁芯开关
magnetic core 磁芯
magnetic coupled multivibrator 磁耦合多谐振荡器
magnetic coupling flowmeter 磁耦合式流量计
magnetic coupling 磁力耦合，磁性耦合，电磁耦合，磁性联轴器，电磁联结器
magnetic crack detection 磁力探伤
magnetic crane 磁力起重机
magnetic creeping 磁滞（现象），磁蠕变
magnetic cross valve 磁阀
magnetic currentmeter 磁力流速仪
magnetic current 磁流，磁通
magnetic cutter 磁力切割机
magnetic cycle 磁化循环
magnetic damper 磁阻尼器
magnetic decision element 磁判定元件
magnetic declination 磁偏角
magnetic defect detector 磁粉探伤仪
magnetic deflection 磁偏转
magnetic degree 磁角度
magnetic delay line 磁延迟线
magnetic demodulation 磁解谐，磁解调
magnetic detent 磁性制动器
magnetic deviation 磁偏移，磁偏差
magnetic dial indicator bracket 磁性千分表架
magnetic difference of potential 磁位差，磁性电势差
magnetic dipole moment 磁偶极矩
magnetic dipole 磁偶极子
magnetic dip 磁倾角
magnetic direction indicator 磁向指示器
magnetic disc 磁盘
magnetic disk memory 磁盘存储器
magnetic disk storage 磁盘存储器
magnetic disk unit 磁盘机
magnetic disk 磁盘
magnetic displacement recorder 磁位移记录器
magnetic displacement 磁位移
magnetic disturbance 磁扰
magnetic domain device 磁畴器件
magnetic domain memory 磁畴存储器
magnetic domain 磁畴，磁域
magnetic doublet 磁偶极子
magnetic drag tachometer 磁感应式转速计
magnetic driller 磁力钻
magnetic drum computer 磁鼓计算机
magnetic drum module 磁鼓存储模件
magnetic drum receiving equipment 磁鼓接收设备
magnetic drum recorder 磁鼓记录器
magnetic drum storage 磁鼓存储器
magnetic drum 磁鼓
magnetic electric drill 磁力电站
magnetic element 磁元件
magnetic elongation 磁致伸长
magnetic energy-density 磁能密度
magnetic equator 地磁赤道
magnetic eraser 消磁器，去磁器
magnetic field balance （探测用的）磁秤
magnetic field clutch 电磁离合器，磁性离合器
magnetic field density 磁场强度
magnetic field dependent resistor 磁敏电阻器
magnetic field distribution 磁场分布
magnetic field indicator 磁场指示器
magnetic field intensity 磁场强度
magnetic field interference 磁场干扰
magnetic field line 磁力线
magnetic field meter 磁场计
magnetic field method 磁场法
magnetic field oscillation 磁场振荡
magnetic field reversal 磁场反向
magnetic field strength 磁场强度
magnetic field vector 磁场矢量
magnetic field winding 磁场绕组
magnetic field 磁场
magnetic figure 磁力线图
magnetic film memory 磁膜存储器
magnetic film parametron 磁膜参变元件
magnetic film storage 磁膜存储器
magnetic film unit 磁膜单元
magnetic film 磁膜
magnetic filter 磁性过滤器
magnetic flaw detector 磁粉探伤仪，磁探伤器，磁力探伤机
magnetic float gauge 磁浮式水位计
magnetic float type level meter 磁浮液位计
magnetic flow measuring device 电磁流量测量仪表，磁流量测量仪表
magnetic flowmeter 磁性流量表，电磁流量计，磁流计，磁通计
magnetic flow tube 磁通管
magnetic flow 磁流
magnetic fluid seal 磁性液体密封
magnetic fluid 磁性流体
magnetic flux density 磁通密度，磁感应密度
magnetic flux inspection 磁力探伤
magnetic flux leakage 磁漏，漏磁，磁链
magnetic flux path 磁路
magnetic flux wave 磁通波
magnetic flux 磁通，磁通量
magnetic follower 磁跟踪器

magnetic force 磁力，磁场强度
magnetic freezing 磁黏
magnetic friction clutch 磁摩擦离合器
magnetic friction 磁摩擦
magnetic fusion energy 磁聚变能
magnetic gap 磁气隙
magnetic gate 磁门
magnetic gear 电磁摩擦联轴器，电磁离合器
magnetic hard steel 硬磁钢
magnetic head 磁头
magnetic history 磁化史
magnetic hologram 磁性综合衍射图，磁性全息照相
magnetic hum 交流哼声
magnetic hysteresis cycle 磁滞循环
magnetic hysteresis loop 磁滞回线
magnetic hysteresis loss 磁滞损耗
magnetic hysteresis 磁性滞后，磁滞，磁滞现象
magnetic hysteretic angle 磁滞角
magnetic ignition 磁电机点火
magnetic impulse counter 磁脉冲计数器
magnetic impulser 磁脉冲发送器
magnetic inclination 磁倾，磁倾角
magnetic induction flowmeter 磁感应式流量计
magnetic induction intensity 磁感应强度
magnetic induction strength 磁感应强度
magnetic induction 磁感应
magnetic inductive capacity 导磁率
magnetic inductivity 导磁率，磁感应率
magnetic inspection 磁性探伤，磁力探伤
magnetic insulation 磁绝缘
magnetic intensity 磁场强度
magnetic interaction 磁相互作用
magnetic iron oxide 磁性氧化铁
magnetic iron remover 除铁器
magnetic iron 磁铁
magnetic jack type drive mechanism 磁力提升式驱动机构
magnetic jack type mechanism 磁力起重式机构【控制棒驱动机构】
magnetic jack 磁力提升器
magnetic lag 磁化滞后，磁滞
magnetic latching relay 磁保持继电器，磁性闩锁继电器
magnetic leakage coefficient 漏磁系数，磁漏系数
magnetic leakage field 漏磁场
magnetic leakage flux 漏磁通
magnetic leakage 磁漏，漏磁
magnetic leg 铁芯柱
magnetic lens spectrometer 磁透镜分光计
magnetic lens 磁透镜
magnetic level sensor 磁水位传感器
magnetic limb 铁芯柱
magnetic line of force 磁力线
magnetic linkage 磁通匝连数，磁链
magnetic links 磁钢片【用于雷电流测量】
magnetic loading 磁负载
magnetic locator 磁性探测器
magnetic lock 电磁锁
magnetic logger 磁测井仪
magnetic loss angle 磁损失角
magnetic loss 磁损
magnetic material 磁性材料
magnetic medium 磁性材料，磁介质
magnetic memory material 磁存储材料
magnetic memory matrix 磁存储器矩阵
magnetic memory plate 磁性存储板
magnetic memory 磁存储器
magnetic mercury cutoff 汞磁力开关
magnetic meridian 磁子午线
magnetic metal 磁性金属
magnetic mirror system 磁镜系统
magnetic modulator 磁性调制器
magnetic moment of particle or nucleus 粒子或原子核的磁矩
magnetic moment 磁力矩，磁矩
magnetic motive force 磁动势
magnetic mouse 磁老鼠
magnetic multiaperture element 多孔磁元件
magnetic needle declination 磁针偏角
magnetic needle inclination 磁针倾角
magnetic needle 磁针，指南针
magnetic neutral state 磁中性点
magnetic noise 磁噪声
magnetic observatory 地磁观察台
magnetic Ohm's law 磁路欧姆定律
magnetic oscillograph 电磁示波器
magnetic overload relay 电磁式过载继电器
magnetic oxide 四氧化三铁，磁性氧化铁
magnetic oxygen cutting machine 磁力氧气切割机
magnetic oxygen meter 磁力测氧计，磁性氧量计
magnetic painting inspection 磁性涂料检查
magnetic particle examination equipment 磁粉探伤检查设备
magnetic particle examination 磁粉检验，磁粉探伤，磁性试验
magnetic particle fluorescent test 磁粉荧光检验
magnetic particle indication 磁痕，磁粉显示
magnetic particle inspection machine 磁粉探伤机
magnetic particle inspection 磁力探伤，磁粒检测
magnetic particle pattern 磁（粉）痕迹
magnetic particle test 磁粉探伤试验，磁粉探伤检验，磁粉检测，磁粉试验，磁粉探伤
magnetic particle 磁性粒子，磁粉
magnetic path 磁路
magnetic permeability curve 磁导率曲线
magnetic permeability measurement 磁性渗透测量
magnetic permeability 磁导率，磁导系数
magnetic permeameter 磁导计
magnetic permeance 磁导
magnetic permibility 磁导率
magnetic perturbation 磁性干扰
magnetic phase shifter 磁控移相器
magnetic pick-up 磁拾声器
magnetic plug 电磁铁，柱形线圈
magnetic point pole test 点磁极检测法
magnetic polarity 磁极性

magnetic polarization intensity 磁极化强度
magnetic polarization 磁极化，体积磁偶极距
magnetic pole gap 磁极间隙
magnetic pole strength 磁极强度
magnetic pole 磁极
magnetic potential difference 磁位差，磁势差，磁压
magnetic potential drop 磁位降，磁压降
magnetic potential 磁势，磁位
magnetic powder flaw detector 磁气粉探伤机
magnetic powder inspection 磁粉探伤
magnetic powder pattern 磁粉图案，磁粉图样
magnetic powder 磁粉
magnetic power-factor 磁功率因数
magnetic power 磁功率
magnetic premaster 磁导计
magnetic printing 磁转印，磁打印
magnetic print-out 磁打印输出
magnetic probe 磁探针，磁探头，探磁圈
magnetic property 磁性
magnetic prospecting 磁法勘探
magnetic pulley 磁力分离滚筒，磁铁分离器，磁性滚筒
magnetic pull 磁铁引力，磁引力
magnetic-pulse 磁脉冲
magnetic pump 电磁泵，磁性泵，磁力泵
magnetic quantum number 磁量子数
magnetic quenching 磁吹灭弧
magnetic reactance 磁抗，感抗
magnetic read-write head 磁读写头
magnetic recording medium 磁性记录材料
magnetic recording system 磁记录系统
magnetic recording technique 磁记录技术
magnetic recording 磁性记录
magnetic reed 磁簧
magnetic regulator 磁力调节器
magnetic relay 磁性继电器，磁继电器
magnetic reluctance 磁阻
magnetic reluctivity 磁阻率
magnetic remanence 剩磁，顽磁
magnetic repulsion 磁斥
magnetic resistance 磁阻
magnetic response 磁反应，磁响应
magnetic resultant 磁力合力
magnetic retardation 磁滞（现象）
magnetic retentivity 顽磁性，剩磁
magnetic reversal 反向磁化，磁性反转，逆磁化
magnetic Reynolds number 磁雷诺数
magnetic rigidity 磁刚性，磁刚度
magnetic roasting 磁化焙烧
magnetic saturation flux density 磁饱和通量密度
magnetic saturation method 磁饱和法
magnetic saturation starter 磁饱和启动器
magnetic saturation 磁饱和
magnetic screening action 磁屏蔽作用
magnetic screening effect 磁屏蔽效应
magnetic screen 磁屏蔽
magnetic seat drill 磁座钻
magnetic-sensing 磁敏的
magnetic sensitive element 磁敏元件
magnetic sensor 磁敏检出器
magnetic separation filter 磁性过滤器
magnetic separation 磁力分离，磁选
magnetic separator 磁力分离器，磁选机
magnetic sheet 硅钢片，电工钢片
magnetic shell 磁壳
magnetic shield 磁屏蔽罩，磁屏蔽
magnetic-shift register 磁位移寄存器，磁移位寄存器
magnetic shunt 磁分路，磁分路器
magnetic skin effect 磁趋肤效应
magnetic slot-seal 磁性槽楔
magnetic sound carrier 磁性音频载波
magnetic sound recording 磁录音
magnetic space constant 真空的磁导率
magnetic spectrometer 磁谱仪，磁分光仪
magnetic spot recorder 磁点记录器
magnetic starter 磁力启动器
magnetic starting switch 磁力启动开关
magnetic steel 磁钢，磁性钢
magnetic stepping motor 磁步进电动机
magnetic stirrer 磁力搅拌器
magnetic storage drum 磁鼓
magnetic storage medium 磁存储介质
magnetic storage register 磁存储寄存器
magnetic storage 磁存储器，磁贮存器
magnetic storm monitor 磁暴监视器，磁暴记录器
magnetic storm 磁暴
magnetic strainer 磁性滤网
magnetic stray field 杂散磁场
magnetic strength 磁场强度
magnetic stripe card 磁条卡
magnetic substance 磁性材料
magnetic surface storage 磁表面存储器
magnetic surface 磁鼓面，磁带面
magnetic survey 磁测，地磁测量，磁法调查，地磁查勘图，磁力调查
magnetic susceptibility type oxygen analyzer 磁化型氧量分析仪
magnetic susceptibility 磁化率，磁化系数
magnetic suspended bearing 磁悬浮轴承，磁浮轴承
magnetic switching system 磁力开关系统
magnetic switching transformer 磁力开关变压器
magnetic switch 磁开关，磁力开关，磁性开关
magnetic-synchro 磁同步
magnetic system 磁系统，磁路
magnetics 磁学，磁性元件
magnetic tachometer 磁流速计，磁转速计，磁力转速计
magnetic tape buffer 磁带缓冲器
magnetic tape controller 磁带控制器
magnetic tape control unit 磁带控制器
magnetic tape control 磁带控制
magnetic tape equipment 磁带装置，磁带机
magnetic tape file operation 磁带文件操作
magnetic tape file unit 磁带文件部件，磁带外存储器部件
magnetic tape function generator 磁带函数发生器

magnetic tape librarian 磁带库管理程序
magnetic tape loading program 磁带记录程序
magnetic tape loading 磁带装入，磁带记录
magnetic tape master file 磁带主文件
magnetic tape memory 磁带存储器
magnetic tape module 磁带模件
magnetic tape parity 磁带奇偶性
magnetic tape reader 读带机，磁带机，磁带读出器
magnetic tape recorder 磁带记录器
magnetic tape record start 磁带记录开始
magnetic tape record 磁带记录
magnetic tape station 磁带机
magnetic tape storage 磁带存储器
magnetic tape system 磁带系统
magnetic tape terminal 磁带终端
magnetic tape transfer 磁带传送
magnetic tape unit 磁带机，磁带装置
magnetic tape verifier 磁带校验机
magnetic tape 磁带
magnetic test coil 探测线圈，检测线圈
magnetic test 磁力探伤，磁探伤，磁试验
magnetic thickness gauge 磁测厚计
magnetic time relay 磁延时继电器
magnetic tools 磁性工具
magnetic torque 电磁转矩
magnetic track 磁道
magnetic transducer 磁换能器
magnetic transfer 磁转印，磁打印
magnetic transient recorder 瞬态磁记录器
magnetic transition temperature 磁性转变温度，居里点
magnetic trigger 磁性启动装置，磁力启动器
magnetic trip breaker 磁脱扣断路器
magnetic type voltage regulator 磁性电压调整器
magnetic valve 电磁阀，磁阀
magnetic variation 磁性变化，磁偏角
magnetic vector potential 磁矢位，磁矢势
magnetic vector 磁场矢量
magnetic vibration 磁振动
magnetic viscosity 磁黏性，磁黏滞性
magnetic voltage regulator 磁性电压调节器
magnetic water treatment 磁水处理
magnetic wave 磁波
magnetic wedge 磁性槽楔
magnetic wire storage 磁性钢丝存储器
magnetic wire 磁性钢丝，磁线，电磁线，磁导线
magnetic writing 磁写
magnetic yoke 磁轭
magnetic 磁的，磁性的，有磁性的，磁铁的，磁学的，有吸引力的，磁
magnet induction density 磁感应密度
magnet ink 磁悬液
magnet inspection 磁粉探伤
magnet intensity 磁场强度
magnetise 磁化，起磁，励磁
magnetising current 磁化电流
magnetising inrush current 磁化合闸电流，磁化冲击电流
magnetising inrush 磁化冲量
magnetism crack detector 磁性探伤仪
magnetism 磁力，磁性，磁学，磁
magnetite 磁性氧化铁，磁铁矿
magnetitite 磁铁岩
magnetizability 磁化强度，磁化能力，可磁化性
magnetizable medium 可磁化介质
magnetizable 可磁化的
magnetization characteristic curve 磁化特性曲线
magnetization characteristic 磁化特性
magnetization curve 磁化曲线
magnetization cycle 磁化循环
magnetization intensity 磁化强度
magnetization loss 磁化损耗
magnetization strength 磁化强度
magnetization 磁化强度，磁化作用，磁化
magnetized spot 磁化点
magnetizer 导磁体，磁化体，磁化器
magnetize 励磁，磁化，起磁
magnetizing ampere-turns 磁化安匝，励磁安匝
magnetizing coil 磁化线圈，励磁线圈
magnetizing current 激磁电流，磁化电流，励磁电流
magnetizing EMF 磁化电动势
magnetizing field 磁化磁场，励磁场
magnetizing force 磁化力
magnetizing inductance 励磁电感
magnetizing loss 励磁损失，磁化损失
magnetizing power 磁化功率
magnetizing reactance in direct axis 直轴磁化电抗
magnetizing reactance 磁化电抗，激磁电抗
magnetizing roasting 磁化焙烧
magnetizing susceptance 磁化电纳，激磁电纳
magnetizing winding 励磁绕组，磁化绕组
magnetizing 磁化的，起磁的，磁化，起磁
magnet lag 磁滞
magnet leakage field 漏磁场
magnet limb 磁极铁芯，电磁铁芯，铁芯柱
magnet line of force 磁力线
magnet loading 磁负载
magnet loop 磁回路
magnet metal 磁性金属
magnet meter 磁通计
magnet-motive potential 磁位，磁动势
magnetoaerodynamics 磁空气动力学，磁性空气动力学
magneto alternator 磁石发电机，永磁同步发电机，永磁交流发电机
magneto bell 磁石电铃
magnetocaloric effect 磁热效应，磁卡效应
magneto central office 磁石式电话总局
magnetochemistry 磁化学
magnetoconductivity 导磁率，磁导率，导磁性
magneto coupling 电磁联轴器
magneto-dynamo 永磁发电机，磁发电机
magnetoelectrical generator 永磁发电机，手摇磁石发电机
magnetoelectricity 磁电学
magnetoelectric machine 磁电机，永磁（式）电机
magnetoelectric relay 磁电式继电器

magnetoelectric tachometer 磁电式转速计
magnetoelectric 磁电的
magneto field scope 磁场示波器
magneto fluid dynamics 磁流体力学
magnetofluidization 磁场流态化
magnetogasdynamics 磁(性)气体动力学
magneto generator 手摇发电机，永磁发电机，永励机
magnetogram 磁力图，磁强记录图
magnetography 磁力探伤法，磁记录法，磁摄影术
magnetograph 地磁记录仪，磁强记录仪
magnetohydrodynamically driven vortex 磁流体动力驱动涡
magnetohydrodynamic energy conversion 磁流体能量转换
magnetohydrodynamic engine 磁流体动力发动机
magnetohydrodynamic generator 磁流体动力发电机，磁流体发电机
magnetohydrodynamic power generation 磁流体发电
magnetohydrodynamic power station 磁流体(发)电站
magnetohydrodynamics 磁流体发电学，磁流体动力学，磁流体力学
magnetohydrodynamic turbulence scale 磁流体动力紊流度
magnetohydrodynamic 磁流体动力(学)的
magneto ignition 永磁发电机点火
magneto inductor 磁电机，磁石感应器
magneto-mechanical effect 磁机械效应，磁力学效应
magnetometer 磁强计，地磁仪，磁力计，测磁仪
magnetometric 测磁的
magnetometry 磁力测定，测磁学，测磁法
magnetomotive force 磁通势，磁动力，磁动势
magnetomotive potential 磁动势，磁位
magnetomotive 磁力作用的，磁动力的，磁势
magneton 磁子
magneto-ohmmeter 兆欧表，摇表
magneto-optical device 磁光器件
magneto-optical laser 磁光激光器
magneto-optical memory material 磁光存储材料
magneto-optical memory technique 磁光存储技术
magneto-optic effect 磁光效应
magneto-optics 磁光学
magnetoplasmadynamic power generator 磁等离子体发电机
magnetoresistance material 磁致电阻材料，感磁场电阻材料
magneto resistance 磁致电阻，磁阻
magnetoresistivity 磁致电阻率，磁阻效应
magnetoresistor 磁敏电阻器，磁控电阻器，磁阻器
magnetoscope 验磁器
magnetoscopic dye penetrate test 磁性探伤检验
magneto spanner set 扳手套装
magnetosphere 磁圈
magnetostatic field 静磁场
magnetostatic 静磁学的
magnetostriction alloy 磁致伸缩合金
magnetostriction echo sounder 磁致伸缩回波检测器
magnetostriction effect 磁致伸缩效应
magnetostriction level meter 磁致伸缩液位计
magnetostriction oscillation 磁致伸缩振荡
magnetostriction oscillator 磁致伸缩振荡器
magnetostriction phenomenon 磁致伸缩现象
magnetostriction transducer 磁致伸缩换能器
magnetostrictive coupling 磁致伸缩耦合
magnetostrictive delay-line storage 磁致伸缩延迟线存储器
magnetostrictive delay line store 磁致伸缩延迟线存储器
magnetostrictive effect 磁致伸缩效应
magnetostrictive motor 磁致伸缩电机
magnetostrictive oscillator 磁致伸缩振荡器
magnetostrictive relay 磁致伸缩继电器
magnetostrictive storage 磁致伸缩存储器
magnetostrictive transducer 磁致伸缩换能器
magnetostrictive 磁致伸缩的
magneto switch 磁电机开关
magneto-telephone 磁石式电话机
magneto-telluric 大地电磁场的
magneto test set 磁力测试仪器
magneto-thermoelectric 磁热电的
magneto turbulence 磁性湍流
magneto 永磁发电机，永磁电机，磁发电机，磁石的，永磁的
magnet pole 磁极
magnetrol 磁放大器
magnetron oscillator 磁控管振荡器
magnetron 磁控管
magnet space 磁场
magnet steel 磁钢
magnet strip 电磁扁线
magnet switch 磁力开关
magnet valve surge arrester 磁吹阀式避雷器，磁吹避雷器
magnet valve 电磁阀
magnet wheel 磁轮，转子
magnet winding 电磁铁线圈，电磁铁绕组
magnet wire coating 电磁线漆膜
magnet wire 电磁线
magnet yoke 磁轭，(直流电机的)定子机座
magnet 磁体，磁铁，磁石
magnification factor 放大因数，放大因子，放大倍数，倍率系数
magnification ratio 伸缩比，放大比，放大率
magnification 增加，增强，放大，放大率
magnifier for reading 读数放大镜
magnifier 放大镜，放大器，放大尺
magnifying glass 放大镜
magnifying power 放大率，放大倍率
magnify 放大，增强
magni-scale 放大比例尺
magnistorized 应用磁变管的
magnistor 磁变管，磁开关
magnitude contours 等值线
magnitude-frequency characteristics 幅频特性
magnitude-intensity correlation 震级烈度相关

(关系)
magnitude margin 幅值裕度，幅值裕量
magnitude of current 电流强度，电流值
magnitude of the earthquake at 8.0 on the Richter scale 里氏8.0级的地震
magnitude of thermal conductivity 热导率的大小
magnitude of traffic flow 车流量，交通流量
magnitude order 数量级
magnitude-phase characteristics 幅相特性
magnitude ratio 增益率，量值比率，幅值比，震级比
magnitude 幅值，大小，尺寸，量，量级，程度，重要，积量，数量，数值，巨大，广大，(地震)级数，量值，振幅，震级
magnon 磁量子，磁性材料中自旋波能量子
Magno 镍锰合金
Magnus effect rotor 马格努斯效应转子
Magnus effect type wind turbine 马格努斯效应式风力机
Magnus effect 马格努斯效应
Magnus force 马格努斯力
magslip resolver 无触点自整角机解算器
mag tape 磁带
MAHBF(mean available hours between failures) 平均无故障可用小时
mahogany russet 红木褐色
mahogany 桃花心木，(硬)红木
maiden trial 初次试验
maid's room 保姆室
mail box 信箱
mail contract 邮务合同
mail day 邮件截止日
mail exploder 邮件分发器，函件分发器
mailing address 通信地址，邮寄地址
mailing list 邮件发送清单，发函清单，通信名单，邮件列表
mail liner 定期邮船
mail order 函购，邮购
mail remittance 信汇，邮汇
mail transfer 信汇
mail 通信
MAI(multiple address instruction) 多地址指令
main access to power house 主进厂通道
main access walkway 主要通道
main access way 入厂主干道
main accounts 主要账户
main air cleaning system 主空气净化系统
main air duct 主风道
main alternator 主发动机
main amplifier 主放大器
main arithmetic processor 主运算处理机
main assembly drawing 主装配图，主要总成图
main axis 主轴
main bang 探测脉冲，放射脉冲，领示脉冲，主脉冲信号
main bar 主钢筋
main base station 主要基本站
main basin 主港池
main battery 主电池
main beam 主梁
main bearing box 主轴承箱
main bearing structure 主承载构件，主要承重结构
main bearing wrench 主轴承扳手
main bearing 主轴承
main bed 主床【沸腾炉】
main body system 主体
main bottom ash control panel 除渣主控制屏
main bottom 基座
main bracing 主腹杆
main branch 支干线
main breakwater 主防波堤
main brush 主电刷
main building 主楼，主厂房，正屋，主要建筑物
main burner 主喷燃器，主火嘴
main busbar 主母线，主汇流排，工作母线
main bus 主母线
main cable 总输出电缆，干线电缆，主索
main chain 主锚链
main channel 主波道，主通路，主信道，主槽
main characteristics 主要特征
main circuit-breaker 主断路器
main circuit diagram of converter station 换流站主接线图
main circuit 主电路，主磁路，主回路，主管路，干线
main circulating water pump 主循环水泵
main circulation pump 主循环泵
main coal stream 主煤流
main communication center 主通信中心
main condensate pump 主凝结水泵
main condensate system 主凝结水系统
main condensate water system 主凝结水系统
main condenser air ejector 主冷凝器空气喷射器
main condenser 主凝汽器
main conditions 主要工况
main connection 电气主接线，主接线
main construction work 建筑物的主要组成部分
main construction 主体工程
main contactor 主接触器，主开关
main contact 主触头，主触点，主接点
main contents 主要内容
main contract 主合同
main control board 主控制盘
main control building 主控制楼，主控楼
main control gate 主控闸门
main controller 主调节器，主控制器
main control panel 主控制板，主控制盘
main control room emergency habitability system 主控制室应急收容系统
main control room human factor engineering 主控制室人因工程
main control room 主控制室，集中控制室
main control valve 主配压阀，主控制阀
main coolant pump 主冷却剂泵，主泵
main coolant 主冷却剂
main cooling flow 主冷却流
main cooling water intake pipework 主冷却水进水口管线
main course 主航道
main crane 主吊车
main current 主电路的电流，主电流，主流

main dam 主坝
main data 主要数据
main deck plate 主平台板
main deck 主甲板
main diagonal 主斜杆
main dike 干堤，主堤
main distributing frame 主配线板
main distributing point 主配电点
main distributing valve 主配汽阀
main distribution board 总配电盘
main distribution frame 总配线架
main ditch 主沟
main divide 主分水界
main drainage 主要排水
main drain 主排水沟
main drive shaft 主动轴，主驱动轴
main driving wheel 主动轮
main duct （通风）总管，主风道，通风干管，总风道
main dyke 干堤，主堤
main economic index 主要经济指标
main electrical connection diagram 电气主接线图
main electrical connections 电气主接线
main electrode 主电极
main elevation 主要立视面，主立视面
main entrance 正门，主要入口
main equipments of power plant 主机
main equipment 主设备
main exciter response ratio 主励磁机电压升速，主励磁机反应系数
main exciter 主励磁机
main exhaust port 主排气口
main factor 主要因素
main fault 主断层
main feature 主要特征
main feedback path 主反馈通道，主反馈通路
main feeder 总馈线，主馈线
main feedwater pump 主给水泵【由电动机或汽轮机驱动】
main feedwater supply 主给水【蒸汽发生器】
main feedwater system 主给水系统
main feedwater 主给水，正常给水
main feed 母线，干线，主馈线
main field circuit 主励磁回路，主磁场回路
main field winding 主磁场绕组，主极绕组
main field 主磁场
main fill line 主充水管
main flow air 主气流
main flow 主流程，主流量
main flue 主烟道
main flux 主磁通
main fly ash control panel 飞灰主控制盘，飞灰主控制屏
main frame computer 主计算机
main frame 主机架，主框架，主龙骨，总配线架，底盘
main fuel trip 主燃料跳闸
main fuel 主燃料
main fuse 主保险丝，主熔断器，电源熔断器
main gap 主间隙
main gate 主闸门
main gating pulse 主控制脉冲，主选通脉冲
main gear box 主变速箱
main generation system 主发电系统
main generator breaker 发电机主断路器
main generator room 主发电机室
main generator 主发电机
main girder lifting and mounting techniques 大板梁吊装技术
main girder lifting and mounting 大板梁吊装
main girder 大板梁，主梁，主桁架
main governor valve 主调节阀
main grounding network 主接地网
main hall of power station 主厂房【电站】
main header 主管道，主联箱
main head 总水头
main heater control switch 主加热器控制开关
main heating system 主供暖系统
main heat sink 主热井
main highway 公路干线，主要公路
main hydraulic ram 主液压活塞
main injection valve 主喷射阀
main inlet control valve 主进汽控制阀，主调节阀
main inlet throttle-stop valve 主进汽节流停汽阀，主汽门
main instruction buffer 主指令缓冲器
main insulation 主绝缘
main intercept valve 主截止阀
main internal memory 主存储器，内存储器
main isolating valve 主隔离阀
main jet 主喷嘴
main joint 主节理
main joist of suspended ceiling 吊顶主龙骨
mainland climate 大陆气候
mainland 大陆
main laying 干管敷设
main lead 主引线，母线，电源线，出线
main leakage 主漏磁
main leg 主材，主管段
main levee 主堤
main line locomotive 干线机车
main line of drainage 排水干线
main line road 主要道路
main line track 干线轨道
main line watermeter 干线水表
main line 主干线，总线，母管，干线，主线
main logic 主逻辑
main loop cooling 主回路冷却
main loop 主回路
main machine building arrangement drawing 主厂房布置图
main machine hall 主厂房，主机房
main machine system 主机系统
main machine 主机
main magnetic path 主磁路
main manifold 主集汽管
main manual operator 主要手动操作器
main material flow 主料流
main material store 主材料贮存库
main maximum 主峰
main mechanical plants 主设备，主机械设备

main members 主材
main memory priority 主存优先级
main memory system 主存储器系统
main memory 中央存储器，主存储器
main merit 主要优点
main motor 主电动机
main negotiator 主谈人
main network 主网
main nozzle for vapor 蒸汽总管
main office 主营业地
main oil lift pump 主油提升泵
main oil pump 主油泵
main oil reservoir 主油箱
main oil tank level alarm test 主油箱油位报警试验
main oil tank 主油箱
main parameter 主要参数
main part 主要部件
main peak 主峰
main pile 主桩
main pipe auto-welding machine 主管道自动焊机
main pipe 干管，母管，主管道，总管，总水管
main piping 母管，主要管路
main point 主点
main pole piece 主极靴，主磁极铁芯
main pole winding 主极绕组
main pole 主杆，主磁极，主电柱
main power block configuration 主厂房布置
main power block 主厂房
main power building arrangement 主厂房布置
main power building 电站主厂房
main powerline 电力干线
main power source 主电源
main power station 主发电厂，主电站，主力电厂
main power transformer 主要电力变压器
main power 主电源，主动力
main primary system 主系统【反应堆】
main processing system 主处理系统，主系统
main program 主程序
main project 主体工程
main propulsion gas turbine 主燃气轮机
main protection 主保护
main pull rod 主拉杆
main pulse 主脉冲
main pump 主泵
main raceway 主电缆管道
main rail line 主干线，干线，主线【铁路】
main reheat steam pipework 主再热蒸汽管道
main reheat stop valve 主再热截止阀
main reinforcement 主要钢筋，主钢筋，主筋
main relay valve 主继动阀，主继电器控制管
main ring 主滑环
main riser 主立管，总立管
main river 干流
main road 主干道
main rotor blade 主旋翼叶片
main rotor 主风轮
main routine 主程序
main runner 吊顶主龙骨
main run 主焊缝【支管焊缝】，主管段
mains antenna 电源天线，主干天线
mains breakdown 电网解联，电网崩溃
mains current 馈路电流
main sea water system 主海水系统
main service fuse 进户线熔断器，总火线保险器，总火线熔断器
main servomotor 主伺服电机
main sewer 排水干管，排水总管，污水干管，污水管干线
mains filter 电源滤波器
mains frequency synchronous motor 工频同步电机，工频同期电动机
mains frequency 电源频率，电网频率，市电频率，工业频率
main shaft bearing 主轴轴承
main shaft flange 主轴法兰
main shaft indicator 主轴位移指示器
main shaft seal 主轴密封
main shaft 主井，主竖井，主轴，传动轴，主动轴
main shelterbelt 主防护林带
main shield 主屏蔽，主体屏蔽
main ship channel 主航道
main shipping track 主航线
main shock 本震，主震
main shutdown chain 主停堆电路
main shutdown system 主停机系统
main site service item 主要现场服务项目
main slant 主斜井
main slope 主斜井
main soil group 主要土类
main solenoid valve 主电磁阀
mains-operated instrument 交流电动仪表
mains-operated set 用交流电源的机组
mains outage 电网断电事故，电源故障，电网转入解列，电网崩溃
main span 主跨
main spar 主梁
main specifications 主要规格
main speed 基本速度
main spillway 主溢洪道，正常溢洪道
mains plug 电源插头
mains power supply 系统供电，电力网供电，主电源
main spray valve 主喷淋阀
mains side 电源侧，馈电侧
mains supply 干线供给，交流电源
mains switch 电源开关，馈路开关
mains synchronization 与电源同步
main stair flight 主要楼梯段
main station building 主厂房
main station 主要观测站，主站，中心站
main steam and feed system 主蒸汽与给水系统
main steam blow-off station 主蒸汽排放站
main steam bypass system 主蒸汽旁路系统
main steam condenser 主凝汽器
main steam condition 进汽参数，主蒸汽参数
main steam control valve 主蒸汽调节阀，调速汽门
main steam flow 主蒸汽流量
main steam header 主蒸汽联箱

main steam isolation valve 主蒸汽隔离阀
main steam lead 主蒸汽管系，主蒸汽进汽管
main steam line rupture accident 主蒸汽管道破裂事故
main steam line 主蒸汽管道
main steam maximum pressure control loop 主蒸汽最高汽压控制回路
main steam maximum pressure limiter 主蒸汽最高汽压限制器
main steam outlet nozzle 主蒸汽出口接管
main steam outlet 主蒸汽出口
main steam penetration isolation valve 主蒸汽贯穿隔离阀
main steam pipe 主蒸汽管（道）
main steam piping system 主蒸汽管道系统
main steam piping 主蒸汽管道
main steam pressure regulator 主蒸汽压力调节器
main steam rate 主蒸汽流量
main steam stop valve leakage test 主汽门严密性试验
main steam stop valve 主汽门
main steam system 主蒸汽系统
main steam temperature control system 主蒸汽温度控制系统
main steam turbine gear unit 主汽轮机齿轮机组
main steam valve bunker 主蒸汽阀隔间
main steam valve 主汽阀
main steam 主蒸汽
main stop valve 主汽阀，主汽门，主截止阀
main storage unit 主存储器装置
main storage 主存储器
main store 主存储器
mains transformer 电源变压器，电力变压器
main stream wind 主流风
main stream 干流，主流
main structural member 主结构构件
main supply duct 总送风道
main-supply radio set 交流接收机，交流收音机
main-supply 主电源，供电干线，交流的，主供水管
mains voltage 电源电压，线电压，厂外供电电压
main switchboard room 主配电盘室
main switchboard 总开关盘，主配电盘，总控制盘
main switching contacts 主通断触头
main switching station 总变电站
main switch 总开关，主开关
mains 总输电线，总管道【main 的名词复数】，输电干线，（建筑物的）污水总管道，（电力）干线，电源，电力网
maintainability 可维护性，可保养性，易保养性，可维修性，可维护度
maintained-action 保持的动作
maintained command 保持命令
maintained load test 维持荷载法
maintainer 检修人员，维护人员，维修人员
maintaining independence and keeping the initiative in one's own hands 保持独立性和主动性
maintaining-level operation 恒水位运行
maintaining voltage 维持电压，保持电压
maintain 保持，维持，维护，保养
main tank 主油箱
maintenance access walkway 维修通道
maintenance alert network 维护警报网
maintenance analysis review technique 维修分析检查技术
maintenance analysis 维护分析
maintenance and increase of the state-owned assets 国有资产保值增值
maintenance and overhaul 维护与检修，保养和检修，维护检修
maintenance and repair shop 维修车间
maintenance and repair 维护修理，维修
maintenance and service platform 维修平台
maintenance approach 维修方法
maintenance area 维修区
maintenance bond 维修保函
maintenance box 检修箱
maintenance building 维修车间，维修间
maintenance bypass 维修旁通
maintenance certificate 维修证书
maintenance charge 维护费用，维修费
maintenance check 维护检查
maintenance contract 维护合同
maintenance control center 维护控制中心
maintenance control data register 维护控制数据寄存器
maintenance control program 维修控制程序
maintenance control retry register 维护控制再试寄存器
maintenance cost 维护费用，维修费用，维护费，维修费
maintenance crew 检修
maintenance cycle 维修周期
maintenance department 维修部门
maintenance depot （机械）修配厂，维修站
maintenance derated hours 维修降负荷小时数
maintenance derated 维护降低出力的
maintenance down time 修理停歇时间，维修停用时间
maintenance engineering analysis record 维护工程分析记录
maintenance engineering and management 维修工艺技术和管理
maintenance engineering 维护工程
maintenance engineer 维修工程师
maintenance equipment 维修（用的）设备
maintenance expense of boiler plant 锅炉设备维修费用
maintenance expense of electric plant 发电设备维修费用
maintenance expense of structure 构筑物维修费用
maintenance expense 维护费用，维修费
maintenance facilities 维护设备，维修设施
maintenance factor 维修系数
maintenance-free battery 免维护蓄电池
maintenance-free operation 不需维护的运行
maintenance-free 免维护的，不需维修的，无需维护的
maintenance frequency 维修次数

maintenance gate　检修闸门
maintenance improvement program　维修改进大纲
maintenance instructions　维护说明书，保养说明，维护规程，维修手册
maintenance interval　检修间隔
maintenance item　维修项目
maintenance lock　维修舱
maintenance manager　维修经理，设备维修经理
maintenance man-hours　维修人时，维修工时
maintenance manual　保养维修手册，维护说明书，维护手册
maintenance man　维修工，维修人员
maintenance margin　维持保证金
maintenance of concentration gradient　浓度梯度的保持
maintenance of equipment　设备保养
maintenance of negative pressure　维持负压
maintenance of the equipment　设备维护
maintenance of true bearing　真实方位保持
maintenance operating instructions　维修操作细则
maintenance outage factor　维护停运因子
maintenance outage hours　维修停用时数，维护停运小时
maintenance outage　检修停机，维修停运
maintenance overhaul　维护检修，经常性检修，年度大修【或两年一度的大修】
maintenance panel　检修盘
maintenance people　维修人员
maintenance period　维修周期，维修期
maintenance personnel　维护人员
maintenance philosophy　维修理念
maintenance platform　维修平台，维修台，检修平台
maintenance point　维修点，维修站
maintenance prevention　维修预防，预防性维修，维修安全措施，维修安全设施
maintenance procedure　维护程序
maintenance program　维护程序，维修程序，维修计划
maintenance proof test　检修保证试验
maintenance record　维修记录
maintenance regulations　维护规程，检修规程
maintenance,repair and operating　维护、修复及运行
maintenance & repair facility　机车、车辆的保养维修设备
maintenance room　维修间
maintenance routine　维护程序，常规小修
maintenance rules　检修规程
maintenance schedule　检修进度表，检修计划，维修计划
maintenance shop　维修车间
maintenance spider　维修星形爬行器
maintenance staff　检修人员，维护人员
maintenance state　检修状态
maintenance station　检修站，巡线站
maintenance superintendent　维修车间主任
maintenance supervision and engineering expenses　设备维修的监督和技术管理费用
maintenance support performance　维修保障性
maintenance system　检修系统，维护系统
maintenance test　维护试验
maintenance time　维修时间
maintenance tool　维修工具
maintenance transformer　检修变压器
maintenance way and stairway　维修通道和楼梯间
maintenance way　检修通道
maintenance worker　维修工
maintenance work　维护工作，维修工作，日常维护
maintenance　维护，日常维修，保养，维持，保持
main terminal　主端子，原边绕组端子
main throttle stop valve　主节流截止阀
main throttle valve　主汽门，主汽阀
main tie　主拉杆
main transformer bay　主变压器区
main transformer station　主变电所
main transformer　主变，主变压器
main transmitter　主发射机
main transmitting station　主发射台
main trap　总疏水器
main traverse line　主导线
main treatment　主处理
main turbine and generator lube oil system　主汽轮机与发电机润滑油系统
main turbine control and diagnostics system　主汽轮机控制与诊断系统
main turbine lube oil reservoir　主汽机润滑油箱
main turbine　主汽轮机，主涡轮机
main turning gear　主盘车装置
main unit　铁芯线圈组
main valve stem　主阀杆
main valve　主阀
main vent　通气立管，通气主管
main vessel void　主容器穴，反应堆室
main voltage　厂外供电电压，电网电压
main vortex　主旋涡
main water supply piping　主供水管
main water table　主潜水面
main waterway　主航道
main wheel　主轮
main winding　主绕组，主线圈
main wiring diagram　电气主接线图，主接线图
main workshop　总厂
main works　主体工程
main wring　电气主接线
main yard conveyor　煤场主皮带机
main　主要的，最重要的，主要部分，主要管道，电源【复数】
maitainability　可维护性，可维修性
major accident　重大事故，主要事故
major alteration　重大变更
major alternative　主要备选方案
major ancillary equipment　主要的辅助设备
major axis　主轴，长轴
major beam　主梁
major behaviorur　主要性状
major break　大破口
major calorie　大卡

major change 主要变更
major characteristics 主要特征
major component 主要部件
major critical component 主要关键部件
major crossing span 重要跨越档
major cycle 主循环，大循环
major dam 主坝
major defective 主要缺陷
major defect 主要缺点，主要缺陷
major development and popularization project 重点开发和推广项目
major discrepancy 主要差异
major divide 主分水界
major drainage basin 大流域
major equipment 重要设备
major executive 高级主管人员
major factor 主要因素
major flow 主流
major fold 主褶皱
major function 强函数，优函数
major grid 主格网
major harbour principal port 主要港口
major importance 至关重要
major individual investment project 重大单项投资项目
major industry 重点工业，大型工业
major inspection 大检查［检验］，主绝缘
majority decision logic 多数表决逻辑
majority gate 多数逻辑门
majority interest 多数股权
majority logic decodable code 多数逻辑可解码，大数逻辑可译码
majority logic decoding 择多逻辑解码
majority logic 多数逻辑，大数逻辑
majority opinion 多数意见
majority 大多数
majorization 优化
majorizing sequence 优化顺序
major key 主关键字
major leak 主泄漏
major maintenance 大型保养，大修
major network 主网
major oscillation 主振荡
major overhaul 大修
major parameter 重要参数
major part 重要零部件
major pieces of equipment 设备主件
major port 主要港口
major principal plane 主平面
major principal strain 大主应变
major principal stress 第一主应力，最大主应力
major problem 主要问题
major project 主跨，重点工程，重点项目
major repair of heat-supply network 热网大修
major repair 大修
major river bed 大河床
major route 主要途径
major specification 主要技术规范，主要规格
major station 主测站，主站
major steam line 主蒸汽管路
major system 重要事件
major trend 重要趋势
major wave 主波
major 长的（轴），主要的，专业
make a bid 报价，递价，投标
make a cable offer 以电报报盘
make a claim on sth. 就某事提出索赔
make a claim with（/against）sb. 向某方提出索赔
make a claim 索赔
make a commitment 承诺
make a contract 订立合同，签订合同
make-action contact 闭合接点
make a down payment 预付现款
make a draft of money 提取款项
make a draft on a bank 从银行提款
make a good price 卖得好价钱
make agreement 签订协定［协议］
make allowance 留余量
make a loss 亏损
make a match 撮合
make a monopoly of 独占，垄断
make an appearance 显露，出现
make an award 判定，裁决
make-and-break capability 闭开［通断］能力
make-and-break contact 闭开触点，断续触点，通断触点
make-and-break cycle 通断周期
make-and-break device 断续器
make-and-break 开合，通断
make an investigation 调查研究
make an investment 投资
make an offer 报价，出价，发价，发盘，还价
make a permanent requisition on the urban and rural land 永久征用城乡土地
make a profit 获得利润
make a quotation 开价，报价
make a request for 请求
make a separate arrangement 另行安排
make a temporary requisition on the urban and rural land 临时征用城乡土地
make available 获取
make award on each group basis 分组决标
make award on item by item basis 分项决标
make award on whole lot basis 总体决标
make-before-break contact 先合后断触点，先合后开触点
make-before-break switch 先合后断开关
make before-break 先闭后开，先接后离，先通后断
make-break capacity 通断能力
make-break contact 合断触点
make-break-make contact 合断合触点
make-break system 断续系统
make-break time 通断时间
make-break transfer switch 通断转换开关
make-break 通断，合断
make-bush key 闭塞电键
make-busy 闭塞，占线
make check payable to 请写支票抬头
make compensation 赔偿，补偿
make concessions 让步

make contact 闭路触点，闭合触点，闭合接点，接通触点，接触点，接通
make correction for 做某方面的修正，做……方面的校正
make dead 切断
make delay 闭合延迟，慢动作
make delivery of the goods 交货
make flush 使埋入，整平
make for 有利于，导致，有助于，走向
make great efforts 努力
make hand-tight 手压致密
make headway 有进展
make impulse 接通电流脉冲，闭路脉冲
make influence on 对……有影响
make inroad into market 打入市场
make it a condition that 以……为条件
make-make-break contact 合合断触点
make-make contact 合合触点
make money 赚钱，挣钱
make no sense 没有意义，讲不通
make out a cheque 开支票
make out bills 算账
make payment beforehand 提前付款，预付款
make payment 支付
make-position 闭合位置，接通位置
make profit 获利
make quotation 给予报价
make reference to 提到，涉及，提及，参考，查阅，加附注
make remittance 汇款，开汇票
make reservations （合同）附保留条件
maker 制造者，制造厂，制造商，接合器，接通器，出票人
make sense 有意义，有道理，讲得通
make snug 使贴紧
make some concession 做某些让步
make spark 闭合火花
make sth. available to sb. 向某人提供……
make terms with 与……达成协议
make the plant area green 厂区绿化
make-up air 补充空气
make up a loss 弥补损失
make up an account 清算，结账
make-up and discharge system 补排水系统
make-up crane 配料起重机
make-up feed pump 补水泵
make-up feed water 补充水，补给水
make-up feed 补给水，补充水补给，补足进料
make-up flow 补充量，补给流量
make-up for coolant gas losses 气体冷却剂损耗补给
make up for 弥补……【不足】
make-up fresh feed water 新鲜补给水
make-up gas 补充气体
make-up heat 补充加热
make-up holding tank 补给贮存箱，补充贮存罐
make up 40% of the total 占总量 40%
make-up oil 补给油
make-up paint 补漆
make-up pipe 补给水管
make-up pump 补给泵，补给水泵，供水泵
make-up shielding 补给屏蔽，临时屏蔽
make-up system 补给系统
make-up tank 补给水箱
make up to 补偿，偿还
make-up valve 补水阀门
make-up water controller 补水控制器
make-up water control 补水控制
make-up water percentage 补水率
make-up water preheater 补水预热器
make-up water pump of heat-supply network 热网补给水泵
make-up water pump 补充水案，补给水泵
make-up water storage tank 补给水贮存箱
make-up water supply control system 补水控制系统
make-up water system 补给水系统
make-up water tank 补水箱，补给水箱，补充水箱
make-up water transfer pump 补给水输送泵
make-up water 补给水，补充水
make-up 补充【原来不足或有损失的】，补给，构成，修理，装配，组成，接通
make use of a credit 使用信用
make use of 使用，利用
make wrench-tight 用扳手拧紧
make 做，接通，闭合，构成，制定，生产，通电
making alive 加电压
making a price 开价，定价
making and breaking capacity 通断能力，关合容量
making capacity 闭合容量，接通能力
making concession 让步
making current release 接通电流脱扣器
making current 接通电流，闭合电流，关合电流
making time 闭合时间，投入时间，接通时间
making 加工，性质，闭合，接通，制造，要素
malachite green 碱性孔雀绿
malachite 孔雀石
maladjustment 失调，失配，不适应，不一致
malalignment 轴线不对中，失中，中心偏移，不重合，失调，不成直线
Malan loess 马兰黄土
malcomising 不锈钢表面热氮化处理
maldistributed airflow 不均匀空气流
maldistribution 分配失调，分配紊乱，分布破坏，不均匀分布
male and female face 凹凸面
male connector 插头，接头
male contact 插塞接点，刀头触片，插头，插塞
male face 凸面
male-female angle 阳阴角套
male-female facing flange 凹凸面法兰
male-female facing 阳阴对面套
male-female globe 阳阴球面
male-female 阳阴
male fitting 凸扣管件，外螺纹连接件
male flange 凸口法兰，凸法兰，凸面法兰，凸缘法兰
maleic acid 马来酸

maleic anhydride 马来酸酐
male plug 插头，插塞
male rotor 显极转子，凸形转子，凸极转子
male thread 阳螺纹，外螺纹
male wrench 套筒扳手
male 突口，突缘，插入式配件，榫面
malfeasance 渎职（罪），非法行为
malfunction analysis 误动作分析
malfunction detection system 故障检测系统
malfunction index 故障索引
malfunction indication 误动作指示器
malfunction probability 误动作概率
malfunction rate 故障率
malfunction routine 定错程序，查找故障程序
malfunction test 故障实验
malfunction time 故障时间
malfunction 误操作，误动作，失灵，故障，不正确动作，不正常工作，异常运行
malleability 可锻性，韧性，展性，恶性，延展性
malleable casting 可锻铸件
malleable cast iron 韧性铸铁，可锻铸铁，马铁
malleable iron 可锻铸铁
malleable 可锻的
malleablizing 可锻化退火
mallet 锤
mall 购物中心，林荫路，草地广场，林荫道
malm brick 白垩砖
malodorness 恶臭
malodorous gas 恶臭气体
malodorous substance 恶臭物质
malodorous 恶臭的
maloperation 不正确运行，误操作，维护不当，误动作
malposition 错位
malthene 软沥青质，石油脂
malthoid 油毛毡，油毡
MA(maintenance analysis) 维护分析
M/A(manual/automatic) 手动/自动
MA(milliampere) 毫安
mA＝milliamps 毫安
mammock 碎块碎片
mammoth blasting 大爆破
mammoth pump 巨型泵
mamometric thermometer 压力式温度计
mamometric 压力式的
MA(municipal authority) 市政当局
managed currency system 通货管理体制
management agreement 经营管理协议
management analysis reporting system 管理分析报告系统
management by objectives 目标管理
management command system 管理指令系统，管理命令系统
management computer branch 管理计算机处
management consultant 管理顾问
management contract 管理合同
management control system 管理监控系统
management cost 管理费
management cycle 管理循环
management data processing system 管理数据处理系统
management decision 管理决策
management efficiency 经营效率
management engineering 管理工程
management engineer 管理工程师
management experience 管理经验
management fee 管理费
management function 管理职能
management information and control system 管理信息和控制系统
management information network system 管理信息网络系统
management information system 厂级管理信息系统，管理信息系统
management level 管理级，经理部层
management of construction area 施工现场管理
management of customer's load 用电负荷管理
management of design change 设计变更管理
management of documents 文件的管理
management of parallel 并网管理
management of technical documents 技术文件管理
management personnel 管理人员
management planning and control system 管理设计和控制系统，管理计划及控制系统
management plan 管理计划
management procedure 管理程序
management program 管理方案
management quality 管理质量
management responsibility 管理职责
management review 管理评审
management science 管理科学
management standard 管理标准
management support utility 管理后援应用程序
management system 管理系统，经营管理系统
management technique 管理技术
management 经营管理职权，管理，经营
managerial ability 管理能力
managerial criteria 管理准则
managerial method 管理办法
managerial skill 管理技能［技巧］
manager of construction service 施工服务经理
manager of expediting 催货经理
manager of the work 工作负责人
manager on duty 值班经理
manager responsibility system 经理负责制
manager's office 经理室
manager 经理
manage 经营
managing director 总经理，总裁，常务董事，执行董事【MD】
managing partner 经营合伙人
man-auto 手动自动，手控自控
man-carried 便携的
man-computer interaction 人机联系
mandate 命令，托管，授权书，委托书，指定委托
mandatory clause 强制条款
mandatory erection tool 必备安装工具
mandatory language 强制性语言
mandatory law 强制性法律

mandatory measures 指令性措施，强制性措施
mandatory period 法定期限
mandatory planning economy 指令性计划经济
mandatory plan 指令性计划
mandatory provision 强制性规定，约束性规定
mandatory requirement 强制性要求
mandatory reserves 法定储备基金
mandatory sign 指示标志
mandatory spare parts 强制性备件，随机备件
mandatory 必须遵守的，强制（性）的，命令的，指示的，委托的，委任的，托管的，义务的，必备的，受委托的，指令性的，代理人，受托人
mandator 命令者，命令着
man-day in absence 缺勤工日数
man-day 劳动日，人日，人工日
mandrel collar 心轴垫圈
mandrel rod 芯棒
mandrel wrench 心轴扳手
mandrel 卡盘【弯曲试验】，顶杆，芯轴，芯棒，紧轴，半导体阴极金属心
maneuverability 操纵性，机动性，灵活性，灵敏性
maneuvering apparatus 控制设备，操纵设备
maneuvering band 操作频带
maneuvering device 控制设备
maneuvering performance 操纵性能
maneuvering space （船舶所需）机动空间
maneuvering test 操纵试验
maneuvering valve 控制阀，操纵阀
maneuver 操纵，运用，调度，机动，演习
MANF＝manifold 母管
manganate 锰酸盐
manganese bacteria 锰细菌
manganese bronze 锰青铜
manganese dichloride 二氯化锰
manganese sand 锰矿砂，锰砂
manganese sesquioxide 三氧化二锰
manganese steel liner 锰钢衬板
manganese steel 锰钢
manganese trioxide 三氧化锰
manganese type electrode 锰型焊条
manganese 锰
manganic oxide 三氧化二锰
manganin wire 锰铜线
manganin 锰铜线，锰铜，锰镍铜合金，锰铜电阻合金
mangano-manganic oxide 四氧化三锰
manganous oxide 一氧化锰
manganous sulfate 硫酸锰
manger of traffic and freight forwarding 交通运输经理
mangle 碾压机
manhead 开人孔的封头
manhole cover 检查井盖，人孔盖，进人孔盖
manhole cross bar 人孔横杆
manhole door 检查井盖
manhole head 检查井盖，人孔盖架
manhole lid 人孔盖
manhole 工作井，人孔，检修孔，检查井
man-hour in attendance 出勤工时
man-hour quota 工时定额
man-hour 工时，人时，人工时，工时数
manifest 报关单，舱单
manifold air oxidation 歧管空气氧化
manifold assembly 集合总管
manifold condenser 多管凝汽器
manifold crane 多用起重机
manifold flow 分叉管水流，簇流
manifolding 复印，复写
manifold penstock 多叉压力水管
manifold pipe 分叉管
manifold system 多头管系
manifolds 集合管，联箱
manifold tunnel 多叉隧洞
manifold valve assembly 歧管阀总成
manifold valve 汇集阀，汇流阀
manifold 母管，集箱，联箱，集管，总管，汇流管，集汽管，歧管，复式接头，多支管联箱，多岔管，多叉管
manila rope 粗麻绳
man-induced erosion 人为冲蚀
man-induced event 人为事件
man-induced 人为的
manipulated soil 重塑土
manipulated variable 操纵量，控制变量，操纵变量
manipulated range 操纵范围
manipulated variable 操纵（变）量
manipulate 操作，操纵，使用，打键，键控
manipulating crane 操作起重机
manipulating grab 操作抓具
manipulating hoist 机械手升降机
manipulating variable 调节量
manipulating 搬运，装卸，调节的
manipulation 操作，操纵，打键，键控，控制，处理
manipulative valve 操作阀
manipulator bridge 机械手桥架
manipulator cell 有机械手的热室
manipulator finger 机械手抓手
manipulator hole 机械手孔道
manipulator joint 机械手关节
manipulator-operated 用机械手操作的
manipulator sleeve 机械手套筒
manipulator traversing gear 机械手水平传动
manipulator trolley 机械手运输车
manipulator turning gear 机械手转动机构
manipulator 操纵者，控制器，操纵装置，键控，机械手，操纵器，焊接操作机
manjak 纯沥青
manlid 检查井盖
man-machine communication 人机通信，人机联系
man-machine control system 人机控制系统
man-machine control 人机控制
man-machine interaction 人机对话，人机联系
man-machine interface device 人机接口装置
man-machine interface 人机接口，人机界面，人机对话，人机交接，人机交换界面［接口］
man-machine interrogation technique 人机询问技术

man-machine simulation 人机模拟
man-machine system 人机系统，人机对话系统，人机通信系统
man-machine 人机的，人与机器的，人机
man-made atmospheric contaminant 人为大气污染物
man-made earth satellite 人造地球卫星
man-made environmental contaminant 人为环境污染物
man-made environment hazard 人为环境公害
man-made environment 人为环境
man-made erosion 人为侵蚀
man-made fibre 人造纤维
man-made hazard 人为危害
man-made land 人造地
man-made noise 人为噪声，工业噪声
man-made pollutant source 人为污染源
man-made pollutant 人为污染物
man-made pollution 人为污染
man-made source 人工放射源
man-made 人为的，人工的，人造的
MAN＝manual 手动
man-months 人月数
manned cable way 载人缆道
manned substation 有人值班变电站
manned 人工操作的，有人操作的，有人驾驶的，载人的
manner of execution 实施方法
manner 方法［式］
Manning roughness coefficient 曼宁糙率系数
Manning roughness factor 曼宁糙率系数
manning 人员配备
manoeuvreability 机动性，灵活性
manoeuvre margin 机动限度
manoeuvre 操纵，运用，调度，机动，演习
manoeuvring test 操纵试验
manoeuvring valve 调节阀
manograph 压力记录器
manometer liquid level indicator 气压计式液位指示器
manometer pressure 表压力
manometer tube 测压管，压力管，压力计连接管
manometer 压力计，压力表，压强计，气压表，汽压表，测压计，风压表，流体压力计，流体压强计，深度压力计
manometric measurement 压力测定
manometric method 测压法
manometry 测压法
manoscope 气体密度测定仪
manostat 稳压器，恒压器，压力继电器
manpower column chart 劳动力柱状图
manpower development 人力开发
manpower resources 人力资源
manpower schedule 人力计划表
manpower training 人才培训
manpower 人力，劳动力
man-rated 适合于人使用的
man-rem cost 人体雷姆费用
man-rem 人体雷姆
man ring 手摇发电机振铃
mansard roof truss 折线形桁架
mansard roof 折线形屋顶，复折屋顶
mansion line 膨胀线
man-time 人次
mantissa 假数，（对数的）尾数，数值部分
mantle friction 表面摩擦
mantle rock 表层岩
mantle 外壳，套，罩，外罩，地幔，风化覆盖层，机壳
manual adjustment 手调，手工调节，手动调整
manual auger 手摇麻花钻
manual-automatic station 手动自动操作站
manual-automatic switch 手动自动开关
manual-automatic 人工自动的，手动自动的，手动自动
manual auto run-up selector 手动自动启动选择器
manual back-up control system 手动备用控制系统
manual barring facility 手动盘车装置
manual batcher 人工配料拌和机
manual blowdown 手动排污
manual booster 手动升压机
manual boring 人力钻探
manual brick 手工砖
manual card 手动插件
manual cleaning device 人工清扫设备
manual cleaning 手工清洗
manual clock 手控时钟
manual compensation 手动补偿
manual computation 笔算，手算
manual control chlorinator 手动控制加氯机
manual control device 人工控制设备
manual control interlock logic unit 手动控制联锁逻辑单元
manual controller 手动控制器
manual control screw 手调螺旋
manual control system 手控系统
manual control unit station 手动操作站
manual control 手动控制，人工控制，手动操作，人工操作，人力控制，手工调整
manual crane 手动起重机
manual cutting 手工切割
manual damper 手动调节风门
manual data entry module 手动数据进入组件
manual data input equipment 人工干预输入装置
manual decode system 人工译码系统
manual discharge station 手动排放操作站
manual drive system 手动系统
manual drive 手动驱动，手操作，手传动，手动
manual examination 手动检查
manual exchanger 人工电话交换机
manual film developing machine 手工洗片机
manual fire alarm call point(/button) 手动火灾报警按钮
manual frequency control 手动频率调节
manual friction brake 手动摩擦制动器
manual gear shifting 手动变速装置
manual hoist 手动起重机，手摇卷扬机
manual hold 人工保持
manual hydraulic handing trolly 手动液压搬运车
manual input program 人工输入程序
manual input 人工输入，手工输入

manual intervention　人工干涉，人工干预
manual loader　手动操作器
manually controlled winch　手动卷扬机
manually locked door　人工闭锁式车门
manually-operated breaker　手动断路器
manually-operated plotting board　人工操作的图形显示面板
manually-operated switch　手动开关
manually-operated turning gear　手动盘车装置
manually-operated valve　手动阀
manually-operated　手工操作的，人工操作的，手动操作的，手操作的，手动的
manual manipulation　手动操作，人工控制
manual megohmmeter　手摇式绝缘电阻表
manual method　手工方法
manual monorail hoist　手动单轨起重机
manual number generator　人工数字发生器
manual of engineering procedures　工程设计程序手册
manual operated control　手动控制
manual operated pump　手压泵
manual operation　手工操作，手动操作，人工操作，手操作，手动
manual operator　手操作器
manual output card　手动输出卡，手动输出件
manual output station　手动输出操作站
manual output　人工输出
manual override　手动控制，人工控制
manual panel　手操作仪表板，操作盘，控制盘
manual plugging meter　手动堵塞计
manual press　手压机
manual program　人工程序
manual pump　手摇泵
manual quasi-synchronization　手动准同步
manual regulation　手调，人工调节，手动调节
manual release　人工释放，手动排放
manual remote handling tool　远距离手工操作工具
manual reset adjustment　手动复位调节，可调区手控
manual reset relay　人工复归继电器
manual reset　手动复位，人工复位
manual ringer　手摇发电机振铃
manual sample collection　人工采样
manual sampling　人工采样，人工取样，手动扫描
manual scram　手动紧急停堆
manual selector switch　手控选择开关
manual setting　手调，人工置定，手动设定
manual shaft turning device　手动盘车装置
manual shutdown　手动停堆，手动停机
manual speed adjustment　手动调速，人工速度调节
manual speed changer　手动同步器
manual starting　手动启动，人工启动
manual station　手动操作站
manual steering　手操纵，手动转向
manual switch group　人力操作组合开关
manual switching　人工切换，手动转换，人工交换
manual switch　手动开关
manual synchronization　人工同步，手动同步
manual system　人工系统
manual tachoscope　手持转速计，手控转速计
manual take-over drive　手动传动装置
manual TIG tack welding　手工钨极惰性气体定位焊
manual titration　人工滴定
manual transfer-switch　手动转换开关
manual transmission　人工传送
manual trigger　人工触发器
manual trip level　手动脱扣杆
manual tripping device　手动跳闸装置
manual trip　手动事故停堆［机］，手动跳闸，手动脱扣
manual tungsten electrode argon arc welding machine　手工钨极氩弧焊机
manual valve positioner　手控阀位置控制器，手动阀门定位器
manual valve　手动阀门，手动阀
manual voltage regulator　手动电压调整器
manual welding　手工焊接，手工焊
manual weld　手工焊
manual word generator　人工字符发生器，人工输入设备
manual word　人工输入字
manual　手控的，手动的，人工的，手工的，手册，指南，人工，手控，手动，说明书
manufactory system　生产体系
manufactory　制造厂，加工厂
manufacture cost　制造成本
manufactured gas　人造煤气
manufacture district　工业区
manufactured product　制成品
manufactured sand　人造砂
manufacture location　制造地点
manufacturer margin　制造厂余量
manufacturer's agent　制造商代理人
manufacturer's brands　制造商厂牌
manufacturer's inspection certificate　制造厂商检验证明书
manufacturer's invoice　制造厂商发票
manufacturer's rating　铭牌额定出力，铭牌出力
manufacturer's representative　厂商代表
manufacturer's certificate　厂商证明
manufacturer's starting curve　制造厂提供的启动曲线
manufacturer　制造厂，制造厂家，制造商，建造者，制造者，生产者
manufacture's brand　制造厂商标
manufacture　制造，制品，制造工厂，生产，制作
manufacturing allowance　制造公差
manufacturing automatic process　自动生产过程
manufacturing capacity　制造能力，生产能力
manufacturing consignment　委托加工
manufacturing cost　制造成本
manufacturing cycle　制造周期
manufacturing deficiency　制造缺陷
manufacturing district　工厂区
manufacturing expense budget　制造费预计
manufacturing expense standing order　制造费用标准通知单

manufacturing expense 制造费用
manufacturing experience 制造业绩
manufacturing failure 制造失效
manufacturing industry 制造工业
manufacturing message specification 制造报文规范
manufacturing operations and surveillance checklist 加工监督查核表
manufacturing period 制造周期
manufacturing plant 制造厂
manufacturing process 制造过程，制造工艺
manufacturing schedule 制造进度
manufacturing sequence 制造顺序，生产顺序
manufacturing shop 生产车间
manufacturing's working curve 制造厂提供的工作曲线
manufacturing technology 制造工艺
manufacturing tolerance 制造公差
manufacturing waste 工业废物
manufacturing 制造
manure 肥料
manuscript base 稿图
manuscript signature 亲笔签字
manuscript 手稿，手抄本，加工图，数字控制指令表，原稿
MANU 手动（方式）【DCS 画面】
manway cover plate 人孔盖板
manway 人孔，人通道
many dimensions 多维
many-group calculation 多群计算
many-stage 多级的，多段的
many-turn 多匝
many-valued logic 多值逻辑
many variable system 多变量系统
map accuracy 地图精度
map appearance 地图整饰
map archive 地图档案
map board 图板
map border 图廓
map code 地图符号
map coordinates system 地图坐标系
map coordinates 地图坐标
map crack 龟裂
map dimension 图幅尺寸，图幅大小
map face 图面
map grid 地图方格网
map legend 地图符号
MAPL(multiple address processing unit) 多地址处理部件
map-making satellite 测图卫星
map-making 绘制图表，制图，制图录
map manoeuvre 图上作业
map margin 图廓
map measurer 量图仪
map measure 图上量算
MAP(message acceptance pulse) 信息接收脉冲
MAP(model and program) 模型和程序
map of groundwater table 地下水分布图
map of navigation system 航运系统图
map of seismic intensity zoning 地震烈度区划图
map of water table 水位地图
mapped real-time disk operating system 实时映射磁盘操作系统
mapping array 映像阵列
mapping chart 测绘图
mapping control 测图控制
mapping fault 变换误差
mapping from photograph 照片测图
mapping of cascade 叶栅映像，叶栅保角交换
mapping unit 图示土类【土壤调查】，(土壤) 图标单位
mapping 绘图，映像，映射，变换，绘制……的地图，计划，测绘，绘制图表，制图，制图录
map projection 地图投影
map relationship 图幅接合表
map revision 地图修订
map sheet 图幅
map symbol 地图符号，图例
map 映像，图像，图表，编录图，水系图，绘制，地图
maraging steel 马氏体时效钢
marasmus 消耗，消瘦
marble fracture 石板状断口
marble slab 大理石板
marble switchboard 大理石开关板，大理石配电盘
marble 大理岩，大理石
marcasite 白铁矿
marching forward method 步进法
marching problem 步进式问题
marconi antenna 马可尼天线
marconigram 无线电报
marco-reticular resin MR 型树脂，巨孔型树脂
margarite 珍珠云母
margin account 保证金账户
marginal adjustment 边界调整
marginal analysis 边界分析
marginal area 边缘地区
marginal bank 边岸，路堤岸
marginal beam 边缘梁
marginal benefit 边际效益
marginal bund 沿水库边缘堤岸
marginal capacity 备用容量
marginal check 边缘检验，边界检查
marginal circuit 备用电路
marginal cost pricing 边际成本制定电价，边际成本定价
marginal cost 边际成本，边际费用
marginal crevasse 边缘裂缝
marginal definition 边缘清晰度
marginal deposit 开发信用证的保证金
marginal discharge 边缘放电
marginal ditch 边缘沟
marginal effect 边际效应，边缘效应，边缘影响
marginal enterprises 边际企业
marginal fuel assembly 周边燃料组件
marginal illumination 最低照明
marginal information 图廓注记
marginal investment 边际投资，最低投资
marginal note 边注，旁注
marginal plateau 缘边台地

marginal power supply capability 备用电力
marginal price 边际价格，最低价格
marginal project 边际项目
marginal relay 定限继电器
marginal revenue 边际收益
marginal soil 贫瘠土壤
marginal stability 临界稳定性
marginal supply capability 备用电力
marginal supply capacity 备用电力
marginal swell 沿岸涌浪
marginal tax rate 边际税率
marginal test 边缘校验，边缘试验
marginal trough 边缘凹陷
marginal type wharf 沿岸式码头
marginal utility 边际效用
marginal value 临界值，边缘值
marginal vortex 边界涡
marginal wharf 堤岸码头
marginal words 旁注
marginal 边际的，边缘的，边界的，临界的，富裕的
margin capacity 富裕容量，备用容量
margined relay 定限继电器
margin knot 边节
margin money 保证金
margin of commutation 切换界限
margin of energy 能量储备，后备能量
margin of error 误差范围，允许误差
margin of power 备用功率
margin of safety 安全储备，安全裕度，安全裕量，安全系数
margin of stability 稳定系数，稳定群限
margin of temperature rise 温升裕度
margin-punched 边缘穿孔的
margin rule 保证金制度
margin switch 限位开关
margin test 边缘校验
margin tolerance 公差，允差，容差
margin to saturation 对饱和的裕度
margin to trip 跳闸裕度，跳闸界限
margin voltage 容限电压
margin 页边距，边缘，极限，利润，（版心外）的空白，边际，保证金，边界，边沿，差价，裕度，界限，余量，裕量，安全裕量，容许极限
MARG＝margin 裕量
marigram 潮汐曲线
marine abrasion 浪蚀
marine accident 海损事故
marine acoustics 海洋声学
marine aerosol 海洋气溶胶
marine air mass 海洋气团
marine air 海洋空气
marine atmosphere 海洋大气
marine bench 海台
marine bill of lading 海运提单
marine biological pollution 海洋生物污染
marine biota 海洋生物（群）
marine boiler 船用锅炉
marine brass 海军铜，锡黄铜
marine cargo insurance 海洋运输货物保险
marine chemical resource 海洋化学资源
marine climate 海洋性气候
marine corrosion 海水腐蚀
marine-cut bench 海台
marine data 海洋水文资料
marine deposit 海相沉积
marine design 船舶设计
marine disposal 海洋处置【指废物】
marine ecology 海洋生态学
marine electrochemistry 海洋电化学
marine energy 海洋能
marine engineering 海事工程
marine engine 船用蒸汽机，船用发动机，船用引擎
marine environment quality 海洋环境质量
marine environment 海洋环境
marine erosion 海蚀，浪蚀
marine facies 海相
marine fouling 海洋污垢
marine gas cooling reactor 船用气冷反应堆
marine geodesy 海洋测地学
marine geomorphy 海洋地貌
marine glue 防水胶
marine hydrographical survey 海洋水文测量
marine insulation 船用绝缘，水下电缆绝缘
marine insurance conditions 海洋运输保险条件
marine insurance policy 海上保险单
marine insurance premium 海上保险费
marine insurance 海损保险，海运险，海险，水险，海上保险，海运保险
marine machine 船用电机
marine measuring instrument 海洋测量仪
marine microbial pollution 海洋微生物污染
marine mineral resource 海洋矿物资源
marine oil spill 海上漏油
marine organism corrosion 海洋生物腐蚀
marine organism 海洋生物
marine petroleum exploitation 海洋石油开发
marine plain 海蚀平原
marine pollutant 海洋污染物
marine pollution monitoring 海洋污染监测
marine pollution prevention law 海洋防污法
marine pollution prevention 防止海洋污染
marine pollution 海洋污染
marine premium 海运保险（费）
marine propulsion reactor 船舶推进反应堆
marine protection 海洋保护
marine resource 海洋资源
marine risk 海上险，水险，海运险
mariner 水手
marine science 海洋科学
marine soil mechanics 海洋土力学
marine steam turbine condenser 船用汽轮机的凝汽器
marine steam turbine 船用汽轮机
marine surveying 海洋测量
marine survey positioning 海洋测量定位
marine swamp 海湿地
marine terminal 海运终点站
marine thermodynamics 海洋热力学
marine thermo energy 海洋热能，海洋温差能

marine transportation insurance 海洋运输保险
marine transportation 海上运输，海运
marine type dynamo 船用直流发电机
marine waste disposal 废物倾海处理
marine water 海水
marine weather observation 海洋天气观测
marine wind regime 海洋风况
marine works 海事工程
marine 船舶的，海生的，海洋的，海运业
maritime air mass 海洋气团
maritime air 海洋性空气
maritime climate 海洋性气候
maritime loss 海损
maritime plant 海生植物
maritime port 海港
maritime provinces 沿海各省
maritime transportation 海运
maritime vegetation 海生植物
mark check 标记检查
mark counting check 符号计数校验
mark current 传号电流，符号电流
mark down 记下，记录
marked capacity 额定容量，额定生产率
marked fluid 示踪流体，标记流体
marked freight prepaid 注明运费已付
marked pellet 示踪粒子，有记号的粒子
marked price 标明价格，标明价目
marked transfer 标明转让
marked weight 标明重量
marked 显著的，有标记的，醒目的，加标志的
marker bed 标志地层
marker control 标识控制
marker gate 标志脉冲门
marker light indicator 标志指示灯
marker pulse conversion 标记脉冲变换，标识器脉冲变换
marker pulse 标志脉冲，指向脉冲
marker register 时标寄存器
marker ring 标志环
marker selector 标志选择器，指向选择器
marker 指向标，指示器，标识，划线，标明，标志，路标
markeshift 权宜之计
marketable reserves 可销售的储量
marketable securities 上市证券
market adjustment 市场调节
market analysis subsystem 市场分析子系统
market analysis 市场分析
market appraisal standard 市场作价标准
market clearing price 市场出清电价，市场结算价，保证市场供求平衡的价格
market clearing 市场出清
market competition rate 市场竞争度指标
market competition 市场竞争
market demand 市场需求
market-developing management 经营开拓型管理
market-directed economy 市场经济
market economy 市场经济
market efficiency 市场效率
market entity 市场主体
market entry certification system 市场准入制度
market equilibrium 市场（供求）均衡
market failure 市场失灵
market fluctuations 市场波动
market force 市场力
market forecasting 市场预测
market information 市场信息
marketing assessment 市场评价
marketing requirement 营销要求
market intervention 市场干预
market life 供销期间，市场寿命
market mechanism 市场机制
market operator 市场运营机构
market orientation 市场导向，市场取向
market-oriented management 经营型管理
market participant 市场参与者
market potential 市场潜力
market power 市场力
market price level 市场价水平
market price 市场价格，市价
market rate 市价
market report 市场调研报告，市场报告
market requirement 市场需求
market research 市场调查
market rigging 操纵市场，操纵市场行为
market risk 市场风险
market saturation 市场饱和
market structure 市场结构
market survey 市场调查
market suspension 市场中止
market tendency 市场趋势，市场趋向
market value 市场价值，市价
market 市场
mark graduation 分度号
marking and preerection of the boiler 锅炉的标记与预组装
marking by sand blasting 喷砂打印
marking circuit 标记电路
marking contact 传号接点，标识触点
marking current 符号电流
marking device 记录装置，压印器，标记设备，压印器
marking gauge 划线规
marking off 标出（刻度），划线，用界线隔开
marking of insulated conductor 绝缘导线的标记
marking out set 划线设备
marking out 标出，规划，划线，标记，定线
marking pencil 标记铅笔
marking pin 测（标）钎，记号钉
marking strip 标识条
marking tree 打印采伐木
marking wave 传号波，符号波，记录波
marking 标记，标识，划线，作记号，记号，打印，识别，辨认，鉴定，加标记
markite 导电性塑料
mark number 标号
Markov process 马尔可夫过程
mark plate 标识牌
mark post 标杆，标记柱
mark reading 标记读出
mark scraper 划线器
mark sensing 符号读出

mark-space multiplier 标号空号乘法器，时间脉冲乘法器
markstone 标石
mark up 标高价格，提高标价
mark 标记，符号，标志，记号，标号，标明，唛头
marlaceous 泥灰质的
marline 细索
marly clay 泥灰质黏土
marly limestone 泥灰质石灰岩
marly sandy loam 泥灰砂质壤土
marly shale 泥灰质页岩
marly soil 泥灰土
marl 泥灰岩
MAR(memory address register) 存储地址寄存器
MARM(magnetic random access memory) 磁性随机存储器
marquench 热浴淬火，间歇淬火，马氏体等温淬火，马氏体分级淬火，分级淬火，等温淬火
marriage problem 匹配问题
marshaling area 调度区，待机区域
marshalling box 编线盒，总配线箱
marshalling service 编组作业
marshalling station 铁路编组场，编组站
marshalling track 编组线，调车线
marshalling yard 铁路编组场，铁路编组站，编组车场，集装箱编组场，货柜汇集场
marshal 调度，整理，安排，引导，带领
marsh area 沼泽区
marsh gas power station 沼气电站
marsh gas 沼气，甲烷
marsh land 沼泽地，沼地
marsh muck 沼泽腐殖土
marsh peat 沼地泥炭
marsh soil 沼土
marshy soil 沼泽土
marshy 沼泽的，湿地的
marsh 湿地，沼泽，海滩沼地，沼地
martemper 分级回火
martensite steel 马氏体钢
martensite strengthening 马氏体强化
martensite 马氏体
martensitic chromium steel 马氏体结构铬钢
martensitic heat resistant steel 马氏体耐热钢
martensitic stainless steel 马氏体不锈钢
martensitic transformation point 马氏体相变点
martensitic 马氏体
martingale 弓形接线
Martin steel 平炉钢，马丁钢
Marubeni Corporation 丸红株式会社【日本】
Marvin recorder 马文记录仪
mar 破损，损坏，擦伤，划痕
mascaret 涌潮，涛
maser action 微波激射作用
maser 微波量子放大器，微波激射器，激射器，脉塞
mashalling cabinets 编组柜，集线柜
mash seam weld 滚压焊
mash 泥浆，灰浆
Maskell method 马斯克尔（风洞阻塞）修正法
mask for radiography 射线照相用的防护面罩
mask index register 屏蔽变址寄存器
masking agent 隐蔽剂，掩蔽剂
masking compound 掩蔽化合物
masking effect 掩蔽效应
masking liquid for radiography 射线照相补偿液
masking method 掩蔽法
masking of odor 恶臭的掩蔽
masking reagent 掩蔽剂
masking solution 隐匿溶液
masking 掩蔽，遮蔽，伪装
mask register 屏蔽寄存器
mask 面罩，面具，罩，遮蔽，屏蔽，掩码
Masoneilan 梅索尼兰公司【美国】
masonite 绝缘纤维板
masonry arch 圬工拱
masonry bit 砖石钻头
masonry block 砖工，砖体，砌筑块
masonry body 圬工体
masonry buttress 圬工支墩
masonry cement 砌筑水泥
masonry check dam 圬工拦沙坝
masonry construction 砌筑结构，砖石结构，砖工构造，圬工结构，砖石建筑
masonry culvert 石圬工涵洞
masonry dam 圬工坝
masonry drill 石钻
masonry fill 圬工填充
masonry flue 砖砌烟道
masonry foundation 圬工基础
masonry grouting 圬工灌浆
masonry joint 圬工砌缝
masonry-lined tunnel 圬工衬砌隧道
masonry materials 砌筑材料
masonry mortar 圬工灰浆，砌筑砂浆
masonry nail 水泥钉
masonry panel wall 圬工格板墙
masonry pier 圬工墩
masonry plate 座板，基础板
masonry reinforcing 圬工加筋
masonry reservoir 圬工贮水池
masonry revetment 圬工护岸
masonry sluice 圬工节制闸
masonry stack 砖砌烟囱
masonry structure 砖石结构，圬工结构，砌体结构
masonry vault 圬工穹窿
masonry wall 石砌护墙，砌石墙，砌筑墙
masonry work 石工，砖工
masonry 圬工，砖工，石工，石工工程，石工行业，砖瓦工工程，砖石建筑，砌（筑）体，石造建筑，砖体，炉衬，砌体结构
mason's float 瓦工刀
mason 泥水工，泥瓦工，石匠，砖石工
mass absorption coefficient 质量吸收系数
mass absorption 质量吸收，全体积吸收
mass acceleration force 质量加速力
mass action law 质量作用定律
mass action 质量作用
mass air flow 空气质量流量
mass analysis 质量分析
mass analyzer 质量分析器，质谱仪

mass-area ratio 质面比
mass attenuation coefficient 质量衰减系数
mass avalanche 大规模崩坍
mass axis 主惯性轴，质量轴
mass balance equation 物料平衡方程
mass balance 质量平衡【聚变反应】，物料平衡
mass-basis sampling 质量基采样
mass blower 大流量吹灰器
mass burning 团块燃烧方式
mass center 质量中心
mass-change rate of solids 固体质量变化率
mass coefficient of reactivity 反应性质量系数
mass concentration of M M的质量浓度
mass concentration 质量浓度
mass concrete dam 大体积混凝土坝
mass concrete invert 大体积混凝土底板【干船坞的】
mass concrete wall 大体积混凝土岸壁
mass concrete 大块混凝土，大体积混凝土
mass conservation equation 质量守恒方程
mass conservation 质量守恒
mass continuity 质量连续性
mass curve of water 累计水量曲线
mass curve 累积曲线
mass damper spring system 质量阻尼器弹簧系统
mass data storage 大容量数据存储
mass data 大量数据
mass decrement 体量减缩量
mass defect 质量亏损
mass density 质量密度
mass depletion rate 质量耗尽率
mass detection limit 质量监测限
mass diagram of earthwork 土方累积曲线图，土方累积图
mass diffusion 质量扩散
mass dispersion coefficient 质量弥散系数
mass dissipation 质量消耗
mass distribution 质量分布，群分布
mass doublet 质量双线
mass dump 大量信息转储
mass eccentricity 质量偏心率
mass effect 质量效应
mass-energy absorption coefficient 质量能量吸收系数
mass-energy conversion 质量能量转换
mass-energy equivalence 质能当量
mass-energy equivalent 质量能量当量
mass-energy transfer coefficient 质量能量转换系数
mass excess 质量过剩
mass float 惯性浮标
mass flow coefficient 质量流量系数
mass flow constant 质量流量常数
mass flow continuity 质量流量连续性
mass flow density 质量流密度
mass flow distribution 质量流分布，质量流量分布
mass flowmeter 质量流量计
mass flow parameter 质量流量参数
mass flow rate perturbation 质量流量率的扰动
mass flow rate （流经管道横截面的）质量流量【单位时间内通过的物质质量】
mass flow ratio 质量流量比
mass flow 整体流动，质量流量，质量流
mass flux 质量流密度，质量通量，质量流量
mass force 质量能，质量力
mass foundation 大块式基础
mass fraction of M M的质量分数
mass fraction 质量份额，质量比
mass heat transfer 质热交换
massic activity 质量活度
massic energy imparted 比授能
massic energy 质量能
massic enthalpy 质量焓
massic entropy 质量熵
massic heat capacity at constant pressure 定压质量热容
massic heat capacity at constant volume 定容质量热容
massic heat capacity at saturation 质量饱和热容
massic heat capacity 质量热容
massic optical rotatory power 质量旋光本领
massic thermodynamic energy 质量热力学能
massic volume 质量体积
massif 地块
mass-impregnated insulation 整体浸渍绝缘
mass-impregnated non-draining cable 不滴流电缆
mass-impregnated paper insulated cable 黏性浸渍纸绝缘电缆
mass-impregnation 整体浸渍
mass inertia 质量惯性
mass inventory 蓄质量
massive exposure 强照射，大剂量照射
massive flange 厚法兰
massive footing 整体式底脚
massive-head buttress dam 大头坝
massive lava 块状溶岩
massive layer 大体积浇筑层
massive quay wall 大体积岸壁
massive rock 块状岩
massive rotor 整体转子，整锻转子
massive structure 大体积结构，整体结构
massive texture 块状结构
mass law effect 质量定律效应
mass law 质量定律
mass load 惯性负载，惯性负荷，惯性力
mass loss rate 质量损耗率，质量损失率
mass loss 质量损失
mass manufacture 大量制造
mass matrix 质量矩阵
mass median diameter 质量中位直径
mass media 媒体
mass memory 大容量存储器
mass moment of inertia 质量惯性矩
mass mortality 大量死亡
mass number 质量数
mass of atom of a nuclide X 核素X的原子质量
mass of molecule 分子质量
mass of UO_2 二氧化铀的质量
mass optical memory 大容量光存储器
mass output 累计出力
mass pile cap 大型桩帽

mass point 质点
mass-produced item 批量生产的项目
mass production 大量生产，成批生产，批量生产
mass quality 含汽率，蒸汽干度
mass range of particles 粒子群射程
mass rate of flow 质量流量率
mass ratio of fission 裂变碎片质量比，裂变质量比
mass ratio 相对流量，质量比
mass resistivity 质量电阻率，比阻，比电阻，体积电阻率
mass sink 质量汇
mass spectrographic analysis 质谱分析
mass spectrographic method 质谱分析法
mass spectrometer 质谱仪
mass spectrometric analysis 质谱测量分析
mass spectrum 质谱，质谱仪
mass spread 质量分布
mass-spring-dashpot system 质量、弹簧、阻尼器体系
mass stiffness 质量刚度
mass stopping power 重量制动能力，质量制动能力，质量阻止本领
mass storage control system 大容量存储管理系统
mass storage facility 大容量存储器设备
mass storage module 集成储存模块
mass storage system 大容量存储系统
mass storage 大容量存储器
mass surface density 表面质量密度，表面密度
mass(/time)-based sampling 以量［时间］为准的取样
mass transfer boundary layer 质量输运边界层
mass transfer characteristic 传质特性
mass transfer coefficient 传质系数
mass transfer cooling 传质冷却
mass transfer rate 传质速率，质量迁移速率，质量转移率
mass transfer 质量交换，质量转移，质量传递，传质，质量输运
mass transportation system 公共交通系统
mass transportation 大量运输
mass transport 大量运输，质量输运
mass unbalance 质量不平衡
mass velocity 质量流速，质量速度
mass 大量的，密集的，总的，堆，团，团块，块，多数，大量，质量
mast arm 支架
M/A station(manual/automatic station) 手动/自动操作站
mast base 电杆底座
mast crane 桅杆式起重机，桅式吊
master agreement 主协定
master air waybill 航空主运单
master alloy 主合金，母合金
master arm 主动臂
master body 标准样件
master change record 主要更改记录
master characteristics 校准特性，主特性曲线
master clock-pulse generator 母钟脉冲发生器，主同步脉冲发生器
master clock 母钟，主钟，母时钟，主时钟
master clutch 主离合器
master connection 主结线
master contract 主约
master control and monitoring 主控和监视
master control console 主控制台
master controller 主控制器，中心控制器，主令控制器，主控器
master control panel 主控制盘，主控制屏，主控制板
master control program 主控程序
master control room 中央检制室，主控制室
master control routine 主控程序
master control system 主控系统
master control valve 主控制阀
master control 总控制，主控制，集中控制，中心控制
master copy 原稿，底图，原件
master credit 原始信用状
master cylinder 主液压缸，主缸
master data control console 主数据控制操作台
master data recorder 主数据记录器
master data 主要数据，基本数据，主数据
master drawing 司令图
master driver 主驱动电机，主激励器，主驱动器
master drive 主拖动
master drum sender 主磁鼓传送器
master electronic control panel 主电子控制箱，主电子控制板
master end 机械手的主动端
master fault 主断层
master file 主文件，不常变文件
master form 靠模，仿形模
master frequency meter 积量频率计，总频率计
master frequency 主振频率，主频
master fuel trip 总燃料跳闸，主燃料跳闸
master gauge 标准量计，校正用仪表
master gear 主齿轮
master governor 主调速器
master gully 集水主沟
master gyroscope 主陀螺仪
master information and data acquisition system 主信息与数据获取系统
master instruction tape 主指令带，主程序带
master instruction 使用说明书
master instrument 主控仪表，校准用仪器，校准仪表
master joint （机械手的）主动端关节，主要节理
master letter of credit 主信用证
master link 主联杆
master list 总清单
master melt 主熔炼
master meter 标准表
master mode 主态，主方式，监督方式
master monitor 主监视器，主监视程序
master motor 主驱动电动机
master operational controller 主操作控制器
master operation system 主操作系统
master oscillator 主振荡器
master overhaul 大修

master phase 基本相位
master phasing chart 基本相位图
master phasing schedule 基本相位表
master plan 司令图，总图，总（体）规划，总计划，总平面图，蓝图
master power switch 主控功率开关，主功率开关
master pressure controller 主压力控制器
master processor 主处理器
master procurement plan 总的采购计划
master program clock 主程序控制装置
master program 主程序
master project schedule 主要项目计划
master pulse 主控脉冲
master regulating control valve 主控调节阀
master relay 主继电器，主控继电器
master routine （执行程序）主程序
master scanner 主扫描器
master schedule 主要图表，综合图表，设计任务书，主要作业表，主进度
master section （电站的）发电机部分，发电机组主段
master set 校对调整，校正调整
master sheet 原图
master side （机械手的）主动端
master signal 主信号
master slave discrimination 主从鉴别
master slave electric manipulator 电传动主从机械手
master slave flip-flop 主从触发器
master slave manipulator 随动机械手，主从机械手
master slave mechanical manipulator 机械传动主从机械手
master slave mode 主仆方式，主从方式
master slave operation 主从运转
master slave system 主从系统
master slave 主从的
master spare positioning resolver 主备用定位分解器
master station 主控台，主台，总台，主控制站，主站
master stream 干流
master switch 主控开关，总开关，主开关
master synchronizer 主脉冲同步器
master system tape 主系统带
master tape 主程序带
master timer 主时间延时调节器，定时装置，主计时器
master-to-slave tape drive 主从机械手的带式传动
master trip 主停车装置，主脱扣器
mastery 控制
master 征服，控制，精通，主导装置，硕士，控制者，主导的，主要的
mast head light 桅杆顶灯
masthead 桅顶，电杆顶，发行人栏，升于桅顶，爬到桅顶
mastic asphalt surfacing 沥青砂胶脂面层
mastic asphalt 石油沥青玛蹄脂，沥青胶泥
mastic pavement 沥青砂胶路面，玛蹄脂路面
mastic weatherproof coating 玛蹄脂保护层
mastic 胶黏剂，油灰，胶泥，玛蹄脂，嵌缝料，树脂
MAST(magnetic annular shock tube) （风动研究用的）磁性环形激波管
mast straightness 棒的平直度
mast switch 柱上开关，柱上闸刀开关
mast timber 桅木
mast-type stationary hoist 柱式固定起重机
mast 柱，杆，电杆，桅，桅杆，杆塔，天线塔，塔桅
masut 重油，残油
matarial buckling factor 材料曲率因子
matarial damage 重大损坏，材料损伤
mat base 垫层
match admittance 匹配导纳
matchboard door 实拼门
match boarding machine 企口镶板机
match boarding 铺企口板，镶拼花板
matched attenuator 匹配衰减器
matched data 匹配数据
matched filter 匹配滤波器
matched impedance 匹配阻抗
matched line 匹配线
matched pair transistor 配对晶体管
matched resistance 匹配电阻
matched termination 匹配终端
matched transformer 匹配变压器
matched transmission line 匹配传输线，匹配输电线
matched 匹配的，适配的
matcher-selector-connector 匹配选择连接器，匹配选择器
match gate “同”门
match impedance 匹配阻抗
matching circuit 匹配电路
matching control 自动选配装置
matching error 匹配误差
matching facilities 配套设施
matching filter 匹配滤波器
matching network 匹配网络
matching of load 负载匹配
matching of pulses 脉冲刻度校准，脉冲匹配
matching of stages 级间匹配
matching of waveguide 波导匹配
matching pad 匹配线路板
matching pillar 匹配柱
matching point 匹配点
matching stub 匹配短线
matching transformer 匹配变压器
matching 匹配，配合，选配，对准
match joint 企口接合
match line 接续分界线
matchmark 模缝线【玻璃制品缺陷】，配合符号，装配标记，为……装配标记
match plane 边刨，槽刨
match termination 终端匹配
match 使相符合，使匹配，使配合，使协调，匹配，相配
mat coat 罩面，面层
material accountability 材料衡算量
material accountancy 材料衡算

material accounting 材料核算
material alteration 实质性变更，重大修改
material analysis 材料分析
material and equipment schedule 设备材料表
material appropriation 材料调拨
material asset 有形资产
material balance accounting 材料收支量衡算
material balance and accounting 材料平衡与核算
material balance area 物料平衡区【核燃料加工】，材料平衡区
material balance component 材料平衡分量
material balance equation 物料平衡方程
material balance period 材料平衡周期
material balance report 材料平衡报告
material balance statement 材料平衡报告
material balance 物料平衡，材料平衡
material balancing calculation 物料平衡计算
material bill 材料清单，材料单，材料表
material breach of contract 严重违约
material buckling 材料曲率【反应堆】，材料压曲，材料的下垂
material budget 材料预算，原料预算
material certificate 材料（检验）合格证
material certification 实物校验，材料认证，材质证明
material change 重大变更
material compatibility 材料相容性
material composite 复合材料
material contract 购料合同
material cost 材料费
material couple 材料配合
material damage 材料损失
material damping 材料阻尼
material defect 材料缺陷
material depreciation cost 材料折旧费
material deviation 重大偏差，重大偏离
material economy 材料经济性
material engineering 材料工程
material engineer 材料工程师
material evaluation 材料评价
material for filling the joint 嵌缝材料
material handling fan 物料输送风机
material handling valve 物料控制阀
material handling 物料搬运，物料操作
material identification 材料鉴定
material index of reservoir inundation 水库淹没实物指标
material industry 材料工业
material inspection 材料检查
material inventory 物料总存量，材料总存量
material investigation 材料调查
material issuing warehouse 发料仓库
materiality 实质性，重要性
material list 材料单，材料表
material lock 材料进出闸
material management system 材料管理系统
material nonlinearity 材料非线性
material of paint 油漆材料
material of tubes 管材，管的材质
material packing list 材料装箱单
material parameter 材料参数
material particle 质点，质粒
material pit 储料坑
material point 质点
material preparation 下料
material processing reactor 材料辐照反应堆
material procurement 材料采购
material property 有形资产
material qualification 原材料测试认证，材料合格鉴定
material radiation effect 材料辐照效应
materials acceptance 材料验收
materials accounting method 材料衡算方法
material safety data sheet 材料安全性数据表
material sample 材料样品
materials assessment 器材评价
materials balance system 材料平衡制度
materials control 器材管理
material selection 材料选择
materials engineering laboratory 材料工程实验室
material separation 物料分离[分层]
materials fatigue 材料疲劳
materials handling system 物料输送系统
materials hauling 搬运材料
material shed 材料棚
material shifting 物料偏移
material shortage 材料缺乏
material shuffling 物料扰动
material slippage 物料下滑
materials management 器材管理
materials manager 器材经理
material specification 材料规格，原材料明细表，材料规范
material spillage 物料溢出
material spring 地表水源
materials-producing reactor 制备新同位素的反应堆【核电】，材料生产反应器
materials strength 材料强度
material standard 材料标准
materials testing reactor 材料试验反应堆
material storage 材料储量，储料场，物料贮存
material storehouse 材料仓库，材料库
material stream 物流
material supplier 材料供应商
material survey 材料调查
materials weldability file 材料可焊性档案资料
materials 器材，设备
material take-off 汇料，材料统计，材料提取，材料表【MTO】
material terms 实质性条款
material testing device 实物试验装置
material test 材料试验
material trampling 物料颠簸
material transfer 物质传递
material type 材料型号
material unaccounted for 不明材料（量）
material use factor 材料使用率
material 资料，材料，原料，物料，物质，器材，物质的，重大的，实际性的
maternal balance accountancy 材料收支衡算
mate's receipt 大副收据，大副据，货运收据，收货单

mate 匹配
mat footing 板式基础，平板底脚
mat foundation 筏基，板式基础，席形基础，地板基础
mat glass 毛玻璃
mathematical analysis 数学分析
mathematical approximation 数学近似法
mathematical computation 数学计算
mathematical definition 数学定义
mathematical equipment 数学计算装置
mathematical expectation 数学期望
mathematical expression 数学表达式
mathematical formulation 数学公式
mathematical function 数学函数
mathematical induction 数学归纳法
mathematical interpretation 数学解释
mathematical linguistics 数学语言学
mathematical load model 负荷数学模型
mathematical logic 数理逻辑
mathematical manipulation 数学运算
mathematical modal 数学模型
mathematical model of controlled plant 控制装置数学模型
mathematical model of excitation system 励磁系统数学模型
mathematical model of prime mover and governor 原动机和调速系统数学模型
mathematical models of electric power network 电网数学模型
mathematical models of electric power system 电力系统数学模型
mathematical models of synchronous machine 同步电机数学模型
mathematical model 数学模式，数学模型
mathematical module 数学模，数学模块
mathematical programming system 数学规划系统
mathematical relation 数学关系
mathematical simulation 数学模拟，数学仿真
mathematical software 数学软件
mathematical statistics of hydrology 水文数理统计
mathematical statistics 数理统计学，数理统计
mathematical stimulation 数学模拟
mathematical subroutine 数学子程序
mathematical treatment 数学处理
mathematical 数学的
mathematic expectation 数学期望
mathematics 数学
mathematic 数学的
mating annular face 环形配合面，环形接配面
mating flange 配合法兰
mating groove 配合槽
mating materials 配合接触的材料
mating plug 连接插头
mating points 交接点，触合点
mating surface 接合面【机械】，啮合面，配合面，接触面
mating 配套的，配合的，接合的，相连的
MAT(memory access time) 存储器取数时间
mat metal 未抛光金属
MAT(microalloy transistor) 微合金晶体管
mat reinforcement 钢筋网
matrix adder 矩阵加法器
matrix addition 矩阵加法
matrix-addressed display 矩阵寻址显示
matrix addressing 矩阵寻址
matrix analysis 矩阵分析
matrix calculation 矩阵计算
matrix calculus 矩阵计算
matrix circuit 矩阵变换电路
matrix column 矩阵的列
matrix compiler 矩阵编码器，矩阵编译程序
matrix computation 矩阵计算
matrix connection 矩阵连接
matrix converter 矩阵变流器
matrix differential equation 矩阵微分方程
matrix element 矩阵元，矩阵元件，矩阵元素
matrix equation 矩阵方程
matrixer 矩阵变换电路
matrixes of drilled holes 钻孔基岩
matrix fraction 矩阵分数
matrix fuel 基体燃料
matrix gate 矩阵门，译码器
matrix graphite 基体石墨
matrix-hardening 基体硬化的
matrixing function 矩阵函数
matrix inversion 矩阵求逆，矩阵反演
matrix iteration 矩阵迭代法
matrix lens 矩阵透镜
matrix management 矩阵管理，矩阵运算
matrix matching 阵列匹配
matrix mechanics 矩阵力学
matrix memory 矩阵存储器
matrix method 矩阵法
matrix operation 矩阵运算
matrix organization 核心组织，矩阵组织
matrix pin board 矩阵插接板
matrix plane output 矩阵板输出
matrix plate 矩阵板
matrix printer 矩阵式打印机，字模打印装置
matrix printing 矩阵式打印
matrix reduction 矩阵化简，矩阵简化
matrix representation 矩阵表示法
matrix ring 矩阵环
matrix row 矩阵的行
matrix storage system 矩阵存储系统
matrix storage 矩阵存储器
matrix suction 填质吸力
matrix switch 矩阵开关
matrix theorem 矩阵定理
matrix transpose 矩阵换行，矩阵转置，矩阵换列
matrix unit 矩阵单元，换算设备
matrix 矩阵，基体，母体，母式，真值表，矩阵变换电路，基质，模型，脉石
matrizant 矩阵积分级数
matsushita pressure diode 压敏二极管
matted vegetation 铺地植被
matter constant 材料常数
matter energy 物质能量
matters needing attention 注意事项
matters not included herein 该协议未尽事宜

matters not settled by this agreement 该协议未尽事宜
matters subject to administrative approval 须经行政审批的事项
matter wave 物质波
matter 物质，材料，物料，事项
matte surface 毛面
matte 冰铜
matting （焊前）清洗工序，无光泽表面，铺层，铺面
mattress array 多列天线阵
mattress ballasting 柴排压沉
mattress 沉排，床垫，空气垫，柴排
matt surface finish 亚光表面光洁度
matt surface 亚光表面，无光泽面，粗面
matt 暗淡的，不光滑的，无光泽的，粗糙的
maturation 熟化，成熟
mature consideration 成熟的考虑
matured bill 到期票据
mature height of tree 树木生长高度
mature hurricane 成熟飓风
mature plan 周密的计划
mature storm 成熟风暴
mature technology 成熟技术
mature topography 壮年地形
mature （票据等）到期，成熟，期满，成熟的，使……成熟
maturing 成熟，凝固【混凝土】，老化
maturity date 到期日，偿还日
maturity value 到期值，终值
maturity （票据等）到期，成熟
mat 板，层，护面，垫板，垫子，席，垫，大木槌
MAU(medium attachment unit) 介质附属单元
mA/V(milliamperes per volt) 毫安/伏
MAWB(master air waybill) 航空主运单
max ambient temperature over the years 多年极端最高温度
max breaking time 最大全分闸时间
maxed output power 最大输出功率
maximal clearance 最大间隙
maximal efficiency 最大效率
maximal time between failures 最大无故障工作时间
maximal value 极大值，最大值
maximal work 最大功
maximal 最大的，最高的
maximization 极限化
maximized voltage response window 最大化电压响应窗口
maximize 使达到最大值
maximizing power output 最大功率输出
maximum absorbed dose 最大吸收剂量
maximum acceptable concentration 最大容许浓度
maximum accident pressure 最大事故压力
maximum accuracy 最高精确度
maximum admissible concentration 最大容许浓度
maximum admissible dimension 最大容许尺寸，最大容许尺度
maximum admitted diameter 最大许可直径
maximum air concentration 最大空气浓度
maximum allowable concentration 最大容许浓度，最高容许浓度
maximum allowable dose 最大容许剂里
maximum allowable draft loss 最大容许风量损失，最大容许通风损失
maximum allowable draft 最大容许风量
maximum allowable force 最大容许力
maximum allowable level 最大容许水平
maximum allowable linear heat generation rate 最大容许线热功率
maximum allowable metal temperature 最高许用金属温度，最高许用壁温
maximum allowable operating temperature 最大许可操作温度
maximum allowable operating time 最大容许运行时间
maximum allowable pressure 最高允许压力
maximum allowable speed 最高容许转速，最大容许速度
maximum allowable temperature 最大容许温度，最高允许温度
maximum allowable velocity 容许最大流速，最大容许流速，最大容许速度
maximum allowable working pressure 最大容许工作压力，最大许用工作压力
maximum allowable 允许的最大限的，最大可允许的
maximum amount limitation 最高额限制
maximum amount 最大量，最高额
maximum and minimum relay 极值继电器
maximum and minimum thermometer 最高最低温度计
maximum annual flood 最大年洪水
maximum annual silt content 年最大含沙量
maximum annual times of freezing-thawing cycle 年最多冻融循环次数
maximum assumed frequency 最大设想频率
maximum asymmetric short-circuit current 最大不对称短路电流
maximum available gain 最大有效增益，最大可用增益
maximum available power 最大可用功率
maximum available storage 最大有效蓄水量，最大有效库容
maximum available time 最大有效（工作）时间
maximum average power output 最大平均功率输出
maximum average wind speed 最大平均风速
maximum average wind velocity 平均最大风速
maximum bare table acceleration 空载最大加速度
maximum bed height 最大床高
maximum beta particle energy β最大能量
maximum blockage 最大阻塞
maximum blowdown pressure 最大排放压力
maximum breaker 最高限（电压或电流）断路器
maximum burn rate 最大燃煤量
maximum capability operation 最大出力运行
maximum capability 最大功率

maximum capacity rating 最大额定容量，最大出力
maximum capacity standard 最大容量标准
maximum capacity 最大出力，最大容量，最高出力，最大称量
maximum capillary capacity 最大毛管持水量
maximum cladding temperature 最高包壳温度
maximum cleaning efficiency 最大清洗效率
maximum clearance 最大间隙
maximum coefficient of heat transfer 最大传热系数
maximum coil power dissipation 线圈最大功耗
maximum compacted dry unit weight 最大压实干容重
maximum computed flood 最大计算洪水
maximum conceivable accident 最大设想事故
maximum condensate flow 最大凝结水量
maximum continuous capacity 额定输送容量，最大连续功率
maximum continuous duty 最大连续工作负载
maximum continuous load 最大持续负载，最大连续负荷
maximum continuous output 最大连续出力，最大连续功率
maximum continuous power 最大连续功率
maximum continuous rating 最大持续额定功率［出力］，最大连续出力［功率］【电机，旋转机械】，最大连续蒸发量
maximum continuous speed 最高持续转速
maximum conversion rating 最大转换率
maximum cooling load 最大冷却负荷
maximum country amount limitation 国家最高额限制
maximum credible accident 最大可能事故，最大可信事故
maximum crest 最大峰值
maximum critical heat flux 最大临界热通量
maximum critical void ratio 最大临界孔隙比
maximum current 最大电流
maximum curvature 最大曲度，最大曲率
maximum cutout 自动切断能力，最大电流自动断路器
maximum cycle temperature 最高循环温度
maximum daily consumption 最高日耗
maximum daily discharge 最大日流量
maximum daily rainfall 单日最大降雨量
maximum daily temperature 日最高温度
maximum daily water consumption 最大日用水量，最高日用水量
maximum deflection 摆幅，最大偏转
maximum demanded power 最大需求功率
maximum demand meter 最大需量计
maximum demand 最大需量，最大需电量，所需最大负荷
maximum density 最大密度
maximum dependable capacity 最大可靠出力
maximum depth of frozen ground 最大冻土深度
maximum depth of snow cover 最大积雪深度
maximum designed wind speed 最大设计风速
maximum design flood 设计最高洪水位
maximum design pressure 最高设计压力
maximum design voltage 最高设计电压
maximum design wave height 最大设计波高
maximum deviation 最大偏差
maximum difference 最大差值
maximum dimension 最大尺寸
maximum discharge of turbine 机组过水能力
maximum discharge principle 最大流量定理
maximum discharge 最大流量
maximum dose equivalent 最大剂量当量
maximum dose 剂量限值，最大剂量
maximum draft 最大吃水深度
maximum draught 最大吃水深度
maximum drop of closure 截流最大落差
maximum dry density 最大干密度，最大干容重
maximum dry unit weight 最大干容量，最大干容重
maximum duty of water 最大灌水率
maximum dynamic error 最大动态偏差
maximum ebb 最大落潮流
maximum economic benefit 最高经济效益
maximum economic rating 最经济出力
maximum effectiveness 最大有效性，最大效用
maximum efficiency point 最高效率点
maximum efficiency 最大效率，最高效率
maximum electric surge flow capacity 最大电涌通流能力
maximum elevation 最大高程
maximum entropy spectral analysis 最大熵谱分析
maximum envelope curve 最大包络曲线
maximum equilibrium current 最大平衡电流
maximum erratum 最大误差
maximum error 最大误差
maximum evaporation 最大蒸发量
maximum excess reactivity 最大过剩反应性
maximum excitation 最大激励
maximum exhaust wetness 最大排汽湿度
maximum failure rate 最高事故率
maximum fall of water level 水位最大落差
maximum fault rate 最大故障率
maximum feasible pressure 最大可能压力
maximum feed size 最大给料粒度
maximum field carrying capacity 田间最大持水量
maximum flood level 最高洪水位
maximum flood 最大洪水
maximum flow capacity 最大过水能力
maximum flow of hot-water supply 热水供应最大流量
maximum flow rate of turbine 涡轮机最大流量
maximum flow rate 最大流量
maximum flow velocity of closure 截流最大流速
maximum flow 最大流量
maximum frequency deviation 最大频率偏差
maximum frozen depth 最大冻结深度
maximum frozen soil depth 最大冻土深度
maximum fuel centerline temperature 最高燃料中心温度
maximum fuel center temperature 最高燃料中心温度
maximum fuel temperature 最高燃料温度，最高燃烧温度
maximum gain 最大增益

maximum gas temperature 最高燃气温度
maximum gradient 最大梯度，最大坡度
maximum ground concentration 最大地面浓度
maximum guaranteed capability 最大保证出力
maximum gust lapse interval 阵风最大递减时段，阵风最大递减时距
maximum headwater 最高上游水位
maximum head 最大水头，最高水头，最大压头
maximum heat flux 最大热流密度，最大热通率
maximum heating load 最大供热负荷
maximum heat load 最大热负荷，最大热负载
maximum heat release rate 最大放热速率
maximum heat-up rate 最高加热速率
maximum high-water 最高水位
maximum high-water level 最高水位
maximum historical flood level 历史最高洪水位
maximum historical flood 历史最大洪水
maximum historically probable earthquake 历史最大可能地震【烈度】
maximum horsepower 最大马力
maximum hot-water heating load 热水供应最大热负荷
maximum hourly outdoor temperature 小时最高室外温度
maximum hourly steam consumption 最大小时耗汽量，最大时用水量
maximum humidity 最大湿度
maximum hypothetical accident 最大假想事故
maximum hypothetical earthquake 最大假想地震
maximum indicating thermometer 最高指示温度计
maximum individual span 极限挡距
maximum input voltage 最高输入电压
maximum instantaneous power 最大瞬时功率
maximum instantaneous wind speed 最大瞬时风速
maximum intensity 最大强度
maximum latency 最大等待时间
maximum lateral shrinkage 最大横向收缩率
maximum length code 最大长度码
maximum lift coefficient 最大升力系数
maximum lifting load 最大提升荷载
maximum likelihood decoding 最大可能性译码
maximum likelihood estimate 极大似然估计，最大可能估计
maximum likelihood 最大可能性
maximum limit 最大极限
maximum linear heat generation rate 最大线热功率
maximum load indicator 最高负载指示器
maximum load power 最大用电功率
maximum load 最大负荷，最高负载，最高负荷，最大载重，最大荷载
maximum loss-of-coolant accident 最大冷却剂丧失事故，最大失水事故
maximum mast height 最高船桅顶高度
maximum mean temperature 最大平均温度
maximum mean wind velocity 最大平均风速
maximum measured power 最大测量功率
maximum mining yield 最大开采水量
maximum mixing depth 最大混合层厚度
maximum moisture capacity 最大持水量
maximum molecular water content 最大分子吸水量
maximum momentary speed variation 瞬时最大升速率
maximum momentary speed 瞬时最大转速，飞升转速，瞬时最大速度
maximum moment 最大力矩
maximum nail holding power 最大握钉力
maximum navigable stage 最高通航水位
maximum net head 最大净水头
maximum nip diameter 最大切断直径
maximum nozzle load 最大管口负荷
maximum observed precipitation 实测最大降水量，最大观测降水量
maximum once flood volume 一次最大洪水总量
maximum one day precipitation 一日最大降水量
maximum one-day rainfall 最大日雨量
maximum operating efficiency 最大运转效率
maximum operating frequency 最高运行频率，最高工作频率
maximum operating pressure differential 最大运行压差
maximum operating temperature 最高运行温度
maximum operating voltage 最高运行电压
maximum operating wind speed 最大运行风速
maximum operational mode 最大运行方式
maximum operation condition 最大工况
maximum output current 最大输出电流
maximum output power 最大输出功率
maximum output voltage 最大输出电压
maximum output 最大输出，最大出力，最高出力，最大输出功率
maximum overall efficiency 最高总效率
maximum overload capability 最大过负荷容量[功率]
maximum overload capacity 最大过载功率
maximum payload ratio 最大有效载荷比
maximum peak forward voltage 最高顺向峰值电压
maximum peak inverse voltage 最大反向峰值电压
maximum peak value 最大峰值，极限峰值
maximum peak 最大峰值
maximum permissible accumulated dose 最大容许累积剂量
maximum permissible body burden 许可的最大人体荷载，体内最大允许剂量
maximum permissible concentration of unidentified radionuclide 未知放射性核素最大容许浓度
maximum permissible concentration value 最大容许浓度值
maximum permissible concentration 最大容许浓度
maximum permissible cumulated dose 最大容许累积剂量
maximum permissible daily intake 最大容许日摄入量
maximum permissible deviation 最大允许偏差
maximum permissible dissipation 最大允许损耗
maximum permissible dose equivalent 最大可接

受剂量当量，最大容许剂量当量
maximum permissible dose rate 最大容许剂量率
maximum permissible dose 最大容许剂量
maximum permissible error 最大允许误差
maximum permissible exposure 最大容许照射量
maximum permissible intake 最大容许摄入量
maximum permissible integral dose 最大容许累积剂量
maximum permissible level 最大容许水平
maximum permissible limit 最大容许限值，最大允许极限
maximum permissible linear rating 最大允许线功率
maximum permissible load 最大允许负荷
maximum permissible pressure 最大允许压力，允许压力
maximum permissible radiation dose 最大容许辐射剂量
maximum permissible release 最大容许排放(量)
maximum permissible revolution 最大许可转数，最高允许转速
maximum permissible speed 最高允许转速
maximum permissible stalled time at rated voltage 在额定电压下的最大允许失速时间
maximum permissible standard 最大容许标准
maximum permissible stress 最大许用应力
maximum permissible temperature rise 最高允许温升
maximum permissible temperature 最高允许温度
maximum permissible velocity 最大允许流速，最大允许速度
maximum permissible void fraction 最大允许空泡份额
maximum permissible weekly dose 最大容许周剂量
maximum permissive power 最大容许功率
maximum pickup current 最大闭合电流，吸合电流最大值
maximum pool level 最高蓄水位
maximum pore size 最大孔径
maximum possible operating factor 最大可能利用率
maximum potential earthquake 最大潜在地震
maximum potential 最高电位，最高电势
maximum power consumption 最大功耗
maximum power dissipation 最大功耗
maximum power of wind turbine 风力机最大功率
maximum power output 最大发电功率，最大功率输出，最大输出功率
maximum power point tracking 最大功率点跟踪，最大功率点追踪，最大功率跟踪
maximum power point 最大功率点
maximum power test 最大功率试验
maximum power tracking 最大功率跟踪法
maximum power transfer theorem 最大功率传输原理
maximum power voltage 最大电源电压
maximum power 最大功率
maximum precipitation 最大降水量
maximum pressure 最高压力
maximum price 最高价格
maximum principal stress theory 最大主应力理论
maximum principal stress 最大主应力
maximum probable error 最大可能误差
maximum probable flood 最大可能洪水
maximum production 最高产量
maximum productivity 最高生产率
maximum propulsive efficiency 最大推进效益
maximum pulling force 最大牵引力
maximum rainfall intensity 最大降雨强度
maximum rainfall 最大降雨量
maximum rated load 最大额定负荷，最大额定荷载
maximum rate of filtering 最大滤速
maximum rate of flow 最大流量速率
maximum rate of rainfall per hour 每小时最大降雨量
maximum rate 最高定额，最大速率，最大流速
maximum rating 最大出力，最大额定值
maximum recorded flood peak 最大实测洪峰
maximum relative cluster capacity 束棒最大相对容量
maximum relay 极限继电器，过载继电器
maximum reliability 最高可靠性
maximum reservoir capacity 水库总库容
maximum resistance 极限电阻，最大电阻
maximum revolution 最大转数
maximum rod deflection 棒最大挠度
maximum rotational speed 最大转速，最大旋转速率
maximum rotative speed 最大旋转速度
maximum rotor blade envelope 风轮叶片最大包络线
maximum routine inspection effort 最大例行视察量
maximum runoff rate 最大径流速度
maximum runoff 最大径流
maximum safe capacity 最大安全容量
maximum safe concentration 最大安全浓度
maximum safe speed 最大安全转速
maximum safety load 最大安全负荷
maximum safety service temperature 最高安全使用温度
maximum safety setting 最大安全定值
maximum safety temperature 最高容许温度，最安全温度，最高安全温度
maximum sag of conductor 导线最大弧垂
maximum salinity layer 最大盐度层
maximum salinity 最大含盐度
maximum scale value 标度终点值，最大标度值
maximum scour depth 最大冲刷深度
maximum secondary discharge 最大次级排放量
maximum second discharge 最大秒流量
maximum sediment concentration 最大沉积物浓度，最大含沙量，最大含沙浓度
maximum seismic intensity 最大地震烈度
maximum sensitivity 最高灵敏度
maximum service life 最大工作寿命
maximum service pressure 最大使用压力
maximum shaft power 最大轴功率
maximum shear strain 最大剪切应变

maximum shear stress theory　最大剪应力理论
maximum shear theory　最大剪力理论
maximum size　最大尺寸
maximum snow depth　最大积雪深度
maximum span　最大跨距，极限挡距，最大挡距【输电线】
maximum specific discharge　最大比流量
maximum spectral luminous efficacy　最大光谱光视效能
maximum speed rise　最高升速
maximum speed　最高速度，最高转速
maximum spoutable bed depth　最大可喷射床深
maximum stable gain　最大稳定增益
maximum stage mark　最高水位痕迹
maximum starting current　最大启动电流
maximum static deflection　最大静挠度
maximum stepping rate　最大步进率
maximum strain energy　最大应变能
maximum strain theory　最大变形理论
maximum stress in bend　最大弯曲应力
maximum stress limit　最大应力极限
maximum stress theory　最大应力理论
maximum stress　最大应力，极限应力
maximum suction point　最大吸力点
maximum sum of hourly cooling load　逐时冷负荷综合最大值
maximum surge current　最大涌浪流
maximum surge　最高浪涌
maximum survival time　最高存活时间
maximum susceptibility　最高磁化率
maximum system voltage　最高线电压，最高系统电压
maximum temperature-rise　最高温升，极限温升
maximum temperature　最高水温，最高温度
maximum tension stress　最大拉伸应力
maximum tension　最大张力
maximum theoretical generation　最大理论发电量
maximum theoretical torque　最大理论转矩
maximum thermometer　最高温度计
maximum thickness of plastic layer　胶质层最大厚度
maximum thrust　最大推力
maximum tide flood　最大潮汛
maximum-to-average power　最大功率与平均功率之比
maximum torque coefficient　最大力矩系数
maximum torque test　最大转矩实验［试验］
maximum torque　最大扭矩，最大力矩，最大转矩，极限转矩
maximum total sag　（输电线的）最大总垂度
maximum transfer time　最大传送时间
maximum treatment efficiency　最大处理效率
maximum tripping current　最大脱扣电流
maximum turning speed of rotor　风轮最高转速
maximum upstream water level　最高上游水位
maximum upsurge　最大上涌浪
maximum usable discharge　（水电站的）最大使用流量
maximum usable frequency　最高可用频率，最大可用频率
maximum valence　最高原子价，最高化合价
maximum value indicator　限值指示器，最高值指示器
maximum value　最高值，最大值
maximum velocity in closure-gap　龙口最大流速
maximum velocity　最大流速，最高流速
maximum vertical shrinkage　最大竖向收缩率
maximum visibility　最大能见度
maximum voltage　最高电压
maximum volume change　最大体积变化
maximum volumetric shrinkage　最大体缩率
maximum water consumption　最大耗水量
maximum water content　最大含水量
maximum water demand　最大需水量
maximum water-holding capacity　最大持水量
maximum water level of waterlogging　内涝最高水位
maximum water level　最高水位
maximum wave　最大波浪
maximum wheel load　吊车最大轮压，最大轮压
maximum wind speed　最大风速
maximum wind velocity of definite period observation　定时的最大风速
maximum wind　最大风力
maximum working load　最大工作负荷
maximum working pressure　最大操作压力，最大工作压力
maximum working temperature　最大工作温度，最高工作温度
maximum working tension　最大工作张力
maximum working voltage　最大工作电压，最高工作电压
maximum　最大的，最高的，最大值，极大值，最大，最大化，最大限度，极限
max longitudinal slope of rails　轨道最大纵坡
MAX＝maximum　最大的，最大值的，最大
maximum thunder days over the years　多年最大雷暴日数
Maxwell distribution　麦克斯韦尔分布
Maxwell field equation　麦克斯韦场方程式
maxwellmeter　磁通计
Maxwell's formula　麦克斯韦尔公式
Maxwell's hypothesis　麦克斯假设
Maxwell　麦克斯韦【磁通量单位】
mayday　求救信号，无线电话中的求救信号
mayor　市长
maze domain　迷路形磁畴
maze　迷宫，曲径
mazout　重油
mazut　重油
MBD(million barrels per day)　每天百万桶
MBE(material balance equation)　物料平衡方程
M-BFP(motor driven boiler feed water pump)　电动锅炉给水泵
M-block tridiagonal matrix　M 块三对角矩阵
MB＝megabyte　兆字节
MBONE(multicast backbone)　多播主干网
MBP(material balance period)　材料平衡周期
MBPS(megabits per second)　兆位每秒
MBQ(modified biquinary code)　改进的二五混合进制码
MBR(material balance report)　材料平衡报告

MBR(memory buffer register) 存储缓冲寄存器
MBS(megabyte per second) 兆字节每秒
mb=millibar 毫巴
MCA(material control area) 材料控制区
MCA(maximum credible accident) 最大可信事故
MCB(miniature circuit breaker) 微型断路器
MCCB(moulded case circuit breaker) 塑壳式断路器
MCC(main communication center) 主通信中心
MCC(main control console) 主控制台
MCC(maintenance control center) 维护控制中心
MCC(motor control center) 电动机控制中心，马达控制中心
MCC(multi-chip carrier) 多片式载体
McCollum terrestrial ammeter 麦科卢姆接地电流计
MCCU(multiple communication control unit) 多路通信控制设备
MCD(manual control device) 人工控制设备
MCDR(maintenance control data register) 维护控制数据寄存器
MCES(main condenser evacuation system) 主冷凝器抽真空系统
MCF(million cubic feet) 百万立方英尺
MCHF(maximum critical heat flux) 最大临界热流密度
MCHFR(minimum critical heat flux ratio) 最小临界热流密度比
MC(management contract) 管理合同模式【工程建设模式】
MC=megacycles 兆周
MCM(magnetic core memory) 磁芯存储器
MCM(micro-circuit module) 微电路模块
MCM(minimum critical mass) 最小临界质量
MCM(Monte-Carlo method) 蒙特卡罗法
MCO(miscellaneous charges order) 旅费证，杂费证
MCO(moisture carryover) （蒸汽中）夹带水分
MCP(main circulation pump) 主循环泵，主泵
MCP(main control panel) 主控盘
MCP(master control program) 主控程序
MCP(measure correlate predict) 测试相关预测
MCP(multiple chip package) 多芯片组件
MCR(main control room) 主控室
MCR(master control routine) 主控程序
MCR(maximum capacity rating) 最大额定容量，最大额定出力
MCR(maximum continuous rating) 最大连续出力【DCS 画面】，最大持续功率，最大持续额定功率［出力］，最大连续出力［功率］【电机，旋转机械】，最大连续蒸发量
MCR(maximum continuous revolution) 最高连续转速
MCR(mean conversion ratio) 平均转换比
MCR(multi-channel receiver) 多信道接收机
MCRR(maintenance control retry register) 维护控制再试寄存器
MCR steam flow 最大蒸汽流量
MCS(magnetic core storage) 磁芯存储器
MCS(main control system) 主控系统
MCS(management command system) 管理命令系统
MCS(master control system) 主控系统
MC/S (megacycles per second) 兆周每秒，兆赫
MCS(modulating control system) 模拟量控制系统【DCS 画面】
MCS(multi channel scaling) 多路（道）定标
MCS(multiple console support) 多操作台支援
MCU(memory control unit) 存储控制器
MCU(microprogram control unit) 微程序控制器
MCV(main control valve) 主控制阀
MCV(minimum critical volume) 最小临界体积
MDA(minimum detectable activity) 最低可测活度
MDA(multi-dimensional access memory) 多维存取存储器
MDA(multi-dimensional analysis) 多维分析
MDBFP(motor driven boiler feed pump) 电动给水泵，电泵
MDCC(master data control console) 主数据控制操作台
MDCF(maximum dependable capacity factor) 最大可靠容量因子
MDC(maximum dependable capacity) 最大保证容量，最大可靠容量
MDC(medium-speed data channel) 中速数据通道
MDD(magnetic-domain device) 磁畴器件
MDE(magnetic decision element) 磁判定元件
M-derived type filter M 导出型滤波器
MDF(main distributing frame) 主配线板
MDFP(motor driven feed water pump) 电动给水泵
MDI(magnetic direction indicator) 磁向指示器
MD(managing director) 总经理
MD(man-day) 人日，人工作日
MD(modulation damper) 调节挡板
MD(modulation-demodulation) 调制解调
MDNBR(minimum departure from nucleate boiling ratio) 最小偏离泡核沸腾比
MDR(master data recorder) 主数据记录器
MDR(memory data register) 存储数据寄存器
MDR(multichannel data register) 多通道数据寄存器
MDSGOP(motor drive speed governing oil pump) 电动调速油泵
MDS(maintenance data system) 维修数据系统
MDS(malfunction detection system) 故障检测系统，故障监测系统
MDS(master drum sender) 主磁鼓传送器
MDS(microprocessor development system) 微处理机开发系统
MDS(minimum discernible signal) 最小识别信号
MDTL(modified diode transistor logic) 改进的二极管晶体管逻辑
MDT(mean down time) 平均停机时间，平均停用时间
meacon 虚造干扰设备，干扰信号发出设备
meadow bog 草地沼泽，水湿地
meadow grass rotation 草地轮作
meadow moor 草地沼泽
meadow soil 草甸土
meadow 牧场，草地，草甸
meager coal 贫煤

meager lean coal 贫煤，贫瘦煤
meagre lime 贫石灰
mean absolute deviation 平均绝对偏差
mean absolute erratum 平均绝对误差
mean absolute error 平均绝对误差
mean accumulation 平均积累
mean aerodynamic center 平均空气动力中心
mean aerodynamic chord 平均气动力弦，气动平均叶弦
mean air temperature 平均气温
mean annual climate temperature 年平均气象温度
mean annual concentration 年平均浓度
mean annual discharge 年平均流量
mean annual efficiency 全年平均效率
mean annual flood 平均年洪水流量
mean annual humidity 年平均湿度
mean annual precipitation 多年平均降水量，年平均降水量，年平均降雨量，平均年降水量
mean annual rainfall 多年平均雨量，平均年降水量，平均年雨量
mean annual range of temperature 平均温度年较差
mean annual range 多年平均变幅
mean annual runoff 多年平均径流，年平均径流(量)，平均年径流量
mean annual temperature 多年平均温度，年平均温度
mean annual water level 多年平均水位，年平均水位
mean annual wind power density 年平均风能密度
mean availability 平均可用度
mean available hours between failures 平均无故障可用小时
mean axial value 轴向平均值
mean basin height 平均流域高度
mean basin slope 平均流域坡降
mean basin width 流域平均宽度
mean beam length 平均射线长度
mean blade chord 叶片平均弦长，叶片平均翼弦
mean camber line 中弧线
mean carrier velocity 平均载气线速度
mean chord length 平均弦长
mean concentration 平均浓度
mean continuous service time 平均连续工作时间
mean current 平均电流
mean curvature 平均曲率
mean daily discharge 日平均流量
mean daily humidity 日平均湿度
mean daily maximum temperature 日平均最高温度值
mean daily stage 日平均水位
mean daily temperature 日平均气温，日平均温度
mean daily 平均每日
mean decade humidity 旬平均湿度
mean decade temperature 旬平均温度
mean depth 平均深度
meander amplitude 蜿蜒幅度
meander belt 蜿蜒带
meandering movement 蜿蜒运动
meandering of river 河道的弯曲
meandering plume model 蛇行羽流模式
meandering 弯曲（化），曲折的
meander ratio 蜿蜒比
meander wave 蜿蜒波
meander 曲流，弯液面，蜿蜒
mean deviation 平均偏差，平均偏移，均差
mean-diameter blade length ratio 叶片平均径高比
mean diameter of grain 平均粒径
mean diameter 平均直径
mean difference 平均差
mean discharge 平均流量
mean distance 平均距离
mean dose 平均剂量
mean down time 平均故障停机时间
mean dust concentration 粉尘平均浓度
mean dynamic head 平均动压头
mean Earth-Sun distance 日地平均距离
mean effective horsepower 平均有效马力
mean effective load 平均有效负荷
mean effective power 平均有效功率
mean effective pressure 平均有效压力
mean effective stress 平均有效应力
mean effective temperature 平均有效温度
mean effective value 平均有效值
mean effort 平均作用
mean environmental wind 平均环境风
mean erratum 均方误差
mean error 平均误差
mean external wind loading 外部平均风荷载
mean failure rate 平均故障率，平均失效率
mean filling factor 平均装填因子
mean flow 平均流量
mean fluctuation power 平均峰谷率，平均涨落功率值
mean fluid temperature 工质平均温度
mean forced outage duration 平均强迫停运延续时间
mean free error time 平均无故障时间
mean free path length for conduction 导热平均自由程长度
mean free path length 平均自由程长度
mean free path of electrons 电子平均自由程
mean free path of phonons 声子平均自由程
mean free path 平均自由行程，平均自由通路，(分子）平均自由程
mean generation time 平均每代中子寿期
mean geometric chord of airfoil 平均几何弦长
mean geometric chord 平均几何弦
mean grain size 平均粒径
mean gross head 平均毛水头
mean heat capacity 平均热容量
mean heat flux 平均热负荷，平均热流，平均热流密度
mean heat input 平均热量输入
mean heat transfer coefficient 平均传热系数
mean height difference 平均高差
mean height of apparent sea 平均视浪高
mean higher high water 平均高高潮
mean highest discharge 平均最大流量
mean highest temperature 平均最高温度

mean high tide 平均高潮
mean high water neap 平均小潮高潮面
mean high water spring 平均大潮高潮面
mean high water 平均高水位
mean hourly wind speed 平均小时风速
mean hydraulic radius 平均水力半径
mean incidence 平均入射角，平均攻角，平均冲角
mean indicated pressure 平均指示压力
meaningless order 无意义指令
meaning 含义，意义
mean intensity 平均强度
mean interdiurnal variation 平均日际变化
mean internal wind loading 平均内部风荷载
mean in vertical 沿垂线平均
mean ionization energy 平均电离能量
mean land level 大陆平均高度
mean length of turn （线）匝的平均长度
mean lethal dose 半数存活剂量，平均致死剂量
mean life rate 平均寿命率
mean lifetime 平均寿命，平均寿期
mean life 平均使用时间，平均年限，平均寿命（期）
mean linear power density 平均线功率密度
mean linear range 平均直线射程
mean line 中弧线
mean load 平均负载，平均负荷
mean logarithmic energy loss 平均对数能量损失
mean logarithmic temperature difference 平均对数温差
mean lower low water 平均低低潮
mean lowest discharge 平均最小流量
mean lowest low water 平均最低枯水位
mean lowest temperature 平均最低温度
mean lowest water level 平均最低水位
mean low tide 平均低潮
mean low water level 平均低水位
mean low water neap 平均小潮低潮面
mean low water spring 平均大潮低潮面
mean low water 平均低潮水位
mean magnetization curve 平均磁化曲线
mean maintenance outage duration 平均维护停运延续时间
mean map 平均值图
mean mass range 平均质量射程
mean maximum air velocity 最大平均空气速度
mean maximum 平均最大
mean meridional circulation 平均径向环流
mean minimum flow 平均枯水流量
mean minimum 平均最小
mean molecular velocity 平均分子速度
mean monthly air temperature 月平均气温
mean monthly discharge 月平均流量
mean monthly humidity 月平均湿度
mean monthly maximum temperature 月平均最高温度
mean monthly minimum temperature 月平均最低温度
mean monthly stage 月平均水位
mean monthly temperature 月平均温度
mean monthly water level 月平均水位
mean monthly wet-bulb temperature 月平均湿球温度
mean neap 平均小潮
mean net head 平均净水头
mean neutron lifetime 中子平均寿命
mean nine-month flow 平均低水流量
mean noon 平正午
mean of valve loop curve 平均调节阀组合特性曲线
mean or maximum wind velocity 平均或最大风速
mean outage duration 平均停运持续时间
mean output 平均出力
mean parallax 平均视差
mean parameter 平均参数
mean particle size 平均粒度
mean planned outage duration 平均计划停运延续时间
mean point of group 点群平均点
mean potential water power 平均水力蕴藏量
mean power 平均功率
mean pressure 平均压力
mean principal stress 平均主应力
mean proportional 比例中项
mean radiant temperature 平均辐射温度
mean range 平均射程，平均自由程，平均变幅
mean recession curve 平均退水曲线
mean recurrence interval 平均重现期
mean refueling rate 平均换料率
mean relative deviation 平均相对偏差
mean relative humidity 平均相对湿度
mean relief ratio of basin 流域平均起伏比
mean repair rate 平均修复率
mean residence time 平均停留时间
mean response 平均响应
mean-root-square error 均方根误差
mean runoff 平均径流量
mean salinity 平均含盐度
mean sampling accuracy 平均抽样精确度
mean sea level datum 平均海面基点
mean sea level 平均海平面，平均海拔，平均海面
mean slope 平均坡度
means of egress 疏散设施
means of escape 避难措施，疏散设施
means of evacuation 疏散设施
means of security 保全措施
means of transportation 运输工具
mean solar day 平太阳日
mean solar time 平（均）太阳时
mean specific heat 平均比热
mean speed 平均速度，平均速率
mean spring range 平均大潮差
mean spring rise 平均大潮高度
mean spring 平均大潮
mean square amplitude 均方根幅值
mean squared departure 均方偏差
mean square deviation 均方差，均方偏差，方差
mean-square error criteria 均方误差准则
mean square error 均方差，均方误差【反映估计量与被估计量之间差异程度的一种度量】
mean square response 均方响应

mean square value 均方值
mean square velocity 均方速度
mean square wind speed 均方风速
mean square 均方
mean steepness 平均陡度
mean stress component 平均应力分量
mean stress 平均应力
mean subarea velocity 部分断面平均流速【两侧流垂线间】
mean sun 平太阳
mean surface temperature 表面平均温度
means 手段，方法，措施，工具，设备，方式
mean temperature difference 平均温差
mean temperature 平均温度
mean-term storage 中期贮存
mean throat 中间喉［颈］部
mean tide level 平均潮位
mean tide rise 平均潮升
mean time before failures 故障前平均时间
mean time between critical failures 平均严重故障间隔时间
mean time between defects 平均故障间隔时间
mean time between degradations 平均衰变间隔时间
mean time between detections 平均（故障）检测时间
mean time between errors 平均错误间隔时间，误差码间平均时间，误差信号间隔平均时间
mean time between failures of auxiliary equipment 辅助设备平均无故障可用小时［工作时间］
mean time between failures （机组的）平均无故障可用小时，平均故障间隔时间，平均无故障时间，平均失效间隔时间
mean time between forced outage 故障停用平均间隔时间，强制停运平均间隔时间
mean time between incidents 平均异常现象间隔时间
mean time between interruption 平均中断间隔时间
mean time between maintenance actions 平均维修间隔时间
mean time between maintenance 平均维修间隔时间
mean time between malfunction 平均误动作间隔时间
mean time between outages 平均停运间运行时间
mean time between overhauls 平均检修间隔时间
mean time between removals 平均拆换间隔时间
mean time between repairs 平均修理间隔时间
mean time between scrams 紧急停堆平均间隔时间
mean time to failure 平均失效时间，平均故障时间，平均故障间隔时间，平均无故障时间，故障前平均（工作）时间，平均失效前时间，平均初次出故障时间
mean time to first failure 平均首次失效前时间，首次故障前平均时间，最初故障平均时间，平均首次故障时间
mean time to forced outage 平均强迫停运间隔时间
mean time to planned minor repair outage 平均计划小修停运间隔时间
mean time to planned outage 平均计划停运间隔时间
mean time to planned overhaul outage 平均计划大修停运间隔时间
mean time to repair 平均修复时间，平均维修时间，平均故障修理时间
mean time to restoration 平均恢复前时间，复原平均时间
mean time to restore 平均故障修复时间，平均修复时间，平均故障修理时间
mean time to unplanned outage 平均非计划停运延续时间
mean time 平均时间
mean turns 平均匝数
mean unavailability 平均不可用度
mean unplanned outage duration 平均非计划停运小时，平均非计划停运延续时间
mean value of periodic quantity 周期量的平均值
mean value process 平均值法
mean value theorem 均值定理
mean value 平均值
mean variation 平均偏差
mean velocity curve 平均流速关系曲线
mean velocity defect 平均速度亏损
mean velocity in section 断面平均流速
mean velocity on a vertical 垂线平均流速
mean velocity point 平均流速点
mean velocity profile 平均速度廓线
mean velocity 平均流速
mean water level 平均水位
mean water stage 平均水位
mean wave height 平均波高
mean wetted length 平均浸湿长度
mean wind profile 平均风速廓线
mean wind speed 平均风速
mean wind velocity 平均风速
mean wind 平均风速
mean winter temperature 冬季平均温度
mean yearly discharge 年平均流量
mean yield 平均产量，平均产出
mean 平均值，平均数，平均，平均的，中间，中间的，中数，中项，手段
MEAS＝measure 测量
measurability 可测性
measurable quantity 可测量
measurable 可测量的，可计量的，可测的，适度的
measurand analogue value x 测量模拟量值 x【功能约束】
measurand 被测对象，被测量，被测状态，可测量
measure correlate predict 测试相关预测
measured accuracy 测量精度
measured amount 测得的数量
measured deviation 测量偏差
measured discard 测定的废弃物
measured discharge 实测流量
measured drawing 竣工图
measured hole 测量孔，测压孔

measured overshoot 测定超调量
measured power curve 测量功率曲线
measured pressure 测定压力
measured quantity 被测量，实测量
measured relieving capacity 被测的泄放容量
measured resources 确定资源
measured runoff 实测径流
measured signal converter 测量信号转换器
measured signal 测定信号，被测信号，测量信号
measured stage 观测水位
measured stress 测定应力
measured ton 体积吨，丈量吨
measured value of a quantity 量的测得值
measured value 实测值，测定值，测量值，被测值，测得量，测得值
measured variable 被测（变）量，可测变量
measured 测定的，量过的，几经推敲的，有分寸的，被测的，测量的，标准的
measure equation 测量公式
measureman 测量工
measurement accuracy 量测精度，量测准确度
measurement adjustment 测量平差
measurement and assessment of power quality characteristics of grid connected wind turbines 并网风电机组功率质量特性测试与评价
measurement and instrument 测量与仪表
measurement arrangement 测量布置
measurement at several different places 在几个不同地点测量
measurement cable 测量电缆
measurement cargo 容积货物
measurement chamber 测量室
measurement channel 测量通道
measurement control system 测量控制体系
measurement control 测量控制
measurement cross-section 测流断面
measurement device 测量设备，量测装置，测量装置
measurement equipment 测量设备
measurement error 测量误差，量测误差
measurement hardware 测量硬件
measurement indicator 测量指示仪
measurement line 测量线
measurement mast 测风杆
measurement mechanism 测量机构，测量装置
measurement of DC resistance 直流电阻测定
measurement of flow rate 流量测量，流量量测
measurement of high impulse voltage 冲击高电压测量
measurement of humidity 湿度测定
measurement of insulation resistance 测量绝缘电阻
measurement of precise height difference 精密高差测量
measurement of size of aerosol 气溶胶粒径测定
measurement of soot and dust concentration 烟尘浓度测量
measurement of speed 速度测量
measurement of time 时间测量
measurement of tube inner diameter 管子的内径测量
measurement of water pollution 水污染测定
measurement parameters 测量参数
measurement period 测量周期
measurement point 测点
measurement precision 量测精度
measurement processor system 测量处理器系统
measurement process 测量过程
measurement range 量程，测量范围，量测范围
measurement result 测量结果
measurement seat 测量位置
measurement sector 测量扇区
measurement set 测量装置
measurement signal 测量信号
measurements reactor 测量用反应堆
measurement standard 计量标准
measurement system 测量系统
measurement technique 量测技术
measurement tonnage 载货容积吨数
measurement transducer 量测变换器
measurement unit 测量设备，测量仪器，测量单位，计量单位
measurement update 测量校正
measurement value 测定值
measurement 测定，测量，测量方法，测量结果，量度，尺寸，度量，量测，丈量
measure of effectiveness 有效性测量
measure of precision 精确测量
measure of spread 概率分布测度
measure of stability 稳定度测量
measurer 量度器，测量器，测量员
measures and weights 权度，度量衡
measures expense 措施费
measures 办法，措施，举措
measure the time by the second 以秒为单位计算时间
measure 测量，测定，计量，手段，办法，措施，测，测度，量度，量度方法，量器，量算，权宜之计
measuring accuracy 测量精度，测量精确度，量测精度
measuring amplifier 测量放大器
measuring and test equipment 测试设备
measuring and testing technique 测试技术
measuring and weighing room 测量与称重室
measuring apparatus 测量工具，测量装置，测量器械，测量仪器，测量仪表，量测器具
measuring appliance 测量器具，测量设备
measuring assembly 测量组件
measuring basis 测量基准
measuring bin 量斗
measuring bolt 量测用螺栓
measuring bridge 测量电桥，测试电桥
measuring buoy 测验浮标
measuring buret 量液滴定管
measuring chain 测量链，测链
measuring chamber guide tube 测量电离室导管
measuring chamber 测量室
measuring channel 测量通道，测量回路，量水槽
measuring circuit 测量电路，测量回路

measuring coil 测量线圈，测试线圈
measuring column 测量仪表柱
measuring cup 量杯
measuring current 测量电流，测试电流
measuring cylinder 量筒
measuring dam 量水坝
measuring data 测量数据
measuring device 测量仪器，测量装置，测量设备，量测装置，量具
measuring diaphragm 测量膜片
measuring dimensions 测量尺寸，测量范围
measuring electrode 测量电极
measuring electronics 测量电子学
measuring element 测量机构，测量元件
measuring equipment 测量设备，测试设备，计量设备，量测装置
measuring error 测量误差，量测误差
measuring flask 容量瓶，量杯，量瓶
measuring flume 测流槽，量水槽
measuring gauge 量规
measuring glass 玻璃量杯，量杯
measuring graduate 刻度量筒
measuring grid 量测格网
measuring head 测量头
measuring hole 测量孔
measuring impulse 测量脉冲
measuring installation 测量设备，测量仪表，测量仪器，计量设备，量测仪器，量具
measuring instrument 测量仪表
measuring junction 测量端，测量接点
measuring lag 测量滞后
measuring line 测量线，测绳，测线
measuring loop 测量回路
measuring mark 测标
measuring mast 测量桅杆，测风塔
measuring means 测量方法
measuring mechanics 测量机构
measuring meter 测量仪表，测量表计
measuring method 测量方法
measuring microscope 量度显微镜，测量显微镜，测微镜
measuring nozzle 测量喷嘴
measuring of air gap between generator stator and rotor 发电机空气间隙测量
measuring of phase sequence 测量相序
measuring orifice 测量孔板，测量用孔口
measuring oscilloscope 测量示波器
measuring panel 仪表屏
measuring peg 测桩
measuring pipette 量液吸液管
measuring pitch 量距
measuring plane 测量面
measuring point 量测点，测点，测试点，测量点
measuring potential 测量电势
measuring potentiometer 测量用电位计，测定电位计，测定电势计
measuring probe insertion and removal tool 测量探头的插入和取出工具
measuring probe 测量探头
measuring pump 剂量泵
measuring rain station 雨量站
measuring range higher limit 测量范围上限值
measuring range lower limit 测量范围下限值
measuring range 测量范围，量程，量测范围
measuring reach 量测河段
measuring reactor 测量反应堆
measuring reel 量测卷尺
measuring relay 测量继电器
measuring resistance 标准电阻，测量电阻
measuring rod 测量杆，测杆
measuring roller 量测鼓轮
measuring rope 测绳，测索
measuring scale 量尺，标尺，刻度尺
measuring section 测验断面，量测断面
measuring sensitivity 测量灵敏度
measuring sequence 测量顺序
measuring signal 测量信号
measuring slide rheostat 测量用滑线变阻器
measuring slide wire 测量用滑线电阻
measuring span 量程
measuring sparkgap 测量用火花放电器
measuring sphere gap 测量球隙
measuring staff 测杆
measuring station 测量站，测位
measuring system 测量系统，测量装置
measuring table 测量台
measuring tank 计量箱，测箱
measuring tape 测尺，卷尺，测带
measuring technique 测量技术，测试技术，量测技术
measuring tool 测量工具，量具
measuring transducer 测量传感器
measuring transformer 仪用变压器，测量用变压器，仪用互感器，测量用互感器
measuring unit 测量装置
measuring voltage 测量电压，测试电压
measuring weir flume 量水堰槽
measuring weir 测量溢流，量水堰
measuring well 量测井
measuring wheel 测轮
measuration 测定，测量，测量方法，求积法
meat 要点，内容，实质，(燃料元件的)燃料部分
MEB(moisture extracting blade) 去湿叶片
mecarta 胶木
Meccalli scale 麦氏地震烈度
mechanical accelerated clarifier 机械加速澄清器
mechanical acceleration clarifying basin 机械加速澄清池
mechanical acoustical coupling 机声耦合
mechanical actuator system 机械执行系统
mechanical admittance 机械导纳，力导纳
mechanical admixture 机械杂质
mechanical advantage 机械效益
mechanical aeration basin 机械曝气池
mechanical aeration system 机械曝气系统
mechanical aeration 机械曝气，机械通气
mechanical aerator 机械曝气器
mechanical ageing test 机械老化试验
mechanical agitation 机械搅拌
mechanical agitator 机械搅拌器
mechanical air separation 机械空气分离

mechanical air supply system 机械送风系统
mechanical air supply 机械送风
mechanical analog computer 机械模拟计算机
mechanical analogue computer 机械模拟计算机
mechanical analogy 机械模拟
mechanical analysis curve 机械分析曲线，颗粒分析曲线，粒径分析曲线
mechanical analysis of heat-supply pipes 供热管道强度计算
mechanical analysis 机械分析，力学分析
mechanical and electrical protection 机电保护
mechanical and hydraulic combined dust 机械、水力联合除尘
mechanical anemometer 机械式风速仪
mechanical angle 机械角
mechanical arm 机械臂
mechanical ash handling system 机械除灰系统
mechanical atomization 机械雾化
mechanical atomizer burner 机械雾化燃烧器，机械雾化喷燃器
mechanical atomizer 机械雾化器
mechanical atomizing oil burner 机械雾油燃烧器
mechanical attraction 机械引力
mechanical axis 机械轴
mechanical balance 机械平衡
mechanical bank 机械存储器
mechanical behavior 受力性能，机械性能，力学性能
mechanical bias 机械偏置
mechanical blower 机械鼓风机
mechanical blowpipe 自动割炬，自动焊炬
mechanical bolt 螺钉
mechanical bond 机械接头，机械结合，机械砌合
mechanical brake 机械制动器，机械测功器
mechanical braking system 机械制动系统
mechanical breakdown 机械故障
mechanical calculation of conductor 导线力学计算
mechanical calculation 机械计算
mechanical cam type programmer 凸轮式机械程序执行机构
mechanical carryover 机械携带
mechanical-centrifugal speed governor 机械离心式调速器
mechanical centring 机械对中
mechanical characteristic of motor 电动机机械特性
mechanical characteristic 机械特性，机械特性曲线，机械性能
mechanical charge 装料机
mechanical check 机械检查
mechanical chief engineer 机务主工
mechanical clarification 机械澄清法
mechanical cleaning device 机械清污设备
mechanical cleaning off dust 机械除尘
mechanical cleaning sedimentation basin 机械清淤沉沙池
mechanical cleaning 机械清理，机械清洗，机械清扫
mechanical clutch 机械离合器【盘、鼓等】
mechanical collector 机械除尘器
mechanical commutation 机械换向，机械整流
mechanical commutator 机械转换器，机械换向器
mechanical completion certificate 安装验收证书，机械安装竣工证书
mechanical compliance 力顺，机械顺从性
mechanical components of PWR nuclear islands 压水堆核岛机械部件
mechanical computer 机械计算机
mechanical connection 机械连接
mechanical contractor 机械合同商
mechanical control and protection 机械控制和保护
mechanical control 机械控制
mechanical convection 机械对流，强制对流
mechanical converter 机械变换器
mechanical cooling tower 机力通风冷却塔，机械冷却塔
mechanical counter wheel 机械数字轮，机械计数轮
mechanical counter 机械计数器
mechanical coupling 机械接头，机械耦合
mechanical current meter 机械传动流速仪
mechanical damage 机械损伤
mechanical damper 机械减震器，机械阻尼器
mechanical damping 机械阻尼，机械减震
mechanical decladding 机械脱壳，机械除皮，机械去壳
mechanical defect 机械缺陷，机械损坏
mechanical dehydration 机械脱水
mechanical dejacketing 机械去壳
mechanical deposit 动力沉积，机械沉积物
mechanical design flow 机械设计流量
mechanical deterioration 机械性损坏
mechanical device 机械装置
mechanical dewatering 机械脱水
mechanical differential 机械差动装置
mechanical draft air cooled condenser 机力通风空冷系统
mechanical draft cooling tower 机力通风冷却塔，机力塔
mechanical draft wet cooling tower 机械通风湿式冷却塔
mechanical draft 机械排风
mechanical draught cooling tower 机械通风冷却塔
mechanical draught 机械通风
mechanical drawing 机械图【对应于电气或土建图】，机械制图
mechanical-driven turbine 机械驱动用透平
mechanical drive system 机械传动系统
mechanical durability 机械寿命，机械强度
mechanical dust collector 机械除尘器
mechanical dust extractor 机械除尘器
mechanical dust removal 机械除尘
mechanical dust separator 机械除尘器
mechanical efficiency 机械效率
mechanical efficient 机械效率
mechanical-electrical transducer 机电变换器
mechanical endurance 机械寿命
mechanical energy 机械能

mechanical engineering 机械工程学，机械工程
mechanical engineer 机械工程师
mechanical equilibrium 机械平衡
mechanical equipment for building 建筑的机械装备
mechanical equipment schedule 机械设备表
mechanical equipment 机械设备
mechanical equivalent of heat 热功当量
mechanical equivalent 机械功当量，热功当量
mechanical erosion 机械磨损
mechanical excitability 机械灵敏性
mechanical exhaust system 机械排气系统，机械排风系统
mechanical exhaust 机械排风，机械排气
mechanical extensometer 机械引伸计
mechanical failing load 机械破坏负荷
mechanical failure 机械故障
mechanical fastener 机械皮带扣
mechanical fatigue test 机械疲劳试验
mechanical fatigue 机械疲劳
mechanical fault 机械故障
mechanical features 机械性能
mechanical feedback 机械反馈
mechanical filter 机械过滤器，机械滤波器
mechanical final control element 机械末控元件
mechanical firing 机械锅炉点火
mechanical flywheel energy storage 机械飞轮储能
mechanical friction coefficient 机械摩擦系数
mechanical generator 机械力发电机
mechanical governor 机械调速器
mechanical grate 机械化炉排
mechanical grit arrester 机械除尘器
mechanical ground 机械基础，机械接地
mechanical handling system 机械搬运系统
mechanical handling 机械搬运，机械操作
mechanical hot-channel factor 机械热通道因子
mechanical hydraulic control system 机械液压调节［控制］系统
mechanical hydraulic control 机械液压式控制
mechanical hydraulic governor 机械液压联合调速器
mechanical hysteresis 机械滞后
mechanical impact 机械冲撞
mechanical impedance 机械阻抗，力阻抗
mechanical impurity 机械杂质
mechanical industry 机械工业，机械行业
mechanical injury 机械损伤
mechanical integrator 机械积分器
mechanical interlocking 机械联锁，机械联动
mechanical interlock relay 机械联锁继电器
mechanical joint 机械接头，机械接合，机械连接，外线端钮
mechanical kinetic energy 机械动能
mechanical latching relay 机械自锁继电器，机械闩锁继电器
mechanical-levertype scale 机械杠杆式皮带秤
mechanical life 机械寿命
mechanical line-up 机械对中，机械找正
mechanical linkage 机械连接，机械联动
mechanical linking 机械连接，机械耦合
mechanical loader 装载机
mechanical load measurements 机械载荷测量
mechanical load 机械载荷
mechanical locker relay 机械闭锁继电器
mechanical locks 机械保险装置，机械锁
mechanical loss 机械损耗，机械损失
mechanically-cleaned rack 机械清除的格栅
mechanically expanded antivibration bar 机械膨胀的防震条
mechanically fired boiler 机械化燃烧锅炉
mechanically-held contactor 机械保持接触器
mechanically laminated member 机械层板构件
mechanically operated bin gate 机动料仓闸门
mechanically operated controller 机械操作控制器
mechanically operated 机械操作的
mechanically switched capacitor 机械开关电容器
mechanically switched shunt capacitor 机械投切并联电容器
mechanically timed relay 机械定时式继电器
mechanically 机械的，机械地
mechanical magnification factor 机械放大因子
mechanical maintenance 机械维修保养
mechanical mass 力质量
mechanical meter 机械（传动）仪表
mechanical mixer 机械混合器
mechanical mixing 机械拌和，机械混合
mechanical monochromater 机械单色器
mechanical movement 机械运动
mechanical multiplication 机械乘法
mechanical multiplier 机械乘法器
mechanical neutron velocity selector 中子速度机械选择器
mechanical noise 机械噪声
mechanical operation 机械操纵，机动
mechanical oscillation 机械振荡，机械振动
mechanical parameter 机械参数
mechanical part 机械零件，机械配件
mechanical penetration 机械贯穿件
mechanical penthouse 屋顶机房
mechanical performance test 机械性能试验
mechanical pile press machine 机械式压桩
mechanical pipe joint 管道的机动接头
mechanical pitching mechanism 机械调桨机构
mechanical pitting 麻点，蚀点
mechanical plug disintegration 机械塞碎裂
mechanical plug expander 机械塞扩管
mechanical plugging 机械堵管，机械封堵
mechanical plug 机械塞
mechanical poisoning （离子交换）机械中毒
mechanical powder 机械喷粉
mechanical power control 机械功率控制
mechanical power 机械功率
mechanical precipitator 机械除尘器
mechanical predictor 机械预报器
mechanical property test 机械性能试验
mechanical property 机械性能，力学性能，力学性质
mechanical protection 机械保护
mechanical pump 机械泵
mechanical quality factor 机械品质因数

mechanical rake 机械清污耙
mechanical rammer 机械夯
mechanical reactance 力抗
mechanical rectification 机械整流，机械换向
mechanical rectifier 机械式整流器，机械整流器
mechanical register 机械记录器，机械寄存器
mechanical relay 机械继电器
mechanical remote control 机械遥控
mechanical resistance 机械阻力，力阻
mechanical responsiveness 力导
mechanical restraint 机械阻尼器，机械约束
mechanical rollers 机械芯辊
mechanical rotor brake 机械风轮制动器
mechanical sample collection 机械采样
mechanical sampling and preparing system 机械采制样装置
mechanical sampling device 机械采煤样装置
mechanical sampling 机械采样
mechanical sand control 机械固沙，机械防砂
mechanical scanning 机械扫描
mechanical seal circulating system 机械密封循环系统
mechanical seal circulation 机械密封循环
mechanical seal cover plate 机械密封盖板
mechanical seal cover 机械密封压盖，机械密封盖
mechanical sealed main coolant pump 机械密封的主冷却泵
mechanical seal flushing system 机械密封冲洗系统
mechanical seal flushing 机械密封冲洗
mechanical sealing 机械密封
mechanical seal ring 机械密封环
mechanical seal section 机械密封段
mechanical seal shaft sleeve 机械密封轴套
mechanical seal water 机械密封水
mechanical seal 机械密封
mechanical sedimentary deposit 机械沉积矿床
mechanical sediment 机械沉积物
mechanical seismograph 机械式地震计
mechanical servo 机械伺服
mechanical sewage treatment 污水机械处理法
mechanical shaft seal sodium pump 机械轴封式钠泵
mechanical shock resistance 抗机械冲击性能
mechanical shock 机械冲击，机械碰撞
mechanical shop 机械车间
mechanical shovel 机械铲
mechanical sifter 机动筛
mechanical similarity 力学相似性
mechanical smoke control 机械控烟
mechanical sodium pump 机械钠泵
mechanical speed governor 机械式调速器
mechanical splice 机械拼接【钢筋】，机械接头
mechanical spreader 撒布机
mechanical stability 机械稳定性
mechanical stabilization 机械加固法
mechanical stabilizer 机械稳定器
mechanical stiffness 力劲
mechanical stirrer 机械搅拌器
mechanical stirring 机械搅拌
mechanical stoker 机械化炉排，机械加煤机
mechanical stop 机械制动，机械阻塞
mechanical storage 机械存储器
mechanical strength 机械强度，力学强度
mechanical stress relieving 消除机械应力
mechanical stress 机械应力
mechanical stretching 机械张拉
mechanical supervisor 机械监理
mechanical switching devices 机械开关装置
mechanical system 力学体系
mechanical test 力学试验，机械试验
mechanical thermal conversion 机械热转换
mechanical thickener 机械浓缩器
mechanical timer 机械计时器，机械定时器
mechanical torque transient 机械转矩暂态
mechanical torque 机械力矩
mechanical transfer function 力传递函数
mechanical translation 机械翻译
mechanical trap 机械式疏水阀【疏水器】
mechanical treatment 机械加工，机械处理
mechanical trowel 抹灰机
mechanical turbulence 机械湍流
mechanical-type exhauster 机械式排气风机
mechanical variable speed drive 机械变速拖动装置
mechanical ventilating system 机械通风系统
mechanical ventilation 机械排风，机械校验
mechanical vibration 机械振动
mechanical wave filter 机械滤波器
mechanical wear 机械磨损，机械磨耗
mechanical weathering 机械风化
mechanical work 机械功
mechanical zero adjuster 机械零位调整器
mechanical zero 机械零位，机械零点
mechanical zinc deposit 机械镀锌
mechanical 机械学的，力学的，机械的，机械
mechanic cutting product line 机械切割生产线
mechanician 机械工人，机械员，技师，机工
mechanic impurity 机械纯度
mechanics of elasticity 弹性力学
mechanics of frozen soil 冻土力学
mechanics of landslide 滑坡力学
mechanics of materials 材料力学
mechanics of metals 金属力学
mechanics of plasticity 塑性力学
mechanics of rigid bodies 刚体力学
mechanics 机械工程学，力学，结构
mechanic 机械工人，机械员，机修工，技工，机械师，机械的
mechanism braking system 机械制动系统
mechanism of cavitation 气蚀机理
mechanism of combustion 燃烧机理
mechanism of creep 蠕变机理
mechanism of landslide 滑坡机理
mechanism of poisoning 中毒机理
mechanism of reaction 反应机理
mechanism of self-purification 自净机理
mechanisms of delivery 实施办法
mechanism 机械装置，机构学，机构，装置，机械，机制，机理
mechanistic model 机理模型

mechanist 机械师
mechanization 机械化
mechanized grate boiler 机械化炉排锅炉
mechanized welding 机械化焊接
mechano-caloric 机械制热的，热功的
mechano-electric 机电的
mechano-electronic transducer 机电变换器
mechatronic system 机电一体化系统
mechatronics 机电一体化，机械电子学，机械电子装置，机械电子技术
mechatronic 机电整合的，机电一体化的，机械电子的
medal 奖（章）
media access control 介质访问控制
media center 媒介中心
medial camber line 中弧线
media 媒质【medium 的复数】
median diameter 中数粒径，中值粒径，中值直径
median discharge 中数流量，中值流量
median elevation of basin 流域中数高程
median fatigue life 中值疲劳寿命
median fatigue strength at N cycles N 次循环的中值疲劳强度
median lethal concentration 中间致死浓度
median lethal dose 半数致死剂量，中间值致死剂量
median lethal time （辐射）平均致死时间，半数致死时间，中间值致死时间
median monthly runoff 中值月径流
median of sounds level 噪声声级的中间值
median panicle size 中数粒径
median size 中值粒径
median stream flow 中值流量
median tolerance level 中间容许水平
median water level 中水位
median wind speed 中值风速
median year 平均年，中常年
median 中值，中位数，中数
media-pressure feed water heater 中压加热器
Mediaterranean climate 地中海气候
mediation decision 调解书
mediation 调解，调停，仲裁
mediator 调解人
medical care 医疗服务
medical control 医疗管理
medical exposure 医疗辐照，医疗照射
medical facility 医疗设施
medical research reactor 医疗研究用反应堆
medical service 医疗服务
medical supervision 医务监督，医学监督
medical surveillance 医疗监督，医学监视
medical therapy reactor 医疗用反应堆
medium access control 介质访问控制，通信介质通路控制
medium active waste 中放废物，中等放射性废物
medium-activity 中等放射性
medium-alloy steel 中合金钢
medium and long term plan 中长期计划
medium and small enterprises 中小企业
medium-arrangement of outdoor switchgear 中型屋外配电装置
medium attachment unit 介质附属单元【收发器】
medium caking coal 中等黏结煤
1/2 medium caking coal 1/2 中黏煤
medium-capacity plant 中容量电站
medium-carbon steel 中碳钢
medium clayloam 中黏壤土
medium consistency 中等稠度
medium construction item 中型建设项目
medium construction project 中型建设项目
medium current 中强度电流
medium dam 中坝
medium clay 中黏土
medium density 中密度
medium diameter 中值粒径，中粒径
medium duty 中级负荷，中型，中等负载
medium-energy lightwater-moderated nuclear reactor 轻水慢化中能核反应堆
medium energy 中等能量
medium expansion foam extinguishing system 中倍数泡沫灭火系统
medium expansion foam 膨胀泡沫材料
medium-fast sweep 中速扫描
medium flow direction arrow 介质流向箭头
medium force fit 中级压紧配合
medium frequency changer 中频变换器
medium frequency generator 中频振荡发生器，中频发电机
medium frequency high voltage generator 中频高压发生器
medium frequency induction furnace 中频感应炉
medium frequency rotary converter 中频旋转换流机
medium frequency transformer 中频变压器
medium frequency 中频
medium fuel oil 中级燃料油
medium grade 中等，中级
medium grained 中等粒度的
medium granular sand 中颗粒砂
medium granular 中等粒度的
medium gravel 中砾石
medium head hydraulic turbine 中水头水轮机
medium head hydroelectric power station 中水头水电站
medium head power plant 中水头电钻
medium head safety injection pump 中等压头安全注入泵
medium head turbine 中水头水轮机
medium-heavy loading 中等加载
medium-height trestle 中等高度支架
medium-high frequency 中高频率
medium hydraulic turbine 中型水轮机
medium hydroelectric power station 中型水电站
medium icing area 中冰区
medium level waste 中等放射性废物，中放废物
medium level work 中（等）放（射性）操作
medium-lift pump 中等扬程泵，中压泵
medium line 中线
medium-lived （放射性）中等寿命的
medium maintenance 中修
medium of international payment 国际支付手段

medium or long term credit 中长期信用
medium power reactor 中等功率反应堆
medium power 中等功率
medium pressure acetylene generator 中压乙炔发生器
medium pressure auto close gear 中压自动关闭器
medium pressure automatic closing gear 中压自动关闭器
medium pressure boiler 中压锅炉
medium pressure element 中压缸【汽轮机】，中压元件
medium pressure governor valve 中压调速汽门
medium pressure hydraulic actuator 中压油动机，中压液压传动装置
medium pressure section 中压缸【汽轮机】，中压段
medium pressure servo-motor 中压油动机，中压伺服马达
medium pressure turbine 中压汽轮机
medium pressure 中压
medium-profile layout 中型布置
medium range forecast 中期预报
medium range 中程，中距离
medium repair of heat-supply network 热网中修
medium repair 中修
medium reservoir 中型水库
medium rutile covering 中钛型药皮
medium sand 中砂
medium scale integrated circuit 中规模集成电路
medium scale integration 中规模集成【电路】
medium scale wind energy conversion system 中型风能转化系统
medium scale wind turbine 中型风电机组，中型风力机
medium series 中系列
medium short wave 中短波
medium-sized coal 中块煤
medium-sized wind turbine 中型风电机组，中型风力机
medium-sized 中型的，中号的，中等的
medium size motor 中型电动机
medium size turbine generator 中型汽轮发电机
medium size 中粒
medium-speed data channel 中速数据通道
medium-speed film 中速胶片
medium-speed mill 中速磨煤机
medium-speed planetary gear 中速行星齿轮
medium-speed pulverizer 中速磨煤机
medium-speed rotor 中速风轮
medium-speed warming 中速暖机
medium-speed 中速
medium standard frequency 标准中频
medium temperature tempering 中温回火
medium tension bushing 中压套管
medium term investment 中期投资
medium term loan 中期放款
medium transmission line 中长输电线路
medium volatile bituminous coal 中挥发分烟煤
medium voltage bushing 中压套管
medium voltage distribution line 中压配电线路
medium voltage distribution network 中压配电网
medium voltage network 中压网络，中压供电网
medium voltage switchgear cubicle 中压开关柜
medium voltage system 中间电压系统
medium voltage winding 中压绕组，中压线圈
medium voltage 中压
medium wave 中波
medium 媒体，媒质，介质，媒介，工质，介质环境，媒介物，中等的，中间的
meet a bill 支付到期的票据
meet commitment 履行承诺
meet demand 满足需求，满足要求
meet expenses 偿付开支
meeting agenda 会议日程
meeting corrosion 接触腐蚀
meeting in camera 秘密会议
meeting of minds 意思表示一致，意见一致，合意
meeting room 会议室
meeting sight distance 回车视距
meeting 会议
meet one's engagements 履行契约，偿债
meet or surpass the standard 达到或超过标准
meet the condition of 满足……条件，适合……条件
meet the draft on presentation 凭汇票付款，凭票即付，出示汇票即付款
meet the ever-increasing export need 满足日益增长的出口需要
meet the requirements of specifications 适应规范的要求
meet the requirement 满足要求，满足需要
meet the specification 合乎规格
meet with 和……会晤
meet 如期付款，偿还，相遇
megabit chip 兆位芯片
megabits per second 兆位每秒
megabit 兆比特，兆位，百万比特
megabusiness 特大企业
megabyte per second 兆字节每秒
megabyte 兆字节
megacycles per second 兆周每秒，兆赫
megacycle 兆周，百万周
megahertz 兆赫
megajoule 兆焦
megaline 兆力线【磁通单位】
megalopolis （人口高度稠密的）都市区
megameter 摇表，高阻表，兆欧表
megampere 兆安，百万安
megaohm 兆欧姆
megarad 兆拉德
megarelief 大地形，宏观地形
megaripple 大波痕
megascopic method 宏观法
megaseism 大地震
megasiemens 兆西门子
megasweep 摇频振荡器
megatectonics 巨型构造
megathermal climate 高温气候
megathermal zone 高温带
megatherm 高温植物
megatonnage 兆吨数，兆吨级

megaton 百万吨级，兆吨
megatron 塔形管
megavar 兆乏，百万乏
megavolt-ampere reactive 兆无功伏安，兆乏
megavolt-ampere 百万伏安，兆伏安
megavolt 兆伏，百万伏，百万伏特
megawatt-day per metric ton of uranium 兆瓦日每吨铀
megawatt-day 兆瓦日
megawatt-hour 兆瓦时，百万瓦（小）时，千瓦时
megawatt size wind turbine 兆瓦级风轮机
megawatt thermal 兆瓦热
megawatt 兆瓦，百万瓦，兆瓦特
MEGC＝megacycles 兆周
meggers lamp 高频电源水银灯
megger 高阻表，兆欧计，绝缘摇表，摇表，兆欧表，测高阻计高阻计
megnetic tape 磁带
megohm bridge 兆欧姆电桥
megohmit （换向器的）云母片，绝缘物质
megohmmeter 兆欧计，高阻计，摇表，兆欧表
megohm 兆欧，百万欧
megomit 绝缘物质，换向器云母片
MEGS＝megacycles 兆周
MEGV＝megavolt 百万伏特
MEGW＝megawatt 兆瓦，百万瓦
meg 兆周，兆欧，小型绝缘试验器
MEH(micro electro hydraulic control system) （锅炉给水泵小汽机）电液控制系统，小型汽轮机电液调节【DCS 画面】
meizoseismal area 极震区，强震区
meizoseismic zone 强震带
mekapion 电流计
melabasalt 暗色玄武岩
meladiabase 暗色辉绿岩
meladiorite 暗色闪长岩
melamine alky varnished glass tape 玻璃漆布带
melanosome 暗色体
melasome 暗色体
MEL(materials engineering laboratory) 材料工程实验室
melon shaped dome 瓜形圆屋顶
meltability 可熔性
meltable 可熔的，易熔的
meltdown accident 熔化事故
meltdown 熔化【燃料】，熔毁
melted iron 铁水
melt-flow index 熔流指数
melting accelerator 熔化加速器
melting and solidification 熔解和凝固
melting coefficient 熔化系数
melting condition 熔化条件
melting core catcher 熔化堆芯收集器
melting curve 熔化曲线
melting decladding 熔融脱壳，熔融去壳
melting degree 上缘熔化度
melting down detector （燃料）熔化探测器
melting equipment 熔化设备
melting furnace 熔化炉，熔炼炉
melting heat 熔化热
melting-in cyclone 熔化旋风筒
melting loss 烧损，熔损
melting of the fuel cladding 燃料包壳熔化
melting point 熔化温度，熔点，熔解点
melting pot 坩埚，熔化锅
melting rate 熔化速度
melting snow 融雪
melting temperature 熔化温度，熔点，溶化温度
melting time 熔化时间
melting vacuum gauge 麦氏真空计
melting zone 熔化区
melting 熔化，熔化状态，溶化
melt-off 熔耗的，消耗掉，熔化掉
melt-out 熔融
melt point 熔点
melt spinning method 熔融纺丝法
melt-through accident （堆芯）全部熔化事故
melt-through 整体熔化，全部熔化
meltwater 熔化水
melt 熔解，熔化，熔炼，融化，（逐渐）减少，融合，融化物，熔融物
member association 成员协会，会员协会
member body 团体会员
member in bending 受弯杆件，受弯构件
member in compression 受压杆件，受压构件
member in flexure 受弯构件
member in shear 受剪杆件，受剪构件
member in tension 拉杆，受拉杆件，受扭构件，受扭杆件
member of consortium 联合体成员，财团成员，集团成员
member of working group 工作组成员
membership 会员资格
members of the public 公众成员
member structure 杆系结构
member 元件，部分，部件，项，成员，构件
membrance wall 膜式壁
membrance 膜，隔膜
membrane analogy 薄膜模拟法，膜类比，薄膜模拟，皂膜比拟法
membrane barrier 隔膜
membrane compressor 隔膜压缩机
membrane-covered amperometric sensor 覆膜式电流测量【余氯分析仪】
membrane curing （混凝土的）薄膜养护
membrane diaphragm seal pressure gauge 隔膜式压力表
membrane equilibrium 膜渗平衡
membrane equipment 薄膜设备
membrane filter technique 膜过滤法
membrane filter 膜滤器
membrane filtration 膜过滤，膜过滤法，膜过渡
membrane grouting process 薄膜灌浆法
membrane isolation valve 薄膜隔离阀
membrane like structure 薄膜型结构
membrane method of waterproofing 防水隔膜法，止水薄膜法
membrane module 膜组件
membrane of electrodialysis 电渗析膜
membrane pack 膜堆
membrane panel 模式壁

membrane permeability　膜的渗透性
membrane poisoning　膜中毒
membrane potential　膜电位
membrane pressure gauge　薄膜式压力计
membrane process　薄膜法
membrane pump　薄膜泵，隔膜泵，膜式泵
membrane resistance　薄膜电阻，膜电阻
membrane roof　薄膜屋顶卷材防水屋面
membrane-sealed tank　薄膜密封箱
membrane-sealed　薄膜止水的
membrane seal mechanical pump　膜密封式机械泵
membrane selectivity　膜的选择性
membrane separation device　膜分离装置
membrane stack　膜堆
membrane stress　薄膜应力，膜应力
membrane tension　膜片张力
membrane theory　薄膜理论
membrane treatment　膜处理
membrane wall　膜式水冷壁，膜式壁
membrane waterproof roof　卷材防水屋面
membrane water wall　模式水冷壁
membrane　膜片，膜，薄膜，隔膜，隔板，振动片，涂层，止水墙
membranous waterproofing　薄膜防水
MEMC Electronic Materials Inc.　美国休斯电子材料公司
memistor　电解存储器
memnescope　瞬变示波器
memoir　论文集，研究报告
MEMO＝memorandum　备忘录
MAE(monitoring and evaluation)　监测和评估
memorandum book　备查簿
memorandum clause　备忘条款，附注条款
memorandum entry　备查记录
memorandum ledger　备查分类账
memorandum of deposit　存款单
memorandum of exchange　兑换水单
memorandum of interview　会谈备忘录
memorandum of meeting　会议纪要，会议备忘录
memorandum of satisfaction　赔偿或还债备忘录
memorandum of understanding　理解备忘录，谅解备忘录
memorandum record　备查记录
memorandum trade　备忘录贸易
memorandum　通知单，备忘录，便条
memorizer　存储器
memorize　存储，记忆
memory access time　存储器存取时间，存储器取数时间
memory access　存取，存储访问
memory address counter　存储地址计数器
memory address register　存储地址寄存器
memory address　存储器地址
memory allocation　存储器配置，存储分配
memory and routing function　存储定向作用
memory bank　存储器
memory block　存储区，存储块
memory buffer register, odd　奇数存储缓冲寄存器
memory buffer register　存储缓冲寄存器
memory buffer　存储缓冲器
memory bus　存储器总线
memory capacitor　存储电容器
memory capacity　存储容量
memory cell　存储单元，存储元件
memory circuit　记忆电路
memory contention　存储器争用
memory control unit　存储控制器
memory core handler　存储磁芯测试键控器
memory cycle　存储周期
memory data register　存储数据寄存器
memory decoder　存储器译码器
memory density　存储密度
memory device　存储装置
memory dump routine　存储器转储程序
memory dump　存储器转储，存储转储
memory effect　记忆效应
memory element　存储元件
memory function　存储作用，存储功能
memory gap　存储间隙，存储间隔
memory guard　存储保护
memory hierarchy　存储器分级体系，存储层次
memory information register　存储信息寄存器
memory in metal　金属存储器
memory input register　存储输入寄存器
memoryless channel　无记忆通道
memoryless transmission channel　无记忆传输通道，无记忆信道
memory location　存储单元，存储位置
memory lockout register　存储封锁寄存器，存储器封锁寄存器
memory magnetic core　存储磁芯
memory management　存储器管理
memory map list　存储变换表
memory map　存储器映像，存储变换，存储器布局图，内存（分配）图
memory matrix　存储器矩阵
memory module　存储模件
memory multiplexor　存储器多路转换器
memory net structure　存储器网结构
memory operating characteristics　存储器操作特性
memory overlay　存储器重叠
memory page　存储页面
memory port　存储器出入口
memory power　存储能力
memory print　存储打印
memory property　存储性能
memory protection　存储保护
memory-reference instruction　存储器访问指令
memory register　存储寄存器
memory relay　剩磁型磁性闩锁继电器，存储器式继电器
memory-scope　记忆式示波器，长余辉同步示波器，存储式同步示波器
memory search routine　存储搜索程序
memory-segmentation control　存储器分区控制
memory self-locked relay　记忆自锁继电器
memory source　内存资源
memory storage　存储器存储
memory swapping　（计算机用）存储交换

memory switch 存储器切换
memory system 记忆系统，存储系统
memory technology 存储器技术
memory test system 存储测试系统
memory time-delay relay 存储延时继电器
memory timer 记忆定时器
memory timing chain 存储器定时链
memory transfer （内存信息的）转储
memorytron 存储管
memory tube 存储管
memory under test 被测存储器
memory unit 存储装置，存储部件，记忆单元
memory 记忆装置，存储器，积累器，记忆，存储
memoscope 存储管式示波器
memotron 阴极射线式存储管
memo 备忘录
Mendeleev's law 门捷列夫元素周期律
mend 修补，修理，恢复
meniscus angle 新月形角
meniscus correction 弯液面校正
meniscus dune 新月形沙丘
meniscus 弯月面，弯液面
menstruum 溶剂，溶媒
mensurability 可测性
mensurable 可测量的，可度量的，有固定范围的
mensural 关于度量的
mensuration 测定法，测量法，求积法
mental attitude 精神面貌
mental 记忆的，思维的
mention 提及
menu 菜单
MEP(mean effective pressure) 平均有效压力
meq. (milligram-equivalent) 毫克当量
Mercalli intensity scale 麦加利地震烈度
Mercalli scale 麦加利地震烈度表
mercantile credit 商业信用
mercantile firm 商行
mercaptan 硫醇
mercaptobenzothiazole 巯基苯并噻唑
merchandise firm 商行
merchandise inventory 商品盘存
merchandise 货物，商品
merchant bank 商业银行
merchant-ship reactor 商船用反应堆
merchant 商人
mercoid switch 水银开关
mercoid 水银转换开关
mercurial barometer 水银气压计
mercurial pyrometer 水银高温计
mercurial thermometer 水银温度计
mercurial 水银的
mercuric chloride 氯化汞
mercuric oxide 氧化汞
mercuric sulfate 硫酸汞
mercurous chloride 氯化亚汞
mercurous oxide 氧化亚汞
mercury-arc converter 汞弧逆变器，汞弧换流阀
mercury-arc DC transformer 汞弧直流变换机
mercury-arc invertor 汞弧逆变器，汞弧逆变阀
mercury-arc lamp 水银弧光灯，汞弧灯
mercury-arc power converter 汞弧变换器
mercury-arc rectifier 汞弧整流器，汞弧整流阀，汞弧管
mercury-arc valve 汞弧阀
mercury-arc 汞弧
mercury arrester 水银避雷器
mercury barometer 水银气压计
mercury breaker 水银断续器，水银断路器
mercury cell 水银电池
mercury column 水银柱，汞柱
mercury complex 汞的络合物
mercury contact relay 水银接点继电器，水银触点继电器
mercury contact 水银开关，水银接点，汞触点
mercury converter 水银换流器
mercury cooled reactor 汞冷却反应堆，汞冷堆
mercury DC transformer 水银直流变换器
mercury discharge lamp 汞弧灯，水银放电灯，水银放电器
mercury distant-reading thermometer 远读式水银温度计
mercury electrode 汞电极，水银电极
mercury filled thermal system 充水银式感温系统
mercury-filled U-tube manometer U形水银管压力计
mercury frequency changer 水银变频器
mercury gauge 水银压力计
mercury-in-glass thermometer 水银温度计，玻璃管水银温度计，玻璃壳水银温度计
mercury intrusion test 压汞法试验
mercury inverter 水银逆变器，水银反向换流器
mercury-jet rectifier 汞流整流器
mercury lamp 水银灯
mercury manometer 水银压力计
mercury memory 汞存储器
mercury mist cooled fast reactor 汞雾冷却快中子反应堆
mercury motor meter 水银电动式安培小时表
mercury motor type 水银电动机式
mercury-needle relay 汞针继电器
mercury plunger relay 水银插棒式继电器
mercury porosimeter 水银测孔仪
mercury power rectifier 大功率水银整流器
mercury-pressure safety valve 汞压安全阀
mercury protoxide 氧化亚汞
mercury recording thermometer 水银记录式温度计
mercury rectifier 水银整流器
mercury regulating unit 水银调节装置
mercury relay 水银继电器
mercury resistor 水银电阻器
mercury safety seal 保安装置汞封
mercury seal 汞封
mercury storage system 汞存储系统
mercury storage 汞存储器
mercury switch 水银开关，水银继电器
mercury tank 水银槽
mercury telethermometer 遥测水银温度计
mercury-thallium thermometer 汞铊温度计
mercury thermometer 水银温度计
mercury tilt switch 水银偏转开关

mercury-tungsten lamp 水银钨丝灯
mercury turbine 水银汽轮机
mercury unit 汞阻单位，西门子电阻单位
mercury vacuum gauge 水银真空计
mercury-vapor frequency changer 汞汽变频器
mercury-vapor rectifier 汞弧整流器，水银整流器
mercury-vapor tube 水银充气整流器，汞汽整流器
mercury well 汞池
mercury-wetted contact 水银触点
mercury-wetted relay 湿式水银继电器，汞接继电器，水银继电器
mere 池，三角湾，沼地
merged company 合并公司，合组公司
merged fault tree 合并的故障树
merged transistor logic 合并晶体管逻辑
mergence 合并，兼并
merger clause 合并条款
merger diagram 状态合并图
mergers and acquisitions by foreign investors 外资并购
merger 合并，兼并
merge search 归并搜索
merge sort 归并分类
merge 兼并，合并，归并，组合，合流，吻合
merging routine 归并程序
merging unit 合并单元
merging 汇合
meridian line 子午线
meridian plane 子午面
meridian 正午，中天，子午线
meridional flow 子午面流
meridional stream line 子午流线
meridional stress 径向应力
meridional velocity 子午面速度，径向速度
meridional weld joint 子午线焊缝
meridional wind 经向风
merit attention 值得注意
merit consideration 值得考虑
merit figure 灵敏值
merit order rank 优先次序等级
merits and fault 优缺点
merit 价值，优点
meromorphic function 半纯函数
merry-go-round type bunker 走马灯式料仓，旋转料仓
merry-go-round windmill 转塔式风车
merry-go-round 旋转木马，循环直达
merry groined train 循环直达列车【英】
mesa 高台，台面，台面晶体管，台式晶体管，平顶山，台地
mesh analysis 筛分分析，网筛分析，筛分，孔分析法，筛分析
mesh aperture 筛孔
mesh boundary 网格边界
mesh cell 网眼
mesh-connected circuit 多边形连接电路，网接电路
mesh connection 网状结线，多角形接法，网形接法
mesh-controlled storage digisplay 网控存储数字显示管
mesh coordinates 格网坐标，网格坐标
mesh current 网络电流，网孔电流
mesh description 网格图形
mesh distortion 格网畸变
mesh division 分格【用于有限单元法】
meshed network 网状网络，环形网络
meshed strainer 网式过滤器
meshed system 多网系统
mesh enclosure 网罩，框架
meshes 铁丝网
mesh filter 筛网过滤器
mesh-form earthing device 接地网，网形接地器
mesh geometry 网格形状
mesh impedance matrix 网络阻抗矩阵
meshing gear 啮合正齿轮
meshing interference 啮合干涉
meshing 衔接，钩住，啮合，结网，建网
mesh inserts 金属丝网插入件
mesh line 网格线
mesh of a system 单网系统
mesh of cable network 电缆网络分布图
mesh point 网格点
mesh power 匹配功率
mesh range 筛目范围
mesh region 网格区域
mesh reinforced brickwork 网状配筋砌体
mesh reinforced cement 钢筋网水泥
mesh reinforcement 钢筋网
30 mesh screen 30目的筛子
mesh screen 筛目，筛网，筛，网筛
mesh sieve 网筛
mesh size 筛号，筛眼大小，网目大小，目径，筛孔尺寸
mesh-star connection 三角形星形接法
mesh structure 格状构造，网状构造
mesh system 网格系统
mesh type separator 网式汽水分离器
mesh voltage 三角形接法线电压，边电压
mesh winding 环形绕组
meshwork 网织品，网络，网状物
meshy 网状的，多孔的
mesh 网眼，网孔，筛孔，筛号，筛目，网格，网络，筛眼，网目，目
mesmorphic state 液晶状态，介晶态
mesoclimate 中尺度气候，局部气候，中期气候，中气候
mesometeorology 中尺度气象学
meson 介子，重电子
mesopause 中层顶，散逸层顶，中间期，中间相
mesorelief 中地形，中起伏
mesosaprobic zone 中等污染区
mesoscale circulation 中尺度环流
mesoscale terrain 中尺度地形
mesoscale 中尺度，中尺度的
mesosphere 中间层，散逸层
mesothermal climate 中温气候
Mesozoic era 中生代
Mesozoic group 中生界
mesozone 中带
message acceptance pulse 信息接收脉冲

message authentication code　信息鉴定码，信息证实码
message buffer　消息缓冲区
message coding　信息编码
message data　信息数据
message digit　信息位，信息符号
message display console　信息显示控制台
message exchange　信息交换
message processing program　信息处理程序
message queue　信息排队
message source　信息源
message switching center　信息转接中心
message switching multiplexer　信息多路转换器
message switching　消息交换，信息交换，信息转接，数据交换
message transfer part　消息传递部分
message-waiting indicator　信息等待指示器
message-waiting　信息等待
message　信息，消息，电报，电信，报文
messenger bottle　浮标瓶
messenger cable　悬吊缆绳，吊线缆
messenger　信使，通信员，电缆吊绳
mess hall　职工食堂，餐厅，旅馆
messmotor　积分马达，积分电动机
mess room　内部食堂
meta-acid　偏酸
meta-anthracite　次无烟煤
metabituminous coal　肥煤，中烟煤
metabolic stage　代谢期
metabolism　新陈代谢
metacenter　稳心，外心点，定倾中心，稳定中心
metacentre　定倾中心，外心点，稳心，稳定中心
metacentric diagram　稳心曲线，稳心图
metacentric height　定倾中心高度，稳心高度
metacentric radius　定倾半径，稳心半径
meta-compound　间位化合物
meta cresol purple　甲酚红紫
metacryst　次生晶
metadyne converter　微场扩流放大机，交磁放大机
metadyne generator　交磁旋转扩大机
metadyne　微场发电机，交磁放大机，磁场放大机
meta-element　过渡元素，母体元素
meta-jacket gasket　金属包垫片
metal air duct　金属风管
metal-alumna-semiconductor　金属氧化铝半导体
metal arc cutting　金属极电弧切割
metal arc welding　金属极电弧焊接，金属电弧焊
metal armouring　金属铠装
metal backing　金属敷层
metal-bath dip brazing　金属浴浸渍硬钎焊
metal bit　金属钻头
metal can　金属罐，金属包壳
metal card storage　金属卡片存储器
metal casting　金属铸件
metal ceramics　金属陶瓷
metal chassis　金属底盘
metal chelate compound　金属螯形化合物
metal-clad cubicle　金属外壳配电箱【开关设备】
metal-clad drawout type switch gear　抽出式金属柜开关装置
metal-clad fuel element　金属包壳燃料元件
metal-clad SF_6 substation　全金属封闭六氟化硫变电站
metal-clad substation　金属外壳升压站
metal-clad switchboard　金属外壳配电盘，铠装配电板
metal-clad switchgear　铠装开关装置，铁壳开关装置
metal-clad　金属包覆，金属铠装，装甲，金属壳体的，铠装的
metal containment　金属安全壳，钢安全壳
metal covering　金属蒙皮
metal crossarm　金属横担
metal crystal　金属结晶
metal cutting band saw　带锯床
metal damage　金属损伤
metal deck floor　金属铺板楼盖，压型钢板楼盖
metal decking　金属护板
metal deck roof　金属薄板平屋顶
metal deck　金属铺板，压型钢板
metal degradation　金属降解
metal deposit　金属沉积，焊着金属
metal desk floor　金属薄板平屋顶
metal detector　金属探测器
metal disintegration machine　金属破碎机
metal disk coupling　金属盘联轴节
metal dust　金属粉尘
metal electrode　金属电极
metal-enclosed apparatus　金属封闭型电器
metal-enclosed switchboard　金属外壳的配电盘
metal-enclosed switchgear　金属封闭开关设备
metal-enclosed　金属外壳的，铠装的
metalepsis　取代
metaler　钣金工
metal expansion steam trap　金属膨胀式蒸汽疏水阀［疏水器］
metal fibre brush　金属纤维电刷
metal fibre　金属纤维
metal filler　填隙金属，填充金属，焊条
metal-film resistance thermometer　金属薄膜电阻温度计
metal finishing waste　金属抛光废料
metal flashing　金属泛水，金属抛光
metal foil detector　金属箔探测器
metal foil window　金属箔窗
metal foil　金属箔
metal frames　金属框架
metal framework　金属构架
metal fuelled reactor　金属燃料反应堆
metal-gas battery　金属气体电池
metal grade　金属品级
metal graphite brush　金属石墨电刷
metal graphite contact　金属石墨触点
metal halide lamp　金属卤化物灯
metal hardness　金属硬度
metal hose　金属软管
metal hydride moderated reactor　金属氢化物慢化反应堆
metal hydride　金属氢化物
metal inclusion　金属夹杂物

metal indicator　金属指示剂
metal inert gas arc welding machine　熔化极惰性气体保护弧焊机
metal inert gas arc welding　熔化极惰性气体保护电弧焊，金属极惰性气体保护焊，惰性气体金属电弧焊
metal inert gas welding machine　熔化极惰性气体保护弧焊机
metal inert gas welding　熔化极惰性气体保护焊，金属焊条电极惰性气体掩弧焊
metal inlaying　镶金属，金属嵌入物
metal inside　金属衬里
metal inspection　金属检验
metal insulator materials　金属绝缘体材料
metal ion　金属离子
metalizing　金属涂覆
metal laboratory　金属试验室
metal lagging　（汽轮机的）金属装饰板，（炉墙外壁）金属外护板，金属保护层
metal lath and plaster　钢板网抹灰
metal lathing　金属网
metal lath partition　钢丝网隔墙
metal lath and plaster ceiling　金属网抹灰顶棚
metal lath　钢丝网板条，钢板网
metalled road　碎石路
metallegraph　金相学
metallic bellows gauge　金属膜盒压力计
metallic brush　金属电刷
metallic carbon brush　金属炭刷
metallic cementation　渗金属法
metallic character　金属特性
metallic circuit　金属回路，不用大地作回路的电路
metallic coating　金属保护层，金属覆盖层，金属护膜
metallic compound　金属化合物
metallic conduit　（布线用的）金属管
metallic contact　金属接触
metallic contamination　金属污染
metallic copper　金属铜
metallic corrosion　金属腐蚀，金属锈蚀
metallic crystal structure　金属晶体结构
metallic crystal　金属晶体
metallic-cup rotor　金属杯形转子
metallic diaphragm gauge　金属薄膜式压力计
metallic dust　金属粉尘
metallic filament　金属灯丝
metallic film　金属薄层，金属薄膜
metallic flexible hoses　金属软管
metallic fuel element　金属燃料元件
metallic fuelled gas-cooled reactor　金属燃料气冷反应堆
metallic fuel　金属燃料
metallic gasket　金属垫片
metallic grounded　金属性接地的
metallic-hydrogenous shield　金属氢屏蔽体
metallic inclusion　金属水垢，金属夹杂物，金属杂质
metallic insulation　金属隔热，金属保温层，金属保温
metallic insulator　金属绝缘子
metallicity　金属性
metallic material storage　金属材料库
metallic material　金属材料
metallic matrix　金属基体
metallic mesh　钢板网，金属网，拉网板
metallic oxide film　金属氧化膜
metallic packing ring　金属填密环
metallic rectifier　干片整流器，金属整流器
metallic return circuit　金属回路电路，金属回路
metallic return　金属回线
metallic ring　金属环
metallic sheath of cable　电缆金属套
metallic shielding　金属屏蔽
metallic short circuit test　金属短路试验
metallic structure installation　金属结构安装
metallic stuffing　金属填料
metallic tape core　金属带形磁芯
metallic tape　金属卷尺
metallic tile　金属瓦片，金属挂片
metallic uranium　金属铀
metallic vapor　金属蒸汽
metallic waveguide　金属波导管
metallic　金属的
metal liner　金属衬里
metalling　碎石层
metallization　包镀金属，金属喷涂，金属喷镀，敷金属
metallized brush　金属石墨电刷，金属电刷
metallized graphite brush　金属石墨电刷
metallized paper capacitor　敷金属纸电容器
metallized paper dielectric capacitor　金属化纸介电容器
metallized　金属涂覆的
metallize　金属喷涂，给……包裹金属层，用金属涂层或处理，使金属化
metal locator　金属探测器
metalloceramics　金属陶瓷
metallographical sand paper　金相砂纸
metallographic analysis　金相分析
metallographic examination　金相分析，金相检查，金相检验
metallographic hot cell　金相热室
metallographic laboratory　金相实验室
metallographic microscope　金相显微镜
metallographic specimen　金相试片
metallographic structure　金相组织
metallographic test　金相试验
metallography　金属学，金相学
metallograph　金相照片，金相显微镜
metalloid　准金属，类金属
metalloscope　金相显微镜
metal louver　金属百叶窗
metallurgical accountability　冶金衡算
metallurgical analysis　金相分析
metallurgical bond　冶金结合，熔结
metallurgical bonding　冶金焊合，冶金黏结
metallurgical cement　熔渣水泥
metallurgical coal　炼焦煤
metallurgical coke　冶金焦
metallurgical component　金相成分
metallurgical defect　冶金缺陷

metallurgical examination　冶金检验
metallurgical fume　冶金烟气
metallurgical microscope　金相显微镜
metallurgical plant　冶金厂
metallurgical product　冶金产品
metallurgical property　冶金性质
metallurgical spectrum analyzer　金相光谱仪
metallurgical structure　冶金结构
metallurgical weldability　冶金可焊性
metallurgical　冶金的
metallurgic microscope　金相显微镜
metallurgic replica　金相复型技术
metallurgic strengthening　冶金强化
metallurgy of non-ferrous metals　有色金属冶金
metallurgy　冶金学，冶金术，冶金
metal mask　金属膜片，金属屏蔽，金属掩模
metal matching　金属温度调节，金属气温匹配
metal mechanic property　金属力学性能
metal mesh　金属网
metal-metal battery　金属-金属电池
metal-metal oxide battery　金属-金属氧化物电池
metal moderated reactor　金属慢化反应堆
metal nitride oxide semiconductor　金属氮氧化物半导体
metalogic　元逻辑
metal-O-ring　金属O形环
metal-oxide arrester　金属氧化物避雷器
metal-oxide catalyst　金属氧化物催化剂
metal-oxide planar transistor　金属氧化物平面晶体管
metal-oxide resistor disc　金属氧化物电阻片
metal-oxide-semiconductor field-effect transistor　金属氧化物型半导体场效应晶体管
metal-oxide-semiconductor for large-scale integration　金属氧化物半导体大规模集成
metal-oxide-semiconductor integrated circuit　金属氧化物半导体集成电路
metal-oxide-semiconductor random access memory　金属氧化物半导体随机存取存储器
metal-oxide-semiconductor　金属氧化物半导体
metal-oxide surge arrester　金属氧化物避雷器
metal-oxide　金属氧化物
metal parts　金属块
metal pipe　金属管
metal plate path　金属板电镀槽
metal plate　金属板
metal plating　金属电镀
metal power cutting　氧熔剂切割
metal probe　金属试样
metal raceway　电线保护用铁管，金属线槽
metal radiant panel heating　金属辐射板采暖
metal radiant panel　金属辐射板
metal rectifier　干片整流器，金属整流器
metal-ring　金属圈，金属环，短路环
metal scraper　金属刮刀
metal-semiconductor contact　金属半导体接触
metal separator　金属分离器
metal shavings　金属切削，金属刨花
metal sheath　金属外皮
metal-sheet cladding　金属板包层
metal-sheet for expansion joint　伸缩缝金属条
metal sheet pile　钢板桩
metal sheet shearing machine　剪扳机
metal sheet wall　金属板墙
metal sheet　薄金属板
metal-shielded wire　金属屏蔽线
metal silicon　金属硅
metal spray gun　金属喷枪
metal spray　金属喷镀，金属喷涂
metalster　金属膜电阻器
metal sulfide　金属硫化物
metal supervision　金属监督
metal tag　金属标签
metal temperature　金属温度
metal-to-metal seating　金属对金属的落座
metal transfer　（电弧焊）熔滴过渡
metal-tube rotameter　金属管转子流量计，金属管转子式测速仪
metal ultrasonic delay line　金属超声波延迟线
metal vessel　金属容器
metal-water reaction　金属水反应
metal wedge　金属楔
metal wiper　金属接触刷
metal wire net　金属丝网
metalwork　钳工作业，金属加工
metal　金属，碎石料
metamagnetism　变磁性
metamic　金属陶瓷材料
metamorphic differentiation　变质分异作用
metamorphic diffusion　变质扩散作用
metamorphic granite　变质花岗石
metamorphic limestone　变质灰岩
metamorphic petrology　变质岩石学
metamorphic rock　变质岩
metamorphic water　变质水
metamorphic　变质的
metamorphism zone　变质带
metamorphism　变质，变质作用
metamorphosis　变质
metanometer　甲烷指示计
metaphenylene diamine　间苯二胺
metaphosphate　偏磷酸盐
metaphosphoric acid　偏磷酸
metaphosphorous acid　偏亚磷酸
metaphrase　直译，逐字翻译
metapole　等角点
metaprogram　元程序，亚程序
metaripple　不对称波痕
metascope　红外线指示器
metasilicate　偏硅酸盐
metastability　亚稳定性，亚稳度
metastable critical flow　亚稳态临界流
metastable equilibrium　亚稳定平衡，亚稳平衡
metastable state　亚稳状态
metastable steam　亚稳态蒸汽
metastable　准稳定，亚稳定，相对稳定，暂时稳定，亚稳的，准稳的，相对稳的
metasynchronism　亚同步
metathesis　双分解，复分解
metathetical reaction　复分解反应
metathetical　复分解的，置换的
metatrophic　腐生的

metavanadate 偏钒酸盐
meteoric water 大气水
meteorite 陨石
meteorogical department 气象局
meteorogram 气象（记录）曲线，气象图解
meteorograph 气象自记仪
meteorological chart 气象图表
meteorological condition 气象条件
meteorological convection 气象对流
meteorological data 气象资料
meteorological department 气象部门
meteorological depression 气象低压（区）
meteorological documentation 气象文件
meteorological elements over years 历年气象数据
meteorological element 气象要素
meteorological essential 气象要素
meteorological factor 气象要素
meteorological map 气象图
meteorological mast 气象塔，气象桅杆
meteorological model 气象模型
meteorological observation 气象观测
meteorological observatory 气象台，气象站
meteorological parameter 气象参数
meteorological range 标准视距，气象视距
meteorological reference wind speed 气象参考风速
meteorological report 气象报告
meteorological satellite 气象卫星
meteorological service 气象局
meteorological site 气象台（站）址
meteorological situation 气象情况，天气形势
meteorological standard condition 标准气象条件
meteorological station 测候站，气象台，气象站
meteorological survey 气象测量，气象考查
meteorological symbol 气象符号
meteorological tide 气象潮
meteorological tower 气象测量塔，气象塔
meteorological wind tunnel 气象风洞
meteorological 气象气球，气象的，气象学的
meteorologist 气象学家
meteorology 气象，气象学
meter ampere 米安
meter board 仪表板
meter bridge 滑线电桥
meter candle 米烛【光的单位】
meter case 仪表外壳
meter constant 表计常数
meter creeping 表的指针爬行现象
meter dial 仪表刻度盘，仪表刻度
metered blow-cast pump 计量吹气箱式泵
metered measurand 计量值
metered reading 表计值，计量值
meter full scale 最大量程，刻度范围
metergate 量水门
meter glass 量杯，刻度烧瓶
meter-gram resistivity 米克电阻率
meter-incircuit 入口节流式回路
metering and control section or cubicle 测量和控制舱或室
metering circuit 测量电路
metering contact 测量触点
metering device 测量装置
metering equipment 测量仪表，测量装置，测量设备
metering error 测量误差
metering fluid 测量的流体
metering gallery 量测廊道
metering hole 定径孔
metering nozzle 测量喷嘴，计量喷嘴，测量喷管，测流管嘴
metering orifice 测量孔板，测流量孔
metering plug 计量插塞
metering point 量测点
metering pump 计量泵，计量录
metering relay 计数继电器，记录继电器
metering section 量测断面
metering system 测试方式，测量系统，计量系统
metering tank 计量箱
metering transformer 表用互感器，仪用变压器，仪用互感器
metering valve 计量阀
metering 计量，测量，计数，量测
meterman 读表人，仪表调整者
meter mercury column 米汞柱
metermultiplier 仪表量程倍增器
meter of water head 米水柱
meter-out circuit 出口节流式回路
meter panel 仪表盘，仪表屏，仪表板
meter pulse 表计脉冲
meter reading expenses 抄表费用
meter reading 仪表读数
meter relay 记录继电器，计数继电器
meter sensitivity 仪表灵敏度
meters per second 米每秒
meter system 测试装置，测试系统，公制
meter box 量水箱，仪表箱
meter flume 测流槽，量水槽
meter rate 按表收费，电费计费率
meter rod 量杆
meter transformer 仪用互感器，仪用变压器，表用互感器
meter type relay 仪表式继电器
meter 米，表，计，仪，测量仪器，计量器，计数器，测量，表计，测量仪表，仪表
methanation 甲烷化
methane cooker 沼气灶
methane derived coating 甲烷热解碳包覆层
methane flow counter 甲烷流量式计数表
methane generator 沼气发电设备
methane level 甲烷水平
methane pond 沼气池
methane-producing bacteria 产沼气菌
methane purifier 沼气净化器
methane store 甲烷贮存库
methane 甲烷，沼气
methanoic acid 甲酸
methanol 甲醇
methenamine 六次甲基四胺，乌洛托品
method and capacity of water purification 净水方式及规模
method by trials 试探法

degree-day method 度日法
method for analysis of water quality 水质分析方法
methodical 有方法的，有条不紊的，按部就班的
method of actuation 驱动方法，执行方式
method of application 申请方法
method of approach 渐近法
method of approximation 近似法
method of average 平均法
method of backward intersection 后方交会法
method of bins 比恩法
method of blade root attachment 叶根连接方式
method of calibration 检验方法，标定方法
method of characteristics 特性线法，特征线法
method of charging out 报销方法
method of charging 计费方法
method of check-off 注销法
method of column analogy 柱类比法
method of coning and quartering 堆锥四分法
method of connection 连接方式
method of continuous flow calorimetry 连续流动量热法
method of corner points 角点法
method of cost accounting 成本计算方法
method of deformation 变形法
method of depreciation 折旧法
method of difference 差分法
method of dimensions 量纲法，因次法
method of direct cooling 直接冷却法
method of direction observation 方向观测法
method of discharge measurement 流量测验法
method of division into groups 分组（平差）法
method of double refraction 双折射法
method of double sight 复觇法
method of elastic center 弹性中心法
method of electric analogy 电模拟法，流变电模拟法
method of electro-mechanical analogy 机电模拟法，机电比拟法
method of energy dissipation 消能方法
method of erection 安装方法
method of estimation 预算法，估计方法
method of false position 试位法，假位置法
method of fictitious load 虚拟荷载法
method of finite difference 有限差分法
method of finite element 有限元法
method of finite increment 有限增量法
method of fixed points 定点法
method of force 力法
method of full voltage starting 全压启动法
method of images 图像法，镜像法
method of impregnation 浸渍法
method of induction 归纳法
method of installation 安装方法
method of interpolation 插值法，内插（法）
method of intersection 前方交会法
method of iteration 迭代法
method of joint 节点法
method of lattice 网格法
method of least square 最小平方差法，最小二乘法
method of levelised costs 平准化成本计算法
method of limit equilibrium 极限平衡法
method of measurement 测量方法
method of minimum square 最小二乘法
method of modulation 调制方法
method of moment distribution 力矩分配法
method of operation 操作法，运转方式，运行方式，操作方法
method of payment 支付方法
method of plane cascade 平面叶栅法
method of plotting position 定点法
method of radiolocation 无线电定位法
method of redress 补救方法
method of redundant reaction 超静定反力法
method of repetition measurement 复测法
method of representative blade element 代表叶素法
method of resection 后方交会法
method of residues 消减法，剩余法
method of reverse osmosis 反渗透法
method of river diversion 导流方法
method of sampling 取样方法，抽样法，抽样方法
method of section 截面法
method of seismic prospecting 地震勘探法
method of sieving 筛分法
method of singularities 奇点法，有限基本解法
method of slope-deflection 倾角变位法
method of small disturbance 小扰动法
method of sound level 声级测量法
method of starting 启动方法
method of statement 施工组织方案
method of substitution 替代法，代替法，置换法，代入法
method of successive approximation 逐次渐进法，逐步近似法，逐次渐近法
method of successive interval 连续间隔法，毗连区间法
method of superposition 迭加法，叠加法
method of surface coating 涂面法【研究边界层流态用】
method of telemetering 遥测方法
method of testing shaft alignment 假轴校直法，假轴找中法
method of test 试验方法
method of trail and error 尝试法
method of transformed section 折算截面法
method of transprtation 运输方式
method of trial and error 逐次逼近法，逐步逼近法，试误法，尝试法
method of trial 试探法
method of undetermined coefficients 待定系数法
method of undetermined parameter 待定参数法
method of unit loads 单位荷载法
method of virtual displacement 虚位移法
method of virtual work 虚功法
method of weighted mean 加权平均法
methodology of interface management 接口管理法
methodology 方法学，方法论，分类法，一套方法

methods for insulation coordination　绝缘配合方法
methods of fusion　熔化方法
method statement on construction　施工方法，施工方案，施工说明书
method statement　（项目）施工说明书，施工组织方案，方法说明
method　方法，手段，方式，规律，顺序
methyl alcohol　甲醇，木精
methylene chloride　二氯甲烷
methyl group　甲基基团
methyl iodide　甲基碘
methyl orange alkalinity　甲基橙碱度，全碱度
methyl orange end-point alkalinity　以甲基橙作指示剂测得的碱度【甲基橙为指示剂的滴定终点pH约为4.5】
methyl orange indicator　甲基橙指示剂
methyl orange　甲基橙
methyl phenyl ether　苯甲醚，茴香醚
methyl purple　甲基紫
methyl red　甲基红
methyl violet　甲基紫
methyl　甲基
meticulous construction　精心施工
meticulous design　精心设计
meticulous diagnosis　精密诊断
MET mast　气象桅杆
MET(meteorological/meteorology)　气象学
metrechon　双电子枪存储器
metrical information content　测度用信息量
metrical information　测度用信息
metric data　量测数据
metric horse power　公制马力
metric machine　米制电机，公制电机
metric scale　米制比例尺
metric screw pitch gauge　国际螺距规
metric system　公制，米制
metric temperature scale　公制温标
metric thread　公制螺纹
metric ton　吨，公吨
metric unit　公制单位
metric wave　米波
metric　计量的，度规的，米制的
metrohm　带同轴电压电流线圈的欧姆计
metrological characteristics　计量特性
metrological regulation　计量规程
metrological requirement　计量要求
metrological supervision　计量监督
metrology　度量衡学，计量学
metropolis　中心城市，首府，大都市
metropolitan area　首都行政区，大都市地区，市中心，大城市区，首都地区
metropolitan atmosphere　大城市大气
metropolitan district　首都行政区，大城市中心区
metropolitan region　大都市地区
metropolitan　大城市的
metro　地下铁道
met stable phase　亚稳相
MEUR　百万欧元
MEV(million electron volts)　兆电子伏，百万电子伏特
Meyer hardness　迈耶硬度
Meyerhof's formula　梅耶霍夫公式
mezzanine floor for cable　电缆夹层
mezzanine floor oil piping　中间层流管
mezzanine floor　高压加热器平台，中间楼层，夹层楼面
mezzanine　夹层，夹层楼面
MEZZ＝mezzanine　高压加热器
MFC(manual frequency control)　手动频率控制
MFC(microfunctional circuit)　微功能电路
MFC(multifunction control)　多功能控制
MFCS(main feedwater control system)　主给水控制系统
MFGCR(mixed flow gaseous core reactor)　混合流气体堆芯反应堆
MFIC(mechanical hydraulic control)　机械液压式控制
MFIS(main feed isolating system)　主给水隔离系统
MFIV(main feed water isolating valve)　主给水隔离阀
MFKP(multifrequency key pulsing)　多频键控脉冲调制
MF(maintenance factor)　维护系数
M/F＝manifest　载货清单，舱单
MF＝microfilm　缩微胶片
MFM(modified frequency modulation)　变频调制
MFNT(most favoured nation treatment)　最惠国待遇
MFOD(mean forced outage duration)　平均强迫停运时间
MFP(main feed pump)　主给水泵
MFP(mean free path)　平均自由行程
MFR(mean failure rate)　平均故障率
mfr＝manufacturer　制造者，制造厂
MFSK(multiple frequency-shift keying)　多频移键控
MFS(magnetic fluid seal)　磁性液体密封
MFS(minimum function specification)　最低功能规范
MFTF(molten fuel test facility)　熔化燃料试验装置
MFT(main fuel trip)　主燃料跳闸，总燃料跳闸
MFT　主燃料失去保护【DCS画面】
MFWLB(main feedwater line break)　主给水管线破裂
MFWP(main feedwater pump)　主给水泵
MGC(maximum guaranteed capacity)　最大保证功率
MGCR(marine gas cooling reactor)　船用气冷反应堆
MGD(million gallons per day)　百万加仑每天
MG(maximum generation)　最大（理论）发电量
mg＝milligram　毫克
M-G(motor-generator)　电动机发电机
M-G set(motor-generator set)　电动机发电机组
$MgSiO_3$(magnesium silicate)　硅酸镁
MGU(motor governor unit)　电动调节单元
MHA(maximum hypothetical accident)　最大假想事故
MHC(mechanical hydraulic control)　机械液压控制，机械液压式控制

MHD effect 磁流体动力学效应
MHD generator duct 磁流体发电机通道
MHD generator 磁流体动力发电机，磁流体发电机
MHD loop 磁流体动力装置回路
MHD(magnet hydro dynamic process) 磁流体动力过程
MHD(magnetohydrodynamics) 磁流体动力学
MHD magnet 磁流体磁体
MHDM(materials handling demonstration module) 材料处理示范单元【美国】
MHD open-cycle 磁流体开环
MHD plant cycle 磁流体动力装置循环
MHE(materials handling equipment) 材料动载设备，材料处理设备
MHF(mounting height above floor) 距地面安装高度
MHI(Mitsubishi Heavy Industries, Ltd.) (日本)三菱重工业公司
MHI(Mitsubishi Heavy Industries) 三菱重工
MH(man-hours) 工时数
mho-impedance relay 姆欧阻抗继电器
mhometer 姆欧计，电导计
mho relay 电导继电器，姆欧继电器，电导型阻抗继电保护装置
mho 姆欧
MHTR(moduled high temperature reactor) 模块式高温气冷堆
MHz＝megahertz 兆赫
mian hall 正厅
mica based resin varnish 云母基树脂漆
mica board 云母板
micabond 米卡邦德绝缘材料
mica cambric 云母布
mica capacitor 云母电容器
micaceous iron oxide phenolplastics paint 云母氧化铁酚醛漆料
micaceous sandstone 云母砂岩
micaceous schist 云母片岩
micaceous shale 云母页岩
micaceous 云母的
mica condenser 云母电容器
mica dielectric capacitor 云母介质电容器
mica dielectric 云母电介体
mica disc 云母盘
micadon 云母电容器
mica flake asphalt tape 沥青云母带
mica flake tape 片云母带
mica flake 云母薄片
mica foil 云母箔
micafolium asphalt paper 沥青云母纸
micafolium shellac paper 虫胶云母纸
micafolium 云母箔
mica-glass-fibre tape 玻璃云母丝带
mica gneiss 云母片麻岩
mica insert 云母插片
mica insulation 云母绝缘
mica lamination 云母片
micalex 云母石，云母玻璃
mica mat 云母垫，云母板
mica molded insulator 粉云母模压绝缘
micanite pipe 云母管
micanite plate 云母板，云母片
micanite sleeve 云母板套筒，云母板套管
micanite 云母板
mica paper foil 粉云母箔
mica paper insulation 全粉云母绝缘
mica paper micanite 粉云母板
mica paper tape 粉云母带
mica paper 云母纸，粉云母纸
mica plate 云母板，云母片
mica resistance 云母电阻
micarex 云母玻璃，云母石，云母板
mica ring 云母环，云母圈
micarta 米卡他绝缘板，胶合云母纸板
mica segment between bars 换向器云母片
mica segment 云母片
mica sheet 云母片
mica spacer 云母垫片，云母隔片
mica splitting 云母剥片，剥制云母，薄片云母
mica strip 云母条
mica tape 云母带
mica tube 云母管
mica undercutter (换向器的)云母下刻机
mica V-ring V形云母环，V形云母圈
mica 云母
micellae 胶态离子，胶束，胶粒
micom 微型计算机
microadjuster 精密调节器，微调节器
micro-aggregate 微集聚体，微团粒
micro-alloy diffused-base transistor 微合金扩散晶体管
microalloy transistor 微合金晶体管
microammeter 微安计
microampere meter 微安计
microanalysis 微量分析，微观分析
microanalytical chemistry 微量分析化学
microarchitecture 微体系结构
microbalance 微量天平
microbarograph 自记微气压计
microbarometer 微气压计，微压计，微气压表
microbe 微生物
microbial attack 微生物侵蚀
microbial degradation 微生物降解
microbial insecticide 微生物杀虫剂
microbial reaction 微生物反应
microbicide 杀菌剂，杀微生物剂
microbiological aerobic corrosion 需氧微生物腐蚀
microbiological contaminant 微生物污染物
microbiological control 微生物控制
microbiological corrosion 微生物腐蚀
microbiological determination 微生物测定法
microbiological factor 微生物因素
microbiological fouling 微生物污染
microbiological induced corrosion 微生物引起的腐蚀
microbreak 微隙，微裂缝
microbubble 微气泡
microburette 微量滴定管
microburst 微猝发风，微爆气流
microcache 微程序缓存，微程序缓冲存储器
microcallipers 百分表，千分尺

microcanonical ensemble 微正则系综
microcanonical partition function 微正则配分函数
microcapacitor 微型电容器
microcard 缩微卡片
micro-catchment 微型集水沟，小集水沟
microchannel plate 微型通道板
microchemical analysis 微量化学分析
microchemical pollutant 微量化学污染物
microchemical pollution 微量化学污染
microchemistry 微量化学
microcircuit module 微电路模块
microcircuit 微型电路，微电路
microclimate heat island 微气候热岛
microclimate 小气候，实验室气候，微气候
microclimatic effect 小气候效应
microclimatology 微气候学，小气候学
microcoding 微编码
microcolorimeter 微量比色计
microcolorimetric titration 微量比色滴定
microcomponent 微型元件
microcomputer 微（型）计算机
microconstituent 显微成分
microconvection 微对流
microcook 用超级电脑高速处理
microcopy 缩微复制，缩微照片
microcorrosion 显微腐蚀，微观腐蚀
microcoulomb 微库仑
microcoulometric titration 微量库仑滴定
microcoulometry 微库仑分析法
microcrack 微裂缝，微观裂纹，微裂纹，细微裂缝
microcurie 微居里
microdensitometer tracing 微密度计的测定线
microdensitometer 微量密度计，显微黑度计
microdepression 小低气压
microdetermination 微量测定，微量测量
microdial 精密标度盘
microdosimetry 微剂量学
microearthquake 微小地震
micro-economics 微观经济学
microeffect 微观效应
micro electro hydraulic control system （锅炉给水泵小汽机）电液控制系统
microelectronics 微电子学
microelement 微量元素，微型元件
microetching 显微侵蚀
microexamination 显微检验，微观检验
microfarad meter 微法拉计
microfarad 微法拉
microfiche 缩微胶片
microfilmer 缩微摄影机
microfilm 缩微胶片
microfilter 微量过滤器，微滤机，微孔微滤器，微型滤波器
microfine coal 超细煤粉
microfinishing 精滚光，精密磨削
microfissure 显微裂缝，微观裂纹，显微裂纹，微裂隙
micro-flaw 发纹，发裂纹，显微裂纹，微裂纹，微观缺陷
micro floc 微小絮状物
microfluorometry 微量荧光分析法
microfluorophotometry 显微荧光照相技术
microflux 微观通量
microfractography 断口显微镜检验
micro-fracture mechanics 微观断裂力学
microfunctional circuit 微功能电路
micro-fuse 微型保险丝
microfusion 微量熔化
micro-galvanometer 微量检流表，微量电流计
micro-gap model 微隙模型
micro-gap switch 微动开关
micro-gasification 微气化
microgenerator 微型发电机
micro-geomorphology 微地形学
microgram 微克【μg】
micrographic examination 显微检验
micrography 显微照片
microgroup 微观群
microhardness testing 显微硬度试验
microhardness 显微硬度
microhenry 微亨利
microhm gauge 微欧计
microhm 微欧姆
microholography 显微全息术
micro-hydraulic station 小型水力发电站
micro-hydraulic turbine 微型水轮机
micro-image storage 缩微存储
microinch finishing 精加工，光制
microinching 慢速运转，微动
microinhomogeneity 微观不匀性
microinstability 微观不稳定性
microinstruction 微程序指令
micro-ionic resin 微离子树脂
micro-irradiation 微束辐照
microjet 微射流
microlamp 微灯，小型人工光源
microlayer evaporation 微层蒸发
microleak 微泄漏
microlevel gauge 微水位计
microlock 微波锁定
micrologic element 微逻辑元件
micrologic 微逻辑
micromachine 微型机械，微电机
micromag 一种直流微放大器
micromanipulator 微型机械手，微型操纵设备，精密控制器，微型操作钳
micromanometer 微压计，微气压计，微压力计
micromation 微型化，微型器件制造法
micromatrix 微矩阵
micromechanism 微观机理
micromeritics 微粒学，粉末工艺学
micromesh screen 微孔筛
micrometeorological data 微气象资料
micrometeorological environment 微气象环境
micrometeorology 微气候学
micrometer adjustment 微量计调整
micrometer calipers 螺旋测径器，千分尺，千分卡尺
micrometer compensator 测微补偿器
micrometer depth gauge 千分深度尺
micrometer drum 测微鼓

micrometer eyepiece 测微目镜
micrometer gauge 测丝规
micrometering 微测，微量测量，测微
micrometer mechanism 测微机构
micrometer microscope 测微显微镜
micrometer run 测微器行差
micrometer screw 测微螺旋
micrometer sextant 测微六分仪
micrometer theodolite 测微经纬仪
micrometer 测微计，千分尺，百分表，千分表
micrometric screw 测微螺旋
micrometry 测微法
micromho 微姆欧，微西门子
micro-microcurie 皮居里，10^{-12} 居里
micromicrofarad 皮法，微微法
microminiature circuit 微型电路
microminiature 微小型的，超小型的
microminiaturization 微小型化，超小型化
micro mini wind turbine 微型或者迷你型风电机组
micromodule 微模块，微模件，微型组件，微型器件
micromotor 微电机，微型电动机
microm 微程序只读存储器
micron efficiency curve （除尘器的）微米效率曲线
microneutronography 显微中子放射照相法
micronormal 微电位曲线
micron 微米
micro-ohm 微欧姆
microoptic level 光学测微水准器
microorganic contaminant 微生物污染物
microorganism identification 微生物鉴定
microorganism treatment 微生物处理
micro organism 微生物
microoscillograph 显微示波器，显微示波仪
microosmometer 微渗压计
micro-packed column 微型填充柱
microparticle 微粒
microphone amplifier 传声器放大器，送话器放大器
microphone 麦克风，微音器
microphonism 颤噪声，颤噪效应
microphotodensitometer 显微光密度计
microphotoelectric 微光电的
microphotogram 显微照相图
microphotography 显微照相术
microphotograph 显微照片【照相】，显微照像
microphotometer 测微光度计
microphotometry 显微光度测量法，浓度计，测微光度计
microphototransistor 微型光敏晶体管
micropile 微型桩
micropipet 微量移液管
micro Pitot tube 微型毕托管
micro-plasma arc 微束等离子弧焊
micropluviometer 微雨量器
micropollutant 微量污染物
micropollution 微量污染
micropore calcium-silicate 微孔硅酸钙
micropore film 微孔滤膜
micropore permeability 微孔孔隙渗透率，细孔孔隙渗透率
micropore 微孔
microporosity 微孔率，显微疏松
microporous calciumsilicate insulation product 微孔硅酸钙保温制品
microporous calcium 微孔硅酸钙
microporous film 微孔膜
microporous membrane technique 微孔膜技术
microporous 微孔的
microporphyritic 微斑状
micropowder 超细粉
microprecessor-based protection 微机保护
micropressure gauge 微压计
micropressure 微压
microprinted circuit 微型印刷电路
microprinted 微型印刷的
microprobe 微型探针，微探针
microprocessor-based protection 微机保护，以微机为基础的保护系统
micro-processer electro-hydraulic control system 小汽机数字电液控制系统，（给水泵小汽机）电液控制系统
microprocessing unit 微处理部件，微处理单元
microprocessor-based protective relay 微机继电保护
microprocessor development system 微处理机开发系统
microprocessor electro-hydraulic system for feed water pump turbine 给水泵小汽机数字电液控制系统【电站】
microprocessor system 微处理机系统
microprocessor totalizer 微处理器累加器
microprocessor unit 微处理机装置
microprocessor 微处理机，微处理器
microprogram control unit 微程序控制器
microprogram data register 微程序数据寄存器
microprogrammable instruction 可编微程序指令
microprogramming language 微程序设计语言
microprogramming 微程序控制，微程序设计，微编程
microprogram storage 微程序存储器
microprogram 微程序
micropunch plate muffler 微穿孔板消声器
micropunch 微穿孔
microquantity 微量
microradiography 显微放射显影法，显微射线照相法
microray 微波，微射线
micro-regionalization 微区划
microrelay 微动继电器
microrelief 小起伏，微地形，微起伏
microresponse 微小反应
microscale circulation 微尺度环流
microscale turbulence 微尺度湍流
microscale 微尺度
microscope with photography attachment 带摄影装置的显微镜
microscope 双目立体显微镜，显微镜
microscopical analysis 显微（镜）分析

microscopic bubble 微气泡
microscopic crack 微裂纹，显微裂纹
microscopic cross section 微观截面
microscopic dust analysis 尘粒尺寸微观分析
microscopic examination 微观检查，微观组织检查，显微镜检验
microscopic fission cross section 微观裂变截面
microscopic fraction 微观粒级
microscopic inspection 微观检查
microscopic matter 显微物质
microscopic particle counter 显微粒子计数器
microscopic stress 微观应力，显微应力
microscopic 微观的，显微的
microscopy 显微镜检查
micro-screen-filter 滤网式精密过滤器
micro-screening 微筛选
microsecond 微秒
microsection 显微截面，显微磨片，显微断面，金相切片
microseepage 微渗
microsegregation 显微偏析，枝晶间偏析
microseismic forecasting 微震预测
microseismograph 微震仪
microseism 强震，微震
micro setting for WTG 风电机组微观选址
microshrinkage 微观缩孔，枝晶间缩孔，显微缩孔
micro siemens 微母，微西门子
micro siting 微观选址
microsize 微小尺寸
microsound scope 微型示波器，小型测振仪
microspec function 特定微功能，特殊操作的微指令
microsphere 微球体
microsphericat catalyst 微球催化剂
micro-stepping motor 微型步进电机
microstep 微步
microstorage 微存储器
microstrainer 微滤机，微过滤器
microstrain 微应变，微过滤
microstress 微观应力，显微应力
microstrip 微波不对称传输线，微波传输带，微带
microstructure 微观结构，金相组织，显微组织，显微结构，高倍组织
microsubroutine 微子程序
microswitch 微动开关，微型开关
microsyn 微动自动同步器
micro syringe 微量注射器
microsystem 微型系统
microtechnic 精密技术
micro-tectonics 微构造
microtest 精密试验
microtexture 微观结构，显微组织，显微地质
microthermal climate 低温气候
microthermometer 精密温度计，微温度计
microtherm 低温植物
microtopography 小地形
micro tremor 微震
microtriangulation 小三角测量
microtronics 微电子学
microtron 电子回旋加速器
micro turbine 小型向心透平
micro turbulence theory 微湍流理论
micro turbulence 微湍流，微湍
micro-vibrograph 微振示振仪，微振图示计
microviscosity 微黏性
microvoid filtration 微孔过滤
microvoid 微孔
microvoltmeter 微伏计，微伏特计，微伏表
microvolt 微伏
micro water power plant 微型水电站
microwattmeter 微瓦计
microwatt 微瓦
microwave band-pass filter 微波带通滤波器
microwave band 微波段
microwave channel 微波通道
microwave communication 微波通信
microwave correlator 微波相关器
microwave delay equalizer 微波延时均衡器
microwave device 微波器件
microwave diagnostics 微波诊断法
microwave frequency band 微微波频带，波波段
microwave frequency counter 微波频率计
microwave generator 微波发生器
microwave heating 微波加热
microwave holography 微波全息术
microwave integrated circuit 微波集成电路
microwave interferometer 微波干涉仪
microwave moisture apparatus 微波含水量测定仪
microwave motor 微波电动机
microwave network 微波通信网，微波网络
microwave oscillator 微波振荡器
microwave oven 微波炉
microwave pilot protection 微波保护
microwave power transmission 微波输送电能
microwave pulse generator 微波脉冲发生器
microwave radiation 微波辐射
microwave radio relay link 微波中继通信线路
microwave ranging measurement 微波测距
microwave reflector 微波反射器
microwave relay communication 微波中继通信
microwave remote sensing 微波遥感
microwave scattering 微波散射
microwave spectrum 微波频谱
microwave station 微波站
microwave telecommunication tower 微波通讯塔
microwave testing 微波检测
microwave transmission 微波通信，微波传输
microwave tube 微波管
microwave 微波
microweigh 微量称量
microweld 微型焊缝
micro-wind filter cartridge 绕线式精密过滤器滤芯
microwire 超精细磁线，微导线
microzonation 微区划
micro 微【10^{-6}】
MICS（model implementation conformance statement）模型实现一致性陈述
mid and long term hydrological forecasting 中长期水文预报
mid board 间壁

mid channel 中央航道
mid chord 弦线中点
mid coil 半线圈
mid diameter 平均直径
mid discharge orifice 泄水中孔
middle and long term investment 中长期投资
middle beam 腰梁
middle bearing bracket 中间轴承座
middle bearing 中间轴线
middle course 中游段
middle-cut file 中纹锉
middle electrode 中间电极
middle layer 中间层，中层
middle leak-off point （轴封）中间抽汽点
middle load power plant 中间负荷电站
middle load thermal power plant 中间负荷火电厂，中间负荷电厂
middle load 中间负荷
middle management 中层管理
middle of cycle 循环周期的中间，循环中期
middle of life 寿命中期，寿期中
middle part of the turbine house 汽轮机房的中部
middle point 中点
middle pole 中间杆
middle power 中间功率
middle reaches 中游
middle reach 中段
middle-shot water wheel 中射式水轮
middle speed mill 中速磨
middle tap （丝锥的）第二锥
middle thunderstorm region 中雷区
middle tooth 主齿
middle water 层间水
middle wire 中线
middle 中间的，当中的
middling coal 中煤
middling duty 中级工作制，中级
middlingly abrasive 中等磨蚀性
middlings 中等燃料，中级品，中煤
midfeather 挡板，隔墙，间壁，隔板
mid-frequency band 中频带
mid-frequency induction heating pipe-bender 感应电热弯管机
mid-frequency noise 中频噪声
mid-frequency of octave band 倍频带中心频率
mid-frequency 中频
midget electric motor 微型电动机，小型电动机
midget fuse 小型熔断器
midget plant 微型电站
midget relay 小型继电器
midget 小型物，小照片，小型的，微型的
mid height plane 半高度面【堆芯】
mid height 一半高度，中高度
midland 内陆的
mid-lethal dose 中等致死剂量
midline capacitor 对数律可变电容器
midline exposure 中线照射量
midline tissue dose 中线组织剂量
midline 中线
mid-merit plant 中等指标工厂
midocean 外海
midplane 中平面
midpoint connection 中点连接
midpoint-grounded socket 中点接地的插座
midpoint method 中点法
mid-point rate 平均汇率
midpoint 中点
mid position contact 中位触点
mid project evaluation 项目中期评价
mid rail 护腰
mid-range forecast 中期预报
mid-range load 中间负荷，腰荷
mid-series termination 半T端接法
mid-series 半串联，串中剖
mid-shunt termination 半π端接法
mid-shunt 并中剖，半并联
midspan joint 接续管
midspan load 跨中荷载
mid span shrouded blade （有减震）凸台的叶片，凸扇叶片
midspan 挡距中间，开度中间，挡距中央，跨距中点，跨中，中跨
midstream depth 中泓水深
midstream of channel 中泓线
midstream 河流正中，中游，中流
midsummer 仲夏
mid-tap coil 中心抽头线圈
mid-term evaluation 中期评估
mid-term report 阶段报告，中期报告
mid-term review 中期审查
mid thickness 半厚度
mid value 中间值，中值
mid width 半宽轴线
mid wing 中翼
midwinter 仲冬
mid 中部，中间
migmatite 混合岩
migmatization 混合岩化
migrate 移动，迁移，迁徙
migrating bar 移动沙洲，游移沙洲
migrating dune 移动沙丘
migration area 位移面积，移动面积，徙动面积
migration constant 迁移常数
migration current 徙动电流
migration length 徙动长度，迁移距离
migration of divide 分水界迁移
migration of fission products 裂变产物迁移
migration process 迁移过程
migration velocity 牵移速度
migration zone 迁徙区
migration 迁移【放射性核素】
migratory bird 候鸟
MIG(metal inert gas) welding 熔化极惰性气体保护电弧焊
miking 测微
mild base 弱碱
mild carburizing 亚共析渗碳
mildew and fungus inductive environment 易霉的环境
mild scale 软垢
mild slope 平缓边坡
mild steel arc welding electrode 低碳钢焊条

mild steel checkered plate 花纹钢板
mild steel yoke 低碳钢磁轭，软钢磁轭
mild steel 低碳钢，软钢，热轧钢
mild wear 中度磨损
mileage indicator 里程指示器
mileage speed 英里风速
mileage stone 里程碑，路标
mileage 按里计算的运费，里数，里程，英里数
mile of wind anemometer 英里风速计
milestone date 里程碑日期
milestone schedule 里程碑计划，里程碑式进度表，一级进度计划
milestone 里程碑，转折点
mile wind speed 英里风速
mile 英里
mil-foot 密尔英尺
military compact reactor 军用小型紧凑反应堆
milker 电池充电用低压直流发电机
milk-glass scale 乳白玻璃刻度盘，毛玻璃标度盘
milk-glass 毛玻璃，乳白玻璃
milk of lime 石灰液，石灰浆，石灰乳
milky glass 乳白玻璃
milky solution 乳状溶液
milky surface of glass 乳白面玻璃
milky way 银河
milk 牛奶，乳剂，蓄电池个别单元充电不足
millammeter 毫安表
mill and coal bunker bay 磨煤机及煤仓间
mill-annealed tube 工厂退火的管子
mill-annealed 轧制退火
mill ball 研磨球，磨煤机钢球
mill classifier 磨煤机的粗粉分离器
mill control system 磨煤机控制系统
millcut saw 磨切锯机
mill cutter 铣刀
mill drying 磨煤机内干燥
milled blade 铣制叶片
milled peat 铲采泥煤
miller 铣床，铣
mill exhauster （磨煤机的）排粉机
mill fan 排粉风机，排粉机，磨煤机风机
mill file 扁锉
mill finish 轧制表面光洁度
mill fire 磨煤机着火
mill housing 磨煤机罩壳
milliammeter 毫安表，毫安计
milliampere man 弱电工程师
milliamperes per volt 毫安每伏
milliampere 毫安
milliard 毫拉德
millibar-barometer 毫巴气压表
millibar 毫巴
millicoulomb 毫库仑
millicurie hour 毫居里小时
millicurie 毫居里
milliequivalents per liter 每升毫克当量
milliequivalent 毫当量
milligamma 毫微克
milligram equivalent 毫克当量
milligram radium equivalent 毫克镭当量
milligrams per liter 毫克每升
milligram 毫克
millihenry 毫亨利【mH】
millilambda 毫微升，10^{-9} 升
milliliter 毫升
millimeter wave communication 毫米波通信
millimeter 毫米
millimetric wave 毫米波
millimicrofarad 毫微法，纳法拉
millimicromicroammeter 飞安计，毫微微安计
millimicron 纳米，10^{-9} 米
millimicrosecond 纳秒，毫微秒
millimole 毫克分子
milling cutter travel 铣刀行程
milling cutter 铣刀
milling head 铣削头，铣头
milling machine 铣床
milling of ores 选矿，矿石处理
milling plant 煤粉制备装置
milling system 制粉系统
milling 碾磨，轧制，破碎，铣，铣平，捣碎，碾碎，磨碎，压碎，磨，制粉
mill inlet pressure control 磨煤机入口负压［压力］控制
millinormal 毫克当量的
milli-ohm 毫欧
million electron volts 百万电子伏特
million gallons per day 百万加仑/天
million operations per second 百万次运算/秒
millions of instructions per second 百万条指令/秒
million 百万
millioscilloscope 小型示波器
millipore filter 微孔滤纸，微孔过滤器
millipore 微孔
millisecond delay blasting 毫秒迟发爆破
millisecond electric blasting cap 毫秒电爆雷管
millisecond 毫秒
millivalve voltmeter 电子管毫伏计，电子管毫伏表
millivoltammeter 毫伏安计
millivoltampere 毫伏安
millivolt analog-digital data encoder 毫伏模拟数据编码器，毫伏模拟数字测试系统
millivoltmeter pyrometer 毫伏计式高温计
millivoltmeter 毫伏表，毫伏计
millivolt 毫伏
milliwatt 毫瓦
milli 毫【词头】，千分之一
mill load ratio 磨煤机负荷率
mill motor 轧机电动机，压延用电动机
mill rejects hopper 石子煤斗
mill rejects system 磨煤机石子煤处理系统
mill room 磨煤机房
mill-run 刚从机器里生产出来的
mill scale 热轧钢锭的氧化，轧制氧化皮，氧化轧皮，轧屑，轧铁鳞，密尔刻度
mill separator 粗粉分离器，磨煤机分离器
millstone grit 磨石砂砾
millstone 磨石
mill tempering damper 磨煤机调温挡板
mill tempering 磨煤机调温

millvoltmeter 毫伏表
mill weir 小水堰
mill 密尔【反应性】，磨，磨煤机，工厂，铣，铣刀，密耳，磨碎机，磨细，磨机，磨粉机
mils per year 密耳/年
mil 密耳
MI(master instruction) 使用说明书
mimesite 粒玄岩
mimetic diagram 模拟图
mimetism 模仿性，拟态
mimic board 模拟图，模拟盘，模拟屏
mimic bus 模拟母线，模拟电路，系统主接线单线图
mimic diagram 模拟图【DCS组态】，示意图
mimic disconnecting switch 模拟断路开关，模拟断路器
mimic display 模拟显示，模拟盘
mimic panel position indicator lamp 模拟盘位置指示器灯【十字形的】
mimic panel 模拟表盘，模拟盘，模拟屏，模拟图
mimic 模拟的，模仿的，拟态的，仿造的，伪造物
MIM(metal insulator materials) 金属绝缘体材料
minable 宜于开采的
min ambient temperature over the years 多年极端最低温度
MINDC(mass-impregnated non-draining cable) 不滴流电缆
mine age 可采年限【矿】
mine coal 原煤，矿煤
mine dump 矿渣废料堆成地
mine dust 矿尘
mine field 煤田
mine mill 竖井式磨煤机
mine motor 矿用电动机
mine mouth plant 矿口电厂，坑口电厂
mine mouth power plant project 坑口电站工程
mine mouth power plant 坑口发电厂，矿口电站，坑口电厂
mine mouth site 矿口【电厂】
mine mouth 坑口
mine-railway motor 矿用牵引电动机，矿用机车电动机
mineral acid 矿物酸，无机酸
mineral aggregate 矿物集料
mineral-bearing water 矿质水，含矿物质的水
mineral board 矿物板
mineral coal 矿物煤，烟煤
mineral compound 无机化合物
mineral constituent 矿质成分
mineral cotton 矿棉
mineral dust 矿尘
mineral extraction 采矿
mineral fiber 矿渣棉，矿物纤维
mineral fibre 矿物纤维，矿渣棉
mineral fuel 矿物燃料
mineral grain 矿物颗粒
mineral hardness (水的)含盐硬度
mineral identification 矿物鉴定
mineral-insulated cable 矿物绝缘电缆，无机绝缘电缆
mineral-insulated coil 矿物绝缘线圈，无机物绝缘线圈
mineral-insulated plastic-sheathed cable 无机物绝缘的塑料包壳电缆
mineral-insulated power cable 无机物绝缘的电力电缆
mineral insulation cable 矿棉绝缘电缆
mineralization 矿化作用
mineralized degree 矿化度
mineral-matter-free basis 无矿物质基
mineral-matter-free coal 无矿物质的煤，无灰煤
mineral matter 矿物质
mineralogical analysis 矿物分析
mineralogical composition 矿物成分
mineralogical 矿物学的
mineral oil 矿物油，石油
mineral particle 矿物颗粒
mineral pitch 地沥青
mineral pollution 矿物污染
mineral reserves 矿产资源
mineral resources 矿物资源
mineral rock 矿物岩
mineral sludge 矿质污泥
mineral soil 矿质土
mineral sulfur 无机硫
mineral tar 软沥青
mineral void 矿质填料孔隙
mineral water 矿水，矿质水
mineral wax 地蜡，石蜡，矿物蜡
mineral wool board 矿渣棉板
mineral wool paper felt 矿棉纸油毡
mineral wool 矿渣棉，石纤维，矿棉，矿渣绒
mineral 矿物的，矿物，无机物，无机的
mine refuse 矿渣
miner's coal ton 煤吨
mine run bin 原煤仓
mine run coal 原煤
mine wastewater 矿山废水
mine water 矿井水，矿坑水，矿水
mine wool 矿棉
mine 矿，矿井，坑道
mingle 混合
mini-amplifier 小型放大器
miniature boiler 小型锅炉
miniature brush 微型电刷
miniature busbar 小母线
miniature capacitance probe 小型电容探头
miniature circuit breaker 微电路开关，微型断路器
miniature current meter 小型流速仪
miniature digital display 小型数码显示管
miniature fission chamber 微型裂变室，小型可移动裂变室，小型裂变室
miniature fuse 小型熔断器
miniature heater 小型加热器
miniature instrument 小型仪表
miniature ionization chamber 微型电离室
miniature lamp 小型灯泡，指示灯
miniature led lamp 小型发光二极管
miniature motor 微型电动机，小型电动机
miniature nuclear battery 小型核电池

miniature probe　微型探针
miniature relay　微型继电器，小型继电器
miniature resistor　小型电阻器
miniature switchboard　小型开关板，小型配电盘
miniature switch　微型开关，小型开关
miniature tractor　小型拖拉机
miniature transformer　小型变压器
miniature valve　小型电子管
miniature　微型的，小型的，小型，缩影，缩样，小型物
miniaturisation　小型化，微型化
miniaturized capacitance　小型化电容
miniaturized ionization chamber　微型电离室
miniborer　小型隧洞掘进机
minibreaker　小型断路器
minicomputer program control　微机程控
minicomputer program　微机程序
minicomputer　微型计算机，小型计算机
minicom　小电感比较仪
miniflow line　小流量管线
miniflow　小流量【再循环】
mini-hydropower station　小（型）水电站
mini-log　单元式晶体管封装电路
minimal access programming　最快存取程序
minimal access　最快存取，最快访问
minimal cut set equation　最小割集方程
minimal cut set　最小割集
minimal detectable activity　最低可探测活度
minimal-latency coding　最快存取编码
minimal path　最短路径
minimal value　最小值，极小值
minimal　最小的，极小的，最低的
minimax approximation　极大极小逼近
minimax principle　极大极小原理
minimax　极小化最大，极大极小，最大最小，扩亮扩暗，极小化极大算法，使对方得点减到最低以使自己得最高分的战略
minimization　化为最小值，化为极小值，求最小值，最小化
minimize　减至最低程度，求最小值，减到最少，缩至最小
minimizing　最小化
minimum accelerating torque　最小加速力矩
minimum acceptability requirement　最少可接受要求
minimum access coding　最快存取编码
minimum access programming　存取时间最短的程序设计，最快存取程序设计
minimum access routine　最快存取程序
minimum access time　最小访问时间
minimum access　最快存取，最优存取
minimum actual tooth spacing　最小实际齿距
minimum allowable cross-section　最小允许截面
minimum allowable oil pressure　最低允许油压
minimum allowable temperature　最低允许温度
minimum amount　最小金额
minimum arc voltage　最小起弧电压
minimum area　最小面积
minimum bend radius　最小弯曲半径
minimum bid　最低展现价格，出价的起价
minimum bubbling velocity　最小鼓泡速度
minimum burn-out margin　最小烧毁裕量
minimum burn-out ratio　最小烧毁比
minimum capacity　最小容量
minimum chance of cladding failure　包壳破损最小概率
minimum charge　最低限度费用
minimum clearance　最小间隙
minimum conductance function　最小电导函数
minimum contact current　最小触点电流
minimum controllable power level　最低可控功率
minimum coolant flow　冷却剂量小流量
minimum cooling surface　最小冷却面
minimum corresponding water level　相应最低水位
minimum cost estimating　最小代价估计，最低价格估计
minimum creep distance　最小爬电距离
minimum critical heat flux ratio　最小临界热流密度比
minimum critical infinite cylinder diameter　无限圆柱最小的临界直径
minimum critical infinite slab dimension　无限平面的最小临界厚度
minimum critical infinite slab thickness　无限平面的极限临界厚度
minimum critical mass　最小临界质量
minimum critical volume　最小临界体积
minimum current　最小电流
minimum curve radius　最小弯曲半径
minimum cut-out　最小电流自动断路器
minimum cut set fault tree　最小割集故障树
minimum cut set　最小路程安排，最小割集【事故树】
minimum daily discharge　最小日流量
minimum dead load　最小静负荷
minimum decision limit　最低判定限
minimum-delay coding　最小延迟编码
minimum demand　最小需量
minimum departure from nucleate boiling ratio　最小偏离泡核沸腾比，最小 DNB 比
minimum detectable activity　最低可测活度
minimum detectable amount　最小可检出量，最小检测量
minimum detectable distance　最小可检距离
minimum detection limit　最低探测限
minimum determination limit　最低测定限
minimum deviation angle　最小偏向角
minimum deviation method　最小偏差法
minimum diameter　最小直径
minimum-differential code　最小差编码
minimum discernible signal　最小识别信号
minimum disruptive voltage　最低击穿电压
minimum distance　最小距离
minimum disturbance　最小干扰
minimum DNB ratio　最小偏离泡核沸腾比
minimum DNBR limiter　最小偏离泡核沸腾比限值器
minimum drag coefficient　最小阻力系数
minimum drag position　最小阻力姿势
minimum drawdown level　最低泄降水位
minimum dropout　（继电器的）最低释放值

minimum dry weather flow 最小枯水流
minimum duration 最短持续时间，最短期限
minimum effective temperature 最低有效温度
minimum emergency cooling system 最小应急冷却系统
minimum emergency planning distance 最小应急计划距离
minimum energy line 最小能量线
minimum energy requirement 最低能量要求
minimum envelope curve 最小包络线
minimum excitation limiter 最低励磁限制器
minimum excitation limiting 最低励磁限制
minimum exclusion distance 最小禁允距离，最小排斥距离
minimum factor 最小因素
minimum fire-protection covering 最小防火保护层
minimum flashover voltage 最低闪络电压
minimum flood 最小洪水
minimum flow bypass line 最小流量旁路管线
minimum flow line 最小流量管线
minimum flow recirculating system 最小流量再循环系统，最小流量装置
minimum flow recirculation control system 最小流量再循环控制系统
minimum flow valve 最小流量阀
minimum flow velocity 最小流速
minimum flow 最小流量【再循环】
minimum fluidization state 临界流态化状态
minimum fluidization 临界流态化
minimum fluidizing velocity 临界流化速度
minimum fresh air requirement 最小新风量
minimum gap 最小间隙
minimum grade 最小坡度
minimum ground clearance 最小对地距离
minimum headway above railway 铁路上部界限
minimum head 最低水头，最小水头
minimum heat flux 最小热流密度
minimum impedance relay 最低阻抗继电器
minimum induced loss windmill 最小诱导损失风车
minimum interference 最小过盈，最小干涉
minimum latency programming 最小等数时间程序设计，最快存取程序设计
minimum latency routine 最快存取程序
minimum latency 最快存取，最小等数时间
minimum leakage door 最小泄漏门
minimum lethal dose 最小致死剂量
minimum level 最低液面
minimum life section 危险截面
minimum liquid level 最低液位
minimum loading of power plant 发电厂最小出力
minimum load operation 最低负荷运行
minimum load 最低负载，最小负荷，最低负荷
minimum low population distance 最小低密度人口距离，最低密度人口距离
minimum moisture capacity 极低持水量
minimum moisture content 最小含水量
minimum navigation depth 最低通航水深
minimum number 最小数
minimum of subsistence 最低生活费
minimum operate voltage 最低工作电压，最低吸合电压，最低操作电压
minimum operating level 最低运行水位
minimum operating pressure 最低操作压力，最低运行压力
minimum operating temperature 最低运行温度
minimum operational mode 最小运行方式
minimum outdoor air 最小室外空气量
minimum output current 最小输出电流
minimum output voltage 最低输出电压
minimum output 最小出力，最低出力
minimum percentage of reinforcement 钢筋混凝土最小配筋率
minimum permissible velocity 最小允许流速
minimum phase shift function 最小相位移函数
minimum phase-shift network 最小相位移网络
minimum phase system 最小相位系统
minimum pool level 最低池水位，死水位
minimum potential water power 最小水力蕴藏量
minimum power factor 最小功率因数
minimum power voltage 最小电源电压
minimum premium 最低保险
minimum price 最低价格
minimum principal stress 最小主应力
minimum profit 最低利润
minimum pulse duration 最小脉冲持续时间
minimum reactance function 最小电抗函数
minimum recirculation flow rate 最小再循环流量
minimum recorded flood peak 最小实测洪峰
minimum redundance code 最小冗余码，最小余度码
minimum relay 低值继电器，低载继电器
minimum requirement 最低要求
minimum resistance of heat transfer 最小传热热阻
minimum response concentration 最低响应浓度
minimum running current 最小工作电流
minimum runoff 最小径流，最小径流量
minimum sample length 最小取样长度
minimum saturation water 最小饱和水量
minimum scale value 标度始点值，最小标度值
minimum signal method 最小信号法
minimum spacing 最小间距
minimum speed at zero stroke 死行程最低转速，零行程时的最低转速
minimum speed 最低速度
minimum stable load 最低稳定负荷
minimum stage 最低水位
minimum standard 最低标准
minimum storage 最小库容
minimum stress limit 下限应力
minimum susceptance function 最小电纳函数
minimum temperature 最低温度
minimum thermometer 最低值温度计
minimum trip duration 最小脱扣时间
minimum turning radius 最小转弯半径
minimum turning time 最短盘车时间
minimum usable frequency 最低可用频率
minimum useful signal 最小可用信号
minimum value 最小值，极小值
minimum velocity 最低速度
minimum voidage 最小空隙度

minimum void ratio　最小孔隙比
minimum voltage　最低电压
minimum water level　最低水位
minimum water storage level　最小蓄水量
minimum wave duration　最小波浪历时
minimum wind to yaw　最小调向风
minimum working current　最小工作电流
minimum yield point at elevated temperature　高温时最低屈服强度
minimum　最小值，最低值，最小的，最低的，极小的，最低限度
mining area　采动影响区，采矿区
mining concession　煤矿开采特许权
mining cost　开采费用
mining motor　矿用电机
mining subsidence　采空坍陷，矿穴沉陷
mining transformer　矿用变压器
mining water use　采矿用水
mining　采矿，开采，矿业，开挖
mini-oil breaker　少油断路器
minioscilloscope　小型示波器
minipad　小垫片
mini-pile　小型桩
mini-plant　小型工厂，（实验室规模）试生产用小型设备
miniplate　小片
mini-steam operation of turbo-generator　汽轮发电机无蒸汽运行
mini-steam operation　少蒸汽运行
ministerial examination　部级考核
ministerial medal　部级奖（章）
minister　部长
ministry-graded quality product　部优产品
Ministry of Coal Industry　煤炭工业部
Ministry of Commerce　商业部
Ministry of Economic Relation with Foreign Countries　对外经济联络部
Ministry of Electric Power Industry　电力工业部
Ministry of Finance　财政部
Ministry of Foeign Trade　外贸部
Ministry of Labour　劳动部
Ministry of Overseas Development　海外开发部
Ministry of Railway　铁道部
Ministry of Water Resources and Electric Power　水利电力部
Ministry of Water Resources　水利部
ministry standard　部颁标准
mini-supercomputer　迷你超级电脑
mini-switch　小型开关
minitrack　电子跟踪系统
minitransistor　小型晶体管
minitype　微型，小型
minivalence　最低价
miniwatt amplifier　小功率放大器
miniwatt　小功率
min index　最小市场份额比
min＝minimum　最小的，最小
MIN(monitoring information notice)　监督信息通知
minometer　充电测读仪，充电计数仪
minor accident　小事故
minor axis　（椭圆的）短轴
minor break　小破口
minor change　小变更
minor constituent　次要成分
minor control change　次要控制变化
minor control data　次要控制数据
minor control point　低等控制点，次等控制点
minor cycle counter　短周期计数器
minor cycle pulse generator　小周期脉冲发生器
minor cycle　小循环，小周期，短周期
minor defect　局部缺陷，次要缺陷
minor determinant　子行列式
minor disaster　小事故，小灾害
minor fault　小断层
minor fold　小褶皱
minor insulation　纵绝缘，次绝缘，局部绝缘
minority interest　少数权益
minor loading condition　次要载荷情况，小加载条件
minor lobe　副波
minor loop　局部磁滞回线，小磁滞回线，小回路
minor losses in pipe　管道局部阻力损失
minor loss　次要损失
minor maintenance　小修
minor overhaul　小修
minor principal plane　小主应面
minor principal strain　小主应变
minor principal stress　第三主应力，最小主应力
minor repair　小修
minor shock　副震
minor switch　小型选择器，小型开关，小型寻线器
minor triangulation　低等三角测量
minor　次要的，较小的，子式，反射镜
MINP(metal insulator n/p) solar cell　MINP太阳电池【一种改进的n/p结高效率硅太阳电池】
mint par of exchange　法定汇兑平价
minuend　被减数
minus allowance　负容差
minus effect　副作用，不良效果
minus electricity　阴电，负电
minus material　低劣材料
minus phase　负相位
minus pole　负极，负端子
minus sign　负号，减号
minus tapping　负抽头，负分接头，负分接
minus terminal　负号端子
minus tolerance　负公差
minus　减去，扣除
minute adjustment　精调，微调
minute break　瞬时断流
minute crack　发状裂缝，细裂缝，细裂纹
minute quantity　微量
minutes of meeting　会议记录，会议纪要
minutes of pre-contract discussion meeting　签约前讨论会的会议纪要
minutes of proceedings　议事记录
minutes of talks　会谈纪要
minutes of the pre-bid meeting　标前会议纪要
minutes　会议纪要，备忘录【minute的复数】，分钟

minute　分，分钟
MIN　最小【DCS画面】
Miocene clay　中新世黏土
Miocene epoch　中新世
mio＝million　百万的缩写
MIPS(millions of instructions per second)　百万条指令/秒
mirabilite　硫酸钠，芒硝
mire　泥沼
MIR(memory input register)　存储输入寄存器
mirror coating　镜子涂层
mirror effect　镜像效应
mirror extensometer　反光伸长计
mirror finish　镜面光洁度
mirror galvanometer　反射式检流计，镜式检流计
mirror hybrid reactor　磁镜混合反应堆
mirror image switch　镜像开关
mirror image　镜像，映像，反像
mirror instrument　转镜式仪表
mirror materials　反射镜材料
mirror method　镜像法
mirror oscillograph　镜式示波器
mirror pyrometer　镜高温计
mirror reactor　磁镜反应堆
mirror reflector　镜面反射器
mirror stone　白云母
mirror symmetry　镜对称，假象对称
mirror torsiograph　镜式扭力记录仪
miry soil　淤泥土
misadjustment　误调整，失调，误调节
misaligned switch and instrument　错位开关和仪表
misaligned　未经找正的，未对中的，线向不正的，方向偏离的，不重合的，未对准的
misalignment　未对中，未对准，非线性，中心偏移，轴线不重合度，不正，未对准，错列，定线不准，偏离，误差
miscalculation　不准确的计算，误算，计算错误，计算误差，算错
miscellaneous building　杂项建筑
miscellaneous business　杂务
miscellaneous charge order　杂费支付通知
miscellaneous clause　杂项条款
miscellaneous cost　杂项费用，零星费用
miscellaneous customer accounts expenses　用户账务其他费用
miscellaneous customer service and informational expenses　其他用户服务和信息费用
miscellaneous distribution expenses　配电其他费用
miscellaneous expense　其他费用，杂项开支，杂费
miscellaneous exposure　零散照射量
miscellaneous fill　杂填土
miscellaneous general expenses　其他行政管理费用
miscellaneous hoisting equipment　其他起重设备
miscellaneous hydraulic power generation expenses　水力发电站其他费用
miscellaneous investments　各项投资
miscellaneous load　其他荷载
miscellaneous maintenance expense of steam generation plant　蒸汽（发电）厂其他设备维修费用
miscellaneous maintenance expenses　其他设备维修费用
miscellaneous nuclear power expenses　核发电其他费用
miscellaneous operation　杂操作
miscellaneous parts　辅助部分，其他部件，附件
miscellaneous payment　杂项支出
miscellaneous provision　杂项规定
miscellaneous pump　杂用水泵
miscellaneous sales expenses　其他销售费用
miscellaneous steam power expense　蒸汽发电其他费用
miscellaneous taxes　杂税
miscellaneous transmission expense　输电其他费用
miscellaneous waste　杂类废物
miscellaneous　其他的，杂项的，各种各样的
miscelleneous floor plan　各层平面图
mischance　障碍，故障，损害，不幸，灾难
mischief　损害，伤害，故障
miscibility　可混性，可混物
miscible fluid　可混流体
miscible liquid　易混溶液
miscible　可混合的，易混合的
misclosure　（测量）闭合差
misc.＝miscellaneous　其他的，杂项的
misconnection　误接，错接
misconvergence　无收敛，不会聚
miscount　错算，误算
misdelivery　误投，发货错误，交付错误
mis-discharged cargo　误卸货物
Mises transformation　米塞斯变换
misfeed　误传送，误馈送，传送失效
misfire handling　瞎炮处理
misfire　不点火，不着火，不正常点火，失燃，空炮，熄火，瞎炮，哑炮
misfit river　不相称河
misfit　不合适，不吻合，配合的零件，不符值，误差
misfocus　散焦
misformation　误传，传错信息
mishandling failure　误操作失效
mishandling　不正确运用，误操作，处理不当
mishap　意外事故，损坏，故障，灾难，不幸
misidentification　错误判读，错误判断
misinterpret　误解译，误判断
misjudge　误判断，看错
misleading　使人误解的，引入歧途的
MIS(management information system)　（厂级）管理信息系统，管理层信息处理系统
mismatched temperature　失配温度
mismatched termination　失配终端
mismatch loss　失配损耗
mismatch　不匹配，不重合，零件错配，错匹配，失配，失谐，失调，解谐，错位，不平衡
M is N times as large as R　M是R的N倍，M比R大［多］(N－1)倍
M is N times as many as R　M是R的N倍，M比

R大［多］（N－1）倍
M is N times greater than R　M比R大N倍，M比R大［多］（N－1）倍
M is N times larger than R　M比R大N倍，M比R大［多］（N－1）倍
misoperation　误操作，误动作，动作异常，工作异常
misplaced winding　失位线圈，配置错误的线圈
misplacement　错改，错位
misplug　（插塞）插错
M is preceeded by N　M以前是N，N在M之前
misprinted　印刷错误
misrepresentation　误述，虚报
misrepresent　误传，误说，不如实地叙述
misrouting　错误指向，不正确指向
misrun casting　有缺陷铸件
miss description　虚报货名
missfire shot　不爆发炮眼，瞎炮眼
misshielding cylinder　飞射物筒状屏蔽
mis-shipped cargo　误装货物
missile barrier penetration　飞射物屏障贯穿件
missile barrier structure　飞射物屏障结构
missile barrier　飞射屏障，飞射物屏障
missile impact effect　飞掷物撞击效应
missile proof　防飞射物
missile protection criteria　飞射物防护准则
missile protection shield　飞射物屏蔽
missile protection slab　防飞射物护板
missile protection　飞射物防护，防飞碎片装置
missile roof　防飞射物
missile shielding concrete　飞射物屏蔽混凝土
missile shielding cylinder　飞射物筒状屏蔽
missile shield　飞射物屏蔽
missile slab　防飞射物板
missile　（风卷）飞掷物，飞射物
missing line　遗漏线划
missing part　差缺件，丢失件
missing　故障，失误，遗漏，失踪的，缺少的，损失
mission failure rate　任务失效率
mission life　工作寿命，工作年限
mission　公司的长期目标和原则，代表团，使团
MIS solar cell　MIS太阳电池
miss operation　误操作，拒绝动作，误动作
Missouri University Research Reactor　（美国）密苏里大学研究堆
misstatement　错报，虚报
miss trip　未跳闸，拒跳闸
miss　错过
mistake in programming　程序差错
mistaken judgement　错误判断
mistake　错误，误差，事故，失误
mist atomizer　喷雾器
mist blower　喷雾器
mist cooling　喷雾冷却
mist eliminator　烟雾消除器，脱湿器，去雾器，除雾器
mistermination　终接失配，端接错误，失配，失谐
mist flow　雾流，雾状流
mist light-signal　雾灯光信号
mist lubrication　油雾润滑
mist of fine sand　细沙雾
mis-trip　解扣失误，误脱扣
mist separator　去湿装置，湿气分离器，油污分离器
mist spray　喷雾
mist trap　除雾装置，捕雾器
mistuning　失谐
mist　霭，薄雾，轻雾，雾凇，烟云，（烟）雾
misunderstanding　误解
misuse failure　错用故障，误用失效
Mitchie test　米奇试验
miter bearing　斜接支撑
miter cutting　斜切割
mitered inlet　水道斜坡进口
miter gate leaf　人字门门叶
miter gate　人字闸门
miter gear　等径斜齿轮
mitering　斜接
miter joint　斜面接合
miter saw　斜截锯，斜切锯
miter sill　人字槛
miter valve　锥形阀
miter wall　人字墙
miter weld　斜接角焊缝
miter　退潮，落潮，衰退，斜接缝
mitigate　缓和，减轻，缓解
mitigation of consequences　减轻后果
mitigation of reactor accidents　减轻反应堆事故措施
mitigation　缓和，减轻，和缓，调节，缓解，治理
MIT(master instruction tape)　主指令带
mitre bend　斜接弯管
mitred core　45°斜接式铁芯
mitred elbow　斜接弯管【肘管】
mitre drain　斜接排水沟
mitre elbow　斜接弯头，虾米腰弯头
mitre gear　等径伞齿轮
mitre joint　斜面接合，斜接头
mitre welding　斜接焊接
mitre　斜撑，斜角，斜面接合，斜接，斜接缝，斜接面，斜角缝，成45°接合
mitron　米管，宽带磁控管
Mitsui & Co　三井物产公司【日本】
mix design　配料设计，混合料配合比设计，配合比设计
mixed acid　混合酸【如硫酸和硝酸混合】
mixed air　混合气体
mixed amplifier　混合放大器
mixed arch furnace　混合拱炉膛
mixed avalanche　混合崩坍
mixed base notation　混基记数法，混合进位制表示法
mixed bed column　混合床离子交换柱，混合柱
mixed bed demineralization　混合床除盐
mixed bed demineralizer　混合床除盐装置，混合床脱矿质器
mixed bedding　复合层理
mixed bed filter train　混合床过滤器管系
mixed bed filter　混合床过滤器，混床过滤器
mixed bed ion exchanger　混合床离子交换器
mixed bed ion exchange　混合床离子交换

mixed bed resin 混床树脂
mixed bed system 混合床系统【水处理】
mixed bed water treatment 混合床水处理
mixed bed 混合床，混床
mixed boundary condition 混合边界条件
mixed car 混合车皮
mixed cement 混合水泥
mixed coal silo 混煤仓
mixed coal 混煤【0～50mm】
mixed column 混合柱
mixed complex 混合配合物
mixed concrete 拌好的混凝土
mixed condensation 混合凝结
mixed construction 混合式建筑，混合结构
mixed consumer 混合用户
mixed convection 混合对流
mixed cooling tower 混合式冷却塔
mixed cooling 混合式冷却
mixed core 混合堆芯
mixed credit 混合信贷
mixed currency 混合货币
mixed current 混合潮流
mixed cycle 混合循环
mixed-dielectric capacitor 复合电介质电容器
mixed-dielectric film capacitor 混合介质薄膜电容器
mixed dislocation 混合位错
mixed electrode 混合电极
mixed fission product 混合裂变产物
mixed flow Francis turbine 混流式法兰西斯水轮机
mixed flow impeller 混流（式）叶轮
mixed flow model 混合流模型
mixed flow pump 混流泵，离心螺旋泵，混流式水泵
mixed flow reactor 混流反应器
mixed flow sewage pump 混流式污水泵
mixed flow turbine 混流式透平，混流式涡轮机，混流式水轮机
mixed flow wheel 混流式水轮
mixed flow 混流式，混合流，混流
mixed fluidized bed 混合流化床
mixed free and forced convection 自由与强迫混合对流
mixed fuel burning ratio 燃料混烧率，混烧率
mixed fuel 混合燃料
mixed gamma spectrometry 混合γ能谱法
mixed gas welding 混合气体保护焊
mixed gas 混合煤气，混合气体
mixed governing 混合调节
mixed-grained coal 混粒煤
mixed-grained 混粒的
mixed heat exchanger 混合式换热器
mixed highs system 高频混合制，混高频系统
mixed indicator 混合指示剂
mixed inhibitor 混合缓蚀剂，混合抑制剂
mixed-in-place 现场拌和
mixed integer programming 混合整数程序设计
mixed interface 混合接口
mixed layer height 混合层高度
mixed linear programming 混合线性程序设计
mixed loop 混合环
mixed lump coal 混块煤
mixed magnetic Reynolds number 混合磁雷诺数
mixed medium-sized coal 混中块煤
mixed module damper 多模数挡板
mixed mortar 水泥石灰砂浆，混合砂浆
mixed number density constant 带分数密度常数
mixed oxide fuel fabrication plant 混合氧化物燃料厂，混合氧化物燃料制造厂
mixed oxide fuel 混合氧化物燃料
mixed oxides 混合氧化物
mixed packing column 混合填充柱
mixed paint 调合漆
mixed pea coal 混粒煤
mixed phase layout 混相布置
mixed power plant 混合式电站
mixed pressure steam turbine 混压式汽轮机
mixed pressure turbine 多压式汽轮机
mixed progressive flow 混合行进水流
mixed pumped-storage plant 混合式抽水蓄能电站
mixed Pu-U oxide 钚与铀混合氧化物
mixed radiation field 混合辐射场
mixed radiation 混合辐射
mixed radix 混基
mixed rags 组合
mixed ratio 混合比例
mixed risk 水陆联运险
mixed rotor winding 混合转子绕组
mixed sample 混合试样
mixed sand-and-shingle spit 砂砾混合砂嘴
mixed sand 混合砂
mixed sea and land risk 水陆联合保险
mixed size particles 混合粒度颗粒
mixed small coal 混末煤【粒径 25mm】
mixed source supply 混合电源
mixed spectrum reactor 混合谱反应堆
mixed-strand bar （电机）交叉股绞线棒，换位线棒
mixed superheater 混合式过热器
mixed support 混合支架
mixed tariff 混合关税
mixed test coupon 混合试样
mixed tube 混频管，混波管，混合潮
mixed type heat exchange 混合式热交换器
mixed UO_2-PuO_2 fuel 二氧化铀-二氧化钚混合燃料
mixed uranium-plutonium carbide 铀-钚混合碳化物
mixed U-Th oxide kernel 混合铀-钍氧化物芯核
mixed video 混合视频
mixed years money 混合年货币
mixer amplification by variable reactance 低噪声微波放大，可变电抗混频放大
mixer crystal 混频器晶体
mixer lorry 移动式混凝土拌和机
mixer plant 拌和楼，配料装置
mixer-settler 混合器沉降槽，混合澄清器
mixer 搅拌器，混合器，混频器，混频管，搅拌机，拌和机
mixing air 混合用空气

mixing amplifier 混合放大器
mixing at site 工地拌和
mixing basin 混合池
mixing box section 混合段
mixing bunker 混煤仓
mixing burner 混合式喷燃器
mixing cell 混合单元
mixing chamber 混合分配器，搅拌室，混合室
mixing channel 混合沟渠，混合槽
mixing characteristics 混合特性
mixing circuit 混频电路，混合电路，或电路
mixing coefficient 混合系数，湍流扩散率，配合比，交混系数
mixing column 混合柱
mixing condensation 混合冷凝
mixing condenser 混合式凝汽器
mixing cup temperature 平均温度
mixing cycle 拌和周期
mixing depth 混合厚度
mixing device 混合装置
mixing drum 拌和鼓
mixing effect 混合作用
mixing fog 混合雾
mixing grid 混合格架
mixing header 混合联箱，混合集箱
mixing heater 混合式加热器，混合加热器
mixing height 混合高度
mixing in place 工地拌和，工地搅拌
mixing layer 混合层，混流层
mixing length theory 混合长度理论，混合长理论
mixing length 混合长度，混合长
mixing of coal sample 煤样掺合
mixing of concrete 混凝土搅拌
mixing orifice 混合孔板
mixing parameter 拌和参数
mixing plant 拌和厂，搅拌站，混合机
mixing platform 拌和平台
mixing plenum 混合腔
mixing plough 混合犁
mixing process 混合过程
mixing pump 混合泵
mixing rates 混合汇率
mixing ratio 混合比
mixing space 混合室
mixing sphere 混合球
mixing Stanton number 交混斯坦顿数
mixing tab 混合片
mixing tank 混合箱
mixing tee T形三通，T形混合管
mixing time 拌和时间
mixing tube 混频管，混合管，混波管
mixing-type heat exchanger 混合式热交换器
mixing-type water heater 混合式加热器
mixing unit 搅拌器，混合器
mixing valve 混合物调节阀，混合阀
mixing vane grid 搅混叶格架
mixing vane （燃料组件的）搅混叶，混流翼，交混翼
mixing velocity 拌用速度，混合速度
mixing vessel 混合容器
mixing waste 混合废料
mixing water requirement 拌和需水量
mixing water （混凝土的）拌和用水
mixing zone 掺冷区，混台区，掺混区
mixing 混合，混频，拌和，加水搅拌混合
mixometer 拌和计时器
mix proportion 混合比【混凝土】，配合比，混配比例
mix ratio 配合比，混合比，搭配比率
mixture chamber 混合室
mixture control valve 混气调节阀
mixture control 混合物成分控制
mixture design 配料设计
mixture distribution （沿汽缸）混合物分布
mixture indicator 混合物成分指示器
mixture of gases 气体混合物【屏蔽气体】
mixture pressure type steam turbine 混压式汽轮机
mixture ratio （组分）混合比，（混合物）组成比，燃料空气比，配合比，混合比例
mixture region 混合区，双相区
mixture water 混合水
mixture 混合气，混合物，配合料，拌和物，掺合，混合剂，混合体
mix 拌和，搅拌，混合，拌和物【混凝土】，混合物，混频，掺和，配合
MKUP(make-up) 补充
MLB(miniland bridge) 小陆桥运输
MLD(minimum lethal dose) 最小致死剂量
mL＝milliliter 毫升
ML＝mill 磨煤机
Mln＝million 百万的缩写
M＝mass 质量
MMCS(man-machine control system) 人-机控制系统
MMD(maximum mixing depth) 最大混合层厚度
M＝mechanical 机械
M＝mega 兆
mmethod of successive correction 逐步校正法
mmf harmonics 磁动势谐波
MMF(magneto motive force) 磁动势
MMHTR(mean manhours to repair) 平均维修工时
MMI(man-machine interface) 人机交换界面［接口］，人机接口
MMI(modified Mercalli intensity) 修正麦加利地震烈度
30mm mesh screen 30mm 筛孔的筛子
mm＝millimeter 毫米
MMM(monolithic main memory) 单片主存储器
MM(modified Meralli scale) 修正的麦卡利地震烈度表
MMOD(mean maintenance outage duration) 平均维护停运延续时数
M＝module 模数，模件
M＝mole 摩尔
M＝moment 力矩
M＝motor 马达
MMS(mannutacture message specification) 制造报文规范
MND(mixed number density) constant 带分数密度常数

mnemonic code　助记码
mnemonic diagram　记忆图
mnemonic symbol　记忆符号
mnemonic　记忆的，记忆符号，记忆存储器
MN＝main　主要的
MNOS(metal nitride oxide semiconductor)　金属氮氧化物半导体
MNTL(modified nonthreshold logic)　改进的非阈值逻辑
moat　壕，护城河
mobile air compressor　移动空压机
mobile anticyclone　移动性高气压
mobile barrage　活动堰
mobile bed model test　动床模型试验
mobile bed model　动床模型
mobile bed　不稳定河床，动床
mobile belt　活动带
mobile blade　动叶片【汽轮机】
mobile closed circuit TV camera　移动的闭路电视摄像机
mobile coal belt　移动式输煤带
mobile conveyor　移动式皮带机
mobile crane charge　移动式其重机费
mobile crane　汽车起重机，汽车吊，移动式吊车，移动式起重机
mobile data acquisition system　可移式数据采集系统
mobile digital computer　可移式数字计算机，可移动数字计算机
mobile drill　轻便钻机
mobile electron　流动电子
mobile equilibrium　动态平衡
mobile equipment garage　车库
mobile equipment replacement cask　移动式设备更换箱
mobile equipment　可移动设备
mobile $FeSO_4$ filming wheel　移动硫酸亚铁涂膜小车
mobile fire extinguisher　移动式灭火器
mobile fire-fighting equipment　移动式消防设备
mobile gas turbine　移动式燃气轮机
mobile generator　移动发电机
mobile generator set　移动电站，移动发电机组
mobile haulage system　移动式牵引系统
mobile heating station　移动式供热站
mobile hydraulic jack　移动式液压千斤顶
mobile ion　流动离子
mobile laboratory　移动式试验室，流动试验室
mobile lamp　行灯
mobile load　活动荷载，活动载荷，活荷载，易变荷载
mobile monitoring system　移动监测系统
mobile nondestructive assay laboratory　移动式无损检验实验室
mobile nuclear power plant　流动核电厂
mobile office　办公箱
mobile oil　机油
mobile phone　手机，移动电话
mobile platform　移动式平台
mobile pollution source　移动污染源，运动污染源
mobile power station　移动式电站，移动电站
mobile power unit　移动式动力装置
mobile remote-control intervention operator　可移动远控干预控制器
mobile sand　流沙
mobile source emission　运动源排放
mobile source of pollution　移动污染源
mobile stage　活动平台
mobile state　运动状态
mobile substation　移动变电站
mobile support　移动支架
mobile telephone　移动式电话，移动电话
mobile test load　可动试验荷载
mobile transformer　移动式变压器
mobile wharf　活动码头
mobile wind speed unit　流动式风速测量车
mobile　可移动的，可动装置，悬挂饰物，移动，运动物体
mobility edge　迁移率边
mobility of particle　颗粒流动性
mobility of sand dune　沙丘流动性
mobility ratio　迁移率比
mobility-type analogy　导纳型模拟，一种电声-机械动态模拟
mobility　机动性，流动度，流动性，迁移率，变动性
mobilization degree　流动度
mobilization fee　启动费，准备费
mobilization schedule　动员计划，进（现）场计划
mobilization　动员，准备
mobilize enthusiasm of staff and workers　调动职工积极性
mock-up experiment　模拟实验
mock-up furnace　模化炉膛
mock-up lattice　模拟栅格
mock-up reactor　模拟反应堆
mock-up test　样机试验，模型试验，模拟试验
mock-up-type measurement　模拟型测量
mock-up　1比1模型，全尺寸模型，实物模型，制造模型，制造样机，实体模型，样机
mock　假的，模拟的，模拟，模仿，制造模型
MOC(main operational controller)　主操作控制器
MOCS(mechanism operated cell switch)　机械操作式开关
modal amplitude　模态幅值
modal analysis　模态分析
modal balancing theory　振型平衡理论，模平衡理论
modal damping　模态阻尼
modal displacement　模态位移
modal force　模态力
modal frequency　模态频率
modal mass　模态质量
modal method　模态法
modal response　动态反应，模态响应，振型反应特性
modal stiffness　模态刚度
modal superposition　模态叠加，振型叠加
modal wind load　模态风载
modal　模型的
mode action　控制作用

mode amplitude 波幅
mode change 运行方式变换
mode control 状态控制，模式控制
mode conversion 波型转换【超声】
mode converter 模变换器
mode-expansion method 模式展开法，模项展开法
mode field concentricity error 模场同心度误差
mode field diameter 模场直径
mode field non-circularity 模场不圆度
mode filter 振荡型滤波器，波形滤波器
mode interference 振荡模干扰
mode I stress intensity factor I型应力强度因子
model adaptive control 模型自适应控制
model agreement 示范协定
model analysis 模拟分析，模型分析
model and program 模型和程序
model atmosphere 标准大气，模式大气
model basin 试验基地，模型试验池
model B30 cylinder B30 型缸
model blocking effect 模型阻塞效应
model casing 模型箱
model coil 模型线圈
model contract 标准合同，示范合同，格式合同，合同范本
modeled on 模仿，仿效，仿制
model efficiency 模型效率
model equipment 模拟设备，模型装置
model experiment 模拟实验，模型实验，模型试验
model fluid 模拟液体
model for quality assurance 质量保证模式
model implementation conformance statement 模型实现一致性陈述
modeling design 模型设计
modeling result 模式预测结果
model law 模拟律，模型律
modelling criteria 模拟准则，建模标准
modelling technique 模拟试验技术，模型技术
modelling verification 模型验证
modelling 模型制作，模型设计，模仿，造型，做模型，模型化，模型试验，模拟（试验）
model machine 样机
model motor 模型电动机，样机电动机
model network 模拟网络
model No. 型号
model number 牌号，型号
model of plastics 塑料模型
model of structure 结构模型
model parameter 模态参数
model power plant 示范电厂，示范电站
model project 示范工程
model-prototype comparison test 模型与原型比较试验
model-prototype relationship 模型与原型关系
model rating curve 模型率定曲线
model reference adaptive control system 模型参考自适应控制系统
model-reference adaptive control 模型参考自适应控制
model-reference 模型参考，模型基准
model rotor 模型转子，模拟转子
model scale 模型缩尺
model scope 模型范围
model similarity 模型相似性
model stator 模拟定子
model stream 典型河流
model study 模型研究
model support 模型支架
model symbol 模型符号
model testing basin 模型试验池
model test 模型试验，模拟试验
model turbine 模型水轮机
model voltage 模型电压
model wake blocking 模型尾流阻塞
model water turbine 模型水轮机
model 样式，模型，样品，样机，型号，式样，原型，模式，类型
MODEM（modulator-demodulator） 调制解调器，调制反调制器
modem pooling 调制解调器系统
mode of combined rotor 组合转子振动波型
mode of connection 接线方法，接线方式
mode of control 控制方式
mode of fluidization 流化形态
mode of interaction 相互作用的模式
mode of operation 运行方式，操作方式，工作方式，工作原理
mode of oscillation 振荡方式
mode of payment 支付方式
mode of requalification 再鉴定的波型
mode of resident relocation 库区移民安置方式
mode of speed regulation 调速方式，调速方法
mode of transportation 输送方式
mode of vibration 振动模态，振型，振动方式
mode of water intake and discharge 取水排水方式
moderate belt 温带
moderate breeze 和风，四级风
moderate concentration 中度浓度
moderate corrosion 中等腐蚀
moderate density 中等密度
moderated reactor 慢化反应堆
moderated 慢化的
moderate erosion 中度侵蚀
moderate fixed duty 低率固定关税
moderate fog 中等的雾
moderate frequency event 中等频率事件
moderate gate 疾风，七级风
moderate gust 中等阵风
moderate head 中水头
moderate heat cement 中热水泥
moderately coarse texture 中等粗质地
moderately enriched uranium 中等浓缩铀，中浓缩铀【5%～80%】
moderately strong wind 中强风
moderate Mach numbers 中等马赫数
moderate odor 中等臭味
moderate rainfall 适度降雨量
moderate rain 中雨
moderate swell 中等涌浪
moderate temperature 中等温度

moderate transient 中速瞬变
moderate tropical storm 中等热带风暴【最大风速 34～47 海里/时】
moderate visibility 中常能见度
moderate wind 和风
moderate 慢化，减速，缓和，适度的，中等的，温和的，稳健的
moderating heat exchanger 省热器【硼热再生辅助系统】
moderating material 慢化材料
moderating medium （中子）减速介质，慢化介质，减速剂
moderating power 慢化能力
moderating property 慢化特征，慢化性能
moderating ratio 慢化比
moderating reflector 慢化反射层
moderation cross section 慢化截面
moderation effect 慢化效应
moderation time 慢化时间
moderation with absorption 有吸收的慢化
moderation 慢化，缓和，减轻，减速
moderator assembly 慢化剂组件
moderator brick 慢化剂块
moderator circuit valve compartment 慢化剂回路阀室
moderator circulation system 慢化剂循环系统
moderator control 慢化剂控制
moderator coolant gas stream 慢化剂的冷却气流
moderator coolant 慢化冷却剂，减速冷却剂
moderator cooler 慢化剂冷却器
moderator cooling system 慢化剂冷却系统
moderator density coefficient 慢化剂密度系数
moderator density feedback effect 慢化剂密度反馈效应
moderator density feedback 慢化剂密度反馈
moderator density fluctuation 慢化剂密度起伏
moderator density reactivity coefficient 慢化剂密度反应性系数
moderator drain shutdown system 慢化剂排放停堆系统
moderator drain system 慢化剂排放系统
moderator drain valve 慢化剂排放阀
moderator dumping safety mechanism 慢化剂倾泻安全机构
moderator dumping 慢化剂倾泻
moderator dump tank 慢化剂倾泻箱
moderator fast drain system 快速排慢化剂系统
moderator-fuel ratio 慢化剂燃料比
moderator inlet 慢化剂入口
moderator lattice 慢化剂栅格
moderator level control system 慢化剂液位控制系统
moderator level control 慢化剂液位控制
moderator logging 慢化剂浸入
moderator loop 慢化剂回路
moderator material 慢化剂材料
moderator merit 慢化指数
moderator poisoning 慢化剂中毒
moderator pressure coefficient 慢化剂压力系数
moderator pressure reactivity coefficient 慢化剂压力反应性系数
moderator pump 慢化剂泵
moderator structure 慢化剂砌体
moderator system 慢化剂系统
moderator tank 慢化剂箱
moderator temperature coefficient of reactivity 慢化剂反应性温度系数
moderator temperature coefficient 慢化剂温度系数
moderator temperature lowering 慢化剂温度下降
moderator to fuel ratio 慢化剂燃料比
moderator void reactivity coefficient 慢化剂空泡反应性系数
moderator water 水慢化剂，慢化水
moderator 减速剂，慢化剂，调解人，阻滞剂，缓和剂
modern architecture 近代建筑
modern control system 现代控制系统
modernization 改进型，现代化，改装
modernize 改进，现代化，改装
modern management 现代化管理
modern nodal method 现代节块法
modern technology 现代技术
modern vegetation 现代植被
modern 现代的，近代的
moder 脉冲编码装置
mode scramble 扰模器
mode selector switch 工作状态选择开关
mode separation 振荡型分离，振型频差，波形间隔
mode shape 波的形状，模态形状，振型
modes of intelligence transmission 信息传递方式
modes of operation 工作状态，操作方法
modest 适度的
mode-superposition procedure 振型叠加法
mode synthesis method 模态综合法
mode synthesis 模态综合
modes 方式
mode transducer （振荡）模变换器
mode transfer switch 转换开关
mode velocity 模态速度
mode voltage 模电压
mode 模式，方法，方式，形式，模，型，模态，振形，模拟，标本
modification factor 改正系数
modification notice 变更通知
modification of bid 标书修改，修改标书
modification of design 设计变更
modification of document 文件更改
modification of drawing 图纸更改
modification of orders 指令改变，指令修改
modification of topography 地形改造
modification of wind 风场改造，风控制
modification to the contractual condition 修改合同条件
modification works 整修工程，改造工程
modification 改进，改型，修改，变化，缓和，变形，变更，更改
modified aeration process 改良曝气法
modified aeration 改进通气法
modified binary code 反射二进码，循环码
modified biquinary code 改进的二五混合进制码

modified complement 变形补码
modified constant potential charge 准恒压充电法
modified diode transistor logic 改进的二极管晶体管逻辑
modified frequency modulation 改进的调频制，变频调制
modified impedance relay 变形阻抗继电器
modified index 修正系数
modified integration digital analog simulator 改进的积分数字模拟仿真器
modified inverted T blade root 改进的倒T形叶根
modified line 变更线
modified loess 次生黄土
modified Mercalli scale 修正的麦卡利（地震）烈度表
modified nonthreshold logic 改进的非阈值逻辑
modified objective function 修改目标函数
modified potential flow 改进位流法
modified shape 改型
modified sliding-pressure 改良滑压运行
modified spacer factor 定位架修正因子
modified velocity 改正流速
modifier register 变址寄存器
modifier 改良剂，改变装置，变址数
modify address 改变地址
modifying agreement 修改协议
modifying factor 变形系数
modify 更改，修改，变更，改进
moding 模度，波模的，振荡模的，传输模的
modulability 调制能力，调制本领
modular air handling unit 组合式空气调节机组
modular brick 符合模数尺寸的砖
modular code system 模块式编码系统
modular concept 模块概念
modular connector 组合式插件
modular construction 模块结构，模件结构，积木式结构，组件结构，单元结构
modular co-ordination 模数协调
modular design method 定型设计法
modular design 积木式设计，标准设计，模件设计，典型设计，定型设计，模块化设计
modular gas turbine power plant 组装式燃气轮机电厂
modular high-temperature gascooled reactor 模块式高温气冷反应堆
modular high temperature reactor 模块式高反应堆
modular information processing equipment 模块信息处理装置
modular integrated housing 模块化集成房屋，打包箱式安置房
modularity 积木性，模块化，调制性
modularization 模件化，模块化
modularized computer 积木式计算机
modularized hardware 模块结构的硬件，模块化硬件
modularized program 模块化程序
modular pile 定形的煤堆
modular programing 模块化程序设计
modular ratio design 模量比法设计
modular ratio method 模量比法
modular ratio 当量系数【钢筋混凝土】，弹性模量比，弹率比，模量比
modular reactor 模块反应堆
modular reclaiming 定式取料
modular structure 模块化结构，模块结构
modular system 模块化系统，单元组合系统，模数制
modular unit 单元组合元件
modular 模（数）的
modulated amplifier 受调放大器
modulated amplitude 受调振幅
modulated beam microbalance 已调制微量天平
modulated carrier channel 已调制载波通道，调制载波通道
modulated carrier 已调载波
modulated continuous wave 调制的连续波，已调连续波
modulated current 调制电流
modulated dimension 模数化的尺寸
modulated oscillator 受调振荡器
modulated pulse amplifier 受调脉冲放大器
modulated quantity 调制量
modulated voltage 调制电压
modulated wave 调制波
modulated 调制的，已调制的，受调的，被调制的
modulate 调制，调谐，调幅，调整
modulating action 调节作用，调制作用
modulating actuator 调节机构
modulating amplifier 调制放大器
modulating choke 调制扼流圈，调制抗流圈
modulating control system 模拟量控制系统，调制控制系统
modulating control 调整控制
modulating frequency 调制频率
modulating oscillator 调制振荡器
modulating valve 调节阀【闭路控制】
modulating voltage 调制电压
modulating wave 调制波，调幅波
modulation bandwidth 调制带宽
modulation capability 最大调制范围，调制能力
modulation code 调制码
modulation controller 调制控制器
modulation degree 调制度
modulation-demodulation 调制解调
modulation depth 调制深度
modulation device 调制设备
modulation distortion 调制畸变，调制失真
modulation divider tube 分频管
modulation eliminator 解调器，调制消除
modulation envelope 调制包线
modulation factor 调制因子，调制系数
modulation frequency ratio 调制频率与载频之比
modulation function of HVDC power transmission system 高压直流输电系统调制功能
modulation generator 调制振荡器
modulation index 调制指数
modulation meter 调制测试计，调制度测试器
modulation noise 调制噪声
modulation rate 调制率

modulation transformer 调制变压器
modulation valve 调制管
modulation 调谐，转变，变换，调节，调制，调幅度
modulator band filter 调幅器带通滤波器
modulator-demodulator 调制反调制器，调制解调器
modulator 调辐器，调制器，调整器，调节器
module efficiency 组件效率
module isolation 模块隔离
module mounting unit 模件安装单元
module of torsion 扭转弹性模数
moduler circuit 一种微型混合集成电路
module type steam generator 组件型蒸汽发生器
module 模，模量，模数，组件，微型组件，模件，插件，模块，系数，太阳电池组件
modulo 2 counter 模数为 2 的计数器，模 2 计数器，二进制计数器，二元计数器
modulo-n adder 模 n 加法器
modulo 2 sum 模 2 和，模数为 2 的和
modulous subgrade reaction 地基系数
modulo 模，模数，模量，组件，按模计算
modulus attenuation 衰减模数
modulus in shear 剪切模数
modulus in tension 拉伸模数
modulus of admittance 导纳模
modulus of aeroelasticity 气动弹性模量
modulus of attenuation 衰减系数，衰减模数
modulus of complex number 复数的模
modulus of compressibility 压缩弹性模数，压缩模量
modulus of compression 抗压弹性模数，压缩模量
modulus of compressive elasticity 压缩弹性模数
modulus of continuity 连续模数
modulus of deformation 变形系数，变形模量
modulus of dilatation 膨胀模量
modulus of discharge 流量模数，流量系数
modulus of drainage 排涝模数，排水模数
modulus of elastic compression 弹性压缩模量
modulus of elasticity in direct stress 拉伸弹性模数
modulus of elasticity in shear 剪切弹性模量，剪切弹性模数
modulus of elasticity in tension 抗拉弹性模数
modulus of elasticity of soil 土壤的弹性模量
modulus of elasticity 弹性模量，弹性模数，弹性系数，杨氏模量
modulus of flexibility 挠性模数
modulus of foundation 基础模量
modulus of impedance 阻抗模
modulus of longitudinal elasticity 纵弹性模数
modulus of machine 机械效率
modulus of periodicity 周期的模
modulus of reaction of soil 土壤弹性均压系数
modulus of resilience 回弹系数，回能系数，回弹模量
modulus of resistance 抗力模量，阻力系数
modulus of rigidity 刚性模数，刚性模量
modulus of rupture in bending 弯曲断裂模数
modulus of rupture 断裂模量，破裂模数，弯曲极限强度，抗裂系数，断裂模数
modulus of section 截面模量
modulus of sodium silicate 水玻璃模数
modulus of soil reaction 土反力模数
modulus of subgrade reaction 地基反力系数，基床反力模量
modulus of transverse elasticity 剪切弹性模数，横向弹性模数
modulus of volume compressibility 体积压缩模量
modulus of volume expansion 体积膨胀模数
modulus 模数，模量，模件，模，振幅，系数，刚性模量
modul 模，模数，模量
modus 方式，方法，程序
MOE(measure of effectiveness) 有效性测量
mog-ohm 兆欧
mogote 灰岩残丘
mogul base 大型电子管底座
mogul lamp-holder 大型灯座
MOH(maintenance outage hours) 维修停用小时数
mohole 超深钻
Mohr circle 摩尔圆
Mohr concentration 摩尔浓度
Mohr-Coulomb criteria 摩尔库仑准则
Mohr-Coulomb criterion 摩尔库仑准则
Mohr-Coulomb soil model 摩尔库仑土的模型
Mohr's circle 摩尔圆
Mohr's envelope 摩尔包络线
Mohr's hardness 摩氏硬度
Mohr's rupture envelope 摩尔破裂包络线
Mohr's stress circle 摩尔应力圆
Mohs scale of hardness 摩氏硬度分度法
moiety 半个，半股
moiling gradation 模拟级配
moirepattern 水纹图样
moire 波动光栅，波纹
moist adiabatic lapse rate 湿绝热递减率
moist adiabatic process 湿绝热过程
moist air 潮湿空气，湿空气
moist ash-free basis 恒湿无灰基，无灰湿基
moist basis （燃料）工作基
moist closet 湿润室
moist curing 湿润养护，雾室养护
moistening 浸湿，润湿，湿润
moisten 变湿，增湿，润湿，弄湿
moist material 湿物料，湿料
moist mineral-free basis 恒湿无矿物质基
moist mineral matter free basis 恒湿无矿物质基
moistness 湿气，湿度，水分
moist plume 水汽羽流
moist sep moisture separator 汽水分离器
moist steam 湿蒸汽
moisture absorbent 吸湿剂
moisture absorber 吸湿器，呼吸器，空气过滤器
moisture-absorbing material 吸湿材料
moisture absorption piping 吸潮管
moisture absorption 吸湿，湿气吸收，水分吸收
moisture adjustment 湿度调整，水分调整
moisture alarm 潮气警报器

moisture-and-ash-free basis 无水无灰基，可燃基
moisture balance 水分平衡
moisture barrier 防潮层，防水层
moisture capacity 含水量，湿度，水分，含湿量
moisture-carrying capacity 携带水分能力，除湿能力
moisture carry over 水分携带，机械携带
moisture catcher 水滴捕集器，水分分离器，去湿槽，汽水分离器，去湿装置
moisture chamber 湿润室
moisture circulation 水分循环
moisture collector 除湿器，集湿器
moisture concentration 水分浓度，潮气冷凝
moisture-conditioned 调湿的
moisture conservation 水分保持
moisture content meter 含水量测定计
moisture content of SF_6 六氟化硫含水量
moisture content of steam 蒸汽湿度
moisture content value 湿度，含水率
moisture content 水分，湿量，湿度，含水量，湿分，含湿量
moisture control 湿度调节，湿度控制，水分控制
moisture deficit 含水量不足，缺少水分
moisture-density curve 含水量密度关系曲线，击实曲线
moisture-density relationship 含水量密度关系
moisture detector 湿度探测器，水分探测器
moisture determination 含水量测定，湿度测定
moisture diffusion controlled 水分扩散控制的
moisture drainage 去湿疏水装置
moisture eliminator 脱湿器，干燥器
moisture entrainment （蒸汽中）夹带水分
moisture equation 水汽量平衡方程
moisture equilibrium 湿度平衡
moisture equivalent 持水当量，含水当量
moisture excess 过湿
moisture excluding efficiency 水分排除效率
moisture extracting bucket 除湿叶片，去湿叶片
moisture-film cohesion 薄膜水内聚力，薄膜水凝聚性
moisture film 水分膜，湿膜，水膜，水汽薄膜，
moisture flux 水汽通量
moisture-free basis 干燥基
moisture-free coal 干燥级煤，干燥煤
moisture-free sample 干试样
moisture-free 无水的，干的，不含水分的
moisture gauge 含水量测定计
moisture gradient 湿度梯度，水分梯度
moisture-hardening varnish 湿固化漆
moisture holding capacity 保水量，持水量，持水能力，最高内在水分
moisture holding 保水
moisture in analysis 分析水分
moisture index 含水指数，湿润指数，水分指数
moisture indicator 湿度计，湿度指示器
moisture in steam 蒸汽湿度
moisture instrument 湿度计
moisture integrator 湿度积分器
moisture in the general analysis test sample 一般分析试验煤样水分
moisture-laden aggregate 饱水骨料
moisture load 湿负荷
moisture loss 湿汽损失，水分损失
moisture maximization 水汽放大
moisture measurement 湿度测量
moisture measuring instrument 水分测量仪
moisture meter 湿度计
moisture monitor 水分指示仪
moisture movement 水分运动
moisture penetration 水分渗入深度
moisture permeability 渗湿性，渗潮性
moisture pickup 带水
moisture prevention 防潮
moisture probe 湿度计
moisture profile 湿度喷线
moistureproof adhesive 防水胶黏剂
moistureproof barrier 防潮层，防湿层
moistureproof construction 防潮结构
moistureproof film 防潮膜
moistureproof generator 防潮发电机
moistureproof insulation 防潮绝缘
moistureproof liner 防潮垫层
moistureproof material 防潮（温）材料
moistureproofness 耐湿性
moistureproof paper 防潮纸
moistureproof type 防潮型
moistureproof 防湿，防潮，耐湿性，耐湿度，防潮的，防温的，不透水的
moisture protection 防潮
moisture regain 回潮，吸湿性
moisture regime 水分情况，水分状况
moisture removal device 去湿装置
moisture removal equipment 去湿设备
moisture removal 除湿
moisture repellent insulation 防潮绝缘层
moisture repellent 防水的，憎水的
moisture resistance 湿敏电阻
moisture resistant coating 防潮涂层
moisture resistant grade 防潮等级
moisture resistant insulation 耐潮绝缘，防潮隔层，防潮层
moisture resistant motor 耐潮电动机
moisture resistant 耐湿，防潮的
moisture retention capacity 持水量
moisture retention curve 持水曲线
moisture retention 保水
moisture seal 防潮密封
moisture separation equipment 汽水分离装置【蒸汽发生器】
moisture separation 汽水分离，去湿，水分分离，除水
moisture separator drain pump 汽水分离器疏水泵
moisture separator drain tank 汽水分离器疏水箱
moisture separator dryer 汽水分离干燥器【蒸汽发生器】
moisture separator reheater system 汽水分离再热系统
moisture separator reheater 汽水分离再热器
moisture separator 汽水分离器，去湿装置
moisture source 湿源，水分来源
moisture status 水分现状
moisture storage capacity 储水量

moisture stress　潮湿应力，水分应力
moisture teller　水分测定仪
moisture-temperature index　温湿指数
moisture tension　水分张力
moisture tester　湿度试验器
moisture-tight　防湿的，防潮的
moisture transfer　湿气传递
moisture trap　疏水器
moisture vapor　湿蒸汽
moisture volume percentage　容计含水率
moisture weight percentage　干重含水率
moisture　潮湿，湿度，潮气，湿气，水分，含水量，湿气水分，降雨量，温度，水汽
moist　潮湿的，湿的，潮湿
molal concentration　克分子浓度，质量摩尔浓度
molality of solute B　溶质 B 的质量摩尔浓度
molality　重量克分子浓度，重模
molal solution　重模溶液
molal volume　克分子体积
molal weight　克分子量
molal　克分子的，摩尔数
molar absorption coefficient　摩尔吸收系数
molar attenuation coefficient　摩尔衰减系数
molar concentration　克分子浓度，容模，摩尔浓度
molar conductivity　摩尔电导率
molar density　克分子密度
molar enthalpy　摩尔焓
molar entropy　摩尔熵
molar fraction　克分子份数
molar gas constant　摩尔气体常数
molar gas flowrate　克分子气体流量
molar heat capacity at constant pressure　摩尔定压比热
molar heat capacity at constant volume　摩尔定容比热
molar heat capacity　摩尔比热
molarity　克分子浓度，容模
molar mass　摩尔质量
molar optical rotatory power　摩尔旋光本领
molar ratio　克分子比
molar solubility　体积克分子溶解度，容模溶解度
molar solution　容模溶液
molar thermodynamic energy　摩尔热力学能
molar volume　摩尔体积
molar weight　克分子量
molar　克分子的，摩尔数
mold and die components　模具单元
mold changing systems　换模系统
mold chillers　模具冷却器
mold core　模芯
molded assembly　模制组件
molded breadth　型宽
molded-case circuit breaker　模制盒式开关
molded coil　浇注线圈
molded commutator　塑料换向器，浇注绝缘换向器
molded depth　型深
molded epoxy insulated coil　浇注环氧绝缘线圈
molded fixed-cord connection　模制固定软线连接
molded insulation　模制绝缘，浇注绝缘
molded plastic　模制塑料
molded relay　模制继电器，浇注绝缘继电器
mold heaters　模具加热器
molding die　模压
molding machine　造型机，制型机
molding micanite　塑型云母板
molding　成型，饰线，线脚装饰
mold polishing　模具打磨
mold rain　梅雨
mold repair　模具维修
mold texturing　模具磨纹
mold　型，模型，模具，造型，铸造，浇注，塑造，发霉，霉菌
mole channel　地下排水道，鼠道
molectron　集成电路，组合件
molecular abundance　分子丰度
molecular attraction　分子引力
molecular compound　分子化合物
molecular concentration of M　M 的分子浓度
molecular conduction　分子传导
molecular constitution　分子结构
molecular crystal　分子晶体
molecular degrees of freedom　分子自由度
molecular diameter　分子直径
molecular diffusion　分子扩散
molecular diffusivity　分子扩散系数
molecular dissipation rate　分子耗散率
molecular energy　分子能
molecular far-acting forces　分子聚合力
molecular filter　分子过滤器
molecular free volume　分子自由容积
molecular gauge　分子压力计
molecular heat　分子热
molecular interaction　分子相互作用
molecular iodine　分子碘
molecularity　分子状，分子性
molecular mass　分子量
molecular mean free path　分子平均自由度
molecular model　分子模型
molecular momentum exchange　分子动量交换
molecular network　分子网
molecular pressure　分子压力
molecular reflection　分子反射
molecular shear stress　分子剪切应力
molecular sieve drying　分子筛干燥
molecular sieve sweetening　分子筛脱硫
molecular sieve　分子筛
molecular spectrum　分子光谱
molecular speed ratio　分子速度比
molecular structure　分子结构
molecular vacuum gauge　分子真空计
molecular velocity　分子速度
molecular viscosity　分子黏性
molecular water　分子水
molecular weight　分子量
molecular　分子的，克分子的
molecule partition function　分子配分函数
molecule　分子，克分子
mole density　克分子密度
mole drainage　暗沟排水，地下排水，鼠道排水
mole drain　地下排水道，地下排水沟，排水洞

mole-electronics 分子电子学
mole fraction of M M的摩尔分数
mole fraction 摩尔分数，克分子分数，克分子份额
mole head 突堤堤头
mole neutron 摩尔中子
mole-pipe drainage 暗沟管道排水
mole plough 暗沟塑孔犁，挖沟犁
mole ratio of solute M 溶质M的摩尔比
mole 克分子量，克分子，摩尔，防波堤，隧洞掘进机，突堤
moling machine 暗沟塑孔机
molion 分子离子
mollification 软化
mollisol 松软土
MOL(middle of life) 寿命中期
molor shaft 电动机轴
molten alloy 熔融合金
molten bath gasification 熔融床气化，熔浴床气化
molten bath 熔化浴
molten clinker 熔渣
molten core material 熔化的堆芯材料
molten fuel 熔化燃料
molten magma 熔融岩浆，岩浆
molten metal 熔融金属
molten plutonium fast reactor 熔融钚快中子反应堆
molten pool 熔穴【堆芯】，熔池，熔渣池
molten salt breeder 熔盐增殖反应堆，熔盐增殖堆
molten salt converter reactor 熔盐转换反应堆
molten salt coolant 熔盐冷却剂
molten salt cooled reactor 熔盐冷却反应堆
molten salt electrolyte battery 熔融盐蓄电池
molten salt fueled reactor 熔盐燃料反应堆
molten salt fuel 熔盐燃料
molten salt mixture 熔盐混合物
molten salt reactor experiment 熔盐反应堆实验装置
molten salt reactor 熔盐反应堆
molten salt receiver tube 熔盐吸热管
molten salts receiver 熔盐吸热器
molten salt tower receiver 塔式太阳能熔盐吸热器
molten salt 熔盐
molten slag 熔渣
molten 熔化的，熔融的
mol. wt. 分子量
molybdate 钼酸盐
molybdenum-reinforced nickelbase alloy 加钼镍基合金
molybdenum steel 钼钢
molybdenum 钼
MO(manual operation) 手操
moment admittance 力矩导纳
momental 惯量的
moment area method 力矩面积法，弯矩面积法
moment arm 力臂，力矩臂
momentary-action switch 转矩动作开关
momentary connection 快速连接，瞬时接通
momentary contact pushbutton 可复归按钮
momentary contact switch 快速开关，暂触开关
momentary current 瞬时电流
momentary discharge 瞬时流量
momentary disturbance 瞬时扰动
momentary duty electromagnet 瞬时工作的电磁铁
momentary duty 瞬时负荷
momentary excess current 瞬时过电流
momentary fall of pressure 瞬时压力降落
momentary fluctuation 暂态波动，瞬时波动
momentary interruption 瞬时断路，瞬时停电
momentary load 短时间负荷，瞬时负荷，瞬时负载
momentary maximum wind velocity 瞬时最大风速
momentary minimum discharge 瞬时最小流量
momentary output 瞬时出力
momentary overload capacity 瞬时过载能力
momentary overload 瞬时过载
momentary power failure 瞬时断电
momentary power 瞬时功率
momentary pressure variation 瞬时压力变化
momentary reserve 短时备用，瞬时备用
momentary rise of pressure 瞬时压力升高
momentary speed adjusting device 瞬时速度调节装置
momentary speed variation 瞬时转速变化
momentary speed 瞬时速度
momentary value 瞬时值
momentary water level 瞬时水位
moment at fixed end 固定力矩，固端弯矩
moment at support 支座弯矩
moment axis 力矩轴
moment balance 力矩平衡
moment block 力矩块
moment center 力矩中心
moment coefficient 力矩系数，弯矩系数
moment-curvature characteristics 弯矩曲率特性
moment curve 力矩曲线，力矩图，弯矩曲线，弯矩图
moment derivative 力矩导数
moment diagram 弯矩图，力矩图
moment distribution method 力矩分配法，弯矩分配法
moment distribution 弯矩分配，力矩分配
moment equilibrium 力矩平衡，弯矩平衡
moment generating function 矩量生成函数，矩量母函数
moment load chart 弯矩荷载图
moment matrix 矩量矩阵
moment method 力矩法
moment of a couple 力偶矩
moment of deflection 弯矩，挠矩
moment of dipole 偶极子矩
moment of force 力矩
moment of friction 摩擦力矩
moment of impulse 冲量矩
moment of inertia in yaw 偏航惯性矩
moment of inertia of section 截面惯性矩
moment of inertia 惯性矩，转动惯量，惰性惯量
moment of magnetic couple 磁偶矩
moment of magnet 磁力矩

moment of momentum equation 动量运方程
moment of momentum theorem 动量矩定理
moment of momentum 动量矩
moment of overturning 倾覆力矩
moment of resistance 抵抗力矩
moment of rotation 转矩
moment of span 跨矩
moment of spectral density function 谱密度函数
moment of stability 稳定惯量，稳定矩，稳定力矩
moment of statics 静矩
moment of torsion 扭矩，扭转力矩
moment redistribution 力矩再分配
moment reference 空气动力焦点，空气动力矩基准点
moments equilibrium 力矩平衡
moments method 力矩法
moment test 弯矩试验
momentum amplifier 动量放大器
momentum balance equation 动量平衡方程
momentum balance 动量平衡
momentum boundary layer 动量边界层
momentum budget 动量收支
momentum coefficient 动量系数
momentum conservation 动量守恒
momentum correction factor 动量修正系数
momentum curve 动量曲线
momentum deficit thickness 动量亏损厚度，动量损失厚度
momentum deficit 动量亏损，动量损失
momentum diffusion 动置扩散
momentum diffusivity 动量扩散率
momentum equation 动量方程
momentum flow vector 动流量矢量
momentum flux 动量通量
momentum gradient 动量梯度
momentum interchange 动量交换
momentum law 动量定律
momentumless buoyant plume 无动量浮力羽流
momentum mixing coefficient 动量交混系数
momentum mixing length 动量混合长
momentum mixing theory 动量混合理论
momentum moment 动量矩
momentum plume 动量羽流
momentum principle 动量原理
momentum range 信号间隔
momentum spectrum 动量分布
momentum theorem 动量定理
momentum theory 动量理论
momentum thickness of boundary layer 边界层动量厚度
momentum thickness 动量厚度
momentum transfer 动量交换，动量转移，动量传递
momentum transport 动量传递
momentum 动量，冲量
moment unbalance 力矩不平衡
moment vector 力矩矢量
moment 力矩，时机，因素，时刻，瞬时，动量，转矩，动差
monthly flow duration curve 月流量历时曲线
MOM(mass optical memory) 大容量光存储器
MO(money order) 汇款单
monadic operation 单值操作，一元运算
monadnock 残丘，残山
monad 单轴，一价元素
monatomic 单质的，单原子的，一价的
monazite 独居石
Monel metal 蒙乃尔镍及铜合金
monentary load 瞬时荷载
monetary base 货币基数
monetary crisis 金融危机
monetary policy 货币政策，金融政策
monetary reward 酬金
monetary terms 货币条款
monetary unit 货币单位
monetary 货币的，金钱的，财政的
money advanced 垫款
money back guarantee 退款保证
money circuit account 资金周转账户
money of account 账面值
money of payment 支付，付款
money order 汇款单，汇票
money paid in advance 预付款
money raising method 筹款方法
money received in advance 预收款
money remittance 汇款
money supply 货币发行量
money transaction 即付交易
money value 货币价值
money 货币，钱，金钱，现金
monic method 随机搜索法
monic polynomial 首一多项式
Monin coordinate 莫宁坐标
Monin-Obukhov length scale 莫宁奥布霍夫长度尺寸
Monin-Obukhov (stability) length 莫宁奥布霍夫（稳定）长度
Monin-Obukhov universal function 莫宁奥布霍夫通用函数
monitor chamber 监察室，控制室
monitor control dump 监督程序控制转储
monitor counter 监视计数器，检验计数器
monitor desk 监测仪控制台
monitor-drain system 疏水检测系统，液体监排系统
monitored area 受监测区域
monitored control system 监控系统
monitored information 监视信息
monitored parameter 监视参数
monitored zone 被监测区
monitored 监视的
monitor equipment 监测设备
monitor index 监控指标
monitoring amplifier 监听放大器
monitoring and hold-up tank （废水）检测和暂存箱
monitoring apparatus 监视装置，监控装置，监控设备
monitoring centre 监视中心，监控中心
monitoring circuit 监听电路，监控电路
monitoring device 监察装置，监控装置

monitoring direction　监视方向
monitoring element　监控元件
monitoring equipment　检测设备，检测装置
monitoring feedback　监控反馈
monitoring hardware　监视硬件
monitoring instrument　监控仪器，控制仪器，监测仪表
monitoring leak-off tube　检测引漏管
monitoring of dam safety　大坝安全监测
monitoring of inconsistency　不一致监视
monitoring point for level　水准监测点
monitoring point　检测点，监测点，监视点
monitoring recorder　监控记录仪
monitoring station　监测站
monitoring store　监测信号存储器
monitoring survey　监测
monitoring system　监视系统，监测系统，监测网
monitoring tap　检漏接头，监测孔
monitoring team　监测组
monitoring test　监听试验，监控试验
monitoring turbidimeter　浊度监测仪，浊度检测仪
monitoring　监视，监控，监督，控制，监测，剂量测定
monitor ionization chamber　监测电离室
monitor light　指示灯
monitor mode　监督方式
monitor on the roof　屋面上的通风楼，屋面上的检修孔
monitor operating system　监控操作系统，监督操作系统
monitor out of service　停止工作监视器，非服务监视器
monitor printer　监控打印机
monitor roof　采光屋顶
monitor routine　监督程序
monitor scope　监视显示器
monitor station reports　监测站报告书
monitor system components　监督系统组成程序
monitor system　监视系统，监控系统
monitor tank drain pump　监测水箱排水泵，检测箱疏水泵
monitor tank　水箱监测仪，检测箱
monitor　监视器，监控器，检测仪器，监测，监听员，监督程序，监测仪，检测，检查
monkey adjustable wrench　活动扳手
monkey-chatter　交叉调制，交叉失真，邻道干扰
monkey drive　锤打机
monkey engine　锤式打桩机
monkey screw wrench　活扳手
monkey wrench　活动管钳，活扳手，活动扳手，管子钳，螺旋钳
MON＝monitor　监视器
monobasic potassium phosphate　磷酸二氢钾
mono-beam bridge crane　单梁桥式起重机
mono-beam hung crane　单梁悬挂起重机
mono-bed　混合床
monobloc forging　整锻
monoblock engine　单排发动机，整体发动机
mono block rotor　整体转子
monoblock unit　单元机组
monoblock　单元机组，整体铸造
monobloc rotor　整体转子，整锻转子
monobloc unit　单元机组
monobloc　整体铸造
monoboard microcomputer　单板微型计算机
monobrid circuit　单片混合电路
monobucket excavator　单斗挖土机
monocarbide　一碳化物
monochord　单软线，单塞绳
monochromatic absorptivity　单色吸收率
monochromatic anaglyph　单色立体图
monochromatic brightness　单色亮度
monochromatic directional emittance　单色定向发射率
monochromatic emissivity　单色放射
monochromatic emittance　单色黑度
monochromatic energy　单色能
monochromatic intensity　单色强度
monochromatic light　单色光
monochromatic measurement　单色测量
monochromatic radiation　单色辐射
monochromatic reflectance　单色反射率
monochromatic scattering coefficient　单色散射系数
monochromatic　单色的
monochromator　单色仪
monochrome camera　单色照相机
monochrometer　单色仪
monochrome　单色的，黑白的
monoclinal block　单斜断块
monoclinal fault　单斜断层
monoclinal fold　单斜褶皱
monoclinal ridge　单斜脊
monoclinal rising wave　单斜上升波
monoclinal river　单斜河
monoclinal stratum　单斜地层
monoclinal valley　单斜谷
monocline　单斜褶皱，单斜
monoclinic form　单斜晶形
monocoil　单线圈的
monocontrol　单一控制，单一调节
monocoque construction　硬壳式结构
monocoque structure　硬壳结构
monocrystalline　单晶体的，单晶形的
monocrystal　单晶体
monocular hand level　手水准
monocular　单筒望远镜
monocyclic alternator　单周同步发电机
monocyclic motor　单周电动机
monocyclic start　单周期启动
monocyclic　单循环的，单周期的，单环的
monodirectional　单向的
monodisperse aerosol　单分散气溶胶
monodispersed pollutant　单分散的污染物
monodisperse fluidized system　单分散性流化系统
monodisperse system　单分散系统
monodrome function　单值函数
monodrome　单值
monoenergetic　单能的
monofier　振荡放大器
monofonically increasing function　单调递增函数

monofonically 单调地
monofonic function 单调函数
monoformer 光电单函数发生器
monofunctional exchanger 单功能离子交换剂
monofunctional extractant 单功能基萃取剂，单官能团萃取剂
monogene rock 单成岩
monogeosyncline 单地槽
monographical study 专题研究
monograph 专题文章，专题著，专论
monokinetic electron 单能最电子
monokinetic 单能量的
monolayer winding 单层绕组
monolayer 单层，单层的
monolithic baffle 整块折流板，整块浇注耐火材料挡墙
monolithic block element 整块状颗粒燃料元件
monolithic block 整体坝块，整块
monolithic computer 单片计算机
monolithic concrete construction 整体式混凝土建筑，整体式混凝土结构
monolithic concrete 整体混凝土
monolithic construction 整体结构
monolithic dam 整体式坝
monolithic hybrid circuit 单片混合电路
monolithic ice 整块冰
monolithic IC 单片集成电路
monolithic integrated circuit 单片集成电路
monolithic main memory 单片主存储器
monolithic masonry lock chamber 整体式圬工闸室
monolithic pouring 整块浇筑
monolithic processor 单片处理器
monolithic quartz crystal filter 单片石英晶体滤波器
monolithic refractory setting 整体耐火材料炉墙
monolithic reinforced concrete structure 整体式钢筋混凝土结构
monolithic roadbed 整体道床
monolithic stability 整体稳定性，整体稳定
monolithic storage 单片存储器，整块存储器
monolithic structure 整体式结构
monolithic system technology 单片系统技术
monolithic 单块的，整块的，整体的，整体式的，单片的
monolith quay wall 整体式码头岸墙
monolith 坝段【两条伸缩缝间的】，独块巨石，整体，(水电) 坝块，整块体
monomeric 单体的
monomer 单体，单元结构
monometallic 单金属
monomodal 单形的
monomolecular film 单分子膜
monomolecular layer 单分子层
monomotor 单发动机，单电动机
monophase 单相的
mono-pitched roof 单坡屋面
monopolar generator 单极发电机
monopolar HVDC system 单极高压直流输电系统
monopolar 单极的
monopole asynchronous motor 单相异步电动机
monopole planned outage times 单极计划停运次数
monopole unplanned outage times 单极非计划停运次数
monopole 单极，单极的
monopolistic competition 垄断性竞争
monopolistic competitive market 垄断竞争市场
monopolistic market 独占市场
monopolist 专利者
monopoly market 垄断市场
monopoly pricing 垄断定价
monopoly 专利权，专利品，垄断，独占
monopulse 单脉冲
monorail bucket hoist 单轨抓斗起重机
monorail hoister 单轨吊
monorail hoists and lifting equipment 单轨吊和起吊设备
monorail hoist 单轨吊，单轨滑车，单轨吊车
monorail trolley 单轨行车
monorail 单轨的，单轨铁道，铁路单线
mono-rotor 整体转子，整段转子
monoscope tube 简单静像管，单像管
monoscope 存储管式示波器，单像管，简单静像管
monosize-distribution 单一颗粒分布
monospar construction 单梁结构
monosphere ion exchange resin 均粒型离子交换树脂
monosphere resin 均粒树脂
monosphere-type resin 均粒型树脂
monostable circuit 单稳电路
monostable flip-flop 单稳态触发器
monostable multivibrator 单稳多谐振荡器，单稳触发器
monostable polarized relay 单稳态极化继电器
monostable trigger 单稳态触发器
monostable 单稳态的，单稳的
monosymmetrical 单轴对称的
monotone function 单调函数
monotone 单调【单一音调】
monotonic function 单调函数
monotonic loading 简单负荷，单调加载
monotonic quantity 单调量
monotonic 单调的
monotron 直越式速调管，无反射极的速调管
monotropic function 单值函数
monotube boiler 直流锅炉，单管锅炉
monotube pile 单管柱桩
monotube 单管
monovalence 单价，一价
monovalent 一价的，单价的
mono-valve filter 单阀滤池
monoxide carbon 一氧化碳
monoxide 一氧化物
monox 氧化硅
monsoon climate 季风气候
monsoon precipitation 季风降水（量）
monsoon prevalent 季风盛行
monsoon weather patterns 季风气候类型
monsoon wind 季风
monsoon 季风，（印度等地的）雨季，季候风，

信风，贸易风
montan wax 褐煤蜡，蒙旦蜡，地蜡
Monte Carlo method 蒙特卡罗法
Monte Carlo model 蒙特卡罗模型
monthly amplitude 月振幅
monthly average wind speed 月平均风速
monthly average 月平均
monthly consumption 月耗量
monthly correlation 月相关
monthly distribution of precipitation 月降水量
monthly distribution 月分配
monthly dosimetry 逐月剂量测定
monthly earnings 每月工资
monthly electricity output 月发电量
monthly estimate 月预算
monthly financial statements 月份结算表
monthly flow hydrograph 月流量过程线
monthly index 月指数
monthly instalments 按月分期付款，按月摊付款
monthly interest 月利息
monthly load curve 月负荷曲线，月负载曲线
monthly load-duration curve 月负荷历时曲线
monthly load factor 月负载率，月负荷率，月负荷因素
monthly load forecasting 月负荷预测
monthly log 月报记录
monthly lowest temperature 月最低温度
monthly maximum load 月最高负荷
monthly maximum temperature 月最高气温
monthly maximum 月最大值，月最大
monthly mean runoff 月平均径流量
monthly mean 月平均值
monthly minimum temperature 月最低气温
monthly minimum 月最小值，月最小
monthly payment 按月付款
monthly peak load 月峰荷
monthly plan 月度计划
monthly precipitation 月降水量
monthly premium 逐月保险费
monthly progression rate 月进度表
monthly progress report 月进度报告
monthly progress 月进度
monthly regulating reservoir 月调节水库
monthly report 月报
monthly runoff 月径流
monthly schedule 月度计划
monthly statement 月报表
monthly summary 月汇总表
monthly tide 月周潮
monthly variation graph of heat consumption in one year 每年的月度热负荷图
monthly variation of windspeed 月风速变化
monthly variation 月变化
monthly wind direction and frequency 逐月风向及频率
monthly 月刊，按月的，月度的，逐月
month to month 逐月
montmorillonite 蒙脱石，高龄石，蒙脱土
monumented station 埋石点
monumented survey point 标志点，埋石点
monument 标石，纪念物
monzonite 二长岩
monzonitic texture 二长结构
moonscape 单象管【显示字符用的】
moorage 停泊处，停泊费
moor coal 疏松的泥煤，疏松的褐煤
mooring area 锚泊区
mooring basin 泊船池，泊地
mooring facility 系泊设施
mooring pile 锚定桩
mooring post 停泊站，系船流柱
mooring stall 浮台
mooring 系留，停泊，系泊，锚泊，下锚
Moorish arch 马蹄形拱，摩尔式拱
moorland 摩尔高沼地
mop basin 拖把洗涤池
mo-permalloy 镍铁钼导磁合金
MOP(main oil pump) 主油泵
MOPS(million operations per second) 百万次运算每秒
mop-up equalization 干线均衡，全程均衡
mop-up 擦干，扫除，结束，(线路)全程
moraine deposit 冰碛沉积
moraine soil 冰碛土
morality 道德，道德准则
moral right 精神权利
moral wear 无形磨损
moral 道德，品德，品行
morass 泥沼，湿地，沼泽
morbidity 发病率
mordant 腐蚀剂，酸洗剂，媒染剂
more or less 大致地，或多或少地，大致
more sales at a lower profit 薄利多销
more thunderstorm region 多雷区
morghology 形态学
MOR(modulus of rupture) 断裂模数
morning-glory shaft spillway 喇叭形竖井式溢洪道
morning-glory spillway 喇叭形溢洪道
morphogenetic force 地貌发生力
morpholine 吗啉
morphological characteristics 地貌特征
morphological identification 形态鉴定
morphological 形态学的
morphology 形态学
morphometry 形态测量学，形态测量
morph 在屏幕上变换图像
Morse nozzle pit 莫式喷嘴坑
mortality 死亡率
mortar admixture 灰浆或水泥砂浆掺合料，砂浆掺合料，灰浆附加剂，砂浆外加剂
mortar bed course 砂浆底层
mortar bedding 砂浆垫层
mortar bed 灰浆层，砂浆层
mortar bond 灰浆砌合
mortar-chalked joint 砂浆勾缝
mortar finish 灰浆抹面
mortar for concrete small hollow block 用于砌筑小型混凝土空心砖的砂浆
mortar joint 灰浆接缝，灰缝
mortarless wall 干砌墙
mortar levelling 砂浆找平
mortar mill 砂浆拌和厂，砂浆拌和机

mortar mixer 灰浆拌和机，砂浆搅拌机，灰浆搅拌机
mortar setting 灰浆凝固
mortar sprayer 喷浆机
mortar trough 灰浆槽
mortar void method 灰浆孔隙法
mortar void ratio 灰浆孔隙比
mortar 灰泥，灰浆，研钵，砂浆
mortgage assets 抵押资产
mortgage bond 抵押债券
mortgage clause 受抵押权条款，抵押条款
mortgage debt 抵押负债
mortgagee 受押人
mortgage insurance 不动产抵押借款保险
mortgage loan 抵押借贷
mortgage of company assets 公司资产抵押
mortgage 抵押，抵押权
mortgagor 出押人
mortice for beam 墙内梁座
mortice 榫眼，卯眼，(榫)接合
mortise and tenon joint 榫槽接合
mortise chisel 榫凿
mortise gauge 划榫线具
mortise hole 榫眼
mortise joint 榫接
mortise lock with lever handle 执柄门锁
mortise & tenon joint 铆榫接头
mortise 榫眼
mortising slot machine 凿槽机
MORV(motor operated regulating valve) 电动调节门
mosaic building blocks 镶嵌块件，拼件
mosaic circuit diagram 镶嵌式电路图，电路镶嵌图
mosaic membrane 镶嵌
mosaic pavement 马赛克铺面，拼花地面
mosaic structure 嵌镶块结构，嵌镶结构
mosaic surface control panel 马赛克面控制屏
mosaic texture 镶嵌结构
mosaic tile flooring 马赛克地面
mosaic 镶嵌，镶嵌的，拼成的，镶嵌结构，马赛克，拼成的瓷砖，镶嵌图，水磨石
MOS controlled thyristor MOS控制晶闸管
MOSFET (metal-oxide-semiconductor field effect transistor) 金属氧化物半导体场效应晶体管
MOSFET power 金属氧化物半导体场效应晶体管功率
MOS flip-flop 金属氧化物半导体触发器
MOSIC (metal-oxide-semiconductor integrated circuit) 金属氧化物半导体集成电路
moss land 沼泽
moss 苔类
most dangerous slip circle 最危险滑动圆
most economical continuous rating (汽机的)最经济连续功率
most economic conductor 最经济的导线
most favorable price 最优惠价格
most favoured nation clause 最惠国条款
most favoured nation treatment 最惠国待遇
most frequent wind direction 盛行风向，主导风向
most important problem 头等重要问题
most probable distribution 最大可能分布
most probable error 最可能误差
most probable molecular velocity 最大概率分子速度
most probable speed 最概然速率，最可概速率
most probable value 最可概值，最可能值，最或然值，最可期值
most reactive condition 最大后备反应性状态
most significant bit 最高有效位，最高位字符，最高二进制数位
most significant character 最高有效字符
most significant digit 最高有效数字，最高有效位
most-spent assembly 最大燃耗组件
most unfavorable circuit 最不利用户环路
most unfavorable steam supply mains 最不利供汽主管路
motar proportioning 灰浆配合
mote 微尘
mothballing 封存，涂防腐油
motherboard 母板，母插件
mother liquor 母液
mother plate 母模，样板，模板
mother rock 母岩
mother solution 母液
motional admittance 动态导纳
motional impedance 动态阻抗
motional 动态的
motion centre 运动中心
motion characteristic 运动特性
motionless ice 静止冰
motionless mixer 固定混合器
motionless 不动的，静止
motion of translation 平动，平移，线性运动
motion pattern 流线图
motion rate 移动速率，插入［抽出］速率【控制棒】
motion register 运动寄存器
motion round wing 绕机翼的环流运动
motion speed 移动速率，提或插棒速率【控制棒】
motion transducer 位移传感器，移动传送器
motion 运动，运转，移动，输送，位移，动作，议案
motivated 有积极性的
motivate 激励，激发……的积极性
motivation 积极性，动力
motivator 操纵装置，操作机构
motive air 气流
motive fluid 原动流体
motive force (原)动力
motive gas 原动气体
motive power machine 动力机械
motive power (原)动力
motive pressure 推动压力，驱动压力
motive steam condenser 驱动蒸汽冷凝器
motive steam generator 驱动蒸汽发生器
motive steam strainer 主蒸汽滤网
motive water 驱动水
motive 动机

motivity 原动力，发动力，储能
MOT(main oil tank) 主油箱
motometer 转数计，电动机型仪表
motomixer 混凝土搅拌汽车
motor actuator 马达执行机构
motor alternator 同步电动发电机
motor amplifier 电动机放大器
motor armature 电动机电枢
motor assembly 马达组件
motor bearing 电动机轴承
motor bed-plate 电动机底板，电动机机座
motor blower 电动鼓风机
motor board 电动机配电盘
motor-boating 低声频或次声频干扰振荡
motor booster 电动升压机
motor capacity 电动机容量
motor carbon 电动机碳刷
motor case 电动机外壳，电动机机座，电动机壳
motor casing 电动机机座，电动机外壳，马达外壳
motor characteristic 电动机特性曲线
motor circuit switch 马达开关，电动机开关
motor circuit 动力电路，电动机电路
motor combination 电动机组合
motor commutator 电动机换向器
motor compressor 电动压缩机
motor control center 电动门控制中心，马达控制中心，电（动）机控制中心
motor controller 电动机控制器
motor control relay 电动机控制用继电器
motor control switch 电动机控制开关
motor control 电动机控制
motor converter 串级变流器，电动变流机
motor cooling 电动机冷却
motor crane 自行式起重机
motor current transformer 电动换流机
motor cutout switch 电动机切断开关
motor damper 电动调节风阀
motor-driven auxiliary feedwater pump 电动辅助给水泵
motor-driven blower 电动鼓风机
motor-driven boiler feed water pump 电动锅炉给水泵，电动给水泵组【带前置泵】
motor-driven compressor 电动压缩机
motor-driven control valve 电动调节阀
motor-driven fan 电动风扇组
motor-driven feed water booster pump 电动给水升压泵
motor-driven feedwater pump system 电动给水泵系统
motor-driven feedwater pump lubrication system 电动给水泵润滑系统
motor-driven feedwater pump 电动给水泵
motor-driven generator 电动发电机
motor-driven hoist 电动起重机
motor-driven interrupter 电动断续器
motor-driven main feedwater pump 电动主给水泵
motor-driven oil pump 电动油泵
motor-driven pump 电动泵，电动泵组
motor-driven relay 电动继电器
motor-driven skylight 电动天窗
motor-driven starter 电动启动器
motor-driven stepper 电动分挡器
motor-driven tensioning winder 电动张紧绞车
motor-driven valve 电动机拖动阀，电动阀
motor-driven variable speed feed water pump 电动调速给水泵
motor driven winch 电动卷扬机
motor-driven 电动的，电（动）机驱动的，电（动）机拖动的
motor drive 电动机驱动装置，电力传动，电动，电动机拖动，电动机驱动
motor driving actuator 电动执行机构
motor dynamometer 电动测力计，电动机测功机
motor dynamo 电动直流发电机
motoreducer 带减速器的电动机
motored 有发动机的，有电动机的，联动的
motor effect 相邻载流体间的作用力
motor efficiency 电动机效率
motor element 电动机部件
motor engine 发动机，马达
motorette （线圈）绝缘寿命试验用模型
motor excitation 电动机励磁
motor exciting current 电动机励磁电流
motor exhaust 发动机排气
motor fan assembly 电动风机组
motor fan 电动机风扇
motor fault 电动机故障，电动机缺陷
motor field control 电动机励磁控制
motor field failure relay 电动机磁场故障继电器
motor field 电动机磁场
motor foot 电动机底脚
motor for tropical use 热带电机
motor for vehicle 车辆用电动机
motor foundation 电动机基础
motor frame 电动机机座
motor fuse 电动机熔断器
motor gain 电动机增益
motor-generator output breaker 电动发电机输出断路器
motor-generator panel 电动发电机仪表盘
motor-generator set 电动发电机组
motor-generator welder 电动发电电焊机
motor-generator 电动发电机
motor grader 机动平地机
motor hoist 电动葫芦
motor hood 电动机罩
motor industry 电机工业
motoring friction 空转摩擦力
motoring test 空转试验，（发电机的）电动回转试验
motoring 电动回转，电动机运行，电动机驱动
motor in industry 工业用电动机
motor integrating meter 电动式积算仪
motorised tipper 电动卸料车
motorised valve 电动阀
motorised winch 机动绞磨
motorization 机动化，电动化，机械化
motorized crab 电动抓斗起重机
motorized gag 电动夹持器
motorized gate 电动闸门，电动阀门
motorized hanging crane 电动悬挂起重机

motorized hoist 电动卷扬机
motorized hydraulic testing pump 电动试压泵
motorized pulley 电动滚筒
motorized regenerant valve 再生剂电动阀
motorized rod insertion 机械控制插棒，电动插棒
motorized valve 电动调节阀，电动阀
motorized 3-way valve 电动三通阀
motorized 装有发动机的，电动的，自动的
motorize 使机动化，使电动化
motorless 无电动机的，无发电机的
motor load 电动机负载，电机负荷
motorlorry 载重汽车
motor lower radial bearing for primary coolant pump 一回路主泵马达下径向轴承
motor magnet coil 电动机励磁线圈
motor manufacturer 电动机制造商
motor maximum torque 电动机最大转矩
motor meter 电磁式仪表，电动式电度表
motor-mount pump 电动泵
motor noise 电动机噪声
motor oil 动机燃料油，马达油，车用机油，电机用油，电动机润滑油
motor-operated barring gear 电动盘车装置
motor-operated extraction valve 电动抽汽门
motor-operated fixed coal plough 电动固定式犁式卸煤机
motor-operated gate 电动闸门，电动阀门
motor-operated main gate valve 电动主闸门
motor-operated movable coal plough 电动移动式犁式卸煤机
motor-operated rheostat 电动操作变阻器
motor-operated steam extraction valve 抽汽电动门
motor-operated switch 电动操纵开关，电动断路器
motor-operated test chain 电动链码
motor-operated valve test 电动门试验
motor-operated valve 电动阀，电动门，电动阀门
motor-operated winch 电动绞车
motor-operated 电动机驱动的，电动机操作的，电动的，发动机驱动的
motor-operating circuit 电动机操作电路
motor-operation 电动机方式运行
motor operator 电动执行器
motor output power 电动机输出功率
motor output 电动机输出功率，电动机出力，电机输出
motor panel 电动机配电盘
motor power input at base load 在基本负荷下电机的功率输入
motor power 电动机功率
motor primary coil 电动机初级线圈
motor protection 电动机保护
motor pulley 电动机皮带轮
motor pump assembly 电动泵组
motor pump set 电动泵组
motor pump 电动泵，机动泵
motor rating 电动机额定值
motor-reducer coupling set 电动机、减速器联合装置
motor-reducer unit 电动机减速箱单元
motor reducing gear 电动机减速器
motor reduction unit 带减速箱的电动机，降速电动机
motor road 公路
motor roller 机动压路机
motor rotor 电动机转子
motor secondary coil 电动机次级线圈
motor shaft power 电动机轴功率
motor sheath 电动机机壳，电动机机座
motor shell 电动机机座，电动机机壳
motor side shaft 电动机侧的轴
motor single-track bridge crane 电动单梁桥式起重机
motorspeed controller 电动机转速控制器
motorspeed control 电动机转速控制
motorspeed 电机转速
motor spirit （车用）汽油
motor stand 电动机架
motor starter 电动机启动器，电动机起动器，电动起动机
motor-starting relay 电动机启动继电器
motor-starting torque 马达启动转矩
motor-stirrer 电动搅拌器
motor support flange 电动机支承法兰
motor support stand 电动机支架
motor switch 电动机开关，电动开关
motor synchronizing 自同步
motor terminal voltage 电动机端子电压
motor tester 电动机检测仪
motor time constant 电机时间常数
motor torque 电动机转矩
motor transformer 电动机发电机组
motor transport 汽车运输
motor type insulator 马达形绝缘子，防震绝缘子
motor type relay 电动机式继电器，电动继电器
motor type 电动机类型
motor upper radial bearing for primary coolant pump 一回路主泵电动机上部径向轴承
motor valve actuator 电动阀执行机构
motor valve 电动阀
motor vehicle liability insurance 机动车责任险
motorway 汽车道
motor winch 电动绞车，机动绞盘，机动卷扬机
motor winding 电动机绕组
motor wiring 电动机布线，电动机嵌线
motor with air cooling 空冷电动机
motor with aluminium winding 铝线电机
motor with compound characteristic 复励特性电动机
motor with direct on-line starting 直接起动的电动机
motor with series characteristic 串励特性电动机
motor with shunt characteristic 并励特性电动机
motor with water cooling 水冷电动机
motor wrench 管子钳，机动管子钳，电动机扳手
motor 发动机，电动机，马达，电机，原动机，原动力，机动的
mottled iron 麻口铁
mottled sandstone 杂色砂岩
mottle 斑点，表面麻孔，树林

mould board　型板
moulded block fuel element　模制块状颗粒燃料元件
moulded case circuit breaker　塑壳式断路器
moulded coal　成型煤
moulded concrete　模制混凝土
moulded core　模压铁粉芯
moulded fuel element　模制颗粒燃料元件
moulded fuel sphere　模制颗粒燃料
moulded-in-place pile　现场模制桩，就地灌注桩
moulded-in-stress　内应力
moulded insulation　模制绝缘塑料
moulded insulator　模制绝缘子
moulded mica　人造云母，模制云母
moulded piece　模制件【预制混凝土的】
moulded resin　模铸树脂
moulde's bellow　皮风箱，皮老虎
moulding board　模板
moulding machine　线条机
moulding micanite　模压云母板
moulding　模板，模制，造型，压制，成型，铸模，模塑，压条，塑造，铸型，塑造物，铸造物
mould loftsman　放样工
mould-making　造型
mould oil　拆模油，脱模油
mould pressing　模压
mouldproof insulation　防霉绝缘
mould rain　梅雨
mould removal　拆模
mould vibrator　模板振动器
mould　制模，模型，样板，造型，铸模，模具
MOU(memorandum of understanding)　理解备忘录
mound breakwater　堆石防洪堤，抛石防波堤，斜坡式防波堤
mound crown　堆石堤顶
mounding　堆土法
mound-type breakwater　倾斜防波堤
mound　墩，护堤，护堤，山丘，堆，高地
mountain air　山地空气
mountain and valley breeze　山谷风
mountain apron　山麓冲积平原
mountain barrier　山地障碍物
mountain belt　山岳地带
mountain bog　山地沼泽
mountain breast　山腰
mountain breeze　山风
mountain chain　山脉
mountain climate　山地气候，山岳气候
mountain effect　山地效应
mountain gap wind　山口风
mountain land　山地
mountain-making force　造山力
mountain-making process　造山运动过程
mountain meteorology　高山气象学
mountain mud flood　泥石流，山洪
mountainous area　多山地区，高山区，山地，山区
mountainous district　山区
mountainous region　多山地区，山区
mountain pass　隘口，山口，山顶
mountain pediment　山麓侵蚀平原
mountain railway　山区铁路
mountain rain　山雨
mountain range　山脉
mountain region　山岳地区
mountain reservoir　山区水库
mountain ridge　山脊，山岭
mountain side　山麓，山腰，山崩
mountain slope　山坡
mountain topography　山岳地形
mountaintop　山顶
mountain torrents　山洪
mountain tunnel　穿山隧洞
mountain-valley wind　山谷风
mountain valley　山谷
mountain vegetation　高山植被
mountain wind　山风
mountain　山，山岳，山区
mountanious region　山地
mounted abrasive wheel　固定式砂轮
mounted-ahead type　前悬挂式
mounted position　装配地点，安装地点
mounting adjustment　安装调整
mounting base　安装基础
mounting bedplate　装配架，装配底板
mounting bolt　安装螺栓，装配螺栓，组装螺栓
mounting bracket　装配架，装配牛腿
mounting claw　安装猫爪
mounting dolly　推力头
mounting fittings　组装配件
mounting flange　安装用法兰盘，安装法兰
mounting hardware　组装金属构件
mounting height above operation floor　距工作面安装高度
mounting height　安装高度
mounting holes and openings in the floors　楼板上的安装孔洞
mounting hole　安装孔
mounting list　安装清单
mounting location　安装位置
mounting lug　悬挂吊耳
mounting of equipment　设备的装配，组装设备
mounting panel　安装盘【用于灯、开关等】，安装板，接线板
mounting plate　安装板，座板
mounting position　安装位置
mounting rack　装配架，工作台，机架，安装架
mounting rail　固定横条，安装用型材
mountings and fittings　管道附件，管件总称
mounting shaft　安装井
mounting strain error　安装应变误差
mounting structure　安装平台
mountings　配件，零件，附件
mounting version　安装方式
mounting yoke　安装架
mounting　装配，安装，架，座，底座，支架，上升，组装
mountor　安装工
mount support　装配支架
mount　安装，底座，山峰，架，架置，安放，装设，装配

mouth of inlet 进水口
mouth ring 密封环
mouth 出口，喉口，短管，进口，口，坑，洞
M-out-of-N code N中取M码
movable air compressor 移动式空压机
movable and immovable property 动产和不动产
movable arm 可动臂
movable bearing 活动支承
movable bed model test 动床模型试验
movable bed stream 动床河流
movable bed 不定河床，动床
movable belt （风洞）移动带，活动地板
movable blade propeller turbine 转桨式水轮机
movable blade turbine 动叶片式水轮机
movable blade 活动轮叶，可动叶片
movable blocking 可移动挡车板
movable block 活动滑车，动滑轮，可移动块体
movable bridge 活动桥
movable coil 移动线圈【控制棒驱动机构】，可动线圈
movable contact 可动触点
movable core cage 可移动堆芯笼
movable core transformer 可动铁芯变压器
movable core 可动铁芯，可移动堆芯
movable counter weight 活配重
movable counter 活配重
movable crane 移动式起重机
movable crest device 坝顶活动调节设备【水电】
movable crusher 移动式轧碎机
movable dam 活动坝
movable detector 移动检测仪
movable distributor 活动配水器
movable filter 移动式过滤器
movable fit 动配合
movable flap 活动襟翼
movable flume 活动水槽
movable form 活动模板
movable gate 活动闸门
movable gripper coil 移动棘爪线圈【控制棒驱动机构】，活动夹持线圈
movable gripper latch arm 移动棘爪闩【控制棒驱动机构】
movable gripper latch carrier 活动夹持销爪支架
movable gripper latch 移动式抓具闩，活动夹持销爪
movable gripper pole 移动棘爪杆【控制棒驱动机构】
movable gripper 移动棘爪【控制棒驱动机构】
movable house 活动板房
movable ladder 可移动扶梯
movable load 动荷载，动载
movable mast 移动式桅杆
movable out-of-pile detector assembly 移动式堆外探测器组件
movable partition 活动隔墙
movable pilot valve 滑动油阀，活动滑阀
movable platen 活动台板
movable platform 活动平台
movable property 动产
movable rack 活动栅格
movable rail 活动护栏
movable roof 活动屋顶
movable span 活动桥跨
movable sprinkler 活动喷灌机
movable support 活动支承，活动支架，活动支座
movable tail-weir 移动式尾水堰
movable target 活动觇标
movable trash rack 活动式拦污栅
movable trestle 活动栈桥，活动支架
movable type 移动式
movable weir 活动堰
movable 可移动的，可动的，活动的，可拆卸的
movable part 可动部分【抽屉、门等】
move down-right 鼠标指针向屏幕右下角移动
move down 插入，将……移至低处，向下移
move-in reflector 向内动反射体
move in 插入，移入
movement joint 变形缝，活动缝
movement of air 空气流动，气体运动
movement of earth crust 地壳运动
movement of front 锋面运动
movement of ice 浮冰
movement of water vapour 水汽移动
movement 位移，运动，移动，活动范围，动作，动程，活动，运转
move mode 传送方式
move-out 抽出
mover 原动机，发动机，马达，动力，推进器，推动力
moves 举措
move-up 上涨
move 移动，推动，措施，步骤，运转
moving average 动态平均值，滑动平均，流动平均数，移动平均（数）
moving axle 动轴
moving bed filter 移动床过滤器，动床滤器
moving bed filtration 活动床过滤，移动床气化
moving bed 移动床
moving belt （风洞）移动带，活动地板
moving blade loss 动叶损失
moving blade 动叶片，动叶【如旋转的，转子的】叶动片【汽轮机】
moving boat method 动船法
moving boundaries 运动边界
moving boundary analysis 界面移动分析
moving cascade loss 动叶栅损失
moving cascade 动叶栅
moving-coil ammeter 动圈式电流表
moving-coil galvanometer 动圈式检流计
moving-coil indicator 动圈式指示仪表
moving-coil instrument 磁电式仪表，动圈式仪表，动圈仪表
moving-coil meter 动圈式仪表
moving-coil motor 动圈式电动机
moving-coil regulator 动圈式调整器，动圈式调压器
moving-coil relay 动圈式继电器
moving-coil type relay 动圈式继电器
moving-coil 动圈式，动圈，可动线圈

moving contact 活动接点，移动接触，动触头，动触点，活动触点
moving conveyor 移动皮带
moving coordinate system 运动坐标系
moving core type relay 活动铁芯继电器
moving curtain filter 卷绕式空气过滤器
moving element 可动部分，活动元件
moving expenses 搬迁费
moving force 动力
moving gaseous medium 气流
moving grate 行进式炉排，移动炉排
moving ground （风洞）活动地板
moving heat sources 运动热源
moving index measuring instrument 固定标度测量仪表
moving iron oscillograph 动铁式示波器
moving iron relay 动铁式继电器
moving iron 动铁芯，动铁式
moving load 移动荷载，活动荷载，动荷载
moving-loop regulator 动圈式调整器
moving magnet galvanometer 动磁式检流计
moving magnet instrument 动磁式仪表
moving operation 动作，移动【控制棒】
moving packed bed 移动填充床
moving particle 运动颗粒
moving partition wall 活动隔墙
moving part 运动部件【如齿轮、轴、皮带、轮等】
moving reference system 运动参考系
moving reflector 移动式反射层
moving ring 动环
moving scale measuring instrument 活动标度测量仪表
moving scale 移动标尺
moving staircase 自动扶梯
moving stairs 自动楼梯
moving trihedral 活动三面形，动标三面形
moving trihedron 流动三面形
moving tunnel 可移动风洞
moving uniform load 均布动荷载
moving urban pollution source 城市运动污染源
moving vane instrument 磁铁式仪表，动叶式仪表
moving vane 动瓣，动叶片，活动轮叶
moving walkway 自动走道
moving wave 行波，行进波
moving weighing scale 动态称
moving winding 动线圈
moving window motor 动窗式电动机
moving 运动，移动，位移，移动的
moviola 音像同步装置
MOV（motor operated valve） 电动门，电动（阀）门
mowing machine 割草机
MOX(mixed oxide) 混合氧化物燃料
moya 泥溶岩
moyle 十字镐
M/O(mandatory/optional) 数据对象必须或可选
MPa＝megapascal 兆帕
MPBB(maximum permissible body burden) 体内最大容许负荷量
MPC(materials preparation center) 材料制备中心
MPC(maximum permissible concentration) 最大容许浓度
MPDE(maximum permissible dose equivalent) 最大容许剂量当量
MPD generator duct 磁等离子体发电机通道
MPD＝magnetoplasmadynamic 磁等离子体
MPD(maximum permissible dose) 最大容许剂量
MPDR(microprogram data register) 微程序数据寄存器
MPF(maximum probable flood) 最大可能洪水
MPG(microwave pulse generator) 微波脉冲发生器
M-phase circuit M 相电路
MPI(maximum permissible intake) 最大容许摄入量
MPLHGR(maximum planar linear heat generation rate) 最大平面线热功率
MPL(maximum permissible level) 最大容许水平
MPL(maximum permissible limit) 最大容许限度
MPOD(mean planned outage duration) 平均计划停运延续时间
MPO(maximum power output) 最大功率输
MPOOD(mean planned overhaul outage duration) 平均计划大修停运延续时间
MPPT(maximum power point tracking) 最大功率点跟踪，最大功率点追踪
MPR(materials processing reactor) 材料辐照反应堆
MPS(mathematical programming system) 数学规划系统
MPS(measurement processor subsystem) 测量处理机子系统
MPS(meters per second) 米每秒
MPS(microprocessor system) 微处理机系统
MPTA(materials open test assembly) 材料开口试验组件
MPT(main power transformer) 主（电力）变压器
MPU(microprocessor unit) 微处理机装置，微处理部件
MPX-CCU(multiplexed communication control unit) 多路转换通信控制器
MPX(multiprogramming executive system) 多例程序执行系统
MQP(master quality plan) 主质量计划
M-Q(multiplier-quotient) register 乘商寄存器
MQR(multiplier-quotient register) 乘商寄存器
mrad 毫拉德
M rather than N 宁可 M 也不 N，是 M 而不是 N，与其 M 不如 N
MRDOS(mapped real-time disk operating system) 实时映射磁盘操作系统
MRF(material request form) 材料申请单
MRG(medium range) 中程，中距离
MRIE(maximum routine inspection effort) 最大例行视察量
MR(maintenance review) 维修检查
M. R. (masking ratio) 掩蔽比
M/R(mate's receipt) 收货单，大副收据
MR(memory register) 存储寄存器
MROD(mean planned minor repair outage duration)

平均计划小修停运延续时间
MRO(maintenance, repair and operating) 维护、修复及运行
MRP(main reactor pump) 主泵【反应堆】
MSB(most significant bit) 最高二进制数位，最高有效位，最高位字符
MSBR(molten-salt breeder reactor) 熔盐增殖反应堆
MSC(most significant character) 最高有效字符
MSD(most significant digit) 最高有效位
MSDS(maritime shipping document of safety) 危险货物安全资料卡
MSE(mean-square error) 均方误差【反映估计量与被估计量之间差异程度的一种度量】
MSG(maximum stable gain) 最大稳定增益
MSG(modular steam generator) 组合式蒸汽发生器
MSGWTG(message waiting) 信息等待
MSHA(Mine Safety and Health Administration) 美矿业安全和卫生管理局
MSIC(medium-scale integrated circuit) 中规模集成电路
MSI(medium-scale integration) 中规模集成
MSIP(mechanical stress improvement process) 机械应力改善过程
MSIV(main steam isolation valve) 主蒸汽隔离阀
MSL(main steam line) 主蒸汽管
MSL(mean sea level) 平均海平面，平均海拔
MSLP(mean sea level pressure) 平均海平面气压
MS(main steam) 主蒸汽【DCS 画面】
MS(main storage) 主存储器
MS(manufacturing specifications) 制造规范
MS(margin of safety) 安全系数
MS(master scanner) 主扫描器
MS(maximum stress) 最大应力
MS(memory system) 存储系统
MS(microprogram storage) 微程序存储器
MS(mild steel) 低碳钢
MSM(message switching multiplexer) 报文多路转换器
MS(moisture separator) 汽水分离器
MSP(main steam pipe) 主蒸汽管
MSPR(master spare positioning resolver) 主备用定位分解器
MSR(main steam rate) 主蒸汽流量
MSR(moisture separator reheater) 汽水分离再热器
MSRV(main steam relief valve) 主蒸汽释放阀
MSS(main steam system) 主蒸汽系统
MSS(material submittal sheet) 材料提交表
MSS(multispectral scanner) 多光谱扫描仪
MSSR(milestone schedule and status report) 里程碑进度和情况报告
MSSV(main steam stop valve) 主汽门
MSTBR(molten-salt thorium breeder reactor) 熔盐钍增殖反应堆
MST(maintenance surveillance test) 维修监视试验
MST(monolithic system technology) 单片系统技术
MSTR＝moisture 湿度
MSU(main storage unit) 主存储器装置
MSV-BV(main stop valve bypass valve) 主汽门旁通阀
MSVC block 多播采样值控制块
MSVC(multicast sample value control) 多播采样值控制
MSV(main steam valve) 主汽阀，主汽门
MSV(main stop valve) 主截门，主汽门，主汽阀
MSV servomotor 主汽阀油动机
MSW(machine status word) 机器状态字
MTBCF(mean time between critical failures) 平均严重故障间隔时间
MTBD(mean time between defects) 平均故障间隔时间
MTBD(mean time between degradation) 平均衰变间隔时间
MTBD(mean time between detection) 平均（故障）检测时间
MTBE(mean time between errors) 平均错误间隔时间，误差码间平均时间，误差信号间隔平均时间
MTBFA(mean time between failures of auxiliary equipment) 辅助设备的平均无故障可用小时，平均无故障工作时间
MTBF(mean time between failures) （机组的）平均无故障可用小时，平均故障间隔时间【可修复系统】，平均无故障（工作）时间，平均失效间隔时间
MTBFO(mean time between forced outage) 故障停用平均间隔时间，强制停运平均间隔时间
MTBI(mean time between incidents) 平均异常现象间隔时间
MTBI(mean time between interruption) 平均中断间隔时间
MTBMA(mean time between maintenance actions) 平均维修间隔时间
MTBM(mean time between maintenance) 平均维修间隔时间
MTBM(mean time between malfunction) 平均误动作间隔时间
MTBO(mean time between outages) 平均停运间隔时间
MTBO(mean time between overhauls) 平均检修间隔时间
MTBR(mean time between removals) 平均拆换间隔时间
MTBR(mean time between repairs) 平均维修间隔时间
MTBSF(mean time between significant failures) 严重故障平均间隔时间
MTBSF(mean time between system failures) 系统故障平均间隔时间
MTBSI(mean time between system incident) 系统异常现象平均间隔时间
MTBS(mean time between scram) 紧急停堆平均间隔时间
MTC(magnetic tape controller) 磁带控制器
MTC(maintenance contract coordination group) 维修处合同协调组
MTC(maintenance training center) 维修培训中心
MTC(memory timing chain) 存储器定时链

MTC(multiply time chain) 乘法时间链
MTCU(magnetic tape control unit) 磁带控制器
MTD(magnetic tape and magnetic drum) 磁带及磁鼓
MTD(maximum tolerated dose) 最大容许剂量
MTD(mean temperature difference) 平均温差
MTE(magnetic tape equipment) 磁带装置，磁带机
MTE(measuring and test equipment) 测量和试验设备
MTL(magnetic tape loading) 磁带记录，磁带装入
MTL(merged transistor logic) 合并晶体管逻辑
MTLP(magnetic tape loading program) 磁带记录程序
MT(magnetic tape) 磁带
MT(Magnetic Test) 磁粉探伤
M/T(mail transfer) 信汇
MT(maximum torque) 最大扭矩，最大转矩
MTMD(multiple tuned mass damper) 多重调谐质量阻尼器
MT(mean time) 平均时间
MT(metric tons) 公吨【1000 千克】
MT(multiple transfer) 多路传送
MTO(multimodal transport operator) 多式联运经营人
MTRE(magnetic tape record end) 磁带记录器终端
MTRG=metering 测量，计量，记录，统计
MTR(magnetic tape record) 磁带记录器
MTR(materials testing reactor) 材料试验反应堆
MTR(material transfer request) 材料转移申请
MTR(mean time to repair) 修理前平均时间
MTRS(magnetic tape record start) 磁带记录器始端
MTS(magnetic tape system) 磁带系统
MTS(main turbine system) 主汽轮机系统
MTS(master trip solenoid) 主跳闸电磁阀
MTS(memory test system) 存储测试系统
MTTFF(mean time to first failure) 平均初次出故障时间，首次故障前平均（工作）时间
MTTF(mean time to failure) 平均无故障时间，故障前平均（工作）时间
MTTFO(mean time to forced outage) 平均强迫停运间隔时间
MTTMO(mean time to maintenance out) 平均维修停运间隔时间
MTTOO(mean time to planned overhaul outage) 平均计划大修停运间隔时间
MTTPO(mean time to planned outage) 平均计划停运间隔时间
MTTR(mean time to repair) 平均修复时间
MTTRO(mean time to planned minor repair outage) 平均计划小修停运间隔时间
MTTUO(mean time to unplanned outage) 平均非计划停运延续时间
MTU(magnetic tape unit) 磁带机
MTU(multiplexer and terminal unit) 多路转换器与终端装置
M-type skylight M形天窗
mu-antenna μ形天线
mu-beta measurement μ-β 同时测试，增益和相角同时测试
much the same 大致一样
muck bottom 软泥底
muck bucket 出碴斗，挖泥斗
muck car 出碴车
mucker 出碴机，装渣机，软土挖运机，挖泥工
muck flat 沼泽平原
muck haulage 运碴，运土
mucking machine 出碴机
mucking method 出碴方法
mucking 出碴，挖土，装岩
muck land 淤泥地
muck loading 装碴
muck soil 腐殖土
mucky soil 淤泥质土、腐殖土
muck 残渣，碴屑，腐泥，软泥，岩碴，淤泥
mu constant μ常数，放大系数μ
mucus 黏液
mud and rock flows 泥石流
mud-apron shield 挡泥板
mud-apron 挡泥板
mud avalanche 泥石流，泥崩
mud baffle 挡泥板
mud ball 泥球
mud bank 泥滩，泥洲
mud bar 泥滩
mud belt 淤泥带
mud boil 泥浆翻腾，泥涌
mud cake 泥饼
mud capping 封泥【爆孔用】
mud chamber 泥浆池，泥箱
mud circulation 泥浆循环
mud coal 煤泥，泥渣煤
mud column 泥浆柱
mud cone 泥火山，泥丘
mud discharge pipe 排泥管
mud dredge 挖泥器
mud drum 泥包，（锅炉）泥鼓，排泥管
muddy creek 泥溪
muddy ground 淤泥地
muddy material 淤泥
muddy soil 淤泥土
muddy wastewater 多泥废水
muddy water 浑水，混浊水，泥浆水
mud flap 挡泥胶皮
mud flat 泥滨，泥滩，淤泥滩
mudflow fan 泥石流（冲积）扇
mud-flow soil 泥流土
mud flow 泥流，混浊水流，浑流，泥石流
mud foreshore 泥前滨
mud hole 出泥孔
mud injection 泥浆灌注
mud jacking 压浆，水泥压浆
mud jack 压浆泵，压浆
mud kibble 挖泥吊桶
mud layer 淤泥层
mud leg 沉泥箱
mud line 泥线
mud lump 淤泥滩
mud pipe 排泥管道，排泥管

mud pump　泥浆泵，污泥泵
mud rake　挖泥耙
mud replacement by blasting　爆破排淤法
mud residue　泥渣
mud ring　（火管锅炉）下脚圈
mud-rock flow　泥石流
mud scow　泥驳
mud scraper　刮泥机
mud scum　浮泥
mud seam　淤泥夹层
mud shovel　泥铲
mud stone　泥岩
mud sump　沉泥池，泥箱
mud trap　泥箱，沉泥池，沉泥井
mud valve　排泥阀
mud volcano　泥火山
mud wall　土墙
mud wave　泥波
mud wing　翼子板
mud　泥浆，滤泥，泥渣，沉淀物，泥，泥沙
mu-factor　μ数，放大系数μ
MUF expected value　最大可用频率预期值
muff coupling　套筒联轴节
muff-joint　套管接头，套管连接
muffle burner　点火喷燃器
muffled glass　遮光玻璃
muffle furnace　点火炉，膛式炉，回热炉，马弗炉
muffler piling　消声打桩
muffler plate　消弧板，消声板
muffler section　消声段
muffler　（熔断器的）消弧片，消声器，套筒
muffling　深孔爆破
muff　套筒，燃烧室外筒
MUF(maximum usable frequency)　最大可用频率
mulched ground　覆盖地
mulching soil　覆盖土
mulching　铺覆盖料
mulch　覆盖物，护根物
mulde　凹地，舟状槽
mule traveler　爬行吊机
mulit-pressure condenser　多压凝汽器
muller　滚轮，混砂机，研磨机
mullion　（门窗）中竖框，竖框
mullite　富铝红柱石，麻来石
multgroup approximation　多群近似
multiaccess computer　多路存取计算机
multiadapter　多用附加器
multiaddress code　多地址码
multiaddress computer　多地址计算机
multiaddress instruction　多地址指令
multiaddress message　多地址信息
multiaddress　多地址的，多地址
multiamplifier　多级放大器
multianalysis　多方面分析
multiannual storage plant　多年调节电站
multianode mercury-arc rectifier　多阳极汞弧整流器
multianode　多阳极的
multi-antireflection coating　多次减反射膜
multiaperture core device　多孔磁芯器件
multiaperture core logic　多孔磁芯逻辑（电路）
multiaperture core　多孔磁芯，多孔铁芯
multiaperture device　多孔磁芯，多孔器件
multiaperture logic element　多孔逻辑元件
multiaperture　多孔的
multiar circuit　多向鉴幅电路，多向振幅比较电路
multiarmature DC motor　多电枢直流电机，多枢直流电动机
multiaxial state　多轴状态，多向状态
multiaxial stress　多向应力，多轴应力
multibag filter　复式袋滤器
multiband antenna　多频带天线，宽频带天线
multiband gap a-Si solar cell　多带隙非晶硅太阳电池
multiband remote sensing　多波段遥感
multiband　多频带，多频段，宽频带
multibank engine　多排发动机，多列发动机
multibank　复台式，复接排
multibar printer　多杆打印装置，多杆印刷机
multibarrier system　多重屏障系统
multibarrier　多重屏障
multibar　多杆的
multibeacon　组合信标，三重调制指点标
multi-beam sounding system　多波束测深系统
multibed system　（水处理）多级系统，多床系统
multi-bed　复床
multiberak　多断口【断路器】
multibit code　多进制码
multi bladed rotor　多叶片风轮
multi bladed windmill　多叶片风车
multiblade　多叶片的
multi-block bidding　多段报价
multi-body system　多体系统
multibranch network　多支路网络，复式网络
multibrator　多谐振荡器
multibreak　多重开关，多断点的
multi-bucket trencher　多斗挖沟机
multibulb　多管的，多灯泡的
multi burner cutting machine　多头切割机
multi-button sequence　多按钮操作程序
multicapacity control system　多容量控制系统
multicapacity process　多容量过程
multicapacity　多容量
multicarrier transmitter　多载波发射机
multicast application association　多播应用关联
multicast sampled value control　多播采样值控制
multicast sampled value control block　多播采样值控制块
multicast　多播
multicavity klystron　多腔速调管
multicavity magnetron　多腔磁控管
multicavity PCRV　多腔预应力混凝土反应堆容器
multi-cell battery　并联蓄电池
multi-cell boiler　多室锅炉
multi-cell fluidized bed　多室流化床
multi-cellular mechanical precipitator　多管式机械除尘器
multi-cellular　多孔的，多网络的，多室的，（除尘器）多管式的
multicenter DC manostat　多通道直流稳压电源

multi-centered arch 多心拱
multichamber drier 多室干燥器
multichamber fluidized bed 多室流化床
multichannel amplifier 多路放大器
multichannel analyser 多道分析器，多波（通）道分析仪
multichannel boiling core 多通道沸腾堆芯
multichannel carrier 多路载波
multichannel coincidence system 多路重合系统，多路符合系统
multichannel communication system 多道通信系统
multichannel correlator 多路相关器
multichannel data register 多通道数据寄存器
multichannel monitoring system 多道监测系统
multichannel on-line noise analysis system 多通道的线噪声分析系统
multichannel oscillograph 多路示波器，多回线示波器
multichannel receiver 多信道接收机
multichannel recorder 多路记录器，多路记录仪，多（通）道记录器
multichannel recording head 多路记录头
multichannel recording oscillograph 多路记录示波器
multichannel revolver 多通道旋转变压器
multichannel telephony 多路电话
multichannel time analyser 多道时间分析器
multichannel transmission 多路传输
multichannel voice frequency 多路话音频率
multichannel 多通道的，多频道的，多信道的，多通道，多孔道
multicharge 混合装药
multichip carrier 多片载体，多切片载体
multichip integrated circuit 叠片集成电路，多片集成电路
multichip 多片，多切片
multicircuit control 多回路控制，多路控制
multicircuit power transmission 多回路输电
multicircuit relay 多路继电器
multicircuit switch 分路开关，多电路转换开关
multicircuit transformer 多绕组变压器
multicircuit winding 多匝绕组
multicircuit 多路，多回路
multiclone 多管旋风除尘器
multicoil 多线圈的
multicollar thrust bearing 多块式推力轴承
multicolour map 多色地图
multicolour recorder 多色记录仪
multicombustion chamber 多腔燃烧室
multicompartment fluidized bed 多层流化床
multicompartment 多隔室
multicomponent balance 多分量天平，多分力天平
multicomponent gas 多成分气体
multicomponent mixture 多元混合物
multicompression 多级压缩
multicomputer system 多计算机系统
multicomputer 多计算机
multiconductor bundles 多根分裂导线
multiconductor cable 多芯电缆，多股电缆，多导体电缆
multiconductor plug 多触点插塞，多线插头，多线插塞
multiconductor 多触点，多线，多导体，多导体的，多导线的
multiconstant speed motor 多速电动机
multicontact auxiliary relay 多触点辅助继电器
multicontact gang switch 多触点联动开关
multicontact relay 多触点继电器
multicontact switch 多触点开关
multicontact 多触点的，多接点的
multicore cable 多芯电缆
multicore combined 多芯组合的
multicore flat cable 多芯扁形电缆
multicore transformer 多芯式变压器
multicore 多芯的
multicountry financing 多国投资，多国集资
multi-coupler 多路耦合器
multicrank engine 多缸复胀式发动机
multi-curie processing 多居里级处理
multi-currency 多种货币
multicycle 多周期的，多循环的
multicyclone collector 多管除尘器
multicyclone dust collector 多旋风除尘器
multicyclone separator 多管式旋风分离器
multicyclone 多级旋风分离器，多管式旋风子，多管旋风除尘器，旋风分离器组
multicylinder stack 多筒烟囱
multicylinder steam turbine 多缸汽轮机
multicylinder turbine 多缸汽轮机
multicylinder 多汽缸
multideck 多层的
multi degree of freedom 多自由度
multi-die press 多模压制
multidigit shift 多位移位
multidigit 多位
multidimensional access memory 多维存取存储器
multidimensional analysis 多维分析
multidimensional code 多维码
multidimensional normal distribution 多维正态分布
multidimensional 多维的，多因次的
multidirectional wind 多向风，不定向风
multidisciplinary group 跨专业小组，多科性小组
multidisciplinary 多种学科的，多学科的
multielectrode counting tube 多电极计数管
multielectrode 多电极
multi-element control system 多冲量控制系统
multi-element model 多段模型
multi-element oscillograph 多振子示波器
multi-emitter transistor 多发射极晶体管
multi-ended circuit protection 多端线路保护
multi-exhaust blade 复式排汽叶片
multi-exhaust turbine 多排汽的汽轮机
multi-exhaust 多排汽
multi-failure 多重故障
multifeed 多点供电的
multifee 复式收费
multifibre cable 多纤光缆
multifilament lamp 多灯丝白炽灯

multifilm 多层胶片
multifinger contactor 多点接触器
multifinned body 多翅体
multi-flash distillation 多级闪蒸法蒸馏，多重闪蒸法，多重急骤蒸馏法
multiflow cooling system 多流程冷却系统
multiflow turbine 多流式汽轮机
multiflow 复流
multi-flue chimney 多管式烟囱，多筒式烟囱，多管烟囱，集束烟囱
multi-flue smoke stack 多烟道烟囱
multi-flue stack 多烟道烟囱
multifluid theory 多流体理论
multiflux method 多通量方法
multiflux 多通量
multifolded plate 多折板
multifold 多倍的，多重的
multi force transducer 多力传感器
multiform function 多值函数
multiframe core 多框式铁芯
multifrequency alternator 多频同步发电机
multifrequency code signaling method 多频编码传信法
multifrequency dialing 多频拨号
multifrequency electronic multiplier 多频电子倍增器
multifrequency generator 多频振荡器，多频发电机
multifrequency key pulsing 多频键控脉冲调制
multifrequency remote control system 多频遥控系统，多频遥控制
multifrequency sender 多频记发器，复频发送器
multifrequency signalling 多频信令
multifrequency 复频的，多频的，宽频带的
multifuel burner 多燃料复合喷燃器，多燃料复合燃烧器，多种燃料燃烧器
multifuel fired boiler 混烧锅炉，燃用多种燃料的锅炉
multifuel fired power plant 混燃火力发电厂，混烧火力发电厂
multifuel firing 多种燃料燃烧
multifunctional nip machine 多功能咬口机
multifunctional 多功能的
multifunction architecture 多功能建筑
multifunction control 多功能控制
multifunction detection and recording system 多功能检测及记录系统
multifunction device 多功能器件
multifunction tool 多用途工具
multifunction 多功能
multigang potentiometer 多联电位器
multigang switch 多联开关
multigap arrester 多隙避雷器，复隙式避雷器
multigap arrestor 复隙式避雷器，多隙式避雷器
multigap discharger 多隙放电器
multigauge 多用规，多用测量仪表
multigrounded common neutral 多接地公共中性点
multigroup calculation 多群计算
multigroup constant 多群常数
multigroup cross section 多群截面
multigroup diffusion equation 多群扩散方程
multigroup equation 多群方程
multigroup flux 多群通量
multigroup method 多群法
multigroup model 多群模型
multigroup neutron diffusion equation 多群中子扩散方程
multigroup theory 多群理论
multigroup transport equation 多群输运方程
multigun 多枪的
multiharmonograph 多谐记录仪
multihole block 多孔型燃料块，蜂窝煤状燃料块
multihole distributor 多孔分布板
multihole grid 多孔栅板
multi-input comparator 多端输入比较器
multi-input switch 多输入开关
multi-input 多端输入，多输入
multi-jet element 多喷嘴管
multi-jet spray 多喷口，喷雾嘴
multi-jet turbine 多喷嘴冲击式水轮机
multi-jet tuyere type gas distributor 多喷口风嘴式气体分布器
multi-jet type sootblower 多喷嘴吹灰器
multi-jet 多射流，多喷嘴
multi-job operation 多作业操作，多作业运算
multijunction solar cell 多结太阳电池
multi-labyrinth seal 多迷宫密封
multi-lane road 多车道道路
multilateral agreement 多边贸易协议，多边协定
multilateral aid 多边援助
multilateral assistance 多边援助，多边支援
multilateral commitment 多边承诺
multilateral cooperation 多边合作
multilateral payments 多边支付
multilateral tax treaty 多边税务条约
multilateral technical assistance 多边技术援助
multilateral trade negotiation 多边贸易谈判
multilateral trade 多边贸易
multilateral treaty 多边条约
multilateral 多边的
multilayer cladding 多层包壳，多层覆盖
multilayer coating 多层覆盖，多层包覆
multilayer coil 多层线圈
multilayered aquifer 多层含水层
multilayered flow 多层流动
multilayered fluid 多层流体
multilayered thermal insulation system 多层隔热系统，多层保温系统
multilayer film 多层
multilayer filter 多层过滤器
multilayer fluidized bed 多层流化床
multilayer insulation 多层绝缘
multilayer interconnection 多层连线，多层布线
multilayer particle 多层颗粒
multilayer printed circuit board 多层印刷电路板
multilayer printed circuit 多层印刷电路
multilayer sandwich 多层夹层结构
multilayer structure 多层结构
multilayer vessel 多层容器
multilayer welding 多层焊
multilayer winding 多层绕组

multilayer 多层的，多层
multilead 多引入线的
multilegged 多支路的
multilength number 多倍长数
multilength 多倍
multilevel address 多级地址
multilevel circuit 多级电路
multilevel control system 多级控制系统
multilevel control theory 多级控制理论
multilevel control 多级控制
multilevel converter 多点式变流器
multilevel indirect addressing 多级间接寻址
multilevel inlet 分层取水式进水口
multilevel interconnection generator 多电平互连式信号发生器
multilevel logical circuit 多级逻辑电路
multilevel store 多级存储器
multilevel 多级的，多层的，多电平的，多水平的，多级
multiline controller 多路控制程序，多路控制器
multiline 多线的，多生产线的
multi-lingual 多种语言的
multiload 多负载
multi-lock 多线船闸
multiloop control system 多环控制系统，多回路控制系统
multiloop control 多回路控制
multiloop feedback system 多环反馈系统
multiloop integral system test 多环路综合系统试验
multiloop safety system 多环路安全系统
multiloop servosystem 多回路伺服系统，多环伺服系统
multiloop superheater 蛇形管过热器
multiloop system 多回路系统
multiloop 多环的，多回路的，多匝的，多回路
multiple bay frame 多跨框架
multimachine controller 多机组控制器
multi manometer 多管压力计
multi media projector 多媒体投影仪
multi media 多媒体
multi-megawatt wind turbine 多兆瓦级的风电机组
multimetering 多次测量，多点测量，多量程测量
multimeter 多用途计量器，万用表，多量程测量仪表，通用测量（万用）仪表
multimodal distribution 多峰分布，多重模态分布
multimodal 多峰
multimode distortion 多模畸变
multimode laser 多模激光器
multimode optical fibre 多模光纤
multimode waveguide 多模波导
multimode 多模态，多方式，多波型
multimotor 多电动机的
multinational company 多国公司
multinational corporation 跨国公司
multinational enterprises 多国企业
multinational fuel centre 多国燃料中心
multinodal 多节点的
multinode 多节的
multinomial distribution 多项式分布
multinomial 多项的，多项式，多项式的
multinormal distribution 多维正态分布，多项分布
multinozzled plate 多喷嘴分布板，多喷嘴布风板
multi-nozzle grid 多喷管叶
multi-nozzle sootblower 多喷嘴吹灰器
multi-nozzle turbine 多喷嘴水轮机
multiobjective planning 多目标规划
multi-operating mode automatic conversion 多工况自动转换
multi-operating mode control system 多工况控制系统
multi-order lag 多阶滞后
multi-orifice plate 多孔布风板，多孔分布板
multioutlet 多引线
multipactoring 次级发射倍增，次级电子倍增
multipair cable 多对电缆
multiparameter analyzing assembly 多参数分析装置
multiparameter regulation circuit 多参数调节电路
multiparameter self-optimizing system 多参数自寻优系统
multiparameter water quality instrument 多参数水质监测仪器
multiparameter 多参数
multipart bidding 多部投标
multi-party arbitration 多方仲裁
multipart 多元件的，多部件的
multi-pass arrangement （受热面）多流程布置
multi-pass boiler 多回程锅炉
multi-pass compiler 多次扫描编译程序
multi-pass cooling 多路冷却
multi-pass core 多流程堆芯
multi-pass flow 多程流
multi-pass welding 叠层焊接，多道焊接，多道焊
multi-pass weld 多道焊，多道焊缝
multi-pass 多程的，多回程，多回路，多次通过，多道，多程
multipath arrester 多火花隙避雷器
multipath cooling 多路冷却
multipath core 多路磁芯，多孔铁芯
multipath ferrite core 多孔铁氧体磁芯
multipath signal 多程信号
multipath transmission 多路输送
multipath 多路的，多孔
multipetticoat insulator 多裙式绝缘子
multiphase circuit 多相电路
multiphase current 多相电流
multiphase diffusion 多相扩散
multiphase flow model 多相流模型
multiphase flow 多相流
multiphase fluid dynamics 多相流体动力学
multiphase generator 多相发电机
multiphase inverter 多相逆变器
multiphase motor 多相电动机
multiphase power transmission 多相输电
multiphase project 多阶段项目，多期工程
multiphase-purpose project 多目标开发工程
multiphase reclosing 多相重合闸
multiphase system 多相系统
multiphase transformer 多相变压器
multiphase 多相，多相的

multi-piece rotor 组合转子，多段转子
multipin connector 多脚插头连接器，多柱（角）连接插头
multipin socket 多脚插座【与多脚插头配合】
multipipe hot-water heat-supply network 多管制热水供热网
multipipe steam heat-supply network 多管制蒸汽热网
multi-pipe 集装管，套装管
multiplane balancing 多面平衡
multiplane-multispeed balancing 多平面多转速平衡法
multiplate fission chamber 多板裂变室
multiple access computer 多路存取计算机，多重存取计算机
multiple access device 多路存取设备
multiple access sequential selection 多路存取顺序选择法
multiple access system 多路存取系统
multiple access time-sharing 多路存取的分时
multiple access 多路存取
multiple accumulating registers 多次累积寄存器，多重累加寄存器
multiple address code 多地址代码
multiple address computer 多地址计算机
multiple address instruction 多地址指令
multiple address message 多地址信息
multiple address processing unit 多地址处理部件
multiple aerial 复合天线
multiple agreement 复合协定
multiple-alternative evaluation 多方案评价
multiple aperture core 多孔磁芯
multiple aperture logic element 多孔逻辑磁元件
multiple aperture reluctance switch 多孔磁阻开关
multiple arch bridge 连拱桥
multiple arch dam 连拱坝，多拱坝
multiple arch retaining wall 连拱式挡土墙
multiple arch 连拱
multiple arm relay 多杆继电器
multiple back reflection 多次底面回波
multiple barrier containment 多层安全壳
multiple bay frame 多跨框架，多跨排架
multiple bay portal 多跨门架
multiple belt 多层皮带
multiple-blade grader 多刃平地机
multiple-blade screwdriver 多刀头螺丝起子
multiple box girder 多室箱梁
multiple box siphon 复式箱形虹吸管
multiple break contact 多断开触点
multiple break switch 多断式开关
multiple bridge 多挡电桥
multiple cable 复电缆
multiple-cage motor 多鼠笼式电动机
multiple-cage rotor 多鼠笼转子
multiple capture 多次俘获
multiple cascades 多列叶栅，多排叶
multiple casing turbine 多缸汽轮机
multiple casing 多缸
multiple chamber lock 多厢船闸
multiple channel indicator 多点指示仪
multiple channel recorder 多点记录仪
multiple channel tray 多槽托架
multiple channel 多通道
multiple check 多重校验
multiple chimney 复式烟囱
multiple chip package 多片组件
multiple circuit line 多回路线路
multiple circuit winding 叠绕组，多匝绕组
multiple circuit 倍增电路，复接电路，多回路
multiple coated fuel particle 多层包覆燃料颗粒
multiple coating 多层包覆
multiple coil 多线圈
multiple coincidence magnetic memory 多度符合磁存储器，多重符合磁存储器
multiple coincidence 多次符合
multiple-coincident magnetic storage system 多重符合磁存储系统
multiple collision calculation 多次碰撞计算
multiple collision probability 多次碰撞概率
multiple collision 多次碰撞
multiple combustor 多燃烧室
multiple communication control unit 多路通信控制设备
multiple compartments （沸腾炉）多床结构
multiple computer operation 多计算机操作［处理］
multiple conductor cable 多导线电缆
multiple conductor concentric cable 多导线同轴电缆
multiple conductor 多重导线
multiple cone insulator 叠锥体绝缘子
multiple connection 并联接法，复接，并联
multiple console support 多操作台支援
multiple consumables vacuum melting 并联自耗真空熔炼
multiple contact regulator 多触点调压器
multiple contact relay 多触点继电器
multiple contact switch 多触点开关
multiple contact 多接点，多触点
multiple containment 多层安全壳
multiple control 平行调节，复式控制，并列调节
multiple converter 并接换流器
multiple convolution 多次卷积式，多重卷积
multiple core 多心的
multiple correlation coefficient 多重相关系数
multiple correlation 多重相关，复相关
multiple current generator 交直流发电机，多电流发电机
multiple-cut trench excavator 多刃挖沟机
multiple cyclone 多管除尘器
multiple damper 百叶窗式挡板
multiple data terminal 多路数据终端
multiple decade counter 多路十进制计数器
multiple deviation 多次偏差
multiple digit decimal adder 多数位的十进制加法器
multiple-dimensioned double head wrench 多尺度双头扳手
multiple disc clutch 多片式离合器
multiple distribution system 并联配电制
multiple dome dam 双曲连拱坝
multiple dose 多剂量，多剂量型，倍剂量，多

次剂量
multiple driver 多滚筒驱动器，多点驱动
multiple drum boiler 多汽包锅炉
multiple drum steam generator 多汽包蒸汽发生器
multiple duct conduit 多导管套管
multiple duct 多导管
multiple dump processing 多重转储处理
multiple earthed system 多处接地系统
multiple earth 多重接地
multiple echo method 多次反射法
multiple echo 多重回声，多次反射回波
multiple effect distillation plant 多效蒸馏厂
multiple effect evaporator 多效蒸发器
multiple electrode process 多电极工艺
multiple entry visa 多次往返的签证，多用入境签证，多次入境签证
multiple error correcting code 多重误差校正码
multiple error 多重错误，多重误差，多次误差
multiple evaporator 多级蒸发器
multiple events 多次事件
multiple examination stand 多种检验台架
multiple excitation 多激，多励，复励，多次激发，多励磁，复励磁
multiple exhaust turbine 多排汽口汽轮机
multiple expansion engine 多次膨胀式发动机，多级膨胀蒸汽机
multiple expansion 多级膨胀
multiple exposure 多次曝光
multiple fault 复式故障，多重故障，复故障，迭断层，复断层
multiple feeder 多电源馈电线，复式并联馈路
multiple feed 多电源供电，多回路供电
multiple fender piles 丛桩缓冲设备
multiple filter plate 多重过滤板
multiple filter 多节滤波器，复式滤池
multiple flow turbine 多排汽口汽轮机
multiple flue chimney 多烟道烟囱
multiple fork type blade root 叉形叶根
multiple frequency-shift keying 多频移键控
multiple frequency 多频，倍频，谐波频率
multiple gamma ray event 多γ射线事件
multiple gap lightning arrester 复隙式避雷器，多隙式避雷器
multiple gated spillway 多闸门溢洪道
multiple glazing 多层窗玻璃
multiple grid method 多重网格法
multiple grid tube 多栅管
multiple groove bearing 多油槽轴承
multiple-harmonic current 多谐波电流
multiple head bonder 多焊接头压接机
multiple hop transmission 多次反射传输
multiple-impulse welding 脉冲点焊
multiple indeterminate 多次超静定的
multiple induction loop 多次归纳循环
multiple instruction multiple data 多指令多数据
multiple instruction single data 多指令单数据
multiple integral 多重积分
multiple ionization 多重电离
multiple jack 复式插孔
multiple jet nozzle 多孔喷嘴
multiple jet 多孔喷射
multiple laccolith 叠岩床
multiple lance （核测量仪的）多头长管
multiple lane road 多车道道路
multiple lap winding 复叠绕组
multiple leaf gate 多叶闸门
multiple-length arithmetic 多倍字长运算
multiple-length number 多倍字长数
multiple-length working 多倍字长处理，多倍字长工作
multiple level line 复测水准线
multiple-lever system 复式杠杆系统
multiple lightning stroke 多隙雷击
multiple linear regression 多重线性回归
multiple-loop control system 多回路控制系统
multiple-loop system 多环路系统，多回路系统
multiple-loop 多环路
multiple manometer 组合压力计
multiple-mercury-column manometer 多级水银柱压力计
multiple mode microwave network 多模微波网络
multiple mode operation 多种方式的操作
multiple modulation 多级调制，复调制
multiple module access 多模块存取
multiple motor drive 多电动机传动，多机传动
multiple motor 多电动机的，复频电动机
multiple multipole circuit breaker 多路多极断路器
multiple objective development 多目标开发，综合开发
multiple objective function 多重目标函数
multiple of unit 倍数单位
multiple operation 并联运行，平行工作
multiple order 多阶的，多级的，多次的
multiple outage occurrence 多重停运事件
multiple outlet 多管出水口，多孔出水口
multiple output circuit 多输出电路
multiple output network 多端输出网络
multiple package contract 多分包合同
multiple parallel winding 复并联绕组，复并励绕组，复叠绕组
multiple pass atmospheric braking 多次气动力减速
multiple pass rotary mixer 多行程转轴式拌和机
multiple-peaked hydrograph 多峰式水文过程线
multiple pen recorder 多笔记录仪
multiple pin plug 多脚插头
multiple pipe inverted siphon 多管式倒虹吸管
multiple plug 复式插头
multiple point borehole extensometer 多点钻孔伸长计
multiple pointer indicator 多针指示仪
multiple point loading 多点集中加载
multiple point recording potentiometer 多点记录式电位计
multiple point to point configuration 多个点对点配置
multiple point 多点的
multiple pole switch 多极开关
multiple pontoon landing stage 多浮筒式登岸栈桥
multiple position shift 多位移位
multiple priming 并联起爆

multiple project　多期工程
multiple pulse coding　多脉冲编码
multiple punch　多滑块压力机，复式冲模
multiple purpose dam　多目标坝
multiple purpose project　多目标工程，综合利用工程，综合开发计划
multiple purpose reservoir　多目标水库，多用途水库，通用蓄水池，综合利用水库
multiple purpose tester　万用测试仪
multiple rate of exchange　复式汇率
multiple rating curve　复式水位流量关系曲线
multiple reaction　多次反应
multiple reactor　多区反应堆
multiple reclosing breaker　多次重合闸断路器
multiple re-entrant multiplex winding　多闭路多重绕组
multiple re-entrant winding　多闭路绕组
multiple reflection　多次反射
multiple region reactor　多区反应堆
multiple regression analysis　复回归分析
multiple regression coefficient　复回归系数
multiple resonance technique　多重共振技术
multiple resonance　复共鸣，复共振
multiple retort underfeed stoker　多槽下饲式炉排
multiple roll crusher　多滚筒碎石机
multiple rotor mixer　多转轴拌和机
multiple runner　复式转轮
multiple sampling　多次抽样法
multiple scattering　多重散射
multiple selection　多次选择
multiple sensor　复合传感器
multiple series connection　串并联，混联
multiple series　双重级联，混联，复联，串并联
multiple shaft　多轴透平
multiple sheet drawing　多张图纸
multiple-shielded high-velocity thermocouple　多层屏蔽高速热电偶
multiple shot reclosing　多次重合闸
multiple signal　复式信号
multiple socket　多用插座
multiple sound insulation　多层隔音
multiple sound insulator　多层隔音装置
multiple source supply　多种电源
multiple source　复合源
multiple-span frame　多跨框架
multiple spark discharge　多火花隙放电器
multiple-speed floating action　多速无定位作用
multiple-speed floating controller　多速无定位控制器
multiple-speed floating control system　多速无静差控制系统
multiple-speed motor　多速电动机，多速马达
multiple spot welding machine　多点焊接机
multiple spot welding　多点焊
multiple-stable-state storage　多稳态存储器
multiple stage amplifier　多级放大器，级联放大器
multiple stage antenna　多级天线
multiple stage compressor　多级压缩机
multiple stage expansion　多级膨胀
multiple stage planetary gear train　多级行星齿轮系
multiple-stage treatment　多级处理
multiple stage triaxial test　多级三轴试验
multiple stage　多级的，多段的，多级式
multiple streamtube model　多流管模型
multiple stroke　多次放电
multiple of a unit of measurement　测量的倍数单位
multiple surge tank　复式调压井
multiple-sway frame　多侧移框架
multiple-swing check valve　多行程止回阀
multiple switching relay　重合闸继电器
multiple switch　复联开关
multiple taril meter　多种电费电度表
multiple telegraph　多路电报机
multiple telephony　多路电话通信
multiple tempering　多次回火
multiple thermocline　复斜温层
multiple tier tray　多层托架
multiple tire roller　多轮胎压路机
multiple-track　多信道的，多路的，多车道的
multiple transfer　多路传送
multiple transmission　复合式传输
multiple tray aerator　多盘曝气
multiple tube counter　多管计数器
multiple-tuned mass damper　多重调谐质量阻尼器
multiple-tuned　多重调谐的
multiple-turn winding　多匝线圈，多匝绕组
multiple twin cable　扭绞多心电缆
multiple twin quad　双股四心电缆
multiple U gauge　多管 U 形压力计
multiple unit control　多元控制
multiple unit running　多机组运行
multiple unit tube　复合管
multiple unit valve　多路阀
multiple unit　多重驱动，多元
multiple upflow filter system　多级上流过滤器系统
multiple-use reservoir　多目标水库
multiple utilization　综合利用
multiple value　多值
multiple vibrating feeder　复合振动给料机
multiple-walled envelope　多层外壳
multiple water sampler　复式取水样器
multiple watershed method　多断并列流域法
multiple wave winding　复波绕组
multiple way switch　多路开关
multiple well pumping　多孔抽水
multiple winding transformer　多绕组变压器
multiple winding　多绕组，并联绕组，复绕组，叠绕组
multiple-wire ionization chamber　多丝电离室
multiple-wire submerged arc welding　多丝埋弧焊
multiple-wound armature　简单并联绕组电枢，多绕组电枢
multiplex-automatic error correction　多路转换自动误差校正，多路自动误差校正
multiplexed communication control unit　多路通信控制器，多路转换通信控制器
multiplexer and terminal unit　多路转换器与终端装置

multiplexer 多路扫描装置，多路调制器，倍增器，多路转换器，信号连乘器，多路调剂器
multiplexing amplifier 多路放大器，倍增放大器
multiplexing equipment 多工设备
multiplexing 多路传输，多路复用
multiplex interface 复用接口
multiplex lap winding 复叠绕组
multiplex link 复用链路
multiplex mode 多路转换方式
multiplexor channel operation 多路转换通道操作
multiplexor communications 多工通信，多重通道通信，多路调制器连通
multiplexor terminal unit 多路转换器终端单元
multiplex pulse modulation 多路脉冲调制
multiplex pulse system 多路脉冲系统
multiplex pump 多路泵【活塞或柱塞】
multiplex switch 复联开关，复式开关，复接开关，多路开关
multiplex telemetering 多路远距离测量，多路遥测
multiplex telephone 多路电话
multiplex transmission 多路传输
multiplex wave winding 复波绕组
multiplex winding 复绕组
multiplex 多路转换，多路传送，多工操作，多路传输，多工，多路（的）
multiple 多倍的，复式的，多重的，多的，多，倍数，多路系统，倍率
multiplicand gate 被乘数门
multiplicand register 被乘数寄存器
multiplicand 被乘数
multiplication by series 级数乘法
multiplication constant 核燃料再生系数，增殖系数，倍增系数
multiplication-division unit 乘法除法器，乘法除法部件
multiplication factor 增殖系数，倍增系数，乘数，放大系数，放大因子，增殖因子
multiplication neutron 倍增中子
multiplication of series 级数乘法
multiplication point 相乘点
multiplication rate constant 倍增率常数
multiplication rate eigenvalue 倍增率本征值
multiplication time 乘法时间
multiplication 增殖，倍增，放大，乘，乘法
multiplicative overall hot spot factor 相乘的总热点因子
multiplicity factor 多重因子
multiplicity 重复度，多样性，多重性，多倍，集
multiplier digit 乘数的数字，乘数的数位
multiplier-divider unit 乘法器除法器部件
multiplier-divider 乘除器
multiplier gain 倍增器增益
multiplier of the contract price 合同价格的乘数
multiplier potentiometer 乘法器电位计
multiplier-quotient register 乘数商寄存器，乘商寄存器
multiplier register 乘数寄存器
multipliers 乘数，倍增器
multiplier zero error 乘法器零点漂移误差
multiplier 乘数，乘法器，倍增器，扩程器，放大器，光电倍增管，系数，因数
multiply and divide 乘与除
multiplyfabric 多层蒙皮
multiplying factor 放大因数，乘系数，复用因子，倍率
multiplying gear 增速齿轮（传动）
multiplying instruction 相乘指令，乘法指令
multiplying medium 倍增介质
multiplying order 乘法指令
multiplying potentiometer 乘法电势计
multiplying power 倍率，放大率
multiplying region 增殖区
multiplying signal 相乘信号
multiplying system 放大系统
multiplying unit 乘法器，乘法部件
multiply periodic 多周期的
multiply reentrant winding 多重闭路绕组
multiply solid woven 多层实心织物
multiply time chain 乘法时间链
multiply type woven 多层织物
multiplywood 多层板
multiply 乘，倍增，繁殖，多层的，多股的
multipoint connector 多点接头
multipoint earthing 多点接地
multipoint indicator 多点指示器
multipoint line 多点线路
multipoint logger 多点（温度）巡测仪
multipoint method 多点法【测流的】，多点测流法
multipoint-partyline configuration 多点共线配置
multipoint priming 多点起爆
multipoint recorder 多点式记录器，多信道记录器，多点记录仪
multipoint-ring configuration 多点环形布局，多点环形配置
multipoint-star configuration 多点星形布局，多点星形配置
multipoint strip chart recorder 多点带状记录器
multipolar dynamo 多极直流发电机，多级电机
multipolar electrical generator stator 多极发电机定子
multipolar external stator 多极外部定子
multipolar generator 多极发电机
multipolar machine 多极电机
multipolar motor 多极电机
multipolar 多极的
multipole breaker 多极断路器
multipole circuit breaker 多极断路器
multipole common flame breaker 共架多极开关
multipole direct drive generator 多极直接驱动型发电机
multipole direct drive variable speed synchronous generator 多极直接驱动变速同步发电机
multipole internal rotor 多极内转子
multipole large diameter stator 多极大直径定子
multipole machine 多极电机
multipole permanent magnet synchronous generator 多极永磁同步发电机
multipole relay 多极继电器
multipole ring generator 多极环式发电机
multipole switch 多刀开关，多极开关

multipole 多级的【电机】，多接点的，多极
multipoplar dynamo 多极发电机
multiport burner 多喷口燃烧器
multiport memory 多端口存储器，多出入口存储器
multiport network 多口网络，多端网络，多端口网络
multiport relief valve 多路溢流阀
multiport sampling nozzle 多头式取样器
multiport valve 多通路阀，多进口阀
multiposition controller 多位式调节器，多位式控制器
multiposition element 多位置元件
multiposition relay 多位置继电器
multiposition 多位置
multi-power source chain network 多电源链形网
multi-power source looped network 多电源环网
multi-pressure condenser 多背压凝汽器，多压式凝汽器
multi-pressure measuring system 多点测压系统
multi-pressure operation 多压运行
multi-pressure turbine 多压式汽轮机，补汽式汽轮机，混压式汽轮机
multi-pressure 多压的
multiprobe 多头探针
multiprocessing system 多处理系统
multiprocessing 多重处理，多道处理
multiprocessor interleaving 多处理器交错
multiprocessor system 多处理机系统
multiprocessor 多处理机，多处理器
multiprogrammed computer 多道程序计算机
multiprogrammed 多程序的
multiprogramming executive control 多道程序执行控制
multiprogramming executive system 多道程序执行系统
multiprogramming 多道程序，多道程序设计
multipuff atmospheric dispersion model 多烟团大气扩散模型
multi-purpose architecture 多功能建筑
multi-purpose building 多功能楼，综合楼
multi-purpose center 多功能中心
multi-purpose computer 通用计算机
multi-purpose crane 多用途起重机
multi-purpose dam 综合利用坝，多目标坝
multi-purpose development 综合利用工程，多目标开发
multi-purpose engineering system 多目标工程系统
multi-purpose fuel reprocessing facility 多用途燃料后处理工厂
multi-purpose house 综合用房
multi-purpose instrument 多功能仪表，万用表，全能仪器
multi-purpose mast 多用途支柱
multi-purpose meter 多功能仪表，万用表
multi-purpose nuclear power plant 多用途核电厂
multi-purpose nuclear station 多用途核电站
multi-purpose project 多目标工程，综合利用工程
multi-purpose reactor 多用反应堆，多用堆
multi-purpose research reactor 多用研究反应堆
multi-purpose reservoir 多目标水库
multi-purpose spent fuel examination facility 综合乏燃料检验装置
multi-purpose tester 多用途测试器，万用表
multi-purpose tractor 万能拖拉机
multi-purpose transformer 多用变压器
multi-purpose use 综合利用
multi-purpose utilization 综合利用
multi-purpose water utilization 水利资源综合利用，多目标水（利）资源利用
multi-purpose 多种用途的，多功能的
multiqueue dispatching 多路排队调度
multiradix computer 多基数计算机
multiradix 多基数的
multirange indicating instrument 多量程指示仪表
multirange instrument 多量程仪表
multirange meter 多量程仪表
multirange 多量程，多量程的，多刻度的，多范围的
multi-reactor nuclear power plant 多堆核电站，多反应堆核电厂
multireflector 多区反射层
multiregion core 多区堆芯
multiregion diffusion theory 多区扩散理论
multiregion lattice 多区栅格
multiregion loading 多区装料
multiregion reactor 多区反应堆
multi-reservoir regulation 水库群调节
multiresonant circuit 多谐振荡电路，多调谐电路
multiresonant 多谐振荡的
multirestraint relay 多持线圈继电器
multi-ribbed plate 密肋板
multirib construction 多肋结构
multirow 多列
multirunning 多道运行
multirun welding 多道焊
multiscale rule 多标度计算尺，多刻度计算尺
multiscaler 通用换算线路
multiscale 多刻度的，多标度
multi-seasonal storage reservoir 多季调节水库
multi-section car retarder 多段式车辆缓行器
multisection coil 多元件线圈，多匝线圈
multisection construction 多段结构
multisection flap 多段襟翼
multisection pile 多节桩
multisection 多段【堆外探测器】，多节
multisegmented rotor 多极转子，多段转子
multisequencing 多机工作，多序列执行
multisequential system 多时序系统
multishaft arrangement 多轴布置
multishaft chimney(/stack) 多管烟囱
multishaft gas turbine 多轴（式）燃气轮机
multishaft type combined cycle 多轴型联合循环
multishell condenser 组合式冷凝器
multishift operation 多位移运算
multishift 多班制的
multisized material 多粒径物料
multisize particles 多粒径颗粒
multislot winding 多槽绕组
multislot 多槽的

multispan beam 多跨连续梁，多跨梁，连续梁
multispan elastic rotor system 多跨弹性轴系
multispan rotors balancing 多跨转子平衡
multispan structure 多跨结构
multispan 多挡，多跨
multispectral photography 多谱线照相术
multispectral remote sensing technique 多光谱遥测技术
multispectral scanner 多光谱扫描，多光谱扫描仪
multispectral survey 多光谱测量
multispeed control action 多速控制作用
multispeed floating mode 多速无静差模式
multispeed motor 多速电动机，多速电机，多速马达
multispeed power unit 多速执行部件
multispeed revolver 多速旋转变压器
multispeed 多速的
multispud burner 多喷嘴喷燃器
multistability 多稳定性
multistable circuit 多稳态电路，多稳电路
multistable storage 多稳态存储器
multistable system 多稳定系统
multistable 多稳态的，多稳的
multistage amplification 多级放大
multistage amplifier 多级放大器，级联放大器
multistage axial flow compressor 多级轴流式压气机
multistage axial-flow fan 多级轴流通风机
multistage blower 多级鼓风机
multistage centrifugal pump 多级离心泵
multistage circuit 多级电路
multistage collector 多级收集器
multistage compression 多级压缩
multistage compressor driverwind tunnel 多级压气机驱动风洞
multistage compressor 多级压缩机
multistage deaerator 多级除氧器
multistage feed water heating 多级给水加热
multistage filtration 多级过滤
multistage fluidized bed 多层流化床
multistage furnace 多级串联布置炉膛
multistage horizontal centrifugal pump 多级卧式离心泵
multistage lock 多级船闸
multistage mixed flow pump 多级混流式水泵
multistage operation 多级操作
multistage power turbine 多级动力透平，多级动力汽轮机
multistage pressure reducing valve 多级减压阀，多级降压阀
multistage pump 多级泵
multistage reactor 多级反应器
multistage separation 多级分离
multistage storage pump 多级蓄能水泵
multistage stressing 多阶段施加应力
multistage transformer 多级变压器
multistage turbine 多级透平，多级汽轮机
multistage volute pump 多级螺旋泵
multistage water power stations 多级水电站
multistage well point system 多级井点系统
multistage 多级的，多级，多段的
multi-stand conductor 多股绞线
multi-station cyclic service 多站循环服务
multistep action 多位作用，多步作用
multistep controller 多位控制器，多步控制器
multistep control servomechanism 多级控制伺服机构
multistep control 多步控制，多级控制
multistep pressure regulator 多级压力调节器
multistep sampling 多段抽样，多级取样
multi-storage system 多级蓄水系统
multi-storey building 高层建筑
multistory block 多层大厦
multistory box building 多层盒式建筑
multistory building 多层房屋，多层建筑
multistory factory 多层厂房
multistory frame 多层框架
multistory structure 多层建筑物
multistream heater 多管加热器
multistud tensioning machine 多螺杆拉伸器
multiswitch 复接开关，复接机键
multi-synchronizing display 多频率同步显示器
multisystem automatic test equipment 多系统自动测试设备
multitap 转接插座，多抽头的
multitask operation 多重任务运行
multitask 多重任务
multi-terminal control 多端（点）控制
multiterminal HVDC power transmission 多端高压直流输电
multiterminal HVDC transmission system 多端高压直流输电系统
multiterminal line 多端线路
multiterminal network 多端网络
multiterminal 多端的，多接线端子的
multitip burner 多头燃烧器
multi-tired tractor 多轮拖拉机
multitone 多频音
multitooth coupling 多齿联轴器
multitorch cutting machine 门式多头切割机
multitrace oscilloscope 多线示波器
multitron 甚高频脉冲控制的功率放大器
multitube chimney 多管式烟囱，多管烟囱，集束烟囱
multitube manometer 多管式测压计，多管压力计
multitube pressure gauge 多管压力计
multitube 多电子管，复用真空管，多管的
multiturn coil 多匝线圈
multiturn winding 多匝绕组，多匝线圈
multiturn 多匝的，多圈的
multituyere distributor 多风嘴分配器
multiunit machine 多部件计算机
multiunit nuclear power plant 多机组核电厂
multiunit plant 多机组电站
multiunit tube 复合管
multi-use bit 多用钻头
multiuser system 多用户系统
multivalence 多价
multivalent cation 多价阳离子
multivalued logic 多值逻辑
multivalued 多值的

multivalue function 多值函数
multivalve 多管的，多阀的
multivariable control system 多变量（多冲量）控制系统
multivariable control 多变量控制
multivariable feedback control system 多变量反馈控制系统
multivariable system 多变量系统
multivariable 多变量
multivariant 多自由度的，多变的
multivariateanalysis and prediction of schedule 进度的多元分析和预测
multivariate analysis 多变量分析，多元分析
multivariate distribution 多维变量分布，多元分布，多变量分布
multivariate normal distribution 多元正态分布
multivariate polynomial 多变元多项式
multivariate statistical analysis 多元统计分析
multivariate 多变量
multivector 多重矢量
multivelocity stages 多列速度级
multivertor 复式变换器
multivibrator circuit 多谐振荡器电路
multivibrator 多谐振荡器
multiview drawing 多视图
multivirtual storage 多虚拟存储器
multivoltage control 多电压控制
multivoltage transformer 多电压互感器
multivoltage 多电压的
multivolt meter 多量程伏特计，多量程电压表
multivolume file 多媒体单元的文件
multivortex mechanical collector 多管式除尘器
multi-wagon tippler 多车翻车机
multiwash scrubber 多段洗涤器
multiwash 多层洗涤
multiwave 多波的
multiway cable 多分支电缆
multiway communication system 多路通信系统，多向通信系统
multiway switch 多路开关，多向开关
multi-wheel drive 多轮传动
multi-wheel roller 多轮碾
multiwinding machine 多绕组电机，多相电机
multiwinding transformer 多绕组变压器
multi-window 多窗口
multiwire strand 多股线，多股绞线
multiwire 多线的，多股的，多股绞线
multiwound relay 多绕组继电器
multizone configuration 多区布置
multizone core 多区堆芯
multizoned reactor 多区反应堆
multi 多【词头】
municipal administration 市政
municipal authority 市政当局
municipal drainage 城市排水
municipal effluent 城市污水及废气，城市废水
municipal engineering 市政工程
municipal environment 城市环境
municipal government 市政府
municipal heating systems 城市热网
municipal heat-supply 城市供热
municipality 市（区），市政府
municipally administered county 市管县
municipally affiliated county 市管县
municipal noise 城市噪声
municipal ordinance 市政条例，城市法令
municipal pollution 城市污染
municipal power plant 市营发电厂，公用发电厂
municipal power utility 城市电力公司
municipal program 城市规划
municipal public facilities 市政公用设施
municipal refuse incinerator 城市垃圾焚化炉
municipal sewage composition 城市污水组成
municipal sewage plant 城市污水处理厂
municipal sewage 城市污水
municipal sewers 都市下水道
municipal sludge 城市污泥
municipal solid waste 城市固体废物
municipal tax 地方税
municipal transportation 城市运输
municipal utilities 市政公用设施
municipal waste disposal 城市废物处理
municipal waste water plant 城市污水处理厂
municipal waste water 市政废水，城市废水
municipal waste 城市废物，城市垃圾
municipal water facilities 城市供水设施
municipal water supply 城市供水，城市给水
municipal water use 城市用水
municipal water 自来水
municipal works 市政工程
municipal 市政的
muniment room 档案室
muniment 档案馆
muntz metal 孟兹合金，蒙次黄铜，蒙氏铜锌合金
MUOD(mean unplanned outage duration) 平均非计划停运小时，平均非计划停运延续时间
mural arch 壁拱
mural column 壁柱
murexide 骨螺紫，紫脲酸铵，红紫酸胺
muriatic acid 盐酸
MURR(Missouri University Research Reactor) （美国）密苏里大学研究堆
muscovite-granite 白云母花岗岩
muscovite-schist 白云母片岩
muscovite 优质云母，白云母，钾云母
museum 博物馆
mush coil 软线圈，散下线圈
mushing error 干扰误差，颤造误差
mushroom diffuser 蘑菇形散流器
mushroom floor 无梁楼板
mushroom-head buttress dam 蘑菇头式支墩坝，大头坝
mushroom insulator 蘑菇式绝缘子，宽裙式绝缘子
mushroom reinforcement 环辐钢筋
mushroom roof 伞形屋顶
mushroom shaped push-button 蘑菇形按钮
mushroom shell 伞形壳
mushroom slab construction 蘑菇形楼板建筑，无梁楼板结构
mushroom slab 带蘑菇形柱头的无梁楼板
mushroom ventilator 伞形风帽

mushroom 蘑菇头，扩流锥
mush winding 散下绕组
mush-wound coil 散绕线圈
mushy consistency of concrete 混凝土流态稠度
mushy consistency 水泥软练法
mush 噪声，干扰，分谐波，糊状物
musical slide rule 计算尺型调音尺
music hall 音乐厅
muskeg 稀淤泥
muslin 细棉布
mussel filter 水生物过滤器
mussel trap 水生物捕集器
mussel 贻贝
must-run ratio 强制运行率
must-run unit 强制运行机组
mutation of water level 水位突变
mutation 变换，更换，变化，变流，变异，突变，转变
mutator 水银整流器，汞弧整流器
mute control 静噪控制，无噪声调整
mute 静噪，噪声抑制，无声的，哑的
muting circuit 镇静电路，噪声抑制电路
muting valve 无噪声管
mutiple arch dam 连拱坝
mutual action 相互作用
mutual admittance 互导纳
mutual advantage 互惠
mutual agreement 互相同意，双方协定，相互协议
mutual anchorage 钢筋搭接
mutual bearing 相互定位
mutual benefit and collaboration 互惠合作
mutual benefit 互利
mutual calibration 相互校正
mutual capacitance 互电容，交互电容
mutual capacity 交互容量，互电容
mutual check 相互核对
mutual compatibility 互容性，兼容性
mutual-complementing code 互补码
mutual complements in economy 经济互补
mutual conductance 互导
mutual confidence 相互信任
mutual consent 相互同意
mutual deadlock 相互锁定
mutual effect 相互作用，互作用
mutual-exchange of needed products 互通有无
mutual excitation 互激励，交换激励
mutual flux linkage 互磁链
mutual flux 气隙磁通，互感磁通
mutual fund 互惠基金
mutual impedance 互感阻抗，互阻抗
mutual-inductance bridge 互感电桥
mutual inductance 互感系数，互电感，互感
mutual induction 互感应，互感
mutual inductor 互感应器，互感线圈
mutual interaction 相互作用
mutual interest 双方的利益
mutual interference 相互干扰
mutual investment company 合股投资公司
mutual investment 相互投资
mutuality of contract 合同上的相互关系
mutuality of obligation 相互义务
mutual joint-sign 相互会签
mutually exclusive alternatives 互斥方案
mutual magnetic field 互磁场
mutual magnetic flux 互磁通
mutual modulation 互调制
mutual noninterference 互不干涉
mutual partial capacitance 互部分电容
mutual perpendicular 互相垂直，互相垂直的
mutual reactance 互电抗，互电阻
mutual stiffness 互逆电容
mutual-supply electricity price 互供电价
mutual surge impedance 冲击互阻抗，互波阻抗
mutual understanding 相互理解
mutual 共同的，相互的，相互
muzzle 喷嘴，嘴管，缩孔板
MVAR(mega-volt-ampere reactive) 兆无功伏安，兆乏
MV bus fast-break protection 中压母线速断保护
MV,LV switcher 中、低压开关装置
mV＝millivolt 毫伏
MWB(master air waybill) 航空主运单
MWD/kgHM(megawatt-day/kilogram of heavy metal) 兆瓦日每公斤重元素【燃耗单位】
MWD(megawatt day) 兆瓦日
MWD/MT(megawatt day/metric ton) 兆瓦日每吨(重金属)【燃耗单位】
MWD/MTM(megawatt day/metric ton of metal) 兆瓦日每吨金属【燃耗单位】
MWD/T(megawatt day per ton) 兆瓦日每吨
MWe(megawatt electric) 兆瓦【电】
MWH(megawatt-hour) 兆瓦小时
MWI(message-waiting indicator) 信息等待指示器
MW-kilometer method 兆瓦公里法
mWL(millwatt logic) 毫瓦逻辑
MW＝megawatt 兆瓦
mW＝milliwatt 毫瓦
MWP(maximum working pressure) 最大操作压力，最大工作压力
MWS(maximum wind speed) 最大风速
MWt(megawatt thermal) 兆瓦热
MW 兆瓦【DCS 画面】
mycalex 云母块，云母玻璃
mylonite 糜棱岩
mylar film negative drawing 聚酯薄膜底图
mylar reproducible copy 透明复印拷贝
mylar 聚酯树脂，聚酯薄膜
mylonitic 糜棱状
MY(man-year) 人年
myriad 无数，无数的
myriametric wave communication 超长波通信
MYR 马来西亚林吉特

N

N/A(not applicable) 不适用，不适当的
NAB control room 核辅助厂房控制室
NAB(nuclear auxiliary building) 核辅助厂房
NACA airfoil NACA 翼型

NACA(National Advisory Committee for Aeronautics) （美国）国家航空咨询委员会
nacelle canopy 机舱罩
nacelle complete lifting 机舱整体吊装
nacelle cover 机舱盖
nacelle electronic controller 机舱电子控制器
nacelle housing 机舱外壳
nacelle main frame 机舱主框架，机舱主机架
nacelle overall 机舱（整体）
nacelle 机舱，发动机舱，连接舱
NAC(net avoidable cost) 净可避免成本
Na contamination of the water 水的钠污染
Na-cooled fast breeder 钠冷快增殖反应堆
nacreous cloud 珠母云
NACR(Nuclear Auxiliary Control Room) 核辅助厂房控制室
nadir distance 天底距
nadir point 天底点
nadir radial line 天底点辐射线
nadir telescope 天底仪
nadir 最低点，最下点，天底
NAFTA(North American Free trade Agreement) 北美自由贸易协定
nager 钎子
Na-H_2O reaction 钠水反应
nail bearing 针形轴承
nailed joint 钉接合
nailed truss 钉接桁架
nail extractor 起钉钳，拔钉器
nail hammer 拔钉锤
nail-holding power 握钉力
nailing strip 可钉条，受钉条
nailing 打钉
nail making machines 造钉机
nail puller 拔钉器，开箱钳，起钉钳
nail punch 深钉冲头
nail-type crack 指甲状裂纹
nail 钉，钉子，爪，钉住，指甲
NaK cooled reactor 钠钾冷却反应堆
naked contract 无担保合同，无偿契约
naked electrode 裸焊条
naked eye observation 肉眼观测
naked eye 肉眼
naked hills and mountains 荒山
naked radiator 裸露辐照器
naked reactor 裸堆，无反射层反应堆
naked source 裸源
naked transfer of technology 无偿技术转让
naked wire 裸线
naked 裸的，无绝缘的，无保护的，无遮盖的
NAK(negative acknowledge character) 否定信号，否定字符
name and type of protection 保护名称及类型
name card 名片
named consignee 指明收货人
name list 名单
name of article 货名，品名
name of commodity 商品名称，货名
name of goods 货物名称，商品名称
name of user 用户名称
name of vessel 船名
nameplate capacity 铭牌容量，铭牌出力
nameplate current 铭牌电流
nameplate drawing 铭牌图
nameplate load 额定功率
nameplate pressure 铭牌压力
nameplate rating 铭牌（上的）额定值，铭牌出力，铭牌数据，铭牌定额
nameplate specification 铭牌说明
nameplate voltage 铭牌电压
nameplate 铭牌，厂名牌
name register 名字寄存器
name 名称
NAM(network analysis model) 网络分析模型
NAND circuit “与非”电路
NAND driver “与非”驱动器
NAND element “与非”元件
NAND gate “与非”门，逻辑“与非”门
NAND logic gate “与非”逻辑门
NAND logic “与非”逻辑
NAND(not and) “与非”
nanoammeter 毫微安计，纳安计
nanoampere 毫微安
nanoanalysis 纳米（级）分析
nanocircuit 超小型电路，毫微电路
nanocurie 毫微居里，十亿分之一居里
nanofarad 毫微法
nanogram 毫微克
nanohenry 毫微亨
N/A(non-acceptance) 拒绝承兑，拒绝接受
nanoprogram 毫微程序
nanoscope 超高频示波器，毫微秒示波器
nanosecond pulser 毫微秒脉冲发生器
nanosecond pulse 毫微秒脉冲
nanosecond 毫微秒，纳秒，十亿分之一秒
N/A(not applicable) 不适用
NA(not assigned) 不指定的，不赋值的
N/A(not available) 没有，不可用，暂缺
nanotechnology 毫微技术，纳米技术
nanowatt integrated circuit 毫微瓦集成电路
nanowatt 毫微瓦
nano 毫微【10^{-9}】
NaOH solution tank 氢氧化钠溶液罐
Na-photocell 纳光电池
naphthalene 萘
naphtha 粗挥发油，石脑油，粗汽油
Napierian logarithm 自然对数
napier 奈培
NAP(network access processor) 网络存取处理机
nappe contraction 水舌收缩
nappe interrupter （溢流堰顶的）水舌掺气齿坎
nappe outlier 飞来峰
nappe over spillway 溢流水舌
nappe over weir 过堰水舌
nappe profile 水舌截面，水舌轮廓线
nappe separation 水舌脱离，水舌分离
nappe-shaped crest of spillway 溢洪道水舌形堰顶
nappe 水舌，外层，蓄水层，溢流水舌
NAR(numerical analysis research) 数值分析研究

narrow band allocation 窄频带分配
narrow band antenna 窄频带天线
narrow band axis 狭通带轴，窄通带轴
narrow band controller 窄频带控制器，窄范围控制器
narrow band correlation 窄带相关
narrow band cross-correlation 窄带互相关
narrow band excitation 窄带激励
narrow band filter 窄频带滤波器
narrow band frequency modulation 窄带调频
narrow band noise 窄带噪声
narrow band proportional control 窄范围比例控制，窄范围比例调节
narrow band reception 窄（频）带接收
narrow band spectrum 窄带谱
narrow band transmission 窄频带发射，窄频带传输，窄频带发送
narrow band 窄频带
narrow base tower 窄基底杆塔
narrow beam condition 窄束条件
narrow beam 窄射束，窄束
narrow cuts of solids 窄粒固体级
narrow cut 窄馏分
narrow entrance slit 窄入射缝隙
narrow gap welding 狭缝焊接，窄间隙焊
narrow gate 窄闸门，窄选通脉冲，窄的电路
narrow gauge railway 窄轨铁路
narrow gauge track 窄轨铁路
narrow gauge 窄轨
narrow gorge 狭谷
narrow gradation 均匀级配
narrow groove 窄槽
narrowing 变窄
narrow-mouthed bottle 细口瓶
narrow-mouth flask 细口烧瓶
narrow-necked bottle 细颈瓶
narrow orifice 狭口
narrow-pulse generator 窄脉冲发生器
narrow-pulse 窄脉冲
narrow range throttling controller 窄范围轭流控制器
narrow range 狭小的范围【蒸汽发生器】
narrow-size distribution 窄筛分
narrow spacing 小节距
narrow span 狭小的范围【蒸汽发生器】
narrow tee 小口径三通
narrow water 狭水道
narrow 狭窄的，山峡
N-ary code N元代码
n-ary relation n元关系
NASA(National Aeronautics and Space Administration)（美国）国家航空与航天局
nascence 起源，发生
nascent action 初生作用
nascent hydrogen 初生（态）氢
nascent neutron 新生中子
nascent oxygen 初生（态）氧
nascent state 初生态，新生态
Nash equilibrium 纳什均衡
Nash pump 纳氏泵，纳希液封型真空泵
National Ambient Air Quality Standard 国家环境空气质量标准
national assembly 议会
National Bank 国家银行
national bond 公债
National Bureau of Standards （美国）国家标准局
national class 国家级
National Committee of Atomic Information 全国原子情报委员会
national competitive bidding 国内竞争性招标
national control survey net 国家控制测量网
national corporation 国营公司
national debt 国债
national depository 国家（废物）贮藏库
national economic plan 国民经济计划
national economy 国民经济
National Electrical Code 国家电气规范，全国电气规程
National Electrical Manufacturers Association 全国电气制造商协会【美国】
National Electric Code （美国）国家电气规范，国家电工标准
National Electric Saftey Code （美国）国家电气安全规范
national enterprise 国营企业
National Environmental Policy Act 国家环境政策法
national excellent design 国家优秀设计
National Fire Protection Association （美国）国家防火协会，（美国）全国消防协会
National Geodetic Net 国家大地测量网
National Grid Square 国家方格网
national grid 国家电网
national high-way 国家公路，国道
national income 国民收入
national industry standards 国家行业标准
nationality 国籍
nationalization of enterprise 企业国有化
nationalization 国有化，国家化
national-level appraisal 国家级鉴定
national mainline track 国家铁路干线
national materials accountability system 国家核材料衡算系统
national medal 国家级奖（章）
national nuclear data centre 国家核数据中心
national power system 国家电力系统
national primary standard 国家一级标准
national property management system 国有资产管理体制
national reference standard 国家参照标准
National Renewable Energy Laboratory 美国国家可再生能源实验室
National Science Board （美国）国家科学委员会
National Security Council 国家安全委员会【美国】
national situation and policy 国情国策
National Society of Professional Engineers （美国）国家专业工程师协会
National Standard Association （美国）国家标准协会
national standard grade Ⅱ high-way 国家二级公路
National Standard Part Association 全国标准零件制造业协会
National Standard Straight Pipe Thread 美国标准直管螺纹

National Standard Taper Pipe Thread 美国标准锥管螺纹
national standard 国家标准
national supergrid 全国超高压电力网，国家超级电力网
national tax 国税
national unified power system 国家统一电力系统
national uniform price 全国统一价格
national welfare and people livelihood 国计民生
National Wind Power Association 全国风电协会
national wind resource 国家风力资源
nation's economic budget 国家经济预算
nation 国家
native asphalt 天然沥青
native correlation 负相关
native language 特种语言，本机语言
native rock 原生岩石
native substrate 同质衬底
native system demand 系统每月一小时的峰荷需量
native 本地的，当地的，土著的
Na-to-steam heat exchanger 钠蒸汽热交换器
natrium hydroxydatum 氢氧化钠
natrolite 钠沸石
natron 氧化钠，泡碱
natural abundance （同位素）天然丰度
natural accretion 自然淤积
natural activity 天然放射性
natural admittance 固有导纳
natural aerodynamics 自然空气动力学
natural afforestation 天然造林
natural aggregate 天然骨料
natural aging 自然老化，自然时效
natural air circulation 空气自然循环
natural air cooling 空气自然冷却，自然通风冷却
natural ambient radiation level 自然环境放射性水平
natural and mechanical combined ventilation 联合通风
natural angle of repose 自然休止角
natural angle of slope 自然安息角，自然休止角，自然倾斜角
natural area 自然地区
natural asphalt 天然沥青
natural atmosphere 天然大气
natural atmospheric dispersoid 天然大气分散胶体
natural attenuation quantity of noise 噪声自然衰减量
natural attenuation 自然衰减
natural background radiation 辐射的天然本底
natural background 天然本底
natural base 自然对数底，天然地基
natural bed circulation 床内自然循环
natural bending frequency 弯曲固有频率
natural binary coded decimal system 自然二进制码的十进制系统
natural binary-coded decimal 自然二进制编码的十进制
natural binary-coded hexadecimal 自然二进制编码的十六进制
natural biological method 自然生物法
natural bluff 天然陡坡
natural boundary condition 自然边界条件
natural boundary layer wind 自然边界层风
natural bow 静挠度，自然挠度
natural bridge 天生桥
natural calamity 自然灾害
natural capacitance 固有电容
natural capacity 固有容量
natural cavity 自生空穴
natural cement 天然水泥
natural characteristic 自然特性，固有特性
natural circular frequency 自然周期，自然圆频率
natural circulating system 自然循环系统
natural circulation boiler reactor 自然循环沸腾反应堆
natural circulation boiler 自然循环锅炉
natural circulation boiling water reactor 自然循环沸水反应堆
natural circulation capacity 自然循环能力
natural circulation cooling system 自然循环冷却系统
natural circulation cooling 自然循环冷却
natural circulation evaporator 自然循环蒸发器
natural circulation heating 自然循环供暖
natural circulation operation 自然循环运行
natural circulation reactor 自然循环反应堆
natural circulation 自然循环，自然环流
natural cleavage lines 自然裂纹线
natural coke 天然焦炭
natural colloid 天然胶体
natural commutation 固有换向，自然换向
natural complex 自然综合体
natural condition 自然条件
natural conservation 自然保护，自然资源保护区
natural consistency test 天然稠度试验
natural consistency 天然稠度
natural construction materials investigation 天然建筑材料勘察
natural control 自然控制
natural convection flow 自然对流流动
natural convection heat transfer 自然对流换热
natural convection loop 自然循环回路
natural convection range 自然对流范围
natural convection within cells 蜂窝结构内自然对流
natural convection 自然对流
natural coolant circulation 冷却剂自然循环
natural cooling system 自然冷却系统
natural cooling transformer 自冷变压器
natural cooling water 自然冷却水
natural cooling 自然冷却，自冷式
natural cover 天然覆盖
natural crack 自然裂纹，自然裂缝
natural curing 自然养护
natural current 自然电流
natural curvature 自然曲率
natural cutoff 天然裁弯段
natural cycle solar heater 自然循环式（太阳）热水器
natural cycle 自然循环
natural damping 自然阻尼
natural defence 自然防护
natural density 天然密度
natural deposit 天然沉积物
natural detail 自然要素

natural deterioration 自然劣化
natural disaster 自然灾害
natural draft air cooled condenser 自然通风空冷系统
natural draft boiler 自然通风锅炉
natural draft condensation tower 自然通风冷却塔
natural draft cooling tower 自然通风冷却塔
natural draft indirect dry cooling system with surface condenser 表面式凝汽器自然通风间接空冷系统，哈蒙系统
natural draft indirect dry cooling system 自然通风间接空冷系统
natural draft machine 自然通风电机
natural draft transformer 自然通风变压器
natural draft ventilation 自然通风
natural draft wet cooling tower 自然通风湿式冷却塔
natural draft 自然透风，自然通风，自然抽风
natural drainage system 自然排水系统
natural draught air cooler 自然通风空气冷却器
natural draught cooling tower 自然通风冷却塔
natural draught transformer 自然风冷变压器
natural draught 自然通风
natural dust 天然粉尘
natural earth electrode 自然接地体
natural earthquake 自然地震
natural ecology 自然生态学
natural economy 自然经济
natural elimination 自然淘汰，自然消除
natural enemy microorganism 天敌有害微生物
natural energy cycle 自然能量循环
natural environment deterioration 自然环境恶化
natural environment protection 自然环境保护
natural environment 自然环境
natural erosion 自然侵蚀
natural escape 天然溢洪道【小水库的】
natural evaporation 自然蒸发
natural excitation 固有励磁，自然励磁
natural exhaust system 自然排风系统
natural existing soil 原土
natural exposure 自然辐照
natural ferrite 天然铁氧体
natural film 自然氧化膜
natural filtering material 天然滤料
natural fission reactor 天然裂变反应堆
natural flow station 自流发电站
natural flow 天然水流
natural foundation 自然基础，天然地基
natural free convection 自然对流
natural frequence vibration test 固有频率振动试验
natural frequency 自然频率，固有频率，自振频率
natural fuel 天然燃料
natural-function generator 自然函数编辑程序，解析函数发生器
natural gangue 自然煤矸石
natural gas fired 燃用天然气的
natural gas fuel cell 天然气燃料电池
natural gas liquid 液化天然气
natural gas pressure-regulating station 天然气调压站
natural gas 天然气
natural-graded cobble ballast 天然级配卵石道碴
natural grade 天然地坪
natural grading 天然级配
natural graphite 天然石墨
natural ground line 天然地面线
natural groundwater level 天然地下水位
natural ground 自然地面，天然地基
natural growth 自然增长
natural gusty wind 天然阵风
natural harbour 天然港
natural hardness 自然硬度
natural hazard 自然风险，自然危害，自然灾害
natural head 天然水头
natural illumination factor 自然采光照度系数
natural illumination 昼光照明
natural impedance 固有阻抗，特性阻抗
natural impoundment 天然蓄水
natural impurity 天然杂质
natural inflow 天然入流量，天然进水量
natural isotopic abundance 天然同位素含量
naturalization 归化
natural landscape 自然景观
natural language 自然语言
natural latex 天然胶乳
natural law 自然规律，自然法则
natural leak 自漏，自然泄漏
natural levee 自然堤，冲积堤，天然堤，天然冲积堤
natural lighting by means of side windows and skylights 借助于侧窗和天窗的天然采光
natural lighting 天然采光，天然照明
natural load of line 线路自然负荷
natural logarithm 自然对数
naturally circulated condition 自然循环工况
natural magnet 天然磁铁
natural mechanical frequency 固有机械频率
natural mica 天然云母
natural mode of vibration 固有振动模式
natural mode shape 固有振形
natural moisture content 天然含水量
natural monopoly 自然垄断
natural neutron 天然源中子
natural noise 自然噪声，天然噪声
natural nuclear reactor 天然核反应堆
natural oil circulation 自然油循环
natural oil cooling 自然油冷却
natural oil 天然油
natural oscillation 自然振荡，本征振荡，固有振荡，自由振动，自由振荡
natural period of vibration 固有振动周期，自振周期
natural period 固有周期，自然周期
natural phenomena 自然现象
natural pollutant 天然污染物
natural pollution 天然污染
natural power （线路的）自然功率
natural productiveness 自然生产力
natural protection area 自然保护区
natural purification process 自然净化过程
natural purification 天然净化，自净作用，自然

净化
natural radioactive isotope 天然放射性同位素
natural radioactive element 天然放射性元素
natural radioactive nuclide 天然放射性核素
natural radioactivity background 天然放射性本质
natural radioactivity 天然放射性
natural radionuclide 天然放射性核素
natural reactor 天然反应堆
natural recharge 天然再充水
natural region 自然区
natural regulation 天然调节，自然调节
natural reserve 自然保护区
natural reservoirs 天然水库
natural resonance frequency 自然共振频率，固有共振频率
natural resonance 固有共振
natural resonant frequency 固有共振频率
natural resources distribution 自然资源分布
natural resources 自然资源，天然资源
natural risk 自然的风险
natural river 天然河流
natural roadbed 天然路基
natural rockfill dyke 天然堆石堤
natural rolling 自然横摇
natural roughness 自然糙率
natural rubber 天然橡胶
natural sag 自然垂度，自然弛度
natural sandstone material 天然砂石料
natural scale model 天然比尺模型
natural scale 自然比例尺，自然量，实物大小
natural scenic area 自然景物区，天然风景区
natural science 自然科学
natural seasoning 自然干燥法，自然时效
natural selection 自然淘汰
natural separation limit 自然分离限值
natural size 原尺寸，天然尺寸
natural slope 自然坡度，天然坡度
natural smoke control 自然控烟
natural soil deposit 天然土沉积物
natural soil stratum 天然土层
natural soil 原土
natural source of pollution 天然污染源
natural stability limit （输电系统的）自然稳定极限
natural state of soil 土的天然状态
natural steam power plant 地热发电厂
natural steam separation 自然蒸汽分离
natural steam 天然蒸汽
natural strain 自然应变
natural stress 自然应力
natural synoptic period 自然天气周期
natural system 自然系统
natural tail water 天然尾水
natural terrain feature 自然地貌
natural terrain 自然地形
natural tide 自然潮
natural torsion vibration 固有扭转振动
natural turbulence 自然湍流
natural turbulent wind 自然湍流风
natural undamped frequency 无阻尼固有频率
natural uranium fueled reactor 天然铀燃料反应堆
natural uranium fuel gas-graphite reactor 天然铀燃料石墨气冷反应堆
natural uranium fuelled heavy-water reactor 天然铀燃料重水反应堆
natural uranium fuel reactor 天然铀燃料反应堆，天然铀堆
natural uranium fuel 天然铀燃料
natural uranium reactor 天然铀反应堆
natural uranium 天然铀
natural value 原始值
natural vegetation 自然植被
natural vegetative cover 天然植被
natural velocity 固有速度
natural ventilation 天然通风，自然通风
natural vibration frequency of the pedestal 底座的自然振动频率
natural vibration frequency 自振频率，固有振动频率
natural vibration intensity 固有振动强度
natural vibration period 自振周期
natural vibration 固有振动，自然振动
natural void ratio 天然孔隙比
natural water content 天然水含量
natural water course 天然水道
natural water resources 天然水利资源，自然水利资源
natural water table 天然地下水位
natural waterway 天然水道，天然水系
natural wavelength 固有波长
natural wave 固有波
natural well 天然井
natural wind boundary layer 自然风边界层
natural wind environment 自然风环境
natural wind field 自然风场
natural wind gust 自然阵风
natural wind hazard 自然灾害
natural wind 自然风
natural 自然的
nature balance 自然平衡
nature conservation area 自然保护区
nature conservation 自然环境保护
nature moisture content 天然含水量
nature of coal 煤质
nature of thermal conductivity 导热系数的性质
nature remaking 改造自然
nature remodeling 改造自然
nature reserve 自然保留地，自然保护区
nature resonance 自然共振
nature sanctuary 自然保护区
nature 自然，性质，特性，本性，特征，本质
naught line 零位线
naught 零，无
NAU(network addressable unit) 网络可访问设备，网络可访问部件
nautical chart 航海图
nautical 海上的
naval brass tube 海军铜管
naval brass 海军铜
naval reactor organic experiment 海军用有机冷却试验反应堆
naval reactor 舰艇用反应堆

naval research reactor　船用研究反应堆
nave　衬套，毂，轮毂，中心
Navier-Stokes equation　纳维斯托克斯方程
navigability　可通航性
navigable channel　航道
navigable dam　可通航坝
navigable depth　通航水深，通航深度
navigable period　通航期
navigable power canal　通航发电渠道
navigable water way　通航河流
navigable　可航行的，可通航的
navigating bridge　驾驶桥楼
navigation aid　助航设备
navigation buoy　导航浮标
navigation canal　通航运河
navigation condition　通航条件
navigation dam　航运坝
navigation head　水陆转运站，航运终点
navigation lamp　航行灯
navigation light buoy　航标灯浮标
navigation light　航行灯
navigation lock　船闸
navigation mark　航标
navigation pass　航道
navigation reservoir　通航水库
navigation signal　航行标志
navigation station　导航台
navigation structure　通航建筑物
navigation-way administration　航道管理局，航道设计院
navigation　航运，航行
navigator　导航仪
navvy pick　十字镐，挖土镐
navvy　挖土工人
navvy barrow　运土手推车
navy motor　船用电动机，海军用电动机
NAWAS(national warning system)　国家警告系统
NBA(narrow band allocation)　窄频带分配
NBCD(natural binary-coded decimal)　自然二进制编码的十进制
NBCH(natural binary-coded hexadecimal)　自然二进制编码的十六进制
NBC(noise balancing circuit)　噪声平衡电路
NBET(Nigeria Bulk Electricity Trader)　尼日利亚大宗电力交易商，尼日利亚大宗电力购买机构
NBFM(narrow band frequency modulation)　窄带调频
n-bit memory　n位存储器
N_2 blanket　氮覆盖层
NB(nominal bore)　标称孔径
NB(nota bene)　注意
N_2 buffer tank　氮气缓冲箱
NCAI(National committee of Atomic Information)　全国原子情报委员会
NCA(non-control area)　非控制区域
NCC(network control center)　网络控制中心
NCC(nuclear circuit cleaning)　核回路清洗
NC contact　常闭触点
NC-curve　噪声评价NC曲线
NC cutting machine　数控切割机
NC cutting　数控切割
NCE(normal calomel electrode)　标准甘汞电极
NCGA(neutron capture gamma-ray analysis)　中子俘获γ射线分析
n channel MOS　n-沟道金属氧化物半导体
N-characteristics　N形特性曲线【电压电流曲线】
N_2 charging system　充氮系统
NCM(network control module)　网络控制模块
NC(natural convection)　自然对流
NC(network controller)　网络控制器
NCN(non-conformance notice)　不符合项通知
NC(non-classified)　无级
NC(normally closed)　常闭
NC(novation contract)　NC模式，新项目管理模式【工程建设模式】
N_2 connection nozzle　氮气接管
NC plate cutter　数控切板机
NCP(network control program)　网络控制程序
NCP(Network Control Protocol)　网络控制协议
NC processing system　数控处理系统
NC program processing　数控程序处理
NCR(non-confirmation record)　未批准记录【尤其指对图纸的确认】
NCR(non-conformance report)　不符合项报告
NCS(net control system)　网络监控系统
NCS(network control station)　网络控制站
NCS(normal cold shutdown)　正常冷停堆
NC torch cutter　数控火焰切割机
N_2 cushion　氮缓冲垫，氮气垫
NCV(no commercial value)　无商业价值
NDACC(natural draft air cooled condenser)　自然通风直接空冷系统
NDAC(not data accepted)　不接受数据
NDA(non disclosure agreement)　保密协议
NDCT(natural draft cooling tower)　自然通风冷却塔
NDE(non-destructive evaluation)　无损评价
NDE(non destructive examination)　无损检验，无损检查
NDE(nonlinear differential equation)　非线性微分方程
NDE(normal de-energized)　正常失电
second grouting layer of equipment foundation　设备基础二次浇灌层
n-dimensional coden　n维代码
n-dimensional cube　n维立方体
n-dimensional normal distribution　n维正态分布
n-dimensional　n维
n-dimensional space　n维空间
n-dimensional vector　n维矢量，n维向量
NDI(non destructive inspection)　无损检验
N-display　N型显示器
NDI　德国工业标准委员会
NDL(network definition language)　网络定义语言
NDM(normal disconnected mode)　正常断开方式，正常拆线方式
ND(nominal diameter)　标（公）称直径
ND(not detect)　不检测，未发现，未检出
NDRC(National Development and Reform Commission)　国家发展改革委

NDR(network data reduction) 网络数据简化
NDR(normalized deficiency rate) 质量缺陷当量值【QA用语】
NDRO(non-destructive read-out) 非破坏读出
NDRW(non-destructive read-write) 非破坏读写
NDS(Nuclear Data Section) (国际原子能机构的)核数据科
NDT(nil ductility temperature) 脆性温度，无延性温度
NDT(nil ductility transition) 非塑性转变，无延性转变
NDT(non-destructive testing) 无损检测，无损检验，无损试验，无损探伤，非破坏性测试
NDT-RT(non-destructive test-radiographic testing) 无损检验-射线照相检验
NDTT(nil ductility transition temperature) 无延性转变温度
NEANDC(NEA Nuclear Data Committee) 核能机构核数据委员会
NEA(Nuclear Energy Agency) (经济合作与发展组织的)核能机构
neap range 小潮涨落差
neap rise 小潮升
neap tide 低潮，小潮，最低潮
near bottom layer 沿底层
near-bottom temperature 底层温度
near breeder 近增殖堆，近增殖反应堆
near-by interference 近区干扰
near-critical assembly 近临界装置
near-critical regeneration 近临界再生
near-critical 近临界的
near delivery 近期交货
near-end cross talk 近端串扰，近端串音
near extinction animal 濒临灭绝动物
near extinction 濒临灭绝
near field 近场，近域【放射性烟云】
near future 近期
near gale 疾风，七级风
near-natural uranium reactor 近天然铀反应堆，稍加浓铀反应堆
near-neutral regime 近中性状态
near perfect matching 准完全匹配
near point correction 近点校正
near-prompt 近瞬时的
near shore zone 沿岸带
nearshore 近岸，近滨，近岸水域，(研究)近岸水域的
near-site inspection 近地视察
near-sonic 近声速的，跨声速的
near surface defect 近表面缺陷
near surface flaw 近表面缺陷
near surface layer 近地面层
near surface wind 近地风
near term decision 近期决策
near terminal fault 近端故障
near-thermal neutron spectrum 近热中子谱
near-thermal neutron 近热中子
near-thermal reactor 近热反应堆
near-thermal spectrum 近热中子谱
near-thermal 近热能的
near-town site 近城镇厂址
near wake region 近尾流区
near wake 近尾流
near zero release 几乎不排放，近零排放
near 近的，靠近
neat-bottom current 沿底流
neat line 内图廓线，细线，准线
neat plaster 纯灰浆
neat Portland cement 纯波特点水泥
NEB(noise equivalent bandwidth) 噪声等效带宽
nebula 喷雾剂
nebulization 喷雾，喷雾作用
nebulizer 雾化器，喷雾器
necessaries 必需品
necessary condition 必要条件
necessary demonstration 必要的示范
necessary equipment 必要设备
necessary expenses 必要经费
necessity 必需品，必要性，必然性，必需
neck bearing 颈轴承，内轴承，中间轴承，弯颈轴承
neck breaking speed 危险速度
neck bush 轴颈套，内衬套
neck collar 轴颈环，轴承环
neck current 束狭急流
neck-down riser 易割冒口
neck-down 颈缩
necked-in (边缘)向内弯曲
necked-out (边缘)向外弯曲
necked part 轴颈部分
neck groove 线槽
neck gutter 天沟
necking down 颈缩，试样局部断面收缩
necking tool 开槽刀
necking 缩颈，断面收缩
neck journal bearing 轴颈轴承
neck journal 轴颈
neck of rip current 离岸裂流颈
neck 颈，颈状物
NEC(National Economic Council) (美国)全国经济委员会
NEC(National Electric Code) 美国国家电气规程，美国国家电工标准
NEC(New Engineering Contract) 新工程合同条件【英国】
NEC(Nippon Electric Company) 日本电气公司
NEC(no error check) 核对无错，无错误检验
needle aberration 磁针偏差
needle beam scaffold 轻型脚手架
needle beam 小横梁，簪梁
needle bearing 针形轴承，滚针轴承
needle contact method 针触法
needle-controlled weir 针阀控制堰
needle control rod 针阀控制杆
needle coupling 针状联轴器
needle density 针入密实度
needle drift chute 散木流放槽
needle electrode 针形电极
needle file 针形锉，针锉
needle frost 针状霜
needle galvanometer 磁针电流计
needle gap 针隙

needle gate 针形闸门
needle gypsum 针状石膏
needle mesh 针孔筛目
needle nozzle 针形喷嘴
needle number 针孔数目
needle penetration test 针入法试验
needle penetrometer 水泥凝结时间检测仪，针式贯入器
needle-plate gap 针板间隙
needle plug 针状插塞
needle point spark gap 针尖火花隙
needle point valve 针形阀
needle probe 探针
needle-regulating nozzle 针形调节喷嘴
needle seating 针阀座
needle servomotor 针阀接力器
needle-shaped structure 针状结构
needle-slot screen 细缝筛
needle stem 喷针杆，针阀杆
needle thermocouple 针形温差电偶，针形热电偶
needle traverse 罗盘施测导线
needle tube 针形管
needle valve seat 针阀座
needle valve 针阀，针形阀，针状活门
needle vibrator 针形振捣器
needle weir 栅条堰
needle 横撑木，针，探针，磁针，指针
needling 横梁支托，横撑木支托
Neel temperature 奈耳温度
negacyclic code 负循环码
negaohm 负电阻材料
negarive characteristic 负特性，下降特性
negater 倒换器，非门
negate 否定，拒绝，使无效，求反，取消，否认，无效
negation 非，否定
negative acceleration 负加速度
negative acknowledge character 否定信号，否定字符
negative acknowledgement 否定确认，否认，否定应答
negative aerodynamic damping 负气动阻尼
negative aerodynamic stiffness 负气动刚度
negative allowance 负公差
negative altitude 负高
negative and gate 与非门
negative angle of attack 负迎角
negative artesian head 负承压水头
negative balance 负平衡
negative base number 负基数
negative battery 负电池
negative bias 负偏压
negative booster 降压器，降压机
negative boosting transformer 减压变压器
negative brush 负电刷
negative buoyancy 负浮力
negative bus 负汇流排
negative charged ion 阴离子
negative charge 负电荷
negative coefficient of fuel temperature 燃料负温度系数
negative conductor 负导线
negative confining bed 负承压底层
negative converter 反向变流器
negative corona 负极电晕
negative correlation coefficient 负相关系数
negative correlation 负相关性
negative current 负电流
negative cutoff 倒截水墙
negative delta 喇叭口，三角湾，河口湾
negative deviation 反向偏差
negative direction 反方向，反向，逆向
negative displacement 负位移
negative electricity 负电
negative electrode 负极，阴极，负电极
negative elevation 负迁移
negative energy 负能量
negative exponent 负指数
negative factor 负（反馈）系数
negative feedback amplifier 负反馈放大器
negative feedback pitch control 负反馈桨距控制
negative feedback 负反馈
negative feeder 负馈电线
negative frequency 负频，负频率
negative friction pile 负摩擦桩
negative friction 负摩擦力
negative gain factor 负增益系数
negative glow 第二阴极辉光，阴电辉，负辉
negative-going reflected pulse 负向反射脉冲
negative-going 负向的
negative gradient 负梯度
negative half-cycle 负半周
negative head 负水头
negative image 负像
negative immittance converter 负导抗变换器
negative impedance 负阻抗
negative incidence 负冲角，负攻角
negative increment 负增量
negative indication 负显示
negative input, positive output 负输入正输出
negative investment 停止投资，减少投资
negative ion source 负离子源
negative ion 阴离子，负离子
negative kilocalorie 负大卡【热量单位】
negative lead （电刷）后移
negative lens 负透镜
negative lift device 负升力装置
negative lift wing 负升力翼板
negative limiting 负向限幅，负值限制
negative logic 负逻辑
negative loop （指示图的）负值部分
negatively buoyancy plume 负浮力羽流
negatively charged ion 阴离子
negatively progressive flow 反向行进水流
negative mass 负质量
negative match 负匹配
negative modulation 负调制
negative moment reinforcement 负弯矩钢筋，负矩钢筋
negative moment 负力矩，负弯矩
negative-order component 负序分量

negative-order current 负序电流
negative or gate 或非门
negative output 负输出
negative peak 最大负值，负峰值
negative phase relay 反向继电器，负相位继电器
negative phase sequence component 负序分量
negative phase sequence current 负序电流
negative phase sequence impedance 负序阻抗，负相序阻抗
negative phase sequence reactance at rated current 在额定电流下的负相序电抗
negative phase sequence reactance 负序电抗
negative phase sequence relay 负序继电器
negative phase sequence resistance 负序电阻
negative phase sequence voltage 负相序电压
negative phase sequence 负相序，逆相序
negative plate 负极板，阴极板，底片，负片
negative plume rise 羽流负抬升
negative pneumatic conveying system 负压气力输送系统
negative polarity 负极性
negative pole 负极，阴极
negative pore water pressure 负孔隙水压力
negative potential 负电位，负位
negative power coefficient of reactivity 反应性负功率系数
negative power coefficient 负功率系数
negative power 接负电，阴极电源接头
negative pressure containment 负压安全壳
negative pressure firing 负压燃烧
negative pressure screen analysis machine 负压筛析仪
negative pressure system 负压系统
negative pressure valve 负压阀
negative pressure 负压，低于大气压的压力
negative pulse 负脉冲
negative quantity 负量，负值
negative reactance 负电抗
negative reaction 负反作用，负反馈，阴性反应，负反动度
negative reactivity coefficient 负反应性系数
negative reactivity margin 负反应性裕量
negative reactivity 负反应性
negative reactor 负反应堆
negative reinforcement 负弯矩钢筋
negative release wave 负放水波
negative resistance bridge 测量负电阻电桥
negative resistance element 负电阻元件
negative resistance relay 负阻继电器
negative resistance 负电阻
negative result 否定结果
negative segregation 负偏析，反偏析
negative self-regulation 负的自动调整
negative sequence braking 负序制动
negative sequence component 负序分量
negative sequence current protection 负序电流保护
negative sequence current 负序电流
negative sequence field impedance 负序磁场阻抗
negative sequence filter 负序过滤器
negative sequence impedance 负序阻抗
negative sequence loss 负序损耗
negative sequence network 负序网络
negative sequence power 负序功率
negative sequence reactance 负序电抗
negative sequence relay 负序继电器
negative sequence resistance 负序电阻
negative sequence symmetrical component 负序对称分量
negative sequence torque 负序转矩
negative sequence voltage relay 负序电压继电器
negative sequence voltage 负序电压
negative sequence 负序，负相序，逆序
negative shear 负剪力
negative sign 负号
negative skin friction 负摩擦力【桩的】，表面负摩阻
negative space charge 负空间电荷
negative spot 负的电荷点
negative suction head 负吸入头，负压头，负吸水头
negative suction 吸入负压
negative supply voltage 负电源电压
negative surge 负涌浪
negative synergism 反协同作用，反协同
negative temperature coefficient feedback 负温度系数反馈
negative temperature coefficient of reactivity 反应性负温度系数
negative temperature coefficient resistor 具有负温度系数的电阻
negative temperature coefficient 负温度系数
negative temperature factor 负温度系数
negative temperature 负温
negative terminal 负极端子，负端子，负线端
negative test 否定测试
negative thread 阴螺纹
negative throat nozzle 渐缩喷嘴
negative torque signal 反扭转信号
negative transmission 负极性转送，负调制转送
negative value 负值
negative virtual mass 负虚质量
negative void coefficient 负空泡系数
negative void reactivity coefficient 反应性负空泡系数
negative water hammer gradient 负水锤梯度
negative wave 负波
negative well 倒渗井
negative wire 负线
negative 负片，底片，否定，负值，阴极，负数，负的，反面的，负电的，阴性的
negator 非门，非元件，倒换器
negatron 负电子，阴电子，双阳极负阻管
negentropy 负平均信息量，负熵
neglect of duty 失职
neglect 遗漏，疏忽，忽略
negligence of operation 操作疏忽
negligence 疏忽，过失，粗心大意
negligible erratum 可忽略误差
negligible error 可忽略误差

negligible quantity 可忽略的量，可略去量，可略量
negligible residue 忽视残留
negligible 可不计的，可忽视的，可以忽略的
negotiability 流动性，流通能力，可流通性，可转让性
negotiable amount 可议付的金额
negotiable bill of lading 可流通提单，可转让提单
negotiable bill 可流通票据
negotiable B/L 流通提单
negotiable CDs 可转让的定期存款单，可转让的存单
negotiable certificate of deposit 可流通存单，可转让存单，可流通定期存单，可转让定期存单
negotiable certificate of time deposit 可转让定期存款证
negotiable check 可流通来人支票，可流通支票
negotiable credit instrument 可转让证券
negotiable credit 可议付的信用证
negotiable document 可转让单据，可转让文件，可转让证券，可流通单据
negotiable instruments 可流通证券，可流通票据，可过户证券，可转让证书，可转让票据
negotiable letter of credit 可转让信用证，可流通信用证，可议付的信用证，可兑信用证
negotiable multiple transport documents 可转让复合运送单据
negotiable note 可流通期票
negotiable on the stock exchange 可在证券交易所流通
negotiable order of withdrawal account 可转让提款指令账户
negotiable original copy 议付正本
negotiable paper 可流通票据，可转让票据
negotiable securities 可流通证券，可转让证券，证券
negotiable warehouse receipts 可转让仓单，可转让仓库收据
negotiable 可通过谈判解决的，可协商的，可转让的，可流通的，(票据)可兑现的
negotiate a contract 协商合同，谈判合同，缔结契约
negotiate a new contract 商议[谈判]一新合同
negotiate business 洽谈业务
negotiated bidding 谈判招标
negotiated contract 议付合同，议价发包，协商合约，议标合同，磋商成交的合同
negotiated price 议价，协商价格
negotiated settlement 经协商的解决方案，协商解决
negotiate exportation 出口结汇
negotiate the amount 商量价钱，议付金额
negotiate 谈判，商谈
negotiating bank 押汇银行，议付银行，购票银行
negotiating body 谈判机构
negotiating condition 议价条件
negotiating date 汇票议付期限
negotiating guidelines and procedures 谈判准则和程序
negotiation against guarantee 凭担保转售，凭保证议付
negotiation attendant 谈判参加人
negotiation bill 议付票据
negotiation commission 议付手续费
negotiation contract 议标合同，协商合同
negotiation credit 让购信用证，议付信用证
negotiation date 议付期限
negotiation of business 交易洽谈，交易磋商
negotiation of contract terms 磋商合同条款
negotiation participant 谈判参加人
negotiation procedure 谈判程序
negotiation room 谈判室
negotiation under letter of credit 按信用证的磋商，按信用证的议付
negotiation under reserve 保结押汇
negotiation with foreigners 涉外谈判
negotiation 洽谈，谈判，协商，磋商，(票据)议付，转让
negotiator 谈判人，磋商者，交涉者
nehyung 沙嘴滩
neighbourhood noise 环境噪声
neighbourhood sheet 邻接图幅
neighbourhood 邻里，街坊，临近地，邻近
neighbouring drainage basin 相邻流域
neighbour 邻近值
NEMA(National Electrical Manufacturers Association) 全国电气制造商协会【美国】
NEMA(National Emergency Management Association) 国家应急管理联合会【美国】
nematic liquid crystal 向列型液晶
nematogen 向列型液晶
NE(normal energized) 正常带电
NE(normal exhaust) 正常排[送]风
neocupróin 新铜试剂
neodymium 钕【Nd】
neo-loess 新黄土
neomagnal 铝镁锌耐蚀合金
neon arc lamp 氖光灯，霓虹灯，氖灯
neon glow lamp 氖辉光灯，氖辉光放电管
neon indicator 氖管指示器，氖管指示灯
neon lamp 氖灯，霓虹灯
neon pulse trigger 氖灯脉冲触发器
neon tester 试电笔，氖测电器
neon-tube transformer 氖管变压器
neon-tube 氖管，霓虹灯，氖光灯管
NEO(nuclear equipment operator) 核设备操纵员
neon voltage regulator 氖电压调整器
neon 氖【Ne】
neoprene compression gasket 氯丁橡胶垫圈
neoprene 氯丁橡胶
neotectonic movement 新构造运动
neo-volcanic rock 新火山岩
NEPA(National Environment Policy Act) (美国)国家环境政策法
NEPA(National Environment Protection Administration Agency) 国家环保局
NEPA(Nigeria Electric Power Authority) 尼日利亚(国家)电力局
NEPA(nuclear energy for propulsion of aircraft) 飞机推进用核能【美国】

NEPC(Northeast Electrical Power Construction Co.) 东北电力建设公司，东北电建
nepermeter 奈培计
neper 奈培
nepheline-syenite 霞石正长岩
nephelinite 霞岩
nephelometer （散射）浊度计，能见度测定器
nephelometric analysis 浊度分析
nephelometric method 比浊法
nephelometric turbidity units 比浊（测量）法浊度单位
nephelometry 浊度测定法
NEP(new equipment practice) 新设备试制
Neptex process 镎萃取法
neptunate 镎酸盐
neptunium 镎
NERC(Nigerian Electricity Regulatory Commission) 尼日利亚电管会，尼日利亚电力管理委员会
NERC(Northern American Electric Reliability Council) 北美电力可靠性协会
neritic zone 浅海区，浅海带
NERO(North America Electric Reliability Organization) 北美电力可靠性机构
nerve tissue 神经细胞组织
nervure 叶脉
nesacoat 氧化锡薄膜电阻
Nesa glass 奈塞玻璃【一种透明导电膜半导体玻璃】
Nesa 奈塞
NESC(National Electrical Safety Code) 全国安全用电代码
nesister 双极场效应晶体管，一种负阻半导体器件
NES(not elsewhere specified) 不另规定，不另作说明
nestable pipe 套管
nest assembly 组件箱，组件箱装置
nested loop 嵌套循环
nested pipe system 套管式【管中管】
nested 内装式的
nesting level 嵌套级
nesting storage 后进先出存储器
nest of pipes 管组
Nestor number 奈斯特数
nest 套，组，束，窝，结网，嵌套，组件箱，机箱
net actual generation 实际净发电量
net aerial production 净空气生产量
net amount 净额
net amplitude 合成振幅，净振幅
net aperture area 净采光面积【聚光集热器】
net area 净面积
net assets 净资产
net assured capability 净保证（发电）能力，净保证出力
net available head 净有效水头
net avoidable cost 净可避免成本
net benefit 净效益，净收益
net book value 账面净值
net buoyance 净浮力
net buoyant weight 净浮重
net caloric value 净热值
net calorific power 低热值
net calorific value at constant volume 恒容低位发热量
net calorific value 低位发热量，低热值
net calorific 净热值，低位发热量
net capability 净（装机）容量，净能力，（最佳运行工况下的）最大供电能力
net capacity 净（装机）容量，净能力
net capital 资本净额
net cash flow 净现金流量
net citizen 网络公民
net coal consumption rate 供电煤耗率
net concentration 净浓度
net control station 网控台
net control system 网络监控系统
net cost 成本净额，净价，净成本
net current assets 流动资产净额
net current 净电流
net cut 纯开挖
net damping 净阻尼
net days 净天数
net deficiency 网路缺电，净缺电量
net dependable capability 净可靠供电能力，平均运行工况下的最大供电能力
net dredger 网式挖泥机
net efficiency 净效率
net electrical output 净输出电功率【电厂】
net electric generation 净发电量
net electric power output 净电功率输出
net enthalpy 净焓，可用焓降
net extraction 净抽气量
net fan requirements 风机设计运行工况
net field 净场
net fill 纯填方
net flux 净通量
net foreign assets 国外资产净额
net foundation pressure 基础净压力
net freeboard 坝顶净超高【水电】
net function 网格函数
net gain 净增益
net generation 净发电量
Net Gen(net generation) 净发电量
net gravitational force 静重力
net head 有效压头，净压头，静压头，净水头
net heat exchange 净热交换
net heating value 净热值，低位热值
net heat input 净输入热量
net heat rate 净热效率，净热耗，净热耗率
net heat 有效热，净热，净耗热量
net height 净高
net horsepower 净马力，有效马力
net income 净收入
net incoming radiation 净入射辐射
net investment 净投资
net irradiance 净（辐射）辐照度
netizen 网民
net leakage 净漏泄
net length of crest 堰顶净长
net line 准线
net load 有效负荷，净损失，净负荷，净负载

net longwave irradiance　净长波辐照度
net longwave radiation　净长波辐射
net loss　净损，净损失，净损耗，纯损
net margin　净利润
net material product　物质生产净值
net maximum capacity　净最大容量
net mean effective pressure　净平均有效压力
NET(network analysis program)　网络分析程序
net of control points　控制点网
net output at terminal　出线端净出力
net output value　净产值
net output　净出力，净输出，净输出功率
net pay　实付工资
net-plant-heat rate　电厂净热耗
net positive suction head requirement　必需的汽蚀余量
net positive suction head test　汽蚀试验
net positive suction head　汽蚀余量，有效吸入高度，净正吸入水头，净正吸入压头
net positive suction　净正吸入
net power output　净输出功率
net power station efficiency　电站净效率
net power　净功率，有效功率
net present value rate　净现值率
net present value　净现值
net pressure head　净压头
net pressure loss　净压力损失
net pressure suction head　临界汽蚀裕量
net price　净价格
net proceeds　净收入
net profit　净利润
net pyranometer　净总日射表，净全天空辐射计
net pyrgeometer　净地球辐射表，净地面辐射计
net pyrradiometer　净全辐射表，辐射平衡表
net radiation flux　净辐射通量
net radiation method　净辐射方法
net radiation　净辐射
net radiometer　净辐射计
net rated capacity　净额定容量
net rate of return on investment　投资净回报率
net rate　净速率
net rating　有效出力
net room area　房间净面积
net safety distance　安全净距
net sectional area　净截面积
net section　净截面
net selling price　净销售价格
net shortwave irradiance　净短波辐照度
net shortwave radiation　净短波辐射
net sieve　网筛
net standard coal consumption rate　供电标准耗煤率
net station heat rate　电站净热耗
net station output　电站有效功率，电站净输出功率
net structure　网络构造
netsuke　悬锤
net sum　净和
net supply interval　地表径流形成期
net tare　净皮重
netted　网状构造的，网状的
net temperature drop　有效温度降
net thermal efficiency　净热效率，送电热效率，输电端热效率
net thrust　净推力
netting　网，网格，结网
net ton　短吨，净吨
net total irradiance　净全（辐射）辐照度
net total power output of a combined cycle power plant　联合循环电厂的净总功率输出
net total power output of a open cycle power plant　开式循环电厂的净总功率输出
net total radiation　净全辐射
net turbine power　水轮机净出力
net value　净值
net volume　净容积
net water head　净水头
net weight　净重
network access processor　网络存取处理机
network address　网络地址
network adjustment　测量网平差
network admittance matrix　网络导纳矩阵
network analog　网络模拟
network analyser　网路分析计算机，网络分析器
network analysis model　网络分析模型
network analysis program　网络分析程序
network analysis　网络分析
network and trunk control　网络及总线控制
network and trunk equipment　网络及总线设备
network calculation　网络计算
network calculator　网络计算机
network capacitance　网络电容
network characteristic　网络特性，网络特性曲线
network chart　网络图
network computer　电网计算机，电网计算台
network connection point　电网连接点
network connection　网络接线
network constant　网络常数
network control board　网络控制盘，电力网控制盘
network control building　通信楼，网络控制楼
network control center　网络控制中心
network controller　网络控制器
network control module　网络控制模块
network control program　网络控制程序
network control room transformer　网控楼变
network control room　网络控制室，网控室
network control station　网络控制站
network control　电力网控制
network conversion　网络变换
network data reduction　网络数据简化
network definition language　网络定义语言
network design　网络设计，电力网设计
network distribution system　电网配电系统
network element　网络元件
network equation　网络方程式
network fault　网络故障，电网故障
network feeder　电力网络馈电线
network flow model　网流模型
network flow routine　网络流程程序
network frequency　网络频率，电网频率
network function　网络函数

network harmonic 网络谐波
network impedance phase angle 电网阻抗相角
network impedance 电网阻抗
network information center 网络信息中心
network information service 网络信息服务
network information system 网络信息系统
networking 联网
network in telecontrol 远动网络
network job processing 网络作业处理
network layout 网络布置，电力网布置
network load characteristic 电网负载特性
network loading statistics 电力网负荷统计
network load 电网负荷
network loss 电力网损失，网络损失
network management unit 网络管理部件
network map 网络图，电力网图
network matching 网络匹配
network matrix 网络矩阵
network measurement center 网络测量中心
network monitoring system 网络监控系统
network of the transportation 交通运输网
network operating system 网络操作系统
network operation and maintenance center 网络操作和维护中心
network parameter 网络参数
network phasing relay 网络相位继电器
network planning 网络计划
network plan 网络布置图，电力网布置
network processing unit 网络处理部件
network protector 网络变换装置，开关站
network relay 电力网继电器，网络继电器
network schedule 网络进度
network security secret system 网络安全保密系统
network sensitivity 网络灵敏度
network server 网络服务器
network splitting 电网解列
network stability 电力网稳定度，网络稳定度
network status display 网络状态显示
network structure 网状构造
network subtransmission 网络状二次输电
network supply 电网供电
network synthesis 网络综合
network system 网络系统，供电网系统，电网系统
network technique 网络技术
network terminating unit 网络终端装置
network test 网络试验
network theorem 网络定理
network theory 网络理论
network topology 网络拓扑
network transfer table 网络传送［转移］表
network transformation 网络变换
network transformer 网络变压器，电网变压器
network-type analog computer 网络型模拟计算机
network vibration 网络振动
network virtual terminal 网络虚拟终端
network voltage 电网电压
network with an earthed neutral 中性点接地电力网
network with arc suppression coil 中性点经消弧线圈接地电力网
network with isolated neutral 中性点绝缘电力网
network 网，网络，电网，管网，网状物，系统
net worth 净值，资本净值
Net Wt(net weight) 净重
net 网，网络，电网，净的，纯的，纯粹的
neurocomputer 神经（式）电脑
neut=neutralizing 中和，中性，平衡
neutral armature 中性衔铁
neutral atmosphere boundary layer 中性大气边界层
neutral atmosphere 中性大气，中性气氛
neutralator 中性点接地电抗器
neutral autotransformer 中性点接地自耦变压器，中性点补偿器
neutral axis 中性轴，中性线，中和轴
neutral balloon 中性气球
neutral body 中性体
neutral boundary layer 中性边界层
neutral brought out 中性引出
neutral bushing （变压器的）中性点套管
neutral bus 中性母线
neutral coil 中性线圈
neutral compensator 中性点补偿器
neutral conductor 中性导体，中性导线
neutral contact 中性接点
neutral current relay 中线电流继电器
neutral current 中线电流
neutral density plume 中性密度羽流
neutral displacement transformer 中性点位移变压器
neutral drag coefficient 中性风阻系数
neutral earthed through an impedance 中性点经阻抗接地电力网
neutral earthing equipment 中性点接地设备
neutral earthing resistor 中性接地电阻器
neutral earthing 中点接地，中线接地
neutral effectively grounded system 中性点有效接地系统
neutral end 中性端
neutral equilibrium 中性平衡，随遇平衡
neutral feeder 中性馈线
neutral flame 中性焰
neutral flow 中性流动
neutral grounded system through an arc suppression coil 中性点消弧线圈接地系统
neutral grounded system through a resistor 中性点电阻接地系统
neutral grounding resistor 中性线接地电阻
neutral grounding state of transformer 变压器中性点接地方式
neutral grounding 中性点接地
neutral gyroscope 自由陀螺仪
neutral impedor 中性点接地二端阻抗元件
neutrality 中性，中和，平衡
neutralization basin 中和池
neutralization chamber 中和槽
neutralization of effluent 排水中和
neutralization pit 中和坑，中和井

neutralization pond 中和池，中和罐
neutralization station 中和站
neutralization tank drain pump 中和箱疏水泵
neutralization tank 中和箱
neutralization 抵消，抑制，中和，平衡，使失效
neutralized series motor 有补偿的串激电动机
neutralized water basin 中和水池
neutralized water pump 中和水泵
neutralizer 中和剂，中和池，中和设备
neutralize 使中和，平衡，抵消
neutralizing agent 中和剂
neutralizing amine 中和胺
neutralizing boil-out 中和煮炉
neutralizing bridge circuit 平衡电桥电路
neutralizing circuit 中和电路
neutralizing coil 补偿线圈，中和线圈
neutralizing condenser 中和电容器，平衡电容器
neutralizing indicator 中性指示器
neutralizing pit 中和坑
neutralizing solution 中和液
neutralizing tank 中和箱
neutralizing treatment 中和处理
neutralizing zone 中和段
neutral lapse rate 中性递减率
neutral layer 中和层
neutral lead 中性点引出线
neutral level 中和界
neutral line 中性线
neutral loading 中性线加载
neutrally buoyancy plume 中性浮升羽流
neutrally stability 中性稳定性
neutrally stable atmosphere 中性稳定大气
neutrally steer 中性转向
neutrally stratification 中性层结
neutrally stratified air 中性分层大气
neutrally stratified flow 中性分层流
neutrally thermal stability 中性热稳定性
neutrally turbulent flow 中性湍流流动
neutral mode of operation 中性运行方式
neutral oil 中性油
neutral packing 中性包装
neutral party bill of lading 中性提单
neutral party 中间方
neutral performance curve 空档特性曲线
neutral plane 中和面，中性面
neutral point arrangement 中性点分布，中性点排列
neutral point displacement voltage 中性点位移电压
neutral point displacement 中性点位移
neutral point earthing 中性点接地
neutral point reactor 中性点电抗器
neutral point solid ground 中性点直接接地
neutral point 中和点，中性点，滴定终点
neutral position 中性位置，中和位置，中性线上（电刷）位置
neutral potential 中性电位
neutral pressure level 中和界
neutral pressure plane 中性面
neutral pressure 中和压力
neutral reaction 中和作用，抵消，中和反应
neutral reactor 中和电抗器，中性线接地电抗器
neutral relay 非极性继电器，中性继电器
neutral resin 中性树脂
neutral rock 中性岩
neutral salt 中性盐
neutral shift 中性点位移
neutral shock wave 中性冲击波
neutral shoreline 中性滨线
neutral soil 中性土壤
neutral solidly grounded system 中性点直接接地系统
neutral solution 中性溶液
neutral stability 中性稳定性，随遇稳定性
neutral stratified flow 中性分层流
neutral stress 中和应力
neutral switch 中性线开关，空挡开关
neutral system grounding 中性系统接地
neutral terminal 中性点接线端，中性线端
neutral-tongue relay 中簧继电器
neutral turbulence flow 中性湍流流动
neutral uneffectively grounded system 中性点非有效接地系统
neutral ungrounded solidly system 中性点不直接接地系统
neutral ungrounded system 中性点不接地系统
neutral water operation 中性水运行工况，中性水运行
neutral water treatment 中性处理
neutral wire 中线
neutral zone control 中间带调节［控制］
neutral zone 中间带，中间区，中性地带，中性区，中和界
neutral 中和的，中间的，中性的，中性点，中性线，中相，零点，零线
neutrator 中性点补偿器
neutrodon 平衡电容器，中和电容器
neutrodyne circuit 中和电路，平衡电路，平差电路
neutrodyne receiver 中和接收机，衡消接收机
neutrodyne 中和接收法，衡消接收法
neutron absorber fluid 中子吸收液
neutron absorber object 中子吸收体
neutron absorber rods 中子吸收棒
neutron absorber solution 中子吸收液
neutron absorber 中子吸收剂，中子吸收体
neutron absorbing liquid 中子吸收液
neutron absorbing material 中子吸材料，中子吸收剂
neutron absorbing 吸收中子的，中子吸收
neutron absorption cross section 中子吸收截面
neutron absorption loss 中子吸收损耗
neutron absorption rate 中子吸收率
neutron absorption 中子吸收
neutron absorptivity 吸收中子的能力
neutron accident dosimetry 事故中子剂量测定法
neutron activated 中子激活的
neutron activation analysis 中子活化分析
neutron activation detector 中子激活探测器
neutron activation kit 中子活化箱
neutron activation method 中子活化法
neutron activation 中子激活，中子活化

neutron activity 中子活化
neutron age 中子年龄
neutron albedo 中子反照率
neutron angular current 角中子流，中子角流量
neutron angular density 中子角密度
neutron angular flux 中子角通量
neutron attenuation 中子流衰减
neutron background 中子本底
neutron balance equation 中子平衡方程
neutron balance sheet 中子平衡表
neutron balance 中子平衡
neutron bath method 中子浴法
neutron beam attenuation 中子束衰减
neutron beam collimation 中子束准直
neutron beam 中子束
neutron belt scale 核子秤
neutron booster 中子增强器，中子倍增装置
neutron burst 中子脉冲
neutron capture cross section 中子俘获截面
neutron capture gamma ray 中子俘获γ射线
neutron capture probability 中子俘获概率
neutron capture radiation 中子俘获辐射
neutron capture 中子俘获
neutron chain reaction 中子链式反应
neutron chain reactor 中子链式反应堆
neutron converter 中子转换器
neutron convert screen 中子转换屏
neutron cooling coefficient 中子冷却系数
neutron counter 中子计数器
neutron count rate 中子计数率
neutron cross section library 中子截面库
neutron cross section 中子截面
neutron current density vector 中子流密度矢量
neutron current density 中子流密度
neutron current vector 中子流矢量
neutron current 中子流
neutron curtain 中子流切断屏
neutron cycle 中子循环
neutron damage 中子损伤
neutron decay 中子衰变
neutron-deficient 缺中子的
neutron DE meter 中子剂量当量计
neutron density disadvantage factor 中子密度不利因子
neutron density distribution 中子密度分布
neutron density fluctuation 中子密度涨落
neutron density 中子密度
neutron detection system 中子探测系统
neutron detection 中子探测
neutron detector system 中子探测器系统
neutron detector well 中子探测器孔道
neutron detector 中子探测器
neutron diagnostics 中子诊断
neutron diffraction 中子衍射，中子衍射术
neutron diffusion area 中子扩散面积
neutron diffusion coefficient 中子扩散系数
neutron diffusion current 扩散中子流
neutron diffusion equation 中子扩散方程
neutron diffusion length 中子扩散长度
neutron diffusion 中子扩散
neutron distribution 中子分布
neutron dose rate 中子剂量率
neutron dose 中子剂量
neutron dosimeter 中子剂量计
neutron dosimetry 中子剂量学，中子剂量测定法
neutron economy 中子节省，中子经济
neutron effective lifetime 中子有效寿期
neutron emergency dosimeter 中子事故剂量计
neutron emission 中子发射
neutron emitting solution 中子发射溶液
neutron energy distribution 中子能量分布
neutron energy group 中子能群
neutron energy spectrum 中子能谱
neutron energy 中子能量
neutron escape 中子泄漏
neutron event 中子事件
neutron excess 中子过剩
neutron exposure 中子曝光，中子照射量，中子照射
neutron film badge 中子胶片剂量计
neutron fission-scintillation detector 中子裂变闪烁探测器
neutron fission 中子裂变
neutron flow density 中子流密度
neutron fluctuation 中子涨落
neutron fluence gradient 中子注量梯度
neutron fluence rate 中子注量率
neutron fluence 中子注量
neutron flux chamber 中子通量室
neutron flux control system 中子通量控制系统
neutron flux converter 中子通量转换器
neutron flux density meter 中子通量密度计
neutron flux density monitor 中子通量密度监测器
neutron flux density scanning assembly 中子通量密度扫描装置
neutron flux density space-time characteristic 中子通量密度时空特性
neutron flux density 中子通量密度
neutron flux detector 中子通量探测器
neutron flux distribution measurement 中子通量分布测量
neutron flux distribution 中子通量分布
neutron flux flattening 中子通量展平
neutron flux gradient 中子通量梯度
neutron flux instrumentation 中子通量测量仪表
neutron flux intensity 中子流强度，中子通量强度
neutron flux level 中子通量水平
neutron flux limiter 中子通量限制器
neutron flux measurement 中子通量测量
neutron flux measuring channel 中子通量测量通道
neutron flux modifying device 中子通量修正设备
neutron flux monitoring system 中子通量监测系统
neutron flux monitoring 中子通量监测
neutron flux monitor 中子通量监测器
neutron flux pattern 中子通量分布图
neutron flux peak 中子通量峰值
neutron flux range 中子通量量程
neutron flux sensor 中子通量敏感元件

neutron flux tilting　中子通量倾斜
neutron flux　中子注量率，中子流，中子通量
neutron generation time　中子每代时间
neutron generation　中子代
neutron generator　中子发生器
neutron group　中子组
neutron hardening　中子谱硬化
neutron hodoscope　中子辐射计数器
neutron howitzer　中子准直器，中子发射器
neutronic behavior　中子性能【堆芯】
neutronic reactor　中子反应堆
neutronic study　中子的研究
neutronics　中子学
neutronic-thermohydraulic model　中子热工水力学模型
neutronic　中子的
neutron importance function　中子价值函数
neutron importance　中子价值
neutron induced damage　中子引起的损伤
neutron induced embrittlement　中子引起的脆化
neutron induced fission reaction　中子诱发裂变反应
neutron induced thermal stress　中子引起的热应力
neutron induced　中子诱发的
neutron instrument　中子仪器
neutron inventory　(反应堆内)中子总量
neutron irradiated　受中子辐照的
neutron-irradiation-induced embrittlement　中子辐照诱发的脆化
neutron irradiation monitor　中子辐照监测器
neutron irradiation　中子辐照
neutron kinetics　中子动力学
neutron leakage spectrum　中子泄漏谱
neutron leakage　中子泄漏
neutron level gauge　中子液位计
neutron life cycle　中子寿命循环
neutron lifetime　中子寿命
neutron loss rate　中子损失率
neutron migration　中子迁徙
neutron moderation　中子慢化，中子减速
neutron monitoring channel　中子监测通道，中子通量测量通道
neutron monitoring　中子监测
neutron monitor　中子监测器
neutron multiplication constant　中子增殖系数
neutron multiplication factor　中子增殖系数
neutron multiplication　中子增殖，中子倍增
neutron multiplying facility　中子增殖装置
neutron noise　中子噪声
neutron nonleakage probability　中子不泄漏概率
neutron nuclear data　中子核数据
neutron number density　中子数密度
neutron number　中子数
neutron period　中子通量时间常数
neutron per second　每秒钟放出中子数
neutron physics　中子物理
neutron pile　中子反应堆
neutron poison　中子中毒
neutron population fluctuation　中子数的涨落
neutron population　中子总数，中子数
neutron powder profile　中子粉末剖面
neutron probe　中子探测器，中子探头
neutron production rate　中子产生率
neutron-proton ratio　中子与质子比
neutron-proton scattering　中子质子散射
neutron pulse　中子脉冲
neutron radiation capture　中子辐射俘获
neutron radiation damage　中子辐射损伤
neutron radiation　中子辐射
neutron radioactivation analysis　中子活化分析
neutron radiobiology　中子放射生物学
neutron radiograph　中子射线照相，中子照相，中子辐射照相，中子射线透照术
neutron reaction　中子反应
neutron reflecting　中子反射
neutron reflector　中子反射层
neutron release rate　中子释放率
neutron release　中子释放
neutron-removal cross-section　中子移出截面
neutron resonance absorption　中子共振吸收
neutron resonance level　中子共振能级
neutron resonance　中子共振
neutron scattering　中子散射
neutron scatter plug　中子散射芯棒
neutron sensitive chamber　中子灵敏（电离）室
neutron sensitive differential thermocouple　中子敏感微分热电偶
neutron sensitive sensor　中子敏感元件
neutron sensitive　中子敏感的
neutron sensor　中子敏感元件
neutron separation energy　中子分离能
neutron shielding　中子屏蔽
neutron shield pad　中子屏蔽块
neutron shield paint　中子防护用涂料，中子屏蔽用涂料
neutron shield panel　中子盾板，抗中子盾板
neutron shield plug　中子屏蔽塞
neutron shield　中子屏蔽体
neutron sink　中子阱，负中子源
neutron slowing down theory　中子慢化理论
neutron source assembly for PWR　压水堆中子源组件
neutron source assembly　中子源组件
neutron source box　中子源柜
neutron source calibration　中子源刻度
neutron source density　总中子源密度
neutron source strength　中子源强度
neutron source thermal reactor　中子源热反应堆
neutron source　中子源
neutron spectrum measurement　中子谱测量
neutron spectrum　中子能谱，中子谱
neutrons per absorption　每次吸收一个中子后所放出的中子数
neutrons per fission　每次裂变所放出的中子数
neutron stream　中子流
neutron strength function　中子强度函数
neutron temperature　中子温度
neutron thermalization theory　中子热化理论
neutron thermalization　中子的热能化，中子热化
neutron threshold　中子阈
neutron-to-gamma sensitivity ratio　中子 γ 灵敏

度比
neutron total cross section　中子总截面
neutron transport cross-section　中子输运截面
neutron transport equation　中子输运方程，中子迁移方程
neutron transport theory　中子输运理论
neutron transport　中子运输，中子迁移
neutron velocity　中子速度
neutron width　中子辐射的幅度，中子宽度
neutron worth　中子价值
neutron yield per absorption　每次吸收的中子产额
neutron yield per fission　每次裂变的中子产额
neutron yield　中子产额
neutron　中子，中子束，中子的
neutropause　中性层顶
neutrosphere　中性层
never lay flat　不可平放
nevertheless　虽然如此
new achievement　新成就
new and high-tech development zone　高新技术开发区
new core component transfer machine　新堆芯部件传送机
new element storage drum　新燃料元件贮存容器
new element storage　新（燃料）元件贮存
newel post　楼梯栏杆柱
newel stairs　盘旋扶梯
newel wall　楼梯隔墙
newel　楼梯栏杆柱
new energy generation　新能源发电
new energy resource　新能源
new energy　新能源
new fold　新褶皱
new fuel assembly storage rack　新燃料储存架
new fuel assembly transfer station　新燃料组件转运站
new fuel elevator　新燃料升降机
new fuel facility　新燃料组装室
new fuel rack　新燃料架
new fuel storage rack　新燃料贮存架
new fuel storage room　新燃料储藏室
new fuel storage vault　新燃料元件贮存库，新燃料储藏间
new fuel store　新燃料元件贮存处
new fuel　新燃料
new jet　新生射流
new-line character　换行字符
newly built power generation project　新建发电工程
newly-built power plant　新建电厂
newly-commenced project　新开工项目
newly laid concrete　新浇混凝土
newly-placed concrete　新铺混凝土
new moon spring tide　新月大潮
new moon　朔
new pin installation　新棒装配
new plume　新生羽流
new products development　新产品开发
new subassembly transfer position　新（燃料）组件输送位置
new technology　新工艺
Newtonian cooling　牛顿流体冷却
Newtonian flow　牛顿流动
Newtonian fluid flow　牛顿型流体
Newtonian fluid　牛顿型流体，牛顿流体
Newtonian friction law　牛顿摩擦定律
Newtonian liquid　牛顿液体
Newtonian mechanics　牛顿力学
Newtonian viscosity　牛顿黏滞度，牛顿黏性
Newton interpolation　牛顿插值
Newton iteration method　牛顿迭代法
Newton iteration　牛顿迭代（法）
Newton's equation　牛顿方程
Newton's interpolation formula　牛顿内插公式
Newton's law of fluid resistance　牛顿流体阻力定律
Newton's law　牛顿定律
Newton　牛顿
next address assembly register　下一地址汇编寄存器
next contingency set　下一个预想事故集
next-event file　下一事件的文件，待处理文件
next to nothing　几乎没有，几乎为零，微乎其微，极少
nexus　互连，网络，节，段，连接，连杆
NFB(negative feedback)　负反馈
NFC(notified for construction)　通知施工
NFC(Nuclear Fuel Complex)　核燃料综合体【印度】
NFC(nuclear fuel contract)　核燃料合同
NFD(no fixed date)　无固定日期
NFI(Nuclear Fuel Industnestries Co.)　核燃料工业公司【日本】
NFM(neutrino flux monitoring)　中微子通量监测
nf=nanofarad　毫微法拉，纳法
NF(Normes Francaises)　法国标准
n-fold multiple integraln　阶重积分
n-fold　n 阶，n 倍，n 重
NFPA(National Fire Protection Association)　全国消防协会【美国】
NFQ(night frequency)　夜间频率
NFU(notified for use)　可供使用
NFW(notification for walkdown)　现场巡查通知
NGN(Nigerian Naira)　尼日利亚奈拉
NGO(non-government organization)　非政府间国际组织
NGR(neutral grounding resistor)　中性点接地电阻
Ångstrom pyrheliometer　埃斯屈朗直接日射表
NHE(normal hydrogen electrode)　标准氢电极
N_2 hold-up tank　氮气贮存槽，氮气滞留槽
NHPE(maximum historically probable earthquake)　历史最大可能地震
NHR(net heat rate)　净热耗
NHSD(Nuclear Health and Safety Department)　（美国）核保健与安全部
NHV(net heat value)　净热值
NHWL(normal high water level)　正常高水位
nibble　（电脑）半字节
nib　模孔，突边
Nicalloy　一种高导磁率铁镍合金，镍锰铁合金

Nichicon　电容器
nichrome wire　镍铬线，镍铬电热丝
nickalloy　一种镍铁合金
nick bend test　刻槽弯曲试验
nick-break test　双缺口破断试验，凹口断裂试验，刻痕断裂试验
nicked fracture test　缺口弯曲试验
nickelage　镀镍
nickel alloy　镍合金
nickel-aluminum pair　镍铝对【热电偶引线】
nickel and nickel-base alloys　镍和镍基合金
nickel brass　镍黄铜
nickel bronze　镍青铜
nickel-cadmium battery　镍镉电池
nickel-cadmium storage battery　镍镉蓄电池
nickel-chrome steel　镍铬钢
nickel-chromium cast iron　镍铬铸铁
nickel chromium-nickel silicon thermocouple　镍铬镍硅热电偶
nickel electrode　镍电极
nickel-electroplating　表面镀镍
nickel-hydrogen battery　镍氢电池，氢镍电池
nickelin　铜镍锰高阻合金，铜镍锌合金
nickel iron accumulator　铁镍蓄电池
nickel iron battery　镍铁电池
nickelizing　镀镍
nickel-molybdenum thermocouple　镍钼热电偶，镍钼温差电偶
nickel-plated steel　镀镍的钢
nickel-plated　镀镍的
nickel steel wire　镍钢线
nickel steel　镍钢
nickel-zinc battery　锌镍电池
nickel-zinc cell　镍锌电池
nicking tool　刻刀
nick　槽口，划痕，伤纹，刻痕，裂痕，裂口
NIC(network information center)　网络信息中心
NIC(normal input cause)　正常输入条件
NIC(not in-contact)　非接触
NIC(Nuclear Information Centre)　核情报中心【日本】
NIEMS(NI Erection Management System)　核岛安装工程管理系统
NIE(nuclear island erection)　核岛安装
Nife accumulator　镍铁蓄电池
Nife cell　镍铁电池
night cooling　夜间冷却
night dip　夜间低谷【负荷曲线】
night duty　夜班
night illumination　夜间照明
night key　夜铃电键
night load　夜间负荷，夜间负载
night observation　夜间观测
night shift room　夜间值班室
night shift　夜班，夜班工人
night signal　夜间航标
nigrometer　黑度计
nil ductility temperature　无延性温度，脆性温度
nil ductility transition temperature　无延性转变温度
nil ductility transition　无延性转变，无塑性转变，脆性转变
nile　奈耳
nil-load　空转
nilometer　水尺，水位计，水位表
niloscope　水位表
nil　零，零点，无
NIM(nuclear instrument module)　尼姆【核仪器系统名称】，核仪器模块，核仪器标准块
ninefold　九倍，九次，九重
nine-grade progressive taxation　九级累进税制
nine point picking out　九点取样法
nine's complement　十进制的反码，九的补码
ninety degree stem turn　90°阀杆转动
NI(non-inductive)　无感应的
NI(nuclear island)　核岛，核电站
niobium capacitor　铌电解电容器
niobium stabilization　（钢）铌稳定
niobium-stabilized　铌稳定的
niobium　铌【Nb】
NIOSH(National Institute For Occupational Safety & Health)　美国国家职业安全与健康研究所
nip angle　（压力加工）咬入角
nip machine　咬口机，辘骨机，咬缝机，（风管）咬口机［辘骨机］
NIP(non-impact printer)　非击打式打印机
nipolet　短管支管台
NIPO(negative input positive output)　负输入正输出
nippers　尖嘴钳，钳子
nipping machine　压平机
nipple joint　螺纹接头
nipple　接头短管，短接管，螺纹接管，喷嘴，接管，短管，短管接头，螺纹接头
nip　海岸低悬崖，浪蚀洞，狭缩
NIS(network information service)　网络信息服务
NIS(network information system)　网络信息系统
NIS(nickel steel)　镍钢
NIST(National Institute of Standards and Technology)　国家标准与技术局【美国】
nitosol　强风化黏磐土
nitra-lamp　充气灯泡
nitrated steel　氮化钢
nitrate nitrogen　硝酸盐氮
nitrate　硝化，硝酸盐，硝酸酯，硝酸根
nitration　氮化作用，硝化，渗氮
nitric acid fume　硝酸烟雾
nitric acid　硝酸
nitric oxide　氧化氮
nitridation　氮化，渗氮
nitrided steel　氮化钢
nitride fuel　氮化物燃料
nitride　氮化物
nitriding steel　氮化钢
nitriding　氮化，渗氮（处理）
nitrification process　硝化过程
nitrile base　叔胺
nitrile-chloroprene rubber　氯丁腈橡胶
nitrile rubber　丁腈橡胶
nitrile　腈
nitrilotriacetic acid　氨川三醋酸，次氨基三乙酸
nitrite nitrogen　亚硝酸盐氮

nitrite 亚硝酸盐
nitro acid 硝基酸
nitro amine 硝胺
nitro bacteria 硝化细菌，氮细菌
nitrobenzene 硝基苯
nitrodope 硝化涂料
nitroenamel 硝基磁漆
nitroexplosive 硝化炸药
Nitrofluor process 硝基氟法
nitrogelatine 硝化明胶炸药
nitrogen blanketing protection 充氮保护
nitrogen blanketing valve 充氮管，氮气管
nitrogen blanketing 氮气覆盖，充氮
nitrogen blanket system 氮气覆膜系统
nitrogen blanket 氮气覆盖层，氮气覆盖
nitrogen bottle battery 氮气瓶组
nitrogen case hardening 渗氮，氮化
nitrogen content 含氮量
nitrogen-cooled closed-cycle gas turbine 氮冷闭环燃气轮机
nitrogen-cooled reactor 氮冷堆
nitrogen cushion 氮气垫
nitrogen cycling 氮循环
nitrogen cylinder 氮气瓶
nitrogen-deficient waste 缺氮废水
nitrogen dioxide index 二氧化氮指数
nitrogen dioxide 二氧化氮
nitrogen flow temperature 氮气流温度
nitrogen gas barrel 氮气瓶
nitrogen gas bottle 氮气瓶
nitrogen gas cylinder 氮气瓶
nitrogen gas seal equipment 氮气密封装置
nitrogen hardening 渗氮硬化处理，氮化
nitrogen heating system 氮气加热系统
nitrogen heating 氮气加热
nitrogen lamp 氮气灯
nitrogen manifold 氮气歧管，氮气总管，氮气汇流管
nitrogen monoxide 一氧化氮
nitrogenous effluent 含氮污水
nitrogenous waste 含氮废物
nitrogen oxide 氧化氮
nitrogen purge line 氮气吹洗管线
nitrogen recondenser 氮气再冷凝器
nitrogen recycle cooler 氮气再循环冷却器
nitrogen return temperature 氮气回流温度
nitrogen-sealed transformer 充氮密封变压器
nitrogen sealing 氮封
nitrogen-strengthened steel 渗氮强化（的）钢
nitrogen supply manifold 氮气总管，供氮联箱
nitrogen supply system 氮气供应系统，供氮系统
nitrogen supply 氮气供应
nitrogen tetroxide cooled reactor 二氧化氮冷却反应堆
nitrogen trap 氮气捕集器，氮气阱
nitrogen trichloride 三氯化氮
nitrogen 氮
nitroglycerine explosive 硝化甘油炸药
nitrohydrochloric acid 王水，硝基盐酸
nitrometal 硝基化金属
nitrometer 氮量测定计
nitrosonitric acid 发烟硝酸
nitroso-R salt 亚硝基 R 盐
nitro-substitution 硝基取代
nitrosyl 亚硝酰
nitrous acid 亚硝酸
nitrous oxide 氧化亚氮，一氧化氮
nitrovarnish 硝基清漆
nivation 霜蚀
nixie decoder 数码管译码器
nixie light 数字管，数码管
nixie readout 数码管读出装置
NJP(network job processing) 网络作业处理
n-level address n 级地址
n-level logic n 级逻辑
NLG 荷兰盾
NLM(noise-load ratio) 噪声负载比
NL(net loss) 净损失
NL(new-line character) 换行字符
n-loop reactor n 环路反应堆
NLP(nonlinear programming) 非线性规划
NLR(no-load ratio) 空载比
NLS(no-load speed) 空载速度
NLS(non linear system) 非线性系统
NMAC(nuclear maintenance assistance center) 核维修援助中心
NMC(network measurement center) 网络测量中心
NMCS(nuclear material control system) 核材料控制系统
NME(noise-measuring equipment) 噪声测量设备
NM(not measured) 未测量的
NMOS(n-channel metal-oxide-semiconductor) n 沟道金属氧化物半导体
NMR(nuclear magnetic resonance) 核磁共振术，核磁共振
NMU(network management unit) 网络管理部件
NMV(normal mode voltage) 串模电压
N＝negative 负的
N＝Newton 牛顿
NNI(noise and number index) 噪声和数值指标
NNI(non-nuclear instrumentation) 非核测量仪表
N_2＝nitrogen 氮气
nn(non-negotiable) 不可转让
NNPC(Nigerian National Petroleum Corporation) 尼日利亚国家石油公司
NNSA(National Nuclear Safety Administration) 国家核安全局【中国】
NNS(non-nuclear safety) 非核安全
NOAC(Nuclear Operations Analysis Center) 核运行分析中心
no-address computer 无地址计算机
no-address instruction 无地址指令
no-address 无地址
NOA(notice of award) 中标通知
no audible warning 禁鸣喇叭
Nobatron DC stabilizer 一种直流稳压器
no bias relay 非偏置继电器，非极化继电器
noble fission gas 惰性裂变气体
noble gas activity 惰性气体放射性

noble gas adsorption to activated charcoal 惰性气体的活性炭吸附
noble gas converter 惰性气体转换器
noble gas fission product 惰性气体裂变产物
noble gas isotope 惰性气体同位素
noble gas 惰性气体，稀有气体
noble metal thermocouple 贵金属温差电偶，贵金属热电偶
noble metal 贵金属
no-bore rotor 无中心孔转子
no-bounce switch 无反跳开关，不跳动开关
no-break emergency power supply 不停电紧急电源
no-break power supply 不能停电的电源，不间断电源
no-break transfer switch 不能停电的转换开关
no-carry 无进位
no charge 免费，未充电，未装载
NOC(network operation center) 网络运行中心
NOC(notice of commencement) 开工通知
NOC(Nuclear Oversight Committee) 核监视委员会【美国】
no code 无代码
no commercial value 无商业价值
no connection 不连接
no-contact pickup 无触点传感器，无触点检波器
N. O.(normal open) contact 常开触点
no-core reactor 空心扼流圈，无铁芯扼流圈
noctirsor 暗视器
NOCT(nominal operating cell temperature) 组件的电池额定工作温度
nocto television 红外线电视
noctovision 红外线电视，暗视
noctovisor 红外线望远镜，红外线摄像机
nocto 红外线
nocturnal boundary layer 夜间边界层
nocturnal cooling 夜间冷却
nocturnal inversion breakup fumigation 夜间逆温破坏下熏
nocturnal radiation cooling 夜间辐射冷却
nocturnal stability 夜间稳定度
nocturnal thermal boundary layer 夜间温度边界层
nocturnal 夜的
nodal admittance matrix 节点导纳矩阵
nodal analysis （反应堆动力学中的）节块分析
nodal circle 节圆，波节圈
nodal diagram （叶轮）节圆振动图，（叶片）节点振动图
nodal diameter 节径
nodal diffusion-depletion calculation 节块的扩散贫化计算
nodal discrete coordinates method 节块离散坐标法
nodal displacement 结点位移
nodal equilibrium equation 节点平衡方程
nodal expansion method 节块展开法
nodal force 节点力
nodal imbedded assembly calculation 节块嵌入组件计算
nodal impedance matrix 节点阻抗矩阵
nodal integration method 节块积分法
nodalization 节块化
nodalizer 波节显示器
nodal line 节线，交点线，波节线
nodal method of analysis 节点分析法
nodal pattern 振动节型
nodal plane 波节面
nodal point keying 波节点键控
nodal point of vibration 振动节点
nodal point 节点，结点
nodal potential 节点电位，波节电位
nodal pricing 节点电价，节点电价法
nodal section 节面
nodal transport theory method 节块输运理论方法
nodal type （流态的）波节式
nodal 结点的，交点的，节的，波节的，中心的，关键的
nodding action 摆动
node-admittance matrice 节点导纳矩阵
node-average flux 节块平均通量
node-average group fluxes 节块平均群通量
node junction method 节点连接法
no-delay base 立即接通制【电话】
node method 节点电位法
node number 节点号
node of separation 分离节点
node of tide 波节
no detect 不检测，未发现
node voltage 节点电压
node 结，结点，波节，中心点，交点，叉点，分支
noding diagram 节点图
no dropping 切勿坠落【包装标志】
nodular cast iron 球墨铸铁
nodular cementite 粒状渗碳体
nodular corrosion 节块状腐蚀
nodular graphite cast iron 球墨铸铁
nodular graphite 球状石墨
nodular iron casting 球墨铸铁件
nodular iron 球墨铸铁
no dumping 切勿抛掷【用于包装外表标志】
NOE(normal operation earthquake) 正常运行地震，设计基准地震
no entrance 禁止入内
no entry 非入口
no exchange surrendered 不结汇
no-excitation detection relay 无激励检测继电器
no-fines concrete 无细骨料混凝土，无砂混凝土
no-flow condition 无流动状况，无流条件
no-flow position 中间位置【液压调节系统中的滑阀】
no-fuse breaker 无熔断器断路器
no-fuse switch 无熔丝开关，无保险丝开关
nogging 木架间水平短撑
no-go gauge 不通过规，不过端量规
no-go 不通过
no honking 禁鸣喇叭
no hooks 不许用钩
no horn 禁鸣喇叭
noise abatement valve trim 低噪声阀结构
noise abatement 噪声抑制，消声

noise absorber 噪声吸收体
noise absorption 噪声吸收
noise-air-pollution 噪声空气污染
noise amplitude 噪声振幅，噪声干扰
noise analysis 噪声分析
noise background 背景噪声，本底噪声
noise balancing circuit 噪声平衡电路
noise barrier 噪声滤波器，噪声声垒，隔声屏障
noise characteristics 噪声特性
noise coefficient 噪声系数
noise compensation 噪声补偿，噪声校正
noise control law 噪声管制法
noise control legislation 噪声管制立法
noise control program 噪声控制规划
noise control 噪声控制
noise criteria curve 噪声标准曲线
noise criteria 噪声标准，噪声判据
noise criterion curve 噪声标准曲线
noise criterion 噪声标准
noise current 噪声电流
noise damper 消声器
noise deadening foam 隔声泡沫材料
noise detector 噪声检测器
noise diagnosis system 噪声诊断系统
noise diagnostics 噪声诊断
noise diode 二极管噪声发生器
noise distortion 噪声畸变，噪声失真
noise dose 噪声剂量
noise effect 噪声效应
noise elimination 消声，减声，噪声消除
noise enclosure 隔声罩
noise environment 噪声环境
noise equivalent bandwidth 噪声等效带宽
noise equivalent flux 噪声等效通量
noise equivalent power 噪声等效功率
noise equivalent resistance 噪声等值电阻
noise exposure forecast 噪声暴露预报
noise exposure meter 噪声暴露计
noise exposure 噪声暴露
noise factor 噪声系数，噪声因子
noise-field intensity 噪声场强
noise figure 噪声指数，噪声系数，噪声数值，噪声因数，噪声图
noise filtering 噪声过滤
noise filter 噪声滤波器
noise footprint 噪声污染的地面区域
noise-free receiver 无噪声接收机
noise-free 无杂声的，无噪声的
noise frequency 噪声频率
noise function 噪声函数
noise generator 噪声发生器
noise immunity 抗噪声度，抗扰度
noise immunization 消除噪声
noise insulation factor 噪声隔离因素
noise insulation 噪声隔离，隔声
noise intensity 噪声级，噪声强度
noise interference 噪声干扰
noise killer 噪声消除器，噪声吸收器，噪声抑制器，降噪器
noiseless channel 无噪声通道
noiseless motor 无噪声电动机
noiseless pavement 无噪声路面
noiseless reactor 无噪声反应堆
noiseless running 无噪声运行
noiseless 无噪声的，无杂声的
noise level analyzer 噪声级分析仪
noise level indicator 噪声电平指示器
noise level measurement 噪声电平测
noise level meter 声级计，噪声级电平表
noise level test 噪声级试验
noise level 噪声级，噪声度，噪声水平，噪声电平
noise-load ratio 噪声负载比
noise margin 噪声容限
noise measurement 噪声测量，干扰测量
noise-measuring equipment 噪声测量设备
noise meter 噪声计，噪声测试器，噪声级表
noise modulated 噪声调制的
noise modulation 噪声调制
noise monitoring 噪声监测
noise muffler 噪声衰减器，消声器
noise network 噪声网络
noise nuisance 噪声公害
noise peak 噪声峰值
noise pollution 噪声污染
noise potential 噪声电压
noise power spectrum 噪声功率谱
noise power 噪声功率，干扰功率
noise precaution 噪声预防
noise prevention 噪声防止
noise-producing equipment 噪声发生装置
noise-producing source 噪声发生源
noise-producing surface 发噪声地面
noise-proof device 防噪声装置
noise propagation 噪声传播
noise protection 噪声防护
noise protector 噪声防护器
noise pulse 噪声脉冲
noise radiation 噪声辐射
noise range 噪声区，干扰源
noise rating curve 噪声分级曲线
noise rating number 噪声评价数
noise ratio 噪声比
noise-reducing telephone 抗噪声电话机
noise reduction device 消声装置
noise reduction system 降低噪声系统
noise reduction 噪声衰减，噪声消减，减少噪声，减声，消声，声衰减，降噪
noise-rejection 噪声抑制
noise resistance 噪声等效电阻
noise shielding 噪声屏蔽
noise sickness 噪声引起的病
noise silencer 噪声吸收器，降噪器，噪声抑制器，消声器
noise source distribution 噪声源分布
noise source 噪声源，干扰源
noise spectrum 噪声频谱
noise squelch 噪声消除器
noise standard 噪声标准
noise suppression circuit 噪声抑制电路
noise suppression 噪声抑制，噪声控制

noise suppressor 噪声抑制器
noise technique 噪声技术
noise temperature ratio 噪声温度比
noise temperature 噪声温度
noise test 噪声试验
noise tolerance 噪声容许量
noise transmission impairment 噪声传输影响，噪声造成的传输质量降低量
noise transmission 噪声传递
noise trap 降噪器
noise use 噪声利用
noise variance 噪声离散
noise vector 噪声向量，噪声矢量
noise voltage generator 噪声电压发生器
noise zoning 噪声分区
noise 杂音，噪声，干扰，声音，噪音
noisiness 喧闹，噪声特性，噪声量
noisy running 噪声运转
noisy 有噪声的，有干扰的
NOK 挪威克郎
no-lag motor 一种补偿式感应电机
no latch 未挂闸
no later than 不迟于
no-leakage magnetic circuit 无漏磁磁路
no-leakage 无漏泄
no-leak 不漏泄，不漏气
no-lines 无空线，全部占线
NOL(normal oil level) 正常油位
NOL(normal operating loss) 正常运行损失
no-load characteristic 空载特性，空载特性曲线
no-load cogging torque 空载齿槽转矩
no-load compensation 空载补偿
no-load condition 空载工况
no-load consumption 空载消耗，空转消耗
no-load current 空载电流
no-load cutout 空载断路器
no-load discharge 空载流量，无负荷流量
no-load excitation 无载励磁，空载励磁
no-load factor 空载系数
no-load field voltage 空载励磁电压
no-load flow 空负荷流量，空载耗汽量
no-load impedance 空载阻抗
no-load inductance 空载电感
no-load loss 空载损失，空转损失，开路损失
no-load magnetization curve 空载磁化曲线
no-load mode 空载模式
no-load opening 空载开度
no-load operating test 空载运行试验
no-load operation control 空负荷运行控制
no-load operation test 卸载运行试验
no-load operation 空负荷运行，空载运行，卸载运行
no-load point 空载点，空负荷点
no-load power consumption 空载能耗
no-load power 空负荷功率，空载功率
no-load pressure 空载压力，空负荷压力
no-load relay 空载继电器
no-load release 空载释放，空载跳闸
no-load resistance 空载电阻
no-load running 空负荷运行，空转，空载运转，空载运行
no-load run 空载运转，空转，无载运转
no-load speed changing 空载变速
no-load speed governor 空载调速器
no-load speed 空载速度，空载转速
no-load starting 空载启动，卸载启动
no-load state 空载状态
no-load steam consumption 空负荷汽耗
no-load switching in 空载合闸
no-load switching off 空载切断
no-load tap-changer of transformer 变压器无载调压开关
no-load tap changer 空载分接开关
no-load temperature 空载温度
no-load terminal voltage 空载端电压
no-load test 空载试验，空负荷试验，空负荷调试
no-load torque 空载转矩
no-load trip 空负荷解列
no-load voltage ratio （变压器的）变比，空载电压比
no-load 空载，空负荷，无负荷，无载，空转
no-man control 无人控制
no-man operation 无人值班操作
no marks 无标记
nomenclature 管道命名表
nomenclature 术语，命名法，专有名称，专用名称
nominal activity 额定活度
nominal air flow rate 额定空气流速率，额定风量
nominal area of screen 标称筛面面积
nominal attenuation ratio 标称衰减率
nominal bore 标称口径
nominal capacity 标称容量，额定蒸发量，额定容量，额定出力
nominal capital account 名义资本账户
nominal capital 名义资本
nominal channel 名义通道
nominal circuit voltage 标称电路电压
nominal condition 名义工况
nominal contract price 名义约定价格
nominal cross section 标称截面
nominal current density 标称电流密度
nominal current 额定电流，标称电流
nominal cut-off frequency 标称截止频率
nominal deflection 名义转折角
nominal deviation 名义偏角
nominal diameter 标称直径，公称直径，名义直径
nominal dimension 公称尺寸，标称尺寸
nominal discharge current 标称通流容量，标称放电电流
nominal electric field strength 标称场强
nominal error 公称误差，名义误差
nominal escalation rate 名义浮动率
nominal frequency 标称频率，额定频率
nominal horsepower 额定马力，标称马力
nominal hydrogen pressure 标称氢压
nominal impedance 标称阻抗
nominal incidence 名义冲角
nominal induced electromotive force 标称感应电

动势
nominal inlet size 标称入口尺寸
nominal insulation voltage 标称绝缘电压
nominal interest rate 名义利率
nominal length 公称长度，名义长度
nominal lifetime 名义寿命
nominal load-bearing capacity 额定承载力
nominal load 标称负载，额定负荷
nominally steady load 名义上的稳定负载，标称稳定负载
nominally 标称，有名无实地，名义上
nominal mix 标称配合比
nominal net head 额定净水头
nominal operating conditions 额定运行工况，设计运行工况
nominal outlet size 标称出口尺寸
nominal output 额定输出，标称输出，额定出力，名义出力，标称出力，额定产量，标称功率，标称生产率
nominal parameter 标称参数
nominal partner 名义合伙人
nominal pipe diameter 标称管径
nominal pipe size 标称管道尺寸
nominal pipe thread 标称管道螺纹
nominal power of windturbine 风电机组的额定功率
nominal power wind speed 额定风速
nominal power 标称功率，标定功率，额定功率，名义功率
nominal pressure 标称压力，名义压力，额定压力
nominal price 名义价格，标称价格
nominal productive head 额定发电水头，毛发电水头
nominal pull-in torque 标称牵入转矩，名义牵入转矩
nominal-rated capacity 额定出力，额定蒸发量
nominal-rated load 额定负载
nominal rate of rise （冲击波的）标称上升率
nominal rate 额定出力
nominal rating condition 额定工况
nominal rating 标称额定值，名义额定值，额定出力，名义出力，标称出力，额定功率
nominal regime 名义工况
nominal reluctance 标称磁阻
nominal rotational speed 额定转速
nominal running load 标定运行负荷
nominal running water pressure 额定运行水压
nominal screen aperture 筛的名义孔径
nominal shares of stock 记名股票
nominal short-circuit capacity 额定短路容量
nominal short-circuit voltage 标称短路电压
nominal situation 正常工况
nominal size of pipe 管材的公称尺寸
nominal slip 额定转差率
nominal speed of water turbine 水轮机额定转速
nominal speed 额定转速，标称转速
nominal steam condition 额定蒸汽参数
nominal steam pressure 额定蒸汽压力
nominal steam temperature 额定蒸汽温度
nominal strain 标称应变
nominal strength 标称强度
nominal stress 标称应力
nominal system voltage 系统标称电压
nominal tension 标称张力，标称电压
nominal terms 名目条款，名义条款
nominal throat diameter 标称喉部直径
nominal tip speed 额定叶尖速度
nominal top size 标称最大粒度
nominal torque 额定扭矩，额定转矩，标称转矩
nominal transformation ratio 标称变换系数
nominal value of geometric parameter 几何参数标准值
nominal value 标称值，公称值，标幺值，额定值
nominal valve size 标称阀门尺寸
nominal voltage of a system 系统标称电压
nominal voltage 标称电压，额定电压
nominal volume 公称容积
nominal wind speed 额定风速
nominal withstand voltage 标称耐压值
nominal working conditions 标称工况，标称工作条件
nominal working pressure 额定工作压力，公称工作压力
nominal 标称的，公称的，名义上的，额定的，名义的
nominated contractor 指定承包商
nominated subcontractor 指定分包商
nominated supplier 指定供应商
nominate 提名，推荐
nomination cargo 指定［指派］货
nomination committee 提名委员会
nomination 任命，指定，提名
nominator 提名者，指定人
nominee 被提名者，被指定人
no-mobile reactor 固定反应堆
nomogram 诺谟图，列线图，计算图表，图解
nomographic chart 诺谟图，列线图，计算图
nomography 列线图解，图算法
nomograph 诺谟图，列线图，算图，计算图表，图解
nomotron 开关电子管
non-abrasive 无磨蚀性
non-absorbing reflector 不吸收反射层
non-absorbing 不吸收的
non-accelerating flow 非加速流
non-acceptance 拒绝承兑
non-activated silica 非活性硅
non-active drains hold-up tank 非放射性疏水暂存箱
non-adiabatic condition 非绝热状态
non-adiabatic cooling 非绝热冷却
non-adiabatic operation 非绝热操作
non-adiabatic process 非绝热过程
non-adiabatic 非绝热的
non-adjustable 不可调节的
non-adjusted relay 无调整继电器
non-aerated flow 不掺气水流
non-aeronautical aerodynamics 非航空空气动力学
non-aggressive 非侵蚀性的
non-aging insulation 不老化绝缘，抗老化绝缘

non-aging steel 无时效钢
non-aging 无时效的，不老化的，未老化的
non-air-entrained concrete 非加气混凝土
non-algebraic adder 非代数加法器，算术加法器
non-algebraic 非代数的，算术的
non-alignment diagram 非列线图
non-alkaline hardness 非碱性硬度
non-analytic function 非解析函数
non-aqueous fluid fuel reactor 非水溶液核燃料反应堆
non-aqueous reprocessing 干法后处理
non-arcing arrester 无弧避雷器
non-arcing 无弧的，无火花的
non-arithmetic shift 循环移位，非算术移位
non-artesian water 非承压地下水
non-articulated arch 无铰拱
non-articulated blade 非铰接叶片
non-assignable 不可转让的
non-association cable 非标准电缆
non-attached check valve 不相连的单向阀
non-attendant operation 无人值班运行
non-attended substation 无人值班变电所
non-attenuated power oscillations 不衰减功率振荡
non-attenuating wave 等幅波
non-autocatalytic reactor 自稳定反应堆，负温度系数反应堆
non-autocatalytic 非自催化的
non-automatic control 手控，非自动控制
non-automatic extraction turbine 非调节抽汽式汽轮机
non-automatic self-verification 非自动的自检验
non-automatic switching 非自动转接
non-automatic tripping 非自动跳闸，非自动断开
non-automatic 非自动的，手控的
non-autonomous system 非自控系统，非自治系统
non-autonomous 非自治的
non-azeotropic mixture refrigerant 非共沸溶液制冷剂
non-bearing partition 非承重隔墙
non-bearing structure 非承重结构
non-bearing wall 非承重墙
non-bidding unit 非竞价机组
non-binary code 非二进制代码
non-binary 非二进制的
non-binding 无约束力的
non-black control element 非黑体控制元件
non-bleeding cable 无泄漏电缆
non-bleed steam rate 不抽汽时汽耗率
non-boiling equilibrium 无沸腾平衡
non-boiling height 不沸腾段高度
non-boiling reactor 非沸腾反应堆
non-boiling region 不沸腾区
non-bore monoblock rotor 无孔实心转子
non-break AC power plant 不中断交流电源设备
nonbreaking wave 未破波
non-breeder reactor 非增殖反应堆
non-breeding 非增殖的
non-bridging contact 非桥接接点
non-bridging wiper 非桥接弧刷
non-bridging 非桥接的，无桥接现象的
non-bubbling fluidization 非鼓泡流态化
non-buoyant plume 非浮力羽流
non-burned block brick 免烧砌块，免烧块砖
nonburning 阻燃
non-business expenditure 营业外支出
non-caked 不黏结的
non-caking coal 不黏结煤，不黏煤
non-cancelable lease term 不可解除的租约条款
non-canonical path 非正则轨线
non-canonical 非正则的
non-capacitive 非电容的，无电容的
non-capillary porosity 非毛管孔隙率
non-capturing medium 非俘获中子介质
non-carbonate hardness 非碳酸盐硬度
non-catalytic process 非催化过程
non-catalytic reaction 非催化反应
non-catalytic single pellet gas-solid reaction 非催化单粒气固反应
non-cavitation flow 无空蚀水流
non-ceramic insulator 非瓷质绝缘子
non-chaotic attractor 非混沌性吸引子
non-circuital field 无旋场
non-circular channel 非圆通道
non-circularity 不圆度
non-circular opening 非圆形口
non-circulatory flow 非环状流
non-classical radix 非经典基数
non-classical 非经典的
non-clogging impeller 不阻塞的叶轮，不堵叶轮
non-clog type pump 不阻塞水泵
non-clog type 无堵塞型
non-cohesive material 非黏性材料
non-cohesive soil 无黏聚性土，无黏性土，非黏性的
non-coincident demand 非同时需要量
noncoking 不结焦的
noncolloidal silica 非胶态硅，非胶态二氧化硅
non-combustible construction 非燃烧结构
non-combustible materials 非燃烧材料
non-combustible 不可燃的
non-commercial agreement 非商业性协议
non-common section 非共同段
non-commutative 不可换的，非交换的
non-compensated length 未补偿长度
non-compensated machine 无补偿电机
non-competition clause 竞业限制条款
non-competition energy 非竞争电量
noncompetitive bid 非竞争性投标
noncompliance with contract term conditions 不符合合同条件事项
noncompliance with specifications 不符合技术规范事项
noncompliance with the notice period 不遵守通知期限
noncompliance 不服从，不履行，不遵守，违约行为，不符规定，不符合，不一致
non-concentrating collector 非聚光型集热器
non-condensable gas purger 不凝性气体分离器
non-condensable gas 不凝性气体，不溶气体
non-condensed gas 不凝结气体

non-condensible gas 不凝结气体
non-condensing bleed turbine 背压式抽汽汽轮机
non-condensing gas 不冷凝气体
non-condensing operation 排汽运行
non-condensing turbine 背压式汽轮机
non-condensing 不冷凝的
non-conducting covering 绝热盖板
non-conducting furnace 绝缘底电炉
non-conducting material 非导电材料
non-conducting particle bed 绝缘颗粒层
non-conducting state 非导通状态，关断状态，阻断状态
non-conducting transistor 不导通晶体管
non-conducting voltage 截止电压
non-conducting 不导电的，绝缘的，不传导的
non-conductor 非导体，绝缘体，电介体
non-conformance report 不符合项案卷，不符合项报告【质保】，不符合报告，不合格报告
non-conformance term 不符合项
non-conformance 不符合，不符合项【质保】
nonconforming goods 不符合规定的货物
nonconforming load 不一致负荷
nonconforming use 不符规定，使用不符合
nonconforming works 不符合同工程
nonconforming 非一致性的
nonconformity 不合格，不整合，非一致性
non-conjunction gate “与非”门
non-connection side span 非连接边节距
nonconservative waste effluent 非保存性废水，非保存性废气
nonconservative 非守恒的
noncontact attemperator 表面式减温器
noncontact desuperheater 表面式减温器
noncontact hardness tester 无触点硬度试验器
noncontacting electrode 非接触电极
noncontacting pickup 无触点传感器
noncontacting switch 无触点开关
noncontacting thickness gauge 无触点厚度计
noncontacting 无触点的
noncontact longitudinal recording 无触点纵向记录
noncontact low-voltage electrical apparatus 无触点低压电器
noncontact nuclear scale 无触点核子皮带秤
noncontactor relay 无触点继电器
noncontact recording 无触点记录
noncontact thermometer 非接触式温度计
noncontact type speed sensing switch 非接触速度传感开关
noncontact 无触头，无接触，无接点，无触点的
noncontaminated atmosphere 未污染的大气
noncontinuous electrode 不连续电极，短电极
noncontracting party 非缔约方
noncontractual claims 非合同规定的索赔，非合同索赔，合同外索赔
noncontractual works 非合同工程
non-contributing area 不产流面积
non-controlled area 非控制区
non-controlled edition 非受控版本
non-controlled rectifier 无控制整流器
nonconvective ponds 无对流太阳池
non-convergent series 非收敛级数
non-cooled distributor 无冷却布风板【沸腾炉】
non-cooperative game 非合作博弈
non-coordinated control 独立控制
non-coplanar force 非共面力，空间力
non-core choking coil 无铁芯扼流圈，空心扼流圈
non-core reactor 空心电抗器，无铁芯电抗器
noncorresponding control 无静差控制，无差调节，无静差调节
noncorresponding 无静差的
noncorrodibility 耐腐蚀性，不腐蚀性
noncorroding 抗蚀的，耐腐蚀的，不锈的
non-corrosive chemical 非腐蚀性化学物质
non-corrosive metal 不锈金属
noncorrosiveness 无腐蚀性
noncorrosive pipe 防腐管
noncorrosive 不引起腐蚀的，不锈的
noncritical circuit 非临界电路
noncriticality 非临界
noncritical phase matching 非临界相位匹配
noncritical pressure drop 非临界压降
noncrop area 非耕作区
noncrystalline solid 非晶化固体
noncumulative letter of credit 不可累积使用的信用证
non-cumulative revolving letter of credit 不可累积使用的循环信用证
non-current-carrying metal parts 不载流金属部分
non-current liabilities 非流动负债，长期负债，非流动欠债
noncurtailing equipment outage 不影响出力停机
non-cultivable land 不可耕地
noncyclical field 无旋场，有位场
noncyclical 非循环的，非周期的
noncyclic memory 非循环存储器
noncyclic storage 非循环存储器
non-cylindrical coil 非圆柱形线圈
non-damage test 无损探伤试验
non-damping vibration 无阻尼振动
non-decimal system 非十进制系统
non-deforming steel 不变形钢
nondelivery 未交付
non-depositing velocity 不淤流速
non-destructive activation analysis 无损活化分析
non-destructive analysis 无损分析
non-destructive assay 无损检验，无损分析
non-destructive compositional analysis 无损成分分析
non-destructive evaluation 无损评价
non-destructive examination group 无损检验组
non-destructive examination 非破坏性检查，无损探伤，非破坏性检验
non-destructive flaw detection 无损探伤
non-destructive inspection 无损检验，无损探伤，不破坏检验，无损检查
non-destructive measuring 不破坏测量
non-destructive read element 无损读出元件
non-destructive reading 无损读数
non-destructive readout 无损读出
non-destructive read-write 无损读写

non-destructive test 非破坏性测试［试验］【混凝土】，无损探伤［检验，测试］
nondeterministic polynomial 不确定性多项式
nondiagonal term 非对角项
nondimensional coefficient 无量纲系数，无因次系数
nondimensional factor 无量纲因子，无因次系数
nondimensional form 无量纲形式
nondimensional frequency 无量纲频率，无因次频率
nondimensionalization 无量纲化
nondimensional parameter 无因次参数
nondimensional velocity 无量纲速度
nondimensional wind shear 无量纲风切变
nondimensional 无量纲的，无因次的
non-dimension 无量纲
nondirectional antenna 不定向天线
nondirectional counter 不定向计数器
non-directional echo sounding 非定向回声测深法
nondirectional 无方向的，不定向的，无方向性的
nondirective 无方向性的，不定向的
non-disclosure agreement 保密协议
non-disclosure clause 保密条款
nondisconnecting fuse 不能隔离的熔断器
non-dished portion 非碟形部分
nondisjunction gate “或非”门
nondisjunction 不分离
non-disperse aerosol 非分散气溶胶
nondissipative network 无耗网络
nondistinct image 不清晰图像
nondivergent field 非发散场
nondivergent vorticity equation 无散量旋度方程
non-drainable superheater 非疏水型过热器
non-drawout relay 固定安装继电器，非抽出式继电器
non-drive end shield 非驱动端端盖
non-drive end 非传动端
non-ductile fracture 无塑性破坏
non-dusting brush 耐磨电刷
non-earthed 不接地的
non-eddying flow 无涡流，无旋流
noneffective lime 失效石灰
noneffective storage 死库容，无效库容
noneffective 无效的，不起作用的
non-elastic closure 非弹性闭合
non-elastic cross section 非弹性截面
non-elastic interaction 非弹性相互作用
non-elasticity 非弹性
non-elastic 非弹性
non-electric 非电的
nonelementary reaction 非基元反应
non-emergency supplied 非应急电源
non-emergency 非应急的
nonenergy-limiting transformer 非限制性恒压变压器
non-engineered safeguard feature 非设计的安全特性
nonequality gate “异”门
non-equilibrium conditions 不平衡状态，不平衡条件
non-equilibrium diffusion equation 非平衡扩散方程
non-equilibrium distribution 不平衡分布
non-equilibrium effect 不平衡效应
non-equilibrium environment 非平衡条件
non-equilibrium flow 不平衡流动，非平衡流
non-equilibrium MHD generator 非平衡磁流体发电机
non-equilibrium process 不平衡过程
non-equilibrium pumping test 不平衡抽水试验
non-equilibrium state 非平衡状态
non-equilibrium thermodynamics 不平衡过程热力学
non-equilibrium 不平衡
non-equiproportional hydraulic misadjustment 不等比水力失调
nonequivalent circuit 非等效电路
nonequivalent 非等效的
non-erasable optical disk 不可擦光盘
non-erasable storage 只读存储器，不可擦存储器，固定存储器
non-erasable 只读的，不可擦除的，固定的
nonerodible channel 不冲刷河槽
nonerodible 不冲刷的
noneroding velocity 不冲刷流速
noneroding 不冲刷的
non-erosive 非侵蚀性的
non-ESF auxiliaries 非设计的安全特性辅助系统
nonessential auxiliaries 非主要的辅助系统
nonessential service chilled water system 非反应堆级冷冻水系统
nonessential variable 不重要的变量
non-excited synchronous motor 无励磁绕组的同步电机，反应式同步电动机
non-exclusive license 非独占许可证
nonexecutable statement 非执行语句
non-executive directors 非执行董事
non-existence 不存在
non-expansive soil 非膨胀性土
non-explosion-proof motor 非防爆型电动机
nonexponential 非指数的
non-exposed installation 非裸露安装
non-extraction operation 不抽气运行
nonextraction throttle flow 无抽汽时的汽轮机入口蒸汽量
nonex 铅硼玻璃
nonfatal error 非致命错误，非严重错误
non-feasible state 不可行状态
non-ferrous alloy 非铁合金
non-ferrous magnetic 非铁磁性的
non-ferrous metal industry 有色金属工业
non-ferrous metal 有色金属，非铁金属
non-ferrous 有色金属的，非铁的
non-fertile material 不可转换材料
non-fertile reflector 非转换材料反射层
non-filtrable residue 悬浮残渣
non-fireproof construction 非防火建筑
non-firm power 备用功率，特殊功率
non-fissile capture 非易裂变材料俘获，非裂变俘获
non-fissile irradiation 非易裂变材料辐照

non-fissile material 不易裂变材料，非易裂变材料
non-fissionable component 不可裂变组分
non-fissionable material 非可裂变材料
non-fissionable 不可分裂的，无裂变的
non-fission absorption 非裂变吸收，无裂变吸收
non-fission capture 非裂变俘获
non-fission interaction 无裂变相互作用
non-fission neutron absorption 无裂变中子吸收，无裂变吸收中子，中子辐射俘获
non-fixed contamination 非固定污染
non-flameproof enclosure 不防爆机壳，不防爆机座
nonflammable 不易燃的，不可燃的，非自燃的，难燃的
non-flood season 非洪水季节，非洪水期
nonflow boiling 不流动沸腾
nonflow system 不流动系统
non-fossil energy 非化石能源
non-foundation rail weigher 无基坑轨道衡
non-free flowing material 非自流物料
non-free flowing 非自由流动
non-freezing stream 不冻河流
non-frequency dependent damper 非频率阻尼器
nonfrontal precipitation 非锋面降水
non-fuel-bearing assembly 非燃料支承组件
non-fuel 非燃料，不包括燃料
non-fulfillment 不履行
non-functional 无功能的，不作用的，非函数的
nonfuse breaker 无熔断器断路器
nonfusible 不熔的
non-galvanized pipe 非镀锌钢管
non-generating period 非发电时期
nongeostrophic flow 非地转流动
nongovernmental institution 非政府机构
nongovernmental investment 社会投资
non-graded sediment 不均粒沉积，非级配沉积物
non-grounded 非接地的，不接地的
non-ground neutral system 中性点不接地系统
non-hangup base 立即接通制
non-harmonic transformer 无谐波变压器
non-hazardous 无危险的，安全的
non-heaving soil 非冻胀土，未隆起土
non-hierarchical relationship 非层级关系
non-homeing tuning system 不复位的调谐系统
non-homing switch 不归位开关，自锁开关
non-homogeneity 不均一性，不同性质，非均匀性
non-homogeneous equation 非齐次方程
non-homogeneous expansion 膨胀不均
non-homogeneous fissured rock 非均匀裂隙岩石
non-homogeneous fluidization 不均匀流态化
non-homogeneous soil 非均质土
non-homogeneous 非均匀的，不均匀的，不同质的，混杂的，非齐次的，不均衡的，非稳态的，非均质的
non-homogenous boundary condition 非齐次边界条件
non-homogenous rock 非匀质岩
nonhumic substance 非腐殖物质
nonhydrogenous shield 无氢屏蔽体，非氢屏蔽体
non-hydrostatic consolidation 非静水压力固结
nonhygroscopic solid 非吸湿性固体
nonhysteretic 无滞后的
nonideal flow 非理想流
nonignitable 不着火的，不可燃的，耐火的
nonignition 不着火，耐火性
non-imaging collector 非成像集热器
nonimpact printer 非击打式打印机
nonimpregnated 未浸透的，未饱和的
non-incentive equipment 非可燃性设备
nonindicating controller 无刻度控制器
nonindicating 无刻度的
non-inductive capacitor 无感电容器
non-inductive circuit 无感电路
non-inductive coil 无感线圈
non-inductive load 无感负载
non-inductive resistor 无感电阻器
non-inductive shunt 无感分流器
non-inductive surge 非感应性电涌，无感冲击电压
non-inductive winding 无感线圈，无感绕组
non-inductive 无感应的，无感的
noninertial coordinate system 非惯性坐标系
noninjurious 无害的
non-instrumented cladding rig 无测量装置的包壳（辐照）试验装置
non-integral connection 机械连接
non-integral quantity 非整数量
non-integral slot winding 分数槽绕组，非整数槽绕组
non-integrated concept 非一体化概念
non-intended product 非预期产品
noninteracting conditions 不相关条件，自治条件，不相关联的条件
noninteracting control system 无互作用控制系统
noninteracting control 无互作用控制，非相关控制，不相关控制，自治调节
noninteracting 不相互影响的，自治的，不相关联的
non-interchangeable fuse 不能互换的熔断器
non-intercooled cycle 无中间冷却循环
non-interference 不相互干扰，不干涉
noninterruptable instruction 不可中断的指令
noninverting amplifier 非倒向放大器，非逆转放大器
non-ionic compound 非离子化合物
non-ionic polymer 不电离聚合物
nonirradiated 未受辐照的
non-irrigable land 不可灌土地
non-isentropic flow 非等熵流，非等熵流动
non-isentropic 非等熵的
non-isolated analogue input 非隔离模拟量输入
non-isothermal flow 非等温流，非等温射流
non-isothermal pressure drop 非等温压降
non-isothermal reactor 非等温反应器
non-isothermal 非等温的
non-isotopic tracer 非同位素示踪
non-isotropic materials 各向异性物质
non-isotropic medium 非均质介质
non-isotropic turbulence 非均匀性湍流

non-isotropic 各向异性的，非各向同性的
non-isotropy 各向异性
non-iterative 非迭代的
nonius zero 游标零点
nonius 游标，游标尺
nonkinetic analysis 非动态分析法
non-latching relay 自动复位继电器
non-lead covered cable 无铅包电缆
non-leakage factor 不泄漏因子
non-leakage probability 不泄漏概率
non-leakage 不泄漏
non-licensed operator 无执照操纵员
non-linear acoustic effect 非线性声效应
non-linear acoustics 非线性声学
non-linear aerodynamic force 非线性气动力
non-linear aerodynamics 非线性空气动力学
non-linear analysis 非线性分析
non-linear autonomous system 非线性独立系统，非线性自治系统
non-linear bridge circuit 非线性桥接电路
non-linear bridge stabilizer 非线性电桥稳压器
non-linear capacitance 非线性电容
non-linear characteristic 非线性特性
non-linear circuit 非线性电路
non-linear computation 非线性计算
non-linear conductance 非线性电导
non-linear conductor 非线性导体
non-linear controller 非线性控制器，非线性调节器
non-linear control system 非线性控制系统
non-linear coordinate transformation 非线性坐标变换
non-linear detector 非线性检测器
non-linear differential equations 非线性微分方程（组）
non-linear diffraction 非线性衍射
non-linear distortion 非线性失真，非线性畸变
non-linear effect 非线性影响，非线性效应
non-linear elasticity 非线性弹性
non-linear element 非线性元件
non-linear equation 非线性方程
nonlinear extraction 非线性提取
non-linear feedback control system 非线性反馈控制系统
non-linear feedback relay servomechanism 非线性反馈继电伺服机构
non-linear feedback 非线性反馈
non-linear field theory 非线性场论
non-linear friction 非线性摩擦
non-linear function 非线性函数
non-linear heat conduction 非线性热传导
non-linear interaction of noisy waves 噪声波的非线性相互作用
non-linear interaction 非线性相互作用
non-linear interpolator 非线性内插器
non-linearity of capacitor 电容器的非线性
non-linearity 非线性度，非线性
non-linear load 非线性负荷
non-linear modulation 非线性调制
non-linear network 非线性网络
non-linear optimizing 非线性最优化
non-linear oscillation equation 非线性振荡方程
non-linear oscillation 非线性振荡，非线性振动
non-linear phase shift 非线性相移
non-linear pneumatic output 非线性气动输出
non-linear potential-divider 非线性分压器
non-linear potentiometer 非线性电势计，非线性电位计
non-linear pricing 非线性定价
non-linear process 非线性过程
non-linear programming 非线性规划
non-linear regression analysis 非线性回归分析
non-linear resistance 非线性电阻
non-linear-resistor-type arrester 非线性电阻器型避雷器
non-linear resistor 非线性电阻器
non-linear resonance 非线性谐振，非线性共振
non-linear response 非线性响应，非线性特性，非线性响应曲线
non-linear sampled-data system 非线性采样数据系统
non-linear saturation 非线性饱和
non-linear scale 非线性标度
non-linear series resistor 非线性串联电阻
non-linear servomechanism 非线性伺服机构
non-linear servomotor 非线性伺服电动机
non-linear simultaneous equations 非线性联立方程组
non-linear system 非线性系统
non-linear theory 非线性理论
non-linear transformer 非线性变压器
non-linear two-port 非线性双口的，非线性两端口
non-linear vibration 非线性振动
non-linear 非线性的
non-line optimum theory 非线性最优化理论
non-liner resistor 非线性电阻
non-load-bearing structure 非承重结构
non-load current 空载电流
non-load curve 空载曲线
non-loaded cable 无负载电缆，非加感电缆
non-loaded 空载的，无载的，空负荷的，空转的，非加感的
non-load 空载负荷，空载，无载
non-locking key 非锁定按钮，自动还原电键，自复按钮
non-locking press-button 非锁定按钮，自动还原按钮，自复按钮
non-locking pushbutton 不闭锁的按钮
non-locking relay 非锁定继电器
non-loss circuit 无损耗电路
non-loss condenser 无损耗电容器
non-loss line 无损耗线路
non-lubricated piston compressor 无润滑的活塞式压缩机
nonluminous flame 不发光火焰
nonluminous gas 不发光气体，不发光烟气
nonluminous 不发光的
non-magnetic alloy 非磁性合金
non-magnetic body 非磁体
non-magnetic clamp 无磁线夹
non-magnetic core 无磁铁芯

non-magnetic flow 无磁流
non-magnetic material 非磁性材料
non-magnetic pressure plate 非磁性齿压片
non-magnetic pressure ring 非磁性压圈
non-magnetic ring 非磁性环
non-magnetic screen 非磁性屏蔽，非磁性罩，
non-magnetic shim 非磁性垫片
non-magnetic substance 非磁性材料，非磁性物质
non-magnetic 无磁的，无磁性的
non-matched data 不匹配数据
non-mathematical program 非数学程序
non-mathematical 非数学的
non-measuring relay 非测量用继电器
non-measuring 非测量的
non-mechanical printer 非机械打印机，非机械的印刷装置
non-mechanical 非机械的
nonmetal air channel 非金属风管
non-metallic element 非金属元素
non-metallic gasket 非金属垫片
non-metallic impurity 非金属杂质
non-metallic inclusion 非金属夹杂物
non-metallic material 非金属材料
non-metallic resistor 非金属电阻器
non-metallic sheathed cable 非金属铠装电缆，非金属包皮电缆
non-metallics 非金属夹杂物
non-metallic 非金属的
non-metal 非金属
non-metering offtake regulator 无量水设备的分水节制闸
non-metering 无读数的，不计数的
nonmicrophonic 无颤噪效应的
non-minimum-phase system 非最小相位系统
non-navigable dam 不通航坝
non-navigable river 不通航河流
non-navigable 不通航的
non-negative number 非负数
non-negative 非负的
non-negotiable bill of lading 不可转让的提单，非流通提单
non-negotiable B/L 非流通提单，不可转让提单，提单抄本
non-negotiable instrument 不可转让的票据
non-negotiable letter of credit 不可转让的信用证
non-negotiable 不可转让的
non-Newton flow 非牛顿流
non-Newtonian flow 非牛顿流
non-Newtonian fluid 非牛顿流体
non-Newtonian viscosity 非牛顿黏性
non-noble-gas nuclide 非惰性气体核素
non-normal distribution 非正态分布
non-normal 异常的，非正规的，不垂直的
non-nuclear component 非核部件
non-nuclear effect 非核效应
non-nuclear environment 一般环境条件，非核环境
non-nuclear hot operation 非核热态运行
non-nuclear instrumentation 非核测量仪表
non-nuclear properties 非核特性
non-nuclear safety 非核安全级，非核安全
non-numerical character 非数字字符
non-numerical data processing 非数字数据处理
non-numeric character 非数字记号
non-numeric information 非数字信息
non-numeric 非数值的，非数字的
non-occupational exposure 非职业性照射
non-occupational noise exposure 非职业性噪声暴露
non-ohmic resistor 非欧姆电阻器，非线性电阻器
non-ohmic 非欧姆的
N/O(no order) 无指示，无定单，不指定人，无抬头
non-operating ampere turn 不吸安匝
non-operating pad 非工作瓦块
non-operating state 不工作状态，停用状态
non-operating time 不工作时间
non-operational cost 非业务费用，非运行费用
non-operational load 非运行负荷
non-operation expenditure 营业外支出
non-operation 非工作态
non-operative cable credit 不可凭使用电报信用证，暂不生效的电开信用证
non-operative part 导入部分，（合同的）序言部分
non-operative 不动的，不工作的，无效的
nonoptimal value 非最佳值
nonoptimal 非最佳的
non-orientable 不能定向的
non-oriented 非取向的
NO(normally open) 常开的
non-oscillating 不振荡的，不摆动的
non-oscillatory circuit 无振荡电路
non-oscillatory control response 非振荡控制响应
non-oscillatory divergence 非振荡发散，静力发散
non-oscillatory transformer 无振荡变压器
non-oscillatory 非振荡的
NO(not operational) 不工作的
non-overflow dam 不溢流坝
non-overflow groin 不过水丁坝，非溢流丁坝
non-overflow section 非溢流（坝）段
non-overflow spur dike 不过水丁坝
non-overlapping energy spectra 不相重叠的能谱
non-overlapping winding 非重叠绕组
non-overlapping 不相重叠的
nonoxidizable 不可氧化的
non-payment 拒不付款，未付款，不支付
non-perennial canal 非常年使用渠道
non-perfect fluid 非理想流体，实际流体，不完全流体
non-periodical repair 不定期维修
non-periodic excitation 非周期性激振
non-periodic function 非周期函数
non-periodic overhaul 不定期检修
non-periodic source 非周期性电源
non-periodic variation 非周期性变化
non-periodic wave 非周期波
non-periodic 无周期的，非周期性的，非振荡的
non-permanent storage 非永久贮存【质保】

non-phantom circuit 非幻象电路
non phase related harmonics 相别无关谐波
non-physical loss 无形的损失
non-planar network 非平面网络
non-plastic silt 无塑性粉土
non-plastic soil 非塑性土，非塑性土壤
non-point pollution 非点源污染
non-point source pollution 不定源污染
non-polarity 无极性
non-polarized relay 非极化继电器，无极化继电器，中和继电器
non-polarized 非极化的
non-polar 非极性的，无极的
non-polluting coating 无污染涂层
non-polluting or pollution-reducing processes 无污染或少污染工艺
non-porous absorber 无孔吸热器
non-porous catalyst 无孔隙催化剂
non-porous oxide film 密实氧化膜
non-porous 无孔的
non-positive 非正的，负的
non-potable water 非饮用水
non-pressure drainage 无压排水
non-pressure treatment 常压处理，无压处理
non-pressurized reactor 不加压反应堆
non-priced factors 非价格因素
non primitive code 非本原码，非原始码
nonprint code 禁止打印码
non-priority interrupt 非优先中断
non-priority objective 非优先目标
non-prismatic channel 变截面河槽
nonprocedural language 非过程语言
nonproduction reactor fuel 非生产堆燃料
non-production staff 非生产人员
non-production wage 非生产工资
nonproductive capture 非生产性俘获
nonproductive construction investment 非生产性建设投资
nonproductive fixed assets 非生产性固定资产
nonproductive investment 非生产性投资
nonproductive operation 非生产操作，非有效操作，辅助操作
nonproductive 不生产的，非生产的，非生产性的，非有效的
nonprogrammed halt 非程序停机
Non-Proliferation Treaty 核不扩散条约
non-propagating crack 不扩展裂纹，停止开裂
non-propagating fatigue crack 不扩展疲劳裂缝
non-propagating 不扩散的
non-protected machine 非防护型电机
non-protective magnetite 非保护性磁性氧化铁
non-quadded cable 对绞电缆
non-radiation decay 无辐射衰变
non-radioactive sodium loop 非放射性钠回路
non-radioactive 非放射性的
non-random 非随机的
non-reactive circuit 无电抗电路
non-reactive filter 无电抗滤波器
non-reactive impedance 无抗阻抗
non-reactive load 非电抗负载，无电抗负荷
non-reactive power 无电抗功率，有功功率
non-reactive resistance 无电抗电阻，纯电阻
non-reactive silica 非活性氧化硅，非反应性氧化硅
non-reactive 无电抗的，非电抗的
non-reciprocal network 非互易网络
non-reciprocal observation 单向观测
nonrecourse financing 无追索融资
nonrecourse 无追索权
non-recoverable storage 不可回收贮存，废弃贮存，永久贮存
non-recuperative system 无回热系统
non-recurrent meeting 非经常会议
non-recurrent wave 非周期波
non-redundant 非多重的，非冗余的
non-reflecting 非反射的
non-reflection attenuation 固有衰减，非反射衰减
nonreflective 无反射的
nonrefundable payment 不予退还的款项
non-regenerative heat exchanger 非再生热交换器
non-regenerative reactor 非再生反应堆
non-regulated area 非控制区
non-regulated discharge 未调节流量
non-regulated transformer 不可调变压器
non-reheating cycle 无再热循环
non-reheat regenerative steam cycle 无再热再生蒸汽循环
non-reheat turbine 非再热式汽轮机
non-reheat unit 非再热机组
non-reimbursable assistance 无偿援助
non-reinforced concrete 素混凝土，无筋混凝土
non-reinforced spread foundation 无筋扩展基础
non-relevant failure 与核安全无关故障
non-reliance clause 完整合约条款
non-renewable buffering 不可再生缓冲
non-renewable energy source 不可再生能源
non-renewable fuse 不能更新的熔断器
non-renewable natural resources 不能更新的自然资源
non-renewable resources 不可恢复的资源，不可更新资源，非再生资源
non-renewable source of energy 非再生能源
non-repetitive 非重复的
nonreproducing code 非复制代码
non-required time 无需求时间
non-resettable counter 加法计算器，不重调计算器
nonresidential area 非居民区
non-residential district 非居民区
nonresident routine 非常驻程序
non-resistive load 非电阻负载
non-resonance absorber 非共振吸收体
non-resonant circuit 非谐振电路
non-resonanting transformer 非谐振变压器，防冲击变压器
non-resonant response 非共振响应
non-resonant standing wave system 非谐振驻波系统
non-resonant 非调谐的，非谐振的
non-respirable 不可吸入的
non-restoring 不复原，不恢复

non-retention 非自保持
nonretentive alloy 软磁性合金
nonretentive material 软磁性材料
nonretentive 非保持
non-returnable container 一次性包装
non-returnable packing 不返回包装
non-return damper （通风）止回阀
non-return flap 逆止门，止回瓣
non-return-flow water tunnel 开路式水洞
non-return-flow wind tunnel 开路式风洞
non-return foot valve 止回底阀
non-return recording 数字间无间隔记录
non-return-to-zero change 不归零制变化
non-return-to-zero magnetic recording 不归零磁性记录
non-return-to-zero mark 不归零标记
non-return-to-zero recording 不归零制记录
non-return-to-zero 不归零，不归零的
non-return trap 止回汽水阀
non-return valve 逆止阀，逆止门，止回阀，单向阀
non-return 单向的，不返回的，止回的，无间隔的
non-reverse assembly 单转向组件，不可逆装配
non-reverse coupling 单转向联轴节
non-reversibility 不可逆性
non-reversible control 不可逆控制
non-reversible machine 不可逆式电机
non-reversible reaction 不可逆反应
non-reversing starter 不倒转的启动器
nonrevolving credit 非循环信用证，非周转信用证
nonrigid connection 非刚性节点
nonrigid coupling device 弹性连接装置
nonrigid joint 非刚性节点
nonrigid pavement 非刚性路面
nonrigid plastics 非刚性塑料
nonrigid system 非刚性体系
nonrigid 非刚性的，弹性的
non-rising stem 非升杆式阀杆【暗杆】，非抬升式阀杆
non-rotary valve 非旋转式阀
non-rotating cable 不旋转的电缆
non-rotating part 不旋转部件
non-rotating 不旋转的，非转动的
non-rotational 无旋的
non-routine operation 非常规操作
non-rusting solution 防锈溶剂
non-rust steel 不锈钢
non-safeguard 非保证安全的
non-safety function 非安全功能
non-safety grade 非安全等级的
non-safety start-up feedwater system 非安全启动给水系统
non-salient pole alternator 隐极同步发电机
non-salient pole generator 隐极式发电机，隐极发电机
non-salient pole （电机的）隐极
non-sampling 非采样的
non-saturable reactor 不饱和电抗器
non-saturated switching circuit 非饱和开关电路
non-saturated 非饱和的，不饱和的
non-scale-forming condition 无垢工况
non-scanning antenna 固定天线，不转动天线
non-scattering gas 非分散性气体
non-scheduled maintenance 不定期维修，非正规维修，计划外维修
non-scouring velocity 不冲流速
non-segregated phase bus duct 共相母线
non-segregated phase bus 不分相（金属铠装）母线
non-segregated phase enclosed bus duct 共箱封闭母线
non-seismic region 不震区，非地震区，无震区
non-selective carryover 非选择性携带
non-selective surface 非选择性表面
non-selective 无选择性的，不选择的
non-selfexcited discharge 非自持放电，非自励放电
non-self-regulating canal 非自动调节渠进
non-self-restoring insulation 非自恢复绝缘
non-self revealing fault 非自显故障
non-separated flow 非分离流
non-separation agreement 不可分割协议书
nonsequential stochastic programming 无顺序随机规划
non-settleable matter 不沉降物质
non-settleable solids 非沉降性固体颗粒
non-set-up job 非准备作业
nonshared control unit 非公用控制单元
non-shatterable glass 不碎玻璃
nonshorting 非短路的，无短路的
non-shrinking cement 不收缩水泥
non-shrinking concrete 不收缩混凝土
non-shrinking 非收缩性，不收缩的
non-sifting tuyere 无筛风嘴
nonsignificant zero 无意义的零
nonsignificant 无意义的
non-silting velocity 不淤流速
nonsimilar solutions 非相似性解
non-sine-wave 非正弦波
non-singular matrix 非奇异矩阵，满秩矩阵
non-sinusoidal current 非正弦电流
non-sinusoidal curve 非正弦曲线
non-sinusoidal voltage 非正弦电压
non-sinusoidal wave 非正弦波
non-sinusoidal 非正弦的
non-skid 防滑装置，防滑的
non-slam check valve 无撞击声止回阀
non-slewing crane 非回转式吊机
non-slip emery insert 金刚砂防滑条
non-slip flooring 防滑地板
non-slip floor 防滑地面
non-slip flow 无滑移流
non-slip insert 防滑条
non-slip nosing 楼梯防滑踏步条
non-slip protection 防滑护面
non-slip strip 防滑条
non-slip 不滑动，防滑的
non-softenable ion 不易软化的离子
nonsoluble 不溶解的
non-sparking motor 无火花马达
non-sparking 无火花的

non-spherical particle 非球形颗粒
non-spinning reserve 非旋转备用
non-square matrix 非方形矩阵
non-square pulse 非矩形脉冲
non-staining 不污染的
non-standard motor 非标准型电动机
non-standard propagation 反常传播
non-standard scheduled outage 非标准的计划停运
non-standard size 非标准尺寸
non-standard 非标准的
non-static 不产生干扰，无静电荷的
non-stationary aerodynamic derivative 非定常气动导数
non-stationary behavior 非稳定工况，非稳定状态
non-stationary flow 不稳定流，非恒定流，非稳定流动
non-stationary method 不定常法
non-stationary motion 非常定运动
non-stationary random walk model 不稳定随机走动模型
non-stationary 不稳定的，非稳定的
non-steady aerodynamics 非定常空气动力学
non-steady current 不稳定流
non-steady flow 非稳定流，不稳定流，非定常流，变量流
non-steady lifting surface theory 非定常升力面理论
non-steady motion 非定常运动
non-steady state model 非稳态模型
non-steady state 瞬态，暂态，非稳态
non-steady supercritical flow 不稳定超临界流
non-steady 不稳定的
nonsteaming economizer 非沸腾式省煤器，不沸腾式省煤器
nonsticking 灵活的，无黏附的，不卡住的
non-stick 不黏结的，不黏附的
non-stochastic effect 非随机效应
non-stoichiometric 非化学比的，非化学计量的
non-stop concreting 混凝土连续浇筑
non-stop switch 不停开关
non-stratified rock 非成层岩石
non-streamlined 非流线型的
non-stressed bar reinforcement 无应力钢筋
non-stress meter 无应力计
non-structural element 非结构部件
non-structural module 非结构性模块
non-sweating 不渗漏的，不结露的
non-swelling 不膨胀的
non-symmetrical adjustment 不对称调整
non-symmetrical network 不对称网络
non-symmetrical winding 不对称绕组
non-symmetrical 不对称的，非对称的
nonsymmetric bending 不对称弯曲
non-synchronizing 异步的，非同步的，不同期的
non-synchronous computer 异步计算机
non-synchronous conduction 非同步传导
non-synchronous motor 异步电动机
non-synchronous 异步的，非同步的，不同期的
non-systematic code 非系统码，非系统化代码
non-systematic error 偶然误差，非系统性误差
non-systematic 非系统化的
non-tapered aerofoil 等弦翼面
non-technical 非技术性的
non-tectonic fault 非构造断层
nontension joint 无张力接头
nonterminal symbol 非终结符号
non-thermal distribution 非热力分布
non-thermal energy 非热能
non-thermal neutron 非热能中子
non-thermal reactor 非热反应堆
non-thermal region 非热能区
non-thermal 非热能的，非热力的
non-threshold logic 无阈逻辑，非阈值逻辑
non-thrust quay 无侧推力岸壁
non-tidal compartment 无潮区
non-tidal river 无潮汐河流
non-tidal section 无潮段
non-tight door 非密封门
nontilting drum mixer 非倾斜鼓式拌和机
non-tilt mixer 非颠倒式拌和机
non-time-delay 无时间延迟
non-transferable L/C 不可转让的信用证
non-transferable letter of credit 不可转让的信用证
non-transferable license contract 不可转让许可合同
non-transferable 不可转让的
nontransferred arc 非转移弧
nontransferred plasma arc 不转移的等离子弧
nontreatment plant 非处理工厂
non-tunable 不调谐的，不可调的
No. =number 编号，数字
non-uniform beam 变截面梁
non-uniform bridge 非均匀电桥
non-uniform burn-up 不均匀燃耗
non-uniform column 变截面柱
non-uniform combustion 不均匀燃烧
non-uniform corrosion 不均匀腐蚀
non-uniform deformation 不均匀变形
non-uniform dielectric 非均匀介质
non-uniform distribution factor 非均匀分布系数
nonuniform distribution 分布不均匀
non-uniform dose distribution 非均匀剂量分布
non-uniform dose 非均匀剂量
non-uniform field 非均匀场
non-uniform flow 非均匀流，变速流，非等速流，不均匀流，紊流
non-uniform gap 不均匀气隙
non-uniform heat flux 非均匀热流密度
nonuniformity coefficient 不均匀系数
nonuniformity 不均一性，不均匀性，异质性，不一致性
non-uniform lift 不均匀分布升力
non-uniform magnetic field 不均磁场
non-uniform magnetization 不均匀磁化
non-uniform medium 不均匀介质
non-uniform motion 不均匀运动
non-uniform pressure 不均匀压力
non-uniform scale 不等分标尺
non-uniform settlement 不均匀沉降

non-uniform shrinkage 不均匀收缩
non-uniform soil 不均质土
non-uniform topography 不均匀地形
non-uniform 不均匀的，非均质的，非均匀的
non-unit system 非单元系统
nonuple 九重的，九倍的，九个一组成的
non-urgent alarm 非紧急报警
nonvariant 不变的，恒定的
nonventilated machine 无风扇电动机
nonventilated tray 不通风的托架
non-vibrating 不振动的
non-virgin neutron 非原生中子，经碰撞的中子
nonviscous flow 非黏滞流
nonviscous fluid 非黏滞流体
nonviscous instability 非黏性梁稳定性
nonviscous liquid 非黏滞液体
nonviscous 非黏性的，无黏性的，不黏的
non-vacuum solar receiver tube 非真空太阳能吸热管
nonvoided concrete beam 实心混凝土梁
non-volatile matter 不挥发物
non-volatile memory 永久性存储器件，非易失性存储器
nonvolatile oil 非挥发性油
non-volatile random-access memory 永久性随机存储器
non-volatile residue 非挥发性残渣
non-volatile storage 非易失性存储器，永久存储器
non-volatile store 永久存储器
non-volatile 不挥发的，永久的，非易失（存储器）
non-volatility 不挥发性
nonvortex 无涡流的，无旋的，无旋涡的
non-waiver 非违约弃权（条款）
non-waste technology 无废技术
non-water tight 非水密的，透水的，不隔水的
non-weeping 不扫除的
nonweighted code 非加权码，无权码
non-wettable film 不可润湿膜
non-wettable 不浸润的，不可浸润的
nonwetting phase 非湿润相
non-wire-wound resistor 非线绕电阻器
non-working flank 非工作齿面
nonzero digit 非零位，非零数
nonzero load 非零荷载
nonzero value 非零值
nonzero 非零
nook 角落，转角处
no-operater control in workshop 车间无人值班控制
no-operation instruction 空操作指令
no-operation 停止操作指令，无操作
NOP＝no-operation 无操作，停止操作指令
no pollution 无污染
no powder-filled cartridge fuse 无填料封闭管式熔断器
no-pressure relay valve 无压继动阀
no-raster 无光栅，无扫描
NOR circuit “或非”电路
NOR element “或非”元件
no responsible 无责任
NOR gate “或非”门
noria 戽斗车
no risk after discharge 不保卸货后的风险，卸货后解除责任
normal acceleration 法向加速度
normal aeration 正常掺气
normal air 标准空气
normal alarm 标准报警
normal and upset conditions 正常和翻转条件
normal annual runoff 正常年径流
normal area of screen 标称筛分面积
normal atmosphere 标准大气，常态大气，正常大气
normal axis 法线轴，垂直轴
normal balance 正常平衡
normal band 基带
normal base line 正常基线，正常基流
normal beam testing 垂直探伤
normal bend 法向弯管，法线弯管，直角弯头
normal binary 标准二进制，普通二进制
normal braking system 正常制动系统
normal bridging 正常桥接
normal butane 正丁烷
normal but infrequent operations 正常但不常进行的作业
normal calomel electrode 标准甘汞电极
normal capacity 正常容量，额定容量，额定功率，标准容量
normal carryover 正常携带
normal cathode fall 正常阴极降
normal cell 标准电池
normal change 正常变化
normal cladding thickness 包壳公称厚度
normal clearance 正常清除
normal close contact 常闭接点
normal close 常闭
normal coldest month 累年最冷月
normal cold startup procedure 正常冷却启动程序
normal combustion 正常燃烧
normal component of force 法向分力
normal component 法向分量
normal concentration 当量浓度
normal concrete 普通混凝土
normal condition 标准状况，标准条件，正常条件，正常工况，正常状态，常态
normal consistency 标准稠度，正常稠度
normal consolidated clay 正常固结黏土
normal consolidation line 正常固结线
normal consolidation soil 正常固结土
normal consolidation 正常固结
normal contact 正常触点，静接点
normal coordinates 法向坐标
normal correlation function 正态相关函数
normal correlation 正态相关
normal critical slope 正常临界比降
normal cross-section 正常截面，正断面
normal current 正常电流
normal curve of error 正态误差曲线
normal curve 正态曲线
normal cut 正切割

normal daily mean 正常日平均值
normal de-energized 正常失电
normal density 正常密度
normal depletion curve 正常退水曲线，正常亏水曲线
normal depreciation 正常折旧
normal depth 正常深度，正常水深
normal derivative 法向导数
normal deviation 常态离差
normal dimension 标定尺寸
normal direction of rotation 正常转向
normal direction 法线方向，正常方向
normal direct solar irradiance 法向直（接日）射辐照度【En】
normal discharge curve 正常流量曲线
normal discharge 额定流量，正常放电，正常流量
normal disconnected mode 正常断开方式
normal displacement 法向位移，正常位移
normal distribution curve 正态分布曲线
normal distribution function 正态分布函数
normal distribution 正规分布，正态分布，正常分布，常态分布
normal divide 正常分水岭
normal dropout voltage 正常释放电压
normal duty 正常工作状态
normal earthquake 浅震
normal electrode 标准电极
normal emittance 法向发射率，法向黑度
normal energized 正常带电
normal equation 法方程式
normal erosion 正常冲蚀，常态侵蚀
normal error integral 正规误差积分
normal excitation 正常激磁，正常励磁
normal fall method 正常落差法
normal fall of water level 正常消落深度
normal fault 正常断层，正断层
normal feedwater supply 主给水
normal feedwater 正常给水
normal flow 正常流量，正常水流
normal fluidization 正常流态化
normal fold 正褶皱
normal force 法向力
normal fracture 正断，正裂
normal frequency 额定频率，正常频率
normal full load 正常满载，正常全负荷
normal gravity 正常重力
normal head 正常水头
normal high water level 正常高水位
normal hottest month 正常最热月
normal hydrogen electrode 标准氢电极
normal hydrostatic pressure surface 正常静水压力水面
normal illumination 正常照明
normal incidence calibrating method 准直标定法
normal incidence pyrheliometer 直接辐射计，直射日射强度计
normal incidence 法向入射
normal induction 正常磁感应
normal input cause 正常输入条件
normal inspection 正常检查
normal insulation 正常绝缘
normality 标准状态，标准态，常态，正规性，正常
normalization condition 归一化条件
normalization constant 归一化常数
normalization factor 归一化因子
normalization of fission source 裂变源归一化
normalization point 归一化点
normalization 归一化，标准化，规格化，正常化，正规化，规范化，正火
normalized admittance 归一化导纳，标准化导纳
normalized apparent impedance plane 归一化视在阻抗平面
normalized autocorrelation function 标准化自相关函数
normalized auto-power spectral density 归一化自功率谱密度
normalized axial power distribution 归一化轴向功率分布
normalized coefficient 归一化系数
normalized correlation coefficient 归一化相关系数
normalized cross-power spectral density 归一化互功率谱密度
normalized curve 标准化曲线，正常曲线
normalized distance 归一化距离
normalized dose 归一化剂量
normalized filter 标准化滤波器
normalized flow distribution 归一化流量分配
normalized force 标准化作用力
normalized frequency 归一化频率，标准化频率
normalized impedance 归一化阻抗，标准化阻抗
normalized output 归一化输出
normalized power distribution 归一化功率分布
normalized power spectral density 归一化功率谱密度
normalized power spectrum function 归一化功率谱函数
normalized reactance 归一化电抗
normalized resistance 归一化电阻
normalized response 标准响应
normalized root-mean-square value 归一化均方根值
normalized spectral density 标准化谱密度
normalized spectrum 归一化谱
normalized steel 正火钢
normalized unit 归一化单位，标准单位
normalized 标准化的
normalizer 规格化装置，标准化部件
normalize 规范化，标准化，归一化，校正，正规化，正火
normalizing effect 正常效应【冶金】
normalizing factor 归一化因子
normal law of error 常态误差定律
normal lift 正常提升高度
normal lighting box 工作照明箱
normal lighting 正常照明
normal line 法线，正垂线
normal loading condition 正常荷载工况
normal load 额定负荷，正常负载，正常荷载
normal low water level 正常低水位

normally closed auxiliary contact 常闭辅助触点
normally closed contact 常闭触点
normally closed control element 常闭控制元件
normally closed interlock 休止辅助触头，常闭触头，常闭联锁装置
normally closed pneumatic diaphragm valve 常闭型气动隔膜阀
normally closed valve 常闭阀，常闭阀门
normally closed 常闭的，静合的
normally close 正常关闭，常闭
normally consolidated clay 正常固结黏土
normally consolidated soil 正常压密土
normally distributed random variable 正态分布随机变量
normally open contact 常开触点
normally open control element 常开控制元件
normally opened contact 常开触点
normally opened pneumatic diaphragm valve 常开型气动隔膜阀
normally open interlock 常开联锁装置
normally open valve 常开阀
normally open 常开，常开的
normally 正常地，通常地
normal magnetization curve 标准磁化曲线
normal mean level 正常平均料位
normal mode field 简正波场
normal mode interference 正常方式干扰，常模干扰
normal mode rejection ratio 常模抑制比
normal mode rejection 正常模抑制，常模抑制，常模排斥力，正态减弱系数
normal mode voltage 常模电压
normal mode 正常方式，正规模态，串模，常态
normal moisture capacity 正常持水量
normal monthly mean 正常月平均值
normal on-load operation 正常带载操作
normal open contact 常开接点，常开触点
normal operating condensate drain-off connections 经常疏水装置
normal operating conditions 正常工作状态，正常工作条件，正常运行工况
normal operating life 正常工作寿命
normal operating limit 正常运转范围
normal operating loss 正常运行损失
normal operating pressure 正常工作压力
normal operating transient 正常运行暂态
normal operational mode 正常运行方式
normal operation 正常操作，正常运行，正常运算
normal orifice 标准孔板
normal osmosis 法向渗透
normal output 正常出力，额定出力，正常输出
normal overload 正常超载
normal path 正常轨线
normal peneplain 正常准平原
normal permeability 正常磁导率
normal phase 正常相
normal phenomenon 正常现象
normal pin 立销
normal pitch 标准行距，标准距，法向齿距，标准间距
normal plane shock 正冲波
normal plane 垂直面，法面
normal point load 法向集中荷载
normal pool level 正常蓄水位
normal Portland cement 普通硅酸盐水泥，普通波特兰水泥
normal position of telescope 正镜
normal position 正常位置，工作状态
normal potential 标准电势
normal power supply 正常电源
normal power 正常功率
normal pressure and temperature 常温常压，标准温度与压力，额定压力和温度
normal pressure angle 正压力角
normal pressure distribution 法向压力分布
normal pressure gradient 法向压力梯度
normal pressure surface （地下水的）常压面
normal pressure 法向压力，常压，正压，标准（大气）压力
normal price 正常价格
normal probability curve 正态概率曲线
normal probe 直探头
normal quantities 正态量
normal quantity of cooling water supplied to condenser 供凝汽器的正常冷却水量
normal rainfall 正常降雨量
normal rated load 标准额定负荷
normal rated power 额定功率
normal rate filtering 正常滤速
normal rating 正常额定值，额定出力
normal reactor operation 反应堆正常运行
normal recession curve 正常退水曲线，正常亏水曲线
normal reflection 法向反射，垂直反射
normal resistance 正常电阻
normal response 正规响应
normal retirement 正常报废
normal revolution 正常转速，额定转速，正常转数
normal ripple 正常波痕
normal river bed 正常水位河床
normal rock pressure 岩石法向压力
normal running load 正常运行负荷，正常运行荷载
normal running speed 正常运行转速
normal running 正常运行
normal runoff 正常径流
normal saturation curve 正常饱和曲线
normal section azimuth 法截面方位角
normal section 法向截面
normal segregation 正偏析
normal sensibility 标准灵敏度，正常灵敏度
normal sequence of operation 正常操作顺序，正常动作程序
normal service 正常服务
normal shock diffusion 正冲波扩散
normal shock wave 正激波，正冲波
normal shock 正向冲击
normal shutdown for wind turbine 风力机正常关机
normal shutdown situation 正常停机状态
normal shutdown 正常关机

normal size 标定尺寸，标准尺寸
normal slip 正常转差率
normal soil erosion 正常土壤冲蚀
normal solution 当量溶液
normal speed 额定速度
normal stacking 常规堆料法
normal stage 正常水位
normal start-up 正常启动
normal start 正常启动
normal state （电力系统）正常状态，常态，正常条件，标准状态
normal storage water level 正常蓄水位
normal strain 法向应变，正应变
normal stratigraphic sequence 正常层序
normal stress difference 法向应力差
normal stress 法向应力，正压力，正应力，垂向应力
normal surface 法面
normal synchronous speed 额定同步速度，正常同步速度
normal system 正常系统
normal temperature and pressure 常温常压，标准状态
normal temperature of oil in reservoir 油罐内油的正常温度
normal temperature 标准温度，正常温度
normal ternary 标准三进制
normal torque 额定转矩，正常转矩
normal uranium reactor 天然铀反应堆
normal uranium 标准铀，天然铀
normal valence compound 正常价化合物
normal value 正常值，额定值，标准值
normal velocity distribution 正常流速分布，法向速度分布
normal velocity 法向速度
normal vision 正常视觉
normal-voltage rating 正常电压下的额定值
normal-voltage 正常电压
normal water depth 正常水深
normal water level 正常水位，标准水位
normal wear 正常磨损
normal weather 正常气候
normal-weight concrete 常规重量混凝土
normal wind year 正常风年
normal working conditions 正常工作条件，正常工况
normal working hours 正常工作时间
normal working point 标准运行点，正常工作点
normal working pressure 标准工作压力，正常工作压力
normal working 正常工作，正常运行
normal year 平水年，正常年
normal 正常，常态，标准，法线，正交，正常的，正规的，法线的，标准的，额定的
normative document 规范性文件
normative 标准的
normatron 模型计算机
Normes Francaises 法国标准
norm for detailed estimates 预算定额
norm management 定额管理
norm of material consumption 材料消耗定额
normol stress 正应力
norm reducing method 减模法
norm 标准，定额，规格，规范
NOR(notice of readiness) 装卸准备就绪通知书
no rough handling 小心轻放【用于包装外表标志】
northeast trades 东北信风
northeastward 东北方向
north elevation 北立面
northern hemisphere 北半球
northern latitude 北纬
north latitude 北纬
north-light roof 锯齿形屋顶
north light 北极光
north-northeast 东北偏北，北东北
north-northwest 西北偏北，北西北
north west elevation 西北立面图
nose angle of spiral casing 涡壳的包角
nose angle （涡壳的）包角，鼻断夹角
nose bearing 抱轴支承
nose bit 手摇扁钻
nose cap 燃烧室前部
nose cone 头锥，整流罩，导流锥，整流帽，鼻锥体，头锥体
nose downmoment 低头力矩
nose end 空端，管口端
nose fairing 头部整流（罩）
nose line 梯段坡度线
nose nacelle 前舱
nose of wing 翼前缘
nose pile 定位桩
nose-pipe 放汽管口
nose slice 机头晃动
nose spar 前缘梁
nose-to-nose separation （气泡）头对头分开
nose up moment 抬头力矩
nose 鼻状物，鼻子，翼前缘，折焰角，突出部，叶片前缘，喷嘴，鼻端，墩端，凸头
nosing 踏步突缘，踏步凸边，头部，机头
no-slip conditions 无滑移条件，无滑流条件
no-slip strip 防滑条
no-slotted armature 无槽电枢
no-slotted turbogenerator 无槽汽轮发电机
no-slump concrete 干硬性混凝土，无坍落度混凝土
no-spark 无火花
no-steam consumption 空负荷汽耗
no-steam operation 无蒸汽运行
no swell 零级浪，零级涌浪
nota bene 注意
notable 值得注意的，显著的
not affecting critical path 非关键路径的时间
NOT-AND element 非与元件
not applicable 不适用，不适当的
notarial certificate 公证证书
notarial deed 公证证书
notarial document 公证文件
notarization 公证书
notarized contract 已办妥公证手续的合约
notarized 公证的
notarize 确认，证明

notary public name change application 公证人姓名变更申请
notary public office 公证处
notary public 公证人，公证员，公证机关
notary 公证人，公证员，公证行
not assigned 不指定的，不赋值的
notation 标记，符号，记号，计数法，符号表示法，批注，图解符号
notch acuity 缺口烈度，锐度
notch amplifier 标记信号放大器，标度信号放大器
notch-bar impact test 切口棒冲击试验
notch brittleness 缺口脆性，切口脆性
notch connection 槽连接
notch drop 凹口跌水
notch ductility 缺口延性，缺口试样断面收缩率，缺口韧性，切口韧性
notched column 有槽柱
notched sill 齿槛，齿形槛
notched specimen 切口试样
notched undulated preheater 凹槽波纹板预热器，凹槽波纹板回转式空气预热器
notched weir 缺口（最水）堰
notched 有缺口的，有槽的
notch effect 刻槽影响，切口效应，切口影响，缺口效应
notch filter 陷波滤波器
notching curve 下凹曲线
notching relay 脉冲次数继电器，分级继电器，加速继电器
notching 穿孔，切痕
notch maker 拉槽机
notch plate 缺口（最小）堰板
notch sensitivity 切口敏感性
notch 槽口，凹口，切口，开槽，豁口，刻槽，缺口
NOT circuit “非”电路
NOT core “非”磁芯
note amplifier 音频放大器
not easily weathered 不易风化的
noted public figures 社会名流
noted 注明（的）
note frequency 声频
not elsewhere indicated 不另详述，在别处未加说明
not elsewhere provided 不另详述，在别处未加说明
not elsewhere specified 不另详述，在别处未加说明
not elsewhere stated 不另详述，在别处未加说明
notes in explanation 注解，注释
notes on talks 会谈纪录
no test link 不需测试链路
note to order 记名票据
note 注解，注，笔记，注意，记录，票据，加注
NOT function “非”作用，“非”功能
NOT gate “非”门
no thoroughfare for vehicles 禁止车辆通行
notice board 布告牌
notice by fax 传真通知
notice date 通知日期
notice in written 书面通知
notice of abandonment 放弃通知，委付通知
notice of acceptance 承兑通知
notice of appeal 上诉通知书
notice of arrival 到法通知书
notice of award 中标通知，中标函
notice of cancellation 撤销通知
notice of change 更改通知
notice of claim clause 索赔通知条款
notice of claim 索赔通知
notice of commencement 开工通知
notice of damage 损坏通知
notice of default 违约通知
notice of defect 缺陷通知
notice of delay 延迟通知
notice of delivery paid 已付款交货通知单
notice of intent 意向书
notice of loss 损失通知
notice of non-payment 拒付通知
notice of readiness 准备就绪通知
notice of shipment 装船通知
notice period 通知期限
notice to bidders 招标通知
notice to commence 开工通知书
notice to proceed 开工通知
notice to suspend payment 暂停支付通知书，中止支付通知
notice 通知，通告
no-tide point 无潮点
notification of approval 批准通知
notification of award 中标通知，签约通知
notification of bidding 招标公告
notification of tender award 中标通知，中标函
notification point 通知要点
notification 通知数据，通知书
notified body 认证机构
notified party 被通知方，通知关系人
no time-delay converter 零延时转换器
no time-delay instantaneous current protection 无时限电流速断保护
no time-delay relay protection 无时限继电保护
not in-contact 非接触
not later than 不迟于
not measured 未测量的
NOT operator 求反算子，“非”算子
NOT OR element “非或”元件，“或非”元件
no touch relay 无触点继电器
no transmission 无传输
no transshipment permitted 不准转口装运，不准转运
no trouble found 未出故障
not-specified 不规定的，不指定的
NOT sum “非”和
not supplier scope 非供应商范围
not to be laid flat 切勿平放
not to be tipped 切勿倒置【包装标志】
not to scale 超出量程，不按比例尺
not water-filled 不充水的
notwithstanding anything to the contrary herein 尽管在本协议中可能有相反（的规定）

notwithstanding the above provision 虽然（规定）有前述条款
notwithstanding 即使，尽管
NOT “非”门
nought 无，零
no upside down 切勿倒置【包装标志】
nourishment source 补给水源
no value declared 没有声明价值
novel design 新颖设计
novel tube 标准九脚小型管
novenary 九进制的
novendenary 十九进制的
no visitors allowed 谢绝参观
Novokonstant 标准电阻合金
novolac epoxy 酚醛环氧树脂
novolac resin 酚醛清漆树脂
no-voltage circuit-breaker 无电压自动断路器
no-voltage coil 无压线圈
no-voltage protection 无电压保护，失压保护
no-voltage relay 失压继电器，无电压继电器
no-voltage release 无压释放，失压脱扣器
no-voltage 无电压
no-volt contact 零压接点
no wind period 无风期
now that 既然，由于
now therefore 特此，因此【与 whereas 连用，后常跟 hereby 译成“兹”，“特此”】
now these presents witness that 兹特立约为据【用于 whereas 条款之后】
NO_x detector 氮氧化物探测器
noxious gas 有毒气体，有害气体
noxious 有害的
NO_x(nitrogen oxides) 氮氧化合物
NOZ=nozzle 喷嘴
nozzle aerator 喷嘴通气器，喷嘴曝气器
nozzle angle 喷嘴角，喷嘴扩散角
nozzle area 接管通道截面，喷嘴面积
nozzle atomizer 喷嘴喷雾器，喷嘴雾化器
nozzle blade cascade 喷嘴叶栅
nozzle blade 喷嘴导叶
nozzle block 喷嘴组
nozzle boss 安装喷嘴用突出部，喷杆突出部
nozzle box assembly 喷嘴室部套
nozzle box chamber 喷嘴室
nozzle box （汽轮机）喷嘴室，喷嘴阀箱
nozzle bridge 喷管板加强筋
nozzle bucker 喷嘴叶片
nozzle button （沸腾炉）风帽
nozzle cage 喷嘴套
nozzle cap nut 喷油嘴紧帽
nozzle carrier 喷嘴座
nozzle cascade 喷嘴叶栅，喷嘴环
nozzle chamber 喷嘴室【汽轮机】，喷嘴内腔
nozzle champ 喷嘴夹
nozzle chest 喷嘴室，喷嘴箱
nozzle clearance 喷嘴出口空隙
nozzle closure 喷口盖，喷口隔板
nozzle coefficient 管嘴系数，喷嘴系数
nozzle configuration 喷口形状
nozzle contour 喷管外形，喷嘴型线
nozzle control valve 喷嘴调节阀
nozzle control 喷嘴调节
nozzle coolant 冷却剂接管
nozzle cover 管嘴盖
nozzle curvature 喷嘴曲率
nozzle cut-out governing 喷嘴调节，断流调节
nozzle dam 管嘴堵板，管嘴闸
nozzle diaphragm 喷嘴隔板，喷嘴装置，喷嘴环
nozzle divergence 喷嘴扩张度
nozzle edge 喷嘴出汽边，喷嘴出气边
nozzle effect 喷嘴效应
nozzle efficiency 喷嘴效率
nozzle element 水帽
nozzle end face 管嘴端面
nozzle enrichment process 喷嘴浓缩法
nozzle expansion ratio 喷嘴扩张比
nozzle filter 喷嘴过滤器
nozzle flame 喷管火焰
nozzle flange 管口法兰
nozzle flapper 喷嘴挡板
nozzle flowmeter 喷管式流量计，喷嘴式流量计
nozzle for arc spraying 电弧喷涂喷嘴
nozzle for flame spraying 火焰喷涂喷嘴
nozzle friction 喷嘴摩擦
nozzle governing 喷嘴调节
nozzle group control 喷嘴调节
nozzle group valve 喷嘴阀
nozzle group 喷嘴组
nozzle guide vane 导向叶片
nozzle header 喷嘴联箱
nozzle head 喷嘴头
nozzle holder 喷枪体
nozzle inspection 管嘴检查，接管检查
nozzle jet separation 喷嘴喷射分离
nozzle leg grinding facility 管嘴段研磨装置
nozzle limit 管嘴边界
nozzle loss 喷嘴水头损失，喷嘴损失
nozzle meter 管嘴式流速计
nozzle mixing burner 引射式喷燃器
nozzle mouth-ring 喷嘴口环
nozzle of sprinkler 喷水管嘴
nozzle opening ratio 接管开孔比，孔径比
nozzle orifice 喷孔，喷嘴口，喷嘴
nozzle outlet air supply 喷口送风
nozzle passage 喷嘴通道
nozzle passing frequency 喷嘴激振频率
nozzle pitch 喷嘴节距
nozzle plate 水嘴板，喷嘴板
nozzle pressure relief valve 喷嘴卸压阀
nozzle pressure 喷嘴压力
nozzle process 喷嘴法【浓缩】
nozzle reaction force 喷嘴反作用力
nozzle regulation 喷嘴调节
nozzle regulator 喷嘴调节阀，喷嘴调节器
nozzle resonance 喷嘴共振
nozzle ring rib 喷嘴环肋板
nozzle ring 涡轮导向器，喷嘴环，环形喷嘴，导叶环
nozzles and tips 接管及管头
nozzle seat face 喷嘴座面
nozzle seating face 喷嘴座面

nozzle section　喷嘴组，接管段，喷嘴剖面，接管部分
nozzle segment　喷嘴组，喷嘴弧段
nozzle shell course　压力壳接管环段
nozzle spray angle　喷嘴喷射角
nozzle stack　喷嘴烟囱
nozzle stand pipe　（沸腾炉的）立管式风帽
nozzle stub　喷嘴短管
nozzle support ring　喷嘴支承环，压力壳支撑环段
nozzle test pilot valve　喷油试验滑阀
nozzle throat area　喷嘴喉部面积
nozzle throat　喷管喉口，喷嘴喉部，喷管喉道
nozzle tip　喷嘴尖，喷头
nozzle tube　喷管
nozzle-type water wheel　喷嘴式水轮（机）
nozzle valve　喷嘴阀
nozzle vane　静叶片
nozzle velocity coefficient　喷嘴速度系数
nozzle wake resonance　喷嘴尾流共振
nozzle wake　喷嘴尾迹，喷嘴尾流
nozzle wearing ring　喷嘴耐磨环
nozzle wrench　喷嘴扳手
nozzle　管接头，喷嘴，喷管，接管，管座，接头，管头，喷口，喷头，（风洞）试验段
NPAR(nuclear plant aging research)　核电厂老化研究
NPB(non-segregated phase busduct)　共箱母线
NPCIL(Nuclear Power Co. of India Ltd.)　印度核电有限公司
NPD(nuclear power demonstration)　示范性核电厂，模式核电厂
n percent duration flow　频率为 n% 的径流量
NPF(nozzle-passing frequency)　喷嘴激振频率
NPG(normal planning group)　日常计划组
NPHR(net plant heat rate)　电厂净热耗
NPIC(Nuclear Power Institute of China)　中国核动力研究院【核一院，乐山】
NPI(Nuclear Power International)　国际核动力公司
NPLA(nuclear plant life assurance)　核电厂寿命保证
Np237-lined fission chamber　镎 237 裂变室
NPL(noise pollution level)　噪声污染级
N-plus-one address instruction　N 加 1 地址指令
NPM(counts per minute)　每分钟计数
Np＝neper　奈培
NPOC(Nuclear Power Oversight Committee)　核动力监视委员会
NPO(nuclear plant operator)　核电厂操纵员
NPP(nuclear power plant)　核电厂
NPPTS(nuclear power plant training simulator)　核电厂培训模拟机
NPRDS(nuclear plant reliability data system)　核电厂可靠性数据系统
NPR(new production reactor)　新生产堆
NPSHA(net positive suction head available)　有效气蚀余量，可用的净正吸入压头
NPSH available in system　系统可用的净正吸入压头
NPSH curve　净正吸入压头曲线
NPSH(net positive suction head)　净正吸入压头，净吸压头，汽蚀余量，有效吸入高度【泵】
NPSH test　净正吸入压头试验
NPSM　美国标准机械接头直管螺纹
NPS(negative phase sequence)　负相序
NPS(neutrons per second)　每秒钟放出中子数
NPS(nuclear power station)　核电站
NPS　美国标准直管螺纹
NPT(National/American Pipe Thread)　美国标准锥管螺纹
NPTC(Nuclear Power Training Center)　核电培训中心【苏州】
NPT(nominal pipe thread)　公称管道螺纹
NPT(normal pressure and temperature)　常温常压，标准压力和温度
NPT(Nuclear Non-proliferation Treaty)　核不扩散条约
NPU(network processing unit)　网络处理部件
NPV(net present value)　净现值
NQR(non quality related)　与质量无关的
NRC(National Research Council)　国家研究理事会【美国】
NRC(noise reduction coefficient)　噪声降低系数
NRC(Nuclear Regulation Committee)　核管会【美国】
NRC(Nuclear Regulatory Commission)　核管理委员会【美国】
NRD(non-return damper)　单向阀
NREL(National Renewable Energy Laboratory)　美国国家可再生能源实验室
NRE(negative resistance element)　负阻元件
NR(name register)　名字寄存器
NRN(noise rating number)　噪声分级值，NR 数
NR(noise ratio)　噪声比
NR(nonredundant)　非冗余的
N/R(notice of readiness)　装卸准备就绪通知书
NRO(numeric read out)　数字读出
NRP(normal rated power)　额定功率
NRTF(naval reactor test facility)　海军反应堆试验装置
NRT(net registered tonnage)　净登记吨，净吨
NRV(non return valve)　止回阀，逆止阀
NRZ-C(non-return-to-zero change)　不归零制变化
NRZ logic　不归零逻辑
NRZ-M(non-return-to-zero mark)　不归零标记
NRZ(non-return-to-zero recording)　不归零制记录
NRZ(non-return-to-zero)　不归零制
NSA(National Standard Association)　（美国）国家标准协会
NSB(National Science Board)　（美国）国家科学委员会
NSC(noise suppression circuit)　噪声抑制电路
NSC(Nuclear Safety Center)　（美国）核安全中心
NSD(network status display)　网络状态显示
NSD(Nuclear Safety Department)　核安全部
NSEA(nuclear safety environmental analysis)　核安全环境分析
N-semiconductor　N 型半导体，电子导电型半导体
NSIC(Nuclear Safety Information Center)　核安全信息中心【属美国原子能委员会】
NS(nuclear safety)　核安全

NSPA(National Standard Part Association) 全国标准零件制造业协会
NSPE(National Society of Professional Engineers) (美国)国家专业工程师协会
NS permeameter NS磁导计
NSRDS(national standard reference data system) 美国国家标准参照数据系统
NSRI(nuclear safety research index) 核安全研究指南
NSSC(Nuclear Safety Standards Commission) (德国)核安全标准委员会
NSSS manufacturer 核蒸汽供应系统制造商
NSSS(network security secret system) 网络安全保密系统
NSSS(nuclear steam supply system) 核蒸汽供应系统，核供汽系统，核锅炉，原子锅炉
NSSS performance 核蒸汽供应系统性能
NSSSS(nuclear steam supply shutdown system) 核蒸汽供应停堆系统
NSV(negative supply voltage) 负电源电压
NSV(nonautomatic self-verification) 非自动的自检验
N. T. C(negative temperature coefficient) 负温度系数
NTC thermistor 负温度系数热敏电阻
NTDC(National Transmission & Despatch Company Limited) 巴基斯坦国家输配电公司
NTE(network and trunk equipment) 网络及总线设备
n-terminal network n端网络
NTF(Nigeria Trusty Fund) 尼日利亚信托基金
NTF(no trouble found) 未出故障
n-th difference n次差分
n-th harmonic n次谐波
n-th order n次，n阶
n-th power n次幂，n次方
n-th root n次方根
NTL(non-threshold logic) 非阈值逻辑
NTL(no test link) 不需测试链路
NT(no transmission) 无传输
NT(numbering transmitter) 编号发送器
NTPC(National Technical Processing Center) (美国)国家热电公司
NTPC(National Thermal Power Corporation) (印度)国家电力集团公司
NTP(network time protocol) 网络时间协议
NTP(normal temperature and pressure) 标准温度和压力
NTP(notice to proceed) 开工令，开工通知
NTR(nothing to report) 无可报告
NTS(not to scale) 不按比例
NTT(network transfer table) 网络传送表
NTU(Nephelometric Turbidity Units) 浊度单位【比浊测量法】
NTU(network terminating unit) 网络终端装置
NTWT(net weight) 净重
N-type semiconductor N型半导体
nuclear absorption 核子吸收
nuclear accident 核事故
nuclear acoustic resonance 核声共振
nuclear activity 核活性
nuclear aftermarket service 核售后服务
nuclear aftermarket 核第二市场
nuclear aircraft 核动力飞机
nuclear amplitude 核振幅
nuclear angular momentum 核角动量，核动量矩
nuclear architect engineering 核建筑工程
nuclear art 核工程，核技术
nuclear atom 核形原子
nuclear auxiliary building 核辅助厂房
nuclear auxiliary power 辅助核动力
nuclear average temperature 核平均温度
nuclear bank 核燃料库
nuclear blinding energy 核的结合能
nuclear boiler 核锅炉，核蒸汽发生器
nuclear boiling 核态沸腾
nuclear breeder 核增殖堆
nuclear breeding 核(燃料)增殖
nuclear calculation 核计算，堆物理计算
nuclear capture 核吸收，粒子捕获，核俘获
nuclear cell 核电池
nuclear chain reaction 核链式反应
nuclear charge 核电荷
nuclear chemical engineering 核化学工程
nuclear chemistry 核化学
nuclear chief engineer 核电总工
nuclear chilled-water system 核冷冻水系统
nuclear cleanliness 核清洁度
nuclear cogeneration plant 核热电厂
nuclear community 核社区
nuclear complex 核联合企业
nuclear component drain and vent system 核设备疏水和排气系统
nuclear component 核部件
nuclear control 反应堆功率调节，核反应控制
nuclear controversy 核争论
nuclear conversion 核转换
nuclear converter 核转换器，转换反应堆
nuclear core monitoring and prediction system 堆芯监测与预报系统
nuclear corrosion 辐照引起的腐蚀，核腐蚀
nuclear cost 核发电成本
nuclear critical accident 核临界事故
nuclear criticality safety 核临界安全
nuclear cross section 核截面
nuclear damage 核损害
nuclear data library 核数据库
nuclear debris 核碎片
nuclear decay 核衰变
nuclear delay 控制棒延迟，核信号延迟
nuclear desalination 核能海水淡化
nuclear desalting plant 核能海水淡化工厂
nuclear design calculations 核设计计算
nuclear design 核设计
nuclear disaster 核事故
nuclear disintegration 核分裂，核衰变，核蜕变
nuclear ecology 核生态学
nuclear effective temperature 有效核温度
nuclear electricity for marine operation 海上核电站
nuclear electricity generation 核能发电
nuclear electricity 核电

nuclear-electric propulsion 核电推进
nuclear emergency brigade 核应急队
nuclear-emulsion film 核乳胶片
nuclear energetics 核动力学，核能技术
Nuclear Energy Agency 核能机构【属经济合作与发展组织】
nuclear energy cost 核能成本
nuclear energy level 核能级
nuclear energy plant 核电厂，原子能发电厂
nuclear energy 原子能，核能
nuclear engineering 核工程学，核工程
nuclear engineer 核工程师
nuclear engine 核发动机
nuclear enthalpy rise factor 核焓提升因子
nuclear enthalpy rise hot channel factor 核焓提升热因子
nuclear environment 核环境，核辐射环境，核环境条件
nuclear equation 核反应方程
nuclear equipment testing laboratory 核设备试验室
nuclear era 核时代，核纪元
nuclear establishment 核设施，核装置
nuclear event 核事件，核试验
nuclear excitation 原子核激发
nuclear excursion 核功率剧增，反应堆工况偏离
nuclear explosion 核爆炸
nuclear facilities 核设施，核装置，设施，设备
nuclear fission power generation 核裂变发电
nuclear fission 核裂变，核分裂
nuclear floating island 浮动核岛
nuclear flux 核通量
nuclear force 核力
nuclear fragment 核碎片
nuclear fuel carbide 碳化物核燃料
nuclear fuel conversion 核燃料转换
nuclear fuel cycle cost 核燃料循环费用
nuclear fuel cycle 核燃料循环
nuclear fuel element 核燃料元件
nuclear fuel expenses 核燃料费用
nuclear fuel fabrication facility 核燃料加工设备，核燃料加工厂
nuclear fuel fabrication plant 核燃料制造厂
nuclear fuelled boiler 核锅炉，核蒸汽发生器
nuclear fuel management 核燃料管理
nuclear fuel pellet 核燃料芯块
nuclear fuel reprocessing plant 核燃料后处理工厂
nuclear fuel reprocessing 核燃料后处理，核燃料再加工
nuclear fuel utilization 核燃料利用率
nuclear fuel 核燃料
nuclear fusion power generation 核聚变发电
nuclear fusion reaction 核聚变反应
nuclear fusion 核聚变，核合成
nuclear gas turbine 核能燃气轮机
nuclear generating cost 核发电成本，核电成本
nuclear generator 原子能发电机，核动力发电机
nuclear geometrical cross-section 核几何截面
nuclear grade ion-exchange resins 核级离子交换树脂
nuclear grade 核纯级
nuclear graphite 核石墨
nuclear grid 核电站电网
nuclear hazard 核危害
nuclear heat and power plant 核热电厂
nuclear heat damage 核（事件）的热损伤
nuclear-heater propulsion 核加热推进
nuclear heat flux hot point factor 核热通量热点因子
nuclear heating 核采暖，核加热
nuclear heat 核热能，反应堆
nuclear hot channel factor 核热通道因子
nuclear hot spot factor 核热点因子
nuclear housekeeping 放射性清理工作，核清理
nuclear ice-breaker 核破冰船
nuclear importance function 中子价值函数
nuclear incident 核事件，核事故
nuclear instability 核反应不稳定性
nuclear installations inspectorate 核设施检查局
nuclear installation 核装置［设施，设备］
nuclear instrumentation channel 核测量线路
nuclear instrumentation system 核功率测量系统
nuclear instrumentation 核测量仪表
nuclear instrument landing system 核仪表控制着陆系统
nuclear instrument thimble （反应堆中）核仪器管道，核仪器孔道
nuclear insurance 核保险
nuclear island chilled water system 核岛冷水系统
nuclear island chiller unit room 核岛冷水系统间
nuclear island demineralized water distribution system 核岛除盐水分配系统
nuclear island fire protection system 核岛消防系统
nuclear island hydrogen distribution system 核岛氢分配系统
nuclear island liquid radwaste monitoring and discharge system 核岛废液监督排放系统
nuclear island nitrogen distribution system 核岛氮气分配系统
nuclear island nonradioactive ventilation system 核岛非放射性通风系统
nuclear island vent and drain system 核岛排气及排水系统
nuclear island 核岛
nuclear know-how 核知识
nuclear legislation 核立法，核法律
nuclear level detector 核煤位探测器
nuclear level device 核子料位计
nuclear level switch 核料位开关
nuclear level 核能级
nuclear liability 核责任
nuclear lifetime 核寿期
nuclear liquid-air cycle engine 液态空气循环核发动机
nuclear lubricant 核级润滑剂
nuclearly safe spacing 核安全距离
nuclear magnetic resonance spectrometer 核磁共振波谱仪
nuclear magnetic resonance 核磁共振
nuclear magneton 核磁子

nuclear makeup 核模拟
Nuclear Management and Resources Council （美国）核管理与资源理事会
nuclear manpower 核能人力资源
nuclear material accountancy 核材料衡算
nuclear material accounting system 核材料衡算制度
nuclear material accounting 核材料的衡算活动
nuclear material control system 核材料控制系统
nuclear material control 核材料控制
nuclear material management 核材料管理
nuclear material safeguard system 核材料保障体系，核材料保障系统
nuclear material safeguard 核材料保障
nuclear material 核材料
nuclear measurement 核测量
nuclear measuring channel 核测量线路
nuclear merchant ship 核商船
nuclear-MHD plant 核磁流体电站
nuclear-MHD power system 核磁流体发电系统
nuclear mockup 核反应堆模拟
nuclear network 核网络
Nuclear Non-proliferation Treaty 核不扩散条约
nuclear nuisance 核公害
nuclear parent 母核
nuclear park 核能联合体
nuclear photoeffect 核光电效应
nuclear physics 核物理
nuclear plant analyzer 核电厂分析器
nuclear plant event report 核电厂事件报告
nuclear plant life extension 核电厂寿命延长
nuclear plant operator 核电厂操纵员
nuclear plant security 核电厂安全保卫
nuclear plant unit 机组，单元【电站】
nuclear plant 核工厂，核电厂，核装置，核设施，核设备
nuclear poison 核毒物，核残渣
nuclear pollution 核污染
nuclear power complex 核动力联合企业
nuclear powered cogeneration power plant 核能热电厂
nuclear powered gas turbine 核动力燃气轮机
nuclear powered reciprocating engine 核动力往复式发动机
nuclear powered tanker 核动力油船
nuclear powered train 核动力火车
nuclear powered urban district heating station 核动力市区供热站
nuclear powered 核动力装置的，核能的，原子能的，核动力的
nuclear power facility 核动力设施
nuclear power fluctuation 核功率涨落
nuclear power generation 核能发电
nuclear power industry 核动力工业，核能工业
nuclear power measurement 核功率测量
nuclear power phase-out 核电逐步停用
nuclear power plant management 核电厂管理
nuclear power plant overall layout 核电站总体布置
nuclear power plant simulator 核电厂模拟机
nuclear power plant site 核电厂址
nuclear power plant 核动力厂，核电站，核电厂，核动力装置，原子能发电厂
nuclear power production 核能生产
nuclear power program 核电发展规划
nuclear power station 核电厂，核电站
nuclear power technology 核动力工艺学，核动力技术
nuclear power unit 核电机组单元，核动力装置
nuclear power 核动力，核电，原子能，核能，核能发电的
nuclear precession angular frequency 核进动角频率
nuclear process heat 核供热
nuclear project 核工程项目
nuclear-propelled 核推进的
nuclear propulsion system 核推进系统
nuclear purity 核纯度
nuclear quadrupole moment 核四极矩
nuclear quadrupole resonance thermometer 核四极共振温度计
nuclear radiation detector 核辐射检测器
nuclear radiation 核辐射
nuclear radius 核半径
nuclear reaction analysis 核反应分析
nuclear reaction 核反应
nuclear reactor art 反应堆技术
nuclear reactor containment 安全壳
nuclear reactor monitoring 核反应堆监测
nuclear reactor primary coolant pump 核反应堆一回路冷剂泵
nuclear reactor safeguard 核反应堆保障
nuclear reactor safety 核反应堆安全
nuclear reactor theory 核反应堆理论
nuclear reactor 核反应堆
nuclear research and development facility 核科研设施
nuclear risk 核危险
nuclear rocket reactor 核火箭反应堆
nuclear runaway 反应功率或反应性的失控上升
nuclear safeguard regulation 核安全保障规章
nuclear safeguard 核安全保障
nuclear safety category 核安全类别
nuclear safety criterion 核安全规范，核安全准则，核安全类别
nuclear safety culture 核安全素养
nuclear safety function 核安全功能
nuclear safety inspection 核安全检查
nuclear safety limit 核安全限值，临界安全限额
nuclear-safety-related system 核安全相关系统
nuclear safety standard 核安全标准
nuclear safety 核安全
nuclear sampling system 核取样系统
nuclear sampling 核取样
nuclear service building 核服务楼
nuclear service business 核售后服务
nuclear shielding 核辐射屏蔽
nuclear ship 核推进船
nuclear site licence 核厂址执照
nuclear siting policy 核电厂选址方针
nuclear soil moisture meter 同位素含水量测定仪，核子土壤湿度计

nuclear spectrum 核能谱，核辐射谱
nuclear spin quantum number 核自旋量子数
nuclear station for district heating and power 用于发电和区域供热的核电站，核热电站
nuclear station 核电站
nuclear steam electric plant 核蒸汽电站，核蒸汽电厂
nuclear steam generating plant 核蒸汽发生装置
nuclear steam generator 核蒸汽发生器
nuclear steam station 核蒸汽站
nuclear steam superheating 核蒸汽过热
nuclear steam supply system 核供汽系统，核蒸汽供应系统，核锅炉，原子锅炉
nuclear steam turbine 核汽轮机
nuclear stockpile 核贮备，核资源
nuclear study 中子学
nuclear superheating 核过热
nuclear superheat reactor 核过热反应堆
nuclear superheat 核过热
nuclear technology 核技术，核工艺学
nuclear temperature coefficient 核（反应性的）温度系数
nuclear temperature 核温度
nuclear test reactor 核试验反应堆
nuclear test 核试验
nuclear thermionic conversion 核热离子转换
nuclear third party liability 第三方核责任
nuclear threat 核威胁
nuclear trade 核贸易
nuclear trainer 核电厂培训设备
nuclear transformation 核转换
nuclear transients calculation 核瞬态计算
nuclear transportation safety 核材料运输安全，核运输安全
nuclear turbine 核能透平，核汽轮机，核电汽轮机
nuclear turbo jet power plant 核透平喷气发动机装置
nuclear uncertainty factor 核不确定因子
nuclear valve 核级阀门
nuclear waste auxiliary building 核废物辅助厂房
nuclear waste disposal 核废物处置
nuclear waste repository 核废物处置库
nuclear waste storage 核废物贮存
nuclear waste treatment 核废物处理
nuclear waste 核废物
nuclear weightment 核子秤
nuclear work site 核施工现场
nuclear work 放射性操作
nuclear 有核的，原子核的，核的，核动力的
nucleate boiling heat transfer coefficient 泡核沸腾传热系数
nucleate boiling suppression factor 泡核沸腾抑制因子
nucleate boiling 核态沸腾，气泡状沸腾，泡核沸腾
nucleate point 汽化核心，汽化中心
nucleate 气泡核，有核的，成核，形成泡
nucleation center 起泡中心
nucleation point 起泡点
nucleation process 核化过程
nucleation rate 起泡速率
nucleation site 起泡点
nucleation temperature 起泡温度
nucleation 成核过程，成核观象，晶核，成核作用
nuclei 核【复数】
nucleonics 核子学
nucleon number 核子数
nucleon 核子
nucleus fission 核裂变
nucleus formation 核生成
nucleus of condensation 冷凝核
nucleus of crystallization 结晶核
nucleus of crystal 晶核
nucleus screening 核屏蔽
nucleus 核，原子核，核心，中心
nuclide mass 核素质量
nuclide 核素
nude cargo 裸装货，裸包装物
nude packs 裸装
nugget size 溶核直径
nugget 熔核
NUG(non-utility generator) 非公用事业发电厂
nuisance analysis 公害分析
nuisance effect 公害效应
nuisance free 无公害
nuisanceless closed loop technique 无公害闭环式工艺
nuisance organism 公害组织，公害机体
nuisance parameter 多余参数，有碍参数，多余参量
nuisance trip 误动作，乱真跳闸
nuisance 谬误的，计划外的误动作，公害，危害，损害，麻烦事，讨厌的东西
null adjustment 零位调整
null and void 无效的，（协议）无法律效力，失效，作废
nullator 双线无损单口网络
null-balance device 零平衡装置
null-balance indicator 平衡零点指示器
null-balance 零点平衡，零平衡
null character 空字符，零字符
null circuit 零电路
null condition 零的条件，平衡条件
null detector 零值检验器，检零器
null device 零型装置
nullification 作废，抑制，取消
nullified 取消的，作废的
null-indicating oscilloscope 零指示示波器
null indicator 零位指示器，零点指示器
null instrument 平衡点测定器
nullity 独立网孔数
null junction 零位连接
null matrix 零矩阵
null method 指零法，零点法，消除法，零位法
null off-set 零点漂移，零点偏移
null point 零点
null position 零位
null potentiometer 零值电位计，零值电势计
null reference system 零基准制
null sequence reactance 零序电抗

null sequence 零序
null-set 零集，空集
null space 零化空间
null-type bridge circuit 零型电桥电路，指零式电桥电路
null-type bridge 指零式电桥
null-type instrument 零型仪表
null valence 零价
null 零（的），零值（的），零位，空的，无效的，无用的，无法律效力的，失效的
nulvalent 零价的，不活泼的
NUMARC（Nuclear Management and Resources Council）（美国）核管理与资源理事会
number address code 数地址码
number and timing of installation 分期付款的次数和时间
number base 数基
number bus 数字总线
number comparing device 数值比较装置
number concentration 计数浓度
number conversion 数转换
number converter 计数制变换器
number density of molecules 分子数密度
number density of particles 粒子数密度
number density 数量密度，数率密度
numbered axis 数字轴线
numbered 已编号的，已达到限定值的
number generator 数发生器
numbering transmitter 编号发送器
numbering 识别，辨认，鉴定，标记，编号，数字，号码，号数
number language 数字语言
number marking 号码标志
number of account 账号
number of air changes 换气次数
number of ampere turns 安匝数
number of blades 叶片数
number of blows per 40cm 每贯入 40 厘米击数
number of blows 击数
number of breaks 断口数
number of cells for battery 每组蓄池组单体电池数
number of coils 线圈数
number of complete air changes 全面换气次数
number of computations 计算量
number of crosslinks 交联数
number of cycles of cracking 开裂循环数
number of cycles of overstress 过应力循环数
number of cycles per second 频率，每秒周期数
number of degree-day of heating period 采暖器度日数
number of divisions 分度数，刻度数
number of field per second 每秒场数，场频
number of fill cycles 充排（熔盐）循环的次数
number of frames per second 每秒帧数
number of lines 行数，线数
number of mode shapes to be considered for dynamic analysis 动力分析所考虑之振型数
number of mode shapes 振型数
number of molecules or other elementary entities 分子或其他基本单元数
number of neutrons per fission 每次裂变放出的中子数
number of orifice 孔数
number of packages 包装件数，件数
number of passes in water space 水区的管程数
number of pass 流程数
number of phases 相数
number of piles to tested 试验桩数
number of pole pairs 磁极对数
number of poles 极数
number of raining days 降雨天数
number of revolution 转数
number of roller passes 碾压遍数
number of safe operation days 安全运行天数
number of sagging plates in each shell 每个凝汽器壳侧的隔板数量
number of scanning lines 扫描行数
number of seats 座位数
number of slots 槽数
number of starts 启动次数
number of start-ups 启动次数
number of stator turn 定子匝数
number of steps 挡数，级数
number of stress cycles 应力循环数
number of teeth 齿数
number of track-lines 车道数
number of transfer units 传热单元数
number of turns in a winding 绕组的匝数
number of turns 匝数
number of utilization hours 利用小时数
number of windings 线圈数，绕组数
number of wires 线数
number of working hours based on maximum load 最大负荷利用小时数
number of yaw drives 偏航驱动器数量
number period 数的周期
number plate 编号牌
number representation system 记数制，数表示制，记数系统
number scale 记数法
number sequence 数的序列，数序
number storage 数字存储器
number switch 号控机，小型交换机，数字开关
number system 数系
number tape 数字带
number theory 数论
number-to-frequency converter 数字频率变换器
number transfer bus 数字传送总线
number-unobtainable tone 空号音
number 号码，编号，数，数字，计数，数值，数量
numeral 数的，数字的，数字，数码
numeration table 数字表
numeration 计算法，读数法，编号，命数法
numerator 计算者，分数的分子，计数器，信号机
numerical analysis research 数值分析研究
numerical analysis 数值分析，数字分析
numerical aperture 数值孔径
numerical approximation 近似数值，数值近似
numerical calculation 数值计算

numerical check　数值校验
numerical code　数字代码，数码，数值码，数字码
numerical computation　数值计算
numerical constant　数值常数
numerical control device　数控设备，数控装置
numerical control　数字控制，数控，数值控制
numerical data　数值数据，数字数据，数字资料
numerical differentiation　数值微分
numerical digit　数字
numerical display device　数字显示器件
numerical display　数字显示
numerical error　数字错误
numerical experimentation　数值试验
numerical fault　数值误差
numerical forecast　数字预报
numerical-graphic method　数值图解法
numerical indication　数字指示
numerical information　数字信息
numerical integral　数值积分
numerical integration　数字积分，数值积分
numerical invariant　不变数
numerical inversion　数字转换
numerically coded　数字编码的
numerically controlled machine tool　数字控制的工作母机
numerically controlled　用数字指令控制的
numerically control　数字控制
numerically　数字的
numerical methods of integration　数值积分法
numerical method　数值法
numerical modeling　数值模拟
numerical model　数值模式，数值模型
numerical notation　数值符号
numerical part　数值部分
numerical positioning control　数字走位控制，数字定位控制
numerical prediction　数值预报
numerical procedure　计算方案，计算程序
numerical protection　数字式保护
numerical readout　数字读出，数值读出
numerical representation　数值表示，数的表示法
numerical selector　号码选择器
numerical simulation analysis　数值模拟分析
numerical simulation　数值模拟
numerical solution　数值解
numerical switch　号控机，数字开关
numerical tabular　数值表
numerical value of a quantity　量的数值
numerical value　数值
numerical weather prediction　数值天气预报
numerical　用数表示的，数字的，数值的
numeric character　数字符号
numeric control system　数字控制系统
numeric control　数字控制
numeric data　数值数据
numeric display　数字显示
numeric-field data　数字域数据
numeric keypad　数字袖珍键盘，数字键区
numeric printer　数字打印机
numeric punch　数字穿孔
numeric ratio　数值比
numeric readout　数字读出
numeric　数，数字，数字的
numeroscope　数字记录器，示数器，数字显示器
numerous handling　多次搬运
numerous　许多的，数量庞大的，大批量的
NUMEX(Nuclear Maintenance Experience Exchange)　核维修经验交流协会
Nu(Nusselt number)　努塞尔数，努谢尔特数
nuplex(nuclear complex)　核联合企业
nurse room　哺乳室
nursery　育婴室
NUSS(Nuclear Safety Standards)　核安全标准
nutating disc flow measuring device　圆盘流量测量仪表
nutating-disk flowmeter　盘式流量计
nutating-disk　盘式的
nut bolt　带帽螺栓，带螺母的螺栓
nut buoy　胡桃形浮标
nut cap　螺帽盖
nut coal　小块煤【30～50毫米】
nut collar　螺母迹圈
nut driver　螺帽起子
nut installation　螺帽装配
nut locking device　螺帽锁紧装置
nut locking　螺母锁紧
nut lock plate　螺帽锁紧板
nut lock washer　锁紧垫圈，止松垫圈
nutrient medium　养分，培养基
nutritive disturbance　营养障碍
nutritive enrichment　营养浓缩
nutritive equilibrium　营养平衡
nutty slack　核状煤屑，核状渣煤
nut wrench　螺帽扳手
nut　螺帽，螺母
nuvistor　小型抗震电子管
NVD(no value declared)　没有声明价值
NVOCC(non-vessel operations common carrier)　无船公共承运人或无船承运人
NVO(non vessel operator)　无船经营人
NVT(network virtual terminal)　网络虚拟终端
NWF(nozzle wake frequency)　喷嘴尾流激振频率
NWH(normal working hours)　正常工作时间
NWL(normal water level)　正常水位
NWL(normal working level)　正常工作水平
NWP(normal working pressure)　正常工作压力
NWT(neutral water treatment)　中性处理
n-year flood water level　n年一遇洪水位
n-year higher high water level　n年一遇最高高潮位
n-year peak flood and its corresponding level　n年一遇洪峰流量和相应洪水位
n-year wind speed　n年一遇风速
nylon brush　尼龙刷
nylon bushing　尼龙轴套，尼龙套筒
nylon dam　尼龙坝
nylon dielectric coating　尼龙绝缘涂层
nylon net　尼龙网
nylon paper　尼龙薄膜
nylon pin coupling　尼龙柱销联轴器
nylon plastics　尼龙塑料

nylon screen 尼龙纱
nylon 尼龙
Nyquist plot 奈奎斯特图
NZD 新西兰纽元
N＝Naira （尼日利亚）奈拉

O

OAC(origin accessory charge) 始发港杂费
OAC(oxyacetylene cutting) 氧乙炔切割
oakum 麻絮
oak 橡木，橡树
OA(office automation) 办公自动化
oar 浆，浆状物
oasis 绿洲
oath 誓言
OAW(oxyacetylene welding) 氧乙炔焊接
OA 操作员自动【DCS画面】
OBE(operating basis earthquake) 运行基准地震
object attribute 对象属性
object beam 物体光束，物体声束
object calibration device 实物校验装置
object clause 宗旨条款
object computer 执行计算机，目标计算机
object distance 物距
object function 目标函数
objectionable constituents 有害成分
objectionable 有害的，不好的，有异议的
objection 障碍，缺陷，缺点，反对，异议
objective analysis 客观分析
objective appraisal 客观的评价
objective assessment 客观评价
objective best-fit 客观最佳匹配
objective cause 客观原因
objective data 客观数据
objective definition 目标清晰度
objective evidence 客观证据
objective factor 客观因素
objective fact 客观事实
objective forecasting 客观预报
objective function 目标函数
objective lens 物镜
objective management 目标管理
objective method 客观法
objective planning 目标计划
objective sample 客观样品
objective table 载物台
objective value 客观价值
objective 目标
objectivity 客观性
object language 目标语言，结果语言
object library 目标程序库
object matter clause 标的物条款
object name 对象名
object program library 目标程序库
object program 目标程序，目的程序，结果程序
object wave 物波
object 对象，物，目标
oblateness 扁率，扁圆形
oblate spheroid 扁球面
obligate 专性的
obligation incurred 承担的债务，已发生的待付债务，承付款项
obligation principle 责任原则
obligation to serve 法定强制供电，供电责任
obligation 义务，责任，职责
obligatory arbitration 强制仲裁
oblige 债主
obligor 债务人
oblique angle 斜角
oblique arch 斜拱
oblique axis 斜轴
oblique bedding 斜层理
oblique crossing 斜交叉路，斜交叉
oblique distance 斜距离，斜距
oblique eccentric loading 斜偏心荷载
oblique exposure length 斜接近段长度
oblique exposure 斜接近，斜曝光
oblique fault 斜断层
oblique flow 斜向水流，斜流
oblique hydraulic jump 斜水跃
oblique incidence 斜入射，倾斜角
oblique joint 斜节理，斜接
oblique ladder 斜梯
obliquely slotted motor 斜槽式电动机
obliquely slotted pole 斜槽磁极
oblique photography 倾斜摄影
oblique plane 斜面
oblique shock wave 斜激波
oblique shock 斜冲波，激波，斜激波
oblique slot 斜槽
oblique string Y形串
oblique valve 倾斜阀
oblique wave 斜波
oblique weir 斜堰
oblique wind 斜向风，斜风
oblique 倾斜的，斜交的，无诚意的
obliquity angle 倾斜角
obliquity 倾角，倾斜，斜度
OBL(ocean bill of lading) 海运提单
oblong and round hole 长方孔和圆孔
oblong bolt hole 椭圆形螺栓孔，长圆螺栓孔
oblong cooler 长方形冷却器
oblong oval 长椭圆形
oblong 长方形的，椭圆形的
obnoxious gas 秽气，恶臭气体
obnoxiousness 讨厌，可恶
obround hole 长方圆孔
obscured glass 磨砂玻璃，毛玻璃，图案玻璃
obscure 暗的，不清楚的
obscurity 不明确，偏僻，含糊，模糊
obsequent flow 逆向流动
obsequent valley 逆向谷
observability 视野性
observance 惯例，遵守
observation accuracy 观测精度
observation adjustment 观测平差
observational data 观测资料

observational record 观测记录
observation apparatus 观测设备
observation balloon 观测气球
observation boat 测量船
observation borehole 观察孔
observation circuit 监听电路，监视电路
observation data 观察数据
observation desk 试验台，观测台
observation door 观察孔，看火孔，观察门
observation equation 观测方程
observation error 观测误差，观察误差
observation failure rate 观察失效率
observation frequency 观测次数，观测频率
observation gallery 观测廊道
observation hole 观测孔，观察孔，看火孔，人孔
observation instrument 观测仪器
observation line 观测线
observation mean life 观察平均寿命
observation monument 测站标石，观测标志
observation network 测站网，观测网
observation of density current 异重流观测
observation of evaporation from water surface 水面蒸发观测
observation of water temperature 水温观测
observation pipe 观测管
observation place 观测地点
observation plate 观测板
observation platform 观测架，观测平台，观测台
observation point 观测点
observation port 观察窗，屏蔽窗，观察孔，看火孔
observation post 观测哨，观测站
observation procedure 观测步骤
observation report 观测报告，观察报告，考查报告
observation series 观测系列
observation station 观察位置，观测站，测位
observation system 监测系统
observation table 观测台，检查台
observation tower 观测塔，瞭望台
observation well 观测井
observation window 观测窗，观察窗
observation 监视，遵守，注视，观察，观测，检查，实测
observatory 观测台，天文台，观象台，观测站
observed azimuth 观测方位角
observed current 实测流
observed data 实验数据，观测数据，实测资料，被观察的数据
observed direction 观测方向
observed maximum wind velocity 实测最大风速
observed mean time 平均观察时间
observed object 观测目标
observed parameter 观测参数
observed pulse shape 被观察的脉冲波形
observed quantity 观测值
observed reading 测量值
observed reliability 观察可靠性
observed stage 观测水位
observed temperature rise 观察温升
observed value 实测值，观测值
observer errors 观察者误差
observer 观察人员
observe's handbook 观测员手册
observe 观察，遵守
observing point 观测点
observing program 观测纲要
obsidian 黑曜岩
obsolescence equipment 陈旧设备
obsolescence technique 陈旧技术
obsolescence 报废，过时，陈旧
obsolete price 过期价格
obsolete project 过时项目
obsolete 废弃的，过时的，不能用的，已废的
obstacle avoidance 排除故障，障碍规避，避障
obstacle detection 故障探测
obstacle indicator 故障探测器，故障指示器
obstacle lighting 障碍照明
obstacle 障碍物，障碍，干扰，阻碍，雷达目标
obstruction-free flow area 有效流动截面
obstruction signal 事故信号
obstruction to vision 视程障碍
obstruction 阻塞，妨碍，阻碍，障碍，障碍物，干扰
obstruct 妨碍，阻挠
obtainable accuracy 可达精度
obtainable burn-up 能达到的燃耗
obtaining a stake in the company 获取公司的股份
obtain 取得
obturating ring 密封圈
obtuse angle blade 钝角叶片
obtuse angled triangle 钝角三角形
obtuse angle 钝角
obtuse arch 钝拱
obtuse 钝的，迟钝的，钝角的，抑制声音
obvious benefit 显著效益
obvious defect 明显的缺陷
obvious objection 明确地反对
obvious 明显的
occasional load 偶然荷载
occasional maintenance 临时维修，偶然维修
occasional peak loads 峰值荷载
occasional storm 稀有暴雨
occasional 临时的，偶然的，特殊的
occasion 时机
occluded cyclone 锢囚气旋
occluded front 锢囚锋
occluded 闭塞的，堵塞
occlude 吸留，吸收，关闭
occlusion 闭塞，阻塞，夹杂，闭合，封闭，吸留，锢囚（作用）
occulting light 明暗灯，隐显灯
occupancy expenses 使用费，占用费
occupancy factor 存在因子，占有因数
occupancy permit 占用许可证
occupancy rate 出租率，占用率，使用率
occupancy right 占用权
occupancy 占有（率、期间），占用，占领，居住（期间）

occupant comfort 居住者舒适性，乘坐者舒适性
occupant 占有人
occupational chemical erosion 职业性化学侵蚀
occupational disease 职业病
occupational environment 职业环境
occupational exposure 职业性照射量，职业辐照
occupational hazard 职业危害
occupational health 职业保健，职业卫生
occupational illness 职业病
occupationally exposed person 受职业性照射的人员
occupationally exposed 职业上的受辐照危险
occupational noise exposure 职业性噪声暴露
occupational noise source 职业性噪声源
occupational poisoning 职业性中毒
occupational radiation dose 职业性辐射剂量
occupational radiation exposure 职业性辐射照射（量）
occupational radiation hazard 职业性辐射危害
occupational risk 职业性危害，职业性危险
occupational safety 职业安全
occupation coefficient 占用率，使用率
occupation exposure 职业性照射
occupation-period 使用时间，占有时间
occupation standard 行业标准
occupation 职业，业务，占有，占据
occupiable area 可居留区
occupied area within power plant's enclosing wall 厂区用地面积
occupied area 占地面积
occupied condition 忙状态，占线
occupied frequency bandwidth 占有频率宽度
occupy market 占领市场
occupy 占用，占有
occurrence frequency 出现频率
occurrence time 出现时间
occurrence 事件，发生，出现，存在，事故，故障，现象
occur 出现，发生，存在，想起
OCCW(open cycle cooling water system) 开式循环冷却水【DCS画面】，开式循环冷却水系统
ocean anticyclone 海洋反气旋
ocean beach 海滩
ocean bill of lading 海运提单
ocean burial 海洋排放，海洋埋葬
ocean cable 海底电缆
ocean circulation 海洋环流
ocean climate 海洋性气候
ocean current meter 海流计，验流计
ocean current 海流，洋流
ocean disposal 海洋处置
ocean energy 海洋能
ocean engineering 海洋工程
ocean environment 海洋环境
ocean freight 海运，海运费（率）
ocean geotechnique 海洋土工学
ocean-going vessel 海轮，远洋（轮）船
oceanic climate 海洋性气候
oceanic condition 海洋状况
oceanic crust 海洋地壳
oceanic tide 大洋潮汐，海洋潮
oceanic troposphere 海洋对流层
oceanic vessel 海洋船
oceanic 海洋的
oceanity 海洋性
ocean marine cargo insurance 海运货物保险，货运保险，海上货运保险，海洋货运保险
ocean observation apparatus 海洋观测装置
oceanographical current meter 海洋流速仪，海流计
oceanographic data 海洋水文资料
oceanographic hydro-logical data 海洋水文资料
oceanographic measuring system 海洋测量系统
oceanographic observation 海洋观测
oceanographic station 海洋观测站
oceanography 海洋学
oceanology 海洋学
ocean pollution 海洋污染
ocean resources 海洋资源
ocean sediment 海洋沉积物
ocean shipping agency 外轮代理公司
ocean shipping 远洋航运
ocean soil 海洋土
ocean structure 海洋结构物
ocean surveying 海洋测量
ocean survey ship 海洋测量船
ocean temperature differential power generation 海水温差发电
ocean temperature 海水温度
ocean thermal energy conversion 海洋热能转换，海洋温差发电
ocean thermal energy 海洋热量
ocean water 海水
ocean wave field 洋浪场
ocean waybill 海运提单
ocean weather station observation 定点观测
ocean wind power generation 海洋风能发电
ocean wind 海洋风
ocean 海洋，洋
oce steel 无渗碳表面硬化用钢
OC(oil cooler) 油冷却器
OCO(open-close-open) 开闭开
OCO(open-close-open contact) 开闭开触点
O&C(operation and checkout) 操作和检查，运算和检查
OC(over current) 过流
OCP(output control pulse) 输出控制脉冲
OCP(overland common point) 内陆公共点或陆上公共点运输
OCR(optical character reader) 光符阅读机
OCS(on-off control system) 开关量控制系统
OCS(operation and control center) 运行与控制中心
OCS(optical character scanner) 光符扫描器
OCS(optical character system) 光符系统
octadic 八个一组的，八进制的
octad 八价物，八个一组
octagonal motor 八角形电动机
octagonal pierhead 八角形堤头
octagonal ring gasket 八角环形垫片，八角形金属环垫
octagonal steel 八角钢

octagon wind tunnel　八角形试验段风洞
octagon　八边形，八角的，八角形
octahedral cleavage　八面解理
octahedral normal stress　八面体法向应力
octahedral shear stress　八面体剪应力
octahedron　八面体
octal base　八脚管底
octal debugger　八进制调试程序
octal debugging technique　八进制调试技术，八进制排错法
octal digit　八进制数制，八进制数字
octal loading program　八进制调入程序，八进制装入程序
octal multiplication　八进制乘法
octal notation　八进制计数法
octal number　八进制数
octal socket　八脚管座
octal system　八进位制，八进制
octal-to-binary converter　八进制到二进制变换器
octal　八进制的，八面的，八角的
octamonic amplifier　倍频放大器
octane rating　辛烷值
octant　八分区，八分仪
octavalence　八价
octave band　倍频程【八度】，倍频程带
octave center frequency　倍频程中心频率
octave　倍频程，八度，八音度
octette　八偶体，八位位组，八角体
octodenary　十八进制的
octode　八极管
octonary number system　八进制
octonary　八进制的，八进制系统
octual sequence　八进制序列
octuplex telegraphy　八路电报
octuple　八重的，八路的，八极的，八倍的
ocular angle　视角
ocular estimation　目测
ocular micrometer　测微目镜
ocular　目镜
oculus　窥视孔
OCV(open-circuit voltage)　开路电压
ODAF(oil directed air forced cooling of transformer)（变压器的）强迫导向油循环风冷
ODA(Official Development Assistance)（外国）官方发展援助
odd-and-even logic　奇偶逻辑
odd-charge nucleus　奇电荷核
odd-controlled gate　奇数控制门
odd-even check　奇偶检验，奇偶校验
odd-even counter　奇偶计数器
odd-even　奇偶
odd-frequency motor　奇频电动机
odd-function　奇函数
odd harmonic function　奇调和函数
odd harmonic　奇次谐波
odd integer　奇整数
odd job　临时工，零工
odd location　奇数存储单元
odd lot　零星货物
odd-numbered line　奇数行
odd number　奇数
odd-odd　奇—奇，奇—奇的
odd-parity check　奇数的奇偶校验
odd parts　多余零件
odds and ends　零星物件，残余
odd symmetric　奇对称的
odd symmetry　奇对称
odds　可能机会，可能性，差别，区别，优势，概率，胜算，不平等
odd winding　奇数绕组
odd　临时的，奇怪的，奇数的，奇的
odeum　音乐厅，剧场
odevity　奇偶性
ODF(optical distribution frame)　光配线架
ODL(output data line)　输出数据线
ODM(original design manufacture)　原始设计制造商
odometer　测距计
o/d(on demand)　即付
odontoid　齿形的
OD(operational directive)　执行导则
OD(operations department)　生产部
odorant　有气味的
odor compound　恶臭化合物
odor concentration　臭气浓度
odor control　臭气控制
odor destruction　恶臭去除
odor dilution　臭气稀释
odor emission　臭气发散
odor free　无气味
odorimeter　气味计
odor inhibitor　臭气清除剂
odor intensity　气味强度，臭味强度
odorless　无臭的，无气味的
odor nuisance　恶臭污染，恶臭公害
odorousness　恶臭浓度，气味浓度，臭气浓度
odorous substance　恶臭物质
odor pollution　臭气污染
odor source　臭气源
odor test　气味试验
odor treatment by ion-exchange　离子交换除臭处理
odor unit　臭气单位
OD(outside diameter)　外径
OD(outside dimension)　外部尺寸
O/D(over draft)　透支
ODP(original document processing)　初始文件处理
ODT(octal debugging technique)　八进制排错法
ODU(output display unit)　输出显示器
ODWF(oil directed water forced cooling of transformer)　强迫导向油循环水冷
oedometer curve　固结曲线，压缩曲线
oedometer　压缩仪，固结仪，膨胀计
oedometric test　固结试验
OEFC(Operation Experience Feedback Committee)　运行经验反馈委员会
OEM(original equipment manufacturer)　原始设备制造商
oeolotropic　各向异性的
OE(operation engineer)　运行工程师
OE(owner's engineer)　业主工程师

oerstedmeter 磁场强度表
oersted 奥斯特【磁场强度单位】
OFAF(oil forced air forced) 强迫油循环风冷
OFA(optimized fuel assembly) 最佳燃料组件
OFA 过燃风【DCS画面】
OFC(oxyfuel gas cutting) 氧燃料气割
of design conditions of condenser 凝汽器变工况
off and on 继续的，不规则的
off-angle drilling 钻斜孔法
off-axis holography 离轴全息图
off-axis 离轴的，轴外的
off-balance 失去平衡
off-bottom rod 可向下伸的燃料棒
off-center condition 偏心度
off-centering of the gasket 密封垫的偏心
off-centering winding 偏心绕组，偏心线圈
off-centering 中心偏移，偏心的
off-center loading 偏心装料
off-center 跑偏，偏心
off-centre bubble 不对称气泡
off-centre coil 偏心线圈
off-centre 偏心
off-contact 开路触点，触点断开
off-critical 偏离临界的
off delay relay 断电延迟继电器
off delay unit 断开延时单元
off delay 断电延时，延迟断开，断电延时
off-design behaviour 变工况性能
off-design conditions 非设计条件，变工况
off-design efficiency 变工况效率
off-design equilibrium running 变工况运行
off-design performance 变工况性能
off-design point 非设计点，非设计工况
off-design 非设计工况，偏离设计工况，偏离计算工况，变工况
off dimension 尺寸不合格
off-duty 不值班的，停用的，备用的
off emergency 紧急断开
offensive odor 臭气，恶臭
offer and acceptance 要约与承诺
offered price 出售价格，标价，报价
offered product 外供产品
offeree 接受报价者
offerer 报盘人，发价人
offer firm 报实盘
offering date 报价日期
offering period 报价有效期
offering 提出，给予，插入，填入，嵌入
offer letter 报价书
offer subject to sample approval 看样后报价
offer subject to seller's confirmation 须经卖方确认的报价
offer with engagement 实盘
offer without engagement 虚盘，不受约束的发价
offer 发盘，报价，出价，开价，卖价，提供，提出，提议，给予
off flavour 臭气
off-gas activated charcoal column 废气活性炭柱
off-gas arisings 气态废物产生，排气量
off-gas buffer tank 废气缓冲罐，废气波动箱
off-gas circuit 废气回路
off-gas cleaning equipment 废气净化设备
off-gas cleaning system 废气净化系统
off-gas compressor 废气压缩机
off-gas condenser 除气凝汽器，废气冷凝器
off-gas control station 废气控制站
off-gas delivery 排气量
off-gas filter 废气过滤器
off-gas hold-up system 废气暂存系统
off-gasing 除气【电缆】，有害气体排放
off-gas monitor 废气监测器
off-gas pipe 废气管
off-gas piping system 废气管道系统
off-gas pump 抽气泵，排气泵
off-gas scrubber 废气洗涤器
off-gas storage tank 废气贮存箱
off-gas system holdup tank 废气系统暂存箱
off-gas system 废气系统
off-gas treatment 废气处理
off-gas vapour trap 废气蒸汽捕集器
off-gas 尾气，出口气，排出的气，废气，排烟，气体排放物
off-gauge 不按规格的，不标准的，非标准的，不合规格的，厚薄不均的
off-grade metal 等外金属
off-grade 劣质的，低级的
off grid energy 下网电量
off grid operation 离网运行
off grid 离网
off-grounded 未接地的，与地断开的
off-ground 中断接地
off-hand 无人管理的，无准备的，自动的，放手的，马上，立即
off-highway truck 越野卡车
off-hook signal 摘机信号
office automation 办公室自动化
office building of fabrication division 建造分公司办公楼
office building 办公楼
office cable 局内电缆
office computation 室内计算
office engineer 内业工程师
office expenses 办公费
office for incoming & outgoing mail 收发室
office lighting 办公室照明
office machine 事务用计算机
office manager 行政主管
office of entry 入境办事处
Office of Saline Water （美国）咸水淡化局
officer 公务员，官员，行政官员
office supplies 办公用品
office treatment 内业处理，室内处理
office worker 办事员
office work 办公室工作，内业，室内工作
office 办公室，办事处，事务所
official acceptance test 正式验收试验
official approval 官方批准，正式批准
official classification 正式分类
official contract 正式合同
official document 官方文件，正式文件
official exchange rate 官方汇率，官方外汇牌价

official fixed price　官价
official holiday　法定假期
official invoice　正式发票
official language　官方语言
official letter　公函
official price　官方价格，正式价格
official recognition　正式认可
official record　正式记录
official representative　官方代表
official seal　公章
official submission　正式提交
official technical document　正式技术文件
official test　正式试验，验收试验
official traffic　公务通信
official visit　正式访问
officiate　行使
off-impedance　断路阻抗
off interval　停电间隔
offlet　放水管，放水口
off-limit alarm　越限报警
off-line analysis　离线分析，非流线分析
off-line computer　离线计算机
off-line control　离线控制，间接控制
off-line data processing　间歇数据处理，离线数据处理，脱机数据处理
off-line data transmission　离线数据传送
off-line electrical consumption　（太阳能发电装置）夜间停运期间的厂用电耗量
off-line equipment　间接装置，离线设备，脱机设备，离线装置
off-line mode　离线方式，脱机方式
off-line operation　独立操作，离线运算，脱机操作，离线操作
off-line output　间接输出，脱机输出，离线输出
off-line printer　离线打印机，间接打印装置
off-line processing　离线处理，脱机处理
off-line processor　离线数据处理装置
off-line process　脱机处理
off-line sensor　管外传感器
off-line storage　离线存储器
off-line test　离线试验，间接试验
off-line unit　脱机单元，离线单元
off-line wash duration　离线清洗时间
off-line working　离线工作，间接工作
off-line　机外的，线外的，离线的，脱机的，非流线，管外的
off-load deaeration　给水在停运装置上除氧
off-loader　卸料机
off-load excitation voltage　空载励磁电压
off-loading　卸负荷，甩负荷，卸料
off-load period　无负荷期，卸载期
off-load refuelling　卸负荷换料，停堆换料
off-load regulation　无载调节，卸荷调节
off-load　卸载，减载，卸货，甩负荷，减荷，无负荷
off-normal condition　不正常情况，异常工况，反常工况
off-normal contact　离位触点，离位接点
off-normal loading　不正常加负载，不正常加负荷
off-normal loss　非正常亏损
off-normal lower　下限越界
off-normal occurrence　不正常现象
off-normal position　非正常位置，不正常位置
off normal upper　上限越界
off-normal　偏位，不正常的，离位，异常
OFF(oldest fuel first)　最先的燃料优先
off-on afterburner　不可调节的燃尽室
off-on control　开关控制
off-on operation　断开接通操作，断开闭合操作
off-on wave generator　键控信号发生器，键控信号振荡器
off-on　断开接通，开关
off-path　反正常路径的，不正常通路的
off-peak demand　峰外需电量，峰外需量
off-peak hours　非峰荷时间，非峰荷小时
off-peak load　低谷负荷，非高峰负荷
off-peak period　非峰荷期间，谷荷期间
off-peak power　非尖峰功率，非最大功率，峰外功率
off-peak system　非高峰系统
off-peak time　离峰时间
off-peak　正常的，偏离高峰的，非峰值的，非峰荷的，峰值外的
off period　断开时期，停运时间，停运周期，闭塞时间
off-plant ash transportation system　厂外输灰系统
off-position　不工作状态，断路状态，关闭位置
off quality　品质不合格
off-rating　非标准状态，不正常状态，非额定工况，超出额定值
off-ratio firing　偏离比例的燃烧
off-resonance　离开共振，非共振
off-return　不复位
off-road vehicle　越野车
off-road　越野的
offscum　废渣
offset angle　偏斜角
offset behavior　补偿性能
offset bend　偏置弯管，迂回管
offset carrier　偏移载波，偏置载频
offset characteristic　偏移特性
offset coefficient　静差系数，偏移系数
offset component　偏差分量，位移分量
offset correction　偏差校正
offset current　失调电流，补偿电流
offset deviation　偏移差
offset disc butterfly valve　偏心（阀板）蝶阀
offset electrode　偏心电极
offset elongation　持久延伸
offset factor　偏移因子
offset fault current　故障偏移电流
offset feed　偏置馈源
offset footing　大方基础，宽阶底脚，踏步式基础，大方脚
offset limit　屈服极限
offset mho relay　偏置姆欧继电器
offset pattern socket wrench　弯脖套筒扳手
offset pipe　迂回管，补偿管，偏置管，乙字弯
offset profile　边线断面
offset reducer　偏心减径管
offset rod　支距杆

offset screwdriver 偏心螺丝起子
offset staff 偏距尺
offset strip fins 错置的带状肋片
offsets 偏置管
offsetting 偏心距，支距测法，位移，偏移
offset voltage 补偿电压，偏移电压
offset wave 偏移波
offset wrench 偏头扳手，斜口扳手
offset yield strength 偏移弹性极限，条件屈服强度，补偿屈服强度，残余变形屈服强度
offset 偏移，抵消，抵扣，补偿，位移，分支，冲账，不重合，偏心的，残留误差，静差，余差，失调，支距
offshore area 近海区，离岸区
offshore bank 境外银行
offshore bar 岸外坝，沿岸洲，沿岸沙坝，离岸沙洲
offshore boring 滨外钻孔，海洋钻探
offshore breakwater 岛式防波堤，岛堤，离岸防波堤
offshore current 离岸流
offshore deposit 近海沉积
offshore discharge 近海排放
offshore disposal 近海处理
offshore drilling 滨外钻探，近海钻井，海洋钻探
offshore engineering 近海工程
offshore environment 近海环境
offshore facies 远岸相
offshore fund 国外资金
offshore mooring 离岸停泊
offshore nuclear power plant 海滨核电站，近海核电厂
offshore office 境外办事处
offshore oil field 滨外油田
offshore oil rig 近海石油平台
offshore power plant 海上电站
offshore power systems floating nuclear plant 海上动力系统浮动核电厂
offshore procurement 国外采购
offshore repairing and inspecting devices 海上维修检测仪器
offshore spread 海上船只
offshore structure 海洋钻架，离岸结构，近海结构物
offshore wind energy 海上风能
offshore wind farm 海上（离岸）风电场
offshore wind power 海上风电
offshore wind 离岸风，陆风
offshore 离岸的，海外的，近海的，境外的，近海，离岸，向海的，滨外，滩外
offsite accident 厂区外事故，厂外事故
offsite auxiliary power source 厂外辅助电源
offsite emergency plan 厂外应急计划，场外应急计划
offsite environmental surveillance 厂外环境监视，场外环境监视
offsite power source 厂外电源
offsite power supply 厂外电源
offsite power system 厂外供电系统
offsite power 厂外电源
offsite protective measures 场外防护措施
offsite supply voltage 厂外供电电压
offsite surveillance 非现场监视，场外监视
offsite ultimate storage （放射性废物）最终厂外贮存，最终场外汇存
offsite 厂区外的，现场外的，工地外，厂区外，场外
off-specification material 不合规格的材料
off-specular peaks 偏离镜面尖峰，偏振辐射
offspeed 转速偏离，比正常（或预料）速度慢的
off-staff people 编外人员
off stall speed 无动力失速速度
off-state current 异常电流，开路电流
off-state leakage 断路漏电
off-state time 断开时间
off-state voltage 断路状态电压
off-state 关闭状态，截止状态，断开状态，切断状态，断开的
off-stream storage 非河道蓄水
off-stream unit 停用设备，备用设备，停运设备
off-stream 停用的
off-street parking lot 路外停车场
off-take agreements 包销协议
off-take contract 承购合同
off-take hood 排风罩
off-take main 排出管道
off-take pipe 排水管，排气管，排出管
off-take point （变电站与输电线路之间的）并网点，取出点，（沸腾炉）排料点
off-taker 承购商，包销买家，（购电协议的）买方，购电方
off-take structure 分水建筑物
off-take target 开采目的层
off-take 出口，（商品）出货，购买，承购，移走，排水管，排气管，支管，分叉，分支，出风口，分接头，购电，取去，移去，扣除，商品销售，（一次）购买量，排出管，泄水处
off-the-air signal 停播信号
off-the-line 解列，停运
off-the-record 非正式的，不留记录的，不许发现的
off the shelf item 非架上的项目
off the shelf 现货供应的，现成的，常备的，成品的，非专门设计的，现役的，流行的
off-the-wall switchboard 离墙配电盘
off-the-wall 离墙的
off-time 停机时间，间歇时间，断开时间，休止时间
off transistor 截止晶体管，不导通晶体管
off-tube 带有断开电子管的闭锁管，断开管
off-tune 失调的
off wind 顺风
off 断开
of good repute 有名誉的，名誉好的，优质的，上等的
of great extent 范围很大
OFLD＝offloaded 卸下，拉货
of one's own accord 自行，自愿
OF(raw water filtration plant) 生水过滤站
OFT(oil fuel trip) 燃料油跳闸，燃油切断

of top priority　最优先的
OFWF　强制油循环水冷，强迫油循环导向水冷
ogee attachment　曲线连接
ogee dam　反弧形断面溢流坝
ogee diversion dam　反弧形引水坝
ogee flange　S形板
ogee ring　S形截面的环
ogee spillway　反弧面溢洪道，反弧形溢流道
ogee　反弧形【即溢流面曲线】，双弯形
ogive　尖顶拱
ohmage　欧姆数，欧姆电阻，欧姆阻抗
ohm ammeter　欧安计
ohmer　欧姆计
ohm gauge　欧姆表
ohmic bridge　（集成电路）欧姆电桥，电阻电桥
ohmic component　电阻性元件
ohmic contact　电阻性接触，欧姆性接触，欧姆接触
ohmic drop　电阻性电压降，欧姆电压降
ohmic heating　欧姆加热，电阻加热
ohmic junction　欧姆（非整流）结
ohmic leakage　漏电阻
ohmic loss　电阻损失，欧姆损失，铜损
ohmic resistance test　欧姆电阻试验，电阻测试，直流电阻损耗，欧姆电阻损耗
ohmic resistance　欧姆电阻，直流电阻
ohmic value　欧姆值，直流电阻值
ohmic　欧姆的，电阻性的
ohmmeter　欧姆计，欧姆表，电阻计
ohm metre(-er)　欧姆米
OHM＝ohmmeter　欧姆计，电阻表
ohm relay　欧姆继电器，电阻继电器
Ohm's law analogy　欧姆定律类比
Ohm's law wheel　欧姆定律轮
Ohm's law　欧姆定律
ohm type reactance relay　欧姆型电抗继电器
ohm　欧姆【电阻单位】
OHO(Overall Hand-over)　全面移交
OH&S(occupational health & safety)　职业健康安全，职业健康与安全，职业健康和安全
OHV(observed heat value)　实测热值
oil absorption　吸油
oil-actuated check valve　油动逆止阀
oil additives　油的添加剂
oil ageing　润滑油的氧化，油老化
oil and chalk process　油和白垩粉工艺
oil and grease storage　油和脂的贮存
oil and water trap　油水收集器
oil asphalt　油沥青
oil atomization　油的雾化
oil atomizer　油雾化喷嘴，油雾化器
oil baffle　挡油板
oil-base paint　油底漆
oil bath lub　油浴润滑
oil bath　油浴，油槽，油浴锅
oil-bearing circulator　油轴承循环风机
oil-bearing sewage drain　含油污水排水管
oil-bearing　含油的
oil berth　油轮泊位，油码头
oil-blast switch　油灭弧开关
oil booster pump　增压油泵
oil bottle　油杯，油瓶
oil bowl　油杯
oil box　油箱
oil breaker　油开关，油断路器
oil-break fuse　油熔断器
oil breather　吸潮器，油枕，油箱
oil buffer　油缓冲器，油减振器，油压缓冲器
oil bumper　油缓冲器，油减震器
oil burner gun　燃烧器油枪
oil burner nozzle　油雾化器，油喷嘴
oil burner　燃油喷嘴，油燃烧室，油燃烧器，燃油烧嘴
oil-burning boiler　燃油锅炉
oil can　油罐，油壶
oil cap　油杯盖
oil casing　油箱
oil catcher　油收集器，盛油器，储油箱，集油器
oil catch-ring　挡油环，集油环
oil catch　集油器
oil cavity　油室
oil cell　油箱
oil chamber　油室，储油箱，油腔
oil change　换油
oil channel　润滑系统，油槽
oil charge　加油
oil chuck　液压卡盘
oil circuit-breaker　油断路器，油开关
oil circuit recloser　油路自动重合闸
oil circuit　油路
oil circulating pump　循环油泵
oil circulating system　循环油系统
oil circulation cooling　油循环冷却
oil circulation　油循环
oil cistern　油杯
oil clarifier　油澄清器
oil cleaner　滤油器
oil-cleaning system　油净化系统
oil-cleaning tank　净油箱，油净化箱，滤油箱
oil clearance　油膜间隙，油隙
oil cloth　绝缘油布，油布
oil-coal slurry　油煤浆
oil cock　油旋塞
oil coke　油焦
oil collecting basin　集油池
oil collector　储油箱，集油装置
oil compression cable　高压充油电缆
oil condenser　油浸电容器
oil condition device　油净化装置
oil conditioner　油调节器，油净化器
oil conditioning system　油净化系统
oil conditioning　油处理
oil conduit　油管路
oil cone　油雾化锥
oil conservation　石油资源保护，节油
oil conservator　油枕，贮油柜
oil consumption rate　油耗率
oil consumption　油耗，耗油量
oil contactor　油接触器
oil container　贮油器，油室
oil containment sump　事故油坑
oil-contaminated water　含油污水

oil-contaminated 油污染的
oil contamination 油污染
oil content 含油量
oil-controlled valve 液动阀，油动阀
oil-controlled 液压传动的，液压控制的
oil control ring 油环
oil cooled stator winding 油冷定子绕组
oil cooled transformer 油冷变压器
oil cooled turbogenerator 油冷汽轮发电机
oil cooled 油冷却的，油冷的
oil cooler 油冷却器，冷油罐，冷油器
oil cooling radiator 油冷散热器
oil corona 油电晕
oil cover 油孔盖
oil crisis 石油危机
oil cup 油杯
oil cut-off valve 油截止阀
oil cylinder 油缸
oil damper 油阻尼器，油缓冲器，液压缓冲器
oil dash pot relay 缓冲油壶继电器
oil dashpot 油缓冲罐，油减震器，油阻尼器
oil deflector 挡油板，导油器
oil degasifier 油除气器
oil dehydrating plant 油脱水车间，油的脱水装置
oil delivery pipe 供油管
oil depot 油库
oil-depth gauge 油位表
oil deterioration 油质劣化，油变质
oil development 石油开发
oil dilution 油的稀释
oil dip rod 量油尺
oil discharge outlet 排油口
oil discharge platform 卸油站台
oil discharge valve 泄油阀
oil dispenser 油分配器
oil dispersant mixture 油分散剂混合液
oil drain cock 放油旋塞，疏油开关
oil drain line 排油管线，疏油管线
oil drain pipe 卸油管
oil drain platform 卸油栈台
oil drain pump 排油泵，卸油泵
oil drain valve at star-up 启动排油阀
oil drain valve 放油阀
oil drain 放油，放油口
oil drilling waste 石油钻井废液
oil-drowned 油浸的，浸于油中的
oil drum 油枕，贮油柜
oil drying method 油干燥法
oil drying 油干燥
oil duct 油沟
oiled paper 油纸，油浸绝缘纸
oiled sleeper 注油木枕
oiled 加油的，油浸的
oil ejector for pump suction 主油泵吸入管的抽油器
oil ejector 注油器，射油器
oil-electric engine 柴油发电机
oil eliminator 除油器，油分离器
oil emulsion adjuvant 油乳化佐剂
oiler 油杯，油壶，加油壶，加油器，注油器，加油员
oil expansion chamber 油枕，储油柜
oil expansion vessel 油枕，储油柜
oil extraction cost 原油开采成本
oil failure trip 油系统故障截断，断油脱扣
oil feed disc 给油盘
oil feeder 加油器，进油器
oil feeding reservoir 给油箱，补油箱
oil feeding system 给油系统
oil feed pump 供油泵，给油泵
oil-filled air-blast cooled 充油鼓风冷却的
oil-filled bushing 充油套管，充油衬套
oil-filled cable installation 充油电缆安装
oil-filled cable 充油式电缆，充油电缆，油浸电缆
oil-filled condenser bushing 充油电容器式套管
oil-filled manometer 油封压力表
oil-filled naturally cooled 充油自然冷却的
oil-filled semistop 充油半截止头
oil-filled system 充油系统
oil-filled transformer 充油式变压器，充油变压器，油浸式变压器
oil-filled 充油的，油浸的
oil filler point 注油孔
oil filler 注油器，给油器
oil filling test 充油试验
oil filling 注油，充油
oil film loss 油膜损耗
oil film oscillation 油膜振荡
oil film rigidity 油膜刚度
oil film stiffness 油膜刚度
oil film technique 油膜（流动显示）技术
oil film thickness 油膜厚度
oil film vibration 油膜振动，油膜振荡
oil film visualization 油膜显示
oil film wedge 油楔，油膜楔
oil film whip 油膜振荡
oil film whirl 油膜旋涡
oil film 油膜
oil-filtering machine 滤油机
oil filter paper 滤油纸
oil filter 滤油机，滤油器，除油器，油过滤器
oil-fired boiler 燃油锅炉
oil-fired furnace 燃油炉膛
oil-fired power plant 燃油电厂，燃油发电厂，燃油电站
oil-fired torch 燃油点火火炬
oil-fired unit heater 燃油热风器
oil-fired 燃油的
oil-firing 燃油
oil fitting 加油器
oil flow detection relay 油流检测继电器
oil flow indicator 油流量指示器，油管上的观察窗
oil flow line 油流管线
oil flow picture 油流谱
oil flow regulating 油流量调节
oil flow visualization 油流流动显示
oil flow 油流量
oil flushing 油冲洗，充油，注油
oil fogger 油雾化器
oil fog 油雾

oil fouling 油污
oil-free air compressor 无油空气压缩机
oil-free air 无油空气
oil-free condensate 无油冷凝水，无油凝结水
oil-free lubrication air compressor 无油润滑空气压缩机
oil-free 无油的
oil fuel trip 燃料油跳闸，燃油切断
oil fuel 油燃料
oil-fume extractor 排油烟机
oil-fume separator 油烟分离器
oil fuse cutout 油浸熔燃断路器
oil fuse 油浸式熔断器
oil gas gathering and transportation 油气集输
oil gas shielded type transformer 充油封闭式变压器
oil gas storage 油气储存
oil gas 石油气
oil gauge 油压计，油标，油量计，油位表
oil grinding machine 油磨机
oil grooves 润滑油槽，油槽
oil guard 防油器，挡油器，护油圈
oil gun torch 油枪
oil gun 机油枪，油枪
oil hardening 油淬火
oil heater storage tank 油加热器的油箱
oil heater 油加热器
oil heating and pumping set 燃油加热及泵送装置
oil hole 油孔
oil hydraulic motor 油马达，液压马达
oil hydraulic stud tensioner 油压螺栓拉伸机
oil ignitor 油点火器
oil-immersed air-blast cooling 油浸风冷，油浸强迫风冷
oil-immersed apparatus 油浸式电器
oil-immersed arc-control device 油浸式电弧控制设备
oil-immersed breaker 油浸式断路器
oil-immersed current transformer 油浸式电流互感器
oil-immersed natural cooling 油浸自然冷却
oil-immersed regulating transformer 油浸式调整变压器
oil-immersed starter 油浸式启动器
oil-immersed type transformer 油浸变压器，油浸式变压器
oil-immersed water cooled transformer 油浸水冷变压器
oil-immersed water cooling 油浸水冷
oil-immersed 油浸的
oil immersion test 油浸没试验
oil impeller 油涡轮，旋转阻尼
oil-impregnated cable 油浸渍电缆
oil-impregnated paper dielectric capacitor 油浸纸介电容器
oil-impregnated paper insulated cable 油浸纸绝缘电缆
oil-impregnated 油浸渍的
oiliness 润滑性，油性
oiling of transformer 变压器注油
oiling system 充油系统
oiling 加油法，加油，注油，注油法
oil injection header 注油总管
oil injection nozzle 喷油嘴
oil injection pump 注油泵，射油泵
oil injection 喷油
oil injector 注油器
oil inlet 加油口
oil insulated terminal 油绝缘端子
oil insulation 油绝缘
oil interceptor 油污截流井，集油器，油收集器
oil-in-water emulsion 水包油乳化液
oil-in-water type coolant 乳化冷却液
oil-in 油入口，进油口
oil-isolated reciprocating pump 油隔离泵
oil jack 油压千斤顶
oil jet pump 射油泵
oil lead 输油管，引油管线
oil leakage sump 泄油池，事故油坑
oil leak 漏油
oil less bearing 无油轴承
oilless circuit breaker 无油断路器
oilless 无油
oillet 孔眼，视孔
oil level alarm test 油位警报试验
oil level gauge 油位表
oil level glass 油位观察窗
oil level indicator 油位指示器
oil level regulator 油位调节器
oil level 油位，油面
oil-lift bearing 高压油顶起轴承
oil lighter 油点火器
oil lighting burner 油点火喷燃器
oil line for turbine proper 本体油管路
oil-line 油路，油管路，油管，油道
oil lock 油栓，油塞
oil-lubricating bearing 油润滑轴承
oil lubrication 油润滑
oil make-up 补油
oil manometer 油压力计，油压表
oil measurement 油的测定
oil-minimum breaker 少油开关，少油断路器
oil mist 油雾
oil motor 液压马达
oil nozzle 喷油嘴，油管嘴
oil-operated voltage regulator 油压操作电压调整器
oil outlet 出油口，排油口
oil-out line 油流出管道
oil-out 出油口
oil overflow valve 溢油阀
oil pad 油垫
oil paint 油漆
oil pan 油盘
oil partition wall 挡油墙
oil petroleum 石油沥青
oil pipeline 输油管道
oil pipe trench 油管沟
oil pipe 油管道，油管
oil pit 油坑，油槽
oil-pneumatic rotary drive 油气动旋转驱动
oil pocket 储油箱，油袋

oil pollution control 油污染防止
oil pollution 石油污染，油污染
oil port 油港，油口，油槽，油池
oil pressure above pilot valve 滑阀上油压
oil pressure adjustment 油压调整
oil-pressure atomizing burner 油压雾化喷燃器
oil pressure check valve 油压止回阀
oil-pressure control relay 油压控制继电器
oil pressure line 压力油管
oil pressure low trip device 油压降低脱扣器
oil pressure regulator 油压调节器
oil pressure relay 油压继电器
oil pressure relief valve 油压安全阀
oil pressure selector 油压选择器
oil pressure supply system 油压装置
oil pressure system 油压系统
oil pressure under pilot valve 滑阀下油压
oil pressure 油压
oil proof rubber 耐油橡皮
oil-proof 不透油的，防油的，耐油的，不漏油的
oil protection wall 挡油墙
oil pump assembly 油泵总成
oil pump capacity 油泵流量
oil pump cover 油泵盖
oil pumping ring 油泵环
oil pump motor 油泵电动机
oil pump pit 油泵坑
oil pump station 油泵房
oil pump 油泵
oil puncture 油击穿
oil purification device 油净化装置
oil purification plant 油净化车间，油净化装置
oil purification room 油处理室
oil purification 油净化
oil purifier 净油器，滤油器，净油装置，油净化器，油净化装置
oil purifying pump 净化油泵
oil quenching 油淬火
oil radiator 滑油散热器，油散热器
oil reclaimer 油再生装置
oil rectifier 油分离器
oil refinery 炼油厂
oil regeneration 油再生
oil-relay governor 油继动调节器
oil-relay system 油动伺服机构
oil-relay 油继动器
oil release valve 油压安全阀，油压调节阀
oil relief valve 放油安全阀
oil removal basin 除油池
oil remover 除油器
oil reservoir 油池，油箱，储油器，油槽
oil residue 油残渣
oil resistance 耐油【皮带面层】，耐油性
oil resisting paronite 耐油石棉橡胶
oil resisting rubber 耐油橡皮
oil retainer collar 润滑油保持环，挡油环，油环挡圈，护油圈
oil retainer ring 润滑油挡油环
oil retaining wall 防油壁，防油堤
oil return line 回油管线
oil return pipe 回油管
oil return valve 回油阀
oil ring guide 油环导槽，油环套管
oil ring retainer 油环挡圈
oil ring 油环，挡油环，甩油环
oil scavenger 排油器，分油器
oil scraper ring 刮油环
oil scraper 刮油刀，刮油器
oil scraping ring 刮油环
oil-sealed 带油封的
oil seal housing 油封套
oil sealing device 油封装置
oil sealing ring 油封环
oil seal 油封，油密封
oil segregating shurry pump house 油隔离灰浆泵房
oil segregating shurry pump 油隔离灰浆泵
oil separate chamber 油分离槽
oil separator 分油器，油分离器
oil servomotor 油动机，油伺服马达
oil shale recovery 油页岩开采
oil shale rock 油母页岩，油密的
oil shale 油页岩
oil shield 护油罩，防油罩
oil shock absorber 油压缓冲器，油压减震器
oil skin 油布
oil sleeve 油套
oil slick 油膜
oil slinger 抛油圈，甩油环
oil sludge 油泥，油渣
oil smoke 油烟
oil snubber 油阻尼器，油减振器，油缓冲器
oil-soluble 溶于油的
oil soot 油烟炱
oil sprayer 油喷雾器，油喷嘴
oil spray 喷油，喷雾
oil stain 油污
oil starter 油启动器
oil stone 油石
oil storage tank foundation 储油罐基础
oil storage tank 储油箱，贮油槽，贮油箱
oil storage 储油
oil store 油库
oil strainer 滤油器，滤油网
oil-submersible electric equipment 潜油电气设备
oil sump 油池，回油箱，油槽，油盆
oil supply bar 供油主管，操作油管
oil supply line 供油管
oil supply mains 供油母管，供油系统
oil supply pump 供油泵
oil supply system 供油系统
oil supply valve 补油门
oil supply 供油
oil switch 油浸开关，油开关
oil syringe 油枪，注油器
oil system failure trip 油系统故障跳闸
oil tank breather 油箱吸潮器，油箱呼吸器
oil tank car 油罐车，油槽车
oil tanker 加油车，油槽车，油轮
oil tank gas exhauster 油箱排气装置
oil tank lorry 油槽车，油罐车

oil tank radiator 油箱散热器，油箱散热管
oil tank wagon 油箱车
oil tank zone 油罐区
oil tank 油槽，油箱，储油罐，油罐
oil tar 焦油
oil temperature gauge 油温度计
oil temperature regulator 油温调节器
oil temperature rise trip device 油温上升脱扣器
oil terminal 石油港，转运油库，石油码头，油轮泊位
oil tester 试油机，试油器
oil thrower 抛油圈，挡油圈，抛油环
oil throw ring 抛油圈，甩油环
oil-tight 油封的，油密的，不透油的，不漏油的
oil-to-air lube cooler 油气式润滑油冷却系统
oil transfer pump 输送油泵，输油泵
oil transportation 输油
oil trap 集油池
oil treater 油净化器
oil treatment room 油处理室
oil treatment 油处理
oil trip test 油跳闸试验
oil trip valve 油压断流阀
oil trip 断油跳闸
oil trough 油槽
oil tube cooler 管状油冷却器
oil turbine 油透平
oil unloading pier 卸油码头
oil unloading platform 卸油站台
oil unloading pump house 卸油泵房
oil unloading station 卸油站
oil unloading system 卸油系统
oil unloading wharf 卸油码头，油码头
oil unload pump 卸油泵
oil valve 补油门，油阀
oil vapor backstreaming 油蒸汽反流，油烟回流
oil vapor diffusion pump 油烟扩散泵，油蒸汽扩散泵
oil vapour 油雾，油气，油烟
oil-varnished cambric 油漆布
oil varnish 油漆
oil warm-up torch 燃油升火火炬
oil waste drainage system 废油排放系统
oil way 油路，润滑油槽
oil wedge 油楔
oil well wastewater 油井废水
oil well 油槽，油杯，油井
oil whip 润滑油起泡，油膜振荡，油沫
oil whirling 油涡
oil wick 油绳，油芯
oil window 油口，油窗
oil wiper 拭油器，刮油器，刮油板，刮油环
oily matter 油垢物，油性物质
oily pollutant 含油污染物
oily sewage 含油污水
oily waste water 含油污水，含油废水
oily water 混油水
oily 油的，油质的，多油的
oil 油，加油，上油
ointment 油膏
OIS(operator interface station) 操作员接口站
OIV(open interceptor valve) 开中压调门
OJT(on the job training) 在岗培训
older quantity 定货量
Oldham coupling 欧氏联轴节，滑块连接，双转块机构，十字滑块联轴器
oleaginous 润滑的
olefin 烯烃，烯
oleoresin 含油树脂，油松脂
oleum 发烟硫酸
oligodynamic 微动力的
oligohaline water 低盐分水
oligohalobic lake 贫营养湖
oligohalobic waters 低污水域
oligohalobic 淡水的
oligomerics 齐聚体，低聚物
oligomer 低聚物，齐聚体
oligopoly market 寡头市场
oligopoly price 卖主垄断价格，寡头统销价格
oligopsony price 垄断价格
olive brown 橄榄棕色
olive green 橄榄绿
olivine rock 纯橄榄岩
olivine 橄榄石
OLO(off line operation) 离线操作
OLO(on line operation) 在线操作
O/L(operation/logistics) 运算逻辑
OL(operator license) 操作员执照
OL＝overload 过载
OLR(overload relay) 过载继电器
OLRT(on-line real-time) 在线实时的
OLTC(on line tap changing) 在线轴头变换，有载轴头变换
OLTC(on load tap changer) 有载调压变压器，加载抽头变换器，有载换接器，有载分接开关
O,M&A(operation, maintenance and administration) 运行、维护与管理
ombrograph 雨量计，雨量记录仪
ombrometer 雨量计
O&M cost(operation and maintenance cost) 运行维护费用
omega rail 欧米茄轨
omega Ω【希腊字母】，远程导航系统
omen 先兆，预兆
OMG(Operations Manager) 生产部部长
OMIS(operation maintenance information system) 检修计算机数据库管理系统
OMIS(ops computer management information system) 生产部计算机管理系统
omission 省略，忽略，遗漏，不作为
omni antenna 全向天线
omnibearing selector 全方位选择器
omnibearing 全方位的，全向的
omnibus bar 汇流排，汇流条，母线
omnibus configuration 总线配置，总线布局
omnibus speaker circuit 公共通话电路
omnibus 公共汽车，多用的，公用的，总的
omnidirectional aerial 全向天线
omnidirectional antenna 非定向天线，全向辐射天线
omni directionality 不定向性，全向性
omnidirectional pressure transducer 全向压力传

感器
omnidirectional solar cell 全向太阳能电池
omnidirectional 无定向的，全向的
omnidistance 全程，至无线电信标的距离
omnigraph （发送电报电码的）自动拍发器，缩图器
omnilateral pressure 全侧向压力
omnimate 简化的自动生产设备
omnirange 全程，全向无线电信标，全方向
omni 全方位
OM(operating memory) 操作存储器
O & M(operationand and maintenance) 运行与维护，"委托运营"模式，"运营—维护"模式【工程建设模式】
O/M ratio(oxygen to metal ratio) 氧（重）金属比
OMR(open market recycle) 露天市场废物回收，公开市场再循环
O & MR (operations & maintenance remainders)（核电厂）运行与维修备忘录
OMR(optical mark reader) 光标阅读器
OMR(optical mark recognition) 光标识别
OMS(occupational medical service) 职业医疗中心
OMS(outage management system) 停电管理系统
OMS(overpressure mitigation system) 超压缓解系统
OMS(ovonic memory switch) 双向存储开关
OMT(operation, maintenance and test) 运行、维护与试验
on a...basis 在……的基础上
on a batch basis 批式的，分批的，间歇的
on a case-by-case basis 按个案（处理），逐个地，视情况而定
on account of 因为，由于，基于
on account 赊账
on a crash basis 紧急的，应急的
ONAF(oil natural air forced) 油浸风冷
on a large scale 大规模的［地］，广泛地
on a national scale 在全国范围内
on and after 从（某日）起
on-and-off switch 通断开关
ONAN(oil natural air natural) 油浸自冷
on a year-by-year basis 按每年的具体情况，按年，一年一年的
on a yearly basis 每年
on board bill of lading 已装运提单，装船提单
on board B/L 已装船提单
on-board computer 船上计
on-board date 装船日期，装运日期
on-board gantry crane 机载龙门起重机，车载龙门起重机
on-board memory 板上内存
on-board notation 已装船批注
on-board ocean bill of lading 已装船的海运提单
on-board 已装船的，船载的，板上的，车载的，机载的
ONB(onset of nucleate boiling) 泡核沸腾起始点
on-bottom rod 不可向下伸的燃料棒，插到底的棒
on call personnel 随叫随到人员
on call 随叫随到的
once forever statistical survey 一次性统计调查
once in one hundred year 百年一遇地［的］
once-in-50-year return wind 50 年一遇的风
on centers 中心距
once-through boiler 直流锅炉
once-through circulation 一次循环，直流流动
once-through condenser cooling 直流式冷凝器冷却
once-through cooling system 开式冷却回路，单程冷却系统，直流冷却系统
once-through cooling water system 直流冷却水系统
once-through cooling 一次通过冷却，单程冷却，直流冷却，非循环冷却
once-through cycle 开式冷却回路，单程冷却系统，单循环
once-through flowsheet 一次循环流程图，一次通过流程图
once-through flow 贯穿流，通流
once-through fuel cycle 一次通过式燃料循环
once-through heat exchanger 单程热交换器，直流式热交换器
once-through operation 单流程操作，一次运算，一次循环
once-through principle 直流原理，非循环原理
once-through reactor 直流冷却反应堆，燃料一次循环反应堆
once-through refueling 直通换料
once-through steam generator 直流式蒸汽发生器
once-through system 一次通过系统，非循环系统，直流（太阳热水）系统
once-through then out 一次通过即卸料
once-through type steam generator 直流式蒸汽发生器
once-through uranium-fuel cycle 一次通过铀燃料循环
once-through water supply system 直流供水系统
once through 直流式，开环，直流的，单流程的，一次的，一次通过式
on 3cm centers 中心距为 3 厘米
oncoming flow 来流，迎面流
oncoming neutron 入射中子
oncoming particle 入射粒子
oncoming turbulence 来流湍流（度）
oncoming wind 来流风，迎面风
on condition that 条件是
on deck bill of lading 船上交货提单
on deck cargo 舱面货
on deck risk 舱面险
on-delay unit 接通延时单元
on-delay 通电延时
on demand guarantee "见索即付"保函，即期担保
on demand performance guarantee 即期履约保证
ondograph 电容式波形记录仪，高频示波器
ondometer 波长计，波形测量仪，测波仪
ondoscope 示波器，辉光管振荡指示器
ondulation 波动，波浪式振荡
on-duty 值班
one addition time 一次加法时间

one-address code 单地址码
one-address instruction 单地址指令
one-address 单地址的
one after another 相继
one and a half breaker arrangement 一个半断路器接线
one and a half breaker 一个半断路器
one and half group model 一群半模型
one-and-two pipe combined heating system 单双管混合式采暖系统
one and zero code 一与零编码
on earth disposal （废物）陆地处置
One Belt One Road Alliance of International Chambers of Commerce 一带一路国际商会协作联盟
One Belt One Road initiative 一带一路倡议
one bladed wind turbine 单叶风电机组，单叶风力机
one-body model 单体模型
one-bolt fastening 单螺栓固定
one-brick wall 单砖墙
one caliper disc brake 一卡钳盘式制动器
one-centered arch 单心拱
one-coil transformer 单圈变压器
one-color indicator 单色指示剂
one-column adder 一位加法器
one-column binary add circuit 一位二进制加法电路
one-column radiator 单柱散热器
one-column 一位的
one component balance 单分量天平
one contracting party 缔约一方
one-course concrete pavement 单层混凝土铺面
one degree of freedom 一次自由度
one-digit adder 半加法器
one-digit delay 一个数位延迟
one-digit substractor 半减法器
one-digit time 一个数字的时间
one-digit 一个数位的
one-dimensional consolidation 单向固结，一维固结
one-dimensional current 一维流
one-dimensional diffusion codes 一维扩散编码
one-dimensional flow 一维流动，一元流
one-dimensional lattice 一维点阵
one-dimensional space 一维空间
one-dimensional 一维的，一元的
one-dimension 一维，一元，单量纲
one direction thrust bearing 单向推力轴承
one-drum type boiler 单汽包锅炉
one-effect evaporator 单效蒸发器，单级蒸发器
one-electron bond 单电子键
one-electron theory 单电子理论
one-energy-storage network 单储能元件网络
one-figure number 单位数
one-figure 单位的，单值的
one-flow core 单流程堆芯，单值数
one fluid model 单流体模型
one gate 或门
one-group approximation 单群近似法
one-group calculation 单群计算
one-group method 单群法
one-group model 单群模型
one-group perturbation theory 单群微扰理论
one-group theory 单群理论
one-group transport equation 单群输运方程
one-group treatment 单群处理
one-half load 半载
one-half period 半周期
one-hinged arch 单铰拱
one-hour duty 一小时工作方式
one-hourly precipitation 一小时降水量
one-hour rating 一小时额定出力
one-kick 单次的，一次有效的
one-layer winding 单层绕组，单层绕法
one-layer 单层的
one leg 单端
one-level address 直接地址，一级地址
one-level code 一级代码，直接码
one-level subroutine 一级子程序
one-line adapter 单线适配器，单线转接器
one-line diagram 单线图，电气主接线图
one-line 单线
onemeter 组合式毫伏安计
one-minute insulation test 一分钟绝缘试验
one number time 一个数的时间，数的周期
one-of-a-kind operation 唯一运行方式
one-of bidding 单次投标
one-off operation 唯一运行方式
one-on-one unit 一拖一机组
one-on-one 一拖一【燃气轮机电厂】
one-out-of-ten code 十中取一码
one-out-of-two system 二取一系统
one-out-of-two taken twice 两次二取一
one pair brush multiphase series motor 单对电刷式多相串激电动机
one-parameter model 单参数模型
one-parameter 单参数的
one-part price 单一制电价
one-pass compiler 一遍编译程序
one-pass operation 一遍操作
one-pen recorder 单笔记录器
one-pen 单笔的
one-phase relay 单相继电器
one-phase short-circuit 单相短路
one-phase 单相的
one piece blade 整叶片
one piece casting 整体铸件，整体铸造
one piece construction 整体结构
one piece crankshaft 组合式曲轴，群体式曲轴，整体式曲轴
one-piece-forged runner 整锻叶轮，整锻转子
one-piece-forged shaft 整锻轴
one-piece-forged 整锻的
one piece frame 整体机座
one piece overalls 联合工作服
one piece ring 单片环
one piece runner 整体式转轮
one piece stator 整体定子
one piece structure 整体结构
one piece support clamp 整体夹持器
one piece 一体的，单片的，单块的，上下身相连的衣服

one-pier elbow draught tube 单墩肘形尾水管
one-pipe circuit cross-over heating system 单管跨越式采暖系统
one-pipe heating system 单管采暖系统
one-pipe hot-water heat-supply network 单管制热水热网
one-pipe loop circuit heating system 水平单管采暖系统
one-pipe series-loop heating system 单管顺序式采暖系统
one-pipe steam heat-supply network 单管制蒸汽热网
one-pipe system 单管系统
one-plus-one address instruction 二地址指令，一加一地址指令
one-plus-one address 一加一地址
one-plus-one instruction 一加一指令
one-point continuous method 单点连续法
one-point method of stream gauging 单点测流法，一点测流法
one-point mooring 单点系泊
one-point wavemeter 定点波长计
one-port network 单口网络
one-position winding 单层绕组
one-pulse delay 一个脉冲延迟
one-pulse one-address 单脉冲单地址
one-pulse time delay 单脉冲延时，单脉冲时间延迟
one-pulse time 单一脉冲时间
one-pulse 单脉冲
one-quadrant multiplier 一象限乘法器，单象限乘法器
one-quadrant 一象限的，单象限的
one-range winding 单平面绕组
one-region reactor 单区反应堆
one risk 一切风险
onerous 承担（过重）义务的，有偿的，繁重的
one's complement 一的补码
one-second theodolite 一秒读经纬仪
one-section reactor 单区反应堆
one setting 一次调整，一次整定
one-shift operation 一班制运行
one shipment 一次装船
one-shot circuit 单触发电路，冲息触发电路
one-shot method 一步法，一步发泡法
one-shot multivibrator 单稳态多谐振荡器，冲息多谐振荡器
one-shot operation 单步操作，单发操作
one-shot request 单发请求【计算机】
one-shot trigger circuit 单稳触发电路
one-shot 只有一次的，一次使用的，一次通过的，单稳的，单触发
one side argument 片面之词
one sided spectral density function 单边谱密度函数
one side opening 单面风口
one side water pressure 单侧水压力
one-speed neutron conservation 单速中子守恒
one-speed neutron 单速中子
one-speed transport equation 单速输运方程
one-speed transport model 单速输运模型
one-speed transport theory 单速输运理论
one-spool engine 单路式涡轮发动机，单路式发动机
one-stage by-pass system 一级旁路系统
one state “1”状态
one station solution service 一站式解决方案服务
one station type service 一站式服务
one-step majority-logic decoding 一步大数逻辑译码
one-stop service 一站式服务
one-storey house 平房
one-storey structure 单层结构
one-tenth period 十分之一衰减期
one-tenth second theodolite 十分之一秒精度经纬仪
one-terminal-pair network 二端网络，单口网络
one-third octave band 三分之一倍频程
one-through 单程
one-time duty 特殊运行状态
one-time recovery 一次性收回
one-to-one assembler 一对一汇编程序
one to one control mode 一对一控制方式
one-to-one control 单个操作，一对一控制
one-to-one correspondence 一一对应
one-to-one transformer 防雷变压器，隔离变压器，一比一变压器
one-to-one translator 一对一翻译程序
one-to-zero ratio 一与零输出比
one-turn magnetic head 单匝磁头
one-turn stairs 单转楼梯
one-turn 单匝的
one-two exchanger 单壳程双管程热交换器
O network O型网络
one-unit plant 单机组电站
one-unit power plant 单机组电站
one-valued function 单值函数
one-valued 单值的
one-variable function 单元函数
one-velocity diffusion theory 单速扩散理论
one-velocity neutron transport equation 单速中子输运方程
one-velocity transport equation 单速输运方程
one-velocity transport model 单速输运模型
one-wall sheet-piling cofferdam 单排板桩围堰
one-wall vessel 单层壁容器
one-wall 实心墙的，单层壁的，单层墙的
one-wattmeter method 单瓦特计法
one-way circuit 单向电路
one-way clutch 单向离合器
one-way cock 单向阀门
one-way communication 单向通信
one-way continuous slab 单向连续板，单向配筋连续板
one-way fare 单程费
one-way feed 单向馈电，单路馈电
one-way lock 单向船闸
one-way radiation 单向辐射
one-way reinforced 单向配筋的
one-way reinforcement 单向钢筋，单向配筋
one-way repeater 单向帮电机，单向增音机，单向中继器

one-way slope 单面坡
one-way switch 单路开关，单向开关
one-way system 单向系统
one-way traffic 单向通信业务，单向联络
one-way transmission line 单路输电线，单向传输线路
one-way transmission 单路输电，单向传输
one-way valve 单向阀
one-way 单向的，单路的
one-writing system 写系统
on flow 流入，进气，支流，涨水，湍流
ongoing evaluation 进行中的评价
ongoing project 进行中的项目，正在实施的项目
on-going qualification test 运行中合格试验
ongoing result 现有成果
on-going 正在运行的
on-grid energy 上网电量
on-grid power tariff 上网电价，标杆电价
on-grid price 上网电价
on hook signal 挂机信号
on hook 挂机
on-impedance 接通阻抗
omnidirectional wind 全向风
on installment basis 分期付款法
on-interval 接通间隔
on-job training 岗位培训，在岗培训
on-job worker 在职人员
on-job 在岗位的，在工地上，在职的
on leave without pay 停薪留职
onlending terms 转贷条件
on-line adjustment 联机调整
on-line alarm 在线报警
on-line analog input 在线模拟量输入
on-line analog output 在线模拟量输出
on-line analyser 在线分析器，在线分析仪
on-line analysis 在线分析，流线分析
on-line central file 联机中央文件
on-line change-over 在线切换
on-line charging 不停堆加料
on-line circuit analysis 联机线路分析
on-line cleaning system 运行中清洗装置，在役清洗装置
on-line cleaning 运行清洗，在役清理
on-line coal-quality detecting analyzer 在线煤质分析仪
on-line computer control 在线计算机控制
on-line computer system 联机计算机系统，在线计算机系统
on-line computer 在线计算机，联机计算机
on-line control system 在线控制系统
on-line control 在线控制，直接控制
on-line coolant clean-up 在线冷却剂净化
on-line data handling system 直接数据处理系统，在线数据处理系统
on-line data processing 在线数据处理，直接数据处理
on-line data reduction 联机数据简化
on-line debugging technique 联机调试技术
on-line debug 联机程序的调试
on-line diagnostic and report system 在线诊断和报告系统
on-line digital input 在线数字量输入
on-line digital output 在线数字量输出
on-line disk file 联机磁盘文件
on-line electrical consumption （太阳能发电装置）白天正常停运期间的厂用电耗量
on-line elimination of oil whip 油膜振荡的在线消除
on-line equipment 在线装置，联机设备，直接装置，在线设备
on-line file system 在线文件系统
on-line inquiry 联机询问
on-line instrument 在线仪表
on-line insulation diagnosis of electrical motor 电机绝缘在线诊断
on-line insulation diagnosis of transformer 变压器绝缘在线诊断
on-line-job assembly 现场装配
on-line load flow 在线电力潮流
on-line loading 不停堆加料
on-line logic simulation system 在线逻辑仿真系统，在线逻辑模拟系统
on-line maintenance 不停机检修，在线维修，使用中维修，不停堆维修
on-line measurement 在线测量
on-line mode 联机方式
on-line monitoring of life 寿命在线监测
on-line monitoring of residual life 寿命在线监测
on-line monitoring of the reactor 反应堆的在线监测
on-line monitoring 在线监测
on-line monitor printer 在线监视打印机
on-line operation 在线运算，联机操作，在线操作
on-line printer 在线打印装置，直接打印装置
on-line processing 连续处理，在役处理
on-line processor 在线数据处理装置
on-line reactivity meter 在线反应性计
on-line real-time branch information transmission 联机实时分路信息传输
on-line real-time 在线实时的
on-line refueling 不停堆换料
on-line sampling bottle 在线取样瓶
on-line storage 联机存储器
on-line switch-over 在线切换
on-line system 联机系统，联机装置，在线系统
on line tap changing transformer 在线抽头变换式变压器，有载抽头变换式变压器
on line tap changing 在线抽头变换，有载抽头变换
on-line test 在线试验，直接试验
on-line treatment 连续处理，在役处理
on-line typewriter 联机打字机
on-line UPS 在线式不间断电源
on-line wash duration 在线清洗时间
on-line working 在线工作，直接工作
on-line 在线的，联机的，直线控制的，并汽，并机，并入电网，在线，在役
on-load access 不停堆进出通道
on-load brush 负载电刷，工作电刷
on-load corrosion 运行腐蚀
on-load disconnecting switch 带负荷隔离开关，

带载隔离开关
on-load factor 通电持续率，负载因数
on-load fuel handling 不停堆燃料装卸
on-load fueling 不停堆换料
on-load indicator 有载指示器
on-loading 换燃料，换料
on-load maintenance 不停堆维修
on-load operation 带负荷运行
on-load refueled reactor 不停堆换料反应堆
on-load regulating transformer 带负荷调压变压器
on-load regulation 带荷调节，有载调节
on-load regulator 带负荷分接开关，带负荷调节器
on-load switch 负荷开关
on-load tap changer 有载调压变压器，加载抽头变换器，带负载抽头变换开关，变压器有载调压开关，带负荷分接开关，有载换接器，有载分接开关
on-load tap-changing transformer 有载分接变压器，带负载抽头变换式变压器
on-load tap-changing 带载抽头变换
on-load tapping switch 带负荷分接开关，有载分接开关
on-load temperature measurement 带电测温
on-load test 带负荷试验
on-load voltage regulating transformer 有载调压变压器
on-load voltage regulation 有载调压
on-load voltage regulator 带负荷电压调整器
on-load voltage 加载电压
on-load washing 带负荷清洗
on-load 带负荷的，带负荷
only in the case that 除非
only 整【表示金额时】
on-mike 靠近话筒，正在送话
ON(observation notice) 观察意见通知
on-off action 开闭式作用，开关作用，通断作用，开闭动作
on-off actuator 开关制动器
on-off circuit 通断电路
on-off controller 双位控制器，双位调节器，通断控制器，二位式调节器，开关型控制器
on-off control system 继电控制系统，开关（量）控制系统，开关式控制体系
on-off control 双位置控制，双位控制，开断控制，开关控制，启停控制，通断控制，继电器式控制
on-off element 开关元件
on-off error detector 继电式误差检测器
on-off feedback control 开关反馈控制
on-off heater 开关式加热器【稳压器】
on-off input 开关型信号，开关量输入
on-off instrumentation 开关仪表，通断仪表
on-off keying 开关键控，振幅键控
on-off measurement 开关测量，离合测量
on-off mechanism 通断机构
on-off operation 接通断开操作，启停运行方式，闭合断开操作
on-off output 开关量输出
on-off push button 合分按钮，通断按钮
on-off relay 通断继电器
on-off service 开停系统
on-off servomechanism 开关随动系统，开关伺服系统，继电伺服机构
on-off servo 继电伺服系统，继电随动系统
on-off signal 开闭信号，二进制信号
onoff sprinkler 开闭式洒水喷头
on-off station 开关量操作器
on off switch 启动停车开关，通断开关，双位开关
on-off system 开关系统【开关控制】
on-off time ratio 开断时间比，开关时间比
on-off timing 通断定时
on-off type control system 双位控制系统
on/off valve 或全开或全关型阀门，全开全关阀，双位阀，截止阀
on-off valve 全开全关阀
on-off variable 开关量，开关变量
on-off 开关，离合，开闭式的
on one's account 为了某人
on one's own account 自负风险，自行负责，依靠自己
on or before 不迟于（某日）
on-peak demand 高峰负荷时的供电要求，峰荷需量
on-peak electric energy 峰值电能
on-peak load 尖峰负荷
on-peak power 峰荷电力，尖峰功率
on-peak 最大的，尖峰的
on period 接通时期，接通周期，闭合周期
on-position 工作状态，“接通”位置
on-power refueling 不停堆换料，带功率换料
on principle 根据原则
on probation 见习，试用
on range 母管供汽
on record 公开发表的，留有记录的
on request 应要求，经要求，承索（即寄等）
on resistance 接通电阻
on richter scale 里氏（地震）级
on-rid energy 上网电量
on-schedule arrival/departure 准班抵离
onset galloping velocity 驰振初始速度
onset of boiling 沸腾起始点
onset of fluidization 起始流化点
onset of nucleate boiling 泡核沸腾起始点
onset velocity 初始速度
onset wind speed 起始风速
onset 开始，着手
on-shift operator 值班操作员
onshore area 海岸区
onshore current 向岸流
onshore gale 向岸大风
onshore wind 向岸风
onshore 向陆地，在岸上，陆上
on-site accident 厂区内事故
on-site arrival certificate 到达工地证书
on-site concrete 现浇混凝土
on-site construction 就地建造
on-site disposal （核电厂废物）就地处置
on-site emergency plans 场内应急计划
on-site energy storage 就地储能
on-site facility 就地设施，现场设施

on-site guidance 现场指导
on-site inspection 就地检验，现场检查
on-site measurement 现场测量，实测
on-site observation 现场观测，实测
on-site plant 工地发电厂
on-site plan 场内应急计划
on-site power source 厂内电源
on-site power system 厂内供电系统
on-site power 厂内电源
on-site repair 现场检修
on-site reprocessing plant （核电厂）就地后处理厂
on-site reprocessing 就地后处理
on-site runoff 涝水，内水
on-site spent fuel storage 乏燃料就地贮存
on-site standby power system 现场备用电力系统
on-site storage 就地贮存
on-site survey 现场调查
on-site test 现场试验
on-site training 现场培训
on-site vehicle 当地车辆，现场车辆
on-site verification 现场核查
on-site waste disposal 废物就地处置
on-site weld 现场焊接
on-site 在工地上，场区内，厂区内，在现场，现场，就地，当地，在施工工地
on standby 待命，备用，待命状态，严阵以待
on-state loss 通态损耗
on-state 开态，通路状态，导通状态，接通的
on-step operation 单步操作
on-stream cleaning 运行中清洗
on-stream inspection 不停工检查，在运转中检查，流水线上的检查
on-stream period 连续开工期间
on-stream pressure 操作压力
on-stream time 连续开工时间
on-stream 在生产中
Ontario Hydro 安大略水电公司【加拿大】
on-the-air 在播音中，正在发射的
on the basis of 根据，按照，依据，以……为基础
on the busiest condition 最繁忙工况，最繁忙工作情况
on-the-fly printer 飞击式打印机
on the gas side 气侧，气程
on the grounds of 基于……理由，根据，依据
on-the-job forming 现场架模
on-the-job professional training 在职专业训练
on-the-job training 在职培训，岗位培训
on-the-job 岗位的，在职的
on the one hand,on the other hand 一方面，另一方面
on the principle of 以……原则为基础
on the request of 在……的要求下
on-the-road mixer 路上拌和机
on the safe side 为了可靠起见
on the shelf 缓行的，废弃的，退役的
on the site labour 工地劳动力
on the site 在施工工地
on the south of ……以南【指相邻并且接壤】
on-the-spot audit 现场审计
on-the-spot investigation 实地调查，现场调查
on-the-spot survey 踏勘
on-the-spot training 就地培训
on the water side 水侧，水程
on the whole 大致上
on this condition 在这条件下
on time 按时，接通时间，工作时间
on(/upon)condition that （只有）在……条件下，条件是，如果
on upstream of 在……之前，在……上游
on voltage 接通状态电压
onyx marble 条纹大理石
OOC(out of commission) 在调试范围之外
oolite 鲕状岩
oolitic limestone 鲕状灰岩
OOM(overall operation manual) 全面运行手册
OON(operating occurrence notice) 操作事件通知
OOS(out of service) 退役
ooze 渗漏，泄漏，泥浆，淤泥，渗出，缓缓流出
opacimeter 暗度计，浊度计
opacity coefficient 不透明系数，遮光率
opacity detector 浊度探测器
opacity technique 隔声技术
opacity tester 不透明度仪
opacity 不透明度，不透明性，不透明，浑浊度，不传导，暗度，浊度
opaco 背阳坡
opal bulb 乳白灯泡
opal glass 乳白玻璃
opal 蛋白石
op-amp(operational amplifier) 运算放大器
opaque body 不透明体
opaque flame 不透明火焰
opaque glass 不透明玻璃
opaque layer 不透明层
opaque mask 不透明屏，不透明屏幕
opaque medium 不透明介质
opaqueness 不透明度
opaque particles 不透明的微粒
opaque white coating 不透明的白色涂层
opaque 浊的，不透明的，不透充的，不传导的
opcode 操作码
OPC(overspeed protect controller) 超速保护控制系统【DCS画面】，超速保护控制
open account 银行开户
open air arrangement 露天布置
open air breather 开启式呼吸器
open air climate 露天气候
open air condenser 大气式凝汽器
open air facilities 露天设施
open air hydroelectric station 露天水电站
open air model 露天模型
open air plant 露天装置，户外装置
open air seasoning 露天干燥
open air station 露天式电站
open air 露天，户外
open amortisseur winding 开路阻尼绕组
open amortisseur 开路阻尼绕组
open and shut action 开闭动作，开闭式动作
open antenna 室外天线，露天天线
open arch 明拱

open arc welding 明弧焊
open area （沸腾炉布风板）开孔面积，广场，开口面积，开阔地带，开放区，开阔场地，露天场地，过流面积
open area 筛面作用面积，筛面有效面积
open barge 敞仓货船
open basin 开放流域
open belt conveyor gallery 露天式栈桥
open bid 公开投票，公开招标
open-bottomed well 套管底开式井
open bucket trap 浮桶式疏水阀，浮桶式疏水器
open burning coal 非结焦性煤，长焰煤
open burning 露天焚烧
open caisson foundation 沉井基础
open caisson 开口沉箱，沉井
open car 敞篷车，敞篷汽车
opencast coal 露天开采的煤
open catwalk 通道
open cavity 明洞
open-center display 空心显示器
open channel diversion 明渠导流
open channel drainage 明渠排水
open channel flowmeter 开口流量计
open channel flow 明沟流动，明渠流，明槽流
open channel spillway 开敞式溢洪道
open channel water model 开式水模
open channel 开式通道，明沟，阳沟，明槽
open circle 开圆
open circuit arc 开路电弧
open circuit cell 开路电池
open circuit characteristic curve 开路特性曲线
open circuit characteristic 开路特性，空载特性
open circuit contact 断路触点，开路触点
open circuit cooling 开启式冷却
open circuit core loss 空载铁耗，开路铁耗
open circuit curve 空载曲线，开路特性曲线
open circuit excitation 空载励磁，开路励磁
open circuit fault 开路故障
open circuit impedance 开路阻抗
open circuit input impedance 开路输入阻抗
open circuit line 开路线路
open circuit loss 空载损耗，开路损耗
open circuit operation 开路运行
open circuit oscillations 开路振荡
open circuit output impedance 开路输出阻抗
open circuit power 无负荷功率
open circuit saturation curve 空载饱和曲线，开路饱和曲线
open circuit tap changer 空载分接开关
open circuit termination 开路终端
open circuit test 空载试验，开路试验，开路测试
open circuit time constant 开路时间常数
open circuit ventilation 开式通风
open circuit voltage ratio 开路电压比
open circuit voltage 开路电压【馈电】，空载电压
open circuit water 直流水
open circuit winding 开路绕组
open circuit wind tunnel 开路式风洞
open circuit 开路，断路，切断电路，开式回路
open circulating cooling water 开式循环冷却水
open circulating make up water pump 开式循环补充水泵
open circulation cooling system 开式循环冷却系统
open circulation 开式循环
open-close-open contact 开闭开接点
open coil winding 开圈绕组
open collector 集电极开路输出门
open competition 公开竞争
open competitive tender 公开竞争投标
open condition 开敞状态
open conductor 明线
open conduit wiring 明管线
open conduit 明流管道
open construction 无蒙皮结构，骨架
open contact leakage current 接点断开漏泄电流
open contract 开启触点，开路接点，敞口合同，开口合同，未结合同
open core transformer 磁路有气隙的变压器，开心式变压器
open core 开口铁芯
open country 开阔地
open courtyard 开口场院
open cut excavation 露天开挖
open cut method 明挖法
open cut mine 露天矿
open cut mining 露天开采法
open cut tunnel 露天坑道，明挖隧道
open cut volume 明挖方量
open cut 大开挖，露天开采，明挖，明堑，露天挖掘的，露天矿
open cycle controller 开环调节器，开环控制器
open cycle control system 开环控制系统
open cycle cooling water system 开式循环冷却水系统
open cycle gas turbine 开式循环燃气轮机
open cycle magnetohydro-dynamic power generation 开式循环磁流体发电
open cycle MHD generator 开环磁流体发电机，开式循环磁流体发电机
open cycle plant 开式循环发电机
open cycle 开环，开环的，开式循环，开路循环，直流循环
open damper winding 开路阻尼绕组
open-delta connection 开口三角接法，开口三角形连接，开三角连接
open district 开阔区域，空旷地区，未建区
open ditch drainage 明沟排水
open ditch 明沟
open-door policy 对外开放政策，门户开放政策
open drainage 明排水，明沟排水
open drain 明沟，开式疏水，排水明沟
open dredging process 明挖法
open dredging 敞开疏浚
open dump body 敞口翻斗车身
open economic region 紧急开放区
open economic zone 开放经济区
opened ended tunnel 直流式风洞
opened inland city 内陆开放城市
open-end barrel lug 开口圆筒接片

open-end concrete block 敞口的混凝土浇筑块
open-end contract 敞口合同，开口合同
open-ended coil 开口线圈
open-ended design 可扩展设计
open-ended line 终端开路长线
open-ended project 敞口项目
open-ended slogging spanner 开口敲击扳手
open-ended spanner with ring ratchet set 梅花棘轮开口扳手套装
open-ended spanner with ring ratchet 梅花棘轮开口扳手
open-ended system 可扩充的系统，开端系统
open-ended transmission line 终端开路输电线，终端开路传输线
open-ended wind tunnel 开路式风洞
open-ended 两端开口的，开环终端的，开口的
open-end investment company 股份不定投资公司
open-end pipe pile 敞口管桩
open-end wrench 开口扳手【扁平型】
opener 开启器
open estuary 不冻河口，开敞河口湾
open excavation 明开挖，明挖
open expansion tank 自由膨胀水箱
open fail 露明栏杆
open feeder 开路馈线
open fiduciary loan 公开信用贷款
open file 打开文件
open filter 开式过滤器
open fissure 张开裂缝
open flame 明火
open float type steam trap 浮桶式疏水器
open floor and-roof wind tunnel 无上下壁风洞
open floor 露明搁栅楼板
open flow duct 明沟
open flow 开放流
open fluidized bed 自由流化床，开式流化床
open flume setting 明槽式装置
open flume turbine 明槽式水轮机，开（敞）式水轮机
open fold 敞开褶皱
open frame girder 空腹大梁，无斜杆桁架梁
open frame motor 开启型电动机
open frame structure 开敞式框架结构
open frame transformer 开启型变压器
open frame 开启型机座
open freight storage 露天货场
open-front 前开式
open fuel assembly 无盒燃料组件
open fuel cycle 一次通过燃料循环
open furnace 开式炉膛
open furrow 明沟
open fuse 敞开型熔断器，明保险丝
open-gap-cooled rotor 开启气隙冷却转子
open gap cooling 开启气隙冷却
open gap 开启气隙
open gate 通门
open gear lubricant 开式齿轮润滑油
open-graded aggregate 开式级配骨料
open grate area 炉排通风面积，炉排活截面
open grate surface 炉排通流截面
open ground storage 开放式地面贮存
open harbour 无掩护港口
open hearth furnace 平炉，马丁炉
open hearth process 平炉炼钢法
open hearth steel 平炉钢
open heater 敞口加热器，开式加热器
open impeller 开式叶轮，无前后盘的叶轮
opening and closing mechanism 车门自动开关机构
opening-and-closing time 开启和关闭时间
opening angle 开敞角
opening bank 开立信用证的银行，开证银行
opening by electric signal 由电信号开启
opening by system pressure 由系统压力开启
opening ceremony 开幕式
opening charge 开发信用证手续费，开立信用证手续费
opening contact 断路接点，断路触点
opening controller 开度控制器
opening force 开启力【机械密封】
opening gap 缺口
opening inventory 初始投料量
opening meeting 首次会议
opening of bids 开标
opening of tender 开标
opening of tuyere 风口
opening policy 开放政策
opening pressure of safety valve 安全阀开启压力
opening pressure （安全阀）起座压力
opening prices 开标，开盘价
opening quotations 开盘
opening session 开幕会议
opening speech 开幕词
opening time 断电时间，断开时间
opening torque 开启力矩
opening to the outside world 对外开放
opening 开放，开口，孔，开孔，开度，开通路，职位空缺，风口，孔口，孔洞，开启
open inlet 不冻河口
open intake structure 开敞式取水建筑物
open intake 开敞式进水口
open into 通向
open issue 未解决的问题
open jetty 穿通码头，高架桥
open jet wind tunnel 开口试验段风洞
open jet 自由射流
open joint 凹缝，松接头，张开节理，露缝接头，明缝
open ladle method 浇桶置冲法
open lap winding 右行叠绕组，开式叠绕组
open lattice core （压水堆）无盒栅格堆芯
open-lattice 无盒栅格
open lend aperture 发散透镜孔径
open lest pit 露天试验坑
open letter of credit 无条款信用书，开放信用证，无特殊条件信用证
open letter 公开信
open line 明线，断开线路
open-loop characteristic 开环特性
open-loop circuit 开环电路
open-loop control circuit 开环控制电路

open-loop control system 开环控制系统
open-loop control 开环控制，无反馈控制，开环系统
open-loop frequency response 开环频率响应
open-loop gain amplifier 开环增益放大器
open-loop gain 开环增益
open-loop head circuit 前部的开环电路
open-loop series circuit 开环串联电路
open-loop stepped motor 开环式步进马达
open-loop system 开环控制系统
open-loop test 开环试验
open-loop transfer function 开环传递函数
open-loop 开回路，开式回路，开路，开环，开路的
open machine 开启型电机
open magnetic circuit 开口磁路
open meeting 公开会议
open mole 开口突堤
open motor 开启型电动机
open mounted 明装
open oil storage 露天油库
open oscillating circuit 开路振荡电路
open-package inspection at site 现场开箱检验
open-package inspection 开箱检查
open parking ground 露天停车场
open phase closing of breaker 断路器非全相合闸，非全相合闸
open phase fault 断相故障
open phase operation 非全相运行
open phase protection 断相保护
open phase relay 断相继电器
open phase running 非全相运行，断相运行
open phase tripping of breaker 断路器非全相跳闸
open phase 开路相，断相
open pier 开口防波堤
open pile surface crusting agent 露天料堆表面黏结剂
open pipe piezometer 竖管式测压计
open pipe ventilated motor 开启型管道通风电动机
open piping 明装管道
open pit excavation 露天开挖
open pit mine 露天矿
open pit mining 露天开采
open pit 露天矿场
open policy insurance 预约保险
open policy 预定保单
open-pool reactor 游泳池型反应堆，水池型反应堆
open pore structure 开孔结构
open port firing 开式燃烧
open port 通商口岸
open position 打开位置
open pot 开放港
open protocol 开放式协议，开放协议
open reactor 空床
open refrigeration 开式制冷
open relay 开启式继电器
open reservoir 露天水池
open return bend U形管
open return 开式回水
open ring wrench 开口扳手
open river discharge 畅流河流泄量
open river navigation 畅流河道航运
open river 畅流河道
open routine 开型程序，直接插入程序
open sea 外海，公海，远海
open set 开集，开放集
open shell and tube condenser 开式壳管式冷凝器
open shop 开放的数据处理中心，开放式计算站
open-shut 开启关闭
open slot 开口槽
open spanner 开口扳手
open spillway 开敞式溢洪道
open stadium 露天运动场
open stairway 敞开楼梯
open-steel 沸腾钢，不完全脱氧钢
open storage pile 露天煤堆
open storage yard 露天堆场
open storage 露天堆料场，露天堆场，露天仓库
open storm channel 雨水明沟
open-string stairs 露明梁楼梯
open structure 开链结构
open subprogram 开型子程序
open subroutine 开型子程序，开放型子程序，直接插入子程序
open surface defect 通向表面的缺陷
open surface method 明挖法
open switchyard 露天开关站
open systems interconnection 开放式系统互联
open system 开放系统，开式系统，敞开系统，开环系统，外通系统，非封闭系统
open tank treatment 冷热槽法
open tank 开式水箱
open tender 公开招标［投标］
open terminal box 敞开式接线盒
open terrain 开阔地形
open the breaker 打开断路器
open the grounding knife-switch 拉开接地刀闸
open the isolator 拉开隔离开关
open throat wind tunnel 开口风洞
open-to close 从开到关
open top culvert 透顶涵洞【指顶部留孔供路面排水的涵洞】
open top hopper 无盖漏斗车
open top mixer 敞顶拌和机
open topped railway wagon 铁路敞车
open top shielded enclosure 上部开启的屏蔽室
open top tray 开顶托架
open transition 开路瞬变
open traverse 不闭合导线
open-trestle unloading system 露天栈桥卸煤装置
open truss bridge deck 开式桁架桥面
open tube spacing 疏管距布置
open tubular column 毛细管柱
open type apparatus 开启型电路，开启型设备
open type composite apparatus 敞开式组合电器
open type condensate return system 开式凝结水回收系统
open type condensate tank 开式凝结水箱
open type feed water heater 开式给水加热器

open type feed water system 开式给水系统
open type frame 开启型机座
open type hot-water heat-supply network 开式热水供热网
open type hot-water heat-supply system 开式热水供热系统
open type motor 开启型电机
open type slot 开口槽
open type substation 敞开式变电站
open type system 开式系统
open type transformer 开式变压器
open type wharf 顺岸栈桥式码头
open type 开启式
open uranium fuel cycle 一次通过铀燃料循环
open vessel functional test 压力壳打开功能试验
open vessel 敞开式容器
open water 地表水
open web beam 空腹梁
open web girder 空腹大梁
open web joist concrete floor 空腹小梁混凝土楼板
open web steel door 空腹钢门
open web steel window 空腹钢窗
open web truss 空腹桁架
open-web 空腹的
open weir 开敞式溢流堰
open weld 开焊
open winding 右行绕组，开路绕组，开口线圈
open-window displaying 开窗显示
open wire circuit 明线线路，架空线路
open wire fuse 开启型熔断器
open wire line 架空电线，明线线路
open wire transposition 明线交叉
open wire 明线，架空线
open wiring 明线布线，明线
open working section 开口试验段
open work iron gate 花铁门，铁栅栏门
open work 户外作业，露天作业
open 开，开启，公开的，开的，敞开的，断路的，打开，断路，断开，分闸
OPE(operation engineer) 运行工程师
operability 可用率，可使用性，可操作性
operable 可操作的，可运算的
operand channel 操作数通道
operand fetch 取操作数
operand-precision register 精度操作数寄存器
operand 运算数，操作数，运算量
operant 操作数，运算对象，操作的，有效的，自发反应，操作性制约，发生作用之人或物
operate current 工作电流，操作电流
operated by oil pressure 油压操作的
operated in three-phase 三相操作，分相操作
operated stack register 操作组号寄存器
operated 开动的，启动的
operate miss 操作误差，控制误差
operate power 操作功率，运行功率
operater action monitor 操作员操作监视
operate time 动作时间，工作时间，操作时间，吸合时间，作用时间
operate voltage 操作电压，工作电压
operate 计算，工作，运转，开动，动作，操作，经营，运行，起作用
operating account 营业账户
operating actions 运行措施
operating ambient temperature 工作环境温度
operating and maintenance cost 运行及维护费
operating apparatus 操作装置
operating area 操作区
operating arm 操作杆
operating assets 经营资产
operating at idling condition 空载运行
operating availability 可能利用率，运行可利用率，运行可用率，时间运行效率
operating bar 操作杆
operating basis earthquake 正常运行地震，运行基准地震
operating bellows 操纵波纹管
operating board 工作台，操作盘，控制盘，操作台
operating bridge 工作桥
operating building 运行办公楼
operating capacity 工作容量，运行容量
operating cast 生产操作费用
operating characteristic curve 运行特性曲线
operating characteristic function 运算特征函数
operating characteristic 运行特性，操作特性，工作特性
operating charge 运行费用，运转费用
operating chart 作业图，操作图表，运行图表
operating circuit 操作电路
operating code 运行规程
operating coil stack assembly 操纵线圈堆组件
operating coil 动作线圈，工作线圈，操作线圈，操纵线圈【控制棒驱动机构】
operating compartment air recirculation system 操作室空气再循环系统
operating compartment 操作隔间【安全壳】
operating condition chart 工况图
operating condition curve 工况曲线
operating conditions 操作条件，运行工况，运行条件，工作条件，运行状态
operating console 操作台
operating contact 工作触点
operating control 运行控制
operating corridor 操作走廊
operating cost 操作费用，运转费用，运行费用，运行费
operating crew 运行人员，作业班人员
operating current range 工作电流范围
operating current 工作电流，操作电流
operating cycle 操作节拍，运行循环，工作循环，操作周期，运行周期，运转周期
operating data 操作数据，有用的资料，运行数据，运行资料
operating deck 操作平台，操作层
operating delay time 操作延迟时间
operating delay 操作延迟
operating device 操作装置
operating diagnostics 运行诊断
operating distance of solar cooker 太阳灶操作距离
operating distance 有效距离，作用距离

operating disturbance 运行干扰
operating duty cycle 操作循环，运行工作循环
operating duty test 运行方式试验，工作方式试验
operating duty 有效方式，运行方式，运行负荷，有效负载，工作制度，动作负载
operating economically 运行经济性
operating efficiency 运行效率
operating electrician 运行电工，值班电工
operating error 偶然误动作，误操作，运行误差，工作误差
operating expense 运行费用
operating experience 运行业绩，运行经验
operating exposes 经营费用
operating face 操作平台
operating factor 工作系数，利用率，运行系数，负载率
operating failure 运行失效，工作破坏，工作特性
operating flexibility 运行灵活性
operating floor of boiler house 锅炉房运转层
operating floor plan of main power building 主厂房运转层平面布置图
operating floor plan 运转层平面图
operating floor 运转层，控制层，操作层
operating flow chart 操作流程图
operating fluid loop 工作流体回路，传热流体回路
operating fluid 工作流体，工作介质，操作液
operating flux 运行通量
operating forced outage rate 运行故障停用率
operating force 作用力
operating frequency 工作频率，操作频率，运行频率
operating gallery 操作廊道
operating handle 操纵手柄，操作杆，操作柄，操纵柄
operating head of filter 滤池工作水头
operating height of solar cooker 太阳灶操作高度
operating history 运行史
operating holding company 经营股权（控股）公司
operating impulse 操作脉冲，执行脉冲
operating incident 运行故障，运行事故
operating income 营业收入，业务收入，营业利润，营业收益
operating index 运行指标
operating influence 工作条件影响
operating instructions and procedure 运行规程
operating instruction 操作规程，调度命令，运行规程，使用说明，操作指令，操作须知
operating leasing 经营租赁
operating level 运行料位，运转层，运行水位
operating lever 操纵杆
operating license 运行许可证，运行执照
operating lifetime 运行寿命
operating life 工作寿命，使用寿命，运行寿命，运行期限
operating limit 运行极限，使用极限
operating line 运行管线，工作运行线
operating load 工作负荷，运行负载，运行负荷，运转荷载
operating location 操作单元
operating loop 运行回路，运行环路
operating loss 运行损耗
operating maintenance 日常维修，小修，维护，运行维护，运行检修
operating management 经营管理
operating manual 运行手册，操作手册，操作规程
operating margin 运行安全裕量，运行裕量
operating mechanism of disconnecting switch 隔离开关操动机构
operating mechanism with stored energy device 带蓄能装置的操作机构
operating mechanism 操纵机构，策动机构，经营机制，操作机构
operating medium 操作介质
operating memory 操作存储器
operating method 操作方法
operating mode of centralized control 集控运行方式
operating mode of unit minimum output 机组最低出力运行方式
operating mode of unit 机组运行方式
operating mode 工作状况，工况，工作状态，运行方式，工作制度
operating motor 操作电动机，执行电机，运行电动机
operating occurrence notice 运行事件情报
operating occurrence 运行事件
operating order 操作命令
operating organization 营运单位
operating overload 运行过负荷，运行过载
operating pad 工作瓦块
operating panel 操纵板，操作屏，操纵盘，控制屏
operating parameter 工作参数，运行参数
operating performance income statement 营业实绩收益表
operating performance 运行特性
operating period 运行期，运行时间
operating permit 运行许可
operating personnel 运行人员，操作人员
operating pipe 工作管道，操纵管，操作管
operating plan 运行计划
operating platform 操作平台，运行平台
operating point 动作点，工作点，运行点，吸合点，工况点
operating position 动作位置，操作位置，工作位置，吸合位置
operating power 运行功率
operating pressure 操作压力，工作压力，运行压力，控制压力
operating principle 作用原理
operating procedure 操作步骤，运行方法，操作程序，运行程序
operating process 操作过程
operating program 操作程序
operating quality cost 运行质量成本
operating range of burners 燃烧器负荷调节比
operating range 运行区段【堆功率】，运转范

围，工作范围，操作范围，作用半径，运行范围
operating rate 开工率，运行速率，工作速率，运算速度，运行率，利用系数
operating reactivity change 运行反应性变化
operating record 操作记录，运行记录
operating region 工作范围，运行区
operating reliability 运行可靠性，操作可靠性
operating repair 日常维护检修，日常维修
operating reserve capacity 运行备用容量
operating reserve 运行备用
operating resistance 运行阻力
operating revenue 营业收入
operating rod 操纵杆
operating room 工作室，操作室，交换室
operating rope 起重索
operating rotational speed range 运行转速范围
operating routine 运行程序，操作过程
operating rule curve 水库操作规程曲线
operating rules 操作规程，运行规程
operating safety evaluation 运行安全评价
operating safety factor 工作安全系数，运行安全系数
operating schedule 运行时间表，操作程序，运行表
operating sequence 操作程序，运行顺序，操作顺序
operating shaft 运转轴
operating source 工作电源
operating specifications 使用规范，操作规程
operating speed 运行速度，运行转速，工作速度
operating staff living quarters 运行人员生活区
operating staff retraining 运行人员再培训
operating staff training 运行人员培训
operating staff 运行人员，操作人员
operating standard 运行标准
operating state 工作状态
operating station 操作站
operating status 运行状态
operating steady-state stability 静态稳定度
operating storage 运用库容
operating stress 工作应力
operating superintendent 运行主任
operating switch 工作开关，操作开关，控制开关
operating system description 操作系统说明
operating system 操作系统，控制系统
operating table 操作台
operating temperature 工作温度，运行温度
operating tension 工作电压，运行电压，运行张力
operating test equipment 操作测试设备
operating test 操作试验，运行试验
operating thermal efficiency 运行热效率
operating threshold 工作阈，工作限
operating time fully backseated to fully mainseated 全开到全关的操作时间
operating time ratio 时间可利用系数
operating time 工作时间，运行时间，动作时间，操作时间
operating torque 操作力矩
operating training unit 操作训练组织，操作训练机构
operating transformer 工作变压器
operating transient 运行瞬变，运行暂态
operating trouble 运行事故
operating unit status report 运行机组的状态报告
operating value 动作值
operating valve 动作阀
operating velocity 操作速度
operating voltage 工作电压，运行电压
operating walk way 操作通道
operating winding 工作线圈，工作绕组
operating wind speed 运行风速
operating with reduced vacuum 恶化真空运行
operating 操作，操作运行，运算，在役，运行，控制，管理，工作的，运行的，操作的
operation activity 维护，保养
operation-address register 运算地址寄存器
operation against rule 违章操作
operational acceptance CFB Boiler 循环流化床锅炉
operational acceptance 运行验收
operational amplifier 运算放大器，运行放大器
operational announcing system 操作通知系统
operational area 运行面积，操作面积
operational bypass 运行旁通
operational calculus 符演算，运算微积分
operational characteristic 运行特性
operational character 操作符
operational check list 运算检查表
operational check 运行检查
operational code 操作码
operational condition 运行工况
operational data analysis 操作数据分析
operational data 工作数据，运行数据，运行资料，运算数据
operational development 运行研究
operational diagnostics 运行诊断
operational digital technique 数字运算技术
operational directive 执行导则
operational earthing 工作接地
operational envelope 运行包络线
operational environment 工作环境
operational exchange rate 业务汇率
operational experience 运行经验，操作经验
operational factor 运算因数，运行率，运转因数
operational failure 运行失效，事故
operational flexibility 运行灵活性
operational flow chart 操作流程图
operational forced outage rate 运行故障停用率
operational forecasting 运行预测
operational function 运行函数，经营职能
operational grounding 工作接地，操作接地
operational impedance 运算阻抗
operational incident 运行事故
operational information system 操作信息系统
operational inspection 运行巡视
operational integrity 运行完整性
operational interval 转速变化范围
operational key 操作键

operational label 操作符号
operational life proof cycle 使用寿命
operational limits and conditions 运行限值和条件
operational loss 运行损失
operational mathematics 运算数学
operational method 运算方法
operational mishanding 操作不当
operational mode 运行方式
operational monitoring 运行监测，操作监测
operational multiplier 运算乘法器
operational order 操作指令，运算指令
operational parameter 运行参数
operational plan 运行计划
operational power supply 操作电源
operational procedure 工作程序
operational processing function 操作处理功能
operational programming 运算程序设计，操作程序设计
operational protection system 运行保护系统
operational range 工作范围，运行范围
operational readiness inspection 操作就绪检查
operational record 运行记录
operational relay 工作继电器，运算继电器，操作继电器
operational reliability 运行可靠性，工作可靠性
operational requirement 操作要求
operational research 运筹学
operational risk 运营风险
operational safety evaluation 运行安全评价
operational safety indicator program 运行安全指标大纲
operational safety 运行安全
operational scanning recognition 操作扫描识别
operational settings 运行整定值
operational shut down 停止运行
operational specification 操作规范
operational standby program 运算等待程序，操作准备程序
operational state 运行工况
operational status indicator 操作状态指示器
operational storage site 操作存储位置
operational stroke 工作行程
operational support directive 操作支援指令
operational support equipment 操作支援设备
operational symbol 运算符号
operational technical manual 操作技术手册
operational test 运行试验
operational time 运行时间，动作时间
operational-transient 运行瞬态
operational trouble 运行故障
operational unit 运算部件
operational voltage 运行电压
operational wind speed 运行风速
operational 运行的，工作的，操作的，运算的
operation analysis 运算分析
operation and checkout 操作和检查，运算和检查
operation and design conditions 运行和设计条件
operation and guardianship system 操作监护制度
operation and logistics management 操作及后勤管理
operation and maintenance cost 运行及维护费
operation and maintenance expenses 运行和维修费用
operation and maintenance manual 运行维护手册
operation and maintenance of substation 变电站运行维护
operation and maintenance 运行与维护
operation area 作业区
operation at elevated pressure 加压物操作
operation at full load 满负荷运行
operation at idling condition 空载运行
operation at part load 部分负荷运行
operation at power 带功率运行
operation at rating 额定状态下运行
operation availability 运行系数，运行可用率
operation board 控制盘，操作盘
operation building 运行大厅
operation chart 工况图，作业图，操作图
operation code 操作码，运算码
operation condition 运行状态，操作条件，运行工况
operation control center 操作控制中心
operation controlled by back pressure 背压控制运行
operation controlled by extraction 调节抽气式运行，抽汽调节运行
operation control switch 操作控制开关
operation control 运行控制
operation cost 运行费，操作费，管理费
operation counter 操作计算器
operation curve 运行曲线
operation cycle 操作周期，运行周期，工作周期
operation data 运行数据
operation decoder 操作译码器
operation desk 操作台
operation disturbance 运行故障
operation duration 运转时间
operation engineering 运行工程学
operation exception 异常运行
operation experience 运行经验
operation factor 运算因素，可用率，运行率，运行因素，运行因子，使用率，投运率
operation floor elevation 运转层标高
operation floor 运行层，运转层
operation flow chart 工序图，作业流程图
operation fuel charge 运行装载
operation guidance 操作指导，运行导则
operation guide 操作指导，导向，制导
operation hydraulic regime 运行水力工况
operation in crippled mode 降级运行
operation indicator 动作指示器
operation in parallel 并行工作，并联运行，并行运行
operation in perfect synchronism 完全同步运行
operation instruction 操作规程，使用说明书
operation licence 运行许可证，运行执照
operation life test 工作寿命试验
operation life 工作寿命，使用期限，使用寿命
operation limit 运行限制值
operation/logistics 运算/逻辑

operation log printout 运行日志打印输出
operation log 运行日志
operation, maintenance and administration 运行、维护与管理
operation management 运行管理
operation manager 运行经理
operation manual 运行手册，操作手册，使用手册，操作说明书
operation margin 营业毛利
operation mistake 操作失误
operation mode 运行状况，运行方式
operation monitoring 运行监测，操作监测
operation number 运算数，操作数，操作号码
operation of gas-steam combined cycle unit 燃气蒸汽联合循环机组运行
operation on distillate fuel oil 靠馏出的燃料油运行
operation on house load 带厂用电运行
operation on light crude oil 靠轻原油运行
operation on natural gas 靠天然气运行
operation part 操作码部分
operation pattern 运行方式
operation performance log 运行记录
operation period 工作周期，运行周期，运行期
operation permit 运行许可，运行执照
operation phase 运行阶段
operation pipe 操作管
operation plane 运行计划
operation platform 操作平台
operation power supply system 操作电源系统
operation pressure diagram 运行水压图
operation pressure line of return pipeline 回水管动水压线
operation pressure line of supply pipeline 供水管动水压线
operation pressure line 动水压线
operation pressure 工作压力
operation procedure 作业程序
operation quality assurance 运行质量保证
operation range 行车范围，工作范围，作用距离
operation register 操作寄存器，操作码寄存器
operation regulation 运行调节，运行规程
operation reliability 运行可靠性
operation research 运筹学
operation response speed 操作响应速度
operation rule 操作规程，运行规则
operations and maintenance personnel 运行维修人员
operation schedule 工作进度表，运行计划，操作程序，运行程序，操作日程，运行步骤
operation sequence control 操作程序控制，操作工序控制，运算程序控制
operation sheet 操作卡片
operations manager 生产组织经理
operation specification 操作规程
operations quality 运行质量
operations related outage 运行相关停运
operations research 运筹学，运用研究
operations superintendent 运行负责人，运行部主任
operation staff 运行人员
operation statement 损益表，营业表
operation state 运行状态
operation station 操作站
operation test 操作试验，运行试验，工作试验
operation transient 运行瞬态
operation tube 操作管
operation voltage in system 系统的运行电压
operation waste 运行弃水
operation 工作，运转，操作，作业，运算，计算，运行，动作，控制，处理，经营
operative attenuation 工作衰减
operative cable credit 可凭使用信用证【信用证的一种】
operative check 运行检查
operative clause 生效条款
operative limits 极限工作条件，运行极限
operative part 正文
operative 工作的，运行的，有效的，操作的，运算的
operator action log 运行人员操作记录
operator aid 助理操纵员，副操作员
operator auto mode 运行人员监控下的自控运行方式
operator call-in signal 呼叫话务员信号
operator-computer interactive communication 人机对话
operator console 操作控制台，操作台
operator control of HVDC power transmission system 高压直流输电运行人员控制
operator error recording 操作错误记录
operator error 操纵员错误，人为误差
operator guidance display 操作指导显示
operator in charge 领班人员
operator induced failure 操作员引起的故障
operator interface station 操作员接口站
operator interface unit 人机接口单元
operator interface 操作员接口
operator interrupt 操作员中断
operator intervention 运行人员干预
operator license 操纵员执照
operator manual mode 运行人员手控方式
operator on duty 值班人员
operator panel 操作板
operator-process communication 生产过程中操纵员的通信联络系统
operator's cab 吊车操作员小室，操作室，司机室
operator's connector 司机控制台
operator's console 操作员控制台
operator's house 操作员宿舍楼
operator's platform 操纵员平台，操作台
operator's request control panel 操作员请求控制台［盘，屏］
operator station 操作员站，操作站
operator training 司机培训
operator 操纵员，操作机构，操作器，操作员，操作人员，值班员，执行机构，算符，运算子，运行人员，接线员，经营者
OPER＝operation 运行
OPF(optimal power flow) 最优的电力潮流

OPG(outage planning group) 大修计划组
OPGW fittings 光缆金具，光缆配件
OPGW(optical fiber composite ground wire) 光缆，光纤复合地线
OPGW tools 光缆工具
ophitron 微波振荡管
opinion 观点，意见，看法
OPI(open package inspection) 设备开箱检查
opisometer 曲线计
opitmum velocity ratio 最佳速比
OP magnet OP磁铁，强顺磁性磁铁
OPM (operations per minute) 操作/分，运算/分
opography and landforms 地形（与）地貌
OP(operation permit) 运行许可证
OP(orifice plate) 节流孔板
OP (oxidation-reduction potential) electrode assembly 氧化还原电位电极装置
opportunity cost of capital 资金的机会成本
opportunity cost 机会成本，选优获益
opportunity 机会，时机
opposed blade damper 双阀瓣风门，相对叶片风门
opposed firing 对冲燃烧
opposed impellers 相对叶轮
opposed multiblade damper 对开式多叶阀
opposed-voltage protective system 启动保护系统，反电压保护系统
opposed wall firing 前后墙对冲燃烧
opposed 对面的，对置的，相反的
opposing coil 反作用线圈，反接线圈
opposing connection 对接，反接
opposing current 逆流
opposing electromotive force 反电动势
opposing field 反向场
opposing ion 反离子
opposing jet 对喷射流
opposing reaction 对抗反应
opposing steam flow 蒸汽回流
opposing torque 反力矩，阻力矩
opposing voltage 反向电压
opposing water mass 对抗水体，交锋水体
opposing winding 反向绕组
opposing wind 逆风，逆面风
opposing 相反的，反作用的，反向的，反对的
opposite acting force 反向作用力
opposite board 组合板
opposite drive end 非驱动端
opposite electricity 异性电
opposite-flow 逆流，逆流的
opposite force 反力
opposite gradient 反坡降
opposite in phase 反相的
opposite in sign 符号相反
opposite joint 对接
oppositely charged 反向充电的
oppositely phased voltages 相位相反的电压
opposite-operating dampers 逆向操作调节器
opposite phase 反相
opposite polarity 反极性
opposite pole 异性极
opposite pressure 反压力
opposite sequence 逆序列
opposite side 对边
opposite signs 异号，反号，相反符号
opposite stress 相反的压力
opposite vertex 对顶
opposite vorticity 反向涡量
opposite 对面的，相对的，对立的，相反的，相反，对立面
opposition method 反接法，对接法，对抗法
opposition test 对组试验，互馈法试验，背对背试验
opposition 反对，对立，反接，反相，反作用，对抗
OPR(organization peer review) 组织同行评审
optical absorption coefficient 光学吸收系数
optical absorption 光学吸收
optical aid 光学仪器
optical air mass 大气光学质量
optical alignment 光学找中
optical altimeter 光学测高计
optical amplifier 光放大器
optical automatic ranging 光学自动测距
optical axis 光轴
optical beam deflection 光束偏转
optical beam 光束
optical cable assembly 光缆组件
optical calibration 光学仪器检定
optical cavity 光学共振腔
optical characteristics 光学性能
optical character reader 光符阅读器，光符阅读机，发光字母读出器
optical character recognition 光学字符识别，光符识别
optical character scanner 光符扫描器
optical character system 光符系统
optical coating 光学涂层
optical combiner 光合路器
optical communication supervisory system 光通信监测系统
optical comparator 光学比较仪，光学比测器
optical concentrating system 光学聚光系统
optical continuum 光连续区
optical contour grinder 光学外形磨床
optical control system 光控制系统
optical coupler 光耦合元器件
optical current meter 光学流速仪
optical data bus 光纤数据总线
optical density 光密度
optical depth 光学厚度
optical detector 光学探测器，光检测器
optical device 光学装置
optical disc 光盘
optical disk drive 光盘驱动器
optical disk storage 光盘存储器
optical disk 光盘，光碟
optical distance measurement 光学测距
optical distribution frame 光配线架
optical dividing head 光分度头
optical efficiency 光学效率
optical fiat 光学平面
optical fiber composite cable 光纤复合电缆

optical fiber composite overhead ground wire 光纤复合架空地线
optical fiber joint box 光纤接续盒
optical fiber wave guide 光学纤维波导
optical fiber 光导纤维，光纤
optical fibre axis 光纤轴
optical fibre bundle 光纤束
optical fibre cable 光纤光缆，光缆
optical fibre coating 光纤涂覆
optical fibre communication 光纤通信
optical fibre connector 光纤活动连接器
optical fibre controller 光纤控制器
optical fibre coupler 光纤耦合器
optical fibre ferrule 光纤套管
optical fibre identification 光纤识别
optical fibre link 光纤链路
optical fibre signal 光纤信号
optical fibre terminal device 光纤终端装置
optical filter 滤光片，滤光器
optical flame fire detector 感光火灾探测器
optical focus 光学焦距
optical generation 光波振荡
optical glass probe 光学玻璃探头
optical glass 光学玻璃
optical grating 光
optical heat efficiency 光热效率
optical hologram 光全息图
optical immersed detector 光浸后式探测仪
optical index 光指数
optical indicator 光学指示器，光指示器
optical information processing 光学信息处理
optical instrument 光学仪器
optical integrator 光学积分器
optical interaction 光学相互作用
optical interface 光接口
optical interferometer 光学干涉仪
optical isolation 光隔离
optical laser torch 光学激光手电筒
optical lime-domain reflectometer 光时域反射测试仪
optical line terminal equipment 光纤线路终端设备
optical line terminal unit 光端机
optical link pilot protection 光纤保护
optically coupled device 光耦合器件
optically roughness 光学粗糙
optically thick limit 光学厚极限
optically thickness 光学厚度
optically thin limit 光学薄极限
optical mark reader 光学指示读出器，光标阅读器，光学标准读出器
optical mark recognition 光标识别
optical maser 激光，激光器
optical measurement 光学法测量
optical measuring method 光学测量法
optical memory 光存储器
optical meteor 大气光学现象
optical method 光学方法，光学法
optical micrometer theodolite 光学测微器经纬仪
optical micrometer 光学测微器，光学显微镜
optical microscopic structure inspection 光学金相显微分析，光学（金相）显微结构检测
optical observation 光学观测
optical outline method 光学轮廓法
optical path length 光学行程长度
optical performance 光学性能
optical period 临界周期
optical phase conductor 光纤复合相线
optical plumbing instrument 光学垂准器
optical plumb 光测垂线，光测铅垂线，光测悬锤，光测悬线，光学垂准线
optical plummet centring 光学垂线对中
optical plummet 光学垂准器
optical position encoder 光位置编码器
optical power composite cable 光纤电力线组合缆
optical probe 光学探头
optical projection system 光学投影装置
optical pulse-counting 光脉冲计数器
optical pulse transmitter using laser 采用激光的光脉冲发送器
optical pyrometer 光学高温计，光测高温计
optical reading balance 光学读数天平
optical reading theodolite 光学读数经纬仪
optical read-out 光示值读数，光读出（数据）
optical recording 光记录
optical regenerative repeater 光再生中继器
optical repeater 光中继器
optical resolution 光学分辨率
optical retardation 光减速
optical scanner 光扫描器［仪］
optical scanning 光学扫描
optical seismograph 光学式地震计
optical sensor 光学传感器
optical signal 光信号
optical spatial filtering 光学空间过滤
optical spectroscopy 分光法
optical spectrum instrument 光谱仪器
optical spectrum 光谱
optical square 光学直角器
optical storage 光存储器
optical strain gauge 光学应变计
optical strainmeter 光学应变计
optical system 光学系统
optical telemetry 光学远距离测量法
optical test 光学试验
optical theodolite 光学经纬仪
optical time domain reflectometry 光时域反射法
optical-to-electrical transducer 光电转换器
optical transmission equipment 光纤传输设备
optical voltmeter 光学电压计
optical warning sign 灯光警告信号
optical waveguide 光波导
optical 光学的
opticator （仪表的）光学部分
optics 光学，光学系统
optimal allocation 最优分配，最优配置
optimal configuration of wind farm 风电场优化
optimal control system 最优控制系统
optimal control 最佳控制，最优控制
optimal damping 最佳阻尼
optimal decision rule 最佳决定律
optimal design 最佳设计

optimal dose for protection 防护最适剂量
optimal efficiency 最佳效率
optimal estimate 最优估算，最佳估算
optimal filtering method 最佳滤波方法
optimality analysis 最佳性分析，优化分析
optimality criterion 最佳判据
optimal load flow 最佳潮流，最优潮流
optimally coded program 最佳编码程序，最优编码程序
optimally sensitive system 最优灵敏系统
optimal multi-variable excitation 最佳多变量励磁控制
optimal parameter 最佳参数
optimal power flow 最优潮流，最佳潮流，最优的电力潮流
optimal reactor shutdown 最佳停堆
optimal regulator 最佳调整器
optimal rotational frequency 最佳旋转频率
optimal solution 最优解
optimal specific frictional head loss 经济比摩阻
optimal stochastic control 最优随机控制
optimal switch system 最佳开关系统
optimal system reliability 最佳系统可靠性
optimal temperature of return water 最佳回水温度
optimal temperature of supply water 最佳供水温度
optimal temperature 最佳温度
optimal transient response 最佳暂态响应
optimal tuning 最佳调谐
optimal type-B2 convolutional code 最佳 B2 型卷积码
optimal value 最佳值
optimal 最适宜的，最佳的，最优的，最理想的
optimatic 一种光电式的光，光电式高温计
optimat 晶体管控制系统
optimeter 光电比色计，光学比较仪
optiminimeter 光学测微计
optimistic prediction 光最优预测，乐观预测
optimization condition 光最优条件
optimization design 优化设计
optimization management 优化管理
optimization method 光最佳化方法，最优化方法，优选法
optimization of parameter 光参数优化，最佳参数选择
optimization of radiation protection 辐射防护最优化，光辐射防护最优化
optimization structure 优化结构
optimization study 最优化研究
optimization technique 最优化技术
optimization 最优化，最佳化，最佳条件选择，最佳特性确定，最佳条件选配，优化，最佳效率阶段
optimize arrangement of production buildings 优化生产建筑物的布置
optimized composite 优化组合
optimized dispatching for reservoir 水库优化调度
optimized fuel assembly 最佳燃料组件
optimized iteration strategy 最优化迭代策略
optimize 使完善，最优化，优选
optimizing controller 最优控制器
optimizing control system 最优控制系统
optimizing control 最优控制，最佳值调节，最佳控制，最佳值控制
optimizing input drive 虽佳状态的输入给定器
optimizing peak-holding 最佳峰值保持
optimizing 最优化，最佳化
optimum allocation of resource 资源最佳分配
optimum allocation 最佳配置
optimum amplitude response 最佳幅值响应
optimum angle 最佳角
optimum bed height 最佳床高
optimum bidding price 最佳报价
optimum capacity 最佳容量
optimum code 最佳代码
optimum coding 最佳编码
optimum compromise 最佳折中
optimum condition 最佳工况，最佳条件，最有利条件，最佳状态
optimum cone configuration 最佳锥体形状
optimum consumptive use 最佳耗水【水电】
optimum control 最佳控制
optimum coupling 最佳耦合
optimum curve 最佳曲线，最优曲线
optimum design in consideration of reliability 可靠性优化设计
optimum design 最佳设计，优化设计
optimum detecting filter 最佳检出滤波器
optimum development plan 最优开发计划
optimum development 最优开发
optimum distribution 最佳分布
optimum efficiency 最佳效率
optimum electrical power 最佳电功率
optimum environmental quality 最佳环境质量
optimum environment 最佳环境
optimum error detection code 最佳误差检验码
optimum feed temperature 最佳给水温度
optimum filter 最佳滤波器
optimum final feed temperature 最佳最终给水温度
optimum fitting method 最佳符合法
optimum flow rate 最佳流率
optimum flow 最佳流量
optimum formula 最佳公式
optimum gas velocity 最佳气体速度
optimum gradation 最佳级配
optimum ground elevation 最佳地面高程
optimum growing condition 最优生长条件
optimum growth temperature 最佳生长温度
optimum investment 最适宜的投资
optimum lattice 最佳栅格
optimum length 最佳长度
optimum life 最佳使用期限
optimum linear system 最优线性系统
optimum load factor 最优负荷因数
optimum load flow calculation 优化潮流计算
optimum load-sharing 最佳载荷分配
optimum load 最佳负载
optimum matching 最佳匹配
optimum method 最佳方法
optimum mix 最佳配合比

optimum moisture content 最优含水量，最佳含水量
optimum moisture 最适水分
optimum number 最佳数
optimum operating condition 最佳运行条件，最佳工况，最佳操作条件
optimum operating current 最佳工作电流
optimum operating frequency 最佳工作频率
optimum operating voltage 最佳工作电压
optimum operation of heat-supply network 热网优化运行
optimum operation 最佳操作，最佳运行，最佳工况运行
optimum output 最佳出力，最佳输出
optimum parameter 最优参数
optimum performance 最佳性能
optimum planning 最优规划
optimum prediction 最优推估，最优预测
optimum Proctor density 最优普氏密度
optimum programming 最佳程序设计，最优规划
optimum program 最佳程序
optimum proportioning 最优配合比
optimum proposal 优化方案
optimum reactor shape 最佳堆形状，最佳反应堆形状
optimum reheat pressure 最佳再热压力
optimum reinforcement 最佳配筋
optimum relay servomechanism 最佳继电伺服机构
optimum reliability 最佳可靠度
optimum response 最佳响应
optimum result 最佳结果
optimum rotational speed 最优转速
optimum section 最优断面
optimum seeking method 优选法
optimum selection 最佳选择
optimum services 最佳服务
optimum site 最佳场址
optimum size of power station 电站的最优尺寸
optimum slip 最佳滑差
optimum solution 最优解
optimum speed to windward 最佳迎风速度
optimum speed 最佳速度
optimum start-up 最佳启动
optimum steam quality （冷却剂中的）最佳含汽率
optimum structural design 结构优化设计
optimum structure 最佳结构
optimum switching function 最佳开关函数
optimum switching line 最佳开关曲线
optimum system design 系统优化设计
optimum temperature gradient 最佳温度梯度
optimum temperature 最佳温度
optimum traffic frequency 最优通信量频率
optimum value 最佳值
optimum velocity 最佳流速，最佳速度
optimum water content 最优含水量
optimum weathering 最适度的磨损
optimum working frequency 最佳工作频率
optimum 最佳状态，最佳值，最佳条件，最佳状况，最适宜的，最优的，最佳的
optional clause 任选条款，选择条款
optional device 选用设备
optional equipment 备选设备，附加设备
optional goods 选货物
optional halt instruction 任选中止指令
optional items 选择项目，暂定项目，备用项目，可选择项目
optional port 选择港
optional project 后备项目，可供选择项目
optional proposal 选择投标，任择议定书
optional suppression 随意消除法，随意消除
optional 任意的，随意的，可选的
option buyer 期权的买方
option contract 期权合同
option money 期权费，期权定金
option price 选购价格
option scheme 选择方案
option switch 选择开关
option trading 期权交易
option 备选方案，选料，选择，选择权，选择物，方案
optiphone 特种信号灯
opto-coupler 光耦合器件
optoelectronic cell 光电池，光电管
optoelectronic device 光电子装置，光电子器件
optoelectronic isolator 光电隔离器
optoelectronic switch 光电开关
optoelectronics 光电子学
optoelectronic 光电子的
opto-isolator 光隔离器，遮光器
optomagnetic 光磁的
optotransistor 光晶体管
OPT(over-speed protection trip) 超速跳闸保护
optronic 光导发光的
optron 光导发光元件
opus latericium 嵌砖混凝土墙
OQA(operation quality assurance) 运行质量保证
OQAP(operations quality assurance programme) 运行质保大纲
oral agreement 君子协定，口头协议，口头修正
oral approval 口头同意
oral contract 口头合同，口头契约
oral examination and explanation 考问讲解
oral instruction 口头指示，口头介绍
oral invitation 口头邀请
oral notice 口头通知
oral presentation 口头说明
orange lac 虫漆片
orange oxide 三氧化铀
orange peel bucket 多瓣抓斗
orange peel excavator 多瓣抓斗式挖土机
orange peeling 油漆皱皮
orange-yellow 橙黄色
orbital angular momentum quantum number 轨道角动量量子数
orbital electron 轨道电子
orbital GTAW welding head 轨道式钨极电弧焊头
orbital GTAW welding 轨道式钨极电弧焊
orbitally stable 轨道稳定的
orbitally 轨道的
orbital motion 轨道运动

orbital sander 轨道喷砂装置
orbital solar power station 轨道太阳能电站
orbital welding 轨道式焊接
orbital 轨道的，范围的
orbit welding 管座旋转焊，环形轨道自动焊，全位置
orbit 轨道
orchestrapit 音乐池
OR circuit “或”电路
OR core “或”磁芯
ORC(origin receiving charge) 始发接单费
ORC(outside reactor containment) 反应堆安全壳内部
order B/L 记名提单
order check 抬头支票，记名支票
order cheque 记名支票，抬头支票
order code 指令码
order contract 订货合同
order controller 顺序控制装置
orderer 订货人
order form 订货单
order good 订货
ordering by merging 并项成序过程，合并排序，并入过程
ordering scheme 程序图
ordering solid solution 有序固溶体
ordering vector 次序矢量
ordering 调整，整理，排次序，有序化
order list 订货单
orderly index 市场有序性指标
orderly shutdown 顺序停堆
orderly 整齐的，有秩序的，守纪律的
order note 购货确认书
order number 序号，指令编码，订单
order of accuracy 正确度，精确度
order of commencement of work 开工令
order of connection 接线顺序
order of deactivation 失活反应级数
order of equation 方程的阶
order of harmonics 谐波次数
order of magnitude 绝对值的阶，数量级
order of merit 优劣顺序
order of precedence （文件效力）优先顺序
order of reaction 反应级数，反应堆级
order of reflexion 反射级
order of stream 河流等级
order quantity 订购量
order register 指令记录器，指令寄存器
order set 指令组
order sheet 订单
orders per second 每秒钟指令数
order structure 指令结构
order tank 顺序存储器
order tape 指令带
order writing 指令写出
order 订货，订货单，指令，命令，次序，顺序，序列，工况，正常状态，规程，制度
ordinal number 编号，序数
ordinal 序数，顺序的，依次的
ordinance bench mark 法定水准标点
ordinance datum 法定基准面
ordinance load 规定负荷
ordinance 法令，法规，条例，管理条例
ordinarily 普通地
ordinary bill 普通支票
ordinary cinder aggregate 普通炉渣骨料
ordinary concrete 普通混凝土
ordinary construction 普通结构
ordinary differential equation 常微分方程
ordinary fault rift 正断层地堑
ordinary fault 常见性故障
ordinary flow 常遇流量，正常流
ordinary hydrogen 普通氢
ordinary lime mortar 普通石灰砂浆，普通灰浆
ordinary load 日常负荷，普通荷载
ordinary maintenance 日常维修
ordinary masonry 普通圬工
ordinary partner 普通合伙人
ordinary pitch 常用屋面坡度
ordinary Portland cement 普通波特兰水泥，普通硅酸盐水泥
ordinary pressure 常压
ordinary steel bar 普通钢筋
ordinary valve type surge arrester 普通阀式避雷器
ordinary water level 常水位
ordinary wave 寻常波，常波
ordinary welder 普通焊工
ordinary 普通的
ordinate scale 纵坐标尺度
ordinate 纵坐标，有规则的，正确的，纵距，纵坐标轴
ordnance datum 规定基准，水准基准
ordnance map 军用地图
ore body 矿体，矿床，沉积【矿】
ore content meter 矿石含量测定器
ore dressing 选矿
ore floatation 矿石浮选
ore industry sewage 采矿工业污水
OR element “或”元件，“或”门
oreliminary examination 粗探伤
ore mill 磨矿机
ore mined in this year 本年度开采量
ore roaster 煅矿炉
ore roasting 矿石焙烧
ore-smelting electric furnace 矿热电炉
ore 矿，矿物，矿石
OR function 逻辑“或”函数，“或”操作，“或”功能
organ burden 器官积存量
organic acid 有机酸
organic additive 有机添加剂
organic analysis 有机分析
organic-base 有机碱
organic basis 有机质
organic carbon 有机碳
organic chemical pollutant 有机化学污染物
organic chemicals 有机药品
organic chloramine 有机氯胺
organic coating 有机护层
organic composition solid potentiometer 有机实芯电位器

organic composition solid resistor 有机实芯电阻器
organic compound 有机化合物
organic contaminant 有机污染物
organic content 有机质含量
organic coolant 有机冷却剂
organic-cooled deuterium-moderated reactor 有机物冷却重水慢化反应堆
organic-cooled reactor 有机物冷却反应堆
organic decomposition product 有机物分解产物
organic deposit 有机物沉积
organic dust 有机灰尘
organic electrochemistry 有机电化学
organic film 有机薄膜
organic floc 有机絮凝物
organic fluid vaporizer generator 有机液体蒸汽发生器
organic foulant 有机污染物
organic fuel 有机燃料
organic glass 有机玻璃
organic-gram 构造示意图
organic impurities test 有机杂质检验
organic inhibitor 有机缓蚀剂
organic insulation 有机绝缘材料
organic iodine 有机碘
organic liquid coolant 有机液体冷却剂
organic matrix 有机基体
organic matter basis 有机基
organic matter in coal 煤中有机质
organic matter 有机物【水分析】，有机物质，有机质
organic moderated and cooled reactor 有机物慢化冷却反应堆
organic moderated reactor 有机物慢化反应堆
organic moderator 有机慢化剂
organic nitrogen 有机氮
organic phosphate 有机磷酸盐
organic phosphorus 有机磷
organic plastics 有机塑料
organic pollutant analysis 有机污染物分析
organic-quaternary compound 有机四元化合物
organic reaction 有机反应
organic reactor 有机反应堆
organic-removal pretreatment 除有机物预处理
organic rock 有机岩
organic sediment 有机沉淀物
organic semiconductor solar cell 有机半导体太阳电池
organic semiconductor 有机半导体
organic sequestrant 有机多价螯合剂
organic sewage 有机污水
organic silicon compound 有机硅化合物
organic silicon paint 有机硅漆
organic silt 有机质泥沙
organic slime 腐泥
organic soil 有机质土，有机土
organic solvent preservative 有机溶剂防腐剂
organic solvent 有机溶剂
organic sulfur 有机硫
organic synthesis 有机合成
organic trap 有机物捕集器
organic treatment 有机处理
organic vapor 有机蒸汽
organic waste water 有机废水
organic waste 有机废水，有机废物
organic 有组织的，有机的，有系统的，有机
organism 有机体，有机组织，有机物体，生物体
organizational freedom 独立的组织机构
organizational goal 组织目标
organizational instruction 管理指令
organizational structure 组织结构
organizational 管理的
organization and administration 组织管理
organization chart 组织关系（机构）系统图，组织机构图表
organization cost amortization 开办费摊销
organization cost 筹备费
organization expense 开办费，筹备费
organization for economics 经济合作开发组织
organization in charge of construction 施工单位
organization of construction 施工单位
organization of workers'position 工作岗位责任制
organization peer review 组织同行评审
organization scheme 组织方案，组织机构系统图
organization table 组织表，结构表
organization 组织，团体，机构，构造，机关，体制
organized air supply 有组织进风
organized exhaust 有组织排风
organized natural ventilation 有组织自然通风
organized ventilation 有组织的通风
organized vortex trail 规则涡迹
organizer 创办人，发起人
organizing committee 组织委员会
organo-boron 有机硼
organochlorine pesticide 有机氯农药
organ 机关，机构，元件
OR gate 逻辑“或”门
orientate 东方的，定方向，定方位，定向，排列方向
orientation analysis 定向分析，方位分析
orientation by backsight 后视定向
orientation control 调向，朝向控制
orientation curve 标定曲线
orientation diagram 定向图，方位图
orientation flange assembly 定向法兰装置，定向翼缘装置
orientation line 标定线
orientation mark 取向标记，参考点
orientation measurement 定向测量
orientation mechanism 定位机构，定向机构，迎风机构
orientation of plane table 平板仪定向
orientation of project 项目方向
orientation point 定向点
orientation 方位，方向性，定位，定方位，排列方向，方向，朝向
oriented array 定向方阵
oriented-core barrel 定向岩芯管
oriented research 定向研究
oriented steel sheet 方向性钢片

oriented 取向的，定向的，与……有关的
orient 东方的，定方向，定向，排列方向，定位
orifice box 孔口量水箱
orifice coefficient 测量孔板系数，孔板系数，开孔率，孔口系数
orifice diameter 压缩喷嘴孔径
orifice discharge 孔流量，孔板流量
orificed lateral 带孔支管
orifice fitting 安装孔板，孔板法兰
orifice flowmeter 孔板流量计
orifice gas 等离子子气
orifice jet 小孔射流
orifice measuring device 孔口测流设备
orifice metering coefficient 孔口流量系数
orifice metering 孔板流量测定法
orifice meter 孔板流量计，孔口流量计，量水孔
orifice of spring 泉眼
orifice outflow 孔口出流
orifice plate flowmeter 孔板流量计，孔板式流量计
orifice plate unit 孔板单元
orifice plate 节流孔板，孔板，节流圈，调压孔板，量孔板
orifice plug 节流圈，孔板
orifice restriction 缩孔
orifice set 节流装置
orifice size （阀门）开口尺寸，节流孔通径，节流面积，孔口大小，孔口尺寸，孔径
orifice spacing 孔间距
orifice spillway 孔口式溢洪道
orifice static tap 静压测量点
orifice throat length 孔道长度
orifice throat ratio 孔道比
orifice-type spillway 孔口式溢洪道
orifice viscosimeter 孔口黏度计
orifice with full contraction 完全收缩孔口
orifice with suppressed contraction 不完全收缩孔
orifice 孔板，节流孔，孔，口，喷嘴，测流孔，管孔
original asphalt 天然沥青
original bill of lading 正本提单，正式提单
original capital 原投资本，原始股本
original cleavage 原生劈理
original coal sample 原始煤样
original cohesion 初始凝聚力，原始内聚力
original contract 合同正本，原合同
original cost 原价
original credit 原信用证
original data 原始数据，原始资料
original design 原设计，初始设计，原始设计，初步设计
original document processing 初始文件处理
original document 原始单据，单据正本，原始文件，正本单据
original drawing 原图，底图
original equipment manufacturer 原设备制造商
original evidences 原始凭据，原始凭证
original ground level 原地面高程
original ground 原地面
original interstice 原生间隙
original investment 原始投资
original letter of credit 原始信用证
original map 原图
original material 初始材料
original package 原装
original part 原始部分
original picture 原图
original plan 原图
original plot 实测原图，原图
original position 初始位置【翻车机】
original price 原价
original print routine 原始打印程序
original program 原始程序
original quotation 报价单正本
original reconnaissance 初次勘测
original record 原始记录
original size 原始尺寸
originals of the contract 合同正本
original soil 原土
original state 原始状态，初始状态
original surface 起始面
original text 原文
original vortex 启动涡流
original vouchers 原始凭单
original wave 基波，固有波
original winding 原绕组
original 最初的，正本，原始的
originate from(/in) 发源于，来自，产生
origin centre 发源中心
origin coordinates 起始坐标
origin depth 震源深度
origin of building coordinate 建筑坐标原点
origin of control net-work 控制网原点
origin of coordinate 坐标原点
origin of force 力的作用点
origin of groundwater 地下水源
origin of height 高程原点
origin stress 原应力
origin 起源，起点，原点，起因，起始地址，成因，发源地，坐标起始点，坐标原点
O-ring flange O形环法兰
O-ring handling fixture O形环装卸工具
O-ring O形环【密封垫】，密封圈，密封环，垫圈
O-ring packing O形密封圈
O-ring seal O形圈密封，O形密封环，O形环密封，O圈密封
O-ring storage rack O形环存放架
ornamental lighting 装饰照明
ornamentation 装饰
ornament 修饰，装饰
orogenesis 造山作用
orogenic disturbance 造山扰动
orogenic force 造山力
orogenic movement 造山运动
orogeny 造山运动
orographic anticyclone 地形反气旋
orographic character 地形特性，山势特性
orographic condition 地形条件
orographic disturbance 地形扰动
orographic divide 地形分水界
orographic effect 地形影响

orographic factor 地形因子
orographic fault 山形断层
orographic influence 地形影响
orographic lifting 地形抬升
orographic model 地形模式
orographic precipitation 地形降水
orographic rain 地形雨
orographic storm 山岳暴雨
orographic upward wind 地形性上升风
orographic 山岳的
orohydrographic model 高山水文地理模型，山水地形模型，山水立体地图
orohydrography 地形水文学，山地水文学，高山水文地理学
OR(oil refinery) 炼油厂
O/R(on request) 应请求，根据需求
OR operator 或算子
orotectonic divide 山岳分水界
O&R(overhaul and repair) 大小检修
OR(over ride) 超控
ORP meter 氧化还原表
ORP(oxidation-reduction potential) transmitter 氧化还原电位变送器
ORP(oxidation-reduction potential) 氧化还原电位
Orsat analysing apparatus 奥尔萨特（烟气）分析仪
Orsat gas analyzer 奥萨特气体分析器
OR switch 或开关
orthicon 低速摄像管，正析摄像管
ortho-acid 正酸
orthocenter 垂心
orthoclase 正长石
orthodox construction 传统结构
orthogonal axes 正交轴
orthogonal base line 直交基线
orthogonal component 正交分量
orthogonal coordinates 正交坐标，直角坐标
orthogonality relation 正交关系
orthogonality 正交性，相互垂直，正交
orthogonalizable convolution code 可正交卷积码
orthogonalization 正交化，相互垂直
orthogonal joint 正交接合
orthogonal lattice 矩形栅格
orthogonal matrix 正交矩阵
orthogonal parity check 正交奇偶校验
orthogonal polarization 正交级化
orthogonal projection 正交投影
orthogonal system of functions 函数的正交系
orthogonal trajectories 正交轨迹线
orthogonal vector 正交矢量
orthogonal 正交，直交，矩形的，正规化，标准化，正交的，垂直的，直角的
orthographic 正视图
orthojector circuit-breaker 高压微量喷油断路器
orthokinetic flocculation 同向絮凝作用
orthometric 正交的，垂直的
orthonol 具有矩形磁滞环线的铁芯材料
orthonormal 标准化的，规格化正交的
orthophosphate 正磷酸盐
orthopole 正交极，垂极
orthopositronium 正阳电子素
orthorhombic crystal system 斜方晶系
orthoscopic 无畸变的
orthoselection 定向选择
orthosilicate 原硅酸盐
orthotelephonic response 正交电话响应
orthotronic error control 正交误差控制
orthotropic body 正交各向异性体
orthotropic slab 正交各向异性板
orthotropic 正交各向异性的
orthotropy 正交各向异性
OR tube “或”门管
OR “或”
Osakatron 大阪管
OSART(operational safety review team) 运行安全评审组
OSBL(outside the battery limit) 界区外
oscillate 摆动，摇摆，舞动，振动，振荡
oscillating agitator 振荡搅拌机
oscillating armature motor 振动电枢式电动机
oscillating center 振荡中心
oscillating circuit high power testing station 振荡回路试验站
oscillating circuit 振荡回路，振荡电路
oscillating coil 振荡线圈
oscillating component 振动部分，振荡分量，振荡部分
oscillating contact 振动触点
oscillating control servomechanism 振荡控制伺服机构
oscillating conveyor 振动式输送机
oscillating crystal 振荡晶体
oscillating current 振荡电流
oscillating curve 振荡曲线
oscillating cylinder viscometer 振动柱体黏度计
oscillating discharge 振荡放电
oscillating electromotive force 振荡电动势
oscillating feeder 振动给料机，振动给煤机，振荡加料器
oscillating flow 脉动流
oscillating force 振动力
oscillating frequency 振动频率，振荡频率
oscillating function 振荡函数
oscillating grate 振动炉排
oscillating impulse 振荡脉冲
oscillating jet 摆动射流
oscillating jump 摆动水跃
oscillating linear motor 振荡式线性电动机
oscillating load 振动荷载
oscillating motion 振动
oscillating movement 振荡，振动
oscillating neutral 振荡中性点
oscillating nozzle 摆动喷嘴
oscillating period 振荡周期
oscillating phase 振荡相
oscillating potential difference 振荡电位差，振荡电压
oscillating pressure field 振荡压力场
oscillating quantity 振荡量
oscillating range 振荡范围
oscillating regulator 振荡式调节器
oscillating relay 振动继电器

oscillating sieve 摇摆筛
oscillating system 振荡系统
oscillating transformer 高频变压器，振荡变压器
oscillating tube 振荡管
oscillating-type blower unit 振动除灰装置
oscillating unsteady flow 波动不稳定流
oscillating voltage 振荡电压
oscillating wave 振荡波
oscillating 振荡的
oscillation absorber 减振器
oscillation amplitude 振动幅度，摆幅，振幅
oscillation bearing 摆动轴承
oscillation boundary 振荡边界
oscillation centre 摆动中心，振荡中心
oscillation circuit 振荡电路
oscillation coil 振荡线圈
oscillation constant 振荡常数
oscillation damping 振动阻尼，摆动阻尼，振荡阻尼，消振，减振
oscillation effect 振荡效应
oscillation frequency 摆动频率，振荡频率
oscillation impulse 振荡脉冲
oscillation indicator 振荡指示器
oscillation intensity 振动强度，振荡强度
oscillation limit 回荡限度
oscillation loop 振荡波腹
oscillation mode 振荡方式，振型
oscillation orbit 振荡轨道
oscillation period 振荡周期
oscillation pickup 测振器
oscillation response spectrum 振荡响应谱
oscillation response 振荡响应
oscillation ripple 摆动波痕
oscillation spectrum 振荡谱
oscillation test 振动试验
oscillation transformer 高频变压器，振荡变压器
oscillation valve 振荡管
oscillation 摆动，波动，振荡，振动，舞动
oscillator coil 振荡器线圈
oscillator frequency 振荡器频率
oscillator response spectrum 震荡响应谱
oscillator tube 振荡管
oscillator voltage 振荡器电压
oscillator wavemeter 振荡式波长计
oscillatory circuit 振荡电路
oscillatory degrees of freedom 振荡自由度
oscillatory discharge 振荡放电
oscillatory flow 振荡流
oscillatory instability 振荡不稳定性
oscillatory motion 摆动
oscillatory occurrence 振荡现象
oscillatory period 振荡周期
oscillatory power surge 功率振荡
oscillatory power 振荡功率
oscillatory reactivity perturbation 振荡反应性扰动
oscillatory surge 振荡过电压
oscillatory system 振动系统
oscillatory torque 振动力矩，振荡力矩
oscillatory wave 摆动波
oscillatory 摆动的，振荡的，振动的，舞动的
oscillator 振荡器，摆动物，激振器，振动器，振子
oscillight 显像管，电视接收管
oscillion 三极振荡管，振荡器管，
oscillistor 半导振导器，半导体振荡管，半导体振荡器，振荡晶体管
oscillogram 波形图，示波图
oscillograph galvanometer 示波器式检流计
oscillographic balancing 示波器平衡
oscillographic integrator 示波积分器
oscillographic 示波的
oscillograph recording system 示波器记录装置
oscillograph trace 示波图
oscillography 示波术，示波法
oscillograph 示波器，波形图，录波器，示波图
oscillometer 示波器，示波计，录波器，示波仪
oscilloprobe 示波器测试头
oscilloscope 录波器，示波器，示波管，录波管，示波仪
oscillosynchroscope 同步示波器
oscillotron 电子射线示波管，阴极射线示波管
oscitron 隧道二极管振荡器
OSC＝oscillograph 示波器
osculating circle 密切圆
osculating plane 接触面
osculation point 密接点
Oseenflow 奥新流
OSHA(Occupational Safety and Health Administration) 美职业安全和卫生管理局
osier 柳条
OSI(open system interconnect) 开放系统互联
OSL(safety & license branch) 安全与执照申领处
osmole 渗透压克分子
osmometer 渗透压力计，渗透计
osmometry 渗压测定法
osmoregulation 渗压调节
osmose 渗透作用，渗透，渗析
osmotaxis 渗透性
osmotic coefficient 渗透系数
osmotic concentration 渗透浓度
osmotic cycling （树脂）渗透周期性
osmotic effect 渗透作用
osmotic equivalent 渗透当量
osmotic factor of solvent A 溶剂 A 的渗透因子
osmotic membrane 渗透薄膜，渗透隔膜
osmotic movement 渗透运动
osmotic potential 渗透势
osmotic pressure 渗透压，渗透压力，渗透压强，渗析压力，渗压
osmotic shock attrition test 渗透冲击磨损试验
osmotic shock 渗透冲击
osmotic 渗透的
OSN(output switch network) 输出转换网络
OS(operation service contract) 生产服务合同
OSOR(on site organization regulation) 现场组织条例
OS(over speed) 过速
osram lamp 钨丝灯
osram 钨丝
OSR(output shift register) 输出移位寄存器

OSR(outstanding reservation) 未完成的保留项
OSS(operational storage site) 操作存储位置
ostensible agent 名义代理人
ostensible partner 名义合伙人
OSURR(Ohio State University Research Reactor) 俄亥俄州立大学研究堆【美国】
OS/VS(operating system/virtual storage) 操作系统/虚拟存储器
OSW(Office of Saline Water) (美国)咸水淡化局
OTBD＝outboard 表盘外形
OTC(training center) 培训中心
OTDR(optical time domain reflectometry) 光时域反射仪
OTEC(ocean thermal emergy conversion) power plant 海洋热能转换发电站
OTE(operating test equipment) 操作测试设备
other expenses 其他费用
other power supply expenses 其他供电费用
other referenced document 其他有关文件
other side 另一面【反义为焊接面】
other than 除了……外，……除外，不同于，(而)不是
otherwise stipulated by law 法律另有规定
OTLT＝outlet 出口
OTM(operational technical manual) 操作技术手册
o-tolidine dihydrochloride 邻联甲苯胺盐酸盐
OT(operating temperature) 工作温度
OT(outstanding task) 遗留工作[任务]，未完成项
OTSG(once-through steam generator) 直流蒸汽发生器
OTTO(once through then out) 一次通过即卸料
ouble row four point contact ball bearing 双列四点接触球轴承
2-ouk-of-3 link 三取二逻辑元件
oultet water temperature of deaerator 除氧器出水温度
ouncer transformer 小型变压器，袖珍变压器
O/U ratio(oxygen to uranium ratio) 氧铀比
OUSR(operating unit status report) 运行机组的状态报告
outage capacity 降额容量
outage cost 停电费用，停电损失
outage factor 停机率
outage for overhauling 大修停运
outage frequency 停运频繁程度
outage management system 停电管理系统
outage management 停运管理
outage occurrence 停运事件
outage of generation 断电，发电中断
outage planning 停运计划
outage rate 停机率，停电率，事故停机率，事故率，停运率
outage repair 停机检修
outage state 停运状态
outage time 发生故障，停运时间，停机时间，停役时间，发生故障时间
outage 停机，停电，停止，停堆，停运，断电，排出口，运行中断，事故，停役，排出量
out-band signaling 带外信令
outboard bearing 外伸轴承，外接轴承，辅助轴承，(泵)外侧轴承
outboard end (泵)外侧端
outboard motor 外装电动机
outboard recorder 外部记录器
outboard support 外伸支架，外侧支架，外支承
outboard 外侧，表盘外形，外装电动机，外端的，外部的
outbound 射出的，输出的，引出的
outbreak 爆发，破裂，中断，冲破，断裂，决口
outburst 尖头信号，爆发，脉冲，闪光，溃决，飞溅，破裂，气喷
outcome 开始，出发，结果，输出，输出量，效果，出口
outcoming electron 离轨电子，逸出电子，出射电子
out-condensing (快堆夹带钠气)凝出
out conductor 外导线，外侧线
outconnector 外接符，改接符
outcrop line 露头线
outcrop of fault 断层露头
outcrop of groundwater 地下水出露
outcrop on surface 地面露头
outcropping 出露地表，露出，露头
outcrop water 渗出水
outcrop 地面露头，露头，露出，岩石露头，露出地面的岩层
outcut 切口
outdated data 陈旧资料
outdevice 输出装置
out-distance 领先，越过，超越
outdoor activity 户外活动
outdoor air design conditions 室外空气计算参数
outdoor air load 室外空气负荷，新鲜空气负荷
outdoor and substruction grounding 厂外和地下构筑接地
outdoor apparatus 户外设备，露天设备
outdoor area 户外区
outdoor arrangement 露天布置
outdoor boiler 露天锅炉
outdoor coal pile 露天煤堆
outdoor condition for designing 设计室外条件
outdoor controlgear 户外控制设备
outdoor corrosion 在敞开空气中的腐蚀，大气腐蚀，室外腐蚀
outdoor critical air temperature for heating 采暖室外临界温度
outdoor current transformer 户外型电流互感器
outdoor design temperature for heating 采暖室外计算温度
outdoor design (电站)露天布置
outdoor disconnecting switch 户外型隔离开关
outdoor distribution board 户外型配电盘
outdoor engineering 室外工程
outdoor environment 户外环境
outdoor equipment 户外设备，露天设备
outdoor exposure test 室外暴露实验
outdoor fire hydrant 室外消火栓
outdoor fossil fuel power plant 露天火电厂
outdoor illumination design 室外照明设计

outdoor installation 露天装置，户外装置
outdoor insulator 户外绝缘子
outdoor landscape design 室外景观设计
outdoor lighting 户外照明
outdoor location 户外场所
outdoor mean air temperature during heating period 采暖期室外平均温度
outdoor motor 户外型电动机，露天电动机
outdoor oil circuit-breaker 露天油断路器
outdoor operation 户外作业
outdoor paint 室外用油漆
outdoor piping 室外管道
outdoor plant 露天发电厂，露天电站
outdoor potential transformer 户外型电压互感器
outdoor power house 露天式厂房
outdoor power plant 露天电站，户外式电站
outdoor power station 露天电站
outdoor regulator 户外型调整器
outdoor reverser 露天型反向器
outdoor service 露天使用
outdoor sol-air temperature 室外空气综合温度
outdoor sound pressure level 露天声压级
outdoor station 露天发电厂，露天变电所
outdoor stockpile 露天料堆
outdoor storage 户外贮存
outdoor substation 露天变电所，室外变电站，户外变电站
outdoor switchgear 屋外配电装置
outdoor switching station 露天开关站
outdoor switchyard 露天开关场
outdoors 户外，野外，露天，在户外
outdoor temperature 室外温度
outdoor thermal power plant 露天火电厂，露天式火力发电厂
outdoor transformer station 露天变压器站，露天变电所
outdoor transformer 露天变压器，户外变压器
outdoor trench 室外管沟
outdoor turbine 露天布置式透平
outdoor type generator 户外型发电机，露天型发电机
outdoor type switchgear 屋外配电装置，屋外开关装置，户外型配电装置
outdoor wiring 户外布线
outdoor work 室外作业，露天作业，户外作业
outdoor 户外的，露天的，露天，室外的，户外，室外，野外
outdo 高出，超过，优于
out-drum mixing 桶外混合
outer air circuit 外部风格
outer annulus 外（部）环状空间
outer atmosphere 外大气层，外层大气
outer bank 外岸，外堤
outer bar 外沙洲，外滩
outer bearing cover 外轴承盖
outer bearing 外轴承
outer bellows seal 波纹管外密封
outer berm 外戗道，迎水面马道
outer border 外图廓
outer brush 外电刷
outer cage （鼠笼电机的）外笼，外机壳，外机架
outer casing 外壳，外机壳，（炉墙外壁）外护板，外汽缸，外缸，外层
outer coil 外层线圈
outer concrete shell 外混凝土壳
outer conductor （三线制的）外侧线，外导线，边相导线
outer containment 外安全壳
outer continental shelf 大陆棚外半部，外侧大陆棚，外大陆架
outer cover ring 外覆环
outer cover 外罩
outer cylinder 外缸
outer dead point 外死点
outer diameter 外径
outer dimension 外形尺寸，外廓尺寸
outer door 外门
outer envelope 外层火焰【氧油气焊接火焰】，包膜外层，外膜，外层信封，包围层
outer enveloping profile 外包络断面
outer frame 外基座，外机座
outer fuel assembly 外部燃料组件
outer fueled zone 外加料区
outer gas pressure tank 外气压供油箱
outer gland 外汽封，外侧汽封
outer harbour 外港
outer housing 外壳，外罩
outer inspection 外观检查，外观测孔
outer insulation 外部绝缘，外绝缘，保护绝缘
outer layer 外层
outer levee 外堤
outer mixing vane 外混流叶片
outermost cage 外鼠笼
outermost 最外面的，最远的
outer motor cooler 电动机外冷却器
outer oil sump 外油槽
outer orientation 外方位
outer packaging 外包装
outer packing 外包装
outer race （滚动轴承）外圈，轴承外部轨道，外环，外轴承圈，外滚道
outer rear cover 外后盖
outer ring of diaphragm 隔板外环
outer ring （轴承）外环，外定位环
outer rotor 外转子，外齿轮
outer sheath 外护套
outer shell bellows compensator 外壳波纹管补偿器
outer shell 外壳层，外缸，外壳，外电子层
outer shield 外屏蔽
outer side 外侧
outer skirt 外裙筒
outer sleeve 外套筒，外套管
outer slope 外坡
outer stator casing 外定子机座，外定子机壳
outer steel shell 钢外壳，外层钢壳
outer strap 外面条带
outer stroke 外行程
outer surface inspection 外表面检查
outer surface 外表面
outer-sync 不同步，失调，不协调

outer toroidal suppression chamber　外侧环形弛压室
outer tube　外管，套管，外套管
outer zone　表面层
outer　外部的，外面的
out-expander　输出扩展电路
out-fade　信号衰减
outfall ditch　排水沟
outfall fan　河口扇形地
outfall losses　跌水损失
outfall pipe　下水管，排放管
outfall sewer　下水管道，下水道，排放管道，出水总管，出水下水道
outfall structure　排水结构，排水站
outfall　排水口，渠口，污水出口管，排泄口，落口，出口，抛下，河道出口，排水结构，排水站，阴沟出口
outfan　输出端，风扇
outfire　灭火
outfit　成套装备，成套备用工具，装置，配备，器械，工具
outflow area　出流面积
outflow channel　放水渠道
outflow conditions　流动出口条件
outflow diagram　出流曲线
outflow duration　出流历时
outflow effluent　出流
outflow from reservoir　水库出流量，水库放水量
outflow hydrograph　出流过程线
outflow rate　泄漏率，流出速率，出流率，出流量
outflow section　（汽轮机）抽汽段
outflow velocity　流出速度，出流流速
outflow vortex　流出旋涡
outflow　流出，外流，流动，流出口，流出物，流出量，出流量，出流率，放水
out frame　外机座，外壳，定子机座
outgassing　除气
outgas　去气，除气，排气
out gate　输出门
outgoing air　排出的空气
outgoing breaker　引出线断路器
outgoing cable　引出电缆，出局电缆
outgoing circuit　输出电路
outgoing condensate water　凝汽器出口循环水
outgoing contactor　引出线接触器
outgoing cubicle　引出小室
outgoing data　输出数据
outgoing direction　出线方向
outgoing feeder gantry structure　出线门形架构
outgoing feeder transformer　出线输电变压器
outgoing feeder　引出馈线，输出馈路，输出馈线，对外馈电线，出线
outgoing group selector　出局选组器
outgoing line circuit　输出线路
outgoing line of overhead line　架空线出线
outgoing line　引出线，出线，引出线起点
outgoing radiation　出射辐射
outgoing repeater　输出增音机
outgoing section　引出隔间
outgoing signal　输出信号
outgoing tide　退潮
outgoing traffic　发端通信业务，发端话务
outgoing trunk　发话中继专用线
outgoing voltage grade　出线电压等级
outgoing water　出水
outgoing wire　出线
outgoing　费用，支出，出发，外出的，输出的，引出的，即将离职的
outgo track　发车线
outgrowth　增生的，结果，产物
outgush　涌出，喷出，流出量
out-in fueling　外区向内区倒换燃料
out-in loading　从外向里装载
out-in refueling　由外往内换料方式，外层燃料内移换料方式，由外向内换料
out-in scatter refueling　从外向内插花式换料
outlaw　取缔
outlay curve　费用曲线，支出曲线
outlay　支出，费用
outleakage　外泄，外漏，泄漏，向外泄漏，向外部的泄漏
outler sluice　泄水闸
outlet air velocity　出口风速
outlet angle　出口角
outlet area of nozzle　喷嘴出口面积
outlet area　出口面积
outlet arm　出口河汊
outlet blade angle　叶片出口角
outlet box　出线盒，引出箱，接线盒，出口盒，出口箱
outlet bucket　出口反弧段，出口消力戽
outlet building　出口建筑物
outlet capacity　泄水能力
outlet chamber　引出室，出风室
outlet channel　出口渠，出水渠
outlet chute　排料口溜槽
outlet conditions　出气参数，出口工况
outlet cone　排料锥
outlet contact　出口接触
outlet control　出口（水位流量）控制
outlet corridor　出口走廊
outlet device　出口装置
outlet discharge　泄水孔流量
outlet duct　出风道，排水道，出线沟
outlet entrance　泄水孔进口
outlet flow angle　出气角
outlet gas temperature　出口烟温
outlet gate　放水闸门，泄水闸门，排料门
outlet geometric angle　出口几何角
outlet guide vane　出口导叶
outlet header　出口集流管，出口联箱，出口部管
outlet head loss　出口水头损失
outlet hole　泄水孔，排出孔，放水孔，出水孔
outlet hose nozzle　水龙带接嘴
outlet housing　出风罩
outlet in dam body　坝身泄水孔
outlet jumper　出口跨接管
outlet line　引出线路，出口管线
outlet loss　出口损失
outlet nozzle dam　出口接管挡板
outlet nozzle　出口管嘴

outlet of tile drain 排水瓦管出口
outlet opening 排料口开度，排除口
outlet pipe 出口管，排出管，出水管，放出管，流出管，出口管道，泄水管，引出管
outlet plenum 出口联箱，出口腔
outlet portal 出口门
outlet port 出口
outlet pressure 出口压力，出水压力
outlet section 出流断面
outlet signal 输出信号
outlet silencer 出口消声器
outlet size 出口尺寸，出料粒度
outlet sluice 放水闸，排水闸
outlet sound absorber 送风口消声器
outlet structure 出口建筑物，泄水建筑物
outlet subcooling 出口欠热度，出口过冷度
outlet temperature profile 出口温度分布
outlet temperature 出口温度，输出端温度
outlet terminal 出线端，引出端
outlet transformer 输出变压器
outlet transition 泄水孔渐变段
outlet valve 出口阀，排出阀，排泄阀，泄水阀
outlet velocity 流出速度，出口速度，排出速度，出口流速
outlet voltage 出线端电压，出口端电压
outlet volute 出口蜗壳
outlet water-head 出口压力，出口水头
outlet works 出口建筑物，排水工程，泄水工程
outlet 出口管，排出口，排水口，排汽口，排气口，电源插座，出口，出水孔，放水口，通风口，排料口，泄水孔，引出线
outline agreement 纲要协定
outline border 外框
outline contour 外形
outline diagram 外形图，略图
outline dimension drawing 外形尺寸图
outline dimension 外形尺寸，总尺寸
outlined procedure 概述的程序
outline drawing 外形图，轮廓图
outline map 轮廓图，外形图
outline of construction organization engineering 施工组织设计纲要
outline of project 项目概况
outline plan 粗略规划
outline shape 外形
outline sketch 轮廓草图
outline 概要，概括，概述，概况，大纲，草图，外形，轮廓（线），提纲，外形线，纲要
outlying area 边远地区
outlying structure 外围构筑物
outmost island 最外岛屿
outmost point 最外点
out of action 不再运转，失去效用
out-of alignment 未对中的
out-of-balance current 不平衡电流
out-of-balance force 不平衡力
out-of-balance load 不平衡负荷，不平衡负载
out of balance 失去平衡
out-of-calibration 刻度以外的
out of center 偏心
out of commission and in commission 停役，服役
out of commission 停用，退役，停役
out of control 失控
out-of-core flux instrumentation 堆外通量测量仪表
out-of-core fuel management 堆外燃料管理
out-of-core neutron flux measurement 堆外中子通量测量
out-of-core neutron flux monitoring 堆外中子通量监测
out-of-core neutron monitoring 堆外中子监测
out-of-core thermionic converter 堆外热离子换能器
out-of-core thermionic reactor 堆外热离子反应堆
out-of-core 堆芯的，堆外的，反应堆外，铁芯外，堆芯外
out-of-date equipment 过时的设备
out-of-date 过期的，过期
out of door hydrodynamic 室外流体动力学
out-of-flatness 平面形缺陷，平坦性缺陷
out-of-frame 帧失调
out-of-gearing 脱开的，不啮合的
out of level 不平坦，不在一水平上
out-of-limit condition 越界，超过定值的条件
out of limit （参数）越限
out-of-line coding 线外编码，越线编码
out of line 不在一直线上
out of operation 停止运行，不能工作，不工作的，不运转的，断开的
out of order 异常工况，不正常状态，失控，无次序，失灵，有毛病，出故障，次序混乱
out-of-phase breaking capability 失步开断能力
out-of-phase breaking current 失步开断电流
out-of-phase breaking test 失步开断试验
out-of-phase current 相位移电流，不同相位电流，反相电流
out-of-phase signal 不同相位信号
out-of-phase voltage 失相电压，不同相位电压
out-of-phase 失相，脱相，异相，不同相，不同相的，异相的，失相的
out-of-pile instrumentation 堆外测量仪表
out-of-pile inventory 堆外（燃料）存量
out-of-pile loop 反应堆外回路
out-of-pile rig 反应堆外台架
out-of-pile test 反应堆外试验
out-of-pile 反应堆外的
out of plane bending 面外弯曲
out of position 脱位，错位，非正确位置
out of proportion 不成比例的，不成比例
out of repair 失修，破损
out-of-roundness 不圆度，不合圆形，椭圆度
out-of-round 不圆
out of scope of supply 非供货范围
2-out-of-3 selection circuit 三取二选择电路
out-of-service bunker 停用的煤仓
out-of-service silo 停用的煤仓
out of service 停止运行的，不能工作的，未投入运行的，停运，停机，退役，不能使用，退出运行，不工作，失效
out-of-size 尺寸不合规定的
out-of-squareness 不垂直度
out of step blocking 失步闭锁

out of step operation 失步运行
out-of-step protection 失步保护
out-of-step relay 失步继电器
out-of-step switching 失步切换，失步开合闸
out-of-step trapping 失步解列，失步跳闸
out-of-step 不同步的，不同步，失步的，失步
out of stock 缺货
out-of-synchronizing 不匹配，不同步，失步，失调
out-of-system unit （测量的）制外单位
1 out of 2 system 二取一系统
2 out of 3 system 三取二系统
2 out of 4 system 四取二系统
out of time 不合时宜，不合拍，太迟，不及时
out of work 不工作，切断，故障，失效
out of 在……范围以外
out-phase component 异相分量
out-phase 相位不重合，异相位，不同相位
outphasing modulation system 反相调制方式，移相调制方式
outphasing modulation 相调，反相调制
outport 外港
out-primary 初级线圈端，原绕组出线头
output admittance 输出导纳
output amplifier stage 输出放大级
output amplifier 输出放大器
output amplitude 输出振幅
output block 输出缓冲区，输出部件，输出信息块，输出存储区
output buffer 输出缓冲器
output bus driver 输出总线驱动器
output capability diagram 出力曲线
output capacitance 输出电容
output capacity 全出力，净出力，输出功率
output carry 输出进位
output characteristic of WTGS 风力发电机组输出特性
output circuit relay 出口继电器
output circuit 输出回路，输出电路，出口回路
output coefficient 输出率，比转矩系数，利用系数，效率
output conductance 输出电导
output constant operation 等功率运行（方式），定负荷运行，恒功率运行
output constant 输出常数
output controller 输出调节器，出力调节器
output control line 输出控制线
output control pulse 输出控制脉冲
output control 输出控制
output cost 出厂价格
output coupling 输出连接
output current 输出电流
output data line 输出数据线
output data 输出数据
output device 输出设备，输出装置
output display unit 输出显示器
output distribution function 输出分布函数
output disturbance 输出扰动
output due to input 输入作用引起的输出，由于输入作用的输出，零状态响应
output element 输出元件
output end 输出端
output error 输出误差
output factor 输出因子，利用率，能量利用系数，出力因数，出力系数，输出系数
output filter 输出滤波器
output flyback 输出回描
output function 输出函数
output governor 输出功率调节器
output hunting loss 输出的搜索损失
output impedance 输出阻抗
output increase block 增负荷
output increase 功率上升
output indicating voltage 输出端指示电压
output indicator 输出指示器
output information 输出信息，输出资料
output lead 输出引线
output level 输出电平
output limit 输出极限，输出范围
output linear group 输出线性部分
output line 输出线
output load resistor 输出负载电阻
output load 输出负荷，输出负载
output loss 出力损失
output mechanism 输出机构
output medium 数据输出装置
output meter 输出计
output monitor interrupt 输出监视中断
output network 输出网络
output nominal speed 额定输出转速
output of desalted water 淡化水产量
output of hydropower station 水电站出力
output of power station 电站出力
output of spring 泉出水量
output order 输出指令
output parameter 输出参数
output performance diagram 出力性能曲线
output plane 输出面
output potentiometer 输出电势计
output-power meter 输出功率计
output power spectrum 输出功率谱
output power 输出功率
output printer 输出印刷装置，输出打印机
output program 输出程序
output pulse amplitude 输出脉冲幅度
output pulse 输出脉冲
output rate 产量，生产率
output rating 额定输出功率，输出功率，产量
output reactance 输出电抗
output recorder 输出记录器
output reduction 降负荷
output register address 输出寄存器地址
output register 输出寄存器
output regulator 输出功率调节器
output resistance 输出电阻
output resonator 输出谐振器
output routine 输出程序
output screen 输出屏
output shaft 输出轴
output shift register 输出移位寄存器
output signal with inert tracer 惰性示踪物输出信号

output signal 输出信号
output speed 输出转速，输出速度
output stability 输出稳定性
output stage 输出级
output station 输出操作站
output subroutine 输出子程序
output switch network 输出转换网络
output system 输出系统
output table 输出函数表
output tax 销项税
output terminal 输出端子，输出端
output test 出力试验
output transfer function 输出传递函数
output transformerless circuit 无输出变压器电路
output transformerless 无输出变压器
output transformer 输出变压器
output transmission 输出传动
output turbine 动力透平，输出功率透平
output unit 输出元件，输出器，输出设备，输出装置，输出部件
output uprating 提高输出功率，提高出力
output value table 输出值表
output value 输出值
output variable 输出变量
output vector 输出矢量
output voltage regulation 输出电压调整率
output voltage stabilization 输出电压稳定
output voltage 输出电压
output winding 输出绕组，输出线圈
output window 输出窗口
output word 输出字
output 输出，输出功率，出力，功率，输出信号，生产量，产量，输出量
outreach program 延伸计划
out reach 起重机臂伸出长度
outrigger jack 悬臂式千斤顶，支撑式起重器
outrigger scaffold 挑出脚手架
outrigger 承力外伸支架【运输集装箱】，叉架，悬臂支架，悬臂梁，舷外支架，外伸叉架，伸出梁
outright electrical failure 纯电气故障
outscriber 输出记录机
out seal 外部密封
out-secondary 次级线圈端，副绕组出线头
outset 开端，最初，开头，开始
outsid air intake duct 新风风道
outside air supply 室外空气供应
outside air 外部空气，室外空气
outside appearance 外观
outside atmospheric pressure 室外大气压力
outside auditor 外部审核员
outside axis line A A列外的（设备等）
outside bank slope 河岸外坡
outside cable 室外电缆
outside calipers 外卡尺，外卡钳
outside cap 外盖，外轴承盖
outside casing 外门窗框
outside circulation 外循环
outside column 外柱
outside condition 室外条件
outside contracting approach 采用外界承包的方法
outside corrosion 外部腐蚀
outside cylinder 外壳，外缸，外径
outside dimension 外形尺寸
outside dovetail root 外包燕尾型叶根
outside envelope 外壳，外包络线
outside exergy dissipation 外部损耗
outside financing 外部筹资
outside flow 外部水流
outside interference 外来干扰
outside its normal operating envelope 在正常运行范围之外
outside leader 外水落管
outside lighting 室外照明
outside marking 外部标记
outside micrometer calipers 外径千分尺
outside-mixing burner 扩散式燃烧器，外混式燃烧器
outside of the field of mathematics 几何生殖线【管道，容器】
outside-packed 外部密封包装的
outside power 外电源
outside primary 初级线圈外端
outside relationship 外部关系
outside run 室外布线，外线装置
outside screw and yoke 外部螺纹和轭架
outside screw yoke 外螺纹阀
outside screw 外螺纹
outside sleeve 外套筒
outside slope 外斜坡
outside specialist 外来专家
outside stairs 室外楼梯
outside steam 外汽源
outside surface 外表面
outside test 外部试验
outside the external building wall 建筑物外墙外侧
outside vertical riser 室外垂直立管
outside washing 体外清洗
outside water 外来水，外水
outside wiring 室外布线，屋外布线
outskirts 市郊，郊区，郊外
out-slot signaling 隙外信令
outsole 脚，基线
outsourced items 外委件
outsource 外包（工程），外购，外购产品或由外单位制作产品
outstanding balance 未偿还余额，未结余额，未付差额，未清余额，待结款项，未用余额
outstanding bonds 未清偿债券
outstanding contract 尚待执行的合同，尚未完成的合同
outstanding contribution 杰出贡献
outstanding personality 杰出人物
outstanding problem 未解决的问题，遗留问题，悬而未决的问题
outstanding talent 优秀人才
outstanding work 突出的工作
outstanding 待完成的，待决的，待付的（款），未付款的，未清偿的，显著的，突出的，未完成的
out station 被控站

outstrip　超前，超过，超出
outsurge　向外膨胀（液体），负波动
outswinger　外弧球
out-swinging casement　外开窗扇
out-sync　失调，不协调，不同步，失步
out-to-out distance　外包尺寸，轮廓尺寸
out-to-out　全长，总宽，总尺度，最大尺度
outward diffusion　向外扩散
outward flange　外凸缘【法兰】
outward-flow turbine　外流式水轮机
outward-flow　离心流动，辐射式流动，外向流，外泄流
outward fuel transfer　乏燃料运出
outward leakage　向外泄漏
outward leaktightness　外向密封，向外密封
outward radial shuffling　径向向外倒料
outward seepage　向外渗出
outwards　向外地
outward thrust　外向推力
outward transfer bridge　向外运输桥吊
outward transfer compartment　向外输送小室
outward transfer position　向外输送位置
outward transfer procedure　向外输送程序，向外输送步骤
outward transfer　向外输送［转运］
outward　向外，向外的
outwash　沉积物，冰水沉积，冲刷
out wire　外线
outworn　用完的，用旧的，磨坏的
out yawing　外偏航
out　外部的，断开的，在外，熄，灭
ouvala　灰岩盆，岩溶谷
ouvered air outlet　百叶式出气口
oval bearing bush　椭圆形轴衬
oval-clearance bearing　椭圆间隙轴承
oval-clearance　椭圆形间隙
oval cross-section　蛋形截面，椭圆截面
oval head screw　椭圆头螺钉
ovality　椭圆度
ovalization　椭圆度
oval nail　椭圆钉
oval orbit　椭圆轨道
oval ring gasket　椭圆环形垫片，椭圆形金属环垫
oval scale　椭圆尺
oval slot　椭圆形槽
oval-type bearing　椭圆形轴承
oval-type oil tank　（变压器的）椭圆形油箱
oval-type slot　椭圆形槽
oval wheel flow measuring device　椭圆齿轮流量测量仪表
oval　椭圆形的，椭圆形
OVCD(over-carried)　漏卸
oven-dried aggregate　烘干骨料
oven wire mesh blanket　金属丝网层
oven　电炉，炉，箱，恒温箱，烘箱，干燥器，干燥箱，烘干炉，烤箱，烤炉，火炉
overachieve　超过（预期目的）
overage and shortage on inventory taking　盘存亏盈
overage　逾龄的，过老化的，超出，过多
overaging　过时效处理，超龄的，老朽的，旧式的
overall accuracy　总的精确度，总精度
overall adiabatic efficiency　总绝热效率
overall amplification　总放大系数
overall analysis of economic benefits　经济效益综合分析
overall appraisal of project　项目的总评价，项目综合评价
overall area　总面积
overall arrangement　总体布置
overall attenuation level　总衰减电平
overall balance　总平衡
overall benefit　总效益
overall breeding rate　总增值率
overall building for production and administration　生产行政综合楼
overall bypass　大旁路
overall characteristic　综合特性
overall coefficient of heat transfer　总传热系数
overall coefficient　总系数，综合系数，总利用系数
overall concentration　总浓度
overall consideration　综合考虑
overall construction duration　总工期
overall construction schedule　总施工进度
overall construction site plan　施工总平面图
overall coolant flow rate　冷却剂总流量
overall cost　总费用，总成本
overall cycle time　总循环时间
overall dead time　总空载时间
overall design peaking factor　设计总峰值因数，总热点因子
overall design　总体设计
overall development　综合开发
overall diameter　全径，机座外径，外包直径
overall differential protection　整体差动保护
overall diffusivity　总扩散率
overall dimension　全部尺寸，总尺寸，轮廓尺寸，外形尺寸，外包尺寸
overall drag　总阻力
overall drawing　总图
overall economic accounting　全面经济核算
overall economic effect　全面经济效果
overall economic efficiency of power station　电厂总经济效率
overall efficiency of separation　除尘效率
overall efficient　总效率
overall effluent　排出物
overall elasticity modulus　综合弹性模量
overall energy　总能量
overall error　总误差
overall evaluation　综合评价
overall expansion ratio　总膨胀比
overall expansion　总膨胀
overall floorage　总建筑面积
overall fog　全面雾翳
overall form factor　总形状因子
overall fuel cost　总燃料费用
overall gain　总增益
overall generator efficiency　放大级总效率

overall geometry 总体尺寸
overall head allowance 总水头容许值，总水头允许值
overall heat balance 全面热平衡，总热平衡
overall heat consumption 总热耗
overall heat loss coefficient 总热损系数
overall heat loss 总热损失
overall heat transfer coefficient 总体传热系数，总传热系数
overall heat transfer 传热过程
overall height 总高度
overall hot-channel factor 总热通道因子
overall hot spot factor 总热点因子
overall information transmission efficiency 总信息传送效率
overall inspection 全面检查
overall layout drawing 总布置图
overall leakage 总泄漏
overall length 全长，总长度
overall load 总负荷
overall loan 总贷款
overall loss coefficient 总热损失系数
overall magnetization 总体磁化
overall mean velocity 全断面平均流速
overall neutron transport 总中子输送
overall objective 整体目标
overall operational cost 全部使用费，总运行费
overall optimization 整体优化
overall output criteria 总体成果标准
over allowance 尺寸上偏差
overall payment agreement 总支付协定
overall performance 综合性能
overall perspective 全景，全貌
overall planning drawing of power plant 全厂总体规划图
overall planning management 全面计划管理
overall planning 全面规划，总体规划，总计划
overall plant design description 电厂设计总说明
overall plant efficiency 电站总效率
overall plant net efficiency combined cycle 联合循环电厂总的净效率
overall plant net efficiency open cycle 开式循环电厂总的净效率
overall plant process flow diagram 整个电厂的工艺流程图
overall plot plan 总平面图，全厂总体规划
overall pressure drop 总压降
overall pressure ratio 总增压比，总压比
overall price index 综合物价指数
overall price 全价
overall program 总进度，总计划
overall project capital 项目计划总资金
overall project 全面计划方案
overall rate 综合出力
overall reaction 总反应
overall regulation 总调节
overall reliability 总可靠性
overall response time 总响应时间
overall responsibility 全部责任
overall risk 总风险，综合险，总风险度
overall safe allumination 总体安全照明
overall schedule of construction 施工综合进度
overall schedule 总进度
overall separation efficiency 总分离效率
overall separation factor 总分离因数
overall shield copper wire 全屏蔽铜线
overall shield 全屏蔽，总屏
overall size 外部最大尺寸，外形尺寸，总尺寸，外包尺寸
overall solids circulation pattern 全部颗粒循环模式
overall soundness of project 项目总体合理性
overall span 全翼展
overall stability constant 总稳定常数
overall stability 总稳定性
overall steam cost 蒸汽总费用
overall strength 总强度
overall structure 整体结构
overall system adjustment 全程调整，综合调整
overall system performance 系统综合性能
overalls 罩衣，工作服
overall theoretical power 理论总功率
overall thermal conductance 总热导
overall thermal efficiency 总热效率
overall thermal resistance 总热阻
overall transfer coefficient 总传热系数
overall transfer function 总传递函数
overall transfer time 总传送时间
overall unit protection 单元总体保护
overall viewer 全貌突击视窗
overall view 全景
overall wear characteristics 总磨损特性
overall width 总宽度，全宽
overall work 全面工作
overall 总的，全部的，全面的，所有的，全的，综合的
overamplification 放大过度，过分放大
over-and-over addition 逐次相加，重复相加，反复相加
over-and-over 反复的
over-and-under controller 自动控制器
over-and-under design 上下型设计
overarch 上架拱圈，上设拱圈
overbalance 超平衡，超出平衡，过平衡，失衡，过重，过量
overbank flow 溢岸流
overbank superheater 位于管束上的过热器
overbank 河滩，溢岸，漫滩
over beam 过梁，悬梁
overbending 过度弯曲
overboots 套靴
overbreakage 超挖度
over break 超挖
over bridge 天桥
overburden depth 覆盖深度
overburden layer 覆盖冰层，覆盖层
overburden pressure 覆盖层自重压力，超载压力，积土压力，覆盖地层压力，表土层压，地静压力
overburden soil 覆盖土
overburden stress 过度应力
overburden stripping 覆盖层剥除

overburden 超负荷，覆盖，过载，超载，覆盖层，（放射性）过量积存
overburning 过烧，烧损
over-burnt brick 过火砖
over capacity design 过容量设计
over capacity 过负荷，过容量
overcapture 多种俘获
over-carried cargo 漏卸货物
over-carried 漏卸
overcast sky 阴天
overcharge 超额装载，超载，过量充电，过负荷
overchorded winding 长距绕组
overclimb 失速，气流分离
overcoated particle 外包覆燃料颗粒
overcoating （燃料颗粒）包覆，涂饰
overcommutation 过换向，过度整流，超前换向
overcompaction 过度压实
overcompensated attenuator 过补偿衰减器
overcompensate 补偿过度，过补偿
over compensation 过补偿
overcompound dynamo 过复励电机
overcompounded motor 过复励电动机
overcompounded 过复励的
overcompound excitation 过复励磁，过复励
overcompound generator 过复励发电机
overcompound 过复励
over consolidated clay 超固结黏土，过度固结黏土，超压密黏土
over consolidated ratio 超固结比
over consolidated soil 超固结土，超压密土
overconsolidation ratio 超固结比，超压密比
overconsolidation 超压密，过度固结，超固结
overcontrol 过分操纵，操纵过量
overcook 过烧
overcooling accident 过冷事故
over cooling 过冷（却）
overcoring 掏心钻进
overcorrection 过校正，过调
overcount 过计数
over-coupled transformer 过耦合变压器
overcouple 过耦合
overcrimping 过分折边的
overcritical 超临界的
overcrossing point 上交叉点
overcrossing 上跨交叉，人行天桥
overcunent release 过流脱扣器
overcurrent cutting-off 过电流切断
overcurrent ground relaying system 过电流接地继电保护方式
overcurrent protection 过流保护，电流过载保护，过电流保护装置，过电流保护
overcurrent protective device 过电流保护装置
overcurrent relay 过流继电器，过电流继电器，过载继电器
overcurrent release 过电流脱扣器，过流释放
overcurrent test 过电流试验
overcurrent trip coil 过电流脱扣线圈
overcurrent tripping 过电流脱扣，过流跳闸，过电流断开
overcurrent trip 过电流脱扣，过电流脱扣器
overcurrent 过电流
overcutting 过度切削，过调制
overdamped harmonic motion 过阻尼谐振荡
overdamped 过阻尼的，过阻尼，强衰减，过度阻尼
overdamp 超阻尼，过阻尼
overdeck superheater 布于管层上的过热器
overdeeping 过量下蚀
over-demand 求过于供
over-design 有裕量的设计，超裕度设计，保险设计，安全设计，过于安全的设计，过度设计
over-development 显影过度，过度开发，超采
over-dimensioned 超过尺寸的，尺寸过大的
overdischarge 过放电
overdosage 超剂量
overdraft facility 透支限额
overdraft 透支，超采，超量取水，过度抽汲，过度抽取，过度通风，上部通风
overdrainage 过度排水
overdraw 透支
overdriven pile 过深桩
overdriven 过载的，过激的
overdrive pile 打过头的桩
overdrive 超速传动，增速传动，过激励，过策动
overdue tax payment 滞纳税款
overdue 过时的，过期的，过期未付
over-duty motor 过载电动机
over-engineering 过度设计
overestimate 估算过高
overexcavation 超挖
overexcitation limiter 过励磁限制器，过励限制器
overexcitation protection 过励磁保护
overexcitation 过励磁，过激磁
overexcited generator 过励磁发电机
overexcited synchronous machine 过励同步发电机
overexcite 过励
overexpansion 过度膨胀，过膨胀
overexposure 超剂量照射，过分辐照的，过曝光，过暴露
overfall crest 溢流堰顶
overfall dam 溢流坝，溢洪道坝
overfall dike 溢流坝，溢洪道坝
overfall spillway 坝顶溢洪道【水电】
overfall weir 溢流堰
overfall with air admission 进气溢流堰
overfall 流出，溢出，溢水坝，溢流
over-fault 上冲断层
overfeed firing 上饲式燃烧，上给料式燃烧
overfeeding 过量供给
overfeed of water into drum 汽包满水
overfeed stoker 上饲式炉排
overfeed 上饲，过量进料，过量供给
overfiling 过分锉磨的
overfire air compartment 上二次风风室
overfire air fan 上二次风风机
overfire air jet 二次风喷嘴
overfire air port 二次风喷口
overfire air 二次风，上二次风
overfire 过烧，烧损

overfiring 过烧，超温
overflash 飞弧，闪络
overflow alarm 溢出报警
overflow arch dam 溢流拱坝
overflow bellmouth 喇叭形溢流口，钟形溢流口
overflow bit 溢出位
overflow chamber 溢流闸室
overflow check indicator 溢出指示器
overflow chute 溢流陡槽
overflow cofferdam 过水围堰
overflow conduit 溢流管沟
overflow contact 满溢触点，全忙触点
overflow control indicator 溢出控制指示器
overflow crest 溢流堰顶
overflow cutting machines for aluminium wheels 铝轮冒口切断机
overflow dam 溢流坝
overflow detector 溢出检测器
overflow dike 溢流堤
overflow discharge 溢流量
overflow earth dike 溢流土堤
overflow earth rock dam 溢流土石坝
overflow edge 溢流堰顶缘
overflow electric power 溢流等效电力，溢流损失电力
overflow error 溢出错误
overflow gate 溢流式闸门
overflow governing 溢洪道调节
overflow groin 溢流式丁坝
overflow indicator 满溢指示器，溢流指示器
overflowing frequency 溢流频率
overflowing of spillway 溢洪道溢流
overflowing rate 溢流速率
overflowing sheet 溢流水层
overflowing spilling water 溢流
overflowing 溢出，溢洪，对流，外溢
overflow installation 溢流设备
overflow land 漫水地
overflow level 溢流水位
overflow line 溢流管线
overflow meadow 泛滥草甸
overflow meter 溢呼次数计，全忙计数器
overflow nappe 溢流水舌
overflow passage 溢水道
overflow pipe 溢流管，溢流道，溢水管
overflow position 溢出位置
overflow prevention 漫溢预防
overflow pump 溢流泵
overflow rate 溢流率，溢流速率
overflow regulation 溢流调节
overflow reservoir 溢流式水库
overflow river 泛滥河
overflow rockfill dam 堆石滚水坝，堆石溢流坝
overflow route 溢呼路由
overflow seal box 溢流密封箱
overflow section 泄水坝段，溢流段，溢流段断面，溢流断面
overflow siphon 溢流式虹吸管
overflow spillway 溢洪道
overflow spur dike 溢流式丁坝
overflow stand pipe 竖井溢洪管，溢流竖管，竖向溢流管
overflow stand 溢流挡墙
overflow station 溢流式电站
overflow structure 溢流建筑物
overflow sump 溢流水池
overflow surge chamber 溢流式调压室
overflow tank 溢流箱，溢流槽
overflow tower 溢流塔
overflow training wall 溢流挡墙
overflow trough 溢流槽
overflow type power house 厂顶溢流式厂房，溢流式厂房
overflow type power plant 溢流式电站
overflow valve 溢流阀
overflow velocity 溢流速度
overflow water pipe 溢流水管
overflow weir 明流堰，溢流堰
overflow well 溢流井
overflow 溢出，溢流，溢流管，计算机溢出，溢出管，自由溢流，漫出，漫流，上溢，溢洪道，溢口
overflux 过励磁，超通量，过激励
overfold 倒转褶皱
overfrequency protection 过频率保护，超频率保护
overfrequency relay 超频继电器，过频继电器
overfrequency 过频率，超频率
over full spring 自流泉
overgagging 过分关紧的【阀门】
overgate capacity （水轮机的）超开度容量
overgauge 超过规定尺寸的，等外的
over generation condition 发电能力超量状况
overgrate air 二次风
overgrate blast 二次风
overgrinding 研磨过细
overground cable 架空电缆
overhang bracket 端部支承，翅托
overhang crane 悬臂起重机
overhanging arm 外伸臂
overhanging bank 陡岸
overhanging beam 挑梁，悬臂梁
overhanging cliff 悬崖，陡壁
overhanging crane 悬臂起重机
overhanging eaves 飞檐，挑檐
overhanging eave timber 挑檐木
overhanging impeller 外伸叶轮
overhanging length 悬挑长度，伸出长度
overhanging pile driver 悬臂式打桩机
overhanging roof 挑出屋顶
overhanging siphon spillway 悬挑式虹吸管溢洪道
overhanging stairs 悬挑式楼梯
overhanging structure 悬臂结构
overhanging support 外伸支架，悬臂支架
overhanging 悬伸的
overhang insulation 端部绝缘
overhang packing 端部衬垫，端部（绝缘）垫块
overhang part 外伸部
overhang 倒悬，挑出，突出，外伸，悬垂，悬垂物，线圈端部，突出物
overhaul check 大修检查
overhaul cost 大修费用

overhauling equipment　大修设备
overhaul inspection　大修，检修，拆检
overhaul life　大修间隔，大修周期
overhaul of boiler　锅炉检修
overhaul period　大修期，电机工作寿命
overhaul reserve capacity　检修备用容量
overhaul reserve　检修备用
overhaul shop　修理车间
overhaul yard　检修场
overhaul　大检修，大修（间隔期），彻底检查，仔细检查，检修
overhead belt conveyor　地面带式输送机
overhead bit　附加位
overhead bridge crane　桥式吊车
overhead bridge　天桥
overhead bundled conductor　架空组合导线
overhead busbar　架空母线
overhead cable line　架空电缆线路
overhead cableway　架空索道
overhead cable　高架缆线，架空电缆
overhead charges　管理费，经常性费用
overhead coal bunker　高位煤斗
overhead coal feeding bridge　高架运煤栈桥
overhead communication line　架空通信线路
overhead conductor rail　架空导电轨
overhead conductor　架空导线，架空电力线
overhead construction　架空线路建设
overhead contact network of electrical railway　电气化铁路接触网
overhead contact system　架空接触网
overhead conveyor　高架轨道，高架运送机，高架输送机
overhead cost　附加开支，杂项开支，（一般）管理费，间接成本，间接费用
overhead crane　高架起重机，行车，桥式起重机
overhead crossing　架空线交叉，立体交叉，高架交叉道
overhead distribution line　架空配电线路
overhead door　高架门
overhead eccentric-jaw crusher　高架偏心颚式破碎机
overhead electric traction line　电气化铁路牵引线
overhead expenditure　经常（性）开支，间接费开支
overhead expenses　经常费用，日常开支
overhead fillet weld　仰焊角焊缝
overhead ground wire　避雷线，架空地线，架空避雷线
overhead GSW　架空（镀锌）钢绞线
overhead insulated conductor　架空绝缘导线
overhead inversion　高空逆温
overhead junction　架空线路线岔，架空线网并线器
overhead laying　架空敷设
overhead lead　架空引线
overhead lighting　天花板上的灯，吊灯
overhead line circuit　架空线路
overhead line expenses　架空输电线路费用
overhead line fitting　架空线路配件
overhead line mechanical characteristic　架空线机械特性
overhead line pole and tower　线路杆塔
overhead line sag　架空线弛度
overhead line　架空电线，架空线，架空线路
overhead loader　高架装载机
overhead main　上分式干管道，架空干管道，架空干管
overhead manipulator　大型万能机械手，高架（重型）机械手
overhead pipeline　架空管道
overhead pipe　高架管道
overhead position welding　仰焊
overhead position　架空的部位【焊接】，仰焊位置，仰位
overhead power cable　架空电力电缆
overhead power line　高架输电线
overhead power transmission line　架空输电线路
overhead power transmission　架空输电
overhead projector　高架探照灯，高射投影仪
overhead railway　高架铁道［路］
overhead rail　高架轨道
overhead rate　管理费分摊率，管理费
overhead ring circuit　架空环形线路
over-head ropeway　架空索道
overhead side deck　外伸侧柄面
overhead stable layer　高空稳定层
overhead static cable　（内包通信电缆的）架空地线，架空避雷线
overhead storage tank　高架贮槽
overhead suspension　悬空架设的
overhead system　架空系统
overheads　（企业等的）日常管理费用，杂项开支，一般经费
overhead tank　压力槽，压力罐
overhead telephone line　架空电话线
overhead transmission line　高架输电线路，架空输电线路，架空送电线路
overhead traveling crane　桥式吊车，桥式起重机，桥式起重，高架行车
overhead traveling crane　天车
overhead traveller　桥式吊车，桥式起重机
overhead travelling crane with crab　绞车式桥式起重机
overhead travelling crane with electric hoist　电动葫芦桥式起重机
overhead travelling crane　移动式高架起重机，桥式吊车，桥式起重机，天车，行车
overhead travelling tripper conveyor　高架卸料车皮带机
overhead travel　天车，行车
overhead tripper conveyor　带卸料车高架输送机
overhead tripper　高架卸料车
overhead trolley　高架行车，架空行车
overhead-underground network　架空埋地混合网络
overhead valve engine　顶阀发动机，顶置气门发动机
overhead valve　顶阀
overhead water tank　高架蓄水箱
overhead welding　仰焊，仰脸焊
overhead weld position　仰焊部位

overhead weld 仰焊
overhead wire locomotive 桥式电线机车
overhead wire 高架电线
overhead 上面的，高架的，头顶上的，（费用等）经常的，管理的，架空的，管理费用，经常费用，船舱的顶板，天花板，在头顶上，在空中，在楼上
overheat control 过热控制
overheated economy 过热经济
overheated structure 过热组织
overheated zone 过热区
overheater 过热器
overheating fault 过热故障
overheating protection 过热保护
overheating 过热，不正常温升，超温
overheat relay 过热继电器
overheat steam 过热蒸汽
overheat 过热
overhung construction 悬吊结构，悬式结构
overhung portion of winding 绕组的端部
overhung rotor 外悬型转子
overhung turbine 悬臂式透平
overhung type hydro-generator 悬式水轮发电机
overhung type motor 外悬型电动机
overhung 悬臂的，悬空的
overhydration 水中毒
over investment 过分投资，投资过度
overirradiation 过度辐照
overland belt conveyor 越野带式输送机
overland common point 内陆公共点或陆上公共点运输
overland drainage 地表排水
overland flow hydrograph 地面漫流过程线，地表径流过程线
overland flow 表面径流，地面漫流，坡面流，漫地水流
overland freight 陆运货运
overland insurance 陆运保险
overland runoff 表面径流，坡面漫流，地表径流，地表漫流，地面径流
overlap angle 重叠角
overlap coefficient 重叠系数
overlap control 重叠控制
overlap-fault 超覆断层
overlap joint 搭接
overlap of channels 通道的重叠
overlapped glass plate air heater 重叠玻璃平板空气加热器
overlapped image 重叠像
overlapped joint 搭接接头
overlapped memories 重叠存储器
overlapping areas 重叠区域
overlapping group technique 群重叠技术
overlapping investment 重复投资
overlapping length 搭接长度
overlapping of order 阶叠加
overlapping outage 重叠停运
overlapping phase 交叠相位
overlapping placement 错缝浇筑
overlapping seam 搭接缝
overlapping surge 重叠涌浪
overlapping tile 搭接瓦
overlapping weld 搭接焊接
overlapping 搭接，重叠，交叠，叠加，跳过
overlap reservoir capacity 重叠库容
overlap span 重叠间隔，分段间隔
overlap technique 覆盖技术
overlap test 重叠试验法
overlap welding 搭焊，搭接焊
overlap zone 搭接区（焊接）
overlap 搭接，重叠，重叠度，覆盖，覆盖度，焊瘤，交叠
overlay cladding 堆焊层
overlaying welding 堆焊，补焊，熔焊
overlaying 堆焊，涂覆，共用存储区，程序段落，重复占位
overlay program 重叠程序
overlay supervisor 覆盖管理程序
overlay tolerance 覆盖容差，交叠容差
overlay 覆盖，覆盖式，涂层，涂，镀，交叠，涂覆层，覆盖层，镀金，覆盖物，外罩，堆焊
overlength goods 超长货物
overlength 超长
overlimed cement 多石灰水泥
over-limited 超过阈值的，越限的
overline bridge 天桥
overload allowance 容许超负荷
overload and rupture test 超负荷与断裂试验
overload breaker 过载断路器，超负荷能力，超载能力，过载容量，过载能量
overload characteristic 过载特性，过载特性曲线，超负荷特性
overload circuit-breaker 过载断路器
overload coefficient 超载系数
overload coupling 超载安全联轴节
overload crack 超载裂缝
overload detector 过载探测器，过载检测器
overload device 过载装置
overloaded cargo 超载货物
overloaded phase 过载相
overload factor 过载系数，超负荷能力，过负荷系数
overload imitation 超量模拟
overload indicator 过载指示器，过负载指示器
overloading 过负荷的【油漆】，超额装载，超载，过载，超负荷
overload light 过载信号灯
overload limit 过负荷极限，过载限
overload margin 过载裕度，过载范围
overload meter 过载测定仪
overload operation 超载运行，超负荷运行
overload pressure 过载压力
overload protection 过载保护，过负荷保护，过载保护装置
overload protective device 防过载装置
overload protective relay 过载保护继电器
overload protector 过载保护器
overload rating 额定过载
overload regulation 超载规定
overload relay 过负荷继电器，过载继电器
overload release 过载脱扣器，过载释放
overload relief valve 过载安全阀

overload rupture 过载断裂
overload safeguard 过载保护装置
overload scram 超负荷紧急停堆
overload signal 过负荷信号
overload strength test 超载强度试验
overload switch 过载开关
overload test 超负荷试验，过载试验
overload time relay 过载限时继电器
overload torque 过载转矩
overload trip 超载切断，过载跳闸
overload value 过载值
overload valve 过负荷阀门，过载阀
overload 超负荷，超载，过负荷
overlooker 监视者，监视员
overlying aquifer 上覆含水层
overlying inversion layer 覆盖逆温层
overlying strata 覆层，上覆地层，上覆层，盖层
overlying stratum 覆盖冰层，覆盖层
overlying 叠加，覆盖
overmeasure 估计过高，余量，高估，裕量
over moded waveguide 过模波导
over mode microwave filter 过波型微波滤波器
overmoderated 过慢化，过慢化的
overmodulation 过调制
overnight construction cost 基础价，一夜建成价
overnight cost 一夜建成价，基础价
overnight load 过夜负荷
overnight shutdown 停机过夜
over-notching relay 过脉冲继电器
overnutrition 富营养化，营养过分
overpacking 外包装
overpass bridge 上跨路桥
overpass 立体交叉，路线桥，上跨路
overpay 多付（款）
overperssure protection device 超压保护装置
overpotential test 过电压试验
overpotential 过电势，过电压，过电位
overpour gate 顶溢式闸门
overpower design condition 设计超功率工况
overpowered 有剩余动力的，有剩余电力的，被压倒的
overpower factor 超功率因数，功率峰值因数
over power for wind turbines 风力涡轮机的过载功率
overpower protection 过功率保护
overpower relay 过负荷继电器
overpower scram 超功率紧急停堆，超功率事故停堆
over power transient 超功率瞬态
overpower trip 超功率事故保护停堆
overpower 过功率，超功率，过负荷，打败，克服
overprediction 过高预计，过高预报
overpressure containment rupture 超压安全壳破裂
overpressured containment 过压安全壳
overpressured protection 超压保护
overpressure test 超压试验
overpressure valve 过压阀
overpressure 超压，过压
overpressurization 过量增压，过压
overpriced 定价过高的，价格过高的
overproduction 生产过剩
overproof 超标准的
overpulling 溢出，溢洪
overpumping 超量取水，过量抽取
overrange limit of short duration 短时过范围极限
overrange limit 过范围极限
overrange protection 过量程保护
overrate 估计过高，超过定额，定额过高，过量程，过范围，超出
over reach 超范围
overreinforced 配筋过多的，超配钢筋的，超筋的
overrelaxation factor 超松弛因数
overrelaxation 过度松弛，超松弛法，超松弛
over … relay 过……继电器
over-rich mixture 过浓混合物，过富混合物，过富拌和物，水泥用量过多的拌和物
override control 越权控制，超驰控制，优先控制
override high excess reactivity 补偿高后备反应性
override index 超控索引
override of xenon oscillations 氙振荡补偿
override switch 过载开关
override the xenon effect 补偿氙效应
override trip 超越跳闸
override 超弛，过载，压倒，过量负荷，超越，超过，过负荷，占优势，超控，盖过
overriding alarm 优先报警
overriding commission 代办佣金
overriding 超越，占优势的，基本的，主要的
overroad stay 跨路拉线
overrolling 过度碾压
overrun error 超限差错，超限校验，溢出误差
overrunning clutch 过速离合器
overrunning of the drive 驱动超速
overrunning 超限运行，超速
overrun 超过，漫出，超限运行，超过额定界限，超限，过度运行
oversailing bricks 挑砖
oversanded mix 多砂拌和物
oversaturated vapour 过饱和蒸汽
oversaturated 过饱和的
oversaturation 过饱和
overseas branch 国外分部[分行]，海外办事处
overseas-funded enterprises 外资企业
overseas investment 海外投资，外资
overseas investor 国外投资者
oversea small hydro-power 国外小水电
overseas market 海外市场
overseas project 国外工程
oversea subsidiary 海外子公司
oversea transportation 远洋运输
oversea 海外
overseer 监视者
oversensitiveness 过分灵敏
oversensitive 过敏的，过于敏感的
overset 翻倒
overshoes 套鞋
overshoot a threshold 越界，超过极限值，超过阈值

overshoot in angle of attack 迎角急增量
overshooting a set point 超越整定点
overshooting a threshold 超出阈值
overshoot 过冲（量），超越，超出规定，过调节，超越度，超调（量）
overshot loader 上倾式装载机
overshot spillway 坝顶泄槽溢洪道【水电】
overshot water wheel 上射式水轮
overshot 超调量，超出规定，从上注入的
oversight 失察，督察，漏失，误差，疏忽
oversize aggregate 超径骨料
oversize and large component transportation 大件运输
oversize brick 大型砖
oversize cobbles 过大的鹅卵石【大于150mm的】
oversized coal 粒度过大的煤
oversize fraction 筛上物份数，限上率
oversize gravel 超径砾石
oversize product 筛余物
oversize 尺寸过大，超过尺寸，带余量的尺寸，加大的尺寸，超径，筛上物
oversizing 余度，尺寸余量
overspeed control device 超速控制装置
overspeed control 超速控制
overspeed detector 过速检测器
overspeed device 超速保护装置，过速装置
overspeed emergency governor 危急保安器，危急截断器
overspeed governor 过速调节器，过速限制器，危急切断器，危机保安器
overspeeding 超速运行，超速的，超速，过速
overspeed limiter 超速限制器
overspeed limiting gear 超速限制器
overspeed monitor 超速监视器，超速监测器
overspeed power 超速功率
overspeed protection controller 超速保护控制器
overspeed protection control 超速保护，超速保护控制
overspeed protection device 超速保护机构
overspeed protection trip 超速跳闸保护
overspeed protection 超速保护，过风速保护
overspeed protective device 防止超速装置，超速保护装置
overspeed spoiler 超速保护扰流板
overspeed stop experiment 过速停机试验
overspeed switch 超速开关
overspeed test of steam turbine 汽轮机超速试验
overspeed test pilot valve 超速试验滑阀
overspeed test 超速试验
overspeed trail 超速试验
overspeed trip device 危急截断器
overspeed trip gear 超速脱扣装置，超速危急保安器
overspeed tripping device 超速保护装置
overspeed tripping 超速脱扣停机
overspeed trip pin 危急保安器脱扣销
overspeed trip setting level 超速脱扣定量值
overspeed trip test 超速跳闸试验
overspeed trip 超速跳闸，超速脱扣，超速停机，超速断流
overspeed 过速，超速，超转速，过速的
overspilling 溢出，溢洪
overspill 上旋
oversplash 飞溅
overspread 漫布，满布，蔓延，铺满
oversquare engine 超宽发动机
overstability 超稳定性
overstable 超稳定的
overstaff 为……配备人员过多
overstaggered 过参差失调的
oversteepening 削峭作用
oversteer 过度转向
overstep 超出
overstock safety device 行程限制器
overstock 过度储备，过裕量
overstrain 过度应变
overstressing technique 超载应力技术
overstress 超应力，过度拉伸，过度应力，应力超限，过应力
overstretch 延伸过长
overswing 超出规定，摆动过头，过调节，过摆，尖头信号，超越
oversynchronous speed 过同步转速
oversynchronous 超同步的，过同步的
overtaking flow 来流，迎面流
overtemperature control device 超温控制装置
overtemperature control 超温控制
overtemperature detector 超温检测装置，过热检测器
overtemperature protection device 超温保护装置
overtemperature protection 过热保护
overtemperature protective device 超温保护装置
overtemperature relay 过热继电器
overtemperature trip device 超温限制器
overtemperature 超温，过热，过热温度
overtension 电压过高，过电压
overtest 超限试验，过量试验
over-the-coupler arm 加工在车钩上的夹持臂
overthrust anticline 逆掩断层背斜
overthrust fault 逆掩断层
overthrust 上冲断层，逆掩断层
overtightening 过分拧紧
over-tile 盖瓦
overtime pay 加班费
overtime usage charge 租箱费
overtime work 加班
overtime 加班时间，超出时间，超时，规定之外的工作时间
overtone 泛音，谐波
overtopped cofferdam 过水围堰
overtopped dam 漫水坝，溢流坝
overtopping discharge 漫顶流量
overtopping flow 越顶流量
overtopping quantity 越顶量
overtopping water stage 漫顶水位
overtop 超出，超出……之上，漫顶，漫溢，越顶
overtravel limit switch 过行程限位开关，行程开关
overtravel switch 超程保护开关
overtravel 过调，超程，再调整
overturned anticline 倒转背斜

overturned beds 倒转地层，倾覆地层
overturned fold 倒转褶皱
overturned tower assembling 倒装组塔
overturning effect 倾覆作用
overturning moment （地震）倾覆力矩，倾覆弯矩，倾翻力矩，颠覆力矩
overturning stability 倾覆稳定性
overturning wind speed 倾覆风速
overturning 翻倒，倾覆，倾翻，倾倒
over twisting 过度扭转
over-under containment 上下型安全壳
overvibration 振捣过分
overview display 概貌显示【DCS 组态】，总貌画面
overview 观察，概观，综述，概述
over voltage alarm 过压报警
overvoltage arrester 过电压保护器
overvoltage computation 过电压计算
overvoltage cutoff 过电压截止，过电压切除
overvoltage device 过电压装置
overvoltage due to potential transformer saturation 电压互感器饱和过电压
overvoltage due to system splitting 解列过电压
overvoltage fuse 过电压熔断器
overvoltage in HVDC transmission system 高压直流输电系统过电压
overvoltage in incomplete phase operation 非全相运行过电压
overvoltage modeling 过电压模拟
overvoltage monitor 过电压监测装置
overvoltage protection 过电压保护
overvoltage protective device 防过电压装置，过电压保护装置
overvoltage relay 过电压继电器，过压继电器
overvoltage release 过电压释放器，过电压脱扣器
overvoltage sensor 过电压传感器
overvoltage test 过电压试验
overvoltage times 过电压倍数
overvoltage tripping 过电压跳闸
overvoltage 超额电压，过电压，超电压
over-wall manipulator 越墙式机械手
overwash 洪积土壤，跃堤冲岸浪，冲刷过，水浪拍打
over weight freight 超重货物
over weight 超重，超量
over weld 过焊
overwinding precaution （提升钢丝绳）过卷预防措施
overwrite 改写
over year regulation 多年调节
over year storage 多年调节库容，多年蓄水
over years 历年，隔年
over 过，越过
ovonics 双向开关半导体元件，交流控制的半导体元件
OWCC（outstanding work completion certificate）遗留工作完工证书
owing degree-day 植物生长所需热量的计算单位
owing to 因为，由于，多亏
own code 扩充工作码，固有编码，自有码
own demand 自用电量，厂用电量
owned capital 自有资金
owned cars 自备车
owned program 专用程序
owner of patent 专利所有人
owner's engineer 业主工程师
owner's equity 所有者权益
ownership cost 所有权费
ownership of fixed assets 固定资产所有权
ownership of trade mark 商标权
owner ship system 所有制
ownership 所有权，所有制
owner 雇主，业主，所有人，所有者，建设单位
own fund 自有资金
own quantity 固有量
own type 固有型，自身型
own weight 自重，（船舶）总载重量
own 拥有
OW(outstanding work) 遗留项工作
oxalate 草酸盐
oxalic acid 草酸，乙二酸
oxbow lake 牛轭湖
oxbow 牛轭形弯道
oxhyhdrogen welding 氢氧焊接
oxic 好氧的，含氧的，氧化的
oxidability 可氧化性
oxidable 可氧化的
oxidant smog 氧化剂烟雾
oxidant 氧化剂，氧化
oxidation and corrosion inhibitor 抗氧抗腐剂【助剂】
oxidation bed 氧化床
oxidation behavior 氧化行为，氧化状态
oxidation ditch 氧化槽
oxidation-inhibited oil 抗氧化油
oxidation kinetics 氧化动力学
oxidation medium 氧化介质
oxidation phase 氧化状态
oxidation pond effluent 氧化塘流出水
oxidation pond 氧化塘
oxidation product 氧化产物
oxidation-reduction indicator 氧化-还原指示剂
oxidation-reduction method 氧化还原法
oxidation-reduction potential 氧化还原电位
oxidation-reduction potentiometer 氧化还原电位计
oxidation-reduction reaction 氧化还原反应
oxidation-reduction system 氧化还原系统
oxidation-reduction titration 氧化还原滴定法，氧化-还原滴定
oxidation-reduction 氧化还原作用
oxidation resistance 抗氧化性
oxidation resistant steel 抗氧化钢
oxidation treatment 氧化处理
oxidation type semiconductor 氧化型半导体
oxidation 氧化，氧化作用，氧化层
oxidative assimilation 氧化同化作用
oxidative chemical decomposition of graphite 石墨氧化化学分解
oxidative decanning 氧化脱壳，氧化去壳
oxidative decladding 氧化脱壳，氧化去壳
oxide coating 氧化物层

oxide core 氧化铁铁芯，氧化物磁芯
oxide deposit 氧化物沉积
oxide film arrester 氧化膜避雷器
oxide film condenser 氧化膜电容器
oxide film insulation 氧化膜绝缘
oxide film lightning arrester 氧化膜避雷器
oxide film 氧化膜，氧化层
oxide-fueled breeder 氧化物燃料增殖反应堆
oxide fuel element 氧化物燃料元件
oxide fuel 氧化燃料
oxide inclusion 氧化物杂质
oxide magnet 氧化物磁铁
oxide semiconductor 氧化物半导体
oxide skin 氧化层
oxide thickness 氧化膜厚度
oxide wear 氧化磨损
oxide 氧化物，氧化
oxidic particle 氧化粒子
oxidizability 可氧化性
oxidizable matter 可氧化物质
oxidization 氧化
oxidized sludge 氧化污泥
oxidized wastewater 氧化废水
oxidizer-cooled 氧化剂冷却的
oxidizer 助燃物，氧化剂
oxidizing action 氧化作用
oxidizing agent 氧化剂
oxidizing atmosphere 氧化气氛
oxidizing covering 氧化层
oxidizing flame 氧化焰
oxidizing reaction 氧化反应
oxido-indicator 氧化物指示剂
oxisol 氧化土
oxyacetylene flame 氧乙炔焰
oxyacetylene welder 氧炔焊机
oxyacetylene welding outfit 氧炔气焊机
oxyacetylene welding 氧乙炔焊接
oxyacetylene 氧乙炔的
oxychloride cement 氯氧化水泥
oxydiacetate 氧二醋酸盐
oxyfuel gas cutting torch 割炬
oxyfuel gas cutting 氧燃料气割
oxyfuel gas welding torch 气焊炬
oxyfuel gas welding 气焊
oxy-gas cutting 氧炔切割，火焰切割
oxygen-acetylence welding 气焊，氧乙炔焊接
oxygen activated sludge system 氧活化污泥系统
oxygen activation 氧活化
oxygen analyzer 氧分析器，氧量分析仪
oxygenated aqueous solution 充氧水溶液
oxygenated water 充氧水
oxygenate 充氧，充氧作用，氧化作用
oxygen attack 氧侵蚀
oxygen concentration cell 氧浓差电池
oxygen conditioning 氧调节
oxygen consumed 耗氧量
oxygen consumption 耗氧量
oxygen-containing compound 含氧化合物
oxygen content of condensate 凝结水含氧量
oxygen content of outlet water 出水含氧量
oxygen content of water 水中含氧量
oxygen content 含氧量，氧量
oxygen correction 氧量校正
oxygen corrosion 氧腐蚀
oxygen cutting machine 氧气切断机
oxygen cutting 气割，氧气切割，火焰切割
oxygen cylinder 氧气瓶
oxygen-deficient combustion 缺氧燃烧
oxygen-deficient condition 缺氧条件
oxygen-deficient environment 缺氧环境
oxygen-demanding pollution 缺氧污染
oxygen demand 需氧量
oxygen detection 氧气探测
oxygen detector 氧气探测器
oxygen dissolved 溶解氧
oxygen electrode 氧电极
oxygen-enriched combustion 富氧燃烧
oxygen enrichment 富氧，空气
oxygen-excess combustion 过氧燃烧
oxygen extractor 除氧器
oxygen-free 无氧的
oxygen gas 氧气
oxygen generation plant 氧气站
oxygen generation station 制氧站
oxygen gouging 火焰气刨
oxygenic biological fluidized bed reactor 含氧生物流化床反应器
oxygen inleakage 氧气漏入
oxygen-isotope ratio 氧同位素比率
oxygenized air 富氧空气
oxygen lance cutting 氧矛切割
oxygen lance 氧气切割器，氧气烧枪
oxygenous 氧的，氧气的
oxygen pipe 氧气管
oxygen pitting 氧蚀麻点
oxygen point 氧点
oxygen propane cutting 氧气丙烷炬切割
oxygen removal by combustion 燃烧除氧
oxygen requirement 需氧量
oxygen-rich area 富氧区
oxygen-rich layer 浓氧层
oxygen saturation capacity （溶解）氧饱和量
oxygen scavenger （水的）化学除氧剂，除氧器
oxygen scavenging 除氧
oxygen station 氧气站
oxygen supply 供氧
oxygen uptake 摄氧量
oxygen welding 氧气焊接
oxygen 氧，氧气
oxyhydrogen blowpipe 氢氧吹管
oxyhydrogen flame 氢氧焰
oxyhydrogen gas monitoring 氢氧气体检测
oxyhydrogen welding 氢氧焊
oyerspeed control 超速控制
oylet 视孔
ozokerite 地蜡，石蜡
ozonation 臭氧化作用
ozonator 臭氧发生器
ozone attenuation factor 臭氧衰减系数
ozone content 臭氧含量
ozone layer 臭氧层
ozone 臭氧

ozonidate 臭氧剂
ozonidation 臭氧化作用
ozonization method 臭氧化法
ozonization 臭氧消毒
ozonizer 臭氧消毒机
ozonolysis 臭氧分解
ozonometer 臭氧计
ozonopause 臭氧层顶
ozonosphere 臭氧层
O 操作员【DCS 画面】

P

PABX(private automatic branch exchange) 专用自动交换分机，专用自动交换机
PABX system 专用自动交换机，程控交换机系统
PABX telephone system 内线电话系统
pace method 步测法
PACE(performance and cost evaluation) 性能成本估价
pacer 步测器，定速装置，调搏器
pace-setter 带头人
pace voltage 跨步电压
pace 步，步速，步伐，步调，速度，步距，一步，定速
pachimeter 弹性切力极限测定计
pachometer 测厚计
Pacific Basin Nuclear Conference 太平洋沿岸地区核能会议
pacing factor 决定性因素，主要因素，主要条件，基本条件，基本因素
pacing 调步，整速，定速，主要的，决定性的，步测
package assembly 组装结构
package bid 一揽子递盘，组合投标
package boiler 快装锅炉
package case 包装箱
package contract 一揽子合同
packaged air conditioner 整体式空调机，整体式空气调节器
packaged boiler 整装锅炉，快装锅炉
packaged circuit 封装电路，浇注电路
packaged component 整装部件
packaged concrete 预填骨料混凝土
packaged design 紧凑设计，装配式结构，组装结构，装配式设计
package deal contract 一揽子承包合同，整套承包合同
package deal 一揽子交易，整批交易，总价交易
packaged economic-type boiler 快装式经济锅炉
packaged electronic circuit 电子线路程序包
packaged electronic instruments 电动单元组合仪表
packaged gas turbine 快装式燃气轮机，组装式燃气轮机
packaged generating set 快装发电机组
packaged heat pump 组装式热泵
packaged investment 一揽子投资
packaged modularised feed unit 组装式给水装置
packaged pneumatic instruments 气动单元组合仪表
packaged reactor 装配式反应堆
packaged source 封装源，密封源
packaged steam turbine 快装式汽轮机
packaged substation 成套式变电所，整装式变电所
packaged technology 成套技术
packaged terminal air conditioners 末端整体空调器
packaged unit 整装机组，可移动装置
packaged water-tube boiler 快装水管锅炉
packaged 成套的，组装的，单元组合的，预先组装成套的
package filling monitor 包装物品监测器
package job contract 整套承包合同
package lead （集成电路）外壳引线，封装引线
package monitor 包装物监测仪
package mortgage 总括抵押，一揽子抵押
package plant 快装电厂
package plan 一揽子计划
package proposal 一揽子建议
package reactor 快装式反应堆，小型轻便反应堆，装配式反应堆
packager 打包机
package system 组装式仪表
package tender 一揽子招标
package treatment plant 成套（移动式污水）处理装置
package turbine 快装式涡轮机，快装式透平，组装式透平
package type 整体式
package unit 成套设备，整体机组，移动式装置
package 快装的，组装的，成套设备，一揽子交易，包装，装箱，打包，机组，套件，部件，机件，标准件，批量，货物，密封装置
packaging cost 包装费
packaging density 封装密度
packaging list 装箱单
packaging machine 打包机
packaging of waste 废物封装
packaging 包装，包装袋，封装，装箱，组件，标准部件
packaging 包装，密封【装箱前】
pack carburizing 固体渗碳
pack cementation 装填水泥固化
packed bed filter 滤清器，填充层过滤器，填充床反应器
packed bed heat transfer 填充床传热
packed bed scrubber 填充床洗涤器
packed bed 填充床，填密床，固定床，压实床，紧装床
packed cargo 包装货物
packed cell 积层电池
packed column for absorption 吸收填充塔
packed column 填充柱，填充塔，填料塔
packed decimal 压缩十进制
packed density 堆积密度
packed expansion joint 填料式补偿器
packed-fluidized bed heat transfer 填充流化床

传热
packed-fluidized bed 填料流化床
packed height 填充高度
packed joint 填塞接缝
packed-particle fuel 装填颗粒燃料
packed-piston pump 填料活塞泵
packed-piston 密封活塞
packed soil 捣实土，素土夯实
packed space 填充容积
packed stone revetment 填石护坡
packed tower 填充塔，填料塔
packer grouting 分段灌浆，栓塞灌浆
packer setting 垫板调整
packer 包装工人，压土机，打包机
packet assembling and disassembling 包装配和拆卸，包装配和解装
packet assembly 分组装配
packet disassembly 分组拆卸
packet multiplexing 包多路转换
packet of laminations 叠片组，铁芯段
packet switch equipment 包交换设备
packet switching 包交换，信包交换，分组交换
packet switch network 包交换网
packet switch 包交换，分组换接
packet type switch 组合开关
packet voidage fraction 颗粒团空隙度
packet 包，束，组，盒，传输组，数据包，信息包
pack-house 堆栈
pack ice 大块浮冰，流冰群
packing and shipping guidelines 包装海运导则
packing block 垫板，衬层，填塞木
packing blowout 密封裂口
packing bolt 填密螺栓
packing box housing 填密函外套
packing box leak-off recovery system 填密函泄漏回收系统
packing box seal 填密函密封
packing box 填料盒，填料函，填料箱，密封盒，包装箱
packing bushing 填料衬套
packing case 汽封套，密封箱壳，填料函
packing chamber 填料盒，填料函
packing charges 包装费（用），包装价
packing check 包装检查
packing coefficient 密实系数
packing collar 垫圈
packing concrete in forms 模内捣实混凝土
packing cost 包装成本
packing course 填层
packing density 装填密度，组装密度，存储密度，充填度，堆积密度
packing depth 填塞深度
packing efficiency 组装效率
packing expenses 包装费
packing factor 充满系数，填充系数
packing filler 充填料
packing flange 填料法兰，止水法兰，填密凸缘，填料涨圈
packing follower 密封衬圈
packing fraction 沉降系数，敛集率
packing gland 密封垫，填料箱，密封压盖，汽封装置，密封装置，填料函盖，填料压盖
packing grease 密封润滑脂
packing head 汽封箱盖
packing inspection 装箱检验
packing instructions 包装须知，包装说明
packing iron plate 垫铁
packing joint 密封连接
packing-leakage 汽封漏汽
packing leak-off flows 汽封漏气量
packing list 包装清单，装箱单，包装单
packing lubrication 填充密封润滑
packing mark 包装标记，包装标识，包装标志
packing material 密封材料，包装材料，填密材料，密封填料
packing micanite 衬垫云母板
packing nut wrench 填密螺母扳手
packing nut 密封螺帽
packing of orders 指令组合
packing of sleeper 轨枕垫衬
packing paper 包装纸
packing piece 衬片，衬垫物，垫块，垫片，隔叶块，密封件
packing plate 垫板，密封垫，密封垫圈
packing puller 填料拉出器
packing removal tool 填料拆除工具
packing ring 填料环，密封环，密封圈，填密环，汽封环，垫圈，胀圈
packing seal 填料密封
packing segment 汽封弧块
packing sheet 填密片，垫片，包装单
packing sleeve 填料轴套，轴封衬套
packing spring 密封弹簧
packing stack 填料堆
packing strip 汽封片
packing tool 捣实工具
packing washer 密封垫圈，垫圈
packing water 密封水
packing 包装，填充，填料，填充物，密封，盘根，填函，包装材料，填密，装箱，打包
packless seal 无填料密封
packless stuffing box 无填料的密封槽【密封环和滑槽】
packless valve 无填料阀门，无封套阀
packless 未包装的，未加封的
pack unit 部件，组件，装箱部件
pack 包装，包，捆，打包，填料，密封，组合件，包装材料
PAC(plasma arc cutting) 等离子弧切割
PAC(polymeric aluminum chloride) 聚合氯化铝【絮凝剂】
PAC(pre-acceptacnce certificate) （电厂）预验收证书
PAC(provisional acceptance certificate) 临时验收证书【合同用语】
P-action(proportional action) 比例作用
pact 公约，协定
pad and chimney foundation 斜柱式基础
pad bearing 轴瓦式轴承，瓦块轴承
pad block 垫衬，垫块，垫铁
pad control 衰耗器控制，垫整调节

padder 垫整电容器，微调电容器
padding capacitor 微调电容器，垫整电容器
padding 衬垫，填料，填充，垫整，调整，充填
paddle aerator 叶轮式曝气池，桨式曝气器
paddle blade 平叶片
paddle coal feeder 叶轮式给煤机，叶轮给煤机
paddle feeder 叶片式给料机，叶轮给料机
paddle-type mixer 桨式混合器
paddle-type pill 锤击式磨煤机
paddle wheel flowmeter 转子流量计
paddle wheel 明轮
paddle 桨，桨叶，轮桨，搅拌桨，叶片，阀门，踏板，轮叶，桨叶板
paddy field 水稻田
pad foundation 衬垫基础
padlocking 寄存，锁住
padlock room 锁住的房间
padlock 挂锁，扣锁，关闭
pad mounted transformer 带隔间的底座安装型变压器
pad mounted 底座安装型
padmount transformer 组合式变压器
pad of a foundation 基础垫层
PAD(product application data sheet) 产品应用资料
pad-reinforced 补强圈加强的
PADS(personnel automatic data system) 人事自动数据系统
pad stone 垫石
pad type flange 盘座式法兰
pad type thermocouple 垫板式热电偶
pad welding 堆焊，垫块焊接
pad 定位凸体【燃料组件】，衬垫，轴承瓦块，台，垫，垫圈，垫片，托架，支座，垫块，护具，便笺簿，填补，滑板，车辆底盘
PAF(primary air fan) 一次风机【DCS 画面】
PAF(procurement application form) 采购申请单
page effect 薄膜效应
page fault 页面出错
page feed-out 纸页送出
page feed 页间走纸【打印机】
page formatter 页面格式标识符，页面格式化程序
page number 页码
page printer 页式印刷机，页式打印机
page-printing telegraph 页式打字电报
page reader 页式阅读机
pager 传呼机，对讲机
page table 页面表，页面调度
page teleprinter 页式印字电报机，页式电传印刷
page turning 页转换
paging dispatching unit 扩音调度机
paging system 分页系统，呼唤系统，寻找呼叫系统，传呼系统，音响播叫系统，对讲系统
paging 分页，调页，页式，分页法，划分页面，翻页
pagoda 塔，塔状结构
PAH(pressure alarm high) 高压报警
PAH(primary air heater) 一次风暖风器
paid duty 已付关税
paid-in capital 实收资本，已收资本，已缴清资本，已付清资本
pail 吊桶，水桶，桶
painted steel 涂漆钢板
painter 油漆工，油漆工具
paint film thickness measuring meter 漆膜测厚仪
paint film visualisation 涂膜流动显示
paint finish 油漆罩面
paint grinder 涂料研磨机
painting gun 喷漆枪
painting machine 喷漆机
painting pump 喷漆泵
painting shop 油漆车间
painting specification 涂漆规范
painting spray 喷漆
painting system with control device of humidity 带湿度控制装置的喷漆系统
painting 涂漆，油漆工作，油漆施工
paint mixer 涂料混合器
paint on plaster 抹灰面油漆
paint primer 涂漆底，底层油漆
paint remover 去漆剂，涂料消除剂
paint sprayer 喷漆器，喷涂料枪
paint spraying machine 喷漆机
paintwork care 油漆养护
paintwork preservation 油漆保护
paintwork 油漆作业，油漆工作，涂上的油漆，油漆的表面
paint 描绘，涂层，涂，涂料，油漆，熟漆，颜料，着色，上油漆，漆
pair-annihilation 正负电子对湮没
paired brushes 配对电刷，电刷对
paired cable 对绞电缆，双线电缆，对组电缆
paired column 双排柱
paired comparison 对偶比较
paired multiplier 乘二乘法器
paired observation 对比观察
paired pulse 成对脉冲
paired running 并列运行
paired 成对的
pairing method 对比法
pairing of lines 叠行现象，并列现象
pairing 成对，成双，配对，电缆心线对绞
pair of poles 极对
pair of vortices 涡对
pair of wheels 轮对
pair of windings 绕组对，一对绕组
PAIR(pipe activity inspection record) 管线检查记录
pair production coefficient 对产生系数
pair production 对产生
pair-to-pair capacity 线对间电容
pair transistor 配对晶体管，双晶体管
pair tube 对偶管，配对管
pair twist 对绞
pair 成对的，配对，对，偶，线对，对绞，双线敷设
pale fencing 栅栏
paleo loess 老黄土
palingenesis 再生作用
palisade 栏栅，围篱，桩

P-alkalinity 酚酞碱度
palladium catalyst 钯催化剂
palladium point 钯点
palladium 钯【Pd】
pallador 铂钯热电偶
palleo-climatology 古气候学
pallet conveyer 板式输送机
palletizer 敷设机，码垛堆积，用货盘装运，堆垛机
pallet truck 码垛车
pallet 平板架，托灰板，衔铁，集装箱，（链式输送机的）托盘，平台，抹子，制模板
palm 掌状物，支承耳孔
palpable result 具体成果
PAL(pressure alarm low) 低压报警
PAL(process assembler language) 过程汇编语言
PAL(programmed application library) 已编好的应用程序库
palstance 角速度
paludal 沼泽的
PAM(partitional access method) 区分存取法
pampas 大草原
PAM＝polyacrylamide 聚丙烯酰胺【助凝剂】
PAM(post-accident monitoring) 事故后监测
PAM(pulse-amplitude modulation) 脉冲调幅
PAMS(post accident monitoring system) 事故后监测系统
PAM winding 极幅调制绕组
panadaptor 扫描附加器，影像接收器
panalarm 报警设备
panalyzor 调频发射机综合测试仪
pancake coil 盘形线圈，饼形线圈，盘管，扁平线圈
pancaked core 扁平堆芯
pancake engine 平置发动机
pancake helix 扁平螺旋线圈
pancake motor 扁平型电动机，短铁芯电动机
pancake reactor 扁平反应堆
pancake shaped motor 扁平型电动机
pancake synchro 扁平型同步机
pancake winding 盘形绕组，扁平绕组，饼形绕组
pan coefficient 蒸发皿校正系数，蒸发器系数
pan conveyor 槽形板式输送机
pane hammer 斧锤
panel bed filter 板层过滤器
panel board 配电盘，仪表板，控制屏，面板，配电箱，镶（嵌）板，配电板
panel box 配电箱，配电盘，分线箱
panel brick 砖砌块
panel connector 面板插座
panel construction 仪表板结构
paneled door 镶板门
panel face arrangement 盘面布置
panel flutter 壁板颤振
panel framework 配电盘框架，屏框架
panel girder 格子梁
panel heating 辐射板供暖，辐射采暖
paneling 镶板
panel interface 平板接口
panelist 专门小组成员
panelized construction 屏式结构
panelize 屏化
panel joint 节点
panel lamp 仪表盘灯，仪表屏灯
panel length 节间长度
panelling 机壳，覆盖物，镶板，门心板
panel meeting 专家小组会议
panel meter 配电板式仪表
panel method 板块法，面元法
panel mounting 配电盘装配
panel of arbitrators 仲裁小组
panel of consultants 顾问小组
panel of experts 专家审议会，专家小组
panel planer and thicknesser 单面木工压刨床，刨板机与厚度刨
panel point 节点
panel processing welding machine 膜式壁成排焊机
panel radiator 板式散热器
panel room 配电盘室
panel strip 嵌条
panel switch 控制盘开关，面板开关
panel system 分组安装制，升降自动电话制
panel test 实验台试验
panel-type construction 框格式建筑，框格式结构
panel-type house 预制墙板式房屋
panel-type instrument 配电盘型仪表
panel-type steel radiator 钢制板式散热器
panel-type switchboard 面板式开关板
panel wall 幕墙，嵌板墙幕墙，水冷壁，隔墙，水冷屏，镶板隔墙
panelwiring one-line diagram 仪表盘配线单线图
panelwiring 仪表盘配线
panel 仪表板，配电板，配电盘，底板，盘，屏，镶板，控制盘，屏蔽，面板，护墙板，表盘，安装板，专门小组
panemone 全风向阻力型竖轴风车
pan evaporimeter 盘式蒸发器
pan factor 蒸发皿校正系数
pan head screw 截头螺钉
panic bolt 太平门栓，紧急保险螺栓，太平门
panic button 报警按钮
panic door 保险门，带有保险栓的门
panlite 聚碳酸酯树脂
pan mixer 盘式拌和机
pannier condenser 侧向式凝汽器
pannier 驮篮，肩筐，背篓，石笼，篾筐
pannoranic 全景的，周向的
panorama 全景
panoramic receiver 侦察接收机，扫调接收机
pans 盘状凹地【尤指盆地】
pantelegraphy 传真电报学
pantelegraph 传真电报
pantelephone 灵敏度特高的电话机
pantiled roof 波形瓦屋顶
panting 晃动，振动，脉动，波动，烧振，拍击，冲击
pantograph isolator 双臂折架式隔离开关
pantograph operation 受电弓操作
pantograph 比例绘图仪，缩放仪，导电弓，受电弓，集电弓，比例画图仪器，伸缩绘图器，放大器，电杆架，比例画图器

pantry　餐具室，食品储存室，配膳室
pan type bed load sampler　盘式推移质采样器
pan type sampler　盘式采样器
pant　整流罩，脉动，晃动
pan　槽板，平底锅，盘，盆，皿，槽，盘状器皿，盆地
Pa＝pascal　帕
PAPE(pending activity prior EESR)　未完工作
paper air filter　纸过滤器，纸空气过滤器
paper-and-cotton insulated cable　纸棉绝缘电缆
p. a. (per annum)　每年
paper audit　书面审计
paper bushing　纸套管
paper cable　纸绝缘电缆
paper capacitor　纸电容器，纸介质电容器
paper condenser　纸介质电容器
paper-core enameled type cable　纸包漆皮绝缘电缆
paper covered wire　纸包线
paper clay　薄层黏土
paper dielectric capacitor　纸介电容器，纸介质电容器
paper drain　纸板排水
paper feed mechanism　送纸机构
paper filler　纸充填物，纸过滤器
paper-foil method　纸箔法
paper for mica tape　云母带纸
paperhanging　贴墙线
paper hat　纸帽【放射性防护】
paper industry　造纸工业
paper-insulated cable　纸绝缘电缆
paper-insulated enamel wire　纸绝缘漆包线
paper-insulated lead sheathed cable　纸绝缘铅护层电缆
paper-insulated　纸绝缘的
paper insulation　纸绝缘
paper-pattern image　纸纹图像
paper shale　薄层页岩
paper spacer　纸绝缘隔片，纸垫片
paper speed　纸带速度
paper strainer　滤纸
paper strip mixed lime mortar　纸筋灰
paper tape coder　纸带编码器
paper tape code　纸带码
paper tape decoder　纸带译码器
paper tape punch　纸带穿孔机
paper tape reader　读带机，纸带输入机，纸带读出器
paper tape transcriber　纸带读数机
paper tape unit　纸带机
paper tape　纸带，纸条，纸质磁带
paper　大样图纸，论文，文件，记录，票据
PA＝polyamice　聚酰胺
PA(primary air)　一次风
PA(process automation)　连续型流程生产自动化
PA(public address)　共有地址，播音装置
para-acid　对位酸，仲酸
parabasalt　普通玄武岩
parabola　抛物线，抛物面
parabolic arch dam　抛物线拱坝
parabolic arch　抛物线拱
parabolic collector　抛物面集热器
parabolic concentrator　抛物面聚光器
parabolic conoid shell　抛物线锥形壳体
parabolic curve　抛物线
parabolic cylinder coordinates　抛物柱面坐标
parabolic cylinder　抛物柱
parabolic detection　平方率检波
parabolic differential equation　抛物线形微分方程
parabolic diffusion equation　抛物线形扩散方程
parabolic-dish collector　旋转抛物面集热器，抛物盘式集热器，蝶式集热器
parabolic dish　抛物面反射镜
parabolic distribution of pressure　压力抛物线分布
parabolic distribution　抛物线分布
parabolic dune　新月形水下沙洲
parabolic girder　抛物线形梁
parabolic mirror　抛物面镜，抛物面反射镜
parabolic plug　抛物线形插塞
parabolic reflector　抛物线反射镜
parabolic trajectory　抛物线轨道
parabolic trough collector　槽型抛物面集热器
parabolic trough solar thermal power generation　槽式太阳能光热发电
parabolic trough　抛物面反射槽
parabolic weir　抛物线形堰
parabolic　抛物线的
paraboloidal aerial　抛物面天线
paraboloidal concentrator　旋转抛物面聚光器
paraboloidal mirror　抛物面镜
paraboloid antenna　抛物面天线
paraboloid condenser　抛物面聚光器
paraboloid of revolution　回转抛物面
paraboloid　抛物面，抛物体
parachor equivalent　原子等张比容
parachor　克分子等张比体积，等张比容，等张体积
parachute jumping　跳伞
parachuting　跳伞运动
paraconductivity　顺电导性
paraconformity　准整合
para-curve　抛物线
paraelectric Curie temperature　仲电态居里温度
paraffined paper　蜡纸
paraffin　石蜡
paraffin wax　石蜡
parafoil　伞翼，翼形伞
parageosyncline　副地槽
paragneiss　水成片麻岩，副片麻岩
paralell operation　平行作业
paralichic contact　异元母质层
parallactic base　视差基线
parallactic polygon　视差导线
parallax angle　视差角
parallax bar　视差杆
parallax computation　方位差计算
parallax correction　视差校正
parallax difference　视差差数，视校差数
parallax displacement　视差移动
parallax elimination　视差消除
parallax error　视差

parallax instrument 视差量测仪
parallax measurement 视差测定
parallax 视差，方位差的
parallel access 并行存取
parallel action 平行作用，并行作用
parallel activity 平行活动
parallel ADC 并行模数转换器
parallel adder 并行加法器
parallel addition 并行加法
parallel admittance 并联导纳
parallel algorithm 并行算法
parallel arithmetic operation 并行的算术运算
parallel arithmetic unit 并行运算装置，并行运算器
parallel arrangement of moving stairs 自动楼梯的平行布置
parallel arrangement 并行布置，并联布置，并联装置，并列式布置
parallel baffled boiler 纵向冲刷锅炉
parallel baffled 纵向冲刷
parallel beam 平行光束
parallel binary accumulator 并行二进制累加器
parallel binary computer 并行二进制计算机
parallel bisection 并联中剖
parallel branch line 并联分支线，并联支路
parallel branch 并联支路
parallel breaking test 并联开流试验
parallel buffer 并行缓冲器
parallel-by-bit 位并行
parallel by character 字符并行处理
parallel canal 平行渠道
parallel capacitance 并联电容
parallel capacitor 并联电容器
parallel cascade action 并联串级动作
parallel centrifuge 单向流动（无逆流）式离心机
parallel chord truss 平行弦桁架
parallel circle 纬圈
parallel circuits per phase 每相并联支路数
parallel circuit 并联电路，平行电路，并行电路，并联管网
parallel coil 并联线圈
parallel compensation （线路的）并联补偿
parallel computation 并行计算
parallel computer 并行计算机
parallel-connected thermocouples 并联热电偶，并联温差电偶
parallel connection 并联
parallel controller 并联控制器
parallel control 并行控制
parallel correction element 并联校正元件
parallel coupling 并联耦合，平行连接
parallel-cross operation method 平行交叉作业法
parallel-cross operation 平行交叉作业
parallel-current condenser 并流凝汽器
parallel current 平行流，并联电流
parallel cut 平行切割
parallel data controller 并行数据控制器
parallel data transmission 并行数据传输
parallel designation 并行标记
parallel detection system 多道同时探测系统
parallel determination 平行测定
parallel dike 平行堤
parallel ditch system 平行沟系
parallel downflow 同方向的下流［顺流］
parallel drainage 平行水系
paralleled alignment 对中，平行线对准
paralleled path 平行通路，并联支路
parallel element-processing ensemble 并行元处理机
parallel entry 并行输入
parallelepiped reactor 平行六面体反应堆
parallel equalizer 并联均衡器
parallel equivalent fourpole network 并联等效四端网络
parallel excitation motor 并激电动机
parallel excitation 并励
parallel fault 平行断层
parallel feedback integrator 并联反馈积分器
parallel feedback 跨步反馈，并联反馈，并励回授
parallel feeder protection 平行馈线保护
parallel feeder 并联馈路，平行馈路
parallel feed 并励馈电，并联馈电
parallel filter 并联滤波器
parallel financing 平行贷款
parallel-flange beam 平行双拼梁
parallel-flange girder 平行双拼梁
parallel flow burner 直流式燃烧器
parallel flow condensation （蒸汽和水）并流凝结
parallel flow condenser 平行流式凝汽器
parallel flow gas turbine 分流式燃气轮机
parallel flow heat exchanger 平行流式热交换器
parallel flow reheater 顺流再热器
parallel flow superheater 顺流过热器
parallel flow turbine 分流透平，平行流汽轮机
parallel flow 平行流，直流，顺流，并流，同向流
parallel gas passes 并联烟道
parallel gate 并联门
parallel-generator theorem 米尔曼定理，并联发电机定理
parallel generator 并联发电机
parallel H＋ and Na＋ ionic exchange 并联氢-钠离子交换
parallel impedance 并联阻抗
parallel inductor 并联电感线圈
paralleling to the grid 与电网同步
paralleling 并列，并行，并联，并车
parallel input 并行输入
parallel interface element 并行接口部件
parallel in 并联输入，并列，并网
parallelism 平行度，并行性，二重性
parallelizing 并列
parallel joint 平行节理
parallel laying 并行敷设，并行架设
parallel light photoelastisity 平行光光弹性
parallel line 并联线路，并行线路
parallel logic 并行逻辑
parallel machine 并行计算机
parallel memory 并行存储器
parallel mirror cavity 平行反射镜共振腔
parallel motion 平行运动，并行动作

parallel mounted 平行布置
parallel multiblade damper 平行式多叶阀
parallel multiplier 并行乘法器
parallel off 解列
parallelogram of velocities 速度平行四边形
parallelogram 平行四边形
parallel-operated electric power plant 并网电厂
parallel operation 并联运行，并列运行，平行操作，并列运转，并车
parallel or right angle horizontal shaft 水平或垂直出轴
parallel oscillatory circuit 并联振荡电路
parallel overlapping work 平行交叉作业
parallel padding 并联调整
parallel-parallel logic 并行并行逻辑
parallel path 并联支路，平行通路
parallel phase 并联相
parallel plane cavity 平行平面共振腔
parallel plate condenser 平板式电容器
parallel plate domain structure 平行板磁畴结构
parallel plate interferometer 平行板干涉仪
parallel plate micrometer 平行板测微仪
parallel port 并行口
parallel processing 并行处理
parallel programming 并行程序设计
parallel reaction 平行反应
parallel reactor 并联电抗器
parallel register 并行寄存器
parallel representation 并行表示
parallel resistance 并联电阻
parallel resonance frequence 并联谐振频率
parallel resonance 并联谐振
parallel resonant circuit 并联谐振电路
parallel ring-type register 并行环形寄存器
parallel ring winding 并联环形绕组
parallel-roller bearing 径向滚珠轴承
parallel route 并行线路
parallel running of generators 发电机并联运行
parallel running 并联运转，并联运行，并列运行，并车，并行运行
parallel search storage 并行存储器，并行搜索存储器，并行检索存储器
parallel seat gate valve 平行座闸阀
parallel seat 平行阀座
parallel section 平行段【抗拉试验】，等截面段
parallel serial conversion 并行串行变换
parallel serial mode 并串联方式，并串方式
parallel serial 并串联，并串行，复联，混联
parallel series connection 并串联接法，混联
parallel series 复联，混联，并串联
parallel shaft reduction gear 平行轴减速齿轮
parallel shafts 平行轴
parallel side notch 边平行槽
parallel slices （气流）平行层
parallel slide gate valve 平行滑移闸阀
parallel slide valve 平行滑阀
parallel slit 平行缝隙
parallel slot 平行槽
parallel stay 平行拉线
parallel storage 并行存储器
parallel stream 平行流
parallel system 并列系统，平行式布置
parallels 垫铁，平行线
parallel table 对照表
parallel tap 并联抽头，平行丝锥
parallel throat 平行喉管
parallel-tile drainage system 平行瓦管排水系统
parallel-to-serial converter 并串变换器，并联串联变换器
parallel to the flow direction 顺水流方向
parallel training wall 顺岸导墙
parallel transductor 并联饱和电抗器
parallel transfer 并行转移，并行传递
parallel transformer 并联变压器
parallel transmission 同时输送，并联输送，并行传送
parallel-travel in cableway 平行移动式缆道
parallel tunnel 平行地槽
parallel unconformity 平行不整合
parallel ventilation branch 并联风路，平行风路
parallel watersheds method 并行流域法
parallel wave 平行波
parallel welding 并联电阻点焊
parallel winding 叠绕组，并联绕组，并绕
parallel wire resonator 平行线谐振器
parallel wire unit 平行钢丝束张拉设备
parallel with 平行，与……比较
parallel work-flow 并行操作
parallel wound coil 并绕线圈，叠绕线圈
parallel 并联的，并列的，并行的，平行的，平行线，并列，并行，对比，使……与……平行
paraloc 参数器振荡电路
paramagnetic absorption 顺磁吸收
paramagnetic material 顺磁性材料，顺磁性物质，顺磁材料
paramagnetic oxygen analyzer 磁性氧量剂
paramagnetic resonance 顺磁共振，顺磁谐振
paramagnetic substance 顺磁物质
paramagnetic susceptibility 顺磁磁化率
paramagnetic 顺磁性的
paramagnetism 顺磁性
paramagnet 顺磁体，顺磁物质
parameter adjustment controls 参数调整装置
parameter checkout engineer 参数检查工程师
parameter checkout 参数检查
parameter correlation 参数相关，参数关联
parameter detection 参数检测
parameter display 参数显示
parameter of electric motor 电机参数
parameter of HVDC power transmission main circuit 高压直流输电主回路参数
parameter optimization 参数优化，参数最优化
parameter plane 参数平面
parameter potentiometer 参数电势计
parameter setting instruction 常数调整指令
parameters of force and energy 力能参数
parameters of heating medium 供热介质参数
parameter storage 参数存储器
parameter transformation 参数变换
parameter variation 参数变化，参数变值法
parameter 参数，参量，指标，系数，特性
parametric amplifier 参量放大器

parametric converter 参量变频器
parametric correlation 参数相关
parametric curve 参数曲线
parametric damping 参数阻尼
parametric diode 参量二极管
parametric down-converter 参量下变频器
parametric equation 参数方程
parametric excitation 参数激励，参数激振
parametric frequency converter 参量变频器，参量频率变换器
parametric frequency divider 参量分频器
parametric frequency multiplication 参量倍频
parametric frequency tuning 参数频率调谐
parametric hydrology 参数水文学
parametric measurement 参数测定
parametric method of hydrology 水文参数法
parametric mixer 参量混频器
parametric model 参数模式
parametric oscillation 参数振荡，参变振荡
parametric oscillator 参数振荡器
parametric resonance overvoltage 参数谐振过电压
parametric sensitivity 参数灵敏度
parametric transformer 参量式变压器
parametric up-converter 参量上变频器
parametric variation 参数变化
parametric 参数的
parametron logic circuit 参数管逻辑电路
parametron 变感元件，参变元件，参变管
paramount clause 首要条款
paramount consideration 首要考虑
paramount 极高的【权力】，最高的，头等的，高过，优于
parapet drainage 女儿墙排水
parapet flashing 女儿墙泛水
parapet skirting 女儿墙泛水，女儿墙脚泛水
parapet wall 防浪墙，女儿墙，压檐墙，胸墙，护墙，矮墙
parapet 防浪墙，女儿墙，护墙，矮墙，胸墙，栏杆，低墙，扶手
paraphase amplifier 倒相放大器，分相放大器
paraphase 倒相
pararock 水成变质岩
paraschist 副片岩，水成片岩
parasite current 寄生电流
parasite drag 寄生阻力，有害阻力
parasite load 寄生载荷
parasite 寄生虫，寄生物，寄生反射器，寄生振荡
parasitic absorber 寄生吸收器
parasitic antenna 无源天线
parasitic bacteria 寄生菌
parasitic capacitance 寄生电容
parasitic capture 寄生俘获
parasitic current 寄生电流
parasitic disturbance 寄生干扰，电磁干扰
parasitic echo 干扰回波
parasitic effect 寄生效应
parasitic element 无源元件，寄生元件
parasitic energy 附加能量
parasitic feedback 寄生反馈，寄生回授
parasitic inductance 寄生电感
parasitic loss 附加损失，寄生损失
parasitic neutron absorption 寄生中子吸收
parasitic noise 寄生噪声
parasitic oscillation 寄生振荡
parasitic pressure drop 附加压降
parasitic resonance 寄生谐振
parasitic stress 次应力
parasitics 寄生效应，寄生现象
parasitic thermoelectromotive force 寄生温差电动势
parasitic torque 寄生转矩
parasitic 附加的，寄生的
parasitism 寄生
parcel of air 气球
parcel of goods 零批货
parcel plan 分部计划
parcel post insurance 邮包保险
parcel speed 气块速度【风电机组】
parental dosage 亲代剂量，接受剂量
parent body 上级机构
parent company 母公司，总公司
parent exchange 母交换机
parenthesis-free notation 免括号记法，无括号标序法
parent material 亲本材料，母材料，基体材料，母质
parent metal （焊接的）母材，基体金属，基本金属，母材金属
parent phase 母相
parent pollutant 原污染物
parent rock 母岩
parent 原始的，起始的，母体，根源
Pareto diagram 帕累托图
pargeting 涂灰泥
parget 粗涂灰泥，涂灰泥，灰泥涂层，石膏粉饰
parging 涂灰泥
Paris Agreement 巴黎协定
Paris fatigue crack growth law 帕里斯疲劳裂纹增长率
parity bit 同位数位，奇偶检验仪，奇偶校验位，一致校验位
parity-check code 奇偶检验码
parity-check digit 奇偶校验数字，一致校验数字
parity-check equation 一致校验方程，奇偶校验方程
parity check 一致校验，奇偶校验，均等核对
parity code 奇偶校验码
parity error 同位误差，奇偶错误，奇偶检验出错，奇偶检验误差，奇偶校验误差
parity matrix 一致校验矩阵，奇偶校验矩阵
parity of treatment 平等待遇
parity switch 奇偶检验开关，奇偶开关
parity 平价，等价，同等，均等一致性
parked location 停放位置【起重机，吊车，行车等】
parked wind turbine 风力机停机
park efficiency 停机效率
Parker screw 派克螺丝
parking and maintenance room in the bulldozer gar-

age 推土机库中的停机和维修间
parking apron 停车道
parking area 停车场
parking brake for wind turbine 对于风电机组停机制动
parking for control rods 控制棒停放处
parking hole 调整孔
parking lane 停车道
parking place 停车场
parking position 停放位置
parking sign 停车标志
parking space 停车场，燃料元件装载区
parking 停车场，停车，停机，停止
parkway cable 公园大道电缆
park 停放，园林
parliament hinge 长脚铰链
parol agreement 口头协议
parol contract 口头合同
parol evidence rule 口头证据规则
parol 口头言词
paronite 石棉橡胶板
PAR(processor address register) 处理机地址寄存器
PAR(project audit report) 设计检查报告
parquet block 镶木地板条块
parquet floor 席纹地板，镶木地板
parse of sentential form 句型分析
parser 分析的程序【语法】
parse 分析【语法】
Parshall flume 巴歇耳量水槽
part area 港区
part assembling diagram 部件装配图
part depth simulation （大气边界层）部分厚度模拟
part flow 部分流量，分流量
partial adjustment 局部平差，部分平差
partial admission degree 部分进汽度
partial admission turbine 部分进汽汽轮机，部分进汽透平，非整周进水式水轮机，局部进水式水轮机
partial admission 部分进汽
partial agreement 局部协定
partial annealing 不完全退火
partial-arc admission degree 部分进汽度
partial-arc admission 部分进汽
partial assembly drawing 零件装配图，装配分图
partial automatic 半自动的，部分自动的
partial bed technique （沸腾炉）分床技术
partial body irradiation 部分身体辐照
partial capacitance 部分电容
partial carry 部分进位
partial cash in advance 预付部分货款
partial coal sample 分样
partial coefficient for action 作用分项系数
partial coefficient for property of material 材料性能分项系数
partial coefficient for resistance 抗力分项系数
partial coefficient 分项系数
partial combustion 不完全燃烧
partial condensation 部分浓缩，部分冷凝
partial containment 局部安全壳
partial correlation coefficient 偏相关
partial correlation 部分相关
partial crack 非贯穿裂纹
partial crit 次临界质量
partial cross-section 部分断面
partial current density 偏中子流密度
partial damping winding 部分阻尼绕组
partial decay constant 部分衰变常数
partial deionization 局部除盐，部分除盐
partial delivery 分批交货
partial denunciation 部分废弃
partial depth rolling 部分深度滚压
partial derivative 偏导数
partial differential equation 偏微分方程
partial differential 偏微分的
partial differentiation 偏微分
partial discharge inception test 电晕放电起始试验，局部放电起始试验
partial discharge pulse 局部放电脉冲
partial discharge test 局部放电试验
partial discharge 部分流量【两测流垂线间的】，部分卸料，局部放电，不完全放电
partial dismantling 部分拆除
partial drawing 零件图
partial drop of pressure 局部压降
partial duration series 部分历时系列
partial earth fault 不完全接地故障
partial earth 部分接地，不完全接地
partial enclosure 局部密闭罩
partial energy competition 部分电量竞争模式
partial equilibrium theory 局部（供求）均衡理论
partial exposure 局部辐照，局部照射
partial failure 部分失效，局部故障，局部破坏
partial film boiling 部分模态沸腾
partial flow 部分流量
partial fluidization 部分流化的
partial ground 不完全接地，部分接地
partial halflife 部分（衰变）半衰期
partial half-time 部分半衰期
partial heat treatment 局部热处理
partial hood 半罩
partial hydrolysis 部分水解
partial increment 偏差，偏增量
partial integration 部分积分
partial interception ratio 部分截流率
partial internal recirculation 局部内部再循环，部分内部再循环
partial internal water recirculation 部分内部水再循环
partial irradiation 局部辐照
partial leveling 重点式平整
partial load heat rates 部分负荷热耗
partial loading 部分负载，部分加载
partial load operation 非满负荷运行，部分负荷运行
partial load rejection 部分甩负荷
partial load steam consumption 部分负荷汽耗
partial load 部分负荷，部分荷载，低负荷
partial loss of flame 部分火焰消失
partial loss 部分损失，局部损失
partially automatic 半自动的，部分自动

partially carbonized lignite　褐煤半焦
partially closed cycle　半闭式循环
partially depleted surface barrier detector　部分耗尽区的表面势垒检测器
partially enclosed apparatus　半封闭设备
partially energized　部分激励的，半激励的
partially exempt obligation　部分免除责任
partially-exposed basement　半地下室
partially fixed-end beam　部分嵌固梁
partially inserted control rod　部分插入的控制棒
partially loaded　部分荷载
partially penetrating well　不完全渗水井
partially prestressed column　部分预应力柱
partially reflected reactor　有部分反射层的反应堆
partially reinforced concrete masonry　少筋混凝土圬工
partially segregated flow　局部分离流
partially separate system　（排水系统的）部分分流制
partially stabilized conductor　局部稳定导体
partially withdrawn control rod　部分提出的控制棒
partial motor　残缺电动机
partial node　不全节，准结点
partial nucleate boiling　局部泡核沸腾
partial opening　部分打开，部分开度
partial operating time　部分运行时间
partial outage state　部分停运状态
partial outage　部分停运
partial oxidation　部分氧化
partial package boiler　半快装锅炉，部分组装锅炉
partial payment on construction contract in process　建设合同执行期间部分付款
partial payment　部分付款，部分支付
partial penetration weld　局部焊透焊缝，部分焊透焊缝，部分贯穿焊缝
partial penetration　部分贯穿
partial plant shutdown　电厂局部停闭
partial plan　部分计划
partial plugging　部分堵塞
partial polarized light　部分偏振光
partial potential temperature　分位温
partial pressure of M in a gaseous mixture　在气体混合物中M的分压力
partial pressure of water vapour　水蒸气分压力
partial pressure　分压，分压力，局部压力
partial prestressing　部分预应力
partial product register　部分积寄存器
partial product sum　部分积和
partial product　部分乘积
partial project　部分工程
partial radiation pyrometer　部分辐射温度计
partial-read pulse　部分读出脉冲
partial refacing of the seal surface　密封表面的局部重磨
partial reflection　部分反射
partial reflux　部分回流
partial refuelling　局部换料，部分换料
partial regression coefficient　偏回归系数
partial regression equation　偏回归方程
partial regulation　不完全调节
partial replacement of defective threads　缺陷螺纹的部分更换
partial resonance　部分谐振
partial response　部分响应
partial risks guarantee　部分风险担保
partial sample　分样
partial saturation　不完全饱和，部分饱和
partial scram　部分快速停堆
partial scroll case　不完全蜗壳
partial-select output　部分选择输出
partial shielding　局部屏蔽法
partial shield　局部屏蔽体
partial shipment　分批交货，分批装船，分批装运
partial short circuit　部分短路
partial shroud　局部遮蔽
partial shutdown　部分停堆，部分停运
partial similarity　部分相似
partial simulation　部分模拟
partial span pitch control　部分跨度桨距控制
partial storage　部分蓄水，分量蓄水
partial sum register　部分和寄存器
partial sum　部分和
partial switching　部分翻转
partial system test　局部系统试验
partial termination of contract　部分终止合同
partial-throttle operation　部分节汽运行
partial transparent mirror　半透镜
partial trip　局部脱扣，局部跳闸
partial vacuum　部分真空，半真空
partial vapor pressure　蒸汽分压
partial waiver clause　部分豁免条款，部分放弃条款
partial-write pulse　部分写入脉冲
partial　局部的，部分的，零件的，部件的
participant　参加人员，参加活动的人员，参加者
participating parties　参加方
participation factor　分享因子
participation in market competition　参与市场竞争
participation　参与，股份份额
particle absorption　粒子吸收
particle accelerator　粒子加速器
particle activity mapping　粒子放射性分布图
particle bed height　颗粒层高度
particle board　木屑（碎片）板，碎屑胶合板
particle chain　颗粒链
particle characteristics　颗粒特性
particle classification　颗粒分级
particle cloud　粒子聚集
particle collection　颗粒收集
particle collector　颗粒计数器，集尘器
particle content　磁悬液浓度，磁粉含量
particle convective component　颗粒对流分量
particle counter　粒子计数器
particle current density　粒子流密度
particle current　粒子流
particle cycle time　颗粒循环时间
particle density　颗粒密度，粒子密度

particle deposition 颗粒沉积
particle detector 粒子检测器
particle diameter 颗粒直径，粒径
particle diffusivity 颗粒扩散系数
particle disintegration 颗粒粉碎
particle displacement 质点位移，颗粒位移
particle emission 粒子发射
particle failure test 颗粒破坏试验
particle fluence ratemeter 粉尘粒子率计，粒子注量率计
particle fluence rate 粒子注量率
particle fluence 粒子流，粒子注量
particle flux density 粒子通量密度
particle flux 颗粒通量
particle generator 造粒器，颗粒发生器
particle growth factor 颗粒生长因子
particle growth 颗粒长大
particle history 颗粒历程，颗粒随时间变化
particle in suspension 悬浮质点
particle interlocking 颗粒联锁
particle kernel 包覆颗粒芯核
particle knifing 颗粒剖切
particle mechanics 颗粒力学
particle motion 粒子运动
particle movement 质点运动
particle number concentration 计数浓度
particle number density 粒子数密度
particle packing density 颗粒聚集密度
particle pathway 颗粒迹线
particle path 颗粒通道，质点路径，质点轨迹
particle pattern 磁粉图形
particle phase 颗粒相
particle radiance 粒子辐射度
particle radiation damage 粒子辐射损伤
particle seepage 颗粒渗漏
particle segregation 颗粒分离，颗粒离析
particle shape factor 颗粒形状因数，颗粒形状系数
particle size analysis curve 粒径分析曲线
particle size analysis 粒度测定，粒度测定术，粒径分析
particle size distribution curve 颗粒大小分布曲线，粒径分布曲线
particle size distribution 粒度分布，粒径分布，粒度分配
particle size of dust 灰尘粒度
particle size of fume 烟气粒度
particle size range 粒径范围
particle size 颗粒尺寸，粒度，粒径，粒子大小，粒径大小
particles of changing size 粒度变化的颗粒
particles of unchanging size 粒度不变的颗粒
particle surface 颗粒表面
particle swarm 颗粒群
particle terminal velocity 颗粒终端速度
particle trajectory 颗粒运动轨迹
particle velocity distribution 颗粒速度分布
particle velocity 质点速度
particle volume 颗粒体积
particle-wall interaction 颗粒与壁间相互作用
particle （燃料）包覆颗粒，粒子，颗粒，质点，微粒
particular cargo 特定货物
particular conditions 专用条款，特别条件
particular integral 特殊积分
particular situation 特殊状况
particulars of goods 货物情况，货物内容
particulars of main pipe 主管道详细资料［数据］
particular solution 特解
particulars 细目，细节，详细情节，详细情况
particular value 特值
particular 个别的，特别的，特殊的
particulate activity level 粒子放射性水平
particulate activity 微粒放射性
particulate chemistry 微粒化学
particulate cloud 粒子云
particulate collector 集尘器
particulate expansion 散式膨胀
particulate filter 颗粒过滤器，微粒过滤器，粒子过滤器
particulate fluidization 颗粒均匀分布的流化，散式流化
particulate fouling 颗粒污垢，微粒污垢
particulate fuel 颗粒燃料，微粒燃料
particulate loading 微粒负荷
particulate matter 粒状物料，颗粒物质
particulate phase 散式相
particulate plume 粒子羽流
particulate pollutant 微粒污染物，粒状污染物
particulate radiation 微粒辐射
particulate region 散式区域
particulate removal 去除微粒，悬浮微粒的去除
particulates 大气尘
particulate waste 粒子状废物
particulate 粒子，质点，微粒，细粒
particuology 颗粒学
parties concerned 当事人
parties insured 被保险方
parties to an agreement 签订协议各方
parties to project 项目有关各方
parties 合同方，合同当事人
parting flange 分离法兰
parting line 分界线【铸造】
parting tool 分离工具
parting 夹矸，夹层【煤矿】，分离，裂理
partitional access method 区分存取法
partition between busbar phases 母线相间隔板
partition block 隔墙块体
partition board 隔板
partition capacitance 部分电容
partition chromatography 分配色层法
partition coefficient 分配系数，分割系数，分配比
partition column 分配柱
partitioned access method 区分存取法
partitioned data set 分区数据集
partitioned logic 分块逻辑
partitioned 分布的，分配的，隔离的
partition effect 分配效应
partition fluidized bed 隔板流化床
partition function of a molecule 分子配分函数
partitioning cycle 分离循环
partitioning matrix 分块矩阵

partitioning of matrix 矩阵分块
partition insulator 绝缘隔板，绝缘套管，绝缘垫
partition liquid 分配液
partition noise 分布噪声，电流分配噪声
partition of load 负荷分布
partition plate 分配板【蒸汽发生器】，隔墙板
partition programming 分块规划
partition stub 管板定位短柱
partition wall 间壁，隔墙，双面露光水冷壁
partition 隔板，间壁，隔墙，划分，分割，隔开，分开，离开，分布
part land and part sea transit 部分陆运与部分海运运输
part length absorbed rod 短吸收棒
part length rod control system 短棒控制系统
part length rod （压水堆用的）短控制棒
part load behavior 低负荷性能
part load diagram 部分负荷图
part load efficiency 低负荷效率
part load operation 部分负荷运行，低负荷运行
part load performance 部分负荷性能
part load 部分负荷
partly-adjustable weir 半调节堰
partly closed slot 部分闭口槽
partly-filled pipe-flow 非满管流
partly full flow 半满流
partly integrated concept 部分一体化方案
partly integrated forced circulation 部分一体化强制循环
partly lifting of nacelle 机舱分拆吊装
partly preformed winding 部分成型绕组
partly pulsating stress 部分脉动应力
partly reducing atmosphere 部分还原性气氛
partly submerged orifice 局部淹没孔口，部分淹没孔口
partly turbulent flow 部分湍流
partnering 合伙模式【工程建设模式】
partnership agreement 合伙契约
partnership at will 自愿合伙
partnership dissolution 解除合伙，散伙
partnership enterprise 合伙企业
partnership 合伙关系，合伙，合股，合伙企业
partner's loan 合伙人贷款
partner 伙伴
part number 零部件编号
part of condensation 密部
part of destination 到货港
part of rarefaction 稀部
part payment 部分付款
part plan 局部平面图
part remote from discontinuities or openings 距不连续或开口处很远的部件【应力分析】
part replica simulator 部分复制型模拟机
part section 局部剖面
parts life expectancy 部件预期寿命
parts list 零（部）件名称表，零部件清单
parts of electrical product 电工产品构件
parts per billion 十亿分率
parts per million 百万分率，百万分之几
parts per thousand 千分率
parts purchased abroad 外购件
part-task simulator 部分任务模拟机
part-through crack 未穿透裂纹
part-through thumbnail crack 半椭圆形表面裂纹
part-time application 短时使用，部分时间使用
part-time staff 兼职人员
part-time 部分时间工作的，业余的
part winding starting 部分绕组启动
party at interest 权益当事人
party A 甲方
party B 乙方
party concerned 当事人
party in breach 违约当事人，违约方，违约的合同当事人
party in default 违约的一方，未履约方
party in delay 迟延方
party line bus 合用总线，并行总线
party line 合用线，同线电话，共用电话线路
party not in default 未违约的一方
party to be notified 受通知人
party 团体，党派，聚会，当事人，（法律协议或辩论中的）一方
part 部件，元件，零件，部分，成分，配件
par value 面值，票面价值
par 同价，平价
Pascals law of fluid pressure 帕斯卡液压定律
Pascal's law 帕斯卡定律
Pascal 帕斯卡
Paschen's law 帕申定律
PAS(planning administration section) 计算机管理科
PAS(power application software) 电力应用软件
PAS(project advisors) 工程顾问
Pasquill-Gifford curve 帕斯奎尔吉福特曲线
Pasquill stability 帕斯奎尔稳定度
passable conduit 可通行管沟
passable trench 通行地沟，通行沟
passage aisle 小通道
passage area 通道面积，通流面积
passage for heat 热通道
passage in transit through another country 经由第三国过境
passage of current 电流通路
passage section 通流截面
passage time 驻留时间，停留时间
passage ventilating duct 通过式风管
passage-way 通道，流道，管路，通路，走廊，过道
passage 通道，通过
passameter 外径指示规
pass band utilization factor 通频带利用系数
pass band 通带，传输频带，通频带
pass by name 按名传送
pass by value 按值传送
pass-by 旁通，旁路
pass capacitor 旁路电容器
passenger and cargo liner 定期客货船
passenger-cargo bus 客货两用车
passenger-cum-goods elevator 客货电梯
passenger elevator 载人电梯，载客电梯
passenger terminal 客运码头
pass horse 渡槽支架

passimeter 内径指示规
passing contact 滑过触点
passing over 从上方穿过
passing place 让车道
passing-screen size 过筛粒径
passing 通过的，经过的，通过，经过
passivant 钝化剂
passivate 钝化
passivating agent 钝化剂
passivating solution 钝化液
passivating （酸洗后）钝化，钝化作用
passivator 钝化剂
passive absorption 被动吸收
passive activation analysis 无源活化分析
passive aerial 无源天线
passive AND gate 无源与门
passive assay equipment 无源分析设备
passive assay 无源分析
passive augmentation methods 无外动力源的强化换热方法
passive bus 无源母线
passive circuit elements 被动电路元件
passive circuit 无源线路，无源电路
passive component 钝化部分，无源设备，非能动部件，无源元件
passive condition 钝态
passive contact 无源接点
passive containment system 非能动安全壳系统
passive contaminant 被污染物
passive control 被动控制
passive cooling 被动式冷却，被动式制冷
passive defects 原因不明的故障
passive diffusion 被动扩散
passive earth pressure coefficient 被动土压力系数
passive earth pressure 被动土压力
passive earth thrust 被动土推力
passive electric network 无源电网络
passive electrode 收集电极，接地电极
passive equivalent network 无源等效网络
passive failure system 故障消极防护系统
passive failure 被动破坏，被动损毁，虚性故障，无效损坏，非能动故障
passive film 钝化膜
passive filter 无源滤波器
passive four-terminal network 无源双口网络，无源四端网络
passive load 无源负载
passively safe reactor 固有安全反应堆
passive method 被动法
passive network 无源网络
passive one-port 无源单口
passive plum 被动羽流
passive pollution 被动污染
passive probe 无源探测器
passive Rankine pressure 被动朗肯土压力
passive reactor zone 反应堆非活性区
passive reflector 无源反射器
passive regulation 被动监管
passive relay station 无源中继站
passive remote sensing 无源遥感
passive resistance 无源电阻
passive resistive load control 无源电阻性负载控制
passive resistor 无源电阻器
passive safeguards 非能动安全装置
passive safe storage 被动安全贮存，封存
passive safety feature 非能动安全设施
passive safety system 非能动式安全系统
passive safety 被动安全，被动安全性，非能动安全
passive solar house 被动式太阳房
passive solar system 被动式太阳能系统
passive source 被动源
passive spread 被动扩散
passive state of plastic equilibrium 被动塑性平衡状态，被动状态的塑性平衡
passive surface of sliding 被动滑动面
passive surface 钝化面
passive system 被动系统
passive trajectory guidance 被动寻的制导
passive transducer 无源换能器，无源传感器，无源变换器
passive two-port 无源双口
passive two-terminal network 无源单口网络，无源二端网络
passive yaw 被动偏航，（风轮）被动调向，被动对风
passive 被动的，消极的，无源的，钝态的
passivity 无源性，被动性，钝性，钝态
pass length average 传送长度平均值
pass length maximum 传送长度最大值
pass length minimum 传送长度最小值
pass length 传送长度
passometer 步测计，计程器
pass-out control valve 抽气控制阀
pass-out regulating valve 排气调节阀
pass-out steam turbine 旁路汽轮机
pass-out steam 抽汽，旁通蒸汽
pass-out turbine 抽汽式汽轮机
pass-out 抽汽，支管，旁通的
pass partition 通道分配【蒸汽发生器】，（热交换器）纵向隔板，管程挡板，分配板
passport 合格证书，护照
pass sequence 焊道顺序
pass sheet 复印清样
pass-test 合格测试【质保】，合格试验
pass the test 通过测试，通过考试，试验合格，经受住考验
pass trestle 渡槽支架
pass valve 直通阀
password 密码
pass 通道，烟道，焊道，经过，通过，道路，合格，传递，传达
pastagram 温高图
past-arc current 弧后电流
past due 过期
paste adhesive 糊状胶黏剂
paste board 层压纸板，绝缘纸板，胶纸板
pasted insulation 涂抹式保温
paste explosive 糊状炸药
paste fuel 糊状燃料

paste reactor　糊状燃料反应堆
paste　浆料，张贴，粘贴
pasting machine　抹灰机
past precipitation　前期降水
pasturage　牧场，畜牧业
pasture land　牧场，草场
pasture　牧草，牧场
pasty material　糊状物料
patch bay　接线架，插头安装板
patch board　接线板，转插板，配电盘，接线盘
patchcord　（配电盘）软线，连接电缆，插接线
patching cord　连接电缆，插接线，调度塞绳
patching-in　临时引线
patching material　修补材料
patching panel　接线板
patching　修补，（道路）补坑，临时性接线
patch-in　临时接入，临时接线
patch out　临时撤线
patch panel　编排板，接线板，中央控制单元
patchplug　转接插头，接插头
patch-program plugboard　变程板
patch rods　补修条
patchy corrosion　斑点腐蚀
patch　补片，连接板，鱼尾板，搭板，临时性线路，插入码，修补
patent agent　专利代理人
patent agreement　专利协议
patent application　专利申请书
patent cooperation treaty　专利权合作条约
patentee　专利权所有人
patent fee　专利费
patent for invention　发明专利权
patent holder　专利持有方
patent in force　生效的专利
patent law　专利法，专卖特许
patent licence contract　专利许可合同
patent licence　专利许可证，专利权特许使用权
patent license agreement　专利特许协议
patent No.　专利号
patent right　专利权，专利证书
patent royalty　专利权使用费，专利使用费，专利费
patent specification　专利说明书
patent　专利权，专利品，专利，特许
patera　插座，接线盒
paternoster elevator　链斗式提升机
paternoster pump　链斗式泵
paternoster　链斗式升降机
path allocation　通道分配
path clearance　通道间隙
path control　通路控制
path curve　轨线
pathfinder　导引装置
path function　过程函数【热力学】
path-gain factor　通路增益系数
path generation method　通路形成法
path increment　行程量
path information unit　通路信息单元
path length　路径长度，行程长度，迹线长度，路程
path line　迹线，流线，轨迹线
path of devil　尘暴路径
path of flow line　流线轨迹
path of percolation　渗漏路径
path of plume　羽流路径
pathogenic bacteria　病菌
pathogen　病原体
path selector　通路选择器【堆芯测量仪表】
path setting　通道设定
path transfer unit　选择通向装置【堆芯测量仪表】
path way　通道，路径
path　通路，路线，途径，轨道，轨迹，路程，路径，道路，走道
patrix　母字模，阳模，上模
patrol and inspection　巡回检查
patrol boat　巡逻艇
patrol inspection　巡回检查
patrolling of line by helicopter　直升机巡线
patrolling of line　线路巡视，巡线
patrol maintenance　巡视修理
patrolman　巡线员
patrol room　巡检室
patrol telephone station　巡线电话站
patrol　巡视，巡查，巡线
patronage　惠顾
pattern allowance　模型余量
pattern analysis　模式分析，图形分析
pattern approval　形式上批准
pattern classification　图样分类
pattern correspondence index　模式符合指标
pattern cracking　规则裂缝，网状裂纹
pattern design　定型设计，图案设计
pattern displacement　光栅位移，图像变位
pattern distortion　图像畸变，像差
patterned conveyor belt　花纹输送带
patterned glass　图案玻璃
patterned plate　网纹钢板
pattern generator　直视装置信号发生器，侧视图案发生器
pattern information retrieval system　模式信息检索系统
patterning　制作布线图案，图案形成
pattern ironwork　铁花饰
pattern layout　图案设计
pattern maker　模型工
pattern of flow　流态，流型
pattern of heat absorption　吸热分布
pattern of scatter load　插花式装料方式
pattern of water flow　水流模型
pattern plate　模板
pattern recognition programming　模式识别程序设计
pattern recognition system　模式识别系统
pattern recognition　模式识别，图形识别
pattern sensitive fault　特殊数据组合故障，特定模式故障
pattern shop　模型车间，木模车间
pattern winding　标准绕组，模型线圈
pattern　样式，图案，图形，图样，形态，模型，铸模，模式，标准图形，帧面图像
paul　爪，棘爪，卡子

pause period 停机时间
pause 暂停，间歇，间隔，停息
paved gutter 铺砌的马路边沟
paved inlet 铺砌进水口
pavement breaker 路面破碎机
pavement concrete 铺面混凝土
pavement marking 路面标线
pavement of clinker 缸砖路面
pavement of cobble stone 大卵石路面
pavement of riprap 乱石护面
pavement of stone block 块石路面
pavement roughness 路面平整度
pavement structure 护面结构
pavement 路面，铺道，铺面，铺砌层
paver 铺料机，铺路机，铺面机
pave with tiles 瓷砖铺面
pave 铺盖，铺设，铺道路，铺地，铺砌
paving beetle 铺面夯具
paving block 铺面块
paving breaker 路面破碎机
paving brick 铺路砖，铺面砖
paving expansion joint 铺面伸缩缝
paving in stone blocks 石块铺面
paving machine 铺路机，铺面机，铺砌机
paving material 铺面材料
paving slab 铺路板
paving stone 铺路石，铺石
paving tile 铺路砖
paving with pebble 卵石铺面
paving 铺路【混凝土或沥青】，铺面，铺砌
pawl 制轮爪，爪，棘爪
PAW(plasma arc welding) 等离子弧焊
paw support 爪式支座
payable account 应付账目，应付账款
payable at a fixed date 定期付款
payable at destination 在目的地付款
payable at maturity 到期付款
payable at par 面值支付
payable at sight 见票即付
payable at usance 远期付款，惯例付款
payable by installment 分期支付
payable in account 赊账付款，应付账款
payable interest 应付利息
payable on demand 见索即付
payable payment period 应付账款的付款时间
payable to bearer 见票即付持票人
payable to order 应付予记名受款者，支付指定人
payable 可付的，应支付的
pay-as-bid settlement 按报价结算
pay as you earn 预扣所得税
pay attention to 重视
pay-back method 偿付方法，回收法
payback period method 回收年限法
payback period 偿付期限，(贷款) 偿还期，还款期，(投资) 回收期，还本期，归本年期
payback time 偿还时间
payback 偿付，付还，回收 (贷款)
pay by installments 分期付款
pay by TT 电汇付款
payee 收款人，受款人
PAYE(pay as you earn) 预扣所得税
payer on commercial instrument 票据支付人
payer 付款人，付款单位
pay for import 进口费用
pay in advance in quarterly installments 分季预缴
pay in advance 预付
paying-off 放线，开卷
paying out drum 放线盘
paying-out machine 放线机
paying out reel 放线盘
paying out 支付，补偿，放电缆
paying the bill 结账
payload capability 有效负载能力
payload capacity 起重量，载重量，最大有效荷载，有效负载容量
payloader 运输装载机
payload fraction 有效载荷部分
payload volume 有效负载体积
payload weight 有效载荷重量
payload 有效负载，有用荷载，有效负荷，净载重量
payment after due date 到期后付款
payment after termination 终止后的付款
payment against arrival of documents 单到付款
payment against arrival 货到付款
payment against documents 凭单付款
payment against document through collection 凭单托收付款
payment against presentation of shipping document 凭运单付款
payment agreement 付款协定，支付协定
payment at maturity 到期付款
payment at regular fixed time 定期付款
payment at sight 见票即付
payment by bill 汇票付款
payment by cash 现款支付
payment by check 支票付款
payment by draft 汇票付款
payment by installment 分期付款
payment by remittance 汇款支付
payment certificate 付款证书
payment declaration 付款通知
payment due 已到期的应付款
payment guarantee 付款保证
payment in advance 提前付款，预付款，预付
payment in cash 现金支付，付现金
payment in due course 按期付款，及时付款，正当付款，到期支付
payment in foreign currencies 外币支付
payment in full 付讫，全部付讫，全额支付
payment in kind 实物支付，以实物支付
payment instrument 支付凭证
payment maturity 到期付款，到期日付款
payment of claims 索赔款的支付
payment of loans 归还借款
payment on arrival 货到付款
payment on delivery 交货付款
payment on demand 即期付款，索取时付款
payment on invoice 凭发票付款
payment on terms 定期付款
payment order 付款通知

payment provisions 支付条款
payment refused 拒付
payment schedule 付款计划表，付款进度，付款日程，支付计划
payment side 付方
payments in advance 预付款，款项先付
payments in arrears 拖欠款项，款项后付，后付款项
payments 拖欠税款
payment terms 付款条件，付款条款，支付条件
payment to contractor 给承包商付款
payment 支付，支付方式，付款
pay off a debt 偿清债务
pay-off bobbin retainer 放线盘架
pay-off method for capital investment 资本投资回收年限法
pay-off period 偿还期，偿还周期，回收期
pay-off reel 放线盘，开卷机
pay-off spool 放线盘
pay-off table 结算表，工资表，支付矩阵
pay-off 偿还，付清，偿清，支付，工资等级表，放线，放线装置，收效，报酬，结果
pay on a consolidated basis 汇总缴纳
pay on delivery 货到付款
payout 偿还，补偿，释放
pay per view 即用即付
payroll payable 应付工资
payroll service agencies 薪资服务中介
payroll tax 工资（所得）税，工薪税
payroll 雇员名单，工资表，工资单，工资总额
pay station 公用电话亭
pay the utmost attention to 优先考虑
pay 薪水，工资，付款
PBFBB(pressurized bubbling fluidized bed boiler) 增压鼓泡流化床锅炉
PBL(planetary boundary layer) 行星边界层
PBNC(Pacific Basin Nuclear Conference) 太平洋沿岸地区核能会议
PB＝polybutylene 聚丁烯
PB(push button) 按钮【DCS 画面】
PC-based 基于个人计算机的，使用个人计算机的
PCB(polychlorinated biphenyl) 多氯联苯
PCB(printed circuit board) 印刷回路板，印刷电路，印刷电路板
PC card enclosure 印刷电路板箱
PCC(packing condition certificate) 包装状态证书
PCC(part completion certificate) 部分完工证书
PCC(particular conditions of contract) 合同的专用条款，合同的特殊条款
PCC(point of common coupling) 公共耦合点，公共连接点，公共供电点
PCC(project consultancy contract) 工程顾问合同
PCFBB(pressurized circulating fluidized bed boiler) 增压循环流化床锅炉
p-channel metal oxide semiconductor P 沟道金属氧化物半导体
PCI(pulse counter input) 脉冲输入
PCM multiplex system 脉码调制多工系统
pcm(pour cent mille) 反应性单位
PCM time multiplex system 脉码调制分时多工系统
PCM transmission line 脉码调制传输线
pcm 10^{-5}【反应性单位】
PCN(plant change notice) 电厂变更通知单
P controller(proportional controller) 比例调节［控制］器
PC(personal computer) 个人计算机
PC(pith circle) 节圆
PC(polycarbonate) 聚碳酸酯
PC(power centre) 动力中心
PCP＝precipitator 除尘器
PCP(primary control program) 基本控制程序，主控程序
PCP(printed-circuit patchboard) 印刷电路接线盘
PC(pressure controller) 压力控制器
PC(pressure control) 压力控制
PC(project controlling) PC 模式，项目总控模式【工程建设模式】
PC(pulverized coal) 煤粉
PCRV(prestressed concrete reactor vessel) 预应力混凝土反应堆容器
PCS(passive containment cooling system) 非能动安全壳冷却系统
PCU(process control unit) 过程控制单元
PCU(program control unit) 程序控制单元
PCV(pressure control valve) 压力调节阀【DCS 画面】
PC 动力中心【DCS 画面】
PD-action(proportional plus derivative action) 比例微分作用
PD controller(proportional plus derivative controller) 比例微分调节（控制）器
PDI(pressure differential indication) 差压指示
PD(potential drop) 电压降，压差【DCS 画面】
PDP(project directing plan) 工程指导计划
PD(proportional plus derivative) 比例微分
PDR(post disturbance review) 事故追忆
PDS(proposal data sheet) 报价数据表
4PDT(four-pole double-throw) 四极双掷
PDT(pressure differential transmitter) 差压变送器
4PDT switch 四极双掷开关
pea coal 粒煤
peacock coal 发亮煤
pea gravel 豆砾石，小砾石
pea-in-pod structure 豆荚结构
peak absorbed does 峰值吸收剂量
peak accelerometer 峰值加速度计
peak amount 高峰量
peak amplitude 振幅，峰值
peak and cyclic load operation 调峰运行
peak arc voltage 电弧电压峰值
peak bandwidth 峰带宽
peak blocked voltage 正向电压峰值，正向峰值电压
peak broadening 峰加宽
peak capacity （卸船机的）峰值出力【最大小时卸煤出力】，尖峰功率，尖峰负荷，尖峰出力，尖峰容量，高峰容量，最大容量
peak-charge effect 峰值充电效应
peak choke 峰值扼流圈，峰值抗流圈
peak chopper 波巅限幅器

peak clipper circuit　削峰电路
peak clipper　峰值限幅器，削峰器
peak clipping　削峰，调峰
peak consumption　高峰汽耗，高峰热耗
peak contact　齿顶啮合
peak corona intensity　峰值电晕强度
peak critical void ratio　最大临界孔隙比
peak current　峰电流，最大电流
peak curve　尖顶曲线
peak-day-load curve　日峰荷曲线
peak demand capacity　峰荷容量
peak demand forecasting　峰荷需量预测
peak detector　峰值检波器
peak dipping　峰值限幅器
peak discharge　高峰流量，最大流量，峰值流量，洪峰流量
peak dose　峰值剂量
peak duration curve　峰值持续时间曲线
peaked roof　尖屋顶
peaked wave　尖峰波
peak efficiency　最高效率
peak electrical load　高峰电力负荷
peak energy　峰值能量
peak envelope power　包络峰值功率
peaker　脉冲整形器，微分电路
peak factor　峰值因子，峰荷因数，峰值系数，振幅系数
peak firing rate　最大燃料耗量，最高燃烧速度
peak flattening　峰展平
peak-flood interval　洪峰间歇
peak flow　洪峰流量，尖峰流量，峰值流量，高峰流量
peak flux density　峰值磁通密度，峰值通量密度
peak forward anode voltage　正向峰值阳极电压
peak forward voltage　正向峰值电压
peak frequency　峰频数，高峰频率
peak ground acceleration　峰值地面加速度
peak gust　最大阵风
peak half width　峰半宽度
peak heat flux　峰值热流密度
peak heating load　尖峰热负荷
peak height　峰高
peak-holding　极值保持
peak hour　峰荷时间，高峰时间，高峰负荷时间
peak indicator　峰值指示器
peaking capacity　最大容量，尖峰容量，尖峰能力，顶值容量，峰值容量
peaking coil　校正线圈，建峰线圈
peaking factor　热点因子，峰值因数，峰值因子
peaking flux　通量峰
peaking installation　尖峰负荷装置
peaking load hours　尖峰负荷时间
peaking load station　尖峰负荷电站
peaking network　峰值网络，建峰网络
peaking plant　尖峰负荷发电厂
peaking power　（局部）功率峰
peaking service　尖峰运行，尖峰负荷运行
peaking transformer　峰值变压器
peaking turbine　尖峰负荷透平
peaking unit　峰荷发电机组，尖峰机组
peaking　最大峰值，剧烈增加，脉冲尖锐化，调峰
peak inrush current　最大峰值电流，最大涌流
peak inverse voltage　反向峰值电压
peak let-through current　峰值通过电流
peak limiter　峰值限制器
peak line　峰线
peak load boiler　尖峰负荷锅炉
peak load calorifier　尖峰加热器
peak load canal　峰荷渠道
peak load capacity　峰荷
peak load condition　尖峰负荷工况
peak load generating set　尖峰负荷机组
peak load generator　峰荷发电厂
peak load heater　尖峰负荷加热器
peak load heat source　峰荷热源
peak load hours　峰荷时间，尖峰负荷时间
peak load hydropower plant　峰荷水电站
peak loading station　峰值负荷电站
peak load operation　高峰负荷运行，尖峰负荷运行
peak load output　峰载出力
peak load period　峰荷期间，高峰负荷时间
peak load power plant　调峰电厂，峰荷（发）电厂，尖峰负荷发电厂
peak load power station　调峰电站，峰荷（发）电站，尖峰负荷发电站
peak load rating　峰载出力
peak load regulating operation　调峰运行
peak load regulation　尖峰负荷调整，调峰
peak load station　调峰电厂，调峰电站，尖峰负荷电站，峰负载发电厂
peak load thermal power plant　尖峰负荷火力发电厂
peak load time　高峰负荷时间
peak load turbine　尖峰负荷汽轮机
peak load unit　峰荷机组
peak load　峰负载，峰荷，峰值负荷，峰值负载，高峰负荷，尖峰负荷，最大负荷，最大荷载，最大短时负载
peak-lopping operation　尖峰负荷运行
peak-lopping station　尖峰负荷电站
peak making-current　最大接通电流，峰值接通电流
peak material-loading rate　最大物料装载率
peak meter　峰值电度表
peak nucleate boiling heat flux　临界热流密度，泡核沸腾峰值热流密度
peak of curve　曲线峰值
peak of load　负荷峰值
peak of negative pressure　负压峰
peak of noise　噪声峰值
peak-of-the-curve　曲线幅值
peak output of pulse　脉冲峰值输出功率
peak output torque　最大输出转矩
peak output　尖峰出力，峰值输出功率，最大输出
peak-peak　峰峰，峰间的
peak performance　最佳工作状态，最大生产率
peak period of construction　施工高峰期
peak point current　峰值电流
peak point　峰值点，最高点
peak power coefficient　峰值功率系数

peak power limitation 峰值功率限制，最大功率限制，最大功率极限，峰值功率极限
peak power output 峰值功率输出
peak power 峰值功率，尖峰功率，最大出力，尖峰能力
peak preheater 峰载加热器
peak pressure indicator 最大压力指示器
peak pressure 峰压，峰值压力，最大压力
peak rate of flood discharge 洪峰流量
peak rate of flow 高峰流量
peak rate of runoff 径流峰值
peak rating 最大额定功率，最大定额
peak regulation turbine 调峰汽轮机
peak response 峰响应，最大响应
peak responsibility 峰值负载能力，尖峰负荷能力，峰值负载量
peak-responsive rectifier 响应峰值的整流器
peak reverse voltage 反向峰值电压，逆向峰值电压，峰值反向电压
peak ripple current 脉动电流峰值
peak runoff 峰值径流，洪峰水量，最大径流
peak sand content hydrograph 沙峰过程线
peak season surcharge 旺季附加费
peak season 旺季
peak separation 脉冲间距
peak sharing station 峰值分担电站
peak shaving 高峰调节，高峰消减，调峰
peak shift 峰值移动，峰荷班次
peak shock pressure 最大冲击压力
peak stage 波峰阶段
peak strength 峰值强度
peak stress 峰值应力，最大应力，最大值应力
peak strip 绕线式脉冲传感器，金属线脉冲传感器
peak switching-current 合闸峰值电流
peak temperature 峰值温度
peak-to-average flux 峰-均通量比，最大通量与平均能量比，峰值通量与平均通量比
peak-to-average ratio 最大值与平均值之比
peak-to-peak amplitude 峰-峰幅度，峰间幅值，峰-峰振幅，峰间振幅，正负峰间幅值［振幅］
peak-to-peak current ripple 电流脉动峰-峰值
peak-to-peak fluctuation of amplitude 振幅峰-峰值起伏
peak-to-peak noise 峰间噪声
peak-to-peak separation 峰间幅值，峰距，峰峰间隔
peak-to-peak value 峰-峰值，峰间幅值，峰间值，峰-峰差值
peak-to-peak voltage 电压峰-峰值
peak-to-peak 峰间的，峰-峰值，峰间幅值，信号总幅值，最大电压波动，峰间值
peak torque 最大转矩，幅值转矩，峰值转矩
peak-total ratio 峰-总比
peak to valley ratio 峰谷比
peak transformer 峰值变压器
peak triaxial acceleration recorder 三轴峰值加速记录仪
peak unit 调峰机组，峰荷机组
peak use rate 最大耗用率
peak-utilization curve 峰荷利用曲线
peak-valley price 峰谷电价
peak-valley pricing policy 峰谷电价政策
peak value measuring equipment 峰值测量设备
peak value 峰值，幅值，振幅，最大值
peak voltage 峰压，幅值电压，尖峰电压，峰值电压，高峰电压
peak voltmeter 峰值电压表，峰值伏特计，幅值电压表
peak volt 幅值伏特
peak volume change 最大体积变化
peak white 白色信号峰值
peak width 峰宽
peak wind speed 最大风速
peak withstand current 峰值耐受电流
peaky 有峰的，尖的
peak 峰值，尖峰，波峰，高峰，最大值，最高点，顶部，尖端，顶点，山顶，顶峰，峰
peamafy 皮马非高导磁率合金
peanut tube 花生管
pearlite concrete 珍珠岩混凝土
pearlite 珠光体，珍珠岩
pearlitic heat-resistant steel 珠光体耐热钢
pearlitic 珠光体的
pearls 珍珠，珍珠状的
pear-push 悬吊式按钮
pear-shaped centrifuge tube 梨形离心管
Pearson type Ⅲ distribution curve 皮尔逊Ⅲ型分布曲线
pear switch 梨形拉线开关，悬吊开关
pea-size coal 豌豆级无烟煤，豆煤
peastone 豆石，豆砾石
peat coal 泥煤，泥炭
peat hag 泥煤地
peatifleation 泥炭化作用
peat inclusion 泥炭夹层
peat land 泥炭地
peat slime 泥炭沉渣
peat soil 泥炭土，沼土
peat water 泥炭水
peaty soil 泥炭土
peat 泥煤，泥炭，泥炭土
pebble bar 卵石滩
pebble-bed core 球床堆芯，球堆芯
pebble-bed filter 卵石层过滤器
pebble-bed gas-cooled reactor 球床气冷反应堆，煤球炉式气冷反应堆
pebble-bed high-temperature reactor 球床高温气冷反应堆
pebble-bed mechanics 球床力学
pebble-bed reactor 球反应堆
pebble-bed rheological behavior 球床流变性能
pebble-bed sagging 球床下陷
pebble-bed shield 球床屏蔽
pebble-bed 卵石床，球床
pebble concrete 小豆石混凝土
pebble-filled trench 填砾石排水沟
pebble flow experiment 球流实验
pebble flow surface 球床表面
pebble fuel element 小球状燃料元件
pebble heater 卵石加热器，粒状载热体加热器
pebble shoal 卵石浅滩

pebble wall 卵石墙
pebble 砾石，水石子，卵石，绿豆砂，小砾石，小粒石
pebbly 砾石覆盖的
PE cable 聚乙烯电缆
PEC controlled variable external resistance PEC控制的外部可变电阻器
pecked line 点虚线
pecking motor 步进电机，步进电动机
peck 琢孔
PEC(packaged electronic circuit) 电子线路程序包
PEC(photo electric cell) 光电管，光电池
PEC(power electronic converter) 电力电子变流器
PEC(printed electronic circuit) 印刷电路
PEC(program element code) 程序单元代码
pectolite 针钠钙石
peculation 挪用，贪污
peculiarity 特点，特性，特质
peculiar test equipment 特殊测试设备
peculiar to 限于，仅由……使用［采纳，实施］，为……所特有的
pecuniary penalty 罚金
pedal-dynamo 脚踏发电机
pedal 踏板，垂线
pedestal abutment 墩座
pedestal base 轴承座底板
pedestal bearing insulation 轴承座绝缘
pedestal bearing motor 座式轴承电动机
pedestal bearing 架座轴承，托架轴承，架座
pedestal body 支座
pedestal cap 轴承座盖
pedestal foot 轴承座
pedestal generator 基准电压发生器
pedestal insulator 座式绝缘子
pedestal pile 爆破桩，扩脚桩，支座桩
pedestal post insulator 针式支柱绝缘子
pedestal 底座，机座，基座，轴承座，台，台座，台式，支座，支架，轴架，基础
pedestrian bridge 人行桥
pedestrian guardrail 人行道栏杆
pedestrian island 安全岛
pedestrian mall 人行林荫路
pedestrian tunnel 人行隧道
pedestrian volume 人流量
pedestrian walk 人行道
pedestrian way 人行道
pediment arch 三角拱
pediment 三角楣饰，山前侵蚀平原【常用复数】，人字墙，麓原，山墙，山形墙，三角墙
pedimetry 步测法，步测计，步程计
pedogenesis 成土作用
pedology 土壤学
pedometer 步测计，步数器，步数计，步程计
pedosphere 表土层，土界
PED＝pedestal 轴承座
ped 土团
peek-a-boo card 同位穿孔卡片
peelable type 可剥型
peeler 剥皮器
peel film 剥落的漆皮
peeling off 剥离
peel 推杆，脱壳
peener 喷丸装置
peening 喷丸硬化【焊缝】，轻敲
peen 喷丸加工，锤尖，锤平
peep door 观火门，窥视门
peep hole 窥视孔
peep-sight alidade 测斜照准仪
peer observation 同行评定
peer review 同行评审，同行评议
peformance number 特性数
peg adjustment 两点校正法【水准仪】
peg board 小钉板
peg fin 分段鳍片
pegging out 标桩定线
pegging steam 启停备用汽源
pegmatite structure 文象结构
peg method 标桩校正法，两点校正法
peg switch 标记转换开关，记次转换开关
peg 标桩，小木桩，桩，钉，钉子，栓，木栓，插塞，销
pelagic zone 水层带，远洋区
pelagos 水层生物
pelite-gneiss 泥片麻岩
pelite 泥质岩
pellet acceptance lot 芯块验收批
pellet arrester 丸形避雷器
pellet center temperature 芯块中心温度【燃料棒】
pellet chips 芯块渣屑
pellet-clad chemical interaction 芯块包壳化学作用
pellet-clad gap 芯块与包壳间隙
pellet-clad interaction 芯块和包壳间的相互作用
pellet cocking 芯块翘起【倾斜】
pellet diameter 芯块直径
pellet dishing 芯块凹陷
pelleted-floc 粒状絮凝物
pellet extrusion 挤压成型
pellet fabrication 芯块制造
pellet fuel element 芯块燃料元件
pellet fuel 芯块燃料
pelleting 芯块压制
pelletization 造球，制粒
pelletized zinc blende 粒装闪锌矿
pelletized 颗粒状的
pelletizer 制粒机
pelletize 造粒，制丸，造渣，粒化，制团，制芯块，芯块制造
pelletizing facility 制芯块设备
pelletizing plant of soot 烟灰粒化装置
pellet layer 芯块层
pellet lightning arrester 丸形避雷器
pellet moisture lot 芯块湿度批
pellet restructuring 芯块重新组织
pellet stack continuity 芯块堆叠连续性
pellet stack （燃烧棒中）芯块柱，芯块堆，芯块堆叠
pellet surface temperature 芯块表面温度
pellet temperature 芯块温度

pellet to can radial gap 芯块与包壳间的径向间隙，冷间隙
pellet-to-clad eccentricity 芯块对包壳偏心率
pellet tracks 芯块跟踪
pellet 丸，片，小球，弹丸，芯块，小块
pellicle 膜，薄膜
pellicular electronics 薄膜电子学
pellicular moisture 薄膜水分
pellicular resin 活化膜树脂
pellicular water 薄膜水，吸附水，黏结水
pellicular zone 薄膜水层
Peltier coefficient for substances M and N 物质 M 和 N 的珀耳帖系数
Pelton turbine 培尔顿水轮机，水斗式水轮机，冲击式水轮机
Pelton water wheel 冲斗式水轮
pelyte 泥质岩
PEMP(provisional environmental management plan) 临时环境管理计划
penal institution 监管机构
penalizing 处罚性的
penal terms 处罚条款
penalty clause 违约罚款条件
penalty factor for rod bow 燃料棒弯曲损失因子
penalty for delayed delivery 延迟交货罚金
penalty for delay 误期罚款，逾期罚款，延误罚款
penalty for nonperformance of contract 违约罚金
penalty function 补偿函数，罚函数
penalty method 补偿法
penalty of breach of contract 违约罚金
penalty punishment 处罚
penalty rate 惩罚率
penalty terms 处罚条款
penalty 罚款，罚金，损失
pen arm assembly 笔杆组件
pencil beam 尖向束，锐方向性射线
pencil grinder 笔状磨头
pencil guide 记录销子导轨
pencil holder 笔尖支架
penciling 铅笔痕，细线
pencil of curves 曲线束
pencil of rays 光束
pencil sketch 铅笔草图
pencil tube 笔形管，超小型管
pencil-type thermocouple 铅笔式热电偶，铅笔式温差电偶
pencil 记录头，射束，束
pendant boiler 悬吊式锅炉
pendant chain 吊链
pendant control station 悬架式操纵台
pendant control 吊灯控制，悬吊控制
pendant cord 吊灯线
pendant drop 垂悬液滴
pendant-fittings 悬吊件
pendant lamp 吊灯
pendant mounted 悬挂
pendant platen 悬挂式管屏
pendant pull switch 拉线开关
pendant push 悬挂按钮
pendant reheater 悬吊式再热器
pendant signal 吊灯信号
pendant superheater 悬吊式过热器，悬式过热器，立式过热器
pendant switch 拉线开关
pendant tube superheater 悬吊管式过热器
pendant tube 悬吊管
pendant 下垂的，悬吊的，悬垂物，悬架，吊架，吊灯，三角旗
pendent-platen superheater 悬吊屏式过热器
pendent 下垂的，悬吊的，悬垂物，悬架，吊灯
pending contract 尚未完成的合同
pending problem 悬而未决的问题，遗留问题
pending question 未解决的问题
pending 待定的，悬而未决的，待解决的（问题）
pen-driving mechanism 记录装置的笔尖传动装置
pendular dynamometer 摆动式测力计
pendular oscillation 摆动
pendulate 摆动，摆振，振动
pendulation period 摆动周期
pendulation 摆动
pendulosity 摆性
pendulum bearing 摆式轴承
pendulum clinometer 摆式测斜仪
pendulum conveyor 摆式运送机
pendulum damping gear 飞摆阻尼机构
pendulum damping tank 摆动阻尼器，摆动阻尼箱
pendulum flap 摆动舌阀
pendulum generator 飞摆发电机
pendulum governor 摆式调速器，飞锤式调速器
pendulum impact test 摆锤式冲击试验
pendulum inclinometer 摆式倾斜仪
pendulum length 摆长
pendulum magnetometer 摆式磁通计
pendulum manometer 摆式压力计
pendulum motion 摆动
pendulum motor 飞摆（调速）发电机
pendulum pile-driver 摆式打桩机
pendulum rectifier 振动式整流器
pendulum relay 振动子继电器
pendulum shaft 摆轴，测锤竖井
pendulum spring 飞摆弹簧
pendulum stability 摆稳定性
pendulum-type impact machine 摆式冲击试验机
pendulum-type tachometer 摆式转速计
pendulum-type 摆式的
pendulum valve 摆动式锁气阀
pendulum 钟摆，摆，铅垂线
peneplain 准平原
penetrability 可穿透性，可渗透性，渗透性，穿透性，穿透能力，穿透率
penetrable 可渗透的，可穿透的，可贯穿的
penetrameter sensitivity 透度计灵敏度
penetrameter （射线）透度计，穿透计，透光计，像质计，图像质量显示器
penetrance 穿透性，贯穿，透射
penetrant action 浸透作用
penetrant aid 渗透辅助手段
penetrant compatibility 渗透计相容性
penetrant dependence 渗透计相关性
penetrant dwell time 渗透计停留时间
penetrant flaw detection 渗透检测

penetrant indication 渗透计显示
penetrant inspection unit 渗透检验单元
penetrant inspection 着色检查，着色探伤，渗透探伤
penetrant method 渗透法
penetrant removal method 渗透计去除方法
penetrant remover 渗透计去除计
penetrant station 渗透检验站
penetrant testing 渗透检验，渗透探伤
penetrant test （液体）浸透探伤试验
penetrant 渗透剂，穿透剂，渗透的，浸透的，贯穿的，穿透的
penetrated crack 穿透裂纹
penetrate 渗透，透过，穿透，贯穿
penetrating capacity 渗透能力
penetrating component 贯穿部件，穿透成分，贯穿成分【辐射】
penetrating cone 贯入锥
penetrating dye 渗透染色，渗透色液
penetrating fluorescent oil test 荧光油浸透试验
penetrating inversion 渗透逆温
penetrating neutron radiation 中子穿透辐射
penetrating power 穿透力，贯穿本领，渗透性
penetrating radiation 穿透性辐射，贯穿辐射
penetrating time 渗透时间
penetrating well 透水井
penetration bead 熔透焊道
penetration chamber 贯穿件室
penetration coefficient 渗透系数
penetration concrete 贯入混凝土
penetration cooling 渗透冷却
penetration corrosion 贯穿腐蚀，渗透性腐蚀
penetration course 贯入层
penetration depth 透入深度，渗透深度，穿透深度，贯入深度
penetration distance 渗透距离
penetration factor 贯穿率，渗透率
penetration flaw detector 渗透探伤
penetration heat 贯穿热
penetration index 贯入指数
penetration isolation valve 贯穿隔离阀
penetration length 穿透长度
penetration liner 贯穿衬套
penetration-load curve 贯入度荷载曲线
penetration method 穿透法，贯入法，灌装法
penetration needle 贯入针
penetration of current 有效肤深透入深度
penetration of pile 桩贯入度
penetration of water 水浸入
penetration per blow 每击贯入度【打桩】
penetration pile 贯入桩
penetration probability 穿透概率
penetration rate 穿透率，焊透率
penetration resistance curve 贯入阻力曲线
penetration resistance 贯入阻抗，贯入阻力
penetration sleeve 穿孔套管，穿墙套管
penetration sounding 贯入触探
penetration surface treatment 贯入式表面处理
penetration test 贯入度试验，针入度试验
penetration time 渗透时间【液体渗透检验】
penetration welding 熔透型焊接法
penetration （焊接母材）熔深，贯穿，穿透，穿过，渗透，穿透度，贯入度，针入度
penetrative probe 可透探头
penetrative 贯穿的，穿透的
penetrator 贯穿件
penetrometer 针穿硬度计，射线透度计，透光计，触探仪，针入度仪，穿透计，贯入器，贯入仪，检测仪，透度计
penetron 射线测厚仪，射线穿透仪
peninsular quay 半岛式码头
peninsula 半岛
pen loading 记录笔压力
pen movement assembly 笔移动组件
penny crack 饼状裂纹
pen recorder 笔式画线记录仪，笔式记录仪
pensions 退休金
pen stiffness 记录笔稳定性
penstock bracing 压力管道支撑
penstock bypass chamber 压力管道旁通管室
penstock drain header 压力管道排水总管
penstock gate 压力管道闸门
penstock manifold 压力管道分岔管
penstock on downstream face of the dam 坝后背管，坝下游面管
penstock section 压力管道断面，压力管道段
penstock valve 压力钢管阀
penstock 救火龙头，给水栓，高压管道，压力管道，水渠，压力水管，水门，消防栓
pentachlorophenol 五氯苯酚
pentad 五价元素，五价物
pentane 戊烷
pentatron 五极电子管
pentavalence 五价
pen tension 记录笔张力
pen tester 试电笔
penthouse on boiler roof 炉顶小室
penthouse 披屋，炉顶小室，炉旁小室，屋顶房间，楼顶房屋
pentice 屋顶房间
pentode flip-flop 五极管触发器
pentode transistor 晶体五极管
pentode 五极管
pen travel 记录笔行程
pent-roof 单坡屋顶
pen-type dosimeter 笔式个人剂量计
penultimate stage blade 次末级动叶片
penultimate stage 倒数第二
pen-writing recorder 笔录器
pen 笔，记录笔
people outside authorized personnel quota 编外人员
people quality 人员素质
people 人员
PEPCO(Pakistan Electric Power Company) 巴基斯坦国家电力公司，巴基斯坦国家电力局
Pe(Peclet number) 贝克列数【等于雷诺数乘以普朗特数】
PE＝polyethylene 聚乙烯
PE(power exchange) 电力交易（所）
PEP(peak envelope power) 包络峰值功率
peptize 胶溶

perambulation 查勘，巡视，测距仪
per annum 每年，按年计算，按年
per capita dose equivalent 人均剂量当量
per capita dose 人均剂量
per capita income 人均收入
per capita 人均
perceived noise level 感觉噪声级
perceive 意识到，认识到，察觉，发觉，理解
percentage absolute humidity 绝对湿度百分数
percentage auxiliary power 厂用电率
percentage by volume 容积百分率，体积百分比
percentage by weight 重量百分率
percentage composition 百分含量，含量百分比，成分百分比
percentage concentration by volume 容量百分比浓度，重量百分比浓度
percentage concentration 百分浓度
percentage coupling 耦合系数
percentage depth dose 百分深度剂量，纵深剂量百分数
percentage differential current 比率差动电流
percentage differential protection 比率差动保护
percentage differential relay 比率差动继电器，百分数差动继电器
percentage duty cycle 工作周期百分数
percentage elongation after fracture 残余延伸百分率，断后伸长率
percentage elongation of stress-rupture 持久断后伸长率
percentage elongation 相对伸长，延伸率，伸长率，延伸率百分数
percentage error 百分比误差，百分误差，误差百分数
percentage extraction 萃取百分率
percentage fee 百分比费用
percentage harmonic content 谐波含量百分数
percentage humidity 湿度百分数
percentage line drop 线路压降率
percentage loss 损失率，损失百分数
percentage method 百分率法
percentage modulation 调制深度，调制百分数
percentage of air space 通风截面比
percentage of articulation 清晰度
percentage of block 阻塞度
percentage of capacitance variation with temperature 电容温度变化率
percentage of consolidation 固结百分率，压缩百分率
percentage of core recovery 岩心采取百分率
percentage of economizer evaporation 省煤器蒸发率，省煤器沸腾率
percentage of elongation 延伸率，伸长率
percentage of excess air 过剩空气系数
percentage of fines 煤粉含量百分比
percentage of maximum momentary speed variation 升速率，最大瞬时速度变化率
percentage of moisture 含水率
percentage of pole embrace 极弧系数
percentage of possible sunshine 日照率
percentage of practical completion 实际完成百分比
percentage of reinforcement 配筋率，钢筋百分率，含筋率【混凝土】
percentage of retention 预留额
percentage of return air 回风百分比
percentage of saturation 饱和率
percentage of shale content 含矸率
percentage of speed rise 转速上升率
percentage of sunshine 日照百分率
percentageof turbulence-intensity 湍流度
percentage of voids 空隙率
percentage reduction of area of stress rupture 持久断面收缩率
percentage reduction of area 断面收缩率
percentage reduction 去除百分数
percentage relay 百分比继电器
percentage reserve 备用功率百分数，备用率
percentage ripple 脉动百分数
percentage slip 转差率
percentage slope 坡度百分比
percentage speed variation 速度变化率
percentage test 含量百分数测定
percentage 百分比，百分率
percent break 脉冲断开时间百分率
percent by volume 体积百分数
percent by weight 重量百分数
percent coarser 超径的百分数
percent conductivity 导电率百分数
percent consolidation 固结百分数，固结度
percent defective 废品率
percent extraction 萃取百分率
percent fines 细骨料百分比，细粒含量百分率
percent harmonic distortion 谐波畸变率
percent make 脉冲持续时间百分率
percent moisture content 含水率
percent moisture 含水量百分数
percent of accuracy 准确度，精度
percent of pass 合格率，通过率
percent passing 过筛百分率
percent pitch 节距系数
percent pole-embrace 极弧系数
percent ratio error 变比误差率，变比误差百分数
percent reactance 电抗百分数
percent regeneration 再生度，百分再生度
percent ripple 波纹百分比
percent sand 含沙率
percent shear fracture 结晶率，塑性断口百分率
percent test 抽查，抽样试验
percent transient deviation 暂态差百分数
percent transmission 透光百分率
percent unbalance of phase voltage 相电压不平衡百分数
percent 百分比，百分率，百分数
perceptibility 感觉力，觉察力
perceptible 可感觉的
perception threshold 可感觉阈
perception 视觉过程，感知过程，知觉，感觉，觉察（力），观念
perceptual 感性的
perched aquifer 栖留含水层，表层含水层
perched groundwater 滞水

perched subsurface stream 地下栖留河
perched water table 上层滞水潜水位
perched water 上层滞水，滞水
perchlorate 高氯酸盐［酯］
perchloric acid 高氯酸
perchloride 高氯化物
perchloronated vinyl paint 过氯乙烯漆
perchlorovinyl lacquer 过氯乙烯树脂清漆
perchlorovinyl resin 过氯乙烯树脂
perchlorovinyl 过氯乙烯
perch 连杆，主轴，栖息
percolate 滤过液，渗出液，渗沥液，过滤，渗出，浸透
percolating bed 渗滤床
percolating filter 渗透滤层
percolating gas 渗滤气体
percolating water 渗水
percolation amount 渗入量
percolation basin 渗滤池
percolation bed 渗滤床
percolation coefficient 渗透系数
percolation filter 渗滤器
percolation friction 渗透摩阻力
percolation gauge 测渗计，渗漏计
percolation line 渗流线
percolation path 渗透途径
percolation pit 渗井，坑
percolation pressure 渗透压力
percolation rate 渗漏率，渗透速率
percolation test 渗漏试验
percolation treatment 渗滤处理
percolation water 渗滤水，滤过水，渗漏水
percolation well 渗滤井
percolation zone 渗漏带
percolation 渗，渗透，渗滤，渗流，渗滤作用，渗出液
percolator 渗滤器
PERCOS(performance coding system) 性能编码系统
percussion action 冲击作用，撞击作用
percussion borer 冲击钻，风钻
percussion boring 冲击钻孔，冲击钻探
percussion centre 撞击中心
percussion drilling 冲击钻探
percussion drill 冲击钻机，冲撞钻
percussion hammer 冲击锤
percussion riveting machine 风动铆钉机
percussion rotary drilling 冲击回旋钻【钻探用】
percussion tool 冲击工具
percussion wave 冲击波
percussion weld 冲击焊接
percussion 冲击，撞击，震动
percussive action 冲击作用
percussive force 撞击力
percussive rig 冲击钻机
per diem 按日计
perdurability 持久性，延续时间
peremptory provision 强制性规定
peremptory regulation 强制性规章
peremptory rule 强制性规则
perennial base flow 常年基流
perennial overdraft 常年过量抽取【地下水】
perennial plant 多年生植物
perennial river 长流河，常年河
perennial stream flow 常年流量
perennial stream 常年河流
perennial yield 常年产水量【地下水】
perennial 常年的，终年的，长久的，多年生的，不断生长的，四季不断的
perfect alignment 精确对准
perfect black body 理想黑体，完全黑体
perfect code 完备码，理想码
perfect combustion 完全燃烧，完全竞争
perfect competitive market 完全竞争市场
perfect condition 理想状态
perfect conductor 理想导体，全导体
perfect crystal 完整晶体
perfect dielectric 理想电介质
perfect differential 全微分
perfect diffusion 完全扩散，完全漫射
perfect dust collection 完全除尘
perfect ejection 完全卸料
perfect elasticity 完全弹性
perfect fluid 非黏性流体，理想流体，完全流体
perfect frame 完整的框架
perfect gas aerodynamics 理想气体动力学
perfect gas 理想气体
perfect integrator 理想积分器
perfect liquid 理想液体，完全液体
perfect living facilities 完善的生活设施
perfectly elastic body 完全弹性体
perfectly mixed reactor 全混反应器
perfect magnetic conductor 理想导磁体，全导磁体
perfect matching 完全匹配
perfect medium 理想介质
perfect mixing 充分混合，完全混合
perfect plasticity 完全塑性，理想塑性
perfect reflection 理想反射
perfect solution 理想溶液
perfect steam turbine 理想汽轮机
perfect transformer 理想变压器
perfect vacuum 理想真空，绝对真空
perfect weir 完全溢流堰
perfect 完善的，完美的
perforated baffle plate 多孔消力板
perforated baffle 穿孔挡板
perforated brick 多孔空心砖，多孔砖
perforated caission breakwater 带孔沉箱防波堤
perforated cap 有孔帽罩
perforated ceiling air supply 孔板送风
perforated cellular brick 多孔空心砖
perforated concrete tube 多孔混凝土管
perforated cover 多孔盖【通风型，即自然通风】
perforated distribution plate 多孔板
perforated distributor 多孔布风板，多孔分配板，多孔分配器
perforated dry pipe 多孔集汽管
perforated fins 穿孔肋片
perforated grill 多孔网格
perforated head 有孔封头
perforated orifice plate 多孔节流板

perforated panel air outlet　穿孔板送风口
perforated panel　多孔板
perforated pipe　多孔管，花管
perforated plate　多孔板，孔板
perforated sheet　多孔板
perforated shrouds of chimney　多孔烟罩
perforated stone　多孔石
perforated strainer　多孔滤管
perforated tape reader　穿孔带读出器
perforated tape　穿孔带
perforated throat wind tunnel　开孔壁风洞【试验段壁有孔】
perforated tray　多孔搭板，多孔托架
perforated wall　孔壁，有漏窗墙，花格墙
perforated wing flap　穿孔襟翼
perforate　穿过，贯穿，穿孔，打眼，凿孔
perforating machine　穿孔机
perforating tool　穿孔工具
perforation rate　穿孔速率
perforation　打孔，冲孔，穿孔，孔眼
perforator　穿孔机，打孔机，穿孔器
perform a contract　履行合同，执行合同
performance adjustment　性能调试
performance analysis and surveillance system　性能分析与监督系统
performance analysis　性能分析
performance and cost evaluation　性能价格比，性能成本估价，性能与成本估价
performance assurance　性能保证
performance-based regulation　绩效挂钩监管法
performance bond　履约保证（金），履约担保，履约保证书，履约保函
performance build-up　性能提高
performance buydowns　履约不足赔偿金
performance calculation　特性计算，性能计算，工况计算
performance capability　（系统或装置的）工作能力
performance certificate　性能证书
performance characteristics of turbine plants　汽轮机组的动力特性
performance characteristic　性能特性，工作特性，运行特性，性能表征参数，操作特性
performance chart　工作特性图，性能图，特性曲线，性能曲线
performance coding system　性能编码系统
performance coefficient　性能系数
performance criterion　特性准则，性能指标，性能标准
performance curve　特性曲线，性能曲线，运行特性曲线
performance data　工作数据，功能数据，性能数据，运行数据
performance degradation assessment　设备性能退化评估
performance degradation data　性能退化数据
performance degradation model　性能劣化模型
performance degradation prediction　设备性能退化预测
performance degradation　性能退化
performance deterioration alarm　性能恶化报警
performance deviation　性能偏差
performance diagram　工况图，性能图，特性图，性能曲线
performance engineering　政绩工程，性能工程，（核电厂）运行性能分析
performance envelope　性能包络线
performance equation　特性方程
performance estimation　性能估计，性能估算
performance evaluate　性能估算，性能评价，性能评定
performance figure　性能数据，质量指标
performance function　功能函数
performance goal　性能目标
performance guarantee conditions　履约保函的条件
performance guarantee test　性能保证试验，性能测试
performance guarantee value　性能保证值
performance guarantee　履约保证（金），履约保证书，履约保函，履约担保，性能保证，性能担保
performance improvement engineering　性能改进工程
performance index　性能指标，性能指数，效能指数
performance indicator　性能指标
performance inspection　性能检查
performance line　特性线
performance measurement　性能测定，工况测定
performance monitoring　性能监察
performance monitor　性能监测器
performance number　特性数，功率值
performance of contract　履行合同
performance optimization　性能最佳化
performance parameter　性能参数
performance period　运行时间，运行期，执行周期
performance rate　操作速度
performance rating　运行定额
performance report　性能报告
performance security　履约保证（金），履约担保，履约保证书，履约保函
performance specification　性能规范，运行特性
performance standard　实施标准
performance test center　性能测试中心
performance test　特性试验，性能测试，性能试验，运行试验
performance warranty　性能保证【电站】
performance　工作性能，运行特性，运行，实行，性能，表现，履行，工况，实施
performed spiral line　预制螺旋形阻尼线
performer　执行器，执行者
performeter　自动调谐的控制谐振器，工作监视器
perform switch check　执行开关检查
perfo-rockbolt　多孔岩栓
per hour　每小时
perhydrol　强双氧水
periclase　方镁石
periclasite　方镁石
peridinetic flocculation　异向絮凝作用
peridotite　橄榄岩

perihelion 近日点
perikon detector 红锌矿检波器，双晶体检波器
peril clause 危险条款
perimeter beam 圈梁
perimeter effect 周边影响，周界影响，周界效应
perimeter grouting 周界灌浆
perimeter shear 环向剪力，周界剪力
perimeter trench 围护沟
perimeter 视野计，圆周，周边，周长，周缘
perimetric joint 周边缝
period amplifier 周期放大器，周期测量计放大器
period-and-level start-up control 按周期和功率的启动控制
period change 周期变化
period channel 周期计通道
period commitment fee 期间承诺手续费
period control 周期控制
period demand 给定周期
period from synchronisation to the completion of commissioning 从并网到调试结束期间
period guaranteed free maintenance 免维修保证期
period hours 统计期间小时数
periodical flood 周期性泛滥
periodical inspection 定期检查
periodically operated switch 周期动作开关
periodical peak load 周期性峰荷
periodical 循环的，周期的，定期的，间歇的，断断续续的
periodic analysis 定期分析
periodic attenuation 周期衰减
periodic blowdown flash tank 定期排污扩容器
periodic blowdown flash vessel 定期排污扩容器
periodic blowdown 定期排污
periodic boundary condition 周期性边界条件
periodic calibration 定期校准
periodic cavitation 周期性汽蚀，间断汽蚀
periodic cavity 周期性空蚀
periodic change 周期变化
periodic chart 周期表
periodic check 定期校验
periodic chemical deactivation 定期化学去活化
periodic circuit 周期性电路
periodic clean 定期清理，定期清洗
periodic component 周期分量
periodic current 周期电流，周期流，周期性流
periodic curve 周期曲线
periodic damping 周期性制动，周期性阻尼，欠阻尼，弱阻尼，周期阻尼
periodic density pattern 周期密度图形
periodic discharge 周期性放电
periodic drift 定期漂移
periodic dust dislodging 定期除灰
periodic duty service 定期运行
periodic duty 周期运行方式，周期性，周期性负载，周期性负荷
periodic electromotive force 周期电动势
periodic envelope 周期包迹
periodic examination 定期检查
periodic external force 周期性外力
periodic filter 周期频率特性滤波器
periodic flow 周期性流动
periodic fluctuating windspeed 周期性脉动
periodic force 周期力
periodic function 周期函数
periodic heat transfer 周期传热
periodic inspection 定期巡视，定期检查
periodicity 周期性，周期，频率，周波
periodic law 周期律
periodic line 梯形网络，链路
periodic load 周期性负载，周期性负荷
periodic log 定时打印，周期性记录
periodic maintenance 定期检修，定期维修，中修
periodic meeting 定期会议
periodic meteorological element 周期性气象因素
periodic motion 周期（性）运动
periodic observation 定期观测
periodic operation 周期（性）运行，定期运行
periodic oscillation 周期振荡
periodic overhaul 定期检修
periodic phenomena 周期性现象
periodic potential difference 周期电压，周期电位差
periodic potential 周期电势
periodic pulse train 周期性脉冲序列，周期性变化，周期脉冲列
periodic quantity 周期量
periodic rating 间断出力，周期性额定值
periodic repair 定期检查，定期检修，定期修理
periodic report 定期报告
periodic resynchronization 周期同步法，周期再同步法，循环同步法
periodic review 定期审查
periodic sampling 周期采样，周期抽样
periodic settling 周期性下沉
periodic shutdown 定期停用，间断停用
periodic solution 周期解
periodic swinging 周期性摆动，周期性振荡
periodic table 周期表，元素周期表
periodic temperature transient 周期性温度变化
periodic test 定期试验，周期性试验
periodic time 周期
periodic treatment 定期处理
periodic turbulent fluctuation 周期性湍流
periodic undulation 周期性波动
periodic unsteady flow 周期性不稳定流
periodic variation 周期（性）变化
periodic vibration 周期（性）振动
periodic visit to site 定期下工地
periodic voltage 周期电压，周期电动势
periodic vortex shedding 周期性旋涡脱落
periodic vortex street 周期性涡街
periodic wake 周变尾流
periodic wave 周期波
periodic 定期的，周期性的
period indicator 周期指示器
periodization 周期化
period load 周期性负荷
period meter 周期计，周波表
period monitor 周期监测器，周期计【反应堆】
period noise 周期计噪声
period of acceleration 加速度周期

period of advance notice 预先通知期限
period of amortization 折旧期
period of analysis 分析期限
period of commutation 换向周期
period of construction 建设期
period of cost recovery 成本回收期
period of credit 贷款期限
period of design 设计年限，设计进度
period of duty 运行时间，运行期
period of element 元素周期
period of flaming 火焰期
period of flood and ebb 涨落潮周期
period of full drawdown 全消落时期
period of guarantee 保证期
period of load development 负荷增长期
period of oscillation 振动周期，振荡周期，舞动周期
period of production 生产期限
period of quality guarantee 质量保证期
period of recovery 恢复时期
period of regulation 调节周期
period of responsibility 责任期限
period of retardation 滞洪期，滞留期
period of rotation 转动周期
period of service 使用期限，使用寿命
period of start-up 启动周期，启动时期
period of time 时段，时间段
period of validity 有效期
period of vibration 振动周期
period of wave 波动周期
periodogram 周期图
period protection 周期保护
period range 周期区段
period-reactivity conversion factor 周期-反应性转换因子
period regulation 周期调节
period regulator 周期调节器
period resonance 周期谐振，周期共振
period run 周期运行
period safety 周期保护
period shutdown 周期保护停堆
period trip 周期保护停堆
period 期间，时间间隔，周期，时期，循环，阶段，期
periophery velocity 圆周速度
periophery 圆周
peripheral air-gap leakage flux 周边气隙漏磁通
peripheral area 边缘区域，周边区域，外围区【电厂动力部分以外】
peripheral break 圆周破裂
peripheral buffer 辅助缓冲器，外围缓冲器，外围设备缓冲器
peripheral building 边缘厂房，周边厂房，外围建筑【电厂动力部分以外】
peripheral charging cone 周边加料锥体
peripheral clearance 周围间隙
peripheral command indicator 外围命令指示器
peripheral component of velocity 切向分速度，圆周分速度
peripheral configuration 外围配置
peripheral contact area （轴承）圆周接触面积
peripheral control program 外围控制程序，外部设备控制程序
peripheral devices 外部设备，外围设备
peripheral direction 圆周方向
peripheral equipment 外部设备，外围设备
peripheral flow 周缘流，圆周绕流，圆柱绕流
peripheral force 周边力，圆周力
peripheral gradient distribution 沿周梯度分布
peripheral gradient （输电线的）沿周梯度
peripheral interchange program 外围交换程序
peripheral interface adapter 外围接口适配器，外围接口转接器
peripheral interface channel 外围接口通道
peripheral joint of arch dam 拱坝周边缝
peripheral joint 周边缝
peripheral limited 外围限制
peripheral loading 周边荷载
peripheral pressure 周边压力
peripheral processing system 外围处理系统
peripheral processor memory 外围处理机存储器
peripheral processor system 外围处理机系统
peripheral processor unit 外围处理机组
peripheral processor 外围处理机
peripheral pump 涡流泵
peripheral ratio 缘速比
peripheral refueling location 周边换料位置
peripheral rooms 外围设备间
peripheral speed 周边速度，圆周速度，圆周速率
peripheral storage control element 外存储器控制部件
peripheral stress 周边应力
peripheral support computer 外围支持计算机
peripherals 外部设备，外围设备
peripheral transfer 外围设备信号传输
peripheral unit fault recognition 外围设备故障识别
peripheral unit processor 外围设备处理机
peripheral units 外部设备，外围设备
peripheral velocity 外缘速度，圆周速度
peripheral weir 周边出水堰
peripheral 边缘的，外部的，外围的，圆周的，周围的，周界的，周边的，次要的，外部设备，外围设备
peripheric velocity 圆周速度
periphery beam 圈梁
periphery of armature 电枢圆周
periphery 周边，周围，圆周，外围，圆柱表面
periscope range-finder 潜望测距仪
periscope 潜望镜
perishable 易腐
perisphere 球形建筑物
peristaltic pump 蠕动泵
peritectic crystal 包晶
peritron 荧光屏可轴向移动的阴极射线管
peri-urban area 城市周边地区，市区外围地带，半城市化地区
perking switch 快动开关，速断开关
perk 详细检查，渗透，动作灵敏，过滤，小费，额外待遇，额外收入
perlite concrete 珍珠混凝土

perlite products 珍珠岩制品
perlite 珍珠岩，珠光体
perlitic structure 珍珠构造，珠光体结构
perlitic texture 珍珠构造
permafrost area 永冻地区
permafrost ground 永冻地
permafrost layer 永冻层
permafrostology 冻土学
permafrost soil 永冻土（壤）
permafrost table 永冻层表面
permafrost 多年冻土，永冻，永冻土
permaliner 垫整电容器
permanence 永恒性，持久性，不变性
permanent action 永久作用
permanent address 永久地址
permanent agreement 永久性协定
permanent alkalinity 永久碱度
permanent architecture 永久建筑
permanent backing strip 永久性垫条
permanent backing 保留垫板【焊接】
permanent back water region 常年回水区
permanent bank protection 永久性护岸
permanent bench mark 固定水准点，永久水准点
permanent-capacitor motor 永久电容器式电动机，电容运行电动机
permanent cave 永久屏蔽室
permanent change 永久性更改
permanent circuit 固定电路
permanent circulation 永久环流
permanent closed contact 常闭接点，动断触点
permanent colour 耐久颜色
permanent connection 不可拆卸的连接
permanent construction 永久建筑
permanent control 永久控制
permanent current 常流，恒定电流，恒定流，恒流
permanent dam 固定坝，永久性坝
permanent deactivation 永久失活
permanent deformation 残余变形，永久变形
permanent delegate 常驻代表
permanent depository 永久处置库
permanent disposal 永久处置，最终处置
permanent distortion 永久变形
permanent droop 持久转差率
permanent earth 永久接地，固定接地
permanent elongation of overhead line 架空线永久性伸长
permanent elongation 持久延伸，永久伸长
permanent establishment 常设机构
permanent fault 永久性故障，长时事故
permanent feedback 固定反馈，刚性反馈
permanent filter 永久过滤器
permanent fission-product poisons 永久性裂变产物毒物
permanent fission-product 永久性裂变产物毒物
permanent flow 稳定流动，稳定流，定常流，稳流，定常流动，常年流量，常定流
permanent forced outage rate 持续强迫停运率
permanent formwork 永久模板，耐耗模板
permanent gauge station 永久水位站
permanent gauge 永久水尺
permanent geodetic beacon 永久大地觇标，永久性测量觇标
permanent groundwater runoff 永久性地下径流
permanent ground 永久接地的，固定接地的
permanent hardness （水的）永久硬度
permanent hard water 永久硬水
permanent increase of depth of indentation 残余压痕深度增量
permanent investment 长期投资，永久投资
permanent joint 不可拆卸连接，固定接点，固定连接，永久缝
permanent lining 永久性衬砌
permanent load 固定负载，恒定负载，持久负荷，永久荷载，连续负荷，恒载
permanently absorbed water 永久吸收水
permanently frozen soil 永冻土
permanently grounded 永久接地的，固定接地的
permanently installed dose rate meter 固定装置的剂量率计
permanently magnetic stator 永磁式定子
permanently occupied areas （人员）持久停留区域
permanent-magnet-actuated brake 永久磁铁驱动的制动器
permanent-magnet alloy 永磁合金
permanent-magnet direct current motor 永磁直流电动机
permanent-magnet excitation 永久磁铁励磁
permanent-magnet flowmeter 永磁型流量计
permanent-magnet generator 永磁发电机
permanent-magnetic field 永磁场
permanent-magnetic motor 永磁式电机
permanent-magnetic pole 永磁极
permanent-magnet instrument 永磁式仪表
permanent-magnetism 永磁
permanent-magnetization 永久磁化
permanent-magnet machine 永磁电机
permanent-magnet material 永磁材料
permanent-magnet meter 永磁式仪表
permanent-magnet motor 水磁电动机
permanent-magnet pole 永磁极
permanent-magnet synchronous generator 永磁同步发电机
permanent-magnet torque 永磁转矩
permanent-magnet 永磁的，磁性耐久的
permanent magnet 永久磁铁，永磁体，永久性磁石
permanent marking 永久性标志
permanent memory 永久存储器，固定存储器
permanent mould 金属模
permanent no-load speed （甩全负荷后）规定的空转转速
permanent outage 永久性停运
permanent output 恒定输出功率
permanent overspeed 终态超速
permanent perched groundwater 永久性地下上层滞水
permanent permeability curve 磁导率曲线
permanent plant 固定设备
permanent power 连续功率
permanent record 永久记录

permanent repair 大修
permanent repository 永久处置库
permanent reserve 永久储量
permanent sash 固定窗扇
permanent set strength 规定残余延伸强度
permanent set 永久变形，永久凝固，残留变形
permanent shuttering 永久性模板
permanent signal 永久标识
permanent site 永久场地
permanent software 固定软件
permanent speed change 整定速度变动率
permanent speed droop 固态转差系数，常态转速下降
permanent speed regulation 静态调速调节
permanent speed variation 整定速度变动率，稳态转速变化
permanent speed 正常运转速度，整定转速，稳定速度，不变的速度
permanent-split-capacitor motor 固定分相的电容器式电动机
permanent sprinkler system 固定式喷灌系统
permanent station services 全厂永久性共用设施
permanent steel shuttering 永久性钢模板
permanent storage 永久性存储器，固定存储器，永久储存，永久保存，永久贮存
permanent store 永久存储器，长久存储
permanent strainer 永久滤网
permanent strain 永久变形，残余变形，永久应变
permanent stream 常流河
permanent strength 持久强度
permanent structure 永久性建筑物
permanent support 永久支护，永久支架
permanent surveying marker 永久性测量标志
permanent varied flow 稳定变速流
permanent vegetation 永久植被
permanent water table 永久潜水位【指地下水位】
permanent wave 驻波
permanent wilting coefficient 永久枯萎系数
permanent works 永久性工程
permanent 不变的，永久的，持久的，恒定的，永恒的
permanganate 高锰酸盐
perma-temp 长期临时工
permatron 磁控管
permeability bridge 测量磁导率电桥
permeability coefficient 渗透系数
permeability curve 磁导率曲线
permeability measurement 渗透性测量
permeability of free space 真空磁导率
permeability of heat 导热性
permeability of permanent magnet 永磁铁磁导率
permeability of vacuum 真空磁导率
permeability rate 渗透率
permeability-reducing admixture 防渗剂，防渗外加剂
permeability-reducing agent 防渗剂
permeability test 渗透试验
permeability-tuned inductor 磁导率调谐电感线圈
permeability 导磁率，导磁性，透磁率，穿透性，透气性，渗透，渗透性，渗透率，透过性，磁导率
permeable breakwater 透水防波堤
permeable confining bed 透水隔层
permeable dam 透水坝，透水堤
permeable groin 透水丁坝
permeable joint 渗透缝，透水缝
permeable layer 透水层
permeable material 透水材料
permeable medium 渗水介质
permeable membrane 渗透膜
permeable pile dike 透水桩堤
permeable rock 透水岩石
permeable slab 渗透性平板
permeable soil 渗透性土，透水土壤
permeable strata 渗透层
permeable stratum 透水层
permeable structure 透水结构
permeable wave absorber 透水消波器
permeable works 透水工程
permeable 能透过的，可渗透的，透水的
permeameter 磁导计，渗透仪
permeance factor 磁导系数，磁导率
permeance 磁导率，磁导，渗透，透过
permeate 渗入
permeation flux 透水量
permeation tube 渗透管
permeation velocity 透水速度
permeation 渗透，渗入，渗气，渗透作用，浸透，透光，透过，充满
permeator 渗透器，渗透膜，渗透层，连通器
Permian period 二叠纪
Permian system 二叠系
permillage 千分率
permissible blasting material 安全爆破器材
permissible compression stress 许用压应力
permissible contamination 容许污染
permissible criterion 容许标准
permissible current 容许电流，允通电流，允许电流
permissible deviation 容许偏差
permissible discharge 容许排放
permissible dose 允许剂量，容许剂量，可接受剂量
permissible draft head 允许吸出高度
permissible draught head 容许吸出水头
permissible drawdown 工作深度【指水库正常蓄水位至最低水位的深度】，容许降落水位
permissible dynamite 安全炸药
permissible erratum 容许误差
permissible error 允许误差，容许误差，公差
permissible exposure 容许照射，容许照射量
permissible heat loss 允许热损失，允许水头损失
permissible limit 容许限度，容许极限
permissible load 许用负荷，容许载荷，允许负荷，容许负荷
permissible motor 密闭型电动机，防爆型电动机，防爆电动机
permissible negative phase sequence current 允许的负相序电流
permissible operating temperature 允许运行温度
permissible overload 容许过负载

permissible revolution　容许转速
permissible speed　容许速度
permissible storage battery locomotive　防爆式蓄电池电机车
permissible stress　许用应力，准许应力，许可应力，容许应力
permissible temperature rise　容许温升
permissible temperature　容许温度
permissible tensile stress　许用拉伸应力
permissible tension　许可张力
permissible time interval between placing layers　直接铺筑允许时间
permissible time interval between placing　加垫层铺筑允许时间
permissible tolerance　容许公差，可容许公差，容许耐力
permissible value　容许量，允许值，容许值
permissible variation　容许误差，容许偏差
permissible velocity　许可流速，安全流速
permissible voltage　允许电压
permissible　容许的，许可的，可接受的
permission flexibility　容许挠性
permission for leave　准假
permission　批准，许可，同意，准允
permissive condition　允许条件
permissive control device　允许式控制装置
permissive control　许可控制
permissive make contact　容许闭合触点
permissive protection　允许式保护
permissive provision　非约束性规定
permissive signal　允许信号
permissive　启动的必要条件，允许条件【复数】，允许的
permit issure　许可人，许可证颁发机构
permittance　准许，许可，容性电纳，电容
permitted assignee　允许受让人
permitted revenue　准许收入
permittivity of medium　介质的介电常数
permittivity of vacuum　真空介电常数
permittivity　介电常数，电容率
permit　许可，认可，允许，批件，许可证，执照，容许，核准，同意
permometer　连接雷达回波谐振器用的设备
permosyn motor　永磁同步电动机
permselectivity　选择渗透性
permutability　可置换性
permutation code　互换码
permutation decoding　置换解码
permutation matrix　置换矩阵
permutation　置换，互换，交换，排列
permutator　转换开关，变换器，机械换流器
permuted code　交换码，变换码，置换码
permute　变更，置换，交换，排列
permutite　人造沸石，滤砂，沸石
peroxidation　过氧化反应
peroxide compound　过氧化物化合物
peroxide of hydrogen　过氧化氢
peroxide　过氧化物
peroxyuranate　过铀酸盐，过铀酸根
perpendicular bisector　垂直等分线，中垂线
perpendicular deviation　垂直偏差
perpendicularity　垂直性，正交，直交
perpendicular layout　垂直式布置
perpendicular line　垂（直）线，正交线
perpendicular magnetization　垂直磁化
perpendicular of velocity　速度变动
perpendicular plane　垂直面
perpendicular recording　垂直记录
perpendicular to the principal direction of working　横向垂直于工作的主方向
perpendicular　垂线，垂直，正交，垂直的，正交的，成直角的
perpetual cycle　永恒循环
perpetually frozen soil　永冻土（壤）
perpetual mobile　永动机
perpetual storage　最终贮存
perpetual　永恒的，永久的，不间断的
perrotation　（气流）预旋
per season　每季的
Persian windmill　波斯（竖轴）风车
persistence of energy　能量守恒
persistence of light radiation　余辉
persistence　坚持，持续，持久性，稳定性，余辉，持久，住留
persistent-cause forced outage　持久性强迫断电
persistent command　持续命令
persistent current memory cell　持续电流存储元件
persistent current　持续电流
persistent fault　持久故障
persistent forced outage duration　持久强迫停运持续时间
persistent forced outage　持久强迫停运
persistent information　持续信息
persistent oscillation　持续振荡
persistent phosphor　余辉荧光体，余辉磷光体
persistent pollutant　持久性污染物
persistent radiation　持久照射
persistent regulation command　持续调节命令
persistent substance　持久性物质
persistent wave　等幅波
persistent wind　持续风
persistent　持续
persistron　持久显示器
personal air lock　人员气闸【反应堆厂房】，人行通道阀
personal air sampler　个人空气取样器，工作人员空气取样器
personal attributes and competencies　个人特质与能力方面的要求【职位招聘】
personal attributes　个人素质，个性特质，个人属性
personal communication net　个人网或个人通信业务
personal communication system　个人网或个人通信业务
personal communications　个人通信
personal computer　个人计算机
personal decontamination　人员去污
personal dose equivalent　个人剂量当量
personal dosemeter　个人剂量计
personal dose monitoring　个人剂量监测

personal dosimeter 个人剂量计，个人剂量仪
personal dosimetry 个人剂量学，个人剂量测量
personal equation 人差方程，人为误差，个人在观察上的误差，个人倾向，个性
personal error 人为误差，个人误差
personal exposure meter 个人曝光表，个人照射量计
personal exposure 个人照射量
personal external exposure monitor 个人外照射监测器
personal factor 人为因素
personal holding company 私有股控公司
personal identification code 个人识别码
personal income tax 个人所得税，个人进口税
personal injury 人身事故
personal insurance 人身保险
personality 品格
personal monitoring 人工监测，个人监测
personal monitor 人工监测仪，人员监视，个人监测器
personal on call 值班应名人员【随叫随到】
personal photographic dosimeter 个人摄影计量仪，胶片计量计
personal property 个人财产，动产
personal proprietor 独资企业，私人业主
personal protection 个人保护，个人防护
personal radiation monitor 人员辐射监测仪
personal radio 小型收音机
personal requisition 人员申请
personal selection 人员选择
personal subsidy 个人津贴
personal television 小型电视机
personal understanding 切身体会
person directly engaged in radiation work 直接参与放射性工作的人员
person in charge of project 工程主管
person in charge of the work 工作负责人
person in charge 负责人
personnel access hatch 人员进出舱
personnel access lock 人行通道闸
personnel air lock 人行通道气闸
personnel automatic data system 人事自动数据系统
personnel basket 吊篮
personnel checkpoint （控制区进出口）人员检验点，人员检查站
personnel decontamination 人员去污
personnel dose meter 个人剂量计
personnel dosimeter 个人剂量计
personnel hatch 人员进出舱口
personnel hazard 对工作人员的危害
personnel information 人事资料
personnel management 人事管理
personnel protection insulation 防烫伤隔热
personnel quota 职工定员
personnel recruitment 人才招聘
personnel training 人才培训，人员培训
personnel 工作人员，全体人员，人员
person not directly engaged in radiation work 不直接参与放射性工作的人员
person on duty 值班人员
person on probation 试用人员
person-time 人次
person who signs the work-sheet 工作票签发人
perspective axis 透视轴
perspective centre 透视中心
perspective drawing 透视图
perspective projection 透视，投影
perspective rendering 透视渲染
perspective representation 透视图表示
perspective sketch 透视草图
perspective view 透视图
perspective 透镜，透视，透视图，立体透视图，远景
perspectivity centre 透视中心
perspex sheet 有机玻璃板
perspex 防风玻璃，有机玻璃
perspiration 渗出，出汗
persuade 说服，劝说
persuasive 有说服力的
persulfate 过硫酸盐
pertaining to 关于，与……有关，适合，属于
pertaining 附属的，与……有关的
pertain to 属于
pertinax 胶纸板，酚醛塑料
pertinence 适当
pertinent boundary curve 对应相分界曲线
pertinent certificate 有关证件
pertinent detail 有关的细节
pertinent to 与……相关的，与……有关的
pertinent 相关的，相干的，中肯的，切题的，有关的，恰当的，关于……的
PERT(program evaluation and review technique) 计划评审技术，程序评价和审查技术，统筹法
pert time schedule 关键路线进度表
perturbable 易被扰动的
perturbance 扰动，干扰
perturbation analysis 小扰动分析，摄动分析
perturbation approximation 微扰近似
perturbation calculation 微扰计算
perturbation coefficient 扰动运动方程系数
perturbation equation 微扰方程，扰动方程
perturbation frequency 微扰频率
perturbation function 扰动函数
perturbation method 小扰动法，摄动法
perturbation of temperature 温度波动
perturbation of velocity 速度变动
perturbation theory 微扰理论，摄动理论，扰动理论，干扰理论
perturbation velocity 扰动速度
perturbation 干扰，波动，扰动，摄动，微扰，不安，扰乱，紊乱，小扰动
perturbed flow 受扰流动
perturbed flux 受扰动的通量
perturbed neutron fluence 微扰中子注量率
perturbed periodic potential 受扰周期势场，受扰周期位场
perturbed 受扰的
perturbographic system 扰动记录系统
perturb 扰乱，烦扰，干扰，使……混乱，使……不安
per unit admittance 每单位导纳，导纳标幺值

per unit calculation 标幺值计算
per unit overvoltage 过电压标幺值
per unit quantity 标幺值
per unit reactance 电抗标幺值
per unit slip 滑差，滑差率
per unit synchronous reactance 单机同步电抗
per unit system 相对值系统，标幺值制，标幺值系统
per unit value 标幺值，相对值，单位值，每单位，每定位
pervade 扩大，蔓延，充满，渗透，普及
perveance 导流系数，电子管导电系数
pervibration 内部振捣
pervious base 透水地基，渗透层，透水层
pervious blanket 透水铺盖
pervious concrete 透水混凝土
pervious course 透水层
pervious foundation 透水地基
pervious material 透水材料
perviousness 渗透性，透水性
pervious overburden 透水覆盖层
pervious rock stratum 透水岩层
pervious sand gravel 透水砂砾石
pervious shell 透视坝壳，透水坝壳
pervious soil 透水性土壤，透水土壤
pervious stratum 透水层，透水地层
pervious subsoil 透水下层土
pervious zone 透水带，透水区
pervious 透水的，可渗透的，能通过的，有孔的
per-word error probability 单字误差概率
pessimistic 悲观的，保守的
pesticide 杀虫剂
pestle 研杵
pestling 研磨
pest 害虫，灾害
pet cock 小龙头，泄放阀，小旋塞，空气阀
petitioner 申请人
petition in bankruptcy 破产申请
PET(physical equipment table) 物理装置表
PET(position-event-time) 位置、事件、时间
petrifaction 化石
petrochemical industry 石油化学工业，石油化工
petrochemical 石化的
petrochemistry 石油化学
petrofabric analysis 岩组分析
petrofabric diagram 岩组图
petrofabrics 岩组学
petrogenesis 岩石成因论
petrographical examination 岩石鉴定
petrographic analysis 灰渣特性分析
petrographic classification 岩石分类
petrographic composition 岩石成分
petrographic facies 岩相
petrographic microscope 岩石显微镜
petrography 岩类学，岩相学
petrolene 沥青脂
petrol engine 汽油内燃机，汽油发动机
petroleum and natural gas development 石油及天然气开采
petroleum asphalt 石油沥青
petroleum bitumen 石油沥青
petroleum coke 石油焦炭
petroleum ether 石油醚
petroleum fuel 石油燃料
petroleum gas oil 石油气体油，石油气
petroleum industry 石油工业
petroleum jelly 矿脂，凡士林
petroleum microorganism 石油微生物
petroleum oil flux 重油
petroleum oil 石油油料
petroleum pith 石油沥青
petroleum products 石油产品
petroleum refinery 石油加工工厂
petroleum spirit 汽油，石精油，石油溶剂，汽油
petroleum waste 石油废水
petroleum 石油，原油
petroliferous shale 油页岩
petroliferous 含石油的
petrolift 油泵，燃料泵
petrol level gauge 汽油油位表
petrology 岩石学
petrol pressure gauge 汽油压力表
petrol station 加油站
petrol 汽油
petrotectonics 岩石构造学
petticoat insulator 裙状绝缘子
petticoat of insulator 绝缘子外裙
petticoat 裙状物，裙状绝缘子，外裙【绝缘子】
PEV(pyroelectric vidicon) 热电光导摄像管
PFBC boiler 增压流化床燃烧锅炉
PFBC-CC (pressurized fluidized bed combustion-combined circulation) 增压流化床联合循环
PFBC-CC unit (pressurized fluidized bed combustion-combined circulation unit) 增压流化床联合循环机组
PFBC(pressurized fluidized bed combustion) 增压流化床燃烧（技术）
PFCs 全氟化合物
PFD(probability of failure on demand) 要求故障概率
PFD(process flow diagram) 工艺流程图
PFI (private finance initiative) 私人融资计划，PFI模式，私人主动融资模式【工程建设模式】
PFI pulse frequency input 脉冲频率输入
PFI purchase mode 主动融资采购模式
PFK(programmed function keyboard) 程序控制的操作键盘
PFM(power factor meter) 功率因数计
PFM(pulse frequency modulation) 脉冲频率调制，脉冲调频
PF(power factor) 功率因数
PFR(power fail recovery system) 电源失效恢复系统
PFR(prototype fast reactor) 原型快中子反应堆
PGAC(professional group-automatic control) 自动控制专业组
PGCC(power generation control complex) 发电综合控制
PGCS (professional group-communication system) 通信系统专业组
PGD(pulse generator display) 脉冲发生显示
PGDS(pressurized gas distribution system) 加压

气体分配系统
PGEC(professional group-electronic computer) 电子计算机专业组
PGM(path generation method) 通路形成法
PGM(project gross margin) 项目毛利
PG(pulse generator) 脉冲发生器
PGR(precision graph recorder) 精密图示记录器
PGS(plant gas systems) 电厂供气系统
PGS(program generation system) 程序生成系统
PGT(performance guarantee test) 性能测试
pH adjustment tank pH 调整槽
pH analyzer pH 表，pH 分析仪
phanerite 显晶岩
phanotron 热阴极充气二极管
phantastron 幻象多谐振荡器，幻象延迟电路，延迟管
phantom circuit 幻象电路
phantom connection 模拟电路接线，幻象电路接线
phantomed cable 幻象线路电缆
phantom loading method 虚负荷法，虚加载法
phantom load 虚负载，假想负载
phantom powering （电容话筒的）幻路供电
phantom 幻象，假象，模型，体模，空幻的
phantophone 幻象电话
pharmaceutical industry 制药工业
pharmaceutical powder 药粉
pharmaceutical 药物的，药物
pharmacy 药房
phase adapter 相位变换适配器，换相器
phase address system 阶段地址系统
phase adjuster （换向器式）进相机，相位调整器
phase adjusting circuit 相位调整电路
phase adjustment 相位调整
phase-advance network 相位超前网络
phase advancer 进相机，相位超前补偿器
phase advance schedule 向前一班倒班
phase-advance 相位超前
phase analysis 相分析
phase angle corrector factor 相角校正系数
phase angle difference 相角差
phase angle error 相角误差
phase angle indicator 相角指示器
phase angle measuring or out-of-step protective relay 相角测量或失步保护继电器
phase angle regulator 相角调节器
phase angle synchronizing 相角同步
phase angle voltmeter 相角电压表
phase angle 相角，相位差，相移角，相位角
phase area integral 相面积分
phase area law 相面积定律
phase axis 相轴线
phase-balance current relay 相平衡电流继电器
phase-balance relay 相平衡继电器
phase balancer 相位平衡器，相平衡
phase bank （变压器的）相组
phase belt 相带
phase bit 定向位
phase boundary 相界
phase bus 相母线
phase cancellation 相消效应
phase change materials 相变材料
phase change-over switch 相转换开关
phase changer 变相器，变相机，换相器
phase change section 相变区
phase change solar system 相变太阳能系统
phase change 相变，换相，相位变化，变相
phase characteristic 相位特性，相频特性，相特性
phase checkup 相量检查
phase code modulator 相码调制器
phase coefficient 相位系数
phase-coherent 相位相参的，相位相干的
phase coil insulation 相线圈绝缘，相间绝缘
phase coil 相位线圈
phase coincidence 相位一致
phase comparator 相位比较器
phase comparison angle-selection 相位比较的角差选择器
phase comparison carrier differential system 相位比较式载波差动保护系统
phase comparison carrier 相位比较载波
phase comparison circuit 相位比较电路
phase comparison protection system 相位比较保护系统
phase-comparison protection 相位比较保护，相差保护
phase comparison relay 相位比较式继电器，相差继电器
phase comparison system （继电保护的）相位比较制
phase comparison 相位比较
phase compensation battery 相位补偿电池
phase compensation 相位补偿
phase compensator 相位补偿器
phase computation 相位计算
phase conductor 相导线，相导体，相线
phase connector 相间连接线
phase constant 相位常数，周相常数，相移常数
phase contacting area 相分界面
phase contours 等相角线，等相线
phase contrast technique 相衬技术
phase contrast 相差衬托
phase control apparatus 相控电器，相控设备
phase control circuit 相控电路
phase control factor 相控因数
phase-controlled motor 相控电动机
phase-controlled rectifier 相位可控整流器
phase-controlled rectifying circuit 相控整流电路
phase control refractometer 相位对比折射计
phase control 相位控制，相位调整，相控
phase conversion 相态转换，换相
phase converter 变相机，相位变换机，相位变换器，换相器
phase correction 相位校正
phase corrector 相位校正器，相位校正电路
phase cross-over frequency 相位交叉频率，相位交越频率
phase crossover 相位交点
phase current 相电流
phase curve 相位曲线

phased array radar　相控天线阵雷达
phase defect　相位差，相位亏损
phase delay schedule　顺序倒班
phase delay　相位延迟
phase demodulator　相位解调器，鉴相器
phase-detecting element　相位检测元件，相敏元件
phase-detecting　检相
phase detection　相位检波
phase detector　鉴相器，相位检波器，同步指示器，相探测器
phase deviation　相偏差，相位移
phase diagram　相图，平衡图，金相图
phase-difference indicator　相差指示器，相位差指示器
phase difference　相差，相位差
phase discriminator　鉴相器
phase disengagement　相分离
phase-displaced current　相位移电流
phase-displacement angle　相位移角，失配相角
phase-displacement error　相移误差
phase-displacement regulation　相位移调整
phase-displacement　相位差，相位移
phase distortion　相位畸变，相位失真
phase-down　停止，关闭
phase drift　相位漂移
phased　调相的，定相位的
phase encoding　相位编码
phase equality control　同相位控制
phase equalization　相位改正，相位均衡，相位稳定
phase equalizer　相位均衡器
phase equilibrium computer　相平衡计算机
phase equilibrium　相位平衡，相平衡，相平衡态
phase error　相角误差，相位误差
phase extension　一期扩建
phase factor　相位因子，相位因数
phase-failure protection　相故障保护，断相保护
phase failure　相故障
phase fault relay　相故障继电器
phase fault　相间短路故障，相故障
phase flag　相位标志
phase flow　（液体或气体的）相流
phase fluctuation　相位波动
phase-frequency characteristic　相位-频率特性
phase-frequency distortion　相频畸变
phase-frequency response　相位-频率响应
phase-frequency　相频
phase front　相位波前，波阵前
phase grounding　相接地
phase hologram　相位全息图
phase hunting　相位摆动
phase imbalance　相间不平衡
phase indicator　相位表，相位指示器
phase-in period　准备阶段
phase-insensitive　相钝的，对相位变化不灵敏的
phase insulated busbar　相绝缘母线，分相绝缘母线
phase insulation　相间绝缘
phase interchange rate　相间交换率
phase inversion theory　相变理论
phase inversion　倒相，相位倒置，相变化
phase inverter　倒相器
phase-in　（分阶段）引入，投入，启用，导入，同步，逐渐采用
phase jitter　相位抖动
phase lag control　相位滞后控制
phase lag　相位滞后
phase lamp　相位指示灯
phase leading　相位超前
phase-lead network　相位领先网络
phase-lead　相位领先，相位超前
phase limitator　相位限制器
phase line　相线，相位线
phase load　相负载，相负荷
phase lock circuit　锁相电路
phase-locked loop　锁相环，锁相回路，锁相环路
phase-locked oscillator　锁相振荡器
phase-locked　锁相的
phase lock　相位同步，锁相，锁相的
phase locus　相位轨迹
phase loss　缺相
phase margin　相补角，相位余量，相角裕量，稳定界限
phase mass　相位量
phase match　相位匹配
phase measuring equipment　相位计，测相设备
phase meter　相位表，相位计
phase method　相位法
phase modifier　整相器，调相器，调相机
phase-modulated signal　调相信号
phase-modulated　调相的，相位调制的
phase modulation　调相，相位调制
phase modulator　相位调制器，调相器，调相机
phase monitor　相位指示器，相位计
phase null detector　相位零值检波器，零相位指示器
phase number　相数
phase of echo　回波相位
phase of exploration　勘察阶段，勘探阶段
phase of line　线相位
phase of reflection coefficient　反射系数的相位
phase of wave　波相
phase-only hologram　纯相位全息图
phase opposition　反相
phase-order impedance　相序阻抗
phase-order indicator　相序指示器，相序表
phase order　相序
phase-out　中止，逐渐停止，不再使用
phase partition　相间隔板
phase pattern　相位图
phase plane　相平面，相位面
phase plate　相序牌，相移片
phase position　相位
phase pushing figure　相位偏移值
phase quadrature　90°相位差，90°相移，转像相差，转象相差
phase regulated rectifier　可控相位整流器
phase related harmonics　相别有关谐波
phase relation　相位关系
phase relay　相继电器

phase reserve 相位余量
phase resistance 相电阻
phase resolution 相位分辨力
phase resolver 分相器
phase resonance 相位共振
phase response 相位响应
phase reversal protection 反相保护，倒相保护
phase reversal transformer 倒相变压器
phase reversal 倒相，反相
phase reverser 倒相器，反相器
phase-reversing connection 反相连接
phase-reversing switch 倒相开关
phase-rotation circuit 相位回转电路
phase-rotation indicator 相序指示器
phase-rotation relay 相序继电器
phaser-reference alternator 参考相位发电机，基准相位同步发电机
phase rule 相律
phaser 相位计，相位器
phase segregated terminal box 隔相接线盒
phase segregation 相离析，相分离
phase-sensitive keyed detector 开关鉴相器
phase-sensitive rectifier 相敏整流器
phase-sensitive switch 相敏开关
phase-sensitive system 相敏装置
phase-sensitive 对相位改变灵敏的，相敏的
phase sensitivity 相位灵敏度
phase separator 相间线圈垫块
phase sequence component 相序分量
phase sequence current 相序电流
phase sequence detector 相序检测器
phase sequence direct order 相序正序列
phase sequence indicator 相序指示器
phase sequence inverse order 相序反序列
phase sequence meter 相序表
phase sequence relay 相序继电器
phase sequence reversal protection 反相保护
phase sequence sensor 相序传感器
phase sequence test 相序试验
phase sequence voltage relay 电压相序继电器
phase sequence 相序
phase shift angle 相位角差
phase shift circuit 移相电路
phase shift circulator 移相式环形器
phase shift coder 相移编码器
phase shift control motor 移相控制电动机
phase shift decoder 相移译码器
phase shift delay 相移延迟
phase shifter 移相器
phase shift frequency curve 相移频率特性曲线
phase shifting capacitor 移相电容器
phase shifting circuit 移相电路
phase shifting network 相移网络
phase shifting transformer 相移变压器，移相变压器
phase shifting 相移
phase shift keying 相移键控，键控移相
phase shift magnetometer 相移磁强计
phase shift modulation 相移调整
phase shift network 相移网络
phase shift transformer 移相变压器
phase shift trigger 移相触发器
phase shift 相位位移，相位移转，相角程，相移，移相，相位移
phase signal 相位信号
phase sign 相序牌
phase space density 相空间密度
phase space volume 相空间体积
phase space 相空间，相宇
phase spacing 相间距离
phase spectrum 相谱
phase speed of electromagnetic wave 电磁波的相平面速度
phase splitter 移相器，分相器
phase splitting capacitor 分相电容器
phase splitting 分相，分相法
phase spread 相带
phase stability 相位稳定度，相稳定性
phase state 相态
phase step 相位跃变
phase swinging 周期性速度波动，相位摆动
phase switch 移相器，移相开关
phase symbol 相位符号
phase synchronism 相位同步
phase synchronization 相位同步
phase system 相位系统
phase terminal voltage 相端电压
phase terminal 相端子，相引出线
phase to earth breakdown 相对地击穿
phase to earth clearance 相对地净距
phase to earth voltage 线对地电压
phase to earth 相对地
phase to ground fault 单相接地故障，相-地故障
phase to ground voltage 相电压，相对地电压
phase to ground 相对地
phase tolerance 相位容限
phase-to-neutral voltage 相电压，相对中性点电压
phase to phase arrester 相间避雷器
phase to phase clearance 相间净距
phase to phase distance protection 相间距离保护
phase to phase fault 相间故障，两相故障
phase to phase spacing 相间距离
phase to phase volatge 相间电压
phase to phase 相间，相间的
phase transformation 相位跃迁，相变，相位变换，相转变
phase transformer 相位变换器
phase transition temperature 相变温度
phase transition 相位变化，相位移动，相变
phase-undervoltage protection 低相电压保护
phase-unstable 相位不稳定的
phase vector 相矢量
phase velocity of electromag-netic wave 电磁波的相平面速度
phase velocity 相（位）速度
phase voltage regulator 相电压调整器
phase voltage 相电压
phase wave 相波
phase winding 相绕组，相线圈
phase wire 相线
phase-wound rotor 绕线式转子

phase zone　相带
phase　阶段，时期，相（数），相别，单相，相位，相位角，定相，调相，状态
phasing adjustment　定相调整，相位调整
phasing back　反相
phasing circuit　定相电路
phasing contact　定相触点
phasing control　相位调整，相位控制
phasing device　定相器
phasing-in　同步
phasing lamp　定相灯
phasing line　定相线
phasing operation　调相运行
phasing switch　调相开关
phasing synchro　相位同步机，变相同步机
phasing transformer　移相变压器
phasing voltage　定相电压
phasing　调整相位，定相，逐步采用
phasitron　调频管，调相管
phasmajector　发出标准视频信号的电视测试设备
phasometer　相位计
phasor amplitude　相量幅值
phasor argument　相量的幅角
phasor current　相量电流
phasor diagram　相量图，矢量图
phasor difference　（平面）相量差
phasor equation　相量方程式
phasor function　相量函数
phasor impedance　复量阻抗，复数阻抗
phasor power factor　基波功率因数，相量功率因数
phasor power　相量功率，复数功率
phasor product　相量积
phasor quantity　相量，复量
phasor quotient　相量商
phasor representation　相量表示
phasor sum　相量和
phasor voltage　相量电压
phasor　相矢量，相位复（数）矢量，相图，相量，彩色信号矢量，相向量
PHCN(Power Holding Company of Nigeria)　尼日利亚电力有限公司
pH control agent　pH 控制剂
pH control　pH 值控制
pH control system　pH 值控制（调节）系统
PHC pipe pile　PHC 管桩，高强度预应力混凝土管桩
PHC(pre-stressed high-strength concrete)　高强度预应力混凝土
pH electrode assembly　pH 电极，pH 电极组件
pH electrode　pH 电极
phenanthrene　菲
phenethylene　苯乙烯
phene　苯
phenocryst　斑晶
phenol acetaldehyde resin　苯酚乙醛树脂
phenol aldehyde plastics　酚醛塑料
phenol derivative　苯酚衍生物
phenol formaldehyde lacquer　苯酚甲醛清漆
phenol formaldehyde resin　苯酚甲醛树脂
phenolic aldehyd　酚醛
phenolic paint　酚醛类漆
phenolic polyamine　苯酚聚胺
phenolic resin binder　酚醛树脂黏结剂
phenolic resin　酚醛树脂
phenolics　酚醛树脂，酚醛塑料，酚醛
phenolic　苯酚的，酚的，石碳酸的
phenology　物候学
phenolphthalein alkalinity　酚醛碱度
phenolphthalein end-point alkalinity　酚酞终端含碱度
phenolphthalein indicator　酚醛指示剂
phenolphthalein　酚酞
phenolplast　酚醛塑料
phenol resin laminate　酚醛树脂叠层板
phenol-resin paint　酚醛聚酯漆
phenol resin　酚醛树脂
phenol sulfonic acid　苯酚磺酸
phenol wastes　石碳酸废液
phenol　苯酚，石碳酸
phenomenological coefficient　唯象系数
phenomenological relation　表象关系
phenomenon　征兆，现象
phenoxin　四氯化碳
phenylethylene　苯乙烯
phenylthiourea　苯基硫脲
phenyl　苯基
Philips gauge　菲利普斯真空计，菲利普斯真空管
Philips rectifier　菲利普斯整流管
philosophy of measurement　测量方法，测量原理
philosophy　策略，哲学，原理
pH indicator　pH 计，pH 值指示器
phlogopite　金云母
pH measurement system　pH 值测量系统
pH meter　氢离子计，pH 计，酸度计
pH monitor　pH 值监测计
phoenix effect　“凤凰”效应
phonautograph　声波计振仪
phonemeter　通话计数器，测声器
phoneme　音素，语音
phone set　电话机
phonetic typewriter　口授打印机，语音打印机
phonetic　语音的，语音学的
phonevision　电话电视
phoney　假的，伪造的
phone　电话机，电话，音素
phonic motor　蜂音电动机
phonics　声学
phonic wheel　音轮
phonic　语音的，声音的，有声的，声学的
phonmeter　测声计
phonodeik　声波显示仪
phonogram　唱片，录音片，话传电报
phonometer　声强计，测声机，测音计
phonon　声子
phonophore　报话合用机
phonoplug　信号电路中屏蔽电缆用插头
phonovision　电话电视
phonozenograph　声波定位器，声波测向器
phoresis　泳动，电泳现象

phosphate alkalinity 磷酸盐碱度
phosphate analyser 磷酸盐分析器
phosphate analyzer 磷表
phosphate esters 磷酸酯
phosphate glass dosemeter 磷酸盐玻璃剂量计
phosphate-hydroxide treatment 磷酸盐苛性碱处理
phosphate of lime 磷酸钙
phosphate powder 磷酸盐干粉
phosphate treatment of water 水的磷酸基处理
phosphate treatment 磷酸盐处理
phosphate wastage 磷酸盐耗蚀
phosphate 磷酸盐
phosphatic 磷酸盐的
phosphating treatment 磷酸盐处理
phosphating 磷酸盐
phosphide 磷化物
phosphine 磷化氢
phosphite 亚磷酸盐
phosphomolybdic acid 磷钼酸
phosphonate 磷酸盐
phosphonium compound 磷化合物
phosphor bronze 磷青铜
phosphor-coated drum 复磷层鼓
phosphor-coated 复磷层的
phosphoric acid 磷酸
phosphoric anhydride 磷酐，五氧化二磷
phosphoric 含磷的
phosphorous acid 亚磷酸
phosphorous anhydride 三氧化二磷，亚磷酐
phosphorous fixation 磷的固定
phosphorous 磷，磷的，亚磷的
phosphor storage 磷光屏存储器
phosphorus pentoxide 五氧化二磷
phosphorus trioxide 三氧化二磷
phosphorus 磷
phosphor 磷光体，荧光物质
phosphurized primer 磷化底漆
photicon 高灵敏度摄像管，光电摄像管
photic zone 透光层
photion 充气光电二极管
photistor 光敏晶体三极管
photoabsorption coefficient 光吸收系数
photoabsorption 光吸收
photo-actinic action 光化作用
photoactive 光敏的
photoactor 光电变换器件
photoaging 光致老化
photo-amplifier 光电放大器
photoassisted production 光辅助生产
photo-audio generator 光电音频信号发生器
photobiological effect 光生物效应
photocartograph 摄影测图仪
photocatalyst 光催化剂
photocathode 光电阴极
photocell pick-off 光电管传感器，光电管发送器
photocell 光电池，光电管，光电元件
photochemical cell 光化学电池
photochemical oxidant 光化学氧化剂
photochemical phenomenon 光化现象
photochemical pollutant 光化学污染物
photochemical pollution 光化学污染
photochemical polymerization 光化聚合
photochemical process 光化学加工，光化学处理，光化学过程
photochemical reaction 光化学反应
photochemical sensitivity 光化学敏感性
photochemical smog model 光化学烟雾模式
photochemical smog 光化学烟雾
photo chemical vapor deposition 光化学气相沉积
photochemistry 光化学
photochopper 光线断路器，遮光器
photochromatic micro-image 彩色照相缩微图像
photochromic 光敏材料，光致变色的
photochromy 彩色照相术
photo-communication 光通信
photocomposition engraving 光学制版
photocomposition 光学排版
photoconduction 光电导，光电导性
photoconductive cell 光敏电阻，光电导管，光导元件
photoconductive detector 光导电探测器
photoconductive driver 光导激励器
photoconductive effect 光电导效应，内部光电效应
photoconductive element 光敏元件
photoconductive fiber 光导纤维
photoconductive 光电导的
photoconductivity 光电导率，光电导性
photoconductor 光电导体，光敏电阻
photocon 光导元件，光导器件
photocopy machine 复印机
photocopy 影印副本
photocurrent 光电流
photodechlorination 感光去氯
photodecomposition 光分解
photodetection 光电探测
photodetector 光检测器，光电检测器，光电探测器，光度感应器
photodiode 光电二极管，光敏二极管
photodissociation 光致分解，光解作用
photodynamic 在光中发荧光的
photod 光电二极管
photoeffect 光电效应
photoelastic analysis 光弹性分析
photoelastic biaxial gauge 光弹性双向应变计
photoelastic coating method 光弹涂层法
photoelastic coating 光弹性涂层，光弹涂层
photoelastic constant 光弹常数
photoelastic effect 光弹效应
photoelastic freezing method 光弹性冻结法
photoelastic fringe pattern 光弹条纹图形
photoelasticity birefringence 光弹性双折射
photoelasticity isochromatic fringe 光弹性等色条纹
photoelasticity 光测弹性学，光弹性，光弹性力学
photoelastic material 光弹性材料
photoelastic method 光弹性法
photoelastic model method 光弹模型法
photoelastic model 光弹模型
photoelastic polariscope 光弹偏光镜
photoelastic reflection method 反射光弹法

photoelastic sensitive 光弹灵敏度
photoelastic strain measuring method 光弹应变测量法
photoelastic stress analysis 光弹性应力分析
photoelastic stress gauge 光弹性应力计
photoelastic surface coating 光弹表面涂层
photoelastic technique 光弹性技术
photoelastic testing 光弹试验
photoelectric absorption coefficient 光电吸收系数
photoelectric absorption 光电吸收，光电吸收效应
photoelectrically balanced potentiometer 光电平衡电位计
photoelectric analysis 光电分析法
photoelectric apparatus 光电装置
photo electric cell 光电管，光电池，光电元件
photoelectric colorimeter comparator 光电比色计
photoelectric colorimeter 光电比色计
photoelectric comparator 光电比较器
photoelectric compensator 光电补偿器
photoelectric controller 光电控制器
photoelectric control 光电控制
photoelectric conversion efficiency 光电转换效率
photoelectric conversion 光电转换，光电变换
photoelectric conversion process （硅晶）光电转换工艺
photoelectric converter 光电转换器
photoelectric coupling 光电耦合
photoélectric current 光电流
photoelectric densitometer 光电式显像密度计
photoelectric detection 光电探测
photoelectric device 光电器件
photoelectric effect 光电效应
photoelectric emission 光电发射
photoelectric eye 光电眼
photoelectric focusing 光电随动系统
photoelectric galvanometer 光电电流计
photoelectric imaging device 光电成像装置
photoelectric inspection 光电检查
photoelectric instrument 光电仪器
photoelectric insulation 光电隔离
photoelectric integrator 光电积分器
photoelectricity 光电学，光电现象
photoelectric measuring instrument 光电测量仪
photoelectric measuring system 光电测量系统
photoelectric measuring 光电式互感器
photoelectric meter 光电表
photoelectric micrometer 光电测微光度计
photoelectric multiplier 光电倍增管
photoelectric nucleus counter 光电核计数器
photoelectric peak 光电峰
photoelectric photometer 光电光度计
photoelectric potentiometer 光电电位计
photoelectric pyrometer 光电高温计
photoelectric reader 光电读出器，光电读数器
photoelectric reading 光电读数
photoelectric relay 光电继电器
photoelectric scanning 光电扫描
photoelectric seismometer 光电地震仪
photoelectric sensing 光电传感，光电读数，光电读出
photoelectric smoke and flame detector 光电烟火探测器
photoelectric smoke detection 光电法检烟
photoelectric spectrophotometer 光电式分光光度计
photoelectric switch 光电开关
photoelectric threshold 光电阈
photoelectric translating system 光电读数系统
photoelectric tube 光电管
photoelectric 光电的
photoelectrochemical cell 光电化学电池
photoelectroluminescence 光致电发光，光控电致发光，光控场致发光
photoelectromagnetic effect 光电磁效应
photoelectromagnetic 光电磁的
photoelectrometer 光电计
photo electromotive force 光电电动势
photoelectronics 光电子学，光电装置
photoelectron 光电子
photoelectrostatic image 静光电影像
photoelectrostatic 静光电的
photoelement 光电管
photoemission 光电发射
photoemissive detector 光电发射检测器
photoemissive effect 外部光电效应
photo emissive power conversion 光电子发射功率转换
photoemissive relay 光电发射继电器
photoemissive 光电发射的，光子发射的
photoengraving 光刻，照相版
photoetching technique 光刻技术
photo-fabrication 光加工，光刻法
photofission product 光致裂变产物
photofission 光致核裂变
photoflash 闪光灯
photoflood （摄影用）超压强烈溢光灯
photofluorogram 荧光屏图像照片
photofluorography 荧光屏图像摄影
photofluoroscope 荧光屏照相机，荧光屏
photoformer 光电函数发生器
photo-fraction 光峰比
photogalvanic effect 光电效应
photogalvanic power conversion 光电流功率转换
photo-generated current 光生电流，光电流
photogenerator 光电讯号发生器
photogeology 摄影地质学
photoglow tube 辉光管
photogrammetric apparatus 摄影测量设备
photogrammetric mapping 摄影测图
photogrammetric survey 摄影测量
photogrammetry 摄影测量（法）
photogram 传真电报，照片，图片
photographical 照相的，摄影的
photographic borehole surveying 钻孔照相测绘
photographic current meter 照相海流计
photographic data recording 数据摄影记录
photographic diagnostics 照相诊断
photographic dosimeter 照相剂量计
photographic dosimetry 摄影计量仪
photographic film dosimeter 照相胶片剂量计
photographic film 胶片，照相底片

photographic finder　探像器
photographic image contrast　照相图像衬度
photographic mapping　摄影测图
photographic negative　底片
photographic observation　摄影观测
photographic plate　照相感光板
photographic process　照相法
photographic pyrometry　摄影高温测定法
photographic recording oscillograph　照相示波器
photographic recording　摄影记录，照相记录
photographic standard irradiance　照相标准辐射度
photographic storage　照相存储器
photographic sunshine recorder　摄影式日照计
photographic survey　摄影测量
photographic tape　摄影带
photographic thermal neutron image detector　热中子成像照相探测器
photographic typesetter　摄影排字机
photographic waste　照相废液
photograph interpretation　相片判读
photography attachment　摄影装置，照相装置
photography　照相术，摄影术
photograph　相片，照片照相，摄影
photogun　光电子枪
photohead　光电传感头
photoheliograph　太阳照相仪
photohomolysis　感光均解
photoimpact　光控脉冲，光电脉冲
photo-induced　光学感生的，光诱导的
photoinduction　光诱导
photo-interpretation　相片辨认，照片判读
photo-interpreter　照片识别器，相片判别装置
photoionization　光致电离，光电离，光化电离，光电离作用
photoisomerization　光异构化
photoluminescence　光致发光，荧光，光激发光
photoluminescent dosimetry　荧光剂量测定法
photoluminescent personal dosimeter　荧光个人剂量计，光点计量笔，光致发光个人剂量计
photolysis　光分解
photomagnetic coupling　光电耦合
photomagnetic effect　光磁效应
photomagnetic power conversion　光磁功率转换
photomagnetic　光磁的
photomagnetoelectric　光磁电的
photomapping　空中摄影测量
photomask　光掩模
photomechanical copy　照相拷贝
photometer　测光仪，分光计，光度计，曝光表，大气光学现象，蒸腾计
photometrical　光测的，光度的
photometric-cube optical pyrometer　带有光度块光学高温计
photometric determination　光度测定
photometric method　光度测定法
photometric pyrometer　光度高温计，光学高温计
photomicrography　缩微照相，显微摄影
photomicrograph　显微镜照相的照片，显微照片
photomission　光电发射
photomixer　光电混频器
photomodulator　光调制器
photomontage　集成照片
photomultiplier counter　闪烁计数器，光电倍增管计数器
photomultiplier tube　光电倍增管
photomultiplier　光电倍增器，光电倍增管
photon counter　光子计数器，光子记数管
photon detector　光子探测器
photonephelometer　光电浊度计
photoneutron emission　光中子发射
photoneutron source　光中子源
photoneutron　光激中子，光致中子，光中子
photon excited X-ray fluorescence analysis　光子激发的X光荧光分析
photon exitance　光子出射度
photon exposure　光子外照射
photon flux density　光子通量密度
photon flux　光子通量
photon intensity　光子强度
photon irradiance　光子照度
photon luminance　光子亮度
photon-neutron coupling calculation　光子中子耦合计算
photon number　光子数
photon radiance　光子亮度
photon radiation　光子辐射
photonuclear activation analysis　光核活化分析
photonuclear cross-section　光致核反应截面
photonuclear reaction　光致核反应
photon　光量子，光电子，光子
photo-optical recorder tracker　光电记录跟踪装置
photooxidation　感光氧化作用，光氧化
photoperiodism　光周期现象，光周期性
photoperspectograph　摄影透视仪
photophone　光音机，光通话，光电话机
photophoresis　光致迁动，光泳现象，光致漂移
photophysical process　光物理过程
photopolarimeter　光偏振表
photopolygonometry　照片导线测量
photopositive　电导率与照度成正比的，正光电导的，正趋光性的
photoprobe　光探头
photoptometer　光敏计，光度计
photoradiogram　无线电传真电报，无线电传真照相
photoreaction　光致反应
photoreader　光度读出器，光电读数器
photoreading　光电读数
photoreceptor　光感受器
photorectifier　光电二极管，光度检波器
photorehucer　照片缩小仪
photorelay　光控继电器
photoresistance relay　光敏电阻继电器
photoresistance　光敏电阻
photoresistor-cell relay　光敏元件继电器
photoresistor　光敏电阻器
photoresist　光阻材料
photo scale　照片比例尺
photoscanning　光扫描
photoscope　透视镜荧光屏

photoscreen method 光屏法
photosensitive film badge 感光胶片
photosensitive 光敏的
photosensitivity 光敏性
photosensitization 光敏作用
photosensor 光电传感器，光敏器件，光敏元件
photo-signal 光电流信号
photostat 复印件
photostereograph 立体测图仪
photoswitch 光电开关，光控继电器
photosynthesis 光合作用
photosynthetical 光合成的
photosynthetic bacteria 光合细菌
photosynthetic cycle 光合循环
photosynthetic efficiency 光合效率
photosynthetic process 光合过程
photo-tape reader 光电穿孔带读出器
photo-tape 光电穿孔带
photo-telegram 传真电报
phototelegraphy 电传真，传真电报，光通信
phototelegraph 传真电报，相片传真
phototelephone 光线电话机
phototelephony 光线电话学
photo theodolite 摄影经纬仪
photothermal effect 光热效应
phototiming 光同步
phototransformer 照片纠正仪
phototransistor 光转换器，光电晶体管，光敏晶体三极管
phototriode 光电三极管
phototron 矩阵光电管
phototropic 向光性的
phototube 光电管，光电池
photounit 光电元件
photovalve 光电元件，光电管
photovaristor 光敏电阻
photovision 电视
photovoltage 光生电压，光电压
photovoltaic array 光伏阵列
photovoltaic cells 光电池
photovoltaic concentrator array field 聚光太阳电池方阵场
photo-voltaic concentrator array 聚光太阳电池方阵
photovoltaic concentrator module 聚光太阳电池组件
photovoltaic conversion 光电转换
photovoltaic driver 光电激励器
photovoltaic effect 光生伏打效应，光电效应，光伏效应
photovoltaic panel 光伏电池板
photovoltaic photon detector 光电光子探测器
photovoltaic plant 光伏电站
photovoltaic power conversion 光生伏打功率转备
photovoltaic power generation 光伏发电
photovoltaic principle 光电原理
photovoltaic silicon pyrheliometer 硅光电池直接日射计
photovoltaic solar module 光伏太阳能组件，太阳能光伏模块
photovoltaic system 光伏发电系统
photovoltaic 光致电压的，光电的，光生伏打的，光伏的
photovoltoics 光电池
photo 光【词头】，光敏，光电，光致，照相，与光有关的
photran 光通可控硅元件，光控管
photronic cell 光电池
photronic 光电池的
phot 辐透，厘米烛光
PH(period hours) 统计期间小时数
PhPh＝phase-phase 相间
PHP 菲律宾比索
phreatic aquifer 地下水含水层
phreatic cycle 地下水位变化周期，潜水位变化周期
phreatic decline 地下水位降低
phreatic discharge 地下水出流量，地下水排放，潜水排出
phreatic divide 地下分水线，地下分水界
phreatic fluctuation 地下水面波动，潜水面升降，地下水位变动，地下水位升降变化
phreatic line 地下水位线，渗水水面线
phreatic low 低地下水位，地下水凹面，潜水低水位，地下水下部含水层
phreatic rise 地下水上升，地下水位上升
phreatic surface of groundwater 地下（静止）水位
phreatic surface 地下水面，潜水面
phreatic water discharge 地下水出流
phreatic water level 地下水水面，地下水水位，地下水位，潜水位
phreatic water surface 地下水位
phreatic water 地下水，潜水，井水
phreatic zone 饱和层，潜水带，地下水区，地下水层
phreatic 地下水的，潜水层的
pH recording pH 值记录
PHR(plant heat rate) 电站热耗
PHT(postweld heat treatment) 焊后热处理
PHT(primary heat transport) 一次热输送
pH transmitter pH 变送器
PHTR＝preheater 预热器
PHT system 一次热输送系统，主回路系统，一次回路系统
phugoid motion 长周期运动，起伏运动
phugoid oscillation 浮沉振荡，长周期振荡
phugoid 长周期运动，长周期的，低频自振动，长周期振动，浮沉的，起浮的
pH value control pH 值控制
pH value pH 值，酸碱度
PHWR（pressurized heavy-water-moderated and cooled reactor） 加压重水慢化和冷却反应堆
phyllite aggregate 千枚岩骨料
phyllite 千枚岩
physical absorption 物理吸收
physical acoustic 物理声学
physical address 物理地址，实际地址
physical adsorption 物理吸附
physical analogue 物理模拟
physical analysis 物理分析

physical and financial progress　工程进度和付款进度【项目实施过程中】
physical assets　实际资产，有形资产
physical atmosphere　物理大气压
physical attenuation　物理衰减
physical attrition　机械磨损
physical behavior　物理属性
physical burnout　物理烧毁
physical capital　实物资本
physical characteristics　物理特性，外形特性
physical circuit　实在电路
physical climate　物理气候
physical climatology　物理气候学
physical connection　物理连接
physical constant　物理常数
physical contact　直接接触
physical control　实际控制，物理控制
physical damage　物理破坏
physical degradation　物理性剥蚀
physical depreciation　有形磨损
physical design　外形设计，结构设计
physical deterioration　有形损坏，物理性损坏，物理性恶化
physical device　物理设备，物理装置
physical dimension　物理尺寸
physical disintegration　物理性崩解
physical disturbance　物理扰动
physical drawing　实体图
physical environment　物理环境，自然环境
physical equipment table　物理装置表
physical evaporation　自然蒸发
physical examination of water　水的物理检验
physical examination　身体检查，健康检查，物理检验，体格检查
physical explosion　物理爆炸
physical features of a basin　流域的自然地理特点
physical features of river basin　流域自然地理条件
physical fidelity　物理逼真度
physical geography　地文学，自然地理学
physical geology　物理地质学
physical half life　物理半衰期
physical hydrological factor　物理水文因素
physical hydrology　物理水文学
physical identification　实物标记
physical incidence　物理迎角
physical index　物理指数
physical inventory　实物丰收量
physical I/O　实际输入/输出
physical layer　物理层
physical layout editor　物理布线编辑器
physical length　实际长度，本体长度
physical life　实际寿命，工作寿命
physical location drawing　实际位置图
physical location　实际位置
physical loss　有形损失，无理损失，物质损失
physical modelling　物理模拟
physical model　实体模型，物理模型
physical node　物理节点
physical parameter　物理参数
physical plastic limit　物理塑限
physical power source　物理电源
physical power trading　电力实物交易
physical process　物理过程
physical properties of rock　岩石物理性测定
physical property　物理参数，物理性质，物理特性，物理性能
physical prospecting　物理勘探
physical protection　实体保护，实物保护
physical quantity　物理量
physical record　实际记录，物理记录
physical resin loss　树脂机械损耗
physical segregation　实体分隔
physical separation　实体分隔
physical similarity　物理相似
physical simulation　物理模拟
physical simulator　物理仿真装置
physical size　实际尺寸，外形尺寸，物理尺寸
physical space　实际空间
physical stack height　烟囱实际高度
physical start-up　物理启动
physical strength　力量
physical stress　物理应力，实际应力
physical system of units　物理单位制
physical system　物理系统
physical time　物理时间，实际时间
physical transmission right　物理输电权
physical treatment　物理处理
physical value　重置价值
physical verisimilitude　物理模拟，物理逼真
physical wear　有形磨损
physical weathering of rock　岩石的机械风化
physical weathering　物理风化
physical withholding　物理持留，容量囤留
physical-yield limit　自然产水量极限【指地下水】
physical　物理的，物质的，实际的，自然（界）的，身体的
physicochemical method　理化方法
physicochemical　物理化学的
physico-geographical factor　自然地理要素
physics of failure　故障机理学，失效机理
physics　物理（学）
physiochemical　生物化学的，生理化学的
physiognomy　地貌
physiographical demarcation　自然区划
physiographic balance　地文平衡
physiographic condition of river basin　流域自然地理条件
physiographic condition　地文条件，自然地理条件
physiographic factor　自然地理要素
physiographic geology　地文地质学
physiographic province　地文区，自然地理区
physiography　地文学，自然地理学
physiological dose　生理剂量，生理效应，生理危害
physiological　生理学的，生理的
phytotron　人工气候室
PI-action(proportional plus integral action)　比例积分作用
piano-conformity　平行整合
PIA(peripheral interface adapter)　外围接口转接器，外围接口适配器

pibal 测风气球
PIC cable 聚乙烯绝缘电缆
PICC(People's Insurance Co. of China) 中国人民保险公司
PICE(programmable integrated control equipment) 积分程序控制设备
pick axe 十字镐
pick breaker 齿式破碎机，锤式破碎机
pick-cell 传感器室
picker 采摘者，挖掘者，采摘机，采摘工具
picket gate 导水栅门【水工试验用】
picking up the telephone 摘机
pickle 酸浸，酸蚀，酸洗，淡酸液，酸洗液
pickling acid 酸洗用酸
pickling agent 酸洗剂
pickling brittleness 酸洗脆裂
pickling plant 酸洗站
pickling test 浸酸试验，酸洗试验，酸洗检查
pickling wastewater 酸洗废液
pickling 酸浸，酸洗，酸处理，浸蚀，化学清洗
pick-off diode 截止二极管
pick-off gear 可互换齿轮
pick-off 传感器，发送器，拾取，脱去，敏感元件
pickout 分辨出，选择，分辨，分类
pick pulse 触发脉冲，启动脉冲
pick-test 取样试验，抽查
pick-up brush 集电刷，集流刷
pick-up calibration 传感器校准
pick-up camera 摄像机
pick-up coil 电动势感应线圈，拾波线圈，检测线圈，耦合线圈
pick-up current 启动电流，接触电流，拾音器电流，始动电流
pick-up fraction 拾取份额
pick-up head （元件盒）抓头，卡头
pick-up loop 耦合圈，拾波环
pick-up plate 信号板
pick-up point 起吊点
pick-up probe 传感器探头，接收探头
pick-up pump 真空泵
pick-up time 动作时间
pick-up truck 清运货车
pick-up value 始动值，（继电器）吸合值
pick-up velocity 起动流速
pick-up voltage 拾取电压，始动电压，接触电压
pick-up 传感器，拾音器，拾波器
pick 镐，十字镐，拾起，采集，挑选，接收到……信号
piclear unit 图像清除器
picofarad 皮法，微微法
picohenry 皮亨，微微亨
picologic 微微秒逻辑，皮逻辑，微微逻辑
PI-controller(proportional plus integral controller) 比例积分调节［控制］器
pico 皮可
PIC(personal identification code) 个人识别码
PIC(person in charge) 具体负责操作人员
PIC(pocket ion chamber) 袖珍电离室
PIC(polymer-impregnated concrete) 聚合物浸渍混凝土
PIC(program interrupt control) 程序中断控制
picric acid 苦味酸，黄色炸药
PICS(Production Information and Control System) 生产信息控制系统
PICS(protocal implimentation conformance statement) 协议实现一致性陈述，协议执行一致性声明
pictorial display 图像显示
pictorial infrared photography 红外成像照相术
pictorial 图示的
picture data 图像数据
picture element 像元
picture facsimile 图片传真
picture measuring 照片量测
picture monitor 图像监视器
picture moulding 挂镜线
picture output 图像输出
picturephone 电视电话
picture plan 照片平面图
picture scale 照片比例尺
picture tube 摄像管
picture 画，图像，图画，影像，照片
PICU(priority interrupt control unit) 优先中断控制器
PID-action(proportional plus integral plus derivation action) 比例积分微分作用
PID controller(proportional plus integral plus derivation controller) 比例积分微分调节［控制］器
P&ID(piping and instrument diagram) 管道及仪表流程图，管道及仪表布置图
P&ID(process and instrumentation diagram) 工艺仪表流程图
PID(proportional-integral-differential) 比例、积分、微分【DCS画面】
piece number 件号
piece of equipment 装备，配备，设备
piece of information for communication 通信信息片
piece quantity 件数
piecewise analytic function 分段解析函数
piecewise approximation method 逐段逼近法
piecewise linear function generator 分段线性函数发生器
piecewise linear interpolation 分段线性插值
piecewise linearity 分段线性
piecewise 分段的，片段的
piece 块，片，部件，零件
piedmont alluvial deposit 山麓冲积物
piedmont alluvial plain 山麓冲积平原
piedmont scarp 山麓断层崖
piedmont 山麓（的），山前地带（的）
pi-electron π电子
PIE(parallel interface element) 并行接口部件
PIE(post-irradiation examination) 辐照后检验
pier-arch system 墩拱系统
pierce circuit 一种晶控自激振荡电路
piercer 冲床，冲头，冲孔器，穿孔机，锥子
pierce 穿孔，渗透，贯穿，刺穿
piercing error （隧道的）贯通误差
piercing 冲孔，穿孔，刺穿
pier-contraction coefficient 闸墩收缩系数

pier drainage 桥墩排水设施
pier foundation 墩式基础
pier head line 港口建筑线
pier head power house 闸墩式厂房
pier head power station 闸墩式电站
pier head 突堤堤头，闸墩头部
pier nose 桥墩尖端，闸墩首部墩尖
pier shaft 墩身
pier-type hydroelectric station 闸墩式水电站
pier with approach trestle 引桥式码头
pier 墩，墩石，桥墩，闸墩，桥柱，码头，柱，防波堤，凸码头，突堤
pie-shaped strand conductor 馅饼形多股线
piestic interval 等压线间距，地下水位等深线间距
piestic water 承压水
pie winding 盘式绕组，饼式绕组
piezocoupler 压电耦合器
piezocrystal 压电晶体
piezodielectric 压电介质的
piezo-effect 压电效应
piezoelectric activity 压电效应
piezoelectrical relay 压电继电器
piezoelectrical 压电的
piezoelectric ceramics 压电陶瓷
piezoelectric constant 压电常数
piezoelectric converse effect 反压电效应
piezoelectric coupling coefficient 压电耦合系数
piezoelectric crystal balance 压电晶体天平
piezoelectric crystal 压电晶体
piezoelectric direct effect 正压电效应
piezoelectric disc 压电圆片
piezoelectric effect 压电效应
piezoelectric element 压电元件
piezoelectric exciter 压电式激振器
piezoelectric gauge 压电计
piezoelectric generator 压电振荡器，压电发电机
piezoelectric high polymer 压电高聚物
piezoelectric inverse effect 逆压电效应
piezoelectricity 压电现象，压电，压电学，压电性
piezoelectric manometer 压电式压力计
piezoelectric modulus 压电系数，压电模量
piezoelectric motor 压电电动机
piezoelectric oscillator 晶体振荡器，压电振荡器
piezoelectric oscillograph 压电示波器，晶体示波器
piezoelectric pick-up 压电拾声器，压电传感器
piezoelectric pressure gauge 压电压力计，压电式压力计
piezoelectric pressure indicator 压电式压力指示器
piezoelectric pressure sensor 电压压力敏感元件
piezoelectric relay 压电继电器
piezoelectric resonator 压电晶体谐振器
piezoelectric sender 压电变换器，压电发送器
piezoelectric stiffness constant 压电劲度常数
piezoelectric strain gauge 压电应变仪
piezoelectric stress 压电应力
piezoelectrics 压电体
piezoelectric transducer 压电传感器，压电式传感器
piezoelectric transformer 压电变压器，压电变换器
piezoelectric vibrator 石英振动片，压电振动器
piezoid 石英片，石英晶体
piezo-luminescence 压电发光
piezomagnetic effect 压磁效应
piezomagnetic 压电磁的
piezometer tip 孔隙水压测头
piezometer tube 测压管
piezometer （流体）压力计，测压计，测压管，微压管，压力表，压强计
piezometric elevation 测压管高程
piezometric head line 测压管水头线
piezometric head 测压管水头，压力水头
piezometric level 测压管水位，水压面
piezometric pipe 测压管
piezometric pressure 测压管压力
piezometric ring 测压环
piezometric surface 测压管水面，测压管液面
piezometric tube 测压管
piezometry （流体）压力测定，压力测量
piezo-oscillator 压电振荡器，晶体控制振荡器
piezophony 压电晶体送话器
piezo pick-up 压电晶体拾音器
piezoquartz 压电石英，压电晶体
piezoresistance 压电电阻
piezoresistor 压电电阻器，压敏电阻器
piezoresonator 压电晶体谐振器
piezo 压电的，压力的【词头】
pie 饼式线圈，饼式绕组
pig and ore process 生铁矿石法
pigeons 杂波，干扰，噪扰，寄生振荡
piggyback connector 背面接头
piggyback container 集装箱
piggyback twistor memory 复合磁扭线存储器
piggyback 背面的
pig iron 生钢，生铁，铸铁
pig lead 铅锭，生铅
pigliner 生铁衬垫，生铁衬套
pigmented glass 有色玻璃
pigment 颜料，色料
pig metal 金属锭
pigtail connection 软辫线连接
pigtail connector 软辫线接头
pigtailing 挠性连接
pigtail resistor 有抽头的电阻器
pigtail splice 编接，绞合
pigtail 引出端，进口跨接管，连接管，刷瓣，软辫线，抽头，进料管
pig 生铁，（金属）锭
pike peak 山巅，山顶
pike 尖头
pilaster block 半露柱块，壁柱块
pilastered wall 带壁柱墙
pilaster 半露柱，壁柱
pile-activated 反应堆内激活的
pile and cylinder jetty 桩筒式码头
pile and waling groin 桩及横撑丁坝
pile assembly 反应堆装置

pile band 桩箍
pile beam 反应堆射线束
pile bearing 桩支承
pile bent 桩排架
pile body 桩身，桩体
pile break test 桩破坏试验，机械荷载试验
pile bulkhead 桩式驳岸
pile capacity formula 桩承载公式
pile capacity 桩承载量，桩承重力，桩的承载能力
pile cap beam 桩顶联系梁
pile capping beam 桩承台梁，桩帽梁
pile capping 桩帽
pile cap 桩承台，反应堆顶盖，桩帽，桩台
pile cluster 系船柱，护墩桩，桩群，桩组
pile cofferdam 桩式围堰
pile collar 桩箍
pile core 桩芯
pile cover 桩帽
pile cross-dike 桩式横堤
pile cross-dyke 桩式横堤
pile crown 桩头
pile cushion 桩垫
pile cutoff 桩截段
piled dike 桩式堤
piled dyke 桩式堤
piled foundation 桩基础，人工基础
pile discharger 料堆排料器
piled jetty 桩式防波堤，桩式码头，桩式栈桥码头
pile-down （反应堆）逐渐停堆
piled quay 桩承驳岸，桩承码头
pile drawer 拔桩机
pile-drawing machine 拔桩机
pile driver barge 打桩船
pile driver hammer 打桩锤
pile driver lead 打桩机导向柱
pile driver tower 打桩机架
pile driver 打桩机
pile driving analyzer 打桩分析仪
pile driving barge 打桩船
pile driving equipment 打桩设备
pile driving formula 打桩公式
pile driving frame 打桩架
pile driving hammer 打桩锤
pile driving machine 打桩机
pile driving method 打桩方法
pile driving record 打桩记录
pile driving resistance 打桩阻力
pile driving rig 打桩设备
pile driving 打桩
piled wharf 桩式码头
pile eccentricity 桩偏位
pile engine 打桩机
pile envelope 反应堆外壳
pile extension 接桩
pile extraction resistance 拔桩阻力
pile extractor 拔桩机，拔桩器
pile fabricating yard 制桩场
pile factor 积累因子，堆积因子
pile fender 防冲桩，护桩
pile foot 桩承底脚，桩脚，桩基
pile for longer term storage 长期贮煤堆
pile foundation integrity tester 桩基完整性测试仪
pile foundation 桩基，桩基础
pile frame 打桩架
pile grade 反应堆级
pile group 群桩，桩群，桩组
pile guide housing 导桩套，桩柱导向套
pile gun 堆测试枪
pile hammer 打桩锤，打桩机，桩锤
pile head chipping 剃桩头，去桩头
pile head 桩顶，桩头
pile height 料堆高度
pile holder 桩架
pile hold 桩的打入深度
pile hoop 桩箍
pile integrity test 桩完整性试验
pile-irradiated 反应堆辐照的
pile irradiation 反应堆照射
pile jetting 射水沉桩，水冲沉桩，水冲法沉桩
pile layout 桩布置，桩的定线，桩位布置
pile loading test 桩的承载试验，桩的载荷试验，桩荷载试验
pile-made source 反应堆中制备的放射源
pile make-up yard 接桩场地
pile material 反应堆材料
pile method constructional equipment 桩法施工设备
pile-neutron physics 反应堆中子物理学
pile-neutron spectrometer 反应堆中子谱仪
pile-oscillator method 反应堆振荡器法
pile oscillator 反应堆振荡器
pile penetration 桩贯入度
pile pier 桩墩
pile pitching 桩的斜度
pileplacing method 打桩方法
pile-planking 打板桩墙，桩上铺板
pile point capacity 桩端承受力
pile point 桩端
pile pressing machine 压桩机
pile puller 拔桩机，拔桩器
pile pulling test 拔桩试验
pile pulling 拔桩
pile radiation 堆辐射，反应堆辐射
pile reactivity 反应堆反应性，反应堆反应能力
pile refusal 桩的抗沉
pile ring 桩箍
piler 堆垛机
pile shaft 桩身
pile sheathing 打板桩
pile shell 桩壳
pile shoe 桩靴
pile-sinking 打桩，沉桩
piles in row 排桩
pile-soil modular-ratio 桩土模量比
pile spacing 桩距
pile spectrum 反应堆中子谱
pile splice 桩的拼接，桩接头
piles resistance 桩群的抗力
pile stem 桩身
piles thrusted-expanded in column hammer 柱锤冲

扩桩法
pile stoppage point 桩止点
pile structure 桩基建筑物，桩式结构
pile-supported footing 桩承基脚
pile-supported platform 支柱平台，桩承平台
pile-supported raft 桩承筏基，桩承浮筏基础
pile-supported 桩支承的
pile target 反应堆靶
pile test 试桩
pile tip resistance force 桩底阻力
pile tip 桩端，桩尖，桩头
pile toe 桩尖，桩头
pile top 桩顶
pile trestle 排桩栈道，桩构排架
pile-type foundation 桩式基础
pile-type selection 桩型选择
pile-type wharf 桩式码头
pile-up effect 堆积效应
pile-up process 堆积过程
pile-up welding 堆焊
pile-up 堆积，积存
pile weir 桩堰
pile winding （分层）叠绕组，分层叠绕
pile with anchors 锚筋桩
pile works 打桩工程
pile 堆，堆积，热电堆，核反应堆，反应堆，料堆，桩，桩柱
pilgering 轧管
piling and piling equipment 打桩和打桩设备
piling bar 钢桩
piling compaction 打桩挤密
piling equipment 打桩设备
piling height 堆积高度
piling machinery 桩工机械
piling machine 打桩机，压桩机，堆垛机
piling plan 桩位布置图
piling rig 打桩机具
piling-up corrosion 叠加溶蚀
piling-up irradiation 堆码辐照
piling-up 汽油积聚，堆积
piling wall 桩墙
piling with straight trench method 打桩直槽法
piling works 打桩工程，桩基工程
piling 打桩，排桩，桩材，打桩工程
pillar buoy 柱形浮标
pillar crane 塔式起重机，转柱式起重机
pillar drain 柱状排水
pillar hydrant 桩式消防栓
pillar industry 支柱产业
pillar insulator 柱式绝缘子
pillar jib crane 转柱悬臂起重机
pillar stone 奠基石，柱石
pillar support 柱基，柱支座
pillar switch 柱式开关
pillar 墩，桩，柱，支柱
pillot wheel 导轮
pillow block 轴承座，轴台，垫台
pillow structure 枕状结构
pillow 枕块，轴承座，轴衬，轴枕，垫座，托瓦，衬，垫，托木
pill transformer 匹配变换器
pilot actuated regulator 间接作用调节器，导阀动作调节器，辅助能源调节器
pilot adit 导洞
pilotage 领港，引航，领港费，操纵，操作
pilot area 超前试验区，示范区
pilot balloon 测风气球
pilot bearing 导（向）轴承
pilot bell 监视铃
pilot bobbin 导柱
pilot boiler 试验锅炉
pilot brush 测试刷，控制刷，副电刷，选择器电刷
pilot burner 点火喷燃器，点火燃烧器
pilot cable 引导电缆，导航电缆
pilot calculation 试算，估算
pilot carbon-pile regulator 辅助碳堆调整器
pilot carrier 导频
pilot cell 监视电池，控制元件
pilot channel 导流河槽
pilot circuit 控制电路，导频电路
pilot combustion chamber 引燃室
pilot controller 辅助控制器，导频控制器
pilot current 控制电流
pilot cut-off valve 先导式截止阀
pilot cut 裁弯导槽
pilot device 先导装置
pilot differential relay 带辅助导线的差动继电器，引线式差动继电器
pilot drift 导洞
piloted reducer valve 导阀控制的减压阀，减压式先导阀
piloted reducing valve 先导型减压阀
piloted relief valve 先导式泄压阀
piloted valve 预启阀，先导式阀，操纵阀，导向阀
pilot engine 辅助发动机，启动发动机
Pilot error correction 皮托管的误差修正
pilot exciter 导频激励器，副励磁机
pilot fire 升炉火炬
pilot fit 定位配合，导向配合，定位孔
pilot flame 升炉火炬，引燃火焰
pilot frequency 领示频率，导频
pilot gas 辅助气
pilot generator 辅助发电机
pilot heading 超前导洞
pilotherm 双金属片控制的恒温器
pilot hole 导向孔，导洞，定位孔，装配孔，辅助孔
pilot ignition 引燃
piloting 操纵，调节
pilot insulator set 跳线绝缘子串
pilotis 重型柱
pilot lamp 信号灯，指示灯，表盘灯，监视灯，点火炬
pilot motor 辅助电机，油动机，伺服电机
pilot oil pressure 控制油压
pilot oil regulating valve 控制油调节阀
pilot oil valve 错油门
pilot oil 操纵油，调节油，控制油
pilot-operated control 导阀控制调节
pilot-operated pressure relief valve 先导式卸

压阀
pilot-operated regulator　液位调节器
pilot-operated relief valve　先导式卸压阀，液压控制溢流阀
pilot-operated safety valve　引导阀操纵的安全泄气阀
pilot-operated valve　导阀操作阀，导阀控制阀，液压控制动作阀
pilot-operated　导阀控制的
pilot oscillator　导频振荡器，主控振荡器
pilot pile　排障桩
pilot plant reactor　试验性反应堆，原型反应堆，中间规模反应堆
pilot plant　半工业性装置，中间试验工厂，中间试验装置，实验装置，试验装置，试验工厂，示范性电站（装置），试验性设备
pilot poppet　液压先导提升阀，液压控制提升阀
pilot project　试点项目，试点工程，样板工程，（小规模）试验计划，试验项目
pilot protection with indirect comparison　间接比较式纵联引线保护
pilot protection　纵联保护，引线保护
pilot reactor　试验性反应堆，中间试验规模的反应堆
pilot reclamation project　试验性改造工程
pilot relay　引示继电器，控制继电器，辅助继电器
pilot relief valve　液压安全阀
pilot-scale　半工业规模，试验性规模
pilot scheme　小规模计划，试验计划
pilot servomotor　中间接力器，引导伺服阀
pilot shaft　导井
pilot shorting relay　辅助导线短路继电器
pilot signal　控制信号，指示信号，导频信号，监控信号
pilot sleeve　导向套筒
pilot solenoid valve　辅助电磁阀
pilot stabilization period　引燃稳定期
pilot streamer　雷电导波，雷电初始低电流放电
pilot study　探索性研究
pilot switch　操作开关，控制开关，辅助开关
pilot tape　引导带
pilot test　小规模试验，半工业试验，中间试验
pilot torch　点火炬
pilot trench　导水沟
pilot tube　指示灯
pilot tunnel method　导洞法
pilot valve bushing　错油门套筒，导阀阀套
pilot valve push stem　先导阀顶杆
pilot valve　操纵阀，导向阀，错油门，滑阀，导阀，控制滑阀，伺服阀
pilot voltmeter　馈线末端电压指示表，送电端电压表
pilot wheel　手轮，导轮
pilot wind power plant　风力发电试验场
pilot wire protection　带辅助导线的纵联保护，纵联差动保护，导引线保护
pilot wire relay　线路纵差保护，辅助线继电保护
pilot wire　领示线，操作线，控制线，辅助导线，导引绳
pilot working　试运行
pilot zone project　试验区项目
pilot　试验性的，调节装置，伺服阀门，控制，导向器，引示信号，引导，指示，驾驶
pimping depression cone　抽水降落锥面
pimple　疙瘩，肿鼓
pimpling　局部隆起
PIM(Project Interface Management)　工程接口管理（程序）
pin-and-bushing coupling　销和套筒连接
pin axis joint　销轴节点
pin bearing　滚柱轴承，销轴承
pinboard　插销板，接线板，插针板
pin bolt　销钉，紧配螺栓
pinborad programming　插接板程序设计
pin bush　定位销套
pincer　钳，钳子
pincette tong　小镊子，小钳子
pinch clamp　弹簧夹
pinch compression　等离子体匝缩
pinch design　夹点设计
pinch effect　夹紧效应，电磁收缩效应
pinchers　木板支撑【沟槽的】
pinch off voltage　夹断电压
pinch off　夹断
pin-chopping machine　细棒切断机
pinch-point　夹点，窄点
pinch-roll forming　压滚成形
pinch valve　夹紧式胶管阀，夹持阀
pinch　箍缩，管脚，夹具，压紧，夹紧，收缩效应，抢风夹点，芯柱，缩紧，挤压（薄板）
pin-connected joint　铰节点
pin-connected truss　铰接桁架
pin-connected　销连接的，栓接的
pin connection joint　铰节点
pin connection　（电子管）管脚连接，铰接，销钉连接
pin coupling　柱销联轴器，销连接
pincushin distortion　（光栅的）枕形失真，正畸变
pin down　把……固定住，使动弹不得
pine board　松木地板
pin-end column　铰端支柱
pin-ended portal frame　铰接式柱脚门形钢架
pine pole　松木电杆
pine resin　松脂
pine-tree crystal　树枝状晶体
pine-tree dovetail　枞树形叶根
pine-tree　水平偶极子天线阵，松树式天线阵，松树
pi network　π形四端网络
pin extraction　销钉拔出
pine　松树
pin face wrench　叉形带销扳手
pin-fin tube　销钉管
pin-fin　针形肋
pin fork　销接式叉形叶根
pinger　声脉冲发送器，声波发射器
pinging noise　颤噪声，传声器效应噪声
ping-pong　往复转换工作
ping　声脉冲，声呐设备的脉冲信号
pinhole failure　针孔状损坏

pin-hole of sight vane 视准孔
pinhole test 针孔试验
pinhole 针孔，引线孔，销孔，塞孔，针眼
pin identification 销钉标记
pin insertion 销钉插入
pin insulator 针式绝缘子，针脚绝缘子
pinion drive 小齿轮传动
pinion gauge 小齿轮，游星齿轮
pinion rack 齿条，齿臂，齿板
pinion 小齿轮，柱销，轴齿轮
pin-jointed 销接的
pin joint 铰链结合，枢结合，销连接
pink noise generator 带有随机噪声的发电机
pink noise 随机噪声，突发性噪声
pin loading 销钉插入
pin lug 针形接线片
pinnate drainage 羽状水系
pinnate tension joint 羽状张拉节理
pinned joint 铰接
pinned shaft 销轴
pinned 铰接的，销接的
pinning 销连接
PIN a-Si solar cell PIN 非晶硅太阳电池【由 P 型非晶硅，本征非晶硅和 N 型非晶硅构成】
PINO(positive-input negative-output) 正输入负输出
pinpoint accuracy 高精确度，高精度，高准确度，高度精确性
pinpoint arrester 针尖放电避雷器
pinpoint 针尖，极精确的，准确定位，高精确度的
pin punch 除销器，冲头
pin replacement 销钉更换
pin spanner 带销扳手
pin splice 销钉拼接
pin terminal 管脚端子，插头端子
pin timbering 锚杆支护法
pintle 开口销，扣钉，枢轴
pint opening 缝张开度
pin type insulator 针式绝缘子
pint 品脱
pin valve 针形阀
pin winding 拉入式绕组
pin with hole 带孔销
pin 销钉，插销，销子，销，插头，引线，钉住，针，支杆，枢轴，栓，管脚，钉
pioneer heading 隧洞导洞，隧洞导坑
pioneering spirit 创新精神，首创精神
pioneering works 先行工程
pioneer well 导井
piont to point interface 点对点接口
PIO(precision iterative operation) 精确迭代操作
PIO(process input/output) 过程输入/输出【通道】
PIO(processor input-output) 处理机输入/输出设备
pipage 管道，管子，管系，管道系统，管道运输
pip amplifier 脉冲放大器
pipe accessories 阀件
pipe adapter 管接头，连接套
pipe aligner 对中器
pipe alignment 管道定线
pipe alley 管道沟
pipe anchorage 管道支座，管道固定
pipe anchor 管道固定点
pipe and duct snips 管子和导管剪
pipe and wire penetration 管道和电缆贯穿件【安全壳】
pipe angle 管道弯头
pipe arrangement 管道布置，管配置，管系
pipe barrel 管筒
pipe bayonet fixing 管脚固定
pipe bell 管子承接口
pipe bender 弯管机
pipe bending machine 弯管机
pipe bending radius 弯管弯曲半径
pipe bending 弯管
pipe bend 管子弯头
pipe bevel end cleaner 管割口清洁器
pipe beveller 坡口机
pipe blowing and cleaning device 管道吹扫设备
pipe blowing silencer 吹管消声器
pipe blowing with steam 蒸汽冲管
pipe blowing 管道吹扫，吹管
pipe blow-through 清洗管道
pipe bracket 附墙支架，管道支吊架
pipe branch 管子分叉
pipe break exclusion region 管道破损排除区
pipe break 管道破裂
pipe bridge 管道桥，管架，管桥
pipe cable 管道电缆
pipe canal 管沟
pipe capacity 管道通流能力
pipe cap 管帽
pipe casing 套管
pipe cavity 缩孔
pipe chair 管托
pipe chase 管道沟槽
pipe chute 斜管
pipe clamp 管卡，管夹，管箍，管架
pipe clip 管夹，管卡，管架
pipe closer 管堵头，管盖，管塞
pipe coil 盘管，蛇形管，管圈，光面管散热器
pipe column 管形支柱【蒸汽发生器】，钢管柱，管柱
pipe connection 管接头，接管
pipe constriction 管道收缩
pipe cooler 管式冷却器
pipe cooling system 冷却水管系统
pipe cooling 水管冷却
pipe cross section 管道断面
pipe culvert 涵洞，管道，涵管，管涵，管式涵洞
pipe cutter 割管机，管子切断机，切管机
pipe cutting machine 管子切断机，切管机
piped compression cable 管内充气的高压电缆
pipe design 管道设计
pipe diameter 水管直径
pipe drainage 暗沟排水，管道排水
pipe drain 管式排水，排水管
pipe duct 管沟，管渠
pipe effect 管道效应
pipe elbow 管道弯头

pipe enamel 管壁涂料
pipe end beveling machine 管端坡口机
pipe end slope-mouth 管端坡口
pipe erector 管道工
pipe expander 胀管器，扩管机
pipe expansion force 管道膨胀力
pipe expansion 管道膨胀
pipe filter 管道过滤器
pipe finder 水管探测器
pipe fitter 管道工，管道装配工
pipe fittings 管道配件，管子配件
pipe flange joint 管子法兰接头，管子凸缘，管子法兰，管法兰
pipe flaring tool 胀管器，管子扩口器
pipe flow 管流，管状流
pipe forceps 管钳
pipe forcing method 管道压入法
pipe friction 管道摩擦阻力，管道摩擦
pipe gallery 敷管廊道，管廊
pipe girth joint operator 管道环缝焊接操纵机
pipe grid 管
pipe-hanger assembly 管道支吊架组件
pipe-hanger load 管道支吊架载荷
pipe-hanger plan location 管道支吊架设计位置
pipe-hanger support 管道支吊架
pipe-hanger 管道（支）吊架，管架
pipe-hanging hook 管道悬吊支架
pipe head 管接头
pipe heat radiator 光管散热器
pipe heat tracing 管道伴热
pipe holder 管道托架，管支架
pipe in series 串联管路
pipe installation 管道敷设
pipe insulating layer 管道保温层
pipe insulation 管道保温
pipe invert level 管内底标高
pipe-in 输进，接通
pipe jacking method 顶管法
pipe jacking 管子推顶法
pipe-joint compound 管子接缝填料
pipe joint 管接头
pipe layer 埋管机，铺管机
pipe laying depth 埋管深度
pipe laying machine 铺管机
pipe laying 管道敷设
pipe lead 管子引入线
pipe leakage 管道渗漏
pipe leg 管式料腿
pipeline booster pump 管线增压泵
pipeline coal transportation 管道输煤
pipeline coal 管道输送（的）煤
pipeline concrete anchor 混凝土管道支座
pipeline gallery 管廊
pipeline gas 管道气
pipeline heat loss 管道热损失
pipeline jetty 管道堤
pipeline layout 管路布置
pipeline manhole 管线入孔
pipeline muffler 管道消音器
pipeline pressure 管道压力，管线压力
pipeline reactor 管道式反应器
pipeline right-of-way 管道通行权，管道走廊权
pipe liner 管套
pipeline strainer 管道过滤器
pipeline support 管架
pipeline survey 管线测量
pipeline telemetering 管路远距离测量，管路遥测
pipeline transportation 管道输送，管道运输
pipeline trench 管线沟
pipeline trestle 管道栈桥，管道支架
pipeline turbine 输送管线用燃气轮机
pipeline warming process 暖管工艺
pipeline with fins 带鳍片管道
pipeline 管线，管道，管路，导管，流水线
pipe lining 管道衬里
pipe locator 探管仪
pipeloop 管圈
pipe manhole 管道入孔
pipe manifold 管道汇集器
pipe measuring thickness instrument 管壁测厚仪
pipe mixer 管道混合器，管式混合器
pipe-mounted sensor 管道安装的传感器
pipe-mounted 管道安装的
pipe network system 管网系统
pipe network 管道网，管网，管道系统
pipe nipple 短节，短管接头
pipe nozzle 管接头
pipe orifice 管形孔口
pipe outlet 管式出水口
pipe penetration 管道贯穿件
pipe pile foundation 管桩基础
pipe pile joint method 管桩接头法
pipe pile joint 管桩接头
pipe pile 管桩
pipe pitch 管子斜度
pipe plugged 管道堵塞
pipe plug 管堵，管塞
pipe pressure 管道压力
pipe racks, brackets, supports or hangers 管道支吊架
pipe rack 管道支架，管架，管道支座，管桥，管托板
pipe radiator 光面管散热器
pipe reaction 管线反力
pipe reducer 变径管，大小头，渐缩管
pipe resistance 管道阻力
pipe roller 管道辊子支座
pipe routing 管道走向
pipe rupture 管道破裂
pipe rust remover 管子除锈机
piper whip protection 管道甩动保护
pipe saddle 管鞍座，管座
pipe sampler 管状取样器
pipe scaffolding 管子脚手架
pipe scale 管垢
pipe scraper 刮管器，刮板式清管器
pipe section 管段
pipe sewer 管沟
pipes for the main steam, the main feedwater, the cold section and the hot section of reheaters （电厂）四大管道
pipe shell 管壳

pipe shoe　管托
pipe sinking method　沉管法
pipe sleeper　管墩
pipe sleeve　管径套管，管套，连接管套
pipe socket　管座
pipe spanner wrench　管（子）钳
pipe-stabilizing pile　稳管桩
pipe stanchion　管支柱
pipe stem　管径，管柱
pipe still　管式蒸馏釜
pipe straightener　管子矫直机
pipe strap　管环，管卡，管子吊架
pipe supports and hangers　管道支吊架
pipe support　管道支座，管道支架
pipe sweeper　扫管机
pipe tap　管分接头，管螺攻，管子螺丝攻，分管接头，堵头
pipe tee　T形管，三通管
pipe test coupon　管道试样
pipe threader　管螺纹机，套丝机
pipe threading and cutting machine　管子套丝切断机
pipe threading machine　管套丝机
pipe thread tap　管螺纹丝锥
pipe thread　管端螺纹，管螺纹
pipe thrusting　顶管法
pipe tongs　管钳，管子钳
pipe tracing　管道加热保温
pipet rack　吸液管架
pipe trench　管沟，管缆沟
pipet stand　吸液管架
pipette analysis　吸管分析法，移液管分析
pipette method　吸管法
pipette　吸量管，吸移管，吸出管，滴管
pipe & tube making machines　管筒制造机
pipe tunnel　管式隧道，管沟
pipe twist　修管器，管钳
pipe type cable　钢管电缆，管装电缆，管式电缆
pipe-type combustion chamber　管形室
pipe-type gas compression cable　钢管压气电缆
pipe-type oil-filled cable　钢管充油电缆
pipet　移液管，滴管，吸液管
pipe union　管接头，联管节，管子接头
pipe-ventilated motor　管道通风式电动机
pipe vice　管虎钳
pipe vise　管子钳
pipe wall tap　管壁分接头
pipe warming-up　暖管
pipe welding machine　焊管机
pipe weld　管道焊缝
pipe well　管井
pipe whip restraint　管道防甩击装置
pipe whip　管道振荡，管道甩动
pipeworks　管道工程，管道系统
pipe wrench　管扳钳，管子钳，管钳，管道扳手，管钳子，管扳手
pipe　管材，管道，管线，管子，管状物，用管子输送，管，测压管头
piping and cable penetration　管道和电缆贯穿件
piping and instrumentation diagram　管线和仪表图，工艺流程及表计图，管道和仪表布置图
piping arrangement plan　管道布置平面
piping attachment　管道附件
piping breaking　管道破裂
piping by heaving　隆起管涌
piping chamber　管道室
piping chase　管道沟，管道槽
piping class　管道等级
piping contractor　管道合同商
piping diagram　管系图，管路图，管道系统图
piping drawing　管系图，管路图
piping effect　管涌
piping element　管道元件
piping engineer　管道工程师
piping erection　管道安装
piping erosion　管道侵蚀
piping fabricator　管道制造商
piping failure exclusion region　管道破损排除区
piping failure　管道破损
piping geometry　管道布置，管道排列
piping layout　管道布置
piping line　管系，管路线
piping load　管道负荷
piping mark band　管道标志带
piping material　管道材料
piping network　管网
piping penetration　管道穿墙
piping plan　管道平面图，管道详图，管路图，管道图
piping pressure drop　管道压力损失
piping pump　柱塞泵
piping ratio　管涌比
piping reaction　管道反作用力
piping rolling shop　卷管车间
piping run　管段，管路装置
piping shop　管工车间
piping support　管道支架，管架
piping symbol　配管图例
piping system　管道，管系，管道系统，管路系统，管汇系统
piping work　铺管工程
piping　管路输送，铺设管线，管道布置，导管，管道，管系
pip integrator　脉尖积分器，脉冲积分器
PIP(peripheral interchange program)　外围交换程序
PI(proportional-integral)　比例积分
pip-squeak　不重要的东西，控制发射机的钟表机构
PI(pulse input)　脉冲量输入，脉冲输入
pip　尖头信号，报时信号，峰值，筒，导管
piracy　侵犯专利权，盗版行为，海盗，袭夺
Pirani vacuum gauge　皮拉尼真空计
pi-section　（梯形网络的）π形结
pisolite　豆石
pisolitic limestone　豆状灰岩
pisolitic structure　豆状构造
pissasphalt　软沥青
piston accumulator　活塞式蓄能器
piston actuator　活塞执行机构
piston attenuator　活塞式衰减器

piston-balanced control valve　活塞平衡的控制阀
piston barrel　活塞筒
piston blast engine　活塞鼓风机
piston check valve　活塞式止回阀
piston compressor　活塞压缩机
piston corer　活塞取芯器
piston cylinder arrangement　活塞气缸部件，活塞气缸排列
piston damping chamber　活塞阻尼室
piston displacement　活塞排量，气缸工作容积
piston drill　活塞式冲击钻
piston flow reactor　活塞流反应器
piston flow　活塞流
piston gauge　活塞（式）测压计，活塞（式）压力表
piston groove　活塞槽
piston lift check valve　活塞抬升单向阀
piston locking　活塞闭锁环
piston motor　活塞液压马达
piston-operated valve　活塞动作阀
piston pin boss　活塞销毂，活塞销孔，活塞销座，活塞销衬套
piston pin　活塞销
piston pump　活塞泵，柱塞泵
piston sampler　活塞取样器
piston seal　活塞密封圈
piston slide valve　活塞滑阀
piston stroke　活塞冲程，活塞行程
piston throttle　活塞式节流阀
piston-type flowmeter　活塞式流量计
piston-type oil pump　柱塞式油泵
piston-type pressure gauge　活塞式压力表，活塞式压力计
piston-type soil sampler　活塞式取样器
piston-type valve　柱塞阀
piston-type variable area flowmeter　活塞式变截面流量计
piston-type wave generator　活塞式生波机
piston valve engine　活塞阀配汽蒸汽机
piston valve steam engine　活塞阀蒸汽机
piston valve　活塞阀，活塞滑阀
piston wrench　活塞扳手
piston　活塞，柱塞
PISW(process interrupt status word)　过程中断状态字
pit barrel pump　地坑筒式泵
pitchabletip　可调桨叶尖
pitch actuation　调桨驱动机构
pitch and catch technique　一发一收探伤法
pitch-and-roll buoy　自动测波浮标
pitch angle controller　桨距角控制器
pitch angle　桨距角，坡度角
pitch axe　鹤嘴斧
pitch bead weld　斜坡堆焊焊缝
pitch bearing　调桨轴承
pitchchange mechanism　变桨距机构
pitch changing　变桨距
pitch-chord ratio　相对栅距，节弦比
pitch circle diameter　节距圆直径，节圆直径
pitch circle　节圆
pitch coal　沥青煤
pitch coil　整距线圈
pitch controller　调桨控制器
pitch control mechanism　桨距控制机构，防纵摇机构
pitch control system　调桨控制系统
pitch control　色调调节，节距调节，音调控制
pitch curve　分度曲线
pitch cylinder　桨距调节气缸
pitch damping　俯仰阻尼，纵摇力矩
pitch ratio　栅距比
pitch diameter　节径，中径，平均直径，节圆直径，分度圆直径
pitch dimension　梯段尺寸
pitched roof　坡屋面，斜屋面
pitched work　块石护坡工程
pitch face　斜凿面
pitch factor　节距系数，节距因数，短距系数
pitch gear　径节声轮
pitch gyroscope　俯仰陀螺仪
pitch indicator　螺距指示器
pitch inertia　纵摇惯量，俯仰惯量
pitching anticline　伏背斜
pitching coefficient　调桨系数，俯仰系数
pitching equation　调桨方程
pitching gear　调桨齿轮
pitching lever　调桨杠杆
pitching moment balance　俯仰力矩平衡
pitching moment coefficient　俯仰力矩系数
pitching moment linearity　俯仰力矩的线性变化
pitching moment　俯仰力矩
pitching motion　俯仰运动
pitching oscillation　纵向振动，俯仰振动，相对横轴振动
pitching pile　插桩
pitching resolution　调桨分辨率
pitching syncline　伏向斜
pitching vibration　俯仰振动，纵摇振动
pitching　俯仰，砌石护坡，倾斜的
pitch interval　音程
pitch macadam　沥青碎石路
pitch mastic　沥青砂胶
pitchmotor　调桨马达
pitch of arch　拱高度
pitch of cascade　叶栅节距
pitch of stabilizer　稳定器的节距
pitch of stairs　楼梯坡度，楼梯斜坡度
pitch of strand　绞距，扭距
pitch of teeth　齿距
pitch of thread　螺距
pitch of transposition　换位节距
pitch of turn　匝距
pitch of winding　线圈节距，绕组节距
pitch paint　沥青漆
pitch pipe　沥青煤
pitch play　齿隙
pitch point　节点
pitch regulated rotor　变距限速风轮
pitch regulated teetering rotor　调桨摇摆风轮
pitch regulated wind turbine　桨距调节风电机组
pitch regulated　桨距调节
pitch regulation of WTG　风电机组变桨距调节

pitch regulation 桨距调节
pitch serve motor 调节伺服马达
pitch setting 桨距调节
pitch-shortening factor 短距系数
pitch spread 节距，间距
pitch system 调桨系统，变桨机构
pitch teeter coupling 变桨摇摆连接器
pitch to diameter ratio 距径比
pitch to feather 顺风调桨
pitch to stall 调桨失速
pitch up 上仰
pitch vibration 俯仰振动，纵摇振动
pitch 变桨，调桨，桨距，齿距，高跨比，间距，节距，螺距，齿节，栅距，斜度，沥青，行距，孔距，管距
pit coal 地坑煤，垃圾煤，矿产煤
pit corrosion 麻点腐蚀
pit drainage pump 坑内排水泵
pit excavation 基坑开挖
pit foundation 坑基
pit furnace 地坑炉
pith-ball electroscope 木髓球验电器
pithead power plant 坑口电站，坑口电厂
pit liner 凹坑衬垫，机坑里衬
pit mill 竖井磨煤机
Pitometer measurement 毕托管量测
Pitometer 毕托管，毕托压差计【测量流速】
Pitot curve 总压曲线，全压曲线
Pitot hole 皮托孔，全压孔，总压孔
Pitot line 皮托管，空速管，风速管
Pitot loss 全压损失，总压损失
Pitot meter 测速管
Pitot pressure gauge 全压测量器，皮托压力计
Pitot pressure inlet port 皮托管压力进口
Pitot pressure 皮托管压力，皮托压力，全压
Pitot probe 皮托管，总压力测量管
Pitot rake 梳状皮托管，全压管
Pitot sphere 毕托球体
Pitot static difference 全静压差，动压头，全压力与静压力差，总压静压差
Pitot static drain 全静压管排水口
Pitot static head 皮托静压水头，全静压探测管头，全静压头
Pitot static rake 梳状动静压管
Pitot static traverse 皮托管静压测定
Pitot-static tube 皮托静压管
Pitot support rod 皮托管支杆
Pitot-traverse method 皮托管移动测量
Pitot tube flowmeter 皮托管流量计
Pitot tube installed rod 皮托管安装杆
Pitot tube method 皮托管法，皮托管入口
Pitot tube 皮托流速测定管，皮托流速管，空速管，皮托管，总压管，流速管
PIT(personal income tax) 个人进口税
pit recharge 坑槽补给
pit-run aggregate 天然级配骨料，未筛选骨料
pit-run gravel 天然料坑砂砾石
pit sampler 槽探取样器
pit sampling 探坑取样
pit skin 金属表面气孔
pits 凹痕，井
pitted area 凹坑部位
pitting attack 点蚀，点状腐蚀
pitting coefficient 汽蚀系数
pitting corrosion resistance 耐点蚀性能，抗点蚀性能
pitting corrosion test 点腐蚀试验
pitting corrosion 麻点腐蚀，孔蚀
pitting factor （局部）腐蚀系数，点蚀系数
pitting guarantee 汽蚀保证
pitting of contact 触点烧坏
pitting 麻点腐蚀，局部腐蚀，点蚀，麻点，小孔，凹痕，锈斑，锈痕，凹陷
pit-type turbine 竖井式水轮机
pit 凹坑，水坑，窖，池，槽，凹痕，井，坑，洞，矿井，矿坑
PIU(path information unit) 通路信息单元
pivotal center （轴颈）回转中心
pivotal point （轴颈）回转中心点
pivot anchor bolt 支枢锚定螺栓
pivot bearing 枢轴承
pivot cell 轴支枢，称重传感器
pivoted bearing 自位轴承
pivoted bucket carrier 翻斗式运料机
pivoted bucket conveyer 斗式运输机
pivoted door 转动门
pivoted end column 铰端支柱
pivoted jib crane 旋转式悬臂吊车，旋转式悬壁起重机
pivoted-pad bearing 自调轴承
pivoted-pad guide bearing 刚性支承式导轴承
pivoted-pad thrust bearing 支枢式推力轴承
pivoted shoe bearing 可倾瓦轴承，自位轴承，立式止推轴承
pivoted shoe 可倾瓦式，轴瓦
pivoted side arm 侧向转臂
pivoted weir 翻板堰，活动堰
pivoted window 转动窗
pivoted 旋转的，铰链接合的
pivot gate 转动式闸门
pivoting stacker 绕中心旋转的堆料机
pivoting support 铰接支座
pivoting 铰链，枢轴
pivot pin 枢轴销钉，枢销
pivot point layout 测深垂线定位，转点定位法，垂直定位法，支点布置
pivot point line 固定起点线【用于水文测验断面】
pivot point 支点，支枢点
pivot ring 支承环
pivot system 枢轴系统
pivot-valve bushing 滑阀套
pivot 枢轴，枢纽，中心点，中枢，要点，旋转中心，主元，在枢轴上转动，转轴，支点
pixel resolution 像元分辨率
pixel 图素，像素
PIXIT(protocol implementation extra information for testing) 协议实施附加信息，协议实现的附加信息，测试用的协议实现附加信息
PKR 巴基斯坦卢比
place a contract 订合同
place and date of issue 签发地点与日期

place an order 订货
placed-in-situ 现浇
placed material 填料
placed on 2mm centers 中心距为2毫米，按2毫米中心距布置
placed rockfill 砌石
placed stone facing 砌石护面
place in circuit 接入电路
place in operation 投入运行
place in service 投入运行
place into commercial operation （电站）投入工业运行
place into requisition 征用
place limits on 限制
placement of fuel channels 燃料组件盒就位
placement 位置，配置，排列，布局，布置，放置，方位，部位，填筑，定位
place of arbitration 仲裁地点
place of delivery 交货地点
place of departure 始发地
place of dispatching 发货地点
place of loading 装货地
place of origin 原产地
place of payment 支付地点
place of production 产地
place of registration 注册地点
place of transshipment 装运地点
place under negative pressure 置于负压
place 地点
placidity 平稳
placing concrete against natural ground 地模混凝土浇筑
placing concrete 浇灌混凝土
placing into commercial operation 投入商业运行
placing machine 混凝土浇筑机
placing nozzle 喷嘴，可置喷嘴
placing of concrete 混凝土浇筑
placing of fuel channels 燃料元件盒就位
placing of the facing 护面铺筑
placing of turbine pedestal 汽轮机台板［基座］就位
placing reinforcement 放置钢筋
placing temperature 混凝土浇筑温度，入仓温度
placing 灌浇（混凝土），浇筑，放置
plagiarism 剽窃，抄袭
plagioclase granite 斜长花岗岩
plagioclase 斜长石
plain arch 平拱
plain balance differential protection 简单平衡的差动保护
plain bar reinforcement 无节钢筋，光面钢筋
plain bars 普通钢筋，光（面）钢筋，无节钢筋（混凝土）
plain bearing 普通轴承，平面轴承，滑动轴承
plain bevel 简单坡口
plain brick wall 清水砖墙，清水墙
plain carbon steel 普通钢，普通碳钢
plain cement 纯水泥，清水泥
plain circular furnace 平直圆形炉胆
plain climate 平原气候
plain concrete pipe 素混凝土管
plain concrete structure 素混凝土结构
plain concrete 素混凝土，无筋混凝土，普通混凝土【没有钢筋或预应力】
plain conductor 普通导线，裸导线
plain cutout 熔丝，保险丝
plain deflector 平面折流器
plain dressing 光面修整
plain end 平端面，平端
plain face 光滑表面
plain fill 素填土
plain fin 平直的肋片
plain flange 对接法兰，平面法兰
plain flap 简单襟翼
plain friction bearing 滑动轴颈轴承
plain furnace 平炉胆
plain gage 平面量规
plain girder 实腹梁
plain glass 防护白玻璃，平板玻璃
plain grate 平炉排
plain hammer 光锤
plain journal bearing 滑动支持轴承
plain labyrinth 平齿迷宫汽封
plain lead-covered cable 光皮铅包电缆
plain masonry 素圬工，无筋圬工
plain mitre joint 平斜接合
plain paper 空白页，空白纸
plain plate 光面钢板
plain presedimentation basin 水平式预沉淀池
plain rebar 光面钢筋，无节钢筋【混凝土】
plain region 平原地区
plain reinforcement 光面钢筋
plain reinforcing bar 光面钢筋
plain resistance 欧姆电阻
plain roller bearing 平行滚珠轴承
plain round bar 光面圆钢筋，光圆钢筋
plain sedimentation 普通沉积，自然沉淀
plain sequence 正序
plain settling tank 自然沉淀池
plain sheet iron 无楞铁皮
plain soil cushion 素土垫层
plain squirrel-cage motor 正常结构的鼠笼电机
plain steel 普通钢
plain surface 光面
plain tile 黏土瓦，平瓦
plain tract 冲积河段，下游河段，平原河段
plain transit 普通经纬仪
plain tube bank 光管束
plain tube 光管
plain vane 柱面叶片，圆柱形叶片
plain washer 平面垫圈
plain water fire extinguisher 清水灭火器
plain water 淡水
plain 平坦的，光滑的，平的，简单的，普通的，平素的，朴实无华的，平原
plan and prepare 筹划
plan and profile drawing （杆塔位）平断面图
planar air-gap generator 平面气隙发电机
planar array 平面阵列
planar body 准二度体，平薄物体
planar defect 平面缺陷
planar diode 平面二极管

planar epitaxial transistor　平面型外延晶体管
planar flow　准二维流动，准平面流动
planar location　平面定位，二维定位
planar motor　平面电动机
planar network　平面网络
planar-shear deformation　平面剪切形变
planar stator motor　平面定子电动机
planar wave system　平面波系
planar　平面的
planation surface　夷平面，侵蚀面
Planck's constant　普朗克常数，普朗克恒量
Planck's mean absorption coefficient　普朗克平均吸收系数
Planck's quantum　普朗克量子
plancton　浮游生物
plan diagram　平面图
plane angle　平面角
plane array　平面排列，平面网络
plane cascade　平面叶栅
plane cell　翼组
plane conductor　平板导体
plane cross section　平断面图
plane curve　平面曲线，平曲线
plane deformation　平面变形
plane drawing　平面图
plane electrode　平面电极
plane face flange　平法兰
plane flow　平面流动，二维流动
plane front　平面波阵面
plane gate　平面闸门
plane geodesy　平面测量学
plane girder　平面大梁
plane hologram　平面全息图
plane interface　平面分界面
plane joint　平面接头
plane lattice　平面网络
plane motion　平面运动
planeness　平整度，平面度
plane neutron source　平面中子源
plane of max　最大剪应力面
plane of polarization　极化面
plane of principal shearing stress　主剪应力面
plane of reference　参考面，参照面，假定面
plane of shear　剪切面
plane of symmetry　对称面
plane of unbalance　不平衡面
plane of weakness　软弱面
plane pad　平垫
plane plastic deformation　平面塑性变形
plane polarization　平面极化
plane-polarized radiation　平面偏振辐射
plane-polarized wave　平面极化波，平面偏振波
planer drilling machine　龙门钻床
plane reference system　平面坐标系统
planer machine　龙门刨床，刨边机
planer　龙门刨床，牛头刨床，刨机，刨工，刨床
plane seal housing　平板密封套
plane separation　平面距
plane shock　平面冲波，正冲波
plane source　平面源
plane stagnation flow in transpiration cooling　发散冷却中的二维滞止流动
plane state of stress　平面应力状态
plane strain fracture toughness　平面应变断裂韧性
plane strain instability　平面应变不稳定性
plane strain toughness　平面应变断裂韧性
plane strain　平面应变
plane stress toughness　平面应力断裂韧性
plane stress　平面应力
plane surface　平面
plane surveying　平面测量
plane table alidade　平板照准仪
plane table fixing　平板仪定位
plane table measurement　平板仪测量
plane table method　平板仪法
plane table photogrammetry　平板仪摄影测量
plane table plate　平板仪测图板
plane table station　平板仪测站
plane table surveying　平板仪测量，平板仪测量法
plane table survey　平板仪测量
plane table tachy-metric survey　平板仪视距测量
plane table tachymetry　平板仪视距法
plane table topographic survey　平板仪地形测量
plane table traverse　平板仪导线
plane table triangulation　平板三角测量
plane table　平板仪
plane-tabling　平板仪测量
planetary atmosphere　行星大气
planetary boundary layer　行星边界层
planetary circulation　行星环流
planetary electron　轨道电子
planetary gearbox yaw drive　行星齿轮偏航驱动器
planetary gearbox　行星齿轮变速箱，行星齿轮箱
planetary gear drive mechanism　行星齿轮传动机构
planetary gearing　行星齿轮传动
planetary gear train　行星齿轮系
planetary gear　行星齿轮
planetary roll mill　行星辊式磨煤机
planetary timing gear　行星定时齿轮，行星齿轮定时装置
planetary train　行星齿轮系
planetary transmission　变速器，齿轮传动，行星齿轮传动，行星传动装置
planetary wave　行星波
planetary wind　行星风
planetary　行星的
planet carrier　行星齿轮架，行星架
planet gearbox　行星齿轮箱
planet gear　行星齿轮
plane triangulation　平面三角测量
planet　行星
plane view　鸟瞰图
plane waves interaction　平面波相互作用
plane wave technique　板波技术
plane wave　平面波
plane　平面，程度，水平，板
plan figure　平面图
planimegraph　面积比例规，缩图器

planimeter 面积仪，侧面器，求积仪
planimetrical 平面的，测面积的，地形平面投影法的
planimetric control point 平面控制点
planimetric coordinates 平面坐标
planimetric line 轮廓线
planimetric method 平面测量法
planimetric position 平面位置
planimetric rectangular coordinates 平面直角坐标
planimetric survey 平面测量，平面图测量
planimetry 测面（积）法
planing boat 滑行艇
planing machines vertical 立式刨床
planing machines 刨床
planing 滑行，刨的
planishing 碾平
planitron 平面数字管
plank check dam 木板谷坊，木板拦沙坝
plank flume 木板槽
plank pile 厚板桩
plank sheathing 木板护壁，木模板
plankton 浮游生物
plank truss 木板桁架
plank 板材，垫板，厚板，板条，木板，板
plan layout 平面布置
plan map 规划图
plan meter 测面仪
plan model 规划模型
plannar structure 成面构造
planned capacity 规划容量
planned commodity economy 计划商品经济
planned derating hours 计划降低出力小时数
planned derating 计划降低出力
planned economy 计划经济
planned emergency exposure 计划中的应急辐照，计划应急照射量
planned hours 计划时数
planned idle time 计划内停运时间
planned investment 计划投资
planned maintenance 计划检修，计划维修
planned management 计划管理
planned minor repair outage hours 计划小修停运小时
planned obsolescence 计划报废
planned outage factor 计划停运系数
planned outage for inspection 计划检查停堆
planned outage hours 计划停运小时
planned outage rate 计划停堆率
planned outage state 计划停运状态
planned outage 计划停役，计划停堆，计划停运，计划停机，计划停车
planned overhaul outage hours 计划大修停运小时
planned payment 计划付款
planned price 计划价格
planned profit 计划利润
planned progress 计划进度
planned project 规划项目
planned shutdown 计划停运
planned special exposure 计划中的特殊辐照，有计划的特殊照射量
planned task 拟定任务
planned 拟定的
planner 计划员
planning and scheduling 制订计划于进度，计划与调度，规划与调度
planning and site supervision 设计和现场管理
planning area 规划区
planning bureau 规划局
planning capacity 规划容量
planning chart 计划图，计划图表
planning commission 规划委员会，计划委员会
planning control sheet 计划控制表
planning for resident relocation 库区移民安置规划
planning group 规划小组
planning idea 规划设想
planning index 计划指标
planning level 规划水平
planning management 计划管理
planning map 规划图
planning of connecting power plant to grid 发电厂接入系统设计
planning of power network 电网规划
planning of power sources 电源规划
planning of river basin 流域整体规划
planning section 计划科
planning site-selection 规划选厂址
planning siting 规划选厂
planning stage 规划阶段
planning supervisor 计划主管人
planning survey 规划测量，设计测量，规划调查
planning target 规划目标
planning 规划，计划的拟定，活动计划
plan of capital construction 基本建设计划
plan of column grid 柱网平面
plan of construction machinery equipping 施工机械装备计划
plan of energy saving 节能计划
plan of measures against accident 反事故措施计划
plan of power demand and supply 电力供求计划
plan of redemption 分期偿还计划
plan of sewerage system 污水管道系统布置，下水道系统布置
plan of site 场地总平面图，总布置图
plan of wiring 线路图，接线图
planometer 测面仪，测平仪，求积仪
plan outline 计划大纲
plan position indicator scope 平面位置显示器
plan position indicator 平面位置指示器
plan section 水平截面
plansifter 平面筛
plant air compressor 厂用空（气）压（缩）机
plant arc cutting 等离子弧切割
plant-area underground piping and culverts drawing 厂区地下沟道图
plant arrangement 工厂布置，工厂布局
plant attenuation 被控对象中的振荡衰减
plant auxiliaries 电厂辅助系统
plant availability 设备可利用率，设备利用率，

电厂可利用率
plant battery 厂用电池
plant bulk 装置外形尺寸
plant capacity 电厂容量，电站容量，装置功率，装机容量，设备功率，设备能力
plant cavitation factor 装置汽蚀系数
plant characteristic curve 全厂特性曲线
plant communication 厂内通信
plant compartment （安全壳内）设备室
plant consumption 厂用电消耗
plant contractor 工厂承包商
plant cooldown 电厂冷却，设备冷却
plant cost 电厂费用
plant cover 植被
plant cycle 理想动力装置循环
plant demand 自用量
plant design capacity 电厂设计容量
plant design output 电厂设计输出功率
plant design 电厂设计
plant director office 厂长室
plant director 厂长
plant discharge 电站泄流量
plant distribution system 电厂配电系统【中低压】
plant districting 工厂划分
plant drainage cooler 电厂疏水冷却器
plant drainage heat exchanger 电厂疏水热交换器
plant drainage system 设备疏排水系统，电厂疏排水系统
plant drainage 工厂排水
plant drains 电厂疏排水，设备疏排水
plant efficiency 发电厂效率，装置效率，电站效率，设备效率
plant effluent 工厂废水，工厂污水
plant electrical consumption 厂用电
plant emergency cooldown system 电厂紧急冷却系统
plant executive 工厂负责人
plant expected capacity 电站预期功率
plant external feedback 电厂外反馈
plant factor 电厂利用率，设备使用系数，设备使用率，设备容量因数，电站容量因数，电站容量因子，电厂利用系数
plant failure 电厂故障，装置故障
plant feed conveyor system 工厂供料输送系统
plant fence 电厂围墙
plant front area 厂前区
plant general office 厂办公室
plant ground elevation 厂区地面标高
plant heat rate 全厂热耗率，电厂热耗，电站热耗
plant inspection program 电厂检查计划
plant lag 被调对象滞后
plant layout drawing 电厂总平面布置图
plant layout 工厂布置，设备布置，厂房布置，动力装置布置图
plant licensing life extension 核电厂批准寿命延长
plant life extension 核电厂寿命延长
plant lighting system 电厂照明系统
plant load demand 电厂负荷需求
plant load factor 电厂负荷因子，设备负载系数，电厂负载系数，电厂容量因数
plant local power plant 地方电厂
plant location programming 厂址规划
plant location 厂址，电站位置
plant loss 电站损失
plant management 电厂运行管理
plant manager 电厂经理
plant metabolism 植物新陈代谢
plant mix 厂拌和物
plant models software 电厂模型软件
plant network 厂内网路，厂内电力网
plant nuclear safety review committee 电厂核安全审评委员会
plant operating reliability 电厂运行可靠度
plant operational capabilities 电厂的运行能力
plant operational quality assurance procedure 电厂运行质量保证程序
plant operation and betterment department 电厂运行及改进部
plant operations review committee 电厂运行审评委员会
plant operator 电厂操纵员，电厂营运者
plant ordering 发电设备排列次序
plant-own power 厂用电
plant parameter 设备参数
plant performance figure 电厂性能指标，质量指标，运行指标
plant performance 设备性能
plant power raising 电厂功率提升
plant power setback 电厂功率下降
plant process display instrumentation 电厂过程显示仪器
plant property line 厂区界线，厂址边界
plant protection system 电厂保护系统
plant quality release 电厂质量证书
plant radiation monitoring system 电厂辐射监测系统
plant radiation monitoring 电厂放射性监测，电站辐射监测
plant rubber 天然橡胶
plant-scale equipment 工厂用设备，生产用设备
plant schedule 设备明细表
plant section 设备零配件
plant security system 电厂保安系统
plant security 电厂保卫
plant service building 生产办公楼
plant service power rate 厂用电率
plant service railway line 铁路专用线
plant service steam header 厂用蒸汽母管
plant service steam piping 厂用蒸汽管道
plant service transformer 厂变
plant shutdown 电厂停运
plant sigma 装置汽蚀系数
plant site coal handling facility 厂内输煤设备
plant site scheme 厂址方案
plant site terrain 厂区地形
plant site 厂址
plant size 电厂规模
plant status 电厂状况
plant steam exhauster 厂用排汽集汽箱

plant superintendent 电站负责人，厂长
plant supervisory information system 厂级监控信息系统
plant switchboard 厂用交换台，厂用开关板
plant system 电厂系统
plant thermal efficiency 全厂热效率
plant time system 电厂时间系统
plant unit 电厂单元机组
plant utilization factor 装置利用系数
plant utilization period 装置利用期，设备利用期
plant waterway 电站输水道
plant without storage 径流式水电站，无库容的水电站，无库容电站
plant 工厂，制造厂，装置，设备，装备，车间，站，植物
plan view 平面图，俯视图
plan 平面图，计划，设计，规划，轮廓，草案，方案
PLA(programmed logic array) 程序控制的逻辑阵列
plash 积水潭
plasma arc cutting 等离子弧切割，等离子体弧切割
plasma arc surfacing 等离子堆焊
plasma arc welding 等离子弧焊
plasma chemical vapor deposition 等离子体化学气相沉积
plasma current 等离子体电流，等离子体流，等离子流
plasma cutter 等离子切割机
plasma cutting machine 等离子切割机
plasma display 等离子显示器
plasma dynamics 等离子体动力学
plasma electric conductivity 等离子体导电率
plasma etching 等离子体刻蚀
plasma flow 等离子体流
plasma frequency 等离子体频率
plasma furnace 等离子炉，等离子熔炼炉
plasma generator 电桨发电厂
plasmaguide 等离子体波导管
plasma heating 等离子加热，等离子体加热
plasma jet 等离子流
plasma loading characteristics 等离子体负载特性
plasma oscillation 等离子区振荡
plasma spray coating 等离子喷涂
plasmas praying 等离子弧喷涂，等离子喷涂
plasma torch and torch carriage 等离子炬和炬架
plasma torch 等离子炬
plasmatron 等离子流发生器，等离子管
plasma welding 等离子焊接
plasma 等离子，等离子体，等离子区
plaster base 粉刷打底，涂层底层
plaster board 石膏板，粉饰板
plaster brick 石膏砖
plastered brickwall 混水砖墙，混水墙
plastered metal-lath ceiling 抹灰金属网顶棚
plastered wall 抹灰的墙，粉刷的墙
plasterer 抹灰工
plaster ground 抹灰找准用的木条，抹灰用的靠尺，抹灰准木
plastering machine 抹灰机
plastering mortar 抹灰砂浆
plastering 抹灰泥，抹面工作，烧石膏，泥水工作，抹灰，粉制
plaster lath wall 板条墙
plaster mineral wool slab 石膏矿棉板
plaster model 石膏模型
plaster of Paris 熟石膏，巴黎石膏
plaster on expanded metal lath 钢板网抹灰
plaster on metal lath 钢丝网抹灰
plaster plate wall 石膏板墙
plaster putty 石膏腻子
plaster relief model 石膏地形模型
plaster rendering 抹灰打底
plaster shooting （孤石的）复土爆破
plaster slab gypsum lath 石膏板
plaster slab 石膏板
plaster sprayed with oil paint 抹灰喷油浆
plaster stone 石膏，生石膏
plaster 灰泥，灰浆，熟石膏，石膏，抹面，涂层，粉饰，涂以灰泥，打上石膏，粘贴
plastic adjustment 塑性调整
plastic analysis 塑性状态分析
plastic asphaltic material 塑性沥青材料
plastic behaviour 塑性
plastic belt 塑料带
plastic bending 塑性弯曲
plastic blade 塑料叶片，塑性叶片
plastic blunting of crack tip 裂纹尖端塑性钝化
plastic body-like behaviour 类塑性体特征
plastic body 塑性体
plastic bottle 塑料瓶
plastic buckling 塑性翘曲
plastic bushing 塑料套管
plastic cable end box 塑料电缆终端盒
plastic cable 塑料电缆
plastic capacitor 塑料电容器
plastic clamp 塑料夹具
plastic clay 塑性黏土，塑性土
plastic-coated EMT 塑料涂层的电气金属管道
plastic-coated metal gasket 塑料涂层的金属密封垫
plastic-coated sheet 塑料护面钢板
plastic-coated tube 有塑料涂层的管子，塑料被覆管
plastic coefficient 塑性系数
plastic collapse 塑性破坏
plastic commutator 塑料换向器
plastic composite 复合塑料
plastic concrete 塑性混凝土
plastic construction 塑料结构
plastic-covered cable 塑料护套电缆
plastic cracking 塑性开裂
plastic deformation 塑性变形
plastic design 塑性设计
plastic dielectric capacitor 塑料介质电容器
plastic drain 塑料排水管
plastic-elastic stress 弹塑性应力
plastic encapsulated coil 塑料绝缘线圈，塑料包封线圈
plastic equilibrium 塑性平衡

plastic failure 塑性破坏
plastic fatigue 塑性疲劳
plastic fibre 塑料纤维
plastic film 塑料薄膜
plastic filter 塑料过滤器
plastic flow limit 塑性流变极限
plastic flow loss 塑流损失
plastic flow phenomenon 塑流现象
plastic flow slide 塑流性滑动
plastic flow 塑性流动，塑性流，塑流，塑性变形，黏（滞）流（动）
plastic fluid 塑性流体
plastic foam 泡沫塑料
plastic foundation 弹性基础
plastic fracture 塑性断，塑性断裂
plastic friction 塑性摩擦
plastic hinge 塑性铰
plastic hysteresis 塑性滞后
plastic incinerator 塑料焚化炉
plastic index 可塑指数
plastic instability 塑性失稳
plastic insulated cable 塑料绝缘电缆
plastic insulated wire 塑料绝缘线
plastic insulation material 塑料绝缘材料
plastic insulation 塑料绝缘
plasticity analysis 塑性分析
plasticity chart 塑性图
plasticity coefficient 塑性系数
plasticity index 塑性指数
plasticity modulus 塑性模量
plasticity retaining 保塑
plasticity stability 塑性稳定性
plasticity theory of failure 塑性破坏理论
plasticity theory of limit design 塑性极限设计理论
plasticity 塑性（力）学，可塑性，塑性，成形性，适应性
plasticized elastomer 增塑弹性材料
plasticizer 可塑剂，增塑剂，塑化度
plasticizing agent 塑化剂，增塑剂
plastic limit tests 塑限试验
plastic limit 塑性限度，塑限
plastic lined pipe 衬塑管
plastic lining 塑料衬里，衬塑料
plastic loss 塑性损失
plastic-made plate pack 塑料制叠板
plastic map 立体地图
plastic material 塑性材料
plastic mechanics 塑性力学
plastic media filter 塑料滤料
plastic membrane 塑料涂层
plastic-moderated reactor 塑性慢化反应堆
plastic moment 塑性弯矩
plastic mortar 塑性砂浆
plasticon 聚苯乙烯薄膜
plastic pipe 塑料管
plastic potential 塑性势
plastic property 可塑性，塑性
plastic range 塑性范围
plastic ratchet 塑料棘轮扳手
plastic read-only disc 塑料只读磁盘
plastic refractory 耐火涂料
plastic resistance 塑性阻抗
plastic rock 塑性岩石
plastics board 塑料板
plastics ceiling 塑料顶棚
plastic scintillator detector 塑料闪烁体探测器
plastic sealing compound 塑料密封剂
plastic seal 塑料密封
plastics film 塑料薄膜
plastic shakedown 弹性稳定【力分析】
plastics handrail 塑料扶手
plastic-sheathed 外面涂塑料的，包塑料的
plastic sheath 塑料护皮，塑料套
plastic slag 塑性渣
plastic soil-cement lining 塑性灰土护岸
plastic soil 塑性土壤，塑性土
plastic sphere 塑料球
plastics refractory 塑性耐火材料
plastics roofing tile 塑料瓦
plastics tape 塑料带
plastic state 塑性状态
plastic strain ratio 塑性应变比
plastic strain 塑性应变，塑性变形，塑性应力
plastic substance 塑料物质
plastics 塑料制品，塑料，塑胶
plastic tape 塑料带
plastic theory of failure 塑性破坏理论
plastic theory of limit design 塑性极限设计理论
plastic theory 塑性理论
plastic tile 塑料瓦
plastic tube 塑料管
plastic veneer 塑料饰面
plastic-viscous flow 塑性黏滞流（动）
plastic volumetric strain 塑性体积应变
plastic wall cladding 塑料壁纸
plastic waste 塑料废料
plastic wire 塑料线
plastic work 塑料加工
plastic yield deformation area 塑性屈服变形区
plastic yield point 塑性屈服点
plastic yield strain 塑流应变
plastic yield stress 塑性屈服应力
plastic yield surface 塑性屈服面
plastic yield temperature 塑性流动温度
plastic yield test 塑流试验，塑料变形测试
plastic yield 塑性变形，塑性屈服
plastic zone 塑性区
plastic 可塑的，塑性的，塑料的，塑料制品，塑料
plastisol 塑料溶胶
plasto-elastic body 塑弹性体，弹塑性物体
plasto-elasticity 弹塑性力学
plastograph 塑性变形记录仪
plastomer 塑料，塑性体
plastometer index 胶质层指数
plastometer indices 胶质层指数
plastometer test 塑性计检验，塑性计试验
plastometer 塑性计
plastometric shrinkage 最终收缩度
plate air preheater 板式空气预热器
plate and angle column 角钢和板组合柱

plate arch 平板拱
plate asphalt 沥青板
plateau characteristic 坪特性
plateau climate 高原气候
plateau section 高原地区
plateau slope 坪斜
plateau 曲线的水平部分，平稳时期，停滞时期，稳定水平，平稳状态，平顶，平直段，坪，海台，高原，台地，托盘，达到平衡，达到稳定时期
plate baffle 金属折焰板
plate beam 板梁
plate bearing test 板承试验，承载板试验
plate bearing 盘式轴承
plate bender 弯板机
plate bending machine 弯板机
plate bending roll 卷板机
plate blade 平叶片
plate bubble 图板水准管
plate casing 护板
plate-circuit detector 板极电路检波器
plate column 板式柱，板式塔
plate cooler 板状冷却器
plate current 板极电流
plate cut-off 阳极截止
plated beam 叠板梁
plate-detection 板极检波的
plate discharger 板形放电器
plated wire memory 镀膜线存储器
plate-edge bending press 压头机
plate electrode slag pool welding 板极电渣焊
plate electrode 板极，屏极
plate feeder 圆盘式给煤机
plate filter 板式过滤器
plate fin coil 板翅式空气加热器，散热片式盘管
plate fin exchanger 板翅式换热器，板肋式换热器
plate finned tube 翅片管，大套片管，肋片管，套片式翅片管
plate flange 平板式法兰
plate flattening machine 板材矫正机
plate foundation 板式基础
plate-fuel assembly 板状燃料组件
plate fuse 片状熔丝，片状熔断器
plate girder upper beam 大板梁
plate girder 钢大板梁，大板梁，板梁
plate glass 平板玻璃
plate grid capacitance 阳极栅极间电容
plate grid capacity 板栅电容
plate grid （蓄电池的）栅板
plate-hammer crusher 锤击式破碎机
plate heat exchanger 板式热交换器，板式换热器
plate height 塔板高度
plate insulation 绝热板
plate lead 阳极引线，屏极引线
plate loading test 平板载荷试验，平板承载试验
plate member holding tension insulator string 耐张挂线板
plate mica 云母板
platen airheater 板式空气预热器
platen aligning hock 平台对中锁钩
plate neutralization 阳极板中性化
platen heat exchanger 屏式热交换器
platen hook 翻车机平台锁钩
platenize 屏化，做成屏状
platen locking device 滑块锁紧装置
platen lock 平台止挡定位装置
platen reheater 屏式再热器
platen superheater 屏式过热器
platen 管屏，屏，平台，台板
plate of accumulator 蓄电池板极
plate orifice 盘式孔板，盘形孔板
plate-pulsed transmitter 板极脉动调制发射机
plate punching machine 平板冲孔机
plate rectifier 极板整流器，板极检波器
plate regenerator 板式回热器
plate roller 卷板机
plate screen 孔板拦污栅
plate shear 剪板机
plate spacing gage （燃料）板间距量规
plate spacing 塔板距
plate spring 板弹簧
plate-steel liner 钢板衬垫，钢板衬砌
plate stiffness 加劲板
plate strainer 板式过滤网
plate superheater 屏式过热器
plate tectonics 板块构造学说
platetest coupon 平板试样
plate theory 塔板理论
plate thickness 筛板厚度，板块厚度，金属板厚度
plate tower 板式塔
plate transformer 阳极变压器，屏极变压器
plate type airheater 板式空气预热器
plate type contraction joint 金属片伸缩缝
plate type cooler 板式冷却器
plate type electrostatic precipitator 板式静电除尘器
plate type heat exchanger 板式热交换器
plate type radiator 板式散热器
plate type reactor 板状元件反应堆
plate type recuperative air heater 板式空气预热器
plate vibrator 平板振动器
plate voltage 屏极电压，板极电压
plate web spar 腹板式（翼）梁
plate-welded spiral case 钢板焊接蜗壳
plate winding 屏极线圈
plate wire 镀线
plate-work shop 金工车间，铆焊车间
plate 板，盘，片，镀，平板，平台，电容器板，电镀，极板，钢板，面板，牌子
platform around a column 塔设备平台
platform balance 台秤，地秤，磅秤
platform bridge 天桥
platform car 平板车箱，平板汽车
platform crane 台车起重机
platform drill 台式钻机
platform for crane repair 吊车修理平台
platform hatch 平台短门，平台舱口
platform hoist 平台式起重机

platform link 翻车机平台连杆
platform lorry 平板车
platform railing 平台栏杆
platform roof 月台棚，站台棚
platform scale 磅秤
platform trailer 平板拖车，平台拖车
platform truck scale 汽车地磅，地秤
platform truck 平板拖车，平板货车，平板车
platform type vibrator 振动台，平台式振动器
platform weigher 地中衡，地磅
platform weighting machine 台秤
platform 桥，平台，地台，台地，人行道，站台，月台，脚手架，装卸台
plating crack 电镀裂纹
plating shop 电镀车间
plating 电镀，喷镀，镀层，镶衬，装甲
platinite alloy 高镍钢合金
platinium resistance thermometer 铂电阻温度计
platinoid 合金
platinum black 铂黑
platinum crucible 白金坩埚，铂坩埚
platinum dish 铂皿
platinum electrode 铂电极
platinum resistance temperature detector 铂热电阻
platinum resistance temperature needle 铂电阻测温探针
platinum resistance thermometer 铂电阻温度计铂阻温度计
platinum-rhodium thermocouple 铂铑热电偶
platinum RTD(platinum resistance temperature detector) 铂热电阻
platinum thermistor 铂热电阻
platinum thermostat 铂恒温器
platinum wire anemometer 铂丝风速计
platinum wire fabric 白金网
platinum 白金，铂
platometer 测面仪，求积仪
plattening 压平
platy joint 板状节理
platykuitic 低峰态
platy limestone 板状灰岩
platy structure 板状结构
plat 地区图
play adjustment 余隙调整
play a leading role 起主导作用
play a … role 起到……作用
playback head 读出头
playback （从磁带上）读出记录，反演，放音，放录像，再现
player 博弈者
play host to 招待，接待
play system 数字数据处理和显示系统
play 间隙，闪动（火焰），游动，游隙，窜动，浮动，余隙
plaza 广场，露天停车场
PLC coal handling system 输煤程控系统
PLCC(plastic leaded chip carrier) 有引线塑料芯片载体【塑封】
PLCC(plastic leadless chip carrier) 无引线塑料芯片载体【塑封】
PLCC(power line carrier communications) 电力线载波通信
PLC failure 可编程逻辑控制器故障，PLC故障
PLC(power-channel carrier channel) 电力线载波通道
PLC(power line carrier) 电力线路载波机
PIC(pressure indication and control) 压力指示和控制
PLC(Programmable Logical Controller) 可编程逻辑控制器，可编程控制器
PLC(public limited company) 股份有限公司，股票上市公司
PLCS(program logic control system) 程序逻辑控制系统
PLD＝payload 有效负载
plead 辩护
pleated 弄皱的，折叠的
pleating 皱纹，折叠
pleats filter 百褶过滤器
pledged assets 质押资产
pledge of movables 动产抵押
pledge 抵押，抵押品，誓言
plenary capacitance 全电容
plenary meeting 全体会议
plenary session 全体会议
plenipotentiary 全权代表，授全权委托的人，有全权的
plentiful runoff 丰水径流
plenum box 充气箱，静压箱
plenum chamber 增压室，充气室，送气室，静压箱，进气增压室
plenum duct 送气道
plenum flow 腔室中的流动
plenum gauging 充气测量
plenum plenum 膨胀空间【燃料棒】，（流化床锅炉）流化风室，水冷风室，压力通风系统，风室，正压室，腔室，充满，送气，增压室，风箱，全体会议，增压的
plenum process of tunnellin 隧洞压气掘进法
plenum space 稳压层
plenum subfactor 腔室分因子
plenum system 送气系统，压力通风系统
plenum ventilation 充气通风，压力通风
plenum zone 空腔区，气体收集区
pleuston 水漂生物
plexicoder 错综编码器
plexiglass 一种聚甲基丙烯酸甲酯有机玻璃，耐热有机玻璃，有机玻璃
PLEX(plant life extension) 核电厂寿命延长
pliability 柔韧性，可挠性
pliable material 软材料
plication 细褶皱
pliers entry 钳子式接入
pliers 老虎钳，克丝钳，台钳，钳子，钳
plier-type current meter 钳形电流表
plinth block 底座木块
plinth 接头座，底座，柱基，勒脚，柱础，基石，踢脚板，趾板，柱脚
Pliocene epoch 上新世
Pliocene series 上新统
pliofilm 胶膜，氯化橡胶薄膜

pliotron 功率电子管，空气过滤器
plished feedwater 净化给水，纯化给水
PLLEX(plant licensing life extension) 核电厂批准寿命延长
PLN 波兰兹罗提【货币】
PLN 印尼国家电力公司
plomatron 栅控汞弧管
plot experiment 小区试验
plot observation 小区观测
plotomat 自动绘图机
plot pen 绘图笔
plot plan 厂址位置图，平面布置图，总平面图，基址位置图，平面图，地区图
plot point 展点
plot runoff 小区径流
plotted line 规划线
plotter 绘图机，绘图仪，标图板，图形显示器，计划者，描绘器
plotting apparatus 绘图仪
plotting board 图形显示板，曲线板，测绘板，绘图板
plotting device 绘图器
plotting equipment 绘图设备
plotting instrument 绘图仪器
plotting machine 绘图机
plotting of points 展点
plotting paper 比例纸，方格纸，绘图纸，坐标纸，绘图比例尺，方格绘图纸
plot 绘图，测绘，描绘，标绘，标绘图，计划，图表，图，画曲线，曲线图，区分，小区
plough-black 利润再投资
plough groove 沟槽
plough scraper 犁式卸料器
plough tripper 犁式卸料器
plough unloader 犁式卸料器
plough 耕作，耕地，犁，刮板
plow feeder 犁式给料机，园盘给料机
plow 犁，刮板，刨，犁煤器，拨煤机
PL(parts list) 零部件清单，零部件明细表
PLSH＝polish/polishing 精处理
PLSHR＝polisher 精处理器
PLS(plant control system) 电厂控制系统
plucking 剥蚀
plug adaptor 转接接头，插塞，插头转接器
plug and feather hole 插楔开石孔
plug and feathers 裂石楔，插楔开石工具
plug and socket connection 插销联结
plug assembly 阻力塞组件
plugboard 插头板，插接板，插件板，插线盘，转接板，转插板，插座式转接板
plug braking 反接制动，反相制动
plug calorimeter 塞形测热计
plug cartridge 熔线塞，插塞式熔线
plug cock 旋塞，旋阀，转阀
plug connection 插塞连接，插头连接
plug contact 插头
plug control valve 塞形控制阀
plug detector 管塞探测器【蒸汽发生器管子】
plug dezincification 栓状脱锌
plug disk globe valve 柱塞阀瓣球阀
plug disk 柱塞阀瓣
plug drain valve 栓塞式排水阀
plug expansion cylinder 管塞膨胀缸
plug expansion mandrel 管塞膨胀心轴
plug expansion 管塞膨胀
plug extraction by metal disintegration machine 由金属粉碎机把管塞拔出
plug feed system 管塞供应系统
plug fit 管塞装配
plug flow model 活塞流模型
plug flow 阻塞流，塞状流，塞状流动
plug for radiographic inspection access hole 射线照相检查用塞子，射线探测孔塞
plug fuse 插塞式保险丝
pluggable unit 插件
pluggable 可插的，可反接制动的
plug gate valve 塞型闸阀
plug gauge 圆柱塞规，塞规，量规塞子，标准插销，孔塞
plugged chute switch 落煤管堵煤开关
plugged-in places 充电场所
plugged-in software 软插件
plugged-in 插上的【熔丝】，插上插头以接通电源，紧跟时代的
plugged program computer 插接程序计算机
plugged program 插入程序
plugger 压塞装置，压塞器
plugging above tubesheet 管板以上堵管
plugging degree 堵塞程度
plugging device assembly 堵管装置组件
plugging indicator 堵塞指示器
plugging inhibitor 防堵塞剂
plugging machine 堵管机
plugging meter 堵塞表【用于测量钠温】
plugging of the diversion opening 导流孔封堵
plugging parameters 堵塞参数
plugging relay 防逆转继电器
plugging-up device 闭塞装置
plugging 反相序制动控制，反向制动，闭塞装置，闭塞，堵塞，阻塞
plug hole 塞孔
plug-in board 插件，插接板
plug-in breaker 插入式断路器
plug-in card 插件
plug-in circuit 插入式电路
plug-in coil 插入式线圈，插换线圈
plug-in coupling transformer 插入式耦合变压器
plug-in device 插入式设备
plug inductor 插换感应线圈，插入式感应线圈
plug-in fuse 插入式熔丝
plug inlet check valve 柱塞入口单向阀
plug-in manipulator 插入式机械手，组合式机械手
plug-in module 插件【电子回路】
plug-in package 插入部件
plug-in printed circuit board 插入式印刷电路板
plug-in relay 插入式继电器
plug-in system 插换制，插塞连接式系统
plug interlock 插塞联锁
plug-in transformer 插入式变压器
plug-in type fuse 插入式熔断器
plug-in type 插入式

plug-in unit （电子回路）插件，插入单元，插入件，插入部件
plug-in 插入，插上，插入式的，嵌入的，带插头的，嵌入，可换插的
plug loader 管塞装填器
plug loading position 管塞装填位置
plug loading sensor 管塞装填传感器
plug loading 管塞装填位置
plug magazine 管塞储放箱
plug mill 芯棒轧管机
plug of cock 旋塞的塞子
plug of the diversion opening 导流孔塞
plug pressing 管塞冲压
plug program patching 插入程序修正
plug puller 管塞拉出器
plug receptacle 插座，插孔板
plug reversal 反接倒转，反接制动
plug selector 塞绳式选择台，塞绳式交换台
plug servicing facility 栓塞维修设施
plug servicing 栓塞维修设施，栓塞维修
plug shell 管塞外壳
plug socket 塞孔，插座，插孔
plug switch 插接开关，插头开关
plug tap 二道螺丝攻，柱形丝锥
plug-to-connector test lead 插头对插座的试验引线
plug-to-receptacle test lead 插头对插孔的试验引线
plug torquing test 管塞转矩试验
plug two-phase flow 团状两相流，塞状两相流
plug-type disc globe valve 塞型阀盘截止阀
plug-type resistance bridge 插塞式电阻电桥
plug-type variable-area flowmeter 插入式可变面积流量计
plug unit 栓塞单元，插塞单元
plug valve 旋塞阀，旋塞活门，塞阀
plug welding 塞焊
plug weld 管塞焊缝，塞焊，塞焊缝，电铆焊
plug wire 插线
plug wrench 火花塞专用扳手
plug 插头，插销，堵塞，接头，丝堵，柱塞，塞子，栓塞，填充物，消防栓，管堵，插塞，插入式连接器，堵住，插上插头，塞住，接插头，插，塞，栓
plumbago 石墨，铅笔图
plumb bob 测锤，垂球，铅锤，线铊，悬锤
plumber block 止推轴承，轴承台
plumber 水暖工，管道工，水管工，管子工，（防止泄密的）堵漏人员
plumbing and drainage system 上下水系统
plumbing and drainage 上下水
plumbing discipline 给排水专业
plumbing drawing 给排水图纸
plumbing engineer 给排水工程师
plumbing fittings 卫生设备，波导管设备，卫生洁具，管配件，管件接头
plumbing fixture 管子附件【复数】，卫生洁具，卫生器具，厕、浴间设备
plumbing inspector 水暖督察，水暖监察，管道检查员
plumbing system 管道系统，水暖设备系统，卫生管道系统，上下水管系统，给排水系统
plumbing tool 管子工具，管扩工具
plumbing works 给排水工程，水管工程，敷设水管工程
plumbing 管工作业，水暖工作，管件的总称，波导管，上下水管道工程，反应堆管道阀门系统，水管设施，铅垂度检查，铅管工程，室内管道工程，铅管工，用铅锤测量，探究
plumb line deflection 垂线偏差
plumb line deviation 垂线偏差
plumb line 铅直线
plumb point 垂准点
plumb rule 垂规，垂线尺
plumbum 铅【Pb】
plumb 使垂直，用铅锤测量，探索，测锤，吊线，铅锤，悬锤，恰恰，正好，垂直地［的］
plum coupling 梅花联轴器
plume advection 羽流平流
plume angle 羽流展角
plume bifurcation 羽流分岔
plume buoyancy 羽流浮力
plume centerline concentration 羽流轴线浓度
plume centerline height 羽流轴线高度
plume coalesce 羽流集聚
plume concentration 羽流浓度
plume depletion 羽流耗损
plume diffusion 羽流扩散
plume dispersion 羽流弥散
plume exhalation 羽流消散
plume exit velocity 羽流出口速度
plume expansion 羽流碰撞
plume geometry 羽流几何形状
plume growth 羽流增长
plume height 烟缕高度，排气高度
plume impingement 羽流拍撞
plume lifetime 羽流生存期
plume model 卷流模型
plume moisture 烟缕湿度
plume of bubbles 气泡卷流，气泡羽流
plume of warm air 热空气羽流
plume opacity 烟缕不透明度
plume optical depth 羽流光学厚度
plume oscillation 羽流振荡
plume passage 羽流行程
plume pollution 烟雾污染
plume profile 羽流廓线
plume rise equation 羽流抬升方程
plume rise formula 羽流抬升公式
plume rise prediction 羽流抬升预报
plume rise 烟羽上升，烟缕上升，羽流抬升
plume smoke visualization 羽流烟显示
plume spread 羽流扩展
plume temperature 羽流温度
plume trajectory 羽流轨迹
plume transport wind speed 羽流输运风速
plume trapping 羽流陷落，羽流收集
plume travel distance 羽流运行距离
plume volume flux 羽流体积通量
plume width 羽流展宽
plume （烟囱）羽状排气，羽状物，羽毛，烟缕，烟羽，羽流，烟云

plummer bearing 止推轴承
plummer block 止推轴承
plummer 轴承座，轴承架
plummet method 测锤法
plummet 铅锤，坠子，测锤，垂球，线铊
plump 使鼓起，膨胀，丰满的，肥胖的，丰富的，充裕的
plum rain 梅雨
plunge cut grinding 切入磨削
plunge pool 跌水池，跌水坑，消力池，水垫塘
plunger armature 吸入式衔铁，插棒式衔铁
plunger half 插入一半【控制棒驱动机构】
plunger motor 柱塞式油马达
plunger pump 柱塞泵
plunger relay 螺管式继电器，插棒式继电器
plunger slurry pump house 柱塞型灰浆泵房
plunger switch 插棒式开关
plunger-type wave generator 冲击式生波机，冲击式造波机
plunger 插棒【控制棒驱动系统】，(加煤机）往复柱塞，插棒式铁芯，柱塞，活塞，可动铁芯，撞针
plunge 浸入，插入，倒转，急降，跌水池，沉入
plunging anticline 倾伏背斜
plunging fold 倾伏褶皱
plunging nappe for the spillway 溢洪道下泄水舌
plurality 多元，复数
plural service 不同源供电，多源供电
plural 复数的
pluriennal regulation 多年调节
pluriennal regulation reservoir 多年调节水库
plus end 正端子
plus grade 上坡，正坡
plusminus screw 调整螺丝
plusminus switching circuit 加减分接开关电路
plusminus switch 加减分接开关
plusminus 正负，加减，调整
plus phase 正相位
plus pressure 正压力，过剩压力
plus tapping 正抽头，正分接
plus 加
plutonate 钚酸盐
plutone 深成岩
plutonia molybdenum cermet 二氧化钚钼金属陶瓷
plutonia 二氧化钚
plutonic acid 钚酸
plutonic earthquake 深成地震
plutonic rock 深成岩
plutonic water 深成水
plutonie plug 岩柱
plutonium-and-power reactor 产钚发电两用反应堆
plutonium balance 钚平衡
plutonium-bearing fuel assembly 含钚的燃料组件
plutonium-bearing 含钚的
plutonium breeding 钚增殖
plutonium buildup 钚累积量，钚累积
plutonium burner 钚反应堆，钚堆
plutonium-containing fuel rod 含钚的燃料棒
plutonium core 钚燃料芯体
plutonium credit 钚收益
plutonium cycle 钚循环
plutonium enriched fuel 富钚燃料，加钚燃料
plutonium factory 钚厂，制钚工厂
plutonium-fuelled reactor 钚燃料反应堆
plutonium hexafluoride 六氟化钚
plutonium hydride 氢化钚
plutonium hydroxide 氢氧化钚
plutonium inventory determination 钚总量［库存］测定
plutonium inventory 钚总量
plutonium load-out vessel 钚卸料容器
plutonium metallurgy 钚冶金学
plutonium-only reactor 产钚专用反应堆
plutonium oxalate 草酸钚
plutonium oxide 氧化钚
plutonium peroxide 过氧化钚
plutonium phosphate 磷酸钚
plutonium pile 钚堆
plutonium poisoning 钚中毒
plutonium producer 钚生产堆
plutonium-producing reactor 产钚反应堆
plutonium production reactor 钚生产堆
plutonium reactor 钚反应堆，生产钚的反应堆
plutonium reclamation facility 钚回收设施
plutonium recovery 钚回收
plutonium recycle capability 钚再循环能力
plutonium recycle critical facility 钚再循环临界装置
plutonium recycle programme 钚再循环程序
plutonium recycle reactor 钚再循环反应堆
plutonium recycle test reactor 钚再循环试验反应堆
plutonium recycle 钚循环，钚再循环
plutonium-regenerating reactor 钚再生反应堆
plutonium-separation plant 钚分离装置，钚分离工厂
plutonium 钚
pluvial denudation 多雨的剥蚀，雨水冲蚀
pluvial erosion 洪水侵蚀
pluvial index 降雨指数，雨量指数
pluvial period 多雨期
pluvial 多雨
pluviogram 雨量曲线
pluviograph 雨量计
pluviometer 雨量表
pluviometric coefficient 雨量系数
ply adhesion 层间黏合力
ply area 芯层
plying 通过
ply separation （输送带）布层分离
plywall 多层壁
plywood door 夹板门
plywood 层压木板，多层夹板，夹板，胶合板
PMC(process management and control) 过程管理和控制
PMC(project management contracting) PMC 模式，项目管理承包模式
PMC 核燃料装卸贮存系统【核电站系统代码】
PMD(post mortem dump) 算后检查转储
p-m(permanent-magnetic) erasing head 永磁式消磁头
PMF(probable maximum flood) 可能最大洪水

PMG(permanent magnet generator) 永磁发电机
PMMA(polymethyl-methacrylate) 聚甲基丙烯酸甲酯
PMN(program management network) 程序管理网络
PMO(privileged memory operation) 特许存储器操作
PM(particulate matter) 颗粒物
PM(permanent magnet) 永久磁铁，永磁体，永久性磁石
PMP(probable maximum precipitation) 可能最大降水（量），概然最大降雨量
PMP＝pump 泵
PM(predecessor matrix) 先趋矩阵
PM(preventive maintenance) 预防性维修
PM(processing module) 处理模块
PM(program memory) 程序存储器
PM reactor portable medium power reactor 可移动式中等功率反应堆
p-m rotor 永磁转子
PMR(post mortality review) 事故追忆
PMSG(permanent magnet synchronous generator) 永磁同步发电机
PMS notation(processor-memory-switch notation) 处理器-存储器-开关表示法
PMS(probable maximum seiche) 可能的最大假潮
PMS(process monitor system) 过程监督系统
PMS(processor-memory-switch) 处理机-存储器-开关
PMS(project management system) 计划管理系统
PMS(protection and safety monitoring system) 保护与安全监控系统
PMSS(probable maximum storm surge) 可能的最大风暴潮
p-m step motor 永磁式步进电机
PMV(multi-stage volute pump) 多级螺旋泵
PNC-curve 噪声评价 PNC 曲线
pneumatic accumulator 气力蓄能器
pneumatic actuator 气动执行机构，气动执行器，气动驱动器，气动装置
pneumatic aeration 压气曝气
pneumatically applied mortar 喷浆用灰浆
pneumatically controlled multistage fluidized bed 气控式多层流化床
pneumatically controlled valve 气动控制阀
pneumatically operated switch 风动开关，气动开关
pneumatically operated 气动操作的
pneumatic amplifier 气动放大器
pneumatic analog computer 气动模拟计算机
pneumatic analog station 气动模拟操作站
pneumatic ash conveyer 干式除灰装置
pneumatic ash conveying plant 气力输灰装置
pneumatic ash handling system 气力除灰系统
pneumatic ash removal system 气力除灰系统
pneumatic ash transmission system 气力输灰系统
pneumatic atomization 气力雾化
pneumatic barrier 压气屏障
pneumatic brake 气动闸，空气制动器，风闸，气闸
pneumatic breakwater 空气防波堤，压气式防波堤
pneumatic buffer 气压缓冲器
pneumatic building 充气建筑物
pneumatic caisson method 气压沉箱法
pneumatic caulking 压气堵缝，气力捻缝
pneumatic chisel 风动凿，压气凿，气錾
pneumatic chute 气力输送槽
pneumatic circuit 气动回路，压缩空气回路
pneumatic classification 气力分级
pneumatic cleaner 气动吸尘器
pneumatic cleaning 风力清除，气力清洗，压气吹洗
pneumatic coal arch breaking system 气力破拱系统
pneumatic coal handling 气力输煤
pneumatic compactor 风动夯，气夯
pneumatic concrete placer 压气式混凝土浇筑机
pneumatic connection 压气连接法
pneumatic controller 气动控制器，气动调节器
pneumatic control relay 气压控制继电器
pneumatic control valve 气动控制阀，气控阀，气动调节阀
pneumatic control 风动控制，气动控制，气力控制，压缩空气控制，气控
pneumatic conveyance 气力输送
pneumatic conveyer 风力输送机，气力输送机
pneumatic conveying system 风动输送系统，气力输送系统，压气输送系统
pneumatic conveying 气力输送
pneumatic conveyor 风动输送机，气力式输送机，压气输送机
pneumatic core sampler 压气钻芯取样器
pneumatic crane 风动起重机
pneumatic cushioning 气压减震
pneumatic cushion 气垫
pneumatic cut-off reversing valve 气动截止换向阀
pneumatic dashpot 风动消振器
pneumatic delivery capability 气动输送能力
pneumatic differential transmitter 气动差动传感器，气动差动变送器
pneumatic digger 风铲
pneumatic dipleg 气控料腿
pneumatic dispatch 气力输送
pneumatic distributor 气力抛煤机，气力分配器
pneumatic drill 风动钻机，风钻
pneumatic drive 气力传动
pneumatic dual-piston linear drive 气动双活塞直线驱动装置
pneumatic duct system 气力管道系统
pneumatic ejector 风动喷射器，气动喷射泵，压气喷射器
pneumatic-electrical converter 气电转换器
pneumatic element 气动元件
pneumatic elevator 气动升降机，压气升降机
pneumatic excavator 风动挖掘机
pneumatic exhaust capability 排气能力
pneumatic feeder 气力送料器
pneumatic feed 风动给料
pneumatic gauge 气动测量仪表

pneumatic gauging 气动测量
pneumatic grinder 风砂轮
pneumatic gripping device 气动抓手，气动夹持装置
pneumatic hammer 风动锤，空气锤，气锤
pneumatic hoist 气动起重机，气动卷扬机，气动绞车，风动绞车，气动葫芦
pneumatic hydraulic clamps 气油压虎钳
pneumatic hydraulic controller 气动液压控制器
pneumatic hydraulic converter unit 气动液压变换单元
pneumatic jack hammer 气动锤
pneumatic jack 压气千斤顶
pneumatic knockout 气动清砂
pneumatic limit operator 气动限幅器
pneumatic loading diaphragm actuator 气动薄膜执行机构
pneumatic local control device 气动基地式调节装置，气动基地式调节仪表
pneumatic lock seamer 风动封口机
pneumatic logic element 气动逻辑元件
pneumatic motor 风动马达，风动机
pneumatic nebulization 气动雾化
pneumatic nut wrench 风动螺母扳手
pneumatic-operated valve 气动阀
pneumatic-operated 气动的
pneumatic operating mechanism 气动操作机构
pneumatic operation 压气操作
pneumatic operator 气动操作器，气动执行机构
pneumatic outlet 气体出口
pneumatic perforator 气压凿孔机
pneumatic pick 风镐
pneumatic pile driver 风动打桩机
pneumatic pile 气压桩
pneumatic piling 压气打桩
pneumatic placer 风动浇筑机
pneumatic positioner 气动位置控制器
pneumatic positioning relay 气动调位继电器
pneumatic post 气动管，跑兔管
pneumatic power tools 气动工具
pneumatic pressure ash handling 负压［正压］气力除灰，干式除灰，气力除灰
pneumatic pressure fluctuation 气压波动
pneumatic pressure test 气压试验
pneumatic pressure 气压
pneumatic pressurizing 气压密合
pneumatic press 气动压力计
pneumatic pump for pressure test 气动试压泵
pneumatic pump 风动泵，气动泵，压气泵
pneumatic pyrometer 气动高温计
pneumatic quick isolation damper 气动快速隔离阀
pneumatic rabbit 气动速送器，气动跑兔
pneumatic ram 压气夯
pneumatic refractometer 气体折射计
pneumatic regulating valve 气动调节阀
pneumatic relay 气动继电器
pneumatic remote control 远距离气动控制
pneumatic restriction 气流限制
pneumatic riveter 风动铆接机
pneumatic rock drill 风动凿岩机
pneumatic scrubber 气动洗涤器
pneumatic seal 气力密封
pneumatic separator 气动分离器
pneumatic servo 气动伺服机构
pneumatic sewage pump 压气污水泵
pneumatic shaft sinking 压气舱沉井
pneumatic shock absorber 压气减震器
pneumatic signal 气动信号
pneumatic sounder 气压测深仪
pneumatic spade 风动铲
pneumatic spanner 气动扳手
pneumatic spring 压气弹簧
pneumatic starter 气力启动器，风动启动器
pneumatic stepper drive 气动步进驱动
pneumatic stiffness 充气刚度
pneumatic storage 充气能
pneumatic structure 充气结构
pneumatic summing unit 气动加法器
Pneumatic surge chamber 气压室调压室
pneumatic switch 气动开关
pneumatic system 气动系统，压气系统
pneumatics 气动装置，气体力学，气动力学
pneumatic tamper 气夯，风动捣固机，风动打夯机
pneumatic tank 压气箱
pneumatic test 气压试验
pneumatic three-way solenoid valve 气动三通电磁阀
pneumatic tide generator 压气式潮汐发生器
pneumatic time delay relay 气动延时继电器
pneumatic-tired roller 气胎碾
pneumatic to current converter 气电转换器
pneumatic tongs 机械手抓手，气动钳
pneumatic tool 充气工具，风动工具，气动工具
pneumatic to voltage converters 气压电压转换器
pneumatic transmission system 气动传输系统
pneumatic transmitter 气动传感器，气动变送器
pneumatic transmitting rotameter 气动传动转子流量计
pneumatic transporting equipment 气动输送装置
pneumatic transport 风动输送，压气传送，气力输送
pneumatic trough 压气槽
pneumatic tube conveyor 风管输送器
pneumatic tube roller 气动胀管器
pneumatic tube 压缩空气管道
pneumatic type wave generator 气压式生波机
pneumatic tyre 充气轮胎
pneumatic valve 气动阀
pneumatic vibrator 风动振动器
pneumatic water barrel 抽气排水筒
pneumatic water supply 压气供水
pneumatic wave recorder 压气波浪计
pneumatic 3-way valve 气动三通阀
pneumatic wrench 气动扳手
pneumatic 风力的，气动的，气力的，气体的，空气的，风动的
pneumatology 气体学
pneumatolysis 气化
pneumodynamics 气动力学
pneumo-electrical convertor 气电转换器

pneumo-hydraulic operation 风动水力操作
pneumohydraulic 气动液压的
pneumonics 射流学，流体学
pneumotransport 气力输送
PNEU＝pneumatic 气动的
pneutronic controller 电子气动调体器
pneutronic level controller 电子气动位面控制器
pneutronic 电控气压的，电子气动的
p-nitrophenol 对硝基苯酚
p-n-p-n switch PNPN 式开关
PNP(prototype nuclear process heat plant) 原型核工业供热厂
p-n-p semiconductor triode PNP 半导体三极管
p-n-p transistor PNP 型晶体管
p-n-p triode PNP 型晶体三极管
PNSRC(Plant Nuclear Safety Review Committee) 电厂核安全审评委员会【美国】
PNTR(permanent normal trade relation) 永久正常贸易关系
PO box(post office box) 邮政信箱
pocket alarm dosimeter 袖珍报警剂量计
pocket beach 袋状滩
pocket boundary 气穴边界
pocket calculator 袖珍计算器，便携式计算器
pocket compass 袖珍罗盘仪
pocket computer 袖珍计算机
pocket dose meter 袖珍剂量计
pocket dosimeter 袖珍剂量计，携带式剂量计
pocket exposure meter 袖珍辐射仪
pocket folding rule 折尺
pocket instrument 袖珍仪表
pocket monitor 袖珍监测仪，袖珍检测仪
pocket of air 气眼，砂眼
pocket rot （木材的）腐孔
pocketscope 轻便示波器
pocket sextant 袖珍六分仪
pocket-sized soil testing instrument 袖珍土壤试验仪
pocket tape 卷尺
pocket telephone 携带式电话机
pocket transit 袖珍经纬仪
pocket type switch 袖珍型开关，组合开关
pocket voltmeter 小型电压表，袖珍伏特计
pocket 烟流死角，囊，矿袋，袋，腔室，凹处，熔蚀坑，壳，气穴，穴，窝，坑，矿穴，凹穴，贮器，储备
pockhole 缩孔
POC(processor operator console) 处理机操作台
Poctor compaction test 葡氏击实试验
Poctor dynamic test 葡氏动力击实试验
pod boiler 荚式蒸汽发生器
podium 墩座，矮墙
POD(port of destination) 目的港
POD(probablity of detection) 探测几率
POD(proof of delivery) 交付凭证
podsol 灰化土
podwer gun 喷粉枪
podwer iron core 铁粉芯
podzolic soil 灰化土壤，灰化土
podzol 灰化土
POF(planned outage factor) 计划停运系数
POH(planned outage hours) 计划停运小时数
poidometer 重量计
poid 形心曲线，腰正弦线
poikilothermic 不定温的
poikilothermy 变温性
point approximation 点近似
point bar 点沙坝
point bearing type pile 支承桩
point-by-point computation 逐点计算
point-by-point integral 逐点积分
point-by-point method 逐点测定法，逐点计算法
point-by-point 逐点，逐项
point charge 点电荷
point-contact diode 点接触二极管
point-contact germanium diode 一点接触式锗二极管
point-contacting temperature measuring device 点接触式测温装置
point-contact rectifier 点接触整流器
point-contact transistor 点接触晶体器
point-contact varactor 点接触变容二极管
point-contact 点接触，一点接触式的
point corrosion 点腐蚀，点蚀
point design 符合要求设计，要点设计，定点设计
point diffusion kernel 点源扩散核，点扩散核
point discharge 尖端放电
point doublet 点偶极子
point drift 点漂
pointed arch 哥德式拱
pointed joint 勾缝
pointed wing tip 尖角翼尖
pointed with cement mortar 水泥勾缝
point electrode 尖端电极
point elevation 点标高
pointer counterbalance 指针平衡器
pointer galvanometer 指针式检流计
pointer-stop meter 有针挡仪表
pointer-type indicating instrument 指针式指示仪表
pointer-type 指针式的
pointer 指示器，指针
point focus collector 点聚焦（型）集热器
point focus transducer 点聚焦传感器
point focus 点聚集
point function （状态函数的）参数，点函数
point gauge 测针，针形水位计
point grid 测点网
point group 点群
point heat theory 点热学说，点热理论
point image 点象
pointing chisel 点凿
pointing error 瞄准误差
pointing joints 勾缝
pointing stuff 勾缝料
pointing trowel 勾缝镘刀
pointing 削尖，瞄准，勾缝，磨尖，嵌缝
point-integrating sediment sampler 积点式（泥沙）采样器
point-integration sampling 积点法采样
point-junction transistor 点结型晶体管，点接触晶体管

point lattice　点阵
point lightning protector　尖端型避雷器，避雷针
point load test instrument　点荷载试验仪
point load test　点荷载试验
point load　集中负荷，点负荷，集中荷载，点载荷
point location　定位
point model　点模型
point mortar joint　勾灰缝
point of application　施力点，作用点
point of attachment　联结点
point-of-best-efficiency operation　最优效率点运行
point of blade　叶片尖端
point of common coupling　公共耦合点，公共连接点，公共供电点
point of connection　连接点，施力点
point of contact　接触点
point of contract　合同要点
point of contra-flexure　反挠点，拐点，反弯点，转变点
point of cultural interest　文物遗址
point of discharge　排放点
point of discontinuity　不连续点，突变点
point of ebullition　沸腾点
point of entry　输入点，进线点
point of failure　破坏点，破损点，故障点
point of fixity　固定点
point of fluidity　流限，屈服点
point of force application　力作用点
point of force concurrence　力交会点
point of fusion　熔点
point of gradient　坡度变点
point of incipient fluidization　起始流态化点
point of inflection　拐点，回折点，反弯点
point of interconnection　互联点
point of intersection　交叉点，交点
point of load　载荷点
point of measurement　测量点
point of onset fluidization　开始流态化点
point of origin　起点，原点
point of reference　水准点，参考点，基准点，控制点
point of zero flow　断流点，流量零点
point of zero moment　零力矩点
point of zero voltage　电压零点
point on wave switching　波形定点分合
point particle　点粒子
point pole　点极
point precipitation　点降水量
point pressure　点压力
point reactor kinetic equation　点反应堆动力学方程
point reactor model　点堆模型，点反应堆模型
point recorder　单点记录仪，点反应堆
point reflection　点反映
point representation　点表示
point resistance pressure　端头阻力
point sampler　单点取样器
point sample　点样
point sampling　点采样法
point scatterer　点散射体
point scattering　点散射
point set　点集
points for attention　注意事项
point single step method　点单步法
points of slinging　起吊点
point source discharger　排放源点
point source dose　点放射源剂量，点源剂量
point source lamp　点光灯
point source model　点源模式
point source turbulent diffusion　点源的湍流扩散
point source　点污染源，点放射源，点源，点热源
point spread function　点分布函数
points scatter　点子分散
point successive over-relaxation method　点逐次过度松弛法
point symmetry　点对称
point tipping method　立堵
point-to-point analysis　逐点分析
point-to-point configuration　点对点配置
point-to-point control　点到点的控制
point-to-point data link　定点数据传输线
point-to-point financial transmission right　点对点金融输电权
point-to-point line　闭合线
point-to-point method　逐点计算法，逐点测定法
point-to-point radio communication　点与点间无线电通信
point-to-point　点对点，点至点的，点位控制的，逐点
point total step method　点全步法
point to　显示……的位置［方向］，表明，指
point transistor　点式晶体管
point value　点值
point velocity　点速度
point vortex　点涡
point welding machine　点焊机
point weld　点焊
pointwise uniformity　点态一致性
pointwise　逐点，逐点的
point　逐点的【模拟】，点，尖端，指向，小数点
POI(point of interconnection)　互联点
poised state　平衡状态
poise　保持平衡，平衡，砝码
Poise　泊【黏度单位】
poishing filter　净化过滤器
poising action　平衡作用
poisohing effete　中毒效应
poison assembly　毒物组件
poison cross section　毒物截面
poison curtain　吸收屏，毒物屏
poison decay　毒物衰变
poison drum control　毒物鼓控制
poison drum　毒物（控制）鼓
poison element　毒物元件
poison finger　细毒物棒，指状毒物棒
poisoning effect　中毒效应
poisoning feedback loop　中毒反馈回路
poisoning feedback　中毒反馈
poisoning override　中毒补偿
poisoning overshoot　中毒过调量

poisoning predictor 中毒预测计
poisoning temperature effect 中毒温度效应
poisoning yield 中毒产额
poisoning 中毒
poison injection system 毒物注射系统
poison material 中子吸收剂，毒物材料
poisonous effect 中毒效应
poisonousness 毒性
poisonous 有毒的，有害的，恶臭的
poison-out 因氙毒不能启动【反应堆】，因氙毒而停堆
poison-prevent system 防（氙）中毒系统
poison-prevent 预防中毒
poison range （反应堆可继续运行的）毒物范围
poison rod 强烈吸收中子的棒，毒物棒
poison section 吸收段【控制棒】，毒物段
poison skirt 毒物围筒
poison sparger 毒物分配器，毒物喷淋环
poison worth 毒物价值
poison 污染，弄坏，毒物，毒化，中毒，毒
Poission's distribution diagram 泊松分布图
Poission's ratio 泊松比
Poisson data distribution 泊松数据分布
Poisson distribution 泊松分布
Poisson equation 泊松方程
Poisson number 泊松数
Poisson probability distribution function 泊松概率分布函数
Poisson ratio 泊松比
Poisson's modulus 泊松模量
Poisson's ratio of soil 土壤泊松比
Poisson's ratio 横向变形系数，泊松比
poke door 拨火孔，拨火门，打焦孔
poke hole 拨火孔，打焦孔
poke rod 通焦杆
poker vibrator 插入式振捣器，插入式振动器
poker 拨火铁棒，火钳，拨火棍，火钩
poke-through wiring 戳通布线
poke welding 手点焊，手动焊
polar air mass 极地气团
polar angle 极角
polar arc 极弧
polar axis 极轴
polar bridge crane 旋转桥式起重机
polar capacitor 极性电容器
polar climate 极地气候
polar control 极坐标法控制
polar coordinate system 极坐标系
polar coordinates 极坐标
polar crane supporting skirt 环形桥中央电视台支承墙
polar crane 环形吊车，回转式吊车，旋转式桥吊
polar curve 极曲线
polar diagram 极线图，极坐标图
polar distance 极距
polar easterlies 极地东风带
polar form coordinates 极坐标
polar front 极锋
polar gantry crane 旋转龙门起重机
polarimeter 偏光计，偏振计，极化计
polar indicator 极性指示器
polar invasion 寒潮
polarising voltage 极化电压
polarity-directional relay 极性方向继电器
polarity effect 极性效应，极性影响
polarity finder 极性测定器
polarity indicator 极性指示器
polarity inspection 极性检查
polarity inversion 极性变换
polarity inverting amplifier 极性反转放大器，倒相放大器
polarity lamp 极性检测灯
polarity mark 极性标记
polarity relay 极化继电器，极性继电器
polarity reversal switch 极性开关
polarity reversal 极性变换，极性颠倒，极性反向器
polarity reversing switch 极性转换开关
polarity selector 脉冲极性选择器
polarity switch 极性开关
polarity test 极性试验
polarity 极性，正相反，偏光性
polarizability 极化性，极化率，极性化，极性率
polarizable 可极化的
polarization cell 极化电池
polarization current 极化电流
polarization degree 偏振度
polarization effect 极化效应
polarization index test 极化率试验
polarization in insulator 绝缘体的极化
polarization microscope 偏光显微镜
polarization plane 偏振面
polarization ratio 极化比
polarization vector 极化矢量
polarization voltage 极化电压
polarization 极化，偏振
polarized capacitor 电解电容器，极化电容器
polarized disk 极化盘
polarized induction 极化磁感应
polarized-latching type relay 闩锁型极化继电器
polarized light 偏光，偏振光
polarized magnet 极化磁铁，永久磁铁
polarized monostable type relay 边稳定极化继电器，单稳态型极化继电器
polarized plug 固定极性插头
polarized radiation 偏镜峰
polarized reflectance 偏振反射率
polarized relay 极化继电器
polarized socket 极化插座
polarized 偏振的，极化的
polarizer 起偏镜，起偏器，偏振器
polarize 使极化，使偏振
polarizing angle 偏振角
polarizing eyepiece 起偏振目镜
polarizing magnetizing force 极化磁力，极化磁强
polarizing optical pyrometer 偏振光测高温计
polar leakage 磁极漏磁
polar line 极线
polar molecule 极性分子
polar moment of inertia 极惯性矩
polar moment 极矩

polarogram 极谱，极谱图
polarographic analysis 极波分析，极谱分析
polarographic analyzer 极谱分析仪
polarography 极谱法，极谱学
polarograph 极谱仪，极谱
polaroid film 偏振胶片
polaroid polarizing filter 波拉偏振滤光镜
polaroscope 偏振光镜
polar pitch 极距
polar plot 极坐标曲线图，极点绘图
polar relay 极化继电器
polar system 极坐标系
polar tip 极尖
polar vector 极矢量
polar wave 极波
polar 磁极的，极地的，极性的，极性，极的，极线，极线图
polatlzing 极化
polder dike 围堤
polder 围海洼地，围堤造地，围圩，围堤
pole agreement 极性协议
pole ampere-turns 极安匝
pole amplitude modulation motor 极幅调制电动机
pole amplitude modulation 极幅调制
pole amplitude 极幅
pole and tower construction 杆塔结构
pole and tower deflection 杆塔挠度
pole and tower design 杆塔设计
pole and tower erection 杆塔组立
pole and tower foundation 杆塔基础
pole and tower of an overhead line 杆塔
pole and tower spotting survey 杆塔定位测量
pole and tower spotting 杆塔定位
pole and tower 杆塔
pole and wire mattress 铁丝木杆沉排
pole arc 极弧
pole armature 凸极电枢
pole arm 横担，线担
pole arrangement 电极布置
pole axis 极轴
pole base 电杆底板
pole body insulation 极身绝缘
pole body 极身
pole bolt 磁极螺栓
pole bore 定子内径，磁极内圆
pole brace 电杆撑杆，电杆支撑
pole bracket 磁极支架，支架，悬臂，支臂，电杆角尺
pole cell insulation 极身绝缘
pole center 磁极中心
pole-change motor 变极电动机
pole changer 换向开关，换极器
pole-change starter 变极启动器
pole changing constant speed induction generator 换极恒速异步发电机
pole changing control 极变换控制，变极控制
pole changing generator 变极发电机
pole changing induction machine 变极异步电机
pole changing winding 变极绕组
pole changing （多速电动机的）极数转接，极变换
pole clearance 极间间隙
pole climber 脚扣
pole coil 磁极线圈
pole control （高压直流）极线控制
pole core width 极身宽度磁极阻尼条
pole core 磁极铁芯
pole count 极数
pole damper 磁极阻尼笼，阻尼器
pole distance 极距，杆距
pole earthing 电杆接地
pole edge 磁极边缘
pole effect 磁极效应，电极效应
pole element 磁极件
pole embrace 极弧系数，极弧空间范围
pole end-plate 磁极压板，磁极端板
pole erecting 架电杆，立电杆
pole erection using helicopter 直升机立杆
pole extension 极靴，极延伸部分
pole-face bevel 极面斜角，极靴表面削斜
pole-face compensating winding 极面补偿绕组
pole-face damper winding 磁极表面阻尼绕组
pole-face loss 极面损失
pole-face shaping 极面整形
pole-face shoe 极靴
pole-face winding 极面绕组
pole-face 极面
pole field 极磁场
pole figure 极象图
pole finder 极性试验器，极靴试验器
pole finding paper 相对端点试纸
pole fixture 电杆支架
pole float 浮标杆，桅式浮标
pole flux 磁极通量
pole form 磁极形状
pole-foundation 电杆基础
pole guy 电杆拉线
pole-hanging transformer 柱上变压器
pole head 极靴
pole height 极高，电杆高度
pole horn 极尖
pole insulation 极身绝缘，磁极绝缘
pole joint 磁极连接，磁极引线
pole lamination 磁极叠片
pole leakage 磁极漏磁
poleless 无电杆的，无极性的
pole line hardware 架空线路金具
pole line switch 杆上线路开关
pole line （架空线路）杆塔群
pole-mast assembly 电杆组件
pole-mounted circuit breaker 柱上断路器
pole-mounted disconnector 柱上隔离开关
pole-mounted load-breaking switch 柱上负荷开关
pole-mounted oil switch 杆装油开关
pole-mounted regulator 柱上电压调整器
pole-mounted substation 杆上变电所，柱上变电站，杆上配电台，柱上配电台
pole-mounted 安装在电杆上的
pole-mounting disconnecting switch 杆装式隔离开关
pole-mounting lightning arrester 杆装式避雷器
pole-mounting transformer vault 杆上变电

pole-mounting type 柱上式
pole oil switch 杆上油开关
pole pair 极对，极对数
pole piece 磁极，极心，极靴，磁极片，极片
pole pitch 磁极距，极距
pole pit 电杆基坑
pole plate 磁极钢板，极板
pole region 极区
pole retriever 自动降压器
pole reverser 换极开关
pole ring 磁极环
pole scaffolds 木杆脚手架
pole shank 极心
pole sheet 磁极叠片
pole shim 磁极与轭间的垫片
pole shoe face 极靴面
pole shoe factor 极靴系数
pole shoe leakage 极靴漏磁
pole shoe 极靴，极片
pole shore 单独立柱
pole shunting 磁极分路
pole sight 标杆
pole slip operation 滑差运行
pole slipping protection 磁极滑动保护
pole slipping 极滑差率，磁极滑动
pole spacing 极间距离，极矩，杆距
pole span 极距，挡距
pole spur 靴檐，靴沿
pole step 电杆爬梯
pole strength 磁极强度
pole structure 杆式结构
pole switch 杆上开关，杆装开关
pole tip 极尖
pole-tool method 杆钻冲击钻进法
pole top switch 杆上开关
pole top transformer 杆上变压器，柱上式变压器
pole-type substation 杆装变电站
pole-type support 柱式管架
pole-type transformer 架杆式变压器，杆装变压器
pole winding 磁极线圈，磁极绕组
pole with syringe 带喷射器的杆
pole 电极，磁极，极，极点，杆，电杆，相（数），柱，桩
policy bank 政策银行
policy clause 保险单条款
policy depression 保险单保险金额贬值
policy fee 保险单签发手续费
policy holder 保险持有人，投保人，保险客户
policy issued basis 以签发保险单为基础
policymaker 决策人，决策者
policy making body 决策机构
policy of insurance 保险单
policy package 一揽子政策，承保多种内容的保险单
policy proof of interest 保险单权益证明
policy reserves 保险单责任准备金
policy year 保险单年度
policy 方针，政策，保险单
poling 立杆，架线路，支撑
polished condensate 净化凝结水
polished feedwater 纯化给水，净化给水
polished finish of stone 石面抛光
polished glass 抛光玻璃，磨光玻璃
polished 抛光的
polisher 抛光剂，抛光机，高纯度水处理装置，精处理器，精处理机［器］
polishing agent 抛光剂
polishing belt 打磨带
polishing control system 精处理除盐控制系统
polishing demineralizer control system 精处理除盐控制系统
polishing filter 净化过滤器
polishing powder 研磨粉
polishing the condensate 凝结水精处理
polishing varnish 抛光漆
polishing 净化，纯化【给水】，给水连续净化处理
polish 打磨，抛光，磨光，磨料，（水）精处理，抛光剂，高纯度水处理
political position 政治地位
political standing 政治地位
polje 灰岩盆地
pollinator 受花粉器
polling and accepting data 定时询问式接收数据【主站端，调度端】
polling controller 轮询控制器
polling list 轮询表
polling mode 轮询方式
polling program 轮询程序
polling system 轮询系统，查询系统
polling telecontrol system 问答式远动系统
polling 对终端设备的定时查询，探询，轮询，查询
pollutant burden 污染物负荷
pollutant chemistry 污染物化学
pollutant concentration 污染物浓度
pollutant diffusion 污染物扩散
pollutant discharge 污染物排出
pollutant dispersal 污染物散布
pollutant dispersion 污染物散布，污染物弥散
pollutant disposal 污染物处理
pollutant distribution 污染物分布
pollutant downwash effect 污染物下洗效应
pollutant effect 污染物影响
pollutant emission 污染物排放
pollutant equilibrium 污染物质平衡
pollutant flux 污染物通量
pollutant identification 污染物鉴定
pollutant index 污染物指数
pollutant pathway 污染物传播途径
pollutant persistency 污染物持久性
pollutant plume 污染物羽流
pollutant recipient 污染物接收者
pollutant recirculation 污染物回流
pollutant release height 污染物释放高度
pollutant source 污染源
pollutant surveillance 污染物监测
pollutant susceptibility 污染物感受性
pollutant target 污染物目标
pollutant 沾污物，污染物，污染物质
polluted air 受污染空气，污染空气

polluted atmosphere 受污染大气
polluted environment 受污染环境
polluted waterway 污染的水道
polluted water 受污染水，污染水
polluted 被污染的
polluter pays principle 污染肇事者付款原则
pollute 污染
polluting substance 污染物
pollution abatement 减少污染，污染消除，减轻污染，污染抑制，污染治理
pollution accretion 污染堆积
pollutional condition 污染条件
pollutional contribution 污染为害部分
pollutional effect 污染效应
pollutional equivalent 污染当量
pollutional index 污染指数
pollutional load 污染负荷
pollution atlas 污染地图集
pollution biology 污染生物学
pollution blackmail 污染威胁
pollution budge theory 污染预算理论
pollution capacity 污染能力
pollution concentration 污染浓度
pollution control function 污染控制作用
pollution control index 污染控制指数
pollution control in hydraulic system 液压系统污染控制
pollution control 污染防治，污染控制
pollution criterion 污染标准
pollution cycle 污染循环，污染周期
pollution degree 污染度
pollution deposit 污染沉积物
pollution discharge test of insulator 绝缘子污秽放电试验
pollution disease 污染引起的疾病
pollution effect 污染的后果
pollution environment 污染环境
pollution exhaust criteria 排污标准
pollution flashover test 污秽闪络试验
pollution flux 污染通量
pollution-free energy resource 无污染能源
pollution-free energy source 无污染能源
pollution-free plastic 无污染塑料
pollution-free 无污染的
pollution hazard 污染危害
pollution horror 污染恐怖
pollution index 污染指数
pollution in limited area 局部污染
pollution intensity 污染强度
pollution layer （绝缘子）污层
pollution level 污染程度，污染度
pollution liability 污染应负的责任
pollution load 污染负荷
pollution management 污染管理
pollution measurement 污染测量
pollution mechanism 污染机理
pollution meteorology 污染气象学
pollution microbiology 污染微生物学
pollution monitoring 污染监测
pollution of groundwater 地下水污染
pollution of natural water 天然水污染
pollution parameter 污染参数
pollution plume 污染的烟缕，污染烟羽
pollution potential 潜在污染
pollution prediction 污染预测
pollution prevention 防污染，污染防治，预防污染
pollution reduction 减轻污染
pollution risk 污染危险频率，有污染危险的
pollution source 污染源，污源
pollution survey 污染调查
pollution tax 污染税
pollution toxicity 污染毒性
pollution type 污染类型
pollution zone 污染带
pollution 污染，污秽，弄脏，污染物
polonium-beryllium neutron source rod 钋铍中子源棒
polonium 钋【Po】
POL（provisional operating license） 临时运行许可证
polution 污染，沾污
polyacrylamide 聚丙烯酰胺【助凝剂】
polyacrylate 聚丙烯酸盐
polyacrylic acid 聚丙烯酸
polyamide plastics 聚酰胺塑料
polyamide slide bearing 聚酰胺滑动轴承
polyamide 聚酰胺，尼龙
polyanode flip-flop tube 多阳极触发管
polyanode 多阳极
polyatomic 多原子的
polyatron 多阳极计数放电管
polyaxial stress 多重（元）轴向应力，多轴向应力
polybasic acid 多价酸
polybutadiene rubber 聚丁二烯橡胶
polycarbonate 聚碳酸酯
poly cell approach 多元近似法，多单元法
poly cell 多晶电池片
poly-centered arch 多心拱
polychlorinated biphenyl 多氯联苯
polychlorlprene 氯丁橡胶，聚氯丁烯
polychromatic 多色的
polycondensation 缩聚作用
polycore cable 多芯电缆
polycrystalline silicon ingot 多晶硅锭
polycrystalline silicon reduction furnace 多晶硅还原炉
polycrystalline silicon reduction 多晶硅还原
polycrystalline silicon solar cell 多晶硅太阳能电池
polycrystalline silicon solar panel 多晶硅太阳能板
polycrystalline solar cell 多晶太阳电池
polycrystalline UO_2 多晶二氧化铀
polycrystalline 多晶的
polycrystal 多晶
polycyclic aromatic hydrocarbons 多环芳香烃
polycyclic 多相的，多环的
poly-dimensional 多维的，多量纲的
polydirectional core loss 多方向性铁芯损耗
polydirectional 多向的

polydisperse aerosol 多相分散气溶胶
polydispersed pollutant 多相分散的污染物
polydisperse fluidization 聚式流化
polydisperse fluidized bed 聚式流化床
polydisperse material 多分散物料
polydisperse system 多相分散系统，多分散系统
polydispersity 多分散性
polyelectrolyte coagulant aid 聚电解质助凝剂
polyelefine 聚烯烃
polyenergetic neutron system 多能中子系统
polyenergetic 多能量的
polyergic source 多能量源
polyester coating wire 聚酯漆包线
polyester fiber(-re) 聚酯纤维
polyester-glassfibre sheet 玻璃钢板，聚酯胶结玻璃丝板
polyester paint 聚酯漆
polyester resin 聚酯树脂
polyesters woven 聚酯织物
polyesters 聚酯，多元酯
polyether-chlorinate 氯化聚醚
polyether 聚醚，多醚
polyethylene bottle 聚乙烯瓶
polyethylene cable 聚乙烯绝缘电缆
polyethylene chloride 聚氯乙烯
polyethylene fiber 聚乙烯纤维
polyethylene foam insulation 聚乙烯泡沫绝缘
polyethylene high-voltage insulation 聚乙烯高压绝缘
polyethylene insulated cable 聚乙烯绝缘电缆
polyethylene insulated wire 聚乙烯绝缘电线
polyethylene insulation 聚乙烯线绝缘
polyethylene pipe 聚乙烯管
polyethylene sheath 聚乙烯绝缘外套
polyethylene tube 聚乙烯管
polyethylene wrapper 聚乙烯包装
polyethylene 聚乙烯
poly foam 泡沫塑料
polyfunctional exchanger 多功能离子交换剂
polygonal annular foundation 多边环基础
polygonal coil 多角形线圈
polygonal column 多角柱
polygonal connection 多边形接线，多角形接线
polygonal furnace 多角形炉膛
polygonal method 导线测量法，导线法
polygonal network 导线网
polygonal orifice 多角形孔口
polygonal overhead contact system 多边形架空接触网
polygonal point 导线点
polygonal roof truss 折线形屋架
polygonal rubble masonry 多角毛石砌体
polygonal traversing 多角测量
polygonal truss 线形桁架，折线形桁架
polygonal voltage （多相系统的）边电压
polygon circuit 多边形电路
polygonization crack 多边化裂纹
polygon leg 导线边
polygon method 多边形法
polygon misclosure 导线闭合差
polygon of voltage 电压多边形
polygonometric network 导线测量网
polygonometric point 导线点
polygonometry 导线测量
polygon 多边形
polygram 多边形，多字母组合，多角形
polygraph 多种波动描记器，复写器
polyhalide water 高盐分水
polyhedron 多面体
polyhybrid 多混合电路
polyhydric alcohol 多元醇
polyisobutylene 聚异丁烯
polylight 多灯丝灯泡
polymer-cement concrete 聚合物水泥混凝土
polymer film 高分子
polymeric aluminium 聚合氯化铝【絮凝剂】
polymeric aluminum chloride 聚合氯化铝【絮凝剂】
polymeric cable 橡塑电缆
polymeric coagulant 聚合凝聚剂
polymeric material 聚合材料，聚合物材料
polymeric silicicacid 聚合硅酸
polymerizable components of monomer system 单体系统的聚合组分
polymerization chemical 聚合剂
polymerization inhibitor 阻聚剂
polymerization 聚合
polymer modified cement mortar 聚合物改性水泥砂浆
polymer semiconductor solar cell 聚合物半导体太阳电池
polymer stabilization 聚合物加固
polymer suspension 聚合物悬浮液
polymer variable condenser 有机薄膜可变电容器
polymer waste 聚合物废料
polymer 聚合物
polymetamorphism 多相变质
polymetaphosphate 聚偏磷酸盐
polymethyl methacrylate 有机玻璃，聚甲基丙烯酸甲酯
polymorphic programming language 全型程序设计语言
polymorphism 多形现象
polynome 多项式
polynomial computer 多项式计算机
polynomial equation solver 多项方程式解算器
polynomial hill 多项式形山丘
polynomial interpolating function 多项式插值函数
polynomial 多项式，多项的，多项式的
polynuclear complex 多核络合物
polynuclear 多核的，多环的
polyolefin insulation 聚烯烃绝缘
polyolefin 聚烯烃
polyoxide 多氧化物
polyoxy acid 多缩含氧酸
polypayre process 一种新型的管道绝热法
polyphase AC power transmission 多相交流输电
polyphase alloy 多相合金
polyphase alternator 多相同步发电机
polyphase armature 多相电枢
polyphase asynchronous motor 多相异步电动机
polyphase circuit 多相电路

polyphase commutator motor 多相换向器式电动机
polyphase converter 多相换流机
polyphase current 多相电流
polyphase directional relay 多相方向继电器
polyphase distance relay 多相距离继电器
polyphase generator 多相发电机
polyphase hysteresis motor 多相磁滞电动机
polyphase induction motor 多相感应电动机
polyphase machine 多相电机
polyphase meter 多相计数器，多相电度表
polyphase motor 多相电动机，多相电机
polyphase phase comparator 多相相位比较器
polyphase power factor meter 多相功率因数计
polyphase power supply 多相供电
polyphase rectifier 多相整流器
polyphase reluctance machine 多相磁阻电机
polyphase rotary converter 多相旋转换流机
polyphase rotating machine 多相旋转电机
polyphase series commutator motor 多相串激换向器电动机
polyphase series motor 多相串激马达
poly phase shunt commutator motor 多相并激换向器电动机，多相并励换向器式电动机
polyphase source 多相电源
polyphase symmetrical system 多相对称系统
polyphase synchronous generator 多相同步发电机
polyphase system component 多相系统元件
polyphase system 多相制，多相系统
polyphase transformer 多相变压器
polyphase winding 多相绕组，多相线圈
polyphase 多相的，多相
polyphone 多音符号，多音字母
polyphosphate treatment 聚磷酸盐处理
polyphosphate 多磷酸盐
polyphosphoric acid 多聚磷酸
polyplexer 天线收发转换开关，天线互换器
polypropylene lining tube 衬聚丙烯管
polypropylene lining 衬聚丙烯
polypropylene 聚丙烯
polyrod antenna 介质天线
polysaprobic 多污水腐生的，重污水的
polysleeve 多路的，多信道的
polyslot winding 多槽绕组
polystage amplifier 多级放大器
polystage 多级的
polystyrene ceiling board 聚苯乙烯天花板
polystyrene chain 聚苯乙烯链
polystyrene foam plastics 聚苯乙烯泡沫塑料
polystyrene paint 聚苯乙烯漆
polystyrene polyamine 聚苯乙烯聚胺
polystyrene resin 聚苯乙烯树脂
polystyrene 聚苯乙烯
polystyrol 高频绝缘材料的聚苯乙烯，聚苯乙烯
polysulfone 聚砜
polysulphide rubber 聚硫橡胶
polytetrafluoroethylene 聚四氟乙烯，特氟隆
polythene 聚乙烯
polytrifluorochloroethylene 聚三氟氯乙烯
polytropic atmosphere 多元大气
polytropic compression 多变压缩
polytropic efficiency 多变效率
polytropic equation 多变方程
polytropic expansion cycle 多变膨胀循环
polytropic expansion 多变膨胀
polytropic exponent 多变指数
polytropic process 多变过程
polytropic 多方的，多项的，多变的
polyurethane foam plastics 聚氨酯泡沫塑料
polyurethane paint 聚氨酯漆，尿烷漆，聚氨基甲酸乙酯漆
polyurethane 聚氨酯，聚氨基甲酸乙酯
polyvalence 多价
polyvalent retreatment plant 多用途再处理设备，多用途后处理工厂
polyvinyl acetate 聚醋酸乙烯酯
polyvinyl alcohol 聚乙烯醇
polyvinyl chloride film 聚氯乙烯薄膜
polyvinyl chloride flange 聚氯乙烯法兰
polyvinyl chloride hose 聚氯乙烯软管
polyvinyl chloride insulated cable 聚氯乙烯电缆
polyvinyl chloride mastic 聚氯乙烯胶泥
polyvinyl chloride pipe 聚氯乙烯管
polyvinyl chloride plastic welding rod 降氯乙烯塑料焊条
polyvinyl chloride resin 聚氯乙烯树脂
polyvinyl chloride sheet 聚氯乙烯板
polyvinyl chloride 聚氯乙烯
polyvinyl fluoride 聚氟乙烯
polyvinyl resin 聚乙烯树脂
poly water 聚合水
polyzonal spiral fuel element 多区螺旋肋燃料元件
polyzonal 多区的
Poncelet wheel 庞塞莱特水轮机
pondage action 调节作用，调蓄作用
pondage correction 蓄量订正，蓄量改正
pondage factor 库容调节系数
pondage power plant 抽水蓄能电站
pondage station 日调节水力发电厂，日调节水力发电站
pondage 水池蓄水量，蓄水，贮水量
ponderomotive force （电磁场的）有质动力
ponding method of curing concrete 混凝土泡水养护法，混凝土养生池养护法
ponding test 浸水试验
ponding 积水，蓄水，泡水，挖池蓄水
pond let 小水池
pond storage pond 蓄水池
pond 池，塘，槽，池塘，水池
pontoon bridge 浮桥
pontoon causeway 浮筒栈桥
pontoon crane 浮式起重机，水上起重机浮吊
pontoon dock 浮码头
pontoon pierhead 趸船码头
pontoon shape body 浮筒式车身
pontoon wharf 趸船码头
pontoon 趸船
Pontriagin's maximum principle 庞特里亚金最大值原理
pony motor 启动电动机，辅助电动机，伺服电

动机
pony-size 小型的，小尺寸的
pony truss 矮桁架
POOH(planned overhaul outage hours) 计划大修停运小时
pool cathode rectifier 汞弧阴极整流器
pool cathode 液体阴极，汞弧阴极
pool concept （快中子增殖堆）池式方案
pool cushion 消能池
pooled inventory management 备品备件集中总存量管理
pool-film boiling 池内膜态沸腾
pool fire 池式起火
pooling 联营
pool rotor 泳池式转子
pool skimming weir 水池溢流堰【反应堆】，水池溢流坝
pool storage 集中贮存
pool test reactor 池式试验反应堆
pool-type facility 池式核装置
pool-type fast breeder reactor 池式快增殖反应堆
pool-type pressure suppression system 池式弛压系统
pool-type reactor 池型反应堆，池式反应堆，游泳池反应堆，池式堆
pool-type rector 池式反应堆
pool vent 燃料水池排气孔
pool water 池水
pooly drained 排水不畅的
pool 池，潭，塘，水池，水潭，水塘，乏燃料贮存水池，深槽段，深渊，联合，联营，统筹，联合电力系统，组合
poop shot 风化层爆炸
poop 尖锐脉冲，喇叭声，情报材料，消息
poor alignment 未对准【焊接】
poor coal 劣质煤
poor combustion 不完全燃烧
poor commutation 不良换向，不良整流
poor concrete 少灰混凝土
poor conductor 不良导体
poor contact 不良接触，接触不良
poor fitup 准备不足【焊接】
poor fusion 未熔合
poor gas 低热值可燃气
poor-grade coal 低级煤，劣质煤
poor in carbon 低碳的
poor in quality 劣质的
poor insulation 不良绝缘
poor insulator 不良绝缘子
poorly drained stream basin 排水不畅的流域
poorly drained 排水不畅的，排水不良的
poorly fluidized bed 不良流化床
poorly-graded aggregate 级配不良的骨料
poorly-graded soil 级配不良土
poorly rounded gravel 磨圆度不良的砾石
poorly-run enterprise 亏损企业
poor management 管理不善
poor mixture 贫燃分混合物
poor penetration 未焊透
poor quality concrete 劣质混凝土
poor quality 不良质量
poor regulation 电机转速宽范围的自调节，不良调节
poor restart 引弧不当【焊接】
poor starting 启动不良
poor subgrade 软弱地基
poor subsoil 劣质底土，软弱底土，软弱地基
poor workmanship 工程质量低劣
PO＝pass-out 抽汽
pop-in 爆裂，初始裂纹增长
poplar 白杨
pop lift 突然打开
PO＝polyolefin 聚烯烃
PO(postal order) 邮政汇票，邮政汇单
PO(post office) 邮政局
PO(power output) 功率输出
poppet air valve 提升式通气阀
poppet injector 菌状油门喷嘴
poppet type extraction valve 提升式抽汽阀
poppet valve 提升阀，圆盘阀，直通阀，活瓣阀
popping point tolerance 突发点容许极限
popping pressure 起跳压力，起座压力，突发压力【安全阀】，提升压力
popping rock 岩石爆裂
popping safety valve 安全阀
popping 阀口打开【阀门】，间歇振荡，爆裂
PO(primary output) 初级输出
pop safety valve 紧急安全阀，弹簧保险阀
popularity 名声
popularize 普及
populated area 居住区
population census 人口普查
population center 居住中心，人口中心
population density 群体密度，人口密度
population distribution 人口分布
population dose 群体剂量
population effect 人口影响
population estimate 人口估计
population exposure 群体照射（量）
population mean 总体平均值，总体均值
population size 群体大小
population 总体
populous 人口稠密的
PO(pulse output) 脉冲量输出
pop-up window display 弹出窗口显示，画面开窗显示
pop-up 弹出
PO(purchase order) 采购订单
pop valve 突开阀
pop 爆裂声，间隙振荡，突然打开
POQAP(plant operational quality assurance procedure) 电厂运行质量保证程序
porcelain bushing shell 瓷套管
porcelain bushing 瓷套，瓷套管
porcelain capacitor 陶瓷电容器
porcelain casserole 瓷勺皿
porcelain clay 瓷土
porcelain cleat 瓷夹
porcelain condenser 陶瓷电容器
porcelain crossarm 瓷横担

porcelain crucible 瓷坩埚
porcelain cup 瓷盘绝缘子
porcelain cutout 瓷断流器
porcelain elbow 弯瓷管
porcelain evaporating dish 瓷蒸发皿
porcelain glaze 瓷釉
porcelain insulator 瓷瓶，瓷（质）绝缘子
porcelain knob 瓷柱，鼓形绝缘子
porcelain lining 瓷衬
porcelain pipe 陶瓷管
porcelain plug 瓷插头
porcelain radiator 陶瓷散热器
porcelain screw holder 瓷质螺丝灯头
porcelain screw socket 瓷质螺丝灯座
porcelain shell 瓷套
porcelain triangle 瓷三角
porcelain tube 瓷管
porcelain type circuit breaker 瓷柱式断路器
porcelainware 瓷器
porcelain 瓷器，陶瓷，瓷件，瓷的
porcess related station level functions 与过程有关的站层功能
porch chamber 船屋，门厅
porch column 门廊柱
porch 门廊
PORC（plant operations review committee） 电厂运行审评委员会
pore air pressure 孔隙气压力
pore diffusion control 空穴扩散控制
pore diffusion 微孔扩散
pore former 微孔形成剂，造孔剂
pore forming agent 造孔剂
pore forming 造孔的
pore-free mass 无孔隙体
pore migration 孔隙徙动
pore moisture 孔隙水分
pore pressure cell 孔隙压力盒，孔隙压力计
pore pressure dissipation 孔隙压力消散，孔隙压力消失
pore pressure gauge 孔隙压力计
pore pressure meter 孔隙压力计
pore pressure parameter 孔隙压力系数，空隙压力参数
pore pressure 孔隙压力
pore size distribution 孔径分布，细孔分布
pore size 孔径，孔尺寸，孔径大小
pore space 孔隙
pore tension 孔隙张力
pore velocity 孔隙流速
pore volume 空隙度，空隙量，空隙率，孔隙度，孔隙率，孔隙容积
pore wall 孔隙壁
pore water content 孔隙水含量
pore water head 孔隙水头
pore water pressure cell 孔隙水压力盒
pore water pressure meter 孔隙水压力计
pore water pressure 孔隙水压力
pore water tension 孔隙水张力
pore water 孔隙水
pore 孔隙，气孔，细孔，毛孔，注视，缝隙
poriness 多孔性，疏松性，孔隙率
poroplastics 多孔塑料，泡沫塑料
pororoca 涌潮
porosimeter 孔率计，孔隙率计
porosimetry 孔隙度测量法
porosity chart 气孔计算图
porosity measurement 孔隙率测定
porosity of particles 颗粒疏松度
porosity 孔隙率，气孔率，多孔性，孔隙度，开比度，松度，气孔，疏松，疏松度
porous absorber 多孔吸热器
porous aggregate 多孔骨料
porous alumina 多孔氧化铝
porous bed 疏松层
porous block type arrester 多孔绝缘体型避雷器
porous breakwater 透水防波堤
porous brick 多孔砖
porous bronze distributor 多孔青铜分布板
porous bronze plate 多孔青铜板
porous brush 多孔电刷
porous carbon 多孔碳
porous ceramics 多孔陶瓷
porous ceramic tile distributor （沸腾炉）多孔陶瓷布风板
porous-clay pipe drain 多孔黏土排水管
porous concrete drain 多孔混凝土排水管
porous concrete pipe 多孔混凝土管
porous concrete 泡沫混凝土，多孔混凝土
porous cooling 发汗冷却，发散式冷却，蒸发冷却
porous dam 透水坝
porous element 多孔元件
porous fibre tube 多孔纤维管
porous film 多孔
porous filter element 多孔滤元
porous foundation 多孔性地基
porous glass membrane 多孔玻璃
porous graphite 多孔石墨
porous ion-exchange resin 多孔离子交换树脂
porous ion exchanger 多孔性离子交换剂
porous layer （岩石的）多孔层，疏松层
porous limestone 多孔灰岩
porous material 多孔材料
porous media flow 多孔介质流
porous media 多孔介质
porous medium 多孔介质
porous membrane 多孔膜
porous metal filter 多孔金属过滤器
porous plastics 泡沫塑料
porous plate 素烧瓷板
porous porcelain evaporimeter 多孔瓷蒸发器
porous reactor 多孔堆芯反应堆
porous rock 多孔岩石，透水岩石
porous scale 多孔垢
porous shroud 多孔罩
porous soil 多孔土
porous solid 疏松固体，多孔固体
porous stone 多孔石，透水石
porous structure 多孔结构
porous thermal insulation 多孔热绝缘
porous tip piezometer 多孔探头式测压管
porous type strongbase resin 多孔型强碱树脂

porous-walled breakwater 多孔墙式防波堤
porous wall wind tunnel 多孔壁风洞
porous wind screen 多孔风障
porous 有孔的，多孔的，疏松的，多孔渗水的
porphyrite 玢岩
porphyritic structure 斑状结构
porphyry 斑岩
porpoise 前后振动，波动
portability 轻便性，可移植性，可携带性
portable agitator 移动式搅拌机
portable air compressor 移动式空气压缩机
portable ammeter 便携式电流表
portable amplifier 移动式放大器
portable antenna 轻便天线，便携式天线
portable appliance 携带式电具
portable arc welding machine 便携式弧焊机
portable automatic tide gauge 轻便式自记潮位计
portable belt conveyor 移动式皮带输送机
portable boiler 移动式锅炉
portable boom conveyor 移动式长臂输送机
portable cement pump 移动式水泥泵
portable compressor 移动式压缩机
portable computer 手提式计算机，便携式计算机
portable counter 携带式计数器，便携式计数器，手提式计数器，轻便计数器
portable crane 轻便起重机
portable derrick crane 移动式人字起重机
portable digital different pressure meter 便携式数字压差表
portable dose rate meter 携带式剂量率计
portable drill 移动式钻机
portable elevator 移动式升降机
portable fire extinguishers 便携式灭火器
portable fire pump with engine 手提机动消防泵
portable hardness tester 便携硬度测试仪，携带式硬度计
portable hydraulic pipe bending machine 手提式液压弯管机
portable instrument 携带式仪表，轻便仪表
portable intelligent terminal 手持式智能终端
portable interphone 手持式对讲机
portable lamp 行灯，手灯
portable low power reactor 移动式低功率反应堆
portable magnetic flaw detector 手提式磁力探伤仪
portable mixer 移动式拌和机
portable monitor 手提式监测仪，轻便监测仪
portable plough car 移动犁式卸料机
portable plough unloader 移动犁式卸料机
portable power source 移动电源
portable pump 可移式泵
portable pyrometer 便携式高温计
portable radiation monitor 携带式辐射监测器
portable radio-equipment 手提式无线电设备
portable railway 工地轻便铁道
portable reactor 可移动反应堆
portable return tube boiler 移动式回火管锅炉
portable sampler 轻便取样器
portable shield 可移动屏蔽体，活动屏蔽
portable shunt 移动式分流器
portable source 可携带放射源，手提源
portable speedmeter 便携式转速表
portable sprinkler system 移动式喷灌系统
portable sprinkler 移动式喷灌机
portable staff gauge 活动水尺
portable staff 轻便式标尺
portable substation 流动变电所，流动变电站
portable switch 轻便开关
portable telephone set 携带式电话机，巡房话机
portable telephone 携带式电话
portable testing set 携带式试验设备
portable tool 手提式工具
portable tower 活动觇标
portable trailer-mounted drill 移动式轻便钻机
portable voltage transformer 便携式电压互感器
portable voltmeter 便携式电压表
portable whole-body counter 携带式全身计数器
portable winch 轻便绞车
portable yard crane 移动式起重机，轻便吊车
portable 可携带的，可移动的，轻便的，轻型的，手提式的，便携的
port administration bureau 港务（管理）局
port administration 港务局
portage 水陆联运，搬运
portal bracing between columns 柱间支撑
portal bracing 桥门联杆，门撑
portal crane 龙门起重机，门式起重机
portal frame construction 门式框架结构
portal frame 门式刚架，门架
portal grab ship unloader 门座式抓斗卸船机
portal hanger 门式吊架【电缆支架】
portal jib crane 龙门吊车，门式旋臂起重机
portal monitor 出入口监测器
portal reclaimer 门式取料机
portal rigid frame 门式刚架
portal scraper 门式刮料机
portal structure 门式结构
portal strut 门式钢架撑杆
portal type bucket wheel stacker-reclaimer 门式斗轮堆取料机
portal type rotary-wheel stacker-reclaimer 门式滚轮堆取料机
portal type screw unloader 门式螺旋卸车机
portal type 门式螺旋卸车机
portal 门形杆，门型架，入口，洞门，门架，门座，桥门，隧道门
port anchorage 港内锚地
port and harbour 港口和海港
port area 港区
port authority 港务（管理）局
port charge 港口费用
port congestion charge 疏港费
portcullis 吊门
port current 端口电流
port debarkation 目的港
ported cylinder 气动直立落煤筒
ported travelling crane 门式移动起重机
port engine 滑阀配汽蒸汽机
porterage 搬运费，搬运业，搬运
port facility 港口设施，港口设备
portfolio copies 公文复件

portfolio 代表作
port glass 观察窗，观察孔玻璃
port-guided 定向的通孔
port handling 港口装卸
porthole 观察孔，口，门
portico （有圆柱的）门廊，柱廊
portion 部分
Portland cement 波特兰水泥，硅酸盐水泥
portlandite 氢氧钙石
Portland-pozzolana cement 火山灰质硅酸盐水泥
Portland-slag cement 矿渣水泥
port of arrival 到达港
port of call 沿途停靠港
port of delivery 卸货港，交货港
port of departure 启运港，出发港
port of destination 到岸港，目的港，目的口岸
port of discharge 卸货港
port of dispatch 发货港
port of distress 遇难港
port of embarkation 发航港，装运港
port of entry 进口港，入境港，报关港
port of exit 出口港
port of last call 最后停泊港
port of loading 装货港
port of option 选择港
port of sailing 起航港
port of shipment 装货港，装运港
port of transhipment 转运港
port of transit 过境港，转口港
port of transshipment port 中途转运港
port of transshipment 转船港，中转港
port of unloading 卸货港
PORT(photo-optical recorder tracker) 光电记录跟踪装置
port power plant 港口电厂
port railway 港区铁路
port risk 港口险
port road 港区公路
port side boiler 左侧锅炉
port size （阀门）通孔尺寸
port structure 港口建筑物
port surcharge 港口附加费，港杂费
port surging 港口涌浪
port terminal facilities 水陆联结设施
port voltage 端口电压
port 港口，港，端口，通口，阀口，门，孔，入口，气门，开槽，孔板，注孔，开口
posiode 正温度系数热敏电阻
posistor 正温度系数热敏电阻
positional accuracy 定位精度
positional control 位置控制
positional deviation 位置偏差
positional error coefficient 位置误差系数
positional information 位置信息
positional notation 位置记数法
positional punch 定位穿孔
positional representation 位置表示法
positional tolerance 位置容差，允许位置偏差
positional variation 位置偏差，位置变化
positional 位置的
position-balance 动平衡，位置平衡
position buoy 定位指示浮标
position coder 位置编码器
position code 位置码
position control servomechanism 位置控制伺服机构
position control system 位置控制系统
position control 位置控制，开度控制
position detection system 位置指示系统
position detector 位置探测器
2-position diverter gate 2位置转向［倒向］闸门
positioned haulage system 定位机牵引装置
positioned weld 定位焊，暂焊
positioned 焊接变位机
positioner arm 定位机臂
positioner carriage 定位机车架
positioner housing 定位器壳套
position error 位置误差
positioner stroke 定位机行程
positioner system 定位机系统
positioner valve 定位阀
positioner 定位器，定位装置，定位监控器，定位机，夹具，反馈装置，位置控制器
position-event-time 位置、事件、时间
position feedback 位置反馈
position finding 定位
position fixing 定位
position head 位置水头，位置压头
position index 位置指示
position indicating system 位置指示系统
position indication 位置指示
position indicator assembly 位置探测装置
position indicator coil 位置指示器线圈
position indicator 位置指示器
positioning accuracy 定位精度
positioning adapter 定位支架，定位衬套
positioning bracket 定位托架
positioning controller 定位控制器
positioning control system 定位控制系统
positioning control 定位控制
positioning device 定位装置
positioning dowel 定位销
positioning drive 定位驱动
positioning process 定位过程
positioning error 定位误差
positioning hole 定位孔，找正孔
positioning indicator 状态指示器
positioning key 定位销
positioning motor 位置控制电机
positioning nut 定位螺帽
positioning of tower leg 塔腿定位
positioning pin 定位销，直线校正杆
positioning plate 定位板
positioning power unit 位置执行部件
positioning rate 定位装置
positioning screw 定位螺钉
positioning sequence 定位程序
positioning servo system 定位伺服系统
positioning 定位，调位，校正，位置控制，布置，定位装置，定位控制
positioning 就位
position limitation 位置限制

position limit switch　限位开关
position line　定位线，位置线
position meter　通话计时器，通话计次器
position motor　位置电动机
position paper　论证书
position recorder　位置记录器
position relay　位置继电器
position selector valve　换向阀，位置选择阀
position selector　位置选择器
position sensor　位置传感器
position servomechanism　位置伺服机构
position signaling　位置信号
3-position splitter gate　3 位置分流闸门
position switch　定位开关，行程开关
position-to-number converter　位置数字变换器
position-to-number　位置数字
position transducer　位置传感器
position transmitter　位置变送器，位置发送器
position-type telemeter　位置式遥测装置
position vector　方位向量，位置矢量
position welding　定位焊
position　定位，位置，状态，职务，安插
positive acknowledgement　肯定确认
positive aerodynamic damping　正气动阻尼
positive allowance　正公差
positive analysis　实证分析
positive approach　主动方法，积极方法
positive artesian head　正承压水头
positive back seat　刚性连接后座
positive biased　正偏的
positive bias　正偏压，正偏置
positive booster　增压机，增压器，升压机
positive brush　正电刷
positive buoyancy　正浮力
positive bus-bar　正汇流排，正母线
positive bus　正汇流排，正母线
positive carrier　正调制载波，正电荷载流子
positive charge　正电荷，阳电荷
positive circulation　强制循环
positive clock pulse　正的时钟脉冲，正的同步脉冲
positive closure　完全闭合
positive coefficient　正系数
positive conductor　正导线，正导体
positive confinement　无泄漏保藏，绝对密封
positive confining bed　上承压层
positive converter　正向变流器
positive coordinates　正坐标
positive corona　正极电晕
positive correlation　正相关
positive coupling　同向耦合
positive current　正电流
positive cutoff wall　正截水墙
positive damping　正阻尼
positive definite matrix　正定矩阵
positive digging force　正面挖掘力
positive displacement flow measuring device　容积式流量测量仪表
positive displacement flowmeter　容积（式）流量计
positive displacement grout pump　排液灌浆泵
positive displacement meter　正位移液体计量器，容积式流量计，变容计量器
positive displacement pump　正排量泵，排液泵，容积式泵
positive displacement slurry pump　容积式灰浆泵
positive displacement type　正向位移式
positive distortion　正畸变，正失真
positive divisor　正除数
positive draft　人工通风，强制通风
positive drift　正阻力
positive effect　正效应
positive electricity　阳电，正电
positive electrode　正极，阳极，正电极
positive electron　正电子
positive elevation　正迁移
positive elongation　正延性
positive energy balance　正能平衡
positive exponent　正指数
positive feedback　正回授，正反馈
positive feeder　正极馈线
positive feed forward control　正前馈控制
positive frequency　正频率
positive function　正函数
positive gearing　直接传动
positive-going　正向的，朝正向变化的
positive grounded battery　正极接地电池
positive impact　积极的影响
positive impedance　正阻抗
positive incidence　正冲角，正攻角
positive infinitely variable gear box　无级变速齿轮箱
positive-input negative-output　正输入负输出
positive integer　正整数
positive ion　阳离子
positive limiting　正值限制
positive logic　正逻辑
positively charged　带正电的
positive modulation　正极性调制，正调制
positive moment reinforcement　正矩钢筋
positive moment　正力矩，正弯矩
positive-negative action　正负作用
positive-negative control　正负控制
positive-negative three position action　正负三位作用
positive-negative three step action　正负三位作用
positive number　正数
positive order component　正序分量
positive order　正指令
positive output　正输出
positive peak　最大正值，正峰值
positive period method　正周期法
positive period　正周期【反应堆】
positive phase-sequence reactance　正序电抗
positive phase-sequence relay　正序继电器
positive phase-sequence symmetrical component　正序对称分量
positive plate　阳极，阳极板，正极板
positive polarity　正极性
positive pole　阳极，正极
positive pressure mill　正压式磨煤机
positive pressure pneumatic ash handling　正压气

力输灰
positive pressure ventilation　正压（机械）通风，正压通气，正压送风
positive pressure　正压，正压力
positive pulse　正脉冲
positive quantity　正量
positive reactivity　正反应性
positive regeneration　正反馈，正回授
positive relationship　积极的关系
positive release wave　正放水波
positive result　肯定结果
positive rotary pump　正向旋转泵
positive scram　正停堆
positive sealing system　可靠止水系统
positive sequence component　正序分量
positive sequence coordinate　正序坐标
positive sequence current　正序电流
positive sequence damping　正序阻尼
positive sequence filter　正序过滤器
positive sequence impedance　正序阻抗
positive sequence network　正序网路
positive sequence power　正序功率
positive sequence reactance　正序电抗
positive sequence resistance　正序电阻
positive sequence voltage　正序电压
positive sequence　正序的，正序
positive shutoff device　绝对切断装置
positive shutoff　完全关闭
positive sign　正号
positive slope　顺坡
positive suction head　正吸入头
positive surge　正涌浪
positive temperature coefficient　正温度系数
positive terminal　正端子，正线端
positive test　肯定测试
positive throat nozzle　缩放喷嘴
positive transmission　正极性传送，正调制传送
positive value　正值
positive ventilation　强制通风，正压通风
positive water hammer gradient　正水锤梯度
positive wave　正波
positive wire　正极引线
positive　正片，正的，正数，确定的，可靠的，肯定，正面的，正向的，正序的
positonging control system　位置跟踪系统，定位控制系统
positoning　定位
positron　正电子
POS(point of sale)　销售点
possession of site and access　现场占有权及其通道
possession of site　占用现场
possession　财产，所有权，占有
possessor　持有人，持有者
possibility of trouble　故障率，事故率
possibility　可能性
possible combination　可能的组合
possible effectiveness　可能效力
possible erratum　可能误差
possible error　可能误差
possible precipitation　可能降水（量）
possitive pole　正极
Posson's ratio　泊松比，横向变形系数
post-aceleration　后加速
post-accident decommissioning　事故后退役
post-accident heat removal　事故后冷却，事故后排热
post-accident instrumentation system　事故后仪表系统
post-accident monitoring instrumentation　事故后监测仪表
post-accident monitoring system　事故后监测系统
post-accident operation upgrading　事故后操作升级
post-accident sampling system　事故后取样系统
post-accident simulator　事故后仿真机
post-accident study　事故研究分析
post-accident　事故后的
postage stamp method　邮票法
postage　邮费
post alloy diffusion transistor　柱状合金扩散晶体管
postal order　邮政汇票，邮政汇单
post amplifier　后置放大器
post analysis　事后分析
post and beam construction　梁柱结构，骨架结构
post and block fence　柱板围墙
post and gird　立柱围梁
post and paling　木栅围篱，木栅栏
post and panel structure　立柱镶板式结构，砖木结构，镶板式结构
post and stall method　房柱法
post audit meeting　监察后会议
post audit　事后审核
post-award meeting　授标后会议
post-beam construction　梁柱结构
postboiler corrosion　炉后腐蚀
post-buckling behaviour　后期压屈特性
postburnout heat transfer coefficient　烧毁后传热系数
postburnout heat transfer　烧毁后传热
post cap　柱帽
post card　明信片
post-Chernobyl era　切尔诺贝利事故后时代
post code　邮政编码
post-commissioning cleaning　运行后的清洗
post-condenser　后置凝汽器
post-cooling method　后期冷却法
postcritical region　过临界区
postcritical vortex shedding　过临界旋涡脱落
postcritical　临界后的
post-detection integral　检波后积分
post-detection noise　检波后噪声
post-detection summation　检波后相加
post-detection　后检波，检波后的
post disturbance review　事故追忆
post-DNB region　DNB 后区
post-dose　照后剂量
post drill　支架式钻机
post dryout heat transfer coefficient　干涸后传热系数
post dryout heat transfer　干涸后传热

post echo 后反射波
post edit 后编辑，后置编辑
posted price 标明价目
post-emulsifiable dye penetrant testing method 后乳化着色探伤法
post-emulsifiable penetrant 乳化后渗透剂
posterior distribution 后验分布
posterior probability 后验概率
posterior 后验的
poster session 展示会，招贴会
post evaluation 后评估
post fence 柱式护栏
post-filter 过置过滤器，后过滤器
post fire hydrant 地上消火栓，柱式消防栓
postheat current 焊后加热电流
postheating temperature 后热温度
postheat 跟踪回火，随后加热，焊后加热，后热
postheat temperature 后热温度
post-hole auger 钻孔取样器，柱孔螺旋钻
post-hole sampler 钻孔取样器
post-impregnated winding （嵌线）后浸渍绕组
postincident analysis 故障分析
post-incident cooling system 应急冷却系统，事故后冷却系统
post-inspection treatment 检验后的处理
post insulator 柱状绝缘子，柱式绝缘子，装脚绝缘子
postirradiation examination 辐照后检验
postirradiation therapy 辐照后治疗
postirradiation 已辐照，辐照后
post light 柱灯
post-LOCA refill 冷却剂丧失事故后再充
post-LOCA-reflooding 冷却剂丧失事故后再淹没
post loss-of-coolent accident protection 冷却剂丧失事故后保护
post mortality review 事故追忆
postmortem analysis 事后分析
postmortem program 算后检查程序
postmortem review 故障分析
postmortem routine 算后检查程序
postmortem 程序完成后的分析，算后检查，事后析误，事后剖析
post-mounting flushing 安装后清洗
post office box 邮政信箱
post office 邮政局
post-operation cleaning 运行后的清洗
post-operative radiotherapy 术后放射治疗法
postponed outage 第 3 类非计划停运【可延迟至 6 小时以后的停运】
postponed payment 延期支付
postponement of payment 延期支付
postponement 搁置
postpone 延缓，延期，搁置，推迟
post-precipitation 继沉淀作用
post processing 后处理，继续（加工）处理
post processor 后信息处理指令，后信息处理机
post-project evaluation 项目后评价
post purge （炉膛）熄火后吹扫
post qualification stage 资格审查后期
post qualification 资格后审
post responsibility system 岗位责任制
post responsibility 岗位专责
post sales services 售后服务
post scriptum 又及【信件正式结尾后附加的字句】，邮政原本，附言，后记，附录，附注
post script 附言
post selection 后选择，补充拨号
post-selector 有拨号盘的电话机
post-service cleaning 运行后的清洗
post shore 单独立柱
post-shutdown period 停堆后时期
post-skill wage system 岗位技能工资制
post spacing 柱距
post stall control 过失速控制
post stall gyration 失速后旋转
post stall maneuverability 过失速机动性
post stall maneuver 过失速机动
post stall performance 过失速性能
post stall recovery 失速后改出
post stall regime 失速后状态
post stall transient 失速后瞬态
post stall 过失速（的）
post storminspection 风暴事后调查
post-stressed concrete 后张法混凝土
post-stressing 后加应力
post stretching 后张法，后张，后张拉，后加拉伸
post technical ability wage system 岗位技能工资制
post-tensioned beam 后张梁
post-tensioned cable 后张钢丝索
post-tensioned concrete beam 后张法预应力混凝土梁，后张混凝土梁
post-tensioned concrete pile 后张混凝土桩
post-tensioned concrete slab 预应力混凝土楼板
post-tensioned concrete structure 后张预应力混凝土结构
post-tensioned concrete 后张混凝土
post-tensioned construction 后张法施工，后张拉结构
post-tensioned method 后张法
post-tensioned preaffirmation concrete beam 后张法预应力混凝土梁
post-tensioned preaffirmation 后张法预应力
post-tensioned prestressed concrete girder 后张法预应力梁
post-tensioned prestressed concrete structure 后张法预应力混凝土结构
post-tensioned prestressed concrete 后张预应力混凝土
post-tensioned prestressed system 后张预应力体系
post-tensioned prestressed 后张预应力的
post-tensioned prestressing 后张法预应力
post-tensioned reinforced concrete 后张应力钢筋混凝土
post-tensioned system 后张拉设备
post-tensioning anchorage 后张锚固
post-tensioning duct 后张钢筋导管
post-tensioning method 后张法
post-tensioning 后张（法）
post Three Mile Island 三里岛事故后的

post title 职务
post training 岗位培训
post treatment 后处理，后续处理，处理以后
post-trip logging 追忆打印
post-trip review log 跳闸追忆记录
post-trip 快速降功率后，保护停堆后的
post-type insulator 柱式绝缘子
postulated accident 假想事故，假设事故
postulated initiating event 假设始发事件，假想始发事件
postulated of induction 归纳法公设
postulated 预计的，假定的，假设的
postulate 先决条件，假设，公设
post wage system 岗位工资制
post-weld heat treatment 焊后热处理
post-write disturb 写后干扰
post 杆，邮政，立柱，柱，标杆，后，站，所，接线柱，端子，岗位
potability 可饮性
potable and sanitory water system 饮用水和卫生水系统
potable water system 生活水系统，饮用水系统
potable water tank 生活水箱，饮用水箱
potable water 饮用水
potamic 江河的
potamogenic deposit 河口沉积
potamology 河流学
potash-magnesia mica 钾镁云母
potash mica 白云母，钾云母
potash 钾碱，碳酸钾
potassa 氢氧化钾，苛性钾
potassium acid phthalate 邻苯二甲酸氢钾
potassium bicarbonate powder 碳酸氢钾干粉
potassium biphthalate 邻苯二甲酸氢钾
potassium chloride 氯化钾
potassium chromate 铬酸钾
potassium cooled reactor 钾冷堆
potassium dichromate 重铬酸钾
potassium dihydrogen phosphate 磷酸二氢钾
potassium ferricyanide 铁氰化钾
potassium ferrocyanide 亚铁氰化钾
potassium hydroxide 氢氧化钾
potassium iodide starch test paper 碘化钾淀粉试纸
potassium manganite 高锰酸钾
potassium mercuric thiocyanate 硫氰酸汞钾
potassium monohydrogen phosphate 磷酸氢二钾
potassium permanganate 高锰酸钾
potassium persurfate 过二硫酸钾
potassium photocell 钾光电池
potassium silicate binder 钾水玻璃，钾水玻璃黏结剂
potassium sodium tartrate 酒石酸钾钠
potassium thiocyanate 硫氰化钾
potassium 钾
potation density 人口密度
potato masher 干扰雷达天线，产生无线电干扰的天线
potcal pot calcination 坩埚煅烧，罐煅烧
pot clay 陶土
pot configuration （钠冷快堆）池式布置【一次回路】
pot core 壶形铁芯
potential ability 潜在能力
potentialality 潜能【用复数】，潜力，势能，可能性
potential antinode 电压波腹
potential attenuator 电位衰减器，电压控制器
potential barrier 势垒
potential capacity of land 土地潜力
potential capillarity 毛管潜能
potential channel 潜在水道
potential circle 电势圆，电位圆
potential circuit 电压电路
potential coil 电压线圈
potential competition 潜在竞争
potential correction 电势校正，电位校正
potential curve 电势曲线，分位曲线
potential damage region 潜在危害区
potential device 电容器式高压装置
potential diagram 电势图，电位图
potential difference 磁位差，电位差，势差，位差
potential distribution 电势分布，电位分布，势分布，位分布
potential divider 分压器
potential drop compensator 电压补充装置
potential drop 电压降，磁压降，电势降，电位降，势降
potential electrical field 位电场
potential emergency 事故酝酿
potential emission 潜在排放
potential energy curve 势能曲线，位能曲线
potential energy of deformation 变形势能
potential energy of twist 扭转势能
potential energy storage system 势能储能系统
potential energy 势能，位能
potential environmental impact 环境的影响
potential evaporation 潜在蒸发，蒸发能力
potential evapotranspi-ration 蒸散发能力
potential failure surface 潜在破坏面
potential fall 电位降，电势降，电压降
potential fault 隐伏断层
potential field 势场，位场
potential flow field 势流场
potential flow in two dimentions 平面势流
potential flow model 势流模型
potential flow theory 势流理论
potential flow 位流，势流
potential fraction of internal energy 内部的势能部分
potential function 势函数，位函数
potential fuse failure relay 故障熔丝电位继电器
potential galvanometer 电位检流计
potential gradient 势能梯度，势梯度，位势梯度，位梯度
potential groundwater yield 地下水供水蕴藏量，地下水蕴藏量，可能地下水产水量
potential hazard 潜在危害
potential head 位置水头，势头
potential heat 潜热
potential hill 势垒，位垒，电位障壁
potential hydroenergy 水能蕴藏量

potential hydropower 水电蕴藏量
potential impulse 电压脉冲
potential indicator 电位指示器
potential instability 位势不稳定
potentiality and actuality 潜能与现实
potentiality of cooperation 合作潜力
potentiality 可能性，潜力，蕴藏量
potential logging 电位测井
potential loop 电压波腹
potentially active 具有潜在活性的
potentially available wind power 潜在可用风能
potentially-radioactive waste 潜在的放射性废物
potentially 可能地，潜在地
potential market 潜在市场
potential measurement 电势测量，电位测量
potential meter 电位差计
potential node 电势波节，电位波节
potential of wind energy 风能潜力
potential phasor 电势相量，电位相量
potential plateau 势坪
potential pollutant 潜在污染物
potential polygon 电位多边形，电势多边形
potential power 水力蕴藏量，动力蕴藏量，潜在功率
potential pressure 静压
potential regulator 电位调节器，电势调节器，调压变压器
potential relay 电压继电器
potential resources 潜在资源
potential rise 电势升，电位升
potential risk 潜在风险
potential site for wind machines 风力机潜场址
potential site 潜在坝址
potential slide area 潜在滑坡区
potential sliding surface 潜在滑动面
potential social and economic impact 社会经济学影响
potential source rectifier 电压源整流器
potential source 势源，位源，电势源，电位源，潜在源
potential stabilizer 电位稳定器，电势稳定器，电压稳定器
potential stress energy 应力势能
potential stress 静电应力，静电强度
potential surface of weakness 潜在软弱面
potential temperature 位势温度
potential terminal 电压端子
potential tester 电压试验器
potential to ground 对地电势，对地电位
potential transformer 测量用变压器，电压互感器，仪用变压器
potential vector 电势矢量，电位矢量
potential vortex 有势旋涡，位涡
potential water power resources 水力资源蕴藏量
potential water power 水力蕴藏量
potential well 势阱
potential winding 电压绕组，电压线圈
potential wind power site 潜在风能场址
potential 电势，电位，势，位，潜力，潜在的，可能的，有可能性的，电压的，蕴藏量，潜能，可能性，势的
potentiodynamic analysis 电位动力分析法
potentiometer braking （串激电机的）分压器式制动
potentiometer card generator 电位计卡片控制发生器
potentiometer circuit 分压器电路，电压电路，电位计电路
potentiometer control 电位器控制
potentiometer error-measuring system 电势计式误差测量系统
potentiometer function generator 电位计函数发生器
potentiometer method 电势计法，电位计法
potentiometer multiplier 电势计式乘法器
potentiometer pick-off 电势计式传感器
potentiometer recorder 电势计式记录器，电位计式记录器
potentiometer resolution 电位计的分辨能力
potentiometer resolver 电势计式解算器
potentiometer-type field rheostat 电位计式磁场可变电阻
potentiometer-type resistor 电势计式电阻
potentiometer-type rheostat 电势计式变阻器，电位器式变阻器
potentiometer 电位器，电势计，电位差计，分压器，电位计，分压计
potentiometric surface 测压管液面，测压管水面
potentiometric titration 电位滴定
potentiometric titrimeter 电势计式滴定计
potentiometric 电势计式的
potentiometry 电位测定法
potentiostat 稳压器，恒势器，电压稳定器，恒电位仪
pothead insulator 套管绝缘子
pothead type transformer 带电缆套管变压器
pothead 端套，（电缆的）终端套管，电缆头
pothole erosion 涡流侵蚀
pothole 壶穴，锅穴，凹坑
pot insulator 罐式绝缘子
pot magnet 罐形磁铁
pot motor 高转速电机，离心罐电动机，两极三相鼠笼电机
potometer 散发仪
potted capacitor 封闭式电容器
potted coil 封闭线圈
potted 罐装的，密封的
pottery 陶制品，陶器，陶器厂，陶器制造术
potting compound 陶瓷体芯块与包壳间相互扩散形成的化合物
potting material 封装材料
pot type reactor 罐式反应堆
pot 罐，筒，坩埚，电位计，装入罐内
poundal 磅达
pounder 夯锤，夯具
pounding 捣碎，捣实
pound mole 磅分子
pounds per cubic foot 磅每立方英尺
pounds per square inch 磅每平方英寸，每平方英寸的磅数
pound 磅，井号
pour cent mille 反应性单位

pour cold 低温浇注
pour concrete 浇灌混凝土，浇注混凝土
pour density 松散密度，松装密度
poured-in-place concrete 现浇混凝土
poured-in-place pile 灌注桩
poured-in-place structural foam 现浇发泡结构，浇模发泡结构
poured-in-place 就地浇筑，现场浇筑，现浇
poured insulation 灌注式保温
poured in two operation 分两次浇灌
poured joint 浇灌缝，浇制接头，浇筑缝
poured monolithically with 与……浇成整体
pouring of concrete 混凝土浇筑
pouring of thermal insulation material 保温层浇注施工
pouring test 浇注试验
pouring 灌注，浇注【混凝土】，流注，注入
pour out 倾倒
pour point 浇注点，流动点，凝点
pour 浇注，灌注，灌，倾注
poverty-stricken area 贫困地区
powder bag 炸药包
powder blend 粉末掺混
powder carbon 粉末炭
powder carburizing 固体渗碳
powder-coal 粉煤
powder coating 粉末涂料
powder core 压粉铁芯
powder coupling 粉末联轴器
powder cutting 氧熔剂切割
powdered absorbent 粉状吸收剂
powdered anion exchange resin 粉末状阴离子交换树脂
powdered asphalt 粉状沥青
powdered carbon 粉状活性炭
powdered catalyst 粉状催化剂
powdered cation exchange resin 粉末状阳离子交换树脂
powdered coal burner 煤粉喷燃器
powdered coal cyclone separator 细粉分离器
powdered coal 粉煤
powdered coke 焦粉
powdered flux 粉状焊剂
powdered fuel 粉状燃料，煤粉
powdered ion exchange resin 粉末状离子交换树脂
powdered iron core 铁粉芯，粉末铁芯
powdered ore 矿粉，粉状矿
powdered resin precoat filter 粉末树脂预涂层过滤器
powdered resin 粉末树脂
powdered synthetic material 粉状合成材料
powder extinguishing agent 干粉灭火剂
powder extinguishing system 干粉灭火系统
powder-filled cartridge fuse 有填料封闭管式熔断器
powder filler 粉料，粉末填充剂
powder fire branch 干粉灭火枪
powder fire extinguisher 干粉灭火器
powder fire monitor 干粉灭火炮
powder fire truck 干粉消防车
powder flame cutting 粉末火焰切割
powder impoverishment 粉末贫化
powder-laden gas 含尘气体
powder-like 粉状的
powder lime 石灰粉
powder magazine 炸药库
powder magnet 压粉磁铁
powder man 炮工
powder material 粉末材料
powder-metallurgical granulation process 粉末冶金粒化法
powder metallurgic forming machines 粉末冶金成型机
powder metallurgy 粉末冶金
powder space 炸药腔
powder stick 炸药棒
powder suppressant 干粉抑爆剂
powder technology 颗粒技术，粉末工艺
powder type insulation 粉末绝缘
powdery coal 粉煤
powdery snow 粉状雪
powdery 粉状的
powder 粉，粉末，粉剂，火药
powdex filter 粉末树脂过滤器
powdex process 粉末树脂法
powdex resin 粉末树脂
powdex 粉末离子交换器，粉末树脂过滤器，粉末离子覆盖过滤器，粉末树脂覆盖过滤器
power absorption 功率吸收
power accumulator 储能器，蓄能器
power-actuated relief valve 动力制动卸压阀
power-actuated relieving valve 动力式排放阀
power-actuated setting device 射钉枪
power-actuated 动力制动的，动力作用的
power actuation 动力制动
power actuator 动力制动器
power allowance 功率余量
power amplification ratio 功率放大系数
power amplification 功率放大系数，功率放大
power amplifier 功率放大器
power and control cable penetration 电力及控制电缆贯穿件
power and miscellaneous resources supply on site 施工力能供应
power angle characteristic （同步电机的）功角特性
power angle measuring device 功角检测装置
power angle measuring system 功角测量系统
power angle 功率角，功角
power application software 电力应用软件
power arc 电弧
power arm 动力臂
power ascension 功率提升，功率上升，功率增加
power-assisted pressure relief valve 动力辅助的卸压阀
power-assisted 动力辅助的
power at zero flow rate 零流率时的功率
power-auger boring 电动螺钻
power auto-transformer 电力自耦变压器
power balance diagram 电力平衡图
power balance 动力平衡，电力平衡，功率平衡

power bank 功率组【控制棒】
power bar bender 动力钢筋弯曲机
power bay 电源间隔，动力间
power block 常规岛，发电区，动力区，动力岛，成套电站设备，成套动力装置，动力滑轮
power board 接线板，电源板，配电板
power bogie 动力转向架
power bracket 功率范围
power brake 电力制动器
power breeder 动力增殖（反应）堆
power breeding reactor 动力增殖反应堆
power bridge 动力桥，功率桥
power broker 电力经纪人
power buggy 动力牵引车，电动小车
power buildup 功率提升，功率增加
power burst facility 功率突增试验装置
power burst 功率猝发，功率突增
power bus-bar 电力母线，动力汇流排
power bus 电源母线，电力母线动力汇流排
power cabinet 电源盘【控制棒驱动装置】
power cable insulation 电力电缆绝缘
power cable line 电力电缆线路
power cable 动力电缆，电力电缆
power calibration 功率标定
power canal 动力渠道，发电渠道，引水渠
power capability 功率容量，可能功率，供电能力
power capacitor room 电力电容室
power capacitor 电力电容器
power capacity of line 线路的容量
power capacity performance curve 功率性能曲线
power capacity 功率容量，动力能量，功率电容
power center 电力中心，动力中心
power change 功率变化
power channel carrier channel 电力线载波通道
power channel 发电渠道
power characteristic 功率特性，幂特性
POWERCHINA（Power Construction Corporation of China） 中国电力建设集团有限公司，中国电建
power circle diagram （输电线路的）功率圆图
power circuit 电力线路，电源线路，电源电路，动力线路，电力网
power coastdown 功率惰降，功率下降
power coefficient of reactivity 反应性功率系数
power coefficient 功率系数，功率因素
power cogeneration 热电联产
power collection system 电力汇集系统【对于风力发电机组】
power collection 电力汇集
power comparator 功率比较器
power component 电阻部分，实部，有功部分，有功分量
power condenser 电力电容器，功率调节器
power conditioning and control 电力调节与控制
power conduit 动力管道，电力管道
power connection 接电源，电源接头
power connector 电源接头
power conservation 节电
power construction project plan 电力建设计划
power consumer 电力用户
power-consuming equipment 耗电设备
power consumption of auxiliaries 厂用电
power consumption 动力消耗量，能耗，电能消耗，电力消耗，耗电量，用电量，功率消耗，功耗
power contactor 电力接触器
power control box 动力箱
power control center 电力控制中心
power controller 功率控制器，功率调节器
power control lever 功率操纵杆
power control member 功率控制元件
power control principle 功率控制原理
power control rod 功率控制棒【反应堆】
power control step change 功率控制阶跃变化
power control unit 动力控制单元
power control valve 动力控制阀
power control 功率调节，功率控制
power conversion efficiency 功率转换效率
power conversion loop 动力转换回路
power conversion unit 功率变换装置
power conversion 能量转换
power converter 功率变换器，功率变流器，电力变换器，电力变换机
power convertor 电力变换器，电力变换机，变流器
power-cooling-mismatch 功率冷却失配（事故）
power corporation 电力公司
power cost 电费，电价，动力费用
power crane 动力起重机，动力吊车
power current 电力电流，强电流
power curve 出力曲线，功率曲线
power cutback system 事故下插系统，降功率系统
power cutback 功率下降
power cut-off 断电，停电，切断电源
power cut 动力切断
power cycle 动力装置循环，热电站循环动力装置系统
power cycling 功率循环
power cylinder 动力缸，动力油缸
power dam 发电坝
power decay 功率递减，功率衰减
power defect 功率亏量
power delay product 功率时延乘积
power demand estimate 电力需要估计
power demand of system 系统的功率需量
power demand 电力需求（量），功率需求（量），动力需求（量），能的需要量
power density distribution form factor 功率密度分布形状因子
power density distribution 功率密度分布
power density monitoring 功率密度监测
power density profile 功率密度分布图
power density spectrum 功率密度谱
power density 功率密度，风能密度，能量密度，释能密度
power department 动力车间，动力部门
power detection 功率检波，强信号检波
power detector 功率检波器，强信号检波器
power developed 实发功率
power development 动力开发
power device packaging 功率器件封装

power diagram 功率图，功率曲线
power diode 功率二极管
power dip 功率骤降【反应堆】
power direction relay 功率方向继电器
power discharge 发电流量
power disconnecting switch 电力隔离开关
power dissipation 功率耗散，功率损耗，功率消耗，耗散功率
power distortion 功率畸变
power distributing cabinet 配电箱
power distributing meter 配电表
power distribution center 配电中心
power distribution component 电源分配组件
power distribution control system 功率分配调节系统
power distribution equipment 配电设备
power distribution flattening 功率分布展平
power distribution line 配电线路
power distribution map 功率分布图形
power distribution network 配电网
power distribution panel power drill 机械钻
power distribution panel 配电盘，电力配电箱
power distribution skewed toward the top 功率分布向上扭曲
power distribution switchgear 配电装置
power distribution system 配电系统，配电方式
power distribution unit 配电部件，配电装置
power distribution 电力分配，配电，能量分布，功率分布
power distributor 配电商
power disturbance 功率失调，功率干扰
power divider 功率分配器
power drain 耗用功率，耗用动力
power drill 动力钻
power-driven rivet 机铆铆钉
power-driven scram 动力驱动快速停堆，动力驱动紧急刹车
power-driven 电动的，动力驱动的，动力传动的
power drive 电力传动，机械传动，动力传动
power dump 切断电源，切断功率供给
power-duration curve 功率特性曲线，功率历时曲线，功率持续时间曲线
power earth auger 动力土钻
power economics 动力经济
power economy 动力经济，电能经济，动能经济
powered actuator 动力执行机构
powered controls 动力控制机构，功率控制机构
powered device 受电设备
powered mechanism 传动机构
powered operator 动力操作器
powered 机动的，供电的，带有发动机的，有动力的，有动力装置的
power effect 功率效应
power efficiency 出力效率，功率效率
power electronic capacitor 电力电子电容器
power electronic circuit 电力电子电路
power electronic control 电力电子控制
power electronic converter 电力电子变流器
power electronic device 电力电子器件
power electronic equipment 电力电子器件
power electronic interface 电力电子接口
power electronic switch 电力电子开关
power electronics 功率电子学，电力电子学
power element 动力元件
power engineering 电力工程，动力工程，动力学
power engineer 动力工程师，电力工程师
power equalizer 功率均衡器
power equation 功率方程式
power equipment factory 发电设备制造厂
power equipment 电力设备，动力设备
power escalation 功率提升，功率上升，功率增加
power exchange 交换容量，功率交换
power excursion accident 功率剧增事故
power excursion 反应堆工作状况偏差，功率偏离（额定值），功率偏差，功率偏移，（反应堆）功率剧增
power export 能量输出，电力输出，功率输出
power extraction coefficient 获能系数，功率系数
power extraction 获能
power factor adjustment 功率因数调整
power factor angle 功率因数角
power factor capacitor 提高功率因数电容器，补充电容器
power factor compensation 功率因数的补偿，相位的补偿
power factor control 功率因数控制，功率因数调节
power factor correcting reactor 功率因数纠正电抗器
power factor correction capacitor 功率因数补偿电容器，功率因数矫正电容器
power factor correction 功率因数校正，功率因数补偿，功率因数调整
power factor improvement 提高功率因数，功率因数改进
power factor indicator 功率因数指示器，功率因数表
power factor lead 功率因数超前
power factor measurement 功率因数测量
power factor meter 功率因数计，功率因数表
power factor regulating relay 功率因数调节继电器
power factor regulator 功率因数调节器
power factor tip up 电机绝缘损失角增量
power factor 功率系数，功率因数，功率因子
power fail recovery system 电源失效恢复系统
power failure 电力故障，停电，断电，电源中断，电源故障
power feeder 馈电线，供电线，电力馈线
power feed 供电
power flattening 功率展平
power flow calculation 电力潮流计算
power flow control 潮流控制
power flow distribution 电力潮流分布，能流分布
power flow model 潮流模型
power flow ratio 功率流量比
power flow tracing method 潮流跟踪法
power flow 功率潮流，电力潮流，功率通量，发电流量
power flux-density 功率通量密度

power form factor 功率形状因数
power frame 电源机架
power frequency arc test 工频电弧试验
power frequency breakdown 工频击穿特性
power frequency characteristic 功率频率调节特性
power frequency control device 功频调节装置
power frequency controller 电源频率控制器
power frequency control loop 电源频率控制回路
power frequency control 电源频率控制，功率频率控制
power frequency current 工频电流
power frequency induction furnace 工频感应炉
power frequency limitation 功率频率限度
power frequency overvoltage 工频过电压
power frequency recovery voltage 工频恢复电压
power frequency sparkover voltage of an arrester 避雷器工频放电电压
power frequency sparkover voltage 工频跳火电压
power frequency test for electric equipment 电气设备工频试验
power frequency testing transformer 工频试验变压器
power frequency test 功率频率试验
power frequency voltage test 工频电压试验
power frequency voltage 工频电压
power frequency welding 工频焊接
power frequency withstanding voltage test 工频耐压试验
power frequency withstand voltage 工频耐受电压，工频耐压，工频耐压水平
power frequency 电源频率，工业频率，市电频率，工频
power-frontal area ratio 功率迎风面积比
powerful spark 强火花
powerful transmitter 强力无线电发射机
powerful 有力的，强力的，强大的，大功率的
power function 幂函数，功率函数
power fuse 动力熔丝【冲出式或限流式】
power gain 功率放大系数，功率增益
power gas 动力气体
power generating hours 发电小时数
power generation control complex 能源生产管理综合体
power generation cost 发电成本，发电费用
power generation department 发电分场
power generation layout 电源布局
power generation planning 发电规划
power generation turbine 发电用汽轮机，电厂汽轮机
power generation using solar pond 太阳池发电
power generation 电力生产，发电，发电量
power generator 发电机
power grader 动力平土机
power grade 功率等级
power grid detector 栅极功率检测器，栅极功率指示器
power grid technology 电网技术
power grid 电力网，动力网，电网
power group corporation 电力集团公司
power hacksaw 动力钢锯
power hammer 电锤
power history 功率史，功率历程
power hoist 动力吊车，机动起重机
power house at dam-toe 坝后式厂房
power house complex 厂房
power house in river channel 河床式厂房
power house machine 主厂房
power house of hydropower station 水电站厂房
power house on river bank 岸边式厂房
power house substructure 电站厂房下部结构，发电厂地下结构
power house superstructure 电站厂房上部结构，发电厂地上结构
power house within the dam 坝内式厂房
power house 动力室，主发电机室，动力车间，发电间，动力间，主厂房，变电站
power house 发电厂房
power hunting 功率摆动，功率晃动
power increase 功率提升，功率增加
power increment 功率增量，功率增长值
power indicator 功率指示器，功率指针
power industry 动力工业，电力工业
power input compressor 压气机轴端输入功率
power input factor 输入功率系数
power input to gearbox input shaft 对齿轮箱输入轴的功率输入
power input 电力输入，输入功率，功率输入(量)，动力输入
power installation 动力装置，电力装置，发电装机
power integrated circuit 功率集成电路
power interchange 能量交换
power interruption 供电中断，停电
power invariance 功率不变性
power inverter 功率逆变器
power ionization chamber 功率区段电离室
power island 动力岛
power jet 强射流，动力喷口，主射流，动力射流
power klystron 功率速调管
power law constant 幂函数规律常数
power law exponent 幂律指数
power law fluid 幂律流体，幂律流
power law for wind shear 风切变幂律
power law index 幂函数规律指数，幂律指数
power law model 幂函数规律模型
power law profile 幂律廓线
power law 指数定律，幂次法则，幂次定律，幂定律
power lead tube 电源引出软管
power lead 电力引线，电源引（入）线
power level channel 功率测量通道
power level control 功率电平控制
power level detector 功率检测器，功率指示器
power level diagram 功率电平图
power level indicator 功率电平指示器
power level monitor 功率监测
power level overshooting 功率过调量
power level safety setting 功率保护定值
power level safety system 功率保护系统
power level 功率级，功率电平，功率水平，功

率位准
power limitation 功率限制
power limiter 功率限制器
power limit factor 功率极限因数
power limit 功率极限，稳定极限
power line bundle 集束输电线
power line carrier channel 电力线载波通道
power line carrier machine 电力线载波机
power line carrier pilot protection 电力线载波保护
power line carrier telephone 电力线载波电话
power line carrier 动力线路载波，电力线载波
power line filter 交流噪声过滤器，电源滤波器，线路滤波器
power line terminal 电力线路终端，电源线接线端子
power line voltage 电源电压，供电电压
power line 电力线，动力线（路），动力网，电源线，输电线，火线
power loading 动力负荷，电力负荷，电负荷
power loss 功率损耗，功率损失，失电
power louvers 电动百叶窗
power magnetic amplifier 功率磁放大器
power magnetics 工业磁学，强磁学
power magnification 功率倍数，功率放大
power mains 输电线，电力输送线，输电干线
power-making 发电的，产生动力的
power maneuvering 电力调度
powerman 动力工作者，发电专业人员
power map 功率曲线分布图
power margin 功率裕度
power measurement terminal 电量测量端子，功率测量终端
power meter 功率表，瓦特表
power missing 能量遗失
power modulation 功率调制
power monitor 功率检查装置，功率监察器
power MOSFET 功率金属氧化物半导体场效应晶体管，功率场效应晶体管
power network at receiving side 受端电网
power network configuration 电网结构
power network frequency 电力网频率
power network 电力网，电网，动力网
power noise 功率噪声
power NOR 大功率或非电路
power of attorney clause 授权书条款
power of attorney 授权函，（法人）授权书，委托书
power-off relay 电源切断继电器，失电继电器
power-off 切断电源，停机，断电
power of interpretation 解释权
power of motor 电动机功率
power of number 幂数
power oil 传动油，动力油，工作油
power only reactor 动力反应堆
power-on time 接入计算机电源的时间，供电时间
power-on 接通电源，加电，接通电源的
power-operated control 动力操纵
power-operated relief valve 功率操作卸压阀
power-operated shearing machine 电动剪断机
power-operated valve 机械驱动阀
power-operated 电动的，机动的，动力操作的，机力操纵的
power operation range 功率运行区段
power operation relief valve 动力操作释放阀
power operation 功率运行，带功率运行，动力操作
power operator 动力操作器【热手动】
power oscillation 功率波动，功率振荡
power oscillator 功率振荡器
power outage 停电
power outlet 电源输出端
power output of hydropower station 水电站出力
power output shaft 动力输出轴
power output 输出功率，功率输出，发电量
power over Ethernet 以太网供电
power override 功率超过给定值
power overshoot 功率超调
power package reactor 装配式动力反应堆
power package 发电装置，动力机组，整装电源机组
power pack 动力单元，单元组，供电部分，电源组，动力箱
power panel 电源板，配电板，配电盘
power park 发电厂区
power peaking factor 功率峰值因数，功率尖峰系数
power peak 功率峰值，最大功率
power penstock 发电管道
power pentode 功率五极管
power performance 功率系数，功率特性
power perturbation 功率微扰
power per unit 单机容量，单机功率
power pipeline 动力管系
power planning 电力规划
power plant accessories 电厂附属设备，电厂辅助设备
power plant air cooling technology 发电机空冷技术
power plant boundary wall 电厂围墙
power plant capacity 发电厂容量
power plant construction 电站的建设
power plant cooling water 发电厂冷却水
power plant decommissioning 发电厂退役
power plant detection device 电厂检测装置
power plant discharge 电站发电流量
power plant emission 发电厂排放
power plant engineering 发电厂工程，动力厂工程
power plant fence 电厂围栅
power plant life extension 电厂超期服役
power plant load factor 电厂负荷因数
power plant main controller 电厂主控制器
power plant make-up 发电厂补充水，电厂补给水
power plant operation 发电厂运行
power plant operator 电厂运行员，电厂操作工
power plant preparatory office 电厂筹建处
power plant regulation and control 电厂调节控制
power plant security 电厂保安
power plants for peak load regulation 调峰电厂
power plant station 电厂站

power plant waste 发电厂废物，发电厂废水
power plant with concentrated fall 集中落差式电站
power plant 动力厂，动力装置，发电设备，电站，电厂，动力设备，动力反应堆，电厂反应堆
power plug 电力插头
power plunger pump 电动柱塞泵
power-plus-plutonium reactor 发电产钚两用反应堆
power pool system 联合电力系统
power pool 联合电力系统，有储备容量的电力网，电力库
power positioner 功率位置控制器
power potential 动力势
power pressure （调节系统中的）工质压力
power producer 动力发生器，动力源
power-producing reactor 动力反应堆
power production building 生产建筑物
power production cost 发电成本，能量生产成本
power production 电力生产，产生功率，发电量
power profile 功率分布图
power project 电力计划，发电工程，动力工程
power protection 功率保护
power pump 动力泵，电动泵，机械驱动力泵
power purchase agreement 购电协议，购电合同
power quality standard 电力质量标准
power quality 电力质量
power rail 传电轨
power raising test 功率提升试验
power rammer 动力夯
power ramping test 功率倾斜试验
power ramp 功率直线上升
power range channel 功率量程通道，功率区段通道
power range high neutron flux trip 功率区段高中子能量保护停堆
power range 功率范围，功率区段，功率量程
power rate 电力费率，电费，电价
power rating 标称功率，标定功率，额定功率，额定容量
power ratio 动力比
power reactor core 动力反应堆活性区，动力反应堆堆芯
Power Reactor Information System 动力反应堆情报系统
power reactor inherently safe module 动力反应堆固有安全模件
power reactor noise 动力反应堆噪声
power reactor package 装配式动力反应堆
power reactor 动力堆，核反应堆，动力反应堆
power receiving ability 受电能力
power receiving and distribution facilities 受配电设备
power receiving box 受电箱
power receiving capability 受电能力
power receiving equipment 受电设备
power receiving point 受电点
power receiving ratio 受电比率
power receptacle assembly 电源插座组件【控制棒驱动机构】
power receptacle 电源插孔，电力插座，电源板，电插座
power recovery turbine 功率回收透平，膨胀透平
power rectifier 整流电源，工业整流器，大功率整流器
power redistribution 功率再分布
power reduction 功率降低
power regulation agency 电力监管机构
power regulation 功率调节
power regulator 功率调节器
power relay 功率继电器，电力继电器
power release 功率释放
power required to drive pump 驱动泵所需功率
power reserve 后备功率，备用功率
power resource 动力资源
power response 功率响应
power retailer 电力零售商
power riveter 电动铆接机
power roof ventilator 屋顶通风机
power runaway accident 功率失控事故
power rush 功率冲击，功率骤增
power saving 节电
power scheme 发电工程
power scraper 动力铲运机
power screen 机械筛
power SCR 大功率可控硅整流器
power sector 电力部门，电力行业
power selector 功率选择器
power selsyn 功率自动同步机，电力自动同步机
power series 幂级数
power setback 降低功率，功率下降，控制功率下降
power set 发电机组，动力装置，电源装置
power shafting 传动轴系
power shaft 动力轴
power sharing 功率份额
power shift pressure oil 动力换挡压力油
power shift pressure port 动力换挡压力孔
power shift pressure solenoid 动力换挡压力电磁阀
power shift pressure 动力换挡压力
power shortage 功率短缺
power shortfall 功率短缺
power shovel 动力铲
power signal 功率信号
power silt 功率倾斜
power site 电站站址
power socket 电源插座
power source for arc gouging 气刨电源，碳弧气刨电源
power source transfer circuit 电源转换电路
power source 电源，能源，动力源
power spectral density 功率谱密度
power spectrum density 功率谱密度
power spectrum 能谱，功率谱
power split 功率分区
power square wave generator 方波功率发生器，方波电源发生器
power stage 功率级
power station at dam toe 坝后式水电站
power station attendant 电厂值班人员

power station auxiliaries 电厂辅助设备
power station boiler 电站锅炉
power station commissioning 电厂调试
power station control system 电厂控制系统
power station environment 电厂环境
power station for hump modulation 调峰电站
power station load curve 电站负荷曲线
power station planning 电站规划
power stations in cascade 梯级水电站
power station voltage 电厂电压
power station 发电厂，动力厂，电站，电厂
power stepping motor 功率步进电机
power storage 发电库容，（水电厂）供发电的贮水（量），能量储存，储电，蓄能
power stretch 超额发电
power stroke 工作冲程，动力冲程
power sub-distributed panel 电源子分配盘
power supplement 动力源，电源
power supply board 电源屏
power supply box 动力箱，动力供应箱，配电箱
power supply bureau 供电所，供电局
power supply cabinet for electric-drive valve 热工配电柜［箱］
power supply cable 供电电缆
power supply circuit 供电回路
power supply contract 供电合同
power supply device 电源装置
power supply distributed panel 配电盘
power supply equipment 电源设备
power supply for civil construction and equipment installation 土建施工和设备安装用电源
power supply for construction 施工供电
power supplying to the station from grid 反送电，倒送电
power supply interruption 供电中断
power supply monitor 电源监控器，电源监视器
power supply network 供电网
power supply of automatic system 自动化系统电源
power supply of electric traction 电力牵引供电
power supply of large enterprise 大型企业供电
power supply panel 电源屏
power supply point 供电点
power supply quality 供电质量
power supply rack 电源架，动力供给架
power supply reliability 供电可靠性
power supply ripple 电源脉动，电源波纹
power supply section 供电区段
power supply set 供电设备，发电机组
power supply source 供电电源
power supply system in communication 通信供电电源系统
power supply system of electric traction 电力牵引供电系统
power supply system 电力系统，供电系统
power supply to subway 地铁供电
power supply transformer 电源变压器
power supply unit 电源部件，供电单元，供电机组，供电设备，发电机组，发电机
power supply vibrator 电源振动器，振动式逆变器
power supply voltage 供电电压，电网电压
power supply with isolated output 带隔离输出的供电
power supply 电力供应，动力供应，电源，供电，功率源
power surge 功率波动，功率冲击
power surplus 功率富裕量
power suspension 停电
power swing 功率摆动，功率波动，功率摇摆
power switchgear 电力开关设备，动力开关设备
power switch 电力开关，电源开关
power system analysis 电力系统分析
power system capacity 系统容量
power system computation 电力系统计算
power system control 电力系统控制
power system drawing 电力系统图，电源系统图
power system harmonic 电力系统谐波
power system interconnection 电力系统联网
power system operation 动力系统运行
power system operator 系统运行机构
power system oscillation 电力系统振荡
power system overload 电力系统过负荷
power system overvoltage 电力系统过电压
power system planning 电力系统规划
power system protection 电力系统保护
power system reliability 电力系统可靠性
power system separation 电力系统解列
power system splitting 电力系统裂解
power system stability 电力系统稳定度
power system stabilizer 电力系统稳定器
power system swing 电力系统功率摆动
power system technology 电网技术
power system tie line 电力系统联络线
power system 电网，动力网，动力系统，电力系统
power-take-off gearing 辅助功率付出装置
power-take-off 动力生产装置，功力输出装置，分出功率
power tamper machine 电动捣固机
power tamper 动力夯，机动夯，蛙式夯
power tansformer 电源变压器，电力变压器
power tariff 电价
power termination 功率负载，吸收头
power tester 功率计，功率测试器，瓦特表
power test 功率实验，动力实验
power tools 动力工具
power-to-weight ratio 功率-重量比，动力-重量比
power tower （太阳能）发电塔，电力塔
power traction tamper 动力牵引式打夯机
power traction 动力牵引
power trading 电力交易
power train boiler car 列车电站锅炉车箱
power train efficiency 功率传输效率
power train 动力齿轮系，动力传动轮系，动力传动系统，列车电站
power transfer relay 故障继电器，电源切换继电器，电力传输继电器
power transfer 功率传输，电力传送
power transformation 变电

power transformer bay 电力变压器间隔
power transformer 电力变压器，电源变压器
power transient 功率瞬变
power transistor 大功率晶体管，功率晶体管，晶体功率管
power transmission and distribution calculation 输配电计算
power transmission and distribution project 输变电工程项目
power transmission and distribution 输配电
power transmission coil 功率传输线圈
power transmission conductor 输电导线
power transmission efficiency 输电效率
power transmission line anchor 输电线地锚
power transmission line hardware 输配电线路金具
power transmission lines corridor 电力出线走廊
power transmission line tower 输变电线路铁塔
power transmission line 输电线
power transmission network 电力传输网络
power transmission protection system 输电保护系统
power transmission system 输电系统
power transmission technique 输电技术
power transmission theory 输电理论
power transmission tower 输电塔，输电塔架，输电铁塔
power transmission 电力传输，输电，电力传送，电力输送
power triangle 功率三角形
power trimming rotor 功率调节风轮
power tube 功率管
power tunnel 发电隧洞
power turbine 动力透平
power type relay 功率式继电器
power unit using solar pond 太阳池发电装置
power unit 动力部件，功率单位，动力单位，动力机组，动力设备，发电机组，执行部件
power uprating 增加功率，提高出力
power-up 带电
power utility 公用电力事业，发电站，电力公司
power utilization 电能利用，用电
power valve 增力阀
power vector 功率矢量
power voltage 电源电压
power water boiler 动力水锅炉
power water 压力水
power-weight ratio 单位重量的功率，功率-重量比
power wheeling 电力转运
power winch 电动绞车，电动卷扬机
power winding 功率绕组
power wiring 电力布线
power wrench 动力扳手
power yield 功率产额
power 功率，能，动力，能力，电源，幂，电力，权，能源，电能，权力
pozzolana cement 火山灰水泥
pozzolana 火山灰
pozzolan cement 火山灰水泥
pozzolanic admixture 火山灰质掺合料
pozzolanic reaction 火山灰反应
pozzolan 火山灰
PPA(power purchase agreement) 购电合同，购电协议
PPB(parts per billion) 十亿分率，十亿分之几
PP(charges prepaid) 运费预付
PPDS(parallel plate domain structure) 平行板磁畴结构
PPFF(priority program flip-flop) 优先程序触发器
PPI(programmable peripheral interface) 可编程外围接口
PPL(polypropylene lined) 聚丙烯内衬，衬塑
PPL(Public Procurement Law) 公共采购法【欧盟】
ppm(parts per million) 百万分率，百万分之几
PPM(peak power meter) 尖峰出力表
PPM(peripheral processor memory) 外围处理机存储器
PPM(procedure preparation manual) 程序准备手册
PPM(pulse-position modulation) 脉冲位置调制
PPM(pulses per minute) 脉冲每分钟
P＝poise 泊【黏度单位】
P＝power 功率
PP(peak power) 峰值功率
P-P(peak-to-peak) 峰间值，峰对峰
PPPOE(point to point protocol over ethernet) 基于以太网的点对点通信协议
PP＝polypropylene 聚丙烯
PPP(point-to-point protocol) 点对点协议
PPP(polluter pays principle) 污染肇事者付款原则
PPP(public private partnership) PPP模式，政府公共部门与私人合作模式，公私伙伴关系【工程建设模式】
PPP(purchasing power parity) 购买力平价
PPS(peripheral processing system) 外围处理系统
PPS(peripheral processor system) 外围处理机系统
PPS(plant protection system) 电厂保护系统
PPS(primary protection system) 一次保护系统
PPS(probability proportionate to size sampling) 抽样调查法
PPS(pulses per second) 脉冲每秒
PPTH(parts per thousand) 千分率
PPU(peripheral processor unit) 外围处理机
PQR(procedure qualification record) 程序鉴定记录
practical application 实际应用
practical capacity 实际容量
practical duty 实际运行方式，实际功率
practical efficiency of solar array 方阵的实际效率
practical electrical unit 实用电单位
practical experience 实际经验，实践经验
practical grain size 实际晶粒大小
practical hydraulics 实用水力学，应用水力学
practicality 实用性
practical life 实际使用寿命，实际寿命
practical module efficiency 组件实际效率

practical personnel 应用型人才
practical position 具体位置
practical proposal 切实可行的建议
practical question 实际问题
practical situation 实际情况
practical standard 实用标准
practical unit 实用单位
practical value 实际值
practical width （共振峰的）实验宽度
practical working model 实用模型
practice code 实用法规
practice fraud 弄虚作假
practice of train dispatching 调车方式
practice 惯例
practicing independent accounting 执行独立核算
prairie 草原
PRAISE(piping reliability analysis including seismic event) 管系可靠性分析
Prandtl-Glauert correction 普朗特葛劳渥修正
Prandtlpitot tube 普朗特风速管，普朗特皮托管
Prandtl's number 普朗特数
Prandtl's tube 普朗特管
Prandtl type wind tunnel 普朗特型回路式风洞，普朗特型式风洞
PRA(probabilistic risk analysis) 概率风险分析
PRA(probabilistic risk assessment) 概率风险评价
praseodymium 镨【Pr】
prata 草甸植被
pravity 障碍，故障
PRBS(pseudo-random binary sequence) 伪随机二进制序列
PRC(pressure record and control) 压力记录和控制
PRCS(primary reactor coolant system) 反应堆一次（主）冷却剂系统
PRDH(pressure reducer desuperheater) 减温减压器
PRD(printer dump) 打印机打印
PRD(proposal review board) 标书审查委员会
pre-acceleration 预加速，前加速
pre-acceptacnce Certificate （电厂）预验收证书
pre-action system 预作用灭火系统
preact time 预动作时间
preact 超载，预作用，超前，提前量
pre-adaptation 预先适应
preaeration tank 预曝气箱
preaeration 预曝气，预通气
preaging 人工老化，预老化
preamble 序，前言，序言
preamplification 预放大，前置放大
preamplifier 前置放大器
preamp＝preamplifier 前置放大器
preanalysis 事前分析
preappraisal 预评估
pre-arcing characteristic 熔化特性熔断器
pre-arcing time 熔化时间【熔断器】
prearranged power interruption 计划停电
preassembled 预装配的，预安装的
preassemble 预装
preassembly method 预制装配法
preassembly 预汇编，预组合，预装配
preassigned 预先指定的，预先分配的
pre-audit meeting 监察前会议
pre-award meeting 标前会议
prebend 预弯【薄板】
pre-bid conference 招标前会议
pre-bid meeting 标前会议
preboiler corrosion 炉前腐蚀
preboiler system 锅前系统，锅前流程
prebox 前置组件
prebreakdown current 预击穿电流，击穿前电流
prebreakdown threshold voltage 预击穿临界电压
prebreakdown 击穿前的，预击穿的
prebreeder 预增殖反应堆，前置增殖反应堆
precalculated 预先计算好的
precarious 不稳定的，不安定的，危险的
precast beam 预制梁
precast block 预制块
precast box 预制箱
precast concrete block 混凝土预制块
precast concrete caisson 预制混凝土沉箱
precast concrete element 预制混凝土构件
precast concrete flume 预制混凝土渡槽
precast concrete form 预制混凝土模板
precast concrete inverted siphon 预制混凝土倒虹吸
precast concrete penstock 预制混凝土压力水管
precast concrete pile 预制混凝土桩
precast concrete pipe 预制混凝土管
precast concrete sheet-pile 预制混凝土板桩
precast concrete shell 预制混凝土薄壳
precast concrete tower 预制混凝土塔架
precast concrete unit 预制混凝土构件
precast concrete 预制混凝土（构件）
precast element 预制构件
precast floor 预制地板
precasting 预制（混凝土）
precast joint filler 预制填缝料
precast joint 预制接缝
precast member 预制构件
precast panel construction 预制板结构
precast panel 预制墙板
precast parts 预制构件
precast pattern block 预制水泥花格
precast pile 预制桩
precast prestressed beam 预制预应力梁
precast products of diabase 辉绿岩制品
precast product 预制件，预制品
precast ribbed roof slab 预制带肋型屋面板，预制大型屋面板
precast segmental sewer 预制沟管段
precast slab form 预制板的模板
precast slab 预制（厚）板
precast terrazzo flooring 预制水磨石地面
precast unit 预制构件，预制设备
precast wall-panel 大型墙板
precast 预浇筑的，预制的，预铸的，预制，预浇制
precautionary approach 预防方法
precautionary measure 预防措施
precautions for use 注意事项

precaution to be observed for storage 存储注意事项
precaution 小心，注意，预防措施
precede M with N 在M前加上N
precedence call 优先呼叫
precedence 领先，优先，优先权，优越性
precedent condition 先决条件
precedent 领先的，先例
precede over 优先于
preceding month 上月，前月
preceding seismic activity 历史地震活动性
preceding settlement 前期结算
preceding tide 前一潮水
preceding wave 前一波浪
preceding 先前的
precessing track 先期存储道
precession 前行，先行，动力，旋进
prechamber 预燃室，前室
precharge 预先充电
precheck 预检，预先检验，预先校验，预校验
pre-chlorination 预氯化（处理），预加氯
precipice 陡壁，峭壁，悬崖，崖
precipitability index 沉降性指数
precipitable water vapour content 可沉积的水蒸气含量
precipitable water 可降水
precipitant pipe 沉淀剂管
precipitant 沉淀剂，沉淀，沉降，沉淀物，降水，脱溶物
precipitating agent 沉淀剂
precipitation amount 降水量
precipitation analysis 沉淀分析
precipitation bed 沉淀层
precipitation deficiency 同期降水量差数
precipitation discharge 沉积放电
precipitation duration 降水持续时间
precipitation efficiency 降水效率，沉降效率
precipitation fouling 析晶污垢
precipitation gage 雨量筒
precipitation gauge network 雨量站网
precipitation hardening 沉淀硬化，弥散硬化，析出硬化，时效硬化
precipitation heat treatment 沉淀硬化热处理，人工时效
precipitation icing 降水成冰
precipitation intensity 降水强度
precipitation observation 沉降观测
precipitation process 沉淀过程，（废水处理）沉淀工艺，降水过程
precipitation rate 降水率，沉降率
precipitation reaction 沉淀反应
precipitation record 降水量记录，雨量记录
precipitation rime 冬雨覆冰
precipitation-runoff relationship 降水径流关系
precipitation station 雨量站
precipitation strengthening 沉淀强化
precipitation tank 沉淀器，沉淀箱
precipitation wind rose 降水风玫瑰
precipitation 沉淀，沉淀池，沉淀作用，沉降，凝结，脱溶，降雨量
precipitator ash handing 除尘器灰斗除灰系统
precipitator ash hopper 除尘器灰斗
precipitator chamber 沉降器室
precipitator 沉淀器，除尘器，电滤器
precipitous sea 八级风浪
precise access block diagram 精确存取框图
precise drawing 精确图
precise examination 精密探伤
precise measurement 精密测量
precise regulation 精密调节
precise 精密的，精确的，合格的
precision accuracy 精度
precision adjustment 精调，精密调节
precision balance 精密天平
precision brazing 精密钎焊
precision casting 精密铸件，精密铸造，精铸
precision counter 精密计数器
precision DC stabilizer 精密直流稳压器，精密直流电源
precision delay line 精密延迟线
precision encoding and pattern recognition 精确编码和模式识别
precision equipment 精密器材
precision express scale 快速精密天平
precision filter 精过滤器
precision finishing machining 精密精加工
precision finishing process 精密精加工（工序）
precision finishing 精密加工，精密光整加工
precision graph recorder 精密图示记录器
precision in products 产品精度
precision instrument 精密仪表，精密仪器
precision iterative operation 精确迭代运算
precision level 精确度【测量】
precision machine element 精密机械元件
precision measurement 精密测量，精确测定
precision meter 精密计量器
precision of analysis 分析的准确性
precision potentiometer 精密分压器，精密电位计
precision prescribed 要求精度
precision pressure gauge 精密压力表
precision regulator 精密调节器
precision resistor 精密电阻器，精密电阻
precision scale 精密天平
precision selection 精度选择
precision standard 精度标准
precision sweep 精密扫描
precision test 精度试验
precision waveform 精确波形
precision work 精密加工，精密工作
precision 精密度，精确度，精确，准确度，精细
precleaner hot precipitator 一级除尘器
precleaning operation 预清洗操作
precleaning 预清理，预清洗
preclude 排除，消除，预防，阻止，妨碍
precoated layer 覆盖层，铺料层
precoat film 预铺
precoat filter 前置式过滤器，覆盖过滤器，预涂层过滤器
precoat filtration 预涂层过滤
precoating cartridges 预涂层滤芯
precoating 打底漆，预涂，底漆，预浇，预敷，

预涂层，预敷层，（在过滤器表面涂敷的）滤料层，铺料，覆盖
precoat pump 预涂层泵
precoat tank 预涂层槽
precoat 预涂，预涂层
precolumn 前置塔【离子交换】，预置柱
precombustion chamber 预燃室，预燃炉
precombustion 预燃，预燃烧
precommercial operation 半商业运行，投产前试运行
precommissioning check 投运前检查
precommissioning test 调试前试验
pre-compacted 预先压实的，预先压紧的
precompaction 预压，初步压块
precompiler program 预编译程序
precompiler 预编译程序
precompressed air 预压缩空气
precompressed zone 预压区
precompression 压缩前，预压缩
precomputed 预先计算好的，预先计算的
preconcentration 预浓缩
precondenser 预冷凝器
preconditioning length （风洞气流）预处理长度
preconditioning 预处理，预调节
precondition 前提，先决条件，预处理
preconduction current 预传导电流
pre-cone angle 预置锥角
preconsolidated clay 预固结黏土
preconsolidation load 预压固结荷载
preconsolidation pressure 先期固结压力
preconsolidation 预先固结
preconstruction activities 前期工作
preconstruction period 施工前阶段
preconstruction planning 施工准备规划
preconstruction review 施工前审核
preconstruction safety analysis report 建造前安全分析报告
preconstruction stage 施工前期，施工准备阶段
preconstruction work 建设前期工作
pre-contract assessment 合同前的评价
pre-contract obligation 先合同义务
pre-contract phase 合同前阶段
pre-contract responsibility 先合同责任
pre-contract risk 前期风险【签约前】
pre-contract works 订立合约前的准备工作
precontrol 预控制
pre-converter 前置变换器
precooled aggregate 预冷骨料
precooler 前冷却器，预冷器，前置冷却器
precooling of aggregate 骨料预冷
precool 预冷，提前冷却
precorrection 预校正，预先校正
precracking by fatigue 疲劳预裂
precritical cold test 临界前冷态试验
precritical hot test 临界期热态试验
precriticality 临界前状态，亚临界状态
precritical test 临界前试验
precritical 亚临界的，临界前的
pre-crushing 预破碎，预压碎
precured period 预养护时期
precuring 早期养护
precursor fission product 前驱裂变产物
precursor time 先行时间
precursor 前驱波【放射性衰减】，初级粒子，产物母体，前任，前兆现象，先驱核，预兆
pre-cut 预切割
PREC 除尘器【DCS 画面】
predecessor matrix 先趋矩阵
predeparture inspection 发车前的检查
predesigned order 预定次序
predesigned 预先确定的，预定的
predesign 前期设计，初步设计，预设计，草图设计
predetection integral 检波前积分
predetection recording 预检记录
predetection 检波前的，检验前的，预检
predetermined bid value 预定标价
predetermined camber 预先起的拱度
predetermined counter 预置计数器
predetermined minimum level 预定的最低料位
predetermined orientation 预先定向
predetermined policy 预定政策
predetermined purge level 预定的料位
predetermined time 预定时间
predetermined value 预定值
predetermined 预定的，预置的
predetermine 预定
predetermining impulse counter 预置脉冲计数器
predicate 断言，断定，使基于……
predication 断定，判断，推算，预测
predictability 可预计性
predictand 预报量
predicted failure rate 预计故障率
predicted high water 预报高水位
predicted performance curve 预计性能曲线，设计性能曲线
predicted performance 预计性能
predicted settlement 预计沉降量
predicted tide curve 推算潮汐曲线
predicted tide 预报潮
predicted value 预测值
predicting filter 前置滤波器，预报滤波器
prediction cost 预测成本
prediction failure 事故预想
prediction method 预测法
prediction of collector performance 集热器性能的预测
prediction of groundwater regime 地下水动态预测
prediction of performance 特性估计
prediction prognosis 预测
prediction 预言，预告，预测，预报，预示，预计
predictive coding 预测编码
predictive control 预测控制
predictive maintenance 预测性检修，预知性检修
predictor circuit 预测电路
predictor control 前置控制
predictor-corrector method 预示校正法，预测校正法
predictor formula 预测公式
predictor servomechanism 前置伺服机构，预测

伺服机构
predictor 预测器，预测机，预报器，预测值，预报装置
predict 预测，前置
predischarge 预放电，预排气，预先卸载
pre-dispatching schedule 预调度计划
pre-disposal treatment 处置前的处理，预处理
predissociation 预分离，预分解
predistorter 前置补偿器，预修正电路
predistortion 预失真，频应预校，预畸变
predominance 优势
predominant current 盛行流，优势流
predominant diameter 主要粒径
predominant direction 主方向
predominant frequency 主频率
predominant vertical flow stream 主要的垂直流动
predominant wave 优势波
predominant wind direction 主风向
predominant 主要的
predose 初始剂量，辐照前
predrier 预干燥器
predrilled 预钻孔的
predrive 前级激励，预激励，预驱动
pre-dry 预干燥
pre-echo 预试反射波
pre-edited interpretive system 预编辑的解释系统
pre-emergency governor 超前危急保安器，辅助危急保安器
pre-emergency speed governor 超前动作危急保安器
pre-emergency 备用，应急，辅助
pre-emphasis circuit 预加重电路，预校电路
pre-emphasis network 预加重网络
pre-emphasis 预加重，预校，预修正，预先加强
pre engineered building 预经工程设计建筑
preentrained jump 预掺气水跃
pre-equalization 预校
pre-equalizer 前置均衡器
preetching 预腐蚀
pre-evaporator 预蒸发器
pre-expansion 膨胀前的
pre-exposure 辐照前的
prefabricated asphaltic blanket 预制沥青铺盖
prefabricated bracing frame 预制加固构架，围笼
prefabricated coil 预制线圈
prefabricated concrete panel 预制混凝土护墙板
prefabricated concrete structure 预制混凝土结构，装配式混凝土结构
prefabricated construction 预制装配式结构
prefabricated formwork 预制模板
prefabricated foundation 预制基础
prefabricated house 预制装配式房屋
prefabricated insulation 预制式保温
prefabricated pipe 预制水管
prefabricated reinforced concrete structure 预制钢筋混凝土结构
prefabricated slab 预制板
prefabricated structure 预制装配式结构，预制结构
prefabricated substation 预装式变电站
prefabricated tie 预制拉杆
prefabricated unit 预制构件
prefabricated 预制加工的，预制的，预建的
prefabrication factory 预制构件厂
prefabrication shop 预制车间
prefabrication 工厂预制，预制（作）
preface 引言，前言
prefade 预衰落
prefault 故障前的
pre-feasibility study 预可研，初步可行性研究，前期可行性研究
prefect site 工程地址
prefecture 专区
preferably 更好地，更适宜，更可取地，宁可，较好
preference for domestic contractors 国内承包商优先
preference for domestic manufacturers 国内制造厂的优惠待遇
preference to adaptability order 适用优先顺序
preference value 特选值，优先值，用户喜好度
preference 优待
preferential agreement 优惠协定
preferential clause 优惠条款
preferential dividend 优先红利
preferential duty 特惠关税
preferential loan 优惠贷款
preferential margins 优惠差额［幅度］
preferential measure 优惠办法
preferential orientation 择优定向，择优取向
preferential policy 优惠政策
preferential price 优惠价格
preferential rate 优惠利率
preferential system 特惠制度
preferential tariff 特惠关税
preferential trade 特惠贸易
preferential treatment 优惠待遇
preferential 优惠的，优先的
preferred noise criteria curve 噪声评价 PNC 曲线
preferred numbers 优先数，从优数
preferred orientation 从优取向，择优取向
preferred power supply 优先供电
preferred process 择优方法，最佳工艺
preferred system 优先使用的系统
preferred temperature 最适温度
preferred value 优先值
prefill surge valve 预先充油补偿阀
prefilter bank 预过滤器
prefiltering 预过滤
prefilter 前置过滤器，前级过滤器，预过滤器
prefiltration package 预过滤器
prefiltration 预过滤
prefired 预燃
prefiring cycle 点火前的准备工作
prefix operator 前置运算符
prefix register 前缀寄存器
pre-flawed specimen 预制裂纹试样
preflex girder 预弯梁

pre-flock chamber 预絮凝室
pre-flock 预絮凝
preflood decrease of storage 汛前库容减少，汛前水库放水
preformed asphalt joint filler 预制沥青缝条
preformed coil 成型线圈
preformed foam 预成泡沫
preformed hole 预留孔
preformed winding 成型绕组
preformed 预制的
prefracture deformation 断裂前变形
preframe 预装配
pregroup modulation 前波群调制
pre-hardening 初凝，预硬化
preheat current 预热电流
preheated air 预热空气
preheater 预热器
preheat flame 预热火焰
preheating aggregate 预热骨料
preheating cell 预热小室
preheating chamber 预热室
preheating coil 预热管盘
preheating of welding pieces 焊前预热
preheating 预热
preheat line 预热管线
preheat oxygen 预热氧
preheat section 预热区，预热段
preheat stage 预热级
preheat-starting 预热启动
preheat temperature 预热温度
preheat time 预热时间
preheat 焊前预热，预热
prehydrolysis 预水解
pre-IF amplifier 前置中频放大器
preignition chamber 预燃室，预燃炉
preignition zone 预燃区
preignition 提前着火
preimpregnated coil 预浸渍线圈
preimpregnated material 预浸渍材料
preimpregnated paper insulation 预浸渍纸绝缘
preimpregnated varnish 预浸渍漆
preimpregnated 预浸渍的
preimpregnating insulation 预浸渍绝缘
pre-inlet valve 预启阀
preinsertion resistor 预先插入电阻器
preinstallation check 安装前的校核
preinstallation test 安装前试验，装配前试验
preinsulated 预绝缘的
preinvestment activity 投资前活动
preinvestment project 投资前项目
preinvestment 投资前
pre-irradiated 预辐照的
pre-irradiation treatment 辐照前处理，预辐照处理
pre-irradiation 照射前，预照射，预辐照，辐照前
prejudice 偏见，歧视，损害
prelatic cycle 地下水循环
prelatic line 地下水线
prelicensed standard design 经过预审批的标准设计
prelicensed 事前许可的，经过预审批的
pre-lifting valve 预起阀
preliminary acceptance certificate 临时移交证书
preliminary adjustment 预调，初步协议，初步协定
preliminary appraisal 初步鉴定
preliminary appreciation 初步评价［评估］
preliminary approval 初步批准
preliminary assessment 初步评价
preliminary audit 初步审计
preliminary budget 初步预算
preliminary calculation 初步计算，概算
preliminary calibration 初步校准
preliminary checkout 初步测试
preliminary computation 初步计算，概算
preliminary contact 预动触点
preliminary correction 初校正
preliminary crusher 初轧碎机，粗碎机
preliminary crushing 预碎
preliminary data report 初步数据报告
preliminary description 初步说明
preliminary design closeout document 初设收口文件
preliminary design drawing 初步设计图
preliminary design phase 初步设计阶段，初设阶段
preliminary design review 初步设计审查
preliminary design stages 初步设计阶段，初设阶段
preliminary design 初步设计
preliminary dimension 预定尺寸，初步尺寸，原始尺寸
preliminary discussion 初步讨论
preliminary disposal technique 初步处置技术
preliminary draft 初稿
preliminary drawing 初（步）设（计）图
preliminary effect 预效应
preliminary elongation 预拉伸
preliminary estimate 初步概算，初步估计，初步估算，概算
preliminary evaluation of technical economy 初步技术经济评价
preliminary evaluation 初评
preliminary examination 初步检验，预备试验
preliminary feasibility study 初步可行性研究
preliminary heating zone 预热层
preliminary heating 预热
preliminary investigation 初步调查［调研］，初勘
preliminary measures 初步措施
preliminary operation 试运行，试运转，初试转
preliminary planning 初步规划［计划］
preliminary plan 初步计划，初步方案
preliminary project 初步设计方案
preliminary reading 预先读出，初步读数
preliminary reconnaissance 初步查勘
preliminary report 初步报告
preliminary result 初步结果
preliminary review 初步审查
preliminary safety analysis report 初步安全分析报告
preliminary scheme 初步方案

preliminary sedimentation tank 初步沉淀池
preliminary sketch 初步设计图，方案草图，草图
preliminary stage 初始阶段
preliminary study 初步调查［调研］，初步研究，预备性研究
preliminary survey 初步调查，初步勘测，初测，初勘
preliminary system design description 初步系统设计描述
preliminary technical development plan 初步技术发展计划
preliminary test 初步试验，预试验，预操作试验
preliminary thermal and hydraulic design 热工水力初步设计
preliminary treatment 预处理，初步处理
preliminary tremor 初期微震
preliminary work for construction 施工准备
preliminary working 初加工
preliminary 预备的，初步的，预先的
prelimit switch 预限开关
preloaded packing 预加载密封件
preloaded spring storage unit （阀门驱动机构）预载弹簧储能器
preloading consolidation 预压加固
preloading method 预压法
preloading of packing gland 填料密封套的预加载
preloading of the bolt 螺栓预紧
preloading 预负载，初载，预加荷载，预压，预先加料
preload spring 预紧弹簧
preload tension 预载张力
preload 预加载，预载，预先加料，预加负荷，预加应力
prelubrication 预先润滑
premagnetization 预先磁化
premature bubbling 过早鼓泡
premature failure 早期破坏，早期损伤，早期故障
premature hardening 早期硬化
premature removal 提前拆修
premature separation 提前分离
premature setting 早凝
premature stiffening 瞬间凝结，（混凝土的）过早硬化，早凝
premature transition 过早过度
premature 不成熟的，不到期的，过早的
prematurity payment 未到期付款，到期前付款
premeditated simulation of accident 事故预想
premises heating 房屋采暖
premises 房产，房屋，前提，根据
premium fuel 优质燃料
premium insulation 优质绝缘，高级绝缘
premium on insurance 保险费，保险金
premium rate 保险费率
premium 保险费，奖金，贴水，优质的
premixed burner 预混式燃烧器
premixed concrete 预拌混凝土
premixed plaster 水泥砂预拌料，预掺料灰泥
premix polymerizer 预拌聚合剂
premix 水泥砂预拌料，预混合
premodulation 预调制
premonitory phenomenon 前兆现象
premultiplication 自左乘
prenatal 出生以前的
preoperational radiological survey 运行前放射性调查
preoperational test 运行前试验，预操作试验
preoperational 运行前的
preoperation test of electrostatic precipitator 电除尘器投入前试验
preoperation 预运行，试运行
preoperative 操作前的，运行前的
pre-OSART mission 运行安全检查先遣组
preoscillation current 起振电流
preoxidation 预氧化
prepackaged concrete 预填骨料压浆混凝土，预压骨料混凝土，预填集料混凝土，作填料用的混凝土
prepacked aggregate concrete 预填集料混凝土
prepacked concrete pile 预垒混凝土桩
prepack method （混凝土骨料）预装法
prepaid insurance 预付保险费
prepaid lump sum 预付总额
prepaid method 压浆法
prepaid shipment 预付装运
prepaid 预付费，预付
prepakt concrete 压浆混凝土
prepakt method 预压骨料法
preparation expense 筹备费
preparation of a document 文件的准备
preparation of aligning lines for steel structure on boiler foundation 锅炉基础上钢结构划线
preparation of bid 准备投标书
preparation of cable conduit 电缆管加工
preparation of cable joints 电缆接头制作
preparation of cable terminal 电缆终端制作
preparation of coal slurry 煤浆制备
preparation of document 文件编制，编制
preparation of procedures 程序的准备
preparation of rigid busbar 硬母线加工
preparation of tender 准备投标书
preparation procedure 编制程序
preparations for construction 施工准备
preparation staff 运行前试运转
preparation 准备，预备，制备，配制，制备品，制剂，编制，处理
preparative treatment 预加工
preparatory committee 筹备委员会
preparatory cost 预备费
preparatory examination 预设计
preparatory measures 初步措施
preparatory meeting 预备会议
preparatory works 预备工程，初期工程，筹备工作，准备工作
preparatory 预备的
prepare budget 编制预算
prepared by 由……准备的
prepared gravel 备制的砾石
prepared hole 预留孔
preparedness 有准备的

prepared refuse 处理过的废物
prepare statement 编制报表
prepare 准备
prepassivation 预钝化
prepayment 提前还款，预付款
prepay-set 投币式公用电话机
preplaced aggregate concrete 预填骨料混凝土，预填集料混凝土
prepost tensioning （预应力的）半后张法
prepreg lay-up 预浸布铺层
prepreg 半固化片，预浸材料
PRE(preliminary for comments) 征求意见稿
prepressurization 预加压，初始内压
prepressurized fuel element 预加压燃料元件
preprocessing 预处理，预加工，加工前的
preprocessor 预处理机，预处理器
preproduction model 试制模型，试制样品，样机
preproduction test 正式投入运行前的试验
preproduction type test 生产前的试验
preproduction 生产前的，试生产的，预制造的，试制
preprogram 预先编程序
pre-proposal site visit 标前踏勘
preprototype 预样机，预原型，预制样品
prepulsing 预馈脉冲，发出超前脉冲
prepump 前置泵
prepunched 预冲孔的
pre-purge 预吹扫
prequalification document 资格预审文件
prequalification evaluation 资格预审，预资格审查
prequalification of bidder 投标人资格预审
prequalification 资格预审
prequalified tendering 资格预审招标
prequalify 资格预审
preread disturb pulse 读前干扰脉冲
preread head 预读头
preregeneration 预再生
pre-repairing of steam turbine 汽机预检修
prerequisite conditions 先决条件
prerequisite event 起始事件
prerequisite 先决条件，必要条件，前提，必须具备的
pre-resinated 用树脂预处理过的
prerogative 特权
prerotation vane 导流叶片，预旋叶片
prerotation velocity 预旋速度
prerotation 预扭，预旋，预先转动，预旋度
prescribed limit 规定限值，给定极限
prescribed velocity 特征速度
prescribed 给定的
prescription balance 药剂天平
prescription water rights 法定水权
prescription 时效
prescriptive period 规定期，时效期
prescrubber 预洗涤塔
presedimentation 预沉降
preselected project 既定项目，预选工程
preselection 预选，前置选择
preselector control 预选控制，预定位控制
preselector relay 预选器继电器
preselector 前置选择器，预选器，高频预选滤波器
preselect 预选，预定，预选送
presence bit 存在位，内存指示位
presence of mind 沉着，镇定
presence 出席，仪表，存在
present amount factor 现值因子
present an amendment 提出修正案
presentation layer 表示层
presentation of receipt 出示收据
present contract 本合同
present insert 凝固前镶嵌件
present period 本期
present position 现任职
present price 市价，现行价格
present staff 现有人员
present status 现状
present value 现值，折现值
present-worth cost 现值成本
present-worthed 现值的
present-worth evaluation 现值估价法
present-worth factor 现值因子
present-worth method 现值法，现值方法
present-worth 现值【即贴现后的钱】，折现值
preservation of water sample 水样保存
preservation treatment of timber 木材防腐处理
preservation 储藏，防腐，保存，保护，防护，贮藏
preservative substance 防腐剂
preservative-treated lumber 防腐处理的木材
preservative 防腐材料，防腐剂，防腐的
preserved storage 保留库容
preserve 保护，保持，保存，保养，留存
preservice cleaning 投运前清洗
preservice inspection 投产前检验，役前检查
preservice training 就业前培训
preserving timber 防腐木材
pre-session documentation 会前文件
preset adjustment 预调，预调整，预调准
preset capacitor 微调电容器，预调电容器，半可变电容器
preset control 程序控制，预调控制
preset counter 预置计数器
preset cut-out speed 给定截断转速
preset decimal counter 预置十进制计数器
preset digit layout 预先给定的数字［数位］配置，预先给定的数位格式
preset flow limit valve 前置流量限制阀
preset high level 预先设定的料位
preset level 给定电平，起始电平
preset limit 预先整定值范围
preset parameter 给定参数，预定参数，固定参数，预置参数
preset position 预调位置，预定位置
preset sequence 给定程序
preset sequential malfunctions 预设序列故障
presetting circuit 预调整电路
presetting period 初凝时间
presettling 预沉积
preset value 预定值

preset 预定的，预置的，预调，预置，预定，预设定
preshoot 预冲，倾斜，前冲，下垂，前置尖头信号
president 总经理
preside over a meeting 主持会议
preslugging 预压制
preslug 片状毛坯
presoftening 预软化
prespective dose 事前计划剂量
presplit blasting 预裂爆破
presplitting 预裂法
prespringing （焊件的）预弯，弹性反应变
pressboard 压制板
press bonding 压焊
press-button control 按钮控制
press-button switch 按钮开关
press-button system 按钮操纵系统
press-button 按钮
press-connection machine 压接机
press controller 按钮控制器
press die 压模
pressed air 压缩空气
pressed and sintered pellet 压制和烧结芯块
pressed block-shaped fuel element 压块状燃料元件
pressed compact 压块
pressed concrete 压制混凝土
pressed film 压膜
pressed fuel 压制煤砖，压制燃料
pressed graphite powder 压制的石墨粉
pressed-in seat ring 压入的阀座环
pressed out boss 压制毂
pressed pile 压入（式）桩
pressed sampler 压入式取样器，压入式取土器
presser cell 压敏元件
presses cold forging 冷锻冲压机
presses crank 曲柄压力机
presses eccentric 离心压力机
presses forging 锻压机
presses hydraulic 液压冲床
presses knuckle joint 肘杆式压力机
presses pneumatic 气动冲床
presses servo 伺服冲床
presses transfer 自动压力机
press filter 压滤机
press filtration 压过滤
press fit 压入配合，压力密合，压配合，紧配合
press for delivery 催货
press formed steel sheet 压型钢板
press forming 模压成型
pressing crack 压裂，压碎
pressing die 压模
pressing-in 注入
pressing 冲压，压紧
press-on ring 压紧环
pressostat 恒压器，稳压器
press plier 压接钳
press-powder magnet 压粉磁铁，粉末磁铁
pressproof 清样
press regulator 压力调节器
press roll 压辊
press switch 压力开关
press-talk system 电话的按讲制，按键通话方式
press transmitter 压力变送器
pressure above atmospheric （对大气压的）表压力
pressure accumulator 蓄力器，蓄压器
pressure adjustment 压力调节，压力调整
pressure aerator 加压曝气器
pressure alarm 压力报警，压力报警器
pressure amplifier 压力放大器，增压器
pressure and temperature reducing 减温减压
pressure anemometer 压力风速表，压力风速计
pressure angle 压力角
pressure arch 压力拱
pressure at expulsion 排气压力
pressure atomization mechanical atomization 压力雾化，机械雾化
pressure atomization 压力雾化
pressure atomizer 机械式雾化器，压力雾化器
pressure atomizing burner 压力雾化燃烧器，机械雾化燃烧器
pressure at right angle 正交压力
pressure at velocity stage 调节级压力
pressure balance 压力平衡
pressure barrier 压力坝
pressure belt 受压带
pressure berme 反压马道护坡道
pressure blasting machine 压送式喷丸机
pressure blower 压力鼓风机
pressure boiler 微正压锅炉
pressure boundary 压力边界
pressure breakdown bush 压力减低衬套
pressure broadening 致压谱线增宽
pressure build-up evaporator 升压蒸发器
pressure build-up 升压
pressure bulb 压力气泡
pressure cable 充油电缆，充气电缆，加压电缆
pressure calibrator 压力校验仪
pressure cell 压敏元件，压力传感器，压力盒，测压计
pressure center 压力中心
pressure chamber 压力室
pressure change rate 压力变化率
pressure change 压力变化
pressure characteristic 压力曲线，压力特性
pressure chart 压力曲线图，压力图
pressure closure 承压封头，上封头
pressure coefficient of reactivity 反应性压力系数
pressure coefficient 压力系数
pressure coil 电压线圈
pressure combustion 微正压燃烧
pressure-compensated flowmeter 压力补偿式流量计
pressure compensating aerator 水压补偿起泡器
pressure compensating control 压力补偿控制
pressure compensator 压力补偿器
pressure component 耐压元件，压力分量
pressure compounded turbine 多级式汽轮机
pressure conduit 有压管道，压力水管
pressure connection 压力联结，有压衔接

pressure connector 压接端子
pressure controlled 按压力调节的
pressure controller 压力调节装置，压力调节器，压力控制器
pressure control loop 压力控制回路
pressure control valve 压力调节阀，压力控制阀
pressure control 压力控制，压力调节
pressure correction factor 压力校正系数
pressure correction 压力修正
pressure cushion 压力枕
pressure dampener 压力阻尼器
pressure damping 压力减震
pressured circulating fluidized bed combustion boiler 增压循环流化床燃烧锅炉
pressure decay rate 降压速度
pressure decay test 风压试验【炉膛，烟道】
pressure decay 压力降低，压力衰减
pressure decrease 压力降低
pressure dehydrater 加压脱水机
pressure detector 压力探测器
pressured fluidized bed combined cycle 增压流化床联合循环
pressured fluidized bed combustion 增压流化床燃烧
pressured gauge 压力表
pressure diagram during nonheating period 非供暖期水压图
pressure diagram in abnormal operation condition 事故工况水压图
pressure diagram of condensate pipeline 凝结水管水压图
pressure diagram 水压图，压力图
pressure die away 压力消失
pressure difference bed load sampler 压差式推移质取样器
pressure difference receiver 压差传感器
pressure difference switch 压差开关
pressure difference type sampler 压差式采样器
pressure difference 压降，压差，电压差
pressure differential meter 压差式流量计，差压式流量计
pressure differential 压差，差压
pressure diffusion 压力扩散
pressure dilatation 受压膨胀系数
pressure discontinue 压力不连续
pressure discontinuity 压力突跃
pressure distribution diagram 压力分布图
pressure distribution 压力分布，电压分布
pressure dividing control valve 压力分配控制阀
pressured oil filter 压力滤油机
pressure drag coefficient 压力阻力系数
pressure drag effect 压阻效应
pressure drag fluctuation 压差阻力波动
pressure drag 压差阻力，压力阻力
pressure drainage system 压力排水系统
pressure drop across the reactor core 堆芯压降
pressure drop at design flow 在设计流量下的压降
pressure drop history 压降历程曲线，压力降随时间变化曲线
pressure drop meter 差压流量计
pressure drop model 压降计算模型
pressure drop sensor 压差传感器
pressure drop 压力降，压降【用于负压燃烧】，压力损失，汽水阻力，通风阻力，烟气阻力，压差
pressure eddy 压力旋涡
pressure effect 压力效应
pressure efficiency 压力效率
pressure element 受压元件，测压元件，压力元件
pressure enclosure 压力套
pressure energy 压力能
pressure-enthalpy chart 压焓圈，pH 图
pressure-enthalpy curve pH 曲线，压焓曲线
pressure-enthalpy diagram 压焓图
pressure equalization 压力均衡
pressure-equalizing device 均压装置，均压线
pressure-equalizing pipe 压力平衡管
pressure excursion 压力突增，压力偏离额定值，压力快速上升
pressure expanded joint 板边胀接接头
pressure exponent 压力指数
pressure feed 正压进料
pressure field 压力场
pressure filter 压力（式）过滤器，压滤机，压力滤池，加压过滤池
pressure filtration 加压过滤，压滤
pressure fired boiler 微正压锅炉
pressure flotation 加压浮选
pressure flow 压力流，有压流
pressure fluctuation 压力波动，压力脉动
pressure fluid 受压流体
pressure flush 加压冲洗
pressure gage assembly 压力计组件
pressure gage 压力计
pressure gas welding 气压焊
pressure gas 压缩气体
pressure gauge tester 压力表校正仪
pressure gauge 测压计，压力表，压力计
pressure generator 压力发生器
pressure governing 压力调节
pressure governor 压力调节器
pressure grade line 压头线
pressure gradient force 气压梯度力
pressure gradient 压力梯度
pressure grouting 压力灌浆（法）
pressure grout pipe 压力灌浆管
pressure gun 压力枪
pressure head loss 压力水头损失
pressure head 压力水头，压力头，压头，压力落差，压位差，扬程，压力传感器
pressure heater 压力加热器
pressure-height curve 气压高度曲线
pressure hole 通气孔
pressure hose 高压软管
pressure housing 承压套罩【控制棒驱动机构】，耐压壳
pressure increase 压力增加
pressure increment 压力增量
pressure indicator 压力指示器，压力表
pressure-induced dipole 压力诱发偶极子

pressure-injected pile 压注桩
pressure in pores 孔隙压力
pressure instrumentation 压力测量仪表
pressure instrument 压力仪表压力计
pressure intake port 压力进水口
pressure intensity 压力强度，压强
pressure jet burner 压力喷射喷燃器
pressure jet 加压射流
pressure jump 气压涌升，压力突跃，压力跃变
pressure keeping line 稳压管路
pressureless turbine 无压式水轮机
pressure let-down standpipe 泄压竖管
pressure level measuring device 压力液位测量仪表
pressure line 压力管路，等压线
pressure-loaded 压力加载的
pressure load 压力载荷，压力荷载
pressure lock 压力栓，压力锁气器
pressure loss coefficient 压力损失系数，压损系数
pressure loss 压力损失
pressure lubrication 压力润滑，强制润滑
pressure manifold 承压集管，排出总管
pressure manometer 压力计
pressure mark 压痕
pressure measuring instrument 压力测量仪表
pressure membrane 压力薄膜
pressure metamorphism 压力变质
pressure meter calibration room 压力表校验室
pressuremeter limit pressure 旁压仪极限压力
pressuremeter modulus 旁压仪模量
pressure meter 测压计，压力表，压力计，测压仪
pressure meter type wave recorder 压力计式波高仪
pressure mill 正压式磨煤机
pressure-momentum curve 压力动量曲线
pressure monitor 压力监测器
pressure node 压力波节
pressure nozzle 机械喷嘴，压力喷嘴
pressure of extracted steam from turbine 汽轮机抽汽压力
pressure of return water 回水压力
pressure of saturated vapour 饱和蒸汽压
pressure of sound wave 声波压强
pressure of supply steam 供汽压力
pressure of supply water 供水压力
pressure oil filter 压力式滤油机
pressure oiling 压力加油，压力注油
pressure oil piping 压力油管道
pressure oil 工作油，压力油
pressure on foundation 基础承压
pressure-operated furnace 微正压炉膛
pressure-operated switch 压力作用开关，压力操作开关
pressure operation 增压运行，微正压运行
pressure orifice 测压孔
pressure oscillation 压力波动，压力脉动
pressure packing 加压填充法
pressure part 受压元件，受压部件
pressure pattern 气压分布型式
pressure pick-off 压力传感器，压力换能器
pressure pickup 压力传感器
pressure pipeline 压力管道［管线］
pressure pipe 承压管，压力水管，压力引水管
pressure piping 压力管道，压力管系
pressure plate anemometer 压板风速计
pressure plate terminal 压力板接线柱
pressure plate 承压板
pressure plotting 压力分布图
pressure pocket 承压腔
pressure port 测压孔
pressure pot feeder 压力排挤加药罐
pressure probe 测压管
pressure profile 压力变化曲线，压力分布图，压力场，压力廓线
pressure propagation 压力传播
pressure pulsation 压力脉冲
pressure pulse 压力脉冲，电压脉冲，压力冲击
pressure pump 加压泵，增压泵，压力泵
pressure rating 额定压力，压力级
pressure ratio regulating valve 压力比例调节阀
pressure ratio 压比，增压比，压力比
pressure receiver 压力传感器
pressure recorder 压力记录器，压力记录仪
pressure recovery factor 压力恢复系数，压力保持系数
pressure recovery 压力恢复
pressure reducer and attemperator 减压减温器
pressure reducer assembly 减压器组件
pressure reducer 减压阀，减压器
pressure reducing and desuperheating valve 减温减压阀
pressure reducing bushing 减压衬套
pressure reducing orifice 降压孔板
pressure reducing regulator 减压调节器
pressure reducing sleeve 减压套
pressure reducing station 减压站
pressure reducing valve assembly 减压阀组件
pressure reducing valve 减压阀
pressure reducing 膨胀减压
pressure reduction effect 降压效应
pressure reduction 压降，电压下降，压力降低
pressure regulating device 调压机构
pressure regulating valve 调压阀，压力调节阀
pressure regulation exhaust 调节排气口
pressure regulation valve 调压阀，压力调节阀
pressure regulation 压力调节
pressure regulator 调压器，压力调节器
pressure release door 泄压门
pressure release window 泄压窗
pressure release 压力释放
pressure relief capability 压力释放能力
pressure relief damper 卸压挡板
pressure relief device 泄压装置【超压保护】，防爆筒
pressure relief facility 压力释放装置
pressure relief opening 压力释放孔
pressure relief system 卸压系数
pressure relief valve discharging into a closed system 卸压阀排放至闭路系统
pressure relief valve discharging to atmosphere 卸

压阀排放至大气
pressure relief valve with auxiliary actuating devices 带辅助制动装置的卸压阀
pressure relief valve with power-assisted lift and blowdown 带动力辅助抬升和喷放的卸压阀
pressure relief valve with power-assisted lift 带动力辅助抬升的卸压阀
pressure relief valve 减压阀，卸压阀，安全阀，泄压阀，压力释放阀
pressure relief vent 防爆筒，（变压器的）安全气道
pressure relief 卸压，压力释放，泄压，消除压力
pressure-relieving vessel 扩容器
pressure repeater 压力传送器
pressure resistance 抗压强度，压力阻力，耐压强度，压差阻力
pressure resistant spot welding 压力电阻点焊
pressure responsive device 感压装置，压敏装置
pressure retaining boundary 承压边界
pressure retaining component 承压部件［设备］
pressure retaining parts 承压零件
pressure retaining vessel 承压容器，压力壳，密闭壳
pressure reversal 压力梯度反向
pressure rise 压升，升压
pressure roller 压紧辊
pressure safeguard 压力监测器，压力防护装置
pressure sand pot 压送式喷砂罐
pressure scanning switch 压力扫描开关
pressure seal bonnet 压力密封阀盖
pressure seal 加压密封
pressure selector switch 测压选择开关
pressure sensing device 压敏元件
pressure sensing line 压敏管线，传压管
pressure-sensitive relay 压敏继电器
pressure sensor 测压传感器，压力传感器
pressure setting test of safety valve 安全门压力整定试验
pressure shell 承压壳体，安全壳壳体
pressure shock 压力激波，压力冲波
pressure side of the blade 叶片正面，叶片工作面
pressure side 受压侧，内弧，压力侧，受压面
pressure slope 压力曲线斜率
pressure span 压力范围，压力区域
pressure sphere anemometer 压力球风速计
pressure spike 压力尖峰
pressure-spring thermometer 弹簧管压力式温度计，压力表式温度计，带有空心弹簧压力计式温度计
pressure stabilizing tank 稳压槽
pressure stage impulse turbine 压力级冲动式汽轮机
pressure stage 压力级
pressure strength 抗压强度
pressure stress 压应力
pressure stroke 压缩冲程
pressure suppression chamber 压力抑制腔，弛压腔
pressure suppression pool 弛压水池
pressure suppression system 压力抑制系统，弛压系统
pressure suppression type containment 弛压型安全壳，压力抑制型安全壳
pressure suppression vent 弛压放气管
pressure suppression zone 弛压区
pressure suppression 压力抑制，弛压
pressure surface 内弧，压力面
pressure surge tank 压力波动箱
pressure surge 气压涌升，压力跃变，压力剧变，水击，压力波动
pressure sustaining valve 恒压阀
pressure switch 压力继电器，压力开关，压力感受器，压力操纵开关
pressure system pulverizer 正压系统磨煤机
pressure system 气压系统
pressure tank reactor 压力罐式反应堆
pressure tank 高压箱，压力供油箱，压力水箱
pressure tapping line 测压管线
pressure tapping point 取压点
pressure tapping tube 取压管
pressure tapping 压力表接头
pressure tap 测压孔［管］，压力计接口，取压分接管，取压口［孔］，压力龙头
pressure-temperature rating 额定热力参数，额定压力温度，压力-温度等级
pressure-temperature safety device 压力温度安全器
pressure-temperature saturating curve 压力温度饱和（曲）线
pressure test report 压力试验报告
pressure test 水压试验，耐压试验，压力试验，试压
pressure thermometer 压力式温度计
pressure tide gauge 水压验潮仪
pressure tight casing 密封护板
pressure-tightness 压力严密性
pressure tight 耐压的，密封的
pressure-time history 压力-时间历程
pressure transducer 压力换能器，压力变送器，压力传感器
pressure transformer 电压互感器
pressure transient 压力瞬变
pressure transmission 压力传递
pressure transmitter 传压器，压力传送器，压力传感器，压力变送器
pressure traverse 压力横向分布，压力剖面
pressure trough 气压槽
pressure tube anemometer 压力管风速计
pressure tube design 压力管设计
pressure tube extension 压力管延伸段，端管段
pressure tube reactor 压力管式反应堆
pressure tube 压力管，排出管，承压管，测压管
pressure tunnel 压力隧洞，有压隧洞
pressure turbine 压力式水轮机
pressure type capacitor （高压用）充高压惰性气体的电容器
pressure type oil burner 机械雾化油燃烧器
pressure vacuum gauge 压力真空表，压力吸力计，真空压力计
pressure vacuum relief gate 压力真空释放门

pressure valve 压出阀，压力阀，增压阀
pressure variance 压力变化（量）
pressure velocity compounded 压力-速度级复合式【汽轮机】
pressure ventilation 压力通风
pressure vessel base blowdown 压力容器排放，压力容器排污
pressure vessel base material 压力容器母材
pressure vessel blowdown 压力容器排污
pressure vessel bottom head 压力容器底封头
pressure vessel cladding 压力容器覆盖层
pressure vessel closure head spray system 压力容器封头喷淋系统
pressure vessel conveyer 压力容器输送机，压力罐输送器，仓式泵
pressure vessel fabrication on site 压力容器现场制造
pressure vessel head installation guild 压力容器封头安装导杆
pressure vessel head lifting beam 压力容器封头提升梁
pressure vessel head 压力容器顶盖，压力容器封头
pressure vessel inlet 压力容器进入口
pressure vessel outlet 压力容器出口
pressure vessel penetration 压力容器贯穿件
pressure vessel pneumatic test 压力容器气压试验
pressure vessel reactor 压力容器式反应堆
pressure vessel shell 压力容器壳体
pressure vessel site fabrication 压力容器现场制造
pressure vessel support lug 压力容器支耳
pressure vessel test 压力容器试验
pressure vessel wall 压力容器壁
pressure vessel 受压容器，压力容器，压力壳
pressure-void ratio curve 压力孔隙比关系曲线，压缩曲线
pressure volume chart 压容图
pressure-volume diagram 压容图，PV 图
pressure warning unit 压力报警装置
pressure washing 压力冲洗
pressure water ash removal 水力除灰
pressure water 加压水，压力水
pressure wave 地震纵波，P 波，压力波
pressure welding machine 压接机
pressure welding 压焊，压力焊
pressure winding 电压绕组，电压线圈
pressure window 压力窗
pressure wind tunnel 增压风洞
pressure zone 压力区，内弧正压区
pressure 压力，压强，加压
pressurization by bypass pipe 旁通管定压
pressurization by compressed air 压缩空气定压
pressurizationbycontinuouslyrunning make-up water 补给水泵连续补水定压
pressurization by make-up water pump 补给水泵定压
pressurization by nitrogen gas 氮气定压
pressurization by steam cushion in boiler drum 蒸汽锅筒定压
pressurization installation 定压装置
pressurization method 定压方式
pressurization point 定压点
pressurization 升压，压紧，气密，增压，压力输送，加压
pressurized air 压缩空气
pressurized boiler 正压锅炉，微正压锅炉，加压锅炉
pressurized bubbling fluidized bed boiler 增压鼓泡流化床锅炉
pressurized by elevated expansion tank 膨胀水箱定压
pressurized circulating fluidized bed boiler 增压循环流化床锅炉
pressurized combustion 微正压燃烧，微负压燃烧，微增压燃烧，加压燃烧
pressurized electrical instrument 过压型电动仪表
pressurized enclosure motor 增压防爆型电动机
pressurized enclosure 过压外壳，压力外壳
pressurized firing 压力燃烧，微正压燃烧
pressurized fluid insulation 密封绝缘油，（密封）加压绝缘液
pressurized fluidized bed boiler 增压流化床锅炉
pressurized fluidized bed combustion 增压流化床燃烧，加压流化床燃烧
pressurized fluidized bed （沸腾炉）增压床，加压流化床，增压流化床
pressurized fluid nozzle 加压喷嘴
pressurized fluid 加压流体
pressurized fuel element 加压型燃料元件
pressurized furnace 微压炉膛
pressurized gap cooling 绝缘气隙冷却
pressurized gas flow ionization chamber 加压气流电离室
pressurized gas method 加压气冷方法
pressurized gas reactor 绝缘气冷反应堆，加压气冷反应堆
pressurized gas 加压气体
pressurized heavy-water-moderated and cooled reactor 加压重水慢化和冷却反应堆，加压重水堆，加压重水慢化冷却堆
pressurized heavy water reactor 加压重水反应堆
pressurized helium gas 加压氦气
pressurized hopper 正压料斗
pressurized H_2O reactor 压水反应堆
pressurized-hydrogen cooling 高压氢气冷却
pressurized hydrogen 加压氢气
pressurized lancing door 充压打焦孔
pressurized light-water moderated and cooled reactor 加压轻水反应堆，加压轻水慢化冷却堆
pressurized motor 充高压气体的密封电动机
pressurized nitrogen 加压氮气
pressurized operation 微正压运行，增压运行
pressurized poke door 加压插孔门
pressurized prototype 正样
pressurized reactor 加压反应堆
pressurized seal 加压密封
pressurized station 加压站
pressurized subcritical experiment 加压次临界试验反应堆【美国】
pressurized suit 气衣，加压服

pressurized test 加压试验
pressurized water graphite reactor 压水冷却石墨慢化反应堆
pressurized water natural uranium reactor 天然铀压水反应堆
pressurized water plant 压水堆电厂
pressurized water reactor power plant 压力堆核电站
pressurized water reactor 加压水冷反应堆，压力水冷反应堆，压水反应堆，压水堆
pressurized water 压力水
pressurized wind tunnel 增压风洞
pressurized （电机）加压防护式，微正压的，增压的，过压的，加压的
pressurizer auxiliary spray line 稳压器辅助喷淋管
pressurizer bunker 稳压器隔间
pressurizer delta connection box 稳压器加热器三角形连接盒
pressurizer discharge line 稳压器排放管
pressurizer heat controller 压力调节器，稳压器热量调节器
pressurizer heater control system 稳压器加热器控制系统
pressurizer heaters group 稳压器加热器组
pressurizer heater 稳压器加热器
pressurizer high level trip 稳压器高水位保护停堆
pressurizer insurge 稳压器正波动
pressurizer level controller 稳压器水位控制回路
pressurizer level control 稳压器水位控制
pressurizer outsurge 稳压器负波动
pressurizer relief line 稳压器泄压管线
pressurizer relief tank 稳压器卸压箱
pressurizer safety valve 稳压器安全阀，稳压器卸压阀
pressurizer spray line 稳压器喷淋管线
pressurizer steam bubble 稳压器蒸汽气泡
pressurizer support skirt 稳压器支撑裙板
pressurizer surge line 稳压器波动管线
pressurizer surge nozzle 稳压器波动管接管
pressurizer system 稳压（器）系统
pressurizer vent valve 稳压器排气阀
pressurizer vent 稳压器放气管
pressurizer water level control loop 稳压器水位控制回路
pressurizer water level 稳压器水位
pressurizer 稳压器，稳压容器，增压装置，保持压力装置
pressurize 增压，密封，（燃料元件）压进，加压
pressurizing system 加压系统，稳压系统
pressurizing tank 加压箱，增压箱
pressurizing vessel 稳压容器，加压容器，压力容器
pressurizing window （电缆）密封封口，气密口
press wheel 压土轮
press 挤压，压，按，压缩，冲压，印刷机，夹具，压床，压力机，压制
PRESS 压力【DCS画面】
prestage 前置级
pre stall regime 失速前状态
prestarting inspection 启动前检查
prestarting 启动前的
prestart operation 启动前操作
pre-start-up check 启动前检查
prestart 预启动
prestige first 信誉第一
prestige 声誉
Preston tube （壁面剪应力测量用）普雷斯顿管
prestore 预先存储，预存储
prestress concrete hollow-cores slab 预应力钢筋混凝土空心板
prestressed anchorage 预应力锚固
prestressed anchor bar 预应力锚杆
prestressed anchors with bond 有黏结预应力锚杆
prestressed anchors without bond 无黏结预应力锚杆
prestressed anchor 预应力锚杆
prestressed bar tensioner 预应力钢筋拉伸机
prestressed beam 预应力梁
prestressed bridge 预应力桥
prestressed cable 预应力钢索
prestressed castiron vessel 预应力铸铁压力容器
prestressed concrete bar 预应力混凝土芯棒
prestressed concrete cable stayed bridge 预应力混凝土斜张桥
prestressed concrete cylinder 预应力混凝土管柱，预应力混凝土圆筒
prestressed concrete pile 预应力混凝土桩
prestressed concrete pipe pile 预应力混凝土管桩
prestressed concrete pipe 预应力混凝土管
prestressed concrete pressure vessel liner tube 预应力混凝土压力容器衬管
prestressed concrete pressure vessel 预应力混凝土压力容器
prestressed concrete process 预应力混凝土
prestressed concrete reactor pressure vessel 预应力混凝土反应堆压力容器
prestressed concrete structure 预应力混凝土结构
prestressed concrete tubular pile 预应力混凝土管桩
prestressed concrete vessel cooling system 预应力混凝土压力容器冷却系统
prestressed concrete vessel top slab 预应力混凝土压力容器顶板
prestressed concrete vessel 预应力混凝土压力容器
prestressed concrete 预应力混凝土
prestressed-glass cable penetration 预应力玻璃电缆贯穿件，预应力玻璃密封贯穿件
prestressed gravity dam 预应力重力坝
prestressed pipe pile 预应力管桩
prestressed precast concrete pile 预应力预制混凝土桩
prestressed reinforcement 预应力钢筋
prestressed rock anchor 预应力岩石锚栓
prestressed rock bolt 预应力岩石锚栓
prestressed steel wire strand 预应力钢绞线

prestressed structure　预应力结构
prestressed tendon　预应力锚索
prestressed wire　预应力钢丝
prestressed　预应力的
prestressing anchorage　预应力台座
prestressing bar　预应力钢筋
prestressing bed　预应力台座
prestressing cable　预应力钢缆，预应力钢索
prestressing equipment　预应力设备
prestressing force　预加应力
prestressing loss due to creep　蠕变引起的预应力损失
prestressing loss due to friction　摩擦引起的预应力损失
prestressing loss due to shrinkage　收缩引起的预应力损失
prestressing loss due to slip at anchorage　锚固滑动引起的预应力损失
prestressing loss　预应力损失
prestressing steel strand　预应力钢绞线
prestressing system　预加应力系统
prestressing tendon　预应力钢筋，预应力钢筋束，预应力锚索
prestressing　预应力的
prestress in the radial direction　径向预应力
prestress transfer　预应力的传递
prestress without bond　无黏结力的预应力
prestress　预应力
prestretched strand　预拉钢索
prestretching　预拉伸
presumption error　预定误差
presumption　（作为推论的）根据，可能性，假定，推测
presumptive error　预定误差
presuppression　预抑制
pre-swirl　预旋转，预转动
presynchronization　准同期
pre-tax margin　税前毛利
pre-tender estimate　标底
pretensioned beam　先张预应力梁
pretensioned binding　（电机）预张绑线
pretensioned member　先张预应力构件
pretensioned pile　先张预应力桩
pretensioned prestressed concrete structure　先张法预应力混凝土结构
pretensioned prestressed concrete　先张法预应力混凝土
pretensioned prestressing　先张法预应力
pretensioned tendon　先张拉钢丝束，预应力钢丝束
pretensioned wire　先张拉钢丝
pretensioning bed　张拉台
pretensioning　先张法，预张紧，先张，预拉伸
pretension load　预张力荷载
pretension loss　预张拉损失
pretension　初张力，预张力，预张拉
pretenvsioned wire　预应力钢丝
pretersonic　超声波的
pretest inspection　试验前检查
pretest pile　预试桩
pretest　预试
pretranslator　前置译码器，预译器
pretravel　预开度，预行程
pretreatment system　预处理系统
pretreatment　预处理，预加工，粗加工
pretrigger　前置触发器，预触发，预触发器
pre-TR tube　前置保护放电管
pretwist angle　（桨叶）预扭角
pretwist　预扭
prevailing climate condition　主要气候条件
prevailing condition　当时条件，主要条件
prevailing current　盛行流
prevailing gas rate　有效气速率
prevailing party　胜诉一方，优势方
prevailing power direction　风能盛行方向
prevailing system　现行系统
prevailing trend　大趋势，主要倾向
prevailing wage　普遍工资，现行工资标准
prevailing westerlies　盛行西风带
prevailing westerly　盛行西风带
prevailing wind direction in spring　春季主导风向
prevailing wind direction of the whole year　全年主导风向
prevailing wind direction　盛行风向，主风向，主导风向
prevailing wind　主导风，盛行风，主风
prevailing　流行的，占优势的，占主导地位的，盛行的
prevail price　现行价格，时价
prevail　流行，（季风）盛行，占上风，占优势，支配，主导
preventative　预防法，预防物，预防性的，防止的
prevent condensation　防止冷凝
prevent device　保护设备
prevented party　受事故影响的一方
preventer　防护设备，防护器
prevention cost　预防成本
prevention measure　保护措施
prevention of furnace explosion　防止炉膛防爆，炉膛防爆
prevention of furnace implosion　炉膛防内爆
prevention of heat loss　防止热损失
prevention of marine pollution　海洋污染防止
prevention of scale　防垢
prevention　防护，防止，阻止
preventive action　预防措施
preventive choke coil　保安扼流线圈
preventive control　预防性控制
preventive maintenance program　预防维修方案
preventive maintenance time　预防性维修时间
preventive maintenance work order　预防性维修工作单
preventive maintenance　预防性保养，计划维修，定期检修，预防性检修，预防性维护
preventive measures　预防措施，预防性措施
preventive overhaul　预防性大修
preventive resistance　保安电阻
preventive routine maintenance　预防性维修，常规维修
preventive test　预防性试验
preventive　预防的（措施），预防性的

prevent or control flood 防洪
preventor 防护装置
preview monitor 预检监视器
preview 预检，预演，试映
previous carry 从前位进位的
previous client 老客户
previous decade 前一个十进位
previously confirmed equipment 早先确认的设备
previous project 以往项目
previous question 先决问题
previous year 上年度
prewarning 预警告，预警报
preweld interval 焊接预热时间
pre-wetting 预湿法
prewhirl valve 预旋阀
prewhirl vane 预旋叶片
prewhirl 预旋（气流），预旋度
prewired instruction 预先排好的指令
prewired program 保留程序
prewired raceway 预先穿线的电缆管道
prewired 预先连线了的，预先排好的
prezenta 黏胶纤维素
prezoelectric crystal 压电晶体
PRF(pulse repetition frequency) 脉冲重复频率
PRG＝purge 吹扫
PRHRS(passive residual heat removal system) 非能动余热排出系统
price adjustment formula 价格调整公式
price adjustment 价格调整
price agreed upon 议定价格
price analysis 价格分析
price bargain 讨价还价
price bidding strategy index 申报价格策略指标
price breakdown 价格细目表，分开列明价格，分项价格表
price cap regulation 价格上限监管法
price comparison 价格比较
price contract 定价合同，价格合同
price control 价格管制
price cost margin index 价格边际成本指数
price cutting 减价
priced bill of quantities 标价工程量清单
price discrimination 价格歧视
priced proposal 报价建议书
price duty paid 已付税价格
price elasticity of demand 需求的价格弹性
price elasticity of supply 供给的价格弹性
price escalation 价格增长
price fluctuation 价格波动
price for prompt delivery 立刻交货价格
price including commission 含佣金价格
price including tax 含税金价格
price indexation 价格指数化，按指数调整价格
price index regulation 价格指数监管法
price index 价格指数，物价指数
price-inflexible demand 不灵活电价需要
price level 价格水平
price list 价格表，价格单，价目表
price negotiation 价格洽谈
price of electricity sent into grid 上网电价
price quotation 报价单
price regulation 价格监管
price review form 价格审查表格
price ruling today 当日有效的价格
prices are subject to change 价格可能有变动
price schedule 价格表，报价表
price scissors 价格剪刀差
Price's guard wire 普赖斯保护线
price sheet 价目单
prices of separate items 分项价格
price stability 价格稳定
price standard 价格标准
price system 价格体系
price tag 价格标签
price term 价格条件，价格条款
price variance 价格差异
price 价格，标价
pricing 计价，定价
prick current meter 旋杯式流速仪
pricker 刺孔针
pricking up 抹灰打底
prick punch 冲孔器中心冲头
prilling tower with densified concrete 增密混凝土的造粒塔
prilling tower with reinforced plastic lining 玻璃钢衬里的造粒塔
primacord 传爆索，导爆索
primage （汽包中随蒸汽带走的）水分带出量，蒸汽带水量
prima humic acid 原生腐殖酸叮
primaral forest 原始森林
primary air fan-coil system 风机盘管加新风系统
primary air fan 一次风机
primary airflow 主气流
primary air heater 第一级空气预热器，一次风暖风器
primary air piping 一次风管道
primary air pollution 同源空气污染质
primary air system 新风系统
primary air 一次风，一次空气，主气流
primary amine 伯胺
primary argon system 一回路氩气系统
primary assets 主要资产
primary auxiliary building to annulus transfer air-lock 一回路辅助厂房至环室的空气闸门
primary auxiliary building 主回路辅助厂房，核辅助厂房
primary battery 原电池
primary beam 主梁
primary benefit 直接利益，直接效益，主要受益
primary blasting 初始爆破
primary branch 主支管
primary breaker 初碎机，粗碎机
primary calorifier 基本加热器
primary cation exchanger 第一级阳离子交换器
primary cell 一次电池，原电池，原生电池
primary chamber 主燃烧室，一次风室
primary circuit of NPP 核电厂一回路
primary circuit pressure boundary 一回路压力边界
primary circuit pressure relief equipment 一回路卸压设备

primary circuit 一次电路，一次回路，前置回路，初级电路，原电路，一级管网，一回路
primary circulation pump 主循环泵，一次循环泵
primary circulation 一级环流
primary circulator 一次风机
primary clarification 初次澄清
primary clearance 初始间隙
primary clock 母钟，主钟
primary coat 第一道抹灰，抹灰底层，抹灰打底，底涂层，底漆层，底漆
primary coil 一次线圈，初级线圈，原线圈
primary coking coal 焦煤
primary colour 底漆色，基本色
primary combustion zone 一次燃烧区域
primary combustion 一次燃烧
primary compensator 初级补偿器
primary component cooling water system 核岛设备冷却水系统
primary component 一回路设备
primary compress 主压缩
primary computation 初算
primary condensate 一回路冷凝水，主冷凝水
primary condenser 初级冷凝器
primary console 主控制台
primary consolidation 初步固结，初期固结，初始固结，原始固结，主固结
primary constant 一次常数
primary constituent 主要组分
primary construction 主结构
primary containment cooling system 一次安全壳冷却系统
primary containment 一次安全壳，内层安全壳
primary control element 灵敏元件，初级检测元件
primary controllable losses 主要可控损耗
primary control network 一等控制网
primary control point 一级控制点
primary control program 主控程序，基本控制程序，主控制程序
primary coolant circuit 初级冷却介质回路，一次冷却（剂）回路
primary coolant contamination 一回路水的污染
primary coolant degradation 一回路水的污染
primary coolant inlet nozzle 一回路水进水管嘴
primary coolant inventory 一次冷却剂总量
primary coolant letdown 一回路水下泄
primary coolant loop 一次冷却剂环路，一次冷却剂回路，一回路
primary coolant outlet nozzle 反应堆冷却剂出口管嘴
primary coolant outlet 反应堆冷却剂出口
primary coolant pipe 反应堆冷却剂管道
primary coolant pressure 反应堆一回路压力
primary coolant pump 一次冷却剂泵，一回路（主）泵，主冷却剂泵
primary coolant system 一次冷却剂回路系统，一次冷却剂系统
primary coolant temperature 一回路温度
primary coolant water 一次冷却水
primary coolant 一次冷却剂，反应堆冷却剂
primary cooling circuit 初级冷却回路，一次冷却回路
primary cooling system 主冷却系统，反应堆冷却系统，一次冷却系统
primary cooling 一次冷却
primary creep 初期蠕变，初始蠕变，第一阶段蠕变
primary crusher 初级碎石机，初碎机，粗碎机，初破碎机
primary crushing 初级破碎
primary cubicle 一回路隔间
primary current 初级电流，一次电流，原电流
primary cut-out 高压断路器
primary cyclone 初级旋风分离器
primary data 原始数据，原始资料，主要数据
primary datum 主基准面
primary deionization system 第一级除盐系统，初级去离子系统
primary demineralization system 一级除盐系统
primary depression 主低压
primary design criteria 主要设计准则
primary desuperheater 一级减温器
primary detecting element 初级检验元件
primary detector 一次探测器
primary device 一回路装置，一次仪表，一次元件
primary distribution main 初级配电网，一次配电干线
primary distribution network 一次配电网
primary distribution system 一次配电系统
primary distribution 一次配电
primary dune 初期沙丘
primary earthquake 初始地震，初震
primary electrical circuit equipment 电气一次设备【中国电气用语】
primary electron 初级电子，原电子，一次电子
primary element 主要内容，一次元件，基本元件，原电池
primary energy resource 一次能源
primary energy 一次能源，初始能量
primary envelope 一次安全壳，主要外壳
primary environment 原生环境
primary equipment 一次设备
primary event 初始事件
primary excavation 初次开挖
primary exchanger 初级离子交换器
primary failure 初级故障，原发性失效
primary fault current 一次故障电流
primary fault test 一次电故障试验
primary feedback 主反馈，主回授
primary filter 粗过滤器，初级过滤器
primary fission-neutron energy 原始裂变中子能量
primary fission product 原始裂变产物
primary fission yield 原始裂变产额
primary flourescent screen 初级荧光屏
primary flow rate 一次流量，主流量
primary forest 原始森林
primary frequency regulation 一次频率调节
primary frequency 一次频率，主频率
primary furnace 首级炉膛，熔渣室
primary fuse 原边熔断器，一次侧熔断器

primary gas filling　充入一次气体
primary gas flow in the steam-generators　蒸汽发生器内的一次气流
primary gas　一次气体【高温气冷堆】，一回路气体
primary grade water　核级水
primary grinding　初步研磨
primary grout hole　一期灌浆孔
primary guard vessel system　一回路保护容器系统
primary gyratory crusher　回转式初碎机
primary harmonic　一次谐波
primary heater　第一级加热器，主加热器
primary heat transfer loop　一次传热回路
primary heat transfer system　一回路传热系统
primary heat transport　一次热输送
primary humic acid　原生腐殖酸
primary hydriding　一次氢化
primary impedance starting　原边串联阻抗启动法
primary increment　初级子样
primary inductance　初级线圈电感，一次绕组电感
primary inlet nozzle　一回路进口接管
primary input　初级输入
primary instrument　一次仪表
primary interstice　原生间隙
primary ionization　一次电离
primary jaw crusher　颚式初碎机
primary jet　喷入炉膛的气粉混合物
primary leakage flux　原边漏磁通
primary leakage intercepting system　一回路泄漏收集系统
primary leakage reactance　原边漏抗，初级漏抗，一次侧漏抗
primary leakage　原边漏泄，一次侧漏泄
primary level line　主水准线
primary limit　初级限值
primary line　一次线
primary lining　底层衬砌
primary loess　原生黄土
primary loop　一回路，主回路
primary magnetomotive force　原边磁势
primary make-up system　一次回路补水系统
primary make-up water tank　一回路补水箱
primary make-up water　一次回路补给水
primary manway　一次侧下部人孔【蒸汽发生器】
primary measuring element　初级测量元件
primary measuring instrument　一次测量仪表
primary member　主要构件，主材
primary membrane stress　一次膜应力
primary meter　一次测量仪表
primary mineral　原生矿物
primary mixture　气粉混合物
primary model　一次模型
primary moisture separator　第一级汽水分离器
primary network　一次电力网，一次电力网络
primary neutron source　初始中子源
primary nozzle dam　一次侧管嘴堵板
primary nozzle　一级喷嘴
primary nuclear fuel　一次核燃料
primary oil　一次油
primary outlet nozzle　一次回路出口接管
primary outlet　一次侧引出线
primary output　初级输出，初期出力，主要出力
primary paint　底漆
primary parameter　一次参数，原边线圈参数
primary phase　初相
primary pilot valve　调速器主滑阀
primary pipe rupture　主管道破裂，一回路管道破裂
primary plant liquid waste　主设备废液，主回路废液
primary plant　一回路设备，主设备
primary pollutant　初级污染物，原生污染物
primary pollution　初级污染，一次污染
primary power distribution　一次配电
primary project benefit　工程初期效益，工程基本收益
primary protection　一次保护，主保护
primary protective barrier　初级防护屏障
primary pump　一次泵，主泵，一次启动油泵
primary purification circuit　主净化回路
primary purification　一次净化，初步净化，炉外处理
primary radiation　一次辐射，原辐射，初级辐射
primary radiator　初级辐射器
primary rate interface　主速率接口
primary reactor　主反应堆
primary recirculation pump　一次再循环泵，主再循环泵
primary record　原始记录
primary reduction gear　初级减速齿轮
primary reheater　再热器冷段，第一级再热器
primary relay　初级继电器，一次继电器
primary relief system　一回路卸压系统
primary relief tank　一回路卸压凝汽箱
primary responsibility　首要责任，主要责任
primary return air　一次回风
primary road　一级干道
primary root　（植物的）初生根，原生岩石
primary sampling equipment　初级取样设备，主要的取样设备
primary sampling system　一回路取样系统
primary scale　主要比例尺
primary seal assembly　初级密封组件
primary seal unit　初级密封单元
primary section　第一蒸发段，净段，一回路隔间，主切面，基本断面
primary sedimentation tank　初级沉淀地
primary sensing point　一次测点
primary sensitive element　主灵敏元件
primary sensor　直接传感器
primary separation　一次分离，初级分离
primary separator　第一级分离器，一次分离器，主分离器
primary settling　初级沉降
primary shaft　主轴，原动轴
primary shield concrete　一回路屏蔽
primary shield cooling system　一次屏蔽体冷却系统
primary shield cooling　一次屏蔽体冷却

primary shielding 第一屏蔽，主屏蔽，一回路屏蔽
primary shield tank 一次屏蔽水箱
primary shield wall 一回路屏蔽墙
primary shield 一次屏蔽体
primary shutdown element 主停堆元件
primary shutdown system 主停堆系统
primary side 初次侧，一次侧，一回路侧
primary silver-zinc battery 一次锌-银电池
primary sludge 原始污泥
primary sodium auxiliary system 一次回路钠辅助系统
primary sodium carrying pipe 一次回路载钠管道，主载钠管道
primary sodium loop 钠一回路
primary sodium make-up pump 一回路钠补给泵
primary sodium overflow vessel 一回路钠溢流容器
primary sodium pump 主钠泵
primary sodium trap 一次钠阱
primary sodium 一回路钠，一次钠
primary soil 原生土
primary source rod 一次（中子）源棒，初始源棒
primary source 一次中子源，主电源，一次电源，初始源，一次源
primary stage 初级阶段
primary standard pyrheliometer 基准直接日射表，一等标准直接日射表
primary standard solar cell 一级标准太阳电池
primary standard 一级标准，基本标准，原始标准，基准，原标准器
primary station 启动站，一等测点，一级测站
primary steam bypass system 主蒸汽旁路系统
primary steam circuit 主蒸汽回路
primary steam flow 主蒸汽流量
primary steam isolation valve 主蒸汽隔离阀
primary steam line 主蒸汽管线
primary steam safety valve 主蒸汽安全阀
primary steam separator 第一级汽水分离器，主汽水分离器
primary steam 主蒸汽，一次蒸汽
primary steelwork 一次钢结构
primary storage 主存储器
primary stratification 原生层理
primary stress 一次应力，基本应力
primary structure 原生构造
primary substation 一次变电所，升压变电所
primary superheater 第一段过热器，第一级过热器
primary supply air system 一次送风系统，主送风系统
primary system blowdown 主系统排放，一回路系统设备排放
primary system component 主系统设备，一回路系统设备
primary system coolant outlet temperature 一回路冷却剂出口温度
primary system diagram 一次系统图
primary system pipe rupture 一回路管道破裂
primary system rupture 一回路系统破裂
primary system 主系统，一次回路，主回路，主要系统，一回路系统，主冷却系统
primary tank 主容器，（快堆）钠池
primary telescope 主望远镜
primary terminal 一次电路端子
primary test board 基本测试台，主测试台
primary test 初始试验
primary thermometer 原始温度计
primary tide station 一级测潮站，主验潮站，主要潮汐观测站
primary tide 原潮汐，主潮
primary time effect 主时间效应
primary-to-secondary leak 一回路向二回路泄漏
primary-to-secondary phase shift 初级对次级的相变换
primary transmission feeder 一次输电的馈电线
primary treatment 初步处理，一级处理，初级处理
primary triangulation point 一等三角点
primary triangulation 一等三角测量
primary trough 主水槽
primary truss 主桁架
primary turbine 前置式汽轮机，第一轴汽轮机
primary turns 一次线匝
primary unit 基层单位
primary valve 一次阀，一次门，根阀，第一道阀
primary voltage 原边电压，一次电压，初级电压
primary wastewater treatment 污水一级处理，污水预处理
primary water pump 主泵，一次泵
primary water purification plant 一回路水净化装置，一次水净化装置
primary water storage tank 一回路水贮存箱
primary water 一次水，一回路水
primary wave 初波，初始波，地震纵波，纵波
primary winding voltage 初级绕组电压，一次绕组电压，原绕组电压
primary winding 初级绕组（变压器），一次绕组，原绕组
primary X-ray beam 原X射线束
primary 原始的，基本的，初级的，一次的，主要的，一回路的
prima 初波
prime bidder 首选投标商
prime coat 底漆，底漆层
prime contract award 总承包合同签订
prime contractor 主承包单位，主要承包商，主合同单位，总承包人（商），主要承包人
prime contract 基本合同，主要承包合同
prime cost 原始成本，成本费，主要成本，原价
primed 打底的
prime energy 一次能源，原始能源，原始能
prime fuel 主燃料
prime investment 优等投资
prime lacquer 打底漆，底漆
prime material 底层材料，打底材料
prime meridian 本初子午线
prime motor 原动机

prime mover governor 原动机的调速器
prime mover 原动机，发动机
prime paint 底层漆
prime plaster 底层灰
prime power 基本功率，原动力
prime pump 启动注油泵，启动泵
prime rate 头等贷款最低利率，头等贷款利率，最优惠利率，优惠利率
primer base 打底涂料，底漆涂料
primer cap 起爆雷管
primer coat （油漆）底层，底漆层
primer pump 启动注油泵，起动注油泵，启动泵，引水泵，初给泵，初级泵
primer valve 启动注水阀，启动注油阀
primer 入门，引子，底漆，底子，底层，底层涂料，始爆器，发火机，启动注油器，点火器，起爆剂，油漆打底层，点火剂，引水泵
primeval forest 原生林
prime 质数，第一阶，主要的，原始的，灌注，使准备好
priming chamber 注水室
priming charge 点火充电，启动注水，起爆药包，雷管药包
priming coat 涂层，底漆，底色，油漆打底层，底面涂层
priming connection 启动注水管道，启动注油管道
priming cup 注水斗
priming depth 启动水深
priming device 启动辅助装置，起爆装置
priming effect 启动效应
priming ejector 启动抽气器
priming level （虹吸管的）启动水位
priming method 启动方法
priming nozzle 点火喷嘴
priming of lube system 润滑充注系统
priming painting 底漆
priming potential 引燃电压，发光电压
priming pump 启动注液泵，启动泵，注水泵，注油泵
priming reservoir 抽水池，吸水池
priming stage 充注阶段
priming syphon 启动虹吸管
priming system 充注系统【泵，箱等】
priming tank 充注箱
priming time 启动时间
priming tube 爆破管
priming vacuum pump 启动真空泵
priming valve 启动注油阀，启动注水阀，启动阀
priming 灌水（入泵），汽水共（沸）腾，蒸汽带水，启动注水［注油］，点火，发动，起爆药，装雷管
primitive area 原始地区
primitive groove 原沟
primitive machine 原型电机
primitive period 基周，原周期
primitive plant 原始植物
primitive polynomial 本原多项式
primitive water 原生水
princess post 小柱
principal accumulator 主蓄能器
principal and interest 本金及利息
principal axes of inertia 主惯性轴
principal axes of strain 应变主轴，形变主轴
principal axes of stress 应力主轴
principal axes 各向异性材料导热系数分布的主轴
principal axis of ellipse 椭圆主轴
principal axis of inertia 惯性主轴
principal axis of stress 应力主轴
principal axis of symmetry 主对称轴
principal axis 主轴线
principal beam 主梁
principal channel 主河道
principal component 主要元件，主要构件
principal concept 基本概念
principal current 主电流
principal debtor 主债务人
principal deformation 主变形
principal designer 主要设计人
principal direction 主方向
principal drafter 主要起草者
principal earthquake 主震
principal element 主要因素
principal engineer 专业主要设计人
principal factor 主要因素
principal feedback 主反馈
principal focus 主焦点
principal fold 主褶皱
principal frequency 主频率
principal horizontal plane 主水平面
principal inertia axis 主惯性轴
principal joint 主节理
principal layout 总布置图
principal market 主要市场
principal meridian 主子午线
principal mode 基型，主型，主模态，主振型
principal moment of inertia 惯性主矩，主惯性矩
principal motor 主电动机
principal normal 主法线
principal nuclear facility 主要核设施
principal officer 首席官员
principal office 主营业地，总部
principal of image 镜像原理
principal of least work 最小功原理
principal of mass conservation 质量守恒定律
principal of momentum and energy 动量与能量原理
principal of operation 工作原理，操作原理
principal of superposition 迭加原理
principal optimality 最优原理
principal organ 主要机构
principal oscillation 主振荡，主振动
principal part 本体部分
principal plane of bending 弯曲主平面
principal plane of symmetry 主对称面
principal plane 主平面
principal point method 主点法
principal post 主柱
principal power 主电源
principal quantum number 主量子数
principal radius of curvature 主曲率半径

principal rafter 上弦，人字木
principal reactance 主电抗
principal reinforcement 主钢筋
principal river 干流
principal section 主截面
principal sector of economy 主体经济
principal size 主要尺寸
principal strain 主形变，主应变
principal stress change 主应力变化
principal stress circle 主应力圆
principal stress trajectory 主应力线
principal stress 主应力
principal tangent 主切线
principal tapping 主抽头，主分接头
principal terminal 主端子
principal theorem 基本定理
principal triangulation 一等三角测量
principal underwriter 主要担保人
principal volcano 主要火山
principal voltage 主电压
principal wave 主波，基波
principal winding 主线圈，主绕组
principal 委托人，当事人，负责人，本金，原理，定律，因素，主要的
principle diagonal 主对角线
principle diagram 原理图
principle of charge compensation 电荷补偿原理
principle of conservation of energy 能量守恒定律
principle of conservation of momentum 动量守恒原理
principle of continuity of electric current 电流连续性原理
principle of continuity of magnetic flux 磁通连续性原理
principle of detailed balancing 细致平衡原理，详细平衡原理
principle of diversity 多样性原则
principle of duality 二象性原理，对偶原理
principle of equipollent load 等效荷载原理
principle of equivalence 等效原理，当量原理
principle of good organization 良好的组织原则
principle of hydraulic similarity 水力相似原理
principle of increase of entropy 熵增原理
principle of least squares 最小二乘法原理
principle of least work 最小功原理
principle of marginality 边际原则
principle of measurement 测量原理
principle of minimum stress 最小应力原理
principle of momentum and energy 动量与能量原理
principle of operation 操作原理，工作原理，运行原理
principle of optimality 最优原理
principle of parity and reciprocity 同等和对等原则
principle of reciprocity 互易原理，互惠原则，对等原则，互为相反作用原则，互易定量
principle of relativity 相对性原理
principle of reversibility 可逆原理
principle of similitude 相似原理
principle of superimposed stresses 应力叠加原理
principle of superposition 叠加原理
principle of virtual displacement 虚位移原理
principle of virtual work 虚功原理
principle of water-turbine engine 水轮机原理
principle of work 工作原理
principles of electric engineering 电工原理
principle 原则，原理，根本方针
print a seal 盖印
printed circuit board 印刷回路板，印刷电路，印刷电路板
printed circuit cable 印刷导线，印刷电路引线
printed circuit card 印刷电路板，印刷电路插件
printed circuit lamp 印刷电路灯
printed circuit motor 印刷电路电（动）机
printed circuit patchboard 印刷电路接线盘
printed circuit relay 印刷电路继电器
printed circuit rotor 印刷电路转子
printed circuit 印刷电路，印刷电路板
printed coil 印刷绕组，印刷线圈
printed conductor 印刷导线
printed electrical conductor 印刷导线
printed electronic circuit 印刷电路
printed inductor 印刷电感线圈
printed matrix wiring 印刷矩阵布线
printed motor 印刷线路电动机
printed rotor 印刷电路转子
printed substrate 印刷线路基板
printed wire 印刷线路
printed wiring board 印刷导线板，印刷线路板
printed 印刷体的
printer controller 打印机控制器
printer dump 打印机打印，打印机转储
printer 打印装置，印刷装置，打印机，印刷机
print format 打印格式
printing and dyeing industry 印染工业
printing calculator 打印式计算器
printing density 印刷密度，打印密度
printing device 打印装置，印刷装置
printing equipment 打印装置
printing head 打印头
printing integrator 印刷积分器
printing mechanism 打印机构
printing-out tape 打印输出带
printing position 打印位置
printing reader 打印读出器，印刷阅读器
printing reperforator 打印复穿孔机
printing station 打印站
printing 打印，印刷
print order 打印指令
print-out rate 印出速度
printout 打印出的计算结果，印出，打印输出，打印出，复制出
print position 打印位置
PRINT(pre-edited interpretive system) 预编辑的解释系统
print room employee 印刷人员
print routine 打印程序
print server 打印服务器
print speed 印刷速度
print subroutine 打印子程序
print suppression 打印封锁指令

print through 打印传送，复印效应
print wheel 印字轮，打印轮
print 复印，印刷，打印，晒图
prionotron 调速管
prior approval 预先批准，预先认可，事先批准，事前批准，事前许可
prior assessment 先期评定
prior-check button 预试按钮
prior checking 预先校验
prior condition 先决条件
prior consent 事先同意
prior consultation 事先协商
prior distribution 先验分布
prior estimate 预先估计
prior ignorance 先前不知
prior incertainty 先验不定性
prior interval 先验间隔
priority check 优先检查
priority circuit 优先电路，优先次序电路
priority claim to general average 共同海损优先权
priority construction 重点建筑，优先建筑
priority indicator 优先指示器，优先权指示符
priority interrupt control unit 优先中断控制器
priority interruption 优先中断
priority interrupt 优先中断
priority level 优先级
priority logic 优先逻辑
priority phase 优先相位
priority processing 优先处理
priority program flip-flop 优先程序触发器
priority project 优先工程
priority switch 优先次序开关
priority use of water 用水优先权
priority 重点，先，前，优先，优先控制，优先次序，优先级，优先权，优先考虑的事情
prior probability 先验概率
prior review 事先审查
prior stress-strain history 先期应力应变历史
prior to 在……之前
prior 优先的，先验的，先前的
PRI(primary rate interface) 主速率接口
prismatical storage 槽蓄容量，动库容，棱柱体槽蓄
prismatical strength 棱柱体强度
prismatic astrolabe 棱镜等高仪
prismatic beam 等截面梁
prismatic bed 等截面河床
prismatic blade 直叶片
prismatic block fuel 棱柱块状燃料
prismatic block reactor 棱柱块高温气冷反应堆
prismatic building 棱柱形建筑
prismatic channel 等截面河槽
prismatic compass 棱镜罗盘仪
prismatic configuration 棱柱形布置
prismatic eyepiece 棱镜目镜
prismatic fuel element 棱柱形燃料元件
prismatic joint 柱状节理
prismatic layer 柱状层
prismatic reactor building 棱柱形反应堆厂房
prismatic soil structure 棱柱状土壤结构
prismatic storage 水库动库容
prismatic structure 柱状构造，折板，棱柱状结构，折板结构
prismatic type high temperature reactor 棱柱形燃料元件高温反应堆
prismatic 等截面的
prism drainage 棱体排水
prism glass 棱镜玻璃，起棱玻璃
PRISM(power reactor inherently safe module) 动力反应堆固有安全模件
prism 棱镜，棱柱，棱柱体
privacy system 保密通信制
private and confidential clause 保密条款
private automatic branch exchange 专用自动交换分机
private automatic exchange 专用自动交换机
private branch exchange 专用小交换机，用户交换机
private branch switchboard 专用交换台，用户总机，用户交换台
private circuit 专用线路
private code 专用代码
private data 专用数据
private enterprise 私人企业，私营企业
private exchange 专用小交换机，用户交换机
private house 私房
private information 私有信息
private library 专用程序库
private line 专用线路
private memory 专用存储器
private network 专用通信网
private project 私人项目
private property 私产
private right 专用权，民事权利，私有权
private section 专用段
private sector military 私营军事公司
private sector 私营部门，私营企业，私营成分
private sewer 专用下水道
private siding 专用铁路（线）
private station 专用电站，自备电站
private water supply 专用供水
private waterway 非公用水道
private wharf 专用码头
private wire 专用线，测试线
privileged instruction 特许指令
privileged memory operation 特许存储器操作
privilege of bidder 报价员权限
privilege 特权，优惠，特许的
prize 奖（章）
PRM(power range monitor) 功率区段监测器
probabilistic analysis 概率分析
probabilistic automat 随机自动机
probabilistic criteria （电力系统可靠性）概率性准则
probabilistic decoding 概率译码
probabilistic design method 概率设计法
probabilistic event 概率性事件
probabilistic logic 概率逻辑
probabilistic machine 概率机，随机元件计算机
probabilistic method 概率论法
probabilistic risk analysis 概率风险分析

probabilistic safety analysis 概率安全分析
probabilistic safety assessment 概率安全评价
probabilistic safety information management system 概率安全情报管理系统
probabilistic study 概率研究
probabilistic technique 概率论法
probabilistic uncertainty 概率的不确定性
probability computer 概率计算机
probability correlation 概率相关
probability curve 概率曲线，或然率曲线
probability density distribution 概率密度分布
probability density function 概率密度函数
probability density 概率密度
probability detector 概率检验器
probability deviation 概率偏差
probability distribution analyzer 概率分布分析器
probability distribution function 概率分布函数
probability distribution 概率分布
probability function 概率函数
probability integral 概率积分
probability interpretation 概率解释
probability law 概论定律
probability method 概率方法
probability multiplier 概率倍增器
probability of busy 占线概率
probability of correcting a single error 改正单个误差概率
probability of erroneous decoding 错误译码概率
probability of error per digit 单位符号误差概率，符号误差概率
probability of failure to close on command 拒合闸概率，拒分闸概率
probability of failure to operate on command 拒动概率
probability of failure 失效概率
probability of flashover 闪络概率
probability of incorrect trip due to fault outside protection zone 越级误分闸故障概率
probability of information loss 信息丢失概率
probability of loss （电话）呼损概率
probability of malfunction 故障概率，误动作概率
probability of occurrence 事故概率，出现概率
probability of stability 稳态概率
probability of subsequent severe core damage 接着发生的严重堆芯损伤概率
probability of survival 可靠概率，幸存概率，试验通过概率
probability paper 概率坐标纸
probability per unit volume of waveform space 振荡空间单位容量的概率
probability theory 概率论
probability 概率，可能性，或然率，几率
probable distribution 概率分布
probable erratum 概率误差
probable error 概差，概然误差，可能错误，概率误差，或然误差，可能误差
probable maximum extratropical storm 可能最大温带风暴
probable maximum flood 可能最大洪水
probable maximum hurricane 可能最大飓风
probable maximum seiche 可能最大假潮
probable maximum storm surge 可能最大风暴潮
probable maximum tropical cyclone 可能最大热带气旋
probable reliability 可靠性概率值
probable value 可能值
probable 可几的，概率的
probablity of detection 探测几率
probate 查验
probationary period 见习期，试用期
probationer 见习人员，实习人员
probation 试行，试用，试用期，预备期，见习期，见习，检验，鉴定
probe carrier unit 探头运送机构
probe coil 探头线圈
probed 终点计体内取样处
probe gas 示踪气体，探头气体，测气体
probe head 探头
probe inlet 探头插孔
probe insertion 探头插孔
probe material 示踪物质
prober 探测器
probe to weld distance 探头焊缝距离
probe unit 测试装置，检测器，测头
probe with long water column coupling 长水柱耦合探头
probe 探测器，传感器，探极，探头，测针，探针，电极，探测，探杆，探查，调查
probing medium 探测介质
probing pin 探测销，探针
probing 探测
problem board 解题插接板
problem-oriented language 面向问题的语言
problem program 问题程序，解题程序
problem-solving language 解题语言
problem time 仿真时间，问题时间
problem 难题，问题
probolog method 电测定器法
probolog 电子探伤仪，电测定器
procedural control 程序管理
procedural language 过程语言
procedural representation 过程表示
procedural test 程序试验
procedure “beyond-design” “超设计准则”程序
procedure control unit 程序控制单元
procedure declaration 过程说明
procedure division （程序语言的）过程部分
procedure document 程序文件
procedure for appeal 上诉程序
procedure for claims 索赔程序
procedure for obtaining approval to invite tenders 获准招标的程序
procedure identifier 程序识别器，过程标识符
procedure of analysis 分析程序
procedure of approval 批准程序
procedure of arbitration 仲裁程序
procedure-oriented language 面向过程的语言
procedure qualification record 程序鉴定记录
procedure qualification test 程序鉴定试验
procedures for import and export clearance of goods 进出口报关手续

procedures of electric power construction 电力基建程序
procedure 步骤，程序，手续，过程，方法，顺序
proceeding 会议录，论文集，会刊，科研报告集，事项，议程，诉讼，行动，程序，进程，过程，进行
proceed of loan 贷款收入
proceeds of consortium 集团收益
proceeds 收入，收益
process air conditioning 工艺性空气调节
process analysis 工艺分析，过程分析
process analyzer 过程分析仪表
process and instrumentation diagram 工艺仪表流程图
process and supervisory system 数据采集与监控系统
process and tooling design status 加工过程和工具设计状况
process annealing 中间退火
process approach 过程方法，过程研究法
process assembler language 过程汇编语言
process automation 过程自动化，工序自动化
process branch indicator 过程转移指示器
process camera 复照仪
process capability 工序能力，过程能力
process change 工艺更改
process channel 工艺管
process characteristics 过程特性
process chart 工艺流程图，工艺卡片
process chromatograph 过程色谱仪，工业色谱仪
process compatibility 工艺适用性
process computer system 数据处理用计算机系统
process computer 工业控制机，（数据）处理计算机，过程（控制）计算机
process condition 工艺操作条件
process control computer system 过程控制计算机系统
process control computer 过程控制计算机
process control diagram 工艺流程控制图
process control instrumentation 工艺控制检测仪表
process control language 过程控制语言
process control level 过程控制级
process control operating system 过程控制操作系统
process control relaying 工艺过程继电控制
process control relay rack 工艺过程控制继电器架
process control relay 工艺控制继电器
process control setting 过程调节器的参数设置
process control software 过程控制软件
process control system 过程控制系统
process control unit 过程控制单元
process control 生产过程控制，过程控制，工艺过程控制
process data 过程数据
process design 工艺设计
process display 过程画面
process disturbance 过程扰动
process drain holdup tank 残液排放贮存箱，工艺排水贮水箱
process drain 工艺疏水，工艺排水，排废水
process drawing 工艺图
processed condition 加工状态
processed information 处理过的数据，处理过的信息，被处理的信息
process engineer's console 过程控制工程师控制台
process engineer （化工生产的）工艺工程师
process environment 过程环境
process equipment 工艺设备，过程设备
processer 处理程序，信息处理机，加工机
process evaluation 工艺过程评价
process feed 运行供料
process flow diagram 工艺流程图
process flow 工艺流程
process fluid compatibility 工艺适用性
process fluid 工作介质，工艺流体
process furnace 加热炉
process gas chromatograph 工艺气体色层（分离）谱
process hazards 工艺危险源
process-heating load 生产工艺热负荷
process heat reactor 工业供热反应堆，供热堆
process heat 工艺热，工业用热
process industry 加工工业
process info 监控进程信息
processing bath 洗相槽
processing behavior 加工性能
processing charges 加工费
processing circuit 处理电路，接通电路
processing computer 数据处理计算机
processing condition 加工条件
processing cycle 操作程序
processing element 处理部件，处理机
processing equipment 加工设备，数据处理装置
processing function 处理功能
processing interrupt 处理中断
processing logic 处理逻辑图
processing method 处理方法加工方法
processing module 处理模块
processing of data 数据处理，资料整理
processing of non conformance 不符合项的处理
processing plant 处理装置，处理工厂，炼油厂
processing program 处理程序
processing room 洗相间
processing section 处理部件，处理部分
processing solution 洗相溶液
processing tax 加工商品税
processing unit 处理部件，处理机，运算器，数据处理装置
processing waste 生产废料
processing 加工，处理
process inherent ultimate safety reactor 固有安全反应堆，超安全反应堆
process input/output channel 过程输入输出通道
process input/output 过程输入输出
process inspection 工序检查，流程检查
process instrumentation and control 工艺仪表和控制

process instrumentation system　工艺仪表系统
process instrumentation　工艺测量仪表，生产过程用检测仪表，热力过程仪表
process instrument calibrator　过程仪表校验仪
process instrument diagram　工艺仪表图
process instrument rack　工艺测量仪表架
process interface schedule　工艺接口一览表
process interrupt status word　过程中断状态字
process inventory estimation　过程中存量估算
process lag　生产过程滞后
process level functions　过程层功能
process-limited　过程受限的
process line　生产线
process management and control　过程管理和控制
process management　过程管理
process measurement　热工测量，过程测量
process media　工艺介质
process monitoring instrument　工艺监测器，过程监测仪
process monitoring　工艺监测，过程监测，工艺监视，工艺（流程）监控
process monitor system　过程监督系统
process monitor　工艺监测器
process multimeter　过程万用表
process of coal-forming　成煤作用
process of compaction　压实过程
process of production　生产流程
process of sedimentation　沉积过程
process of self-excitation　自励过程
process optimization　过程优化
process optimizing　过程最优化
processor address register　处理机地址寄存器
processor controller　处理机控制器
processor control program　处理机控制程序
processor error interrupt　处理机出错中断
process-oriented sequential control　工艺过程的顺序控制，过程顺序控制
processor input-output　处理机输入输出（设备）
processor-limited　受处理机限制的
processor-memory-switch notation　处理器存储器开关表示法
processor-memory-switch　处理机—存储器—开关
processor operator console　处理机操作台
processor stack pointer　处理器栈指针
processor status register　处理机状态寄存器
processor status word　处理机状态字
processor　处理程序，处理系统，信息处理机，加工机，处理机，处理器
process pipeline　工艺管道
process piping drawing　工艺管线图
process piping systems　工艺管线系统
process point　过程点
process pressure　过程压力
process pump　流程泵
process qualification report　工艺过程鉴定报告
process qualification　工艺过程鉴定
process quality control　过程质量控制
process reaction curve　过程反应曲线
process regulation　工艺规程
process relay control　工艺过程继电控制
process requirement　工艺要求
process rules　工艺规程
process scale　工业规模，生产规模
process self-regulation　过程自动调节
process simulation　过程仿真，过程模拟
process sodium level　工艺钠位，钠工艺液位
process specification　工艺规范书
process station　过程站
process status　过程状态
process steam　工艺用汽，生产用汽
process-stream monitor　工艺物流监测器
process stream　工艺流程，工艺物流
process switch　过程开关
process system　工艺系统
process tap　工艺接头
process technique　程序加工技术
process temperature　过程温度
process time lag　过程时滞
process tube elevator　工艺管升降机
process tube　工艺管【如脉冲管线】
process unit　工艺设备
process variable　可调变量，过程变量
process vessel　工艺容器，加工容器
process waste　工艺废物
process water　工艺用水
process　工艺流程，工艺过程，过程，步骤，程序，工序，方法，处理，手续，加工
proclaim　宣布，宣告，公布，声明，表明
proclamation　公布，公告，宣告，宣布，声明
PROC(programmed computer)　程序控制计算机
Proctor compaction test　普氏击实试验，普罗克特击实试验
proctor method　葡式压实法
procuration signature　代签
procuration　代理（权），获得
procuratorial organ　监察机构
procurator　代理人
procure an agreement　达成协议
procurement authorization　采购授权书
procurement contract　订货合同
procurement control　采购控制
procurement division　采购部门
procurement document　采购文件
procurement item　采购项目
procurement method　采购方式
procurement of goods　货物采购
procurement of items and services　采购物项和服务
procurement packages　一揽子采购
procurement price　采购价格
procurement program　采购计划
procurement specification　采购规范书
procurement　采购
procure　采购，取得
procuring goods　采购货物
PRODAC(programmed digital automatic control)　程序数字自动控制
prod-contact magnetization　触头磁化
prod magnetization method　触头（通电）磁化法
produce influence on　对……有影响

producer gas 发生炉煤气
producer surplus 生产者剩余
producer 产生器，发生器，煤气发生炉，振荡器，发电机，生产者
produce's quality cost 生产者质量成本
producing reactor 生产反应堆
product accumulator 乘积累加器
product analysis 成分分析
product application data sheet 产品应用资料
product buy back 产品返销
product catalogue 产品目录
product cycle 积循环
product degradation 产品降低粒度等级
product development 产品开发
product generator 乘积发生器
product identification 产品标识
product in short supply 短线产品
product inspection record 产品检验记录
product inspection 产品检验
product integrator 乘积积分器
production analysis control technique 生产分析控制技术
production and construction funds 生产建设资金
production area 生产区
production base 生产基地
production branch of electric power industry 电力生产部门
production building 生产厂房
production calibration 生产检定
production capability 生产能力
production capacity investment 生产能力投资
production capacity 生产能力
production control 生产控制，生产管理，产品检查
production cost budget 生产成本预算
production cost 生产费用，生产成本
production counter 产品计数器
production department 生产部，生产部门
production effectiveness 生产效益
production engineer 工艺工程师
production engine 正常系列发动机，标准生产的发动机
production expense 生产费用
production facilities 生产设施
production foreman 生产工长，值班长
production line 生产线
production log 生产记录
production management personnel 生产管理人员
production management 生产管理
production-oriented management 生产型管理
production permit 生产许可
production planning 生产规划
production plan 生产计划
production process 生产过程，生产流程
production prototype testing 生产前试验
production prototype 投产样机
production rate 产量，生产率
production reactor 生产反应堆，生产堆
production record 生产记录
production reserve funds 生产预备费
production run equipment 成批生产设备
production run 产生式程序，生产过程
production schedule 生产计划
production-size 生产规模
production specifications 生产技术要求［条件］
production staff 生产人员
production supervisor 生产工长
production technique 生产技术
production technology section 生技科
production test coupon 产品试样
production test 产品试验
production time 运转时间，工作时间
production type wind tunnel 生产性风洞
production volume 产量
production water supply 生产供水，生产用水供应
production water 生产用水
production weld data sheet 制造焊缝数据单
production weld test assembly 制造焊缝试验组件
production weld test coupon 制造焊缝试样
production weld 制造焊缝
production yield 生产量
production 产品，出产，生产
productive capacity 生产能力，生产率
productive capital project 生产性投资项目
productive construction investment 生产建设投资
productive force 生产力
productive head 发电水头
productive investment 生产投资
productive 生产的，生产性的，多产的
productivity 生产率，生产力，生产能力
product liability 产品责任
product licensing file 产品执照申请档案
product modulator 乘积调制器
product moment correlation 积矩关联
product of combustion 燃烧产物
product of generating functions 生成函数积
product on sale 产品销售税
product quality certificate 产品质量证明
product quality test 产品质量检验
product quality 产品质量
product register 乘积寄存器
product relay 乘积继电器
product requiring high-precision techniques 精加工制成品
product rule 乘法定则
products catalogue 产品目录，产品样本
product size 产品粒度
product spare 产品备件
product specification sheet 产品说明书
product specification 产品规格
product standard 产品标准
product supervision 产品监督
product test 产品试验
product top size 产品最大粒度
product verification 产品验证
product 产品，成品，产物，生成物，积，乘积，结果，成果
prod 触头，刺激，激励，热电偶，温差电偶
proeutectic 先共晶（的）
pro-eutectoid phase 先析相
professional accountant 专业会计师

professional advisor 专业顾问
professional and technological post 专业技术职务
professional auditor 职业审计师
professional certification 专业证书
professional competence 业务能力
professional consultant 专业咨询人员
professional design 专业设计
professional engineer 专业工程师，职业工程师
professional ethics 职业道德
professional experience 专业经验
professional exposure 职业性照射（量）
professional group-automatic control 自动控制专业组
professional group-communication system 通信系统专业组
professional group-electronic computer 电子计算机专业组
professional integrity 职业操守，诚信敬业
professional knowledge 专业知识
professional liability insurance 职业责任（保）险
professional management 专业管理，专业经营
professional organization 行业组织
professional personal integrity 职业操守
professional personnel 专业人员
professional practice 专业操作，专业实践，专业经验
professional qualification 专业资格，业务能力
professional resources exchange 专业人才交流
professional senior engineer 教授级高级工程师
professional service 专业性服务
professionals floating 科技人员流动
professional skill 专业技术
professional staff 专业职员，专门人员，专业人士
professional standard 职业标准，专业标准
professional title assessment 职称评定
professional title 职称
professional vocabulary 专门名词
professional 内行，人才，职业，专业
professorial translator 译审
professor of engineering 教授级高级工程师
professor of translation 译审
proficiency 熟练，精通
profile analysis 线型分析
profile angle 齿形角，翼型角
profile cavitation （叶片）断面汽蚀，叶型汽蚀，断面空蚀
profile chart 轮廓图，透视图，剖面图
profile chord 翼弦
profile component 翼剖面分量
profile correction 齿廓修行
profile curve 轮廓曲线
profile cutting 成形切削，定型切削
profiled bar rectification machine 型钢矫正机
profile diagram 断面图，轮廓图，纵断面图
profile distortion 线型畸变
profiled outline 外形轮廓
profile drag 型阻，叶型阻力，翼型阻力，外形阻力
profile drawings 纵断面图
profiled steel sheet 压型钢板
profile error 齿形误差
profile flow 翼型绕流，叶型绕流
profile gauge 轮廓量规
profile in elevation 高程纵断面
profile in plan 平剖图
profile levelling 纵断面水准测量，纵断面测量
profile lift 翼型升力，叶型升力
profile loss 叶型损失，型面损失
profile machine 仿形机床，仿形铣床
profile meter 表面测量仪，（表面粗糙度）轮廓仪
profile modeling 仿形，靠模
profile modification 齿形修整
profile of ground-water level 地下水位纵剖面
profile of road 道路纵断面
profile of teeth 齿形
profile plotting 断面测图
profile pressure distribution 沿叶型的压力分布
profile recorder 表面轮廓记录器
profile resistance 翼型阻力
profiler 仿形工具机
profile set 翼型系列
profile shape 翼型
profile shifted gear 变位齿轮
profile steel 型钢
profile survey 剖面测量，断面测量，纵断面测量
profile testing instrument 表面光洁度检测仪
profile thickness 翼剖面（最大）厚度
profile tracer 靠模
profile 型线部分【叶片截面】，型，翼型，剖面图，纵断面图，剖面，轮廓，外形，表面光洁度，分布图，侧面图，断面（图），廓线，剖视图，纵断面，叶型，协议集，侧面，简况，描……的轮廓，扼要描述
profilograph 轮廓曲线仪，表面光度计
profilometer 表面光度计
profilooks 发电负荷及预期负荷指示器
profitability 盈利能力
profit and loss statement 损益表，盈亏报表
profit and loss 损益，盈亏
profit appointment 利润分摊
profit capitalized on return of investment 利润归还投资
profit drawing 利润提成
profiting 盈利
profit management 利润管理
profit of enterprise 企业利润
profit rate of investment 投资利润率
profit statement 利润表
profit submitted to the state 上缴利润
profit tax 利得税
profit 利润，收益，红利，盈余
proforma invoice 估价单，形式发票
proforma 形式上的，预计的
profound contradiction 深刻的矛盾
profound impact 深远影响
profound observation 深入观察
profound zone 深海底，深水层
prognosis 预报
prognostic chart 预报图

prognostic wave chart 波浪预报图
prognostic 前兆，预测，预兆
progradation 延伸作用
prograding shoreline 推进岸线
program address counter 程序地址计数器
program analysis 程序分析
program appraisal and review 程序鉴定及检查
program authorization 程序审定
program baby 程序体
program block diagram 程序框图
program block 程序块
program board 程序控制台
program card 程序卡
program change package 程序交换包
program chart 程序框图，程序图
program check interrupt 程序检查中断
program check 程序校验
program chooser 节目选择器，程序选择器
program compatibility 程序兼容性
program compiler 程序编译器，编译程序
program compiling 程序编译
program composition 程序设计，程序编制
program control device 程序控制装置
program-controlled computer 程序控制计算机
program-controlled interruption 程序控制中断
program-controlled network 程控网
program-controlled telephone exchanger 程控电话交换机
program-controlled telephone 程控电话
program-controlled 程序控制的
program controller 程序控制器
program control register 程序控制寄存器
program control system 程序控制系统
program control unit 程序控制部件，程序控制装置
program control 程序控制
program cost estimate 程序价格估计
program counter store 程序计数存储器
program counter 程序计数器
program damage assessment and repair 程序故障的估定及修复
programdation 推进作用，进积作用，冲蚀作用
program debugging 程序调试
program design 程序设计
program development time 程序编制时间
program development 程序编制
program device 程序装置
program display 程序显示
program editor 程序编辑器
program element code 程序单元代码
program element 程序单元
program error 程序误差
program evaluation and review technique 统筹法，统筹安排法，程序估计和检查技术，计划评审法，计划评审技术
program evaluation procedure 程序鉴定过程
program event record 程序事件记录
program exception code 程序异常代码
program-exit hub 程序输出集线器，程序输出插孔
program extension 程序扩展
program flowchart 程序流程图
program flow diagram 程序流程图
program generation system 程序生成系统
program generator 程序生成器
program index register 程序索引寄存器
program information department 程序信息部
program interrupt control 程序中断控制
program interrupt 程序中断
program lending 计划贷款
program library 程序库
program list 程序表
program loading 程序装入
program logic 程序逻辑
program loop 程序周期，程序循环
programmable by software 软件设置
programmable calculator 可编程计算器
programmable communication interface 可编程通信接口
programmable controller control system PC 控制系统
programmable controller 程控器，可编程控制器
programmable control 可编程序控制
programmable counter 可编程计数器
programmable instrumentation 可编程仪表
programmable integrated control equipment 积分程序控制设备
programmable limit switch 程控程位开关
programmable logical controller 可编程逻辑控制器
programmable logic array 可编程逻辑阵列
programmable logic controller 可编程逻辑控制器
programmable peripheral interface 可编程外围接口
programmable power supply 可编程电源
programmable read-only memory 可编程序的只读存储器
programmable ROM 可编程只读存储器
programmable stimulus 可编激励
programmable 可编程（序）的
program maintenance 程序维护
program management network 程序管理网络
programme chart 进度表
programme check 程序校验
programme clock 程序钟
programme composition 程序编制
programmed action safety assembly 按程序动作的安全装置
programmed application library 已编好的应用程序库
programmed automatic circuit tester 程序控制的自动电路测试器
programmed blowing 程序吹灰
programmed checking 程序检验
programmed computer 程序控制计算机
programmed control computer 程序控制计算机
programmed control in time 时（间程）序控制
programmed controller 程序控制器
programmed control 程序控制
programmed digital automatic control 程序数字自动控制
programmed dump 程序转储

programmed electron beam 程序控制电子束
programmed function keyboard 程序控制的操作键盘
programmed guidance 程序控制程序控制制导
programmed halt 程序控制停机
programmed heating 顺序加热，程序控制加热，程序控制升温
programmed inspection 程序控制检查
programmed instruction 程序指令，程序教学，循序渐进的教学
programmed-interconnection pattern 编程连接图
programmed logic array 编程逻辑阵列，程序控制的逻辑阵列
programmed logic 程序控制的逻辑
programmed operator 程序控制的算子
programmed protection 程序控制保护
programmed teaching 程序教学
program memory 程序存储器
programme parameter 程序参数
programmer-defined macroinstruction 程序员定义的宏指令
programmer's console 程序员控制台
programmer 程序设计器，程序设计员，程序编制员，程序器，程序装置，程序员
programme 操作程序，程序，计划，大纲，纲要，进度表，方案，说明书
programming check 程序校验
programming computer 程序设计计算机
programming controller 程序控制器
programming control panel 程序控制盘［板，台］
programming device 程序编制装置
programming documentation 编程资料
programming language 程序设计语言
programming linguistics 程序设计语言学
programming module 程序模块
programming program 自动编程，编程序的程序，编译程序
programming relay 顺序继电器，程序继电器
programming support system 程序设计支援系统
programming system 程序设计系统
programming technique 编程技术
programming tool 程序工具
programming unit 程序装置
programming 程序设计，程序编制，计划，规划，编程，设计
program modularity 程序模块化
program module 程序模块
program of work 工作日程
program operation 程序操作
program order address 程序指令地址
program-output hub 程序输出插孔
program package 程序包
program parameter 程序参数
program priority 程序优先级
program pulse 程序脉冲
program reference table 程序引用表
program regeneration 程序再生
program register 程序寄存器
program regulation system 程序调节系统
program relay 程序继电器
program ring 程序环
program run 程序运行
program segment 程序段
program-sensitive error 程序过敏故障，程序敏感性误差
program-sensitive fault 特定程序故障
program set station 程序设定操作站
program sheet 进度表
program simulator 程序模拟器
program specification 程序要求，程序说明
program stage 程序阶段
program statement 程序语句
program status word 程序状态字
program step 程序步
program stop 程序停止，程序停止指令
program storage unit 程序存储单元，程序存储器
program storage 程序存储器
program structure code 程序结构代码
program support and advanced system 程序支援和改进［先进］系统
program support document 程序支援文件
program switch 程序开关
program tape 程序带
program test group 程序检验组
program testing time 程序检验时间
program testing 程序检验
program test system 程序检验系统
program test 程序测试，程序检验
program time analyzer 程序时间分析器
program time 程序时间
program tracking 程序跟踪
program translation 程序翻译，程序转换
program transmitter 程序发送机
program unit counter 程序单位计数器
program validation 程序验证
program word 程序字
program 程序，大纲，图表计划，说明书
progress chart 施工进度（图）表，进度图表，进度表
progress control 进度控制
progressing pilot streamer 雷电前导波
progress in science and technology 科技进步
progression 级数，数列
progressive approximation 渐近近似法，渐次逼近法，渐进近似法
progressive concentration 逐步提浓，递增浓度
progressive-conversion model 逐级转化模型
progressive corrosion 发展中的腐蚀
progressive deformation 累积变形，连续变形
progressive die 顺序冲模，连续冲模
progressive error 累积误差，累进误差
progressive failure 渐进故障，逐渐破坏，逐步破坏
progressive flow 行进水流
progressive fracture 渐进疲劳断裂，逐渐断裂
progressive grinding 逐步研磨【金相检验】
progressive lap winding 右行叠绕组，前行叠绕组
progressive payment 分期付款，累计支付［付款］，按进度付款
progressive plastic yield 渐进塑性屈服

progressive scanning　顺序扫描
progressive settlement　累积沉陷，逐渐沉陷
progressive slide　递增滑动，逐渐滑动
progressive surface wave　前进表面波
progressive type of tidal oscillation　前进式潮汐波动
progressive wave winding　波绕法，前行波绕组
progressive wave　行波，前进波，行进波
progressive winding　右行绕组，前进绕组
progressive　渐进的，累进的，递增的
progress of technology　技术进步
progress of welding　焊接方向
progress of work　工作进程
progress payment　按进度付款，按进度分期付款，分阶段付款，工程进度款
progress plan　进度安排
progress record　进度记录
progress report　进展报告，进度报告
progress schedule　进度计划，时间表
progress sheet　进度表
progress　进步
PROGR＝program　程序
prohabit　严禁
prohibited area　禁区
prohibited goods　禁运品
prohibited product　违禁物品
prohibited　禁止的
prohibition　禁止，禁令
project acceptance certificate　工程验收证书
project agreement　项目协定
project all risks insurance　工程一切险
project alternative　项目备选方案
project analysis　项目分析
project appraisal　工程评价，项目评估
project approval　项目批准
project area　工程面积，项目区
project assessment　工程评定
project assistance　项目援助
project audit report　方案检查报告，设计检查报告
project benefit　工程效益
project bid　项目投标
project brief　项目概况，项目简介
project budgetary estimate　项目概算，工程预算估计，工程投资概算
project budget　工程投资预算，工程预算，项目预算
project change notice　项目变更
project chief engineer　设计总工程师，项目总工
project classification　项目分类
project communication management　项目沟通管理
project completed　已完成的项目
project completion report　项目完成报告
project completion settlement　工程竣工结算
project components　项目组成部分
project comprehensive evaluation　项目综合评价
project consortium　项目集团
project construction manager　工程施工经理，项目施工经理
project construction schedule　工程建设进度
project consultation　项目咨询
project contracting　工程承包
project contractor　成套设备承包公司，项目承包人，工程承包商
project contract　工程承包，项目合同
project controller　项目管理员
project control　项目管理
project conviction report　竣工报告
project coordinator　工程协调人
project cost estimate　工程费用估算，工程投资估算
project cost management engineer　项目费控工程师
project cost management　项目费用管理
project cost　工程造价，工程费，项目费
project cycle　项目周期
project description　工程说明（书）
project design organization　工程设计组织
project design　项目设计
project detail　项目细节
project development fee　项目开发费
project director　工程负责人
project economic evaluation　项目经济评价
projected area of blade　叶片投影面积
projected area　投影面积
projected cost　预计完工费用
projected curve　投影曲线
projected depth　设计水深
projected display　投影显示器
projected dose　预期剂量
projected flood　设计洪水
projected scale instrument　投影刻度仪表
projected　投影的
project engineering　工程设计，计划工程
project engineer system　项目工程师制度
project engineer　项目工程师，设计工程师，总设计师，规划工程师，设总
project evaluation　工程评定，项目评估
project file　项目文档，项目档案
project financer　项目融资方
project financing　项目融资，项目投资，项目集资
project focus spot　投影焦点
project for equipment updating　设备更新项目
project formulation　项目拟定
project for technical renovation　技术改造项目
project funds　项目资金
project group　项目小组
project human resource management　项目人力资源管理
project identification　方案鉴定，项目确定，项目选定
projectile　抛射体，飞射物
project implementation and supervision　项目执行与监督
project implementation　项目执行，工程实施
project information　项目信息
projecting bar　伸出［突出］的钢筋【混凝土】
projecting figure　投影图
projecting pole rotor　凸极转子
projecting pole　凸极

projecting scaffold　悬挑式脚手架
projecting scoop　（汽轮发电机转子上的）凸出风斗
projecting　投射，凸出的，隆起的，悬臂式，外伸的
project in preparation　筹建项目
project in progress　在建工程［项目］
project integration management　项目综合管理
project interface　项目接口
projection angle　投射角，投影角
projection axis　投影轴
projection centre　投影中心
projection distance　投影距离
projection grid　投影格网
projection lamp　投光灯，投影灯
projection oscillograph　投影示波器
projection rectification　投影纠正
projection room　放映室
projection screen　投影屏幕
projection welding　多点凸焊，凸焊
projection weld　凸焊焊接
projection　规划，投射，投影【如管子对管板的】，投影图，设计，抛射物，突出，突出物，估算，预测
project item　工程项目
projective grid　投影格网
projective welding　凸出焊接
project kick-off meeting　项目启动会，项目开工会
project layout　总体布置
project leader preset director　工程负责人
project leader　项目负责人
project loan　项目贷款
project management system　计划管理系统
project management unit　项目（管理）部
project management　项目管理，工程管理，工程监理
project manager　项目主任，项目经理，工程经理，工程负责人
project manger system　项目经理制度
project manual　工程手册
project master plan　工程总计划［规划］
project master schedule　项目主进度
project milestone　工程里程碑
project miscellaneous materials　工程散料，项目的杂项材料
project negotiation　项目谈判
project objective　项目目标
project of direct electricity supply from power plant to end-user　直供电发电工程
project of highest priority　最优先项目
project organization　工程组织机构
projector type filament lamp　集光型白炽灯
projector　投影仪，聚光灯，投光灯，幻灯，放映机，投射器，设计者
project owner　项目业主
project parameters　工程参数
project participant　项目参加人
project performance detail　项目实施细节
project planning and scheduling　工程计划与进度安排
project planning　工程规划，项目计划
project plan　项目拟定
project precipitation　设计降水（量）
project preparation section　筹建处
project preparation　项目准备
project preparatory department　工程筹建处
project price　工程造价
project procurement management　项目采购管理
project procurement supervisor　设备采购负责人
project procurement　项目采购
project programming　制定项目计划
project progress　工程进度
project promoter　项目发起人
project proposal　项目建议书
project quality management　项目质量管理
project quality　工程质量
project quantity sheet　工程数量表
project reconciliation　项目最终结算
project reporting manual　工程报告手册
project reserve funds　工程预备费
project review　项目审查
project revision　项目修订
project risk management　项目风险管理
project salt vault　盐井计划，盐库计划【放射性废物处置用】
project scale　工程规模
project schedule　工程进度，项目进度，工程进度表，项目进度表
project scope management　项目范围管理
project scope　项目范围
project settlement　工程结算
projects for renovation and expansion　改建扩建工程
project site appraisal　工程地址鉴定
project site inspection　工程地址调查
project site-selection　工程选厂
project site　工地，现场
project siting　工程选场，工程选址
project sponsor　项目主办者
project staff　项目人员
project state　工程状态
project strategy　项目策略
project superintendent　项目主任
project supervision department　项目监理部
project supervision　工程监理，项目监督
project survey　工程测量
project team　工程队，项目组，项目工作组
project time management　项目时间管理
project to project　项目有关各方
project type assistance　项目型援助
project under construction　在建工程［项目］，正在实施的项目
project undertaker　项目建设单位，项目实施单位
project work breakdown structure　工程工作项目分级体系
project　投射，投影，发射，计划，设计，设想，凸出，规划，工程，项目，放映，伸出
pro-line run-off switch　沿线跑偏开关
pro-line safety pull switch　安全拉线开关
pro-line speed switch　沿线速度开关
pro-line tilt switch　沿线倾斜开关

pro-listed companies　准上市公司
prolongation　延长，延期
prolonged agitation　延时搅拌，持续搅动
prolonged erosion test　长期耐蚀试验
prolonged humidity cycling test　长期防潮周期性试验
prolonged low dose rate exposure　长期低剂量照射，长时间低剂量照射
prolonged outage　长期停机
prolonged suspension of project　工程持续停工
prolonged test　连续负荷试验
prolong the period of validity　延长有效期
prolong　延长，外延
promenade　堤顶大路【防波堤上的】，宽廊，通道，走廊，散步广场
prometal　一种耐高温铸铁
prominent　显著的，突出的
promiser　订约人，立约人，许诺者，作出诺言的人
promise to do sth.　承诺做某事
promise　承诺，许诺，允诺
promisor　立约人，订约人，许诺者，契约者
promissory notes　本票，借据，期票
promoters for dropwies condensation　珠状凝结的助聚剂
promoter　促进剂，助催化剂，助聚剂，激发器，催化剂，发起人，主办单位
promote　改进，促销，促进
promotion period　优惠期
promotion　升级，提升，晋升
PROM(programmable read-only memory)　可编程序的只读存储器，可编程只读存储器
PROM programmer　可编程只读存储器的编程器
prompt capture gamma radiation　瞬发俘获γ辐射
prompt cash payment　即期付现，立即付款
prompt coefficient　瞬发系数
prompt critical burst　瞬发临界中子猝发
prompt critical condition　瞬发临界条件
prompt criticality　瞬间临界性，瞬时临界，瞬发临界
prompt critical reactor　瞬发临界反应堆
prompt critical transient　瞬发临界瞬态
prompt critical　瞬发临界，瞬发临界的
prompt delivery　立即交货
prompt dismantling　电站的立即退役
prompt dose　瞬时剂量
prompt drop of reactivity　反应性速降
prompt drop　速降
prompt fission gammas　瞬发裂变γ射线
prompt fission product　瞬发裂变产物
prompt fission　瞬发裂变
prompt flux tilt　瞬发通量斜变
prompt gamma energy　瞬发（裂变）γ能量
prompt gamma radiation　瞬发（裂变）γ辐射，瞬态伽马辐射
prompt gamma ray analysis　瞬发（裂变）γ射线分析
prompt gamma ray　瞬发（裂变）γ射线
prompt gammas　瞬发（裂变）γ射线，瞬发伽马射线
prompt generation time　瞬发（中子）每代寿命，瞬发（中子）每代时间
prompt jump approximation　瞬跳变近似
prompt jump of reactivity　反应性瞬发跳变
prompt jump　瞬跳变，瞬跳
promptly born neutron　瞬发中子
prompt multiplication　瞬发增殖
prompt negative temperature coefficient　瞬发负温度系数
prompt neutron activation analysis　瞬发中子活化分析
prompt neutron area　瞬发中子面积
prompt neutron decay constant　瞬中子衰变常数
prompt neutron fraction　瞬发中子份额，瞬发中子率
prompt neutron generation time　瞬发中子每代时间
prompt neutron lifetime　瞬发中子寿期
prompt neutron multiplication　瞬发中子增殖
prompt neutron reactor　瞬发中子反应堆
prompt neutron spectrum　瞬发中子能谱
prompt neutron　瞬发中子
prompt NO_x　快速温度型氮氧化物
prompt nuclear analysis　瞬发核反应分析
prompt nuclear reaction analysis　瞬发核反应分析
prompt payment discount　立即付款折扣
prompt payment　立即付款
prompt period accident　瞬发周期事故
prompt period　瞬发周期
prompt poisoning　瞬发中毒
prompt radiation　瞬发辐射
prompt reactivity jump　瞬发反应性跳变
prompt reactivity　瞬发反应性
prompt reactor　瞬发中子反应堆
prompt release　即时放行
prompt Rossi alpha　瞬发罗西α
prompt sampling system　瞬时取样系统
prompt shipment　即期装船
prompt shutdown coefficient　瞬发停堆系数
prompt shutdown　瞬发停堆
prompt subcriticality　瞬发次临界
prompt subcritical reactor　瞬发次临界反应堆
prompt supercriticality　瞬发超临界
prompt-supercritical　瞬发超临界的
prompt supercrititcal reactor　瞬发超临界反应堆
prompt temperature coefficient　瞬发温度系数
prompt　敏捷的，迅速的，瞬发的，促使，促进，增长，提升
promulgate　发布，公布，传播
promulgation　颁布，公布，发布
prone to　易于……的，很可能……的，易于遭受……的，有……倾向的
pronged-root blade　叉形叶根式叶片
prong　齿尖，管脚，叉形叶根，叉股
pronounced stall　严重失速
prontour chart　高空预报图
proofed cloth　防水布
proofing　证明，防护
proof load　试验负荷［荷载］，保证负荷［荷载］，检验［验证］荷载，标准（保证）负荷
proof of concept machine　方案验证机
proof of delivery　交付凭证

proof of loss　损失证明
proof of performance　合同已履行的证明
proof-plane　验电板
proof pressure　耐压压力
proofreader　校对员
proof reading　校读
proof sample　样品，试样，标准试样
proof strength of non-proportional elongation　规定非比例延伸强度
proof strength of total extension　规定总延伸强度
proof strength　极限强度
proof stress　试验应力，屈服点，检验应力，弹性极限应力
proof test　验证试验，安全试验
proof voltage　耐电压
proof　证据，证明，校验，校样，试验，论证，检验，凭证
propaganda fire vehicle　宣传消防车
propagate　传导，传播，蔓延，繁殖，普及
propagating shock wave　行冲击波，冲击行波，行激波
propagating wave　传播波，行波
propagation characteristics　传播特性
propagation coefficient　传播系数
propagation condition　传播条件，传播情况
propagation constant　传播常数
propagation distance　传播距离
propagation factor　传播因数，传播系数
propagation loss　传播损耗，传播损失
propagation of sound　声传播
propagation rate of elastic wave　弹性波传播速率
propagation stall　旋转失速，传播失速
propagation vector　传播矢量
propagation velocity　传播速度
propagation　传送，传播，普及，繁殖，扩散，蔓延
propane dichloride　二氯丙烷
propane gas cutting　丙烷气切割
propane storage tank　丙烷储罐
propane torch　丙烷火炬
propane　丙烷
propanone　丙酮
propargyl alcohol　炔丙醇
propellant　发生剂，推进剂，推进物，推进燃料，推进的
propeller advance　螺旋桨进程
propeller agitator　螺旋桨式搅拌器，螺旋搅拌器
propeller blade current meter　旋桨（式）流速仪
propeller blade　螺旋桨式叶片，螺旋桨叶
propeller boss　桨毂
propeller control switch　推进器控制开关
propeller current meter　旋桨流速仪
propeller discontinue　桨叶旋转面上气流不连续性
propeller fan　螺旋桨式风扇，旋桨式扇风机，轴流式风机
propeller generator　风车发电机，螺旋桨风力发电机
propeller governor　推进器调速器
propeller hub　螺旋桨毂，螺旋桨桨毂
propeller inserts　螺旋桨式插入件
propeller mixer　螺旋桨搅拌器
propeller motor　螺旋桨（驱动用）电机
propeller pump　轴流式泵，螺旋泵，螺旋桨式水泵，旋桨式水泵
propeller release　旋桨排气装置
propeller shaft turning wrench　螺旋桨轴扳手
propeller turbine　螺旋桨式水轮机，螺旋涡轮机
propeller type anemometer　螺旋桨风速计
propeller type current meter　叶轮流速计
propeller type fan　轴流式风机
propeller type impeller　螺旋桨式叶轮，推进器式叶轮，旋桨式叶轮，轴流叶轮
propeller type mixer　螺旋桨式拌和机
propeller type pump　螺旋桨式泵，轴流泵
propeller type turbine　螺旋桨式水轮机，螺旋桨式涡轮机，轴流定桨式水轮机
propeller type wind machine　螺旋桨型风力机
propeller vane anemometer　螺旋桨风速计，螺旋桨叶片风速计
propeller vane　螺旋桨式叶片
propeller water turbine　定桨式水轮机
propeller　螺旋桨，推进器，螺旋桨推进器，旋桨，叶轮
propensity to invest　投资倾向
propensity　倾向，习惯
proper alignment　同心度
proper authorities　有关方面，相关当局
proper brown coal　正褐煤
proper conduction　固有电导
proper equation　特征方程
proper field　固有场
proper function　本征函数，特征函数，正常函数
proper law　准据法
properly posed problem　适定问题
proper node　正常结点
proper operation　正确操作，合适的运行，合乎规程的运行，正常运行
proper oscillations　固有振动
proper phasor　特征相量
proper response　适当反应
proper semiconductor　固有半导体，本征半导体
proper shock　本震
proper shutdown　最后停堆，退役
properties division　财产分割
property assets　不动产
property boundary　厂界
property insurance contract　财产保险合同
property insurance　财产保险
property line　建筑红线，地界线
property mechanism　产权机制
property ratio method　物性比法
property right　财产所有权，财产权，产权
property tax rate　财产税率
property tax　财产税
property test　性能试验
property　特性，性能，性质，资产，财产，所有权
proper use factor　适当利用因子
proper value　本征值，固有值，恰当值
proper vector　特征矢量
proper　本体，本征的，适当的

prophylactic repair 定期检修，预防性检修，预防检修
propionaldehyde 丙醛
propionic acid 丙酸
propogation delay 传播延迟
proponent 建议人
proportional accumulator 比例积算器
proportional action coefficient 比例作用系数
proportional action 比例作用
proportional band 比例区，比例带，线性范围，比例范围，比例区域
proportional component 比例元件
proportional control action 比例调节作用
proportional control factor 比例调节系数
proportional controller 比例控制器，比例调节器
proportional controlling means 比例控制方法
proportional control relay 可变控制继电器
proportional control 比例控制，比例调节
proportional counter tube 正比计数器，正比计数管，比例计数管
proportional counter 比例计数器
proportional distribution 比例分配
proportional divider 比例分配器，比例规
proportional elastic limit 弹性比例极限
proportional error 相对误差，比例误差
proportional feedback 比例反馈
proportional flow controller 比例流量调节器
proportional flow control unit 比例流量调节器
proportional flow counter 气流式正比计数管
proportional function 正比例函数
proportional gain 比例增益
proportional governor 比例调节器
proportional heater 比例加热器，可调式加热器
proportional-integral controller 比例积分控制器，PI 控制器
proportional-integral control 比例积分调节
proportional-integral-derivative control 比例积分微分控制，PID 控制
proportional-integral 比例积分
proportionality coefficient 比例因数
proportionality constant 比例常数
proportionality factor 比例因子，比例常数，比例系数
proportionality limit 比例极限
proportionality range 比例范围
proportionality 比值，比例，相称，均衡
proportional limit 比例极限
proportional liquid sampler 比例液体采样器
proportional part 比例部分
proportional plus derivative action 比例微分作用
proportional plus derivative control 比例微分控制，PD 控制
proportional plus derivative rate control action 比例微分调节作用
proportional plus derivative rate controller 比例微分调节器，比例微分控制器
proportional plus derivative 比例微分
proportional plus floating control 比例加无静差调节，比例无差调节，比例加无静差控制
proportional plus integral action 比例积分作用
proportional plus integral control action 比例积分调节作用
proportional plus integral controller 比例积分调节器
proportional plus integral control 比例积分控制，比例积分调节，PI 控制
proportional plus integral plusderivation controller 比例积分微分调节［控制］器
proportional plus integral plus derivative action 比例积分微分作用
proportional plus integral plus derivative control action 比例积分微分调节作用
proportional plus integral plus derivative controller 比例积分微分调节器
proportional-plus-integral-plus-derivative control 比例积分微分控制，PID 控制
proportional plus integral plus derivative 比例积分微分
proportional plus integral 比例积分
proportional power unit 比例执行部件
proportional pressure reducing valve 定比减压阀
proportional pressure reducing 定比减压
proportional quantity 比例量
proportional region 比例区域，正比区
proportional sampler 比例采样器
proportional sampling device 比例取样装置，比例取样器
proportional scale 比例尺
proportional speed floating controller 比例速度浮动控制器，比例速度浮动调节器
proportional tax 比例税
proportional totalizer 比例积算器
proportional weir 比例式堰
proportional 成比例的，相称的，比例数，比例项
proportion band of a controller 控制器的比例带，调节器的比例带
proportioner 比例装置，比例调节器，配量设备，比例器，配料器，定量（给料）器
proportioning by absolute volume 按绝对体积配合（法）
proportioning by arbitrary assignment 按经验配合法
proportioning by grading chart 按级配图配合（法）
proportioning by mortar void method 灰浆孔隙配合
proportioning by trial method 试配法
proportioning by volume 按容积配合法
proportioning by weight 按重量配合（法）
proportioning controller 比例控制器
proportioning damper 分配挡板
proportioning device 配料设备
proportioning meter （水处理加药）剂量计
proportioning motor 比例电动机
proportioning of waste 废料配比
proportioning plant 配料车间
proportioning pump 配料泵，比例泵，剂量泵
proportioning tank 计量箱，加药箱
proportioning valve 比例调节阀
proportioning vessel 配比容器
proportioning wheel （装料机）配料轮

proportioning 按比例配合，成比例，配量
proportion, integral, differentiation 比例、积分、微分
proportionment 比例，按比例分配，相称
proportion of equipment in good 设备完好率
proportion of ingredient （混合物的）成分比例
proportion of resin present 树脂含量
proportion 比例，比率，比，部分，相称
proposal documentation 投标文件，报价文件
proposal for project 项目建议书
proposal preparation 方案准备
proposal version 报价版
proposal 投标书，报价（书），方案，建议（书），提案，议案，用户技术建议
propose an amendment 提修改建议
proposed access road 拟建入口道路
proposed dam axis 推荐坝轴线
proposed life 拟用年限
proposed price 投标价
proposed project 拟建工程
proposed site 待选坝址，推荐的场址
proposed 拟定的，拟建的
propose 提议
propositional logic 命题逻辑
proposition 主张，提议，建议，命题，定理，论题，提要
propping 支撑，支柱，支架，设支撑，加固
proprietary name 专利商标名
proprietary product 专利产品
proprietary program 专有程序，特许程序
proprietary rights 所有权，产权
proprietary system 企业自备系统
proprietary technique 专利技术
proprietary 所有的，专有的
proprietorship 所有权
proprietor 所有人，东道主
propriety 适当
PRO=protection 保护
prop stay 支架
prop test rig 支柱试验装置
propulsion efficiency 推进效率
propulsion force 推进力
propulsion machinery 推进机械
propulsion motor 推进电动机
propulsion power 推进功率
propulsion reactor 推进用反应堆
propulsion wind tunnel 推进试验风洞
propulsion 推进器，推进，推力，发动机，驱动
propulsive coefficient 推力系数，推进系数
propulsor 推进器
propyl alcohol 丙醇
propylbenzene 丙苯
propylene 丙烯
prop 支撑，支柱
prorata basis 按比例
prorata freight 比例运费，按里程比例计费
prorata gratuity 按比例领取酬金
pro rata 按比例（分配的），成比例，按比例的，成比例的
prorata 按比例分配，摊派，按比例分配的，成比例地
prorate average 按比例均摊
prorated joint rate 按比例分摊联合运费
prorated section 分截，分段
prorated unit 分截
prorate tax rate 比例税率
prorate 按比例分配，按比例分派，摊派
proration 按比例分配
prosecution 诉讼
prospect for cooperation 合作前景
prospect hole 探孔
prospecting by boring 钻探
prospecting shaft 探井
prospecting 勘察，勘探
prospection 勘探，探测
prospective bidder 有希望投标人，潜在投标人，预期的投标人
prospective current 预期电流，远景电流
prospective design 勘察设计
prospective earnings 预期收益，远景收益
prospective peak value 远景峰值，预期峰值
prospective study 前瞻性调查，前瞻性研究，远景调查，远景研究
prospective survey 前瞻调查，远景调查
prospective yield 预期收益
prospective 预期的，未来的，可能的，有希望的
prospect pit 探坑
prospectus 计划任务书，计划书，任务书，说明书
prospect 展望，远景，前景
pross heating installation 、生产工艺热用户
pross heating load 生产工艺热负荷
prosthesis 取代，置换
protactinium absorption 镤吸附
protactinium poisoning 镤中毒
protactinium 镤【Pa】
protected against corrosion 防腐【电机防护等级】
protected against dripping water 防滴，防滴式【电机防护等级】
protected against explosion 防爆【电机防护等级】
protected against splashing 防溅【电机防护等级】
protected against spraying 防淋水【电机防护等级】
protected against the effects of immersion 防浸水
protected against water heavy seas 防海浪【电机防护等级】
protected against water jet 防喷水【电机防护等级】
protected area 保护区（域）
protected data 保护数据
protected grounding network 保安接地网
protected location 保护存储单元
protected machine 防护式电机，防护型电动机
protected monitor 避风天窗
protected motor 防护型电动机，密封马达，防火马达
protected probe 有保护膜的探头
protected stairway 疏散楼梯
protected switch 盒式开关，防护型开关
protected type 屏蔽式
protected zone 保护区，保护区域，防护带
protected 受保护的
protecting cage 防护梯笼

protecting coating　保护涂层
protecting cover　防护罩
protecting device　保护设备，保护装置，保安装置
protecting equipment　保护装置
protecting mask　保护面具
protecting net　保护网
protecting planting　防护造林
protecting plate　护板
protecting rack　保护格栅，保护格
protecting screen　防护屏，保护屏
protecting sleeve　防护套管
protecting transformer　保护用互感器
protecting tube　保护管
protecting wall　防护墙，挡土墙
protection action　保护动作
protection activation information　保护启动信息
protection actuation value　保护动作值
protection against corrosion　防腐
protection against lightning　防雷保护，防雷
protection against radioactivity　放射性防护
protection against rotor ground　转子接地保护
protection against rust　防锈
protection against single-phasing　单相保护
protection against unsymmetrical load　不对称负载保护
protection and indemnity risk　保赔责任险
protection-apron　护坦
protection cabinet　保护柜
protection cap　防护罩，保护帽
protection channel　保护通道
protection circuit　保护回路，安全回路，保护电路
protection coat　防护层
protection code　保护码
protection course　保护层
protection cover　保护罩，保护壳
protection device against fault switching　开关误操作保护装置
protection device test　保安装置试验
protection device　防护装置，防护设备，保护设备，保安装置，安全装置
protection diagram　保护原理图
protection earthing　保护接地
protection embankment　防护堤
protection equipment　保护设备，专设安全设备，安全装置
protection factor　防护因子，防护因数
protection fault relay　故障保护继电器
protection forest　防护林，保护林
protection for interturn short-circuits　匝间短路保护装置
protection for large generator-transformer units　大型发变组保护
protection guide　保护导管
protection and interlock　保护与联锁
protectionism　保护贸易政策
protection level at steep current　陡波冲击保护水平
protection lever　保护等级
protection of water induction　防进水保护，防进水
protection of water resources system　保护水资源制度
protection panel　保护盘，保护屏
protection period　保护期
protection ratio　保护率
protection relay panel　继电保护屏
protection relay　保护继电器【超电压，频率等】
protection sleeve　保护套管
protection spectacles　护目镜
protection status　保护现状
protection survey　保健物理监测，辐射防护监测
protection switch　保护开关
protection system　保护系统
protection test　保护试验
protection valve　安全阀
protection wall　防护墙
protection works　防护工程
protection　保护，防护，保护措施，防御，保护装置
protective action　预防性行动，保护作用安全作用
protective agent　防护剂，抗氧化剂
protective apparatus　防护器械
protective apron　防护围裙
protective atmosphere　保护气氛
protective band　保护箍
protective barrier　防护屏障，防护栅障，保护屏障，防护屏
protective breaker　保护断路器
protective cap　保护帽，护罩
protective casing　保护外壳，保护罩
protective choke　扼流线圈，抗流线圈，防干扰线圈
protective circuit breaker　保护断路器
protective circuit for thyristor　晶闸管保护电路
protective circuit　保护电路
protective clause　保护性条款
protective clothing　防护罩，防护衣具，防护衣
protective coating　保护层，防护涂层，护面
protective colloid　保护胶体
protective coloration　保护色
protective covering of cable　电缆外护层
protective covering　保护层，保护涂层，护面
protective cover of vegetation　防护性植被
protective cover　防护盖，防护罩，保护层，保护盖，保护罩，覆盖
protective device test　保护装置试验
protective device　安全装置，保护装置，防护装置
protective duty　保护关税
protective earthing　保护接地
protective effect　保护效应
protective enclosure　屏蔽室
protective end cap　保护端盖
protective envelope　防护外壳
protective equipment　保护设备
protective facing　护面
protective film　保护层，保护膜
protective filter　防护滤层
protective forest belt　防护林带，护田林带
protective gap　保护间隙，保护性间隙，安全

间隙
protective gas 保护气体
protective gear 防护装置，保护装置
protective glove 防护手套
protective goggles 护目镜
protective groin 护岸丁坝
protective ground 保护性接地
Protective groyne 护岸丁坝
protective head gear 保护的顶端齿轮
protective helmet 安全帽，防护帽，钢盔
protective hood 防护罩
protective housing 保护罩，防护式外壳
protective insulation 外绝缘，保护绝缘
protective interlocks 防护性联锁，保安互锁装置，保安联锁装置
protective jacket 保护套【围绕绝热的管子】
protective layer 保护层，防护层
protective lead-glass 防护用铅玻璃
protective lighting 夜间防护照明
protective link 保险连杆，保护线路，保险丝
protective magnetite layer 磁性氧化铁保护层
protective margin 保护裕度
protective mask 防护面罩
protective measure 防护措施
protective multiple earthing 保护性多点接地
protective oxide film 氧化物保护层
protective pack 包装
protective policy 保护政策
protective potential 保护电位
protective principle 保护原理
protective ratio （避雷器的）保护比
protective redundancy 保护冗余度
protective rejector circuit 保护性抑制器电路，保护性带阻滤波器电路
protective relaying 继电保护
protective relay panel 保护继电盘
protective relay 保护继电器
protective ring 保护环
protective screen 防护屏，防护屏蔽，保护屏，保护屏蔽
protective sea wall 护岸海堤墙
protective service 防护设备，预防设施
protective sheath （变压器一、二次绕组间的）接地屏蔽
protective shelter 防护罩
protective shield 防护屏蔽，防护屏，防护屏蔽体
protective shoe cover 保护鞋套
protective shroud 护罩
protective sleeve 防护套管，防护套轴
protective storage 监护封存，（退役第一阶段）封存
protective structure 防护建筑物，防护结构
protective suit 防护衣，潜水服，防护服，防护衣具
protective system false operation rate 继电保护误动作率
protective system 保护装置，保护系统
protective tariff 保护关税
protective thickness 防护厚度
protective trade 保护贸易
protective transformer 保护变压器
protective valve 安全阀
protective varnish 防护漆
protective zone 保护范围，保护区，防护区
protective 防护的，保护的
protect market 受保护市场
protector protective gear 保护装置
protector tube 管形避雷器
protector 防护罩，保护装置，防护剂，保护器，防护器
protect relay 保护继电器
protect 保护，防止
protein foam concentrate 蛋白泡沫液
protein nitrogen 蛋白质氮
protein 蛋白质
protest a bill 拒付票据
protest 抗议
protoclase 原生解理
protoclastic structure 原生碎屑结构
protoclastic texture 原生碎屑结构
protocol analysis system 协议分析系统
protocol conversion 协议转换
protocol converter 协议转换器
protocol data unit 协议数据单元
protocol implementation conformance statement 协议实现一致性陈述
protocol implementation extra information for testing 测试用协议实现附加［额外］信息
protocol machine 协议机
protocol （数据通信的）规约，议定书，协议，约定，草案，礼仪，会谈备忘录
proto magma 原始岩浆
protomalenic acid 原烯酸
proton capture 俘获质子，质子俘获
proton detection 质子探测
proton gas model 质子气体模型
protonic solvent 质子性溶剂
proton number 质子数
proton separation energy 质子分离能
proton 质子
prototype and first part tests 原型和第一次部件试验
prototype atmospheric wind 原型大气风
prototype core 原型堆芯，模式堆芯
prototype filter 原型滤波器
prototype furnace 原型炉
prototype gradation 天然级配
prototype machine 样机，原型电机
prototype nuclear process heat plant 原型核工业供热厂
prototype observation for concrete dam 混凝土坝原型观测
prototype observation for underground structure 地下建筑物原型观测
prototype observation 原型观测
prototype plant 原型工厂，样板厂，实验装置
prototype project 原型项目
prototype reactor 原型反应堆
prototype Reynolds number 原型雷诺数
prototype rotor 原型转子
prototype structure 原型结构

prototype test　原型试验，样机试验
prototype turbine　原型水轮机
prototype wind speed　原型风速
prototype　原型，原型物，模型，样品，样机，典型，试制品，实验性的，试制的
protoxide　氧化亚物，低价氧化物
protozoan　原生动物
protracted test　疲劳试验
protraction dose　迁延剂量
protractor　量角器，分度规，分度器，分规
protrude　突出，伸出，凸出，使凸出
protruding nozzle　插入式接管，内伸式接管
protruding plug　凸出塞
protruding tube　凸出管子
protruding　凸出的，隆起，浮雕
protrusion　伸出，凸出，凸起，凸出物
protuberance　突出部，突起，节疤，突出物，凸度
provenance　起源
proven components　经考验的部件
proven design　成熟的设计，经验证的设计，已有的设计
proven equipment　经过验证的设备
proven experience　成熟经验
proven reactor type　成熟堆型
proven reserves　勘定储量
proven technique　成熟技术
prover of ball flow measuring device　球形流量测量仪表校验装置
prover of flow measuring device　流量测量仪表校验装置
provided hole　预留孔
provided that　只要，如果，假如，倘若，以……为条件
provided with　装有，配有，备有，设有
provide for　保证，规定，供给
provide M with N　给［向］M 提供 N，给 M 装备 N，把 N 装到 M 上
provider　提供方
provide　提供
provincial capital　省城，省会
provincial civilized unit　省级文明单位
provincial construction commission　省建委
provincial electric power bureau　省电力局
provincial government　省政府
provincially administered municipality　省辖市
provincial medal　省级奖（章）
provincial people's government　省人民政府
provincial planning commission　省计委
provincial state-owned enterprises　省级企业
proving program　验证程序
proving ring　测力环，量力环
proving run　试车前检查，试运行
proving trial　检验
provisional acceptance report　临时验收报告
provisional acceptance　临时验收，暂时验收
provisional agenda　临时议程
provisional agreement　临时协议
provisional clause　临时性条款
provisional contract　临时合同
provisional coordinates　资用坐标
provisional cost　暂列金，临时列入的成本
provisional environmental management plan　临时环境管理计划
provisional government　临时政府
provisional invoice　临时发票
provisional licence　临时许可证
provisional list　临时性清单
provisional measures for the administration　管理暂行办法
provisional measures　临时措施
provisional method　暂定方法
provisional point　临时点
provisional regulation　暂行条例
provisional rules of procedure　临时程序规则
provisional subcommittee　临时小组委员会
provisional sum　（合同价里的）暂定金（额），备用款，不可预见费用，暂列款项
provisional take-over certificate　临时移交证书
provisional value　初步值
provisional version　临时版本，临时文本
provisional wiring　临时布线
provisional　临时的，暂时的，暂定的，初步的
provision as to payment　付款规定
provision in the contract　合同规定
provision of agreement　协议条款
provision of water，electricity，gases，heat，communication and machinery for construction　力能供应
provisions as to arbitration　仲裁规定
provision to salvage reusable oil　可再用油回收装置
provision　规定，条款，提供，供应，供应品，储备，预备，设备
provisory clause　除外条款，限制性条款词句，附带条款
proviso　限制性条款
provitization programme　私有化方案［计划］
provocation　激发
provoke　引起，诱发
proximate analysis　工业分析，（燃料）近似分析，组分分析，实用分析
proximate and ultimate analysis　组分和元素分析
proximate　近似的，最接近的，即将发生的
proximitor　前置器
proximity detector　渐近指示仪
proximity effect　邻近效应，邻线影响
proximity meter　近似测量仪
proximity of zero order　零阶逼近
proximity sensor　接近度（防撞）传感器，近距离传感器
proximity switch　接近开关，极近开关，临近开关
proximity type　接近型
proximity　逼近，近似，接近，临近，前牵
proxy　代理人，代理权，委托书
PR(performance report)　性能报告
PRP(pseudo-random pulse)　伪随机脉冲
PR(pressure ratio)　压力比
PR (pressure recorder)　压力记录计
PR(program register)　程序寄存器
PR(progress report)　进展报告
PRR(pulse-repetition rate)　脉冲重复频率

PRS(pattern recognition system) 模式识别系统
PRS(prototype retrieval system) （美国）原型废物可回取系统
PRTD(platinium resistance temperature detector) 铂电阻温度检测器
PRT(platinium resistance thermometer) 铂电阻温度计
PRT(program reference table) 程序引用表
prudent investment 审慎的投资
prudent resource utilization 资源节约利用
PRV(peak reverse voltage) 峰值反向电压
PRV(pressure reducing valve) 减压阀
PRV(pressure relief valve) 过压保护阀，泄压阀
pry bar 撬棒
PSAFC(polysilicate aluminium ferric chloride) 聚硅酸氯化铝铁【絮凝剂，混凝剂】
psammite 砂屑岩
psammitic structure 砂屑构造
PSA(probabilistic safety analysis) 概率安全分析
PSA(probabilistic safety assessment) 概率安全评价［评估］
PSAR(preliminary safety analysis report) 初步安全分析报告
PSAS(program support and advanced system) 程序支援和先进系统
PSCD(probability of severe core damage) 严重堆芯损伤概率
PSCE(peripheral storage control element) 外存储器控制部件
P-scope 平面位置显示器，P型显示器
PSC(program structure code) 程序结构代码
PSDD(preliminary system design description) 初步系统设计描述
PSD(position-sensitive detector) 位置灵敏探测器
PSD(power spectral density) 功率谱密度
PSD(program support document) 程序支援文件
P-semiconductor P型半导体，空穴导电型半导体
PSE(packet switch equipment) 包交换设备
psephicity 磨圆度
psephite 砾质岩
psephitic structure 砾状结构
psephitic 砾状的
PSE(Pressurized Subcritical Experiment) 加压次临界试验反应堆【美国】
pseudo adiabatic convection 假绝热对流
pseudo adiabatic expansion 假绝热膨胀
pseudo adiabatic lapse rate 假绝热直减率
pseudo adiabatic process 假绝热过程
pseudobinary 伪二元的
pseudo boiling 似沸腾
pseudo-burnup 准燃耗
pseudo-code 伪代码
pseudo-complement 伪余
pseudoconformity 假整合
pseudocritical temperature region 似临界温度区
pseudocritical temperature 似临界温度
pseudocritical 似临界的
pseudocrystalline 假晶的
pseudodielectric 赝电介质，准电介质
pseudo-dynamic analysis 准动力分析
pseudo-equilibrium 准平衡，假平衡
pseudofilm boiling 似膜态沸腾
pseudofluidization 拟流态化
pseudo fluid 假流体，赝流体
pseudo instruction 伪指令
pseudo-language 伪语言
pseudo-linear 伪线性的，假线性的
pseudonoise sequence 伪噪声系列
pseudo-norm 伪模
pseudo operation 虚拟操作，伪运算，伪操作
pseudo order 伪指令
pseudo-particulate expansion 假散式膨胀
pseudo-period 伪周期，赝周期
pseudo-polymetric structure 假聚合结构
pseudo potential flow 假位流，赝势流
pseudo-primary thermometer 准原始温度计
pseudo-program 伪程序
pseudo-radial equilibrium 假径向平衡，准径向平衡
pseudo-random binary sequence 伪随机二进制序列，伪随机二进制数列
pseudo-random code 伪随机码
pseudo-random force 假随机力
pseudo-random noise signal 伪随机噪声信号
pseudo-random noise 伪随机噪声
pseudo-random number 伪随机数
pseudo-random perturbation 伪随机扰动
pseudo-random process 伪随机过程
pseudo-random pulse 伪随机脉冲
pseudo-random sequence 伪随机序列
pseudo-random 伪随机的，伪随机，赝随机，虚拟随机，伪随机数
pseudoscalar quantity 假标量，伪标量
pseudoscalar 伪标量
pseudo-shock 伪冲击波
pseudo-static method 拟静力法
pseudostationary flow 准稳定流动
pseudostationary state 似稳状态
pseudostationary 伪稳态的，假稳的，准稳定的，伪正常的
pseudo-steady state 假稳态
pseudo-turbulent bed 假湍流床
pseudovector 伪矢量，轴矢量
pseudo-viscous flow 拟黏滞流
pseudo 伪的
PSH 屏式过热器【DCS画面】
psia(pounds per square inch absolute) 磅每平方英寸【绝对压力】
psig(pounds per square inch gauge) 磅每平方英寸【表压】
psi(pounds per square inch) 磅每平方英寸
PSI(preservice inspection) 役前检查
PSK(phase-shift-keyed) 相移键控，移相键控
PSMS(power shape monitor system) 功率形状监测系统
P-S-N curve P-S-N曲线
PSN(packet switch network) 包交换网
psophometer 噪声电压计，噪声计，电噪声计，测听计，声级计，杂音计
psophometric electromotive force 杂音计电动势
psophometric voltage 估量噪声电压

PS＝polystyrene　聚苯乙烯
PS(price sheet)　价目单，价目表
PS(project service)　工程服务
PSRP(power sector restructuring plan)　电力行业重组计划
PSR(processor status register)　处理机状态寄存器
PSRV(pressurizer safety relief valve)　稳压器安全释放阀
PSSCD(probability of subsequent severe core damage)　接着发生的严重堆芯损伤概率
PSS(peak season surcharge)　旺季附加费
PSS(power system stabilizer)　电力系统稳定器
PSS(primary sampling system)　主取样系统
PSS(product specification sheet)　产品说明书
PSS(programming support system)　程序设计支援系统
PSU(program storage unit)　程序存储器，程序存储单元
PSV(pressure safety valve)　压力安全阀
PSW(program status word)　程序状态字
PSWS(potable and sanitory water system)　饮用水和卫生水系统
psychrometer effect　干湿效应
psychrometer　干湿球湿度计，湿度表，湿度计
psychrometric chart　湿度计算图，焓湿图
psychrometric difference　干湿球差
psychrometric table　干湿球湿度表
psychrometry　测湿学，空气湿度测定法
P. S.　邮政原本，附言，又及【信件正式结尾后附加的字句】
PTA(program time analyzer)　程序时间分析器
PT balance protection　电压互感器平衡保护
PT circuit broken　电压互感器回路断线
PTC(pulse time code)　脉冲时间码
PTC(positive temperature coefficient) thermistor　正温度系数热敏电阻
PTE(peculiar test equipment)　特殊测试设备
PTE　葡萄牙埃斯库多【货币单位】
PTFE envelope gasket　聚四氟乙烯包复垫片
PTFE(polytetrafluoroethylene)　聚四氟乙烯
PTFE sliding plate　聚四氟乙烯滑动板
PTG(program test group)　程序检验组
PTM(pulse time modulation)　脉冲时间调制
pto coupling　动力输出轴联轴节
pto driven　动力输出轴驱动的
pto power　动力输出轴功率
peak to peak　从峰值到峰值
PT(potential transformer)　电压互感器
PTP(point-to-point)　点对点
PT(pressure transmitter)　压力变送器
Pt－Rh electrode　铂铑合金电极
PTR(pool test reactor)　游泳池式试验性反应堆
PTR　反应堆水池和乏燃料水池的冷却和处理系统【核电站系统代码】
PTSA(pure time-sharing assignment)　纯分时指定
P-T saturating curve　压温图饱和线
PTS(pressurized thermal shock)　加压热冲击
PTS(program test system)　程序检验系统
PTT(push-to-talk)　按键通话
P-type　P型的
P-type semiconductor　P型半导体
P-type transistor　P型晶体管
PTZ(pan tilt zoom)　平移倾斜变焦【摄影机】
public acceptance　公众认可，公众接受性
public accumulation　公共积累
public address system　有线广播系统
public address　公共地址
public aid　公共补助金
publication　出版
public authority　公共管理机构
public bidding　公开投标，公开招标
public bid opening　公开开标
public bid　公开招标［投标］
public budget　公共预算
public data network　公用数据网
public disaster　公害
public domain　公众领域
public electricity supply　公用电气事业
public enterprise　公营企业
public facility　公共设施
public figure　知名人士
public goods　公共物品
public hazard　公害
public health survey　公共卫生情况调查
public health　公共卫生
public information　公开信息
public investment　公共投资
publicity　名声
public land　公共用地
public law　公法
public liaison office　公众联络处
public library　公用程序库
public lighting circuit　路灯线路
public lighting　公共照明，街道照明
public limited company　股份有限公司，股票上市公司
public meeting　公开会议，公众会议
public money　公款
public nuisance　公害
public offering　公开发售
public offer　公开报价
public participation　公众参与
public place　公共场所
public power station　公用发电厂
public price hearings　价格听证
public pricing　公共定价
public procurement law　（欧盟）公共采购法
public procurement　公共采购
public reservoir　公用水库
public road　公共道路
public safety　公共安全
public sanitary engineering　公共卫生工程
public security bureau　公安局
public security　公安
public service facilities　公共服务设施
public space　公共场所
public telegram　公用电报
public telephone card　公用电话磁卡
public telephone network　公用电话网
public telephone system　公用电话系统
public tender procedures　公开招标程序

public tender 公开招标，公开投标
public transportation system 公共交通系统
public use of water 水的公用
public utilities pipeline and cable 公用事业设施管道及电缆
public utilities regulatory commission 公共事业管委会
public utility and service 公用事业和服务机构
public utility network 公用电网
public utility plant 公用事业电站
public utility power plant 公用事业发电厂
public utility 公用事业公司，公用事业，公共设施
public water consumption 公用耗水量
public water supply 公用给水
public waters 公共水域
public waterway 公用水道
public water 公用水
public welfare facilities 公用福利设施
public welfare fund 公益金
public welfare 公共福利
public wharf 公用码头
public works 公用工程，市政工程
public 公众
Pu burner 钚燃料反应堆，全钚堆芯
puckle 一种锯齿波振荡电路
puddle cofferdam 夯土围堰
puddle cut-off 夯土截水墙
puddled backfill 捣实回填，夯实回填
puddled clay 夯实黏土
puddled core 夯实胶土心墙，夯实土质心墙
puddled wall 夯实胶土墙
puddle dyke 黏土堤
puddler 捣拌机，捣实器
puddle welding 堆焊
puddle 稠黏土浆，捣拌，水坑
puddling 捣成泥浆，黏土夯实
puffback 逆喷吹，反吹
puff diffusion 喷团扩散
puffer breaker 六氟化硫吹弧断路器
puffing plume 喷团型羽流
puff model 喷团模式
puff of smoke 烟团
puff release rate 烟团释放率
puff rope 拉绳
puff source 喷团源
puff superposition model 喷团叠加模式
puff trajectory 喷团轨迹
puff 爆燃，爆音，扑灭，喷团，烟团
PUFR(peripheral unit fault recognition) 外围设备故障识别
pugger 耐火泥塞
pugging 隔声材料，隔声层
pug-mill type mixer 捏拌机
pugmill wet unloader 湿灰搅拌卸载装置
pugmill 搅土机，搅拌机，拌泥机
pulcom 小型晶体管测微指示表
pull angle 牵引角
pull blade 牵引板【犁电缆沟用】
pull box 引线盒，分线箱，引线箱
pull button 拉钮
pull cord switch 拉线开关，拉绳开关
pull-down current 反偏电流
pull-down menu 下拉式菜单
pull down 拆除（建筑物），下拉，拉开
pulled-out control rod 抽出的控制棒
pulled-type coil 穿入式线圈，插入式线圈
puller and tensioner 张力牵引机
puller hoist 拉出器起重机械，手动葫芦
puller 牵车机，拉出器
pulley assembly 滑轮总成
pulley bearing resistance 滚筒轴承阻力
pulley block 滑车，滑轮组
pulley class gate lifting device 滑车类闸门启闭机
pulley cleaner 皮带轮清扫器
pulley lagging 滚筒包胶
pulley motor 皮带轮电动机
pulley shaft 提升井，传动滑轮轴
pulley tackle 滑车组，辘轳
pulley torque 牵入扭矩
pulley 皮带轮，滑车，滑轮，转向轮
pull grader 拖拉式平地机
pulling compound 拉延膏
pulling eye （电缆牵引用）引线孔，拉环，牵引环
pulling in parallel operation 牵入并联运行
pulling into step 牵入同步
pulling into synchronism 牵入同步
pulling-in wire 牵拉线
pulling out of step 牵出同步，失步
pulling out of synchronism 失步，牵出同步
pulling tension 牵引拉力
pulling test 张拉试验
pulling 拉拔，牵引
pull-in range 同步区，牵入范围，同步范围
pull-in step 使同步，牵入同步
pull-in test 整步试验，牵入试验
pull-in torque 整步转矩，牵入转矩，牵引力矩
pull-in value 吸入值，动作值
pull-in winding 插入式绕组，穿入式绕组
pull-in 引入，同步引入，进入同步频率，接通
pull-lift 轻便提升设备，神仙葫芦，牵拉线
pull-off cable 拖出的电缆
pull-off 使失步，使不同步，拐弯架线拉张器
pull-on value 吸合值，动作值
pull-out bond test （钢筋）拔拉握裹试验
pull-out force 拔力，拉出力
pull-out fuse 插入式保险丝，插入式熔断器
pull-out load 拔脱负载，拉拔力，拉拔载荷
pull-out power 牵出功率，失步功率
pull-out pump 抽出式泵
pull-out resistance 拔拉阻力
pull-out rotor angle 转子失步角
pull-out slip 失步转差率
pull-out speed 失步转速
pull-out test for bundle of bars 并筋拉拔试验，钢筋束拉拔试验
pull-out test 牵出试验，失步试验，拔桩试验
pull-out time 动作时间，吸动时间
pull-out tool 拔出工具
pull-out torque 失步力矩，失步转矩，脱扣力

矩，临界过载力矩，牵出扭矩，拔拉转矩，最大力矩，最大转矩
pull-out type fracture 剥落破坏
pull-out 使失步，使不同步，牵出，拔出，拉出，抽出，落下，释放，断电
pull-rise curve 拔升曲线
pull rod 拉杆，拉棒
pull shovel 反铲，拉铲，拖拉铲运机
pull string switch 拉线开关
pull switch 拉线开关
pull-through positioner 前牵式定位器
pull-through winding 穿入式绕组，插入式绕组
pull-type fuse 插入式熔丝
pull-up circuit 工作电路，负载电路
pull-up cold 冷态拉紧
pull-up position （衔铁）吸持位置
pull-up test 最低启动试验
pull-up time 动作时间，吸动时间
pull-up torque （电机）最低启动转矩，最小启动转矩，启动过程最小力矩，最小启动扭矩
pull-up 拉起，停止，停车，刹住，吸引，吸动
pull 拉力，拉，拖，牵引力，牵力，牵引，拖曳，曳
pulmonary respiration 肺呼吸
pulmonary silicosis 矽肺
pulpit 操纵台，控制室，控制台
pulp mill 湿浆磨，碎浆磨
pulp pump 纸浆泵
pulp 淤泥，泥浆，矿浆，浆，纸浆
pulsafeeder 脉动供料机，脉动电源
pulsatance 角频率
pulsate 脉动，波动
pulsating burner 脉动喷燃器
pulsating combustion 脉动燃烧
pulsating-correction factor 脉动校正系数
pulsating current 脉动电流
pulsating electromotive force 脉动电动势
pulsating energy 脉动能源
pulsating extractor 脉冲萃取器
pulsating field 脉动磁场，脉动场
pulsating firing 振动炉排
pulsating flow measurement 脉动流测定，脉动流量测量
pulsating flow 脉动流，非均匀流，脉动流量
pulsating fluidized bed 脉动流化床
pulsating load 脉动荷载，脉动负载
pulsating motor 往复运动电动机
pulsating oil piping 脉动油管道
pulsating power control 脉冲功率控制
pulsating pressure 脉动压力，脉动电压
pulsating rotating field 脉动旋转磁场
pulsating sampler 脉动采样器
pulsating stress 脉动应力
pulsating torque 脉动转矩
pulsating voltage 脉动电压
pulsating wave 脉动波
pulsating 脉冲的，脉动的，片断的
pulsation dampener 脉动阻尼器
pulsation damper 脉冲减震器
pulsation interference 脉动干扰
pulsation loss 脉动损耗
pulsation of current 电流脉动
pulsation period 脉动周期
pulsation source 脉动源
pulsation welding 脉冲焊接
pulsation 脉动，波动，振动，跳动，间断，（交流电的）角频率
pulsator clarifier 脉冲澄清池，脉冲澄清器
pulsator jig 凿岩机
pulsatory 脉动的
pulsator 振动器，断续器，脉冲澄清池，水锤泵
pulsatron 阴极充气脉冲管，双阴极充气三极管
pulscope 脉冲示波器
pulse accumulator 脉冲存储器，脉冲累加器，脉冲计数器
pulse-actuated circuit 脉冲激励电路
pulse-actuated 脉冲激励的
pulse advancing 脉冲提前
pulse amplifier 脉冲放大器
pulse amplitude modulation 脉冲幅度调制
pulse amplitude 脉冲振幅
pulse attenuation 脉冲衰减
pulse averaging circuit 脉冲平均电路
pulse bag filter 脉冲布袋除尘器
pulse bandwidth 脉冲频带宽度
pulse blocking 脉冲阻塞
pulse bucking adder 脉冲补偿加法器
pulse bus 脉冲总线
pulse capacitor 脉冲电容器
pulse carrier 脉冲载频，脉冲载波
pulse channel 脉冲通道
pulse circuit 脉冲电路
pulse cleaning 脉冲吹扫
pulse-code demodulator 脉冲电码解调器，脉冲编码解调器
pulse-code modulation telemetry system 脉冲编码遥测系统
pulse-code modulation 脉冲编码调制，脉冲码调制
pulse-code modulator 脉冲码调制器
pulse-code processing system 脉冲编码处理系统
pulse-coder 脉冲编码器
pulse-code signaling system 脉冲码信号系统
pulse-code transmission 脉冲码传输
pulse-code 脉冲码，脉冲编码，脉冲电码
pulse-coincidence detector 脉冲重合检测器
pulse column 脉冲柱
pulse command 脉冲命令
pulse communication 脉冲通信
pulse compression ratio 脉冲压缩比
pulse controller 脉冲控制器
pulse control 脉冲控制
pulse converter 脉冲变换器
pulse corrector 脉冲校正器
pulse counter 脉冲计数器
pulse-counting system 脉冲计数系统
pulse-counting 脉冲计数
pulse count input 脉冲计数输入
pulse crest 脉冲顶峰
pulse current amplitude 脉冲电流幅值
pulsed arc 脉冲电弧，脉冲弧
pulsed argon arc welding 脉冲氩弧焊

pulsed attenuator 脉冲分压器
pulsed beam 脉冲束
pulsed cavitation 脉动空蚀
pulse DC motor 脉冲式直流电动机
pulsed counter 脉动计数器
pulsed dielectric laser 脉冲式电介质激光器
pulsed discharge 脉冲放电
pulsed drive 脉冲驱动
pulse decay time 脉冲后沿持续时间，脉冲衰落时间
pulse decoder 脉冲译码器
pulse delay circuit 脉冲迟延电路
pulse delay nomogram 脉冲延时列线图
pulse delay time 脉冲后沿持续时间
pulse delay 脉冲延迟
pulse demagnetization 脉冲退磁
pulsed fast reactor 脉冲快中子反应堆
pulsed field 脉动场
pulsed filter 脉冲滤波器
pulsed fluidized bed 脉冲流化床
pulsed gaseous core reactor 脉冲气体堆芯反应堆
pulse discriminator 脉冲甄别器，脉冲鉴别器
pulse distortion 脉冲畸变
pulse distributor 脉冲分配器
pulsed magnetic field 脉动磁场
pulsed neutron source 脉冲中子源
pulsed neutron technique 脉冲中子技术
pulsed neutron 脉冲中子
pulsed operation 脉冲状态工作
pulsed-plasma arc welding 脉冲等离子弧焊
pulsed-power supply 脉冲电源
pulsed reactor 脉冲反应堆
pulse driver 脉冲驱动器
pulsed ruby laser 脉冲式红宝石激光器
pulsed servo 脉冲随动系统，脉冲伺服系统
pulsed sieve-plate column 脉冲筛板塔，脉冲筛板柱
pulsed signal 脉冲信号
pulsed source 脉冲源
pulsed spray transfer 脉冲喷射过渡
pulsed superconducting magnet 脉冲超导磁铁
pulsed tower 脉冲塔
pulse duct 冲压管
pulse duration controller 脉宽控制器
pulse duration counter 脉冲时间计数器
pulse duration modulation 脉冲宽度调制，脉冲持续时间调制
pulse duration ratio 脉冲占空系数，脉宽周期比
pulse duration 脉冲宽度，脉冲持续时间
pulse duty factor 脉冲占空因数
pulsed water jet 脉冲水力喷射
pulsed wire anemometer 脉冲热线
pulsed 脉冲的
pulse-echo-overlap 脉冲回波重叠
pulse-echo 脉冲反射
pulse eddy current testing 脉冲涡流检测
pulse encoder 脉冲编码器
pulse envelope 脉冲包线
pulse excitation 脉冲激励
pulse fall time 脉冲下降时间，脉冲下沿时间，脉冲降落时间
pulse former 脉冲形成电路，脉冲形成器
pulse-forming amplifier 脉冲形成放大器
pulse-forming delay line 脉冲形成延迟线
pulse-forming network 脉冲形成网络
pulse-forming 脉冲形成
pulse form 脉冲波形
pulse frequency divider 脉冲分频器
pulse frequency modulation 脉冲频率调制，脉冲调频
pulse frequency 脉冲频率
pulse gate 选通脉冲，门脉冲
pulse generating means 脉冲发生装置
pulse generator display 脉冲发生显示
pulse generator 脉冲振荡器，脉冲发生器
pulse height discriminator 脉冲幅度鉴别器
pulse height selector 脉冲幅度选择器
pulse height 脉冲高度，脉冲幅度
pulse inhibit 脉冲封锁，脉冲禁止
pulse input 脉冲量输入
pulse integration 脉冲积分法
pulse interval 脉冲间隔，脉冲周期
pulse inverter 脉冲倒相器
pulse-jet collection 脉冲式除尘器
pulse-jet filter 脉冲式过滤器
pulse keyer 脉冲键控器
pulse length modulation 脉冲宽度调制
pulse length 脉冲长度，脉冲宽度
pulse level 脉冲电平
pulse matching 脉冲刻度校准
pulse measuring oscilloscope 脉冲测量示波器
pulse memory circuit 脉冲存储电路
pulse moder 脉冲编码电路，脉冲编码装置
pulse modulated wave 脉冲调制波
pulse modulated 脉冲调制的
pulse modulation 脉冲调制
pulse modulator 脉冲调制器
pulse monitored 有脉冲监控的，有脉冲信号的
pulse motor 脉冲电动机
pulse multiplex transmission 多路脉冲传输
pulse narrowing system 脉冲压缩系统
pulse number check 脉冲数目校验
pulse number 脉动数，脉波数
pulse-on 启动
pulse oscillator 脉冲发生器，脉冲振荡器
pulse output 脉冲量输出，脉冲输出
pulse packet 脉冲群
pulse pattern generator 标准脉冲发生器，脉冲信号发生器
pulse peak 脉峰
pulse period 脉冲周期
pulse-phase modulation 脉冲相位调制
pulse-phase 脉冲相位
pulse-position modulation 脉冲位置调制
pulse-position 脉冲位置
pulse power 脉冲功率
pulse rate indicator 脉冲速率指示仪
pulse rate 脉冲速率，脉冲频率，
pulse-recurrence rate 脉冲重复频率
pulse regeneration 脉冲再生，脉冲恢复
pulse relaxation amplifier 脉冲张弛放大器

pulse relay　脉冲继电器
pulse-repetition frequency　脉冲重复频率
pulse-repetition period　脉冲重复周期
pulse-repetition rate　脉冲重复率
pulse-repetition　脉冲重复
pulse response　脉冲响应，脉冲响应技术
pulse reverse jetting air compressor　脉冲反吹空（气）压（缩）机
pulse ripple　脉冲波纹，脉动
pulse rise time　脉冲上升时间，脉冲上沿时间，脉冲升起时间
pulser　脉冲源，脉冲发生器，脉冲发送器
pulsescope　脉冲示波器
pulse separator　脉冲分离器
pulse-sequence detector　脉冲次序检测器
pulse-series generator　脉冲序列发生器
pulse-series　脉冲序列
pulse servosystem　脉冲伺服系统
pulse shaper　脉冲形成器，脉冲整形器，脉冲形成电路，脉冲整形电路
pulse shape　脉冲整形电路，脉冲整形器，脉冲波形
pulse-shaping circuit　脉冲形成电路
pulse-shaping　脉冲成形，脉冲整形
pulse signal generator　脉冲信号发生器
pulse signals　脉冲量，脉冲信号
pulse solar simulator　脉冲式太阳模拟器
pulse source　脉冲源
pulse spike　脉冲尖峰
pulse steepness　脉冲陡度
pulse stretcher　脉冲展宽器
pulse stretching　脉冲展宽
pulse switching circuit　脉冲开关电路
pulse switch　脉冲开关
pulse synchroscope　脉冲同步示波器
pulse system　脉冲制，脉冲系统
pulse time code　脉冲时间码
pulse time demodulater　脉冲时间解调器
pulse time modulation　脉冲时间调制
pulse time　脉冲时间
pulse timing　脉冲计时，脉冲同步，脉冲定时
pulse-to-pulse correlation　脉冲间相互关系
pulse-to-pulse　脉冲间的
pulse train generator　脉冲序列发生器
pulse train input conversion　脉冲输入转换
pulse train　脉冲序列，脉冲列，脉冲串，脉冲群，一串脉冲
pulse transformer　脉冲变压器
pulse transmission　脉冲传输，脉冲发射
pulse transmitter　脉冲发送机，脉冲发射机
pulse triggering　脉冲触发
pulse trigger　脉冲触发器
pulse tube　脉冲管
pulse-type detector　脉冲型探测器
pulse-type MHD generator　脉动式磁流体发电机
pulse-type signal　脉冲信号
pulse unit　脉冲部件
pulse value　脉冲量
pulse velocity　脉动速度
pulse washing　脉冲清洗
pulse-width coding　脉宽编码
pulse-width decoder　脉冲宽度译码器
pulse-width discriminator　脉冲宽度鉴别器
pulse-width encoder　脉冲宽度编码器
pulse-width keyer　脉宽键控器
pulse-width modulated converter　脉宽调制变流器
pulse-width modulated　脉冲宽度调制的，脉宽调制的
pulse-width modulation　脉宽调制，脉冲宽度调制
pulse-width modulator　脉冲宽度调制器
pulse-width　脉冲持续时间，脉冲宽度
pulse　脉量，脉冲，脉动，冲量，波动
pulsimeter　脉冲计
pulsing relay　脉冲继电器
pulsing　脉动，脉冲，发送脉冲，脉冲发送
pulsometer pump　蒸汽抽水泵
pulsometer　蒸汽抽水机，气压抽水机，脉冲计
pulstance　角速度
pultruded aluminium blade　挤拉铝叶片
pultrusion　拉挤成型
pulvation action　尘化作用
pulveriser　粉碎机，喷雾机，雾化器
pulverization　雾化，磨成粉，研末，粉碎，喷雾
pulverized asbestos　石棉粉
pulverized burner　煤粉燃烧炉
pulverized coal as fired　入炉煤粉
pulverized coal ash　粉煤灰
pulverized coal boiler　粉煤锅炉，煤粉锅炉
pulverized coal bunker　粉煤仓，煤粉仓，粉仓
pulverized coal burner　煤粉燃烧器，煤粉喷燃器
pulverized coal chute　落粉管
pulverized coal collector　煤粉收集器
pulverized coal conveyer　输粉机
pulverized coal deposits　积粉
pulverized coal distributor　煤粉分流装置，煤粉分配器
pulverized coal duct　煤粉管道
pulverized coal exhauster　排粉（风）机
pulverized coal exhaust fan　排粉（风）机
pulverized coal feeder　给粉机
pulverized coal feed piping　给煤粉管道，送粉管道
pulverized coal fired boiler　煤粉锅炉
pulverized coal firing　粉煤燃烧
pulverized coal injection　高炉喷吹
pulverized coal piping　粉煤管道
pulverized coal preparation system　煤粉制备系统
pulverized coal preparation　煤粉制备
pulverized coal screw conveyer　螺旋输粉机
pulverized coal self-ignition　煤粉自燃
pulverized coal silo　煤粉仓
pulverized coal spontaneous combustion　煤粉自燃
pulverized coal storage system　煤粉仓储系统
pulverized coal system　制粉器
pulverized coal　煤粉，粉煤
pulverized fuel ash　粉煤灰
pulverized fuel boiler　煤粉锅炉
pulverized fuel combustor　煤粉炉膛
pulverized fuel exhauster　排粉机
pulverized fuel feeder　给粉机
pulverized fuel fired boiler　煤粉锅炉
pulverized fuel line　煤粉管道

pulverized fuel 煤粉，粉状燃料
pulverized lime 石灰粉
pulverized ore 粉矿
pulverized 粉状的，磨成粉的
pulverizer air 送粉风
pulverizer coal feeder 给粉机
pulverizer control system 磨煤机控制系统
pulverizer exhauster 排粉机
pulverizer temperature control 煤粉温度控制
pulverizer 粉碎机，磨煤机，磨粉机，磨碎机，喷雾机，雾化器
pulverize 磨碎，粉化
pulverizing machinery 制粉设备
pulverizing mill 磨煤机
pulverizing mixer 粉碎拌和机
pulverizing 压碎，磨碎
pulverous 粉的，粉状的
pulverulent 粉的，粉状的
pulveryte 细粒沉积岩
pulv＝pulverizer 磨煤机
PULV 磨煤机【DCS 画面】
pumice block 浮石块
pumice concrete 浮石混凝土
pumice sand 浮石砂
pumice slag brick 浮石渣砖
pumice 泡沫岩
pumpability 泵的输送能力，泵送能力，可用泵输送，可泵性，输送量
pumpable 可泵的
pump adapter 泵接合器
pumpage rate 扬水率
pumpage 泵抽送量，泵的抽运能力，泵的出力，泵的抽运量，抽水（量），抽送能力
pump air chamber 泵的气室
pump assembly 泵组合，泵装置，水泵组件
pump-back system 泵回系统
pump-back test （电机）反馈试验
pump barrel 泵筒
pump bay （厂房内的）水泵区，给水泵区
pump-body water drain valve 泵体放水门
pump bowl flange 泵壳法兰
pump capacity 泵流量，泵容量，泵的抽水量，抽水机出水率，水泵出水率
pump casing 泵壳
pump chamber 泵室
pump characteristic curve 泵特性曲线
pump characteristics 泵特性
pump circuit 激励电路，泵用电源电路
pump circulation 强制循环
pump concrete 泵浇混凝土，泵送混凝土
pumpcrete machine 混凝土泵
pumpcrete 泵浇混凝土，泵送混凝土
pump cylinder 泵缸，泵筒
pump delivery 泵供水量，泵排出量，泵排量
pump discharge manifold 泵排出总管
pump discharge pipe 泵出口管
pump discharge shutoff valve 泵出口截止阀
pump discharge 泵抽送量，泵出口，泵的流量，泵的排出管，水泵出水量，水泵排水量
pump discharging acid vehicle 泵卸式酸槽车
pump discharging caustic vehicle 泵卸式碱槽车
pump displacement 泵排水量
pump drainage 机泵排水，抽排
pump drain 抽水，水泵排水沟
pump dredger 抽泥机，吸泥机
pump driveline 泵传动线
pump drive motor 泵用电动机
pump drive 泵的传动
pump duty 泵出力
pumped drainage requirement 排水泵排放要求
pumped drainage 泵站排水
pumped fluid 泵送流体
pumped hydro energy storage 抽水储能
pumped power generation 扬水发电
pumped storage aggregate 抽水蓄能机组
pumped storage development 抽水蓄能开发
pumped storage generating set 抽水蓄能机组
pumped storage generation 抽水蓄能发电
pumped storage hydroelectric plant 抽水蓄能水电站
pumped storage hydro-plant 抽水蓄能水电站
pumped storage hydropower station 抽水蓄能水电站
pumped storage hydro-unit 抽水蓄能机组
pumped storage power generation 抽水蓄能发电
pumped storage power plant 抽水蓄能电站，抽水蓄能发电厂
pumped storage power station 抽水蓄能电站，抽水蓄能发电厂
pumped storage project 抽水蓄能工程
pumped storage schedule 抽水蓄能计划
pumped storage scheme 抽水蓄能计划
pumped storage station 抽水蓄能电站
pumped storage unit 抽水蓄能机组
pumped storage 泵储能【抽水或压气】，抽水蓄能
pumped vacuum system 抽真空系统
pumped well drain 水井抽排
pumped 用泵送的
pump efficiency 泵效率，水泵效率
pump end bearing 泵端轴承
pump end 泵端
pumper 带泵消防车，司泵员，泵式油井
pump failure 泵故障
pump flow 泵流量，水泵流量
pump for heating network 热网水泵
pump for scrubber 除尘水泵
pump frequency 泵激频率，激励频率
pump gallery 水泵廊道
pump governor 泵调节器
pump head 泵扬程，泵压头，水泵扬程，水泵压头
pump horsepower 泵马力
pump house ventilation system 泵房通风系统
pump house 水泵间，泵房，水泵房
pump impeller 泵轮【主动轮】，泵叶轮
pumping action 抽水作用
pumping appliance 排水设备
pumping back method 反馈负载法
pumping back 反馈
pumping capacity 抽水能力，抽气能力，泵流量
pumping depression area 抽水降落范围

pumping duration 抽水历时
pumping during construction 施工期抽水
pumping engine 蒸汽泵
pumping head 抽水扬程，水泵扬程
pumping level 抽水高程
pumping lift 水泵扬程
pumping line 水泵出水管，扬水管道
pumping main 泵站母管
pumping-out system 抽水系统
pumping plant capacity 抽水站抽水能力
pumping plant 泵站，抽水站
pumping power station 抽水蓄能电站
pumping power 泵送功率，抽运功率
pumping rate 抽气速率，抽运速率，泵送率
pumping relief well 抽水减压井
pumping ring 泵环
pumping room 泵房间
pumping shaft 抽水竖井，扬水竖井
pumping signal 参数激励频率信号
pumping speed 泵送速度，抽水速度
pumping station （水）泵站，扬水站，抽水站
pumping system 抽水系统
pumping test 泵压试验，抽水试验
pumping unit 泵设备
pumping-up power plant 抽水蓄能电厂
pumping-up power station 抽水蓄能电站
pumping water 扬水
pumping well 抽水井
pump intake tunnel 泵的吸水管道
pump intake 水泵进水口
pump internals storage stand 泵内构件贮存架
pump internals 泵内件
pump jacket 泵体护套
pump lever 泵杆
pump lift （水）泵扬程
pump motor 泵用电动机
pump nozzle 泵接管，喷头
pump operating curve 泵运行特性曲线
pump operating duty 泵工况
pump output 抽水（量）
pump-out vanes 泵出口叶片，背叶片
pump performance 水泵性能，泵性能
pump pit 水泵坑，泵坑，抽水机坑
pump power input 泵功率输入【马力，制动马力】
pump power output 泵功率输出
pump power 泵耗电力，泵耗功率
pump preheating system 暖泵系统
pump pressure gauge 泵压力计
pump primer 水泵启动泵，泵的启动装置
pump priming 泵启动注水，泵启动注油，泵充水
pump protection filter 泵保护过滤器
pump protection strainer 泵保护滤网
pump rod 泵杆
pump room 泵房，泵舱，泵室
pump screen 泵用滤网
pump sets running 泵站机组运转
pump set 泵组
pump shaft seal water 泵轴封水
pump shaft 泵轴
pump shutoff head 泵的关闭压头
pumps in parallel 并联泵
pumps in series 串联泵
pump's operating range 泵的运行范围
pump specific speed 水泵比速
pump speed 泵速度
pump stage 泵级
pump station of recirculation water 循环水泵站
pump station （水）泵站，扬水站
pump storage group 抽水蓄能机组
pump storage power plant 抽水蓄能电站
pump strainer 泵进口端拦污网，泵口拦污网
pump stroke 泵冲程
pump submergence 水泵淹没深度
pump suction well 水泵集水井
pump sump 泵池，泵井，水泵集水井
pump surging 泵脉动
pump system 泵系
pump torque 泵转矩
pump train 泵系列
pump-turbine 水泵水轮机，可逆式水轮机，泵用透平，涡轮泵
pump unit 泵装置，水泵机组，抽水机组
pump up 用泵把……抽上来，泵送
pump valve cage 泵阀盖座
pump valve 水泵阀
pump warming valve 暖泵门
pump works 水泵工程
pump 泵，水泵，抽水机，用泵打，抽，注入，抽吸，抽运，泵送，抽水
punch card machine 穿孔卡片机
punch card reader 穿孔卡片读入机，穿孔卡片读出器
punch card 穿孔卡片，卡片穿孔
punch-dressed masonry 凿毛圬工
punched bend 冲压弯头，压制弯头
punched card computer 穿孔卡计算机
punched card processing equipment 穿孔卡片数据处理装置
punched card processing 穿孔卡片数据处理
punched card reader 穿孔卡片输入机
punched card recorder 穿孔卡记录器
punched card 穿孔卡片，打孔卡片
punched paper tape reader 穿孔纸带读出器
punched paper tape 穿孔纸带
punched tape program 穿孔带程序
punched tape storage 穿孔带存储器
punched tape 穿孔带
punched 穿孔的
puncheon 半圆木料，短柱
puncher 穿孔器，穿孔机
punch format 穿孔格式
punch hole 冲孔
punching drilling 冲击钻井
punching machine 穿孔机，冲床
punching position 穿孔位置
punching press 冲孔机
punching shear 冲剪应力，冲剪力，冲剪
punching 冲孔，冲压，冲片，穿孔
punch list 消缺单，问题清单，剩余工作清单，工程将竣工前做最后检查的清单
punch load 冲剪荷载
punch machine 冲孔机

punch operator 穿孔员
punch out 冲孔输出
punch press 冲床
punch pulse 穿孔脉冲
punch rivet hole 冲铆钉孔
punch shear test 冲剪试验
punch speed 冲压速度，穿孔速度
punch-through 击穿现象，穿通现象
punch 冲头，冲床，穿孔机，冲压，冲孔，冲压机，打孔机
punctiform source 点状源
punctured code 收缩码
punctured element 针孔元件
puncture field intensity 击穿场强
puncture impulse voltage 击穿冲击电压
puncture lightning arrester 击穿保险器
puncture of insulation 绝缘击穿
puncture place 击穿位置
puncture resistance 击穿阻力
puncture strength 击穿强度
puncture tester 耐（电）电压试验器
puncture test 打孔试验，击穿试验，耐压试验，冲孔试验
puncture time 击穿时间
puncture voltage 击穿电压
puncture-withstand test 耐击穿电压试验
puncture 刺穿，击穿
punish according to law 依法惩办
punish corruption 惩治腐败
punish 惩罚
punitive damages 惩罚性损害赔偿金
punitive measure 惩罚措施
punitive sanction 惩罚性制裁
pun 捣，打夯，夯，夯捣
p. u. (per unit) 标幺值
pupin coil 加感线圈
pupinization 加感，加负荷
pupinize 加感
PU(poly urethane) 聚氨酯
PUP(peripheral unit processor) 外围设备处理机
purchase agreement 购货协定，购买合同
purchase confirmation 购货确认书
purchase contract 购货合同，订购契约，订购合同
purchased power expenses 外购电力费用
purchase expenses 购置费
purchase guideline 采购导则
purchase in advance 预购
purchase list 订货单
purchase note 购货确定书
purchase order number 订货单号
purchase order 订单，定单，购货单，定货单
purchase price of asset 资产采购价
purchase price 采购价
purchaser 采购人员，购买人，买方，招标人，招标方
purchase specification 采购说明书，采购规范，采购标准
purchase 采购，购买
purchasing agent 采购代理人，代购商
purchasing by invitation to bid 招标采购
purchasing chief 采购科长
purchasing department 采购部门
purchasing equipment 采购设备
purchasing power parity 购买力平价
purchasing procedure 采购程序
purchasing schedule 订货采购进度
purchasing specification summary sheet 采购规范汇总表
PURC(Public Utilities Regulatory Commission) 公共事业管委会
pure air 纯洁空气，纯空气
pure aluminum wire 纯铝线
pure bending instability 纯弯不稳定性
pure bending 纯弯（曲）
pure binary number base 纯二进制数基
pure binary number 纯二进制数
pure binary 纯二进制的，纯二进制
pure carrier network 纯载波网络
pure chemical control 纯化学控制
pure circulation round wing 纯机翼环流
pure compound 纯化合物
Pu recycle （热中子堆中）钚再循环
pure diffusion control 纯扩散控制
pure gas turbine cycle 纯燃气轮机循环
pure generator 产生程序的程序
pure laminar flow 纯层流，纯片流
pure lime 纯石灰
purely automatic 全自动化的
purely random process 纯随机过程
purely random 纯随机的
purely viscous fluid 纯黏性流体
pure neutron shielding 纯中子屏蔽
pure oscillation 纯振荡，正弦振荡
pure profit 纯利润
pure reactance network 纯电抗网络
pure reaction turbine 纯反动式汽轮机
pure scattering 纯散射
pure semiconductor 无杂质半导体，纯半导体
pure shear 纯剪（切）
pure sound 纯声
pure stress 纯应力，单种应力
pure time delay 纯时延
pure time-sharing assignment 纯分时指定
pure tone 纯音
pure torsion flutter 纯扭颤振
pure umbrella type generator 全伞式发电机
pure variable-pressure turbine 纯冲动透平
pure water 纯净水，纯水
pure wave 正弦波
pure zinc anticorrosion sleeve 防腐锌套
pure 纯的
purge air exhaust fan 安全壳的排气风机【核电】
purge air filter 吹洗空气过滤器
purge air 吹洗空气，吹扫空气
purge and drain system 吹洗和疏排水系统
purge and wash system 吹洗系统
purge device 清扫设施
purge gas flow 吹洗气流
purge gas supply system 吹洗供气系统
purge gas supply 吹洗供气
purge gas system 吹洗气体系统

purge gas　吹扫气，清洗气，吹洗用气体
purge gate　放（空）气阀
purge interlock　吹扫联锁
purge loop　吹洗回路
purge operation　吹洗，吹扫操作
purge rate　吹扫风量
purge rotometer　气动转子流量计
purger　吹洗装置，吹扫设备
purge stream　清除气流，吹洗气流
purge system　吹洗系统，吹扫系统，清洗系统，净化系统
purge tank　冲洗箱，冲洗槽
purge unit　抽气器
purge valve　抽气阀，排气阀
purge　放出，排空，净化，吹扫，吹，清洗，使清洁，吹洗
purging efficiency　排净效率
purging N_2　吹洗用氮气
purging of heat-supply pipeline　供热管道清洗
purging of turbine oil system　汽轮机油循环系统清洗
purging the pipeline　管线吹扫
purging　净化，清洗，吹洗，清除
purification and letdown system　净化和疏排系统
purification bypass outlet　净化旁通管出口
purification circuit　净化回路，清洗回路
purification constant　净化常数
purification efficiency　净化效率，纯化效率
purification flow rate　净化流量
purification index　净化指数
purification loop　净化回路
purification plant　净化装置
purification rate　净化流量，净化速率
purification system　净化系统
purification tank　净化槽
purification tower　净化塔
purification train　净化系统
purification　纯化，净化，提纯，精制，精炼
purified-gas buffer tank　净化气体缓冲罐
purified-gas compressor　净化气体压缩机
purified-gas supply system　净化气体供应系统
purified helium aftercooler　净化氦气后冷却器
purified helium compressor with integral intercooler　带一体化中间冷却器的净化氦气压缩机
purified helium gas　净化氦气
purified helium product cooler　净化氦气产品冷却器
purified helium storage tank　净化氦气贮存箱
purified helium　净化氦气
purified water　净化水，提纯水
purifier　清洗装置，净化装置，净化器
purifying equipment　净化设备
purifying pond　净化池
purifying separator　净化分离器
purifying station　净化站
purifying　净化
purity　洁净，纯度，纯净，洁净度，含量
purlin brace　檩条撑
purling　下洗流，涓流
purlin support　檩托
purlin　（平行的）桁条，檩桁条，梁，檩，檩子，檩条
purportedly　据称
purport　声称是……，意味着，意图
purpose loan　目的贷款
purpose-made brick　特质砖
purpose　目的
purpurate　紫尿酸盐，红紫酸盐
pursuance　追求，履行，执行
pursuant to law　依法，按照法律
pursuant to　依照，按照，依据，根据
pursuing a deflationary policy　执行通货紧缩政策
push and pull brace　推拉杆，推拉把手
push and pull switch　推拉开关
pushbutton actuator　按钮执行机构，按钮开关
pushbutton bank　按钮组
pushbutton box　按钮盒
pushbutton contact　按钮触点
pushbutton control station　按钮控制站
pushbutton control system　按钮控制系统
pushbutton control　按钮控制
pushbutton guard　按钮保护罩
pushbutton interface　按钮接口
pushbutton matrix　按钮矩阵
pushbutton or pull box　按钮或分线盒
pushbutton panel　按钮控制面板
pushbutton switch　按钮开关
pushbutton telephone set　按钮电话机
pushbutton timer　按钮定时器
pushbutton　按钮
push cam　推杆凸轮
pushcart　手推车，运料车
push contact　按压触点，按钮开关
push control rod　推杆
push down list　后进先出表
push down stack　下推栈
push down storage　下推存储器
push-down-to-close　按下致闭
push-down-to-open　按下致开
push down　推下，下推，叠加，叠式存储器
push dozer　推土机
push effective address　下推有效地址
pusher block　推杆，撞杆
pusher carriage　推车器，铁牛
pusher dog　（链式输送机的）椎头，拨爪
pusher-helper locomotive　后推辅导机车
pusher leg drill　气腿式凿岩机
pusher puller plate　推进拉出器板
pusher-puller　推进器拉出器
pusher relay　推杆式继电器
pusher tractor　后推机
pusher　四面体，推进器，顶推机，推车器，铁牛，推出器，推动器，推送机车，推杆
push fit　推入配合
push-in positioner　后推式定位器
push in　推进，闯入
push-on connector　推进连接器
push-on push-off button　可定位按钮，开关键
push on starter　按钮启动器
push onto the market　推向市场
push out　推出

push pedal　推板
push plate with finger push　带按指的推板
push plate　推板
push pole　推杆
push pull amplifier　推挽放大器
push-pull cable　推拉电缆【用于远距离控制的阀门】
push pull circuit　推挽电路
push-pull connector　推拉连接器【便于快断】
push pull critical experiment　推挽式临界实验
push pull driver　推挽式驱动器，推挽式激励器
push-pull drive system　推拉传动系统
push pull drive　推挽激励
push-pull gun　推拉式焊枪
push pull hood　吹吸式排风罩
push pull modulation　推挽调制
push pull oscillator　推挽振荡器
push pull rectifier　推挽式整流器
push pull rod　（机械手）推挽棒
push pull running　可逆运行
push-pull torch　推拉式焊枪
push pull transformer　双程变压器，推挽变压器
push pull　推挽的
push rod actuator　推杆执行机构
push rod　推杆
push stem power unit　直进式执行部件
push switch　按钮开关
push through loading system　推换装料系统
push through winding　穿入式绕组，插入式绕组
push through　向中心推进式
push to talk switch　按钮操纵的传话开关
push to talk　按键通话
push to type operation　按钮操纵的电报操作，按钮启动打印操作
push tube　推管
push up list　先进先出表，上推表
push-up-to-close　上推致闭
push welding　手点焊，钳点焊，手压点焊
push well　平推管井
push　推，推进，推力，按钮
put aside　搁置，撇开，排除，储存……备用
put … back in service　恢复运行
put forward　提出，促进
put in a (/one's) claim　提出索赔
put in circuit　接入电路
put in series　串联接入
put in service　投入使用，使生效
put into effect　付诸实施，生效，使生效
put into excitation　投入励磁
put into practice　实行
put into production　投产
put into service　投入生产运行，投入运行，启用，投运
put in touch with　与……联系
putrefaction　腐烂，腐败物
put in motion　带动
put statement　放置语句，送出语句
put the plant into operation　将工厂［装置］投入运行
putting into parallel operation　投入并联运行
putting stator onto location　发电机静子吊装就位
putting the hydrogen cooling system into service　投氢，氢冷系统投用
putting the upper casing onto the lower of the turbine　汽机扣缸，汽轮机本体扣盖
put to earth　接地
put to test　试验
puttying　嵌油灰，上腻子
putty joint　油灰缝
putty knife　油回收系统
putty up　上腻子
putty　油灰，腻子，封泥，油腻子
put word in string　成串放字
PUTWS (put word in string)　成串放字
puzzolana cement　火山灰水泥
PVA (polyvinyl alcohol)　聚乙烯醇
PV array　光伏方阵，光伏阵列，太阳电池组列
PVC cable　聚氯乙烯绝缘电缆
PVC (polyvinyl chloride)　聚氯乙烯
PVCR (polyvinyl chloride resin)　聚氯乙烯树脂
PVDF (polyvinylidene fluoride)　聚偏氟乙烯，聚偏二氟乙烯
P-V diagram　压力比体积图，P-V 图
PV effect　光电效应
PVF (polyvinyl fluoride)　聚氟乙烯
PV industry　光伏产业
PV＝photovoltaic　光伏，光电的
PV (pneumatically operated valve)　气动门
PVQAB (Pressure Vessel Quality Assurance Board)　压力容器质量保证局【英国】
PVR (polyvinyl resin)　聚乙烯树脂
PV string　光伏组件串
PVSV (pressure vacuum safety valve)　压力真空安全阀
PV system　光伏系统，光伏发电系统
PVT (pressure-volume-temperature)　压力体积温度
PVTS (pressure vessel thermal shock)　压力容器热冲击
PV wire　太阳能模组用电线，太阳能光伏线缆
PWB (power water boiler)　动力水锅炉，水均匀动力堆
PWB (printed wiring board)　印刷导线板
PWC (pulse width coding)　脉宽编码
PWD (post write disturb)　写后干扰
PWD (pulse width discriminator)　脉冲宽度鉴别器
PWE (pulse width encoder)　脉冲宽度编码器
PWGR (pressurized water graphite reactor)　压力石墨反应堆
PWHT (post-weld heat treatment)　焊后热处理
PW method (present worth method)　现值方法，现值法
PWM (pulse width modulation)　脉宽调制，脉冲宽度调制
PWM switching circuit　脉冲宽度调制开关电路，脉宽调制开关电路，PWM 开关电路
PWN (plant work notification)　追加工作通知
PW (plant water)　厂用水
PWP (peak working pressure)　最大工作压力
PW (present worth)　现值（即贴现后的钱）
PW (program word)　程序字
PW (pulse width)　脉冲宽度

PWR emergency core cooling system 压水堆应急堆芯冷却系统
PWR fuel assembly 压水堆燃料组件
PWR internals 压水堆堆内构件
PWR nuclear power plant 压水堆核电厂
PWR(pressurized water reactor) 加压水冷反应堆，压力水冷反应堆，压水反应堆，压水堆
PWR primary coolant pump 压水堆一次冷却剂泵
PWR steam generator 压水堆蒸汽发生器
PWR system 压水堆系统
PWR vessel and its internal 压水堆本体
PWS(potable water system) 饮用水系统
PWTCR(production weld test coupon report) 产品焊缝试件报告
PWT(production weld test coupon) 产品焊缝试件
PWT(propulsion wind tunnel) 推进风洞
PXS(passive core cooling system) 非能动堆芯冷却系统
pycnometer 比重计，比重瓶，密度瓶
pycnometry 测比重法
pygmy current meter 微型电流表，微型流速仪
pygmy lamp 微型灯
pygmy 微型的，微小的
pyknometer 密度瓶
pylon antenna 铁塔天线
pylon tower 耐张力铁塔，桥塔
pylon 铁塔，电杆，塔门，铁塔式电杆，高压电缆塔，桥塔，立柱
pyod 热电偶
PR(pressure relay) 压力继电器
pyramidal bin 角锥形料仓
pyramid balance 塔式天平
pyramid carry 锥形进位
pyramid hopper 锥形灰斗
pyramid 金字塔，棱锥，四面体，锥形，角锥体
pyranograph 总日射计
pyranometer 日射强度计，辐射强度表
pyrex flask 派热克斯烧瓶，硬质烧瓶
pyrex glass tube 钛硼硅酸硬玻璃管
pyrex glass 硬质玻璃，派热克斯玻璃
pyrex sphere 耐热硬质玻璃球
pyrex 硼硅酸玻璃，耐热硬质玻璃
pyrgeometer 大气辐射表，地面辐射计
pyrheliometer 直接日射表，直接日射计，太阳热量计，日温计
pyridine 吡啶
pyrite handling system 石子煤处理系统
pyrite roasting 黄铁矿焙烧
pyrites dewatering bin 石子煤脱水仓
pyrites hopper 石子煤斗
pyrite 黄铁矿，二硫化铁，石子煤
pyritic sulfur 黄铁矿硫，硫铁矿硫
pyro acid 焦酸
pyrocarbon-coated particle 热解碳包覆颗粒燃料
pyrocarbon coating 热解碳包覆
pyrocarbon deposition on guide tuber 导向管上热解碳沉积
pyrocarbon 高温碳，热解碳，高温石墨
pyrocatechol 儿苯酚，苯磷二酚，焦儿苯酚
pyrochemical processing 高温化学处理
pyroclastic rock 火成碎屑岩
pyroclastic 火成碎屑物
pyroconductivity 高温导电性，热电导
pyrodigit 一种数字显示温度指示器
pyroelectric effect 热电效应
pyroelectricity 热电学，热电，热电现象
pyroelectrics 热电体
pyroelectric thermometer 热电温度计
pyroelectric 热电的，热电物质
pyroferrite 热电铁氧体
pyroflow system 热流动系统
pyrogallate 焦酸盐
pyrogallic acid 焦酸，焦性没食子酸，邻苯三酚
pyrogallol 焦培酸，焦酚
pyrogenetic rock 火成岩
pyrogenetic 热解的
pyrogenic deposit 火成矿床
pyrogenic effect 发热效应
pyrogenic rock 火成岩
pyrogenous 火成的
pyrohydrolysis 高温水解
pyrohydrolytic process 高温水解法
pyrolith 火成岩
pyrology 热工学
pyrolysis 高温分解，热解，加热分解
pyrolytic carbon-coated particle 热解碳包覆颗粒
pyrolytic carbon 热解碳
pyrolytic cerbon-coated fuel 热解碳包覆颗粒燃料
pyrolytic coating 热解包覆
pyrolytic damage 热损伤，高温分解损伤
pyrolytic graphite 热解石墨
pyrolytic layer 热解碳层
pyrolytic silicon carbide 热解碳化硅
pyrolyzer 热解器
pyromagnetic generator 热磁发电机
pyromagnetic motor 热磁电动机
pyromagnetic 热磁的
pyrometallurgical processing 高温冶金处理
pyrometallurgy 高温冶金学
pyrometer 高温计
pyrometric cone 测温锥
pyrometry 高温测定法，高温测定，高温测定学
Pyromic 一种镍铬耐热合金
pyrophoricity accident 自燃事故
pyrophoric waste 自燃废物
pyrophoric 自燃物，引火物，自燃的，引火的
pyrophosphate 焦磷酸盐
pyrophosphite 焦亚磷酸盐
pyroscan 一种红外线探测仪
pyroscope 辐射热度计，高温计
pyroshale 可燃性油母页岩
pyrostat 高温调节器，高温自动调节仪，恒温器，恒温槽
pyrosulfate 焦硫酸盐
pyrosulfite 焦亚硫酸盐
pyrosulphate 焦硫酸盐
pyrotechnic projector 信号发射器，信号枪
pyrotomalenic acid 焦麻酸
pyroxene-dacite 辉英安岩
pyroxene 辉石
pyroxenite 辉岩

pyroxylin　火棉
PYR＝pyrometer　高温计
pyrradiometer　全辐射表
pyrron detector　黄铁矿检波器
Pythagorean theorem　毕德哥拉定理，勾股定理
PZT(piezoelectric transducer)　压电换能器
PZT(piezoelectric transition)　压电跃变
P　各种坑和池【核电站系统代码】
P　压力【DCS 画面】

Q

QA data package　质量保证数据包，质量保证文件袋
QADP(quality assurance data package)　质量保证文件数据包
QADS(quality assurance data system)　质量保证数据系统
QAM(quality assurance manual)　质保手册
qanat　暗渠
QAP(quality assurance planning)　质量保证计划
QAP(quality assurance program)　质量保证大纲
QA program　质量保证大纲
QA(quality assurance)　质量保证
QAS(quality acceptance standard)　质量验收标准
QASR(quality assurance surveillance report)　质量监督报告
QA system　质量保证系统
QBS(quality-based selection)　根据质量选择
QCB(queue control block)　队列控制块
QCD(quality control data)　质量控制数据
QCE(quality control engineer)　质检工程师
QC(quality control)　质量控制，质量管理，质检
QCR(quality control report)　质量控制报告
QCS(quality control system)　质量控制系统
QDC(quick dependable communication)　快速可靠通信
Q demodulator　Q 信号解调器
QDS(technical qualification data sheet)　技术评定数据表
QEL(queue element)　队列元素
Q factor(quality factor)　品质因数，质量因数
QF(quality factor)　质量因数，品质因数
QIC(quality insurance chain)　质量保险系统
QI(quality index)　质量指标，质量指数
QI(quality inspection)　质量检查
QIT(quality information and test system)　质量信息及试验系统
Q-matching　Q 匹配，四分之一波长匹配
QMB(quick make-and-break contact)　快速通断接触
QMI(qualification maintainability inspection)　合格维修检验
QMQB(quick-make, quick-break)　快通快断
QMS(quality management system)　质量管理体系
QNC(quality non-class)　非质量级别，质量无级
QOL(quality of life)　生活水平，生活质量，基本生活条件
QOM(quality organization manual)　质量管理手册
QOV(quick open valve)　排污阀，快开阀
QPM(quality plan management)　质量计划管理
QP(quality plan)　质量计划
QRA(quality reliability assurance)　质量可靠性保证
QR(quality related)　与质量有关（的）
QR(quick reaction)　快速反应
QSDR(quality surveillance deficiency report)　质量监督缺陷报告
QSDR(quality surveillance deviation report)　质量监督偏差报告
Q-section　四分之一波长线段
QSP(quality safety plan)　质量安全计划
QS(quality surveillance)　质量监督
QSR(quality and safety related)　与质量及（核）安全有关的
QTCWPDS(welding procedure data sheet for qualification test coupon)　用于评定试件的焊接工艺数据单
Q-terminal　Q 信号输出端
quad array　方阵，方阵列
quadbundle　四分裂导线
quadded cable　扭绞四心电缆，四线电缆
quadergy　无功能量，千乏小时
quad hopper car　四段式漏斗车
quad-in-line package　四列直插式组件
quad location　四角定位
quad pair cable　扭绞八心电缆
quadrangle frame　四边形机座
quadrangle　四角形，四边形，四方院子，四合院
quadrangular　四角形的，四边形的，四棱柱
quadrantal deviation　象限偏差
quadrantal error　象限误差
quadrantal　象限的，扇形的
quadrant angle　射角，象限角
quadrant-edged orifice plate　方边的孔板
quadrant electrometer　象限静电计
quadrant iron　方钢
quadrant opener　开启机构【扇形齿轮】
quadrant strand conductor　扇形多股线
quadrant　象限仪，扇形体，象限，扇形齿轮，四分仪
quadrate　四等分，正方形，正方形的，方块物
quadratic confidence limit　二次置信界限
quadratic constraint　二次约束
quadratic damping　平方阻尼，按平方律衰减
quadratic equation　二次方程式，二次方程
quadratic expression　二次式
quadratic form　二次型，二次方程式，二次式
quadratic function　二次函数
quadratic lag　平方滞后
quadratic magneto-optic effect　二次磁光效应
quadratic mean deviation　二次平均偏差
quadratic performance index　二次性能指标
quadratic potentiometer　平方电位计
quadratic programming　二次规划
quadratic sum　平方和
quadratic　平方的，二次的，正方形的，方形的，象限的，四方的
quadrat major　主平方区
quadratron　热阴极四极管

quadrature amplitude modulation　正交调幅，正交幅度调制
quadrature axis component　交轴分量，横轴分量
quadrature axis magnetic flux　横轴磁通，交轴磁通
quadrature axis magnetizing reactance　交轴磁化电抗，横轴磁化电抗
quadrature axis reactance　横轴电抗，交轴电抗
quadrature axis reluctance　横轴磁阻，交轴磁阻
quadrature axis subtransient impedance　交轴次瞬态阻抗，横轴次瞬变阻抗
quadrature axis subtransient reactance　交轴次瞬变电抗，横轴次瞬变电抗
quadrature axis subtransient voltage　横轴次瞬变电压，交轴次瞬变电压
quadrature axis synchronous impedance　横轴同步阻抗，交轴同步阻抗
quadrature axis synchronous reactance　交轴同步电抗，横轴同步电抗
quadrature axis transient impedance　横轴瞬变阻抗，交轴暂态阻抗
quadrature axis transient reactance　交轴瞬态电抗，横轴瞬态电抗
quadrature axis transient voltage　交轴瞬变电压，横轴瞬变电压
quadrature axis voltage　交轴电压，横轴电压
quadrature axis　交轴，横轴
quadrature booster　正交增压器
quadrature bridge　正交电桥，90°相移电桥
quadrature channel　正交通道
quadrature circuit　横轴电路，交轴电路
quadrature component　横轴分量，正交分量，电抗分量
quadrature detector　积分检波器
quadrature-drop compensation　正交下降补偿
quadrature exciting watt　激磁乏，正交励磁伏安
quadrature factor　无功因数
quadrature field　正交场
quadrature impedance　横轴阻抗，交轴阻抗
quadrature information correlator　平方律相关器
quadrature information　平方律
quadrature lagging　滞后 90°，后移 90°
quadrature modulation　直角相位调制，正交调制
quadrature phase　正交相位，90°相位移
quadrature spectrum　正交谱
quadrature subroutine　平方子程序
quadrature tide　弦潮
quadrature transformer　正交变压器
quadrature tube　直角管，电抗管
quadrature voltage　正交电压
quadrature winding　横向绕组，正交绕组，鼠笼绕组
quadrature　方照，90°相移，90°相位差，求面积，求积分，正交
quadrat　平方区
quadra　勒脚
quadrel　方瓦，方砖
quadric equation　二次方程
quadricorrelator　自动调节相位线路
quadric　二次的
quadrilateral characteristics　四边形特性
quadrilateral directional distance relay　四边形方向性继电器
quadrilateral　四边形的，四边形，四角的
quadriphase system　四相制
quadripolar　四极的，四端的
quadripole　四端电路，四端网络，四极
quadrivalence　四价
quadrode　四极管
quadroxide　四氧化物
quadruple compound turbine　四缸复式涡轮机
quadruple direct shear apparatus　四联直剪仪
quadruple dumper　四联翻车机
quadruple error detection　四重误差检测
quadruple-flow turbine　四排汽口汽轮机
quadrupler　四倍器，四频器
quadruplex cable　四芯电缆
quadruplex system　四路传输制，四路多功制
quadruplex telegraph　四路多功电报
quadruplex telephony　四路多功电话术
quadruplex　四重的，四倍的，四路多功的
quadruple　四重的，四倍式，四路，四倍，翻两番
quadruplication　四倍，一式四份
quadrupling　四倍
quadrupole　四极
quad word　16 字节长的字
quad　四心线组，象限，四心导线，四边形，四倍的
quagmire　颤沼，泥沼，软泥地
quag＝quagmire　泥沼
quake-proof structure　耐震构造
quake-proof　防震
quake protection of line　线路防（地）震
quake　地震，震动
quaking bog　颤沼，跳动沼
qualification and capability　资格与能力
qualification certificate　资格证明，资格证书，鉴定合格证
qualification document　资格文件，资格证明文件
qualification evaluation　资格考评，资格评定，资格评价，合格判定
qualification information　资质信息
qualification maintainability inspection　合格维修检验
qualification of bidder　投标商资格
qualification rate of welds　焊口合格率
qualification renewal　重新鉴定
qualification specification　鉴定规范书
qualification test coupon　鉴定试样
qualification test　鉴定试验，验收试验，合格（性）试验，质量鉴定试验，资格考试
qualification welder　合格焊工
qualification weld　鉴定焊缝
qualification　鉴定，审定，合格，合格证书，资格，资质，限制条件
qualified acceptance　有条件承付
qualified certificate　附条件证明书，合格证
qualified goods delivery　合格交货
qualified life　合格寿期
qualified managerial personnel　合格管理人员
qualified notary　合格公证人

qualified personnel 合格的人员，有资质的人员
qualified product 合格产品
qualified scientist and technician 合格科技人才
qualified technician 合格技术员
qualified 合格的，经过鉴定的，有限制的，有资格的
qualifier 限定词
qualify as 通过考试等取得当……的资格，把……描述为……
qualify 合格，限制，使具有资格，证明……合格，取得资格，有资格
qualitative analysis 定性分析
qualitative collection 定性采集
qualitative control for heat supply 供热的品质调节
qualitative data 定性资料
qualitative description 定性描述
qualitative filter paper 定性滤纸
qualitative forecast 定性预报
qualitative information 定性信息
qualitative interpretation 定性判读，定性说明
qualitative investigation 定性分析，质量集中分析，定性研究，定性调查
qualitative measurement 定性测量
qualitative method 定性方法
qualitative operational requirement 定性操作要求
qualitative safety goal 定性安全目标
qualitative test 质量试验
qualitative 品质的，定性的，合格的，质量的，性质上的
quality activity 质量活动
quality analysis 质量分析
quality appraisal 质量鉴定
quality assessment system for electronic components 电子元器件质量评定体系
quality assessment 质量评定
quality assurance acceptance standard 质量保证验收标准
quality assurance branch 质量保证处
quality assurance data package 质量保证档案
quality assurance data system 质量保证数据系统
quality assurance examination 质量保证检验
quality assurance inspection 质量保证检查
quality assurance organization 质量保证组织
quality assurance planning 质量保证计划
quality assurance procedure 质量保证程序
quality assurance program 质量保证大纲，质量保证计划
quality assurance system 质量保证体系，质量保证系统
quality assurance 质量保证
quality auditor 质量审核员
quality audit 质量监查，质量检查，质量审核
quality award 质量奖
quality awareness 质量意识
quality bonus 质量奖
quality certificate 质量证明书，质量证书
quality certification organization 质量认证机构
quality certification system 质量保证体系
quality change 质量变化
quality characteristic curve of X ray film X射线胶片质量特性曲线
quality characteristic 质量特性，质量特征
quality check 质量检查
quality class 质量等级
quality coefficient 蒸汽干度，质量系数
quality committee 质量委员会
quality competition 质量竞争
quality concrete 优质混凝土
quality control department 质量控制部门
quality control for construction 施工质量管理
quality control group 质量管理小组
quality control manual 质量控制手册
quality control procedure 质量管理程序
quality control program 质量控制大纲
quality control specialist 质量管理专家
quality control standard 质量控制标准
quality control system 质量控制系统
quality control 质量检查，质量控制，质量检验，质量管理
quality cost report 质量成本报告
quality criteria 质量标准
quality defect 质量缺陷
quality determination 质量鉴定
quality discrepancy 品质差异
quality document 质量文件
quality education 质量教育
quality engineering 质量工程
quality engineer 质量工程师
quality evaluation 质量评定，质量评价
quality factor 品质系数，品质因数，质量因素
quality first,customers supreme 质量第一，顾客至上
quality governing 质量调节
quality guarantee 质量保证
quality identification function 质量鉴别功能
quality ideology 质量意识
quality improvement 质量改进
quality information and test system 质量信息及试验系统
quality in marketing 营销质量
quality in procurement 采购质量
quality in production 生产质量
quality inspection fee 质量检验费
quality inspection 品质检验，质量检查，质量检验
quality inspector 质量检查员
quality judgement of process point 过程点质量判断
quality judgment 质量判断
quality levels for acceptance 验收质量等级
quality level 质量标准，质量水平，质量等级
quality loop 质量环
quality loss 质量损失
quality management committee 质量管理委员会
quality management medal 质量管理奖（章）
quality management system 质量管理体系
quality management 质量管理
quality manager 质量经理
quality manual 质量手册
quality measurement 质量测定
quality measure 质量措施，质量度量

quality monitoring　质量控制，质量监督
quality objective　质量目标
quality of air environment　空气环境质量
quality of appearance　观感质量
quality of balance　平衡质量，平衡度
quality of commutation　换向质量
quality of design　设计质量
quality of documentation　文件质量
quality of electric energy　电能质量
quality of enterprise　企业素质
quality of fit　配合等级
quality of fluidization　流态化质量
quality of lighting　照明质量
quality of operations　操作质量
quality of power supply　电能质量
quality of professional services　专业服务质量
quality of radiation　辐射质量
quality of raw water　原水性质
quality of regulation　调节质量
quality of tolerance　公差等级
quality of treated water　处理水质（量）
quality of work　工作质量
quality planning　质量策划
quality plant　高品质（核）电厂
quality plan　质量规划，质量计划
quality policy　质量方针
quality prize　质量奖
quality procedure　质量程序
quality product　优质产品
quality program　质量大纲
quality proposal　质量建议书
quality record　质量记录
quality region　含汽区
quality regulation　质量监管
quality-related cost　质量成本
quality-related requirements specification　质量有关要求说明书
quality release　质量证明，质量合格证明
quality reliability assurance　质量可靠性保证
quality requirement　质量要求
quality restriction　质量限制
quality review　质量审查
quality specification　质量标准，质量说明书，质量技术规格，质量规范
quality standard　质量标准
quality supervision　质量监督
quality surveillance　质量监督
quality system elements　质量体系要素
quality system registration　质量体系注册
quality system review　质量体系评审
quality target　质量目标
quality test　质量检验
quality tolerance　品质公差
quality verification　质量验证
quality　品质，质量，性质，等级，含汽率，特性
quango　特殊法人
Quantenary sediment　第四纪沉积物
Quanternary period　第四纪
Quanternary　第四纪
quantifiable non-material deviation　可度量的非实质性误差
quantifier　计量计，量词，配量计
quantify　定量，量化
quantile　分位点，分位数
quantitative analysis　定量分析
quantitative assessment　定量评价
quantitative attribute　量的属性，数量特征，品质的属性
quantitative classification　定量分类
quantitative data　定量资料，数字资料
quantitative description　定量描述
quantitative determination　定量测定
quantitative estimation　定量估算
quantitative filter paper　定量滤纸
quantitative forecast　定量预报，数值预报
quantitative hydrology　定量水文学
quantitative measurement　定量测量，定量测定
quantitative metallography technique　定量金相技术
quantitative method　定量法
quantitative morphology　定量形态学
quantitative reaction　定量反应
quantitative safety goal　定量安全目标
quantitative system　定量分类系
quantitative test　定量试验
quantitative　数量的，定量的
quantity certificate　数量证书
quantity delivered　交付数量
quantity determination　数量确定
quantity flow　流量
quantity galvanometer　电量检流计，测量电荷量的电流计
quantity governing　量调节
quantity measurement　数量测量
quantity meter　电量表，电量计
quantity of cooling water per cooler　每个冷却器的冷却水量
quantity of current　流量
quantity of electric charge　电量
quantity of electricity　电量，电荷
quantity of energy　能量
quantity of flow　流量
quantity of heat　热量
quantity of illumination　光照量，曝光量
quantity of information　信息量
quantity of light　光量
quantity of radiant energy　辐射能量
quantity of refuse　垃圾量
quantity of seepage　渗漏量
quantity of vegetative cover　植被覆盖量
quantity of work　工程量，工作量，劳动量，功量
quantity overrun　工程量差额
quantity production　大量生产
quantity restriction　数量限制
quantity sheet　工程量表，工程数量表，土方表
quantity surveyor　工程量估算员，工程量统计员，数量勘查员
quantity survey　工程量估算
quantity underrun　工程量差额
quantity　量，数量，定量，分量
quantization error　量化误差，整量化误差

quantization noise　分层声噪，量化噪声
quantization signal　量化信号
quantization　量化，量子化，整量化
quantized pulse modulation　分层脉冲调制，量化脉冲调制
quantized signal　量化信号
quantized system　量化系统
quantizer　脉冲调制器，量化器，数字转换器
quantize　数字转换，量化，分层，量子化
quantizing encoder　整量化编码器，量化编码器
quantizing　整量化，取整，量化
quantometer　光子计数器，光谱分析器，辐射强度测量计，光量计，冲击电流计
quantum constant　量子常数
quantum counter　量子计数器
quantum effect　量子效应
quantum efficiency　量子产额，量子效率
quantum law　量子定律
quantum mechanics　量子力学
quantum number　量子数
quantum yield　量子产额，量子效率
quantum　定额，定量，份额，量子，量子的
quaquaversal fold　穹形褶皱
quaquaversal structure　穹形构造
quaquaversal　穹形，穹形圆顶
quarantine fee　检验检疫费
quarantine office　检疫所
quarantine　隔离
quarle　异形耐火砖，大块耐火砖
quarrier　采石工人
quarry blasting　采石爆破
quarry brick　缸砖，方砖
quarry car　采石场斗车
quarry drill　采石钻机
quarry dust　采石场粉尘
quarry engineering　采石工程
quarry equipment　采石场设备，采石设备
quarry face　采石面
quarrying method　采石方法
quarrying rock breakwater　块石防波堤，石块防波堤
quarrying　采石
quarryman　采石工
quarry material　采石场石料
quarry refuse　采石场废石，采石场弃石
quarry rock　采石场毛料，毛石料
quarry rubbish　石碴
quarry run rockfill　毛石填筑
quarry run rock　毛石料
quarry run stone　毛石料
quarry-run　采石量
quarry stone bond　毛石砌合
quarry stone　粗石
quarry tile　缸砖
quarry water　石坑水，石矿水，石窝水
quarry　采石场
quarter bend　90°弯管
quarter chord point　四分之一弦点
quarter circle orifice plate　四分之一圆孔板
quarter deck　后甲板
quarter-diurnal tide　四分日潮
quartering process　四分法
quartering wind　斜风，后侧风，尾舷风，侧风
quartering　四分法
quarterly schedule　季度计划
quarterly　25%，百分之二十五，四分之一，季度的
quarterly　每季的
quarter phase circuit　两相回路，两相电路
quarter phase system　二相制
quarter phase　两相的，四分之一，四等分，相互垂直
quarter-point deflection　四分点挠度
quarter-point loading　四分点荷载
quarter squares multiplier　四分之一乘方乘法器，开四次方乘法器
quarter squares　四分之一乘方，开四次方的
quarter-turn rotary valve　四分之一转动阀【柱塞、球或蝶阀】
quarter-turn valve　四分之一转动阀
quarter-turn　四分之一转动
quarter wavelength line　四分之一波长线
quarter wave stub　四分之一波长短线
quarter wave transformer　四分之一波长变换器
quarter wave　四分之一波长的
quarter　街区，四分之一
quartic　四次
quarts sand　石英砂
quartz-andesite　石英安山岩
quartz bulb　石英灯泡
quartz crystal oscillator　石英晶体振荡器
quartz crystal stabilizer　石英晶体稳定器
quartz crystal　石英晶体
quartz-diorite　石英闪长岩
quartz fiber dose meter　石英丝剂量计
quartz fiber microbalance　石英丝微量天平
quartz fiber　石英丝，石英纤维
quartz fibre dosimeter　带有石英丝的剂量计
quartz fibre electrometer　带有石英丝的静电计
quartz fibre electroscope　带有石英丝的验电器，石英丝验电器
quartz filter　石英过滤器，石英滤波器
quartz glass　石英玻璃
quartz grain　石英颗粒
quartzite　石英岩
quartzitic intrusion　石英侵入脉
quartz lamp　石英灯，水晶灯
quartz mercury lamp　石英水银灯
quartz-monzonite　石英二长石
quartzose shale　石英页岩
quartzous　含石英的，石英质的
quartz particles　石英颗粒
quartz-porphyrite　石英玢岩
quartz porphyry　石英斑岩
quartz powder　石英粉
quartz resonator　石英谐振器
quartz sand　石英砂
quartz spiral-torsion microbalance　石英弹簧扭力秤
quartz tube　石英管
quartz　石英，水晶
quart　夸脱【等于四分之一加仑】

quasi albedo approach method 准反照率近似方法
quasi anisotropic layered medium 准各向异性分层介质
quasi annular flow 准环状流
quasi-attached flow 准附着流
quasi breeder reactor 准增殖反应堆
quasi chemical method 似化学方法
quasi-cleavage rupture 准解理断裂
quasi coherent 准相干光
quasi conductor 准导体，半导体
quasi continuous operation 似连续运行，准连续工作
quasi coordinate 准坐标
quasi critical damping 准临界阻尼
quasi critical 准临界的
quasi cyclic code 准循环码
quasi diagonal matrix 准对角矩阵
quasi diagonal 准对角的
quasi diffusion 准扩散
quasi elastic dipole 准弹性偶极子
quasi elastic vibration 准弹性振动
quasi electric field 准电场
quasi equilibrium 准平衡
quasi ergodic 拟遍历性的，准各态历经的
quasi-geostrophic approximation 准地转近似
quasi-harmonic oscillation 准谐振
quasi homogeneouse wave 准均匀波
quasi homogeneous reactor 准均匀反应堆
quasi-homogeneous turbulence 准均匀湍流
quasi homogeneous 准均匀的
quasi-incompressible fluid 准不可压流体
quasi instruction 拟指令，伪指令
quasi insulator 准绝缘体
quasi linear approximation 准线性近似
quasi linear equation 拟线性方程
quasi linear feedback control system 拟线性反馈控制系统
quasi linearization method 准线性化方法
quasi linearization 准线性化
quasi linear 拟线性（的）
quasi longitudinal wave 准纵波
quasi maximum value 准最大值
quasi monoenergetic neutrons 准单能中子
quasi-normal flow 准正常水流
quasi peak detector 准峰值检测器，类峰值检测器，准峰值检波器
quasi peak meter 准峰值测量仪
quasi peak output 准峰值输出，类峰值输出
quasi peak （电场强度的）准峰值
quasi perfect code 准完备码
quasi periodicity 准周期性
quasi-periodic vibration 准周期振动
quasi permanent combination 准永久组合
quasi permanent value of an action 作用准永久值
quasi permanent value 准永久值
quasi-preconsolidation pressure 似先期固结压力，准先期固结压力
quasi punctuated probe 拟点状探头
quasi random access memory 准随机存取存储器
quasi random access 准随机存取
quasi random 拟随机的，准随机
quasi resonance 准共振，准谐振
quasi-simple wave 拟简波
quasi sine-wave 准正弦波
quasi-smooth flow 准平滑水流
quasi stability 拟稳定性，准稳定性，拟稳定
quasi stable state 准稳态
quasi-state cycling loading 准静态周期载荷
quasi state 准静态
quasi-static assumption 准静态假设
quasi-static core 准静态堆芯
quasi-static deflection 准静态挠度
quasi-static derivative 准静导数
quasi static process 准静态过程
quasi-static response 准静态响应
quasi-static wind load 准静态风载
quasi-static 准静力的，准静态的
quasi stationary current 似稳电流，准稳电流
quasi stationary state 似稳状态，拟稳态
quasi stationary 准静止的，准固定的，似稳的
quasi-steady aerodynamics 准定常空气动力学
quasi-steady approach 准定常方法
quasi-steady flow 准稳定流，准稳定流动，准稳态流，准定常流
quasi-steady state 准稳态，似稳态
quasi-steady 准定常的，准稳定的
quasistellar radio sources 类星射电源
quasi-synchronization 准同步，准同期
quasi-synchronous sampling 准同步采样
quasi-synchronous 准同步的，准同步
quasi-three-dimensional solution 准三元解
quasi-transverse wave 准横波
quasi-two-dimensional flow 准二维流，准二元流
quasivariable 准变数
quasi-viscous 似黏滞性的
quasi 准，即，似，类似的，外表的
quaterdenary 十四进制的
quaternary 四进制的，四元的
quaternionic 四元的
quaternion wall 堤岸墙
quaternion 四元数
quayage 码头总长
quay berth 码头泊位
quay crane 港岸起重机，码头起重机
quay pier 岸壁式突码头，防波堤，突堤码头
quay shed 码头仓库，前方仓库
quay structure 岸壁结构
quay surface 码头地面
quay wall foundation 岸壁基础
quay wall on pile foundation 桩基岸壁
quay wall on piles 桩承岸壁
quay wall 岸壁，岸壁型码头，岸墙，堤岸
quay 岸壁，码头，岸墙，顺岸码头
queen post truss 双桩屋架，双柱桁架，双柱大梁，双柱上撑式桁架，双柱架
queen truss 双柱桁架
quefrency 类频率，拟频率，逆频
quench aging 淬火（后自然）时效
quench alloy steel 淬硬合金钢
quench circuit 灭弧电路，熄灭电路
quench cooling phase 骤冷阶段

quench crack 淬火裂纹
quench drum 骤冷箱
quenched and tempered steel 调制钢
quenched and tempered 淬火并回火的
quenched gap 淬熄火花隙，灭弧隙
quenched steel 淬硬钢
quenched 淬火的
quench effect 冷激效应
quencher 灭弧器，熄灭器，阻尼器，冷却池，冷渣设备，灭火器
quench front 急冷面，淬火面【失水事故-堆芯再淹没】，骤冷前沿
quench hardened case 淬硬层
quench hardening 淬火，淬硬化
quench hot 高温淬火
quenching and high temperature tempering 调质，淬火及高温回火
quenching and tempering 调质处理，冷淬及回火
quenching bath 淬火浴
quenching brittle crack 淬硬脆化裂纹
quenching circuit 淬熄电路，灭弧电路，消火花电路
quenching compound 冷却化合物，冷却介质，冷却液，淬火剂
quenching crack 淬致裂纹，淬火裂纹
quenching degree 淬透性
quenching effect 淬灭效应
quenching frequency 淬熄频率，辅助频率
quenching gas 淬熄气体
quenching heat-exchanger 急冷废热锅炉，急冷热交换器
quenching medium 淬火剂，冷却介质，冷淬液
quenching moment 淬熄时刻，灭弧时刻，灭弧点
quenching of excitation 突然灭磁
quenching pot （开关的）灭弧室
quenching pulse 熄灭脉冲，消隐脉冲
quenching speed 淬火速度
quenching system 冷风系统
quenching time 淬熄时间，灭弧时间
quenching transformer 淬熄变压器，灭弧变压器
quenching water 冷却水
quench oil 淬火油，急冷油
quench tank 骤冷槽
quench time 间歇时间
quench water 骤冷水
quench 淬火，熄灭，骤冷，硬化，（金属）淬火处理，抑制，阻尼，熄灭，断开，遏制，冷浸
queried access 询问存取
query in writing 书面质疑
query language 查询语言
query program 查询程序
query （数据传输的）问号，疑问，问题回答系统，查询，询问
questionable 有争议的
question answer 问题回答
questionnaire 调查表，书面调查，征求意见表
question of paramount importance 头等重要的问题
question of principle 原则性问题
question of substance 实质性问题
question under discussion 正在讨论的问题
question 问题，疑问
queue algorithm 排队控制算法
queue control block 队列控制块
queued access 排队存取
queued telecommunication access method 排队通信存取法
queue element 队列元素
queueing problem 排队问题
queueing process 排队过程
queueing theory 排队论
queueing 排列
queue request 排队请求
queue 排队，队列
queuing delay 排队延时
quibinary code 五二码
quibinary 五二进制的，五二码
quick access drum 快速存取磁鼓
quick access memory 快速存取存储器，快速存储器
quick access 快速访问，快速存取，快速存取的
quick-acting actuator 快速执行机构
quick acting automatic switch 高速自动开关，高速自动断路器
quick acting charging 快速充电
quick acting coupling 速动联轴节
quick acting operator 快速操作器
quick acting regulator 速动调节器，快动作调节器
quick acting relay 速动继电器
quick acting shutoff 快速关闭
quick acting starter 速动启动器
quick acting switching-off 快速分闸，快速切断
quick acting switching 快速切换，速动合闸
quick acting voltage regulator 速动电压调整器
quick acting 快速动作
quick action contact 快动作触点
quick action fuse 速动熔丝
quick action switch 速动开关
quick action valve 快动阀门，速动阀
quick action 快动，速动，快作用
quick adjusting 快速调整的
quick assets 速冻资产
quick break cut-out 快动断路器
quick break fuse 速断熔断器
quick break knife-switch 速断闸刀开关
quick break switch 高速断路器
quick break tester 快速熔断丝试件
quick break 速断，高速断路器
quick cement 快凝水泥，速凝水泥
quick change of characteristic 特性的快速改变
quick change of trim 特性的快速改变
quick change orifice plate 快速变换的孔板
quick charger 快速充电机，快速充电器
quick clip 快速线夹
quick close down 快速关闭
quick closing damper control 快关风门控制
quick closing damper 速关气门，快关风门，速关挡板
quick closing isolating damper 速关隔离挡板
quick closing isolating valve 快关隔离阀
quick closing valve 快闭阀，快关阀

quick-connect coupling 快速离合器接合
quick-connect fitting 快速连接件
quick connection 快速接头
quick-connect lug 快速连接夹
quick-connect receptacle 快速连接插座
quick-connect terminal block 快速连接端子板
quick-connect terminal 快速连接端子
quick coupling 速动联轴节，速接联轴节，快速接头
quick deexcitation 快速灭磁
quick demagnetization 快速去磁
quick dependable communication 快速可靠通信
quick disconnect coupling 离合器快速分离，速断连接，速断耦合
quick disconnect lug 快速断开夹
quick disconnect 快速拆卸，速断，迅速断开
quick dismountable connector 快速拆卸接头
quick-drying varnish 快干漆，快干清漆
quick-dump car 速卸式货车
quick exhaust valve 快速排气阀
quick fitting 快速连接的，快装的
quick freezing 快速冻结
quick-hardening cement 快硬水泥
quick-jellying property 速凝性
quick liabilities 短期债务，速动负债
quick lime 氧化钙，生石灰
quick load change capacity 快速变负荷能力
quick lock in 快速同步
quick lock 快速闭锁
quick look record 速见记录
quick magnetic saturated current transformer 速饱和电流互感器
quick make-and-break contact 快速通断接触
quick make,quick break 快通快断
quick make 快速接通，快速闭合
quick oil classification 土壤简易分类法
quick open flow characteristic 快开流量特性
quick-opening gate valve 快动阀门
quick-opening valve characteristic 快开阀门特性
quick-opening valve 快开阀门，快开阀，速开阀
quick-opening 快动的，快开的，快速开启
quick operating relay 速动继电器
quick operating valve 旋阀门，快开阀门
quick operating 快动的，快速的
quick-priming siphon spillway 快速动作的虹吸式溢洪道
quick ratio 速动比率
quick reaction 快速反应
quick release grip 快卸夹具
quick release mechanism 快速脱扣机构
quick release 速释，速断，快速释放的
quick releasing relay 速放继电器
quick response excitation control 快速励磁调整
quick response excitation system 强励系统，快速励磁系统
quick response excitation 快速励磁，强励磁
quick response exciter 快速反应励磁机
quick response governor 快速反应调速器
quick response thermocouple 小惯性热电偶
quick response transducer 小惯性传感器
quick response voltage control 快速电压调整，快速电压控制
quick response 快速响应，快速响应的，快速反应的
quick restarting after a scram 紧急停堆后的快速再启动
quick-return flow 快速回流
quick sand 流沙，飞沙
quick-selling product 畅销品
quick set cement 速凝水泥
quick-setting additive 快凝掺和剂，快凝剂
quick-setting agent 速凝剂
quick-setting asphalt 快凝沥青
quick-setting cement 快凝水泥，速凝水泥，快干水泥
quick-setting curing asphalt 快凝沥青
quick-setting instrument 快速整定仪表
quick-setting level 速调水平仪
quick-setting mortar 快凝灰浆
quickset 快速设置，快速键，树篱
quick shear strength 抗快剪强度，快剪强度
quick shear test 快剪试验
quick silver 水银，汞
quick start air ejector 快速启动空气喷射器
quick start capability 快速启动能力
quick start lamp 瞬时启动灯，快速启动灯
quick start 快速启动
quick static loading 快速静荷载
quick steaming unit 快速蒸发装置
quick steaming 快速蒸发
quick test 快剪试验
quick turnover and small profit 薄利多销
quick 快速的
quiescent carrier telephony 载波抑制电话
quiescent carrier 抑制载波，静载波
quiescent condition 静止状态
quiescent current 静态电流
quiescent fluidization 平稳流态化
quiescent fluidized bed 平稳流化床
quiescent load 静载荷，固定载荷
quiescent operation point 安定运行点，静态运行点
quiescent plasma 静等离子体
quiescent point 静点，静态工作点
quiescent settling 静止沉降
quiescent state 静止状态，静态
quiescent telecontrol system 静态远动系统
quiescent value 无载运行值，静态值，开路值
quiescent 静止的
quiet circuit 无噪声电路
quieting sensitivity 静噪灵敏度
quiet motor 无噪声马达
quiet run 平稳运转
quiet-type switch 无声开关【如水银开关】
quiet 平静的
quill shaft 挠性短轴，套管轴，中空轴
quill 衬套，套管，套管轴
quilted insulation 内填式隔热，填塞绝缘
quinary digit 五进制数字
quinary notation 五进制计数法
quinary 五倍的，五进制的，五的
quincuncial piles 梅花桩

quindenary 十五进制的
quinhydrone electrode 氢醌电极
quinine sulfate 硫酸奎宁
quinquevalence 五价
quintessence 实体
quintic 五次
quintillion 10^{18}【美国、法国用法】，10^{30}【英国、德国用法】
quintupler 五倍倍压器，五倍倍频器
quintuplicate 一式五份
quirk 沟，深槽
quit claim deed 放弃权利的契约，放弃索偿契约，放弃合法权利转让契据
quit claim 放弃索赔权利，放弃要求权，放弃权利，放弃产权
quitforcing fabric 钢筋网
quit office 退职
quit work 停工
quit 退出
qunty＝quantity 量，数量
quod vide 参见该条，参照，参阅
quoin culvert 闸隅涵洞
quoin 墙角，屋角石，楔子，打楔子夹紧，隅角
quorum 法定人数
quota agreement 定额协议
quota-free product 非配额产品
quota management 定额管理
quota remuneration 定额计酬
quota restriction 配额限制
quota system 定额分配制，配额制，配额制度
quotation clarification meeting 报价书澄清会议
quotation document 报价书
quotation list 行情表，报价表
quotation of price 报价（单）
quotation scope 报价范围
quotation 报价，标价，报价单，引证，引文，估价单
quotative discharge of hydropower station 水电站引用流量
quota 定额，限额，分摊额，配额
quoted price 报价
quoted securities 挂牌证券
quotient-difference algorithm 商差算法
quotient relay 比例继电器，商数继电器
quotient 比，商【数】，系数
quotiety 率，系数
Q value Q值【核反应能值，单位为 MeV】
qwerty key 字符数字键

R

rabbet joint 半槽接合，企口接合
rabbet plane 半槽边刨，槽刨，凸边刨
rabbet 缺口，止口，镶口，刨刀，榫接，插孔，槽，凹部，凹凸榫接，半槽企口，槽口，槽口接缝处，企口缝，嵌接，在……上开槽口，由槽口接合
rabbit tube 跑兔管
rabbit 跑兔，气动速送器
rabble aggregate 毛石骨料
rabble catchwater drain 毛石截水沟
rabble 铁耙，搅拌棍，拨火棒，高低缝，铲口，刮刀，搅拌器，耙，司炉
race condition 竞态条件
race pulverizer 中速球磨机
race rotation 空转，逸转
race running 空转
racer 轴承环
race time 惰走时间，空转时间
racetrack winding 跑道形绕组
racetrack 跑道形电磁分离器，跑道形放电管
raceway guide drawing 电缆管道指导图
raceway layout drawing 电缆管道布置图
raceway system （电缆）走向系统
raceway wire 槽板布线
raceway 电缆管道，布线槽，输水道，电缆管，电缆通道，轨道，轴承座圈，水沟，走向
race （轴承的）滚道，（滚珠轴承钢球的）座圈，环，空转，皮带槽，急流，座圈
rachet 棘轮
racing speed 逸转速度
racing 空转，逸转，飞逸
rack-and-lever jack 齿条杠杆千斤顶
rack and panel accessories 支架和配电盘附件
rack and panel connector 机柜连接器，转接插头座
rack and pinion drive （小）齿轮齿条传动装置
rack and pinion gearing 齿条与齿轮联动，齿条小齿轮传动装置
rack and pinion gear 齿条与小齿轮传动装置
rack and pinion jack 齿条齿轮千斤顶
rack and pinion mechanism 齿条-小齿轮机构
rack angle 机架角
rack bar 齿条，齿杆
rack-cleaning machine 拦污栅清理机
rack drive 齿条传动
rack earth 机壳接地
racked-in and connected 装进架子并连接的
racked-in 装进架子的【推进就位，但不连接】
racked-out【drawer】 从柜中抽出（抽屉）
racked timbering 斜角支撑
rack for common piping 综合管道支架
racking 用齿条移位，企口接缝，架子，湿法富选，阴阳榫，放到架上
rack insulator 架式绝缘子
rack mounted instrument 架装仪表
rack railway 齿轨铁道
rack rail 齿轨
rack rake 拦污栅耙
rack seat 拦污栅栅座
rack soot blower 伸缩式吹灰器
rack system 托架系统
rack type jack 牙条式千斤顶
rack 导轨，支架，齿条，机架，架，台架，栅格，拦污栅，格架
RAC(random access control) 随机存取控制器
RAC(read address counter) 读地址计数器
radar aerial 雷达天线
radar altimeter 雷达测高仪
radar altimetry 雷达测高法

RADA(random access discrete address) 随机存取离散地址
radar antenna 雷达天线
radar band 雷达波带
radar beacon 雷达信标
radar begin 雷达波速
radar-directed 雷达操纵的，雷达定向的
radar echo 雷达回波
radar illumination swathe 雷达照射条带
radar mapping 雷达测图
radar measurement 雷达测量
radar meteorological observation 雷达气象观测
radar meteorology 雷达气象学
radar photograph 雷达航摄照片
radar plotter 雷达量测仪
radar precipitation echoes 雷达降水回波
radar ranging 雷达测距
radar return analysis 雷达回波分析
radarscope 雷达显示器，雷达示波器
radar sounding 雷达探空
radar station 雷达站
radar surveying 雷达测量
radar weather observation 雷达天气观测
radar wind sounding 雷达测风
radar 无线电探测器
RADAS(random access discrete addressing system) 随机存取离散编址系统
RADA system (random access discrete address system) 随机存取离散位址系统
radcrete 经辐射处理的混凝土
radex(radiation exclusion) 放射性物质的排除
radgas 放射性气体
radiac instrumentation 放射性检测仪表，放射性探测仪器
radiac instrument 辐射仪表
radiacmeter 核辐射测定计，剂量计，辐射剂量计
radiacwash 放射性去污液的商名
radiac 放射性的，核辐射的测定，辐射仪
radial acceleration 径向加速度
radial admission 径向进汽
radial air gap 径向气隙
radial and axial cooling 径轴向通风冷却
radial armature 凸极电枢，径向电枢
radial arm stacker-reclaimer 旋臂堆取料机
radial arm stacker 辐射臂堆料机
radial axial compressor 径轴向混流压气机
radial axial flow turbine 辐向轴流式水轮机，混流式水轮机，轴向辐流式水轮机
radial ball bearing 径向滚珠轴承
radial bar 辐射形钢筋
radial basin 辐射状流域
radial bearing 径向轴承，横力轴承
radial bladed impeller 辐射式叶轮，叶片径置式叶轮
radial blade element 径向叶片元素
radial blade-fan 径向叶片风机
radial blanket （水轮发电机的）辐射式机架，径向转换区
radial breeder 径向增殖区
radial brush holder 辐式刷握
radial brush 辐向电刷，径向电刷
radial buckling 径向衬套，径向曲率
radial carbon reflector 径向碳反射层
radial centre 辐射中心
radial check gate 弧形节制闸门
radial clearance detector 径向间隙检测装置
radial clearance 径向间隙
radial collector well 辐射状集水井
radial commutator 辐向排列整流子，径向排列整流子
radial component 辐向部分，辐向分量，径向分量，径向部分
radial compressor 径向压缩机，离心式压缩机，辐流式压气机
radial concentration profile 径向浓度分布
radial cooling slot 径向冷却槽
radial cooling 径向冷却
radial coordinate 极坐标
radial core barrel support 心轴径向支承，吊篮径向支承
radial crack 辐射形裂缝
radial cylinder 径向排列汽缸
radial cylindrical-roller bearing 径向滚柱轴承
radial density 径向密度
radial deviation 径向偏差
radial diffusion coefficient 径向扩散系数
radial diffusion 径向扩散
radial direction 径向
radial displacement 径向位移
radial distance 径向距离
radial distribution feeder 辐射式配电馈线，径向配电馈线
radial distribution system 放射形配电制
radial distribution 径向分布
radial drainage pattern 放射状水系型
radial drainage system 辐射形排水系统
radial drainage 辐射形排水，辐射状水系，径向排液
radial drill 摇臂钻床
radial eccentricity 径向偏心率，径向跳动，径向圆跳动
radial-entry blade 径向装入式叶片
radial equilibrium 径向平衡
radial expansion 径向膨胀
radial extension vibration mode 径向伸缩振动模
radial fault 放射断层
radial feeder 辐射式配电馈线，径向馈线，单馈线
radial field cable 分相屏蔽电缆
radial fin 径向鳍片，径向汽封片
radial fissure 径向裂缝
radial flow back pressure turbine 辐流式背压汽轮机
radial flow compressor 辐流式压缩机，径流式压气机
radial flow fan 径向流风机
radial flow impeller 离心式叶轮，辐流式叶轮，径流叶轮
radial flow impulse turbine 辐流冲动式汽轮机
radial flow model 径向流模型
radial flow noncondensing turbine 辐流背压式汽

轮机
radial flow pump 辐流泵
radial flow rotor 径向通风转子
radial flow settling basin 辐射流沉淀池
radial flow steam turbine 辐流式汽轮机
radial flow tank 径流式水池
radial flow tube bundle （凝汽器中）有辐向蒸汽通道的管束
radial flow turbine 辐流式汽轮机，径流透平，辐流式水轮机
radial flow 径向流动，径流，辐流，辐向流
radial flux distribution 径向通量分布
radial flux flattening 能量径向展平
radial flux permanent magnet synchronous generator 径向磁通永磁同步发电机
radial flux 径向通量，径向流
radial force 辐射力，径向力，辐向力
radial form factor 径向形状因子
radial gate 弧形闸门
radial gland 径向汽封
radial gradient 径向梯度
radial guide 径向导向
radial highway 辐射式公路
radial ice 径向覆冰
radial impeller pump 径向叶轮泵
radial inflow turbine 径向流透平，向心辐内流式水轮机
radial inflow 向心式
radial inlet impeller 径向入口叶轮
radial instability 径向不稳定性
radial intersection method 辐射交会法
radial inversion 径向倒料
radial inward admission 向心进汽
radial inward flow 辐射内向流，径向内流
radial inward turbine 向心式透平
radial journal bearing 轴颈轴承，径向支承轴承，径向轴承，支持轴承
radial junction 径向接头
radial labyrinth gland 径向迷宫汽封
radial lead assembling tool 径向引导组装工具
radial lead resistor 径向引线电阻器
radial line plot 辐射线图
radial line triangulation 辐射三角测量
radial line 径向线，径向沟，辐射线
radial load 径向负荷
radially bare （反应堆）径向无反射层的
radially inward steam flow 向心蒸汽流
radially non-uniform heat flux density profile 径向非均匀热流密度分布图
radially split casing pump 径向拼合泵壳的泵
radially split casing 径向拼合壳体
radially split pump 径向剖分泵
radially split type 径向剖分式
radially split 径向对分的，径向拼合
radially uniform heat flux density profile 径向均匀热流密度分布图
radially 径向地
radial drill 摇臂钻
radial mixing model 径向混合模型
radial mode 径向（振动）模式，径向模态
radial network 径向网络，辐射状网络，辐射形网
radial node 径向波节
radial operation 辐射运行
radial outflow turbine 离心式透平
radial outward admission 离心进汽
radial outward flow 辐射外向流，径向外流
radial packing 径向汽封，径向密封
radial passage 径向风道，径向通道
radial pattern 放射型，辐射型
radial peaking factor 径向峰值因数，径向峰因子，径向峰值因子
radial peaking（hot spot）factor 径向峰值热点因子
radial photometer 径向光度计
radial pin coupling 径向销连接
radial pin 径向销
radial play 径向游隙
radial plunger pump 径向柱塞泵
radial pole piece 径向磁极块，径向极靴
radial position 径向位置
radial power density distribution 径向功率密度分布
radial pressure gradient 径向压力梯度
radial pressure 径向压力
radial projection 放射型投影，放射投影
radial ray 辐射线，放射线
radial rearrangement of fuel 径向倒料，燃料径向倒换，燃料径向再分布
radial reflector saving 径向反射层节省
radial rigidity 径向刚度
radial road 辐射式公路
radial rolling-contact bearing 径向滚动接触轴承
radial rotor machine 凸极电机
radial runout 径向跳动
radial scan 径向扫描
radial scheme 辐射式系统布置【如道路、管路、线路等】
radial seal 径向密封
radial section 径向截面
radial self aligning roller bearing 径向自定位滚柱轴承
radial sewer system 辐射式下水道系统
radial shaft seal 径向轴封
radial shake 辐裂
radial shuffling of fuel 径向倒料，燃料径向倒换
radial sleeve bearing 径向套筒轴承，径向滑动轴承
radial slot 径向槽，辐射槽
radial solid mixing 径向固体混合
radial spherical-roller bearing 径向球面滚柱轴承
radial spill strip 径向汽封片
radial spoke fan 径向叶片式风机
radial stacker 辐射式堆料机
radial stay 径向拉撑
radial strain 径向形变，径向应变
radial stress 径向应力
radial subtransmission 放射状二次输电
radial supply line 辐射形供电线路
radial support key 径向支撑键【反应堆压力壳】
radial support pad 径向支承块
radial support saddle 径向支座【反应堆压力壳】
radial support system 径向支承系统

radial support 径向支撑
radial symmetry 径向对称
radial system 放射系统，径向配电制，辐射状配电制，径向系统
radial tapered-roller bearing 径向锥形滚柱轴承
radial tension 径向张力
radial thermal gradient 径向温度梯度
radial thrust bearing 径向推力轴承，径向止推轴承
radial thrust 径向推力
radial tie-rod 径向拉杆
radial tilt factor 径向不均匀系数，径向倾斜系数
radial tilt 径向不平衡【通量】
radial tower stacker 塔式辐射堆料机
radial transfer 内外传送
radial transmission feeder 放射形输电的馈电线
radial transmission line 径向传输线，辐射式输电线路
radial-traveling cableway 辐射移动式缆索起重机
radial triangulator 辐射三角仪
radial tunnel 辐射式地槽
radial turbine 辐流式透平，径向流透平
radial type core 辐射式铁芯
radial type unloader 旋臂式卸煤机
radial vane 径向叶片
radial vector 径向向量
radial velocity 径向速度
radial ventilation 辐向通风，径向通风
radial vibration 径向振动
radial viewing head 径向瞄准头
radial whirl 辐向旋涡，径向涡流
radial winding 辐式绕法，辐式绕组
radial wind sounding 径向风测深
radial wobble 径向摆动
radial xenon oscillations 径向氙振荡
radial 放射的，辐射的，辐向的，半径的，径向的，放射状的
radiance exposure 曝辐照度
radiance 辐射亮度，辐射度，辐射率
radian frequency 角频率
radian measure 弧度，弧度法
radiant absorption surface 辐射受热面
radiant absorption 辐射吸收
radiant arc furnace 辐射电弧炉
radiant boiler 辐射锅炉，辐射式锅炉
radiant burner 无焰喷燃器
radiant density 辐射密度
radiant drying 辐射法干燥
radiant efficiency 辐射效率
radiant emittance 辐射发射度
radiant emmisivity 辐射率
radiant energy density 辐射能量密度
radiant energy fluence rate 辐射能流率，辐射能流
radiant energy flux 辐射能通量
radiant energy intensity 辐射强度
radiant energy thermometer 辐射温度计
radiant energy 辐射能量，辐射能【Q】
radiant exitance 辐射出射度【M】
radiant exposure 辐照量，曝辐（射）量
radiant flow 辐射流
radiant flux density 辐射通量密度
radiant flux 辐（射能）通量【Φ】
radiant gain 辐射增益
radiant heat-absorbing surface 辐射受热面
radiant heat exchanger 辐射热交换器
radiant heat exchange 辐射热交换
radiant heat flux 辐射热流，辐射热流量，辐射热通量
radiant heating surface 辐射受热面
radiant heating 辐射采暖，辐射供暖
radiant heat lamp 辐射热灯
radiant heat-transfer coefficient 辐射传热系数，辐射放热系数
radiant heat transfer 辐射传热
radiant heat 辐射热
radiant intensity 辐射强度【I】
radiant interchange 辐射换热，辐射热交换
radiant panel heating 辐射板供暖
radiant power 辐射功率
radiant reflectance 辐射反射比
radiant reheater 辐射式再热器
radiant section 辐射区
radiant sensitivity 辐射灵敏度
radiant superheater 辐射式过热器
radiant surface 辐射面
radiant-target pyrometer 辐射高温计
radiant temperature 辐射温度
radiant transmittance 辐射透射比
radiant zone 辐射区
radiant 辐射的
radian 弧度
radiated interference 辐射干扰
radiated noise 辐射噪声
radiate dose meter 射线剂量仪
radiated power 辐射功率
radiated wave 辐射波
radiate 发光，发热，散热，辐射，放射
radiating angle 辐射角
radiating body 发射体，辐射体
radiating capacity 辐射能力
radiating circuit 辐射电路，发射电路
radiating crack 辐射裂纹
radiating element 辐射元
radiating fin 散热片
radiating gill 散热片
radiating guide 辐射波导
radiating heat 辐射热，散热
radiating particle 辐射粒子
radiating plasma 辐射等离子体
radiating power 辐射功率，辐射本领，辐射能，辐射强度，发射能力
radiating rib 散热肋，散热片
radiating slot 辐射槽
radiating source 辐射源
radiating surface 散热面，辐射面
radiating vein 辐射状岩脉
radiating 辐射的
radiation ablation 辐射消融
radiation absorption analysis 辐射吸收分析

radiation absorption 辐射吸收
radiation absorptivity 辐射吸收率
radiation accident 辐射事故
radiation ageing 辐射老化
radiation alarm system 辐射报警系统
radiational tide 辐射潮
radiation amount 辐射总量
radiation analysis 辐射分析
radiation analyzer 辐射分析器
radiation angle 发射角
radiation appliance 辐射装置，辐照设备
radiation attenuation 辐射衰减
radiation balance concentration X 辐射平衡聚光度 X
radiation balance meter 辐射平衡表
radiation balance 辐射平衡，辐射差额
radiation bank count 辐射储备量
radiation barrier 射线防护屏，防辐射屏，辐射屏蔽
radiation beacon 辐射报警器
radiation beam 辐射束
radiation biological effect 辐射生物学效应
radiation breakdown 辐射损伤，辐射杀伤
radiation budget 辐射收支
radiation burn 辐射烧伤
radiation capacity 辐射能力
radiation capture 伴生辐射俘获，辐射俘获
radiation carcinogenesis 辐射致癌
radiation cavity 辐射空间
radiation chemical yield 辐射化学产额
radiation chemistry of macromolecules 高分子辐射化学
radiation chemistry 辐射化学
radiation conductivity 辐射传导率
radiation constant 辐射常数
radiation-control area 辐射管理区，辐射管制区
radiation cooled tube 自然冷却电子管，辐射冷却电子管，气冷管
radiation cooling 辐射冷却
radiation correction 辐射修正
radiation corrosion 辐射腐蚀
radiation counter tube 辐射计数管
radiation counter 辐射计数器
radiation coupling 辐射耦合
radiation damage criteria 辐射损伤准则
radiation damage product 辐射损伤产物
radiation damage reaction 辐射损伤反应
radiation damage susceptibility 辐射损伤敏感性
radiation damage 辐射损害【非生物】，辐射线损伤，辐射损伤，辐射杀伤，放射致伤
radiation damping 辐射阻尼，辐射衰减
radiation decomposition 辐解，辐射分解
radiation density 辐射密度
radiation detection instrument 辐射探测仪
radiation detection technology 辐射探测技术
radiation detection 辐射探测
radiation detector 辐射检测器，射线检测仪，辐射探测器，放射线探测器，放射性检测仪
radiation dosage indicator 辐射剂量指示器
radiation dose distribution 辐射剂量分布
radiation dose meter 辐射剂量计
radiation dose monitoring 辐射剂量监测
radiation dose standard 辐射剂量标准
radiation dose unit 辐射剂量单位
radiation dose 辐射剂量
radiation dosimeter 辐射剂量计
radiation dosimetry 辐射剂量学，辐射剂量测定法
radiation ecology 辐射生态学
radiation effect on man 人的辐射效应
radiation effect 放射效应，辐射效应
radiation efficiency 放射效率，辐射效率
radiation embrittlement 辐射脆化
radiation emissivity 辐射发射率
radiation emitter 辐射体
radiation energy 辐射能
radiation equilibrium 辐射平衡，放射平衡
radiation exchange between surface 表面间的辐射换热
radiation excitation 辐射激发，辐射激励
radiation exposure of the environment 环境的射线照射量
radiation exposure 射线照射辐照，辐射照射辐照，辐射照射量，辐射照射
radiation factor 辐射率，辐射系数，放射率
radiation field 辐射场，辐照场，照射场
radiation fin bonnet 带散热片阀盖
radiation fin 散热片
radiation flux density 辐射通量密度
radiation flux 射线通量，辐射流，辐射通量
radiation fog 放射雾，辐射雾
radiation furnace 辐射炉
radiation genetics 辐射遗传学
radiation hardening 辐射硬化，增强抗辐射，辐照硬化
radiation hazard 辐射危害，辐射危险，辐射伤害，放射危害
radiation heat exchange 辐射热交换
radiation heat flux 辐射热通量
radiation heating criteria 辐射加热准则
radiation heating surface 辐射受热面
radiation heating 辐射加热，辐照热，辐射增温
radiation heat transfer 辐射传热
radiation heat 辐射热
radiation hygiene 放射保健学，辐射卫生学
radiation image contrast 辐射图像衬度
radiation impact 辐射影响
radiation impedance 辐射阻抗
radiation indicator 射线指示器
radiation induced cancer 辐射致癌
radiation induced creep effect 辐射感生蠕变效应
radiation induced creep 辐照蠕变
radiation induced diffusion effect 辐射感生扩散效应
radiation induced genetics 辐射遗传学，辐照遗传学
radiation induced oxidation of the graphite 石墨的辐照氧化
radiation induced shrinkage 辐照收缩
radiation induced swelling 辐射肿胀
radiation injury 放射损伤，辐射损伤
radiation instability 辐射不稳定性

radiation instrument 辐射仪，剂量仪
radiation insult 辐射伤害，辐射侵害
radiation intensity 辐射强度
radiation inversion 辐射逆温
radiation ionization 辐射电离
radiation ionizing 辐射电离
radiation jump 辐射突跃
radiation leakage at the Fukushima Nuclear Power Plant （日本）福岛核电站辐射水泄漏事件
radiation leakage 辐射漏泄
radiation length 辐射长度
radiation level meter 辐射式料位计，核辐射式料位计
radiation level 辐射能级，辐射水平
radiation load 辐射负荷
radiation loss 辐射损耗，辐射热损失，辐射损失
radiation maze 防辐射用的迷宫式入口，辐射防护曲径入口，辐射迷宫网，迷宫式辐射屏蔽
radiation measurement assembly 辐射测量装置，辐射测量组件
radiation measuring instrument 辐射测量仪，辐射测量仪器
radiation measuring room 辐射测量室
radiation meter 射线计，辐射计，伦琴计，辐射测量仪
radiation method 放射测量法（平板测量用的），放射法，辐射法
radiation mode 辐射模式
radiation modification 辐射改性
radiation monitoring and alarm assembly 放射性探测报警器，辐射监测报警装置
radiation monitoring equipment 辐射监测装置
radiation monitoring film 辐射监测胶片
radiation monitoring instrument 辐射监测仪，辐射监测器
radiation monitoring system 辐射监测系统
radiation monitoring 辐射监测
radiation monitor 辐射监视器，辐射监测器，辐射监测仪
radiation opacity 辐射不透明度
radiation passage monitor 通道辐射监测器
radiation pattern 辐射方向图，辐射图
radiation physics 射线物理学，辐射物理学，辐射物理
radiation polymerization 放射聚合
radiation potential 辐射电势，辐射电位
radiation power 辐射功率
radiation precaution sign 辐射警告标志
radiation pressure 辐射压力
radiation probe 辐射探测仪
radiation processing reactor 辐射处理反应堆
radiation processing 辐射加工，辐射处理
radiation proof electric equipment 耐辐射电气设备
radiation proof 防辐射的
radiation protection assessment 辐射防护评价
radiation protection device 辐射防护装置
radiation protection dosimetry 辐射防护剂量学
radiation protection guide 辐射防护指南
radiation protection officer 辐射防护官员
radiation protection regulation 辐射防护管理
radiation protection standard 辐射防护标准
radiation protection superviser 辐射防护监督员
radiation protection survey 辐射防护测量
radiation protection technique 辐射防护技术
radiation protection window 辐射防护窗
radiation protection 辐射防护，辐照防护装置，放射防护
radiation purity 辐射纯度
radiation pyrometer temperature measurement system 辐射高温计温度测量系统
radiation pyrometer 辐射高温计
radiation quality 辐射质量，辐射品质
radiation quantity 辐射量
radiation rate 辐射强度
radiation ray 放射线
radiation receiver 辐射接收机
radiation regime 辐射状态
Radiation Research Association （英国）辐射研究协会
radiation resistance property 耐辐射性
radiation resistance 抗辐射性，辐射阻力，辐射电阻，耐辐射性，抗辐射力，耐辐射力
radiation resistant cable 耐辐照电缆
radiation resistant low-wear material couple 耐辐射与耐磨材料的配合
radiation resistant relay 耐辐照继电器
radiation resistant 耐辐射的，抗辐射的
radiation risk 辐射风险，辐射危险
radiation safety control 辐射安全控制
radiation safety measure 辐射安全措施
radiation safety 辐射安全
radiation scale 辐射标准值
radiation sensor 辐射传感器，辐射探头
radiation shielding sheet 辐射屏蔽板
radiation shielding 辐射屏蔽
radiation shield 辐射防护屏，辐射遮热屏，辐射屏蔽层，辐射护罩，防辐射屏
radiation sign 辐射符号，辐射标记
radiation slip method 辐射滑动方法
radiation source container 辐射源容器
radiation source guide fixture 辐射源导向架
radiation source 放射源，辐射源
radiation specimen guide 辐照样品导向管
radiation spectrum 辐射谱
radiation stability 辐照稳定度
radiation standard 辐射标准
radiation stress 辐射应力
radiation superheater 辐射式过热器
radiation surveillance capsule 辐射监视舱
radiation survey meter 辐射测量计
radiation survey 辐射调查
radiation swelling 辐照肿胀
radiation thermocouple 辐射热电偶，辐射温差电偶
radiation thermometer 辐射温度计
radiation thickness gauge 射线测厚仪
radiation thimble 辐照孔道
radiation transfer 辐射传递
radiation transport equation 辐射迁移方程，辐射输送方程

radiation trap 辐射阱
radiation treatment 照射处理，放射疗法，放疗，放射线处理，射线防腐
radiation type timbering 辐射式支撑
radiation value 辐射值
radiation warning assembly 辐射报警装置
radiation warning symbol 辐射警告符号
radiation width 射线辐射的宽度
radiation window 辐射窗
radiation worker 辐射工作人员，辐射工作者，直接从事放射性工作的人员
radiation work 放射性工作
radiation yield 辐射产额
radiation zone 活性区【反应堆】，辐照区，辐射区
radiation 放射，辐射，辐射热，辐射线
radiative absorption 辐射吸收
radiative capture cross-section 辐射俘获截面
radiative capture 辐射俘获
radiative collision 辐射碰撞
radiative diffusivity 辐射扩散率
radiative effect 辐射效应
radiative equilibrium 辐射平衡
radiative excitation 辐射激发
radiative heat exchange 辐射热交换
radiative heat transfer 辐射传热
radiative inelastic scattering cross section 辐射非弹性散射截面
radiative 发光的，辐射的，发射的，发热的
radiator case 散热器箱
radiator false front 散热器护栅板，散热器前护栅
radiator fan 散热器风机，散热器风扇
radiator fin 散热片
radiator gill 散热器散热片
radiator grid 散热器格栅
radiator hanger 散热器吊钩
radiator header tank 散热器上水箱
radiator heating system 散热器采暖系统
radiator machine 带散热器的电机
radiator nipple 散热器螺纹连接短管
radiator of sound 声辐射器，声源
radiator pedestal 散热器角，散热器支座
radiator screw nipple 散热器螺纹连接短管
radiator section 散热器片
radiator shield 散热器罩
radiator shutter 散热器百叶窗，散热器百叶罩
radiator tank （变压器的）带散热器的油箱
radiator trap 散热器疏水器
radiator tube 散热管
radiator type transformer 散热管式变压器，散热器式变压器
radiator valve 散热器调节阀
radiator vent valve 散热器放气阀
radiator vent 散热器排放口
radiator with thin fin 翼形散热器
radiator 辐射体，散热片，辐射器，散热器，辐射电热器
radical contradiction 根本矛盾
radical sign 根号
radical 根部，基础
radicand 被开方数
radication 开方，生根
radices 根，基，基数，底
radicle 基，根，原子团
radio-acoustic method 电声法
radio-acoustic ranging 电声测距法
radio acoustics 无线电声学
radio acoustic 无线电声学的
radioactivation analysis 放射活化分析
radioactivation 辐射活化
radioactive accident 放射性事故
radioactive aerosol 放射性气溶胶
radioactive age determination 放射性测年，放射性年代测定性
radioactive age 放射性寿命
radioactive airborne particulates 放射性尘埃【大气中】
radioactive air sampler 放射性空气取样器
radioactive ash 放射性尘埃，放射性灰（渣）
radioactive assay 放射性检验，放射性验定，放射性分析
radioactive atom 放射形原子
radioactive background 放射性本底
radioactive blowdown treatment line 放射性排污处理管线
radioactive cemetery 倾倒放射性废料的场所
radioactive change 放射性变化
radioactive charging 装放射源
radioactive cloud 放射性（烟）云
radioactive compound 放射性化合物
radioactive concentration 放射性浓度
radioactive constant 放射性常数，放射常数
radioactive contaminant 放射性污染物，放射性沾染物
radioactive contamination of system water （电站）热网水的放射性污染
radioactive contamination regulations 放射性污染规章
radioactive contamination 放射性污染
radioactive content 放射性含量
radioactive cooling 活性冷却，放射性冷却
radioactive corrosion product carryover 放射性腐蚀产物夹带
radioactive corrosion product 放射性腐蚀产物
radioactive damage 放射性物质引起的伤害，辐射伤害，放射性损伤
radioactive dating 放射性年代测定性，放射性测年
radioactive daughter product 放射性子体产物
radioactive debris 放射性碎片
radioactive decay constant 放射性衰变常数
radioactive decay law 放射性衰变定律
radioactive decay 放射（性）衰变
radioactive decontamination 放射性钝化，消除放射性，放射性去污
radioactive density gauging 放射性密度测量
radioactive density 放射性密度
radioactive deposit 放射性沉积物，放射性沉降物
radioactive detector 放射性探测器
radioactive discharge 放射性排弃物

radioactive disintegration 放射性蜕变
radioactive drain header 放射性排污集管
radioactive drug 放射性药物，放射性制剂
radioactive dry fall-out 放射性干沉降物
radioactive dust 放射性灰尘，放射性尘埃
radioactive effect 放射性效应，放射性影响
radioactive effluent disposal 放射性排放物处置
radioactive effluent drain pipe 放射性液流排放管道
radioactive effluent plant area 放射性液流处理厂区
radioactive effluent 放射性流出物，放射性废水，放射性排放物
radioactive element in coal 煤中放射性元素
radioactive element 放射性元素
radioactive emission 放射性排放
radioactive energy 放射能
radioactive equilibrium 放射性平衡
radioactive evaporation concentrates 放射性蒸发浓缩物
radioactive evaporator bottoms 放射性蒸发浓缩物，放射性蒸发底渣
radioactive fallout 放射性沉降物，放射性落下灰，放射性落尘，放射性散落物
radioactive fission gas 放射性裂变气体
radioactive fission product release 放射性裂变产物释放
radioactive fission product 放射性裂变产物
radioactive gas mixture 放射性气体混合物
radioactive gas 放射性气体
radioactive half-life 放射性的半衰期
radioactive heating surface 放射性（辐射）加热面
radioactive heat 放射性热，放射性衰变热
radioactive hood 放射性通风柜
radioactive inclusion 放射性夹杂物
radioactive indicator 放射性示踪物，放射性指示器，放射性指示剂
radioactive iodine 放射性碘
radioactive isotope dilution method 放射性同位素稀释法
radioactive isotope thermoelectric generator 放射性同位素热电发生器
radioactive isotope 放射性同位素
radioactive krypton isotope 放射性氪同位素
radioactive leak 放射性泄漏
radioactive level device 放射性料位计
radioactive level gauge 放射性液面计
radioactive level transmitter 辐射式液位变送器
radioactive level 放射性水平
radioactive liquid waste concentrate 放射性废液浓缩物
radioactive liquid waste disposal 放射性液体废物处置
radioactive liquid waste 放射性废液
radioactive material 放射性物质，放射性材料
radioactive mineral deposit 放射性矿物
radioactive nature 放射特性
radioactive nuclide 放射性核素
radioactive particles 放射性微粒
radioactive pebble 放射性卵石
radioactive phosphorus 放射性磷
radioactive poisoning 放射性中毒
radioactive poison 放射性毒物
radioactive pollutant 放射性废物，放射性污染物，放射性污物
radioactive pollution 放射性污染
radioactive processing 放射性物质加工
radioactive purity 放射性纯度
radioactive radiation 放射性辐射
radioactive rainout 辐射尘
radioactive rain 放射性雨
radioactive ray 放射线
radioactive release 放射性释放
radioactive residue method 放射性残余法
radioactive salt 放射性盐
radioactive sample 放射性样品
radioactive sand 放射性沙
radioactive series 放射（性）系列
radioactive-solution gauging 放射性溶液测流速法
radioactive source term 源项，放射性源项
radioactive source 放射（性）源，辐射源
radioactive standard 放射性标准，放射标准
radioactive substance 放射性物质
radioactive surface contamination 放射性表面污染
radioactive thickness gauge 放射性测厚计
radioactive tracer gas 放射性示踪气体
radioactive tracer method 放射性示踪法
radioactive tracer salt 放射性示踪盐
radioactive tracer 放射性指示剂，放射性同位素示踪器，放射性示踪物，放射性示踪剂
radioactive transfer 放射性迁移
radioactive ventilation system 放射性通风系统
radioactive waste disposal 放射性废物处置
radioactive waste drain system 放射性废渣排放系统
radioactive waste liquid 放射性废液
radioactive waste management 放射性废物的管理
radioactive waste material 放射性废物
radioactive waste processing and disposal 放射性废物处理和处置
radioactive waste processing 放射性废物处理
radioactive waste release 放射性废物释放
radioactive waste repository 放射性废物处置库
radioactive waste residue 放射性废渣
radioactive waste storage 放射性废物贮存
radioactive waste store 放射性废物贮存
radioactive waste treatment and storage system 放射性废物处理及贮存系统
radioactive waste treatment 放射性废物处理
radioactive wastewater disposal system 放射性废水处置系统
radioactive wastewater 放射性废水
radioactive waste 放射性废物，放射性排放物
radioactive water 活化水，放射性水
radioactive xenon isotope 放射性氙同位素
radioactive 放射性的
radioactivity concentration 放射性浓度
radioactivity content 放射性含量

radioactivity decontamination agent 放射性去污剂
radioactivity decontamination 放射性去污
radioactivity determination 放射法测定
radioactivity level 放射性程度，放射性水平［能级］
radioactivity logging method 放射测井法
radioactivity logging 放射性测井
radioactivity log 放射性测井记录
radioactivity measuring point 放射性测量点
radioactivity meter 放射性测量计，放射性测定计
radioactivity protection 放射性防护
radioactivity release 放射性（物质）释放
radioactivity resistant 耐放射性的，抗辐射的
radioactivity standard 放射性标准
radioactivity strength 放射性强度
radioactivity surveillance 放射性监护
radioactivity 放射性，放射（现象），放射学
radioaerosol 放射性气溶胶
radio alarm signal 无线电报警信号
radio altimeter 无线电测高计
radio amplifier 无线电放大器，高频放大器
radioanalysis 放射分析，放射性分析
radio antenna 无线电天线
radioassay 放射性分析，放射性测量，放射性检验
radio astronomy 无线电天文学
radioautocontrol 无线电自动控制
radio autogram 无线电传真
radio beacon 无线电（航空）信标，无线电指向标
radiobioassay 放射生物测定
radiobiological 放射生物学的
radiobiology 放射生物学
radio broadcasting 无线电广播
radiocable 高频电缆
radio call letters 无线电呼号
radio carrier frequency 无线电载波频率
radio cartogram （全身）放射性统计图
radiocartography 放射统计法
radio center 无线电中心
radio ceramics 高频陶瓷，射频陶瓷
radio channel 无线电信道，无线电波道
radiochemical analysis 放射化学分析
radiochemical centre 放射化学中心
radiochemical contamination 放射化学污染
radiochemical laboratory 放射化学实验室
radiochemical method 放射化学法
radiochemical synthesis 放射化学合成
radiochemical 放射化学的
radiochemistry 放射化学
radiochromatography 放射色谱法
radiocircuit 高频电路
radiocobalt 放射性钴
radio code 无线电通信电码，无线电缩语
radio coil 无线电线圈
radiocolloid 放射性胶体
radio communication engineering 无线电通信工程
radio communication 无线电通信
radio component 无线电零件
radio contact 无线电呼叫，无线电联络
radiocontaminant 放射性污染物
radiocontamination 辐射性污染
radiocontrast agent 放射性对比剂
radio control box 无线电操纵箱
radio control relay 无线电控制继电器
radio control 无线电操纵，无线电控制
radio current meter 无线电测流速仪
radio detection and ranging 无线电定向与测距【雷达】
radio direction finder 无线电定向仪，无线电测向计
radio distance finder 无线电测距器
radio duct 无线电波道
radio-echo detector 无线电回波探测器
radio-echo 无线电回波
radioecology 放射生态学，辐射生态学
radioelement 放射性元素
radio emission 无线电发射
radio environment 无线电环境
radio equipment 无线电设备
radio field strength 电磁波场强，高频场强
radio fix 无线电定位
radio-frequency presentation 射频显示，不检波显示
radio-frequency alternator 高频振荡器，高频发电机，高频发生器
radio-frequency amplification 射频放大
radio-frequency amplifier 射频放大器
radio-frequency cable 射频电缆，无线电频率电缆，高频电缆
radio-frequency carrier shift 射频载波漂移
radio-frequency channel 射频通道
radio-frequency choke 射频扼流圈
radio-frequency converter 射频变频器
radio-frequency current 射频电流
radio-frequency interference 无线电射频干扰，射频干扰
radio-frequency transformer 射频变压器，射频变量器
radio-frequency transmission 无线电频率传输
radio-frequency 射频，高频，无线电频率
radiogoniometer 无线电测角器，无线电定向仪
radiogoniometry 无线电测向术，无线电定向仪，无线电方位测定法
radiographic appearance of discontinuity 缺陷的射线照相显示
radiographic contrast 射线透明对比度
radiographic examination 射线照相探伤，射线照相检验，射线检验，射线照相检查法
radiographic exposure 射线曝光
radiographic inspection 放射线检查，射线照相检查，射线探伤［检验］，射线照相探伤（法）
radiographic instrument 无线电图示仪
radiographic projection technique 射线照相投影技术
radiographic quality 射线照相质量
radiographic screen 射线照相屏
radiographic technique 射线照相技术，射线探伤，投照技术，摄片技术
radiographic term 射线照相术语

radiographic testing　射线检验
radiographic test　放射线探伤，放射线照相试验，放射性检查，射线检验，射线探伤
radiographic　射线照相术的，放射照相的，放射线的，射线照相的
radiograph test　射线试验，(放）射线检查
radiography examination　射线探测
radiography　射线照相（术)，射线照片，射线照相探伤
radiograph　射线照相，伦琴射线照相，伦琴射线照片，放射照相
radiohazard　射线危害
radioimmunoassay　放射免疫分析法
radioimpulse　高频脉冲，无线电脉冲
radioinduced　辐射诱发的，辐射导致的
radio influence field　高频干扰场
radio influence voltage　无线电干扰电压
radio influence　无线电干扰，射频感应
radio instrument　无线电测量仪器
radio interference voltage　高频干扰电压
radio interference　无线电干扰
radio interferometer　无线电干扰仪，无线电干涉仪
radioiodine　放射性碘
radioisotope all or nothing relay　放射性同位素继电器
radioisotope battery　放射性同位素电池
radioisotope pollution　放射性同位素污染
radioisotope production reactor　同位素生产反应堆
radioisotope release　放射性同位素排放
radioisotope smoke alarm　放射性同位素烟雾报警器
radioisotope submersible engine　放射性同位素动力水下浸没式发动机
radioisotope technique　放射性同位素技术
radioisotope tracer　放射性同位素示踪物
radioisotope　放射性同位素
radio jammer　无线电干扰发生器
radio jamming　无线电干扰
radiokrypton　放射性氪
radiolesion　放射性损伤
radio-linked tide gauge　无线电电传感潮位计
radio link protection　无线电保护装置
radio location　无线电定位
radio locator　雷达，无线电定位器
radiological contamination　放射污染
radiological detriment　放射损害
radiological examination　放射学检查
radiological exclusion area　放射性禁区
radiological hazards　放射（性）危害，放射线风险
radiological inspection　放射性检查
radiologically controlled area ventilation system　放射控制区通风系统
radiological physics　放射物理学
radiological program　放射学规划
radiological protection monitor　辐射防护监测仪
radiological protection　放射防护装置，放射（性）防护，辐射防护
radiological reference state　自然环境放射性水平
radiological safety control　放射性安全管理
radiological safety protection　放射性安全防护
radiological safety　放射安全
radiological study　放射性研究
radiological survey　放射性调查
radiological technique　放射技术
radiological　放射学的
radiology　放射学
radiolysis gas　辐解气体，辐射分解气体
radiolysis of methane　甲烷的辐解
radiolysis　射解（作用)，辐解，辐射分解，放射性分解
radiolytic breakdown　辐射分解
radiolytic damage　辐射损伤
radiolytic decomposition　辐射分解
radiolytic dissociation rate　辐射离解率
radiolytic oxidation　辐射氧化
radiolytic　辐射分解的
radio magnetic indicator　无线电磁指示器
radio measurement　无线电测量
radio metal locator　无线电金属探测器
radiometeorograph　无线电气象仪
radiometer gauge　辐射计式压力计，辐射测压计，辐射真空计
radiometer manometer　辐射计式压力计
radiometer　射线探测仪，辐射计，辐射表
radiometric age　放射性年代
radiometric analysis　放射分析法
radiometric inspection　辐射检查
radiometric measurement　辐射测量
radiometric temperature measurement　辐射温度测量
radiometry　辐射测量学，辐射测量术
radio micrometer　微量辐射计，辐射微热计，无线电测微计
radionecrosis　辐射致坏死，放射性坏死
radio network　无线电通信网
radio noise source　无线电噪声源，无线电干扰源
radio noise　无线电噪声
radionuclide absorption　放射性核素吸收
radionuclide contamination　放射性核素污染
radionuclide distribution　放射性核素分布
radionuclide elimination　放射性核素排出
radionuclide kinetics　放射性核素动力学
radionuclide metabolism within human body　放射性核素在人体内的代谢
radionuclide metabolism　放射性核素代谢
radionuclide migration　放射性核素迁移
radionuclide retention　放射性核素滞留
radionuclide transfer in environment　放射性核素环境转移
radionuclide transfer in organism　放射性核素体内转移
radionuclide turnover　放射性核素更新
radionuclide　放射性核素
radio-operated remote-recording system　无线电遥测系统
radiopharmaceutical　放射性药物
radiophobia　辐射恐惧
radio phone　无线电话，无线电收发话机

radio pill 放射性小球
radio pilot 无线电控制气球，无线电测风气球
radioquiet 不产生无线电干扰的
radio range 无线电测距，等信号区无线电信标
radio ranging measurement 无线电测距
radio receiver 无线电接收机
radio receiving set 无线电接收设备，无线电接收装置
radio reception 无线电接收
radio relay communication 无线电中继通信
radio relay station 无线电中继台，无线电转播台，转播台
radio relay 无线电中继的，无线电中继
radio remote-control 无线电遥控
radio reporting river gauging station 无线电报汛站
radioresistance 抗辐射性，耐辐照性
radioresistant 耐辐射的，抗辐射的
radioresonance method 射频共振法
radioscope 放射探测仪，放射镜，剂量测定用验电器，X射线检试法
radioscopy 射线检查法
radiosensitivity 辐射灵敏的，辐射灵敏性
radio shielded 高频屏蔽的，防高频感应的
radio shielding 射频屏蔽
radio signal 射频
radiosity 有效辐射
radiosonde observation 无线电探空仪观测
radio sonde 无线电探空仪，无线电测距仪，无线电高空测候器
radio spectroscopy 无线电频谱学，电磁能全景接收技术
radio spectrum 射频频谱
radio technics 无线电技术
radio technology 无线电工艺，无线电技术
radio telecontrol 无线电遥控
radio telegraph 无线电报机，无线电报术
radiotelemetering system 无线电遥测系统
radiotelemetering 无线电遥测
radio telemetry 无线电遥测学，无线电测远术，无线电远距离测量术
radio telephone network 无线电电话网
radio teletypewriter 无线电传打字机
radiotherapy 放射疗法
radiotolerance 耐辐照度，抗辐射性
radiotolerant 抗辐射的，耐辐照的
radiotoxicity 放射性毒性，辐射中毒
radiotoxicology 放射毒理学
radiotoxin 放射性毒素
radiotracer （放射）示踪物
radio transmission 无线电波传播
radio transmitter 无线电发射机
radio transmitting station 无线电发射台
radiotron 真空管，三极电子管
radio tube performance chart 电子管特性曲线
radio tube 电子管，真空管
radio warning 无线电报警
radio-wave frequency 射频
radio-wave propagation 无线电波传播
radio-wave sounding 无线电波测深
radio-wave spectrum 无线电频谱
radio-wave 无线电波
radio wind 无线电测风仪，测风雷达
radioxenon 放射性氙
radio 无线电通信
radius at bend 弯曲半径
radiused area 辐射状面积
radius of channel's curvature 航道曲度半径
radius of convergence 收敛半径
radius of curvature 曲率半径
radius of extrados 外拱圈半径
radius of gyration 回转半径，转动惯量半径，旋转半径
radiusofinertia 惯性半径
radius of influence 影响半径，有效半径【低水头泵】
radius of intrados 内拱圈半径
radius of plume 羽流半径
radius of rounding 圆角半径
radius of soffit 内拱圈半径
radius of turn 旋转半径
radius tapping 径距取压
radius vector 矢径，幅，位置矢量
radius 半径，幅度，辐射线
radix complement 补码
radix converter 计数制变换器
radix-minus-one complement 反码，基数减一补码
radix sorting 基数分类
radix system 基数制，计数制基，基数系统
radix two counter 多位二进制计数器
radix 根，基，基数，底，语根
radome 天线罩，雷达罩
radon gas 氡气
radon method 氡测法【用氡测定水中镭含量】
radon 氡【Rn】
rad=radian 弧度
rad(radiation absorption dose) 拉德【吸收辐射剂量单位】
rad=radius 半径
RAD(rapid access data) 快速存取数据
RAD(rapid access device) 快速存取设备
RAD(rapid access disc) 快速存取磁盘
radsafe 辐射安全
radsalts 放射性盐
raduction of area 断面收缩率
radux 远距离双曲线低频导航系统，计数制的基数
radwaste building HVAC system 废料厂房高压交流系统
radwaste building sump 放射性废物厂房集水坑
radwaste building 放射性废物厂房
radwaste disposal 放射性废物处置
radwaste management 放射性废物管理
radwaste store 放射性废物贮存（库）
radwaste system 放射性废物处理系统
radwaste transport 放射性废物运输
radwaste treatment 放射性废物处理
radwaste 放射性废物，放射性排放物
raffinate 残液
raft bridge 木筏浮桥
raft chute 筏道，过筏斜槽
rafter and strut-framed dam 木框架式坝

rafter 椽子，桶，屋面梁
raft foundation of nuclear island 核岛筏基
raft foundation 筏基，筏式基础，浮筏基础，联合式基础【架空线路】
raft harbour 木筏港
rafting canal 放筏运河，放排运河
rafting channel 筏运水路
rafting operation 筏运作业
rafting reservoir 筏运水池，筏运水库
rafting 木筏流放
raft of land vegetation 沿岸植被漂流块
raft passageway 筏道
RAFT(radiological assessment field team) 放射性评价现场队
raft-wood 流放的木材
raft 筏基，层，垫板，筏，托，座板，木筏
Raga index 罗加指数
raggle block 防漏盖块
raggle 承水槽【防雨板】
raglin 平顶搁栅【防雨板】
RPG(report program generator) 报告程序生成程序
rag rubble 粗毛石，粗面块石
rag works 石板砌筑
rag 石板瓦，条石
railage 铁路运输，铁路运费
rail-barge combinations 铁路海船联运
rail base 轨底
rail-bender 弯轨机
rail bond 导轨夹紧器，轨端接续线
rail cambering machine 钢轨弯曲机
railcar coupling 车钩
rail carrier 钢轨搬运机
railcar 铁道用的汽车
rail clamp 夹轨器
rail classification yard 铁路编组场
rail crane 轨道起重机
rail current 轨流线
rail flat lorry 有轨平板车
rail gauge 铁道轨距，轨距
rail haulage 铁路运输，卡车托动
rail hauler 铁路拖车
railing around mounting hole 吊装孔栏杆
railing 栅形干扰，栏杆，扶手，栏栅，围栏
rail intake plant 铁路受煤设备
rail jack 起道机
rail joint gap 轨缝
rail joint 钢轨接头
rail layer 铺道机
rail mountd reclaimer 轨道式取料机
rail-mounted bucket wheel S/R 轨道式斗轮堆取料机
rail-mounted crane 轨行起重机
rail-mounted excavator 轨行挖掘机
rail-mounted feeder 轨道式给煤机
rail-mounted feed hopper 轨道式给煤斗
rail-mounted overhead crane 轨行高架起重机
rail-mounted stacker 轨道式堆料机
rail of square steel 方钢轨
rail or truck sampling 火车或卡车取样
rail pile 钢轨桩
rail post 栏杆柱
rail rate 铁路运价率
rail resistance 钢轨电阻，干线电阻
rail return current 钢轨回流
railroad car 铁路车辆
railroad clearance 铁路限界
railroad engineering 铁路工程
railroad overcrossing 铁路跨线桥，铁路天桥
railroad siding 铁路岔线
railroad spur line 铁路专用线
railroad station 火车站
railroad tariff 铁路运价
railroad trestle 铁路栈桥
railroad 铁道，铁路，铁路部门
rail scale 轨道衡
rail scraper 轨道刮板
rail shipment 铁路运输
rail splice 联接板，鱼尾板，连接板
rail spur 铁路岔线，铁路支线
rail square 准轨尺
rail steel 轨钢
rail sweep 轨道清扫
rail table 轨道平台
rail top 轨顶
rail track for trash rack cleaner 拦污栅清理机轨道
rail track gauge 铁路轨距
rail transportation 铁路运输
rail transport 铁路运输
rail unit scale 轨道衡
rail unloading hopper 铁路卸煤斗
rail upright 栏杆立柱
rail voltage 电源线，电源，干线电压
rail-water route 铁路水路
railway administration 铁路局
railway and highway combined bridge 铁路公路两用桥
railway bill of loading 铁路运货提单
railway convertor 铁路用变流机
railway crane 铁路起重机
railway crossing 跨铁路
railway design institute 铁道设计院
railway distribution system 铁路配电系统
railway freight 铁路运费
railway junction 接轨
railway loading 火车装车
railway main line 铁路干线
railway motor 铁路用电动机
railway-side power plant 铁路边电厂，路口电厂
railway siding 专用铁路线，铁路专用线，铁路侧线
railway spur line 铁路专用线
railway station 火车站
railway substation 铁路变电所
railway track 铁路线
railway traffic volume 铁路运输量
railway transportation 铁路运输，铁道运输
railway turbine 机车燃气轮机
railway-yard 调车场
railway 铁路设施，铁道部门，铁路，铁道
rail weigher 轨道衡

rail 铁路，铁轨，导轨，钢轨，栏杆，扶手，轨道，横木
rain arc-over test 湿飞弧试验
rain area 降雨面积
rainbow sprinkler 彩虹喷水器
rain channel 雨水沟
rain chart 雨量图
rain day 下雨日
rain down-pipe 水落管
raindrop spectrometer 雨滴谱仪
rain erosion damage 雨蚀损伤
rain erosion 雨冲蚀
rain factor 雨量因子
rainfall amount 降雨量，雨量
rainfall area 降雨面积，降雨区，雨区
rainfall curve 雨量曲线
rainfall distribution coefficient 雨量分布系数
rainfall distribution 雨量分布
rainfall duration 降雨持续时间，降雨历时
rainfall-evaporation balance 降雨蒸发平衡
rainfall frequency 降雨频率
rainfall in 24 hours 24小时降雨量
rainfall intensity curve 雨量强度曲线，降雨强度曲线
rainfall intensity frequency 雨量强度频率，雨强频率
rainfall intensity meter 雨量强度计
rainfall intensity recurrence interval 雨量强度重现期
rainfall intensity 降雨强度
rainfall inversion 雨量逆增
rainfall loss 雨量损失
rainfall map 雨量分布图，雨量图
rainfall mass curve 雨量累积曲线
rainfall over the years 多年平均降水量
rainfall penetration 雨水渗透
rainfall probability 降水概率
rainfall quantity 雨量
rainfall rate 降雨率
rainfall recorder chart 自记雨量图
rainfall-runoff correlation 降雨径流相关（关系）
rainfall runoff 地面径流，降雨径流
rainfall station 雨量站
rainfall trend 雨量趋势
rainfall variation 雨量变化
rainfall 降雨（量），降水量，降雨
rain gauge network 雨量站网
rain gauge station 雨量站
rain gauge 雨量计
rain guard 防雨设施
rain gush 降雨，骤雨
rain gust frequency 暴雨频率
rain gust 暴雨
rain gutter 雨水槽
rainhead 花洒顶喷
raininess 降雨强度
raining packed bed 淋雨填充床
raining slag 淋渣
raining works 整治工程
rain intensity 降雨强度
rain in torrents 暴雨
rain leader 水落管，雨水管
rainless period 无雨期，枯水期
rainless 无雨的
rain-like condensation 滴状冷凝
rainmaking 人工降雨
rain of electrons 电子流
rain-out wash-out 雨水冲洗，大气中的尘埃被雨水清除
rainout 凝雨散落物，凝雨沉降物，放射性坠尘沉降，雨散落物，雨沉降物
rain pipe 雨水管
rain precipitation 雨量
rain-proof casing 防雨罩
rain-proof test 防雨试验
rain rate 雨强
rain season 雨期
rain-shield insulator 防雨绝缘子
rain shower 雨淋花洒
rain simulator 雨量计
rain spell 多雨期，阴雨期
rainstorm 暴风雨
rainsuit 雨衣
raintight 不漏雨的，防雨的
rain visor 遮雨板
rainwash 雨冲洗，雨水冲蚀，雨水冲刷（物），被雨冲走的东西
rainwater channel 雨水明沟
rainwater drainage pit 雨水排水坑
rainwater drain 雨水排水管
rainwater head 水落斗，雨水斗
rainwater leader 水落管
rainwater outlet 雨水出口
rainwater pond 雨水池
rainwater pump house 雨水泵房
rainwater run-off 排雨水
rainwater 雨水
rainy day 雨天
rainy period 雨期
rainy season 雨季
rainy tropics 多雨热带
rainy year 多雨年
rain 电子流，雨
raise a claim 提出（所有权的）要求，提出索赔要求
raise capital 筹措资金
raised arch 突起拱
raised beach platform 上升滨岸平滩
raised bench 上升阶地
raised bog 高地沼泽
raised-chord truss 起拱桁架
raised countersunk head screw 凸出钻孔头螺丝，半沉头螺钉
raised face flange 凸面法兰
raised face 凸台面，凸面，突面
raised floor 高架地板，提升地板，活地板，假地板
raised pattern safety flooring 凸出花纹安全地板
raised skylight 隆起天窗
raised 凸出的
raise funds to build a power plant 集资建电站
raise interlock relay 上升联锁继电器

raise limit switch 升高限位开关
raise one's claim 提出索赔
raise overtravel limit switch 提升超限开关
raiser 上升烟道，上升管，升管，提升器
raise scaffold 升降式脚手架
raise statement 引发语句
raise steam 锅炉点火
raise 提高，上升，升高
raising and removal of reactor internals 堆内构件的提升和移走
raising funds 集资，筹措资金
raising of dam 坝的加高
raising of water level 水位壅高
raising plate 顶升板
raising platform 提升式平台
raising pressure 升压
raising speed 升速
raising voltage from zero 零起升压
raising 提升，上升，加高
rake angle 桨叶倾斜角，偏斜角
rake bond 斜纹砌合
rake classifier 倾斜式筛分机
rake conveyer 耙式输送机
rake dimension 梯段尺寸
raked joint 剔缝
raked pile 斜桩
raked wing tip 斜翼尖
rake forward 前斜
rake hoist platform 清污耙起重机平台
rake of blade 桨叶斜角，桨叶倾斜度
rake of bow 斜度
rake of pole 电杆的倾斜度
rake of sampling tube 梳状取样排管
rake probe 排管，梳状探针
raker 耙路机
rake 耙，用耙子耙，倾斜，倾斜度，火钩，倾角，前倾面，齿耙，耙子
raking balustrade 斜面栏杆
raking equipment 清污机
raking out 冲刷，擦净
raking pile 斜桩
raking shore 斜撑，斜撑木
raking strut 斜支撑
raking trestle 人字排架
ralative density of material 散料相对密度
RALU(register, arithmetic and logic unit) 寄存器、运算及逻辑部件
ram-air pipe 冲压管
ram-air turbine 冲压式空气涡轮
ram-air 冲压空气
ram and retort grate 柱塞进煤的下饲式炉排
Raman spectroscopy 拉曼光谱学
ram compression 冲压压缩，动压头
ram compressor 冲压式压气机
ram cylinder 柱塞式油缸，活塞式油缸
ram drag 冲压阻力
ram effect 冲压效应
ram feeder 活塞给煤机，推煤柱塞
ramification 支线，支流，分支，分叉，分派，门类，支脉
ramified pipe system 树枝状管网
ramify 分支
ram machine 打桩机
rammed clay 夯实黏土
rammed concrete 夯实混凝土
rammed-earth house 干打垒
rammed-earth 夯实土，素土夯实
rammed ground 夯实地基
rammed lime earth 夯实灰土
rammed plain soil 素土夯实
rammed soil-cement pile 夯实水泥土桩法
rammed-soil pile 夯实土桩
rammer 锤体，夯锤，夯具，夯实机，夯土机
ramming machinery 夯实机械
ramming 打夯，夯实
ram pack 基础垫铁
rampant arch 跛拱，高低脚拱
rampart 防御，防御物，壁垒，波蚀残丘
ramp bridge 斜坡引桥
ramp characteristic 加速特性曲线
ramp delay circuit 斜坡延时电路
ram penetrometer 锤式贯入硬度计
ramp function 直线上升函数
ramp generator 斜坡发生器
ramp grade 斜坡度
ramping 主汽压跃升过程
ramp input signal 瞬变输入信号
ramp input 瞬变输入
ramp insertion of reactivity 反应性的倾斜输入，反应性线性引入
ramp load change 倾斜负荷的变化，负荷线性变化
ramp load variation 倾斜负荷的变化
ramp load 连续负荷
ramp power change 功率线性变化
ramp rate 倾斜率，缓变率，等变率
ramp reactivity insertion 反应性线性引入
ramp reduction 坡式下降
ramp response time 斜坡响应时间
ramp response 斜坡响应
ram pressure 冲压，冲压压力，速度头，全压力
ramp voltage 斜坡电压
ramp 倾斜，斜面，滑台，倾斜端头，挑坎，斜道，斜坡，(斜) 坡道，匝道，对冲断层
RAM(random access memory) 随机访问存储器，随机存取存储器
RAM(random access method) 随机存取法
RAM(relational access method) 相关存取法
RAM(reliability-availability-maintainability) 可靠性、可利用率和可维修性
RAM(repair and maintenance) 检修与维修
ramsonde 锥式硬度计
ram steam pile driver 汽锤打桩机
ram the earth 把土夯实
RAM 控制棒驱动机构电源系统【核电站系统代码】
ram 柱塞，撞头，撞，夯，捣实，撞击子，锤体，动力油缸，液压缸，夯锤，夯实，夯土机，水轮泵，推出机，桩锤，撞击装置，(水压机的) 活塞，(压力泵的) 柱塞
ranch 牧场

RANDAM (random-access nondestructive advanced memory) 随机存取非破坏性先进存储器
random access and correlation for extended performance 扩展性能的随机存取及相关性
random access card equipment 随机存取计算机装置
random access control 随机存取控制器
random access device 随机存取装置
random access discrete addressing system 随机存取离散编址系统
random access discrete address 随机存取离散地址
random access disk file 随机访问磁盘文件，随机存取磁盘文件
random access file 随机存取文件
random access input/output 随机存取输入输出
random access memory 随机访问存储器，随机存取存储器，随机存储器
random access method of accounting and control 计算和控制的随机存取法
random access method 随机存取法
random access nondestructive advanced memory 随机存取非破坏性先进存储器
random access parallel tape 随机存取并行带
random access programming 随机访问程序设计，随机存取程序设计
random access storage and control 随机存取存储器和控制器
random access storage and display 随机存取存储器和显示器
random access storage device 随机存取存储装置
random access storage 随机取数存储器，随机存取存储器
random access time 随机存取时间
random access 随机存取，随机访问
random addressing 随机寻址，随机编址
random analysis 随机分析
random array 随机阵列
random ashlar 乱砌料石
random bond 杂乱砌合
random car 杂车，异型车，混编车辆
random check 随意抽查，随机查核
random-coding bound 随机编码限
random coincidence 随机叠合（重合），偶然符合
random correlation method 随机相关法
random counting error 无规计数误差
random crack 不规则裂缝，不规则裂纹
random defect 不规则缺陷，杂乱缺陷
random diffusional motion 无规则扩散运动
random distribution 随机分布，随机分配
random disturbance 随机骚扰，随机扰动
random embankment 任意填方
random-error-correcting 纠正随机错误
random error 偶然误差，随机误差，无规误差
random event 偶然事件，随机事件
random excitation 随机激励，随机激振，无规则扰动
random failure 偶然故障，随机故障，随机损坏
random fatigue 随机疲劳
random fault 偶然故障
random fluctuation 不规则起伏，随机起伏，随机变动，随机脉动，随机涨落，无规则变动
random forcing function 随机强制函数
random function 随机函数
random Gaussian number 随机高斯数
random gust 随机阵风
random incidence 随机入射
random increment 随机增量
random inhomogeneous medium 随机不规则介质
random inspection 抽查，随机抽查
random item 无规则件
randomization 不规则化，无规则化，随机化
randomizer 随机数发生器
randomize 使随机化，使不规则化
random jump motion 随机跳跃运行
random jump 不规则跳动
random load 随机荷载
randomly fluctuating data 随机涨落数据
randomly fluctuating 随机起伏
randomly oriented flaw 随机取向缺陷
randomly packed bed 随机堆积床
randomly time-varying linear system 随机时变（参数）线性系统
randomly 任意地，随机地
random masonry （随石头不同大小堆砌的）堆砌泥瓦工
random material 任意料
random modulated signal 随机调制信号
random motion 不规则运动，随机运动
random multiple access 随机多路存取
randomness 不规则性，随机性
random neutron density wave 随机中子密度波
random noise background 随机噪声背景
random noise level 随机噪声级
random noise 无规则噪声，杂乱噪声，随机噪声
random-number generator 随机数发生器
random-number 随机数
random order 任意顺序，任意状态，随机位
random oscillation 随机振动
random packing 随机填料，无规则填充
random paralleling 自同期，非同期并车，不规则并联
random pattern 不规则型式
random phase 随机相位
random policy 随机策略
random processing 随机存取数据处理
random process 随机过程，随机处理
random pulse sequence 随机脉冲序列
random pulse 杂乱脉冲，随机脉冲，无规脉冲
random quantity 随机量
random radiography 随机射线照相检查
random reactivity 随机反应性
random response 随机响应
random rockfill 任意石料堆筑，乱石堆筑
random rubble fill 乱石堆填
random rubble wall 乱毛石墙，乱石墙
random sampling method 随机采样法
random sampling 随机取样，随机采样，任意取样，随机抽样

random scanning 随机扫描
random search 随机搜索
random sequence 随机序列
random shock 随机冲击
random signal analysis technique 随机信号分析技术
random superimposed coding 随机迭加编码
random surface renewal 无规则表面更新
random synchronizing 不规则同步，自同期
random thermal creep stress 随机徐变应力
random train 车型不一的列车
random turbulence 无规则湍流，随机湍流
random variable 随机变数，随机变量
random vector 随机矢量
random vibration 随机振动，无规振动，无规则振动，不规则振动
random vortex shedding 随机旋涡脱落
random walk model 随机行走模型，随机走动模式
random walk 随机游动
random winding 散下线圈，不规则线圈，散嵌绕组，散绕组
random wound coil 散绕线圈
random 随机的，随便的，无规则的，机遇的，任意的，偶然的，随机，随意
rangeability 幅度变化范围，量程，范围度，变化幅度
range accuracy 测距精度
range adjustment 范围调整，量程调整
range arithmetic 区间运算
range calibration 距离校正
range change-over switch 量程转换开关
range changing switch 量程变换开关
range changing 变量程
range coal 块煤
range control switch 量程控制开关
range conversion 距离变换
range correction 距离校正
range coverage 覆盖区段，覆盖量程
range determination 距离测定
range extension 量程扩展，范围扩大
range finder 测距器，测程器，测距计，测距仪
range finding 遥测术，测距技术，测距
range-gate generator 距离选通脉冲发生器
range indicator 距离指示器
range interval 间距
range light 导灯
range line 叠标线
range masonry 整层圬工
range measurement 测距
range of adjustment 调节区域，调节范围
range of allowable error 容许误差范围
range of equipment 成套机具
range of equivalence of parameters 参数的等效范围
range of error 误差范围
range of fluctuation 变动范围
range of heat-supply service 供热半径
range of indication 指示范围
range of instability 不稳定范围，不稳定区
range of instrument 仪表量程
range of linearity 线性范围，线性区域
range of load fluctuation 负荷波动范围
range of load 负荷范围
range of measurement 测量范围
range of nominal tension 额定电压范围
range of operation 工作范围，操作范围，运行范围
range of price 价格范围
range of regulation 调节范围
range of scatter 离散范围，分散范围，扩散范围
range of sensitivity 灵敏度范围
range of speed adjustment 调速范围
range of speed control 转速控制范围，变速范围
range of stability 稳定阶段，稳定范围
range of stress variation 应力变动范围
range of stress 应力范围，应力幅度
range of temperature 温度范围
range of transmission 发送范围，传输范围
range of use 使用范围
range of values 范围值
range of variation 变化范围，偏差范围，变化区域
range of visibility 视程
range of wind tide 风潮差
range pole 标杆
range resistor 量程电阻器
ranger 模板横撑，板桩横挡，测程器，测距仪
range scope 距离指示器
range selection 量程选择
range selector 波段开关，距离转换开关
range spacer 隔块，隔片
range span 量程范围
range switch 量程开关，波段开关
range system 母管制
range unit 测距装置
range works 层砌工程
range 变程，草原，测程，范畴，山脉，射程，调整，幅度，跨度，间隔，量程，区间
ranging bond 排列式砌筑
ranging data 测距数据
ranging line 测程线
ranging 测程，测距，测量，定界，定线，距离调整，广泛搜索
rank first in the world 占世界第一
Rankine cycle engine 朗肯循环发动机
Rankine cycle 朗肯循环
Rankine scale 朗氏温标
Rankine's earth pressure theory 朗肯土压力理论
Rankine state 朗肯状态
Rankine's theory of earth pressure 朗肯土压说
Rankine's vortex 朗肯涡流
rank of matrix 矩阵的秩
rank smell 难闻气味
rank 等级，横排，煤阶，排，列，级，分级，入列，煤的品位
RAN(read-around numbers) 周围读数
rapeseed 油菜子
rapid access data 快速存取数据
rapid access device 快速存取设备
rapid access disc 快速存取磁盘

rapid access loop　快速存取回路
rapid access management information system　快速存取管理信息系统
rapid access memory　快速存取存储器
rapid access storage　快速取数存储器，快速存取存储器
rapid access　快速存取
rapid action valve　快动阀
rapid-activity loss　快速活性损耗
rapid aging　快速老化
rapid analysis method　快速分析方法
rapid analysis　快速分析
rapid automatic checkout equipment　自动快速检查装置
rapid circuit etch　快速电路蚀刻
rapid circulation　快速循环
rapid condensation　快速凝结
rapid construction　快速施工
rapid control rod insertion　控制棒快速插入
rapid cooling　快速冷却
rapid corrosion　快速腐蚀
rapid curing asphalt　快干沥青
rapid curing cutback　速干，速凝
rapid curing　快速养护
rapid current　急流
rapid discharge bottom dump hopper　快卸底开门漏斗车
rapid discharge hopper car　快速卸式（漏斗）货车
rapid discharge hopper　快卸式漏斗车
rapid emergency insertion of the control rods　控制棒的紧急快速插入
rapid erosion　快速侵蚀
rapid excavation　快速开挖
rapid field assessment　快速现场检定
rapid filter　高速过滤器，快滤池
rapid filtration　快滤
rapid flow　急流，湍流
rapid fluctuation of water level　水面急剧波动
rapid fluctuation　急剧波动
rapid forcing voltage　快速强励电压
rapid-hardening cement　快干水泥，快硬水泥，快凝水泥
rapid infiltration　快速渗滤
rapid information technique for evaluation　快速信息鉴定技术
rapid insertion　快速插入
rapid internal storage　快速内存储器
rapidity　迅速，速度
rapid loading　快速加载
rapidly extensible language　可快速扩充语言
rapid mix and flocculation basin　快速混合絮凝池
rapid permeability　快速渗透
rapid printer　快速打印装置，快速打印机，快速印刷装置
.rapid random access memory　快速随机存取存储器
rapid reconnaissance　快速勘测
rapid record oscillograph　快速记录示波器
rapid replacement　快速置换
rapid response motor　快速响应电动机
rapid response　迅速响应
rapid-rise test　快速升压试验
rapid river　陡峻河流，湍急河流
rapid-sand filter strainer　快速沙滤器
rapid sand filter　快速沙滤池，快速砂滤器
rapid screw　大节距螺钉
rapid shutdown　快速停堆【反应堆】
rapid spoil removal　快速除渣法
rapid start lamp　快速启动灯，瞬时启动灯
rapid storage　高速存储器，快速存储器
rapid survey　快速测量
rapids　险滩，急流，湍流
rapid tunnel driving　隧洞快速掘进
rapid tunnelling　隧洞快速掘进
rapid　迅速的，急促的
raplot　等点绘图法
rapper　振动器
rapping allowance　铸型尺寸增量
rapping device　振打装置
rapping gear　锤打装置，振动装置
rapping　振动，（除尘器）集尘电极的定期振打
RAP(reliability assurance program)　可靠性保证程序
RAP(resettlement action plan)　移民行动计划
RAP(resource allocation processor)　资源分配处理机
RAPS(remote area power supply)　偏远地区供电
RAPTAP(random access parallel tape)　随机存取并行带
rapture envelope　破裂包线
rapture strength　破坏强度
rapture　破裂
rare animal　珍稀动物
rare earth element　稀土元素
rare earth metal　稀土金属
rare earth　稀土，稀土的，稀土族
rarefaction state　稀疏状态
rarefaction wave　稀疏波
rarefaction　稀疏，稀化，稀薄
rarefied gas　稀薄气体
rare flood　稀遇洪水
rarefy　抽空，抽稀，稀薄化，抽真空
rare gas　稀有气体
RA(reliability analysis)　可靠性分析
rare metal couple　稀有金属温差热偶
rare metal　稀有金属
R/A(revision appendix)　修订/附录
rare　稀有的
RARP(reverse address resolution protocol)　逆地址解析协议
RAR(risk assessment report)　风险评估报告
Rasching ring　拉希环
rashing　煤下页岩，煤层底板的软碳质页岩
rasp cut file　木锉
RASP(reactor analysis support package)　反应堆分析支持包
rasp　（粗）木锉
RAS(reliability, availability and serviceability)　性能最佳性【可靠性、可用性及可维修性】
RASTAC(random access storage and control)　随机存取存储器和控制器

RASTAD(random access storage and display) 随机存取存储器和显示器
raster scope 光栅式阴极射线管
raster 光栅，(荧光屏) 扫描区
ratchet-and-pawl mechanism 棘爪机构
ratchet blocking system 棘轮闭锁装置
ratchet coupling 爪形联轴节，棘轮联结器
ratchet cyclometer 棘轮回转计
ratchet demand 计费基准最大需量
ratchet drill 棘轮扳钻
ratcheting deformation 棘轮效应变形
ratcheting 离合器，棘轮效应，联轴节
ratchet pawl 棘轮掣子，棘轮卡子
ratchet relay 棘轮式继电器
ratchet screwdriver 棘轮螺丝起子
ratchet spanner 齿轮扳手，棘轮扳手
ratchet strain 棘齿应变
ratchet wheel backstop 棘轮逆止器
ratchet wheel 棘轮
ratchet with T-handle T柄棘轮扳手
ratchet wrench 棘轮扳手
ratchet 棘轮，棘爪，单向齿轮
rate action 比率作用，速率作用，微分作用
rate aided signal 定标信号
Rateau turbine 压力外级冲动式汽轮机
rate control action 微分调节作用，比率调节作用
rate controlling regime 速度控制系统
rate controlling step 速度控制步骤
rate control 速率控制，比例控制，速度控制，按被调量变化率调节，微分控制
rated accuracy 额定精确度
rated apparent power 额定视在功率
rated back-to-back capacity bank breaking current 额定背靠背电容器组开断电流
rated base impedance 额定基本阻抗，额定基极阻抗
rated bleedoff temperature 设计的抽汽温度
rated boost pressure 额定吸入压力，额定升压
rated brake power 额定制动功率
rated breaking current 额定断开电流
rated breaking 额定截断功率
rated burden 额定负载，额定负荷，额定功耗
rated cable-charging breaking current 额定电缆充电开断电流
rated capability 额定出力
rated capacitance 标称电容量
rated capacity 额定出力，额定能力，装机容量，额定容量，额定载荷，额定功率，额定蒸发量，设计能力，标称功率
rated closing capability 额定关合能力
rated coil current 额定线圈电流
rated coil voltage 额定线圈电压
rated condition 额定工况，额定参数，额定状态，额定条件
rated consumed power 额定消耗功率
rated consumption 额定消耗，额定消耗量，额定耗量
rated contact current 额定触点电流
rated contact voltage 额定触点电压
rated continuous capacity 额定持续功率
rated continuous current 额定持续电流
rated continuous working voltage 额定连续工作电压
rated current of generator 发电机额定电流
rated current of switch 开关额定电流
rated current 额定电流
rated duration of short-circuit 额定短路持续时间
rated duration 额定工作时间
rated duty 额定功率，额定工作制，额定负载，额定工况，额定税率
rated efficiency 额定效率
rated energy 额定能量
rated excitation 额定励磁
rated filling pressure 额定充气压力
rated final resistance 额定的最终阻力【空气过滤器】
rated fire-resistive period 额定耐火时限
rated flow coefficient 额定流量系数
rated flow 额定流量
rated frequency range 额定频率范围
rated frequency 额定频率
rated fuel power 额定燃料比功率
rated governor 调速器
rated head 额定水头
rated heating capacity 额定供热量
rated high voltage 额定高压
rated hoisting capacity 额定起重量
rated horsepower 额定马力，额定功率
rated impedance 额定阻抗
rated impulse protective level 额定冲击保护水平
rated impulse withstand voltage 额定耐冲击电压
rated input 额定输入，额定输入功率
rated insulation class 额定绝缘等级
rated insulation level 额定绝缘等级
rated kilovolt-ampere 额定千伏安，额定容量
rated kVA tap 额定千伏安抽头
rated kVA 额定千伏安
rated life time 额定寿命
rated life 额定寿命，额定使用期限
rated line-charging breaking current 额定线路充电开断电流
rated load condition 额定负载状态，额定负载运行工况
rated load excitation 额定负载励磁
rated load operation 额定负荷运行
rated load test 额定负荷试验
rated load torque 额定负荷转矩，额定转矩
rated load 额定负载，额定负荷，额定荷载
rated normal current 额定正常电流
rated operating condition 额定运行工况，设计运行工况，额定运行条件
rated operational current 额定工作电流
rated operational voltage 额定工作电压
rated operation condition 额定工况
rated out-of-phase breaking current 额定失步开断电流
rated output capacity 额定输出容量
rated output power 额定输出功率，额定功率
rated output 额定出力，标准产量，额定产量，额定功率，额定输出
rated peak withstand current 额定峰值耐受电流

rated performance 额定性能
rated power coefficient 额定力矩系数
rated power density 额定功率密度
rated power factor 额定功率因数
rated power operation 额定功率运行
rated power output 额定功率输出，额定动力输出
rated power 标称功率，额定功率
rated pressure 额定压力
rated primary current 一次侧额定电流
rated primary voltage 一次侧额定电压
rated quantity 额定量
rated reactive power 额定无功功率
rated reliability 额定可靠性
rated resistance 标称电阻，额定电阻
rated revolution 额定转数，额定转速
rated ripple current 额定纹波电流
rated ripple voltage 额定纹波电压
rated rotation speed of generator 发电机额定转速
rated rotation speed of rotor 风轮额定转速
rated rupturing capacity 额定截断容量
rated secondary voltage 二次侧额定电压
rated short-circuit breaking current 额定短路开断电流
rated short-circuit current 额定短路电流
rated short circuit duration time 额定短路持续时间
rated short circuit making current 额定短路关合电流
rated short-line fault breaking current 额定近区故障开断电流
rated short time current 额定短时（耐受）电流
rated short time withstand current 额定短时耐受电流
rated single capacitor bank breaking current 额定单个电容器组开断电流
rated slip 额定滑差，额定转差率，额定滑差率
rated specific speed 额定比速，额定比转速
rated speed of rotation 额定转速
rated speed 额定速度，额定转速，额定速率
rated steam conditions 额定蒸汽参数
rated steam 额定蒸汽
rated temperature coefficient 标称温度系数
rated temperature-rise 容许温升
rated temperature 标称温度，额定温度，规定温度
rated tension 额定张力
rated thermal current 额定发热电流
rated thermal power 额定热功率
rated thermal stability current 额定热稳定电流
rated throughout 额定生产能力，额定出力
rated tip speed ratio 额定叶尖速度比【标准高速性系数】
rated torque coefficient 额定力矩系数
rated torque 额定转矩
rated travel 额定行程
rated tube current 额定管电流
rated turning speed of rotor 风轮额定转速
rated uninterrupted current 额定持续电流
rated value 额定值
rated valve capacity 阀门额定容量
rated valve discharge pressure 额定的阀门出口压力
rated VA 额定伏安（数）
rated voltage drop 额定压降
rated voltage 额定电压
rated watts input 额定输入功率
rated weight 额定重量
rated wind speed 额定风速【风力机】
rated wind velocity 额定风速
rated withstand voltage 额定耐压
rated 标称的，额定的，计算的，设计的
rate equation 速率方程
rate feedback 速度反馈
rate gain 微分增益，比率增益
rate generator 比率发电机
rate gyroscope 速率陀螺仪
rate indicator 速率指示器
rate integrating gyroscope 速度积分陀螺仪
rate integrating gyro 速度积分陀螺
rate integrating 速度积分
rate limitation 速度限制
rate limit 速度限制，速率限制
ratemeter 速率表，测速计，速率计，测速表，计数率计，辐射量计
rate of absorption 吸收率
rate of air circulation 空气循环率，换气次数
rate of annual depletion 年平均耗水率
rate of attendance 出勤率
rate of blowdown 排污率
rate of boiling 蒸发率，沸腾强度
rate of capacity utilization 开工率
rate of capillary rise 毛细管上升速度，毛管水上升率
rate-of-change indicator 变率指示器
rate-of-change limiting control 变化率限值控制
rate-of-change of circulation 环量变化率
rate-of-change of frequency 频率变率
rate-of-change relay 变率继电器，微分继电器
rate-of-charge 充电时间，充电速率
rate of coastal erosion 海岸侵蚀速率
rate of combustion 燃烧速度
rate of concrete placement 混凝土浇筑速率
rate of consumption 耗量
rate of control-rod insertion 控制棒插入速率
rate of conversion 转化速率
rate of conveyance loss 输水损失率
rate of cooldown 冷却速率
rate of cooling-air flow 冷却空气流量
rate of cooling-air required 冷却空气需要量
rate of correct actuation of electric protection system 电气保护正确动作率
rate of correct actuation of protection system for thermal auxiliary equipment 热工辅机保护正确动作率
rate of correct actuation of thermal process protection system 热工保护正确动作率
rate of correct operation 正确动作率
rate of corrosion 腐蚀速率
rate of crack growth 裂纹扩展率，裂纹生长速度
rate of crack propagation 裂缝扩展速度，裂缝

扩展速率
rate of creep 蠕变速率，徐变速度
rate of crystal growth 晶体增长速度
rate of curve 曲线斜率
rate of curving 弯曲率
rate of damaged cargo 货损率，损坏货物率
rate of damping 阻尼率
rate of decay 衰减率，衰变速率，衰减速度，衰减速率
rate of deceleration 减速率
rate of decent 下降速度
rate of deformation 变形速率
rate of deposition rate 熔敷速度
rate of deposition 沉积速率，熔敷速度
rate of depreciation 折旧率，贬值率
rate of diffusion 扩散速度，扩散速率，扩散率，稀释率，稀释比
rate of disappearance 消失速率
rate of discharge 出口流量，流量，排出率，放电率，排水量
rate of discount 贴现率
rate of dissipation 耗散率
rate of distortion 畸变率
rate of divergence 发散速率
rate of dividend 股利率
rate of downward gradient 下降速度
rate of duty 关税率
rate of effusion 流出速度
rate of emergency water make-up 事故补水量
rate of emission 排放率
rate of energy loss 能量损失比
rate of enthalpy increase 焓增加率
rate of equipment in good condition 设备完好率
rate of equipment utilization 设备利用率
rate of evaporation 蒸发量，蒸发率，蒸汽负荷
rate of exchange 兑换率
rate of expansion 膨胀率
rate of extension 延伸率
rate of fall-out 衰减率
rate of fall 沉降速率
rate of feeding 投料速度
rate of filtration 滤速
rate of flattening of flood wave 洪水波的平复率
rate of flow increase 流量增加速率
rate of flow indicator 流量指示器
rate of flow 流率，流速，流量，流量率
rate of frequency change 频率变率
rate of frost heave 冻胀速率
rate of gas flow 气体流速
rate of groundwater discharge 地下水流量率
rate of groundwater flow 地下水流量率
rate of grout absorption 吸浆率
rate of growth 增长率，增长速度
rate of heat exchange 热交换率
rate of heat flow 热流强度，热流率
rate of heat generation 释热强度，热负荷
rate of heating 暖机速度，加热速度
rate of heat release 热量释放速度【火灾】，放热率
rate of heat transfer 传热强度，热流率
rate of house power 厂用电率
rate of incidence 入射率
rate of increment 增长率
rate of infiltration 入渗率
rate of information loss 信息丢失率
rate of information throughout 信息传送速率
rate of load change 负荷变化速率
rate of load growth 负载增长率，负荷增长率
rate of loading 加载速度，负载率，加负荷速度，负荷率，加荷速率
rate of load up 负荷上升率
rate of loss of coolant 冷却剂丧失率
rate of loss 损耗率
rate of mass transfer 传质速度，质量转移速度
rate of metal consumption 金属消耗率
rate of metal corrosion 金属腐蚀速率
rate of occurence of opening without proper command 误分闸故障率
rate of occurrence of closing without proper command 误合闸故障率
rate of occurrence of short-circuit events 短路故障率
rate of oxidation 氧化速率
rate of penalty 罚款率
rate of perforation 穿孔速度
rate of power change 功率变化速率
rate of preassembled pieces in boiler erection 锅炉安装组合率
rate of premium 保险费率
rate of pressure change 压力变化速率
rate of pressure reduction 降压速率
rate of profit and tax of investment 投资利税率
rate of profit and tax on investment 投资税利率
rate of profit 利润率
rate of pulse repetition 脉冲重复频率
rate of qualification of boiler water quality 炉水品质合格率
rate of qualification of condensate quality 凝结水品质合格率
rate of qualification of feed water quality 给水品质合格率
rate of qualification of steam quality 蒸汽品质合格率
rate of radiation 辐射强度
rate of rainfall 降雨率
rate of reaction 反应速率
rate of reactivity change 反应性变化率
rate of refunded 退税率
rate of regulation 调节速度
rate of replacement 更换率
rate of reservoir silting （水库的）淤积率
rate of residual information loss 残留信息漏失率，信息丢失漏检率
rate of residual value 残值率
rate of return on investment 投资回收率，投资收益率
rate of return on total assets 总资产利润率
rate of return regulation 投资回报率监管法
rate of return （投资）回收期限，利润率，收益率，资本收益率
rate of revolution 转速
rate of rise detector 增长速率测试仪

rate of rise of TRV 瞬态恢复电压上升率
rate of rise of voltage 电压上升率
rate of rise restriking voltage 暂时恢复电压上升率，再击穿电压上升率
rate of rise temperature detector 温升速率探测器
rate of rise 上升率，（曲线）上升斜率，陡度，增长速度，增长速率
rate of rotation 转速
rate of runoff 径流模数
rate of salvage value 残值率
rate of sand movement 移沙率
rate of scale formation 生垢率
rate of secondary consolidation 次固结速率
rate of sediment yield 产沙率
rate of self-purification 自净速度
rate of selfregulation 自调节速度
rate of settling 沉降速率
rate of shear 切变速率
rate of speed 速度变化率，速率
rate of strain tensor 应变率张量
rate of strain 材料应变率，应变速度，应变率
rate of stress application 施加应力的速率
rate of surface runoff 地表径流率，地面径流率
rate of swelling 膨胀速率
rate of telegraphic transfer 电汇汇率
rate of temperature change 温度变化速率，温度变化率
rate of temperature rise 温度上升速率【传感器】
rate of tidal current 潮流速率
rate of transmission losses 线损率
rate of travel of flood peak 洪峰行径速度，洪峰移动速度
rate of travel speed 行走速度
rate of travel 控制棒移动速率
rate-of-turn gyroscope 转度陀螺仪
rate-of-turn recorder 转速记录器
rate-of-turn 转动速度，转速
rate of unloading 减负荷速度
rate of upward gradient 上升速度
rate of utilization of electric automation systems 电气自动投入率
rate of utilization of electric protection systems 电气保护投入率
rate of utilization of HP heater 高加投入率
rate of utilization of measuring points for data acquisition system 数据采集系统测点投入率
rate of utilization of protection systems for thermal auxiliary equipment 热工辅机保护投入率
rate of utilization of thermal automation systems 热工自动投入率
rate of utilization of thermal interlock systems 热工连锁投入率
rate of utilization of thermal process protection systems 热工保护投入率
rate of utilization of thermal sequence control systems 热工程序控制投入率
rate of vacuum down 真空下降率
rate of voltage rise 电压上升速度
rate of volume flow 体积流量，体积流动速率
rate of water level rise 水位上升速率
rate of water make-up 补水量
rate of wear 磨损率
rate process 速率过程
rate regulator 流量调节器
RATE(remote automatic telemetry equipment) 自动遥测装置
rate response 导数反应，微商作用
rate-sensitive 对速度变化灵敏的
rate servo 速度随动系统，速度伺服系统
rate signal 与速度成比例的信号
rate system of electricity 电费收费制
rate time 微分时间，预调时间，比率时间
rate 比率，评定，等级，价格，费用，功率，流量，频率，速度，率，速率
rather M than N 宁可 M 也不 N，是 M 而不是 N，与其（说）N 不与（说）M
rather than （要）……而不……
rathole flow 管状流动
rathole scabbard 鼠洞管
rathole 细长流道，鼠穴，挂料，挂煤
ratholing 鼠洞流，深孔流，挂料，挂煤
ratification of agents'contract 批准代理人签订的合同
ratification 批准，确认，正式批准，认可
ratifier 批准者
ratifying date 报告审批日期
ratify 批准
rating canal 率定槽
rating channel 检定槽
rating curve 流量特性曲线，率定曲线，标定曲线，水位流量关系曲线
rating data 额定数据
rating designation 参数标志
rating discharge curve 率定流量曲线
rating factor 额定值系数，额定因数
rating flume 率定槽
rating method 额定法，鉴定法，率定法
rating of coolers 冷却器的额定功率
rating of current meter 流速仪率定
rating of heater elements 加热元件的额定功率
rating of machine 电机定额，电机规格，电机额定值
rating output 额定输出
rating plate 标牌，铭牌
rating power consumption 额定功耗
rating power 额定功率
rating region table 分级区域表
rating station 率定站
rating system 定额制，定额制度，评级系统，评分系统，配给制
rating table 评级量表法，率定表
rating tank 率定池，率定柜，率定箱
rating test 评级试验，规格试验
rating under air conditioning condition 空调工况制冷量
rating value 额定值，标称值，标准值
rating 出力，承载能力，定额，规格，定额值，额定功率，额定值，额定参数，出力，评价，率，流量，标称值，等级，参数，测定，检定，率定
ratio adjuster 抽头变换位置，（变压器的）分接开关

ratio arm box 电桥电阻箱，比例臂电阻箱
ratio arm 比例臂，电桥臂
ratio balance relay 比率平衡继电器
ratio by volume 体积比
ratio calculator 比例计算机
ratiocination 推理，推论
ratio controller 比率控制器，比率调节器，比值控制器，比例调节器
ratio control system 比率控制系统，比率调节系统，比例控制系统
ratio correction factor 比例校正系数，变比校正系数
ratio detector 比例检波器，比值检波器，比例鉴频器
ratio-differential relay 比率差动继电器
ratio dissipation 比率功耗
ratio error 比率误差，(变压器的）变比误差
ratio factor 比率系数
ratio flow controller 流量比控制器
ratio flow 流量比
ratio interference voltage 无线电干扰电压
ratio meter 比率表，(电流）比率计
ratio method for telemeasurement 比率遥测法
rational adjustment 合理调整
rational analysis of hydrological data 水文资料合理性分析
rational analysis 合理性分析，有理分析
rational depth of well 合理井深
rational design 合理设计
rational electrical unit 合理化电单位
rational exploitation 合理开发
rational expression 有理式
rationale 基本原理，基础理论，理由，根据
rational function 有理函数
rational integer 有理整数
rational interpolating function 有理插值函数
rationalization proposal 合理化建议
rationalization 有理化，合理化
rationalized MKSA unit 合理化的米、千克、秒、安制单位
rationalized unit 合理化单位
rationalize 有理化
rationalizing denominator 有理化分母
rational method 合理方法，推理法
rational number 有理数
rational runoff formula 推理径流公式
rational spacing between wells 合理井距
rational structure 合理结构
rational unit 合理化单位
rational 合理的
ration of coolant to fuel volume 冷却剂与燃料的体积比
ration 定量，限额，定额，配额，配给量，合理的量，正常量，限量供应，配给供应
ratio of attenuation 衰减比
ratio of blade thickness to chord 叶片厚度与叶弦之比
ratio of closing error 闭合比
ratio of compression 压缩比
ratio of current assets to total assets 流动资产对资产总值比率
ratio of damping 阻尼系数，减幅系数
ratio of distance 截距比
ratio of elongation 伸长率
ratio of expansion 膨胀比
ratio of fluctuation 起伏比【沸腾床表面】
ratio of gains 增益比
ratio of generating functions 生成函数比
ratio of heat to electricity 热电比
ratio of height to sectional thickness of wall or column 砌体墙、柱高厚比
ratio of longitudinal reinforcement 纵向配筋率
ratio of “off” to “on” impedance “断开”与“闭合”的阻抗比
ratio of over load 过载度
ratio of principal stresses 主应力比
ratio of reduction 收缩率，衰减率，还原率
ratio of reinforcement 钢筋比率，配筋率
ratio of remaining value 残值率
ratio of revolution 转数比
ratio of slope 坡度，坡度比，坡率，坡度率
ratio of specific heat 热容比，比热比
ratio of the massic heat capacity 质量热容比
ratio of the specific heat capacity 比热容比
ratio of tidal range 潮差比
ratio of tip-section chord to root-section chord 叶片根梢比
ratio of transformation of current transformer 电流互感器的变流比
ratio of transformation of potential transformer 电压互感器的变压比
ratio of transformation 变换系数，变压比，变换比，变比
ratio of transformer 变压器变比
ratio of transmission 传动比
ratio of tunnel nozzle 风洞喷管面积比
ratio of utilization 利用系数
ratio of varying capacitance 电容变化比
ratio of viscosity 黏度比
ratio of winding 匝（数）比，绕组比
ratio Poisson's 泊松比
ratio relay 比例继电器
ratio set 比率装置，比值设定
ratio station 比率控制站，比值操作站
ratio switch 比例开关
ratio table 比率表
ratio test 比率检验法，(变压器的）变比测定
ratio transformer 比例变压器
ratio-turn 匝数比
ratio-voltage 电压比，比例电压
ratio 比例，速度，速率，变化率
ratproof electric installation 防鼠型电气装置
rattail file 鼠尾锉【圆的和锥状的】
rat tail 鼠尾形，鼠尾，天线水平部分与引下线的连接线
rattan 藤条
rattler loss 磨损
rattler 猛烈雷暴，磨损试验机
rattle 颤动
rattling 拍击
ravage 岸
ravine stream 溪流

ravine wind 峡谷风
ravine 冲沟，沟壑，谷沟，山涧，峡谷
raw air 未处理的空气，未过滤的空气
raw and processed material 原料
raw brick 生砖，砖坯
raw coal bin 原煤仓
raw coal bunker 原煤仓，原煤斗
raw coal handling 原煤处理，原煤手选
raw coal piping 原煤管道
raw coal separation system 原煤分选系统
raw coal silo 原煤斗
raw coal sizing screen 原煤分级筛
raw coal 原煤
raw data electronic acquisition system 原始数据电子采集系统
raw data 原始数据，原始资料
raw effluent 原废水，未经处理的废水
raw energy consumption 毛电能消耗量
raw gas 粗煤气，未净化气体，未经净化的气体，原料气
raw gypsum 生石膏
raw information 原始信息
rawinsonde 无线电探空测风仪
raw lacquer 生漆
raw materials for routine consumption 日常消耗用原材料
raw materials industry 原材料工业
raw material storage 原料库
raw material 原材料，原料
raw mica 未加工云母，原云母
raw mix 生混合料，未加工混合料
raw paint 生漆
raw sample 原样品
raw sanitary water 未经处理的生活废水
raw sewage screening 原污水过滤
raw sewage 未经处理的污水，未净化污水，原污水
raw sludge 未经处理的污泥
raw slurry 粗制泥浆
raw subgrade 天然路基
raw wastewater 未经处理的废水
raw waste 生水，原水，未处理水，原废物，待处理的废物
raw water heater 生水加热器
raw water pumping station 生水泵站
raw water pump 生水泵
raw water system 生水系统
raw water tank 生水箱
raw water 生水，原水，未净化的水
ray acoustic 几何声学
ray-control electrode 射线控制电极
ray displacement 射线位移
ray effect 射线效应
ray filter 射线过滤器
Rayleigh criterion 瑞利判据
Rayleigh distribution 瑞利分布
Rayleigh number 瑞利数
Rayleigh scattering 瑞利散射
rayl 雷耳【声阻率的值】
ray modulation 射线调制
Raymond Concrete pile 雷蒙式混凝土桩
rayon 人造纤维，人造丝
ray-proofing device 防射线设备
ray-proof 防射线的，防辐射的
ray＝radiation 射线
Raytheon tube 雷通管，全波整流管
ray tracing 射线轨迹
ray 放射线
RAZ 核岛氮气分配系统【核电站系统代码】
RBCW(reactor building chilled water system) 堆厂房冷冻水系统
RBD(reliability block diagram) 可靠性方块图
RBE dose 相对生物效应剂量，生物学效应剂量
RBE(relative biological effectiveness) 相对生物效应
RBE(remote batch entry) 远程成批输入
RBH(relative biological hazard) 相对生物危害
RBMK 石墨反应堆，石墨水冷堆，沸水冷却式石墨慢化反应堆，压力管式石墨慢化沸水反应炉
RB(reactor building) 反应堆厂房
RB(read backward) 反向读出
RB(read buffer) 读缓冲器
R/B(release to birth ratio) 释产比
RB(return to bias) 归零制
RB(roller bearing) 滚柱轴承
RB(run back) 快速减负荷，快速降负荷，快速返回，辅机故障减负荷
RBT(resistance bulb thermometer) 变阻泡温度计
RB 快速返回【DCS 画面】
RCA(Regional Co-operative Agreement) 地区合作协定
RCB(reactor containment building) 反应堆安全壳建筑物
RCCA alignment 控制棒束组件对准
RCCA(rod cluster control assembly) 控制棒组件
RCCA(root cause corrective action) 根本原因纠正行动
RCC assembly 棒束控制组件，控制棒束组件
RCCA withdrawal accident 提棒事故
RCC changing fixture 控制组件置换装置
RCC drive shaft 束控制驱动轴
RCCD(roller compacted concrete dam) 碾压混凝土坝
RCC element actuation 控制元件动作
RCC element insertion limit 控制元件插入限度
RCC element 棒束控制元件，控制元件
RCC guide tube removal 除去控制棒束导向管
RCC guide tube 控制棒束导向管
RCC handling station 控制棒装卸站
RCCM 法国核电标准【核电站系统代码】
RC coupling 阻容耦合
RCC(radio chemical centre) 放射化学中心
RCC(read channel continue) 读通道继续
RCC(remote communication complex) 远程通信复合系统
RCC(remote communication console) 远程通信操作台
RCC(rod cluster control) 棒束控制
RCC(roller compacted concrete) 碾压混凝土
RCC(rotor current controller) 转子电流控制器
RCC spider body 控制棒束星形连接体

RCC storage rack 控制棒贮藏架
RCC thimble plug changing fixture 阻力塞置换装置
RCC thimble plug tool 控制棒束阻力塞抽出工具
RCD(residual current device) 剩余电流装置
RCDTL(resistor-capacitor diode transistor logic) 电阻电容二极管晶体管逻辑
RCDT(reactor coolant drain tank) 反应堆冷却剂排放箱
RCE(rapid circuit etch) 快速电路蚀刻
RCFC(reactor containment fan cooler) 反应堆安全壳风扇冷却器
RCGM(reactor cover gas monitor) 反应堆覆盖气体监测器
RCIC pump turbine 反应堆堆芯隔离冷却水泵汽轮机
RCIC pump 反应堆堆芯隔离冷却水泵
RCIC(reactor core isolation cooling system) 反应堆堆芯隔离冷却系统
RCIC(reactor core isolation cooling) 反应堆堆芯隔离冷却
RCI(read channel initialize) 读通道开始，读出信道初始化
RCIC system 反应堆堆芯隔离冷却系统
RCMP(reactor coolant motor pump) 一回路电动泵
RCM(reliability-centered maintenance) 以可靠性为核心的维修
RCO(remote control oscillator) 遥控振荡器
RCO(representative calculating operation) 典型计算操作
RCPB(reactor coolant pressure boundary) 反应堆冷却剂压力边界
RC(reactor coolant) 反应堆冷却剂
RCP(reactor cooling pump) 堆冷却泵，主泵，一回路泵，反应堆冷却剂泵
RC(product of resistance times capacitance) 电阻乘电容之积
RCP 反应堆冷却剂系统【核电站系统代码】
RC(ray-control electrode) 射线控制电极
RC(read and compute) 阅读和计算
RC(reader code) 阅读器代码
RC(reinforced concrete) 钢筋混凝土
RC(remote control) 遥控，远距离控制
RC(resistance-capacitance) 电阻电容
RC(resistor-capacitor) 电阻电容
RC(Rockwell hardness) 洛氏硬度
RCR(reader control relay) 阅读器控制继电器
RCR(reservation clear request) 保留项消除申请
RCS(main coolant system) 主冷却剂系统
RCS-NAA(radiochemical separation-neutron activation analysis) 放射化学分离中子活化分析
RCS(reaction control system) 反应控制系统
RCS(reactor coolant system) 反应堆冷却剂系统
RCS(rearward communications system) 后向通信系统
RCS(reloadable control storage) 可写控制存储器
RCS(remote control system) 遥控系统
RCTL(resistor-capacitor transistor logic) 电阻电容晶体管逻辑
RCTL(resistor-coupled transistor logic) 电阻耦合晶体管逻辑
RCU(recovery control unit) 恢复控制器
RCV＝recovering 回收
RCV(removal-chemical and volume control system) 化学容积控制系统
RCV 化学和容积控制系统【核电站系统代码】
RCW(return control word) 返回控制字
RDCHK(read check) 读出检查
R & D cost amortization 研究与开发成本的分摊
RDC(remote data collection) 远程数据收集
RDE(rotating disc electrode) 旋转圆盘电极
RDF(relational data file) 关系数据文件
RDF(resource data file) 资源数据文件
RDH(radioactive drain header) 放射性疏水总管
RDM(rod drive mechanism) 控制棒驱动机构
RDO(rebars delivery order) 钢筋发料单
R & D(research and development) 研究与开发
RD(rotating disc) 叶轮
RD(run down) 降负荷，自动减负荷，负荷追降
RDS(reactor diagnostic system) 反应堆诊断系统
RDTE(research, development, test and evaluation) 研究、研制、试验与鉴定
RDT(remote data transmitter) 远程数据发送器
reaccelerate 再加速【飞轮等】
reacceleration 再加速
reach an agreement 达成协议
reach an identity of views 取得一致看法
reach of protection 保护范围
reach the standard 达标
reach 外伸幅度【起重机】，达到，延伸，到达，段，河段，范围
REAC(radiological emergency assessment center) 放射性应急评价中心
reactance amplifier 电抗耦合放大器
reactance arm 电抗支路，电抗臂
reactance bond 电抗搭接，电抗耦合，接合扼流圈
reactance breaker 有电抗线圈的分段开关
reactance bridge method 电抗桥接法
reactance capacity 无功功率
reactance coil 电抗线圈，扼流圈
reactance compensation 电抗补偿
reactance coupling 电抗耦合
reactance curve 电抗曲线
reactance-drop compensation 电抗电压降补偿
reactance drop 电抗压降
reactance EMF 电抗电势
reactance factor 电抗因数，无功功率因数
reactance frequency multiplier 电抗式倍频器
reactance function 电抗函数
reactance grounded 电抗接地的
reactance in direct axis 直轴电抗，纵轴电抗
reactance leakage 电抗漏泄
reactance meter 电抗计
reactance modulation system 电抗调制方式，直接调制方式
reactance network 电抗网络
reactance of opposite phase sequence 逆序电抗，反相序电抗
reactance power 无效功率

reactance protection 电抗保护装置
reactance relay 电抗继电器，电抗型继电保护装置
reactance resistance ratio 电抗电阻比
reactance room 电抗器室
reactance starter 电抗启动器
reactance synchronizing method 电抗同步法
reactance theorem 电抗原理
reactance-to-resistance ratio 品质因数，电抗电阻比
reactance transformer 电抗变换器
reactance triangle 电抗三角形
reactance-tube modulator 电抗管调制器
reactance-tube 电抗管
reactance voltage 电抗电势，电抗电压
reactance 电抗，反应性
reactant ratio 反应物比
reactant 成分，反应物，组分，反应剂，试剂
reactatron 低噪声微波放大器，一种晶体二极管
reacted auxiliary system 反应堆辅助系统
reactimeter 反应性测定仪
reacting bar 承力杆
reacting field 反应场，反作用场
reacting fluid 反应流体
reacting space 反应间隔
reacting system 反应系统，反应体系
reaction alternator 无功发电机，反应式同步发电机
reaction balance 反作用平衡，反动平衡，反应平衡
reaction beam 反应梁
reaction blade 反动式叶片
reaction brazing 反应钎焊
reaction classification 反应分类
reaction coefficient 反馈系数
reaction coil 反作用线圈，反馈线圈
reaction component 无功分量，电抗分量，复阻抗的虚部
reaction control 反馈控制
reaction coupled 电抗耦合的
reaction coupling 电抗耦合
reaction current 反作用流
reaction curve 反应曲线
reaction cycle 反应周期
reaction degree 反动度
reaction energy 反应能（量）
reaction exergy 反应热，反应有效能
reaction force from thermal compensator 热补偿器反力
reaction force 反作用力，反力
reaction impact crusher 反击式破碎机
reaction locus 反力轨迹【用于拱结构】
reaction motor 反作用式电动机，反应（式）电动机，磁阻电动机
reaction of combustion 燃烧反应
reaction of support 支点反力，支座反力，支点反酌
reaction period 反应周期，反应期
reaction pit 反应槽
reaction principle 反作用原理
reaction product 反应产物
reaction rate constant 反应速度常数
reaction rate 反应速率，反应速度，反应率
reaction rim 反应边
reaction ring 止推环
reaction stage 反动级【汽轮机】，反应级，反应阶段
reaction steam turbine 反动式汽轮机
reaction tank 反应池
reaction thrust 反推力
reaction timer 反应计时器，反应速度测定器
reaction time 反馈时间，反应时间
reaction turbine 反击式涡轮，反动式汽轮机
reaction type stage 反动级
reaction type turbine 反动式汽轮机
reaction velocity 反应速度
reaction vessel 反应器
reaction water turbine 反动式水轮机，反击式水轮机，反动水涡轮机
reaction wheel 反动叶轮
reaction winding 反作用绕法，反作用绕组
reaction zone 反应区，反应堆活性区
reaction 反力，反应，感应，反作用
reactivation of catalyst 催化剂再活化
reactivation 恢复活性，再生，复原，重激活
reactive capability 无功容量，无功功率
reactive circuit 反馈电路，电抗电路
reactive coil 电抗线圈，扼流圈
reactive compensation equipment 无功补偿装置
reactive compensation 无功补偿
reactive compensator 无功补偿装置
reactive component of current 电流无功分量
reactive component 电抗部分，无功分量，虚数部分，无功部分
reactive current compensation 无功电流补偿
reactive current compensator 无功电流补偿器
reactive current meter 无功电流计
reactive current 电抗性电流，无功电流，感性电流
reactive DC voltage drop 电抗性直流电压降
reactive drop 电抗电压降，无功电压降
reactive electromotive force 无功电动势
reactive element 无功元件，电抗元件
reactive EMF 无功电动势
reactive energy 反应能
reactive factor meter 无功功率因数表
reactive factor 电抗系数，无功因数
reactive force 反作用力，反力
reactive ground 电抗性接地
reactive in respect to 相对……呈感性
reactive kilovoltampere-hour meter 无功电度表
reactive kVA 无功功率，无功千伏安
reactive line 无功线路，电抗线路
reactive-load compensation equipment 无功补偿设备
reactive-load surge 无功负载浪涌
reactive-load 电抗性负载，无功负载，电抗性负荷，无功负荷
reactive loss 无功损耗，无功损失
reactive material 反应物质
reactive metal 活性金属
reactive muffler 抗性消声器

reactive output 无功输出
reactive-peak limiter 无功峰值限制器
reactive power absorption 无功功率吸收
reactive power capability 无功能力，无功功率
reactive power capacity 无功容量
reactive power compensation converter station 换流站无功补偿装置
reactive power compensation device 无功补偿装置
reactive power compensation system 无功补偿系统
reactive power compensation 无功补偿，无功功率补偿
reactive power compensator 无功功率补偿器
reactive power consumption 无功消耗
reactive power control of converter station 换流站无功功率控制
reactive power control 无功功率控制
reactive power factor 无功功率因数
reactive power loading 无功负载
reactive power loss 无功功率损耗
reactive power meter 无功伏安计，无功功率计，反应功率计
reactive power regulator 无功功率调节器
reactive power relay 无功功率继电器
reactive power service 无功调节服务
reactive power set-point 无功功率设定点
reactive power supply capacity 无功电源容量
reactive power 无功功率，无效功率，反应功能，无功，无功能力
reactive radiator 无源辐射器，无功辐射器
reactive synchro-transformer 无功同步控制变压器，电抗性自整角变压器
reactive thrust 反作用力
reactive type coal crusher 反击式碎煤机
reactive value 无功量，无功值
reactive voltage component 无功电压分量
reactive voltage 电抗性电压，无功电压
reactive volt-ampere hour 无功伏安小时
reactive volt-ampere meter 无功功率表，无功伏安计
reactive volt-ampere 无功伏安，乏
reactive 反作用的，反应的，活性的，无功的，电抗性的
reactivity accident 反应性事故
reactivity addition rate 反应性添加率
reactivity addition 反应性增加
reactivity alteration 反应性改变量
reactivity balance 反应度平衡，反应性平衡
reactivity behaviour 反应性行为，反应性状态
reactivity binding 反应性约束
reactivity change 反应性变化
reactivity charge 剩余反应性，反应性余额
reactivity coefficient of the coolant temperature 冷却剂温度反应性系数
reactivity coefficient of the fuel temperature 燃料温度反应性系数
reactivity coefficient 反应性系数
reactivity compensation 反应性补偿
reactivity computer 反应性测定仪
reactivity consumption rate 反应性消耗率
reactivity contribution 反应性贡献
reactivity control component 反应性控制部件
reactivity control system 反应性控制系统
reactivity control 反应性调节，反应性控制
reactivity decrease 反应性降低
reactivity disturbance 反应性扰动
reactivity effectiveness 反应性的有效性
reactivity effect 反应性影响，反应性效应
reactivity efficacy 反应性的有效性
reactivity equation 反应性方程
reactivity equivalent 反应性当量
reactivity excursion 反应性剧增，反应性急速上升，反应性闪变
reactivity feedback 反应性反馈
reactivity fluctuation 反应性涨落
reactivity forcing function 反应性驱动函数
reactivity gain 反应性增益
reactivity holddown 反应性抑制
reactivity increase 反应性增加
reactivity increment 反应性增量
reactivity in dollars 以“美元”计反应性
reactivity in inhour unit 以倒时单位计反应性
reactivity insertion accident 反应性引入事故
reactivity insertion rate 反应性引入率，反应性插入率，反应度插入率
reactivity insertion 反应性的输入
reactivity interval 反应性范围
reactivity jump 反应性跃变
reactivity lifetime 连续运行时限【反应堆】，反应性寿命，反应性寿期
reactivity life 反应性寿命［寿期］
reactivity limitation 反应性约束
reactivity loss 反应性损失
reactivity margin 反应性裕度，剩余反应性，反应裕量
reactivity meter 反应性测定仪，反应性测量计，反应性仪，反应度计
reactivity mismatch 反应性失配
reactivity noise 反应性噪声
reactivity of a nuclear reactor 核反应堆的反应性
reactivity oscillator 反应性振荡器
reactivity penalty 反应性亏损
reactivity perturbation 反应性微扰
reactivity power coefficient 反应性功率系数
reactivity pressure coefficient 反应性压力系数
reactivity ramp rate 反应性等变率
reactivity ramp 反应性线性增加
reactivity rate 反应性变化率，反应度（变）率
reactivity reactor behaviour 堆反应性性能
reactivity regulation 反应性调节
reactivity release 反应性释放
reactivity reserve 反应性储备
reactivity shimming 反应性补偿
reactivity shutdown margin 负反应性余量
reactivity sinusoidal variation 反应性正弦变化
reactivity spectral density 反应性谱密度
reactivity spectrum 反应性谱
reactivity step change 反应性阶跃变化
reactivity surge 反应性急变
reactivity temperature coefficient 反应性温度系数

reactivity testing 反应性试验
reactivity tramp 反应性线性增加
reactivity transient 反应性瞬变，反应性瞬态
reactivity unit 反应性单位
reactivity variation 反应性变化
reactivity worth 反应性价值
reactivity 反应度，反应性，反应能力，电抗性，活动性，反应率
reactor accident mitigation analysis 反应堆事故减轻分析
reactor accident 反应堆事故
reactor advanced manoeuverability package 反应堆改进的灵活操纵部件
reactor alarm system 反应堆报警系统
reactor appurtenances 反应堆辅助（附属）设备
reactor art 反应堆建造技术
reactor assembly 反应堆装置，反应堆组合体
reactor at shutdown 停下的反应堆
reactor autocatalytic behavior 反应堆自动催化特性
reactor auxiliaries cooling water system 反应堆辅助设备冷却水系统
reactor auxiliaries 反应堆辅助设备
reactor auxiliary building normal ventilation system 反应堆辅助厂房正常通风系统
reactor auxiliary building sump 反应堆辅助厂房污水坑
reactor auxiliary building switch gear room ventilation system 反应堆辅助厂房开关间通风系统
reactor auxiliary building 反应堆辅助厂房
reactor auxiliary equipment 反应堆辅助设备
reactor auxiliary system 反应堆辅助系统
reactor availability factor 反应堆使用效率，反应堆时间利用率
reactor availability 反应堆（时间）可利用率
reactor based activation analysis 反应堆活化分析
reactor batch loading 反应堆分批装料
reactor beam 反应堆射束
reactor behaviorur 反应堆性能，反应堆运转状态
reactor blanket 反应堆转换区，反应堆再生区
reactor block 反应堆体，反应堆室，反应堆区，反应堆舱，反应堆本体
reactor boron and water makeup system 反应堆硼和水补充系统
reactor boundary 反应堆边界
reactor breakdown 反应堆运行事故
reactor bridge 反应堆顶桥式起重机，反应堆顶过桥
reactor building crane 反应堆厂房吊车
reactor building drain system 反应堆厂房疏水系统
reactor building equipment components 反应堆厂房设备部件
reactor building exhaust air system 反应堆厂房排气系统
reactor building gantry and peripheral rooms 反应堆厂房龙门吊及其外围设备间
reactor building spray system 堆安全壳喷淋系统，反应堆厂房喷淋系统，主厂房喷淋系统
reactor building sump 安全壳污水坑，反应堆厂房污水坑
reactor building supply air system 反应堆厂房供气系统
reactor building track 反应堆厂房轨道
reactor building 反应堆厂房
reactor burner 非增殖反应堆
reactor cavity cooling system 堆腔冷却系统
reactor cavity flooded 反应堆空腔充满水
reactor cavity level 反应堆腔水位
reactor cavity roof （反应）堆换料腔顶盖
reactor cavity wall 反应堆腔墙
reactor cavity 反应堆换料腔，反应堆空腔，反应堆腔，堆腔，反应堆换燃料池
reactor cell 反应堆隔间，反应堆栅元
reactor chamber 反应堆室
reactor channel 反应堆孔道
reactor charge 反应堆装料
reactor charging 反应堆装料
reactor charging machines 反应堆装料机
reactor chemistry 反应堆化学
reactor circuit 反应堆回路
reactor coil 电抗线圈，扼流圈
reactor compartment 反应堆室，反应堆舱
reactor component 反应堆部件
reactor compressor 反应堆压缩机
reactor concept 反应堆概念
reactor construction 反应堆结构
reactor containment boundary 反应堆安全壳边界
reactor containment building 反应堆安全壳厂房
reactor containment equipment cooling water heat exchanger 反应堆安全壳设备冷却水热交换器
reactor containment shell 反应堆安全壳
reactor containment 反应堆安全壳
reactor control material 反应堆控制材料
reactor control station 反应堆操纵器
reactor control system 反应堆（功率）调节系统，反应堆控制系统
reactor control 反应器控制，反应堆控制
reactor converter 转换堆，转换反应堆
reactor coolant and pressurizer system 反应堆冷却剂与稳压器系统
reactor coolant bypass filter 反应堆冷却剂旁路过滤器
reactor coolant chemistry 一回路化学
reactor coolant circulating pump 反应堆冷却剂主循环泵
reactor coolant clean-up demineralizer 反应堆冷却剂主净化除盐器
reactor coolant clean-up filter holding pump 反应堆冷却剂净化过滤器压力保持泵
reactor coolant clean-up filter precoat pump 反应堆冷却剂净化过滤器预涂层泵
reactor coolant clean-up filter 反应堆冷却剂净化过滤器
reactor coolant clean-up system 反应堆冷却剂净化系统
reactor coolant design basis activity 反应堆冷却剂设计基准放射性
reactor coolant drains pump 一回路泄水泵
reactor coolant drain tank pump 一回路排水箱泵

reactor coolant drain tank　反应堆冷却剂疏水箱，一回路排水箱
reactor coolant filter　一回路水过滤器，堆冷却剂过滤器
reactor coolant gas circulator seal gas supply system　反应堆气体冷却剂循环风机的密封气体供应系统
reactor coolant gas　反应堆气体冷却剂，反应堆冷却气体
reactor coolant inlet nozzle　反应堆冷却剂出口接管
reactor coolant leakage　反应堆冷却剂泄漏
reactor coolant letdown　一回路水下泄，堆冷却剂下泄
reactor coolant loop　反应堆冷却剂环路，反应堆冷却剂环路，一回路冷却剂系统，一次冷却回路
reactor coolant material　反应堆冷却剂材料
reactor coolant outlet nozzle　反应堆冷却剂出口接管
reactor coolant outlet　反应堆冷却剂出口
reactor coolant pipe break　反应堆冷却剂管道破裂
reactor coolant pipe　一回路管道，反应堆冷却剂管道
reactor coolant piping section　反应堆冷却剂管段
reactor coolant piping　反应堆冷却剂管，主管道
reactor coolant pressure boundary　反应堆冷却剂压力边界
reactor coolant pressure control system　反应堆冷却剂压力控制系统
reactor coolant pressure　反应堆冷却剂压力
reactor coolant pump bunker　（反应堆）主泵隔间，主泵舱
reactor coolant pump can　主泵屏蔽套
reactor coolant pump set　一回路泵组合，堆冷却剂泵组合
reactor coolant pump underspeed trip　反应堆冷却剂泵低速停堆
reactor coolant pump　反应堆冷却剂泵，一回路冷却剂泵，一回路泵，主泵
reactor coolant recirculation pump　反应堆冷却剂再循环泵
reactor coolant supplementary system　反应堆冷却剂补给系统
reactor coolant system cold leg isolation valve　反应堆冷却剂系统冷段隔离阀
reactor coolant system cold leg　反应堆冷却剂系统冷段
reactor coolant system component　一回路设备，堆冷却剂系统设备
reactor coolant system depressurization accident　一回路失压事故
reactor coolant system drain tank pump　反应堆冷却剂系统疏水箱泵
reactor coolant system drain tank　反应堆冷却剂系统疏水箱
reactor coolant system hot leg　反应堆冷却剂系统热段
reactor coolant system instrumentation　反应堆冷却剂系统仪表
reactor coolant system isolating valve　反应堆冷却剂系统隔离阀
reactor coolant system　反应堆冷却剂系统，主冷却剂系统，主回路系统
reactor coolant temperature control system　反应堆冷却剂温度控制系统
reactor coolant temperature　反应堆冷却剂温度
reactor coolant water　反应堆冷却水，一次水
reactor coolant　反应堆冷却剂，反应堆载热剂，一回路水
reactor cooling system　反应堆冷却系统
reactor cooling water　反应器冷却水
reactor core assembly　堆芯组合体
reactor core baffle　堆芯围板
reactor core barrel　堆芯吊篮
reactor core design　堆芯设计
reactor core disassembly　堆芯解体
reactor core flooding system　堆芯淹没系统
reactor core internals　反应堆堆内构件
reactor core inventory control system　（高温堆）堆芯燃料总量控制系统
reactor core isolation cooling system　堆芯隔离冷却系统
reactor core isolation cooling　堆芯隔离冷却
reactor core lifetime　堆芯寿期
reactor core meltdown　堆芯熔化
reactor core power distribution　堆芯功率分布
reactor core pressure drop　堆芯压降
reactor core refueling　反应堆堆芯重新装料，反应堆堆芯换料
reactor core restraint　堆芯限制器，堆芯抑制
reactor core structure　反应堆堆芯结构，堆芯结构
reactor core support structure　堆芯支承结构
reactor core　反应堆堆芯，堆芯，反应堆活性区
reactor cross section　反应堆截面
reactor cubicle　反应堆室，反应堆舱
reactor damage　反应堆损坏
reactor dead time　反应堆死时间【即停堆后因中毒而不能启动的时间】
reactor debris　反应堆裂变碎片，反应堆的裂变物，反应堆的裂变产物
reactor decommissioning　反应堆退役
reactor design　反应堆设计
reactor disaster　反应堆事故，反应堆严重事故
reactor discharging　反应堆卸料
reactor dismantling　反应堆拆除
reactor dome　安全壳圆顶建筑，反应堆球形安全壳
reactor doubling time　反应堆倍增时间
reactor-down　反应堆功率下降
reactor dry containment　反应堆干式安全壳，干井
reactor dynamics　反应堆动力学
reactor effluent　反应器流出物
reactor emergency shutdown　反应堆应急停堆
reactor engineering　反应堆工程
reactor enthalpy rise　反应堆焓升
reactor equation　反应堆方程
reactor excursion accident　反应堆功率剧增事故
reactor excursion　功率偏差，反应堆工作状况偏

差，反应堆功率剧增
reactor experimental facility　反应堆实验装置
reactor family　堆体系，堆串列
reactor feedwater flow　反应堆给水流量
reactor feedwater pump　反应堆给水泵
reactor feedwater　反应堆给水
reactor feed　反应堆装料
reactor flux shield　电抗器磁屏蔽
reactor follow mode operation　反应堆跟随运行方式
reactor fuel element　反应堆燃料元件
reactor fuel grid position　反应堆燃料格架位置
reactor fueling　反应堆装料
reactor fuel　反应堆燃料
reactor geometry　反应堆几何
reactor grade fluid　反应堆质量级流体
reactor grade　反应堆等级，核等级
reactor grid position end closure　反应堆栅格定位端封头
reactor-grounded neutral system　小电流接地系统，中性点电抗接地系统
reactor group　电抗器组
reactor hatch　反应堆舱口
reactor head　反应堆顶盖
reactor housing　反应堆安全壳
reactor inherent protection　反应堆固有保护
reactor inlet pressure　反应堆进口压力
reactor in shutdown condition　停堆工况反应堆
reactor instrumentation and control　反应堆仪表监测和控制
reactor instrumentation monitoring system　反应堆仪表监测系统
reactor instrumentation　反应堆检查控制仪表，反应堆检测仪表
reactor instrument　反应堆仪表，反应堆仪器
reactor integrated reprocessing　堆厂一体化后处理
reactor internal feedback　反应堆内反馈
reactor internal recirculation pump　反应堆内循环泵
reactor internals lifting device　堆内构件提升装置
reactor internals storage area　堆内构件储存区
reactor internals storage pool　反应堆内构件贮存水池
reactor internals　反应堆内部部件，（反应）堆内构件
reactor irradiated　反应堆内辐照的
reactor island　堆岛，反应堆装置，核岛
reactor jacket　快堆外套
reactor kinetics equations　反应堆动力学方程
reactor kinetics　反应堆动力学
reactor lattice parameters　反应堆栅格参数
reactor lattice　反应堆芯栅格，反应堆栅格
reactor licensing　反应堆审批
reactor lid　反应堆顶盖
reactor liner　反应堆衬里
reactor liquid level　反应堆液位
reactor load acceptance　反应堆带负荷
reactor loading　反应堆填料
reactor load pick-up　反应堆带负荷
reactor loop　反应堆环路
reactor maintenance　反应堆维修
reactor manufacturer　反应堆制造商
reactor material　反应堆材料
reactor measurement　反应堆测量
reactor meltdown accident　反应堆燃料熔化事故
reactor meltdown　反应堆燃料熔化
reactor metallurgy　反应堆冶金学
reactor moderator material　反应堆慢化剂材料
reactor moderator　反应堆慢化剂，反应堆减速计
reactor multiplication　反应堆增殖
reactor neutron spectrum　反应堆中子能谱
reactor noise analysis　反应堆噪声分析
reactor noise　反应堆噪声，反应器噪声
reactor of gas system　反应堆除气系统
reactor operating floor　反应堆操纵平台
reactor operating mode　反应堆运行模式
reactor operating temperature　反应堆运行温度
reactor operation　反应堆操纵，反应堆运行，堆控制
reactor operator chief　反应堆运行班长
reactor operator　反应堆操纵员，反应堆运行人员
reactor optimization　反应堆最优化
reactor oscillation technique　反应堆振荡法
reactor oscillator　反应堆振荡器
reactor outlet pressure　反应堆出口压力
reactor outlet temperature　反应堆出口温度
reactor outlet valve　反应堆出口阀
reactor output　反应堆功率，反应堆出力
reactor overpower　反应堆超功率
reactor performance　反应器性能，反应堆性能
reactor period meter　反应堆周期表
reactor period　反应堆周期
reactor physics calculation　反应堆物理计算
reactor physics constant　反应堆物理常数
reactor physics　反应堆物理学
reactor pipe rupture safeguard device　反应堆管道破裂安全保护装置
reactor pit ventilation system　反应堆坑通风系统
reactor pit　反应堆水池，反应堆坑
reactor plant heat-up and cooldown　反应堆装置加热及冷却
reactor plant layout　反应堆厂房布置
reactor plant　反应堆装置
reactor poisoning　反应堆中毒
reactor poison removal　反应堆毒物排除，反应堆毒物去除
reactor poison　反应堆毒物
reactor pool and cavity　反应堆水池和腔体
reactor pot　反应堆槽
reactor power control system　反应堆功率控制系统
reactor power limiter　反应堆功率限制器
reactor power monitor　反应堆功率监测器
reactor power noise　反应堆功率噪声
reactor power raising　反应堆功率提升
reactor power　反应堆功率
reactor pressure vessel closure head　反应堆压力容器封头
reactor pressure vessel failure accident　压力壳破

裂事故
reactor pressure vessel internals 反应堆压力容器内部构件，堆芯构件
reactor pressure vessel penetration 反应堆压力容器贯穿件
reactor pressure vessel steel liner 反应堆压力容器钢内衬
reactor pressure vessel venting system 反应堆压力容器排气系统
reactor pressure vessel 反应堆压力容器
reactor primary circuit 反应堆一回路
reactor project 反应堆项目，堆计划
reactor protection board 反应堆保护板
reactor protection channel 反应堆保护通道
reactor protection device 反应堆保护装置
reactor protection panel 反应堆保护（仪表控制）盘
reactor protection parameter 反应堆保护参数
reactor protection relay cabinet 反应堆保护的继电器柜
reactor protection system 反应堆保护系统
reactor prototype 原型反应堆，原型堆
reactor pump 反应堆泵
reactor radiation field 反应堆辐射场
reactor recirculation pump 反应堆再循环泵
reactor refueling floor 反应堆换料平台
reactor refueling 反应堆换料
reactor reserve shutdown system 备用停堆系统
reactor residual heat removal system 反应堆余热排出系统
reactor resistance 电抗器电阻
reactor restart 反应堆再启动
reactor RHR system 反应堆余热排出系统
reactor roof 反应堆顶盖
reactor room 反应堆室，反应堆舱
reactor runaway 反应堆失控
reactor rundown （反应）堆功率下降
reactor run 换料间隔期，堆运行周期
reactor safety analysis 反应堆安全分析
reactor safety circuits 反应堆保护系统，反应堆安全电路
reactor safety fuse 反应堆安全保险器，反应堆安全保险装置
reactor safety study 反应堆安全研究
reactor safety system 反应堆安全系统
reactor safety 反应堆安全
reactor scram system 反应堆紧急停堆系统，反应堆维修紧急系统
reactor scram 反应堆紧急停堆
reactor secondary circuit 反应堆二回路
reactor service bridge 反应堆维修桥式起重机
reactor service building 核辅助厂房，反应堆公用厂房
reactor shield analysis 反应堆屏蔽分析
reactor shielding material 反应堆屏蔽材料
reactor shielding 反应堆屏蔽
reactor shield plug positioning 反应堆屏蔽塞定位
reactor shimming 反应堆补偿
reactor shutdown cooling system 反应堆停堆冷却系统，反应堆余热冷却系统
reactor shutdown margin 停堆深度
reactor shutdown period 停堆周期
reactor shutdown 反应堆停堆
reactor shut-off period 停堆周期
reactor shut-off system 停堆系统
reactor simulator 反应堆模拟器，反应堆模拟装置，反应堆模拟机
reactors in series 串联反应器，串联电抗器
reactor source 反应堆启动源，反应堆中子源
reactor sphere 反应堆圆顶
reactor stability 反应堆稳定性
reactor starting 电抗器启动
reactor-start motor 电抗器启动电动机
reactor start-up rate 反应堆启动速率
reactor start-up 反应堆启动，开堆
reactor statics 反应堆静力学
reactor string 堆体系，堆串列
reactor structural material 反应堆结构材料
reactor structure 反应堆结构
reactor support 反应堆支承结构，反应堆底座
reactor surveillance 反应堆监测，（反应）堆监视
reactor system 反应堆系统
reactor tank 堆箱
reactor tap changer 电抗器抽头切换装置
reactor technology 反应堆工艺
reactor theory 反应堆理论
reactor thermal power 反应堆热功率
reactor time constant 反应堆时间常数
reactor transfer function 反应堆传递函数
reactor transient behavior 反应堆瞬态行为
reactor trip breaker 反应堆跳闸断路器
reactor trip signal 反应堆事故保护停堆信号
reactor trip system 反应堆停堆系统，反应堆跳脱系统，紧急停堆系统
reactor trip 事故保护停电，反应堆事故保护停堆，停堆，反应堆紧急停堆
reactor tunnel 反应堆屏蔽走廊
reactor type tap-change mechanism 电抗器型抽头切换装置
reactor type 反应堆类型
reactor upper/lower internals storage stand 堆内上部与下部构件贮存架
reactor-up 反应堆功率增大，反应堆功率增高
reactor variable 反应堆变量
reactor vault 反应堆坑室
reactor vessel attachment members 反应堆压力壳附属构件
reactor vessel body 反应堆容器本体
reactor vessel bolting flange 反应堆容器螺栓紧固法兰
reactor vessel clevis 反应堆压力壳键槽
reactor vessel closed circuit TV inspection unit 反应堆压力壳闭路电视检查装置
reactor vessel design 反应堆（压力）容器设计
reactor vessel flange seal resurfacing 反应堆压力壳法兰密封重修表面
reactor vessel flange 反应堆压力壳法兰
reactor vessel guide stud 反应堆压力壳导向螺杆
reactor vessel head center disc 反应堆容器封头中心碟形盘
reactor vessel head external thermal insulation 反

应堆容器封头外部保温
reactor vessel head laydown area 堆压力壳顶盖储存场地
reactor vessel head lifting device 反应堆压力壳顶盖起吊装置
reactor vessel head lifting rig 反应堆压力壳顶盖起吊设备
reactor vessel head storage ring 反应堆容器顶盖贮存环
reactor vessel head 反应堆压力壳顶盖，反应堆压力容器封头，反应器槽顶盖
reactor vessel heating jacket 反应堆容器加热套
reactor vessel sealed by dummy closure head 由模拟顶盖密封的反应堆压力壳
reactor vessel sealed by vessel head 由顶盖密封的反应堆压力壳
reactor vessel seal ring 反应堆压力壳密封环
reactor vessel shipping skid 反应堆容器运输托架，反应堆压力壳运输导轨
reactor vessel support structure 反应堆容器支承构件
reactor vessel support 压力壳支承
reactor vessel-to-cavity seal ring 压力壳与水池间密封环
reactor vessel to refueling cavity seal 反应堆容器与换料腔间的密封
reactor vessel torus ring support 反应堆容器环形支承
reactor vessel water level measurement 压力壳水位测量
reactor vessel 反应堆壳体，反应堆（压力）容器[外壳]
reactor waste clean-up filter 反应堆废水净化过滤器
reactor waste 反应堆废物，反应器废料
reactor water clean-up heat exchanger 反应堆水净化热交换器
reactor water clean-up system 反应堆水净化系统
reactor water purity 反应堆水质纯度
reactor water recirculation 反应堆水再循环
reactor water temperature 反应堆水温
reactor well closure concrete slab 反应堆井坑混凝土盖板
reactor well cover 反应堆井盖
reactor well drain pump 反应堆井疏水泵
reactor well fill line 反应堆井注水管道
reactor well flooding 反应堆井注水
reactor well water 反应堆井水
reactor well 反应堆井，干井
reactor winding 电抗器线圈
reactor with a nonlinear feedback 非线性反馈的反应堆
reactor with external coolant recirculation 冷却剂堆外再循环反应堆
reactor year 堆年
reactor 反应堆，反应器，电抗器，电抗线圈，扼流圈
readability 清晰度，可读性，读出能力，可读度
read accumulator 读累加器
read address counter 读地址计数器
read-after-write scheme 写且读出方法
read amplifier 读出放大器
read-around number 周围读数，多次读取数，环读次数，读出次数
read-around ratio 读数比，读出比，多次读取比
readatron 印刷数据读出和变换装置
readback signal 读回信号
read backward 反向读出
read-back 回读，读要写回的信息
read buffer 读缓冲器
read channel continue 读通道继续
read channel initialize 读通道开始
read check 读出检查
read cycle 读数周期
read digital input 读数字输入，数字量读入
readdressing routine 重新编址程序，改变地址程序
readdressing 再寻址地址变换，改变地址
read driver 读数策动器，读数驱动器
reader code 阅读器代码
reader control relay 阅读器控制继电器
reader-interpreter 阅读翻译器，读入解释
reader-printer 读出打印机
reader-punch equipment 阅读穿孔设备
reader-punch 读出穿孔机
read error 读取误差，读数误差
reader-sorter 读出分类器
reader stop 阅读器停机
reader-typer 读数打字机
reader unit 读数器，读数装置
reader 读本，阅读程序，读出器，阅读器，读者，抄表员
read forward 正向读出
readily available water 速效水
read-in cryotron 写入冷子管
readiness review 启用前检验，安装完检验，待用检查
readiness test 准备状态测试
readiness 备用状态
reading accuracy 读数精度
reading board 读板
reading circuit 读出电路
reading device 读数装置，示读装置，阅读器阅读设备
reading dial 读数盘
reading error 读数误差，读数错误
reading glass 阅读放大镜，读数放大镜
reading head 读（磁）头
reading instrument 指示式仪器，指示仪表
reading microscope 读数显微镜
reading-off position 读数位置
reading plate 读板
reading precision 读数精度
reading pulse 读出脉冲
reading room 阅览室
reading signal 读出信号
reading speed 读出速度
reading station 读数站，输入站
reading system 读数系统
reading time 读时间

reading 读数，（仪表）指示
readjustment 再调整，修正，重新调整，微调，重调，校准，重调整
readjust 再调整，重调，微调，修正，核正
read memory 读存储器
read-mostly memory 主读存储器
read-only memory 只读存储器，永久性存储器，固定存储器
read-only procedure 只读过程
read-only storage 只读存储器
read-only store 只读存储器
read-only 只读的，只读
read-out cryotron 读出冷子管
read-out gate 读出门
read-out time 读出时间
read-out-tube 读出管
read-out winding 读出绕组
read-out 示值读数，数显装置，结果输出，数字显示装置，读出（数据），读数示出
read period 读数周期
read printer 读出打印机
read-punch unit 读出穿孔部件
read-punch 读出穿孔
read-record head 读数记录头
read-record 读数记录
read routine 读数程序
read speed 读出速度
read statement 读语句
read-while-writing 同时读写，读的同时进行写
read-write channel 读写通道
read-write check indicator 读写检验指示器
read-write continue 继续读写
read-write cycle 读写周期
read-write equipment 读写装置
read-write head 读写头
read-write memory 读写存储器
read-write random-access memory 读写随机存储器
read-write storage 读写存储器
ready coal storage 日用煤堆
ready delivery 即期交货
ready for insertion into the reactor （燃料元件）准备插入反应堆
ready for operation 操作就绪，运行准备就绪，运算就绪
ready-for-use 随时可用的，备好待用的
ready-made hinge 普通合页
ready market 现有市场
ready-mixed concrete 搅拌好的混凝土，预拌混凝土
ready-mixed paint 调和漆
ready-mix plant 预拌混凝土厂，预拌混凝土机
ready-mix truck 预拌混凝土运送车
ready-money price 现金价格
ready-packaged 快装式
ready pile 周转煤堆
ready to start condition 可随时启动状态
reaeration in river 河流再曝气作用
reaeration rate 再曝气率
reaeration sludge 污泥复氧
reaeration 重新充气，再充气，再掺气作用，再曝气，通风
reaffirming 重新确认，重申，再次确认
reafforestation 重新造林
reagent blank correction 试剂空白校正
reagent bottle 试剂瓶
reagent injection plant 加药间，试剂注入装置
reagent method of watersoftening 化学试剂软化法
reagent paper 试纸
reagent removal 试剂去除
reagent 反应力，试剂，反应物
reaging 反复老化
reak in relay 插入继电器
real accumulator 实数累加器
real aperture 实际孔径
real assets 实际资产
real capital 实际资本
real component 实部，有功分量
real constant 实常数，有效常数
real cost 实际成本
real density 真（实）密度
real domain 实域
real-dump truss 耳座
real economy 实体经济
real escalation rate 实际浮动率
real estate agent 地产商
real estate title deed 房地契
real estate 不动产，地产，房地产
real flow 真实水流
real fluid 黏性流体，实际流体，真实流体
real flux 有效通量
real frequency axis 实频率轴
real function 实函数
real-gas effect 真实气体效应
real gas 实际气体
real heat efficiency 实际热效率
realignment 改线，重定线
real income 实际收入
real intention 真正意图
real interest rate 扣除通货膨胀影响后的利率
real investment 实际投资
realistic model 仿真模型
realization and liquidation 变产清盘
realized load curve 实际负荷曲线
realized price 实际价格
realize 实现
real load curve 实际负荷曲线
real load 有效负载
reallocation of land 土地重新规划
reallocation 重新配置，重新规划，重新划拨，再分配
realm 领域
real negative coefficient 实负系数
real operating time 实际运转（操作）时间
real output 有效输出
real overshoot 实超调量
real-part condition 实部条件
real-part （复数的）实部，实数部分，有功分量
real power analysis 有效功率分析
real power 实际功率，有效出力，实际出量，有效功率，有功功率

real price 实价
real quantity 实数
real resistance 实电阻，欧姆电阻
real root 实根
real size 实际尺寸
real slot 实槽
real specific gravity 真比重
real storage 实存储器
real-time ampacity 实时载流量，实时安培容量
real-time basic executive 基本实时执行程序
real-time clock 实时计时器，实际时钟，实时时钟
real-time communication 实时通信
real-time computation 实时计算
real-time computer programming 实时计算机程序设计
real-time continuous-wave holography 实时连续波全息术
real-time control system 实时控制系统
real-time control 实时控制
real-time cost analysis 实时成本分析
real-time counter 实时计数器
real-time data processing 实时数据处理
real-time data server 实时数据服务器
real-time data system 实时数据处理系统
real-time data 实时数据
real-time date reduction 实时数据缩减，实时数据预变换
real-time digital governor 实时数字控制
real-time executive routine 实时执行程序
real-time executive 实时执行，实时执行程序
real-time exposure testing 实时寿命试验
real-time graphic system 实时制图系统
real-time information retrieval system 实时信息检索系统
real-time I/O control system 实时输入输出控制系统
real-time logging 随机打印
real-time machine 快速计算机，实时计算机
real time market 实时市场
real-time monitoring 实时监测，实时监控
real-time monitor 实时监视器
real-time operating system 实时操作系统
real-time operation 快速操作，实时操作，实时运算
real-time power system simulator 电力系统实时仿真装置
real-time price 实时电价
real-time process control system 实时过程控制系统
real-time process control 实时过程控制
real-time processing 实时处理
real-time processor 实时数据处理装置
real-time process 实时过程，实时处理
real-time programming 实时编程，实时程序设计
real-time servicing 实时维护，实时服务
real-time simulation 实时模拟
real-time simulator 实时模拟装置
real-time system 实时系统
real-time trade subsystem 实时交易子系统
real-time trading 实时交易
real-time trend display 实时趋势画面
real-time working 实时操作，实时工作
real-time 实时，实时的，快速的
real-valued process 真实过程
real value 实际价值，有效值
real variable 实变量
real velocity 实际速度
real wage 实际工资
real work 有效功
real 实际的，真实的
reamer 铰床，铰刀，锪钻，扩孔器，钻孔器
reaming machine 铰孔机
reaming 铰孔，扩孔，铰削
reamplifying 再放大
reappear arc 再燃弧
reappraise stocks and assets 清产核资
rear apron 闸后护坦
rear arch 后拱
rear axle drive 后轮传动
rear beam 后部梁
rear bearing housing 后轴承箱
rear bearing 后轴承
rear channel 后水道
re-arcing 再燃弧
rear connection 背面接线，盘后接线，屏后接线
rear damper 后阻尼环，后阻尼器
rear diaphragm 后光阑
rear discharge stoker 后出渣炉排
rear-dump truck 后卸式卡车
rear-dump wagon 后卸式（货）车
rear edge 后缘
rear elevation 背面立视图，背视图，后视图
rear end-plate 后端盖
rear fire box head 后火箱封头
rear-fired boiler 后墙燃烧锅炉
rear girder 后梁
rear heat recovery surface 尾部受热面
rear hub 后轮轴承
rear intercept valve 后截门
rear mounted aerofoil 后翼板
rear node 后方节点
rear of core 铁芯轭部，铁芯背部
Rear oil retainer ring 后挡油环
rear-panel mounting 后部盘板装配
rear panel 后部盘板
rear pass 后烟道
rear plate superheater 后屏过热器
rear pusher 后推式推车器
rearrangement of fuel 倒换燃料
rearrangement （燃料元件的）倒换，重新布置，重新排列，变位，重新配置，重新安排
rear row velocity stage 复速级的后列【汽轮机】
rear shaft seal 后轴封，后汽封
rear span 外形后宽
rear spar 后（翼）梁
rear stagnation point 后驻点
rear suspension wrench 后悬挂螺母扳手
rear tipper 后倾自卸车
rear turbine bearing 透平后轴承
rear view 背面立视图，背视图，后视图
rear wall enclosure superheater 后包墙过热器

rear wall firing 后墙燃烧
rear wall 后墙
rearward communications system 后向通信系统
rearward face 叶片凸面，背弧面
rear water wall 后墙水冷壁
rear-wheel bearing nut wrench 后轮轴承螺母扳手
reasonable comparability 合理的可比性
reasonable disposition of buildings 建筑物的合理布局
reasonable judgement 合理判断
reasonable level 正确的标准
reasonable price 合理价格，公平价格
reasonable profit 合理利润
reasonable proposal 合理化建议
reasonable time 适当的时候
reasonable use of water 合理用水
reasonable 合理的
reasonably priced 价格合理，定价合理
reason 原因
reassemble 重新装配，复装
reassembling of rotating machine after inspection 转动机械解体后组装
reassembling of valve after inspection 阀门解体后组装
reassembly 重汇编，重（新）装配，再组装，重新组装，二次组装
reassign 重新指定，再分配
reattaching flow 再附着气流中的流动
reattachment line 再附线
reattachment point 重附着点，再附点
reattachment zone 再附区
reattachment 再附（着）
REA 反应堆硼和除盐水补给系统【核电站系统代码】
rebanding 重新绑扎，重扎钢丝
rebar bending machine 钢筋弯曲机
rebar butt-welding machine 钢筋碰焊机
rebar cutting machine 钢筋切断机
rebar detector 钢筋探测器
rebar schedule 钢筋规范表【用于钢筋混凝土】
rebar straightening machine 钢筋调直机
rebar 圆钢筋，加固筋，钢筋
rebated joint 半槽接合
rebate tariff 折扣计费率
rebate 半槽企口，槽口，抵扣，回扣，折扣，减少，打折扣
rebidding 重新招标
rebind 重捆，重绑，重新装订
reboiled heating valve 再沸腾加热门
reboil effect 再汽化作用
reboiler 再沸腾器，重沸器，再沸器，单鼓式蒸汽发生器
reboiling of the coolant 冷却剂再沸腾
reboiling 再沸腾
rebound apparatus 回弹仪
rebound curve 回弹曲线
rebound elasticity 回弹性
rebound gauge 回弹计
rebound hammer 回弹锤
rebound hardometer 回跳硬度计
rebound method 回弹法
rebound of a pile 桩的回弹
rebound rate （喷射混凝土）回弹率
rebound valve 逆止阀，止回阀
rebound 回弹，回跳，弹回
rebuild 改建，再建，重建，修复
reburning chamber 再燃烧室
recalibrate 重新校准，再校准
recalibration screw 重新校准螺钉
recalibration 重新校准，重新校正，重校，复校，再校准，再标定，重核准
recall factor 检索率
recalling of bid 重新招标
recall signal 回答信号，回叫信号
recall 追忆
recede 退却
receding hemicycle 退却半周期
receding line 退缩线
receding side of belt 皮带松边
receding water table 退落地下水位
receding 回流，逆流
receipt form 回执表格
receipt of goods 收货凭据
receipt of licence clause 收到许可证条款
receiptor 收货单位，收货人
receipt signal 回答信号，接收信号
receipt title 收据抬头
receipt 接收，收到，收据
receive a patent 取得专利
received for shipment B/L 收货待运提单，备运提单
received heat 受热量，传热量
received power 接收功率
received signal 接收信号
received wave 接收波
receive-only device 只收设备
receive-only equipment 只收设备
receive-only monitor 只收监听器
receive-only 只接收
receiver drum 集汽包
receiver element 接收机单元
receiver of remote-control system 遥控信息接收器
receiver outlet-voltage 接收机输出端电压
receiver protective device 接收机保护装置
receiver relay 接收机继电器
receiver sensibility 接收灵敏度
receiver tank pump 废液滞留箱泵
receiver tube 吸热管
receiver 吸热器【太阳能】，接收器，集汽包，容器，接收机［器］，贮液器，收报机，话筒，电话听筒，收件人，收款人，受话器
receive-send keyboard set 收发键盘装置
receive 接收
receiving aerial 接收天线
receiving agent 收货代理人
receiving area 验收地点
receiving capacity 接收能力，接收容量
receiving center 接收中心
receiving chemical waste fluid 接收化学废液
receiving circuit 接收电路

receiving collecting electrode　集尘电极
receiving core　接收铁芯
receiving crystal　接收晶体
receiving department　验收部门
receiving-end impedance　受电端阻抗，接收端阻抗
receiving-end transformer　接收端变压器
receiving-end voltage　终端电压，接收端电压，受电端电压
receiving-end　接收端，受电端，受端
receiving equipment　接收设备
receiving flammable liquid　承接易燃液体
receiving frequency　接收频率
receiving goods　收货
receiving hood　接受式排风罩
receiving hopper　受料漏斗，受煤斗
receiving inspection　接收监督，进货检验
receiving instrument　受信机，接收器，接收仪表
receiving loop　环形接收天线
receiving magazine　接收装置，接收箱
receiving note　收货通知
receiving paraboloid　抛物面接收天线
receiving party　收货单位，收货人
receiving point　（放射性废气）接受点
receiving power　接收功率，输入功率
receiving quality　接收质量
receiving record　到货记录，收货记录
receiving region　容泄区
receiving register　接收寄存器
receiving relay　接收继电器
receiving report　验收报告
receiving selsyn　同步系统接收装置，自动同步接收机，接收的自整角机
receiving signal　接收信号
receiving station hopper　受料斗平台
receiving station　接收台，收信台，接收站
receiving store　接收仓库
receiving stream　容泄水流
receiving substation　降压变电站
receiving tank　贮槽
receiving target　接收目标
receiving transducer　接收传感器，接收变换器，受波器
receiving trough　接受槽
receiving tube　接收管，收信管
receiving voltage　接收电压
receiving water body　吸收电站废热冷却水
receiving　接收，接收监督
recently deposited soil　新近堆积土
receptacle box　插孔盒
receptacle plug　插头
receptacle　插孔，容器，（电器）插座，塞孔，插口，贮槽，插入式连接器
receptance　敏感性
reception basin　受水区
reception channel　接收通道
reception level　接收电平
reception of pilot frequency　导频接收
reception poor　接收不良
reception room　接待室，传达室，会客室，会见室，接见室
reception tank　受水箱，尾水箱
reception voltage　接收端电压
reception　接收法，接收
receptivity　可接收度，吸收力，感受性，接收性，吸收能力，容量，接纳能力
receptor location　接受点
receptor　接闪器，接收性，感受性，接收器
recessed bottom gondola car　凹底无门敞车
recessed plug　开槽塞
recessed surge tank　埋藏式调压井
recessed tube　开槽管
recessed　开槽的【管子对管板焊接】
recess in brick wall　砖墙凹处
recessing machine　刨槽机
recession curve　退水曲线
recession equation　退水方程
recession hydrograph　退水过程线
recession limb　过程线下降段
recession of level　水面下降
recession time　地面水消退时间，退水时间
recession velocity　退水流速，退水速度
recession　凹坑，凹陷，后退，经济衰退，退水
recessive　隐性的
recess　凹槽，凹进部分，凹部，凹进处，凹座，使凹进
rechain　重新丈量
rechange by rainfall penetration　雨水渗入补给
rechargeable battery　可再充电电池
rechargeable primary cell　可再充电原电池
rechargeable　可再充电的
recharge area of aquifer　含水层补给区
recharge area　补给区
recharge by seepage of stream　渗漏补给
recharge cone　回灌锥
recharge of basin　盆地补给
recharge period　再灌注期
recharge rate　补给率，回灌率
recharge well　水回灌井
recharge　再充电，再装填，补给，再次装料，再次加料，更换燃料元件，回灌
recheck　复查，复核
reciever horn　喇叭形接收器
recieving antenna　接收天线
recipher　译成密码，密码文件
recipience　接受，容纳
recipient channel　受主通道
recipient country　接受国，受援国
recipient　接受的，容纳的，信息接收器，容器，接收器，收货方，接收方，接收者
reciprocable generator　往复式发电机
reciprocable machine　往复式电机
reciprocable motor　往复运动电动机，往复式电动机
reciprocal account　往来账户
reciprocal action　交互作用
reciprocal agreement　互惠协定
reciprocal and mutually advantageous arrangement　互惠互利协议
reciprocal avoidance of double taxation　相互避免双重收税

reciprocal bilinear form 互逆双线性型
reciprocal capacity 逆电容
reciprocal circuit 可逆电路
reciprocal compressor 活塞式压气机，往复式压气机
reciprocal contract 互惠合同
reciprocal depth 对应水深
reciprocal diffusion length 扩散长度倒数
reciprocal dose theorem 剂量互易定理
reciprocal duty 互惠关税
reciprocal function 反商函数
reciprocal inductance 逆电感
reciprocal lattice 倒易点阵
reciprocal L/C 对开信用证
reciprocal letter of credit 对开信用证
reciprocal levelling 对向水准测量
reciprocal linear dispersion 线性扩散倒数
reciprocal multiplication 倒数相乘，倍增倒数
reciprocal network 互易网络，倒易网络
reciprocal observation 对向观测
reciprocal of ohm 欧姆的倒数
reciprocal period 堆周期倒数，倒周期
reciprocal polynomial 反多项式
reciprocal relationship 互反关系
reciprocal sight line 双向测线
reciprocal sight 对向照准
reciprocal space 倒易空间
reciprocal square root 反平方根
reciprocal stress 往复应力
reciprocal tariff 互惠关税，互惠税率
reciprocal temperature 温度倒数
reciprocal theorem 互易定理，倒易定理
reciprocal trade 互惠贸易
reciprocal treatment 互惠待遇
reciprocal vector 互易矢量，倒易矢量
reciprocal velocity 速度倒数
reciprocal 往复的，可逆的，互易的，可易的，互惠的，相互的，相反的，交互的，往复的
reciprocating air compressor 往复式空气压缩机
reciprocating air pump 活塞式气泵，往复式空气压缩泵，往复式空气泵
reciprocating charging pump 往复式上充泵
reciprocating compressor 活塞（式）压缩机，往复式压缩机
reciprocating condensing unit 往复式冷冻机
reciprocating diaphragm pump 往复式隔膜泵，往复隔膜泵
reciprocating drill 往复式钻机
reciprocating duplex pump 往复式双缸泵
reciprocating engine 往复式发动机
reciprocating feeder 往复式给煤机，往复式给料机
reciprocating grate bar 往复式炉条
reciprocating linear motor 往复式直线电动机
reciprocating machine 活塞式机械，往复式机械，往复式电机
reciprocating membrane pump 往复式活塞隔膜泵
reciprocating motion 往复运动
reciprocating motor 往复电动机
reciprocating piston pump 往复式活塞泵，往复活塞泵，来去式活塞泵，往复泵活塞
reciprocating pump with counterrunning piston 对向活塞泵
reciprocating pump 往复泵【活塞，柱塞，薄膜等】，往复式水泵
reciprocating ram feeder 往复式给料机
reciprocating screen 振动筛，往复式振动筛
reciprocating sieve 振动筛，往复式浆料泵
reciprocating sliding 往复滑动，往复滑移
reciprocating type 往复式
reciprocating vacuum pump 往复式真空泵，往复真空泵
reciprocating 往复的【活塞或柱塞泵】
reciprocity calibration 互易校准
reciprocity clause 互惠条款
reciprocity failure 反比定律失效
reciprocity law 反比定律，倒易律，互易律
reciprocity rules for exchange factor 交换系数互换率
reciprocity theorem 互易定理，可逆定理
reciprocity 交互作用，互易性，可逆性，相关性，倒易，互惠待遇
RECIRC＝recirculation 再循环
recirculated air cooler 回风冷却器，再循环空气冷却器
recirculated air 回风，再循环风，再循环空气
recirculated flue gas 再循环烟气
recirculating air damper 再循环空气挡板
recirculating ball screw 循环滚珠螺杆，循环滚珠轴杆
recirculating bed 再循环床【沸腾炉】
recirculating cooling water system 循环冷却水系统
recirculating current 环流
recirculating fan 再循环风机
recirculating flow 回流
recirculating flue gas damper 再循环烟气挡板
recirculating fluidized bed 循环流化床
recirculating gas damper 再循环烟气挡板
recirculating jet 环形射流
recirculating line 再循环管路
recirculating loop frequency 闭合回路频率
recirculating loss 环流损失
recirculating of ash sluicing water 灰水再循环系统
recirculating pipe 再循环水管
recirculating piping 再循环管
recirculating pump 再循环泵
recirculating ratio 循环比，再循环倍率
recirculating steam generator 再循环蒸汽发生器
recirculating storage 动态存储器，循环存储器
recirculating store 重复循环存储器，再循环存储器
recirculating system 循环系统，再循环系统
recirculating wake flow 回流尾流
recirculating water piping 再循环水管道
recirculating water supply system 循环供水系统
recirculating water 循环水，再循环水
recirculating zone 回流区
recirculation capacity 再循环容量
recirculation cavity 回流空穴

recirculation control 再循环控制
recirculation cooler 再循环冷却器
recirculation cooling water 再循环冷却水
recirculation current 再循环流
recirculation flow control 再循环流量控制
recirculation flow rate 再循环流量
recirculation flow 再循环流量，回流
recirculation heat exchange 再循环热交换器
recirculation inlet nozzle 再循环水进口接管
recirculation inlet 再循环入口
recirculation line 再循环管线
recirculation loop 再循环回路
recirculation of cooling air 冷却空气回流
recirculation outlet 再循环水出口
recirculation phase （失水事故后）再循环阶段
recirculation pipe 再循环管
recirculation pump house 循环泵房
recirculation pump 再循环泵
recirculation ratio 再循环率，再循环比
recirculation reactor line 再循环管（线）
recirculation reactor 循环反应器
recirculation region 回流区
recirculation system 再循环系统
recirculation valve 再循环阀，再循环门
recirculation water flow 再循环水流量
recirculation water pump 再循环水泵
recirculation water 循环水，回流水，再循环水
recirculation zone 回流区
recirculation 二次循环，再循环，回流，信息重记，重复循环
recirculator tube 再循环管
recirculator 再循环器，再循环管，再循环系统管道
recital （合同）陈述部分，说明条款
reckoner 计算器，计算表，计算者
reckoning 算账，推算
reckon 计算，算作，计数，认为
reclaim area 取料范围
reclaimed ground 填筑地
reclaimed land 开垦地，填筑地
reclaimed water 回收水
reclaimer 斗轮机，回收程序，回收设备，无用单元收集程序，扒料机，回收装置，取料机
reclaim hopper 取料机
reclaiming facility 取料设备
reclaiming machine 斗轮堆取料机
reclaiming rate system 取料装置
reclaiming 回收，取料
reclaim point 取料点，受煤点
reclaim tunnel 取煤地槽
reclaim 收复，回收，再生，矫正，改正，收回，开垦，要求恢复，要求归还，复归
reclamation of land 土地填筑
reclamation 收回，矫正，改正，废物利用，回收，开垦，填筑，再生
reclocking 重复计时
reclose circuit-breaker 重合闸断路器
recloser 自接入继电器，自动重合开关，自动重接器，自动重合闸，自动开关，重合闸，再次接通，重合闸装置
reclosing breaker 自动重合闸，再闭合断路器
reclosing fuse 重合熔丝
reclosing relay 自接入继电器，重合闸继电器
reclosing time 重合闸时间，再闭路时间
reclosing 重合，重合闸，再投运，再合闸，再次接通，重新接通，重闭
reclosure 再次接入，重合闸，自动接入
recluse 重合
RECMARK(record mark) 记录标记
recoal tunnel 取煤地槽
recoating of precoat filter 覆盖过滤器的重新铺料
recoating （用油漆等）再涂，重新铺料
recoding medium 记录介质
recognised overload 认可过载
recognition device 识别装置，识别系统
recognition network 识别网络
recognition program 识别程序
recognition system 识别系统
recognizable state 可识别状态
recognizer 识别程序，识别器，测定器，识别算法
recoil effect 反冲效应
recoil electron 反冲电子
recoiler 卷取机，重绕机
recoil line 回复线
recoil motion 反冲动作，反冲运动
recoil nucleon 反冲核子
recoil proton ionization chamber 反冲质子电离室
recoil proton 反冲质子
recoil 反冲，后坐，重绕，再绕，弹回
recolonize 再殖，再度移民到
recombination coefficient 复合系数
recombination efficiency 复合效率
recombination rate 复合率，再化合率
recombination train 复合管系
recombination 复合物，再化合，复合
recombiner loop 复合回路
recombiner system 复合器系统，复合装置系统
recombiner unit 复合器，复合单元，复合装置
recombiner 复合器，接触器
recommendation is made that 建议，值得推荐的是，最好是
recommendation letter 推荐信
recommendation 建议，推荐
recommended comment 推荐意见
recommended frequency 推荐频率
recommended limit 推荐限值
recommended operating speed 推荐运行速度
recommended parameter 推荐参数
recommended practice 建议实践
recommended product 推荐产品
recommended proposal 推荐方案
recommended spare part list 推荐的备件清单
recommended spare parts 推荐备件
recommended value 推荐值
recommend 建议，推荐
recommision 复役
recompense 补偿，报偿，酬金，赔偿（金）
recompilation 再编译，重新编译
recomplement 再补码，再补数
RECOMP＝recomplement 再补码，再补数

recompression curve 再压缩曲线
recompression 再压实
reconcentrating （重水）提浓，再浓集
reconciliate （使）和解，（使）和好，调解，调停
reconciliation agreement 调解协议，调解书，和解书
reconciliation of the project 项目的最终结算
reconciliation process 账目查对进程
reconciliation statement 对账表，调整表，核对表
reconciliation 调节，对账表，对账，和解，调和，和谐
recondenser plant 再凝结设备
reconditioning 检修，重新调整，再处理，重建，修复，恢复，重调节，修改，重整
reconfiguration 重新组合，重配置，结构变换，再组合，重新装置
reconfirmation 再确认
reconnaissance diagram 选点略图
reconnaissance fire vehicle 勘察消防车
reconnaissance map 草测图，踏勘图
reconnaissance report 踏勘报告
reconnaissance satellite 侦察卫星
reconnaissance survey 草测，查勘测量，概查，普查，踏勘，踏勘测量，选线测量
reconnaissance 查勘，调查，踏勘
reconnection to power 重新接上电源
reconsideration 复议，重新考虑，重新审议，再固结
reconstituent assembly 可拆式燃料组件
reconstitutable assembly 可拆式燃料组件
reconstitutable fuel assembly 可重新组成的燃料组件
reconstituted mica 改制云母，粉云母
reconstructed district 改建区
reconstructed image 重现（的）像
reconstructed project 改建工程
reconstruction of a company 重整公司
reconstruction of acoustic hologram 声全息图的再现
reconstruction of company 公司重组
reconstruction of local flux 局部通量的重建
reconstruction of pin power distribution 芯棒功率分布的重建
reconstruction project 改造工程
reconstruction structure 再现结构
reconstruction works 改建工程
reconstruction 改建，重建
reconstructive policy 改建政策
reconvergent node 再收敛节点
reconvergent path set 再收敛通路集，再收敛通路组
reconvergent variable set 再收敛变量集
recooler 再冷却，二次冷却器
recool 再冷却，二次冷却，循环冷却
recordable condition 可录记的条件
record amplifier 记录放大器，录音放大器
record and process input table 记录及进程输入表
record circuit 记录电路
record conversion system 记录变换系统
record conversion 记录变换
recorded flood 实测洪水
recorded hydrograph 实测过程线，实测水文过程线
recorder adjustment 记录装置调整
recorder chart 自动记录器记录纸，记录纸
recorder controller 记录控制器，记录器控制器
recorder float 自记浮标
recorder motor 记录装置电动机
recorder panel 记录器仪表盘
recorder pen 记录笔，自动记录器笔
recorder reception 记录接收
recorder scale 记录器标度
recorder sensitivity 记录器的灵敏度
recorder 记录器，记录员，记录装置，录音机，记录仪，记录仪表
record in extensor 详细记录
recording accelerometer 记录式加速度计
recording ammeter 自记安培计，记录式电流表
recording anemograph 自记风速计
recording apparatus 记录仪器
recording barometer 记录式气压计
recording buoy 自记浮标
recording calorimeter 记录式热量计
recording controller 记录控制器
recording counter 记录计数器
recording curve 记录曲线
recording demand meter 记录式需量计
recording densitometer 记录式显像密度计
recording density 记录密度
recording device 记录装置
recording differential manometer 记录式差示压力计
recording drum 记录滚筒，记录磁鼓
recording duration 记录时间
recording echo sounder 自记回声测深仪
recording electrometer 记录静电计
recording equipment 记录装置
recording flowmeter 记录式流量计，流量自动记录仪
recording galvanometer 记录式检流计
recording head 记录头，录音头
recording hygrometer 记录式湿度计
recording impulse 记录脉冲
recording instrument 记录仪表
recording kilo-watt-hour meter 记录电度表
recording level 记录水平
recording liquid level gauge 记录式液位计
recording manometer 压力记录仪
recording mechanism 记录机构
recording medium 记录介质，记录载体，记录纸
recording meter panel 记录表盘
recording meter 记录仪，记录式仪表
recording microbalance 记录式微量天平
recording microphotometer 记录式测微光度计
recording oscillometer 记录示波器
recording paper of sounder 测深仪记录纸
recording paper 记录纸
recording pen linkage 记录笔尖的传动装置
recording pen stiffness 记录笔稳定性
recording pen tension 记录笔张力

recording pen travel 记录笔行程
recording pen 记录笔，自记笔
recording pointer 自记笔
recording potentiometer 记录式电势计，记录式电位计
recording pressure gauge 记录式压力计
recording psychrometer 记录式干湿球湿度计
recording ratemeter 记录速率仪
recording rotameter 记录式转子流量计
recording spectrophotometer 记录式分光光度计
recording strip 记录带
recording stylus 记录笔尖，记录笔，录音针
recording system 记录系统
recording tachometer 自动记录流速仪，自动记录转速仪，记录式转速计
recording technique 记录技术
recording thermometer 记录式温度计，自计温度计
recording tide gauge 自动潮位计
recording unit 记录装置
recording vacuum gauge 记录式真空计
recording vacuum tube voltmeter 记录式真空管伏特计
recording viscometer 记录式黏度计
recording voltmeter 记录式伏特计，自记电压表，记录式容积计
recording watt and varmeter 自动记录式瓦乏计
recording wattmeter 自记电力表，记录瓦特计
recording Wheatstone bridge 记录式慧斯通电桥，记录式慧斯登电桥
recording wind vane 自记风向计
recording 记录，录音，录像，录制
record 记录
record of acceptance 验收记录
record of inspection 检查记录
record of performance qualification test 性能鉴定试验的记录
record of welder or welding operator qualification test 焊工或焊接操作者鉴定试验记录
record retrieval 记录恢复，记录检索
record selection expression 记录选择表达式
record separator 记录分隔符
record sheet 记录单
record system 记录制度
record 记录，读数示出
recoup 补偿，扣除，赔偿（金）
recourse action 追索诉讼
recourse to arbitration 提请仲裁
recourse 求援，追索，追索权，依靠，求助
recoverable coal reserves 煤可采储量
recoverable coal 可回收煤，可恢复煤
recoverable heat 可再生热，可回收热
recoverable pile 可回取煤堆
recoverable reserve 可采储量，探明储量
recoverable scrap 可回收的碎燃料芯块【燃料制备】
recovered rate 回收率
recovered solution 回收溶液
recover the cost 收回成本
recovery boiler 废热锅炉
recovery coefficient 复原系数，恢复系数，回收系数
recovery control unit 恢复控制器
recovery control 恢复控制
recovery creep 回复蠕变
recovery curve 恢复过程曲线，过渡过程曲线
recovery cycle 回升周期
recovery device 回收装置
recovery factor 恢复系数，恢复因子
recovery furnace 回收炉
recovery gypsum 石膏回收
recovery information set 恢复信息组
recovery interrupt 恢复中断
recovery management support 恢复管理支援
recovery method 回收法
recovery of ash slurry water 灰水回收
recovery of chemical energy 化学能恢复
recovery of critical 重返临界
recovery of deformation 变形恢复
recovery of loss 弥补亏损
recovery of principal and interest 收回本息
recovery of waste heat 余热回收
recovery pan 接油盘
recovery peg 重复用测标
recovery plant 回收装置，再生装置
recovery processing 恢复处理，恢复加工
recovery rate 回收率，再生速度，恢复速度
recovery ratio 恢复系数比，恢复比，回收率
recovery scrubber 回收洗涤器
recovery signal 返回信号
recovery stability time 恢复稳定时间
recovery system 恢复系统
recovery tank 回收箱
recovery temperature 复原温度，恢复温度
recovery test 回收试验
recovery time constant 恢复时间常数
recovery time 复归时间，恢复时间，过渡过程持续时间
recovery unit 回收装置
recovery vessel 回收容器
recovery voltage 恢复电压，还原电压
recovery 复原，恢复，回收
recover 收回，抵扣，恢复，弥补，重新获得
RECP＝receptacle 插座，插孔，容器
recreated neutron 再生中子
recreational facility 再生设施，文娱设备
recreational water 文娱场所用水，再生加工用水
recreation area 休养区
recreation complex 综合旅游区
recreative facilities 娱乐设施
recriticality 重返临界
recruitment examination 征聘考试
recruitment 招聘
recrushing 二次破碎，重轧碎
recrystallization annealing 再结晶退火
recrystallization 再结晶
recrystallized graphite 再结晶石墨
recrystallized Zr 再结晶的锆
recrystillation 再结晶
rectanglar comb filter 矩形梳式滤波器
rectangle 矩形，长方形
rectangular air duct 矩形风管

rectangular array　矩阵列
rectangular axes　直角坐标轴
rectangular axis　直交轴，正交轴
rectangular bar blade　矩形条叶片
rectangular bar copper　矩形铜条
rectangular bed　矩形床
rectangular bus bar　矩形汇流条，矩形母线
rectangular card potentiometer　矩形卡片状电势计
rectangular channel　矩形通道
rectangular conductor　矩形导体，矩形导线
rectangular coordinate system　直角坐标系统
rectangular coordinate　直角坐标
rectangular copper wire　矩形截面铜线
rectangular core　矩形铁芯
rectangular distribution　（概率）矩形分布
rectangular duct cross-section　矩形通道横截面
rectangular duct　矩形导管，矩形风道
rectangular electrode　矩形电极
rectangular equation　直角坐标方程
rectangular fin　矩形肋
rectangular fluidized bed　矩形流化床
rectangular foundation　矩形基础
rectangular function　矩形函数
rectangular guide　矩形波导管
rectangular header　矩形联箱
rectangular hoops　矩形环箍
rectangular hysteresis loop　矩形磁滞回线
rectangular impulse　矩形冲击波，矩形脉冲
rectangular lug　矩形接线片
rectangular mesh　长方网络
rectangular orifice　矩形风口
rectangular panel　矩形电屏，矩形电盘
rectangular parallelepiped　长方体
rectangular pulse　矩形脉冲
rectangular register　矩形调节风口
rectangular rib　矩形肋
rectangular section tube　矩形截面管
rectangular section　矩形截面
rectangular-shaped pulse　矩形脉冲
rectangular sky-light　矩形天窗
rectangular slot　矩形槽
rectangular tank　矩形油箱
rectangular throttling slide damper　矩形节流挡板
rectangular timber　方木
rectangular tower body　扁塔身
rectangular type core　矩形铁芯
rectangular voltage　矩形电压
rectangular vortex ring　矩形涡环
rectangular waveguide　矩形波导
rectangular wave inverter　矩形波逆变器
rectangular wave　矩形波
rectangular wind tunnel　矩形试验段风洞
rectangular wing tip　矩形翼尖
rectangular wing　矩形翼
rectangular working section　（风洞）矩形试验段
rectangular　直角的，成直角的，矩形的，长方形的，正交的，直角，矩形
rectiblock　整流片
rectification capacitor　整流电容器
rectification characteristic　整流特性曲线
rectification circuit　整流电路
rectification coefficient　整流系数
rectification column　（重水）精馏塔，分馏塔
rectification factor　整流系数
rectification of channel　河槽（道）整治
rectification of river　河道整治
rectification process　精馏工艺，精馏过程
rectification　整流，检波，调整，修正，精馏，提纯，校正，改正，矫正，纠正
rectified action　整流作用
rectified current　整流后的电流，已整电流，整流电流
rectified feedback　整流反馈
rectified output　整流输出
rectified power　整流功率
rectified value　整流值，已整值
rectified voltage　已整流电压
rectified waveform　整流波形
rectified　检波的，已整流的，精馏过的
rectifier assembly　整流器装配
rectifier bridge comparator　整流电桥式比较器
rectifier bridge relay　整流电桥式继电器
rectifier bridge　整流桥，整流器电桥
rectifier cell　整流元件
rectifier characteristic　整流器特性
rectifier circuit　整流（器）电路
rectifier conduction time　整流管导电时间
rectifier doubler　倍压整流器
rectifier-driven motor　整流器供电的电动机
rectifier efficiency　整流器效率
rectifier equipment　整流器设备，整流阀设备
rectifier filter　平滑滤波器，整流器滤波器
rectifier instrument　整流仪器，整流式仪表，有整流器的仪表
rectifier inverter　整流换流器
rectifier lamp　整流管
rectifier load　整流器负载
rectifier operation　整流器运行
rectifier phase comparator　整流型相位比较器
rectifier photocell　整流光电管，整流光电池
rectifier photoelectric cell　整流光电管
rectifier plate　整流器片
rectifier power source　整流电源
rectifier protection　整流器保护
rectifier relay　整流器继电器
rectifier ripple factor　整流波纹系数，整流脉动系数
rectifier stage　整流级
rectifier station　整流站
rectifier thyristior　整流器闸流管
rectifier transformer　整流变压器，整流器用变压器
rectifier type differential relay　整流式差动继电器
rectifier type distance relay　整流式距离继电器
rectifier type meter　整流式仪表
rectifier type motor　整流式电动机
rectifier type relay　整流式继电器
rectifier unit　整流设备，整流机组，整流器组件
rectifier welding set　整流焊机
rectifier　整流管，整流器，检波器，精馏器，矫正器，整流装置

rectiformer 整流变压器
rectify and reform 整改
rectify economic order 整顿经济秩序
rectifying action 检波作用，整流作用，检测作用
rectifying device 整流装置，检波装置
rectifying film 阻挡层
rectifying method 调流方法
rectifying screw 调整螺旋
rectify 整流，矫正，检波，提纯，精馏，调准，纠正，整顿
rectilinear flow 直线流，直匀流
rectilinearity 直线性
rectilinear manipulator 直线式机械手，坐标控制器
rectilinear motion 直线运动
rectilinear propagation 直线传播
rectilinear vortex 直线涡流，直涡丝，直线旋涡
rectilinear vorticity 直线涡流
rectilinear 线性的，直线的
rectiplex 多路载波通信设备
rectistack 整流器
rector 氧化铜整流器
rect＝rectifier 速流器，检波器
RECT＝rectifier 整流器
rectron 电子管整流器
recumbent fold 伏褶皱
recuperate 复原，恢复，回收，再生
recuperation 恢复，蓄热，余热利用，反馈，再生，回收，换热
recuperative airheater 间壁式空气预热器
recuperative gas turbine 间壁回热式燃气轮机
recuperative heater 同流加热器
recuperative heat exchanger 同流热交换器，间壁式热交换器，同流换热器，间壁式换热器
recuperative regenerator 表面式回热器，间壁式回热器
recuperative type attemperator 间壁式减温器
recuperative type desuperheater 间壁式减温器
recuperative 复原的，同流换热的，间壁换热的，复热的
recuperator 同流换热器，同流换热室，间壁式换热器，表面式回热器，间壁式回热器，废油再生装置
recurrence equation 递推方程
recurrence formula 迭代公式，递归公式，循环公式，递推公式
recurrence frequency 脉冲重复频率，重复频率
recurrence interval 重复间隔，脉冲周期，重现间隔，重现期
recurrence period 重现周期
recurrence rate 重复率，重复速度
recurrence relation 循环关系，递推关系，重现关系
recurrence 复现，递归，递推，再现，循环，重复，再发生，重现
recurrent appropriation 经常性拨款
recurrent code 连环码，卷积码
recurrent frequency 重发频率
recurrent item 经常性项目
recurrent network 重复网络，链形网络
recurrent procurement 经常性采购，周期性采购，反复采购
recurrent pulse 周期脉冲，复现脉冲
recurrent signal 重发信号
recurrent subvention 经常资助金
recurrent 再现的，周期的，递归的，重复的，经常发生的，周期性发生的，循环的，复发的，周期性的，反复发生的
recurring cost 续生成本，经常成本
recurring deficiency 反复出现的缺陷
recurring 循环的，再发的，复发的，经常的
recursion equation 递归方程
recursion function 递归函数
recursion 递归，递推，递归式
recursive computation 递归计算
recursive function 递归函数
recursive macro call 信息宏呼叫
recursive procedure 递归过程
recursive routine 递归程序
recursive 递归的，循环的
recurved parapet 挡水墙，防浪墙
recurved spit 弯曲沙嘴
recur 再现，循环，回想，回到，复现，递归
recyclable coagulant 可再循环凝结剂
recycle compressor 循环压缩机
recycle cooler 再循环冷却器
recycled air 再循环气体
recycled backwash water 循环回洗水
recycled coagulant 循环凝结剂
recycled fuel 回烧燃料，再循环燃料
recycled gas mixing method 再循环气体混合法
recycled material 回用材料，循环物质，循环使用材料
recycled off-gas 循环尾气
recycled waste-process 废水回流过程
recycled water 循环水
recycle fuel plant （后处理厂）核燃料再循环工厂
recycle gas 循环气体
recycle ratio 循环比
recycle system 循环系统
recycle unit 循环设备
recycle valve 再循环阀
recycle-water 回收水，回用水
recycle 再循环，回收，再生，（核燃料）回烧，重复利用
recycling chromatography 循环色谱法
recycling of plutonium 钚的再循环
recycling process 循环过程
recycling pump 再循环泵
recycling reactor （核燃料）再循环反应堆
recycling 再生，再循环
redactor 减压器
red and yellow podzolic soil 红黄色灰化土
red bed 红色岩层
red brass 红铜
red clay 红（色）黏土
red clinker tile 红缸砖
red copper weld 紫铜焊缝
red earth 红土
redeemable preferred stock 可赎还优先股

redeem 偿还，弥补
redemption loan 偿还贷款
redemption date 偿还（贷款）日期
redeposited loess 次生黄土，再积黄土
redeposition 再沉淀，再沉降
redeposit 再沉积
redesign 重新设计
redetermination 再测定
redevelopment 改建，再开发
red hardness 红硬性，热硬度
red heat 红热
red-hot 赤热的
redial 重拨
redirecting surface 改向面
re-dispatch 再分配
redissolve 再溶解
redistillation 再蒸馏
redistilled water 再蒸馏水
redistribution of stress 应力重分布
redistribution prestress 预应力重分布
redistribution 再分布，再分配
red lead paint 红丹漆，红丹涂料
red lead primer 铅丹底漆
red lead 红丹漆，红丹，红铅
REDOPS(ready for operation) 操作就绪，运算就绪
redox flow cell 氧化还原液硫电池
redox potential 氧化还原电位
redox reaction 氧化还原反应
redox resin 氧化还原树脂
redox 氧化还原作用
red particle 红磁粉
red pine 红松
redressed current 已整流电流
redress 矫正，再平衡，调整，纠正，使再平衡，补正，重新穿上，修整，赔偿（金）
redriving （桩工）复打
red rust 红锈
Red Sea reservoir 红海蓄水池
red-shortness 热脆性，热脆
red-short 热脆
red soil 红土
red-tape operation 程序修改，红带运算，辅助操作
reduce a vacuum 减低真空
reduced air 减压风
reduced capacity tap 降容量分接头
reduced capacity valve trim 减容阀内件
reduced capacity 降低容量
reduced current 折算电流
reduced damping 折算阻尼
reduced-density effect 约化密度效应
reduced density 对比密度
reduced discharge 折减流量，折算流量
reduced equation 简化方程，对比方程
reduced flow 收缩水流，折算流量
reduced frequency 折算频率
reduced gravity 折合重力
reduced group 缩减能群
reduced head 折减水头，折算水头
reduced height 折算高度
reduced impedance 折算阻抗，归一化阻抗
reduced instruction set computer 精减指令集计算机
reduced insulation 降低绝缘
reduced iron 还原矿
reduced kilovolt-ampere tap 降低千伏安出力的抽头
reduced kVA tap 低负荷抽头，降负荷抽头
reduced length of the geodetic line 大地线改化长度
reduced length 折合长度，折算长度
reduced mass-flow 折合流量
reduced mass velocity 对比质量流速
reduced mass 折算质量
reduced matrix element 约化矩阵元素
reduced modulus of elasticity 折算弹性模数
reduced moment of inertia 折算惯性矩
reduced natural frequency 折算固有频率
reduced neutron width 约化中子宽度
reduced ore 还原铁
reduced output 降低出力
reduced parameter 折算参数
reduced port valve 缩径孔道阀门
reduced power tapping 降低功率分接
reduced power 降低功率，折算功率
reduced pressure coefficient 折算压力系数
reduced pressure 折算压力，减压，对比压力，降压
reduced reserve 裁减储备
reduced resistance 折算电阻
reduced scale 缩尺，缩小比例尺
reduced scope high realism simulator 缩小范围、高逼真度电厂仿真机
reduced scope 缩减范围
reduced section test specimen 缩减段试样
reduced section 拉细部位，减缩部位，折算断面
reduced section 缩减段【张拉试验】
reduced spectrum 折算谱
reduced-speed operation 减速运行
reduced speed 折合转速
reduced temperature difference 归一化温差
reduced temperature 对比温度，折算温度
reduced time unit 折合时间单位，对比时间单位
reduced value 折算值
reduced viscosity 对比黏度
reduced voltage starter 减压启动器，降压启动器
reduced voltage starting 降压启动
reduced voltage start 降低电压启动
reduced voltage 下降电压，折算电压
reduced volume 对比体积，缩小体积
reduced width 折算宽度
reduced winding diagram 简化绕组图
reduced wind speed 折算风速
reduced 变换后的，减少的，简化的
reduce earthwork and stonework 减少土石方量
reduce first cost 压缩初期投资
reduce frequency 降低频率
reducer mechanical power rating 减速机额定功率

reducer 减压器，减压阀，减速器，节流器，减振器，管子缩节，扼流圈，大小头，减速机，渐缩管，异径管，变径管
reduce tee 异形三通
reduce the benching and revetment of the hilly area 减少山区的阶梯式挖土和护坡
reduce the staff 裁员
reduce 减去，扣除，降低
reducibility 还原性
reducible 可缩减的，可还原的，可简化的
reducing agent 还原剂
reducing-and-cooling plant 减温减压装置
reducing atmosphere 还原气氛
reducing bend 变径弯头，异径弯头
reducing bore valve 缩腔阀门
reducing capacity 还原能力
reducing coupling 大小头，异径管箍，异径管接头
reducing cross 异径四通
reducing elbow 变径肘管，渐缩弯头，变径弯头，异径弯头
reducing environment 还原气氛
reducing excitation 减励磁
reducing flame 还原焰
reducing flange 异径法兰，缩口法兰
reducing gas 还原性气体，还原气
reducing gear 减速齿轮，减速装置，减速箱
reducing head 缩径弯头
reducing joint 缩接套，异径接头
reducing lamp holder 灯头缩节
reducing nipple 异径短管
reducing nozzle 异径管接头
reducing of frequency 频率降低
reducing pipe 异径管，渐缩管
reducing socket 大小头承插管，异径管套
reducing swage 锻制异径管
reducing tee 渐缩T形管，缩径三通管，异径三通
reducing transformer 降压变压器
reducing turbine 抽汽式汽轮机
reducing valve 减压阀
reducing waste 还原废料
reducing 减少，减缓，简化，还原，还原的，减低的，折合的，减低
reductant 还原剂
reduction admittance matrix 简化导纳矩阵
reduction agent 还原剂，脱氧剂
reduction coefficient 折减系数，折合率，折算系数
reduction crusher 次轧碎石机
reduction device 缩分装置
reduction factor of pile group 群桩折减系数
reduction factor 换算系数，折减系数，减缩因数
reduction gear box 减速（齿轮）箱
reduction gear housing 减速齿轮箱
reduction gear ratio 减速比
reduction gear unit 齿轮减速设备
reduction gear 减速齿轮，减速装置
reduction in area 断面收缩率，断面收缩，缩颈
reduction in degree of protection afforded by enclosures 机壳的保护程度的降低
reduction in design quality 设计质量下降
reduction in visibility 能见度减低
reduction of area 断面收缩率
reduction of availability 可用率降低，利用率降低
reduction of coal sample 煤样的缩分
reduction of CO_2 emission 二氧化碳减排
reduction of data 信息简化，数据还原
reduction of dissolved oxygen 溶解氧的还原
reduction of evaporation from water surface 抑制水面蒸发
reduction of gradient in curve 曲线折减坡度
reduction of gradient 坡度折减
reduction of longitudinal slope 纵坡折减
reduction of porosity 孔隙度降低
reduction oxidation potential 氧化还原电位
reduction oxidation 还原氧化作用
reduction partition 缩分
reduction potential 还原电位
reduction rate 减速率，减速比，还原速率，传动比，破碎比，缩小比例，缩小率
reduction test on tubes of metal 金属管缩口试验
reduction type semiconductor 还原型半导体
reduction voltage 降压
reduction with calcium 钙热法
reduction with carbon 固碳还原
reduction zone 还原带
reduction 减少，缩减，变换，简化，还原，下降，折合，折减，折算
reductive water 还原水
reductor 减压器，减速器，伏特计附加电阻
redundance 多余度，冗余（度）
redundancy and diversity 多重性和多样性
redundancy bit 冗余位
redundancy character 冗余字符
redundancy check 剩余信息校验
redundancy computer system 冗余计算机系统
redundancy design 冗余设计
redundancy determination 多余测定
redundancy device 冗余设备
redundancy digit 冗余位【十进制】
redundancy factor 冗余系数
redundancy principle 冗余性原则，多重性原则
redundancy reduction 冗余简化
redundancy test 冗余试验
redundancy 冗余，冗余度，剩余度，冗余码，冗余位，冗余性，多重性
redundant architecture 冗余结构【仪表】
redundant bracing member 辅助支撑件，冗余支撑件
redundant channel 冗余通道
redundant check 冗余校验
redundant circuit 多重回线路【反应堆保护系统】，备用电路，冗余电路
redundant code 冗余码
redundant configuration 冗余配置
redundant constraint 多余约束
redundant design 多重设计，冗余设计，多样设计
redundant digital signal 冗余数字信号

redundant digit　多余位，多余数字
redundant equipment　多重设备，冗余设备
redundant information　冗余信息
redundant logic circuit　冗余逻辑回路
redundant member　多余杆件
redundant scram　多重性紧急停堆，冗余事故停堆
redundant state　冗余状态
redundant structure　超静定结构
redundant subsystem　多重子系统，冗余子系统
redundant support　多余支承，多余支座
redundant system　冗余系统，备援系统，多重系统，备用系统
redundant trip coil　双跳闸线圈
redundant unknown　多余未知数
redundant　冗余的，多余的，过量的，多重的，冗余码，冗余位，冗余，多余部分
reduplicate sampling　多份采样
red water　铁锈水，锈水
reed blade　舌簧叶片
reed bog　芦苇沼泽
reed capsule　簧片膜盒
reed contact　簧片触点，舌簧触点
reeded ramp　有凹槽的斜坡
reed fence　芦苇栅
reed frequency meter　振簧式频率计
reeding　防滑条
reed march　芦苇沼泽
reed mat　芦席，苇席
reed relay　衔铁继电器，舌簧继电器
reed switch　舌簧开关，舌簧触点
reed tachometer　簧片式转速表
reed valve　针阀，簧片阀
reed　簧片，舌簧，衔铁，簧，芦苇
reefer cargo　冷冻货
reefer goods　冷冻货
reefing　缩帆
reef　暗礁
re-electrolysis　二次电解，再电解
reel fiber concrete　钢纤维混凝土
reel flange　电缆盘
reeling machine　绕线机，卷取机
reel length　线盘长度
reel stand　卷线轴架
reel-up　绕上，绕线
reel　卷轴，卷，盘，卷筒，绕，缠，线轴，线盘，绕线架，绞车，卷线，绞盘
re-embrittlement　再脆化，重新脆化
ree mineral acidity　游离无机酸酸度
re-emission　再发射，二次辐射，再排出，二次发射，次级辐射
reenergize　使又通电流，重新激励
re-engage　再啮合，重新接入
reenterable routine　可重入程序
reentering angle　凹角
reenter　再进入
reentrainment of dust　二次扬尘
reentrance　（程序）可重入性，重入
reentrant angle　凹角
reentrant cavity resonator　凹型空腔谐振器
reentrant comer　凹角
reentrant flow of gas　再入气流
reentrant inlet　凹入式进水口
reentrant intake　凹入式进水口
reentrant program　可重入程序
reentrant trunking　中继再进入
reentrant winding　闭路绕组，闭合绕组
reentrant　凹腔的，（微波管结构）凹腔型，再入的，凹角
reentry communication　（卫星通信的）重返大气通信
reentry system　重入系统
reentry turbine　回流式汽轮机
reentry　再记入，再进入
reevaporation　再汽化，再蒸发
reeve　穿结，穿（绳）入孔，结牢
reeving thimble　穿绳套环
re-export　转口，再出口
re-extract　再萃取物
refacing　重磨，重修表面
referee method test　仲裁法试验
referee test　抽样试验，仲裁试验
referee　仲裁员
reference accuracy　基准精确度
reference address　参考地址，基准地址，访问地址，引用地址，基本地址
reference area　基准面
reference axis　参考轴（线），基准轴（线）
reference black level　黑色信号参考电平，黑色信号基准电平
reference block　对比试块，参考试块
reference book　参考书
reference chord　基准弦
reference circle　参考圆
reference colour　基准色，参考色
reference concentration　参比浓度
reference conditions　基准条件，参考条件
reference coordinate　基准坐标
reference count　检验读数，基准计算，参考读数
reference critical stress intensity factor　参考临界应力强度因子
reference current meter　标准流速仪
reference cycle　参考循环，标准循环
reference data　参考资料，参考数据
reference datum　参考基准面
reference debugging aids　调试辅助程序
reference defect　参考缺陷
reference design　参考设计，标准设计
reference device　基准器件
reference dimension　参考尺寸
reference distance　基准距离
reference document　参考文件，引用文件，参考凭证
reference drawing　参考图
reference effective pressure　基准有效压力
reference electrode　参考电极，参比电极
reference ellipsoid　参考椭圆体
reference enthalpy　参考焓
reference feedback　基准反馈
reference file　参考档案
reference form　引用形式
reference frame　参考标架，参考格网，读数系

统，空间坐标
reference frequency 基准频率，参考频率
reference fuel 参考燃料，标准燃料
reference gauge 标准水尺，参考水尺
reference generator 基准信号发生器，参考信号发生器
reference grid 参考格网
reference ground 参考接地
reference height 基准高度
reference-input element 基准输入元件
reference-input signal 基准输入信号
reference-input 参考输入，标准输入量，基准输入，基准输入量
reference inspection 原始检查，基准检查
reference instrument 标准仪器，校准用仪器
reference junction compensation 基准结补偿，(温差电偶）输出端补偿，温差电偶的冷端补偿
reference junction compensator 基准结补偿器，冷端补偿器
reference junction 基准接点，参考接点，基准结，参考结
reference kilowatthourmeter 标准电度表
reference language 标本语言，参考语言
reference length 参考长度
reference letter 推荐信，介绍信
reference level 基准水位，参考水准面，基准电平，参考水平，参考高度，海图基准面
reference line 参考线，基准线，起读线，零位线
reference list of installations （关于同样的）装置的业绩表
reference location 基准位置
reference manual 参考手册
reference man 参考人
reference mark 标准点，参考标记，基准点，参考点，参证点，基准标记，假定水平基点，控制点
reference material 参考资料，参考材料，参比物质
reference nil ductility temperature 参考无延性温度
reference nil ductility transition temperature 参考无延性转变温度
reference nuclear power plant 参照核电厂，参考核电站
reference number of document 文件编号
reference operating condition 参考工作条件，参比工作条件
reference order 转接指令，控制指令
reference organ 参考器官
reference performance characteristic 参比性能特性
reference performance 参考性能，参比性能
reference plane 参考面，参证面，基准面，参考平面，基准平面
reference plant design 参考电厂设计
reference plant 参照电厂，被仿真电厂，参考电站，参照电站
reference point 参考点，基准点，依据点，参证点，控制点
reference position 基准位置，参考位置，原位
reference potential 基准电势，参考电位，标准电位，参考电势
reference power meter 基准功率计
reference power plant design 参考电厂设计
reference power plant 参考电厂
reference power supplement 基准动力源
reference power 参考功率，基准功率
reference pressure 参考压力
reference price 参考价格
reference-quality level 参考质量等级
reference quota 参考指标
reference radiation source 放射性标定源
reference radiation 参考辐射
reference radiography 参考射线照片
reference radiometer 标准辐射表
reference record 参考记录
reference response 参考源响应
reference room 资料室
reference roughness length 基准粗糙长度
reference safety analysis report 参考安全分析报告
reference section 参考断面
reference signal 参考信号，参比信号，标准信号
reference snow pressure 基本雪压
reference solar cell 参考太阳电池，标准太阳电池
reference solution 参比溶液
reference source 标准源，参考源，校准源
reference square wave 基准方形波
reference standard 参照标准
reference station 参考港，参证站
references to similar works 对类似工程的业绩
reference structure 参考结构
reference system 参考系（统）
reference temperature method 参考温度法
reference temperature 定性温度，参考温度
reference test piece 参考试块，参考试片，标准试片，标准试块
reference test 基准试验
reference time 基准时间
reference to storage 访问存储器
reference transition temperature 参考转变温度
reference value scale （量的）基准值标度，（量的）参考值标度
reference value standard 参考标准值
reference value 参考值，基准值
reference variable 基准变量，参比变量，参考变量，参比量，设定值
reference velocity 参考速度，表征速度
reference voltage 基准电压，参考电压，标准电压
reference water gauge 辅助水尺
reference wave 参考波
reference wedge 参考楔块
reference white level 白色信号参考电平，白色信号基准电平
reference winding 参考绕组，基准绕组
reference wind pressure 基本风压
reference wind speed 参考风速，基本风速
reference wind 参考风

reference 参考，参照，基准，参考点，标记，参数，基准端
referencing of films 底片的参照【射线照相】
referencing 连测
referring factor 折合系数
referring impedance 折算阻抗
referring of rotor quantity 转子量的折算
referring reactance 折算电抗
referring to detail drawing 参见详图
refer to 参见，参考，参照
refer 参考，涉及，提到，查阅，委托，归诸于，使……求助于
refilling process 再充水过程
refilling 回填土，重新蓄水
refill stage of LOCA 失水事故的再充水阶段
refill 再注满，再灌满【失水事故】，再注，再填
refinancing 融通，重新筹集资金
refined asphalt 精致沥青
refined oil 精制油
refined 精制的
refinement 提纯，精炼，精制，改善，改进，净化
refinery gas 炼油气
refining 精炼，精制，提炼
refiring of fly ash 灰再燃
refit 重新装配，改装，修理，修改，翻新
reflash alarm 再闪报警
reflatten （通量）再次展平
reflectance 反差【电】，反射，反射比，反射系数，反射率
reflected binary code 反射二进码
reflected binary 反射二进位制
reflected code 反射码，循环码
reflected core （反应堆中的）带反射层的堆心，有反射层的活性区
reflected energy 反射能量
reflected harmonics 反射谐波
reflected heat 反射热
reflected impedance 反射阻抗
reflected impulse 反射脉冲
reflected light 反射光
reflected power 反射功率
reflected pulse 反射脉冲
reflected ray 反射线
reflected reactor 有反射层反应堆
reflected shock front 冲击波反射波面
reflected shock wave 反射振荡波
reflected signal 反射信号
reflected solar irradiance 反射日辐照度【Er】
reflected solar radiation 反射日射，反射太阳辐射
reflected surface 反射表面
reflected tension wave 反射张力波
reflected wave 反射波
reflecting barrier 反射隔板
reflecting boundary 反射边界
reflecting galvanometer 镜式电流计，反射电流计，反射镜式检流计
reflecting layer 反射层
reflecting level 反射水准仪
reflecting loss 反射损失
reflecting material 反光材料
reflecting mirror 反射镜
reflecting objective 反射物镜
reflecting power 反射率，反射能力，反射系数，反射功率
reflecting prism 反射棱镜
reflecting region 反射区
reflecting stereoscope 反光立体镜
reflecting surface 反射面
reflecting wave 反向波
reflection angle 反射角
reflection coefficient of sound intensity 声强反射率
reflection coefficient of sound pressure 声压反射率
reflection coefficient 反射系数，振幅反射率
reflection crack 对应裂缝，反射裂缝
reflection factor 反射系数，反射因数，反射因子
reflection interaction 反射波相互作用
reflection law 反射定律
reflectionless waveguide 无反射波导
reflection loss 反射损耗
reflection material 反光材料
reflection method 反射法
reflection mode filter 反射滤波器
reflection modulation （储存管的）反射调制
reflection of light 反光
reflection plane 反射板，反射平面
reflection polariscope 反射偏振镜
reflection prospecting 反射法勘探
reflection shield 反射屏蔽层
reflection 反光，反映，反射，反射波
reflective boundary condition 反射边界条件
reflective coating 反射（涂）层，反射涂料
reflective insulation 反射隔热，反射性绝热
reflective material 反光材料
reflective mirror 反光镜
reflective probe 反射探针
reflectivity 反射，反射性，反射度，反射率，反射能力
reflectogage 超声波测厚仪
reflectogram 反射图形
reflectometer 反射比表，反射仪
reflectometry 反射体（法）
reflector blanket 反射层转换区
reflector bowl 抛物面形反射器
reflector control 反射层控制
reflector economy 反射层节省
reflector element 反射层元件
reflector lamp 反光灯
reflector liner 反射层衬里
reflector material 反射层材料，反光材料
reflector moderated reactor 反射层慢化反应堆
reflector panel 反光板
reflector region 反射层区
reflector rod 反射层棒
reflector saving 反射层节省，反射层节省中子
reflector thickness 反射层厚度
reflector 反光镜【太阳能】，反射器【电磁】，

反射体，反光片，反光板，反射层，反射板，反射镜
reflectoscope 反射仪，脉冲式超声波探伤仪，超声波探伤仪
reflects element 反射器单元
reflex amplification 来复式放大
reflex amplifier 来复式放大器
reflex circuit 来复电路
reflex condenser 回流凝汽器
reflexion 反光，反映，反射，反射波
reflex klystron 反射速调管
reflex reception 回复反射
reflex valve 止回阀，单向阀，逆止阀
reflex 反射的，反射，回复
refloating operation 打捞工程
reflooding 再淹没
reflood stage of LOCA 失水事故的再淹没阶段
reflood 再淹没
reflow 回流，逆流，反流
refluence 逆流，回流，退潮
refluidization 再流化
refluidize 重新流化【沸腾炉压火后】
reflux condensation 冷凝回流
reflux condenser 回流冷却器，回流凝汽器，回流冷凝器
reflux extraction flowsheet 回流萃取流程
refluxing 回流
reflux liquid （重水提纯柱）回流液
reflux pipe 回流管
reflux pump 回流泵
reflux rate 回流速率
reflux ratio 回流比
reflux tank 回流槽
reflux tube 回流管
reflux valve 回流阀
reflux 倒流，落潮，逆流，回流，退潮，反射（作用，现象），反射的，反流
refooding phase 再淹没阶段
reforestation 重新造林，再造林
reform and open up 改革开放
reformation 重整，矫正，重作，改善，改造
reformed gas 重整气
reforming temperature 重整温度
reforming 重整，恢复，改良，还原，换算，变换
reform in professional title 职称改革
reform of financial system 金融体制改革
reform of organization structure 机构改革
reform of system 体制改革
reform of the education system 教育体制改革
REFORM(reference form) 引用形式
reformulated program 重订方案
reform 变换，换算，改革，改造
refracted beam 折射波束
refracted flow 折向流
refracted pulse 折射脉冲
refracted ray 折射线
refracted wave 折射波
refracting angle 折射角
refracting prism 折射棱镜
refraction angle 折射角
refraction coefficient 折射系数
refraction diagram 折射图
refraction effect 折射效应
refraction error 折射误差
refraction index 折射率
refraction law 折射定律
refraction of sound 声折射
refraction seismograph 折射式地震仪
refraction shock wave 折射冲击波
refraction survey 折射法勘探
refraction wave 折射波
refraction 折射
refractive index 折光指数，折光系数，折射率，折射指数
refractive medium 折射介质
refractivity 折射系数，折射性
refractoloy 镍基耐热合金，里弗雷克达洛依耐热合金
refractometer 折射计，折射仪
refractoriness 耐熔性，耐火性，耐熔度
refractory alloy 耐热合金，耐火合金，高温合金
refractory ash 难熔灰
refractory backing 炉衬
refractory baffle 卫燃带
refractory belt 卫燃带
refractory block 耐火垫
refractory brickwork 耐火砖墙
refractory brick 耐火砖
refractory cement 耐火水泥，耐火胶黏材料
refractory ceramics 耐火陶器
refractory chamber 耐火炉衬燃烧室
refractory clay 耐火泥，耐火黏土
refractory coating 耐火涂层
refractory concrete 耐火混凝土
refractory cover 耐火盖板
refractory curing 烘炉【锅炉安装完，使用前】
refractory erosion 炉衬侵蚀，炉墙侵蚀
refractory-faced 涂覆耐火材料的
refractory fibre 耐火纤维
refractory fireclay block 黏土耐火砖
refractory furnace 耐火材料砌的炉膛
refractory glass 耐火玻璃
refractory lined fire-box boiler 砖砌火筒锅炉
refractory liner 耐火材料衬里
refractory lining 耐火（材料）衬里，耐火衬砌，耐火炉衬
refractory material 耐火材料
refractory metal-oxide-semiconductor 耐热的金属氧化物半导体
refractory-metal 耐熔金属的
refractory mortar 耐火砂浆，耐火灰浆
refractory nozzle blade 耐火材料喷嘴叶片
refractory organics 耐热有机物
refractory period 不应期，不起反应期
refractory property 耐火性，难熔性
refractory protection 耐热涂层，耐热保护层
refractory steel 耐热钢，耐强钢
refractory surface 重辐射面，绝热表面
refractory test 耐火试验
refractory wall 耐火砖墙
refractory 耐火，耐火材料，耐火的，难熔的，

耐熔的，高熔点的，不易处理的
refract 使折射
REF＝reference 参考
refresher training 进修培训
refresh time 更新时间
refresh 刷新，更新，再生
refrigerant pump 冷剂泵
refrigerant 冷冻剂，制冷剂，冷却介质，冷却液，冷却剂，制冷的
refrigerated wind tunnel 低温风洞，冷却风洞
refrigerated 冷冻的，冷却的
refrigerating capacity 产冷量，制冷量，制冷能力
refrigerating compressor 制冷压缩机
refrigerating cycle 制冷循环
refrigerating effect 制冷效果，产冷量
refrigerating engineering 制冷工程
refrigerating engineer 制冷工程师
refrigerating equipment 冷冻设备
refrigerating machine 冷冻机，制冷机
refrigerating medium 制冷剂，载冷剂
refrigerating plant room 制冷机房
refrigerating plant 制冷装置
refrigerating station 冷冻室，制冷机房，制冷站
refrigerating storage 冷藏库
refrigerating system oil separator 制冷系统油分离器
refrigerating system 制冷系统，制冷装置，冷冻系统
refrigeration chamber 冷冻室
refrigeration compressor 制冷压缩机
refrigeration cycle 制冷循环
refrigeration plant 制冷站
refrigeration station 制冷站
refrigeration system 制冷系统
refrigeration ton 制冷吨
refrigeration unit 制冷机组
refrigeration 冷冻，致冷，冷藏，制冷
refrigerator room and its control room 制冷机房及控制室
refrigerator （电）冰箱，制冷机，冷冻机，冰柜，冷藏室，制冷装置
refueling accident 换料事故
refueling canal 换料通道
refueling cavity 反应堆空腔，反应堆换燃料池
refueling pool 换料水池
refueling scheme 换料方案
refueling shutdown 换料停堆
refueling water storage tank 反应堆换料水贮存箱
refueling 换料，换装核燃料，装卸料
refuelling accident 换料事故
refuelling and shuffling scheme 燃料倒换方案
refuelling batch 换料量
refuelling canal 换料输送管道
refuelling cavity drain pump 反应堆井疏水泵，换料腔疏水泵
refuelling cavity fill line 反应堆井注水管线，换料腔注水管线
refuelling cavity fill pump 换料水腔注水泵，反应堆水腔注水泵
refuelling cavity flooding 换料腔注水，反应堆井注水
refuelling cavity floor 换料腔底面
refuelling cavity lining 换料腔内衬
refuelling cavity 换料腔，反应堆井
refuelling cell 换料室
refuelling cycle 换燃料周期
refuelling flask 换料罐
refuelling hatch cover 换料舱口盖板
refuelling hatchway 换料舱口通道
refuelling hatch 换料舱口
refuelling machine cooling and purge system 换料机冷却和吹洗系统
refuelling machine maintenance compartment 换料机维修间
refuelling machine repair compartment 换料机修理间
refuelling machine 换燃料机，装燃料机，装卸料机
refuelling outage time 换料停堆时间
refuelling outage 换料停堆
refuelling pattern 换料模式
refuelling plan 换料计划［方案］
refuelling platform 换料平台
refuelling pool 换料水池
refuelling port 换料舱口，换料孔
refuelling procedure 换料程序，换料步骤
refuelling region 换料组，换料区，换料单元
refuelling replacement batch 换料量
refuelling schedule 换料日程表
refuelling scheme 换料方案
refuelling seal ledge 换料密封突环
refuelling shutdown 换料停堆
refuelling slot gate 换料闸门
refuelling slot 换料槽
refuelling sodium level 换料钠液位
refuelling space 换料空间，换（燃）料场地
refuelling system 换料系统
refuelling tank 换料箱
refuelling test tower 换（燃）料试验塔
refuelling time 换（燃）料时间
refuelling tool 换料工具
refuelling tube 换料管
refuelling water pump 换料水泵
refuelling water storage tank system 换料水储存箱系统
refuelling water storage tank 换料水贮存箱
refuelling water 换料水，反应堆井水
refuel 加燃料，换燃料
refugee capital 游资，投机性短期资本
refuge harbour 避风港
refuge 安全岛，安全地带，避车台
refund of duty 退税
refund 偿还，归还（款），退还，退款
refurbishing 大修，彻底检查，修配
refurbishment 再刷新，整修
refurnish or replace the supplies 重新供货或换货
refusal of consent 拒绝同意
refusal point 桩的止点
refusal pressure 回抗压力
refusal to accept 拒收

refusal 拒绝
refuse an order 拒绝命令
refuse burner 垃圾焚化炉
refuse-burning plant 燃烧垃圾发电厂
refuse chute 垃圾道，排矸溜槽
refuse coal 煤渣，下脚煤
refuse combustion gas 垃圾燃烧气
refuse-content 含矸率
refuse destructor 垃圾处理厂，垃圾焚化炉
refuse disposal 垃圾处理，垃圾处置
refuse dump 垃圾堆
refuse-fueled boiler 废料锅炉
refuse fuel 废物燃料
refuse furnace 垃圾焚化炉
refuse-incinerated boiler 垃圾焚烧炉
refuse incineration furnace 燃垃圾炉膛，燃废料炉膛
refuse incinerator 垃圾焚烧锅炉，垃圾焚化炉
refuse in coal 煤矸石
refuse power generation 垃圾发电
refuse power plant 垃圾电厂
refuse to grant 拒绝承认
refuse to indemnify 拒绝赔偿……
refuse treatment 垃圾处理
refuse truck 垃圾车
refuse utilization plant 废料利用装置
refuse wagon 垃圾车
refuse 残渣，废料，垃圾，渣滓，灰渣，矸石，拒绝，反驳
refute 驳斥
regarded alkali soil 再生碱土
regarding 关于，就……而论
regavolt transformer 一种柱式接触调压器
regenerable candle-type filter 可再生烛形过滤器
regenerant level 再生水平
regenerant proportioning pump 再生剂计量泵
regenerant pump 再生剂泵
regenerant utilization 再生剂利用率
regenerant 再生，再生的，反馈，再生剂
regenerated capacity （离子交换树脂的）再生容量
regenerated cellulose 再生纤维素
regenerated fuel 再生燃料，堆后料
regenerated oil tank 再生油箱
regenerated resin 再生好的树脂
regenerated water 再生用水
regenerate 还原，再生，回热，正反馈，回收，反馈，更新
regenerating circuit 再生电路
regenerating gas cooler 再生气体冷却器
regenerating gas heater 再生气体加热器
regenerating heat treatment 再生热处理
regenerating reagent 再生剂
regenerating solution 再生液
regenerating unit 再生装置
regenerating 再生
regeneration air heater 回转式空气预热器
regeneration column 再生塔
regeneration counter 再生计数器
regeneration curve 水位回升曲线
regeneration cycle 再生循环
regeneration efficiency of ion-exchange resin 离子交换树脂再生效率
regeneration efficiency 再生效率
regeneration equipment 再生设备
regeneration factor （燃料）再生因子，（堆燃料的）再生系数
regeneration level 再生水平
regeneration loop 再生回路
regeneration method 再生方法
regeneration of catalyst 催化剂再生
regeneration of current 电流再生
regeneration of ion-exchange resin 离子交换树脂的再生
regeneration of ion-exchange softener 离子交换软化剂的再生
regeneration period 再生阶段，再生周期
regeneration sequence 再生程序
regeneration system 再生系统
regeneration temperature 再生温度
regeneration test 再生试验
regeneration tower 再生塔
regeneration water 入渗水
regeneration 再生（作用），还原，回收，更新，回热，正反馈，交流换热，后反馈放大
regenerative air heater 再生式空气加热器，回转式空气加热器，蓄热式空气加热器
regenerative air preheater 回转式空气预热器，再生式空气预热器
regenerative brake 再生制动（器），再生制动
regenerative braking contactor 正反馈制动接触器
regenerative braking performance 再生制动性能
regenerative braking system 再生刹车系统，再生制动系统
regenerative braking 反激制动，回馈制动，反馈制动，再生制动，再生发电制动
regenerative chamber 回热室
regenerative circuit 再生电路
regenerative converter 再生转换堆，再生转换器
regenerative coolant 再生式冷却剂
regenerative cycle 再生周期，回热循环，再生循环
regenerative detector 再生检波器
regenerative efficiency 再生效率
regenerative extraction cycle 抽汽回热循环
regenerative extraction steam 回热抽汽，调整抽汽
regenerative feedback loop 再生反馈电路，正反馈电路，正反馈回路
regenerative feedback 再生反馈，正反馈
regenerative feedwater heater 回热给水加热器
regenerative feedwater heating system 给水回热加热系统
regenerative fuel cell system 再生燃料电池系统
regenerative fuel cell 再生式燃料电池
regenerative gas turbine plant 回热循环燃汽轮机装置
regenerative gas turbine 回热式燃气轮机
regenerative heater 回热加热器
regenerative heat exchanger 回热式热交换器，

回热式换热器，再生式热交换器
regenerative heating 回热加热
regenerative integrator 正反馈积分器
regenerative loop 再生电路，正反馈电路
regenerative noise 再生噪声
regenerative pump 旋涡泵
regenerative reactor 再生反应堆
regenerative reception 再生接收法
regenerative regenerator 再生式回热器
regenerative reheat cycle 回热再热循环
regenerative repeater 再生中继器，再生式转发器
regenerative stage number 回热级数
regenerative stage 回热循环燃气-蒸汽联合装置，再生阶段
regenerative steam turbine 回热式汽轮机
regenerative storage 再生存储器
regenerative system 回热系统，再生系统
regenerative turbine 回热式汽轮机
regenerative voltage 再生电压
regenerative waste water 再生废水
regenerative water heater 给水回热加热器
regenerative water heating 给水回热加热
regenerative 回热式的，再生的，正反馈的，更生的，再生，反馈
regenerator effectiveness 回热度
regenerator repeater 再生中继器
regenerator 蓄热式回热器，再生器，回热器，再发器，蓄热器，再生装置
regeulating reservoir 再调节水库
regime analysis 情况分析，状况分析
regime canal 稳定河槽
regimen 水情
regime of runoff 径流情势
regime 工况，状态，方法，范围，方式，情况，状况
regional agreement 区域协定
regional analysis of hydrological data 水文资料区域分析
regional analysis 区域分析，区域性分析
regional bench mark 地区水准点
regional channel 区域性通道，区域性波道
regional climate 区域气候
regional climatology 区域气候学
regional cooperation 地区合作
regional damage index 区域性损失指数
regional development 地区开发，区域开发
regional diagnosis 局部诊断
regional diffusion 区域扩散
regional dispatching 地区间调度
regional distribution network 区域配电网
regional economic integration 地区经济一体化
regional electricity market 区域电力市场
regional factor 地区性因素
regional flood forecast 区域洪水预报
regional forecast 区域预报
regional geography 地志学，区域地理学
regional geology 区域地质（学）
regional group 区域集团
regional heating plant 区域供热锅炉房
regional heating 区域供暖，区域供热
regional hydro-logical forecast 区域水文预报
regional hydrology 区域水文学
regionalism 区域主义，地方主义
regionality 地区性，区域性
regionalization 区划
regional layered electricity market 区域共同市场
regional map 地区地图
regional metamorphic rock 区域变质岩
regional network 区内电力网
regional office 地区办事处
regional planning commission 区域规划委员会
regional planning program 区域规划方案
regional planning 地区规划，区域计划
regional pollution 地区性污染，局部污染
regional power plant 区域（发）电厂
regional power station 区域性发电厂
regional power system 地区电力系统
regional price difference 地区差价
regional research 区域（性）调查
regional scale storm 区域性风暴
regional seismology 区域地震学
regional sewerage system 区域排水系统
regional soil 地带性土壤
regional strike 区域走向
regional substation 区域变电站
regional survey 区域勘测，区域性调查
regional tectonic stability 区域构造稳定性
regional trade 区域性贸易
regional transmission group 地区性输电集团
regional uniform electricity market 区域统一市场
regional velocity control 区域速度控制
regional water authority 地区水利管理局，区域河道管理局
regional water balance 区域水量平衡
regional water budget 区域水量平衡
regional water supply planning 地区给水规划
regional water supply 区域给水，区域供水
regional wind climate 地区性风气候
regional wind regime 地区风况
regional wind resources 地区风力资源
regional wind 地区性风
region discharge burnup 卸料区燃耗
region loading 分区装料
region of convection 对流层
region of disturbance 干扰区，干扰范围
region of fracture 断裂区域
region of intake 补给区
region of limited proportionality 有限正比区
region of non-operation 不动作范围
region of operation 动作区，动作范围，工作区
region of outflow 排泄区
region of power supply 供电区
region of stability 稳定区域
region of variation 变化区域
region of weakness 弱区
region without heating 非供暖地区
region zone 区域
region 地带，地区，地域，范围，区，区域，部位
register allotter 寄存器分配程序
register a claim 提出索赔

register arithmetic and logic unit （微处理器中具有）寄存器的运算和逻辑部件，寄存器、运算及逻辑部件
register chooser 记发器选择器，记发器寻线器
register circuit 寄存电路
register company 注册公司
register control 对准控制，寄存器控制，定位控制
registered accountant 注册会计师
registered airmail 航空挂号信
registered bond 记名债券
registered brand 注册商标
registered capital amount 注册资本额
registered capital 注册资本，注册股本，登记资本额
registered certificate of stock 记名股票
registered corporation 注册公司
registered design 注册设计
registered equipment management system 寄存器管理系统
registered letter 挂号信
registered mail 挂号信，挂号邮件
registered office address 注册办公地址
registered participant 已注册的参与者
registered trademark 注册商标
register finder 记发器寻线机，记发器选择器
register function 记发器功能
register information of bidder 报价员注册信息
register information of bidding unit 竞价机组注册信息
register information of market participant 市场成员注册信息
registering beam 重量标示器
registering instrument 记录仪器，自记器
registering voltmeter 记录伏特计
register in switch 交换中寄存器
register length 寄存器长度
register pin 定位销，固定销
register ratio （记录器）机械传动比
register relay 记录继电器
register rotation 寄存器循环移位
register signaling 寄存器信令
register storage 寄存存储器
register trade-mark 注册商标
register transfer language 寄存器传送语言
register transfer module 寄存器传送模件
register translator 寄存译码器
register 记录器【数据，信息处理】，登记，调风器，计量装置，计数计［器］，寄存器，挡板，音区
registrar of companies 公司注册局
registration number 注册号
registration of change 变更登记
registration of consumer 立户（用电）
registration 登记，读数，定位，对齐，记录，配准，注册
reglet in concrete 混凝土中的狭长槽
reglette 基线尺端点分划尺
reglet 平嵌线【防雨板】，平条，墙上凹槽，扁条，木嵌条，平嵌饰线
regolith 表层岩，表土，风化层，浮土，疏松母质岩，土被
regrade 重整坡度
REG=regulation 调节，规程，规则
REG=regulator 调节器
regression analysis 回归分析
regression coefficient 回归系数
regression curve 回应曲线，回归曲线
regression equation 回归方程
regression estimate 回归估计
regression line 回归线
regression 海退，回归，退步，倒退，消落
regrinding 重磨，再磨，修磨
regula-falsi method 试位法
regular analytical function 正则分析函数
regular budget estimate 经常概算
regular budget 经常预算
regular castable 常规浇注料
regular change 有规律变化
regular check 定期检查，定时检查
regular coast 平直岸
regular communication 定期交流
regular coursed rubble 整齐层砌毛石
regular discharge 正常放电
regular disk 标准的阀瓣
regular dividend 固定股息，正常股息
regular forecast 定期预报
regular function 正则函数，正常函数
regular inspection of equipment 设备定期检查
regular inspection 定期检查，定期审查
regular load change 正常负荷变化
regular maintenance 定期检修，定期维护，日常维修，定期检查
regular meeting 例会
regular movement 有规律移动
regular observation 定时观测
regular operation 正常运行
regular order 定期订货，正规订单，正规定单
regular overhauling 定期维修，大修
regular pit dewatering 经常性排水
regular practice 常规做法，习惯做法
regular pressure distribution 有规律的压力分布
regular price 正常价格
regular procedure 正规手续
regular project 正式项目
regular reflection factor 规则反射系数
regular reflection 单向反射，习惯做法，镜面反射，规则反射，常规反射，正反射
regular screwdriver 普通的螺丝起子
regular service condition 正常工作条件
regular service 正常操作，正常运行
regular shutdown 正常停机
regular signaling link 主用信令链路，常规用的信号链路
regular stream pattern 正常水系
regular survey 正规测量
regular vortex shedding 规则旋涡脱落
regular vortex street 规则涡街
regular vortex trail 规则涡迹
regular 定期的，有规律的，合格的，整齐的
regulated agreement 调节协议
regulated area 控制区，节制区

regulated attenuator 调节衰减器
regulated coefficient 调节系数
regulated drain valve 疏水调节阀
regulated extraction steam 调节抽汽，调整抽汽
regulated extraction turbine 调节抽汽式汽轮机
regulated flow 调节后流量，调节流量
regulated power supply 稳定电源
regulated-set cement 控凝水泥
regulated stay area 管制停留区
regulated stream flow 调节流量
regulated stream 受调节水流
regulated value 调节值
regulated variable 被调变量
regulated water stage 调节后水位
regulated work area 管制工作区
regulated year 调节年度
regulate flow 调节流量
regulate 规定，控制，调整，调节，稳定，调准，校正
regulating action 调节作用
regulating apparatus 调节设备，调节器
regulating auto-transformer 调压自耦变压器
regulating baffle 调节挡板
regulating capacity 调节容量
regulating chamber 调节汽室，调节室
regulating characteristics 调节特性
regulating circuit 调节电路
regulating cock 调整旋塞
regulating coil 调节线圈
regulating command 调节指令，调节命令
regulating control rod 调节棒
regulating course 路面整平层
regulating current 调节电流
regulating damper 调节挡板，调节风门
regulating dam 调节坝，分水坝
regulating device 调节装置
regulating element 调节机构，调节元件
regulating error 调节误差，调节温差
regulating extraction steam valve 调节抽汽阀
regulating fault 调节误差，调节故障
regulating frequency 调节频率
regulating gate 调节闸门，控制闸门
regulating gear 调节机构
regulating gradient 调节梯度，调节陡度
regulating impulse 调节脉冲
regulating instrument 调整仪器
regulating loop 调节回路
regulating mechanism by adjusting the pitch of blade 变桨距调节机构
regulating mechanism of rotor out of the wind side-ward 风轮偏侧式调速机构【使风轮轴线偏离气流方向的调速机构】
regulating mechanism 调节机构，调速机构
regulating member 调节元件
regulating motor 调节电动机
regulating needle 调节针
regulating pondage 调节容积，调节容量，调节蓄水量
regulating pond 调节池
regulating property 调节特性
regulating pull rod 调节拉杆，推拉杆
regulating range 调节范围
regulating relay 调节继电器
regulating reserve 调节储量
regulating reservoir 调节池，调节水库，调节蓄水池
regulating resistor 调节电阻
regulating rheostat 调节变阻器
regulating ring 调节环，调整环，调速环
regulating rod 控制棒，调节棒
regulating screw 调节螺旋
regulating set 调节机组
regulating slide valve 调节滑阀
regulating speed 调节速度
regulating stage 调节级
regulating starting rheostat 启动调节变阻器
regulating station 调节发电厂
regulating step command 步进调节命令，增量命令
regulating storage capacity 调节库容
regulating storage 调节库容
regulating structure 整治建筑物
regulating switch 调节开关
regulating system 调节系统
regulating transformer 调节变压器，调压变压器
regulating tube 调节管
regulating unit movement 调节机移动
regulating unit 调节部件
regulating valve bonnet 调节阀帽
regulating valve gland 调节阀压盖
regulating valve seat bush 调节阀座衬套
regulating valve seat 调节阀座
regulating valve stem 调节阀杆
regulating valve 调节阀，调整阀
regulating variable 调节变量
regulating voltage 调节电压，调整电压
regulating winding 调节绕组
regulating works 整治工程
regulating 调节，控制，调整
regulation and metering station 调压计量站
regulation capacity 调节能力
regulation characteristics 调节特性
regulation circuit 调节电路
regulation cock 调节栓
regulation constant 调节常数
regulation curve 负载特性曲线，电压调整曲线，调整曲线
regulation degree 调准度
regulation discharge 调节流量
regulation drop-out 停止调节，失调负荷
regulation energy of system 系统的调节能量
regulation factor 调整系数
regulation for safe operation 安全（操作）规程
regulation for safety 安全章程
regulation for technical operation 技术操作规程
regulation in steps 分级调整，分档调整
regulation line 整治线
regulation loss 调节损失
regulation of line voltage 电压变动范围，线路电压调整
regulation of output 输出调整
regulation of river 河道整治，治河

regulation of runoff 径流调节
regulation of sediment-laden stream 多沙河流整治
regulation of tidal river 潮汐河流整治
regulation of tidal water 潮汐控制
regulation of water level 水位调节
regulation oil pump 调速油泵
regulation on liability of checking and autograph 专业会签制度
regulation on liability of design and checking 设计校审责任制度
regulation on loans 贷款规章条例
regulation on prevent from radiation harzard 辐射防护法规
regulation period 调节期
regulation pull-out 停止调节，失调负荷，失调负载
regulation quality 调节品质
regulation resistance 变阻器，调节电阻，电位器，分压器
regulation stage 调节级
regulation table 调节表
regulation voltage 调节电压
regulation works 整治工程
regulation 电压调整率，转速调整率，（稳压管的）稳压度，调节，守则，条例，规则，规章，调整，校准，控制，法规，法则，规程，规定，监管，细则，整治，治理
regulator capacity 调节器容量
regulator coil 调节器线圈
regulator limiter 调节器限制单元
regulator potentiometer 调节电位器，调节器电压计
regulator rod 调节棒
regulator storage 调节库容
regulator tube 稳压管，稳流管
regulator works 整治工程
regulatory agency 管理机构，管理机关，审批部门，监督管理部门，主管机关，管制机构，监管机构
regulatory control function 调节控制作用
regulatory control 管理机构
regulatory guide 管理导则
regulatory inspection 管理检查
regulatory requirement 规章要求
regulatory text 法规原文，规定原文
regulatory works 规管工程，整治设施
regulatory 管理的，法定的
regulator 调节装置，调整器，调节器，调速器，控制器，操纵器
regulex 电机调节器，磁饱和放大器
rehabilitation cost 修复成本
rehabilitation redevelopment 改建
rehabilitation works 修复工程，整治工程
rehabilitation 改建，复原，修复，重建
reheat air conditioning system 再热式空气调节系统
reheat and bypass steam system 再热及旁路系统
reheat boiler 再热锅炉
reheat combustion chamber 再热燃烧室
reheat condensing turbine 再热凝汽式汽轮机
reheat control valve 中压调节阀，再热调节阀
reheat control 再热汽温控制
reheat cracking 再加热裂纹
reheat cycle 再热循环
reheated steam flow 再热蒸汽流量
reheated steam safety valve 再热蒸汽安全门
reheat emergency valve 再热汽门，中压汽门
reheat engine 带加力燃烧室的发动机
reheater bypass 再热器旁路
reheater furnace 布置再热器炉体的炉膛
reheater tube 再热器管
reheater 灯丝加热器，再热器，回热器，中间过热器，加力燃烧室
reheat factor 重热系数，再热系数
reheating crack 再热裂纹
reheating cycle 再热循环
reheating hot well 回热式热井
reheating regenerative cycle 再热回热循环
reheating steam turbine 中间再热汽轮机
reheating type steam turbine 中间再热式汽轮机
reheating 再加热，（中间）再热，重热
reheat in stage 级中再热
reheat interceptor valve 中压调节阀，再热截流阀
reheat pressure zone 再热压力段
reheat pressure 再热汽压，再热压力
reheat regenerative steam cycle 再热回热蒸汽循环
reheat-return point 再热蒸汽引入点
reheat section 再热部分，再热段
reheat stage 再热级
reheat steam conditions 再热蒸汽参数
reheat steam flow 中间再热蒸汽流量
reheat steam pipe 再热蒸汽管
reheat steam system 再热蒸汽管道，再热蒸汽系统
reheat steam temperature control 再热汽温控制
reheat steam 再热蒸汽
reheat stop interceptor valve 中压联合汽门，再热联合汽门
reheat stop valve 再热截止阀，再热主汽门，中压汽门，再热汽阀，再热汽门
reheat system 再热系统
reheat temperature 再热温度
reheat turbine 再热（式）汽轮机
reheat type （中间）再热式
reheat vapor cycle 再热蒸汽循环
reheat vernier bypass valve 再热微调旁路阀
reheat 再加热，重热，再热
rehypothecate 再抵押
rehypothecation 再抵押，再抵押权，转抵押
reignite 重燃
reignition device 延弧装置
reignition voltage 再引弧电压
reignition 逆弧，二次点燃，反点火，重燃
reimbursable expenditure 可补偿支出
reimbursable technical assistance 有偿技术援助
reimbursable 可偿还的，可赔还的，可赔偿的
reimbursed time 补偿时间
reimbursement clause 补偿条款
reimbursement credit 偿付信用证

reimbursement 报销，偿付，偿还，补偿，赔偿，陪还，退还，支付
reimburse 偿付，赔款，付还，补偿，偿还（款项），报销
reindexing 改变符号
re-inerting 再充惰性气体
reinforced bar 钢筋
reinforced bituminous layer 加强的沥青层
reinforced brick lintel 钢筋砖过梁
reinforced brickwork 配筋砖砌体
reinforced chemical churning pile 植筋旋喷桩
reinforced concrete base 钢筋混凝土地基
reinforced concrete bridge 钢筋混凝土桥
reinforced concrete buttressed dam 钢筋混凝土支墩坝
reinforced concrete canopy 钢筋混凝土顶盖
reinforced concrete chimney 钢筋混凝土烟囱
reinforced concrete construction 钢筋混凝土建筑，钢筋混凝土结构
reinforced concrete containment 钢筋混凝土安全壳
reinforced concrete dam 钢筋混凝土坝
reinforced concrete draft tube 钢筋混凝土尾水管
reinforced concrete embedded steel cylinder pipe 埋有钢衬里的钢筋混凝土管【彭纳型】
reinforced concrete facing rockfill dam 钢筋混凝土面板堆石坝
reinforced concrete flume 钢筋混凝土渡槽
reinforced concrete footing 钢筋混凝土底座，钢筋混凝土基脚
reinforced concrete foundation 钢筋混凝土基础
reinforced concrete gate 钢筋混凝土闸门
reinforced concrete lined tunnel 钢筋混凝土衬砌隧洞
reinforced concrete lintel 钢筋混凝土过梁
reinforced concrete masonry shear wall structure 配筋砌块砌体剪力墙结构
reinforced concrete missile shielding cylinder 飞射物钢筋混凝土防护筒
reinforced concrete outer shield 钢筋混凝土外屏蔽体
reinforced concrete penstock 钢筋混凝土压力水管
reinforced concrete pier 钢筋混凝土闸墩
reinforced concrete pile 钢筋混凝土桩
reinforced concrete pipe 钢筋混凝土管
reinforced concrete poles 钢筋混凝土电杆
reinforced concrete pressure pipe 钢筋混凝土压力水管
reinforced concrete radial gate 钢筋混凝土弧形闸门
reinforced concrete retaining wall 钢筋混凝土挡土墙
reinforced concrete rigid frame 钢筋混凝土钢架
reinforced concrete road 钢筋混凝土路
reinforced concrete sector gate 钢筋混凝土扇形闸门
reinforced concrete sheet pile 钢筋混凝土板桩
reinforced concrete sleeper 钢筋混凝土轨枕，钢筋混凝土枕木
reinforced concrete spiral casing 钢筋混凝土蜗壳
reinforced concrete structure 钢筋混凝土结构
reinforced concrete surge tank 钢筋混凝土调压塔
reinforced concrete wall 钢筋混凝土墙
reinforced concrete works 钢筋混凝土工程
reinforced concrete 加强孔，补强孔，钢筋混凝土
reinforced core 强化磁芯，加强线芯
reinforced cover plates of the trenches crossing roads 跨越道路沟的加强盖板
reinforced earth dam 加筋土坝
reinforced elastomer 增强的弹性材料
reinforced flange fitting 加强的法兰管接件
reinforced hose 加强的软管
reinforced insulation 加强绝缘，补强绝缘
reinforced masonry structure 配筋砌体结构
reinforced plastic plate 玻璃钢板材
reinforced plastics 增强塑料
reinforced rib 加强肋
reinforced seam 加强焊缝
reinforced square set 加固方框
reinforced structure 加固结构
reinforced welding 补强焊
reinforced weld seam 加强焊缝
reinforced 补强的，加强的，加固的，加钢筋的，配钢筋的
reinforcement arrangement 钢筋布置，钢筋排列
reinforcement bar straightening machine 钢筋调直机
reinforcement bar 加强筋，钢筋（混凝土）
reinforcement brick lintel 钢筋砖过梁
reinforcement brickwork 配筋砖砌体
reinforcement by thickened nozzle 接管加厚补强
reinforcement cage 钢筋骨架
reinforcement concrete 钢筋混凝土
reinforcement corrosion 钢筋腐蚀
reinforcement displacement 钢筋位移
reinforcement drawing 配筋图，钢筋图
reinforcement exposed 露筋
reinforcement limitation 配筋限度
reinforcement mat 钢筋网
reinforcement meter 钢筋测力计
reinforcement of openings 开孔补强
reinforcement of weld 加强焊缝
reinforcement pad 补强垫板
reinforcement plate 补强板，增强板
reinforcement prefabrication 钢筋预加工
reinforcement protective cover 钢筋保护层
reinforcement range 补强范围
reinforcement ratio 配筋比，配筋率，含筋率，钢筋比
reinforcement ring 补强圈，补强环
reinforcement rod 钢筋
reinforcement summary 钢筋明细表
reinforcement yard 钢筋存放场
reinforcement 钢筋，加强，补强，加厚，加固，增强，加强物，强化物，（焊缝）余高，加筋，加强件，配筋，配筋率，强化材
reinforce packing 加固包装
reinforce 加固
reinforcing agent 增强剂

reinforcing bar spacer　钢筋隔块
reinforcing bar spacing　钢筋间距
reinforcing bar　加强筋，钢筋，加固筋【混凝土】
reinforcing cage　钢筋笼
reinforcing collar　补强圈，加固圈
reinforcing element　钢筋【混凝土】
reinforcing fabric　钢筋网
reinforcing fiber　增强纤维
reinforcing girder　加力梁
reinforcing gusset　增强结点板
reinforcing material　加强材料
reinforcing mesh　焊接金属网，承重金属网
reinforcing pad　补强垫，补强板，增强衬板
reinforcing personnel　增员人员
reinforcing post　辅助支柱
reinforcing relay　加强继电器
reinforcing rib　加强肋
reinforcing ring　补强环，加强环
reinforcing rod　钢（筋）条
reinforcing saddle　鞍形补强板
reinforcing steel area　钢筋截面积
reinforcing steel bender　钢筋弯曲机
reinforcing steel cutter　钢筋切断机
reinforcing steel　钢筋（混凝土）
reinforcing strap　补强带
reinforcing system　钢筋系，补强体系
reinforcing wire　钢筋钢丝
reinforcing　配筋，圆钢筋，增强
reinjection system　飞灰复燃装置
reinjection　回送，再喷入，飞灰回收
reinserted fuel　重新插入的燃料
reinserted subcarrier　还原副载波
reinsertion of carrier　载波恢复，载波的重插入
reinsertion　直流量的再生，直流成分的恢复，（元件）再插入
reinsert　重新插入，重新引入，重新埋入
reinspection　复查，重复检验，重新检查
reinstallation test　安装前试验
reinstallation　再安装
reinstatement cost　重安装成本
reinstatement value insurance　重置价值保险
reinstatement　复原，修复，恢复，恢复原状
reinsurance　再保险
reinvest　再投资
re-irradiation　再照射，重复辐照
reiteration　反复，重复，迭代
rejacket　再装套
reject a bid　拒绝投标
reject an appeal　驳回上诉
reject a tender　拒绝投标
reject bit　废料
reject chute　废料槽
rejected coal sample　弃样
rejected heat　废热，散失热量，放出热量，排出热
reject failure rate　抑制故障率
rejection amplifier　带阻放大器
rejection filter　抑制滤波器，拒波滤波器
rejection gate　或非门
rejection level　抑制电平
rejection of full load　抛满负荷，甩满负荷
rejection of heat　热的排出，散热
rejection of load　甩负荷，抛负荷，弃负荷，甩负载
rejection of tender　拒绝投标，废标
rejection surge　弃荷涌浪
rejection to bid　拒标
rejection trap　拒波器，带阻滤波器
rejection　报废，拒绝，排斥，排除，抑制，障碍，衰减，废弃，废品，废料，尾料，抛弃
reject load　甩负荷
rejector-acceptor circuit　（复合滤波器的）拒斥接收电路
rejector circuit　抑制器电路，带阻滤波器电路
rejector　抑制器，反射器，反射体，带阻滤波器
rejects ejection　排出废品
reject steam　旁路蒸汽
reject　废品，废料，排放，排弃物，质次品，驳回
rejuvenated water　再生水
rejuvenation of tube　电子管复活，电子管再生
rejuvenation　复活，再生，复原，恢复活力
rekindling　重新点火
related communication　有关来文
related company　联营公司
related department　有关部门
related interpolator　相关内插器
related standard　二级标准
related terminal　相关线端
related to　与……相关的
related　相关的
relate　叙述
relational access method　相关存取法
relational database　关系数据库
relational data file　关系数据文件
relational operator　关系运算子
relational system　关系系统
relation curve　关系曲线，相关曲线，关系式
relationship　关系，联系，关系曲线
relation　联系，叙述，比例关系，关系（式），关系曲线，方程（式），关联，相关
relative acceleration　相对加速度
relative accuracy　相对准确度，相对精度
relative activity method　相对活度测定法
relative activity of soil water　土壤水相对活动性
relative activity of solute B　溶质 B 的相对活度
relative activity of solvent A　溶剂 A 的相对活度
relative address programming implementation device　相对地址程序设计装置
relative address　相对地址
relative age　相对年代
relative air density　相对空气密度
relative airstream　相对气流
relative amplitude　相对振幅
relative anchor point　相对死点
relative angle of attack　相对迎角
relative aperture　相对孔径
relative atomic mass　相对原子质量
relative attenuation　相对衰减
relative biological effectiveness dose　相对生物效应剂量

relative biological effectiveness 相对生物效应
relative biological effect 相对生物效应
relative blade height 相对叶高
relative bucket cascade flow 叶栅相对流
relative burnup 相对燃耗
relative capacitivity 相对电容率
relative centrifugal force 相对离心力
relative change of speed 相对速度变化
relative coding 相对编码
relative concentration 相对浓度
relative consistency 相对稠度
relative conversion ratio 相对转换比
relative coordinates 相对坐标
relative correction 相对校正
relative cost factor 电机成本因数
relative cost 比较成本
relative current 相对流
relative damage factor 相对损害因子，相对损伤因数
relative damping factor 相对阻尼系数
relative damping 相对阻尼
relative dead center 转子相对于静子膨胀时的基准点，相对死点
relative deficit 相对亏损
relative deformation 相对变形
relative degree of oxidation 相对氧化度
relative delay 相对延迟
relative density of air 空气相对密度
relative density 相对密（实）度
relative detuning 相对失调，相对解调
relative deviation 相对离差
relative dielectric constant 相对介电常数
relative diffusion 相对扩散
relative discharge 相对流量
relative displacement 相对位移
relative divergence of parameter 参数相对误差
relative eccentricity 偏心率
relative eddy 相对涡流，相对旋涡
relative efficiency 相对效率
relative elevation 相对标高，相对高程
relative elongation 相对伸长
relative enthalpy 相对焓
relative equilibrium 相对平衡
relative erratum 相对误差
relative error 相对误差
relative evaporation 相对蒸发量
relative exit angle 相对流出角
relative exposure factor 相对曝光系数
relative film speed 相对胶片速度
relative flow coefficient 相对流量系数
relative flow 相对流动，相对流量
relative frequency distribution 相对频率分布
relative fundamental content 相对基波含量
relative gain 相对增益
relative harmonic content 相对谐波含量
relative height 相对高度
relative humidity 相对湿度
relative importance 相对（中子）价值，相对重要性
relative inclinometer 相对倾斜仪
relative informational capacity 相对信息容量
relative kinetic energy 相对动能
relative level 相对标高
relative light path retardation 相对光程延迟
relative limit of error 误差相对极限
relative log exposure 相对对数曝光量
relative loss 相对损失
relative magnitude 相对值
relative mass defect 相对（单位）质量亏损
relative mass density 相对质量密度
relative mass excess 体量减缩量，相对（单位）质量过剩
relative molecular mass 相对分子质量
relative motion 相对运动
relative movement 相对运动
relative overshoot 相对超调量
relative permeabilities 相对渗透率
relative permeability 相对磁导率，相对渗透率，相对透气性
relative permittivity 相对电容率，相对介电常数
relative pitch-shortening value 短距系数
relative pitch 相对跨距，相对节距
relative power-density distribution 相对功率密度分布
relative power level 相对功率水平
relative pressure coefficient 相对压力系数
relative pressure 相对压力
relative price 比价，相对价
relative programming 相对程序设计
relative radiometer 相对辐射表
relative range of control 相对控制范围
relative refractive index 相对折射率
relative regulation 相对调节，调整率
relative restraint 相对约束
relative retardation 相对减速
relative rigidity 相对刚度
relative risk 相对危险度
relative rod effectiveness 控制棒相对效率
relative rotor displacement 转子相对位移
relative roughness 相对粗糙度，相对糙率
relative sensitivity 相对灵敏度
relative settlement 相对沉陷
relative spectral response 相对光谱响应【相对光谱灵敏度】
relative speed drop 相对转速降低【降落】，转速降落率
relative speed rise 转速升高率，相对转速升高
relative speed variation 转速变化率
relative speed 相对速率，相对速度
relative stability 相对稳定性
relative stiffness ratio 相对刚度比
relative stiffness 相对刚度
relative strength 相对强度
relative thickness of airfoil 翼型相对厚度
relative toxicity 相对毒度
relative to 和……有关，关于，相对于
relative transpiration 相对蒸腾
relative travel 相对行程
relative velocity 相对速度
relative viscosity 相对黏度
relative voltage drop 相对电压降
relative voltage response 相对电压响应

relative volume mass 相对体积质量
relative water content 相对含水量
relative water depth 相对水深
relative wind direction 相对风向
relative wind speed 相对风速
relative wind 相对风，迎面气流，相对来流
relative 成比例的，相对的，有关系的，相互有关的，比较而言的，亲属，亲戚，相关物
relativistic mass 相对论质量
relativistic speed 相对论速度
relativity principle 相对性原理
relativity 相对论，相关性
relaxational oscillation 张弛振荡
relaxation circuit 张弛电路
relaxation effect 松弛效应，张弛效应
relaxation function 松弛函数
relaxation generator 张弛振荡器
relaxation inverter 张弛逆变器
relaxation length 张弛长度，衰减距离
relaxation loss 张弛损失
relaxation method 张弛法，松弛法，逐次近似法，卸载法
relaxation model 松弛模型
relaxation of steel 钢筋的松弛
relaxation of stress 应力（的）松弛
relaxation oscillation 松弛振荡，张弛振动，张弛振荡
relaxation oscillator alarm 张弛振荡器报警器
relaxation oscillator 弛张振荡器
relaxation period 弛豫时间，松弛周期
relaxation phenomenon 松弛现象
relaxation point 松弛点
relaxation process 张弛过程，弛豫过程，张弛工艺
relaxation scanning 张弛扫描
relaxation technique 叠弛技术
relaxation test 松弛试验，张弛试验
relaxation time 缓和时间，松弛时间，张弛时间，弛豫时间
relaxation velocity 松弛速度
relaxation 平衡的自动恢复，张弛，衰减
relaxed rock 松弛岩体，卸载岩体
relaxed stress 松弛应力
relaxor 张弛振荡器
relay act trip 继电器操作跳闸
relay actuation time 继电器动作时间
relay actuator 继电器操作机构，继电器启动装置
relay adjustment 继电器调节
relay air valve 继动空气阀
relay amplifier 中继放大器，继电器放大器
relay armature hesitation 继电器衔铁瞬时延时
relay armature travel 继电器衔铁行程
relay armature 继电器衔铁
relay automatic telephone system 继电器式自动电话系统
relay back-up 后备继电器，继电器后备保护
relay base 无线电中继站
relay bias winding 继电器偏置绕组
relay board 继电器屏
relay bridging 继电器触点桥接
relay cabinet 继电器柜
relay center 中继中心，数据转送中心
relay characteristic 继电器特性
relay circuitry 继电回路
relay circuit 继电器电路
relay-coil resistance 继电器线圈电阻
relay comparator 继电器式比较器
relay computer 继电器计算机，继电器式计算机
relay construction 继电器结构
relay contact combination 继电器触点组合
relay contact gauging 继电器触点间距调测
relay contact network 继电器触点网络
relay contactor 继电器接触器
relay contact system 继电器接触系统
relay contact 继电器触点
relay control cable 继电控制电缆
relay controller 继电器控制器
relay control system 继电控制系统，继电器控制系统
relay control 继电（器）控制
relay core 继电器铁芯
relay counter 继电器式计数器
relay cycle timer 周期定时继电器，继电器式周期定时器
relay cylinder 继动油缸，错油门壳体
relay damping ring 继电器阻尼环
relay distribution system 中继配电系统
relay driver 继电器激励装置，继电驱动器
relay driving spring 继电器驱动弹簧
relay dump valve 继动切断阀
relay duty cycle 继电器工作周期
relayed capacity 继电保护所容许的功率
relay effect 继电效应
relay electromagnetic system 继电器电磁系统
relay element 继电器构件，继电器元件
relay equipment 继电设备
relay finder 中继寻线器
relay for position indication 指示位置的继电器
relay frame 继电器座子
relay friction effect 继电器摩擦效应
relay fritting 继电器触点熔合
relay gear 继动装置
relay governing 间接调节，继动调节
relay governor gear 继动调速器
relay governor 继动调速器，继电器调节器
relay group 继电器群，继电器组
relay housing 继电器室，继电器壳
relaying current transformer 继电器用电流互感器，继电器用变流器
relaying panel 继电器屏
relaying sounder 中继发声器，中继音响器
relaying station 中继站，中继台
relaying voltage transformer 继电器用电压互感器
relay interrupter 继电器式断路器，继电器式断续器
relay inverse time 继电器反时限特性
relay just-release value 继电器始可释放值
relay line 中继线
relay logic card 继电器逻辑卡
relay logic 继电器逻辑
relay module 继电组件，继电单元

relay mounting plane 继电器安装平面
relay multiplier-divider unit 继电器乘除部件
relay must-operate value 继电器保证启动值
relay must-release value 继电器保证释放值
relay non-operate value 继电器未动作值
relay normal condition 继电器正常状态
relay oil （调节系统中的）调速油，继动油
relay-operated accumulator 继电器操作的累加器
relay-operated controller 继电器操作的控制器
relay-operated interlocking 继电器联锁装置
relay-operated 继电器操作的［操纵的］，错油门操作的［操纵的］
relay operate time characteristic 继电器工作时间特性曲线
relay operating time 继电器动作时间，继电器吸合时间
relay output 继电器输出
relay outside lead 继电器线圈外端引线
relay overrun 继电器超限运行
relay panel 继电器盘，继电器屏
relay piston valve 继动活塞阀
relay piston 继动活塞，调节器一次滑阀
relay pointer 继电器指针
relay point 中继站，继电器触点
relay preselector 中继预选器
relay protection planning 继电保护设计
relay protection 继电保护
relay pull curve 继电器引力曲线
relay pulse sender 中继脉冲发送器
relay pump 中继泵
relay rack room 继电器室
relay rack 继电器架
relay receiver 中继接收机
relay register 继电器式寄存器
relay regulator 继电调节系统，断续式调节器
relay reoperate time 热继电器释放时间，继电器再动作时间
relay repeater 中继转发器
relay retractile spring 继电器复位弹簧，继电器回缩弹簧
relay return spring 继电器复位弹簧
relay room 继电器间，继电器室
relay satellite 中继卫星，卫星中继站
relay saturation 继电器磁饱和
relay seating time 继电器入位时间
relay seating 继电器衔铁就位，继电器衔铁入位
relay selecting circuit 继电器选择电路
relay selector 继电器式选择器
relay selsyn motor 中继自动同步电动机
relay selsyn 中继自动同步机
relay sequence 切换顺序
relay servomechanism 继电伺服机构
relay setting 继保整定，继电保护，整定计算，继电器整定值，继电器整定
relay set 继电器组
relay sleeve 继电器铁芯套管
relay slug 继电器的延时套管
relay socket 继电器插座
relay spring curve 继电器弹力曲线
relay spring stop 继电器弹簧制动件
relay spring stud 继电器弹簧推杆
relay stack 继电器触点组合
relay stagger time 继电器不同触点组启动时差，继电器参差时间
relay starting switch 继电器启动开关
relay static characteristic 继电器静态特性
relay station 中继站，传播台
relay storage 继电存储器，继电器存储器
relay switchboard 继电器盘，继电器屏
relay switching circuit 继电器开关电路
relay switch 继电器接线器，继电器开关
relay system 中继制，全继电器制，中继系统
relay testing equipment 继电器测试设备
relay testing technique 继电器测试技术
relay test 继电器试验
relay theory 继电器理论
relay timing 继电器时延，继电器定时
relay transmission 中继传输
relay transmitter 中继发射机
relay tree circuit 继电器树状电路
relay tube 替续管，电子继电器
relay-type recorder 继电器式记录器
relay-type recording instrument 继电式自记仪表
relay-type 继电器式的
relay unit 继电器组，继电器单元
relay valve 继动阀，减压阀，继动滑阀
relay winding 继电器线圈
relay yoke 继电器轭
relay 继动器，继电器，错油门，替续器，中继，转播
release alarm 释放报警
release bid bond 退还投标保证金
release brake position 制动器松开位置
release clause 豁免条款
release coil 跳闸线圈，释放线圈
release condition 释放状态
release current 开断电流，释放电流
release curve 泄放曲线
released energy 释出能量
released fission product 释放的裂变产物
released heat 释放热，放出热
released neutron 释（放）出中子
released structure 释放结构
releasee 受让人
release for shipment 运输许可证
release-free relay 自由释放继电器
release gate 泄水闸门
release gear 释放装置
release guard 释放监护
release hold 保持松开
release lever 释放杆
release line of tracer 示踪物释放线
release magnet 释放电磁铁
release of gas 排气，放气
release of guarantee 解除担保
release of mortgage 解除抵押
release of radioactivity 放射性（物质）释放
release of stored water 蓄水泄放
release order 释放指令
release permit 排放许可
release point 释放点

release position 放松位置，（偏置开关的）释放位置
release rate limit 释放率限值
release rate 释放率，排放率，（裂变气体）释放量与产生量之比，释产比
release ratio 释放比
release relay 复归继电器，释放继电器，话终继电器
releaser 溢水口，上升管，立管，排气门，升降器，释放器，释放装置
release signal 复原信号，释放信号
release sluice 泄水闸
release ticket 发放证书
release time 释放时间
release to birth ratio 释产比
release to the environment 向环境排放
release valve 放泄阀，排气阀，排出阀，安全阀，释放阀，卸压阀，泄放阀
release voltage 释放电压
release wave 放水波
release works 泄水建筑物
release 发布，放出，放弃，解除，免除，释放，释放装置，释放机构，断开，脱扣，豁免，（放射性）排放，脱扣器，断路器
releasing coil 释放线圈
releasing contact 释放触点
releasing current 释放电流
releasing of relay 继电器释放
releasing 泄放
relending terms 分贷条件
relevance 相关性
relevant authorities 有关当局
relevant clause 有关条文
relevant experience 有关经验
relevant failure 相关故障
relevant party 相关方
relevant part 相关部分
relevant 相关的
relevelling 复测水准
relevent indication 相关显示，相关指示
relevent vocoder 相关声码器
reliability accounting 可靠性计算
reliability allocation 可靠度分配，可靠性分配
reliability analysis 可靠性分析
reliability and maintainability assurance 可靠性和可维修性保证
reliability and maintainability audit 可靠性和可维修性审计
reliability and maintainability control 可靠性和可维修性控制
reliability and maintainability plan 可靠性和可维修性计划
reliability and maintainability program 可靠性和可维修性大纲
reliability and maintainability surveillance 可靠性和可维修性监察
reliability apportionment 可靠度分配，可靠性分配
reliability assessment 可靠性评价
reliability assurance program 可靠性保证程序
reliability,availability,serviceability 性能最佳性【即可靠性、可用性及可维修性】
reliability block diagram 可靠性框图
reliability-centered maintenance 以可靠性为中心的维修
reliability coefficient 可靠性系数
reliability compliance test 可靠性验证试验
reliability criteria 可靠性准则
reliability criterion 可靠性准则
reliability data 可靠性数据
reliability design 可靠性设计
reliability determination test 可靠性测定试验
reliability economics of an electric power system 电力系统可靠性经济学
reliability engineering 可靠性工程，可靠性工程学
reliability estimation 可靠性估计
reliability evaluation of electric power system 电力系统可靠性评估
reliability evaluation of generating capacity 发电容量可靠性评估
reliability evaluation planning 可靠性评价规划
reliability evaluation 可靠性评估
reliability for electric power industry 电力工业可靠性
reliability function 可靠性函数
reliability improvement factor 可靠性提高系数
reliability improvement 可靠性改进
reliability index system 可靠性指标系统
reliability index 可靠性指标，可靠度指数
reliability index β 可靠指标 β
reliability management 可靠性管理
reliability maturity index 可靠性老化指标
reliability model 可靠性模型
reliability of forecast 预报可靠性
reliability of fuel supply 燃料供应的可靠性
reliability of generating system 发电系统可靠性
reliability of heat-supply system 供热系统可靠性
reliability parameter data bank 可靠性参数数据库
reliability performance measure 可靠性测量
reliability prediction 可靠性预测，可靠性预计
reliability price 可靠性电价
reliability running 可靠性运行
reliability screening 可靠性筛选
reliability service 可靠运行
reliability study 可靠性研究
reliability test assembly 可靠性测试装置
reliability test 可靠性试验
reliability theory 可靠性理论
reliability trials 耐久试验，可靠性试验，强度试验
reliability worth assessment 电力系统可靠性价值评估
reliability 可靠性，确实性，可靠度
reliable operation 可靠运行，可靠操作，运行可靠
reliable 可靠的，安全的，确实的
reliance 信赖
reliction 出水土地，水位渐消退，新生地，海退，陆进
relict mountain 残山
relict sediment 残留堆积物

relict texture 残余构造
relict 残余，遗迹
relief amount of elastic energy 弹性能释放量
relief angle 后角
relief by-pass valve 安全旁通阀
relief cock 减压旋塞
relief data 地貌数据，地貌资料
relief device 减压装置
relief ditch 泄水沟
relief drawing 地形图
relief energy 起伏量，地势起伏量
relief feature 地形要素
relief gas 排出气，废气
relief intensity 地势起伏强度
relief line 泄压管路【稳压器】
relief map 地势图，地形图，立体地图
relief model 地形模型
relief nozzle 泄压接管，释放嘴，膨胀管嘴，扩容管嘴
relief poppet valve 泄压提动阀
relief port 放气口
relief pressure valve cartridge 减压阀芯
relief pressure valve 卸压阀，安全阀
relief representation 地形表示
relief sewer 泄流排水管，溢流道
relief shift system 交接班制度
relief siphon 安全虹吸管，放水虹吸管
relief spring 保险弹簧，平衡弹簧，减压弹簧
relief system 卸压系统
relief tank 卸压箱，释放箱，卸载箱
relief track 避车线
relief valve effluent 卸压阀排出液
relief valve opening 释放阀启开
relief valve 安全阀，减压阀，调压阀，释放阀，泄压阀，保险阀，溢流阀，自动气阀
relief vent 安全通风管，释压阀，减压通气管
relief well 减压井，排水井，泄水井
relief worker 急救员
relief 地貌，地势，地形，起伏，浮雕，卸载，泄压，减荷，降压，减轻，释放，溢流，卸压，（债务）免除，解除
relieve a vacuum 减低真空，降低真空
relieved load 甩负荷，卸载
reliever 减压装置，辅助炮眼
relieve stress 消除应力
relieve valve 安全阀，减压阀，溢流阀
relieve 减轻，免职，释放，卸载，解除，替换
relieving anode 辅助阳极
relieving arch 辅助拱
relieving capacity 排放量，排汽能力，通流能力，释放量
relieving device 释放装置
relieving gear 卸荷机构，解脱机构
relieving of internal stress 消除内应力
relieving shot 卸载爆破
relieving timber 辅助支架
relieving 减轻
religion building 宗教建筑
religious customs 宗教习惯
re-line 重新划线，更换衬套，修理炉衬，换填料，重浇轴瓦
reloadable control storage 可重写控制存储器，可写控制存储器
reload batch （一次）换料量
reload core design 换料堆芯设计
reload core nuclear design 换料堆芯核设计
reload enrichment 换料的浓度
reloader 复载机
reload fuel 换（燃）料，再填装燃料
reloading curve 重新加载曲线
reloading modulus 再压模量
reloading of the reactor core 反应堆堆芯换料，反应堆堆芯重新装料
reloading pattern 换料图形
reloading procedure 换料步骤，换料程序，重装步骤，重装程序
reloading split 换料分区
reload 换燃料，换料，重复荷载，重新装料，重新加载，重装载
relocatable code 浮动码
relocatable emulator 可置换的仿真器
relocatable loader 可再定位输入程序，浮动输入程序
relocatable program 浮动程序，可再定位程序
relocatable 可再定位的，浮动的
relocate 改线，移位，程序安置，变换，浮动，迁移，重新安置
relocating loader 浮动装入程序
relocation dictionary 重新配位表
relocation hardware 可重新配位的硬件
relocation register 可重分配的寄存器
relocation 改线，迁移，重新安置，再定位，再分配，移位
relocator 定位设备，定位程序
REL(rate of energy loss) 能量损失比
reluctance generator 磁阻发电机
reluctance linear motor 磁阻式直线电动机
reluctance machine 磁阻电机，反应式同步电机
reluctance motor 磁阻马达，反应式同步电动机，磁阻电动机
reluctance power 磁阻功率
reluctance speed sensor 磁阻式转速传感器
reluctance stepper 磁阻式步进电机，反应式步进电机
reluctance stepping motor 磁阻式步进器，反应式步进电动机
reluctance synchronizing 磁阻同步
reluctance-synchronous machine 磁阻式同步电机，反应式同步电机
reluctance torque 磁阻转矩，反应转矩
reluctance-type synchronous motor 反应式同步电动机，磁阻式同步电动机
reluctance 磁阻，阻抗，勉强，不情愿
reluctancy 比磁阻，磁阻系数，磁阻率，磁阻，阻抗，嫌恶，不愿
relying on one's own efforts 自力更生
remainder register 余数寄存器
remainder term 余项
remainder 残渣，残余，剩余物，剩余货，余数，余项，剩余
remaining as provided in this 其余按这样的规定
remaining benefit 剩余效益

remaining load 剩余负载，遗留负载
remaining problem 遗留问题
remaining strain 残余应变
remaining stress 残余应力，剩余应力
remaining vibration 残余振动
remains 残余物
remake 改造
remanence permeability 剩磁磁导率
remanence relay 剩磁继电器
remanence type relay 剩磁感应式继电器
remanence 剩余磁通密度，剩余磁感应，顽磁，剩磁
remanent core 带剩磁铁芯
remanent field 剩余磁场
remanent flux density 剩余磁密
remanent induction 剩余磁感应，剩余感应
remanent magnetism 剩余磁性，剩磁
remanent magnetization 剩磁，顽磁性，残留磁气，剩余磁化强度
remanent polarization 剩余极化强度
remanent strain 残余应变
remanent 剩余的，残余的
remapping 重测图
remarkable achievement 巨大成就
remarkable 显著的，卓越的
remarks column 备注栏
remarks 备注，附注，加注，批注
rematching relay 再匹配继电器
remedial action 补救措施
remedial maintenance 补救维修，检修，故障后维修
remedial measure 补救办法，补救措施
remedial treatment 补强处理
remedying defect 修补缺陷
remedying fault 排除故障
remedy measures 补救措施
remedy the default 纠正违约
remedy work 修复工程
remedy 补救，纠正，制裁
remembering 存储
remesh 重啮合，再啮合
remission 减免
remittance 汇款（额）
remittee 汇款的收款人
remitter 汇款人
remitting bank 托收银行
remit 汇款
remixed concrete 二次搅拌的混凝土
remixing 再拌和，重拌和【混凝土】
remnant magnetism 剩磁
remnant 残余物，残余，残余的
re-mobilization 重新进场，重新准备现场
remodulation 二次调制，再调制
remolded soil sample 重塑土样
remote access 远程存取
remote adjustment 遥控，远程调整
remote aftereffect （辐射损伤的）远期后效
remote alarm 远方报警，遥警，远方警报
remote-antomatic control 遥控操作
remote arc striking device 遥控引弧装置【钨极惰性气体电弧焊】
remote area power supply 偏远地区供电
remote area 边远地区
remote automatic telemetry equipment 自动遥测装置
remote auto 远程自动
remote back-up protection 远后备保护
remote back-up 远距离后备保护
remote batch access 远距群组存取，远控信息存取，按链式顺序存取
remote batch computing 远程成批计算
remote batch entry 远程成批输入
remote batch processing 远程成批处理
remote batch terminal 远程成批处理终端
remote calculator 远程计算器，解算装置
remote communication complex 远程通信复合系统
remote communication console 远程通信操作台
remote computing system language 远程计算系统语言
remote computing system log 远程计算系统记录
remote computing system 远程计算系统
remote concentrating system 遥控集中系统
remote concentrator 远端集线器
remote console 远程操纵台
remote continual verification system 远距离连续核实系统
remote control apparatus 遥控仪器，远距离控制设备
remote control assembly 遥控装置
remote control at audiofrequency 音频控制
remote control ball valve 遥控球阀
remote control board 遥控盘
remote control butterfly valve 遥控蝶阀
remote control channel 远程控制信道
remote control command 遥控指令，遥控命令
remote control device 遥控装置
remote control equipment 遥控设备，遥控装置，远程控制装置
remote control fuel valve 遥控燃料阀
remote control gear 遥控操作机构
remote control hydroelectric station 遥控水电站
remote-controlled apparatus 遥控装置
remote-controlled automated GTAW welding 遥控自动钨极惰性气体保护焊
remote-controlled diesel locomotive 遥控内燃机车
remote-controlled lens 遥控镜头
remote-controlled operation 遥控操作，遥控运行，遥控人员
remote-controlled power station 遥控电站
remote-controlled revolving turret 遥控转动架
remote-controlled switching locomotive 遥控调车机车
remote-controlled torquing machine 遥控扭矩机
remote-controlled valve 远距离控制阀，遥控阀
remote-controlled 远距离操纵的，遥控的，远距离控制的
remote control lighting switch 遥控照明开关
remote-controlling 遥控，远距离操纵
remote control manipulator 遥控人员，遥控操作器，遥控操作机，远距离控制器

remote control mode （协调）手动方式
remote control operation 遥控操作
remote control oscillator 遥控振荡器
remote control panel 遥控盘
remote control receiver 遥控接收
remote control relief valve 远程控制溢流阀
remote control station 遥控站，远距离控制站
remote control substation 遥控变电所
remote control switch 遥控开关
remote control system 遥控系统
remote control unit 遥控装置
remote control valve 遥控阀
remote control water level indicator 遥控水位指示器
remote control 遥控，远程控制，远方控制，远距离操作［操纵，控制］
remote cut-off tube 遥截止管
remote cut-off 遥控开关，遥截止，远距离截止
remote-data collection 远程数据收集
remote-data indicator 数据遥示器
remote-data processing 远程数据处理
remote-data station 远程数据站
remote-data transmitter 远程数据发送器
remote debugging 远程排错，远程调试
remote device 远程设备，远程终端设备
remote display 远程显示
remote error sensing 远距离误差传感
remote float system 浮子遥控系统
remote float 遥控浮子
remote handler 远距离操作设备
remote handling device 远距离装卸装置，远距离操作装置
remote handling system 远距离操作系统
remote handling 远距离装卸，远距离操作，远距离操纵
remote hand-operated valve 远距手动操作阀
remote head pump 远距离给料泵
remote ignition 远方点火
remote-indicating instrument 远距离指示仪器，遥测仪器，远距离指示仪表
remote-indicating rotameter 远距离指示转子流量计
remote-indicating thermometer 遥示温度计
remote-indicating 远距离指示，遥示的，远距离指示的
remote indication manometer 远距离指示（式）压力计，遥示压力计
remote indication telemeter 远距离指示遥测器
remote indication 远距离指示，遥远显示
remote indicator 遥控指示器，遥示器，远距离指示器
remote inquiry 远程询问
remote job entry protocol 远程作业输入协议
remote job entry 远程作业输入
remote job processing 远程作业处理
remote job service 远程作业服务
remote level indicator 电接点液位计，液面遥示器，遥读水位表，远距离水位指示器
remote liquid level indicator 远距离水位指示器
remote load dispatch control 远距离负荷分配控制
remote loader 远方操作器
remote loading 远距离装料
remote set 远方设定
remote location 偏僻地区
remotely controlled object 远距离控制的对象
remotely controlled plant 远距离控制装置
remotely maintained plant 遥控维修的装置，远距离维修的车间
remotely operated circuit 遥控电路
remotely operated 遥控的，远距离操作的
remotely sensed scanner 遥感扫描器
remotely 遥远地，偏僻地
remote maintenance 远距离操作维护，远距离维修
remote manipulating equipment 远距离操作设备
remote manipulation 远距离操作
remote manual control station 远距离手动控制站
remote manual control 手动遥控
remote manual loader 手动操作器【仅有手动操作输出，操纵一个或多个远程仪表的装置】
remote measurement/metering 遥测，遥测技术，遥感勘测，自动测量记录传导
remote measurement 遥测，远距离测量
remote measuring element 遥测元件，远距离测量元件
remote measuring equipment 遥测设备
remote measuring system 远距离测量系统，遥测系统
remote measuring 遥测
remote mechanical operation 远距离机械操作
remote metering 遥测，远距离测量
remote monitoring equipment 远距离监视设备
remote monitoring system 远距离监控系统
remote monitoring 远距离监控，远距离监测
remote monitor 远距离监测器，遥控装置
remote operated balance 遥控操作天平
remote operated substation 远距离操作变电站
remote operated valve 远距离操作阀
remote operated 遥控的，远距离操作的
remote operation hoist 遥控起重机
remote operation 遥控操作，远距离操作，遥控运行
remote operator panel 远程操作板
remote panel loader 遥控加载盘
remote patch system 遥控补入编码系统
remote pick-up 电视实况录像，远距离电视摄像
remote pipetting 遥控移液
remote pipet 遥控吸移管，遥控移液管
remote position control 远程位置控制，位置遥控
remote position indicator 远距离位置指示器
remote processing 遥控处理，远程处理
remote protection 远方保护
remote-reading gauge 遥测仪表
remote-reading strainometer 遥测式应变计
remote-reading tachometer 遥读流速计，遥读转速计，远程读数转速计
remote-reading tank gauge 远距离液位计
remote-reading water level indicator 远距离显示水位表
remote-reading 远距离读数，远程读数，遥测

显示，遥控读数
remote readout　远距离读数显示
remote recorder　远距离记录器
remote recording　远距离记录
remote reencapsulation　遥控分装，远距离重新包装
remote reservoir　偏僻地区水库
remote sampling　远距离取样，遥控取样
remote sensing equipment　遥感设备
remote sensing material　遥感资料
remote sensing platform　遥感平台
remote sensing technique　遥感技术
remote sensing technology　遥感技术
remote sensing　遥感，遥测，远距离读出
remote sensor　遥感器
remote servicing　远距离操作，远距离维护
remote set controller　远方设定控制器，远方设定调节器
remote set point adjuster　远方设定点调整器
remote set　远方设定
remote shutdown panel　紧急遥控停堆控制盘，遥控停机盘，远程紧急停堆控制盘【备用】
remote signaling plant　远距离信号装置，遥信装置
remote signaling　遥信，远距离信号装置，远距离发信号
remote station alarm　远程站报警
remote station data terminal　远程站数据终端
remote station　远程站，遥控站，对方站，他站，子站，远处工作站
remote storage (solar water heating) system　分体式（太阳热水）系统
remote supervision system　遥测系统，远距离监督系统
remote supervisory control　远距离监视控制
remote terminal support　远程终端辅助设备
remote terminal unit　远程终端设备，远程终端装置，远动终端，远动终端机，远程终端单元，站内远方终端
remote terminal　遥控终端，远方终端
remote testing　远程调试，远程测试，远程排错
remote transfer point　远程传送点
remote transmission　远距离传送，远距离传输
remote transmitter　远距离变送器
remote trip control　远动跳闸系统
remote trip protection with carrier current　载波（电流）远方跳闸保护
remote trip　远方跳闸
remote tuning　遥远整定，遥调
remote underwater manipulation　水下遥控，远距离水下操纵，遥控水下操作
remote underwater manipulator　遥控水下机械，遥控水下操作器
remote valve control　远距离阀控制
remote valve　遥控阀，远距离控制阀
remote velocity　远方速度
remote viewing equipment　远距离观察设备，遥视设备
remote viewing system　远距离观察系统
remote water level indicator　遥测水位指示仪，远距离水位指示计［器］
remote　遥控的，远程的，遥远的，偏僻的，远方的
remoulded fatclay　重塑肥黏土
remoulded sample　重塑土样
remoulding degree　重塑度
remoulding index　重塑指数
remoulding loss　重塑后的强度损失，重塑减弱
remould　改造，重塑，重新塑造
remounting　再安装
removable contaminant　可除去的污染物
removable core basket　可拆除堆芯吊篮
removable deck　活动平台
removable disk storage　可换磁盘存储器
removable floor grating　活动楼面格栅
removable fuel assembly　可拆装的燃料组件
removable fuel rod assembly　可拆装燃料棒的组件
removable heavy concrete block　可拆卸重混凝土屏蔽块
removable in locked and unlocked position　在闭锁和不闭锁位置之间移动
removable irradiation specimen access plug　活动辐照样品进入塞
removable jumper　可移跳线
removable patch panel　可移动式接线板
removable-pin hinge　活动销铰链
removable-pin　可移动销，松动销
removable plugboard　可拆换的插接板
removable plug　可除去的管塞
removable railing　活头栏杆
removable reinforced concrete cover　活动钢筋混凝土（沟）盖板
removable shaft adapter　可拆卸的联轴短管
removable shape memory plug　可除去的形状记忆塞
removable shielding plug　可拆式防护塞
removable-skin forming varnish　漆皮可揭的保护漆
removable slab　可更换的板
removable strainer　活动的过滤器
removable structure　可拆结构
removable unit of insulation and canning　可拆卸隔热层和罩壳组合件
removable　活动的，可移动的，可［易］拆卸的，可更换的，可撤换的，可置换的
removal and rebuilding　拆迁
removal carriage　移动滑架
removal-chemical and volume control system　化学容积控制系统
removal cost　拆除费用，拆迁成本
removal cross section　分出截面，迁移截面
removal diffusion method　分出扩散法
removal diffusion theory　分出-扩散理论
removal efficiency　清除效率，排出效率，去除效率，脱除率
removal flux　分出通量
removal from storage　从仓库取出
removal gripper　采掘抓斗
removal of ABS from effluent　污水中 ABS 的去除
removal of administrative lockout　取消行政禁令
removal of bolted link　除去栓接的连杆

removal of cable jacket　除去电缆套管
removal of closure head　除去顶盖
removal of faults　排除故障
removal of impurities from station circuit　电厂回路中杂质的去除
removal of internal stress　消除内应力
removal of jacketing　除去外套管
removal of lagging　除去外套，除去绝缘层材料
removal of material　物料的清除
removal of oxygen　除氧
removal of silica　除硅
removal of sulfur　脱硫
removal of the reactor vessel head　拆除反应堆压力容器顶盖
removal of upper internals　除去上部堆内构件
removal ratio　除砂比【沉砂池】
removal shield　活动防护屏，移动式屏蔽
removal time　（核素后处理）去除时间
removal tool　拆卸工具
removal with a hammer　用锤除去【临时附件】
removal　拆卸，移去，除去，迁移，迁走，搬迁，免职，去除量，排除
remove burrs　消除毛刺
removed position　移开位置
remove from service　将……退出运行
remove ground wire　拆除接地线
remove interlock　除去联锁（保护）
remove iodine　除碘
remove residual heat　余热排出
remover　排除装置，清除器
remove　排出，移走，拆除（建筑物），摘除
removing indemnity　搬迁费
removing iodine　除碘
removing of a back layer　去底层
removing　清除
REMS(registered equipment management system) 已注册设备管理系统
remunerate　酬报，酬劳，补偿
remuneration calculation　报酬计算
remuneration　报酬，酬金
rem　雷姆，人体伦琴当量【与1拉德X射线的生物学效应相同的任何电离辐射量】
RENDA(requests for neutron data measurement) 对中子数据测量的要求
rendering coat　抹灰底层
rendering　建筑透视图，建筑渲染，抹灰，涂层，包壳，覆盖，外粉刷，渲染
render　（墙壁）初涂，（抹在墙上的）底灰，打底，抹灰，粉刷，提炼，给予，初涂
renegotiation　重新谈判［协商］
renewable contract　可（延）续的合同，可续签的合同
renewable energy resources　可再生能源，可持续性能源
renewable energy　可持续性能源，可再生能源
renewable fuse-link　可更换熔断器，可更换熔断链
renewable fuse　可更换熔断器
renewable natural resources　可再生自然资源
renewable parts　更新部件
renewable resource　可再生资源
renewable seat ring　可更换的阀座环
renewable　可再生的，可持续的，可继续的，可续订的，可更新的，可翻新的
renew a contract　重定合同
renewal expense　更新费用，修补费用
renewal filter　再生过滤器
renewal of contract　合同续订，合同展期
renewal of equipment　设备更新
renewal of facility　设施更新
renewal of qualification　重做合格鉴定
renewal part　更换部分，更换部件
renewal-penetration concept　表面更新穿透概念
renewal　备件，补充，更新，再生，重做，恢复，续借，续签，续约，续订，续期
renewed fault　复活断层
renewed recourse　再追索人
renewing contract　续签合同
renew product variety　产品创新创优
renew the agreement on expiry　协议期满后到期续订
renew　使更新，使恢复，重新开始，补充
renormalizability　可重正化，可再归一化
renormalized surface function　重正化面函数
renounce　丢弃，放弃，拒绝，否认
renovated water　再生水
renovate technology and equipment　更新技术及设备
renovation and reformation fund　更新改造资金
renovation cost　修缮费
renovation plan　更新计划，改造计划
renovation project　改造项目
renovation works　改建工程
renovation　改建，改造，革新，修复，恢复，修理，更新，整修
rental　租费，租金额
rent deposit　租赁时支付的保证金
renter　承租人，租客，租赁人
rent out　租出，出租
rent　出租，租金，裂纹，租用
renunciation　弃权
REN　核取样系统【核电站系统代码】
re-observation　重测
reoperate time　再动作时间
reordering　重新排序，改组
reorder signal　重排信号
reorganization　改组，重整
reorganized assets　重组的资产，资产重组
re-oxidize　再氧化
repackable stuffing box　可更换填料的填料函
repackable　可换填料的
repacking and storage　重新包装和贮存
repacking　改装，重包装，再包装
repack　换填料，再注
repairability　可修理性，可检修性
repair and maintenance of heat-supply system　热网维修
repair and maintenance　修理和维护
repair and supply workshop　修配分场，修配车间，机修车间
repair cavity　补焊凹坑
repair chart　维修工艺卡片

repair crew 检修班
repair cycle 检修周期，维修周期
repair delay time 修理延误时间
repair end plug 修理端塞【节流塞组件】
repair expense 修理费
repairing 修复
repair load 检修荷载
repair machines for closure head flange seal resurfacing 对顶盖法兰密封表面整修的修理机
repairman 检修工
repair of crack on pressurized component 承压件裂纹处理
repair order 修理通知单
repair outage 停堆修理，检修停堆
repair part 备件，备品，检修备件
repair piece 备品，配件，备用件
repair rate 返修率，修复率
repair record 检修记录，维修记录
repair replication 修复复制，修补复制
repair reserve of capacity 检修备用容量
repair schedule 检修计划［进度］
repair shop 检修间，修理车间，修配分厂，修配车间
repair time 维修时间，修复时间
repair tool 修理工具
repair weld 补焊，返修焊
repair with disassembling 解体检修
repair with disassembly 解体检修
repair worker 修理工
repair workshop 修配厂，机修车间
repair work 修理工作
repair 检修，维修，修补，修复，修理，修配
reparations 赔款
repatriated income 从国外汇回本国的收入
repatriate 遣返，遣送回国
repatriation of proceeds 从国外汇回收入
repatriation of profits 利润汇回本国
repatriation 汇回本国，遣返，送回
repayment of credit 偿还贷款
repayment of loan 偿还贷款
repayment period of loan 贷款偿还年限，贷款偿还期
repayment period 偿还期
repayment 偿还，报答，还款，赔偿（金）
repay 偿还，付还
repeal 撤销，废止
repeatability error 重复性误差
repeatability of measurement 测量的重复性
repeatability 反复性，可重复性，再现性，重复性，可复现性
repeat a contract 重复合同
repeat analysis 重复分析
repeat an order 复诵命令
repeat-back 指令应答装置，回复信号发送装置
repeat circuit 转发电路，中继电路
repeat cycle timer 重复周期计时器
repeated action 多次重复作
repeated addition 叠加，重复相加
repeated bending test 弯曲疲劳试验
repeated capacity 重复容量
repeated compression test 压缩疲劳试验
repeated cycle of freezing and thawing 冻结及融解的重复周期，重复的冻融循环
repeated cycle 重复周期
repeated differentiation 多次微分
repeated-flow turbine 回流式汽轮机
repeated fluctuating stress 反复应力
repeated hardening 多次硬化
repeated impact test 重复冲击试验
repeated impact 反复冲击
repeated loading 反复负载，循环负载，重复荷载
repeated load test 重复荷载试验
repeated load 交变荷载，重复负载，重复荷载
repeated normalizing 重复正火
repeated observations 多次观察
repeated reflection 重复反射
repeated shock 反复冲击，重复冲击
repeated signal 重复信号
repeated start 重复启动
repeated strain 疲劳应变
repeated stress failure 疲劳断裂，疲劳失效
repeated stress 重复应力，反复应力
repeated tempering 多次回火
repeated tensile stress test 反复拉伸试验
repeated tension and compression test 拉压疲劳试验
repeated test 重复试验
repeated torsion test 扭曲疲劳试验，反复扭力试验
repeated utilization 重复利用
repeated 再三的，反复的
repeater circuit 增音机电路，中继电路，转接电路
repeater group 中断组
repeater lamp 指令应答灯，复式灯，应示灯，控制信号灯，监视灯
repeater panel 指令应示屏，复示屏板，增音机屏
repeater relay 中继继电器，转发继电器
repeater section 中继站，增音段，转发站
repeater spacing 增音机间隔，中继站间隔
repeater system 循环回复系统，重发系统
repeater-transmitter 中继发射机
repeater 重发器，增音器，中继器，转变器，中间继电器，中间继动器
repeating center 转播中心，枢纽站
repeating coil 转发线圈
repeating current meter 复测流速仪
repeating installation 帮电装置，中继装置
repeating optical transit 复测光学经纬仪
repeating relay 转发继电器
repeating section 增音段
repeating signal 重复信号，重述信号
repeating sounder 转发音响器，中继音响器
repeating station 增音站，广播转播台，中继站
repeating theodolite 复测经纬仪
repeating timer 重复定时器
repeating 转播，转发，中继
repeat offer 重新发价
repeat operator 重复操作指令
repeat production 重复生产
repeat test 复核试验

repeat 转发，中继，重发
repellent 防护剂，排斥的，弹回的，防水的
repeller mode 反射极振荡模
repeller 反射极，推斥极，导流板
repelling board 防水板
repelling groin 挡水丁坝
repel 排斥，弹回
reperforator 收报穿孔机，复式穿孔机
repertory 仓库，库存，代码指令表
repetency 波率
repetitional load 反复荷载
repetition instruction 重复指令
repetition interval 重复周期
repetition measurement 复测
repetition period 重复周期
repetition-rate divider 重复频率分频器
repetition-rate 重复频率，重复率
repetitive addressing 重复寻址
repetitive analog computer 重复模拟计算机
repetitive computer 重复运算计算机
repetitive cycle 重复循环
repetitive error 重复误差
repetitive form 反复使用的模板
repetitive load 反复载荷
repetitively pulsed reactor 脉冲反应堆
repetitive operation 重复操作，重复运算
repetitive power reactor 系列动力堆
repetitive process 迭代法
repetitive pulse 重复脉冲
repetitive routine 重复程序
repetitive shock 重复冲击
repetitive statement 重复语句
repetitive stress 重复应力
repetitive surge test 反复冲击试验
repetitive 重复的
repipe 调换管段，换管子
replaced by B 用B替代
replaced with B 用B替代
replacement assets 重置资产
replacement capital 重置资本
replacement cost 维修费，更新费用
replacement investment 更新投资
replacement material 代用材料
replacement method （地基处理的）换土法，换土施工法，置换法，置换施工法
replacement of parts 更换部件
replacement parts 更换件，替换零件，备用零件，更换用的零件，代用件
replacement reaction 置换反应
replacement value 重置价值
replacement 代替，更换，重置
replacer 拆装器，替换器
replace 取代，替换，替代，更换
replacing the small capacity unit with a large capacity one 上大压小【电厂改造，上马大机组，淘汰小机组】
re-planning 重新规划
replating 重镀
repledge 转抵押
replenishable source of energy 能量补给源
replenishable sources 可重复利用能源［资源］，可补充能源［资源］，可再生能源［资源］
replenisher powder 粉末补充剂
replenisher 补充器，充电器，感应起电机
replenishing period 再补给期，再灌注期
replenishment of groundwater 地下水的再补给，地下水的重蓄
replenishment supply 补充，补给
replenishment 补充，补给，回灌，再装满
replenish 补充【对原来不足或有损失的】，再补给，再装满，装足
replica aeroelastic model 仿样气动弹性模型
replica examination 复制品检验
replica impedance 重复阻抗
replica temperature relay 仿形温度继电器
replicate plant 同样的电厂，翻版电厂
replication 复制品，重复，重现，复制，套用
replica 复印，复制品，拷贝，仿形，复制物
replying to your telegram 回复电报
reply to an offer 复报价
reply to general enquiry 答复一般询问
reply 回答
repointing 重勾缝，重嵌灰缝
REP-OP(repetitive operation) 重复操作，重复运算
report about equipment modification 设备改装报告
report and electronic magazine server 报表及电子杂志服务器
report on starting construction 开工报告
reporter 指示器
report generator 报告生成器
reporting at customs 报关
reporting contract 申报式保险契约
reporting station 水情站
reporting year 报告年度
report of earnings 收入报告
report of plant site selection 选厂址报告
report of tests 试验报告
report on a special topic 专题报告
report on final settlement of account 决算书
report on item condition at receiving 物项接收状态报告
report on item condition before shipment 发运前对物项状态的报告
report on operating occurrence 运行事件报告
report on the closed accounts 决算（报告）书
report on the loss 报损
report period 报告期
report program 报告程序
report ratifier 报告审批单位
report to dispatcher 向调度汇报
report 证书，鉴定书，报告
repose angle 休止角，安息角，自然休止角
repose slope 休止坡度
repose 静止
repositioning rate 移动速率
repositioning 调节，倒换，移动，移位
reposition of redundant personnel 分流富裕人员
reposition 复原，复位，移动，位移，贮存，回位
repository system 处置库系统
repository 贮藏所，地下贮藏，贮藏室，（放射

性）废物最终处置设施库
repowering （电站通过改造后）扩大容量，改造增容，重新接上电源
REP(reactor establishment permit) 堆装置许可
reprecipitation 再沉淀，再沉降
REP＝repair 维修
representation 表示，表示法，代表，代表性，申述，显示
representative area 特征面积
representative basin 代表性流域，典型流域
representative calculating operation 典型计算操作
representative calculating time 典型计算时间
representative data 典型资料
representative dew point 代表性露点
representative length 特征长度
representative method of sampling 抽样表示法，代表性抽样法
representative model 表现模，典型代表，代表性模型
representativeness 代表性
representative number 代号
representative observation 代表性观测，典型观测
representative office 代表处
representative of legal entity 法人代表
representative of legal person 法人代表
representative property 典型性质
representative sample 代表性试样［样品］，典型试样［样品］
representative sampling 特征取样
representative scale 惯用比例尺，图示比例尺，代表性比例尺
representative section 代表性剖面，等效截面
representative soil sample 代表性土样
representative speed 特征速度
representative station 代表性测站
representative temperature 代表性温度
representative value of an action 作用代表值
representative values of a load 荷载代表值
representative value 代表值
representative year for wind energy resource assessment 风能资源评估代表年
representative 代表，典型，有代表性的
represent dynamically 动态模拟
representee 被陈述人
represent 主张
repressing 补加压力
repressurization 再加压
reprocessed fuel 堆后料，再生燃料
reprocessing analysis 后处理分析
reprocessing center 后处理中心
reprocessing cycle 再加工循环，再生循环，后处理循环
reprocessing loss 后处理损失
reprocessing of the depleted fuel 乏燃料后处理
reprocessing plant （核燃料）后处理工厂
reprocessing procedure 后处理程序
reprocessing rate 后处理率
reprocessing waste 后处理废物
reprocessing 后处理【燃料】，再加工
reprocessor 后处理器
reprocess 后处理，再加工，再处理，再生
reprod(receiver protective device) 接收机保护装置，天线转换开关
reproduced drawing 复制图
reproducer 复制器，再生器，复制机，再现设备
reproducibility of measurement 测量的再现性
reproducibility 再现性，复现性，再生性，还原性，重复性，重现性，复制性，重复能力
reproducible copy 复印拷贝
reproducible intermediate 复制媒体
reproducible 可复制的，能再生的，能再现的
reproducing camera 复照仪
reproducing head 复制头，拾音头，唱头
reproducing punch 复穿孔机
reproducing unit 复制组件，再现装置
reproduction constant 再生常数
reproduction factor 再生系数，再生常数，转换因子，重现因数
reproduction 复现，复制，再生，重现
reproductive property 再生性能
reproductivity 再现性
reprogrammable read-only memory 可重编程只读存储器
reprogrammable 可重新编程的
reprogramming 改编程序
reprogram 重编程序
REPROM(reprogrammable read-only memory) 可重编程只读存储器
reprom 重编程序永久只读存储器
reptile 爬虫
repudiate a contract 否认合同有效，拒绝执行合同
repudiation of claims 拒绝赔偿要求
repudiation of the contract 不履行合约
repudiation 拒付
repulsion and induction type motor 推斥感应式电动机
repulsion motor 推斥电动机
repulsion start induction motor 推斥启动感应电动机
repulsion 斥力，排斥
repulsive force 推斥力，排斥力
repumping house 二级泵房
repurchase 再次采购，重复间隙
REPU(reprocessed uranium) 后处理铀，堆后料
repurification 再净化
reputation first 信誉第一
reputation of firm 厂商信誉
reputation product 名牌产品
reputation 名声，声誉，信誉
requalification 复审
request for approval 请求批准
request for bid 招标，要求承包
request for comments 请求注解
request for proposal 征求报价，招标
request for quotation 询价书
request for supplier's approval 请求供应商的认可
request of clarification 澄清请求
request on line 请求在线连接，请求接通线路

request-send circuit 请求发送线路
request signal 请求信号，查询信号
request to extend time-limit of claim 要求延长索赔期
request 请求，申请
required capacity 必需容量
required function 规定功能
required head 所需要的水头
required information 需要的信息
required length of life 耐用年限
required NPSH curve 必需净正吸压头曲线
required NPSH 必需汽蚀裕量
required personal attributes （招聘）要求的个人特质
required power 需用功率
required rate of discharge 所需的排放速度
required system 必要系统
required time 需求时间
required value 待定值，特定值，预期值
required 必需的，要求的
requirement contract 有限制条款的合同
requirement for control of operational pollution 控制操作污染的规定
requirement for installation 安装要求
requirement of written form 书面形式要求
requirement 技术条件，必要条件，需要，要求
require performance of an obligation 要求履行义务
require 要求
requisite document 必备文件
requisite quality level 必需的质量水平
requisite 需要的，必不可少的，必备的，必需品
requisition land 征用土地
requisition of land 征用土地
requisition plan 申请计划
requisition 必要条件，征用
REQ 请求，要求【DCS画面】
reradiate 反向辐射，再辐射【吸收热量】，转播
reradiation 再辐射，反辐射，转播
re-read 重读
RE(real number) 实数
reregulating reservoir 反调节水库，水库反调节，再调节水库，平衡水库
re-regulation 再调节
rering signal 再呼叫信号，重发振铃信号
rering 重发振铃信号，再呼叫
reroute 重设路径
rerouting 路由再选定
RER(radiation effects reactor) 辐射效应反应堆
RERTR(reduced enrichment for research and test reactors) 降低研究堆和试验堆的浓缩度
rerun point 重新运行点，重算点，重复运行点
rerun routine 重复运行程序，恢复程序
resale contract 转卖合同
resale value 转让价值
resample 重新取样
rescaling 改变尺度，改比例
rescanning 二次检索，重新扫描
rescind an agreement 取消和约
rescind contract 取消合同
rescind the contract 撤销合同
rescission of the contact 解除合同
rescission 撤销，废除，（合同）解除
rescreening 再次筛分，重新筛分
rescue dump （计算机）检验点清除，拯救性信息转储，重入点信息转储
reseal 再封
research and test reactor 研究与试验堆
research center 研究中心
research department 研究部门
research & development tax concession 研发税优惠
researcher 研究人员
research facilities 研究设备
research institute 研究所，研究院
researchist 研究人员
research project 研究项目
research reactor 研究堆，研究性反应堆，试验性反应堆
research report 研究报告
research turbine 试验透平
research work 研究工作
research 研究，调查
reseating machine 修整阀座机械
reseating pressure （安全阀）回座压力
reseating 复位，修整阀座，修正
reseat 回座
reseau 网状物，滤屏，栅网，晶格，网络，线路，电路
resected point 后方交会点
resection method 后方交会法
resection 后方交会
resequent valley 再顺谷
reservation characteristic of heat-supply system 供热备用性能
reservation clause 保留条款
reservation park 自然保护区
reservation 预定，预约，保留，限制，备用，储备
reserve battery 储备电池
reserve bus bar 备用母线，备用汇流条
reserve cable 备用电缆
reserve capacity 备用库容，功率储备，备用容量，预备容量
reserve coal storage 贮备煤堆
reserve cost 预备费
reserved coal sample 留样
reserved hole 预留孔
reserved opening 预留孔洞
reserved space 预留场地
reserved storage location 保留存储单元
reserved word 保留字，预定字
reserve energy 储备能量
reserve equipment 备用设备
reserve factor 备用系数
reserve feed tank 备用给水箱
reserve filter 备用过滤器
reserve free 备用
reserve fund of project 工程预备费
reserve fund 备用基金，储备基金，准备基金，后备基金，准备金
reserve generating capacity 后备发电容量

reserve generator turbine 备用发电机
reserve hours 备用小时
reserve machine 备用机
reserve oil tank 辅助油箱
reserve opening 储备
reserve parts 备件
reserve plant 备用装置
reserve power circuit 备用电力电路
reserve power of system 系统的备用容量
reserve power plant 备用电厂
reserve power station 备用发电厂
reserve power supply 备用电源
reserve power 备用容量，备用功率
reserve protection 后备保护
reserve pump 备用泵
reserve service 备用服务，预备役
reserve set 备用机组
reserve shutdown availability time 备用停堆可利用时间
reserve shutdown hours 备用停堆时数，备用停机持续小时
reserve shutdown system 备用停堆系统
reserve shutdown 备用停机
reserve signaling link 备用信令链路
reserve state 备用状态
reserve storage pile 备用煤堆
reserve storage 备用仓库，备用存储，贮备性贮存
reserve supply 备用电源
reserve the right to claim 保留索赔权
reserve the right to 对……保留权利
reserve time 备用时间
reserve unit 备用机组
reserve well 备用孔道
reserve 保留，预定，备用，预备，备品，储备，预约，储备量，储量，蕴藏量，贮量，保留的，预备的
reservoir accretion survey 水库淤积测量
reservoir action 调蓄作用
reservoir area-capacity curve 水库面积库容关系曲线
reservoir area survey 库区调查
reservoir area 水库面积
reservoir axis 水库轴线
reservoir bank 水库库岸
reservoir bed 储水层
reservoir capacitor 充电电容器
reservoir capacity 库容，水库库容，水库容量
reservoir cleaning 水库净化，水库清理
reservoir conditions 滞止参数
reservoir depleting curve 水库放水曲线
reservoir deposits 水库沉积物
reservoir design flood 水库设计洪水
reservoir design 水库设计
reservoir draught 水库实际供水量
reservoir drawdown 水库水位泄降
reservoir-empty condition 库空情况
reservoir emptying 水库放空
reservoir enthalpy 滞止焓
reservoir evaporation 水库蒸发
reservoir experiment station 水库实验站
reservoir factor 库容系数
reservoir filling 水库蓄水
reservoir for irrigation 灌溉用水库
reservoir for overyear storage 多年调节水库，年际调蓄水库
reservoir for power generation 发电水库
reservoir group system 水库群系统
reservoir group 水库群
reservoir immersion 水库浸没
reservoir impoundment 水库蓄水
reservoir induced earthquake 水库诱发地震
reservoir inflow hydrograph 水库入流过程线
reservoir inflow 入库流量，水库来水量
reservoir inspection record 水库观测记录
reservoir inundated area 水库淹没区
reservoir inundation line survey 水库淹没界线测量
reservoir investigation 水库调查
reservoir leakage 水库渗漏
reservoir length 水库长度
reservoir level 水库水位
reservoir life 水库使用年限
reservoir lining 水池衬砌
reservoir live storage 水库有效蓄水量
reservoir loss 水库损失
reservoir observation record 水库观测记录
reservoir of clean water 清水池
reservoir operation chart 水库运用图表
reservoir operation curve 水库调度曲线
reservoir operation dispatching 水库调度
reservoir operation guide curve 水库运用指导曲线
reservoir operation 水库调度，水库运用
reservoir outlet 水库放水口
reservoir pollution 水库污染
reservoir regulation 水库调度，水库调节，水库调节作用
reservoir release rate 水库放水流量，水库泄放速度
reservoir retention capacity 水库滞洪能力
reservoir routing 水库调洪演算，水库演算
reservoir runoff regulation 水库径流调节
reservoir sedimentation 水库淤积
reservoir sediment washout 水库冲沙
reservoir seepage 水库渗漏
reservoir shoreline 水库岸线
reservoir silting 水库淤积
reservoir-site selection 库址选择
reservoir stage 水库水位
reservoir storage 库容，水库储水量，水库库容，水库容量，水库蓄水量
reservoir stripping 水库表土剥除，水库清理
reservoir survey 水库测量，水库调查
reservoir system 水库系统
reservoir tank 储水池［箱］
reservoir temperature 滞止温度
reservoir type power plant 蓄水式发电厂
reservoir utilization 水库利用
reservoir value 滞止值
reservoir works 水库工程
reservoir yield 水库供水量

reservoir 水库，容器，贮槽，油箱，水箱，储存器，蓄水池，罐
resetable 可复位的
reset action 复位动作，重新调整动作
reset attachment 复位装置
reset bar 复位杆
reset button 复位按钮
reset coil 复归线圈
reset condition 复位条件，清除条件，复原状态，重设状态
reset control action 积分调节作用
reset controller 积分调节［控制］器，再调调节［控制］器，复归控制器
reset control 复位控制
reset current 翻转电流，复位电流，移位电流
"reset failure" push button 事故复位按钮
reset flip-flop 归位触发器，置"0"触发器
reset gate 复位门
reset input 复位输入
reset key 复归键，复位键，置"0"键，清除键
reset magnet 复归磁铁
reset mode 复位方式，重置方式
reset pulse 清除脉冲
reset rate 再调率，积分率
reset relay 恢复继电器
reset-set flip-flop 置位复位触发器
reset signal 复位信号
reset spring 复位弹簧
reset state 复位状态，置零状态
reset switch 复位开关
resettability （振荡的）再调谐能力，可重调性
reset time 重调时间，复位时间，再调时间，恢复时间，积分时间，还原时间
resetting current 复归电流
resetting device 复归装置，再整定装置
resetting key 置"0"开关，复位开关【计算机】，复位键
resetting magnet 复归磁铁
resetting method 复解法
resetting of zero 零复位，调回零位
resetting price 重新定价
resetting ratio 返回系数，复归系数，恢复系数
resetting shaft 反馈轴
resetting spring 复归弹簧
resetting time 复归时间，复位时间
resetting value 复归值
resettlement action plan 移民行动计划
resettlement 重新安置，拆迁
resettling 二次沉降，二次沉淀
reset value 复归值
reset valve 微调阀，重调阀，再调阀
reset voltage 翻转电压，复位电压，移位电压
reset wind-up 积分饱和
reset 复归，返回，重调，复位，复原，回零，置零，置闲，再整定，重新接入，重安装
reshaper 整形器
reshuffle 改组，重配置，重安排，转变
reshuffling 转换【燃料】，对调，倒换燃料
residence half-life （放射性灰尘从平流层落下的）半落下期，半停留期
residence time distribution 停留分布时间
residence time （燃料元件在堆内）停留时间
residence 滞留【在反应堆中】，居住，停留
resident adviser 常驻顾问
resident advisor 常驻顾问
resident engineer 现场工程师，工地工程师，驻工地工程师
resident executive 常驻执行程序
resident field personnel 驻现场人员
residential area 居民点，居民区
residential customer 住宅用户
residential district 住宅区
residential load 住宅负荷
residential population 居住人口
residential quarter 住宅区
residential standard 住宅标准
resident monitor 常驻监视器
resident representative office 常驻代表机构
resident routine 常驻程序
residents'living 居民生活
resident time 停留时间，驻留时间
resident 驻留，常驻部分
reside 常驻，驻留，常留
residing time in furnace 炉内停留时间
residual aberration 剩余像差
residual activity 残余活性，剩余放射性，剩余活性，残余放射性活度
residual air 余气
residual angle of internal friction 残余内摩擦角
residual attenuation 剩余衰减，净衰减
residual austenite 残余奥氏体
residual background 剩余本底
residual bias 剩余，偏差
residual burning equipment 残余燃烧设备
residual capacitance 剩余电容
residual charge 剩余电荷
residual chlorine analyzer 余氯测定仪，余氯分析器
residual chlorine 余氯
residual coal （加工后的）残渣煤
residual cohesion 残余黏聚力
residual contamination 残留污染
residual content 残留（含）量，含渣量
residual current circuit breaker 漏电断路器，剩余电流断路器，漏电开关
residual current device 剩余电流装置
residual current relay 零序电流继电器，剩余电流继电器
residual current 残余电流，剩余电流，零序电流，余流
residual cycle life 剩余循环寿命
residual defect rate 残留缺陷率
residual deflection 残余偏转
residual deformation 残余变形，剩余变形
residual deposit 残积
residual discharge 剩余放电
residual displacement 剩余位移
residual drying （汽轮机湿蒸汽）残余干燥
residual effect 残效
residual elasticity 剩余弹性，残余弹性
residual electricity 剩电
residual element 残余元素

residual elongation 残余延伸率
residual error probability 残留差错概率
residual error rate 残留差错率
residual error 残差，剩余误差
residual evaporator 余热蒸发器
residual excitation 剩磁励磁
residual exciting current 剩余励磁电流
residual field 剩余磁场，剩余域
residual fission product 剩余活度裂变产物，剩余裂变产物
residual flow 余流
residual flux density 剩磁通密度
residual flux 剩磁通
residual force 残余力
residual fuel oil contamination 残余燃料油污染
residual fuel oil 残余燃料油，残渣燃料油
residual fuel 剩余燃料
residual gas analyzer 残余气体分析仪
residual gas 残余废气，（裂化后的）残余气体
residual generation 残留物生成
residual geosyncline 残余地槽
residual hardness 残余硬度，残留硬度
residual heat exchanger 余热换热器，余热热交换器
residual heat removal and suppression pool cooling system 余热排出及弛压水池冷却系统
residual heat removal chain 余热排出管系
residual heat removal condenser 余热排出凝汽器
residual heat removal control valve 余热排出控制阀
residual heat removal control 余热排出控制
residual heat removal exchanger 余热排出热交换器
residual heat removal facility 余热排出装置
residual heat removal feedwater pump 余热排出给水泵
residual heat removal heat exchanger 余热导出热交换器
residual heat removal loop 余热排出回路
residual heat removal pump 余热导出泵，余热排出泵
residual heat removal reducing station 余热排出降压站
residual heat removal system heat exchanger 余热排出系统热交换器
residual heat removal system line 余热排出系统管线
residual heat removal system 停堆余热冷却系统，余热冷却系统，余热排出系统
residual heat removal 余热导出，余热排出【反应堆停堆冷却】
residual heat 剩余热，残余热，衰变热，余热
residual hydrogen 残余氢
residual image 残余图像
residual impulse 剩余脉冲，残留脉冲
residual induction 残余感应，剩磁电感，剩余磁感应，剩余电感，剩磁感应
residual life 残效期，剩余寿命
residual liquid 残余岩浆
residual loss 其他损失，剩余损耗
residual magma 残余岩浆
residual magnetic field 剩磁场
residual magnetic flux density 剩余磁通密度，剩余磁感应强度
residual magnetic flux intensity 剩余磁通密度
residual magnetic flux 剩磁通量
residual magnetism measurement 剩磁测定
residual magnetism 剩磁，顽磁
residual magnetization 剩余磁化强度
residual mass curve 差积曲线
residual noise spectrum 剩余噪声谱
residual nonwetting fluid saturation 残留的不润湿流体的浸润度
residual oil 残油，渣油
residual oxygen 残余氧
residual packing （颗粒）残留堆积，残留填充物
residual penetrant 剩余渗透剂
residual permeability 剩余导磁率
residual polarization 剩余极化
residual pollution product 残余污染产物
residual pore pressure 剩余孔隙压力
residual power 剩余功率
residual pressure 残余压力
residual product 残余产物，副产品
residual radiation 剩余辐射，残余辐射
residual radioactive dust 残余放射性尘埃
residual radioactivity 残余放射性
residual rain 剩余雨
residual reactivity effect 剩余反应性效应
residual resistance 剩余电阻，残余电阻，余阻力
residual resistivity 剩余电阻率
residual sag 剩余挠度
residual salt content 残余含盐量
residual settlement 残留沉降，残余沉降
residual shear strength 残余抗剪强度
residual shim （继电器的）防黏垫片
residual shrinkage 残余收缩
residual soil 残积土，残余土
residual static loading 残余张力
residual strain 残余延伸率，残余应变
residual strength 残余强度
residual stress 残［剩］余应力，焊接残余应力
residual supply index 供给剩余系数
residual time constant 剩余时间常数
residual velocity loss 余速损失
residual velocity 余速
residual vibration 剩余振动
residual voltage of arrester 避雷器残压
residual voltage 残压，剩余电压，零序电压，残余电压
residual water head 剩余水头
residual water 残留水，废水，污水
residual welding stress 残余焊接应力
residual 残渣的，剩余的，残余的，留数，残数，残余，残留的，剩余，余量
residuary water 残留水，污水
residue code 剩余码
residue coke 残留焦炭
residue oil 渣油
residue on evaporation 蒸发残余
residue slime 残渣
residues 剩余物，残渣，留数，残数，剩数，

残渣量，剩余，残余，渣滓
residuum 残渣，灰烬，残渣油，残余物，残油，渣油
resignation 辞职
resign 放弃（权利），辞职
resilience coefficient 回弹系数
resilience modulus 回能模量
resilience 回弹
resiliency 弹回，弹性，弹力，冲击值
resilient coupling 弹性联轴节，弹性联轴器
resilient material 弹性材料
resilient mounting 弹性安装
resilient ring 弹性环
resilient seal 弹性密封
resilient seating 弹性落座
resilient sleeper bearing 弹性垫板
resilient sleeve 弹性套筒
resilient support 弹性支架
resilient 弹回的，有弹性的，有弹力的，有伸缩性的
resilimeter 回弹仪
resin add tank 树脂添加箱
resin anchor bar 树脂锚杆
resin and filter media sluice waters 树脂和过滤材料洗涤水
resin bead technique 树脂颗粒法
resin bed regeneration 树脂床再生
resin bed 树脂床
resin-bonded mastic 树脂胶泥
resin-bonded mortar 树脂水泥砂浆
resin capacity 树脂交换容量
resin catcher 树脂收集器，树脂捕集器
resin-cement concrete 树脂混凝土
resin cleaning tank 树脂清洗罐
resin column 树脂柱，树脂塔
resin deuterization plant 树脂氘化装置
resin exhaustion 树脂耗尽
resin-filled joint 树脂填充电缆接头
resin fill pump 树脂充填泵
resin fill tank 树脂充填箱
resin fill 树脂充填
resin flushing pump 树脂冲洗泵
resin hold-up tank 树脂滞留暂存箱
resin hopper 树脂装料斗
resinification 树脂化
resin impregnant 树脂浸渍剂
resin impregnated cloth 上胶布，浸树脂布，浸胶布
resin-impregnated tape 浸树脂带，胶布带
resin-insulated winding 树脂绝缘绕组
resin-insulated 树脂绝缘的
resin life 树脂寿命
resin loading 树脂吸附量，树脂操作容量
resin loss 树脂损耗
resin mixed tank 树脂混合罐
resin molded transformer 树脂浇注变压器
resin operating capacity 树脂工作交换容量
resinous 负电性的，阴电性的，含树脂的，像树脂的
resin proportioning vessel 树脂配料容器
resin regeneration vessel 树脂再生罐
resin regeneration 树脂再生
resin-rich mica tape 多树脂云母带
resin separation column 树脂分离塔
resin separation tank 树脂分离罐，树脂分离塔
resin separation vessel 树脂分离器
resin shipping container 树脂运输容器
resin sluice pump 树脂冲洗泵
resin sluice water 树脂冲洗水
resin sluicing line 树脂冲洗管
resin sluicing pump 树脂冲洗泵
resin solidification 树脂固化
resin splicing 树脂接合
resin storage tank 树脂贮存罐
resin storage 树脂储存
resin support bed 树脂支持层
resintering test 再烧结试验
resin transfer container 树脂输送容器，树脂转运容器
resin trap 树脂捕捉器，树脂收集器
resin wearing 树脂磨损
resin 树脂，松香
resist abrasive wear 耐磨
resistance alloy 电阻合金
resistance amplifier 电阻耦合放大器
resistance arc furnace 电阻电弧炉
resistance attenuator 电阻衰减器
resistance box yoke 电阻壁盒偏转器
resistance box 电阻箱
resistance braking 电阻制动
resistance brazing 电阻钎焊，接触电阻加热钎焊
resistance bridge circuit 电阻电桥电路
resistance bridge 电阻测量电桥
resistance bulb thermometer 变阻泡温度计，电阻球温度计
resistance bulb 变阻灯泡，电阻灯泡，测温电阻器
resistance-capacitance coupling 阻容耦合
resistance-capacitance divider 阻容分压器
resistance-capacitance network 电阻电容网络
resistance-capacitance oscillator 阻容振荡器
resistance-capacitance transistor logic 阻容晶体管逻辑
resistance-capacitance 电阻电容
resistance-capacitor coupled transistor logic 阻容耦合晶体管逻辑
resistance-capacity filter 阻容滤波器
resistance coefficient for wind 风阻力系数
resistance coefficient of soil 土壤电阻系数
resistance coefficient 阻力系数，电阻系数
resistance coil 电阻线圈
resistance commutation 电阻换向
resistance comparison circuit 电阻比较电路
resistance comparison relay circuit 电阻比较继电器电路
resistance compensation 电阻补偿
resistance convertor 电阻转换器
resistance-coupled amplifier 电阻耦合放大器
resistance-coupled 电阻耦合的
resistance coupling 电阻耦合
resistance curve 阻力曲线

resistance divider 电阻分压器
resistance drop compensation 有功压降补偿，电阻压降补偿
resistance drop 有功电压降，电阻性电压降
resistance element 电阻元件
resistance erodibility 抗蚀性
resistance factor 阻力系数，电阻系数，阻力因数
resistance furnace 电阻炉
resistance grading 电阻防晕层，电阻梯度平滑法
resistance grounded neutral system 中性点电阻接地系统
resistance grounding 电阻接地
resistance head 阻力损失
resistance-heated high temperature fluidized bed 电阻加热高温流化床
resistance heater 电阻加热器
resistance heating 电阻加热，电阻加热法
resistance in parallel 并联电阻
resistance in series 串联电阻
resistance in the dark 暗电阻
resistance lamp 电阻灯
resistanceless 无电阻的
resistance loss 电阻损耗
resistance material 电阻材料
resistance measurement 电阻测量
resistance method of temperature determination 电阻测温法
resistance moment 阻力矩
resistance net work analogue for seepage 渗流模拟阻力网
resistance network 电阻网络
resistance of an earthed conductor 接地导线的电阻，接地电阻
resistance of ducting 管道压力损失，管道阻力
resistance of fluid friction 流体摩擦阻力
resistance of friction 摩擦阻力
resistance of ground connection 接地电阻
resistance of heat transfer 传热阻
resistance of materials 材料力学
resistance paper analogy 电阻纸模拟
resistance per square 方电阻
resistance pressure transmitter 电势计式压力传感器
resistance probe 电阻测温器插头
resistance protection 电阻保护装置
resistance pyrometer 电阻式高温计
resistance ratio 电阻比
resistance-reactance-ratio 电阻电抗比
resistance relay 电阻继电器
resistance ring 端环，短路环
resistance sensitivity 电阻灵敏度
resistance-shunt method 电阻分流法
resistance soldering 电阻钎焊
resistance spot welding 电阻点焊
resistance spot weld 电阻焊点，电阻点焊
resistance-stabilized oscillator 电阻稳频振荡器
resistance-start motor 电阻启动电机
resistance strain gauge 电阻应变片，电阻应变仪
resistance switchgroup 电阻切换组合开关
resistance temperature bulb 热电阻元件
resistance temperature coefficient 电阻温度系数
resistance temperature detector 电阻测温计，热电阻，电阻（式）温度检测器，电阻温度检查器，电阻式温度感应装置
resistance test 电阻试验
resistance thermal shock 抗热冲击性
resistance thermometer assembly 热电阻组件
resistance thermometer 电阻（式）温度计，热电阻温度计
resistance thermometry 电阻测温技术
resistance to abrasion 耐磨性
resistance to air flow 气流阻力
resistance to alternating current 交流电阻
resistance to bending strain 抗弯曲应变
resistance to bending 抗弯能力
resistance to breakage 抗碎强度
resistance to buckling 抗挠强度，抗弯强度
resistance to case 对机壳电阻
resistance to chemical attack 抗化学腐蚀能力
resistance to cold weather 耐寒性
resistance to compression 抗压缩性，压缩变形阻力
resistance to corona 耐电晕性
resistance to corrosion fatigue 耐腐蚀疲劳
resistance to deformation 变形抗力
resistance to direct current 直流电阻
resistance to driving 打桩阻力
resistance to earth 接地电阻
resistance to effect of heat 耐热性能
resistance to electronic transmitter 带阻尼的电子变送器
resistance to fire 耐火性
resistance to fouling 抗腐蚀性，防污着性
resistance to ground 对地电阻，接地电阻
resistance to heat 耐热性
resistance to impact 抗冲击，抗冲击性，耐冲击性，冲击抗力，碰撞阻力
resistance to laminar tearing 耐层状撕裂能力
resistance to overflow 流出阻力【气、液体】
resistance to pneumatic transmitter 带阻尼的气动变送器
resistance to pressure 抗压性，耐压
resistance to shock 冲击阻力，抗震动，耐震，耐冲击，抗冲击
resistance to sliding 摩擦阻力
resistance to sparking 耐击穿性，击穿电阻
resistance to tearing 抗撕裂，抗撕裂性，耐撕裂性，扯裂强度
resistance to tracking 表面漏泄电阻，爬电电阻
resistance to vibration 动态强度，抗振刚度
resistance to water vapour permeability 蒸汽渗透电阻
resistance to water vapour permeation 蒸汽渗透阻
resistance to wear 耐磨性
resistance to weathering 耐风化强度
resistance to weather 抗风化能力，耐风化能力，耐受天气自然作用的能力，气候稳定性
resistance to yield 屈服抗力
resistance to 带阻尼的
resistance transducer 电阻传感器

resistance transformer 电阻变换器
resistance transistor logic 电阻晶体管逻辑
resistance transmitter 电阻传感器
resistance tube 管形电阻，电阻管
resistance-type flowmeter 电阻式流量计
resistance-type level transducer 电阻式液位传感器
resistance upset-butt welding 电阻锻压对顶焊接
resistance voltage divider 电阻分压器
resistance voltage 电阻电压
resistance welding time 焊接通电时间【电阻焊】
resistance welding 电阻焊（接），接触电阻焊，欧姆线圈，线绕电阻
resistance wire extensometer 电阻丝引伸仪
resistance wire wave gauge 电阻丝测波仪
resistance wire 电阻线，电阻丝
resistance 电阻，阻力，抗力，流阻，阻尼，抵抗力，对抗力
resistant spot welding 电阻点焊
resistant welding 电阻焊
resisted rolling 阻尼横摇
resister 电阻器
resisting force 阻力，抗力
resisting medium 黏性介质，阻尼介质
resisting moment 抗力矩，阻力矩，抵抗力矩，阻抗力矩
resisting shear 抗剪力
resisting torque 抗转矩，抗力矩，反力矩，阻力矩，抵抗力矩
resisting 抵抗，忍住
resistin 一种锰铜电阻合金
resistive attenuator 电阻式衰减器
resistive-capacitive 阻容的
resistive component 有功部分，实数部分，电阻分量
resistive conductor 电阻导线，电阻性导体
resistive DC voltage drop 电阻性直流电压降
resistive divider 电阻分压器
resistive ground 电阻性接地
resistive heating 电阻加热
resistive load 电阻性负载，有功负载，有效负载
resistive muffler 阻性消声器
resistive thermal detector 电阻式热探测器，电阻式温度检测器
resistive 电阻的，有阻力的，有抵抗力的
resistivity exploration 电阻查勘法
resistivity meter 电阻率计
resistivity method 电阻法【用于物探】
resistivity profile 电阻率剖面
resistivity prospecting log 电阻率测井记录，电阻率勘探日志
resistivity prospecting 电阻勘探
resistivity 抵抗（能）力，电阻系数，电阻率，抵抗性，比电阻
resistojet 电阻加热电离式发动机
resistor-capacitor diode transistor logic 电阻电容二极管晶体管逻辑
resistor-capacitor transistor logic 电阻电容晶体管逻辑
resistor-capacitor 电阻电容
resistor-coupled transistor logic 电阻耦合晶体管逻辑
resistor disc 电阻片【避雷器】
resistor divider 电阻分压器
resistor matrix 电阻矩阵
resistor tap-change method 电阻器式抽头切换法
resistor-transistor logic 电阻晶体管逻辑
resistor 电阻器，电阻
resist-valve arrester 阀型避雷器
resist 抵抗，承受，耐得住，抗拒，忍耐，忍住，抗蚀剂，防染剂，保护层，抗蚀层
resite 丙阶酚醛树脂，不溶酚醛树脂
resolution error 解算误差，分辨误差
resolution of board 董事会决议
resolution of force 力的分解
resolution of polar to cartesian 极坐标-直角坐标转换
resolution of stress 应力分解
resolution of time tagging 时间标记的分辨率
resolution of vector 矢量的分解
resolution ratio 分辨率
resolutions of shareholders'meeting 股东大会决议
resolution threshold 分辨度
resolution time 分辨时间
resolution 分辨（能力），辨别，分解，甄别，分辨度［率］，解决，分辨力，决议，图形分辨率，溶解，清晰度
resolvent 分解物
resolver 解算器，分解器
resolve 分辨，分解
resolving ability 分辩能力
resolving potentiometer 解算电势计
resolving power 分辨能力，分辨率，鉴别能力
resolving time 分辨时间
reso-meter 谐振频率计
resonance absorber 共振吸收体
resonance absorption cross section 共振吸收截面
resonance absorption of neutrons 中子共振吸收
resonance absorption 共振吸收，谐振吸收，共振吸收反应
resonance amplifier 调振放大器，调谐放大器
resonance blocking 谐振闭锁
resonance bridge 谐振测量电桥
resonance build-up delay 谐振起建时滞
resonance capture of neutrons 中子共振俘获
resonance capture 共振俘获，共振吸收
resonance characteristic 谐振特征
resonance circuit 谐振电路
resonance condition 共振条件
resonance cross-section 共振截面
resonance current 谐振电流
resonance curve 反应（特性）曲线，谐振曲线
resonance detector 共振探测器
resonance effect 谐振现象，共振现象，共振效应
resonance energy 共振能
resonance escape probability 共振逃脱率，逃脱共振概率［几率］
resonance fission integral 共振裂变积分
resonance flux 共振中子通量
resonance frequency meter 谐振式频率计
resonance frequency 共振频率，谐振频率，共

鸣频率
resonance hump 共振峰
resonance indicator 谐振指示器
resonance instrument 谐振式仪表
resonance integral 共振积分
resonance interference 谐振干扰
resonance isolator 共振隔离器
resonance level 共振能级
resonance line 共振线，谐振线
resonance matching 谐振匹配
resonance method 共振法，谐振测定法
resonance modulus 共振模量
resonance neutron 共振中子，谐振中子
resonance oscillation 共振振荡，谐振振荡，谐振，共振荡
resonance overvoltage due to broken line 断线谐振过电压
resonance overvoltage 调谐造成的过电压，谐振过电压
resonance parameter 共振参数
resonance peak 谐振波峰，谐振曲线峰值，共振峰
resonance phenomena 谐振现象
resonance point 共振点
resonance pulsator 共振脉动器
resonance radiation 谐振辐射
resonance range 谐振范围
resonance region 共振区，共鸣区，共振区域
resonance resistance 谐振电阻
resonance scattering cross section 共振散射截面
resonance scattering 共振散射
resonance screen 共振筛
resonance sharpness 谐振锐度
resonance shielding factor 共振屏蔽因子
resonance speed 共振速度
resonance state 共振态
resonance testing 谐振检查法
resonance transformer 调谐变压器，谐振变压器
resonance tube 共振管
resonance type transformer 谐振式变压器
resonance type voltage regulator 谐振式电压调整器
resonance vibration 共振
resonance voltage 谐振电压
resonance wavemeter 谐振式波长计
resonance wave 共振波，谐振波
resonance width 共振幅度
resonance zone 谐振区
resonance 共振，谐振，共鸣
resonant absorber 共振吸音体
resonant angular frequency 共振角频率
resonant cavity 共振腔
resonant circuit 谐振电路，共振电路，谐振回路
resonant column method 共振柱法
resonant column triaxial test apparatus 共振柱三轴试验仪
resonant earthed neutral system 中性点谐振接地系统
resonant earthed system 谐振接地系统
resonant eddy 共振涡旋
resonant excitation 共振激励
resonant feeder 谐振馈线
resonant-flip-flop 谐振触发器
resonant four-terminal network 谐振四端网络，共振四端网络
resonant frequency testing 共振频率试验
resonant frequency 共振频率，谐振频率
resonant grounded system 共振接地系统，谐振接地系统，中性点谐振接地系统
resonant grounded 谐振接地的
resonant interaction 共振干扰
resonant iris switch 谐振膜转换开关
resonant iris 谐振膜片，谐振窗
resonant line 谐振线
resonant peak value 共振峰值
resonant pile driver 共振打桩机
resonant pitching 共振俯仰，共振纵摇
resonant reed relay 谐振簧片继电器
resonant relay 谐振继电器
resonant response factor 共振特性系数
resonant response 共振响应
resonant shunt 谐振分路
resonant sound absorber 共鸣消音器，共振吸音器
resonant speed 共振转速，临界转速
resonant transformer 谐振变压器，调谐变压器
resonant type instrument 谐振式仪表
resonant type transformer 共振式变压器
resonant vibration 共振，谐振
resonant vortex excitation 共振旋涡激励
resonant vortex shedding 共振旋涡脱落
resonant 共振的，谐振的，共鸣的
resonate 共振，谐振，共鸣
resonating circuit 谐振电路
resonator gap 谐振器间隙
resonator grid 谐振器栅极
resonator-tron 谐振电子管，谐腔四极管
resonator 共鸣器，共振器，谐振器，谐振腔，集音盒
resonatron 谐振管
resonoscope 共振示波器
resorption 熔蚀，再吸收，再吸收循环
resource advantages 资源优势
resource allocation and network scheduler 资源分配及网络调度程序
resource allocation processor 资源分配处理机
resource appraisal 资源评价
resource base 资源基础
resource conservation 资源保护
resource-conserving 资源节约的
resource data file 资源数据文件
resource development 资源开发，手段研究
resource enhancement 提高资源质量
resource exploitation 资源开采
resource management 资源管理
resource map 资源分布图
resource material 核原料，源材料
resources appraisal 资源评价
Resources Association （美国）水资源协会
resources development 资源开发
resources equilibrium 资源平衡
resources evaluation 资源评价

resources exploitation 资源开发
resource-sharing executive 资源共享执行
resource-sharing 资源共享
resources planning and scheduling method 资源计划与调度法
resource 实际电源，实声源，物资，资源，手段，方法
respect 方面，注意
respirable dust content 可呼吸的粉尘含量
respirable dust 可呼吸的粉尘
respiraiton 呼吸
respiration calorimeter 呼吸热量计
respirator filter 呼吸面罩，口罩
respiratory protection equipment 呼吸保护设备
respiratory protection 呼吸面罩的过滤器
respiratory system 呼吸系统
respiratory tract 呼吸系统
respirator 口罩，防毒面具
respondent 相反应的
responder 响应器，应答机，回答器
respond in damages 承担损失赔偿，对损失负责
response ability index 响应能力指数
response bandwidth 响应带宽
response characteristic 响应特性曲线，响应特性，灵敏度特性曲线
response curve 反应曲线，灵敏度特性曲线，响应曲线，应答曲线
response excursion 扰动偏移
response factor 响应系数，反应系数
response frequency 响应频率
response function 响应函数
response gust factor 响应阵风因子
response index 响应指数
response lag 响应延迟，反应滞后
response matrix method 响应矩阵法
response matrix 响应矩阵
response period 反应期，雨水集流时间
response pressure 开启压力
response pulse duration 响应脉冲宽度
response rate 反应速度，响应速度
response ratio 响应比，响应系数
responser 响应机，应答机
response spectrum analysis 反响应谱分析方法，反应谱分析，响应谱分析
response spectrum annunciator 响应谱报警器
response spectrum recorder 响应谱记录仪
response spectrum 响应谱，反应谱
response speed 响应速度
response surface analysis 表面效应分析
response team 应急响应工作队
response temperature 响应温度
response time history analysis 动力反应时程分析
response time index 响应时间指数，响应时间系数
response time 反应时间，应答时间，动作时间，响应时间
response to unit impulse 对单位脉冲的响应
response voltage pulse characteristics 感应电压脉冲特性
response 反应，应答，响应，反响，答复，灵敏度，频率特性，响应曲线
responsibility accounting 责任会计
responsibility audit 责任审计
responsibility range 职责范围
responsibility system of factory director 厂长负责制
responsibility 责任，职责，经济责任
responsible approach 响应方法，可靠的方法
responsible department 责任部门
responsible for 对……负有责任
responsible governmental department 政府主管部门
responsible organization 负责单位，负责机构，负责的部门
responsible 负有责任的，负有义务的
responsive bid 符合性投标，响应性投标
responsiveness to bid 对标书的响应
responsiveness 响应度，响应性，灵敏度
responsive tender 响应标
responsive to bidding document 符合招标文件要求
responsive 响应的，敏感的，灵敏的，应答的
responsor 响应器，应答机
RE STAGE(regenerative stage) 回热循环燃气蒸汽联合装置，再生阶段
rest angle 休止角，安息角
restarting 再启动
restart routine 程序启动程序
restart time 再启动时间
restart up 再启动
restart zone 再引弧区
restart 再启动，重新启动，重新开始
restate 重申，再声明，重新陈述
rest bucket 工具吊架
rest condition 静止状态
rest contact 静止触点，静触头
rest energy 粒子静态能
resting basin 休息池
resting contact 静止触点，静止接点
resting pool 休息池
resting position 静止位置
resting rung （铁爬梯的）休息梯级
resting support 支承结构
resting type spring constant support 弹簧恒力托架
resting type spring support 弹簧托架
restitution coefficient （桩的）回弹系数
restitution 偿还，复原，恢复，还原，解调，归还，回弹
rest mass of electron 电子静质量
rest mass of neutron 中子静质量
rest mass of proton 质子静质量
rest mass 静质量
restorable change 可逆变化
restoration drawing 修复图
restoration force 恢复力
restoration heat treatment 恢复热处理
restoration of cladding 堆覆层的修复
restoration of hydro-logical data 水文（逻辑）资料还原，水文（逻辑）资料修复
restoration of storage 重新蓄水

restoration process （电力系统的）恢复过程
restoration time after system failure 电力市场运营系统故障恢复时间
restoration voltage 恢复电压
restoration 复原，恢复，还原，复位，修复，重建
restorative control 恢复控制
restorative state 待恢复状态
restorer diode 恢复二极管
restorer 复位器，恢复器，还原器，恢复设备
restore the status quo ante 恢复原状
restore voltage 恢复电压，重建电压
restoring component 还原元件
restoring couple 恢复力偶
restoring current 恢复电流
restoring force 复原力，恢复力
restoring moment 恢复力矩
restoring relay 恢复继电器
restoring spring 复原弹簧
restoring time 恢复时间
restoring 重新启动，恢复，回复，再接通，还原
rest point 平衡点
rest potential 静态电位，静态电势
restrained beam 约束梁，固端梁
restrained deformation 约束变形
restrained end 固定端，约束端
restrained pile 约束桩
restrained system 约束系统
restrainer 抑制器，阻制器，定位器
restraining coil 阻尼线圈，制动线圈，吸持线圈，保持线圈
restraining force 抑止力
restraining moment 约束力矩，固端力矩
restraining pressure 约束压力
restraining resistance 制动电阻
restraint coefficient 约束系数
restraint condition 约束条件
restraint cylinder 耐震补强筒
restraint degree 约束度
restraint garter 限制箍，约束箍
restraint intensity 拘束度
restraint location 紧固位置
restraint plane 约束平面
restraint relay 牵制式继电器，制约式继电器
restraint ring 约束环，固紧环
restraint stress 约束应力
restraint structure barrel （堆芯）阻尼构件筒
restraint system 监督系统，约束系统
restraints 约束件，阻尼件
restraint tank 堆容器
restraint zone 紧固带，限制区
restraint 限定，约束，抑制，限制，收缩，阻止，制止，限制器，阻尼器，限动器
restricted accessibility space air recirculation system 限制接近室空气再循环系统
restricted access 限制进入
restricted article 受限制商品，限制物品
restricted-bore 限制的孔径
restricted connected-speech recognition system 有限连接语言识别系统
restricted distribution 内部分发
restricted earth fault protection 限制接地故障保护
restricted earth protection 限制接地保护
restricted hour maximum demand 限制时间内最高需量
restricted hour rate 定时电价
restricted information 内部保密资料
restricted lag network 有限滞后网络
restricted linear collision stopping power 定限线碰撞阻止本领
restricted message 密电，密件
restricted orifice surge chamber 阻抗式调压室
restricted orifice surge tank 阻力孔式调压塔
restricted problem 约束问题，限制问题
restricted water level in flood period 汛期限制水位
restricted waterway 限制航槽
restricted zone 禁区
restricting element 节流元件
restriction control 节流控制
restriction efficiency 限定效率，节流效率，限制效率
restriction loss 节流损失
restriction orifice 限流孔板
restriction 节汽门，扼流圈，收口，节流口【落煤管】，限定，限制，约束
restrictive clause 限制性条款
restrictive valve 节流阀，限制器
restrictor by-pass valve 节流旁通阀
restrictor valve 限流阀
restrictor 限流器，节流阀，节流器，扼流圈，节气门，限制器
restrict 制约
restrike of arc 电弧再触发，再点火
restriking voltage 再击穿电压，再闪击电压
restriking 电弧再触发
rest room （公共建筑物内的）公用厕所，洗手间，休息室
restructure the power sector 电力行业重组
restructure 改组
restructuring trust corporation 重组信托公司，改造信托（资金）
rest time correction factor 间歇校正系数
rest time 停机时间，积分作用时间常数
rest upon the price 视价格而定
rest with the buyer 由买方决定的
rest 架，刀架，盈余
result address 结果地址
resultant accuracy 总精度
resultant action 总作用
resultant amplitude 合成振幅
resultant compressive stress 合成压应力
resultant couple 合成力偶
resultant current 合成电流，合成流
resultant curve 合成曲线
resultant displacement 合成位移
resultant drag 总阻力
resultant drive factor 总传动系数，合成传动系数
resultant error 合成误差
resultant field 合成磁场
resultant force 合力

resultant gear ratio　总传动比
resultant impedance　合成阻抗，总阻抗
resultant load　合成荷载
resultant motion　合成运动
resultant mutual flux　合成互磁通
resultant of forces　合力
resultant of velocities　速度的合成
resultant phasor　合成相量
resultant pitch　合成节距
resultant pressure　总压力
resultant strain　合成应变，合应变
resultant stress　合成应力
resultant vector　合成矢量
resultant velocity　合成速度
resultant vibration　合成振动
resultant wave　合成波
resultant wind direction　合成风向
resultant wind　合成风
resultant　合成的，结果的，复合的，合力，合量，组合，合成（矢）量
result function　目标函数
resulting display　最终显示
resulting stress　合成应力
resulting　致使，产生
result in science and technology　科技成果
result in　导致，引起
result of measurement　测量结果
results of inspection　视察结果
results of rating　考核成绩
result　成果，结果，结论，效果
resume　概述，恢复，重新开始，简历，自传
resumption of concreting　恢复浇灌混凝土
resumption　恢复，重新开始，继续，取回
resuperheater　再过热器
resuperheat　再过热
resurface　整修表面，修理炉衬，磨锐工具
resurvey　重测
resuspension coefficient　再悬浮系数
resuspension model　再悬浮模式
resuspension rate　再悬浮率
resynchronization of electric power system　电力系统再同步
re-synchronization　二次同步，再同步
resynchronize　再同步
retail competition　零售竞争模式
retailer　零售商
retail load　零售负荷
retail price　零售价格，销售电价
retail sales　零售
retainable　拦蓄的
retained amount　筛余量
retained austenite　残余奥氏体，残留奥氏体
retained cost　保留成本
retained earnings　留存收益，未分配利润
retained income　留存收益
retained item　保留项目
retained percentage　残留百分率
retained profit　利润留成，留存利润，保留利润，净利润
retained surplus　留存盈余
retained wage　保留工资
retained wall　挡土墙
retained waste　保留废物
retained water　持面水
retainer plate spacer　止动定位板条
retainer plate　止动挡板
retainer shoe　护套，托板
retainer sleeve　夹持套筒
retainer　保持架，定位器，止动器，护圈，承盘，制动，抵住物，保留物，制动装置
retaining and protecting for foundation excavation　基坑支护，基础开挖的支护
retaining bar　夹持杆，防震条
retaining circuit　保持电路
retaining coil　吸持线圈
retaining current　维持电流，吸持电流，制动电流
retaining cylinder　护筒
retaining dam　拦水坝，蓄水坝
retaining grid　挡板
retaining nut　锁紧螺母
retaining plate　支撑板【可燃毒物棒束】
retaining power　截污能力，截污容量
retaining ramp　护坡
retaining ring bar　扣环钢条
retaining ring pliers　扣环钳
retaining ring　挡圈，固定环，护圈，扣环，持环，定位环，支撑环，护环
retaining spring　止动弹簧
retaining structure　支挡结构
retaining time of control process　控制过程持续时间
retaining time　持续时间，保持时间
retaining valve　单向阀，止回阀，逆止阀
retaining wall　挡墙，挡土墙
retaining washer　止动垫圈
retaining wedge　槽楔
retaining works　挡水工程，蓄水工程
retain　保持，保留，保护，留存
retaliate　打击报复
retapping　桩工的复打，重攻螺纹
retardant agent　缓凝剂
retardant　缓凝剂，耐火的，阻化剂，抑制剂，缓凝剂，阻燃
retardation angle　减速角
retardation coefficient　扼流系数
retardation coil　扼流圈，抗流线圈
retardation factor　阻滞因数
retardation field　减速场
retardation network　时延网络
retardation of phase　相位滞后
retardation of the tide　潮时后延
retardation of wind　风力减弱
retardation reservoir　滞洪库容
retardation time　推迟时间，延迟时间，滞后时间，制动时间
retardation　迟滞，延迟，延缓，阻力，障碍物，减速，推迟，缓凝，缓凝作用，放慢，阻止，阻滞，迟延，阻滞作用
retarded admission　延迟进气
retarded caving　滞后崩落
retarded cement　缓凝水泥
retarded combustion　延迟燃烧

retarded control 推迟控制，迟延调节
retarded elasticity 推迟弹性
retarded flow 拦滞水流，阻滞水流
retarded ignition 延迟着火
retarder 阻凝剂【混凝土】，阻滞剂，限制器，缓凝剂，减速剂，制动器，延迟器，减速器，延迟剂，缓行器，抑制剂
retarding action 迟延动作
retarding admixture （水泥）缓凝剂
retarding agent 阻滞剂，缓凝剂
retarding disk 制动盘
retarding electrode 减速电极
retarding field 减速电场
retarding force 制动力，减速力
retarding reservoir 拦洪水库，滞洪水库
retarding torque 制动转矩
retarding transmitter 延时发送器
retard 迟滞，减速，缓解，推迟，延迟，制动，阻碍
retempering （混凝土的）重拌和，重新混合
retensioning wrench （装料机）拧紧扳手，再拉伸扳紧器
retention ability 持水力
retention area 保存区
retention basin 贮水池，贮留槽，贮液池
retention capability 持留能力，滞留容量
retention capacity 贮留能力【过滤器】，迟缓率，调节能力，泄洪能力，滞流量
retention category 保存期分类
retention characteristics 滞留特性
retention duration 保存期限
retention effect 滞留效应
retention efficiency 捕渣率，除尘效率
retention factor 保持系数，持水系数，保留因子，滞留因子
retention fraction 保留份额
retention index 持着系数
retention level 壅水水位
retention money 保留金，保留款额
retention of activity 放射性滞留
retention of nutrients by reservoir 水库养分保留
retention of radioactivity 放射性滞留
retention of records 记录的保存
retention period 保存期，保留期
retention pin 定位销
retention pit 滞留罐
retention pond 澄清池
retention rate 保水率，持水率
retention reservoir 洪水调节池，滞洪水库
retention screen 持留网，贮留筛
retention structure elevation 拦洪高程
retention system 贮留系统，保留系统
retention tank 贮存箱，贮存槽
retention time 保留时间，滞留时间，保持时间，停留时间，保存时间，阻滞时间
retention wafer level 正常高水位
retention wall 挡水墙
retention 保持（力），抑制，保存，保留，自保持，滞留，贮留，保留金，停留，滞留量
retentive alloy 硬磁性合金
retentive material 硬磁性材料
retentive memory relay 保持记忆继电器
retentiveness 保持力，持水性
retentive soil 持水土壤
retentive 保持，保持的，有保持力的
retentivity of vision 视觉暂留
retentivity 保持性，保持力，顽磁性，剩磁
retesting 重新试验
retest 重复试验，复核试验
reticle 刻线，分度线，标线，十字线
reticular 网状的
reticulation 网，网状（结构），起皱
reticule alignment 十字线标定，十字线，十字线校准，网线对准
reticule 标度线，分度线，十字线
retiform 网状的，有交叉线的
retimbering 重新支撑
retired cadres administration office 老干部退休办公室
retirement age 退休年龄
retirement annuity 退休年金
retirement funds 退休基金
retirement of generating unit 发电机组退役
retirement pay 退休金
retirement pension 退休金，养老金
retirement （电站）退役，退休，退职
retort gas 甑中产生的气体，蒸馏气体
retort grate 下饲式炉排
retorting 干馏，甑馏
retort-type furnace 坩埚式液态排渣炉膛
retort-type slag tap furnace 坩埚式液态排渣炉膛
retort-type stoker 下饲式炉排
retort 曲颈瓶，杀菌釜，蒸馏炉，反驳
retotrol exciter 一种旋转放大励磁机
retraceline 回描线，回扫时间
retracement line 后视线
retrace ratio 回描率
retrace time 回描时间
retractable-blade safety knife 可缩回刀片的安全刀
retractable gripping arm 可伸缩的夹持臂
retractable head pulley of belt conveyor 带式输送机头部伸缩装置【三位置】
retractable oil gun 伸缩式油枪
retractable soot blower 伸缩式吹灰器
retractable thimble 可伸缩的套管
retractable type soot blower 伸缩式吹灰器
retractable 可伸缩的，可收缩的
retract bid 撤回出价
retracted thimble 缩回的套管
retractile cable 蜿蜒的电缆
retractile cord 可伸缩的电线【电话架空线等】
retractility 伸缩性
retracting motion 回缩运动
retracting stroke 回程，回行冲程
retracting type soot-blower 伸缩式吹灰器
retraction stress 收缩应力
retraction 回缩，缩进
retractive tube expander 收缩式胀管器
retract 拉回，收缩，取消
retraining 再培训
retransmission 转播，转发，中继

retransmitted signal 中继信号
retransmitter 中继发射机，转播发射机
retransmit 转移（信号，信息）
retreatment 再处理，再加工
retreat of valley sides 谷坡后倾
retreat velocity 排放流速，退却速度
retreat 退却
retrenchment of expenditure 紧缩开支
retrenchment policy 紧缩政策
retrenchment 节省，紧缩，删除
RET＝return 返回
retrievability 可回取性
retrievable storage 可回取贮存
retrievable surface storage facility 可回取地面贮存设施
retrieval by on-line search 联机检索
retrieval conduit 补偿导管
retrieval facility 检索设备［设施］
retrieval language 检索语言
retrieval routine 检索程序
retrieval system 检索系统，返回系统
retrieval time 恢复时间
retrieval 查找，检索，（信息）恢复，收回，挽回，补偿，弥补，取回，重新得到
retrimming 重新调整，重新平衡
retrim 再调整，再平衡，再配平
retroaction 再生，反作用，反馈
retroactive amplification 再生放大，反馈放大
retroactive audion 再生检波管
retroactive 再生的，反馈的
retroact 回动，倒行，反作用
retrofit 翻新，改造，改建，旧机组改装，改型，更新，小修小改，返回改变，返回更改
retrogradation 衰退，退化，衰减，退减，下降状态
retrograde rotation 逆向旋转
retrograde wave 反向波，退行波，逆行波，后退波
retrogression 倒退，逆向运动，反向运动，衰退，退化
retrogressive erosion 向源侵蚀
retrogressive landslide 牵引式滑坡，向源逆行滑坡
retrogressive lap winding 倒退叠绕组，左行绕组
retrogressive wave 退行波
retrogressive winding 倒退绕组，左行绕组
retrogressive 倒退的，退行性，逆行性，逆行的
retrospective dose 事后实计剂量，事后实受剂量
retrospective search 逆检索法，追溯检索法
retrovert 使翻转，使后倾
retry 重试
retube 换管子
retune 再调谐
returnable heat 可返回热，可再用热
returnable 可再使用的，可再回收的
return action 回复作用
return address 返回地址
return air condenser 回气式凝汽器
return air course 回风巷道
return air fan 回风机
return air grille 格栅回风口
return air inlet 回风口
return air ticket 往返机票
return airway 回流风道
return air 回风，回流空气，回气，循环气体
return albedo 再进入的反射率
return and restore 返回和复原
return a visit 回访
return bend 180°倒转，180°弯头，反向弯管，回转弯头
return branch of radiator 散热器回水支管
return brush 回流电刷
return cable 回流线，回流电缆
return call 回答呼叫
return channel 返回流道
return cinder fan 飞灰复燃风机，飞灰再燃送风机
return circuit rig 反向导流器
return circuit wind tunnel 回路式风洞
return circuit 返回线路，回流电路，返回管路，回路
return code 返回码
return conductor 回流导线，回路导线
return control word 返回控制字
return current 反流，回流
return curve 回复曲线
return difference 回差
return drain pump 疏水泵
returned goods 退回货物
returned oil 回油
returned sludge ratio 污泥回流比
return fan section 回风机段
return feeder 负馈路，回流馈路，负馈线
return flange tray 曲折边缘托架
return flange 曲折边缘
return flow atomizer 回油雾化器
return flow oil burner 回油式燃烧器
return flow tunnel 回流式风洞
return flow wind tunnel 回流式风洞
return flow zone 回流区
return flow 回流，倒流，逆流，反流
return flume 回水槽
return from interrupt 中断返回
return from subroutine 从子程序返回
return fuel fan 回料风机
return guide vane 回流导向叶片，回流导叶
return header 回流联箱，回流总管
return idler 返回托辊，回空侧托辊，回程托辊
return information 返回程序，返回信息
returning air mass 回归气团
returning current from earth wire 地线返回电流
returning grid 可转动分布板，翻板
returning wave 回波
return instruction 返回指令
return interval 回描间歇，回描周期
return leg 回料管
return line corrosion 回水管路腐蚀
return line 回线，回流管路，回扫线，回描线，回流管（线）
return loss 回波损失，回波损耗，反射波损耗
return main 回流主管，回水干管，回水总管
return motion 回复运动
return of dial 拨号盘回复原位

return offset 回归偏置管
return oil piping 回油管道
return on assets 资产收益率
return on equity 股东资产净值的盈利，资本收益率，资本权益回报率，资本金净利润率，净资产收益率，净资产回报率，股本回报率，股权收益，股票收益
return on investment 投资回报率
return on positive 为正值返回【指令】
return on tangible assets 有形资产收益率
return on total assets 资产总额收益率
return path 归路，返归路线，回流
return period 重现期，回复周期
return pipe 回浆管，回流管，回水管
return pressure of safety valve 安全阀回座压力
return pump 反馈泵
return rate price 经营期电价
return recording 数字间带有间隔的记录
return register 返回寄存器
return relief line pipe 回油管道
return riser 再循环上升管
return run 回程，(管道的）折回段
return seepage 回归渗流
return side 回行侧【输送带】，回空侧
return signal 返回信号，回答信号
return sludge flow 污泥回流
return sludge 回流污泥
return speed 复位转速
return spring 返回弹簧【控制棒驱动机构棘爪组件】，复原弹簧，回位弹簧，回动弹簧
return statement 返回语句
return steam header 回汽联箱
return strand 回空分支
return streamer 回流
return stroke 回程，回行冲程
return tank 回流箱
return ticket 来回票
return time 回复时间
return to bias 归零制
return to normal 回复正常
return to nucleate boiling 返回沟核沸腾
return to original conditions 恢复原状
return to the scale 规模收益
return to the status quo 恢复原状
return-to-zero mark 归零标记
return-to-zero method 归零法
return-to-zero position 返回零位
return to zero recording 归零记录
return-to-zero record 归零记录
return-to-zero 归零点，复零，归零制
return transfer function 返回传递函数，回路传递函数
return trap 回水盒，疏水器
return trip 回程，返回
return tube boiler 回火管锅炉
return tube 回流管，溢流管
return type wind tunnel 回路式风洞
return valve 回流阀，回水阀，回油阀
return vane 返导叶
return voltage 回复电压
return wash 回转冲刷
return water collecting header （回水）集水器
return water header 回水联箱
return water pipe 回水管
return water pump 回收水泵
return water temperature 回水温度
return water 回水
return wire 回线
return yoke 旁轭
return 返回，回程，收益
reunion （断裂）复合
reusability 重新使用的可能性
reusable component 重复使用的部件
reusable containers 可再用的容器
reusable resources 可多次使用的资源
reused water 复用水，再用水
reuse of plastic wastes 废塑料的利用
reuse water 再用水
reuse 重复利用，再使用
reusing industrial effluent 工业废水再利用
reusing water pool 复用水池
re-utilization 重复利用
revaluation of data 数据换算
revaluation 升值，增值，重新估计［估价］
revaluate 对……重新估价
revalve 更换阀门，更换电子管
revamping of exiting enterprise 改组现有企业
revamp 翻新，更新，修理，修补，整修，改进，修改
revaporization 二次蒸发，再汽化
revaporize 再蒸发
revasal 倒转
revcur(reverse current) 反向电流
REVD＝reversed 反向的
revealment 显露
reveal 揭露，显露，展现，揭示，呈现，泄漏，外露，显示
revegetation 再生植被
reventing system 再排气系统
revenue and cost 收入与成本
revenue authority 税务局
revenue cap regulation 收入上限监管法
revenue expenditure 收益性支出
revenue requirement ·必要年收入
revenue risk 收入风险
revenue stamp 印花税票
revenue tax 营业收入税
revenue 税收，收入，收益，营业收入
reverberate 弹回，反射
reverberation meter 混响时间测量
reverberation-suppression filter 混响抑制滤波器
reverberation time 混响时间
reverberation 反响，回响，反射
reverberatory furnace 反射炉
reverberator 反射炉，反射器
reverberometer 混响计，混响时间测量计
reversal chamber 转向室
reversal control of HVDC power transmission system 直流输电潮流反转控制
reversal loss 反向损耗
reversal of current 反流，电流反向，倒流
reversal of diode 二极管反接

reversal of machine 机器反向，机器倒转
reversal of magnetism 磁极变化，磁性反转，反磁化
reversal of phase sequence 相序倒转
reversal of phase 相位反向，倒相，反相，相位改变 180°
reversal of pole 极性变换
reversal of stress 应力交变
reversal point 反向点
reversal rotation 反转，倒转
reversal shuttle belt 可逆移动皮带
reversal valve 换向阀
reversal washing 反冲洗
reversal 反向，回复，倒向，（极性）变换，变号，转向，换向，反转，倒转
reverse acting actuator 反作用执行机构
reverse-acting control element 反作用调节元件，反作用控制元件
reverse-acting controller 反作用控制器
reverse-acting control valve 反作用控制阀
reverse-acting valve 反作用阀
reverse-acting 反作用的
reverse action 反作用，相反动作
reverse-air collection 反向气流式除尘器
reverse-air filter 逆流式空气过滤器
reverse azimuth 反方位角
reverse bearing 反方位
reverse bend of metals 金属反复弯曲试验
reverse bend property 反复弯曲性
reverse bend 反向弯曲
reverse biased 反偏置
reverse bias 反向偏压，逆向偏压，反偏压，反向偏置
reverse blocking interval 反向关断期间，反向闭锁期间
reverse blocking state 反向关断状态，反向闭锁状态，反向阻断状态
reverse blocking thyristor 反向阻断晶闸管
reverse blocking 反向闭锁
reverse breakdown 反向击穿
reverse buckling rupture disk 可逆爆破安全膜
reverse Carnot cycle 逆卡诺循环
reverse characteristics 反向特性
reverse compound-wound motor 差复激电动机
reverse-conducting thyristor 逆导型晶闸管
reverse construction 反向设定
reverse counting rate 计数率倒数
reverse counting 反向计数
reverse count rate 计数率倒数
reverse creep 反向蠕变
reverse current braking 反向电流制动
reverse current breaker 逆流断路器
reverse current filter 逆流过滤器
reverse current metering 反向电流测量，逆流测量
reverse current protecting equipment 逆电流保护装置
reverse current relay 逆电流继电器
reverse current release 反向电流脱扣器，反向电流释放器，逆电流脱扣器
reverse current switch 逆流开关
reverse current time-lag relay 逆流延时继电器
reverse current trip 逆流自动切断
reverse current 逆序电流，反向电流，反流，倒流，逆流
reverse curve 反向曲线
reverse cycle 逆循环
reversed bending stress 交变弯曲应力
reversed bending 反向弯曲
reversed bias 反偏压
reversed capital flow 资本流向逆转
reversed control 反控制
reversed current 逆流
reversed-directional element 倒相单向元件，反向元件
reversed filter 反滤层
reversed flow condenser 反流冷凝器，逆流式凝汽器
reversed flow turbine 反流涡轮机，回流式汽轮机
reversed flow 变向流动，混合冲刷，逆流，回流
reverse direction 反方向，逆方向，反向，逆向
reversed-loop winding 逆行叠绕组，左行叠绕组
reversed pendulum 倒垂线【观测大坝挠度用】
reversed phase coil 反相线圈
reversed phase 倒相，反相
reversed polarity 反极性，异极性，反接
reversed projection 逆投影
reversed return system 同程式系统
reversed river 反向河
reversed stratigraphic sequence 倒转层序
reversed stream 反向河
reversed stress 交变应力，反向应力
reversed tickler 负反馈线圈
reversed wind 反向风
reverse fault 逆断层
reverse feedback 反相反馈，负反馈
reverse field 反向场
reverse flattening test 反向展平试验
reverse flow cooling system 逆流冷却系统
reverse flow cooling tower 逆流式冷却塔
reverse flow nozzle 反向水流喷嘴，回流喷嘴
reverse flow region 逆流区
reverse flow valve 换向阀
reverse flow 倒流，反向流，逆流，逆向流动
reverse flush pipe 反冲洗管
reverse gear shift 倒车调挡
reverse gear 倒车挡
reverse-graded media 逆粒度滤料
reverse grid current 反向栅极电流
reverse loss 反向损耗
reverse motion 返回运动，反向运动
reverse of excitation 励磁反向，激励反向
reverse operation 倒转
reverse osmose membrane 反渗透膜
reverse osmosis apparatus 反渗透设备
reverse osmosis desalination 反渗透脱盐
reverse osmosis desaltinating 反渗透脱盐
reverse osmosis filter 反渗透过滤器
reverse osmosis membrane 反渗透膜
reverse osmosis module 反渗透组件
reverse osmosis permeator unit 逆渗透渗漏装置
reverse osmosis permeator 逆渗透渗漏计

reverse osmosis process 反渗透法，反渗透过程
reverse osmosis separation 逆渗透分离器
reverse osmosis system 反渗透系统
reverse osmosis treatment 反渗透处理
reverse osmosis unit 反渗透装置
reverse osmosis 反渗，反渗透，反渗透法
reverse overcurrent relay 反向过载继电器，逆过电流继电器
reverse peak voltage 反向峰压
reverse-phase current relay 反相电流继电器
reverse-phase protection 反相保护
reverse-phase relay 反相继电器
reverse-phase 反相
reverse polarity protection 反极保护
reverse polarity 相反极性
reverse position 反常位置，反转位置
reverse potential 反相电位
reverse power flow 反向潮流
reverse power protection 逆功率保护
reverse-power relay 反相功率继电器，逆功率继电器
reverse-power tripping 逆功率动作，逆功率跳闸
reverse-power 逆功率
reverse reaction 逆反应
reverse reactive flow 反相无功电流
reverse relay 逆流继电器
reverse remittance 逆汇
reverse return system 回水系统，同程系统
reverse rotation of the pump 泵反转，泵倒转
reverse rotation 反转，倒转
reverser 换向开关，反向器，换向器，反演机构
reverse saturation-current effect 反向饱和电流效应
reverse saturation current 反向饱和电流
reverse signal 反转信号，反向信号
reverse sign 反号
reverse slip-face 反滑落面
reverse slip fault 逆断层
reverse slope 逆坡
reverse spoon cutter 反向式料勺截样器【出口煤样】
reverse surge voltage 反向浪涌电压，反向冲击电压
reverse telescope 倒镜
reverse torsion machine 扭转疲劳试验机
reverse torsion test 反复扭转试验
reverse turbine 可逆式水轮机
reverse valve 可逆阀，双向阀
reverse voltage 反向电压
reverse 背面，反向，反转，颠倒，换向，倒转，使反向，反向的，相反的，逆向的，倒的
reversibility 可反向性，可逆性，可反转性
reversible absorption current 可逆吸收电流
reversible action 可逆动作，可逆作用，可逆运行
reversible adiabatic change 可逆绝热变化
reversible adiabatic compression 可逆绝热压缩
reversible adiabatic expansion 可逆绝热膨胀
reversible adiabatic process 可逆绝热过程
reversible binary counter 可逆二进制计数器
reversible boom conveyor stacker 可逆悬臂皮带堆料机
reversible boom conveyor 可逆悬臂皮带机
reversible booster 可逆增压机
reversible cell 可逆电池
reversible controller 可逆控制器，双向控制器
reversible counter 可逆计数器
reversible crusher 可逆破碎机
reversible cycle 可逆循环
reversible deformation 可逆变形
reversible drill 双向钻
reversible drive 可逆传动
reversible electrode 可逆电极
reversible expansion 可逆膨胀
reversible fan 可逆转风扇
reversible flow 可逆流
reversible gear train 可逆齿轮系
reversible generating set 可逆发电机组
reversible generator-motor 可逆式发电电动机
reversible globe valve 双向球阀
reversible hammermill 可逆锤式破碎机
reversible HVDC system 双向输电高压直流系统
reversible hydraulical machine 可逆式水力机械
reversible impact hammer crusher 可逆反击锤式破碎机
reversible level 回转式水准仪，可倒水准器，可逆水准仪，活镜水准仪
reversible magnetic process 可逆磁化过程
reversible motor 可逆电动机
reversible operation 可逆运转
reversible permeability 可逆磁导率
reversible process 可逆过程
reversible pumped storage station 可逆式抽水蓄能电站
reversible pump-storage unit 可逆式抽水蓄能机组
reversible pump-turbine 可逆式水泵水轮机
reversible ratchet 双向棘轮套筒扳手
reversible reaction 可逆反应，对行反应
reversible series starting rheostat 可逆串联启动电阻器
reversible speed control 可逆调速
reversible stepping motor 可逆式步进电动机
reversible switch 换向开关
reversible turbine 可逆式水轮机
reversible variable speed motor 可逆式变速电动机
reversible water meter 可逆式水表
reversible yard conveyor 可逆煤场皮带机
reversible 可反向的，可逆的，可反转的，可倒的，可转换的
reversing belt conveyer 逆带式输送机
reversing belt feeder 双向带式给料机
reversing blade 转向导叶片
reversing chamber 转向室
reversing commutator 电流方向转换器
reversing controller 可逆控制器，双向控制器
reversing device 反向装置
reversing engine 可倒转发动机
reversing gear 换向齿轮，回行机构
reversing key 换向电键
reversing load 反向负荷
reversing mill motor 可逆轧机电动机

reversing motor 向电动机，可逆电动机
reversing operation 可逆运行
reversing starter 可逆启动器，双向启动器
reversing switch contact 换接开关触点
reversing switch 反向开关，换向开关，换接开关
reversing thermometer 颠倒温度表
reversing turbine 可逆转式透平，倒顺车透平
reversing valve （凝汽器）反冲洗阀，逆洗阀，换向阀，可逆阀
reversing 反转，反向，颠倒，换向，倒转，逆转，回动
reversion 颠倒，返原，恢复，复原，反转
reverting call （同线用户间）相互呼叫
revertive impulse 反脉冲，回送脉冲，倒脉冲
revertive impulsing circuit 反脉冲电路
revert statement 回复语句
revert 恢复原状，回复
revetment-dike 护堤
revetment project 护岸工程，防洪工程
revetment wall 护墙
revetment 衬里，金属内衬，保护层，覆盖层，防浪堤，护坡，铺面，堑壕，护墙，护岸
revibration 再振动
review board 审查委员会
review conference 审查会议
reviewer 审核者，审阅人，审查人
review form 校审单
review of document 文件审查
review of tender 评审投标书
review personnel 评价人
review procedure 评审程序，审查程序
review 回顾，检查，探伤，观察，评论，评审，审查，审核
revised design method 校正设计法
revised design 改进设计，修正设计，已修改的设计，修改后的设计
revised edition 改版，修订版
revised price 修改后的价格
revised text 修订文本
revised version 修订版
revise 修订，校订，校正
revision block 修订框，修订记录栏
revision number 修改号
revision status 修订状况
revision survey 修测
revision 修订（版、本）【文件】，复核，修改，修正
revitalize 使恢复元气，使新生，使复兴
revived fault 复活断层
revivification 再生，复活
revivifier 再生器
REVLN＝revolution 转数
revlolving window 旋转窗
REV. No.（revision number） 修订版号
revocable connector 可拆接头
revocable letter of credit 可撤销信用证
revocable offer 可撤销的报价
revocable unconfirmed banker's credit 可撤销无保兑银行信用证
revocation of contact 解除合同
revocation 撤回，撤销，废除
revoke 撤销，废除，取消，吊销
revoluting field induction motor 旋转磁场式感应电动机
revoluting speed 转速
revolution axis 回转轴，转轴，旋转轴
revolution body 回转体
revolution counter 旋转计数器，转数表，计数器，转数计，转速计
revolution door 旋转门
revolution drop 转速下降
revolution ellipsoid 回转椭圆体，旋转椭圆体
revolution indicator 转数计，转数表，转数指示器
revolution meter 转数计，转速表
revolution of polar to Cartesian 极坐标-直角坐标转换
revolution period 公转周期
revolution per minute 转每分，每分钟转数
revolution speed transducer 转速传感器
revolution speed 旋转速度，转速
revolutions-per-minute indicator 每分钟转数指示器
revolutions per minute 每分钟转数，转每分
revolutions per second 转每秒
revolution surface 回转面
revolution 回转，转动，旋转，转数，转速，运行
revolved representation 回转图示法
revolved section 回转剖面
revolver crane 旋转起重机
revolver 溶剂，旋转器，解算器，快速访问磁道
revolve 旋转，循环
revolving antenna 旋转天线
revolving-armature type machine 旋转电枢电机
revolving-blade mixer 转叶式拌和机
revolving coil 旋转线圈
revolving crane 旋臂吊车，旋转式起重机
revolving credit 循环信用证
revolving door 旋转门，转动门，转门
revolving drum concrete mixer 转筒式混凝土搅拌机
revolving drum screen 转筒筛
revolving drum stroboscope 带有旋转鼓轮的频闪观测仪
revolving drum 转鼓，回转（圆）筒
revolving electro-motive force 旋转电动势
revolving excavator 旋转式挖土机
revolving field theory 旋转场理论
revolving field type machine 旋转磁场式电机
revolving field type motor 旋转磁场式电动机
revolving field 旋转场
revolving fish screen 旋转式鱼栅
revolving fuel manipulator crane 旋转式装料机
revolving fund 周转资金
revolving gantry crane 旋转移动门式起重机
revolving gate 旋转式闸门
revolving grate 转动炉排，旋转炉篦
revolving magnetic field 旋转磁场
revolving magnetic flux 旋转磁通量
revolving mirror stroboscope 带有旋转镜的频闪观测仪
revolving pillar jib support 旋转柱起重杆支座

revolving pipet stand　盘式吸液管架
revolving screen　旋转式拦污栅，旋转式筛，转动式滤网，转筒筛
revolving shovel　旋［回］转式挖土机，旋转式机铲
revolving shutter　卷帘百叶窗，卷筒百叶窗，转动闸门，回转断路器
revolving sprinkler head　回转式喷头
revolving sprinkler　旋转式洒水器
revolving stacker　旋转堆料机
revolving steam shovel　回转式汽铲，旋转蒸汽挖土机
revolving storm　热带风暴，旋转风暴
revolving vane　旋转式轮叶
revolving window　旋转窗
revolving　周期的，旋转的，循环的
REV=reverse　反向
REV=revision　修改，修改版
REV=revolutions　转数
reward distribution system　奖金分配制度
reward payment　奖金
reward　奖励，补焊，重焊，返修焊
rewetting　再湿
rewinding coil　重绕线圈
rewinding motor　重绕电动机
rewind switch　倒带开关，反绕开关，卷带开关
rewind　重绕，反绕，倒带，卷带
rewiring　重新布线，重新接线
rework cost　返工成本
reworking　返工【如焊接返工】
rework of flange seal faces　法兰密封表面的返工
rework of tapping　攻丝返工
rework　返工，重做，返修，再加
rewriting circuit　重写电路
REWR(read and write)　读写
REX(real-time executive routine)　实时执行程序
Reynolds analogy　雷诺相似，雷诺比拟，雷诺模拟，雷诺类比
Reynolds approach　雷诺方法
Reynolds averaged Navier-Stokes equations　雷诺平均纳维埃-托克斯方法
Reynolds averaging　雷诺平均
Reynolds criteria　雷诺数，雷诺准则
Reynolds criterion　雷诺数，雷诺准则
Reynolds critical velocity　雷诺临界流速
Reynolds equation　雷诺方程
Reynolds factor　雷诺因子
Reynolds law of similarity　雷诺相似律
Reynolds number based on chord　弦长雷诺数
Reynolds number based on diameter　直径雷诺数
Reynolds-number correction　雷诺数校正
Reynolds number effect　雷诺数效应
Reynolds number of turbulence　湍流雷诺数
Reynolds number range　雷诺数范围
Reynolds number similarity　雷诺数相似
Reynolds number　雷诺数
Reynolds similarity law　雷诺相似定律
Reynolds stress tensor　雷诺应力张量
Reynolds stress　雷诺应力
RF amplification　无线电频率放大，高频放大
RF bandwidth　射频带宽
RFB(response for bidding)　标书响应
RF channel synchronization　高频信道同步，射频信道同步
RF choke　射频扼流圈，高频扼流圈
RFC(return for correction)　退回改正【工程文件用语】
RFC(test refused, to be reperformed)　试验不合格，需重做
RFDH(reserve shut down forced derated hour)　强迫降低出力备用停机小时
RF dielectric heating　电介质射频加热
RF energy　射频能量
RF excited ion laser　射频激励离子激光器
RFF(reset flip-flop)　复位触发器，置“0”触发器
RF generator　高频发生器，射频发生器
RFID(radio frequency identification)　射频识别
RFI-immune　不受射频干扰的
RFI(radio frequency interference)　射频干扰
RFI(ready for implementation)　已准备好可以实施
RFI　运行结束报告
RF-PMSG(radial flux permanent magnet synchronous generator)　径向磁通永磁同步发电机
RFP(reactor feed pump)　反应堆给水泵
RFP(registered financial planner)　美国注册财务策划师学会
RFP(request for proposal)　征求报价，报价邀请书，招标书
RF presentation　射频显示，不检波显示
RFQ(request for quotation)　报价邀请书，招标书
RF(radial flow)　径向流动
RF(radio frequency)　无线电频率，射频
RF(raised face flange)　突面法兰，凸面法兰
RF(rate of flow)　流速
RF(read forward)　正向读出
RF(reserve free)　备用
RFR(reject failure rate)　抑制故障率
RF standard signal generator　射频标准信号发生器
RF switching relay　高频转换继电器，射频转换继电器
RF transformer　高频变压器，射频变压器
RFW(reserve feed water)　后备给水
RGER　再生式【DCS 画面】
RGL(full length rod control)　长棒控制系统
RGL　控制棒控制系统【核电站系统代码】
RG(reduction gear)　减速齿轮
RG(regulatory guide)　管理导则
RG(reset gate)　复位门
RHB(reheater by-pass)　再热器旁路
rheobase　基本电流强度，稳定的阴极电流强度
rheochord　滑线电阻器
rheograph　流变记录器，示波器
rheological behaviour　流变性状
rheological character　流变性
rheological measurement　流变测量
rheological model　流变模型
rheological properties　流变性质
rheologic theory　流变理论
rheology　流变学，液流学

rheometer 流速计，电流计，黏质流速计
rheometry 流变测定法
rheonome 电流强度变换器
rheopectic fluid 触变震凝流体
rheopexy 振凝
rheo＝rheostat 变阻器，电阻箱
rheoscope （电流）检验器，验电器
rheostan 一种高电阻合金
rheostat alloy 变阻器合金
rheostatic braking 电阻制动
rheostatic control 变阻控制，变阻调节
rheostatic excitation control 变阻器励磁控制
rheostatic regulator 变阻调节器
rheostatic starter 电阻启动器
rheostatic voltage regulator 变阻式电压调整器，变阻式调压器
rheostatic 电阻的，变阻器的
rheostat loss 变阻器损耗
rheostat slider 变阻器滑动触头
rheostat 可变电阻，变阻器，电阻箱，电阻器
rheostriction 箍缩效应，夹紧效应，紧缩效应，流变压缩
rheotome （周期）断流器，中断电流器
rheotron 电磁感应加速器
rheotrope 电流转换开关
rhexistasy 破坏平衡
rhe 流值【流度单位】
rhizosphere 根层区
rhodanine 绕丹宁，绕丹酸
rhodium 铑
rhogosol 粗骨土
rholite 熔剂
rhombic antenna 菱形天线
rhombic form 正交晶型
rhombic shingle 菱形瓦
rhombic symmetry 斜方对称
rhombic transformer 菱形变压器
rhomboid chain 菱形连锁
rhomboid 长斜方形，长菱形，平行四边形，长斜方形的，长菱形的
rhombus 菱形
rhometal 镍铬硅铁磁合金
rhone 排水槽
RH＝reheater 再热器
RH＝reheat 再热
RH(relative humidity) 相对湿度
RHR heat exchanger 停堆冷却器，余热排出热交换器
RH(Rockwell hardness) 洛氏硬度
RHR(residual heat removal) （反应堆停堆冷却时的）余热排出
RHRS(residual heat removal system) 余热冷却系统，余热排出系统
RHR system line 余热排出系统管线
RHR system pump 余热排出泵，停堆冷却泵
rhumbatron 环状共振器，空腔共振器
rhumb line 等角线
rhumbus algorithm 菱形算法
rhyolite 流纹岩
rhyometer 电流计
rhyotaxitic texture 流纹状构造
rhysimeter 流体流速测定计
rhythmic sedimentation 韵律沉积
rhythmic succession 韵律层序
rhythmic unit 韵律层，韵律单位
rhythmic 有节奏的，合拍的
rhythm 周期性的变动
RH 再热器【DCS 画面】
ria 沉溺河
riband stone 条纹砂岩
rib arch 肋拱，扇形拱
ribbed arch bridge 肋拱桥
ribbed arch 肋拱
ribbed bar reinforcement 竹节钢筋
ribbed bar 竹节钢，钢筋
ribbed bolt 起棱螺栓
ribbed floor slab 肋形楼板
ribbed floor 密肋楼板，肋构楼面，肋形楼板
ribbed frame 有散热筋的机座
ribbed motor 散热片型电动机
ribbed pipe 加肋管，肋片管
ribbed radiator 肋片对流散热器
ribbed sheet metal 带肋的钢板
ribbed shell 带肋薄壳
ribbed stiffener 加劲肋
ribbed surface machine 散热筋型电机，翅面电机
ribbed tube 内螺数管，肋片管，内壁螺纹管
ribbed 肋片的，加肋的
ribbing 肋条
ribbon antenna 带形天线
ribbon blender 螺旋叶片式混合器
ribbon bond 带状连接器
ribbon burner 带式（开孔）燃烧器
ribbon cable 带状电缆
ribbon clay 条带状黏土
ribbon coil 带绕线圈
ribbon conductor 带形导体，带形导线
ribbon course 带状瓦层
ribbon-cutting ceremony 剪彩仪式
ribbon-cutting 剪彩
ribbon fuse 熔线片，带状熔线
ribbon gauge 花带状应变片
ribbon iron 扁铁，带钢
ribbon mixer 螺条混合器
ribbon panel 回带管屏
ribbon resistance 金属带电阻，带状电阻
ribbon structure 带状结构
ribbon tape 皮带尺
ribbon winding 带绕组
ribbon-wound core 钢带绕的铁芯，卷铁芯
ribbon-wound pole 扁绕磁极
ribbon 带，条，带状电阻，散热片，饰条，卷尺
rib-cooled motor 散热筋冷却型电动机
rib flange 肋凸缘
rib member 肋杆
rib mesh 肋条钢丝网
rib of slab 板肋
rib reinforcement at the hole 洞口加筋
rib reinforcement 加强肋，加强筋
rib roughness of cooling tower 冷却塔加强肋粗糙度

rib snubber 外围掏槽眼
rib strip 肋条
rib vaulting 肋形拱顶，扇形肋穹顶
rib web 加强肋腹板
rib 加强筋，加强肋，肋片，加肋片，肋条，肋状物，筋，肋
rice 一种粒度非常细的无烟煤
rich coal 肥煤，化质煤，长焰煤
rich concrete 多水泥混凝土，富混凝土
rich experience 丰富的经验
rich gas 高热值煤气
rich mixture 富燃分混合物
rich mix 稠浆拌和物
Richter magnitude scale 里氏震级【地震】
Richter scale 里氏地震烈度，里克特震级，里氏震级【共分 10 级】
rich 丰富的，富有的，肥沃的，昂贵的
ricker 堆垛机
rick-filling 堆石护趾
RIC 堆芯测量系统【核电站系统代码】
ridding bubble chamber 消泡箱
riddle 粗筛，筛，筛子，筛滤器
riddlings hopper （炉排下）灰斗
riddlings return 漏煤回送
riddlings 漏落物，（炉排下）漏煤
rideograph 平整度测定仪
rider cap 承台
rider 跟车工，游码，制导器，横架，斜撑
ride-up 涌浪
ridge beam 栋梁，脊梁，屋脊梁木
ridgecap 脊瓦
ridge covering 脊盖
ridge crest 脊顶
ridged profile 起伏断面
ridge line 山脊线，屋脊线
ridge purlin 脊檩
ridge roof 有脊屋顶
ridge tile 脊瓦
ridge waveguide 脊形波导管
ridge 分水岭，山脊，脊，隆起线，垅，垄，屋脊，背，波峰，山岭，起皱纹
ridging 围埂
riding comfort 乘坐舒适性
Riemann space 黎曼空间
Rieman surface 黎曼面
rieselikonoscope 移像式光电稳定摄像管
riffled 有槽的
riffle sampler 试样划分器，分格取样器，格槽式缩样器，分割采样器
riffle 漕，沟，二分器，急流，刻痕，浅石滩，浅滩，沙沟，水面微波，压花
rifled tube water wall 内螺纹水冷壁
rifled tube 内螺纹管【来复线管】
RIF(reliability improvement factor) 可靠性提高系数
rift crack 心裂
rift valley 地沟，地堑，断缝谷，裂谷
rift 长狭谷，断陷谷，裂缝，裂痕，裂口
rig atmosphere composition 辐照装置内气体成分
rig base 钻架支撑座
rig boom 钻机臂杆
rig column 钻机柱架
rig for model test 模型试验台
rigger 装配工，束带滑车
rigging 索具，装配，安装，装备，绳索，调整，传动装置，悬挂
right and obligation 权利与义务
right-angled bend 直角弯管
right-angled elbow 直角形肘管
right angle drive 直角传动
right angle impulse 直角脉冲
right angle 直角
right arch 正拱
right ascension circle 赤经圈
right bank tributary 右岸支流
right circular cylinder 规正的圆柱体
right cylindrical reactor 正圆柱形反应堆
right elevation 右立面图，右视图
rightful notice 正式通知
right-half plane 右半面
right-hand adder 低位加法器，右移加法器
right-hand component 右侧数
right-hand coordinate system 右手坐标系统
right-hand derivative 右导数
right-hand door 右手门
right-handed wound 右向绕组
right-hand engine 右旋发动机
right-hand loose joint hinge 右边活节合页
right-hand rotating fun 右旋风机
right-hand rotation 右旋
right-hand rule 右手定则
right-hand screw rule 右手螺旋定则
right-hand screw 右旋螺丝
right-hand thread 右螺纹
right-hand winding 右向绕组
right-hand 右向的，顺时针方向的
righting force 还原力，恢复力
righting moment 扶正力矩
right of access 查阅权，知情权，接近权，近岸权，通行权，出入权
right of appeal 上诉权，索赔权
right of approval 批准权
right of claim 索赔权，追索
right of eminent domain 土地征用权
right of entry 进入权
right of monopoly 专利权，独占权
right of passage 通行权，通过权
right of patent 特许权，专利权
right of power use 用电权
right of priority 优先权
right of recourse 追索权
right of rescission 解约权
right of subrogation 代位求偿权
right of way clearing for pipeline 管线通廊的清障
right of way clearing for utility line 输电线路通廊的清障
right of way width 线路走廊宽度
right of way 通行权，路权，（管道的）先行权，杆线权，有架线权的地区，线路走廊
right-running characteristic 右伸特征线
right-running Mach wave 右伸马赫波
rights and interests 权益

right scale integration 适当规模集成，适当规模集成电路
right shift 向右移位
right side up 勿倒置
right side wall 右侧墙
rights reversion 权利复原
right to access 出入权
right to interpret 解释权
right to vary 变更权
right 垂直的，权利，右，正确
rigid alignment 精确对准
rigid analysis scheme 刚性方案
rigid arch dam 刚性拱坝
rigid arch 刚性拱
rigid armouring 刚性钢筋
rigid axle 刚性轴
rigid balancing 刚性平衡
rigid base 刚性基层
rigid beam 刚性梁
rigid bearing 刚性轴承
rigid bent frame 刚性排架
rigid body 刚体
rigid boundaries working section （风洞）刚性壁试验段
rigid boundary 刚性边界
rigid bracing 刚性撑杆
rigid busbar 刚性母线，硬母线
rigid casing 刚性外壳
rigid composite material 硬质复合材料
rigid conduit 刚性管道，刚性导管，刚性套管
rigid connection 刚性连接，刚接
rigid copper bus 硬铜母线
rigid core earth-rock dam 刚性心墙土石坝
rigid core wall 刚性心墙
rigid coupling 刚性联轴节，刚性联轴器，刚性连接
rigid design 刚性设计
rigid distortion 刚性扭曲
rigid-elastic analysis scheme 刚弹性方案
rigid element aeroelastic model 刚性组件气动弹性模型
rigid facing 刚性护面
rigid fastening 刚性固定，刚性连接
rigid fixing （电缆）刚性固定
rigid foam 硬质泡沫
rigid foundation 刚性基础
rigid frame bent 刚性排架
rigid framed structure 刚架结构
rigid frame with honeycomb-web 空腹刚架
rigid frame with plate-web 实腹刚架
rigid framework 刚性构架
rigid frame 刚构，刚架，刚性架，刚性构架
rigid girder 刚性梁
rigid grid 刚性网格
rigid hanger 刚性吊架
rigidimeter 刚度计
rigid insulation 硬质绝缘
rigidity coefficient 糙率系数，刚度系数，刚性系数
rigidity condition 刚性条件
rigidity criterion 刚度准则
rigidity factor 刚度系数，（同步电机的）整步功率
rigidity gear 刚性齿轮
rigidity index 刚度指数［标］
rigidity modulus 刚性模量，刚性模数
rigidity 刚度，刚性，强度，硬度，严格，刻板，僵化，坚硬
rigid joint 刚性接头，刚性节点，刚性结点，刚性连接
rigidly mounted blade 刚性安装的叶片
rigid magnet 硬磁铁
rigid material 刚性材料
rigid member 刚性构件
rigid membrane 刚性膜片
rigid mesh 刚性网
rigid metal conduit 刚性金属套管
rigid metal girder 刚度大的金属梁
rigid model on elastic base 弹性底座刚性模型
rigid model 定床模型，刚性模型
rigid motion 刚体运动
rigid nonmetallic conduit 刚性非金属套管
rigid particle 硬颗粒
rigid pavement 刚性路面，刚性铺面
rigid penstock 刚性压力水管
rigid plastic material 刚塑性材料
rigid plastic 硬质塑料
rigid plate 刚性底板
rigid polyvinyl chloride 硬聚氯乙烯
rigid PVC pipe 硬聚氯乙烯管
rigid PVC 硬聚氯乙烯
rigid reinforcement 刚性钢筋
rigid retaining wall 刚性挡土墙
rigid return 刚性回复
rigid roll 固定辊【辊碎机】
rigid rotator 刚性转动体
rigid rotor vane pump 刚性转子叶片泵【旋转泵型】
rigid rotor 刚性转子
rigid shaft 刚性轴
rigid steel conduit 钢制电线管，硬钢管
rigid structure 刚性结构
rigid support 刚性支架，刚性支座
rigid transverse wall 刚性横
rigid trestle 刚性支架
rigid tube 刚性管
rigid tunnel boundary 刚性风洞壁
rigid wake model 刚性尾流模型
rigid wall （风洞）刚性壁
rigid 严密的，坚硬的，刚性的，严格的
rigorous adjustment 严密平差
rigorous similarity 严格相似性
rig test 台架试验
rig 台架，装置，装配，装备，台，试验台，钻机，钻具
RI level 无线电干扰电平
rill drainage 毛沟排水，细流排水
rill erosion 细沟冲刷，细沟侵蚀
rill washing 细沟冲刷
rill 溪流，细沟，小溪
RIL(research information letter) 研究信息通信
rim bearing 环承，周缘支承
rim clutch 凸缘离合器

rim cooling 轮缘冷却
rim deposit 边沿沉积，边沿矿床
rime deposit 雾凇
rim effect 轮圈效应
rime fog 雾凇，雾
rime ice 霜冰
rim exhaust 槽边排风罩
rime 霜凇，白霜
riming 结凇
rim keying （水轮发电机）磁轭打键
rim lamination （水轮发电机）磁轭叠片
rimmed glass cylinder 带边玻璃量筒
rimmed steel 净缘钢，未静钢，沸腾钢，不脱氧钢
rimming steel ingot 沸腾钢钢锭
rim of guide blading 导向叶片环
rim shaft 主轴
rim speed 轮缘速度
rim stacking 磁轭叠压，（水轮发电机）磁轭堆叠
rim stress 轮缘应力
rim velocity 轮缘速度，轮周速度
rim ventilation 槽边通风
rim wrench 轮缘扳手
rim 轮缘【枞树型叶根槽】，边框，边缘，轮，边，缘，齿圈，垫环，磁轭，边沿，齿环
ring-and-ball apparatus 环球仪
ring-and-ball method 环球试验法【用于测定沥青软化点】
ring-and-ball point 沥青软化点
ring armature 环形电枢
ring-back key 回铃键，回叫键
ring-back tone 回铃音
ring badge 胶片剂量计
ring balance manometer 环秤压力计
ring-ball mill 中速球磨机，E型磨煤机
ring beam 环梁，圈梁
ring burner 环形燃烧器
ring busbar 环形母线，环形汇流排
ring-bus 环形母线
ring cable circuit 环形电缆回路
ring canal 环管
ring casing pump 环壳泵
ring casing 环形泵壳
ring circuit operation 环路运行
ring circuit 环路，环形电路，回路
ring coil 环形线圈，电铃线圈
ring collector 集电环
ring compression 环压力
ring connection 环形接法，环联
ring connector 联结环
ring core reactor 环状堆芯反应堆
ring core 环形铁芯
ring counter 环形计数器
ring course 拱圈层
ring cover 环形盖
ring crusher 环式破碎机
ring current-transformer 环形电流互感器
ring current 环流，环形电流
ring-cutting 环切
ring dam 环形坝
ring discharge 环形放电
ring distribution system 环形配电网
ringdown signaling 振铃信号，传送信号，低频监察信号
ringdown system 振铃信号制
ring drilling 环向钻孔法
ring earth external 杯形接地体
ringed network 环形网络
ring electrode 环形电极
ring electromagnet 环形电磁铁
Ringelman chart 林格曼煤烟浓度测定表，标准烟色图，林格曼烟气图，烟色比较图，烟气比较图
Ringelman's smoke chart 林格尔曼排烟浓度表，林格尔曼烟浓度图
ring embankment 月牙堤
ringer bay 振铃机架
ringer bell 电铃
ringer oscillator 铃流发生器，振铃信号振荡器
ringer test panel 振铃器测试盘
ringer 电铃，振铃机，信号机
ring-face gasket 环形垫片
ring feeder 环形干线，环式馈路，环形馈线
ring filament 环形灯丝
ring filler 垫环
ring forging 锻环，环状锻件
ring form coil 环形线圈
ring form turn 环形线匝
ring foundation 环形基础
ring gauge 环规
ring gear diameter 被动齿轮直径
ring gear 内齿圈，齿圈，大牙轮，环形齿轮
ring generator 环式发动机
ring girder 环形主梁
ring groove 环形槽
ring ground 环形接地装置
ring hammer 环锤
ring header 环状集管，环形集箱，环形联箱，汇流环【液体】
ring housing 环套
ring infiltrometer 环式测渗仪
ringing cut-off relay 铃流切断继电器
ringing dynamo 铃流发电机
ringing experiment 振铃实验
ringing fail alarm 振铃信号故障报警
ringing generator 铃流机，电动磁石发电机，铃流发电机
ringing key 振铃键，呼叫键
ringing period 振铃时间，呼叫时间
ringing pilot lamp 呼叫指示灯，振铃指示灯
ringing position 振铃位置，呼叫位置
ringing relay 呼叫继电器，振铃继电器
ringing set 铃流机组，振铃装置
ringing signaling 电铃信号装置
ringing test 呼叫试验，振铃试验
ringing tone 回铃音，振铃音，呼叫信号
ringing transformer 振铃变压器
ringing-trip relay 呼叫电流切断继电器，振铃切断继电器
ringing wave 呼叫波
ringing wire 振铃线，呼叫线

ringing 环绕，环形接线片，振铃，阻尼振荡，减幅振荡
ring insert 环形插入件
ring joint face 环（连）接面
ring-joint facing 环接密合面
ring-joint flange 环接法兰
ring joint metal gasket 环形连接金属垫片
ring joint seal O形密封
ring joint 环接，围缘接合，环结件，环型接头
ring levee 月堤
ring-like foundation 环形基础
ring line 环行管路
ring liquid cooler 环形液体冷却器
ring liquid heat exchanger 环形液体热交换器
ring liquid strainer 环形液体过滤器
ring liquid tank 环形液体箱
ring load 环路
ring lug 环形接线片
ring magnetic circuit 环形磁路
ring magnet 环形磁铁
ring main unit 环网开关柜
ring main 环形干线，环形管线，环形主管
ring manometer 环形压力计
ring micrometer 圆径测微计
ring modulator 环形调制器
ring motor 环形电动机
ring nozzle chest 叶轮全周进汽喷嘴室，环形喷嘴室
ring nozzle 环状管嘴
ring-off button 拆线按钮，话终按钮
ring-off lamp in pair 话终指示对灯，拆线指示对灯
ring-off lamp 拆线指示灯，话终指示灯，挂机指示灯
ring-off signal 话终信号
ring off 话终振铃
ring of ten circuit 十进位环形电路，十进制环形电路
ring-oiled bearing 油环润滑轴承
ring-oiler 润滑油杯
ring-operated network 环形运行的电网
ring operation 环式运行
ring-out 呼出振铃
ring pellet 环状芯块
ring piston servomotor 环形活塞继动器，环形活塞接力器
ring plate 环形板
ring reinforcement furnace 环形加强炉
ring reinforcement 增强环，增强圈
ring rim type rotor 环形磁轭转子
ring road 环路，环形道路
ring-roller mill 中速辊式磨煤机
ring-roll mill 中速平盘磨煤机
ring route 环状道路
ring sampler 环刀取样器
ring seal gate 环形支水闸门
ring seal 密封环
ring section feed pump 环段式给水泵
ring shake 环裂
ring-shaped core 环形铁芯
ring-shaped heat-supply network 环状供热管网
ring-shaped shield 环形屏蔽，环形防护屏
ring shear apparatus 环形剪力仪
ring shear test 环剪试验
ring shielding 环形屏蔽
ring slogging spanner 梅花敲击扳手
ring spanner 梅花扳手，环形扳手
ring sparger 喷淋环，配水环，环形喷淋器
ring stiffened semi-monocoque construction 加筋半硬壳结构
ring stoker 环形旋转炉排
ring stone 楔块
ring strain gauge 环式应变计
ring support 环形支座［支架］
ring switch 环形开关
ring tank 环形水池
ring tension 周边拉力，周边应力
ring-to-ring voltage 滑环间电压
ring transformer 环心变压器，环形变压器
ring-type adder 环形加法器
ring-type coal crusher 环式碎煤机
ring-type crusher 环式破碎机
ring-type element 环形元件
ring-type hammer mill 环锤式破碎机
ring-type head 环形头
ring-type nozzle chest 全周进汽喷嘴室
ring-type scraper 环形刮泥器
ring-type transformer 环心变压器，环形变压器
ring-type 环形的
ring valve 环形阀
ring vortex 环形涡，涡环
ring wale 环形横梁【支撑用】，环撑
ring waste 淋洗废液
ring water supply inlet 清洗供水入口
ring wear 环形磨损
ring whirl 环形涡旋
ring winding 环心线圈，环心绕组，环形绕组，环形绕法
ring 环，环形，环形物，环形电路，环绕，计数环，圆形环，包围，吊环，环段，圈
RINPO(Research Institute of Nuclear Power Operation) 核电运行研究院【武汉105所】
RIN(reference identification number) 基准标识号
rinse bed 淋洗床
rinsed resin 冲洗好的树脂
rinse period 清洗周期
rinse water supply inlet 升液管，上升管，提升管，溢流管
rinse 冲洗，漂洗
rinsing water 冲洗水
rinsing 漂洗，淋洗，清洗
riot and civil commotion insurance 暴动及民变险
riot and civil commotion 暴乱及民变，骚动及内乱
riparian land 沿岸地带，沿岸地
riparian pump house 岸边泵房
riparian right 沿岸使用权
riparian state 沿岸国家
riparian water right 沿岸地用水权
riparian works 河岸工程，治水工程
riparian 河岸的，水边的
rip channel 裂流水道

rip cord　气囊拉索，线头拉索
rip current　岸边回流，离岸流，裂流
ripe sludge　熟污泥
ripe wood　成熟木材
rip head　离岸流头
rip neck　离岸流颈
ripper dozer　松土推土机
ripper rooter　犁土机
ripper　松土机，耙路机，粗齿锯，裂具
ripping device　释放装置
ripple bedding　波痕层理，流纹层理
ripple contain　波纹系数
ripple current motor　脉动电流电动机
ripple current　弱脉动电流，波纹电流
rippled　波动的
ripple electromotive force　轻微脉动电动势
ripple eliminator　灭波器
ripple factor　波纹因数，脉动系数，波纹系数，脉动因子，波纹因子
ripple filter　脉动滤波器，平滑滤波器，平流滤波器
ripple flow　波纹流
ripple frequency　脉动频率，脉冲频率，波纹频率
ripple in output　输出脉动
ripple mark of current　水流波浪
ripple mark　波痕，波迹，涟痕
ripple noise　波纹电压噪声，电源交流噪声
ripple percentage　波纹百分数，脉动百分数
ripple quantity　脉动量
ripple ratio　脉动系数
ripple through carry　行波传送进位
ripple voltage　波纹电压，脉动电压
ripple waveform　脉动波形
ripple wave length　纹波的波长
ripple　焊缝波纹，波纹，使成波浪形，焊波，波动，脉动，鳞纹
rippling through　行波传送
rippling　跃移沙波
riprap breakwater　抛石防波堤
riprap protection of slope　乱石护坡，抛石护坡
riprap works　抛石工事
riprap　防冲乱石，防冲抛石，海漫，乱石护坡，抛石，石块
rip rooter　犁土机
rip surf　离岸流
rip tide　浪潮
rip　急流河段，裂浪
RI(radio influence)　射频感应
RI(radio interference)　无线电干扰
RI(reliability index)　可靠性指标
RISC(reduced instruction set computer)　精减指令集计算机
rise above grade　提高精度等级
rise amount　潮升量
rise-and-fall pendant　升降式吊灯
rise and run ratio　踏步级高与踏步宽比
rise factor　增长因子
rise in price　涨价
rise of a flight　梯段高度
rise of flood　洪水上涨
rise of ground-water level　地下水位上升高度
rise of span　跨高
rise of watertable　地下水位上升高度
rise ratio　升高比
riser characteristic　上升特性
riser circuit　上升回路
riser diagram　立管图
riser flow　上升流
riser height　踏步高度
riser leg　上升管支管
riser pipe　立管，上升管，直立管
riser reactor　提升管反应器
riser shaft　主井，提升井，提升器轴
riser tube　上升管，引出管
riser type cyclone　上升管型旋流分离器
riser　垂直井【坝内的】，冒口【铸件的】，垂直管，挡步板，（锅炉）上升管，立管，惯性重块，楼梯踢板，起步板，升降器，溢水口，提升器，90°立管
rise-span ratio　高跨比
rise time constant　（反应性扰动）上升时间常数
rise time indicator　上升时间指示器
rise time jitter　上升时间跳动
rise time　升压时间，升起时间，上升时间【冲击波】，建立时间，前沿时间
rise velocity　上升速度
rise　楼梯级高，上升，上涨，升高，踏步高度，梯段高度，提升，斜坡，增长，矢高
rising air　上升气流
rising bubble　上升气泡
rising characteristic　上升特性
rising coast　上升岸
rising edge　前沿，上升边
rising flow　上升流
rising gust　上升阵风
rising height of plume　羽流抬升高度
rising height of smoke plume　烟上升高度
rising limb of hydrograph　过程线上升段
rising limb　过程线上升段，涨水线段
rising magnetization curve　基本磁化曲线，上升磁化曲线
rising main　上升总管，竖管
rising performance curve　上升性能曲线
rising period　上升周期
rising pipe　竖管，出水管
rising plume　抬升羽流
rising stem　上升的阀杆，升杆式【明杆】
rising tide　涨潮
rising transient　上升瞬态
rising void　空隙度增大
rising whirl　上升旋涡
rising　上升的
risk allocation　风险分担，风险分配
risk analysis　风险分析
risk assessment　风险评估，风险分析，风险评价
risk bearing　承担风险的
risk consideration　风险的考虑
risk cost　保险费
risk coverage note　投保单
risk estimate　危险度估价
risk evaluation　风险评价，危险度评价

risk factor　危险系数，危险度
risk framework　危险模式
risk identification and allocation　风险的辨识和分配
risk investment　风险投资
risk level　风险水平，危险水平
risk management　风险管理
risk note　暂保单，保险凭条
risk of breakage　破损（风）险
risk of contamination　污染险
risk of damage　损坏风险
risk of error　过失险
risk of leakage　泄漏风险，泄漏险
risk of loss　损失风险，损失险
risk of oil damage　油渍险
risk of rust　生锈险
risk of shortage　货差险
risk of sling damage　吊索损险
risk of sweat damage　汗潮损失险，吊索损险
risk of warehouse to warehouse　仓库到仓库险
risk premium　风险费
risk prophecy　风险估计
risk quantification　风险定量评估
risk reduction factor　风险降低因数
risk reserve funds　风险准备基金
risk upper bound　危险度上限，危险度上界
risk　冒险，风险，危险（率）
RIS(recovery information set)　恢复信息组
RIS　安全注入系统【核电站系统代码】
RITE(rapid information technique for evaluation)　快速信息评价技术
RIT(rate of information throughput)　信息传送速率
rival　竞争者，对手
rivel varnish　皱纹清漆
river abstraction　导流
river and forest conservation　水利及森林资源保护
river authority　内河管理局
river-bank erosion　河岸冲蚀
river-bank spillway　河岸式溢洪道
river bank storage　河岸储水
river bank　河岸，河堤
river barge　河驳
river barrage　拦河坝，拦河建筑物，拦河堰
river basin balance　流域（水量）平衡
river basin development　流域开发
river basin model　流域模型
river basin planning　流域规划
river basin　（河流）流域
river beach terrace　河滩台地
river bed sorting　河床分级
river beheading　夺流
river bluff　河边陡岸
river board　河流管理局
river boil　涡流
river bottom deposit　河底沉积物
river bottom protection　护底
river bottom　河床，河底
river branch development　支流开发
river branch　支流
river cable　过河电缆
river capture　夺流
river closure　施工截流
river course　河道
river crossing facility　跨河设施
river crossing levelling　跨河水准测量
river crossing section　跨河段
river crossing　跨河线，过河
river deposit　河流沉积
river diversion arrangement　导流布置
river diversion tunnel　导流隧洞
river diversion　导流
river drainage system　河道排水系统
river driving　木材流运
river dynamics　河流动力学
river fan　河流冲积扇
river fork　汊河
river gauge　水标尺
river gauging　河川水文测验
river improvement　河道整治
river junction　汇流点
river marsh　河边沼泽
river of movable bed　动床河流
river outlet works　河口工程
river outlet　汇流处
river passing　支流
river piracy　夺流
river plain　河流平原，冲积平原
river potential analysis　河势分析
river profile　河流纵剖面
river robber　夺流河
rivers confluence　汇流点
river side face　临水面
river side power house　岸边式厂房
river side slope　临水岸坡，外坡，迎水坡
river side wharf　临水码头
river side　河岸，河边
river site　河边厂址
river span　河跨
river station　河道测站
river suspended drift　河流漂浮物质
river system　水系
river type pumping plant　河岸式油水站
river valley development project　流域开发规划
river valley reclamation project　流域垦殖规划
river wall　河堤
river water treatment plant　河水处理厂
river water treatment　河水处理
river weir　拦河坝，拦河堰
river width　河宽
rivet allowance　铆孔留量
rivet buster　铆钉铲，铆钉截断器，一种铲除铆钉头的工具
rivet connection　铆钉连接，铆接
rivet driver　铆钉机
riveted boiler　铆接锅炉
riveted bond　铆接
riveted butt joint　铆钉对接
riveted-drum boiler　铆接汽包锅炉
riveted girder　铆接梁
riveted joint　铆接
riveted seam　铆钉接缝，铆接缝

riveted steel structure 铆合钢结构
riveted truss 铆接桁架
riveted weld joint 铆焊边接
riveter 铆钉枪，铆机，铆钉机，铆工
rivet gun 铆钉枪
rivet holder 铆钉夹具
rivet hole 铆钉孔
riveting clamp 铆钉夹具
riveting gun 铆钉枪
riveting hammer 铆钉撑锤
riveting machine 铆钉机
riveting press 压铆机
riveting punch 铆接冲孔机
riveting 铆，铆合，铆接，铆接法
rivet in single shear 单剪铆钉
rivet joint 铆接
rivet length 铆钉杆长度
rivet line 铆钉线
rivet list 铆钉表
rivet pitch 铆钉间距，铆距
rivet snap 铆头模
rivet spacing 铆钉间距
rivet test 铆钉试验
rivetting 铆，铆接
rivet 铆钉，铆接
rive 碎片
RIV level(radio interference voltage level) 无线电干扰电平
RIV measurement 无线电干扰电压测量
rivulet 溪流
RJEP(remote job entry protocol) 远程作业输入协议
RJE(remote job entry) 远程作业输入
RJP(remote job processing) 远程作业处理
RJS(remote job service) 远程作业服务
RLC network 阻感容网络
RLD(relocation dictionary) 重新配位表
RL(relay logic) 继电器逻辑
RL(return loss) 回波损失
RMA(random multiple access) 随机多路存取
RMB 人民币
RMCC(remote control channel) 远程控制信道
R. M. C. (reverse mud circulation) 反向泥浆循环
RMDH (reserve shut down maintenance derated hours) 维护降低出力备用停机小时
RMF(reactivity measurement facility) 反应性测量设备，反应性测量装置
RMI(reliability maturity index) 可靠性老化指标
RMM(read-mostly memory) 主读存储器
RMOS(refractory metal-oxide-semiconductor) 耐热的金属氧化物半导体
RM(radiation monitor) 辐射监测仪
RM(read memory) 读存储器
RM(reference material) 参考材料，标准样品
R&M(regulation and metering station) 调压计量站
RM(research memorandum) 研究备忘录
RMS current 有效电流，电流有效值，均方根电流
RMSE(root mean square error) 均方根误差
RMS error 均方根误差
RMS-horsepower method 均方根马力法
RMS meter 均方根测量仪
RMS power 均方根功率，有效功率
RMS pressure coefficient 均方根压力系数
RMS(radiation monitoring system) 核辐射监测系统
RMS(recovery management support) 恢复管理支援
RMS(remote monitoring system) 远距离监测系统
RMS(root-mean-square) 有效值，均方根，均方根的，均方根值
RMS sound pressure 均方根声压
RMS value 有效值，均方根值，均方根，均方值
RMS voltage 有效电压，电压有效值，均方根电压
RMWS(reactor makeup water system) 反应堆补水系统
RNBK＝runback 快速减负荷
rneasurement circuit 量测线路
rnechanical stoker 机动加煤机
RN meter(radio noise meter) 无线电干扰仪
RN(reconvergent node) 再收敛节点
RN(Reynolds number) 雷诺数
RNS(normal residual heat removal system) 常规余热排放系统
road and railroad crossing 公路铁路交叉
road approach 桥梁引道，桥头路，引桥路
road asphalt 筑路沥青
road axis 道路中心线
road barricade 路障
road-bed 路基，明道床
road bend 道路弯道
road border 路缘
road capacity 道路容量，道路通行能力
road construction machine 筑路机械
road crossing 道路交叉点，跨公路，道路交叉
road crown 路拱
road curb 路缘（石）
road curve 道路曲线，道路弯道
road diversion 绕行道路
road drag 路刮
road drainage 道路排水
roaded catchment 道路型集水区
road embankment 路堤
road engineering 道路工程
road excavator 筑路挖掘机
road finisher 路面整修机
road frost-heave 道路冻胀
road grader 平路机
road grade 道路坡度
roadheader 综掘机【巷道掘进机】
road heater 路面加热器
road intersection 道路交叉
road junction 道路交叉点
road lamp 路灯
road levelness 路面平整度
road lighting fixture 路灯
road-making plant 成套筑路机
road metacentre 筑路碎石料
road network 道路网
road oil 铺路油【指慢凝沥青】

roadpacker 夯路机
road relocation 道路改线
road rerouting 道路改线
road roller 压路机，轧道机，（压）路碾
roads and utility networks 道路和电力网
road scraper 刮路机，筑路铲运机
road shoulder 路肩
road side ditch 路边明沟
roadside erosion control 路侧防冲措施
roadside power plant 路边电厂
road sign 路标
road sprinkler 道路洒水车
roadstead 碇泊区，锚泊区，锚地，停泊所
road subgrade 路基
road surface roughness 路面平整度
road surface 路面，面层
road tamping roller 路面填压滚筒
road tar 筑路用柏油
road vibratory roller 振动压路机
roadway drainage 路面排水系统
roadway light 路灯
roadway maintenance 路面养护
roadway side 平洞侧墙
roadway 道路，路幅，车行道
road works 道路工程
road 道路，公路，路
roak 表面缺陷
roaster 焙烧炉
roasting in air 氧化焙烧
roasting of sulfide ores 硫化矿焙烧
roasting test 焙烧试验
roasting 焙烧
roast 焙烧，焙烧生成物
roattion 旋转
robber ball distributor 胶球分配器
robber 抢夺
Robinson bridge 鲁宾逊电桥
Robitzsch bimetallic actinography 罗比兹双金属片总日射计
Robon glass 一种防热玻璃
robot device 自动装置
robotic arm 机械臂
robotics 机器人学，自动化学
robotic tool carrier arm 自动工具支架臂
robotization 自动化
robotized plant 自动化工厂
robotized 自动化的
robot scaler 自动计算装置，自动换算装置
robot vehicle 机器人车
robot 机器人，机械手，自动机，遥控设备
robust control 鲁棒控制，强健控制
roches moutonnee 羊背石
rock analysis 岩石分析
rock anchor foundation 岩锚基础
rock anchor 岩地锚
rock and earth engineering 岩土工程
rock and soil erosion 岩土侵蚀
rock asphalt pavement 石沥青路面
rock asphalt 石沥青，岩沥青，天然沥青
rock attitude 岩态
rock avalanche 岩石崩塌，岩崩
rock bank 石岸
rock bar 石梁，岩坝
rock basin 石盆地，岩盆
rock beam 石梁，岩槛
rock bed 岩床
rock bench 岩台
rock bit 凿岩机钻头
rock blasting 岩石爆破
rock block 块石，岩块
rock body 岩体
rock bolt 岩石锚栓
rock borer 钻岩机
rock boring device 岩石钻孔设备，钻岩装置
rock bottom price 最低底价
rock breaker 碎石机，岩石破碎机
rock bridge 岩桥
rock burst 岩爆
rock cave 石窟
rock chamber 岩石洞室，渣室
rock chunk 石碴，石块
rock classification 岩石分类
rock cleavage 岩石劈理
rock cliff 悬岩，岩崖
rock compression 岩石压缩
rock core barrel 岩芯管
rock core bit 取（岩）芯钻头
rock core recovery 岩心获得率
rock core sampler 岩芯取样器
rock core 岩芯，岩心，岩石核心
rock cover 岩石覆盖
rock creep 岩石蠕动
rock crosscut 石门
rock crusher 岩石破碎机，轧石机
rock-crushing plant 碎石厂
rock crystal 岩晶
rock-cut building 石窟
rock-cut job 凿石工作
rock cutter 割岩机
rock debris 岩屑
rock deformation 岩石变形
rock discontinuity structural plane 岩体结构面
rock drilling machine 钻岩机
rock drill 凿岩机
rock dyke dam 堆石坝
rock dyke 岩脉
rock embankment 堆石堤，石堤
rocker arm of gearbox 齿轮箱摇臂
rocker arm 摇臂
rocker bar 摇杆
rocker car 侧翻料车，翻转料车
rocker mechanism 摇杆机构
rocker-mounted ring girder support 摆柱支座
rocker shovel 翻斗铲
rocker switch 摇杆开关
rockery 假山
rocker 摆杆，摇移器，摇杆，刷架，摇轴
rocket engine 火箭发动机
rocket meteorograph 火箭气象计
rock excavation 采石工程，石方工程，石方开挖，挖石工程
rock excavator 岩石挖掘机

rock explosion 石方爆炸
rock exposure 裸露岩石，岩石露头
rock fabric 岩石结构
rock-faced dam 堆石护面坝
rock fall 岩崩
rock fan 石质扇形地
rockfill breakwater 堆石防波堤
rockfill cofferdam 堆石围堰
rockfill dam with asphaltic concrete core wall 沥青混凝土心墙堆石坝
rockfill dam with concrete facing 混凝土斜墙堆石坝
rockfill dam with vertical clay core 垂直黏土心墙堆石坝
rockfill dam 填石坝，堆石坝
rockfill dike 堆石堤，填石坝
rockfill diversion weir 堆石导流堰，堆石引水堰
rockfill drain 堆石排水体，填石排水沟
rock-filled crib weir 木笼填石堰
rock-filled jetty 堆石突堤
rock-filled timber crib dam 木笼填石坝
rock fill embankment 填石坝
rock fill foundation 填石基础
rock fill groin 堆石丁坝
rock fill groyne 堆石丁坝
rock-filling 堆石
rock fill platform 堆石平台
rock fill revetment 填石铺面
rock fill riprap 堆石护坡
rock fill spur dike 堆石丁坝
rock fill timber crib 叠木石笼
rock fill 填石，堆石（体），岩块填料，堆石填方
rock fissure 岩石裂缝
rock flour 岩粉
rock-flowage zone 岩石流动带
rock foreshore 岩滩
rock formation 岩层，岩系
rock foundation 岩石基础，岩石地基，岩基
rock fracture zone 岩石裂隙带
rock fragment 岩屑
rock gap 岩隙
rock gas 天然气
rock glacier 石流
rocking-contact speed regulator 振动接触式速度调节器
rocking-contact 振动接触的
rocking grate 摇动炉排
rocking-sector regulator 扇形摆动式电压调整器
rocking 翻转【机械】，倾卸，倾翻
rock ledge 岩礁
rock-magma 岩浆
rock mass factor 岩体系数
rock massif 山丘，整体岩块
rock mass structure 岩体结构
rock mass 岩体
rock material 石料
rock mechanics 岩石力学，岩土力学
rock melting 岩石溶化
rock mole 堆石突堤
rock-mound breakwater 堆石防洪堤，抛石防波堤
rock oil 石油
rock outcrop 岩石出露面
rock perviousness 岩石透水性
rock picker 碎石镐，碎岩风镐
rock pillar 岩柱
rockpinning 岩石插筋锚固
rock pin 岩石插筋
rock plug blasting 岩塞爆破
rock plug 岩塞
rock pocket 岩穴
rock pore 岩石孔隙
rock powder 岩粉
rock pressure 山岩压力，岩石压力
rock prestressing 岩石预加应力
rock product 石制品
rock property 岩石特性，岩土性质，岩性
rock quality designation 岩石质量标识，岩石质量指标，岩性符号，岩性指标
rock quarrying 石料开采
rock rake 耙岩机，清岩耙
rock rating 岩石分级
rock removal 除去岩石
rock revetment 块石护岸
rock ridge 岩脊
rock riprap 乱砌石块，抛石，填石
rock salt mine 岩盐矿
rock salt 岩盐
rock series 岩系
rock shovel 岩铲
rock slide 塌方，岩滑，岩石滑坡
rock-socketed bored pile 钻孔嵌岩桩
rock stratification 岩石层
rock stratum 岩层，岩石层，石流
rock stream 石流
rock strength 岩石强度
rock stress measurement 岩石应力量测
rock tar 石油，原油
rock test 岩石试验
rock toe 块石坝趾
rock tunnel 石隧道
rock underdrain 填石暗沟
rock waste 岩碴
rock weathering 岩石风化
Rockwell apparatus 洛氏硬度计
Rockwell A scale 洛氏硬度A标，洛氏硬度A级
Rockwell hardness number 洛氏硬度值
Rockwell hardness scale 洛氏硬度标尺
Rockwell hardness tester 洛氏硬度计
Rockwell hardness test 洛氏硬度试验
Rockwell hardness 洛氏硬度
Rockwell hardometer 洛氏硬度计
Rockwell number 洛氏硬度值
Rockwell superficial hardness number 表面洛氏硬度值
Rockwell superficial hardness test 表面洛氏硬度试验
rock wool packing 矿石棉包装
rock wool 矿石棉，矿毛绝缘纤维，矿物棉，矿渣棉，岩棉
rocky beach 岩滩

rocky bottom 岩底，岩石基底
rocky coast 岩岸
rocky harbour 岩岸港，岩滨港
rocky reef 岩礁
rocky shallows 石滩
rocky shore 岩岸
rocky soil 石质土
rocky subsoil 石质底土
rocky 石质的，石，尾矿，岩石，基岩，底岩
ROC(rate of change) 变化率
rod array 棒布置
rod assembly （燃料）棒组件
rod bank overlap 棒组重叠
rod bank position control system 控制棒组位置控制系统
rod bank position value 棒组位置值
rod bank 控制棒组
rod bottoming 棒端触底
rod bow penalty 燃料棒弯曲损失因子
rod bow 棒弯曲
rod bundle 棒束
rod calibration 测杆校准
rod centre pitch 棒中心节距
rod chopping 棒截断
rod cluster control assembly 控制棒束，棒束控制组件
rod cluster control changing fixture 控制组件抽插机构，控制组件转换装置，控制棒束装卸装置
rod cluster control handling and changing fixture 控制束棒装卸与更换固定件
rod cluster control handling station 控制棒装卸站
rod cluster control storage rack 控制棒贮藏架
rod cluster control thimble plug tool 束棒控制阻力塞抽出工具
rod cluster control 棒束控制
rod cluster type fuel assembly 棒束型燃料组件
rod cluster 棒组，（燃料元件）棒束
rod configuration 棒布置
rod control system 棒控制系统
rodded end 钢筋加固端
rodded system 棒状系统
rod deviation 棒位置偏差
rodding （混凝土的）通条，用棒捣实，砂心骨，使小圆棒通过钢管内以检查钢管内径
rod drive mechanism 控制棒驱动机构
rod drive screw 棒驱动螺杆
rod drop accident 落棒事故
rod drop experiment 落棒实验
rod drop method 落棒法
rod drop time 落棒时间
rod drop 落棒，紧急停堆，控制棒插入
rod ejection accident 弹棒事故，控制棒弹出事故
rod ejection transient 弹棒瞬态
rod ejection 棒弹出，控制棒弹出
rod end cap 棒端帽
rodent-proof cable 防啮电缆
rodent-proof 防蚀的
rod fabrication 棒生产
rod float 测流杆，杆式浮标
rod follower （压水堆）随动棒
rod for stilling water 稳水杆
rod gap 棒间隙，棒状放电器
rod geometry 棒几何位置，棒几何形状
rod guide plate 棒导向板
rod guide tube 控制棒导管
rod guide 控制棒导管，棒导向装置
rod holding plate 棒夹持板
rod holding 控制棒保持
rod insertion 棒插入，控制棒插入
rod insulator 棒型绝缘子
rod interval 标尺间隔
rod lattice pitch 燃料棒栅距
rod lattice 燃料棒栅格
rod loading 棒装填
rodman 标杆员，司尺员
rod mill 棒磨机，轧辊，辊轧机，辊碾磨机
rod motion 棒运动，棒过程
rod mounting adapter 棒支承接头
rod oscillation experiment 控制棒振荡实验
rod outside diameter 棒外径
rod pattern 棒布置形式
rod-plane gap 棒板间隙
rod position annunciation 棒位置信号通知
rod position detector coil 棒位置指示器线圈
rod position indicating system 棒位置指示系统
rod position indicator 棒位置指示器
rod position measurement detector 棒位置探测器
rod position measurement system 棒位置测量系统
rod power 棒线功率
rod pulling fixture 拉棒夹具
rod reading 标尺读数
rod removal 提棒
rod retention force 棒夹持力
rod-rod gap 棒棒间隙
rod seizure accident 卡棒事故
rod shadow effect 控制棒的荫蔽作用
rod shadowing 棒荫蔽
rod shaped fuel element 棒状燃料元件
rod shoot out accident 弹棒事故
rod sounding 杆测法【水深探测】
rod spacing 钢筋间距
rod stop command 控制棒停止命令
rod stop 控制棒停止
rod stroke 棒行程
rod support 水准尺垫
rod surface temperature 棒表面温度
rod suspended current meter 悬杆式流速仪
rod suspension 杆式悬装
rods with roughness 带有粗糙度的燃料棒
rod thermistor 棒形热敏电阻器
rod travel assembly 棒组行程装置
rod travel housing 棒行程套筒【控制棒驱动机构】，棒行程壳，棒行程罩
rod travel 棒行程
rod type fuel element 棒状燃料元件
rod withdrawal rate 提棒速率
rod withdrawal 提棒
rod worth 控制棒价值，反应性价值
rod 杆，棒，拉杆，连杆，避雷针，标尺，条
roentgendefectoscopy X射线探伤法

Roentgen meter 伦琴射线计，X射线计
roentgenodiagnosis X射线诊断
roentgenofluorescent analysis X射线荧光分析
roentgenography X射线照相法
roentgenospectral analysis X射线光谱分析
roentgen 伦琴【照射量单位】
ROE(return on equity) 股东资产净值的盈利，资本收益率，资本权益回报率，资本金净利润率，净资产收益率，净资产回报率，股本回报率
ROF(reactor operating experience) 反应堆运行经验
Roga index 罗加指数
Rogowski coil 罗戈夫斯基线圈
ROI(return on investment) 投资利润率［报酬率］
rollable 可卷的
roll angle 滚转角
rollback point 反转点
rollback routine 反转程序，重新运行程序
roll back to 反转
rollback 反转，反绕，重绕，重新运行，重算
roll bending 滚压弯曲
roll bonded clad 滚轧结合覆盖层
roll bonded composite plate 滚轧结合复合板
roll-bonding process 辊压接合工艺
roll chart drive unit 卷纸机构
roll chart 卷筒记录纸，卷纸
roll crusher 辊式破碎机，滚筒破碎机，滚轴式碎石机
roll damping 滚转阻尼
roll dovetail 球形叶根
rolled asphalt 碾压沥青
rolled beam 轧制钢梁
rolled compacted concrete dam 碾压混凝土坝
rolled condenser 卷筒电容器
rolled core type transformer 卷铁芯式变压器
rolled dam 碾压土坝
rolled-earth fill dam 碾压土坝
rolled-earth fill 碾实填土，碾压填土，压实填土
rolled earth-rock dam 碾压式土石坝
rolled-fill dam 碾压土坝
rolled hardening 轧制硬化
rolled-in tube end 轧入管头，胀接管端
rolled iron 轧制钢，钢材
rolled joint 辘口，胀接，胀接接头，辘口连接
rolled plate 轧制钢梁
rolled rockfill dam 碾压堆石坝
rolled roofing material 屋面卷材
rolled section steel 轧制型钢
rolled section 轧制型钢，轧制叶型，辗压断面
rolled shape 滚压的形状，轧制型材
rolled sheet metal 轧制金属板
rolled sheet steel 轧制钢板
rolled steel beam 热轧工字钢梁
rolled steel channel 轧制槽钢
rolled steel section 热轧型钢，轧制型钢
rolled steel 辊轧钢，轧制钢，钢材
rolled strip roofing 屋面卷材
rolled tube 胀过的管子
rolled up one's sleeves 撸起袖子（加油干），卷起袖子
rolled up vortex 卷起涡
rolled-water-proofer 防水卷材
rolled 滚压的，压制的，轧制的，胀接的
roller actuator 转子开关
roller bearing support 滚轴支座
roller bearing 滚柱轴承，辊柱轴承，滚轴支承
roller bit 碾子锥头，滚子旋锥
roller box 轮箱
roller bucket 消力戽
roller bumper 滚轴防冲设备
roller-carried plunger 滚道柱塞，滚子支承柱塞
roller chain 滚轮链
roller coating 滚涂
roller compacted concrete dam 碾压混凝土坝
roller compacted concrete 碾压混凝土
roller compaction of earth dam 土坝碾压，土坝压实
roller compaction 碾压
roller contact bearing 滚柱轴承
roller contact 滚动触点，滚动接触
roller conveyor 辊式运输机，辊道运输机，（移动）滚柱式传送机
roller coupling 圆柱联轴器
roller crusher 辊式破碎机
roller curtain filter 卷帘式过滤器
roller cutter for pipe 滑轮割管器
roller drum gate 圆管闸门
roller drying plant 滚辗干燥设备，滚筒干燥装置
roller drying 滚辗干燥
roller expander 滚子式扩管器，滚子式胀管器，辘管器
roller extensometer 滚轮式伸长计
roller gate 辊式闸门
roller mill 中速辊式磨煤机
roller mounted ring girder support 滚动支座
roller mounted 滚子支承
roller nut system 螺杆传动系统【控制棒驱动机构】
roller nut 活动螺帽
roller pulverizer 中速辊式平盘磨煤机
roller ring mill 中速辊式磨煤机
roller screen 滚轴筛
roller set 滚筒组
roller shut-off valve 辊轮关断阀
roller shutter door 卷帘门
roller support 滚轴支座，滚动支座，轧辊支承
roller tappet 滚柱挺杆
roller track 滚轮轨
roller tractor 轮式拖拉机
roller-type rotary pump 滚轴式回转泵
roller vibrator 滚轴式振动器
roller weir 圆辊堰
roller 导轮，辊子，滚柱，滚轨机，滚轮，轧辊，滚筒，滚轴，碾压机，压路机，路碾，碾子，托辊，球形滚柱
roll feeder 辊式给料机，旋转式加料机
roll filter 卷绕过滤器
roll-forming machine 滚压成形机械
roll forming 滚压成形
roll grip pipe wrench 链条管子钳

roll-in and roll-out　转入转出
rolling ball plan meter　滚球式求积仪
rolling bearing　滚动支座
rolling-contact type regulator　滚动触点式调整器
rolling country　丘陵区
rolling couple　倾侧力偶，横摇力偶
rolling cylinder gate　滚动式圆筒闸门
rolling dam　滚动式活动坝
rolling direction　轧制方向，滚压方向
rolling disc planimeter　转盘式求积仪
rolling display　滚动画面显示
rolling door　滚动门，卷门
rolling element line bearing　滚动元件线轴承
rolling flow wind tunnel　旋转流风洞
rolling force and energy　轧制力能
rolling friction resistance　滚动（摩擦）阻力
rolling friction　滚动摩擦
rolling gate　滚筒闸门
rolling gear　盘车装置
rolling grade　起伏坡度
rolling ground　起伏地面
rolling hammermill　滚压式破碎机
rolling instability　倾侧不稳定性，横摇不稳定性
rolling land　起伏地
rolling lift bridge　滚动开合桥
rolling load　滚动荷载，运输装载量［载荷］
rolling mark　滚压的标记
rolling mergin and cut pieces　卷边搭接和切割
rolling method　碾压方法
rolling mill gauge　钢板厚度计
rolling moment　滚转力矩，倾侧力矩
rolling plan　滚动计划，逐年扩展计划，波状平原
rolling pontoon　滚动浮筒
rolling region　起伏地区
rolling resistance　滚动阻力
rolling rotor　滚筒式转子
rolling scale　轧制氧化皮
rolling shutter door　卷帘门
rolling shutter　卷升百叶窗，转动百叶窗
rolling speed　滚动速度，暖机转速，转动速度
rolling stall　旋转失速
rolling steam　冲转用的蒸汽
rolling stock gauge for standard gauge railroad　标准轨距铁路机车车辆限界
rolling stock　机车车辆
rolling support　滚动支架
rolling surface　丘陵地表
rolling terrain　丘陵地带
rolling texture　轧制织构
rolling the turbine rotor by steam　汽机的冲转
rolling thrust bearing　滚动止推轴承
rolling topography　起伏地形
rolling track　导轨
rolling　滚动，胀管，滚轧，转动，滚动的，碾压，旋转的，轧制（钢板）
roll-jaw crusher　滚动鄂式碎石机
roll of attorneys　律师名录
roll off monitor　停转监视器
rollout breaker　滚出式断路器
roll out　辊平，卷边，扩口，铺开，滚出，碾平，转出
roll oversteering　倾侧过度转向
roll pass　滚道
roll-pin　滚动销
roll rate indicator　滚速指示器
roll rate　滚动角速度，侧倾角速度，滚转率，滚速
roll resonance　横摇共振
roll-rotor stepper motor　滚动转子式步进电动机
roll spot welding　滚点焊
roll steer　倾侧转向
roll surface temperature element　转动面温度元件
roll-table pulverizer　平盘磨
roll transition zone　滚压过渡区
roll type filter unit　滚筒型过滤器
roll-type filter　滚动式过滤器
roll-up door　卷帘门，卷升门，滑升门
roll-up type solar array　卷式方阵
rollway　滚物道
roll wear test　滚动磨损试验
roll welding　滚焊，热辊压焊接，滚压焊
roll　辊子，滚筒，辊碎机，卷，滚压，滚轧，滚动，卷轴，胀管，转动
ROL(request on line)　请求接通线路
Roman cement　罗马水泥
Roman spectrum　拉曼光谱仪
ROM coal　毛煤
ROMIS(remotely operated miniature inspection system)　远距离操作的小型检查系统
ROMON(receive-only monitor)　只收监听器
ROM(read-only memory)　只读存储器
ROMT(rehabilitate-operate-maintain-transfer)　ROMT 模式，“修复-经营-维修-转让”模式【工程建设模式】
Rontgen ray　伦琴射线
roofage　屋面材料
roof ash load　屋面积灰荷载
roof baffle　炉顶挡板
roof beam　屋面梁
roof boarding　屋面板
roof bolt head　顶板锚头，洞顶锚栓头
roof bolting method　洞顶锚栓法【隧洞】
roof bolt　顶板锚栓，顶板锚杆，顶锚，洞顶锚栓
roof bracing　屋面支撑
roof break　顶板开裂
roof burner　顶棚燃烧器
roof clay　隔水黏土
roof coating　屋面涂料
roof collapse　坍顶
roof construction　屋顶构造，屋面结构，屋盖构造，屋顶结构
roof crack　顶板裂缝
roof deck　屋顶板，屋顶平台，屋面盖板
roof dormer　屋顶采光窗
roof drainage　屋面排水
roof drain　屋顶排水口
roof expansion joint　屋顶伸缩缝
roof extract unit　屋顶排风机
roof fall　（隧洞等的）掉顶，坍顶
roof flashing　屋顶泛水

roof gate　屋顶式闸门
roof girder　顶梁
roof glazing　天窗
roof grinder　屋面梁
roof gutter　天沟，屋顶排水沟，屋檐水槽
roofing accessories　屋面附件
roofing and siding material　屋面和板壁材料
roofing asphalt　屋面沥青
roofing board　屋面板
roofing felt　屋面卷材，屋面（油）毡
roofing material　屋面材料
roofing millboard　屋面麻丝板
roofing paper　屋面纸
roofing plume　屋脊型羽流
roofing sand　屋面铺砂
roofing tile　屋面瓦
roofing　波形板屋面，屋顶，屋面
roof insulating　屋顶保温，屋面绝缘
roof joist　屋顶龙骨
roof level of the heater bay　加热器间的屋面标高
roof light flashing　天窗泛水
roof overhang　挑檐
roof panel　屋面板
roof pitch　屋顶高跨比，屋顶坡度
roof resisting to external fire exposure　防火屋面
roof ridge　屋脊
roof runoff　建筑物顶雨水
roofslab　顶板，屋面板，盖板
roof slate　屋面石板瓦
roof slope　屋顶屋面坡度
roof smoke screen　屋顶挡烟隔板
roofs of houses with low clear height　净高低的房屋的屋面
roof span　屋顶跨度，屋架间距
roof strainer　雨水斗罩
roof structure to falls　屋顶结构找坡
roof structure　屋顶结构
roof superheater　顶棚过热器
roof tank　屋顶水箱
roof terrace　屋顶平台
roof tile　屋瓦
roof-top air conditioner　屋顶式空调器
roof-topping　通量拉平，（通量分布曲线）顶部的削平
roof truss lower chord elevation　屋架下弦标高
roof truss　屋顶架，屋架，桁架
roof ventilation　屋顶通风
roof ventilator　筒形风帽，顶篷通风器，屋顶通风器
roof weir gate　屋顶堰闸门
roof weir　屋顶式堰
roof　顶，顶板，屋顶，屋面
room absorption　房间吸声量
room air plant monitoring system　室内空气装置监测系统
room identification list　房间编号清单
room of higher cleaning standard　较高净度的房间
room ratio line　房间比例线
room-scattered neutron　室内散射中子
room scattering　室内散射
room temperature superconductor　常温超导体，室温超导体
room temperature test　室温下的试验
room temperature　室内温度，室温
room thermostat　室内恒温调节器
room　室，房间
ROO（rehabilitate-own-operate）　ROO 模式，“修复-拥有-经营”模式【工程建设模式】
roost　栖息
root angle　安装角，齿根圆锥角，伞齿轮底角
root bead　根部焊道
root bend test　根部弯曲试验
root cap　根冠
root cause corrective action　根本原因纠正行动
root chord length　翼根弦长
root chord　翼根弦
root circle　齿根圆
root concentration　根的集中度【指每层土层中的根系占总根系的百分比】
root constant　根常数
root-cooled blade　叶根冷却式叶片
root crack　根部裂缝，焊根裂纹
root defect　根部缺陷
root diameter　齿根圆直径
root directory　根目录，主目录
root distance　齿根距
root distribution　根系分布
rootdozer　除根机
rooter plow　除根犁
rooter　拔根器，除根犁，路犁
root face　钝边，（焊件坡口）根面，（焊缝坡口）钝边
root fillet　叶根垫片，根圆角
root fixing　叶根连接
root flexibility　叶根柔度
root fusion　根部未焊透
root gap　根部间隙
root hole　根孔
rooting-in of blade　叶片安装
root locus method　根轨迹法
root locus　根轨迹
root mean square criterion　均方根判据，均方根准则
root mean square current　均方根电流，有效电流
root mean square deviation　均方根差
root mean square error　标准差，均方根误差
root mean square horsepower　均方根马力，有效马力
root mean square load　有效负载，均方根负载
root mean square noise voltage　噪声的有效电压，均方根噪声电压
root mean square quantity　均方根值，有效值
root mean square response　均方根响应
root mean square root extractor　均方根计算器
root mean square value　有效值，均方根值，均方根，均方值
root mean square　均方根，有效值
root of a weld　焊缝根部
root of blade　叶根
root of cavity　空穴根部【补焊】
root of joint　接头根部

root of the notch　切口根部
root of weld　焊根
root opening　根部间隙
root pass　根部焊道，焊道根部，第一焊层，底焊道
root radius　根部半径，齿根半径
root reinforcement　（焊缝）背面余高，根部加强高
Roots blower　螺旋式鼓风机，罗茨型鼓风机
root segment　基本段，（程序的）常驻段
Roots fan　罗茨风机
root side　根面，背面
roots of polynomial　多项式的根
root solver　求根器
root spacing　焊缝根部间隙
root span of foundation　挡距【线路工程】
Roots pump　罗茨泵，机械增压泵
root-squaring method　求平方根法，平方根法
root stall　根部失速
root surface　焊缝背面
root symbol　基本元符号
root tip　根尖
root valve　根阀，一次门，一次阀
root zone　根部区【焊缝】，根系层
root　根部【焊接】
rope and belt conveyor　钢丝绳牵引皮带机
rope cord　绳索
rope-driven barney　钢丝绳传动推车器
rope-driven carriage　钢丝绳驱动的小车
rope drive　钢丝传动，钢索传动
rope guide　导绳器
rope-lay cable　复铰电缆
rope lay conductor　多股绞合电缆，复绞导线
rope packing　绳状填料
rope roll　钢绳卷筒，绞车卷筒
rope sheave　绳轮
rope storage　磁芯线存储器，“编结”存储器
rope stranded wire　绞合线，合股线
rope-suspension bridge　悬索桥，吊桥
rope system　钢丝绳系统
ropeway　缆道，索道
rope　钢索，绳索，绳，系缆
ROP(remote operator panel)　远程操作板
RO(read only)　只读
RO(receive only)　只接收
RO(remote operation)　遥控运行
RO(restriction orifice)　节流孔（板），限流孔板
RO(reverse osmosis)　反渗透法
rose bit　玫瑰状钻头
Rosemount　罗斯蒙特仪表公司【美国】
Rosenberg dynamo　罗森堡直流发电机
Rosenberg generator　罗森堡发电机
Rosenberg starting　罗森堡启动法
ROSE(retrieval by on-line search)　联机检索
rose spraying　喷嘴，喷淋器
rosette　接线盒插座，喷头，玫瑰形饰物，接线盒，插座，三向应变计
rose　玫瑰，玫瑰图，（泵进口的）滤网，灯线盒，接线盒，记录盘
rosin　松香
ROS(read-only storage)　固定存储器，只读存储器
Rossby number　罗斯贝数
Rossby parameter　罗斯贝参数
rotameter　转速表，转子流量计，转子流速计，浮子流量计
rotary acceleration　旋转加速度
rotary actuator　旋转式执行机构
rotary air compressor　回转式空气压缩机
rotary air heater regenerative air heater　回转式空气预热器
rotary air heater　回转式空气预热器
rotary air preheater　回转式空气预热器
rotary air pump　回转式空气泵
rotary amplifier　电力扩大机，电机放大机
rotary atormizer　旋转喷雾器
rotary attenuator　旋转衰减器
rotary baffle　旋转挡板，旋转折流板
rotary balance　转动平衡，动平衡
rotary barrel throttle　旋转筒形节流阀
rotary bin discharger　廻转式料仓排料器
rotary blower　旋转式风机
rotary boring machine　旋转钻孔机
rotary boring　回转钻进，旋转钻孔
rotary bradford breaker　滚筒碎选机，滚筒破碎机，旋转压碎机
rotary breaker　滚筒破碎机
rotary brush　旋转电刷
rotary bucket-type flowmeter　旋转叶片式流量计
rotary bucket-wheel reclaimer　斗轮堆取料机
rotary burner　转子式喷燃器
rotary cam limit switch　旋转凸轮限位开关
rotary car-dumper sampler　转筒取样器
rotary car dumper　转子式翻车机
rotary check valve　旋启式止回阀
rotary circulating pump　旋转式循环水泵
rotary coil　转动线圈
rotary compressor　旋转压缩机
rotary condenser　调相机
rotary continuous filter　旋转连续性过滤器
rotary converter　旋转变流机
rotary conveyer　回转式运输机
rotary core drill　旋转式岩心钻机
rotary counter　旋转式计数器
rotary-coupled car　带旋转车钩的车皮
rotary crane　回转式起重机，旋臂起重机
rotary crusher　回转式碎石机，旋转式碎石机，旋转压碎机，旋转轧碎机
rotary cup atomization　旋杯雾化，转杯雾化
rotary cup atomizer　转杯式雾化器
rotary cup oil burner　转杯式油燃烧器
rotary current　回转水流，旋转电流，多相电流
rotary dehumidifier　转轮除湿机
rotary diaphragm　回转隔板
rotary diode exciter　旋转二极管励磁机
rotary disc feeder　旋转送料器
rotary discrepancy switch　旋转差动开关，转动灯光开关
rotary disc valve with freeze seal　带冻结密封的旋转圆盘阀
rotary disk　回转盘
rotary distribution　旋转配水

rotary distributor　回转配料机，旋转分配器
rotary drill　旋转钻井机
rotary drum concrete mixer　转筒式混凝土搅拌机
rotary drum filter　滚筒（式）过滤器，旋转滤网
rotary drum mixer　回转式鼓形拌和机
rotary drum screen　旋转式鼓形栅网
rotary drum　旋转筒
rotary dump car　翻卸用敞车
rotary dumper receiving hopper　旋转翻车机受料斗
rotary dump gondola car　翻卸用敞车，翻转式卸矿车
rotary dump gondola　供翻卸用的敞车
rotary dump unloading　车辆翻卸
rotary dump　翻卸
rotary eccentric pump　旋转偏心泵
rotary engine　转缸式发动机，旋转式发动机
rotary excavator　旋转式挖掘机
rotary feeder　旋转（式）给煤机［给料机］
rotary field　旋转场，旋转磁场
rotary filter　旋转（式）过滤器
rotary flow indicator　旋转流通指示器
rotary flow　旋转流
rotary frequency converter　旋转变频机
rotary fuel spray　旋转式（分配器的）燃料喷嘴
rotary gap　旋转火花隙，旋转放电器
rotary gate　旋转式闸门
rotary hammer breaker　回转锤式破碎机
rotary heat exchanger　转轮式换热器
rotary inertia　转动惯量
rotary interrupter　旋转式灭弧室
rotary inverter　旋转逆变器
rotary isolating switch　旋转式隔离开关
rotary isolator　旋转式隔离开关
rotary joint　回转接头
rotary kiln　回转窑
rotary machine　旋转机构
rotary magnetic effect　旋转磁效应
rotary magneto　旋转式磁电机
rotary magnet　旋转电磁铁
rotary mill　滚动球磨机，旋转式粉碎机
rotary miner　旋转式开采机
rotary mixer　回转式搅拌机，转动搅拌机
rotary motion　回转运动，旋转运动
rotary motor　旋转电动机
rotary moving blade vacuum pump　旋转刮板真空泵
rotary paddle feeder　旋桨式加料器
rotary paddle type　旋转桨叶式
rotary parts　回转件，旋转部件，回转体零件
rotary penetration　旋转贯穿件
rotary percussion drill　旋转冲击式钻机
rotary piston engine　旋转活塞发动机
rotary piston pump　旋转活塞泵，旋转式活塞泵
rotary piston　旋转活塞
rotary plow feeder　叶轮给料机
rotary plow reclaimer　叶轮取料机
rotary positioner　旋转定位器，翻转架
rotary power unit　回转执行机构
rotary process　旋转法
rotary pump　回转泵，旋转泵，转轮泵，转子泵
rotary pushbutton switch　转动推进型开关
rotary rectifier　旋转整流器，旋转整流机
rotary regenerative heater　回转再生式加热器
rotary regenerative heat exchanger　回转再生式换热器
rotary regenerator　旋转式回热器
rotary relay　旋转式继电器
rotary rig　旋转钻探机
rotary rod head　旋转棒头部
rotary sampler　回转式取样器
rotary sampling switch　旋转采样开关
rotary scraper　旋转式铲运机
rotary screen filter　旋网过滤器
rotary screen　筒形旋转筛，旋转式筛分机，转筛
rotary seal　锁气器，旋转密封，动密封
rotary shaft sealing　旋转轴封
rotary shovel　旋转式铲
rotary shunt-type flowmeter　旋转分路式流量计
rotary side car dumper　翻车机
rotary siphon　旋转虹吸
rotary slide valve　回转滑阀
rotary-sliding-vane flowmeter　旋转滑动叶片式流量计
rotary solenoid relay　旋转螺管式继电器
rotary soot blower　旋转式吹灰器
rotary speed detector　转速探测器
rotary spreader　旋转式抛煤机
rotary stem　旋转阀杆
rotary stepped machine　旋转式步进电机
rotary stepping motor　旋转式步进电机
rotary stepping relay　旋转式步进继电器
rotary substation　（具有旋转电机的）变流站
rotary supply outlet　旋转送风口
rotary switch　旋转开关
rotary swivel coupler　旋转式车钩
rotary synchronizer　指针式同步指示器
rotary synchronouscope　指针式同步指示器
rotary table feeder　圆盘式给煤机
rotary table　转车盘
rotary telephone　拨盘式电话机
rotary tidal current　回转潮流
rotary tipper　转子翻车机
rotary tower crane　塔式回转起重机，回转塔吊
rotary transducer　转速传感器
rotary-transformer　旋转变量器，旋转变压器
rotary tunnel excavator　旋转式隧洞挖掘机
rotary type coupler　旋转式车钩
rotary type sprinkling nozzle　旋转式喷嘴
rotary vacuam filter　旋转真空过滤器
rotary vacuam pump　旋转真空泵，回转（式）真空泵
rotary valve　回转阀，球形阀，旋转阀
rotary vane feeder　转叶式加料器
rotary vane pump　转动叶片泵
rotary voltmeter　高压静电伏特计，旋转式伏特计
rotary wagon tippler　转子式翻车机
rotary　回转式（的），旋转的，回转的
rotaside wagon tipper　侧倾式翻车机
rotatable crossarm　转动横担

rotatable phase-adjusting transformer 调相器，旋转相位调整变压器
rotatable shield 转台式屏蔽
rotatable slide 转动滑盖
rotatable transformer 旋转变量器，可旋转变压器
rotatable walkway 旋转通道
rotatable 可旋转的
rotated car dumper 侧倾翻车机
rotate 旋转
rotating air outlet with movable guide vanes 带有可动导叶的旋转送风口
rotating amplifier excitation system 旋转放大器式励磁系统
rotating amplifier 电机扩大器，旋转放大器，旋转式放大机
rotating annulus 转环转盘
rotating armature generator 转枢式发电机
rotating armature machine 转枢式电机
rotating armature motor 转枢式电动机
rotating armature relay 转动衔铁继电器
rotating armature type exciter 转枢式励磁机
rotating armature 旋转电枢
rotating arm 旋臂
rotating ball nut 滚珠螺母
rotating ball spindle 滚珠轴杆，液球螺杆
rotating beacon 旋转式信标
rotating blade groove 动叶片叶根槽
rotating blade row 动叶列
rotating blade 转动叶片【汽轮机】
rotating cascade 旋转叶栅
rotating choke 旋转扼流圈，旋转抗流圈
rotating coil 转动线圈
rotating commutator 旋转换向器，旋转整流子
rotating control assembly 旋转控制装置
rotating convection 旋转对流
rotating converter 旋转变流器
rotating core type relay 旋转铁芯式继电器
rotating core 转动核心
rotating counter weigher 旋转计数秤
rotating counter weight weightment 旋转计数秤
rotating cup atomizer 转杯式雾化器
rotating cylinder 转筒
rotating damper 旋转阻尼
rotating diaphragm 旋转隔板
rotating differential transformer 差动式旋转变压器
rotating diode 旋转二极管
rotating disc electrode 旋转圆盘电极
rotating disc flow measuring device 圆盘流量测量仪表
rotating dishpan 转盘【全球气候物理模型】
rotating double-sphere dumbbell-type viscometer 旋转双球哑铃式黏度计
rotating drum relay 转鼓式继电器
rotating drum screen 旋转滤网
rotating drum store 转筒贮存器
rotating eccentric mass excitation test 偏心块起振试验
rotating electrical machine 旋转电机
rotating equipment data 转动设备数据
rotating equipment 旋转设备
rotating excitation generator 旋转励磁发电机
rotating excitation 旋转励磁
rotating exciter 旋转励磁机
rotating farthrowing nozzle 旋转远射式喷嘴
rotating field exciter 旋转磁场式励磁机
rotating field generator 旋转磁场式发电机
rotating field instrument 旋转磁场式测量仪表
rotating field machine 旋转磁场式电机
rotating field power 旋转磁场功率
rotating field transformer 旋转磁场变压器
rotating field type alternator 旋转磁场同步发电机
rotating field 旋转场，旋转磁场
rotating flow 有旋流
rotating fluidized bed 旋转流化床
rotating flux 旋转磁通
rotating frame antenna 转动框形天线
rotating frame 旋转框形天线
rotating gate 转动式闸门
rotating generator 旋转发电机
rotating guide vane 可转导叶
rotating heat exchanger 旋转式热交换器
rotating induction motor 旋转感应电动机
rotating inertia 转动惯量
rotating-insulator switch 旋转绝缘子开关
rotating machinery performance test 旋转机械性能试验
rotating machine 旋转电机，旋转机械
rotating magnetic field 旋转磁场
rotating magnet magneto 转极式磁电机
rotating mass 旋转质量
rotating mercury tachometer 旋转式水银转速计
rotating mixer 回转式拌和机
rotating motor 旋转电机
rotating nature frequency 旋转固有频率
rotating paddle 旋转叶片
rotating pay-off 旋转放线装置
rotating phasor 旋转相量
rotating pilot valve 旋转滑阀
rotating plate type regenerative air heater 受热面转动型再生式空气预热器
rotating plug system 旋转屏蔽塞系统
rotating plug 旋转屏蔽塞
rotating pole type magneto 转极式磁电机
rotating reactor shield plug 旋转式反应堆屏蔽塞
rotating reading head 旋转读头
rotating rectifier alternator 旋转整流器发电机
rotating rectifier exciter 旋转整流器励磁机
rotating rectifier 旋转整流器
rotating resistance box 旋转式电阻箱
rotating ring-disc electrode 旋转环盘电极【RRDE】
rotating seal ring 旋转密封环
rotating shaft 转动轴
rotating shield plug system 旋转屏蔽塞系统
rotating shield plug 旋转屏蔽塞
rotating shield 旋转屏蔽体
rotating silicon rectifier 转动式硅整流器
rotating sleeve type magneto 转套式磁电机
rotating speed 转速

rotating stall 旋转失速
rotating standard 标准电度表【千瓦时表】
rotating stiffness 旋转刚度
rotating support 转动支架
rotating synchronous exciter 旋转同步励磁机
rotating synchro 旋转同步的，旋转同步机
rotating thrust collar 转动推力环
rotating thyristor excitation system 旋转式可控硅整流器励磁系统
rotating thyristor 旋转式可控硅整流器
rotating torque 旋转力矩
rotating tower crane 旋转塔式起重机
rotating transformer 旋转变压器
rotating type filter 旋转过滤器【空气】
rotating type screen 旋转滤网
rotating type soot-blower 旋转式吹灰器
rotating union 旋转接头
rotating valve 回转阀
rotating variable resistor 旋转式变阻器
rotating vector 旋转矢量
rotating vibration test 旋转振动试验
rotating wave 旋转波
rotating 旋转，旋转的
rotational balance 转动平衡
rotational component 角位移分量，旋转分量
rotational constant 旋转常数
rotational direction 旋转方向
rotational electromotive force 旋转电动势
rotational energy 旋转能
rotational fault 旋转断层
rotational field 旋转场，有旋矢量场，旋转磁场
rotational flow 旋转（水）流，有旋流，旋流
rotational fluid motion 旋转流体运动
rotational frequency 旋转频率
rotational hysteresis 转动磁滞，旋转磁滞，循环磁滞
rotational inertia coefficient 转动惯量系数
rotational inertia of hydro-generator 水轮发电机转动惯量
rotational inertia 转动惯量
rotational loss 空载旋转损耗，旋转损失
rotationally sampled wind velocity 旋转采样矢量
rotational motion 旋转运动，转动
rotational resistance of idlers 托辊回转阻力
rotational restraint 旋转约束
rotational slide 圆弧滑动
rotational slip 圆弧滑动
rotational speed of hydro-generator 水轮发电机转速
rotational speed 转速
rotational stiffness factor 转动刚度系数
rotational stiffness 旋转刚度，旋转劲度
rotational stream surface 旋转流面
rotational symmetry 轴对称，旋转对称
rotational velocity control system 转速控制系统
rotational viscometer 旋转式黏度计
rotational voltage 旋转电压，旋转电势
rotational wave 有旋波
rotational 旋转的，旋转式的，转动的
rotation angle 旋转角
rotation axis 旋转轴，转动轴
rotation belt conveyer 旋转带式输送机
rotation cooling 循环冷却
rotation direction 旋转方向
rotation distribution of water 轮流配水
rotation energy 转动能
rotation gate 旋转式闸门
rotation indicator 转动指示器
rotation moment 转动力矩，转矩
rotation motor 转动马达
rotation period 自转周期
rotation plate 转向指示牌
rotation recorder 旋转记录仪
rotation source 旋转源
rotations per minute 每分钟转数
rotation vector 旋转矢量
rotation velocity 旋转速度，转速
rotation 轮回，旋转，旋度，转动
rotative component 转动分量
rotative flow 旋转水流
rotative moment 转矩
rotative power 转动功率
rotative 回转的，循环的，旋转的，转动的
ROTATN＝rotation 转动
rotatory condenser 同步调相机
rotatory current 旋转电流，三相电流
rotator 旋转器，转子，翻转架，转动体
rotatrol 电动放大器
rothemuhle type air heater 风罩回转式预热器
rotodynamic pump 回转动力泵
rotometer 旋转流量计，转子流量计
rotopeening 旋转敲击
rotoplug 旋塞
rotor ampere turns 转子安匝
rotor angle detector 转子角检测仪
rotor angle indicator 转子角位指示器
rotor angle （发电机）转子角
rotor angular momentum 转子角动量
rotor assembling 转子装配
rotor assembly rotating parts 转子组件转动部件【转子，叶轮，旋转密封环等】
rotor assembly 转子总成，转子装配
rotor axial thrust 转子轴向推力
rotor axis 转子轴线，转子轴
rotor balancing test 转子平衡试验
rotor balancing 转子平衡
rotor bandage 转子绑线
rotor banding 转子绑扎，转子护环装配
rotor bar 转子条，转子线棒，转子导条
rotor bearing 转子轴承
rotor blade 动叶（片），风轮叶片，转子叶片
rotor-blocked test 转子止转试验
rotor-body slot 转子本体槽
rotor bore 转子中心孔
rotor bow 转子挠度，转子弯曲
rotor brake 转子刹车
rotor bushing 转子套管，转子衬套
rotor cage bar 转子笼条
rotor cage 转子笼，鼠笼，转子铁芯
rotor can （全密封泵）转子包壳，转子密封套，转子屏蔽套
rotor cap 转子护环，转子护盖

rotor center part 转子中心体
rotor center 转子中心部
rotor circuit 转子电路
rotor coil connection 转子线圈连接
rotor coil 转子线圈
rotor column 风轮柱
rotor conductor slot 转子线槽
rotor conductor 转子导体
rotor contraction 转子收缩
rotor cooling 转子冷却
rotor copper loss 转子铜损，转子铜耗
rotor copper 转子铜线
rotor core assemble 转子铁芯装配
rotor core lamination 转子铁芯冲片
rotor core 转子铁芯
rotor critical speed 转子临界转速
rotor current controller 转子电流控制器
rotor current 转子电流
rotor diameter 风轮直径，转子外径
rotor disc 风轮桨盘，叶轮，转子轮盘
rotor displacement angle 转子相角差
rotor displacement 转子偏移
rotor dynamic balancing 转子动平衡
rotor dynamics 转子动力学
rotor efficiency 风轮效率，转子效率
rotor end bell 转子护环
rotor end-cap 转子端箍，转子端帽，转子护环
rotor end ring 转子端环
rotor erection pedestal 转子装配台
rotor expansion 转子膨胀
rotor exterior ring 转子外环，磁轭
rotor feed type motor 转子馈电式电动机
rotor field spider 转子支架
rotor field 转子磁场
rotor flux 转子磁通
rotor frequency 转子频率
rotor ground 转子接地
rotor heating 转子发热
rotor hub 风轮桨毂，转子轮毂
rotor induced EMF 转子感应电动势
rotor interface 风轮界面
rotor iron loss 转子铁损
rotor iron 转子铁，转子铁芯
rotor leakage field 转子漏磁场
rotor leakage flux 转子漏磁通
rotor leakage reactance 转子漏磁电抗，转子漏抗
rotor leakage 转子漏磁
rotor length 转子长度
rotor loading 转子负载
rotor locking disc 转子锁止盘
rotor locking ring 风轮锁定环
rotor lock 风轮锁
rotor loss 转子损耗
rotor magnetic field 转子磁场
rotor meter 转子流量计
rotor nominal speed for constant wind turbines 恒速风电机组风轮的额定转速
rotor nominal speed for dual speed wind turbines 双速风电机组风轮的额定转速
rotor nominal speed for variable speed wind turbines 变速风电机组风轮的额定转速
rotor nominal speed 额定转速，转子的标称转速
rotor of condenser 电容器转片，可变电容器的动片
rotor oscillation 转子振动
rotor outer diameter 转子外径
rotor overhang 风轮悬突体
rotor overload protection 转子过载保护
rotor peripheral speed 转子圆周速度
rotor phase 转子相位
rotor plate 动片
rotor pole 转子极
rotor position limiting device 轴向位移保护装置
rotor position protection 串轴保护，轴向位移保护
rotor position sensor 转子位置传感器
rotor power coefficient 风能利用系数
rotor pre-cone 风轮预锥角
rotor punching 转子冲片
rotor reaction 转子反应
rotor resistance starter 转子电阻启动器
rotor resistance starting 转子串电阻启动
rotor resistance 转子电阻
rotor retaining ring 转子护环
rotor rheostat 转子变阻器
rotor rim 转子磁轭，转子轮缘
rotor ring 转子环，磁轭
rotor sagging 转子挠度
rotor seal 转子密封
rotor shaft hatch 转子轴孔
rotor shaft tilt angle 主轴倾斜角度
rotor shaft 转子轴
rotor sheet 转子冲片
rotor short-circuit 转子短路
rotor side converter 转子侧变流器
rotor slip inducted potential 转子滑差感应电势
rotor slip ring 转子滑环
rotor slip 转子转差率
rotor slot-pitch 转子槽
rotor slot wedge 转子槽楔
rotor solidity 风轮实度
rotor speed range 转速范围
rotor speed 风轮转速，转子转速，转速，转子速度
rotor spider 转子支架，转子支臂
rotor spoke 转子轮福
rotor spool 转子线圈
rotor starter 转子启动器
rotor static balancing 转子静平衡
rotor stray field 转子杂散磁场
rotor stress distribution 转子应力分布
rotor stress program 转子应力程序
rotor stress 转子应力
rotor swept area 风轮扫风面积
rotor temperature gradient 转子温度梯度
rotor terminal box 转子出线盒
rotor thrust 风轮推力
rotor tip vortex 旋翼叶尖旋涡
rotor tooth 转子齿
rotor torque 风轮转矩
rotor turning gear 盘车装置

rotor turn 转子线匝
rotor vent 转子风孔
rotor vibration resonance speed 转子共振转速
rotor voltage 转子电压
rotor volume 转子体积
rotor wake 风轮尾流
rotor water box 转子水箱
rotor wave resistance 转子波阻
rotor wedge 转子槽楔
rotor wheel 叶轮，工作轮，转子
rotor windage loss 转子风阻损耗
rotor winding resistance 转子绕组电阻
rotor winding 转子绕组，转子线圈
rotor wire 转子导线
rotor with non-salient poles 隐极转子
rotor without blades 转子体，无叶片转子
rotor without center bore 无中心孔转子
rotor yoke 转子轭，转子铁芯
rotor 风轮，转子，旋翼，回转体，转动体，动片
rotosyn 滑差检测投磁装置，旋转同步装置
rototrol generator 自励电机放大发电机
rototrol 旋转式自励自动调整器，自励电机放大器
rotproof 防腐朽的
ROT(rate of turn) 转动速度
ROT（Renovate-Operate-Transfer） ROT模式，"改扩建—运营—移交"模式【工程建设模式】
ROT(reserve oil tank) 备用油箱
rot-resistant quality 抗磨蚀性
ROT=rotary 转动的
ROT=rotor 转子
rotter 自动瞄准干扰发射机
roturbo 涡轮泵，透平泵
rough adjustment 粗调整
rough and finish grading of the plant area 厂区地面平整工程
rough area 崎岖地区
rough ashlar 粗琢方石
rough bevelling 粗斜切
rough bolt 粗制螺栓
rough burning 不均匀燃烧
rough cast finish 墙面粗涂【灰泥】
rough cast glass 毛玻璃
rough casting 毛坯铸件
rough cleaning 粗净化
rough coal 粗煤，原煤
rough concrete 粗糙混凝土
rough country 丘陵地带，崎岖地区
rough-cut file 粗纹锉
rough cut 粗切，粗纹
rough draft 草图
rough drawing 草图
rough-drilled hole 粗钻孔
roughened can 粗糙表面燃料包壳
roughened surface 毛糙化的表面
roughening concrete surface 混凝土毛面
roughening 打毛，凿毛，琢毛
rough estimate 粗估，粗略估算，概算
rough-faced form board 粗面模板
rough filter bank 预过滤组件
rough filtering package 预过滤组件
rough filter 粗滤器
rough floor 粗糙地面
rough gas 粗净化气体
rough grading 土地粗整
rough grind 粗磨
rough ground 崎岖地面
rough hardware 粗五金
roughing filter 粗过滤器，粗滤池，前置过滤器
roughing machine 粗加工机
roughing pump 前置真空泵
rough lathing 粗车
rough lumber 粗锯材，木材毛料
roughly 大致，大致地
rough machined 粗加工的
rough machining 粗加工
rough milling machine 粗磨机
roughness band 粗糙带
roughness category 糙率分级
roughness cavitation 粗糙性空蚀
roughness class 粗糙度等级
roughness coefficient 粗糙（度）系数
roughness concentration 粗糙度
roughness elements 粗糙元
roughness exposure 粗糙地貌开敞度
roughness factor 粗糙因数，粗糙因素，粗糙系数，粗糙度因子，粗糙度系数
roughness height 粗糙度高，粗糙度值，粗糙高度
roughness index 粗糙指数
roughness layer 粗糙层
roughness length 粗糙长度
roughness of particles 颗粒粗糙度
roughness of surface 表面粗糙度，表面粗度
roughness of terrain 地表粗糙度
roughness parameter 粗糙度参数
roughness Reynolds number 糙率雷诺数【即卡门数】
roughness rose 粗糙度玫瑰图
roughness sample 粗糙度样品
roughness screen 粗糙丝网
roughness similitude 粗糙度模拟
roughness standard 粗糙度标准
roughness （粗）糙度，不平（整）度，糙率，粗度
roughometer 平整度测定仪
rough plank 毛木板
rough sea 四级风浪
rough sketch 略图
rough surfaced blade 表面粗糙叶片
rough-surfaced can 粗糙表面燃料包壳
rough surface flow 粗糙面水流
rough surface 粗糙表面，毛面
rough survey 草测
rough terrain diffusion model 粗糙地形扩散模式
rough vacuum 低真空
rough-wall boundary layer 粗壁边界层
rough wave effect 波浪效应
rough weather works 恶劣气候下作业
rough weather 狂风暴雨天气
rough 粗糙的，粗调，大致的
roundabout drier 回转式干燥器
round-about 间接的，迂回的，迂回线路

round and obround penetrations 圆形和非圆形贯穿件
round angle 周角
round arch 环形拱，圆拱
round bar handle 圆杆门拉手
round bar 圆钢，洋圆，圆棒
round-body motor 圆机座电动机
round bolt 圆形螺栓
round-bottom flask 圆底烧瓶
round brush 圆电刷
round concrete bar 混凝土用圆钢筋
round conductor 圆导体，圆导线
round cowl 筒形风帽
round-crested weir 圆顶堰
round damper 圆挡板
round die 圆板牙，圆螺丝板
round down 把……四舍五入
rounded aggregate 卵石骨料，圆形骨料
rounded analysis 全面分析
rounded corner 圆角
rounded-crest measuring weir 圆顶量水堰
rounded-entrance flow meter 圆口流量计
round-edged orifice plate 圆边孔板
rounded indication 全面显示
rounded off to next higher whole number 四舍五入到下一个整数
rounded roof 圆形屋顶
rounded up 四舍五入，取整
round-end and socket bearing 圆端-凹穴支承
round figure 整位数
round file 圆锉
round head bolt 圆头螺钉，圆头螺栓
round head buttress dam 大头坝，圆头式支墩坝
round header 圆联箱
round head machine screw with grommet nut 圆头机器螺丝带手拧螺母
round head nail 圆头钉
round head rivet 圆头铆钉
round head screw 圆头螺丝
round head 圆头
round holes plate 圆孔板
round hole 圆孔【燃料组件的上管座】
rounding error 取整误差，化整误差，舍零误差，舍入误差
rounding off figure 略去尾数
rounding tool 弄圆工具
rounding 舍入，弯成圆圈，舍去小数
round it off to the nearest whole number 把它化整到最近的整数
round jet 圆形射流
round leading edge airfoil section 顿前缘翼型
roundness 圆度，球度
round nosed airfoil 圆头叶形
round-nosed slug 圆头涌节
round-nosed stamp 圆形冲孔
round-nose pliers 圆嘴钳
round-off accumulating 舍入误差累加
round-off accumulator 舍入误差累加器，舍零误差累加器
round-off constant 舍入常数
round-off error 舍入误差，取整误差，化整误差
round-off number 舍入数，取整数
round-off order 舍入指令
round-off 弄圆，舍入指令，舍入，四舍五入，舍去（零数）【取整数】，化成整数
round of holes 一组炮眼
round orifice 圆形孔口
round pole 圆形磁极
round-robin experiment 周游实验
round-robin scheduling 循环调度
round-robin servicing 循环服务，循环维护
round-robin 循环，循环的，循环法，依次的
round rotor 整圆转子
round section joint ring O形圈
round solid bar 圆形实心棒
rounds per shift 每班操作循环
round steel 圆钢
round strand conductor 圆形多股线
round-table conference 圆桌会议
round-table meeting 圆桌会议
round-the-clock job 昼夜施工
round-the-clock 连续不停的，昼夜不停的
round the world echo 环球回波
round timber pile 圆木桩
round trip air ticket 往返机票
round trip expenses 来回旅费
round trip ticket 来回票
round trip time 周转时间
round trip 往返
round tubular beam 圆管梁
round up 集合，把……四舍五入，使数目恰好（进），使……集拢，去以成整数
round washer 圆形垫圈
round way cock 通海旋塞，通海阀
round 炮眼组，绕的，围绕，圆形，圆形的
route and profile survey 路径与断面的测量
route clearing 清障
route-indicating bottle 流程指示瓶
route inspection 线路常规检查
route key map 路径总图
route map 路径图
route of escape 安全疏散路线
route planning 定线
route relocation 路线改线
router 路由器，定路线程序，刳刨者，刳刨工具，刳刨机
route selection of heat-supply network 供热管网选线
route selection 路径选择
route selector 路由选择器
route survey 路线测量
route 定路线，路线，路程，流程，流水作业，航线，路径，途径，线路
Routh criterion 劳斯准则，劳斯判据
routine analysis 日常分析，常规分析，例行分析
routine attention 日常维护
routine card 程序卡，工艺卡
routine check test 常规检查，例行检查
routine cleaning 定期清理，常规清理
routine coil test 常规线圈试验
routine communication 日常交流
routine control 常规管理，常规检验，例行控

制，常规监督，常规控制
routine determination 常规检测
routine electrical shop test 常规车间电气试验
routineer 定期测量装置，定期操作器
routine experiment 例行试验
routine inspection 例行检查，日常检查，常规检查，定期检查
routine library 程序库
routine maintenance 常规维护［维修］，定期维修，日常维修［检修］，例行维修
routine method 常规方法
routine monitoring 常规监测
routine observation 常规观测，例行观测
routine operation 例行运行
routine outage 常规停机，计划停运
routine overhaul 例行大修，例行维修，日常检修
routine program 操作规程
routine quality control 日常质量管理
routine release 例行脱扣，正常断开
routine repair 小修，日常维修
routine replacement of fuel elements 燃料元件例行更换
routine report 例行报告
routine sampling 常规采样，常规取样
routine serving 例行维修
routine shutdown 计划停机［停堆］，例行停机［停堆］
routine soil test 常规土工试验
routine startup 例行启动
routine test 例行检查，例行试验，常规试验，定期试验，例行测验，程序检验
routine work 常规作业
routine 例行的，程序，常规，常规的，日常工作，例行公事
routing axis 行进轴线
routing card 流动卡片
routing curve 演算曲线
routing decision 常规决策
routing design for cable 电缆敷设设计
routing drawing 路线图
routing inspection 巡检
routing method 定线方法
routing of wire 电路路径
routing sheet 流动单
routing 确定路线，路线，轨迹，路由，通路指示，发送，路径，通道，路程，定线
routinization 程序化，经常化
ROV(remote-operated vehicle) 遥控小车
row-binary card 横式二进制卡片，行式二进制数穿孔卡片
row binary 行式二进制数，横式二进制码
row engine 单列汽缸发动机
Rowland winding 罗兰线圈
rowlock arch 顺砌砖拱
row matrix 行矩阵
row of piles 排桩
row of vortices 涡排
ROW(right of way) 通行权
rows per inch 行每英寸
row-to-row distance 排间距离，节距
row vector 行向量
row 排，列，行，序列，天线阵列
royalty fee 提成费
royalty-free transfer 免费转让
royalty rate 版税率，专利费税率，提成率，提成支付的比例，使用费率
royalty （著作的）版税，特许权，专利（权）税，土地使用费，提成费，专利权使用费，特许权使用费，王权，皇室
RPCP(reinforced plastic composite pipe) 增强树脂复合管
RPC(remote position control) 远程位置控制
RPDB(reliability parameters data bank) 可靠性参数数据库
RPDH(reserve shut down planned derated hours) 计划降低出力备用停机小时
RPD(rupturing disc) 叶片断裂
RPE 核岛排气和疏水系统【核电站系统代码】
RPI(rows per inch) 行每英寸
RPL(remote panel loader) 遥控加载盘
rpluvial period 雨季
rpm gauge 转速表
RPMI(revolutions-per-minute indicator) 每分钟转数指示器
RPM(reliability performance measure) 可靠性测量
RPM(revolutions per minute) 每分钟的转数，转每分【DCS 画面】
RPN(reverse Polish notation) 逆波兰表示法，逆波兰计数法
RPN 核仪表系统【核电站系统代码】
RP(radiation protection) 辐射防护
RP(real property) 固定资产
RP(resettlement plan) 移民规划
RPR(read printer) 读出打印机
RPR 反应堆保护系统【核电站系统代码】
RPSM(resources planning and scheduling method) 资源计划与调度法
RPS(random pulse sequence) 随机脉冲序列
RPS(reactor protective system) 反应堆保护系统
RPS(reconvergent path set) 再收敛通路组
RPS(revolutions per second) 每秒钟的转数，转每秒
RPV bottom head 反应堆压力容器底封头
RPV(reactor pressure vessel) 反应堆压力容器
RPV water level measurement 反应堆压力容器水位测量
R＆QC(reliability and quality control) 可靠性与质量控制
RQD(rock quality designation) 岩厂质量标识，岩石质量指标，岩性符号，岩性指标
R(Rankine degree) 朗氏度，朗肯度，朗肯温标
RRA(Radiation Research Association) （英国）辐射研究协会
r-ray flaw detector r 探伤仪
r-ray source alarmr 源报警器
r-ray source pipe crawler r 源爬行器
RRA 余热排出系统【核电站系统代码】
RRB 硼回路加热系统【核电站系统代码】
RRC 反应堆控制系统【核电站系统代码】
R(Reaumer degree) 列氏温度

RRF(risk reduction factor) 风险降低因数
RRI 设备冷却水系统【核电站系统代码】
RRM 控制棒驱动机构通风系统【核电站系统代码】
RR(repetition rate) 重复频率
RR(research report) 研究报告
RR(running reverse) 反转，反向操作
RSAC(Reactor Safety Advisory Committee) 反应堆安全咨询委员会
RSAI(rules, standards and instructions) 规则、标准和说明
RSA(remote station alarm) 远程站报警
RSARR (Republic of South Africa Research Reactor) 南非研究堆
RSB(reactor service building) 反应堆辅助厂房
RSC(rolled steel channel) 轧制槽钢
RSD(relative standard deviation) 相对标准偏差
RSDT(remote station data terminal) 远程站数据终端
RSEXEC(resource-sharing executive) 资源分享执行
RSH 顶棚过热器【DCS 画面】
RSI(right scale integration) 适当规模集成
RS&I(rules, standards and instructions) 规则、标准和说明
RSK(receive/send keyboard set) 收发键盘装置
RSL(reserved storage location) 保留存储单元
RSM 转子应力监测【DCS 画面】
RSPL(recommended spare part list) 推荐的备件清单
RSP(record selection expression) 记录选择表达式
RS(reader stop) 阅读器停机
RS(record separator) 记录分隔符
RS(register storage) 寄存存储器
RS(remote station) 远程站，遥控台
RS(reset key) 置“0”键，清除键
RS(roll schedule) 滚动计划
RSS(reserve shutdown system) 备用停堆系统
RSST(reserve station service transformer) 备用厂用变压器
RST(readability, strength, tone) 可读性、强度和音调
RST=restart 再启动
RSUDH(reserve shutdown unit derated hours) 机组降低出力备用停机小时
RSV(reheat steam valve) 再热主汽阀
RSV(reheat stop valve) 再热截止阀，中压主汽门
RSV=reserve 备用
RSV 快速降负荷【DCS 画面】
RSWF(radioactive scrap and waste facility) 放射性废物装置【美国】
RS 印度卢比
RTA(reliability test assembly) 可靠性测试装置
RTC(real time counter) 实时计数器
RTC(reheat temperature controller) 再热温度控制器
RTD(resistance temperature detector) 电阻（式）温度检测器［探测器］
RTD(resistive thermal detector) 电阻式热探测器
RTE-B(real-time executive, basic) 基本实时执行程序
RTE(real-time executive) 实时执行程序
R-test 固结不排水剪切试验
RTF(radio telephone) 无线电话
RTG (radioactiveisotope thermoelectric generator) 放射性同位素热电发生器
RTG(regional transmission group) 地区性输电集团
RTGS(real-time graphic system) 实时制图系统
RTIOCS(real-time I/O control system) 实时输入输出控制系统
RTI(relative toxicity index) 相对毒性指数
RTI(rise-time indicator) 上升时间指示器
RTIRS (real-time information retrieval system) 实时信息检索系统
RTL(resistor-transistor logic) 电阻晶体管逻辑
RTM(real-time monitor) 实时监视器
RTM(register transfer module) 寄存器传送模件
RTN-N(return to normal) 复位
RTOS(real time operating system) 实时操作系统
RTP(remote transfer point) 远程传送点
RT(radiographic testing) 射线探伤，射线检验
RT(radiograph test) 射线探伤
RT(radioisotope tracer) 放射同位素示踪物
RT(ratio station) 比率控制站
RT(room temperature) 室温
RTR(resonance test reactor) （美国）共振试验堆
RTR(return and restore) 返回和复原
RTR=rotor 转子
RTS(reactor trip system) 事故保护停堆系统
RTS(return from subroutine) 从子程序返回
RTU(remote telemetry unit) 遥测装置
RTU(remote terminal unit) 远程终端单元，远程终端设备［装置］，站内远方终端，远动终端机
RTV(run-time variable) 运行时间变量
rubbed concrete 水磨石
rubbed finish 磨平修整
rubber asbestos 橡胶石棉
rubber ball collection net 收球网
rubber ball recirculating pump 胶球再循环泵，胶球泵
rubber ball 胶球
rubber belt conveyer 橡胶皮带输送机
rubber belt 胶带，橡胶皮带，胶布带，橡皮带
rubber buffer 橡皮减震垫
rubber bulb 橡皮吸球
rubber bump 减震垫，橡皮缓冲器
rubber cable 橡皮绝缘电缆
rubber cement 橡胶胶浆，橡胶水泥
rubber compression-type seal 橡皮压缩式止水
rubber cord 橡皮软线
rubber-covered braided wire 橡皮绝缘的编包线
rubber-covered wire 橡皮绝缘线，胶皮线
rubber-covered 橡皮绝缘的，橡皮覆盖的
rubber cushion 橡皮绝缘垫
rubber dam 橡胶坝
rubber disc type 胶圈式
rubber disc （缓冲托辊的）橡胶圆盘
rubber expansion joint 橡胶膨胀节

rubber fastening band　橡胶止水带
rubber fender system　橡皮防冲装置
rubber gasket　橡胶［橡皮］垫片，橡皮［橡胶］密封，橡皮［橡胶］垫圈
rubber glue　橡皮胶浆
rubber hammer　橡皮锤
rubber hose　橡胶管，橡皮软管
rubber-insulated cable　橡皮绝缘电缆
rubber-insulated wire　橡皮绝缘线，胶皮线
rubber-insulated　橡皮绝缘的
rubber insulation　橡皮绝缘
rubberized belt　胶皮带
rubberized cloth　橡胶布
rubber lagging　包胶
rubber latex　橡胶浆
rubber layer　橡胶层
rubber lined diaphragm valve　橡胶衬里隔膜阀
rubber lined tube　衬胶管
rubber lined valve　衬胶门
rubber lined　衬橡胶的，橡胶衬里
rubber lining　衬胶，橡胶衬里
rubber membrane　橡皮膜片
rubber mounting　橡皮垫
rubber packing　橡皮填料，橡胶垫圈，橡胶密封垫，胶皮碗，橡胶盘根，橡皮垫（料）
rubber padding　橡胶料垫
rubber pad　橡皮垫
rubber plate　橡胶板
rubber plug　橡胶塞
rubber policeman　橡皮淀帚
rubber ring coupling　胶圈联轴器
rubber sealing ring　橡皮止水环
rubber sealing strip　橡皮止水条
rubber seal　橡皮止水
rubber-sheathed flexible wire　橡皮软线
rubber-sheathed trailing cable　橡皮护皮拖曳电缆
rubber-sheathed wire　橡皮外包线
rubber sheath　橡皮（护）套，橡皮外套
rubber shock absorber　橡胶吸振器
rubber stopper　橡皮塞
rubber strip　橡皮防水条
rubber-tired compactor　轮胎式压实机
rubber-tired dozer　轮胎式推煤机
rubber-tired excavator　轮胎式挖掘机
rubber-tired mobile crane　轮胎移动式起重机
rubber-tired roller　轮胎式碾压机
rubber-tired tractor　轮胎式拖拉机
rubber tube　橡胶管，橡皮管，橡皮绝缘管
rubber-tyre crane　轮胎式起重机
rubber-tyred bulldozer　轮胎式推土机
rubber-tyred mounted reclaimer　轮式取料机
rubber vibration insulator　橡皮隔振器
rubber washer　橡皮垫
rubber waterstop　橡皮水封，橡皮止水
rubber　粗石，橡胶，橡胶制品，橡皮
rubbing contact　摩擦触点，滑动触点
rubbing face seal　紧密密封
rubbing seal　滑动密封
rubbing strip　（轻水堆元件）定位带
rubbing surface　摩擦面
rubbing　摩擦（机械），研磨，擦伤
rubbish-burning fuelled power plant　垃圾电站
rubbish　垃圾
rubble aggregate　粗骨料，毛石骨料
rubble ashlar　方块毛石，毛方石
rubble catchwater drain　毛石截水沟
rubble concrete　块石混凝土，毛石混凝土
rubble drain　块石排水沟，石砌排水沟
rubble drift　碎石堆
rubble flow　碎石流
rubble foundation　抛石基床
rubble masonry dam　毛石圬工坝
rubble masonry footing　毛石基础
rubble masonry　毛石砌体，乱石圬工，毛石圬工，毛石砖体
rubble mound structure　抛石建筑物
rubble mound　堆石斜坡堤，毛石堆，毛石护坡
rubble pitching　乱石护面
rubble retaining wall　毛石挡土墙
rubble soling　乱石基底
rubble stone　毛石，石渣，乱石
rubbles　毛石
rubble wall　毛石墙，毛石圬工墙
rubble work　毛石工程
rubble　粗石，乱石，毛石，瓦砾
rub-out signal　指示错误的信号
ruby laser　红宝石激光器
rub　摩擦，磨，障碍，不平坦
rudaceous rock　粗碎屑岩
rudaceous　砾质的，砾状的
rudder pit　舵槽，舵穴
rudder post　舵柱
rudder　舵
RUDH(reserve shut down unplanned derated hours)　非计划降低出力备用停机小时
rudite　砾状岩
rudyte rudite　砾质岩
rudyte　砾状岩
ruff　轴环
rugged area　恶劣的地区（气候），崎岖的地区
rugged duty　磨损量过大的工作状态
ruggedized computer　耐震计算机
ruggedized instrument　抗震仪表
ruggedized　抗震的，加固的
ruggedness number　粗度数
ruggedness　粗糙度，强度，刚性，紧固性
rugged terrain　畸形地形，起伏地面
rugged topography　崎岖地形
rugosity coefficient　糙度系数，糙率
rugosity factor　糙率系数
rugosity　粗糙度，糙度
rug　毯
RUHS(reserve ultimate heat sink)　备用最终热阱
ruination　没落
rule and procedure　规则和程序
rule base　规则库
rule of thumb　近似计算法，右手法则，经验法则
ruler　尺，直尺，直规
rules and regulations in enterprise　企业规章制度
rules and regulations　规则和条例，规章制度
rules of administration　管理规章
rules of arithmetic　算术定则

rules of implementation 施行细则
rules of penalties 处罚规则，惩罚措施
rules of safety operation 安全操作规程
rules 规定，规则，守则，条例，标准，常规，尺，比例尺，法则
ruling condition 控制条件
ruling grade 限制坡度
ruling language 支配语言，主导语言，合同语言，裁决语言，支配性语言
ruling point 控制点
ruling span 代表挡距
ruling 支配，刻度，划线，管理
rumble 隆隆声，低频音，噪声
rummage 搜查，海关检查
RUM(remote underwater manipulator) 遥控水下机械手
run a line 敷设管道，放线
runaround track 越行线
runaway chain reaction 失控连锁反应
runaway corrosion （氧化膜）脱落型腐蚀，剧增腐蚀
runaway curve 飞逸曲线
runaway governor 超速调节器，失控调节器
runaway reaction 失控核反应
runaway reactor 失控反应堆
runaway speed of water turbine 水轮机飞逸转速
runaway speed 飞车转速，极限转速，飞逸速度，飞逸转速，发电机超速
runaway 超速运行【汽轮机】，失去控制，飞车，失控
runback test 辅机故障减负荷试验
runback 辅机故障减负荷，回流管，溢出管，快速减负荷，快速返回，（汽轮机）减速
run coal bin 卸煤仓，日用仓
run coal 软烟煤，炉前煤，原煤
run curve 运行曲线
rundown pipe 溢流管
rundown tank （冷凝液的）接收贮槽
rundown （泵）减速，负荷（指令）追降，流出，减少，下降，降负荷，停止，停机惰走，功率降低，控制棒插入，自动减负荷
run down 撞倒，走下坡路，往下流，变弱，用尽，慢下来，减少，停止运转，惰转，耗尽，停产，下降，裁减，破败，浏览【注：本词条为动词性质，不同于 rundown】
run duration 风速持续时间，运转时间
run empty 空转
run-free 空转，无载运行，开路
rung ladder 攀梯（指坡度大于75°的），直爬梯
run gravel 冲积砾石
rung 轮辐，梯级，车辐，横梁，直梯踏步，（梯子）横档
run idle 空转
run-in period 先期运行阶段，试运转
run-in step 同步运行
run-in 试运转，试车，（控制棒）插入
run-length 扫描宽度
runlet 细流
run location routine 在程序带上检测记录的程序，运行定位程序
run-mine coal 毛煤
runnel 隧道，溪沟
runner blade 转轮轮叶
runner casing 转轮外壳，转轮罩
runner chamber 转轮室
runner cone 转轮轮锥
runner crown 转轮轮冠
runner disc 转轮轮盘
runner housing 转轮壳
runner hub 转轮轮毂
runner inlet 转轮进水口
runner level 转轮安装高程
runner removal access 转轮检修道
runner seal ring 转轮密封环，转轮止水环
runner series 转轮系列
runner vane 转轮叶片
runner 导滑车，流道【铸造】，叶轮，转子，运转体，转轮
running ability 运行能力
running angle support 悬垂转角杆塔
running attention 运行维护
running backwards of the drive 驱动倒转
running balance 动平衡
running block 传动滑车
running characteristic 工作特性，运行特性
running charge 运行费用
running clearance 动静件间的间隙
running condition 工作条件，运行工况，运行状态，运行情况，工作状态
running cost 日常管理费用，运行费用，运转费
running current 工作电流
running curve simulator 运转曲线模拟计算机，工作曲线模拟计算机
running curve 运行曲线，工作曲线
running days 连续日数
running experience 运转经验
running fit （转）动配合
running gear 传动装置，工作机构，操纵机构，传动机构，运行装置，行走机构
running-hot 运行发热
running hour meter 运转小时计
running hours 运行时数
running idle 空转，惰走
running-in period 试运行期
running-in test 试运行试验，空转试验
running-in wear 跑合磨损
running-in 试运转，试车，跑合
running-light test 轻载试验，空载试验
running light 空转试验，空载运行
running load 运行负载
running maintenance 日常修理，日常保养，小修
running no-load 空负荷运转
running of trains 行车量
running operation 运转操作
running order 动作次序，运转次序
running out block 放线滑车
running-out period 运行结束期
running out system 自然到期制
running performance 运行特性，运行性能
running period （设备）运转周期，运转时间，工作时间
running position 运转位置，运行位置，工作位置

running program　操作程序，运算程序，运行工况，工作程序，运转程序，运行程序
running rabbit　（显示器）跑动的干扰信号，串动干扰信号
running rate　运转率
running repair　日常维修，临时修理，小修，运行中检修
running resistance　运行阻力
running reverse　反转，反向操作
running sand　流沙
running saturation test　运行饱和度试验
running slag　散流渣
running smoothness　运转平稳性
running speed　运转速度，运行速度，工作速度
running state　运行状态
running status　运行状态
running temperature　工作温度
running test　常规试验，运转试验
running threshold　活动门限
running time　操作时间，运行时间，工作时间，运转时间，行车时间
running track　行走线
running trial　运行试验
running turbine components　透平旋转部件
running-up test　启动试验
running up the turbine　汽机的升速
running-up　启动，升火
running voltage　工作电压，运行电压
running water　活水，流水
running wheel　工作轮
running without load　空转
running　运转，工作，流动，行程，控制
runoff coefficient　径流系数
runoff computation　径流计算
runoff depth　径流深度
runoff distribution curve　径流分布曲线
runoff formula　径流公式
runoff generating precipitation　产流降水
runoff hydroelectric power station　径流式水电站
runoff hydrograph　径流过程线
runoff in depth　地下径流
runoff of dry season　旱季径流
runoff of low water　枯水径流
runoff plot　径流区域
runoff process　径流工程
runoff producing precipitation　产流降水
runoff regulation of hydropower regulation　水电站径流调节
runoff regulation　径流调节
runoff-river power station　径流式电站
runoff-river station　径流式水电厂，河流式水电厂
runoff source　补给水源
runoff station　径流站
runoff tab　（焊缝）引出板
runoff volume　径流体积，径流量
runoff weld tab　引出板
runoff yield　产流量
run-off　泄出，流出，流量，溢出，径流（量），漏失量，流走之物，耗量，跑偏
run-of-mine coal　原煤，原矿，毛煤
run-of-river hydroelectric station　径流式水电站
run-of-river plant without pondage　无调节池的径流式电站
run-of-river plant with pondage　有调节池的径流式电站
run-of-river plant　径流式水电站，（无调节水库的）河流式水电厂
run-of-river power station　径流式水电站
run of the bank filling　岸土流失
run of the wind　风程
run-on tab　（焊接）引弧板
run-on time　运转时间，连续时间
run out a contract　合同满期
run out capacity　失配容量，失载容量
run-out check　（汽轮机）转动检查
runout flow rate　最大流量
run-out flow　抽出流量
run-out key　输出电匙，输出电键
run out load　失载负荷
run-out of shaft　轴的摆度
run-out　惰走，滑行，输出，满期，流出，惰转，跑偏
runout　偏心率，用尽，消退，用尽，总径流量
run-round system　折返系统
run schedule　运行调度表，程序的运行图表
runtime storage　运行时存储器，存放程序变量的存储器
run time variable　运行时间变量
run time　试验时间，运行时间
run-to-failure　（燃料）辐照考验到破坏为止
run-up time　启动时间，起转时间
run-up　起转，试车【汽轮机】，迅速增大，上坡，负荷追升，预备阶段，急剧增长，升负荷，升速，负荷指令回升，试运行，启动，抽出（控制棒），提升功率，爬高，上涨
runway girder　行车大梁
runway level　走道标高
runway rails　轨道
runway track　吊车导轨
runway　（机场）跑道，滑道，导轨
run　操纵，流动，行程，流程，运行，运转，路线，试车，进行（试验等），试验，一个堆芯周期，路程，走向，管道，通道，梯级踏步
rupture angle　破坏角，破裂角，缺角
rupture cross section　断裂截面
rupture deformation　破裂变形
rupture diaphragm　破裂盘，破裂片，安全膜
rupture disc nozzle　破裂圆盘接管
rupture disk　破裂片［盘］，安全膜［盘］，爆破片［膜］，破裂圆盘【高压安全阀】
rupture elongation　断裂伸长，断裂延伸率
rupture envelope　破坏包线
rupture factor　断裂系数
rupture failure　断裂破坏
rupture life　持久强度
rupture limit　破坏极限
rupture line　破裂线，断裂线
rupture load　破断荷载
rupture modulus　断裂模量，断裂模数，破裂系数，断裂力矩，破坏力矩
rupture plane　破裂面
rupture point　断裂点，破裂点

rupture pressure 破坏压力
rupture stage 破损阶段
rupture strain 断裂应变
rupture strength 断裂强度，破裂强度
rupture stress 破坏应力，断裂应力，破裂应力
rupture test 断裂试验，破坏试验，持久试验
rupture time 破裂时间
rupture 断裂，裂口，断口，管道断裂，破裂，破损
rupturing capacity test 截断容量试验
rupturing capacity 截断容量（开关的）
rupturing current 切断电流
rupturing duty （短路）切断功率，（触点的）灭弧工作
rupturing operation 截断操作，切断操作
rupturing voltage 击穿电压
rural and township enterprises 乡镇企业
rural architecture 乡村建筑
rural area 乡村，乡村地区
rural boundary layer 乡村边界
rural diffusion coefficient 乡村扩散系数
rural distribution network 农村配电网，农网
rural distribution wire 郊区配电线
rural distribution 郊区配电
rural electricity consumption 农村用电
rural electric power network 农村电力网
rural electrification 农村电气化
rural enterprise 乡镇企业
rural high voltage distribution network 农村高压配电网
rural line 郊区线路
rural low voltage distribution network 农村低压配电网
rural medium voltage distribution network 农村中压配电网
rural network 郊区电力网
rural party line 郊区电话合用线
rural power distribution 农村配电
rural power network 农村电力网
rural power service 农村供电
rural power source 农村电源
rural service 郊区供电，农村供电
rural small hydroelectric power generation 农村小水电
rural small thermal power generation 农村小火电
rural subscriber 郊区用户
rural substation 农村变电站
rural type transformer 农用变压器
rurban area 靠近都市之农村地区
RU(run up) 升负荷，负荷迫升
rush channel 冲沟
rush current 冲击（电）流
rush letter of credit 速开信用证
rush of current 流水冲击
rush out of the low valley 冲出低谷
rust grease 防锈脂
rusticated joint 粗接缝，凸形构缝
rustic brick 粗面砖
rustic masonry 粗面圬工
rustiness 生锈
rusting 锈蚀
rust-inhibiting paint 防锈漆
rust-inhibiting 防锈的
rust-inhibitive primer 防锈底漆
rust inhibitor 防锈剂
rust preventer 防锈剂
rust preventing agent 防锈剂
rust preventive 防锈漆
rust-proof oil 防锈油
rust-proof paint 防锈漆
rust-proof 不锈的，防锈的，抗锈的
rust protection 防锈
rust removal 除锈
rust remover 除锈剂，除锈器
rust-resistance 防锈
rust-resistant 防锈的
rust-resisting material 防锈材料
rust-resisting paint primer 防锈底漆
rust-resisting property 抗腐蚀性能，防锈性能
rust-resisting steel 不锈钢
rust spot 锈斑
rust stains 锈斑
rust surface 锈蚀面
rusty scale 锈鳞，锈垢
rusty surface 锈蚀表面
rust 锈，生锈，锈蚀
ruthenium 钌【Ru】
rutile type electrode 钛型焊条
rutile 金红石
rut 皱纹
RVLIS(reactor vessel level instrumentation system) 堆容器水位测量仪表系统
RV(rated voltage) 额定电压
RV(reactor vessel) 反应堆容器
RV(relief valve) 安全阀，减压阀
RV(self-regulating valve) 自动调整门
RVS(reconvergent variable set) 再收敛变量集
RV to cavity seal ring 反应堆容器与堆腔间的密封圈
RWC(read,write and compute) 读写和计算
RWC(read write channel) 读写通道
RWC(read write continue) 继续读写
RWCS(reactor water cleanup system) 反应堆一回路水净化系统
RWCU(reactor water cleanup) 堆冷却水净化
RWL(remaining work list) 剩余工作清单
RWM(read write memory) 读写存储器
RWM(rod worth minimizer) 控制棒价值最小化装置
RWP(radiation work permit) 辐射工作许可
RWP(reusing water pool) 复用水池
RW(read/write) 读写
RW(regular word) 正规字
RWS(raw water system) 生水系统
RWST(recirculating water storage tank) 再循环水贮存箱
RXS(reactor system) 反应堆系统
Ryznar stability index 雷兹纳稳定指数
RZ logic(return-to-zero logic) 归零制逻辑
RZM(return-to-zero mark) 归零标记
RZ(return-to-zero record) 归零记录
RZ(return-to-zero) 归零制

RZ system(return-to-zero system) 归零制系统
R 反应堆【核电站系统代码】

S

SAA(surface active agent) 表面活性剂
sacked cement 袋装水泥
sacked concrete revetment 袋装（水下）混凝土护岸
sack gabion 袋形铅丝石笼
sack 粗布带，麻袋
sacred right 神圣权利
sacrificial anode protection 牺牲阳极保护法
sacrificial anode 牺牲阳极
sacrificial electrode 牺牲电极
sacrificial formwork 损耗模板
SAC(semi-automatic coding) 半自动化编码
saddle axis 鞍轴
saddleback gutter 鞍形天沟
saddle back roof 鞍形屋顶
saddle back 鞍背
saddle bend 鞍顶
saddle coil magnet 鞍形线圈磁体
saddle fold 鞍形褶皱，鞍形接合，咬口接头
saddle pier 鞍形支墩
saddle-point method 鞍点法
saddle point 鞍点
saddle reef 鞍状矿脉
saddle scaffold 鞍形脚手架
saddle shell 鞍形壳
saddle siphon 鞍形虹吸管
saddles 鞍形填料
saddle type 鞍形【指流态等】
saddle vein 鞍状矿脉
saddle 鞍，鞍状物，鞍形支座，鞍状构造，鞍座，支座，阀座，座板，滑动座板
safe alarm system 安全报警系统
safe allowable load 安全容许负载，安全许可负荷，安全负荷
safe allowable stress 安全许可应力
safe anchorage 安全锚泊地
safe bearing capacity 安全承载能力
safe bearing load 安全荷载，安全（容许）荷载
safe by geometry 安全的几何形状，几何安全
safe by mass 安全质量
safe carrying capacity 安全承载能力，安全过水能力，安全载流（量）
safe clearance 安全净空
safe code 安全规程
safe coefficient 安全系数
safe concentration 安全浓度
safe containment 安全封隔
safe corridor 安全走廊
safe current carrying capacity 安全载流能力，安全载流量
safe current 安全电流，容许电流
safe-deposit vault 安全储存室
safe diameter 安全直径
safe distance 安全距离
safe dose 安全剂量
safe enclosure 安全封入，安全封存，封存【退役第一阶段】
safe end weld 不同合金焊，异金属焊
safe end 安全端，(异金属的) 过渡段
safe escape 安全出口
safe exit 安全出口
safe failure 安全故障
safe gear 安全设施
safe geometry shipping container （核燃料）几何安全运输容器
safe geometry 安全几何条件，安全形状
safeguard agreement 保障协定
safeguard auxiliaries 安全保障系统
safeguard auxiliary building 安全保障辅助厂房
safeguard circuit 安全保护电路
safeguard clause 保障条款
safe guard construction 安全防护结构
safeguard equipment room 保安设备间
safeguarding 保护
safeguard measure 保障措施
safeguard provisions 保障规定
safeguard regulation 安全管理
safeguards agreement 保障协定
safeguards analysis 保障分析
safeguards approach 保障方案
safeguards breaker 安全保障断路器
safeguards containment 保障封隔
safeguards inspector 保障视察员
safeguards material control system 保障（核）材料控制系统
safeguards measures 保障措施
safeguards switchboard 安全保障配电盘
safeguards system 保障制度，安全系统，外设安全保护系统
safeguards technique 保障技术
safeguard 核保障【国际原子能机构】，安全装置，保安措施，护罩，保护，保安，保险器，安全设备，保安装置，保护装置，防护装置
safe handling 安全操作
safe health level 安全保健水平
safe heating limit 安全加热限度
safe in operation 安全操作，安全运行
safe integral reactor 安全一体化反应堆
safe life 安全寿命
safelight screen 安全灯滤光片
safe limit 安全极限
safe load-carrying capacity 安全承载能力
safe load 安全负荷，容许负荷，安全荷载，容许荷载
safe low power critical experiment reactor 安全低功率临界试验堆
safely test 安全试验
safe margin 安全裕度
safe mass 安全质量
safe measure 安全措施
safe operating area 安全工作区
safe operating temperature 安全运行温度，容许运行温度
safe operation life 安全运行寿命
safe operation 安全运行，安全操作

safe port 安全港，避风港
safe pressure 安全压力
safe program 安全规程
safe range of stress 应力安全范围
safe range 安全距离，安全范围
safe reliability 安全可靠性
safe rules 安全准则
safe running 安全运行
safe separation criterion 安全分隔准则
safe service life 安全使用年限
safe shower 安全淋浴器
safe shutdown earthquake 安全停堆地震
safe shutdown 安全停堆
safe sight distance 安全视距
safe spacing 安全间距
safe standstill 安全停机
SAFE（State Administration of Foreign Exchange）（中国）国家外汇管理局
safe steel wire 防坠钢丝绳
safe stress 许用应力，安全应力，容许应力
safe temperature 安全温度
safety accessory 安全装置
safety action 安全措施，安全操作，安全动作，安全作用，保护作用，屏蔽作用
safety actuation system 安全动作系统
safety against fracture 防断裂余量
safety aid 安全设备，安全援助
safety alarm annunciating system 安全报警系统
safety alarm device 安全报警装置
safety alarm system 安全报警系统
safety alarm 安全警报器
safety allowable stress 安全容许应力
safety allowance 安全补偿，安全宽限
safety analysis report 安全分析报告
safety analysis 安全分析
safety and control scheme 安全控制策略
Safety and Reliability Directorate （美国）安全与可靠性管理局
safety and reliability test 安全性与可靠性试验
safety apparatus 安全设备
safety appliance 安全设备
safety arch 分载拱，安全拱
safety assessment 安全评价
safety authorities 安全管理当局
safety barrier 安全屏障
safety basin 安全泊地，避风港地
safety belt buckle 安全带扣环
safety belt 安全带，保安带
safety block 安全保护单元，安全保护部件
safety bolt 安全螺栓，保险螺栓
safety boots 工作鞋
safety brake 安全制动器，保险闸
safety breaker 安全开关，安全断路器
safety cage ladder 安全笼罩爬梯
safety cap 安全帽
safety catch 安全挡
safety certification 安全认证
safety chain 安全链
safety channel （反应堆）安全保护通道，安全通道
safety circuit 安全电路
safety class barrier 安全级屏障
safety classification 安全等级，安全分级
safety class structure 安全级构筑物
safety class 安全等级
safety clearance 容许间隙，安全间隙
safety clutch 安全离合器
safety cock 安全旋塞
safety code 安全规程，安全规范，安全码
safety coefficient 安全系数
safety color 安全色
safety component 安全部件
safety concept 安全方案
safety condition 安全条件，安全状态，安全工况
safety control equipment 安全控制设备
safety control 安全管制，安全控制，安全管理，安全监测
safety copy 备用原图，第二底图
safety coupling 保险连接器，安全联轴节，安全离合器
safety cover 安全罩，防护罩
safety criterion 安全准则
safety critical equipment 安全关键设备
safety culture 安全素养
safety curtain 安全屏障，防火幕
safety cut-off device 安全切断装置
safety cutout 保安器，熔丝断路器，安全断流器，安全断路器
safety design 安全设计
safety device 安全装置，保安装置，保护装置，安全设备，保险装置
safety discharge capacity 安全过水能力
safety discharge in river 河道安全泄量
safety discharge 安全泄量
safety distance 安全距离
safety dose 安全剂量
safety drill 安全演习
safety earthing 安全接地
safety element 安全元件
safety engineering 安全工程（学），保安工程
safety engineer 安全技术工程师，安全工程师
safety equipment 安全设备
safety estimation 安全估价
safety evaluation report 安全评价报告
safety evaluation 安全评估，安全评价
safety exercise 安全演习
safety exhaust 安全排汽
safety exit of pipe duct 管沟事故人孔
safety experiment 安全试验
safety explosive 安全炸药
safety facilities 安全设施
safety factor for drop-out 返回安全系数
safety factor for holding 吸持安全系数
safety factor 安全因素，安全系数，保险系数，安全率，安全因数
safety failure fraction 安全失效分数
safety feature 安全装置，安全设施，安全故障
safety-first engineering 保安技术
safety flange coupling 安全凸缘联轴器
safety for overspeeding 过速安全性
safety fuel 防爆燃料

safety function 安全功能
safety fuse 安全保险丝，安全熔断器，保险丝，熔丝
safety fuunel tube 安全漏斗管
safety gap 安全隙，保安放电器
safety gate 安全闸门
safety gland 安全密封套，安全填料盖
safety glasses 安全镜，安全眼镜，防护眼镜
safety goal 安全目标
safety goggles 护目镜
safety governor 危急保安器，危急截断器
safety guards 护栏，安全罩，安全护栅，保险板
safety guide 安全导则
safety handling 安全操作
safety hardware 安全硬件
safety harness 安全导线
safety hat 安全帽
safety helmet 安全盔，安全帽
safety impedance 安全阻抗
safety injection process 安全注射过程
safety injection pump 安全注射泵
safety injection system 安全注射系统，安全注入系统
safety injection tank 安全注射箱
safety injection 安全注射，安全注入
safety in production 安全生产
safety inspection 安全检查
safety installation 安全装置
safety instruction 安全规定，安全要求
safety instrumented function 安全仪表功能
safety instrumented system 安全仪表系统
safety integrity level 安全完整性等级，安全综合水平
safety interlock system 安全联锁系统
safety interlock 安全闭锁装置，安全联锁
safety island 安全岛
safety isolating transformer 安全隔离变压器
safety isolating valve 安全隔离阀
safety lamp 安全灯【井道用】
safety lighting 安全照明【火灾出口信号等】
safety light 安全信号灯
safety limit 安全极限，安全限值，安全限度，安全范围
safety load 安全负载，安全负荷，安全荷载，容许荷载，容许负载
safety logic assembly 安全逻辑装置
safety logic circuit 安全逻辑电路
safety loop 安全保护回路，安全圈，保险圈
safety management 安全管理
safety manager 安全经济
safety margin 安全极限，安全界限，安全余度，安全范围，安全边缘，安全裕量
safety marking 安全标志
safety measures 安全措施
safety member 安全棒，事故棒，（反应堆）事故控制元件
safety monitor for steam turbine 汽轮机安全监测装置
safety monitoring assembly 安全监测装置
safety monitoring 安全监测
safety monitor 安全监测器
safety net 安全网
safety nozzle 安全阀接管，安全管嘴
safety officer 安全员
safety-operating life 安全运行寿命
safety operation condition 安全运行工况
safety operation specification 安全操作规程
safety operation 安全操作，安全运行
safety panel 安全盘
safety parameter display console 安全参数显示控制台
safety passage and escape stair 安全通道和太平梯
safety passage 安全通道
safety performance 安全性能
safety philosophy 安全原则，安全原理
safety pin 定位销，安全销
safety plug socket 安全塞座，熔丝塞座
safety plug 安全塞，熔丝塞，熔线塞
safety power consumption 安全用电
safety power cutback system 安全功率下降系统
safety practice 安全措施
safety precautions 安全措施，安全保护，预防措施，安全防范
safety pressure margin 富裕压力
safety procedure 安全规程，安全程序，安全操作程序
safety protection 安全防护
safety provision 安全措施
safety recommendation 安全规则，安全建议
safety regulations 安全规程，安全守则
safety-related auxiliaries 安全相关的辅助系数
safety-related component 安全相关部件
safety-related electrical equipment 安全级电气设备
safety-related equipment 安全相关的设备
safety-related event 安全相关事件
safety-related feed and bleed function 安全相关的给排功能
safety-related function 安全相关功能
safety-related incident 安全相关事件
safety-related system 安全相关系统
safety relay 安全继电器，断路继电器，保护继电器
safety release spring 安全放松弹簧
safety relief valve 安全泄放阀，安全泄压阀
safety requirement 安全规定，安全要求
safety resin catcher 树脂安全收集器，树脂安全阱
safety resin trap 树脂安全阱，树脂安全收集器
safety rod channel 安全棒管道
safety rod 事故棒，安全棒，停堆棒【反应堆】
safety rules 安全规程，安全规则，安全规定，安全守则
safety screen 安全筛，安全网，保护挡板，安全隔层
safety seal 安全密封
safety setting 安全整定值
safety shoes 安全靴
safety shower 安全喷淋，应急喷淋
safety shutdown system 安全停堆系统
safety shutdown 事故保护停堆，安全停堆

safety shutoff valve 燃油快速关断阀
safety siding 安全避车道，安全线
safety signal 安全信号
safety significant incident 重要安全事件
safety slipping coupling 安全滑动（补偿）联轴器
safety source 保安电源
safety standard 安全标准
safety stop 安全制动，安全停止
safety supervision section 安监科
safety supervision 安全监督
safety switch 保险开关，安全开关，安全按钮，紧急开关，快速停堆按钮
safety system functional inspection 安全系统功能检查
safety system support features 安全系统辅助设施
safety system 安全系统
safety tape 安全带
safety technique 保安技术
safety temperature 安全温度
safety terminal box 保险接线盒
safety test box 安全试验盒
safety test 安全试验
safety train 安全系列
safety trip lever 安全触摆杆，紧急触摆杆
safety trip valve 自动断路阀，燃油快速关断阀，安全跳闸阀
safety trip 事故保护停堆，安全脱扣装置，紧急断开装置，安全脱扣器
safety tube 安全管，安全漏斗
safety valve blowback 安全阀回座
safety valve easing gear 安全阀松动装置
safety valve easing 安全阀松动，安全阀疏水
safety valve escape 安全阀泄漏
safety valve muffler 安全阀消声器
safety valve of deaerator 除氧器安全门
safety valve operating pressure 安全门动作压力
safety valve operation test 安全阀动作试验
safety valve relief 安全阀，保险阀
safety valve reseating pressure 安全门回座压力
safety valve seat 安全阀座
safety valve set pressure 安全阀设定压力
safety valve stand 安全阀支架
safety valve 安全阀，安全门，保险阀
safety voltage 安全电压
safety water tube boiler 安全水管锅炉
safety window 安全窗
safety wire 保险丝
safety zone protective zone 安全区
safety zone 安全区
safety 安全，安全性，稳定，可靠，保护，保险，安全设备，保险装置
safe wet mass （临界）安全湿质量
safe wind speed for lifting 吊装安全风速
safe working load 安全工作负荷，安全资用荷载
safe working pressure 安全工作压力，设计压力
safe working stress 安全工作应力
safe yield 安全给水量，安全开采量，可靠产量
safe 安全的，可靠的，保险箱
SAFP(shop assembled fabricated piece) 车间组装的制作件
SAFR(sodium advanced fast reactor) 先进钠冷快堆
SAFSTOR(safe storage) 安全贮存
safty clearance for power transmission lines 输电线的安全间隙
SAF 密封风机【DCS 画面】
sag and tension chart 弧垂-张力曲线
sag and tension curve 弧垂-张力曲线
sag and tension table 盘面的倾斜和拉紧
sag calculation 弛度计算，弧垂计算
sag correction 垂度校正
sag curve 挠度曲线
sagging moment 下垂力矩，正弯矩
sagging of drum 汽包下垂度
sagging plate （凝汽器的）隔板，中间隔板
sagging price 极度下跌价格
sagging 下垂，垂度
sag of the span 架空明线的垂度，挡间弛度
sag ratio 垂度和跨度比，垂跨比
sag rod 系杆，拉杆，吊杆
sag roller 托辊
sags and crests 凹凸不平
SAG(standard address generator) 标准地址发生器
SAG(system analysis group) 系统分析组
sag template 弧垂模板
sag 弧垂【架空线路】，弛度，松弛，凹陷，下垂，（下）垂度，挠度，下降，下沉，沉降，下陷，（物价的）下跌，萧条
SAH drain and vent system 暖风机疏水及排气系统
SAH(steam air heater) 暖风机，暖风器【DCS 画面】
sailaba 泛滥平原
sail cloth 帆布
sailer 帆船
sail fabric 风帆蒙布
sail handing 帆操纵
sailing boat 帆船
sailing course 航道，航线
sailing date 开航日期，启航日
sailing schedule 船期表
sailing ship 帆船
sailing stage 通航期
sailing vessel 帆船
sailing yacht 帆艇
sailing 航海，航行
sailor 水手
sail tip 帆尖
sail wind generator 帆翼风力发电机
sail wing rotor 帆翼风轮
sailwing wind generator 帆翼式风力发电机
sail wing windmill 帆翼风车
sail 帆片，风帆
salability 畅销，销路，可售性，适销性
salable coal sample 商品煤样
salable goods 畅销货
salable item 可售产品
salable 畅销的，适于销售的
salamander 焙烧炉

sal-ammoniac cell 氯化氨电池
salary 薪金
sale by instalments 分期分批出售
sale of bidding document 出售标书
sales allowance 销售折让
sales clause 销售条款
sales confirmation 销售确认书
sales contract 销售合同
sales expenses 销售费用
sales invoice 销售发票
sales manager 销售经理，营业主任
salesman 销售员
sales price to network 上网电价
Sales price to power grid 上网电价
sales profit 销售利润
sales program 推销计划
sales tax and extra charges 销售税金（及）附加
sales tax 销售税，营业税
salicylaldehyde 水杨醛
salicylate 水杨酸盐
salicylic acid 水杨酸
salient edge 凸缘，凸角
salient feature 主要特征
salient instrument 凸装型仪表
salient point 凸起点
salient pole armature 凸极电机
salient pole electric rotor 凸极转子
salient pole generator 凸极式发电机，凸极发电机
salient pole machine 凸极电机
salient pole motor 凸极电动机
salient pole rotor 凸极转子
salient pole stator frame 凸极定子框架
salient pole stator winding frame 凸极定子绕组框架
salient pole synchronous generator 凸极同步发电机
salient pole 显极，凸极
salient trait 特征
salient 显著的，突出的，凸出的，显现的，凸角，扇形地背斜轴
salification 积盐
salina 盐水蒸发槽，盐田，盐碱滩
saline area 盐渍地区
saline concentration 含盐度，含盐量
saline land 盐地
salineland 盐碱池
saline solution 盐溶液
saline tolerant plant 耐盐碱植物
saline water conversion power reactor plant 海水淡化核动力厂
saline waters 咸水域
saline 含盐的，盐水，盐场
salinity anomaly 盐度反常
salinity circulation 盐度环流，盐度循环
salinity determination 盐度测定
salinity diference power generation 海洋盐浓度差发电
salinity gradient 含盐量梯度，盐度梯度
salinity indicator 盐量计
salinity intrusion 盐分侵入
salinity meter 测盐计，含盐量测定计
salinity tolerance 耐盐度
salinity 含盐量，含盐度，盐浓度，盐分，咸度，盐度
salinization 盐碱化
salinometer 盐分计，盐量计，盐度计，盐液密度计，测盐计，含盐量测定计
SAL(save address latch) 保存地址门闩
SAL(symbolic assembly language) 符号汇编语言
salt accumulation 盐分蓄积
salt affected soil 盐碱土壤
salt age value 残值
saltating soil particle 跃移土粒
saltation friction velocity 跃移摩擦速度
saltation Froude number 跃移弗劳德数
saltation load 跃移质
saltation mode 跃移模型
saltation phase 跃移相位
saltation transport 跃移输送
saltation velocity threshold 跃移启动速度
saltation velocity 跳跃速度
saltation （泥沙的）跃移，跳跃，跳动，突然变动
salt bath brazing 盐浴钎焊
salt bath carburizing 盐浴渗碳
salt bath dip brazing 盐浴浸渍硬钎焊
salt bath 盐浴槽
salt bearing liquid wastes 含盐废液
salt bearing rock 含盐类岩石
salt carryover 盐分携带
salt concentration gradient 盐浓度梯度
salt concentration 盐浓度，含盐量
salt content 盐分，含盐量
salt cote 盐泉，盐矿（井）
salt deficient dehydration 缺盐性脱水
salt deposit 盐垢，积盐
salt detector 盐分探测器
salt-dilution method 盐分稀释法
salted water 盐水，咸水
salt elimination 除盐
salt endurance 耐盐性
salt entrain 盐分携带
saltern 盐场
salt exclusion 除盐
salt film 盐膜
salt filter 滤盐器
salt fog test 盐雾试验
salt fog 盐雾
salt haze 盐霾
salt index 含盐指数
saltiness 含盐性
salting agent 盐化剂
salting-out agent 盐析剂【混凝土】
salting-out 盐析，加盐分离
salt intrusion 盐分侵入
salt laden air 盐雾
salt lake 盐水湖
salt meter 盐量计
salt mine disposal 盐矿处置
salt mine repository 盐矿处置库
salt mine 盐矿

salt mist resistant 防盐性，耐盐雾
salt pan 盐场
salt penetration 渗盐
saltpeter 硝石
salt pillar 岩盐柱
salt rejection ratio 脱盐率，盐排除率
salt-resistive insulator 耐盐绝缘子
salt screen 盐类屏
salt section 锅水盐段
salts of strong anions 强阴离子盐
saltsplitting capacity 盐劈容量，盐劈，中性盐分解容量
saltsplitting 盐劈
salt spray fog corrosion 盐雾腐蚀
salt spray test 盐雾腐蚀试验
salt spray 盐雾
salt tolerance 耐盐性
saltus 跃幅，振幅，急变，飞跃
salt water barrier 咸水屏障
salt water circulating pump 海水循环泵
salt water cooling system 海水冷却系统
salt water intrusion 咸水入侵
salt water lock 防咸水闸
salt water stratified towing tank 分层盐水托槽
salt water underrun 咸水潜流
salt windmill 盐场风车
salt wind 盐风
salty soil 盐渍土
salty 含盐的
salt 盐
salutatory 欢迎词
salvageable tool 可抢救的工具
salvageable 可抢救的
salvage boat 打捞船
salvage value 残值
salvage 废物利用，废品处理，废物处理，修补，抢救，利用废料
SAMA diagram(scientific apparatus manufacturer association diagram) SAMA 图，（美国）科学仪器制造商协会图例
samarium build-up 钐积累
samarium poisoning 钐中毒
samarium valley override 钐谷补偿
samarium 钐，Sm
SAMA(Scientific Apparatus Manufacturer Association) （美国）科学仪器制造商协会
same assignment 同一批货
same educational level 同等学历
same-phase 同相，同相位
same-size ratio 等量比
samica 粉云母
samlped hologram 取样全息图，取样全息照片
sampe system miniaturization 取样系统小型化
sample analysis 样品分析
sample-and-hold amplifier 取样与同步放大器
sample-and-hold circuit 采样和保持电路
sample bias 样品误差
sample bottle 取样瓶
sample bulb 取样球管
sample carrier 试样容器，样品容器
sample car 样车
sample chamber 样品室
sample changer 样品更换器
sample changing mechanism 样品更换器
sample chute 取样落煤管
sample collecting tube 收集样品的管子，联样管
sample collection 煤样收集
sample collector 取样器
sample conditioning equipment 取样调节装置
sample connection 采样接口，连接试样
sample contamination 样品污染
sample cooler 取样冷却器
sample coupon （反应堆材料）试样件，试样块
sample covariance 样本协方差
sample cutter 截样装置
sample cutting device 截样装置
sampled analog data 抽样模拟数据，采样模拟数据
sampled analogue value 模拟值采样值
sampled data analog computer 采样数据模拟计算机
sampled data computer 抽样数据计算机，采样数据计算机
sampled data control system 采样数据控制系统，数据采样控制系统
sampled data system 数据取样系统，抽样数据系统，数据采样系统
sampled data 取样数据，样本数据，采样数据
sample division 试样缩分
sampled measured value control 采样测量值控制
sampled measured value 采样测量值
sampled offer 附样报盘
sample drain tank 试样排放箱
sample drawn 抽样
sampled signal 采样信号，取样信号
sampled value control 采样值控制
sampled value 采样值
sample for commercial coal 商品煤样
sample for reference 参考样品
sample frequency 采样频率
sample heat exchanger 样品热交换器，样品冷却器
sample holder 样品盒，样品夹
sample hole 取样孔
sample hopper 标准漏斗车
sample house 取样器，取样间
sample importance function 取样价值函数
sample in measurement 样品测量
sample interval 采样时间间隔，取样间隔，试样间距
sample investigation 抽样调查
sample irradiation hole 样品照射孔
sample line 取样管线
sample log 测井取样剖面，取样记录
sample mixing 试样混合
sample of commercial coal 商品煤样
sample operating system 样本操作系统
sample panel 取样盘
sample period 取样周期
sample plot 样图
sample preparation 制样
sample program 抽样程序，取样程序

sample reduction 试样破碎
sample reject chute 剩余煤样的落煤管
sample room 样品室
sampler spoon 取样勺，采样勺
sampler 取样器，采样机，抽样转换器，采样器，样板，抽样
sample sink 样品处理小池，取样槽，样品槽
sample size 样本大小
sample space 取样空间，取样间
samples per second 每秒抽样数
sample splitting 煤样分割
sample standard deviation 样板标准差
sample statistics 样本统计（学）
sample strength 样品（放射性）活度，样品强度
sample strip 采样带
sample survey 抽样检查
sample taker 取样器，采样器
sample-taking 采样
sample tap 取样口
sample thief 取样器，采样器
sample to background activity ratio 样品本底活度比
sample treatment 样品处理
sample tube 管样
sample variance 采样离散，样品离散，采样方差，样品方差
sample 样品，取样，试样，抽样，样本，标本，试块，从……取样，采样
sampling action 取样，采样，抽样，取样动作，采样动作，采样作用
sampling analyzer panel 取样分析仪表盘
sampling and decontamination rooms 取样及去污室
sampling aperture 取样孔
sampling apparatus 采样装置，取样器，取样装置
sampling area 取样区
sampling bag 采样袋
sampling bench 取样台
sampling bottle 取样瓶［管］
sampling building 取样楼
sampling by electrostatic precipitation 静电沉淀法采样，静电集尘法采样
sampling by filtration 过滤采样
sampling cabinet 取样箱
sampling chamber 取样室
sampling circuit 采样电路，幅度脉冲变换电路
sampling controller 采样控制器
sampling control 采样控制，取样控制
sampling cooler 取样冷却器
sampling cross-section 采样断面
sampling device 取样设备，取样装置
sampling distribution 采样分布
sampling disturbance 取样扰动
sampling element 采样元件
sampling equipment 取样设备
sampling error 抽样误差，样品误差，取样误差
sampling frequency 取样次数，取样频率，采样频率
sampling function 采样函数
sampling hole 测孔
sampling inspection 抽样检查，抽样检验，取样检查，取样检验
sampling instant 采样时间
sampling instrument 取样器，取样仪表，采样器，采样仪表，采样工具
sampling interval time 取样间隔时间
sampling interval 采样间隔，取样间隔
sampling leader 采样管线
sampling line 取样管路
sampling method 取样方法，取样法
sampling network 信号取样网络
sampling normal distribution 抽样正态分布
sampling nozzle 取样头，取样水嘴，取样接管，取样管嘴
sampling observation 取样检查
sampling of aerosol 气溶胶采样
sampling oscilloscope 取样示波器
sampling parametric computation 抽样参数计算，采样参数计算
sampling period 采样周期
sampling pipe 取样管（路）
sampling plan 取样计划
sampling point 取样（地）点
sampling port 采样口，取样孔，测孔
sampling position 采样地点
sampling probe 采样探头，采样探针
sampling process 抽样过程，采样过程
sampling pulse 抽样脉冲，选通脉冲
sampling rake 取样排管
sampling rate 取样率
sampling room 取样室
sampling rule 抽样规则
sampling servomechanism 采样伺服机构
sampling servosystem 采样伺服系统
sampling signal 采样信号
sampling sink 取样水槽
sampling spoon 取土筒，取样环刀，取样勺
sampling station 采样站，监测站，取样站
sampling survey method 抽样调查法
sampling survey 抽样调查
sampling system 采样系统，取样系统
sampling tap 取样管咀
sampling technique 采样技术，取样技术，取样方法
sampling test 样品试验，样机试验，抽样试验
sampling theorem in the frequency domain 频域采样定理
sampling theorem 抽样定理，采样原则，采样定理
sampling theory 采样理论
sampling thief 采样器
sampling time 取样时间
sampling tool 取样工具，取样器械
sampling tube 取土筒，取样管
sampling turntable 取样转盘
sampling unit 采样单元，取样装置
sampling valve 取样阀
sampling variance 抽样方差
sampling waste 取样废水
sampling well 取样井，样品槽

sampling zone　取样区
sampling　采样，取样，抽样，抽样试验
samploscope　取样示波器
SAMS(sampling analog memory system)　抽样模拟存储系统
Samsung　(韩国)三星公司
sanatron　窄脉冲多谐振荡器
sanction number　批准文号
sanction　许可，制裁
sand-aggregate ratio　砂骨比
sandal-wood　檀香木
sand and coarse aggregate ratio　砂石比
sand and gravel alluvium　砂砾石冲积层
sand and gravel extraction　砂石精选
sand and gravel overlay　砂砾覆盖层
sand and gravel trap　沉砂砾池，沉砂砾井，沉砂砾槽，沉砂砾段
sand and gravel wash　洗涤砂及砾石
sand and gravel　砂砾石，砂石料
sand asphalt　砂质沥青
sand avalanche　砂崩
sand backfill　砂土回填
sandbag cofferdam　砂袋围堰
sandbag revetment　砂袋护坡
sandbag walling　砂袋筑墙
sandbag wall　砂袋护墙
sandbag　砂包，砂袋
sand bailer　拦砂筒
sand bank　沙坝，沙滩
sand barrier　砂埂
sand bar　砂坝
sand basin　沉砂池
sand bath　喷砂清洗，砂浴锅，砂浴
sand beach　沙滩
sand bearing wind　带沙风，含沙风
sand bedding course　砂垫层
sand bed filter　砂滤器，砂滤床，砂床过滤器
sand bed　砂层，砂床
sand-bentonite fill　砂膨润土充填
sand bitumen core　沥青砂土心墙
sand blanket　砂垫层
sand blast apparatus　喷砂机，喷砂装置
sand blast cleaning　喷砂清理
sand blaster　喷砂机
sand blast finish　喷砂饰面，喷砂修整
sand blasting & dust removing system　喷砂与除尘系统
sand blasting gun　喷砂枪
sand blasting machine　喷砂机
sand blasting method　吹砂清理法
sand blasting　吹砂，喷砂，喷砂冲毛，喷砂处理，喷砂清除法
sand blast unit　喷砂装置
sand blast　喷砂（处理）
sand boil　喷砂，砂沸，涌砂，冒水翻砂
sand borrow area　砂料场
sand-break　防沙林
sand by-pass　旁通输沙道
sand carrier with grab bucket　带抓斗运砂船
sand casting　砂型铸造，砂型铸件
sand catcher　水中集砂器
sand classifier　砂子选分机
sand-clay pavement　砂质黏土路面
sand cleaning machine　清沙机
sand cloud　沙云
sand-coarse aggregate ratio　砂与粗骨料比
sand compaction pile　挤密砂桩
sand control dam　防沙堤
sand cracker　砂子裂解炉
sand cushion　砂垫层
sand deposit　砂层
sand desert　纯沙沙漠，沙漠
sand devil　沙旋风，沙暴
sand disk　砂轮
sand drain　砂井，砂井排水，填沙排水井
sand drifting　流沙
sand drift　风沙，沙丘
sand-driving wind　风沙流，挟沙风
sand dryer　烘砂器
sand dune area　沙丘地区
sand dune fixation　沙丘固定
sand dune stabilization　沙丘固定
sand dune　沙丘
sanded bitumen felt　铺砂的沥青毡
sand ejector　喷砂器
sand eliminator　除沙设备
sand equivalent test　含沙当量试验
sand equivalent　含沙当量
sand erosion　砂蚀
sander　喷砂机，撒沙器
sand escape　排沙道，排砂道
sand falls　砂瀑（布）
sand-filled drainage well　填沙排水井
sand filter bed　沙滤层，沙滤床
sand filter layer　砂质过滤层
sand filter　砂滤器，砂滤池
sand filtration　砂滤法，砂滤
sand-flash valve　排沙阀
sand flat　沙滩，砂坪
sand flood　沙暴
sand flow　沙流
sand flushing canal　冲沙渠（道）
sand flushing channel　冲沙渠（道）
sand flushing　冲沙
sand foundation　砂基
sand fraction　沙粒径
sandglass　沙漏【用于计算时间】
sand grading　砂的级配
sand grain　砂粒
sand-gravel aggregate　砂砾石骨料
sand-gravel cushion　砂石垫层
sand-gravel foundation grouting　砂砾石地基灌浆
sand-gravel mixture　砂子砾石配比【骨料】
sand-gravel pile　砂石桩法
sand grout　砂浆
sand-guide channel　导沙槽
sand-guide sill　导沙坎
sand haze　沙霾
sand heap analogy　堆沙模拟
sand hedge　沙丘围护
sand hill　沙阜，沙冈，沙丘
sand hole　砂眼

sandier spoon 取样勺
sand inclusion 夹砂
sanding machine 带式抛光机，喷砂机，撒沙器
sanding 喷砂，用砂纸打磨，砂纸打光
sand island method 筑砂岛法
sand jack 砂箱千斤顶
sand layer 砂层
sand lens 透镜状砂层
sand levee 沙堤
sand-like 砂状的
sand lime brick 灰砂砖
sand line 砂层线
sand mat 砂垫层
sand mill 碾砂机
sand mist 沙雾
sand mound 沙堆
sand moving wind 起沙风
sand paper 砂纸
sand pile 砂桩
sand pit 采砂场，采砂坑
sand plant 采砂场
sand prevention 防沙
sand production 砂料开采
sand-protecting dam 防沙坝，防沙堤
sand-protecting plantation 防沙林
sand pump 抽沙泵，排沙泵，沙泵，吸沙泵
sand ridge 沙脊，沙丘线
sand ripple 砂纹
sand rock 砂岩
sand sampler 采砂器
sand screen 砂筛
sand sediment trap 沉沙槽，沉沙井
sand shadow 沙影，背风积沙区
sand sheet 沙片
sand shoal 浅沙滩
sand size 砂粒大小
sand sluicing 冲沙
sand snow 沙性雪
sand soil 砂土
sand specimen 砂样
sand spit 砂嘴
sandstone 砂岩，砂石
sand storage pit 储砂坑
sand storm 暴风砂，沙尘暴，风沙，沙暴，尘暴
sand stratum 砂层
sand streak 起砂【混凝土表面】
sand stream 沙流
sand strip 沙带，截沙设施
sand sucker 泥泵
sand tornado 沙龙卷
sand-total aggregate ratio 砂与骨料总量之比
sand trap scour gate 砂阱式冲砂闸门
sand trap 集砂器，分砂器，沙坑障碍，拦沙坑，拦沙阱，沙挡，沙槽
sand-washing machine 洗砂机
sand-washing plant 洗砂厂
sand-water mixture 砂水混合物
sand-water separator 砂水分离器
sand wave 沙浪
sand whirl 沙旋
sandwich arrangement 交错重叠布置
sandwich beam 夹层梁，层结梁，多层组合梁，多层叠合梁
sandwich brush 分层电刷
sandwich cell 夹层电池
sandwich construction 夹层结构
sandwich digit 中间数字，中间位
sandwiched packed bed 层状填充床
sandwiched 夹在当中的
sandwich frame 双构架
sandwich irradiation 夹层辐照
sandwich isolater 剪切型防松器
sandwich panel 夹层板，多层板，复合板，夹心板
sandwich plate type fuel element 夹心板型燃料元件
sandwich plate 多层板
sandwich structure 夹层结构，多层结构，多层叠合结构
sandwich tape 多层带
sandwich-type element （晶体管的）多层结构元件
sandwich type wall panel 夹心墙板
sandwich type 夹层型的
sandwich wall panel 复合墙板，夹芯墙板
sandwich winding 交错多层绕组，叠层绕组，饼式绕组，交叠式线圈
sandwich wound coil 交错多层绕组线圈，盘式线圈，饼式线圈
sandwich-wound 叠层绕的，分层绕的
sandwich 夹层（结构），多层（结构），夹心，多层的
sand yard 堆砂场，砂料场
sandy clay foundation 砂质黏土地基
sandy clay loam 砂质黏壤土
sandy clay 含沙黏土，砂质黏土，砂黏土
sandy gravel stratum 沙砾层
sandy gravel 含砂砾石，砂砾石
sandy limestone 砂质石灰石
sandy loam 砂质壤土，砂壤土
sandy loess 砂质黄土
sandy marl 砂质泥灰岩
sandy shale 砂质页岩
sandy silt 砂质粉砂，砂质粉土，砂质淤泥，砂质黏土淤泥
sandy slate 沙质板岩
sandy soil 含砂土，砂质土，砂类土
sandy 砂质的
sand 砂，沙（土），沙滩，沙地
sanitarium 养老院
sanitary drainage 生活污水排放，生活污水排泄系统
sanitary effluent 生活排水
sanitary engineering 卫生工程（学）
sanitary equipment 卫生设备
sanitary facilities 卫生设备
sanitary fittings 卫生设备配件，洁具，卫生设施
sanitary fixture 卫生器具
sanitary lock 卫生舱
sanitary pipe 污水管
sanitary pump 生活污水泵，污水泵
sanitary sewage 厕所污水，生活污水，生活污

水管，污水管
sanitary standard for drinking water 饮用水卫生标准
sanitary ventilation 卫生通风
sanitary venting system 卫生通气系统
sanitary wastewater 厕所污水，生活污水
sanitary waste 生活污水
sanitary works 卫生工程
sanitary 卫生的
sanitation （环境）卫生，卫生设备，下水设备
SANTA（systematic analog network testing approach）系统化模拟网络测试方法
SAON(semi-auto on) 半自动开
SAO(stand alone operation) 独立操作
saphe 拟位相，同态相位
saponaceous liquid wastes 皂质废液
saponification number 皂化值
saponification 皂化，皂化
sapphire 蓝宝石
saprocol 灰质腐泥
saprolith 残余土
sapromixite 藻煤
sapropelic coal 腐藻煤
sapropel 腐泥，煤泥，腐殖泥，腐殖质
saprophytic bacteria 腐生细菌
saproplankton 污水浮游生物
sapropolization 腐泥化作用
SAP(share assembly program) 共享汇编程序
SAP(symbolic assembly program) 符号汇编程序
SAP(system assurance program) 系统保险程序
sap 渗碳钢棒的中心未渗碳部分
SAP 压缩空气生产系统【核电站系统代码】
saran 萨冉树脂，莎纶
sarking board 屋面衬板
SAR(safety analysis report) 安全分析报告
SAR(storage address register) 存储地址寄存器
SAR 仪表用压缩空气分配系统【核电站系统代码】
SASAC(State-owned Assets Supervision and Administration Commission)（国务院）国有资产管理委员会，国资委
SASD(single-access single-distribution) 单存取单分配
SA(secondary air) 二次风
SA(semi-auto off) 半自动关
sash-bar rotor 倒丁字形线棒的鼠笼转子
sash bar 窗框条
sash bolt 窗插销
sash chain 吊窗链
sashcord 吊窗绳
sash frame 窗框
sash handle 窗拉手
sash pulley 吊窗滑轮
sash window 框格窗，上下拉动的窗子
sash 窗框，吊窗，窗框格，门框，窗扇
SAS installation 变电站自动化系统设备
SAS parameter set 变电站自动化系统参数集
SAS product family 变电站自动化系统产品系列
SAS(safety assessment system) 安全评价系统
SAS(self-adaptive system) 自适应系统
SAS(substation automation system) 变电站自动化系统
SASSY(small aperture separator system) 小隙缝分离系统
SAST(safety standard) 安全标准
SA(substation automation) 变电站自动化
SATCOM(satellite communication) 卫星通信
SATD＝saturated 饱和的
satellite altimetry 人造卫星测高法
satellite altitude determination 卫星测高
satellite based radio equipment 卫星无线电设备
satellite borne sensing system 卫星上传感系统
satellite borne sensor 卫星上传感器
satellite city 卫星城
satellite communication 卫星通讯
satellite computer 辅助计算机，卫星计算机
satellite equipment 卫星装置，中继装置
satellite geodesy 卫星大地测量学
satellite hole 辅助炮眼
satellite image 卫星图像，卫星影像
satellite line 卫线，伴线
satellite meteorology 卫星气象学
satellite nuclear power station 卫星核动力站
satellite observation 卫星观测
satellite peak 伴峰
satellite photograph 卫星图像，卫星照片
satellite processor 卫星处理机
satellite solar power station 卫星太阳能电站
satellite station 小型中继台，卫星电台，星际站
satellite-switched multiple-access system 卫星转接多路通信系统
satellite town 卫星城镇
satellite travelling wave tube 卫星行波管
satellite 卫星，人造卫星，卫星楼，附属物，伴线，附属的，辅助的
satisfactory service 服务周到
satisfactory stall 良好失速
satisfactory 满意的
satisfy 令人满意，令人满足，满足，说服，使相信，使满意
SAT＝saturate 饱和的
SAT(simulator acceptance test) 模拟器验收试验
SAT(site acceptance test) 工地验收测试
SAT(site availability test) 现场可利用率试验
saturability 可饱和性，饱和度，饱和能力
saturable choke 饱和抗流圈，饱和扼流圈
saturable core generator 饱和铁芯发电机
saturable core inductor 饱和铁芯线圈
saturable core 饱和铁芯，饱和磁芯
saturable inductor 饱和电感线圈
saturable reactor 饱和电抗器，饱和扼流圈
saturable servosystem 饱和伺服系统
saturable transformer 饱和变压器
saturable 饱和的
saturated activity 饱和放射性
saturated adiabatic lapse rate 饱和绝热直减率
saturated adiabatic process 饱和绝热过程
saturated adiabatic 饱和绝热的
saturated air 饱和空气
saturated asynchronous motor 饱和异步电动机
saturated belt 饱和带
saturated blowdown stage 饱和喷放阶段

saturated blowdown 饱和喷放，饱和水急剧流失
saturated boiling 饱和沸腾
saturated brine 饱和食盐溶液，饱和盐水
saturated calomel electrode 饱和甘汞电极
saturated Carnot vapor cycle 饱和蒸汽卡诺循环
saturated clay 饱和黏土
saturated core 饱和铁芯
saturated diode valve 饱和二极管
saturated diode 饱和二极管
saturated dissolved oxygen 饱和溶解氧
saturated homopolar alternator 饱和单极同步发电机
saturated humidity 饱和湿度
saturated hysteresis 饱和滞后
saturated index 饱和指数
saturated layer 饱和层
saturated liquid curve 饱和液态线
saturated liquid 饱和液
saturated logic circuit 饱和型逻辑电路
saturated machine 饱和电机
saturated magnetic induction density 饱和磁感应强度
saturated moist air 饱和湿空气
saturated permeability 饱和透水性
saturated pH value 饱和 pH 值
saturated polarization 饱和极化
saturated pole 饱和磁极
saturated pressure 饱和压力
saturated reactance 饱和电抗
saturated reactor compensator 饱和电抗补偿器
saturated reactor-type stabilizer 饱和电抗器型稳压器
saturated reactor 饱和电抗器
saturated recovery time 饱和恢复时间
saturated region 饱和区
saturated release time 饱和释放时间
saturated resin 饱和树脂
saturated rock 饱和岩层
saturated sample 饱和试样
saturated sand 饱和砂
saturated soil 饱和土（壤）
saturated solution 饱和溶液
saturated specimen 饱和试样
saturated-stabilizer transformer 饱和稳压器的变压器
saturated state 饱和状态
saturated steam cycle 饱和蒸汽循环
saturated steam generator 饱和蒸汽发生器
saturated-steam phase 饱和蒸汽状态
saturated steam pipe 饱和蒸汽管
saturated steam safety valve 饱和蒸汽安全门
saturated steam supply line 饱和蒸汽供应管线
saturated-steam temperature 饱和蒸汽温度
saturated steam turbine 饱和蒸汽汽轮机，湿蒸汽汽轮机
saturated steam 饱和蒸汽
saturated surface 饱和面
saturated transductor 饱和磁放大器
saturated unit weight 饱和单位重量，饱和幺重，饱和容重
saturated vapor pressure 饱和蒸汽压，饱和蒸汽压力
saturated vapor state 饱和蒸汽状态
saturated vapor temperature 饱和蒸汽温度
saturated vapor 饱和水汽
saturated-water phase 饱和水状态
saturated water 饱和水，饱和温度时的水
saturated zone 饱和区
saturated 饱和的
saturate 使饱和，使浸透，饱和
saturating current transformer 饱和电流互感器
saturation adiabatic lapse rate 饱和绝热递减率
saturation-adiabatic process 饱和绝热过程，湿绝热过程
saturation adiabat 饱和绝热线，湿绝热
saturation boiling 饱和沸腾
saturation capacity 饱和能力，饱和容量
saturation carrying capacity 饱和夹带能力
saturation characteristics 饱和特性（曲线）
saturation choke 饱和阻流圈
saturation coefficient 饱和度，饱和系数
saturation coil 饱和线圈
saturation concentration 饱和浓度，饱和浓缩，饱和沸腾
saturation condition 饱和状态
saturation constraint 饱和限制
saturation core regulator 饱和铁芯调压器，磁饱和稳压器
saturation current density 饱和电流密度
saturation current 饱和电流
saturation curve 饱和曲线
saturation deficiency 饱和差
saturation deficit 饱和差
saturation degree 饱和度
saturation disintegration rate 饱和衰变速率
saturation effect 饱和效应
saturation extract 饱和浸出液，饱和提取液
saturation factor 饱和系数
saturation field 饱和场
saturation flux density 饱和磁通密度，饱和通量密度
saturation flux 饱和磁通，饱和通量
saturation gain 饱和增益
saturation gradient 饱和梯度
saturation humidity ratio 饱和湿度比
saturation inductance 饱和电感
saturation induction intensity 饱和磁通密度
saturation induction 饱和感应，饱和磁感应强度
saturation level 饱和度
saturation limit 饱和极限
saturation line 饱和线，浸润线
saturation magnetic circuit 饱和磁路
saturation magnetic flux intensity 饱和磁通密度
saturation magnetic induction 饱和磁感应，饱和磁密
saturation magnetization 饱和磁化强度，磁饱和
saturation magnetometer 饱和磁强计
saturation mixing ratio 饱和混合比
saturation moisture content 饱和含水量
saturation of excitation system 励磁系统的饱和
saturation of iron 铁磁饱和，铁芯饱和
saturation of photo-electric current 光电流饱和

saturation of pole 磁极饱和
saturation percentage 饱和百分率
saturation pH value 饱和 pH 值
saturation plane 饱和面
saturation pressure 饱和压力
saturation range 饱和范围
saturation rate 饱和速率
saturation ratio 饱和系数，饱和率
saturation reactance 饱和电抗
saturation region 饱和区
saturation signal 饱和信号，极限信号
saturation specific humidity 饱和比湿
saturation starter 饱和启动器
saturation state 饱和状态
saturation steam pressure 饱和蒸汽压力
saturation steam table 饱和蒸汽表
saturation swelling stress 饱和湿胀应力
saturation temperature effect 饱和温度效应
saturation temperature 饱和温度，饱和点
saturation threshold 饱和阈
saturation transformer 饱和变压器
saturation type acid cell 饱和式酸性电池
saturation value 饱和值
saturation vapour pressure 饱和蒸汽压力，饱和水汽压
saturation vapour tension 饱和蒸汽张力
saturation voltage drop 饱和压降
saturation voltage 饱和电压
saturation water 饱和水
saturation zone 饱和层，饱和带，饱和区
saturation 浸润度，饱和（作用），饱和度，饱和水，渗透
saturator 饱和剂
saturex 饱和器
SAT 公用压缩空气分配系统【核电站系统代码】
sausage antenna 圆柱形天线
SAU(storage access unit) 存储存取部件
savanna climate 热带稀树草原气候
Savathe cycle 定容等压循环，双燃循环
save address latch 保存地址门闩，保存地址寄存器
save area 保存区
save file 副本文件
save instruction 保存指令
save on food 节约粮食
saving account 储蓄账户
saving bank 储蓄银行
saving deposit 储蓄存款
savings 滤屑
saw blade 锯片，锯刀
sawdust 锯末
sawing machines-band 带锯床
sawing machine 锯床
sawing unit 锯开装置
sawn lumber 锯材
saws band 带锯锯带
saw-shaped impulse 锯齿形脉冲
saws horizontal band 卧式带锯
SAW(submerged arc welding) （用焊丝或焊条的）埋弧焊工艺，埋弧焊接，埋弧焊
saws vertical band 立式带锯
saw-tooth antenna 锯齿形天线
saw-tooth attester 锯齿形避雷器
saw-tooth current 锯齿形电流
saw-tooth curve 锯齿形曲线，锯齿波曲线
saw-toothed 锯齿形的
saw-tooth fin 锯齿形鳍板
saw-tooth fouling 锯齿型污垢
saw-tooth generator 锯齿波发生器
saw-tooth impulse 锯齿形脉冲
saw-tooth oscillation 锯齿形振荡
saw-tooth oscillator 锯齿波振荡器
saw-tooth pulse 锯齿形脉冲
saw-tooth roof 锯齿形屋顶，锯齿形屋面
saw-tooth test signal 锯齿形试验信号
saw-tooth trip 锯齿形挡板
saw-tooth type 锯齿形
saw-tooth voltage 锯齿形电压
saw-tooth wave 锯齿波
saw-tooth 锯齿形
saw trace 锯痕
saw 锯
say 大写【表示金额时】
SB alloy 低温电阻合金
S-band S 波段
SBC(single board computer) 单板计算机
S bend S 形弯管（头）
SBE 热洗衣房清洗去污系统【核电站系统代码】
SBFR(settled bed fast reactor) 沉积床快中子反应堆
SBGTS(standby gas treatment system) 备用气体处理系统
SBI(storage bus in) 存储器总线输入
SBLC(standby liquid control system) 备用液体控制系统
SBLWR(soot blower) 吹灰器
SBO(storage bus out) 存储器总线输出
SBP(small bore pipe) 小口径管
SB(selected bit) 被选位
SB(sponge ball) 海绵球
SBT(surface barrier transistor) 表面势垒晶体管
SBU(station buffer unit) 站缓冲部件
SBWR(small boiling water reactor) 小型沸水堆
S by E(south by east) 南偏东
S by W(south by west) 南偏西
scabbled dressing 粗琢面
scabbling hammer 粗琢锤
scab land 崎岖地
scab 表面铸瘤，疵点，铸件表面黏砂
SCADA interface SCADA 接口
SCADA(supervisory control and data acquisition) 监督控制与数据采集，监控与数据采集
SCAE(State Commission for Atomic Energy) （美国）州原子能委员会
scaffold board 脚手板
scaffold building and removing cost 脚手架搭拆费用
scaffolder 架子工，脚手架工
scaffolding 搭脚手架，脚手架材料，搭脚手架的材料

scaffold pole 脚手杆
scaffold 搭脚手架，跨越架，脚手架
SCAF 火检冷却风机【DCS 画面】
scagliola 仿云石
scalability 可量测性，可伸缩性
scalant 结垢物（质）
scalar conductivity 标量电导
scalar control variable 标量控制的变量
scalar control 标量控制
scalar field 标量场
scalar flux 标量通量
scalar function 标量函数
scalar impedence 标量阻抗
scalar matrix 标量矩阵
scalar point function 标量点函数
scalar potential field 标位场，标量势场
scalar potential 标电位，标量势，标量位，标势
scalar product cycle 数量积循环
scalar product 数积，点积，标量积
scalar quantity 标量
scalar 标量的，标量，无向量
SCALD（structured computer-aided logic design）结构化计算机辅助逻辑设计
scale base 比例基数，标尺基线
scale beam 秤杆
scale benefit 规模效益
scale buildup 结垢
scale calibration device 秤校验装置
scale coefficient 水垢（热阻）系数
scale crust 垢层
scale deposit 水垢，积垢
scaled hydraulic model 水力比例模型
scale disk 刻度盘
scale distortion 比例尺变形
scale division 标度分格，分格，刻度，分度
scale down model 缩尺模型
scale down 缩小比例，按比例减少
scale drawing 缩尺图
scaled 薄片状的，有刻度的，成比例的
scale economy 规模经济
scale effect 比尺效应，规模效应，缩尺影响，比尺影响，尺度影响
scale error 比例尺误差，标度误差，定标误差，刻度误差
scale factoring 标度选配
scale factor 比例因子，标度因子，标度因数，比例系数，换算系数，换算因数，污垢系数，比例尺因子，刻度因子，定标因数，缩尺因数
scale formation 结垢，垢的生成，生成水垢
scale-forming material 结垢物质
scale-forming salt 形成垢的盐
scale-forming tendency 结垢倾向
scale-free heating surface 无垢受热面
scale height 均质大气高度
scale inhibitor 阻垢剂
scale interval 格值，分格值，刻度间隔
scale layer 垢层
scale length 标度长度
scale-like particles 类似鳞片状颗粒
scale load 标准载荷，标准负荷
scale lofting 比例放样
scale mark base （标度）基线
scale mark 标度标记，分度，刻度，刻度线，分度标记
scale merit 规模效益
scale model study 缩尺模型研究
scale model test 物理模型试验
scale model turbine 缩尺模型水轮机
scale model 比例模型，刻度模型，缩尺模型，物理模型
scale numbering 标度数字
scale of aerial photograph 航空照片比例尺
scale of eddies 涡旋尺度
scale of enterprise 企业规模
scale off 分层，剥落
scale of hardness 硬度
scale of living 生活水平
scale of pollution 污染标度
scale of ten circuit 十分标电路，十进制换算电路
scale of turbulence 湍流尺度，紊流度
scale-of-two circuit 二进制电路，二进计数器
scale-of-two 二进制的
scale of wind damage 风害规模
scale paper 比例纸，方格纸，坐标纸
scale parameter 尺度参数，标度参量
scale plate 标度盘
scaleplate 刻度盘，刻度板，标度盘，标盘，表盘
scale platform 台秤
scale-production water 生垢水
scale range 标度范围，刻度范围
scale ratio 标度比，尺度比
scale removal 去除氧化皮，除垢，除锈，除鳞，除锅垢，除氧化
scale resistance 水垢热阻
scale resistant steel 抗氧化钢
scale-resisting 抗锈蚀
scaler-printer 打印计算装置，打印换算装置，定标打印机
scaler 除垢器，计数器，换算器，定标器，定标电路，换算电路，脉冲分频器
scale spacing 标度分格间距
scale-up effect 比例放大效应
scale-up factor 放大比例系数
scale-up problem 放大问题
scale up 放大比例，按比例增加
scale value 标度值，刻度值
scale 结垢，水垢，水锈，结盐，氧化皮，刻度（尺），换算，比例尺，标度，秤，规模，天平，鳞片，削皮，剥落，分层，起鳞，定标，（缩放）比例，数值范围，生水锈，测量
scaling action 剥落作用
scaling air fan 密封风机
scaling chip 剥落碎屑
scaling circuit 定标电路，校准电路，计数电路，刻度，标度
scaling factor 放大系数，定标系数，比例系数，换算系数，缩尺因子，缩尺比，定标（比例）因数，污垢因子
scaling law 尺度律
scaling loss 结垢损失

scaling parameter　缩尺参数
scaling rate　结垢速率
scaling ring test　密封环试验
scaling tool　清水垢工具，去垢工具
scaling unit　换算器，换算电路，分频器
scaling zero point　标尺零点
scalloping　开扇形孔
scalped anticline　削峰背斜
scalping grizzly　粗料格筛，粗筛
scalping　铲除草皮
scaly coating　鳞片状涂层
scaly structure　鳞片状构造
scaly texture　鳞片状结构
scan angle　扫描角
scanatron　扫描管
scan converter tube　扫描转换管
scan converter　扫描转换器
scan flyback interval　回描时间
scan-fold chart drive unit　折纸机构
scan-fold chart　折叠式记录纸
scan frequency　扫描频率
scan gate number　扫描门数
scan-in　移位寄存器装入，扫描输入
scan line　扫描线
scanned area　扫描场，扫描范围
scanned point　扫描点
scanner recorder　多点记录器
scanner　扫描仪，扫描天线，扫描设备，析像器，扫描程序，扫描装置
scanning acoustic holography　扫描声全息术
scanning aperture　扫描孔
scanning beam　扫描射束
scanning control register　扫描控制寄存器
scanning device　扫描装置
scanning electron microscope　扫描电子显微镜
scanning electron microscopy　扫描电子显微镜检查，扫描电子显微术，扫描电镜法
scanning element　扫描元件
scanning in switching　交换中扫描
scanning interferometer　扫描干涉仪
scanning linearity　扫描线性
scanning monitor　扫描监测器
scanning motor　扫描电动机
scanning pattern　扫描图形，扫描模式
scanning period　扫描周期
scanning point　扫描光点
scanning projection system　扫描投影系统
scanning rate　扫描率
scanning speed　扫描速度，扫掠速度
scanning system　扫描系统【超声波检验】
scanning voltage　扫描电压
scanning yoke　扫描线圈
scanning　扫描的，观测的，展开，搜索，监测，扫描【超声波检验】
scan period　扫描周期
scan rate　扫描速率，扫描率
scan-round　循环扫描
scansion　扫描，扫掠，图像分析
SCANS(scheduling and control automation by network system)　用网络系统实现的调度和控制自动化
SCAN(switch circuit automatic network)　开关电路自动控制网络
scantling　小块木料，小方材
scan valve　扫描阀
scan　扫描，搜索，扫查
scarce currency　稀缺通货，硬通货
scarcity　缺乏，不足，短缺，缺货
scarfing　火焰表面清理
scarf joint　斜对接接头，嵌接
scarf splice　镶嵌拼接
scarf welding　斜面焊接
scarf　火焰清理，嵌接，切口，斜面，斜角
scarifier　翻土机，松土机
scarify　翻松，翻挖，凿毛，琢毛
scarp zone　悬崖带
scarp　陡坡，陡崖，悬崖，崖，峭壁
SCAR(significant corrective action request)　重大纠正措施要求单
scar　疤痕，伤痕
SCA(secondary control area)　次要控制区，二级控制区
SCA(selectivity clear accumulator)　选择清除累加器
scatter absorption coefficient　散射吸收系数
scatter diagram　散布图，散点图
scattered band　分散带
scattered beam　分散线
scattered data　分散的数据，分散数据
scattered directive radiation intensity ratio　散射比
scattered dose　散射剂量
scattered echo　散射回波
scattered light method　光散射法
scattered light photoelastic analysis　散射光光弹性分析
scattered radiation　散辐射，散射辐射
scattered ray　散射光线，散射线
scattered refueling　分散式换料，插花式换料
scattered wave　散射波
scattered　散射的，耗散的
scatterer　散射体，分散体
scatter factor　分散因子【辐射】
scattering amplitude　扩散幅度
scattering angle　扩散角，散射角
scattering area　散射面积，扩散面积
scattering attenuation coefficient　散射衰减系数
scattering coefficient　散射系数，分散系数，漏损系数
scattering collision　散射碰撞
scattering cross-section　（有效）散射截面
scattering efficiency　散射效率
scattering factor　散射系数
scattering frequency　散布频度
scattering function　散射函数
scattering integral　扩散积分
scattering-in　内部散射
scattering law　散射定律【辐射】
scattering layer　散射层
scattering loss　散射损失
scattering matrix　散射矩阵【辐射】
scattering of sound　声散射
scattering phenomena　散射现象

scattering process 散射过程
scattering region 散射区域
scattering solar irradiance 散射（日射）辐照度
scattering solar radiation 散射日射
scattering volume 散射体积
scattering 散射
scatter loading 分散装入，（反应堆堆芯）插花式装料
scatter range 分散范围
scatter read-write 分散读写
scatter read 分散读入
scatter reloading pattern 插花换料方式
scatter storage technique 分散存储法
scatter write operation 分散写操作
scatter 散射【核反应】，散布，分散，扩散
scavenge oil 废油，（轴承）回油
scavenge port 换气口，扫气口
scavenger pump （轴承）回油泵
scavenger 清除剂，净化剂，换气管
scavenge 除垢，清除，换气，扫气箱
scavenging air 清净空气，除垢空气
scavenging blower 换气鼓风机，清除鼓风机
scavenging coefficient 清除系数
scavenging efficiency 扫气效率
scavenging engine 扫气式发动机，换气式发动机
scavenging precipitation ion exchange process 清除沉淀离子交换法
scavenging process 净化过程，清除过程
scavenging pump 换气泵，清除泵
scavenging ratio 换气比
scavenging stroke 扫气冲程
scavenging 扫气【内燃机】，检查存储器中剩余数据，清除无用数据，清除，打扫，排除废气，吹洗，（通风）清扫
SCAV＝scavenge 吹扫
SCBR（steam-cooled breeder reactor） 蒸汽冷却增殖反应堆
SCC（security sur-charge） 安全附加费
SCC（sequence control counter） 时序控制计数器
SCC（set conditionally） 按条件设置
SCC（single chip computer） 单片计算机
SCC（special conditions of contract） 专用条款，特殊条款
SCC（storage connecting circuit） 存储连接电路
SCC（stress corrosion cracking） 应力腐蚀龟裂，应力腐蚀裂纹
SCC（supervisory computer control） 监督计算机控制（系统），管理计算机控制
SCC（system control centre） 系统控制中心
SCC system 监督计算机控制系统
SCDSB（suppressed-carrier double sideband） 抑制载波的双边带
SCD（severe core damage） 严重堆芯损伤
scenario 对白【人机对话】，假设，假定，情形，（行动的）方案，剧情概要
scenic area 风景区
scenic spot 风景区
scenography 透视图法
scenograph 透视图
SCE（single cycle execute） 执行单行（打印）
SCFBB（solids circulation fluidized bed boiler） 固体循环流化床锅炉
SCFBR（steam-cooled fast-breed reactor） 气冷快中子增殖反应堆
SCFM（subcarrier frequency modulation） 副载波调频
SCF（stress-concentration factor） 应力集中系数
schalstein 辉绿凝灰岩
schedule are subject to change without prior notice 如有变更将不作事先通知
schedule commitment 进度保证
schedule control of program executive 程序执行的调度控制
scheduled completion date 计划完工日期
scheduled contract energy 计划合同电量
scheduled date 计划日期
scheduled decommissioning 计划退役
scheduled discharge burn up 计划卸料燃耗
scheduled downtime 计划停工时间
scheduled frequency offset 规定频率偏移
scheduled frequency 规定频率，计划频率
schedule diagram 进度计划图
scheduled inspection outage 计划检查停堆
scheduled interruption 计划断电
scheduled investment 计划中的投资
scheduled maintenance 计划检修，进度维护，计划维修，预定检修
scheduled net interchange 规定交换量，计划交换量，计划交换功率
scheduled operating time 规定操作时间，规定运转时间
scheduled outage duration 计划停电持续时间
scheduled outage rate 计划停役率，计划停运率，计划检修停机率
scheduled outage times 计划停机次数
scheduled outage 计划停运，计划停炉，计划停机，计划停电，计划停堆，检修停机
scheduled overhaul 计划性检修，计划检修，定期大修
scheduled power consumption 计划用电
scheduled purchasing 计划采购
schedule drawing 工程图，工序图
scheduled reactor outage period 计划停堆周期
scheduled repair reserve 计划内维修备用
scheduled repair 计划修理，定期修理
scheduled shutdown 计划停堆，计划停机，计划停车，计划停运，计划停炉
scheduled ventilation 程序通风
schedule imbalance deviation 计划不平衡差额
schedule number 表号，系列号
schedule of bid price 报价表
schedule of construction 建设进度，工程进度表
schedule of delivery 交货进度
schedule of part 零部件明细表
schedule of payment 付款计划表
schedule of prices 价格表，价格明细表［清单］
schedule of quantity 数量表
schedule of requirements 需求一览表
schedule outage rate 计划停机率
schedule outage 计划停机
schedule planning 日程计划

schedule requirement 进度要求
scheduler 制表人，计划员，调度器，排程器，调度程序，程序机
schedules of rates and prices 价格与费率表
schedule speed 规定速率，规定速度
schedule to an agreement 合同附表
schedule 时间表，资料表，进度表，调度表，表，目录，图表，日程表，安排，计划（表），进度计划，明细表，清单，一览表，系列
scheduling algorithm 调度算法
scheduling and control automation by network system 用网络系统实现的调度和控制自动化
scheduling and the control of production 生产的调度控制
scheduling control automation 自动程序控制
scheduling coordinator 计划协调人
scheduling management display 调度管理显示
scheduling market 计划市场
scheduling rule 调度规则
scheduling system 调度系统
scheduling 编制时间表，编制进度表，计划的拟订，活动计划，进度安排，时序安排
schematic arrangement 简要布置
schematic design phase 初步计划阶段，方案设计阶段
schematic design 方案设计，原理（图设）计，草图设计
schematic diagram 草图，简图，（电路）原理图，概略图，工序图，示意图
schematic drawing 方案图，示意图，略图，简图
schematic layout 示意图，设计原理图，原理图，流程图
schematic map 略图
schematic piping diagram 管道系统示意图
schematic section 示意剖面
schematic wiring diagram 接线原理图
schematic 图解的，简图的，示意的，简略的，原理的，简图，概略，接线图，（电路）原理图，示意图
schematization 规划，设计
schema 大纲，概要，（衰变）纲图，图解，略图，方案示意图
scheme comparison 方案比较
scheme competition 方案竞赛
scheme diagram 方案图
scheme drawing 方案草图，方案图
scheme for construction organization measures 施工组织措施计划
scheme of electric power supply 供电方案
scheme of power supply 供电方案
scheme table 图表
scheme 方案，线路图，电路，图解，计划，设计图，规划，图，流程图，示意图
schist gneiss 片状片麻岩
schistose clay 片岩质黏土
schistose granular 片粒状
schistose rock 层状岩石，片状岩石
schistose structure 片状构造
schistose subbase 片岩基础
schistose 片状的
schistosity plane 片理面
schistosity 片理
schist 片岩
schlieren apparatus 条纹仪
schlieren device 声影法观察装置
schlieren method of photography 纹影法照相术
schlieren method 纹影法
schlieren system 纹影法系统
schlieren technique 纹影术
schlieren 纹影仪，纹影
Schmidt number 施密特数
Schneider Electric S. A. 法国施耐德电气股份有限公司
Schottky diodes 肖特基二极管
Schottky solar cell 肖特基太阳电池
schriftgranite 文象花岗岩
SCHWR(steam cooled heavy water reactor) 汽冷重水反应堆
sciagraph 房屋纵断面图，投影图
SCIB(selective channel input bus) 选择通道输入总线
SCIC(semi-conductor integrated circuit) 半导体集成电路
science and technological management system reform 科技体制改革
science and technological personnel 科技人员
science and technology appraisal 科技评价
science and technology award 科技奖励
science and technology commission 科学技术委员会
science and technology information 科技信息
science and technology investment 科技投资
science and technology progress in enterprises 企业科技进步
science and technology section 科技科
science and technology 科学技术
science of the sea 海洋科学
science research management 科研管理
scientific and technical funds 科技经费
scientific and technical research 科技研究
scientific and technical training 科技培训
scientific and technological achievement 科技成果
scientific and technological cooperation 科技合作
scientific and technological key achievement 科技攻关成果
scientific and technological progress prize 科技进步奖
scientific and technological progress 科技进步
Scientific Apparatus Maker's Association （美国）科学仪器制造商协会
Scientific Apparatus Manufacturers Association （美国）科学仪器制造商协会
scientific hydrology 理论水文学
scientific invention 科学发明
scientific management 科学管理
scientific manpower 科研人才
scientific payoff 科研成果
scientific research equipment 科研装备
scientific researches reserve force 科学研究后备力量
scientific research institute for atomic reactors 核

反应堆科学研究院【苏联】
scientific research mechanism 科研机制
scientific research project 科研项目
scientific research 科学研究
scientific symposium 科学（专题）讨论会
scientific terminology 科学名词
scientific worker 科学工作者
scientist 科学家
SCIF（standard thermal column irradiation facility） 标准热柱辐照设施
SCIG（squirrel cage induction generator） 鼠笼异步发电机
scinticounting 闪烁计数，用闪烁的方法测量放射性
scintillating material 发光材料，闪烁材料
scintillation analysis 闪烁分析
scintillation chamber 闪烁室
scintillation converter 闪烁转换器
scintillation counter 闪烁计数器
scintillation counting 闪烁计数，用闪烁的方法测量放射性
scintillation crystal 闪烁晶体
scintillation detector 闪烁检测器，闪烁计数计，闪烁探测器，闪烁式检测仪
scintillation duration 闪烁持续时间
scintillation effect，random varation in capacitance 电容量闪变
scintillation effect 闪烁效应
scintillation method 闪烁记数法
scintillation noise 闪烁噪声
scintillation probe 闪烁探针
scintillation scanner 闪烁扫描器
scintillation spectrometer 闪烁谱仪，闪烁分光计
scintillation 闪烁，火花，闪烁现象
scintillator fast neutron fluxmeter 闪烁体快中子通量计
scintillator radiation meter 放射性闪烁探测器，辐射闪烁探测器
scintillator 闪烁体，闪烁器，闪烁仪
scintillometer 闪烁计数器
SCISRS（sigma center information storage and retrieval system） 西格玛中心数据储存和检索系统
scissors crossing 交叉跨越【输电线】
scissors crossover 双渡线
scissors 剪刀
SCI（stress corrosion index） 应力腐蚀指数
sclerometer 回跳硬度计，肖式硬度计，硬度计
scleroscope hardness test 回弹硬度试验，肖氏硬度试验
scleroscope hardness 肖氏硬度，回跳硬度
scleroscope 回跳硬度计，肖氏硬度计，硬度计，验硬器
scler（scleroscope hardness） 肖氏硬度，回跳硬度
SCL（speed control logic） 速度控制逻辑
SCL（system control language） 系统控制语言
SCM（small core memory） 小磁芯存储器
scoll 涡形
scoop channel 斗式槽
scoop condenser 自流式凝汽器
scoop dredge 斗式挖泥船
scooper 斗式升运机
scooping machine 铲斗挖土机
scoop shovel 铲斗挖土机
scoops 勺，水舀
scoop tube 勺管
scoop wheel pump 斗轮泵
scoop wheel 斗轮，戽斗水车
scoop 煤斗，铲子，收集器，吸入口，取样勺，通风斗，掘，舀取，水斗
scope and scale of industries 工业规模
scope drawing 范围图
scope of a convention 条约的范围，公约的范围
scope of application 使用范围，适应范围
scope of bid 招投标范围
scope of business 经营范围
scope of cover 承保范围，责任范围
scope of design 设计范围
scope of employment 聘用职责
scope of responsibility 职责范围
scope of service 服务范围
scope of supply 供应范围，提供范围
scope of woks 工程范围，工作范围
scope qualification 应用范围
SCOPE（schedule control of program executive） 程序执行的调度控制
scope 范围，范畴，目标，指示器，显示器，阴极射线管，观测设备，风斗，水斗
scorched earth 焦土
score 刻痕，线痕
scoria 火山渣，炉渣，熔渣
scoring 划痕，刻痕，划线，分格，擦蚀作用
SCO（senior control operator） 控制室高级操纵员
scotch 刻痕
scour apron 防冲护坦
scour basin 冲刷盆地
scour below dam 坝下淘刷
scour criterion 冲刷标准
scour culvert 冲沙涵洞
scour depth 冲刷深度
scour forecast 冲刷预测
scour gallery 冲沙廊道
scour hole 冲蚀穴，冲刷坑，冲刷孔
scouring abrasion 冲刷
scouring action 冲刷作用
scouring and silting 冲刷及淤积
scouring capability 冲刷能力
scouring effect 冲刷作用
scouring equipment 冲沙设备
scouring force 冲刷力
scouring of river banks 河岸冲刷
scouring sluice 冲沙闸，冲沙孔，冲沙阀，冲沙道
scouring test 冲刷试验
scouring velocity 冲刷速度
scour off 冲刷掉
scour outlet channel 冲沙泄水道
scour outlet 冲沙孔
scour pad 防冲衬垫
scour pipe 冲沙管道

scour preventer 防冲刷装置
scour prevention 防冲，防止冲刷
scour protection 防冲刷，防冲刷保护
scour rate 冲刷率
scour sluiceway 冲沙泄水道
scour tunnel 冲沙隧洞
scour valve 冲沙阀
scourway 冲刷水道
scour 擦掉，洗去，（渣对炉衬的）侵蚀，冲刷，擦净，洗净，淘刷，洼地，冲蚀，擦，腹泻，擦亮，洗涤，冲洗，清除
SCPA(solar cell panel assembly) 太阳能电池板组件
SCP(symbolic conversion program) 符号转换程序
SCP(system control program) 系统控制程序
scram accumulator isolating valve 紧急停堆蓄压箱隔离阀
scram accumulator 紧急停堆蓄压箱
scram aiding spring 紧急停堆弹簧
scrambler 扰频器，倒频器，编码器，保密器
scramble time 零星机动时间，编码时间，量化时间
scrambling matrix 译码矩阵
scrambling 译码
scram breaker 紧急停堆断路器
scram button 紧急停堆按钮
scram circuit 紧急刹车电路，快速停堆电路
scram control valve 紧急停堆控制阀
scram criterion 紧急停堆准则
scram dump tank 紧急停堆排放箱
scram initiation spring 触发紧急停堆的弹簧
scram initiation 引起紧急停堆，触发紧急停堆
scram limit value 紧急停堆极限值
scram magnet 紧急停堆磁铁
scram mechanism 紧急停堆机构
scrammed rod 紧急停堆棒，插入的事故棒
scrammed 紧急停堆的
scramming mechanism 紧急停堆机构
scram position 快速停堆位置
scram protection 快速停堆事故防护装置
scram rate 快速停堆速度
scram rod drive mechanism program 紧急停堆棒传动机构程序
scram rod 紧急停堆棒
scram runback 紧急停堆复位
scram signal 快速停堆信号，紧急停机信号，紧急停堆信号
scram system 快速停堆系统，紧急停机系统，紧急停堆系统
scram tank 紧急停堆蓄压箱
scram time 快插时间【控制棒】，紧急停堆时间
scram valve 紧急停堆阀，快关阀
scram water valve 紧急停堆水阀
scram 快速停堆，紧急停堆【反应堆】，刹车，快速断开
scrap build 设备改装
scrap collecting tank 废料收集箱
scrap equipment 报废设备
scraper bar 刮板
scraper bowl 铲运机斗
scraper bucket excavator 铲斗挖土机
scraper bucket 铲斗
scraper chain conveyor 链板输送机，链式铲运机
scraper coal feeder 刮板式给煤机
scraper conveyer 刮板（式）运输机，刮板输送机，链板传送带
scraper conveyor 刮板（式）运输机，刮板输送机，链板传送带
scraper dredger 耙泥船
scraper excavator 铲斗挖掘机
scraper feeder 刮板式给煤机，刮板式给料机
scraper flight 刮板带
scraper for land levelling 平地用铲运机
scraper haulage 铲运机拖运
scraper loader 刮斗装载机，耙斗装料机
scraper plow reclaimer 刮板取料机
scraper pulverized coal conveyer 刮板输粉机
scraper reclaimer 刮板取料机
scraper ring 刮油环
scraper shoveling machine 铲土机
scraper stacker 刮板式堆煤机
scraper 刮板，刮刀，铲运机，铲土机，刮煤机，犁煤器，平土机，刮料机，刮削器
scrape type of settling tank 刮板沉淀器
scrape 擦伤，刮，擦，刮削
scraping blade 犁刀
scraping edge 刮刀，刮口
scraping grader 刮土平地机
scraping 刮痕，擦伤，刮平，刮削加工
scrap iron 废铁
scrap rate 废品率
scrap recovery plant 废料回收厂
scrap recovery 切屑回收
scrap recycle 废料利用，碎屑再循环
scrap separator 废料分离器
scrap tire 废轮胎
scrap value 废品残值
scrap 废弃物，碎片，屑
scratch hardness 刮痕硬度
scratching 损伤
scratch-pad memory 暂时存储器，中间结果存储器，草稿存储器
scratch-pad storage 暂存器，中间结果存储器
scratch pad 高速暂存储存器
scratch tape 暂存带，废带
scratch 擦痕，擦伤，刮伤，刻痕，划痕，擦除
SCR bank 可控硅功率电桥【控制棒驱动】
screaming 啸叫声，振荡，振动
scree cone 碎石锥
screed board 样板，平泥板
screed-coat 找平层
screed finish 饰面找平
screeding 用样板刮平，找平，抹平【混凝土】
screed 抹面用的刮板，匀泥尺，准条，找平，拉平
screen action 屏蔽作用
screen aftergrow 荧光屏余辉
screen analysis 筛分分析，筛分析
screen angle 筛面倾角
screen aperture 筛孔
screen bar 筛条

screen bed 筛面
screen blinding 筛子堵塞
screen brightness 荧光屏亮度
screen cages 筛笼
screen chamber 拦污栅室
screen classifier 筛分机
screen cleaner 拦污栅清污器
screen coil motor 屏蔽线圈电动机
screen coil 屏蔽线圈
screen collector 幕式集尘器，筛式除尘器
screen constant 屏蔽常数
screen deck area 筛面面积
screen deck 筛面
screen dike 隔堤，围堤
screen display 屏幕显示
screen distributor 筛网分布器
screen door 纱门
screen drier 百叶窗（水分）分离器
screened air intake 有防尘网的进汽口
screened cable 屏蔽电缆
screened coal 筛分煤，筛选煤
screened collector 包网集水器
screened electrode 屏蔽电极
screened feeder 屏蔽馈线
screened grading 筛分级
screened indicator 遮蔽指示剂
screened intake 有栅的进水口
screened material 筛过的材料
screened pipe 包网管
screened probe 包网取样管
screened type cable 屏蔽型电缆
screened well 过滤井
screen-factor 屏蔽系数
screen filter 筛滤板
screen grid current 屏栅电流
screen grid dissipation 屏栅耗散
screen grid input 帘栅极输入
screen grid 帘栅极，屏栅极
screen hole 筛眼，筛孔
screen hood 筛罩
screening ability 筛分能力
screening agent 掩蔽剂
screening aggregate 筛分骨料
screening analysis 筛分析
screening and crushing station 筛分破碎装置
screening angle 筛面倾角
screening capacity 筛选能力
screening characteristics 筛分特性
screening cobble ballast 筛选卵石道碴
screening coil 屏蔽线圈
screening curve 筛分曲线，筛选曲线
screening device 筛分设备
screening dewatering 污水排水除污
screening drum 旋转滤网
screening effect of rail 铁轨屏蔽效应
screening effect 屏蔽效应，掩蔽效应
screening efficiency 筛分效率
screening equipment 筛分设备
screening facility 筛分设备
screening factor 屏蔽系数
screening function 屏蔽功能
screening hopper 筛分料斗
screening machine 筛分机，筛选机
screening material 屏蔽材料
screening plant 筛分厂，筛分设备，筛选装置
screenings 筛余物，筛上物
screening test 筛选试验，甄别试验
screening wire 屏蔽线
screening 筛分，筛选，过筛，屏蔽，隔离，遮蔽，筛屑，筛渣，网眼，遮护，防火隔墙
screen lock 锁屏，筛子定位器，死机
screen marker 荧光屏标记
screen mesh wire diameter 筛目钢丝直径，筛眼钢丝直径
screen mesh 筛孔，筛目，筛网，筛眼，细孔，丝网
screen of supply 供货范围
screen of work 分工
screen opening 筛孔，筛眼
screen out 筛出，筛分
screen overflow 筛上物
screen-packed bed 网状填充床
screen packing 网状填料
screen pipe 包网管
screen plate 筛板，过滤板，（碎煤机的）底筛
screen potential （电子射线管的）屏电势
screen printing contact 丝网印刷电极
screen printing process 丝网印刷工艺
screen-protected cubicle 屏蔽板保护的小室【开关设备】
screen-protected machine 网罩防护型电机
screen rack 拦污栅栅条，格栅除污机
screen rake 拦污栅清理机，拦污栅清理耙
screen-reflector 金属网反射器
screen residue 筛剩物
screen sash 纱窗
screen separator 钢丝网分离器
screen size gradation 筛分级配
screen sizer 筛式分级机
screen slot size 筛网网眼尺寸
screen tailings 筛余物
screen test 筛分试验
screen thermometer 百叶箱温度表
screen touching 触摸屏幕（式）
screen tray 筛盘
screen tube 凝渣管，垂帘管
screen-type superheater 屏式过热器
screen underflow 筛下物
screen unsharpness 荧光屏不清晰度
screen wall 花格墙
screen wash pump 滤网冲洗泵
screen well point 滤网式井点
screen window 纱窗
screen 阻尼网【风洞】，格网，屏幕，荧光屏，滤网，筛，筛子，屏蔽，防护网罩，拦污栅
scree 山麓碎石，岩屑堆
screw anchor foundation 螺旋锚基础
screw anchor 螺钉锚固
screw and worm gate lifting device 螺旋蜗杆式闸门启门机
screw auger 螺旋钻
screw axis 螺旋轴线

screw bar anchorage 螺丝端杆锚具
screw block-lifter 螺旋升高机
screw block 千斤顶
screw bolt 螺栓，方头螺
screw bonnet 螺纹阀盖，螺纹阀帽
screw-cage reclaimer 绞笼取料机
screw capstan head 螺旋式绞盘杆
screw cap 螺纹（阀）帽
screw car unloader 螺旋卸车机
screw centrifugal pump 螺旋式离心泵
screw clamp 螺旋夹
screw compressor 螺杆式压缩机，螺杆式压气机
screw connector 螺纹接头
screw conveyer(-or) 螺旋（式）运输机
screw conveying unloader 螺旋卸车机
screw coupling 螺纹接头
screw decanter type centrifuge 螺旋倾析型离心分离机
screw depth regulator 螺旋式入土深度调节器
screw die 螺丝板
screw differential 螺旋式差动装置
screw dislocation 螺纹错位
screw-down bolt 压紧螺栓
screw-down valve 螺旋阀
screw driver bit 螺钉旋具头，螺丝起子
screw drivers set 成套螺丝刀
screw driver 改锥，螺丝刀，螺丝起子
screwed connection 螺纹连接，丝扣连接
screwed-end connection 拧紧的端部连接
screwed fittings 螺纹连接件
screwed flange 拧紧的法兰
screwed gland 拧紧的密封套
screwed joint 螺纹接头，螺纹连接
screwed pipe 螺纹管
screwed piping joints 螺丝状的管接头
screwed pressed-in seat ring 拧紧的压入阀座环
screwed seat ring 拧紧的阀座环
screwed 螺丝状的，螺旋的，拧紧的
screw elevator 螺旋升降机
screw extractor 起螺丝器，拉马，断螺钉联出器，螺杆旋出器，螺旋挤压机
screw feeder 螺旋给煤机，螺旋给料机
screw feed stoker 螺旋加煤炉排，下饲式炉排
screw feed 螺旋给料，螺旋进给
screw gate-lifting device 螺旋式闸门启门机
screw gauge 螺纹规
screw gearing 螺旋传动装置
screw hoist 螺旋起重机
screw holding screwdriver 夹持螺钉的起子
screw hole 螺孔
screw impeller pump 螺旋叶轮泵
screw impeller 螺旋叶轮
screw jack 螺旋千斤顶
screw locking device 螺纹锁紧装置
screw lock plate 螺纹锁紧板
screw luffing 螺杆变幅
screw micrometer caliper 螺旋千分卡尺
screw mixer 螺旋搅拌机
screw mounted 螺钉安装
screw nail 螺钉
screw nipple 丝对
screw nut 螺母
screw pile dolphin 螺旋带缆桩
screw pile wharf 螺旋桩码头
screw pile 螺旋桩
screw pitch gauge 螺距规，螺纹样板
screw pitch 螺距
screw plate 板牙
screw plug fuse 旋入式保险丝
screw plug 安全塞，螺纹接头，螺旋塞，螺纹接线柱
screw press 螺旋压力机
screw pump 螺旋泵，螺旋桨泵，螺杆泵
screw retention （螺纹接线的）防自松装置
screw rolling machine 滚丝机
screw scraper deasher 螺旋捞渣机
screw setting 定位螺旋
screw shackle 螺旋拉紧设备，螺旋钩环
screw ship unloader 螺旋卸船机
screw shutdown 传动停堆【沸水堆】
screw slag conveyor 螺旋式捞渣机
screw-socket 螺口接座，螺丝灯头
screw stairs 盘梯
screw stem 螺杆
screws 螺丝
screw take-up 螺旋拉紧（装置）
screw tap 螺丝攻，丝锥，套丝板
screw terminal 螺丝接线端，螺纹接线柱，螺栓端子
screw thread gauge 螺纹规
screw thread lubricant 螺纹润滑剂
screw thread 螺纹
screw-type air compressor 螺杆式空（气）压（缩）机
screw-type centrifugal nozzle 螺旋离心式喷嘴
screw-type coal distributor 螺旋给煤机
screw-type fuse 螺旋式熔断器
screw-type rotary pump 螺旋式回转泵
screw-type strainer 旋转滤网
screw unit 螺旋拉紧装置
screw unloader 螺旋卸车机
screw valve 螺旋（式）阀
screw wagon unloader 螺旋卸车机
screw-wheel gearing 斜齿轮传动装置
screw worm 螺杆
screw wrench 螺钉扳手
screw 螺杆，螺旋，螺丝，螺旋桨，螺，螺丝钉，旋紧螺钉
SCR excitation system 晶闸管励磁器，可控硅整流器励磁系统
scribe mark 划线刻痕
scriber 划线器，描绘标记的用具，划线针，划针
scribe 划线
script 手稿，正本
scroll case drain 蜗壳排水管
scroll case vent 蜗壳通气管
scroll case 螺旋形箱，蜗壳
scroll casing 涡形壳，蜗壳
scroll housing 涡室外壳，蜗壳
scroll meander 内侧堆积曲流，涡形曲流
scroll 卷动，上卷，滚动，屈卷法，卷轴，画

卷，名册，卷形物，(使) 成卷形，蜗壳，涡管
SCR(scanning control register) 扫描控制寄存器
SCR(selective catalytic reduction) 选择性催化还原法【用于烟气脱氮的一种手段】，触媒式脱硝工艺
SCR(semiconductor controlled rectifier) 半导体可控整流器
SCR(silicon controlled rectifier) 可控硅，可控硅整流器
SCR(specific commodity rate) 指定商品运价
SCR(state control register) 状态控制寄存器
SCR switch 可控硅开关
scrubber baffle 洗汽挡板，(汽包内的) 分离器
scrubber collector 洗涤集尘器，洗涤收集器
scrubber precipitator 水膜除尘器
scrubber tower 洗涤塔
scrubber wash tower 洗涤塔
scrubber water pump 除尘水泵
scrubber 洗涤器【汽包】，洗涤塔，湿式除尘器，除尘器，洗气器，刮管器，涤气器
scrubbing column 洗涤柱
scrubbing tower 洗涤塔
scrubbing 洗气【指废气处理】，洗涤
scrub solution 洗涤剂，洗涤液
scrub 气体洗涤，灌木，擦洗
scrutiny 详尽研究，仔细检查，仔细审查
SC(safety class) 安全级
SC(safety coefficient) 安全系数
SC(semi-conductor) 半导体
SC(sequential circuit) 时序电路
SC(service contract) 服务合同
SC(shift control) 移位控制器
SC (shift counter) 移位计数器
S/C(short circuit) 短路
SC(sine-cosine) 正弦余弦
SC(single casing) 单缸【汽轮机】
SCSM (specific communication service mapping) 特定通信服务映射
SCS(sequence control system) 顺序控制系统【DCS 画面】，程序控制系统
SCS(shutdown cooling system) 停堆冷却系统
SCS(silicon controlled switch) 硅可控开关，硅控开关，可控硅开关
SCS(single-channel simplex) 单路单工
SCS (system completion schedules) 系统完工计划
SC(standard condition) 标准状态
SC(stop-continue register) 停续寄存器
SC(supercharge) 增压
SC(supervisory control) 管理控制
SCTL(short-circuit transmission line) 短路传输线
SCTL(sodium component test loop) 钠设备试验回路
SCT(step control table) 分段控制表
SCT(subroutine call table) 子程序调用表
SCUAE(State Committee on the Utilization of Atomic Energy) (苏联) 国家原子能利用委员会
scuffing 刻痕，磨损，划伤，拉伤，磨蚀
sculpture 雕刻
sculpt 造型
scum baffle 浮渣挡板
scum barrier 除渣栅
scum board 浮渣挡板，刮渣板
scum cock 放浮沫的旋塞
scum collector 浮渣收集器
scum gutter 放渣槽
scum line 浮垢条纹
scummer 水面清污器
scum pump 污水泵
scum trough 浮渣槽
scum weir 浮渣排除堰
scum 浮垢，泡沫，水垢，浮渣，浪花，渣滓，糟粕，去除浮渣，产生泡沫
S-curve 单位线相应的径流积分曲线，总和曲线
SCU(station control unit) 站控制部件
SCU(system control unit) 系统控制部件
scuttle 使 (船) 沉没，煤斗，天窗
SDA(source data automation) 源数据自动化
SDAT(symbolic device allocation table) 符号设备分配表
SDA 除盐水生产系统【核电站系统代码】
SDB(store data buffer) 存储数据缓冲器
SDC(secondary distribution center) 辅助配电中心，二次配电中心
SDC(shutdown cooling) 停堆冷却
SDC(signal data converter) 信号数据转换器
SDD(system design description) 系统设计说明
SDE(storage distribution element) 存储分配部件
SDF(seasonal derating factor) 季节性降低出力系数
SDF(separator demonstration facility) 分离器示范装置
SDF(standard distribution format) 标准分配格式
SDH(seasonal derating hours) 季节性降低出力小时
SDI(selective dissemination of information) 信息选择传播
SDI(silt density index) 淤泥密度指数
SDI(system diagram index) 系统图标志，系统图索引
SDLC(synchronous data link control) 同步数据链路控制
SDM(system design manual) 系统设计手册
SDO(source data operation) 源数据操作
SDPS(signature data processing system) 特征数据处理系统
SDP(standard depth of penetration) 标准穿透深度
SDR(sodium D_2O reactor) 钠冷重水慢化堆
SDR(special drawing right) 特别提款权【国际货币基金组织】
SDR(statistical data recorder) 统计数据记录器
SDR(storage data register) 存储数据寄存器
SD(sample delay) 抽样延迟
SD(seasonal derating) 季节性降低出力
SD(sectional drawing) 断面图
SD＝shutdown 停机
SD(shut-off damper) 关断挡板
SDS(sanitary drainage system) 卫生排水系统
SDS(simulation data subsystem) 模拟数据子系统
SDS(standard data set) 标准数据集
SD(standard deviation) 标准差

SD(steam lead drain valve) 蒸汽疏水阀
SD(superintendent of document) 文件主管人
SD(surface charge destination) 目的站地面运输费
SDU(station display unit) 站显示部件
sea ability 续航力
sea bank 海堤
sea based reactor 海上反应堆
sea beach 海岸
sea bed disposal (放射性废物)海床处置
sea bed 海底
sea bench 海滩
sea book 航海图
sea breach 破坏性碎波
sea breeze front 海风前锋
sea breeze 海风
sea burial (放射性废物)海洋埋置，海洋埋葬
sea chart 海图
seacoast 海岸，海滨
sea damage 海损
sea defense works 海岸防护工程
sea disposal (放射性废物)投海处置，海洋处置
sea dumping (放射性废物)倾倒入海洋
sea embankment 海堤
sea fog 海雾
sea freight 海运费(率)
sea front 滨海区，海边
sea going capability 适航性
sea-going vessel 海轮，远洋(轮)船
sea-going 航海的
sea inlet 海水吸入口
seal air fan 密封空气风机
seal air system 密封风系统
sealant compound 嵌缝料
sealant 封闭剂，封口料，密封剂，密封胶，填缝料
seal assembly removal tool 密封组件拆卸工具
seal assembly 密封装置，密封组件
seal block 汽封块
seal box 密封盒，密封箱
seal bushing 汽封套
seal cage 填料环
seal chamber 密封室
seal coat 封闭层，止水层
seal course 密封层
seal diaphragm 密封隔板
seal disc 密封盘
seal disk 密封盘
sealed bid 限制性投标，密封投标书，非公开招标，秘密投标
sealed cabin 密封舱
sealed chamber 气封室，密封室
sealed coil relay 线圈密封式继电器
sealed compartment 密封舱
sealed contact 密封触点
sealed cooling 封闭式冷却，闭式冷却法
sealed double glazing 密封双层玻璃
sealed-dry-type transformer 密封型干式变压器
sealed earth 填封土
sealed enclosure 密封机壳，密封外壳
sealed envelope 密封信封
sealed fuel element 密封式燃料元件
sealed germanium diode 密封的锗二极管
sealed in fuel element 封装的燃料元件
sealed-in nozzle 密封喷嘴
sealed insulation 密封绝缘
sealed in unit 封装设备，封入装置
sealed joint 止水逢
sealed off counter 密封计数管
sealed proposal 密封投标
sealed pump 密封泵
sealed relay 密封(式)继电器
sealed sample container 密封煤样箱
sealed source 密封的中子源，密封放射源
sealed system 封闭系统
sealed tank 密封油箱，密封箱
sealed tender 密封投标书，限制性投标，非公开招标，秘密投标
sealed transformer 密封型变压器
sealed transistor 封闭式晶体管
sealed winding 密封绕组，密封线卷
sealed 密封的
seal element 密封件
sealer 封闭剂，密封层
sea level datum 水深基准面
sea level departure from normal 潮位偏差
sea level elevation 海拔高度
seal face 密封面
seal fitting 密封接头，密封配件
seal flange 密封法兰
seal gas extraction duct 密封抽气管
seal gas flow feed and shut-off control system 密封气体供给和关闭控制系统
seal gas flow 密封气体流量
seal gas for the primary gas circulators 供一次气体循环用的密封气体
seal gas injection 密封气体注入
seal gasket 密封垫片
seal gas pressure 密封气体压力
seal gas-supplied seal 密封气体提供的密封
seal gas system 密封气体系统
seal gate valve 密封闸阀
seal gland 密封盖
seal housing 密封壳，密封罩
sea line 海岸线
sealing air fan 密封风机
sealing air 密封风，密封空气
sealing barrier 密封膜，外壳
sealing behaviours of vessel 容器的密封性状
sealing bellows 密封弹簧箱，密封波纹管
sealing block 密封块
sealing box 电缆终端套管，密封箱
sealing bucket 封水戽斗【自动虹吸管的】
sealing bushing 汽封套
sealing by bush 衬套密封
sealing casing 密封盖，密封外壳
sealing clearance 密封间隙
sealing coat 路面沥青涂层，密封层，止水涂层
sealing collar 止水圈
sealing compound 密封膏，封口胶，封口料，封闭剂，密封黏胶剂

sealing concrete 封混凝土
sealing cover 密封压盖
sealing damper （气体管道）密封闸板
sealing device 止水设备，止水设施
sealing diaphragm 密封隔板
sealing efficiency 密封效率
sealing end 封端，电缆封端，封口，密封端
sealing face 密封面
sealing fin 汽封片
sealing flange 密封法兰
sealing fluid 密封流体【密封水等】
sealing gasket 密封衬片，止水衬垫
sealing gland 汽封装置，密封装置，密封盖
sealing glue 密封胶
sealing joint 密封接头
sealing key 密封键
sealing layer 防水层
sealing mastic 填塞料
sealing material 封填材料，止水材料
sealing medium 密封用的介质
sealing of soil 土壤板结
sealing oil pump 密封油泵
sealing oil system 密封油系统
sealing oil tank 密封油箱
sealing pad 密封瓦
sealing pipe 封闭式管套，密封管
sealing piston 密封活塞
sealing plate 密封板，包覆板
sealing plug 密封塞
sealing pot 密封槽
sealing pressure 密封压力
sealing reliability 密封可靠性
sealing ring bushing 密封环衬套
sealing ring cap 密封环盖
sealing ring 汽封环，密封环，止漏环
sealing rope 密封条
sealing run 封底焊道
sealing sleeve handling pallet 密封套装卸托板
sealing steam box 汽封蒸汽室
sealing steam desuperheater 汽封蒸汽冷却器
sealing steam 密封蒸汽，汽封蒸汽
sealing steel shell 密封钢壳
sealing strip 密封条，止水条，汽封片，填缝片，填缝条
sealing surface area 密封面面积
sealing surface of seat 阀座密封面
sealing surface 密封表面，密封面
sealing system 密封系统，汽封系统
sealing test 密封试验
sealing tooth 汽封齿
sealing up 止水，防漏
sealing washer 密封垫圈
sealing water pump 轴封水泵，密封水泵
sealing water system 密封水系统
sealing water 密封水，密封用水
sealing weld 封焊
sealing wire 封装用线，焊接线，密封引线
sealing with fines 细粒料填缝，细粒料止水
sealing works 填缝工作，止水工程
sealing 密封，封接，封口
seal-in relay 自保持继电器，密封继电器保护装置
seal joint 密封接头
seal lac 封口漆，止水虫胶
seal leakage 汽封泄漏
seal leakoff 密封泄漏
seal ledge 密封凸缘【反应堆压力壳】
seal lip 密封突缘【顶盖，欧米茄焊】
seal loop 封环
seal membrane 密封膜片
seal of corporation 公司印章
seal oil back up test 密封装置反馈试验
seal oil head tank and transfer barrier 密封油压力槽及传输隔板
seal oil head tank 密封油压力槽
seal oil piping 密封油管道
seal oil pressure 密封油压
seal oil system 密封油系统
seal oil 密封用油，密封油
seal packing 密封填料
seal plate 密封板，止水板
seal plug 密封塞【堵头】
seal pot 密封水箱
seal ring O形环，密封圈，密封环，止水环，垫圈
seal runner 密封滑槽
seal sleeve 汽封套筒
seal spacer 密封垫圈
seal tank 密封槽
seal tooth 汽封齿
seal trough 密封槽
seal unit 密封单元
seal up 查封
seal washer 密封垫圈
seal water booster pump 轴封水升压泵
seal water filter 轴封水过滤器
seal water heat exchanger 轴封水热交换器，密封水换热器，密封水热交换器
seal water injection filter 轴封注水过滤器
seal water injection 轴密封水的注入（主泵）
seal water leakoff line 密封水泄漏管线
seal water leakoff pump 轴封水引漏泵
seal water pipe 水封管
seal water pump 轴封水泵
seal water reflux filter 轴封水回流过滤器
seal water storage tank 密封水虹吸水箱
seal water system 密封水系统【阻止空气漏入处于真空的阀门】，轴封水系统
seal water tank 水封水箱
seal water 密封水，密封排水，轴封水
seal wax 封蜡，火漆
seal-welded taper pipe threaded joint 锥管螺纹密封焊连接
seal-welded 密封焊接的
seal welding 密封焊，密封焊接
seal weld machining unit 密封焊机单元
seal weld （燃料棒的）密封焊缝，封闭焊
seal well 密封井
seal 汽封，密闭，封接，封口，嵌塞，封记，焊封，密封（面），密封垫，盖印，图章，印章
seaman 水手

sea map 海图
sea mark 航海标志
seam blast 缝隙爆破
seam clamp 接缝夹板
seamed sheet metal 焊接钢板
seamed steel pipe 有缝钢管
sea mile 海里
seaming 焊缝，卷边接缝
seam inundation 充水岩层
seamless pipe 无缝管
seamless steel pipe 无缝钢管
seamless steel tube 无缝钢管
seamless tube 无缝管
seamless 无焊缝的，无缝的，无焊的
seam locker for round elbow 圆弯头咬口机
seam-sample of coal 煤层煤样
seam welding machine 缝焊机
seam weld 滚焊，缝焊，缝焊接，缝熔接，缝焊缝
seam 生裂缝，焊缝，缝，接缝，接口
seaport 港口
sea puss 沿岸急流
seaquake intensity 海震强度
seaquake 海震
search coil test 探测线圈测试法
search coil 探测线圈，探查线圈
search cycle 检索周期
searcher 探测器
search flood light 探照灯
search for 搜查
searching current 探测电流
searching lamp 探照灯
searching storage 相联存储器
searching tube 测针，探头
search key 检索键，检索关键字
searchlight generator 探照灯发电机
searchlighting 雷达搜索，探照灯搜索，探照，照射
searchlight signal 探照灯信号
search light 探寻灯，探照灯
search method 搜索法
search operation 检索操作，觅数操作
search sweep 搜索扫掠，搜索扫描
search tube experiment 探管试验
search unit tracking and recording 探头跟踪记录
search unit 搜索器，探测器，探头
search variable 研究变量
search 调查，搜索
sea reach 近海河段
sea risk 海险
s-earth 信号地线
sea scale 风浪等级
sea shore breeze 海岸风
sea shore coast 海滨
sea shore industrial reservation 沿海工业地带
sea shore 海岸，海滨
seaside thermal power plant 沿海火电厂
seaside 海滨的
seasonal correlation 季（节）相关
seasonal depletion 季节性枯竭
seasonal energy efficiency ratio 季节能效比
seasonal energy 季节性电能
seasonal heating load 季节性热负荷
seasonal incidence 季节影响
seasonal interruption of coal shipment 煤炭运输的季节性中断
seasonality 季节性
seasonal load curve 季节性负荷曲线
seasonal load 季节性负荷，季节性负载
seasonal peak load 季节性尖峰负荷
seasonal plan 季度计划
seasonal price 季节性电价
seasonal recovery 季节性回复
seasonal regulating reservoir 季调节水库
seasonal river 季节性河
seasonal runoff 季径流（量）
seasonal tariff 季节性电费率
seasonal variation of rainfall 雨量季变化
seasonal variation 季节变化
seasonal wind 季风
season cracking 风干裂纹，应力腐蚀裂纹，季裂，（木材的）干燥开裂
seasoned timber 干材
seasoned wood 风干木材
seasoning 气候处理，不稳定性，时效，老化，陈化，干燥，风干
season timber 风干木材
season 时期，季节，老化，陈化，变得成熟，变干燥
seat cage 阀座
seated armature 已闭合的衔铁，入位衔铁
seated gas generator 固定式燃气发生器
seated position 落座位置【翻车机】
seated 落座的
sea temperature chart 海水温度图
sea-thermal power generation 温差发电
seating contact 落座接触
seating of the vessel head 压力容器顶盖就位
seating ring 座环
seating surface 座表面
seating 落座【球、锥等】，基础，支座
seat leakage measurement 阀座泄漏测量
seat leakage test 阀座泄漏试验
seat leaktightness （阀）支座防漏严密性
seat of settlement 沉降影响范围
sea-tossed 波浪颠簸的
seat pad 垫块，坐垫
sea transportation 海上运输，海运
seat reservation system 预定坐位系统【计算机系统】
seat ring 阀座环，座环
seat seal 阀座密封
seat-type breather （变压器的）座式换气器，座式吸潮器
seat wear compensation 阀座磨损补偿
seat 使固定，阀座
sea valley 海底谷
seawall 防波堤，防浪堤，海堤
sea water concrete 海水拌制混凝土
sea water cooling system 海水冷却系统
sea water corrosion 海水腐蚀
sea water desalination reactor 海水淡化堆

sea water desalination 海水淡化
sea water desalinator 海水淡化器
sea water desalinization 海水淡化
sea water electrolysis 电解海水
sea water intrusion 海水入侵
sea water pump 海水泵
sea water scrubbing method 海水洗涤法
sea water supply 海水供给
sea water thermometer 海水温度表
sea water 海水
seaway 航路
seaweed 海草，海藻
sea wind 海风
sea works 海岸（防护）工程
seaworthiness 耐波性，适航性
seaworthy packing 适合海运包装
sea 海，海洋，海域
SEA 生水系统【核电站系统代码】
SEB(Source Evaluation Board) （英国）原始资料评价委员会
SE by E(southeast by east) 东南偏东
secant galvanometer 正割检流计
secant modulus 割线模量
SECD (self-regulatingerror-correctcoder-decoder) 自调节误差校正编码-译码器
secohmmeter 电感表
secohm 秒欧【电感单位】
second addition time 二次加法时间
second addition 二次加法
second approximation 二次近似
secondary action 副作用，二次作用，二次操作，二级操作
secondary air fan 二次风风机，二次风机
secondary air heater 第二级空气加热器
secondary air piping 二次风管道
secondary air pollutant 次生空气污染物
secondary air 二次风，二次空气，二次气流
secondary battery 二次电池组，二次蓄电池组，二次电池
secondary beam 次梁，二次射线束，付波瓣
secondary benefit 附属效益，间接效益
secondary blasting 二次爆破
secondary breakdown 二次击穿
secondary brush 副电刷
secondary cell 二次电池，蓄电池，副电池
secondary chamber 二次风室
secondary circuit 次级电路，二次回路，二次电路，二次管网，二回路
secondary circulation 二次循环，次级环流，二次环流，二级环流
secondary clarifier 二次澄清池
secondary clay 次生黏土
secondary clearing 次级澄清，次级分层
secondary cleat 次生割理
secondary cleavage 次生劈理
secondary coil 次级线圈，二次线圈，副线圈
secondary cold front 副冷锋
secondary collimator 辅助准直器
secondary combustion zone 二次燃烧区
secondary combustion 再燃烧，二次燃烧
secondary compensator 次级补偿器
secondary computer 副计算机，辅助计算机
secondary concentrator 二次聚光器
secondary connection diagram 二次接线图
secondary consequent stream 次顺向河
secondary consolidation settlement 次固结沉降
secondary consolidation 二次固结，次固结
secondary constant 二次常数
secondary construction work 抹光，（建筑工程）收尾工作
secondary containment 二次安全壳
secondary controller 二次控制器
secondary control 二等控制
secondary converter 二次转换器
secondary coolant circuit 二次冷却剂回路，二回路
secondary coolant loop 二次冷却剂回路
secondary coolant 二次冷却剂，二回路水，二回路冷却剂
secondary cooling circuit 二次冷却回路
secondary cooling system 二次冷却系统，辅助冷却系统，二回路
secondary cooling water loop 二次冷却水回路
secondary copper loss 二次铜耗，次级绕组铜耗，副边铜耗
secondary core support 堆芯二次支承系统，辅助堆芯支撑，堆芯辅助支撑
secondary creep 附加蠕变，蠕变稳定区，第二阶段蠕变，稳态蠕变，二期蠕变，中期蠕变
secondary crevice 再生裂隙
secondary crusher 二次破碎机，中碎机，次级轧石机
secondary current of CT 电流互感器的二次电流
secondary current 次级线圈电流，二次电流，次级电路电流，次级电流
secondary cyclone 次生气旋，副气旋
secondary damage 连带损坏，二次损坏，继发损坏
secondary dam 副坝
secondary defect 二次缺陷
secondary deposit 次生沉积，次生矿床
secondary depression 副低压，次（生）低压
secondary desuperheater 二级减温器
secondary device 二次装置，二次仪表，二次元件【如变送器】
secondary dike 副堤
secondary discharge 次级放电
secondary distribution feeder 二次配电馈电线
secondary distribution mains 二次配电干线
secondary distribution network 二次配电网
secondary distribution system 二次配电系统
secondary distribution trunk line 二次配电干线
secondary distribution 二次配电
secondary distributor 次级分布器
secondary drag 二次流阻力
secondary earthquake 次地震
secondary earth 次级线圈接地，二次线圈接地
secondary effect 二次效应，二次流影响，副作用
secondary efficiency 二次效率
secondary electrical equipment 电气二次设备【中国】
secondary electrode 副电极

secondary element　二次元件【如变送器】
secondary emission characteristic　二次发射特性
secondary emission curve　二次发射曲线
secondary emission multiplier　二次放射电子倍增器
secondary emission photocell　二次发射光电管
secondary emission ratio　二次发射系数
secondary emission　二次发射，二次排放
secondary energy source　二次能源
secondary energy　次级能量，二次能源
secondary enlargement　次生加大，次生扩大
secondary enrichment　次生富集
secondary environment　次生环境
secondary equipment in convertor station　换流站的二次设备
secondary equipment　二次设备
secondary exchanger　二级离子交换器
secondary excitation system　次级励磁系统
secondary extinction　次级消光
secondary failures　二次故障，从属性失效
secondary fault　二次（绝缘子击穿）故障
secondary feeder　备用馈路
secondary feedwater inlet　二次回路给水进口
secondary feedwater　二次回路给水
secondary fiber　次级纤维
secondary field　副磁场，次级场
secondary filter　细过滤器，二次过滤器，精滤器，二次滤网
secondary fire　二次燃烧
secondary flow loss　二次流损失
secondary flow pattern　二次流型
secondary flow　二次流，副流
secondary fold　次生褶皱
secondary forest　再生林
secondary front　副锋
secondary fuel filter element replacement　次级燃油滤清器滤芯的更换
secondary fuel filter element　次级燃油滤清器滤芯
secondary furnace　燃尽室
secondary gamma ray　二次伽马射线
secondary gas chromatograph　二次气相色谱仪
secondary gear　超前动作危急保安器
secondary generation neutron　第二代中子
secondary geosyncline　次生地槽
secondary governor　辅助调节器，超前动作危急保安器
secondary group　中生界
secondary grout hole　二期灌浆孔
secondary grouting　二次灌浆
secondary growth　次生加大
secondary gyratory crusher　二次回转轧碎机
secondary gyroscope　二自由度陀螺仪
secondary hardening　二次硬化
secondary heat exchanger　二次热交换器
secondary humic acid　次生腐殖酸
secondary indicating instrument　二次指示仪表
secondary industry　二次工业
Secondary inflow　二次入流
secondary inlet nozzle　二次进口接管
secondary input　副边输入功率，二次输入功率
secondary instrument　二次仪表，辅助仪器
secondary interference　副干扰
secondary interstice　次生裂隙
secondary ionization　二次电离
secondary iteration　副迭代，第二迭代
secondary jack truss　次桁架
secondary lattice group　次级网络群，二次网络群
secondary leakage flux　副边漏磁通
secondary leakage reactance　次级线圈漏抗，副边漏抗
secondary leakage　二次侧漏泄
secondary levee　副堤
secondary lift　二次抬升
secondary load　次要荷载，二次负载，副载
secondary loess　次生黄土，再积黄土
secondary loop　二次回路，二回路
secondary loss　二次损耗，二次侧损耗，二次损失，端部涡流损失
secondary low　次生低压，副低压，副气旋
secondary main　副总线，副干线，二次干线
secondary manway chinning bar　二次侧人孔装卸杆
secondary manway　蒸汽发生器二次侧人孔【二回路】
secondary martensite　二次马氏体
secondary material　再用材料
secondary measuring instrument　二次测量仪表
secondary measuring system　二次测量系统
secondary member　次要部件，次要构件，次要杆件，副构件
secondary memory　副存储器，外存储器
secondary meter　二次测量仪表，二次仪表，二次表计
secondary mineral　次生矿物
secondary model　二次模型
secondary moment　附加力矩
secondary network　二次侧电力网，二次网络
secondary neutral grid　二次中心点接地网，配电线中性网络
secondary neutron　次级中子
secondary occupation　第二职业
secondary oil　二次油
secondary operating sleeve　二次操作套筒
secondary order correlation　二介相关
secondary orogeny　次生造山运动
secondary outlet nozzle　二次出口接管
secondary output　（电流互感器的）二次侧容量，次级输出
secondary parameter　二次参数，次级线圈参数，副边参数
secondary plant　再加工的技术设备，二次装置
secondary pole　副极
secondary pollutant　二次污染物，二级污染物，次级污染物，再生污染质
secondary pollution　二次污染，二级污染，次级污染
secondary port　次等港，副港
secondary power distribution　二次配电
secondary power　超出保证功率部分，辅助电源
secondary principal stress　第二主应力
secondary product　次级产物，二次产物

secondary pump house 二级泵房
secondary pump 二次泵，二回路泵
secondary pusher dog 副推杆
secondary raceway 二次电缆管道
secondary radial distribution 次级辐射形网络
secondary radiation 次级辐射
secondary radio interference effect 二次波干扰效应
secondary rainbow 霓
secondary reaction 二次反应
secondary recording instrument 二次记录仪表
secondary reference fuel 副标准燃料
secondary reference point 次级参考点，次级基准点，二级参考点
secondary reference solar cell 二级标准太阳电池
secondary refrigerant 载冷剂，冷媒
secondary regulating pond 副调节池
secondary reheater 第二级再热器
secondary reinforcement 辅助钢筋
secondary relay 次级继电器，二次继电器
secondary resistance 电极线圈电阻，二次侧电阻
secondary return air 二次回风
secondary ridge 次分水岭
secondary saline soil 次生盐土
secondary sample cutter 二次采样器
secondary screen 二道筛
secondary sealing unit 二次密封单元
secondary seal source assembly 二次密封源组件
secondary section cubicle 二次舱（室）
secondary sedimentation basin 二次沉降池
secondary sedimentation tank 二次沉淀池
secondary sedimentation 二次沉降
secondary separation 二次分离
secondary settlement 二次沉降
secondary settling tank 二次沉淀池，二次沉降池
secondary set 副环设定
secondary sewage treatment 二级污水处理，污水二级处理
secondary shield concrete 二次混凝土屏蔽
secondary shielding 次级屏蔽，第二屏蔽，二次屏蔽
secondary shield 二次屏蔽体
secondary shutdown system 辅助停堆系统，第二停堆系统
secondary shutdown unit 第二停堆装置
secondary side inspection 二次侧检查
secondary side 蒸汽发生器二次侧，二次侧
secondary sodium loop 二次钠回路
secondary sodium pump 二次钠泵
secondary sodium 二回路钠，二次钠
secondary source rod 二次源棒，二次中子源棒
secondary source 二次源，二次中子源
secondary stair flight 次要楼梯段
secondary standard pyranometer 副基准总日射表，二等标准总日射表
secondary standard pyrheliometer 副基准直接日射表，二等标准直接日射表，副基准全辐射表，二等标准全辐射表
secondary standard solar cell 二级标准太阳电池
secondary standard 二级标准，副标准，二次标准，次级标准
secondary station secondary steam flow rate 二次蒸汽流量
secondary station 二次站，二等站
secondary steam generator 二次蒸汽发生器
secondary steam pipe 二次蒸汽管
secondary steam-to-steam heat exchanger 二次蒸汽蒸汽热交换器
secondary steam 二次蒸汽
secondary steelwork 二次钢结构
secondary storage 二级存储器，辅助存储器
secondary stratification 次生层理
secondary stream 二次流
secondary stress 二次应力，次级应力，次应力
secondary structure 次生构造
secondary subroutine 二次子程序
secondary substation 二次变电所
secondary superheater 第二段过热器，第二级过热器
secondary support column 辅助支撑柱
secondary swelling 二次肿胀
secondary switchboard 辅助配电盘
secondary system chemistry 二回路化学
secondary system diagram 二次系统图
secondary system hydrotest 二回路水压试验
secondary system 二次系统，二次回路
secondary test 二回路试验，二次试验
secondary thermal relay 二次热敏继电器
secondary thermometer 派生温度计
secondary tidal wave 副潮波
secondary time effect 次时间效应
secondary transformer 次级变压器
secondary treatment process 二级处理过程
secondary treatment 二次处理，二级处理
secondary triangulation 二等三角测量
secondary turbine 后置式汽轮机
secondary ventilation 辅助通风
secondary voltage 次级电压，二次电压
secondary vortex after cascade 叶栅后的二次涡
secondary vorticity 二次涡量
secondary waste 二次废物
secondary water pollution 次生水污染
secondary water 二次水，二回路水，次生水
secondary wave 次波，次生波，次级波，地震横波
secondary weir 辅助堰，副堰
secondary wind direction 次级风向
secondary wind flow effect 次级风效应
secondary winding 二次绕组，次级绕组，二次线圈，次级线圈，副绕组
secondary 二次的
second axial moment of area 截面二次轴距
second breakdown 二次击穿
second carrier （第）二程船
second channel interference 副通道干扰，图像干扰，镜像干扰
second circuit 次级电路，二次回路
second class conductor 第二类导体，电介质
second class pyranometer 二级（工作）总日射表
second class pyrheliometer 二级（工作）直接日射表

second class pyrradiometer　二级（工作）全辐射表
second coat of paint　第二道漆
second coat　二道抹灰
second coefficient of viscosity　第二黏性系数
second controllable losses　次要可控损耗
second convert component　二次转换元件
second critical field　第二临界磁场
second critical speed　第二临界转速，二阶临界转速
second-cut file　二次刻纹锉
second-degree price discrimination　二级价格歧视
second derivation control　按二次导数调节
second derivative action　二次微分作用，二阶微分作用，二阶导数作用
second derivative control　二次微商控制，二次导数控制
second derivative　二次导数
second differential　二次微分
second evaporator　二次蒸发器
second generation computer　第二代计算机
second generation neutron　第二代中子
second half-value layer　第二半值层
second half year　下半年
second harmonic magnetic modulator　二次谐波磁调制器
second harmonics　二次谐波
second harmonic　二次谐波的
second law of thermodynamics　热力学第二定律
second level controller　第二级控制器
second limit theorem　二次极限定理
second liquid　两次流【次级液体】
second moment of area　截面惯性矩
second moment　二次矩
second mortgage　第二抵押，再次抵押
second-motion engine　带减速器的发动机
second order activity　二次放射性
second order bench mark　二等水准标点
second order correction　第二级校正
second order difference　二次差分
second order differential equation　二阶微分方程
second order dynamic system　二阶动力系统
second order lag　二阶滞后
second order levelling　二等水准测量
second order linear system　二阶线性系统
second order polynomial　二阶多项式
second order reaction　二级反应
second order servo　二阶随动系统，二阶伺服系统
second order subroutine　二级子程序
second order system　二级系统，二阶系统
second order triangulation　二等三角测量
second order　二阶的
second pitch of coil　线圈前节距，线圈第二节距
second polar moment of area　截面极惯性矩
second principal stress　第二主应力
second prize　二等奖
second probability distribution　第二概率分布
second radiation constant　第二辐射常量
second ranked firm　排名第二的公司
second remove subroutine　二级子程序
seconds counter　秒数计数器
second shelterbelt　副防护林带
second sort tempering brittleness　第二类回火脆性
second stage cofferdam　第二期围堰
second stage concrete　二次浇灌混凝土
second stage cooler　（抽气器的）第二级冷却器
second stage cooling　二期冷却
second stage development　二期开发
second stage ejector　第二级抽气器
second stage excavation　二期开挖
second stage steam bypass　二级蒸汽旁路
second tap　二道丝锥
second　次要的，二次的，附加的，第二的，支持，临时调派，附议，赞成提案
SECO(sequential control)　程序控制，顺序控制
secrecy and confidentiality clause　保密条款
secrecy system　语言编码电路加密器
secretariate　秘书处，书记处，秘书
secret code　密码
secret communication　保密通信
secret dovetailing　暗楔接合，暗榫接合
secret gutter　暗檐槽
secret joint　暗榫接合
secret meeting　秘密会议
secret screwing　暗螺丝拼接
secret　秘密
SEC(simple electronic computer)　便携式电子计算机
SEC(single-entry single-exit circuit)　单入口单出口电路
sectional area of reinforcement　钢筋截面积
sectional area　横截面积，截面积
sectional coil　分段线圈
sectional construction　预制构件拼装结构
sectional-continuous function　分段连续函数
sectional-continuous　分段连续
sectional conversion　分段转化
sectional core　拼合型芯，分段铁芯
sectional detail　剖面，大样
sectional diagram　断面图
sectional dimension　截面尺寸
sectional-divided furnace　分隔式炉膛
sectional drawing of main power building　主厂房横剖面布置图
sectional drawing　断面图，剖面图，截面图，剖视图
sectional drive　多电机驱动，分段驱动
sectional elevation　截视立面图，立剖面图，立剖面
sectional formwork　工具模板，活用拼装模板
sectional form　断面形状
sectional-header boiler　分联箱锅炉
sectional inspection　分段巡线
sectionalization　分组，分段，分节
sectionalized bus-bar　分段母线
sectionalized configuration　单母线分段接线
sectionalized economizer　多级省煤器
sectionalized reheater　分段再热器
sectionalized single-bus with auxiliary busbar　单

母线分段带旁路母线接线
sectionalized superheater 多级过热器，分段过热器
sectionalized transformer 分节变压器，分组变压器
sectionalized winding 分组绕组，分段绕组
sectionalized 分段的，分区的，分组的
sectionalizer 分段器
sectionalize 分段，分组，分区
sectionalizing air-break switch 分段空气断路开关
sectionalizing of network 电力分区，电力网分段
sectionalizing oil-circuit breaker 分段油断路器
sectionalizing valve 分区截流阀
sectional-line 断面线，分区线
sectional observation 断面观测
sectional plan 分区平面图
sectional pump 节段泵
sectional shaft 分段轴，组合轴
sectional side elevation 纵剖图
sectional steel 型钢
sectional switch 分段开关
sectional terminal block 分段接线板
sectional trash rack 分节式拦污栅
sectional view 截面图【机】，断面图，剖面图，剖视图
sectional winding 分段绕组
sectional 分级的，部分的，可拆卸的，剖断的
section blocking 分段闭塞
section box 电缆交接箱
section breaker 分段开关，分段断路器
section chief of a speciality 专业科长
section chief 科长
section chord technique 截面弦法
section chord 剖面弦，叶片剖面弦长
section configuration 断面外形，断面形状，截面轮廓
section construction 分段施工，拼装结构
section contour 剖面外形，截面轮廓
section crack 断面裂纹
section disconnecting switch 分段隔离开关
section drag 截面阻力，翼型阻力
section drawing 断面图，剖面图
section elevation 剖面图
section engineer 工段工程师，主任工程师
section factor 断面系数
section foreman 分段工长
sectioning valve 分段阀，切断阀
section insulation 段间绝缘
section insulator 分段用绝缘子
section iron 型钢，型铁
section length 耐张段长度
section lift 截面升力，翼型升力
section line 断面线，剖面线
section manager 部门经理
section model 截段模型
section modulus 断面模数，截段模量，剖面系数［模数］，截面模数［惯性］，断面系数［模量］
section moment 截面力矩
section of an overhead line between two strain towers 耐张段
section of communication 通信段
section of construction project 工段
section of highest hydraulic efficiency 最高水力效率断面
section of line 线段
section of rotor 转子区段
section paper 方格纸
section plan 平面剖视图，剖面图
section profile 横断面，横截面
section steel 型钢
section switch 分段开关，区域开关
section valve 隔离阀
section view 断面图，剖面图
section 剖面，截面，断面，剖视图，断面图，剖面图，划线，分割，切片，分节，分段，工段，区段，切面，部门，型钢
sector box model 扇形箱模式
sector cable 扇心电缆
sector coal-yard 扇形煤场
sector conductor 扇形导线，扇形引出线
sector display 分区显示，分段显示，扇形显示
sector gate 弧形闸门，扇形门
sectorial wave 扇形波
sectorial 扇形的
sectoring 扇形扫描，分扇区
sectorized feeder 扇形加料器
sector liner 扇形瓦里衬
sector magnetic field mass spectrometer 扇形磁场质谱计
sector-pattern instrument 扇形仪表
sector pivot gate 扇形旋转式闸门
sector regulator 扇形节制闸门
sector-rim type rotor 扇形叠片磁轭转子
sector scan indicator 扇形扫描显示器
sector scanning 扇形扫描
sector scan 扇形扫描
sector-shaped conductor 扇形导线
sector switch 分段开关，扇面开关
sector-type relay 扇形继电器
sector valve 扇形阀
sector weir 弧形堰
sector 扇面，扇形，部分，部门，段，区段，象限，扇形片，扇形轮
sectrometer 真空管滴定计
sect 宗派，教派，部分
secular change 长期性变化，多年变化
secular cycle 百年周期，长年周期
secular drift 缓慢漂移
secular effect 长期效应
secular equation 特征方程，长期方程
secular stability 长期稳定度，长期稳定性
secular trend 长期趋势
secular variation 长期性变化，多年变化
secundum usum 根据惯例
secure and economical operation 安全经济运行
secure an obligation 保证负债义务
secured loan 有担保贷款，有抵押贷款
secure normal state 安全正常状态
secure one's agreement 征得……的同意
securing key 定位销

securing 卡紧，固定
securities company 证券公司
securitron 电子防护系统
security analysis 安全分析
security area 保卫区
security assessment 安全性评估，安全性评价，可靠程度估计
security assessor 可靠性鉴定器
security automatic equipment 安全自动装置
security building control desk 保安楼控制台
security building 保安楼
security control 安全措施，安全控制，保护措施，安全监督，安全管理
security coupling 安全联轴器
security deposit 交易保证金
security detector 安全检测计
security evaluation of vibration 振动安全判别
security glass 安全玻璃
security lock 安全锁
security manager 治安经理
security margin 安全裕度
security monitoring 安全监视
security of an electric power system 电力系统的安全性
security office 保安值班室
security of loan 贷款担保
security rod 安全棒
security seal 安全密封
security system 保安系统，保卫系统
security wall 护墙
security wind speed 安全风速
security zone 安全区，设防区
security 有价证券，证券，债券，安全，可靠（性），保安，保证（金），担保，抵押品，保函，投标保证（金）
SEC 重要厂用水系统【核电站系统代码】
sedentary deposit 原生沉积
sedentary soil 残积土，原地土
sedentary 原地的
sedimental 沉淀的，沉积的
sedimentary band 沉积带
sedimentary basin 沉积盆地
sedimentary clay 沉积黏土
sedimentary cover 沉积盖层
sedimentary cycle 沉积循环，堆积轮回
sedimentary deposit 沉积层，沉积物，成层沉积
sedimentary differentiation 沉积分异作用
sedimentary dyke 水成岩墙
sedimentary facies 沉积相
sedimentary fault 沉积断层
sedimentary formation 沉积层
sedimentary intrusion 沉积侵入
sedimentary peat 沉积泥炭
sedimentary petrography 沉积岩相学
sedimentary resistance 沉积阻力
sedimentary ripple 沉积波纹
sedimentary rock 沉积岩，堆积岩，水成岩
sedimentary soil 沉积土（壤）
sedimentary structure 沉积构造
sedimentary tectonics 沉积构造作用
sedimentary 沉积物
sedimentated dust 沉降的灰尘，降尘
sedimentation analysis 沉淀分析，沉降分析法
sedimentation balance 沉降天平
sedimentation basin 沉淀池，淤积池，沉降池，沉砂池
sedimentation chamber 沉淀池，澄清器，沉降室
sedimentation coefficient 沉降系数
sedimentation cycle 沉积循环
sedimentation diameter 沉降粒径
sedimentation equilibrium 沉积平衡
sedimentation mechanism 沉积机制
sedimentation method 沉降法
sedimentation mode 沉积方式
sedimentation pit 沉降池
sedimentation process 沉淀过程
sedimentation rate 沉积速率，（水库的）淤积率
sedimentation reservoir 沉淀池，沉沙池，沉沙水库
sedimentation storage 淤积库容
sedimentation survey 淤积测量，淤积调查
sedimentation tank 沉淀池，澄清池，沉积池，沉降槽
sedimentation time 沉淀时间
sedimentation velocity 沉降速度
sedimentation 沉降，沉积（作用），淤积（作用），沉积法，沉淀，淤积指数，沉积学
sediment box 沉淀箱
sediment-carrying capacity 挟沙能力
sediment charge 含沙量，泥沙相对含量【泥沙重量对流水重量之比】
sediment classification 冲积物分类
sediment concentration 含沙量，沉积物的浓度
sediment content 沉积量
sediment control dam 拦沙坝
sediment diffusion 泥沙扩散
sediment discharge curve 水位输沙量关系曲线，输沙量曲线
sediment discharge measurement 输沙量测定
sediment discharge 输沙量，输沙率，土沙流量，沉积物流量，沉淀物排放
sediment dislodging 清除泥沙
sediment ejection 排沙
sediment escape 冲沙泄水道
sediment flushing at low water level 低水位泥沙冲刷
sediment flushing 排沙
sediment-free water 无沉积物水
sediment grade size 沉积物分级尺寸
sedimenting system 沉降系统
sediment-laden river 多泥沙河流，多沙河流
sediment-laden water 挟沙水
sediment longitudinal transport 沉积物纵向迁移
sediment measurement 泥沙测验
sedimentography 沉积岩相学
sedimentology 沉淀学，沉积学，沉积岩石学
sedimentometer 沉积测定仪
sediment percentage 含泥量
sediment runoff curve 固体径流曲线
sediment runoff 固体径流，泥沙径流，输沙量
sediment sampler 采泥器，沉积物采样器

sediment source 泥沙来源
sediment storage basin 蓄沙池
sediment storage capacity of reservoir 水库储沙能力，水库堆沙容积
sediment tank 沉淀箱，泥箱
sediment transport 泥沙运输
sediment trap 沉积物收集器，沉淀物捕集器，沉积区
sediment-water system 沉积物水系统
sediment yield 固体径流量，产沙量
sediment zone 沉淀区，（直流炉的）沉盐区
sediment 沉淀，沉积，沉渣，沉淀物，泥沙
SED(static electricity discharge) 静电放电
SED 核岛除盐水分配系统【核电站系统代码】
seebeck coefficient for substances M and N 物质M和N的塞贝克系数
see copy 阅副本
seed and blanket arrangement 点火区转换区布置
seed and blanket core 点火区转换区堆芯
seed area 点火区
seed blanket breeder reactor 点火区转换区增殖堆
seed blanket reactor 点火区转换区反应堆，点火区再生区反应堆
seed blanket 点火区转换区
seed cluster 点火区棒束
seed core reactor 强化堆芯反应堆
seed core 强化堆芯
seeded plasma 籽晶等离子体
seed element 点火区燃料元件
seeding 强化，点火，播种，放入晶种
seed loading 点火区装料
seed material 点火材料
seed recovery 点火区燃料回收
seed region 点火区【反应堆】，点燃区
seed subassembly 点火区组件，点火区元件盒
seed unit 点火区
seed 导火管组件，点火燃料组件，点火区元件，点火区，颗粒，晶粒
SEE-IN(significant event evaluation and information network) 重要事件评价和通报网络
seek a negotiated settlement 寻求和平谈判解决
seeking overseas funds 寻求，引进外资
seeking overseas investment 寻求外资，引进外资
seepage analysis 渗透分析
seepage apron 防渗护坦
seepage area 渗流区，渗漏区，渗漏面积
seepage bed 渗滤床
seepage boil 渗流翻腾
seepage calculation 渗透计算
seepage channel 渗水通道
seepage coefficient 渗流系数
seepage control wall 防渗墙
seepage control 防渗，渗流控制
seepage curve 渗透曲线
seepage deformation 渗透变形
seepage discharge 渗透流量，渗流量
seepage distance 渗透距离
seepage face 渗流面
seepage failure 渗流破坏，渗透破坏
seepage flow 渗出水流，渗流
seepage force 渗流力，渗透力
seepage gauge 渗流计
seepage gradient 渗透坡降，渗透梯度
seepage gutter 渗水沟
seepage head 渗流水头
seepage line 渗流方向，渗水线
seepage loss 渗流损失，渗漏损失
seepage measurement 渗流测定
seepage meter 测渗仪，渗透计，渗透仪
seepage model 渗流模型
seepage path 渗径，渗透途径
seepage pattern 渗流型式
seepage pit 渗水坑
seepage pressure 渗流压力，渗透压力
seepage prevention 防渗（漏）
seepage quantity 渗流量，渗透量
seepage rate 渗透率
seepage test 渗透试验
seepage uplift 渗透扬压力
seepage velocity 渗流速度，渗透速度
seepage water 渗流水，渗透水，渗漏水，渗水
seepage zone 渗流区
seepage 渗出（量），渗透，漏出，渗漏，渗流，渗流线，渗液
seep holes 渗水孔
seep off 渗出
seep tar 渗出焦油
seep water 渗出水，渗流水
seep 渗出（量），渗漏现象，渗漏
SEER(seasonal energy efficiency ratio) 季节能效比
seesaw circuit 反向放大器，反相放大电路
seesaw motion 跷板式运动
seesaw rotor 跷板式风轮
seesaw switch 转换开关
see-through power plant 透明模型电厂
S-effect S效应，表面电荷效应
Segent bearing 多片瓦轴承，轴瓦块轴承
seger cone 温度锥，测温锥
segistration 叠加，重叠，重合
segmental arc 弧段，弧形段
segmental barrel vault 弧筒形穹顶
segmental blading （扇形）叶片组
segmental conductor 扇形导线，弧形导线
segmental girder 弓形大梁
segmental lamination 扇形冲片，扇形迭片
segmental nozzle group control 喷嘴调节
segmental orifice plate 圆球孔板，圆缺孔板，扇形孔板
segmental punching 扇形冲片
segmental-rim rotor （水轮发电机的）扇形片磁轭转子
segmental roof truss 折线形屋架
segmental rotor 分瓣转子，扇形转子
segmental sluice gate 弧形泄水闸门，扇形泄水闸门
segmental stamping 扇形冲片
segmental thrust bearing 分瓣推力轴承，扇形止推轴承

segmental 弓形的，部分的，扇形的
segment arch 弓形拱
segmentation 程序分段，分段，段式，分块，换向片，分裂
segment bearing 瓦块式轴承
segment display 分段显示
segmented bed slumping 部分床压火【沸腾炉】
segmented-electrode channel 分段电极通道
segmented fin （螺旋）切片式的肋片
segmented fuel rod 分段的燃料棒
segmented roller 分段滚压机
segmented-rotor 分瓣转子
segmented thermocouple 扇形热电偶
segment gate 弧形堰，扇形闸门
segment hydrograph 输沙过程线
segmenting-faraday generator 分段法拉第发电机
segment mark 分段标记
segment mica 换向器云母片
segmentor （程序）分段装置
segment pitch 换向片节距，片距
segment shoe 扇形瓦
segment table entry 段表，段表条目，段表项
segment table origin register 段表起始寄存器
segment table 段表
segment-tooth roll 节段齿辊【辊碎机】
segment valve 弓形阀
segment voltage 换向片间电压
segment weir 弓形堰
segment （回转式预热器）舱格，部分，段，节，轴瓦，扇形瓦，喷嘴弧段，扇形片
segregability 离析性
segregated collection system 分流集水系统，分凝收集系统
segregated equipment 分批提供的设备
segregated flow 分离流
segregated-phase bus 分相母线
segregated-phase common enclosure bus 隔相共箱封闭母线
segregated spot 偏析区
segregated 分开的，被隔离的
segregate 分隔开【功能，物理上】，分凝，分析，分离，分隔，偏析
segregating concrete 离析的混凝土
segregating unit 分离装置，分离机
segregating valve 隔离阀
segregation ash-deposit 选择性积灰
segregation degree 离析度
segregation map 偏析图，熔析图
segregation of items 项目的分类，项目划分
segregation of loss 损耗分离
segregation 分凝，离析，分隔，分离，骨料分离，偏析，分开，隔离，划分，熔析【冶金】
segregator 分离器，分配器，分凝器
SEH 废油和非放射性水排放系统【核电站系统代码】
SEIA（strategic environmental impact assessmen）战略环境影响评价
seibt rectifier 低压真空管全波整流器
seiche amplitude 水面波动幅度
seiche nodality 驻波节数
seiche 湖涌，湖震

seismic acceleration 地震加速度
seismic accident analysis 地震事故分析
seismic activity 地震活动，地震活动性
seismically induced liquefaction 地震诱导液化
seismic amplifier 地震放大器
seismic analysis 地震分析
seismic area 地震区
seismic belt 地震带，地震区
seismic calculation 地震计算
seismic category 地震等级
seismic center 地震中心
seismic coefficient 地震系数
seismic concept design of buildings 建筑抗震概念设计
seismic conditions 地震条件
seismic cross-section 地震剖面
seismic data 地震数据，地震资料
seismic design decision analysis 抗震设计决策分析
seismic design intensity 地震设计烈度
seismic design provision in building code 建筑抗震设计规范
seismic design 抗震设计，地震设计，地震计算
seismic detection 地震探测
seismic detector 地震测波器，地震仪
seismic duty 抗震能力
seismic effect 地震效应
seismic exploration 地震勘探，地震探测
seismic factor 地震系数
seismic floor joint cover 地面抗震缝盖板
seismic focus 地震震源，震源
seismic force 地震力
seismic fortification criterion 抗震设防标准
seismic fortification measures 抗震措施
seismic gap 地震活动空白地带
seismic geophone 地震测波器
seismic geophysical method 地球物理地震法，地震物探法
seismic head wave 地震首波
seismic instrumentation 测震仪表［设备］
seismic instrument 地震仪
seismic intensity expectance map 预期地震烈度图
seismic intensity micro-regionalization 地震烈度小区域划分
seismic intensity scale 地震烈度等级，地震烈度表
seismic intensity 地震烈度，地震强度，震度
seismic investigation 地震探测
seismicity activity map 地震活动分布图
seismicity chart 地震烈度图
seismicity degree 震度
seismicity gap 地震活动空白地带
seismicity map 地震区域图
seismicity 地震等级，地震活动性，地震活动度，地震活动
seismic joint 抗震缝，地震缝
seismic load factor 地震载荷因子
seismic load 地震荷载
seismic location 地震定位
seismic logger 地震测井仪

seismic logging 地震测井
seismic map 地震图
seismic measurement 地震测验
seismic method of exploration 地震法勘探
seismic method of prospecting 地震法勘探
seismic model 地震模型
seismic motion 地震运动
seismic origin 地震成因，震源
seismic plate 测震板
seismic profile 地震剖面
seismic property 地震性质
seismic prospecting system 地震探查装置
seismic prospecting 地震勘探，地震探查，地震探测
seismic record 地震记录
seismic reflection method 地震波反射法，地震折射法
seismic regionalization map 地震区划图
seismic regionalization 地震区划分
seismic region 地震区
seismic resistance 抗地震
seismic response analysis 地震反应分析
seismic response spectra 地震反应谱
seismic sea wave 地震津波
seismic shock 地震，地震冲击，地震震动
seismic signal 地震信号
seismic sounding 地震测深法
seismic spread 地震传播，地震扩散
seismic stability of electric equipment 电气设备抗震
seismic stability 抗震稳定性
seismic stop 防震缓冲器
seismic stress 地震应力
seismic support platform 抗震支撑台【控制棒驱动机构】
seismic survey 地震测量，地震调查，地震探查
seismic switch 地震开关
seismictectomic line 地震构造线
seismic test 抗震试验
seismic travel time 地震波传播时间
seismic trigger 地震启动装置
seismic wave path 地震波路径
seismic wave velocity test 地震波速测试
seismic wave 地震波
seismic zone of intensity ……度地震区
seismic zone 地震带，地震区
seismic zoning map 地震分区图
seismic zoning 地震区划分
seismic 地震，地震的
seismocope 地震仪
seismogenesis 地震成因
seismogeological map 地震地质图
seismo-geology 地震地质学
seismogram 地震波曲线
seismographic observation 地震观测
seismographic record 地震记录
seismographic station 地震台
seismography 地震学
seismograph 测震仪
seismological bureau 地震局
seismological observation 地震观测
seismological observatory 地震观测站
seismologieal evidence 地震实迹
seismologist 地震学家
seismology 地震学
seismolog 测震仪【附有摄影设备的】
seismomagnetic effect 地震地磁效果
seismometer station 地震测站
seismometer 地震仪
seismoscope 地震波显示仪，验震器
seismostation 地震台
seismotectonic line 地震构造线
seismotectonic province 地震构造区，地震活动范围
seismotectonic 地震构造
seized screw 卡住的螺钉
seized stud removal tools 咬住螺杆的清除工具
seized stud removal 清除咬住的螺杆
seize 没收，抢夺，咬住，卡住，占有，抓住
seizing signal 约定信号，约束信号，占用信号
seizure evacuation 强迫撤离
seizure wear 卡滞磨损
seizure 卡住，捉住，夺取
SEK 常规岛废液排放系统【核电站系统代码】
SEK 瑞典克朗
SELCH(selector channel) 选通器通道
SELD＝selected 已选好的
select address and contract operate 选择地址及对比操作
selectance 选择度，选择系数
select and execute command 选择和执行命令
select before operate 操作前选择
select best alternative 选择最优方案
select bit 选择位
selected bidder(/tenderer) 入选的投标人，预选的投标人
selected bidding 选择性招标
selected bit 被选位
selected cell 寻址单元，选址单元
selected core 被选铁芯
selected filling 选择的填料
selected tenderer 选定的投标人，被选择的投标人
select finger 选择指针
selecting circuit 选择电路
selecting magnet 选择磁铁
selection check 选择校验，选择检验
selection coefficient 选择性系数
selection command 选择命令
selection control 选线控制，选择控制
selection core matrix 选择磁芯矩阵
selection core 选择磁芯
selection criteria 选择标准
selection element 选择元件
selection factor 选择因素
selection index 选择性指标
selection measurement 选线测量
selection mechanism 选择机构
selection of auxiliary electrical equipment 厂用电设备选择
selection of cable type and size 电缆选型
selection of conductor cross-section 导线截面

选择
selection of converter station site 换流站站址选择
selection of dam type 坝型选择
selection of equipment 设备选型
selection of hydro-logical data 水文资料选择
selection of overhead ground wire 架空地线选择
selection of plant site 厂址选择，站址选择
selection of tender 选择投标人
selection of water source 水源选择
selection plugboard 选号插头板，选择插头板
selection procedure 选择程序
selection sort 选择分类
selection wire 选择线
selection 选择，（自动电话）拨号，挑选，筛选
selective absorbing surface 选择性吸收面
selective absorption 选择吸收，选择性吸附，选择性吸收
selective-access display 随机显示
selective assembly 选配
selective bidding 选择性招标
selective calling system 选择呼叫系统
selective calling 选择呼叫
selective carryover 选择性携带
selective catalytic reduction 选择性催化还原（法）【烟气脱硝技术】
selective channel input bus 选择通道输入总线
selective circuit 选择电路
selective coating 选择性涂层
selective coefficient 选择性系数
selective-collective type 群控自动
selective combustion method 选择性燃烧法
selective control system 选择控制系统
selective control 选择操作
selective corrosion 选择腐蚀，选择性腐蚀
selective crusher 选择性破碎机
selective differential relay 选择性差动继电器
selective digging 分类挖土
selective dissemination of information 信息选择传播
selective dump 选择转储
selective electrode 选择电极
selective enrichment 选择性富集
selective erasing 选择擦除，选择性侵蚀
selective exam 抽查
selective excavation 选择开挖法
selective fading 选择性衰落
selective feeder 选择给料器，选择进料器
selective firing 换层燃烧
selective ground relay 选择接地继电器
selective headstock 变速箱
selective heating 局部加热，选择性加热
selective identification feature 选择识别特点
selective information retrieval 选择性情报检索
selective inspection 抽查
selective interrogation command 选择查询命令
selective ion exchange 选择性离子交换
selective leaching 选择沥滤
selective mark insertion 选择标记的插入
selective mark 选择标记
selective measuring system 选择测量系统
selective measuring 选择测量
selective network 选择性网络
selective-preferential type 优择方式，选择性优先方式
selective printer programmer 选择性打印机的程序编制器
selective protection 选择性保护
selective radiation 选择性辐射
selective radiometer 选择辐射计
selective relay 选择性继电器，谐振继电器
selective sequence calculator 选择程序计算机
selective signaling 选择信号装置
selective surface 选择性表面
selective switch 选择开关
selective tender 选择性招标
selective trace 选择跟踪
selective transformer 选择性变压器
selective withdrawal 选择性抽取
selective writing 选择写入，选择记录
selectivity characteristic 时限特性，选择特性
selectivity clear accumulator 选择清除累加器
selectivity coefficient 选择性系数
selectivity control 选择性控制
selectivity curve 选择性曲线
selectivity factor 选择性因子
selectivity order 选择性顺序
selectivity 选择性能力，选择性，选择能力
select magnet 选择磁铁
select material 选用材料
selector busbar 选择器母线
selector button 选择按钮，选择器按钮
selector channel 选通器通道，选择通道，选择器通道
selector circuit 选择器电路
select order 选择指令
selector key 选择电键
selector magnet 拨号电磁铁
selector relay 选择开关，选择继电器，步进开关
selector stepping magnet 选择器步进磁铁
selector switch 切换开关，选路开关，（自动/旁路/就地）选择开关，波段开关
selector valve 选择阀，多道阀，换向阀
selector 调谐旋钮，选择器，波段开关，选数器，选择开关
select read numerically 数字选择读出
selectron storage 选数管存储器
selectron 选数管，聚酯树脂
select switch 选择开关，选线器，选路开关
select unit 选择电源
select 分选，选择
selenite cement 石膏水泥
selenite 透明石膏
selenium cell relay 硒光电池继电器
selenium cell 硒光电池，硒整流片
selenium conductive cell 硒光敏电池
selenium diode 硒二极管
selenium layer 硒层
selenium photometer 硒光度计
selenium rectifier 硒整流器
selenium resistor 硒电阻
selenium stabilizer 硒稳定器
selenolite 石膏岩

seletron 硒整流器
self-absorption factor 自吸收系数，自吸收因子
self-absorption 自吸收
self-acting control automatic control 自行控制
self-acting controller 自作用控制器
self-acting control 自行控制，自动控制，直接调节，直接控制
self-acting discharge valve 自动泄水阀
self-acting pump station 自动抽水站
self-acting thermostat 自调恒温器，自动恒温器
self-acting valve 自调阀，自动阀
self-acting 自作用的，自动的，自作用，自动式
self-action 自动，自动作，自作用
self-activating 自激活的
self-actor 自动机
self-actuated controller 自作用控制器，自给能控制器，自激控制器
self-actuated 自激的，自行的，直接的，自作用的
self-actuating pressure relief valve 自致动卸压阀
self-adapting filter 自适应滤波器
self-adapting 自适应的，自适应
self-adjoint system 自伴随系统
self-adjoint 自共轭的，自伴性，自共轭性，自伴的，自伴随
self-adjustable 可自动调节的
self-adjusting bearing 自调整轴承
self-adjusting gland 自调整汽封
self-adjusting weir 自动调整堰
self-adjusting 自动调整的，自动调整，自调整
self-adjustment 自调节
self-admittance 固有导纳，自导纳
self-aerated flow 自动渗气水流
self-agglomerating fluidized bed 自团聚流化床
self-aligning bearing 自调轴承，自动调整轴承，自位调整，自位轴承，自找中轴承
self-aligning belt training idler 自动调心托辊
self-aligning roller bearing 调心滚子轴承，球面滚柱轴承
self-aligning system 自动调准系统，自定位系统
self-aligning 自找正定位，自动就中，自行调整，自动调准
self-anchorage 自锚
self-anchored suspension bridge 自锚式悬索桥
self-assessment 自我评定
self-baking electrode 自焙干电极
self-balance protection 自平衡保护，差动保护
self-balancing bridge 自平衡电桥
self-balancing capacitance 自动平衡电容
self-balancing device 自动平衡装置
self-balancing instrument 自平衡仪表
self-balancing measuring equipment 自平衡测量仪器
self-balancing potentiometer 自平衡电势计，自平衡电位计
self-balancing pump 自平衡泵
self-balancing receiver-recorder controller 自平衡接收记录式控制器
self-balancing stress 自平衡应力
self-balancing system 自平衡系统
self-balancing 自平衡，自动平衡的
self-ballasted lamp 自镇流灯
self-bearing steel pipe 自承式钢管
self-bearing wall panel 自承重墙板
self-bias cut-off effect 自偏压截止效应
self-biased off 自动偏压截止
self bias resistor 自偏置电阻
self-bias 自给偏压，自动偏移
self-blocking 自动联锁，自中断，自断路，自动阻塞
self-boiling waste storage tank 自沸废液贮罐
self-bonding 自动焊合【利用热量结合在一起】
self-boring pressure meter 自动钻入的压力仪表
self braking 自动制动
self breakdown 自击穿
self-bursting fuse 自爆熔断器
self-calibrating 自校正的，自校准的
self-capacitance 自电容，固有电容
self-capacity 自电容，固有电容
self-catalyzed reaction 自催化反应
self-centering idler 自动调心托辊
self-centering mandrel 自动定心芯轴
self-centering vibrating screen 自定中心振动筛
self-centering 自动定心，自定中心
self-checking behavior 自检行动，自检动作
self-checking code 自校验码，自校代码，自检验码
self-checking number system 自校验数系统
self-checking number 自检验数
self-check program 自检程序
self-check system 自检系统
self-check 自动检验，自检验，自检，自校，自校核
self-circulation 自循环
self-cleaning belt magnetic separator 自清带式磁力分离器
self-cleaning contact 自清洁触头
self-cleaning drilling 水冲钻探
self-cleaning grate 自清灰炉排
self-cleaning hammermill 自清式破碎机
self-cleaning screen 自动清污栅网，自净拦污栅
self-cleaning velocity 自净流速
self-cleaning wagon 自动卸料车
self-cleaning 自动清洗，自动清灰，自清扫
self-climbing formwork 自升模板
self-climbing tower crane 自升式塔式起重机
self-clocking 自计时，自同步
self-closing butter-fly valve 自动关闭蝴蝶阀
self-closing door 自闭式防火门
self-closing oil adsorption filter 自封式吸油过滤器
self-closing stop valve 自动断流阀
self-closing valve 自闭阀，自动关闭阀
self-closing 自关闭，自动关闭，自动闭合，自接通，自闭合
self-collapse 自动倒塌，自溃
self-collapsing gate 自动翻倒式闸门，自溃闸门
self-combustion 自燃
self commutated inverter 自换流逆变器，自动换向逆变器
self-commutation （可控硅的）自整流
self-compensated machine 自补偿电机

self-compensated motor 自补偿电动机
self-compensating system 自补偿系统
self-compensating 自补偿的，自动补偿的，自动补偿
self-compensation balance 自补偿天平
self-compensation of pipes 管道的自补偿
self-compensation 自补偿，自然补偿
self-complementing code 自补代码
self-condensation 自冷凝
self-conductance 自电导
self-confinement 自限制，自制约，自约束
self-consistent field 自调和场，自励磁
self-consolidation 自动固结
self contained battery 自备电池
self-contained boiler 设备齐全的锅炉，整装式锅炉，整装锅炉
self-contained breathing apparatus 自备呼吸装置
self-contained closed-circuit TV system 自备闭路电视系统
self-contained complete 配套的
self-contained cooling unit 自备冷却机组
self-contained data-base management 独立的数据库管理系统
self-contained emergency lighting unit 自备应急照明装置
self-contained engine 独立发动机
self-contained equipment 成套设备，独立设备
self-contained generator 整装发电机
self-contained instrument 整装仪器，机内仪表
self-contained navigation aid 自载导航设备
self-contained oil-filled cable 自容式充油电缆
self-contained pneumatic type 自保持气动式
self-contained pressure cable 自容式压力电缆
self-contained unit 整装组件，自持装置
self-contained with 自成一体
self-contained 自持的，自备的，自带的，机内的，配套的，设备齐全的，整装的，独立的，自治的，自保持的，自给的
self-control 自动调整，自动控制
self-cooled machine 自冷式电机
self-cooled transformer 自冷式变压器
self-cooled 自冷式的
self-cooling vane 自冷式叶片
self-cooling 自冷式的
self-correcting code 自校正码
self-correcting 自校正，自改正
self-correction code 自校正码，自校编码
self cost 成本
self-coupled voltage-down starting 自耦降压启动
self-coupling separable contacts 自动联结分离触头
self-coupling 自动连接
self-damping conductor 自阻尼导线
self-damping crystal 自阻尼晶体
self-damping property 自阻尼特性，固有减振性能
self-damping 自阻尼
self-demagnetization 自去磁，自动退磁，自动去磁
self-demagnetizing field 自退磁场
self-descripiton 自描述，自我描述
self-developing film 自显影照片
self-diagnosis 自诊断
self-diagnostic alarm display 自诊断报警画面
self-diagnostic function 自诊断功能
self-diagnostic of fault 事故自诊断
self-diffusion 自扩散
self-discharge 自放电，自卸
self-discharging coal barge 自卸煤驳船
self discharging 自动泄水
self dosing 自动关闭
self-draining superheater 可疏水式过热器
self-draining valve 自动疏水阀
self-draining 自疏水
self-drill grouted anchor bar 自钻式注浆锚杆
self-drilling anchor bolt 自钻式锚杆
self-drilling screw 自钻孔螺钉
self-driven 自动的
self-drive 自动步进，自启动，自激励
self-dual 自对偶的
self-dumping car 自卸车
self-dumping skip 自卸吊斗
self-dumping truck 自卸车
self-dumping wagon 自卸车
self elastance 自倒电容
self-elevating platform 自升式平台
self-elevating pontoon 自升浮筒
self-employed individual 个体户
self-emptying 自动放空
self-energized 自供能量的，自激的
self-energizing gasket 自加力垫圈
self-energizing seal 自咬口密封
self-energizing 自紧的（密封）
self-energy interrupter 自能灭弧室
self-energy 固有能量【粒子】，自身能量，内能量，自具能
self-equalizing expansion joint 自均衡膨胀节
self-etching primer 自腐蚀涂料
self-evaporation 自蒸发
self-excitaion winding 自励绕组
self-excitation of electrical motor 电机自励磁
self-excitation oscillation 自激振荡
self-excitation process 自励过程
self-excitation transductor 自激磁变换器
self-excitation 自激，自励，自励磁
self-excited AC generator 自励交流发电机
self excited aerodynamic moment 自激励空气动力矩
self-excited booster 自激式升压机
self-excited circuit 自激电路
self-excited discharge 自激放电
self-excited dynamo 自激电机，自激发电机，自励式发电机
self excited force 自激力
self-excited generator 自励发电机，自激振荡器
self-excited machine 自励电机
self-excited MHD generator 自激式磁流体发电机
self-excited motor 自励电动机
self-excited multivibrator 自激多谐振荡器
self-excited oscillation 自激振荡
self-excited oscillator 自激振荡器
self-excited phase advancer 自激式进相机

self-excited regenerative braking 自激再生式制动方式
self-excited series winding 自励串激绕组
self-excited shunt field 自励并激磁场
self-excited transductor 自激磁变换器
self-excited vibration 自激振动
self-excited vortex shedding 自激涡脱落
self-excited 自激的，自励的
self-exciter 自励发电机，自励励磁机
self-exciting condition 自激条件
self-exciting 自励的，自激的，励磁的
self-exited oscillation 自振
self-exited vibration 自激振动
self-extensible language 自扩充语言
self-extinction 自熄
self-extinguishing circuit-break 自灭弧断路
self-extinguishing 自动熄火的，自熄火的
self-feedback 自反馈，自回授
self-feed conveyer 自装料输送机
self-feeder 自动给料器，自动加料器，自给器，自馈器
self-feeding 自激，自馈，自进给，自送料，自动供给，自动给料，自动进给
self-feed 自给，自馈
self-filling microburet 自充满微量滴定管
self finance 自筹资金
self-flocculation 自絮凝
self-fluxing brazing filler metal 自钎剂硬钎料
self-forced ventilation 自强力通风
self-forming 自成模板
self-gating 自选通
self-generated plutonium recycle 钚的自持再循环，自产钚再循环
self-hardening 自行硬化
self-healing breakdown 自复性击穿
self-healing fuse 自复式熔断器
self-healing soil 自愈性土
self-healing 自动愈合
self-heating 自加热
self-heterodyne 自差，自拍
self-holding contact 自保持触点
self-holding push button 自保持按钮，自持按钮
self-holding 自保持的
self-hold 自锁，自保持
self-homing device 自动寻的装置
self-hooped penstock 自箍式压力水管
self-hunting 自动寻线
self ignition engine 柴油机，压燃式发动机
self ignition fuel 自点火燃料
self ignition point 自燃点
self ignition 自发火，自动点火，自点火，自燃
self-impedance 固有阻抗，自阻抗，自动阻挠抗
self-induced EMF 自感电动势
self-induced magnetic flux 自感磁通
self-induced oscillation 自激振荡
self-induced 自感应的
self-inductance coefficient 自感系数
self-inductance 自感，固有电感，自感系数
self-induction coil 自感线圈
self-induction type coil 自感型线圈
self-induction 自感应
self-inductive 自感的
self-inductor 自感应器，自感线圈，自感应线圈
self-inhibition 自抑制
self-inspection 自检，自查
self-instructed carry 自动进位
self-interrupter 自行断续器
self irradiation damage 自辐射损伤
self-irradiation 自辐照
self-jetting wellpoint 自射式井点
self-leveling paint 自动找平漆
self-levelling instrument 自动找平水准仪
self-limiting model 自限模型
self-limiting oxidation 自限制氧化
self limiting parallel heater 自限制式并联加热器
self-limiting property 自限性
self-limiting reactor 自调节反应堆，自限制反应器
self-limiting 自限【负荷】，自限制，制约本身的
self-loading scraper 自动装载铲运机
self-loading 自动装卸，自行加载，自动装载
self-lock device 自锁装置
self-locking nut 自锁螺母，防松螺帽
self-locking screw 自锁螺钉，自锁螺旋
self-locking worm gear 传动装置自锁蜗轮
self-lock 自锁，自同步，自动制动
self-lubricated bearing 自润滑式轴承
self-lubricated （被泵液）自润滑的
self-lubricating bearing 自润滑轴承
self-lubricating brush 自润滑电刷
self-lubricating 自润滑
self-magnetic field 自激磁场
self-magnetic 自感的，自磁的
self-maintained discharge 自持放电
self-maintained push-button 自持按钮
self-mending fuse 自复熔断器
self-mixing 自混合
self-mobile 自动的
self-modeling 自建模
self-modulation 自调制
self-monitoring 自监控的
self-mulching soil 自植被土
self-multiplying chain reaction 自增殖链式反应
self-mutual inductance 自互感
self-mutual induction 自互感
self-noise 固有噪声
self-oiling bearing 含油轴承
self-operated controller 自动控制器，自给能控制器，自力式控制器，自作用控制器，自操作控制器
self-operated control valve 自力式调节阀，自动阀，自作用阀
self-operated control 自行控制，自动操作控制，自力式控制
self-operated differential pressure control valve 自力式压差控制阀
self-operated flow control valve 自力式流量控制阀
self-operated measuring unit 自动测量装置
self-operated pressure control valve 自力式压差控制阀
self-operated regulator 自力式调节器
self-operated thermostatic controller 自作用的恒

温控制器
self-operated 自行操作的
self-operating construction pattern 自营方式
self-optimizing control 最佳自动控制，自寻最佳控制
self-optimizing decision system 自寻优决策系统
self-optimizing filter 自寻最佳化滤波器
self-optimizing system 自寻优系统
self-optimizing 最佳自动的，自动最佳的，自寻优
self-organizing computer 自组织计算机
self-organizing function 自组织功能
self-organizing machine 自组织机
self-organizing system 自组织系统
self-organizing 自组织（的）
self orientating （风轮）自动调向，向动对风，自动定向的，自动定位的
self-orientation 自动定向
self-orthogonal block code 自正交块码，自正交分组码
self-orthogonal convolutional code 自正交卷积码
self-orthogonal 自成正交
self-oscillating 自激振荡的
self-oscillation method 自振荡法
self-oscillation 自激振荡，自震荡，自振荡，自激振荡器
self-oxidation 自动氧化
self partial capacitance 固有部分电容
self-passivating metal 自钝化金属
self-perpetuating 自生自存，自保持，自保持的
self-polarizing relay 电流极化继电器，自极化继电器
self-potential 自位，自势
self-powered detector 自给能探测器
self-powered neutron detector 自给能中子探测器
self-powered 自备电源的，自己供电的，自给能的
self-priming pump 自充水泵，自启动水泵，自吸泵，自吸式泵
self-priming siphon 自动虹吸管
self-priming 自启动注水
self-programming computer 自编程计算机
self-programming 程序自动化，自动编程序，自动程序设计
self-propelled concreting plant 自行式混凝土浇筑机
self-propelled crane 自行式起重机，自走式起重机，移动式起重机
self-propelled electric locomotive 自备电源式机车
self-propelled grader 自动式平地机
self-propelled mobile reactor 自行式反应堆
self-propelled pneumatic roller 自动式气胎压路机
self-propelled scraper 自动式铲运机
self-propelled segmented roller 自动分段滚压机
self-propelled shaking table 自动式振动台
self-propelled 自行的，自动推进的，自走式
self propulsion test 自航试验
self-protected transformer 自保护变压器
self-protecting 自保护的，自行保护的
self-protective 自行保护的，自保护的
self-pulse modulation 自生脉冲调制
self-purification ability 自净能力
self-purification activity 自净活性
self-purification capacity 自净能力
self-purification characteristic 自净特性
self-purification constant 自净常数
self-purification of waters 水体自净作用
self-purification parameter 自净参数
self-purification waterbody 自净化水体
self-purification 自纯化，自净化，自净，自净作用，自澄清
self-quenched 自动熄灭的
self-quenching 自动熄灭（的）
self-raise fund 自筹资金
self-raising tower crane 自升式塔式起重机
self-reactance-to-resistance ratio 自有品质因数，固有电抗电阻比
self-reactance 固有电抗，自电抗
self-reading instrument 自动读数仪器
self-reading level rod 自读水准尺
self-reading staff 自读水准尺
self-recorder 自动记录器
self-recording anemometer 自动风力仪，自记风速计
self-recording apparatus 自动记录器
self-recording barometer 自计气压计
self-recording deflection gauge 自记挠度计
self-recording evaporimeter 自记蒸发计
self-recording flow-meter 自记流量计
self-recording gold-leaf electroscope 自动记录式金箔验电器
self-recording instrument 自动记录器
self-recording mechanism 自动记录机构
self-recording micro-meter 自记测微器
self-recording precipitation station 自记雨量站
self-recording pressure gauge 自记压力计
self-recording rain gauge 自记雨量计
self-recording strain gauge 自记应变仪
self-recording unit 自动记录器
self-recording wattmeter 自动记录瓦特计
self-recording 自动记录，自记
self-recovery 自动复位，自动恢复
self-rectified circuit 自整流电路
self-rectifying 自整流
self-registering anemometer 自计风速计
self-registering instrument 自动记录仪器
self-registering 自动记录的
self-regulated machine 自调节电机
self-regulating alternator 自调节交流发电机
self-regulating canal 自动调节渠道
self-regulating characteristics 自动控制特性，自调（节）特性
self-regulating dam 自动调节坝
self-regulating error-correct coder-decoder 自调节误差校正编码-译码器
self-regulating heating strips 自动调节的加热条
self-regulating process 自调节过程
self-regulating windmill 自动调节风车
self-regulating 自调节的，自行调整的，自调整的，自行调节的
self renewal resource 自更新资源
self-reparative 自动修复的，自动补偿的

self-replenishing contact 自补偿触点
self-reproduction 自再生，自复制
self-reset manual release control 自动复原手动释放控制
self-reset relay 自动复位式继电器，自行复归式继电器
self-resetting timer 自回复定时器
self-resetting 自复位，自动复位，自动复归
self-resistance 固有电阻，自电阻
self-resonant frequency 自谐振频率
self-restoration 自然更新，自恢复
self-restoring insulation 自恢复绝缘
self-restoring relay 自复继电器，自复归继电器
self-restoring 自复的
self-restraining 自抑
self-revealed failure 本身显露的故障
self-revealing fault 自显故障
self-running 不同步的，自行启动的
self-saturation 自饱和
self-scattering 自动扩散的
self-screening 自屏蔽
self-sealing capacitor 自封闭电容器
self-sealing 自封闭，自动密封，自密封，自密封的，自动封接
self-shadowing 自荫蔽
self-shielded burnable poison 自屏蔽可燃毒物
self-shielding coil 自屏蔽线圈
self-shielding effect 自屏蔽效应
self-shielding factor 自屏蔽系数，自屏蔽因子
self-shielding 自屏蔽的，自屏蔽
self-shifting synchronous clutch 自动同步离合器
self-similar 自相似
self-spillway dam 自由溢流坝
self spring 自拉
self-stabilisation 自稳定性
self standing tower 独立塔架
self-starter 自动启动器，自启动器，自动启动机
self-starting motor 自启动电动机
self-starting property 自启动特性
self-starting relay 自启动继电器
self-starting reluctance motor 自启动磁阻式电动机，自启动反应式电机
self-starting rotary converter 自启动旋转变流器
self-starting synchronous motor 反应式同步电动机，自启动同步电动机
self-starting 自启动，自启动的
self-stiffness 固有刚度，自反电容，自逆电容
self-stressing cement 自应力水泥
self-stressing concrete 自应力混凝土
self-stressing 自应力
self-sufficiency 自足性，独立性
self-supply power plant 自备动力厂，自备发电厂，自备电厂
self-supply power project 自备电厂工程
self-supporting aerial cable 自支持架空电缆，自承式架空电缆
self-supporting cable 自撑式电缆
self-supporting chain reaction 自持链式反应
self-supporting coil 自持线圈，自立式线圈
self-supporting combustion 稳定燃烧
self-supporting partition 自承重隔断
self-supporting property 自承能力
self-supporting scaffold 自承脚手架
self-supporting stack 自承式烟囱
self-supporting steel chimney 自立式钢烟囱
self-supporting tower 自立式杆塔，独立塔架，自立塔
self-supporting wall 自承重墙
self-supporting 自立式的，自支持，自支承，自支撑，自保持
self-surge impedance 涌阻抗，行波阻抗，浪自冲击阻抗
self-sustained discharge 自持放电
self-sustained oscillation 自激振动，自持振荡
self-sustained 自激的，自持的，自给的
self-sustaining chain reaction 自持链式反应
self-sustaining combustion 稳定燃烧
self-sustaining discharge 自持放电
self-sustaining reactor 自持反应堆
self-sustaining speed 自持转速
self-sustaining 自激的，自立的，自驱动的，自励的，独立的，自持的
self-switching 自开关，自动开关的
self-synchronism probability screen 自同步概率筛
self-synchronizing device 自同期装置
self-synchronizing 自同步，自同期
self-synchronous device 自同步装置
self-synchronous motor 自动同步机，自同步机，自同步电动机
self-synchronous repeater 自同步中继器
self-synchronous 自动同步的
self tapping screw 自攻螺钉，自攻螺纹的
self-temperature compensated element 温度自补偿元件
self tempering 自热回火，自身回火
self-test circuit 自试电路
self-test factor 自测系数
self-testing and repair 自检修
self-testing 自检验，自检验的
self-timer 自动计秒表，照相机自拍装置
self-timing 自动分时，自动同步，自动定时
self-tipping 自动翻落式，自动倾倒式
self-trapping 自陷
self-triggering program 自启动程序
self-triggering 自激励的，自引发的，自启动的，自触发的
self-trig 自触发，自适应
self-tuned filter 自调谐滤波器
self-tuning control 自校正控制
self-tuning regulator 自调谐调节器，自校正调节器
self-tuning 自调整，自动调谐
self-unloading barge 自卸煤驳
self-unloading carrier 自卸船
self-unloading pump 自动卸载泵
self-unloading ship 自卸船
self-unloading trailer 自卸拖车
self-unloading type hopper wagon 自卸式底开车
self-unloading 自卸，自卸载
self vaporation 自汽化
self-ventilated machine 自通风电机

self-ventilated motor 自通风电动机
self-verifying 自检，自动校验
self-virtualizing 自虚拟式，自虚拟
self-weight non-collapse loess 非自重湿陷性黄土
self-weight of pipeline 管道自重
self-weight stress 自重应力
self-weight 自重
self-welding of structural materials in liquid sodium 钠液中结构材料自焊接
self-wending 自动焊接
self-winding 自卷的，自绕的，自动上发条的
sell by retail 零售
seller-consignor 卖方发货人
seller's credit 卖方信贷
seller's obligation 卖方义务
seller's responsibility 卖方责任
seller 卖方
selling bidding document 售标
selling price 售价
selling rate of telegraphic transfer 电汇卖出汇率
selling short 卖空
selling space 营业厅
SELS(selsyn motor) 自同步马达
selsyn control 自动同步机控制
selsyn differential 自整角机差动装置
selsyn-drive electric motor 自整角机驱动电动机
selsyn generator 自同步发电机，自整角发电机
selsyn motor 自整角电动机，自同步电动机，自同步马达
selsyn system 自同步机系统，自整角系统
selsyn train 自动同步传动，自整角传动装置
selsyn transformer 自同步变压器
selsyn-type electrical motor 自整角型电动机
selsyn-type electric machine 自整角型电机
selsyn 自整角机，自动同步机，自同步机，自动同步，自整角器
seltrap （利用半导体二极管反向特性的）变阻器
SEL 常规岛废液贮存排放系统【核电站系统代码】
semantic error 符号含义误差，语义误差
semantic information 语义信息
semantic network 语义网络
semaphore signal 杆上信号
semaphore 信号灯，信号标，信号，横杆信号，信号量
semaphorne signal unit 壁板信号机
semiactive homing guidance 半主动寻的制导
semiactive homing 半主动寻的
semiactive 半主动的
semi aerodynamic shape 半气动形状，半流线型
semi-angle 半角
semiannual period 半年期
semiannual report 半年度报告
semiannual 半年的
semi-anthracite coal 贫煤，半无烟煤
semi-anthracite 半无烟煤
semiaquatic 半水生的
semi-arch 半拱
semiarid area 半干旱地区
semiarid climate 半干旱气候
semiarid 半干旱的
semi-auto cutting machine 半自动切割机
semi-automatic arc welding machine 半自动电弧焊机，半自动弧焊机
semi-automatic arc welding 半自动弧焊，半自动电弧焊
semi-automatic batcher 半自动化拌和楼
semi-automatic call 半自动呼叫
semi-automatic checkout equipment 半自动检查装置
semi-automatic coding 半自动化编码
semi-automatic control 半自控制
semi-automatic crimping machine 半自动压接机
semi-automatic cycle 半自动循环
semi-automatic gas shielded arc welding 半自动气体保护焊
semi-automatic hydro-electric station 半自动化水电站，半自动水电站
semi-automatic message switching center 半自动信息转接中心
semi-automatic metal arc welding 半自动金属电弧焊
semi-automatic operation 半自动操作
semi-automatic programming 半自动程序设计
semi-automatic quality control 半自动质量控制
semi-automatic rectifier 半自动整流器
semi-automatic sprinkler system 半自动喷水系统
semi-automatic starter 半自动启动器
semi-automatic submerged arc welding machine 半自动埋弧焊机
semi-automatic submerged arc welding 半自动埋弧焊
semi-automatic substation 半自动变电所
semi-automatic switching 半自动交换
semi-automatic system 半自动系统
semi-automatic trimming machine 半自动截锯机
semi-automatic welder 半自动焊机
semi-automatic welding 半自动焊，半自动焊接
semi-automatic 半自动的
semi-automation 半自动化
semi-axial-flow pump 半轴流式水泵
semi-axis （椭圆的）半轴
semi-base load power station 半基荷电站
semi-batch operation 半间歇操作，半分批操作
semi-bituminous coal 半烟煤，贫煤
semi-circle 半圆形
semi-circular arch 半圆形拱，半圆拱
semi-circular basin 半圆形港池
semi-circular drain 半圆形排水管
semi-circular electric meter 半圆形仪表
semi-circular fluidized bed 半圆形流化床
semi-circular sheet steel flume 半圆形钢板渡槽
semi-circular surface crack 半圆形表面裂纹
semi-circular surface flaw 半圆形表面缺陷
semi-circular trash-rack 半圆形拦污栅
semi-circular vault 半圆形穹顶
semi-circular weir 半圆形堰
semi-circular 半圆形
semi-circumferential flow 半圆周绕流
semi-closed cycle gas turbine 部分废气再循环的燃气轮机，半闭式循环燃气轮机，半闭式燃气

轮机装置
semi-closed cycle 半闭式循环
semi-closed feed system 半闭式给水系统
semi-closed jet wind tunnel 半闭口式风洞
semi-closed slot 半闭口槽
semi-closed 半封闭的
semicoke 半焦化
semi-coking stoker 半焦化炉排
semicompiling 半编译
semiconducting ceramics 半导体陶瓷
semiconducting coating 半导体漆层，半导体涂层
semiconducting crystal 半导体晶体
semiconducting glassed relay 玻璃半导体继电器
semiconducting glass 半导体玻璃
semiconducting jacket 半导体护套
semiconducting layer 半导电防晕层
semiconducting 半导体的，半导电的
semiconduction 半导电性
semiconductive suspension 半导体悬浮
semiconductive varnish 半导体漆
semiconductive 半导体的，半导电的
semiconductor amplifier 半导体放大器
semiconductor bolometer 半导体辐射热测量器
semiconductor cold lens 半导体冷镜
semiconductor controlled rectifier 半导体可控整流器
semiconductor detector 半导体探测器
semiconductor device 半导体器件
semiconductor diode 半导体二极管
semiconductor electronics 半导体电子学
semiconductor film 半导体薄膜
semiconductor gauge 半导体应变片
semiconductor glazed insulator 半导体釉绝缘子
semiconductor hot-wire anemometer 半导体热线风速仪
semiconductor insulating varnish 半导体绝缘漆
semiconductor integrated circuit 半导体集成电路
semiconductor laser 半导体激光器
semiconductor memory 半导体存储器
semiconductor-metal-semiconductor 半导体-金属-半导体
semiconductor motor 半导体电动机
semiconductor on thermoplastic 热塑性塑料半导体
semiconductor pressure sensor 半导体压力传感器
semiconductor probe 半导体探头
semiconductor rectifier cell 半导体整流器元件
semiconductor rectifier 晶体管整流器，半导体整流器
semiconductor resistor 半导体电阻器
semiconductor storage unit 半导体存储器
semiconductor strain gage 半导体应变计
semiconductor switching device 半导体开关器件
semiconductor switch 半导体开关
semiconductor tape 半导体带
semiconductor valve 半导体换流阀
semiconductor varnished glass cloth 半导体玻璃漆布
semiconductor 半导体，半导体器件
semi-container ship 部分集装箱船
semicontinuous-duty motor 半连续工作制电动机
semicontinuous operation 半连续操作
semi controlled device 半控型器件
semicon 半导体
semi-counter-current regeneration 半逆流再生
semicrystalline 半晶质
semi-desert 半沙漠，半荒漠
semi-digital read-out 半数字示值
semidirect fined system 半直吹制制粉系统
semidirect illumination 半直接照明
semidirect lighting 半直接照明
semi-direct maintenance 接触维修，半直接维修
semi-distributed winding 半分布绕组
semidiurnal wave 半日波
semidiurnal 半日的，半天的
semidry state 半干状态
semiebonite 半硬橡胶
semi-elastic impact 半弹性冲击
semielectronic 半电子的
semi-ellipsoidal head 半椭球形封头
semielliptic flaw 半椭圆形缺陷
semielliptic surface crack 半椭圆形表面裂纹
semiempirical criteria 半经验判据
semi-empirical equation 半经验方程
semi-empirical formula 半经验公式
semi-enclosed impeller 半封闭式叶轮
semi-enclosed machine 半封闭型电动机
semi-enclosed motor 半封闭式电动机
semi-enclosed slot 半闭口槽
semi-enclosed water body 半封闭水体
semi-enclosed 半封闭式
semiexciting type regenerative generator 半励磁再生发电机
semiexciting 半激励式的
semi-finished material 半成品，半制品
semi-finished product storage 半成品库
semi-finished product 中间产品，半成品
semi-fireproof building 半防火建筑
semi-fixed penstock 半固定式压力水管
semi-flexible coupling 半挠性联轴节，半挠性联轴器
semi-fluidized bed 半流化床
semi-fluidized system 半流化系统
semi-fluid mass 半流动体，半流质
semi-fluid 半流体
semi-fused slag 半熔状渣
semi geared wind turbine 半齿轮驱动风电机组
semi girder 悬臂梁
semigloss enamel 半光磁漆
semigloss paint 半光漆
semigraphical method 图解法，半图解法
semigraphical 半图解的
semigraphic method for solving storage equation 蓄量方程半图解法
semi-graphicpanel 半图形板，半图解式面板
semigroup 半群
semiguarded machine 半防护型电机
semi-high arrangement of outdoor switchgear 半高型屋外配电装置
semi-high-profile layout 半高型布置
semi-hot laboratory 中等放射性水平实验室，半热室

semi-humid climate 半湿润气候
semi-humid soil 半湿土
semi-humid 半潮湿，半湿润的
semihydrate 半水合物
semi-hydraulic fill dam 半水力冲填坝
semi-hydraulic fill method 半水力冲填法
semi-implicit method 半隐式方法
semi-indirect illumination 半间接照明
semi-indoor power plant 半露天式电厂，半露天式电站
semi-industrial gas turbine 半工业燃气轮机装置
semi-infinite body 半无限体
semi-infinite cascade 半无限叶栅
semi-infinite cloud model 半无限烟云模式，半无限云模式
semi-infinite half-space 半无限空间
semi-infinite plane reactor 半无限平面反应堆
semi-infinite shield 半无限屏蔽体
semi-infinite type 半无限型
semiinfinite 半无限
semi-insulating 半绝缘的
semi-insulator 半绝缘体
semi-iterative method 半迭代法
semi-killed steel 半镇静钢
semi knock down 半散件组装
semi-laminar flow 半层流
semi length scale 半长尺度
semi-liquid waste 半液体废物
semi-logarithmic coordinate 半对数坐标
semi-logarithmic paper 半对数（坐标）纸
semi-logarithmic plot 半对数曲线图
semi-logarithmic scale 半对数标度
semi-logarithmic 半对数的
semilog coordinate 半对数坐标
semi log paper 半对数（坐标）纸
semilongitudinal fault 偏斜走向断层
semi-luminous flame 半发光火焰
semilumped parameter 半集中参数
semilumped 半集中的，半集中参数的
semilunar dune 新月形沙丘
semi-magnetic controller 半磁控制器
semi-magnetic 半磁的
semi major axis 长半轴
semimanufacture 半成品
semi matt 半亚光
semimetallic gasket 半金属垫片
semimetal 半金属
semi-microbalance 半微量天平
semi-mimic panel 半模拟盘［屏］
semi minor axis 短半轴
semi mobile dune 半流动沙丘
semi mobile sand 半流动沙
semimonocoque construction 半硬壳式结构
seminar 研究班，（专题）研讨会，讲习班
semi-open-air arrangement 半露天布置
semi-open-air installation 半露天装置
semi-open fuel assembly 半开式燃料组件
semi-open furnace 半开式炉膛
semi-open impeller 半开式叶轮，无前盘的叶轮
semi-open jet wind tunnel 半开口式风洞
semi-open slot 半开口槽
semi-open type electrical equipment 半开启式电气设备
semi-open type 半敞开式，半封闭式
semi-open 半开敞，半开启式
semi-orthogonal 半正交的
semioscillation 半周期振荡
semiotics 形式语言论，符号学，符号语言学，形式语言学
semi-outdoor arrangement 半露天布置
semi-outdoor power house 半露天厂房
semi-outdoor power plant 半露天式电厂
semi-outdoor thermal power plant 半露天式火力发电厂，半露天火电厂
semi-outdoor 半露天式
semi-packaged boiler 组装锅炉，半快装锅炉
semi-peak station 中间负荷电站
semi-perched groundwater table 半栖留地下水位，半滞水地下水位
semi-period 半周的，半周期
semipermanent cave 半永久屏蔽室
semipermanent damage 半永久性（辐射）损伤
semipermanent depression 半永久性低压
semipermanent structure 半永久性建筑，半永久性结构
semipermeable material 半透水材料
semipermeable membrane 半渗透膜
semipermeable 半渗透性的
semi-pervious material 半透水材料
semi-pervious zone 半透水区
semi-physical simulation 半实物仿真
semi-plant 中间试验工厂，试验装置
semi-plastic stage 半塑性阶段
semi-polar 半极性的
semi-portable engine 半移动式发动机
semi-portal crane 半门式起重机
semipotentiometer 半电位计，半电势计
semi-private circuit 半专用线路
semiproduct 半成品
semiprotected enclosure 半保护型护罩
semiprotected harbour 半防护港
semiprotected motor 半防护型电动机
semiprotected 半保护的
semi-public company 公私合营公司
semi-rational formula 半经验公式
semiremote handling 半远距离操纵
semi-reversibility 半可逆性
semi-revolution 半周
semi-rigid approach 半刚性方法
semi-rigid connection 半刚性接合，半刚性连接
semi-rigid framed structure 半刚性框架结构
semi-rigid model 半刚性模型
semi-rigid penstock 半刚性压力水管
semi-rural environment 半乡村环境
semi-saturation 半饱和
semi-silica brick 半硅砖
semisintered 半烧结的
semi-skilled worker 半熟练工人
semi-solid asphalt 半固态沥青
semi-solidified liquid 半固化液
semi-solid state 半固态
semispan 半翼展

semi-spherical 半球形
semispiral case 不完全蜗壳
semistalled condition 半失速状态
semistall （气流）局部分离，半失速
semisteel 半钢，钢性铸铁
semistop barrier （电缆的）半截止挡头
semistop （电缆的）半截止头
semistor 正温度系数热敏电阻
semi-straight-blow type powdering system 半直吹式制粉系统
semi-symmetric 半对称的
semi-synchronizing device 准同期装置
semitight 半密封的
semitrailer 半拖车
semi-transless 半无变压器式
semi-transverse fault 偏斜倾向断层
semi-truss （四面坡屋顶的）半桁架
semi-tubular turbine 半贯流式水轮机
semiturbulent 半紊流的
semi-turnkey contract 半统包合同
semi-umbrella type generator （水轮发电机的）半伞式发电机
semi-underground power house 半地下厂房，半地下式厂房
semiuniform 半均匀的
semi-up position 倾斜位置【焊接件】
semi-urban environment 半城市环境
semi-vertical configuration 半垂直排列
semi-weathered state 半风化状态
semi 半
SEM(scanning electron microscopy) 扫描电镜法
SEMS(severe environment memory system) 苛求环境的存储系统
send back 遣送回国，退还
send-break 传输中断
sender link frame 记发器触排，发射机线弧
sender （信号）发送器，发射机，引向器，记发器，发货人，发信人
sending allowance 传输损耗
sending end transformer 送电端变压器
sending end 送电端，发送端，送端
sending power 发送功率，发射功率
sending relay 发送继电器
sending station 发信台，发送台，发射台
sending stress 弯曲应力
sending substation 送电变电站
send only 只发送
sendout 出力，功率，烟囱
send-receive switch 发送接收开关
sendust core 铝硅铁粉磁芯
send 发送，派遣
SENET(slotted envelope network) 时间片分割法网络
senior accountant 高级会计师，资深会计员
senior consultant 资深咨询专家
senior engineer 高级工程师【相当于副教授级】
senior management department 高级管理部门
senior management 高级管理
senior manager 高级经理
senior personnel 高级人员
senior reactor operator 反应堆高级操纵员
senior representative 高级代表
senior vice president of engineering and construction 设计施工高级副总裁
senior vice president of nuclear power 核电高级副总裁
sense amplifier 读出放大器
sense antenna 辨向天线
sense-digit line 位读出线【计算机】
sense direction finding 定向，测向
sense finder 定向器，无线电罗盘
sense finding 测向
SEN=sensor 传感元件，传感器，探测器
sense of current 电流的方向
sense of rotation 旋转方向
sense switch 敏感开关
sense 检验，读出，方向，意义
sensibility reciprocal 灵敏度倒数
sensibility 敏感性，灵敏度，感觉，识别力
sensible cooling 等湿冷却，显热制冷
sensible heating 等湿加热
sensible heat meter 热量表
sensible heat storage 显热贮能
sensible heat transfer 显热传递
sensible heat 显热
sensible plan 切合实际的计划，明智的计划，合乎情理的计划
sensible 灵敏的，敏感的，明显的，明智的
sensing component 敏感元件
sensing dement elevation 敏感元件高度
sensing dement 敏感元件
sensing device 传感器，感受设备，传感装置
sensing element elevation 敏感元件高度
sensing element 传感元件，敏感元件，灵敏元件，传感器，传感器（元件）变（发）送器，探测器，检出器
sensing equipment 传感设备
sensing head 敏感头，灵敏头
sensing line assembly 检测管线组合【检测管线，阀，泵，取样小室等】
sensing line 取样［采样］管路
sensing relay 传感式继电器
sensing station 读出站
sensing switch 敏感开关【压力开关，限位开关，恒温器等】
sensing system 感应系统，测读系统
sensing transducer 传感器
sensing unit 传感器，敏感元件，传感装置
sensing 敏感的，感光的，敏感，感觉，读出，信号传感，方向指示
sensistor 正温度系数热敏电阻，硅电阻
sensitiser 激活剂，感光剂，敏化剂
sensitive analysis 敏感性分析
sensitive clay 灵敏性黏土
sensitive day 灵敏黏土
sensitive drilling machine 高速钻床
sensitive element 传感器，敏感元件，灵敏元件
sensitive film 感光胶片
sensitive galvanometer 灵敏检流计，灵敏电流计
sensitive instrument 灵敏仪表
sensitive lining 敏感衬里
sensitive manometer 灵敏压力计

sensitiveness of the governor 调速器灵敏度
sensitiveness 灵敏度，灵敏性，敏感性
sensitive paper 感光纸
sensitive polarized relay 灵敏极化继电器
sensitive relay 灵敏继电器，灵敏继电保护装置
sensitive resistor 敏感电阻器
sensitive sensor 传感器
sensitive static relay 灵敏静态继电器
sensitive switch 微动开关，敏感开关
sensitive thermometer 敏感温度计
sensitive volume thickness 核磁敏感体积厚度
sensitive volume 感应容量，（辐射）敏感区，敏感体积，（辐射）敏感范围
sensitive 灵敏的，敏感的
sensitivity analysis 灵敏度分析，敏感性分析
sensitivity characteristic 灵敏度特性
sensitivity coefficient 敏感系数
sensitivity compensator 灵敏度补偿器
sensitivity controller 灵敏度控制器，灵敏度调节器
sensitivity control 灵敏度调整，灵敏度控制
sensitivity correction 灵敏度校正
sensitivity curve 灵敏度曲线
sensitivity data 灵敏度数据
sensitivity drift 灵敏度漂移
sensitivity of deflection 偏转灵敏度
sensitivity of following wind 调向灵敏性【表示随风向的变化，风轮迎风是否灵敏的属性】
sensitivity of instrument 仪器灵敏度
sensitivity of protection relay 保护装置灵敏度
sensitivity of radiation 辐射灵敏性
sensitivity ratio 灵敏度比
sensitivity reading 灵敏度读数
sensitivity region 灵敏区
sensitivity setting 灵敏度调整
sensitivity study 灵敏度研究
sensitivity test 灵敏度试验
sensitivity-time control 灵敏度时间调整，灵敏度时间控制，敏感时间控制（器）
sensitivity to decarburization 脱炭敏感性，易脱炭性
sensitivity to light 感光性
sensitivity to radiation 辐射灵敏性
sensitivity 灵敏度，灵敏性，敏感性
sensitization 灵敏，感光，敏感，敏化，激活
sensitized 激活的，敏化的
sensitizing heat treatment 强化热处理
sensitometer 感光度测定计，感光计，曝光表，曝光表
sensitometry 感光度测定法
sensor-based system 基于传感器的系统
sensor coil 传感线圈
sensor detector 传感探测器
sensor for monitoring expansion parameters 监测膨胀参数的传感器
sensor information 传感器信息，遥感信息
sensor measuring mechanical quantity 测机械参量的传感器【震动探测器，位移探测器等】
sensor measuring physical quantities 测物理参量的传感器
sensor-mounting beam 传感器支架梁
sensory control 感觉控制
sensor 传感器，变送器，发送器，探测器，敏感元件，感应头
sentence length count 句长计算
sentence of bankruptcy 破产判决
sentence 判决
sentinel pyrometer 高温计
sentinel 标记，传送器，发送器，发讯器
sentron 防阴极反加热式磁控管
SEN 辅助冷却水系统【核电站系统代码】
SEO 电厂污水系统【核电站系统代码】
separability 可分性
separable coupling 可拆式联轴器
separate adjustment 分部平差分别调整
separate aeration 独立通风
separate and joint references for incorporation of the consortium 由联合体组成的公司的单独的和共同的业绩
separate baseplate pump 分座式泵
separate bill 拆单【提单】
separate blade 分离叶片，薄根叶片
separate chamber 隔离室
separate contactor 分项承包商
separate contract 分项承包
separate cut-off valve 独立停气阀
separated aggregate 分级骨料
separated boundary layer 分离边界层
separated economizer 可分式省煤器，独立式省煤器
separated flow model 分相流动模型，分流模化
separated flow vortex 分离流的旋涡
separated flow 分离流，分离水流，分离流动
separated furnace 独立炉膛，分隔炉膛
separated joint venture 松散型联营体
separated phase layout 分相布置
separate drawing 单体图
separate drive 个别驱动，单独驱动
separated shear layer 分离剪切层
separated 分开，隔开
separate earth electrode 独立接地体
separate excitation 他励磁，他励，他激
separate exciter 他励励磁机，独立励磁机
separate-from-fuel absorber 与燃料分离的吸收体
separate grade crossing 立体交叉道
separate gravity structure 分离式重力结构
separate ground electrode 独立接地体
separate handling of ash and slag 灰渣分除
separate heterodyne 独立本机振荡外差法
separate inverter unit 独立逆变器
separate lead sheathed cable 分铅型电缆
separate-lead-type cable 分开铅包电缆
separate loop 切断环路，隔离环路
separately-cooled machine 外冷式电机
separately cooling 外冷，他冷
separately driven circuit 他激励电路
separately excited dynamo 他励发电机，他励电机，分激电机
separately excited generator 他励磁发电机，他励发电机
separately excited motor 他励电动机
separately excited multivibrator 他激多谐振荡

器，分激多谐振荡器
separately excited oscillator 他激振荡器，分激振荡器
separately excited rotor winding 他励磁转子绕组
separately excited 他励的，他激的
separately fired superheater 布置在单独炉膛内的过热器
separately-leaded cable 铅皮分包电缆
separately nozzle chest 分离式喷嘴室
separately-ventilated machine 外通风电机
separately-ventilated 外通风的，单独通风的
separate negotiation 个别协商
separate network operation 电网分列运行
separate self-excitation 间接自励磁
separate sewerage system 单用污水系统，分区污水系统，分流制污水系统［排水系统］
separate sewer system 分流排污系统，分流污水系统，分流下水道系统
separate sewer 单用污水管，单用下水道
separate sludge digestion tank 单独的污泥消化槽
separate sludge digestion 分离式污泥净化法
separate source 不同来源
separate system （雨水污水）分流系统，离散系统
separate tank 分离池
separate ventilation 独立通风，外通风
separate water supply system 分区供水系统
separate winding 单独绕组
separate 分离
separating agent 分离剂
separating calorimeter 分离式量热计
separating chamber 分离室，沉淀室
separating character 分隔字符
separating column 分离柱
separating drum 分离汽包
separating funnel 分液漏斗
separating layer 间隔层
separating nozzle 分离喷嘴
separating screen 分类筛，分选筛，分离筛
separating sewer 分流污水管
separating system 分离系统
separating tank 分离槽，澄清槽
separating wall 分隔墙
separation angle 分离角
separation bubble 气流离体区，气流分离，分离气泡
separation by diffusion 扩散分离
separation cavity 分离空穴
separation chamber 分离槽，展开槽
separation coal 富选煤，精选煤
separation coupling 可拆联轴节
separation distance 隔距
separation efficiency 除尘效率，分离（除尘）效率
separation energy 分离能
separation factor 分离因数，分离系数，分离因子
separation flow 分离流
separation fracture 分离剥裂
separation groin 导水丁坝，分水丁坝
separation levee 分流堤，隔堤
separation line 分界线，分离线
separation of groundwater flow 地下径流分割，分割地下径流，区分地下径流
separation of hydrograph components 水文过程线各分量的分割
separation of losses （电机损失测试中的）损耗分离
separation of particle mixture 颗粒混合物分离
separation of two frequencies 两频率间的间隔
separation of two-phase fluid 汽水分层，双相流体分层
separation of variables 变量分离
separation partition 分隔墙
separation pier 隔墩
separation plane 分离面
separation point 分离点
separation principle 分离原理
separation ratio 分离比
separation region 离散区【指水流的】，分离区
separation regulator 煤粉分离器，（煤粉）细度调节器
separation screen 分离筛
separation streamline 分离流线
separation structure 分隔结构
separation tank 分离罐
separation vortex 分离涡
separation wake 分离尾流
separation works 分隔工程
separation 分隔，分开，分离，离析，间距，间隔，导线间距，分流，流体的分离
separative element 分离元件
separative excitation system 他励系统
separative power 分离功率，分离本领
separative unit 分离装置，分离单元
separative work content 分离功量值
separative work unit 分离功单位，分离功单元
separative work 分离功【浓缩】
separator flask 分离器容器
separator insulation 层间绝缘
separator regulator （煤粉）分离器的，（细度）调节器，分流式导流轮
separator tube 分离器管
separator vessel 分离器容器
separator 分隔标志，分离器，气液分离器，隔板，区分符号
sepiolite 海泡石
SEPP(standard erection procedure package) 标准安装程序包
SEP＝separate 分离
SEP＝separator 分离器
SEP(simulated echo pattern) 模拟回波图形
SEP(standard engineering practice) 标准工程惯例
septavalence 七价
septenary 七进制的
septendecimal 十七进制的
septic sewage 腐化污水
septic tank 腐化池，化粪池
septum 隔板，隔膜
SEP 分割开的【DCS 画面】
SEP 饮用水系统【核电站系统代码】

sequence calling 顺序调用
sequence-checking routine 顺序校验程序，序列校验程序
sequence-checking 顺序校验
sequence circuit 时序电路，程序电路
sequence command 顺序指令
sequence component 序分量
sequence control counter 时序控制计数器，程序控制计数器
sequence-controlled computer 程序控制计算机
sequence-controlled contact 程序控制触点
sequence-controlled sootblower system 顺序控制吹灰系统
sequence-controlled system 程序控制装置
sequence-controlled 顺序控制的，程序控制的
sequence controller 顺序控制器，程序控制器
sequence control of HVDC power transmission system 高压直流输电顺序控制
sequence control plaque 程序控制板
sequence control register 顺序控制寄存器，程序控制寄存器
sequence control system 顺序检验系统，顺序控制系统，程序控制系统
sequence control test 顺序控制试验
sequence control 程序控制，顺序控制
sequence counter 指令计数器
sequenced computer 序列计算机
sequence diagram 工序图
sequence display system 程序显示系统
sequence equipment 程序装置
sequence event recorder 事故顺序记录仪
sequence filter 相序滤波器，相序过滤器
sequence generator 程序发生器
sequence initiation switch 程序启动开关
sequence in time 时序
sequence monitor 顺序控制器［监控器，监测器］，巡回监测器，顺序监督程序
sequence number indicator 顺序号指示器
sequence number 顺序号
sequence of connection 接线顺序
sequence of events recorder 事件顺序记录器
sequence of events 事故顺序记录，事件的顺序，大事记，事件序
sequence of operation 操作程序，操作顺序
sequence program 次序程序
sequence reactance 相序电抗
sequence record 顺序记录
sequence register 顺序寄存器
sequence relay 顺序动作继电器
sequence representation of hydrological data 水文资料系列代表性
sequencer （控制棒）定序器，序列发生器，程序装置
sequence scheduling 按序调度
sequence signal 时序信号
sequence starting 顺序启动
sequence switching mechanism for control rod movement 控制棒动作程序开关机构
sequence switch 时序开关，顺序开关
sequence table 程序表
sequence test 连锁顺序试验，运行顺序试验
sequence tripping 顺序跳闸
sequence valve 顺序阀
sequence 序号，次序，数列，程序，序列，顺序，后处理，排序
sequencing blow-down valve 顺序排污阀
sequencing computation 顺序计算
sequencing timer 顺序定时器
sequencing unit 命令发送器，控制器
sequential access display 顺序扫描显示
sequential access memory 按序存取存储器
sequential access storage 顺序存取存储器
sequential access 按序存取，按序访问
sequential advanced control 顺序先行控制，按顺序推进的控制
sequential automatic control 顺序自动控制
sequential batch operating system 按序批操作系统
sequential bidding 分次竞价
sequential carry 顺序进位
sequential circuit 时序电路，序列电路
sequential coding 序列编码，连续编码
sequential color television system 顺序制彩色电视系统
sequential computer 顺序计算机
sequential construction 序列施工法
sequential controller 顺序控制器
sequential control station 顺序控制站
sequential control system 顺序控制系统，顺控系统
sequential control 顺序控制，时序控制，程序控制，连续控制
sequential decoding 时序译码，有序译码
sequential decomposition 顺序分解
sequential fault 二次故障
sequential filtering 序列滤波
sequential flash 连续闪光
sequential image 连续图像
sequentialization of design 设计顺序化
sequential least squares estimation 序贯最小二乘估计
sequential logic 时序电路
sequential machine 时序机
sequential measurement 顺序检测
sequential network 时序网络
sequential operation 时序操作，按序运算
sequential operator 顺序算子，顺序运算符
sequential optimal control algorithm 序列最优控制算法
sequential optimization 顺序优化
sequential order of the phase 相序
sequential organization 时序组织
sequential phase control 顺序相位控制，顺序相控
sequential probability ratio 序列概率比
sequential processing 按序处理
sequential program 顺序程序
sequential radiograph 连续射线照片
sequential regeneration 程序再生
sequential relay 顺序动作继电器，顺序继电器
sequential sampling inspection 逐次抽样检查
sequential sampling 序列取样，顺序抽样，顺序

取样
sequential scanning 顺次扫描，顺序扫描
sequential scheduler 按序调度程序
sequential scheduling system 按序调度系统
sequential search 顺序检索
sequential selection 顺序选择
sequential short-circuit 连续短路
sequential starting control 顺序启动控制
sequential switching system 顺序开关系统
sequential switching 顺序开关
sequential time delay 连续时延
sequential transducer 顺序电路，顺序变换器
sequential tripping 顺序跳闸，顺序脱扣
sequential 顺序的，连续的，序列的，时序的
sequent 结果的，继续的，连续的，结果
sequestering agent 螯合剂
sequester 螯合，查封，扣押财产，没收
sequestrant 多价螯合剂
sequestrate 查封
SEREP（system environment recording，editing and printing） 系统现场记录、编辑和打印
serial access storage 串行存取存储器
serial access 串行存取
serial accumulator 串行累加器
serial ADC 串行模数转换
serial adder 串行加法器
serial arithmetic operation 串行的算术运算
serial arithmetic unit 串行运算装置
serial arithmetic 串行运算
serial batch system 串行成批处理系统
serial bit 串行位
serial by bit 按位串行的
serial by character 按字符串行的
serial by word scan 按字串扫描
serial by word 按字串行的
serial capacitor compensation device 串联电容补偿装置
serial computer 串行计算机
serial computing machine 串行计算机
serial digital computer 串行数字计算机
serial digital decoder 串行数字译码器
serial distribution （渗水系统的）串联布设
serial dot character printer 串行点字符打印机
serial file 串行文件
serial flow 串行流
serial input register 串行输入寄存器
serial interface 串行接口
serial I/O 串行输入输出
serialization 串行
serialized scheduling 串行调度
serializer 串行转换器
serial line internet protocol 串行线路网际协议
serial logic 串行逻辑
serial memory 串行存储器
serial mode 串行方式
serial multiplier 串行乘法器
serial number 编号，序串联号，串联数，序数，连续编号，顺序号，系列号
serial operation 串行操作
serial-parallel conversion 串行并行变换，串并行转换
serial-parallel multiplication 串并行乘法
serial-parallel-serial configuration 串并串行结构
serial-parallel 串并行，串行并行
serial photograph 连续照片
serial port 串行口
serial printer 串行打印机
serial processing 串行处理
serial processor 串行处理器
serial production 成批（系列）生产，批量生产
serial programming 串行程序设计，串行编程
serial register 串行寄存器
serial representation 串行表示
serial sampling 连续取样
serial storage 串行存储器
serial time sharing system 串行分时系统
serial-to-parallel converter 串行并行变换器
serial-to-parallel 串行并行
serial transfer 串行传送，串行转移
serial transmission 串行传输
serial tube 系列电子管
serial 串行的，串联的，连续的，依次的，系列的
series and parallel 串并联，混联
series arc lamp 串联弧光灯
series arc regulator 串联电弧调节器
series arc welding 串联电弧焊
series arrangement 串联，配套
series bisection 串联中剖
series booster 串联式升压器
series camera 连续摄影机
series capacitance 串联电容
series capacitor compensation device 串联电容补偿装置
series capacitor compensation 串联电容器补偿，串联电容补偿
series capacitor 串联电容器，附加电容器
series characteristic motor 串激特性电动机
series characteristic 串联特性，串激特性
series choke 串联抗流圈，串联扼流圈
series circuit 串联电路
series coil 串联线圈
series commutator motor 串励换向器式电动机
series compensating capacitor 串联补偿电容器
series compensating device 串联补偿装置
series compensating winding 串联补偿绕组
series compensation 串联补偿
series compounding excitation 串复励
series condenser block 串联电容器组
series condenser 串联电容器
series conduction motor 交流串励换向器电动机，单相串励换向器式电动机
series conductor 串联导体
series-connected starting-motor 串联启动电动机
series-connected system 直流（太阳热水）系统
series-connected 串联的
series connection 串联
series controller 串联控制器
series control 连续控制
series development 级数展开
series distribution system 串联配电制
series dynamo 串激发电机，串励电动机

series element 串联元件
series excitation coil 串励线圈
series excitation 串励，串激，串联励磁
series excited generator 串励发电机
series excited motor 串激电动机，串励电动机
series excited 串励
series expansion 级数展开
series feedback 串联反馈
series feed oscillator 串联馈电振荡器
series feed 串励馈电
series field diverter 串激磁场分流电阻
series field loss 串激磁场损耗
series field resistance 串激磁场（绕组）电阻
series field shunt 串激磁场分流器
series field winding 串励绕组，串激绕组
series field 串励磁场，串激磁场
series filtration 连续过滤
series flow turbine 单轴多缸汽轮机
series gap 串接气隙，串联间隙
series generator 串激发电机，串励发电机
series impedance 串联阻抗
series inductance 串联电感
series lighting 串联照明
series loading 串联负荷
series loop （电机线圈的）鼻端，串联回路
series machine 串行计算机，串励电机
series mode interference 串模干扰
series mode rejection ratio 串模抑制比
series mode rejection 串模抑制
series mode signal 串模信号
series mode voltage 串模电压
series modulation 串馈式屏极调制
series-mounting 串接方法，串联安装
series-multiple connection 串并联，混联，串接装配
series negative feedback 串联负回授，串联负反馈
series of observations 观测系列
series of power stations 梯级电站
series of tubes 管束，管组，串列管束
series operation 串联运行，串联运转
series-opposing connection 反向串联
series-opposing （线圈等的）反向串联，反向串联线圈
series-parallel circuit 串并联电路
series-parallel connection 串并联，混联，串并联连接
series-parallel controller 串并联控制器
series-parallel control 串并行控制，串并联控制
series-parallel-coupled buzzer 串并联蜂鸣器
series-parallel motor 串并激电动机，串并联电动机
series-parallel network 串并联网络
series-parallel reaction 串并联反应
series-parallel starter 串并联启动器
series-parallel starting 串并联启动
series-parallel switch 串并联转换器
series-parallel winding 串并联绕组，多重绕组
series-parallel 混联的，串并联（连接）的
series pipe 串联管路
series position 串联位置
series-produced item 系列产品，批量生产
series-produced power reactor 系列发电反应堆
series production 系列产品，大批量生产，批量生产
series reactance 串联电抗
series reactor 串联扼流圈，串联电感器，串联电抗器，附加电感器
series regulation （电源的）串联调整率
series relay 电流继电器，串联继电器
series-repulsion motor 串励推斥电动机
series resistance 串联电阻
series resistor 附加电阻器，串联电阻器，分压器
series resonance circuit 串联谐振电路，串联共振电路
series resonance 电压谐振，电压共振，串联共振，串联谐振
series resonant device 串联谐振装置
series rheostat in rotor circuit 转子串电阻
series ring winding 串联环式绕组
series shot-firing 串联爆破
series shunt 复激电机串激绕组的分流电阻
series spark-gap 串联火花隙
series stage 串联级
series starting rheostat 电动机串联启动变阻器
series stepping motor 步进电机
series transductor 串联饱和电抗器，串联磁放大器
series transformer 串联变压器
series transmission 串联输送，串行传输
series tripping 串联跳闸，电流跳闸
series turn 串联线匝
series type excitation 串激励磁，串联励磁
series type limiter 串联式限幅器
series type of peaking circuit 串联式高频补偿电路
series type phase advancer 串联式相位补偿器
series voltage 串联电压
series welding 串联电阻点焊
series winding 串联绕组，串激绕法
series-wound motor 串激电动机，串励电动机
series-wound 串绕的，串励的
series 串联，系列，级数，批，序，次序
serioparallel 串并行的，串并联的，混联的
serious accident and failure 重大事故
serious accident 重大事故
serious consequence 严重后果
SERI(Solar Energy Research Institute) 美国太阳能研究所
serpentine coil 蛇旋管
serpentine pipe 蛇形管
serpentine-tube header-type heater 蛇形管联箱式加热器
serpentine 盘旋管道，盘曲水管，蛇形管，螺旋形的，蛇根碱，蛇纹石
serrasoid 锯齿波
serrated belt 齿形三角带，风扇带
serrated coupling 齿型联轴器
serrated disc 缺口圆盘，缸口耙片
serrated fin 锯齿形的肋片
serrated overflow weir 锯齿形溢流堰
serrated pulse 顶部有切口的帧同步脉冲，缺口

脉冲
serrated ratchet plate　齿状棘轮板，固定齿套【主泵】
serrated ring　（冲裁用）齿圈
serrated root　（叶片的）锯齿形根，锯齿形叶根
serrated shaft　细齿轴
serrated topography　齿形山脊
serrated vertical synchronizing signal　交错垂直同步信号
serrated　锯齿形的
serrate-type steam seal　梳齿式汽封
serrate　锯齿形的，使成锯齿状
serration　细齿，锯齿
serrodyne　线性调频转发器
SER(sequence event recorder)　事故顺序记录仪
SER(Significant Event Report)　重大事件报告
SERS(surface-enhanced Raman spectroscopy)　表面增强拉曼光谱
SER(system environment recording)　系统现场记录
servant　公务员
served lead-covered cable　黄麻皮铅包电缆
server class　服务器类
server-confromance requirement　服务器一致性要求
server　服务器
serviceability criterion　适用性准则
serviceability limit state　使用极限状态，正常使用极限状态
serviceability limit　正常使用极限（状态）
serviceability　操作上的可靠性，使用可靠性，维护保养方便性，操作性能，可用性，可服务性，适用性，供给能力，耐用期限，耐用性，使用能力，可维修性
serviceable　合用的，耐用的
service access point　服务访问点
service after nuclear sale　核售后服务
service after sale　售后服务
service age　服务期限，使用期限
service aid　维修工具
service air dispersion model　大气扩散模式
service air system　厂用压缩空气系统
service air　厂用气，工业用气，压缩空气
service and admin building　服务设施及行政楼
service area　供电区，有效工作区，无线电服务范围，辅助面积，服务范围，有效作用区，生活服务区，公用设施区
service auxiliaries　服务性辅助系统
service auxiliary transformer　辅助变压器
service availability　供电可用率
service basin　工作水池，维修间，装配段，装配间
service behaviour　工作性能，运用性能，服务行为，使用状态，运转情况
service bin　日用仓，接线盒，供电箱
service breaker　辅助设施断路器
service bridge　工作桥
service building　辅助厂房，运行厂房
service cable　用户电缆，引入电缆
service capacity　电机铭牌容量，蓄电池工作容量
service center　服务中心
service channel　公务通道
service charge　服务费用，手续费
service check　运行检查
service circuit　业务通信电路
service column　工作塔
service compressed air distribution system　公用压缩空气分配系统
service compressed air system　厂用压缩空气系统
service compressed air　工作用压缩空气，厂用压缩空气
service condition　使用条件，工作条件，运行工况，运行状态，工作状态，工况
service conductor　接户线，供电导线，用户引入线，辅助设施导线
service connection　用户连接
service contract　劳务合同
service corridor　操作廊，巡回走廊
service corrosion　运行腐蚀
service crane　维修用的吊车，维护用吊车
service damage　使用损伤
service data　业务资料，运行数据，（设备）维护数据
service dead load　有效死荷载，工作恒载
service deck　服务平台
service demand availability factor　需求可用系数
service demand factor　需求系数
service designation　用途代号
service discharge　供用量
service drain　生活服务排水
service drop　架空引入线，架空接户线
service duty test　工作能力试验
service earth electrode　工作接地电极
service effluent tank　公用废水排水箱，公用废水箱
service effluent　生活服务排出物，强迫排放物
service engineer　服务工程师
service entrance conductor　进户线；用户引入线
service entrance　进户线，进线口
service equipment　供电设备，辅助设备，服务设施，维修设备
service experience　运行经验
service facilities　服务机构，公用设施，辅助设施，服务设施
service factor　服役率，利用系数，运行系数，使用系数，服务系数，运行率，负载系数，工作系数，运行因子，与电网连接投运率，时间利用系数，保险系数，劳务因素
service failure　供电中断，运行故障，使用故障
service fee　手续费
service floor area　辅助面积
service floor　工作面，操作面，修理间，服务间
service gallery　维修廊道
service gate　工作闸门
service generator room　厂用发电机室
service ground　工作接地
service hatch　备用舱口，使用闸门，维修孔
service head　（电缆）终端套管
service hoist　维修起吊设备，维修用吊车
service hours　运行时数，运行小时

service inspection 业务检查
service instruction 维护规程，运行规程，操作规程，使用规程，使用说明书
service intermittent 间断运行，间歇式工作
service interruption 停电，供电中断
service kid 维修（工具）箱
service length 工作周期
service level 服务级别
service liability 服务责任
service lifetime 使用寿命，工作期限
service life 使用寿命，使用期限，营运寿命，工作寿命
service lift 送货电梯
service limit 使用限制
service line 接户线，供电线
service link antennas 业务链路天线
service live load 实际工作活荷载
service load 供电负载，工作负荷，工作荷载，使用荷载，作用荷载
service machine 工作机，操作机械，维修机械
service mains and service entrance 接户线与进户线
service main 分配总管
service man-months 服务人月数
service manual 使用和维护手册［规程］，维修手册，使用手册
servicemen 技术服务人员
service meter 用户水表
service note 服务记录
service package 综合服务
service period 使用期限，运行期，运行周期
service pipe 给水管，工作管道，用户水管
service platform 工作台，操作平台
service position 工作位置
service power rate 厂用电率
service power source 厂用电源
service power 厂用电，供电功率
service pressure rating 额定工作压力
service pressure 运行压力，工作压力，使用压力
service primitive 服务原语
service processor 服务器，业物处理机
service program 服务程序，运行程序，操作程序
service property 使用特性，服役性能
service provider 服务提供商
service pump 辅助泵，备用泵，厂用水泵，伺服泵
service rating 使用等级，运行功率，运行额定值
service regulations 操作规程，维修规程
service regulator 工作调节器
service reliability of customers 用户供电可靠性
service reliable 使用可靠性
service requirement 运行要求
service reservoir 备用水池，给水池，供水水库，配水池
service restoring relay 恢复运行继电器
service road 厂用道路
service room 运行房间
service routine 服务程序，辅助程序
service run 工作行程
service sector 第三产业
service set 厂用机组
service shop 检修间，维修车间
service sink 工作室洗涤池
service skid 工作滑机
service sleeve 预埋套管
service speed 运行转速
service spillway 常用溢洪道，正常溢洪道，主溢洪道
service stairs 旁门楼梯，勤务楼梯
service switchboard 公用服务设施配电盘
service switch 控制器，业务寻线器，维修开关
service system 服务系统
service tariff of exclusive transmission project 专用输电工程服务价
service tax 劳务税
service test 操作试验，运行试验，性能试验，工作试验，使用检查，使用试验，运转试验
service time 服务时间
service tool 维修工具
service trade 劳务贸易
service trial 运行试验
service tunnel 工作隧洞
service-type test 使用状态试验，移交试验，使用试验
service valve 工作阀门
service voltage 工作电压，供电电压，额定工作电压，运行电压
service wage 劳务费
service walkway 维修走道，检修通道
service water collected tank 厂用水［生活服务水］回收水箱
service water piping 工业水管道
service water pump 工业水泵
service water supply pipe 生产水管线
service water supply system 厂用水供给系统
service water system 厂用水系统，杂用水系统，工业水系统，公用水系统
service water 工业用水，生产用水，厂用水，（处理过的）生活服务水，家用水，杂用水
service way 服务通道
service wear index 实用耐磨指数
service weight 使用重量
service weldability 使用焊接性
service wire 用户进线，引入线，入户线
service year 使用年限
service 业务，技术维护，检修，服务，操作，工作，运行，使用，维护，保养
servicing area 维修区
servicing time 维修时间，供电时间，工作时间
serving company chairman 现任公司董事长
serving hatch 送饭口
serving jute 黄麻被覆，（电缆的）黄麻包皮
serving of cable 电缆外皮
serving 侍候，服务，被覆物
servlet 小服务程序
servo action 伺服作用
servo-actuated control 伺服控制，从动控制
servo-actuated 伺服的，从动的
servo-actuator 伺服做功器，伺服执行机构，伺服驱动装置，伺服电机

servoamplifier　伺服放大器，伺服放大机，跟踪系统放大器
servo-analog computer　伺服模拟计算机
servo-analog　伺服模拟的
servo brake　伺服制动器
servo component　伺服元件
servoconnection　伺服连接
servo contact　伺服接触器
servo-controlled rod　伺服传动控制棒，伺服棒
servocontrol system　伺服控制系统
servocontrol　伺服控制机构，伺服控制，伺服补偿机，随动控制
servo-cylinder　伺服油缸
servo-driven　伺服拖动的
servodrive　伺服传动装置，伺服传动，伺服拖动
servodyne　伺服系统的动力传动装置，随动系统的动力传动装置
servo element　伺服系统元件，随动系统元件
servo function generator　伺服函数发生器
servogear　伺服机构
servo generator　伺服发电机
servohydraulic　伺服液压控制的
servoing　同步装置，伺服系统，随动装置，辅助设备
servo integrator　速度伺服机构，伺服积分器
servo-link　伺服传动装置
servo loop　伺服回路
servomagnet　伺服电磁铁
servomanometer　伺服压力计
servomechanism tester　伺服机构试验器
servomechanism unit　伺服机构部件
servomechanism　伺服机构，继动机构
servomotor actuator　伺服马达驱动器，伺服马达执行机构
servomotor capacity　接力器容量，伺服电动机容量
servomotor monitor　油动机监视保护仪
servomotor position indicator　油动机行程指示器
servomotor setting　伺服马达调整
servomotor　伺服电动机，伺服电机，伺服马达，伺服机构，继动器，接力器，油动机
servomultiplier　随动乘法器，伺服乘法器
servo-operated control　伺服调节，具有伺服装置的调节
servo piston　油动机活塞，伺服活塞
servo positioning system　伺服定位系统
servo potentiometer　伺服电势计
servo power drive　伺服动力驱动装置
servo pressure　伺服压力
servopump　伺服泵
servo recorder　带传动系统的记录装置，伺服记录器
servorudder　伺服舵
servo-selsyn system　随动自动同步机系统，追踪自动同步机系统
servosimulator　伺服模拟器，伺服模拟装置，伺服模拟机
servo speed control　随动速度控制
servo stability　伺服系统稳定性
servosupply　伺服供电
servo swap　磁带交替连续工作，交替换磁带
servo system　伺服系统，随动系统，辅助系统
servo table　带有随动系统的记录架
servotab　伺服调整片，伺服补偿机
servotron　高压启动式汞弧整流器
servo-type mechanism　伺服式机构
servo-type　伺服式的
servounit　伺服机构
servo valve　随动滑阀，伺服阀，从动阀
servo　伺服系统，伺服装置，随动系统，伺服机构，伺服的，随动的
SER　常规岛除盐水分配系统【核电站系统代码】
sesquioxide　倍半氧化物
sessile drop　座固液滴
sessile organism　附着生物
session layer　会话层
session　对话时间，会议闭幕，预约时间，学期，会期，会议，开庭
SES(Society of Engineering Science)　（美国）工程科学学会
SES(socioeconomic survey)　社会经济调查
SES(supplier evaluation sheet)　供应商评审表
SE(state estimate)　状态估计
SES　热水生产和分配系统【核电站系统代码】
set a limit to　（对……加以）限制
set an example　示范
set angle of blade　叶片安装角
set a price on(/upon)　对……预定价格
set aside　留出，拨出（款）
setback building　台阶式建筑
setback system　反插系统，功率下降系统
setback type　台阶式
setback　逆转，后退，减速，下降，阻碍，将指针拨回，（控制棒）下插，挫折，退税，台阶式
set bolt　固定螺栓
set conditionally　按条件设置
set-controlling admixture　（添加于混凝土的）控凝剂，调凝剂
set-deformation　永久变形
set-down area　搁置区
set-down location　放置处，搁置处
set-down pod　搁置架
set-down position　搁置位置
set-down　平均波面下降，水位下降
set flip-flop　置位触发器
set forth facts and reasons out　摆事实，讲道理
set forth facts　摆事实
set forth for　出发前往
set forth　阐明，宣布，提出，详尽地解释，陈述，说明，规定，把（会议等）提前，起程，出发，动身，展示
set forward one's proposition　提出……建议
set forward　提出，阐述，促进，提前，拨快，往前移动
set free　释放
set frequency　流向频率
set gate　置“1”门，门置位，随模浇口
set gauge　定位规
set-going　启动，开动

set hammer　击平锤
set hard　凝固，结硬
set-in branch pipe　嵌入的支管
set in motion　启动，开动
set light　照明灯具，照明设备，背景照明
set-maker　收音机制造者，接收机制造者
set-meter　沉降仪
set noise　机内噪声，本身固有噪声
set of brushes　电刷组
set of coils　线圈组
set of combs　梳齿组
set of current meter　流速仪装置
set of current　流向
set of curves　曲线族
set of holes　炮眼组
set of instructions　指令组
set of nozzles　喷嘴组
set of posts　接线柱组
set of spare parts　成套备件
set of spare unit　成套设备
set of symbols　符号组
set of tubes　管组
set-on branch pipe　安放的支管
set-on　安置，就位
set out angle position　确定转角位置
set out　定线，放线，放样，陈述，规定
set period　规定时间
set pin　定位销
set point adjuster　定值器，给定点调节器
setpoint controller　整定点控制器
set point control　整定点控制
set point generator　定值器
set point list　设定点清单
set point signal　参考信号，标准信号
setpoint station　整定点站
set point　设定点［值］，整定点［值］，给定值，整定工况，定点，定值，参比（变）量
set pressure of safety valve　安全阀设定［许用］压力
set pressure test　整定压力试验【安全阀或减压阀】
set pressure　调定压力，给定压力，设定压力，整定压力
set price　固定价格，预定价格
set pulse　置位脉冲
set recording　沉降记录，调整记录
set-reset pulse　置复位脉冲
set screw wrench　止动螺钉扳手，固位螺钉扳手
set screw　定位螺，固定螺钉，定位螺钉，定位螺丝
SET＝setting　设定
set sieves　筛组【指一套筛子】
setter　安装器，给定装置，给定器，调节器，安装人员，安装工，装订器
set the terms of payment　指定支付条款
set the world record　创世界纪录
set thrust plate　推力挡板
set time　凝固时间，凝结时间，定时
setting accelerator　促凝剂
setting accuracy　整定值精确性
setting agent　凝固剂【用于混凝土】
setting angle of blade　叶片安装角
setting angle　安装角，装配角度
setting basin　澄清池
setting button　调整钮
setting chamber　沉淀室
setting clearance　调整间隙
setting coat　罩面层
setting command　整定指令，存储指令
setting depth　电杆埋入深度
setting device　整定机构
setting drawing　装配图
setting encoder　设定值编码器
setting error　调定误差
setting for automatic protective device　自动保护装置整定值
setting group conrtolc lass　定值组控制类
setting group control block　定值组控制块
setting group editable　定值组可编辑
setting group　定值组
setting height　安装高度
setting interval　整定间隔
setting mechanism　整定机构
setting of bridge location　桥位测设
setting off　断流，关闭
setting of safety valves　安全阀整定
setting-out detail　定线详图，放样详图
setting-out diagram　放样草图
setting-out line　测定的线，划定的线
setting out of building　建筑放线
setting-out rod　定线杆，定线桩
setting-out work　定线工作，放样工作
setting out　放线，放样
setting pad　安装瓦块【轴承】
setting piece　调整片
setting plate　定位板
setting point　（水泥等）凝固温度，凝固点
setting pressure　给定压力，整定压力，设定压力，安装压力，初撑力
setting process　凝固过程，硬化过程
setting rail clamp　定位夹轨钳
setting range　调整范围，整定范围
setting reservoir　沉淀水池
setting retarder　缓凝剂
setting screw　止动螺钉，定位螺
setting shrinkage　凝固收缩
setting stake　定线桩
setting strength　（混凝土的）凝固强度
setting test　整定试验
setting time test　凝固时间试验【混凝土】
setting time　凝固时间【混凝土】，建立时间，整定时间，稳定时间，调整时间
setting to work　投入运行
setting up　设立，建立，装配，调整，快速凝固，快速干燥，调定，编排
setting value of the differential protection relay　差动保护继电器整定值
setting value of the time relay　时间继电器整定值
setting value　设定值，整定值
setting　调节，校正，校准，匹配【混凝土】，装配，调整，炉墙，支座，整定值，设置，定值，设定，凝固

settle a bargain 达成交易
settleable solids 可沉降固体（粒子）
settle accounts 结算
settle a claim 解决索赔（问题）
settled bed fast reactor 沉积床快中子反应堆
settled bed 沉降层
settled density 沉降密度
settled depth 澄清深度
settled dust 沉降尘埃
settled layer 沉积层
settled sewage 澄清的污水
settled sludge 沉积的污泥
settled water 澄清水
settle freight at actuals 按实际支出结算运费
settlement account 结算账户
settlement allowance 安置补偿费
settlement analysis 沉降分析
settlement area 住宅区
settlement bank 结算银行
settlement based on system marginal price 按边际价格结算
settlement and billing subsystem 结算管理子系统
settlement by arbitration 以仲裁方法解决
settlement cell 沉降盒
settlement coefficient 沉降系数
settlement contour 沉降等值线，等沉降线
settlement crack 沉降裂缝
settlement crater 沉陷坑
settlement curve 沉降曲线
settlement device 沉陷测定装置
settlement due to compression 压缩沉降
settlement failure 沉陷破坏
settlement gauge 沉降计，沉陷计
settlement inquiry 结算质疑
settlement interval 结算周期
settlement isoline 沉降等值线，等沉降线
settlement joint 沉降缝，压紧密封
settlement measurement 沉降测量
settlement method 结算方法
settlement meter 沉陷计，沉降（测定）计
settlement of accounts 结算
settlement of claim abroad 在海外理赔
settlement of claim according to accounting value at loss 按损失时账面理赔
settlement of claim 理赔，索赔的处理
settlement of disputes 争端的解决
settlement of exchange 结汇
settlement of insurance claim 保险理赔
settlement on account 账面结算
settlement plate 沉降板
settlement platform 沉降平台
settlement prediction 沉降预测，沉陷预测
settlement process 沉降过程
settlement rate 沉降速率，沉陷率
settlement slope 沉陷斜度
settlement stress 沉降应力
settlement-time curve 沉降时间曲线
settlement 沉降，沉陷，沉淀（物），地基下沉，结算，交割，清算，解决，处理，聚落
settler 澄清器，沉淀器，沉降器
settle through consultation 通过洽商解决
settle 沉淀，澄清【水处理】，固定，调整，沉降
settling accelerator 沉降加速器
settling baffle 沉降挡板
settling bank 结算银行
settling basin 澄清池，沉淀池，沉降池，沉沙池
settling chamber （风洞）稳定段，沉降室，沉淀槽
settling characteristics 沉降特征
settling compartment 分隔沉降室
settling efficiency 沉降效率
settling matter 沉淀物，沉降物
settling method 沉淀法，沉降法
settling of ground 地面沉降
settling of supports 支座下沉
settling pit 沉淀池，沉降坑
settling pond 澄清池，沉淀池，沉降池
settling process 沉降法，沉降过程
settling rate 沉降速度，沉降速率，下沉速度
settling ratio 沉降系数
settling solids 沉淀固体（颗粒）
settling speed 下沉速度，沉降速度
settling tank 沉降槽，沉淀池，澄清箱，澄清池，沉淀箱，沉渣箱，沉箱
settling time 沉淀时间，稳定时间，还原时间，建立时间，沉降时间
settling vat 沉降缸
settling velocity 沉淀速度
settling zone 沉淀区，沉淤地带
settling 安置，固定，沉淀，下沉，沉降
set to zero 调到零，零调整
setup amplitude 顶托幅度
setup diagram （计算机系统的）准备工作框图
setup error 调整误差
setup procedure 准备过程，准备程序，装配程序，装配过程
setup right 竖立
setup-scale instrument 无零位仪表
setup scale 无零点标度
setup sheet 装配图表
setup the structure 架构组立
setup the support with a herringbone pattern 倒落式人字抱杆整体立杆塔
setup time 建立时间，安装时间
setup unit 给定装置
setup 装置，装备，构成，设立，机构，安装，建立，调定，调整，整定
set value control 给定值控制
set value 整定值，给定值，变定值，标准值
set 调整，安装，组，套，固定，集，置位，置值，设定，机组，凝结，（混凝土）凝固
seven-heater cycle 给水七级回热加热循环，有七个回热加热器的热力系统
seven put-throughs of electricity, water supply, drainage, communication, heating, gas and road and site leveling 七通一平
seven-unit code 七单位电码
severability clause 可分离条款，合同中止条款，部分有效条款，可分割性条款
severability doctrine of arbitration clause 仲裁条款独立原则

severability of arbitration clause　可分离的仲裁条款
severability of the contract　合同条款的可分割性
severability　合同条款的可分割性
several-layer solenoid　多层圆筒形螺线管，多层螺线管
several separate winding　复分绕法，复分绕组
several　个别的，不同的，若干
severance type break　切断破裂
sever cut　切断
severe accident　严重事故
severe cold　严寒的
severe environment memory system　苛求环境的存储系统
severe exposure　严重暴露
severe flood　严重洪水
severe frost　严重霜冻
severe-fuel-damage　燃料严重损坏
severe gale　厉风
severe gust　强阵风
severe pressure　严重压力
severe salinization　严重盐碱化
severe storm　大暴雨，强风暴，狂烈风暴，猛烈风暴
severe stress　危险应力
severe tropical storm warning　强热带风暴警报
severe tropical storm　强热带风暴
severe weather　恶劣天气
severe　剧烈的，严峻的，严厉的，苛刻的
severity factor　硬度系数
severity level　严格程度
severity stall　失速严重
severity　严重，严格
sewage aeration　污水曝气
sewage analysis　污水分析
sewage basin　污水池
sewage charge　污水费
sewage chlorination　污水氯化作用
sewage composition　污水成分
sewage conduit　污水管道
sewage discharge standard　污水排放标准
sewage discharge　污水排放
sewage disposal plant　污水处理厂
sewage disposal process　污水处理过程
sewage disposal system　污水处理系统
sewage disposal works　污水处理厂
sewage disposal　污水处置
sewage ditch　污水沟
sewage drainage standard　污水排放标准
sewage drain　排污管，下水道
sewage ejector　污水压气喷射器，污水射流泵
sewage evaporator plant　污水蒸发器装置
sewage examination　污水检查
sewage filter　污水过滤器，污水过滤池
sewage final settling basin　污水最终沉降池
sewage final settling tank　污水最终沉降槽
sewage flow　污水流
sewage gas　垃圾沼气
sewage grit chamber　污水沉沙室
sewage grit　下水淤渣
sewage inflow　污水流入量
sewage loading　污水负荷量
sewage outfall　污水流出口
sewage oxidation pond　污水氧化塘
sewage oxidation　污水氧化
sewage particulate　污水颗粒
sewage pipeline　污水管线
sewage pipe　污水管
sewage pit　污水坑，污水井
sewage plant effluent　污水厂出水
sewage plant　污水处理厂，污水厂
sewage pollution　污水污染
sewage preliminary basin　污水初次沉降池
sewage preliminary tank　污水初次沉淀槽
sewage pressure　渗流压力
sewage pumping plant　污水泵站
sewage pumping station　污水泵站
sewage pump　排污泵，污水泵
sewage purification　污水净化
sewage purifier　污水净化设备
sewage rate　污水流量
sewage reservoir　污水池，污水库
sewage screen　污水筛网
sewage sink　污水槽，污水池
sewage sludge disposal　污水污泥处理
sewage sludge drying bed　污水污泥干燥床
sewage sludge drying　下水污泥干化
sewage sludge gas　污水污泥气体
sewage sludge incineration　污水污泥焚化
sewage sludge treatment　污水污泥处理
sewage sludge washing　污水污泥洗涤
sewage sludge　污水污泥，污泥浆
sewage stream　污水流
sewage system　污水系统
sewage tank　污水槽，化粪池，污水池
sewage treatment installation　污水处理设施
sewage treatment plant effluent　污水厂出水，排放污水，污水排水口
sewage treatment plant　污水处理装置，污水处理厂
sewage treatment process　污水处理过程
sewage treatment station　污水处理站
sewage treatment structure　污水处理构筑物
sewage treatment system　污水处理系统
sewage treatment tank　污水处理池
sewage treatment works　污水处理工程
sewage treatment　污水处理
sewage water　污水
sewage works effluent　污水厂流出水
sewage works　污水工程
sewage　下水道系统，污水，下水道，污物
sewerage dredger　挖泥机
sewerage law　污水工程法
sewerage system overflow　下水道溢流
sewerage system　排污水系统，下水道系统，污水系统，排水系统
sewerage　污水工程，下水道，下水（道）工程，排水系统，污水处理
sewer arch　排水管拱顶，污水管拱顶
sewer capacity　下水道容量
sewer catch basin　沉泥池，集泥井，阴沟沉泥井，下水道沉泥井

sewer cleaner 清沟机
sewer culvert 污水涵管
sewer gas 污水气体
sewer inlet 污水管进口，污水管入口，阴沟进口
sewer line 下水管道
sewer manhole 下水道检查井，污水道检查孔，下水道人孔
sewer ordinance 污水条例
sewer outfall 污水管出口，污水排放口，阴沟出口
sewer outlet 污水管出口
sewer pill 清理污水井管的木球
sewer pipe drain 污水排泄管
sewer pipe 排水管，下水道管
sewer rod 清污杆
sewer sluice valve 污水管冲洗阀，污水管闸阀
sewer system 污水系统，污水管系统，下水道系统
sewer tile 污水瓦管
sewer tunnel 排水通道，污水道隧洞，下水道
sewer utility 污水公用事业
sewer 阴沟，下水管，下水道，污水管，排水沟，污水道
SEW＝sewer 下水道
sexadecimal external number base 十六进制的外数基
sexadecimal 十六进制的，十六近位的，十六分之一
sexivalence 六价
sexpartite vault 六肋拱穹顶
sextant 六分仪
sextic equation 六次方程
sextuple-flow turbine 六排汽口汽轮机
sextuple 六倍的，六重的
SF_6 circuit breaker 六氟化硫断路器，六氟化硫开关
SFC(specific fuel consumption) 比燃料消耗，燃料消耗率，比燃耗油
SFCS(spent fuel cooling system) 乏燃料元件冷却系统
SFD(severe fuel damage) 燃料严重损坏
sferics network 天电观测网
sferic 远程雷电，天电，天电学，大气干扰
SFF(safety failure fraction) 安全失效分数
SFF(set flip-flop) 置位触发器
SF_6 gas-insulated 六氟化硫绝缘的
SF_6 gas leakage detector 六氟化硫气体泄漏检测器
SF_6 gas reclaiming equipment 六氟化硫气体回收设备
SF_6 gas 六氟化硫气体
SF_6 insulated substation 六氟化硫开关室
SFL(substrate feed logic) 衬底馈电逻辑
SFPCCS(spent fuel pool cooling and cleanup system) 乏燃料水池冷却及净化系统
SF_6 retrieving 六氟化硫气体回收
SFR(safety range) 安全距离
SFR(sweden's final repository for reactor wastes) 瑞典反应堆废物最终处置库
SFR(system failure rate) 系统事故率
SF(safety factor) 安全系数
SF＝self-feeding 自动给料的，自动进料的，自供料的，自激，自给
SF(service factor) 运行系数
SF(shearing force) 剪力
SF(shift forward) 前移
SF(shrinkage factor) 收缩系数
SF(signal frequency) 信号频率
SF(skip flag) 跳过标记
SF(spontaneous fission) 自发裂变
SFS(spent fuel pool cooling system) 乏燃料池冷却系统
SF(standard frequency) 标准频率
SF(steam flow) 蒸汽流量
S/F(store and forward) 存储转发
SF(stowage factor) 货物积载因数
SFT＝shaft 轴
SFU(standby filter unit) 备用过滤器单元
SF_6 六氟化硫
SGAFWS(steam generator auxiliary feedwater system) 蒸汽发生器辅助给水系统
SGAHRS(steam generator auxiliary heat removal system) 蒸汽发生器的辅助排热系统
SG cast iron(spheroidal graphite cast iron) 球墨铸铁
SGCB 定值组控制块
SGCI(special grade cast iron) 特级铸铁
SGD 新加坡元
SGHWR(steam generating heavy water reactor) 产汽重水反应堆，蒸汽发生重水反应堆
SGN(scan gate number) 扫描门数
SGRI(second general rate increase) 第二次运价上调
SGR(sodium-cooled graphic moderated reactor) 钠冷石墨慢化堆
SGR(sodium-graphite reactor) 钠石墨反应堆
SG(screen grid) 帘栅极
SG(set gate) 置“1”门
SGS(steam generator system) 蒸汽发生器系统
SG(standard gauge) 标准线规
SG(standby generator) 备用发电机
SG(steam generator) 蒸汽发生器，锅炉
SG＝switchgear 开关装置
SG(symbol generator display) 符号发生显示器
SG(symbol generator) 符号发生器
SGTR(steam generator tube rupture) 蒸汽发生器管破裂
SGTS(standby gas treatment system) 备用气体处理系统
SGV(slide gate valve) 插板阀
SGWH(steam generator water hammer) 蒸汽发生器水锤
SGZ 厂用气体贮存和分配系统【核电站系统代码】
shackhole 沉陷灰岩坑
shackle bolt 卡环螺栓销子
shackle joint 钩环接头
shackle pin 卡环销子
shackle U形挂环，U形吊环，钩环键，卸扣，锁扣，钩环，卸钩，U形环，卡环

shack 窝棚，棚屋，简陋小屋
shade and railing 遮拦
shade angle 保护角
shaded area 阴影面积
shade deck 遮阳甲板
shade disk 遮光片
shaded pole motor 罩极电动机
shaded pole relay 遮极盘式继电器，屏蔽磁极式继电器
shade holder 灯罩座，灯罩点
shade ring 遮光环
shade 挡板遮光罩，遮篷，遮蔽
shading band 短路环
shading circuit 补偿电路
shading coefficient 遮阳系数
shading coil 屏蔽线圈，罩极线圈，校正线圈，短路环
shading facility 遮阴设备
shading generator 黑点补偿信号发生器
shading-pole 罩极，屏蔽磁极
shading ring 短路环，校正线圈，罩极环
shading wedge 屏蔽楔
shading 描影法【制图】，遮阳
shadow area 阴影区，遮蔽区
shadow band 遮光带
shadow column instrument 阴影指示式仪表
shadow cone 荫蔽锥，影锥
shadow effect 荫蔽效应，阴影效应，阴影影响
shadow formation principle 阴影形成原理
shadowgraph scale 阴影标度，投射标度天平
shadowgraph X射线照片，影像图，阴影像片，阴影图像，阴影仪，阴影图
shadow method 阴影法
shadow picture 影像
shadow price 影子价格
shadow scattering 荫蔽散射
shadow shield 局部屏蔽，阴影屏蔽，遮光板
shadow 静区，(雷达）盲区，荫蔽，阴影，遮蔽
shaf vibration monitor 轴承振动监视器
shaft alignment gauge 轴对中规
shaft alignment 轴对中
shaft angle 轴线夹角
shaft vibration amplitude 轴承振动值
shaft bearing 轴承
shaft bending 轴弯曲
shaft butt 轴端
shaft collar 轴肩挡圈
shaft concentricity 轴同心度
shaft cooling shroud 轴冷却管套
shaft counter 轴转数计
shaft coupling 联轴节，联轴器
shaft crook 轴弯曲
shaft current of turbogenerator 透平发电机轴电流
shaft current 轴电流
shaft-cylinder differential 轴与汽缸的差胀
shaft deflection 轴挠度，轴偏摆，轴摆度，轴位移
shaft-driven exciter 同轴励磁机，轴传动励磁机
shaft-driven oil pump 主油泵
shaft-driven 轴驱动的，轴传动
shaft eccentricity 轴偏心度，大轴晃动（值）
shaft encoder 轴角编码器
shaft end output 轴端出力，轴端功率
shaft end play 轴端游隙
shaft end seal 轴封
shaft end 轴端
shaft excavation 竖井开挖
shaft extension 轴外伸部
shaft fatigue 轴疲劳
shaft flange 轴法兰
shaft generator 轴传动发电机
shaft gland steam cooler 轴封冷却器
shaft gland 轴封
shaft governor 轴速调速器，轴向调速器
shaft headworks 竖井首部建筑
shaft height 轴高（发动机）
shaft horsepower 轴输出功率，轴马力
shafting stability 轴系稳定性
shafting vibration 轴系振动
shafting 轴，轴系
shaft installing sleeve （密封用）轴套
shaft intake 竖井式进水口
shaft journal 轴颈
shaft jumbo 竖井掘进盾构，竖井钻车
shaft key 轴键
shaft kiln 竖式
shaftless motor 无轴电动机
shaft line alignment 轴线调整
shaft line 轴线【泵、电动机、联轴节和中间轴】
shaft lining 轴衬，竖井衬砌
shaft load 轴负载，轴载
shaft misalignment 轴装配误差
shaft mounted drive unit 轴装式驱动单元
shaft mounted unit 轴安装设备
shaft neck 轴颈
shaft nut 轴螺帽
shaft of column 柱身
shaft output 轴输出功率，轴功率
shaft packing leakage 轴封漏气
shaft packing 轴封，轴封填料
shaft plumbing wire 竖井垂直钢丝，预应力高强度钢丝
shaft position digitizer 转角数字转换器
shaft position encoder 轴位编码器
shaft position indicator 轴向位移指示器
shaft position mechanism 轴位置调节机构
shaft potential 轴电位，轴电压
shaft power coefficient 轴功率系数
shaft power curve 轴功率曲线
shaft power source 同轴功率，同轴电源
shaft power station 竖井式电站
shaft power 轴端功率，轴功率
shaft pumping station 竖井式抽水站
shaft raising gear 顶轴装置
shaft raising 向上打井
shaft revolution indicator 轴转速指示器
shaft revolution 轴转速
shaft runout 轴跳动
shaft seal assembly 轴封组件
shaft sealed pump 限漏流泵，轴封泵

shaft seal extraction cooler　轴封抽气冷却器
shaft seal extraction fan　轴封抽气风机
shaft sealing pump　轴封泵
shaft sealing　井筒密封
shaft seal leakage heater　轴封漏气加热器
shaft seal leakage steam piping　轴封漏气管道
shaft seal pump　轴封泵
shaft seal steam supply system　轴封供汽系统
shaft seal system　（汽轮机）轴封系统
shaft seal　轴封，轴封装置
shaft shoulder　轴肩
shaft sleeve nut　轴套螺帽
shaft sleeve　轴套，轴衬
shaft spillway　竖井式溢洪道
shaft stationary sleeve　固定轴套
shaft stem　柄
shaft straightening　直轴
shaft strain　轴变形
shaft swinging value　大轴晃动值
shaft torque　轴扭矩
shaft-type rotary encoder　轴式旋转号码机
shaft unlatching checkout tool　轴的拆解和检验工具
shaft vibration　轴振动
shaft voltage　轴电压
shaft well　竖井
shaft work　轴功
shaft　井道，竖井，通风井，筒身，轴杆，矿井，轴，烟囱筒身
shaitan　尘旋
shakedown analysis　（蠕变疲劳）安定性分析
shakedown operation　试运行
shakedown　（蠕变疲劳）安定，试用，试运转，调整，试运转的
shakeproof　防震的，防振，抗振
shaker filter　振打式过滤器
shaker screen　摇动筛
shaker table　振动台
shaker-type collection　机械振动式除尘器
shaker　振动试验机，振动式传送机，振动器，振动筛，振荡器，振动机，振动筛
shake table　震动台
shake　裂纹
shaking chute　振动溜槽
shaking feeder　振动送料器，振动加料器
shaking grate　振动炉排
shaking machine　振动机
shaking screen　摇动筛，振动筛
shaking sieve　摇摆筛，振动筛
shaking table　振动台
shaking　手摇式，摇动
shale clay　页岩黏土
shale content　泥质含量，黏土含量
shale fragment　页岩碎片
shale oil　页岩油
shale pit　页岩坑
shale planer　刨土机
shale pyrolysis　页岩热分解
shale refuse　矸石
shale tar　页岩焦油
shale　页岩，板岩，泥板岩
shallow beam　浅梁
shallow bed　浅床，浅层，薄层
shallow bin　浅仓
shallow cast　浅水施测
shallow channel tray　浅的槽形托架
shallow cut　浅挖
shallow depression　芯块端面碟状陷
shallow ditch　浅沟，小沟
shallow dose equivalent index　浅层剂量当量指数
shallow-draft navigation　浅水航行
shallow draught waterway　浅水航道
shallow dredging　浅水疏浚
shallow earthquake　浅层地震，浅发地震
shallower fundament　浇成的基础
shallow estuary　浅水河口
shallow fluidized bed　浅层流化床
shallow footing　浅基脚
shallow foundation　浅基础
shallow freezing　表面冻结
shallow granitic formation　浅层花岗岩
shallow ground disposal　浅地层处置
shallow ground grid　浅层接地网
shallow groundwater　浅层地下水
shallow grouting　浅层灌浆
shallow-hole blasting　浅孔爆破
shallow karst　浅层岩溶
shallow land burial　浅地层埋葬
shallow layer　浅床，浅层，薄层
shallow manhole　浅入孔
shallow open defect　浅的开口缺陷
shallow percolation　浅层渗透
shallow pond　浅水池
shallow reach　浅滩段
shallow river bed　浅水河床
shallow sampling with soil auger　麻花钻浅层取样
shallow soil　浅层土
shallow sounding apparatus　浅水测深设备，浅水回声测深仪
shallow stratum profile instrument　浅地层剖面仪
shallow tube well　浅管井
shallow water area　浅水区
shallow water clapotis　浅水驻波
shallow water deposit　浅水沉积
shallow water effect　浅水效应
shallow water tide　浅水潮汐
shallow water wave　浅水波
shallow water　浅水
shallow well　浅井
shallow　表面的，浅水，浅的，薄层的，浅滩，变浅
shaly sand　页岩质砂
shaly　页岩状
shank　螺杆【无螺纹部分】，螺栓的无螺纹部分，小轴，柄，叶身
shape coefficient　形状系数，体型系数
shape constant　形状常数
shape cutting　仿形切割
shaped channel　定形河道
shaped-conductor cable　（三相）特形铁芯，导体电缆
shaped conductor　非圆形导体，异形导线

shaped graphite brick 成形石墨砖
shaped nozzle 成形喷嘴
shaped orifice 型线孔口
shaped pipe 异形钢管
shaped plate 型板
shaped steel 型钢
shaped wire 型线【电线】
shape elastic scattering cross section 形状弹性散射截面
shape factor algebra 形状因子代数
shape factor 波形系数［因子］，形状系数［因子］，波形因数，形状因数
shape function 形状函数
shape information 波形信息，形状信息
shape memory alloy 形状记忆合金
shape memory plug 形状记忆塞
shape of bottom roughness 底部糙率型
shape of envelope 包络线线型
shape parameter 形状参数
shape resistance 形阻，形状阻力
shaper （脉冲）形成电路，整形器，成形机，脉冲整形器，牛头刨（床），造型者
shape steel 型钢
shape 型砖，型铁，型钢，形态，形状，模型，外形
shaping circuit 成形电路，整形电路
shaping loop 脉冲整形电路，脉冲形成电路
shaping machine 牛头刨床，成型机
shaping milling cutter 异形铣刀
shaping network 整形网络
shaping unit 信号形成器，整形部分，整形器
shaping wave 成形波
shaping 成形
share alike 均摊
share assembly program 共享汇编程序
share capital 股本
shared channel 复用信道，共用信道
shared control unit 共用控制单元
shared control 共享控制，共用控制
shared display system 共享的显示系统
shared file 共用文件
shared frequency station 同频台
shared logic 共享逻辑
shared memory 共享存储器
shared path 共用通路
shared resource 共用资源
shared routine 共用程序
shared storage 共用存储器
shared store 共用存储
shared subchannel 共用分通道
shared time control 分时控制
shared watershed 共用的流域
shared waters 共有水域
shared 分享的，共享的，共用的
share equally 均摊
shareholder structure chart of company 公司股东结构图
shareholder 股票持有人，股东
shareholding system 股份制
share-issuing enterprise 股份制企业
share of resources 资源共享
share operating system 共享操作系统，分时操作系统
share ownership enterprises group 股份制企业集团
share responsibility 承担责任
share the expenses equally 均摊费用
share 分享，股份，股票
sharing of structures 共用构筑物
sharing system 划分制
sharing 共用的，分时的，划分的
sharp aggregate 有棱角的骨料
sharp angle 锐角
sharp bend 小半径变弯管，急弯头，陡弯
sharp-crested weir 薄壁堰，锐顶堰，锐缘堰
sharp crest 锐顶
sharp cutoff pentode 锐截止五极管
sharp cutoff 锐截止
sharp doil method 二极管电容充电峰值电压测量法
sharp draft 强力通风
sharp-edged attack 锐缘侵蚀
sharp-edged body 尖锐物体
sharp-edged building 尖缘建筑
sharp-edged crest 锐缘堰顶
sharp-edged groove 锐缘（腐蚀）槽
sharp-edged gust 突发阵风
sharp-edged orifice 锐缘孔口
sharp-edged pelledized fuel 锐边芯块燃料
sharp-edged separation 尖缘分离
sharp-edged strip 锐缘条
sharp-edged wheel 锐缘轮
sharp-edged 锐缘的
sharp edge gauging weir 锐缘量水堰
sharp edge 陡缘，陡沿
sharpener 锐化器，刃磨机，锐化电路
sharpening machine 磨钻机
sharpening 削尖
sharp front 陡锋
sharpness of regulation 调节正确度，调节精密度
sharpness of resonance 共振峰的尖锐度，谐振锐度
sharpness of selection 选择锐度
sharpness 清晰度，锐度
sharp nosed blade 尖头叶片
sharp pop 突然发生
sharp pulse 尖脉冲
sharp starter crack 尖锐起始裂纹
sharp transition 突变过渡，突变段
sharp-tuning 锐调整
sharp turn 陡弯
Sharpy impact test 夏比冲击试验
shatter cut 龟裂掏槽
shattering 粉碎
shatter-proof glass 防碎玻璃
shatter strength 落下强度
shatter 破碎，碎片
shave 刹，刮，刨
shaving board 刨花板
Shaw hardness 肖氏硬度
shear beam 抗剪梁，剪切梁
shear bolt 安全螺栓，保险螺栓，剪力栓

shear box 剪切盒
shear center 剪切中心
shear cleat （预应力混凝土压力容器）剪力固着楔
shear coefficient of viscosity 黏滞剪切系数
shear connector 剪切连接，抗剪切连接
shear core structure 剪力核心（抗风）结构
shear crack 剪切裂缝
shear deformation 剪切变形，切变
shear diagram 剪切图，剪力图
shear displacement 剪切位移
shear effect 剪切作用，剪切效应
shear elasticity 剪切弹性，剪切弹性模量
shear failure 由剪应力引起的断裂，剪切破坏
shear fillet 受剪贴角焊缝
shear flow theory 切变流理论
shear flow turbulence 切变流紊动，剪切流紊流度
shear flow 剪切流，切变流，黏性流，旋流
shear force 剪切力，切断力
shear fracture resistance 剪切破坏抗力
shear fracture 剪切断口，切断
shear generator （风）切变发生器，剪切发生器
shear-gravity wave 切变重力波
shearing current 切变流
shearing force 剪切力，切断力，剪力
shearing movement 切变运动
shearing pin 受剪销钉
shearing strain 剪应变
shearing strength 抗剪强度
shearing stress 切向应力，剪切应力，切应力，剪应力
shearing stress 最大剪应力面
shearing wall structure 剪力墙结构
shearing 剪断，剪切
shear instability 剪切不稳定性
shear joint 受剪节点
shear layer reattachment 附面层再附着
shear layer 剪切层，切变层
shear leach 切断浸出法
shear leg derrick crane 合撑式起重机
shear legs 起重机梃杆，起重机支架
shear line 断层线
shear loading 剪切荷载
shear loss 剪切损失
shear lug 抗剪键，剪力吊耳，止动挡块
shear mode 切变模
shear modulus of elasticity 剪切弹性模数，剪切弹性模量
shear modulus 剪变模量，抗剪弹性模量，切变模量
shear of wind 风切度
shear pin 安全销
shear plane 剪切面
shear rate 切变速率
shear reinforcement 抗剪钢筋，剪刀钢筋
shear stiffness 剪切刚度，抗剪刚度
shear strain hypothesis 剪切应变假说
shear strain 剪应变，切应变
shear strength 剪切强度，抗剪强度，切变强度
shear stress 剪应力，切变应力，切应力，剪切应力
shear surface 剪切面
shears 剪床，剪断机
shear test 剪切试验，抗剪试验
shear theory 切变理论
shear turbulence 剪切湍流
shear velocity 切变速度，剪切速度
shear viscosity coefficient 切变黏滞系数
shear wall structure 剪切墙结构，剪力墙结构
shear wall 剪力墙
shear wave probe 切变波探头
shear wave tester 剪切波测试仪
shear wave velocity 切变波速度
shear wave 剪切波，切变波，剪力波
shear wind 切变风，断层带，切变区
shear 剪，切，切断，切力，剪切力，剪切，剪切机，切变
sheath-bonding transformer 电缆套接地变压器
sheath-circuit eddy （电缆）外皮涡流
sheath current 表皮电流，表皮涡流
sheathed chromel-alumel thermocouple 铠装镍铬镍铝热电偶
sheath eddy （电缆的）表皮涡流
sheathed-pilot system 金属屏蔽引线选择保护系统
sheathed pump 密封泵，屏蔽泵
sheathed thermocouple 铠装热电偶
sheathed 加护套的，铠装的
sheath effect 表皮（涡流）效应
sheathe 包壳覆盖，覆盖，包，装在包壳内
sheathing and shielding cable 铠装和屏蔽电缆
sheathing wire 金属护皮电线，铠装线
sheathing 装套，鞘，外壳，护套，覆盖层，铠装，包端，屋面板，屋顶板，支撑板，覆以包壳
sheath loss 护皮损失，电缆铅耗
sheath protector of single-core cable 单芯电缆护层保护器
sheath protector 护层保护器
sheath 电缆包皮［包壳］，外皮，壳，（合成绝缘子）护套，护层，覆盖，铠装，外壳，包覆层
sheave block 滑轮组
sheave stand 滑轮座
sheave wheel 滑轮
sheave 滑车，滑车轮，滑轮，皮带轮，绳轮，槽轮，凸轮盘
shed damage 绝缘子外裙损坏
shedding of solids 颗粒散落
shed for storage of dry coal 干煤棚
shed line 分界线
shed roof 单坡屋面
shed storage 有棚仓库
shed the load 截断若干线路的电流供应
shed vortex 脱体涡，脱出涡
shed vortices 尾迹涡系
shed wake vortex 脱落涡
shed 棚，车库，干煤棚，工棚，摆脱，流出，泻下，泻掉，车棚，棚库，分水岭
sheep-foot roller 羊脚碾，羊足碾，羊蹄滚筒，羊蹄压路机

sheer leg　三脚体
sheet aluminium　铝板
sheet anchor　船首的副锚，备用锚，备用大锚
sheet and plate work　薄板制作
sheet and strip　板材和带材
sheet assembly　图幅接合表
sheet boiling　膜状沸腾
sheet border　图廓
sheet capacitance　箔电容
sheet cavitation　平面空蚀，片状空蚀，表面汽蚀
sheet conductance　面电导，薄层电导
sheet dielectric　片状电介质
sheet dimension　图幅尺寸，图幅大小
sheeted roof　薄板屋顶
sheet erosion　表面侵蚀，片状冲刷，片蚀
sheeter　压片机
sheet flow　薄层水流，片状流，片流
sheet gasket　密封垫片，密封垫圈
sheeting board　护堤板
sheeting driver　打板桩机
sheeting jack　板桩千斤顶
sheeting pile　板桩
sheeting plank of dyke　护堤板桩
sheeting plank　护堤板
sheeting　衬板，护板
sheet iron gauge　钢板厚度计
sheet iron leader　铁皮水落管
sheet iron　薄钢板
sheet joint　席状节理
sheet lightning　片状闪电
sheet line　图廓线
sheet material　板材
sheet metal brush　金属片电刷
sheet metal casing　锅板蜗壳
sheet metal forming machines　金属板成型机
sheet metal screw　薄板螺钉
sheet metal shop　白铁车间
sheet metal worker　钣金工
sheet metal working machines　金属板加工机
sheet metal works　钣金工程
sheet metal　薄钢板，钢板，金属片，金属薄板
sheet name　图幅名称
sheet number　图幅编号
sheet of vorticity　涡旋面
sheet pavement　整片路面
sheet pile anchorage　板桩锚，板桩锚定
sheet pile breakwater　板桩防波堤
sheet pile bulkhead　板桩岸壁
sheet pile cell　板桩格笼，潜水箱，沉箱
sheet pile cofferdam　板桩围堰
sheet pile curtain　板桩帷幕
sheet pile cut-off　板桩截水墙【水电】
sheet pile enclosure　板桩围堰
sheet pile extracting　板桩拔除
sheet pile hammer　板桩锤
sheet pile levee　板桩堤
sheet pile quaywall　板桩码头
sheet pile retaining wall　板桩挡土墙，板桩式挡土墙
sheet piler　垛板机
sheet pile screen　板桩防护栏，板桩围幕
sheet pile wall　板桩墙，钢板桩墙
sheet pile wharf　板桩码头
sheet pile　板桩，打板桩，防渗板桩
sheet piling bulkhead　钢板桩墙
sheet piling　板桩，打板桩
sheet plate　厚钢板
sheet resistance　薄膜电阻，表面电阻
sheet rubber　橡皮板
sheet shickness meter　薄板测厚仪
sheet stacker　垛板机
sheet steel duct　薄钢板风管
sheet steel flume　钢板渡槽
sheet steel form　钢模板
sheet steel　薄钢板，钢片
sheet stringer construction　板桁结构
sheet structure　片状结构
sheet thrust　席冲断层
sheet tin　马口铁皮
sheet vortices　片涡
sheet wall　薄壁
sheet washing　片状冲蚀
sheet wash　片状冲刷，片蚀，平面冲刷
sheet water-proofing　板材屋面防水
sheet winding　扁线绕组
sheet　板，层，薄板，钢板，涡面，页，图表，表，单
shelf aging　闲置老化，搁置老化
shelf edge　大陆架边缘
shelf life　贮藏期限，搁置寿命，适用期
shelf mounted instrument　盘装仪表，架装仪表
shelf retaining wall　衡重式挡土墙，棚架式挡土墙
shelf　搁板，托架，架子
shellac varnish　虫胶漆
shellac　紫胶，虫胶，虫胶清漆
shell-and-auger boring　冲击与螺旋钻探
shell and coil condenser　壳管式冷凝器，盘管式冷凝器
shell and coil　壳管
shell-and-tube condenser　壳管式冷凝器
shell-and-tube evaporator　壳管式蒸发器
shell-and-tube heater　管壳式加热器
shell-and-tube heat exchanger　表面式热交换器，管壳式热交换器，管壳式换热器（多）管式交换器
shell-and-tube preheater　管板式预热器，壳管式预热器
shell-and-tube type attemperator　管壳式减温器，表面式减温器
shell-and-tube water condenser　壳管式水冷凝器
shell arch　薄壳拱
shell armature motor　管式电枢电动机
shell-bearing clay　含贝壳黏土
shell construction　薄壳建筑，薄壳构造，薄壳结构
shell-cooling annulus　环形冷却套
shell drain nozzle　壳侧疏水接管
shell element　壳体元素
sheller　去皮机，脱壳机
shell expansion indicator　汽缸膨胀指示器

shell flange 筒体法兰
shell-form power transformer 外铁型电力变压器
shell foundation 壳体基础
shell-less pile 无壳桩
shell lime 坝壳石灰
shell manway 容器壁上人孔
shell model 壳层模形
shell molding 壳模法
shell of ferro-cement 钢丝网水泥壳体
shell of heat exchanger 换热器外壳
shell pass 壳程
shell perm process 沥青乳剂灌浆
shell pile 带壳桩
shell plate thickness 壳板厚度
shell plate 锅筒板，筒身板
shell portion 筒身段，筒体部分，圆柱体段
shell pressure 壳侧压力
shell pump 抽沙泵
shell puncture 外壳击穿
shell ring 筒圈，筒节
shell rock 坝壳岩
shell shake 环裂
shell side enhancement technique 壳程强化法
shell side flow 壳程流
shell side performance 壳程性能
shell side pressure drop 壳程压力降
shell side pressure loss 壳侧压力损失，壳侧阻力
shell side 蒸汽发生器二次侧，管际空间，壳侧，壳程
shell slab 薄壳板
shell space 壳层空间，壳体空间
shell structure roof 壳体屋顶
shell structure 薄壳结构
shell thermocouple 壳式热电偶
shell thickness 壳体厚度
shell transformer 外铁型变压器，壳型变压器
shell tube type heat exchanger 壳管式热交换器
shell type attemperator 管壳式减温器，表面式减温器
shell type boiler 锅壳锅炉【指卧式火管锅炉】，筒形锅炉
shell type branched pipe 无梁壳型岔管
shell type core 壳型铁芯
shell type iron core 壳式铁芯
shell type motor 封闭型电动机
shell type rotor 空心转子
shell type servo motor 空心转子伺服电机
shell type surface attemperator 管壳式表面减温器
shell type transformer 外壳型变压器，壳式变压器
shell vibration 蒙皮振动，外壳振动
shelly limestone 坝壳灰岩
shell 壳，罩，壳体，筒身，筒壁，外壳，锅壳，套，皮，管套，绝缘子裙
shelter area 掩蔽区
shelter belt 防风林带
shelter-clad 金属铠装，金属防护罩
shelter deck 遮蔽甲板
sheltered area 掩蔽区【C级】
sheltered bay 避风河湾
sheltered location 掩蔽场所
sheltered station 有遮蔽的观测站
sheltered waters 有掩护的水域
sheltered 提供食宿的【尤指为病弱者】，受庇护的，受保护的，免税的，提供抚养方便的
shelter forest belt 防护林带
shelter forest 防护林
sheltering effect 遮蔽效应，防护作用
shelter 保护，躲避，掩蔽处，棚，屏障，掩蔽所，掩蔽，遮蔽
shelter 车棚
shelve 搁置，放在架子上
shelving insert 置架嵌入件【墙壁】
shelving 搁板材料，搭架子
shematic diagram 原理图
sheradizing 镀锌防锈法，粉锌镀法，渗锌，电镀
Sherwood number 薛伍德数
SHF(super-high frequency) 超高频
SHG(second-harmonic generation) 二次谐波产生
SHG(second harmonic generator) 第二谐波发生器，倍频器
shield analysis 屏蔽分析
shield angle 屏蔽角
shield area 屏蔽区
shield assembly 屏蔽（防护）装置
shield cable 屏蔽电缆
shield cell 屏蔽室
shield coil 屏蔽线圈
shield configuration 屏蔽布置
shield container 屏蔽容器
shield cooler 屏蔽冷却热交换器，屏蔽冷却器
shield cooling air fan 屏蔽冷却风扇
shield cooling duct 屏蔽冷却管道
shield cooling heat exchanger 屏蔽冷却热交换器，屏蔽冷却器
shield cooling pump 屏蔽冷却泵
shield cooling system 屏蔽冷却系统
shield cooling water return tank 屏蔽冷却水回水箱
shield cooling water system 屏蔽冷却水系统
shield cooling 屏蔽冷却系统，屏蔽冷却
shield design 屏蔽设计
shield-driven tunnelling 盾构法隧道施工，盾构隧洞掘进法
shield-driven tunnel 盾构开挖隧道
shield driving 盾构掘进
shield earthing 屏蔽接地
shielded antenna 屏蔽天线
shielded arc electrode 屏蔽电弧电极
shielded arc welding 屏蔽式电弧焊，气体保护焊
shielded area 屏蔽区
shielded bridge 屏蔽电桥
shielded cable 屏蔽电缆
shielded capacitor 屏蔽电容器
shielded cask 屏蔽罐
shielded cave facility 屏蔽地下室设施
shielded cell 屏蔽室
shielded coffin 屏蔽运输容器

shielded compartment 屏蔽的隔间
shielded conductor cable 屏蔽导线电缆
shielded conductor 屏蔽导线，屏蔽导体
shielded container 屏蔽容器
shielded door 屏蔽门
shielded facility 屏蔽设施
shielded flask 屏蔽容器
shielded from interference 防止干扰的
shielded galvanometer 磁屏蔽式检流计
shielded glove box 厚壁手套箱，屏蔽的手套箱
shielded inert-gas metal-arc welding 惰性气体熔化极电弧焊
shielded inspection booth 屏蔽的检查室
shielded joint 屏蔽接头
shielded measuring instrument 铠装型测量仪表，屏蔽型测量仪表
shielded metal arc welding 自动保护金属极电弧焊，熔化极自动保护电弧焊，手工电弧焊
shielded pole instrument 屏蔽极仪表
shielded spent fuel shipping cask 屏蔽乏燃料运输罐
shielded storage area 屏蔽贮存区【堆芯测量仪表】
shielded thermocouple 屏蔽式温差电偶，屏蔽式热电偶
shielded transformer 屏蔽型变压器
shielded weld 保护焊
shielded wire 屏蔽线
shielded 隔离的，屏蔽的，铠装的，屏蔽式的
shield facility 屏蔽（试验）装置
shield factor 屏蔽系数
shield gas flow rate 保护气体流量
shield gas 保护气体
shield heating 屏蔽发热
shield hole 屏蔽孔
shield inert gas metal arc welding 惰性气体保护金属极电弧焊
shielding action 屏蔽效应，屏蔽作用
shielding angle 保护角
shielding arc welding 盾护电弧焊
shielding barrier equipment 屏蔽设备
shielding block 屏蔽块
shielding box 屏蔽盒，屏蔽箱
shielding cabin 屏蔽小室
shielding calculation 屏蔽计算
shielding case 屏蔽壳，屏蔽罩
shielding castle 屏蔽箱，屏蔽容器
shielding coefficient 屏蔽系数，遮阳系数
shielding coil 屏蔽线圈
shielding concrete 屏蔽用混凝土，防护用混凝土，混凝土防护
shielding container 屏蔽容器
shielding effect 屏蔽效应，屏蔽作用，遮蔽效应
shielding equipment 屏蔽设计，屏蔽装置
shielding facility 屏蔽装置，防护装置，防护设施
shielding factor 屏蔽系数，遮蔽因子
shielding failure 雷电绕击
shielding gas 保护气体
shielding gate valve 屏蔽闸阀
shielding material 屏蔽材料
shielding measures 屏蔽措施
shielding of high-voltage laboratory 高电压实验室屏蔽
shielding plug 防护塞，屏蔽塞
shielding properties 防护特性，屏蔽性质
shielding ring 屏蔽环，屏蔽环段
shielding seal 屏蔽的密封终端
shielding slab anchoring 屏蔽板固定
shielding slab 屏蔽块，屏蔽板
shielding system 屏蔽装置，隔离装置
shielding tank 防护箱，屏蔽罐
shielding tape 屏蔽带
shielding transmission rate 屏蔽穿透率
shielding wall 屏蔽墙
shielding window 屏蔽观察窗，观察窗，屏蔽窗
shielding wire 屏蔽线
shielding 屏蔽，屏蔽的，遮蔽的，防护的，防护层
shield jig 保护夹具
shield method 盾构法【用于隧洞工程】
shield motor 屏蔽电机
shield opening 屏蔽孔
shield pit 屏蔽孔，屏蔽坑
shield ring 屏蔽环
shield tank 屏蔽罐，屏蔽水箱
shield test facility 屏蔽试验装置
shield test pool facility 池式屏蔽试验设施
shield test reactor 屏蔽试验反应堆
shield tunnelling 盾构法开挖隧洞
shield window 观察窗，屏蔽窗
shield wire spacing 地线间距
shield wire 屏蔽线
shield 屏，罩，盾构，防护罩，屏蔽，铠装，隔屏，护板，保护装置，掩体，遮护板
shift attendant room for house consumption 厂用电值班室
shift change 位移，转动
shift chief-operator 倒班主操，值班主操作工
shift chief supervisor 倒班主管，值长
shift circuit 位移电路
shift computer 偏移计算器
shift control 偏移调整，移位控制器
shift converter 频移变换器，相移变换器
shift counter 移位计数器
shifted divisor 位移的除数
shifted measurement 换挡测量
shifted 移位的，位移的，变换的
shift engineer room 值长室
shift engineer 值班工程师
shifter factor 移动因子，平移因子
shifter plate 开关板
shifter 转换机构，切换装置，移相器，倒相器，移位器，移位寄存器，移动装置
shift foreman 值班长
shift-freguency modulation 移频调制
shift-in character 移入字符
shifting anchor 移锚
shifting beam 活动梁
shifting bearing 活动支座
shifting camshaft 可调准的凸轮轴
shifting center 稳心

shifting channel 不固定河槽
shifting cultivation 轮垦
shifting current 不定流
shifting of divide 分水界迁移，分水界移动
shifting of fuel 换燃料
shifting of river channel 河道的变迁
shifting order reaction 变级数反应
shifting ring 调整环，调整圈
shifting sand 流沙
shifting spanner 活动扳手
shifting transfer （燃料元件）倒换
shifting wind 变向风
shifting wrench 活动扳手，活口扳手
shifting 分级，移位，移动
shift knob 开关按钮
shift left and count instruction 左移计数指令
shift left logical instruction 向左移位逻辑指令
shift lever 变速杆
shift log 值班记录
shift manager 值班经理
shift matrix 移位矩阵
shift of origin 原点转移
shift on duty 轮班
shift operator 值班员
shift order 移位指令
shift-out 移出
shift personnel 值班人员，轮换班组，换班人员，倒班人员
shift pulse 移位脉冲
shift reaction 转移反应
shift register drive 移位寄存器驱动器
shift-register generator 移位寄存器式发生器
shift register 移位寄存器
shift report 值班报表
shift reverse 反向移位
shift right double 双倍右移，倍右移
shift ring 移动环，控制环，调速环
shift supervisor station 值长操作台
shift supervisor 值班主任，倒班主管，值班经理，值长
shift system 轮班制
shift test coordinator 值班试验协调员
shift theorem 移位定理
shift turnover 位移，转动
shift 变换，调挡，轮班，班，漂移，改变，移动，平移，位移，移位，移数，值班
shim action 粗略调整，垫补作用，填隙作用，粗调
shim block 垫片，垫块，楔块
shim bolt 固定衬垫螺栓
shim control rod 补偿控制棒
shim control 粗调，补偿控制
shim element 补偿元件
shim follow-up rod 补偿随动棒
shim member 补偿元件
shimming coil 补偿线圈
shimming plate 填隙板
shimming valve 补偿阀，微调阀
shimming 用垫片调整，填隙，垫补法，磁场的调整，调节磁场
shimmy 摆振，横摆
shim plate 垫铁
shim rod actuator 补偿棒的传动机构
shim rod bank 补偿棒组
shim rod 补偿棒，调整棒，粗调棒
shim safety rod 补偿安全棒
shim 调整垫片【控制棒驱动装置】，夹铁，垫片，薄垫片，垫板，填隙片，粗调棒，补偿棒
shingle bank 粗砾岸，砾石堤
shingle barrier 粗砾滩
shingle beach 砾滨，砾石滩
shingle roofing 木板屋顶
shingle 屋顶板，砾石（堤岸），砂砾，卵石，石棉瓦，木板瓦，木瓦，瓦，板
shingly 砾石的
ship a contract 装运合同的货物
ship berth 停泊处
ship integrated power system 船舶综合电力系统
shiplap sheet piling 搭叠板桩
shiplap timber-sheet-pile 搭叠缝木板桩
shiplap 搭叠，木板槽口的接合处，用以相接槽口的木板
ship lifter with sloping deck 斜面升船机
ship lift 升船机
shiploader 装船机
shiploading dock 装船码头
shiploading 装船
ship lock 船闸
ship log 航海日记
shipment by installments 分批装运
shipment by railway 铁路运输
shipment clause 装运条款
shipment condition 装运条件
shipment date 装船日期
shipment dimensions 货运体积，货物尺寸
shipment in three lots 分三批装运
shipment port 起运港
shipment release 运输证件
shipments under B/L No.... ……号提单的货物
shipment 运输，装运，发运，发货，装船，装货，发运货物
shipped B/L 已出运的货物提单，已装船提单
shipped on board bill of lading 已装船提单
shipper and carrier 托运人与承运人
shipper box 发货人栏［格］
shipper loaded container 货主自装集装箱
shipper receiver difference 发方与收方计量差，收发差
shippers letter of instruction 空运托运书
shipper 发货方，发货人
shipping advice 装船通知，装运通知，发货通知书
shipping agent 海运代理人，船公司代理，船舶代理，运货代理商，轮船代理人
shipping and traffic 水陆运输
shipping bill 装船单据
shipping capsule 运输（放射性物质的）小盒
shipping cask grid 运输容器格架
shipping cask loading 运输容器装料
shipping cask setdown grid 运输容器搁置架
shipping cask （乏燃料）运输容器，运输铅罐，铅屏蔽罐

shipping clearance 航运净空
shipping company's certificate 船运公司证书
shipping company 航运公司
shipping container unloading station 运输容器卸货站
shipping container 运输容器
shipping contract 海运合同
shipping cylinder 运输用容器
shipping date 装船日期
shipping diameter （压力容器）运输直径
shipping dimension 装船尺寸
shipping document 发运单据，装船文件，运单，货运单据，装运单据
shipping enterprise 航运企业
shipping flask lay-down location 运输容器放置处
shipping flask transfer 运输容器输送车
shipping flask transport route 运输容器运输路线
shipping flask 运输容器，运输铅罐，铅屏蔽罐
shipping inspection 运输检验
shipping instructions 装船须知，装运须知
shipping invoice 装货发票
shipping lane 航道
shipping list 货运单
shipping mark 货运唛头，唛头，货运标记，装船唛头，发货标志，运输标志，装运标记
shipping notice 装船通知
shipping order 发货单，装货单，运货单，发货通知
shipping port 发货港，装运港
shipping quality and weight 离岸品质和离岸重量
shipping quality terms 装船质量条件
shipping quality 装船质量，可运输性，离岸品质
shipping regulations 运输规程
shipping schedule 装船进度
shipping shield 运输用屏蔽
shipping size 装船尺寸
shipping skid 运输滑橇【重件运输】，运输滑轨，运输托架
shipping space 舱位
shipping tag 货运标签
shipping weight 装运重量，装船重量
shipping 船舶，海运，航运
ship power station 船舶电站
ship propulsion reactor 船用推进反应堆
ship reactor 船用反应堆
ship shape line 船舶型线
ship skin 船壳
ship stabilizer 船舶稳定器
ship unloader 卸船机
ship unloading installations 卸船设备
shipworm 蛀木虫
shipyard gantry ciane 船坞门吊，船厂龙门起重机
shipyard 船坞
ship 船
shiver 蓝色板岩，碎块，页岩
SHM(shared memory) 共享存储器
shoal area 浅滩区
shoal head 沙嘴
shoaling coefficient 浅水系数
shoaling 浅滩的淤积，淤浅
shoal water zone 浅水区
shoaly land 浅滩地
shoal 浅滩
shock absorbent material 吸震材料
shock absorber stop 缓冲装置犁子
shock absorber 阻尼器，减震器，消震器，减震器缓冲装置
shock absorption device 减震设备
shock absorption 消震，减震，吸震作用
shock adiabatic curve 冲击绝热线
shock attenuation 减震，震动衰减
shock baffle 防震隔板
shock chamber 骤冷室，激冷室
shock coefficient 撞击损失系数
shock cooling 骤冷
shock countermeasure 防冲击措施
shock current 触电电流
shock damper 减震器
shocked flow 冲波气流，激波流
shocked plasma 受冲等离子体
shock effect 冲击影响
shock eliminator 缓冲器，减振器
shock excitation 震动激励，冲击激励
shock exciter 强行励磁机，冲击励磁机
shock-expansion method 冲波膨胀法
shock-free flow 无激波流
shock-free making 无震动合闸
shock-free 无冲波的
shock front measurement 冲波波前测量
shock generated vorticity 激波产生的涡量
shock hazard 冲击危险性，电击事故
shock heating 骤热
shock-induced boundary layer 冲击边界层
shock isolator 隔震器
shockless entrance 无震动入口段
shockless inlet 无震入口
shock line 冲波线，激波线
shock load 突加负载，冲击负荷，冲击负载，冲击荷载
shock loss 震动损失，冲波损失，激波损失
shock plate 冲击板
shock point 激波点
shock polar 冲波极线
shock position 冲波位置
shock pressure 冲击压力，震动压力
shockproof mounting 防振座
shockproof relay 防震继电装置
shockproof structure 抗震架构
shockproof 防震的，耐震的，防电击的，抗震的，防冲击的
shock pulse 冲击脉冲，震动脉冲
shock recovery 压力恢复，激波恢复
shock reducer 缓冲器
shock reflection 冲击波反射
shock region （流动）突变区，激波区
shock resistance 抗冲击性，抗震，抗冲击
shock ring 减震环
shock speed measurement 冲波速度测量

shock-stall 冲波失速
shock stress 冲击应力
shock suppressor 抑制冲击器，防冲击器，消冲击器
shock table 震动实验台，震动台【混凝土测试用】
shock test 冲击试验
shock thickness 激波厚度
shock tube 冲波管，激波管
shock tunnel simulation 激波风洞模型试验
shock velocity 冲击波速
shock visits 突击视察
shock wave amplitude 冲波强度
shock wave angle 冲波倾角，激波倾角
shock wave cancellation 冲波消失
shock wave decay 冲波衰减
shock wave discontinuity 冲击波不连续性
shock wave drag 冲波阻力
shock wave formation 冲波形成
shock wave front 冲击波前沿，冲击波面
shock wave propagation 冲击波传播
shock wave radiation 冲波延伸
shock wave reflection 激波反射
shock wave rise 冲波出现
shock wave shadow 冲击波痕
shock wave wake 冲波尾流
shock wave 冲击波，冲波，激波
shock zone 冲击区域，冲击区
shock 激波，震动，冲击，冲波，电击，冲击波，振动，激发，碰撞
shoe brake 闸瓦制动器
shoe-button cell 鞋扣电池
shoe connection 箍环接合
shoe-elbow 底弯肘管
shoegear 受电靴
shoe 极靴【电渣焊】，垫板，瓦形物，托，导向板，管头，闸瓦，轴瓦
sholder nipple 长丝
shooting flow 超临界流，射流
shooting trouble 找出错误，查明故障
shooting 发射，射击，找出，查明，放射，照射【射线照相】
shoot through 不灭弧，不熄弧，贯通
shoot 滑槽，斜槽
shop-assembled boiler 组装式锅炉
shop-assembled 工厂整装的，快装式的
shop assembly 工厂装配
shop bill 工厂材料单
shop-built form 厂制模板
shop crane 厂用起重机
shop detail drawing 工厂加工详图
shop detailing 放样
shop drawing 制造图，工作图，生产图，加工图
shop driven rivet 厂铆铆钉
shop erection 厂内安装
shop-fabricated boiler 快装锅炉
shop-fabricated component 车间加工的部件
shop-fabricated equipment 工厂加工的设备
shop-fabricated member 工厂预制构件
shop fabrication 车间（加工）制造
shop flat rail cart 车间运料平车
shop inspection department 车间检验工段
shop inspection 工厂检验
shop-labor cost 工厂装配成本
shop painting 工厂上漆
shop panel 车间油漆
shopping center 购物中心
shop preassembly 厂内预安装，厂内预组装
shop quality release 车间质量证书
shop rivet 工厂铆合的铆钉
shop splice 工厂拼接
shop test assembly 车间试验台
shop test 车间试验，工厂试验，出厂试验
shop traveler 流动卡片（车间），车间行车
shop-welded connection 厂焊接头
shop-welded 车间焊接的，工厂焊接的
shop weld 车间焊接，工厂焊接，厂内焊接
shop 车间，商店
shore beacon 岸标
shore bridge 顺岸栈桥
shore current 岸边流
shore cutting 岸边冲刷，海滨冲刷
shored composite beam 有支撑的叠合梁
shore deposit 岸边淤积，沿岸沉积
shore-end cable 浅海电缆，海底电缆
Shore hardnessn umber 肖氏硬度值
Shore hardness scale 肖氏硬度标度
Shore hardness tester 肖氏硬度计
Shore hardness test 肖氏硬度试验
Shore hardness 肖氏硬度
shore ice belt 岸冰
shore intake 河岸进水构筑物
shoreline denudation 海岸剥蚀
shoreline feature 海岸线特征
shoreline fumigation 海岸线熏烟
shoreline profile 岸线纵断面
shoreline regression 坍岸
shoreline 岸边线，岸线，滨线，海岸线，海滨线
shore material 海岸物质
shore mattress 护岸柴排
shore pier 岸墩
shore platform 海滨平台
shore profile 海岸剖面
shore progress 海岸进展
shore protection embankment 海岸堤防
shore protection facility 海岸防护设施
shore protection structure 海岸防护建筑物
shore protection works 海岸防护工程，护岸工程
shore protection 海岸防护，护岸
shore reef 岸礁
Shore scleroscope 肖氏硬度仪
shore stabilization 海岸稳定，海滨稳定
shore structure 护岸建筑物
shore subsidence 海岸坍陷
shore terrace 岸边台地，海岸阶地，海滨阶地
shore to shore clause 陆运条款
shoreward mass transport 向岸质量运输
shoreward 朝岸（的），向岸的
shore wave recorder 岸用波浪自记仪
shore wind 海岸风
shore 岸，滨，支撑，支柱
shoring layout 支撑布置

shoring method 支撑方法
shoring of foundation 基础支撑
shoring of trench 基坑支撑
shoring sheeting 临时木板支撑
shoring support 支撑
shoring 临时支撑，支撑，支柱，支架
short-access storage 快速存取存储器
short-access 快速存取
shortage in weight 重量不足
shortage of earth 缺土
shortage of manpower 人力短少
shortage 不足，不足额，缺少量，短缺，缺乏，不足，缺少
short barrel lug 短筒接线片
short bar 短路棒
short blow 快速排污
short-bore generator 短铁芯发电机
short brittle 热脆
short-bunker unloading system 短煤斗卸煤系统
short burnout 短焰燃烧
short-chorded coil 短距线圈
short-chord winding 短距绕组
short circuit admittance 短路导纳
short circuit analysis 短路分析
short circuit armature time constant 短路电枢时间常数
short circuit armature 鼠笼电枢，短路电枢
short circuit between conductors 线间短路
short circuit between lines 线间短路
short circuit breaking capacity 短路截断容量
short circuit brush 短路电刷
short circuit calculation 短路电流计算，短路计算
short circuit capability 短路容量
short circuit characteristic curve 短路特性曲线
short circuit characteristic 短路特性
short circuit coil 换向线圈，短路线圈
short circuit condition 短路状态
short circuit contact 短路触点
short circuit current capability 短路电流允许值
short circuit current density 短路电流密度
short circuit current gain 短路电流增益
short circuit current level 短路电流水平
short circuit current ratio 短路电流比
short circuit current rush 短路电流冲击
short circuit current withstand capability 短路电流耐受能力
short circuit current 短路电流
short circuit curve 短路特性曲线
short circuit detector 短路检测器
short-circuited armature coil 短路电转子线圈
short-circuited armature 短路电枢，鼠笼电枢
short-circuited circuit 短路电路
short-circuited feeder 短路馈线
short-circuited impedance 短路阻抗
short-circuited inductance 短路电感
short-circuited loop 短路回线
short-circuited resistance 短路电阻
short-circuited rotor bar 短路转子导条，鼠笼条
short circuited rotor 短路式转子，鼠笼式转子
short-circuited turn 短路线匝
short-circuited winding 鼠笼绕组，短路绕组
short-circuited 短路的
short circuit end 短路端，短路侧
short circuit equivalent test 短路等效试验
short circuiter 短路器，短路装置
short circuit force 短路力
short circuit generator 短路发电机
short circuit impact 短路冲击
short circuit impedance 短路阻抗
short circuit indicator for distribution line 配电线路短路指示器
short-circuiting arc 短路
short-circuiting bar 短路线棒
short-circuiting ring 短路环
short-circuiting terminal block 短路接线板
short-circuiting transfer 短路过渡
short-circuiting 短路，短路循环，发生短路
short circuit in winding 绕组短路
short circuit level 短路电流
short circuit load 短路荷载
short circuit making and breaking test 短路关合和开断性能试验
short circuit method 短路法
short circuit operation 短路运行
short circuit output admittance 短路输出导纳
short circuit period 换向周期，短路持续时间
short circuit plate 短路板
short circuit plug 短接接头
short circuit power 短路功率
short circuit proof 防短路的
short circuit protection device 短路保护装置
short circuit protection 短路保护
short circuit protector 短路保护装置，短路保护器
short circuit rating 短路额定值，开断能力
short circuit ratio 短路比，短路率
short circuit relay 防止短路继电器
short circuit ring 短路环
short circuit spark 短路火花
short circuit step 短路级
short circuit strength 短路强度
short circuit surge 短路涌流
short circuit termination 短路终端
short circuit test 短路试验
short circuit time 短路时间
short circuit torque 短路转矩
short circuit transfer admittance 短路转移导纳
short circuit transition 短路过渡过程
short circuit transmission line 短路传输线
short circuit turn 短路线匝，短路匝
short circuit voltage 短路电压
short circuit winding 短路线圈，短路绕组，鼠笼绕组
short circuit withstand test 耐短路试验
short circuit 短路
short column 短柱
shortcoming 毛病，缺点
short control rod 短控制棒
short cooled fuel 短期冷却的核燃料
short-crested wave 短脊波，短峰波
short-cut calculation 简化计算
short-cut multiplication 简化乘法

short-cut 简化，短路，捷径，省力省时的办法，快捷之法，提供捷径的，有近路的
short cylindrical shell 短圆柱形壳
short damping winding 短阻尼绕组
short delivery 短交（货）
short drive shaft 短传动轴
short duration gust 短期阵风
short duration loading 短期荷载
short duration 短历时，短时
shorted turn fault 匝间短路故障，线匝短路故障
shorted turns indicator 匝间短路指示器
shortened blade 削短叶片
shortened pulse 削短脉冲
shortened river section 裁直河段
shortening of pitch 缩短节距
shorten 缩短
shortest path problem 最短路径问题
shortest processing time 最短的处理时间
shortest route problem 最短途问题
short fall 不足，缺少，亏空，欠缺
short feedback admittance 短路反馈导纳
short flame coal 短焰煤
short-flaming 短焰的
short footed 短柄的
short format instruction 短型指令
short gravity wave 短重力波
short-handled tool 短柄工具
short-handled unlatching tool 短柄拆解工具
short haul aircraft 短程飞机
short haul call 短距离通话
short haul fare 短途票价
short haul ferry 短程渡轮
short haul flight 短途航班
short haul route 短程航线
short haul transit 短途运输
short haul 短距旅行，短程运输，短途托运，短途的，近距离的
short heavy swell 六级涌浪
short hole method 浅孔爆破法
short-hour motor 重复短时工作电动机，短时工作电动机
shorting bank contact （电话的）短路触排接点
shorting out with jumpers 用跨接线短路
shorting switch 短路开关
short input admittance 短路输入导纳
short iron 脆性铁
short irradiated 短期照射过的
short-landed cargo 短卸货物
short lease 短期租赁
short-lever armature 短杆衔铁
short life intermediates 短寿命中间产物
short-life 不耐用的，短寿命的
short line fault breaking capability 近区故障开断能力
short line fault breaking current 近区故障开断电流
short line fault test 近区故障试验
short line fault 近区故障
short line 短线路
short list 供最后选择的候选人名单，人数已缩减的候选人名单，（最终）候选人名单，决选名单，短名单
short lived fission gas 短寿命裂变气体
short lived fission product 短寿命裂变产物
short lived isotope 短寿命同位素
short lived noble gas isotope 短寿命惰性气体同位素
short lived noble gas nuclide 短寿命惰性气体核素
short lived nuclide 短寿命核素
short lived radioactivity 短寿命放射性
short long-lived radionuclide 短寿命与长寿命放射性核素
short low swell 一级涌浪
short moderate swell 三级涌浪
shortness （金属等的）松脆，脆性，短缺，短小，简略，不足
short number 短数
short out 短路，使短路，缩减
short pattern valve 短型阀门
short period fluctuation 短期变动
short period forecast 短期预报
short period motor 短时工作电动机
short period overspeed 短期超速
short period seismograph 短周期地震计
short period wave 短周期波
short pipe 短管
short-pitch factor 短距系数
short-pitch winding 短距绕法，短距绕组
short positive integer 短的正整数
short-proof horn 灭弧角
short-proof ring 引弧环，短路防止环
short purchase order No. 短期购买订货号
short radius elbow 短半径弯头
short radius return 短半径 U 形弯头
short-range communication 短途通信，近距离通信
short-range forecast 短期预报
short-range investment 短期投资
short-range order parameter 短程序参量
short-range sprinkler 短射程喷灌机
short-range weather forecast 短期天气预报
short-run marginal cost pricing 短期边际成本定价
short separation bubble 分离短气泡，短泡分离
short sequence scheme 短程序方案，短程序系统
short series 短系列
short shipment 短装
short-shipped cargo 短装货物
short-shipped 漏［少］装
short-shunt compound machine 短并复激电机
short-shunt compound winding 内并联复励绕组，短并复励绕组
short-shunt 短分（路），短分流，短并
short shutdown 短期停堆，临时停堆，临时停机
short slag 短渣
short stroke engine 短行程发动机
short term concentration 短期浓度
short term contract 短期合同
short term cost 短期费用
short term credit overdraft 短期信用透支
short term cut-out wind speed 短时切出风速

short term dynamics 短时动态特性，短时动态
short term effect 短时效应，瞬时效应，近期效益
short term export credit insurance 短期出口信用险
short term exposure 短期接触，短期暴露，短时间照射
short term forecast 短期预报
short term foreign capital 短期外资
short term goal 近期目标
short term insurance 短期保险
short term investment 短期投资
short term load 短暂荷载
short term loan 短期贷款
short term memory 短期记忆
short term planning 短期计划
short term rate 短期利率
short term reactor power control 短时间反应堆功率控制
short term record 短期观测记录
short term station 短期测站
short term test 短时运行试验
short term trade subsystem 短期交易子系统
short term weather forecast 短期天气预报
short term wind data 短期风资料
short term 近期，短期
short test section wind tunnel 短试验段风洞
short time breakdown voltage 短时击穿电压
short time capability 短时能力
short time climatic variation 短期气候变化
short time constant 短时间常数
short time current 短时电流
short time delay 短延时
short time diffusion factor 短期扩散因子
short time duty 短时工作制，短时运行方式，短时功率，短时负载【旋转机械，马达等】
short time load 短时负载
short time memory device 短暂存储装置
short time MHD generator 短时磁流体发电机
short time overload capacity 短时过载能力
short time overload test 短时过载试验
short time parallel 短期并列
short time rated motor 短时定额电动机
short time rated output 短时额定出力
short time rating 短时负载额定值，短时额定出力
short time storage reservoir 短期调蓄水库
short time test 短期试验，快速试验
short time variation 短期变化
short time washout factor 短期冲洗因子
short ton 短吨，美吨
short transfer path 传递捷径
short wave adapter 短波适配器，短波附加器
short wave communication 短波通信
short wave converter 短波变频器
short wavelength radiation 短波辐射
short wave radiation 短波辐射
short wave reception 短波接收
short wave sharp cut-off colour glass filter 短波端锐截止型有色玻璃滤光片
short wave 短波
short 短的
shot bit 冲击式钻头
shot blast cleaning 喷丸清理，钢珠除灰
shot blast 喷丸处理，喷丸，喷砂
shot break 爆炸时间标志，爆炸信号
shot-cleaning device 钢珠除尘装置
shotcrete gun 喷混凝土枪
shotcrete lining 喷浆混凝土衬砌
shotcrete machine 混凝土喷射机，喷浆机，水泥砂浆喷射机
shotcrete-rock bolt-wire mesh support 喷锚网支护
shotcrete shell 喷混凝土壳体
shotcrete support 喷混凝土支护
shotcrete system 喷浆法，喷浆混凝土支护法
shotcrete 喷射混凝土，喷浆混凝土，喷混凝土
shot depth 爆炸深度
shot distribution and storage tank 弹丸分配和贮存箱
shot effect 散粒效应
shot filter tank 弹丸过滤箱
shot firing cable 引燃电缆，引炸电缆
shot firing wire 引线
shot firing 放炮，引爆
shot hole 爆破孔，炮眼
shot noise 散粒噪声，发射噪声，散射噪声
shot peening mechanics 喷丸加工技术
shot peening parameters 喷丸参数
shot peening 锤击法【冷加工件表面】，喷丸强化，珠击处理，喷丸处理，弹射增韧法，珠击法
shot point 爆炸点
shot retrieval tank 弹丸取回箱
shot well 小球孔道【快速停堆用】
shot 放炮
shoulder bag 挎包
shoulder ditch 坡肩截流沟
shoulder joint （机械手的）肩关节
shoulder line 路基边缘，路肩线
shoulder pivot （机械手的）肩关节
shoulder the responsibility for 承担……的责任
shoulder 路肩，山肩，台肩，凸肩
shoved moraine 推碛
shovel access 挖土机工作半径
shovel blade 铲刀
shovel bucket 挖土机铲斗
shovel-crane 挖掘起重两用机
shovel crawler 履带式单斗挖土机
shovel dipper 挖土机铲斗
shovelling machine 铲土机
shovel loader 支臂式装载机
shovel sprinkler stoker 铲式抛煤机
shovel stoker 铲式抛煤机炉排
shovel truck 汽车挖掘机
shovel-type loader 铲式装料机，斗式自动装载机
shovel 铲，铲车，铁铲，单斗挖土机，铁锹
shower basin 淋浴池
shower head （除氧器）淋盘，淋浴喷头
shower meteors 陨石雨
shower room 浴室

shower rose　淋浴喷头，喷头，花洒头
shower　淋浴，倾注，洒水，阵雨
SHP(standard hardware program)　标准硬件程序
shredder ring　破碎机环锤
shredder　破碎机，切碎机，撕碎机
shred　碎片
shrinkage allowance　收缩容许量
shrinkage and temperature reinforcement　收缩与温度钢筋
shrinkage bar　收缩钢筋
shrinkage cavity　缩孔
shrinkage coefficient　收缩系数，干缩系数
shrinkage crack　收缩裂缝，缩裂
shrinkage curve　收缩曲线
shrinkage degree　收缩度，收缩率
shrinkage factor　收缩因子，收缩因数
shrinkage-fit bearing　热装轴承
shrinkage fit　收缩配合，热套，红套
shrinkage heat　收缩热
shrinkage hole　缩孔
shrinkage index　收缩指数
shrinkage joint　伸缩接头，缩缝
shrinkage limit　收缩极限，收缩限度，缩限
shrinkage loss　收缩损失
shrinkage of concrete　混凝土的收缩
shrinkage of film　膜的收缩
shrinkage pipe　缩管
shrinkage porosity　缩松，松心
shrinkage ratio　收缩率
shrinkage redundance　收缩裕度
shrinkage reinforcement　收缩钢筋
shrinkage strain　收缩应变
shrinkage stress　套合应力，收缩应力
shrinkage tear　收缩拉裂
shrinkage test　收缩试验
shrinkage void　缩孔
shrinkage　收缩，减少，干缩，收缩量【材料】，消融
shrink defect　收缩缺陷
shrink disk　缩紧盘，收缩盘
shrink fit pressure　热套表面压力
shrink-fitted shell　热套筒体
shrink fit　过盈配合，红套配合，冷缩配合，收缩配合
shrinkhole　缩孔，气沟砂眼
shrinking core model　缩核模型
shrinking particle　缩小颗粒
shrinking ratio　收缩比，收缩率
shrink-mixed concrete　缩拌混凝土，细拌混凝土，二次拌和的混凝土
shrink on　红套，热套
shrinkproof　防皱，防缩的
shrink ring　热压轮圈，红套环
shrink stress　收缩应力，过盈应力
shrink-swell potential　潜在胀缩能力
shrink　变小，收缩
shroud band　围带
shroud can　（燃料组件）外套管
shrouded blade　有围带的叶片
shrouded impeller　有盖板的叶轮，闭式叶轮
shrouded rotor　有罩风轮
shrouded wheel　有围带的叶轮
shroud head　堆芯围筒顶盖
shrouding wire　包箍线，屏蔽线，拉筋
shroud ring　包箍，箍带，箍环，围带，复环
shroud segment　围带弧片
shroud stabilizer　堆芯围筒稳定器
shroud support　堆芯围筒支承
shroud tube storage pool　套管贮存池
shroud tube　套管
shroud　覆盖物，罩，套，覆盖，围带，屏蔽，覆板，护罩，围筒，管套，屏，遮板，屏蔽板，屏蔽套筒，（风轮）集风罩，罩盖，燃料元件稳流套，覆环
SHR＝shrinkage　收缩
shrubland　灌丛带
shrub　灌木，灌木丛
shrunk-on coupling　红套联轴节
shrunk-on disc　套装叶轮，红套叶轮
shrunk-on disk　热嵌叶轮【汽轮机】，红套叶轮，套装叶轮
shrunk-on gear ring　热装齿圈
shrunk-on ring carrier　冷缩配合环座，热装环座
shrunk-on ring　红套箍环
shrunk-on rotor　套装转子
shrunk package　压缩包装
SH(service hours)　运行小时
SH(Shore hardness)　硬度，肖氏硬度
SH＝superheater　过热器
SHTC(superheater steam temperature control)　过热器汽温控制，主汽温控制
shuffle　混合，缓慢移动，洗牌，推诿，推卸，搬移，搁置
shuffling device　燃料倒换装置
shuffling of fuel　燃料的调换，转换燃料，倒换燃料
shuffling position　倒燃料位置
shuffling procedure　倒换（燃料）程序
shuffling　搅乱，搅动，倒料
shunt admittance　并联导纳
shunt and series starter　串并联（换接）启动器
shunt bisection　并联中剖
shunt box　分流器箱
shunt capacitance　分路电容，并联电容
shunt capacitive reactance　并联容性电抗
shunt capacitor bank complex　并联电容器组成套装置
shunt capacitor bank connection　并联电容器组接线
shunt capacitor banks　并联电容器组
shunt capacitor　并联电容器，旁路电容器
shunt capacity　并联电容
shunt characteristic　并励特性
shunt circuit-breaker　分路断路器
shunt circuit　并联电路，分流电路，分路
shunt coil　并联线圈，并激线圈，分流线圈
shunt commutator motor　并励换向器式电动机
shunt compensating capacitor　并联补偿电容器
shunt compensation　并联补偿
shunt compensator　并联补偿装置
shunt connection　并联
shunt control　并联控制

shunt current 分路电流
shunt diode 旁路二极管
shunt displacement current 旁路位移电流
shunted capacitor motor 并联电容器电动机
shunted meter 分流电流表
shunted 分流的，分路的，并联的
shunt excitation coil 分励线圈
shunt excitation 并激，并励，并励磁
shunt-excited dynamo 并励（发）电机
shunt-excited machine 并励电机
shunt-excited 并励
shunt factor 分路系数
shunt feed antenna 并联馈电天线
shunt feedback 并联反馈
shunt feed 并联馈电
shunt field coil 并激线圈
shunt field loss 并激损耗
shunt field relay 磁分路式继电器，并励继电器
shunt field resistance 并激绕组电阻
shunt field rheostat 分路变阻器，并激磁场变阻器，磁分路变阻器
shunt field winding 并激绕组，并励绕组
shunt field 并励磁场，并激磁场
shunt generator 并激发电机
shunt impedance 并联阻抗，分路阻抗
shunting action 分流作用
shunting condenser 并联电容器，分路电容器
shunting effect 分路效应，分流作用
shunting engine 调车机车
shunting impedance 并联阻抗，分流阻抗
shunting in marshalling yard 编组站调车作业
shunting line 调车线
shunting point 分流点
shunting service 调车作业
shunting siding 调车线
shunting signal 调车信号
shunting sign 调车标志
shunting track 调车线，岔道
shunting train 调车列车
shunting yard 调车场
shunting 分接，分流，分路，并联
shunt lead 分流器引线
shunt motor 并激电动机
shunt of brush 电刷瓣
shunt off 分路，分流
shunt operation 并联运行，分路操作
shunt peaked circuit 并联建峰电路
shunt phase-advancer 并激进相机
shunt rail line 调车线，岔线
shunt ratio 分路比
shunt reactor 并联电抗器，分流扼流圈，分路电抗器
shunt release 分励脱扣器
shunt resistance 并联电阻
shunt resistor 分流电阻器，并联电阻
shunt resonance 并联谐振
shunt resonant device 并联谐振装置
shunt rheostat 并联变阻器
shunt ring 分流环
shunt transition 分路换接过程
shunt trip coil 并联跳闸线圈【控制棒驱动机构】，分励脱扣线圈
shunt tripping 电压跳闸，并联跳闸，联跳
shunt type current turbo flow meter 分流旋翼式流量计
shunt valve 分流阀，分水龙头，旁通阀
shunt winding 并励绕组，并联绕组，并激绕组，异联绕组，分流线圈，分路绕组，单绕组变压器原副边公用绕组
shunt wound arc lamp 分绕弧光灯
shunt wound exciter 并激励磁机
shunt wound generator 并激发电机
shunt wound motor 并励电动机
shunt wound 并励的，并联的，并绕的，并激的
shunt 并联，并励，分流，分路，分流器，分路器，（电刷的）刷瓣，并联的，调车，转辙器，分道叉，转轨
shutdown amplifier 停堆放大器
shutdown at emergency 紧急停机
shutdown at rating parameters 额定参数停机
shutdown at sliding parameters 滑参数停机
shutdown bank 停堆棒组
shutdown boron concentration 停堆硼浓度
shutdown circuit 停止电路
shutdown condensate pump 停堆凝结水泵
shutdown condition 非工作状况，停运状态
shutdown cooler 停堆冷却器
shutdown cooling pump 停堆冷却泵
shutdown cooling system 停堆冷却系统
shutdown cooling 停炉冷却，停堆后冷却，停机冷却，停堆冷却
shutdown curve at sliding parameters 滑参数停机曲线
shutdown curve 停机曲线
shutdown device 停机装置
shutdown diagnosis 停机诊断
shutdown feature （电流）切断装置，停止装置
shutdown heat 停堆余热
shutdown inspection 停工检查，关闭检查
shutdown margin 停堆深度，停堆裕度
shutdown of boiler 锅炉停运
shutdown of power plant 电厂停运
shutdown oil pressure 停机油压
shutdown operation 停止操作，停机操作，停堆操作，停止运行，停机
shutdown period 停工期，停运期，停机时期，停堆期，停用期
shutdown power 停堆功率，剩余功率
shutdown procedure （反应堆的）停堆步骤，停堆程序，停机程序
shutdown rate 停堆速率，关闭速率
shutdown RCC assembly 停堆棒束组件
shutdown reactivity margin 停堆裕度，停堆（反应性）深度
shutdown reactivity 停堆反应性
shutdown relay 停机继电器，断路继电器
shutdown rod bank 停堆棒组
shutdown rod drive 安全棒驱动机
shutdown rod group runaway 停堆棒组失控
shutdown rod 安全棒【反应堆用】，停堆棒
shutdown signal 停堆信号，断路信号
shutdown switch 断路开关

shutdown system 停机系统，停堆系统
shutdown temperature 停机温度
shutdown test 停机试验
shutdown time 停堆时间，停役时间，停用时间
shutdown unit 停堆装置
shutdown valve 停机阀
shutdown wind speed 刹车风速，停机风速
shutdown with variable parameter 滑参数停运
shut down 关机，切机，停运，停堆，停炉，停机【shot down 作动词词组用】
shutdown 停堆，停运，停车，停炉，停机，关闭，断路【shutdown 作名词用】
shut-in pressure 闭合压力
shutoff contact 断路接触器
shutoff damper 关断挡板，截流闸阀，截止挡板
shutoff device 断流装置
shutoff electromagnet 电磁关闭器
shutoff gate valve 截止阀
shutoff gate 截止门，挡板门，闸门
shutoff head conditions （风机或泵的）空转压头
shutoff head 关闭压头【泵或风机】，全闭压头，关闭扬程，截流水头，截流压头
shutoff mechanism 切断机构
shutoff operation （泵）关死运转，关断运行
shutoff period 停运时间
shutoff pressure 关闭压力
shutoff signal 断开信号
shutoff slide valve 关断滑阀
shutoff valve 断流阀，截止阀，危急截断器，关断阀，切断阀，关闭阀
shutoff wind speed 刹车风速
shutoff 断开，切断，关闭，关闭器，断路，停机，关掉，停堆，使不进入
shut out cargo 退关货物
shutter blind 百叶帘
shutter crack 裂纹
shutter dam 翻板坝
shutter door 百叶门
shutter gate 百叶式闸门，翻板闸门
shuttering project 模板工程
shuttering technology 滑模技术
shuttering works 模板工作，模板工程
shuttering 模板，立模，支架
shutter-proof glass 不碎玻璃
shutter vibration 调节板振动，模板振动，木条板闸门振动
shutter weir 横轴闸门堰，翻板堰
shutter 百叶窗，节气门，风门片，断流器，光闸，快门，闸板，节流门，调节板，翻板门，围堰，闸门，开关，控制板
shutting-down device 停车装置
shuttle armature 梭形电枢，H形截面电枢
shuttle belt conveyor 往复式移动带式输送机
shuttle cableway 往复移动式缆索道
shuttle car 往复式料车
shuttle conveyer 自走式［可逆式/往复式/摆动式］输送机，移动式胶带机，穿梭式运输机
shuttle head 伸缩头【皮带机】
shuttle ropeway 往复移动式索道
shuttle 航天飞机，来回运动，滑闸，梭，往复
SHY 氢气生产和分配系统【核电站系统代码】
sial 硅铝带
siamese joint 分叉接头，二重连接
Siberian land bridge traffic 西伯利亚大陆桥运输
siccative varnish 快干清漆
siccative 催干剂
SiC corrosion 碳化硅腐蚀
sickness leave 病假
sic passim 全书下同
SIC(semiconductor integrated circuit) 半导体集成电路
sidac 硅对称二端开关元件，交流硅二极管
side abutment pressure 边墩压力
side adjustment 边平差
side agreement 附属协议
side air admission （链条炉排）侧向进风
side-aisle 边侧通道
side anchor 边锚
side armature relay 边衔铁继电器
side armature 边衔铁
side arm 单面横臂【输电线】，单面横担
side band capacity 边带容量
side band interference 边带干扰，邻频道干扰
side band 边能带，边频带，边带
side bay （混凝土面板等的）边块
side beam mirror 一阶衍射光镜
side beam 副波束
side bend test 侧面弯曲试验
side-binding 侧面接合
side blocking 侧向封堵
side board 侧面挡板
side break switch 边断路开关
side bunker system 侧面煤仓系统
side by side reaction 并列反应
side by side 肩并肩【指高温堆分置式】
side car dumper 侧倾翻车机
side channel spillway 侧槽式溢洪道
side channel 边渠，侧槽
side-cleaning stoker 侧墙出渣式机械化炉排
side clearance 旁隙，（机）侧间隙
side condenser 侧排汽凝汽器
side contraction 侧边收缩，侧收缩
side cover pump 侧盖泵
side culvert 侧边涵洞
side-cutting pliers 侧切钳
side cutting 堤旁取土
side discharge 旁边排出，侧面排出
side discharging car 侧卸车
side ditch 边沟，侧沟，路边明沟
side door hopper barge 侧开式料斗船
side dozer 侧铲推土机
side drag （自航耙吸挖泥船的）边耙
side drainage 路边排水
side drain 路边排水沟
side drift method （隧洞开挖用的）边侧导坑法
side dump bucket 侧卸式料斗
side dump car 侧翻车，侧卸车
side dumper 侧卸卡车
side dump stoker 侧翻板加煤机
side dump trailer 侧卸拖车
side dump truck 侧卸卡车，侧卸式货车
side dump wagon 侧倾式翻斗车

side effect 副作用，边界效应
side elevation drawing 侧视图
side elevation 侧面图，侧视图，侧立面
side entrance manhole 旁侧进口人孔
side entry blade 侧装叶片
side entry serration root 侧装式锯齿型叶根
side equation 边线方程
side erosion 侧向侵蚀
side falling accident 片帮事故
side fed water wheel 侧射式水轮
side feet 两侧底脚
side-fired 侧墙燃烧的
side firing 侧面燃烧
sideflash 侧击雷，侧面放电
side-flow weir 侧流堰
side flow 旁通流，旁流
side force 侧向力
side form 边模
side frame of gate 闸门（门叶）边端构架
side frequency 旁频率，边频
side friction 侧面摩擦，侧向摩擦
side gear 边齿轮
side guide wheel 侧向导轮
side guide （闸门的）侧向导轨
side gust 侧阵风
side headwind 侧逆风
side hill fill 半路堑，山坡填土
side hill seepage 山坡渗流
side-hinged window 侧铰链窗
side hood 侧吸罩
side hung casement 平开窗
side hung door 开扇门，平开门
side hung window 平开窗
side illumination 侧面照明
side intake 旁侧进水口
side leading wind 侧顺风
side leakage 侧面泄漏
side light 边灯，侧向采光，舷灯
sideline 副业
sideling placed blade 斜置叶片
side loop 侧回路，旁路，侧环
side lurch 侧倾
side mixing nozzle 侧边混合喷嘴
side-mounted condenser 侧装式凝汽器
side of foundation pit 基坑侧壁
side outlet tee 侧向口三通，支流三通管
side outlet 侧向出口，侧泄水孔
side overfeed stoker 侧墙上饲式机械化炉排
side overflow 侧向溢流
side panel 边盘（板），边屏
side pavement 人行道
side plate 侧靠板
side projection 侧向投影
side rail 边梁
side rake 侧斜角，侧前角
side reaction 副反应
sidereal clock 恒星时钟
sidereal day 恒星日
sidereal month 恒星月
sidereal time 恒星时
sidereal year 恒星年
side reflected 带侧面反射层的【反应堆】
side reflector 侧面反射层【反应堆】
side restraint band 侧面箍带
siderite 兰石英，菱铁矿，陨铁
side roller 侧边滚轮，侧面滚子
side scanner 侧扫描器
sidescan sonar 旁侧扫描声纳
side seal 侧水封，边缘止水
side shield 侧面屏蔽体
side shore 侧撑
side shoring 边撑
side slicing 侧切片
sideslip angle 侧滑角
sideslip derivative 侧滑导数
sideslip 侧滑
side slope 边坡
side span 边跨
side spillway dam 旁侧溢流坝
side spillway 岸边溢洪道，旁侧溢洪道
side spin 侧旋
side-stable relay 边稳定极化继电器
side-stepping 径向脉动
sidestream filtration 旁流过滤
sidestream treatment 分流处理
sidestream 分流，旁流
side suction 侧吸
side support 侧面支承
side sway 侧摆，侧移，侧倾
side thermal shield 侧面热屏蔽体
side thrust 侧压，侧向力，侧推力，侧向推力
side-tipping skip 倾卸翻斗车
side-tipping wagon 侧倾式翻斗车
side-tip wagon 侧翻车
side-to-side vibration 左右振动
side track 侧线
side trench 边沟
side tripper 侧倾车
side tube 支管
side valve engine 侧阀发动机
side valve 旁路阀
side vane 侧翼【在风轮侧面利用风压使风轮偏离风向的机构】
side view 侧面图，侧视图
sidewalk 边道，步道，侧道，人行道
sidewall air supply 侧面送风
sidewall box 防焦箱
sidewall enclosure superheater 侧包墙过热器
sidewall insert （风洞）侧壁插件
sidewall 岸墙，边墙，侧墙，风洞侧壁，井壁
side wash （气流）侧洗
side water wall 侧墙水冷壁
side wave 侧波
sideways motion 横向运动
sideways sum 数位叠加和
side wedge 侧向槽楔
side weir 侧堰
side winding 侧面绕组
side wind 侧风，横风
sidewise buckling 侧向翘曲
sidewise restraint 侧向约束
side 边，侧，侧面，方面

siding for ash handling 除灰专用线
siding frame 挡板架
siding wall 板墙
siding 侧线，岔线，专用线，挡板，覆盖层，覆层，围护板
Siemens armature H形截面电枢，西门子电枢
Siemens （德国）西门子公司
SIE(single instruction execute) 单指令执行
sieve-analysis curve 筛分曲线
sieve analysis method 筛选法
sieve analysis of coal 筛分分析
sieve analysis test 筛析试验，筛分析试验，筛分试验
sieve analysis 过筛分析，筛分分析，筛分析，筛析，粒度测定
sieve cloth 筛布
sieve curve 筛分曲线
sieve diameter 筛孔直径
sieved particles 筛余颗粒
sieve mesh 筛目，筛眼
sieve method 逐步淘汰法，筛法
sieve number 筛号，筛孔，筛眼
sieve plate column 筛板塔，筛板柱
sieve plate 筛板
sieve problem 筛问题
sieve residue 筛余物，筛渣
sievert 西弗特【缩略词为 Sv，剂量当量单位】
sieve series standard 筛号标准
sieve series 筛制
sieve size 筛孔尺寸，筛眼尺寸，筛号
sieve test 过筛试验，筛分试验
sieve trays column 筛板塔
sieve tray 筛盘，筛板
sieve 筛，筛子，筛分，过滤器，筛滤器，筛选
sieving test machine 筛分机
sieving 过筛，筛分，粒度测定，粒度测定术
SIF(safety instrumented functions) 安全仪表功能
sifter 筛子，筛，滤波器
sifting 过筛，筛分，筛选，挑选，细分筛，筛屑，炉排漏煤
sift 筛分，过筛，筛，筛选，挑选，细查
sight alidade 视准仪
sight angle 视角，视线角
sight aperture 观测孔
sight axis 视轴
sight bill 即期汇票，即期票据
sight control 目视检查，直观检查，直观控制
sight distance 视距
sight draft 即期汇票
sight flow glass 流量观测镜，玻璃流量计
sight flow indicator 流量指示器
sight for sag 垂度计，垂度仪
sight gauge 观测计
sight glass oiler 窥视玻璃加油器
sight glass window 窥视窗
sight glass 观察窗，观察孔，观察玻璃，窥视玻璃，监视孔，窥视镜，窥镜，视镜
sight hole 窥视孔，观察孔，检查孔，视孔
sighting board 觇板，觇标，测视板
sighting device 照准设备
sighting disc 觇板
sighting error 瞄准误差，视准误差
sighting gear 照准器
sighting line 视线
sighting mark 测标，照准标
sighting pendant 瞄准锤
sighting piece 观测件
sighting point 视准点
sighting rod 花杆
sighting target 测标，觇标，观测目标
sighting tube 水位表管
sighting wire 照准丝
sight inspection 目测，目检，外观检查
sight letter of credit 即期信用证，即期信用状，见票即付的信用证
sight level gauge 可见水位计，透明管水位计
sight level indicator 可见水位指示器
sight level 水准仪
sight line 视线
sight plane 视准面
sight point 瞄准点，视点
sight pole 标杆花杆
sight port 窥视孔，检查孔，观察窗
sight rail 视准轨
sight rule 觇尺
sight size 透光孔口
sight vane alidade 测斜照准仪
sight vane of fore side 前视准板
sight vane 觇板
sight window 观察窗，窥视窗
sight 观测，瞄准器，视线，视域
sigma phase σ相
sigma pile σ反应堆
sigma welding 惰性气体保护金属极电弧焊
sigmoid curve 剂量响应曲线，S形曲线
sign a contract 签合同，签约
sign after receiving 签收
signal ability 信号监测能力
signal alarm bell 信号警铃
signal alarm 信号报警，音响信号
signal amplifier 信号放大器
signal amplitude sequencing 信号幅值顺序控制
signal and display 信号和画面
signal anemometer 信号风速计
signal aspect 信号方式
signal bell 信号铃，警铃
signal board 信号盘
signal booster 信号前置放大器
signal cabin 信号室
signal cable 信号电缆
signal channel 信号通道
signal characterizer card 函数卡
signal characterizer 信号表征器
signal circuit 信号电路
signal clock 信号钟
signal code 信号码
signal common 信号公共点
signal component 信号分量
signal conditioner 信号调制器
signal conditioning 信号处理
signal control relay 信号控制继电器
signal converter 信号转换器

signal correlation 信号相关
signal curbing 信号抑制
signal data converter 信号数据转换器
signal detection and estimation 信号检测和估计
signal direction 信号方向
signal distortion 信号失真，信号畸变，符号失真
signal distribution component 信号分配组件
signal distribution panel 信号分配盘
signal distribution 信号分配
signal duration 信号持续时间
signal element 信号元件，信号元素
signal equipment 信号设备
signaler 信号员，信号设备
signal extraction 信号提取
signal feed-back panel 信号返回屏
signal flag 信号旗
signal flare 闪光信号
signal flow chart 信号流图
signal flow diagram 信号流图，信号流程图
signal for blocking the track 闭塞信号
signal forming time 信号产生时间，信号发生时间
signal forming 信号发生，信号形成
signal frequency 信号频率
signal function 信号函数
signal generator 信号发生器
signal graph 信号流图
signal ground 信号接地
signal identification 信号识别
signal indicator 信号器，信号指示器，信号指示灯
signaling alarm equipment 信号报警设备
signaling alphabet 信号字母
signaling channel 信令通道
signaling equipment 信号装置
signaling information 信令消息
signaling key 信号电匙
signaling light 信号灯
signaling link 信令链路
signaling module 信令模块
signaling network 信令网
signaling path 信令通道
signaling route 信令路由
signaling routing 信令路由选定
signaling security 信令安全性
signaling set 信号设备
signaling speed 信号发送速度，通信速度，发码速度
signaling system 信号系统，信令系统
signaling test 发信试验，振铃试验，信号发送试验
signaling 信号设备，发信号，传信，信号装置，信号化
signal intensity 信号强度
signal interface 信号接口
signal interpretation 信号解释
signal isolation 信号隔离
signal lag 信号延时
signal lamp 信号灯
signal level distribution 信号电平分布
signal level 信号功率，信号级，信号电平
signal light 信号灯
signal-like density distribution 类似信号的密度分布
signal malfunction 信号误动作
signal message 信号消息
signal meter 信号指示器
signal modulator 信号调制器
signal monitor 信号监视器
signal noise ratio 信噪比，信号噪声比
signal optimization program 信号优化程序
signal output 信号输出
signal panel 信号板，信号盘
signal preservation 信号保持
signal processing board 信号处理板
signal processing 信号处理
signal pulse 信号脉冲
signal quality detection 信号质量检测
signal reading 信号读出
signal receiver 信号接收器
signal reconstruction 信号重构
signal relay 信号继电器
signal repeater 信号中继器，信号复示器
signal running time 报警延续时间，信号持续时间
signal saturation 信号饱和
signal scaling 信号标定
signal scanner 信号扫描器
signal selector relay 信号选择继电器
signal selector 信号选择器
signal shielding 信号铠装，信号屏蔽
signal standardization 信号标准化，信号整形
signal status 信号状态
signal-strength meter 信号强度计，信号强度指示器
signal-to-disturbance ratio 信骚比
signal-to-noise ratio 信号噪声比，信噪比
signal tower 信号塔
signal tracer 信号示踪器，信号故障寻找，信号故障寻找器
signal transfer point 信号转接点
signal transferring lag 信号传输滞后
signal transformation 信号变换
signal transformer 信号变换器，信号变压器
signal transmission 信号传输
signal voltage 信号电压
signal winding 信号绕组，控制绕组
signalyzer 电路调整和故障寻找综合试验器，信号分析器
signal 信号器，信号，发信号，记号，标志，标记，显著的，信号的，发信号
sign and issue 签发
sign a referendum contract 草签合同
signatory of a contract 合同签字方
signatory party 签字方
signatory 签约国，签名人，签约方，签字人，签署的，签约的
signature data processing system 特征数据处理系统
signature extraction 特征提取
signature 签名（字），签署，署名
sign bit 符号位

signboard 标价牌，路标
sign changer 符号变换器，正负变换器
sign changing unit 信号变换部分
sign character 符号字符
sign check indicator 符号检验指示器
sign contract 签署合同
sign-control flip-flop 符号控制触发器
sign-controlled circuit 符号控制电路
sign-controlled 符号控制的
sign convention 符号规约，符号规定
sign detection 符号检查
sign digit 符号数字，符号位
signed agreement 签署的协议
signed and sealed 签名盖章
signed contract 签订的合同
signed decimal 带有符号的十进制数
signed field 带符号字段
signed 签过名的，带有符号的，已签署的
signer 签字人
sign flip-flop 符号触发器
significance arithmetic 有效位运算
significance of digit position 数位的有效性，有效位
significance test 显著性检验，显著性试验
significance 有效位，有效数，显著性，重要性，意义
significant achievement 显著成绩
significant condition adverse to quality 对质量不利的重要状况
significant contribution 重大贡献
significant defect 重大缺陷
significant deficiency 重要偏差【质量保证】，重要差错，重大缺陷
significant depth 有效深度
significant digit 有效数字，有效位
significant dimensions 有效尺度，特征尺寸，重要量纲
significant error 显著误差
significant event evaluation and information network 重要事件评价和通报网络
significant event report 重要事件报告
significant figure 有效数字，有效位
significant hydraulic parameter 特征水力参数
significant information pertaining 有关重要资料
significant operating experience report 重要运行经验报告
significant quantity 显著量，重要量
significant wave height 有效波高
significant wave length 有效波长
significant wave period 有效波周期
significant wave 有效波
significant 有效的，显著的，有意义的，重要的，灵敏的，值得注意的，数的有效部分，象征，有意义的事物
signify 表明，符号化
signing ceremony 签字仪式
signing of agreement 签订协定［协议］
signing of bid 标书的签署
signing of contract 签订合同，签约
signless integer 无符号整数，正整数
sign modification 符号改变
sign-off （终端用户）停止工作，去掉符号，符号结束
sign of inequality 不等号
sign of operation 运算符号
sign of rotation 转动符号，旋转方向
sign-on 加上符号
sign position 符号位，符号位置
sign pulse 符号脉冲
sign reversing 反号，符号变换
sign test 符号检验
signum 正负号函数
sign 标志，符号，迹象，牌子，签订，签字，签名，署名，签署
SIGOP(signal optimization program) 信号优化程序
silt-covered 淤泥覆盖的
silane 硅烷
silastic 硅橡胶
silastomer 硅塑料
silchrome steel 硅铬耐蚀耐热钢
silding pressure operation of deaerator 除氧器滑压运行
silence cabinet 隔音室，隔声室
silencer chamber 减声室
silencer cover 消声罩
silencer for gas turbine 燃气轮机消声器
silencer 消声器
silence signal 停机信号
silence zone 平静区
silence 禁鸣喇叭，消音
silent arc 无声电弧，静弧
silent discharge 无声放电
silent motor 无噪声电机
silent pile driver 无声打桩机
silent pile driving 无声打桩法
silent running 无声运行
silent tuning 无噪调谐
silent 无噪声的，安静的
silex glass 石英玻璃
silex 燧石
silhouette 黑色轮廓像
silica analyser 硅分析器，硅表
silica-bearing material 含硅轴承材料
silica block 硅砌块，硅酸盐砌块
silica brick 硅酸盐砖，硅砖
silica cement 硅石水泥
silica coated lamp 涂硅灯泡
silica content 含硅量
silica deposit 硅酸盐沉积，二氧化硅沉积
silica exchange capacity 硅交换容量
silica gel absorber 硅胶吸附剂
silica gel drier 硅胶干燥器
silica gel filter 硅胶过滤器
silica gel 硅胶
silica glass 石英玻璃
silica-laden steam 带二氧化硅的蒸汽
silica lamp 石英水银灯
silica-meter 硅表
silica purge 二氧化硅清洗
silica rectifier 硅整流器
silica refractory 硅土耐火材料

silica removal 除硅
silicarenite 石英砂岩
silica sand 硅砂，石英砂，硅质砂
silica sheet 硅氧片层
silicate brick 矽砖
silicate cement 硅酸盐水泥
silicate fouling 二氧化硅污染
silicate industry 硅酸盐工业
silicate of lime 硅酸钙
silicate scale 硅酸盐水垢
silica tetrahedron 硅氧四面体
silicate 硅酸盐
silication 硅化作用
silica wool 石英棉
silica 硅石，二氧化硅
siliceous aggregate 硅质骨料
siliceous limestone 硅质灰岩
siliceous phyllite 硅质千枚岩
siliceous sandstone 硅质砂岩
siliceous schist 硅质片岩
siliceous shale 硅质页岩
siliceous 含硅的，硅质的，硅酸的
silicic acid anhydride 硅酸酐，二氧化硅
silicic acid （正）硅酸
silicide 硅化物
silicification grouting 单液硅化法
silicification 硅化加固，硅化作用
silicified rock 硅化岩
silicium dust 含矽粉尘
silicium rectifying unit 硅整流装置
silicium 硅
silicized carbon 碳化硅，金刚砂
silicoformer 硅变压整流器
silicon alloy diode 硅合金二极管
silicon alloy 硅合金
silicon barrier detector 硅垒探测器
silicon bronze 硅青铜
silicon carbide coated fuel 碳化硅包覆核燃料
silicon carbide heating rod 硅碳加热棒
silicon carbide lamp 碳化硅灯
silicon carbide varistor 碳化硅压敏电阻
silicon carbide 金刚砂，碳化硅
silicon cell 硅电池
silicon controlled inverter 可控硅逆变器
silicon controlled rectifier element 可控硅元件
silicon controlled rectifier fast acting fuse 可控硅快速熔断器
silicon controlled rectifier 可控硅整流器，可控硅
silicon crystal 硅晶体
silicon diode excitation 硅二极管（整流器）励磁
silicon diode 硅二极管
silicon dioxide 二氧化硅，硅石
silicon earth 硅土
silicone-based insulating oil 硅有机基绝缘油
silicone electrical insulant 有机硅电绝缘材料
silicone foam 硅树脂泡沫
silicone grease 硅酮润滑脂，硅脂
silicone insulation 硅树脂绝缘
silicone lacquer 硅树脂漆
silicone neuron 模拟神经元硅片
silicone rubber insulation 硅橡胶绝缘
silicone rubber tube 硅酮橡皮管
silicone rubber 硅橡胶，硅酮橡胶
silicone 硅有机树脂，硅酮
silicon fusion transistor 硅熔接晶体管
silicon iron lamination 硅铁叠片，硅铁片
silicon iron 硅铁
siliconized plate 硅钢片
silicon-organic lacquer 硅有机清漆
silicon oxide 二氧化硅
silicon photodiode 硅光电二极管
silicon photoresistor 硅光敏电阻器，硅光电晶体管
silicon photovoltaic cell 硅光电池
silicon point-contact diode 点接触硅二极管
silicon precision alloy transistor 硅精密合金晶体管
silicon rectified arc welding machine 硅整流弧焊机
silicon rectifier 硅整流器
silicon ribbon solar cell 带硅太阳电池
silicon sheet varnish 硅钢片漆
silicon solar cell 硅太阳能电池
silicon solid circuit 硅固体电路
silicon stack 硅堆
silicon steel lamination 硅钢片
silicon steel plate 矽钢片，硅钢片
silicon steel sheet 硅钢片，矽钢片
silicon steel 硅钢
silicon surface barrier detector 硅面垒控测器
silicon tetrafluoride 四氟化硅
silicon transistor 硅晶体管
silicon unijunction transistor 硅单结晶体管
silicon unilateral switch 硅单向开关
silicon varistor 硅压敏电阻器
silicon welding machine 硅焊机
silicon 硅
silico-rudite 硅砾岩
silistor 硅电阻器
silit resister 碳硅电阻器
silk covered wire 丝包线
silk insulation 丝绝缘
silk screening 丝网（印刷电路）法
silk streamer 气流观察丝线带
sill anchor 地脚锚栓
sill beam 侧靠车梁
sill course 窗台层
sill for jump control 控制水跃的底槛
sillimanite brick 硅线石砖
sillimanite 硅线石
sill plate 门槛板，木骨架的底木条
sill 窗台，门槛，岩床，基面，基石，底槛，垫底横木
silo bay 煤仓间
silo cell group 圆形群仓
silo cell 仓筒
silo-cone slope 筒仓锥体斜度
silo pressure 筒仓压力，贮仓压力
silo storage 筒仓贮存
silo with wedge-shaped hoppers 具有楔形漏斗的筒仓
siloxicon 氧碳化硅，硅碳耐火材料

silo 仓，筒仓，地坑，地下室，料仓，圆筒形储仓，地下仓库
SIL rating of ESD system 紧急停车系统的安全整体性等级
SIL(safety integrity level) 安全完整性等级，安全综合水平
SIL(speech interference level) 话音干扰电平
SIL(switching impulse level) 操作冲击水平
silt accumulation 泥沙淤积
silt and expanded metal absorber 狭缝和多孔金属网吸热器
silt arrester 拦沙设备，拦淤设备
siltation 淤积，淤塞
silt box 泥沙（拦截）箱
silt-carrying capacity 携沙量，挟沙能力
silt-carrying river 多沙河流
silt charge 含沙量，含泥量
silt concentration 含沙量
silt content 含沙量，含泥量
silt density index 淤泥密度指数
silt deposition 泥沙淤积
silt discharge capacity 排沙能力
silt-discharge gate 冲沙闸
silt-flushing sluice 放淤闸
silt fraction 粉粒粒组
silt gravel 粉砾
silting basin 沉沙池
silting up 淤积
silting 沉沙，淤积
silt-laden flow 含泥水流
silt-laden river 多泥沙河流
silt layer 粉沙层，淤泥层
silt lens 淤泥透镜体
silt loam 粉壤土
silt orifice 排泥沙孔口
silt pressure 淤沙压力
silt-releasing sluice 放淤闸
silt seam 淤泥层，淤泥夹层
siltstone 粉砂岩，泥砂岩
silt storage capacity 拦沙能力
silt stratification 泥层，淤泥层
silt stratum 粉土层
silt-trap dam 拦沙坝
silt trap scour gate 砂阱式冲砂闸门
silt up 淤塞
silty clay loam 粉砂质黏壤土，粉质黏壤土
silty clay 粉砂质黏土，粉质黏土
silty gravel 粉质土砾
silty loam 粉砂壤土，粉砂质垆姆，粉质垆姆
silty sand 粉砂
silty soil 粉砂壤土
silt 泥沙，淤泥，粉砂，淤泥（砂）的
silumin alloy 硅铝明合金，铝硅铸造合金
siluminite 石棉或云母基复合绝缘材料
silver bromide photocell 溴化银光电池
silver-cadmium contact 银镉触点
silver chloride 氯化银
silver cladding 银镀层
silver electrode 银电极
silver frost 银光霜
silver fuse 银熔线
silver grain 银粒
silver halide 卤化银
silver-hydrogen battery 氢银电池
silvering 镀银
silver-jacketed wire 镀银线
silver medal 银质奖
silver microfilm 银盐缩微胶卷
silver paste 银浆
silver-plated copper 镀银铜
silver plated 镀银
silver-silver chloride electrode 银氯化银电极
silverstat regulator 接触式调节器，银触头电压调整器
silverstat relay 银触头继电器
silvertoun testing set 电缆故障测定设备
silvertoun 电缆故障寻找器
silver voltameter 银电解伏打计，银电解式电量计
silver-zinc battery 锌银电池
silver-zinc reserve battery 锌银储备电池
sima 硅镁层
SIMD(single instruction multiple data) 单指令多数据
similar condition 相似条件
similar flow 相似水流
similar geometry wind tunnel 几何相似风洞
similarity condition 相似条件
similarity coordinate 相似坐标
similarity criterion 相似准则
similarity factor 相似系数
similarity law 相似定律，相似性定律
similarity level 相似性水平
similarity measure 相似性测度
similarity parameter 相似参数，相似准数，相似准则
similarity principle 相似原理
similarity relation 相似关系
similarity rule 相似规则，相似率
similarity solution 相似性解
similarity theory of turbulence 紊流的相似理论
similarity theory 相似理论
similarity transformation 相似变换
similarity variable 相似变量
similarity 类似，相似点，相似性
similarly hereinafter 下同
similar matrix 相似矩阵
similar number 相似准则数
similar power plant 同类型电厂
similar project 类似项目，同类工程
similar size power plant 相似规模的电厂
similar system of numbers 相似数系
similars 相似导体，相似导线
similitude law 相似（定）律
similitude method 相似法
similitude principle 相似原理
similitude 比拟，比喻，相似，类似，相似物，相似性
simmer 徐沸，（阀门）前泄，抖动，（安全阀）迸发喷射，即将爆发，即将沸腾的状态
simple alternating current 简谐电流，正弦交流
simple arithmetic expression 简单算术表达式

simple beam 简支梁
simple bending 单纯弯曲，纯弯曲
simple compression 单向压缩
simple contract 简易合同，简单契约，单纯合同，无签名的合同，简约
simple credit 简单信用证，单纯信用证，扣账信用证
simple cycle gas turbine 简单循环燃气轮机
simple cycle 简单循环
simple degenerate kernel 简单退化核
simple design 结构简单
simple dimensional method 单量纲法
simple direct connection 简单直接连接
simple doubling time 简单倍增时间
simple electronic computer 便携式电子计算机
simple engine 单式引擎，单级发动机
simple fan 单级风机
simple feeding system 直接供电系统
simple gas turbine cycle 简单燃气轮机循环
simple gauge relation 单一水位关系
simple grid 简单格架
simple harmonic current 简谐电流
simple harmonic law 简谐定律，正弦定律
simple harmonic motion 简谐运动
simple harmonic quantity 正弦量
simple harmonic vibration 简谐振动
simple harmonic wave 简谐波
simple header boiler 整联箱式锅炉
simple interest 单利
simple key paging 单键翻页
simple lag network 简单相移网络
simple licence contract 普通许可（证）合同
simple majority vote 简单多数票
simple network time protocol 简单网络时间协议
simple one-constituent bed 单组分床
simple open-shut 简单的开关
simple parallel winding 简单并联绕组
simple pendulum 单摆
simple potentiometer 线性电势计，线性电位差计
simple process factor 单一分离因子【浓缩】，单级过程分离系数，简单过程因子
simple program structure 简单程序结构
simple radial equilibrium 简单径向平衡
simple retrofit 简单改装
simple RLC parallel circuit 简单的 RLC 并联电路
simple sedimentation 简单沉降
simple self-excitation system 简单自励系统
simple series winding 单串联绕组，波绕组
simple servomechanism 简单伺服机构
simple shaded pole 普通屏蔽极，普通罩极
simple shear apparatus 单剪仪
simple shear flow 简单剪切流动
simple shear 单剪
simple signal 简单信号，单频信号
simple sinusoidal current （简单）正弦电流，简谐电流
simple sinusoidal electromotive 正弦电动势
simple sinusoidal quantity 正弦量
simple storage system 单独蓄水系统
simple strain 纯应变
simple stress 单轴应力，简单应力
simple support grid 单一支撑格架
simple system structure chart 系统结构简图
simple trestle 独立式支架
simplex lap winding 单叠绕组
simplex method （线性规划中的）单纯形法
simplex motor 感应式同步电动机
simplex pump 单缸泵
simplex reciprocating pump 单缸往复式泵
simplex reciprocating 单筒往复式（泵）
simplex steam pump 单缸蒸汽泵
simplex system 单向通信系统，单工系统
simplex tableau 单纯形表
simplex telegraphy 单工电报学
simplex telephony 单工电话
simplex transmission 单工传输，单向传输
simplex wave winding 单波绕组
simplex wave 单波
simplex winding 单重绕组
simplex 单体，单工，单向通信，单一的，简化的，单体的
simplicity 简易性
simplification 单一化，精简，简化，理想化
simplified diagram 简化图，简图
simplified model 简化模型
simplified periodical inspection （锅炉）系统定期检修
simplified periodical repair work 定期小修
simplified statistical method for insulation coordination 绝缘配合简化统计法
simplified 简化了的
simplify working process 简化工序
simplify 精简
simply connected curve 单连通曲线
simply supported beam 简支梁
simply supported end 简支端
simply supported on four sides 四边简支
Simpson's formular 辛普森公式【利用区间二等分的三个点来进行积分插值】
simulant 模拟装置，伪装的，模拟的
simulated approach 模拟法
simulated assembly 模拟组件，假组件
simulated bubble 模拟气泡
simulated computer 仿真计算机
simulated condition 相似条件，模拟条件
simulated converter 模拟转换器
simulated data 模拟数据
simulated echo pattern 模拟回波图形
simulated environment 模拟环境
simulated fall-out 模拟沉降物，模拟落下灰
simulated flow 模拟流动，模拟水流
simulated fuel assembly 模拟燃料组件
simulated fuel slug 模拟燃料块，假燃料块
simulated gravity 模拟重力
simulated input processor 模拟输入处理程序
simulated in situ method 模拟现场法
simulated line 仿真线路
simulated machine 模拟电机
simulated management operation 模拟管理作业
simulated motion 模拟运动
simulated network analysis program 模拟网络分

析程序
simulated operating procedure　模拟操作过程
simulated output program　模拟输出程序
simulated program　仿真程序，被模拟的程序
simulated remote station　远程模拟站
simulated repair weld　模拟补焊
simulated service test　模拟运行试验
simulated slot　模拟槽
simulated snow　模拟雪
simulated source　模拟源
simulated test　模拟试验
simulated waste　模拟废物
simulated wind　模拟风
simulate　仿真，模拟
simulating material　模拟材料
simulating motor　模拟电动机
simulating test　模拟试验
simulation algorithm library　仿真算法库
simulation analysis　模拟分析
simulation block diagram　仿真（方）框图
simulation center　仿真中心
simulation chamber　模拟室
simulation clock　仿真时钟
simulation condition　模拟条件
simulation data base　仿真数据库
simulation data subsystem　模拟数据子系统
simulation environment　仿真环境
simulation equipment　仿真设备
simulation evaluation　仿真评价
simulation experiment mode library　仿真实验模式库
simulation experiment　仿真实验
simulation expert system　仿真专家系统
simulation graphic library　仿真图形库
simulation hardware　模拟硬件，模拟装置
simulation information library　仿真信息库
simulation input tape　模拟输入带
simulation interrupt　仿真中断
simulation job　仿真作业
simulation knowledge base　仿真知识库
simulation laboratory　模拟实验室，仿真实验室
simulation language　仿真语言
simulation methodology　仿真方法学
simulation model library　仿真模型库
simulation model　仿真模型，模拟模型
simulation process time　仿真过程时间
simulation process　仿真过程
simulation program　仿真程序，模拟程序
simulation result　仿真结果
simulation run　仿真运行
simulation software　仿真软件
simulation support system　仿真支持系统
simulation system　仿真系统
simulation technique　模拟技术，仿真技术
simulation test　模拟试验
simulation training　模拟培训
simulation type　仿真类型
simulation velocity　仿真速度
simulation work station　仿真工作站
simulation　仿真，模拟
simulative cycle　模拟循环
simulative generator　模拟发生器，模拟振荡器
simulative network　模拟网络
simulator control features　仿真机功能，仿真机控制性能
simulator control　仿真机控制
simulator load　模拟器负载
simulator program　模拟程序
simulator software　仿真机软件
simulator　模拟程序，模拟电路，模拟装置，模拟计算机，模拟机，模拟器，仿真机
simultaneity factor　同时系数，同时率
simultaneity　同时性
simultaneous access　并行存取，同时存取，同时访问
simultaneous adaptation　同时适应
simultaneous call　同时呼叫
simultaneous carry　同时进位，并行进位
simultaneous computer　并行计算机，同时操作的计算机
simultaneous conformity phasing　同步符合度调相
simultaneous differential equations　微分方程组
simultaneous equations　联立方程，联立方程式，方程组
simultaneous exposure　瞬时曝光
simultaneous faults　同时故障
simultaneous ground fault　同时接地故障
simultaneous input-output　同时输入输出
simultaneous input pulse　同时输入脉冲
simultaneous level line　双线水准测量
simultaneous linear algebraic equations　线性代数方程组
simultaneous multiplier　同时乘法器
simultaneous observation　同时观测
simultaneous peripheral operations on-line　外围设备同时联机操作
simultaneous reactions　联立反应
simultaneous system　同时系统，同步系统
simultaneous track processor　并行轨道处理机
simultaneous　同时存在的，同步的
SINAD ratio(signal to noise and distortion ratio)　信噪失真比
sine-cosine encoder　正弦余弦编码器
sine-cosine mechanism　正弦余弦机构
sine-cosine potentiometer　正弦余弦电势计，正弦余弦电位计
sine-cosine　正弦余弦，正弦余弦的
sine curve　正弦曲线
sine distribution　正弦分布
sine electrometer　正弦电流计，正弦检流计
sine-forced response　正弦强迫响应
sine function　正弦函数
sine galvanometer　正弦电流计，正弦检流计
sine generator　正弦波发生器，正弦波发电机
sine junction gate　禁止门
sine mode　正弦模态
sine potentiometer　正弦电位计
sine qua non　必要条件，要素
sine series　正弦级数
sine-shaped　正弦波形
sine voltage　正弦电压
sine-wave frequency　正弦波频率

sine-wave generator 正弦波发电机，正弦波振荡器，正弦波发生器
sine-wave inverter 正弦波逆变器
sine-wave modulator 正弦波调制器
sine-wave oscillation 正弦波振荡
sine-wave oscillator 正弦波振荡器
sine-wave power generator 正弦波发电机
sine-wave sharpener 正弦波锐化电路
sine-wave 正弦波
sine 正弦
singal flag 标旗
singing arc 声弧，发声电弧
singing sand 鸣沙
singing signal 振鸣信号，蜂鸣信号
singing 嗡鸣，振鸣
single access single-distribution 单存取单分配
single acting centrifugal pump 单动式离心泵
single acting compressor 单动空压机，单向作用压缩机
single acting cylinder 单作用油缸
single acting door 单向门
single acting pile hammer 单动式桩锤
single acting power unit 单作用的执行部件
single acting pump 单作用泵【活塞，柱塞】，单动式泵
single acting ram 单动式桩锤，单向夯
single acting reciprocating 单作用往复泵
single acting relay 单侧作用继动器
single acting steam hammer 单动汽锤
single acting 单作用的，单动的
single action packed slip expansion joint 单向滑动填料函补偿器
single action printer 单作用打印机
single action 单动，单作用
single active component failure 单一部件故障，单一主动部件损坏
single-address code 单地址码
single-address computer 单地址计算机
single-address instruction 单地址指令
single-address message 单地址信息
single-address 单地址，单地址的
single admission 单侧进汽
single and double layer compound winding 单双层混合绕组
single angle bevel 单角坡口
single-anode rectifier 单阳极整流阀
single-apertured core 单孔铁芯
single-apertured 单孔的
single aperture seal 单孔密封
single-arch dam 单拱坝
single armature convertor 单枢变换机，单枢变流机
single-armature DC generator 单枢直流发电机
single-armature DC motor 单枢直流电动机
single-armored 单层铠装的
single axial stress 单轴应力
single band super-heterodyne 单波段超外差
single band 单波段，单频带
single bank of trays 单组托架
single-bank superheater 单级过热器
single-bank 单组
single barrier containment 单层安全壳，一层安全壳
single-bar winding 单线棒绕组
single base 单独底座
single beam bridge crane 单梁桥吊
single beam cantalever crane 单梁悬臂吊
single beam gantry crane with electric hoist 电动葫芦单梁门式起重机
single beam hoist 单梁起重机
single beam oscillograph 单程示波器
single beam spectrometer 单光分光计
single bearing machine 单轴承电机
single bearing 单轴承
single beat main stop valve 单座式主气门
single bed filter 单介质过滤器，单层滤料滤池
single bed fluidized reactor 单层流化床反应器
single-bevel groove 单斜面坡口，半V形坡口，单斜槽
single biased relay 单偏位继电器
single blade adjustment 单叶片调整
single blade switch 单刀开关
single blade windmill 单叶片风车
single block bidding 单段报价
single block footing 独块底座
single block method 单块试验
single block 单轮滑车
single board computer 单板计算机
single board microcomputer 单板微（型）计算机
single-boiler single-turbine unit 单元机组，单锅炉单汽机机组
single bounce technique 二次波法，一次反射法
single bounce 单脉动
single-box siphon 单箱式虹吸管
single break circuit breaker 单截断断路器
single break contact assembly 单断口触头组
single break switch 单断口开关，单断路开关
single break 单断口【断路器】
single-bucket excavator 单斗挖土机
single buoy mooring 单浮筒系泊
single burst pulsed reactor 一次脉冲反应堆
single busbar system 单母线系统
single busbar 单母线
single bus configuration 单母线接线方案
single bus connection 单母线接线
single bus scheme 单母线接线方案
single bus 单母线
single-button sequence 单钮操作顺序
single-buyer mode 单一购买者模式
single-capacity plant 单容量调节对象
single-capacity process 单容量过程
single-capacity system 单容量系统
single-capacity 单容量的
single-car dumper 单车翻车机
single carry 单位进位
single cascade 单列叶栅
single-casing steam turbine 单缸汽轮机
single-casing structure 单缸结构
single-casing 单缸【汽轮机】
single-cavity klystron 单腔调速管
single-cavity magnetron 单腔磁控管

single centrifugal pump　单进口式离心泵，单离心泵
single chain conveyer　单链板输灰机
single-chamber drier　单室干燥器
single-channel access　单能道进入
single-channel analyzer　单波（通）道分析仪，单道分析器
single-channel control　单路控制
single-channel discriminator　单道鉴别器
single-channel monopulse processor　单信道单脉冲信息处理机，单路单脉冲处理机
single-channel regulator　单通道调节器，单路调节器
single-channel simplex　单路单工制
single-channel　单通道，单路的，单道的
single-chip circuit　单片电路
single-chip computer　单片计算机
single-chip　单片的
single-circuit line　单回路线，单路输电线
single-circuit nuclear power stations　单回路核电厂
single-circuit power supply　单回路供电
single-circuit power transmission　单回路输电
single-circuit reflex receiver　单回路来复式收音机
single-circuit transformer　自耦变压器，单绕组变压器
single-circuit transposition　单路换位
single-circuit　单回路的，单工线路
single-clad board　单面（印刷电路）板
single clamp　单卡头
single coat　单涂层的
single-coiled relay　单线圈继电器
single-coil filament　单圈灯丝
single-coil lamp　单线圈灯
single-coil regulation　单线圈调整
single-coil selector　单线圈选择器
single-coil type winding　单圈式绕组
single-column disconnecting switch　单柱型隔离开关
single column footing　单柱底座
single column manometer　单管式压力计
single command　单命令
single component balance　单分量天平
single component system　单组分系统
single-conductor cable　单导体电缆
single-conductor　单导体的【电缆】，单导线，单导体
single conduit　单孔涵洞
single contact　单触点
single conveyor flight　单路皮带
single cord　单线塞绳
single-core binary counter　单磁芯二进制计数器
single-core cable　单芯电缆【英用法】
single-core　单铁芯，单芯线，单磁芯的，单芯的
single-cotton covered wire　单纱包线
single-cotton covered　单纱包的
single course pavement structure　单层式路面结构
single-crystalline-silicon solar cell　单晶硅太阳能电池
single crystal　单晶
single current method　单向电流法
single-curvature surface　单曲率曲面
single-curvature vane　单扭曲叶片
single-cut file　单纹锉
single cycle boiling water reactor plant　单循环沸水堆电站
single cycle execute　单循环执行，单周执行
single cycle forced-circulation boiling water reactor　单循环强制循环沸水反应堆
single cycle gas turbine power plant　简单循环燃机电厂
single cycle natural-circulation boiling water reactor　单循环自然循环沸水反应堆
single cycle plant　（采用直接燃气轮机循环的）单循环电站
single cycle system　单循环系统
single cycle　单循环
single cyclone dust collector　单旋风除尘器
single cyclone　单级旋风分离器
single cylinder turbine　单缸汽轮机
single cylinder　单缸
single day load curve　单纯日负荷曲线
single day tide　全日潮
single-decade counting unit　单十进计数元件
single-decade　单十进数的
single-deck vibrating screen　单层振动筛
single deflection grille　单偏转栅，单层百叶风口
single degree of freedom gyro　单自由度陀螺仪
single degree of freedom system　单自由度体系，单自由度系统
single differential servosystem　单差动伺服系统
single-digit adder　一位加法器
single-digit　一位的
single dike　单堤
single dipole　单偶极子
single discharge type turbine　单向出流式水轮机
single-disc polyphase meter　单圆盘多相计数器，单盘多相电表
single-disc　单盘的
single disk brake　单圆盘制动器
single disk clutch　单片式离合器
single disk gate valve　单瓣闸阀
single disk parallel-seat gate valve with spring　带弹簧的单瓣平行座闸阀
single disk　单个刹车盘
single-distilled water　一次蒸馏水
single door　单扇门
single-dose　一次剂量，单次剂量
single flow HP turbine　单流高压汽轮机
single drive　单驱动
single-drum boiler　单鼓式蒸汽发生器，单汽包锅炉
single-drum hoist　单筒卷扬机
single-drum steam generator　单鼓式蒸汽发生器
single duct air conditioning system　单风管空气调节系统
single duct conduit　单导管套管
single duct system　单风道系统，单风管系统
single duct　单导管
single earthing switch　单接地刀闸
single effect　单向作用，单效

single-electric conductor 单导电体
single electrode process 单电极工艺
single electrode system 单电极系统
single element cask 单元件运输罐
single element relay 一元继电器，单元件继电器
single-end break 单向破裂
single-ended guillotine break 单端剪切断裂
single-ended open-jawed spanner 单头开口爪扳手，单开口扳手
single-ended resonant converter 单端谐振变换器
single-ended wrench 单端扳手【开口端或套筒】
single-end grounding 单端接地
single-end loading system 单端装料系统
single end neutron 单能中子，单向中子
single-end refuelling 单端换料
single-energy reactor 单能反应堆
single-entry compressor 单侧进气压气机
single-entry impeller 单入口叶轮
single entry pump 单吸泵
single entry single-exit circuit 单入口单出口电路
single entry visa 单次入境签证
single entry 单吸式入口，单一记录
single-error correcting code 单差校正码
single-error correction 单一误差校正
single-error detecting code 单差检验码
single-error 单一误差的，单一误差
single-event 单一事件
single-exposure 一次照射，单次照射
single-extraction turbine 单抽（式）汽轮机
single-extraction 单抽汽
single-faced tape 单面绝缘带
single failure criteria 单一故障准则
single failure criterion 单一事故准则，单一故障准则
single failure 单一故障
single fed asynchronous generator 单馈异步发电机
single fed induction generator 单馈感应发电机
single feeder 单馈线
single flat seam locker 单平咬口机
single-flow filter 单流向滤池
single-flow mechanical filter with dual media 单流双介质机械过滤器
single-flow turbine 单流式汽轮机
single-flow 单流，单流的，直流的
single-fluid cell 单液电池
single-fluid MSBR 单流体熔盐增殖堆
single-fluid process 单液法
single footing 独立底脚，独立基础
single fork type blade root 单叉式叶根
single-frame motor 整体机座电动机
single frequency excitation 单频率激励
single fuel firing 单一燃料燃烧
single fuel system 单燃料系统
single furnace boiler 单炉膛锅炉
single furnace 单炉膛
single-gate lock 单闸门船闸
single gearbox with multiple electrical generators 带有多台发电机的单齿轮箱
single-girder crane 单轨吊
single-girder loader-unloader 单梁装卸桥
single-grab 单抓具
single groove 单面坡口
single helical gear 单线螺旋齿轮
single-hinged arch 单铰拱
single-impeller impact breaker 单叶轮冲击破碎机
single inductive shunt 单线圈感应分流器
single-inductor 单电感线圈
single-inlet fan 单吸风机
single-inlet impeller 单面进气叶轮，单入口叶轮
single-inlet 单吸式
single in-line package 单列直插式组件
single input single output control system 单输入单输出控制系统
single inside elbow 简单内部弯头
single instruction execute 单指令执行
single instruction multiple data 单指令多数据
single instruction single data 单指令单数据
single-instrument control 单仪表控制
single-instrument 单仪表的
single-investor enterprise 独资企业
single isolated bubbles 单个气泡
single-jet nozzle 单孔喷嘴
single-J groove 单 J 形坡口
single ladder dredger 单梯式挖泥船
single lamp transformer 单灯用变压器
single lamp 单面出口标志灯
single-lane road 单行车路
single layer fluidized bed 单层流化床
single layer layout 单层布置
single layer welding 单层焊
single layer winding 单层绕组
single layer 单层
single-leaf type rolling gate 单门叶式滚动闸门
single-leg manometer 单管压力计
single letter of indemnity 单签担保提货书
single level address 单级地址，直接地址
single level circuit 单级电路
single level memory 一级存储器
single level process 单级过程
single level steam turbine 单排式汽轮机
single-level 单级的
single-lift lock 单级船闸
single-line diagram 单线回路图，单线图
single-line feeder 单路馈线
single-line ground 单线接地
single-line highway 单线公路
single-line 单路，单线的
single liquid quenching 单液淬火
single load 单一负荷，集中负荷，集中荷载
single lock 单厢船闸
single-loop controller 单回路控制器
single-loop control system 单回路控制系统
single-loop coordination strategy 单环协调策略
single-loop safety system 不互相连接的安全系统，单回路安全系统
single-loop system 单回路系统
single-loop 单回路，单回路的，单环路
single-machine capacity 单机功率
single-main distribution 单主管配水
single mass system 单体系
single-mitre bend 单斜接面弯管

single mode generator 单模发生器，单模震荡器
single mode optical fiber 单模光纤
single module damper 单模数挡板
single-motion turbine 单动式汽轮机
single-movable out-of-pile detector element 单个移动式堆外探测元件
single negotiating text 单一协商文本
single nozzle burner 单喷嘴燃烧器
single open ended spanner 单头开口扳手，活动扳手
single open end wrench 单开口端扳手
single-order device 单指令装置
single-order subtractor 单指令减法器
single-order 单指令的
single orifice plate 单孔板
single outage occurrence 单一停运事件
single outside elbow 简单外部弯头
single-pair 单对
single pan balance 单盘天平
single-parameter control 单参数控制
single-part bidding 单部投标
single-pass boiler 单烟道锅炉
single-pass condenser 单流程凝汽器
single-pass core 单流程堆芯
single-pass heat exchanger 单流程热交换器，直流热交换器
single-pass operation 单程操作
single-pass water 一次通过冷却水，单流程冷却水
single-pass welding 单道焊，单焊道
single-pass weld 单焊道焊缝
single payment 一次整付
single-peaked 单峰值的
single pellet reaction 单颗粒反应
single phase alternator 单相同步发电机，单机交流发电机
single-phase boundary layer 单相边界层
single-phase bridge rectifier 单相桥式整流器
single-phase circuit 单相电路，单相回路
single-phase critical flow 单相临界流
single-phase current 单相电流
single-phase earth fault 单相接地故障
single-phase fault 单相故障
single-phase flow 单相流
single-phase fluid 单相流体
single-phase generator 单相发电机
single phase grounding 单相接地
single-phase ground short circuit 单相接地短路
single-phase kilowatt-hour meter 单相电能表
single-phase line 单相线路
single-phase machine 单相电机
single-phase magnet 单相电磁铁
single-phase motor 单相电动机，单相电机
single-phase operation 单相运行
single-phase pad-mounted transformer 单相箱式变压器，箱变
single-phase passage 单相通道
single-phase potential transformer 单相电压互感器
single-phase power station 单相发电厂
single-phase power supply 单相供电
single-phase power transmission 单相输电
single-phase reactor 单相电抗器
single-phase rectification 单相整流
single-phase rotor 单相转子
single-phase section 单相段
single-phase series motor 单相串激电动机
single-phase stator 单相定子
single-phase synchro-control 单相同步控制
single-phase synchronous generator 单相同步发电机
single-phase synchronous motor 单相同步电动机
single-phase system 单相制，单相系统
single-phase three-limb core 单相三柱铁芯
single-phase transformer 单相变压器
single-phase voltage regulator 单相调压器
single-phase wattmeter 单相瓦特计
single-phase winding 单相绕组
single-phase 单相，单相的
single-phasing damage 单相故障，单相运行损坏
single-phasing feeding 单相供电
single-phasing protection device 单相保护装置
single-phasing transformer 单相变压器
single-phasing （多相电机的）单相运行，单相接入，单相（的）
single-piece erection 整体吊装
single-piece frame 整体机座，整体机壳
single-piece rotor 整体转子
single-piece 整体铸造
single-pier draught tube 单墩尾水管
single pile 单桩
single-pin type spanner wrench 单销式扳手
single-pipe circuit cross-over heating system 单管跨越式采暖系统
single-pipe heating system 单管采暖系统
single-pipe loop circuit heating system 水平单管采暖系统
single-pipe series-loop heating system 单管顺序式采暖系统
single-pipe system 单管系统
single piston pressure-vacuum gauge 单活塞压力真空计
single-pitch roof 单坡屋顶
single pivot instrument 单轴承仪表
single-place shift 单位移位
single-place 单位的
single plane balancing 单面平衡
single planetary gear train 单级行星齿轮系
single-ply solid woven 单层实芯织物
single-point conrtol 单点控制
single-point controllable status output 单点可控状态输出
single-point earthing 单点接地
single-point information 单点信息
single-point lifting 单点起吊法
single-point mooring 单点系泊
single-point recorder 单点记录器
single-point status information 单点状态信息
single-point 单点的
single polarization 单极化
single-pole circuit breaker 单刀断路器
single-pole double-throw switch with center-off posi-

tion 不对称位置的单极双掷开关
single-pole double-throw switch 单极双掷开关
single-pole double-throw 单刀双掷，单刀双投
single-pole isolator 单极式隔离开关
single-pole knife switch 单刀闸刀开关
single-pole line 单杆线路
single-pole mercury switch 单极水银开关
single-pole reclosing 单相重合闸
single-pole scaffold 单柱脚手架
single-pole single-throw switch 单极单掷开关
single-pole single-throw 单刀单投，单刀单掷
single-pole switch 单刀开关，单极开关
single-pole 单极，单杆，单相，单柱，单刀，单针，单杆的，单极的
single-ported control valve 单座控制阀
single-ported globe valve 单座球阀
single-ported slide valve 单口滑阀，单口滑阀【往复机】
single-post row brush dam 单排桩梢料坝
single-post type disconnecting switch 单柱式隔离开关
single precision arithmetic 单字长运算，单精度运算
single pressure condenser 单压凝汽器
single pressure operation 单压运行
single pressure stage 单压力级
single pressure steam cycle 单压蒸汽循环
single pressure steam turbine 单压式汽轮机
single probe testing method 单探头检测法
single program initiation 单程序启动
single program initiator 单程序初始化程序
single proprietorship 独资，独资经营
single pulley drive 单滚筒驱动
single pulse decatron 单脉冲十进管
single pulse device 单脉冲装置
single pulse voltmeter 单脉冲电压表
single pulse 单脉冲
single purpose computer 专用计算机
single purpose engineering system 单一的工程系统
single purpose reactor 专用反应堆
single purpose reservoir 单用途水库
single random failure 单一随机故障
single-range indicating instrument 单量程指示仪表
single-range instrument 单量程仪表
single-range 单量程的，单波段的
single rate prepayment meter 单率预付费用电表
single rectangular box girder bridge 单矩形箱桥梁
single re-entrant winding 单闭路绕组，单回路绕组
single-region core 单区堆芯
single-region reactor 单区反应堆
single regulation 单调节
single reheat cycle 一次再热循环
single reheat type once-through boiler 一次再热式直流锅炉
single reheat 一次再热
single-resonance level 单共振能级
single responsibility contract 单一责任合同
single retort stoker 单槽下饲式炉排
single return wind tunnel 单回路风洞
single-ring feed 单环输电
single-riveted joint 单行铆接
single rod burst test 单棒破裂试验
single rod lattice 单棒栅格
single roll crusher 单辊破碎机，单滚筒破碎机
single roll eccentric crusher 单辊偏心破碎机
single roll grinder 单辊碎渣机
single roll type idler 单辊式托辊
single-rotatable plug 单旋转防护塞
single-rotor gas turbine 单轴燃气轮机
single row governing stage 单列调节级
single row layout 单列布置
single row pile coffer-dam 单排桩围堰
single row stage 单列级
single row wheel 单列速度级叶轮
single row winding 单排线圈，单排绕组
single row 单排的，单列的
single-runner water turbine 单转轮式水轮机
single-run welding 单道焊
single sampling 单次取样
single-seated control valve 单座控制阀
single-seated main stop valve 单座式主气门
single-seated valve 单座阀
single-seated 单阀座的，单座的
single service 单路供电
single-shaft arrangement 单轴布置
single-shaft chimney 单管烟囱
single-shaft gas turbine 单轴燃气轮机
single-shaft stack 单管烟囱
single-shaft steam turbine 单轴汽轮机
single-shaft turbine 单轴汽轮机
single-shaft type combined cycle 单轴联合循环
single shear 单剪
single sheathing 单层面板
single-shell casing 单层缸
single-shell cylinder 单层缸
single-shot multivibrator 单稳多谐振荡器
single-shot operation 单步操作
single-shot reclosing 一次式重合闸
single-shot 单投的
single shrouded wheel 半闭式叶轮
single side band modulation 单边带调制
single side band modulator 单边频带调制器
single side band receiver 单边带接收机
single side band reception 单边带接收
single side band suppressed carrier 单边带抑制载波法
single side band 单边频带，单边带
single-sided linear motor 单边型直线电动机
single-sided 单侧面的
single side feeding 单边供电
single side stable 单稳态，单侧稳定
single-signal 单信号
single-silk enamel wire 单丝漆包线
single site mechanism 单点机构
single-sized aggregate 单径骨料，一定规定尺寸的骨料
single-skin steel dam 单层钢面板坝
single-skin timber dam 单层木面板坝

single sling 单索吊具
single slope roof 单坡屋顶
single-slot winding 单槽绕组
single socket 单插座
single solar cell 单体太阳电池
single-solenoid valve 单螺线管阀
single-solenoid 单螺线管的
single source model 单源模式
single source procurement 单一源采购
single source solution 一站式服务
single space 单间隔
single span beam 单跨梁，单跨轴
single span flexible support 单跨弹性支承
single span shaft 单跨轴
single span 单跨
single spar construction 单梁结构
single-speed floating action 单速无定位作用
single-speed floating controller 单速无定位控制器
single-speed floating control system 单速无静差控制系统
single-speed synchro data system 单速同步数据系统
single-speed 单速的
single-spool engine 单路式涡轮发动机，单路式发动机
single spring clip 单弹簧夹
single squirrel cage motor 单鼠笼电动机
single squirrel cage rotor 单鼠笼转子
single squirrel cage 单鼠笼
single-stage air cooled turbine 一级气冷式透平
single-stage amplifier 单级放大器
single-stage compression 单级压缩
single-stage compressor 单级压气机
single-stage crushing 一级破碎
single-stage curing 单级养护【预制混凝土】
single-stage fluidized bed 单级流化床
single-stage furnace 单燃烧室炉膛
single-stage modulation 单级调制
single-stage natrium ion exchanger 一级钠离子交换器
single-stage one-envelope bidding procedure 单步单信封投标程序
single-stage pump 单级泵
single-stage recycle 单级再循环
single-stage regenerative pump 单级旋涡泵
single-stage regulator 单级调节器
single-stage separation factor 单级分离系数
single-stage steam bypass 单级蒸汽旁路，蒸汽大旁路
single-stage turbine 单级透平
single-stage two-envelope bidding procedure 单步双信封投标程序
single-stage 单级，单级的
single station analysis 单站分析
single station forecast 单测站预报，单站预报
single-step distance relay 单段距离继电保
single-step method 单步法
single-step operation 单步操作
single-step process 单步法
single-step 一步，单级，单步，单级的
single-storey building 平房
single-strand wire rope 单股钢丝索
single-strand 单股
single string 单联金具串
single-stroke deepwell pump 单冲程深井泵
single suction centrifugal pump 单吸离心泵
single suction impeller 单吸入口叶轮，单吸叶轮
single suction pump 单吸式泵，单吸泵，单吸抽水机
single suction 单吸
single supply 单电源，单电源供电
single-sweep 单相扫描
singles 煤的颗粒级
single-tank breaker 单筒式油断路器
single-tank switch 单筒式开关
single tariff 单一税率
single tax system 单一税制
single terrain 简单地形
single-throw circuit breaker 单掷断路器
single-throw contact 单投接点
single-throw switch 单掷开关
single-throw 单投，单掷
single thrust bearing 简单推力轴承
single ticket 单程票
single-time-lag servo 单时延的随动系统，单时延的伺服系统
single-time-lag 单时延的
single tower 单柱塔，独塔
single track hoist 单轨吊
single track railway 单线铁路
single transducer operation 单探头操作
single trapezoidal box girder 单梯形箱梁
single-triplet 单三重线
single-trolley system （电车的）单触线制
single T slab 单T形板
single tube chimney 单筒式烟囱
single-tube evaporator 单管蒸发器
single-tube installation 单管装置
single-tube oscillator 单管振荡器
single-turbine-single boiler construction 单机-单锅炉结构
single-turn coil 单匝线圈
single-turn core-type current transformer 单匝内铁型电流互感器
single-turn potentiometer 单匝电势计
single-turn transformer 单芯式电流互感器，单匝式电流互感器
single-turn winding 单匝绕组
single-turn 单匝的
single-U groove 单U形坡口，单面U形坡口
single unit hydroelectric station 单机组水电站
single unit nuclear power plant 单堆核电厂
single unit 单一机组，单机，单组
single user computer 单用户计算机
single-valued function 单值函数
single-value nonlinearity 单值非线性
single-value 单值的
single-valve circuit 单管电路
single-valve filter 单阀滤池
single-valve operation 单阀操作
single-variable control system 单变量控制系统

single-variable system 单变量系统
single-variable 单变量的
single variety 单一种类
single V butt weld 单V形对接焊
single-vee groove 单V形坡口
single velocity stage 单速度级
single vernier 单游标
single viewing 简单检查［检验］，单一考察
single-voltage rating 单电压额定值，单额定电压的
single-volute pump 单螺旋泵
single vortex strength 单涡强度
single-wall buttress 单垛支墩
single-wall cofferdam 单墙围堰
single-wall containment 单层安全壳
single-wall tank 单壁容器
single watershed method 单独流域法
single wattmeter method 单瓦特计法
single wave operation 单波运行
single wave rectification 半波整流
single wave rectifier 半波整流器
single-way radio communication 单向无线电通信
single-welded joint 单焊缝【只从一面焊接】
single well pumping 单孔抽水
single wheel impulse turbine 单级冲动式透平
single winding motor 单绕组电动机
single winding multispeed motor 单绕组多速电动机
single winding transformer 自耦变压器，单绕组变压器，单卷变压器
single-wing stacker 单悬臂堆料机
single wire armoured 单线铠装的
single wire circuit 单线电路
single wire system 单线制
single-zone reactor 单区反应堆
single 单一的，单程的
singular attractor 奇异吸引子
singular control 奇异控制
singular cycle 连续循环
singular integral 奇异积分
singularity 奇异性
singular linear system 奇异线性系统
singular line 奇异直线，单回线路
singular ordinal 特异序数
singular perturbation 奇异摄动，奇异摄动法
singular point 奇异点，奇点
singular series 奇异级数
singular 单数的，奇异的
singulizer 分类器，分级器，单一器
singulizing disc 分级盘，分类盘
singulizing （颗粒燃料元件）分级，分类
sink coal sample 沉煤样
sink drawing （管材）无芯棒拨制
sinker bar 冲击钻杆
sinker chain 沉锤锚链
sinker drill 冲钻
sinker （浮标等的）沉锤，沉块，冲钻，向下凿岩机
sink flow 汇流
sink hole erosion 陷穴侵蚀
sink hole 灰岩坑，石灰渗水坑，落水洞，渗坑，阴沟洞
sinking by jetting 射水下沉法
sinking caisson 沉箱
sinking curve 下沉曲线
sinking fund depreciation 偿债基金折旧
sinking fund factor 偿债基金因子
sinking fund 累积基金，偿还基金，偿债基金，零存整取基金，分期提存还本基金，还本准备金，提存金
sinking jumbo 向下掘井钻车
sinking pile by water jet 水冲沉桩
sinking region 下沉区
sinking tube well 下井筒
sinking tubular pile 沉管桩
sink sample of coal 沉煤样
sink streamline 汇流线
sink strength 汇强
sink 沉没，降沉，沉落，坍陷，消退，汇点，汇集，渗坑，盥洗池，漏水池，污水槽
Sino-foreign cooperative design 中外合作设计
Sino-foreign cooperative enterprise 中外合作企业
Sino-foreign joint venture 中外合资企业
sinoidal 正弦形的，正弦波的，波形的，起伏的
SINOMACH（China National Machinery Industry Corporation Ltd.） 中国机械工业集团有限公司，国机集团
SINOPEC（China Petroleum & Chemical Corporation） 中国石油化工集团公司，中国石化
sinor 复量，相量，彩色信息失真
Sinosure 中国出口信用保险公司
sinterability 烧结，粉末冶金
sinter density 烧结块密度，烧结密度
sintered ceramic 烧结陶瓷
sintered deposit 高温黏结灰，烧结沉积
sintered flux 烧结焊剂
sintered fuel 烧结燃料
sintered glass 烧结玻璃
sintered membrane 烧结膜
sintered-metal ultrafine filter 烧结金属超细过滤器
sintered metal 烧结金属
sintered plastics 烧结塑料
sintered plate 烧结分布板
sintered porcelain plate 烧结瓷板
sintered powder 烧结的粉末
sintered UO_2 pellet 烧结二氧化铀芯块
sintering after Aluminium evaporating 蒸铝烧结
sintering aid 烧结助剂
sintering boat 烧结舟
sintering coal 烧结煤
sintering furnace 烧结炉
sintering in a reducing atmosphere 在还原气氛中烧结
sintering point 软化点，烧结点，软化温度
sintering strength （灰）烧结强度
sintering temperature 烧结温度
sinter-roasting 烧结焙烧
sinter 结渣，渣，熔渣，熔结，烧结，粉末冶金
sinthetics 合成产品，合成产物
sinuosity 弯曲率，蜿蜒度，蜿蜒

sinuous coil　波形管圈，蛇形管
sinuous flow　乱流，曲折水流，蛇形流动
sinuous header　波形集箱
sinuous plume　蛇形羽流
sinuous vortex　蛇形涡
sinuous　波状的，蜿蜒的，弯曲的
sinusoidal current　正弦电流
sinusoidal cyclic load　正弦周期载荷
sinusoidal density wave　正弦磁密度
sinusoidal distribution　正弦分布
sinusoidal envelope　正弦包线
sinusoidal excitation　正弦型激励
sinusoidal field　正弦场
sinusoidal function　正弦函数
sinusoidal generator　正弦波发生器
sinusoidal impulse　正弦脉冲
sinusoidal input　正弦输入信号
sinusoidal lock-in excitation　正弦锁定激励
sinusoidal meander　正弦波形曲流
sinusoidal mode shape　正弦模态形状
sinusoidal modulation　正弦调制
sinusoidal motion　正弦运动
sinusoidal oscillation　正弦振荡
sinusoidal oscillator　正弦波振荡器
sinusoidal output　正弦输出
sinusoidal quantity　正弦量
sinusoidal response　正弦响应
sinusoidal rotating flux field　正弦旋转磁场
sinusoidal sender　正弦发送器
sinusoidal sequence　正弦序列
sinusoidal signal　正弦信号
sinusoidal time function　正弦时间函数
sinusoidal variation　正弦变化，正弦曲线变化
sinusoidal vibration　正弦振动
sinusoidal voltage　正弦电压
sinusoidal wareform　正弦波形
sinusoidal wave　正弦波
sinusoidal　正弦波的，正弦的，正弦曲线的
sinusoid generator　正弦波发生器
sinusoid response　正弦响应
sinusoid wave　正弦波
sinusoid　正弦，正弦波，正弦曲线
sinus vibrometer　正弦振动计
sinus　正弦，正弦的
siphon action　虹吸作用
siphon barometer　虹吸式气压计，虹吸气压表
siphon breaker　虹吸截断器，虹吸破坏器，虹吸切断器
siphon culvert　虹吸涵洞
siphon drain　虹吸疏水
siphon filter　虹吸滤池
siphon gauge　虹吸表
siphon head　虹吸压头
siphonic closet　虹吸马桶
siphonic effect　虹吸效应
siphonic outlet　虹吸式出水口
siphonic system　虹吸系统
siphonic water collection　虹吸集水
siphoning drainage of the reactor vessel　反应堆容器虹吸疏排
siphoning　虹吸作用
siphon inlet　虹吸进水口
siphon intake　虹吸式取水
siphon mouth　虹吸管进口
siphon pipe　虹吸管
siphon pit　虹吸坑
siphon pump　虹吸泵，虹吸水泵
siphon recorder　虹吸记录器
siphon suction　虹吸抽水
siphon transition　虹吸管渐变段
siphon trap　虹吸式防气弯管，存水弯
siphon-type flume　虹吸式渡槽
siphon vacuum breaker　虹吸切断器
siphon well　虹吸井
siphon　虹吸，虹吸管，通过虹吸
sipping test container　啜漏试验室【辐照过的燃料检验】
sipping tester　啜吸探漏器
sipping test　啜漏试验【辐照后元件检漏】，啜吸试验，热析检验【燃料组件】
SIP(simulated input processor)　模拟输入处理程序
SIP(single in-line package)　单列直插式组件
SIP(symbolic input program)　符号输入程序
siren　汽笛，警报器
SIR(safe integral reactor)　安全一体化反应堆
SIR(selective information retrieval)　选择性情报检索
SIR(serial input register)　串行输入寄存器
SIR(symbolic input routine)　符号输入程序
sirufer　细铁粉铁芯
SIR　化学试剂注入系统【核电站系统代码】
SI(sample interval)　采样时间间隔
SI(saturation index)　饱和指数
SI(screen-grid input)　帘栅极输入
SISD(single instruction single data)　单指令单数据
SI(serial interface)　串行接口
SI(shift-in character)　移入字符
SI(signal interface)　信号接口
SI(site instructions)　现场指示书
SISO(single-input and single-output)　单输入单输出控制系统
SI(special information)　专用信息
SI(special instruction)　专用说明书
SIS(safety injection system)　安全注射系统，安全注入系统
SIS(safety instrumented system)　安全仪表系统
SIS(supervisory information system)　(厂级)监控信息系统
Si-steel　硅钢
sister block　双滑车
sister company　姊妹公司
sister hook　双钩
SI(system international of units)　公制，国际单位制
site acceptability　厂址的可接受性
site acceptance test　现场验收测试
site access control system　现场出入控制系统
site accommodation　工地生活设施，现场食宿
site analysis　场址分析
site appraisal　坝址评价，厂址评价，工程地址

鉴定
site area 厂区占地，厂区
site-assembled 工地装配的，散装的
site assembly 现场组装
site assessment 场址评价
site auxiliaries 厂区辅助系统
site availability test 现场可利用率试验
site boundary 现场的边界，现场范围
site canteen 工地食堂
site cast concrete pile 现场浇制混凝土桩
site clearance and leveling 现场清理与整平
site clearance （施工前的）场地清理，清理现场，工地清障
site clearing （施工前的）场地清理，清理现场，工地清障
site close to load centre 近负荷中心的厂址
site conditions 建厂条件，厂址条件，现场条件
site-constructed 工地装配的，散装的
site construction schedule 现场施工进度
site construction 现场施工
site control 就地控制，现场控制
site criterion 选址准则
site data 厂址资料，现场数据
site diary 施工日志
site dimension 现场尺寸
site director 工地主任
site disposal 现场处理
site DNI data 现场直接辐射数据
site drainage 现场排水
site electrical facilities 风场电器设备
site elevation 厂址标高
site engineer 工地工程师，现场工程师
site-erected 工地安装的
site error 位置误差
site establishment plan 施工组织设计总平面布置图
site evaluation 厂址评估，场址评价
site exploration 厂址勘查，厂址勘探，现场勘察，现场勘探
site fabrication 现场制造，现场制作
site facility 现场设施
site finish painting 现场最后涂漆
site grade level 厂址地面标高
site grading and drainage 现场平整［地坪处理］与排水
site grading 场地平整
site handover and takeover 现场移交
site hand-over 现场移交
site height 测站高程
site inspection 现场检查
site installations 现场组装，工地设施
site intensity 场地烈度
site investigation report 场地勘察报告
site investigation 厂址调查，现场调查，现场勘测［勘察/勘探］，工程地址调查
site laboratory 厂区实验室
site layout plan 场地平面图
site layout 场地布置
site leveling drawing 场地平整图
site leveling 厂址平整，场地平整，现场平整
site levelness 场地平整度
site management 现场管理
site manager 现场经理
site map 现场图
site measurement 现场实测
site meeting 现场会议
site mobilization 进场，现场动员，现场准备
site monitoring 场区监测
site office 工地办公室，现场办公室
site operation file 现场运行档案
site operation 工地作业，现场施工
site organization 现场施工组织
site permit report 厂址批准报告
site personnel 厂区人员，工地［现场］工作人员
site planning 场地规划，地盘规划，总平面设计，总平面规划
site plant 现场设施
site plan 厂区［工地］总平面图，厂址平面图，工地布置图，工地平面（布置）图
site plot plan 厂区规划
site plot 厂址图
site precast 现场预制
site prefabrication 现场预制
site preliminary works 厂址前期工作
site preparation prior to construction starting 开工前施工现场准备
site preparation program 现场准备计划
site preparation 现场平整，现场准备
site proper 厂区【占地】
site railway 工地铁路
site rated output 现场额定出力
site rating 现场评级，厂址定级
site reconnaissance （工程的）地质勘测，现场踏勘
site reference conditions 现场基准条件，现场参考条件
site related design parameters 厂址相关设计参数
site related extreme event 厂址相关极端事件
site requisition form 现场请求表
site requisition 现场请求单
site resident representative 工地代表
site road 工地道路，现场道路，施工道路
site security 厂区的保安，现场保卫，现场安全
site selection investigation 选址查勘
site selection principles 选址原则
site selection 坝址选择，厂址选择，场地选择，位置选择，站址选择，选址
site simulation test 现场模拟试验
site-specific 现场特有的
site storage 现场保管
site survey plan 现场测量图
site survey 工程地址调查，现场测量，现场勘测
site technical director 工地技术主任
site technical service 现场技术服务
site test 现场试验，工地试验
site touch-up painting 现场修整涂漆
site trainee 现场受训人员
site transfer and access 现场移交
site utilization area 场地利用面积
site utilization factor 场地利用系数

site visit　现场勘察，现场踏勘
site weld(ing)　工地焊，现场焊，现场焊接
site workshop manager　现场车间经理
site works schedule　现场工作计划，施工进度
site work　现场工作，场地工程
site　现场，位置，厂址，厂区，地点，工地，定址，选址，选厂址，布置，(工程)选点
siting criteria　定址准则，选址准则
siting source term　选址源项
siting　定厂址
SI(special information) tone　专用信息音
SIT(simulation input tape)　模拟输入带
situation display　状态显示，状况显示
situation plan　位置图
situation　场所，地点，情况，状况
SIT　给水化学取样系统【核电站系统代码】
SI unit(Standard International Unit)　标准国际单位
SIWT(safety injection water tank)　安全注入水箱
six-break oil breaker　六断点油断路器
six component balance　六分量天平
six-element ring　六合环
six-factor formula　六因子公式
sixfold　六重的，六倍的，六次的
six-hole rock drilling machine　六孔钻岩机
six-line system　六线系统
six-phase connection　六相接法
six-phase generator　六相发电机
six-phase power transmission　六相输电
six-phase rectifier　六相整流器
six-phase ring connection　六相环接法
six-phase star connection　六相星形接法
six-phase system　六相系统，六相制
six-phase voltage　六相电压
six-phase winding　六相绕组
six-phase　六相的
six pole motor　六极电动机
six poles dual feed generator　六极双馈发电机
six series airfoil NACA　六位数系列翼型
sizable　相当大的，颇大的，可观的
size analysis　筛分析，粒度分析，粒径分析
size change　尺寸变化
size characteristics　尺寸特性，粒径特性
size classifier　粒度分级器
size coefficient　粒度系数
size composition　颗粒度组成，粒径组成
sized coal　分级煤，筛选煤，粒级煤，分级煤
sized cut　粒级，规模缩减
sized gypsum　筛分石膏
size discrimination　尺寸鉴别
size distribution curve　粒度分布曲线
size distribution　粒度分布，尺寸分布，粒径分布，粒径分配
sized　按大小分类的，筛过的
size factor　尺寸因素，尺度因数
size fractions of crashed coal　煤破碎的粒级
size fractions of raw coal　原煤自然级
size fractions　粒级，颗粒组，粒度级份额
size frequency distribution　粒度频率分布
size grading　粒径级配
size increment　粒度增加，颗粒长大
size interval　尺寸间隔，粒度间隔
size of electrode　焊条规格
size of memory　存储器容量
size of mesh　筛眼大小，筛号
size of pipe and tubing　管道标称尺寸，管子公称尺寸
size range of separation　分选粒级
size range　粒度范围
size ratio　粒径比
size reduction factor　缩尺因子，破碎比
size reduction　磨细，粉碎
sizer　分粒器，分选机，上胶器，分级器，大小分拣器，筛选器，填料器
size-selective dust-sampler　粒度选择微尘取样器
size separation　按颗粒度分类
size shape-factor　尺寸形状因子
size　尺寸，大小，规模，号码，规格，粒度，测定尺寸，定尺寸，筛选，按大小分类，涂胶水，校正，分选
sizing analysis　筛析
sizing grid　筛格
sizing screen　分级筛，分类筛
sizing test　粒度分析
SJAE(steam jet air ejector)　射汽抽气器
skate　滑动接触片，侧向装置
SKD(semi knock down)　半散件组装
skein winding　分布绕组
skein　一绞，一束(导线)，(水轮机的)轴套
skeletal code　骨架代码
skeletal coding　程序轮廓编码，程序纲要，骨架编码
skeletal density　骨架密度
skeletal end shield　框架式端盖
skeletal material　骨料
skeletal soil　粗骨土
skeletal structure　骨架结构
skeleton assembly　燃料组件骨架
skeleton bay　(厂房)预留机组段
skeleton construction　骨架构造，骨架结构
skeleton diagram　概略图，轮廓图，单线圈，方框图，原理图，结构图，框架图
skeleton framework　(燃料组件)骨架
skeleton frame　框形机座
skeleton line　中弧线
skeleton motor　开敞式电动机
skeleton structure　骨架结构，框架结构
skeleton table　骨架表
skeleton-type bridge　阻抗电桥
skeleton unit　预留机组
skeleton　骨架，构架，框架，结构，轮廓，梗概，概略
skellering　翘曲
skelp　焊管铁条
skeptical　怀疑的
sketch drawing　示意图，素描图，简图，草图，略图
sketching board　绘图板
sketching sketch plan　初步计划
sketch plan　初步设计，草图
sketch　简图，草图，略图，画草图，示意图，设计图，概略，纲要
skew angle　斜交角，歪扭角

skew-back 拱脚，拱脚砖，起拱石，斜块拱座
skew-barrel arch 斜筒形拱
skew bevel gear 斜齿锥齿轮
skew bridge 斜桥
skew bucket 扭曲挑坎
skew coil winding 端部斜接绕组，斜圈绕组
skew coil 斜线圈，不对称绕组
skew curve 不对称曲线，斜曲线
skew direction 斜向
skew distribution 不对称分布
skewed boundary layer 扭曲边界层
skewed pole-shoe 斜极靴
skewed pole 斜极
skewed rotor 斜槽转子
skewed slab 斜板
skewed slot 斜槽
skewed split spacer 歪分离式定位架
skewed storage 错位存储器
skewed streamline 扭曲流线
skewed tooth 斜齿
skew factor 斜槽系数，歪斜系数，槽扭系数，绕组不对称系数
skew frequency curve 偏斜频率曲线
skew inclination 倾斜
skewing allowance 歪扭余量
skewing slot 斜槽
skewing 偏态，歪斜，偏移，时滞，相位差，偏置
skew integral 反称积分
skew intersection 斜交叉
skew leakage flux 斜槽漏磁通
skew matrix 斜对称矩阵
skewness coefficient 偏态系数
skewness 不对称，偏斜，偏态，斜度，偏差度，偏度，偏斜度
skew notch 斜槽口
skew slot factor 斜槽系数
skew symmetrical matrix 反对称矩阵，反号对称矩阵，斜对称矩阵
skew symmetric 斜对称的，非对称的，反号对称的
skew wind 斜风
skew 斜交，歪斜，变形，斜的，歪的，弯曲的，扭曲不对称的，反称的
SKH 润滑油和油脂贮存系统【核电站系统代码】
skiagraph 房屋纵断面图，投影图
skid bar 滑杆
skid beam 滑道梁
skid derrick 滑动式起重机
skidding behavior 滑动特性
skidding 滑痕，曳运
skid-mounted 滑动安装的，装有滑轨的
skid pile-driver 滑动打桩机
skid proof net 防滑网
skid proof 防滑
skid resistance 抗滑，抗滑力
skid shoe 滑靴
skid way 滑道
skid 导轨，滑动，滑动垫木，滑轨，滑道，滑行架，溜滑
ski-jump energy dissipater 挑流式消能工
ski-jump spillway 滑雪道式溢洪道，挑流式溢洪道
skilled labourer 熟练工人
skilled worker 技工，熟练工人
skill testing 技能考核
skill 技能，技巧，技艺，熟练
skimmer circuit 去浮回路，溢流回路，撇浮回路
skimmer cyclone 撇渣旋风筒
skimmer equipment 臂斗式挖掘机
skimmer pump 排渣泵
skimmer shovel 铲土机
skimmer wall 拦污墙
skimmer 分离装置【取水】，撇除器，去浮器，分液器，撇除装置，撇渣器
skimming boom 分离装置【取水口】，撇除装置
skimming circuit 撇浮回路
skimming door 出渣门
skimming filter 撇浮过滤器
skimming of waste water 自废水表面撇取石油
skimming pit 撇油池
skimming pond 撇油池
skimming pump 排渣泵
skimming tank 撇油槽
skimming weir 拦沙撇水堰
skimming 浮渣，撇（渣）
skim 表面薄层，撇去，撇去浮沫，去渣
skin bending stress 表面弯曲应力
skin bonded construction 蒙皮胶合结构
skin casing 隔热材料内套，外护板
skin coat （胶带各衬层间的）薄涂层
skin contamination incident 皮肤污染事故
skin contamination 皮肤污染
skin cut 表面切除
skin depth 有效肤深，集肤深度，透入深度，趋肤深度
skin dissipation 趋肤耗散
skin dose 皮肤剂量
skin dosimeter 皮肤剂量计
skin drying 油漆结皮
skin effect in conductor 导线内的集肤效应
skin effect loss 集肤效应损耗
skin effect rotor 深槽式转子
skin effect winding 深槽式绕组
skin effect 集肤效应，趋肤效应，表面效应
skin exposure 皮肤照射量
skin friction coefficient 表面摩擦系数
skin friction force 表面摩擦力
skin friction tests 蒙皮磨损阻力试验
skin friction 表面摩擦，表面摩擦力，（桩的）表面摩擦阻力，表面磨阻
skin hardness 表面硬度
skin heating 表面加热
skin irritation 皮肤刺激，皮肤发炎
skin lamination 蒙皮表层，表皮分层
skin material 蒙皮材料
skin monitoring 皮肤污染监测，皮肤监测
skinning 剥皮的【导线】
skin patch 蒙皮补片
skin reinforcement 蒙皮加强

skin resistance capacity 摩擦承载力
skin resistance 表面阻力，表面摩擦阻力，（桩的）表面摩擦阻力，表面磨阻
skin rivet 蒙皮铆钉
skin streamline 表面流线
skin stress 表层应力，表面应力
skin stretch forming 蒙皮拉伸成形
skin stringer construction 蒙皮桁条结构
skin temperature 表面温度，表皮温度
skin test 蒙皮试验
skin tolerance dose 皮肤耐受剂量，皮肤容许剂量
skin weld 表面焊接，表面焊
skin 外表，表皮，表面，表面薄层，外皮，蒙皮，皮肤，外壳，外层，表层
skiodrome 波面图
skip bucket 倾卸斗
skip crane 吊斗起重机，翻斗提升机
skip flag 跳过标记
skip gradation curve 不连续级配曲线
skip hoist 倒卸式提升机，吊斗提升机，料斗绞车，料斗升降机，斜斗提升机，斗式运煤机
skip loader 翻斗式装料机
skip multiplexer 跳群复接设备
skip over 跳过【程序步骤】
skipper arm 挖土机斗柄
skip rolling 间断滚压
skip sequence 跳焊
skip track 翻斗轨道，料车轨道
skip welding 跳跃焊
skip zone 跳跃区，静区
skip 吊斗，翻斗【装混凝土用】，翻斗车，料斗，料车，起重箱，“空白”指令，跳，跳跃【进位】，空指令，省略
skirting board 踢脚板，侧壁，侧护板，裙板，挡板，围板，导料挡板，约制槽，导料槽两侧的裙板
skirt plate 围板，导料挡板，导料槽两侧的裙板，翻车机的平台
skirt 侧板，挡板，侧挡板，护墙板，裙板，踢脚板，边缘
skir 绝缘子外裙，边缘，套筒，对称谐振曲线的边下部分，侧缘
skiving machine 切削机
SK＝sink 排水管，排水沟，散热器
sky condition 天空状况
skylab 天空实验室
skylight frame 天窗架
skylight 天窗，天棚照明，屋顶采光窗
sky radiation 天空散射辐射，天空辐射
skyscraper 摩天大楼
sky shine 天空辐照，天空回散照射
skyway 高架公路
slab and column buttress dam 板柱式支墩坝
slab and girder 板梁结构
slabbing cut 分片爆破，分片开挖
slab coil 平线圈
slab culvert 平板式涵管
slab foundation 板式基础，平板基础
slab insulation 板材保温
slab joint 面板接缝
slab lattice 平板栅格
slab loading 板状装料
slab pile 平板反应堆
slab reactor 平板反应堆
slab rectifier 平板整流器
slab reinforcement 面板钢筋
slab robot 电传动车式仿效机械手
slab source 板状源，面源
slab spacer 板内垫块
slab structure 板块构造，平板结构
slab tamper 平板振捣器
slab-type building 板式建筑
slab-type sliding gate valve 平板式滑动闸阀
slab winding 盘形绕组，容性线圈
slab with ribs tuned-up 反梁板
slab 平板，厚板，板，片，铁块，切片，块，板坯，石板，扁平块，楼板
slack adjuster 松紧调整器
slack bus 浮游母线，松弛母线
slack coal 煤屑，末煤
slackening 松弛
slack hopper 细煤斗
slacking index 松散指数
slacking （煤的）风化，风化作用
slackline cableway 松弛的缆索
slackline scraper 拖铲挖土机
slackline 拖铲挖土机
slack quenching 调质，断续淬火，细热化处理
slack season 淡季
slack side （皮带的）松边
slack span 放松（孤立）档
slack tide 憩潮，憩流
slack time 间歇时间，松弛时间
slack tip 松坍
slack turn 松线匝【钢丝绳】
slack variable 松弛变量
slack water time 平潮期
slack water 憩流，死水
slack 松弛【输送带、缆索等】，放松，减弱，减速，碎煤，煤屑，松弛的，缓慢的
SLAC(shipper's load and count) 货主装载、计数
SLACS(shipper's load, count and seal) 货主装载、计数和加封
SLAC(Stanford linear accelerator center) 斯坦福直线加速器中心
slag aggregate 熔渣骨料
slag bath 渣池【电渣焊】
slag-bed 渣床
slag blanket 渣层
slag blob 渣瘤
slag blower 吹灰器，熔渣吹扫器
slag breaker 碎渣机
slag breakthrough 炉底漏渣
slag brick 矿渣砖，灰渣砖
slag bridging 堵渣，渣搭桥
slag buildup 结渣，结焦
slag catcher 捕渣器
slag cement 矿（钢）渣水泥
slag chunk 渣块
slag coating 熔渣覆盖层
slag collecting bunch 捕渣管束

slag concrete block　矿渣混凝土砌块
slag concrete　矿渣混凝土
slag conveyor　捞渣机
slag cork　渣塞
slag corrosion　熔渣腐蚀
slag cotton　矿渣棉
slag cover　渣层
slag crusher　碎渣机
slag crust　渣皮，渣瘤
slag dam　（出渣口的）渣栏
slag deposit　结渣
slag discharge system　排渣系统
slag disposal area　灰场
slag disposal yard　渣场
slag disposal　除渣
slag dragging machine　捞渣机
slag-drip opening　排渣口
slag dump　渣场，堆渣场，废渣堆，渣坑
slag enclosure　夹渣
slag extractor　排渣机，捞渣机
slag fiber　矿渣棉
slag film　渣膜，渣皮
slag flushoff　放渣
slag formation　造渣，渣化
slag-free operation　无灰渣运行
slagging factor　（炉膛）沾污系数
slagging furnace　液态排渣炉
slagging in furnace　燃烧室结渣【锅炉】
slagging operation　有渣运行
slagging property　结渣性
slagging reaction　渣侵蚀
slagging temperature　结渣温度
slagging　结渣，成渣
slag granulation plant　熔渣粒化装置
slag heap　渣堆
slag heat　熔渣热量
slag hole　渣口
slag inclusion　焊缝夹碴，夹渣，炉渣杂质
slag layer　渣层
slag melting point　渣熔点
slag muck　废渣
slag notch　出渣口
slag overflow　放渣
slag particles　焊渣粉粒【重复使用焊药】
slag penetration　熔渣渗透
slag pinup　灰渣泵
slag pit　渣井
slag pocket　渣室
slag pool　渣池
slag pump　灰浆泵，灰渣泵
slag quenching　炉渣骤熄
slag removal equipment　除渣设备
slag removal　除渣
slag remover　出渣机，除渣机
slag resistance　抗渣性
slag sand　熔渣砂
slag screen　费斯顿管，防渣屏，凝渣管，捕渣管束，捕渣网，防渣管，防渣管筛
slag sluice　冲渣沟
slag slurry pond　渣浆池
slag slurry sump　渣浆池
slag spatter　焊渣飞溅
slag spout　出渣口
slag stone　熔渣石
slag stringer　线状夹渣
slag-sulphate cement　石膏矿渣水泥
slag tank　渣池
slag tap boiler　液态排渣炉
slag tap bottom furnace　液态排渣炉膛
slag tap bottom　液态排渣炉底
slag tap firing　熔灰燃烧
slag tap furnace　液态排渣炉膛
slag tapping boiler　液态排渣炉
slag tapping　液态排渣
slag-tap plugging　排渣堵塞
slag trench　灰渣沟
slag welding　夹焊
slag wool　矿渣棉，渣棉
slag yard　渣场
slag　矿渣，熔渣，使成渣，变熔渣，渣，结渣，炉渣，渣化
slaked lime　消石灰，熟石灰
slaker　消石灰器
slake tank　消石灰槽
slake　消除，熄灭，熟化
slaking clay　水解黏土
slaking test　风化试验，湿化试验，水化试验，（土的）崩解试验
slaking time　熟化时间
slaking value　水化度，水化值
slaking　潮解
slamming　使劲关上
SLAM(simultaneous localization and mapping)　即时定位与地图构建
SLAM technology　即时定位与地图构建技术
slant distance　斜距
slant height　斜高
slanting baffle　斜挡板
slanting chute　斜槽
slanting leg manometer　倾斜式差示压力计
slanting strut　斜撑
slanting　倾斜式的
slant range　斜距
slant washer　斜垫圈
slant　斜向
slapping　拍击
slasher　单轴多片圆锯机
slat conveyor　板式输送机，条板式输送机
slate cement　板岩水泥
slate coal　板岩煤，页岩煤
slate hanging　墙面铺石板
slate lath　挂瓦条
slate oil　页岩油
slate　板岩，石板，弄斜，斜面
slaty cleavage　板岩劈理
slat　板条，狭板
slave analyzer　伺服分析器
slave arm　（操作器或机械手的）从动臂
slave clock　子钟
slave drive　从动，随动，随动拖动
slaved system　受役系统
slaved tracking　随从跟踪

slave end 从动端【机械手】
slave flip-flop 自激多谐振荡器，从触发器
slave hand actuator 从动手传动机构【机械手】
slave joint 从动端关节【机械手】
slave manipulator 仿效机械手，从动机械手
slave mode 仆从方式，用户方式
slave module 从模块
slave motion 从动作，从动【机械手】
slave motor 随动电动机
slave relay 辅助继电器，随动继电器，从动继电器
slave station 被控站，从站，子站
slave system 从属系统
slave unit 从属部件，从属单元，从属装置
slave valve 液压自控换向阀，随动阀
slave 从属的，次要的，随动的，从动，随动，从属，从动装置
slaving principle 役使原理，从属原理
slaving voltage 从动电压
slay 芯子，铁芯
SLB(Siberian land bridge traffic) 西伯利亚大陆桥运输
SLC(selector channel) 选择器通道
SLC(shift left and count instruction) 左移计数指令
SLCS(standby liquid control system) 备用液体控制系统
SLD(single line diagram) 单线图
sledge hammer 大锤，大榔头，大铁锤，重磅铁锤
sledge 大铁锤
sleeper beam 垫梁
sleeper block 卧木
sleeper joist 小搁栅，轨枕梁
sleeper plate 垫板，轨枕垫板
sleeper wall 地龙墙
sleeper 轨枕，枕木
sleeping company 挂名公司
sleeping room 卧室
sleet jump of conductor 导线跳跃
sleet jump 覆冰跳跃
sleet layer 冰凌层，覆冰层
sleet load 覆冰负载，冰凌负载，冰雪负载
sleet melting 融冰
sleet-proof 耐冰凌的，防冰凌的
sleet 冻雨，雨夹雪，雨淞，冰雹，下雨夹雪，下冰雹，冻雨拍打，使下霰般落下
sleeve antenna 同轴偶极天线
sleeve armature 套筒式电枢
sleeve barrel 套筒
sleeve bearing 套筒式轴承，套筒轴承
sleeve cock 衬套旋塞
sleeve coupling 套管连接，套管接头，套筒式管接头
sleeve dipole 同轴偶极子
sleeve expansion joint 套管伸缩器，套筒补偿器
sleeve guide 套筒导向
sleeve half bearing 半套筒轴承
sleeve insertion 套筒插入
sleeve joint 套筒接合，套筒接头，套管接头，套管接合
sleeve lower neck 套筒下部颈
sleeve nut 套筒螺母，轴套螺帽
sleeve piece 壳筒，嵌环
sleeve-pipe wrench 套筒扳手
sleeve plugging 套管堵塞
sleeve rotor motor 空心转子电动机
sleeve rotor 空心转子
sleeves and escutcheons 套管和孔罩
sleeve stop 套筒制动
sleeve terminal 套筒端子
sleeve valve 套阀，筒式阀
sleeve 联轴器，套管，管接头，套筒，护套，套管，外套，轴套，外壳
sleeving 装套管
slender beam 细长梁
slender body theory 细长体理论
slender column 细长柱
slender contour 流线型外形
slenderness ratio 长细比，细长比，纵横比，长度与直径比，高宽比
slenderness 细长度
slender pointed hydrofoil 细尖水翼
slender proportion 细长比
slender ratio 细长比
slender tower 细高塔架，细高大厦
slender 细长的
SLE(static load error) 静态负载误差
slewable guide 旋转导轨
slewable 能旋转的
slew angle 堵转角
slew drive 回转驱动装置
slew increment 回转增量
slewing crane 旋臂起重机，回转式吊机
slewing mechanism 回转机构
slewing motor 回转电动机
slewing radius 回转半径
slewing rate 转换速度，回转率，旋转速度，变速范围
slewing reclaimer 旋转取料机
slewing ring 齿环，轴承齿环，回转支承，回转环
slewing speed 旋转速度
slewing stacker 悬臂堆料机
slewing type S/R 回转式斗轮堆取料机
slewing 快速定向，快速瞄准，悬臂
slew 泥沼，回转，使旋转
SLF(sender link frame) 发射机，接线架
slice method 条分法
slicer 切片机
slice 薄片，切片，部分，铲
slickensided hair cracks 擦痕面发状裂缝，擦痕面毛细裂纹
slickenside （岩石的）擦痕面，断层擦痕，擦痕
slicker 叠板刮路机
slidac 滑线电阻调压器
slide angle 滑动角
slide area 滑动面积
slide-back voltmeter 偏压补偿式电压表
slide bearing 滑动轴承
slide block type universal joint spindle 滑块式万向接轴

slide block 滑块
slide bridge 滑线电桥
slide caliper 滑动卡尺，游标卡尺
slide coil 滑触线圈
slide contact 滑动接触，滑动触点
slide control 滑动调整，滑动调节，平滑控制，防滑
slide conveyer 滑轨运送机
slide damper 闸板，滑动挡板，插板阀
slide door 滑动门，拉门
slide down 滑落
slide drive assembly 滑移传动组件
slide escape 逃生滑梯
slide fit 滑动配合
slide gate valve 插板阀
slide gate 滑动阀门，滑动闸门，滑阀
slide gauge 卡尺，游标卡尺
slide mass 滑动体，滑移体
slide multiplier 滑臂式乘法器
slide path 坍滑路径
slide plane 滑动面
slide plate gate 插板门
slide plate 滑板
slide pressure mode 滑压状态
slide rail 滑动导轨，滑轨
slide resistance 滑动电阻
slide rheostat 滑线电阻，滑线变阻器
slide-rule nomogram 计算尺型列线图
slide-rule 计算尺
slider 滑动触点，滑板，滑座，游标，（可变电阻的）滑触头，滑尺，导板
slide sash 推拉窗
slide switch slide switch 滑动开关，拨动开关
slide track 滑道
slide transformer 调节变压器，滑线变压器
slide vacuum pump 滑阀式往复真空泵
slide valve buckle 滑阀套
slide valve bush 滑阀衬套
slide valve chest 滑阀箱
slide valve seat 滑阀座
slide valve spindle 滑阀杆
slide valve thimble 滑阀套管
slide valve 滑阀，闸阀，分流阀
slideway 滑道
slide window 推拉窗
slide wire bridge 滑线电桥
slide wire potentiometer 滑线电位器，滑线式电势计
slide wire resistor 滑线电阻器
slide wire 滑线
slide 滑板，滑动，滑行，滑动片
sliding angle 摩擦角
sliding arm 滑动臂
sliding avalanche 滑动崩坍
sliding barrier 防滑器
sliding bearing 滑动轴承，滑动支座
sliding box-shaped caisson 滑动式箱形沉箱
sliding bracket 滑动支架
sliding caisson 滑动式沉箱
sliding caliper （游标）卡尺，游标卡
sliding cam 滑动凸轮
sliding clutch 滑动离合器
sliding coefficient 滑动系数
sliding contact suction 密封环
sliding contact 滑动触点，滑动接触
sliding coupling 滑块，滑板，离合器
sliding current 滑动电流
sliding curve 滑坡曲线
sliding damper 滑动挡板，闸板
sliding distance 滑动距离
sliding door hardware 拉门五金零件
sliding door rail 推拉门滑轨
sliding door 滑动门，拉门，滑门，推拉门
sliding electrical contact 滑动电接点
sliding expansion joint 滑动式伸缩缝
sliding factor 滑动系数
sliding failure 滑动破坏，滑坍破坏，滑移破坏
sliding floor 滑动地板
sliding formwork 滑动模板，滑模
sliding form 滑动模板，滑模
sliding fracture 滑动断裂，剪切断裂，滑动破裂
sliding frictional force 滑动摩阻力
sliding frictional resistance 滑动摩擦阻力
sliding friction 滑动摩擦
sliding gate pump 滑片泵
sliding hinge support 滑动铰支座
sliding hinge 滑动铰，烟斗铰
sliding housing 滑动阀壳
sliding joint 滑动接合，滑动接头
sliding key pressure 滑销压力
sliding key system 滑销系统
sliding key 滑销，滑键
sliding pad 滑板
sliding panel weir 插板堰
sliding phenomenon 滑动现象，滑移现象
sliding pin system 滑销系统
sliding pin 滑销
sliding plane 滑动层面，滑移面，坍滑面
sliding platform 移动平台车
sliding pontoon 滑动浮筒
sliding pressure air storage reservoir 滑压蓄气库
sliding pressure operation 滑压运行
sliding pressure shut-down curve 滑压停机曲线
sliding pressure shut-down 滑参数停机
sliding pressure starting-up curve 滑压启动曲线
sliding pressure starting 滑参数启动
sliding pressure 滑压【启动】
sliding pulley 滑轮
sliding rheostat 滑动变阻器，滑线式变阻器
sliding ring girder support 滑动支墩
sliding rule 滑尺
sliding rupture 剪切断裂，滑动断裂
sliding saddle 滑动鞍座
sliding scaffold 滑动脚手架
sliding scale 滑尺
sliding screen 活动滤网
sliding seal 滑动密封
sliding shim 滑动垫片
sliding shoe 滑块，滑瓦，滑动制动器，滑靴
sliding shuttering 滑动模板
sliding shutter 推拉百叶窗，活动屏蔽体
sliding sleeve valve 滑动套阀

sliding sluice gate 滑动闸门
sliding snout （重水堆装料机）滑动头部
sliding spanner 活络扳手
sliding stem 滑动杆
sliding support 滑动支架，滑动支座
sliding surface 滑动面
sliding timber weir 插板堰
sliding track 滑道
sliding traveller 滑动吊板
sliding tripod 伸缩三脚架
sliding vane rotary pump 滑片机械泵，滑片转动泵
sliding-vane-type vacuum pump 滑片式真空泵
sliding vector 滑动矢量
sliding wear 滑动磨损
sliding wedge method 滑动楔体法
sliding wedge 滑动三角土楔
sliding window 滑动窗
sliding wire resistance 滑线电阻
sliding 滑移，滑动的，滑行的，活动的，可调整的
slight breeze 轻风【二级风】
slighted enriched fuel 低富集燃料
slight flagging 轻微旗状【二级植物指示风力】
slight glass 视镜
slightly active 低放射性的，低活性的
slightly enriched fuel 低富集燃料
slightly enriched uranium 低加浓铀，低浓缩铀，低富集铀，稍浓缩铀
slightly enriched 低加浓的，低浓缩的，低富集的
slightly moderated 轻度慢化，轻度慢化的
slightly offset longitudinal welds 稍偏的轴向焊缝
slightly radioactive 稍有放射性的，低放射性的
slightly soluble salt 微溶性盐
slight pressure device 微正压装置
slight rain 微雨
slight sea 二级风浪
slight 轻微的，少量的
slime coal 煤泥
slime separator 黏泥分离器
slime 残渣，黏泥，泥渣，淀渣，腐泥，粘液
slimicide 杀黏菌剂
sliminess 稀黏程度
slimness of furnace 炉膛细（长）度
sling bag 背带袋
sling block 吊索滑车
sling dog 吊钩
slinger ring 吊环，吊索，甩油环
sling hygrometer 摆动湿度计
slinging wire 电源线，供电线
slinging 吊起
sling psychrometer 手摇干湿表，悬吊式湿度计，旋转式干湿球湿度计
sling stay 链钩拉撑
sling thermometer 手摇温度表，旋转温度计
sling tube 悬吊管，吊环，吊具，吊索，链钩，钩索，绳扣
slip angle 横偏角侧滑角
slip a pole 滞后一个极，滑差一个极
slip at anchorage 锚固滑移
slip back one pole 滑差一个极，滞后一个极
slip back 倒流，逆流
slip band 滑动带，滑移带
slip bolt 插销
slip circle method 滑弧法
slip circle 滑动圆弧，滑弧
slip cleavage 滑移劈理
slip clutch 滑动离合器
slip counter 转差计
slip coupling 滑动联轴器，滑差联轴节
slip curve 转差曲线
slip energy 转差能量
slip erosion 滑移侵蚀
slip factor 滑动因数，滑动因子，转差系数
slip flow 滑流，滑移流
slipform concrete silo 滑模式混凝土筒仓
slipform construction 滑模施工
slipform lining 滑模衬砌
slip formwork 滑动模板，滑模
slip form 滑动模板，滑模
slip fracture 滑动破裂
slip-free drive 无滑移传动
slip frequency 滑差频率，转差频率
slip-joint pliers 滑接钳，鲤鱼钳
slip joint 滑动连接，滑动接头，伸缩接头
slip-line method 滑移线法
slip line 滑动线
slip loss 转差损耗
slip meter 转差计
slip-off slope bank 冲积坡岸
slip-off slope 冲积坡
slip-on connection 卡套式连接
slip-on coupling 活箍
slip-on flange weld 法兰的平焊
slip-on flange 滑套法兰，平焊法兰，平移法兰
slip-on marker 活动标志
slip-on welding flange 滑套焊接法兰，平焊法兰
slippage force 滑动力【在栅格栅元内部】
slippage 滑动，滑程，滑移，滑距，错动，滑动量，下跌（量），转差率，滑差率，滑移量
slipper bearing 滑动轴承
slipper 游标，滑动部分，滑触头
slipping clutch 安全摩擦离合器，可调极限扭矩摩擦离合器，滑动离合器
slipping of pole 磁极滑移
slipping of wheel 车轮打滑
slipping plate 滑动式支承板【抗震支承】
slipping 转差率，滑动面，滑移面，滑面，滑差率，滑动，空转，打滑
slip power 滑差功率，转差功率
slip rate 滑移率
slip ratio 滑移比率，滑动比，转差比率，汽水两相速度比，滑速比，转差率
slip regime 滑流态
slip regulator 滑率调节器，转差调节器
slip relay 转差继电器
slip ring armature 滑环式电枢
slip ring brush 滑环电刷，滑环套
slip ring contact 转动开关，滑环接触
slip ring cover 滑环罩

slip ring current 滑环电流
slip ring device 滑环装置
slip ring induction motor 滑环感应电动机
slip ring motor 滑环电动机
slip ring rotor 滑环式转子
slip ring side 滑环侧
slip ring spider 滑环支架
slip ring voltage 滑环电压
slip ring 滑环，集电环，滑动环
slip seal 滑动密封
SLIP(serial line internet protocol) 串行线路网际协议
slip speed 滑差速度，转差速度
slip stream 滑流
slip surface of failure 破坏滑动面
slip surface 滑动面
slip test 转差试验，滑差试验
slip velocity 滑动速度
slip water bottle 带阀门取水样器
slipway 滑道，船台
slip-wedge method 滑动楔体法
slip wire bridge 滑线电桥
slip 滑矩【泵，电机】，港池，船台，滑差，滑移，滑动，转差率，空转
SLI(shipper's letter of instruction) 空运托运书
SLI(suppress length indicator) 抑制长度指示器
slit erosion 隙缝侵蚀
slit pipe 纵向切口管
slitting 切口，切开，纵切
slit-type bucket 窄缝式挑坎
slit-type burner 缝隙式燃烧器
slit width 缝宽
slit 细缝，缝，开缝，裂口，缝隙
sliver 裂片，碎片，分层，（条）片，分裂物，碎裂物
SLOE(special list of equipment) 专用设备清单，专用设备表
slope angle 坡度角，坡角，倾斜角
slope-area discharge measurement 比降面积测流法
slope-area method 比降面积法
slope base 坡脚
slope benching 坡级，坡台
slope board 坡板
slope collar(/mouth) 斜井井口
slope compaction 边坡压实
slope-conveyance method 比降输水法
slope correction 倾斜改正
slope current 倾斜流，斜坡流
slope design 斜坡设计
slope detection method 斜率鉴别法，斜率鉴定法
slope detector 坡度测定仪
slope-discharge curve 水面坡降流量曲线
slope distance 斜距离，斜距
slope down-wind 下坡风
slope drainage 斜坡排水
slope drain 山坡排水沟
slope engineering 边坡工程
slope erosion 边坡侵蚀，坡地侵蚀，坡面冲刷
slope face drainage 贴坡排水
slope face 坡面
slope failure 边坡破坏，边坡坍塌，斜坡崩塌，坍坡
slope flow 山坡气流，坡面流
slope foot 坡脚
slope ground 斜坡，倾斜地表
slope impedance 动态电阻，微分电阻
slope inclination 斜率
slope indicator 测斜仪，坡度指示器
slope intake 岸坡式进水口
slope invert 斜井仰拱
slope-lining machine 边坡衬砌机
slope loess 坡地黄土
slope mine 斜井矿
slope of a curve 曲线斜率
slope of formation 地层坡度
slope of ground leveling 场地平整坡度
slope of ground 地面斜坡
slope of lift curve 升力曲线斜率
slope of natural ground 自然地形坡度
slope of river bank 河岸坡度
slope of river 河道坡降，河流比降
slope of stack 堆积坡度
slope of stair 楼梯坡度
slope paving machine 斜坡铺砌机
slope paving 护坡，斜坡护面，斜坡铺砌
slope peg 坡桩
slope potentiometer 跨导调整电位器
slope protection of gabion 边坡土石筐垒墙保护
slope protection of grass 边坡植草保护
slope protection of shotcrete 边坡水泥喷浆保护
slope protection of the retaining wall 边坡挡土墙保护
slope protection 边坡保护，护坡
slope ramp 斜面坡度
slope resistance 微分电阻，动态电阻
slope revetment 边坡护面
slope scale ratio （模型的）坡降比尺
slope factor 倾斜系数，斜率，坡度因子
slopes of chute 落煤管的斜度
slopes of hopper 漏斗斜度
slope stability analysis 斜坡不变性剖析
slope stability evaluation 边坡稳定性评价
slope stability 边坡稳定（性）
slope stake 坡桩
slope station 坡地测站
slope steepness 边坡陡度
slope tamping 边坡夯实
slope tatus 边坡堆积
slope top 坡顶
slope trimmer 斜坡修整机，斜坡修整，修坡
slope up-wind 上坡风
slope-velocity formula 比降流速公式
slope wash 冲积堆，坡积物，坡面冲刷
slopeway 斜坡道
slope weighting 边坡压重
slope weir 斜坡式堰
slope wind 山坡风，坡风
slope 比降，坡度，斜坡，斜坡道，倾斜，斜度，斜率，斜面，山坡，坡降
sloping apron 斜护坦，斜坡式护坦

sloping aquifer 倾斜含水层
sloping blanket drain 倾斜排水铺盖
sloping block 斜块
sloping bottom 倾斜炉底【炉膛】，冷灰斗
sloping breakwater 斜坡堤
sloping channel 斜槽
sloping concrete fill （混凝土）倾斜充注
sloping core earth dam 斜墙土坝
sloping core earth-rock dam 黏土斜墙土石坝
sloping core 斜墙，斜心墙
sloping discharge pipe 倾斜卸料管
sloping faced breakwater 斜坡式防波堤
sloping field 坡地
sloping floor 倾斜底
sloping gauge 斜坡水尺
sloping-grate stoker 倾斜炉排
sloping ground 坡地
sloping land 坡地
sloping of river bank 河岸边坡放缓
sloping ramp with slats 带防滑条的坡道
sloping span length 斜档距
sloping surge chamber 倾斜式调压室
sloping toward open drain 坡向明沟
sloping transfer line 倾斜输送管
sloping valley gutter 斜沟
sloping-wall-type breakwater 斜墙式防波堤
sloping water table 倾斜地下水面
sloping wave 缓波
sloping 有坡度的
slop-on flange 滑套法兰
slop sink 污水池，污水槽，污水盆
slop 污水坑，废油，水坑
slosh 溅泼，溅，泼，搅动，晃动，把……泼溅出，泥泞，溅泼声
slot-and-wedge anchor bar 楔缝式锚杆
slot-and-wedge bolt 楔缝式锚栓
slot armor 槽衬，槽内线圈主绝缘
slot array antenna 缝隙天线阵
slot atomizer 缝隙喷雾器，缝隙式喷油嘴
slot bar 槽部线棒，槽内导线
slot base 槽底
slot breadth 槽宽
slot cell 槽绝缘垫片
slot closing wedge 槽楔
slot coal trough 缝式煤槽
slot coil 槽部线圈，槽线圈
slot combination 槽配合
slot contraction factor 槽缩小系数
slot coordination 槽配合
slot cross-field voltage 跨槽磁场电压
slot cross-field 槽横向磁场
slot cross-section 槽截面
slot diffuser 条缝式雾化器，条缝型风口
slot discharge analyzer 槽部放电分析仪
slot discharge 槽部放电
slot distribution pipe 缝隙式配水管
sloted flip bucket 差动式挑坎
slot effect 槽齿效应【电机】
slot electric field 槽电场
slot exhaust hood 槽边排风罩
slot exhaust on edges of tanks 槽边通风
slot factor 槽系数
slot field-effect transistor 沟道场效应晶体管
slot field 槽磁场，槽漏磁场
slot filler 槽内垫条，槽内填料
slot frequency 信道间插入频率
slot harmonic 齿谐波，槽谐波
slot hole 长孔
slot impedance 槽阻抗
slot inserter 嵌线机
slot insulation 槽绝缘
slot insulator 槽内线圈间的绝缘
slot joint 狭槽接合
slot leakage field 槽漏泄磁场
slot leakage flux 槽漏磁通
slot leakage reactance 槽漏电抗
slot leakage 槽漏磁，隙缝泄漏
slotless air-gap winding 无槽气隙绕组
slotless generator 无槽发电机
slotless motor 无槽电机
slot lining 槽衬
slot magnetron 有槽磁控管
slot mesh 长缝筛孔
slot meter 预付式电度表，算表，投币式电度表
slot number ratio 槽数比
slot opening 槽口，槽口宽
slot outlet 条缝形送风口，条缝形出风口
slot packing 槽内填料
slot permeance 槽部磁导
slot pitch angle 槽距角
slot pitch 槽距
slot pulsation frequency 槽脉动频率
slot pulsation 槽脉动
slot reactance constant 槽漏抗常数
slot reactance 槽电抗，槽漏抗
slot ripple frequency 槽脉动频率
slot ripple 槽脉动
slot scrubber 缝式洗涤器，百叶窗分离器
slot seal 槽楔，槽封口
slot separator 槽内线圈间的绝缘
slot slope factor 槽斜系数
slot slope 槽斜度
slot space-factor 填充系数，槽满率
slot space 槽空间
slots per pole per phase 每极每相槽数
slots per pole 每极槽数
slotted aerofoil 开缝机翼
slotted armature 有槽电枢
slotted base plate 有槽底板
slotted blade 开槽叶片
slotted bucket 齿槽式挑流鼻坎
slotted commutator 有槽换向器
slotted core 开槽铁芯
slotted fin 开缝的肋片
slotted gravity dam 宽缝重力坝
slotted hole 槽孔，长槽孔
slotted line carriage 开槽线架
slotted line 开槽线
slotted nozzle 槽孔式喷嘴，开槽喷嘴
slotted nut 开槽螺母【用开尾销锁住】
slotted punched plate 长眼板
slotted roller bucket 槽齿式戽，差动式鼻坎

slotted screw 有槽螺钉
slotted stator 开槽定子
slotted strap 槽板【燃料组件】
slotted teeth 开槽齿
slotted tube 开缝管
slotted tubular pin 开缝空心销
slotted waveguide 开槽波导管
slotted winding 槽部绕组，槽绕组，嵌入槽内的绕组
slotted working section 开槽壁试验段，开缝壁试验段，开缝式试验段
slotted 开槽的，有槽的，有沟槽的，开缝的
slotter 铣槽机，插床
slotting machine 插床
slot-tube anchor bar 缝管锚杆
slot type bed load sampler 槽孔式推移质采样器
slot type sampler 槽孔式采样器
slot vector method 槽矢量法
slot wall 槽壁
slot wave 间隙波
slot wedge material 槽楔材料
slot wedge 槽楔
slot welding 槽焊
slot weld 槽焊，槽焊缝，切口焊缝
slot width 槽宽
slot 槽，槽口，插槽，狭槽，沟，长槽孔，缝，缝隙，开缝，开槽，裂缝，切口，长煤槽的开口部分
sloughage 脱落，脱
slough 坍塌，滑坍，泥坑，泥沼，绝境，蜕皮，脱落，使陷入泥沼
slow-access memory 慢速存储器
slow-access storage 慢存取存储器
slow-acting relay 缓动继电器
slow-acting switch 延时开关
slow assets 呆滞资产
slow banking method 慢速筑堤法
slow-bend test 慢弯试验
slow-blow fuse 缓慢熔化的保险丝
slow break switch 缓断开关
slow-break 缓断，迟释
slow bubble model 慢速气泡模型
slow-burning construction 耐燃建筑物
slow cement 慢凝水泥
slow convection 弱对流
slow cooling 缓冷却
slow-curing asphalt 慢凝液体沥青
slow death 老化
slow diffusant 缓慢扩散源
slow direct shear test 慢剪直剪试验
slow down 缓解，减速
slower-than-real-time simulation 欠实时仿真
slow-fast reactor 慢快中子双区反应堆
slow filter 慢滤池
slow filtration 慢滤，慢滤法
slow fission 慢裂变
slow-flowing stream 缓慢水流
slow flux density 慢中子通量密度
slow hardening 缓慢硬化
slowing-down age 慢化龄，慢化年龄
slowing-down area 减速区，慢化面积，减速面积，减能区面
slowing-down cross-section 慢化截面
slowing-down density 慢化密度
slowing-down energy distribution 慢化能量分布
slowing-down equation 慢化方程
slowing-down kernel 慢化核
slowing-down length 慢化长度
slowing-down power 慢化效率，减能本领，中子慢化能力
slowing-down process 慢化过程
slowing-down spectrometer 慢化能谱仪
slowing-down 慢化【中子】
slowly variable flow 缓变流
slow match 慢燃导火线，缓燃引信
slow memory 慢速存储器
slow mode 慢变模态
slow-motion screw 微动螺旋
slow-moving depression 慢移低气压
slow-moving 缓动的
slow neutron activation analysis 慢中子活化分析
slow neutron capture 慢中子俘获
slow neutron filter 慢中子过滤器
slow neutron fission cross section 慢中子裂变截面
slow neutron fission 慢中子裂变
slow neutron flux density 慢中子通量密度
slow neutron reactor 慢中子反应堆
slow neutron region 慢中子区
slow neutron scitillator 慢中子闪烁体
slow neutron velocity spectrometer 慢中子速度谱仪
slow neutron 慢中子
slow-operating relay 缓吸合继电器，慢动作继电器，缓动继电器
slow period 长周期
slow powder 缓爆炸药
slow reaction 慢反应
slow release action 缓释动作
slow release relay 慢释继电器
slow release 缓释，缓放
slow replacement 缓慢置换
slow response 慢响应
slow rinse 置换冲洗，慢速冲洗
slow sand filter 慢沙滤池，慢沙过滤器，慢滤池
slow sand filtration 慢砂过滤
slow setting cement 缓凝水泥
slow setting 缓凝，慢凝
slow settling particles 慢沉颗粒
slow shear test 慢剪试验
slow shear 慢剪
slow speed machine 低速电机
slow speed main shaft 低速主轴
slow speed motor 低速电动机
slow speed operation 低速运行
slow speed rotor 低速风轮
slow speed 慢速
slow state 慢变状态
slow storage 慢速存储器
slow subsystem 慢变子系统
slow surface wave 慢表面波
slow-to-operate relay 缓动继电器，缓放继电器

slow-up 慢化，减速
slow-vehicle lane 慢车道
slow wave structure 慢波结构，高频波减速结构
slow 慢的
SLRAP（standard low-frequency range approach）标准低频段法
SLRN（select read numerically） 数字选择读出
SLSF（sodium loop safety facility） 钠回路安全装置【美国】
SLSI（super large-scale integration） 超大规模集成
SL（supply limit） 供货边界
SLT（solid logic technology） 固态逻辑技术
SL type cable（separate lead type cable） 分头电缆
SLT 更衣室通风系统【核电站系统代码】
sludge age 污泥龄
sludge bed 污泥场，污泥床
sludge blanket 泥渣层，渣体
sludge bottom 污泥底
sludge cake 泥饼，渣饼
sludge chute 污泥管，泥煤管，污水管
sludge circulation clarifier 泥渣循环澄清池
sludge collector pipe 泥渣收集器管
sludge collector 集泥器，污泥收集器
sludge concentration 污泥浓缩
sludge concentrator 泥渣浓缩器
sludge density 污泥密度
sludge deposit 污泥沉积，淤泥沉积物
sludge dewatering system 污泥脱水系统
sludge dewatering unit 污泥脱水装置
sludge dewatering 污泥脱水
sludge digester 污泥消化池
sludge digestion tank 污泥消化池
sludge digestion 污泥消化
sludge discharge pipe 淤泥排管，排泥管
sludge disintegration 污泥分解（作用）
sludge disposal 污泥处理
sludge drain tube 排泥管
sludge drying 污泥干化
sludge examination 污泥检查
sludge excess 过剩污泥
sludge extractor 污泥分离池
sludge formation 泥浆形成
sludge gas 污水气体，淤渣气
sludge handling process 污泥处理过程
sludge handling 泥浆处理，污泥处理
sludge hopper 泥斗，渣斗
sludge incineration plant 泥渣焚化炉
sludge inhibitor 泥浆抑制剂
sludge lagoon 淤泥池
sludge lance 泥浆喷水枪
sludge lancing equipment 泥浆冲洗设备
sludge lancing trailer 泥浆冲洗拖车
sludge lancing 高压水冲洗淤渣，泥浆冲洗
sludge moisture content 污泥含水量
sludge moisture 污泥水分
sludge pipe 泥浆管，排泥管，污泥管，污水管
sludge pressing 污泥压干
sludge pump house 排泥泵房
sludge pump 泥浆泵，污泥泵，泥渣泵，排污泵
sludge reaeration 污泥再曝气
sludge recirculation 泥渣再循环
sludger 抽沙泵，吸沙泵，离心式污泥泵
sludge sample 污泥试样
sludge scraper 刮泥机
sludge separator 泥渣分离器
sludge shredder 污泥破碎机
sludge solid concentration 污泥固体浓缩
sludge tank 泥渣箱，残渣箱，蒸发器底渣储箱
sludge test 污泥试验
sludge thickener 污泥浓缩池
sludge thickening 污泥浓缩
sludge treatment 污泥处理
sludge ultimate disposal 污泥最终处置
sludge utilization 污泥利用
sludge volume index 泥渣体积指数【污水生化处理时】
sludge water 污泥水
sludge 泥渣，淤渣，污泥，泥浆，泥，沉淀，沉淀物，沉积物，积垢，杂质，软泥，污泥指数，析出物，油泥，淤泥，泥沼
sluff 雪崩
slug calorimeter 嵌片形测热计
slug flow model 团状流动模型
slug flow （汽）团状流动，活塞流，迟滞流
slugged relay 延时继电器，缓动继电器
slugging bed 腾涌床
slugging fluidized bed 腾涌流化床
slugging frequency 腾涌频率
slugging wrench 敲击扳手
sluggish circulation 缓慢环流
sluggish flow 缓流，缓慢水流，黏滞流
sluggish stream 缓流
slug length 气栓长度
slug matching 短线匹配法
slug nose 气栓顶端
slug of air 空气团
slug press 棒条压制
slug relay 惯性继电器
slug size 气栓尺寸
slug spacing 气栓间隙
slug two-phase flow 团状两相流
slug velocity 气栓速度
slug 重击，在……中插，嵌片，金属片状毛坯，金属块，铁芯，有惯性的，燃料块，短时间内排放的大量浓化学废水，未蒸发的燃料液滴
sluice barrage 泄水闸
sluice board 泄水闸板，闸底板
sluice box 泄水箱
sluice chamber 泄水闸室
sluice collar 泄水道截水环
sluice control mechanism 水闸操作机械
sluice damper 插板阀
sluice discharge 水力除灰，水力除灰管道，泄水闸流量
sluice door 泄水门
sluice entrance 泄水道进口
sluice flow 闸下出流
sluice gate chamber 泄水道闸门室
sluice gate 水闸【水泵站】，泄水闸门

sluice gun 冲洗枪
sluice nozzle 冲洗喷嘴，泄流喷管
sluice opening 水闸孔
sluice pump 冲灰水泵，冲洗水泵
sluice separation 淘析
sluice structure 泄水建筑物
sluice timber 闸木
sluice trough 冲灰沟，长斜水槽
sluice valve 闸门，闸板阀，泄水阀，截止阀，滑动阀门，闸阀
sluiceway 冲砂道，泄水道，排水渠
sluice weir 泄水堰
sluice 冲洗，漂净，闸，水闸，沟，管道，冲洗，开闸放水
sluicing and scouring 泄水冲沙
sluicing canal 泄水渠
sluicing capacity 冲刷量，泄水闸过流量
sluicing chamber 冲沙室
sluicing earth dam 冲填土坝
sluicing nozzle 冲灰喷嘴
sluicing-siltation dam 冲填淤积坝
sluicing-siltation method 冲填淤积法
slump belt 崩坍地带
slump block 崩坍块体
slump cone 坍落锥形，混凝土锥形筒
slumped bed height 坍落层高度
slumped bed 坍落床
slumped condition 塌落条件
slump limitation 坍落度限值
slump test value 坍落度
slump test （混凝土）坍落度试验，（泥浆）调度试验
slump 压火【流化床】，坍塌，坍落，（混凝土）坍落度，暴跌，萧条，衰退，滑动
slum 页岩煤
slurry characteristic 料浆特性
slurry circulation pump 浆液循环泵
slurry circulation type flocculator 浆液循环型絮凝器
slurry cutoff 泥浆防渗墙
slurry drilling method 泥浆钻孔法
slurry equipment 拌浆设备
slurry explosive 水胶炸药，塑胶炸药
slurry feeding pump 给浆泵
slurry fuel 混合燃料浆，煤浆，浆液状态燃料
slurry handling 淤浆处理
slurry hole-boring method 泥浆护壁钻孔法
slurry injection 泥浆灌注
slurry method 泥浆法
slurry pipeline 泥浆管道
slurry pipe 灰浆管，灰渣浆管
slurry pump house 灰浆泵房，灰渣浆泵房
slurry pump 灰浆泵，泥浆泵
slurry reaction kinetics 浆料反应动力学
slurry reactor 悬浮液反应堆，浆料反应器
slurry recycle 浆料循环
slurry-reinforced overburden 泥浆增强的覆盖层
slurry trench cutoff 泥浆截水墙
slurry trench method 泥浆槽法
slurry trench 灰浆沟，泥浆槽，灰渣浆沟
slurry waste 浆状废物
slurry 泥浆，矿浆，煤泥浆，灰浆，淤浆，浆液，浆料，煤泥
slusher 扒煤机，活动拖铲
slush field 雪水聚集处
slushing 抗腐蚀，涂油灰
slush pump 泥浆泵，污水泵
small amplitude disturbance 小（振幅）扰动
small amplitude wave 小振幅波
small angle scattering 小角度散射
small area flood 小面积洪水
small bore 小口径
small break LOCA 小破口失水事故
small capacity wind turbine 小容量风电机组
small capacity 小容量
small catchment 小集水区
small coal 末煤，煤屑，小块煤
small core memory 小磁芯存储器
small cruciform guide 小十字导向（柱）
small-curved meander 小弯度曲流
small decorative zone 小装饰区
small disturbances method 小扰动方法
small enterprise 小企业
small focal spot 小焦点
small freeboard 低稀相区
small gravel 细砾石
small hydraulic turbine 小型水轮机
small hydroelectric power station 小型水电站
small inductive breaking current 小电感开断电流
small internal bypass valve 内旁通阀
small line break 小管道破口
small-lot production 小规模生产
small-mesh sieve 细眼筛
small nonlinearity 弱非线性
small nuclear power plant 小型核电厂
small-oil-volume circuit-breaker 少油断路器
small pane 小镶板
small perturbance theory 小扰动理论
small power motor 小型电动机，小功率电机
small power station 小型发电厂
small profit and quick returns 薄利多销
small puller 小推车器
small reactivity 小反应性
small repair 小修
small reservoir 小型水库
small risk event 危险小的事故，小概率事故
small sample assay system 小样品测定系统
small sample perturbation 微小扰动
small scale construction project 小型基建项目
small scale drawing 小比例图
small scale gust 小尺度阵风
small scale integration 小规模集成
small scale map 小比例尺地图
small scale model 小比例尺模型，小尺度模型
small scale production 小规模生产
small scale survey 小比例尺测量
small scale turbulence 小尺度紊动，小尺度湍流
small scale WECS 小型风能转换装置
small scale wind energy conversion system 小型风能转换系统
small scale wind turbine 小型风电机组
small scale 小规模的，小型的

small shutdown spheres 停堆小球
small-signal equivalent circuit 小信号等效电路
small-size computer 小型计算机
small sized wind machine 小型风力机
small sized wind tunnel 小型风洞
small size eddy 小尺度涡流
small size motor 小型电动机
small specimen 小试件
small submarine reactor 美国小型潜艇反应堆
smalls 煤的颗粒等级，末煤【0～13 毫米】
small tools such as bench clamp 台钳等小型工具
small U-bend region 小U形弯曲区
small water power station 小（型）水电站
small watershed 小集水区
small watersupply system 小型给水系统
small wind turbine safety 小型风电机组安全性
small wiring 二次接线
small 小的
smart-grid technology 智能电网技术
smart grid 智能电网
smart pressure transmitter 智能（式）压力变送器
smart sensor 智能传感器
SMARTS(status memory and real-time system) 状态存储器及实时系统
smart terminal 智能终端
smart transmitter 智能变送器
smart 智能的
smash glass 破碎玻璃
SMAW(shielded metal arc welding) 手工电弧焊，自动保护金属极电弧焊，气体保护金属极电弧焊，熔化极自动保护电弧焊
smaze 烟雾，干性烟，烟霾【smog+haze】
SMD(scheduling management display) 调度管理显示
smearable contamination 可擦抹的污染
smear camera 扫描摄影机
smear density 有效密度
smeared-out boundary 模糊不清边界
smear test 擦拭法检查
smear 弄污，涂抹法，涂，擦，敷，涂片，擦拭法，擦拭法检验
smectite 蒙脱石
smee cell 银锌电池
smelter 冶炼炉，熔炉，冶金厂
smelting point 熔化温度，熔点
smelting waste 熔炼废渣
smelt 熔炼，精炼
SMF(system management facilities) 系统管理设备
S/MH(spacing-to-mounting height ratio) 距高比
SMI(static memory interface) 静态存储器接口
Smith chart 史密斯圆图
Smith-Putnam wind turbine 史密斯普特南大型风轮机
smith welding 锻接
smith 锻造，铁匠
SML(simulator load) 模拟器负载
smock mill 裙形古风车
smog aerosol 烟雾气溶胶
smog alert 烟雾警报
smog chamber 烟雾室
smog control 烟雾控制
smog episode 烟雾中毒事件
smog-free 无烟雾
smog horizon 烟雾层顶，烟雾顶层
smog index 烟雾指数
smog injury 烟雾危害
smog reaction 烟雾反应
smog-sensitive 对烟雾敏感的
smog 烟雾
smog warning 烟雾警报
smoke abatement device 消烟除尘设备
smoke abatement 除烟，消烟
smoke agent 发烟剂，烟雾剂
smoke alarm system 烟气报警系统
smoke alarm 烟气报警器，防烟警报
smoke aloft 高空烟
smoke and dust 烟尘
smoke and soot 烟尘，烟灰
smoke bay 烟分区，烟罩
smoke belching 冒黑烟，冒浓烟
smoke black 烟炱，烟黑
smoke box 烟箱
smoke cloud 烟云
smoke concentration 排烟浓度，烟浓度
smoke condensate 烟凝结物，聚烟器
smoke control door 控烟门
smoke control 烟气控制
smoke curtain 挡烟帘
smoke damage 烟害
smoke damper 烟气挡板，防烟阀
smoke deflector 折烟器
smoke density indicator 烟气浓度计，烟浓度指示器
smoke density meter 烟浓度计，烟尘密度计，烟气浓度测定器
smoke density 排烟密度，烟尘密度，排烟浓度
smoke detection system 检烟系统
smoke duct 烟道，烟尘
smoke eliminator door 带消烟器的炉门，消烟炉门
smoke emission standard 排烟标准
smoke emission 冒烟，排烟
smoke evacuation system 排烟系统
smoke evacuation 排烟
smoke exhaust damper 排烟阀
smoke exhaust 排烟
smoke extraction 排烟，烟尘抽取
smoke extractor 排烟风机
smoke eye 烟雾报警器，烟眼
smoke filter 雾烟过滤器
smoke fire detector 感烟火灾探测器
smoke flow visualization 烟流流动显示，烟流显示，烟流显形
smoke flue 烟道
smoke fog 烟雾
smoke from diesel exhaust 柴油机排烟气
smoke funnel 烟筒，烟囱
smoke generator 烟雾发生器，发烟器
smoke grenade 烟幕弹
smoke haze 烟霾

smoke horizon 烟层顶
smoke indicator 烟气浓度计
smoke injury 烟害
smoke laden air 含烟空气
smokeless coal 无烟煤
smokeless combustion 完全燃烧，无烟燃烧
smokeless fuel 无烟燃料
smokeless powder 无烟火药
smokeless zone 无烟区
smokemeter 烟尘测量计，烟尘计
smoke nozzle 喷烟嘴
smoke ordinance 烟管理条例
smoke pall 厚烟层
smoke particle 烟粒
smoke pattern 烟流谱
smoke pipe 排烟管，烟囱，烟道
smoke plume （排烟）烟柱，羽状烟缕，烟羽
smoke point 发烟点
smoke pollution 烟污染，烟雾污染
smoke prevention and dust control 消烟除尘
smoke prevention well 防烟楼梯间
smoke prevention 除烟法
smoke probe 烟雾探测器
smoke proof damper 防烟阀
smoke proof staircase 防烟楼梯间
smoke protection 烟尘防护
smoke puff 烟迹
smoker detector 烟雾指示器
smoke recorder 烟尘记录器
smokescope 烟尘密度测定计，烟气浓度计，检烟镜
smoke screen 烟幕
smoke shade 烟雾色调
smoke shaft 排烟竖井
smoke spot 烟斑
smoke stack bead 烟囱帽
smoke stack emission 烟囱排放
smoke stack head 烟囱头
smoke stack hood 烟囱帽
smoke stack outlet 烟囱出口
smokestack 烟囱，烟筒，烟囱帽
smoke stone 烟晶
smoke stratification 烟层
smoke suppressant additive 防烟添加剂
smoke suppressant 防烟剂
smoke test cycle 烟尘浓度测试循环
smoke tester 烟浓度检验计
smoke test 火焰高度试验，烟示气流法，通烟试验
smoke trail 烟迹
smoke tube boiler 烟管锅炉
smoke tube 烟管，火管
smoke vent 排烟口，排烟风门
smoke visualization 烟流流动显示
smoke volatility index 烟挥发指数
smoke wind tunnel 烟风洞
smoke wire method 烟丝法
smoke zone 烟雾带
smoke 冒烟，烟尘，烟气
smoking agent 发烟剂
smoking room 吸烟室
smoking 冒烟
monthly discharge 月流量
smooth armature 光滑电枢，无槽电枢
smooth blasting 光面爆破
smooth body ACSR 光面钢芯铝绞线
smooth burning 稳定燃烧
smooth circular bend 平滑圆形弯道
smooth coast 平直岸
smooth conductor 光滑导体
smooth core armature 平滑电枢，无槽电枢
smooth-core generator 隐极发电机，无槽发电机
smooth core motor 无槽电动机
smooth core 光滑铁芯，无槽铁芯
smooth curve 平滑曲线
smooth-cut file 平纹锉
smoother 整平器
smooth-faced wheel 光端面齿轮
smooth flow 平滑流，平滑流动，平直水流
smooth fluidization 平稳的流态化
smooth grate 平炉排
smoothing and dressing of the heater well bevel 铣削热套管的斜边
smoothing choke 滤波阻流圈，平滑扼流圈
smoothing device 平滑装置，滤波器
smoothing factor 平滑因子
smoothing filter 平滑滤波器
smoothing iron 压平铁碾
smoothing reactor 滤波电抗器，平波电抗器，平滑电抗器
smoothing tool 打磨工具
smoothing 磨光，轧光【混凝土】，平滑，校正，滤波，滤除，精加工，使平滑
smoothline 参量均匀分布线，平滑线
smoothness 光滑度，平滑度，平滑
smooth nozzle 平滑管嘴
smooth operation 平稳操作
smooth passage 光滑通道
smooth profile 平滑断面
smooth raised face 光滑突面
smooth roller press 光面辊式压球机
smooth roller 整理机，光辊
smooth running 平稳运行，平稳运转
smooth sea 微浪，小浪，二级风浪
smooth starting 平稳启动
smooth steel pipe 光面钢管
smooth surface 平滑（水）面
smooth-wall waveguide 滑壁波导管
smooth-wheel(ed)roller 光轮压路机
smooth-wheel roller 平轮压路机，平碾
smooth wind 平滑风
smooth-wound armature 光滑电枢，无槽电枢
smooth 光滑的，整平
smothering pan 闷熄箱
smother 冒烟，浓烟，熄火，使窒息
smoulder 冒烟，烟，发烟燃烧，阴燃，闷烧，慢燃，低温炼焦
SMR(solid-moderated reactor) 固体慢化反应堆
SMR(standard mortality ratio) 标准死亡比
SM(slave module) 从模块
SMS(semiconductor-metal-semiconductor) 半导体金属半导体

SMS(special monitoring system) 特殊监测系统
SM(status modifier) 状态修改符
SM(successor matrix) 后继矩阵
smudge 斑点，弄脏
smuggled goods 走私货
smuggled 走私
smut 污垢，煤烟，污物，劣质煤
SMVC(sample measuring value control) 采样测量值控制
SMV(sample measuring value) 采样测量值
SMX(submultiplexer unit) 子多路转换器
snab pulley 增面滚筒，改向滚筒，扼紧滚筒
snagging grinder 粗磨床
snagging 不平整的
snag grinding machine 砂轮机
snag 暗礁，障碍
snake bar 蛇形钢筋
snake （敷线用）牵引线，蛇形
snaking of cable 电缆的蛇形敷设
snaking 蛇形，蜿蜒
snap action contact 瞬动触点，突动触点，速动触点
snap action 突动作用，快动作
snap back test 快速复位试验
snap gauge 卡规
snap-head rivet 圆头铆钉
snap lock 弹簧锁
snap off diode 急变二极管
snap-on mounted 卡轨安装
snap-on socket wrench 有爪套筒扳手
snap-on 搭锁的
snapover 闪弧
snapped bolt 圆头螺栓
snapping sound 气蚀噼啪声
snap ring 开口环
snapshot dump 抽点打印
snapshot 抽点打印，抽点打印程序
SNAP(simulated network analysis program) 模拟网络分析程序
SNAP(structural network analysis program) 结构网络模拟程序
snap switch 瞬动开关，速动开关，快动开关
SNAP(system for nuclear auxiliary power) 核辅助动力系统
snap valve 快动阀
snap 快速的，铆头模，抽点打印
SNA(system network architecture) 系统网络结构
snatch block 有盖滑车，转向滑车，开口滑车，紧线滑轮，扣绳滑轮
snatch 抢夺
SNCR(selective Non-Catalytic Reduction) 选择性非催化还原法
SNCR 烟气脱硝
S-N curve for 50% survival 50%存活率的疲劳试验曲线
S-N curve 疲劳试验曲线
S-N diagram 疲劳试验曲线图，应力周数图
sneak circuit 寄生电路
sneak out current 寄生电流，潜行电流
sneak path 潜通路
sneck 塞缝小石块
SNERDI(Shanghai Nuclear Engineering Research & Design Institute) 上海核工程研究设计院【728院】
SNG(substitute natural gas) 代用天然气
snib 门闩，插销
sniffer test 嗅漏试验
sniffer 吸气探针，探测装置，取样器，压强探针
sniffing check （破损燃烧元件）嗅测
sniffle valve 吸气阀
sniff （检漏）取样，探测，吸气，吸入空气，嗅
SNIF(standard neutron irradiation facility) 标准中子辐照设施
snifter 自动充气器
snifting technique 嗅测技术
snifting valve 排气阀，吸气阀，取样阀
snip 剪，剪切小片
SNI(sequence number indicator) 顺序号指示器
SNM(special nuclear material) 特殊核材料
snook rectifier 旋转机械整流机
snore piece 水泵吸入管头，通气孔
snout 喷口
snow and rain 雨夹雪
snow avalanche 雪崩
snowball 急速发展，快速增长
snow banner 雪旗
snow barrier 雪障
snow bearing wind 裹雪风，风花雪
snow belt 雪带
snowberg 雪盖冰山，雪山
snow blockage 积雪阻塞
snow blower 扫雪机
snowbreak thaw 解冻
snowbreak 雪障，防雪林
snow cave 雪洞
snow climate 雪原气候
snow cloud 雪云
snow cover 积雪层，雪被
snow crystal 雪晶
snow day 雪日
snow density meter 积雪密度计
snow deposit 积雪
snow devil 雪卷风
snow drift control 吹雪控制
snow drift density 雪堆密度，雪空间密度
snow drift field 堆雪场
snow drift intensity 吹雪强度
snow drift station 堆雪站
snow drift 风吹雪，雪堆，吹雪，积雪
snow dune 雪丘
snow erosion 霜蚀
snowfall totalizer 累计雪量计
snowfall 雪量，降雪
snow fence 积雪栅，防雪栅
snowfield 雪原
snowflake 雪花
snow forest climate 雪林气候
snow garland 雪花环
snow gauge 量雪器
snow grain 粒雪
snow guard 防雪棚

snow gully 雪沟
snow layer 雪层
snow level 噪声电平，噪声级
snowline 雪线
snow load 雪荷载，雪载
snow management 堆雪控制
snow mat 雪席
snow measuring plate 积雪板，量雪板
snowmelt runoff 融雪径流，雪水径流
snowmelt 融雪，雪水
snow niche 雪蚀龛
snow particle 雪粒
snow pellet 雪丸
snow plow 扫雪机
snow plume 鹅毛雪，雪羽流
snow precipitation line 降雪线
snow protection plantation 防雪林
snow region 雪区
snow removal device 扫雪器
snow retention 积雪
snow roller 雪卷
snows ampler 雪取样器，取雪器
snow scale 雪尺
snow screen 防雪栅
snow shield 防雪设施
snow simulator 雪模拟器
snowslide 雪崩
snow slip 雪崩
snow slope 雪坡
snow squall 雪飑，暴风雪
snow stake 测雪桩
snow storm center 雪暴中心
snow storm meter 雪暴测定仪
snow storm path 雪暴路径
snowstorm 雪暴
snow trap 集雪器，“雪阱”【指除去四氟化锆蒸汽】
snow tube 集雪管
snow 雪
SNPDRI(State Nuclear Electric Power Planning Design and Research Institute) 国核电力规划设计研究院
SNPTC(State Nuclear Power Technology Corporation) 国家核电技术公司
SN(saturation noise) 饱和噪声
S/N(signal-to-noise ratio) 信号噪声比
SNSS(senior nuclear shift supervisor) 核电厂高级值班主任
SNTP(simple network time protocal) 简单网络时间协议
SNTS(starred nonterminals) 加星非终结符
snubber cylinder 缓冲器，缓冲汽缸
snubber hole 消震炮眼，掏槽眼
snubber piston 缓冲器，缓冲活塞
snubber test 减振试验
snubber type shook absorber 干摩擦式减擦器
snubber valve 滞流阀
snubber 阻尼器，减震器，消震器，减振器，消声器，缓冲区，缓冲器，掏槽眼，整体围带
snubbing post 系缆柱
snub 突然制止
snug fit 滑动配合
snug 舒适的
SNW(spot network) 点式网络
soakage 浸渍性，静电荷，吸水量，填充，渗入
soakaway 渗滤坑
soaking-in 浸入，吸入，电荷透入，电荷渐增
soaking method 浸泡清洗法
soaking test 浸湿试验
soaking （焊接）保温，均匀加热
soak out 漏电，剩余放电
soak period 暖机阶段
soak 渗入，浸泡，吸收，浸液，填充
soap bubble leak detection 皂泡检漏
soap bubble test 皂泡检漏，皂泡试验
soap-film analogy 皂膜比拟法，肥皂膜模拟法
soap solution 肥皂水
soap-stone 皂石
soaring wind 上升气流
SOAR(state of the art reports) 技术现况报告
SOA(safe operating area) 安全工作区
SOA(state of the art) 目前工艺水平，目前工艺条件
social activities 社会生活
social benefit 社会效益，社会受益
social cohesion 社会凝聚力
social contract 社会契约
social cost 社会费用
social cybernetics 社会控制论
social economic benefit 社会经济效益
social environment 社会环境
social infrastructure 社会基础设施
social insurance 社会保险
social investigation 社会调查
social overhead investment 社会间接投资
social participation 社会参与
social phenomenon 社会现象
social position 社会地位
social rate of discount 社会贴现率
social security tax 社会保险税
social status 社会地位
social survey 社会调查
social welfare 社会福利
Society of Engineering Science （美国）工程科学学会
socio-cybernetics 社会控制论
socioeconomic system 社会经济系统
socioeconomic 社会经济学的
socket adapter 灯座接合器，管座接合器，接线板
socket-and-spigot joint 承插接头，插口接合，承窝接合
socket bend 管节弯头
socket cap 球窝形钢帽
socket clevis eye 碗头挂板
socket clevis 双联碗头，碗头挂板
socket-connected footing 杯形基础
socket-contact 插座，插座式接点
socket cover 插座盖
socketed tube 套管
socket end 承接端
socket extraction tool 插口抽出工具

socket eye 单联碗头
socket hexagon 3/8 八分之三六角套筒
socket inlet 插座口
socket off-set wrench 弯头套筒扳手
socket outlet adaptor 分支插头，插座底壳
socket outlet 电器插座，插座
socket pipe 承插管，套接管，承口管
socket power 插接电源，外接电源
socket ratchet wrench 套筒棘轮扳手
socket spanner 套筒扳手
sockets set 套筒套装
socket switch 灯头开关
socket tool （机械手）带插口的抓手
socket type fitting 套节管座
socket welded branch pipe 承插焊接的支管
socket welded joint 承插焊连接［接头］
socket weld end connection 承插焊端接
socket weld fitting 承插焊接装配
socket welding end 承插焊端
socket welding flange 承插焊法兰
socket welding 承插焊
socket weld 插焊，承插焊接，承插焊缝
socket wrench with 3c-handle 3c 把手套筒扳手
socket wrench with T-handle and magnet 带 T 形柄及磁性的套筒扳手
socket wrench 套筒扳手
socket 插孔，插座，管座，灯座，承窝，插口，承口，活接头，管接头，管套，承插焊支管台，插入式连接器，套管，套节，轴孔
socle girder 悬臂梁
socle 座石
SOC(shipper's own container) 货主箱
soda acid extinguishing agent 酸碱灭火剂
soda acid fire extinguisher 酸碱灭火器
soda asbestos 苏打石棉
soda ash 纯碱，无水碳酸钠，苏打灰
sodaclase granite 钠长花岗岩
sodafining 碱洗
soda glass particle 钠玻璃颗粒
soda glass 钠玻璃
soda-lime 苏打石灰，碱石灰
sodalye 氢氧化钠
soda process 苏打软水法，苏打法
SODAR(sonic detection and ranging) 声波探测与测距
soda solution grouting 碱液法
SODA(source-oriented data acquisition) 面向源程序的数据采集
SODAS(structure-oriented description and simulation) 面向结构的描述和模拟
soda wash solution 碱洗溶液
soda 纯碱，苏打，碳酸钠
sod cover 草皮覆被
sod culture 草皮培植
sodded spillway 草皮溢水道
soddy soil 生草土
sodic soil 钠质土，苏打土
sodiium-D_2O reactor 钠冷重水反应堆
sodium aluninate 铝酸钠，偏铝酸钠
sodium analyzer 钠表
sodium-based chemicals 钠化学药物
sodium-base grease 钠基润滑脂
sodium benzoate 苯甲酸钠
sodium bicarbonate powder 碳酸氢钠干粉
sodium bicarbonate 碳酸氢钠
sodium bisulfate 亚硫酸氢钠【还原剂】
sodium boiling noise 钠沸腾噪声
sodium bonding 钠结合
sodium carbonate 碳酸钠
sodium carbon reactor 钠-石墨堆
sodium carrying plant component 载钠设备部件
sodium carrying system 载钠系统
sodium chloride battery 氯化钠电池
sodium chloride 氯化钠
sodium cinnamate 肉桂酸钠
sodium clean-up system 钠净化系统
sodium cold trap 钠冷阱
sodium cooled breeder reactor 钠冷增殖反应堆
sodium cooled decay store 钠冷衰变贮存
sodium cooled fast breeder reactor 钠冷快中子增殖（反应）堆
sodium cooled fast reactor 钠冷快中子反应堆
sodium cooled graphic moderated reactor 钠冷石墨慢化堆
sodium cooled reactor 钠冷中子堆，钠冷反应堆，钠冷堆
sodium cooled thermal reactor 钠冷热中子反应堆
sodium cooled zirconium hydride moderated reactor 钠冷氢化锆慢化反应堆
sodium cycle 钠周期，钠型循环
sodium dehydrogen phosphate 磷酸二氢钠
sodium deposition 钠沉积，钠沉积物
sodium discharge lamp 钠气灯
sodium D_2O reactor 钠冷重水慢化反应堆
sodium draining system 疏钠系统
sodium electrode 钠电极
sodium exchanger 钠离子交换器
sodium exchange 钠交换
sodium experimental circuit 钠试验回路
sodium expulsion 钠逐出
sodium filled transfer pot 充钠输送罐
sodium flow measurement 钠流量测量
sodium flowmetering 钠流量测量
sodium formate 甲酸钠，蚁酸钠
sodium-free water 无钠水
sodium freeze-up 钠冻结
sodium gluconate 葡萄糖酸钠
sodium graphite reactor 钠石墨反应堆
sodium hammer effect 钠锤效应
sodium heat transfer system 钠传热系统
sodium hexametaphosphate 六（聚）偏磷酸钠
sodium hot trap 钠热阱
sodium hydrate 氢氧化钠
sodium hydrogen carbonate 碳酸氢钠
sodium hydrogen phosphate 磷酸氢二钠
sodium hydrogen sulfite 亚硫酸氢钠【还原剂】
sodium hydroxide solution 氢氧化钠溶液
sodium hydroxide 烧碱，苛性钠，氢氧化钠
sodium hypochlorite generator 次氯酸钠发生器

sodium hypochlorite 次氯酸钠【氧化剂】
sodium indigo disulfonate 靛蓝二磺酸钠，酸性靛蓝
sodium inlet pipe 钠进口管
sodium intermediate circuit 钠中间回路
sodium iodide crystal 碘化钠晶体
sodium iodide 碘化钠
sodium ion analyzer 钠离子分析器
sodium ion electrode 钠离子电极
sodium ion sensitive glass electrode 钠离子敏感玻璃电极
sodium lactate 乳酸钠
sodium lamp 钠气灯
sodium leakage flow 钠漏流
sodium leakage tank 钠泄漏箱
sodium leak detector 钠泄漏探测器
sodium level gauge 钠液位表，钠液位探头
sodium level instrumentation 钠液位测量仪器，钠液测量
sodium level measurement 钠液位测量
sodium logging effect 钠浸效应
sodium loop （传热）钠回路
sodium lubricated hydrostatic bearing 钠润滑静压轴承
sodium lubricated hydrostatic floating bearing 钠润滑静压浮动轴承
sodium metaborate 偏硼酸钠
sodium metaphosphate 偏磷酸钠
sodium metasulfite 焦亚硫酸钠
sodium naphthalene sulfonate 萘磺酸钠
sodium nitrite 亚硝酸钠
sodium orthophosphate （正）磷酸钠
sodium pentaborate solution 五硼酸钠溶液
sodium permanganate 高锰酸钠
sodium peroxide 过氧化钠
sodium persulfate 过硫酸钠
sodium photocell 钠光电池
sodium pool 钠池
sodium potassium alloy 钠钾合金
sodium potassium tartrate 酒石酸钾钠
sodium process level 工艺钠位，钠工艺液位
sodium pump 钠泵
sodium purification system 钠净化系统
sodium reactor 钠冷反应堆
sodium refilling system 充钠系统
sodium-saturated clay 钠饱和黏土
sodium separation device 钠分离装置
sodium silicate binder 水玻璃黏结剂，钠水玻璃
sodium silicate cement 硅酸钠水泥
sodium silicate concrete 水玻璃混凝土
sodium silicate mastic 水玻璃胶泥
sodium silicate mortar 水玻璃砂浆
sodium silicate 硅酸钠，水玻璃
sodium sulfate 硫酸钠
sodium sulfite 亚硫酸钠
sodium-sulfur accumulator 钠硫蓄电池
sodium sulphur battery 钠硫电池
sodium thiosulfate 硫代硫酸钠
sodium-to-sodium exchanger 钠钠热交换器
sodium tripolyphosphate 三聚磷酸钠
sodium vapour lamp 钠蒸气灯
sodium vapour trap 钠蒸气收集器，钠蒸气阱
sodium-void coefficient 钠空穴系数
sodium-void effect 钠空效应
sodium-void worth 钠空穴（反应）价值
sodium washing effluent 钠洗涤废水
sodium wash plant 钠洗涤装置
sodium water heat exchange 钠水热交换
sodium water reaction test 钠水反应试验
sodium water reaction 钠水反应
sodium wetted 浸钠的
sodium zeolite 钠沸石
sodium 钠
sod revetment 草皮护岸，草皮铺面
sod strip 带状草皮
sod turf 草皮
sod waterway 草地泄水道
sod 草皮
SOE(sequence of event) 事件的顺序，事故顺序记录
sofa and coffee table 沙发茶几
soffit cusp 拱尖
soffit level 板底标高
soffit lining 拱底衬砌
soffit radius 拱内弧半径，内拱圈半径
soffit scaffolding 拱腹架，砌拱支架
soffit springing 拱内弧起拱线
soffit spring 拱背起拱线点
soffit 拱腹，拱内侧面，涵管内顶面，下端
SOF(statement of facts) 装卸事实记录
soft aggregate 软骨料
soft and free expansion sheet making plant 软板及自由发泡板机组
soft-annealed wire 软金属线
soft annealing 软化退火，不完全退火
soft arc 软电弧
soft asphalt 软质沥青
soft black 碳黑，烟炱
soft braking 软制动
soft chemical decontamination process 软化学去污工艺
soft clay 软黏土
soft component 软部件【辐射】
soft constraint 软约束
soft control 软控制
soft currency 软通货
soft cushioning 软式减震
soft cut in 软切入
soft damping 软阻尼
soft design 软件设计
soft device 软设备
softdog 软件狗
soft-drawn wire 软拉线
soft driven system 软传动系统
soft driven train 软传动系
softenability 可软化性
softenable 可软化的
softened water pump 软水泵
softened water tank 软水箱
softened water 软化水
softener 软化剂，软化器，增塑剂
softening agent 软化剂

softening coefficient 软化系数
softening filter 软化滤水器，软化过滤器
softening of water 水软化
softening point test 软化点试验
softening point 软化点
softening process 软化法，软化过程
softening temperature 软化温度
softening 软化退火【硼硅玻璃】，软化
soften the terms 放宽条件
softest possible terms 最优惠的条款
soft foundation 软基
soft gasket 软密封垫
soft-glare lighting 软眩光照明
soft grid connection 软并网
soft ground 松土地
soft hail 软雹
soft handling 软操作
soft iron armature 软铁电枢
soft iron core 软铁芯
soft iron instrument 软铁芯电磁测量仪表
soft iron oscillograph 电磁示波器
soft keyboard 软键盘
soft landing （经济）软着陆
soft limestone 软石灰岩
soft loan 软贷款，优惠贷款，贴息贷款
soft machine check 机器软件校验
soft magnetic material 软磁材料
soft manual operation 软手操
soft manual 软手操（作），软手动
soft muddy clay 软泥质黏土
soft mud 软淤泥
soft packing 软垫
soft particle 软弱颗粒
soft patch 柔性搭板
soft radiation 软射线
soft rime 雾凇
soft rock 软岩
soft-sealing fasteners 软密封固定器
soft-seated valve trim 软阀座结构
soft seating 软落座【腈等】
soft-sectored 程序分段的，软扇区的
soft self-excitation 软自励，软自激
soft soil base 软土地基
soft soil 软土
soft spot 软点，模糊光点，弱点
soft starter 软启动器
soft start type 软启动型
soft start 软启动，缓启动，平稳启动
soft steel 软钢，低碳钢
soft stratum 软弱层
soft subsoil 软底土
soft superconductor 软超导体
soft terms 特惠条款，优待办法
soft tower 软性塔架
soft transmission system 软传动系统
software compatibility 软件兼容性
software configuration 软件组态
software cost 软件成本
software design procedure 软件设计过程
software development library 软件开发库
software development plan 软件开发计划
software development process 软件开发过程
software documentation 软件资料
software document 软件文件
software engineering 软件工程
software environment 软件环境
software fix 固定软件，软件修复
software flexibility 软件灵活性
software for wind farm design 风电场设计软件
software interface 软件接口
software library 软件库
software maintenance 软件服务，软件维护
software monitor 软件监督，软件监督程序
software multiplexing 软件复用
software package of CAD 计算机辅助设计软件包
software package of computer aided design 计算机辅助设计软件包
software package 软件包
software piracy 软件私自复制
software platform 软件平台
software portability 软件可移植性
software positional limit 限位软件
software product 软件产品
software psychology 软件心理学
software quality 软件质量
software reactive power droop setting 软件无功调差
software rejuvenation 软件恢复【针对软件老化】
software reliability 软件可靠性
software resource 软件资源
software revision 软件版本
software system 软件系统
software technique 软件技术
software testing plan 软件测试计划
software testing 软件测试
software test system 软件测试系统
software tool 软件工具
software 软件，软设备，程序系统
soft water 软水
soft winding 软绕组
soft wood 软木，针叶树材
soft 柔软的
soggy soil 湿润土
SOH(scheduled outage hours) 计划停用时数
SOH(start of header) 标头开始
soil accumulation yard 堆土场
soil aggregate mixture 土集料混合物
soil aggregate surface 碎石土面层，土砂石路面
soil aggregate 土团粒，集料土，碎石土，土混骨材，土料，土壤团聚体
soil alkalinization 土壤盐碱化，土壤改良
soil amelioration 土壤改良
soil analysis 土（粒径）分析
soil anchor 土层锚杆
soil and rock erosion 岩土侵蚀
soil and water conservation engineering 水土保持工程
soil and water conservation 水土保持
soil and water loss 水土流失
soil auger 麻花钻，土样钻，土钻
soil bearing capacity 土壤承载能力

soil bearing test 土的承载力试验
soil bitumen 土沥青
soil borrow area 取土区
soil borrow pit 取土坑
soil borrow 取土
soil bulk density 土壤容重
soil cement 稳定土【基础】，水泥及土混合料，掺土水泥，土水泥
soil classification 土的分类，土壤分类
soil climatology 土壤气候学
soil cohesion 土的黏结力，土壤黏聚力
soil colloid 土胶体
soil compaction pile method 挤密土桩法
soil compaction 土的压实，土壤压实
soil composition 土的组成，土壤成分
soil compressibility 土的压缩性
soil concrete 掺土混凝土
soil conditioner 土壤改良剂
soil conductivity 土的导电性，土壤电导率
soil conservation 土壤保持
soil consistency 土壤稠度
soil consolidation 土的固结，土壤固结
soil constant 土常数
soil core tube 取土样管
soil core wall 土心墙
soil cover 地被物，土被
soil cracking 土裂
soil creep 土的蠕动，土滑
soil damping 基土阻尼
soil data 土壤参数
soild block 实心砌块
soil dike 土堤
soil displacement 换土
soil drifting 土壤风蚀
soil dyke 土堤
soil engineer 土工工程师
soil erosion 土壤侵蚀，土蚀
soil evaporimeter 土壤蒸发仪
soil excavation 挖土
soil expansion 土膨胀
soil exploration 土壤勘察 土壤勘探，土质查勘，土样钻探
soil flexibility 基土柔度
soil flow 泥流
soil-forming rock 成土母岩
soil foundation 土基
soil fraction 土粒
soil friction circle method 按圆弧滑动面分析法【土体稳定计算用】
soil gradation 土的级配
soil grain 土的粒径，土粒
soil group 土类
soil heterogeneity 土的不均一性
soil horizon 土层
soil humidity 土的湿度
soil identification 土的鉴定
soil impermeability 土的不透水性
soil imperviousness 土的不透水性
soil improvement 土壤改良
soil in situ 现场土，原位土
soil investigation 土壤调查，土质勘察，土壤勘察，土壤勘探
soil layer 土层
soil legend 土壤图例
soil levee 土堤
soil lime 石灰土
soil liquefaction 土壤液化，土的液化
soil loss tolerance 容许土壤流失量，土壤流失容许限度
soil lump 土块
soil mantle 土被
soil mass 土体
soil material 土料
soil mechanics laboratory 土工试验室，土力学试验室
soil mechanics 土力学
soil moisture content analyser 土壤水分测定仪
soil moisture content 土壤含水量
soil moisture meter 土壤湿度计
soil moisture 土壤水分
soil mortar 土灰浆
soil nailing wall 土钉墙
soil nail 土锚钉
soil parameter 土壤参数
soil particle size 土粒大小，土粒粒径
soil particle 土粒
soil permeability 土的透水性，土壤渗透性
soil phase 土相
soil piping 土的管涌
soil porosity 土的孔隙率，土壤空隙度
soil pressure cell 土压盒
soil pressure 土壤压力，土压力
soil profile 土的剖面
soil quality 土质
soil reconnaissance 土的查勘
soil replacement method 换土法
soil replacement works 换土工程
soil resistivity 土壤电阻率
soil retaining structure 挡土建筑物
soil retaining 挡土的
soil-rock composite subgrade 土岩组合地基
soil sampler spoon 取土器
soil sampler 采土器，取土器，取土样器
soil sample test 土样试验
soil sample 土样
soil-saving dam 保土坝
soil-saving dike 拦土堤
soil science 土科学
soil sealant 土样密封剂
soil sedentary soil 残积土
soil series 土系
soil shear strength 土壤抗剪强度
soil skeleton 土的骨架
soil slip 塌方，坍方，土崩，土滑
soil specimen 土样
soil spring 土壤弹性
soil stabilization by electro-osmosis 电渗法稳定土体
soil stabilization by freezing 冻结法稳定土体
soil stabilization 土壤稳定，土体稳定
soil stabilizer 土的加固剂，土的稳定剂
soil stratum 地层，土层

soil strength 土的强度
soil-structure interaction 土基与建筑物相互作用
soil-supported structure 土支承结构物
soil surface 土表面
soil survey 地质勘测【勘查，勘察】，土壤调查
soil suspension 土悬液
soil technology 土工工艺学
soil testing laboratory 土壤试验室
soil test 土工试验
soil texture classification 土结构分类，土组织分类
soil texture 土壤粗密度，土质，土结构，土组织
soil thermometer 土温计
soil thrust 土壤压力
soil triangle 土的三角坐标图
soil vibrator 土的振动器
soil water conservation 水土保持
soil water experiment 土壤水分实验
soil water relationship 土水关系
soil water 土壤含水
soil wedge 土楔体
soil weight 土重
soil 土，土地，土壤，地面，大地，泥土
solar-air temperature 太阳作用气温，综合气温
solar air conditioner 太阳能空调器
solar air heater 太阳能空气加热器
solar air-heating system 太阳能空气加热系统
solar altitude angle 太阳高度，太阳高度角
solar altitude 太阳高度（角）【h】
solar and wing climatology 太阳和风气候学
solar appliance 太阳能器具
solar array banket 太阳电池方阵库
solar array 太阳能电池阵，太阳电池板，太阳组
solar-assisted reactor 太阳能辅助反应堆
solar atmosphere 太阳大气
solar azimuth angle 太阳方位角
solar azimuth 太阳方位角【ψ】，太阳方位
solar battery 太阳（能）电池
solar cell area 太阳电池面积
solar cell array 太阳电池方阵，太阳电池组，太阳电池阵，太阳能发电器
solar cell basic plate 太阳电池底板
solar cell module area 太阳电池组件面积
solar cell module surface temperature 太阳电池组件表面温度
solar cell module 太阳（能）电池组件
solar cell motor 太阳能电动机
solar cell panel 太阳能电池板
solar cell temperature 太阳电池温度
solar cell 太阳电池，太阳能电池，太阳能电池片
solar chimney 太阳能烟囱，太阳烟囱
solar climate 数理气候，太阳气候，天文气候
solar collector array 太阳能集热器阵列
solar collector system 太阳能集热系统
solar collector 太阳能集热器
solar component tide 太阳分潮
solar concentrating system 太阳聚光系统
solar concentrator 太阳聚光器
solar constant 太阳常数【E_{sc}】
solar contribution 太阳能供热量
solar cooker 太阳灶
solar cooling（air-conditioning）system 太阳能制冷（空调）系统
solar daily radiation 太阳能日辐射量
solar day 太阳日
solar declination 太阳赤纬【δ】
solar desalination system 太阳能海水淡化系统
solar disk 日面，太阳圆面，太阳圆盘
solar distillation 太阳能蒸馏
solar diurnal tide 太阳全日潮
solar dryer 太阳能干燥器
solar drying system 太阳能干燥系统
solar electrical energy generation 太阳能发电
solar electric panel 太阳电池板
solar elevation angle 太阳高度角
solar energy boiler 太阳能锅炉
solar energy cell array 太阳能电池阵列
solar energy collector 太阳能集热器，太阳能聚热器，太阳能收集器
solar energy conversion facility 太阳能转换设施
solar energy conversion 太阳能转换
solar energy prime mover 太阳能发动机
Solar Energy Research Institute （美国）太阳能研究所
solar energy storage 太阳能的存储
solar energy system 太阳能系统
solar energy water heater 太阳能热水器
solar energy 太阳能
solar evaporation pond 太阳能蒸发池
solar evaporation 曝晒蒸发
solar field 太阳能场，太阳能集热场
solar flux 太阳辐射通量，太阳能通量
solar fraction 太阳能保证率
solar furnace 太阳炉
solar generator 太阳能发电机
solar grade polysilicon production technology 太阳能级多晶硅生产技术
solar heat collector 太阳集热器
solar heat engine 太阳热发动机
solar heater 太阳能加热器
solar heat gain factor 太阳增热因数
solar heating of buildings 建筑物的太阳能取暖
solar heating system 太阳能加热系统
solar heating 太阳能采暖
solar heat 太阳热
solar hot water system 太阳（能）热水系统
solar hour angle 太阳时角【ω】
solar hydrogen energy system 太阳氢能系统
solarigraph 总日射计
solarimeter 日射总量表，太阳辐射强度计，总日射表
solar incident angle 太阳入射角
solar infrared spectrum 太阳红外光谱
solar insolation 太阳辐射，太阳辐照量
solar irradiance simulator 太阳（辐照度）模拟器
solar irradiance 太阳辐照，太阳辐照度
solar irradiation 太阳辐照量
solarization 负感现象，暴晒作用
solar magnetic 太阳磁周期
solar magnetograph 太阳地磁记录仪

solar mirror 太阳能反射镜
solar module array 太阳能电池组件阵列
solar noon 日照午时，中午日照时，太阳正午
solar-only system 太阳能单独系统，太阳能独立系统
solar orientation 朝阳方位
solar panel 太阳能电池板
solar parallax 太阳视差
solar photovoltaic energy system 太阳光伏能源系统
solar photovoltaic module 光伏模组，太阳能光伏模块
solar photovoltaic power generation 太阳能光伏发电
solar photovoltaic 太阳能光伏发电
solar plus supplementary heat-source system 太阳能带辅助热源系统
solar pond 太阳能贮藏设备，太阳能贮藏池，太阳池
solar power corporation 太阳能动力公司
solar power generation 太阳能发电
solar power plant 太阳能发电厂，太阳能发电站
solar power satellite station 太阳能卫星电站
solar power satellite 太阳能电力卫星
solar power station 太阳能电站，太阳能发电站
solar power 太阳能，太阳能动力
solar preheat system 太阳能预热系统
solar pump 太阳能泵
solar radiant energy 太阳辐射能量
solar radiant heat 太阳辐射热
solar radiation flux 太阳辐射通量
solar radiation intensity measuring 太阳辐射强度测量
solar radiation intensity 太阳辐射强度
solar radiation meter 太阳辐射计
solar radiation sonde 日射探空仪
solar radiation spectrum 太阳辐射波谱
solar radiation thermometer 日射温度计
solar radiation 太阳辐射，日光照射
solar radiometer 太阳辐射计
solar radiometry 太阳辐射测量
solar reflection panel 太阳能反射板
solar reflective curing membrane 混凝土养护日光反射膜
solarscope 太阳方位仰角显示器
solar semidiurnal component tide 太阳半日分潮
solar simulator 太阳模拟器
solar spectrum 太阳光谱
solar spot light 太阳聚光灯
solar still 太阳能蒸馏器
solar stove 太阳灶
solar system 太阳能系统
solar terms 节气
solar thermal collector 太阳（能）集热器
solar thermal distributed power system 分散式太阳能热发电
solar thermal energy collecting mirrors 太阳热能收集镜
solar thermal power generation system 太阳能热发电系统
solar thermal power generation 太阳能光热发电
solar thermal power plant 太阳能热电站
solar thermoionic generator 太阳离子发电机
solar thermoionic power generation 太阳能热离子发电
solar tide 太阳潮，太阳潮汐，日潮
solar time 日照时间，太阳时
solar-to-fuel energy conversion efficiency 太阳能与燃料能转换效率
solar total radiation 太阳总辐射
solar tracker with shade disk kit 自动遮光装置
solar tracker 太阳跟踪器
solar tracking device 太阳追踪设施
solar tracking system 太阳追踪系统
solar water heater 太阳能热水器
solar water heating system efficiency 太阳热水系统效率
solar water heating system 太阳（能）热水系统
solar wind energy hybrid system 太阳能风能混合系统
solar zenith angle 太阳天顶角，太阳高度角
solar 太阳的，太阳能的
solderability 钎焊性
solder connector 钎焊接头
solder covered wire 焊料涂覆铜线
solder dipping 浸焊，搪锡
soldered fitting 钎焊接头
soldered joint （低温）焊接接头，软钎焊接头，软钎焊连接
soldered seam 钎焊焊缝
solderer 焊工，钎焊机
soldering blowpipe 钎炬
soldering flux 焊剂，钎焊剂
soldering hammer 钎焊焊接器，焊烙铁
soldering iron 焊铁，（锡焊）烙铁
soldering machine 钎焊机
soldering point 焊接点
soldering seam 焊缝
soldering set 焊具
soldering tag 焊片
soldering terminal 接线柱，焊片
soldering tin bar 焊锡条
soldering torch 钎焊焊炬
soldering wire 焊丝
solder interface 软钎焊面
solder joint 钎焊接头
solderless lug 无焊料接线片
solderless terminal 无焊接线柱【线夹、螺丝快速拆卸等】
solder lug 钎焊接线片
solder metal 软钎缝金属
solder paste 钎焊膏，钎焊剂
solder spike 焊料毛刺
solder strip 钎焊条
solder 钎焊，焊料，焊锡，焊剂，软钎缝，软钎料，焊，锡焊，低温焊料
soldier arch 立砌砖拱
soldier beam 立桩【基坑围护】
soldier course 立砌砖层，排砖立砌
soldier 模板立柱，模板支撑
sole agency 独家代理
sole licence contract 独家许可（证）合同

sole licence 排他许可证，独家许可
solenoid actuated relay 螺线管启动式继电器
solenoid-actuated sampling valve 螺线管采样阀
solenoid actuated stepping motor 螺线管线圈励磁的步进电动机
solenoid actuation 螺线管调节，螺线管驱动
solenoid actuator 电磁阀执行器，电磁执行机构，螺线管执行器，螺线管执行机构
solenoidal field winding 螺线管磁场绕组
solenoidal field 无散场，螺线管磁场
solenoidal magnetization 纵向磁化，螺线管磁化
solenoidal phasor 无散相量
solenoidal vector field 无散矢量场，螺线矢量场
solenoidal 圆筒形线圈的，螺线管形的
solenoid armature 螺线管式电枢，螺线管衔铁
solenoid brake 螺线管制动器
solenoid clutch 螺线管式离合器
solenoid coil 圆筒形线圈，螺线管线圈，电磁线圈
solenoid core 螺线管心
solenoid operated switch 电磁控制开关
solenoid operated valve 电磁控制阀，电磁阀，螺线管阀
solenoid operated 螺线管操作的，电磁操作的
solenoid pilot valve 电磁先导阀
solenoid plunger 螺线管插棒铁芯
solenoid relay 螺纹管继电器，螺线管式继电器
solenoid switch 磁力开关，螺线管开关
solenoid tripping 电磁脱扣停机
solenoid tumbler 电磁转换开关
solenoid valve 电磁阀，螺线管操作阀，螺线管阀，螺线电磁阀
solenoid vibrating feeder 电磁振动给料机
solenoid 螺线管，螺线管线圈，电磁线圈，柱形线圈，筒形电磁线圈，螺线形电导管
sole piece 钢筋填块【模板内】
sole plate 基础板，底板，垫板，基板，台板
sole proprietorship 独资（经营）
sole source investment enterprise 独资企业
sole source 唯一供应商
sole timber 垫木
sole 底脚，底部，单一的
solfatara 硫磺气喷出孔
sol-gel process 溶胶凝胶法
sol-gel technique 溶胶凝胶技术
sol-gel technology 溶胶凝胶工艺
solicitor 律师
solicit 恳求，索要
solid acoustical lens 固体声透镜
solid active waste building 固态放射性废物贮存厂房
solid adsorbent 固体吸附剂
solid amorphous metal 固态无定形金属的
solid angle 立体角
solid arch 实体拱
solid asphalt 固体沥青
solid balustrade 实心栏杆
solid barrier 实体屏障
solid beam 实体梁
solid bearing 整体轴承
solid bed 定床，固料层
solid blockage 固体阻塞
solid block 实体块
solid body 固体
solid-borne noise 固源噪声
solid-borne sound 固体载声
solid bottom cable tray 整底电缆托架
solid bottom gondola car 无底门敞车
solid bottom tray 整底托架
solid boundary （风洞）实壁边界，固体边壁
solid brick 实心砖
solid buttress 实体支墩
solid cable 胶质浸渍的纸绝缘电缆，实心电缆
solid carbon brush 实心电刷
solid carbon dioxide 固体二氧化碳，干冰
solid carbon 固体碳，烟黑，碳黑
solid carburizing 固体渗碳
solid carrier 固体载体
solid casing 整体泵壳【不拼合】
solid-cast insulation 浇铸固化绝缘
solid-cast transformer 干式变压器，浇注固化式变压器
solid circuit 晶体管电路，固体电路
solid coating 固体涂料
solid concentration 固体浓度
solid concrete beam 实心混凝土梁
solid concrete block 整体混凝土块
solid conductor cable 实心导体电缆
solid conductor 实心导线，实心导体，单线
solid constituent of soil 土的固体成分
solid contact softening 固体接触软化
solid content 容积，固体含量
solid copper wire 实心铜线，单铜线
solid core reactor 固态堆芯反应堆
solid core support column 实心（堆芯支撑）柱
solid core 固体堆芯，实心铁芯
solid coupling 刚性联轴器，刚性联轴节
solid cover 整块盖板
solid cylindrical rotor 实心圆柱形转子，隐极转子
solid dam 实体坝
solid debris 固体碎屑
solid delay line 固体延迟线
solid deposit 泥沙淤积
solid dielectric breakdown 固体电介质击穿
solid discharge 固体径流量
solid-discharging dissolver 可排出固体不溶物的溶解器，排渣溶解器
solid door 防火门
solid drawn tube 无缝管，拉制管
solid draw pipe 无缝钢管
solid drum rotor 整锻鼓形转子
solid earth 固定接地，完全接地
solid electrode 固体电极，实心焊条
solid electrolyte battery 固体电解质电池
solid encapsulated transformer 固封变压器
solid enclosure 密闭外壳，密封罩
solid error 固定误差
solid expansion thermometer 固体膨胀温度计
solid expansion 固体膨胀
solid extruded polymeric insulated cable 固体挤压聚合电缆
solid fin 连续鳍片

solid-fired gas turbine 固体燃料燃气轮机
solid flexible wedge gate valve 整体柔性楔闸阀
solid flow 固体径流，固体流动
solid-forged rotor 整锻转子
solid forging 整体锻造，实锻
solid fuel heterogeneous reactor 固体燃料非均匀反应堆
solid fuelled reactor 固体燃料反应堆
solid fuel 固体燃料
solid gravity dam 实体重力坝
solid grounding 整体接地，直接接地，固定接地
solid-head buttress dam 大头坝
solid heterogeneous reactor 固体非均匀反应堆
solidification fouling 凝固污垢
solidification heat 固化热，凝固热
solidification in cement （放射性废物）水泥固化
solidification of radwaste 放射性废物固化
solidification plant 固化装置
solidification point 固化点，凝固点
solidification process 固结过程
solidification temperature 凝固温度
solidification 固化【放射性废物】，凝固，固化作用
solidified slag 焊渣，渣壳
solidified waste 固化废物
solidified water 固化水，固结水分
solidify concentrate 固化浓缩物
solidifying point 凝点，凝固点
solidify 凝固
solid impurity 固体杂质
solid inclusion 固体杂质
solid ingredients 固体组分
solid inorganic insulation 固体无机绝缘
solid insulating material 固体绝缘材料
solid insulation 固体绝缘
solid insulator 实心绝缘子
solid iron core 实心铁芯
solid iron cylindrical-rotor generator 实心圆柱转子发电机，隐极发电机
solid iron machine 实心铁芯电机
solidity coupled shaft 刚性连接轴
solidity factor 刚度，刚性系数
solidity losses 实度损失
solidity ratio 硬度比，充满系数，实积比，实度比
solidity （风轮的）实度，固化性，固体性，完整性，固态，硬度
solid jet 密实射流，连续射流
solid leaching 固料浸出
solid-leg tripod 固定腿三脚架
solid line 实线
solid liquid cyclone 固液分离旋流器
solid liquid equilibrium 固液平衡
solid logic technology 固态逻辑技术
solid lube bearing 黄油润滑轴承
solid lube 黄油，固体润滑剂
solidly grounded neutral system 中性点永久接地制，中性点直接接地系统
solidly ground 直接接地，永久性接地
solid masonry retaining wall 实体圬工挡土墙
solid masonry 实体圬工
solid mass 实体
solid matrix 骨架，固体基质
solid matter 固体物质
solid mechanics 固体力学
solid-medium sonic delay line 固体介质声延迟线
solid metal serrated gasket 整体金属齿形垫片
solid MIG wire 实心的金属惰性气体保护焊丝
solid mineral fuel 固体矿物燃料
solid molding 固体造型
solid neutral switch 整体中性开关【没有中性触点】
solid neutral 整体中性【没有中性开关】
solid nonmetallic impurity 固体非金属夹杂物
solid oil 固体油
solid panel 厚镶板，实体板
solid particles 固体颗粒，固体粒子
solid partition 实体隔墙
solid pellet 固体芯块
solid pier 实心桥墩
solid plate 实心板，未开孔的板
solid pole motor 实心磁极电动机
solid pole piece 实心极块，实心磁极
solid pole 整块磁极，实心磁极
solid porcelain bushing 实心瓷套管
solid portion （废物浓缩物）固体部分
solid precipitation 固态降水
solid pump 固体输送泵，固体吹散泵，杂质泵
solid puncture 固体击穿，完全击穿
solid radioactive waste storage 固体放射性废物贮存库
solid radioactive waste treatment 放射性固体废物处理
solid radioactive waste 固体放射性废物
solid radwaste system 固体废物处理系统
solid rate 固体加料率
solid rectangular beam 实体矩形梁
solid removal and proportioning device 固体清除及配比装置
solid residue 固体残余物，固体残渣
solid-rib arch 实肋拱
solid roller bucket 实体消力戽
solid rotor 实心转子，整锻转子，整块转子
solids backflow 固体回流，固体倒流
solids blending 固料掺和
solids concentration 盐浓度，（锅炉水）含盐量
solids-delivery valve 固料输送阀
solid separator pump 杂质分离泵
solids expansion 固相膨胀
solids feeder 固体加料器
solids feed line 固体加料管线
solids feed rate 固体加料率
solids flow 固相流
solids full out 固料充实量
solid shaft 实心轴，整轴
solid short-circuit 稳定短路
solids inventory 固体存料量
solid slab 实心板
solids leakage 漏料
solid sleeve 密实套筒
solids level 料面
solids makeup feed 补充给料

solids of infinite thermal conductivity 热导率为无限大的固体
solid solubility 固体可溶性，固体溶解度
solid solution cermet 固溶体金属陶瓷
solid solution 固体溶液，固溶体
solid-spandrel arch 实腹拱
solids population balance 固料总量平衡
solids removal line 固料排放管
solids run-back 固体漏料
solids-sliding zone 固料移动区
solids stress 固体粒子应力
solid state analog-to-digital computer 固态模数计算机
solid state capacitor 固态电容器
solid state circuit breaker 固态断路器
solid state circuit 固态电路
solid state commutator 固态整流器
solid state component 固态元件
solid state computer 固态计算机
solid state detector 固体探测器
solid state device 固态装置
solid state diode 固态二极管
solid state dosimeter 固体剂量计
solid state electrical conductivity dosimeter 固体电导率剂量仪
solid state electronics 固态电子学
solid state element 固态元件
solid state excitation 固态励磁，固体励磁
solid state exciter 固态励磁装置
solid state frequency converter 固态变频器
solid state inverter 固态逆变器
solid state logic card 固态逻辑组件
solid state logic circuit 固态逻辑电路
solid state logic timer 固态逻辑计时器
solid state logic 固态逻辑
solid state memory element 固态存储元件
solid state memory 固态存储器
solid state motor 固态电动机
solid state physics 固体物理学
solid state power equipment 固体电力设备
solid state power supply 固体电源
solid state rectifier 固体整流器，固态整流器
solid state relay 电子继电器，固体组件继电器，固态继电器，固体继电器
solid state safety logic assembly test 固体安全逻辑装置试验
solid state safety logic assembly 固体安全逻辑装置
solid state sensor 固体传感器
solid state switch 固态开关
solid state technique 固体电路技术
solid state time relay 固体时间继电器
solid state welding 固态焊
solid-state 固态的，使用电晶体的，不用真空管的，固态电子学的，固态物理学的
solid state 固态，固体状态
solid steel pole 实心钢磁极，整块钢磁极
solid steel rotor 整体钢转子
solid-stem liquid in glass thermometer 玻璃温度计里的棒式液体
solid storage engineering test facility 固体（放射性废物）贮存工程试验设施
solid stream of water 连续水流
solids turn over 固体翻转
solids weeping 固料流落
solid swelling 整体膨胀
solids 固体颗粒
solid transport 固体物质输移
solid type cable 油浸纸绝缘电缆，实心电缆
solid type jetty 实体突码头
solid type rotor 实体式转子
solidus 固相点，固相线，凝固线，斜线
solid walled structure 承重墙结构
solid walled 单层容器壁的，实心容器壁的，实心墙的，单层墙的
solid wall test section 实壁试验段
solid waste baler 固体废物包装机
solid waste burial pit 固体废物埋葬坑
solid waste classification 固体废物分类
solid waste compactor 固体废物压实机
solid waste disposal 固体废物处置
solid waste evacuation 固体废物清除
solid waste handling 固体废物处理
solid waste incinerator 固体废物焚化炉
solid waste press 固体废物压实机
solid waste processing system 固体废物处理系统
solid waste storage 固体废物贮存
solid waste treatment facility 固体废物处理设备
solid waste treatment 固体废物处理
solid waste 固体废物，固体废弃物，固体废料，固体排放物
solid water reactor 水冷固体化反应堆
solid web beam 实腹梁
solid web bracket 实腹牛腿
solid web bridge 实腹式桥
solid web column 实腹柱
solid web girder 空腹梁
solid web steel window 实腹钢窗
solid web 实心腹板
solid wedge gate valve 整体楔闸阀
solid wedge valve 整体楔形闸板阀
solid-welded panel 焊成整体的管屏
solid wharf 实体码头
solid wind break 实壁风障，固体防风
solid wire 整根导线，实心焊丝，单股线，实线
solid wood fire-rated door 实木防火门
solid woven belt carcass 整芯编织带芯
solid woven belt conveyor 织物带输送机
solid wrench 死扳手，呆扳手
solid yoke 实心磁轭
solid 单态的【一回路系统】，整体的，固体的，坚固的，实心的，固体
solifluction 泥流作用，土流
solifluxion 泥流作用
solion 溶液离子管
soliquoid 悬浮体
solitary wave 单波
solodization 脱碱作用
solodyne 只用一组电池组工作的接收机
solod 脱碱土
solo operation 单人值班

solo valve 单向阀
solo 单独布置
solstices 二至点【即夏至点和冬至点】
SOL(system output language) 系统输出语言
solubility coefficient 可溶性系数
solubility gradient 溶解性梯度
solubility product 溶度积
solubility test 溶解度实验
solubility 溶解度，可溶性，溶解能力
solubilization 溶解作用，溶液化
soluble ash content 可溶性灰分含量
soluble boron 可溶硼
soluble chemicals 可溶性化学品
soluble chemistry 溶解化学
soluble contaminant 可溶性污物
soluble glass 溶性玻璃，水玻璃
solubleness 溶解度，溶解性
soluble neutron absorber 可溶性中子吸收体
soluble neutron poison 可溶中子毒物
soluble poison 可溶毒物，可溶性核毒物
soluble salt 可溶盐
soluble silica 可溶性氧化硅
soluble starch 可溶淀粉
soluble substance 可溶物质
soluble 溶解的，可溶的
solum 风化层
solute 溶质，溶解物，溶解，溶解的
solution annealing 溶体化退火，固溶退火
solution basin 溶蚀盆地
solution by electrical analogy 电模拟解法
solution cavern 溶洞
solution channel water 溶洞水，溶蚀槽的水
solution culture 溶液培养
solution deposit 溶液沉淀物，溶质沉积
solution feeder 溶液进料器
solution groove 溶沟
solution hardening 溶液硬化，固溶淬火
solution heat treatment 溶液热处理
solution heat 溶解热
solution hole 溶蚀穴
solution opening 溶洞
solution pipeline 溶液管道［线］
solution poison 溶液毒物
solution pool 溶蚀潭
solution pressure 溶解压力，溶解压强
solution reactor 溶液反应堆
solution strengthening 固溶强化
solution tank 溶解槽，溶解箱，溶液箱
solution treatment 固溶处理
solution type reactor tank 溶液型反应堆槽
solution type reactor 溶液型反应堆
solution valley 溶谷
solution 溶解，溶液
solvable problems 可解问题
solvable 可解的，可溶的
solvation 溶解，溶剂化
solved pipe joint 螺纹管连接
solvency power 溶解能力
solvency 偿付能力，偿债能力，溶解力
solvend 可溶物质
solvent ability 溶解度，溶解力
solvent catch tank 溶剂收集槽
solvent cement 溶剂胶接剂
solvent cleaner 溶剂清洗剂
solvent developer 溶剂显像剂
solvent extraction-AAS procedure 溶剂萃取原子吸收光谱法
solvent extraction column 溶剂萃取柱
solvent extraction contactor 溶剂萃取器
solvent extraction cycle 溶剂萃取循环
solvent extraction monitor 溶剂萃取监测器
solvent extraction plant 溶剂萃取工厂，溶剂萃取装置
solvent extraction process 溶剂萃取过程
solvent extraction tower 溶剂萃取塔
solvent extraction 溶剂萃取，溶剂淬取法
solvent feed 溶剂料液
solvent inventory 投入溶剂量，加入溶剂量
solvent isotope effect 溶剂同位素效应
solventless coil 无溶剂线圈
solvent loading 溶剂萃取负荷
solvent makeup 溶剂补充
solvent phase 溶剂相
solvent power 溶剂能力，溶解能力
solvent purification 溶剂纯化
solvent raffinate 废有机相，反萃后的溶剂
solvent recovery column 溶剂再生塔，溶剂回收塔
solvent recovery 溶剂回收
solvent regeneration process 溶剂再生过程，溶剂再生法
solvent-removal test 溶剂去除试验
solvent remover 溶剂去除剂
solvent stripping 溶剂反萃取
solvent treatment 溶剂处理
solvent vapour 溶剂蒸汽
solvent wash 溶剂洗涤
solvent welding 溶剂焊【用于塑料】
solvent 溶媒，溶剂
solver 解算装置，解算机，解算器
solvolysis 溶剂分解
solvolyte 溶剂化物
solvolyze 溶剂分解
solvus 溶解度曲线
sol 溶胶
somascope 超声波检测仪
somatically significant dose 躯体显著剂量
somatic effect 躯体效应
sommer 地梁
SOM(standard on-line module) 标准联机组件
SOM(start of message) 报文开始
sonar beacon 声呐信标
SONAR(sound navigation and ranging) 水声测位仪
sonar 声波定位仪，水声测位仪，声呐
SONCAP(Standards Organization of Nigeria Conformity Assessment Programme) 尼日利亚标准组织产品合格认证证书
sonde 探测气球，探测装置，探测器，探头，探针
sonic alarm modules 声响报警组件
sonic altimeter 声测高度计

sonic anemometer 声波风速计，声学风速表
sonic barrier 声障
sonic cleaning 声波清洗
sonic delay line storage 声延迟线存储器
sonic delay line 声延迟线
sonic detection and ranging 声波大气探测，声雷达【用于勘测大气层状况】
sonic detector 声波探伤器
sonic device 声学仪器
sonic field 声场
sonic flowmeter 声测流量计
sonic flow 音速流，音速流动，声音流动，声流
sonic frequency 声频
sonic interferometer 声干涉仪
sonic leak detection 声波探漏方法
sonic logger 声速测井仪
sonic log 声波测井，连续声速测井，声测井记录
sonic memory 声存储器
sonic nozzle 音速喷嘴
sonic pressure 声压
sonic scattering layer 声散射层
sonic signal 音响信号，声信号
sonic sounding gear 回声测探仪
sonic speed 声速，音速
sonic storage (/memory) 声存储器
sonic strain gauge 声学应变仪
sonics 声能学
sonic testing 音响试验
sonic thermometer 声学温度计
sonic type 超声波（计量仪器）式，声波式，音叉式
sonic velocity 声速，音速
sonic Ventrui-nozzle 音速文丘里喷嘴
sonic wave 声波
sonic 声波的，音速的，声音的，声的
sonigauge 超声波测厚仪
SONIM(solid nonmetallic impurity) （固体）非金属夹杂物
soniscope 脉冲式超声波探伤仪，声波探测仪
sonoelastic coefficient 声弹性系数
sonometer 振动式频率计，弦音计
soot and dust 烟尘
sootblower nozzle 吹灰器喷嘴
soot blower system 吹灰系统
soot blower 除灰器，吹灰器，烟灰吹除机，吹灰机，吹灰装置
soot blowing 吹灰，烟灰吹除
soot carbon 烟黑，碳黑
soot collector 集灰器
soot concentration 烟灰浓度
soot deposit 积灰
soot door of chimney 烟囱灰门
sootfall 烟灰沉降
sootflake 碳黑片，烟灰薄片
sooting 烟灰沾污，积灰
soot-laden 烟灰沾污了的，含烟灰的
soot lance 吹灰枪，吹灰器
sootless flame 不产生烟灰的火焰
soot particle 灰粒
soot pit 灰斗，灰坑
sooty 烟灰的
soot 烟灰，碳黑，煤灰，煤烟
sophisticated technology 尖端技术
sophisticated 尖端，精致的，有经验的
sophistication 改进，混杂（信号），完善化
Sopoznikov's penetrometer test 索波兹尼科夫（氏）穿透试验，胶质层指数测定
SOP(seal oil pump) 密封油泵
SOP(simulated operating procedure) 模拟操作过程
SOP(simulated output program) 模拟输出程序
SOP(standard operating procedure) 标准操作过程
SOP(symbolic optimum program) 符号优化程序
sopwith staff 塔尺
SOP 密封油泵【DCS画面】
sorbent 吸附剂，吸收剂
sorbite 索氏体，富氮碳钛矿
sorel cement 菱镁土水泥
sorghum 高粱
sorptional capacity 吸附容量
sorptional expansion 吸收性膨胀
sorption capacity 吸附能力
sorption process 吸着过程
sorption 吸收，吸着作用，吸附作用
sorptive material 吸附性材料
SOR(successive over-relaxation) 逐次超松弛
sort description 分类说明
sorted radioactive waste 分类放射性废物
sorter-comparator 分类比较器
sorter-reader 分类阅读器
sorter 分类装置，分类机，分类器，卡片分类器，定序器，分类程序
sorting action 分选作用
sorting bench 放射性污染分类台
sorting circuit 分类电路
sorting coefficient 分选系数
sorting device 分选机
sorting grizzly 拣选筛
sorting machine 分类机
sorting of waste 废物分类
sorting track 编组线，分类线
sort key 分类键
sort package 分类归并软件包
sort 种类，分类，排序，分级，分选，区分，划分
SO(seal oil) 密封油
S/O(send only) 只发送
S/O(shipping order) 装货单，托（运）单，下货纸，关单
S/O=shutoff 断开，关闭
SOS(sample operating system) 样本操作系统
SOS(save our ship) （国际通用的）呼救信号
SOS(share operating system) 共享操作系统，分时操作系统
SOS(symbolic operating system) 符号操作系统
SOTA(state of the art) 目前工艺条件，目前工艺水平
SO test(system-operation test) 运行系统试验
SOT(system operation test) 系统操作试验
sough 排水沟

sound absorber 吸声器，吸音器，消声器
sound-absorbing board 吸音板
sound-absorbing coefficient 吸声系数
sound-absorbing facility 吸声设施
sound-absorbing lining 吸音衬砌
sound-absorbing material 吸音材料
sound-absorbing object 吸声物体
sound-absorbing 吸声，吸声的，消声，消声的
sound absorption characteristic 吸声特性
sound absorption coefficient 吸声系数
sound absorption material 吸声材料
sound absorption mat 声音吸收垫
sound absorption requirement 吸声要求
sound absorption 吸音，吸声
sound analysis 声响分析
sound aperture 声孔径
sound arrester 隔声装置，隔音装置
sound attenuation device 隔音装置
sound attenuation 消声
sound barrier 声垒，声障
sound bearing 声定位
sound burst 猝发声
sound cable 良好的电缆，通信电缆，传声电缆
sound cement 安定性水泥
sound channel 伴音信道，声道
sound conductivity 传声性能
sound deadening 消声，消音，隔音，隔音法
sound deafening capacity 消声量
sound deafening 消声，隔音
sound detection 听音检查
sound detector 测音器，检声器
sound diffraction 声衍射
sound directivity factor 声源指向性因数
sound directivity index 声源指向性指数
sound dispersion 声频散
sound door 消声门
sound emission 声发射
sound emitter 发声器
sound energy density 声能密度
sound energy reflection coefficient 声能反射系数
sounder 测深员，发音器
sound field 声场
sound focuser 聚声器
sound frequency 声频
sound granite 完整花岗岩
sound-hard boundary 声学刚性边界
sound image 声像
sounding alignment 测深定线，水道测深浅
sounding apparatus 测深器
sounding balloon 探测气球
sounding boat 测深船
sounding bob 测深锤，测深重锤
sounding by echo 回声测深
sounding by lead 铅锤测深
sounding chart 测深图
sounding cone 测深锥
sounding datum 测深基准面
sounding device 测深设备，回声测深仪，音响设备
sounding equipment 测深设备
sounding impulse 探测脉冲
sounding lead 测深铅锤
sounding line correction 测深绳偏斜校正
sounding line layout 测深垂线布置，测深线定位
sounding line 测深绳，测深锤绳，测深索，测深线
sounding method 测深法
sounding of cross section 断面测深
sounding pole 测深杆
sounding profile 测深断面，探测断面
sounding protractor 测深量角器
sounding rod 测深杆，触探杆，探杆
sounding signal 音响信号
sounding sinker 测深锤
sounding survey 水深测量
sounding vertical 测深垂线
sounding weight 测深重锤
sounding well 探测井
sounding wire 测深绳
sounding 测深，探测，测高，音响
sound-insulating door 隔音门
sound-insulating glass 隔音玻璃
sound-insulating material 隔声材料，隔音材料
sound-insulating wall 隔音墙
sound insulation board 隔音板
sound insulation casing 隔声罩
sound insulation value 隔声值
sound insulation 隔声，隔音，音绝缘
sound insulator 隔声材料，隔音材料
sound intensity level 声强级
sound intensity 声强
sound isolating layer 隔声层
sound level meter 声级计，噪声计，声强分贝表
sound level 噪声程度，噪声水平，噪声电平，声级，噪声级
sound lever meter 声级计
sound localization 声源定位
sound locator 声波定位器
sound measuring 声响测量，音响测量
sound navigation and ranging 声波导航与测距
soundness test 安定性试验，气密性试验
soundness 无缺点，完整性，坚实性，致密性，整体性，坚固性，完好性，合理性，安定性
sound-on-vision 视频上的音频干扰
sound particle velocity 声质点速度
sound pattern 声图案
sound pickup 拾音器
sound pollution 声响污染
sound power density 声功率密度
sound-powered telephone system 声动力电话系统
sound-powered telephone 声动力电话
sound power level 声功率级，声能级，声能强度
sound power reflection coefficient 声功率反射系数
sound power 声功率
sound pressure level 声压水平，声压等级，声压级，声级
sound pressure reflection coefficient 声压反射系数
sound pressure reflection factor 声压反射因数
sound pressure response 声压响应特性曲线
sound pressure transmission coefficient 声压透射

系数
sound pressure transmission factor　声压透射因数
sound pressure　声压
sound probe　探声器，声探头
sound professional judgement　正确的专业判断
sound-proof box　隔音箱
sound-proof door　隔音门
sound-proof fiber board　隔声纤维板
sound-proofing lagging　隔声外套
sound-proof measures　隔音措施
sound-proof wall　隔音墙
sound-proof　不透声的，隔声的，隔音的，隔音
sound propagation　传声，声传播
sound radar　声雷达
sound radiation pressure　声辐射压
sound radiation　声辐射
sound-ranging　声波测距法
sound-rated door　标定音量的门
sound ray trace　声线轨迹
sound ray　声线
sound reduction factor　声降因素
sound reduction index　隔声量
sound reflection factor　声反射系数
sound reflection　回声，声反射，声音反射
sound-resistive glass　隔声玻璃，隔音玻璃
sound scattering object　声散射物
sound shadow region　声影区
sound signal　音响信号，音频信号，可听信号，声信号
sound source strength　声源强度
sound source　声源
sound spectrograph　音频频谱仪
sound spectrometer　声谱仪
sound speed　声速
sound-transmission coefficient　声透射系数，透声系数
sound transmission loss　传声损失
sound transparent　透声的
sound tube　完好的管子
sound velocimeter　声速仪
sound velocity meter　声速仪
sound velocity　声速
sound wave　声波
sound　声音，噪音，发出声音，回响，测深，健全的，合理的，可靠的，有效彻底的
soup　燃料液
source address　源地址
source alphabet　信源符号集，信源字母集
source and sink method　源汇法
source area　发源地，源区
source assisted fission converter　源辅助转换反应堆
source capsule　（射线）源盒，源管，源的封装
source changing　换放射源
source code control system　源代码控制系统
source compartment　放射源室
source condition　放射源的工作条件
source container　放射源容器，放射源罐
source core　放射性源芯，放射源芯
source data automation equipment　源数据自动化设备
source data automation　自动化源数据，源数据自动化
source data entry　源数据输入
source data operation　源数据操作
source data　源数据，原始数据
source decoder　信源译码器
source density function　源密度函数
source density　源密度
source depletion model　源耗减模式
source destination code　无操作码
source destination system　源目的数据传输系统
source distribution　源分布
source document　原始单据，原始文件
source-drain voltage　源极漏极间电压
source encoder　信号编码器
source error　电源引起的误差
source evaluation　源项评价
source extrapolation　源外推
source flow　源流动
source follower circuit　源极输出电路
source forcing　源激励
source free diffusion equation　无源扩散方程
source free　无源的
source holder　（射线）源夹持器，电源架，放射源座
source image　源像
source impedance　电源阻抗，源阻抗
source index　资料索引
source information　源信息
source inspection　源项检查
source intensity　源激烈性，源强度
source intercalibration　源的相互校准
source interlock　源联锁装置
source introduction method　源引入法
source iteration method　源迭代法
source jerk method　跳源法
source language　源语言
source level flux monitor　源区段通量监测器
source level　源能级，源水平
source library　源库
source loading　装源
source making technique　制源技术
source material　源材料【核】，原始资料，权威性资料
source module　源模块
source mover　源移动装置
source multiplication method　源倍增法
source neutron　源中子
source node　源节点
source normalization　源强度归一化
source of air pollution　空气污染源
source of atmospheric dust　大气尘埃源
source of buoyancy　浮力源
source of carbon monoxide　一氧化碳源
source of coal　煤源
source of disturbance　干扰源
source of dust　尘源
source of energy　能源
source of error　误差源
source of fission neutrons　裂变中子源
source of funds　资金来源

source of goods 货源
source of heat release 散热源
source of ignition 导火线，点燃源
source of moderated fission neutrons 慢化裂变中子源
source of noise 噪声源
source of particulate matter 颗粒物质来源
source of pollution （大气）污染源
source of power 能源，电源，动力源
source of radiation 辐射源
source of resonant excitation 谐振激励源，共振源
source of sound 声源
source of supply 货源
source of trouble 事故源，事故来源
source of water pollutant 水污染物源
source of water 水质检验
source-oriented data acquisition 面向源程序的数据采集
source plate 源板，板状源
source position 源位置
source power 源强度，电源功率
source preparation 源制造，源制备
source program optimizer 源程序优化程序
source program 源程序
source push rod 放射源推杆
source range channel 源量程通道，源区段通道
source range high neutron flux trip 源区段高中子通量保护停堆
source range measuring channel 源区段测量通道
source range monitor 源区段监测器
source range neutron monitoring assembly 源区段中子监测器
source range 源量程，源区段
source reactor 中子源反应堆
source region 发源地，来源地，震源域
source related limit 源相关限值
source rod assembly 中子源棒组件
source rod cladding tube 源棒包壳管
source rod 中子源棒，源棒
source routine 源程序
source scanning 声源扫描
source size 源大小，源尺寸，射线源尺寸，源体积
source skin distance （辐射治疗中）源皮间距
sources of error 误差来源
source statement 源程序语句，源语句，输入语言语句
source streamline 源流线
source strength 点源强度，源强，源强度
source superposition 源叠加
source surface distance 放射源与被照表面距离，源面距离
source surveillance 货源监督，供货监督
source term experimental program 源项试验大纲
source term 源项
source tissue 源组织
source tool 源棒吊装工具【核电】
source transfer 电源切换
source upper bound 源的上界
source verification planning 源验证计划
source voltage 电源电压
source 源，源地，电源，气源，放射源，中子源，来源，资源，能源，信号源，辐射源
sour gas 含硫化氢天然气，酸气，酸性气
sour water 含硫水，酸性水
south by east 南偏东
south by west 南偏西
southeast by east 东南偏东
southeast trade wind 东南信风
southeast trade 东南信风带
southeast 东南，东南的
south elevation 南立面图
south southeast 南东南
south ward 向南（方）
southwest monsoon 西南季风
southwest 西南
south 南
sovereign guarantee 主权担保
SOV(solenoid-operated valve) 电磁阀
sowback 底板隆起，沙丘，山脊
sowing drill 条播机
sowing in drill 条播
sowing in holes 穴播
sowing in strips 条播
SOW(start of word) 起始字
SO_x emission rate 硫化物排放率
space air diffusion 气流扩散
space antenna 空间天线
space based reactor 空间反应堆
space borne reactor 空间反应堆
space calculation 空间计算
space character 间隔符号
space charge density 空间电荷密度
space charge effect 空间电荷效应
space charge field 空间电荷电场
space charge potential 空间电荷电位
space charge region 空间电荷区
space charge tetrode 空间电荷四极管，双栅管
space charge 空间电荷
space communication 空间通信，宇宙通信
space cooling load 房间冷负荷
space coordinate 空间坐标
space correlation 空间相关
spacecraft reactor 空间飞行器反应堆
space current 空间电流
space curve 空间曲线
space demarcation 空间划分
space dependent noise 空间相关噪声
space derivative 空间导数
space-dipole array 分集偶极天线阵
space disposal 宇宙处置，空间处置
space distribution 空间分布
space division switching 空分交换
space dosimetry 空间剂量学
spaced-out loading 分隔装料
spaced tube wall 疏排水冷壁
spaced winding 间绕绕组，疏绕线圈，间绕
space encoding 空间编码
space energy distribution 空间能量分布
space energy flux 空间能量通量
space exploration 空间探索

space factor 占空系数，空间系数，填充系数
space filling factor 占空系数，空间填充因数，槽满率
space for future expansion 为将来扩充的余地，为以后扩建预留的空地
space frame 空间构架
space frequency modulation 空间频率调制
space grid 空间网格，定位格架
space harmonics 空间谐波
space heater 空间加热器，空间电热器，加热防潮设备，机房供热
space heat gain 房间得热量
space heating installation 供暖热装置，采暖热用户
space heating load data per unit building volume 采暖体积热指标，供暖体积热指标
space heating load data per unit floor area 采暖面积热指标，供暖面积热指标
space-heating load 供暖热负荷
space heating 供暖，采暖，空间加热
space intersection 空间交会
space lattice 空间点阵
space-mark 空间测标
space model 立体模型
space moisture load 房间湿负荷
space nuclear auxiliary power 空间核辅助电源
space nuclear power unit 空间核电力装置
space optical communication 空间光通信
space orientation 空间定位
space oscillator 空间振荡器
space permeability 空间磁导率
space phase angle 空间相角
space phase 空间相位
space phasor 空间相量
space polar coordinate 空间极坐标
space power reactor 空间动力堆
space propulsion reactor 空间推进堆
space radio 无线电通信
spacer assembly 定位组件
spacer bar （电机）定位棒，定位钢筋，架立筋
spacer basket 定位框架
spacer block 定位块
spacer car 飞车
spacer damper 阻尼间隔棒
spacer dimple 定位波纹，定位架
space resection 空间后方交会
space reserved for extension 为扩建预留的面积
spacer factor 填充系数，占空系数，槽满率
spacer ferrule 定位环，定位箍，定位短管
spacer fin 定位肋片
spacer flange 中间法兰，过渡法兰，对接法兰
spacer grid 定位格架【燃料组件】
spacer lattice spring clip grid 弹簧夹定位格架
spacer lattice 定位格架
spacer pad 定位垫块，定位垫
spacer plug 定位塞
spacer ring 衬垫环，垫圈，间隔圈，衬圈，定位环
spacer rod 定位棒
spacer sleeve 定位套管
spacer strip 垫片，垫板
spacer tube 定位套接管，间隔管
spacer 垫片，隔离物，定位架【吸收棒组件】，间隔棒【高架线路】，水泥垫块，格架，定位片，间隔器，隔板，定位装置，衬垫，隔片，垫板，（燃料元件的）定位件
space segment 空间段
space structure 空间结构
space telemetry 航天遥测
space temperature variation 区域温差
space thermionic auxiliary reactor 空间热离子辅助反应堆
space-time 时空
space truss analogy 空间桁架模拟法
space utilization factor 空间利用系数
space variation 空间变化
space velocity 空间速度
space vibration （中子通量的）空间振荡
space wave 空间波
space winding 断续卷绕法
space 空地，空白，间隔，场地，间距，距离，齿槽，空间，空间桁架
spacial 空间的
spacing block 隔块
spacing board 定位板
spacing boss of outer shell 安全壳定位止挡
spacing contact 隔离触头，空号触点
spacing container 隔离容器
spacing of bars 钢筋间距
spacing of hangers 吊架的间距
spacing of movable supports 活动支座间距
spacing of reinforcement 钢筋间距
spacing of stirrup legs 钢筋肢距
spacing of stirrups 钢筋间距
spacing piece 定距块
spacing pulse 间隔脉冲【数据传输】
spacing ring 隔环
spacing signal 间隔信号，停息信号
spacing timber 定位木，隔条
spacing washer 间隔垫圈
spacing wave 间隔信号，间隔波，空号信号
spacing 间距，间隔，留间隔，空隙，空白，定距，距离，符号与数字之间的距离
spade lug 铲形接线片
spader 铲具
SPADE(spare parts analysis, documentation and evaluation) 备用元件分析、记录和鉴定
spade tuning 薄片调谐
spade-type vibrator 铲式振动器
spade vibrator 铲式振动器
spade 铲
spaghetti 绝缘套管，漆布绝缘管
spallation fragment 散裂碎片，散变碎片
spallation neutron source 散变中子源
spallation product 散变产物，散裂产物
spallation 散裂，剥落，分裂，蜕变
spall drain 碎石排水沟
spall hammer 碎石锤
spalling force 爆破（破裂）力【混凝土】
spalling resistance 耐热震性，抗裂性
spalling stress 碎裂应力
spalling 爆裂，剥落，散裂【混凝土】

spall 碎石
spam 垃圾邮件
span adjustment 量程调整
span and safety factor 挡距及安全系数
span centre 跨距中点，跨中
span chord ratio 展弦比
span clearance 跨度净空
span-depth ratio 距高比，跨高比
spandrel arch 拱肩拱【指主拱背上的小拱】
spandrel beam 梁托，托架梁
spandrel-braced arch bridge 桁架式拱桥，空腹拱桥
spandrel-filled arch bridge 实肩式拱桥
spandrel-filled arch 实肩拱
spandrel girder trimmer 托梁
spandrel space 拱肩上空间
spandrel wall 窗肚墙，拱上侧墙
spandrel wall 窗裙墙
spandrel 拱肩，拱肩墙
span drift 量程漂移
span efficiency factor 翼展效率因数
span error 量程误差
span length 跨度距离，翼展长，挡距
span load distribution 展向载荷分布，跨向荷载分布
span loading 翼展负载
spanner wrench 活络扳手，开脚扳手，插头［销］扳手，转动扳钳，水带扳手
spanner 扳钳，扳手
span of coil 线圈节距
span of foil 叶片展长，叶片展
span of roof truss 屋架跨度
span of truss 屋架跨度
span of wing 叶栅节距
span reduction factor 风速不均匀系数
span rise 起拱
span roof 等斜屋顶
span shift 量程迁移
SPAN(statistical processing and analysis) 统计处理及分析
span uneven factor 风速不均匀系数
span wire 拉线，吊线
spanwise constant thickness ratio 展向恒定厚度比
spanwise correlation 展向相关，跨向相关
spanwise distance 展向翼展距离
spanwise flow 展向流动，跨向流动
spanwise load change 展向载荷变化
spanwise spar 展向梁
spanwise velocity component 展向分速
spanwise vortex 展向涡，跨向涡
spanwise 叶高方向，叶展方向
span 跨度，跨距，期间，间距，挡距，节距，量程，叶片宽度，间隔【桥】，展长，展宽
spar buoy （海上风车的）圆柱式浮台，柱形浮标，圆柱浮标
spare boiler 备用锅炉
spare capacity 备用容量
spare car 备用车
spare circuit 备用电路，备用回路
spare coil 备用线圈
spare condensate polisher 备用的凝结水净化装置
spare contact 备用触头，备用触点，备用接点
spare details 备件明细表
spare equipment 备用设备
spare exciter change-over test 备用励磁机切换试验
spare exciter 备用励磁机
spare fuel bundle 备用燃料元件束
spare fuel channel （沸水堆）备用燃料通道
spare fuse 备用熔断器
spare generator set 备用发电机组
spare instrument nozzle 备用仪表接管
spare line 备用线路
spare machine 备用机器
spare margin 备用裕度
spare part management 备品管理，备件管理
spare parts storage 备件库
spare parts 备品，备件，备用部件，备品备件，待用件，备用部分
spare power 备用电力
spare quantity 备用数量
spare resin 备用树脂
spare room 备用房
spare set 备用机组
spare transformer 备用变压器
spare unit 备用部件，备用设备，备用机组
spare valve 备用阀
spare wire 备用线
spare 备件，备品，备用的
sparge pipe distributor 分配管式布风板
sparge pipe 再沸腾管，（布风板上的）送风管，（事故冷却水的）配水管，喷水管
sparger ring 配水环，喷淋环，环形喷淋器
sparger 喷雾器，配电器，分布器，淋水器，喷淋器，配水管，起泡（扩散）装置
sparge water 冲洗水
sparging system 喷淋系统
sparging 喷淋，喷布，喷雾，飞溅，鼓泡，充气［汽］到液体中
sparing action 屏蔽效应，防护作用，保护作用
sparingly soluble salt 微溶性盐
spark absorber 消弧器
spark advance 火花提前，点火提前
spark arrester 火花避雷器
spark at breaking 断电火花
spark at make-contact 闭合火花
spark ball 火花球
spark capacitor 灭火花用电容器
spark catcher 火花挡
spark chamber 灭弧腔，火花室，火花熄灭器
spark coil 点火线圈，火花线圈
spark discharge 火花放电
spark emission 散发火花
spark erosion 电火花腐蚀，电火花侵蚀
spark excitation 火花激励，火花激发
spark-extinguishing 灭火花的，消弧的
spark frequency 火花放电振荡频率，火花频率
spark gap generator 间隙火花发生器
spark gap voltmeter 火花间隙电压表
spark gap 火花放电器，火花放电隙，火花间

隙，火花隙，避雷器
spark ignition engine 火花点火发动机
spark ignition 电火花点火，火花点火
sparking contact 断弧辅助接点
sparking distance 火花间隙，跳火距离
sparking limit 换向器发火花的电机极限容量
sparking of brush 电刷火花
sparking on commutator 换向器发生火花
sparking plug 火花塞
sparking potential 闪电电压，闪络电压
sparking voltage 跳火电压，放电电压，击穿放电电压
sparking 发火花，点火，发火
spark killer 火花抑制器，火花熄灭器
spark length 火花长度
sparkle plan 星火计划
sparkless breaking 无火花断路
sparkless commutation 无火花换向，无火花整流
sparkless running 无火花运行，无火花运转
sparkless 无火花的
spark oscillator 火花发生器，火花振荡器
sparkover protective device 击穿保险器，击穿熔断器
spark-over test 火花放电试验
spark-over voltage 火花放电电压，跳火电压
spark-over 火花跳越，火花放电，绝缘击穿
spark plug box socket spanner 火花塞套筒扳手
spark plug box spanner 火花塞套筒扳手
spark plug insulator 火花塞绝缘子
spark plug socket wrench 火花塞（套筒）扳手
spark plug type leak detector 火花塞检漏仪
spark plug wrench 火花塞扳手
spark plug 电嘴，火花塞
spark potential 击穿电压，火花电压
spark-proof 防火花器
spark quenching circuit 消弧电路，消火花电路
spark quenching condenser 灭火花电容器
spark resistance （电极间的）火花电阻
spark signal 火花信号
spark suppressor 火花抑制器
sparks 船上无线电报务员，随机电气技术员
spark test 火花试验
spark through 击穿
sparkwear 火花烧毁，(触点等）烧坏
spark 电火花，火花，火花放电
sparse-data area 资料缺乏地区，资料稀疏地区，未经调查地区
sparsely populated area 人口稀少区
sparse matrix 稀疏矩阵
sparse 稀少的，稀疏的
SPAR(space power advanced reactor) 先进空间动力堆【美国】
SPAR(symbolic program assembly routine) 符号程序的汇编子程序
spartalite 氧锌矿
spar 晶石，翼梁，柱体式平台
spasmodic burning 脉动燃烧
spasmodic variation 间歇性的变化
spate 暴风雨，大水，洪水猛涨，涨大水
spatial aerotriangulation 空中三角测量
spatial angle 空间角
spatial arrangement 空间排列
spatial autocorrelation function 空间自相关函数
spatial behaviour 空间工作性能
spatial change 空间变化
spatial concentration of sediment 单位容积含沙量
spatial coordinates 空间坐标
spatial correlation 空间相关
spatial covariance 空间协方差
spatial discretization 空间离散化
spatial distribution 空间分布
spatial dose distribution 剂量空间分布
spatial flow visualization 空间流动显示
spatial flux distribution 通量空间分布
spatial gust 空间阵风
spatial harmonic 空间谐波
spatial instability 空间不稳定性
spatial load 空间荷载
spatially coherent beam 空间相干光束
spatially varied flow 沿程变量流，空间变化水流
spatial model 空间模型
spatial non-uniformity 空间不均匀性
spatial orientation 立体定位
spatial power distribution flattening 功率空间分布展平
spatial power distribution 功率空间分布
spatial randomness 空间随机性
spatial resolution 空间分辨率
spatial spectral density function 空间谱密度函数
spatial structure 空间结构，立体结构
spatial variation 空间变化
spatial vorticity distribution 空间涡量分布
spatial waveform 空间波形
spatial 空间的，立体的
SPAT(silicon precision alloy transistor) 硅精密合金晶体管
spats 护脚
spatter loss coefficient 飞溅率
spatter loss rate 飞溅率
spatter 飞溅，溅，滴落
spat 流线型罩
spawl 劈开，碎片
spa 矿泉，矿泉疗养院
SPCA(spare plate critical assembly) 备用平板临界装置【美国】
spcctraoradiometric instrument 光谱辐射计
SPCSO(single point control-lable status output) 单点可控状态输出
SPCS(special-purpose computing system) 专用计算系统
SPC(stored-program control) 存储程序控制，存储程序控制器
SPC(system power control) 系统电源控制器
SPD(self-powered detector) 自给能探测器
SPD=speed 速度，转速
SPDT(single-pole double-throw) 单刀双掷
speaker 扬声器
speaking position 通话位置
spearhead attack 标枪头侵蚀
spear pilot valve 针阀导阀
spear servomotor 针阀接力器
spear stroke 针阀冲程

spear tip 针阀尖
special account 专用账户
special administrative region 特别行政区
special agent 特别代理人
special allocation 特别拨款
special allowance 特别津贴
special alloy steel 特种合金钢
special amenities 特别待遇
special analog system 专用模拟系统
special anthracite 专用无烟煤
special appliance 特种设备，专用设备
special appointment contract 指定承包合同
special appointment works 指定承包工程
special appropriation for special use 专款专用
special appropriation 专用拨款
special area 特殊区域
special audit 特殊审计，专项审计
special award 特别奖
special-bands reinforced wye piece 三梁岔管
special bronze 特种青铜
special building 特殊建筑物
special cable 特殊电缆，专用电缆
special capital 专用资本
special carbon-graphite material 特种碳石墨材料
special-carte 胶木纸，电木纸，特制纸片
special case category 特殊案例分类
special casting 专用铸件
special cement 特种水泥
special channel 特殊信道，专用信道
special clause 特别条款
special clothing 特种工作服
special code selector 特殊业务台选择器，特殊编号选择器
special coinbox discrimination tone 专用投币电话鉴别音
special committee 特别委员会
special conditions of contract 合同的专用条件
special conditions 特殊条件
special conference 专题会议
special contactor 特种接触器，专业承包商
special corporation 特殊法人
special cost 专项费用
special credit 特别信用证，特定信用证
special current loading （电机的）线负荷
special damages 特殊损害赔偿
special design door 特设门，特殊设计的门
special detail drawing 特殊详图，特种详图，大样详图，足尺大样图
special development area 特别开发区
special discount 特别折扣
special district 特区
special door lock 专用门锁
special drainage requirement 具体排放要求
special drawing rights 特别提款权
special ductile frame 特种延展性框架，特种延性框架
special-duty motor 特殊工作制电动机
special-duty wagon 专用货车
special economic zone 经济特区
special edition 专辑
special effect amplifier 专门效应信号放大器，特殊效果放大器
special emission 特殊排放
special emplacement technique 特别安置技术
special entity 特设机构
special environment electric equipment 特殊环境用电气设备
special environment 特殊环境
special equipment 特殊设备
special errand fire vehicle 专勤消防车
special expanded diameter conductor 特有扩径导线
special export 专门出口
special fissionable material 特种可裂变材料
special flange 特殊法兰
special form radioactive material 特殊形式的放射性材料
special function 特殊函数
special fund 专款，专项资金
special fuse 特种熔断器
special glass 特种玻璃
special grace period 特殊宽限期
special hanger 专用吊架
special hardware 专用硬件设备，特殊金具
special highway 专用公路
special import 专门进口
special instruction 特殊规定，特殊要求，特殊巡视
special insulation 特殊绝缘
special investigation 特别调查
specialist report 专题报告
specialist 专家
special item 特殊件
speciality 特殊件，特性，特制品，专长，专业
specialization 专业化
specialized agency 专门机构
specialized bank 专业银行
specialized body 专业团体
specialized conference 专业会议
specialized factory 专业厂家
specialized field 专业领域，专业
specialized knowledge 专业知识
specialized production 专业化生产
specialized quality manual 专用的质量手册
specialized skill 专业技术
specialized standard 专用标准
special jaw 特殊卡爪，专用卡爪
special joint 特种接头
special legal person 特殊法人
special licence 特别许可，特殊许可证
special line 专线
special list of equipment 临时设备清单
special load 特殊负载
specially-designated enterprise 定点企业
specially-designated factory 定点厂家
specially designed welded seal 特殊设计的焊接密封【顶盖，欧米茄等】
specially-made brick 特制砖
specially-made 特制的
special machine 特殊电机，专用电机
special main resistance factor 特种主要阻力系数
special mechanical plug 特殊机械塞

special meeting 专题会议
special metal 特殊金属
special modulator 特殊调制器，专用调制器
special monitoring 特殊监测
special nuclear material 特殊核材料
special packaging 特种包装
special permission export 出口特许
special pH test paper 精密 pH 试纸
special physical examination 专项体检
special pipe 特种水管
special power excursion reactor test 专用功率剧增试验
special preference trade agreement 特惠贸易协定
special preference 特惠
special prize 特别奖
special procedure 特殊程序，特殊工序
special processes 特殊工序
special process heat tracing system 特殊工艺伴热系统
special production reactor 专用生产堆
special project 特殊项目
special provisions 特殊条款
special-purpose cable 特种电缆，专用电缆
special-purpose computer 专业计算机，专用计算机
special-purpose computing system 专用计算系统
special-purpose fund 专用基金
special-purpose machine 专用计算机
special-purpose map 专用地图
special-purpose memory 专用存储器
special-purpose motor 专用电动机，特殊功用电动机
special-purpose relay 专用继电器，特殊功用继电器
special-purpose section 异型钢材
special-purpose spanner 专用扳手
special-purpose 专用的，专用目的
special railway line 铁路专用线
special railway 专用铁路，铁路专用线，专线
special rate 特价
special regulation 特殊规定，专项监管
special report 特别报告
special requirement 特殊规定，特殊要求
special risk 特殊风险
special rules 特殊规定
special secondary resistance factor 特种次要阻力系数
special service railroad 铁路专用线
special service railway coupling point 铁路专用线接轨点
special service railway line 铁路专用线
special session 特别会议，专题会议
special-shaped fish plate 异形鱼尾板
special-shaped steel tower 异形铁塔
special shear wall 特种剪力墙
special siding 专用线
special spanner 特种扳手
special star-delta starter 延边星三角启动器
special steel plate 特种钢板
special steel 特种钢
special structural steel 特种结构钢
special subject evaluation 专题评价
special support 特殊管架
special task 特殊任务
special technical skill 专业技术
special terms and conditions 专用条款
special term 特别条款
special test mode 专用测试模式
special test reactor 专用试验堆
special tool 专用工具
special topic 专题
special tower 异形塔
specialty metal 特殊金属
special type 特定类型
specialty steel 特种钢
specialty 专业
special ventilation system of nuclear power stations 核电厂的特别排风系统
special welding equipment 专业焊接设备
special work permit area 特殊工作区
special zone 特区
special 特别的，特殊的，专用的
species certificate 品种证明书
species 物种，种类
specific abrasion 磨损率
specific absorption 吸收系数，吸收率，比吸收
specific accumulation 累积率
specific acoustic impedance 声阻抗率
specific activity 比放射性，（放射性）比活度，放射性比度，活性比
specific address 具体地址，绝对地址
specific adhesion 黏着系数，黏着比
specific area 比面积
specific armature loss 单位电枢损耗，电枢比损耗
specification curve 标准曲线
specification deviation 规格偏差
specification drawing 规范图
specification for materials 材料规格
specification number 规范号
specification of quality 质量规范，质量规格
specification of service 使用说明
specification requirements 规格要求
specifications for material 材料规格
specification standard 规格标准
specifications （技术）规格，明细表，（技术）规范，（技术）规程，说明书，（技术）要求
specific bed load transport 推移质输送比率
specific bed load 推移质比率
specific blinding energy 比结合能
specific brightness 亮度率
specific burnup 比燃耗，燃烧率
specific capacitance 单位电容，比电容
specific capacity 比功率，比容量，单位出力，功率系数，单位蒸发量，单位容量
specific capital outlays 基建单位投资
specific ionization coefficient 比电离系数
specific code 绝对代码，特殊编码，代真码
specific commodity rate 指定商品运价
specific communication service mapping 特定通信服务映射

specific compression 单位压缩量
specific conductance 电导率
specific conductivity analyzer 比导电率表
specific conductivity 导电系数，导电率，电导率
specific consumption of cellulose 纤维素的比耗
specific consumption of fuel 燃料的消耗率
specific consumption of steam 汽耗率
specific consumption 单位耗量，消耗率，比耗
specific coolant activity 冷却剂比（放射性）活度
specific cost 特定成本，特定费用
specific creepage distance 爬电比距
specific creep 比蠕变
specific damping 衰减常数
specific density 比重，比密度，相对密度
specific dielectric strength 比介电强度
specific discharge 比流量，单位流量，流量率
specific duty 单位负载，单位生产量，单位出力，单位蒸发量，单位功率
specific efficacy 功用功能，内在功效
specific electric load （电机）线电负荷
specific elongation 伸长率
specific emission （污染）比排放量
specific energy cost 比能量费
specific energy imparted 比授予能
specific energy of current 水流比能
specific energy of flow 水流比能
specific energy 比能，单位耗能，比授予能
specific enthalpy 比焓
specific entropy 比熵
specific environment 特定环境
specific equipment 特定设备
specific evaporation 单位蒸量
specific exchange capacity 比交换容量
specific exergy costsum optimization 按比成本和的最优化
specific exergy rate 比火用输出率
specific exergy 比火用
specific expansion 膨胀率
specific extinction coefficient 比消光系数
specific flow 比流量，单位流
specific force 比力
specific frictional resistance 比摩阻
specific fuel consumption 比燃耗油，燃料消耗率，单位燃料消耗量
specific fuel power 比燃料费
specific gamma-ray constant 比射线常数，特征射线常数，伽马射线固有常数
specific gas constant 比气体常数
specific gas detector 特色气体探测器
specific gas flow rate 气体流率
specific gravity balance 液压比重计
specific gravity floater 比重浮标，浮标式比重计
specific gravity fraction 比重组分
specific gravity 比重，密度
specific head diagram 比水头曲线，单位水头曲线
specific head loss 水头损失率
specific head 比水头
specific heat capacity at constant pressure 比定压热容
specific heat capacity at constant volume 比定容热容
specific heat capacity at saturation 比饱和热容
specific heat capacity 比热容
specific heat consumption 单位热耗，热耗率
specific heat content 比焓，比热含量
specific heat flow 比热流
specific heat load 散热强度
specific heats ratio 比热比，绝热指数
specific heat 比热，比热容
specific Helmholtz free energy 比亥姆霍兹自由能
specific humidity 比湿，比湿度
specific hydraulic slope 水力坡降率
specific impedance 阻抗率
specific impulse 比冲，比冲量
specific inductive capacity 比电容，电容率，介电常数
specific inertia 比惯量
specific input 比进料量
specific inventory 比投料量，比装载量
specific investment cost 比投资费用，单位投资费用，单位投资成本
specific investment 单位投资，比投资
specific ionization 比电离，单位电离，特殊电离
specific item 具体项目
specificity 特性
specific leakage distance 泄漏比距
specific load 单位负载，单位载荷，比负载，比负荷，负荷率
specific locations of termination points 分界点的位置
specific location 具体位置
specific loss of design head 设计水头的单位损失
specific loss 单位损耗，比损耗
specific magnetic loading 磁负荷，平均磁密
specific magnetising moment 磁化强度
specific maintenance job 专门维修工作
specific mass flow 比质量流量
specific mass 比质量，质量密度，密度
specific material demand 单位材料需要量
specific nuclear fuel power 核燃料比功率
specific objective 特定目标
specific optical rotatory power 比旋光本领
specific order 特别订货
specific output 比出力，比输出，输出率，功率系数，单位出力，单位输出功率
specific parachor 比等张比容
specific penetration resistance 比贯入阻力
specific performance 具体履行【合同义务】，特别履行【通常是法院衡平救济的方式之一】
specific permeability 比渗透率，比通过本领
specific potential power 单位水力蕴藏量
specific power consumption 比功率消耗
specific power 比功率，单位功率，单位出力，功率系数
specific pressure drop 比压降
specific pressure 比压
specific productivity 单位生产率
specific program 专用程序
specific project 特定项目

specific radioactivity 比放射性，特定放射性
specific rate of flow 单位流量
specific reluctance 磁阻率
specific requirement 具体要求，特定要求
specific resistance of ash 灰比电阻
specific resistance 比电阻，电阻率
specific resistivity 比电阻，电阻率，电阻系数，比阻
specific retention 持水度，持水率
specific rod power 线功率
specific routine 专用程序
specific sliding 比滑，滑差系数，滑移系数，有限叶片系数，滑率
specific specification 具体技术要求
specific speed of pump 水泵比速
specific speed of rotation 比转速
specific speed 比转速，折算速度，比速
specific standard 特定标准
specific steam consumption 汽耗率
specific steam enthalpy 蒸汽比焓，比蒸汽焓
specific storage 单位储水量
specific strength 比强度
specific subject 专题
specific surface area 比表面积
specific surface energy 表面比能
specific surface loading 单位表面负荷
specific surface 表面系数，比面【水泥】，比表面，单位表面
specific susceptance 电纳率
specific suspended load 悬移质比率
specific terms of reference 具体职权
specific test reactor 特定试验堆
specific thermodynamic energy 比热力学能
specific thrust 比推力
specific torque coefficient 比转矩系数
specific torque 比转矩
specific unbalance 不平衡率
specific unit capacity purchases 规定卖出容量
specific utilization coefficient 利用系数，单位利用率
specific value 比值
specific viscosity 比黏度
specific voltage 比电压
specific volume 比体积，单位体积，比容（积）【单位质量的容积】
specific water absorption 吸水率
specific water power potential 水力蕴藏量
specific water power resources 水力蕴藏率，特定水力资源
specific water retention 单位持水量
specific weight meter 比重计
specific weight of soil 土壤比重
specific weight 比重
specific yield 出水率，单位产水量，单位给水量
specific 明确的，具体的，特种的，特殊的，特定的
specified administrative department 归口管理单位
specified capacity 额定容量，额定气体量，规定容量
specified characteristic curve 指定特性曲线
specified compressive strength of concrete 规定的混凝土抗压强度
specified condition 额定工况，额定参数，额定条件
specified contractor 指定承包商
specified criteria 明细规范，给定（技术）条件
specified data 确定数据，规定数据
specified dropout （继电器的）规定释放值
specified grading 规定的级配
specified horsepower 额定功率，铭牌功率
specified-in-detail item 详细规定项目
specified limit 规定极限
specified load 规定负荷，额定负荷，设计负荷，标准荷载
specified mix 指定配合比
specified objective(/target) 规定的目标
specified operation conditions 规定运行工况
specified penetration 指定贯入度
specified performance 设计性能，保证性能
specified pump head 规定扬程
specified quality measure 规定的质量度量
specified quality 合格质量
specified rated load 额定负荷
specified rate 额定量
specified requirement 规定的要求
specified sensitivity 规定灵敏度
specified speed 额定转速
specified strength 规定强度
specified temperature 规定温度
specified value 规定值，给定值，某一给定值
specified 规定的，明细的
specifier 区分符，规格制定者
specifies breakaway torque 规定的最初起动转矩
specify task asynchronous exit 特殊任务异步出口，指定任务异步出口
specify 规定，详细说明
specimen blank 试件毛坯
specimen capsule 样品盒，样品夹
specimen copy 样本
specimen grip 试样夹钳
specimen holder 样品夹
specimen machine 样机
specimen mould 试模
specimen rack 试样架
specimen 标本，样本，范本，样品，样机，试样，试件
speckle pattern 斑纹图样
speckle 斑点，斑纹
speck 斑点，瑕疵，灰尘，污点，小颗粒
Spec. No.（specification number） 规范号
SPEC(scram prevention evaluation checklist) 防止紧急停堆评定清单
spectacle blind 管孔盲板，八字盲板，带双圈的盲板，眼圈盲板
spectacle plate 管孔盲板
spectral absorptance 光谱吸收比，单色吸收率
spectral absorption coefficient 光谱吸收系数
spectral absorption factor 光谱吸收因数
spectral absorption 光谱吸收
spectral absorptivity 单色吸收率
spectral analysis 光谱分析法，频谱分析
spectral angular cross-section 能谱角截面

spectral carbon　光谱碳棒
spectral concentration of radiant energy density　辐射能密度的光谱密集度
spectral concentration of vibration mode　晶格振动模式密度，点阵振动模式密度
spectral cross section　谱截面
spectral decomposition　频谱分析，光谱分析
spectral density function　谱密度函数
spectral density　光谱密度，谱密度
spectral detectivity　单色探测率
spectral distribution of black body　黑体的光谱分布
spectral effect　光谱效应
spectral efficacy at a specified wavelength　特定波长的光谱视能
spectral emission rate　光谱发射率
spectral emissivity　光谱发射率，单谱发射率
spectral flux density　分谱通量密度
spectral function　谱函数
spectral gap　谱隙
spectral hardening　能谱硬化，谱硬化
spectral index　光谱系数
spectral intensity　光谱强度
spectral irradiance distribution　光谱辐照度分布
spectral irradiance　光谱辐射，光谱辐照度
spectral limit　光谱范围
spectral line　谱线
spectral luminous efficacy　光谱光视效能
spectral luminous efficiency　光谱光视效率
spectrally smoothed transmissivity　光谱平滑透射率
spectral method　频谱法
spectral mismatch　光谱失配
spectral particle fluence rate　谱粒子注量率
spectral particle flux density　谱粒子通量密度
spectral photon irradiance　光谱光子辐照度
spectral power density　光谱功率密度
spectral pyranometer　分光总日射表
spectral pyrometer　光谱温度计
spectral radiance factor　光谱辐射亮度因数
spectral radiancy distribution curve　谱辐射分布曲线
spectral radiant energy density　光谱辐射能密度
spectral radiant power　光谱辐射功率
spectral-radiation thermometer　单谱辐射温度计
spectral reflectance　单色反射率，光谱反射比
spectral reflection factor　光谱反射因数
spectral refractivity of water　水的光谱折射率
spectral response curve　波谱响应曲线，谱响应曲线
spectral response　光谱响应【光谱灵敏度】
spectral scattering coefficient　单色散射系数
spectral sensitivity　光谱灵敏度
spectral shift control reactor　谱移反应堆，谱移控制反应堆
spectral shift control　谱漂移控制，谱移控制
spectral shift technique　谱移技术
spectral shift　光谱的偏移，谱移
spectral softening　能谱软化
spectral source density　光谱源密度
spectral standard solar cell　光谱标准太阳电池
spectral transmission factor　光谱透射因数
spectral transmittance　光谱透射比
spectral window　谱窗
spectral　光谱的，谱的
spectra　谱，光谱，波谱，频谱，能谱【spectrum 的复数】
spectro-angular cross section　谱角截面
spectro-angular flux density　谱角通量密度
spectrobologram　分光变阻测热图
spectrocalorimetry　光谱色度学
spectrograde reagent　光谱级试剂
spectrograph　光谱仪，摄谱仪
spectrometer　分光计，分光仪，光谱机，光谱仪，频谱仪，摄谱仪
spectrometric analysis　光谱测定分析
spectrometry　光谱测定法
spectrophotometer　分光光度计，分光光度仪
spectrophotometric analysis　分光光度计分析
spectrophotometric cell　分光光度吸收电池
spectrophotometry　分光光度法
spectroradiometer　分光辐射仪
spectroradiometry　光谱辐射度学，分光辐射度量学，分光辐射度学
spectroscope　分光镜
spectroscopic analysis　分光分析
spectroscopic hygrometer　分光湿度计
spectroscopic thermometer　光谱温度计
spectroscopic　分光镜的，光谱的，与分光镜联合的
spectroscopy　分光学，光谱学，频谱学
spectrum amplitude　谱幅
spectrum analyser　光谱分析器，频谱分析器
spectrum analysis　光谱分析
spectrum analyzer　光谱分析仪，频谱分析仪
spectrum-averaged cross section　能谱的平均截面
spectrum infrared radiation system　红外线辐射
spectrum infrared radiation thawing system　红外线解冻装置
spectrum instrument　便携式光谱仪
spectrum leak tester　频谱泄漏测试仪
spectrum level　谱级
spectrum of an ionizing radiation　电离辐射谱
spectrum of frequency　频谱
spectrum of horizontal gustiness　水平阵风谱
spectrum of radiation　辐射光谱
spectrum of turbulence energy　湍流能谱
spectrum of turbulence　湍流谱
spectrum of velocity fluctuation　脉动谱
spectrum of vertical gustiness　垂直陈风谱
spectrum selector　频谱选择器
spectrum stabilizer　能谱稳定器，稳谱器
spectrum stripping　剥谱法，剥谱，差谱法
spectrum synthesis　能谱综合
spectrum　领域，频谱，光谱，谱，波谱，交叉谱，范围
specular coal　亮煤
specularly reflecting surface　镜反射表面
specular reflectance　镜面反射率
specular reflection　镜面反射
specular surface　镜面
speculate　投机

speculation company 皮包公司
speculation 投机
speculative investment 投机性投资
speculative resources 推测资源
speculative 抽象的，推理的，纯理论的，思索的，投机性的
specus 地下水渠，渡槽，高架水渠
speech digit signaling 语音数字信令
speech-enhancement system 语音增强系统，语音校正系统
speech frequency 语频，音频，通话频率
speech interference level 话音干扰电平
speech inversion system 话频颠倒制
speech inverter 倒频器，一种语言保密器
speech modulation 语音调制
speech pattern 语言模式
speech power 通话功率，语言功率
speech processing system 语音处理系统
speech range 话频范围，音频范围
speech recognition system 语音识别系统
speech recognition 语音识别
speech scrambler 扰频器，语言编码器
speech synthesizer 语言综合器
speed-adjusting rheostat 调速变阻器
speed adjustment 速度调整，速度调节
speed brake 气动力减速装置，速度制动器，减速板
speed change box 变速箱
speed change gear 变速箱，同步器，变速齿轮
speed changer governing range 同步器调速的范围
speed changer load-reducing time 同步器的减负荷时间
speed changer motor 同步器电动机
speed changer synchronizing range 同步器的同步范围
speed changer 转速变换器，变速器，同步器
speed characteristic curve 速度特性曲线
speed-coding system 快速编码系统
speed coefficient 速度系数，转速系数，速率系数
speed constant 速度常数
speed control by changing the field current 调磁调速
speed control by electromagnetic slip clutch 电磁转差离合器调速
speed control by pole-changing 变极调速
speed control device 调速装置
speed controlled motor 调速电动机
speed controller with inverter 变频调速器
speed controller 调速器，转速调节器，转速控制器，速度调节器
speed control logic 速度控制逻辑
speed control loop 调速回路，速度控制回路
speed control mechanism 转速控制机构
speed control of commutatorless motor 无换向器电动机调速
speed control of motor 电动机调速
speed control range 速度控制范围
speed control system 调速系统
speed control valve 调速阀
speed control with constant power 恒功率调速
speed control with constant torque 恒转矩调速
speed control 速度控制，转速控制，转速调节，速度调节
speed converter 速度变换器
speed correction signal 转速校正信号
speed counter 转速计数器，转速表，速率计
speed decreaser 降速器
speed decrement 减速
speed detector 测速装置，转速检测装置
speed down 减速
speed drop characteristic 转速下降特性，惰走特性
speed drop 速度降，转速降
speed envelope 超速包络线
speeder gear 调速装置，调速机构，同步器，转速变换器，调速器，增速器
speeder motor 同步器电动机，调速电动机
speed error signal 转速偏差信号
speeder 调速器，增速器，转速变换器，同步器，调速装置
speed factor 速度因子，速率因数
speed fluctuation 转速波动，速度波动
speed frequency 转速频率
speed gauge 速率计
speed governing device 调速装置，调速机构，调速器
speed governing droop 转速不等率
speed governing oil pump 调速油泵
speed governing operation 调速运行，调频运行
speed governing system 调速系统
speed governing 速度调整，调速
speed governor operation 调速运行
speed governor 速度调节器，调速器
speed increaser 增速器
speed increasing gear pair 增速齿轮副
speed increasing gear train 增速齿轮系，增速齿轮副
speed increasing gear 增速齿轮，增速器
speed increasing ratio 增速比
speed indicating generator 调速发电机
speed indicator 转速表，速度计，示速器，示速计，速度指示器
speed-induced voltage 旋转电压，旋转电动势
speed insensibility 速度不灵敏度
speed isopleth figure 等速线图
speedlight 闪光管，频闪放电管
speed limitator 转速限制器，限速器
speed limiter 限速器
speed limiting device 限速装置，转速限制器
speed-load characteristic 转速负荷特性
speed loss 速度损失
speed measurement instrument 测速仪
speed measurement 速度测量
speed measuring instrumentation 速度测量仪
speed meter 速度计
speed monitor 转速监视器
speed of circulation 循环速度
speed of cooling 冷却速度
speed of electromagnetic wave in vacuum 电磁波在真空中的传播速度
speed of flow 流速

speed of insertion 插棒速率【控制棒】
speed of manipulation 键控速度
speed of oil trip 危急截断器动作转速
speed of ratio 速度比
speed of reducer 减速器
speed of reset 再调速度
speed of response 反应速度，响应速度，调节速度
speed of revolution 旋转速度
speed of rotation 转速
speed of transmission 发送速度
speed of turbine trip 跳闸动作转速
speed of variation 速率变化，变速
speed of wind 风速
speed of withdrawal 提棒速率【控制棒】
speedomax 电子自动电势计
speedometer 速度计，转速表，转速计，测速计，速率计
speed on load 负载转速
speed or frequency matching device 速度或频率匹配装置
speed orifice 测速孔【风洞】
speed oscillation 速度振荡
speed overshoot 速度超调量
speed polar 速度极线图
speed programming 速度程序设计
speed pulser 磁阻发生器
speed range 变速范围，转速范围
speed rate 速率
speed ratio control 速率比控制
speed ratio 速比
speed reducer 减速器，减速机，减速装置，减速齿轮
speed reducing gear train 减速齿轮系
speed reducing motor 减速电动机
speed reduction gear 减速齿轮，减速器
speed reduction ratio 减速比
speed-regulating rheostat 调速变阻器
speed-regulating servomotor 调速接力器，调速伺服电动机
speed-regulating 调速的
speed regulation pump 调速泵
speed regulation 调速，速度调节，速度变动率，速度不等率
speed regulator 调速器，速度调节器
speed relay 速度继动器
speed responsive element 速度响应元件
speed ring 控制环，调速环，速度环
speed rise on load rejection 甩负荷引起的转速上升
speed rise 转速升高
speed sensor 转速传感器，速度传感器
speed setpoint 速度设定点［设定值，控制点］
speed setter 速度设定器，转速定值器
speed setting gear 速度设定装置
speed setting 速率整定，转速整定，速度调定
speed stability 速度稳定
speed stage 速度级
speed-time curve 速度-时间曲线
speed timer 速度计
speed tolerance 容许转速偏差
speed-torque characteristic 转速-转矩特性曲线，速度转矩特性
speed torque curve 转速力矩特性曲线
speed transmitter 速度变送器，速度传送器
speed triangle 速度三角形
speed-up capacitor 加速电容器
speed-up construction schedule 加快施工进度
speed-up effect 加速效应
speed-up factor 加速因子
speed-up rate 升速率
speed-up 加速，升速
speed variation rate 速度变化率
speed variator 变速器
speed voltage generator 测速发电机
speed voltage 速度电势，旋转电动势
speed way 高速车道，高速公路，快车道
speedy cableway 快速缆道
speed 速度，速率
spelaeo-meteorology 洞穴气象学
spell 短时间间隔，短时间中断，轮班
spelter solder 锌焊料
spending beach 消波滩
spent acid 废酸
spent-bed cooler 流化床式冷渣器
spent caustic soda 苛性钠废液
spent caustic 废碱
spent clay 废土
spent core 一炉废燃料
spent extractant 用过的萃取剂，废萃取剂
spent fuel assembly storage rack 废燃料元件组件贮存架
spent fuel assembly storage 废燃料元件组件贮存
spent fuel assembly 乏燃料组件，废燃料组件
spent fuel building crane 废燃料贮存库的起重机
spent fuel canal 废燃料沟道
spent fuel cask loading area 废燃料容器装料区
spent fuel cask shipping area 废燃料装运区
spent fuel cask 废燃料容器
spent fuel disposal 乏燃料处置，废燃料处置
spent fuel element handling cell 废燃料（元件）操作小室
spent fuel element 乏燃料元件，辐照过的核燃料元件
spent fuel flask 废燃料罐，乏燃料罐，存放废燃料的容器
spent fuel intermediate storage compartment 乏燃料过渡贮存隔间
spent fuel manipulator 乏燃料操作机
spent fuel pit bridge 乏燃料池桥台
spent fuel pit building crane 废燃料水池房吊车
spent fuel pit building 燃料水池房
spent fuel pit cooling and clean-up system 废燃料贮存水池冷却与净化系统
spent fuel pit cooling loop 废燃料贮存水池冷却回路
spent fuel pit demineralizer 乏燃料池除盐装置
spent fuel pit filter 乏燃料池过滤器，废燃料贮存水池过滤器
spent fuel pit heat exchanger 乏燃料池热交换器，废燃料贮存水池热交换器

spent fuel pit lining　废燃料贮存水池衬里
spent fuel pit pump　乏燃料池泵
spent fuel pit skimmer filter　乏燃料水池撇浮过滤器
spent fuel pit storage rack　乏燃料池组件贮藏架
spent fuel pit　乏燃料池【核电站】，乏燃料冷却水池，废燃料池
spent fuel pool bridge crane　废燃料池桥式吊车
spent fuel pool cooling pump　废燃料池冷却泵
spent fuel pool purification pump　废燃料池净化泵
spent fuel pool　乏燃料池，乏燃料贮存池【核电站】，废燃料池
spent fuel rack　废燃料架
spent fuel reprocessing　乏燃料后处理，废燃料后处理
spent fuel shipping cask　废燃料运输铅罐，废燃料装运容器
spent fuel shipping　废燃料装运
spent fuel storage area　废燃料贮存区
spent fuel storage building　废燃料贮存库房
spent fuel storage pit　乏燃料储存池，废燃料贮存水池
spent fuel storage pool　乏燃料储存池，废燃料贮存水池
spent fuel storage rack　乏燃料组件贮存架，废燃料元件组件贮存架
spent fuel storage　乏燃料储存，废燃料贮存，废燃料贮存库，废燃料元件组件贮存
spent fuel tank　废燃料存放槽
spent fuel water pit　废燃料水池
spent fuel　乏燃料，废燃料，用过的燃料【核电】，辐照过的核燃料
spent gas　废气
spent ion exchange resin concrete incorporation plant　废树脂水泥固化装置
spent ion exchange resin concrete store　废离子交换树脂贮存库
spent ion exchange resin slurry　废离子交换树脂淤浆
spent ion exchanger resin tank　废离子交换树脂箱
spent liquor　废液
spent material　废物，废料，用过的材料
spent residue　废渣，废物
spent resin storage tank　废树脂贮存箱
spent resin　废树脂，用过的树脂
spent scrub stream　废洗涤液
spent water　废水
spent　烧过的，贫化的，浓度降低的
sperone　黑榴白榴岩
SPERT(special power excursion reactor test)　专用功率剧增试验堆
SPE(stored program element)　存储程序部件
SPFP(spent fuel pool)　废燃料贮存池
SPF(storage protect feature)　存储保护特性
SPGR(specific gravity)　比重
sphaerophone　利用可变电容器变频的音频电子仪器
sphalerite　闪锌矿
spheno-conformity　楔形整合
sphere discharge tube　燃料球卸料管
sphere feed tube　燃料球供料管
sphere gap　球状避雷器，球隙避雷器
sphere of action　作用范围
sphere of penetration　穿透范围
sphere path curve　（燃料）球轨迹
sphere-plane gap　球板间隙
sphere pole　球状电极
sphere sequence frequency　球顺序频率
sphere-sphere gap　球球间隙
sphere valve　球形阀
sphere　球，球体，地球的圈层
spherical array　球面阵
spherical azimuth　球面方位角
spherical bearing　球面轴承
spherical branched pipe　球形岔管
spherical buoy　球形浮标
spherical cap bubble　球帽状气泡
spherical cap　球形泡罩
spherical casing　球形汽缸
spherical cavity　球形空腔，球形空穴
spherical configuration　球形构形，球形布置
spherical containment　球形安全壳
spherical coordinates　球面坐标
spherical curve　球面曲线
spherical distance　球面距离
spherical distribution chamber　球形配水室
spherical dome　球形屋顶
spherical explosion　球形爆破法
spherical faced pan　球面盘
spherical free progressive sound wave　球面行波
spherical fuel element　球形燃料元件
spherical fuel handling system　球形燃料装载系统
spherical geometry　球形几何形状
spherical joint　球形接合
spherical lamp　球形灯
spherically dished bottom head　半球形底封头
spherically dished head　球形封头
spherically faced closure washer　球面封头垫圈
spherical mirror　曲面镜
spherical mixing chamber　球形混合器【复合循环锅炉】
spherical particle　球形颗粒
spherical pile　球形反应堆
spherical radiator　球面辐射器
spherical reactor　球形反应堆
spherical reflector　球面反射体
spherical resolver　球形解算器
spherical resonator　球状共振器，球状谐振器
spherical roller bearing　球形滚柱轴承
spherical roller thrust bearing　球面滚动推力轴承
spherical seat　球面座
spherical steel containment vessel　球形钢安全壳
spherical steel containment　球形钢质安全壳
spherical surface　球面
spherical tank　球形水箱
spherical thrust bearing　球面推力轴承
spherical triangle　球面三角形
spherical valve　球阀，球形阀
spherical vault　球形穹顶
spherical vortex　球面旋涡，球形涡
spherical washer-shaped collar　球面垫圈形套环

spherical wave function　球波函数
spherical wave　球面波，球形波
spherical weathering　球形风化
spherical　球面的，球面形的，圆的，球形的
sphericity　圆球度，球状，球性，圆球率，球形度
spheroidal coordinates　球体坐标
spheroidal graphite cast iron　球墨铸铁
spheroidal material　球状材料
spheroidal roller mill　中速钢球磨
spheroidal weathering　球状风化
spheroidal　球的，球状的
spheroidization of pearlite　珠光体球化
spheroidization　球化
spheroidized particle　球化颗粒
spheroidizing annealing　球化退火
spheroidizing graphite iron　球墨铸铁
spheroidizing　球化（处理），球化退火
spheroid tank　球罐
spheroid　球体
spherometer　球径计，球径仪，球面曲率计
SPH(specific heat)　比热
sphygmoborometry　脉压测量
sphygmometer　脉搏计
SPIC(State Power Investment Corporation)　国家电力投资公司
spider armature　带径向支架的电枢
spider assembly　星形连接组件【燃料组件】
spider bar　横杆，横梁
spider gear assembly　星形齿轮总成
spider hub　转子中心体
spider rim　支架轮缘，磁轭
spider's web of cracks　蛛网形裂隙
spider web winding　蛛网形绕组
spider web　支臂
spider wire entanglement　蛛网形铁丝网
spider wrench　星形扳手，星形套筒扳手
spider　蜘蛛形接头【燃料组件】，十字叉，十字头，星形接头，三脚架，星形轮，多角架，多脚撑
spigot and socket pipe　承插管
spigot joint　套筒接合，套管接合，窝接，套筒连接，连接器，套筒接头
spigot ring　套圈，凸圈
spigot　凹凸槽，套管，插头，插孔，插口，塞子，龙头，栓
spiked core　强化堆芯
spike diverter　尖脉冲分流器
spike double roller　双齿辊
spike drawer　拔钉钳
spiked reactor　强化反应堆
spike driver　道钉钉入机
spike eliminator　尖峰脉冲消除器
spike enrichment fuel　点火用富集核燃料
spike fuel element　点火燃料元件
spike hammer　道钉锤
spikeless　非峰值的，非尖锐的
spike nail　道钉
spike-over shoot　上冲
spike pulse　尖脉冲，窄脉冲
spike suppressor　尖脉冲抑制装置
spike-tooth　钉齿
spike voltage　峰值电压
spike　道钉，尖端，峰值，顶点，波峰，尖峰信号，测试信号，尖峰脉冲，点火燃料组件，耙齿
spiking　尖峰信号，峰值形成，强化，尖峰释放
spile hole　小气孔
spile　插管，小塞子
spillage detection test　检漏试验
spillage loss　溢流损失
spillage of coal　溢煤
spillage　溢出，泄漏量，泄漏，溢出物，溢出量，洒落物，撒料，溢水
spill atomizer　回油式雾化器
spill burner　回流式喷燃器，回油式燃烧器
spill current　差电流，不平衡电流，动作电流
spilled feed　溢出的物料
spill index　去污指数
spilling breaker　崩碎波，溢型碎波器
spilling flap　阻尼板
spilling surge tank　溢流式调压室
spilling water　溢水
spilling wind　溢出风能【限速法】
spilling　泄漏
spill loss　轴向漏汽损失
spillover effect　扩散效应，溢出效应，外溢效果
spill over tank　溢流槽
spillover　泄漏放电，溢流管，信息漏失，附带结果，溢出
spill plate　防撒板
spill strip　级内轴封
spill tip burner　回油式喷燃器
spill truss　落料管支架，翻车机
spill valve　溢出阀，溢流阀，溢油阀
spillwater　溢出水
spillway apron　溢洪道护坦
spillway basin　溢水池
spillway bay　溢洪道堰孔，溢洪道闸孔
spillway bridge　溢洪道桥
spillway bucket lip　溢流堰消能鼻坎
spillway bucket　溢洪道消力戽
spillway capacity　溢洪道容量，（溢洪道）溢洪能力
spillway chamber　溢流室
spillway channel　溢流槽
spillway chute　溢洪道陡槽
spillway crest gate　溢流堰顶闸门
spillway crest level　溢洪道堰顶高程
spillway crest　溢洪道顶，溢流堰顶
spillway culvert　溢流涵洞
spillway dam　溢流坝
spillway design flood　溢洪道设计洪水
spillway discharge　溢洪道泄量
spillway face　溢流坝面
spillway gate　溢洪道闸门，溢洪闸
spillway lip　溢洪堰缘，溢洪道前缘
spillway performance　溢洪道运行特性
spillway pier　溢洪道闸墩
spillway section　溢流段
spillway slap　溢流面板
spillway surge tank　溢流式调压井
spillway tunnel　泄洪隧洞，溢洪隧洞

spillway 喇叭形溢洪道，泄洪道，泄水道，溢洪道，溢水孔，溢水口
spill 溢出，流出，溅出，泄出，泄漏，散落，溢流
spilt contract 分项承包
spin angular momentum quantum number 自旋角动量量子数
spin burst test 旋转破坏试验
spin counter 转数计
spindle assembly 心轴组件，主轴总成
spindle holder 心轴托
spindle nut 轴螺母
spindle oil 锭子油，轴润滑油
spindle sleeve 轴套
spindle 轴，杆，蜗杆，主轴，心轴，棒端定位插销，锭子
spin-echo information storage 自旋回波法信息存储器
spin-echo storage 自旋回波法信息存储
spin-echo 自旋回波法，自旋回波的
spine fins 棘状的肋片
spinel 尖晶石
spin gearing 传动装置
spin intensity 旋转强度
spinner blade （分离器中的）旋流叶片
spinner motor 双转子电动机
spinner 整流罩，螺旋桨整流罩，桨毂盖，旋转器，自旋体
spinning ball 旋转球
spinning carriage 绕索小车
spinning cup burner 转杯式燃烧器
spinning disk humidifier 离心式加湿器
spinning electron 自旋电子，旋转电子
spinning machine 离心式旋制机，离心机
spinning mode 自转模式
spinning operation 热备用运行
spinning period 热备用期
spinning reserve capacity 动态备用容量，运转备用容量，旋转备用容量
spinning reserve 热机备转容量，热备用，运转备用，旋转备用，空转备用，动态备用容量，旋转备用容量
spinning rotor 旋转转子
spinning wheel 绕索轮
spinning wind tunnel 螺旋试验风洞【垂直风洞】
spinning 离心法，离心作用，分离作用【成形作业】，旋转
spin off 附带的效果，伴随的，附带物，有用的副产品
SPIN(standard procedure instruction) 标准过程指令，标准程序训令
spin tunnel model tests 螺旋风洞模型试验
spin vector 角速度矢量
spin 旋转，自旋，拔丝，自旋的
spiracore 钢带螺旋绕铁芯，卷铁芯
spiral anchored ring 螺旋锚圈
spiral auger 螺旋钻
spiral bar 螺旋钢筋，螺旋筋
spiral bevel gear 螺旋伞齿轮，弧齿锥齿轮
spiral-bevel 螺旋伞形的
spiral-bladed rotor 螺旋叶片式转子
spiral case access 蜗壳进人孔
spiral case manhole 蜗壳进人孔
spiral case relief sluice 蜗壳安全泄水道
spiral case 蜗壳
spiral casing storey 蜗壳层
spiral casing 蜗壳
spiral chute 螺旋形陡槽
spiral coil 螺旋形线圈，螺旋形盘管
spiral concrete pile 螺旋式配筋混凝土桩
spiral concrete 螺旋式配筋混凝土柱
spiral condenser 盘管凝汽器，蛇形管冷凝器
spiral conveyer(-or) 螺旋输送机
spiral cooling 螺旋形冷却
spiral cross current 螺旋形横向环流
spiral current 螺旋形水流
spiral feeder 螺旋进料器，螺旋给料器
spiral filament 螺旋形灯丝
spiral fin fuel element 螺旋肋燃料元件
spiral-finned tube 螺旋肋片管
spiral fin 螺旋肋片
spiral flow aeration tank 螺旋流式曝气池
spiral flow aeration 螺旋流通气
spiral flow 螺线流，螺旋流
spiral fringe 螺旋形条纹
spiral-gear pump 斜齿轮式泵
spiral groove 螺旋槽
spiral guide 螺旋形导向装置
spiral heater 盘管式加热器
spiral heat exchanger 螺旋板式换热器
spiral hoop 螺旋形箍
spiral housing 涡壳，蜗室
spiral idler 螺旋托辊
spiral induction sodium pump 螺旋式感应电磁钠泵
spiraling 成螺旋形
spiral joint 螺旋接合
spirally coiled tube 盘管，螺盘管
spirally reinforced column 螺旋钢筋柱
spirally reinforced concrete column 配螺旋箍筋的混凝土柱
spirally wound tube 螺旋围绕管圈，水平围绕管圈
spiral mode of motion 螺旋运动
spiral pancake winding 螺旋饼式绕组
spiral pin 螺旋销
spiral plate heat exchanger 螺旋板式换热器
spiral plate 螺旋板
spiral pump 螺旋泵
spiral-quad cable 星铰四心软电缆
spiral-ramp vortex generators 螺旋坡道涡流发生器
spiral reinforcement 螺纹钢筋【预制柱桩】，螺旋钢筋，螺旋环筋
spiral-riveted pipe 螺旋形铆接管
spiral rope 螺旋绳
spiral scanning 螺旋形扫掠，螺旋扫描
spiral slot 螺旋形槽
spiral spring 螺旋形弹簧，螺线弹簧，盘簧
spiral staircase 螺旋式楼梯
spiral stairs 螺旋楼梯，螺旋梯
spiral-threaded roofing nail 螺纹屋面钉

spiral tube 螺旋管，螺旋形弹簧管，盘簧管
spiral turbine 蜗壳式叶轮机，蜗壳式水轮机
spiral type core 螺旋形铁芯，卷铁芯
spiral unloader 螺旋卸载机
spiral vortex 螺旋形涡流
spiral water turbine 螺旋式水轮机
spiral welded pipe 螺旋焊接管，螺旋形焊接管
spiral welded steel pipe 螺旋焊接钢管
spiral welding 螺旋焊接【管道制造】
spiral weld 螺旋形焊缝
spiral winding 螺旋绕组，波绕组
spiral wire insert 螺旋金属丝插入件
spiral wound cable 螺旋缠绕电缆
spiral wound gasket 缠绕式垫片，螺旋缠绕填料
spiral wound module 螺旋卷式组件，涡卷型组件
spiral-wrapped pin 螺旋缠绕销
spiral 螺旋的，螺旋形的，螺线，螺旋，螺旋管，螺旋体，螺旋线
spiratron 螺旋管
spire roughness simulation technique 尖塔粗糙元模拟技术
spire 尖塔【旋涡发生器】
spirit lamp 酒精灯
spirit level 水准仪，水平仪，气泡水准仪
spirit of enterprise 企业精神
SPI(single program initiation) 单程序启动
SPI(single program initiator) 单程序的启动程序
spit growth 沙嘴增长
spitzkasten 锥形选粒器
spit 沙嘴
splash apron 挡泥板
splash bar 淋水板【冷却水塔】
splash block 水簸箕，水落管下导水的砌块
splash board 挡泥板，防泥板
splash erosion 雨淋冲刷
splash fill 点滴式填料
splash-guard ring 防溅环
splash lubrication 飞溅润滑
splash packing 点滴式填料
splash plate 淋水板【冷却水塔】，挡板
splash-proof machine 防溅式机器，防溅型电机
splash-proof motor 防溅型电动机
splash ring 润滑油环
splash shield 防溅护板
splash type lubricating system 飞溅润滑系统
splash wing 挡泥板，翼子板
splash zone 飞溅区，稀相区
splash 溅，飞溅，喷溅
splat cooling 急冷
splayed abutment 八字形桥台，翼形拱座
splayed footing 八字形基脚
splayed jamb 八字形侧墙
splayed joint 八形缝，八字形接合，斜接
splay foot 八字脚
splay 喇叭形，向外张开的
splice angle steel 拼接角钢
splice bar 拼接钢筋
splice box 电缆套管，分编盒，分线盒
splice case 电缆套管，连接盒，分编箱
splice connector 拼合接头
spliced pile 接桩，拼接桩
spliced pole 叠接电杆
splice joint 拼合接头，鱼尾板接合，铰接，接头
splice length 搭接长度
splice loading （电缆）接头加感
splice plate 拼合板，拼接板，鱼尾板，镶接板
splicer 连接工具，接带机，接片机，接片器，交接器，接合器，接线
splice sleeve 连接套筒，连接套管，联轴器
splice 叠接，铰接，接合，接头，编织，铰接处，接续管，拼接，搭接，镶接
splicing attenuation 熔接衰耗
splicing box 电缆分线箱，电缆套管，接线盒，分接盒
splicing chamber 电缆分编室，接头室
splicing fitting 接线夹
splicing sleeve 接续管，拼合套筒
spline bushing 齿槽轴瓦
splined coupling 花键连接
splined hub 花键套
splined joint 花键连接
splined shaft 滑键轴，花键轴，有齿轴
splined slip joint 花键，多槽的伸缩式连接
spline function 样条函数
spline joint 填实缝
spline model 仿样模型，样条模型，
spline 方栓，齿条，止转楔，花键，键槽，塞缝片，样条
splint coal 硬烟煤
splintering 碎裂
splinter of stone 石片
splinter-proof glass 防碎玻璃
splint 夹板，托板，碎片
split air conditioner 分体空调器，分体式空调器
split air conditioning system 分体式空调系统
split anode magnetron 分瓣阳极磁控管
split anode 分瓣阳极
split batch charging 分散配料
split batch fuel management 分区燃料管理
split bearing 对开式（滑动）轴承，剖分轴承
split bed 分层床
split blip 峰，裂峰，分裂尖头信号
split block insulation 拼合的保温块
split body 分体，拼合体
split box 分线箱
split break 断口，破口，裂口，纵向断裂
split bridge 炉门坎，空冷火桥
split brush 分裂式电刷，分块电刷
split burner 缝隙式燃烧器
split bus protection 分相母线保护
split cable 分股电缆
split carbon brush 分块炭刷
split-carrier system 分裂载波制
split casing pump 拼合壳体泵
split casing 拼合泵壳，中分面式汽缸，破裂的套管，拼合外壳，对分外壳
split collector ring 分块式汇流环，分块式集电环
split conductor cable 分裂导线电缆，多心导线电缆
split conductor protection 分裂导线作引线的保

护系统
split conductor 多心线，多股绝缘线，分裂导线
split contact 头接点
split core type current transformer 分裂铁芯式电流互感器，钳形电流互感器
split core 分裂铁芯，分离堆芯
split cotter pin 开口销，开尾销
split cup 分瓣感应环
split delivery 分批交货
split diaphragm 对分隔板
split drum mixer 裂筒式拌和机
split-feed control 分路馈给控制
split-feed 分路馈给的
split-field motor 分串激电动机，串激绕组分段式直流电动机，串行绕组分段式电动机
split fire bridge 炉门坎，空冷火桥
split flange 对开法兰
split flap 分裂式襟翼
split-flow reactor （冷却剂）分流反应堆
split-flow 分离流，分流
split frame 分瓣机座
split gasket 拼合密封垫
split gear 拼合齿轮
split hammer 裂口锤
split jet injector 缝隙式喷嘴
split joint 拼接（头）
split key 开口键，切断电键
split leaf gate 分扇拼合闸门
split-level house 错台式住宅
split lever 分半式拐臂
split package contract 分片承包合同，大分包合同
split-phase capacitor motor 分相电容器式电动机
split-phase induction motor 分相感应电动机
split-phase motor 分相电动机
split-phase relay protection 分相继电保护
split-phase starting 分相启动，辅助相启动
split-phase start motor 分相启动电动机
split-phase winding 辅助绕组，分相绕组
split-phase 分相，分裂相位
split pin 分裂销子，开尾销，开口销
split pole converter 分裂磁极变流机
split pole flux 分裂磁极磁通
split pole motor 分极电动机，分裂磁极电动机
split pole 分裂磁极
split ranging control 分程控制
split reactor 分裂电抗器
split regeneration 分流再生
split ring commutator 分环换向器，分环整流子
split ring key 开口环键
split ring 环键，开口环，开口垫圈，对开挡环，裂环
split rotor 分裂式转子
split runner 拼合式转轮
split Savonius wind turbine 开裂式S型风力机
split saw 粗齿锯
split secondary 分级次级线圈，有抽头的次级线圈
split-series field motor 分串激电动机
split-series motor 分串激电动机，串激绕组分段式直流电动机
split series servomotor 绕组串激伺服电动机
split shaft gas turbine 分轴燃气轮机，分轴式燃气轮机
split sleeve bearing 轴瓦式轴承
split spoon sample 对开式取土样
split spoon 对开式取土勺
split stator condenser 分定片电容器
split stator 分瓣定子
split-stream softening 分流软化，氢钠并联软化
split system 分流系统
split table reactor 可伸缩堆芯反应堆
split tensile strength 劈裂抗拉强度
splitter blade 分流叶片
splitter box 分线盒
splitter gate 分流闸门，分叉阀门
splitter pier 分流墩
splitter plate 导流板，分流板，分割（稳定）板
splitter vane 分水叶片
splitter wall 分水墙
splitter 分解器，分离器，分裂机，劈裂器，分离设备，分流片，分流装置，导流板
split test 劈裂试验
split-throw winding 分叉绕组，异槽绕组
splitting action 分水作用，劈裂作用
splitting resistance 劈裂强度
splitting tensile strength 抗拉裂强度【开缝】，劈裂抗拉强度
splitting tensile test 分裂拉力试验，劈裂抗拉试验
splitting test 劈裂试验
splitting the bill down the middle 均摊费用
splitting 分裂，剖开，劈裂，切口，切开，纵切
split tooth 开小槽的齿，开有窄缝的齿
split tube sampler 对开管式取土器，对开式取样器
split tube 对开管口
split turbine 分轴式透平
split variometer 分档可变电感器
split ventilation 分流通风
split washer 开口垫圈，裂缝垫圈，拼合垫圈
split wedge gate valve 拼合楔闸阀
split wedge inclined-seat gate valve 拼合楔倾斜座闸阀
split wedge valve 对开楔形闸板阀
split winding rotor 分裂绕组转子
split winding type synchronous motor 分裂绕组式同步电动机
split winding type transformer 分裂绕组变压器
split winding 多头线圈，抽头绕组，分裂绕组
split-word operation 分字运算
split year 跨年度
split 拼合的，分体的【非整体泵壳】，裂开的，双绕组的，劈，劈开，劈裂，破裂，分离，裂开，划分，分开，分解，分裂，分割，剥裂，中分面，裂缝
SPL(sound pressure level) 声压级
SPM(set point manual) 整定值手册
spodic horizon 灰化层
spodic 灰化的
spodosol 灰土
spoil area 废渣场，弃土区

spoil disposal 废渣处理
spoil earth 废土石，废物
spoiled material 损毁材料
spoiled product 次品，废品
spoiler brake 阻流减速板，扰流板刹车
spoiler 绕流器，阻流板，扰流片，扰流器，扰流板
spoiling flap 阻尼板【随风速的变化用来阻止风轮转数增加的构件】
spoil removing 废渣清除
spoil 损害，弃土，废土石，抛土，损坏，破坏，弃渣，废物，挖出的泥土和岩石
SPOM(specialized & programmed operational matrix) 矩阵式生产运行模式
sponge ash 海绵状灰
sponge ball cleaner 胶球清洗器
sponge ball cleaning device 胶球清洗装置
sponge ball cleaning system 海绵球清洗系统
sponge ball cleaning 胶球清洗
sponge ball 胶球
sponge iron 海绵铁
sponge plastics 多孔塑料
sponge 海绵，泡沫材料，多孔材料，多孔塑料
spongy porous membrane 海绵状多孔薄膜
spongy rubber cleaning ball 海绵橡胶球【清洁用】
spongy texture 海绵状结构
spongy 海绵状的，多孔的
sponsor 发起者，发起人，资助者，牵头方，主办单位
spontaneous combustion point of oil 燃油自燃点
spontaneous combustion 自燃
spontaneous cracking 自裂
spontaneous decay 自发衰变
spontaneous emission 自发射
spontaneous evaporation 自然蒸发
spontaneous fission 自发核裂变，自发裂变
spontaneous ignition temperature 自燃温度
spontaneous ignition 自燃
spontaneously inflammable 可自燃的
spontaneous magnetic domain 自然磁畴
spontaneous magnetization 自然磁化，自发磁化
spontaneous oscillation 自发振动
spontaneous polarization 自发极化
spontaneous reaction 自发反应
spontaneous splitting （气泡）自行分裂
spontaneous transmission 自发传输
spontaneous 自然地，自生的，自发的
spoofing transmitter 干扰发射器，干扰信号发射器
spool chamber 阀芯腔
spool displacement 阀芯位移
spool drawing 管段图
spool end 阀芯端面
spool face 阀芯表面
spool gear 长齿轮
spooling system （软件的）假脱机系统
spool insulator 圆柱形绝缘子，直脚绝缘子
spool land 阀芯台肩
spool piece 联轴节，短管【主泵/反应堆冷却剂泵】
spool pilot valve 滑阀
spool position 阀芯位置
SPOOL(simultaneous peripheral operations on-line) 假脱机操作
spool stroke 阀芯行程
spool travel 滑阀芯行程
spool valve 滑阀
spool winding 模绕绕组
spool 带圈，短管，卷筒，卷轴，线轴，线圈架，线圈，卷，盘，直管段
spoon dredger 单斗式挖泥船
spoon sampling 勺钻取样
spoon-shaped blade 匙状轮叶
spoon 勺
sporadic trouble 偶然性故障
spot analysis 点滴分析，点渍分析
spot annealing 局部退火
spot beam antenna 点波束天线，区域覆盖天线
spot beam 点波束
spot board 小拌板
spot check 抽样检查，局部检查，抽查，现场抽查
spot check test 抽查试验
spot corrosion 点腐蚀，点蚀
spot dancing 斑点跳动
spot definition 光斑清晰度
spot delivery 当场交付（货）
spot facing 锪端面【在轴上加工平面】，局部整平
spot galvanometer 光点电流计
spot level 标高，高程点
spot lighting 局部照明
spot market 现货市场
spot measurement of flow 单点测流
spot measurement 单点测量
spot network 点式网络
spot pattern 光斑图案
spot plate 滴试板
spot price 现货价格
spot punch 点式穿孔，点穿孔器
spot repair 现场修理
spot sample 定时定点取样
spot temperature 工作地点温度
spotter 定位器，定心钻，测位仪，测位计，搜索雷达
spot test 点滴试验，抽查，当场测试，现场测试，斑点试验
spotting error 定位误差
spotting gain control 增益校正调节
spotting 确定准确位置，对位，配合，定位【在锚固以前】
spot trading 现货交易
spot-type detector 瞬时值探测器
spot-type fire detector 点型火灾探测器
spotty rainstorm patter 雷阵雨型
spot weld-bonding 胶接点焊
spot welder 点焊机
spot welding machine 点焊机
spot welding point 点焊点
spot welding 点焊
spot weld spacing 焊点距

spot weld 点焊，点焊接头，点焊缝
spouse 配偶
spoutable bed depth 可喷射床深
spoutable 可喷射的
spout-delivery pump 无压供水泵
spout diameter 喷嘴直径
spouted bed drier 喷动床干燥器
spouted bed 喷动床
spouting velocity 喷出速度，喷射出口速度
spout radius 喷嘴半径
spout region 喷射范围
spout 排料口，排水口，流槽，缝隙，气龙卷，水龙卷，水落管，水柱，喷射，排水管道，喷口，喷出，出料斜槽
sprage 喷雾
sprag 短支柱
spray additive tank 化学添加剂箱，氢氧化钠罐
spray aerate 喷洒通气器
spray aerator 喷雾曝气器，喷嘴通气器
spray and tray type deaerator 喷雾淋水盘式除氧器
spray angle 雾化角
spray apparatus 喷射设备
spray arc 喷射电弧
spray arrester 喷射避雷器
spray atomizer 喷雾器
spray attemperator 喷水减温器
spray box 喷雾箱，箱型喷雾集管
spray burner 喷嘴，雾化喷燃器
spray burning 喷雾燃烧
spray canal 喷水槽道
spray catcher 滴漏水捕集器，捕雾器
spray chamber 喷水段
spray coating 喷涂
spray column 喷雾柱，喷雾塔
spray concentrate 浓喷雾液
spray condenser 喷雾冷凝器，喷水冷凝器
spray cone angle 喷射锥角
spray cooled reactor 喷雾冷却反应堆
spray cooler 喷淋冷却器，喷淋式冷却器
spray cooling process 喷雾冷却法
spray cooling system 喷淋冷却系统
spray cooling tower 喷淋冷却塔
spray cooling 喷淋冷却，喷雾冷却
spraycrete 喷射混凝土
spray deck 喷淋盘
spray dedusting facility 喷水防尘设施
spray density 淋水密度
spray desuperheater 喷水减温
spray distributor 喷水机
spray dryer method 喷雾干燥法
spray drying process 喷雾干燥法
spray drying 喷雾干燥
spray dust suppression 喷湿抑尘
sprayed cathode 喷涂阴极
sprayed concrete 喷射混凝土
sprayed-on Al layer 喷铝层
spray eductor 喷淋引射器，化学添加剂引射器
spray eliminator 喷淋净化器，喷雾净化器
sprayer head （油雾化器）分油嘴
sprayer plate （油喷嘴）雾化片
sprayer 喷头，喷枪，喷洒机，喷雾器，喷射器
spray fan 喷雾风扇
spray field 喷水范围
spray flow rate 喷淋流率，喷雾流率
spray granulator 喷雾成粒器
spray gun 喷枪
spray header 喷淋水总管，喷淋环管
spray head 喷头
spray heat exchanger 喷淋热交换器
spray humidification 喷雾加湿
spraying agent 喷雾剂
spraying car 洒水车
spraying equipment 喷雾设备
spraying jet 喷射流
spraying nozzle 雾化喷嘴
spraying of thermal insulation material 保温层喷涂施工
spraying painting 喷漆
spraying pistol 喷枪
spraying water pond 喷水池
spray irrigation 喷灌
spray lacquering 喷漆
spray line connection 喷淋管连接
spray line nozzle 喷淋管接管
sprayline 喷灌管道，喷淋管路
spray manifold 喷水减温器，喷水减温集箱
spray metal coating 金属喷涂
spray method 喷射法
spray module 喷水组件
spray nozzle density 喷嘴密度
spray nozzle header 喷淋嘴联箱
spray nozzle ring 喷淋管环
spray nozzle 喷淋管嘴，喷雾嘴，喷嘴，水雾喷头，雾化喷嘴
spray of antireflective coating 喷涂减反射膜
spray-on fire proofing 喷涂防火
spray-on method 喷雾法
spray painting 喷漆
spraypak packing 喷淋网型填料
spray pattern 喷雾流型，喷纹
spray-pipe system 喷水管道系统
spray pipe 喷淋管（线）
spray plating 喷镀
spray pond 喷水池
spray pump 喷淋泵
spray radiation treatment 全身照射治疗
spray rate 喷淋率
spray ring 喷淋环，配水环，喷雾环
spray rinse 喷洗
spray solidification 喷雾固化（过程）
spray stuffing type deaerator 喷雾填料式除氧器
spray system 喷雾系统，喷淋系统
spray test 喷淋试验，喷雾试验，雾化试验，喷淋防水性测试
spray tower 喷雾冷却塔，喷雾塔，喷淋塔
spray transfer 喷射过渡
spray tray deaerator 喷雾淋盘式除氧器
spray treatment 喷雾处理
spraytron 静电喷射器
spray-type air washer section 喷水段
spray-type air washer 喷淋式空气洗涤机，喷淋

式洗涤机
spray-type attemperator 喷水减温器
spray-type cooling 喷淋式冷却，喷雾（式）冷却
spray-type deaerator 喷雾式除氧器
spray-type desuperheater 喷水式过热蒸汽减温器，喷水减温器
spray-type evaporator 喷淋式蒸发器
spray-type heat exchanger 喷淋型热交换器
spray-type separator 喷淋型分离器
spray valve 喷淋阀
spray water flow 喷淋水流
spray water pipe 喷水管
spray water protection shroud 喷雾水保护罩
spray water 喷雾水，喷水
spray zone 浪花带
spray 喷洒，喷淋，喷射，喷涂，喷雾，雾化，飞沫，喷洒液，水雾，喷雾器，喷嘴
spreadability 铺展性
spreader beam 担梁，起重机吊梁，传力吊架
spreader box 铺料箱
spreader feeder 抛煤机
spreader firing 抛煤机炉排燃烧
spreader stoker boiler 抛煤机锅炉
spreader stoker fired boiler 抛煤机锅炉
spreader stoker 抛煤机（炉排）
spreader strip 分流防冲带
spreader tool 扩张工具【防震条】
spreader 散布器，抛煤机，扩张器，碎石撒布机
spread factor （绕组）分布系数
spread footing 扩大式底脚，扩展基础，扩展式基础，桩台
spread foundation 扩大式基础，扩展基础，扩展式基础
spreading and leveling 平仓
spreading area 分洪面积，分水面积
spreading device 铺设装置
spreading draught tube 扩大式尾水管，喇叭形尾水管
spreading machine 铺料机
spreading of spray 喷雾角
spreading-out 漫流，推开
spreading project 推广项目
spreading range 扩展范围
spreading resistance 扩张流动损失
spreading roller 扩张滚筒
spreading thickness 碾压厚度
spread in performance 性能参差
spread in pulse height 脉冲高度参差
spread length 传播长度
spread of bearings 方位角摆动范围
spread of coil 线圈节距
spread of radioactive material 放射性物料扩散
spread of stall area 失速（气流分离）区的扩展
spread of wing 翼展
spread recorder 扩张记录器
spread source 分布源，分散源，涂敷源
spread variable capacitor （频带）展宽可变电容器
spread velocity 扩散速度
spread voltage 分布电压
spread 传播，伸展，扩散，扩展，展开，分布，蔓延，传开，范围，概率散度，射流扩散角
SPRG＝sparge 喷淋，鼓泡
spring abutment 弹簧支座
spring accelerated 弹簧加速的【控制棒】
spring action 弹簧动作
spring assembly 弹簧组件
spring back labyrinth gland 弹簧支撑迷宫汽封
spring back radial seal 弹簧支撑径向密封
spring-back sealing 回弹式汽封
spring back 弹性变形恢复
spring-balanced bell gauge 弹簧平衡式钟形压力计
spring balance 弹性平衡，弹簧秤
spring ball check valve 弹簧球式止回阀
spring barrel 弹簧锁定装置
spring board 跳板
spring bolt 弹簧螺栓
spring bracing 弹簧支撑架
spring break 春季解冻
spring buffer 弹簧缓冲器
spring butt hinge 弹簧铰链
spring catch 弹簧插销
spring circulation 春季环流
spring clamp 弹簧线夹，弹簧夹
spring clip grid 弹簧夹定位格架
spring clip retainer 弹簧保险圈
spring compression adjustment 弹簧压紧装置
spring cone penetrometer 弹簧锥贯入硬度计
spring constant hanger 弹簧恒力吊架
spring constant 弹簧常数
spring-controlled valve 弹簧控制阀
spring-controlled 弹簧控制的
spring core catcher 弹簧式土芯取样器
spring coupler damper 弹簧防震器
spring deflection 弹簧挠度
spring diaphragm reducing valve 弹簧薄膜式减压阀
spring diaphragm 弹簧薄膜式【执行器，阀门等】
spring-disk valve 簧盘阀
spring divider 弹簧量规
spring equinox 春分
springer 拱底石，起拱石
spring fender 弹簧护板
spring finger of the latch mechanism 锁定机构弹簧销
spring governor 弹簧调速器
spring gradient 弹簧变化率
spring hanger 弹簧支架，弹簧吊架
spring head 弹簧头
spring height 弹簧高度
spring hinge 弹簧合页
springing centre 起拱中心
springing course 起拱层
springing height 起拱高度
springing line 起拱线
springing of intrados 拱腹起拱线
springing operation 绿化
springing 弹动

springless valve 无弹簧阀
springless 无弹簧的
spring line 倒缆，拱脚线，起拱线
spring loaded activator 弹簧加载执行机构
spring-loaded anchor 弹簧承载弹顶锚固［锚定］，弹簧锚定
spring loaded brake 加载弹簧制动器，弹簧闸，弹簧制动装置
spring-loaded centrifugal latch 加载弹簧离心闩锁
spring-loaded check valve 弹簧加压逆止阀
spring-loaded clamp brake 弹簧加载闸瓦
spring-loaded membrane 弹簧承力膜片
spring-loaded needle valve 弹簧承力针形阀
spring-loaded packing 弹簧加载的填料密封
spring-loaded pressure relief valve 弹簧卸压阀
spring-loaded regulator 弹簧承力调节器
spring-loaded safety valve 弹簧式安全阀
spring-loaded silent check valve 弹簧式无声止回阀
spring-loaded stuffing box 弹簧加载的填料函
spring-loaded tube suspension 弹簧管子吊架
spring-loaded valve 弹簧阀
spring-loaded 簧压的，弹簧加压的，弹簧加载的，弹簧承力的
spring lock washer 弹簧垫圈
spring lock 弹簧锁
spring low-water line 大潮低潮线
spring manometer 弹簧式压力表
spring motor 发条传动装置，发条驱动用电动机
spring of pilot valve 错油门弹簧
spring-opposed air valve 弹簧承力空气阀
spring-opposed bellows 弹簧承力波纹管
spring-opposed piston 弹簧承力活塞
spring-opposed 弹簧承力的
spring overturn 春季对流
spring pack 弹簧组件
spring pad 弹簧座
spring pin 弹簧销
spring piston 弹簧活塞（式）
spring plate detector 簧板检示器
spring plate 弹簧板
spring plug 弹簧塞【燃料组件上管座】
spring pocket 弹簧套
spring point 弹簧辙尖
spring preload 弹簧预载
spring protrusion 弹簧突出
spring range 大潮潮差
spring rate 大潮流速，弹力比
spring reaction fendering 弹簧缓冲
spring reamer 弹簧铰刀
spring-relief crusher 弹簧式破碎机
spring relief mechanism 弹簧释放机构
spring relief system 弹簧减荷装置
spring relief valve 弹簧安全门
spring reservoir 弹簧组件
spring return limit switch 有弹簧复位的限制开关
spring return switch 弹簧回复开关
spring return 弹簧复位
spring rigidity 弹簧刚度
spring ring 弹簧圈，开口环
spring rise 大潮升
spring rod 弹簧杆
spring runoff 春季径流
spring scale 弹簧秤
spring season 大潮期
spring set 弹簧装置，弹簧组件【燃料组件上管座】
spring shim 弹簧垫片
spring shock absorber 弹簧减振器
spring shovel spreader 铲式抛煤机
spring stack 弹簧柱【控制棒驱动机构】
spring steel 弹簧钢
spring stiffness 弹簧刚度
spring stored-energy operating mechanism 弹簧贮能操作机构
spring strip air preheater 弹簧带式空气预热器
spring support 弹簧支承，弹簧支架，弹簧吊架，弹簧支座
spring suspension 弹簧吊架，弹簧悬吊，弹簧悬置
spring switch 弹簧开关
spring tester 弹簧试验器
spring thrust bearing 弹簧推力轴承
spring tidal current 大潮流
spring tide range 大潮潮差
spring tide 高潮，大潮，潮汛
spring-tube manometer 弹性管式压力计
spring-tube 弹性管
spring-type governor 弹簧式调速器
spring valve 弹簧阀
spring velocity 大潮流速
spring viscous damper 弹簧黏滞阻尼器
spring washer 弹簧垫圈
spring water 泉水
spring wire 弹簧钢丝
spring wood 春材
spring 弹性，弹簧，跳跃
sprinkler device 喷淋装置
sprinkler fire extinguisher 自动喷水式灭火器
sprinkler head system 喷水装置【人工降雨器】
sprinkler head 喷水头，撒水头
sprinkler irrigation 喷灌
sprinkler jockey pump 洒水器操控水泵
sprinkler line 喷灌管道
sprinkler nozzle 喷洒嘴
sprinkler pattern 喷灌方式
sprinkler pipe system 喷管装置
sprinkler stoker 抛煤机炉排
sprinkler system 喷水（灭火）系统，（冷却塔的）喷淋系统，洒水灭火系统
sprinkler 撒煤机【沸腾炉】，洒水车，洒水器，喷水机，喷雾器，洒水灭火器，自动喷水灭火装置，喷洒装置，洒水装置，水喷淋喷头
sprinkle 喷淋，泼洒，喷洒，微阵雨，小雨
sprinkling basin 喷水池
sprinkling filter 喷淋过滤器
sprinkling lawn 喷灌草地，喷灌草地
sprinkling machine 喷洒装置
sprinkling nozzle 洒水喷嘴
sprinkling pond 喷水池
sprinkling system 洒水系统

sprinkling truck 洒水车
sprinkling 喷撒
SPRINT(selective printer programmer) 选择性打印机的程序编制器
sprocket belt pitch drive 链轮齿皮带调桨驱动
sprocket belt 链轮齿皮带
sprocket hole 凿孔，定位孔
sprocket pulse 计时脉冲，时钟脉冲
sprocket wheel 链轮
sprocket winch 链轮绞车
sprocket 链轮，链轮齿，扣链齿，扣链齿轮
SPRT(standard platinum resistance thermometer) 标准铂电阻温度计
spruit 溪，小河道
sprung arch 悬吊拱，扇形拱
sprung mass 悬挂质量，簧载质量
sprung 加载弹簧，支在弹簧上的
SPRY=spray 喷射
SPSB(Shenzhen Power Supply Bureau) 深圳供电局
SP(set point) 设定值
SPS(samples per second) 抽样每秒，每秒抽样数
SPS(secondary protection system) 二次保护系统
SPS(serial-parallel-serial configuration) 串并串行结构
SPS(stored program system) 存储程序系统
SPS(string process system) 字符串处理系统
SPS=supersonic 超音速的
S/P(stowage plan) 货物积载图，也称船图、舱图
SP(structured programming) 结构程序设计
SPST(single-pole single-throw) 单刀单掷
SPSW(swap program status word) 交换程序状态字
SPT blow count 标准贯入击数，标准贯入试验击数
SPT(shortest processing time) 最短的处理时间
SPT(standard penetration test) 标准贯入试验
spudding shoe 钻杆接头
spudding 铲除
spud vibrator 振动棒
spud 煤气喷嘴，煤气枪，夹板，销，销钉，草铲，定位桩
spun concrete 离心浇灌混凝土，旋制混凝土【构件】
spun glass 玻璃纤维，玻璃丝
spun part 离心浇铸部件
spur dike 挑水坝，丁坝
spur gearing 正齿轮传动装置
spur gear operated valve 正齿轮传动阀门
spur gear pump 正齿轮泵
spur gear 直齿圆柱齿轮，正齿轮，圆柱正齿轮，直齿式齿轮
spurious action 虚假动作，乱真动作
spurious alarm 假报警
spurious capacitance 杂散电容，寄生电容
spurious count 虚假的计数，假计数，乱真计数
spurious coupling 杂散耦合，寄生耦合
spurious echo 虚假回波
spurious event 假事件
spurious frequency 寄生频率
spurious high level trip 虚假的高水位紧急停堆，虚设高水位保护停堆
spurious indication （测量仪表）假指示
spurious lift 虚假抬升
spurious noise 寄生噪声
spurious oscillation 寄生振荡
spurious pulse 乱真脉冲
spurious response 假信号响应，无线电干扰
spurious shutdown 假停堆，误停堆
spurious signal 虚假信号，伪信号，干扰信号，寄生信号，乱真信号
spurious trip 谬误跳闸，误跳车，误动作，乱真跳闸
spurious 假的，虚的，虚假的，乱真的，寄生的，伪的
spur line 输出线的短分支，支线
spurnwater 防浪板
spur protection 护岸丁坝
spurted plastic tube 喷塑管
spur track 铁路岔线
spurt 喷出，溅散，喷射
spur wheel 正齿轮
spur 丁坝，支脉，短木支撑
sputtering of electrode 电极溅蚀
sputtering temperature 溅散温度
sputtering 阴极溅镀
sputter vane 分水叶片
sputter 飞溅，喷涂，溅射，喷射，喷镀
SPVOL(specific volume) 比容
SPV(second party verification) 第二方验证
SPV(special purpose vehicle) 专用车辆
spy hole 观测孔，窥孔
squall line 飑线
squall 飑
squamose structure 鳞片状构造
square array 方格排列【堆芯】
square bar 方棒，方钢，方钢筋
square bellmouth transition 方喇叭形渐变段
square body 方塔身
square bolt 方头螺栓
square box wrench 方套筒扳手
square bracket 方格形
square building 方形建筑
square butt weld 方形对接焊
square cell 正方形栅元
square channel tile 方槽形瓦管
square coil 矩形线圈
square configuration 正方形布置，正方形构形
square cowl 方伞形风帽
square crossarm 方横担
square crossing 垂直交叉
square cube law 平方立方原则
square deviation 平方偏差
squared off tip 方形翼尖
squared paper 方格纸
square-drive coupler 四分之一四方驱动连接头
squared timber 方木，方正木材
squared value 平方值
squared 平方的
square-edged broad-crested weir 直角宽顶堰

square-edged inlet 直角进水口
square-edged orifice plate 方边的孔板
square-edged thin orifice (plate) 直角边薄孔板
square-edge timber 方木，方正木材
square elbow 直角肘管
square engine 方形发动机
square file 方锉
square fluidized bed 矩形流化床
square foundation 方形基础
square frame 方框
square fuel assembly 正方形燃料组件
square grid 方格网，方栅
square groove 方形槽，垂直坡口
square head bolt 方头螺栓
square head wrench 方头螺栓扳手
square inserted handle box wrench 四方插柄套筒扳手
square lattice 方形栅格【堆芯】
square law capacitor 平方律可变电容
square law detector 平方律检波器
square law function generator 平方律函数发生器
square law rectifier voltmeter 平方律整流伏特计
square law rectifier 平方律整流器
square-loop core 矩形磁滞回线铁芯
square loop 方形环路
square matrix 方阵，方形矩阵
square measure 面积，面积单位，平方尺度
square mesh sieve 方眼筛
square mesh 方格网
square nail 方钉
square network 方格网
square net 方格网
square normal closed double head wrench 四方环形双头扳手
square-nosed slug 方头涌节
square nut 方螺母，四方螺母，螺帽
square opening 方口，方孔
square or rectangular wave AC voltage 正方形或长方形波交流电压
square pulse 矩形脉冲，方形脉冲
square ring spanner 四方环扳手
square root calculator 平方根计算器
square root compensation 平方根补偿
square root computer 平方根计算器
square root deviation 均方根差
square rooter 平方根程序，平方根计算器
square root extractor module 均方根模块
square root extractor unit 均方根模块
square root extractor 开方器，取平方根器
square root-floating 平方根浮点运算
square root function 平方根函数
square root method 平方根法
square root planimeter 平方根面积计
square root 方根，平方根
square rubble 方块毛石
squarer 平方器，矩形波形成器，方波脉冲发生器
square sail 横帆，直角帆
square shingle 方块瓦
square signal 方波信号
square sine wave 正弦波平方
square single end tubular wrench 四方单头套筒扳手
square socket wrench 方套筒扳手
square S-shaped closed double head wrench S形四方环形双头扳手
square steel 方钢
square-stone masonry 方石圬工
square thread 方螺纹，矩形螺纹
square trench 方形沟槽
square vortex ring 方形涡环
square washer 方垫圈，方形垫圈
square wave-form oscillator 方波振荡器，矩形波振荡器
square wave generator 方波发生器，矩形波发生器
square wave inverter 矩形波逆变器，方波逆变器
square wave response curve 方波响应曲线
square wave 方波，矩形波
square wind tunnel 方形试验段风洞
square wire 方线
square 平方，方块，方形，广场，直角尺，丁字尺，正方形，正方形的，四方的，求平方，形成矩形脉冲，削方
squaring circuit 矩形脉冲形成电路
squaring count 求平方数
squashed delta （变压器的）互联三角形连接
squat pier 矮墩
squatting pan 蹲式大便器
squeeze motion （机械手抓手的）夹取动作
squeeze time 预压时间
squeeze valve 挤压阀
squeeze 挤，压，挤压，压榨
squegger 自动消失振荡器
squegging oscillator 断续振荡器，间歇振荡器
squeg 作非常不规则的振荡
squelch control 静噪控制
squibbing 扩孔爆破
squib 电气引爆器，引火线，点火器
squinch 内角拱，突角拱
squirrel cage AC motor 鼠笼式交流电动机
squirrel cage armature 鼠笼式电枢
squirrel cage asynchronous motor 鼠笼型异步电动机
squirrel cage bar 鼠笼条
squirrel cage heat exchanger 鼠笼式换热器
squirrel cage induction generator 鼠笼异步发电机
squirrel cage induction motor 笼式感应电动机，鼠笼式感应电动机
squirrel cage motor 笼式电动机，鼠笼式电动机，鼠笼式马达
squirrel cage rotor 鼠笼式转子
squirrel cage type rotor 鼠笼型转子
squirrel cage winding 鼠笼绕组，鼠笼式绕组
squirrel cage 鼠笼，鼠笼式，鼠笼式的
squirt pump 注射泵
squirt 射流，水枪
squitter pulse 断续脉冲
squitter 间歇振荡器，断续振荡器
S register S寄存器，存储寄存器，和数寄存器

SRE 放射性废水回收系统【核岛，机修车间，厂区试验室，核电站系统代码】
SR(set-reset) flip-flop 置“0”置“1”触发器
SRH 第二级再热器【DCS 画面】
SRI 常规岛闭路冷却水系统【核电站系统代码】
SRM(stuck rod margin) 卡棒裕度
srob 抢夺
SR(Significant Incident Report) 重大事件报告
SRSS(square root of sum of squares) 平方和开平方组合法
SR(styrene rubber) 苯乙烯橡胶
SR(supply request) 供应要求（申请）单
SSA(shift supervisor assistant) 付值长
SSB(single side band) 单边带
SSC(static series compensators) 静态串联补偿器
SSC(submerged scraper conveyer) 水浸式刮板捞渣机
self-accelerating reaction 自加速反应
self-regulation 自动调整，自调节，自调整，自动调节
SSE(safe shutdown earthquake) 安全停堆地震
S-shaped bar 蛇形钢筋
silica carryover 二氧化硅携带
single-curvature arch dam 单曲率拱坝，单曲拱坝
solar irradiance 太阳辐照度，太阳辐照率，太阳辐照，太阳照度
SSPD(short-shipped) 漏（少）装
SSPE(solid-state power equipment) 固体电力设备
spoke 轮辐
SSP(special surveillance plan) 专门监督计划
SSPS(satellite solar power station) 卫星太阳能电站
SSR(subsynchronous resonance) 次同步谐振
SS(selector switch) 选择开关
SS(shift supervisor) 值长
SSS(secondary sampling system) 次取样系统
SS(stainless steel) 不锈钢
SS(suspended solids) 悬浮固体，悬浮物
S-S(startup-standby) transformer 启/备变
SSU(subsequent signal unit) 后续信号单元
SSU(synchronisation signal unit) 同步信号单元
SSW(solid-state welding) 固态焊
stabilidyne receiver 高稳式接收器
stabilisation 稳定
stabilising condenser 稳定电容器
stabilistor 一种铁磁稳压管
stability against collie 抗倒塌稳定性
stability against sliding 抗滑移稳定性
stability against tilting 抗倾覆稳定性
stability against tipping 抗倾翻稳定性
stability alarm 稳定度警报，稳定受破坏警报
stability analysis 稳定分析，稳定性分析
stability and integrity of the structure 结构的稳定性和整体性
stability boundary 稳定性边界
stability calculation 稳定计算
stability category 稳定性类别
stability class 稳定度等级
stability coefficient of capacitance 电容稳定性系数
stability coefficient 温度系数
stability condition 稳定条件，稳定状态，稳定性条件
stability constant 稳定常数
stability criterion 稳定判据，稳定性判据，稳定准则，稳定性准则
stability curve 稳定特性曲线
stability degree 稳定度
stability derivative 稳定性导数
stability diagram 稳定图
stability disruption 稳定破坏
stability factor 稳定系数，稳定因子，稳定因数
stability index 稳定性指数，稳定度指数，稳定指数
stability in high wind 大风中的稳定性
stability in the large 总稳定度
stability in the small 局部稳定度
stability limit 稳定限度，稳定极限，稳定性极限，稳定度极限
stability margin 稳定界限，稳定裕量，稳定裕度
stability moment 稳定力矩
stability number 稳定系数
stability of a linear control system 线性控制系统的稳定性
stability of arch dam abutment 拱坝坝肩稳定
stability of atmosphere 大气温度
stability of cement grout 水泥浆的稳定性
stability of floatation 浮选稳定性
stability of flow 流量稳定，流动稳定
stability of following wind 调向稳定性【风力发电】
stability of rating curve 率定曲线稳定性
stability of reservoir slope 水库边坡稳定
stability of stage-discharge relation 水位流量关系稳定性
stability of surge tank 调压井的稳定性
stability of unstable rock 危岩稳定性
stability or coal water mixture 水煤浆稳定性
stability parameter 稳定性参数
stability range 稳定范围，稳定区，稳定区域，稳定域
stability series 稳定系列
stability simulation 稳定度模拟
stability test 机械稳定性试验，稳定性试验，安全度试验
stability theory 稳定性理论
stability under irradiation 辐照稳定性
stability zone 稳定区
stability 稳定，稳定度，稳定性
stabilivolt tube 稳压管
stabilivolt valve 稳压阀
stabilivolt 稳压器，稳压管
stabilizability 可稳性
stabilization by cement grouting 水泥灌浆加固
stabilization coefficient 稳定系数
stabilization efficiency 稳定效率
stabilization factor 稳定因子，稳定系数，稳定因素

stabilization lagoon 稳定塘
stabilization network 稳定网络
stabilization of frequency 频率稳定
stabilization of gasoline 汽油的稳定化
stabilization of level 电平的稳定
stabilization of super conducting magnet 超导磁体稳定化
stabilization pocket 稳定，稳定袋
stabilization pond 稳水池
stabilization power 稳定功率
stabilization ratio 稳定比
stabilization time 稳定时间，持续时间
stabilization 加固处理，镇定，稳定，稳定化，稳定（化）作用
stabilizator 稳定剂，稳定器，稳压器
stabilized channel 不冲不淤河槽
stabilized condition 稳定工况，稳定状态
stabilized feedback 稳定反馈
stabilized grade 不冲不淤比降
stabilized master oscillator 稳定主振荡器
stabilized operating temperature 稳定工作温度
stabilized platform 稳定平台
stabilized plume height 羽流稳定高度
stabilized power source 恒定电源，稳定电源
stabilized power supply device 稳压电源装置
stabilized power supply 稳定电源，稳压电源
stabilized pressure water tank 稳压水箱
stabilized shunt-wound generator 稳定并励发电机
stabilized shunt-wound motor 稳定并励电动机
stabilized shunt 稳定分路，稳并
stabilized steel 稳定性钢
stabilized 稳定的
stabilizer＝stabilizer 稳压器，稳定剂
stabilizer 加固剂，稳定剂，稳定器，稳燃器，稳压器
stabilizing agent 稳定剂
stabilizing annealing 稳定化退火
stabilizing baffle 火焰稳定器
stabilizing burner 助燃用燃烧器
stabilizing choke 稳定扼流圈
stabilizing circuit 稳定电路
stabilizing factor 稳定系数
stabilizing feedback network 稳定反馈网络
stabilizing feedback 稳定反馈
stabilizing fin 稳定鳍，稳定翼板
stabilizing load 安定载荷，稳定载荷
stabilizing measures 稳定措施
stabilizing network 稳定网络，镇定网络
stabilizing potential 稳定电位
stabilizing resistance 稳流电阻
stabilizing transformer 稳定变压器，稳压变压器
stabilizing treatment 稳定处理，稳定化处理
stabilizing winding 稳定绕组，平衡绕组
stabilovolt 镇伏器【电压稳定装置】
stabistor 稳定二极管
stable air mass 稳定气团
stable air 稳定空气
stable area 稳定地区
stable atmosphere 稳定大气
stable atmospheric condition 稳定大气条件
stable coal arch 稳定的煤拱
stable combustion 稳定燃烧
stable component 稳定环节
stable condition 稳定工况
stable control 稳定控制
stable current 定常流
stable density gradient 稳定密度梯度
stable density stratification 稳定密度层结
stable element 稳定元素，稳定元件，稳定因素
stable equilibrium point 稳定的平衡点
stable equilibrium 稳定平衡
stable film boiling 稳定膜态沸腾
stable filter 稳定滤波器
stable fission product poison accumulation 稳定裂变产物毒素积累
stable fission product poisoning 稳定裂变产物中毒
stable flame 稳定火焰
stable flow 稳定流
stable fluidized bed 稳态流化床
stable grout 稳定灌浆
stable head curve 压头稳定曲线
stable hydraulic section 稳定水力断面
stable ignition 稳定点火
stable isotope 稳定同位素
stable lapse rate 稳定递减率
stable load test 稳定负荷试验
stable load 稳定负荷
stable local oscillator 稳定的本机振荡器
stable matrix 稳定矩阵
stable nappe 稳定水舌
stable operating condition 稳定运行情况
stable operation 稳定运行，安定运行，稳定操作
stable oversteer 稳定过渡转向
stable period 稳定反应堆周期，稳定周期
stable rathole diameters in silo 筒仓中稳定的细长流道直径
stable reactor period 反应堆稳定周期
stable region 稳定域
stable rock bed 稳定岩基
stable-sliding 稳定滑动
stable slug 稳定气栓
stable state 稳态，稳定状态，平稳状态
stable stratification 稳定层结
stable stratified flow 稳定分层流
stable structure 稳定结构
stable system 稳定系统
stable tracer 稳定指示剂
stable transistor 稳定的晶体管
stable tunnel speed 风洞稳定风速
stable value 稳定值
stable wave 稳定波
stable 稳定的
stab 刺穿
stack air activity release 烟囱空气放射性释放
stack design 烟囱设计
stack desulfurization 排烟脱硫
stack dilution factor 烟囱稀释因子
stack disposal 烟囱处置，烟囱排放
stack draft 烟囱拔风，自生通风压头，烟囱排

烟，烟囱通风
stacked bed 叠床，多层床
stacked Darrieus rotor 重叠式达里厄转子
stacked-disk thermopile 层叠盘式热电堆
stacked-gate avalanche injection type MOS 叠栅雪崩注入金属氧化物半导体
stacked integrated circuit 层叠集成电路
stacked length （纵接核燃料芯块的）叠堆长度
stacked solar cell 叠层太阳电池
stacked strip conductor 由多股扁线组成的导条
stacked tray 堆叠的托架
stack effect pressure 热压
stack effect 烟囱效应，抽吸效应，烟囱压差效应
stack effluent monitor 烟囱排出物监测仪
stack effluent 烟囱排放物
Stackelberg decision theory 施塔克尔贝格决策理论
stack emission 烟囱排放，烟气排放，烟囱排放量，尾气排放
stacker conveyor 堆存用输送机
stacker crane 自动存取机高架吊车，塔式起重机，码垛机
stacker-reclaimer 斗轮堆取料机，堆取料机
stacker 堆料机，堆煤机，堆垛机，堆叠器，卡片积存器，栈式存储器
stack exit velocity 烟囱出口［排烟］速度
stack factor 烟囱因素，烟囱因子
stack flue gas monitor system 烟囱烟气监视系统
stack flue 烟道
stack gas desulfurization facility 烟气脱硫装置
stack gas desulfurization technique 烟气脱硫技术
stack gas desulfurization 烟气脱硫
stack gas monitor 烟囱排气监测仪
stack gas purifier 烟气净化器
stack gas 排烟，排气，烟囱废气，烟道气
stack guy 烟囱风缆
stack height 烟囱高度
stack-in cooling tower technology 排烟冷却塔技术，烟塔合一技术
stack induced downwash 烟囱诱导下洗
stacking angle 堆料角
stacking capacity 堆料出力
stacking coal 堆积煤
stacking conveyor 堆料皮带机
stacking factor （铁芯的）叠压系数，（线圈的）占空系数
stacking fault 层错
stacking height 叠片高度，（电机）铁芯高度
stacking mechanism 叠片机构
stacking and reclaiming system 堆取料系统
stacking site 堆料场【土方、设备】
stacking truck 叉车升降机
stacking 堆积，堆料
stackless nuclear power plant 无烟囱核电厂
stack loss 排烟损失，烟囱损失
stack monitor 烟囱监测仪
stack nozzle 烟囱口，烟囱排烟口
stack of Belleville springs 碟形弹簧柱，盘形弹簧柱
stack of fuel element 释热元件组件
stack of lamination 铁芯段，叠片组，（线圈的）占空系数
stack of pipes 管堆
stack operation 栈操作，栈运行
stack-oriented 面向堆栈的
stackout and reclaim system 堆取系统
stack-out conveyor 堆料皮带机
stack outlet 烟囱出烟孔，烟囱出烟口
stack parameter 烟囱参数
stack plume 烟羽
stack pointer 栈指示字，栈指针
stack pressure 烟囱压力
stack register 栈存储器
stack sampling 烟囱取样
stack shaft 烟囱身
stack tape 组合磁带
stack-tip downwash 烟囱出口烟气下洗
stack-tower 烟塔【烟塔合一】
stack-up 层叠，叠装
stack velocity 出烟速度
stack vent 通风竖管，烟囱排烟道
stack yard 堆料场
stack 烟囱，烟道，堆积，堆栈，通风管，竖管，浪蚀岩柱，草堆
staddle 承架
stadia computing disk 视距计算盘
stadia constant 视距常数
stadia diagram 视距图表
stadia formula 视距公式
stadia hair 视距丝
stadia interval 视距间隔
stadia levelling 视距测高
stadia line 视距丝
stadia method 视距法
stadia plane-table survey 平板仪视距测量
stadia point 视距点
stadia reading 视距读数
stadia rod 视距尺
stadia slide rule 视距计算尺
stadia station 视距测站
stadia survey 视距测量
stadia table 视距计算表
stadia telescope 视距望远镜
stadia theodolite 视距经纬仪
stadia transit 视距经纬仪
stadia wire 视距丝
stadia 视距仪
stadium 体育场
STAE(specify task asynchronous exit) 特殊任务异步出口，特指任务异步出口
staff auditor 雇用审计师
staff coefficient 定员系数
staff error 标尺误差，水尺误差
staff float 长杆浮标
staffing criteria 人员配备标准
staffing schedule 人员配置计划表
staff in post 在职人员
staff living quarters 职工宿舍
staff man 司尺员
staff-on-duty room 值班室
staff quarters 职工宿舍

staff reading 标尺读数
staffs canteen 职工食堂
staff training expense 职工教育经费
staff 标尺，标杆，量杆，杆，测尺，水尺，（全体）工作人员
stage aeration 分段曝气，分级通气【处理活性污泥的一种方法】
stage bleeding （汽轮机）级间抽汽
stage-capacity curve 水位容量（关系）曲线，水位库容曲线
stage casing insert 多级泵壳垫圈【多级泵】
stage casing pump 分段式多级泵，多级套管泵
stage casing with bleed off 有抽头的中段泵壳
stage casing 中段泵壳，中壳【多级泵】
stage characteristic 级的性能，阶段性特点
stage compaction 分级压实法
stage construction 分期施工
staged capacity 分期容量
staged combustion 分级燃烧
staged contactor 分级接触器
staged diversion 分期导流
stage development 分级开发，分期发展
staged fluidized bed 分级流化床
stage-discharge curve 水位流量曲线
stage-discharge formula 水位流量关系公式
stage-discharge record 水位流量记录
stage-discharge relation curve 水位流量关系曲线
stage-discharge relation 水位流量关系
stage diversion 分期导流
stage dredging 分期疏浚
staged transformer 分级变压器，多抽头变压器
stage-duration curve 水位变化曲线，水位历时曲线
stage duration 水位历时
stage efficiency 级效率，级放率，放大级效率
stage evaporation 分段蒸发，分级蒸发
stage filter 分级过滤器
stage fluctuation range 水位起伏幅度
stage fluctuation 水位变化，水位涨落
stage forecasting 水位预报
stage frequency curve 水位频率曲线
stage frequency 水位频率
stage grouting 分段灌浆，分期灌浆
stage head 级压头
stage heat drop 级的焓降
stage heater 回热加热器
stage hydrograph 水位曲线图，水位图
stage I fatigue crack 第一阶段疲劳裂纹
stage improvement 分期改善
stage intercooling 级间冷却
stage load 级负荷，分级加荷，分阶段加荷
stage matching 级间匹配
stage observation 水位观测
stage of completion 完工阶段
stage of decommissioning 退役阶段
stage of design 设计阶段
stage of freezing 封冻期
stage of zero flow 断流水位
stage planning 分期规划
stage pump 多级泵
stage recorder 水平记录仪
stage record 水位记录
stage rise 水位上升
stage routing 水位演算
stage station 水位站
stage structure 分期建筑
stage teeth （汽封）高低齿
stage transfer function 级间传递函数
stage trickling filter 分段滴滤池
stage 工程分期，级，阶段，时期，台
stagflation 滞胀【金融】
stagger angle （汽轮机的）安装角，叶片扭角，交错角
stagger cycle 交错周期
stagger delay line 交错延迟线
staggered arrangement 梅花形布置，错列布置
staggered bank 错列布置管束
staggered brush arrangement 电刷的交错布置
staggered bundles 错叠排管
staggered circuit 参差调谐电路，相互失谐级联电路
staggered gauging 参差调测，交错调测
staggered header 波形集箱
staggered intermittent fillet weld 交错断续角焊缝
staggered joint 错缝，错缝接合，错开接缝，错列接头
staggered labyrinth gland 交错迷宫汽封
staggered perforated plate 错叠多孔板
staggered pile 梅花桩
staggered piling 错列打桩，打梅花桩
staggered rivet joint 错列铆接
staggered tube 错列管，交错管排，叉排管
staggered vortex street 交替涡街，交错涡街
staggered vortex 交错旋涡
staggered welding 错列焊接
staggered 成梅花形，错列布置的（管束），错列的，交错的
staggering advantage 参差调频效果
stagger of stabilizer 稳定器的摆动
stagger time 参差时间
stagger tuning 参差调谐，串联调谐
stagger-wound coil 篮形线圈，笼形线圈
stagger 交错，错列，参差，错列布置，（叶片）安装角，摆动，跳动
staging area 集合场，分级区，集结待命地区
staging 分期，级，级组，台架
stagnancy state 停滞状态
stagnant air 停滞空气，滞止汽流
stagnant area corrosion 停滞区腐蚀
stagnant area 死水区，停滞区，滞流区
stagnant condition 滞止条件
stagnant film 滞止膜
stagnant gas 停滞气体，滞流气体
stagnant groundwater 上层滞水
stagnant inversion 停滞逆温
stagnant layer 滞止层
stagnant pocket 停滞区
stagnant point 临界点，滞留点
stagnant pool 积水池
stagnant sodium layer 停滞钠层，滞流钠层
stagnant steam 不流动的蒸汽，停滞蒸汽

stagnant wake 静区，（气流中的）死区，滞流区
stagnant water 停滞水，不流动水，死水
stagnant zone 滞流区
stagnant 停滞的，不景气的，静水的
stagnated air mass 停滞气团
stagnating tunnel flow 风洞滞止气流
stagnation area 滞止区
stagnation boundary layer 滞止边界层
stagnation condition 滞止状态
stagnation density 滞止密度
stagnation enthalpy 滞止焓
stagnation flow 滞止流动
stagnation line 滞止线
stagnation point flow 滞止点流动
stagnation point heat transfer 驻点传热
stagnation point 驻点，临界点，滞止点
stagnation pressure 滞止压力，全压，驻点压力，临界压力
stagnation properties 滞止参数
stagnation state 滞止状态，停滞状态
stagnation streamline 滞止流线
stagnation surface 驻面
stagnation temperature 滞止温度，驻点温度，临界温度
stagnation value 滞止值
stagnation zone 滞止区
stagnation 停滞，萧条，闷晒，滞止，滞流，层次
stained glass 彩饰玻璃，彩色玻璃，有色玻璃
staining 污染，着色，生锈，锈蚀
stainless cladding steel 不锈钢复合板
stainless clad steel 不锈钢复合钢板
stainless steel cladding aluminum sheet 不锈钢覆铝板
stainless steel cladding guardrail 不锈钢复合护栏
stainless steel cladding nuclear fuel 不锈钢包壳核燃料
stainless steel cladding tube 不锈钢包壳管
stainless steel cladding 不锈钢堆焊层，不锈钢包层
stainless steel electrode 不锈钢焊条
stainless steel foil insulation 不锈钢箔保温（层）
stainless steel foil 不锈钢箔
stainless steel pipe 不锈钢管
stainless steel plate 不锈钢板
stainless steel pump 不锈钢泵
stainless steel sheet 不锈钢板
stainless steel wire 不锈钢丝
stainless steel wool 不锈钢绒
stainless steel 不锈钢
stainless 不锈的
stain 沾污，污斑，斑点，污点，生锈，锈蚀，染（着）色
staircase generator 台阶形波发生器，梯形波发生器
staircase of bifurcated type 合上双分楼梯，双分叉式楼梯
staircase of helical type 螺旋式楼梯
staircase of open-well type 带梯井式楼梯，明梯井式楼梯
staircase tower 楼梯间上的塔楼
staircase with several flights 多梯段楼梯
staircase 楼梯间，楼梯，楼梯井
stair exit 楼梯安全出口
stair flyer 楼梯梯级
stair hall 楼梯休息平台
stair headroom 楼梯净空
stair landing 楼梯平台
stair nosing 梯级突边
stair rail 楼梯扶手
stair riser 楼梯踢板，梯段高度
stairs carriage 楼梯纵梁
stairs landing 楼梯平台，楼梯休息平台
stairs platform 楼梯平台
stairs post 楼梯柱
stairs riser 楼梯的竖板，楼梯级高，楼梯踏步高度
stairs stringer 楼梯梁
stair step 楼梯踏板，楼梯踏步
stairs tread 楼梯踏步，楼梯踏步平板，楼梯踏级板
stairs trimmer 楼梯段端部托梁
stair stringer 楼梯梁
stairs 楼梯
stair tower 楼梯间
stair tread 楼梯踏板
stairway enclosure 有围护的楼梯
stairway 楼梯间，楼梯，楼梯井，阶梯
stake dam 排桩滞流透水坝
stakeholder 股东，利益相关者，利益攸关方，干系人
stake line 标桩线
stake-man 标桩工
stake-resistance 桩极电阻
stake 标桩，树桩，柱杆，桩，竖管
staking out 打桩放样
staking 标桩定线
stalactite 钟乳石
stalagmite 石笋
stale B/L 陈旧提单
stale wastewater 腐臭污水
stalked bacteria 杆状菌
stall angle of attack 失速攻角
stall angle 失速角
stall condition 零速工况【转速比为零时的工况】，失速状态
stall control 失速控制
stalled area 失速区
stalled blade 失速叶片
stalled condition 失速条件
stalled flow 失速流，失速气流，分裂流
stalled flutter 失速颤振
stalled rotor 失速转子
stalled speed 失速速度，失速转速
stalled time 失速时间，停滞时间，停转时间
stalled torque motor 在逆转状态中的电动机
stalled torque 失速转矩，制动转矩，逆转转矩
stall flutter 失速颤振
stall hub 失速轮毂
stall hysteresis excitation 失速迟滞激励

stalling action 失速行为
stalling angle of attack 失速迎角，临界迎角
stalling characteristic 失速特性，分离特性
stalling current 停转电流，制动电流
stalling incidence 失速冲角，失速迎角
stalling limit 失速界限
stalling load 停转负载
stalling moment 失速力矩
stalling relay 失速继电器
stalling speed 失速速度，失速转速
stalling test 失速试验
stalling torque 停转转矩，失速转矩
stall limit 失速极限
stall line 失速线
stall margin 喘振裕度，失速裕度，喘振边界
stallometer 失速信号器，失速指示器，失速仪
stall out 失速，气流分离
stalloy 硅钢，硅钢片，薄钢片
stall pattern 失速流谱
stall performance 失速性能
stall point 失速点
stall profile blade 失速型叶片
stall proof 防失速
stall propagation 失速传播
stall regulated rotor 失速式限速风轮
stall regulated wind turbine 失速调节风电机组
stall regulated 失速调节
stall regulation of WTG 风电机组失速调节
stall regulation 失速式限速
stall severity 失速的严重程度
stall strip 失速条
stall torque 失速力矩
stall-type shower 分隔式淋浴间
stall-warning 失速告警的
stall zone 失速区
stall 停车，（因速度不足而）停转，失速，喘振，阻止，妨碍，厢房
stalpeth cable 钢铝聚乙烯复合铠装电缆
stal turbine generator 辐流式汽轮发电机
stamp duty 印花税
stamped capacity 标明容量
stamped nail 压制钉
stamped pressure rating 标明的额定压力
stamped pressure 标记压力
stamped relieving capacity 标明的释放容量
stamping parts 冲压机
stamping press 冲压机，压印机
stamping 冲压，冲压件，捣碎
stamp marking 冲标记
stamp press 冲压机，压印机
stamp 冲头，锥子，打孔器，戳记，印章，捣碎机，冲压，捣磨，盖印
stanchion mounted 柱上安装
stanchion 窗间小柱，支柱，柱子，承杆，钢桩
stand-alone capability 独立能力
stand-alone mode 独立方式
stand-alone operating time at rated load 在额定负荷下的独立运行时间
stand-alone operation system 独立操作系统
stand-alone operation 独立操作
stand-alone system 独立系统，单机系统
stand-alone （电脑外围）可独立应用的，自备的，独立的，整装的
stand alone 独立运行
standard address generator 标准地址发生器
standard agreement 标准协定
standard air pressure 标准大气压
standard air 标准状态空气，标准大气，标准空气
standard ambient condition 标准大气状态
standard ambient 标准环境温度
standard ammeter 标准安培计
standard and specification 标准和规范
standard antenna 标准天线
standard arbitration clause 标准仲裁条款
standard atmosphere pressure 标准大气压，标准大气压力
standard atmosphere 标准大气（压）
standard atmospheric condition 标准大气状态，标准大气条件
standard atmospheric pressure 标准大气压
standard atmospheric state 标准大气状态，空气的标准状态
standard bidding document 标准招标文件
standard block 标准试件，标准块
standard bolt 标准螺栓
standard brick 标准砖
standard buret 标准滴定管
standard cable 标准电缆
standard calliper gauge 标准测径规
standard candle 标准烛光
standard capacitance 标准电容
standard capacity 标准容量
standard casement window 标准平开窗
standard cell 标准电池
standard chamber 标准计量室，标准电离室
Standard Chartered Bank 渣打银行
standard chart 标准图【晶粒大小确定】
standard circuit 标准电路
standard clause 标准条款
standard coal consumption rate of heat supply 供热标准煤耗率
standard coal consumption rate 标准煤耗率
standard coal equivalent 标准煤当量【单位为MJ/g】，当量煤吨
standard coal 标准煤
standard coil 标准线圈
standard compaction test 标准击实试验
standard component 标准部件
standard condenser 标准电容器
standard condition 标准工况【温度和压力】，标准条件，标准状态
standard conductivity 标准导电率
standard containment vessel 标准安全壳
standard containment 标准安全壳
standard contract form 标准合同格式
standard contract provision 标准合同条款
standard contract 标准合同
standard control module 标准控制模件
standard coordinate system 标准坐标系
standard coordinates 标准坐标
standard copper wire 标准铜线

standard cost 标准成本
standard counter 标准计数器
standard counting error 标准计数偏差
standard coupler 标准车钩
standard cross-section 标准横断面
standard cruciform guide column 标准十字导向柱
standard current meter 标准流速仪
standard current probe 标准电流互感器
standard current transformer 标准电流互感器，故障变流器
standard curve 标准曲线
standard data set 标准数据集
standard density altitude 标准密度高度
standard depth of penetration 标准穿透深度
standard depth 标准水深
standard design 标准设计
standard detail 标准细部图，通用详图，标准详图，标准大样
standard deviation 标准偏差，典型偏差，标准差，均方误差
standard deviation 均方差
standard dial 标准标度盘
standard-dimensioned motor 标准规格电动机
standard dimension 标准尺寸
standard distribution format 标准分配格式
standard draft 标准草案
standard drawing package 标准图包
standard drawing 通用图，标准图
standard drum 标准桶
standard duty 标准负荷，标准功率
standard electrode potential 标准电极电位
standard electrode 标准电极
standard element 标准构件
standard elevation 标准高程
standard engineering practice 标准工程惯例
standard enrichment specimen 浓缩标样，标准浓缩样品
standard equation 标准方程
standard equipment 标准设备
standard erection procedure 标准安装程序
standard error 标准误差
standard explicit method 标准显示法
standard fall velocity 标准降落速度
standard flux assembly 标准通量装置
standard form 标准格式，标准形式
standard free energy 标准自由能
standard frequency signal 标准频率信号
standard frequency 标准频率
standard frost penetration 标准冻深
standard fuel assembly 标准燃料组件
standard fuel management 标准燃料管理
standard fuel rod 标准燃料棒
standard fuel 标准燃料
standard function 标准函数
standard gas bottle including pressure-reducing valve 带减压阀的标准气瓶
standard gas bottle 标准气瓶
standard gauge railway 标准轨矩铁路
standard gauge 标准量规，标准量具，标准线规，标准轨距，标准水尺
standard gravitational acceleration 标准重力加速度
standard gravity 标准重力
standard guideline 标准指南，标准指导路线
standard hardware program 标准硬件程序
standard hook 标准弯钩
standard hydrogen electrode 标准氢电极
standard implicit method 标准隐式法
standard inductance 标准电感
standard installation 标准装配
standard instrumentation 标准仪表装置
standard instrument 标准仪表
standard insulation class 标准绝缘等级
standard international unit 国际标准单位制
standard internet web browser 标准互联网浏览器
standard ionization chamber 标准计量室，标准电离室
standardization of solar radiometer 太阳辐射计的标准化
standardization 标准化【对已知标准的校准】，校正，规格化，标定
standardized building 标准化建筑
standardized design 定型设计
standardized graphics package 标准化图形软件包
standardized house 标准住宅
standardized installation categorization 标准化设施分类法
standardized instruction 标准化指令
standardized normal distribution 标准化正态分布
standardized plant series 标准化电站系列
standardized plant subseries 标准设备的次分类
standardized plant 标准化电站
standardized processing condition 标准化处理条件
standardized product 标准产品，定型产品
standardized program 标准化程序
standardized random variable 标准化随机变量
standardized-shape member 标准构件，标准截面构件
standardized wind speed 标准风速
standardized 标准化的
standardize 标准化
standardizing rheostat 标准化变阻器
standard laboratory atmosphere 标准实验室（大）气压
standard laboratory temperature 标准实验室温度
standard lamp 标准灯
standard layout 标准配置
standard length rail 标准长度钢轨
standard lifting equipment 常用起重设备
standard load test 标准荷载试验
standard load 标准负荷
standard low-frequency range approach 标准低频段法
standard man 标准人【辐射防护】
standard map 标准地图
standard mass 标准质量
standard mean chord 标准平均弦【等于机翼总面积除以全展长】，平均几何弦
standard meter 标准仪表

standard mix (混凝土）标准配合比
standard model 标准样品，样机，标准形式
standard motor 标准电动机
standard neutron source 标准中子源
standard noise generator 标准噪声发生器
standard notation 标准计数法，标准符号
standard nuclear power block 标准核电机组
standard nut 标准螺母
standard observation 标准观测
standard of design flood 设计供水标准
standard of ethics 道德准则，伦理标准
standard of feasibility 可行性标准
standard of fisheries water 水产用水标准
standard of living 生活水平
standard of material 材料标准
standard of performance 实施标准
standard of project appraisal 项目标准评估
standard of reasonableness 合理性标准
standard of USA 美国标准
standard ohm 标准欧姆
standard on-line module 标准联机模件
standard operating conditions 标准工作条件
standard operation procedures 标准操作程序，标准操作过程
standard operator 标准算符，标准操作符
standard orifice 标准孔口
standard oscillator 标准振荡器
standard oxidation potential 标准氧化电位
standard package 标准组合件
standard part 标准（零）件
standard penetration resistance 标准贯入阻力
standard penetration test 标准贯入试验，标贯试验
standard penetration 标准贯入度
standard penetrometer 标准贯入仪
standard periodical inspection 标准定期大修，大修
standard pile 标准反应堆，标准桩
standard pipe support 标准管架
standard pipe 标准管
standard platinum resistance thermometer 标准铂电阻温度计
standard plug-in unit 标准插换部件
standard pole 标准电杆
standard port 标准港，参考港
standard potential transformer 标准电压互感器
standard power curve 标准功率曲线
standard power 标准功率
standard pressure 标准压力
standard price 标准价格，基准价，中间价【银行】
standard procedure instruction 标准过程指令
standard procedures manager 标准与程序经理
standard procedure 标准程序
standard Proctor test 标准普氏压实试验
standard program 标准程序
standard projection 标准投影
standard provision 标准条款
standard pulse 标准脉冲
standard radiation source 标准放射源
standard railway gauge 标准轨距
standard rated output 标准额定功率，标准额定出力
standard rated power 标准额定功率
standard rating cycle 标准循环
standard rating 标准额定出力，标准额定值，标准定额，标准制冷量
standard recommended practice 推荐使用标准
standard reference condition 标准基准条件
standard reference material 标样，标准参考物质
standard resistor 标准电阻，标准电阻器
standard response spectrum 标准响应谱
standard review plan 标准审批计划，标准审批大纲
standard ruler 标准尺
standard safety analysis report 标准安全分析报告
standard sample 标准样品，标准试样
standard sand 标准砂
standard screen 标准筛
standard screw 标准螺钉
standard sea level 标准海平面
standard section 标准断面，标准截面，标准型材，标准剖面
standard security 一般安全性
standard ship size 标准船型尺度
standard ship 标准船
standard short tube 标准短管
standard shunt 标准分流器
standard sieve 标准筛
standard signal generator 标准信号发生器
standard signal 标准信号
standard size 标准尺寸
standard solar cell 标准太阳电池
standard solenoid 标准螺线管
standard solution 标准液，标准溶液
standard source 标准源
standard span 标准跨度，标准挡距
standard specification 标准技术规范，标准规范，标准规格
standard specimen 标准试样
standard state 标准状态
standard steel section 标准型钢
standard submerged orifice 标准淹没孔口
standard subroutine 标准子程序
standard suit 标准工作服
standard symbol 标准符号
standard system of level 标准水准系
standard tails assay 尾料标准浓度【浓缩】，标准尾料丰度测定
standard tape executive program 标准带执行程序
standard taper plug 标准圆锥塞
standard temperature and pressure 标准温度和压力
standard temperature 标准温度
standard tender document 标准招标文件
standard tension 标准拉力
standard terminal program 标准终端程序
standard test block 标准试块，校正试块
standard test condition 标准测试条件
standard test mode 标准测试模式

standard test specimen 标准试件
standard test 标准试验
standard thermocouple 标准热电偶
standard thermometer 标准温度计
standard thyristor 标准晶闸管
standard time clock 标准时钟
standard time zone 标准时区
standard time 标准时间
standard tip screwdriver 标准头螺丝起子
standard tower height 标准塔高，呼称高【线路工程】
standard transformer 标准变压器
standard trickling filter 标准滴滤池
standard tube 标准管
standard turbidity solution 浊度标准液
standard-type insulator 普通型绝缘子
standard-type test 标准型式试验
standard uncertainty 标准不确定度
standard unit 标准单位
standard value 标准值
standard velocity 标准速度
standard visual diffuse density 标准的可见扩散浓度
standard visual range 标准视程
standard voltage gauge 标准电压表
standard voltage oscillator 标准电压振荡器
standard voltage 标准电压
standard water meter 标准量水计
standard wave form 标准波形
standard wave-meter 标准波长计，标准波频计
standard wave 标准波
standard weather bureau pan 气象局标准盘式蒸发器
standard weight 标准重量，标准砝码，称重标准块
standard wire gauge 标准线规
standard 标准，规格，规范，准则，基准，标准的
standby agreement 备用协定
standby analogue controller 备用模拟控制器
standby application 备用，待用
standby arrangement 备用安排
standby auxiliaries 备用辅助系统
standby auxiliary plant 备用辅助发电设备
standby battery 备用电池
standby heat source 备用热源
standby boiler 备用锅炉，备用炉
standby capacity 备用容量
standby circuit 备用电路，附加电路
standby circulating water pump system 循环水补充水泵系统
standby computer 备用计算机
standby condition 备用状态
standby cooling 备用冷却
standby credit 备用贷款安排，备用信用证，保证信用证，备用信贷
standby current 维持电流
standby diesel generator set 备用柴油发电机组
standby electric supply for production 生产备用电源
standby equipment 备用设备
standby exciter 备用励磁机
standby facilities 备用设施
standby gas treatment system 备用气体处理系统
standby generating plant 备用发电厂
standby generator set 备用发电机组
standby generator 备用发电机
standby heating 值班采暖，值班供暖
standby heat source 备用热源
standby heat 储备热，热备用
standby hydro unit 备用水力机组
standby L/C 备用信用证
standby letdown valve 备用下泄阀
standby letter of credit 备用信用证
standby liquid control system 备用液体控制系统
standby load 备用荷载
standby loss 备用容量损耗，备用损失，备用容量损失
standby oil pump 备用油泵
standby period 备用期
standby plant 备用电厂，备用装置
standby power energy 备用能源
standby power plant 备用电厂，备用发电厂
standby power source automatic put-in device 备用电源自投装置
standby power source 事故电源
standby power station 备用电站
standby power supply 备用供电，备用电源
standby power system 辅助供电系统
standby power 备用电源，备用功率，备用电力，备用动力
standby pump 备用泵，备用水泵
standby redundancy 备用冗余（量），备用信息，备用位，备用码，备品备件
standby register 备用寄存器
standby service water 备用厂用水
standby service 备用
standby set 备用机组
standby source of power 备用动力源
standby source 备用电源
standby state 备用状态
standby station 备用电站，尖峰负荷电站
standby supply 备用电源
standby system 辅助系统
standby tank 备用罐
standby time 备用时间，待命时间，待机时间，闲置时间
standby transformer 备用变压器
standby turning gear 后备盘车装置，事故用盘车装置
standby unit 备用机组
standby 备用的，备用，备用设备，后备的，备品，等待，备用品，待机
stand by 处于待命状态，遵守（诺言等），坚持（决议等），做好准备，准备行动
stand for 支持【用于否定句、疑问句】，表示，代替，意味着，主张，容忍，接受
standing balance 静平衡
standing body 常设机构
standing bubble 驻泡
standing cloud 驻云
standing cost 固定成本

standing current 稳定电流，驻流
standing derrick 扒杆
standing eddy 立轴旋涡
standing gravity wave 重力驻波
standing harbour 临时港口
standing internal wave 内立波，内驻波
standing oscillation 驻波振荡
standing pier 独立桥墩，直立桥墩
standing pipe 直立管
standing seam 直立缝
standing surface wave 表面驻波
standing swell 立涛，直立涌浪
standing vibration 稳定振动
standing vortex 驻涡
standing water level 常驻水位
standing water 停滞水
standing wave detector 驻波检波器
standing wave flume 驻波槽
standing wave indicator 驻波检测器
standing wave line 驻波线
standing wave meter 驻波测定器，驻波测量计，驻波指示器
standing wave pressure 驻波波压
standing wave ratio 驻波系数，驻波比
standing wave tide 驻波潮
standing wave 驻波，立波
standing 静止的，固定的，稳定的，长期的，直立的
standoff insulator 支柱绝缘子
standoff terminal 支座端钮
standoff tray 有支座的托架
standoff 定距套筒
stand of tide 停潮，停止潮
standpipe and hose system 立管水龙带消防系统
standpipe closure 竖管封盖
standpipe connection 竖管连接
standpipe cooling system 竖管冷却系统
standpipe head 竖管头部
standpipe piezometer 竖管式测压计
standpipe plug grab 竖管塞抓取装置
standpipe plug 竖管塞
standpipe storage rack 竖管贮存架
standpipe 立管，竖管【主泵密封】，直立圆筒，落煤井，平衡管，圆筒形水塔，立式送风管【沸腾炉布风板上】
stand post hydrant 地上消火栓
stand still corrosion 锅炉停炉腐蚀
standstill current 止转电流
standstill locking （感应电动机）止转
standstill reactance 止转电抗
standstill seal 静止密封
standstill 静止，停止，停滞，停机，停车状态
stand 立场，支架，架，台，台架，框架，机座，停滞，竖立
stank 水沟
stannous chloride 氯化亚锡，二氯化锡
Stanton number 斯坦顿数
Stanton tube 斯坦顿管
staple air sampler 纤维空气取样器
staple joint 夹持接头
staple 钉书钉，环钩，骑马钉，（用书钉）装订
stapling machine 压钉机
star aerial 星形天线
star bit 梅花钻头，十字钻头，星形钻头
starboard boiler 右侧锅炉
starboard （船）右舷，右侧
star box 星形连接电阻箱
star configuration 星形配置
star connected system 星形连接制，Y连接
star connected 星形连接的，星形结线的
star connection 星形连接，Y接法
star coupling 万向联轴节
star crack 星形裂纹
star current 星形电流，Y电流
star-delta changing device 星形三角形切换装置
star-delta connection 星形三角形结线，星形三角形连接
star-delta starter 星形三角形启动器
star-delta starting 星形三角形启动
star-delta switching starter 星形三角形切换启动器
star-delta switching 星形三角形切换，星形三角形转接
star-delta switch 星形三角形转换开关
star drill 梅花钻，星形钻
star electromotive force 星形电动势
star enterprise 明星企业
star feeder 鼓式给煤机
star-grounded 星形中性点接地的
star knob 星形操作柄
starling 桥墩尖端
star network 星形网络，Y形网络
star point 中性点【星形接法】，星点
star-polygon conversion 星形多角形变换
star program 无错程序
star-quad cable 星铰四线电缆
starred nonterminals 加星非终结符
star-shaped slug 星形块（铀）
star-star connection 星形-星形结线，Y-Y接线，星-星形接法
star-star-delta connection 星形星形三角形连接
star system 星形制，Y制
startability 启动性能
start address 起始地址
start and stop control of HVDC power transmission system 高压直流输电启停控制
start and stop of HVDC power transmission system 高压直流输电系统启停
start and stop push button 启停按钮
start a new brand 创牌子
start BFBP 启动锅炉给水前置泵
start bit 起始位【计算机】
start button 启动按钮
start capacitor 启动电容器
start characteristic curve 启动特性曲线
start characteristic test 启动试验
start command 启动命令
started time 开始时间
starter bar 连接钢筋【混凝土】，预留搭接钢筋，启动杆，起始筋，露头钢筋，伸出钢筋
starter gap 起导间隙
starter generator 启动器用发电机

starter motor 启动电动机，起动电动机
starter relay 启动继电器
starter rheostat 启动变阻器
starter switch 启动开关
starter 启动器，启动装置，点火极，点火装置
starting algorithm 起步算法
starting amortisseur （同步电机的）启动阻尼绕组
starting and ending time 起止时间
starting and load-limiting device 启动和负荷限制装置
starting anode （整流器的）启动阳极
starting at sliding parameters 滑参数启动
starting autotransformer 启动自耦变压器
starting box 启动箱，电阻箱
starting burner 点火燃烧器，启动燃烧器
starting button 启动按钮
starting by frequency-sensitive rheostat 频敏变阻器启动
starting capacitance 启动电容
starting capacitor 启动电容器
starting characteristic curve （汽轮机）启动特性曲线
starting characteristics 启动特性
starting characteristic test 启动特性试验
starting circuit 启动电路，触发电路
starting coil 启动线圈
starting command 启动命令
starting compensator 启动补偿器，启动自耦变压器
starting condition 启动条件，启动工况，启动状况
starting contactor 启动接触器
starting current at rated voltage 在额定电压下的启动电流
starting current 启动电流，起动电流，启始电流
starting curve at sliding parameters 滑参数起动曲线，滑参数启动曲线
starting curve 启动曲线
starting damper 启动风阀
starting device 启动装置
starting duty 启动功率，启动负载
starting ejector 启动抽气器
starting electrode 启动电极
starting element 启动元件
starting equipment 启动设备，启动装置
starting failure 启动失败
starting force 启动力
starting forge 始锻
starting from cold 冷态启动
starting fuel 升炉燃料
starting hand-wheel 启动手轮
starting impedance relay 启动阻抗继电器
starting impulse 启动脉冲
starting load 启动负载
starting loss 启动损失
starting magneto 启动磁电机，点火磁电机
starting manostat 启动稳压器
starting material 原材料，起始物料，原料
starting method 启动方法
starting mixture 启动混合气
starting moment 启动力矩，启动转矩
starting motor 启动电动机
starting of oscillation 起振
starting oil 启动油
starting performance test 启动性能试验
starting period 启动期，启动周期
starting plume 初始羽流
starting point 起点，启动点，出发点
starting position 启动位置
starting power 启动动力，启动功率
starting procedure 启动程序
starting pulse 启动脉冲
starting reactance 启动电抗
starting reactor 启动电抗器
starting relay 启动继电器
starting reliability 启动可靠度
starting resistance 启动电阻
starting resistor 启动电阻器
starting rheostat 启动变阻器
starting separator 启动分离器
starting sequence 启动顺序，启动程序
starting signal 启动信号
starting source 启动源
starting steam source 启动汽源
starting success 启动成功
starting switch 启动开关
starting system of gas turbine 燃气轮机启动系统
starting system 启动系统
starting test 启动试验
starting time constant 启动时间常数
starting time 启动时间
starting torque coefficient 启动力矩系数
starting torque 启动力矩，启动扭矩，启动转矩
starting transformer 启动变压器
starting under load 带载启动
starting under no-load 空载启动
starting up equipment 起动设备
starting up period 启动周期
starting up 开机，启动，触发，升炉
starting value 初值，启动值
starting valve 启动阀
starting velocity 启动速度
starting vessel 启动分离器【直流锅炉】
starting voltage 启动电压
starting vortex 启动涡
starting weld tab 引弧板
starting winding 启动绕组
starting wind speed 起动风速
starting wind velocity 启动风速
starting with load 带载启动
starting 启动，起动，投运
start key 启动键
start knob 启动旋钮，启动把手
start lead 线圈起始端，线圈内引线
start of operation 开始运行
start of run-up 开始升速
start of run 开始运转
start of text character （计算机）正文起始符
start of text 正文开始，正文起始
start point 起始点
start pulse 触发脉冲，启动脉冲

start pump 启动泵
start relay 启动继电器
start request 启动要求
start rheostat 启动变阻器
start rolling （汽轮机）开始转动
start signal 启动信号
start-stop control circuit 开停车控制电路
start-stop counter 一次计数器
start-stop multivibrator 启停多谐振荡器，单稳多谐振荡器，延迟多谐振荡器
start-stop synchronism 启停同步，起止同步
start-stop telecontrol transmission 启停式远动传输
start-stop transmission 启停传输
start-stop 开停，启闭，断续，起止，启动停止
start-to-discharge pressure （安全阀的）前泄压力
start-to-discharge pressure 开始排出的压力
start-up accident 启动事故
start-up after a cold shutdown 冷停堆后启动
startup and commissioning of T/G set 汽轮发电机组整套启动试运
start-up BFP 启动锅炉给水泵
start-up boiler house 启动锅炉房
start-up boiler 启动锅炉
start-up capacity 启动能力
start-up circuit 启动回路
start-up condition 启动工况
start-up control valve 启动控制阀
start-up cost 启动费，开办费
start-up date 开工日期
start-up device 启动装置
start-up diagram 启动图
start-up ejector 启动抽气器
start-up experience 启动经验
start-up feed pump 启动给水泵
start-up feed water pump 启动给水泵
start-up filter 启动过滤器
start-up flash tank 启动分离器，启动扩容箱
start-up flow rate 始动流量
start-up flow 启动流量
start-up heater 启动加热器
start-up instrumentation 启动测量仪表
start-up light oil burner 启动用轻油燃烧器
start-up line 启动管路
start-up loss 启动热损失
start-up meeting 开工会议
start-up neutron source 启动中子源，中子启动源
start-up of auxiliary oil pump 辅助油泵启动
start-up of medium-pressure cylinder 中压缸起动
start-up of T/G set 汽轮发电机组启动
start-up operation 试运转启动运行，启动操作
start-up parameters 启动参数
start-up period 启动期，启动阶段，启动周期
start-up phase 起始相
start-up piping 启动管系
start-up power source 启动电源
start-up pressure 启动压力
start-up procedure 启动步骤，启动程序
start-up protection 启动保护
start-up pump 启动泵
start-up ramp 启动匀增率
start-up range 启动区段
start-up rate 启动速率
start-up reducing station 启动降压装置
start-up rod 启动棒
start-up separator 启动分离器
start-up sequence 启动程序
start-up curves 启动曲线
start-up source 启动源
start-up system 启动系统
start-up test engineer 启动试验工程师
start-up test 启动试验
start-up time 启动时间
start-up torque 启动力矩，启动转矩
start-up transformer 启动变压器
start-up valve 启动阀
start-up wind speed 启动风速
start-up with variable parameter 滑参数启动
start-up with variable pressure 滑压启动
start-up zero power test 启动零功率试验
start-up zone 点火启动区
start-up 启动，开动，起动，投运，再启动
start with bypass system 利用旁路系统启动
start working 动工
star type cable 星型电缆
star valve （灰斗用）旋转阀
starved 缺乏的
star voltage 相电压【星形接法】，相线与中性点间电压
star wheel dosing pump 星形计量泵
star wheel 棘轮，星形轮
star-zigzag connection 星形曲折接法，Y-Z接线
star 星形，星形连接
statampere 静电安培
STATCOM(static synchronous compensator) 静止同步补偿器
statcoulomb 静电库仑
state assignment 状态分配
state balance 状态平衡
state bank 国家银行
state border 国界
state capital 国家资本
state coffer 国库
state constraint 状态约束
state control register 状态控制寄存器
state coordinate system 国家坐标系统
state council 国务院
state derivative 状态导数
state development bank 国家开发银行
state diagram 状态图
stated meeting 例会，例行会议
state economic and trade commission 国家经贸委
state enterprise 国营企业，政府企业，国企
state equation model 状态方程模型
state equation 状态方程
state estimation 状态估计
state estimator 状态估计器
state event 事件状态【仿真模拟】
state feedback 状态反馈
state first-level enterprises 国家一级企业
state function 状态函数

state grade 国家级
State Grid Corporation of China 国家电网公司
state highway 国道
state information 状态信息
state in service 运行状态
state insurance 国家保险
state intervention 国家干预
state investment 国家投资
state land administration 国家土地管理局
state loan 国家贷款
statement analysis 财务报表分析，决算表分析
statement at completion 竣工报表
statement balance 账单余额
statement editor 语句编辑器
statement form 报表格式
statement of account 会计报表，对账单
statement of assets and liabilities 资产负债表
statement of cash flow 现金流量表
statement of information 情况说明
statement of loss and gain 损益表
statement of loss and profit 损益表
statement of net guarantee annual energy figures 关于保证的年度净发电量的声明［说明］
statement of opinion 意见陈述
statement of problem 问题的提法
statement of profit and loss 盈亏报表
statement of profit distribution 利润分配表
statement of the defendant 被告人辩解
statement 报表，陈述，声明，报告书
State Nuclear Electric Power Planning Design and Research Institute 国核电力规划设计研究院
State Nuclear Power Technology Corporation 国家核电技术公司
state observer 状态观测器
state of alarm 报警状态
state of capillary equilibrium 毛细管平衡状态
state of control 控制状态
state of disturbance 扰动状态
state of elastic equilibrium 弹性平衡状态
state of flow factor 流动状态因子
state of flow 流态
state of insulation 绝缘状态
state of limit equilibrium 极限平衡状态
state of plastic-elastic equilibrium 塑弹性平衡状态
state of plastic equilibrium 塑性平衡状态
state of rest 静态
state of runtime machine 机器运行状态
state of starting operation 起动工况
state of strain 应变状态，应力状态
state-of-the-art equipment 先进设备
state-of-the-art facility 目前最新技术，目前最现代化设备
state-of-the-art record 最新记录，目前工艺水平记录
state-of-the-art technology 先进技术，先进工艺水平，目前最现代化的技术
state of the art 技术发展最新水平，当前发展状况，目前工艺发展水平，目前技术发展水平
state-of-the-art 使用最先进技术的，体现最高水平的，最新式的，最先进的，最现代化的
state-of-the-theoretical-art 理论发展水平
state organ 国家机关
state-oriented approach 与状态有关的探讨
State-owned Assets Supervision and Administration Commission （国务院）国有资产管理委员会，国资委
state-owned backbone enterprise 国有骨干企业
state-owned corporation 国营公司
state-owned economy 国有经济
state-owned enterprise income 国有企业收入
state-owned enterprise 国营企业
state parameter 状态参数
state planning commission 国家计委
state planning 国家计划
State Power Investment Corporation 国家电力投资集团公司
state register 状态寄存器
state-run economy 国营经济
state-run enterprise 国营企业
state-run large and medium sized enterprise 国有大中型企业
state space description 状态空间描述
state space kinetics 状态空间动力学
state space method 状态空间法
state space model 状态空间模型
state space 状态空间
state-specified standard 国家规定标准
states superior limit 上限
state subsidy 国家补贴
state trajectory 状态轨迹
state-transition diagram 状态转移图
state transition equation 状态转移方程
state-transition matrix 状态传递矩阵，状态转移矩阵
state transition model 状态转移模型
state treasury 国库
state-variable equation 状态变量方程式
state variable method 状态变量法
state variable technique 状态变量法，状态变量技术
state variable 状态变量，状态参数
state-vector equation 状态向量方程
state vector 态矢量，状态矢量，状态向量
state with an adequate legal system 法制国家
state 国家，陈述，申述，态，状况，状态，情况，条件
statfarad 静（电）法拉
stathenry 静电亨利
static accuracy 静态偏差，静态精（确）度
static action 静态作用
static admittance 静态导纳
static aerodynamics 定常流空气动力学
static aeroelasticity 静气动弹性力学
static aerothermoelastic behaviour 空气热弹性静力特性
static air pressure 空气静压
statical head 静压头，静压传感器
statical load 静载
statically balanced 静平衡的
statically broken 静荷载破坏
statically determinate beam 静定梁

statically determinate frame 静定构架
statically determinate structure 静定结构
statically determinate system 静定系统
statically determinate truss 静定桁架
statically indeterminate structure 超静定结构
statically indeterminate system 超静定体系统
statically indeterminate 超静定（的）
statically stable 静态稳定的
statically switching 静态开关的，无触点开关的
statically 静态地
statical model 静态模型
static analog device 静态模拟装置
static analysis scheme of building 房屋静力计算方案
static analysis 静态分析
static arc 静态电弧
static autoclave test 高压釜静态试验
static balancer 静平衡器，静电平衡器
static balance 静平衡，静力天平
static barrier 静止隔板
static batch weigher 静态批量计量装置
static batch weighing device 静态批量计量装置
static bed 静止床【沸腾炉】
static bidding 静态投标
static calculation 静力计算
static calibration 静力校准，静态校准
static capacitance 静态电容
static cascade 静叶
static characteristic curve 静态特性曲线
static characteristic test 静态特性试验
static characteristic 静态特性
static charge 静电荷
static check 静态校验
static condenser 固定电容器
static condition 静态，静态条件
static conducting 导电【皮带面层】
static cone penetration test 静力触探试验
static constant 静态常数
static contact 静触点，静触头
static control 静态控制
static conversion 静态变换
static converter 静止换流器，静止变流器
static corrosion test 静态腐蚀试验
static coupling 静电耦合
static decoupling 静态解耦
static deflection 静挠度
static derivative 静导数
static deviation 静态偏差
static direct reactance 静态纵轴电抗
static discharge 静电放电
static distance relay 静态距离继电器
static draft 静压头
static dump 静态打印，静态转储
static dynamic pressure technique 静动压技术
static efficiency 静态效率，静效率
static electricity discharge 静电放电
static electricity 静电，静电学
static elevation difference 静压头，（水的）提升高度
static elevation 位差
static enthalpy 静焓
static equilibrium 静平衡，静态（力）平衡
static error coefficient 静态误差系数
static error 静态误差
static excitation system 静态励磁系统，静止励磁系统
static exciter room 静态励磁设备间
static exciting equipment 静止励磁装置
static fatigue 静力疲劳
static field 静止场
static flip-flop register 静态触发寄存器
static force divergent 静力发散
static frequency changer 静止变频器，静态变频器
static frequency 静频率
static friction factor 静摩擦因数
static friction 静摩擦
static gain 静态增益
static gauge pressure 静表压
static generator 静电起电机
static hardness test 布氏硬度试验
static harmonic absorber 天电谐波滤除器，天电谐波吸收器
static head 静压头，静压传感器
static hole 静压测量孔
static impedance 静止阻抗，静态阻抗
static indentation test 球硬度试验静凹痕试验
static induction thyristor 静电感应晶闸管
static induction transistor 静电感应三极管，静电感应晶体管
static induction 静电感应
static input-output model 静态投入产出模型
static instability 静态不稳定性，静态不稳定，静力学不稳定性
static investment cost of power generation project 发电工程静态投资成本
static investment 静态投资
staticiser 静态化装置
static level meter 静态水位指示器
static life 静态寿命
static lift 静升力
static load characteristic 负荷静态特性
static load deflection 静荷载挠度
static load error 静态负载误差
static load tester 静载荷试验仪
static load 静态载荷，静负载，静荷重，静力荷载，静载荷，恒载，静载
static machine 静电起电机
static mark 静电感光痕迹
static measurement 静态测量
static medium 静态介质
static memory interface 静态存储器接口
static memory 静态存储器
static micro relay 静电型微继电器
static model 静态模型
static modulus of elasticity 静弹性系数
static modulus of soil 土壤静态模量
static MOS inverter 静态金属氧化物半导体反相器
static multiplication 静态放大
static noise 静电噪声，天电噪声
static nozzle tester 静态喷嘴试验台

static of fluid 流体静力学
static operation condition 静态工况
static oscillograph 静电示波器
static output 静态输出
static overvoltage 静电过电压
static parameter 静态参数
static penetrometer 静力触探仪，静探仪
static performance 静态特性
static phase comparator 静电相位比较器
static phase-shift circuit 静态移相电路
static phase shifter 静止移相器
static pile press extractor 静力压拔桩机
static pile press 静力压桩机
static pipe pile 静压管桩
static point resistance 静力触探探头阻力
static power coefficient 静态功率系数
static precipitator 静电除尘器，电气除尘器
static pressure coefficient 静压系数
static pressure compensation joint 静态压力补偿接头
static pressure compensation 静压力补偿
static pressure correction vent 静态压力校正孔
static pressure difference 静压差
static pressure distribution 静压分布
static pressure drop 静压差
static pressure error 静压误差
static pressure gradient 静压力梯度
static pressure hole 静压孔
static pressure level sensor 静压水位传感器
static pressure line 静水压线
static pressure loss 静压损失
static pressure method 静压法【施工】
static pressure orifice 静压测量孔
static pressure pitot tube 静压皮托管
static pressure probe 静压测针
static pressure ratio 静压比
static pressure tap 静压孔
static pressure vent hole 静压力测定孔
static pressure 静压，静压力
static probe 静力传感器
static random access memory 静态随机存取存储器，静态随机存储器
static reactive compensation 静止无功补偿
static reactivity 静态反应性
static rectifier 静止整流器
static reflector 固定反射层
static register 静态寄存器
static regulation 静调节
static relay 固体组件断电器，静态继电器，无触点继电器
static reserve 静止备用
static ring 静环
static self-excitation method 静止自励方式
static self-inductance 静态自感系数
static sensitivity 静态灵敏度
static series compensators 静态串联补偿器
static settling tank 自然沉降器
static settling 静止沉降
static shielding 静电屏蔽
static shift register 静态移位寄存器
static shunt capacitor 静电分路电容器
static shunt reactor 静电分路电抗器
static slot （风洞）静压补偿缝，静压缝口
statics of fluid 流体静力学
static solution 静态溶液
static stability limit 静态稳定度极限
static stability 静态稳定，静态稳定度，静态稳定性
static status 静态
static stiffness 静态刚度
static storage volume of reservoir 静库容
static storage 静态存储器
static strain 静应变
static subroutine 静态子程序
static suction bead 静吸入水头
static suction of water turbine 水轮机吸出高度
static switching 静态开断，静态转换
static switch 静态开关
static synchronous compensator 静止同步补偿装置
static synchronous condenser 静止同步调相机
static synchronous series compensator 静止同步串联补偿装置
statics 静力学，静态
static temperature 静温
static test stand 静态测试台
static test 静态试验
static thrust 静推力
static time-current relay 静态时间电流继电器，无触点时间电流继电器
static time delay 静时延，静态时间继电器
static touch-probe test 静力触探试验
static track scale 静态轨道衡
static transconductance 静跨导
static transfer switch 静态开关，静态转换开关
static transfer system 静态转移系统
static transformer 静电变压器，静止感应器，感应电压调整器
static trial 静态试验，静压管
static unbalance 静不平衡
static var compensator test 静止无功补偿装置试验
static var compensator 静止无功补偿器，静止无功补偿装置
static variable 静态变量
static vent 静压孔
static vibration 静态振动
static voltage stability margin 静态电压稳定裕度
static voltage stabilizer 静止稳压器【用于远距离交流输电】
static voltage 静态电压
static voltmeter 静电伏特计
static water head 静水头
static water level 静水位
static wind load 静态风载
static 静的，静止的，静态的，静力的，静电的，静电
statimho 静电姆欧
station accuracy 定点精（确）度
station adjustment 测站平差
stationary air compressor 固定式压缩空气机
stationary air pollution source 固定性空气污染源

stationary anode 固定阳极
stationary appliance 固定式仪表，固定式器具
stationary barrier 固定隔板
stationary bar type of screen 固定栅条式栅网
stationary belt conveyor 固定式皮带运输机
stationary blade carrier 静叶持环
stationary blade groove 静叶叶根槽
stationary blade loss 静叶损失
stationary blade row 静叶列
stationary blade 固定叶片，定叶片，静叶，静叶片【汽轮机】
stationary boiler 固定式锅炉，陆用锅炉
stationary breaker contact 断路器静触头
stationary breaker 固定式断路器
stationary bubble 不动气泡
stationary cascade 静叶
stationary cavity 常驻空穴，固定空穴
stationary classifier 稳态分离器
stationary coal unloader 固定式卸煤机
stationary coil 固定线圈【控制棒驱动机构】
stationary combustion source 固定燃烧源
stationary component 静止分量，恒定分量
stationary concrete pump 固定式混凝土泵，混凝土固定泵
stationary condition 固定条件，恒定条件
stationary contact member 静触头零件，静触头
stationary contact spring 静触点弹簧
stationary contact 静触点，固定触点
stationary conveyor 固定皮带机
stationary core components 固定的堆芯部件
stationary crane 固定式起重机
stationary crusher 固定式碎石机
stationary current 稳定流
stationary cyclone 滞留气旋
stationary decanting unit 固定式倾析装置
stationary derrick 固定式起重机
stationary echo 固定回波
stationary electric field 驻立电场，固定电场
stationary equilibrium 稳态均衡
stationary field 驻波场，稳定场，恒定场
stationary film 稳定膜
stationary flow field 平稳流场
stationary flow 稳定流，稳流
stationary front 滞留锋
stationary fuel cluster 固定核燃料棒束
stationary gas turbine 固定式燃气轮机
stationary grate bar 固定炉条
stationary grate 固定炉排，静止炉排
stationary gripper coil 固定棘爪线圈【控制棒驱动机构】，固定夹持线圈
stationary gripper latch arm 固定棘爪闩【控制棒驱动机构】
stationary gripper latch 固定夹持销爪
stationary gripper pole 固定棘爪杆【控制棒驱动机构】
stationary gripper 固定棘爪【控制棒驱动机构】
stationary guide blade 静叶，导叶
stationary guide vane 固定导叶
stationary hopper 固定式漏斗
stationary hysteresis 稳态滞后现象
stationary installation 固定装置
stationary instrument 固定仪表
stationary jump 停滞水跃，稳态水跃
stationary layer 静止层，平稳层
stationary load 固定负荷
stationary magnetic field 恒定磁场
stationary mixer 固定式拌和机
stationary motion 常定运动，稳定运动
stationary neutron detector 固定中子探测器
stationary optical fibre controller 静态光导纤维控制器
stationary part 静止部件
stationary piston sampler 固定活塞式取样器
stationary plane 固定面
stationary plant 陆用装置
stationary plate type regenerative air heater 风罩转动型再生式空气预热器
stationary poisoning 稳定中毒
stationary ported cylinder 直立的带落煤孔的圆筒【即落煤井】
stationary power reactor 固定式动力反应堆
stationary process 平稳过程
stationary random function 平稳随机函数
stationary random input 平稳随机输入
stationary random load 平稳随机荷载
stationary random process 平稳随机过程
stationary reactor 固定式反应堆
stationary rectifier 静止整流器，固定整流器
stationary rotor 静止转子
stationary seal ring face 固定密封环面
stationary seal ring 固定密封环，静止密封环
stationary shock 驻激波，定激波
stationary source 固定污染源，固定源
stationary state 稳定水位，静止状态，稳态，定态，固定状态，平稳状态，稳定状态
stationary steam turbine 电站汽轮机
stationary stochastic process 平稳随机过程
stationary storage battery 固定式蓄电池
stationary stream 稳流
stationary thrust collar 固定推力环，静止止推环
stationary tidal wave 驻潮波
stationary tire 定位轮毂
stationary tripper 固定式倾料器
stationary tube sheet 固定管板
stationary turbine components 透平静止部件
stationary turbine 固定式透平
stationary type 固定式
stationary value 平稳值
stationary vane 静叶
stationary ventilator 固定式通风器
stationary vibration 固定的振动
stationary vortex 固定旋涡
stationary wave front 驻波阵面
stationary wave ratio 驻波比
stationary wave theory of tide 潮汐立波理论，潮汐驻波理论
stationary wave theory 驻波理论
stationary wave 驻波，定波
stationary weighing scale 静态称
stationary winding 定子绕组，静止线圈
stationary window 固定窗
stationary 不动的，固定的，稳定的，静止的，

恒定的，平稳的
station auxiliary motor　厂用电动机
station auxiliary powe rate　厂用电率
station auxiliary power system　厂用电系统
station auxiliary turbine　厂用辅助透平
station battery　厂用电池
station blackout accident　全厂断电事故
station blackout　全厂断电
station board　电厂操纵台
station buffer unit　站缓冲部件
station capacity　电站功率，电站容量，发电厂容量
station consumption　电站消耗
station control unit　站控制部件
station control　测站控制，电厂控制
station display unit　站显示部件
station distribution board　厂用配电盘，厂用配电屏
station efficiency　电站效率
station equation　测站方程
station error　测站误差
stationery　文具用品
station front　厂前区
station grounding　发电厂接地，变电所接地
stationing　布点，设站
station interrogation command　站查询命令
station keeping　保持驻留的
station level function　站层功能
station load factor　发电厂负载率，发电厂负荷因数
station mark　测站标点
station network density　站网密度
station network　测站网，站网，所用电力网，厂内电力网
station of destination　目的站
station output　电站输出功率
station plant factor　电站设备利用率，发电厂设备利用率
station pointer　三杆分度器，三角分度规
station power consumption rate　厂用电率
station rating curve　测站水位流量关系曲线，测站流率曲线
station refixation　重新埋石
station service air system　厂用空气压缩系统，专用空气压缩系统
station service distribution switch gear　厂用配电设备
station service electrical system　厂用电系统
station service electric system initially energized from grid　厂用电初次受电
station service generator cubicle　厂用发电机开关柜
station service load　厂用电负荷
station service power consumption　厂用电（耗）量
station service power source　站用电源
station service power supply　厂用电电源
station service power　厂用动力，厂用电，站用电
station-service transformer　站用变压器，厂用变压器
station service water tank　杂用水箱
station service　厂用，厂用电，所用电
station signal　测站标志
station siting　站址选择，场址
station square　厂前广场
station system efficiency　厂内系统效率
station track　站线
station transformer　站用变压器，厂用变压器
station type　站用型，厂用型
station underground grounding grid　全厂地下接地网
station-year method　站年法
station　站，所，台，局，发电厂
statisitic absorption coefficient　统计吸收系数
statisitic accuracy　统计精度
statisitic analysis　统计分析
statisitic data　统计数据，统计资料
statisitic distribution　统计分布
statisitic error　统计误差
statisitic figure　统计数字
statisitic inhomogeneous medium　统计不均匀性介质
statisitic interpretation　统计解释
statisitic method　统计方法
statisitic noise　统计噪声
statisitic quality control　统计质量控制
statisitic reception　统计接收
statisitics　统计学
statistical analysis　统计分析
statistical approach　统计方法
statistical calculation　统计计算
statistical communication theory　统计通信理论
statistical computer　统计计算机
statistical control　统计控制
statistical correlation　统计相关
statistical data recorder　统计数据记录器
statistical data　统计数据，统计资料
statistical decision method　统计决定法，统计判定法
statistical dependence　统计相关
statistical description　统计说明
statistical discrete gust method　孤立阵风统计方法
statistical distribution　统计分布
statistical error　统计误差
statistical estimate of error　误差的统计估计
statistical estimation　统计评价法
statistical factor　统计因子
statistical figure　统计数字
statistical fluctuation　统计起伏，统计涨落
statistical forecast　统计预测
statistical inference　统计推断
statistical information　统计资料
statistical interpretation　统计解释
statistical machine　统计机
statistical mechanics　统计力学
statistical method for insulation coordination　绝缘配合统计法
statistical method　统计法，统计方法
statistical model　统计模型
statistical parameter　统计参数
statistical prediction　统计预测

statistical problem 统计问题
statistical processing and analysis 统计处理及分析
statistical processing 统计（数据）处理
statistical property 统计性质，统计特性
statistical quality control 统计质量管理，统计质量控制
statistical quality surveillance 质量统计监督
statistical range 统计范围
statistical regression analysis 统计回归分析
statistical relative frequency 统计相对频率
statistical rule 统计规律，统计相似率
statistical sampled-data control system 统计数据采样控制系统
statistical sampling 统计抽样
statistical series 统计系列
statistical simulation 统计模拟
statistical standard 统计标准
statistical surveillance 统计监督
statistical table 统计表
statistical technique 统计技术
statistical test 统计检验，统计试验
statistical theory 统计理论
statistical thermodynamics 统计热力学
statistical tolerance limit 统计容许极限
statistical treatment 统计处理
statistical turbulence theory 湍流统计理论
statistical uniformity 统计均匀性
statistical weight factor 统计权重因数
statistical weight 统计权重
statistical 统计的，统计学的
statistic analysis 统计分析
statistic data 统计数据，统计资料
statistic decision 统计决策
statistic hydrology 统计水文学
statistician 统计人员
statistic parameter 统计参数
statistic pattern recognition 统计模型识别
statistic prediction 统计预测
statistic survey 统计调查
statistics 统计，统计学，统计法，统计资料
statistic test 统计试验
statistic 统计（上）的，统计学（上）的
statist 统计人员
statohm 静电欧姆
stator ampere-turn 定子安匝
stator assembly 定子装配，定子总成
stator bar 定子线棒
stator blade carrier ring 静叶环套
stator blade ring 静叶环
stator blade 静叶（片），（电容器的）静片，导向器叶片，静子叶片，定子叶片，固定叶片
stator bore 定子内孔，定子内径
stator cage 定子机壳，定子机座
stator can 定子屏蔽套
stator cascade loss 静叶栅损失
stator case 定子外壳，定子罩
stator circuit 定子电路
stator coil overhang 定子线圈悬
stator coil pin 定子线圈拉杆
stator coil 定子线圈
stator conductor 定子导线
stator contactor 定子接触器
stator cooling water system 发电机定子冷却水系统
stator cooling water 定子冷却水
stator copper loss 定子铜耗
stator copper 定子铜，定子铜线
stator core lamination 定子铁芯叠片
stator core of generator 发电机定子铁芯
stator core segment 定子铁芯扇形片
stator core 定子铁芯
stator current 定子电流
stator earth-leakage 定子对地泄漏
stator end cover 定子端盖
stator end plate 定子端板
stator end 定子端部
stator field 定子磁场
stator flange 定子突缘
stator flux 定子磁通
stator frame 定子机座
stator groove 定子槽
stator housing 定子机壳，定子机座，定子（外）壳
stator impedance 定子阻抗
stator inductance 定子电感
stator insulation 定子绝缘
stator iron loss 定子铁耗
stator iron 定子铁，定子铁芯
stator lamination 定子叠片
stator leakage field 定子漏磁场
stator leakage flux 定子漏磁通
stator leakage reactance 定子漏抗
statorless advancer 无定子进相机，无定子相位超前补偿
statorless compensator 无定子调相机
stator loss 定子损耗
stator of condenser 电容器定片
stator of hydro-generator 水轮发电机定子
stator outer diameter 定子外径
stator permeance 定子磁导
stator phase 定子相位
stator plate assembly （电容器的）定片组
stator plate （电容器的）定片
stator punching 定子冲片
stator reactance 定子电抗
stator resistance 定子电阻
stator rheostat 定子变阻器
stator-rotor starter 定子转子启动器，全套启动装置
stator shell 定子外壳，机壳
stator slot leakage reactance 定子绕组槽漏抗
stator slot 定子槽
stator spool 定子线圈
stator starter 定子启动器
stator structure 定子结构
stator surface 定子表面
stator tester 定子试验器
stator tooth 定子齿
stator turns 定子匝数
stator vane 静叶
stator vibration 定子振动

stator voltage phasor diagram　定子电压相量图
stator voltage　定子电压
stator winder　定子绕线机，定子嵌线机
stator winding copper　定子绕组铜线
stator winding DC resistance per phase　每相的定子绕组直流电抗
stator winding of generator　发电机定子绕组
stator winding　定子绕组，定子线圈
stator yoke　定子磁轭，定子轭，（直流电机的）机座
stator　定子，静子，（汽轮机）缸体，（电容器的）定片
statoscope　高差仪【测】，微动气压计，变压计，高差仪，灵敏高度表
stato　静（电）【词头】
statunits　厘米克秒静电制单位，CGS 静电制单位
status at air failure　气源故障时状态
status display　状态显示，现场显示
status indication group　状态指示组
status information　状态信息
status input/output contact or channel　状态输入/输出接点或通道
status memory and real-time system　状态存储器及实时系统
status modifier　状态修改符
status of legal person　法人地位
status quo ante　原状，以前的状态
status quo　现状，现状的，维持现状
status recorder　状态，记录仪
status region　状态区
status register　状态寄存器
status report　情况报告
status signal　状态信号
status source document　状态源文件
status　状态，状况，情况，条件
statute law　成文法
statute of frauds　反欺诈法，防止诈欺法
statute　规约，法令，法规，规则，章程
statutory activity　法定活动
statutory address　法定地址
statutory agent　法定代理人
statutory authority　法定机构，法定权力
statutory obligation　法定职责，法定义务
statutory representative　法定代表
statutory right　法定权利
statutory tax rate　法定税率
statutory text　法规原文
statutory　管理的，法定的，规定的，法规的
statvolt　静电伏特
staubosphere　尘层，尘圈
stave flume　板条渡槽
stave pipe　木板条水管
stave　狭板条
staving varnish　烤漆，牢固漆
stay baffle　支承挡板
stay bar　拉条钢筋
stay blade　固定导叶
stay bolt　支撑螺栓，撑螺栓
stay brace　支撑
Staybrite　斯特布赖特镍不锈钢
stay cable　斜拉索
stay collar　拉环，牵环
stayed pole　拉线杆
stayed tower　拉线铁塔
stay guy　固定拉索，拉线
staying wire　拉线
stay insulator　拉线绝缘子
stay line method　标线测流法，缆索测流法
stay pile　拉索桩，锚固桩
stay plate　撑板，垫板，缀合板
stay pole　撑杆
stay-put　原位不动，不返回【动作终止后，停留在该位置上】
stay ring　座环
stay rod　牵条螺栓，撑杆，拉线桩，拉线杆，缀条，拉索
stay thimble　拉线垫环，电杆上拉线的终端环
stay tightener　拉线张紧器
stay time　持续时间
stay tube　支撑管，拉撑管，撑管
stay vane ring　水轮机固定叶环
stay vane　固定导水瓣，固定导叶，水轮机固定导叶
stay　撑条，拉绳，拉条，拉索，拉杆，停留，支撑，固定，支持
STC(said to contain)　内容据称，据称内装
STC(site testing committee)　现场调试委员会
STCS(steam turbine control system)　汽轮机控制系统
STD-BY(stand by)　备用
steadiness　稳定度，稳固性
steady accelerated fluid　稳定加速流
steady acceleration　等加速度
steady air flow　稳定气流
steady alarm　持续报警
steady bearing　导轴承，固定轴承
steady component　稳态分量，稳定部分
steady condition　稳定工况
steady creep　稳定蠕变，等速蠕变，持续徐变
steady current　稳定电流，稳恒电流
steady direct current　恒稳直流电
steady emission rate　定常排放率
steady equation　稳态方程
steady flame　稳定火焰
steady flow aerodynamics　定常流空气动力学
steady flow coefficient　稳定流的流量系数，定常流动系数，稳流系数
steady flow energy equation　稳定流动能量方程
steady flow field　定常流场
steady flow model　定常流模式，稳流模型
steady flow reactor　（冷却剂）稳流反应堆
steady flow system　稳定流系统
steady flow turbine　稳流式透平
steady flow　稳定流，稳定水流，稳流，恒流，恒定流，定常流
steady fluidized bed　稳态流化床
steady galloping response　定常驰振响应
steady gas dynamics　定常流气体动力学
steady gradient　连续坡度
steady heat conduction　稳定热传导
steady heat transfer　稳定传热

steady illumination 稳定照明
steadying baffle 消浪栅，稳水栅
steadying resistance 镇流电阻
steady jump 稳定水跃
steady laminar flow 稳定层流
steady load 稳定负载，稳定载荷，稳定负荷，不变负荷
steady motion 定常运动，稳定运动，稳态运动
steady noise 稳态噪声
steady nonuniform flow 稳定变速流，稳定不均匀流，稳定非均匀流
steady operation 稳定运行，稳态运行
steady point 稳定点
steady pressure 稳定压力
steady price 不变价格
steady rain 连绵雨
steady rate 定常速率
steady resistance 镇流电阻
steady rotation 稳定旋转
steady running condition 稳定运转工况
steady running 稳定运行，稳定运转
steady seepage quantity 稳定渗流量
steady seepage 稳定渗流
steady short-circuit current 稳定短路电流
steady short-circuit 稳定短路，持续短路
steady signal 稳定信号
steady solar simulator 稳态太阳模拟器
steady solution 定常解
steady speed motor 稳速电动机
steady speed 稳定速度，稳定转速
steady state amplitude 稳态振幅
steady state availability 稳态可用度
steady state behavior 稳态工况性能
steady state boundary condition 稳态边界条件
steady state boundary layer 稳恒边界层，稳恒状态边界层
steady state characteristics 稳态特性（曲线），静态特性
steady state commutation 稳态换向
steady state compressible flow 稳态可压缩流
steady state concentration 稳态浓度
steady state condition 稳定条件，稳定工况，稳态工况，稳态条件
steady state constant 稳态常数
steady state core 稳态芯部
steady state creep 稳态蠕变，稳态徐变
steady state current 稳态电流
steady state cycle 稳态循环
steady state density 稳态密度
steady state deviation of the N-th order N阶稳态偏差
steady state deviation 稳态偏差
steady state equation 稳态方程
steady state error coefficient 稳态误差系数
steady state error 稳态误差，静态误差，静差
steady state estuary 定态河口
steady state flow 稳态流，静态流
steady state frequency control 稳态频率控制
steady state fuel conversion ratio 稳态燃料转换比
steady state fusion reactor 稳态聚变反应堆
steady state heating 稳态加热
steady state heat loss 稳态热损失
steady state heat transfer 稳态传热，稳定传热，定态传热
steady state incompressible flow 稳态不可压缩流
steady state load characteristic 稳态负荷特性
steady state measurement 稳态量测
steady state models of performance 集热器性能的稳态模型
steady state noise 稳态噪声
steady state of flow condition 稳定流态
steady state operating condition 稳态运行工况，定态运行工况
steady state operation 稳定状态运行，稳态运行
steady state oscillations 稳态振荡
steady state output 稳态输出功率，稳态输出，稳态出力
steady state performance 稳定工况性能，稳态特性，稳态运行特性，稳态性能
steady state period 稳态周期
steady state plume 定常羽流
steady state poisoning 稳态中毒
steady state pore pressure 稳态孔隙压力
steady state power coefficient 稳态功率系数，稳态电源条件
steady state power limit 稳态功率极限
steady state pressure 稳态压力
steady state process 稳态过程
steady state reactance 稳态电抗
steady state reactor operation 反应堆稳态运行
steady state regulation 静不等率，稳态调节
steady state response 定常响应，稳态响应
steady state short-circuit 稳态短路
steady state signal 稳定信号，稳态信号
steady state solution 稳态解，静态解
steady state speed regulation 稳态速度调节，稳定速度调整率，稳定调速率
steady state speed 稳态速度，稳态速率，整定速度
steady state stability calculation 静态稳定计算
steady state stability factor 静态稳定度系数
steady state stability limit 静态稳定极限，静态稳定限度
steady state stability margin 静态稳定储备系数
steady state stability 静态稳定，静态稳定性，静态稳定度
steady state temperature distribution 稳态温度分布
steady state temperature 稳定状态温度
steady state test technique 稳态试验法
steady state theory 稳恒态学说
steady state transport 稳态输运
steady state unavailability 稳态不可用度
steady state value 稳态值
steady state vibration 稳态振动，稳定振动状态
steady state 稳定状态，稳定工况，稳恒状态，稳态
steady steaming condition 稳定蒸发条件
steady stream 稳定流，稳恒流
steady stress 静应力

steady temperature 稳定温度
steady uniform flow 等速均匀流，稳定等速流，稳定均匀流
steady unsaturated flow 非饱和稳定流
steady vibration 定常振动
steady voltage 稳定电压
steady wave 稳定波
steady wind load 定常风载
steady wind 定常风
steady working condition 稳定工况
steady 稳定的
steam accumulator 蓄汽器，蒸汽蓄热器，蒸汽联箱，蒸汽蓄力器
steam admission side 进汽侧
steam admission slit 进汽口
steam admission valve 进汽阀
steam admission 进汽
steam air ejector 蒸汽抽气器，射汽抽气器
steam air heater coil 暖风机盘管
steam air heater system 暖风器系统
steam air heater 前置预热器，蒸汽（加热式）空气预热器，蒸汽暖风器
steam air humidification system 蒸汽湿润空气系统
steam air ratio 蒸汽空气比
steam and gas combined cycle power plant 蒸汽燃气联合循环电厂
steam and gas turbine 蒸汽燃气轮机
steam and power conversion system 蒸汽发电厂，蒸汽动力转换系统
steam and water piping design 汽水管道设计
steam and water separation 汽水分离
steam atomization 蒸汽雾化
steam atomizer 蒸汽雾化喷嘴
steam atomizing oil burner 蒸汽雾化油燃烧器
steam attemperator 蒸汽减温器，蒸汽调温器
steam bath 蒸汽浴
steam bending force 蒸汽弯曲力
steam bending stress 蒸汽弯应力
steam benefit 蒸汽收益
steam blanket 汽塞，循环停滞，蒸汽垫层
steam blast cleaning 吹汽清洗
steam bleeding system 抽汽系统
steam bleeding 抽汽【汽机，不可调抽汽】，放汽
steam bleedoff 抽汽，放汽
steam blinding 汽塞，蒸汽堵塞，蒸汽粘结
steam blower 蒸汽吹灰器
steam blowing out （蒸汽）吹管［吹扫］
steam blowing pipe 蒸汽吹管
steam blowing 蒸汽吹管，蒸汽吹洗，蒸汽吹扫，蒸汽吹灰，蒸汽喷吹法
steam boiler 蒸汽锅炉
steam borne impurities 蒸汽带杂质，蒸汽带盐
steam borne spray 蒸汽携带的水雾
steam break accident 主蒸汽管道破裂事故
steam bubble 气泡
steam bypass line 蒸汽旁路管线
steam bypass system 蒸汽旁路系统
steam bypass 蒸汽旁路
steam calorimeter 蒸汽热量计
steam can 蒸汽发生器
steam capacity at max condensing duty 在最大冷凝负荷时的蒸汽容量
steam capacity 蒸发量
steam chamber of steam governing valve 调节汽阀蒸汽室
steam chamber 蒸汽室
steam chart 水蒸气图表，蒸汽特性图表，蒸汽焓熵图
steam chest assembly 进汽室组合件
steam chest 进汽箱，蒸汽室【汽轮机】
steam circular motion 蒸汽循环运动
steam circulating pipe 蒸汽循环管
steam circulation 蒸汽流动
steam cleaning 用蒸汽清洗
steam coal blending 动力配煤
steam coal 动力煤，锅炉用煤，蒸汽锅炉用煤
steam coil air heater 蒸汽盘管式暖风机，蒸汽盘管式空气预热器
steam collector 分汽缸，集汽包
steam condensate-feedwater system 蒸汽凝结水给水系统
steam condensate 蒸汽冷凝水，蒸汽凝结水
steam condensation 蒸汽凝结
steam condenser 凝汽器，蒸汽冷凝器
steam condensing temperature 蒸汽凝结温度
steam conditions 蒸汽参数
steam conductivity 蒸汽导电度
steam cone 蒸汽喷嘴
steam consumption diagram 汽耗工况图
steam consumption rate 汽耗率
steam consumption 汽耗，汽耗量
steam contamination 蒸汽污染
steam content 含汽量
steam converter 蒸汽发生器，蒸汽转换器
steam converting valve 减温减压阀
steam cooled fast breeder reactor 蒸汽冷却快增殖堆
steam cooled fast breeder 汽冷快增殖堆
steam cooled heavy water reactor 气冷重水
steam cooled reactor 蒸汽冷却堆，过热蒸汽堆
steam cooled roof 顶棚管过热器
steam cooled wall 包墙管过热器，蒸汽冷却壁，汽冷壁
steam cooler 蒸汽冷却器
steam cooling coil 蒸汽冷却蛇管
steam cooling 蒸汽冷却
steam curing chamber 蒸汽养护间
steam curing of concrete 蒸汽养护混凝土，混凝土的蒸汽养护
steam curing 蒸汽养护【混凝土】
steam cushion 汽垫，蒸汽缓冲垫
steam cycle reactor 蒸汽循环反应堆
steam cycle 蒸汽循环
steam cyclone 蒸汽旋风分离子
steam cylinder 汽缸
steam dago 往复横截锯机
steam deaerator 热力除氧器
steam decontamination cabin 蒸汽去污室
steam department （电业管理局的）热工科，热力设备科

steam discharge pipe 排汽管
steam discharging 蒸汽排空，蒸汽排放
steam-disengaging surface 蒸发面
steam displacement sensor 蒸汽排量传感器
steam distributing gear 配汽机构
steam distributing valve 配汽阀
steam distribution device 配汽机构
steam distribution header 分汽缸
steam distribution internals 蒸汽分配内部构件
steam-distribution plate 蒸汽分配板，多孔板
steam distribution 配汽，蒸汽分配
steam distributor 蒸汽分配器
steam dome 蒸汽空间，汽室，蒸汽区，汽包
steam drain 排汽口
steam draw-off 抽汽，蒸汽取样
steam drier 蒸汽干燥器
steam-driven blowing engine 汽动鼓风机
steam-driven feed water pump 汽动给水泵
steam-driven oil pump 汽动油泵
steam-driven pump 汽动泵
steam-driven riveting machine 汽动铆钉机
steam-driven 蒸汽驱动的，汽动的
steam drive 蒸汽驱动
steam drop 蒸汽压降
steam drum paint 汽包漆
steam drum section 汽水分离装置【蒸汽发生器】
steam drum 汽包，上汽包，汽鼓，蒸汽锅筒
steam dryer 蒸汽干燥器
steam drying 蒸汽干燥
steam dump control 蒸汽排放控制
steam dumping 排汽【到凝汽器】
steam dump system 蒸汽排放系统【汽轮机事故时】
steam dump to atmosphere 蒸汽向大气排放
steam dump to condenser 蒸汽排入凝汽器
steam duty rate 蒸汽负荷率
steam ejector 射汽抽汽器，蒸汽喷射器
steam-electric generating set 汽轮发电机组
steam-electric generating station 蒸汽电站
steam-electric power generation 蒸汽发电
steam-electric power plant 蒸汽发电厂，火力发电厂
steam engine 蒸汽机
steamer 蒸汽发生器，汽锅，轮船，蒸汽机
steam escape valve 放汽阀
steam evaporator 蒸发器
steam excavator 汽动挖掘机，汽动掘土机
steam exhaust hood 排汽室
steam exhaust outlet 排汽口
steam exhaust pipe 排汽管
steam exhaust port 排汽口
steam exhaust system 蒸汽排放系统
steam exhaust 乏汽
steam exit pressure 蒸汽出口压力
steam exit temperature 蒸汽出口温度
steam explosion 蒸汽爆炸
steam extraction capacity 抽汽量
steam extraction header 抽汽母管
steam extraction line 抽汽管路
steam extraction pipe 抽汽管道
steam extraction system 抽汽系统【汽机，可调的】
steam extraction tube 抽汽管【沸水堆汽水分离器】
steam extraction turbine 抽汽式汽轮机
steam extraction 抽汽【汽机，可调的】
steam feed heater 回热加热器
steam feed pipe 供汽管
steam feed pump 汽动给水泵
steam feedwater flow mismatch 水-蒸汽流量不匹配
steam field 蒸汽田
steam film 蒸汽膜，蒸汽层
steam filter 滤汽器
steam fire protection 蒸汽消防
steam flow channel 蒸汽流道
steam flow condition 蒸汽流量
steam flow diagram 蒸汽流程图
steam flow direction 汽流方向
steam flow excited vibration 汽流激振
steam flow meter 蒸汽流量表
steam flow orifice 蒸汽流量孔板
steam flow recorder 蒸汽流量记录仪
steam flow 汽流，蒸汽流量
steam flushing 蒸汽清洗
steam fog 蒸汽雾
steam forging hammer 蒸汽锤
steam free 无蒸汽的，不含蒸汽的
steam from inside 厂用汽源
steam-gas cycle 蒸汽燃气联合循环
steam gate valve 蒸汽闸阀
steam gauge 蒸汽压力表，汽压表
steam generating bank 蒸发管束
steam generating circuit 锅炉循环回路，汽水系统
steam generating heavy water reactor 蒸汽的重水反应堆
steam generating plant 蒸汽发生装置
steam generating rate 产汽率
steam generating reactor 蒸汽发生反应堆
steam generating surface 蒸发段【直流炉】
steam generating tube 蒸汽发生管
steam generating unit 蒸汽发生装置，锅炉
steam generation expenses 蒸汽生产费用
steam generator block 锅炉机组
steam generator blowdown condenser 蒸汽发生器的排污冷凝器
steam generator blowdown demineralizer 蒸汽发生器排污除盐器
steam generator blowdown flash tank 蒸汽发生器排污扩容箱
steam generator blowdown system 蒸汽发生器排污系统，锅炉排污系统
steam generator blowdown tank 蒸汽发生器排污箱，锅炉排污罐
steam generator building 蒸汽发生器厂房，锅炉房
steam generator bunker 蒸汽发生器隔间
steam generator cavity 蒸汽发生器舱
steam generator cell 蒸汽发生器室
steam generator compartment 蒸汽发生器室

steam generator cubicle　蒸汽发生器隔间
steam generator decontamination system　蒸发器放射性去污系统
steam generator isolation and dump system　蒸汽发生器隔离与排放系统
steam generator lead　主蒸汽管道
steam generator module　蒸汽发生器组件
steam generator relief line　蒸汽发生器卸压管
steam generator shell　蒸汽发生器壳体
steam generator system　蒸汽发生器系统，锅炉系统
steam generator tube failure　蒸汽发生器管子破裂
steam generator tube rupture　蒸汽发生器管子断裂
steam generator unit　锅炉，蒸汽发生器
steam generator water level control loop　蒸汽发生器水位控制回路
steam generator water level control valve　蒸汽发生器水位控制阀
steam generator　锅炉，蒸汽发生器
steam governing valve with link lever　调节汽阀及连杆
steam governing valve　调节汽阀
steam hammering analysis　汽锤分析
steam hammer　汽锤，蒸汽锤
steam header　蒸汽集箱，蒸汽联箱，供汽联箱，集汽联箱，分汽缸
steam heated air-heater　蒸汽加热空气预热器，暖风器
steam-heated evaporator　蒸汽加热的蒸发器
steam heater　蒸汽供热装置，蒸汽加热器
steam heating boiler　蒸汽采暖锅炉
steam heating coil　蒸汽加热盘管
steam heating system　蒸汽供暖系统
steam heating　汽暖，蒸汽采暖，蒸汽供暖，蒸汽加热
steam heat-supply network　蒸汽热网
steam heat-supply system　蒸汽供热系统
steam heat tracing　蒸汽伴加热
steam humidification　蒸汽加湿
steam humidifier　蒸汽增湿器，蒸汽湿润器
steam impingement corrosion　汽击腐蚀
steaming capacity　蒸发量
steaming conditions　汽化条件，蒸发条件
steaming economizer　沸腾式省煤器
steaming header　沸腾联箱
steaming point　汽化点，沸点
steaming process　蒸汽法
steaming rate　蒸发率
steaming temperature　汽化温度
steaming zone　蒸发区
steam injection equipment　蒸汽喷射装置
steam injection pump　蒸汽喷射器
steam inlet condition　进汽参数
steam inlet pipe　进汽管
steam inlet　进气口
steam isolation valve　蒸汽隔离阀
steam jacket　蒸汽夹套，蒸汽套管
steam-jacket tracing　蒸汽夹套伴热
steam jet air ejector　射汽抽汽器，蒸汽抽汽喷射泵
steam jet air extractor　射汽抽汽器
steam jet ash conveyer　射汽输灰器
steam jet burner　蒸汽雾化式燃烧器
steam jet deaerator　鼓泡式除氧器
steam jet draft　蒸汽拔风
steam jet ejector　射汽抽汽器，蒸汽喷射器
steam jet exhauster　蒸汽喷射式排气器，蒸汽引射拔风
steam jet heat pump　蒸汽喷射式热泵
steam jet hot water heating system　蒸汽喷射热水采暖系统
steam jet pump　蒸汽喷射泵
steam jet refrigerating machine　蒸汽喷射式制冷机
steam jet refrigeration cycle　蒸汽喷射式制冷循环
steam jet siphon　蒸汽喷射泵
steam jet soot blower　蒸汽吹灰器
steam jet sprayer　蒸汽雾化器
steam jet　蒸汽射流，蒸汽喷射器，蒸汽喷射（嘴）
steam lance　蒸汽吹灰枪
steam lane　汽道
steam lead pipe　蒸汽导管，导汽管
steam lead　蒸汽导管，蒸汽管道
steam leakage test of boiler　锅炉蒸汽严密性试验
steam leakage　漏汽
steam leak-off valve　漏汽阀
steam leak　漏汽
steam line blowing　蒸汽吹管，吹洗
steam line break accident　蒸汽管道破裂事故
steam line break　蒸汽管道破裂
steam line flow　层流
steam line pressure　二回路蒸汽压力【核电】
steam line　蒸汽管道，蒸汽管路
steam liquid type rupture　汽区水区管道破裂
steam load　蒸汽负荷
steam lock　汽封，蒸汽塞
steam locomotive　蒸汽机车
steam loop　蒸汽旁通管路
steam main　主蒸汽管道，蒸汽母管，主蒸汽管
steam manifold　蒸汽集箱，蒸汽联箱，分汽缸
steam manometer　蒸汽压力计，蒸汽压力表
steam meter　蒸汽流量计
steam mist　蒸汽轻雾
steam moisture　蒸汽湿度
steam nitrogen mixture　蒸汽氮气混合
steam nozzle　蒸汽接管，蒸汽喷嘴
steam off　蒸发
steam-operated air ejector　射汽抽气器
steam-operated　蒸汽驱动的
steam outlet nozzle　排气口，排气管嘴
steam outlet plenum　蒸汽出口腔
steam outlet　蒸汽出口
steam output　蒸发量
steam-oxidized　气流氧化的
steam parallel slide stop valve　蒸汽闸阀
steam parameter　蒸汽参数
steam passage drain　蒸汽管线疏水

steam passage 蒸汽通道
steam path 蒸汽流道，蒸汽通道
steam pile driver 蒸汽打桩机
steam pile hammer 汽桩锤，打桩汽锤，蒸汽打桩锤
steam pipe expansion loop 蒸汽管膨胀圈
steam pipeline 蒸汽管路，蒸汽管道，蒸汽管线
steam pipe 蒸汽管道，蒸汽管
steam piping for steam ejector 射汽抽气器蒸汽管路
steam plenum head 蒸汽腔头部
steam plough 汽犁
steam plume 蒸汽柱
steam pocket 汽室，汽袋
steam pollution 蒸汽污染
steam port 蒸汽口
steam power cycle 蒸汽动力循环
steam power engineering 蒸汽动力工程
steam power plant 火力发电设备，蒸汽电站，蒸汽（动力）发电厂，蒸汽动力装置，蒸汽动力厂，火力发电厂
steam power 蒸汽动力
steam pressure chart 蒸汽压力曲线
steam pressure gauge 蒸汽压力表，汽压表
steam pressure temperature reducer 蒸汽减压减温装置
steam pressure test 汽压试验
steam pressure 汽压，蒸汽压力
steam pressurization 蒸汽定压
steam pressurizer 蒸汽稳压器
steam production 蒸汽生产，产汽量
steam projection system 蒸汽喷射系统
steam property 蒸汽参数
steam pump 汽泵
steam purge 蒸汽吹洗
steam purging of piping 蒸汽吹管
steam purging 蒸汽吹洗，蒸汽吹扫
steam purification 蒸汽净化
steam purifier 汽水分离器
steam purifying equipment 蒸汽净化设备
steam purity meter 蒸汽盐量计
steam purity 蒸汽纯度，蒸汽品质
steam quality at minimum heat transfer coefficient 最高壁温处含汽率
steam quality by mass 质量含汽率，重量蒸汽含量，蒸汽干度
steam quality by volume 容积蒸汽含量
steam quality by weight 重量蒸汽含量，蒸汽干度
steam quality 蒸汽质量，蒸汽浓度，蒸汽品质，蒸汽干度，含汽率
steam radiator 蒸汽散热器
steam-raising boiler 蒸汽（发生）锅炉【即工业锅炉】
steam raising unit 蒸汽发生器
steam raising 锅炉升汽，产汽，蒸汽蒸发，汽化蒸发，产生蒸汽
steam rate guarantee 保证汽耗
steam rate 汽耗，汽耗率，汽耗量，蒸汽流量，
steam receiver 集汽管
steam reducing valve 蒸汽减压阀
steam reformer 蒸汽重整器
steam reforming 蒸汽重整
steam regulator 蒸发量调节器
steam reheating cycle 蒸汽再热循环
steam reheating in nuclear reactor 核反应堆内蒸汽再热
steam reheating system 再热蒸汽系统
steam rejection system 蒸汽喷射系统
steam releasing surface 蒸发面
steam relief valve 蒸汽释放阀，安全阀
steam relieving area 蒸发面，蒸发面积
steam-relieving capacity 蒸发量
steam resistance 汽阻
steam return line 蒸汽返回管道，冷凝蒸汽管线
steam ring 环形配汽室
steam roller 蒸汽压路机
steam sampling point 蒸汽取样点【蒸汽发生器】
steam sampling 蒸汽取样
steam-saturation temperature 饱和蒸汽温度
steam screen 汽水分离器
steam scrubber 蒸汽清洗装置，洗汽板
steam scrubbing 蒸汽清洗
steam seal cooler 汽封冷却器
steam seal diverting valve 汽封分流阀
steam seal gland 汽封
steam seal header 汽封联箱
steam sealing line 蒸汽轴封管路
steam sealing 汽封
steam seal regulator 汽封调节器
steam security valve 蒸汽安全阀
steam separator cover 汽水分离器罩壳
steam separator 凝汽罐，蒸汽分离器，汽水分离器
steam shock 蒸汽激波【汽轮机通道内】
steam side 汽侧
steam slug 蒸汽段塞，蒸汽柱，汽团，汽塞
steam space 蒸汽空间，汽空间，汽容积
steam stagnation 蒸汽停滞，（水循环中的）汽塞
steam staitc bending stress 蒸汽静弯应力
steam sterilizer 蒸汽消毒器
steam stop valve 蒸汽截止阀
steam strainer 蒸汽滤网，蒸汽过滤器
steam stripping 蒸汽脱除，汽提，蒸汽脱附
steam superheater 蒸汽过热器
steam supply and power generating plant 热电联产厂，供热火力发电厂，热电厂
steam supply conditions 供汽参数
steam supply for shaft sealing 轴封供汽
steam supply pipe 供汽管（道）
steam supply turbine 供汽式汽轮机
steam supply 蒸汽源，供汽
steam table 水蒸气表
steam tapping 排汽，抽汽
steam temperature control 汽温调节，蒸汽温度控制
steam temperature regulation 汽温调节
steam tension 蒸汽张力
steam thawing equipment 蒸汽解冻设备
steam throttling 蒸汽节流
steam throughput 蒸汽流量
steam tightness 汽密性

steam tight test 汽密性试验
steam tight 汽密，汽密性，汽密性的，蒸汽密封，汽密的，不透蒸汽的
steam to steam heat exchanger 汽汽热交换器
steam trace 蒸汽伴随加热
steam tracing 蒸汽加热保温，蒸汽伴热
steam transformer 蒸汽变换器【换热器】
steam trap connection 疏水装置
steam trap valve 蒸汽疏水阀
steam trap 蒸汽管去水装置，（蒸汽）疏水器，疏水阀，汽水分离器，蒸汽疏水阀，汽阱，阻汽回水阀，阻汽排水阀
steam treatment 蒸汽处理，暖管
steam tug 蒸汽拖轮
steam turbine barring gear 汽机盘车装置
steam turbine by-pass system 汽轮机旁路系统
steam turbine differential expansion 汽轮机胀差
steam-turbine-driven alternator 汽轮发电机
steam-turbine-driven generator 汽轮发电机
steam turbine for cogeneration 供热汽轮机，热电联产汽轮机
steam turbine for feed-water pump 给水泵汽轮机
steam turbine generating set 汽轮发电机组
steam turbine generator 汽轮发电机
steam turbine governing system 汽轮机调节系统
steam turbine of feed water pump 给水泵汽轮机
steam turbine oil system 汽轮机油系统
steam turbine oil whip vibration 汽轮机油膜振荡
steam turbine performance test 汽轮机性能试验
steam turbine power plant 汽轮机发电厂
steam turbine proper 汽轮机本体
steam turbine self-starting 汽轮机无电源启动
steam turbine speed governor 汽轮机调速器
steam turbine stage 汽轮机级
steam turbine starting 汽轮机起动
steam turbine thermal performance 汽轮机热力特性
steam turbine vibration 汽轮机振动
steam turbine with air-cooled condenser 空冷汽轮机
steam turbine 汽机，汽轮机，蒸汽轮机，蒸汽透平
steam turboalternator 汽轮发电机，汽轮发电机组，汽轮交流发电机组
steam turboset 汽轮发电机组
steam type airheater 蒸汽加热式空气预热器
steam type rupture 汽区管道破裂
steam vacuum pump 蒸汽真空泵
steam vacuum system 真空式蒸汽系统
steam valve body 汽阀体
steam valve 汽阀，蒸汽阀
steam vapor 水蒸气
steam vent stack 排汽管
steam void coefficient of reactivity 反应性汽泡系数
steam void fraction 汽泡份额
steam void 汽泡
steam washer 蒸汽清洗设备，洗汽装置，蒸汽清洗装置
steam washing 蒸汽清洗
steam water chemical lab 汽水化验站
steam water cooled reactor 汽水冷却反应堆
steam water dump system 汽水排放系统
steam water experimental loop 汽水试验回路
steam water film 汽水膜
steam water interface 汽水混合物
steam water mixed heat exchanger 汽水混合式换热器
steam water mixture 汽水混合物，蒸汽水混合物
steam water sampling analyze tore room 汽水取样分析间
steam water sampling rack room 汽水取样盘间
steam water sampling room 汽水配样间
steam water separation 汽水分离
steam water separator 汽水分离装置
steam water shock 汽水冲击
steam water supervising room 汽水监督室
steam water system 蒸汽水系统，汽水系统，二回路系统
steam water two phase flow 汽水两相流
steam water type heat exchanger 汽水式换热器
steam wetness 蒸汽湿度
steam whirl 汽流涡动
steam with moisture carryover 夹带水分的蒸汽
steam working pressure 蒸汽工作压力
steam 水蒸气，蒸汽
stearine pitch 硬脂沥青
steatite 块滑石
steel angle 角钢
steel arch-gate 弧形钢闸门
steel area ratio 钢筋面积比
steel area 钢筋面积
steel armored cable 钢铠电缆
steel armoured 钢铠装的
steel ash door 铁板出灰门
steel ball 钢球，钢珠
steel bandage 钢丝绑扎，钢绑线，钢绑环
steel bar bending machine 钢筋弯曲机
steel bar cutter 钢筋切断机
steel bar shearing machine 钢筋切断机
steel bar straightening and shearing machine 钢筋拉直切断机
steel bar 扁钢，钢筋，棒材，条钢，钢条，型钢
steel beam enclosed in concrete 混凝土包钢梁
steel bender 弯钢筋机，钢筋工
steel bending yard 钢筋工场
steel blasting and priming shop 预处理车间【加工】，钢喷砂加底漆车间
steel bottom ash silo 钢制底灰库
steel box pile 盒形钢桩，箱形钢桩
steel bracing plate 钢支撑板
steel bulkhead 钢板驳岸，钢闷头【钢管的】
steel butt-welding seamless pipe fittings 钢制对焊无缝管件
steel cable 钢缆，钢索，钢丝绳
steel caisson breakwater 钢沉箱防波堤
steel caisson 钢沉箱
steel cantilever dam 悬臂式钢坝
steel casement 钢窗，钢窗框
steel casting 钢铸件，铸钢
steel cellular bulkhead 钢板格型岸壁
steel cell 钢小室，潜水箱，沉箱

steel central post 钢中心柱
steel channel 槽钢
steel checkered plate 花纹钢板
steel chimney 钢烟囱
steel cladding tube 钢包壳管
steel-clad switch box 铠装开关箱
steel-clad 包钢的，铠装的，铁甲的，装甲的
steel cofferdam 钢围堰
steel composite construction 劲性钢筋混凝土结构
steel concrete composite girder 钢材混凝土组合梁
steel concrete 钢筋混凝土
steel conical tubular tower 锥形钢筒塔架
steel construction 钢结构
steel consumption of main power building per kW 每千瓦主厂房钢材消耗量
steel consumption of power generation project per kW 发电工程每千瓦钢材消耗量
steel containment shell 钢安全壳壳体
steel containment sphere 钢安全壳球体，球形钢质安全壳
steel cord belt conveyer 钢绳芯胶带机
steel cord rubber belt conveyor 钢绳芯胶带机
steel cord 钢丝绳芯
steel core concrete column 钢管内填混凝土柱，钢芯混凝土柱
steel-cored alluminium conductor 钢芯铝绞线
steel-cored aluminum wire 钢芯铝线
steel-cored copper wire 钢芯铜线
steel core rubber belt 钢丝绳芯胶带
steel core 钢心，钢芯
steel dam 钢坝
steel deck girder dam 钢面板梁式坝
steel deck-plate 钢盖板，钢甲板
steel deck 压型钢板
steel diaphragm 钢隔板，钢膜片，钢心墙【坝内的】
steel dog 钢索钩
steel door buck 钢门边柱
steel door 钢门
steel dowel 插筋，钢暗键
steel facing 钢板护面，钢面板
steel fixer 钢筋定位器，钢筋绑扎工
steel flat 扁钢
steel floral tube pile 钢花管桩
steel foil insulation 钢箔保温，钢箔保温层
steel formwork 钢模板
steel form 钢模
steel frame construction 钢架结构
steel frame dam 钢架坝
steel frame plate girder 锅炉钢架大板梁
steel frame 钢框架，钢结构，钢架
steel grade 钢号
steel grillage type foundation 钢格栅基础
steel guy wire 钢拉线
steel hanger 钢吊架
steel I-beam 工字钢
steel industry 钢铁工业
steel ingot 钢锭，钢坯
steel jacket platform 钢套管架平台【近海建筑物用】
steel ladder 钢梯
steel lagging 钢挡板，钢套
steel lathing 抹灰用钢丝网
steel lattice truss 钢格构桁架
steel lattice work 钢格构
steel-lined concrete pipe 钢板衬砌混凝土管
steel-lined wire winding channel 衬钢钢丝绕线孔道
steel-lined 衬钢的
steel liner 钢衬，钢衬套，钢衬圈
steel lining 钢板衬砌
steel magnet 磁钢
steelmaker 炼钢者
steelmaking 炼钢
steel mark 钢标号
steel member 钢构件
steel membrane 钢膜，钢覆面层
steel mesh lagging 钢丝网衬板
steel mesh reinforcement 网状钢筋
steel mesh 钢丝网
steel mill 钢铁厂
steel mould 钢模
steel movable form 钢滑动模板
steel panel 钢护板
steel penstock 压力钢管
steel pile 钢桩
steel pipe column 钢管柱
steel pipe flange 钢管法兰
steel pipe pile 钢管桩
steel pipe 钢管
steel pit 钢表面凹坑
steel plain splice 钢板平面拼接
steel plate circular washer 圆形钢垫片
steel plate gate 平板钢闸板
steel plate girder 钢板梁
steel plate hanger 钢托兜
steel plate membrane embedded in the foundation 安全壳埋入基础的钢板
steel plate shearer 剪板机
steel plate square washer 方形钢垫板
steel plate stamping 钢板冲压
steel plate 薄钢板，钢板
steel-ply form 折叠式钢模板，钢平台
steel prestressing tendon 预应力钢筋束
steel profile 钢材剖面
steel punch 钢字模，钢冲头
steel pylon 输电铁塔
steel radiator 钢制散热器
steel rail 钢轨
steel ratio 配筋百分率，钢筋百分率
steel reactor pressure vessel 反应堆钢压力容器
steel reflector 钢反射层
steel refueling cavity liner 换料腔钢内衬
steel reinforced aluminium conductor 钢芯铝绞线
steel reinforced aluminum alloy conductor 钢芯铝合金绞线
steel reinforcement 钢筋
steel relaxation 钢筋的松弛
steel rod 圆钢
steel rolling mill 轧钢厂

steel rolling-up door 钢卷门
steel roof joist 屋面钢梁
steel roof truss 钢屋架
steel rope fittings 钢丝绳配件
steel rope 钢丝绳，钢索
steel rotor 钢转子
steel rule 钢尺
steel sash 钢框格
steel section 型钢
steel shape reinforcement 型钢配筋
steel shape 型钢
steel sheathing 钢面板，钢棉面板
steel sheet liner 钢皮内衬
steel sheet pile break-water 钢板桩防波堤
steel sheet pile bulkhead 钢板桩驳岸，钢板桩护岸
steel sheet pile cellular cofferdam 钢板桩格型围堰
steel sheet pile cofferdam 钢板桩围堰
steel sheet pile earth cofferdam 钢板桩填土围堰
steel sheet pile quaywall 钢板桩岸壁
steel sheet pile wharf 钢板桩码头
steel sheet pile 钢板桩
steel sheet piling cut-off 钢板桩截水墙
steel sheet 薄钢板，钢板
steel-shelled concrete pile 钢壳混凝土桩
steel shell leak rate 钢壳漏率，钢衬泄漏率
steel shell 钢壳，钢壳体
steel-shielded isolation box 钢屏蔽隔离箱
steel shoe 钢桩靴
steel shot boring 钢珠钻探
steel shot drilling 钢珠钻探，钢珠钻孔，钢砂钻进
steel skeleton building 钢骨架房屋
steel skeleton 钢骨架
steel slab bridge 钢板桥
steel sleeper 钢枕木
steel sleeve 钢套管，钢套筒
steel spike 钢针
steel split-ring 钢裂环
steel square 钢角尺
steel stack 钢烟囱，钢通风管
steel stranded wire 钢绞线
steel strand 钢绞线
steel strap 扁钢，带钢，钢条，钢片
steel structural member 钢结构构件
steel structure 钢结构，钢架
steel supporting 钢支架
steel support structure 钢支撑结构
steel-tank rectifier 钢壳汞弧整流器
steel-tape armouring 钢带铠装
steel tape 钢卷尺，钢尺
steel tendon 钢丝束
steel tower framing 铁塔构造
steel tower outline 铁塔简图
steel tower swivel clevis 铁塔转动挂板
steel tower test 铁塔试验
steel tower 钢塔，铁塔
steel trestle 钢栈桥
steel-troweled concrete 钢镘抹面混凝土
steel trowel 钢镘刀
steel truss 钢桁架
steel tube economizer 钢管省煤器
steel tube 钢管
steel wall framing 钢墙框架
steel weir 钢堰
steel welding construction 钢焊接结构
steel window 钢窗
steel wire armoured 钢丝铠装的
steel wire brush 钢丝刷
steel wire lashing 钢丝绑扎
steel wire net 钢丝网
steel wire rope 钢丝索，钢丝绳
steel wire sheath 钢丝外皮
steel wire wheel brush 钢丝轮刷
steel wire 钢丝
steel wool 钢丝绒
steelwork component 钢结构部件
steelwork superstructure 上部钢结构
steelwork 钢结构，钢架，钢制品，炼钢厂，钢铁构造
steelyard 秤，吊秤，杆式秤
steel 型钢，钢
steening brick lining 砖衬
steening 石衬，石砌，砖砌
steen 砌砖石内壁，加衬，许许多多的
steep bank revetment 陡坡护岸
steep channel 陡槽，陡坡河槽
steep cone 小张角锥
steep descending gradient 陡下坡道
steep dip 陡倾角
steep down grade 陡下坡
steepen （使）变得陡峭
steepest ascent 最陡上升
steepest descent 最陡下降
steep-fronted wave 陡峭前缘波
steep-front-of-wave test 陡波试验
steep-front wave 前陡波，陡前沿波，雷电波
steep-front 陡峭的前沿
steep grade tunnelling 陡坡隧洞掘进
steep grade 陡坡
steep gradient 陡坡
steeple-compound turbine 上下叠置式汽轮机
steeply 险峻地，大坡度地，
steepness of pulse edge 脉冲边沿陡度
steepness of the lightning current 雷电流陡度
steepness of wave front 波前陡度
steepness 陡度，斜度
steep performance curve 陡斜性能曲线
steep slope 陡坡
steep-walled canyon 陡壁峡谷
steep-walled valley 陡壁河谷
steep wall 陡壁
steep wave 陡波
steep 陡的，陡峭的，急剧升降的，浸，沉浸，浸渍，悬崖
steerable aerial 可控天线
steerable 可控的，可操纵的
steering angle 转向角
steering apparatus 转向机构
steering body 指导机构
steering box 转向机构箱，操能箱

steering brake 转向闸
steering circuit 控制电路，操纵电路
steering clutch 转向离合器
steering column jacket 转向轴套
steering committee 指导委员会，筹备委员会
steering computer 操纵系统计算机，驾驶用计算机
steering current 引导气流
steering engine 转向机
steering gear 转向机构，转向装置，轮向齿轮
steering handle 转向旋盘，舵把
steering instruction 控制指令
steering pole 主梁
steering program 导引程序，操纵程序
steering resistance 转向阻力
steering response 操纵灵敏性，转向响应
steering rod 转向杆
steering routine 执行程序，控制程序
steering sail 转向帆，补助帆
steering wheel switch 操纵轮开关
steering wheel 操纵轮，舵轮，转向轮
steering 操纵，驾驶，控制，转向，转向器
Stefan-Boltzmann constant 斯蒂芬-玻尔兹曼常数
Stefan-Boltzmann law 斯蒂芬-玻尔兹曼定律
Stefan's constant 斯蒂芬常数
Steinmetz coefficient 磁滞系数，司坦麦兹系数
stellar plan 星形布置图
stellate snow flake 星状雪花
Stellite-clad stainless steel journal 司太立覆面不锈钢轴颈【主泵】
Stellited stainless steel journal 司太立不锈钢轴颈【主泵】
Stellite overlap 硬质合金覆盖面
Stellite protection strips 司太立合金防蚀片
Stellite 司太立合金，史太莱合金，司太立钴铬钨耐磨硬质合金
stem bank 主干堤岸
stem bearing 叶柄轴承
stem dike 干堤
stem packing assembly 阀杆密封组件
stem flow 沿茎水流
stem-guided 阀杆导向的
stem guide 阀杆导向
stem leakage 阀杆泄漏
stemless nozzle 无柄水帽
stemming material 炮泥材料
stemming rod 填塞炸药棒
stemming 炮泥，填塞，炸药孔炮泥，装药孔塞柱
stem nut 阀杆螺母【控制阀杆的上下运动】
stem thrust bearing housing 阀杆止推轴承罩
stem thrust bearing 阀杆止推轴承
stem valve 杆阀【截止阀】
stem 传动杆【阀门】，钻杆，杆，柄，茎
stencil 镂花【模板】，蜡纸，用模板印刷，用蜡纸印刷
stenography 速记
step action 位式作用，阶跃作用
step attenuator 步进衰减器
step-back relay 跳返继电器，话终继电器
step bearing 立轴止推轴承，踏板轴承
step block 阶梯试块
step bolt 脚钉
step by step adjusting command 步进调节命令，增量命令
step-by-step approach 逐步逼近法
step-by-step calibration 逐步校验
step-by-step carry 按位进位，逐位进位
step-by-step computation 按步计算
step-by-step control 步进控制，步进调节，步进式控制，逐级控制
step-by-step design 逐步设计
step-by-step dial system 步进拨号制
step-by-step drive 步进传动
step-by-step excitation 逐步激发，逐步励磁，逐级激励
step-by-step feed 间歇进料
step-by-step impulse 步进脉冲
step-by-step method 步进法，逐步求解法，逐步逼近法，逐步测量法
step-by-step motor 步进电动机，步进电机
step-by-step procedure 步进法，按步法，逐步法
step-by-step process 逐步逼近求解过程
step-by-step regulation 逐级调节
step-by-step relay 步进继电器
step-by-step seam welding 步进缝焊
step-by-step selection 步进选择
step-by-step selector 步进式选择器，步进式选线器
step-by-step simulation 步进模拟，逐步模拟
step-by-step solution 逐步求解法
step-by-step spot welding 步进点焊
step by step switch 进步式开关
step-by-step system 步进制系统
step-by-step telegraph 步进式电报
step-by-step test 逐级试验
step-by-step transmitter 步进式传送器，步进式发送机
step-by-step variable gear 多级变速齿轮
step-by-step 步进的，逐步的
step change in reactivity 反应性阶跃变化
step change in wall thickness 壁厚阶跃变化，断面变化
step change reactivity 阶跃变化反应性
step change 有级变速，阶跃变化
step coil 分步线圈，阶梯线圈
step cone 宝塔轮，级轮
step connection 分级连接
step controller 分级控制器
step control table 分级控制表，分段控制表
step control 分级控制，分步控制，步进控制
step core 阶梯形铁芯
step counter 计步器【控制棒】，步进式计数器
step crushing 分级破碎
step current 阶跃电流
step delay 分级延迟
step down gear 减速器，降速齿轮
step down ratio （变压器）降压比
step down side （变压器）低压侧
step down station transformer 低压厂用变压器，降压厂用变压器

step down substation 降压变电所，降压变电站
step down transformation 降压
step down transformer 降压变压器
step down turn ratio 降压匝比【变压器】
step down 降压，下降，降低，减慢，减速
step drawdown test 多级降深抽水试验
step flight 链板
step-forced response 阶跃扰动响应
step-forced 阶跃扰动的
step forward relay 步进继电器
step forward 步进
step function response 阶跃函数响应
step function signal 阶跃函数信号
step function 阶跃函数，阶梯函数
step generator 步进发电机，阶梯信号发生器
step grate 阶梯式炉篦
step hydroelectric power station 梯级水电站
step in attenuation 衰减的差距
step induction regulator 步进感应电压调整器
step input signal 阶跃输入信号
step input 阶跃输入，阶梯输入
step insertion of reactivity 阶跃式引入反应性
step joint 齿槽接头，齿式接合
step junction 阶跃结
step ladder 踏步梯，梯子
step length 步长
stepless control 无级控制，连续控制，无级调节，均匀调节，连续调节
stepless regulation 平滑调节，无级调节
stepless speed change device 无级变速装置
stepless speed control 平滑调速
stepless variable drive 无级变速传动
stepless voltage regulator 无级调压器
stepless 无级的，不分级的
steplike-distribution wave 阶梯形波
step-like joint 阶梯接头
steplike slot 阶梯形槽
step line-voltage change 分档调整线电压，分级线电压变动
step load acceptance 带阶跃负荷
step load change 分级负载变动，分档调整负荷，负荷阶跃变化，阶跃负荷变化
step load 梯级负荷
step measuring 步测
step meter rate 级差电费制
step mode of control 分段调节法
step motion electric drive 步进电气传动
stepmotor 步进式电动机，步进电动机，步进马达
step multiplier 步进式乘法器，阶梯式乘法器
step-on-coil energization 步进式线圈激励
step-on-deenergization 步进式断电，步进式去激励
step-out locking circuit 失步镇定电路
step-out relay 失步继电器，解列继电器
step out 失步，解列，失调
stepped aging process 分级时效工艺
stepped aging 阶段时效，分段时效
stepped charging method 分段充电法
stepped coil 多抽头线圈
stepped column 阶形柱
stepped concrete foundation 混凝土台阶式基础
stepped control 分级控制
stepped curve distance-time protection 阶梯时限距离保护
stepped design 分阶段设计
stepped footing 阶梯形基础
stepped foundation 阶梯式基础
stepped labyrinth gland 高低齿迷宫式汽封
stepped layout 阶梯式布置
stepped pulley 宝塔式滑轮
stepped retaining wall 阶式挡土墙
stepped slot 阶梯形槽
stepped spillway 阶梯式溢洪道
stepped start stop system 分段启停系统
stepped tommy bar 分节旋棒
stepped voltage regulation 分级电压调整
stepped voltage 步进电压
stepped wall 台阶岸壁
stepped wave 梯形波
stepped winding 阶梯绕组，抽头绕组，多头线圈
stepped 阶梯的，步进的，分级的，分阶段的，阶梯式
stepper motor 步进电机，步进电动机，步进马达
stepper piston 步进活塞
stepper 步进电动机，分档器，分节器
steppe 草原，大草原，干草原
stepping control 分级控制
stepping core 阶梯形铁芯
stepping counter 步进式计数器，级进式计数器
stepping distance 步距
stepping-mode 步进方式
stepping motor 步进电动机，步进电机，步进马达
stepping movement 步进位移
stepping piston 步进活塞
stepping rate 步进速率
stepping register 步进式寄存器
stepping relay 步进继电器
stepping sequence 步进程序
stepping switch type relay 步进开关式继电器
stepping switch 步进开关，分档开关
stepping technique 步进法
stepping type relay 分段动作继电器，步进型继电器
stepping 步进式的，使成梯级【如控制棒驱动机构或电缆端】
step position information 步位置信息
step potential 跨步电压，梯级势能
step power change 阶跃功率变化
step pulse 阶跃脉冲
step quenching 分级淬火
step-rate prepayment meter 分级预付费用表
step recovery diode 阶跃恢复二极管
step response curve 阶跃响应曲线
step response time 阶跃响应时间
step response 阶跃响应，阶跃特性，瞬态响应，瞬态特性
step-servo system 步进伺服系统
step signal 阶跃信号
step size 步长
step slot 阶梯形槽

step-switch converter 步进式变换器，步进式转换器
step-switch 分段开关，步进开关，步进的
steps 踏步，台阶
step-taper pile 分段锥形桩
step tracking 步进跟踪
step transformer 升降压变压器
step-type counter 步进式计数器
step-type image quality indicator 阶段型影像质量指示计
step-type IQI 阶段型像质计，阶段型透光计
step-type penetrameter 阶段型穿透计，阶段型透光计
step-type recording wave gauge 分挡式波高仪
step-type voltage regulator 分级式电压调整器
step-up error 设置误差，加速误差
step-up excitation 升压励磁
step-up frequency changer 升频变换器
step up gearbox 增速齿轮箱
step-up gear 增速齿轮，增速传动装置，增速装置
step-up instrument 无零位仪表
step-up pulse transformer 升压脉冲变压器
step-up ratio （变压器）升压比，增速比，升压比率
step-up side （变压器）高压侧
step-up substation 升压变电所
step-up switchyard 升压站
step-up transformer 升压变压器
step-up 升压，升高，加速，增速
step voltage regulator 分级调压器
step voltage test 逐级加压试验
step voltage 跨步电压，阶跃电压
step waveform 阶跃波形
step wearing ring 多级磨损环
step winding 阶梯形绕组，多段绕组，多抽头绕组
stepwise change in power 功率阶跃变化
stepwise cut-out control 逐步切出控制，分段断路器控制
stepwise dissociation 逐步离解，分级离解
stepwise disturbance 逐级干扰
stepwise impulse 阶式脉冲
stepwise refinement 逐步精化
stepwise regeneration 分步再生
stepwise regression 逐步回归
stepwise relaxation 逐级松弛
stepwise repositioning 阶跃式位移，阶跃式移动
step 级，阶，步，阶段，步骤，阶跃，楼梯踏步，踏步，台阶，梯级
steradian 球面度【一种立体角度单位】，立体弧度
sterad 球面度【一种立体角度单位】，立体弧度
stereobate 无柱底基
stereocamera 立体摄影机
stereocartograph 立体测图仪
stereocompara-graph 立体坐标测图仪
stereo display 立体显示
stereogrammetriq method 立体测量法
stereogram 实体图，立体地图，立体照相
stereographic grid 立体格网
stereographic projection 球面投影
stereograph 立体照片
stereology 体视学
stereo-measurement 立体量测
stereometer 立体量测仪，视差测图镜，体积计
stereomethod 立体测量法
stereometrical map 立体测绘地形图，立体测量图
stereometry 立体测量学
stereo microscope 立体显微镜，体视显微镜
stereomotor 一种永磁转子电动机
stereo pair 立体对象
stereophotogrammetric survey 立体摄影测量
stereophotogrammetry 立体摄影测量
stereophotography 立体摄影
stereoplotter 立体测图仪
stereoplotting instrument 立体测图仪器
stereoscan examination 立体扫描鉴定
stereoscope 立体镜，立体显微镜
stereoscopic device 立体观测装置
stereoscopic measurement 立体量测
stereoscopic method 立体成像法
stereotelemeter 立体测距仪
stereotheodolite 立体经纬仪，体视经纬仪
stereotopographic map 立体测绘地形图
stereotyped command 标准指令
stereotyped 老一套的，标准的，固定不变的
stereotype 老框框，旧规矩，固定形式，铅版
stereo wave 立体波
sterile water 无菌水
sterilization 杀菌，消毒
sterilizer 消毒器
sterilize 消毒，灭菌
sterilizing room 消毒室
stern sea 尾浪
stern wave 尾波
stern 尾部
stethoscope 听诊器
stevedore charge 装卸费，码头装卸费用
stevedoring 装卸，装卸工作
1st extraction 一段抽汽
stffening piece 加劲件
STG＝stage 级
STG(steam turbine generator) 汽轮发电机（组），三大主机
stickability 黏着能力
stick circuit 吸持电路，自保电路
stick electrode 焊条
stickiness 黏性，黏着，黏附
sticking coefficient 黏着系数
sticking of contacts 触点烧结
sticking potential 饱和电位，极限电位
sticking voltage 粘着电压
sticking 卡住（阀），黏的，黏住，黏附，黏着，滞附，黏滞，黏结
stick insulator 棒式绝缘子
stick-on 粘贴式
stickout 伸出
stick powder 炸药筒
stick relay 自保持继电器，吸持继电器
stick-slip flow 黏性滑移流动

stick valve spool 斗杆阀阀芯
stick winding 棒绕组
sticky ash 黏性灰
sticky dust 黏性粉尘
sticky limit 黏韧度，黏着界限，黏限
sticky material 黏性物料
sticky matter 黏性物质
sticky paper 黏性纸
sticky particle 黏性颗粒
sticky point 黏点
sticky 胶黏的，黏性的，黏稠的，黏的
stick 杆，棒，手柄，操纵杆，粘贴，胶带，棒状炸药
stiction 静摩擦，静摩擦力，静态阻力
stiff building 刚性建筑
stiff clay 硬黏土
stiff concrete 稠混凝土
stiff consistency concrete 干硬性混凝土
stiff consistency 硬稠性
stiff design 硬件设计
stiff dimple 刚性纹
stiff driven train 硬传动系统
stiffened girder 经加固的大梁
stiffened micropile 劲性微型桩
stiffened panel 加劲板，加强板
stiffened sheet 加强板
stiffener plate 加劲板，补强板
stiffener ring 环形加强肋板
stiffener 刚性筋，加强板，加劲板，硬化剂，加肋板，补强板，刚性元件，加劲角钢，加劲件，支肋
stiffening angle 加劲角钢，加强角铁
stiffening band 刚性筋，刚性带
stiffening effect 加强效果
stiffening girder 加强梁
stiffening piece 加固件
stiffening plate 加强板，补强板，加劲板
stiffening rib 加强圈，加强筋，加劲肋
stiffening ring 加强圈
stiffen 硬化
stiff fissured clay 硬裂缝黏土
stiff flow 黏滞流动
stiff frame 刚架
stiff girder connection 梁的刚节点
stiff joint 刚性接头
stiffleg derrick 刚腿人字起重机，刚性柱架起重机
stiff mix 干硬性拌和物
stiff mud brick 硬泥砖
stiffener 加强件，加固件
stiffness coefficient of fulcrum bearing 支点刚度系数
stiffness coefficient 韧性系数，劲度系数，刚度系数
stiffness constant 刚度常数
stiffness degradation 刚度降低
stiffness driven oscillation 刚度驱动振动
stiffness factor 劲度因数，劲度系数，刚度系数，刚劲因数
stiffness index 刚度指数
stiffness matrix 刚度矩阵，劲度矩阵
stiffness method 刚度法
stiffness modulus 劲度模量
stiffness of cable 电缆的抗挠性
stiffness of system 系统稳定度
stiffness ratio 刚度比
stiffness reactance 劲度力抗
stiffness tester 刚性度试验器，劲度试验器
stiffness 刚性，韧性，硬度，刚度，劲度，稳定性，抗扰性，反电容，逆电容
stiff penalty 严厉的处罚
stiff rotor 刚性转子
stiff-shaft turbine 刚性轴汽轮机
stiff-shaft 刚性轴
stiff soil 硬土
stiff stability 强稳定性
stiff tower 刚性塔架
stiff transmission system 硬传动系统
stiff 刚性的，硬的，坚硬的，黏的
stile 门边木，窗边木，门竖木，竖框
still air conditions 静止空气环境
still air cooling 静气冷却，自然冷却
still air sag 无风弧垂
still air tension 无风张力
still air 静止空气，无风
stiller 消力池
still gas 蒸馏气体
stilling basin of spillway 溢洪道消力池
stilling basin 消力池，消力塘
stilling box 消力箱
stilling cavern 消力孔洞
stilling chamber 消涡室
stilling device 防波装置
stilling pond 消力塘，消力池，静水池
stilling storage 静库容
still photography 静物摄影术
Still's equation 司蒂尔方程
Stillson wrench 斯蒂尔森扳手，管子钳，可调管扳手
still tide 平潮
still water level 静水位
still water navigation 平水航运
still water region 死水区
still water 死水
still 静止的，蒸馏，蒸馏器，蒸馏装置，仍然，更，静止地
stilted arch 上心拱
stilted vault 上心拱
stilt 支材
stimulated emission 诱导发射
stimulated transition frequence 受激跃迁频率
stimulating economy 搞活经济
stimulation tower test 模拟塔试验
stimulator 激励器，激活剂，激活器，激流器，激励装置
stimulus-response technique 激发响应技术
stimulus 激发
sting balance 尾撑天平
sting mounted model 尾撑模型
sting 尾撑，支杆
stipulated damages 规定的违约赔偿金
stipulated in the contract 在合同中予以规定的

stipulated limit　规定的限值
stipulated norm　（合同）规定的定额
stipulation　规定，契约，约定，条文
stirred bed　扰动床【沸腾炉】
stirred fluidized bed　搅动流化床
stirred tank reactor　搅拌槽反应器
stirrer blade　搅拌器叶片
stirrer　搅拌器，搅拌机，搅棒，混合器，拌和器
stirring apparatus　搅拌装置
stirring filter　搅拌过滤器
stirring mechanism　搅拌机构
stirring mixer　搅拌混合器
stirring motion　湍流，涡流，紊流运动
stirring screw conveyor　螺旋拌和输送机
stirring tank　搅拌箱
stirrup-ties　箍缀筋
stirrup wire　箍钢丝
stirrup　钢筋箍【混凝土钢筋】，镫形铁件，镫形夹，钢箍，箍筋
STIR(subject to immediate reply)　立即回答生效，须经立即回答才生效
stir　搅动，搅拌，摇动
stitch bonding　自动点焊，跳焊，针脚式接合
stitch welding　合缝焊接
stitch weld　缝焊
stitch　绑结，缝合，压合
STM=steam　蒸汽
stmt support　（风洞）支架，支柱架
STM　蒸汽【DCS 画面】
stochastical　偶然的，随机的，推测的
stochastic analysis　随机分析
stochastic approach　随机方法
stochastic continuity　随机连续性
stochastic control system　随机控制系统
stochastic differential equation　随机微分方程
stochastic disturbance　随机扰动
stochastic effect　随机性效应
stochastic event　随机事件
stochastic finite automaton　随机有限自动机
stochastic grammar　随机文法
stochastic heating　随机加热
stochastic hydraulics　随机水力学
stochastic hydrological model　随机性水文模型
stochastic hydrology　随机水文学
stochastic loading　随机载荷
stochastic modeling　随机建模
stochastic model　随机模型，随机模式
stochastic nature　随机性
stochastic network　随机网络
stochastic noise　随机噪声
stochastic point reactor kinetic equations　随机点反应堆动力学方程
stochastic process　随机过程
stochastic pushdown automaton　随机下推自动机
stochastic sampling　随机采样，随机抽样
stochastic simulation　随机模拟
stochastic stability boundary　随机稳定边界
stochastic system　随机系统
stochastic variable　随机变量，无规变量
stochastic vibration　随机振动
stochastic wind load　随机风载
stochastic　偶然的，随机的
stockade groin　栏栅丁坝
stockade groyne　栏栅丁坝
stockade　围桩
stock brick　普通砖
stockbridge damper　防震锤，（线路导线）减震器
stock clerk　货存管理员
stock company　股份公司
stock controller　库房管理者
stock dividend　股票股利，股息票
stock equity securities　股权证券
stocker　堆垛机，贮料机
stock fluctuation　储量变化
stock holders equity　股东权益
stockholder　股东，股票持有人
stockholding　库存量
stocking area　堆放场地
stocking out tower　堆煤塔
stock interest　股权
stock item　库存品
stock management　库存管理
stockman　仓库管理员
stockpile area　堆料场
stockpile base　贮料堆底面
stockpile　贮存，（储）料堆，煤堆，堆，（原料等）储备，贮备，库存
stockpiling equipment　储存，堆放设备
stockpiling exception　储存例外
stockpiling facility　堆料设备
stockpiling　堆料，囤积，贮存
stock premium　股票盈价
stock program　仓库规划
stock record card　货存记录卡
stock room　储藏室，库房
stock size　常规尺寸
stock solution　储备液
stock taking　盘存
stock water development　畜牧水源开发
stock water　畜牧用水
stockyard　堆料场，贮料场，货场，煤场
stock　台，贮藏物，毛坯，备料，备品，成品库，原料，存货，股份，股票，库存，坯料，树干，常备的，存货的，陈旧的
stoichiometric calculation　化学计算
stoichiometric coefficient　理论系数，化学当量系数
stoichiometric combustion　按理论空气量燃烧
stoichiometric composition　化学计量组分
stoichiometric equation　化学计算方程式
stoichiometric mixture　按化学反应比例混合的混合物，理论混合物
stoichiometric number of B　B 的化学计量数
stoichiometric ratio　化学计算比，化学计算当量比
stoichiometric relation　化学计量关系
stoichiometric　化学计量的，化学配比的，化学计算的，化学当量的
stoichiometry　化学计量法，化学计算法，化学计量学

stoikiometric ratio　化学当量比
stoke hole control　炉内工况监察
stoke hole　加煤孔，拨火孔
stoker air compartment　（炉排下的）风室
stoker coal　加煤机用煤，炉排煤
stoker combustion rate　炉排燃烧热强度
stoker feed boiler　链条炉
stoker feed　机械加煤
stoker-fired boiler　机械化加煤炉排炉，炉排锅炉
stoker-fired grate　机械炉排
stoker-fired plant　炉排炉电厂
stoker firing　机械化炉排燃烧
stoker grate　加煤机炉篦
stoker rating　炉排热强度
stoker side cooling box　炉排防焦箱
stoker surface　炉排面积
stoker　（翻转）炉排，司炉，层燃炉，机械加煤机，给煤机，自动加煤机
Stokes flow　斯托克斯流
Stokes formula　斯托克斯公式
Stokes law　斯托克斯定律
Stokes number　斯托克斯数
Stokes regime　斯托克斯流态
Stokes stream function　斯托克斯流函数
Stokes wave　斯托克斯波
stoke　斯【运动黏度单位】，烧火，添煤
stomatal transpiration　气孔散发
stoma　小孔
Stone and Webster Standard Plant　斯通韦伯斯特标准（压水堆）核电厂
Stone and Webster Standard Safety Analysis Report　斯通韦伯斯特标准安全分析报告
stone apron　抛石护坦
stone axe　凿石斧
stone ballast　石碴，碎石
stone-block groin　条石丁坝
stone-block groyne　条石丁坝
stone-block pavement　石块铺路面，石块路面
stone bolt　底脚螺栓
stone breakwater　石防波堤
stone bridge　石桥
stone carving　石雕
stone chip　石屑
Stone circuit　斯通电路，抗流圈式电路
stone coal　块状无烟煤，硬煤，石煤
stone column　碎石桩
stone concrete　块石混凝土
stone construction　石构造
stone crusher　碎石机
stone crushing plant　碎石厂
stone curb　石路牙，石井栏，路缘石
stone dam　石坝
stone desert　戈壁，石漠
stone dike　砌石堤，石堤
stone drainage ditch　石棚排水沟
stone drainage　填石排水
stone drain　砌石排水沟，石砌排水沟
stone drill　钻岩机
stone dust　石屑
stone dyke　砌石堤
stone embankment　石堤
stone-faced masonry　石饰面圬工
stone-filed trench　填石排水沟
stone-filled crib dam　填石木框坝
stone-filled timber crib groin　木笼填石丁坝
stone-filled timber crib groyne　木笼填石丁坝
stone-filled trench　填石暗沟
stone-filled works　堆石工程
stone filling　填石
stone fire resister　砾石阻火器
stone forest　石林
stone foundation　石基，石基础
stone fragment　石片
stone generator　车轴驱动恒流发电机
stone insulator　石式粗陶低压绝缘子
stone jetty　石防波堤，石突堤式码头
stone levee　石堤
stone lifting bolt　吊石栓
stone-like coal　石煤
stone lined　块石衬砌的
stone line　碎石带
stone mansonry　石砌体
stone marker　标石
stone masonry construction　石砌结构
stone masonry dam　砌石坝
stone masonry　砌石工程，砌石建筑，石圬工，石砌体，砌石
stonemesh apron　钢丝网填石沉排
stonemesh groyne　钢丝网填石丁坝
stonemesh mattress　钢丝网填石沉排
stone mill　磨石机
stone mulch　石幕
stone pavement　铺石路面，石块铺面，石块路面
stone picker　选石机
stone-picking machine　除石机，选石机
stone pitching surface　砌石护面，砌石护坡
stone pit　采石场，石坑
stone pockets of concrete　混凝土蜂窝状气孔
stone quarry　采石场
stone riprap　乱砌石块
stone roll　石辊
stone sculpture　石雕
stone slab　石板
stone steps　石砌台阶，石级
stone structure　石结构
stone surfacing　石护面
stone transmission bridge　抗流圈式馈电电路
stone wall　料石墙，石壁，石墙，石垣
stoneware pipeline　缸瓷管道
stoneware　瓷器，石器，石制品，陶器
stone weir　堆石坝，堆石堰，石堰
stoneworks　砌石工程，石方工程，石方，石砌体，石圬工
stoney gate　提升式平板闸门
stone　石，砌石，石材，石料
stoning　碎石护岸
stony soil　含石土，石质土
stony stream　石流
stool plate　垫板
stool　托架，托座，座驾【管道支撑】，凳子，

粪便
stoop 门廊
stop-and-go valve 起停阀
stop bank 堤岸
stop-belt sampling 停带取样
stop BFBP 停锅炉给水前置泵
stop bit 停止位
stop block 限位块
stop bushing 止动衬套
stop butterfly valve 截流蝶阀
stop button 停机按钮，停止按钮，制动按钮
stop calculation 停止计算，停止计算指令
stop check valve 切断式止回阀，断流单向阀
stopcock buret 活塞式滴定管
stopcock 管塞，旋塞，活栓
stop command 停止命令
stop-continue register 停续寄存器
stop device 限位装置，止动装置
stop dog 止动器
stoper 向上式凿岩机
stope 采矿场
stop-gap measure 权宜之计，临时措施
stop gate 快速闸门
stop impulse 停住信号脉冲
stoping 顶蚀作用
stop instruction 停止指令，停机指令
stop joint 嵌固接头，电缆接头，塞止接头
stop-leak compound 密封剂
stop log gate slot 迭梁闸门槽
stoplog gate 叠梁门，叠梁闸门
stoplog groove 叠梁门槽
stoplog hoisting equipment 叠梁吊装设备
stop log weir 迭梁堰
stop log 插板，叠梁闸门，叠梁，闸板
stop loop 停止循环，循环停止
stop operation and undergo shake-up 停产整顿
stoppage of work 停工
stoppage period 停工期，停工时期，故障周期
stoppage 闭塞，停止，中止，停机，跳闸，脱扣，断路，阻断，阻塞
stop pawl 制动爪
stop payment notice 止付通知
stop payment 停（止支）付
stopped condition 停车工况
stopper drill 套筒式凿岩机
stopper 限位器，制动器，阻进器，挡圈，塞子
stopping condenser 隔直流电容器
stopping distance 停车滑行距离
stopping equivalent （带电粒子穿过物体时）能量损失当量
stopping potential 遏止电势，遏止电位
stopping power 防辐照能力【辐射】
stopping sight distance 停车视距
stopping signal 停止信号
stopping switch 停机开关
stop-pin 制动销
stop plank 插板，止流板
stop position 停止状态，停止位置
stop pulse 停止脉冲
stop relay 静止继电器
stop ring 定位环，止推环
stop screw 止动螺钉，制动螺旋
stop signal 停车信号，终止信号，停闭信号
stop sign 停止标志，停止信号，停车牌
stop spindle 止动杆
stop standby 停止备用
stop-start unit 启停部件，起止单元
stop statement 停机语句，停止语句
stop-valve pressure 主汽门压力
stop valve 截止阀，关断阀，节流阀，断流阀，停汽阀，停止阀，主汽阀，切断阀
stop watch 停表，秒表
stop wind speed 停止风速
stop with guide spring 带定位弹簧的止挡块
stop work order 停工命令
stop 停车，停止器，停机，停止，阻止，截止，制动器，止挡器，挡板，车挡，止挡块，行程限位器
storage access control 存储存取控制
storage access unit 存储存取部件
storage access 存储器存取
storage address register 存储地址寄存器
storage address 存储器地址
storage allocation 存储器分配，存储器配置，存储分配，库容分配
storage and retrieval system 存储和检索系统
storage and retrieval 存储及检索
storage area 存储区，贮存区，储存场地，蓄水面积
storage assembly 储存组件
storage atmosphere 贮存气氛
storage balance 蓄水量平衡，蓄水平衡
storage barns system 缝式贮存系统
storage barn 贮仓
storage basin 蓄水池
storage battery DC system 蓄电池直流系统
storage battery discharging 蓄电池放电
storage battery locomotive 电瓶机车
storage battery 储能电池，蓄电池，蓄电池组
storage bay 蓄潮湾，蓄水湾
storage bin fire 煤仓着火
storage bin 贮仓，贮料仓，贮料罐
storage block 存储块，存储器部件
storage buffer 存储器缓冲器
storage bunker type coal pulverizing system 贮仓式燃煤制粉系统
storage bunker 贮料仓，贮煤柜
storage burn system 贮槽系统
storage bus in 存储器总线输入
storage bus out 存储器总线输出
storage canal 贮存沟，贮存井
storage canister 贮存密闭容器
storage capacitor 存储电容器
storage capacity curve 库容曲线，贮水量曲线
storage capacity of drainagebasin 流域蓄水量
storage capacity of reservoir 水库库容
storage capacity of watershed 流域的蓄水容量，流域蓄水量
storage capacity 蓄热容【太阳能光热利用】，贮存能力，调蓄能力，存储容量，蓄电池容量，蓄水容量，蓄水库容，存储器容量，积聚电容，库容，储量

storage cask 贮存罐，贮存容器
storage cell 存储单元，蓄电池
storage change 蓄水量变化
storage charge 仓储费，存储费，保管费
storage circuit 存储电路
storage coefficient 存储系数【线圈，电路】，库容系数
storage compacting 存储紧密化
storage compartment for strongly active components 高放射性部件贮存库
storage compartment for weakly active components 低放射性部件贮存库
storage compartment 存储单元，储料室，蓄电池
storage condition 贮存条件
storage connecting circuit 存储连接电路
storage constant for aquifer 含水层蓄水常数
storage content 储存内容，存储量
storage counter 中间结果存储计数器
storage curve 储量曲线，库容曲线，蓄量曲线
storage cycle period 存储周期，最大等待时间
storage cycle time 存储周期时间，存储周期
storage cycle 存储周期，蓄水周期
storage data register 存储数据寄存器
storage decoder 存储译码器
storage density 存储密度，贮存密度
storage device 存储器件，存储设备，存储装置，储能装置
storage diode 存储二极管
storage-discharge curve 库容流量曲线，槽蓄曲线，蓄量出流曲线
storage distribution element 存储分配部件
storage draft curve 蓄泄曲线
storage dump 信息转贮，存储（器）转储，存储器内容打印
storage element 存储元件，蓄电池
storage-elevation curve 高程库容关系曲线，高程-库容曲线
storage energy curve 蓄能曲线
storage environment 贮存环境
storage equation 储量方程，蓄（水）量方程
storage equipment 贮存设备
storage excavation ratio 蓄水量与挖方量之比
storage facilities 拦蓄设施，储存设备，贮藏室，贮藏装置，贮存设施
storage factor 存储因数，品质因数，库容系数
storage fee 保管费，仓储费
storage fill 存储器填充
storage firing system 仓储制燃烧系统
storage for decay 衰变贮存
storage generator 蓄能发电机
storage grid 贮存格架
storage head-end plant 乏燃料处理设施首端车间
storage hierarchy 存储器分级体系，存储体系
storage hopper 储料斗
storage house 仓库
storage-in bus （存储器的）输入总线
storage instruction 保管说明
storage integrator 存储积分器，存储用积分器，记忆用积分器
storage inventory 库存量
storage irrigation system 蓄水灌溉系统
storage key 存储关键字
storage level 水库水位，蓄水位
storage lifetime 贮存期限
storage life 贮存期限，存储寿命，存储时间
storage location 存储位置，贮存位置，存储单元，存储器地址
storage loss 储存损失
storage mark 存储标志
storage materials 储能材料
storage matrix 存储矩阵
storage means 存储方法
storage medium 存储介质，存储媒介，存储体，信息存储体，存储媒体
storage mesh 存储网
storage of stability 稳定储备系数
storage of surplus water 过剩水量的贮存，剩余水量的蓄存
storage operation 蓄水运行
storage oscillograph 存储示波器
storage-out bus （存储器的）输出总线
storage-outflow curve 蓄水量与泄水量的关系曲线，蓄泄曲线
storage parity 有奇偶校验的存储器
storage path 贮存通路
storage percentage 蓄水程度
storage period 蓄水期，贮藏期限
storage pile activator 储料堆活化器
storage pile discharger 贮料堆排料器
storage pile 储料堆，煤堆
storage pit （放射性物质）贮藏窖，贮存坑，贮存槽，贮槽，贮水坑，堆存坑
storage place 贮料场
storage pond 贮存池
storage position detector 贮存位置探测器【螺孔检查】
storage position 贮存架位置，贮存场所
storage power plant 蓄能电站，蓄能电厂
storage power station 蓄能电站
storage protect feature 存储保护特性
storage protection 存储保护
storage pump 蓄能水泵，蓄能泵
storage rack 贮藏架，贮存架
storage rain gauge 蓄水式雨量计
storage ratio 调蓄率
storage receiver 贮存容器
storage register 存储寄存器
storage regulator 调节池
storage relay 存储继电器
storage requirement 需要存储量，需要库容
storage reservoir 蓄水库
storage restoration 重新蓄水
storage ripple 存储器脉动
storage room 贮藏室，贮藏间
storage routing in natural channel 天然河槽的槽蓄演算
storage routing 调洪计算，库容演算
storage scan 存储器扫描，存储单元检查
storage schema 数据存储图
storage selection circuit 存储器选择电路

storage shed 库棚
storage shelter 贮存棚
storage silo 贮料筒仓，储料仓，储料罐，贮仓
storage slip 仓单
storage space 储料场，存储空间
storage stability 贮存稳定性
storage stack 存储体，储煤堆，一组存储器，存储堆栈
storage stand 贮藏架
storage station 蓄能电厂，蓄能电站，蓄能水电站
storage surface 储存表面，存储屏蔽
storage system 存储系统，仓储式制粉系统，蓄能系统
storage tank 贮存槽，贮存箱，储料罐，储箱，贮罐，储槽
storage time 存储时间，蓄水时间
storage tube 存储管
storage-type heat exchanger 蓄热式热交换器
storage unit 存储器，存储装置
storage vault 贮藏室，贮存库
storage vessel 储存容器，贮水箱
storage volume 储存量，蓄水量，蓄量
storage water 储存水，拦蓄水
storage well 贮存井
storage yard 露天堆场，贮煤场，堆场
storage zone 存放区
storage 存储，存储器，仓库，积聚，蓄电，库存，储备，贮藏，保管，贮藏库，贮量
storascope 存储式同步示波器
storatron 存储管
store and clear 存储和清除
store and forward 信息转接，存储转发
store conveyer 煤场用输送机
stored address 被存地址
store data buffer 存储数据缓冲器
stored count 累积计数
stored data 存储数据，内存数据
stored energy system of wind power 风能发电储能系统
stored energy 蓄能，储能，存储的能量，贮存能量
stored heat 贮存热量，存储热量，蓄热，储热
stored magnetic energy 存储的磁能
stored pressure fire extinguisher 贮压式灭火器
stored program computer 存储程序计算机
stored program control 存储程序控制，存储程序控制器
stored program element 存储程序元件
stored program logic 存储程序逻辑
stored program system 存储程序系统
stored program 存储程序，内存程序
stored response 存储响应
stored routine 存储程序
stored word 存储字
store for next year 为下年度存储
storehouse for dangerous articles 危险品库
storehouse 仓库，库房，堆栈
store index in address 地址存储指针，地址的存储变址
storekeeper 保管员，仓库管理员
store location 存储地址
storeman 保管人员，仓库管理员，仓库工人，店主
store register 存储寄存器
store room 贮藏室
storer 存储器
storey height 楼层高
storey 层
store 保存，存储，料场，商店，仓库
storied house 楼房
storing device 存储装置
storing of information 信息存储
storing solution 储备液
storm and tempest insurance 暴风雨保险
storm axis 暴雨轴线，风暴轴
storm berme 风暴浪滩肩
storm centre 风暴中心
storm cloud 风暴云，雷雨云
storm cone 风暴信号
storm decay index 暴雨衰减系数
storm door 防风暴门
storm drainage system 雨水排放系统
storm drum 风暴风量筒
storm duration 风暴期，暴雨历时
storm eye 风暴眼
storm front 风暴锋
storm gale 暴风【十一级风】，强烈风
storminess 风暴度
storm intensity 暴雨强度，风暴强度
storm lane 风暴路径
storm loading （架空线路的）暴风负荷
storm maximization 暴雨极大化，暴雨放大
storm model 风暴模式
storm path 风暴路径
storm pavement 防暴雨护面
storm-proof 防暴风雨的，耐风暴的
storm rainfall 暴雨量，风暴降雨
storm region 风暴区
storm runoff 暴雨径流，雨洪径流
storm-sewage system 暴雨污水合流系统
storm sewerage 暴雨水下水道工程
storm sewer 雨水道，雨水管
storm signal 风暴信号
storm surge forecast 风暴潮预报
storm surge protection breakwater 风暴潮防波堤
storm surge 风暴潮，风暴大浪，风暴汹涌
storm's wake 风暴尾迹
storm-swept 受风暴破坏的
storm tide 风暴波浪，风暴潮
storm track 风暴路径
storm valve 排水口止回阀
storm warning 风暴警报
storm water inlet 雨水进口
storm water overflow 暴雨溢流
storm water system 暴雨排水系统，雨水（排放）系统
storm wave 风暴波，风暴浪，风暴潮
stormy weather 风暴天气
storm 暴风雨，狂风，风暴，十级风
story grate 倾斜炉排
story 楼层

stove bolt 炉用螺栓，槽头螺栓，埋头螺栓，（丝扣到根的）短螺栓，粗制螺栓
stove coal 火炉无烟煤
stove heating 火炉采暖
stove 窑，炉，暖房
stoving varnish 烘漆
stowage fee 理舱费
stowage space 有效空间，装载空间
stowing 理仓
stow motor 可变磁阻变速电动机
STPD＝stopped 已停止
STPG＝stopping 停止的
STP＝stop 停止
straddle inverted T-root 外包倒T形叶根
straddle scaffold 跨立式脚手架
straddle-type root 外包式叶根
straddle wrench 叉形扳手
straggling parameter 偏差参数，误差参数
straight abutment 无翼桥台
straight amplification 直接放大
straight asphalt 纯沥青
straight beam contact unit 接触法直探头
straight beam examination 直探头检验
straight beam faced unit 带保护面的直探头
straight beam inspection 直射法探伤
straight beam method 直射法超声探伤，直射法
straight beam probe 直探头
straight beam technique 直射法，直射波技术
straight beam transducer 直探头
straight beam ultrasonic examination 纵向超声波探伤
straight bevel gear 水平伞齿轮，直齿锥齿轮
straight binary 标准二进制，普通二进制
straight bituminous filler 纯沥青填塞料
straight bit 一字形钻头
straight blade 直叶片，等截面叶片
straight blow 直击强风
straight B/L 收货人记名提单
straight brace 直拉条
straight cable 直线型钢丝束
straight cement mortar 纯水泥灰浆
straight cement 纯水泥
straight centrifugal separation 纯离心分离
straight check valve 水平单向阀
straight circuit 直通线路，直通回路
straight coastline 平直岸线
straight coast 平直岸
straight-compound cycle 平行双轴循环
straight condensing turbine 纯凝汽式汽轮机
straight conical draught tube 直锥形尾水管
straight connector 水平接头
straight cross 等径四通（管）
straight cycle 直接蒸汽循环，纯蒸汽循环
straight dozer 直铲推土机
straight drop spillway 直落式溢洪道
straight dynamite 纯炸药
straightedge 直尺
straight embedment of anchorage 钢筋直端锚固【无弯钩】
straightener blade 整流静叶片
straight energy rate 单一制电价
straightener 拉直机，校直器，矫正器，（风洞）整流段，（风洞风扇）止旋片
straightening cliff 平直崖壁
straightening fixture 矫直夹具
straightening machine 整直机
straightening vane 整流翼，整流导叶，导流叶片
straightening 整直，变形矫正，矫直，校正（使）变直，把……弄直，直线折旧法
straight flight stair 直跑楼梯
straight flow combustion 直流式燃烧室，直流型燃烧器
straight flow pump 贯流泵，直流泵，轴流泵
straight flow system 直流系统
straight flow valve 顺流阀，直流式阀
straight forward experiment 直接实验
straight forward phenomenon 直观现象
straight forward 顺向
straight globe valve 水平球阀
straight grained wood 直纹木材
straight hammer 直锤
straight idler 平托辊
straight joint 无分支连接，直线接头，直线接合
straight line capacitor 直线性可变电容器，电容标度正比电容器
straight line characteristic 直线特性
straight line code 直接式程序，无循环程序
straight line coding 直接式程序编制，无循环程序编制
straight line commutation 直线换向
straight line depreciation 直线折旧法
straight line distribution 直线分布
straight line graticule 直角坐标网
straight line interpolation 直线内插法
straight line motion screen 直线运动筛
straight line relationship 直线关系
straight line section 直线段
straight line support deprecated 直线杆塔
straight line tower 直线塔
straight multiple 直接复接
straight natural circulation operation 直接自然循环运行
straightness 直线度，平直度
straight one-way valve 直通式单向阀
straight-on starter 直接启动器
straight-on starting 直接启动
straight-pattern snip 直型剪
straight pin porcelain insulator 直脚瓷瓶
straight pin （绝缘子）直脚钉，圆柱销
straight pipe heat loss 直线管道热损失
straight pipe thread 直管螺纹
straight pipe 直管
straight polarity direct current 直流正接
straight polarity electrode 正极性电极
straight polarity 正极性【电极】，正接
straight pressure atomizer 简单压力式雾化器
straight quay 直线岸壁式码头，直线岸壁
straight radial blade 径向平叶片
straight reach 直线河段
straight regeneration 直流再生，顺流再生
straight roller 平辊
straight-run bitumen 直馏沥青

straight section 直线段
straight-shank bit 直柄钻头
straight shank socket wrench 直柄套筒扳手
straight shunt-wound motor 直并激电动机
straight-slope conical silo 直坡锥体筒仓
straight splice （套管的）无分支连接，直接铰接
straight steam cycle 非再热循环，直接蒸汽循环
straight-stem butterfly valve 水平杆蝶阀
straight streamline 直线流线
straight system 高放式，直放式
straight tee 等径三通（管）
straight tenon 直榫
straight-through arrangement 直通布置
straight-through connector 直通接头
straight-through cooling 直流冷却
straight-through current transformer 穿心式电流互感器
straight-through perforated tube silencer 直通穿孔管消声器
straight-through reactor 直通反应堆
straight-through safety valve 直通安全阀
straight-through splicing 直通拼合
straight-through tube 直流式风洞【无回路的】，直通管
straight-through wind tunnel 直流式风洞，开路式风洞
straight-through 直通
straight top 平顶
straight track 铁路直线段
straight tube bundle type heat exchanger 直管束型热交换器
straight tube bundle 直管束
straight tube steam generator 直管式蒸汽发生器
straight tube type heater 直管式加热器
straight tube 直管
straight tuner 直接放大式调谐器
straight vane 水平叶片
straightway check valve 直通单向阀，纵贯单向阀
straight way diaphragm valve 直通式隔膜阀，纵贯式隔膜阀
straight way suction piston-pump 水平吸入活塞泵
straight way test 直接试验
straight way type plug valve 直通式旋塞阀
straight way valve 直流阀，直通阀
straight-way 直通的
straight weld 直焊缝
straight wheel 平面砂轮
straight 直的，连续的，水平的【相对于角】
strain acceleration 应变加速度
strain-age brittleness 应变时效脆性
strain aged embrittlement 应变时效脆性
strain ageing 应变时效
strain aging impact absorbing energy 应变时效冲击吸收功
strain aging impact toughness 应变时效冲击韧度
strain aging sensitivity factor 应变时效敏性系数
strain amplitude 应变幅
strain analysis 应变分析
strain cell 应变盒
strain center 应变中心
strain clamp 耐张线夹
strain component 应变分量
strain concentration factor 应变集中因数
strain concentration 应变集中
strain-controlled load test 应变控制荷载试验
strain control shear apparatus 应变控制剪力仪
strain control 应变控制
strain cracking 应变开裂
strain crack 应变裂缝
strain creep 应变蠕变
strain damp 耐张线夹，耐拉线夹
strain deviation 应变偏移
strain deviator 应变偏差器，应变转向装置
strain due to torsion 扭曲应变
strained angled tower 耐张转角塔
strained bond 多股绞线连接
strain energy density 应变能密度
strain-energy method 应变能法
strain energy of distortion 畸变应变能
strain energy theory 应变能理论
strain energy 应变能，变形能
strainer-and-check valve 滤网和单向阀
strainer cap 滤帽
strainer element 水帽
strainer head 过滤器水头
strainer mesh size 滤网的网格大小，滤网的目径
strainer well 过滤井
strainer 粗滤器，（金属）滤网，滤管，筛滤网，过滤器，筛，筛网，除污器，雨水箅子，松紧扣，拉紧装置，张紧器
strain fatigue 应变疲劳
strain field 应变场
strain figure 应变图形
strain force 变形力
strain gage transducer 应变计传感器
strain gage 变形测量传感器，应变计，应变片
strain gauge for general prupose 通用应变片
strain gauge load cell 应变式称重传感器，应变仪式荷重传感器
strain gauge logger 应变记录器
strain gauge measurement 应变测量技术，应变仪测量
strain gauge method 应变仪法
strain gauge rosette 应变片丛
strain gauge scanner recorder 应变计式多点记录器
strain gauge tensionmeter 应变张力计
strain gauge transducer 应变传感器
strain gauge type pressure transducer 应变仪压力传送器
strain gauge 电阻应变计，电阻丝应变仪，拉力计，应变计，应变规，应变仪，应变片，应力仪
strain gauging 应变测量
strain gradient 应变梯度
strain hardenability 金属材料的应变硬化性，受力变硬性
strain hardened zone 加工硬化区【金工】
strain hardened 变形硬化的

strain hardening coefficient　应变硬化系数
strain hardening exponent　应变硬化指数
strain hardening　形变硬化，冷加工硬化，应变硬化，冷作硬化，冷锻
strain hinge　U形拉板
strain increment　应变增傲
strain indication　应变显示
strain indicator　应变指示器
straining element　抑制元件
straining piece　系杆，拉条，拉杆
strain input　应变输入端
strain installation　应变仪
strain insulator-string　耐张串
strain insulator　耐张绝缘子
strain invariant　应变不变量
strainless ring　无张力环
strain limit　应变极限
strain matrix　应变矩阵
strain measurement　应变测量
strain measuring device　应变测定仪
strain measuring instrument　应变仪
strain meter　应变计，应变仪，应变测定仪，
strain-optical constant　应变光学常数
strain path　应变过程，应变路径
strain rate　变形速度，应变速率，应变率
strain-recovery characteristics　应变回复特性
strain relief method　应变释放法
strain relief tempering　消除应力回火
strain relief　应变消余，应变消除
strain response　应变响应
strain rosette　应变片花
strain section　耐张段【高架线】
strain sensibility　应变灵敏度
strain sensing type　应变敏感型
strain sensing　应变检测
strain-slip cleavage　应变滑动劈理，错动劈理
strain-slip folding　滑移褶皱，应变滑动褶皱
strain softening　变形软化，变缓和
strain strengthening　形变强化
strain-stress loop　应力应变环形曲线
strain tensor　应变张量
strain threshold　应变极限
strain-time curve　应变时间曲线
strain tower of double circuit transmission　双回路输电耐张塔
strain tower　耐张（铁）塔
strain transducer　应变传感器
strain vector　应变向量
strain wave　应变波
strain　应变，拉紧，张力，粗滤，变形，延伸
strait　海峡
strake　层，段，侧板，铁箍，箍条，条纹，破风圈
strand conductor　绞线
strand deposit　海滨沉积物
stranded cable　多股绞合电缆，股绞缆索
stranded conductor　多股线，绞线
stranded copper cable　铜绞线电缆
stranded copper　编织铜线
stranded cost　搁浅成本
stranded galvanized steel wire　镀锌钢绞线
stranded steel wire　钢绞线
stranded wire　单股钢丝绳，多股绞合线，绞合线，绞线
stranded　多股的
strander　绞线机，合股机
strand flat　潮间坪
strandflat　海滨滩
stranding effect　（导线）绞合效应
stranding factor　绞线系数
stranding harbour　浅水港
stranding machine　绞线机
stranding　搁浅，绞合
strand insulation　单股导线绝缘，单股绞线绝缘
strand line　岸边线，岸线，滨线
strand plain　潮间平原
strands of cotton yam　面纱簇
strand　股，股线，绞线，细索，绳股，绞，绞合线，导线束
strangler　节流门
strap bolt　扁尾螺栓，带形螺栓
strap brake　带闸
strap clamp　带状卡
strap coil　铜带绕制的线圈
strap conductor　扁线
strap hinge　带式铰链
strap iron　扁铁
strapped magnetron　腔式磁控管，均压环式磁控管
strapped wall　板条墙
strapping bolt　箍紧螺栓【对分式外壳或气缸】
strapping of a multiplecavity magnetron　多腔磁控管的模式分割
strapping wires　双连开关接线法
strap steel　扁钢，带钢
strap stiffener　加劲窄板
strap style coil　带绕线圈，扁绕线圈
strap support　吊架
S-trap　S形曲颈管
strap-wound coil　扁绕线圈
strap-wound winding　带绕绕组，扁绕绕组
strap wrench　带状夹板
S-trap　虹吸曲颈管
strap　燃料组件夹板【可以刚性的，钢线】，条，带，带片，搭接片，皮带，搭板，垫片
strata alternation　层的交互变化
stratabed　双层床
strata behaviour　岩层特性
strata bridge　岩桥
strata dip　岩层倾角
strata displacement　岩层移动
strata gap　岩层间断
strata group　地层群
strata inclination　地层倾斜度
strata movement　地层运动，岩层移动
strata pinching　地层尖灭，地层变薄
strata profile　岩层剖面
strata section　岩层剖面
strata sequence　层理次序，地层次序，层次，层序【指地层】
strata strike　地层走向
strata succession　层序【指地层】

strata tilting 地层倾倒
strategic function 策略函数
strategic goal 战略目标
strategic partner 战略性伙伴
strategic point 关键点，战略点
strategy （控制）策略
strath 平底谷
stratificated air conditioning 分层空气调节
stratification line 分层线
stratification of wind 风层
stratification plane 层理面，层面
stratification 层，分层，层次，层理，层化，分层现象，成层，成层作用，分层作用，层化作用，带状，分层结构
stratified atmosphere turbulence 分层大气湍流
stratified atmosphere 分层大气
stratified bed 分层床，双层床
stratified coal sample 分层煤样
stratified deposit 成层沉积，成层矿床
stratified drift 成层漂碛
stratified flow 分层流动，层流，分层流
stratified foundation 成层地基，层状地基
stratified inner layer 分层内层
stratified material 层状材料
stratified media 成层介质
stratified pavement 成层路面
stratified random sampling 分层随机取土样
stratified rock 成层岩，层状岩，层岩
stratified sampling 分层取样，分层采样
stratified sand 层夹砂
stratified seam-sample of coal 分层煤样
stratified shear flow 分层剪切流
stratified soil 层状土，分层土，成层土
stratified structure 层状构造
stratified subgrade 成层路基
stratified volcano 层状火山
stratified water 成层水
stratified 层状的，成层的，分层的
stratiform 成层，层状
stratify 分层，层化
stratigrapher 地层学家
stratigraphical break 地层断缺
stratigraphical division 地层划分
stratigraphical profile 地层纵断面
stratigraphical time scale 地层年表
stratigraphical time table 地层年表
stratigraphical 地层的，地层学的
stratigraphic analysis 地层分析
stratigraphic column 地层柱状图
stratigraphic correlation 地层对比
stratigraphic cross-section 地层横剖面
stratigraphic gap 地层间断
stratigraphic geology 地层学
stratigraphic heave 地层平错
stratigraphic hiatus 地层间断
stratigraphic information 地层资料
stratigraphic map 地层图
stratigraphic overlap 地层超覆
stratigraphic section 地层剖面，地层剖面图
stratigraphy （某地区、某国家的）地层情况，区域地层，地层学，地层中的岩石组成
stratit element 钼电阻加热元件，钨电阻加热元件
stratjgraphic unit 地层（分层）单位
stratopause 平流层顶
stratosphere 同温层，平流层
stratospheric cloud 平流层云
stratospheric coupling 平流层耦合作用
stratospheric fallout 平流层落尘
stratospheric ozone 平流层臭氧
stratospheric polar vortex 平流层极涡
stratospheric pollution 平流层污染
stratospheric 平流层
stratovolcano 成层火山
stratum lithology 地层岩性
stratum structure 地层结构
stratum water 层状水，地层水
stratum 岩层，阶层，分层，地层，层次，材料层
straw bag 草包，草袋
straw-earth cofferdam 草土围堰
straw mattress 草垫，草席
straw-mud mortar 草泥灰
straw rope 草绳
straw 稻草，麦秸
stray air-current 离散气流
stray capacitance of winding 绕组的杂散电容
stray capacitance 寄生电容，杂散电容
stray capacity 杂散电容，寄生电容
stray copper loss 杂散铜损
stray coupling 杂散耦合，寄生耦合
stray current corrosion 杂散电流腐蚀
stray current 杂散电流，涡流，漏泄电流
stray electron 杂散电子
stray emission 杂散发射
stray field 杂散磁场，漏磁场
stray flux 杂散磁通，杂散磁通量
stray heat 散失热，漫射热
stray inductance 杂散电感，漏电感
stray load loss 杂散负载损耗
stray load 杂散负荷
stray loss factor 杂散损耗系数
stray loss 杂散损耗
stray magnetic field 杂散磁场，漏磁场
stray parameter 杂散参数，补充参数，随机变量，寄生参数
stray pickup 杂散拾波
stray power 杂散功率
stray radiation 杂散辐射
stray reactance 杂散电抗，寄生电抗
stray resonance 杂散谐振
stray torque 杂散转矩
stray transformer 大漏抗变压器，高漏磁变压器
stray voltage tester 寄生电压测试器
stray voltage 寄生电压
stray wave 杂散波
stray （无线电）干扰，天电，杂散电容，杂散的，寄生的
streak camera 扫描照相机
streak line 流动条纹线，条纹线，脉线，染色线
streak stripe 条纹
streak test 刻痕试验
streak 条痕，条纹

stream abstraction 夺流
stream anchor 尾锚
stream angle 气流（偏）角，流线角
stream bank erosion control 河岸防冲
stream bank erosion 河岸冲蚀
stream bank stability 河岸稳定
stream bed intake 河底取水口
stream borne material 河流挟带物
stream centre line 水流轴线，中泓线
stream channel erosion 河槽冲刷
stream channel pattern 水系类型
stream channel slope 河道比降，河道坡降
stream condition 水流情况，气流状态
stream current 海洋急流
stream deflector 折流设施
stream degradation 河道冲刷
stream deposit 河流沉积
stream dimension 水流尺度
stream drainage pattern 河网类型，水系类型
stream enclosure 地下输水道
streamer theory of gas discharge 流注放电理论
streamer 测风带，（测风）飘，彩色纸带，饰带，射束
stream filament 流丝，流线，水流线
stream flow data compilation 汇编流量资料
stream flow data 流量资料
stream flow depletion 河川径流消退，退水
stream flow forecasting 流量预报
stream flow formation factor 径流形成因素
stream flow gauging 水文观测，流速及流量测量
stream flow hydrograph 流量过程线
stream flow measurement station 流量站
stream flow record 测流记录，流量记录，河道流量记录
stream flow regulation 径流调节，流量调节
stream flow separation 水流分离
stream flow 流动，射流
stream form 流型
stream function 流函数
stream gauging operation 水文测验作用
stream gauging section 水文测验断面
stream gauging station 测流站，流量站，水文测量站
stream gauging 测流
stream-head 河源
stream hydrometric station 河流水文站
streaming effect 流束效应，中子流效应
streaming gravity flow 层状重力流
streaming-potential 水流势能
streamlet 溪
streamline angle 流线偏角
streamline body 流线体
streamline coincident 流线重合
streamline contour 流线型外廓
streamline curvature effect 流线弯曲效应
streamlined blade 流线型叶片
streamlined body 流线体
streamline diagram 流线图
streamlined insert 流线型插入件
streamlined nosing 流线型头部整流罩
streamlined production 流水作业生产
streamlined spillway surface 流线型溢流面
streamlined weight 流线型重锤
streamlined 流线型的，层流的
streamline fairing 流线型罩
streamline flow motion ，层流，平流
streamline flow resistance 流线型流动阻力
streamline flow 层流，线流，平滑绕流，流线型流
streamline form 流线型
streamline motion 流线型运动
streamline pattern 流线型
streamline picture 流线图
streamline profile 流线型轮廓
streamline section 流线型截面
streamline shape 流线形，流线型
streamline space 流线间隔
streamline squeezing effect 流线汇集效应
streamline theory 一维流动理论，流线理论
streamline 流（通量）线，流线型，流线型的，制成流线型，精简（机构），使简化
streamlining effect 流线效果
stream load 水流泥沙
stream of electrons 电子流，电子束
stream over airfoil 翼型绕流
stream pattern 流线谱
stream penetrating 喷入射流
stream piracy 溪流截夺
stream power 河流功率，水流功率
stream pressure probe 水流压力传感器
stream purification 河水净化
stream reaeration 河流再曝气
stream runoff 河川径流
stream segment 河道分支，河流分支
stream set 流向
stream sinking river 伏河
stream's self-purification 河流自净能力
stream surface 流面
stream swirl 涡流，气流涡流
stream temperature 流体温度，气流温度
stream time 工作周期，连续开工时间
stream traction 水流挟带作用
stream tube 流管
stream turbulence 气流湍流（度）
stream velocity fluctuation 流速脉动
stream velocity 水流速度，流速
streamwise component 流向分量
streamwise flow condition 沿程水流特征
streamwise pressure gradient 流向压力梯度
streamwise section 水流沿程剖面
streamwise vorticity 流向涡量
streamwise 沿流动方向的，沿气流方向，顺流的
streamy 流动的
stream 流，作业线，气流，溪流，溪，水流
street dust 街道尘埃
street lamp 路灯
street level wind 街道风
street level 街道路面标高
street lighting and signal system 路灯和交通信号系统
street-lighting transformer 街道照明变压器，路灯变压器

street main　沿街干线
street manhole　街道检查井
street　街道
strength across grain　逆纹强度
strength calculation　强度计算
strength contour graph　等强度线图
strength decrease　强度递减
strength-deformation characteristics　强度变形特性
strengthening mechanism　强化机制
strengthening of metal　金属强化
strengthening　补强，加固
strength envelope　强度包络线，强度包线
strength factor　强度系数，强度因数
strength in undisturbed state　未扰动状态强度
strength limit state　强度极限状态
strength limit　强度极限
strength of brick　砖标号
strength of current　电流强度
strength of ebb interval　最大落潮流间隙
strength of electric field　电场强度
strength of field　场强
strength of flood interval　最大涨潮流间隙
strength of magnetic field　磁场强度
strength of materials　材料力学
strength of rain　雨力
strength of shell　磁壳强度
strength of solution　溶液浓度
strength of timber　木材强度
strength parallel to grain　顺纹强度
strength parameter　（材料）强度参数
strength ratio　强度比
strength requirement　强度要求
strength retrogression　强度衰退
strength safety coefficient　强度安全系数
strength test　强度试验
strength theory　强度理论
strength under shock　冲击强度
strength-weight ratio　强度重量比
strength weld　承载焊缝
strength　强度，力，含量【化学】
stress accommodation　拉紧装置，张紧夹具
stress acoustic constant　声应力常数
stress alternation　应力交变
stress amplitude　应力变幅，应力幅
stress analysis　应力分析
stress analyzing error　应力分析误差
stress boundary condition　应力边界条件
stress calculation　应力计算
stress carrying covering　应力蒙皮，受力蒙皮
stress carrying member　受力材，受力件
stress chart　应力表，应力谱
stress checking for non-operation condition　冷态应力验算
stress circle　应力圆
stress coat brittle lacquer　应力涂层脆性漆
stress-compensation beam　应力补偿梁
stress-compensation factor　应力补偿系数
stress-compensation slab　应力补偿板
stress-compensation　应力补偿
stress component　应力分量
stress concentration factor　应力集中系数，应力集中因子
stress concentration　应力集中
stress condition　应力状态
stress-controlled load test　应力控制荷载试验
stress-controlled test　应力控制试验
stress controller　应力控制器
stress-control shear apparatus　应力控制剪力仪
stress-control tubing　应力控制套管
stress-control winding　可调脉冲电压绕组
stress corrosion cracking　应力腐蚀裂纹，应力腐蚀裂缝，应力腐蚀破裂
stress corrosion fracture　应力腐蚀断裂
stress corrosion index　应力腐蚀指数
stress corrosion rupture　应力腐蚀断裂
stress corrosion　应力腐蚀
stress crack　应力裂纹
stress curve　应力曲线
stress cycle diagram　应力循环图
stress cycle　应力循环，应力周期
stress deformation diagram　应力应变图
stress deformation　应力变形
stress deprivation　应力消失
stress deviation　应力偏量，应力偏移
stress diagram　应力图
stress difference　应力差
stress dispersion　应力扩散
stress distribution pattern　应力分布图
stress distribution　应力分布
stress due to torsional vibration　扭振应力
stressed covering　受力蒙皮
stressed part　承力部件
stressed skin construction　薄壳建筑
stressed skin structure　承力蒙皮结构
stressed skin　应力蒙皮
stressed　受应力的
stress ellipse　应力椭圆
stress ellipsoid　应力椭圆面
stress engineer　应力工程师
stress envelope　应力包络线
stress equation　应力方程
stress equilibrium　应力平衡
stresses combination　应力组合
stresses due to friction　摩擦应力
stress evaluation　应力评值
stress factor　应力因子
stress field　应力场
stress fluctuation　应力脉动
stress-free corrosion　无应力腐蚀
stress freezing　应力冻结
stress function　应力函数
stress gradient　应力梯度
stress history　应力历程
stress increment　应力增量
stress in earth crust　地壳应力
stressing cable　预应力钢缆，加应力钢缆
stressing gallery　预应力走廊，加应力走廊
stressing　施加应力
stress intensification factor　应力增大系数
stress intensity factor　应力强度系数，应力强度因子
stress intensity　应力强度

stress invariant 应力不变量
stress locus 应力轨迹线
stress matrix 应力矩阵
stress maximum 应力最大值，应力峰值
stress measurement 应力测量
stressmeter 应力计
stress-number curve 持久曲线，应力周数图，疲劳曲线
stress-number of cycles 应力循环次数
stress of conductor 导线应力
stressometer 应力计
stress optical constant 光应力常数，光弹性常数
stress-orientation line 应力方向线
stress path 应力路径，应力过程
stress pattern 应力条纹图，应力图样，应力型式
stress peak 应力高峰
stress-penetration curve 应力-贯入曲线
stress-raiser 应力集中源
stress range reduction factor 应力范围减小系数
stress range 应力变化范围，应力变化幅度
stress ratio 应力比
stress redistribution 应力重分布
stress relaxation curve 应力松弛曲线
stress relaxation rate 应力松弛速度
stress relaxation test 应力松弛试验
stress relaxation under constant load 恒定荷载下应力松弛
stress relaxation 消除应力，应力松弛
stress release channel 应力解除槽
stress release 应力解除，应力消除
stress relief annealing 消除应力退火，去应力退火，消除内应力
stress relief blasting 应力解除爆破
stress relief core 应力解除岩心
stress relief cracking 消除应力裂缝
stress relief heat treatment （消）除应力热处理
stress relief method 卸载法
stress relief treatment 消除应力处理
stress relief tubing 应力消除套管
stress relief 消除应力，应力释放，应力解除，应力消除
stress relieved type 消除应力型
stress relieved 消除应力的，应力消除的，应力消失的
stress relieving heat treatment 消除应力热处理，退火
stress relieving 去应力退火，消除应力【热处理】，低温回火，应力消除
stress report 应力计算报告
stress reversal 应力反向
stress rupture limit 持久强极限
stress rupture notch sensitivity factor 持久缺口敏感系数
stress rupture plasticity 持久塑性
stress rupture test 应力破坏试验，持久强度试验，持久试验
stress rupture 应力断裂
stress sensitivity 应力灵敏度
stress sheet 应力表
stress solid 应力体
stress space 应力空间
stress state chart 力学状态图
stress state 应力状态
stress-strain creep curve 应力应变蠕变曲线
stress-strain curve 应力应变曲线，应力形变曲线
stress-strain diagram 应力应变图
stress-strain loop 应力应变回线
stress-strain modulus 应力应变模量，形变模量
stress-strain ratio 应力应变比
stress-strain relationship 应力应变关系
stress superposition 应力叠加
stress system 应力体系，应力系统
stress table 应力表
stress tensor 应力张量
stress-to-rupture test 应力与断裂关系试验，断裂应力试验，蠕变断裂试验
stress to rupture 断裂应力
stress trajectories 应力轨迹线
stress transfer 应力传递
stress transmission 应力传递
stress variation 应力变化
stress wave 应力波
stress 应力，重要性，压力，强调
stretchability 拉伸性
stretched pulse 展宽的脉冲
stretcher bond 顺砖，顺砖砌合
stretcher course 露侧层，平砌（砖），顺砖层
stretcher strain line 滑移流线【冶金】
stretcher strain 滑移流线【冶金】
stretcher 伸张器，延伸器，张紧器，拉紧设备，展宽器，露侧砖，拉伸机，拉紧机，伸长器，薄板矫直机
stretching apparatus 张拉设备
stretching bed 张拉台
stretching device 拉紧装置，张拉装置
stretching factor 展宽倍数，伸长（展开）因数
stretching insulator 耐拉绝缘子
stretching mode 变速模式
stretching screw 扩张螺丝，拉紧螺丝
stretching strain 拉伸应变
stretching stress 拉伸应力
stretching vibration 伸缩振动
stretching wire for prestressed concrete 预应力混凝土用张拉钢丝
stretching wire 张拉钢丝
stretching 拉伸，张拉
stretch operation 延长运行
stretch out and draw back 伸缩
stretch-out contract 期限拖长的合同
stretch-out cycle （元件）延长寿期
stretch-out operation 延长运行，延长周期运行
stretch-out phase 延长阶段
stretch out 推迟卸料期【电站】，降功率以延长换料周期，加深燃耗，延长
stretch thrust 引伸冲断层
stretch 拉伸，伸长，伸展，延伸，展宽，段
stria depth 割纹深度
striated bedrock 有擦痕的基岩
striated discharge 成条纹的放电
striated marble 条纹大理石

striated rock surface （岩石的）擦痕面
striation 线状，条纹，光条，擦痕
stria 擦痕
strickle 刮平，刮平器
strickling strikling 斗刮
strict alignment 严格对中
strictest 最严格的
strict inspection 严格检查
strict liability 严格（赔偿）责任
strictly forbid 严禁
strictly prohabit 严禁
strict standard 严格标准
stride scale 步测比例尺
striding level 跨水准（器）
strike board 刮板【用于刮平混凝土面】
strike fault 走向断层
strike joint 走向节理
strike line 走向线
strike off 勾销，删除
strike of rock strata 岩层裂隙走向
striker pin fuse 撞针式熔丝
striker plate （车钩的）冲击座
striker 冲击器
strike shift 走向变位
strike slip fault 平移断层，走向滑动断层，走向滑断层
strike slip 走向滑距
strike valley 走向谷
strike 打击，撞击，触发，闪击，引弧，放电
striking arc （电焊中）引弧
striking current 起弧电流，击穿电流，起弧距离，击距
striking distance 放电距离，击穿距离，
striking end of an electrode 电焊条引弧端
striking formwork 拆除模板
striking job 拆模作业
striking of arc 起弧，闪弧
striking off lines 勾划线
striking point （熔丝）熔断点
striking potential 起弧电位
striking the arc 起弧
striking time 拆模时间
striking voltage 起弧电压，引弧电压，引燃电压
striking winding 起弧线圈，引燃线圈
striking 起弧，引燃，触发，放电，拆除，拆模，引人注目的，显著的
string bead 窄焊缝
string board 楼梯斜梁侧板
string break 串断点
string chart 架线图表，架线垂度
string constant 串常数
string course 腰线，束带层，带状层
string efficiency （绝缘子的）串效率
string electrometer 弦线电流计
stringent requirement 严格要求
stringer bead technique 窄焊道工艺
stringer bead 窄焊缝，线状焊道，线状焊缝
stringer bracing 纵梁联，纵梁支撑，纵梁斜撑
stringer corrosion 线状腐蚀
stringer vein 细脉
stringer wall 梯墙
stringer 楼梯斜梁，燃料柱，纵梁，斜梁，连接杆
string galvanometer 弦线电流计，弦线检流计
stringing block 紧线滑车
stringing of conductor 紧导线，把导线拉紧，把导线拉直
stringing section 紧线段
stringing temperature 紧线气温
stringing 架线
string insulator unit 绝缘子串元件
string manipulation language 串处理语言
string manipulation 串处理
string measurement 绳测
string of block pieces 葡萄串
string of cars 一列车
string of insulators 绝缘子串
string of mixed-bed filters 混合床过滤器管系
string oscillograph 弦线式示波器
string pendulum 线摆
string polygon 索线多边形，索多边形
string processing language 串处理语言
string process system 字符串处理系统
string-wire concrete 钢弦混凝土
string wire 钢弦
string 串，绝缘子串，架线，绳，细绳，细索，线，带子，弦，行，列，排，把……拉直，把……拉紧，把……串起，把……排成一列
strip-area 带形地区
strip chart recorder 条纸记录器，纸带记录器，长图式记录仪
strip coat 可剥离层
strip column 反萃取柱，反萃取塔
strip conductor 带状导线，扁导线
strip copper 铜带，铜条
strip electrode 带极，板极，焊条
stripe of paint 油漆条纹
strip-extraction 反萃取
strip feed 反萃取剂，反萃剂进液
strip footing 条形基础
strip form 条形模板
strip foundation 条形基础
strip fuse 片状保险器，片状熔丝
strip intercepting 条带截取法
strip iodine 除碘
strip iron 带钢
striplight lamp 顶灯，带形照明器天幕灯
striplight 带形照明器，长条状灯
stripline connector 带状线接头
strip load 条形荷载
strip mining 露天开采
strip of fuses 熔线排
strip of keys 按钮片，电建排
strip of tags 接线条组，端子板
strippable coating 可剥离层
strippable film paint 可剥离的薄膜涂料
stripped fraction 贫化部分
stripped joint 露出的接头
stripped magnetron 耦腔磁控管
stripped plasma 完全电离的等离子体
stripped solution 贫化溶液，已用于反萃取的溶液

stripped surface 表层剥除的地面
stripper column 脱气柱【沸水堆废水处理】
stripper 冲孔模板，剥离器，汽提塔，剥绝缘器，贫化器，刨土机，反萃取器
stripping area 开箱区【新燃料组件】
stripping attachment 拆卸器
stripping column 洗提塔
stripping depth 剥离深度
stripping factor 脱气系数
stripping line 清基线
stripping machine （燃料元件盒）脱盒机，起模机，剥皮机
stripping of the boundary layer 边界层剥离
stripping operation 拆模作业
stripping section 汽提段，洗提段，反萃段，贫化段，抽取段
stripping shovel 表土层剥离机铲
stripping time 拆模时间
stripping topsoil 剥除表土
stripping workshop 反萃溶液，解吸溶液
stripping 剥离，拆模，脱模，起模，汽提，洗涤，反萃，拔出
strip radiant panel 带状辐射板
strip recorder 自动展开记录仪，条形记录仪
strip steel 带钢，扁钢
strip strain gauge 条形应变片
strip surfacing 带极堆焊
strip-suspension type 悬片式，轴尖支撑式（仪表）
strip terminal 片接头
strip theory 切片理论
strip tillage 带状整地
strip-type magnetron 耦腔磁控管
strip winding 扁线绕组，带绕绕组
strip-wound armature 扁线绕组电枢，带绕组电枢
strip-wound core 卷铁芯
strip-wound pressure vessel 绕带（钢）压力容器
strip-wound 带绕的，扁线绕的
strip 带，条，条带，压缝条，片，剥，带钢，焊条，地带，剥去，分解，拆卸（端子）
striving goal 奋斗目标
STRNR＝strainer 滤器
strobe line 选通脉冲线路
strobe pulse 选通脉冲，读取脉冲
strobe 闸，闸门，选通脉冲，选通的，读取脉冲
strobing pulse 选通脉冲
strobodynamic balancing machine 频闪观测平衡机
strobodynamic 频闪观测的
stroboscope polarimeter 频闪观测仪式偏振计
stroboscope 频闪观测仪，频闪观测器
stroboscopic image 频闪观测仪像
stroboscopic instrumentation 频闪观测仪式检测仪表
stroboscopic instrument 频闪式仪表
stroboscopic light 高速闪光灯
stroboscopic meter disc 电度表的频闪测圆盘
stroboscopic tachometer 频闪转速计
stroboscopic 频闪观测的
strobotac 闪频转速计，频闪测速计
strobotron circuit 门电路，选通电路，频闪管电路
strobotron 频闪放电管，有控制栅的冷阴极充气管
strobo 频闪观测器，闪光放电管，闪光仪
stroke adjuster 冲程调整器
stroke-bore ratio 冲程内径比，冲程缸径比
stroke current 冲击电流
stroke-incidence rate 雷击事故率
stroke limiter 冲程限制器，行程限制器
stroker 炉排
strokes law 冲量定律
stroke time 行程时间【开启或关闭】
stroke 行程，冲程【活塞】，动程，回程，返行程，冲击
stromwater tank 雨水池
strong acid ion exchange resin 强酸性离子交换树脂
strong acid 强酸
strong add cation exchange resin 强酸阳离子交换树脂
strong add cation exchanger 强酸阳离子交换器
strong aqua ammonia 浓氨水
strong-back （焊接）定位板
strong base anion exchanger 强碱性阴离子交换器
strong base anionic resin 强碱性阴离子交换树脂
strong base group 强碱基，强碱团
strong base 强碱
strong box 保险箱
strong breeze 强风，六级风
strong caustic anion exchange resin 强碱阴离子交换树脂
strong caustic anion exchanger 强碱阴离子交换器
strong coal 硬煤
strong current control 强电控制
strong current 强电流，大电流
strong earthquake 强地震，强震
strong fibre 强力纤维
strong gale 烈风，九级风
strong grid 强电网
strong liquor 浓溶液
strongly absorbing 强吸收
strongly acidic cation-exchange resin 强酸型阳离子交换树脂
strongly acidic hydrogen cation exchange 强酸氢离子交换
strongly acid 强酸
strongly active components store 高放射性部件贮存库
strongly basic anion exchange 强碱阴离子交换
strongly basic 强碱性的
strongly coupled system 强耦合系统
strongly soluble salt test 易溶盐试验
strong-motion seismograph 强动地震仪，强震仪
strong point 优点
strong room 保险库，保险室
strong sewage 浓污水

strong signal 强信号
strong solution 浓溶液
strong solvent 强溶剂，活性溶剂
strong source 强源
strong thunderstorm region 强雷区
strong tide river mouth 强潮河口
strong vibration 强烈振动
strong-weathered rock 强风化岩
strong wind boundary layer 强风边界层
strong wind fumigation plume 强风下垂型羽流，强风下沉型烟羽
strong wind 大风，强风
strong 强烈的，坚固的
strontium isotope 锶同位素
strontium unit 锶单位
strontium 锶【Sr】
Strouhal frequency 斯特鲁哈频率
Strouhal number 斯特鲁哈数
Strowger two-motion switch 史端桥双动开关
STR＝stator 定子
STR＝strainer 拉紧装置，过滤器
STR(system thermal ratio) 系统热比
STRTD＝started 已启动的
STRTG＝starting 正启动
STRT＝start 启动
struck joint 刮平缝
struck nucleus 被轰击核
struck 卡车
structural alloy steel 合金结构钢
structural analysis by matrix method 矩阵法结构分析
structural analysis 结构分析
structural angle 角钢结构
structural area 结构面积
structural attachment 结构附件
structural bar 结构钢筋
structural basin 构造盆地
structural beam 结构梁
structural buckling 结构屈曲
structural build-up 结构累积
structural calculation 结构计算
structural coefficient 结构系数
structural column 构造柱
structural component 结构部件，构件
structural concrete column 混凝土构造柱
structural concrete 结构混凝土【墙、地板、天花板等】
structural controllability 结构可控性/结构能控性
structural coordination 结构协调
structural crack 结构裂缝
structural damage 结构破坏
structural damping characteristic 结构阻尼特性
structural damping 结构阻尼
structural decomposition 结构分解
structural density 结构密度
structural design 结构设计
structural diagram 结构图，状态图
structural discontinuity 结构的不连续性
structural discordance 构造不整合
structural drawing 构造图
structural dynamic similarity 结构动力相似
structural dynamics 结构动力学
structural element 结构部件，结构构件
structural engineering 结构工程
structural engineer 结构工程师，建筑工程师
structural feature 构造细部，结构特征，构造形迹
structural flexibility 结构柔度
structural formula 结构式
structural form 构造形式
structural frame 结构框架
structural geology 构造地质学
structural girder 构造梁
structural grain refining 晶粒细化处理
structural hardware 构件
structural heat sink 建筑冷源
structural information 结构性信息
structural integrity test 结构完整性实验
structural integrity 构造完整性，结构完整程度
structural joint 构造缝
structural landform 构造地形
structural-load-carrying capacity 结构的承载能力
structural loading 结构荷载
structural matereial 结构材料
structural material rod 结构材料棒
structural material swelling 结构材料肿胀
structural mechanics 结构力学
structural member section 结构构件截面
structural member 结构构件
structural model 结构模型
structural mode 结构模态，结构模式
structural modification 结构改变
structural modulus 结构模量
structural network analysis program 结构网络模拟程序
structural noise 结构噪声
structural observability 结构可观测性，结构能观测性
structural of turbulence 湍流结构
structural oscillation 结构振动
structural parameter 结构参数
structural part 结构部件
structural passability 结构可通性
structural pattern of superstructure 上部结构的结构形式
structural pattern 构造型式
structural petrology 岩石构造学
structural prefabrication shop 铆焊车间
structural reliability 结构可靠性
structural response 结构反应，结构响应
structural return loss 匹配回路衰耗
structural rigidity 结构刚度
structural rolled steel 轧制结构板
structural section 构件型材，构造剖面图
structural shape 结构用钢板，型钢，（热轧）建筑型钢
structural shielding 结构防护
structural shield 结构屏蔽
structural sill 结构梁
structural stability 结构稳定性
structural steel erection 钢结构安装
structural steel frame 钢架，钢结构

structural steel grade　型钢钢号
structural steel member　钢构件
structural steelwork drawing　钢结构图
structural steelworks　钢构件【厂房】
structural steel　结构钢，钢结构
structural stiffness　结构刚度
structural strengthening　结构加强
structural structure part　结构件
structural support　支架
structural test　结构试验
structural transfer function　结构传递函数
structural type　结构型式
structural unit　构造单元
structural viscous damping　结构黏性阻尼
structural wall tile　承重墙用空心砖
structural wall　承重墙
structural weld　结构焊缝
structural　构造的，结构的，结构上的，建筑的
structure analogue　结构模拟
structure analysis　结构分析
structure beyond repair　无法维修的建筑物
structure chord　结构弦
structure closure　构造闭合度
structure columnar　柱状结构
structure contour　构造等高线
structure damage　结构损坏
structured data type　结构数据类型
structure diagram　结构图
structured programming　结构化程序设计，结构程序设计
structure drawing　构造图，结构图
structured shell of particles　颗粒薄壳结构
structure engineering　结构工程
structure failure　结构损坏
structure fire protection　建筑防火
structure flexibility　结构挠性，结构弹性
structure for cooling water　冷却水构筑物
structure gauge for standard gauge railroad　标准轨距铁路建筑限界
structure gauge　建筑限界
structure in space　空间桁架，空间结构
structure line　构造线
structure model　结构模型
structure noise　结构噪声
structure of aggregation　聚集态结构
structure of brickwork　砌筑结构
structure of cooling water　冷却水结构
structure of corona protection　电晕保护结构
structure of pavement　路面结构
structure of rock mass　岩体构造
structure of soil mass　土体结构
structure of turbulence　湍流结构
structure-oriented description and simulation　面向结构的描述和模拟
structure origin　坐标原点
structure oscillation　结构振动
structure perturbation approach　结构摄动法
structure plane　构造面
structure reliability and reliability index　结构可靠度及结构可靠指标
structures and improvement　建筑物及改良工程
structure steel fabrication　钢结构制作
structure steel frame　型钢结构，钢框架
structure steel　结构钢
structure water　构造水
structure　架构，结构，构造，组织，构筑物，建筑物，结构物，装置设备
strum　吸入滤网
strung busbar　柔性母线
strut beam　支梁
strut bracing　支撑，压杆
strut-framed beam　撑架式梁，支撑梁
strut-framed bridge　撑架式桥
strut frame　支撑构架
strutted pole　撑杆
strutting of pole　电杆加固
strutting piece　支撑件
strutting　支撑系统，支撑，撑
strut　支撑（杆），支柱，支架
STR　蒸汽转换系统【核电站系统代码】
STS(static transfer switch)　静态开关，静态转换开关
ST tube　直管
stub angle　插入式基础的短柱角钢
stubby screwdriver　木螺钉起子
stub cable　连接电缆，短截电缆
stub card　计数卡片
stub cutoff wall　防渗矮墙
stub-end feeder　短截馈线
stub end　翻边短节，连杆端
stub feeder　直接电源的馈路
stub for transmission tower　组塔用的插铁【输电线路工程】
stub guy　有拉桩的拉线
stub pipe　短管
stub protection　短引线保护
stub setting template　插铁定位模板
stub shaft　短轴，轴颈
stub stack　短烟囱【设在锅炉顶上】
stub station　高峰负载发电厂
stub tenon　短榫
stub tube　短管，管接头
stub tuner　短线调谐器，短柱调谐器
stub wall　矮墙
stub　短管，短柱，轴端，残端，短截线，短轴，管接头
stucco finished wall　粉刷完的墙
stucco stippling　拉毛粉刷
stucco　粉刷灰泥，毛粉饰，灰墁
stuck rod criterion　卡棒准则
stuck rod margin　卡棒裕度
stuck rod　卡棒
stuck subassembly　卡住的分组件
stud bolt　双头螺栓，柱头螺栓，柱螺栓
studded fin　交错嵌钉式的肋片
studded　焊有销钉的
studding of tubes　管子上加销
studding　装销
stud driving　螺杆拧紧
stud elongation measuring tool　螺栓伸长测量工具
student engineer　见习工程师

stud gap 衔铁撑杆间隙，衔铁传动杆间隙
stud gun 销钉枪
stud hole cleaning machine 螺孔清洁机
studhole inspection 螺孔检查
studhole plug handing tool 螺孔塞装卸工具
studhole plug 螺孔塞
stud nut 柱螺栓螺母
stud partition 隔墙架
stud pre-tensioning 螺栓预拉伸
stud remnant 残存螺杆
stud tensioner 双头螺栓拉伸器
stud terminal 螺栓接线柱，接线柱
stud transistor 柱式晶体三极管
stud tube wall 焊销钉的水冷壁
stud tube 销钉管
stud type chain 挡环式链，日字链
stud welding 螺栓焊接，螺柱焊
study group 研究小组
study on counterguarantee system 反担保制度研究
study on type selection of main auxiliary equipment 主要辅机选型论证
study on type selection of main equipment 主设备选型论证
study 设计，调查，研究，学习
stud 大头螺钉，焊钉，螺杆，栓钉，螺柱，模板支柱，双头螺栓，双端螺栓，柱螺栓，柱，销，中间支柱
stuffing box head cover 填料函（帽）盖
stuffing box housing 填料函套
stuffing box leakage 填料函泄露
stuffing box loss 填料函损失
stuffing box shaft sleeve 填料函轴套
stuffing box sleeve 填料函套
stuffing box 密封垫，填料函，密封压盖，密封袋，填充体，填料盒，填料箱
stuffing gland 填料密封，填料压盖
stuffing 填料，填塞，填塞料
stuffy 窒息的
stuff 材料
stuke 灰墁
stull 顶梁
stump puller 拔根机
stump 残桩
style of work 工作作风
style 格式，式样
styling 造型，样式
stylobate 柱座
stylolitic structure 缝合构造，柱状构造，柱状结构，缝合结构
stylus 尖笔，钢针，记录针，笔尖，笔头，触针
S type rotor(Savonius type rotor) S型转子
styrene base resin 苯乙烯基树脂
styrene-dielectic cable 苯乙烯绝缘电缆
styrene-divinylbenzene beads 苯乙烯二乙烯苯白球
styrene polyvinyl cation resin 苯乙烯-聚乙烯阳离子交换树脂
styrene 苯乙烯
styroflex cable 聚苯乙烯软性绝缘电缆
subacid solution 微酸溶液
subacid 弱酸，弱酸的，微酸的
subadiabatic atmosphere 亚绝热大气
subaerial spring 地表水
subangular panicle 略带棱角的颗粒
subaquatic illumination 水下照明
subaqueous cable 过河电缆，水底电缆
subaqueous concrete 水下混凝土
subaqueous concreting 水下浇筑混凝土
subaqueous corrosion 水下侵蚀
subaqueous foundation 水下基础
subaqueous pump 潜水泵
subaqueous reconnaissance survey 水下勘测
subaqueous rock excavation 水下采掘岩石
subaqueous survey 水下测量
subaqueous tunnel 水下隧洞
sub arch 副拱
subarea 部分断面
subarid 半干旱的
sub-array 子方阵【太阳电池子方阵】
subartesian well 地面下自流井，半自流井
sub-article 子条款
subassembly-can （燃料）组件外壳，元件盒外壳
subassembly levitation （燃料）组件漂浮
subassembly shroud 分组件包壳
subassembly storage location 分组件贮存场所
subassembly 部件，分组件，配件，分部件，分部装配，分部组装件，局部装配，组件，部件装配，辅助装置，子配件，分总成
subatmospheric hydraulic model 负压水工模型
subatmospheric pressure system 负压系统，低于大气压系统
subatmospheric pressure ventilation 负压通风
subatmospheric pressure 真空计压力，负压，真空度，真空
subatomospheric steam 负压蒸汽
subaudio oscillation 亚声波振荡，亚音频振荡
sub-bank 分组
subbase course 底基层，垫层
sub-based drainage system 地基排水系统
sub-basement 下层地下室
subbase 粒料路面底层，底基层，底座
subbasin 子流域
subbed 路基
sub-bituminous coal 次烟煤，亚烟煤
sub-bituminous 劣烟煤的，亚烟煤的，次烟煤的
sub-block 小组信息，数字组
sub-borrower 转借人
sub-bottom profiler 浅地层剖面仪
subbranch 次分支
sub-cadmium neutron 次镉中子，亚镉中子
subcapillary interstice 次毛细孔隙
subcapillary opening 亚毛管孔
subcarrier frequency modulation 副载波调频
subcarrier frequency 副载频
subcarrier wave 副载波
subcarrier 副载波，副载频，调制载波的载波
subcatchment 支流集水区
subcentre 主分支点
subchannel analysis 子通道分析
sub channel （棒束中冷却剂流动）子通道，分

流道，支通道
subchord 副弦杆
sub-circuit 分支电路，辅助电路，支路，分路
subclad 堆覆层下面【裂纹等】
subclass 子类
subclause 小条，子条款
subcode 子码
sub-cofferdam 子围堰
subcommittee 分会，分组委员会
subcommutator 副换向器
sub-company 子公司
sub-conductor spacing 分裂间距
subconductor 再分导线，子导线，次导线，分裂导线中的单导线
subconsequent stream 次顺向河
sub-consultant 咨询分包人
subcontinent 次大陆
subcontract change notice 转包合同更改通知
subcontract for the project 分包工程合同
subcontracting agreement 分承包协议
subcontracting arrangement 分包办法
subcontracting cost 转包成本
subcontract item 转包合同项目
subcontractor 分包商，分承包人，转包商
subcontract package 分包合同包
subcontract 分包，转包，分包合同，转包合同
sub control area 准管理区，准监管区
sub-controlling unit 辅助控制装置
sub control 辅助控制器，辅助控制
subcooled blowdown 欠热排放
subcooled boiling 过冷沸腾，欠热沸腾
subcooled section 欠热段
subcooled surface boiling 欠热表面沸腾
subcooled water flowing in downcomer 下降管内欠热水流
subcooled water 欠热水
subcooled 过冷的，欠热的，过冷
subcooler 欠热冷却器
subcooling boiling 欠热沸腾
subcooling degree 过冷度
subcooling line 欠热管
subcooling margin 过冷浴度
subcooling neutron 欠热中子
subcooling 低温冷却，过度冷却，过冷（却）
subcool state 欠热状态
subcrack 皮下裂纹
subcritical behavior 亚临界工况
subcritical boiler 亚临界锅炉
subcritical boiling 亚临界沸腾
subcritical condition 次临界条件，次临界状态
subcritical crack growth 亚临界裂纹增生
subcritical equilibrium 次临界平衡，亚临界平衡
subcritical experiment 次临界实验，亚临界试验
subcritical facility 次临界装置，亚临界装置
subcritical flow regime 亚临界流动状态
subcritical flow 次临界流，缓流
subcritical fossil fuel power plant 亚临界火电厂
subcritical generating unit 亚临界发电机组
subcritical hardening 低温硬化，亚临界温度淬火
subcritical hot shutdown 热停堆
subcriticality measurement 次临界测量
subcriticality 次临界
subcritical limit 次临界限值
subcritical mass 亚临界质量，次临界质量
subcritical multiplication constant 次临界倍增因子
subcritical multiplication factor 次临界倍增因子，次临界增殖因子
subcritical neutron 次临界条件，次临界状态
subcritical pressure boiler 亚临界压力锅炉
subcritical pressure turbine 亚临界汽轮机，亚临界压力汽轮机
subcritical pressure 亚临界压力
subcritical reactivity 次临界反应性
subcritical reactor 次临界反应堆
subcritical Reynolds number 亚临界雷诺数
subcritical rotor 亚临界转子
subcritical state 次临界状态，次临界系统，亚临界状态
subcritical steam conditions 亚临界蒸汽参数
subcritical steam parameter 亚临界蒸汽参数
subcritical test 次临界试验
subcritical turbine 亚临界压力汽轮机
subcritical vortex shedding 亚临界旋涡脱落
subcritical wind speed 亚临界风速
subcritical 次临界的，亚临界的，临界以下的
sub-cutout 分路断流器
subdevice 子装置
sub-diagonal 副斜杆
subdistribution board 分配电盘
subdivided capacitor 可变电容器
subdivided gap 分割间隙
subdivided transformer 带分接头的变压器
sub-divide 支流分水岭
subdividing truss 再分桁架
subdividing 分小块
subdivisional works 分项工程
subdivision control 分段控制
subdivision zone 再分区
subdivision 细分，再分，再分度，细分度
sub-drainage structure 地下排水构筑物
subdrain 地下排水管，排水暗沟
sub-duct assembly 防回流装置
sub-edge connector 片状插座
subexchange 电话分局
subexciter 副励磁机
subface 底面
subfactor 分因子，子因子
subfeeder 副馈线，分支配电线
subfield subcode 子域子码
subfield 子域，分区
subfill 垫层，垫料
subfloor 粗地板，底层地板，毛地板
subfractional horsepower motor 次分马力电动机，微型分马力电动机
subfractional rating 小分数功率
sub-frame 底架，下支架，副架，分支机架，辅助构架
subfrequency 次谐波频率，副谐频
subfunction group level 子功能组级
subgeostrophic wind 次地转风

sub glacial 冰下的
subgrade capacity 路基承载力
subgrade construction 路基施工，路基建筑，路基工程
subgrade drainage 路基排水
subgrade elevation 路基标高
subgrade foundation soils 地基
subgrade modulus 地基模数，路基模量
subgrade of highway 公路路基，道路基层
subgrade of railway 铁路路基，道路基层
subgrade reaction 路基反力，地基反力
subgrade soil 路基土
subgrade test 路基承载试验
subgrade 路基，地基，掩埋的，地下的
subgradient wind 次梯度风
subgrain boundary 亚晶界
subgrain 亚晶粒
subgrid-scale motion 次网格尺度运动
subgroup function control 子功能组级控制
subgroup signaling equipment 子群信号设备，列信号设备
subgroup 亚类，子组，小群，隶属的小组织
subharmonic current 次谐波电流
subharmonic generator 分谐波发生器
subharmonic oscillation 分谐波振荡
subharmonic oscillator 分谐波振荡器
subharmonic resonance 次谐波共振，次调和共振
subharmonic response 分谐波响应
subharmonic voltage 次谐波电压
subharmonic 次谐波，分谐波，分谐波的，亚谐波
sub-header 分联箱
subheading 细目，小标题
sub HP 次高压【DCS画面】
subhumid climate 半湿润气候
subhumid soil 半湿土
subhumid 半湿润的
subindividual 晶片
subintradosal 内拱砌块
sub-item 分项
subjacent bed 下卧层
subjacent waters 下层水域，下方水域，下伏水域
subject contrast 被摄物衬度
subject heading 议题
subjective definition 主观清晰度
subjective error 主观误差
subjective factor 主观因素
subjective fault 人为性故障，主观性故障
subjective noise meter 主观噪声计
subjective photometer 直观光度计
subjective preference scaling value 主观优选标度值
subjective probability 主观概率
subjective reading 主观读数
subjective test 主观测试
subject matter 主题事项，主要事项，标的物
subject program 源程序
subject to approval 需经批准，须报批，以确认为准，以同意为准
subject to change at short notice 一有通知就可能马上改变
subject to change without notice 可不经通知就进行更改
subject to change 可能发生变化，随时可能变更，容易改变
subject to contract 根据合同，以合同为准
subject to immediate replay 立即回答生效，须经立即回答才生效
subject to our confirmation 须经我方确认，以我方确认为准
subject to prior approval 须经预先核准
subject to ratification 须经批准
subject to 服从于，易遭受，须经……，受……管制，以……为准，以……为条件
subject wave 主体波
subject 标的，主题，主语，受制于……的，服从的
subjob 子作业
sub-joint 副接头，辅助接头
subjunction gate “禁止”门
sublateral 暗沟分支，分支管
sublayer laminar 次层流
sublayer 内层，下层，次层，亚表层，底层，层流边界层
sublease 转租
sublessor 转租人
sublethal amount 亚致死量
sublethal damage 亚致死损伤
sublethal dose 亚致死剂量
sublethal exposure 亚致死照射量，亚致死照射
sublethal level 亚致死水平
sublet 转租
sublevel 次层，次级，次能级，副准位
sub-licence contract 从属许可（证）合同
sub-license 可转让许可
sublimation cooling 升华冷却
sublimation drying 升华干燥
sublimation energy 升华能
sublimation pressure 升华压力
sublimation temperature 升华温度
sublimation 纯化，升华
sublime 升华，纯化，升华的
sub-line 副线，辅助线
subloop 子回路
sublot 小批量
submain sewer 支主管下水道
submain 次主管，地下总管，辅助干线，干线的分支
submanual 分册
submarine advanced reactor prototype 改进型潜艇原型堆
submarine advanced reactor 潜艇用改进型反应堆
submarine armour 潜水服
submarine bank 水下沙洲
submarine bar 水下沙堤
submarine bell signal 潜水钟信号，潜水钟
submarine blasting 水下爆破
submarine cable communication 海底电缆通信
submarine cable laying 海底电缆敷设
submarine cable link line 海底电缆连接线路

submarine cable power transmission line 海底电缆输电线路
submarine cable 海底电缆，水下电缆
submarine camera 水下摄影机
submarine core drilling 水下岩心钻探
submarine detonator 水下雷管
submarine earthquake 海震
submarine illumination 水下照明
submarine installation 水底敷设，水底设施
submarine intermediate reactor 潜艇用中能堆，潜艇用中子反应堆
submarine landslide 水下滑坡
submarine photometer 水下光度计
submarine pipeline 水下管路
submarine prototype reactor 潜艇原型堆
submarine slide 水下塌方
submarine summit 海底峰
submarine thermal reactor 潜艇用中子反应堆，潜艇用热堆
submarine thermometer 水下温度计，海水温度计
submarine transmission cable 海底输电电缆，水下输电电缆
submarine transmission 水下输电，海底输电
submarine volcanism 海底火山作用
submarine volcano 海底火山
submarine works 潜水工作，水底工程
submarine 海底，水下的
submeander 次生河曲
submegawatt wind turbine 次兆瓦级的风电机组
submenu 子菜单，子选项单
submerge-arc welding 埋弧焊
submerged antenna 水下天线
submerged arc welding machine 自动埋弧焊机
submerged arc welding process 埋弧焊工艺
submerged arc welding with wire or strip electrode 用焊丝或焊条的埋弧焊
submerged arc welding 埋弧焊，潜弧焊，水下弧焊
submerged area 淹没地区
submerged bar 暗沙洲，淹没沙洲，水下沙洲
submerged centrifugal pump 潜水离心泵
submerged chain conveyor 捞渣机
submerged coast 沉没岸，淹没岸，下沉岸，溺岸
submerged condition 潜水状态
submerged crib 淹没式取水口
Submerged curved surface 淹没曲面
submerged dam 潜坝
submerged delta 水下三角洲
submerged depth 潜航深度，下潜深度
submerged discharge 淹没出流，淹没泄流
submerged displacement 排出量
submerged exhaust 地下排气
submerged filter 淹没式滤池
submerged float 水下浮标
submerged flow 浸没流动，淹没流，淹没水流
submerged gate 淹没式闸门，水下闸门
submerged glandless motor 潜水无压盖电动机
submerged groin 淹没式丁坝
submerged groyne 淹没式丁坝
submerged-head boiler 立式埋头锅炉
submerged heating 浸渍电加热
submerged hydraulic jump 淹没水跃
submerged intake 淹没式进水口，深孔式进水口
submerged isothermal current 等温潜流
submerged jet 深层排水【排水构筑物】，淹没式射流
submerged jump 淹没式水跃
submerged land mass 淹没陆块
submerged land 淹没地
submerged margin 水下边界，淹没边缘，淹没边界
submerged nappe 淹没水舌
submerged obstacle 水下障碍物
submerged operation 水下操作
submerged orifice 淹没孔口
submerged outcrop 水下露头
submerged outlet 淹没式出水口
submerged pier 潜水墩
submerged pipeline 水下管道
submerged plant 沉水植物
submerged pump 潜水泵
submerged reef type breakwater 暗礁式防波堤
submerged reef 沉没礁
submerged-rotor pump 全密封泵，转子浸没泵
submerged scraper conveyer 埋刮板输送机，水浸式刮板捞渣机
submerged shelf 淹没陆架
submerged sill 淹没式底槛
submerged slag conveyor 埋刮板捞渣机
submerged sluice 潜水闸
submerged speed 潜水航速
submerged spillway 淹没式溢洪道
submerged spur dike 淹没式丁坝
submerged spur 淹没式丁坝，潜丁坝
submerged structure 水下建筑物，水下结构
submerged surface 淹没平面
submerged tube 浸没管道
submerged tunnel 沉埋隧道，水下隧道
submerged-type construction joint 暗式施工缝
submerged-type desuperheater 浸泡减温器
submerged unit weight 水下容重，浮容重
submerged valley 沉没河谷，溺谷
submerged vegetation 沉水植被，水下植物
submerged weight 浮重
submerged weir 潜堰
submerged work chamber 下沉式工作舱
submerged 浸没
submergence bubble testing 浸没式气泡检漏
submergence control （泵的）吸入头高度调节
submergence degree 淹没度
submergence depth 淹没深度
submergence factor 淹没系数
submergence ratio 淹没系数
submergence 插入深度【控制棒】，（卸压系统下降管的）浸没深度，沉没，淹没
submerge 浸没，淹没
submergible drainage pump 潜水式排水泵，淹没式排水泵
submergible motor pump 潜液电泵
submergible roller dam 可淹没式圆辊坝
submerse 淹没

submersible bridge 漫水桥
submersible distribution transformer 潜水柜式配电变压器
submersible drainage pump 潜水式排水泵，淹没式排水泵
submersible electric equipment 潜浸式电气设备
submersible machine 潜水式机器，潜水型电机
submersible motors within oil 潜油电机
submersible motor 潜水电机，潜水型电动机，潜浸式电机
submersible pump for active drain tank 放射性疏水箱潜水泵
submersible pump 潜水泵
submersible research vehicle with nuclear propulsion 核动力水下研究船
submersible sensing element 浸没式敏感元件
submersible sewage pump 潜污泵
submersible transformer 潜水型变压器，浸没式变压器，地下式变压器
submersible 淹没
submersion depth 浸入深度
submersion duration 淹没期
submersion 沉没
submeter 分表，辅助计量
submicron aerosol 亚微米气溶胶
submicro particle 亚微米颗粒
subminiature motor 微型电动机
subminiature relay 超小型继电器
subminiature 超小型（元件）
subminiaturization 超小型化
submission date 提交日期
submission of bid 标书的提交
submission of tenders 标书的提交
submission program 提交程序
submission to arbitration 提请仲裁
submission 投标书的提交
submit an amendment 提出修正案
submit a tender 投标
submittal 提交物
submitting a bid 投标
submit to arbitration 提交仲裁
submit to the superior for approval 报请上级批准
submit 提交，提供
submodulation 副调制
submodulator 副调制器，辅助调制器
sub-module 子模件
sub-mortgage 二次抵押
submultiplexer unit 子多路转换器
submultiple 次倍数，分谐波，次谐波
subnetwork 子网
subnitron 放电管
subnormal frequency 亚正常频率的
subnormal-pressure surface 低于饱和层顶面的压力面
subnormal pressure 低于正常压力
subnormal 次法线，低于正常的
subnumber 附加号
suboptimal control 次优控制
suboptimal excitation system 次最佳化的励磁系统
suboptimality 次优性
suboptimal system 次优系统
suborder 亚纲【土壤】
subordinate body 下属机构
subordinate concept 从属概念
subordinated loan 附属贷款
subordinate load 次要荷载，附加荷载
subordinate station 辅助站
subordinate tide station 辅助观潮站
subordinate unit 下级单位
subordination 下属
suboxide 低氧化物
sub-panel 辅助板，安装板
sub-parallel 辅助并行标记
subpermafrost water 冻土底水
subpile control-rod compartment 反应堆下控制棒小室
subpile room 反应堆底室，堆底间
subpolar low-pressure belt 副极地低压带
subpost 副柱，小柱
subpower range 低于额定功率范围
subpower 部分功率，亚功率
subprogram 子程序
sub-project 子工程，子项目
sub-prompt criticality 近瞬发临界
subprovincial administrative region 专区
subpulse 子脉冲，次脉冲
sub-punch and ream 留量冲孔并扩孔
sub-purlin 小桁条，副檩条
sub-quality product 次品
sub-rack 辅助机架
subresonant 亚共振
subrogation 代位，代位偿清，代位求偿权，代位权，取代，债权移转
subroutine analyzer 子程序分析程序
subroutine call table 子程序调用表
subroutine call 子程序调用
subroutine library tape 子程序库磁带
subroutine library 子程序库
subroutine tape 子程序带
subroutine 分程序，子程序
subsalt 碱式盐
sub-sampling 二次抽样
subsattion automation system 变电站自动化系统
subsattion automation 变电站自动化
subsattion configuration description language 变电站配置描述语言
subsattion configuration description 变电站配置描述
sub-scan 副扫描，辅助扫描
subscribed data 预订数据
subscriber branch line 用户分线
subscriber-busy signal 用户忙信号
subscriber cable 用户电缆
subscriber circuit 用户电路
subscriber equipment 用户设备
subscriber line 用户线
subscriber meter 用户计数器，用户表计
subscriber selector 用户选择器，预选器
subscriber set 用户电话机
subscriber's line 用户专用线
subscriber's loop 用户线路

subscribers meter 用户计数器
subscriber's protector 用户保护装置
subscriber trunk dialing 用户干线拨号
subscriber 订户，用户
subscribe's line 用户线
subscribe store 用户存储器
subscript 下标，脚注，注脚
subsequent call 连续呼叫
subsequent core 后续堆芯
subsequent cost 后续费用
subsequent handling 后续工序
subsequent settlement 附加沉降量
subsequent signal unit 后续信号单元
subsequent valley 后成谷
subsequent 后续的，随后的，后来的
subseries of standardized plants 标准电厂的次分类
subsidability 湿陷性，下陷性
subsidence basin 沉陷盆地
subsidence coast 下沉岸
subsidence flow 下沉流，沉降流
subsidence inversion 下沉逆温
subsidence of ground 地面沉降
subsidence rate 沉陷率，下陷率
subsidence ratio 递减比，阻尼比
subsidence settlement 沉积物
subsidence shoreline 下沉岸线
subsidence temperature inversion 下沉逆温
subsidence transient 衰减过渡过程
subsidence zone 下沉区
subsidence 沉降，下沉，沉陷，沉淀，递减，衰减，降落，坍陷，下陷，消退
subside （土壤）下沉，（水处理）沉淀，下陷，消退
subsidiary body 附属机构
subsidiary building 辅助建筑
subsidiary company 附属公司，子公司，附属公司
subsidiary condition 附加条件，辅助条件
subsidiary conveyer 辅助输送机
subsidiary dam 辅助坝，副坝
subsidiary department 辅助部门
subsidiary drain 排水支沟
subsidiary equation 辅助方程
subsidiary loop 辅助回路，子回路
subsidiary occupation 副业
subsidiary organ 附属机构
subsidiary plant 辅助电站，辅助设备
subsidiary point 分支点
subsidiary right 邻接版权
subsidiary spar 辅助（翼）梁
subsidiary station 辅助电站
subsidiary triangulation 辅助三角测量，小三角测量
subsidiary weir 辅助堰，副堰
subsidiary welfare fund 福利补助金
subsidiary 分公司，子公司，附属机构，辅助的，附属的
subsidization rate 优待电价
subsidization 补助（金），津贴
subsidized loan 贴息贷款
subsidize revenue 补贴收入
subsidy appropriate to particular jobs 特殊岗位津贴
subsidy for health 保健费
subsidy income 补贴收入
subsidy of electric rate 电价补贴
subsidy 津贴，补助金
subsoil condition 地基情况，地基条件
subsoil drainage 底层土壤排水
subsoil drain 地下排水沟
subsoil exploration 底土查勘，地基土勘探
subsoil flow 地下径流
subsoil improvement 土层改良
subsoil test 土工试验
subsoil water movement 地下水运动
subsoil water 底土水，地下水
subsoil 底土，下层土，亚层土，地基土，天然地基，底层土壤
subsonic aerodynamics 亚音速气体动力学
subsonic airfoil 亚音速机翼
subsonic flow field 亚音速流场
subsonic flow 亚声速流，亚音速流
subsonic speed 亚音速
subsonic turbine 亚音速涡轮
subsonic wind tunnel 亚音速风洞
subsonic 亚音速的
subspan galloping 次档距舞动，次跨驰振
subspan oscillation 次档距舞动，次档距振荡，次跨振动
subspan wake-induced galloping 尾流诱导次跨驰振
subspan 次档距，次跨
sub-stage 亚阶【地层】，分期
substance for calculation 计算物质
substance modification of bid 标书实质性修改
substance modification 实质性修改
substance of bid 标书的实质性
substance 物质
substandard cement 水泥次品
substandard instrument 次标准仪器
substandard product 等外品
substantial change 实质性改变［变化］，重大变更
substantial completion 基本竣工，基本完工，大致竣工，大体上已完成
substantial damage 重大损害
substantial derivative 物质微商
substantial detriment 实质性的损害
substantial deviation 严重偏离
substantial information 实际资料
substantiality 实质性
substantial modification 根本性变更，重大变更
substantial 实质的，基本的
substantiation of claims 索赔证明
substantive agreement 实质性协议
substantive article 实质条款
substantive clause 实质条款
substantive issue 实质性问题
substantive provision 实质性规定
substantive session 讨论实质性问题的会议
substantive 本质的，实质的，直接的，实在的，

大量的，巨额的
substation automation system　变电站自动化系统
substation automation　变电站自动化，变电所自动化
substation busbar　变电站母线
substation capacity　变电站容量
substation composite automation　变电站综合自动化
substation control room　变电站控制室
substation department　（电业管理局的）变电处
substation fitting　变电金具
substation framework　变电架构
substation gantry　变电站门型架
substation ground　变电所接地
substation island　电气岛
substation layout　变电站布置
substation master　变电站主单元
substation operating cost　变电站运行费
substation relay room　变电站继电保护室
substation structure　变电站构架
substation transformer　配电变压器
substation　变电站，变电所，主配电设备，开关室，配电站，分站，分所，分台
substitute A instead of B　用 A 代替 B
substitute article　代用品
substituted for B　代替了 B
substitute fuel　代用燃料
substitute material　代用材料，替换材料
substitute M for N　用 M 代替/取代 N
substitute N by(/with) M　用 M 代替 N，用 M 取代 N
substitute　代用品，替代品，替换物，取代
substitution method　代替法，代换法
substitution reaction　取代反应，置换反应
substitution theorem　替代原理【网络学】
substitution　置换，代替，取代，替换，变换作用
substochiometric　化学配比的
sub-strata formation　下层地层构造
substrate feed logic　衬底馈电逻辑
substrate leakage　衬底漏电
substrate material　衬底材料
substrate　衬底，底层，底物，基质，基层，基底，基片，水道底层物质
substratum　底土层，下层，下卧层，下伏地层
substructure work　地下工程
substructure　底部结构，底层结构，地下结构，基础结构，下部结构，下层结构，子结构，亚组织，路基
substrut　副撑
sub-subroutine　次子程序
sub-supplier　分供货厂家
subsurface boring　地下钻探
subsurface checking　深层裂纹，表面下的裂纹
subsurface circulation　次表层环流
subsurface correlation　地层对比
subsurface corrosion　表面下腐蚀
subsurface coverage　地下覆盖段
subsurface current　次表层流
subsurface defect　表面下的缺陷
subsurface detention　地下滞留
subsurface disposal　地下排放
subsurface divide　地下分水界，地下水界，地下分水线
subsurface drainage basin　地下流域，地下水流域
subsurface drainage check　地下排水节制闸
subsurface drainage　地下排水，浅地表排水
subsurface echo sounding　水下回声测探
subsurface erosion　潜蚀，水下侵蚀
subsurface exploration　地下查勘，地下探测
subsurface float　深水浮标，水下浮标
subsurface flow　地下水流，潜流
subsurface geology　地下地质学
subsurface investigation　地下勘探，地下探测
subsurface layer　地下层
subsurface level　次表层水位
subsurface mark　地下标点
subsurface pipe drainage system　暗管排水系统
subsurface pipe drainage　暗管排水
subsurface runoff　表层流，地下径流
subsurface sewage disposal　地下污水处理
subsurface soil condition　地下土质条件
subsurface soil　次表土，亚表土
subsurface structure　地下结构，下层结构
subsurface velocity　水面下流速
subsurface water pollution　地下水污染
subsurface water　次表层水，地下水，底土水
subsurface wave　水下波
subsurface　表面下的，地面下（的），地下（的），地下，水面下，表层下，水面下的，路面下层，下层面
subsynchronous oscillation　亚同步振荡，次同步振荡
subsynchronous reluctance motor　次同步磁阻电动机
subsynchronous resonance　次同步谐振，次同步共振，亚同步谐振
subsynchronous speed　次同步速率
subsynchronous　亚同步的，次同步的
subsynoptic scale weather system　次天气尺度系统
subsynoptic scale　次综观尺度
subsystem description　子系统说明书
subsystem function　子系统功能
subsystem simulation model　子系统模拟模型
subsystem　分系统，子系统，分支系统，辅助系统，次要系统
subtangent　次切线，次初距
subtask　子任务
subtended angle　夹角
subtense bar　横测尺
subtense method　视差角法
subtense traverse　视差导线
subterranean cable　地下电缆
subterranean disposal　（放射性物体的）地下处置
subterranean divide　地下分水线
subterranean flow　地下径流
subterranean heat energy　地下热能
subterranean heat　地下热
subterranean layer　地下层
subterranean line　地下线路
subterranean pipe line　地下管路

subterranean railway　地下铁道
subterranean river course　地下水道，地下河流
subterranean stream　地下河，地下河流，伏流河
subterranean tunnel　地下隧道
subterranean water parting　地下水界
subterranean watershed　地下分水界［分水线］
subterranean water　底土水，地下水
subterranean works　地下工程
subterranean　地下的
subthermal energy　亚热能，次热能
subthermal neutron　次热中子
subthermal　次热的
subtidal zone　潮下带
subtie　副系杆
subtitle　副标题，小标题
sub-total price　分类计价
sub-total　小计
subtract counter　减法计数器
subtracter　减法器
subtraction circuit　减法电路，比较电路
subtraction　减法，减去，扣除
subtractive polarity　（变压器的）减极性
subtract M and N　把 M 和 N 相减
subtract M from N　从 N 中减去 M
subtractor　减法器，减数
subtract output　减法输出
subtract pulse　减法脉冲
subtract　减法，减，减法指令，减去，扣除
subtrahend　被减数，减数
subtrain　分系列
subtransient component　次瞬态分量
subtransient current　次瞬变电流
subtransient flux-linkage　次瞬变磁通链
subtransient internal voltage　次瞬变内电压
subtransient period　次瞬变周期，超瞬变周期
subtransient reactance　次暂态电抗，次瞬变电抗，起始瞬变电抗
subtransient time constant　次瞬态时间常数
subtransient　次瞬态的，次瞬态，次瞬间，起始瞬态
subtransmission system　分支输电系统，二次输电系统
subtransmission　二次输电，辅助变速箱
subtropical calm　副热带无风带
subtropical high belt　副热带高压带
subtropical high　副热带高气压
subtropical jet stream　副热带急流
subtropical rain forest　亚热带雨林
subtropical region　副热带地区，亚热带地区
subtropical zone　亚热带，副热带，亚热带地区
subtropical　副热带的，亚热带的
subtroposphere　副对流层
subulate　锥形的，钻状的
suburban call　郊区呼叫
suburban line　郊区线路
suburban power distribution　郊区配电
suburban power supply　市郊供电
suburban switchboard　郊区交换台
suburban terrain　郊区地形
suburban wind exposure　郊区地貌风开敞度
suburban　市郊的
suburb　郊区，郊外，近郊，市郊
subvendor　分供厂商，分供货方，分供货商
subvention　津贴
subvertical member　副垂直杆【桥梁】
subwatering　地下给水
subwatershed　次分集水区
subway cable system　地下电缆系统
subway distribution line　地铁配电线路
subway substation　地下变电站
subway type transformer　浸没式变压器，地下变压器［变电站］，防水型变压器，地道型变压器
subway　地下铁道，地下电缆管道，地道
subzero temperature　零下温度
subzero treatment　冷处理
subzero　零下，负的
subzone network　小区网络，分区网络
subzone　分区，小区
succeeding component　接续部件
succeeding lift　后继浇筑层
succeeding tide　后续潮
success diagram　成果图
successful attempt　成功尝试
successful bidder　中标人，中标者，得标商
successful bid　中标
successful call　呼叫到达
successful experience　成功经验
successful forced line energization　线路强送电成功
successful operation　有效运行，成功的操作
successful re-closure　重合闸成功
succession　接续，连续，顺序性，继承，次序，顺序，地层层序
successive approximate method　逐次趋近法，逐步渐近法
successive approximation　逐步逼近法，逐次逼近法，逐次近似（计算）法，逐次趋近
successive barrier　多道屏障
successive collapse　连续破坏
successive commutation failure　（变流器或电机的）连续换向故障
successive difference　逐次差分，递差
successive displacement method　逐次位移法
successive exposure　连续曝光
successive failure　逐步破坏，逐渐破坏
successive grinding　连续研磨【金相检验】
successive launching method　顶推法
successive layer　逐层
successive line over-relaxation　逐次行超松弛
successive neutron capture　逐次中子俘获
successive overrelaxation method　逐次超松弛法
successive over-relaxation　逐次超松弛
successive reaction　连续反应
successive reduction　逐次简化
successive relaxation　逐次松弛，逐级松弛
successive sedimentation　逐次沉淀
successive slide　分级式滑坡
successive slip　渐进性滑动
successive substitution method　失代法，逐次代入法
successive substitution　递代法，逐次代换法，

逐次代替法
successive surge 连续涌浪
successive transformation 递次变换
successive value 逐次值
successive 连续的，逐次的，逐步的，相继的，
successor matrix 后继矩阵
successors in title 权利继承人
successor 继承人
success rate 成功率
success record 成功记录
success 成功
succinic acid 丁二酸
sucessive approximation method 逐步近似法
suck-back 倒吸
suck down wind tunnel 下吸式风洞
suck dry 吸干
sucker （泵的）活塞，吸入管，吸入器，吸泥管，吸头
sucking action 吸引作用
sucking coil 可动铁芯调节线圈
sucking solenoid 吸持螺线管
sucking tube 吸气管，吸入管
suck in 吸入，吸进
suck out 吸出，抽出
suck up 吸收，吸入
suck 渗入，吸收，吸力
suction adapter 吸入过滤段
suction anemometer 吸管式风速计
suction attachment 吸管接头
suction barrel 吸入筒
suction bay 吸水池
suction bell 台锥形吸水管【取水口】
suction branch 抽汽支管
suction casing 进气缸，吸入壳
suction check valve 进水管逆止阀
suction cleaner 吸尘器
suction coefficient 负压系数，吸力系数
suction condition 吸入口状态
suction conduit 吸入管
suction cone 锥形吸入管【钠冷快堆容器】
suction cover 吸入盖
suction culvert 吸扬涵洞
suction cup 吸盘【电抛光】
suction current 吸入水流
suction-cutter dredge 旋桨吸泥机
suction datum 吸入数据
suction deaerator 真空除氧器
suction double volute casing pump 立式端吸双蜗壳泵
suction drainage 吸扬式排水
suction dredger 吸泥机
suction eddy 负压旋涡
suction effect 抽吸作用，吸气作用，吸入效应
suction elbow 吸入弯道，吸水肘管
suction eye 吸入孔
suction fan 吸风机，引风机
suction filter 吸入侧过滤器，抽吸式过滤器，真空滤池
suction firing 不鼓风燃烧，吸风燃烧
suction flask 吸滤瓶，真空过滤瓶
suction flow 吸流，吸入流
suction force 吸力
suction gauge 负压表，真空计
suction gradient 负压梯度
suction grid 吸水网
suction head cover 吸入头盖
suction head 吸入压头，吸水压头，吸入水头，吸升水头，吸水水头，吸水高度【泵】，虹吸水头
suction height 吸水高度
suction hood 吸气罩
suction hose 吸水软管，真空皮管
suction inlet of pump 泵的吸水口
suction inlet 吸入口
suction jet 抽吸射流，吸水射流
suction level 吸入高度
suction lift of pump 泵吸入管吸上高度
suction lift 吸升水头［压头］，吸引升力，抽吸高度，吸升高度，吸程
suction line 吸入管线
suction load 吸力荷载
suction main 抽吸管道，吸水干管，吸水总管
suction manifold 吸入总管，吸入汇流联箱，吸出集流管
suction mill 负压磨煤机
suction nozzle 入口接管，吸入管，进口管，吸水管口，吸入管嘴
suction of blade 叶片背面
suction opening 吸入口，抽吸口
suction operation 负压运行
suction peak 吸力峰，力峰值
suction-pipe loss 吸水管损失
suction pipe 吸管，吸水管，吸泥管，吸气管，吸入管（道）
suction pit 吸水坑
suction plant 抽吸（式）泵站，吸尘设备，（用吸入式煤气发生炉产生煤气的）煤气厂
suction point 抽吸点
suction port 进汽
suction pressure 抽吸压力，吸入压力，负压（力）
suction pump 抽吸泵，吸入泵，抽气泵，抽水泵
suction pyrometer 空吸式高温计，抽气式高温计
suction scour 波浪退吸冲刷
suction side cover 吸入端盖
suction side 吸力面，进口侧，吸入端，负压侧
suction slot spillway 吸槽式溢洪道
suction slot 吸槽
suction specific speed 汽蚀比转速
suction stage 吸入级
suction strainer 吸水口滤网，吸入粗滤器，吸水过滤器，吸滤器
suction stroke 吸气冲程，吸入冲程
suction sump 吸水槽
suction surface 叶型背弧，负压面
suction system pulverizer 负压制粉系统，负压磨煤机
suction system to remove fission products 用以排除裂变产物的吸入系统
suction tank 吸入箱
suction temperature 吸气温度，抽气温度，入口温度

suction thermometer 吸气热电偶，吸气式温度计，抽气式温度计
suction tube 吸出管，吸水管，吸管，吸筒
suction type airfoil 吸入式机翼
suction type sampler 吸入式取样器
suction type suspension load sampler 抽吸式悬移质采样器
suction valve 吸入阀，吸水阀【活塞泵和柱塞泵】
suction velocity at return air inlet 回风口吸风速度
suction vortex 吸入旋涡，进口旋涡
suction water column 负压水柱，吸力水柱
suction wave 负压波，空吸波，吸入波，稀散波，稀疏波
suction well 吸水井
suction 进气【内燃机】，抽气［水］，吸气［水］，吸入［出］，抽吸，吸引［收］，吸力，入口
sudden catastrophe 突然灾变
sudden-change relay 突变继电器
sudden closure 突然关闭
sudden contraction 突然收缩
sudden drawdown 突降
sudden enlargement 突然扩大
sudden failure 突发故障，突然失效
sudden fault 突发故障
sudden fracture 突然断裂
sudden injection method 突然注入法
sudden load change 负荷突变，负载突变，荷载突变
sudden load increase 负荷突增
sudden load interruption 突然甩负荷
sudden load reduction 负荷急剧下降
sudden load rejection 突然甩负荷
sudden load 突加负载，骤加荷载
suddenly applied load 突加荷载
sudden overload 意外过载
sudden phase anomalies 相位的突变
sudden-pressure relay 液压继电器
sudden reduction 骤然下降
sudden settlement 突然沉陷
sudden short-circuit current 突然短路电流
sudden short-circuit test 突然短路试验
sudden short-circuit 突然短路
sudden shower 突降阵雨
sudden step load increase 负荷阶跃骤增
sudden stress 骤加应力
sudden transition 突变段
sudden water release 突然泄水
sue 起诉，控告，提出诉讼
suffering demurrage 滞留费
sufficient condition 充分条件
sufficient estimate 充分估计量
sufficient material 足够的材料
sufficient receiver 充分接收机
sufficient slope 充分的坡度
sufficient statistic 充分统计量
sufficient 足够的，充分的
suffix signal 词尾信号，后缀信号
suffix 下标
suffocation 熄灭，窒息，窒息作用
suffusion 弥漫
suggested permissible level of noise 参考容许噪声级
suggested specification 建议性规范
suggestions for improvement 改进意见
suggestion 建议，意见，提议
sugrader 路基平整机
suitability test 适用性试验
suitability 适合性，适用性，适当
suitable 适当的
suit and hood unit 头罩式防护服
suite 组，排，套【电气柜或架子】
suit in field 现场决定
suit 适合，适应
Sulan 聚丙烯的商名
sulfamic acid 氨基磺酸
sulfanilic acid 磺胺酸，对氨基苯磺酸
sulfate attack 硫酸盐侵蚀
sulfate-phate 硫酸盐
sulfate soundness test 硫酸盐安定试验，抗硫酸盐侵蚀试验，抗硫酸盐试验
sulfate sulfur 硫酸盐硫
sulfating roasting 硫酸盐焙烧
sulfation rate 硫酸盐化程度
sulfation reaction 硫酸盐化反应
sulfatizing roasting 硫酸盐化焙烧
sulfide 硫醚，硫化物
sulfite 亚硫酸盐
sulfoaluminate cement 膨胀水泥
sulfonated coal 磺化煤【一种软化剂】
sulfonate 磺化，磺酸盐
sulfonation 磺化，磺化作用
sulfone-thiocyanate 硫代氰酸砜
sulfonic acid cation exchange 硫酸阳离子交换
sulfonic acid functional group 磺酸官能团
sulfonic acid group 磺酸基
sulfonic acid 磺酸
sulfonic polystyrene resin 磺化聚苯乙烯树脂
sulfonic resin 磺化树脂
sulfosalicylic acid 磺酸水杨酸
sulfourea 硫脲
sulfur acid 硫酸
sulfur and nitrogen free basis 无硫无氮基
sulfuration 硫化，硫化作用
sulfur bacteria 硫细菌
sulfur-bearing fuel 含硫燃料
sulfur-bearing 含硫的
sulfur cement 硫磺胶合剂
sulfur coal 高硫煤
sulfur content 含硫量
sulfur corrosion 硫腐蚀
sulfur dioxide pollution 二氧化硫污染
sulfur dioxide 二氧化硫
sulfur dust 硫灰尘
sulfur emission 硫的排放
sulfur-free basis 无硫基
sulfur-free fuel 无硫燃料
sulfur hexafluoride cable 六氟化硫电缆
sulfur hexafluoride circuit breaker 六氟化硫断路器
sulfur hexafluoride insulating transformer 六氟化

硫绝缘变压器
sulfuric acid 硫酸
sulfuric 硫磺的
sulfur in ash 灰中硫
sulfur monoxide 一氧化硫
sulfurous acid 亚硫酸
sulfurous 亚硫的
sulfur-removing additives 脱硫添加剂
sulfur-removing sorbent 脱硫吸着剂
sulfur retention 脱硫，固硫
sulfur-rich char 高硫木炭
sulfur sesquioxide 三氧化二硫
sulfur content 硫分
sulfur＝sulphur 硫
sulfur trioxide 三氧化硫
sulphate-carbonate ratio （锅水）碱度
sulphate test 硫酸盐试验
sulphate 硫酸盐
sulphide sulfur 硫化物硫，硫酸盐
sulphide 硫化物
sulphite 亚硫酸盐
sulphoaluminate cement 硫酸盐水泥
sulphonic acid 磺酸
sulphur-bonded mortar 硫磺砂浆
sulphur-cement concrete 硫磺混凝土
sulphur coal 高硫煤
sulphur concrete 硫磺混凝土
sulphur content 含硫量
sulphur dioxide detector 二氧化硫探测器
sulphur dioxide 二氧化硫
sulphur hexafluoride circuit breakers 六氟化硫断路器
sulphur hexafluoride gas 六氟化硫气体
sulphuric acid 硫酸
sulphuric 含硫的
sulphur mastic 硫磺胶泥
sulphurous acid 亚硫酸
sulphur print 硫印，硫印法
sulphur hexafluoride circuit breaker SF_6 断路器，六氟化硫断路器
sulphur water 含硫水
sulphur 硫
Sulzer boiler 苏尔寿锅炉，苏尔寿直流锅炉
sum accumulator 和数累加器
sum digit 和数位，和数数字
sum equation 累加方程
sum in words 大写金额
Sumitomo Corporation 住友商事株式会社
summand 被加数
summarized information 被综合的信息
summarize instruction 求和指令
summarize 汇总
summary card punch 总计卡片穿孔机
summary classification 简要分类
summary counter 累加计数器
summary gang punch 总计复穿孔机
summary list 汇总表，汇总清单
summary of element cost 分项造价汇总
summary of pipe and pipe fitting 管道及配件汇总表
summary punch 总计穿孔机
summary rating 综合评价等级
summary record 简要记录
summary schedule 汇总表
summary sheet for bid opening 开标一览表
summary sheet 汇总表
summary table 汇总表，综合表
summary 简易的，扼要的，总结，概括，概要，摘要，汇总（表），一览（表）
summated output 总和输出，总加输出
summation check 求和检查，总和校验
summation current transformer 总和电流互感器，总加电流互感器
summation curve 累积曲线
summation instrument 总和仪表
summation metering 累积计量法
summation of series 级数求和
summation percentage 累计百分率
summation transformer 总和变压器，总加变压器
summation 求和，相加，累积，总和
summator 总加器，加法器，相加器
summed current 总电流
summer dike 夏堤
summer flood 伏汛
summer full load 夏季满发负荷
summer levee 夏堤
summer season ventilation calculating temperature 夏季通风计算温度
summer solstice 夏至，夏至点
summer time system 夏时制
summer with gain 带增益加法器，增益加法器
summer wood 大木材
summer 加法器，楣石，夏季
summing amplifier 加法放大器
summing box 汇总盒
summing circuit 总和线路，加法电路
summing integrator 求和积分器，加法积分器
summing network 求和网络
summing point 相加点
summing unit 求和装置，相加单元
summing 总计
summit crater 山顶火山口
summit level 峰顶面
summit of a mountain 山顶
summitor 相加器
summit pond 山顶水池
summit talk 高级会晤
summit 顶峰，高峰，山顶，顶点
sum of phasor 相量和
sum-of-years digits depreciation 年数总和折旧
sum out gate 和数输出门
sum output 和数输出
sump cooling 地坑水冷却
sump pit 积水坑，集水井，集水坑，污水坑，污水井，排水坑
sump pump 污水泵，地沟水泵，地坑泵，立式泵，油池泵，井底水窝水泵
sum-product output 和积输出
sum-product register 和积寄存器
sum-product 和积
sump tank pump 污水泵，污水箱泵，地坑水箱泵

sump tank　集水箱，沉淀箱，废水箱，聚水池，污水箱，贮水槽
sum pulse　和数脉冲
sump water hold up tank　污水滞留箱，污水暂存箱
sump water monitoring and storage tank　污水监测贮存箱
sump water train　污水处置系统
sump water treatment　废水处理，排水处理
sump water　污水
sump　地坑【安全壳】，沉淀池，集水井，污水坑，贮槽，贮油槽，水坑，水池
sum readout　和数读出
sum rule　加法定则
SUM(startup manager)　调试经理
sum storage　和数存储器
sum　总数，总和，总额，金额，概要
sun-and-planet gearing　行星齿轮传动
sun-and-planet gear　行星（式）齿轮（装置），太阳行星齿轮系，太阳系型齿轮传动装置
sun-and-planet　太阳系型的
sun axle　中心轴
sunbeam　日光线
sunblind　百叶窗
sunbreaker　遮阳板
sun burst　日现色
sun crack　干裂，晒裂
sunder　切割，割开
sundries　杂物
sundry charges　杂费
sundry　杂货【复数形式】，各式各样的，杂项的
sun following　太阳跟踪
sun gear　太阳齿轮，中心齿轮
sunk cost　沉没成本
sunken evaporation pan　埋置式蒸发器
sunken fascine layer　沉柴埽层
sunken fascine mattress　沉排
sunken fascine works　埽工
sunken fascine　沉柴埽，沉柴排
sunken reef　暗礁
sunken-ship breakwater　沉船防波堤
sunken　水底的
sunk gutter　暗天沟
sunk hydrant　地下式消火栓
sunk key　暗键，埋头键
sunk rivet　埋头铆钉
sunk screw　沉头螺钉，埋头螺钉，埋头螺丝
sunk shaft foundation　沉井基础
sunk shaft　沉井
sunk-type switch　埋装式开关
sunk well foundation　沉井基础
sunk well　沉井
sunlight battery　太阳电池
sunlight intensity　日光强度
sunlight power generation　太阳光发电
sunlight wind machine　日照型风力机
sunlight　日光
sun louver　遮阳板
sun-path diagram　太阳行程图
sun power　太阳能
sun relay　太阳继电器
sun rise effect　（长波通信的）日出效应
sunrise　日出
sun's altitude　太阳高度角
sun's azimuth　太阳方位角
sun screen　遮阳板，百叶窗
sun's disk　日面
sun set effect　（长波通信的）日落效应
sunset　日落，（法律的）自动废止期
sunshade blind　遮阳
sunshade grille　遮阳花格
sun shade　遮阳，遮阳板，百叶窗，遮光罩
sunshading board　遮阳板
sun shading disk　遮光片
sun shield　遮阳板
sunshine duration　日照长度，日照时间，日照时数，日照持续时间
sunshine hours　日照时数，日照小时数
sunshine recorder　日照记录仪，日照计
sunshine record　日照记录
sunshine time　日照时间
sunshine tracking system　阳光跟踪系统
sunshine unit　锶单位
sunshine　日照
sun's meridian altitude　太阳中天高度
sunspot cycle　太阳黑子循环
sunspot　太阳黑子
sunsynchronous　与日光同步的，与太阳同步的
sun test　日照试验
sun tracker　太阳跟踪器
sun-tracking controller　太阳跟踪控制器
sun visor　遮阳板
sun　太阳
supecrooling degree　过冷度
superabundant measure　过多测量
superactivity　超活性
superadiabatic atmosphere　超绝热大气
superadiabatic lapse rate　超绝热递减率
superadiabatic layer　超绝热层
superadiabatic state　超绝热状态
superalloy　超耐热合金，高合金钢，超合金
super-bundle arrangement　多股分裂导线的排列
supercapillary interstice　超毛细孔隙
supercapillary percolation　超毛细管渗透
supercapillary porosity　超毛细管孔隙率
supercapillary seepage　超毛细管渗流
supercapister　超阶跃变容二极管
supercapital　副柱头
supercargo　押运人
supercavitating flow　超空蚀水流
supercavitating hydrofoil　超空蚀水翼
supercavitation impeller　超汽蚀叶轮
super-ceded　不能继续进行的，过时废弃的
supercell　超栅元
super-cement　超级水泥
supercharged boiler　增压锅炉
supercharged engine　增压式发动机
supercharged pressure furnace　增压炉膛
supercharged steam generator　增压锅炉
supercharge loading　超载，过载
supercharger　增压器
supercharge　增压

supercharging boost 增压
supercharging equipment 增压设备
supercharging pressure ratio 增压比
superchlorination 过氯化作用
superchopper 特快断续器，超速断续器
supercode 超码
super coil 超外差线圈
super compaction 密集压实，超压实
superconducting bolometer 超传导辐射热测量计，超导电阻测温计
superconducting cable 超导电缆
superconducting coil 超导线圈
superconducting device 超导装置
superconducting dynamo 超导发电机［电动机］
superconducting energy storage 超导储能
superconducting field winding 超导励磁绕组
superconducting film 超导体薄膜
superconducting generator 超导（体）发电机
superconducting insulation 超导绝缘
superconducting machine 超导电机
superconducting magnetic energy storage system 超导磁贮能装置，超导磁贮能系统
superconducting magnet 超导磁铁
superconducting motor 超导（体）电动机
superconducting state 超导态
superconducting super collider 超导超级对撞机
superconducting transformer 超导变压器，超低温变压器
superconducting wire 超导电线
superconducting 超导，超导体的
superconduction magnet 超导磁铁
superconduction 超导
superconductive cable 超导电缆
superconductive coil 超导线圈
superconductive energy storage 超导能量储存
superconductive generator 超导发电机
superconductive winding 超导绕组
superconductive wire 超导线
superconductive 超导（电）的，超导体的
superconductivity 超导性，超导率，超导电率
superconductor energy gap 超导体能隙参数
superconductor transition temperature 超导体转变温度
superconductor 超导体
supercontrol tube 可变互导管，超控制管
super converter 超级转换反应堆
supercooled air 过冷空气
supercooled Austenite 过冷奥氏体
supercooled droplet 过冷水滴
supercooled electric machine 超冷电机
supercooled field winding 超冷磁场绕组
supercooled rotor coil 超冷转子线圈
supercooled state 过冷状态
supercooled water droplet 过冷水滴
supercooled water 过冷水
supercooled 过冷的
supercooling degree 过冷度
supercooling neutron 过冷中子
supercooling point 过冷却点
supercooling 过冷（却），过度冷却
supercritical accident 超临界事故
supercritical behavior 超临界工况
supercritical boiler 超临界锅炉
supercritical centrifuge 超临界离心机，高于临界转速的离心机
supercritical cycle 超临界循环
supercritical flow regime 超临界流动状态
supercritical flow 超临界流，超临界流动，超临界水流
supercritical fluid 超临界流体
supercritical generating unit 超临界发电机组
supercriticality 超临界状态，超临界
supercritical Mach number 超临界马赫数
supercritical mass 超临界质量
supercritical medium 超临界介质
supercritical power plant 超临界电站
supercritical pressure boiler 超临界压力锅炉
supercritical pressure cycle 超临界压力循环
supercritical pressure power plant 超临界压力发电厂
supercritical pressure steam turbine 超临界压力汽轮机，超临界汽轮机
supercritical pressure 超临界压力
supercritical reactor 超临界反应堆
supercritical region 超临界区
supercritical Reynolds number 超临界雷诺数
supercritical rotor 挠性转子，柔性转子，超临界转子
supercritical state 超临界状态
supercritical steam conditions 超临界蒸汽参数
supercritical steam power plant 超临界蒸汽发电厂，超临界压力火力发电站，超临界压力蒸汽动力装置
supercritical steam pressure 超临界蒸汽压力
supercritical steam 超临界蒸汽
supercritical system 超临界系统
supercritical tower 硬塔架，刚性塔架
supercritical turbine 超临界汽轮机
supercritical unit 超临界机组
supercritical vortex shedding 超临界旋涡脱落
supercritical 超临界，超临界的
supercurrent 超导电流
superdeep holes 超深孔
superdense 超密的
superdiamagnetic 超抗磁的
super-digits 发光二极管的
superdip 超倾磁力仪
superdirective antenna 超锐定向天线
superdirectivity （天线的）超锐定向性
superduty belt conveyor idler 皮带机超载托辊
superduty castable refractory concrete 超高温耐火混凝土，超级耐火混凝土
super economic boiler 超经济式锅炉
superelevated-curve 超高曲线
superelevation of outer rail on curve 曲线外轨超高
superelevation ramp 超高的坡道
superelevation slope 超高顺坡，超高横坡度
superelevation 超高【指高程】，超高值
superexchange 超交换，超交换的
superexcitation 过激，过励，强励
superferromagnetism 超铁磁体

superficial area 表面积
superficial cementation 表面渗碳
superficial charring （木材的）表面碳化处理
superficial compaction 表面压实，表土夯实
superficial corrosion 表面腐蚀
superficial degradation 表层剥蚀，表面损坏
superficial deposit 基岩上的表土
superficial deterioration 表层剥蚀，表面损坏
superficial dimension 表面尺度
superficial expansion 表面膨胀
superficial fault 表面缺陷
superficial flow rate 表观流量
superficial fold 表层褶曲
superficial gas velocity 空塔速度，表观气速
superficial hardness 表面硬度
superficial layer 表面层
superficial measure 表面积计量
superficial molar flux 表面摩尔通量
superficial treatment of timber 木材表面处理
superficial value 表观值
superficial velocity 空床流速，表观速度，折算速度，引用速度，空塔速度
superficial water 表层水，地表水，地面水
superficial 表面的
superfine cement 超细水泥
superfine pulverized coal 超细煤粉
superfine quality 最好的质量
superfines 超细粉末
superfinishing 超精研磨
superflood 特大洪水
super-fluid 超流体
superfluous term 冗余项
superfluous water 过剩水分
superflux reactor 超高通量反应堆
superfrequency 超高频，特高频
supergain antenna 超增益天线
supergain 超增益
super-gel ion-exchange resin 超凝胶型离子交换树脂，超凝胶型树脂
supergene enrichment 次生富集
supergeostrophic wind 超地转风
supergradient wind 超梯度风
supergravity 超重力
supergrid network 超级电力网络
supergrid substation 超高压电网的变电站
supergrid 特大功率电网，超高压输电网
super-group 超群【信道】，大组
superhard 超硬的，超硬度的
superharmonic response 超谐波响应
superharmonic 超谐波的
superheat boiling-water reactor 过热沸水反应堆
superheat control 过热调节，过热汽温调节
superheat degree 过热度
superheated area 过热区，过热蒸汽区
superheated section 过热段
superheated sensitivity 过热敏感性
superheated steam control 过热蒸汽调节
superheated steam cooler 过热蒸汽冷却器
superheated steam discharge pipe 过热蒸汽排放管
superheated steam reactor 核过热反应堆，过热蒸汽反应堆
superheated steam safety valve 过热蒸汽安全门
superheated steam temperature control 过热汽温控制
superheated steam zone 过热蒸汽区
superheated steam 过热蒸汽
superheated tube 过热器管
superheated vapor 过热蒸汽
superheated water station 给水过热站
superheated water 过热蒸汽
superheater assembly cluster 过热器组件束
superheater assembly 过热器组件
superheater bank 过热器管束
superheater cavity 过热器级间气室
superheater chamber 过热器区
superheater coil 过热器管圈
superheater furnace 布置过热器炉体的炉膛【双炉体锅炉】
superheater outlet leg 过热器出口联箱
superheater outlet line 过热器出口管线
superheater steam circuit 过热器蒸汽回路
superheater trimming damper 过热器烟气流量调节挡板
superheater vent 过热器空气阀
superheater 过热器
superheat fuel assembly cluster 过热器燃料组件束
superheat fuel element 过热器燃料元件
superheating boiling reactor 沸腾过热反应堆
superheating curve （蒸汽）过热曲线
superheating heat 过热热量
superheating reactor 过热反应堆
superheating section 过热段
superheating surface area 过热器受热面面积
superheating surface 过热器受热面
superheat 过热
superheat reactor 过热反应堆
superheat region 过热区，过热蒸汽区
superheat section 过热段，过热器段
superheat steam temperature control 过热汽温控制
superheat temperature 过热温度
superheavy hydrogen 超重氢，氚
superheavy water 超重水，氚水
superheavy 超重的
super-high frequency engineering 特高频工程学
super-high frequency oscillator 特高频振荡器
super-high frequency tube 特高频管
super-high frequency 超高频
super-high pressure boiler 超高压锅炉
super-high pressure mercury lamp 超高压水银灯
super-high pressure turbine 超高压汽轮机
super-high pressure 超高压
super-high tension 超高电压
super-high voltage equipment 超高压设备，超高压装置
super-high voltage 超高压
super highway 超级公路【指超高速公路】
super-huge turbo-generator 特大型汽轮发电机，巨型汽轮发电机
superimposed arch 层叠拱

superimposed backpressure　叠加背压
superimposed circuit　重叠电路
superimposed drainage　叠置水系，重叠排水系统
superimposed effect　叠加效应
superimposed fill　加填土
superimposed flow　叠流
superimposed fluidized bed　叠层流化床
superimposed load　超负荷，叠加荷载，附加荷载
superimposed on　叠加到……上
superimposed oscillation　叠加振动，叠加振荡
superimposed soil　上覆土
superimposed stress　附加应力
superimposed wave　重叠波
superimpose unit　（电站的）前置机组
superimpose　叠加，重叠
superimposition　叠加，重叠，叠置
superimposure　叠加，重叠
superinduction　超感应，增加感应，添加感应
superinfragenerator　远在标准外的振荡器，远在标准下的振荡器
superinsulant　超绝缘体，超绝热体
superinsulation　超绝缘，超绝热
superintegrated　高密度集成的
superintendence　指挥，主管，监督
superintendent of construction　安装施工段主任
superintendent of document　文件主管人
superintendent　负责人，监督人，主管人，总裁，总监，主管
superinvar　超级殷钢，超级镍钴钢
superior air　高空下降空气
superior limit　上限
superior quality　优等
superior wave　主波，基波
superjacent waters　上方水域，上覆水域
superjacent　叠加，盖在上面
super large-scale integration　超大规模集成
super-lattice　超点阵
superlethal dose　超致死剂量
superlethal　超致死量的
superload　超载，附加负荷，附加荷载
superloy　超合金
supermalloy　高导磁合金，镍铁钼高导磁合金
supermatic drive　高度自动化传动，完全自动传动的
supermatic　高度自动化的，完全自动化的
supermendur　铁钴钒合金材料
supermicrocomputer　超级微型计算机
superminiature capacitor　超小型电容器
superminiature　超小型的
superminicomputer　超小型计算机
supermumetal　超合金
supernatant liquid　澄清液，上层液体，上层清液
supernatant　浮在表面的，浮于上层的，上层清液，浮在表层的东西
supernetwork　超级线路网，超级道路网
superparamagnetism　超顺磁性
superpass　超过，优于，胜过
superperformance　超级性能
superperiod　超周期
super-permalloy　超级坡莫合金
superphantom circuit　超幻想电路
superphosphate　过磷酸盐
superplasticity　超塑性
superplasticizer　超增塑剂，强塑剂，高效减少剂
superposed bleeder turbine　前置抽汽式汽轮机
superposed circuit　叠加电路
superposed current　叠加电流
superposed drainage　重叠排水系统
superposed field　叠加场
superposed fluid　叠加流体
superposed magnetization　重叠磁化，辅助磁化
superposed plant　前置机组
superposed turbine　前置式汽轮机
superposition integral　重叠积分
superposition law　叠加定律
superposition method　叠加法，重叠法
superposition of directive pattern　方向图叠加
superposition of flow patterns　流型叠加
superposition of signal　信号叠加
superposition of stress　应力叠加
superposition of vortex　旋涡叠加
superposition principle　叠加原理，重叠原理
superposition specific heat　叠加比热
superposition superimposed　迭加的，叠加的
superposition theorem　迭加原理，迭加定理
superposition　迭加，叠加，重叠，叠置，重合，复合
superpotential　过电压的
superpower coefficient　超功率系数
superpower klystron　特大功率调速管
superpower station　大功率发电站
superpower system　超级电力系统
superpower　（一个地区内的）联合发电总量，特大功率的，超功率
super pressure balloon　过压气球
super pressure plant　超高压设备，超高压装置
super profit　超额利润
super-prompt-critical accident　超瞬发临界事故
super-prompt criticality　超瞬发临界
super-purity　高纯度
super-rapid hardening cement　超快硬水泥，高级快硬水泥
super-regenerative circuit　超再生电路
super-regenerative grid detector　超再生栅极检波器
super-regenerative stage　超再生级
super-regenerator　超再生振荡器
super-regulator　高灵敏度调整器，极精确调整器
super safety rod　超安全棒
supersaturated air　过饱和空气
supersaturated solid solution　过饱和固溶体
supersaturated solution　过饱和溶液
supersaturated steam　过饱和蒸汽
supersaturated vapour　过饱和蒸汽，过饱和水汽
supersaturated　过饱和的
supersaturation curve　过饱和曲线
supersaturation water　过饱和水
supersaturation with respect to water　水面过饱和
supersaturation　过饱和（度），过饱和现象
superscript　上标，上角标，标在上角的字，

符号
superseded by 由……代替
superseded 不能继续进行的，过时废弃的
supersede 取代，替代，接替
super-sensitive 高灵敏度的，超灵敏度的，过敏的
supersession head 除尘用的化学喷头
supersilent motor 高无噪声电动机
supersmart card 超灵敏智能卡
supersonic aerodynamics 超音速空气动力学，超音速气体动力学
supersonic contoured nozzle 超音速型面喷嘴
supersonic delay line 超声波延迟线
supersonic detector 超声波检测器，超声波探测仪
supersonic echo sounder 超声波回声测深仪
supersonic flaw detector 超声波探伤仪
supersonic flow 超声流，超音速流
supersonic frequency 超音频
supersonic generator 超声波发生器
supersonic machining 超声波加工
supersonic nozzle throat 超音速喷嘴喉部
supersonic nozzle 超音速喷嘴
supersonic oscillation 超声波振荡，超声波振动
supersonic sounding 超声波测深法
supersonic speed 超音速
supersonic switch 超声波开关
supersonics 超声速空气动力学，超声波学
supersonic test 超声波检验，超声波检查
supersonic turbine 超音速透平，超音速涡轮
supersonic velocity 超声速，超音速
supersonic wave 超声波
supersonic 超声波的，超音速的，超声的，超音频的
super-speed 超速
superstabilizer 超稳定器，超稳定剂
superstation 特大功率电台，特大型发电厂
superstratum 上覆层
superstructure arrangement （锅炉的）塔式布置
superstructure 上部结构，上层建筑，（建筑物、船等的）上面部分，塔头
supersulphated cement 高硫酸盐水泥
supersynchronous braking 超同步制动
supersynchronous motor 超同步电动机
super-synchronous resonance 超同步共振
supersynchronous speed 超同步转速
supersynchronous 超同步的
super-system 超系统，超级系统
supertanker wharf 超级油轮码头
super tanker 超大型油轮
supertension line 超高压线路
supertension network 超高压电力网
supertension 超高压，过压
superterranean 架空的，天上的，地表的
super thermal 超热的
super thermostatic oil bath 超级恒温油浴
super thermostatic water bath 超级恒温水浴
superturbulent flow 超紊流
superturnstile antenna 三层绕杆式天线，蝙蝠翼天线
superuniversal shunt 超万能分流器
supervene 并发
supervised area 监督区
supervised circuit 监控电路
supervised training 监督训练
supervise man-months 监督人月数，监理人月数
supervise 监督
supervising architect 监理建筑师
supervising engineer 总工程师，监查工程师
supervising system 监测系统
supervision afterwards 事后监督
supervision architect 监理建筑师
supervision at site 现场监督
supervision business 监理业务
supervision certificate 监理证书
supervision company 监理公司
supervision engineer 监理工程师
supervision expenses 监理费用
supervision in advance 事前监督
supervision level 监控级
supervision of construction 施工监督
supervision of energy saving 节能监督
supervision work 监理工作
supervision 监督，监视，检查，管理，检测，观察，监控
supervisor call 管理程序调用
supervisor control 遥控监督，监视控制系统
supervisor interrupt 管理程序中断
supervisor mode 管理状态，管态
supervisor's desk 监控台，监视操作台
supervisor workstation 主管工作站
supervisory authority 监督机构
supervisory board 监事会
supervisory body 监督机构
supervisory capability 管理能力
supervisory center 监理中心
supervisory circuit 监控电路，监视回路
supervisory committee assessed expenses 监管委员会下摊费用
supervisory computer control system 监督控制系统
supervisory computer control 计算机监视控制，管理计算机控制，监督计算机控制（系统）
supervisory computer 监控计算机
supervisory control and data acquisition 监视控制和数据采集系统，监控与数据采集
supervisory control relay 监控继电器
supervisory control system 监督控制系统
supervisory control 监视控制，监督控制，管理控制
supervisory device 监控仪器
supervisory equipment 监视装置
supervisory gear 检测监视装置
supervisory information system 厂级监控信息系统，监控信息系统
supervisory instrumentation 监控仪表
supervisory instrument 检测装置，监视装置
supervisory lamp 监视灯
supervisory organ 监察机构
supervisory panel 监视信号盘
supervisory personnel 监督人，监管人员
supervisory processer 监视处理机

supervisory relay 监视继电器
supervisory signal 监视信号，监控信号，监督信号
supervisory system 监督系统
supervisory tone 监视音
supervisory work 监视操作，管理工作
supervisory 监督的，监视的
supervisor 管理者，监督人，控制器，管理器，监控装置，监理，值长，领班
supervoltage radiotherapy 超高压放射治疗
supervoltage transmission 超高压输电
super voltage 超高压
superwide band oscilloscope 超宽频带示波器
super 特级的，极好的
supplemental document 补充文件
supplemental essential variable 补充基本变量
supplemental fired heat recovery combined cycle 排气助燃联合循环
supplemental steel 补充钢结构
supplemental 补充的，附加的
supplementary agreement 补充协定
supplementary air 补充空气
supplementary angle 补角
supplementary benchmark 辅助水准点，水准补点
supplementary budget 补充预算，追加预算
supplementary building 附属建筑物
supplementary condition 补充条件
supplementary contact 附加触点
supplementary contour 补助等高线，辅助等高线
supplementary contract 补充合同，合同补充书
supplementary control point 辅助控制点
supplementary cost 补充费，附加费用
supplementary explanation 补充说明事项
supplementary exploration 补充查勘，补充勘探
supplementary income tax 追加所得税
supplementary insurance 追加保险
supplementary item 补充项目
supplementary keyboard 辅助键盘
supplementary lighting 辅助照明
supplementary list 补充项目表
supplementary load loss 附加损耗
supplementary loss 附加损失
supplementary measure 配套措施
supplementary observation 补充观测，辅助观测
supplementary point 补充测点
supplementary power 补充电力
supplementary pressure 附加压力
supplementary provisions 补充规定，附则
supplementary relay 辅助继电器，中间继电器
supplementary report 补充报告
supplementary statement 补充报表，辅助报表
supplementary station 补充测站
supplementary storage 辅助存储器
supplementary subroutine 辅助子程序
supplementary tax 附加税
supplementary treaty 补充条约
supplementary triangulation 补充三角测量
supplementary water source 供水源
supplementary water 供水
supplementary work 附属作品
supplementary 补充的，补助的，附加的，增补的，追加的，辅助的，补角，辅助建筑物
supplement 补充，增补，补遗，添加物，附件，附录，补角
supplied-air suit 气衣
supplied heat 供热量
supplier data sheet 供货商数据单
supplier evaluation 供方评价
supplier's installation instructions 供应商的安装规程
supplier's invoice 供货方发票
supplier surveillance 供货监督，供货商监督
supplier 供货商，供货方，供应方，供方，卖方，厂商
supply agreement 供货协议
supply air duct 送风管道
supply air outlet 送风口
supply air rate 进风量
supply air system 供气系统，送风系统
supply air temperature difference 送风温差
supply air 供气，空气源，送气，送风
supply and erection 供应与装设
supply and installation 供应与安装
supply area 供电区域，供电范围，补给区
supply bay 电源架
supply boat 供应船
supply boundary 供应边界
supply cell 馈电间隔，供电间隔，电源间隔
supply change over 电源转换
supply circuit 馈电电路，电源电路
supply current 馈电电流，供电电流
supply curve 供给曲线
supply-demand ratio 市场供需比
supply district 供电区，供电范围
supply equipment 供水设备
supply falls short of demands 供不应求
supply fan room 送风机室
supply fan 送风扇，送风机
supply fault recovery circuit 电源故障恢复电路
supply flume 供水槽
supply frequency 电源频率
supply header 进口联箱
supply in full sets of 成套供应……
supplying party 供货方
supply interruption 断电
supply lead 馈电线，电源线
supply-line 供应线，给水管线，馈电线路，供电线路
supply main 供电线路，馈电干线，给水总管
supply meter 供电电表，馈路电度表
supply network 馈电网，供电网
supply of equipment 设备供应
supply of goods 货源，供货
supply of material 材料供应
supply of power 动力供应，供电
supply pipe 供给管
supply point 供货点
supply pressure 供给压力，供水压力，供气压力，供油压力，电源电压
supply price 供货价格
supply rate 供水率

supply regulator 电源调节器
supply requirement 供货要求
supply side 馈电侧，电压侧，供给端
supply source 电源
supply suspension 断电
supply system 供应系统，电源系统
supply tank 供水柜，供水箱
supply terminal 供电点，馈电终端
supply transformer 供电变压器
supply tube 供水管，供给管，输送管
supply unit 电源装置
supply voltage 电源电压，供电电压，动力电压
supply water distribution header 分水器
supply water makeup 补给水
supply water temperature 供水温度
supply waveform 供电波形
supply wire 电源线
supply 供电，供电电源，电源，进给，给料，供给，供货，供应品
supportability 可维护性，可支持，可忍耐
supportable 可支承的，可支持的
support abutment 支座
support and hanger 支吊架
support and lifting trunnion 支承和起吊耳轴
support arm 支臂
support assembly 支撑组件
support bar 支承杆件，支杆
support beam 支撑梁，支臂
support block 支持块，支撑块，撑块
support brace 拉紧板，支承板，支承
support bracket 支托，支架，托架，支承牛腿，支承架
support column 支架，支撑，支承柱，支柱
support concrete 支承混凝土
support cushion 拱坝坐垫
support cylinder 支承筒
support device 支承部件
supported along four sides 四边支承
supported at two edges 两边支承
supported cable 支撑电缆
supported layer 支持层，承托层，垫层
support features 辅助设施
support flange 支承法兰
support foot 支承底部，支承脚，支撑腿
support grid 支承栅格
supporting bearing 支撑轴承，支座轴承
supporting body 承载体
supporting capacity 承载能力，支承能力
supporting course supporting layer 持力层
supporting course 持力层
supporting data 专业间配合资料
supporting documentation 支持性文件，辅助性文档
supporting document 证明文件，支持文件
supporting element 支撑构件
supporting facilities 配套设施
supporting force 支承力，支点力
supporting frame 支承框架，支撑架【管道】，承重构架，支承构架，支架，固定架
supporting girder 支承柱
supporting insulator 支撑绝缘子
supporting layer 承托层，持力层，垫层
supporting machinery 支承机械
supporting mechanism 支撑机构，承重装置
supporting member 支承构件
supporting pad 支撑垫，支承垫块
supporting pier 支墩
supporting point 支点
supporting policy 配套政策
supporting post 支柱
supporting rack 支持架
supporting resistance 支承反力，支架阻力，支承阻力
supporting resources 后备资源
supporting ring 支持环，绑环，端箍，座环，垫环
supporting roller 托辊
supporting rope 支撑缆索
supporting skirt 支承裙筒
supporting spring 支承弹簧，托片，托簧，承簧
supporting structure 支承结构，支撑结构，支护结构，下部结构
supporting strut 支杆，支柱
supporting technique 支撑杆术
supporting tool 支持工具
supporting tube 悬吊管，吊挂，吊拉管，支撑管
supporting wedge 支撑斜钢
supporting 支持，保证，支持的，辅助性的，次要的
support insulator 支持绝缘子
support interference correction 支架干扰修正
support interference 支架干扰
supportion roller 托辊
support item 配套项目
support ledge 支撑柱【环形吊车】，支承凸缘
support leg 支架，支撑
support lug 支承耳，(压力容器) 支承块，支承撑架
support mast 支撑柱
support material 补强材料
support moment 支承力矩，支座弯矩
support of ducts 风管支架
support of motors 电动机支架
support of pipelines 管道支架
support pad 支承垫块，支承架，支撑座
support pattern 支护方式
support pin 支撑销
support plate 支撑板，支承板，支承隔板
support plinth 支承底座
support point 支撑点，支承点，控制点，支点
support pressure 支座压力
support program 支持程序，辅助程序
support rail 承轨
support reaction 支座反力，支承反力，支点反力
support removal 拆除支撑
support ring insulation (电机绕组的) 端箍绝缘
support ring 支撑环，支承环【控制棒驱动机构】
support rod 托轮，支承棒
support roller 支承滚柱，支承轮
support saddle 支座

supports and hangers 支吊架
support settlement 支座沉陷，支架沉陷
support skirt 支撑裙板，支承裙筒，支承裙座
support software 支撑软件
support steelwork 钢架
support structure 支撑结构，支承结构，支承构件
support tower 支撑塔
support tripod 支撑三脚架
support tube 支撑管，支承管
support unit 单体支架，支承元件
support wire 支撑张线
support 基片，基底，支撑，支承，支架，支护，支持，支柱，支座，配套，辅助，后援
suppressant 抑制剂
suppressed alarm signal 清除的警告信号
suppressed-carrier double sideband 抑制载波的双边带
suppressed-carrier 载波抑制式
suppressed contraction 不完全收缩
suppressed range 正迁移范围，提升范围
suppressed scale 无零标度，压缩刻度
suppressed span 正迁移量程，提升量程
suppressed weir 无侧收缩堰
suppressed zero instrument 无零位仪表，无零点仪表
suppressed zero meter 无零点仪表
suppressed zero pressure gauge 无零刻度压力计，无零位压力表
suppressed zero range 正迁移零位
suppressed zero scale 抑零标度，无零点刻度盘，零点正迁移量，零点正迁移范围
suppressed zero 无零点的，刻度不是从零开始的
suppressed 正迁移
suppressing agent 抑尘剂
suppression chamber cooling system 弛压室冷却系统
suppression chamber 弛压室
suppression of harmonics 谐波抑制，消除谐波
suppression pool cooling system 弛压水池冷却系统
suppression pool water filter 弛压水池过滤器
suppression pool water heat exchanger 弛压水池热交换器
suppression pool water 弛压水池水
suppression pool 弛压水池
suppression pulse 抑制脉冲
suppression ratio 正迁移比
suppression resistance 抑制电阻
suppression time 抑制时间
suppression 压缩，抑制，消除，镇压，平定，正迁移，熄灭，遏止，禁止发行，遏制
suppress length indicator 抑制长度指示器
suppressor choke 降低干扰扼流圈，抑制扼流圈
suppressor current 抑制电流
suppressor grid 遏止栅极，抑制栅极，遏止栅
suppressor impulse 抑制脉冲
suppressor of harmonics 谐波抑制器
suppressor 抑制器，消除器
suppress 封锁，取缔
supraconductivity 超电导率
supra-conductor 超电导体
supralethal dose 超致死剂量
supralethal irradiation 超致死辐照
supralittoral zone 上沿岸带，潮上带
supramesh 超细孔网
supratidal zone 潮上带
supremum 上确界，上限
SUP＝support 支承
supurious wave 无效波
SUPVN＝supervision 监视
SUPV＝supervisory 监视的
surcharge charge 附加费
surcharge angle of the handled material 所输送的物料积载角
surcharge angle 积载角
surcharge depth 超高水深【溢流堰顶以上】
surcharge for overdue tax payment 滞纳金
surcharge in import 进口附加税
surcharge load 超载荷重
surcharge storage 超高库容
surcharge 超高水深【溢流堰顶以上】，附加费（用），附加税，充电过度，洪水超高，附加荷载，过载，超载
sure accumulator 蓄能器
surety bond 担保书
surety company 保险公司
surety money 保证金
suretyship 保证人的地位，保证人的资格
surety 保证，担保，保证人，保证金，抵押品
surface ablation 表面消融
surface abrasion 表面磨蚀，表面磨损，表面磨耗
surface absorbed dose 表面吸收剂量
surface absorption 表面吸收
surface accumulation of salt 地表积盐，盐分表聚
surface acoustic wave coder 表面声波编码器
surface acoustic wave convolver 表面声波卷积器
surface acoustic wave correlator 表面声波相关器
surface acoustic wave filter 表面声波滤波器
surface acoustic wave phase shifter 表面声波移相器
surface acoustic wave transducer 表面声波换能器
surface acoustic wave 表面声波
surface action 集肤作用，表面作用
surface activated plant component 电站表面活化部件
surface activator 表面活性剂
surface active agent 表面活性剂，表面活化剂
surface active material 表面活性材料
surface active 表面活化
surface activity 表面活度，表面活性，表面放射性
surface adsorption 表面吸附作用
surface aeration 表面曝气
surface agent 表面试剂
surface air 地面空气，大气底层
surface appearance 表面状况
surface application 敷面
surface arcing 表面电弧

surface area goodness factor 表面积优度因数
surface area of aggregate 骨料表面面积
surface area of soil grain 土颗粒表面面积
surface area 表面面积，曲面面积，表面积
surface averaged partial current 表面平均偏中子流
surface barrier detector 表面检测仪，面垒型探测器
surface barrier diode 表面阻挡二极管，表面势垒二极管
surface barrier transistor 表面势垒晶体管，表面势垒层晶体管
surface barrier 表面势垒，表面位垒
surface bearing 面支承，平面轴承
surface blemish 表面疵病
surface blowdown 表面排污
surface blowhole 表面气孔
surface blowoff 表面排污，表面排污管
surface boiling in a subcooled pool 大容积欠热表面沸腾
surface boiling 表面沸腾
surface boundary layer 表面边界层，近地边界层
surface breakdown 表面击穿
surface brightness 表面亮度
surface cable 表面电缆
surface carburization 表面渗碳，表面渗碳处理
surface channel 明沟
surface charge destination 目的站地面运输费
surface charge 地面运输费【代理人收取此费为SUA】，面电荷
surface checking 表面网状裂纹
surface cleaning 表面净化，表面去污
surface cleanliness 表面清洁度
surface coating 表面包覆，表面涂层
surface coefficient of heat transfer 表面换热系数
surface coefficient 表面系数
surface colour 表面色
surface combustion burner 表面式喷燃器，无焰喷燃器
surface combustion 表面燃烧
surface compaction 表面夯实
surface concentration 地面浓度，表面浓度
surface condensation 表面凝结
surface condenser 表面式凝汽器，表面式冷凝器
surface conditioner 底材表面处理剂
surface condition 表面状态，表面条件，表面状况
surface conductance 表面电导
surface conductivity 面电导率
surface conduct 表面接触
surface configuration 表面形状，地表形态
surface connected flaw 露于表面的缺陷
surface containment 地面安全壳
surface contaminated object 表面污染物体
surface contamination indicator 表面沾污指示器
surface contamination meter 表面污染计量仪，表面沾污测量仪
surface contamination 污染表面，表面污染
surface contour map 等高线地形图
surface contraction 表面收缩
surface coolant 表面冷却剂
surface cooled reactor 表面冷却反应堆
surface cooler 表面冷却器，表面式冷却器
surface cooling 表面冷却
surface corrosion 表面腐蚀
surface course 表层，面层
surface cover 地表被覆
surface crack test 表面裂纹检验
surface crack 表面裂缝，表面裂纹
surface creepage 表面爬电，表面放电
surface creep 表层土蠕动，表面蠕动
surface crust 地表层，地表硬壳
surface current 表层流，表面流，表面流动，表面电流
surface curvature 表面曲率
surface curve 表面曲线
surface damage 表面损伤，机械损伤，表面损坏
surface decontaminstion 表面净化，表面去污
surface defect 表面疵病，表面瑕疵，表面缺陷
surface density of charge 电荷面密度
surface density of mechanical impedance 声阻抗率
surface density 表面密度，地表密度
surface dent 表面凹坑，表面凹痕
surface depletion method 表面损耗法
surface depletion model 表面耗减模式
surface deposit 表层堆积物，表层沉积
surface detention 表面滞留，地表滞流，地面滞流，地面阻滞
surface development 表面开采
surface diffusion 表面扩散
surface discharge in dielectric oil 油中沿面放电
surface discharge 延面放电，表面放电
surface discontinuity 表面非连续性缺陷
surface ditch 明沟，阳沟
surface divergence 表面散度
surface dose equivalent index 表浅剂量当量指数
surface dose 表面剂量
surface drag coefficient 表面阻力系数
surface drag 表面曳力
surface drainage ditch 地表排水沟
surface drainage 表面排水，地表排水，地面排水，地面排出的污水
surface drain 地面排水沟，排涝沟
surface dressing 敷面，表面修整，表面修琢
surface drift bottle 水面浮标瓶
surface drift 表层漂流
surface drying 表面干燥，油漆结皮
surface dump 地面堆放场
surface earthing 表面接地
surface echo 表面回波
surface eddy 表面涡旋，表面涡流
surface efficiency 表面效率
surface elevation 地面高程
surface emissivity 表面辐射，表面发射率，表面发射系数
surface energy 表面能【核子】
surface-enhanced Raman spectroscopy 表面增强拉曼光谱

surface erodibility　地面可蚀性，表面侵蚀度
surface erosion　表面冲刷，表面侵蚀
surface evaporation pan　地面蒸发器
surface evaporation　表面蒸发
surface evaporative condenser　表面蒸发式凝汽器
surface evenness　表面平整度
surface examination　表面检验
surface exploration　地表探测
surface fault　表面缺陷，表面故障
surface feature　表面形态，地貌
surface-fed intermittent stream　表源间歇河流
surface feed heater　表面式给水加热器
surface film of water　水的表面膜
surface film　表面膜
surface filtration　表面过滤
surface finishing　表面精加工，表面抛光
surface finish sample　表面抛光样品
surface finish standard　表面抛光标准
surface finish　表面抛光，表面修整
surface flatness　表面平整度
surface flaw　表面缺陷
surface float method　水面浮标侧流法
surface float　水面浮标
surface flooding　地面泛流
surface flow pattern　面流形态
surface flow visualization　表面流动显示
surface flow　表流，地表径流，地表水流（量）地面径流，地面水流
surface flux　表面流，表面通量
surface folding　表面褶皱
surface force　表面力
surface friction velocity　地面摩擦速度，表面摩擦速度
surface friction　表面摩擦，表面摩擦力，地面摩擦力
surface gauge　画针盘
surface geologic map　表层地质图
surface geology　地表地质
surface geometry　地表几何形状，表面几何图形
surface geostrophic wind　地表地转风
surface gloss　表面光泽
surface gravity wave　表面重力波
surface hardened spacer dimple　表面硬化的定位波纹
surface hardener　表面硬化剂【混凝土】
surface hardening　表面淬火，表面硬化
surface heater　面式加热器，表面式加热器
surface heat exchanger　表面式换热器
surface heat flux　表面热流密度，表面热通量
surface heat-transfer coefficient　表面传热系数
surface heat-transfer rate　表面热传导率
surface heat transfer　表面传热
surface heat treatment　表面热处理
surface horizon　表土层
surface-hydrated cement　表面水化水泥
surface ignition engine　表面点火发动机，热球式发动机
surface imperfection　表面缺陷
surface indicator　表面规，表面找正器
surface inflow　地表入流，地表水流（量），地面来水（量）
surface inspection　表面检查，表面探伤
surface insulation resistance　表面绝缘电阻
surface insulation　表面绝缘
surface integral　面积分
surface integrated pressure　表面积分压力
surface inversion layer　地面逆温层
surface inversion　地面大气逆温
surface irradiation　表面辐照
surface irregularity　表面凹凸不平，表面不平度，表面不平性
surface lap　表面折叠，表面折痕
surface layer　表层，表面层，地表层，近地层
surface leakage current　表面泄漏电流，表面漏电
surface leakage　表面漏泄
surface level　地表水准，表面能级，地面高程，水平度盘水准器
surface load　表面荷载
surface loss　表面损耗
surface mark　地面标志
surface mean size　表面积平均粒度
surface method　水面侧流法
surface mine　露天矿
surface moisture content　表面水分
surface moisture　外在水分，表面水分
surface monitoring　表面（沾污）监测
surface mounted　明装的，吸顶安装的
surface mounting　平面式安装
surface movement　表层移动
surface net current　表面净中子流
surface noise level　表面噪声水平
surface of ballast　道碴面
surface of contact　接触面
surface of density discontinuity　密度不连续面
surface of discontinuity of current　水流不连续面
surface of discontinuity　不连续面
surface of equal density　等密度面
surface of equal dynamic depth　等动力深度面
surface of equal specific volume　等比容面
surface of equi-pressure　等压面
surface of fracture　断裂面，破裂面
surface of friction　摩擦面
surface of no motion　无流动面
surface of revolution　回转曲面
surface of rotation　旋转面
surface oil film　表面油膜
surface paint　面漆
surface penetrability　表面穿透率
surface penstock　露天（式）压力水管，明压力钢管
surface pitting　（混凝土的）表面点蚀，表面凹孔
surface plate　面板，平板，平台
surface porosity　表面气孔
surface-potential gradient　面电位梯度
surface power density　表面功率密度，面功率密度
surface power station　地面电站
surface preparation　表面预处理，表面预加工，表面处理，表面加工

surface prepared for welding 焊接面
surface pressure coefficient 表面压力系数
surface pressure fluctuation 表面压力脉动
surface pressure 表面压力
surface pretreatment 表面预处理
surface process 表面过程
surface profile 水面纵剖面，水面线
surface protection 表面保护
surface pyrometer 表面高温计
surface quality 表面质量
surface raceway 表面电缆管道
surface radiation 表面辐射
surface radiator 表面散热器
surface radioactive contamination monitoring 表面放射性沾污监测
surface radioactive contamination 表面放射性沾污
surface rating 表面光洁度等级
surface ratio 表面积比
surface reactance 表面电抗
surface reaction control 表面反应控制
surface reaction 表面反应
surface reflection 表面反射
surface relief method 表面浮雕法，胶片漂白法
surface relief 地表起伏，地势，地形
surface renewer theory 表面更新理论
surface reservoir 地面水库
surface resistance of heat transfer 表面热阻
surface resistance 表面电阻，表面阻力
surface resistivity 表面电阻率，表面电阻系数
surface retention 地表滞留
surface reverberation 地表混响，海面混响，表面反射
surface ripple 地表波痕，表面波纹
surface roller 表面旋滚
surface roughening 使表面粗糙，表面糙化
surface roughness element 地面粗糙元，表面粗糙元
surface roughness gauge 表面粗度测量仪
surface roughness measurement 表面粗糙度测量
surface roughness microscope 表面光洁度显微镜
surface roughness tester 表面粗糙度测试仪
surface roughness 地面粗糙度，表面粗糙度，地面平整度，表面糙度，表面糙率
surface runoff 地表径流（量）
surface scaling 表面掉皮
surface scouring 表面冲刷
surface sealed system 表面密封系统
surface sharp factor 表面形状因子
surface shear stress 表面剪应力
surface shrinkage 表面收缩
surface slope 地表坡度，地面坡度，水面比降
surface smoothness 表面平整度
surface soil 表土
surface sound pressure level 表面声压强度
surface source 面源
surface speed （磁鼓表面的）线速度
surface spillway 表孔溢洪道
surface state 表面状态
surface storage 地表蓄水，地面贮存
surface strain 表面应变
surface stratum 表层
surface streamline 表面流线
surface stress layer 表面应力层
surface stress 表面应力
surface structure 地表构造
surface sublayer 地表副层
surface subsidence 地面沉降
surface table 大理石桌面
surface tear 表面划伤
surface temperature 地面温度，表面温度
surface temper 表面回火
surface tension balance 表面张力计
surface tension transfer welding 表面张力（过渡）焊
surface tension wave 表面张力波
surface tension 表面张力，表面张力系数，表面压强
surface texture 表面结构，表面组织
surface thermal conductance 表面热导率
surface thermocouple 表面热电偶
surface traverse 地面导线
surface treatment 表面处理
surface tuft method 表面丝线法
surface turbulence 地面湍流
surface type attemperator 表面式减温器，面式减温器
surface type desuperheater 表面式减温器，面式减温器
surface type heat exchanger 表面式换热器，表面式热交换器
surface type intercooler 表面式中间冷却器
surface type meter 面板式仪表
surface undulation 地面起伏
surface velocity 表面流速
surface vibration 表面振动
surface vibrator 表面式振动器，表面振捣器，附着式振动器
surface visibility 地面能见度
surface-volume ratio 面积体积比
surface washing 表面冲洗，表面清洗
surface wash-off 地表冲刷
surface waste 表面损失
surface water drain 地表水排除，地面水排水管
surface water hydrology 地表水水文学，地面水水文学
surface water inlet 地表水进口，地面水入口
surface water intake facility 取地表水设施
surface water-proofer 表面防水剂
surface water-proofing 表面防水
surface water quality 地表水质
surface water resources 地表水资源
surface water sampling 表层采水
surface watershed 地面分水岭
surface water supply 地表水补给，地面水给水，地面水供应
surface water yearbook 地表水年鉴
surface water 表层水，表面水，地表水，地面水
surface wave probe 表面波探头
surface wave 表面波，地面波，地表电波，面波
surface weather chart 地面天气图
surface welding 表面焊接

surface winding　表面绕组
surface wind speed　表面风速
surface wind　地面风，近地风
surface wiring　表面布线
surface wound armature　光滑电枢，无槽电枢
surface wound machine　无槽电机
surface　表面，面，表面的，外表，表面加工
surfacing electrode　堆焊焊条
surfacing machine　表面平整机
surfacing material　铺面材料
surfacing treatment　表面处理
surfacing welding electrode　堆焊焊条
surfacing welding　堆焊
surfacing　表面加工，表面处理，表面修整，覆盖层面，堆焊层面，铺面，堆焊
surfactant　表面活化剂，表面活性剂
surf beat shore　浪击海岸
surf beat　岸边碎波拍击
surf board　防浪板
surfusion　过冷现象
surf　岸边碎波，破波，破浪
surge absorbed capacitor in converter station　换流站冲击波吸收电容器
surge absorber　过电压吸收器，过电压吸收装置，冲击波吸收器，电涌吸收器
surge admittance　冲击导纳，波导纳，电涌导纳
surge amplitude　涌浪振幅
surge arrester for transmission line　线路避雷器
surge arrester　电涌放电器，过电压吸收器，避雷器
surge basin　调压池，调压井
surge bin　缓冲仓，平衡仓
surge capacity　浪涌能力，涌浪速度，过载能力，波动容量，缓冲量
surge chamber　缓冲室，波动室，调压室，（水电站的）调压井，平衡室
surge characteristic　过电压特性，浪涌特性
surge control　喘振防护
surge-crest ammeter　冲击峰值电流表
surge current　冲击电流，浪涌电流
surge damping valve　峰值衰减阀
surge diverter　避雷器，避雷针
surge electrode current　浪涌电极电流
surge excitation　冲击励磁
surge from　涌浪前沿
surge front　冲击波前沿
surge gap　脉冲放电器
surge generator　冲击波发生器，脉冲发生器，冲击发生器
surge height　涌浪高度
surge hopper　缓冲料斗
surge impedance loading　（输电线的）自然功率
surge impedance　冲击阻抗，浪涌阻抗，波阻抗
surge level　冲击电压电平，浪涌电压电平
surge limit　喘振极限
surge line　波动管路【稳压器】，喘振线，波动管线，海边波涛线，浪涌线
surge load　冲击负载
surge margin　喘振裕度
surge nozzle　波动管接口
surge oscillography　脉冲示波器
surge parameter　冲击波参数
surge pile　储料堆，缓冲煤堆
surge pipe　波动管，调压管，平压竖管
surge point　缓冲点
surge power generator　冲击电源发生器，冲击发电机
surge pressure　冲击压力，峰值压力
surge-preventing system　防喘振装置
surge-proof intertripping relay　防冲击联锁跳闸继电器
surge-proof intertrip receive relay　防冲击联锁跳闸接收继电器
surge-proof　防电涌的，防冲击的，非谐振的
surge propagation　喘振扩展，涌浪传播
surge protecter　突波保护器
surge protection　冲击保护，冲击电压保护，浪涌保护
surge relay　冲击继电器，波前陡度继电器
surge resistance　波冲电阻
surge shaft　井式调压室，调压井
surge spray　不均匀喷射
surge stockpile　缓冲料堆
surge storage　暂时过量贮存
surge strength　冲击强度，浪涌强度
surge suppression resistor　抑制冲击电阻器，冲击抑制绕组
surge suppression winding　电涌抑制绕组，冲击抑制绕组
surge suppressor　冲击波抑制器，电涌抑制器，涌波抑制器，浪涌电压吸收器，涌浪消减设备
surge tank drain pump　流动箱疏水泵
surge tank with cooling coil　带冷却盘管的波动箱
surge tank　缓冲罐，缓冲槽，平衡罐，平衡箱，波动水箱，中间贮槽，调压池，调压井，调压水槽，稳压罐，缓冲水池，缓冲箱，波动箱，均压箱
surge tester　脉冲试验仪
surge test　冲击波试验
surge-to-stall margin　喘振失速裕度
surge transformer　浪涌变压器
surge voltage generator　冲击电压发生器，冲击发电机
surge voltage recorder　冲击电压记录器
surge voltage suppressor　冲击电压抑制器
surge voltage　冲击电压，浪涌电压
surge water tank　稳压水箱
surge wave　涌波
surge withstand capability　抗冲击能力，冲击电压承受能力
surge　浪涌，涌浪，波动，湍振，喘振，冲击，冲击波，电涌，汹涌，巨涌
surgical gloves　外科用手套
surging characteristic　喘振特性
surging flow　脉动气流，脉动流
surging force　冲击力，涌浪力
surging limit　喘振边界
surging line　浪涌线
surging pad　推力瓦片
surging shock　涌浪冲击
surging　电涌，冲击，浪涌，脉动，喘振，涌起，起浪

surmounted arch 超半圆拱
surpass 超过
surplus air 过剩空气
surplus capacity 过剩出力，过剩功率，剩余容量
surplus coal 余煤
surplus earth 多余土方，余土
surplus energy 剩余流量
surplus factor 多余因子，剩余因子
surplus flow 剩余流量
surplus fund 盈余资金
surplus generation 富裕发电量，剩余发电量
surplus heat utilization 余热利用
surplus load 超额负荷，超额负载
surplus neutron 过剩中子
surplus or deficit of thermal power 火电盈亏
surplus power 剩余电力，剩余功率
surplus reserve 盈余公积金
surplus value 剩余价值
surplus valve 溢流阀
surplus variable 多余变量，剩余变量
surplus water 弃水，溢水
surplus 超量，过剩，盈余，剩余的
surprise visit 突击视察
surrender 效率系数
surrogate product 替代产品，更换产品
surrogate system 备用系统
surrogate 代理的，代用品，代理人
surrounding air speed 环境风速
surrounding building 周围建筑
surrounding countries 周边国家
surrounding media 周围介质
surrounding obstacles 周围阻碍
surrounding rock 围岩
surroundings around foundation pit 基坑周边环境
surroundings 环境，周围
surrounding wind 周围风
surrounding 周围的，附近的，外界的
surtax 超额累进税，附加税
surveillance and evaluation system 监视与评价系统
surveillance personnel 监督人员
surveillance plan 监督计划
surveillance release 监督证书
surveillance requirements 监督条件
surveillance testing 检查性试验
surveillance 检测【放射性】，监视，监督，监测，检查，跟踪
survey and design expense 勘察设计费
survey and drawing 测绘
survey and exploration for power plant site selection 厂址勘测
survey area 测区面积
survey beacon 测量标志
survey before shipment 发运前检验
survey boat 测量船
survey bureau 测量局
survey camera 测量摄影机
survey clause 检验条款
survey data 测量资料
survey datum 测量基准（面）
survey engineer 测量工程师
survey error 测量误差
survey grid 测量格网，地形勘测网
survey group 测量队，测量组
surveying agent 检验代理人
surveying boat 测验船
surveying calculation 测量计算
surveying computation 测量计算
surveying coordinate grid 测量坐标网
surveying coordinate 测量坐标
surveying equipment 测量设备，测量仪器
surveying guide 调查准则
surveying instrument 测量仪器
surveying line 测线
surveying map board 测图板
surveying peg 测桩
surveying pin 测量针
surveying plane table 测量平板仪
surveying rod 测杆
surveying system 测量系统
surveying vessel 测量艇
surveying 地形学，测量学，测量
survey map 测量图
survey marker 测量标志，测量标记，测量标，方位标
survey mark 基测标记，测标，测量标志，基准标志【地形测量学】
survey meter 巡测仪
survey mission 考察团
survey monument 测量标石，方位标
survey network 测量网，测图网，地形勘测网
survey net 地形勘测网
survey of damage to goods 货损检验
survey of dam breach flood 溃坝洪水调查
survey of water conservancy works 水利措施调查
survey of wind farm 风电场勘察
surveyor's certificate 公证人证明书，鉴定书
surveyor's compass 测量用罗盘仪
surveyor's dial 测量用罗盘仪
surveyor's level 水准仪
surveyor's pole 标杆
surveyor's report 鉴定书
surveyor's table 平板仪
surveyor 鉴定人，勘测员，测量员，检查员，测量人员
survey outline 勘测大纲
survey pin 测钎
survey planning 总体规划
survey platform 观测平台
survey point 测点
survey probe 探针
survey report on quality 品质鉴定证明书
survey report 测量报告书，检验报告
survey route 检查线，勘察路线
survey scale 测量比例尺
survey scope 测量范围
survey service 测绘局，测量局
survey sheet 测图原图，实测原图，测量底图，测量图，外业原图
survey station 测站

survey target 测量觇标
survey team 测量队
survey traverse 测图导线
survey vessel 海底调查船
survey 测量，查勘，视察，设计，调查，研究，地形测绘
survival curve 残存曲线，存活曲线
survival rate 存活率
survival time 存活时间
survival wind speed 安全风速
survive （中子）不被吸收，幸存，活下来
SUR 俄罗斯卢布
susceptance 电纳
susceptibility meter 磁化率计
susceptibility to interference 干扰灵敏度
susceptibility 易受影响或损害的状态，敏感性，灵敏度，感受性，磁化系数，磁化率
susceptible to 易受……的影响
susceptible 敏感的，灵敏的，允许的
susceptiveness 敏感性，灵敏度，灵敏性，敏感度，磁化率
susceptivity 灵敏度，敏感性
SU/SD(startup/shutdown) 开机/停机
suspect assembly 可疑组件
suspend an order for goods 中止订货
suspended acoustic ceiling 悬吊式吸热天棚，悬吊式吸音天棚
suspended arch 悬拱，吊拱
suspended ash 悬浮灰分
suspended boiler 悬吊式锅炉
suspended cable roof 悬索屋盖
suspended cable structure 悬索结构
suspended cable 悬索
suspended catalyst 悬浮催化剂
suspended ceiling system 吊顶体系
suspended ceiling 吊顶，吊型天花板，悬式天花板
suspended coil galvanometer 悬圈式电流计
suspended diode 中断二极管
suspended-drop current meter 滴漏式流速仪
suspended droplet 悬浮小滴
suspended dust 悬浮尘埃
suspended fender 悬挂式护舷木，悬式护弦木，悬木缓冲器
suspended form 悬吊式模板
suspended-frame weir 活动式悬架堰
suspended gypsum board with emulsion paint 涂有乳化油漆的悬吊式石膏板
suspended hydrogenerator 悬式水轮发电机
suspended impurity 悬浮杂质
suspended joint 悬式接头
suspended lamp 吊装照明器
suspended load budget 悬移质平衡
suspended load measurement 悬移质测验
suspended load sampler 悬移质采样器，悬移质取样器
suspended load sampling 悬移质采样
suspended load transport 悬移质搬运，悬移质输送
suspended load 悬荷载，悬荷
suspended matter 悬浮物，悬浮质，悬浮物质，悬移质，悬质
suspended moisture 悬着水分
suspended mud 悬浮淤泥
suspended-net type check dam 悬网式谷坊，悬网式拦沙坝
suspended particles 悬浮颗粒，悬浮粒子，悬浮微粒，悬浮质点
suspended particulate 悬浮粒子
suspended sand 悬浮沙
suspended scaffolding 悬吊式脚手架，悬式脚手架
suspended scintillator 悬浮闪烁体
suspended sediment concentration 悬移质含沙量，悬移质浓度
suspended sediment discharge 悬移质输沙率
suspended sediment 悬浮泥沙，悬浮沉淀物
suspended silt 悬浮泥沙
suspended slab 悬式板
suspended solid contact equipment 悬浮固体接触反应设备
suspended solid contact reactor 悬浮固体接触反应器
suspended solid material 悬浮固体物质
suspended solid 悬浮固体，悬浮固体物，悬浮物
suspended span 悬吊跨度，悬跨
suspended state 悬浮状态
suspended stiffening truss 悬索加劲桁架
suspended structure 悬吊结构，悬挂结构，悬挂式结构物
suspended substance 悬浮物，悬浮体
suspended support 悬索式支架
suspended tray 悬挂托架
suspended truss 悬式桁架
suspended type generator 悬式发电机
suspended type hydrogenerator 悬式水轮发电机
suspended velocity 悬浮速度
suspended water 悬着水
suspended wave velocity logging instrument 悬挂式波速测井仪
suspended 吊挂的
suspender arm load 吊杆臂负荷
suspender 吊材，吊杆，吊筋
suspending insulator 悬挂式绝缘子
suspending magnetic separator 悬挂筒式磁铁分离器
suspending medium 悬浮介质
suspending velocity 悬浮速度
suspend payment of salary 暂停支付工资
suspend payment 暂停付款，宣告破产
suspend work 暂时停工
suspend 悬浮，悬挂，暂停，中止
suspension angle tower 直线转角塔
suspension assembly 悬垂绝缘子串组
suspension box-cofferdam 吊箱围堰，悬浮箱形围堰
suspension bridge 索桥，悬桥，悬索桥
suspension cable 悬缆，吊索，悬索
suspension centre 吊点，悬点
suspension circuit breaker 悬式断路器
suspension clamp 悬垂线夹【带碗头挂板】，悬挂线夹

suspension combustion 悬浮燃烧，火室燃烧
suspension current 含沙流，浑浊流，悬质流
suspension damp 悬垂线夹
suspension electrodynamometer 悬置式电功率计
suspension firing 火室燃烧，悬浮燃烧
suspension fittings 挂线配件，悬挂金具
suspension fuel 悬浮燃料
suspension girder 悬梁
suspension hinge U形挂板
suspension insulator string 悬垂绝缘子串，悬式绝缘子串
suspension insulator 悬式绝缘子
suspension link 悬杆
suspension liquid 悬浮液
suspension load 悬浮荷重
suspension matter 悬浮物
suspension medium 悬浮介质
suspension method 悬吊法【测流用】
suspension of business 停业
suspension of contract 中止合同
suspension of payment 中止付款，中止支付
suspension of snow particle 雪粒子悬浮
suspension of the prescriptive period 时效期中止
suspension of work 工程停工
suspension percentage 悬浮率，悬浊率
suspension pin 悬挂销
suspension pipe 吊管
suspension point 悬挂点
suspension power only pile 悬浮动力堆
suspension ratio 悬浮比
suspension reactor 悬浮液反应堆
suspension reel 悬索卷轴
suspension rod 吊材，吊杆，吊索，悬杆
suspension scaffolding 悬空脚手架
suspension set 悬垂绝缘子串组，悬垂线夹
suspension spotlight 悬挂式聚光灯
suspension straps of a suspension clamp 悬垂线夹的悬垂挂扳
suspension string 悬垂绝缘子串，悬垂式绝缘子串
suspension switch 吊装开关
suspension system 悬置系统
suspension tower 直线塔
suspension transposition tower 直线换位塔
suspension-truss bridge 悬索桁架桥
suspension type alternator 悬吊型发电机
suspension type fuel 悬浮态燃料
suspension type 悬挂式
suspension isolation system 悬吊隔振装置
suspension wire 吊索，吊线
suspension 悬挂，悬浮（物，液），吊，悬置，中止，暂停，悬挂物，悬挂吊架
suspensoids 悬溶胶体
sustainability 可持续性
sustainable development 可持续发展
sustainable energy policy 可持续能源政策
sustainable 可持续的
sustained arc 持续电弧
sustained backfire 持续回火
sustained capacity 持久功率
sustained ceiling current 最大持续电流，持续峰值电流
sustained collector voltage 集电极保持电压
sustained combustion 稳定燃烧
sustained current 持续电流
sustained deviation 持续偏差
sustained fault 持续故障
sustained interruption 持续停电
sustained load 持久负荷，持续负荷，持续荷载，永久载荷，长周期负荷
sustained modulus of elasticity 持久弹性模量
sustained oscillation 持续振荡，持续摆动，等幅振荡，自持振荡
sustained outage 持续停运
sustained overload 持续过载
sustained overvoltage 持续过电压
sustained radiation 持续辐射，等幅波辐射
sustained reactance 稳态电抗
sustained short-circuit test 持续短路试验
sustained short-circuit 持续短路，稳定短路
sustained stress 持续应力
sustained surge 持续涌浪
sustained three-phase short-circuit current 稳态三相短路电流
sustained vibration 持续振动
sustained wave 持续波，等幅波
sustained wind 持续风
sustained yield 持续（生）产量
sustained 持续的，持续不断的，持久的，持久不变的，被支持的
sustaining power 支持能力
sustaining slope 连续斜坡，顺流坡度
sustaining wall 扶墙
sustain 承受
SU＝startup 开机，启动
SU(surface charge) 地面运输费【代理人收取此费为 SUA】
sutficient 丰度，丰富
Sutherland constant 苏斯兰德常数
Sutherland model 苏斯兰德模型
SUT(start up team) 调试启动队
suttle weight 净重
Sutton diffusion parameter 苏通氏扩散参数
Sutton formula 苏通氏（大气污染）公式
Suydam criterion 赛达姆判据
SVA 辅助蒸汽分配系统【核电站系统代码】
SVC(static var compensator) 静止无功补偿器
SVE 运行前试验用蒸汽分配系统【核电站系统代码】
SV(sedimentation velocity) 沉降速度
SV(servo valve) 电液伺服阀【电液转换器】
SVSM(static voltage stability margin) 静态电压稳定裕度
SV(solenoid valve) 电磁阀
SV(stop valve) 截止门，截止阀
swab 擦拭法试验
swaged nipple 异径短节
swage fitting 型模配合（件）
Swagelok fitting 世伟洛克配件
swage nipple 型钢短管
swage 锻造，锻压，铁模，铁型，锻制管，铸模
swaging tool 挤压工具

swallet stream 地下河，伏流
swallowed subsurface water 渗入地下水
swallow hole 落水洞，石灰渗水坑
swallowing capacity （涡轮机的）临界流量
swallowtail 燕尾榫
swampiness 沼泽化
swamping box 补偿箱，浸没箱【热电偶】
swamping resistance 扩量程电阻
swamping 扩程的，沼泽化
swamp land 沼泽地
swampy area 沼泽区
swamp 泥沼，沼地，沼泽，湿地
swan base 卡口灯头，卡口灯座
swan fuse 插入式熔断器
swan neck insulator 弯脚绝缘子
swan neck pin porcelain insulator 弯脚瓷瓶
swan neck 鹅颈弯
swan socket 卡口灯座，卡口管座
swap action valve 交换作用阀，快速作用阀
swap data 交换资料
swap-in 交换入，换入，换进
swap market 交易市场
swap-out 交换出，换出
swap program status word 交换程序状态字
swap time 交换时间
swap 交换，交流，换，（程序的）调动
sward 草地，草土，草坪，草皮
swarm velocity 群集速度
swarm 群，成群
swash bank 波浪爬溅岸
swash channel 冲流水道
swash height 波浪爬高
swash limit 波浪上冲界线
swash mark 冲痕
swash plate 旋转斜盘
swash way 冲流水道
swash 波浪上爬，冲流水道，卫星扫描带
S wave S 波
sway acceleration 摆动加速度
sway angle 摆动角
sway brace 斜支撑
sway response 摆动响应
sway rod 防倾杆，斜撑杆
sway vibration 摆振
sway 摆动，摇动，振摇，摇晃
SWB(seaway bill) 海运单
SWCP(service water closed pump) 闭式水泵
SWC(service water closed) 闭式水
SWCS(service water closed sump) 闭式水箱
sweat cooling 蒸发冷却，流逸冷却，发散冷却
sweating 结露水，渗湿
sweat 熔化【耐火材料】，凝上水珠，水气
SWECS(small scale wind energy conversion system) 小型风能转换系统
swedge 型铁，缩小管径，校直弯管
Swedish National Institute of Radiation Protection 瑞典国家辐射防护研究所
Swedish State Power Board 瑞典国家电力局
sweepage 扫描，扫掠
sweep angle 后掠角【桨叶】
sweep away 冲走
sweep circuit 扫描电路，扫掠电路
sweep-delay circuit 扫描延迟电路
sweeper 扫描振荡器，扫频振荡器，摆频振荡器
sweep frequency 扫描频率
sweep gas 残气，尾气
sweep generator 扫描振荡器，扫描发生器
sweeping coil 扫描线圈，偏转线圈
sweeping 清扫【通风】，清除，除尘，扫除，扫描
sweep interval 扫描时间间隔
sweep method 追赶法
sweep motor 扫描用电动机
sweep of gases 烟气冲刷
sweepolet 镶入式支管嘴
sweep oscillator 扫描振荡器，扫频振荡器
sweep outline 流线型轮廓
sweep tee 弯曲三通管
sweep trace 扫描线
sweep tube 扫描管
sweep unit 扫描装置
sweep voltage 扫描电压，扫频电压
sweep wave form 扫描波形
sweep wood 沉底木
sweep 打扫，清理，清扫，扫描，扫掠，（烟气）冲刷
sweet corrosion 淡水腐蚀
sweetened-sour gas 脱硫化氢天然气
sweeter 频率计，全景分析器
sweet gas 无硫气【天然气处理后】
sweetware 软硬件的合理利用
sweet water 淡水，软水，甜水
swell age 余波龄
swell coal 膨胀的煤
swell conditions scale 涌浪状况等级
swelled bed 膨胀床
swell factor 膨胀系数
swell head 起浪水头
swell height 涌浪高度
swell increment 膨胀量
swelling agent 膨胀剂
swelling capacity 膨胀量，湿胀量
swelling degree 膨胀度
swelling effect 肿胀效应
swelling expansion 肿胀，膨胀
swelling ground 冻胀土，隆起地基
swelling index 回弹指数，膨胀指数，湿胀指数
swelling limit 膨胀极限
swelling material 膨胀性材料，湿胀性材料
swelling of the tidal wave 潮波升高
swelling potential 膨胀潜能，膨胀势
swelling pressure 膨胀压，膨胀压力
swelling property 膨胀性
swelling rate 肿胀速率
swelling rock and soil 膨胀岩土
swelling rock 膨胀性岩石
swelling soil 冻胀土，膨胀土
swelling stress 膨胀应力
swelling test 膨胀试验
swelling zone 膨胀区
swelling （材料）肿胀，膨胀，溶胀，泡胀，鼓

起，河水猛涨
swell scale 涌浪等级
swell-shrink potential 胀缩敏感性
swell-shrink soil 胀缩性土
swell trail 余波
swell 浪涌，涌浪，涨水，膨胀，湿胀，肿胀，隆起
swemar generator 扫描与标志信号发生器
swept area of rotor 风轮扫掠面积
swept area 扫及区，摆动面，扫掠面积
swept back vane 后弯式叶片
swept back 后掠
swept-band 可变波段
swift 急流，湍流，快的，快速的
swimming-pool reactor 泳池式反应堆，水池型反应堆
swing anchor 摆动锚
swing angle 旋角，旋转角
swing bearing 摇座
swing bolt 摆动螺栓
swing boom 起重机旋转臂
swing bridge 平旋桥，旋开桥，转动式桥
swing capacity 冲击性瞬时功率
swing check valve 摆动式止回阀，单摇板止回阀，回旋逆止阀，旋启式止回阀
swing curve 摆动曲线，振荡曲线
swing door 摆动门，双开门，弹簧门，游荡门，旋转门，转门
swing excavator 全回转式挖土机
swing gate 摆动门，旋转门
swing hammer crusher 锤击式破碎机
swing hammer 摇摆锤，回转锤
swinging angle 摆动角，晃动角
swinging arm 摆动臂，摇臂
swinging bracket 悬垂串延长架
swinging choke 摆动扼流圈，变感抗扼流圈，交感扼流圈
swinging damper 旋转挡板
swinging hammer crusher 旋锤式碎石机
swinging hammer 摆锤
swinging-in casement 内开窗扇
swinging load 波动负荷，负荷摆动
swinging mooring 单点系泊
swinging mounting 摆动支架
swinging radius 旋转半径
swinging scaffold 两点悬吊式脚手架
swinging screen 振动筛
swinging sieve 摇筛
swinging single suspension string 单联悬垂串【悬垂转角塔用】
swinging table 回转台
swinging valve 平旋阀
swinging 摆动，摇摆
swing-jib crane 旋臂式起重机
swing link 摆杆
swing loader 转臂式装载机
swing moor 游移沼泽
swing nozzle 摆动喷嘴
swing offset 摆支距，回转支距
swing-out 铰链，枢轴
swing sash 弹簧窗，弹簧窗扇
swing span 平旋跨
swing test 摆动试验，频率不稳定试验
swing-type hammermill 摆锤式破碎机
swing-type rubber lined check valve 旋启式衬胶止回阀
swing valve 止回阀，回转阀
swing wire 摆动缆
swing 摆动，动荡，摆度，振幅，摆幅，回转，旋转，侧移，漂移，悬挂，悬吊，挥舞，摇摆，摇晃，振荡
S-wire S线，塞套引线
swirl angle 涡流角，出汽角的余角
swirl atomizer 旋流式雾化器
swirl diffuser 旋流风口
swirler 旋流器
swirl-flow device 涡流装置，旋流装置
swirl flow 旋流
swirling ascending 旋动上升
swirling combustion chamber 涡流燃烧室
swirling flow 涡流，旋流
swirling vortex 旋涡
swirling wind 旋风
swirling 旋动，打旋，旋涡
swirlmeter 旋进流量计，涡流式流量计，旋涡流量计，涡流计
swirl nozzle type injector 旋流喷嘴型喷射器
swirl nozzle 旋流喷嘴
swirl speed 涡动速度
swirl-stabilized burner 平衡旋流燃烧器
swirl vane moisture separator 旋流叶片汽水分离器
swirl vane 旋流叶片，涡流器叶片，涡流片
swirl 水涡，涡动，涡流，旋风，旋涡
swishing sound 嗖嗖的声音
switchable busbar 有载分段母线
switchable capacitor 可开闭电容器，可换向电容器
switchable 可变换的，可换向的，可用开关控制的
switch adjustment 开关调整
switch-and-fuse combination 开关和熔丝组合
switchback station 车辆调车站
switchback （山区的）之字形路，改变角度
switch bail 开关把手
switch bank 转换触排
switch bar 开关柄
switch base plate 开关板
switch bay 开关安装地位，开关架
switch blade 开关闸刀，闸刀
switchboard attendant 配电盘值班员
switchboard cable 局用电缆，开关板电缆
switchboard cell 配电盘间隔，交换机塞孔单元
switchboard equipment 配电盘设备
switchboard fittings 配电盘配件
switchboard framework 配电盘框架
switchboard frame 配电盘构架
switchboard fuse 配电盘熔断器
switchboard instrument 配电盘测量仪表
switchboard jack 交换机插孔
switchboard lamp 配电盘照明灯
switchboard meter 配电板仪表

switchboard panel 配电盘，配电屏，开关板，转换开关板
switchboard position 交换台坐席
switchboard room 配电间，配电室，开关间，控制室，电话接线室
switchboard type instrument 配电盘式仪表
switchboard wiring diagram 配电盘接线图
switchboard 配电盘，配电板，配电屏，开关板，开关柜，开关装置，交换台，转换器，电话交换机
switch box 开关箱，开关柜，配电箱，转换开关盒，转换开关箱
switch cabinet 开关柜
switch cabin （电缆）汇接室，电缆夹层，电缆配线室
switch cell 开关间，开关间隔
switch characteristic tester 开关特性测试仪
switch circuit automatic network 开关电路自动控制网络
switch clock 开关钟
switch closure 开关闭合，开关闭合信号
switch combination 开关组合
switch contact 开关触点，开关触头
switch control 开关控制
switch cover plate 开关盖板
switch cover 开关盖
switch desk 开关台，控制台
switch-disconnector 隔离开关
switched busbar circuit-breaker 母线分段断路器
switched bus 变换的母线
switched capacitor 开关电容
switched circuit automatic network 开关电路自动控制网络
switched current 开断电流
switched memory 反转存储器
switched network 开关网络，交换网络
switched on 接通（电流）
switched reluctance motor 开关磁阻电动机
switch engine 调车机车
switch equipment 开关设备
switcher 转换开关，转接开关，调车机车
switches and buttons 开关及按钮
switchette 小型开关
switch frame 开关架
switch frequency 切换频率，开关频率
switch-fuse 熔丝开关，开关熔丝
switchgear arrangement 开关柜布置
switchgear cabinet 开关柜，开关间，开关室
switchgear control 开关设备控制
switchgear cubicle for looped distribution network 环网开关柜
switchgear cubicle 开关柜
switchgear and electrical equipment room 配电装置及电气设备室
switchgear framework 开关装置的构架，配电装置的构架
switchgear panel with switching device on truck 手车式开关柜
switchgear panel 开关柜
switchgear room 开关间，开关设备室，室内开关站
switchgear 开关装置，开关设备，配电设备，交换设备，配电装置
switchgroup 组合开关
switch handle 开关把手
switch hook 开关挂钩，开关钩键
switch house 开关间，开关室，控制室，配电室
switch identifier 开关识别器
switching action 开关作用
switching AC voltage 开关交流电压
switching arc 开合闸电弧
switching arrangement 配电装置，电路布置
switching boundary 开关边界
switching capacity 开断容量
switching circuit 开关电路，转换电路
switching command 切换命令
switching compound 开关站
switching computer 开关计算机，开关计算装置
switching configuration 调车线，开关配置，开关组态
switching constant 开关常数
switching criterion 开关准则
switching current 合闸电流，操作电流，开合电流，开关电流
switching curve 开关曲线
switching cycle 开关操作循环，转换周期，整流周期
switching DC voltage 开关直流电压
switching device 开关装置
switching differential 切换差
switching diode 开关二极管
switching duration 开关时间，切换时间
switching element 开关元件
switching endoscope （燃料元件观察用的）旋转内窥镜
switching equipment 开关装置，转换装置，切换装置车设备
switching flow rate 变换流动速率
switching force 合闸力
switching frequency 开关频率
switching function 电路开闭函数，电路开闭作用
switching impulse insulation level 操作冲击绝缘水平
switching impulse sparkover voltage 开关冲击闪络电压，操作过电压
switching impulse test 操作冲击试验，冲击试验
switching impulse withstand voltage 操作冲击耐受电压
switching impulse 开关脉冲
switching input 电路开闭输入，开关输入
switching installation 开关装置
switching in 接入，接通，合闸
switching key 开关操作钥匙
switching kiosk 开关亭，配电亭
switching knob 合闸按钮，切换按钮
switching lever 合闸手柄，合闸杠杆
switching locomotive 调车机车
switching magnet 合闸磁铁
switching manipulation 开关操纵，开关操作
switching matrix 开关矩阵，开闭矩阵，交换矩阵，开合闸矩阵

switching mechanism 开闭机构，切换机构
switching member 开关构件，切换构件
switching module 开关模块，转换模块
switching network 开关网络，转换网络，开闭网络，转接网络，交换网络
switching no-load long line test 分合空载长线试验
switching no-load transformer test 分合空载变压器试验
switching-off arc 开关断路电弧
switching office （电话）中继局
switching-off transient 断路时的瞬变现象
switching off 断开，断路
switching on-off unloaded line 切合空载线路
switching on-off unloaded transformer 切合空载变压器
switching on surge current 合闸涌流
switching on transient 合闸时的瞬变现象
switching on 合闸，接入，接通，连入
switching operation 开关操作，转换操作，切换运行
switching order 开关操作顺序，合闸顺序，操作命令
switching out 断开，断电，断路
switching over of auxiliary power system 自用电源切换
switching overvoltage 操作过电压
switching over 切换，转接，换接
switching performance 开闭性能，开关性能
switching period 切换时间，切换周期
switching point 开关点，开关站，切换点，转接点，（电话）中继站
switching power 切换功率
switching process 转换过程，线路倒换过程，电路倒换过程
switching program （开关）开闭程序
switching pulse 控制脉冲，转换脉冲
switching push button 开关按钮
switching range 开关转换范围
switching rate 转换速度，开关速度
switching relay 转换继电器，开关操作继电器
switching resistor 合闸电阻器
switching rod 开关绝缘棒
switching room 配电室
switching selector repeater （电话）选局中继器，分区换接器
switching sequence 开关顺序
switching shunt reactor test 分合并联电抗器试验
switching signal 转换信号，合闸信号，触发信号
switching stage 交换级
switching starter 直接启动器，开关启动器
switching station 开闭所，开关站，交换站
switching substation 开关站
switching surge absorber 操作过电压吸收器
switching surge overvoltage 操作过电压
switching surge protective level 开关冲击波保护水平，操作过电压保护水平
switching surge test 操作波试验
switching surge 开关冲击，操作过电压，开关峰值，开关浪涌，操作波
switching system 开关系统，转换系统，操作制度
switching test 开合试验
switching time 切换时间，开关时间，转换时间，（磁芯的）翻转时间
switching tone 交换音
switching torque 换接时转矩，（电动机）合闸转矩
switching transient 开关操作过电压，开关瞬态，开关切换时的瞬变现象
switching tube 开关管，转换管，换接管
switching unit 开关模块，转换模块
switching value 开关量，开断值，切换值
switching 通断，切换，开关，转换，整流，配电，（两设备之间）交替，启闭，开闭，转接，换向，交换，调度，调车
switch interlock 开关互锁
switch in 接入，接通，合闸，接上
switch key 开关电键，电话电键
switch knob 开关旋钮
switch lever 开关操纵杆
switch lighting 开关照明
switch lock 道岔锁闭器
switchman （铁路的）扳道员
switch-mode converter 开关方式变换器
switch motion 开关动作，切换动作
switch motor 转辙电动机
switch off 开闸，断路，断开，关闭，关断，切断
switch oil 开关油
switch on 接通，合闸，合上，接入
switch order 开关指令
switch out 断开，关断，关上，开闸
switch over chute 切换下料管，切换下料槽
switch over 转换，（两设备之间）切换，转变
switch panel 控制屏，开关板，控制板
switch pillar 开关柱
switch plant 开关设备，电力开关站
switch position 开关位置
switch quantity signal 开关量信号
switch rail 转辙器，侧线
switch register 开关寄存器
switch resistance box 旋匙式电阻箱，开关电阻箱
switch room 配电室，开关间，电话交换室
switch signal 道岔表示器
switch sleeper 道岔木枕
switch socket 开关插座
switch solenoid drive 开关螺线管传动装置
switch spring 开关弹簧
switch step 切换阶段
switch tank 开关油箱
switch terminal block 开关接线板
switch terminal 开关接线端子
switch through 接转
switch transistor 开关晶体管
switch trip 开关锁链
switch tube 开关管，电子管切换开关
switch turn-off 用开关断开
switch-type function generator 开关型函数发生器
switch-type modulator 开关式调制器
switch-type 开关型的

switchyard compressed air production system 开关站压缩空气生产系统
switchyard structure 变电架构
switchyard 调车场【铁路】，变电站开关室，升压站出线部分，开关站，户外配电装置，露天开关场
switch 开关，交换机，换接，断电器，换向器，转换器，电键，转换，切换，转接，道岔
swivel armchair 转椅
swivel bridge 旋开桥
swivel connectors for pilot wire 抗弯连接器
swivel coupler 固定式车钩，旋转车钩
swivel coupling 旋转式联轴器，转环联轴节
swivel damper 旋转挡板
swivel disc globe valve 可转动的阀盘截止阀
swiveled 回转的
swivel handle with sliding T 活结T型扳手
swivel handle 活动头手柄
swiveling characteristic 转换特性【阻抗继电器】，应变特性
swiveling discharge chute 旋转式卸料槽
swiveling mechanism 旋转装置
swiveling pile driver 旋转打桩机
swiveling vane 回转叶片
swiveling 回转，旋转的
swivel rod 旋转杆
swivel-socket wrench 旋臂套筒扳手
swivel support with ladder boom 附有梯架的旋转支座
swivel support 铰接支座
swivel tightener 旋转紧线器，旋转拉紧器
swivel 防捻器，活节，轮轴，转轴，旋转，转体，旋转体
swiverling pile driver 旋转打桩机
swollen resin 膨胀树脂
SWOP(service water open pump) 开式水泵
swop 交换
SWO(service water open) 开式水
SWR(standing-wave ratio) 驻波比
SW(socket weld) 承插焊，插套焊
SWS(service water system) 辅助水系统
SW(steam water) 汽水
SW=switch 开关
SWU(separative work unit) 分离功单元
syenite 正长岩
systematology 系统学
syllabified code 字节代码
syllable 字节，字段
Sylphone temperature relay 细耳风恒温继电器
sylphon 膜盒，波纹管
sylvite(sylvine) 钾盐，天然氯化钾
symbiont 共存程序
symbol code 符号码
symbol generator display 符号发生显示器
symbol generator 符号发生器
symbolic addressing 符号编址
symbolic address 符号地址
symbolic algebra 符号代数
symbolic assembler 符号汇编程序
symbolic assembly language listing 符号汇编语言表
symbolic assembly language 符号汇编语言
symbolic assembly program 符号汇编程序
symbolic assembly system 符号汇编系统
symbolic circuit 符号电路
symbolic code 符号代码
symbolic coding 符号编码
symbolic conversion program 符号转换程序
symbolic debugging 符号调试，符号排错
symbolic deck 符号卡片组
symbolic device allocation table 符号设备分配表
symbolic expression 符号式
symbolic function 符号函数
symbolic input program 符号输入程序
symbolic input routine 符号输入程序
symbolic language 符号语言
symbolic logic 符号逻辑
symbolic method 符号法
symbolic model 符号模型
symbolic notation 符号表示法
symbolic number 符号数字
symbolic operating system 符号操作系统
symbolic operation 符号运算
symbolic operator 符号算子
symbolic optimal assembly program 符号最优汇编程序
symbolic optimum program 符号优化程序
symbolic program assembly routine 符号程序的汇编子程序
symbolic programming 符号程序设计，符号编程
symbolic representation 符号表示法
symbolic vector 符号矢量，形式矢量
symbol identifier 符号识别器
symbolism 符号体系，符号化，符号系统
symbol language 符号语言
symbol-manipulation technique 符号处理技术
symbol note 符号说明
symbol of unit of measurement 测量单位符号
symbol printing mechanism 符号打印机
symbol processing 符号处理
symbols list 图例
symbol 标志，标记，符号，象征
symeresis 脱水收缩
symmag 对称磁元件
symmetric adjustment 对称调整
symmetrical airfoil 对称翼型
symmetrical alternating quantity 对称交变量
symmetrical arch dam 对称式拱坝
symmetrical balance 对称平衡
symmetrical body 对称体
symmetrical expression 对称式
symmetrical flow 对称流
symmetrical fold 对称褶皱
symmetrical load 对称荷载
symmetrical matrix 对称矩阵
symmetrical mode 对称模态
symmetrical plane 对称面
symmetrical section 对称状分段
symmetrical syncline 对称向斜
symmetrical system 对称系统
symmetrical turnout 对称道岔
symmetrical vein 对称脉

symmetrical wave　对称波
symmetrical　均匀的，对称的
symmetric balance　对称平衡
symmetric binomial distribution　对称的二项分布
symmetric blading　对称叶片
symmetric breaking capacity　（每相间）对称截断容量
symmetric cable　对称电缆
symmetric center　对称中心
symmetric component filter　对称成分过滤器，滤成器
symmetric-component method　对称分量法
symmetric component　对称分量
symmetric current　对称电流
symmetric deflection　对称偏移，对称偏转
symmetric distribution　对称分布
symmetric fault　对称故障
symmetric filter　对称滤波器
symmetric formulation　对称表示法
symmetric function　对称函数
symmetric heterostatic circuit　对称异势差连接电路，对称异电位电路
symmetric impedance　对称阻抗
symmetric induction machine　对称感应电机
symmetric load　对称负载，对称荷载，均衡荷载，均衡负载
symmetric matching　对称匹配
symmetric matrix　对称矩阵
symmetric negative sequence current　对称负序电流
symmetric network　对称网络
symmetric parameter　对称参数
symmetric phase　对称相
symmetric polyphase system　对称多相制
symmetric profile　对称叶型
symmetric short-circuit　对称短路
symmetric stage　对称级
symmetric state　对称态
symmetric system　对称系统
symmetric three-phase load　对称三相负载
symmetric three-phase system　对称三相制
symmetric transducer　对称换能器
symmetric varistor　对称变换器，对称变阻器
symmetric voltage　对称电压
symmetric wave　对称波
symmetrized winding　对称绕组
symmetroid　对称曲面
symmetry axis　对称轴，旋转轴
symmetry centre　对称中心
symmetry law　对称定律
symmetry plane　对称面
symmetry point　对称点
symmetry type　对称型
symmetry　对称（性）
sympathetic vibration　谐振，共振
sympiezometer　双测压管测流计
symposium　（专题）讨论会，专题会议，座谈会，论文集，学术报告会
symptom　迹象，表征，征兆，症状，故障特征
sync circuit　同步电路
synchroclock　同步电钟
synchro control differential transmitter　自动同步控制差动式发送机
synchro control receiver　同步控制接收机
synchro control transformer　同步控制变压器
synchro control　同步控制，同步调整
synchro coupling　同步耦合
synchro data　自整角机数据，自动同步机数据
synchro device　同步机，自整角机
synchro differential receiver　自整角接收机，同步差动接收装置
synchro differential transmitter　自整角发射机，同步差动发射装置
synchro drive　同步传动
synchrodyne circuit　同步电路
synchrodyne　同步机
synchro generator-transformer system　同步发电机变压器系统
synchro generator　同步发电机，自动同步振荡器
synchro indicator　同步指示器
synchro light　同步指示灯
synchrolock　同步保持电路，自动频率控制电路
synchromagslip　自动同步机
synchromesh　同步配合，同步啮合
synchrometer　同步表，同步指示器
synchro motor　同步电动机，自动同步机
synchronal　同步的，同时的
synchronisation signal unit　同步信号单元
synchronisation　同步
synchroniser　同步器，同步装置，整步器
synchronism check relay　同步校验继电器，检同期继电器
synchronism detection　同步检查
synchronism indicator　同步指示器
synchronism or synchronism-check device　同期或同期检查装置
synchronism　同步，同步性，同期，同时性
synchronization character　同步字符
synchronization checking　同步鉴定
synchronization communication　同步通信
synchronization effect　同步效应
synchronization error　同步误差
synchronization factor　同步系数
synchronization generator　同步发电机
synchronization of flow　径流同步
synchronization of station　电站并网
synchronization of turbogenerator　透平发电机并网
synchronization regulator　同步调速器
synchronization system　同步系统
synchronization test　准同期试验
synchronization voltage check　同期电压检查
synchronization zone　同步区
synchronization　并网，同步，同期，整步，使时间相互一致
synchronize and close　并网
synchronized breaker　同步断路器
synchronized hydraulic traction　液压同步牵引
synchronized operation　同步运行
synchronized oscillation　同步振动
synchronized-signal　同步信号
synchronized sweep　同步扫描

synchronizer hand wheel 同步器手轮
synchronizer 同步机，同步器，同步装置，整步器
synchronize 使同步，同步，整步
synchronizing amplifier 同步放大器
synchronizing apparatus 同步设备，整步设备
synchronizing band 同步频带，整步频带
synchronizing bus-bar 同步母线，整步母线
synchronizing capacity 同步能力
synchronizing characteristics 同步特性
synchronizing checking relay 同步检查继电器
synchronizing circuit 同步电路
synchronizing coefficient 同步系数
synchronizing control 同步控制机构，同步调整装置
synchronizing current 同步电流，整步电流
synchronizing cycle 同步周期
synchronizing detection 同步探测
synchronizing device 同步装置，整步装置
synchronizing error 同步误差，整步误差
synchronizing force 同步力，整步力
synchronizing frequency 同步频率，整步频率
synchronizing gear 同步装置，同步机构
synchronizing generator 同步信号发生器，整步信号发生器
synchronizing-impulse generator 同步脉冲发生器，整步脉冲发生器
synchronizing impulse 同步脉冲，整步脉冲
synchronizing lamp 同步指示灯
synchronizing linkage 同步装置，同步联锁
synchronizing moment 同步的瞬间
synchronizing motor 同步电动机
synchronizing of image 图像同步
synchronizing potential 同步电位，同步电平
synchronizing power 同步功率
synchronizing process 同步过程，整步过程
synchronizing pulse 同步脉冲，整步脉冲
synchronizing range 同步范围，整步范围
synchronizing reactor 同步电抗器
synchronizing relay 同步继电器，整步继电器
synchronizing signal compression 同步信号压缩
synchronizing signal generator 同步信号发生器
synchronizing signal level 同步信号电平
synchronizing signal separator 同步信号分离器，整步信号分离器
synchronizing signal 同步信号
synchronizing switch 同期开关，同步开关
synchronizing torque 同步转矩，整步转矩
synchronizing transformer 同步变压器，整步变压器
synchronizing voltage 同步电压，整步电压
synchronizing voltmeter 同步电压表，整步电压表，零值电压表
synchronizing winding 整步绕组
synchronizing 同步，并网，整步，同步的
synchronometer 同步计，同步指示器
synchronoscope 同步指示器，同步示波器，脉冲示波器
synchronous advancer 同步进相机
synchronous alternator 同步交流发电机
synchronous angular velocity 同步角速度
synchronous-asynchronous motor 同步异步电动机
synchronous booster-converter 同步调压换流器
synchronous booster-inverter 同步调压逆变器，同步调压变流器
synchronous booster 同步增压机
synchronous capacity 同步能力
synchronous circuit breaker 同步开关
synchronous circuit 同步电路
synchronous communication satellite 同步通信卫星
synchronous communication 同步通信
synchronous commutator 同步换向器，同步整流子
synchronous compensator 同步调相机，同步补偿器
synchronous computer 同步计算机
synchronous condenser operation 同步调相运行
synchronous condenser 同步调相机，同步电容器，调相器，调相机
synchronous converter 同步变流机，同步换流机，同步转换器
synchronous convertor 同步变流机，同步换流机，同步转换器
synchronous data link control 同步数据链路控制
synchronous data network 同步数据网
synchronous deposit 同时沉积
synchronous detector 同步检测器，同步检波器
synchronous effect 同步作用，整步效应
synchronous electric clock 同步电钟
synchronous electric machine 同步电机
synchronous fading 同步衰减，同时衰落
synchronous field 同步场
synchronous frequency booster 同步频率放大器
synchronous frequency changer 同步变频机
synchronous gate 同步闸门，同步选择
synchronous generator 同步发电机
synchronous hysteresis motor 同步磁滞电动机
synchronous impedance method 同步阻抗法
synchronous impedance 同步阻抗
synchronous induction motor 同步感应电动机
synchronous inverter 同步逆变器
synchronous linear motor 同步线性电动机
synchronous looping-in 同期合环
synchronous machine 同步电机
synchronous motor-generator 同步电动发电机
synchronous motor hunting 同步电动机振荡
synchronous motor 同步电机，同步电动机，同期电动机
synchronous operation 同步运行
synchronous oscillation 同步波动，同步振动
synchronous parallel 同期并列
synchronous phase converter 同步变相机
synchronous phase modifier 同步相位补偿器
synchronous power 同步功率
synchronous pull-in torque 同步牵入功率
synchronous pulse 同步脉冲
synchronous radiation 同步辐射
synchronous reactance 同步电抗
synchronous receiving motor 同步接收电动机
synchronous rectifier 同步整流机，同步变流机

synchronous regulator 同步调节器
synchronous reluctance motor 同步磁阻电动机
synchronous rolling 同步横摇
synchronous RPM 同步转速【转每分】
synchronous running 同步运行，同步运转
synchronous sequential system 同步时序系统
synchronous serial system 同步串联系统
synchronous spark gap 同步火花隙，旋转火花隙
synchronous speed 同步速度，同步转速
synchronous starting 同步启动
synchronous system 同步系统
synchronous tie 同步连接，同步耦合
synchronous time-division multiplexing 同步时分多路转换
synchronous timer 同步计时器，同步秒表
synchronous time selector 同步时间选择器
synchronous torque 同步力矩，同步转矩
synchronous tracking computer 同步跟踪计算机
synchronous transformer 同步变压器，自整角变压器
synchronous transistor logic 同步晶体管逻辑
synchronous transmission 同步传输
synchronous transmit 同步发送
synchronous vibration 同步振动
synchronous vibrator 同步振动器
synchronous voltage 同步电压
synchronous watt 同步功率
synchronous working 同步工作
synchronous 同步的，同期的
synchron 同步
synchro phase shifter 同步移相器
synchrophasing 同步定相
synchro phasometer 同步相位计
synchrophasotron 同步稳相加速器
synchroprinter 同步打印机
synchro reader 同步读出器
synchro receiver 自整角接收装置，同步接收装置
synchro repeater 同步重发器，同步能转发器
synchro resolver 同步分析器，同步解算器，自整角机
synchroscope 同步示波器，同步指示器，同步指示仪，同步测试器，同步表
synchro-shifter 同步转换机构，同步移位器
synchrospeed motor 同步电动机
synchrospeed 同步速度，同步转速
synchrostat regulator 同步调节器
synchrostep remote control 远距离同步控制
synchro-switch 同步开关
synchro system 自整角系统，同步系统
synchros 同步器
synchrotector 同步检波器
synchrotie 同步机联线，同步机馈电线，同步耦合
synchrotimer 时间同步器，同步计时器，同步定时器
synchro-transformer 自整角变压器，同步变压器
synchrotransmitter 同步发送机，同步发送器，自整角发射装置
synchrotrans 同步变压器，同步转换
synchrotron 同步加速器
synchro 同步器【机械变速】，自动同步机，同步的，自动信号同步器，同步机
SYNCH 同步，同期【DCS画面】
synclator 同步振荡器
synclinal axis 向斜轴
synclinal closure 向斜闭合，舟状褶皱
synclinal limb 向斜翼
synclinal valley 向斜谷
synclinal 向斜的
syncline 向斜
synclinore 复向斜
synclinorium 复向斜
sync pulse 同步脉冲
sync section 同步部分
sync separator circuit 同步信号分离电路
sync separator 同步信号分离器
sync signal 同步信号
sync stretch circuit 同步脉冲扩展电路，同步脉冲展宽电路
sync-stretching 同步展宽
syndetic 连接的
syndicated loan 银团贷款
syndicate 银团
syndrome-threshold decoder 校正子阈解码器，校正子门限译码器
syndrome 出错，出故障，(纠错码的)伴随式
synergetics 协同学
synergism 联合作用
synergistic effect 协同效应，协合作用，协合影响
synfuel 合成燃料
syniphase excitation 同相激励，同相励磁
synopsis 概要，摘要，梗概，大纲，提要
synoptic analysis 天气分析，天气图分析
synoptic chart 天气概要图
synoptic correlation 天气相关
synoptic data 天气资料
synoptic forecast 天气预报
synoptic map 天气图
synoptic meteorology 天气学
synoptic model 天气模式，天气模型
synoptic report 天气报告
synoptic scale 天气尺度
synoptics 天气学
synoptic weather chart 天气图
synoptic weather map 天气图
synoptic 天气的
synroc process 合成岩石法【废物处理】
synroc 合成岩石
syntactic analyser 语法分析程序
syntactic analysis 句法分析，语法分析
syntactic class 语法子分类
syntactic pattern recognition 句法模式识别
syntax-control 语法控制
syntax graph 语法图【计算机软件】
syntax language 语法语言
syntax tree 语法树【软件】
synthesis gas 合成煤气
synthesis of hydrograph 过程线合成，水文过程线的构成

synthesis of network 网络综合
synthesis wave 合成波
synthesis 合成，综合
synthesized accelerogram 综合加速图
synthesizer mixer 合成混频器
synthesizer 合成器，综合器
synthetic address 合成地址，综合地址
synthetical 合成的，综合的，人造的
synthetic ammonia 合成氨
synthetic castor oil 合成蓖麻油
synthetic climatology 综合气候学
synthetic curve 综合曲线
synthetic data 综合数据，综合资料
synthetic division 综合除法
synthetic fiber cloth 合成纤维编织布
synthetic fiber 合成纤维
synthetic filter media 合成过滤介质
synthetic flocculant 合成絮凝剂
synthetic foam concentrate 合成泡沫液
synthetic fuel 合成燃料
synthetic gas clean-up process 煤气净化工艺
synthetic gas stream 综合气流
synthetic instruction 合成指令
synthetic insulating rod 合成绝缘杆
synthetic language 人工语言
synthetic load testing 综合负载试验，双频试验法
synthetic method 综合法
synthetic paint 合成涂料
synthetic-pattern generator 复试验振荡器
synthetic piezoelectric crystal 人造压电晶体
synthetic pollution data 综合污染指数
synthetic proof 综合证明
synthetic quartz-crystal 人造石英晶体
synthetic resin filtering material 合成树脂滤料
synthetic resin 人造树脂，合成树脂
synthetic response matrix 综合响应矩阵
synthetic rubber 合成橡胶
synthetic seismogram 合成地震记录
synthetic silicoaluminate 人造硅铝酸盐
synthetic study 综合研究
synthetic system 综合系统
synthetics 合成品
synthetic unitgraph method 综合单位过程线法
synthetic unitgraph 综合单位线
synthetic unit hydrograph method 综合单位线法
synthetic unit hydrograph 综合单位过程线，综合单位线
synthetic varnish 合成清漆
synthetic weather factor 综合气象因数
synthetic 合成的，人造的，综合的，合成物，合成纤维，合成剂
synthon 合成纤维
syntonic circuit 谐振电路
syntonic 谐振的，共振的，调谐的
syntonization 谐振法，同步，同期
syntonize 调谐，谐振，共振
syntony 共振，谐振，调谐
syphonage 虹吸作用
syphon hood 虹吸罩
syphon mouth 虹吸管进口
syphon＝siphon 虹吸管，虹吸道，虹吸
syphon tube 虹吸管
syren 汽笛，警报器，警报机，验音盘
syringe tong 吸管式钳
syringe 喷水器，注射器
system absorbed power 系统吸收功率
system aggregation 系统集结
system allocation document 系统分配文件
system analysis chart 系统分析图
system analysis group 系统分析组
system analysis 系统分析
system and procedure 系统及过程
system approach 系统方法
system assessment(/evaluation) 系统评价
system assurance program 系统保险程序
systematical observation 系统观测
systematic analog network testing approach 系统化模拟网络测试方法
systematic analysis 系统分析
systematic block codes 系统分组码
systematic code 系统码
systematic design method 系统设计法
systematic deviation 系统偏差
systematic drift rate 系统的漂移率
systematic error checking code 系统误差校验码
systematic error 系统误差
systematic evaluation 系统评价
systematic explanation 系统讲解
systematic failure 系统事故
systematic guidance 系统的指导
systematic observation 系统观测
systematic procedure 系统化的程序
systematic random sample 系统随机样本
systematic research 系统研究
systematic sample 系统采样，系统样本
systematic sampling method 系统采样法
systematic sampling 系统采样，系统抽样法
systematic search decoder 系统搜索译码器
systematic seasonal deviation 系统性季节偏差
systematic separation 系统分离
systematic surveillance 系统地监督
systematic 系统的
systematization 系统化，规则化
system black-start capacity 系统黑启动容量
system black-start generator 系统黑启动发电机
system busbar 系统母线
system cabinet 系统柜，系统箱
system centralized control 系统集中控制
system centre 系统中心
system chart 系统框图，系统流程图，系统图
system check 系统校验
system communication 系统通信
system component 系统部件
system configuration 系统组态，系统配置，系统结构图
system connection 系统连接
system constant 系统常数
system control and load dispatching expenses 系统控制和负荷调度费用
system control language 系统控制语言
system control program 系统控制程序
system control room 系统控制室

system control unit 系统控制部件
system control 系统控制
system damping 系统阻尼
system debug 系统调试
system decomposition 系统分解
system demonstration 系统论证
system description 系统说明
system design description 系统设计说明
system design 系统设计
system detailed design 系统详细设计
system development 系统开发
system deviation 系统偏差
system diagnosis 系统诊断
system diagram 系统接线图，系统图
system disconnection and field extinction 解列灭磁
system disconnection 系统解列
system dispatch 系统调度
system display 系统显示
system disturbance 系统扰动，系统干扰
system dynamics model 系统动力学模型
system dynamics 系统动力学，系统动态特性
system earth 系统接地
system efficiency 系统效率
system eigenvalue 系统本征值
system electrical power swing 系统电功率摆动，电力系统振荡
system emergency protection 系统应急保护，系统事故保护
system engineering 系统工程
system environment recording 系统现场记录
system environment 系统环境
system equivalent network 系统等值网络
system error 系统误差
system evaluation 系统评价
system failure recovery 系统故障恢复
system failure 系统故障
system fault 电力网故障，系统故障
system file 系统外存储器，系统文件
system flow chart 系统流程图，系统操作程序图
system flushing 系统冲洗
system for controlling 控制系统
system for coordination of peripheral equipment 外围设备协调系统
system frequency excursion 系统频率偏移
system frequency 系统频率，电网频率
system function 系统函数，系统功能，装置功能
system gain 系统增益
system generation 系统形成，系统生成
system grounding chart 体系接地图
system grounding conductor 系统接地导体
system grounding 系统接地
system halted 死机
system head characteristics 系统阻力特性
system head curve 管网特性，系统压力特性曲线，系统阻力曲线
system head 系统压头，系统阻力
system homomorphism 系统同态
systemic contamination 系统污染
systemic effect 系统效应
system identification 系统识别，系统辨识
system impedance ratio 系统阻抗比
system impedance 系统阻抗
system improvement time 系统的改进时间
system input unit 系统输入部件
system input 系统输入
system installed capacity 系统装机容量
system integrator 系统集成商
system integrity 系统完整性
system interconnection 电力系统相互连接，系统连接
system interface 系统间接口
system international of unit 国际单位制
system International 国际单位制
system interrogation 系统询问
system isomorphism 系统同构
system ladder chart 系统梯形图
system layout 系统布置
system library 系统库
system life cycle 系统寿命周期
system linking 系统耦合
system load control 系统的负荷控制
system load factor 系统负荷系数
system loading condition 系统负载状况，系统负荷状况
system load 系统负载，系统负荷
system logic 系统逻辑
system loss 系统损耗
system main power 电源主动力系统
system maintainability 系统可维护性
system maker 整机制造厂，装配厂
system management facilities 系统管理设备
system management 系统管理
system manual 系统手册
system matrix 系统矩阵
system maximum delay 系统最大延时
system maximum hourly load 系统小时最大负载
system mistake 系统错误
system modeling 系统建模
system model 系统模型
system monitoring and control 系统监控
system network architecture 系统网络体系结构
system noise 系统噪声
system of automatical internal audit 内部自动审计制度
system of centralization 集权制
system of decentralization 分权制
system of deliberation 评议制度
system of director assuming full responsibility 厂长负责制
system of economic accounting 经济核算制度
system of economic cooperation 经济合作制度
system of electrical energy utilization 用电体系
system of fixed quota 定额制（度）
system of job responsibility 责任制
system of liability 赔偿责任制度
system of licensing 许可证制度
system of measurement units 测量单位制
system of nets 网系，网组
system of notation 记数法，符号系统
system of periodical maintenance 定期修缮制
system of plates 电容器片组，极板组

system of promotion 晋级制度
system of quantities 量制
system of rewards and penalties 奖惩制度
system of the economic responsibility within an enterprise 企业内部经济责任制
system of time-sharing system 分时系统的系统，分时系统的体系
system of units （测量）单位制
system of verification 核查制度
system of working hours 工作时间制度
system of working shifts 倒班制
system on-line network for information control 信息控制的系统联机网络
system operating capacity 系统工作容量
system operating department 动力系统运行科
system operation test 系统操作试验
system operation 系统操作，系统运行
system optimization 系统优化
system oscillation 系统震荡
system outline design 系统概要设计
system output language 系统输出语言
system output unit 系统输出部件
system overshoot 系统超调量
system packaging 系统封装
system parameter 系统参数
system pattern 系统连接方式，系统模式
system peak load 电网尖峰负荷，系统尖峰负荷
system performance index 系统性能指标
system performance 系统性能
system planning 系统规划
system power control 系统电源控制器
system power swing 系统振荡，系统震荡
system pressure 系统压力
system processing 系统处理
system programmed operator 系统程序操作员
system programming 系统编程，系统程序设计
system program 系统程序
system queue area 系统排队区
system reactance 系统电抗
system reference library 系统引用库
system related outage 系统相关停运
system relay protection 系统断电保护
system reliability 系统可靠性
system requirement 系统要求
system research 系统研究
system reserve capacity 系统备用容量
system reserve 系统储备
system reset 系统复位
system resistance 系统阻力，系统电阻，电力网电阻
system resolution 系统的分辨能力
system response 系统频率特性，系统反应，系统响应
system safety engineering 系统安全工程
systems analysis technique 系统分析技术
systems analysis 系统分析
system scale 系统规模
system schematic 系统图
system selection 系统选择
system self regulation 系统自调节
systems engineering 系统工程，系统工程学
system sensitivity 系统灵敏度
system separation 系统分列，系统解列
system service control point 系统服务控制点
system service transformer 系统供电变压器
system shape factor 系统形状因子
system short circuit capacity 系统短路容量
system simulation 系统仿真，系统模拟
system software 系统软件，系统程序
system specification 系统规范
system split incident 系统解列事故
system splitting 系统解列，系统分列，系统分离
systems proof test 系统检验试验
systems science 系统科学
system stability 系统稳定性
system start-up 系统启动
system state 系统状态
system statistical analysis 系统统计分析
system status display 系统状态显示
systems test facility 系统测试设备
system stored energy 系统储能
system structure chart 系统结构图
system supervisor 系统监控程序，系统管理程序
system support management 系统支撑管理
system switching 系统切换
system synthesis 系统综合
system test time 系统测试时间
system test 系统测试
system thermal power plant 系统热力发电厂，系统火力发电厂，电力系统火电厂
system thermal ratio 系统热比
system transfer function 系统传递函数
system tuning 系统调试
system under test 被测系统
system unit 系统的单元
system utilization factor 系统利用率
system variable 系统变量
system verification 系统验证
system voltage drop 系统电压降，电力网电压降
system voltage stabilizer 系统电压稳定器，电力网电压稳定器
system voltage 系统电压，电力网电压
system volume including pressurizer system 包括稳压器系统的系统容积
system with effectively earthed neutral 中性点有效接地系统
system with non-effectively earthed neutral 中性点非有效接地系统
system without idle periods 无空转周期的系统
system 系统，体系，装置，网络，方式
SYS 系统【DCS 画面】
syzygy tide 朔望潮
syzygy 大潮期
SZR（sodium cooled zirconium hydride moderated reactor） 钠冷氢化锆慢化反应堆
S 公用系统【核电站系统代码】
S 设定，设置【DCS 画面】

T

tabacosis 烟尘肺，烟末沉着病
tabella 侧板
tabernacle 临时性房屋，帐篷
tabetisol 层间不冻层【在永久冻土之间】，不冻地
tab 接片，标签，制表
tableau 表
table base register 表基址寄存器
tabled fish plate 带扣榫的接合板
table-driven compiler 表格控制的编译程序
tabled scarf joint 叠嵌接合
tabled scarf 叠嵌接合
table earthwork balance 土方平衡表
table feeder 平板式给料机，圆盘给煤机，圆盘式给煤机
table land 高原，台地，高地
table look-aside buffer 表格后备缓冲器
table look-up instruction 查表指令
table look-up 查表，一览表【指令】
table mountain 平顶山
table of clauses 条款目录
table of compound interest 复利表
table of contents 目录
table of difference 差分表
table of drawings 图纸目录
table of errata 勘误表
table of hydro-metric data 水文测验成果表
table of limits 限额表，公差表
table of organization 机构表
table projector 台式投影仪
table simulation 表模拟
table simulator 算表模拟程序
tables of equipment 设备表
table tap 台用插头，台用分接头
tablet coating 压片涂层
tablet computer 平板电脑
table top computer 台式计算机，桌上电脑
tablet 小块，图形输入板
table vibrator 台式振动器
table vice 台式虎钳
table-water 饮用水
table 目录，台架，平板，表，表格，平台，图表
tabular calculation 表格计算法
tabular data 表列数据，表列资料
tabular display 表格显示
tabular figure 表列数据
tabular form 表格形式
tabular gypsum 层状石膏
tabular joint 板状节理
tabular language 列表语言，表列语言
tabular sequence control 列表时序控制器
tabular structure 板状构造
tabular values of Bessel function 贝塞尔函数值表
tabular value 表列值
tabular 表列的，平板状的，薄层的，表格式的，表格
tabulated solution 表解
tabulated value 表列值
tabulating card 卡片
tabulating equipment 卡片设备，卡片系统
tabulating machine 制表机
tabulating system 制表系统
tabulating 列表，制表
tabulation analysis 列表分析
tabulation character 制表字符
tabulation method 列表法
tabulation 编表，制表，制表机
tab washer 止动垫圈，锁片垫圈，有耳垫圈
tab 接片【焊接】，调整片
tacheometer staff 视距标尺
tacheometer 流速计，速度计，准距仪，视距仪，测距器
tacheometrical alidade 视距照准仪
tacheometrical survey 视距测量
tacheometric polygon 视距导线
tacheometric rule 视距计算尺
tacheometric table 视距计算表
tacheometric traverse 视距导线
tacheometry 视距测量，视距测量法
tachometry 转速测量法
tacho-alternator 测速发电机
tachodynamo 测速发电机
tachogenerator 测速传感器，测速发电机，转速发电机，转速表传感器
tachograph 转速表，转速记录器，转速图表，旋转表，速度记录器，转速记录仪，转速记录图
tachometer generator 测速发电机，转速传感器，测速电机
tachometer sender 测速发送机
tachometer stabilized system 转速计的稳定系统
tachometer 转速表，转速计，流速计，测速计，测速仪，视距仪
tachometric relay 转速继电器
tachometry 测速，转速测定法
tachomotor 测速电机
tachoscope 手提转速计
tachoscopy 手提转速表
tacho 转速计
tachymeter 视距仪
tacit acceptance 默示接受
tacit agreement 默契
tacit approval 默认
tacit confirmation 默示确定
tacit consent 默示同意，默许
tacitly approve 默认
tacit ratification 默示批准，默认
tackability 黏着能力
tack and pinion gate lifting device 齿条齿轮式启门机
tack claw 钉爪
tack coat 黏结涂层
tack eliminator 防黏剂
tackifier 增黏剂
tacking 定位焊，定位铆，紧钉
tackle block 滑车组，滑轮组，滑轮，滑车，钢绳滑车，复滑车

tackleburton 辘轳
tackle for paying-off 放线滑车
tackle 滑车，索具，绞车，装备，用具，扭倒，处理，抓住，固定，与……交涉
tackling system 起重系统，起吊系统
tackproducing agent 增黏剂
tack rivet 定位铆钉，平头铆钉
tack welding 定位焊，点固焊，点焊，平头焊接
tack weld 定位焊，固定焊，定位点焊，平头焊，点固焊，点焊，定位焊缝，临时点焊，间断焊，预焊
tack 点焊，平头钉，抢风调向，图钉
taconite 铁燧岩
tactile 触觉的
TAC(total acceptance) 完全接受
tactron 冷阴极充气管
tact system 流水作业
TACT(the air cargo tariff) 航空货物运价
tag attachment 挂牌
tag block 标记端子排，标签接线板
tag-end 末端
tagged atom 标记原子，示踪原子
tagged element 标记元素
tagged item 有特点的物质
tagged particle 示踪颗粒
tagged photon spectrometer 标记光子谱仪
tagged 加标签的
taggers 薄铁皮
tagging method 标记法，跟踪法
tagging 识别，辨认，鉴定，标记，做标记
tag holder 标签夹
tag-line method 牵线法
tag marking 特征记号
tag No. 位号【美国用法】，标志号，编号
tag number 编号
tag of cable 电缆标牌
tag protocoli dentifier 标签协议标识
tag system 特征系统
tag wire 终端线
tag 标签，特征，标志，签条，标号，标牌，标条，金属包头，电缆终端接头，辅助信息，贴以标签
tail bay 尾水池，下游河段
tail beam 端梁
tail bearing 梁尾端支承，尾轴承
tail block 末端滑轮
tail board scraper 尾板铲土机
tail channel 尾水渠【水电站】
tail drum 尾部滚筒
tail effect 尾端效应
tail end booster 线路末端升压器
tail end process 尾端过程
tail end treatment 尾端处理
tail end （核燃料后处理）尾端【仅卧式泵】
tail gas treatment technology 尾气处理技术
tail gas treatment unit 尾气处理装置
tail gas 尾气，废气
tail gate 尾水闸门，下尾闸门，下游闸门
tail head 尾水水头
tail-heating surface 尾部受热面
tail idler 后部托辊
tailing dam 尾矿坝
tailing turbine 后置式汽轮机
tailing 拖尾，跟踪，衰减尾部，尾渣，底渣，残渣，富选残余物，蒸馏残渣，尾材
tail light 尾灯
tail of the peak 峰尾
tail-of-wave impulse test voltage 闪络试验波尾电压，冲击试验波尾电压
tail-of-wave 波尾
tailor-made motor 定制的电动机，专用电动机
tailor-made 量身打造的，定制的
tail piece 墙端短梁，尾管
tail pipe chamber 尾气管燃烧室
tail pipe 排气（尾）管，尾水管，吸入管
tailplane spar 水平尾翼翼梁
tail pulley 尾部滚筒
tail race dam 尾水坝
tailrace platform 尾水平台
tailrace pool 尾水池
tailrace surge basin 尾水调压池
tailrace surge chamber 尾水调压室
tailrace tunnel 尾水隧洞
tailrace （水电厂的）尾水渠，退水渠
tail reach 渠尾段
tail reservoir 后池，下（水）库
tail rope 尾索
tails assay 尾料分析【浓缩装置】，尾料测定率【浓缩度】，尾气分析，尾料丰度
tail section 尾部
tail-sharpening inductor 后沿锐化线圈
tailstock 尾架
tail stream 尾料流，废料流
tail swing 尾摆度
tail tower 尾塔
tail trimmer 墙端短托梁
tail-up on a rope 拉紧装置，收紧绳索
tail-up 拉紧装置，收紧轮
tail vane 尾翼
tailwater canal 尾水渠
tailwater channel 尾水渠
tailwater depth 尾水深度
tailwater elevation 尾水高程
tailwater erosion 尾水冲刷
tailwater gallery 尾水廊道
tailwater level 尾水水位，尾水位，下游水位
tailwater pit 尾水坑
tailwater pond 尾水池
tail water rating curve 尾水水位流量关系曲线
tailwater recovery 尾水回收
tailwater tunnel 尾水隧洞
tailwater velocity 尾水流速
tail water 废水，尾液，尾水，下游水
tail wheel 尾轮【尾舵上的多叶片风轮】
tail wind component 风的顺风分量
tail wind 顺风
tail wire 尾端线，出线
tail 引线【电子管】，末端，尾料，贫料，尾巴，尾部，废料，尾随脉冲
tainter gate 弧形闸门
tainton method 高电流密度锌电解法
take account of 考虑

take action 着手
take advantage of 利用
take a profit 获得利润
take a step 采取措施
take back goods 退货
take charge of 主持
take delivery of goods 收货
take down 拆卸，降下，压下
take effect from(/on) 自……开始生效
take effective measures 采取有效措施
take effect 生效，施行
take exception to 提出异议
take . . . for example 以……为例
take-home pay 实得工资
take . . . in and out of service 将……投运与退运
take into account 考虑，入账
take inventory 盘存
take legal action 起诉
take levels 抄平
take measures 采取措施
take moment about point A 对A点取力矩
taken as final and binding upon 对……均有最终约束力
taken at 2.0m below 取……以下2米
take no account of 不考虑
take notes on 记录
taken over by 由……接管
take off coupling table 摘钩平台
take-off line 引出管
take-off pipe 放出管，放水管，排出管，分支管，取出管
take-off point 引出点，分支点
take-off the line 解列
take-off tower 出线塔
take off valve 输出阀
take-off 起飞，发射，取出，取走，移去，分接头，输出功率，施工测量
take of grout 吃（耗）浆量
take on new order 接受新订单
take on 承担，呈现，雇用
take-or-pay agreement 或取或付协议
take out and maintain the insurance in effect 投保并使保险保持有效
take out a patent for 得到……的专利权，取得……的专利
take out a policy 投保
take out of service 停止使用
take out of storage 出仓
take out 取出，去掉，出发，发泄，把……带出去，邀请（某人）外出，开始
takeover bid 盘进（一个公司）的出价
takeover certificate （电厂）验收合格证书，验收合格证
take overhead 从塔顶取出，从顶接收
take over land for use 征用土地
take over 承担，接管，（电厂）移交，交接，验收
take part in international market competition 参与国际市场竞争
take part with 与……合作
take precautionary measures against 采取措施以防止……
take precedence of 优先于
take precedence over （应）优先于……
take precedence 优先
take remedial action 采取补救措施
take revenge 打击报复
take sights 抄平，瞄准
take sth. seriously 重视
take stock 盘存
take suction from 从……吸取
take the floor 参加讨论
take the form of 取……形式
take the top priority of 优先考虑
take up a bill 付清款项以收回票据
take up a position 担任职位
take-up mechanisms 拉紧机构
take-up pulley 拉紧装置滚筒
take-up travel 拉紧行程
take-up unit 拉紧装置
take-up 卷线装置，张紧装置，拉紧，缠绕，卷线盘，松紧装置，提升装置
taking a wait-and-see approach 采取等看的方式
taking of steam 抽汽
taking-over certificate 移交证书
takktron 辉光放电高压整流器
taktron 冷阴极充气二极管
talbe 无线电应答装置
talcose schist 滑石片岩
talc-schist 滑石片岩
talc slate 滑石板岩
talcum powder 滑石粉
talcum 滑石
talc 滑石，滑石粉
talents market 人才市场
talent 才能，人才
talik （在永久冻土之间的）层间不冻层，不冻地
talk-back circuit 联络电路，内部对讲电路，回话电路
talking on the telephone 电话通话，通话
talk-listen button 通话按钮
talk-listen switch 通话开关
talk position 通话位置
talk-ring key 呼叫通话两用键
talk 谈话，讨论
tall beaker 高形烧杯
tall building 高层建筑，高层建筑物
tall capsule building 高层盒式建筑
tall chimney 高烟囱
tall frame shear wall structure 高层框架剪力墙结构
tall frame structure 高耸框架结构
tall guyed tower 高拉索塔
tall shear wall structure 高层剪力墙结构
tall slender structure 细高结构
tall stack source 高烟囱（污染）源
tall stack 高烟囱
tall suspension building 高层悬挂建筑
tall tube building 高层筒体建筑
tall tube-in-tube building 高层筒中筒建筑，高管

套管式建筑
tally fee 理货费
tallyman 点货（数）员，理货员
tally order 作总结指令，结算指令
tally sheet 计数纸
tally 计算，结算，总计，标签，筹码，运算，标记牌
tall 高的
talus accumulation 坡下碎屑堆积
talus cone 岩屑堆
talus creep 山麓堆积体蠕动
talus deposit 岩屑堆积
talus material 坡积物
talus slide 堆积体滑坡
talus wall 堆石护岸
talus 坝脚抛石（水电），山麓堆积
talweg 河道中泓线，最深谷底线，主航道中心线
tamped backfill 夯实回填土，回填土夯实
tamped cinder 炉渣夯实
tamped concrete 捣实混凝土
tamped finish 夯实整修面
tamped tunnel 折返，调头
tamper-proof 防干扰的
tamper-resistant 不可破坏的，抗干扰的，无掺杂的
tamper 打夯机，捣棒，捣固机，夯具，夯实机，夯实器，夯土机
tamping bar 夯棍
tamping device 捣实工具
tamping drum 碾压滚筒
tamping in layers 分层夯实
tamping machine 打夯机，击实机
tamping plate 夯板
tamping rod 击实杆
tamping roller 夯击式碾，夯实压路机
tamping stick 装药棒
tamping 打夯，夯实，夯
tamp-iron collection box 废铁收集箱
tamp 捣固，夯实，捣实
TAM remote manipulator arm TAM 远距离操作器臂
tandem arrangement 纵列装置，串联装置
tandem axle 双轴，串列轮轴
tandem blade 串列式叶片
tandem-bowl scraper 串联斗铲运机
tandem car dumper 双车翻车机
tandem central office 中心转接站
tandem-completing trunk 转接中继线
tandem-compound doubleflow 单轴双排汽
tandem-compound singleflow 单轴单排汽
tandem compound turbine 单轴多缸汽轮机，串联双轴汽轮机，纵联复式汽轮机，串联复式汽轮机，单轴汽轮机
tandem-compound turbogenerator 串联双轴汽轮发电机
tandem connection 串联
tandem control 串联控制
tandem crescent rotary dumper 新月形串联翻车机
tandem double-flow turbine 单轴双排汽汽轮机
tandem drive 串联驱动，双轴驱动
tandem dumper 双联翻车机
tandem engine 串列式发动机
tandem-junction solar cell 串结太阳能电池
tandem knife switch 级联刀闸开关
tandem mirror experiment 串列磁镜试验
tandem mirror fusion-fission hybrid 串列磁镜聚变裂变混合堆
tandem mirror reactor 串级磁镜反应堆
tandem mixer 复式拌和机
tandem motor 双电枢电动机，级联电动机
tandem office 中继局，转接局
tandem-pulley gear 串联式滑轮传动
tandem road roller 两轮压路机
tandem roller 串联式碾压机，双轮式滚压机
tandem selection 中继选择，转接选择
tandem sequence submerged arc welding 纵列多丝埋弧焊
tandem solar cell 叠层太阳电池【级联太阳电池】
tandem switch 汇接中继机键
tandem transistor 双晶体三极管
tandem tunnel 串联式地槽
tandem turbine 单（串）轴汽轮机，串联汽轮机，单轴汽轮机
tandem type pump 串联式水泵
tandem unit 串联驱动
tandem wheel gripper 夹轮器
tandem wound coil 纵列线圈
tandem 单轴的，串联的，串联，纵列，前后排列，纵列的
tangency condition 相切条件
tangent-angle tower 直线转角塔
tangent angle 正切角
tangent dead end 直线终端
tangent deflection 切线挠距，切向偏移
tangent distance 切线距离
tangent friction force 切向摩擦力
tangent galvanometer 正切电流计
tangential acceleration 切向加速度
tangential admission 切向引进，切向进入
tangential approach principle 切线进场法
tangential approximation method 切线逼近法
tangential beam tube 切向（中子）束管道
tangential blade spacing 叶片栅距
tangential bracing 切向加强筋
tangential burner 角式燃烧器，角式喷燃器，切向燃烧器
tangential channel 切向孔道
tangential component 切向分量，切向部分
tangential corner firing 切向四角喷燃
tangential couple 切向力偶
tangential curvature 切面曲率
tangential deformation 切向变形
tangential dimension 切向尺寸
tangential discharge 切向放电
tangential distortion 切向畸变
tangential fan 贯流式通风机
tangential firing 切向燃烧，切圆燃烧
tangential flow stream turbine 切向回流式汽轮机
tangential flow turbine 切击式水轮机，冲击式水

轮机
tangential flow 切向流动，切向流，切向水流
tangential force coefficient （叶片）切向力系数
tangential force 切向力
tangential gradient probe 切向梯度探头
tangential in-phase （叶片）切向同相位
tangential key 切向键
tangential keyway 切向键槽
tangential line height 切线之间的高度
tangential line load 切向线荷载
tangential line 切线
tangentially fired boiler 切向燃烧锅炉
tangentially fired combustion chamber 有角式燃烧器的炉膛，切向燃烧炉膛
tangentially fired furnace 角式燃烧炉膛
tangential motion 切向运动
tangential nozzle 切向接管
tangential out-of-phase （叶片）切向异相位
tangential partial turbine 切向非整周进水式水轮机
tangential point load 切向点荷载
tangential shearing stress 切向剪应力
tangential shrinkage 切向收缩
tangential stiffness matrix 正切刚度矩阵
tangential strain 切向应变
tangential stress in flange 法兰切向应力
tangential stress 切向应力
tangential thrust 切向推力
tangential turbine 切击式水轮机，冲击式水轮机，切流式水轮机，切向水轮机
tangential velocity distribution 切向速度分布
tangential velocity 切线速度，切向速度
tangential wave 切向波
tangential wheel 切击式水轮机，冲击式水轮机，切向水轮
tangential wind stress 切向风应力
tangential wind 切向风
tangential 切线的，切向的，切面的，切线，正切，正切的，切向
tangent key 切向键
tangent line 切线
tangent method 切线法
tangent modulus martrix 正切模量矩阵
tangent modulus of elasticity 弹性正切模量，切线弹性模量
tangent offset 切线垂距，切线支距
tangent of helix angle 导程角正切，螺旋角正切
tangent of the loss angle （电容器的）损耗角正切
tangent pitch tube furnace 切接管炉膛，密排管炉膛【无间隙】
tangent plane 切面
tangent scale 正切标度，正切尺，切线斜率
tangent strain 直线耐张
tangent suspension tower 直线铁塔
tangent suspension 直线悬垂
tangent tower （输电线的）直线塔，直线塔
tangent tube construction 切接管结构，密排管结构
tangent tube wall 切接管水冷壁，密排管水冷壁，小节距管壁
tangent wedge 切向槽楔
tangent welding 多级平行弧焊
tangent 切线，正切，正切的，切线的
tangible assets 可计资产，实物资产，有形资产
tangible benefit 有形效益
tangible product 有形产品
tangible 可触知的，现实的，有形的
tankage 容积，容量，箱容量，燃料箱，储槽，池容量，贮槽
tank base 箱底
tank block 玻璃浴池耐火砖，箱座
tank boiler 火管锅炉，箱式锅炉
tank bottom heater 箱底加热器
tank capacitor 槽路电容器，振荡回路电容器
tank capacity 储水量，贮水量
tank cap 箱盖
tank cart up-dump 槽车【无盖货物】
tank car 运水车
tank circuit 谐振电路，槽路
tank cooler 油箱冷却器
tank cover 箱盖
tanker vehicle 罐车
tanker 油箱，加油车，油轮，油罐，供水车
tank farm 废物罐场地，（油）罐场
tank furnace 浴槽炉
tank gauge 液位计
tanking clearance 油箱侧壁与器身间的距离
tanking 贮在槽内【基础】，装入油箱
tank level indicator 箱液面指示器
tank life 槽液寿命
tank-mounted 箱体装配的
tank oil cooler 罐油冷却器
tank on tower 高架水箱
tank overflow 水箱溢流
tank railway car 油槽车
tank reactor 罐式反应堆
tank retention 桶（罐）内贮存
tanks and accessories 箱罐和附件
tank sender 液面信号发送器
tank sheet 槽用钢板
tank strainer 水柜滤器
tank surge 调压井涌浪
tank test 水槽试验，水箱试验
tank tower 水箱塔架
tank trailer 油槽拖车
tank transformer 油箱式变压器，油冷变压器
tank truck 槽车，油槽车
tank-type circuit breaker 罐式断路器，箱式断路器
tank-type reactor 箱式反应堆
tank voltage 槽电压
tank wagon 油罐车，油槽车，罐车
tank 水箱，油箱，槽，柜，容器，（储能）电路，罐，筒仓，箱
tannic acid 丹宁酸，鞣酸
tanning bleach 硬化漂白
tannin 丹宁，丹宁酸
tantalate 钽酸盐
tantalum absorber 钽吸收剂
tantalum capacitor 钽电容器
tantalum lamp 钽丝灯

tantalum rectifier　钽整流器
tantalum　钽
tantamount　等值的，等价的，同等的，相等的
tap adjuster　分接头调整装置，分接头切换开关【变压器】
tap bolt　龙头螺栓
tap-change operation　分接变换，分接头换接【变压器】
tap changer control of converter transformer　换流变压器分接头控制
tap changer control　分接头控制
tap changer　抽头转换器【变压器】，抽头开关，调压开关，抽头换接［切换］开关，分接头换接位置，分接开关
tap-change transformer　有分接头切换的变压器，分接头切换变压器
tap changing apparatus　分接开关【变压器】，分接头切换装置
tap changing arrangement　分接开关【变压器】
tap changing device　抽头转换装置
tap changing switch　分接开关【变压器】，分接头切换开关
tap changing transformer　抽头转换变压器
tap changing　抽头切换，分接头切换
tap cock　水管栓
tap density　压实密度
tap-drill　螺孔钻，螺纹底孔钻
tap each face blind　（法兰）每面盲孔攻丝
tape alternation　磁带交替切换
tape-armoured cable　钢带铠装电缆
tape armour　钢带铠装
tape beginning marker　带始标
tape bootstrap routine　带引导程序
tape code　磁带码，带码
tape comparator　带比测器
tape control unit　磁带控制单元，带控装置
tape control　磁带控制
tape covering　带包绝缘层
tape data family　带数据族
taped insulation　带绕绝缘
tape drive　（磁）带驱动器，（磁）带驱动装置，磁带输入机
tape drum　带鼓
tape duplicator　磁带复制机
tape editor　磁带编辑程序
tape feed　磁带卷盘，拖带机构，纸带馈送，磁带馈送
tape file　磁（纸）带外存储器
tape gauge　传送式水尺，活动水尺
tape handler　卷带机
tape handling unit　带处理装置
tape identification unit　带标识器
tape insulated coil　带绝缘线圈
tape insulation　带绝缘，带绕绝缘
tape label　磁带标号
tape layer　扁绕
tape layout　带设计，带配置
tape leader　带首，带头，引导带
tape library　带程序库
tape limit　带限制
tapeline　钢卷尺，卷尺
tape-loop storage　环形带存储器
tape loosing　磁带松弛
tape machine　自动收报机，磁带录音机
tape marker　磁带终端标记，磁带始端标记，磁带标记
tape mark　磁带标记，带标，带端标记
tape measure　卷尺，皮尺，带尺
tape mechanism　卷带机构
tape memory　磁带存储器
tape merge　带合并
tape module　磁带模件
tape operating system　磁带操作系统，带操作系统
tape oriented system　面向磁带系统
tape parity　带奇偶
tape perforator　带穿孔机
tape plotting system　磁带绘图系统
tape pool　磁带组
tape preparation unit　带准备装置
tape preventive maintenance　带预防维护
tape printer　带式打印机，纸条印刷电报机
tape probe　磁带探头
tape-processing simultaneity　带同时处理
tape puncher　纸带穿孔机
tape punch　纸带穿孔
taper angle　尖削角，锥角，楔削角
taper bit　锥形绞刀，锥形钻头
taper bridge reamer　锥桥形绞刀
taper cone　圆锥
taper dowel　锥销
tape read and write library　带读写库
tape reader　读带机，带读出器，带阅读机
tape reading　带读数
tape recorder　磁带记录器，磁带录音机
tape recording density　磁带记录密度，带记录密度
tape recording　磁带记录
tapered aeration　渐减曝气，逐渐减弱通气
tapered aerofoil　不等弦翼面
tapered beam　递变截面梁，楔形梁
tapered bed　锥形床
tapered bit　带锥形接头的钻头
tapered blade　变截面叶片，锥形叶片，楔形叶片
tapered-body pole　梯形极
tapered bolt anchorage　锥形螺杆锚具
tapered cascade　锥形级联
tapered center-column rotameter　带有锥形芯柱的转子流量计
tapered chimney　锥形烟囱
tapered column　递变截面柱，锥形柱
tapered cowl　锥形风帽
tapered deep-bar squirrel-cage rotor　楔形深导条笼型转子
tapered downcomer　斜降液管，斜降液板
tapered draft tube　锥形尾水管
tapered enlargement　锥形扩大
tapered extension　锥形延伸段
tapered-flange beam　楔形工字钢
tapered haunch　斜腋
tapered inlet　锥形进口

tapered-land thrust bearing　斜面式推力轴承
tapered line　阻抗递变线路
tapered-overhang winding　锥端绕组
tapered pin　圆锥销，锥形销
tapered plug valve　锥塞阀
tapered plug　圆锥形插塞，锥形塞
tapered potentiometer　非线性电势计，非线性电位差计
tapered press fit　楔形压配合
tapered reamer　锥形扩孔器，锥形绞刀
tapered ring seal　锥形环密封垫
tapered roller bearing　斜滚柱轴承
tapered roller　锥形辊
tapered-shank bit　锥形柄钻头
tapered sheet covering tapered structure　覆盖锥形结构的锥形薄板
tapered slide valve　锥形滑阀
tapered slot　斜槽，斜沟，梯形槽
tapered square washer　斜方形洗槽
tapered tenon　锥形榫，斜榫
tapered transition joint　锥形过渡接头
tapered tube rotameter　带有锥形管柱的转子流量计
tapered wedge　斜楔
tapered winding　圆锥形线圈
tapered wire rope　变截面钢丝绳，锥头索
tapered　锥形的，斜的
tape reel　磁带盘，色带盘，带圈
tapering gutter　端宽沟，斜沟
tapering spindle　锥形轴
taper joint　锥形接头
taper key　楔键，楔形销，斜键
taper land thrust bearing　斜面式推力轴承
taper-loaded cable　锥形负载电缆，端部负载逐渐变小的电缆
taper or cone hood　锥形排风罩
taper pile　锥形桩
taper pin　圆形锥，圆锥销
taper pipe thread　锥管螺纹
taper pipe　缩节，异径管，锥形管
taper ratio　锥度比
taper seat valve　锥形座阀
taper seat　锥形阀座，锥形座
taper tap　丝锥
taper wedge　斜楔
taper　锥度，锥形，(圆)锥体，楔销，拔销，斜度，锥形物，变尖，逐渐变细［缩减］，尖锥
tape sampler　大气微粒取样器，纸带取样器
tape search unit　带搜索单元
tape selection unit　带选装置
tape shielding　条带屏蔽的
tape skip　跳带指令
tape speed　磁带转速
tape splice　胶带黏结
tape station　磁带机，纸带机
tape storage　磁带存储器，纸带存储器
tape-stored　磁带存储的
tape-to-card converter　纸带到卡片的转换器
tape-to-card　从带到卡片
tape-to-tape conversion　带带转换
tape trailer　带尾
tape transmitter-distributer　带发送分配器，读带机
tape transport　磁带输入机，带传送
tape unit　磁带机，纸带机
tape width　磁带宽度，纸带宽度
tape winding　布带缠绕
tape-wound core　带绕磁芯
tape-wound　带绕的
tape　磁带，带，胶带，卷尺，皮尺，纸带
tap field control　分段励磁电动机速度控制，抽头激磁控制
tap field motor　有分段励磁绕组的电动机
tapgential strain　切向应变
taphole　出口，出渣口
taping machine　（线圈的）包带机
taping　缠带
tap lead　抽头
tap line　分接线路
tap-off box　分线箱【支线连接】
tap-off　抽头，分接头，分接
tapped actuator　偏心轮执行机构
tapped air-gap reactor　分段气隙电抗器
tapped air　抽出的空气
tapped coil　多接头线圈，抽头线圈
tapped control　分接头控制
tapped delay line　抽头延迟线
tapped inductor　抽头感应线圈，多头感应线圈
tapped line　T 接线路，支线，抽头线，多抽头线路
tapped-potentiometer function generator　抽头电势计式函数发生器
tapped potentiometer　带有抽头的电势计
tapped stator winding　抽头定子绕组
tapped transformer　抽头式变压器
tapped-tuned circuit　抽头调谐电路
tapped variable inductor　可抽头式感应线圈
tapped-winding　分组线圈，多抽头绕组
tapped　抽头的，分接的，带向分接头的，带分接头的，攻螺丝的
tappered HP rating　分级功率【调速直流电动机】
tapper　轻击锤，字母打印设备的键，电报键
tappet clearance　挺杆间隙，阀杆与阀座间的间隙
tappet drum　凸轮鼓
tappet gear　挺杆传动装置
tappet roller　止推辊，推杆滚柱
tappet shaft　凸轮轴
tappet switch　制动开关【带导轮的开关】
tappet　挺杆，推杆，随行件，捣杆
tapping adjuster　分接开关调节器【变压器】
tapping attachment　攻丝装置，攻丝夹头
tapping bar　出渣通道的手柄
tapping breast　放流口泥塞，风口铁套
tapping capacity　分接容量
tapping changing mechanism　分接开关调节机构【变压器】
tapping clamp　钻孔夹具
tapping contactor　分接接触器
tapping current　分接电流

tapping factor 分接系数
tapping hole 螺纹孔，螺丝孔，排渣孔
tapping machine 钻孔眼机
tapping point 抽头点，（汽轮机的）抽汽口，抽汽点，分支点，分接头，（地下水的）出水点，（测压管的）取压点
tapping power 分接容量
tapping quantity 分接量
tapping range 分接头调节范围，分接范围【变压器】
tapping ratio 总匝数与抽头匝数之比【变压器】
tapping screw thread 自攻螺纹
tapping screw 攻纹螺钉，自攻螺钉
tapping-selector drum switch 鼓形分接头选择开关
tapping slag 排渣
tapping sleeve 分流套管
tapping spout 出渣口
tapping step 分接级【变压器】
tapping switch 分接开关，抽头开关【变压器】
tapping temperature 出炉温度
tapping transformer 抽头变压器
tapping voltage 分接电压
tapping 抽头，攻丝，分接头，分支，泄放，出渣，开孔，夺流，分接，攻螺丝，攻螺纹，流出口，泄水管
tap position indicator 分接头位置指示器
tap position information 分接头位置信息
tap selector 分接头选择器，分接开关
tap support 螺丝攻支架
tap switch 分线开关
tap the furnace bottom 排除炉底熔渣
tap type transformer 抽头式变压器
tap voltage 抽头电压，分接头电压
tap water works 自来水厂
tap water 自来水，饮用水
tap wrench 丝攻扳手，丝锥扳手，铰杠，螺丝攻扳钳，螺丝纹扳手
tap 抽头，分接头，螺丝攻，（液态排渣炉）排渣，龙头，丝锥，开发，接头，抽头，开关，塞，水龙头，丝攻
tar asphalt 柏油，焦油沥青
tar boiler 柏油熔化炉
tar cement 柏油胶合料
tar coated road 柏油路
tar concrete 焦油混凝土，焦油沥青混凝土，柏油混凝土
tar cracking 焦油裂化
tared filter 配衡过滤器
tare gross 皮重
tare-load ratio 自重载重比
tar epoxy paint 柏油环氧漆
tare ratio 重新校正
tare weight 包装重量，车皮自重，皮重，（集装箱）自重，容器重量
tare 皮重，（风洞）支架阻力皮重，包装重，（集装箱）自重，容器重
tar felt 柏油毡
tar finisher 柏油铺面机
tar from lignite 褐煤焦油
target acquisition 目标显示，目标捕获
target analysis 目标分析
target area 觇板面积，目标区，目标区域
target burnup 目标燃耗【辐照】
target computer 结果程序计算机，特定程序计算机
target cost 目标成本
target cut off voltage 靶截止电压【摄像管】
target date 目标日期，预定日期
target flow characteristics （模拟）目标气流特性
target flow measuring device 靶式流量测量仪表
target flowmeter 靶式流量计
target flow transmitter 靶式流量变送器
target flow 目标流量，预定流量，规定流量，指定流量
target hitting activities 达标活动
target homing 目标自动接近
target image 目标图像
target irradiation 目标燃耗【辐照】
target lamp 灯塔，目标灯
target language 目标语言，译成语言
target load 目标负荷
target machine 目标机，靶机
target material 靶材料，靶材
target nucleus 靶核
target plane 靶面
target plate 靶板
target position indicator 目标位置指示器
target preparation 靶制备
target price 目标价格
target program 目标程序，目标计划，目标语言组成的程序
target purchased price 标定买价
target return method 资本报酬率定价法
target rod 觇板水准尺，标杆
target search 目标检索
target simulator 目标仿真器
target specification 目标规范
target-speed responder 特定速响应仪
target speed setter 目标转速给定装置
target tissue 靶组织
target tracking beam 目标跟踪射束
target variable 目标变量
target word 目标字
target 靶，目标，中间电极，（继电器动作）指示标志，标志，觇标，目标值，指标
tar heater 柏油加热器
tariff agreement 关税协定
tariff compact 关税协定
tariff concession 关税减免
tariff configuration 费率设置【计费表】
tariff for foreign trade 国际贸易税则
tariff of electricity sent to the grid 上网电价，标杆电价
tariff policy 关税政策
tariff preference 关税优惠
tariff rate 关税率
tariff 关税表，关税率，关税，税，价目表，价格表，费，费率，运价，收费表
tar macadam pavement 柏油碎石路面
tar macadam road 柏油碎石路
tar macadam 柏油碎石路

tarmac 铺地用沥青
tar membrane 焦油沥青涂层
tarnish film 锈膜，氧化膜
tarnish 生锈，锈，污点，（表面）变色
tar oil preservative 柏油防腐剂
tar oil 煤焦油，焦油
tar paper 柏油纸
tarpaulin 防水布，防水油布，柏油帆布
tar paving 柏油铺面
tar pollution 焦油污染
tar pump 柏油泵
tarp 柏油帆布，防水油布
tarred felt 焦油浸渍油毡
tarred marline 柏油麻绳
tarred road 柏油路
tarred roofing felt 屋面油毡
tarred rope 柏油麻绳
tarred 涂焦油的
tarring number 焦（油）化值
tarring value 焦（油）化值
tar road 柏油路
tarry hemp fiber 沥表麻丝
tarry valve 调时阀，定时阀
tarry 柏油的，涂柏油的
tar sand 焦油砂
tar still 焦油蒸馏釜
tartaric acid solution 酒石酸溶液
tartaric acid 酒石酸，二羟基丁二酸
tarviated macadam 柏油碎石路
tar 柏油，沥青，焦油
tasimeter 微压计
task abnormal exit routine 任务异常出口程序
task allocation 任务分配
task analysis 任务分析
task attribute 任务属性
task condition 任务状态
task control block 任务控制块
task coordination 任务协调
task cycle 作业周期
task descriptor 任务描述信息，任务描述符，状态矢量
task detach 任务清除，任务终止
task dispatcher 任务调度
task execution memory 任务执行存储器
task family 任务族
task force on the project 项目特别小组
task force 特遣队，专案组，专门工作组，（执行特定任务的）工作小组，特别（行动）小组，突击队，攻关小队
tasking （计算机系统中的）任务分配
task level language 作业级语言
task lighting 工作照明
task list 工作项目表，工作任务清单，工作单
task management 任务管理
task memorandum 任务备忘录
task optimization 任务优化
task performance 工作特性，工作效能
task program 作业程序
task scheduler 任务调度程序
task schedule 任务调度，任务计划
task sheet 工作单
task suspension 停工
task unit 任务单位，任务单元
task 作业，工作，任务
taste 口味
TA(tangent angle) 正切角
TA(technical adaptation) 技术适应性修改
T-attenuator T衰减器，T形接法阻尼器
taut band 紧带，发条
tautline position-reference system 紧绳监测定位标准系统
taut line system 张紧绳装置
tautness meter 拉力计
tautness 拉紧
tautology 重言式
taut strip model 拉条模型，紧带模型
taut strip suspended 紧带悬挂的
taut wire traverse 张绳
taut wire 紧拉钢线，张线
taut 拉紧的，紧张的
tax abatement 减税
taxable income 应税所得，应纳税所得额，须纳税的收入，应税收入
taxable item 税目
taxable 可征税的
tax accrued 应计税金
tax allowance 免减税
tax and profit submitted to the state 上缴税利
taxation law 税收法
taxation rate 税率
taxation 税款，税务，税制，征税
tax authorities 税务当局
tax avoidance 避税
tax bearer 纳税人
tax bureau 税务局
tax collector 收税人
tax concession policy 税收优惠政策
tax concession 税收优惠，税项减免
tax credit 税收抵免
tax day 纳税日
tax-deductible 在计算所得税时予以扣除的
tax delivery 上交税金
tax dispensation 税收制度
tax dodger 偷税漏税人
tax dodge 逃税，偷税
tax due 应付税金
taxes and duties 税费
tax-evader 偷税漏水者
tax evasion 避税，逃税，偷税
tax exclusion 免税
tax exemption 免税
tax fraud 偷漏税
tax-free investment 免税投资
tax-free 免税
taxi cab 出租车
taxi driver 出租车司机
tax income 征收所得税
taxi 出租车
tax jurisdiction 税务管辖权
tax law 税法
tax levied 征税
tax on business 营业税

tax on enterprise　企业税
taxonomy　分类学
tax on　征税
tax payer　纳税人
tax payment receipt　纳税凭证
tax rate　税率
tax rebate　退税
tax reduction　减税
tax refund　退税
tax reimbursement　退税
tax relief　税款减免
tax revenue　税收收入
tax shelter　避税
tax stamp　纳税印花
tax system　税制
tax　税
Taylor expansion　泰勒展开式
Taylor's approximation　泰勒近似法
Taylor's series　泰勒级数
Taylor's stability chart　泰勒土坡稳定图
Taylor's statistical theory　泰勒统计理论
Taylor standard screen　泰勒标准筛
Taylor system　泰勒制
T-bank position control loop　温度控制组位置调节控制回路
T-bar　T形钢，丁字钢
TBD(to bo determined)　待定
T-beam　丁字梁
TBEA　特变电工【一家以生产变压器为主业的公司名称】
T-bend　三通管，三通接头
T-BFBP(turbine-driven boiler feedwater booster pump)　汽动锅炉给水前置泵
TBFP(turbine-driven boiler feedwater pump)　汽泵，汽动锅炉给水泵
TBL(through bill of lading)　全程提单
T-bolt　T形螺栓
T-branch pipe　三通管
TB(transfer damper)　转换挡板
TCC(transmission congestion contract)　输电阻塞合同，传输拥挤合同
TCD(target completion date)　目标完成日期
T-circuit　T形电路，T形网络
TCN(transmission company of nigeria)　尼日利亚输电公司
T-connector　T形头，T形连接器，T形线夹
TCPAR(thyristor controlled phase angle regulator)　晶闸管控制相角调节器
TCP(transmission control protocol)　传输控制协议
T/C ratio(thickness chord ratio)　翼型厚-电抗器
T/C ratio　翼型厚弦比
T-crossing　丁字路口
TCR(thyristor controlled reactor)　晶闸管控制电抗器
TCSC(thyristor controlled series capacitor)　晶闸管控制的串联电容器
TCS(turbine building closed cooling water system)　汽轮机厂房循环冷却水系统
TCS(turbine control system)　汽轮机控制系统
TC＝thermocouple　热电偶
T/C(time charter)　定期租船，简称期租
TC(time constant)　时间常数
TCT(time charter on trip basis)　航次期租
TCV(temperature control valve)　温度控制
TC　高压主汽门控制【DCS画面】
TDCS(total distributed control system)　集散控制系统
TDC(total distributed control)　集散控制系统
TDD(technical design office)　技术设计室
TDEFL(triple diffused emitter-follower logic)　三次扩散射极跟随器逻辑
T&D equipment (transmission & distribution equipment)　输配电设备
TDH(total dynamic head)　总动态压头，总动压头
T-dike　丁字堤
TDM(turbine diagnose monitoring)　汽轮机运行监视、诊断管理
TDM(turbine transient data management system)　汽机瞬态数据管理系统【汽机振动分析】
TDS(tax deducted at source)　从源扣缴的税收，源头税
TDS(total dissolved solids)　总溶解固体物，总溶盐量【水分析】
TDS(turbine Island vents, drains, and relief system)　汽轮机厂房通风、排水和溢流系统
T&D(transmission and distribution)　输(变)电和配电，输配电
TD　汽机负荷指令【DCS画面】
teach and repeat　施教和不断重复
teaching machine　教学机
teaching programming　示教编程
teaching reactor　教学用反应堆，训练用反应堆
teagle　绞车，卷扬机，绞盘机，绞辘
teak wood　柚木
teak　柚木
team chief　队长
team decision theory　队决策理论
team design　成套设计
team leader　队长，领队
team of expert　专家小组
team theory　队论
tear and wear allowance　磨损留量，允许磨耗
tear and wear　磨损，损坏
tear down　拆卸
tear-fault　捩断层
tear foil　防爆
tearing rupture　扯裂
tearing strength　扯裂强度
tear resistance　耐磨力
tear up the contract　撕毁合同
tear　撕裂，磨损，扯裂
teaser coil　梯塞线圈，调节变比的线圈
teaser transformer　梯塞变压器
teaser winding　梯塞绕组，可调绕组，辅助绕组
technetronic　以使用电子技术来解决各种问题为特征的
technetron　场效应高能晶体管，场调管
technial specification　技术规范书
technical advantage　技术优势
technical advice　技术建议
technical adviser　技术顾问

technical advisory group 技术顾问团【美国】
technical advisory personnel 技术咨询人员
technical advisory service 技术咨询服务处
technical advisory 技术咨询
technical advisor 技术咨询人员，技术顾问
technical agreement 技术协议
technical aid 技术援助
technical alternatives of various power plant system process 工艺系统技术方案
technical analysis 工业分析，技术分析
technical and economical advantage 技术及经济优势
technical and economical index 技术经济指标
technical and economical indicator 技术经济指标
technical and economical target 技术经济指标
technical and economic analysis 技术和经济分析
technical and economic appraisal 技术经济评价
technical and economic assessment 技术经济评价
technical application 技术应用
technical appraisement 技术鉴定，技术评估［评价］
technical approach 技术因素分析法，技术途径
technical approval team 技术审议组
technical approval 技术认可，技术审批
technical arbitration 技术性仲裁
technical archives 技术档案
technical assessment 技术鉴定
technical assistance for coordination 协调的技术援助
technical assistance program 技术援助方案
technical assistance 技术援助
technical atmosphere 工业大气压，工程大气压
technical backbone 技术骨干
technical background 技术经历，技术背景
technical backstopping 技术依托
technical basis 技术依据
technical bid tabulation 技术报价对比表，技术评标
technical bid 技术标，技术报价
technical breakthrough 技术突破
technical bulletin 技术通报
technical capacity 技术能力
technical center 技术中心
technical characteristic 技术特性，技术性能
technical chart 技术曲线，曝光曲线
technical check record 技术检验记录
technical check record 技术检验记录
technical check 技术检验
technical circular 技术通报
technical class 技术等级
technical code for ground treatment of buildings 建筑地基处理技术规范
technical code for work and acceptance 施工验收技术规范
technical code 技术规程，技术规范
technical coefficient 技术系数
technical collaboration 技术协调［协作］
technical commission 技术委员会
technical committee 技术委员会
technical competence 技术才能，技术能力
technical condition 技术条件
technical constant 技术常数
technical control board 技术控制板
technical control 工艺控制，技术管理，技术控制
technical coordination 技术协调［协作］
technical corrigendum 技术勘误表，技术勘误
technical criteria 技术性准则，技术标准【电力技术可靠性】
technical data sheet 技术数据表
technical data 技术数据，技术资料
technical delay 技术延迟
technical delegation 技术代表团
technical description 技术说明
technical design phase(/stage) 技术设计阶段
technical design 技术设计
technical deviation 技术偏差，技术指导
technical discussion 技术讨论
technical documentation 技术资料
technical document 技术文件
technical drawing 技术图纸
technical economical comparison 技术经济比较
technical and economical index 技术经济指标
technical economic analysis 技术经济分析
technical-economic efficiency 技术经济效率
technical-economic evaluation 技术经济评价
technical-economic quota 技术经济定额
technical economy 技术经济
technical education 技术教育
technical evaluation criteria 技术评价标准
technical evaluation 技术评价
technical expertise 技术鉴定，技术知识，技术专长
technical expert 技术专家
technical failure 技术故障
technical feasibility 技术可行性
technical feature 技术性能
technical flow diagram 工艺流程图
technical force 技术力量
technical grade 工业级
technical handbook 技术手册
technical identification 技术验证
technical index 技术指标
technical information center 技术情报中心
technical information 技术信息，技术资料
technical innovation investment 技术革新费
technical innovation 技术革新
technical inspection report 技术检验报告
technical inspection 技术检验
technical instructions given to the construction personnel 施工技术交底
technical instructions 技术说明书，技术指导
technical interface 技术接口
technical interpretation of design intention before construction （施工图的）设计交底
technical interrogation and reply 技术问答
technical investigation 技术调查
technicality 技术性，技术细节
technicalization 技术专门化，技术化
technical justification material 技术验证资料
technical know-how contract 技术诀窍合同

technical know-how education 技术诀窍教育
technical know-how 技术秘密，技术诀窍，技术知识，技术专长，专有技术
technical knowledge 技术知识
technical know-why 技术理论
technical leader 技术总负责（人）
technical lettering 工程字
technical levelling 工程水准测量
technical level 技术水平
technical life technical data 技术数据
technical lifetime 技术寿命
technical life （电厂或其他设施）技术上能服役的年限，技术寿命
technical load 工艺负荷，技术负荷
technically pure 工业纯，工业纯的
technical maintenance 技术保养
technical management 技术管理
technical manager 技术经理
technical manual 技术导则，技术手册，技术细则
technical man 技术员
technical measurement 技术测定
technical measures for construction work 施工技术措施
technical measures 技术措施
technical memorandum 技术备忘录
technical merit 技术价值
technical mission 技术代表团
technical monitoring 技术监查
technical monopoly 技术专利
technical negotiation 技术协商
technical notes 技术注释，技术备忘录，技术说明，工艺说明
technical objective document 技术指标文件
technical obsolescence 技术陈旧
technical offer 技术报价
technical open-air climate 技术室外气候
technical operation 技术运行
technical order 技术规范
technical parameter 技术参数
technical part 技术部分
technical patent 技术专利
technical performance 技术性能
technical personnel(/staff) 技术人员
technical person 技术员
technical presentation 技术演示【设计交底会，会审】，技术报告
technical problem 技术难题
technical program presentation 技术方案介绍
technical proposal 技术报价书，技术建议书，工程建议书，技术报价，技术方案
technical provision 技术规定，技术条款
technical-pure 工业纯的
technical ranking of enterprise 企业技术等级
technical reconstruction 技术改造
technical reference 技术参考文献
technical regulation 技术规范，技术规程，技术条例
technical reliability 技术可靠性
technical renovation project 技术改造工程
technical renovation 技术革新
technical report 技术报告
technical requirement 技术要求，技术条件
technical responsibility 技术责任
technical retrofit project 技术改造工程
technical review 技术审查
technical safety appraisal 技术安全评价
technical scrutiny 技术审查
technical service company 技术服务公司
technical service fee 技术服务费
technical service 技术服务
technical skill 技巧，特殊技能
technical solution 技术方案
technical specification for retaining and protection of building foundation excavation 建筑基坑支护技术规程
technical specifications 技术规格，技术条件，技术规范（书），技术说明（书）
technical standard 技术规格，技术标准
technical storage 技术储备
technical superintendent 技术厂长，技术主任
technical supervision 技术监督
technical support 技术配合［支持，支援］
technicals 技术术语，技术细则
technical talks 技术会谈
technical terms 技术术语，工程术语，专业术语，技术名称，专门名词
technical theory 技术理论
technical title 技术职称
technical tolerance 技术公差
technical total loss 推定全损
technical training 技术培训
technical transfer 技术转让
technical transformation 技术改造
technical visa 技术签证
technical viscometer 工程黏度计
technical work 技术工作
technical 技术的，工艺的，学术的，专业的
technician on probation 见习技术员
technician 技术员，技师
technicist 技术人员，技师
technicolor 彩色电影，彩色电视
technic 术语，专门技术，工艺术语，技术，技巧
technique assembly 装配工艺
technique for project evaluation 项目评估技术
technique of display 像重显技术
technique of X-ray topography X射线形貌技术
technique patented by sb. 由某人取得的专利
technique route 工艺路线
techniques of geotextile laying 土工布铺设工艺
technique 工艺，技术
techno-economic analysis 技术经济分析
techno-economic center 技术经济中心
techno-economic forecast 技术经济预测
techno-economic norm 技术经济标准
techno-economic professional 技经人员
techno-economic verification 技术经济论证
techno-economist 技术经济学家
technogovernance 高技术治国，高技术统治
techno-hip 精通现代科技界行话的，精通现代技术行话的

techno-industrial 技术工业的，技术先进工业的
technological advance 技术进步
technological and economic index 技术经济指标
technological characteristics 工艺特性
technological conditions 工艺条件
technological contract law 技术合同法
technological cooperation 技术合作
technological design 工艺设计
technological development 技术开发
technological economics 技术经济学
technological equipment 工艺设备
technological exchange 技术交流
technological factor 技术因素
technological forecasting 技术预测
technological gas 工业气体
technological innovation 技术创新
technological laws and regulations 技术法规
technological layout 工艺流程图
technological parameter 工艺参数
technological performance test 工艺性能试验
technological process 工艺过程，工艺流程
technological progress in enterprises 企业技术进步
technological progress 技术进步
technological radiation protection 工艺辐射保护，工艺辐射防护
technological rating 技术作业定额
technological renovation 技术改造
technological requirement 工艺要求
technological sophistication 技术完善程度，采用尖端技术
technological source 工艺过程（污染）源
technological specifications 技术规格
technological transformation 技术改造
technological 技术的，工艺的
technologist 工程技术人员，科技工作者
technology assessment 技术评价，评估
technology-based solutions 以技术为基础的解决方案
technology contract 技术合同
technology equipment importation 技术设备引进
technology for electricity saving 节约用电技术
technology import 技术引进
technology innovation expense 技术革新费
technology-intensive enterprises 技术密集型企业
technology market 技术市场
technology of operations 经营技术
technology of production 生产技术
technology of waste disposal 废物处置技术
technology service company 技术服务公司
technology structure 技术结构
technology system 工艺系统
technology trade 技术贸易
technology transfer patent 技术转让专利权
technology transfer 技术转让
technology transformation 技术转移
technology 技术，工艺，工艺学
technometrics 技术计量学
technostructure 技术专家控制体制，技术专家体制，技术（专家）阶层
technote 工程符号，工程记号
tectogenesis 构造运动
tectonic analysis 构造分析
tectonic basin 构造盆地
tectonic characteristics 构造特征
tectonic earthquake 构造地震
tectonic element 构造单元，构造要素
tectonic fissure 构造裂隙
tectonic force （地壳）构造力
tectonic geology 构造地质（学）
tectonic instability 构造不稳定性
tectonic level 构造层位
tectonic line 构造线
tectonic metamorphism 构造变质
tectonic movement 构造运动
tectonic stress 构造应力
tectonic system 构造体系
tectonics 大地构造学，地壳构造的，构造地质学
tectonic terrace 构造阶地
tectonism 构造作用
tectonoblastic 构造变晶的，构造碎裂的
tectorium 疏松层
tectosphere 构造圈
tee-beam and slab construction 肋形板结构
tee-beam 丁字梁
tee bend T形接头，三通
tee branch 三通管
tee-coil T形线圈
tee connector T形分线夹，T形分线端子，T形接头，三通，T形接法，T形连接
tee connector T型接头【用于分接头】
teed line T接线路
tee fitting 三通管，T形管，三通接头
tee girder 丁字形大梁，T形梁
tee iron 丁字形铁件，T字钢
tee joint 三通，T形连接套管，T形接头，三通接头
tee junction T形接头，三通
teem 浇铸，铸造，倒出
tee network T形网络
tee-piece T形接头，三通
tee pipe T形管，丁形管
Teepol 阴离子表面活性剂的商名
tee-root T形叶根
tee slot T形槽
teeter angle 摇摆角
teeter bearing 摇摆轴承
teeter bed 搅动床
teeter brake 跷板式制动器
teeter bumper 摇摆防撞器
teeter hinge 跷板铰链
teetering angle 跷板角，摇摆角
teetering bearing type 摇摆轴承类型
teetering hinge 摇摆铰链
teetering hub 摇摆轮毂，跷板式桨毂
teetering rotor 跷跷板风轮，摇摆风轮
teeter restraint model 跷跷板控制模型
teeter restraint 摇摆限制器，摇摆阻尼器
teeter 摇摆器，跷板式结构，跷跷板，摇晃，摇摇欲坠，举棋不定
teeth bar 齿条

teething trouble 早期故障
tee transformer T形变压器
tee tube T形管
tee weld T形焊缝
tee 三通，T字钢，T字形，T形物，分接，丁字管接，三通管
TEFC(totally enclosed with fan cooler) 全封闭风冷型
teflon-coated beads 涂聚四氟乙烯的玻璃珠
teflon hose 聚四氟乙烯软管
teflon insulation 聚四氟乙烯绝缘，绝缘塑料
teflon lined pipe 衬聚四氮乙烯管
teflon O ring 聚四氟乙烯O形环
teflon resin 聚四氟乙烯树脂
teflon rod 聚四氟乙烯棒
teflon seal tape 聚四氟乙烯密封带
teflon sheet 聚四氟乙烯板
Teflon 聚四氟乙烯，特氟纶，特氟隆
tefzel 乙烯-四氟乙烯共聚物
TEG 废气处理系统【核电站系统代码】
telautomatics 遥控力学，远距离自动控制
tele-action 遥控作用
teleadjusting 远程调节
teleammeter 遥测安培计，遥测电流计
teleautomatics 遥控力学，远距离自动控制
telebar 棒料自动送进装置
telebit 二进制遥测系统，遥远比特位
telecamera 电视摄像机
telechirics 遥控系统
telechron motor 电钟用电动机
telecommand 远方指挥，远方调度，遥控，遥令，远程命令
telecommunicaiton environment 远方通信环境
telecommunication access method 电信存取法
telecommunication bureau 电信局
telecommunication cable 远程通信电缆，通信电缆
telecommunication circuit 电信电路
telecommunication data interface 电信数据接口
telecommunication environment 远方通信环境
telecommunication equipment control room 通信控制室
telecommunication equipment 无线电通信设备，电信设备
telecommunication installation 电信装置
telecommunication line 电信线
telecommunication network 电信网
telecommunications interface 远方通信接口
telecommunications trunk line 通信干线
telecommunication system 电信系统
telecommunication terminal 电信终端
telecommunication traffic circuit 电信业务电路
telecommunication trunk network 电信干线网
telecommunication working group 电信工作组
telecommunication 电信，电讯，无线电通信，无线电通讯，远程通信
telecom 电信
teleconference 远距离通信会议，电报会议，电话会议
teleconnexion 远距离联系
teleconrtol intefrace 远方控制接口
telecontrol center 调度中心
telecontrol configuration 远动布局，远动配置
telecontrol frequency 远动信号用频率
telecontrolled power station 遥控电站
telecontrolled 远距离控制的
telecontrolling 遥控，远距离操纵
telecontrol system 遥控系统，远动系统
telecontrol transfer time 远动传送时间
telecontrol 遥控，远距离控制，远动，远程控制
telecon 电报会议，电话会议
telecord 电话机上附加的记录器
Telecoseal 太莱古西尔合金
telecounting 远程累计
telecoupler 共用天线耦合器
teledata processing 远程数据处理
teledial fuel pin 电话拨盘形孔内燃料细棒
teledial 电话拨盘型
telefacsimile 电话传真
telefault 电缆故障位置检测线圈，故障检测电感线圈
teleflex 软套管
telegauge 遥测仪表，测远仪，遥测计
telegoniometer 遥测角仪，无线电测向仪，方向计
telegon 一种自动同步机，无接点交流自整角机
telegram office 电报局
telegraph cable 电报电缆
telegraph equation 电报方程
telegraphic address 电报挂号
telegraphic charge 电报费
telegraphic draft 电汇汇票，电汇票
telegraphic keying 发报
telegraphic money order 电汇单，电汇汇单，电汇
telegraphic money 电汇款
telegraphic report 电汇报表
telegraphic toll 电报费
telegraphic transfer payment 电汇付款
telegraphic transfer rate 电汇汇价
telegraphic transfer reimbursement 电汇付款
telegraphic transfer selling 电汇卖出汇率
telegraphic transfer 电汇
telegraph network 电报通信网
telegraph relay 电报继电器
telegraph repeater 电报帮电机，电报转发器
telegraphy 电报学，电信
telegraph 电报
telehygrometer 遥测湿度计
teleindication 遥信
teleindicator 远距离指示器
teleinstruction 远程指令，远距离指令
tele-irradiation 远距辐照
telemanometer 遥测压力表，遥测压力计
telemechanical apparatus 远动装置
telemechanical device 遥控机械装置，远动装置
telemechanical installation 远动装置
telemechanical receiver 远动接收器，远动接收机
telemechanical transmitter 远动发送机，远动发送器，远动发送装置

telemechanical 遥远机械的，远动的
telemechanic system 远控系统
telemechanics 远动学，遥控机械学，遥控动力学，遥控力学
telemechanic 远控的
telemechanism 远动装置，远动操作机构，远动力学
telemechanization 遥远机械化，远动化，远距离机械化
telemetered signal 遥测信号
telemeteric control 遥测控制
telemetering channel 遥测通路
telemetering communication system 遥测通信系统
telemetering current meter 遥测流速仪
telemetering depth meter 遥测深度计
telemetering equipment 遥测装置
telemetering hydrophone 遥测水下测音器
telemetering installation 遥测装置
telemetering medium 远距离测量装置
telemetering pick-up 遥测传感器
telemetering rain gauge 遥测雨量计
telemetering sender 遥测发送器
telemetering system of frequency division type 频分（制）遥测系统
telemetering system of time division type 时分（制）遥测系统
telemetering system 遥测系统
telemetering tide gauge 遥测验潮仪
telemetering totalizer 遥测加法器
telemetering water level indicator 遥测水位指示仪
telemetering 遥测，远距离测量
telemeter reading system 电能量计量系统
telemeter 遥测装置，遥测器，遥测仪，测距器，遥测计，遥测表，电量计费表，测远仪
telemetric instrument 遥测仪器
telemetry analog to digital information converter 遥测模拟数字信息转换器
telemetry and command system 遥测指令系统
telemetry filter 远测滤波器
telemetry information 遥测信息
telemetry link 遥测线路，远动通道
telemetry 遥测法，遥测技术，遥测装置，遥感勘测，远动，遥测，自动测量记录传导
telemometer 遥测式直读荷重计
telemonitoring interface 远方监视接口【如到工程师工作站】
telemonitoring 远程监视，遥控
telemonitor 遥远监视，遥远监控，遥控
telemotion 无线电操纵
telemotor 遥控马达，遥控电动机，遥控传动装置
telenmetering 远程测量
teleobjective 遥测对象，望远物镜
teleological system 目的系统
tele-operated 遥控操作的
telepancurietherapy 全身性照射疗法
telepantoscope 单向扫描器件
telephone and carrier current equipment 载波电话设备
telephone block 电话闭塞
telephone booth 电话亭
telephone bureau 电话局
telephone cable 电话电缆
telephone call-box 电话通话室，电话间
telephone channel 电话通道，话路
telephone company 电话公司
telephone conference 电话会议
telephone cord 话机绳
telephone data set 用电话线连接的数据终端，电话数据设备
telephone dial key 电话键
telephone duct 电话管道
telephone equipment diagnosis 电话设备诊断
telephone exchange room 电话交换室，电话总机站
telephone handset 电话听筒
telephone harmonic factor 电话干扰系数
telephone harmonics form factor 电话波形系数
telephone influence factor 电话干扰系数，电话影响因子
telephone interference 电话干扰，通话干扰
telephone kiosk 电话亭，电话间
telephone meter 电话计数器
telephone modem 电话调制解调器
telephone number 电话号
telephone office 电话局
telephone relay 电话继电器
telephone ringer 电话振铃器
telephone service desk 话务台
telephone service signal 话务信号
telephone set 电话机
telephone switchboard room 电话交换机室
telephone test 电话试验，电话干扰检查
telephone transmission measuring set 电话传输测量器
telephone trunk zone 长途电话电路段
telephone-type relay 电话式继电器
telephone 电话
telephonic enquiry 电话询价
telephonometer 通话计时器
telephonometry 通话计时
teleplotter 电传绘迹器
teleprinter code 电传打印机代码
teleprinter 电传打印机，电传打字机，电传打字电报机
teleprocessing 远处理，远程信息处理，电传信息处理
teleprotection 远方保护
telepunch 遥控穿孔
telequipment 遥控装翼，远动装置
telerecorder 遥测自动记录仪
tele-recording system 遥测记录系统
teleregulation 遥调，远程调节
telerun 遥控
telescope caliper 光学测微仪
telescoped joint 套管接头，管套连接，套管连接，套管连接装置
telescope-feed joint 套筒连接
telescope-feed 套筒式
telescope for surveying 测量望远镜

telescope-type coal chute 伸缩式落煤管
telescopic axle 套筒轴
telescopic bar 伸缩套杆，伸缩套筒
telescopic blower 伸缩式吹灰器
telescopic boom 伸缩式臂杆，套筒式臂杆
telescopic caisson 套筒式沉箱
telescopic chimney 套筒式烟囱
telescopic chute 伸缩槽，伸缩管，伸缩落煤管，套筒式卸槽，卸料槽，卸料管
telescopic hoist 伸缩吊，套管式起重机
telescopic joint 套管接合，套管连接，套筒式接头
telescopic manipulator 伸缩式机械手
telescopic mast 伸缩套管天线
telescopic pipe 伸缩管
telescopic range finder 望远镜测距仪
telescopic screw 套筒螺丝
telescopic settlement gauge 套筒式沉陷计
telescopic slide 望远镜式滑动
telescopic staff 伸缩标尺，塔尺
telescopic steel shuttering 套筒式钢模板
telescopic tube 伸缩管，伸缩套管
telescopic 望远镜的，套筒式的，套管的，可伸缩的，套筒，套管，伸缩的
telescopiform chute 伸缩（落煤）管
telescopiform 套叠式的，套管式的，可伸缩的
telescoping chute 伸缩的落煤管
telescoping column （机械手）套筒柱，套管柱
telescoping door 套叠门
telescoping downspout 伸缩落煤管
telescoping drive 可伸缩套管式驱动装置
telescoping form 伸缩式模板
telescoping manipulator mast 伸缩式机械手支柱
telescoping mast 可伸缩套筒柱
telescoping 套进
teleseme 信号机
telesignalisation 遥测信号化，远距离信号设备，遥信，远程信号
telesmoke 烟气望远镜
telesounder 遥测器
teleswitching 远程切换
teleswitch 遥控开关
tele-symbiont 远程共存程序
telesynd 遥测设备，远程同步遥控装置
telesyn 遥控同步机
teletachometer 遥测转速计
telethermometer 遥测温度计
telethermoscope 遥测温度计
teletorque 交流自整角机
teletransmission 遥测传送，远程传送
teletransmitter 遥测传送器，遥测发送器
teletron 显像管
teletype printer 电传打字机
teletype tape 电传打字带
teletype writer exchanger 电传打字机交换机
teletypewriter marking pulse 电传打字机信号脉冲
teletypewriter pulse spacing 电传打字机脉冲间隔
teletype writer 电传打字机
teletype 电传打字机，电传打字
television camera 电视摄像机
television control center 电视控制中心
television interference 电视干扰
television mast 电视塔，天线杆
television monitoring 电视监控
television monitor 电视检查装置
television picture 电视图像
television scanning 电视扫描
television signal receiver 电视信号接收器
television system 电视系统
television tower 电视塔
televisual inspection of SG shell side 蒸汽发生器外壳侧的电视检查
televoltmeter 遥测伏特计，遥测电压表
televox 声控机器人
telewattmeter 遥测瓦特计
telewriter 电传打字机，打字电报机
telex channel 专线电信通道
telex fee 电传通讯费
telex 用户电报，电传
tele 远的，远距离的，遥的，遥控的，电信的
tellevel 液面高度计
telltale hole 显示孔【支管焊缝】，警报孔，信号孔
telltale lamp 信号灯
telltale signal 告警信号
telltale 信号装置，警报器，寄存器，计数器，舵位指示器，指示器
telluric 电场效应的
tellurium 碲
tellurometer 接地欧姆表，雷达测距仪，无线电测距仪
telnet 远程登录，远程登陆，远程登录协议，远程登录服务
telpher conveyer 电动缆索输送机，缆车输送机
telpher line 高架索道，电气索道
telpher 高架索道，电动缆车
telstar 通信卫星
TEMA（Tubular Exchanger Manufacturers Association）美国管式换热器制造商协会
temeplate 型板
temoin 挖方土柱
temperate belt 温带
temperate climate zone 温带气候带
temperate climate 温带气候
temperate discontinuity 温度不连续性
temperate field 温度场
temperate gradient 温度梯度
temperate inversion 逆温
temperate lapse rate 温度递减率
temperate latitude 中纬度
temperate low belt 温带低压带
temperate rain forest 温带雨林
temperate rainy climate 温带多雨气候
temperate zone 温带
temperate 适度的，温和的
temperating air 调温风
temperature above zero 零上温度
temperature absolute 绝对温度
temperature acclimation 温度适应
temperature action 温度作用

temperature actuated switch 热动开关
temperature adjustment curve 温度调节曲线
temperature adjustment 温度调节
temperature alarm 过热警报器，过热信号器，过热信号，温度报警，过热报警
temperature and pressure reducer 减温减压器
temperature anomaly 温度异常
temperature balance 温度平衡，温度均衡
temperature band 温度范围
temperature behavior 温度特性，温度变化情况
temperature behind the cooler 冷却器后温度
temperature booster 加热器，预热器，加热器室
temperature boundary layer 温度边界层
temperature build-up 温度升高
temperature bulb 测温包，热敏元件
temperature buzzer 温度蜂鸣报警器
temperature calibration 温度标定
temperature change 温度变化
temperature characteristic 温度特性，温度特性曲线
temperature chart 温度曲线
temperature classification 耐热等级，绝缘等级
temperature coefficient feedback 温度系数反馈【反应性】
temperature coefficient of delay 温度延迟系数
temperature coefficient of reactivity 反应性温度系数
temperature coefficient of resistance 电阻温度系数
temperature coefficient of viscosity 黏度温度系数
temperature coefficient power regulation 温度系数功率调节
temperature coefficient 温度系数
temperature coloration 温度色变，受热色变
temperature-compensated overload relay 温度补偿过载继电器
temperature-compensated 温度补偿的
temperature compensating device 温度补偿装置
temperature compensating 温度补偿
temperature compensation alloy 温度补偿合金
temperature compensation 温度补偿
temperature compensator 温度补偿器
temperature condition 温度条件
temperature conductivity 导热性，热导率，导温率，温度传导率
temperature-constant operation 等温运行方式，等温运行
temperature contour 等温线
temperature contraction stress 温缩应力
temperature contrast 温度不均匀分布，温度差，温度对比，不均匀温度分布
temperature control coating 调温涂层
temperature control device 温度控制装置
temperature control equipment 温控设备
temperature controlled material 温控材料
temperature controlled pane 温度控制板
temperature controlled relay 温控电器
temperature controlled sensor 温度控制传感器
temperature controlled system test set 温度控制系统测试设备
temperature controlled system 温度控制系统
temperature controller 温度调节器，温度控制器，温度调节仪表
temperature control loop 温度控制回路
temperature control point 温度控制点
temperature control relay 温控继电器
temperature control sensor 温度控制传感器
temperature control switch 温度控制开关
temperature control system 温度控制系统
temperature control unit 温度调节装置，温度控制装置
temperature control valve 温度调节阀，温度控制阀
temperature control 温度控制，温度调节
temperature correction factor 温度校正系数
temperature correction 温度改正，温度校正
temperature crack 温度裂缝
temperature curve 温度变化曲线，温度曲线
temperature cycle test 温度循环试验
temperature cycle 温度循环
temperature cycling stress 温度交变应力
temperature cycling test 冷热交变试验
temperature cycling 温度交变，温度循环，温度周期变化
temperature-decreased pressure reducer 减温减压器
temperature decrease 温度降低，温降
temperature defect 温度亏损【反应性】
temperature deformation 温度变形
temperature-density coexistence curve 温度密度图上的分界线
temperature departure 气温距平，温度偏差
temperature dependence of insulation resistance 绝缘电阻温度特性
temperature dependence 温度关系
temperature dependency 温度相关性
temperature-dependent diode 随温度变化的二极管
temperature-dependent resistor 热变电阻
temperature-dependent 热变的，与温度有关的
temperature derated voltage 温度降额电压
temperature detector 测温器，热敏元件，检温器，探温计，温度探测器
temperature deviation 温度偏差
temperature difference between oil outlet and cooling water inlet 油出口和冷却水入口之间的温差
temperature difference correction factor 温差修正系数
temperature difference module 温差组件
temperature difference technique 温差法
temperature difference 温度差，温差
temperature differential filter 温差过滤器
temperature differential 温差
temperature diffusivity 温度扩散率
temperature discontinuity 温度不连接性
temperature distortion 热变形，受热变形
temperature distribution 温度分布，按温度分布
temperature disturbance 温度振动
temperature drift instability 温度漂移不稳定性
temperature drift 温度变差，温度漂移

temperature driving force　温压
temperature drop　温度下降，温降，温差，降温
temperature effectiveness　温度有效度，温度效率
temperature effect　温度效应
temperature efficiency　温度效率
temperature embrittlement　温度脆化
temperature endurance　耐温性
temperature energy　热能
temperature-entropy chart　温熵图表
temperature-entropy coordinate　温熵坐标
temperature-entropy curve　温熵曲线
temperature-entropy diagram　温熵图
temperature-entropy plot　温熵图
temperature environment　温度环境
temperature equalisation system　均温系统
temperature equilibrium　温度平衡，热平衡
temperature error　温度误差
temperature excess　温差【热交换器中】，传热温差
temperature excursion　温度偏差，温度剧增，温度偏离额定值
temperature expansion　温度膨胀，热膨胀
temperature extreme　温度极限
temperature factor　温度因素，温度系数
temperature fall　温降
temperature feedback　温度反馈
temperature field　温度场
temperature fluctuation range　温度波动范围
temperature fluctuation　温度变幅，温度起伏，温度脉动，温度波动
temperature front　温度界面
temperature gauge unit　温度感受元件
temperature gauge　温度表，温度计
temperature gradient　温度梯度，温梯度
temperature hardened spectrum　温度硬化能谱
temperature head　温压
temperature history　温度随时间的变化
temperature hysteresis　温度滞后作用，温度滞后现象
temperature imbalance　温度失衡
temperature-increasing rate　升温率
temperature increment　温度增量，温度升高值
temperature independent　与温度无关的
temperature indicating crayon　热敏元件，温度指示笔，测温元件
temperature indicative and control　温度指示和控制
temperature indicator　温度指示器
temperature in front of the cooler　冷却器前温度
temperature instability　温度不稳定性
temperature instrumentation　测温仪表
temperature instrument　温度计
temperature inversion　逆温，温度逆增，温度转换，逆温层，逆温现象，温度倒布
temperature joint　温度缝
temperature jump coefficient　温度飞跃系数
temperature jump　温度突跃
temperature lag　温度滞后
temperature-limiting criteria　温度要求限制，温度变化限制
temperature-limiting relay　过热继电器，限温继电器
temperature-limiting　温度限制
temperature limit　温度极限
temperature margin　温度裕度，温度限度
temperature measurement instrument　温度测量仪表
temperature measurement system　温度测量系统
temperature measurement　温度测量
temperature-measuring element　温度测量元件
temperature-measuring meter　温度测量仪
temperature-measuring station　温度测位
temperature-measuring thermocouple　温度测量热电偶
temperature-measuring　温度测量
temperature monitoring　温度监控
temperature motion　（原子的）热运动
temperature movement　温度位移
temperature noise　温度噪声
temperature observation　温度观测
temperature of coolant entering the chemical and volume control system　化学容积控制系统的冷却剂进口温度
temperature of coolant returned to the reactor coolant system　返回反应堆冷却剂系统的冷却剂温度
temperature of fusion　熔解温度
temperature of outgoing air　排气温度，排风温度
temperature of saturation　饱和温度
temperature of supply steam　供汽压力
temperature on operation　运行温度
temperature on standby　待机温度
temperature overload relay　测温式过载继电器，温度检测式过载继电器
temperature overshoot　温度超调
temperature pattern　温度场
temperature peak　温度峰值
temperature pickup　测温传感器，温度感受器，温度传感器
temperature plug　测温热电偶
temperature plume　热羽流
temperature potential　温度势
temperature power coefficient　功率温度系数
temperature probe　测漏探针，测温传感器，测温探针
temperature profile　温度场，温度（分布）曲线，温度分布，温度分布图，温度廓线
temperature programmer　程序升温器
temperature progression　温度上升
temperature radiator　温度辐射体，散热器
temperature ramp　温度上升斜率
temperature range　温度范围，稳定范围，温度变化范围
temperature ratio　温比
temperature record and control　温度记录和控制
temperature recorder　温度记录器，温度记录仪，自记温度计
temperature recovery factor　温度恢复系数
temperature reduction　温度订正
temperature-regulated bath　恒温浴
temperature-regulating valve　温度调节阀，恒温器

temperature regulator 温度调节器，温度控制器
temperature reinforcement 温度钢筋
temperature relay 温度继电器，热动继电器
temperature resistance 耐温性
temperature-resistant coating 耐热敷层
temperature-resistant material 耐热材料
temperature-resistant 耐热的，抗热的
temperature response 温度响应
temperature rise curve 温升曲线
temperature rise of deaerator 除氧器温升
temperature rise test 升温试验，温升试验
temperature rise value 温升值
temperature rise 加热，升温，温升
temperature scale 温度标，温度标尺，温度刻度，温标
temperature seiche 温度静振
temperature sending device 传温器
temperature-sensing bulb 感温包
temperature-sensing cable 感温电缆
temperature-sensing device 温度敏感元件
temperature-sensing element 热敏元件
temperature-sensing resistance thermometer 热敏电阻温度计
temperature-sensing system 温度探测系统，感温系统
temperature-sensing 温度变化灵敏的，温度探测，测温，热敏的
temperature sensitive control 按温度冲量控制
temperature sensitive element 热敏元件
temperature sensitive paint 热敏油漆
temperature sensitive switch 温敏开关
temperature sensitive 热敏的，对温度变化灵敏的
temperature sensor 测温传感器，温度传感器
temperature shift 温度变化
temperature shock 热震，热冲击，温度骤变，温度冲击
temperature spread 温度范围
temperature stability 温度稳定性
temperature strain 温度应变，热变形
temperature stratification 温度层结
temperature strength 温度强度
temperature stress 温度应力，热应力
temperature surge 温度波动
temperature switch 温度开关
temperature test 温度试验
temperature threshold 温度定值
temperature-time data 温度时间数据
temperature tolerance 温度容限
temperature transducer 测温传感器
temperature transient type 瞬态温度型的
temperature transient 温度暂态，温度瞬变值
temperature transmitter 温度变送器，温度传送器，温度传感器
temperature traverse 移动测温
temperature uniformity 温度均匀性
temperature up 温升
temperature variation curve 温度变化曲线
temperature variation 温度变化，温度波动，热振动
temperature warping stress 温度翘曲应力
temperature warping 热应力引起的扭曲，温度引起的翘曲
temperature wave 温度波
temperature zone 温度带
temperature 温度
temper brittleness 回火脆性
temper crack 回火裂纹
temper current 回火电流
tempered glass 钢化玻璃
tempered martensite 回火马氏体
tempereds orbite 回火索氏体
tempered steel 淬硬钢，回火钢
tempered troosite 回火屈氏体
tempered 回火（冶金）
temper embrittlement 回火脆性
temper film 回火色的底片，回火氧化膜
tempering-air damper 调温风挡板
tempering air 调温风，添加风
tempering brittleness 回火脆性
tempering moisture 调节用水分
tempering resistance 回火稳定性，抗回火性，耐回火性
tempering temperature 回火温度
tempering 回火，人工老化，混合，调和，回火【冶金】，捏和【黏土等】
temper time 回火时间【电阻焊】
temper 回火，调节，变柔软，韧度，钢的含碳量，硬度
tempest 风暴天气
temping rod 捣棒
template base 模板库
template beam 垫梁
template frame 放样架
template hinge 样板铰链
template matching 模板匹配
template method 样板法
template shop 放样车间
template 模板，型板，足尺模型，定线样板，样板
templet 模板，样板
temporal average 时间平均值
temporal change 瞬时变化
temporal coherence 时间相干
temporal coherent beam 瞬时相干光束
temporal discharge tube 临时泄水管
temporal dose distribution 时间剂量分布，瞬时剂量分布
temporality 暂时性
temporal mean velocity 瞬时平均流速
temporal overvoltage 暂时过电压
temporal randomness 时间随机性
temporal sampling theorem 时间采样定理
temporal 时间的，暂时的，当时的，现世的
temporarily absorbed water 暂时吸收水
temporarily suspend 暂停
temporary absorber 临时吸收体
temporary access road 临时便道
temporary address 临时地址
temporary adjustment 暂时调整
temporary admission 暂时批准进口
temporary aquifer 临时含水层

temporary assembly 临时装配
temporary assistance 临时协助，暂时协助
temporary attachment 临时性附件
temporary backing 临时垫板
temporary bank protection 临时性护岸
temporary base-level 临时基面
temporary bench mark 临时水准点
temporary bridge 便桥，临时桥
temporary buildings and facilities 临建设施
temporary buildings for construction 施工现场临建
temporary buildings and structures 临建
temporary building 临时建筑，临时性房屋，临建
temporary cofferdam 临时围堰
temporary communication 临时交通
temporary connection 临时接线
temporary construction opening 临时安装孔
temporary construction 临时建筑，暂设工程
temporary control curtain （毒物）控制屏
temporary current 短暂流
temporary dam 临时挡水坝
temporary dike 临时堤
temporary discharge tube 临时泄水管
temporary disturbance 瞬态扰动
temporary droop 暂态转差系数
temporary dyke 临时堤
temporary earth 临时接地
temporary employee 临时职工
temporary employment 短期就业，临时工
temporary facilities for construction 施工临时设施
temporary facilities 临时设施
temporary failure 临时故障，暂时故障
temporary fault 临时故障，瞬时故障
temporary fixing method 临锚
temporary flash board 临时挡水板
temporary foreign capital 短期外资
temporary ground 临时接地
temporary guy 临时拉线
temporary hardness （水的）暂时硬度，硅酸盐硬度
temporary hard water 暂时硬水
temporary hearing defect 暂时听力损失
temporary heating 备用供暖
temporary house 临时住房
temporary huts 临时棚屋，临建工棚
temporary instability 暂时不稳定性
temporary interruption 临时停电
temporary investment 短期投资，临时投资
temporary joint 临时接缝，施工缝
temporary laborer 临时工
temporary labour 短工，临时工，临时劳动
temporary lighting 临时照明
temporary liquefaction 暂时液化
temporary loan 短期贷款
temporary magnet 临时磁体，暂时磁体
temporary maintenance 临时检修，临修
temporary marking 临时标志
temporary memory 暂时存储，暂时存储器，暂存器
temporary outage 临时性停运
temporary overvoltage 短时过电压
temporary payment 临时付款，暂时付款
temporary plug 临时插头
temporary poison 暂时性毒物
temporary protection 临时性保护
temporary register 暂时寄存器
temporary repair 临修，小修，临时修理，临时检修
temporary service 临时性业务，临时性工作
temporary set 弹性形变，弹性变形
temporary shelter for workers 临时工房
temporary shelter 临时隐蔽所
temporary shield 临时屏蔽体
temporary signal 临时信号，临时觇标
temporary site facilities 临时厂区设施
temporary speed rise 瞬时升速
temporary standard 暂定标准
temporary stoppage 临时停机，暂时停机
temporary storage pile 临时煤堆
temporary storage 暂时存储器，暂存器，临时蓄水量，临时贮存，暂时库容，暂时容蓄
temporary storing 短暂贮存
temporary strainer 临时粗滤器，临时过滤器
temporary stress 暂时应力
temporary structure 临时性构筑物
temporary support and protection materials 临时支撑及保护材料
temporary support 临时支撑，临时支护
temporary test plug 试验用的临时管塞
temporary toe weight 坡脚临时压重物
temporary torrent 临时性急流
temporary water-bearing layer 上层滞水，暂时含水层
temporary wiring 临时接线【比方为了试验】
temporary word descriptor table 暂时字描述表
temporary worker 临时工
temporary works 临时工程
temporary 暂时的，临时的，瞬时的，短暂的
temporizing measures 权宜措施
tempreture limitation 温度限制
TEMP＝temperature 温度
TEM 温度【DCS 画面】
tenacious 黏的
tenacity 黏性，韧性，韧度
tenancy agreement 租借合约，租赁协议
tenancy condition 租赁条件
tenancy from month to month 按月租赁
tenancy （土地，房屋的）租佃，租用，租赁，租期
tenant farmer 土地租用人
tenantless 无人租赁的，无人居住的
tenantry 承租人，佃农
tenant 承租人，房客，租客，租户，佃户
ten-day mean 旬平均
ten-days average humidity 旬平均湿度
ten-days average temperature 旬平均温度
tendency of stage variation 水位变化趋势
tendency toward sliding 滑动趋势
tendency 倾向，趋势
tender advertisement 招标广告

tender analysis 投标审查
tender bidding 投标
tender bond 投标保函，投标担保，投标保证金，投标保证书
tender closing date 投标截止日期
tender committee 投标委员会
tender control price 招标控制价
tender data sheet 投标资料表
tender design 投标设计
tender document 招标文件，交单，标书
tendered price 投标价
tenderer 投标人，投标者，报价人
tender evaluation 标书评估，投标评估价
tender for the construction of 投标承建……
tender guarantee 提供担保
tender guarantor 投标保证人
tendering data sheet 招标资料表
tendering document 投标文件
tendering form 投标书格式
tendering notification 招标通知
tendering procedure 招标程序
tendering 招（投）标
tender invitation 招标
tender list 投标目录，投标一览表
tender notice 招标公告，招标通知
tender offer 股权收购，投标报价
tender opening date 开标日期
tender price 投标报价，投标价格
tender procedures 招标程序，投标手续
tender sheet 报价单
tender sum 投标总价
tender system 投标制
tender 报价，投标，投标书
tendon buttress 钢筋束支墩
tendon tube 钢索管
tendon 预应力钢筋，（预应力）钢筋束，预应力钢板，锚索
tend to decline 趋势下降
tend to rise 趋势上涨
tend to 易于，往往会，倾向于
tenement 房屋，住户，租房子，房产，公寓
tenon and mortise 雌雄榫
tenon-bar splice 榫条拼接
tenon joint 榫接
tenon （叶片）铆钉头，榫头，榫面
tenorite 黑铜矿
ten-position path group selector 十通道群选择器【堆内测量用】
tens dial 十分度盘
tens digit 十进制数字
tense lines 紧线
tensibility 伸长性
tensile bar 受拉钢筋
tensile bending test 弯曲拉力试验
tensile break 拉伸断裂，拉伸破裂
tensile circumferential stress 圆周拉应力
tensile cracking 张拉破裂
tensile creep strain 拉蠕变应变
tensile deformation 拉伸变形
tensile elongation 拉伸长度，延伸率，拉伸
tensile extension 拉伸延长
tensile failure 拉断，受拉破坏
tensile force transducer 张力传送器
tensile force 拉力，张力，牵引力
tensile fracture 拉伸破裂，拉断
tensile plastic strain 拉塑性应变
tensile property 抗拉性能
tensile region 受拉区
tensile reinforcement 抗拉钢筋，受拉钢筋
tensile reinforcing bars 锚束
tensile section 张拉段【自由段】
tensile splitting test 劈裂抗拉试验
tensile strain 拉伸应变，拉应变
tensile strength limit 拉力强度极限
tensile strength structural coefficient 抗拉强度结构系数
tensile strength 抗张强度，抗拉强度
tensile stress area 张拉应力区
tensile stress-strain curve 拉伸应力应变曲线
tensile stress 抗张应力，抗拉应力，拉伸应力，拉应力，张应力
tensile test by the no-hold method 用不保持法做的拉伸试验
tensile test diagram 拉力试验图
tensile testing machine 拉力试验机
tensile test piece 受拉试件
tensile test specimen 拉应力样品
tensile test 拉力试验，拉伸试验，张力试验，抗拉试验，张拉试验
tensile trajectory 受拉轨迹线
tensile type load cell 张拉式压力盒
tensile vibration test 拉伸振动试验
tensile yield point 拉力屈服点
tensile yield strength 拉伸屈服强度
tensile zone 受拉区
tensile 张力的，拉力的，拉伸的，抗张的，抗拉的
tensility 延性
tension meter 张力计，拉力计，牵引测力计
tensimeter （流体）压力计，拉力计，湿度计，张力计
tensiometer 拉力计，张力计，表面张力计，土壤湿度计
tension anchor bar 张拉锚杆
tension area 受拉面积
tension axis 张力轴
tension bolt 拉紧螺栓，张力螺栓
tension brace 拉条，拉撑
tension carriage 拉力架
tension clamp 耐张线夹
tension coefficient 拉伸系数
tension crack 拉裂，拉伸裂缝，受拉裂缝，张裂，张裂缝
tension diagonal 斜拉杆
tension drum of conveyer 皮带机的拉紧鼓筒
tension drum 拉紧滚筒
tension dynamometer 拉力测力计，张力动力计，张力功率计
tensioner 螺栓张拉机，张紧器，拉伸机，张力机
tension failure 拉伸破坏，受拉破坏
tension fault 张断层
tension fiber 受拉纤维

tension foil 张力薄片，张力箔
tension fracture 受拉断裂
tension-free pore water 无张力孔隙水
tension gauge 张力计
tension graph 张力曲线
tension head 张力水头
tension indicator 张力指示器
tensioning apparatus 张拉设备
tensioning bolt 张紧螺栓，拉伸螺栓
tensioning rod 拉杆
tensioning tool 拉紧工具
tension insulator string 耐张绝缘子串
tension insulator 耐张绝缘子
tension intensity 拉力强度
tension joint 张节理，受拉接头，受拉接合
tension load 张力荷载
tension measurement 拉力测量，张力测验
tension member 拉杆，受拉构件
tension paying-off 张力放线
tension pile 抗拔桩
tension pin 张力销
tension pulley 受拉滑轮，张紧滚筒，张紧轮
tension readjusting spring 张力重新调整弹簧
tension regulator 电压调整器
tension reinforcement 拉力钢筋，受拉钢筋
tension ring （圆屋顶中承受张力的）箍圈，拉力环，张拉环
tension rod 拉杆
tension rupture 拉断
tension screw 张力螺栓
tension section 耐张段
tension set 耐张线夹，耐张串组
tension sleeve 带丝扣套环【钢丝绳用】，拉紧套管
tension spring 伸张弹簧，拉簧
tension state 张拉状态
tension strain 拉伸应变
tension strecting force 拉力
tension stress 拉应力
tension stringing equipment 张力架线设备
tension stringing 紧线，张力放线，张力架线
tension strings 耐张力绝缘子串
tension support 耐张杆塔
tension test by the no-hold method 用不保持法做的抗拉试验
tension test 张拉试验，张力试验，拉力试验，拉伸试验，抗拉试验
tension tower 耐张塔，张力塔
tension transformer 电压互感器
tension type anchor bar 张拉型锚杆
tension wave front 拉力波前沿，张力波前沿
tension 拉力，张力，压力，电压，（气体）压强，紧张，应力，拉伸，张紧，牵力，张拉
tensiostat 张力保持器
tensometer 拉力计
tensor force 张量力【核】
tensor transformation 张量变换
tensor 伸张器，张量
tentation data 试验数据，试用数据，假设数据，预定数据，暂定数据
tentation 假定，假设
tentative conclusion 暂时结论
tentatively 试探性的，试验性的，暂时的
tentative method 暂行方法
tentative program 暂行程序
tentative regulation 暂行规定
tentative site 初选工程地址
tentative specification 暂行规程，暂行技术规范，试行技术规范，暂行规范
tentative standard 试用标准，暂行标准，试行标准，标准草案
tentative technical specification 暂行技术规范
tentative value 试用值
tentative 临时的，试验（性）的，假定的，暂行的，暂时的
ten-tear program 十年规划
tenter chord 中心弦
tenth gearbox 十比一变速箱
tenth-thickness value 十分之一厚度值
tenth-value layer 十分之一值层【健康物理学】
tenth value thickness 十分之一厚度值
TEN(total except noted) 完全接受，但有注明的除外
tent 帐篷，帐幕
tenuity 细，薄，稀薄
tenure 任期，使用期
ten-year flood 十年一遇洪水
ten-yearly outage program 十年一次停运计划
ten-yearly outage 十年一次停运
TEPCO 东京电力公司
TEP 硼回收系统【核电站系统代码】
T-equivalent circuit parameter T形等效电路参数
Teratron 亚毫米波振荡器
terawatt hour 太（拉）瓦时，万亿瓦（特）时，兆兆瓦（特）时
terawatt 太瓦
terdenary 十三进制的
terephthalic acid 对苯二酸
term bill 期票
term bonds 定期债券
term by term differentiation 逐次微分
term contract 定期合同
term deposit 定期存款
terminable contract 有限期的合同
terminal address selector 终端地址选择器
terminal and host interface processor 终端与主机的接口处理机
terminal architecture 终端体系结构
terminal area distribution processing 终端区域分布处理
terminal area real-time program 终端区实时程序
terminal area sequence and control 终端区时序控制
terminal area 焊接区，接点
terminal assembly 接线板，接头装配
terminal bar 端子连接棒
terminal-based conformity 端基一致性
terminal-based linearity 端基线性度
terminal block end plate 接线板端板
terminal block module 接线板模件
terminal block with bonding jumper 带跨接线的端子板

terminal block 端子板，端子排，接线板，接线盒
terminal board 端子板，终端板，端板，接线板
terminal box （接线）端子箱，接线盒，出线盒，端子盒
terminal cabinet 端子柜
terminal cable 终端电缆
terminal changer 线端转换开关
terminal charges 终端码头费，装卸费，堆场费
terminal clamp 终端线夹，出线夹
terminal computer 终端电子计算机
terminal conditions 终端参数
terminal connector 终端线夹，出线夹，设备线夹
terminal control system 终端控制系统
terminal control 终端控制
terminal cover 端盖
terminal decision 最终决策
terminal device 终端设备
terminal diagram 端子接线图
terminal difference 端差
terminal efficiency 发电机端效率
terminal equipment 终端设备
terminal face 端面
terminal facilities 码头设施，车站设施，航空站设施
terminal field 终点场【相轨迹】
terminal group orientation 端子排排列方向，端组定向
terminal handling charge 码头附加费
terminal header 出口联箱
terminal head 端头
terminal hood 接线端盖帽【电机】
terminal impedance 终端阻抗
terminal installation 终端设备
terminal insulation 端子绝缘
terminal insulator 端子绝缘子，终端绝缘子
terminal interface module 终端接口模件
terminal interface processor 终端接口处理机
terminal interface 终端接口
terminal lead 终端引线
terminal line 终线
terminal load 末端荷载，终端负载
terminal lug 端接片，接线端子
terminal marking piece 端子号牌
terminal marking 端子记号，终端标记，线端记号
terminal monitor program 终端监督程序
terminal moraine 尾碛
terminal of circuit 线路终端
terminal office 终端局
terminal pad 接头垫板
terminal pin 引线插头，尾销
terminal plate 接线板，终端片，端子板
terminal points drawing （终）端点图
terminal point 终点，连接点，接线点，分界点，端子
terminal pole 终端杆
terminal port 终点港
terminal post 接线柱，终端柱
terminal pressure 末端压力，终点压力
terminal resistance 端电阻
terminal row 端子排
terminal screw 端子螺钉
terminal service 码头作业，终点站作业
terminal shaft 终端轴
terminal speed 末速度，终端速率，极限速度
terminals per station 每站终端数
terminal station 终端站
terminal strip 端子排，接线盒，端子板，接线板
terminal stud 端接螺杆
terminal substation 终端变电站
terminal supervisory program 终端管理程序
terminal support 终端杆塔，终端支架
terminal temperature difference calculation 端差温度计算
terminal temperature difference of condenser 凝汽器终端温差
terminal temperature difference of heater 加热器端差
terminal temperature difference 温度端差，端差，末端温差，最大温差，端部温差
terminal-to-computer multiplexer 终端计算机多路转换器
terminal tool 端子工具
terminal tower 终端塔
terminal transformer 终端变压器
terminal treatment plant 最终处理厂
terminal treatment 最终处理
terminal unit 终端设备，终端，端子单元
terminal valley point capacitance 谷值电容
terminal velocity 终速，终端速度，末速，极限速度，终点速度
terminal voltage 终端电压，线端电压，端电压
terminal weld 端部焊接
terminal wiring diagram 端子安装接线图
terminal 端子，终点，终端机，终端设备，终结符，端接线柱，接线片，线端，线接头，卸货码头，终点站，接线柱，引线，接头，末端，转运站，中转站，交接点，接线端，终点的，末端的
terminate a contract 解除合同，终止合同
terminate an agreement 终止协议
terminated line 端接线
terminated project 已结束项目
terminated ramp 有限直线上升
terminated two-port 具有端接的双口
terminate the contract 解除合同
terminating plug 终端插头
terminating point 端接点，终点，最终结局
terminating resistance 终端电阻
terminating set 终端装置
terminating 端接，终端，线端加负载，线端绑扎
termination box 终端盒
termination cabinets 端子箱
termination circuit 终端电路
termination clause （合同）解除条款，（合同）终止条款
termination design 终端设计
termination fittings 端接配件
termination flange 接口法兰
termination forthwith clause 立即终止生效条款

termination link 终端连接
termination of a contract 合同终止，终止合同，合同期满
termination of an agreement 协定期满，终止协议
termination of lines 线路终端
termination of prescription 时效终止
termination of risk 风险责任终止，保险责任终止
termination pay 解雇费
termination report 总结报告
termination right provisions 终止权规定
termination schedule 端接表
termination supervisor 终端检查员
termination 终止，终端，结束，终端装置，端接法，终端负载，终止契约
terminator-string 终结字符串
terminator 终端套管，线端套管
terminology 术语，专门名词
terminus 终点，限界，中心站
termipoint connection 端点连接
termite 白蚁
term loan 定期贷款
term mortgage 定期抵押
term of a contract 契约期限
term of an agreement 协议条款
term of delivery 交货条件
term of entrustment 委托期限
term of grace 宽限条款
term of life 使用期限
term of loan redemption 还贷期限
term of loan 贷款条件
term of office 任期
term of payment 付款方法，付款条件
term of redemption 偿还期限
term of residence 居住期限
term of service 使用期限
term of validity 有效期限，有效期间
term of year 年限
termperate zone 温带
terms and conditions of delivery 交货条款
terms and conditions of employment 聘用条件
terms and conditions of the purchase agreement 购货协定条件
terms and conditions 条款
term sheet 合同书，（协议或合同的）条款清单，（协议或合同的）条款说明书
terms of agreement 协议条款
terms of appointment 委任条款
terms of consignment reference 委托权限条款
terms of contract 合同条款
terms of dealing 交易条件
terms of electric power engineering 电力工程术语
terms of employment 雇佣条件，劳动待遇
terms of payment 付款条件，付款条款，支付条件，支付条款
terms of reference （委托）权限，职权范围，职责范围，受权调查范围【委员会等】，委托的规定说明，任务大纲
terms of reimbursement 偿付条款
terms of shipment 装货条款
terms of trade 贸易条件
terms of trust 信托合同条款
term of validity 有效期限
term 术语，要求，条件【复数】，期限，条款，专门名词，期
ternary addition 三进制加法
ternary code 三进制码
ternary counter 三进制计数器
ternary fission 三元裂变，三分裂
ternary logic element 三进制逻辑元件
ternary material 三元材料
ternary mixture 三元混合物
ternary multiplication 三进制乘法
ternary number 三进制数
ternary sediment 三相沉积
ternary sequence 三进制程序
ternary system 三元系统，三进制系统，三元系
ternary 三相，三元素，三进制的，三重的
terne plate 白铁皮，镀铅锡合金薄钢板
tern 三个一套的，三重的
terotechnology 设备综合工程学，设备综合管理学
teroxide 三氧化物
terra cariosa 板状硅藻土
terrace at entrance 入口平台
terrace channel 地埂槽沟
terraced arrangement of buildings and structures 建、构筑物阶梯布置
terraced arrangement of subareas at different elevations 不同标高上分区的阶梯布置
terraced arrangement 阶梯式布置
terraced grate 台阶式炉排
terraced ground 阶梯地面
terraced land 地埂地
terrace grade 地埂坡度
terrace ridge 台地脊线
terrace roof 平台，平屋顶
terrace step 平台踏步
terrace 地埂，地坪，阶地，露台，平台，台，台地，梯田，土堤
terracing grader 筑埂平地机
terra-cotta block protection 陶砖护岸
terra-cotta tile 陶瓦
terra cotta 陶土砖，赤褐色（的），赤褐，赤陶，赤陶土，赤土陶器（的），赤土色的
terra firma 大地，稳固地基
terrain and topography 地形地势
terrain category 地形分类，地形类别
terrain classification 地形分类
terrain clearance 离地高度
terrain coefficient 地面利用系数
terrain configuration 地表外形
terrain details 地面细部
terrain downwash 地形性下洗
terrain environment 地形环境
terrain estimation 地面估测
terrain feature 地面要素
terrain height 地形高度
terrain induced mesoscale system 地形诱导中尺度系统
terrain interpretation 地形判读
terrain model 地形模型

terrain monitoring 地表监测
terrain movement 地层移动
terrain restriction 地形限制
terrain roughness 地面粗糙度，地形粗糙度
terrain 地形，地势，地面，地带
terra ponderosa 重晶石
terra rossa 红色灰土
terra verde 绿土
terrazzo brick 水磨石砖
terrazzo finish grinder 水磨石机
terrazzo finish 水磨石地面，水磨石面层，水磨石饰面
terrazzo floor 磨石子地面，水磨石地面，水磨石地坪
terrazzo 磨石子，水磨石
terrestrial ambient noise 大地环境噪声
terrestrial attraction 地心吸力
terrestrial current 地电流
terrestrial deposition 地表沉积
terrestrial deposit 陆地沉积
terrestrial disposal 地面处置【废物】
terrestrial ecology 陆地生态学
terrestrial environment 陆地环境
terrestrial equator 地球赤道
terrestrial erosion 地表侵蚀
terrestrial eyepiece 正像目镜
terrestrial facies 陆相
terrestrial globe 地球，地球仪
terrestrial gravitation 地球重力
terrestrial heat 地热
terrestrial horizon 地平
terrestrial irradiance 地球（辐射）辐照度
terrestrial magnetic field 地磁场
terrestrial magnetic pole 地磁极
terrestrial magnetism 地磁，地磁学
terrestrial photogrammetric camera 地面摄影测量摄影机
terrestrial photogrammetry 地面摄影测量学
terrestrial photographic mapping 地面摄影测图
terrestrial photograph 地面摄影照片
terrestrial radiation 地面辐射，地球辐射，大地辐射
terrestrial solar cell array 地面太阳电池方阵
terrestrial surface 地球表面
terrestrial transfer model 陆地转移模式
terrestrial triangulation 地面三角测量
terrestrial whirl wind 陆旋风
terrestrial wind system 陆成风系
terrestrial wind test 陆成风试验
terrestrial 陆地的，陆地生物
terre verte 绿土
terrigenous deposit 陆源沉积
terrigenous 陆地的，陆源的
territoriality 大陆性，陆地性
territorial sea 领海
territorial waters 领海，领水
territory lighting 地面照明
territory 领土，领地，领域，范围
tertiary air-type pulverized fuel burner 乏气喷燃器
tertiary air 三次风，三次空气
tertiary amine 叔胺，三级胺
tertiary chamber 三次风室
tertiary circulation 三级环流
tertiary coil 三次线圈，第三级线圈
tertiary coolant circuit 中间冷却剂回路，三回路
tertiary creep 第三阶段蠕变，蠕变加速区，末期蠕变
tertiary grout hole 三期灌浆孔
tertiary industry 第三产业
tertiary interference 第三线干扰
tertiary reheater 第三级再热器
tertiary sampler 三重取样器
tertiary sewage treatment 三级污水处理
tertiary storage 三种存储器，第三存储器
tertiary substation 三次变电所
tertiary super heater 第三级过热器
tertiary system 第三系统，第三系
tertiary treated wastewater 三级处理废水
tertiary treatment （废物）三级处理，三次处理
tertiary voltage 三次电压
tertiary wastewater treating process 三级废水处理过程
tertiary wastewater treatment 污水三级处理
tertiary winding 三次绕组，第三绕组
tertiary 三重的，第三级的，第三位的
tervalent acid 三价酸
tervalent 三价的
terylene 涤纶
TER 废液排放系统【核电站系统代码】
Tesla coil 特斯拉感应圈，特斯拉线圈
Tesla coupling 特斯拉耦合
Tesla current 特斯拉电流
Tesla induction coil 特斯拉感应圈
Tesla transformer 特斯拉变压器，特斯拉空心线圈
tesla 特斯拉【磁通密度单位】
testability 可测试性
test age 试验龄期
test and adjustment 调试
test and diagnostic language 测试和诊断语言
test and inspection of welding joint 焊接接头检验
test and inspection procedure 试验和检查程序
test and repair processor 测试修复处理机
test and set 测试与置位
test and verify program 检验证明程序
test apparatus 试验设备，试验仪器，试验装置
test area 试验场地
test assembly 测试组件
test atmosphere 试验环境
test bar 试验杆，试验棒，试杆，试验间，试车台
test beam 试验梁
test bed 试验装置，试验台，测验台
test block 试块，试验台，试片，试验台架
test board 测试用配电盘，试验台，测试台
test boiler 试验锅炉，调试锅炉
test boring 试验钻探【土壤调查】，试孔
test box 试验盒
test cell 试验间，试验台，试验盘
test certificate 试验鉴定书，试验证书，检验合格证
test chain 链码，链式砝码

test chamber （风洞）试验段，试验间，试验箱，试验室
test check 试验检查，试样检验，抽检，抽查法
test circuit 测试电路，试验电路
test clamp 试验夹钳
test clip 测试夹
test coal 分析用煤，试验用煤
test cock 试水旋塞，试验栓
test code 试验规范，检验规范，测试标准
test coil 试验线圈
test conditions 试验工况，试验条件，试验状况
test conductor 试验员
test connection 试验接头
test coupon 试样
test cube 立方体试块，立方试体
test cylinder 圆柱试体
test data acquisition 试验数据采集
test data 测试数据，试验数据
test department 试验所，（电业管理局的）试验科
test desk 测试台，校验台，试验台
test device 试验设备
test dummy 试验模型
tested capacity （井的）试验出水率
test efficiency 试验效率
test equipment 试验设备，测试设备
test error 试验误差
tester 试验人员，测验员，检查仪，试验仪器，测试器，试验器，测定器，检验器，试验机，试验计，试验装置，分析仪器
test even 偶次谐波测试
test facility 试验装置，试验设备，试验设施，测试机构
test figures 试验数据
test floor 试验台，试验场地
test flume 试验槽
test for balance 平衡试验
test for convergence 收敛检验
test for expansive pressure 膨胀压力试验
test for flow 气流试验
test frequency 测试频率，探测频率
test fuel 试验燃料
test gag 试验夹持器
test gauge 标准表，标准规，检验规
test generator 测试码生成程序
test ground 试验场
test head 试验水头
test hole 试验孔，试孔
test house 试验室，试验站
testimonial 证明书，鉴定书
testimonium clause 认证条款
testimony 表示，证词，宣言，陈述，证据，证言
testing accuracy 测试精度，试验精度
testing and inspection 试验和检验
testing and maintenance 试验与维修
testing and sampling 检查与抽样
testing and setting of turbine regulating and automatic protection system 汽轮机调节系统和自动保护装置的测试与整定
testing apparatus 试验装置
testing block load 试重块
testing car for traction substation 牵引变电设备检测车
testing certificate 检验证书
testing conditions 试验工况，试验条件
testing data 测检数据
testing device 测试装置
testing equipment 测试设备，试验设备
testing facility 测试设备
testing gallery 试验廊道
testing generator 试验发电机
testing instrument 试验仪表，测试仪表，测试仪器，试验仪器
testing jack 试验端子
testing jig 测试架
testing kinetics model 试验动力学模型
testing machine 试验机
testing method 试验方法，测试方法
testing of characteristic 特性试验
testing of pump performance 水泵性能试验
testing panel 试验板
testing point 试验点
testing position 试验位置，试验场地
testing pressure 试验压力，试验原理
testing procedure outline 试验程序大纲
testing procedure 测试程序，试验程序
testing programme 试验大纲
testing pushbutton 试验按钮
testing reactor 试验反应堆
testing rig 试验台
testing set 试验器组，测试器组
testing similar condition 实验相似条件
testing stand 试验台，试验架
testing station 试验站
testing tank 试验池
testing technique 试验技术
testing terminal 测试端子
testing transformer 试验变压器，试验用变压器
testing voltage 试验电压
testing 试验，测试，检验，测验
test in service 生产条件下试验
test inspection 检查，试验，检验
test installation 试验装置
test instrumentation system 测试仪表系统，试验仪表系统
test instrument 试验仪表
test item 测试项目
test jack 测试塞孔，测试插孔，试验插孔
test lamp 试验灯
test leader 试验负责人，试验指挥
test lead 试验引线，测试导线
test level 试验水准
test line 试验线段
test load 试验负载，试验负荷，试验荷载
test loop control 试验回路控制
test loop 测试回路，试验回路
test machine 试验机
test man 测试员，试验员
test material 试验材料
test metal 试验用金属
test method 测试方法
test model 试验模型

test module 试验模数
test object 试件
test observation 试验观测
test odd 奇次谐波测试
test of ability 能力测验
test of automatic turbine start-up control system 汽轮机自启停控制系统启动试验
test of dynamic balance 动平衡试验
test of end 结算结果检验
test of life 寿命试验
test of material 材料试验
test of turbine load trip 汽轮机甩负荷试验
test of welded joint 焊接试验
test on bed 台架试验
test on completion 竣工试验，竣工测试
test on pile 试桩
test on site 现场试验
test operation condition 试验运行工况
test operation 试运行，试运转
test operator 试验操作员
test pack 测试卡片组
test panel 试验盘，预试屏板
test paper 试纸
test pattern evaluator 测试码模式鉴定程序
test pattern 试验模式，试验图像，测视图，测试卡
test pencil 测电笔，试电笔
test performance 试验性能
test period 试验期，试验阶段，试验周期
test phase 测试阶段
test piece 试样，试件，试块，样品，试验品
test pile 试验桩，试桩
test pilot 试验滑阀
test pin 触针
test pit 试验坑，探坑
test plant 试验车间
test plate with broad groove 带宽坡的试板
test plate 试验板，试板
test point 测量点，测试点，试验点，测点，试点
test position 检测位置，试验位置
test pressure gauge 标准压力表
test pressure nozzle 水压试验接管
test pressure 试验压力，水力试验压力
test probe 测试探头，试验检测器
test procedure 试验程序
test program 试验大纲，测试程序，试验程序
test pump 试验泵，水压试验泵
test quality level 测试质量水准
test rack 试验台
test range 试验范围
test reactor 试验反应堆，试验堆
test record 试验记录
test regulation 试验规则
test relay 测试继电器
test report 试验鉴定书，试验报告
test requirement 试验要求
test result 试验结果，试验成果
test Reynolds number 试验雷诺数
test rig 试验装置，试验设备，试验台，试验用具
test rod 探棒
test room 试验间
test routine 试验程序，测试程序，校验程序
test rules for inventory management 库存管理的检验规则
test-run 试车，试运行，测试运行，试运转
tests after completion 竣工（后）试验
test sample 试件，试样，试验样本
test schedule 试验程序
test section （风洞）试验段，试验断面
test sequence 试验顺序
test series 试验系列，检测系列
test setup 检测装置，测试设置，试验装置
test set 测试设备，试验装置，测试机组
test sheet 试验单
test shield 测试护套
test sieve 试验筛
test signal 测试信号
test site 试验场地
test socket 试验插座
test source 试验用放射源
test specification 试验规范
test specimen 试样，试件，试验样品
test stand 试验台，试验架
test status 试验状态
test steam 试验蒸汽
test storage 试验贮存
test stream （风洞）试验（段气）流
test summary report 试验总结报告
test switch 试验开关
test technique of hydraulic machinery 水力机械测试技术
test terminal 测试端子，试验端子
test to destruction 破坏试验
test transistor 测试晶体管
test trench 试槽，视槽，探槽
test tube stand 试管架
test tube 试管
test two bits 测试二位，检验两位
test under load 在负荷下试验
test unit 测试器，试验装置
test value 测试值，测定值
test valve 测试阀，试验阀，检查阀
test verification report 检验证明报告
test vessel 试件容器
test voltage 试验电压
test volume 测试量
test weight 砝码
test weld 试验焊缝，试焊
test well 试验套管，探井
test work group 测试工作组
test 实验，测试，试验
TES 固体废物处理系统【核电站系统代码】
TE(temperature element) 温度元件，温感元件
tethered balloon 系留气球
tethered 范围，系链
tethersonde 系留探测气球
tetrachloride 四氯化物
tetrachloroethane 四氯乙烷
tetrachloroethylene 四氯乙烯
tetrad 四个，四个一组
tetrafluoro-ethylene resin 聚四氟乙烯树脂
tetrafluoroethylene 四氟乙烯，特氟纶

tetrahedral element 四面体单元
tetrahedron block revetment 四面块护岸
tetrahydroxy-quinone 四羟苯醌
tetrapolar 四端的，四极的
tetrapropylenebenzene sulfonate 四丙烯苯磺酸盐，表面活性剂
tetrasodium pyrophosphate 焦磷酸四钠
tetravalence 四价
tetrode transistor 晶体四极管
tetrode 四极管
tetroon 等容气球
tetroxide 四氧化物
TEU(twenty-foot equivalent units) 计算单位，20英尺换算单位
TEU 废液处理系统【核电站系统代码】
TEWAC(totally enclosed with water cooled air cooler) 全封闭水冷式空气冷却器，全封闭水空冷型
tewel 烟囱，烟道，口，通风孔
text box 文本框
text-editing system 文本编辑系统
text file 文本文件，行文文件
textile fibre 纺织纤维
textile industry 纺织工业
textile wastewater 织物废水
text of final act 最终议定书文本
text of the protocol 议定书文本
textolite wedge 胶木槽楔
textolite 层压胶布板，夹布胶木，夹布胶板，布层塑料
textural classification of soil 土的质地分类，土结构分类
texture brick 粗面砖
textured solar cell 绒面太阳电池
texture hardening 基体硬化
texture 构造，纹理，织构，质地，组织
text 文本，正文
TE 汽缸膨胀【DCS 画面】
TFD(Test Fluid Drawing) 水压试验流程图
T flip-flop door 不用活塞制作正反门
T-flip flop T形触发器，触发式双稳多谐振荡器，反转触发器
TFPMSG(transverse flux permanent magnet synchronous generator) 横向磁通永磁同步发电机
TFT(transversal filament tape) 横向纤维带
TF(turbine follow) 汽机跟随方式
TFT winding(transversal filament tape winding) 横向纤维带缠绕技术
TFT winding 横向纤维带缠绕技术
TF 汽机跟随锅炉【DCS 画面】
TGA(thermogravimetric analyzer) 热失重分析仪
TGCU(tail gas cleanup unit) 尾气净化装置
T-G foundation 汽轮发电机基础
T-girder T形梁
TGOP(turbine gear oil pump) 盘车油泵
T-G pedestal 汽轮发电机基座
TGRD(temporary general rate decrease) 临时运价下调
T groove T形叶根槽
TGR(turbine gear) 盘车装置
TG supervisory instrumentation 汽轮发电机监控仪表
TG(turbine generator) 汽轮发电机
TG(turning gear) 汽机盘车
TGTU(tail as treatment unit) 尾气处理装置
thalassothermal energy 海洋热能
thalassothermal plant 海洋热能电站
thalofide cell 铊氧硫光电管
thalofide photocell 硫化铊光电池，铊氧硫光电管，氧硫化铊光电管
thalofide 氧硫化铊
thalweg 河道中泓线，最深谷底线，海谷底线，海谷深泓线，深泓，深泓线，剖面线，河槽各横断面最大水深点的连线
T-handle screwdriver T柄螺钉起子
T-handle socket wrench T柄套筒扳手
T-handle square fixed socket wrench 丁字柄四方固定套筒扳手
thanks to 由于
thatchboard 草结板
thawing and soaking shed 解冻库
thawing area for wagons 车辆解冻区［处］
thawing area of cars 车辆解冻区［处］
thawing heater 加热，融化
thawing house 解冻室
thawing plant 解冻装置
thawing station 解冻装置
thawing 解冻，融化
thaw pit （煤的）解冻房
THB 泰铢
THC(terminal handling charge) 码头操作附加费，码头附加费
T-head hammer 丁字锤
the air cargo tariff 航空货物运价
the atmospheric boundary layer thickness 大气边界层厚度
the axis of rotation of dumper 翻车机迴转轴线
the beginning of the year 年初
the benchmark rate of return 基准收益率
the chairman and managing director 董事长兼总经理
the China Insurance Regulatory Commission 保监会
the Chinese version 中文版本
the dielectric 电介质
the draft centerline of the cars 车辆牵引中心线
the family of airfoil 翼型族
the first stage of creep 蠕变第一阶段
the first stage of stress relaxation 应力松弛第一阶段
the following 下列（的），下述的
theft of electricity 窃电
theft and non-delivery 窃盗遗失
theft risk 盗窃险
the fuel safety and burner control system 燃料安全和燃烧器控制系统
the gate-type frame of the outgoing line 出线门型构架
the greatest casualties of this financial crisis 金融危机的最大牺牲品
the insured 被保险人
thematic session 专题会议

the meeting is posphoned till further notice 会议改期，时间另行通知
theme lecture 专题报告
the minutes of a meeting 会议记录
thenardite 无水芒硝
theodolite goniometer 经纬仪测角仪
theodolite with compass 罗盘经纬仪
theodolite 光学经纬仪，经纬仪，精密经纬仪
Theodorsen's circulation function 西奥多森环量函数
Theodorsen's theory 西奥多森理论
the operating power requirements 所需运行功率
theorem law 定理
theorem of least work 最小功定理
theorem of three moments 三力矩定律，三弯矩定理
theorem of virtual displacement 虚位移原理
theorem proving 定理证明
theorem 定理，法则，原理
theoretical acid 理论需酸量
theoretical air for combustion 燃烧理论空气量
theoretical air fuel ratio 理论风煤比
theoretical air quantity 理论空气量
theoretical air 理论空气，理论空气量
theoretical analysis 理论分析
theoretical annual production 理论年发电量
theoretical boundary 理论边界
theoretical calculation 理论计算
theoretical capacity 理论容量
theoretical captial cost 理论投资成本
theoretical combustion gas volume 燃烧气体理论容量
theoretical combustion temperature 理论燃烧温度
theoretical concentration 理论聚光度
theoretical condition 理论条件，理论工况
theoretical curve 理论曲线
theoretical cutoff frequency 理论截止频率
theoretical cycle 理论循环，理想循环
theoretical discharge 理论计算流量
theoretical dose 治疗剂量
theoretical draft （烟囱）理论拔风压头，理论抽力
theoretical efficiency 理论效率
theoretical equation 理论方程，理论公式
theoretical error 理论误差
theoretical flame temperature 理论火焰温度
theoretical flow coefficient 理论流量系数
theoretical flow rate 理论流量
theoretical fluid dynamics 理论流体动力学
theoretical fluid 理论流体
theoretical H_2O release rate 理论释水率
theoretical horsepower 理论功率
theoretical hydrodynamics 流量流体动力学
theoretical hypothesis 理论假设，原理假说
theoretical investigation 理论研究
theoretically developed profile 理论叶型
theoretically perfect plate 理论（塔）板
theoretical maximum density 理论最大密度
theoretical mechanics 理论力学
theoretical meteorology 理论气象学
theoretical mix of concrete 混凝土理论配合比
theoretical model 理论模型
theoretical occupation density 理论占有密度
theoretical output 理论输出功率
theoretical oxygen demand 理论需氧量
theoretical plate height 理论塔板高度
theoretical plate number 理论塔板数
theoretical plate 理论塔板，理想板
theoretical power 理论功率
theoretical prediction 理论判断
theoretical profile of gravity dam 重力坝基本剖面
theoretical pump displacement 水泵理论排水量
theoretical pump head 理论扬程
theoretical size 理论尺寸
theoretical span 理论跨度
theoretical stress concentration factor 理论应力集中系数
theoretical thermal efficiency 理论热效率
theoretical thermodynamics 理论热力学
theoretical tray 理论塔板，理想板
theoretical vacuum 理论真空
theoretical value 理论值
theoretical velocity ratio 理论速比
theoretical velocity 理论速度
theoretical water power 理论水力
theoretical weld-throat thickness 焊缝计算厚度
theoretical yield 理论产额，理论产率，理论生产率
theoretical 理论的，理论上的
theorization 理论化
theory of ablation 消融理论
theory of aeroelasticity 空气弹性理论
theory of chances 概率论
theory of communication 通信理论，通信学
theory of commutation 换向理论
theory of conjugate beam 共轭梁原理
theory of conjugate 共轭梁原理
theory of consolidation 固结原理
theory of constant energy of deformation 形变能守恒原理
theory of continental drift 大陆移动理论
theory of continental growth 大陆成长理论
theory of cross-flow 横向流动理论
theory of damage accumulation 累积损伤理论
theory of dynamic programming 动态规划理论
theory of elasticity 弹性理论
theory of electric breakdown of gas 气体击穿理论
theory of equivalent slope 等倾度理论
theory of errors 误差论，误差理论
theory of field 场论
theory of function 函数论
theory of induced cleavage 诱生裂隙理论
theory of large scale system 大系统理论
theory of limit 极限理论
theory of models 模型理论
theory of optimal control 最优控制理论
theory of optimum growth 生长最适度理论
theory of plane stress 平面应力理论

theory of plasticity 塑性理论
theory of plastic viscous flow 塑性黏滞流理论
theory of probability 概率论
theory of proportioning 混凝土配合比理论
theory of reliability 可靠性理论
theory of sampling 取样理论
theory of similarity 相似理论
theory of stability 稳定性理论
theory of straight-line distribution 直线分布理论
theory of strength 强度理论
theory of superposition 重叠理论
theory of suspended separation 悬浮分离理论
theory of tidal channel 潮汐沟渠理论
theory of turbulence 湍流理论
theory 理论，原理，定理，学说
the party awarding the contract 发包方
the People's Insurance Company of China 中国人民保险公司
the power of inertia 惯性力，惰性力
the range of stress indensity factor 应力强度因子范围
therapy dose 治疗剂量
therapy model 治疗模型
thereafter 此后，在那之后，据此
thereby 因此，由此，从而
thereinafter 在下文，以下，在下一部分中
thereinbefore 在上文，在上一部分中，前头
therein 在那里，在那点上，在那方面
thereof 它的，其
thereon 在其上，在那上面，……之后立即
the rest may be inferred and so on and so forth 依此类推
thereto 与之，向那里
thereunder 在其下，在那下面，依据
thereupon then 从而
thereupon 关于那，就此，于是，立即，随即，在其上，关于那件事
therewith 以此，此外
thermal abrasion 热蚀，热磨蚀
thermal absorber 吸热器
thermal absorption 热吸收，热量吸收，热中子吸收，吸热
thermal accommodation coefficient 热适应系数，热调节系数
thermal accumulated fatigue index 热积累疲劳指数
thermal activation energy 热激活能
thermal activation 热（中子）活化
thermal activity 热中子导致的放射性
thermal aging property 热老化特性
thermal aging test 热老化试验
thermal aging 加热老化，受热老化
thermal agitation voltage 热扰动电压
thermal agitation 热扰动
thermal air buffer compartment 热空气缓冲室
thermal alteration 热蚀变
thermal analog computer 热力计算用的模拟电脑，热模拟计算装置
thermal analysis 热分析
thermal anemometer 热风速仪
thermal anomaly 热的反常，热反常
thermal automation instrument 热工自动化仪表
thermal automation 热工自动化
thermal baffle 隔热板，绝热隔板
thermal balance 热量平衡，热平衡
thermal barrier coating materials 隔热涂料
thermal barrier coating 热障涂层
thermal barrier flange 热屏蔽法兰
thermal barrier 热屏障【换热器类型】，热障，隔热层，保温层，绝热层，热垒，热屏蔽
thermal beam 热中子束
thermal behavior of solar pond 太阳池的热性能
thermal belt 高温带
thermal bimetal 热双金属
thermal Biot number 传热毕奥数
thermal-blanket insulation 隔热层
thermal bond （元件的）热结合
thermal boundary layer thickness 热边界层厚度
thermal boundary layer 热边界层，温度边界层
thermal breakdown of capacitor 电容器热击穿
thermal breakdown 热击穿，热破坏
thermal breaker 配有热继电器的自动开关
thermal breeder reactor 热增殖堆，热中子增殖堆
thermal breeder 热中子增殖堆
thermal breeding 热（中子）增殖
thermal bridge 热电桥【测无线电干扰用】
thermal brittleness 热脆性
thermal buckling 热皱折，热失稳
thermal budget 热平衡，热收支
thermal buffer tube 热缓冲管
thermal buffer 热缓冲器【热处理】，热缓冲
thermal bulb 热敏元件，测温包
thermal burst 发裂
thermal bus 传热母线
thermal capacitance 热容
thermal capacity 热容量，热值
thermal capture cross section 热中子俘获截面
thermal capture 热中子俘获
thermal characteristics 热工特性
thermal charge 热负荷
thermal chemical reaction 热化学反应
thermal chemical test 热化学试验
thermal chimney 热风筒
thermal choking 热阻现象
thermal circuit breaker 热胀断路器
thermal circuit concept 热回路初步设计
thermal circulation 热力循环，热环流
thermal climate 热型气候
thermal coefficient of conduction 导热系数
thermal coefficient of expansion 热膨胀系数
thermal coefficient 热系数
thermal collector 热力除尘器，集热器
thermal column beam 热柱中子束
thermal column 热柱
thermal comfort 热舒适性
thermal compensation alloy 热补偿合金
thermal compressor blade 热力压气机叶片
thermal computing element 热计算元件
thermal condition 热工况，热状态
thermal conductance 热导，热电导
thermal conduction characteristic 热传导特性，

导热特性
thermal conduction resistance 导热热阻
thermal conduction 热传导，导热，导热性
thermal conductivity analyzer module 热导式分析模块
thermal conductivity bridge 热导电桥
thermal conductivity cell detector 热导元件检测器，热导池检测器
thermal conductivity cell 导热池
thermal conductivity coefficient 热传导系数，导热系数
thermal conductivity detector 热导检测器
thermal conductivity factor 导热系数
thermal conductivity gas analyzer 热导式气体分析仪
thermal conductivity leak detector 热电导检漏仪
thermal conductivity 导热性，热导率，热电导，导热率，导热系数
thermal conductor 导热体
thermal cone 温度锥
thermal confinement 热约束
thermal conformer 温度随变生物
thermal connection 热连接【低温焊，钎焊或焊接】
thermal consolidation 热固结
thermal constant 热常数，温度常数
thermal contact 热触点，热接触
thermal contraction 热收缩
thermal control equipment 热工控制设备
thermal control switch 热控开关
thermal control system 热控制系统
thermal control 热工控制，温度控制
thermal convection loop 热对流回路
thermal convection resistance 对流热阻
thermal convection storm 热对流暴风雨
thermal convection 热对流
thermal converter 热变换器，热电偶式变换器，热转换堆
thermal core 热中子活性区
thermal corrosion rate 热腐蚀率
thermal couple 热电偶
thermal cracking 热裂，热裂解
thermal creep stresses field 温度徐变应力场
thermal creep stress 温度徐变应力
thermal creep velocity 热蠕变速度
thermal creep 热蠕变，热力蠕变，热潜变
thermal critical assembly 热中子临界装置
thermal critical point 热临界点，临界温度
thermal cross section 热截面
thermal current density 热流密度
thermal current rating 热额定电流
thermal current 热流，对流
thermal cutoff energy 热中子截止能，热中子谱切割能
thermal cutout 热熔断路器，热熔保险器，热断流器，热断路
thermal cutting 热切割
thermal cycle 热循环
thermal cycling properties of fuel elements 燃料元件热循环特性
thermal cycling stability 耐热循环性，热循环稳定性
thermal cycling strength 热循环强度
thermal cycling stress 热循环应力
thermal cycling 热交变，热循环
thermal deaeration 热力除氧
thermal deaerator 热力除氧器
thermal decanning 热法去壳，热脱壳
thermal decomposition 热分解，热解作用
thermal deformation 热变形，温度变形
thermal degradation 热降解
thermal delay timer 热力定时器
thermal demineralization of water 水的热力除盐
thermal density current 温差异重流
thermal design flow 热工水力设计流量
thermal design of reactor 反应堆热工设计
thermal design 热工设计
thermal detector 热电探测器，热力探测器，热中子探测器，热探测器
thermal deviation 热偏差
thermal device 热工设备
thermal diffusibility 热扩散性
thermal diffusion coefficient 热扩散率，热扩散系数
thermal diffusion constant 热扩散常数
thermal diffusion effect 热扩散效应
thermal diffusion factor 热扩散因子
thermal diffusion plant 热扩散工厂，热扩散装置
thermal diffusion ratio 热扩散比
thermal diffusion 温度扩散，热扩散
thermal diffusivity 热扩散率，热扩散性，导温系数，热扩散系数，散热率，散热系数
thermal dilatation 热膨胀
thermal disadvantage factor 热中子不利因素
thermal discharge 放热，排出热
thermal dispersion coefficient 热弥散系数
thermal dispersion property 散热性能
thermal dispersion rate 热扩散率
thermal dispersion 热弥散，热扩散
thermal distortion 热扭曲，热变形
thermal distress 热损坏
thermal distribution 热力分布，热分布
thermal dose 热中子剂量
thermal drain 热疏水
thermal drilling 热力钻进
thermal driving force 热推动力
thermal drop 焓降，热降
thermal-dynamic mechanical endurance 耐热机械动应力
thermal dynamics 热能动力学
thermal effectiveness 热力经济性
thermal effect 热效率，热效应
thermal efficiency curve 热效率曲线
thermal efficiency indices 热经济性指标
thermal efficiency of cycle 循环热效率
thermal efficiency of power plant 发电厂热效率
thermal efficiency ratio 相对热效率，热效率比值
thermal efficiency test 热效率试验
thermal efficiency 热效率，发热量
thermal effluent 热废气，热废液，热排出流，热废水，温排水

thermal electrical relay 电热继电器
thermal electric generation 火力发电
thermal electricity 热电
thermal electromotive force 热电动势，温差电动势
thermal electron 热电子
thermal element 热敏元件，热元件，热电偶，熔断器，热偶元件
thermal embrittlement 热脆性
thermal endurance 耐热度，热寿命
thermal energy neutron 热中子，热能中子
thermal energy power engineering 热能动力工程
thermal energy range 热能区
thermal energy region 热能区
thermal energy storage equipment 热能储存装置
thermal energy storage system 储热系统，热储能系统
thermal energy storage 热能
thermal energy yield 热能当量，热能产额
thermal energy 热能
thermal enrichment 核电站废热，热增浓
thermal environment 热环境
thermal equation 热力方程
thermal equator 热赤道
thermal equilibrium 热平衡
thermal equipment 热工设备
thermal equivalent of work 功热当量
thermal equivalent power 热当量功率
thermal equivalent （功的）热当量
thermal etching 热侵蚀
thermalexcel 一种沸腾换热强化管
thermal exchange 热交换
thermal expansion coefficient 热膨胀系数
thermal expansion monitor 热膨胀监测保护仪
thermal expansion reactivity coefficient 热膨胀反应性系数
thermal expansion relay 热膨胀继电器
thermal expansion ring 热膨胀环
thermal expansion stress 热胀应力
thermal expansion 热膨胀
thermal expansivity 热膨胀系数
thermal fatigue cracking 热疲劳裂纹
thermal fatigue creep 热机械疲劳蠕变
thermal fatigue fracture 热疲劳断裂
thermal fatigue life 热疲劳寿命
thermal fatigue 热疲劳，高温疲劳
thermal field 温度场
thermal fisser 热中子裂变物质
thermal fission cross section 热中子裂变截面
thermal fission factor 热中子裂变因子
thermal fission yield 热中子裂变产额
thermal fission 热中子裂变
thermal flash 温度突变
thermal flow diagram 热力系统图
thermal flowmeter 热流量计，量热式流量计
thermal flow 热流
thermal flux density 热流密度
thermal flux 热流，热通量，热流率，热中子通量
thermal follow-up unit 热跟踪单元，热跟踪装置
thermal force 热力
thermal friction 热摩擦
thermal galvanometer 温差电偶电流计
thermal gradient energy 温差能
thermal gradient measuring probe 热梯度探测仪
thermal gradient 温度梯度，热梯度
thermal gradiometer 热量陡度计
thermal graphite reactor 石墨慢化热中子反应堆
thermal growth 热膨胀，热生长
thermal head 热力压头
thermal healing process 透热疗法
thermal history 温度随时间变化历程，温度时间关系
thermal hydraulic analysis 热工水力分析
thermal hydraulics 热工水力学
thermal hydraulic test facility 热工水力试验装置
thermal hydraulic test 热工水力试验
thermal hydraulic 热工水力学，热工水力的
thermal imaging system 热成像系统，热像仪
thermal impedance 热阻
thermal incineration 加热焚化
thermal incinerator 加热焚化炉
thermal inelastic cross section 热中子非弹性散射截面
thermal inelastic scattering 热非弹性散射
thermal inertia 热惯性，热质量
thermal inland 热岛
thermal instability 热不稳定性
thermal instrument 热电偶仪表，热电效应仪表，热工仪表
thermal insulance 热绝缘系数
thermal insulated barrier 保温屏障
thermal insulated metal roofing 保温金属屋面
thermal insulated metal wall 保温金属板墙
thermal insulating disc 隔热盘，绝热片
thermal insulating mortar 保温砂浆
thermal insulating property 绝热性质
thermal insulating roof 保温屋顶
thermal insulating value 绝热指数
thermal insulation blanket 绝热层
thermal insulation block 保温块
thermal insulation cabinet 保温柜
thermal insulation door 保温门
thermal insulation lagging 保温保护层
thermal insulation lining 保温内衬
thermal insulation material 绝热材料，保温材料
thermal insulation of roof 屋顶保温
thermal insulation plug 隔热芯块
thermal insulation structure 保温结构
thermal insulation support brackets 隔热层支承架
thermal insulation system 绝热系统
thermal insulation tape 耐热绝缘带
thermal insulation 保温层，绝热层，绝热，热绝缘，隔热，保温，保温设施
thermal insulator 绝热器，绝热层，绝热体，热绝缘材料
thermal intelligence 热信息
thermal intensity 热强度
thermal internal boundary layer 热力内边界层
thermal inversion layer 逆温层
thermal inversion 逆温
thermal isolation 绝热
thermalization column 热柱

thermalization cross section 热化截面
thermalization range 热化区
thermalization time 热化时间
thermalization （中子的）热能化，热化
thermalizing column 热柱
thermalizing time 热化时间
thermal jump 热落差
thermal lag 热惯性，热滞后
thermal lattice spectra 热栅格谱
thermal leakage factor 热泄漏因子，热中子漏逸率
thermal level 热级
thermal lifetime 热（中子）寿命
thermal life 热寿命
thermal limit 耐温极限
thermal liner 热衬里
thermal load 热负荷，热载荷，热荷载
thermal logger 温度记录仪
thermal loop 传热回路
thermal loss 热损失，热耗
thermally activated creep 热激活蠕变
thermally black 对热中子为黑体的
thermally fissile Pu 热裂变钚
thermally fully developed 热定型流动的
thermally induced flow 热导流（动）
thermally treated 热处理的
thermal machine 热机
thermal-magnetic breaker 热磁开关
thermal-magnetic circuit breaker 热磁保护断路器
thermal-magnetic trip breaker 热磁脱扣断路器
thermal-magnetic trip element 热磁切断装置元件
thermal-magnetic trip 热磁切断装置
thermal-magnetizing material 热磁材料
thermal mass 热质量，热惯性
thermal Maxwell flux distribution 热中子通量的麦克斯韦分布
thermal measurement instrument 热工测量仪表
thermal measurement system 热工测量系统
thermal measurement 热工测量
thermal mechanical fatigue 热机械疲劳
thermal mechanical performance test 热机性能试验
thermal metamorphism 热力变质
thermal misadjustment of consumer heating system 热用户热力失调
thermal mixing 热混合
thermal modification 热改造
thermal monitoring 热力监测
thermal mortar 保温砂浆
thermal motion （原子的）热运动
thermal neutron activation 热中子激活，热中子活化
thermal neutron breeding 热中子增殖
thermal neutron cross section 热中子截面，热中子（通量）校准
thermal neutron detector 热中子探测器
thermal neutron fission cross section 热中子裂变截面
thermal-neutron fission 热中子裂变，热中子分裂
thermal neutron flux amplification 热中子注量放大
thermal neutron flux density 热中子通量密度
thermal neutron flux spectrum 热中子通量谱
thermal neutron flux standard 热中子通量标准
thermal neutron flux 热中子通量
thermal neutron inelastic scattering 热中子非弹性散射
thermal neutron leakage 热中子泄漏
thermal neutron lifetime 热中子寿期
thermal neutron nonleakage probability 热中子不泄漏概率
thermal neutron range 热中子能区
thermal neutron reactor 热中子反应堆，热中子堆
thermal neutron spectrum 热中子谱
thermal neutron 热中子
thermal noise thermometer 热噪声温度计
thermal noise 热噪声
thermal nonleakage probability 热中子不泄漏概率
thermal NO_x 热力型氮氧化合物
thermal ohm 热欧姆
thermal oscillation 热振荡
thermal outfall 热水排放
thermal output of heating surface 加热面热强度
thermal output 热功率输出【电厂、核蒸汽供应系统、堆芯等】，热功率，传热量，热输出量，热出力，热输出
thermal over-current relay 热力式过载继电器
thermal over-current trip 过量电流热跳脱
thermal over-current 热过电流
thermal overload protection 热过载保护
thermal overload relay 热过载继电器
thermal overload release 热过载脱扣器
thermal overload trip 热过载保护
thermal oxidation 热力燃烧
thermal pad 热屏蔽【堆芯】
thermal parameter 热力参数，热力指标，热力指数，热工参数
thermal peak 热峰值
thermal performance calculation 热工性能计算
thermal performance characteristics 传热性能的表征
thermal performance of building 建筑热性能
thermal performance test 热力性能试验
thermal performance 热力特性，热力性能，热工性能
thermal plant 热电厂，火电厂
thermal plume 放射性烟云，热烟羽，热羽流，热卷流，热烟流
thermal pollution 热污染
thermal power engineering 热能动力工程
thermal power generation 热力发电，火力发电
thermal power installation 热能动力装置
thermal power plant 火电厂，火电站，热电厂，热电站，热力发电厂，热能动力装置
thermal power rating 额定热功率
thermal power station 火力发电厂［站］，热力发电厂［站］，热电站［厂］，火电站［厂］
thermal power system 热力系统
thermal power 热能，火力，热力，热能动力，热功率
thermal precipitation 加热沉淀，热力沉降

thermal precipitator 热集尘器，热沉淀器
thermal pressure 热压，热压力
thermal probe （炉膛）热力测针，测热探头
thermal process control 热工过程控制
thermal process curve 热力过程曲线
thermal process 热力过程
thermal properties of materials 材料的热特性
thermal property 热力特性，热性质，热力参数
thermal protection structure 热防护结构
thermal protection 热防护，热屏蔽，热保护
thermal pump 热力泵
thermal quench 热激冷
thermal radiant flux 辐射热通量
thermal radiation detector 热辐射检测仪
thermal radiation 热辐射
thermal range 热（中子）区
thermal ratcheting 热棘轮效应
thermal rating 热功率，热定额，耐热等级
thermal reactor output 热中子反应堆出力
thermal reactor power 热中子反应堆功率
thermal reactor 热堆，热反应堆
thermal recovery factor 热恢复系数
thermal recovery unit 热回收装置
thermal refining 调质热处理
thermal reflux 热回流
thermal regeneration 加热再生，热量回收
thermal regime 热状况
thermal region 热区，热能区
thermal relay 热敏继电器，热式继电器，温度继电器，热继电器
thermal release 热释放器，热脱扣器，热敏释放装置，过电流热保护
thermal resistance of submerged heat sources 埋藏热源的热阻
thermal resistance thermograph 热敏电阻温度计
thermal resistance type 热阻型的
thermal resistance 热阻，热稳定性，热电阻，耐热性能
thermal resistivity 热阻率，热阻系数
thermal resistor 热敏电阻器，热敏电阻
thermal response 热力特性，热响应
thermal reversibility 热力可逆性
thermal runaway 热散发，热散逸，热耗散
thermal saturation property 饱和热力参数，热饱和特性
thermal scanning equipment 热扫描仪
thermal sensing element 热敏元件
thermal sensitive element 热敏元件
thermal sensitivity 热灵敏度
thermal sensor 热传感器
thermal shield assembly of reactor vessel 压力壳热屏蔽组件
thermal shielding 热屏蔽
thermal shield radial support 热屏径向支承
thermal shield support lug 热屏支承耳
thermal shield 热屏蔽，挡热板，热屏蔽体，热屏蔽层
thermal shock crack 热冲击裂缝，热冲击裂纹
thermal shock fracture 热冲击断裂
thermal shock resistance 抗热冲击，抗热冲击能力
thermal shock rig 热冲击试验装置
thermal shock shield 热冲击屏蔽
thermal shock stress 热冲击应力
thermal shock 热震，热冲击，温度骤变，温度冲击
thermal shrinkage temperature 热收缩温度
thermal shrinkage 热收缩，热收缩率，热缩
thermal similarity 热相似性
thermal siphon 热虹吸作用，热虹吸
thermal skin effect 热肤效应
thermal sleeve 热套管，热套管接头，热阱，热套
thermal source 热源
thermal spectrum calculation 热谱计算
thermal spike 最大热值，温度峰值，热峰
thermal spraying oxygen eliminator 热力喷雾除氧器
thermal spraying 热喷涂
thermal stability 热稳定性，耐热性，热稳定度
thermal stabilization 热加固，热稳定
thermal state 热态
thermal static draft 自拔风
thermal station 热力发电厂，火力发电厂
thermal steering 热引导
thermal storage capacity 蓄热能力，热容量，蓄热量，蓄热容
thermal storage device 蓄热器
thermal storage effect 蓄热特性
thermal storage tank 蓄热水池
thermal storage 储热，蓄热
thermal straightening 热校直
thermal strain meter 热应变计
thermal strain 热应变，热变形
thermal stratification 热分层现象，热层结，温度分层
thermal stress analysis 热应力分析
thermal stress cracking 热应力裂纹
thermal stress due to nonuniform temperature 温度分布不均引起的热应力
thermal stress fatigue index 热应力疲劳指数
thermal stress fatigue 热疲劳，热应力疲劳
thermal stress meter 热应力计
thermal stress ratchet 热应力棘轮效应
thermal stress 热应力，温差应力，温度应力
thermal stretching 热伸长
thermal structural coupling 热结构耦合
thermal structure 热结构
thermal subtraction technique 消热俘获法
thermal switch 热控开关，温控开关，热开关，温度开关
thermal testing 热工测试
thermal test 热力试验
thermal time constant 热时间常数
thermal time delay relay 热延时继电器
thermal time delay switch 热时间延迟开关
thermal tracer 热示踪物
thermal transfer 传热
thermal transient facility 热瞬变装置
thermal transmission 热传播，传热
thermal transmittance 热传导，热传递，透热率
thermal transport property 热输运性质，热传递性质

thermal trap 热阱，热收集器
thermal treatment plant 热处理室
thermal treatment 热处理
thermal tripping 热跳闸
thermal trip type breaker 热脱扣型断路器
thermal trip 热敏短路，热效应断路器，热脱口器，热（双金属）断路装置
thermal turbulence 热湍流
thermal type meter 热电式仪表
thermal unbalance 热不平衡
thermal unit 热量单位
thermal utilization factor 热中子利用系数
thermal velocity neutron 热中子，热能中子
thermal viscoelastic properties 热黏弹性特性
thermal wake 热迹
thermal water pollution 热水污染
thermal weldability 热焊接性
thermal well 热井
thermal wind 热成风
thermal writing recorder 热记录仪
thermal yield 热中子产额
thermal zone 高温区
thermal 热的，热力学的，热电温度计
thermic boring 热钻孔法
thermic 热的，热力的
thermionic conduction 热离子导电，热离子转换
thermionic converter 热离子转换器，热离子换能器
thermionic current 热离子电流
thermionic emission 热电子发射，热电子放射，热离子发射
thermionic energy converter 热离子能转换器
thermionic engine 热离子发动机
thermionic fuel element 热离子燃料元件
thermionic generation 热离子发电
thermionic power generation 热离子发电
thermionic reactor 热离子反应堆
thermionic rectifier 热离子整流器，热电子整流器
thermionic relay 热离子继电器
thermionic tube 热电子管
thermionic vacuum gauge 热离子真空计
thermionic work function 热离子功函数
thermionic 热离子的，热电离
thermisopleth 变温等值线，等变温线
thermistor bolometer 热敏电阻测热器，热敏电阻辐射热测量计，热敏电阻辐射仪
thermistor bridge 热敏电阻电桥，热电阻电桥，热阻电桥
thermistor flowmeter 热变电阻流量计
thermistor gauge 热敏电阻式压力计
thermistor heat detector cell 热敏电阻测热器
thermistor nomogram 热敏电阻列线圈
thermistor probe 热敏电阻探针
thermistor thermometer 热敏电阻温度计
thermistor voltage stabilizer 热敏电阻稳压器
thermistor wattmeter 热敏电阻瓦特计
thermistor 热变阻器，热敏电阻，热控管，热电阻，热敏电阻器，热电调节器
thermite welding 铝热焊，铝热焊接
thermit process 铝热剂工艺
thermit reaction 热剂反应
thermit welding 热剂焊
thermoaero-elasticity 热气动弹性力学
thermo-ammeter （测量微电流用的）热电偶安培计［表］，热电流表，温差电偶安培计
thermo anemometer 温差式风速计
thermobalance 热量平衡，热天平
thermobarometer 温度气压计
thermo capillary movement 温度毛管水运动
thermo-compression bonding 热压焊接，热压接合
thermocell 热电池
thermochemical equilibrium 热化学平衡
thermochemical means 热化学方法
thermochemical process 热化学过程
thermochemical reaction 热化学反应
thermochemical test 热化学试验
thermochemical treatment 化学热处理
thermochemical 热化学的
thermochemics 热化学
thermochemistry 热化学
thermocline layer 斜温层
thermocline 斜温层，变温层，温跃层
thermocoax 超细管式热电偶
thermo-compensation 温度补偿
thermocompensator 热补偿器
thermocompression bonder 热压焊接机
thermo compression wire bonder 热压引线接合机
thermo-compressor 热力压气机
thermocontamination of warm water discharge 温排水的热污染
thermocouple alloy 热电偶合金
thermocouple ammeter 温差电偶安培计，热电偶安培计
thermocouple assembly 热电偶组件
thermocouple burnout 热电偶烧断
thermocouple cannula 热电偶保护套管
thermocouple cold junction boxes 热电偶冷端盒
thermocouple cold junction output 热电偶冷端输出线
thermocouple column assembly 热电偶套管柱组件
thermocouple column replacement 热电偶柱更换
thermocouple column seal assembly 热电偶套管柱密封组件
thermocouple column 热电偶套管柱
thermocouple comparator 热电偶比较器
thermocouple conduit 热电偶导管
thermocouple current meter 热电偶流速仪
thermocouple element 热电偶元件，恒温器，温度自动调整装置
thermocouple extension cable 热电偶补偿电缆
thermocouple extension wire 热电偶延长线
thermocouple gauge 热电偶仪表
thermocouple generator 热电偶发电器，温差发电器
thermocouple instrument 温差电偶仪表，热电偶仪表
thermocouple junction box output panel 热电偶连接盒输出盘
thermocouple junction 热电偶接点

thermocouple lead wire 热电偶引出线
thermocouple map 温度分布图
thermocouple material 热电偶材料
thermocouple metal 热电偶合金
thermocouple pile 热电偶柱
thermocouple probe 热电偶测针
thermocouple pyrometer 热电偶高温计
thermocouple relay 热电偶继电器
thermocouple restorer 误差热偶恢复器
thermocouple-sensing system 热电偶探测系统
thermocouple sensor 热电偶传感器
thermocouple thermometer 热电偶温度计
thermocouple wattmeter 热电偶功率计，热电偶瓦特表
thermocouple well 热电偶测温孔，热电偶套管，热电偶管座
thermocouple wire 热电偶线
thermocouple 热电偶，温差电偶
thermocurrent 热电流，温差电流
thermocutout 热保险装置，热断流器
thermo-deaeration 热力除氧
thermodecomposition （燃料）热分解
thermodetector 热检测器，测温计
thermodiffusion coefficient 热扩散系数
thermodiffusion 热扩散
thermodiffusiophoresis 热扩散泳
thermoduric 耐热的
thermodynamic analogy 热力模拟
thermodynamic argument 热力学论证
thermodynamic change 热力学变化
thermodynamic circulation 热力环流
thermodynamic coefficient 热力系数
thermodynamic critical magnetic flux density 热力学超导临界磁通密度
thermodynamic cycle 热力循环
thermodynamic diagram 热力系统图，热力学图
thermodynamic drain trap 热动力式疏水阀（器）
thermodynamic effectiveness 热力经济性
thermodynamic effect 热力效应，热力效果
thermodynamic energy 热力学能
thermodynamic equilibrium 热力平衡
thermodynamic force 热动力
thermodynamic function 热力学函数
thermodynamic gas scale 热力学气体温标
thermodynamic heating 热力供暖
thermodynamic instrumentation 热工仪表
thermodynamic instrument 热工仪表
thermodynamic level 热力能级
thermodynamic loss 热力损失
thermodynamic non-equilibrium 热力不平衡
thermodynamic optimization 热力学优化
thermodynamic performance test 热力特性试验
thermodynamic potential 热势，热力位，热动力势
thermodynamic process 热力过程
thermodynamic property 热力学特性，热力学性质，热力学参数
thermodynamic quality 热力学特性，热力学性质
thermodynamic relation 热力学方程，热力学关系式
thermodynamic scale 热力学温标，绝对温标
thermodynamics engineer 热力学工程师
thermodynamic similarity 热力学相似
thermodynamic stability 热力学稳定性，热力稳定性
thermodynamic state 热力学状态
thermodynamics vapor cycle 热力学蒸汽循环
thermodynamic system diagram 热力系统图
thermodynamic system of conventional island 常规岛热力系统
thermodynamic system 热力学系统，热力系统
thermodynamics 热力学
thermodynamic temperature scale 热力学温标
thermodynamic temperature 热力学温度
thermodynamic test 热力试验
thermodynamic tractive test 牵引热工试验
thermodynamic trap 热动态活门
thermodynamic type steam trap 热动力式（蒸汽）疏水器
thermodynamic yield 热力学产额
thermodynamic 热动力学的，热动态的，热力学的，使用热动力的，热动的
thermoelastic effect 热弹性效应
thermoelastic epoxy 热弹性环氧树脂
thermoelastic insulation 热弹性绝缘
thermoelasticity 热弹性
thermoelastic property 热弹性
thermoelectric and mechanical property 热电和力学性能
thermoelectric cell 热电元件，温差电池
thermoelectric comparator 热电比测器
thermoelectric conversion 热电转换
thermoelectric converter 热电变换器
thermoelectric cooler 热电冷却器
thermoelectric cooling 热电冷却，温差电制冷
thermoelectric couple 热电偶，温差电偶
thermoelectric current 热电电流，温差电流
thermoelectric effect 热电效应，温差电效应
thermoelectric emission current 热电发射电流
thermoelectric energy converter 热电换能器
thermoelectric generating set 热力发电机组，火力发电机组
thermoelectric generator 热电式发电机，温差发电器，温差电势发生器，温差发电机
thermoelectric instrument calibrator 热工仪表校验仪
thermoelectric inversion 热电转换
thermoelectricity 温差电，热电
thermoelectric junction 热电偶，温差电偶接头，热电偶接点
thermoelectric measurement 热电测量
thermoelectric pile 热电堆，热电偶，温差电堆，温差电池
thermoelectric potential 热电势
thermoelectric power generation 热力发电
thermoelectric power plant 火力发电厂
thermoelectric power station 火力发电厂，热电站
thermoelectric power 热电动力，电功率
thermoelectric property 热电性质
thermoelectric pyrometer 热电高温计，热电温度计
thermoelectric refrigeration 热电制冷

thermoelectric standard reference material 热电标准参考材料
thermoelectrics 热电学，温差电流学
thermoelectric thermometer 热电温度计
thermoelectric unit 热电电池，温差电池
thermoelectric 热电的，温差电的，由温差产生电流的，热电势率的，差电势率的
thermoelectrochemistry 热电化学
thermoelectrode 热电电极
thermoelectromagnetic 热电磁的
thermoelectrometer 热电计
thermoelectromotive force between substance M and N 物质M与N之间的温差电动势
thermoelectromotive force 温差电动势
thermoelectronic effect 热电子效应
thermoelectronic 热电子的
thermoelectron 热电子
thermoelement 热电元件，热电偶，温差电偶，热电部件，热偶丝，热电元件
thermoemission reactor 热发射反应堆
thermoendosmosis 热内渗透
thermoexcell 强化换热管
thermo expansion valve 热膨胀阀
thermofication factor 热化系数
thermofor 蓄热器
thermofractography 热裂谱法
thermo-fuse 热熔断器
thermogalvanic corrosion 温差电流腐蚀，热电腐蚀
thermo-galvanometer 热电偶检流计，热偶电流计，温差电偶检流计
thermo-generator 热偶电池，热偶发电器
thermogram 温度过程线，温谱图
thermography 温度记录（法），发热记录，自记温度，热谱
thermograph 温度记录器，温度过程线，温度记录仪，自记式温度计，热录像仪，热图
thermogravimetric analysis 热解重量分析
thermogravimetric analyzer 热失重分析仪，热重分析仪
thermogravimetric curve 温度重量曲线
thermogravimetry 热重分析法
thermoguard motor 一种内装热保护装置的电动机
thermohaline convection 热盐对流
thermohardening 热硬化
thermohydraulic core design 堆芯热工水力设计
thermohydraulics 热工水力学
thermohydraulic 热工水力（的），热工水力学的
thermohydrodynamic instability 热工流体动力不稳定性
thermohydrometer 温差比重计，温差相对密度计，热比重计
thermohygrograph 温湿度过程线
thermohygrometer 温湿表，温度湿度计
thermo-hygrostat 恒温恒湿器，恒温恒湿装置
thermoionic emission 热离子发射
thermoionic 热离子的，热电子的
thermoion 热离子
thermoisopleth 等温线，等温等值线
thermojunction 热电偶接头，热电偶接点，温差电偶
thermolabile 不耐热的，感热的
thermolite 红外辐射用大功率碳丝灯
thermolith 耐火水泥
thermoluminescence 热发光
thermoluminescent detector 热释光剂量计，热释光探测器
thermoluminescent dosimeter 热释光剂量计
thermoluminescent dosimetry 热发光剂量测定法，热释光剂量学
thermoluminescent film badge 热发光的胶片剂量计【保健物理】
thermolysis 热分解的
thermomagnetic conversion 热磁转换
thermomagnetic effect 热磁效应
thermomagnetic generator 热磁发电机
thermomagnetic motor 热磁电动机
thermomagnetic protection 热磁保护
thermomagnetic 热磁的，热磁效应的
thermomagnet 热磁
thermomatic motor 一种内装热保护装置的电动机
thermomechanical behaviour 热力学行为，热变性
thermomechanical curve 热机械曲线
thermomechanical effect 热力学效应，热机械效应，一种热机械效应，热机效应
thermomechanical exergy 热机械火用
thermomechanical generator 热机械发电机
thermomechanical heat treatment 形变热处理
thermomechanical instability 热机失稳
thermomechanical method 热力学方法
thermomechanical noise 热机械噪声
thermomechanical properties 热机械性能，热机械特性
thermomechanical supervisor 热机监理
thermomechanical treating 形变热处理
thermomechanical treatment processing 形变热处理工艺
thermomechanical treatment 形变热处理，热机械处理，热处理工艺周期，加工热处理
thermomechanical 热机（械）的
thermometallurgy 火法冶金
thermometal 双金属
thermometamorphism 热变质作用
thermometer bulb 温度计测温包，温度计泡
thermometer calibration room 温度表校验室
thermometer coefficient 温度系数
thermometer probe 测温计探头
thermometer reading 温度计读数
thermometer resistor 电阻测温器
thermometer screen 温度表百叶箱
thermometer shelter 百叶箱温度表，温度表百叶箱
thermometer support 温度计架
thermometer well 温度计袋，温度计套管
thermometer 温度计
thermometrical 温度计的，测温的
thermometric analysis 测温分析
thermometric coefficient 温度系数
thermometric conductivity 导温系数
thermometric constant 热量常数，温度常数

thermometric depth 测温深度
thermometric device 测温设备，温度计
thermometric element 测温元件
thermometric fluid 温度计液体
thermometric gas 测温气体
thermometric indicator 测温仪表
thermometric instrument 温度仪表
thermometric lag 热滞后
thermometric scale 温标
thermometrograph 温度记录器
thermometry 计温学，测温学，测温技术，测温法
thermomodule 热电微型组件
thermomolecular pressure 热分子压力
thermonatrite 水碱
thermonegative 吸热的
thermonuclear cycle 热核循环
thermonuclear energy 热核能
thermonuclear fusion 热核反应，热核聚变
thermonuclear machine 热核装置
thermonuclear plasma 热核等离子体
thermonuclear power plant 热核电站，核聚变电站
thermonuclear power 热核能，热核动力
thermonuclear process 热核过程
thermonuclear reaction 热核反应
thermonuclear reactor 热核反应堆，聚变反应堆，热核堆
thermonuclear 热核的
thermonucleonics 热核技术，热核学
thermoosmosis 热渗（透）
thermopaint 测温涂料
thermopair 热电偶
thermopause 热成层顶
thermophone 传声温度计，热线式受话器
thermophoresis 热泳现象，热迁移现象
thermophore 载热体，蓄热器
thermophysical property 热物理性质
thermophysical 热物理的
thermopile generator 温差发电机，热堆发电机，温差电堆
thermopile stick 测温色笔
thermopile 温差电堆，温差电池，热电堆，热电偶
thermoplastic behavior 热塑性
thermoplastic-covered wire 热塑绝缘线
thermoplastic hologram 热塑全息图
thermoplastic insulation 热塑性绝缘
thermoplasticity 热塑性
thermoplastic polymer 热塑聚合物
thermoplastic resin 热塑性树脂
thermoplastic 热塑性的，热塑性塑料，热熔塑料
thermopositive 放热的
thermopower automation 热工自动化
thermo power plant 热电厂
thermo power station 火力发电站
thermopower 热电动势
thermopress paint spraying 热压喷漆
thermoprobe 测温探针，热探针
thermo quenching 热浴淬火
thermore flectance 热反射，热反射能力
thermoregulation 温度调节
thermoregulator 温度调节器
thermorelay 热继电器，温差电偶继电器，温差电偶
thermoremanence 热顽磁
thermoresistance 热阻
thermoresistance 温变电阻，热电阻
thermoscope 测温器，验温器
thermoscreen 热屏蔽，隔热屏
thermosensitive resistance 热敏电阻
thermosensitive resistor 热敏电阻器
thermoset plastic 热固塑料
thermosetting insulation 热固性绝缘
thermosetting material 热固材料
thermosetting plastics 热固塑料
thermosetting resin 热固性树脂，热固树脂
thermosetting 热凝固
thermoset 热固性，热凝物，热固树脂，热固（性）塑料，热固的，热固性的
thermosiphon circulation 热对流循环
thermosiphon system 热虹吸系统
thermosiphon 热虹吸，热虹吸管，温差环流系统
thermosistor 调温器
thermosnap 过热保护自动开关
thermosonde 热感探测仪
thermosounder 热探测仪
thermosphere 热层，电离层
thermostability test 热稳定试验
thermostability 热稳定性
thermostable 耐热的，热稳定的
thermostat container for electrode 焊条保温筒
thermostat container 恒温（集装）箱
thermostat controlled ventilation arrangement 恒温控制的通风设备
thermostated container 恒温箱
thermostated 恒温的
thermostat film developing machine 恒温洗片机
thermostat for room temperature control 对室温控制的温度自动调节器
thermostatic bimetal 双金属温度元件，恒温双金属
thermostatic control valve 恒温控制阀
thermostatic control 恒温控制
thermostatic expansion valve 热力膨胀阀
thermostatic steam trap 恒温式疏水器，恒温蒸汽阱，恒温冷凝罐
thermostatic switch 恒温开关
thermostatic trap 恒温式疏水阀［疏水器］，热静态活门
thermostatic type steam trap 热力式疏水器
thermostatic valve 恒温控制阀，恒温器，温度调节阀
thermostatic 恒温的，热静力学的
thermostat relay 恒温继电器
thermostat 恒温调节器，恒温器，恒温箱，热动开关，自动调温器
thermoswitch 热敏开关，热开关，热断路器，热继电器
thermosyphon cooling system 温差环流冷却系统
thermosyphon tube 热虹吸管

thermotank 恒温箱
thermotechnical measurement 热工测量
thermotechnical 热工的
thermotel （变压器的）负载指示器
thermotolerant 耐热的，热稳定的，热稳的
thermovent 散热口
thermoweld 热焊接，熔焊
thermowell 热电偶套管，温度计袋，温度计保护管
THERM＝thermal 热力的
therm 克卡，千卡【煤气热量单位】
thernamastat 中空保温纤维
the said 上述的，前述的
the same below 下同
the secondary shaft 副轴
the second stage of creep 蠕变第二阶段
the second stage of stress relaxation 应力松弛第二阶段
thesis 论文
the span of life 寿命，使用期间
The State Bureau of Seismology 国家地震局
the successful bidder 中标人
the successful tenderer 中标人
the third stage of creep 蠕变第三阶段
the title of a technical post 职称
the tropics 热带
the White Book 白皮书
the words that precede 前面所说的话
thiamine 硫胺（素）【维生素 B_1】
thick airfoil 厚翼型
thick and thin fin labyrinth 厚薄齿迷宫汽封
thick-cylinder theory 厚壁圆筒理论
thickener 增稠器，增稠剂，沉降槽，浓缩池，浓缩器，浓缩设备
thickening agent 增稠剂
thickening vessel 加浓容器
thickening 变厚，变浓，加厚，浓缩，增稠
thick film capacitor 厚膜电容器
thick film circuit block 厚膜电路块
thick film circuit 厚膜电路
thick film hybrid integrated circuit 厚膜混合集成电路
thick film lubrication 厚油膜润滑
thick film power resistance 厚膜电阻
thick film resistor 厚膜电阻器
thick film thermistor 厚膜热敏电阻器
thick film trimmer 厚膜微调电容器
thick fog 浓雾
thickness chord ratio （翼型）厚弦比
thicknesser 划线盘，压刨机，测厚计，刨压机，单面刨床
thickness extension vibration 厚度伸缩振动
thickness function of airfoil 翼型厚度函数
thickness gauge 厚度计，塞尺，测隙规，测厚计
thicknessing machine 压刨机，定厚机
thickness measurement 厚度测量
thickness meter 测厚仪，厚度计
thickness of airfoil 翼型厚度
thickness of earth-fill cover 覆土深度
thickness of electric wire icing 电线覆冰厚度
thickness of half absorption 半吸收厚度
thickness of honeycomb wall 蜂窝壁厚
thickness of oxide layer 表面氧化层厚度
thickness of section 截面厚度
thickness of slab 板厚
thickness of the roof insulation 屋面保温厚度
thickness regulator 厚度调节器
thickness to high ratio of arch dam 拱坝厚高比
thickness tolerance 厚度公差
thickness transducer 厚度传感器
thickness 厚度，稠度，黏度，密度，壁厚
thick plank 厚板，厚木板
thick-shell penstock 厚壁压力水管
thick source 厚源
thick steel plate 厚钢板
thick walled steel tube 厚壁钢管
thick 厚的，浓的
Thiele modulus 蒂勒模量
thigmoreceptor 触觉感受器
thimble assembly 套管组件
thimble cell 套管栅元【格架】
thimble coupling 套筒联轴节，套管联结器
thimble end plug 套管端塞
thimble filter 套管型过滤器，指型过滤器
thimble fire tube boiler 指形管式火管锅炉
thimble guide tube 导向管，导向套管
thimble hole 镶套筒的孔
thimble ionization chamber 顶针式电离室，套管电离室
thimble nut 套管螺帽
thimble plug assembly flow restrictor 阻力塞组件限流器，阻力塞限流组件
thimble plug assembly 阻力塞组件
thimble plug changing fixture 阻力塞装卸固定装置
thimble plug 套管塞，阻力塞
thimble screw 套管螺纹【燃料组件下管座】
thimble storage 套管贮存
thimble tube boiler 指型管式锅炉，套管锅炉
thimble tube 指形套管
thimble 衬圈，套筒，套管，反应堆垂直试验管道，穿线环，电缆接头，顶针，活动套管，壳筒，嵌环，心形环
thin aerofoil 薄翼
thin airfoil theory 薄翼理论，薄翼型理论
thin airfoil 薄翼型
thin-arch dam 薄拱坝
thin consistency 稀黏滞度
thin core dam 薄心墙坝
thin covered electrode 薄药皮焊条
thin-cylinder formula 薄壁圆筒公式
thin film capacitor 薄膜电容器
thin film carrier 薄膜载体
thin film ceramics 薄膜陶瓷
thin film circuit 薄膜电路
thin film conductor 薄膜导体
thin film diode 薄膜二极管
thin film distributed constant MIC 薄膜分布参数微波集成电路
thin film evaporator 薄膜蒸发器
thin film gauge 薄膜测热仪

thin film inductor 薄膜电感器
thin film insulator 薄膜绝缘体
thin film lubrication 薄油膜润滑
thin film memory 薄膜存储器
thin film MIC 薄膜微波集成电路
thin film passive circuit 薄膜无源电路
thin film photocell 薄膜光电池
thin film photovoltaic cell 光生伏打薄膜电池，薄膜光电池
thin film resistor 薄膜电阻器，薄膜电阻
thin film silicon solar cell 薄膜硅太阳能电池
thin film solar cell 薄膜太阳电池，薄膜太阳能电池
thin film storage 薄膜存储器
thin film technology 薄膜工艺，薄膜技术
thin film thermistor 薄膜热敏电阻器
thin film thermo-electron device 薄膜热电子器件
thin film transistor 薄膜晶体管
thin film type resistor 薄膜型电阻器
thin film 薄膜的，薄，薄膜
thin gauge conduit 薄铁管，布线管
thin grout 薄浆
think better of 给予认真考虑
think highly of 重视
thinking-machine 计算机，电脑
thin layer chromatography 薄层色谱法
thin layer gel filtration 薄层凝胶过滤
thin layer 薄层
thin magnetic film memory 磁性薄膜存储器
thin membrane box structure 薄壁箱形结构
thin membrane dam 薄心墙坝
thinner 稀释剂，冲淡剂
thinning 减薄
thin oil 稀油
thin plank 薄板
thin plate element 薄板单元，薄板构件
thin plate orifice 锐孔板
thin plate weir 薄壁堰
thin plate 薄板
thin region theory 薄区理论
thin rim type rotor 薄轮缘式转子
thin section analysis 薄片分析
thin section 薄片
thin sheet flow 薄层水流
thin sheet 钢皮
thin shell barrel roof 筒形薄壳屋顶
thin shell barrel 薄壳圆筒
thin shell bearing 薄壳轴承
thin shell construction 薄壳结构
thin shell cylinder 薄壳圆筒
thin shell element 薄壳单元
thin shell foundation 薄壳基础
thin shell gate 薄壳闸门
thin shell roof 薄壳屋顶
thin shell structure 薄壳结构
thin shell 薄壳
thin-skin calorimeter 薄蒙皮测热计
thin-slab structure 薄板结构
thin source 薄源
thin steel plate 薄钢板
thin target 薄靶
thin wall calorimeter 薄壁测热计
thin wall cased pile 薄壁管桩
thin wall construction 薄壁结构
thin wall counter 薄壁计数管
thin-walled structure 薄壁结构
thin-walled tube 薄壁管
thin-walled 薄壁的
thin wall penstock 薄壁压力水管
thin wall pipe pile 薄壁管桩
thin wall pipe 薄壁管
thin wall sampler 薄壁取土器，薄壁取样器
thin wall structure 薄壁结构
thin wall tube 薄壁管
thin wall tubular inner liner 薄壁管内衬【可燃毒物棒】
thin webbed girder 薄腹梁
thin 薄的
thioacetic acid 硫代乙酸
thiourea 硫脲
third-brush generator 三刷发电机
third-core reload fuel management 三分之一换料的燃料管理
third critical speed 第三临界转速
third-degree price discrimination 三级价格歧视
third filter 第三倍频滤波器
third-generation computer 第三代计算机
third-generation hardware 第三代硬设备，第三代硬件
third-generation software 第三代软件
third harmonic attenuation 三次谐波衰减
third harmonic component 三次谐波分量
third harmonic excitation 三次谐波励磁
third harmonic neutral voltage 三次谐波中性点电压
third harmonic residual voltage 三次谐波剩余电压
third harmonic voltage 三次谐波电压
third harmonic （第）三次谐波，第三谐波
third law of thermodynamics 热力学第三定律
third-order levelling 三等水准测量
third party certification 第三方认证，第三方评核，第三方鉴定制
third party insurance 第三方保险
third party liability insurance 第三方责任险
third party liability 第三方责任
third principal stress 第三主应力
third prize 三等奖
third quarter neap tide 下弦小潮
third rail 输电轨
third subharmonic 第三次（分）谐波
third voltage range 第三级电压范围
third winding 第三绕组
third wire 中线，第三线
this side up 此端向上
this suggests that 这样我们就可以假定……
thistle board 轻质板，石膏板
thixotropic agent 触变剂
thixotropic effect 触变效应
thixotropic fluid 触交流体
thixotropic hardening 触变硬化
thixotropic injection 触变灌浆

thixotropic liquid 触变液体
thixotropic paint 触变漆
thixotropic property 触变性
thixotropic 触变性的
thixotropy 振动液化，摇溶，触变性
THM(top head mass) 机顶质量
Thomas cavitation constant 托马斯气蚀常数
Thompson bridge 汤姆逊电桥
thorat area 喉部面积
thorat opening 炉口，喉宽
thoriated electrode 钍钨电极
thoriated filament 敷钍灯丝
thoriated waste 含钍废物
thoride 钍化物
thorium-base fuel 钍基燃料
thorium-bearing fuel element 含钍的燃料元件
thorium blanket 钍转换区
thorium breeder reactor 钍增殖反应堆
thorium breeder 钍增殖堆
thorium compound 钍化合物
thorium converter reactor 钍转换反应堆
thorium converter 钍转换堆
thorium cycle hybrid reactor 钍循环混合反应堆
thorium cycle 钍循环
thorium fission 钍的裂变，钍裂变
thorium fuel cycle 钍燃料循环
thorium-fuelled pebble-bed reactor 钍燃料球床反应堆
thorium-fuelled reactor 钍燃料反应堆
thorium fuel pebble bed reactor 钍燃料球床反应堆
thorium fuel 钍燃料
thorium high-temperature reactor 钍高温反应堆
thorium oxide sol 氧化钍溶胶
thorium reactor 钍反应堆
thorium-uranium breeder 钍铀增殖堆
thorium-uranium cycle 钍铀循环
thorium-uranium dicarbide particle 钍铀二碳化物颗粒
thorium-uranium fuel cycle 钍铀燃料循环
thorium-uranium fuelled reactor 钍铀燃料反应堆
thorium 钍
thorough analysis 全面分析
thorough burning 完全燃烧
thorough condensation 完全凝固
thoroughfare insulator 穿墙绝缘子，穿越绝缘子
thoroughfare 干道
thoroughly 彻底地
thorough metal construction 全金属结构
thorough overhaul 彻底检修
thorough 彻底的，全面的，充分的
thorugh-flow part 通流部分
thoruranium 钍铀
thrashing 颠簸，因争用主存储器引起的失效
THRBRG(thrust bearing) 推力轴承
thread condition 螺纹状态
threaded anchorage 有螺纹锚具
threaded bonnet 螺纹阀帽
threaded cap 螺纹堵头
threaded closure stud hole （反应堆压力容器）封头螺柱孔
threaded code 串式代码
threaded connection 螺纹连接，螺纹接头，丝扣连接
threaded end 螺纹端
threaded fastener 螺纹紧固件
threaded flange 栓接法兰，螺纹法兰
threaded insert 内螺纹管嵌入件
threaded-in winding 插入式绕组，穿入式绕组
threaded joint 螺纹接头，螺纹连接
threaded nut 螺母
threaded opening 螺孔
threaded pipe connection 螺纹管接头
threaded pipe 螺纹管
threaded rod 螺纹杆【无头螺纹固定器】，螺纹吊杆
threaded sleeve 螺纹套管
threaded spindle 螺杆
threaded stay bolt 螺纹拉撑螺栓
threaded surface 带有螺纹的表面
threaded 拧紧的
threader cutter 套丝机割刀
thread gauge 螺纹规
thread grinder 螺纹磨床
thread insert 螺旋垫圈
thread interval 十字丝间距
thread-like corrosion 线状腐蚀，螺纹状腐蚀
threadolet 螺纹支管台
thread reamer 套丝机铰板头
thread refurbishment 螺纹整修
thread roller 滚丝机
thread root radius 滤网底部半径
thread root 螺纹根部
thread 螺纹，车螺纹，攻丝，线，丝，穿线于，穿过，螺纹加工
three-address code 三地址码
three-address computer 三地址计算机
three-address instruction 三地址指令
three-address 三地址
three-ammeter method 三安培表法
three-armature DC motor 三枢直流电动机
three aspect signal 三位显示信号
three-axis attitude stabilization 三轴姿态稳定
three-axis table 三轴转台
three axle tandem roller 三轴串列压路机
three-batch fuel cycle 三区燃料循环周期
three-bearing machine 三轴承电机
three-bed demineralizer 三床式除盐设备
three caliper disc brake 三卡钳盘式制动器
three-casing double flow turbine 三缸双排汽汽轮机
three-center arch 三心拱
three-centered arch dam 三心拱坝
three-chamber surge tank 三室式调压井
three-circuit receiver 三电路接收机
three-circuit transformer 三绕组变压器
three-circuit 三电路的
three-coil generator 三绕组发电机
three-coil regulation 三线圈调整
three-column core 三柱型铁芯
three-column radiator 三柱散热器
three-column transformer 三心柱变压器

three-conductor 三导体
three core cable 三芯电缆
three core conductor 三心导线
three core 三芯的
three-cornered file 三角锉
three-crystal electron interferometer 三晶体电子干涉仪
three-crystal 三晶体的
three cup anemometer 三杯风速表
three-cylinder turbine 三缸汽轮机
three digit display 三位显示器
three digit group 三数字组
three dimensional air model 三维气模
three-dimensional analog computer 三维模拟计算机
three-dimensional appearence 立体电影，立体显示
three-dimensional consolidation 三维固结，三向固结
three-dimensional design 三维设计
three-dimensional display 三维显示，立体显示
three-dimensional equation 三维方程
three-dimensional fin profile 三维肋片剖面
three-dimensional flow 三元流，三维流，空间流，三维流动，三向水流
three-dimensional fluidized bed 三维流化床
three-dimensional flutter 三维颤振
three-dimensional four wire 四根线的三度重合
three-dimensional holography 三维全息术
three-dimensional image 三维像
three-dimensional lattice 三维点阵，空间点阵
three-dimensional magnetic field 三维磁场
three-dimensional model 三维模型，立体模型
three-dimensional motion 三维运动，空间运动
three-dimensional photoelasticity 三向光弹性
three-dimensional photoelastic stress analysis 三维光弹应力分析
three-dimensional pitot probe 三向毕托管
three-dimensional scope 三维显示
three-dimensional space 三维空间
three-dimensional stagnation flow 三元滞止流动
three-dimensional stress distribution 三维应力分布
three-dimensional stress 三向应力
three-dimensional structure 三维结构
three-dimensional system 空间体系
three-dimensional three wire 三根线的三度重合
three-dimensional velocity log 三维速度测井
three-dimensional vortex street 三维涡街
three-dimensional wing 三维机翼
three-dimensional 三维的，三度的，立体的，空间的
three-dimension design 立体造型设计
three-dimension reinforced 立体配筋
three-drum boiler 三汽包锅炉
three-electrode arc lamp 三电极弧光灯
three-electrode transistor 晶体三极管
three element arch 三分拱
three element control loop 三元控制回路
three element control 三冲量调节，三元调节，三冲量调节系统
three element feedwater control system 三元给水控制系统
three element regulator 三冲量调节器
three element servo system 三元件的伺服系统
three element 三冲量，三元件
three equal divisions 三等分
three equal portions 三等分
three-field generator 三磁场发电机
three-finger gripper 三指棘爪
threefold statically indeterminate 三次超静定的
threefold 三倍的，三重的
three-group theory 三群理论
three-hinged arch dam 三铰拱坝
three-hinged arch 三铰拱
three-hinged braced chain 三案桁链
three-hinged stiffening truss 三铰加劲桁架
three-hole core 三孔铁芯
three-hole 三孔的
three-input adder 全加法器，三输入加法器
three-input switch 三输入开关
three-input 三输入的
three-integrator system 三积分器的系统
three-integrator 三积分器的
three island ship 三岛式船
three jaw chuck 三爪卡盘
three-lane road 三车道道路
three leaved bear-trap 三叶式活动闸门
three-leg sling 三套索
three-legs type core （变压器的）三心柱铁芯
three-letter code 三字母编码
three-level laser 三能级激光器
three-level master 三能级微波激射器
three-level recording 三级记录
three-level 三级的，三位的
three-line diagram 三线图
three lobe bearing 三油叶轴承
three-loop nuclear power plant 三环路核电厂
three-member joint 三杆联结
three moment equation 三弯矩方程
three month flow 丰水量【水力发电】
three-pass boiler 三回程锅炉
three phase alternating current 三相交流电
three-phase armature 三相电枢
three-phase asynchronous induction motor 三相异步电动机
three-phase asynchronous motor 三相异步电动机，三相感应电动机
three-phase autotransformer 三相自耦变压器
three-phase bar-winding 三相条绕组
three-phase breaking time 三相断路时间
three-phase bridge rectification 三相桥式整流
three-phase cable 三相电缆
three-phase circuit 三相电路
three-phase commutator machine 三相换向器式电机
three-phase connection 三相接法，三相连接
three-phase current limiting fuse 三相限流熔断器
three-phase current transformer 三相电流互感器
three-phase current 三相电流
three-phase diode bridge 三相二极管桥
three-phase E-core 三相E铁芯

three-phase electromotive force 三相电动势
three-phase exciter 三相励磁机
three-phase fault 三相故障，三相事故
three-phase five-column potential transformer 三相五柱式电压互感器
three-phase five-limb core 三相五柱铁芯
three-phase five-wire system 三相五线制
three-phase four core cable 三相四心电缆
three-phase four-wire system 三相四线制
three-phase full wave rectifier 三相全波整流器
three-phase generator 三相发电机
three-phase grid 三相电网
three-phase induction regulator 三相感应电压调整器
three-phase insulated copper winding 三相绝缘铜线绕组
three-phase inverter 三相逆变器
three-phase kilowatthour meter 三相电度表，三相千瓦时表
three-phase line 三相输电线
three-phase load 三相负载
three-phase motor 三相电动机，三相电机
three-phase network 三相电力网
three-phase oil immersed transformer 三相油浸式变压器
three-phase operation 三相运行
three-phase potential transformer 三相电压互感器
three-phase power supply 三相供电
three-phase power transmission 三相输电，三相电力输送
three-phase reactor 三相电抗器
three-phase rectifier 三相整流器
three-phase regulator 三相调压器
three-phase series motor 三相串励电动机
three-phase seven-wire system 三相七线制
three-phase short circuit 三相短路
three-phase shunt motor 三相并励电动机
three-phase squirrel-cage induction motor 三相鼠笼式感应电动机
three-phase star-delta mixed winding 三相星形三角混合绕组
three-phase starter 三相启动器
three-phase stator winding 三相定子绕组
three-phase switch 三相开关
three-phase symmetrically distributed stator winding wound 三相对称分布定子绕组
three-phase synchronous generator 三相同步发电机
three-phase system diagram 三相系统图
three-phase three-wire system 三相三线制
three-phase transformer 三相变压器
three-phase variable frequency bi-directional back-to-back four-quadrant PEC 三相变频双向背靠背四象限 PEC
three-phase voltage regulator 三相电压调整器，三相调压器
three-phase winding 三相绕组
three-phase 三相的，三相
three-piece mitre bend 三节斜接弯管
three-piece rotor 三段组合式转子
three-pin plug 三心插塞，三脚插头
three-pipe water system 三管制水系统
three-plus-one address 三加一地址
three-ply wood 三层胶合板，三夹板
three-point comtact ball bearing 三点接触球轴承
three-point difference equation 三点式差分方程
three-point intersection 后方交会，三点交会
three-point method of stream gauging 三点（测流）法
three-point roller bearing 三点滚子轴承，三点式单轴承
three-point support 三点支承
three-pole double-throw knife switch 三刀双掷闸刀开关
three-pole gang operated switch 三极联动开关
three-pole solid-neutral switch 三极整体中性开关
three-pole switch 三极转换开关，三刀开关
three-pole 三极，三极的
three-pool system 三库系统【潮汐发电】
three-position controller 三位式调节器，三位式控制器
three-position polarized relay 三位极化继电器
three-position relay 三位置继电器
three-position switch 三位置转换开关
three-post type disconnecting switch 三柱式隔离开关
three-prong socket 三脚插座
three put-throughs and site leveling 三通一平
three-range winding 三列绕组
three-region charging cycle 三区加料循环
three-region core 三区堆芯
three-region loading cycle 三区加料循环
three-ring cluster 三圈燃料束
three-roller pulverizer 中速三辊磨煤机
three rollers reeling machine 三轴卷板机
three-roll in-line type of idler 三辊直线式托辊
three-roll offset type of idler 三辊偏置式托辊
three-roll rolling machine 三辊滚板机
three-row stage 三列级
three-section coil 三节线圈
three-shift work system 三班制
three space diagram 立体图
three-span continuous bridge 三跨连续桥
three-square file 三面锉
three-square scraper 三角刮刀
three-stable state device 三稳态设备，三稳态器件
three stable state 三稳态
three-stage by-pass system 三级旁路系统
three-stage compound engine 三级复合蒸汽机
three-stage slewing planetary gear unit 三级回转行星齿轮装置
three-stage supercharger 三级增压器
three state controller 三位控制器，三位调节器
three-step controller 三位控制器，三位调节器
three-step control 三位控制，三位调节，三级控制
three-step distance relay 三段距离保护装置
three-step 三级的
three teeth, straight, curved fir-tree blade root 三齿、直线、弧形枞树型叶根
three-terminal contact 三端子触点［节点］

three-terminal 三端的
three-throw pump 三冲程水泵
three-tiered rotor 三层式叶轮
three-tier winding 三列绕组
three valve manifold 三阀组
three-vaned centrifugal pump 三叶片离心泵
three-voltmeter method 三伏特表法
three-wattmeter method 三瓦特计法
three-way ball valve 三通球阀
three-way coal chute 三通落煤管
three-way cock 三向旋塞，三通旋塞
three-way cord 三心塞绳，三线塞绳
three-way elbow 三向弯头
three-way flat slab 三向无梁楼板
three-way pipe 三通管
three-way plug valve 三通旋塞阀
three-way receiver 三用接收机，三调谐电路接收机
three-way reinforcement 三向配筋
three-way shuttle chute 梭式三通落煤管
three-way solenoid type pilot valve 三路通电磁导阀
three-way solenoid valve 电磁三通阀
three-way switch 三路开关，三通开关
three-way tube 三通管
three-way valve manifold 三通阀联箱，三路阀箱
three-way valve 三通阀
three-way wiring 三通配线
three-way 三通的，三路的，三用的
three wedge bearing 三油楔轴承
three-wheeler 三轮卡车
three-wheel roller 三轮路碾
three-winding autotransformer 三绕组自耦变压器
three-winding transformer 三绕组变压器
three-wire balancer 三线均压器
three-wire control 三线控制
three-wire generator 三线发电机
three-wire installation 三线装置
three-wire levelling 三丝水准测量
three-wire line 三线线路
three-wire meter 三线式电度表
three-wire system 三线制，三线式
three-wire 三线的，三线式，三线制
three-zone cycling （燃料）三区循环
threshold adjustment 阈值调整，限界调整
threshold amplifier 定值放大器
threshold circuit 阈电路，门限电路
threshold concentration 阈浓度
threshold current 阈电流，门限电流
threshold crack 极限裂纹
threshold detection capability 极限检测能力
threshold dose 门限剂量
threshold detection logic 极限值探测逻辑
threshold detector 定值式检测仪，阈探测器
threshold dose 剂量阈值，阈剂量，最低致害剂量
threshold effect 阈效应，临界效应
threshold element 阈值元件，临界值元件
threshold energy 阈能
threshold field 临界场，阈场
threshold frequency 临界频率，阈频率
threshold hypothesis 有阈假说
threshold inductance 阈电感，临界电感
threshold in fatigue crack propagation 疲劳裂纹扩展界限
threshold law 阈定律
threshold limit value 阈限值
threshold-logic network 阈逻辑网络
threshold logic unit 阈逻辑元件
threshold neutron 阈能中子
threshold odor number 嗅阈值
threshold of audibility 可闻度阈
threshold of comfort 舒适阈
threshold of danger 危险阈
threshold of detectability 探测阈，检波阈
threshold of discomfort 不舒适阈
threshold of hearing 听觉阈值，听觉极限
threshold of luminescence 发光定值
threshold of nucleation 成核临界温度
threshold of perception 感受阈
threshold of relaxation 松弛阈
threshold of sensitivity 灵敏度界限，灵敏度阈
threshold of unpleasant 不悦感阈
threshold overshoot 超过定值
threshold reaction 定值反应
threshold region 临阈区域
threshold sensitivity 阈灵敏度
threshold signal-to-interference ratio 临界信号干扰比，阈信号干扰比，阈信杂比
threshold signal-to-noise ratio 临界信号噪声比
threshold signal 阈限信号，阈信号，临界信号
threshold size 临界尺寸
threshold speed 阈速度
threshold temperature 临界温度，阈温，阈温度
threshold transmission 阈值传输
threshold treatment 入口处理，阈值处理
threshold value 阈值，临界值，界限值，临阈值
threshold velocity 临界吹动速度，初动速度，起动流速
threshold voltage 阈电压，临界电压
threshold 阈值，阈，临界值，界限，临界，门槛，门口，限值，定值
thria 氧化钍
thribble 三联管
thrift basin 节水池
THRM＝thermal 热力的
throat area 喉部截面积
throat bush 节流套，喉管护套
throat depth 焊口有效厚度
throating 滴水槽
throat liner 尾管直段衬砌
throat of fireplace 壁炉烟道口
throat opening （叶栅的）喉部宽度
throat sheet 喉板
throat tail piece 喉管尾部
throat-tap flow nozzle 长颈式流量喷嘴
throat track capacity 咽喉区通过能力
throat velocity 喉部速度
throat 收缩颈【冷却塔】，喉管，喉口，（风洞）试验段，管颈，混合管段，节流圈，（低速）风洞试段，咽喉，颈前部，滴水管，喉部，临界截面，锁口管
THROT＝throttle 节流

throttle apparatus 节流装置
throttle bushing 节流衬套
throttle bush seal 节流阀衬套密封
throttle bush 节流阀衬套
throttle button 节流阀钮
throttle chamber 节流室，混合室
throttle conditions （汽轮机）入口参数，进口参数
throttle control 节流控制
throttled steam 节流蒸汽
throttled surge chamber 阻尼式调压室，带喉管的圆筒式调压室
throttled upper chamber surge tank 阻抗上室式调压室
throttle flap 节流闸门
throttle flow ratio 节流流量比
throttle flow （汽轮机）入口流量
throttle gate 节流门
throttle-governed turbine 节流调节汽轮机
throttle governing 节流调节，节流调速
throttle governor 节流调节器
throttle loss 节流损失
throttle orifice 节流孔板，节流圈
throttle pilot valve 节流滑阀
throttle plate 节流孔板，节流板
throttle pressure control 主蒸汽压力控制
throttle pressure regulator （新汽）节流压力调节器
throttle pressure setpoint （新汽）节流压力设定点
throttle pressure （汽轮机）入口压力，新汽压力
throttle regulation 节流调节
throttle regulator 节流调节器
throttle setting 节流阀调整
throttle slide valve 节流滑阀
throttle steam pressure 节流阀处蒸汽压力
throttle steam temperature 节流阀处蒸汽温度
throttle steam （汽轮机）入口蒸汽，新汽
throttle temperature （汽轮机）入口温度，新汽温度
throttle type rotor 节流式转子，深槽式转子
throttle valve position indicator 节流阀位置指示器
throttle valve 调节节流阀，阻流活门，节流阀，调节阀
throttle 节流圈，扼流圈，节流阀，节流，调速汽门，扼流，风门
throttling action 节流作用，扼流作用，调节作用，节制作用
throttling and desuperheating unit 减温减压装置
throttling butterfly valve 节流蝶形阀
throttling calorimeter 节流式量热器
throttling cock 节流阀，扼流旋塞
throttling coefficient 节流系数
throttling component 节流件
throttling control element 节流控制元件
throttling controller 节流控制器
throttling control 节流控制
throttling disc 节流孔板
throttling expansion 节流膨胀
throttling governing 节流调节
throttling governor 节流调速器
throttling loss 节流损失【蒸汽】
throttling orifice 节流孔板，阻力孔
throttling process 节流过程
throttling range adjustment 节流区域调整
throttling range 节流范围，调节范围，节流区域
throttling valve 节流阀，减压阀，调节阀【精确控制】
throttling 节流的，调节的，节流
through air duct 通过式风管
through air way bill 空运直达提单，联运空运货单
through beam 连续梁，下承梁
through bias 直通偏流，直通偏压
through bill of lading 联运提单
through B/L 联运提单
through-bolt 贯穿螺栓，双头螺栓
through bridge 下承式桥
through bushing 通过式套管，穿心式套管
through cargo 直达货
through conduit gate valve 直通式闸阀
through corrosion 穿透腐蚀
through crack 贯穿裂纹，穿透裂纹
through current bias 直通电流偏置
through current 通过电流，直通电流
through cut 贯穿开挖
through fault condition 穿越故障状态
through fault current 穿越故障电流
through fault 穿越故障
through flow theory 通流理论
through flow turbine 贯流式水轮机
through flow velocity 通流速度
through flow ventilation 贯穿通风
through flow 通流量，通流，直流，流过量，穿透流，穿堂风
through freight traffic 货物联运
through freight 直达货运
through head refueling 顶端换料
through hold-down bolt 贯穿固体螺栓
through hole for closure stud 封头螺柱直通孔
through hole refueling 贯通换料
through hole 通孔，贯穿孔
through insulator 穿越绝缘子，穿越绝缘管，套管绝缘子
through joint 状态接线装置
through line 直通线
through-manipulator 穿墙式机械手
throughout 贯穿，遍及，全部
through-pass 过道
through penetration technique 渗漏探伤技术
throughput agreement 交货量协议
throughput capacity 生产能力，出力，流通能力，通过容量
throughput rate 通流率，通过速度，生产率
throughput time 解题时间
throughput 生产能力，生产量，通流能力，物料通过量，处理量，通过功率，吞吐能力，工厂的总产量，流量，（输入输出信息的）通过量，生产率，吞吐量
through quenching 淬透

through rate 通过速率，贯空速率
through repair 大修
through repeater 中间增音器，中间帮电机
through screw 穿钉
through-ship average capacity 整船平均出力
through span 穿过式桥跨
through stay 贯通拉撑
through stone 穿墙石，系石
through supervision 状态监控，直通监控
through switching exchange （电话）转接局
through-the-wall handler 穿墙式机械手
through-thickness direction 垂直于板面的方向【金属板】
through-thickness 全厚度的
through traffic 直达交通
through transport by land and water 水陆联运
through transport 联运
through truss bridge 穿过式桁架桥
through truss 下承桁架，下承式桁架
through tube 贯穿管，穿墙管
through-type arrangement 贯通式布置
through-type current transformer 贯通式电流互感器，单心式电流互感器
through-type transformer 穿越式变压器，单心式变压器
through-type 直通型的，穿越式的
through unbalance 穿越不平衡
through-valley 贯通谷
through ventilation 穿堂风
through wall crack 贯穿壁厚的裂纹，贯穿裂纹
through wall defect 贯穿壁厚的缺陷
through wall flaw 穿透性裂纹
through wall periscope 穿墙潜望镜
through wall protection （电缆）穿墙保护
through way type 直通式
through way valve 直通阀
through way 高速公路，直达道路，直通
through 通过
throw-away cartridge 一次性通用过滤器
throw-away cycle 一次通过（燃料）循环
throw-away packing 一次性包装
throw-away 对空排放
thrower ring 油环，甩油环
thrower 抛料机，抛油环，喷射器，投掷器
throwing-off load 甩负荷，卸载
throwing-off 抛，甩（负荷）
throwing-on 负荷剧增
throwing 倒伏状【植物风力指示】
throw in 接入，注入，插入，接通
thrown fault 下落断层
throw of coil 线圈节距
throw-off load 甩负荷
throw-off 断开，切断，断路，关闭
throw-out blasting 抛掷爆破
throw-out 切断，断路，抛出
throw-over relay （固定）双位继电器
throw-over switch 投掷开关，投掷式转换开关
throw-over 转换，转接，换向，换速
throw 冲程，行程，偏心距，摆度，断距，摆幅，槽距，节距，摆度
THR＝thrust 推力
thrust-adjusting gear 推力调整装置
thrust anemometer 推力风速计
thrust at springer 拱端推力
thrust ball 止推滚珠，推力球
thrust bearing bracket 推力轴承座架，推力轴承机架
thrust bearing collar 推力轴承环，止推轴承环
thrust bearing housing 推力轴承壳套，推力轴承箱，推力轴承室
thrust bearing liner 推力轴瓦里衬，推力轴瓦
thrust bearing load 推力轴承负荷
thrust bearing of hydro-generator 水轮发电机推力轴承
thrust bearing pad 推力瓦块，推力轴承瓦块，推力轴承垫座
thrust bearing protection device 推力轴承保护装置
thrust bearing resisting plate 推力轴承支承座
thrust bearing runner 推力轴承滑槽，推力头，转环
thrust bearing safety device 推力轴承安全装置
thrust bearing segment 扇形推力轴瓦
thrust bearing shim 推力轴承垫片
thrust bearing shoe 推力轴瓦，止推轴瓦，推力轴承瓦块
thrust bearing trip test 推力轴承跳闸试验
thrust bearing trip 推力轴承脱扣装置
thrust bearing wear trip 推力轴承磨损脱扣
thrust bearing 推力轴承，止推轴承，止推支座，推力座
thrust block plate 推力轴承底板
thrust block seat 推力轴承座
thrust block 推力瓦，止推座，推力轴承座，止推（承）座，推力轴承，止推支座
thrust borer 冲击钻机
thrust boring 冲击钻孔
thrust built-up 推力增大
thrust coefficient 推力系数，拉力系数
thrust collar 推力盘，止推环，推力环，止推垫圈，止推轴承定位环，
thrust cone 推力头
thrust cylinder 推力缸【摩擦焊】
thrust disc 推力盘
thrust disk collar 推力盘
thruster 顶推装置，推杆，推进器，推力器
thrust fault 逆冲断层，推力断层，逆断层，冲断层
thrust force 推力，轴向力
thrust journal bearing 推力径向轴承
thrust journal 止推轴颈
thrust key 止推盘
thrust limiter 推力限制器
thrust line 推力线
thrust loading coefficient 推力负荷系数
thrust load 推力载荷
thrust metal 推力轴承座
thrust meter 推力计
thrust of arch 拱推力
thrust of pump 泵推力，泵压力
thrust pad temperature 推力瓦块温度
thrust pad 止推轴承衬，推力瓦块

thrust plane 冲断面，冲推面，坍落面
thrust plate 推力板，止推板
thrust reverser 反推力器，反推力装置，制动推力器
thrust ring 推力环，止推环
thrust screw 推力螺旋
thrust segment 推力瓦块
thrust shaft 推力轴，止推轴
thrust slip 冲断滑动
thrust stand 推力试验台
thrust stiffness factor 侧移劲度系数
thrust surface 冲断面，推力面
thrust-time curve 推力时间曲线
thrust vector control system 推力矢量控制系统
thrust washer 推力垫圈，止推垫圈
thrust-weight ratio 推力重量比
thrust 牵引力，推力，冲断层，止推
TH(total hardness) 总硬度
TH(total head) 总水头，总压头
thumber 制动器
thumb nut 蝶形螺帽
thumb rule 安培右手定则，经验法则
thumbscrew 元宝螺钉，蝶形螺钉
thumb tack 图钉
thumbwheel switch 拇指旋转开关，滚轮开关
thumbwheel 带柄滚轮
thumping sound 重击声
thumping 尺码大的，极大的，非常的，极好的
thump 低音噪声，强冲击，撞击筛
thunder and lightning 雷电
thundercloud 雷雨云，雷云
thunderhead 雷暴云砧
thunder protected cable 防雷电缆
thunder protection 防雷
thunder-shower 雷阵雨，雷雨
thunderstorm cloud 雷云
thunderstorm day 雷暴日
thunderstorm downburst 雷暴下击
thunderstorm high 雷暴高压
thunderstorm outflow 雷暴外流
thunderstorm rainfall 雷暴降雨（量），雷暴雨，雷暴降雨量
thunderstorm rain 雷暴雨
thunderstorm wind 雷暴风
thunderstorm 雷暴，雷雨
thundery precipitation 雷雨云降水
thunder 雷，雷击
thus it can be seen that 由此可知，由此可见
thus 从而
thwack 捣实
thymol blue 百里酚蓝
thymolphthalein 百里酚酞
thymotrol 闸流管电动机控制，闸流管电动机
thyratron chatter detector 闸流管振动检测器
thyratron commutation 闸流管换向，闸流管整流
thyratron control characteristic 闸流管控制特性
thyratron controlled generator 闸流管控制发电机
thyratron drive 闸流管拖动
thyratron motor 闸流管电动机
thyratron rectifier 闸流管整流器
thyratron relay 闸流管继电器
thyratron servo 闸流管伺服系统
thyratron stroboscope 闸流管闪光测速仪，闸流管频闪观测器
thyratron 闸流管
thyrector 可变电阻的硅二极管
thyristack 可控硅整流器堆，硅堆
thyristor-assisted commutation 可控硅帮助换向
thyristor bridge circuit 可控硅桥式线路，晶闸管桥式线路
thyristor bridge 可控硅桥，晶闸管桥
thyristor brush 可控硅电刷
thyristor commutation 可控硅换向
thyristor-controlled motor 可控硅控制的电动机
thyristor-controlled phase angle regulator 晶闸管控制相角调节器
thyristor-controlled reactor 晶闸管控制电抗器
thyristor-controlled series capacitor 晶闸管控制串联电容器
thyristor-controlled series compensator 晶闸管控制串联补偿装置
thyristor-controlled transformer 晶闸管控制变压器
thyristor-controlled 闸流管控制的，可控硅控制的
thyristor converter 闸流管换流器，可控硅换流器
thyristor-driver DC motor 可控硅供电的直流电动机
thyristor element 硅整流元件
thyristor excitation 可控硅励磁
thyristor-fed motor 可控硅供电的电动机
thyristor frequency converter 闸流管变频器
thyristor gating circuit 晶闸管触发电路，可控硅触发电路
thyristor inversion 可控硅换流，可控硅逆变
thyristor inverter 可控硅逆变器
thyristorized supply 可控硅整流供电
thyristor low-voltage electrical apparatus 可控硅低压电器
thyristor motor 闸流管电动机，可控硅电动机
thyristor power supply 可控硅供电，可控硅电源，晶闸管供电
thyristor rectifier 可控硅整流器
thyristor soft-starter 晶闸管软启动器
thyristor supply 可控硅电源
thyristor switched capacitor 晶闸管投切电容器
thyristor switched reactor 晶闸管投切电抗器
thyristor switch 可控硅开关，晶闸管开关
thyristor valve block 晶闸管阀组
thyristor valve loss 换流阀损耗
thyristor valve 晶闸管阀，可控硅阀，换流阀
thyristor 半导体闸流管，硅可控整流器，晶体闸流管，晶闸管，可控硅，半导体开关元件，可控硅元件，闸流晶体管
thyrite arrester 非线性电阻避雷器
thyrite exciter 非线性电阻励磁机
thyrite lightning arrester 砂砾陶瓷避雷器
thyrite valve block 非线性电阻阀片
thyrite 砂砾陶，碳化硅陶瓷材料
thyrode 硅可控整流器
thyroid dose 甲状腺剂量
thysyn motor 可控硅励磁同步电动机
TIBL(thermal internal boundary layer) 热力内边界层

ticker 蜂音器，振荡器，断续装置，自动收报机
tickler coil 反馈线圈，回授线圈
tickler 反馈线圈，回授线圈
tickle 反馈，回授
tickling 反馈，回授
tick mark 记号
tidal action 潮汐作用
tidal amplification 潮型放大
tidal amplitude 潮幅
tidal analysis 潮汐分析
tidal atlas 潮汐图
tidal backwater 潮水顶托
tidal ball 报潮球
tidal bank 潮滩
tidal basin 潮港池，潮汐港池
tidal beach 潮滩
tidal bedding 潮流层理
tidal bench mark 验潮水准点，潮汐水准点
tidal bore 潮津波，潮水上涨，潮水涌浪，涌潮，怒潮，潮汐隆起
tidal channel 排洪沟
tidal chart 潮汐图
tidal compartment 潮区，有潮区
tidal component 分潮
tidal constant 潮汐常数
tidal constituent 分潮，潮汐组分
tidal correction 潮汐订正
tidal creek 潮沟
tidal current chart 潮流图
tidal current curve 潮流曲线
tidal current cycle 潮流同期
tidal current diagram 潮流图
tidal current difference 潮流速度差
tidal current observation 潮流观测
tidal current scour 潮流冲蚀
tidal current signal 潮流信号
tidal current survey 潮流测量，潮汐测量
tidal current table 潮汛涨落表，潮流表，潮流预报表
tidal current 潮汐流，潮流
tidal curve 潮候曲线，潮位曲线，潮汐曲线
tidal cycle 潮汐周期
tidal data 潮汐潮流资料，潮汐数据，潮汐资料
tidal datum plane 潮汐基准面
tidal datum 潮位基准面，潮准线
tidal day 潮汐日
tidal delta marsh 潮盐沼
tidal delta 潮汐三角洲
tidal diagram 潮汐图
tidal difference 潮时差，潮高差，潮差，潮汐改正值
tidal divide 潮道分水岭
tidal double ebb 双低潮，双重低潮
tidal double flood 双高潮
tidal drift 潮汐漂移
tidal effect 潮汐效应，潮汐影响
tidal efficiency 潮汐效率
tidal electric power generation 潮汐发电
tidal embankment 挡潮堤
tidal energy resources 潮汐水能资源
tidal energy 潮能，潮汐能，潮汐能量
tidal envelope 潮汐包线
tidal epoch 潮相
tidal estuary 潮区，有潮河口，潮汐河口
tidal excursion 潮程
tidal fall 落潮
tidal flat 潮成平地，潮坪，潮滩，潮汐滩地
tidal flood current 涨潮流
tidal flood strength 最大涨潮流速
tidal flow 潮流
tidal flush 潮流冲刷，潮流冲刷作用
tidal force 潮力，潮汐力
tidal friction 潮汐摩擦
tidal gaging station 潮汐站，验潮站
tidal gate 潮汐闸门
tidal gauge 验潮仪
tidal-generating force 成潮力
tidal generator 潮汐发电机
tidal gravity correction 潮汐重力改正
tidal harbour 有潮港
tidal hat deposit 潮滩沉积
tidal height 潮位
tidal hour 潮时
tidal hydraulics 潮汐水力学
tidal impulse 潮浪冲击
tidal indicator 示潮器
tidal information 潮汐情报，潮汐资料
tidal instability 潮汐不稳定性
tidal interval 潮汐间隙
tidalite 潮棱岩，潮流沉积
tidal lag 潮汐滞后，潮滞
tidal lamination 潮汐分层
tidal land 潮间地，潮浸区，海埔地，潮汐地
tidal level 潮水位，潮位
tidal load 潮压
tidal lockup 潮汐顶托，潮壅
tidal mark 潮标，潮痕
tidal marsh 潮泛沼泽，潮沼
tidal mashland 潮漫滩
tidal mechanism 潮汐发生仪
tidal meter 验潮仪，测潮表
tidal migration 潮汐推移
tidal mill 潮力磨坊
tidal mixing 潮汐混合
tidal model test 潮汐模型试验
tidal model 潮汐模型
tidal movement 潮流运动
tidal mud deposit 潮区淤泥沉积
tidal mud flat 潮区泥泞低地
tidal observation well 验潮井
tidal observation 潮汐观测，验潮
tidal observatory 潮沙观测站，验潮站
tidal oscillation 潮汐振动，潮汐振荡
tidal period 潮汐周期
tidal plane of reference 潮位参考面
tidal plant 潮汐发电站
tidal platform ice foot 潮间台状冰脚
tidal pole 验潮杆
tidal pool 潮间带水坑，潮池
tidal potential 潮汐势，潮汐位，潮汐势能，潮汐位能
tidal power barrage 潮汐发电坝

tidal power generation 潮力发电，潮汐能发电
tidal power generator set 潮汐发电机组
tidal power generator 潮汐发电机
tidal power plant 潮力发电厂，潮汐发电厂，潮汐发电站，潮汐电站
tidal power resources 潮汐水能资源
tidal power station 潮汐电站，潮汐发电站
tidal power 潮汐发电功率，潮汐发电，潮汐力，潮力，潮汐能
tidal pressure ridge 潮压冰脊
tidal progressive wave 潮汐推进波
tidal quay 潮汐岸壁式码头
tidal race 强潮流，潮汐流
tidal range at springs 大潮潮差
tidal range 潮位变幅，潮差，潮汐涨落范围
tidal regime 潮流特性，潮汐状况
tidal region 潮汐地区
tidal-regulating structure 潮汐调节构筑物，潮汐节制结构物
tidal regulation 潮汐控制
tidal report 潮汐报告
tidal resonance 潮汐共振
tidal rip 潮隔
tidal river station 潮水河测站
tidal river 潮水河，潮汐河流
tidal rotary current 旋转潮流
tidal sand ridge 潮沙脊
tidal sand wave 潮沙波
tidal scale 观潮水尺
tidal scheme 潮汐发电方案，潮汐发电工程
tidal scour 潮流冲蚀，潮流冲刷
tidal section 潮汐段
tidal signal station 潮位信号站
tidal signal 潮位信号
tidal stage 潮位
tidal stand 平潮，停潮
tidal station 潮沙观测站，验潮站
tidal stencils 潮型
tidal stirring 潮汐混合
tidal strain 潮汐应变
tidal stratification 潮汐成层，潮汐分层
tidal stream 潮汐河流，潮流
tidal stress 潮汐应力
tidal stretch 潮区
tidal surface 潮汐面
tidal surge 涌潮，潮汐起伏，潮水涌浪
tidal swamp 潮淹沼泽
tidal synthesizer 潮浪合成仪
tidal system 潮系
tidal table 潮汐表
tidal theory 潮汐理论
tidal undulation 潮振动
tidal waterway 潮汐航道
tidal water 潮水
tidal wave theory 潮波理论
tidal wave 潮汐波，潮波
tidal wind 潮沙风
tidal zone 潮间带
tidal 潮汐的
tide age 潮龄
tide amplitude 潮汐振幅，潮幅，潮振幅
tide analysis 潮汐分析
tide box 潮水箱【水工模型用】
tide-control apparatus 潮汐控制仪
tide cycle 潮汐预报，潮汐周期
tide data 潮汐数据
tide duration 潮汐历时
tide embankment 防潮堤
tide epoch 大潮发生时间
tide flat 潮成平地，潮汐滩地
tide force 引潮力
tide forecast 潮汐预报
tide-free 无潮汐的
tide gate 防潮闸，当潮闸
tide gauge well 验潮井
tide gauge 潮位计，测潮计，潮位水尺
tide generating force 起潮力
tide generating mechanism 潮浪发生器
tide generating potential 引潮势
tide generator 生潮机
tide harbour 有潮港
tide height 潮高
tide indicator 示潮器
tide interval 潮流间隙
tide lagging 潮汐迟滞，潮时延后
tide lag 潮时差，潮汐滞后
tide land 潮间地
tide level 潮位
tide limit 潮区界限，潮汐界限
tide lock 潮闸
tide mark 潮标，潮痕
tide marsh 潮沼
tide-meter 潮汐计
tide-motor 潮汐发动机
tide observation 潮汐观测，验潮
tide phase 潮汐位相，潮相位
tide-plant 潮汐发电站
tide pole 验潮杆，测潮杆，测量潮高标尺，潮位水尺，潮水尺，水尺，潮标
tide pool 潮池，潮间带水坑，满潮池
tide predication 潮汐预报
tide predicting machine 潮汐自动推算机，潮汐预报机，潮汐预测机
tide prediction 潮位预报，潮汐推算
tide predictor 潮汐推算机
tide pressure 潮压
tide priming 潮时提前
tide-producing force 引潮力
tide-producing potential 引潮势
tide-raising force 引潮力，涨潮力
tide range 潮位变幅，潮差
tide record 潮汐记录
tide reproduction 潮汐重演
tide retardation 潮时后延
tide rip 潮隔
tide rise 潮升
tide-rock 受潮岩
tide sequence 潮汐序列
tide sluice 挡潮泄水闸
tide staff 潮位水尺，验潮杆
tide stand 停滞潮
tide station 潮汐测站

tide table 潮流表，潮汐表
tide type 潮汐类型
tide variation 潮位变化
tide water level 潮水位
tide wave 潮波
tide way 潮汐水道，潮路
tide zone facies 潮区相
tide 潮，潮汐
tiding 消息，音信
tidology 潮汐学
tie back wall 拉杆挡墙
tieback 横梁，系梁，拉条，内支杆
tie bar 连接铁条，钢筋【混凝土】，系杆，连杆，拉杆，连接杆
tie beam roof 系梁屋架
tie beam 下弦，大料，系梁，圈梁，连系梁
tie bolt 锚杆，拉紧螺杆，系紧螺栓，连接螺栓，拉杆螺栓
tie breaker 连络开关【在开关站输电线头处】
tie bus 连络母线，连系母线
tied-arch 弦系拱
tied expansion joint 带拉杆膨胀节
tie-down anchors 锚定装置
tie-down failure 拉条毛病
tie-down insulator 悬垂形绝缘子
tie-down 系紧，拴系，拉住
tied purchase clause 限制性采购条款
tied retaining wall 拉杆挡土墙
tie feeder 电源联络线，联结馈线
tie hoop 连系钢箍
tie-in transformer 联络变压器
tie jumper 连接跳线
tie line interchange 联网输送
tie line oscillation 联络线振荡
tie line 连接线，联络线，直接连接线
tie member 拉杆，系件
tie plate 系板，固定板，连接板，接点，联络点，连测点
tier building 多层房屋
tiered burner 分层布置的喷燃器
tiering 分层布置，堆叠
tie-rod disposal facility 系杆处置装置，系杆破碎室
tie-rod end 球接头，转向横拉杆球铰接头，横拉杆接头
tie-rod stator frame 拉杆连接的定子机座连接棒
tie-rod 栏杆，连接杆，拉杆，拉条，系杆
tierra helada 高山永冻带
tierra templada 高山温带
tier tube 间隔管
tier 列，层，盘，堆叠，行，排
tie-station 汇接站，通信中心站
tie transformer 联络变压器
tie-wire 钢丝捆带，绑扎用的铁丝，拉筋，扎线，绑线
tie 横拉撑【混凝土钢筋】，捆，扎，结，扣，拉杆，连杆，枕木，联系，关系，连接，绳，束缚，系材，系杆
TIF(heat transfer fluid) 导热油，导热工质【太阳能光热利用】
TIF trough type 导热油槽式【太阳能光热利用】
tight alignment 精确调整
tight bond 紧密耦合，强耦合，紧密键合
tight-closing isolating damper 密闭隔离挡板
tight coupling 紧耦合，强耦合
tight curve 急弯
tighten diagonally opposite 对角拉紧【螺栓】
tightened inspection 紧固检查
tightener 张紧器，紧线器，紧固器，紧带轮，收紧器，张紧轮，张紧装置
tightening force 紧力
tightening ring 紧箍，紧环
tightening screw 扣紧螺钉
tightening torque 拧紧扭矩，紧固力矩，上紧转矩
tightening-up 收紧装置
tightening wire 拉线
tightening 拧紧
tighten in opposing sequence 相对方向顺序拉紧
tighten line for strained tower 耐张塔紧线
tighten using a wrench 用扳手拧紧
tighten 上紧
tight fit 紧密配合，牢配合，紧配合，过盈配合
tight joint 密封接头
tight lattice 稠密栅格
tight-lock coupler 密锁自动耦合器
tight-lock 密锁的
tightly enclosed 紧身封闭
tightly organized group of enterprises 紧密型企业集团
tight manhole cover 密封式人井盖板
tight-money policy 货币紧缩政策，紧缩银根政策
tightness test of piping system 管道系统严密性试验
tightness test of transformers 变压器整体密封试验
tightness test of turbine steam valves 气门严密性试验
tightness test 密封性试验
tightness 气密性，紧密性，紧密度，不可穿透性，密封性，密封度，紧密
tight-packed lattice 稠密栅
tight penetration 密封贯穿件
tight pulley 固定滑轮
tight seal 严密止水，密封
tight setting 气密式炉墙
tight-side tension of pulley 紧边张力滑轮
tight surface 密封面
tight 紧密的
TIG(tungsten inert gas arc welding) 钨极惰性气体保护电弧焊
TIG(tungsten inert gas) 钨极惰性气体
tile baffle 耐火砖挡板
tile drainage system 瓦管排水系统
tile drainage 排水瓦管
tile drain 瓦管排水沟，瓦管排水
tile floor 砖铺地坪，瓷砖地板
tile pipe 瓦管
tile roof 瓦屋面，瓦屋顶
tiler 瓦工
tile subdrain 地下瓦管

tile underdrainage 瓦管地下排水
tile underdrain 地下排水瓦管，瓦管暗沟
tile 花砖，瓷砖，瓦片，铺地砖，贴面砖，瓦，盖瓦，铺瓦，铺砖，贴瓷砖，贴砖
tillage 整地
till further notice 在另行通知以前，需等进一步通知，直到进一步通知
till 泥砾土，漂积土
tilt adjustment 倾角校正
tilt angle of collector 集热器倾角
tilt angle of rotor 风轮仰角
tilt angle of the idler axis 托辊轴线前倾角
tilt angle 倾斜角，倾角
tilt autosyn 倾斜自动同步机
tilt axis 倾斜轴
tilt dozer 斜推式推土机
tilted block 翘起地块，偏斜地块
tilted flume 倾斜水槽
tilted model 斜置模型
tilted pad bearing 可倾瓦轴承
tilted tray 倾斜池
tilted up 向上摆动
tilter 倾倒设备，倾卸车，振动台
tilt finder 倾斜测定器
tilt hopper 倾斜料斗
tilting angle 倾角
tilting bar 翻转炉排片
tilting bearing 可倾轴承，斜垫轴承，自位轴承
tilting bucket rain gauge 倾斜桶式雨量计，倾斗雨量器
tilting burner control 摆动燃烧器调节
tilting burner 摆动式燃烧器，摇摆式燃烧器
tilting cart 倾卸式手推车
tilting dam 堰顶可倾式过水坝
tilting device 翻倒装置
tilting-disc check valve 无冲击止回阀，斜阀瓣止回阀
tilting drum concrete mixer 斜筒式混凝土拌和机
tilting failure 倾倒破坏
tilting float 倾斜式浮标
tilting-insulator switch 斜置绝缘子开关
tilting level 微倾水准仪
tilting load 倾覆荷载
tilting manometer 倾斜压力计，回转式压力计
tilting micromanometer 倾斜式微压计
tilting moment 倾覆力矩
tilting nozzle 摆动喷嘴
tilting-pad bearing 可倾瓦块轴承，斜垫轴承，自位式轴承，自调式推力轴承，可倾瓦轴承
tilting pad thrust bearing 金氏推力轴承
tilting pad 可倾瓦块，斜垫
tilting platform 倾卸平台
tilting ring tachometer 摆环式转速计
tilting screw 倾斜螺丝，倾斜螺旋
tilting skip 倾斜料斗
tilting sluice gate 倾倒式泄水闸门
tilting-tangential burner 角置摆动喷燃器，角置摆动燃烧器
tilting trestle 翻料架
tilting 倾斜的，回转的，摆动式的
tilt-loading trailer 倾斜装卸式拖车
tilt measurement 斜度测量
tiltmeter 倾斜仪
tilt motor 俯仰操纵电动机
tilt stability 倾覆稳定性
tilt switch pile detecting probes 偏转开关料位探头
tilt-trailer 倾卸式拖车
tilt-type mixer 可倾式拌和机
tilt-up door 向上倾斜的门
tilt-up tower 倾斜塔
tilt 倾角，倾斜，倾斜面，天线仰角，侧倾
timber apron 木护坦
timber bearing block 垫木
timber block 木块
timber chute 木滑道
timber construction 木结构
timber crib breakwater 木笼式防波堤
timber crib dam 木框坝
timber crib 木框，木笼
timber culvert 木构涵洞
timber dam 木坝
timber-deck weir 木板堰
timber defect 木材缺陷
timber depot 贮木场
timber dog 骑马钉，扒钉
timber fender 防撞护木
timber floating 木材浮运
timber floor 木地板
timber forest 乔木林
timber form 木模板
timber grade 木材等级
timber guide crib 导水木笼
timber harbour 木材港
timbering gallery 木框支撑式廊道
timbering method 支撑方法
timbering of a cut 挖方支撑
timbering 木模结构
timber jetty 木结构突堤
timber line 森林线
timber partition 木隔板，木隔墙
timber pass 木筏道
timber pickling 木材浸渍【一种防腐处理】
timber piled jetty 木桩式突堤，木桩突码头
timber pile 木桩
timber plain splice 木材平面拼接
timber pond 贮木池
timber preservation 木材防腐
timber roof truss 木屋架
timber scaffold 木脚手架
timber sheet pile bulkhead 木板桩护岸
timber sheet pile groin 木板桩丁坝
timber sheet pile groyne 木板桩丁坝
timber sheet pile 木板桩
timber shelter 木棚
timber slide 木滑道
timber structure 木结构
timber trestle 木栈桥
timber truss bridge 木桁架桥
timber truss 木桁架
timber warehouse 木材仓库
timber work 木工作业

timber 用木材支撑【坑道】，厚木板，木材，坑木，木料，原木
time adjusting device 时限调节装置
time allocation 时间配置
time analyzer 时间分析器
time-angle 时间角
time-anticoincidence circuit 时间反符合电路
time-anticoincidence 时间反符合的
time-area-depth curve 时间面积降水深度曲线，时面深曲线
time-area method 时间面积法
time-averaged value 时均值
time-averaged 按时间平均的，时均化的
time average fringe pattern 时间平均条纹图样
time average hologram 时间平均全息图
time average holography 时间平均全息术
time average interferometry 时间平均干涉量度学
time average measurement 时间平均值测量
time average velocity 时均流速
time average 时间平均，时间平均值
time-averaging operation 按时间求平均值运算
time-averaging 按时间平均的，按时间平均
time axis 时间轴
time base circuit 时基电路，扫描电路
time-based maintenance 定期维修
time base error 时基误差
time base generator 时基发生器
time base integrated circuit 时基集成电路
time base 时基，时基坐标，时间坐标，扫描线，时间基线，时间基准
time-basis sampling 时间基采样
time behavior 时间性能，时间状态
time between overhaul 大修间隔
time between refuellings 换料间隔时间
time bias integrator 时偏积分器
time bias 时偏，时间补偿
time break 爆炸信号，定时爆炸
time card 计时卡片，时间表，考勤卡，工作时间记录卡
time cell 时间间隔
time clock 时钟，计时钟
time closing 延时闭合
time code generator 时间码发生器
time code word 时间代码字
time coherence 时间相干性
time-coincidence circuit 时间符合电路
time-coincidence 时间符合的
time comparator 时间比较仪
time-consolidation curve 时间-固结曲线
time constant of controlled plant 调节对象时间常数
time constant of excitation 激励时间常数，励磁时间常数
time constant of photoresistor 光敏电阻的时间常数
time constant of thermistor 热敏电阻的时间常数
time constant range 可达到链式反应范围，时间常数范围，时间常数区段，周期区段
time constant 时间常数
time constraint 时间限制
time consuming 费时间的
time-contract 定期交货合同
time controller 时间控制器，自动定时仪
time-controlling factor 时间控制因素
time control 时间控制
time correlation analysis 时间相关分析
time correlation function 时间相关函数
time correlation method 时间相关法
time correlation system 时间相关系统
time correlation 时间关联，时间相关
time-count cycle 计时循环
time-count 计时的
time-current characteristic 时间电流特性
time-current threshold 时间电流临界值
time curve 时间曲线
time cut-out 定时断路器，定时断路，定时停车，定时短路器
time cycle controller 时间周期控制器
time cycle 周期
time deformation 随时间而发生的变形，依时变形，依时间而变的变形
time delay characteristic 延时特性
time delay circuit 延时电路，时间延迟回路
time delay contactor relay 延时中断接触器
time delay de-energizing 延时去势，延时释放
time delay energizing 延时加势，延时带电
time delay fuse 延时熔丝
time delay mechanism 延时机构
time delay module 延时模件
time delay of transient recovery 瞬时恢复电压的时延
time delay opening relay 延时断路继电器
time delay push-button 延时复位按钮
time delay relay 延时继电器，缓动继电器
time delay servo 时延随动系统，时延伺服系统
time delay starting or closing relay 延时启动或闭合继电器
time delay stopping relay 延时停机继电器
time delay switch 延时开关
time delay system 时滞系统
time delay undervolt relay 延时低电压继电器
time delay unit 延时单元，延时器
time delay valve 延时操作阀
time delay 时间延迟，时延，延时，时滞，时间滞后
time dependence 时间相关，随时间变化，与时间有关
time dependent deflection 随时间变化的挠度
time dependent flow 非定常流动，随时间变化流动，随时间变化的水流
time dependent load 非定常荷载，随时间变化荷载
time dependent method 时间相关法
time dependent stress analysis 时效应力分析
time dependent 与时间有关的，随时间变化的
time derivative of temperature 温度的时间导数
time derivative 时间导数
time deviation 时间偏差
time device 计时装置
time difference 时差
time dimension 时间因次，时间维度

time discriminator　时间鉴频器
time distance curve　时间距离曲线，走时曲线
time distance relay　时间距离继电器
time distance　时距
time distribution　时间分配
time-divided control　时分控制
time divider　时分器
time-division data link　时分数据链路
time-division multiple access　时分多址
time-division multiplexing　时分多路转换，时分复用
time-division multiplex　时分多路复用，时分多路
time-division multiplier　时分多路复用器，时间脉冲乘法器
time-division switching system　时分开关系统
time-division switching　时分交换
time-division switch　时分开关
time-division system　时间划分制
time-division　时间划分，时间分隔
timed magneto　定时磁电机
time domain analysis　时域分析
time domain concept　时域概念
time domain display　时域显示，时畴显示
time domain electromagnetics　时域电磁学
time domain matrix　时域矩阵
time domain method　时域法
time domain model reduction method　时域模型降阶法
time domain　时间范畴，时域，时畴
timed pulse　时控脉冲，定时脉冲
timed relay　定时继电器
time duration　持续时间，延续期
timed　定时的，同步的，时控的
time edge effect　时间边缘效应
time effect　时间效应，时效
time element relay　时限元件继电器，延时继电器
time element　时间元件，延时继电器
time-event　时间事件
time factor　时间因数，时间利用系数，时间因子
time fitting　时间配合
time for access to the site　进入现场时间
time for completion　竣工时间
time for delivery　交付时间
time for notice to commence　开工通知发出时间
time for submission of programme　提交工作计划时间
time frame　时间框架，时间范围
time free solution　与时间无关的解
time-frequency duality　时间频率对偶性
time-frequency　时间频率
time fuse　定时信管
time gate　时间门
time graph　时间曲线，时间图
time harmonic　（电压和电流波形的）时间谐波
time history analysis method　时程分析法
time history triaxial acceleration recorder　三维时间关系曲线加速度仪
time history　时间关系曲线图，随时间的变化，随时间变化过程，时程，纪事
time horizon　时间范围
time increment　时间增量
time independent　定常的
time in milliseconds　毫秒，单位时间
time integral　时间积分
time integrated concentration　时间积分尺度
time integrated neutron flux　时间积分中子通量，中子注量
time-integrating sampler　积时式取样器
time-integration sampling　积时法采样
time-intensity graph　时间强度关系图，时间-强度关系线
time-interval indicator　时间间隔指示器
time-interval　时间间隔，间隔时间，时段，时间区间
time-invariant channel　非时变信道
time-invariant regulator　时间恒定调整器
time-invariant system　定常系统，非时变系统，时间恒定系统
time-invariant wind speed　稳定风速
time-invariant　时间恒定的，非时变的，时不变的
timekeeper　计时员，钟表，工作时间记录员
timekeeping system　时间记录系统
timekeeping　时间记录，计时，计工，守时
time lag action　延时作用［动作］
time lag effect　时滞效应
time lag fuse　延时熔断器
time lag relay　延时继电器，时限继电器
time lag switch　延时开关
time lag system　时滞系统
time lag　时间延迟，时间滞后，时间间隔，时滞
time lapse indicator　时间经过指示器
time lapse　时间推移
time lead　导前时间
timeless observation　与时间无关的观察
timeless　不定时的，与时间无关的
time limit attachment　限时装置
time limit breaker　限时断路器
time limit failure　时间限制失败
time limit for claims　索赔期限
time limit for notice　发出通知的期限
time limit of insurance　保险期限
time limit relay　限时继电器
time limit　期限，时间限制
timeliness of evaluation　评价的及时性
timeliness of performance　成果的及时性，成果的时间性
timeliness　及时性
time-load curve　时间荷载曲线
time loan　定期贷款
time lock　定时锁，时间锁定
timely completion　按期竣工
timely delivery　及时交货
time magneto　定时磁电机
time mark　时间标记
time mean concentration　时均浓度
time measurement　时间测定
time meter　计时器
time modulation　时间调制

time monitor 时间监视器
time motor 计时电动机
time node 时间节点
time normalization 时间规范化
time of completion 竣工时间，完成期限，完工时间
time-of-day clock 日历钟，日时钟
time-of-day price 峰谷电价
time-of-day rate 分时电价
time of delivery 交货期，交货期限
time of effect 有效期限
time of fall 退水历时
time of flight method 飞行时间法
time of flight neutron spectrometer 飞行时间中子能谱仪
time of flight 粒子飞行时间
time of lag 延迟时间，延滞时间
time of occurrence at the focus 震源震时
time of payment 付款期
time of pumping 抽水时间
time of response 响应时间
time-of-season price 季节性电价
time of shipment 装运期
time of the essence clause 重要时限条款
time of transit 运送时间
time-of-use pricing 峰谷分时电价，分时电价
time of utilization of installed capacity 装机容量利用小时数
time on stream 运行时间
time-operated 定时操作的
time optimal control 时间最优控制，时间最佳控制，快速控制
time-optimal 时间最优的
time-oriented sequential control 定时顺序控制
time overcurrent protection 时限过电流保护
time overcurrent relay 时限过电流继电器
time overcurrent 时限过电流
time payment 定期付款
time period for recovery of investment 投资回收期
time phase angle 时间相角
time phase 时间相位，时相
time phasor 时间相量
time point for payment 付款节点
time priority table 时间优先表
time programmed control 时（间程）序控制，时序控制
time program 分时程序，时间程序
time proof 耐久的，长寿命的
time proportioning controller 分时比例控制器
time proportioning control 时间比例控制
time pulse distributor 时间脉冲分配器
time pulse relay 定时脉冲继电器
time pulse 时间脉冲
time quadrature 90°时间相位差，90°时间相移
time rate of heat capacity 单位时间的热容量
time rate system 计时制
time rate 时变率，时间放电率，时间变率
time rating 额定工作时间，连续（或断续、短时）运行功率
timer clock 定时钟
time recorder 时间记录器
time relay setting value 时间继电器整定值
time relay 定时继电器，时间继电器
time release 限时释放器，延时释放器，定时（小量）排放
time resolution constant 时间分辨常数
time resolution 时间分辨，时间分辨率
time response 时间响应，时间特性，时间反应
time reversal 时间反转，时间反演
time rise 随时间上升，（电动势）随充电时间上升［增长］
timer 计时器，定时器，时间继电器，时间发送器，时钟，时间记录器，计时员，跑表，延时调节器
time sampling 按时采样，时间采样
time-scale factor 时标因子，时间比例尺
time-scale 时间刻度，时间标度，时间比例，时间量程，时间比尺，时标
time schedule controller 时间程序表控制器，时序控制器
time schedule control 时间程序控制，时序控制
time schedule 时间表，进度表，日程，时间进度表
time sequence 时间顺序，日程表，时间序列，时序
time series analysis 时间序列分析
time series of ocean measurement 海洋测量时间序列
time series 时间序列
time setting 时间整定，时间设置，时间设定
time-settlement curve 时间沉降曲线
time shaft 时间轴
time shared control 分时控制
time-shared executive 分时执行程序
time-shared supervisor assembly system 分时管理程序汇编系统
time-shared system 分时系统
time share 时间区分，时间分配，时分的
time sharing bidding 分时竞价
time-sharing control 分时控制
time-sharing operation 分时操作
time-sharing option 分时选择
time-sharing power supply scheme 分时供电方案
time-sharing scheme 时间划分制
time-sharing system command 分时系统命令
time-sharing system 分时体系，时间分配系统，分时系统
time-sharing 分时，时间分配，时间划分
time-signal station 报时站
time signal 时间信号，报时信号
time-slotting 分时间片，时间曲线
time-slot 时隙，时间空挡
time sorting 时间分选
time span 时间间隔，允许的贮存期限
time stationary process 时间平稳过程
time step selection 时段选择
time step 时间步，时间步长，时间间隔
time structure 时间结构
time-supervisor 监视时间定时器
time sweep 时间扫描
time-swelling curve 时间膨胀曲线

time switch 定时自动开关，计时开关，定时器，时间开关，时间接线器
time system 定时系统
times 时代
time table 时间表，时间曲线
time tag 绝对时标，时间标记
time temperature curve 时间温度曲线，温度时间曲线
time temperature-transformation curve 时间温度转变曲线
time tested 经过长期运行试验的，经受过时间考验的
time-to-amplitude converter 时间幅度变换器
time to computation 计算时间
time to crest 波头时间
time to failure 失效时间
time-to-number converter 时间数字变换器
time-to-number 时间数字的
time to rupture 破断时间
time to sparkover 到跳火的时间，到闪络的时间
time totalizer 计时器，时间累加器
time-travel curve 时距曲线
time undervoltage protection 时限低电压保护装置，限时低电压保护
time utilisation factor during availability 可用性期间的可利用小时率
time utilisation factor 时间利用系数，与电网连接投运率
time variable control 程序控制，按时间控制
time variable data 随时间变化的数据
time variable 时间变量
time variation 时间变化
time varying coefficient 随时间变化的系数
time varying field 时变场，随时间变化的磁场
time varying linear system 时变参数线性系统
time varying parameter 随时间变化的参数，时变参数
time varying reactivity 随时间变化的反应性
time varying system 时变系统
time varying time scale 时变时间比例尺
time varying wind speed 随时间变化风速
time varying 时变的，随时间变化的
time-voltage test 定时电压试验
time wage 计时工资
time-weighted average overall efficiency 时间加权平均全效率
time word 时间字
time yield limit 持久延伸极限
time zone 时区
time 年代，时间
timing adjuster 定时器
timing adjustment 定时
timing and control unit 定时及控制装置
timing capacitor 计时电容器
timing chart 时间图
timing circuit 定时电路，时滞电路
timing code 时间码
timing contactor 定时接触器
timing controller 定时器，时序控制器
timing control 定时控制，时间控制
timing current switch 定时电流开关
timing cycle 定时循环
timing data distributor 计时数据分配器
timing data input-output 计时数据输入输出
timing device 定时装置，时限装置，计时器，定时器，定时设备
timing diagram 时间图
timing element 时间元件，定时元件
timing error 定时误差，计时误差
timing extraction 定时抽取
timing float 定时浮标
timing magneto 定时磁电机
timing mark 时标，时间标志，时间标记
timing mechanism 定时机构
timing motor 定时电动机，时计电动机
timing pulse generator 定时脉冲发生器
timing pulse 定时脉冲
timing recovery 定时恢复
timing register 定时寄存器，定时记录器，自动计时器
timing relay driver 定时继电器驱动
timing relay 定时继电器，延时继电器，时间继电器，时限继电器
timing signal 定时信号
timing switch 定时开关
timing synchronization 时间同步
timing system 定时系统
timing track 同步磁道，定时纹
timing unit 定时装置，计时装置，延时装置
timing verifier 时序验证器
timing voltage 定时电压，同步电压
timing wave 定时信号，定时波
timing 定时，计时，时间选择，时间设置，延时（作用）
tinbender 钣金工
tin bronze 锡青铜
tin-coated steel 镀锡钢
tin dichloride 二氯化锡
tin-ferrite core 宽温铁氧体磁芯
tin-foil paper 锡箔纸
tin foil 锡箔
tin galvanized 镀锡
tin-lined lead pipe 衬锡铅管
tinned cable 镀锡的铜
tinned copper wire 镀锡铜线
tinned copper 镀锡铜
tinned wire 镀锡线
tinners tool 白铁工具
tinning 镀锡，包锡
tin oxide film capacitor 氧化锡薄膜电容器
tin oxide film 氧化锡膜
tin-plated copper 镀锡铜
tin-plated 镀锡的
tin-plate 白铁皮，马口器，镀锡铁皮，马口铁
tin-plating 镀锡
tinsel conductor 箔导体
tinsel 箔片，金属丝，金属片，锡铅合金
tin sheet iron 白铁皮，镀锡铁皮，马口铁
tinsmith 白铁工
tin soldering alloy 锡焊料合金
tin soldering paster 焊锡膏
tin soldering wire 焊锡丝

tin soldering　锡焊，焊锡
tin solder　焊锡，锡焊条，锡焊料，软焊料
tinted float glass　有色浮法玻璃，有色平板玻璃
tinted glass　有色玻璃
tinting material　着色剂
tinting pigment　着色颜料
tint　金属受热变色，着色，浮脏，色彩，色调，淡季，浅色，染（发），给……着淡色
tintometer　色度计，比色计
tin wire　焊锡丝
tiny capacitance probe　小电容探头
tiny crack　微裂纹
tiny electric motor　微型电动机
tin　锡
tip angle　顶锥角，顶圆锥角，顶端角，顶尖角
tip barrow　倾卸式手车
tip box　倾卸箱
tip brake system　叶尖制动系统
tip brake　叶尖制动器
tip car　自动卸货车，倾卸车
tip chord　翼梢弦
tip chute　倾卸滑槽
tip circle　齿顶圆
tip clearance loss　叶顶漏汽损失
tip clearance　叶片间隙，叶顶间隙
tip condition　轮缘条件
tip diameter　（叶轮）顶部直径
tip eddy　桨叶尖旋涡，叶梢涡流
tip end　末梢
tip feathering　叶尖顺桨
tip flap control　叶尖襟翼控制
tip flap　叶尖襟翼，叶尖小翼
tip-hub ratio　轮毂比
tip jack　插孔，塞孔
tip leakage　顶部漏汽，顶部漏气
tip lorry　自动卸货车，翻斗卡车
tip loss factor　叶尖损失因子
tip loss　顶部损失，叶尖损失
tip noise　叶尖噪声，端节点
tip of blade　叶片轮缘，叶尖
tip of file　锉尖
tip off　焊下
tip of pole shoe　极靴尖
tip of the pile　桩尖
tip orifice　喷头孔板
tip-over　倾卸，倾翻，翻转（机械），翻倒，翻载
Tipper test　蒂氏金属脆性转变温度试验
tipper　倾卸装置，自卸车
tipping barge　倾卸式甲板驳船
tipping bucket flusher　翻斗式冲洗器
tipping bucket forklift truck　翻斗式叉车
tipping bucket raingguage　翻斗式雨量计
tipping bucket　倾卸斗，翻斗，倾倒式吊灌
tipping cart　倾卸小车
tipping device　翻车机
tipping grate　翻转炉篦
tipping height　仰倾高度，倾卸高度
tipping hopper　倾卸漏斗
tipping over　倾卸，倾翻，翻转【机械】
tipping stone rubble　抛填块石
tipping through mixer　倾卸式搅拌机
tipping truck　自卸车
tipping vehicle　倾卸车
tipping wagon　倾卸车
tipping　俯仰，倾翻，翻，倒，倾斜，倾卸，翻转【机械】
tip plate　顶板
tippler wagon　自卸车
tippler　侧倾式翻车机，翻车机，翻斗车，翻斗机，卸料机，自翻卡车
tipple　筛煤楼，转运站
tip radius　尖端半径【用于粗糙度计的笔】，（风力涡轮机）叶尖半径
tip seal　（叶片）顶封
tip section pitch　叶尖段桨距
tip shroud　围带，环形叶栅外柱面
tip speed ratio　叶尖速度比【高速性系数】
tip speed　叶顶速度，轮缘速度，叶尖速度
tip stall　叶顶失速，叶尖失速
tip truck　翻斗卡车，自卸汽车
tip upright　在垂直方向失去平衡
tip vane augmented wind turbine　叶尖翼增力型风力机
tip vane　叶尖小翼
tip velocity　（风机）叶尖速度，叶顶速度
tip vortex　叶尖旋涡，叶顶涡
tip wagon　倾卸车
tip-wire　塞尖引线，T线
tip　尖端，触点，顶部，（熔化极弧焊机）导电嘴，叶顶，焊嘴，触点，末端，末梢，梢
tire bolt　轮箍螺栓
tired roller　轮胎碾
tired　疲劳
tire-load　轮胎负荷
tire　轮胎
Tirfor　手提拖拉及吊重机，手扳葫芦【输变电项目】
Tirill regulator　梯瑞尔电压调整器
T-iron　丁字铁
tissue activity meter　肌体放射性测量仪表
tissue at risk　受到危险的组织
tissue dose　肌体组织剂量，组织剂量
tissue equivalent dosimeter　组织等效剂量计
tissue equivalent ionization chamber　组织等效电离室
tissue equivalent material　（机体）组织等效物质
tissue equivalent　等效机体组织
tissue paper　薄纸，纱纸
tissue　生物组织，织物
titania calcium type electrode　钛钙型焊条
titania type electrode　钛型焊条
titania　二氧化钛
titanium alloy　钛合金
titanium dioxide paint　二氧化钛漆
titanium dioxide　二氧化钛
titanium gettering system　钛吸气系统
titanium getter　钛吸气器
titanium steel　钛钢
titanium tetrachloride　四氯化钛
titanium tube　钛管
titanium　钛

TI(technical information) 技术资料
titer 滴定度
title block drawing 标题框
title block 标题栏，图标框
title insurance 产权保险
title of drawing 图纸名称
title page 扉页
title 标题，名称，题目
titrant 滴定剂，滴定用的标准液
titratable acidity 可滴定酸度
titrater 滴定计
titrate 被滴定液
titration analysis 滴定分析
titration cell 滴定池
titration curve 滴定曲线
titration detector 滴定检测器
titration end point 滴定终点
titration error 滴定误差
titration stand 滴定架
titration 滴定，滴定法，牵引
titrator 滴定器
titrimeter 滴定计
titrimetric analysis 滴定分析
titrimetry 滴定分析
TI(turbine island) 汽机岛
T-joint T接头，T形接头
T junction T形接合，T形接合器
TK＝tank 箱，容器
TK 水箱【DCS画面】
TLG(torque limiting gearbox) 转矩限制齿轮箱
TLO(total loss only) 全部灭失险
T-matching T形匹配
TMCR(turbine maximum continuous rating) 汽机最大连续出力
TMD(tuned mass damper) 可调质量阻尼器，调谐质量阻尼器
TML(transportable moisture limit) 可输送的水份限值
TMR(triplex modular redundant) 三重模块冗余
TMS(turbine master system) 汽机主控系统
TMT(thermo mechanical treatment) 形变热处理
T-network T形四端网络，T形节，T形网络
TNT 三硝基甲苯
to a certain extent 在一定程度上，多少有点儿
to a considerable extent 在相当大的程度上
to a great large extent 在很大程度上，基本上
to-and-from data 起止数据
to-and-from list 起止清单
to-and-fro motion 往复运动
to-and-fro oscillation 来回摆动
to be against regulations 违规操作
to be put into operation 投入运行
TOB(takeover for blocking) 隔离移交
TOC(technical on call) 技术待召
TOC(truck operated cell switch) 小车操作式开关
TOD(total oxygen demand) 理论需氧量
toe basin 静水池
toe bearing 止推轴承
toe blanket 坝趾铺盖，趾部铺盖
toe circle 坡脚圆，坡趾圆
toe crack 边界裂纹，焊缝边缝裂纹，焊趾裂纹
toe drainage 滤水坡脚
toe drain 背水面坡脚排水，坝趾排水层
toe filter 坝趾反滤层，坝址滤层
toe groove 底部切口
toe-guard 踢脚板
toe nail 斜钉
to end 至终端
toe of slope 坡脚
toe of weld 焊缝边界
toe plate 底盘，踢脚板，趾板
toe protection 坝趾防护，坡脚保护
toe slab 趾板
toe stress 坝趾应力
toe trench of levee 堤基防渗槽
toe wall 坝趾墙（水电），坡脚墙
toe weight 坝趾压重，压重填土
toe 齿顶，端，缘，坡脚，趾，趾部，根部
toft 屋基
toggle actuator 拨转开关，起倒开关
toggle bolt 弯头螺栓，系墙螺栓
toggle circuit 触发电路，触发器
toggle condition 触发条件
toggle flip-flop 反转触发器，计数型触发器
toggle lever 肘节杆
toggle link 肘节杆
toggle mechanism 肘节机构
toggle rate 计时频率
toggle switched capacitor inverter 拨动式开关电容倒相器
toggle switched capacitor 拨动式开关电容
toggle switched differencer 拨动式开关差动器
toggle switched floating four-port 拨动开关浮地四端口
toggle switch 拨动开关，扳扭开关，跳动开关，触发开关，弹簧开关，肘节开关，钮子开关
toggle-type jaw crusher 复杂摆动式颚式破碎机，肘节式颚式破碎机
toggle 触发器，反复电路，紧线钳，紧线套环，肘节
toilet bout 抽水马桶
toilet bowl 抽水马桶，大便器
toilet paper holder 手纸盒
toilet 厕所，盥洗间（室），洗手间
Tokamak fusion fission reactor 托卡马克聚变裂变混合反应堆
token pass system 令牌系统
token passing system 符记传递系统
tolerable backlash 容许齿隙游移
tolerable concentration 可容许浓度，许可浓度
tolerable criterion 许可准则
tolerable level 可容许量，许可水平
tolerable limit 容许极限，容许限度
tolerable out of service time 允许停用时间
tolerable radiation exposure 可容许的射线照射量
tolerable settlement 容许沉降量
tolerable 可以的，可容忍的
tolerance deviation 容许偏差
tolerance domain 容许量范围
tolerance dose rate 耐受剂量率
tolerance dose 容许剂量，耐受剂量，耐药量

tolerance fit 公差配合
tolerance for contaminant 污染容限
tolerance limit 容许界限，有效极限，容许限度，容差限度
tolerance on rated resistance 额定阻值容许偏差
tolerance range 存活界限，公差范围，容差范围
tolerance unit 公差单位
tolerance 公差，容限，允许量，配合公差，忍耐度，容差，容许极限，容许间隙，容许量，耐药性，裕度，允许误差，容许偏差，余度
tolerant species 抗性植物
tolerated stress 极限应力
tolerence on resistance ratio 阻值比容许偏差
tolidine 联甲苯胺
Tolimetron 电触式指示测微表
toll board 长途交换台
toll-cable network 长途通信电缆网络
toll-cable 长途通信电缆
toll call 长途呼叫
toll enrichment 委托浓缩
toll final selector 长途终接器
toll information board 长途查询台
toll line 郊区线，长途电话线
toll message 长途通话
toll switchboard 长途交换台
toll switching stage 长途电话交换台，长途电话转接站
toll telephone system 长途电话系统
toll 通行税，同行费，长途电话费，伤亡人数
toluene 甲苯
tombolo 陆连洲，连岛沙洲，陆连岛
to minimum 最低程度
tommy bar with sliding T T形滑动头手柄
tommy bar 撬棒，螺丝钻，旋棒
tommy wrench 螺旋头部用（活动）杆（如虎钳上）套筒扳手
tommy 螺旋杆
tomography technique 体层摄影术
tomography 层析X射线摄影法，X线断层摄影术，体层摄影术
tomograph 层析X射线摄影机，断层X光摄影装置，射线断层摄影机，三维结构图
TOM(take over for maintenance) 维修移交
tonality 音值
tone-burst generator 单音脉冲发生器
tone generator 音频发生器，音频振荡器
tong-term debt 长期债务
tong-test ammeter 钳式安培计，钳式电流表，夹线式电流表
tongue and groove face 榫槽面，榫舌和企口密合面
tongue and groove joint 企口接缝，舌槽接合
tongue and groove sheet pile 企口板桩
tongue and groove siding 企口侧墙板
tongue and groove 舌榫
tongued and grooved board 企口板
tongued and grooved flooring 企口板地面，企口地板
tongue face 榫面
tongue joint 舌榫接合
tongue piece 舌片
tongue wedge 舌片式槽楔
tongue 衔铁，舌片，语言，旋钮，舌状物，雄榫，岩舌
tong 抓手，钳子，夹具，钳
tonic medicine 营养药品
tonlar 稳定电话电路损耗系统
tonnage calculation 牵引计算
tonnage of bridge 桥梁吨位
tonnage rating 牵引定数，牵引拱额定值，载重率
tonnage throughput 处理量，处理吨数
tonnage 吨数，吨位，登记吨位，（货物的）每吨运费，以吨计算的重量
tonne of coal equivalent 煤的等量吨数
tonne of oil equivalent 石油的等量吨数
8-tonne self propelled vibrating roller 8吨自行式振动碾路机
ton＝tonne 吨
tonns equivalent coal 当量煤吨数
tonometer 张力计
tons per hour 每小时吨数，吨/时
tool box 工具箱
tool carriage 工具车
tool cleaning 机械清理
tool enrichment 委托浓缩，收费浓缩
tool kit 全套工具
tool maker vises 精密平口钳
tool-post wrench 刀架扳手
tool pouch 工具袋
tools and machines for construction 施工机具
tools and machines for overhead line construction 高架输电线路施工机具
tools for cutting valve body-to-bonnet seal welds 阀体对阀盖密封焊缝的切割工具
tools for dismantling and assembling MSV 拆装主汽门工具
tools for measuring thimble reinsertion force 测量套管再插入力的工具
tools for measuring thimble removal force 测量套管移动力的工具
tools for welding valve body-to-bonnet seal welds 阀体对阀盖密封焊缝的焊接工具
tools management 工具管理
tool steel 工具钢
tool 工具，施工设备，加工设备
to one's advantage 有利，对……有好处
to operate a machine contrary to its instructions 违规操作
to order B/L 指示提单
to order of consignee 收货人指示抬头
to order of issuing bank 押汇银行指示抬头
to order 空白抬头【凭开证行指示】，凭（某人）指定，按要求【信用证用语】
tooth addendum 齿端高，齿顶高
tooth axe 齿斧
tooth chisel 齿凿
tooth clutch 齿轮离合器
tooth dedendum 齿根高
tooth depth 齿高

toothed armature 有齿电枢
toothed bar 齿条
toothed belt 有齿传动带，牙轮皮带
toothed coupling 齿轮联轴器
toothed disc crusher 齿盘式碎煤机
toothed hammer 带齿锤
toothed rack 齿轨
toothed ring connector 齿环连接器
toothed ring 带齿环，齿环
toothed roller crusher 齿辊式碎煤机
toothed roll 齿辊
toothed sill 齿槛
toothed synchronous belt drive 齿形同步皮带驱动器
toothed wave 锯齿波
toothed wheel 齿缘轮，齿轮
toothed 有齿的，锯齿状的，锯齿形的
tooth flank 齿面
tooth fracture 轮齿断裂
tooth gauge 齿规
tooth gear 齿轮
tooth harmonic 齿谐波
tooth head 齿顶高
tooth height 齿高
toothing 齿连接，齿轮连接
tooth joint 槽齿接合
tooth kinks （电机）齿致谐波，齿致脉动
tooth lock washer 带齿锁紧垫圈
tooth pitch 齿距
tooth ratio 齿槽宽度比
tooth ripple harmonic 齿脉动谐波
tooth ripple loss 齿谐波损耗，齿脉动损耗
tooth ripple 齿纹波，齿形波纹
tooth roll 齿辊式
tooth saturation 齿部饱和
tooth sector 扇形齿轮，转向齿轮
tooth space 齿槽，齿间，齿距，齿（间）隙
tooth spacing angle 齿间角
tooth spacing error 齿距误差
tooth spacing testing 齿距检查仪
tooth spacing 齿距，齿（间）隙
tooth support 齿压板，压指
tooth thickness 齿厚
tooth tip leakage flux 齿尖漏磁通
tooth tip leakage reactance 齿顶漏抗
tooth tip leakage 齿尖漏磁
tooth tip reactance 齿尖漏抗，齿间漏抗
tooth tip 齿尖
tooth top 齿顶
tooth type coupling 齿式联轴器
tooth type gasket 齿形垫
tooth width 齿宽
tooth 齿状物，啮合，加齿，使成锯齿状，齿
top and bottom chord of the roof truss 屋面桁架的上、下弦
top and bottom guided 顶部和底部导向的
top-and-port guided 顶部和通路导向的
top bar 顶层筋【板或梁的钢筋】，上层线棒，上鼠笼条
top beam 顶梁，冠梁
top boom 上弦杆
top core grid plate 堆芯上栅格板
top neutron shield 上部中子屏蔽体
top reflector 上反射层
top bracing 主机顶部支撑系统，顶撑
top burner 顶部燃烧器
top cap 顶盖
top chamber 上闸室
top chord of the car body 车厢上弦
top chord 上弦，上弦杆
top clearance 顶部间隙，顶部净高度
top coarse 顶层
top coat 外涂层，面层，面漆，顶层
top contraction 顶部收缩
top core grid plate 堆芯上格板
top course 顶层
top cover 正面覆盖层【上胶层】
top diameter （圆木的）梢径，（叶片）顶部直径
top digit 高位数，最高位
top discharge 顶部排出
top down approach 自上而下的方法
top down development 自上而下开发
top down testing 自上而下测试
top draft hood 上吸罩
top dressing 表面处治，敷面，敷面料
top drive tachometer 高速转速计
top dump gondola 卸料敞车
top efficiency 最高效率
top elevation 顶面标高
top end fixture 顶端夹持器
top end plug 上端塞【燃料棒或阻力塞】
top end stud 顶端螺杆
top end 顶端，端梢
top entry 顶部引入
top event 顶端事件
top-fill valve and nozzle 顶部装料阀和接管
top-fired burner 顶置燃烧器
top fitting 上部配件【燃料组件】，上管嘴，上端件
top flange plate 上翼缘板
top flange 顶部法兰，上翼缘，上凸缘，上法兰
top floor 顶层
top fluid 上级工质【二元循环】
top frame of gate 闸门的顶部构架
top grid assembly 顶部格架组件【燃料组件】
top-guided 顶部导向的
top half casing 上缸
top half inner casing 内上缸
top half outer casing 外上缸
top hamper 塔头
top hat shielding 顶帽屏
top heading method 上导洞法
top heading 顶部导坑，上导洞
top head mass 机顶质量
top head 上封头【压力容器】，顶盖
top heater 末级加热器
top-hung window 上悬窗
topic 论题，题目
top lateral bracing 上弦横向水平支撑
top level of flood control 防洪高水位
top light 顶部采光，桅杆顶灯

top loaded aerial 顶负载天线，电容负载天线
top loading 顶部加料，顶部装料
top load 尖峰负荷，最高负荷
top longitudinal bracing 上弦纵向水平支撑
top management 上层管理机构
top manager 高层管理者，高级管理人员
top mounted 顶部安装的
top notch 第一流
top nozzle assembly 上管座组件
top nozzle 上管座【燃料组件】
topo beacon 地形测量标志
topoclimate 地形气候
topoclimatology 地形气候学
top of embankment 堤坝的顶层
top of inactive storage 水库最低运行水位，水库死水位
top of pipe 管顶
top of slot belt 上层绕组区【双层绕组】
top of support 支架顶
top of weir 堰顶
topogram 内存储信息位置图示
topographical deflection 地形垂线偏差
topographical change 地形变化
topographical condition 地形条件
topographical data 地形资料
topographical distribution 地形分布
topographical divide 地表水分水线，地形分水界
topographical drafting 地形绘图
topographical drawing 地形图
topographical effect 地形效应［影响］
topographical element 地形因素
topographical exposure factor 地形开敞度
topographical factor 地形因素
topographical features of drainage basin 流域地形特征
topographical feature 地势，地形特征，地形要素
topographical form 地形
topographical interference 地形干扰
topographical interpretation 地形判读
topographical legend 地形图符号
topographical levelling line 地形测量水准线
topographical manuscript 地形图底稿，地形原图
topographical mapping 测绘地形图，地形测绘，地形绘图
topographical map 地形图
topographical model 地形模型
topographical modification 地形改造
topographical plan 地形图
topographical plotting 地形描绘
topographical point 地形点
topographical profile 地形纵断面
topographical recon-naissance 地形勘测
topographical relief ratio 地形起伏比
topographical relief 地形起伏
topographical survey team 地形测量队
topographical survey 地形测量，地形测绘
topographical symbol 地形图符号
topographical 地形测量的，地志的，地形学上的
topographist 地形测量员
topography channel wind 狭道风
topography of the site 厂区地形
topography 地形【地面起伏的形状】，地貌，地形测量学，地势，地形学，地志
topograph 地形描述
topological diagram of network 网络拓扑图
topological structure 拓扑结构
topology analysis 拓扑分析
topology tolerance （集成电路元件的）布局容差
topology 地志学，拓扑结构，拓扑学
topometry 地形测量
toposequence 地形系列
top overhaul 中修，大修
top panel 顶盖，（风洞）顶壁顶面板，顶部板
topping governor 主控调速器
topping plant 前置电站，前置机组
topping set 前置机组
topping steam turbine 前置式汽轮机
topping turbine 前置式汽轮机
topping unit 前置机组
topping 漫（过坝顶），涂层，覆盖层，前置，顶盖，拉平，溢顶，加顶，超越
top pintle 顶枢【闸门的】
top plan view 俯视图
top plate of foundation 基础的顶板
top plate 顶板【燃料组件】
top pressure closure 密封封头，密封顶盖
top pressure 最高压力
top price 最高价，最高价格
top priority 绝对优先
top property of performance 性能优良
top-quality consultant 高级咨询人员，最有资格的咨询人员
top rail 上冒头
top-ranked firm 排名第一的公司
top reflector sleeve 顶部反射层套管
top reflector 顶部反射层，上部反射层
top region material 顶区材料
top reward 头等奖
top right corner 右上角
top roll 顶部旋滚
top science 尖端科学
top secret 绝密
top section 顶截面
topset bed 顶部沉积层
top shield 顶部防护屏，上部屏蔽，顶部屏蔽
topside fabrication yard 组块场地
top-side 上侧
top size 产品最大粒度
top slab 顶厚板，顶板
top sod 表层草皮
topsoil cover 表土保护层
top soil erosion 表土侵蚀
topsoil stripping 表土剥离
top soil 表层土，表土，表土层，浮土
top speed 最大速度
topspin 上旋
top spray 顶部喷淋
top stepped shielding ring 顶部阶梯式屏蔽环
top strand 上分支
top stratum 地表岩层
top suction 顶部吸入
top-supported boiler 顶部悬吊支承锅炉

top-supported structure　悬吊式锅炉构架，支承式锅炉构架
top surface of the beam　梁顶面
top surface　顶面
top temperature controller　最高温度控制器
top thermal shield　顶部热屏蔽体
top thickness　顶厚
top thread　顶丝
top turbine　前置汽轮机
top undesired event　顶部意外事件
top-view　顶视图，俯视图，顶视
top water level for flood control　防洪高水位
top water level　最高蓄水位
top width　顶宽
top wind speed　最大风速
top wing　上翼
topworks　顶部结构【阀帽及以上】
top yoke　（变压器的）上部磁轭
top　顶，顶部，尖端，上部
torch brazing　分火焰硬钎焊，火焰钎焊
torch corona　火焰式电晕
torch cutting edge　火焰切割边
torch downward movement　焊炬向下移动
torch igniter　火炬点火器
torching　（受热面上积集可燃物）迅速燃烧，喷灯除漆
torch oil gun　点火油枪
torch plan　火炬计划
torch soldering　火焰钎焊，火焰软钎焊
torch welding　吹管焊接
torch　火舌，火炬，焊炬，喷灯，吹管，焊接灯，手电筒
torispherical head　准球形封头
torispherical　准球形
tori　环形铁芯，环状线圈
torkret　喷射混凝土
tornado affected area　龙卷风影响区
tornado alley　龙卷风通道
tornado axis　龙卷风轴
tornado belt　陆龙卷带，龙卷风带
tornado center　龙卷风中心
tornado core　龙卷风核
tornado cyclone　龙卷风气旋
tornado damage area　龙卷风破坏区
tornado dust collector　龙卷风式集尘器，旋风除尘器
tornado effect　龙卷风效应
tornado F-scale　龙卷风 F 等级
tornado funnel　龙卷风漏斗体
tornado hazard　龙卷风灾害
tornado parameter　龙卷风参数
tornado path　龙卷风路径
tornado resistant design　抗龙卷风设计
tornado simulation model　龙卷风模拟模型
tornado simulator　龙卷风模拟器
tornado track　龙卷风轨迹
tornadotron　旋风管，微波亚毫米波转换定子谐振器
tornado type wind energy system　旋风型风能系统
tornado vortex　龙卷涡
tornado wind load　龙卷风载
tornado　飓风，旋风，龙卷风，陆龙风，陆龙卷风，十二级风
torney fee　律师费
toroidal cavity resonator　环形空腔谐振器
toroidal coil　环形线圈
toroidal core　环形铁芯，圆环柱芯
toroidal current-transformer　环形电流互感器
toroidal field winding　环形磁场绕组
toroidal membrane　环形隔板，弹性油箱【水轮发电机】
toroidal memory core　环形存储磁芯
toroidal permeability　环磁导率
toroidal potentiometer　环形电位计
toroidal pressure suppression chamber　环形弛压室，环形压力抑制腔
toroidal reactor　环状反应堆
toroidal transformer　环状铁芯变压器
toroidal winding　环状线圈，环芯绕组
toroidal　环形的，螺旋管形的，圆环的，曲面，环形
toroid　圆环面，环形线，螺旋管
torpedo sand　粗粒砂
torpedo-sinker　鱼雷型测锤
torpedo-type weight　鱼雷形重锤，鱼雷形测锤
torque allowance　转矩允许误差
torque angle　转矩角，负载角
torque arm　支耳，扭力杆，扭力臂，扭矩臂，转矩臂
torque balance converter　扭矩平衡变换器，转矩均衡换流器
torque balance device　扭矩平衡装置
torque balance element　扭矩天平元件
torque balance　扭矩平衡，转矩平衡
torque bar　转矩杆
torque booster　增扭器
torque calculation　转矩计算
torque coefficient　力矩系数，转矩系数
torque compensator　转矩补偿器
torque constant　转矩常数
torque converter　扭矩转换器，转矩换向器
torque convertor transmission　变扭器传动
torque-current constant　转矩电流常数
torque-current　转矩电流
torque curve　扭矩曲线
torque density　转矩密度
torque dependent safety coupling　转矩相依安全联轴节
torque diagram　转矩图
torque-error constant　转矩误差常数，转矩放大常数
torque flow impeller　扭转流叶轮
torque fluctuation　转矩波动
torque force　扭力，转矩力
torque free body　自由旋转体
torque head screw　转矩头螺丝
torque limitation　转矩限制
torque limiter　转矩限制器
torque limiting gearbox　转矩限制齿轮箱
torque magnetometer　转矩磁力计
torque measurement instrument　扭矩测量仪

torque measurement 转矩量测
torque measuring instrument 转矩测量仪表
torque measuring wrench 转矩指示扳手
torque meter 扭矩测量仪，扭矩计
torque moment 力偶矩，转矩
torque motor 转矩电动机，力矩电动机
torque peak 转矩峰值，最大转矩，扭矩峰值
torque pickup 转矩传感器
torque pulsation 转矩脉动
torque rating 额定转矩
torque reaction bar of gearbox 齿轮箱转矩从动杆［反应杆］
torque reaction stand 扭矩测量台
torque release coupling 限矩型联轴器
torque release fluid coupling 限矩型液力联轴器
torque ripple 转矩波动，转矩脉动
torque rod 转矩杆
torque rotary actuator 力矩旋转执行机构
torquer 转矩发送器，转矩发生器，转矩装置，扭矩计
torque screwdriver 转矩螺钉起子
torque sensor 转矩传感器
torque-slip characteristic 转矩滑差特性
torque-slip curve 转矩滑差曲线
torque spanner 力矩扳手
torque speed characteristic 转矩转速特性
torque spike 转矩脉冲，扭矩峰值
torque summing member 转矩相加器
torque switch 转矩开关
torque-synchro generator 同步转矩发电机，旋转磁场发电机
torque-synchro receiver 同步转矩接收机，旋转磁场接收机
torque-synchro transmitter 同步转矩发射机，旋转磁场发射机
torque tester 转矩测试仪
torque-to-inertia ratio 扭矩惯性比，转矩惯性比
torque-to-inertia 转矩惯性
torque transducer 扭矩传感器，转矩传感器
torque-tube force-balance level sensor 扭力管压力平衡水位传感器
torque tube 扭矩管，扭力管
torque type power meter 扭矩式功率计，转矩式功率计
torque variator 转矩变换器
torque viscosimeter 扭矩式黏度计
torque wrench 扭矩扳手，转矩扳手，扭力扳手，定力矩扳手，力矩扳手
torque 转矩，扭矩，力矩，扭矩装置，扭转力矩
torquing and untorquing motor 转矩和反转矩电动机
torquing arm 力矩臂
torquing device 转矩装置
torquing head 转矩头
torquing machine 螺栓紧松器
torquing 扭转
torrential bedding 山峰层理
torrential flood 山洪
torrential rain 暴雨
torrential river 湍急河流
torrent 山洪，急流
Torricellian unit 托里拆利真空单位
torrid zone 热带
torsel 承梁木，漩涡状的
torsiogram 扭矩图，扭振图
torsiograph 扭力记录仪，扭力计，扭振自记器
torsional buckling 扭转屈曲
torsional buffeting 扭转抖振
torsional compliance 扭力柔度
torsional critical speed 扭转临界速度，扭转临界转速
torsional cyclic load triaxial apparatus 扭转动力三轴仪
torsional damper 扭振减振阻尼器，（输电线）防振锤，扭力阻尼器
torsional damping 扭转阻尼，扭转减振
torsional deformation stress 扭应力
torsional deformation 扭曲变形
torsional divergence 扭转发散
torsional elasticity 扭转弹性，抗扭弹性
torsional flexibility 扭转挠性，扭转柔度
torsional flutter 扭转颤振
torsional force 转动力，扭曲力，扭力
torsional fracture 扭转断裂
torsional frequency 扭转频率
torsional galloping 扭转驰振
torsional in-phase （叶片）扭转同相位
torsional load 扭力荷载，扭转负载
torsional mode 扭振形式，扭转模态
torsional moment 扭矩，扭转惯量，扭转力矩
torsional oscillation 扭转振荡，扭转振动，扭振
torsional out-of-phase （叶片）扭转异相位
torsional pendulum viscometer 扭摆黏度计
torsional pendulum 扭摆
torsional reinforcement 抗扭配筋
torsional restraint 扭转约束
torsional rigidity 抗扭刚度，扭转刚度
torsional shear test 扭剪试验
torsional stall flutter 扭转失速颤振
torsional stiffness 扭转刚度，抗扭刚度，扭转刚性
torsional strain 扭曲应变，扭应变
torsional strength 抗扭强度
torsional stress 扭转应力，扭应力
torsional test 扭转试验
torsional vibration of shaft system 轴系扭振
torsional vibration 扭转振动，扭振
torsional wave 扭转波，扭波
torsional 扭转的
torsion angle 扭转角，扭角
torsion balance 扭力秤，扭力天平
torsion bar safety valve 扭杆式安全阀
torsion bar 扭力杆
torsion beam 抗扭梁
torsion constant 扭转常数
torsion couple 扭力偶
torsion dynamometer 扭力功率计，扭转式测力计
torsion failure 扭转破坏
torsion galvanometer 扭转式电流计，扭转式检流计
torsion indicator 扭力指示器，扭力计

torsionmeter 扭力计，扭矩仪
torsion mode 扭转模态
torsion modulus 扭曲模量
torsion moment 扭转力矩
torsion pendulum 扭摆
torsion radius 扭转半径
torsion shear 扭转剪切，扭剪
torsion spring 扭转弹簧
torsion test of metallic wire 金属线材扭转试验
torsion test 扭力试验，扭转试验
torsion 扭矩，扭转，扭力，挠曲
TOR(terms of reference) （委托）权限，职权范围，职责范围，受权调查范围【委员会等】，委托的规定说明，任务大纲
tortional buckling 扭曲
tortuosity factor 弯曲因素
tortuosity 弯曲，曲折，弯曲度，迂曲
tortuous flow 乱流，紊流
tortuous passage 曲折通路
TOR type isolated input board 晶闸管型隔离式输入板
tort 侵权行为
torus air space 环形腔空气空间，弛压室空气空间
torus cooling system heat exchanger 环形弛压腔冷却系统热交换器
torus 环形室，轮环，圆环，环面，椭圆环，环形铁芯，环形线圈
torward circuit 正向电路
torx bit screwdriver 十字刀头螺钉起子
tosimeter 微压计
to (such) an extent that （竟然）达到这样程度以致……，到这样的程度以致……
to sum up 综上所述
total ablation 总消融量
total absorbed dose 总吸收剂量
total absorbing ionization chamber 全吸收电离室
total absorption coefficient 总吸收系数
total absorption detector efficiency 全吸收探测器效率
total absorption of sound 总吸声率
total absorption 全吸收
total absorptivity 总吸收率，全吸收性
total access time 总取数时间
total accumulation rate 总累积率
total accumulation 总累积量
total acidic group 总酸性基
total acidity 总酸度
total acid value 总酸值
total activity discharge to the atmosphere 向大气排放的总（放射性）活度
total activity 总活性，总（放射性）活度
total air for combustion 燃烧总空气量
total air pollution 总空气污染
total alkalinity 全碱度，总碱度
total allowance and deduction 扣除及免税总额
total allowance for machining 加工总余量
total allowance 总留量
total amount of weld fumes 焊接发尘量
total amount 总额，总量，总数
total amplitude 总振幅，全振幅，总幅值，双振幅
total angular momentum quantum number 总角动量量子数
total assets 总资产
total atomic stopping power 总原子阻止本领
total backlash 总的间隙，总间隙
total bed weight 总床层重量
total benefit 总效益
total bid price excluding VAT 不含增值税的总投标价
total bid price including VAT 包含增值税的总投标价
total bid price 投标总价，总报价
total body dose 全身剂量
total body neutron activation analysis 全身中子活化分析
total capability for load 总负载能力，总供电能力
total capacitance 总电容，全电容
total capacity of reservoir 总库容
total capacity 总容量
total capital investment cost 总基建费，总基建投资
total carbon 总碳
total chain yield 总链产额
total charge 总装料量
total chlorine residue 总余氯
total chlorine 全氯
total cohesion 总内聚力
total collapse 总湿陷量
total composite error 总合成误差
total consumption of nuclear fuel 核燃料总耗量
total containment （整个核电厂的）全部安全壳
total content 总含量
total contract price 合同总额［总价］
total cooling capacity 总冷却量
total cooling effect 总冷却效果
total core flow rate 堆芯总流量
total core heat output 堆芯总热功率
total core pressure drop 堆芯总压降
total core thermal output 堆芯总热功率
total corrosion 全面性腐蚀
total cost of power production 发电总费用
total cost 总成本，总费用
total counter 总计数器
total creep 全蠕变
total cross section 总截面
total current 总电流
total curvature 总曲率
total daily flux 总日照量
total damping 总阻尼
total defluidization 完全失流
total deionization 全除盐
total delivery head 供水总水头，总输水水头
total demineralization 全除盐
total departure 总偏离度，总偏差，总离差
total depreciation 折旧总额
total derivative 全导数，全微商
total differential 全微分
total digitalization 全面数字化
total direct digital control 集散控制系统

total discharge head 总排出扬程
total discharge rate 总排放率，总疏散率
total displacement 总流量，总排水量，总位移
total dissolved salt 总含盐量
total dissolved solids 总溶盐量，总溶解固形物，总溶解固体（量），总溶解固体物
total distortion ratio 全失真比
total distortion 总失真
total distributed control system 总体分散型控制系统，集散控制系统
total divisor 总除数，全因子
total drag 总阻力
total dynamic discharge head （水泵的）实际出水扬程
total dynamic head （水泵的）实际扬程，总动态压头，总动压头，总扬程［水头］
total effective collision cross section 总有效碰撞截面
total effect 总效果
total efficiency 总效率，滞止效率
total electric field strength 合成场强
total electric load 总电负荷
total emission control 总排放控制
total emissivity 总辐射率
total emittance 总黑度
total enclosed type 全封闭式
total energy approach 能量综合利用
total energy concept 总能概念，能量综合利用
total energy gas turbine 能量综合利用燃气轮机
total energy head 总能头
total energy peak 全能峰
total energy plant 全能电站
total energy 总能，（燃气轮机）能量综合利用
total enthalpy rise factor 总的焓提升因子
total enthalpy 总焓，总热含量，滞止焓
total entropy 总熵
total environmental control 全环境控制，全环境管理
total environment 总（体）环境
total erosion 总侵蚀
total erratum 总误差
total error allowance 允许总误差
total error 总误差
total evaporation 总蒸发量
total evaporator 全汽化器，全蒸发器
total evapotranspiration 总蒸发散发量，总蒸发蒸腾量
total exchange capacity 全交换容量
total exhaust loss 总排汽损失
total exhaust pressure loss 总的排气［汽］压力损失
total exposure of the body 全身照射，全身照射量
total exposure 总辐照
total failure 完全失效，彻底失败
total feed 总进料
total filter 组合滤波器
total fissile content 易裂变物质总含量
total flooding extinguishing system 全淹没灭火系统
total flooding 全淹没
total floor area 展开面积
total flow expander 全流膨胀机
total flow 总流量
total fluctuation 全起伏
total flux linkage 全磁链
total flux 总通量
total flywheel effect of generator rotating parts 发电机定子部件的总飞轮效应
total force 合力
total fraction of delayed neutron 缓发中子总份额
total freight tonnage （铁路）货运总量
total frequency curve 频率累计曲线
total fuel energy input to gas turbine 燃机的总燃料能输入
total gauge pressure 总计示压力
total gradient current 总梯度流
total half life 总半衰期
total hardness of water 水的总硬度
total hardness （水的）总硬度，（总）硬度
total harmonic distortion 全谐波畸变，总谐波失真，总谐波畸变率
total head efficiency 滞止效率
total head gradient 总水头梯度
total head line 总水头线
total head tube 全压头管，总压头管
total head value 全压头值
total head 总扬程，总压头，全压头，全水头，总落差，总能头，总水头
total heating surface 总受热面
total heating value 高热值，总热值
total heat 总热量，全热
total heave 总胀升量
total height 总高度
total holdup 总存料量
total humic acid 总腐殖酸
total impedance 总阻抗
total inflow of river network 河网总入流量
total information system 全信息系统，总信息系统
totaling meter 求积计数器，计数综合器
total input impedance 总输入阻抗
total input 总输入信号
total installed capacity 总装机容量
total installed power 总装备功率
total investment cost 总投资费
total investment 总投资额，总投资
total ion exchange capacity 全离子交换容量
total ionization by a particle 粒子总电离
total ionization 全电离，总电离
total iron 全铁
total irradiance （倾斜面）总辐照度，（倾斜面）太阳总辐照度，全（辐射）辐照度
total irradiation 倾斜面总辐照量，倾斜面总日照量
totaliser 加法器，累加器，加法求和装置，累计表
totality 总体
totalizator rain gauge 累积雨量计
totalizator 累积计算器
totalizer 加法器，累积计算器
totalizing measuring instrument 累计仪表

totalizing meter 加法计
totalizing relay 综合继电器
totalizing wattmeter 累计功率计，总功率表，综合瓦特计
totalizing 总计，合计，累计
total leakage flux 总漏磁通
total leakage reactance 总漏磁电抗
total leakage 总漏料量
total leak rate test 总泄漏率试验
total length of overflow front 溢流前缘总宽度
total length 总长度
total liabilities 总负债
total light flux 总光通量
total linear stopping power 总线阻止本领
total load demand 总电能需要量，总负荷需求
total load rejection full income 全部甩负荷
total load rejection 甩全负荷，甩满负荷，全部弃荷
total load 总负载，总负荷，总载荷，总载重
total loss in the iron 铁损
total loss of magnetic materials 磁性材料的总损耗
total loss of power 全厂断电，失去所有电源
total loss of prestress 总预应力损失
total loss 总损失，总损耗
totally closed slot 全闭口槽
totally-enclosed apparatus 全封闭式设备
totally-enclosed box 全封闭出线盒
totally-enclosed busbar 全封闭母线
totally-enclosed frame 全封闭外壳，全封闭机座
totally-enclosed machine 全封闭式电机
totally-enclosed motor 封闭式马达
totally-enclosed pipeline busbar 全封闭管道母线
totally-enclosed substation 全封闭变电站
totally-enclosed 全封闭的
totally reflecting resonator 全反射共振腔
total macroscopic cross section 宏观总截面
total macroscopic scattering cross section 宏观总散射截面
total magnetic flux 全磁通
total management 全面管理
total mass airflow 空气质量总流量
total mass stopping power 总质量阻止本领
total maximum permissible individual dose 单个最大容许总剂量
total mean free path 总平均自由程
total mercury analysis 总汞分析
total mineral acidity 总矿物酸度
total moisture coal sample 全水分煤样
total moisture 总水分，全水分
total motion resistance 总运动阻力
total multiplication time 全乘法的时间
total net ablation rate 总净消融率
total neutron cross section 中子总截面
total neutron source density 总中子源密度
total normal stress 总法向应力
total number of welding joints 焊口总数
total number of welding 焊口总数
total occupied area of power plant per kW 单位发电用地面积
total occupied area of power plant 电厂总用地面积
total on-line program and information control system 总的机内联用程序和信息控制系统
total openflow area 总开孔流动面积
total operating expense 总操作费用
total ordering quantity 总订购数量
total organic carbon analyzer 全有碳测定仪
total organic carbon 总有机碳，总有机碳量
total output panel 全出力配电盘
total output value 生产总值
total output 总出力，总功率
total oxidized nitrogen 总氧化氮
total oxygen demand 总需氧量
total parallax 总视差
total peaking factor 总峰值因子
total permeability 总磁导率
total phosphorus 全磷
total plume rise 羽流总抬升
total plutonium 总钚
total porosity 全孔隙率，总孔隙率
total potential 总位势
total power consumption 总功耗
total power inhomogeneous coefficient 总功率不均匀系数
total power loss 总功率损耗
total power of station 电站装机总容量
total power system 全功率方式
total power 总功率
total precipitation 总降水量
total pressure head 总压差，总压力头
total pressure loss coefficient 总压损失系数，全压损失系数
total pressure Pitot tube 全压皮托管，总压皮托管
total pressure probe 总压测针，全压测针，总压探针
total pressure tapping 总压取压孔
total pressure 全压，总压，总压力
total price 总价
total probability 总概率
total profit 总利润
total pumping head 水泵总水头［压头］，水泵总扬程
total quality control 全面质量控制，全面质量管理
total quality management 全面质量管理
total quality manual 总质量手册
total radiation fluxmeter 全辐射通量计
total radiation pyrometer 全辐射高温计
total radiation temperature 总辐射温度，全辐射温度
total radiation thermometer 全辐射温度计
total radiation 总辐射，全辐射
total rail freight tonnage 总（铁路）货运量
total rainfall 总雨量
total reactor core recirculation flow 堆芯再循环总流量【沸水堆】
total reactor feed 反应堆总装料量
total reflected reactor 有完整反射层的反应堆
total reflected 全反射的
total reflection method 全反射法
total reflection 全反射

total reflux operation　全回流操作
total reservoir storage　水库总容量
total resistance　总电阻，总阻力
total risk　全损险
total runoff　总径流量，总水量
total runout　总偏心率
total sag　总垂度
total salary allowance amount　工资津贴总额，总薪金津贴金额，总薪酬津贴额
total salinity　总含盐量，总盐度
total salt content　总含盐量
total scattering　全散射
total sediment load transport　总输沙量
total sediment load　总含沙量
total sensitivity　总灵敏度，综合灵敏度
total separation efficiency　除尘效率
total settlement　总沉降，总沉降量
total short-time duty　全部短暂工况
total silica　全硅
total skin dose　皮肤总剂量
total sodium inventory in primary system　主路系统钠部装量【钠冷快堆】
total solar radiation estimation　太阳总辐射估计
total solar radiation　太阳总辐射
total solid runoff　总固体径流量
total solids concentration　总含盐量，全固形物含量
total solids content　总固体含量
total solids　全固形物，总固体物，总固体
total span　总跨度
total static head　总净扬程
total station coordinate method　全站仪坐标法
total station gyroscope　陀螺全站仪
total station instrument　全站仪
total station theodolite　全测站经纬仪
total station　全站仪
total stem operating force　总的阀杆操作力
total step iteration　整步迭代法
total step method　全步法
total storage capacity　水库总容量，总库容
total storage　总蓄水量，总库容
total stress method　总应力法
total stress path　总应力路径
total stress strength parameter　总应力强度参数
total stress　总应力
total suction head　总吸入扬程，总吸升水头
total suction　总吸水量
total sulfur　总硫量，全硫
total sulphuroxide measuring instrument　氧化硫含量测定器
total sulphur　全硫量
total surface area　总表面积
total surface dose　总表面剂量
total suspended particulate index　悬浮颗粒总量指数
total suspended particulates　总悬浮颗粒物
total suspended solids　总悬浮物，悬浮固体物总量，悬浮固体总含量【水分析】
total swell　总胀升量
total temperature coefficient　总温度系数
total temperature restoration coefficient　总复温系数
total temperature　总温，全温，滞止温度，总温度
total thrust　总推力
total time lag　总时延【继电器】
total transfer capability　输电总容量，总输电容量，总传输容量
total travel　全行程
total treatment　完全处理
total trip　安全停堆
total uncertainty factor　总的不定因子
total unsharpness　总不清晰度
total value of the goods　货物总价
total variation　总变差
total VAT on bid price　投标价格的总增值税
total voltage　总电压
total vorticity　（涡系）总涡量
total water storage　总蓄水量
total weight　全重，总重
total well capacity　井的总出水量
total width　总宽度
total working time　总工作时间
total work　总功
total　总数，合计，全部，全体，总，总的，总和，总计
totameter　流量计
totem pole　图腾柱，推拉输出电路
to terminal　至接线端
to that extent　达到那样的程度
to the advantage of　有利，对……有利
to the best of one's ability　尽量，尽可能
to the extent of　达到……程度，在……方面，到……的程度
to the extent that　达到……的程度以致……，在这样的范围内，从而，就……来说
to the extent　到……程度
to the full extent of　最大限度地，充分地
to the utmost extent　在最大可能范围内，竭尽全力
to the south of　……以南【指相邻不接壤】
to this end　为此，所以，为这个目的
TOTO(take over for temporary operation)　临时运行移交
TOT(transfer of title)　产权转让，所有权转让
TOT(transfer-operate-transfer)　TOT模式，“转让-运营-移交”模式【工程建设模式】
touch down point　接地点
touchdown　（羽流）着地
touching screen　触屏，触摸式屏幕，触摸屏
touch spark　接触火花
touch switch　触摸开关
touch-up　修整，韧度，润色
touch voltage　接触电压
tough copper　韧铜
toughened glass insulator　钢化玻璃绝缘子
toughened glass　钢化玻璃
tough fracture　（材料）韧性断裂
tough metal　韧性金属
toughness-brittleness transition temperature　韧脆转变温度
toughness index　韧性指数
toughness test　韧性试验，冲击负荷强度试验

toughness 韧性，韧度，刚度，耐久性，黏稠性
tough 坚硬的，韧性的
tourist attraction 旅游胜地
tourist 旅游的
touting device 启动装置
towage 拖运，拖运费
tow boat 拖驳，拖船
tow car 拖缆车
tow conveyor 皮带运输机
towed grader 牵引式平土机
towed type vibrating sheep-foot roller 牵引式振动羊足碾
towel rack 毛巾架
tower and pole erection 组塔立杆
tower antenna 铁塔天线
tower base electronic controller 塔基电子控制器
tower base 塔底座，塔基
tower basic body and crossarms 塔身本体与横担【无塔腿】
tower block 塔式建筑，塔形大厦
tower body extension 塔身延伸段
tower body 塔身，塔体
tower boiler 塔式锅炉
tower building 塔式建筑，摩天大楼，摩天楼
tower cable excavator 塔式缆索挖土机
tower center peg 杆塔中心桩
tower climb ladder 塔架爬梯
tower construction 塔的结构
tower crane 塔吊，塔式起重机
tower erection works 组塔工作
tower excavator 塔式挖土机
tower focus solar power station 集中聚热太阳能电站
tower footing resistance 塔基电阻
tower footing 塔基，塔脚
tower foundation 铁塔基础
tower gantry 塔式龙门起重机
tower height 塔架高度
tower hoist 塔吊
tower insulation 铁塔绝缘
tower intake at bank side 岸塔式进水口
tower intake 塔式进水口
tower leg 塔架腿，塔腿
tower like structure 塔形结构
tower line 铁塔线路
tower list 杆塔明细表
tower loading 塔负载
tower load test 铁塔加载试验
tower location 塔位
tower nodding 塔架摆动
tower number sign 塔号牌
tower pad 塔基，塔座
tower pier 塔形墩
tower pond 冷水塔集水池
tower prototype test 塔原型试验，塔的真型试验
tower reclaiming system 塔式回收系统
tower ring foundation 塔环式基础
tower schedule 杆塔明细表
tower section 塔节
tower shadowing effect 塔影效应【风力发电】
tower shadow 塔架阴影
tower shielding reactor 塔式屏蔽反应堆
tower sign 杆塔标志牌
tower slewing crane 塔式悬臂起重机
tower structure 塔式结构，铁塔结构
tower support 塔支架
tower truss 铁塔桁架
tower type boiler 塔式锅炉
tower type building 塔式建筑
tower type stacker with swing boom 旋臂塔式堆料机
tower type with heavy angle 中角度塔型
tower type with medium angle 中角度的塔型
tower type 塔架类型，塔型
tower unloader 塔式卸料机
tower wagon 高架检修车
tower washer 塔式洗涤器
tower window 塔窗
tower （输电线）塔，塔架，天线塔，建筑物
towing basin 拖曳试验池
towing channel 拖曳试验槽
towing force 拉牵力，拖曳力
towing hook 牵引钩
towing machinery 拖拉机械
towing plate 牵引板
towing tank 拖曳水槽
towing 拖曳
tow-line 拖绳
town and country planning 城乡规划
town fog 城市雾
town planning 城市规划，市镇规划
town refuse 城市垃圾
Townsend theory of gas discharge 汤生放电理论
township enterprise 乡镇企业
township line 市镇界线
town water supply 城市给水
town 城镇，市镇
tow path 拉纤道
tow phase flow 两相流
tow pin 牵引销
tow rope 拖缆，拖索
tow tractor 牵引车
tow 拖拉，牵引，拖缆，拖索
toxicant 毒物，毒（物）
toxication 中毒
toxic chemical 有毒化学物质
toxic discharge 有毒排放
toxic dose 中毒剂量
toxic effect 毒性效应
toxic fume 毒烟
toxic gas detector 毒性气体检测器
toxic gas 有毒气体，毒气
toxic industry waste 有毒工业废物
toxicity analysis 毒性分析
toxicity assessment 毒性评价
toxicity of metal dust 金属粉屑的毒性
toxicity 毒性
toxicology 毒物学，毒理学
toxic pollutant 有毒污染物，毒性污染物
toxic pollution 有毒污染物，毒性污染
toxic response 中毒反应

toxic salt reduction 毒性盐类浓度降低
toxic smog 毒雾
toxic smoke 毒烟
toxic substance 有毒物
toxic value 中毒剂量
toxic waste 毒性废料，有毒废水
toxic 有毒的，毒性的，毒剂，毒物
toxin 毒素
toy motor 微型电动机
TPC(throttle pressure control) 主蒸汽压力控制
TPC(transmission power control) 传输功率控制
T-piece T形部件，T形接头，三通管
TP monitor(teleprocessing monitor) 远程处理监控
T-post 丁字形支柱
TPP(Trans-Pacific Partnership) 跨太平洋伙伴协定
TQC(total quality control) 全面质量管理
TQP(typical quality plan) 典型质量计划
TQ(training and qualification) 培训和资格评定
traceability 追踪能力，跟踪能力，可追溯性，溯源性，可描绘性
traceable 可示踪的
trace amount 痕量，示踪量
trace analysis 痕量分析
trace ball 跟踪球
trace chemistry 示踪化学
trace command 跟踪指令，结尾指令
trace element in coal 煤中微量元素
trace element 微量元素，痕量元素
trace heating 管道（电阻）加热保温，伴热
trace impurity 微量杂质，痕量杂质
trace interval 扫描时间
trace metal 微量金属
trace program 跟踪程序
tracer agent 示踪剂
tracer atom 示踪原子，标记原子
tracer dye 示踪染色水
tracer element 示踪元素
tracer experiment 示踪实验
tracer gas 示踪气体
tracer head 跟踪头
tracer input signal 示踪物输入信号
tracer isotope 示踪同位素
tracer material 示踪材料
tracer milling 靠模铣
trace routine 跟踪程序
tracer particle 示踪粒子
tracer technique 示踪技术
tracer test 示踪原子法试验
tracery 窗格
tracer 示踪剂，示踪物，示踪器，故障检寻器，追踪程序，描绘器，跟踪程序，跟踪装置，描图号，指示剂
trace statement 跟踪语句
trace 轨迹，跟踪，微量，示踪，扫描，轨线，痕迹，痕量，踪迹，描绘，描图
trachyte andesite 粗面安山岩
trachyte porphyry 粗面斑岩
trachyte 粗面岩
tracing and insulation 伴热
tracing arm 寻迹臂，跟迹臂
tracing cloth 描图布
tracing experiment 示踪试验
tracing instrument 描图仪器
tracing paper 描图纸
tracing pen 描迹笔
tracing pipe 伴热管
tracing point 轨迹点，描迹点
tracing program 跟踪程序
tracing routine 跟踪程序，检验程序，追踪程序
tracing steam 伴热蒸汽
tracing stiffness 支撑刚度
tracing streamline 示踪流线
tracing 跟踪，示踪，寻找故障，描绘，管道（电阻）加热保温，管道伴热，描图，追踪
track address register 磁道地址寄存器
track adhesive 轨道黏着力
track advancing 轨道接长
trackage 轨道，轨长
track air compressor 拖带式空气压缩机
track and hold amplifier 跟踪与同步放大器
track bed 道床
track cable 载运索
track cant angle 轨道倾角
track chart 航道图，航路图
track circuit 轨道电路
track configuration 轨迹线
track crane 轨道起重机
track curve 弯道
track density 磁道密度
track dozer 履带式推土机
tracked type bulldozer 履带式推土机
tracked vehicle 轨道车
tracker 跟踪系统，跟踪器，跟踪仪
track gauge 轨距，轨距规
track hoist 履带起重机，履带吊
track hopper 卸煤沟
tracking apparatus 跟踪装置
tracking circuit 跟踪电路
tracking collector 跟踪集热器
tracking control 跟踪控制
tracking cross 跟踪十字光标
tracking cursor 跟踪光标，跟踪十字
tracking error 跟踪误差
tracking filter 跟踪滤波器
tracking frequency 跟踪频率
tracking information 跟踪信息
tracking mode 跟踪方式
tracking motion 跟踪运动
tracking network 跟踪网络
tracking number 快递单号
tracking program 跟踪程序
tracking resistance 爬电电阻，耐电痕性
tracking sensibility 跟踪灵敏度
tracking system 跟踪系统
tracking test 漏电痕迹试验
tracking 跟踪，(绝缘子表面的)漏电痕迹，追踪
track jack 起轨器
track-laid bunker 铁路卸煤斗
track-laid platform 铁路卸台
track-laid trestle 铁路栈桥

track-laid tunnel 铁路卸煤沟
track layer 铺轨机
track laying machine 铺轨机
track laying tractor 铺轨式拖拉机
track layout 铁路线平面布置图
track level 轨面
track line 轨道，迹线
track locomotive 干线机车
track lubricator 轨道润滑器
track miles of the industrial siding 专用线长度
track mounted 导轨式
track obstruction 线路故障
track pitch 转向架间距，道距
track profile 轨道剖面，线路纵断面图
track relay 轨道继电器，拨轨用继电器
track return circuit 轨道回流电路，路轨回流电路
track ring of brush 电刷在换向器上留下的遗迹圈
track road 纤道
track scale 称量车，轨道衡
track switch 触轨分段断路器开关，磁道转换
track-to-track access time 道到道的存取时间
track tractor 牵引拖车
trackway 轨道，踏出来的路，行迹
track wheel 履带轮
track-while-scan 扫描跟踪
track width adjustment 轮距调整
track width 轨矩
track wrench 轨道扳钳
track 轨道，轨迹，径迹，线索，路径，跟踪，磁道，导轨，轮距，跨距，路线，铁路线
traction angle 牵引角
traction boiler 牵引式锅炉
traction crusher 牵引式破碎机，牵引式碎煤机，牵引式碎石机
traction current 牵引电流
traction electrical apparatus 牵引电器
traction engine 牵引式发动机，牵引机
traction-free surface 无张力表面
traction generator 牵引发电机，牵引用发电机
traction load 牵引负载，牵引荷重
traction motor 牵引电（动）机
traction network 牵引网
traction rope 牵索，牵引索
traction substation 牵引变电站
traction system 牵引系统，牵引装置
traction test 牵引试验
traction theory 牵引理论
traction thrust 推挽力
traction transformer 牵引（式）变压器
traction-type fluid coupling 牵引型液力联轴器
traction wheel 牵引轮，主动轮
traction 拖拉，牵引
tractive armature type relay 衔铁上升继电器，吸铁式继电器
tractive capability 牵引能力
tractive capacity 牵引能力
tractive efficiency 牵引效率
tractive force 牵引力，拉力，推移力
tractive grade 牵引坡度
tractive resistance 牵引阻力
tractive 牵引的
tractor crane 拖拉起重机
tractor-dozer 拖拉推土机
tractor-driven pump 拖拉机驱动泵
tractor garage 拖拉机库
tractor haulage 牵引机拖拉
tractor-loader 拖拉装料机
tractor-pulled sheep-foot roller 拖拉机牵引的羊足碾
tractor scraper 拖拉机拖曳铲土斗，拖拉机拖曳铲运机
tractor shovel 拖拉铲土机
tractor trailer 拖拉机挂车
tractor truck 牵引车，拖车头
tractor wagon 拖运料车
tractor 牵引车，牵引机，拖拉机
tractrix 等切面曲线，曳物线
tract 通道，土地
trade agency 代理商
trade agreement 贸易协议，贸易协定
trade and economic cooperation 经贸合作
trade and payment agreement 贸易支付协定
trade by agreement 协定贸易
trade clause 贸易条款，贸易中心
trade contract 贸易合同
trade credit 商业信贷
trade deficit 贸易逆差
trade dispute 商务争端
trade fair 交易会，商品交易会
trade investment 商业投资
trade mark law 商标法
trade mark licence contract 商标许可（证）合同
trade mark licence 商标注册证
trade mark registration 商标注册
trade mark right 商标权
trade mark 牌号，商标
trade name 商品名
trade-off analysis 综合比较分析，权衡分析
trade-off evaluation system 综合鉴定系统
trade-off study 权衡分析，折中研究
trade-off 比较评定，折中方案
trade pattern 贸易模式
trade procurator 商务代理人
trade protocol 贸易协定书，贸易议定书
trade refuse 工业废物
trade representative 商务代表
trade secret 商业秘密
trade sewage 工业污水
tradesman 手工工人
trade terms 贸易用语
trade wind belt 信风带
trade wind desert 信风沙漠
trade wind region 信风区
trade wind 信风
trade zone 贸易区
trade 职业，行业，交易，商业，贸易
trading classification 交易类型
trading interval 交易时段
trading manner 交易方式
trading port 通商口岸

trading profit and loss account 销售盈亏账户
trading service charge 交易服务费
traditional biomass 传统生物质能
traditional construction 传统式施工
traditional industry 传统产业
traditional management 传统管理
traditional tunnel reclaim system 老式地槽取料系统
traditional windmill 传统风车
traditional 传统的
tradition 传统，惯例
traffic allowance 车费津贴
traffic and transportation 交通运输用
traffic capacity 通行能力
traffic carrying capacity 输送能力
traffic density 交通密度，(通信) 业务量密度，占线密度
traffic distributor 话务分配器
traffic-free 没有车辆来往的，无通信的
traffic group (记录) 磁道组
traffic lane 行车道，车道
traffic load 交通荷载量
traffic map 交通地图
traffic pan 车压硬土层
traffic vibration 行车振动
traffic volume 交通容量，通车量
traffic 交通 (量)，运输 (量)，交通，传输，通信量
trail builder 拖挂式筑路机
trail car 拖车
trailed fire pump 牵引消防泵
trailer block 随附信息组
trailer 活动板房，轻便房，挂车，拖车，尾部，尾车
trail hole 探井
trailing antenna 下垂天线，拖曳天线
trailing axle 从动轴
trailing bogie 后转向架
trailing brush edge 后刷边
trailing brush holder 后倾式刷握
trailing bubble 尾随气泡
trailing cable 拖动电缆，拖曳电缆
trailing conveyor 拖带式输送设备
trailing edge angle 后缘角
trailing edge cap (叶片) 后缘帽
trailing edge delay (脉冲) 后沿延迟
trailing edge flap 后缘襟翼
trailing edge shock 出口边激波
trailing edge spar 后梁，尾梁
trailing edge thickness 后缘厚度，出汽边厚度，出气边厚度
trailing edge vortex 后缘涡
trailing edge 后缘，后沿，尾边，(脉冲的) 下降边，(叶片) 出汽边，(脉冲) 后沿
trailing grate 活动炉排
trailing load 牵引重量
trailing pole horn 后极尖
trailing pole tip 后极尖
trailing section loss 出口边损失
trailing shock wave 尾冲波，尾激波
trailing vortex-line 后缘涡流线
trailing vortex sheet 尾涡面
trailing vortex 后缘涡流，尾涡，拖曳涡，出汽边涡流
trailing 拖尾
trail mile 线路，里程
trail of the fault 断层痕迹
trail pit 视坑，探坑
trail rope 拖绳
trail 后缘，拖曳物，尾，痕迹，跟踪，追踪，拖曳
train arrangement 串联布置
train compartment 火车车厢
train control 列车控制
train crew 列车检修班
trainee 学员，接受培训人员，受训人员，实习生，见习生
trainer 教练设备，教练员，教员
trainfall intensity-duration curve 雨量强度历时曲线
train indexer 列车对位分车机
train indexing 列车逐次对位
training and qualification 培训和资格认定
training center 培训中心
training courses setup 培训课程设置
training course 培训课程
training dike 导水堤，顺堤
training documentation 培训资料
training duration 培训持续时间
training dyke 导流坝
training expenses 培训费
training facilities 培训设施
training fee 培训费
training for employer's staff 对业主人员的培训
training for sediment 导沙
training for specific posts 定向培训
training idler 调心托辊，对中托辊
training jetty 导水突堤
training levee 导流堤，导水堤，顺坝
training management 培训工作管理
training manual 培训手册
training of employee 职工培训
training period 培训期
training personnel 培训人员
training program 培训计划
training reactor 培训用反应堆，教学用反应堆，培训堆
training, research and isotope production reactor 培训、研究和同位素生产反应堆
training scheme 培训方案
training service 培训服务
training session 培训时期
training setup 培训设置，培训方案
training simulator building 训练模拟机厂房
training simulator 培训仿真机，培训模拟机，训练仿真器
training site 培训场所
training time 培训时间，训练时间
training wall 导流壁，导水墙
training works 整治工程，治导工程
training 培训，训练 (人员)，整治
train inspection service 列检作业

train loading station 铁路装车站
train lock 列车锁定器
train make-up 列车编组
trainmen 列车乘务员，列车车务员，机车司机
train of gears 齿轮系
train of the current pulse for testing core 测试磁芯的电流脉冲序列
train of wave 波列
train positioner 列车定位机，列车定位器
train power generation 列车发电
train power station 列车电站
train program subsystem 系列程序子系统
train split-up service 解列作业
train travel 列车行程
train unloading facility 列车卸料设施
train wave 列波
trainway 电车轨道，索道
train 训练，系列，列，序列，齿轮系，机组，列车牵引车，培养
traipse 拖曳
trait of river channel 河道特性
trait 品质，特性
trajectory bucket-type energy dissipater 挑流鼻坎式消能结构
trajectory bucket 挑流鼻槛，挑流鼻坎
trajectory case 轨道角色
trajectory model 轨道模型
trajectory of principal stress 主应力轨迹
trajectory 流轨，轨道，轨线，轨迹，弹道，抛射线
trake resistance 制动阻力
tramegger 摇表，高阻表，兆欧表
trammel 椭圆规，量规，(旋转)圆筒筛，转筒
tramper 不定期船
tramp iron chute 落铁管
tramp iron container 集铁箱
tramp iron pocket 废铁室
tramp iron separator 杂铁分离器
tramp iron 废杂铁，煤中铁块，杂铁
tramp metal detector 废金属探测器
tramp screen 走动筛子
tramper 不定期船，夯实器
tramway 电车(轨)道，电车道
tram 有轨电车
tranquil flow 平流
tranquilizer 稳流装置，增稳装置，稳流板，镇静剂
tranquilling tank 稳流箱
tranquil regime 平稳状态
tranquil 平静的，稳定的，安静的
transacter 处理装置
transaction at buyer's option 由买主决定的交易
transaction cost 交易成本
transaction data 交易数据，事务数据
transaction processing system 事务处理系统
transaction tape 更新数据带
transaction value 成交价格
transaction 交易，处理，办理，事务，学会，学报，会刊
transactor 计算机输入的问答元件，询答装置
transadmittance 跨导纳，互导纳
transbasin diversion project 跨流域引水工程
transbasin diversion 跨流域引水
transbeam 横梁
transbooster 可调扼流圈【稳定电压用】，带补偿整流器的饱和电抗器
transcalifornium element 超锎元素
transceiver data link 收发机数据链路
transceiver 无线电收发机，收发器
transceiving (数据的)收发
transcendental curve 超越曲线
transcendental equation 超越方程
transcendental function 超越函数
transcendental transfer function 超越传递函数
transcendental 超越的
trans-century project 跨世纪工程
transcoder 代码转换机
transcode 自动译码(系统)
transconductance bridge 跨导电桥
transconductance meter 跨导表
transconductance 跨导
transcorporation 跨国公司
transcriber 复制装置，抄录器，读数器，信息转换器，转录器
transcription device 转抄装置，卡片穿孔装置
transcription 抄本件，记录
transcript 抄本件，副本，正式文本
transcritical flow regime 过临界流动状态，跨临界流动状态
transcritical Reynolds number 过临界雷诺数，跨临界雷诺数
transcritical 跨临界
transcrysraline fracture 穿晶断裂
transcrystalline corrosion 穿晶腐蚀
transcrystalline crack 晶内裂纹，穿晶裂纹
transcrystalline fracture 穿晶断裂，穿晶断口
transcurium element 超锔元素
transcurium 超锔
transcurrent fault 横推断层
transcurrent thrust 横冲断层
transcurrent 横向电流，横向流动的
transducer cabin 变送器小间
transducer face velocity 换能器表面振动速度
transducer insertion loss 换能器插入损耗
transducer response 传感器响应
transducer room 传感器室，变送器室
transducer 传感器，换流器，变换器，变频器，发送器，换能器，变送器，探测器，转换器
transduce 传感，转换，变频，变送
transductance 互导，跨导
transductor amplifier 磁放大器
transductor-controlled generator 磁放大器控制发生器
transductor 饱和电抗器，磁放大器
transect 横断面
transet 动圈式电子控制仪
transferable and divisible letter of credit 可转让分割信用证
transferable and separable letter of credit 可转让与可分割信用证
transferable contaminant 可运送的污染物，可运送的沾污物

transferable credit 可转让信用证，可转账信用证
transferable documentary letter of credit 可转让的跟单信用证
transferable letter of credit 可转让信用证，可转移的信用证
transferable licence contract 可转让许可（证）合同，可转让（专利）许可合同
transferable load 可调负荷
transferable specific delivery contract 可转让的定期交货合同
transferable to be allowed 允许转让
transferable 可转移的，可转让的
transfer admittance 转移导纳
transfer an amount by mail 信汇
transfer area 传递面积，转运站
transfer arm 转移臂
transfer basket 转运吊篮，转运容器
transfer behaviour 输运性能
transfer belt 输送带
transfer bridge 渡桥
transfer building 转运站
transfer busbar 旁路母线
transfer bus 切换母线，换接母线
transfer canal tilting device 输送管道［渠道］倾斜装置
transfer canal 燃料输送管道，输送渠道
transfer capacitance 跨路电容
transfer capital 转让资本
transfer car pit 迁车坑
transfer carriage 输送车，转运车
transfer cart 输送小车，输送车，运输车
transfer case 变速箱，分动箱
transfer cask （燃料）传送罐，中转容器
transfer cell 传送热室
transfer certificate of title 所有权转让证书
transfer characteristic 转移特性，传输特性
transfer charges 转运手续费
transfer check 传送校验，转账支票
transfer chute 切换下料管，切换下料槽，转运站落煤管，转运溜槽
transfer circuit 转移电路
transfer coefficient 传递系数，传热系数
transfer conditionally 有条件传送
transfer constant 传递系数，转移常数，转换常数
transfer contact 转换触点，切换触点
transfer control register 传送控制寄存器
transfer control 转移控制
transfer cost 转运费
transfer crane 运送吊车
transfer current ratio 转移电流比
transfer curve （饱和电抗器的）静特性曲线
transfer device 切换装置
transfer drive screw 传动螺杆
transferee 受让方
transfer efficiency 合金过渡系数【焊接材料】，转换效率，传输效率
transference 移动，转送，传递，传送，转移
transfer equation 传热方程
transfer equipment 转运设备
transfer facility 转运设备
transfer fee 转账费
transfer flask 传送容器，中转容器
transfer fluid 载热剂
transfer from storage to installation area 从贮存场转移至安装区
transfer function amplifier 传递函数放大器
transfer function analyzer 传递函数分析器
transfer function for feedback 反馈传递函数
transfer function matrix 传递函数矩阵
transfer function meter 传递函数测试仪【反应堆】
transfer function model 传递函数模型
transfer function of reactor kinetics 反应堆动力学传递函数
transfer function 传递函数，转移函数，转换函数
transfer gantry 高架起重机，龙门起重机
transfer gas 输送气体
transfer hopper 转运料斗，转运煤斗
transfer house 转运站
transfer impedance 转移阻抗，传输阻抗
transfer in channel 通道转换
transfer instruction 转移指令
transfer instrument 转换式仪表
transfer interpreter 传送翻译器
transfer key 转接电键
transfer lag 传输延迟，转移滞后，传递时滞
transfer line 传输线路，传输管路，输送管，传递线，输送管线
transfer locus 传递函数轨迹图
transfer loss 传递损耗
transfer machine isolation valve 输送机隔离阀，转运机隔离阀
transfer machine （燃料元件）转运机，输送机
transfer matrix 转移矩阵
transfer mechanism 传动机构，传递机构
transfer memory 转移存储
transfermium element 超镄元素
transfer of accountant 转账
transfer of auxiliary supply 备用电源切换，后备电源切换
transfer of axis 轴系转换
transfer of equipment 设备转让
transfer of heat 传热
transfer of knowledge 知识转让
transfer of load 负载转移，负荷变换
transfer of ownership 产业转让，所有权转让
transfer of prestressing force 预应力的传递
transfer of profit 让利
transfer of right 权利转让
transfer of risk 风险转移，保险金的转让
transfer of technology requirement 技术转让要求
transfer of title 所有权的转让，产权转让
transfer oil pump 输油泵
transfer on no flow 无溢出转移
transfer on nonzero 非零转移
transfer operation 转移操作
transfer order 转移指令
transferor 转让方
transfer ownership 过户，转让所有权

transfer payment 转让性付款
transfer penetration 燃料传送穿孔件
transfer pipet 移液管
transfer pipe 传动管
transfer point 转运站
transfer position 输送位置
transfer power 输送功率
transfer pressure 分级压力
transfer price 调拨价格，转让价格
transfer probability 迁移概率，输运概率
transfer pump 运输泵，中继泵，输送泵，转运泵
transfer rate 输送速率，转移速率，转换率
transfer ratio 传递比
transferred arc 转移弧
transferred potential 传递电位
transferred tripping 远方跳闸，运动跳闸
transfer relay 切换继电器
transfer resistance 转移电阻
transferring or assigning copyright 版权转让
transfer scheme 读出和记录电路
transfer station 输送站，转运站，中转站
transfer storage bin 中转仓
transfer strength 传递强度
transfer strip 切换片
transfer surface 传热面，传质面
transfer switch 转接开关，转换开关
transfer system 转移系统，传输系统
transfer tank 扬液器，转运仓
transfer terminal 中转码头
transfer time 传递时间，转移时间，转换时间，传送时间，切换时间
transfer tower 转运塔，转运站
transfer track 输送轨道，转运轨道
transfer tripping system 远方跳闸系统，远动释放系统
transfer trip 远动释放，远方脱扣
transfer trolley 转运车
transfer trough 输送槽
transfer tube （燃料）输送管道
transfer tunnel 转输通道
transfer-type heat exchanger 传递式换热器
transfer valve 输送阀
transfer voltage ratio 转换电压比【线性无源网络】
transfer winding 成形绕组
transfer 交替【两个设备之间】，对调，对换，传输，传导，传递，传送，输送，转换，转移，调动，变换，迁移，进位，切换，移交，转让，转账
transfilter 变压器滤波器
transfinite 超限的
transfluxor storage 多孔磁芯存储器
transfluxor 多孔磁芯，多孔磁芯存储器
transformational criterion 变换分析准则
transformational model 变换模型
transformational 变换分析的
transformation coefficient 变换系数
transformation diagram of super-cooled austenite 过冷奥氏体转变图
transformation effect 转化效果
transformation equation 变换方程
transformation grammar 转换文法
transformation group 交换群
transformation hardening 相变硬化，硬化处理
transformation load 变换荷载
transformation matrix 变换矩阵
transformation of coordinates 坐标变换，坐标系变换
transformation of state variable 状态变量变换
transformation period 变换周期
transformation point temperature 转变点温度，临界点温度
transformation point 转变点，临界点
transformation ratio meter 变换系数计，变比计
transformation ratio 变比【变压器】，变换系数，变换比，变压比，变压系数
transformation temperature 转换温度，相变温度，相变点，变态点
transformation 变换，转换，变化，转化，改变，改造，转变，传递
transformator 变压器，变换器
transformed area 换算面积
transformed flow net 变换流网
transformed section 换算截面，折算断面
transformed value （被测量的）变换值
transformed wave 变相波
transformer action 变压器作用，变压器效应
transformer active power 变压器有功功率
transformer amplifier 变压器耦合放大器
transformer bank 变压器组
transformer bay 变压器间
transformer booster 升压变压器
transformer breaker 变压器断路器
transformer bushing 变压器套管
transformer bus 变压器母线
transformer cabin 变压器室
transformer casing 变压器外壳
transformer characteristic 变压器特性
transformer circuit-breaker 变压器断路器
transformer coil 变压器铁圈
transformer compound 变压器组
transformer converter （变交流为直流的）变压变流机
transformer cooler 变压器冷却器
transformer cooling system 变压器冷却系统
transformer cooling 变压器冷却
transformer core 变压器铁芯
transformer-coupled amplifier 变压器耦合放大器
transformer-coupled load 变压器耦合负荷
transformer-coupled oscillator 变压器振荡器
transformer-coupled pulse amplifier 变压器耦合脉冲放大器
transformer coupling 变压器耦合
transformer DC resistance tester 变压器直流电阻速测仪
transformerd emting factor 变压器额定下降因子
transformer differential protection 变压器差动保护
transformer drop compensator 变压器电压降补偿器
transformer effect 变压器效应，（电磁感应的）

脉动效应
transformer efficiency　变压器效率
transformer electromotive force　变压器电动势
transformer equivalent circuit　变压器等效电路
transformer fan　变压器风扇
transformer-fed system　经变压器的供电系统
transformer feeder protection　变压器馈线保护
transformer fitted with OLTC　有载调压变压器
transformer for hand lamp　行灯变压器
transformer gallery　变压器廊
transformer hall　变压器室
transformer hum　变压器铁芯哼声，变压器噪声
transformer inductance　变压器电感
transformer iron　变压器铁芯
transformer kiosk　变压器亭，变压塔
transformer lamination　变压器叠片
transformer lead　变压器引线
transformerless multiplex switch　无变压器多接开关
transformerless power-supply　发电机直供电源，无变压器电源
transformer load center　变压器负荷中心
transformer load　变压器负荷
transformer loss　变压器损耗
transformer magnetic shunt　变压器的磁分路
transformer maintenance room　变压器检修间
transformer matching　变压器匹配
transformer network　变压器网络
transformer noise　变压器噪声
transformer oil conservator　变压器油枕
transformer oil cooler　变压器油冷却器
transformer oil drum　变压器油枕
transformer oil purification　变压器油净化
transformer oil tank　变压器油箱
transformer oil　变压器油
transformer output　变压器输出
transformer outside axis line A　A 列外（主）变压器
transformer overhaul bay　变压器检修间
transformer overhaul frame　变压器检修架
transformer overload　变压器过负荷
transformer pillar　变压器塔，变压器柱
transformer protection　变压器保护
transformer rated current　变压器额定电流
transformer rated ratio　变压器的额定变比
transformer rated voltage　变压器额定电压
transformer ratio bridge　变压比电桥
transformer ratio　变压比，变流比
transformer reactance　变压器电抗
transformer reactive power　变压器无功功率
transformer read-only storage　变压器（式）只读存储器
transformer rectifier　变压整流器
transformer relay　变压器继电器
transformer rotor　变压器转子【旋转变压器】
transformer section　变压器部分
transformer sheet　变压器硅钢片
transformer stamping　变压器冲片
transformer starter　变压启动器
transformer station for structural fabrication yard　铆焊用变电所，结构制造场用变电所
transformer station　变电站，变电所
transformer steel　变压器硅钢片
transformer substation　变电所，变电站
transformer tank　变压器箱，变压器油箱
transformer tap　变压器分接头
transformer turns ratio　变压器匝比
transformer utilization factor　变压器利用系数
transformer vault　变压器室，变压器间【拱顶室】
transformer voltage regulation range　变压器调压范围
transformer voltage　变压器电压，变压器绕组
transformer with natural cooling　自然冷却式变压器，自然冷却变压器
transformer with ring core　环芯式变压器
transformer with series-parallel connection　串并联变压器
transformer with split winding　分裂变压器
transformer yard　变压器场
transformer yoke　变压器磁轭
transformer　变压器，变换器，互感器
transformeter　电流互感器仪表总成
transforming station　变电站
transform　变压，转换，变换，改造
transgranular corrosion　穿晶腐蚀
transgranular crack　穿晶裂纹，穿晶龟裂
transgranular fracture　穿晶断裂
transgranular　穿晶的，横晶的
transgression for using electricity　违章用电
transgression speed　超速【静电除尘器内】
transgressive intrusion　不整合侵入，贯层侵入
tranship cargo　转运货物
tranship goods　转运货物
transhipment bill of lading　转运提单
transhipment B/L　转运提单
transhipment bond note　转运保证书
transhipment cargo　转船货，转口货
transhipment crane　运输用起重机
transhipment shipping bill　转运单，转船单
transhipment volume　转运量
transhipment　转运，转船
transhipmsnt entry　转运报单
transient ablation area　暂时消融区
transient aerodynamic characteristics　瞬态气动特性，瞬态空气动力学特性
transient aerodynamic heating　瞬时空气动力加热
transient aerodynamic performance　瞬态气动性能
transient aerodynamics　瞬变空气动力学
transient amplifier　瞬态放大器
transient analysis　暂态分析，瞬变分析，瞬态分析
transient analyzer　瞬变分析器
transient availability　瞬时可用率，瞬时可用度
transient behaviorur　瞬变特性，瞬时性态，暂态行为，过渡特性，暂态特性，瞬态
transient blade-flapping behavior　叶片挥舞运动的过渡状态
transient boiling　瞬态沸腾，过渡沸腾
transient-cause forced outage　暂态性故障
transient characteristics　瞬态特性，暂态特性，

过渡特性，动态特性，过渡特性瞬态特性
transient characteristic tracer 瞬态特性图示仪，瞬态特性示踪剂
transient chemistry 过渡化学
transient component 瞬变分量
transient condition 过渡工况，瞬时条件，瞬变工况，瞬变条件，瞬态，瞬态工况，过渡状态，暂态，瞬变值
transient creep 不稳定蠕变，瞬时蠕变，瞬时徐变
transient critical flow 瞬态临界流
transient critical heat flux 瞬态临界热流密度
transient current offset 瞬态电流补偿
transient current 瞬时电流，瞬态电流，暂态电流，过渡电流
transient curve 瞬态曲线，过渡过程曲线
transient data 暂态数据
transient-decay current 瞬衰电流
transient deformation 瞬时变形
transient design pressure 瞬时设计压力
transient deviation 瞬态偏差
transient displacement 瞬时位移
transient disturbance 瞬时扰动
transient drag 瞬时阻力
transient electromotive force 瞬变电动势
transient equilibrium 瞬态平衡，暂时平衡
transient event 暂态事件，暂现事件
transient experiment 瞬态实验
transient fault 瞬息故障，短暂故障，瞬态故障
transient field current 瞬态励磁电流
transient flow 瞬间流动，瞬变流动，暂态流，非定常流，瞬变流
transient flux-linkage 瞬间磁通链
transient forced outage duration 瞬时强迫停运持续时间
transient-forced outage rate 元件瞬时性强迫停运率
transient forced outage 瞬时强迫停运
transient-free 无瞬变现象的
transient gain 暂态增益
transient generator 暂态发电机
transient harmonics 瞬间谐波，瞬变谐波
transient heat conduction equation 瞬时热传导方程
transient heat conduction 非稳态导热，不稳定导热
transient heat flux 瞬变热流
transient heating condition 非定常热状态，过渡热状态
transient heating 瞬间加热，瞬时加热
transient heat transfer 瞬态热传递，不稳定传热
transient heat 瞬态热
transient hot channel factor 瞬态热通道因子
transient impedance 瞬变阻抗
transient information 瞬变信息，瞬间信息，速变信息
transient instability 瞬态不稳定性
transient internal voltage 瞬变内电压
transient interruption 瞬断
transient load characteristics 暂态负荷特性
transient load 瞬时荷载，瞬变负荷，瞬间荷载，瞬时负荷，瞬态荷载
transient magnetic variation 瞬磁变
transient measurement 瞬态量测
transient motion 瞬变运动，瞬时运动
transient network analyzer 暂态网络分析仪
transient neutron slowing down 瞬态中子慢化
transient of RLC circuit RLC 回路的过渡过程
transient operating conditions 过渡工况，瞬态工况
transient operation condition of deaerator 除氧器暂态运行工况
transient operation 瞬态运行，过渡工况运行
transient oscillation 瞬态振荡
transient oscillogram 过渡过程波形图
transient oscillograph 瞬变过程示波器
transient outage 瞬时停运
transient output 瞬态出力，瞬态输出
transient overload current 瞬时过载电流
transient overpower accident 瞬态超功率事故
transient overshoot 过冲，瞬态超调
transient overspeed 瞬态超速
transient overvoltage capability 暂态过电压能力
transient overvoltage counter 瞬时过电压计数器
transient overvoltage 瞬时过电压
transient part 瞬变分量，瞬变态分量，瞬变部分
transient peak loading 瞬时峰荷，暂态峰荷
transient peak 瞬间峰值，合闸尖峰电流
transient performance 过渡特性，过渡性能，瞬态特性，瞬态性能
transient period 暂态期，过渡期，瞬变周期，过渡时期，瞬态时期，瞬态周期
transient phenomena 暂态现象
transient phenomenon on switching 开关暂态现象
transient phenomenon 暂时现象，瞬时现象，瞬变现象，不稳定现象
transient power disturbance 瞬态电源扰动
transient power system 暂态电力系统
transient power 瞬态功率
transient pressure 瞬态压力，过渡工况压力，瞬变压力
transient process characteristic curve 暂态特性曲线
transient process 瞬态过程，暂态过程，过渡过程，瞬变过程
transient property 瞬时特性
transient pulse 瞬态脉冲
transient radiation damage 瞬时辐照损伤
transient radiation effect 瞬时辐照效应
transient reactance 瞬变电抗，暂态电抗
transient reactivity 瞬发反应性，反应度瞬变
transient reactor period 反应堆的瞬变周期
transient reactor test facility 瞬态反应堆试验装置，反应堆瞬态试验装置
transient recorder 瞬态记录器，瞬间记录器
transient recording 瞬态记录
transient recovery voltage 瞬变恢复电压
transient regime 瞬变状态
transient resistance 瞬态电阻
transient response characteristic 瞬态响应特性
transient response 瞬态响应，过渡反应，过渡

特性，瞬态特性，瞬时响应
transient running 过渡工况运行
transient saliency 瞬态凸极性
transient saturation 瞬态饱和
transient seepage flow 不稳定渗流
transient service 临时检修
transient shielding 暂态屏蔽
transient solution 过渡解
transient-speed characteristics 瞬变过程中的转速特性，瞬时转速特性
transient stability calculation 大干扰稳定计算，暂态稳定计算
transient stability limit 瞬态稳定边界
transient stability region 瞬时稳定区
transient stability 暂态稳定度，瞬间稳定，瞬态稳定度，瞬时稳定性，暂态稳定（性）
transient state power limit 暂态功率极限
transient state stability factor 暂态稳定度系数
transient state stability limit 暂态稳定度极限
transient state stability 暂态稳定度，瞬态稳定度
transient state vibration 非定常振动
transient state 过渡状态，非稳态
transient stress 瞬态应力，瞬时压力
transient swing 瞬时摇摆，暂态波动
transient temperature gradient 瞬时温度变化梯度
transient temperature 瞬变温度，瞬时温度，不稳定温度
transient thermal response 瞬时热响应
transient thermal stress 瞬态热应力
transient time-constant 瞬变时间常数
transient time 瞬态时间，过渡时间
transient vibration 瞬态振动
transient voltage regulation 瞬态电压调整
transient voltage 瞬态电压，过渡电压，瞬时电压
transient vortex 瞬变涡，瞬时涡，非稳定旋涡
transient wave 瞬变波
transient wind speed 瞬时风速
transient working area 暂时工作区
transient xenon poisoning 瞬态氙中毒，非稳态氙中毒
transient zone 过渡区
transient 瞬变的，瞬时的，暂态的，短暂的，路过的，瞬态，暂态，瞬变，过渡，非稳态，暂态过程，瞬变过程，过渡过程，瞬变现象，瞬变值，过往旅客，候鸟
transilog 晶体管逻辑电路
transimpedance 互阻抗，互导倒数，跨导倒数
transinformation 转移信息，传递信息
transistance 晶体管作用，晶体管效应
transistor ageing 晶体管老化
transistor amplifier 晶体管放大器
transistor analysis recording equipment 晶体管分析记录设备
transistor and component tester 晶体管及元件测试仪
transistor base 晶体管基极
transistor bias circuit 晶体管偏压电路
transistor chopper 晶体管斩波器
transistor circuit 晶体管电路
transistor collector 晶体管集电极
transistor computer 晶体管计算机
transistor-coupled logic 晶体管耦合逻辑
transistor-coupled transistor logic 晶体管耦合晶体管逻辑
transistor current steering logic 晶体管电流导引逻辑
transistor decade counter 晶体管十进制计数器
transistor demodulator 晶体管解调器
transistor diode logic 晶体管二极管逻辑
transistor diode transistor logic 晶体管二极管晶体管逻辑
transistor display and data processing system 晶体管显示器及数据处理系统
transistor-driven core memory 晶体管驱动的磁芯存储器
transistored automatic control 晶体管自动控制器
transistored bridge 晶体管电桥，装有晶体管的电桥，半导体二极管电桥，半导体电桥
transistored computer 晶体管计算机，晶体管化的计算机
transistored counter 晶体管计数器
transistored 使用半导体的，晶体管化的
transistor electronics 晶体管电子学
transistor equation 晶体管方程
transistor flip-flop 晶体管触发器
transistor ignition 晶体管点火，晶体管触发
transistor inverter 晶体管反相器，晶体管反向换流器
transistorized circuitry 晶体管电路图
transistorized magnetic core memory 晶体管磁芯存储器
transistorized servo amplifier 晶体管伺服放大器
transistorized 装有晶体管的，晶体管化的，用晶体管的
transistor logic circuit 晶体管逻辑电路
transistor measuring oscilloscope 晶体管测量示波器
transistor memory 晶体管存储器
transistor motor 晶体管电动机
transistor multivibrator 晶体管多谐振荡器
transistor-operated magnetic storage 晶体管磁芯存储器，晶体管控磁存储器
transistor operational circuit 晶体管运算电路
transistor oscillator 晶体管振荡器
transistor output 晶体管输出
transistor parameter 晶体管的参数
transistor relay 晶体管式继电器，晶体管继电器，晶体管继电保护装置
transistor-resistor logic 晶体管电阻逻辑
transistor-resistor-transistor logic 晶体管电阻晶体管逻辑
transistor servo preamplifier 晶体管伺服前置放大器
transistor switch 晶体管开关
transistor tester 晶体管试验器
transistor torquemeter 半导体扭矩计
transistor-transistor logic 晶体管晶体管逻辑
transistor trigger 晶体管触发器
transistor unit 晶体管元件

transistor voltage stabilizer 晶体管稳压器，晶体管电压稳定器
transistor voltmeter 晶体管电压表，晶体管伏特计
transistor 晶体三极管，晶体管，半导体管
transit cargo 转载货，过境货
transit clause 转运条款
transit company 中转公司
transit crane 全转式起重机
transit dose 通过剂量
transit dues 转口税
transit duty 过境税，通行税
transit insurance 内陆运输险
transitional area 过渡区
transitional belt 过渡带
transitional boiling 过渡沸腾
transitional climate 过渡性气候
transitional level 过渡电平
transitional measure 过渡措施
transitional profile 过渡段水面曲线
transitional region 过渡区
transitional 过渡的，不稳定的，瞬变的
transition anode 过渡阳极
transition band 过渡频带
transition belt 过渡带
transition boiling 暂态沸腾，过渡沸腾
transition chute 过渡槽
transition coefficient 过渡系数
transition coil 过渡线圈，分流线圈
transition condition 瞬态，暂态，过渡状态
transition cone 过渡锥形体【蒸汽发生器】
transition contact 过渡触头
transition core 过渡堆芯
transition curve 过渡曲线，转变曲线，缓和曲线
transition diagram 转移图
transition energy 转变能，跃迁能量
transition fit 过渡配合
transition flare 喇叭形过渡段
transition flow regime 过渡流态
transition flow 过渡流，过渡流动，不稳流动
transition height （风速廓线）转换高度，过渡高
transition hopper 中继渣斗
transition idler 过渡托辊
transition impedance 过渡阻抗，转移阻抗
transitioning member 过渡构件
transition joint 中间接头，过渡接头
transition layer capacitance 过渡层电容
transition layer （层流与紊流之间的）过渡层
transition loss 转变损失，过渡损失
transition matrix 跃迁矩阵，转换矩阵，状态转移矩阵
transition of height difference 过渡高差
transition period 过渡时期
transition piece 过渡段
transition point 过渡点，转折点，转捩点
transition process 转变过程，过渡过程，跃迁过程
transition range 转变范围
transition rate 转变速率
transition ratio 传动比
transition regime 转变方式
transition region 过渡区，转变区
transition resister 过渡电阻器【变压器抽头转换用】
transition Reynolds number 转捩雷诺数
transition ring （锻件）过渡环【直流炉】
transition section （风洞）过渡段，过渡断面，过渡区
transition stage 瞬时状态，过渡状态
transition strip 转捩带
transition temperature 转变温度
transition time 跃迁时间，转换时间
transition type 过渡型
transition wire 转捩绊线
transition wrapper 过渡段【蒸汽发生器外壳】
transition zone 过渡区【直流炉，焊接】，转换区，过渡层，过渡带，渐变层，跃迁区
transition 过渡，转变，通过，变换，转换，（放射性核素）转移，过渡段，跃迁
transit letter of credit 转口信用证
transit mixer 运送式拌和机
transit operation 倒罐操作
transitory section 过渡段，转换区
transitory 过渡的，不稳定的，跃迁的
transit pass for imports 进口过境单
transit phase angle 过渡相角【带电粒子】
transit port 中继港，中转港口
transitrol 自动调频管，自动频率微调管
transitron 碳化硅发光二极管，负互导五极管
transit shed 转运堆栈
transit station 转接站
transit store 临时仓库，（废物）金属桶储存区
transit time spread 渡越时间的离散
transit-time tube 速调管，渡越时间管
transit time 过渡时间，停留时间，转移时间，切换时间，渡越时间，通过时间，中转时间
transit visa 过境签证
transit with stadia wire 视距丝式经纬仪
transit 运送，渡越，跃迁，通过，飞渡，过境，运行，中转，转口
translating circuit 译码电路
translating gear 变换齿轮
translating system 转换系统，变换系统
translating 变换
translational energy 平动能量，平移能量，直线运动能
translational molecular heat 平移分子热
translational moment 侧移弯矩
translational motion 转移，迁移，平移，平动
translational movement 转移，迁移
translational slide 平移滑动
translational speed of tornado vortex 龙卷涡移动速度
translational speed of vortex street 涡街移动速度
translational vibration 平移振动
translational wave 移动波
translational wind machine 平移型风力机
translational 平移的，直移的
translation and rotation 平动与转动
translation circuit 译码电路，变换电路
translation coordinates 平移坐标
translation exception 转换失效，转换故障

translation lookaside buffer 翻译后援［后备］缓冲器【计算机】，转换旁视缓冲区
translation memory 译码存储器
translation of axes 轴系平移
translation of rigid body 刚体平移
translation of transfer function 传递函数的变换
translation 转移，迁移，转发，中继，变换，转播，移动，翻译，平动，平移，译码
translator-compiler 翻译程序编译程序
translator routine 翻译程序
translator unit 译码器部件
translatory fault 直断层，直移断层
translatory flow 平移流
translatory wave 推进波
translator 翻译者，译码程序，译码器，转发器，转换器，中继器，传送器，翻译程序，翻译器，译员
translawrencium element 铹后元素
translay relay 感应差动继电器，纵差继电器
transless 无变量器的，无变压器的
translocation 改变位置，移位，位置转移
translucence 半透明性，半透明度
translucent glass 半透明玻璃
translucent mask 半透明屏
translucent mosaic 半透明马赛克
translucent PVC panel 半透明聚氯乙烯板
translucent scale 透明标度盘
translucent 半透明的，透明的
translucid glass 半透明玻璃
transmeridional metallogenic belt 东西向成矿带
transmeridional 东西向的
transmissibility coefficient of aquifer 含水层导水系数
transmissibility coefficient 可透率，传输率，传输系数
transmissibility 可传性，可透过性，传输率，传递率，可传输性，透射性
transmissible credit 可转让信用证
transmissible letter of credit 可转让信用证
transmissible pressure gauge 电远传压力表
transmission accuracy 传动精度
transmission adapter 传输转接器，传输适配器
transmission and distribution equipment 输配电设备
transmission and distribution line 输配电线路
transmission and distribution price 输配电价
transmission and distribution substation 变配电站
transmission and distribution tariff 输配电价
transmission and distribution 传输与分配，输电和配电，输配电，输变电
transmission and exchange system of information 信息传输与交换系统
transmission and reception controller 传输及接收控制器
transmission attenuation 传输衰减
transmission bar 传力杆
transmission cable 输电电缆
transmission capacity of tie line 联络线的传输容量
transmission capacity 输电能力，输电量，传输容量，传播能力，传递能力，渗透能力，投射性
transmission case 传动箱，变速箱
transmission channel 传输通道，传输通路，发送信息，传输信道
transmission characteristics 输电特性，传输特性
transmission coefficient of water vapor 水蒸气的传输系数
transmission coefficient 传输系数，透射系数，传动比，传递系数，传导系数，渗滤系数，输送系数
transmission congestion contract 输电阻塞合同，传输拥挤合同
transmission congestion 输电阻塞
transmission constant 传输常数，传递常数，传导常数，渗透常数
transmission control character 传递控制符号
transmission control element 传输控制部件
transmission controller 发送控制器
transmission internet protocol 传输网间协议
transmission control protocol 传输控制协议
transmission control 传输控制
transmission cost 输电费用
transmission current capacity 输送载流量
transmission curve 传输曲线，输电曲线
transmission delay 传输延迟
transmission distance 输送距离，输电距离
transmission and distribution equipment 输配电设备
transmission-distribution price 输配电价
transmission dynamometer 传动式测力计，传输式测力计
transmission efficiency 输电效率，传输效率，传动效率，传送效率
transmission electron microscope 透射电子显微镜，透射电子显微术
transmission end 输电端
transmission equipment 传输设备
transmission equivalent 传输当量
transmission error alarm 传输差错报警
transmission error 传动误差
transmission expense 输电费用
transmission facility 输电设备
transmission factor 传输系数，传导系数，传递系数，透射因数
transmission frequency 输电频率，传输频率
transmission gear 传动齿轮
transmission grid 输电网
transmission interface converter 传输接口转换器
transmission interruption 传输中断
transmission lag 传输滞后
transmission length 传递长度
transmission level 传输水平，传输电平
transmission line capacity 输电线容量
transmission line control room 配电控制室，网络控制室，网控室
transmission line corridor 输电线路走廊，输电线走廊
transmission line discharge class 输电线路放电等级
transmission line inductance 传输线电感
transmission line insulation level 线路绝缘水平

transmission line loss 输电线损耗，传输线损耗
transmission line parameter 输电线路参数
transmission line passage way 线路通道
transmission line protected region 输电线路保护区
transmission line survey 线路测量
transmission line to grid 上网输电线
transmission line tower foundation 杆塔基础
transmission line tower 线路铁塔，输电杆塔，输电铁塔，输电线路铁塔
transmission line 输电线路，输电线，送电线路，传输线
transmission loss coefficient 输电损失系数
transmission loss 传输损耗，传导损失，穿透损失，传递损失，输电损失，透射损失
transmission magnifying machine 传动放大机构
transmission mast 输电线杆
transmission medium 传输媒介，传输装置
transmission method 透过法【超声波探伤】
transmission network 输电网
transmission of change-of-state information chronologically 状变信息按时标顺序传输
transmission of change-of-state information in order of priority 状变信息按优级传输
transmission of electricity 输电
transmission of energy test 能量穿透法试验
transmission of energy 能量输送
transmission of heat by convection 对流传热
transmission of heat 热的传递
transmission of load 荷载传递
transmission of power 动力传送，输电
transmission on demand 按请求传输
transmission overhead line insulation level 架空线路绝缘水平
transmission owner 输电所有者
transmission path 传输路径，传递通路，传输通道
transmission performance rating 传输特性额定值
transmission performance 传输性能
transmission pole 输电线杆
transmission power control 传输功率控制
transmission price 输电电价
transmission protocol 传输协议
transmission pulse 发射脉冲
transmission range 输电范围，传输范围，传输距离
transmission rate 传输速率，传动比
transmission ratio 传动比，速比，传输比
transmission regulator （通信）传输调节器
transmission reliability margin 输电可靠性裕度
transmission right 输电权
transmission route survey 输电线路测量
transmission route 传输路线，输电路线，传输路由
transmission semiconductor detector 穿透式半导体探测器
transmission shaft 传动轴
transmission spare capacity 剩余输电容量，输电备用容量
transmission subsystem interface 传输子系统接口
transmission switch yard and offsite power system 传输开关站与厂外电源系统
transmission system operator 输电系统运营商
transmission system relaying 输电系统继电保护
transmission system 输电系统，传输系统，传动系统，发射系统
transmission target 透射式靶
transmission technique 传输技术
transmission throughput 有效传输速度，发报速度，数据传输吞吐量
transmission tie capacity 输电联络线容量
transmission time 传播时间
transmission tower 输电杆塔，输电（铁）塔
transmission unit 传输装置
transmission voltage class 输电电压等级
transmission voltage grade 输电电压等级
transmission voltage 输电电压
transmission wave 发射波
transmission wire 输电线
transmission with decision feedback 判决反馈传输
transmission with information feedback 信息反馈传输
transmission with two line in parallel 双回路输电
transmission 输送，发送，发射，传递，透射，传动，传输，输电，通过，传动装置，传送，传射
transmissivity 透光度，透射率
transmit clock 发送时钟
transmit data register 发送数据寄存器
transmit dues 过境税，转口税
transmit heat 传热
transmit-receive switch 收发转换开关
transmit-receive unit 发送接收装置
transmit and receiving 发送与接收
transmit station 发信台
transmittal document 传递单据
transmittance meter 能见度测量仪，透射比测量仪
transmittance 透射率【τ】，传递系数，透明性，透射系数，穿透率，透光率
transmitted power 输电功率，发射功率，输送功率
transmitted radiation 透过辐射，漏过辐射
transmitted wave 传输波，发射波
transmitter distributor 发射机配电器
transmitter horn 喇叭形发射器
transmitter rack 变送器架
transmitter-receiver 收发两用机，收发信机
transmitter site error 发射机产生的方位误差
transmitter 变送器，传感器，发送器，传递器，发射机，发报机，传送机，发射器
transmitting capacity 输电能力，传输能力，发送能力
transmitting core 发送磁芯
transmitting crystal 发送晶体
transmitting element 发送元件
transmitting instrument 发送仪表
transmitting level 发送电平
transmitting medium 传导介质，发送装置
transmitting relay 发送继电器
transmitting rheostat 传送变阻器

transmitting selsyn 自动同步传感器，自动同步发送装置
transmitting station 发射台
transmitting system 发送系统，传递系统
transmitting transducer 发送器，传送变换器，发射换能器
transmitting transformer 输电变压器
transmitting voltage response 发送电压响应
transmitted 已发送
transmityper 导航信号发送机
transmit 输电，传输，发射，发送，传导，传送，转移（信号，信息）
transmountain diversion project 穿山引水工程
transmountain diversion 穿山引水
transmultiplexer 复用转换器
transmutations per second 每秒核转变数
transmutation 变形，变质，变化，蜕变，转变
transnational corporation 跨国公司
transnational management 跨国经营
transnobelium element 超锘元素
transnormal 超越常规的
transoceanic communication 越洋通信，海外通信
transom 结构中横向构件，上亮子，气窗，腰头窗，腰窗，(门窗）中横框，横梁，横楣
transonic aerodynamics 跨音速气体动力学
transonic cascade 跨音速叶栅
transonic compressor 跨音速压气机
transonic flow field 跨音速流场
transonic flow 跨音速流
transonic turbine 跨音速透平，跨音速涡轮
transonic wind tunnel 跨音速风洞
transonic zone 跨音速区
transonic 跨音速的，超音速的
transparence 透明性，透明度，透明
transparency clarity 透明度
transparency meter 透明度测定器，透明度计
transparency of atmosphere 大气透明度
transparency 透明度
transparent conduction oxide 透明导电膜
transparent cover plate 透明盖板
transparent cover 透明盖板
transparent flame 透明火焰
transparent glass 透明玻璃
transparent lacquer 透明亮漆，透明漆
transparent plastic tile 透明塑料瓦
transparent side wall 透明边墙
transparent to neutron 对中子透明的
transparent tracing paper 透明描图纸
transparent two-dimensional fluidized bed 透明二维流化床
transparent 透明的，透彻的，明显的
transpiration coefficient 叶面蒸发系数，蒸腾系数
transpiration-cooled blade 发散冷却叶片
transpiration cooling 发散冷却，蒸发冷却
transpiration efficiency 蒸腾效率
transpiration from leaves 叶面蒸发
transpiration loss 蒸腾损失
transpiration rate 蒸腾率
transpiration ratio 叶面蒸发比值，蒸腾比，散发比
transpiration stream 散发流
transpiration wind tunnel 蒸腾风洞
transpiration 蒸腾
transpired honeycomb absorber 发泡蜂窝结构吸热器
transpirometer 蒸腾计
transplutonicum element 超钚元素
transplutonicum isotope 超钚同位素
transplutonicum 超钚
transplutonium element 超钚元素
transponder 发送机应答器，转发机，脉冲收发两用机
transportability 输送能力，可运输性，转用能力
transportable fire extinguisher 推车式灭火器
transportable moisture limit 可输送的水分限值
transportable radionuclide 可转移性放射性核素
transportable reactor 可运输式反应堆，移动式反应堆
transportable substation 可迁移的变电所
transportable 可移动的
transport agreement 运输协定
transport air compressor 输送空气压缩机
transport approximation 输运近似
transportation and storage condition 传输和贮存条件，运输和储存条件
transportation and storage system 储运系统
transportation by leaps and bound 跳跃式输移，（泥沙）跃移
transportation charge 转运费
transportation clearance 运输尺寸
transportation condition 运输条件
transportation cost 运费，拆迁费用，运输成本
transportation department 运输部门
transportation facilities 运输设施
transportation hub 中转港
transportation instruction 运输说明
transportation lag 运输滞后
transportation limit 运输限度
transportation line 运输线
transportation noise 运输噪声
transportation of bed load 推移质输送
transportation of debris 岩屑输送
transportation of ion exchange resin 离子交换树脂的输送
transportation permit 转运许可证，转运许可
transportation rate of suspended load 悬移质输送率
transportation risk 运输险
transportation route 运输路径
transportation velocity 输送流速
transportation volume 运输量
transportation 运输，输送，输移
transport basket 笼式运输容器
transport capacity of bed load 推移质输送量，推移质挟沙能力
transport capacity of suspended load 悬移质输送量
transport capacity 运送能力，输送能力，输送量
transport characteristics of wind 风力输运特性
transport coefficient 输运系数
transport container 运输集装箱
transport correction 输运校正

transport crane 运输用起重机
transport cross section 输运截面
transport delay unit 运输延迟单元
transport delay 传输延迟
transport dimension 运输尺寸
transport disengaging height 输送分离高度
transport distance 运输距离，运程
transport document 运输单据
transported soil 运积土
transport equation 输运方程，迁移方程
transport equipment 运输设备
transporter 运输机，运输装置，运送者，运送机
transport index 输运指数
transporting action of running water 流水搬运作用
transporting capacity 搬运能力，运输量
transporting equipment 运输设备
transporting erosive velocity 输移冲刷流速
transporting power 输送功率
transporting vehicle 运输工具
transporting velocity 搬运速度
transport kernel 输运核
transport lag 传输延迟
transport layer 传输层
transport load 运输负荷
transport loss 输运损失
transport mean-free path 输运平均自由程
transport mode 输运模式
transport nozzle 冲灰喷嘴【水力除灰沟内】
transport number 迁移数
transport of heat 热传输
transport of momentum 动量输运
transport of nuclear materials 核材料的运输
transport of pollutants 污染物运输
transport operator 输运算符
transport pallet for closure nut carrier racks 封头螺母装载架运输托架
transport parameters 传递参数，传输参数
transport process 输运过程，迁移过程
transport profile 传输协议集
transport property 迁移特性，输运性质，输送性质
transport system （仓储式）煤粉输送系统，传递系统
transport theory 输运理论
transport tunnel 运输隧洞
transport weight 载重量
transport wind speed 输运风速
transport 运输，迁移，输送，传输，搬运，运送，转移
transposed conductor 换位导线
transposed integral equation 转置积分方程
transposed matrix 换位矩阵，转置矩阵
transposed storm 移置暴雨，移置风暴
transposed strip 换位导线，换位导条
transposed transmission line 换位的输电线路
transposed winding 换位绕组
transpose matrix 转置矩阵
transposer 换位器，移项器
transpose 换位，（矩阵的）转置，移项，交叉
transposition cycle 线路必需的换位距离
transposition insulator 交叉绝缘子，换位绝缘子
transposition interval 换位间隔，换位节距
transposition method 移置法
transposition of telephone line 电话线交叉换位
transposition of unit hydrograph 单位过程线移置
transposition pole 换位杆，交叉杆
transposition section 交叉区，（线路的）交叉段
transposition support 换位杆塔
transposition tower 换位塔，交叉塔
transposition 换位，移置，移项，交叉，转置，传输，调换
transput （计算机设备之间的）交换，互换
transreactance 互抗
transrectification factor 换流因数，检波系数
transrectification 阳极检波
transrectifier 电子管检波器，阳极掩波器
transresistance 互阻
transship cargo 转运货物
transship goods 转运货物
transshipment 转船【运输】，转运
transsusceptance 互纳，互导纳的虚数部分
transsusceptification 交换整流，多电极管整流
transtage 过渡级，中间级
transtat 可调变压器
transthorium element 超钍元素
TRANS＝transfer 转换
transudate 渗出液
transuranic element 超铀元素
transuranic 超铀的
transuranium element 超铀元素
transuranium processing plant 超铀元素处理厂
transuranium 超铀元素
transversal arrangement 横向布置方式，横向布置
transversal axis 横轴
transversal cofferdam 横向围堰
transversal discrepancy 横向差值
transversal displacement 横向位移
transversal effect 横向效应
transversal erosion 横向侵蚀
transversal filament tape winding 横向纤维带缠绕技术
transversal filament tape 横向纤维带
transversal filter 横向滤波器
transversal flux permanent magnet synchronous generator 横向磁通永磁同步发电机
transversal galloping 横向驰振
transversal level 横轴水准器
transversal line 横轴线
transversal loading 横向荷载
transversal oscillation 横向振动
transversal plunging 横向浮沉
transversal rib 横向肋片
transversal section 横断面
transversal shrinkage 横向收缩
transversal stability 横向稳定性
transversal 横向的，横断的，横断线的，贯线，截线，横截线，横向
transverse alignment 横向调整［找正］
transverse anisotropy 横向各向异性
transverse arch 横向拱
transverse axis 横轴

transverse beam 横向梁，横梁
transverse bending test 横向弯曲试验
transverse bending 横向弯曲
transverse bent 横向构架
transverse brace 横向支撑
transverse buckling 横向变形，横向曲率
transverse circulation 横向环流
transverse component 横向分量
transverse condenser 横向布置凝汽器
transverse conductance 横向导电
transverse correlation 横向相关
transverse crack 横向裂纹，横向裂缝
transverse crevasse 横向裂缝
transverse current 横向电流，横向流，横流
transverse curvature 横向曲率
transverse differential protection 横联差动保护转置，横差保护
transverse diffraction 横向衍射
transverse dike 横堤
transverse direction 垂直于工作的主方向【冶金】，横向
transverse displacement 横向位移
transverse distribution 横向分布
transverse electromagnetic wave 横电磁波
transverse exchange 横向交换【反应堆冷却剂】
transverse expansion joint 横向伸缩缝
transverse extension 横向扩张，横向伸长
transverse fiber belt winding 横向纤维带缠绕
transverse field 横向场
transverse fissure 横向裂缝
transverse flow cooling tower 横流式冷却塔，横流式塔
transverse flow tube bundle 横流管束
transverse flow 横向流，横向气流，横流，横向流动
transverse focussing electric field 横向聚集电场
transverse force 横向力，剪切力
transverse fracture 横向断裂
transverse galloping 横向驰振
transverse gate 横向门
transverse gradient 横向比降，横比降
transverse gyro frequency 横向旋转频率
transverse hatching 横向剖面线
transverse interference 横向干扰
transverse irregularity 横向不规则性
transverse isotropy 横向同性
transverse joint 横向接缝，横缝
transverse leakage 横向泄漏
transverse load 横向荷载，水平荷载
transversely-finned fuel element 带横向肋的燃料元件
transversely isotropic material 横向同性材料
transverse magnetic filed 横向磁场
transverse magnetic wave 横磁波
transverse magnetization 横向磁化
transverse main passage between two units 两台机组之间的横向主通道
transverse maximum-to-average-power ratio 横断面最大功率与平均功率比【堆芯】
transverse member 横构件，横向构件
transverse metacenter 横稳心
transverse mixing 横向混合
transverse movement 横向位移，横向运动
transverse net 导线网
transverse oscillation 横向振荡，横向振动
transverse piezo-electric effect 横向压电效应
transverse pitch 横向节距
transverse plunging 横向浮沉
transverse prestress 横向预应力
transverse profile 横断面
transverse projection 横向投影
transverse reciprocating blade 横向往复式刀片
transverse reinforcement 横向钢筋
transverse resolution 横向分辨力
transverse rupture stress 弯曲强度
transverse section 横截面，横断面，横剖面
transverse shape 横断面形状
transverse shear 横向剪切
transverse shrinkage 横向收缩
transverse slide 横向滑移
transverse slipway 横向滑道
transverse slope 横比降，横坡，横向坡度
transverse slot 横向槽
transverse stability 横向稳定性
transverse stiffener 横向加劲杆
transverse strain 横向应变
transverse strength 横向强度，弯曲强度，抗挠强度
transverse stress 横向应力，弯曲应力
transverse strut 横撑
transverse test 抗弯试验
transverse thrust 横冲断层
transverse to the flow direction 垂直水流方向
transverse vibration 横向振动
transverse wall 横墙
transverse wave testing method 横波探伤法
transverse wave 地震横波，横波
transverse wedge 横楔
transverse weld tensile test 横向焊缝抗拉试验
transverse weld tension test 横向焊缝抗拉试验，横向焊缝拉伸试验，横向焊缝张力试验
transverse weld 横向焊缝
transverse 横的，横向的，横贯的，横断的，横梁
transverter 变频器，换能器，（交直流）变换器，变压整流机，换流器
transwitch 可控硅开关
trap circuit 陷波电路
trapdoor 调节风门
trap dyke 暗色岩脉
trapeze 梯形，吊架，秋千
trapezium 不规则四边形，梯形
trapezoidal flume 梯形水槽
trapezoidal hydrograph 梯形过程线
trapezoidal measuring flume 梯形量水槽
trapezoidal notch weir 梯形堰
trapezoidal notch 梯形缺口
trapezoidal reinforced concrete channel 梯形钢筋混凝土槽
trapezoidal rib 梯形肋
trapezoidal rule 梯形法则，梯形规则
trapezoidal truss 梯形桁架，梯形屋架
trapezoidal wave 梯形波

trapezoidal weir 梯形堰
trapezoidal windrow pile 梯形煤堆
trapezoidal wing tip 梯形翼尖
trapezoidal 梯形的
trapezoid-shaped slot 梯形槽
trapezoid thread 梯形螺纹
trapezoid truss 梯形桁架
trapezoid wind 梯形线绕丝
trapezoid 梯形
trap on overflow 溢出俘获
trapped air （气包中的）滞留空气
trapped vortex combustion chamber 驻涡燃烧室
trapped vortex 脱体涡
trapped wave 陷波
trapper 陷波器，捕提器
trapping centre 陷阱中心
trapping muck 卸碴
trapping snow 挡雪
trapping system （裂变产物）捕集系统
trapping 捕捉，收集，收集器，蹋落
trap rock 暗色岩，黑色火成岩
trap setting （在程序中）设置“陷阱”
trap size 存水弯的管径
trap valve 除污阀，过滤阀，滤阀
TRAPV(trap on overflow) 溢出俘获
trap-water purification 泄放水净化
trap 疏水器，疏水阀，捕集器，汽水分离器，陷波电路，沉沙槽［井］，存水弯，汽水阀，曲颈管，（捕捉动物的）夹子，陷阱，阱，收集，捕获
trash baffle 拦污墙
trash burning 废物燃烧
trash can 垃圾桶
trash chute 排污斜槽，垃圾道
trash disposal 污物处理
trash dump 垃圾堆
trash gate 拦污闸门
trash rack cleaning 清污机
trash rack loss 拦污栅损失
trash rack rake 拦污栅清理耙，拦污栅清污耙
trash rack wash water pump 滤网（栅栏）清洗水泵
trash rack 拦废物栅栏，拦污栅【泵站】
trash rake car 废料耙车
trash rake 拦污栅【冷却水入口】，清污机
trash-removal device 除污设备
trash remover 清污机
trash screen 拦污栅，杂物筛，防污栅
trash sluice 排污闸
trashway 泄污道
trash 垃圾，废料，废物，碎屑，除去废料，漂浮物，污物，杂物
trass cement 火山灰水泥
trass concrete 火山灰混凝土
trass mortar 火山灰砂浆
trass 浮石凝灰岩，火山土，火山灰
TRAS(test report analysis sheet) 试验报告分析单
trauma 伤害
travel charge 差旅费
travel drive 输运装置【吊装设备】
traveler 流动卡片【车间】
traveling bar 横梁
traveling bridge crane 移动桥式吊车，龙门起重机，桥式起重机
traveling cable 电梯用电缆
traveling carriage shuttle conveyor 梭式皮带机
traveling crab 小车起重机
traveling crane 移动式起重机
traveling-field electromagnetic pump 行场电磁泵
traveling-field 行波场，行移场
traveling gantry crane 行走式龙门起重机
traveling-grate boiler 链条炉排锅炉
traveling in-core probe 可移动的堆内探测器
traveling load 动荷载
traveling motor crab 电动移动起重机
traveling motor hoist 移动式电动起重机
traveling overvoltage 行进波过电压
traveling phase 前进相位，移动相位
traveling resistance 行走阻力
traveling screen 移动式滤网，筛带，带式筛
traveling shuttering 移动式模板
traveling wave klystron 行波速调管
traveling wave line 行波线
traveling wave magnetron 行波磁控管
traveling wave maser 行波量子放大器，行波脉塞，行波微波激射管
traveling wave oscilloscope 行波示波器
traveling wave protection 行波保护
traveling wave tube 行波管
traveling wave 行波
traveller （自动弧焊机）行走机构，移动式起重机，吊车活动起重架，起吊小车
travelling beam 行车梁
travelling belt 运输带，传送带
travelling boom stacker 移动式旋臂堆料机
travelling bridge 高架桥，移动桥
travelling cableway 移动式缆道
travelling cathetometer 活动测高计
travelling cavitation 移动空穴
travelling check 旅行支票
travelling coal unloader 移动式卸煤机
travelling crane beam 桥式起重机主梁
travelling crane 行车，桥式起重机，移动式起重机
travelling depression 移动性低压
travelling disturbance 行进扰动
travelling dune 移动沙丘
travelling equipment 起重机移动的装置
travelling expense 差旅费
travelling feed hopper 移动喂料斗
travelling form 移动式模板
travelling gantry crane 移动式门架起重机，移动式龙门起重机，移动式门机
travelling grate spreader stoker 抛煤机链条炉排
travelling grate stoker boiler 链条炉排锅炉
travelling grate stoker 链带炉排，链条炉排
travelling grate 链条炉篦
travelling highs 移动性高气压
travelling hoist 移动式卷扬机
travelling hopper 移动煤斗
travelling jack 吊运小车

travelling lamp transformer 行灯变压器
travelling load 动荷载
travelling lows 移动性低气压
travelling mechanism 行走机构
travelling mixer 移动式混凝土拌和机，移动式拌和机
travelling motor crab 电动移动起重机
travelling paddle feeder 移动式的叶轮给料机
travelling reclaimer 直行式取料机
travelling rolling hammermill 移动式滚压破碎机
travelling-rotary plow feeder 移动螺旋给料机
travelling rotary plow 移动式回转拨煤机【即叶轮给煤机】
travelling scaffold 移动式脚手架
travelling scan 前后扫查
travelling screen 带筛，活动筛
travelling shutter 移动式模板
travelling sootblower 移动式吹灰器
travelling standard 移动标准器
travelling tipper 移动式翻车机
travelling tower 移动式塔【缆道】
travelling scraper 移动式刮板
travelling tripper 移动卸料小车
travelling wave tube 行波管
travelling wave 行波，行进波
travelling wheel 行走轮
travelling winch 移动式绞车
travelling winged stacker 移动式堆料机
travelling 移动的，走行
travel mechanism 走行机构
travel parameter 汇流参数
travel-reversing switch 终点转换开关
travel speed 行走速度【焊接】，运行速度
travel stop 定位装置，行程终端【限位】
travel switch 行程开关
travel time curve 时程曲线，走时曲线
travel time of flood wave 洪水波传播时间
travel time 传播时间，行进时间
travel truck 行走台车
travel 行程【阀杆】，大车运行【起重机】，行进，移行
traversable gallery 横穿廊道
traverse adjustment 导线平差
traverse closure 导线闭合
traverse gauge 坐标架
traverse gear 横向转动齿轮，横动装置，回转装置
traverse layout 导线布设
traverse measurement 导线量测
traverse net 导线网
traverse pin 横销
traverse point 导线点，交点
traverse rib 横肋
traverser （铁路的）转盘，拉线器，转车台
traverse side 导线边
traverse station 导线站
traverse survey 导线测量
traverse 小车运行【起重机】，横的，横梁，横断物，横向测量，横断的，横向移动
traversing breakdown unit 可移动拆卸单元，可移动解体单元
traversing bridge 横移式活动桥
traversing carriage 移动支架
traversing chute 摆动落煤管
traversing crane 桥式吊车，移动小车
traversing equipment 横行机构
traversing in-core fission chamber probe system 可移动堆内裂变室探测系统
traversing in-core ion chamber 移动式堆内电离室
traversing incore probe 可移动的堆芯内探测器
traversing lever 横杆
traversing method 导线法
traversing motion 往返移动【吊运】
traversing probe 横向移动探针，横移探测管
traversing start-up neutron detector probe 移动式启动中子探测器探头
traversing tube 横向移动测量管
traversing 导线测量
travertine 钙华，石灰华，钙质凝灰岩
trawl board 拖网板
trawl head 拖网架
TRAWL（tape read and write library） 带读写库
trawl 拖网
tray aerator 浅盘式通气器
tray bank 托架组合
tray bracket support 电缆托架的牛腿支架
tray cable 托架电缆
tray height 托架高度
tray structure 托架结构
tray type bed load sampler 盘式推移质采样器
tray type deaerating heater 淋盘式除氧器
tray type deaerator 盘式除氧器
tray type of raceway 电缆托架，托架
tray type sampler 盘式采样器
tray （电缆）支架，溅水盘，浅盘，托盘，垫座，电缆托架，盘，盘架，桥架
TRBL＝trouble 故障
TR cell 胞式保护放电管
TRC（turbine rated load） 汽轮机额定负荷
treadle 踏板
tread return 踏板端翼
tread run 踏步宽度，级宽
tread tractor 大型拖拉机
tread width 踏板宽度，踏步宽度，行踏板宽度
tread 轨顶，楼梯踏板，轮距，踏板，踏步板，踏级宽，行轨距
treasure and valuable goods 贵重货物
treasurership 会计员之职
treasurer 司库，财务主管，出纳财务，财务主任
treasure 财富
treasury bonds 国库券
treasury 金库
treated air 处理过的空气
treated sleeper 防腐处理过的木料，防腐枕木
treated timber 已处理木材，已加工木料
treated wastewater 处理过的废水
treated water 除盐水，处理过的水，净化水，二次水
treating agent 处理剂
treating chemicals 水处理药品
treating plant 处理设备

treating process 处理过程
treating water 处理水
treatise 论文，专题论文
treatment effect 处理效果
treatment efficiency 处理效率
treatment facility 处理设备
treatment loop 处理回路
treatment of cracking rotor 转子裂纹处理
treatment of elevation 立面处理
treatment of fault and fracture zone 断层破碎带处理
treatment of garbage 垃圾处理
treatment of refuse 垃圾处理
treatment of sanitary sewage 生活污水处理
treatment of sewage 污水处理
treatment of water 水处理
treatment plant effluent 处理厂出水
treatment plant 处理厂
treatment procedure 处理方法
treatment process 处理过程
treatment rate 处理速率
treatment zone of reservoir inundation 水库淹没处理范围
treatment 处理，加工，待遇
treaty articles 合同［协定］条款
treaty conditions 合同［协定］条件
treaty content 合同［协定］内容
treaty-making capacity 缔约能力
treaty obligation 条约［合同］义务
treaty of establishment 定居条约
treaty of mutual assistance 互助条约
treaty particulars 合同摘要
treaty port 通商口岸
treaty valid 合同生效
treaty 合同，条约，协定，协议
treble block 三轮滑车
trebler 三倍倍频器，频率三倍器
trebles 50～75毫米的块煤
treble 三倍的，三重的，三层的，高频，使成三倍
tree circuit 树形电路
tree code 树码
tree diagram 树形图
treeing breakdown 树枝状击穿
treeing 树枝化【绝缘材料劣化】
tree line 森林线
treenail 木钉
tree ordering 树形排序
tree propagation 电树枝扩展
tree-shaped heat-supply network 枝状管网
tree structure 树状结构
tree system 链式（树）系统【配电方式】，树枝形配电方式
tree wire 跨树保护线
tree 树形网络，树状物
trefoil arrangement 三叶形布置
trefoil cable 三叶形电缆
trefoil 三个相同芯块组成的辐照件，三叶花形【燃料排列】，三叶形
trellis drainage 格形排水系统，格状水系，葡萄藤状水系
trellised drainage 格状排水系统
trellis girder 格（构）梁
trellis 格构，格架，棚架
trembler 蜂鸣器，颤振部件，振动器，颤振
trembling bell 震颤式电铃
tremendous contribution 巨大贡献
tremendous potentiality 巨大潜力
tremie concrete seal 水下混凝土封底
tremie method 导管法
tremie pipe 导管
tremie placing 导管浇筑
tremie seal 水下混凝土止水层，水下封底
tremie 混凝土导管，漏斗管
tremor 颤动
trench backfill 沟槽回填
trench brace 槽撑
trench cover plate 沟盖板
trench cover slab 管沟盖板
trench cover 沟盖板
trench drain 沟槽式排水
trenched 开槽的
trencher style stacker 挖沟式堆料机
trencher type reclaimer 挖沟式取料机
trencher 挖沟机，凿沟机
trench excavation 挖沟工程
trench fault 沟状断层
trenching and pipelaying machine 挖沟铺管机
trenching machine 挖沟机
trenching reclaimer 挖沟式取料机
trenching sampling 沟取法取样
trenching type S/R 挖沟式斗轮堆取料机
trenching 挖沟，下切
trench line 沟道
trench refilling 沟槽还土，堑壕回填
trench-type storage 贮煤沟
trench wall 槽壁
trench 沟，水渠，探槽，槽，沟道，沟槽，壕沟，管沟，渠道，行地槽，地沟
trend analysis 趋势分析
trend curve 曲线板
trend display 趋势显示
trend in data 数据变化趋势
trend method 趋势法
trend recorder 趋势记录仪
trend recording 趋势记录
trend surface analysis 趋势面分析
trend surface 趋势面
trend variable 趋势变量
trend 倾向，趋势，趋向，动态
trepass 侵犯，侵入
Tresca criteria 特雷斯卡准则【强度理论】
trestle bay 栈桥跨距
trestle bent 栈桥排架
trestle bridge 栈桥
trestle crane 高架起重机，门式起重机
trestle dam 栈桥式木坝，支架坝
trestle flume 栈桥式渡槽
trestle stand 栈架
trestle system 栈桥卸煤装置
trestle unloading system 栈桥卸煤系统
trestle works 栈桥工程

trestle　台架，支架，栈桥，高架桥，排架桥，（脚手架等的）栈架结构
TREX（tube reduced extrusions）　管子减径挤压
TRG　目标【DCS 画面】
triac circuit　三端双向可控硅开关线路
triacetate　三醋酸盐
triac relay driver　三端双向可控硅继电器激励装置
triacs　三端双向可控硅元件
TRIAC（tri-electrode AC switch）　三端双向交流半导体开关
triad　三价物
trial-and-error method　逐次逼近法，试凑法，尝试法，试错法，试算法，尝试误差法
trial-and-error procedure　尝试法，试探法
trial-and-error solution　尝试解法
trial and error theory　尝试误差说
trial-and-error　尝试法，尝试法的，逐步逼近，逐步逼近法
trial assembly　试验性组装，试装配，试装
trial balance　试算表
trial batch　试验性拌和
trial borehole　试钻孔
trial boring　初探钻孔，试钻探
trial condition　试车状态
trial control valve　试验控制阀
trial erection　试装配，试装
trial firing　点火试运行
trial fit-up　（可携带的）校验台
trial function　试探函数，尝试函数
trial hole　试孔
trial implementation　试行
trial line charging　试送电
trial line energization　线路试送电
trial load analysis　试荷载分析
trial load method　试载法
trial load　试验荷载
trial-manufacture　试制，试生产
trial mass　试重
trial method　试算法
trial mix　试配合，试拌和
trial operation of heat-supply network　热网试运行
trial operation　试运行
trial order　试销订购单，试销订单，试订
trial period　试用期
trial pile　试验桩
trial piling　试桩（工程）
trial production　试制，试生产
trial run　试运行，试车，调试
trial solution　试探解
trial survey　试测
trial table　试算表
trial test　探索（性）试验，初步试验，预试验
trial value　试验值
trial wedge method　楔形破裂面试算法
trial　试验，试用，试车，检验
triangle belt　三角皮带
triangle closing error　三角锁闭合差
triangle closure　三角形闭合
triangle connection　三角接线法
triangle generator　三角波发生器
triangle gin　三脚起重机
triangle load distribution　三角形荷载分布
triangle mesh　三角网
triangle noise　三角形分布噪声
triangle pole　三角杆
triangle tie　三角撑
triangle to sine wave converter　三角波正弦波转换器
triangle　三角形，角铁
triangular array　三角形阵列
triangular classification chart　三角坐标分类图
triangular compass　三角规
triangular configuration　三角形布置
triangular crib groin　三角形填石木笼丁坝
triangular crib groyne　三角形填石木笼丁坝
triangular dam profile　三角形坝体断面
triangular distribution toad　三角形分布荷载
triangular element　三角形单元
triangular fin　三角形肋片
triangular frame dike　三角框架堤
triangular frame　三角架
triangular function generator　三角函数发生器
triangular hydrograph　三角形过程线
triangular lattice configuration　三角形栅格布置，三角形栅格构型
triangular lattice　三角形栅格
triangular load distribution　三角形荷载分布
triangular load　三角形荷载
triangular noise　三角波噪声
triangular notch weir　三角形缺口堰
triangular pile　三角形煤堆
triangular pitch　三角形节距
triangular plotting scale　绘图用三棱尺
triangular prism　三棱柱
triangular pulse　三角形脉冲
triangular pyramid　三棱锥
triangular rabbet joint　三角形切口槽接合
triangular segmental conductor　三角截面导体
triangular skylight　三角形天窗
triangular support system　三角形支撑体系
triangular support　三角架
triangular symbol　三角符号
triangular truss　三角形屋架
triangular wave　三角形波
triangular weir　三角形堰，三角堰
triangular windrow　三角形煤堆
triangular　三角的
triangulation class　三角测量等级
triangulation height　三角测量高程
triangulation method of ground resistance test　接地电阻试验的三角测量法
triangulation net adjustment　三角网平差
triangulation network　三角网
triangulation net　三角测量网
triangulation order　三角测量等级
triangulation pillar　三角点标石
triangulation point　三角点
triangulation signal　三角测量觇标【塔】
triangulation station　三角点
triangulation survey station　三角网测站
triangulation survey　三角测量

triangulation 三角测量
triatomic 三原子的
triax connector 三芯接头
triaxial accelerometer 三维加速度仪
triaxial apparatus 三轴试验装置
triaxial bore-hole deformation gauge 三轴钻孔变形计
triaxial cable 三线电缆，三芯电缆，三轴电缆
triaxial cell 三向压力传感器，三轴测压仪，三轴压力盒
triaxial chamber 三轴压力室
triaxial compression apparatus 三轴压缩仪
triaxial compression test 三维（空间）抗压试验，三轴压缩试验
triaxial extension test 三轴延伸试验
triaxial loading 三向加载
triaxial photoelastic measurement 三轴光弹量测法
triaxial quick test 三轴快剪试验
triaxial response spectrum recorder 三维响应谱记录仪
triaxial seismic switch 开维地震开关
triaxial shear equipment 三轴剪力仪，三轴仪
triaxial shear test 三轴剪力试验，三轴剪切试验
triaxial strain 三轴应变
triaxial stress 三轴应力，三向应力
triaxial 三轴的
triax 三轴，三柱块，三层床
triboelectricity 摩擦电
triboelectrification 摩擦起电
tribunal 法庭
tributary area 支流流域面积，从属面积，单桩分担地基面积
tributary basin 支流流域
tributary ditch 支沟
tributary 支流，支流的，从属的，辅助的
tricharged 三电荷的
Trichel pulse 特里切尔脉冲
trichloroethane 三氯乙烷
trichloroethylene 三氯乙烯
trichloromethane 三氯甲烷，氯仿
trichlorophenol 三氯苯酚
trichromatic coordinates 三色坐标，色品坐标
trickle bed 滴流床
trickle charger 小电流充电器，浮充电器，弱电持续充电器
trickle charge 涓流充电，点滴式冲电，浮充电
trickle collector 涓流集热器
trickle cooler 淋式冷却器，滴流冷却器
trickle drain 滴流排水管
trickle flow 细流
trickle hydrodesulfurization 滴流加氢脱硫法
trickle tube 滴水管
trickle valve 滴流阀
trickle 小电流，滴流，细流，一滴滴地流
trickling cooling plant 淋水冷却装置
trickling down 淋下，滴下
trickling filter high rate 高速滴滤器
trickling filter process 滴滤过程
trickling filter 水淋过滤器，滴滤池，生物滤池
trickling filtration 滴滤
trickling water 渗漏滴水
trick 把戏，计谋，骗局，恶作剧，诡计
triclene 三氯乙烯
triclinic prism 三斜柱体
tridimensional analogue computer 三维模拟计算机
tridimensional flow 三维流动
tridimensional map 立体地图，三度地图
tridimensional measurement 立体量测
tridimensional 三度的，三维的，立体的
tri-drum boiler 三气包锅炉
tri-electrode AC switch 三端双向可控硅开关
triethanolamine 三乙醇胺
triethylamine 三乙胺
trifilar wire 三股线
triflux 三工质热交换器，气汽热交换器
trifurcating box 三心线终端套管
trifurcating joint 三心电缆与三根单心电缆的接线盒
trifurcating 分成三枝的，分成三叉的
trigatron 触发管，引燃管
trigdatum 三角测量资料
trigger action 触发作用，脉冲触发，激发作用
trigger amplifier 触发放大器
trigger battery 触发电池组
trigger button 触发按钮
trigger circuit for thyristor 晶闸管触发电路
trigger circuit 触发电路
trigger control 触发控制
trigger countdown 降低触发，信号重复率
trigger decoder 触发译码器，触发解码器
trigger-delay circuit 触发（脉冲）延迟电路
trigger-delay 触发延迟，触发脉冲延迟
trigger device 触发器件
triggered gaps 电花隙避雷器
triggered time clock 触发信号钟
triggered 触发的
trigger electrode 触发电极，引燃电极
trigger flip flop 计数触发器，启动型双稳态触发器
trigger gap 触发间隙
trigger gate 触发电闸，触发选通脉冲，触发门
trigger generator 触发脉冲发生器
triggering device 触发装置
triggering disturbance 触发扰动
triggering point 触发点
triggering signal 触发信号
triggering system 触发系统
triggering 触发
trigger lock 触发机构保险器
trigger mechanism 触发机构
trigger option enabled 触发选择允许
trigger option for data-change 数据变化触发选项
trigger option for data update 数据更新触发选项
trigger option for filtered-data change 过滤后数据变化触发选项
trigger option fro quality-change 品质变化触发选项
trigger option 触发选项
trigger pulse 触发脉冲
trigger register 触发寄存器

trigger relay 触发继电器
trigger sensitivity 触发灵敏度
trigger shaper 触发脉冲形成器
trigger sharpener 触发脉冲锐化电路
trigger signal 触发信号
trigger switch 触发开关
trigger tube 触发管
trigger valve 启动阀，触发管，启动开关
trigger voltage 触发电压，启动电压
trigger winding 触发绕组，触发线圈
trigger 触发器，触发，启动器，启动，启动电路，扳机，启动装置，锁定装置，引发
trigistor 三端开关器件，双稳态 pnpn 半导体组件
trigonometrical calculation 三角测量计算
trigonometric function 三角函数
trigonometric levelling 三角高程测量，三角水准测量
trigonometric transformation 三角变换
trigonometric 三角函数的，三角学的，三角法的
trigram 三字字母
trigraph 三字母组
trijunction 三线交点
trilateration method 三边测量法
trilateration 三边测量网，三边测量
trilite 三硝基甲苯
trillion 兆，万亿，10^{12}【美国用法】，10^{18}【英国用法】，大量，无数
trim and grind the welding 修磨焊点
trim angle 纵倾角，俯仰角
trimer 三聚物
trimmed state 配平状态
trimmed surface 修琢面
trimmer beam 承接梁，托梁
trimmer capacitor 微调电容器
trimmer dozer 推扒机
trimmer potentiometer 微调电位计，微调电位器
trimmer pot 微调电位计
trimmer 调整片，微调铁芯，堆煤机，微调电容器，承接梁，裁员，修整机
trimming capacitor 微调电容器，调谐电容器
trimming change 平舱费
trimming cleaning dozer 清仓机
trimming inductance 微调电感
trimming machine 截锯机
trimming moment 纵倾力矩，配平力矩
trimming motor 微调电动机
trimming potentiometer 微调电位器
trimming regulator 微调调节器
trimming 清铲，修整，（铸件）去毛刺，切边，微调，使平衡，刨边
tri-model tape converter 三模型带转换器
trim saw 截锯机
trim tab 配平片，纵倾调整片
trim valve 微调阀
trim 修整【阀盘、阀座、密封圈、杆等】，微调，整理，（仪表）调整，刨边，阀芯，门头线，配平，去毛刺
trinistor 三极晶管
trinitrotoluene 三硝基甲苯
trinodal seiche 三节静振
triode alternating current switch 三极管交流开关，三端双向可控硅开关
triode generator 三极管振荡器
triode-heptode 三极七极管
triode-hexode 三极六极管
triode-mixer 三极混频管
triode-pentode 三极五极管，三极四极管
triode 三极真空管，三极管
trioxide 三氧化物
trip amplifier （事故）保护停堆放大器
trip and monitoring system 跳闸与监测系统
tripartite agreement 三方协定
tripartition 三分裂
trip breaker 紧急停堆断路器
trip bus 跳闸母线，脱扣母线
trip channel 跳脱控道
trip circuit 跳闸电路，脱扣电路，断开电路
trip coil 跳闸线圈，脱扣线圈，解扣线圈，跳脱线圈
trip device 脱扣装置
trip discharger 翻卸装置
trip dog 止动钩，脱扣钩
trip free mechanism 自由脱扣机构
trip free relay 自动跳闸继电器，自动断路继电器，自由脱扣继电器
trip free release 自动脱扣，自动释放
trip free 自动跳闸的，自动脱扣的，自由断路的
trip gear 扳动装置，脱扣机构
triphase 三相的
triple access 三重存取
triple action press 三向压缩机
triple and quad hopper 三段式和四段式漏斗车
triple blade windmill 三叶片风车
triple bundle 三分裂导线
triple busbar system 三母线系统
triple busbar 三母线，三联汇流条
triple cambered aerofoil 三曲翼面
triple-car dumper 三车翻车机
triple chain conveyor 三链式输送机
triple circuit angle tower 三回路转角塔
triple circuit nuclear power station 三回路核电厂
triple-coincidence matrix 三重符合矩阵
triple-coincidence switch 三重符合开关
triple-coincidence 三重符合
triple-concentric cable 三心同轴电缆
triple conductor 三股导线
triple-control-grid gate tube 三控制栅选通脉冲管
triple-core cable 三心电缆
triple-core wire 三心电线
triple cotton-covered 三层纱包的
triple course pavement structure 三层式路面结构
triple diffused emitter-follower logic 三次扩散射极跟随器逻辑
triple dumper 三联翻车机
triple dumping 三联翻车
triple-error correcting code 三差校正码，三重误差校正码
triple-expansion engine 三胀式蒸汽机
triple-expansion steam pump 三次膨胀蒸汽泵

triple flow turbine　三排汽口汽轮机，三排汽汽轮机
triple frequency harmonic　三次谐波
triple-furnace　三炉膛，三炉体炉膛
triple-hinged arch　三铰拱
triple hopper car　三段式漏斗车
triple integral　三重积分
triple-layer winding　三层绕组
triple modular redundancy　三重模件冗余，三重模块冗余
triple modulation telemetering system　三重调制遥测系统
triple motor　横联三电枢电动机
triplen harmonics　三的整数倍次谐波
triplen　三次谐波序列，三重结构
triple petticoat bell shaped insulator　三裙碗式绝缘子
triple-phase　三相的
triple-pin blade　三联叶片
triple point path　三相点轨迹
triple point temperature　三相点温度
triple point　（水的）三相点
triple-pole double-throw switch　三极双投刀开关
triple-pole double-throw　三刀双掷（开关）
triple-pole switch　三极开关
triple-pole　三极的
triple-reduction speed reducer　三级减速器
triple redundance　三冗余
triple redundancy　三冗余
triple-roll crusher　三滚筒破碎机
triple-shed insulator　三檐绝缘子
triple-squirrel-cage rotor　三鼠笼转子
triple-stroke deep-well pump　三冲程深井泵
triple tariff meter　三部收费电度计
triplet　T形接头，三个一组，三重线
triple-tuned transformer　三绕组调谐变压器
triple valve body　三通阀体
triple valve　三通阀
trip level　脱扣连杆，事故保护停堆功率水平
triple-wound transformer　三绕组变压器
triplex cable　三心电缆
triplex coated particle　具有三种涂层的复合包覆颗粒
triplex layer filter　三层滤料滤池
triplex modular redundant　三重模块冗余
triplex pump　三缸式泵
triplex reciprocating pump　三缸往复式泵
triplex stainless steel chain　（不锈钢）滚柱链，三联链
triplex winding　三重绕组，三分绕组
triplex　三倍的，三重的，三联的，三线的，三部分的，由三个组成的
triple　三倍的，三次的，三层的，三重的，三个的
triplicate　一式三份
triply wood　三夹板
trip margin　脱扣余量，停堆裕度
trip mechanism　释放装置，脱钩装置
tripod derrick　三脚起重机，三脚桅杆
tripod mast　三脚桅杆
tripod planting　等边三角形岸边植柳
tripod plate　三脚架顶面
tripod socket　三脚架底座
tripod　三脚架
trip off　脱扣，断开，跳开
trip oil　脱扣油
tripolar　三极的
tripotassium phosphate　磷酸钾
tripotential　三电位的
trip-out coil　跳闸线圈
tripout rate　跳闸率
tripout　跳闸，自动跳闸，甩负荷，切断，跳开
trip paddle　脱扣闸板
tripped breaker　断开的断路器
tripped　事故保护停堆
tripper car　尾车，翻斗车，卸料车
tripper coaling point　卸料车上煤位置
tripper conveyor　带卸料车的皮带机，自动倾卸输送机
tripper discharge slot in floors　地面的卸料车落料槽口
tripper floor　卸料车层
trip period　事故保护停堆周期
tripper stacker　卸料车式堆料机
tripper valve　翻板阀
tripper　解扣装置，自动切断器，断路装置，分离机构，自动卸货车，倾料器，倾卸装置，倾卸车，自动解扣装置，跳开装置，脱扣器，卸料车，尾车，自动翻车机，自动卸料机
tripping-bar　跳闸杆，脱钩杆
tripping battery　脱扣用电源蓄电池
tripping characteristics　脱扣特性
tripping circuit　跳开电路，跳闸回路
tripping coil　跳闸线圈
tripping current　脱扣电流，释放电流，跳闸电流，分闸电流
tripping device　释放装置，跳脱装置，脱扣装置
tripping fence　跳闸栅栏
tripping frequency　跳闸频率
tripping mechanism　脱扣机构
tripping off　断路，跳开
tripping out　脱扣，断开，断路，跳开
tripping pulse　切断脉冲，触发脉冲，跳闸脉冲
tripping relay　跳闸继电器，释放继电器，跳开继电器，切断继电器
tripping rib　跳闸肋
tripping speed　脱扣转速
tripping strip　脱扣带
tripping voltage　跳闸电压
tripping wire　绊线，绊网
tripple-error-correcting code　三差校正码
tripple-error-correcting　三差校正
trip point　脱扣点
trip power　事故保护停堆功率，停堆阈功率
trip pressure　解列压力，启闭（阀门的）压力
trip relay　切断继电器，跳开继电器，脱扣继电器
trip setpoint　紧急停堆整定点【反应堆】，事故保护停堆整定值
trip setting　事故保护停堆整定值
trip signal　事故保护停堆信号，脱扣信号
trip speed　危急截断器动作转速

trip summary log　跳闸一览记录
trip switch　控制开关，跳闸开关
trip test　脱扣试验
trip threshold　停堆阈值【反应堆】
tripton　非生物悬浮细粒
trip value　（电流或电压）脱扣值
trip valve　脱扣阀，应急阀，速闭阀，捶拍阀，切断阀，跳闸阀
trip　（事故保护）停堆【反应堆】，跳闸，脱扣，断开，切断，抛负荷，自动停止机构，（往返）行程，旅程，解扣，断路，绊倒，脱扣器
TRIR(total recordable injury frequent rate)　总可记录伤害频繁率
trisector air heater　三分仓回转式空气预热器
trisector regenerative air preheater　三分仓空气预热器
trisguare　曲尺
trisodium phosphate　磷酸钠，磷酸三钠【加入汽包】
tri-state logic　三态逻辑
tri-states　三状态
trisulfate　三硫化物
trisulphate　三硫化物
tritiated compound　氚化物
tritiated water　氚化水，氚水
tritiate　制取氚
tritiation　氚化，氚化作用
tritide　氚化物
tritium activity　氚活性，氚（放射性）活度，氚活度
tritium air monitor　氚气监测仪
tritium breeding　氚增殖
tritium carrier　氚载体，含氚载体
tritium concentration　氚浓度，氚浓集
tritium generation reactor　产氚反应堆
tritium impregnated titanium　吸氚钛
tritium labelled　氚标记的
tritium level　氚含量
tritium monitor　氚监测器
tritium oxide　氚水
tritium production reactor　产氚反应堆
tritium ratio　氚比
tritium separation　氚分离
tritium target　氚靶
tritium transverse titanium target　氚钛靶
tritium transverse unit　氚单位
tritium unit　氚单位
tritium zirconium target　氚锆靶
tritium　氚【T】，超重氢【H_3】
triton　氚核
TRIT(turbine rotor inlet temperature)　透平转子进口温度
trituration　研成粉，研制（作用）
triturator　研钵
trivalence　三价
TRKG＝tracking　跟踪
TRM(cargo transfer manifest)　转运舱单
trochoidal curve　余摆线
trochoidal form　余摆线形
trochoidal ripple　余摆线涟波
trochoidal wave form　摆动波形
trochoidal wave　余摆线波，摆动波
trochoid　余摆线，摆动线管，长短幅旋轮线
trochotron　摆线磁控管，余摆磁旋管，电子转换器，电子滚线管
troffer　槽箱式照明设备
trolley batcher　装滚轮的配料器
trolley beam　行车梁
trolley busway　滑触母线
trolley bus　电车
trolley car　电车
trolley chain conveyor　悬挂输送机
trolley-frog　接触电线的线岔，电车吊线分叉
trolley head　触轮，受电头，集电头
trolley hoist　电动小吊车，电葫芦，架空单轨吊车
trolley-jib tower crane　小车变幅塔式起重机
trolley-pivot　受电器杆支枢，集电器杆支枢
trolley pole　触轮杆，接电杆
trolley positioning system　桥式吊车定位系统
trolley system　缆车系统
trolley wheel　触轮
trolley wire　电车架空线
trolley　有轨电车【美】，无轨电车【英】，空中吊运车，滑轮，手推车，（电车）滚轮，起吊小车，桥式吊车，小车，行车，运输车，滑架
Trombe Wall　太阳能吸热壁，特隆布墙
trombone　（长度）可调的U形波导节
trommel screen　旋转式圆筒筛
trommel sieve　滚筒筛
trommel　旋转筒筛
troostite　屈氏体，锰硅锌矿
tropical air mass　热带气团
tropical air　热带空气，热带气团
tropical calm zone　热带无风带
tropical climate condition　热带气候条件
tropical climate　热带气候
tropical climatic zone　热带气候带
tropical continental air mass　热带大陆气团
tropical continental air　热带大陆空气，热带大陆气团
tropical curing　高温养护法
tropical cyclone　热带气旋
tropical depression　热带低压
tropical forest　热带森林
tropical hurricane　热带飓风
tropical insulation　（适用于）热带的绝缘，防绝缘
tropicalization　热带气候化，使设备适应热带气候条件
tropical marine air　热带海洋空气，热带海洋气团
tropical month　分至月
tropical rain forest climate　热带雨林气候
tropical rain forest　热带雨林
tropical rain　热带雨
tropical region　热带地区
tropical revolving storm　热带风暴，热带旋转风暴
tropical storm　热带风暴
tropical switch　热带用开关

tropical tidal current 回归潮流
tropical tide 热带潮汐
tropical zone 热带
tropical 热带的
tropic higher high water interval 回归高高潮间隙
tropic higher low water 回归高低潮
tropic lower high water 回归低高潮
tropic lower low water interval 回归低低潮间隙
tropic lower low water 回归低低潮
Tropic of Cancer 北回归线，夏至线
Tropic of Capricorn 冬至线，南回归线
tropic range 回归潮差
tropic tide 回归潮
tropic 热带，回归线
tropopause funnel 漏斗状对流层顶
tropopause inversion 对流层顶逆温
tropopause 对流层顶，对流顶层
troposphere scatter communication 对流层散射通信
troposphere stratosphere exchange 对流层平流层交换
troposphere 对流圈，对流层
tropospheric aerosol 对流层气溶胶
tropospheric dust suspension 对流层悬浮尘
tropospheric fallout 对流层沉淀物，对流层落尘
tropospheric 对流层效应，对流层的
trouble back jack 故障信号插孔，故障返回信号插孔
trouble back signal 故障返回信号
trouble back tone 故障音
trouble blinking 事故指示信号
trouble block 故障部分
trouble check list 故障检查表
trouble diagnosis 故障诊断
trouble duration 故障持续时间
trouble finder 故障寻迹器，故障探测器
trouble free operation 无故障运行，安全运行
trouble free service 安全服务，安全工作，安全维修
trouble free 无故障，无故障的，不发生事故的
trouble lamp 故障指示灯
trouble light 故障信号灯
trouble locating 故障定位，排除故障，故障检寻
trouble-location problem 故障定位问题，故障探测问题
trouble-location 故障点测定，故障探测
trouble-proof 防故障的，安全的，防故障
trouble recorder oscillograph 故障记录示波器
trouble recorder 故障记录器，故障记录仪
trouble removal 事故处理
trouble-saving 预防故障的
trouble sensing element 故障检测元件
troubleshooting data 修理指南，检修指南
troubleshooting diagram 检查及排除故障图
troubleshooting method 故障诊断［排除］方法
troubleshooting oscilloscope 测定故障用示波器
troubleshooting 排除故障，故障检修，消除缺陷，事故分析，查明故障，测定故障
trouble signal 故障信号
trouble-spot 故障点
trouble symptom 故障征兆
trouble zone 损坏区域，故障区
trouble 故障，事故，毛病
trough ability 成槽性
trough angle 槽角
trough-belt conveyer 槽式皮带输送机
trough concentrating unit 槽式聚光单元
trough duct 电缆沟槽
trough factor 槽形系数
trough girder 槽形大梁
trough grate 下饲式炉排
trough idler 槽式托辊
troughing angle 槽角
troughing idler 槽形布置托滚【皮带机用】
troughing side 槽形侧【输送带】
troughing 明电缆槽，波谷，电缆走线架
trough line 槽线
trough of low pressure 低压槽
trough plate 槽形板，槽型板
trough roller 槽型托辊，槽滚
trough section 槽形截面
trough solar collector 槽式聚光器，槽式集热器
trough tip wagon 槽形翻斗车，倾斜车，倾卸车
trough tray 槽形托架
trough urinal 小便池
trough valley 槽形河谷
trough 沟，槽，电缆架，喇叭口，低压槽，水槽，坳槽，电缆架
trowel(l)d stucco 抹子拉毛
trowel finish 抹光面，抹平
trowel(l)d finish 镘抹光面，抹面
trowel(l)d surface 抹光面
trowel(l)d course 找平层
trowel(l)ing machine 镘平机，抹平机，抹灰机
trowel 镘，镘刀，抹子，镘平，抹光，抹平
TRSHRK(trash rack) 拦污栅，滤网
TR(technical report) 技术报告
TR(test report) 试验报告
TR(thermal ratio) 热比
TRTMT＝treatment 处理
truck agent fee 汽代费
truck capacity 卡车载重量
truck chock 转向架止挡块
truck crane 车载起重机，汽车吊，汽车式起重机
truck dump receiving hopper 卡车卸受煤斗
truck entrance 卡车入口
truck grader 汽车平地机
truck hopper 卡车卸煤受煤斗
trucking ball 球标
trucking company 车队，汽车运输公司
truckle 小轮，滑轮
truck lift 车式提升机，自动装卸机
truck load bin 装车料仓
truck loading fee 小件装车费
truck load 卡车载重量
truck lock assembly 转向架止挡器
truck lock 转向架止挡定位器
truck-mixed concrete 拌和车拌制的混凝土
truck mixer 车载伴和机，混凝土搅拌车，汽车式拌和机
truck-mounted concrete pump 汽车载运的混凝土

泵，混凝土泵车
truck-mounted crane 安装在卡车上的起重机，汽车式起重机，汽车吊
truck-mounted dragline 车载索铲
truck-mounted drilling rig 车载钻机
truck-mounted drill 车装钻机，钻车
truck-mounted jumbo 汽车台钻
truck-mounted power shovel 车载动力铲
truck receiving hopper 卡车受煤斗
truck scale 汽车衡
truck shipment 汽车运输
truck-shovel 铲运机
truck side frame 转向架侧架
truck-station （废物库）运出站
truck tractor 汽车拖拉机
truck trailer 载重拖车
truck transport 卡车运输
truck-type switchgear 台车式开关
truck unit scale 汽车衡
truck unloader 汽车卸车机
truck unloading hopper 汽车卸料斗
truck weigher 汽车衡
truck 卡车，运货车，矿车，台车，转向架，货车，载重汽车
true angle of attack 真实迎角
true angle of internal friction 真内摩擦角
true arch 纯拱
true area of contact 实际接触面积
true average value 实平均值
true azimuth 真方位角
true bearing 真象限角，真方位，真实方位
true carry 真实进位
true cleavage 真劈理
true cohesion 真凝聚力，真黏着力
true coincidence correction 真符合校正
true coincidence 理想［准确，真正］的叠合［重合］，真符合
true complement 补码，实补码，补数
true density 实际密度，精确密度，真密度
true deviation 真实偏差
true dip 真倾斜
true erosion 实际侵蚀
true exhaust velocity 实际排汽速度
true fault 真误差，真实故障
true form 真实状态，原码形式
true groundwater velocity 地下水实际流速
true heat capacity 真实热容量
trueing device 整平装置，修正装置
true integral 理想积分
true interest 纯利息
true length 实长
trueness 真实性，正确，准确度
true ohm 实际欧姆，电磁制欧姆
true output 实际出力
true power consumption 有效功率消耗
true power output 有效功率
true power 实际功率，有效电功率
true probability density function 实概率密度函数
true rate of interest 实际利率
true reactor 实际反应堆
true relative density 真相对密度
true resistance 实电阻，直流电阻
true running test 试车
true specific gravity 真比重
true specific heat 真实比热
true stress-strain curve 实际应力变形曲线
true stress 实际应力，真实应力
true threshold 实际阈值
true time operation 实时操作，实时运算
true time 实时
true to line 线条平直的
true-to-scale print 足尺复印
true to size 尺寸准确的
true triaxial apparatus 真三轴试验仪，真三轴仪
true triaxial test 真三轴试验
true up 校准，调准，整形
true value of a quality 量程的实际值，量的真值
true value 真值【量值】，实际值
true velocity 实际速度，真流速
truing 矫直
T-rule 丁字尺
truly linear 完全线性的
trumpet spillway 喇叭形溢洪道
truncated bowstring truss 平端弓形桁架
truncated chimney 截顶烟囱
truncated cone shaped combustion chamber 斜截头圆锥体状燃烧室
truncated cone-shaped 截锥形的
truncated cone 截锥
truncate distribution 截尾分布
truncated normal distribution 截断常态分布
truncated part 截断部分
truncated spectral model 截谱模式
truncated spectrum truncation error 截断误差
truncated value 截断值，舍位值
truncated Venturi tube 截尾文丘里管
truncated 被截的
truncation error 舍项误差，舍位误差，截断误差，公式误差，方法误差
truncation 切去，截去，舍位，舍项，切断，削蚀，截短，切掉顶端
trundle 滚动，移动
trunk cable 干线电缆，中继电缆，长途电缆
trunk call 长途电话，长途通信
trunk circuit 干线，中继电路
trunk concentrator 中继集线器
trunk connection 长途接线，长途通信
trunk cut-off 干线切断
trunk distribution frame 中继线配线架
trunk duct 主干管，总管
trunk exchange 长途电话局
trunk group 中继线群
trunk hunting connector 中继线寻线器
trunking loss 管道损失
trunking matrix 二极管开关输出矩阵
trunking 管道，中继，线槽
trunk line circuit 干线回路，长途线路
trunk line waterway 水道干线
trunk line 干线，中继线
trunk main 大直径管，主干线
trunk multiple 干线复接，出线复接
trunk pipe 总管，干管

trunk piston 筒状活塞
trunk railway 铁路干线
trunk relay 中继线继电器
trunk road 干道，主干道
trunk sanitary sewer 污水管干线
trunk sewer 总阴沟
trunk 本体，总管，主要部分，干线，信息通路，树干，中继线
trunnion air seal assembly 端轴空气密封
trunnion mounted ball valve 装在轴上的球阀
trunnion mounted 耳座支承
trunnion type suspension clamp 中心回转式悬垂线夹
trunnion 耳轴，凸耳，轴耳，轴颈，枢轴，耳座，车轴，支座枢轴
truss action 桁架作用
truss arch 桁架拱
truss bridge 桁架桥
truss-buttress dam 桁架式支墩坝
truss camber 桁架起拱
trussed beam 桁架式梁，桁架梁
trussed-box-girder bridge 箱形桁架桥
trussed bracing 桁架式支撑
trussed member 构架杆件
trussed partition 桁架式隔墙
trussed roof 桁架屋顶
truss framed with angle sections and bolted 螺栓联结角钢组成的桁架
truss frame 桁架
truss girder 桁架梁
truss rod 桁架拉杆
truss spacing 桁架间距
truss span 桁架跨度
truss veticle 桁架竖杆
truss wing 桁梁臂杆
truss with parallel chords 平行弦桁架
truss 桁架，构架，屋架
trust bank 信托银行
trust company 信托公司
trust corporation 信托公司
trusted program 受托程序
trustee 受托管理人，受托人
trust and investment business 信托投资业务
trust investment 信托投资
trust 信任
truth strain 真实应变
truth stress 真实应力
truth table 真值表
TRU waste 超铀废物
TRVL=travel/travelling 运行，旋转
TRVSCN(travelling screens) 旋转滤网
TRX(two region critical experiment) 双区临界实验装置
try-and-error method 逐次逼近法，试探法，尝试法，试错法
try cock 试验旋塞，试水位旋塞
try hard 努力
trying paper 描图纸
try out 试行
try square 方形规，曲尺，验方尺
TSA(technical safety appraisal) 技术安全评价
TSA(technical service agreement) 技术服务协议
TSC(thyristor switched capacitor) 晶闸管投切电容器
TSD(temporary special device) 临时专用设施
T-section roof girder T形薄腹梁
T-section T形截面，丁字钢
T-shaped conductor T形导体
T-shaped current-displacement bar T形电流容量棒
TSI(temporary surveillance instruction) 临时监督指令
TSI(turbine supervisory instrumentation) 汽轮机监测仪表，汽轮机安全监视系统
TSI 汽机本体监视系统【DCS画面】
T-slot T形槽
TSO(transmission system operator) 输电系统运营商
TS/PQR(technical specification for welding procedure qualification report) 焊接工艺评定报告技术规程
TSP(total suspended particulates) 总悬浮颗粒物
TS/PWTCR(technical specification for production weld test coupon report) 产品焊缝试件报告技术规程
T-square 丁字尺
TSR(thyristor switched reactor) 晶闸管投切电抗器
TSS(total suspended solids) 总悬浮物【水分析】
T-stamping T形冲片
TS(technical specification) 技术规范，技术规格
TS(technical support) 技术支持
TS(test supervisor) 试验负责人
tsunami barrier 地震海啸防波堤
tsunami protection breakwater 地震海啸防波堤
tsunami 地震津波，海啸，津浪
TS/WQR(technical specification for welder qualification report) 焊工评定报告技术规程
TTD(terminal temperature difference) 温度端差
t=temperature 温度
T/T(telegraphic transfer) fee 电汇手续费
three-point starting box 三端启动箱
t=ton 吨
TT(temperature transmitter) 温度变送器
turbulent fluctuation 湍流脉动，紊流脉动，湍流起伏，紊动
two-position threeway solenoid valve 二位三通电磁阀
T-type antenna T形天线
T-type pipe 三通管
T-type strainer T形粗滤器，T形过滤器
tube adapter 电子管适配器，管接头
tube anchor bar 管式锚杆
tube and plate collector 管板式集热器
tube arrangement 管子排列
tube array 多排管子，管子网，管排
tube axial fan 轴流式风扇
tube ball mill 筒形球磨机
tube bank 管束，管组，排管
tube bender 弯管机
tube bend region 管子弯曲区
tube beveling 管子坡口

tube brush　管刷，瓶刷
tube building　简体建筑
tube bunch　管束
tube bundle access plate　管束隔板【蒸汽发生器】
tube bundle shroud　管束罩壳，管束套筒
tube bundle wrapper　管束套
tube bundle　管束，管排
tube burst due to creep　蠕胀爆管
tube burst　爆管
tube caisson foundation　管柱基础
tube cap　管帽
tube caulker　管子扳边工具
tube cavitation　管的气蚀
tube cleaner　管子清洁器，管子清洗器，洗管器，管内清垢器
tube clip　管卡管束，管卡管排
tube closure　管盖，管堵头
tube coil cylinder　盘管圆筒
tube coil type heat exchanger　盘管式热交换器
tube coil　盘管，管圈，蛇形管
tube conductor　空心导线，空心导体
tube connection　管接头
tube connector　管接头，(导线)连接管
tube construction　管子结构
tube cooled motor　管道通风式电动机
tube coupling　管子联结
tube cracking furnace　管式裂解炉
tube cutter　切割机
tube delivery gate　管式斗门，管式输水斗门
tube denting　管子凹陷
tube diode　真空二极管
tube drainage system　管道排水系统
tube drainage　管道排水
tube economizer　管式省煤器
tube electrode　管形电极
tube element　管件
tube end　管端
tube erosion　管子侵蚀
tube exchanger　管式热交换器
tube expander　扩管机，胀管机，胀管器
tube expansion roller　胀管芯辊
tube expansion　胀管
tube explosion analysis　爆管分析
tube explosion　爆管
tube failure　管子损坏，管子漏泄
tube fin exchanger　管肋式换热器
tube fin pitch　管子鳍片节距
tube fittings and accessories　管子配件和附件
tube float　管式浮标
tube flow　管流
tube fluorescent lighting fixture　简式荧光灯
tube friction　管道摩擦阻力
tube furnace　管式炉
tube fuse　管形保险器，管形熔断器
tube galvanometer　电子管电流计，电子管检流计
tube generator　电子管振荡器
tube gill pitch　管子鳍片节距
tube hanger　管子吊架
tube heat exchanger　管式换热器
tube hole groove　管孔槽
tube holes coefficient　管孔带减弱系数
tube hollow　空心管坯
tube-inlet orifice　管子进口节流圈
tube-in-sheet heater　管板式加热器
tube in-shell heat exchanger　管壳式热交换器
tube-in-sleeve alidade　转镜照准仪
tube installation　装管，管道安装
tube-in-tube building　简中筒建筑，套管式建筑
tube-in-tube chimney　套筒式烟囱
tube-in-tube condenser　套管式凝汽器
tube-in-tube heat exchanger　套管式热交换器
tube-in-tube structure system　筒中筒结构体系
tube-in-tube structure　筒中筒结构
tube lane　U形管束中间水道【蒸汽发生器】，管间通道
tube leak　管子泄漏
tube manufacturer　装管工
tube mill　钢球磨煤机，球磨机
tube nest　管束
tube of electric induction　电感应管
tube of force　力管
tube of magnetic force　磁感应管，磁力管
tube optical cable　管道光缆
tube outer surface　管子外表面
tube panel　管屏
tube pile　管桩
tube pitch　管节距，管距
tube platen　管屏
tube-plate packing　管子与管板之间的密封
tube plate stay　管板支撑
tube plate　管板
tube precipitator　管式除尘器
tube profilometry　管子表面光洁度仪
tube protrusion　管子凸出
tube radiator　管状辐射器，管式散热器
tube rattler　洗管器，清管器
tubercular corrosion　点腐蚀，结瘤腐蚀
tuberculation　(金属)结瘤，管瘤
tube reactor　管式反应器
tube reduced extrusions　管子减径挤压
tube reduction　管子减轻
tube removal　管子清除
tube resistance　管阻
tube rework to specification　管子按技术要求返工
tube rolling roller　管子滚压辊
tube row　管排
tube rupture　爆管
tube sample boring　薄管取样钻探
tube sampler　管式取样器
tube scale　管垢
tube scraper　刮板式清管器
tube screen　管状筛
tube search　管子通球试验
tube seat　管座
tube setter　管式沉淀器
tube sheet core　管板芯
tube sheet　管板
tube shield　管棚，管罩，电子管屏蔽，射线管套
tube side pressure loss　管侧阻力
tube side　管侧【设计压力】

tube-sinking cast-in-situ pile　沉管灌注桩
tubes in line　顺排管束
tube skin temperature　管子外表污垢层温度
tube sleeving　管子装套管【蒸汽发生器】
tubes of field intensity　场强管
tube spacing　管距，管排配置，管距布置，管子间隔
tube span　管子间距
tube specimen　管子样品
tube-spring manometer　管状弹簧式压力计
tube-spring　管状弹簧
tube support plate　U形管支撑板【蒸汽发生器】，肋板，中间管板
tube swelling　管子过热胀粗，管子变形
tube tester　电子管试验器
tube-to-tube construction　密排管结构
tube-to-tube wall　密排水冷壁
tube transformer　圆柱形变压器【具有同轴绕组】
tube type cooler　管式冷却器
tube type cyclone　管式旋风分离器
tube type lightning arrester　管形避雷器
tube type motor　圆筒形机座电动机
tube valve　管式阀
tube vibration　管子振动
tube vibrator　振管器
tube wall dash pot　管壁缓冲筒
tube wall　管壁
tube wastage　管壁耗蚀，管子磨耗，管壁变薄现象
tube wear　管子磨损
tube well output　单井出水量
tube well　管井，自流井
tube　风洞管，电子管，隧道，包壳，管子，地下道
tubing installation　管道安装
tubing loop　管圈
tubing manufacturer　装管工
tubing stress analysis　管系应力分析
tubing supervisor　管道检查员
tubing tray　穿管托架
tubing　管道系统，管道，管系，管路，安装管道，配管，装管
tubo-annular combustion chamber　环管燃烧室
tubular airheater　管式空气加热器
tubular air preheater　管式空气预热器
tubular bal mill　筒式磨煤机
tubular bridge　管桁，管桁桥
tubular busbar　管形母线
tubular cladding　管形包壳
tubular cluster　管束
tubular coil　管形线圈，空心线圈
tubular column drilling method　管柱钻孔法
tubular combustion chamber　管形燃烧室
tubular concrete section　混凝土圆筒节
tubular condenser　管形电容器，管式凝汽器
tubular conductor　空心导体，空心导线
tubular construction　管状建筑，筒形结构
tubular cooler　管状散热器，管式冷却器，管状冷却器
tubular cork borer　管式（木塞）穿孔器
tubular cutting shop　铜工车间
tubular cyclone　管状旋风分离器
tubular drier　管式干燥器
tubular electrostatic precipitator　管式静电除尘器
Tubular Exchanger Manufacturers Association　美国管式换热器制造商协会
tubular exchanger　管式换热器
tubular field　管状场
tubular filter element　管状滤元件
tubular flow　管状流，管内流
tubular foundation　管状基础
tubular frame　管式框架
tubular fuel element　管状燃料元件
tubular guide　管形导轨，管状导向装置
tubular heater　管式加热器，列管加热器
tubular heat exchanger　管式热交换器，管式换热器
tubular interacting fuel pin　（与内外石墨）相互贴紧的管状包覆颗粒燃料细棒
tubular lacing wire　管状拉筋
tubular level　水准管
tubular linear motor　圆筒形直线电动机
tubular magnet　管形磁铁
tubular membrane　管状
tubular module　管式组件
tubular motor　管状电动机
tubular non-interaction fuel pin　（与内外石墨）不相互紧贴的管状包覆颗粒燃料细棒
tubular pile foundation　管桩基础
tubular pile　管桩
tubular pole arm　管形横担
tubular pole　管形电杆
tubular preheater　管式空气预热器
tubular pressure vessel　管式压力容器
tubular radiator　管式散热器，管状光管散热器
tubular reluctance motor　管形磁阻电动机
tubular resistance　管形电阻
tubular rivet　空心铆钉
tubular scaffolding　管子脚手架
tubular section　管状截面
tubular segment　筒节
tubular sheath　管状护套
tubular slug　管状块
tubular socket wrench　管状套筒扳手
tubular source　管状源
tubular spanner wrench　管状扳钳
tubular strander　管式绞线机
tubular strand　空心股线
tubular structure loader-unloader　管结构装卸桥
tubular structure　钢管结构，管状结构
tubular thermocouple　管式热电偶
tubular tower　圆筒式塔架
tubular turbine generator　贯流式水轮发电机
tubular turbine　贯流式水轮机
tubular type collector　管式除尘器
tubular type tank　（变压器的）管式油箱
tubular well　管井
tubular　管状的，管形的，管式的，有小管的
tubule auto-welding machine　小管自动焊机
tub　盆，槽，桶，浴缸盆
tuck-in　卷入

Tudor plate 都德电池极板，都德阳极板
tufa 石灰华
tuffaceous phyllite 凝灰千枚岩
tuffaceous schist 凝灰片岩
tuff agglomerate-lava 凝灰集块熔岩
tuff-breccia 凝灰角砾岩
tuff cement 凝灰岩水泥
tuffcrete 凝灰岩水泥混凝土
tuff sandstone 凝灰质砂岩
tuff volcano 凝灰火山
tuff 凝灰岩
tufted flow study 丝线法气流研究
tuft method 丝线法
tug boat 拖轮
tugger hoist 拖拉式卷扬机
tugger 拖拉式卷扬机
tug 拉，拖
tumble bay 回旋池，消能池
tumble gate 倒翻闸门
tumble home 舷缘内倾
tumbler switch 快速转换开关，快速翻转开关，翻转开关
tumbler 翻转开关，倒扳开关，转换开关，转臂，转筒，换向齿轮，齿轮换向器，铸件磨光滚筒
tumble 滚动，跃落
tumbling drum 磨光滚筒
tumbling flow 倾泻水流
tumbling mixer 转筒混合器
tumbling motion 滚动
tumbling regime 跌水情况
tumbuckle 花篮螺栓，螺丝扣，套筒螺母
tumor dose 肿瘤剂量
tumult 骚动
tumulus 钟状火山
tunable 可调谐的
tunance 并联谐振，电流谐振
tundra climate 冻原气候
tundra 冻土地带，冻原，苔原
tuned-anode 阳极回路调谐的，已调阳极回路
tuned autotransformer 调谐自耦变压器
tuned blade 调频叶片
tuned circuit 调谐电路
tuned damper 可调阻尼器，调谐阻尼器
tuned detector 调谐检波器
tuned-grid 调栅的
tuned high frequency 调谐的高频
tuned mass damper 可调质量阻尼器，调谐质量阻尼器
tuned radio-frequency amplifier 射频调谐放大器
tuned-reactance grounding system 调谐电抗接地系统
tuned-reed frequency meter 调谐振簧式频率计
tuned-reed indicator 振动指示器
tuned-reed rectifier 调谐振簧式整流器
tuned-reed relay 调谐舌簧继电器
tuned-secondary transformer 次级线圈调谐变压器
tuned transformer 调谐变压器
tune in 调入，调谐
tune out 解谐，失谐，解调，失调
tuner 调谐设备，调谐器
tune-up 调整，调准
tune 调谐，调整
tungar rectifier 钨氩管整流器
tungar 吞加整流管，（二极）钨氩整流管
tung oil 桐油
tungsten arc lamp 钨丝弧光灯
tungsten arc weld 钨极电弧焊
tungsten arc （高强度的）钨电极弧
tungsten carbide 碳化钨
tungsten contact 钨触点
tungsten halogen lamp 钨卤素电灯
tungsten inclusion （焊缝）夹钨
tungsten inert gas arc welding machine TIG 焊机，钨极惰性气体保护弧焊机
tungsten inert gas arc welding 钨极惰性气体保护电弧焊
tungsten inert gas cutting 钨极惰性气体保护切割
tungsten inert gas welding 钨极惰性气体保护焊
tungsten inert gas 钨极惰性气体
tungsten lamp 钨丝灯
tungsten-rhenium thermocouple 钨铼热电偶
tuning capacitor 调谐电容器
tuning circuit 调谐电路
tuning coil 调谐线圈
tuning condenser 调谐电容器
tuning dial 调谐用标度盘
tuning fork oscillator 音叉振荡器
tuning indicator 调谐指示器
tuning inductance 调谐电感
tuning mass 调整质量
tuning motor 调谐电动机
tuning range 调谐范围
tuning scale 调谐标度盘
tuning unit 调谐装置，调谐器
tuning wave 调谐波
tuning 整定
tunnel alignment 隧洞定线
tunnel balance test 风洞天平试验
tunnel blasting 隧洞爆破
tunnel blockage correction 风洞阻塞修正
tunnel blockage 风洞阻塞
tunnel boring machine method 掘进机法
tunnel boring machine 隧道掘进机，隧洞掘进机
tunnel boundary 风洞边界，风洞洞壁
tunnel building clearance 隧道建筑限界
tunnel ceiling 风洞顶壁
tunnel constraint 风洞壁约束
tunnel crack 条状裂纹
tunnel diode transistor logic 隧道二极管晶体管逻辑
tunnel diode 隧道二极管
tunnel diversion 隧洞导流
tunnel drainage 隧洞排水
tunnel drill 隧洞凿岩机
tunnel driving 隧道掘进，隧洞掘进
tunnel effect element 隧道效应元件
tunnel effect 隧道效应
tunnel entrance 隧道口
tunnel erosion 地下洞蚀
tunnel excavation 隧道开挖，隧洞开挖

tunnel excavator 隧洞挖掘机
tunnel face 隧道面
tunnel form 隧洞模板
tunnel heading 隧道工作面，隧洞导洞，隧洞导坑
tunnel induced velocity 风洞诱导速度
tunneling record 隧洞掘进记录
tunneling 隧洞掘进，隧洞开挖
tunnel inlet 风洞进气口
tunnel intake 风洞进气段，引水隧道
tunnel interference correction 洞壁干扰修正
tunnel intersection 隧洞交叉
tunnel invert 隧洞底拱
tunnelled winding 嵌入绕组，下线后的绕组
tunnelless underground laying 无地沟敷设
tunnelling equipment 隧洞掘进设备
tunnelling machine 隧道开挖机，隧洞掘进机
tunnelling method 隧洞掘进方法
tunnelling operation 隧道作业
tunnelling shield 隧洞防护支架
tunnelling 隧道掘进，隧洞掘进，隧洞施工
tunnel lining 隧道衬砌，隧洞衬砌
tunnel measurement 风洞测量
tunnel muck 隧洞弃渣
tunnel plug 隧洞堵塞段
tunnel portal 随道口，隧洞入口
tunnel reclaim conveyor 地槽内取煤皮带机
tunnel reference pressure 风洞参考压力
tunnel return flow 风洞回流
tunnel ring 隧道圈，隧洞衬砌环
tunnel roof 隧洞顶板
tunnel shaped crack 隧道形裂纹
tunnel shape factor 风洞（试验段）形状因子
tunnel shield 隧盾
tunnel-site selection 隧洞位置选定
tunnel soffit 隧洞顶面，隧洞拱顶
tunnel speed control 风洞风速控制
tunnel speed hole 风洞洞壁测速孔
tunnel spillway 隧道排水槽，泄洪隧洞
tunnel spoil 隧洞弃渣
tunnel support 隧洞支撑，隧洞支护
tunnel survey 隧道测量
tunnel technique 风洞试验技术
tunnel test 风洞试验
tunnel transfer trolley 燃料转送车
tunnel transition section 隧洞渐变段
tunneltron 隧道管
tunnel turbulence 风洞湍流（度），风洞紊流（度）
tunnel vault 隧洞顶拱
tunnel ventilation 隧道通风
tunnel wall interference 洞壁干扰
tunnel-wound armature 闭口槽绕组电枢
tunnel 地槽，地道，涵洞，隧道，隧洞
tunoscope 调谐指示器，电眼
tup 撞锤
turbator 带环形谐振腔的磁控管
TURBFLW(turbine flow) 汽轮机流量
turbid flow 浑流，混浊水流
turbidimeter fineness 浊度计细度
turbidimeter 浊度测定计，比浊计，浑浊度仪，浊度表，浊度计
turbidimetric titration 浊度滴定法
turbidimetry 浑浊度测定，比浊法
turbidite 浊流层
turbidity coefficient 混浊系数
turbidity current 浑浊流，混浊流，浊流
turbidity degree 浑浊度
turbidity factor 浑浊因子
turbidity meter 浊度测定计，比浊计，浊度表，浊度计
turbidity of atmosphere 大气的浑浊度
turbidity point 浊度点
turbidity-producing material 制备浊度材料
turbidity removal 浑浊度去除
turbidity titration 浊度滴定
turbidity value 浊度值
turbidity 浑浊度，浑浊性
turbid liquid 浑浊液体
turbid matter 浑浊物质
turbidness 浑浊
turbidometer 浊度计
turbidometry 浊度分析法
turbid water 浑水，浊水
turbid 混浊的，混乱的，浑浊的
turbine accessories 透平辅助设备
turbine AC lube-oil pump 汽机交流主油泵
turbine aerator 水轮机补气装置，涡轮充气器，涡轮搅拌器
turbine air vent pipe 水轮机通气管
turbine and feed heating plant preservation during outage 汽轮机、给水加热器停机保护
turbine area 汽轮机区
turbine automatic control 汽轮机自动控制，涡轮机自动控制
turbine axial clearance adjustment 汽轮机轴间隙调整
turbine back pressure 汽轮机背压
turbine baseplate in position 汽机底板就位
turbine baseplate 透平底盘，涡轮机底板
turbine bearing 汽轮机轴承
turbine biological shield 汽轮机生物屏蔽
turbine blade renewal 汽轮机换叶片
turbine blade velocity diagram 透平叶片速度图
turbine blade 透平叶片，汽轮机叶片，水轮机叶片
turbine bleed point 汽轮机抽汽口
turbine block 汽轮机基座，汽轮机平台
turbine blower 涡轮式鼓风机
turbine-boiler control room 机炉控制室
turbine-boiler coordinated control system 机炉协调控制系统
turbine-boiler paralell control 锅炉汽轮机平行控制，锅炉汽轮机并联控制
turbine brake 涡轮制动器，涡轮测功器
turbine bucket entrance 透平动叶入口
turbine bucket 汽轮机叶片，透平动叶片
turbine building ventilation system 汽轮机厂房通风系统
turbine building 汽机房，汽轮机厂房，汽轮机房
turbine bypass control system 汽轮机旁路控制系统

turbine bypass pipe 汽轮机旁通管，汽轮机旁路管
turbine bypass system 汽机旁路系统，汽轮机旁路系统
turbine bypass valve 汽轮机旁通阀，汽轮机旁路阀
turbine bypass 汽轮机旁通管，汽轮机旁路
turbine case 水轮机机壳，水轮机罩
turbine casing 透平汽缸，涡轮机匣，汽轮机缸
turbine center elevation 水轮机中心高程，转轮中心高程
turbine centrifugal pump 涡轮式离心泵
turbine chamber 水轮机转轮室，水轮机室，水轮机机坑
turbine characteristic curve 透平特性曲线，涡轮机特性曲线
turbine characteristics 水轮机特性，汽机特性，涡轮机特性，透平特性
turbine circuit 透平循环，涡轮机循环
turbine compressor 透平压缩机，透平压气机，涡轮式压气机，涡轮压缩机
turbine condensate polishing 汽轮机凝结水净水
turbine condenser 汽轮机凝汽器
turbine constant 水轮机常数
turbine control fluid system 汽机调节液系统
turbine control panel 水轮机控制盘，透平控制盘
turbine control room 汽机控制室
turbine control system 汽轮机控制系统
turbine control transmission ratio 水轮机控制传递比
turbine control 汽轮机控制，水轮机控制
turbine cover 水轮机盖，水轮机顶盖及底环
turbine crown 透平叶轮轮缘
turbine cycle 透平循环，涡轮机循环
turbine cylinder 汽轮机缸，汽缸
turbine DC lube-oil pump 直流事故油泵
turbine deck 汽机平台
turbine demand 汽轮机给定功率
turbine department 汽机分场
turbine direct drive 透平直径驱动
turbine discharge 水轮机过流量，水轮机流量
turbine discipline 汽机专业
turbine disc 透平轮盘，透平叶轮
turbine disk 透平叶轮，透平轮盘
turbine draft tube 水轮机尾水管
turbine drain 汽轮机疏水
turbine draught tube 水轮机尾水管
turbine drive boiler feed pump 汽动给水泵
turbine-driven auxiliary feedwater pump 汽动事故给水泵，汽动辅助给水泵
turbine-driven boiler feed pump 透平驱动锅炉给水泵，汽动锅炉给水泵
turbine-driven compressor 透平驱动压缩机
turbine-driven feedwaer booster pump 汽动给水升压泵
turbine-driven feedwater pump system 汽动给水泵系统
turbine-driven feedwater pump 汽动主给水泵，汽动给水泵
turbine-driven pump 汽动泵，汽轮机驱动泵
turbine-driven set 汽轮发电机组，涡轮驱动装置
turbine-driven 透平驱动的，汽轮机驱动的
turbine drive 透平驱动
turbine dynamometer 涡轮功率计，涡轮测功计
turbine efficiency 水轮机效率，透平效率
turbine enclosure 汽机罩
turbine end point 汽轮机终态焓
turbine end 汽轮机端，透平端，水轮机端
turbine exhaust spray system 汽机排气口喷淋系统
turbine-exit temperature 排汽温度
turbine expansion line 汽轮机膨胀过程线
turbine fan 涡轮风扇
turbine first stage pressure 汽轮机第一级压力
turbine floatation 叶轮浮选
turbine floor 水轮机层
turbine flow measuring device 涡轮流量测量仪表
turbine flowmeter 涡流流量计，涡轮流量计，转子式流量表，透平流量计
turbine flow section 通流部分
turbine following mode 汽（轮）机跟踪方式
turbine follow-up control 汽轮机随动控制
turbine follow 汽（轮）机跟踪，汽（轮）机跟随
turbine foundation 汽轮机基础，透平基础，汽机基础
turbine gallery 水轮机廊道，水轮机层
turbine gas burner 叶轮式气体喷燃器
turbine gate 水轮机闸门，进水闸门
turbine gear unit housing 涡轮减速器箱体
turbine gear 盘车装置
turbine generator building 涡轮发电机房
turbine generator island 汽轮发电机岛
turbine generator monitor 汽轮发电机监控器，涡轮发电机监控器
turbine generator pedestal 汽轮发电机基座
turbine generator protection system 汽轮发电机组保护系统
turbine generator remote control system 汽轮发电机遥控系统
turbine generator set 汽轮发电机
turbine generator shaft system 汽轮发电机组轴系
turbine generator thermal efficiency 汽轮发电机组热效率
turbine generator unit 汽轮发电机组，涡轮发电机组，透平发电机组
turbine generator 透平发电机，汽轮发电机，水轮发电机，涡轮发电机
turbine gland steam and vent system 汽轮机轴封蒸汽及排气系统，汽轮机汽封系统
turbine gland system 汽机轴封系统
turbine governing system 汽机调节系统
turbine governor 汽轮机调速器，水轮机调速器
turbine guard valve 水轮机保护阀
turbine half coupling 透平端的半联轴器
turbine hall elevator 汽机房电梯
turbine hall oil tank fire protection system 汽机间油箱消防系统
turbine hall overhead crane 汽机房桥吊，汽机房行车，天车
turbine hall 汽轮机厂房大厅，汽机房
turbine hand trip 手动打闸停机

turbine heat rate 汽机热耗
turbine high pressure bypass 高压旁路装置
turbine house bridge crane 汽机房桥式起重机
turbine house 汽轮机房，汽机房
turbine housing 汽缸
turbine idling 水轮机空转
turbine inlet bend 水轮机进水弯管
turbine inlet gas temperature 燃气轮机初温
turbine inlet pressure 透平进口压力
turbine inlet temperature 透平进口温度，汽轮机进汽温度
turbine inlet valve 水轮机进水阀，汽机进汽阀
turbine input 水轮机输入功率
turbine inspection platform 水轮机检查平台
turbine intake capacity 汽轮机最大通流量
turbine island chemical feed system 汽轮机厂房化学药品补给系统
turbine island 汽机岛
turbine jacking oil device 顶轴装置
turbine jacking oil pump 顶轴油泵
turbine load 汽轮机负荷
turbine low pressure bypass 低压旁路装置
turbine lube oil convey pump 小机润滑油输送泵，汽机润滑油输送泵
turbine lube oil treatment system 汽轮机润滑油处理系统
turbine lube oil 汽机润滑油
turbine lubricating oil tank gas exhaust fan 汽轮机润滑油箱排烟风机
turbine lubricating oil tank 汽轮机润滑油箱
turbine lubrication, jacking and turning system 汽轮机润滑、提升和盘车系统
turbine main shaft 水轮机主轴，汽轮机大轴
turbine main steam and drains system 汽机主蒸汽和排水系统
turbine master 汽轮机主控器
turbine maximum continuous rating 汽轮机最大连续功率，汽轮机组（保证）最大连续出力
turbine meter 涡轮流量计，涡轮式流量计
turbine middle layer heating steam header 汽轮机夹层加热进汽联箱
turbine missile 汽轮机飞射物
turbine mounting 透平安装，汽轮机安装
turbine nozzle blade 透平喷嘴叶片
turbine nozzle 汽轮机喷嘴，透平喷嘴
turbine oil 透平油，汽轮机油
turbine operating board 水轮机操作板
turbine outlet temperature 透平出口温度
turbine output 水轮机输出功率，水轮机出力，汽机出力
turbine overspeed testing pit 汽轮机超速试验坑
turbine package 汽机包
turbine pedestal 汽机底座［支座，支架］
turbine pit-lining 水轮机机坑衬砌
turbine pit 水轮机机坑
turbine plant 透平装置，汽轮机装置
turbine power control valve 汽轮机功率控制阀
turbine-powered generator 汽轮发电机，涡轮发电机
turbine pressure ratio 透平压比
turbine proper water drain 汽轮机本体疏水
turbine proper 汽轮机本体
turbine protection against water induction 汽轮机进水保护
turbine protection system 汽轮机保护系统
turbine protective device 透平保护装置
turbine pump 透平泵，涡轮泵，叶轮泵
turbine quick cooling device 汽机快速冷却装置
turbine rated load 汽轮机额定负荷
turbine rating 透平额定功率
turbine reduction gear 汽轮机减速齿轮
turbine regulation, safety protection and oil systems 汽轮机调节保安系统和油系统
turbine regulation 水轮机调节
turbine relief valve 涡轮机空放阀
turbine room basement 汽轮机房底层
turbine room 汽机房，汽轮机房，水轮机室
turbine rotor inlet temperature 透平转子进口温度，涡轮机转子进口温度
turbine rotor 汽轮机转子，涡轮机转轮，水轮机转轮，透平转子
turbine runback module 汽轮机降功率组件
turbine runner 水轮机转轮
turbine run-up 汽轮机启动至全速，汽轮机冲动
turbine scale 透平叶片机垢，透平机垢
turbine security valve 汽轮机安全阀
turbine self-regulation factor 水轮机自调因数
turbine sequence control 透平程序控制，涡轮机程序控制
turbine setting height 水轮机安装高程
turbine setting 汽轮机安装（高程）
turbine set 涡轮机组，透平机组
turbine shaft gland 水轮机［汽轮机］轴封套，水轮机［汽轮机］轴密封
turbine shaft 汽轮机轴，涡轮机轴，水轮机轴，透平主轴
turbine side 机端，透平端
turbine simulator 水轮机仿真器
turbine sitting hight 水轮机的安装高程，涡轮机静置高度
turbine sole plate 汽轮机底板，汽轮机基础板
turbine speed monitor 汽轮机转速监视器
turbine spindle 涡轮机轴，透平轴
turbine sprinkler 涡轮式喷灌器
turbine stage 涡轮机级，透平级
turbine starting characteristic curve 汽机启动特定曲线
turbine starting monitoring 透平启动监控，涡轮启动监控
turbine start-up panel 汽轮机启动盘，透平启动盘
turbine start-up 汽机启动
turbine stay ring 水轮机座环
turbine steam by-pass system 汽轮机蒸汽旁路
turbine steam condition 汽轮机蒸汽参数
turbine steam rate 汽轮机汽耗率
turbine storey 水轮机层
turbine stress analyzer 透平应力分析仪
turbine stress evaluator sensor 透平应力测定传感器
turbine stress supervisory system 汽轮机热应力监控系统

turbine stub shaft （接转速监视器的）透平端轴
turbine supervisory instrumentation 汽轮机（本体）监测仪表，透平（本体）监控仪表
turbine supervisory system 透平监控系统
turbine system 涡轮机系统，汽轮机系统
turbine thermodynamic efficiency 透平热效率
turbine throttle automatic governing 汽轮机节流自动调节
turbine throttle 汽轮机节流阀
turbine transient data management system 汽轮机瞬态数据管理系统【汽机振动分析】
turbine trip 透平跳车，汽轮机事故保护停机，汽轮机跳闸
turbine turning gear motor 汽轮机盘车电动机
turbine turning gear 汽轮机盘车装置，盘车装置
turbine turning system 汽轮机盘车系统
turbine type agitator 汽动搅拌器
turbine type centrifugal pump 透平式离心泵
turbine type deep-well pump 涡轮式深井泵
turbine type wind machine 涡轮式风力机
turbine valve 汽轮机阀，水轮机（进水）阀，涡轮活门
turbine variable pressure operation 汽机变压运行
turbine venting 水轮机补气，水轮机通气
turbine vibration protection 汽轮机振动保护
turbine vibration supervision 汽轮机振动监控（保护）
turbine warming-up 汽轮机暖机，暖机
turbine washing equipment 透平清洁装置
turbine wheel 汽轮机转子，透平叶轮，涡轮机叶轮
turbine wicket gate 水轮机导叶
turbine with axial condenser 轴向排汽汽轮机
turbine with horizontal joint 带有水平中分面的汽轮机
turbine with vertical joint 带有垂直中分面的汽轮机
turbine 透平，汽轮机，水轮机，涡轮机，透平机，涡轮
turblent Schmidt number 紊流什米特数
turbo-alternator 透平交流发电机（组），汽轮发电机，汽轮交流发电机，涡轮交流发电机
turbo blower 透平鼓风机，汽轮鼓风机，涡轮式鼓风机
turbo boiler feed pump 汽动给水泵
turbocharged 透平增压的，涡轮增压的
turbocharger 透平增压器，涡轮增压器
turbo-charging 涡轮增压
turbocompressor 透平空压机，涡轮式压气机
turboconverter 汽轮驱动变流机，汽轮发电变换机
turbodrill monitor 涡轮钻机喷水枪
turbo-drill 涡轮钻机
turbo dynamo 直流水轮发电机，直流汽轮发电机，透平发电机，叶轮发电机，汽轮发电机
turboexciter 涡轮励磁机，汽轮励磁机
turbo exhauster 叶轮排气机，涡轮排气机
turboexpander 膨胀式汽轮机
turbofan 透平鼓风机，涡轮风扇，涡扇发动机，涡轮风机
turbo-feed pump 汽动给水泵
turbofurnace 旋风炉膛，扰动式炉膛
turbo-generating unit 汽轮发电机组
turbo-generator maximum capacity 透平发电机最大容量
turbo-generator rated output 透平发电机额定出力
turbo-generator rotor 透平发电机转子
turbo-generator running at rated speed for syncronization 汽轮发电机组定速
turbo-generator set 透平发电机组，汽轮发电机组，水轮发电机组，汽轮发电机，涡轮发电机组
turbo-generator shaft 水轮发电机轴，涡轮发电机轴
turbo-generator stator 透平发电机定子
turbo-generator 汽轮发电机，涡轮发电机，透平发电机
turbo-jet engine 涡轮喷气发动机
turbojet 涡轮喷气发动机
turbolator 扰流子
turbomachine performance 叶轮机械性能
turbo-machinery 透平机械，涡轮机械，叶轮机械
turbomachine 叶轮机械，涡轮机
turbomat 汽轮机自动控制
turbomic 云母基绝缘材料
turbonator 汽轮发电机
turbopause 湍流层顶，湍流顶层
turbo-power unit 透平动力装置
turboprop 涡轮螺桨，涡桨发动机
turbo-pump 透平泵，涡轮泵，叶轮泵
turbo-rotor 涡轮转子，汽轮机转子
turboseparator 旋风分离器，轴流式分离器，旋风子
turboset 透平机组，汽轮发电机组，涡轮机组，汽轮机组
turbosphere 湍流层
turbosuperchanger 透平增压器，涡轮增压器
turbotrol 汽轮机控制
turbo-unit 汽轮发电机组
turboventilator 透平通风机
turbo-visory equipment 汽轮机监控装置，涡轮机监控装置
TURB SYNC 汽机同期允许【DCS画面】
TURB＝turbine 汽轮机
turbu-flow transport system 紊流输送系统
turbulence admittance factor 湍流导纳因子
turbulence closure model 湍流封闭模型
turbulence coefficient 湍流系数，紊流系数
turbulence component 湍流分量
turbulence damping 紊动性衰减
turbulence decay 湍流衰减
turbulence degree 湍流度，紊动度，紊流率
turbulence-development fetch 湍流发展风程
turbulence distortion 湍流畸变
turbulence effect 湍流效应
turbulence energy content 湍流能量含量
turbulence energy spectrum 湍流能谱
turbulence energy 湍流能，湍流能量，紊动能
turbulence excitation 湍流激励
turbulence factor 湍流因子
turbulence generation mechanism 湍流发生机理
turbulence generation 湍流发生器
turbulence grid 湍流格网

turbulence index 紊流指数
turbulence indicator 紊流计
turbulence induced vibration 湍流致振
turbulence intensity 湍流强度，扰动强度，紊动度，紊动强度，紊流强度
turbulence inversion 湍流逆温，湍流逆增
turbulence length scale 湍流长度尺度
turbulence level 湍流度，紊动度
turbulence loss 紊流损失
turbulence lump 紊流微团
turbulence meter 紊流计
turbulence method 湍流法
turbulence nonuniformity 湍流不均匀性
turbulence number 湍流度，湍流数
turbulence promoter 湍流促进器
turbulence propagation 紊流传播
turbulence report 湍流报告
turbulence resistance 湍流阻力
turbulence response 湍流响应
turbulence Reynolds number 湍流雷诺数
turbulence scale parameter 湍流尺度参数
turbulence scale 湍流尺度，紊流尺度，紊流度，紊流大小
turbulence scatter 紊流散布
turbulence screen 湍流网，（风洞）阻尼网，湍流栅
turbulence separation 湍流分离现象，紊流分离
turbulence spectrum 湍流谱，湍谱
turbulence sphere 湍流球
turbulence statistics 湍流统计学
turbulence stimulator 湍流激发器
turbulence structure 湍流结构
turbulence transition 湍流转捩，湍流转变
turbulence viscosity 湍流黏性
turbulence 骚乱，湍流，涡流，紊流，湍流强度，扰动，旋涡，紊动
turbulent bed 扰动床【沸腾炉】
turbulent boundary flow 紊流边界流动
turbulent boundary layer 湍流边界层，紊流边界层
turbulent buffeting 湍流抖振
turbulent burner 旋流式燃烧器
turbulent burning 紊流燃烧
turbulent bursting 湍流猝发
turbulent condition 湍流工况，紊流工况，紊流状态
turbulent convection 紊流对流，涡旋对流，湍流对流
turbulent core 湍流核心，紊流中心
turbulent current 紊流
turbulent density fluctuation 紊流密度变动
turbulent deposition velocity 湍流沉降速度
turbulent diffusion coefficient 湍流扩散系数
turbulent diffusion 湍流扩散，紊流扩散
Turbulent diffusivity of momentum 动量紊流扩散系数
turbulent diffusivity 湍流扩散率，紊流扩散系数
turbulent dispersion 湍流弥散
turbulent dissipation ratio 紊流耗散率
turbulent drag 湍流阻力
turbulent eddy 湍涡
turbulent Ekman layer 埃克曼湍流层
turbulent energy spectrum 湍流能谱
turbulent energy 湍流能量，紊流能量
turbulent entrainment 湍流卷挟
turbulent exchange 湍流交换，紊动转换，紊流交换
turbulent field 紊流区
turbulent firing burner 扰动式燃烧器
turbulent flame 紊流火焰
turbulent flow burner 湍流燃烧器
turbulent flow contact absorber 湍流接触吸收器
turbulent flow core 湍流堆芯，紊流中心
turbulent flow motion 紊流，湍流
turbulent flow regime 紊流状态
turbulent flow theory 湍流理论，
turbulent flow velocity 湍流速度
turbulent flow vibrating 紊流振动
turbulent flow 紊流，湍流，乱流，涡流
turbulent fluidized bed 湍流流化床，湍动流化床
turbulent friction velocity 湍流摩擦速度
turbulent friction 湍流摩擦，紊流摩擦
turbulent gust 湍流阵风
turbulent heat exchange 湍流热交换，紊流换热，紊流热交换
turbulent heating 湍流加热
turbulent heat transfer 湍流传热，紊流传热
turbulent intensity 湍流强度
turbulent interchange of moisture 水汽湍流交换，水汽紊流交换
turbulent inversion 湍流逆温
turbulent jet 紊动射流
turbulent layer 湍流层，紊流层
turbulent Lewis number 湍流刘易斯数
turbulent mass exchange 湍流质量交换
turbulent mass transfer 湍流传质
turbulent microstructure 湍流微结构
turbulent mixing 湍流混合，湍流交混，紊流混合
turbulent motion 扰动，紊流运动，湍动，紊动，湍流运动，涡旋运动
turbulent pattern 紊流结构
turbulent Peclet number 紊流贝克列数
turbulent plume 湍流羽流
turbulent Prandt number 湍流普朗特数，紊流普朗特数
turbulent process 湍流过程
turbulent property 湍流特性，紊流特性，紊流性
turbulent reattachment 紊流再附着
turbulent regime 湍流状态
turbulent region 紊流区，湍流区
turbulent resistance 紊流阻力
turbulent resistivity 湍性电阻率
turbulent Reynolds number 湍流雷诺数
turbulent ring 紊流环
turbulent Schmidt number 湍流施密特数
turbulent separation 湍流分离（现象），紊流分离，紊流脱离
turbulent shear flow 紊动剪切流
turbulent shear layer 湍流剪切层
turbulent shear 湍流剪切，湍流切变，紊流剪切

turbulent shock wave 湍流冲击波
turbulent single-phase mixing 湍流单相交混
turbulent skin friction 紊流表面摩擦
turbulent stream 紊动水流，紊流
turbulent stress 紊流应力
turbulent thermal conductivity 湍性热导率
turbulent transfer 湍流传递
turbulent transport theory 湍流输运理论
turbulent transport 湍流输运
turbulent velocity field 紊流速度场
turbulent velocity 紊流速度
turbulent viscosity 湍流黏性
turbulent vortex street 湍流涡街
turbulent vortex 紊流涡，紊流旋涡，湍动涡流
turbulent wake 湍流尾流，紊动尾流，紊流尾迹
turbulent wind field 湍流风场
turbulent wind fluctuation 湍流风脉动
turbulent wind 湍流风
turbulent zone 湍流区，紊流区
turbulent 湍流的，激流的，骚乱的，混乱的，吵闹的
turbulivity 湍流度，湍流系数
TURB 汽轮机【DCS 画面】
turf cutter 切草机
turf dam 草土小坝
turf 泥煤，泥炭，泥炭土，草皮，草地
turgescence 膨胀，肿胀，胀起
turgescency 膨胀，肿胀，胀起
turgo impulse turbine 斜击式水轮机
Turgo turbine 土尔戈式水轮机，斜击式水轮机，中水头冲击式水轮机
Turing machine 图灵机
Turing test 图灵实验
turnable bridge 平旋桥
turnable gate 转动式闸门
turnabout pulley 改向滚筒［滑轮］
turnabout the axis 绕轴旋转
turnabout 转盘，旋转，转向，经济转好，180度转弯
turn angle 回转测角
turnaround document 周转文件
turnaround efficient 周转效率
turnaround period 周转期
turnaround sequence 转向序列
turnaround system 周转系统
turnaround time of the train 列车周转时间
turnaround time 换向时间，周转期时间，周转时间，往返时间，检修时间，维持时间【后处理工厂】
turnaround 小修，预防性维修，回车场，换向，预防修理，转向，回车道，转机
turn-back arrangement 折返式布置
turn buckle screw 旋扣螺丝，花兰螺丝
turn buckle 花篮螺栓，花篮螺丝，索具，紧线扣，花蓝螺母，螺栓张力器，螺旋扣，螺丝接头，紧线器，松紧螺套，拉线螺丝，丝扣接头，套筒螺母
turnbutton 旋钮
turndown capability 调节能力
turndown range 调节范围
turndown ratio （燃烧器负荷）调节比，量程可调比，极限负荷比
turndown 关闭
turned bolt 精制螺栓，车制螺栓，车光螺栓，旋制螺栓
turned over 翻倒，翻转
turn from deficits to profits 扭亏增盈
turn indicator 转数指示仪，转弯指示器，转数计，匝数计，转动指示器
turning angle 偏转角，旋转角
turning bridge 平旋桥
turning by every 180° （每次）盘转 180 度
turning circle 回转圈，转车盘
turning crane 旋转式起重机
turning effort 回转力
turning error 回转误差
turning device 盘车装置
turning gear hand wheel 盘车手轮
turning gear motor 盘车装置的电动机
turning gear oil pump 盘车装置油泵，盘车油泵
turning gear on 盘车机合上
turning gear 盘车装置，转动机构，旋转装置，回转装置，盘车齿轮
turning joint 铰接，铰链，活动关节
turning moment 旋转力矩，转矩，倾覆力矩
turning motor 盘车电动机
turning of the tide 转潮
turning oil pump 盘车油泵
turning oil 循环油
turning plate 尺垫
turning point 转向点，转折点
turning power unit 回转执行部件
turning radius 回转半径
turning rolls 焊接滚轮架
turning room 转向室
turnings 切屑
turning vane 转动叶片，导向叶片，导流叶片，（风洞）导流片
turning wheel 转轮
turning 盘车【汽轮机】，转动，转弯
turn insulation of rotor winding 转子绕组绝缘
turn insulation 线匝绝缘，匝间绝缘，匝绝缘
turnkey approach 交钥匙方式，启钥方式
turnkey bid 按交钥匙方式承包的招标，交钥匙投标，一揽子招标，交钥匙方式总承包
turnkey contract 统包式合同，交钥匙合同，总承包，总承包合同
turnkey delivery 立即可用的交付，交钥匙式的交付
turnkey installation 成套设备
turnkey investment 统包式投资，交钥匙投资
turnkey plant 交钥匙电厂
turnkey project 统包式项目，交钥匙工程，交钥匙项目，工程总承包
turnkey services 交钥匙服务
turnkey tender(/bid) 整体承包投标
turnkey terms 交钥匙条件
turnkey 总控钥匙，总承包，交钥匙承包，交钥匙工程
turn-knob 旋钮
turn-light switch 带灯转动开关，灯光旋转开关
turnmeter 转速计，回转计，转速表

turn number （绕组的）匝数
turn-off characteristic 关断特性
turn-off time 断开时间
turn-off transient 关断瞬态
turn-off 岔道，断开
turn of the tide 潮流转向
turn of tidal current 转流
turn of tidal stream 潮流转向，转流
turn-on time 接通时间，接入时间
turn-on transient 接通瞬态
turn-on 开，接通，闭合，接入，合闸
turn-out track 牵出线，转道辙
turn-out 分叉道【铁轨】，道岔，避车道，分水闸，分水处，生产量
turn out 切断，断开，断路，关闭，关掉，生产，结果是，证明是，出席，参加，熄灭（电灯，煤气等），（以某种方式）发生
turnover bonus 年终额外津贴
turn-over door 翻转门
turn-over form 翻转模板
turnover of capital 资本周转率
turnover of current capital 流动资金周转率
turnover of fixed assets 固定资产周转率
turnover tax 周转税
turnover time 周转时间
turnover 成交量，营业额，翻腾，翻转，倒转，倾覆，倒置，回转，移交，交接【所有权的转移】，成交量，周转，周转率，翻倒
turn over 见背面，转入下页，翻转，倒转，倾覆，倒置，回转，移交，交接
turn-push-light switch 转-按-灯光开关，转-按灯光旋钮
turn-push switch 转动推进型开关
turn radius 回转半径，转弯半径
turn ratio 变比，匝数比，线匝比【变压器】
turn separator 匝间间隔片，匝间垫条，匝间绝缘
turns ratio of the rotor-to-stator winding 定子与转子绕组的匝数比
turnstile aerial 绕杆式天线
turnstile 交叉天线，十字形天线，绕杆，旋转式栅门
turn switch 扭转开关，旋转开关
turntable bearing 回转轴承，滚动轴承
turntable （风洞）转盘，回转台，转车台，转台，转盘
turn-to-push discrepancy switch 转-按差动开关
turn-to-turn capacitance 匝间电容
turn-to-turn fault 匝间故障
turn-to-turn insulation 匝间绝缘
turn-to-turn short circuit 匝间短路
turn-to-turn shunt capacity 匝间分路电容
turn-to-turn 匝间，匝间的
TURN＝turning 转动
turn-up loss 倒流损失
turnup 达到一定转速，卷起部分
turn-voltage vector 线匝电压矢量
turn 车削，匝，转向，圈，转，转动，转数
turpentine oil 松节油
turpentine varnish 松节油清漆
turpentine 松节油，松油，松脂
tusk and tenon joint 榫槽接合
tusk tenon 长牙榫
tutorial 教程，导师的，指导的
Tuttle tube-factor bridge 塔特尔电子管参数测量电桥
tutwork 计件工作
tuyere belt 风口圈
tuyere distributor 风帽分布板
tuyere hole 喷口
tuyere notch 风嘴孔，风口
tuyere spacing 风帽间距
tuyere 风口，喷口，炉，喷气管嘴
TV camera 电视录像机
TV monitor 电视监视器
TV remote pickup 遥控电视摄像管
TV system 电视系统
TV＝television 电视
TV(throttle valve) 主汽门，节流阀
TV 主汽门【DCS 画面】
tweezers 夹子，镊子
twinax 屏蔽双线馈线
twin beams 并置梁，双梁
twin-bed 双床【沸腾炉】
twin-blade runner 双叶片转轮
twin boundary 双晶面，孪晶间界，孪晶界
twin box girder 双箱梁
twin branch lines 双分裂导线
twin-bridge gantry 双路栈桥
twin bundled conductor 双分裂导线
twin bundled lines 双分裂导线线路
twin-cable ropeway 双缆索道
twin-cable 双心电缆
twin-car dumper 双车翻车机
twin-channel 双通道的，双路的
twin coil 双线圈
twin columns 并置柱，双柱
twin compressor 复式压缩机
twin conductor 双股导线
twin contact 双触点
twin-core cable 两芯电缆，双心电缆
twin-core current transformer 双心电流互感器
twin-core 双心
twin crankshaft engine 双曲轴发动机
twin-cylinder air-cooled engine 双缸风冷式发动机
twin-delta 三角形-三角形接法，双三角形连接
twin diode 双二极管
twin drive 双传动，双电动机驱动
twin electrode 双芯焊条
twin-engined 双引擎的，双发动机的
twin-engine tractor and scraper 双引擎拖拉机和铲运机
twine 两股绳，双股，盘绕，缠绕，股绳，股线
twin-filter 双联冷油器
twin-fluid atomization 双流体雾化
twin-furnace boiler 双炉膛锅炉
twin-furnace-type boiler 双炉膛锅炉
twin furnace 双炉膛
twin gasket 双垫片
twin I-beam 双工字梁
twin insulator strings 双联绝缘子串

twin-jaw crusher 双颚式破碎机
twin jetties 双边突堤，双导流堤
twinkle 闪烁
twinkling 闪烁，闪光，瞬间，顷刻
twin layout 双岸式布置
twinned grains 双晶粒【结构】
twinning 扭成对，双股绞合，孪生
twin pier 双桥墩
twinplex plotter 双投影器测图仪
twinplex projector 双投影仪
twinplex 四信路制，双路移频制
twin power plant 双岸式电站
twin probe system 双晶探头系统
twin pump 双水泵
twin push-button 双按钮
twin reactor station 双堆核电厂，双堆核电站
twin rectangular box girder 双矩形箱梁
twin-screw extruder 双螺杆挤压机
twin screw 双螺杆
twin-shaft 双轴的，双轴式
twin-shell condenser 双壳凝汽器，双壳冷凝器
twin spark magneto 双发火磁电机
twin-spiral water turbine 双蜗壳水轮机
twin tape transporter 双带运输装置
twin-T network 双T形网络
twin tower 双柱塔
twin triode 双三极管
twin tunnels 双隧洞
twin turbine generator （单堆配的）双汽轮发电机
twin turbine （对称）双流型汽轮机，弯接式水轮机
twin twisted strand 双绞股线
twin twisted 双绞的
twin-type cable 对铰多心电缆
twin type safety valve 复式安全阀
twin volute pump 复式螺旋泵
twin volute 双蜗壳
twin-wagon tippler 双车翻车机
twin water turbine 双流式水轮机，双轮水轮机
twin wharf 突码头
twin wire 双股线
T-wire “T”线塞尖引线
twist aerofoil 扭曲翼剖面
twist angle of blade 叶片的扭角
twist angle （叶片）扭转角
twist axis 扭曲轴，扭转轴
twist blade 扭转叶片
twist centre 扭转中心
twist connectors for pilot wire 旋转连接器
twist drill 螺旋钻头，麻花钻
twisted auger 取土样的麻花钻
twisted bar 螺旋钢筋
twisted blade 扭曲叶片，扭叶片，扭转叶片
twisted column 螺旋形柱
twisted cord 铰合软线，双铰软线
twisted joint 绞接头，扭接头
twisted-lead transposition 扭线换位，端部翻转换位
twisted line 绞线
twisted packing 绞合的填充物
twisted-pair cable 双绞线电缆，双扭线电缆
twisted pair wiring 双心铰合线
twisted pair 双心铰合线，双绞线，铰线对，扭铰双线电缆，双心绞合线
twisted shackle 扭转式U形环
twisted shielded four 编织屏蔽的四芯导线
twisted shielded pair 编织屏蔽的双芯导线
twisted shielded quad 编织屏蔽的四芯导线
twisted shielded triplet 编织屏蔽的三重线
twisted steel 螺纹钢筋
twisted strata 扭曲地层
twisted-tape fuel element assembly 纽带型燃料组件，旋涡型燃料组件
twisted-tape insert 扭转带插入件
twisted wire 绞合线，扭绞钢丝，绞线
twisted 双绞的，绞合的（导线）
twisting bar 扭绞线棒，扭绞导条，换位线棒
twisting channel 摆动河道
twisting force 扭力
twisting inter moment 扭转惯性矩
twisting moment curve 扭矩曲线，弯矩曲线
twisting moment 扭矩，扭转力矩
twisting stiffness 抗扭刚度
twisting strength 抗扭强度
twisting stress 扭转应力，扭应力
twisting 绞合的，扭曲的，扭转，铰合
twist joint 铰合接头，铰接
twist of blade 叶片扭角
twistor 磁扭线，绞扭线
twist outlet 旋流风口
twist rotation （机械手拉手）扭转
twist strain 扭应变
twist stress 扭应力
twist vector 扭矢
twist vibration 扭振
twist warp 扭转翘曲
twist waveguide 扭波导
twist wire 绞线
twist 搓，扭，缠绕，曲解，铰合，编织，扭曲，扭动，旋转，转动，扭转
twi-tier blade 双层叶片
two-address code 二地址码
two-address computer 二地址计算机
two-address instruction 二地址指令
two-address 二地址
two and three dimensions 二维和三维问题
two-armature DC motor 双枢直流电动机
two-armature generator 双枢发电机
two-axis equivalent circuit 双轴等效电路
two-axis stop 二维限位架
two-axis synchronous machine 双轴同步电机
two-bank engine 双列汽缸发动机
two banks opposing spray pattern 对喷模式
two basic valve tee 二位三通基本阀
two-bay spillway 双跨溢洪道
two-bearing computer 双方位计算机
two-bearing machine 双轴承型电机
two-bed demineralizer 双床除盐装置
two bladed wind turbine 双叶风电机组
two-bowl scraper 双斗铲运机
two-breaker arrangement 双断路器接线

two-cage rotor 双鼠笼转子
two caliper disc brake 二卡钳盘式制动器
two casing condensing turbine 双缸凝汽式汽轮机
two casing unit with single shaft 单轴双缸机组
two-centered arch 双心拱
two channel impeller 双流道叶轮
two channel 双通道的
two-circuit boiling water reactor 双回路沸水反应堆
two-circuit nuclear power plant 双环路核电厂
two-circuit prepayment meter 双路预付电度计
two-circuit receiver 双调谐电路接收机
two-circuit transformer 双绕组变压器
two-circuit winding 双路绕组
two-circuit 双回路
two-coil instrument 双线圈仪表
two-coil relay 双线圈继电器
two color laser Doppler anemometer 双色激光多普勒风速计
two color thermometer 双色温度计
two-column core 双柱铁芯
two-column radiator 双柱散热器
two-component heat pipe 双组分热管
two-conductor cable 双导体电缆
two-conductor 双导体，二导体
two-contingency factor 双重意外事故因子
two conveyor flight 双路皮带
two-core cable 双芯电缆
two-core excitation winding 双铁芯励磁绕组
two-core reactor 双芯反应堆
two-core switch 双磁芯开关
two-core transformer 双芯变压器
two-core voltage regulating transformer 双铁芯调压变压器
two-core 双芯【电缆】
two-couple method 双力偶法
two current meter method 双计测流法
two-cycle engine 双冲程发动机
two-cycle generator 双周波发电机，双频发电机
two-cycle steam engine 双循环蒸汽机，双冲程发动机
two-cycle system 双循环系统
two degrees of freedom 二级自由度
two-digital group 二位组
two-dimensional airfoil 二维翼型
two-dimensional bed 二维床
two-dimensional boundary layer 二维边界层
two-dimensional calculation 二维计算
two-dimensional diffusion theory 二维扩散理论
two-dimensional display 二维显示
two-dimensional electric field 二维电场
two-dimensional finite element method 二维有限元法
two-dimensional flow 二维流，平面流，二元流，二维流动，二元水流
two-dimensional Gaussion distribution 二维高斯分布
two-dimensional guidance 二维制导
two-dimensional irrotational flow 二维无旋流
two-dimensional jet 二维射流
two-dimensional magnetic field 二维磁场
two-dimensional model 二维模型
two-dimensional motor 二维电动机
two-dimensional numerical analysis 二维数值分析
two-dimensional parallel flow 二元平行流
two-dimensional photoelastic analysis 二维光弹性分析
two-dimensional plume 二维羽流
two-dimensional positioning system 双向定位系统
two-dimensional puff 二维喷团
two-dimensional rig 二维试验设备
two-dimensional scanning 二维扫描
two-dimensional space 二维空间
two-dimensional state of stress 双向应力状态
two-dimensional storage 二维存储器
two-dimensional surface 二维平面
two-dimensional system 二维系统
two-dimensional transonic cascade 二元跨音速叶
two-dimensional vector 二维矢量，平面矢量
two-dimensional water tunnel 二元水洞
two-dimensional wind tunnel 二元风洞
two-dimensional wing 二维机翼
two-dimensional 二维的，平面的，二度的，二元的
two directional stress crack 双向应力裂纹
two-direction linear stepping motor 双向直线步进电动机
two-direction 双向的
two-drum boiler 双汽包锅炉
two-drum hoist 双滚筒卷扬机
two-duct conduit 双导管套管
two-element electronic system 双元件电子系统
two-element feed control 双元件馈给控制，双冲量进给调节
two-element relay 双元件继电器，双线圈继电器
two-element resistively probe 双极电阻率探针
two-element 双元件
two-envelope evaluation method 双信封评标法
two-envelope method 双信封（评标）法
two-field induction machine 双磁场异步电机
two-film theory 双膜理论
two-flanged butterfly valve 双法兰蝶阀
two-flow core 双流程堆芯
two fluid process 双液法
two fluid reactor 双流体反应堆
two fluid system 双流体系统
twofold door 双折门，双褶门
twofold 两倍的，两重的，双卷的，两层的
two-furnace boiler 双炉体锅炉
two-gang variable capacitor 双联可变电容器
two-gang 双联的，双排的
two-group critical equation 双群临界方程
two-group critical mass 双群临界质量
two-group diffusion theory 双群扩散理论
two-group model 两群模型，双群模型
two-group perturbation theory 双群微扰理论
two-group theory 两群理论，双群理论
two-gun oscillograph 双枪示波器，双线示波器
two-hinged arch 二铰拱，双绞拱

two-in-hand winding 双路叠绕法
two-in-one motor 二合一电动机
two-input adder 双输入加法器，半加法器
two-input gate 双输入门
two-input servo 双输入随动系统，双输入伺服系统
two-input subtractor 双输入减法器
two-input switch 双输入开关
two-input 双输入
two-integrator system 双积分器系统
two-integrator 双积分器
two-jet-type turbine 双喷嘴水轮机
two-lane road 双车道道路
two-layer spiral coil 双层螺旋式线圈
two-layer winding 双层绕组
two-layer 双层的
two-lay sling 双吊索
two-legged transformer 双柱式变压器
two-legs type core 双柱式铁芯
two-level logic 两级逻辑，双电平逻辑
two-level non-return system 双级无零记录系统，双位无回程系统
two-level return system 双级（归零）记录系统，双位有回程系统
two-level subroutine 二级子程序
two-level 双级的，双位的
two-liquid hardening 双液淬火
two-loop servomechanism 双回路伺服机构
two-loop test apparatus 双环路试验设备
two-loop 双回路的
two-mass, spring-connected and subresonant tuned vibrating feeder 双质量、弹簧串联、亚共振调谐振动给煤机
two-metal corrosion 双金属腐蚀
two-motion switch 两级动作开关，两位开关，双向开关
two-neck distilling flask 双颈蒸馏瓶
two-on-one gas-steam combined cycle unit 二拖一燃气蒸汽联合循环机组
two-on-one unit 二拖一机组
two-on-one 二拖一【燃气轮机】
two-out-of-five code 五中取二码
two-out-of-three circuit 三取二电路
two-out-of-three trip 三取二事故保护停堆
two out of three 三取二
two parallel conveyer system 双路皮带系统
two-parameter adaptive control system 双参数自适应控制系统
two-parameter 双参数
two particle model 双颗粒模型
two parties 双方
two-part network synthesis 四端网络综合
two-part network 二对端网络，四端网络
two-part price 两部制电价
two-part tariff 二部电价制
two-party application association 双方应用关联
two-party line 对讲电话线，两户合用电话线
two-pass boiler 双回程锅炉
two-pass condenser 双流程凝汽器
two-pass cooling 双路冷却
two-pass core 双流程堆芯
two-pass 双行程的，双回程，双流程的
two-pen potentiometer continuous recorder 双笔电势计式连续记录仪
two-phase and dry-steam geothermal systems 两相和干蒸汽地热系统
two-phase boundary layer 两相边界层
two-phase bubble assemblage model 两相气泡汇集模型
two-phase circuit 两相电路
two-phase connection 两相接法
two-phase consecutive model 两相连续模型
two-phase cooling 两相冷却
two-phase five-wire system 二相五线系统
two-phase flow loop 两相流回路
two-phase flow regime 两相流动工况
two-phase flow 两相流，双相流，两相流动，两相水流
two phase fluid 双相流体
two-phase four-wire system 两相四线制
two-phase generator 两相发电机
two-phase ground fault 两相接地故障
two-phase induction motor 两相感应电动机
two-phase light water 两相轻水，沸腾轻水
two-phase mixing 两相交混
two-phase mixture 两相混合物
two-phase operation 两相运行
two-phase reactor 两相反应堆
two-phase region 两相区
two-phase relay 双相继电器，双圈继电器
two-phase Schrage motor 两相施拉奇电动机
two-phase section 两相段，两相区
two-phase short circuit and ground fault 两相接地短路故障
two-phase short circuit fault 两相短路故障
two-phase short circuit 两相短路
two-phase state 双相态
two-phase system 两相系统
two-phase theory 两相理论
two-phase three-wire system 两相三线制
two-phase transformer 两相变压器
two-phase 两相的
two-piece ring 双片环
two-piece wedge 分半槽楔
two-piece 双体的
two-pier elbow draught tube 双墩肘形尾水管
two-pinned arch 双钉拱
two-pipe heating system 双管采暖系统
two-pipe hot-water heat-supply network 双管制热水热网
two-pipe steam heat-supply network 双管制蒸汽热网
two-pipe water system 两管制水系统
two-ply 双层的，双股的
two-point extraction cycle 两段抽汽循环
two-point lifting 两点悬吊法
two-point method of stream gauging 两点测流法
two-point suspension scaffold 两点悬吊式脚手架
two-pole circuit 双极回路
two-pole machine 两极电机
two-pole network 二端网络
two-pole switch 双极开关

two-pole voltage meter 双极电压计
two-pole 两极的
two-pool layout 双池式布置
two-pool plan 双池式布置【潮汐发电】
two port network 双口网络
two port 双口的，二端对的，四端的
two-position action 双位置动作，两位作用
two-position controller 二位式调节［控制］器，双位置调节器
two-position control 双位控制，两位控制，双位调节
two-position differential gap control 带中性区的双位控制，双位差隙控制
two-position power actuated valve 两位式电动控制阀
two-position relay 二位继电器，双位继电器
two-position winding 双层绕组
two-position 双位的，二位置的，双位（托板车）
two-pressure steam power cycle 双压力蒸汽动力循环
two-prism square 双棱镜直角器
two-pulse timer 双脉冲定时器，双脉冲定时装置
two-pulse 双脉冲
two-quadrant multiplier 二象限乘法器
two-quadrant 二象限的
two-range winding 两平面绕组
two-rate meter 双额电表，双费率计
two-rate tariff 双电费率
two-reaction machine theory 双反应电机理论
two-reaction method 双反应法，双反应分析法
two redundance 二冗余
two redundancy 二冗余
two region critical experiment 双区临界实验装置
two region reactor 双区反应堆
two-roll crusher 双辊破碎机，双滚轴破碎机
two-row velocity-stage 双列速度级
two runner turbine 双转轮式水轮机，双转轮水轮机
two-scale 双刻度，双标度，二进制记数法，二进制
two's complement 二的补码
two-section coil 二节线圈
two-section 双节的
two-shaft gas turbine 双轴燃气轮机
two-shaft turbine 双轴汽轮机，双轴水轮机
two-shift operation 两班制运行
two-shot method 双管法
two-sided mosaic 两面嵌镶
two-sided reinforced concrete cover plate 双面钢筋混凝土盖板
two sided spectral density function 双边谱密度函数
two-side power supply 双边供电
two-speed motor 两速电动机，双速电动机
two stable position 双稳定状态，双稳位置
two stage by-pass system 二级旁路系统
two-stage combustion 两级燃烧
two-stage compressor 二级压气机
two-stage controller 双级控制器
two-stage cracker 两段裂化反应器
two-stage crushing 两级破碎
two stage demineralization system 二级除盐系统
two-stage demineralization 两级除盐
two-stage ejector 两级抽气器
two-stage filtration 两级过滤
two-stage ionic exchange 两级离子交换
two-stage magnavolt 两级式旋转放大机
two-stage natrium ion exchanger 二级钠离子交换器
two-stage preheating circuit 二级预热系统
two-stage preselector control 两级预选控制，两步预选控制
two-stage pump 两级水泵，双级泵
two-stage reciprocating compressor 二级往复式压缩机
two-stage regulator 双级调节器
two-stage relay 两阶段继电器，两级继电器
two-stage rotary air compressor 二级旋转式空气压缩机
two-stage servo 两级随动系统，两级伺服系统
two-stage sodium exchange 两级钠离子交换
two-stage stationary air compressor 二级固定式空气压缩机
two-stage steam ejector 两级射汽抽气器
two-stage two-envelope bidding procedure 两步双信封法投标程序
two-stage 两级的，双级的
two state controller 二位控制器
two-stator induction machine 双定子感应电机
two-stator-rotor motor 双定子转子电动机
two-step action 双级作用，两步作用，两位作用
two-step controller 二位控制器，二位调节器
two-step control 两级控制
two-step normalizing 二段正火
two-step-relay turntable 两阶段继电器
two-step relay 两阶段继电器，两级作用继电器
two-step shedding 两步卸载的，两步减荷
two-step 两步的，两阶段的，两级的，双级的
two-storeyed intake 分层式取水
two-story sedimentation tank 双层沉降槽
two streams of the coal handling system 双路输煤系统
two-stroke cycle 双冲程循环
two-stroke engine 双冲程发动机
two-stub transformer 双柱式变压器，双芯变压器
two-symbol 二进位的
two-terminal generator 两端发电机
two-terminal HVDC power transmission system 两端高压直流输电系统
two-terminal network synthesis 二端网络综合
two-terminal network 二端网络
two-terminal pair network 四端网络
two-terminal series-parallel network 二端串并联网络
two-terminal switching circuit 二端开关电路
two-terminal switching network 二端开关网络
two-terminal switch 双端开关
two-terminal symmetrical switch 二端对称开关

two-terminal system （直流）两端点制，两站制
two-terminal 二端的
two-tier blade 双层叶片
two-tier winding 两平面绕组
two-time scale system 双时标系统
two-tone modulation 双音频调制
two to one frequency divider 频率减半器
two unit power plant 两单元机组发电厂
two-unit 双机组的
two-value capacitor motor 双值电容式电动机
two-variable computer 二变数计算机，双变量计算机
two-variable matrix 双变量矩阵
two-variable 二变数的，双变量的
two-velocity stage 复速级
two vertical planes system 双竖向平面系
two-wall sheet-piling cofferdam 双排板桩围堰
two-wattmeter method 双瓦特计法
two-way arch bridge 双曲拱桥
two-way beam 双向梁
two-way break-before-make contact 双向先断后合触点
two-way channel 双向通道，双向信道
two-way clamp 双向钳位电路
two-way communication 双向通信
two way configuration 二线制
two-way contact 双投触点，切换触点
two-way feed 两路馈电，双向馈电
two-way flat slab 双向无梁楼板
two-way footing 双向受力基脚
two-way heat-supply 双向供热
two-way make-before-break contact 双向先合后断触点
two-way operation 双向运行
two-way pump 双向泵
two-way radio 无线电对话机
two-way ramp 双向坡道
two-way ram 两路冲压机，双作用活塞
two-way reinforced concrete 双向配筋的钢筋混凝土
two-way reinforced 双向配筋，双向配筋的
two-way reinforcement 双向钢筋
two-way repeater 双向帮电机，双向增音机，双向中继器
two-way slab 双向板，四边支承板
two-way switch 双向开关，双路开关
two-way system of reinforcement 铜筋双向制
two-way telephone 双向电话，双工电话
two-way transmission 双路传输
two-way valve 二通阀，双通阀，双向阀
two-way 双向的，双通的
two-wheel tandem roller 双轮串联式压路机
two-wheel tractor-pulled scraper 双轮拖拉机牵引式铲运机
two-winding transformer 两绕组变压器，双绕组变压器
two-wire circuit 两线制电路
two-wire loop terminal 双线回路端子
two-wire switch 双联开关
two-wire system 二线制，两线制
two-wire-wide strands 双排股线
two-wire winding 双线线圈，双线绕组
two-wire 二线的，双股线的，双线，二线制
two-zoned-core 双区堆芯
two-zone distance relay 两段距离继电器
T-wrench 丁字柄套筒扳手
TWR＝tower 塔
TW(tail wind) 顺风
TW＝thermowell 热电偶套管
TWV 三通阀
twystron 行波速调管
tying coil 绑扎线圈
tying contract 附有条件的合同，搭卖合同
tying tape 绑带
Tyler sieve analysis method 泰勒筛析法
tympanum （电话机的）振动膜
type A container A 型容器
type A insulator A 型绝缘子【不击穿型绝缘子】
type A single-prong tester A 型单叉探测器【测绝缘子用】
type B insulator B 型绝缘子【可击穿型绝缘子】
type B package B 型包装
type B two-prong tester B 型双叉探测器【测绝缘子用】
type design of main building 主厂房的典型设计
type distribution 类型分布
type efficiency 型效率
type-house 定型房屋
type of blade root 叶根型式
type of blade 叶片类型
type of car 车型
type of coal 煤种
type of connection 连接方式
type of contract proposed 建议的合同样本
type of contract 包工方式，合同类型
type of control 控制方式，调节方式
type of coupling 连接形式
type of current 电流类别【交流或直流、极性】
type of dam 坝型
type of disturbance 搅动形式
type of flow 流动方式，流动状态
type of joint 接头型式
type of load 负荷类型
type of locomotive 机车类型，机车形式
type of operation 操作方式【手动/自动】
type of organization 组织形式
type of remuneration 报酬种类
type of stack gas plume 烟柱形式
type of tide 潮型
type section 定型剖面
types of power supply 供电方式
types of roof or floor structure 屋盖、楼盖类别
types of wind site 风场类型
type 1 system 1 型系统
type 2 system 2 型系统
type 0 system 0 型系统
type test 典型试验，定型试验，试验类型，型式试验
type tide graph 典型潮位过程线
type writer output routine 打字机输出程序
type writer 打字机
type 型号，类型，样式，型式

typhonic rock　深源岩
typhoon eye　台风眼
typhoon rain　台风雨
typhoon squall　台风飑
typhoon surge　台风涌浪
typhoon track　台风路径
typhoon warning　台风警报
typhoon　台风
typical analysis　典型分析
typical cell　典型栅元
typical coal handling system　典型输煤系统
typical data　典型数据
typical design　标准设计，典型设计，定型设计
typical detail　典型详图
typical drawing　典型图
typical floor plan　高层建筑中标准层平面
typical grid cell　标准格架栅元
typical hydrograph　典型过程线
typical installation　典型装配
typical layout　典型布置
typical module　典型模式
typical operation condition data　典型运行工况数据
typical pipe support　典型管架，通用管架
typical railroad car positioner　标准铁路车皮定位器
typical sample　代表性试样，典型试样，典型样本，典型样品
typical section　典型断面
typical shipment　标准装货量
typical stream surface　流面模式
typical title option　标头方案
typical value　标准值，典型值，代表值
typical wiring diagram　典型接线图
typical　有代表性的，典型的
typify　代表，象征，说明……特征
typist room　打字机室
typotron　显字管，高速字标管
tyre-wheel loader　轮式装载机
tyre-wheel stacker-reclaimer　轮式堆取料机
tyre　轮胎
tyrolean plaster　喷灰浆，拉毛抹灰
T　三废处理【核电站系统代码】

U

UAM(unit automatic management)　机组自启停管理系统
UAM(unit automatic master)　机组自动系统【主控】
UAT(unit auxiliary transformer)　机组厂用变压器，机组副变压器，厂用变压器
UAV(unmanned aerial vehicle)　无人机
ubac　背阳坡
UBC(Uniform Building Code)　统一建筑规范【美国】
U bend expansion　U形补偿器
U-bend heat treatment　U形弯曲热处理
U-bend specimen　U形弯曲（应力腐蚀）试样
U-bend　U形弯头，U形管，U形弯管
U-bend zone heat treatment　U形弯曲区热处理
ubi infra　参见下文
ubi supra　参见上文
U-bolt　U形耳子，U形螺钉，U形螺栓，
U burner　铀235燃烧堆
UCC(Uniform Code Council)　美国统一代码委员会
UCC(unit completion certificate)　机组完工证书
UCC(unit coordinated control)　机组协调控制
UCD(unit completion date)　机组完工日期
U-clamp　U形夹，U形夹头
U235 content　铀235含量
U core　U形磁铁芯
UCP(uniform customs and practice for documentary credits)　跟单信用证统一惯例
UCP600　国际商会跟单信用证统一惯例第600号出版物
UCR(unit control room)　单元控制室
UCTE(Union for the Co-Ordination of Transmission of Electricity)　欧洲输电联盟
UDC(utility distribution company)　配电公司
udent　湿新成土，湿变性土
Udimet　尤迪麦特镍基耐热合金
UDL(upper destruct limit)　破坏极限上限值
udoll　湿软土
udomograph　自记雨量计
udox　湿氧化土
UD-test(unconsolidated-drained test)　不固结排水剪切试验
udult　湿老成土
UEA(U type Expansion Agent for Concrete)　U型混凝土膨胀剂
UES(unexpected event reports)　意外事件单
U-flame furnace　U形火焰炉膛
U flow　底流，地下水流，下溢
U format　U形记录格式
U-form tube　U形管
U-gauge　U形管压力计
U-groove　U形坡口
UHF ceramic capacitor　超高频瓷介电容器
UHF correlator　特高频相关器
UHMW polyethylene-liner material　超高分子量聚乙烯内衬材料
UH(unavailable hours)　不可用小时
UHVAC system　特高压交流电力系统
UHVAC transmission line　特高压交流输电线路
UHVDC system　特高压直流电力系统
UHVDC transmission line　特高压直流输电线路
UHV line shunt reactor　特高压并联电抗器
UHV power grid　特高压电网
UHV power transmission　特高压输电
UHV transmission line　特高压输电线路
UHV(ultra high voltage)　特高压
UJT　单结晶体管
UKAEA(Atomic Energy Authority UK)　英国原子能管理局
ULD(unit load device)　集装器，集装化设备
ULD　机组负荷指令【DCS画面】
UHV substation　特高压变电站
U235-lined fission chamber　铀235裂变室

ullage 油箱油面上部的空间，漏损，损耗量
U235 load 铀 235 装量
U-loop U 形管圈
ultimate analysis 元素分析【燃料】，极限分析，最后分析
ultimate bearing capacity 极限承载量，极限承载能力
ultimate bending moment 极限弯矩
ultimate bending strength 抗弯强度极限
ultimate bending stress 极限挠曲应力，极限弯曲应力
ultimate beneficiary 最终受益人
ultimate capacity 最大容量，最大功率，极限容量
ultimate composition （燃料的）元素成分
ultimate compressive strength 极限抗压强度
ultimate compressive stress 极限抗压应力
ultimate cost 最终成本，最后价值，最终费用
ultimate creep 蠕变极限
ultimate crushing strength 极限抗压强度
ultimate customer 最终用户
ultimate damping 极限阻尼，极限衰减
ultimate deformation 极限变形
ultimate demand 最终需要
ultimate design resisting moment 极限设计抗力矩
ultimate design 极限设计
ultimate destination 最后目的地
ultimate disposal 最后处置，最终处置，最终处理
ultimate elongation 极限伸长
ultimate expansion 远景扩建
ultimate factor of safety 极限安全系数
ultimate filter 终端过滤器
ultimate fuel burn-up 最终燃料燃耗
ultimate gain 最大增益
ultimate heat sink basin 最终热阱水坑
ultimate heat sink 极限热井，最终热井，最终热阱
ultimate height of plume 羽流最终高度
ultimate installation 最大装机容量
ultimate installed capacity 最终装机容量
ultimate interest 最终利益
ultimate limit state 最大极限状态，极限限制状态，承载能力极限状态
ultimate load analysis 极限荷载分析［设计］
ultimate load design 极限载荷设计
ultimate load 最大负荷，极限负荷，极限荷载
ultimately controlled variable 最终被控变量，最终受控变量
ultimately 最终的
ultimate maximum 极限最大值
ultimate mechanical strength 极限机械强度
ultimate objective 最终目的
ultimate output 最大出力，最大功率，极限容量，极限产量
ultimate oxygen demand 极限需氧量
ultimate pH 极限 pH
ultimate pressure 最终压力
ultimate regenerative cycle of saturated steam 饱和汽极限回热循环
ultimate resistance 极限阻力
ultimate sampling unit 最后取样单位
ultimate sensibility 最高灵敏度，极限灵敏度
ultimate settlement 最终沉降
ultimate set 相对伸长
ultimate shear stress 极限剪切应力，极限剪应力
ultimate stable temperature 最终稳定温度
ultimate storage drum 最后贮存桶
ultimate storage facility （放射性废物）最后贮存装置
ultimate strain 极限应变
ultimate strength design 按极限强度设计，极限强度计算，极限强度设计
ultimate strength of rupture 最终破裂强度
ultimate strength 强度极限，极限强度
ultimate stressed state 极限应力状态
ultimate stress 极限应力
ultimate tensile strength 抗拉强度极限，极限抗拉强度
ultimate tensile stress 极限拉应力，极限张拉应力
ultimate tension 极限张力
ultimate user （燃料元件）最后使用者
ultimate vacuum 极限真空
ultimate value 极限值，最大值，最后值
ultimate waste disposal 废物最终处置
ultimate working capacity 极限工作强度，最大强度
ultimate yield 最终产量，最终收率
ultimate 极限的，最大的，最后的，根本的，基本的，最终的
ultimatum 最后通牒
ultipor 多孔环氧纤维
ultra acoustic 超声学
ultra-audible frequency 超声频率，超音频
ultra-audion 超三极管，回授栅极检波器
ultrabasic rock 超基性岩
ultra-calan 超卡兰
ultra capacitor 超级电容器
ultracentrifugation 超速离心法【浓缩】
ultracentrifuge 超高速离心机，超速离心机
ultra-clay 超微粘粒
ultra-clean air system 超净空气系统
ultraclear environment 超洁环境
ultra-cold neutron 超冷中子，超快中子
ultrafilter membrane 超滤膜
ultrafilter 超滤器
ultrafiltrate 超滤液
ultrafiltration membrane process 超滤膜法
ultrafiltration process 超滤法
ultrafiltration system 超滤系统
ultrafiltration weakly acidic hydrogen cation exchange 超滤弱酸氢离子交换
ultra-filtration 超过滤作用，超滤，超过滤
ultrafine dust 特细粉末
ultrafine enameled wire 特细漆包线
ultrafine glass-coated wire 特细玻璃丝包线
ultrafine ion-exchange resin 超微料树脂
ultrafine particle 超微粒，特细粒子
ultrafine soot 特细煤烟
ultrafine 超细的，超精细的，特细的
ultraharmonics 超高次谐波

ultraharmonic 超调和的
ultra-high-early-strength portland cement 超高早强硅酸盐水泥
ultra-high frequency tuner 超高频调谐器
ultra-high frequency 特高频，超高频
ultra-high power electric arc furnace 超高功率电弧炉
ultra-high-purity helium atmosphere 超高纯度氦气氛
ultra-high purity 超高纯度
ultra-high resolution 超高分辨力
ultra-high speed transient wave form analyzer 超高速瞬时波形分析器
ultra-high speed 超高速
ultra-high strength steel 超高强度钢
ultra-high-temperature reactor 超高温反应堆
ultra-high temperature 超高温
ultra-high vacuum valve 超真空阀
ultra-high vacuum 超高真空
ultra-high voltage 特高压
ultra large coal 特大块煤
ultra light fold out solar-generator 超轻级伸展式光伏发电站
ultralimit 超极限
ultra-low emission 超低污染排放
ultra-low frequency 超低频
ultra-low temperature thermistor 超低温热敏电阻器
ultra-magnifier 超磁放大器
ultrametamorphism 超变质（作用）
ultramicroanalysis 超微量分析
ultramicrobalance 超微量天平
ultramicrochemistry 超微量化学
ultramicrodetermination 超微量测定
ultramicroearthquake 超微（地）震
ultramicrometer 超测微计
ultramicron 超微粒，超细粒
ultramicroscope 超倍显微镜
ultramicroscopic dust 特细微尘
ultramicro-technique 超微量技术
ultraminiature 超小型
ultramylonite 超糜棱岩
ultraoscilloscope 超短波示波器
ultra photometer 超光度计
ultra-porcelain 超高频瓷
ultra pressure turbine 超高压汽轮机
ultrapure water 高纯水，超纯水
ultrapure 超纯，超纯的
ultrapurity 超纯度
ultra-rapid flow 超速水流，高速水流
ultra-rapid hardening cement 超快凝水泥
ultra-red ray 红外线
ultra-red 红外的，红外线的
ultra sensitive clay 超灵敏黏土
ultrasensitive leak testing method 超灵敏检漏法
ultra-short wave 超短波
ultrasonator 超声振荡器，超声波发生器
ultrasonic alarm system 超声波报警系统
ultrasonic anemometer 超声波风速计
ultrasonic atomizer 超声波雾化器
ultrasonic beam 超声波束
ultrasonic bed load sampler 推移质超声波测定器
ultrasonic bonding 超声波焊接
ultrasonic bond meter 超声波式结合强度测定计
ultrasonic cavitation 超声空化
ultrasonic cleaner 超声波清洗机，超声清洗器
ultrasonic cleanliness 超声波净化
ultrasonic coagulation 超声凝聚
ultrasonic corona detector 超声波电晕检测器
ultrasonic crack detection 超声波探伤
ultrasonic current meter 超声波流速仪
ultrasonic damping 超声阻尼
ultrasonic defect detector 超声波探伤仪
ultra-sonic dehydration 超声波脱水
ultrasonic delay line 超声延迟线
ultrasonic depth finder 超声波回声测深仪
ultrasonic detector 超声波探测仪，超声检测器
ultrasonic dicing 超声切割
ultrasonic echo sounder 超声波回声测深仪
ultrasonic echo-sounding device 超声-回声探测器
ultrasonic electrode cleaner 超声波电极清洗器
ultrasonic examination 超声波检验［检查/探测/探伤］
ultrasonic flaw detection 超声检测，超声波探伤
ultrasonic flaw detector 超声波探伤机［器，仪］
ultrasonic flow measurement 超声波测流
ultrasonic flow measuring device 超声流量测量仪表
ultrasonic flowmeter 超声波流量计，超声流量计
ultrasonic frequency 超声频率，超音频，超声波频率
ultrasonic fusion joining 超声熔接
ultrasonic generator 超声波发生器
ultrasonic grinder 超声波打磨机
ultrasonic hardness measurement 超声波硬度测量
ultrasonic holography 超声波全息照相法，超声波全息术
ultrasonic humidifier 超声波加湿器
ultrasonic image tube 超声波显像技术
ultrasonic inspection and measurement 超声波检测
ultrasonic inspection 超声波检查，超声波探伤，超声波检验，超声检查，超声探伤
ultrasonic interferometry 超声波干涉量度法
ultrasonic investigation 超声波探测，超声波探查
ultrasonic leaching 超声波浸出
ultrasonic leak locator 超声波检漏仪
ultrasonic lens 超声波透镜
ultrasonic level measuring device 超声物位测量仪表
ultrasonic level meter 超声波（式）料位计，超声物位计
ultrasonic level sensor 超声波探测器
ultrasonic light diffraction 超声光衍射
ultrasonic liquid level indicator 超声波液位指示器
ultrasonic machining 超声波加工
ultrasonic measurement of stress 超声波应力测量

ultrasonic measurement 超声波测量
ultrasonic modulator 超声调制器
ultrasonic noise 超声波噪声
ultrasonic nonodestruction evaluation 超声无损评价
ultrasonic probe 超声波探头，超声探头
ultrasonic pulse method 超声脉冲法【测混凝土强度】
ultrasonic radiation 超声辐射
ultrasonic scanner 超声波扫描器
ultrasonic sensor 超声波传感器
ultrasonic separation 超声波分离
ultrasonic soldering 超声波软钎焊
ultrasonic spectrum 超声谱
ultrasonic spraying 超音速喷涂
ultrasonic stress meter 超声波应力仪，超声应力计
ultrasonics 超声波学，超声学
ultrasonic tank 超声波（液体）槽
ultrasonic testing method 超声试验法
ultrasonic test 超声波试验，超声波探伤试验，超声波检查
ultrasonic thermometry 超声波测温法
ultrasonic thickness gauge 超声波厚度计，超声波测厚仪
ultrasonic thickness indicator 超声波厚度测定器
ultrasonic thickness measurement 超声波厚度测量
ultrasonic thickness meter 超声波测厚仪，超声厚度计
ultrasonic vibration 超声振动
ultrasonic vibrator 超声波振动器
ultrasonic wave 超声波
ultrasonic welding inspection 超声波焊缝检查
ultrasonic welding 超声波焊，超声波焊接
ultrasonic 超声波的，超音速的，超声波，超声的
ultrasonoscope 超声波探测仪
ultrasound 超声
ultrastability 超高稳定度，超稳定度
ultrastable subsystem 超稳定的子系统
ultrastable system 超稳定系统
ultra-supercritical coal-fired power plant 超超临界火电厂
ultra-super critical pressure turbine 超超临界压力汽轮机
ultrasupercritical turbine 超超临界汽轮机
ultra-temperature 超高温
ultra thermostat 精密恒温器
ultrathin membrane 超薄膜
ultrathin 特薄的，超薄的
ultraudion 超三极管，（三极管）反馈线路
ultra-violet absorption spectrometry 紫外线吸收光谱法
ultraviolet band 紫外线谱带
ultraviolet energy 紫外线能量
ultraviolet filter 紫外线滤色片
ultraviolet lamp 紫外线灯
ultraviolet light source 紫外线光源
ultraviolet photoresistor 紫外光敏电阻器
ultraviolet pyranometer 紫外总日照表
ultraviolet radiation 紫外辐射
ultraviolet ray 紫外线
ultraviolet region 紫外线区
ultraviolet-visible spectrophotometer 紫外和可见光分光光度计
ultraviolet 紫外线，紫外线的，紫外的
ultra vires 越权，超越权限
ultrawhite synchronizing 超白色同步信号
ultra 超……的，超
ultron 波导耦合正交场放大管
umacro 通用宏指令，超宏
umbilical cable 连接电缆
umbilical 操纵用的，控制用的，地面缆线及管道
umbrella aerial 伞形天线
umbrella agreement 一揽子协议
umbrella arch 陡道护拱，隧洞顶拱
umbrella article 总括条款
umbrella gripper 伞形抓斗
umbrella insulator 伞形绝缘子
umbrella term 涵盖性术语
umbrella type alternator 伞形交流发电机
umbrella type generator 伞形发电机
umbrella type hydro-generator 伞式水轮发电机，伞式水力发电机
umbrella type of disc vibration 叶轮的伞形振动
umbrella 烟齿顶罩，保护伞，庇护，伞形结构
umformer 变换器，交流机
umklapp process 翻转过程，倒逆过程，碰撞过程
umpire assay 仲裁分析，校检分析
UMS(unit master sequence) 机组主控顺序
UM(unit master) 机组主控
unabsorbed cost 待摊成本
unacceptable criterion 不可接受性判据
unacceptable 不能接受的
unaccounted-for loss 未计入损失，其他损失
unadjustable speed electric drive 非调速电气传动
unadjusted 未调节的，无调节的
unaflow 单流，直流
unaided eye 肉眼
unaka 残丘
unalienable right 不可剥夺（的）权利
unallocated physical storage 未分配实存储器
unallowable instruction digit 非法字符，非法指令
unalloyed copper 非合金铜
unalloyed steel 非合金钢
unalterable element 系统的不变化部分
unalterable 不变化的
unaltered 不变的，未被改变的
unambiguous determination 单值确定
unambiguous 明显的，明白的，单值的，容易辨认的，不含糊的，清楚的，明确的
unamortized exchange rate 未摊还汇兑率
unanimous decision 一致决定
unanimous verdict 一致裁定
unarmoured cable 非铠装电缆
unary operation 一元操作，一元运算
unary operator 一元算子，一元运算符

unary 一元的
unassembled 未装配的
unasserted claim 未确定的索赔
unassignable letter of credit 不可转让的信用证
Unat(natural uranium) 天然铀
unattackable 耐腐蚀的，耐侵蚀的
unattended automatic exchange 无人值班自动电话交换机
unattended boiler 自动化锅炉
unattended booster pumping station 自动增压泵站，无人（操作）增压泵站
unattended hydroelectric power station 无人值班水电站
unattended operation 无人值班运行，无监督的操作，自动操作，全自动运行
unattended power station 无人值班发电站，无人管理电站
unattended remote-control power station 无人值班的遥控电站
unattended substation 无人值班变电所，无人值班变电站
unattended time 停机时间，（机器的）空闲时间，待修时间，非工作时间
unattended 无人监视的，无人值班的，自动的，无监督的
unauthorized person 未经许可的人员
unauthorized signature 未经授权签字
unauthorized 未经批准的，越权的
unavailability time 停用时间
unavailability 无效，无法利用，不能利用性，停堆，停机，不可利用率
unavailable energy 无用能
unavailable factor 不可用系数
unavailable heat 无效热，不可利用的热
unavailable hours 不可用小时
unavailable state 不可用状态
unavailable time 不可用时间
unavailable water 无效水
unavailable 不可利用的，不在役
unavoidable accident 不可避免的事故
unavoidable cost 无法俭省的成本，不可避免的成本，不可控成本
unavoidable risk 不可避免的风险
unavoidable 不可避免的
unbacked 无衬的，无支持的，无助的
unbaffled bed 无挡板床
unbaffled rod bundle 裸棒束
unbalanced bridge 不平衡电桥
unbalanced circuit 不平衡电路
unbalanced component 不平衡分量
unbalanced current 不平衡电流
unbalanced dynamic 动不平衡
unbalanced end thrust 不平衡轴向推力
unbalanced factor 不平衡系数，不平衡因素
unbalanced fault 不平衡故障，不平衡事故，不对称故障
unbalanced flow 非平衡流，不平衡流动，非稳定流
unbalanced force from internal pressure 管道内压力不平衡力
unbalanced force 不均衡力
unbalanced line 不平衡传输线，不平衡线路
unbalanced load 不平衡负载，不对称负荷
unbalanced magnetic pull 不平衡磁拉力
unbalanced operating condition 不稳定工况
unbalanced operation 不对称运行，不平衡运行
unbalanced phase 不平衡相位，不对称相位
unbalanced polyphase load 不平衡多相负荷，不对称多相负荷
unbalanced ring 不平衡环
unbalanced seal 不平衡密封
unbalanced stator fault 不平衡定子故障，定子失衡故障
unbalanced system 不平衡制，不对称制
unbalanced tension 不平衡张力
unbalanced three-phase voltage 不平衡三相电压，不对称三相电压
unbalanced trench 不通行管沟
unbalanced valve 不平衡阀
unbalanced voltage 不平衡电压，不对称电压
unbalance dynamic 动（态）不平衡
unbalance force 不平衡力
unbalance loading condition 不平衡工况
unbalance mass 不平衡质量
unbalance moment 不平衡力矩
unbalance protection 不平衡保护
unbalance reduction ratio 不平衡减少率
unbalance response 不平衡反应，不平衡特性
unbalance static 静不平衡
unbalance voltage 不平衡电压
unbalance weight 不平衡重量
unbalance 不平衡
UNBAL＝unbalance 不平衡
unbiased error 无偏误差
unbiased estimation 无偏估计
unbiased estimator 无偏估计量，无偏估计值
unbiased fuel element 未结合的燃料元件
unbiased manner 无偏见的态度
unbiased test 无偏压试验
unbiased variance 无偏方差，均方差
unbiased 不偏的，未加偏压的，无偏压的，公正的，无偏见的
unbiassed error 无偏误差
unbiassed estimation 无偏差估计
unbiassed statistic 无偏统计，无偏差统计量
unbiassed test 无偏差检验
unblended 未掺和的
unblocked area 非堵塞截面
unblocked level 开启电平
unblock 解锁，接通，开启
unbolt 卸螺钉，打开
unbonded design 按无握裹效应设计
unbonded prestressed bar 未黏结预应力钢筋
unbonded prestressing tendon 无粘结预应力锚索
unbounded function 无界函数
unbounded lattice 无限大栅格【堆芯】
unbounded medium 无限介质
unbounded stream 无界水流
unbound water 非结合水
unbound 游离的，自由的
unbraced frame 无支撑框架
unbraced length of column 无支撑柱长

unbraced length 无支撑长度
unbraced 不加撑的
unbranched stream 无分叉水流
unbreakable glass 不碎玻璃
unbridled competition 盲目竞争
unbroken wave 未碎波，未破波
unbuffered report control block 不缓存报告控制块
unbuffered report control 不缓存报告控制
unbuffered report 非缓存报告
unbuild 不能自励，去磁，失磁，破坏，拆毁
unbuilt 未兴建的
unbundled rate 分解费率
unbundled service 分解服务，分类服务
unburned carbon in flue dust 飞灰含碳量
unburned carbon in slag 炉渣含碳量
unburned carbon loss 机械不完全燃烧损失，未燃尽碳损失
unburned carbon 未燃烧碳，未燃尽碳，未燃碳，飞灰含碳
unburned combustible in flue dust 飞灰可燃物含量
unburned combustible in sifting 漏煤可燃物含量
unburned combustible in slag 炉渣可燃物含量
unburned combustible loss 未机械不完全燃烧损失，未燃尽碳损失
unburned combustible 未燃尽可燃物
unburned gas 未燃尽气体
unburned hydrocarbon 未燃烃
unburnedness 未燃尽度
unburned part 未燃部分
unburned refuse 未燃垃圾
unburnt brick 欠火砖，未烧透砖
unburnt carbon loss 未烧尽炭损
uncaging 释放，松开
uncanned 无外壳的，不加外壳的
uncased well 未加固水井
uncased 已去外壳的，未罩外壳的
uncertain factor 不确定因素
uncertain region 不确定范围，不可辨区
uncertainty analysis 不确定性分析
uncertainty in measurement 测量误差，测量的不确定度，测量的不精确性
uncertainty principle 测不准原理，不确定性原理，不定性原理
uncertainty range 不确定范围
uncertainty relation 不定关系
uncertainty 不定因素，无把握，不确定，不可靠性，无常，误差
uncertain variability of wind 风的不确定性
uncertain 含糊的
uncharacteristic harmonic 非特征谐波
uncharged 不带电的，未充电的，不带电荷的
unchecked 未校核的
unclad structure 无围护结构
unclad 无包壳的，裸露的
unclaimed cargo 待领货物
unclamp 松开
unclassified 未分类的
unclean B/L 不清洁提单
unclosed contour 不闭合等高线
unclosed porosity 不封闭孔隙率
unclosed switchgear panel 敞开式开关柜
unclosed traverse 不闭合导线，展开导线
uncoated oxide fuel particle 无包覆氧化物燃料颗料
uncoated substrate 无涂层衬底
uncoated 未被包覆的
uncoaxiality 不同轴性
uncoded output 非编码的输出
uncoded word 未编成代码的字
uncoiler 开卷机，拆卷机
uncoil 松开线圈，拉直，伸开，伸展
uncollectible account 呆账
uncollided flux density 未碰撞通量密度
uncollided flux 非碰撞能量
uncollided intensity 未碰撞强度
uncollided neutron flux 未经碰撞的中子通量，初始中子能量
uncollided neutron 未碰撞中子
uncombined form 游离态
uncombined 未化合的，游离的，自由的
uncommitted amount 未支配款项
uncommitted contract 不受约束的合同，不受约束合同
uncommitted cost 未承诺费用
uncommitted surplus 未定用途的盈余，未支配盈余
uncommitted 未支配的，不受约束的
uncompacted rockfill 未压实堆石
uncompacted 不密实的，未压实的
uncompensated amplifier 无补偿放大器
uncompensated attenuator 未补偿衰减器
uncompensated flowmeter 无补偿流量计
uncompensated 未加补偿的，无补偿的
uncompleted construction project 未完成的工程[项目]，在建工程[项目]
uncompleted contract 未完成的合同
unconditional acceptance 无条件接受
unconditional branch 无条件转移
unconditional delivery 无条件交付
unconditional-jump instruction 无条件转移指令
unconditional jump 无条件转移
unconditional L/G 无条件保函
unconditionally stable 无条件稳定的
unconditionally 无条件地
unconditional probability 无条件概率
unconditional stability criterion 无条件稳定判据
unconditional transfer instruction 无条件转移指令
unconditional transfer 无条件转移
unconditional 无条件的
unconditioned zone 非空气调节区
uncondition statement 无条件语句
unconfined aquifer 非承压含水层，非承压水含水
unconfined compression apparatus 无侧限压力仪，无侧限压缩仪，无侧限压缩装置
unconfined compression strength 无侧限抗压强度
unconfined compression test 无侧限压缩试验
unconfined compression 无侧限压缩
unconfined cylindrical sample 无侧限圆柱体试样
unconfined groundwater 非承压地下水

unconfined vortex wind machine 非约束涡型风力机
unconfined 自由的，松散的，无侧限的
unconfirmed credit 不保兑信用证
unconfirmed letter of credit 非保兑信用证，不保兑信用证，未确认信用证
unconfirmed 未经确认的
unconformity surface 不整合面
unconformability 不整合
unconformable strata 不整合地层
unconformity 不整合，不一致
unconsolidated-drained test 不固结排水剪切试验
unconsolidated investment 不合并计算的投资
unconsolidated sediment 未固结沉积
unconsolidated soil material 未固结土料
unconsolidated surface layer 未固结表层
unconsolidated-undrained test 不固结不排水试验
unconsolidated-undrained triaxial test 不固结不排水三轴试验，室内土工试验
unconsolidated 未固结的，未凝固的，松散的，非固结的
unconstrained air 非约束空气
unconstrained body 非约束体
unconstrained trading schedule 无约束交易计划
uncontaminated soil 未污染土
uncontaminated 未沾染的，未污染的，无杂质的
uncontamination 未污染
uncontracted weir 非收缩堰
uncontrollable cost 不可控成本，不易管理成本
uncontrolled acceleration 不可控加速作用
uncontrolled area （核电厂内）非控制区
uncontrolled chain reaction 失控链式反应
uncontrolled crest spillway 无闸门控制的溢洪道
uncontrolled device 不可控器件
uncontrolled fuel-assembly 燃料组件失控
uncontrolled leakage （冷却剂）失控漏泄
uncontrolled levitation 失控漂浮【燃料组件】
uncontrolled natural ventilation 无组织自然进风，自然通风
uncontrolled oscillation 无控振荡，不可控振荡
uncontrolled reaction 失控反应
uncontrolled rise 失控上升
uncontrolled rod withdrawal accident 控制棒失控提升事故，失控提棒事故
uncontrolled runaway 失控上升【堆功率或反应性】
uncontrolled storage 无控制设备的水库
uncontrolled weir 无控制堰
uncontrolled 不受控制的
unconverted reactant 未转化反应物
uncooled motor 无冷却式电动机
uncooled reactor 无冷却剂反应堆
uncorrected data processor 错误数据处理器
uncorrected delay 未校正延迟，未修正结果
uncorrected 未修正的
uncorrelated variable 不相关变量
uncoupled and resting on back seat 脱开并搁置在后座上【反应堆冷却剂泵】
uncoupled car 已摘钩的车辆
uncouple 去偶，去耦，解去联系，拆开，脱开，分开
uncoupling control station 摘钩控制站
uncoupling 解列，解耦，去耦，脱离，脱扣，非耦联，摘钩
uncoursed rubble masonry 不分层乱石圬工
uncovered tray 无盖托架
uncovered wire 无绝缘导线，裸线
uncovering 堆芯裸露，未湮没【堆芯】
UNCTAD（United Nations Conference on Trade and Development） 联合国贸易和发展会议
uncultivated land 荒地，未垦地
uncut grass 未剪草地
undaform 浪蚀地形
undamped control 不稳定控制，不稳定调节
undamped frequency 无阻尼频率
undamped motion 无阻尼运动
undamped natural frequency 非阻尼固有频率，等幅固有频率
undamped natural of vibration 固有模态
undamped oscillation 无阻尼振荡，未衰减振荡，等幅摆动
undamped resonant frequency 无阻尼共振频率，无阻尼谐振频率
undamped vibration 无阻尼振动
undamped wave 无阻尼波，未衰减波
undamped 无阻尼的，未加阻尼的，不衰减的，未受潮的
undated 未定日期的，未注明日期的，无限定日期的
undathem 浪蚀岩层
unda 浪蚀底
undecimal 十一进制的
undecipherable 不可译的
undefined record 未定界记录，不定长记录
undelayed channel 无延迟信道
undelayed pulse 无延迟脉冲
undelete 恢复功能
under adverse pressure condition 在反压条件下
under all circumstances 无论如何
under allowance 尺寸下偏差
under-annealing 不完全退火
underbalance 平衡不足，欠平衡
underbead crack 焊接热影响区裂纹，焊道下裂缝［裂纹］，内部裂纹
under beam girder 梁下纵桁
underbed coal feed port 床底加煤点
underbed 底架，底座
under(/below)ground hopper 地下料斗
underbid 投低价标，出价低于……，出较低的价钱，叫牌偏低，出价较低
under bracing 下支撑，帮桩
undercapacity factor 欠载系数
undercapacity 容量不足，功率不足，出力不足，生产率不足，非饱和容量
undercarriage 底架，机架
undercharge 充电不足，非饱和充电
underclad cold cracking 堆覆层下冷裂
underclad cracking 堆覆层下裂纹
underclad 堆覆层下面
undercoat 底涂层，涂层下面，内涂层
undercommutation 整流不足，换向不足，滞后换向

undercompensated integral control 欠补偿的积分控制
undercompensated 欠补偿的
undercompensation 不完全补偿，欠补偿，补偿不足
undercompound excitation 欠复励磁，欠复励，欠复激
undercompound generator 欠复励发电机
undercompound winding 欠复激绕组，欠复激绕法
undercompound 欠复激的，欠复绕的，欠复励的
under-consolidated clay 欠固结黏土，欠压密黏土
under construction project 进行中的项目
under construction 正在施工，施工中
undercooled steam 过冷蒸汽
undercool 过度冷却，过冷，欠热
under correction 有待改正，不一定完全对
undercoupling 不完全耦合，耦合不完全
undercrimping 卷边不足
undercritical 次临界的，亚临界的
undercroft 穹形地下室
undercrossing 地下通道
undercuring 欠处理，养护不够
undercurrent circuit breaker 欠电流断路器
undercurrent protection 欠电流保护
undercurrent relay 低地低电流继电器，欠电流继电器，欠流继电器
undercurrent release 欠流释放，欠流脱扣，欠流断路器
undercurrent 欠电流，电流不足，暗流，欠电流的，低电流，底流，潜流，下层流
undercut commutator （被电刷）磨损太深的换向器
undercut slope 底切岸坡，淘蚀岸坡
undercutter 下刻机
undercutting 咬边【焊缝】，底部切口，淘空，淘刷，下切，下切作用，下陷
undercut trimmer saw 脚踏截锯机
undercut 咬边【焊缝】，咬齿，根切，下部凹陷，切槽，凹割，下刻，凹口，底切，掏槽，掏挖
underdamped harmonic motion 欠阻尼谐振荡
underdamped 欠阻尼，欠阻尼的，阻尼不足的
underdamp 阻尼不足，欠阻尼，弱阻尼
under-deck cargo 舱内货
underdeposit corrosion 垢下腐蚀
underdesign 安全系数不足的设计
underdeveloped area 未发展地区
underdeveloped country 不发达国家，欠发达国家，低度开发国家
underdeveloped region 不发达区域
underdeveloped 不发达的，未充分开发的
underdevelopment 显影不足，不发达，欠发达（地区）
underdome bell 机内式电铃，罩盖式电铃
underdrainage system 地下排水系统
underdrainage 暗沟排水，地下排水，地下排水沟，阴沟
underdrain distributor 排水配水器
underdrain system 排水系统
underdrain 地下沟道，暗沟，阴沟，地下排水管，用暗沟疏水
under edge 下缘
under-engineering 设计不足
underestimate 低估
underestimation 低估价，低估
underexcitation protection 欠励保护
underexcitation 欠励，励磁不足
underexcited motor 欠励电动机
underexcited operation 欠励磁运行
underexcited 励磁不足的，欠励的
under existing condtions 在现有条件下
underexpand 膨胀不足，弱膨胀
underexpansion 膨胀不足
underexposure 曝光不足
under favourable condtions 在有利条件下
underfeed burning 下给料式燃烧
underfeed combustion 下给料式燃烧
underfeed firing 下饲式燃烧
underfeed fuel bed 下饲式燃料床
underfeed furnace 下饲式炉膛
underfeed 补给不足，地下补给，下饲
underfelt 底层油毡
underfiling 下面锉磨
underfill 未满（缺焊），缺焊
underfilm corrosion 膜下腐蚀
under-fire air 炉排下进入的空气，燃料层下进风
underfired 下部燃烧的，自下燃烧的
underfiring fuel 未烧尽的燃料
underfit river 不相称河，废河
underfloor condenser 地面下凝汽器，底层凝汽器
underfloor heating 地板采暖
underfloor raceway 地板下电缆管道
underflow baffle 导流板，底流挡板
underflow density current 底部异重流
underflow gate 底流式闸门
underflow turbidity current 底部混浊流
underflow water 伏流水
underflow 地下水流，底流，潜流，下溢，槽下水流
underframe 底架，基础构架，机座，底框，托架
under frequency operation 低频率运行
under frequency protection 低频保护，低周波保护
under frequency relay 欠频继电器
under frequency sensor 低频传感器
under frequency trip 低频率事故保护停堆，低频率紧急停堆
under frequency 低频率，频率过低，低于额定值的频率
under given conditions 在特定条件下，在一定条件下
underglaze 上釉前着色的，用于上釉之前的，釉底的，釉下的
undergorund power distribution in residential district 住宅区地下配电
undergorund power house 地下厂房
undergo 经历，遭受，忍受
undergrate air 炉底空气，炉排下进风，（火床燃烧）一次风

undergrate plenum chamber　炉排下风室
underground AC power transmission　地下交流输电
underground air channel　地下风道
underground alignment　地下放线，坑道定线
underground cable channel　地下电缆槽
underground cable distribution network　电缆配电网
underground cable　地下电缆【可以直接掩埋或穿在导管中】
underground canal　地下渠道
underground chamber　地下室
underground channel　地下河道，地下通道，地下渠道
underground coal bucket room　地下煤斗间
underground coal tunnel　地下运煤隧道
underground coal unloading ditch　地下卸煤沟
underground communication　地下通信
underground compressed air storage reservoir　地下蓄气库
underground conduit net　地下水道网
underground configuration　地下轮廓线
underground construction　地下构筑物
underground containment　地下安全壳
underground continuous wall　地下连续墙
underground corrosion　地下腐蚀
underground dam　地下坝
underground DC power transmission　地下直流输电
underground diaphragm wall　地下连续墙
underground disposal　地下处置，地下放置，排入地下
underground distribution cable　地下配电电缆
underground distribution　地下配电
underground drainage　地下排水
underground drain　地下排水管
underground electrical duct　地下电气导管
underground engineering　地下工程
underground erosion　地下侵蚀，潜蚀
underground excavation　地下开挖
underground exploration　地下勘探
underground explosion　地下爆破
underground facilities　地下设施
underground facility arrangement　地下设施布置
underground flow　地下水流，潜水流
underground free way　地下通道
underground garage　地下车库
underground gasification　地下气化
underground hopper　地下煤斗
underground hydraulic power plant　地下水电厂
underground hydroelectric power station　地下（式）水电站
underground hydropower house　水电站地下厂房，地下式水电站
underground injection　（废物的）地下注入
underground installation　地下敷设，地下设施，地下设施布置
underground laying　地下敷设【电缆、管道等】
underground levelling　坑道水准测量
underground line　地下线路
underground mark　地下标点
underground nuclear power plant　地下核电厂
underground nuclear unit　地下核设施
underground oil reservoir　地下储油仓
underground opening　地下孔洞
underground parking　地下停车
underground part　地下部分，地下部件
underground passage way　地下通道
underground penstock　地下埋管
underground pipeline finder　地下管线探测仪
underground pipeline　地下管线，地下管道
underground piping　地下管道
underground plant　地下工厂
underground plastic insulated wire　地埋线
underground pollution　地下污染
underground power distribution　地下配电
underground power house　地下式厂房
underground power plant　地下电站
underground power station　地下电站，地下式电站
underground power supply　地下供电
underground power transmission　地下输电
underground pumped hydrostorage　地下抽水储能
underground railway　地下铁道
underground reservoir　地下水库
underground river　暗河，地下河流
underground runway　地下滑道
underground sprinkler system　埋管式喷灌系统
underground steam generation　地下蒸汽发电，地热发电
underground storage　地下储存，地下储量，地下水库，地下蓄水
underground stream　伏流河
underground structure　地下建筑（物），地下结构
underground substation　地下变电所，地下变电站
underground survey　地下测量，坑道测量
underground system　地下输电系统，地下系统
underground tank　地下罐，地下水箱，地下槽
underground transfer post　地下转运站
underground transformer　地下变压器
underground transmission cable　地下传输电缆
underground transmission line expenses　地下输电线路费用
underground transmission　地下输电，地下传输
underground utilities　地下公共设施，地下公用管道设施【水、电、气等】，地下设施
underground water basin　地下水域
underground watercourse　地下河道，地下水道
underground water divide　地下分水线，地下水分水线
underground water level　地下水面，地下水位
underground water pollution　地下水污染
underground water resources　地下水资源
underground watershed　地下流域
underground water supply　地下水供应
underground water table　潜水面
underground water　地下水，潜水
underground works　地下工程
underground　地下，地下的
underheating　欠热
underhung door　扯门
underhung spring　悬簧

underhung window 扯窗
under-impedance relay 低阻抗继电器
under-land waste disposal 深埋废物
under lap 负重叠度
underlayer 底层，下层，底垫，垫底层，下垫层
underlaying material 衬底材料
underlaying 垫底，衬底材料，基础，垫物
underlay 衬底，从下面支撑，铺在下面
underlie 下伏
underline 强调
underload (circuit)breaker 欠载断路器
underloading 负载不足，轻载
underload relay 欠载继电器，负载不足继电器
underload 欠负载，加载不足，负载不足，负荷不足
underlying bedrock 下伏基岩
underlying bed 下垫层，下伏岩层，下卧层
underlying cause 根本原因
underlying company 附属公司，子公司
underlying concept 基本概念
underlying contract 基础合同
underlying document 原始单据
underlying foundation material 基础垫层材料
underlying metal 底层金属
underlying rock 下垫岩石，下伏岩石
underlying soft soil 下伏软土
underlying soil 下伏土
underlying strata 底板岩石
underlying stratum 下伏地层
underlying surface 伏面，下垫面，下伏面
underlying topography 下垫面地形
underlying 潜在的，下层的，表面下的，下伏的
undermine 侵蚀……的基础，暗中破坏，底部冲刷，潜挖，挖掘道，渐渐破坏，挖掘地基，挖……的墙角，危害
undermining blast 坑道爆破，潜挖爆破
undermining method 暗挖法【地下工程】
undermining 底部冲刷，淘底冲刷，挖坑道
undermoderated blanket 欠慢化转换区
undermoderated lattice 欠慢化栅格
undermoderated mixture 欠慢化混合物
undermoderated reactor 欠慢化反应堆
undermoderated system 欠慢化体系
undermoderated 慢化不足的，欠慢化，欠调制的，调制不足的
undermoderation 慢化不足，欠慢化
undermodulated 调制不足，欠调制
undermodulation 欠调制，调制不足
under no circumstance 无论如何都不
under one's hand and seal 签名盖章
under otherwise equal conditions 其他条件都相同时
under otherwise identical conditions 其他条件都相同时
underpan 底壳，底盘，托盘
underpass 地道，地下通道，下穿道，高架桥下通道，桥下孔道
underpavement 下层路面，下层铺面
underpile feeder 堆下给煤机
underpile hopper 煤堆下料斗，堆下煤斗
underpinned pile 托换桩
underpinning 基础结构，托换基础
underpin 托换基础
underpitch groin 两个不同高度的圆拱接头
under pivoted type suspension clamp 上杠式悬垂线夹
underplaster raceway 灰浆层下电缆管道
underplate 底板，底座，垫板，基础，基座
underpour type gate 下泄式闸门
underpower protection 低功率保护
underpower relay 低功率继电器
underpower 功率不足，动力不足，低功率
underpressure mode 欠压模式
underpressure 欠压，负压，抽空，压力不足，失压，压力降低，空气稀薄
under pressure 在压力下，面临压力，受到压力
underproof 不合标准的，不合格的，被试验的
underrating 出力不足
underreamed anchor 扩体型锚杆
underreamed bored pile 扩底钻孔桩
underreamed foundation 底部扩大式基础
underreamed pile 扩孔桩
underreamed zone 管下扩眼井段
underreamed 扩孔的
underreamer cutter 管下扩眼器刀刃
underreinforced concrete 少筋混凝土
under-reinforced 配筋不足的
under … relay 低（或欠）……继电器
under reserve 保留追索权，保留追偿权
under-roof area 建筑面积【厂房或建筑】
underrun 低负荷运行，欠负荷运行，在……下通过，欠载运行
undersanded mix 少砂拌和物
undersaturation 欠饱和，未饱和
undersea cable 海底电缆
undersea delta 水下三角洲
undersea detection 水下探测
undersea device 潜水装置，水下设备
undersea 海底
undersell competitor 售价比竞争者低
under service conditions 在使用条件下
undershoot 低于额定值，小于额定值，负脉冲信号，负尖峰信号
undershot water wheel 底射式水轮，下击式水轮
undershot 下射式
undershrub 矮灌木
underside （水平管的）下半部，底侧，下侧，下端，下面，阴暗面
undersigned （法律文件末尾的）签名者，签字人【前面加定冠词】，签名的
undersize aggregate 逊径骨料
undersize fraction 限下率
undersize material 筛底物料，筛下物料
undersize 尺寸不足，尺寸过小，筛下物
undersluice 底部泄水道，下泄式水闸
undersoil 下层土
under specified conditions 在特定条件下
underspeed switch 低速开关
underspeed 欠速
underspin 下旋
understable 欠稳定的，不够稳定的

understandability 可理解性
under standard and off-standard condition 在标准和非标准条件下
understand 理解，谅解
understratum 下部地层，下层
understructure works 地下工程
understructure 基础，下部结构
undersupply 供料不足，供电不足，供应不足，缺电
underswing 负尖峰【信号】，（摇摆）幅度不足
undersynchronous braking 低于同步制动，次同步制动
undersynchronous 低于同步的，次同步的
undertake one's security 担保某人的安全
undertake receiving hopper 室内受煤斗
undertake responsibility 承担责任
undertaker 承办人，承办商
undertake to do sth. 承诺（做）……
undertake 承担，从事，保证，同意，答应，承诺，进行，许诺，担保
undertaking contract project 承包工程
undertamping 夯实不足
undertension 电压不足，低于额定值的电压
under the auspices of 由……主办，在……赞助下
under the circumstances 在这些情况下
under the condition that 在……条件下
under the condtion of 在……条件下
under the direction of 在……领导下
under the given conditions 在给定条件下
under the pain of nullity 否则按无效论处
under-the-plug refueling （屏蔽）塞下换料
under the precondition of 在……前提下
underthrust 俯冲断层
undertightening 拧紧不足
undertow current 回卷流，裂流
undertow 回卷流，激流，裂流
undertrack hopper 轨下煤斗
undertracks receiving hopper 铁轨下受煤斗
undervibration 振捣不充分
undervoltage coil 失去电压的线圈
undervoltage condition 欠压状态
undervoltage protection 低电压保护，欠压保护
undervoltage relay 欠压继电器，低电压继电器，低压继电器
undervoltage release 低电压释放，低电压跳闸，低电压脱扣器
undervoltage sensor 低压传感器
undervoltage trip 低压自动跳闸，低电压保护停堆，低电压紧急停堆
undervoltage 电压不足，低电压，欠电压，失压，欠压，低于额定值的电压
underwater apron 水下护坦
underwater bank 水下浅滩，水下沙滩
underwater blasting 水底爆炸，水下爆破
underwater cable 水底电缆，水下电缆
underwater camera 水下摄影机
underwater cartridge chopper 水下元件切断机，水下元件切割机
underwater cast-in-place pile 水下灌注桩
underwater concrete 水下混凝土
underwater concreting 水下灌注混凝土，水下混凝土浇筑
underwater construction 水下施工
underwater contour 水深线
underwater conveyor 水下输送机
underwater crane 水下升降机
underwater current 水中暗流
underwater cutting 水下切割
underwater drilling 水下钻探
underwater driving 水下打桩
underwater earthquake 水下地震
underwater engineering 水下工程
underwater excavation 水下开挖
underwater excavator 水下掘削机
underwater explosion 水底爆炸，水下爆破，水中爆炸
underwater fill 水下填土
underwater floodlamp 水下聚光探照灯
underwater foundation 水下基础
underwater fuel handling 水下燃料操作
underwater illumination 水下照明，水中照度
underwater lamp 水下灯，水底灯，水景灯
underwater lighting system 水下照明系统
underwater lighting 水下照明
underwater listening post 水下收听站
underwater maintenance 水下保养，水下维修
underwater manipulator 水下机械手，水下控制器
underwater motor 潜水电动机
underwater nephelometer 水下浊度计
underwater observation 水下观察
underwater operation 水下作业
underwater outfall 水下出口
underwater photogrammetry 水下摄影测量学
underwater pipeline 水下管道，水下管路
underwater platform 水下平台
underwater power plant 水下发电厂
underwater probing 水下探测
underwater pump 潜水泵
underwater reactor 水下反应堆
underwater resources 地下水源
underwater revetment 水下覆护，水下挡墙
underwater sawing 水下锯割
underwater sound gear 回声测探仪，水声设备
underwater storage tank 水下储罐
underwater storage 水下储存
underwater survey 水下调查，水下勘测
underwater survillance 水下监视
underwater technology 水下工艺学
underwater telemeter 水下遥测器
underwater telemetry 水下遥测
underwater television 水下电视
underwater test 水下试验
underwater topography 水下地形
underwater transparency meter 水下透明度计
underwater tunnel 水下隧洞
underwater TV camera 水下电视摄像机
underwater ultrasonic 水下超声波
underwater viewer 水下潜望镜，水下窥视窗
underwater visibility 水下能见度
underwater wave guide 水下波导

underwater welding 水下焊，水下焊接
underwater works 水下工程
underwater 水下的，在水中的，水面下的，在水下
underway 正在进行中
underweight 重量不足
under-wood 丛林
underwriter 保险人，保险公司，保险商，承诺支付者，承销人，担保人
underwriting the risk 承担风险
underwriting 承担
under 依照，根据，低于，在……之下
undesired signal 无用信号
undesired sound 噪声
undetectably low concentration 探测不出的低浓度
undetected error rate 残留差错率，漏检错误率
undetermined coefficient 待定系数，未定系数
undetermined multiplier 未定乘数，待定乘数
undetermined value 未定值
undeveloped area 不发达地区，处女地，未开发地区，未开发区
undeveloped land 荒地，未开发地区
undeveloped region 不发达地区
undiluted oil 纯油
undiluted 不掺和的
undirected tree 不定向树，双向树
undirectional coupler 单向耦合器
undirectional flow 不定向流
undiscovered resources 未发现资源
undisputed soil sample 无争议的土样
undistorted model 正态模型，无畸变模式
undistorted wave 无畸变波
undistorted 无失真的，不失真的，无畸变的
undistributed profit 未分配利润
undisturbed airflow 未受扰气流
undisturbed boundary layer 未扰动边界层
undisturbed clay 未扰动黏土，原状黏土
undisturbed flow 未扰动气流，无扰动流
undisturbed fluid 无扰动流体
undisturbed ground 未扰动地基
undisturbed jet 无扰动射流
undisturbed one output 未打扰的“1”输出，不干扰“1”输出【计】
undisturbed rock 原状基岩
undisturbed sample boring 原状土样钻取
undisturbed sample 不扰动土样
undisturbed sand 原状砂
undisturbed soil sample 未扰动土样，原状土样
undisturbed soil with uplift resistance 原状抗拔土体
undisturbed soil 未扰动土，无扰动土，原状土
undisturbed stream 无扰流
undisturbed structure 未扰动结构
undisturbed transfer 无扰动切换
undisturbed velocity 未扰动流速
undisturbed water 未扰动水
undisturbed wind speed 未受扰风速
undisturbed wind 未扰动风
undisturbed-zero output 未打扰的“0”输出
undisturbed 未被打扰过的，未扰动的，静止的
undivided profit 未分配利润
undocking 出坞，离码头
undo 拆卸
undrained creep 不排水蠕变
undrained quick shear test 不排水快剪试验
undrained shear strength 不排水抗剪强度
undrained shear 不排水剪切
undrained test 不排水试验
undrained triaxial compression test 不排水三轴压缩试验
undrained triaxial test 无排水三轴试验
undressed 未经处理的，未加工的
undrilled 未钻孔的
undrinkable water 非饮用水
undue debt 未到期债务
undue exposure 超剂量照射
undue 不适当的，不正当的，非法的，过度的，未到期的
undular jump 波状水跃，不完全水跃
undular 波形的，波状的，波纹的
undulated nappe 脉动水舌
undulated series winding 波绕组
undulated 波动的，波形的，起伏的
undulating ground 起伏地，丘陵
undulating surface 波状表面
undulating terrain 丘陵，地势起伏，起伏地形
undulating topography 起伏地形，丘陵地形
undulation period 波动周期
undulation 波浪形，波动，起伏，波纹，波荡
undulatory jump 波状水跃
undulatory motion 波荡运动，波状运动
undulatory theory 波动学说
UND(unit derating) 机组降低出力
unearthed 未接地的，不接地的
uneconomic site 不经济坝址
unelasticity 非弹性
unelastic 非弹性的
unemployment insurance 待业保险费，失业保险
unemployment 失业
unencapsulated 不封装的，未封装的
unenclosed relay 敞开型继电器
unenclosed 敞开的，开启的
unending circulation 连续环流
unequal angle iron 不等边角钢，不等边角铁
unequal angle steel 不等边角钢
unequal extension 杆塔长短腿
unequal foundation settlement 不均匀地基沉降
unequal pressure 不等压力
unequal settlement 不等沉陷，不均匀沉降
unequal subsidence 不均匀沉降
unequal tower leg 高低（塔）腿【线路】
unequivocal 不含糊的，明确的
uneven fracture 不规则断裂，参差状断裂
uneven grain 不均匀颗粒
uneven ice accumulation 不均匀覆冰
uneven irradiation 不均匀辐照
unevenly compressible foundation 不均匀压缩地基
unevenly 不均匀的，不平坦的
uneven melting of electrode covering 焊条药皮的不均匀熔化

unevenness 不均匀性，凹凸度，不平度，不均匀度，不平整度，粗糙度
uneven running 不稳定运行
uneven settlement of foundation 基础不均匀沉陷
uneven settlement 不均匀沉降
uneven surface 粗糙表面
uneven temperature distribution 不均匀温度分布
uneven terrain 不平坦地形
uneven 非偶数的，不平的
unexcited 未励磁的，未激励的，欠激励的
unexecuted agreement 未执行的协议
unexpected expenses 不可预见费
unexpected surveillance 意想不到的监督
unexpired term 未满期限
unexplained loss 不明损失，无名损失
un-exploited 未被利用的
unexposed side 未暴露侧，冷侧
unexposed 未受辐照的，未经照射的，未曝光的
unextracted 不被提取的，不被萃取的
unfair competition 不公平竞争
unfair price 不合理价格
unfair treatment 不公正待遇
unfaulted pole 非故障极
unfaulty conditions 无故障状态
unfavorable balance of trade 贸易逆差
unfavoured factor 不利因子，不利因素
unfavoured 不利的，不适宜的，不良的
UNFCCC(United Nations Framework Convention on Climate Change) 联合国气候变化框架公约
unfilled and enclosed cartridge fuse-link 无填料封闭管式熔断器
unfilled order 未发货订单
unfilled porosity 非饱和孔隙度
unfiltered air 未过滤的空气
unfiltered water 未过滤水
unfiltered 未过滤的
unfinned fuel element 无肋片燃料元件
unfired combined-cycle system 不补燃联合循环系统
unfired drum 不受火气包
unfired heat recovery combined cycle 排热回收联合循环
unfired pressure vessel 非受火压力容器
unfired steam generator 非燃式蒸汽发生器
unfired waste heat boiler 不补燃的余热锅炉
unfired 不受火的，未燃烧的
unfitted 不合格的，未装备的
unfixed sand 未固定沙
unfluidized state 非流化状态
unfolding 展开，显露
unforeseeable event 不能预料的事件，难以预见的情况
unforeseen event 未预见事件
unforeseen item 不可预见项目
unforeseen miscellaneous 未预见项
unformatted read 无格式读
unformatted write 无格式写
unformatted 无格式的
unformed 不用模板的，无模板的
unfractured medium 无裂缝介质
unfree water 固着水
unfreeze 解冻
unfrozen port 不冻港
unfrozen river 不冻河流
unfrozen 不冻的
unfulfilled obligation 尚未履行的责任［义务］
unfull quenching 不完全淬火
unfurling 扬帆
ungated flow 敞开泄流
ungated 闭塞的，截止的
ungauged basin 未设站流域
ungauged drainage 未设站流域
ungauged watershed 未设站流域，未施测流域
ungearing 脱扣
unglazed collector 无透明盖板集热器
ungoverned 未经调节的，不调节的
ungraded aggregate 无级配骨料
ungraded gravel 非标准砂砾，未经拣选的砂砾
ungraded 不分级的，不合格的
ungrounded neutral system 中性点不接地系统
ungrounded supply system 不接地供电系统
ungrounded system 非接地系统
ungrounded 不接地，非接地的，未接地的，不接地的
unguided 非制导的，无控制的，不能操纵的
unhandy 难操纵的，不灵巧的
unhardened concrete 未硬化混凝土
unhealthy tendency 不正之风
unhydrated plaster 干灰膏，未水化灰泥
unhygienic 不卫生的
uniaxial compression test 单轴压缩试验
uniaxial eccentricity 单轴向偏心
uniaxial strain test 单轴应变试验
uniaxial stress 单向应力，单轴应力
uniaxial 单轴的，同轴的
unibus 单一总线，单母线
unicast sampled measured value control 单播采样测量值控制
unicast sampled valueconrtol block 单播采样值控制块
unicast sampled value control 单播采样值控制
unicast 单播
unicellular yeast 单细胞酵母
unichoke 互感扼流圈
uniclinal structure 单斜构造
unicoil type motor 单线圈式电动机
unicoil winding 单线圈绕组
unicoil 单线圈
uniconductor waveguide 单导体波导
uniconductor 单导体，单导线
unicontrol 单向控制，单一控制，单钮调谐
unicrystalline germanium thermistor 单晶锗热敏电阻器
unicrystalline silicon thermistor 单晶硅热敏电阻器
Unidal 铝镁锌系形变铝合金
unidentified leakage 未识别的泄漏【排水】
unidentified reading frame 未鉴定读框
uni-dimensional consolidation 单向固结
unidimensional flow 线性流动
unidimensional stream 单向流
unidimensional 一度空间的，一维的，线性的，

单方向的
unidirectional antenna 单向天线
unidirectional contract for difference 单向差价合同
unidirectional current 单向电流，单方向电流，直流电
unidirectional element 单向元件
unidirectional flow ventilation 单向流通风
unidirectional flow 单向流动，直流，一维流，单向水流
unidirectional HVDC system 单向输电高压直流系统
unidirectional magnetization 单向磁化
unidirectional motor 单向电动机
unidirectional pressure 单向压力
unidirectional pulse train 单向脉冲列，单向脉冲序列
unidirectional pulse 单向脉冲
unidirectional relay 单向继电器
unidirectional rotation 单向旋转
unidirectional stepping relay 单向步进继电器
unidirectional stress 单向应力
unidirectional transducer 单向传感器，单向转换器
unidirectional transmission 单向输电，单向传输
unidirectional wind 单向风
unidirectional working 单翼工作面，单翼开挖面
unidirectional 单向的，不可逆的，单向性的
UNIDO(United Nations Industrial Development Organization) 联合国工业及发展组织
unification administration construction pattern 统建方式
unified atomic mass constant 原子量的统一常数，原子质量常量
unified boiler control 锅炉联合控制
unified bus 一元化总线
unified construction 统建，统一施工
unified distribution 统一分配
unified electrical machine theory 统一化电机理论
unified foreign exchange rate 单一汇率
unified member 统一构件
unified modeling language 统一建模语言
unified planning and overall arrangement 统筹安排
unified plan 统一规划［计划］，综合规划
unified power flow controller 统一潮流控制器
unified precast element 通用预制构件
unified programming 统一规划
unified soil classification system 土的统一分类系统，统一土壤分类法
unified specification 统一规范
unified standard for reliability design of building structures 建筑结构可靠度设计统一标准
unified standard 统一标准
unified structure element 统一构件
unified tax system 统一收税制度
unified 统一的
unifilar winding 单线绕法，单线绕组
unifilar 单丝的，单线的
uniflow boiler 直流锅炉
uniflow cooling 单向流动冷却，直流冷却
uniflow steam engine 单流蒸汽机
uniflow 直流，单向流，顺流，单向水流
unifluxor 匀磁线
Unifont 铝镁锌系铸造铝合金
uniform acceleration 匀加速度，等加速度
uniform-admission turbine 全周进汽汽轮机
uniform-air-gap machine 均匀气隙电机，隐极电机
uniform air-gap 均匀气隙
uniform array 一致阵列，一致数组
uniform attack 均匀腐蚀，均匀侵蚀
uniform automatic data processing system 一致自动数据处理系统
uniform beam 等截面梁
uniform blade adjustment 整体叶片调整
uniform blade 等截面叶片
uniform Building Code 统一建筑规范【美国】
uniform channel 等断面渠道
uniform coal flow 均匀煤流
uniform code 等距码，均匀码
uniform coefficient of sand 砂的均匀系数
uniform coefficient 均匀系数
uniform column 等截面柱
uniform combustion 均匀燃烧
uniform continuity 均匀连续性
uniform conversion model 均匀转化模型
uniform cooling 均匀冷却
uniform corrosion 均匀腐蚀
uniform cross-section 等截面
uniform customs and practice for documentary credits 跟单信用证统一惯例
uniform design 定型设计
uniform dielectric 均匀电介质
uniform distributed load 均匀分布负荷
uniform distribution 均匀分布
uniform efficiency 等值效率
uniform electric field 均匀电场
uniform elongation 均匀伸长，均匀生长
uniform encoding 线性编码
uniform field 均匀场，均强场
uniform flow wind tunnel 均匀流风洞
uniform flow 等流，等流速，均匀流，稳定流
uniform fluidization 均匀流态化
uniform function 单值函数
uniform Gaussian noise 均匀高斯噪声
uniform gradient series 等差系列
uniform invoice 统一发票
uniformity coefficient 均一系数，均匀度，均匀（性）系数，均质系数
uniformity factor 统一系数
uniformity index 均匀性指数
uniformity modulus 均匀模数
uniformity of fluidization 流态化均匀性
uniformity of galvanizing coating 镀锌层均匀度
uniformity ratio 均匀度
uniformity settling 均匀沉降
uniformity 均匀度，均匀性，一致性，无变化，同质性，均质性，统一
uniformization 同质化，均质化，均匀化，单值化
uniform lattice 均匀栅格
uniform lighting 均匀照明

uniform line 均匀线路
uniform loading 均匀加载，均匀加感
uniform load 均匀荷载，均匀负荷，均布载荷
uniformly accelerated motion 匀加速运动
uniformly accessible storage 均匀存取存储器
uniformly asymptotic stability 一致渐近稳定性
uniformly distributed load 均布荷重，均布载荷，等分布荷重，均布荷载
uniformly distributed winding 均匀分布绕组
uniformly-graded aggregate 均匀级配骨料
uniformly loaded cable 均匀加载电缆
uniformly loaded reactor 均匀装载反应堆
uniformly retarded motion 匀减速运动
uniformly rotating fluid 匀速旋转流体
uniformly shaped blade 等截面叶片
uniformly varying load 匀变分布负荷
uniform magnetic field 均匀磁场
uniform magnetization 均匀磁化
uniform medium 均匀介质
uniform mixing 均匀混合
uniform motion 匀速运动，等速运动
uniform plane wave 均匀平面波
uniform pressure 均匀压力
uniform quality 均一性，均质性，均匀性
uniform reporting system 一致报告系统
uniform rules for demand guarantee 见索即付保函统一规则
uniform sampling 等间隔采样
uniform scale 等分标尺，均匀刻度
uniform-section blade 等截面叶片
uniform section 均匀截面
uniform series 等额系列
uniform size aggregate 等粒径骨料，均匀骨料
uniform slope 等坡度
uniform sound distribution 均匀声分布
uniform stability 一致稳定性
uniform steady wind 均匀定常风
uniform stream 均匀流
uniform strength beam 等强度梁
uniform strength disc 等强度叶轮
uniform strength 等强度
uniform stress field 均匀应力场
uniform suspension 均匀悬浮物
uniform temperature distribution 均匀的温度分布
uniform thermal-neutron source 均匀热中子源
uniform thickness 等厚式
uniform velocity 匀速度，等速度
uniform viscous flow 均匀黏性流
uniform vorticity 均匀涡量
uniform wall heat flux density 均匀壁面热流密度
uniform waveguide 均匀波导
uniform wind 均匀风
uniform 均匀的，均一的，均匀，统一，一致，统一的，一致的
unifrequency block 单频阻塞
unifrequency 单频率
unifunctional circuit 单功能电路
unifunctional exchange resin 单官能团交换树脂
unifying composition operation 合复运算
unijunction transistor 单结晶体管
unilateral antenna 单向天线
unilateral conduction 单向导电
unilateral conductivity 单向导电性，单向传导
unilateral contract 单务合同
unilateral decision 片面决定
unilateral denunciation 单方面废除
unilateral impedance 单向阻抗
unilateral surface 单侧曲面
unilateral thickness gauge 单面厚度计
unilateral trade 单边贸易
unilateral winding 单向绕组，单向绕法
unilateral 单侧的，片面的，单向的，单方面的，单向作用的，单向性的
uniline 单一线路，单线，单相线路，单行
unilpolar flux 单极磁通
unimetallic strip thermal relay 单金属片式热继电器
unimeter 伏安表，多刻度电表
unimodal distribution 单峰分布
uninstrumented fuel assembly 未装测量仪表的燃料组件
uninsulated conductor 裸线，非绝缘导线，未绝缘导线
un-insulated pipe 不保温管
uninsulated 未绝缘的，无绝缘的
uninsulation 未绝缘，不绝缘，不绝热，未绝热
uninsurable risk 不可保的险
unintended 无意的
unintentional reactor shutdown 无意停堆，误停堆
unintentional scram 意外紧急停堆
unintentional 无意识的
uninterrupted concreting 混凝土连续浇筑
uninterrupted DC power system 不间断直流电源
uninterrupted duty 不间断工作方式，不间断工作制度，不间断工作制，长期工作制
uninterrupted power supply 不间断（供电）电源，不停电电源
uninterrupted production 不间断生产
uninterrupted raceway 连续的电缆管道
uninterrupted service 不间断运行，不间断服务，无故障运行，无事故运行
uninterrupted 不间断的，不中断的，连续的，不停的
uninterruptible power supply 不间断电源，不停电电源
uninterruptible power system 不可间断的供电系统
union bonnet 活接阀盖
union cap 活接（阀）帽
union connection 管子接头
union coupling elbow 弯管接头
union coupling sleeve 连接衬套，连接套管
union coupling 联轴节，管连接，联管节
union elbow 结合弯头，弯头
Union for Coordination for Transmission Electricity 欧洲输电联盟
union joint 管接头，连接管，活络接头
union link 连环
union melt welding 埋弧焊接
union nipple 接管接头
union nut 联管螺母，管接螺母
union organization 工会组织

union pipe 接管
union reducer 渐缩管接头
union ring 联管节
union screw 对动螺旋，管接头对动螺纹
union socket 联管套节
union swivel end 旋转联管节，转环连接端，联管节旋转端
union tee T形管接头
union wrench 联管节扳手，管接头扳手
union 管套节，活接头，螺纹管接头，管节，接合，结合，连接器，联合会
uniphase output 单相输出
uniphase 单相的
unipivot 单支枢，单枢轴
uniplanar flow 单面流
unipolar armature 单极电枢
unipolar dynamo 单极（发）电机
unipolar generator 单极发电机
unipolar input 单极输入
unipolarity 单极性
unipolar machine 单极电机
unipolar 单极的
unipole antenna 无方向性天线
unipotential 单电位的，等电位的，单势的，等势的
uniprocessing system 单处理系统
uniprocessor 单一处理机
unipump 整体式电动泵
unique identification 单一标志
unique label 特有标记
unique method 独特的方法
uniqueness theorem 唯一性定理，单值定理
uniqueness 单一性，单值性，独特性，均匀性，唯一性
unique number 唯一的数字，唯一的编号，独特的号码，特有标记
unique solution 唯一解
unirradiated fuel store 未照射过的燃料贮存库
unirradiated fuel 未经照射过的（核）燃料，未辐照过的（核）燃料
unirradiated sample 未照射过的样品
unirradiated 未受照射的，未受辐照的
uniselector 单动作选择器，旋转式选择器，旋转式转换开关
uniserial 单排的，单系列的
unisource 单源的
unistrand 单股的
unit amplitude 单位振幅
unit area building cost 建筑物单位面积造价
unit area loading 单位面积载荷
unit area 比面积，单位面积
unitary matrix 单位矩阵，酉矩阵
unitary operator 单式算子
unitary sampling 单次采样
unitary 单一的，单式的，个体的，单位的
unit automatic exchange 小型自动电话交换台，内部自动电话交换机
unit auxiliaries 机组辅助系统
unit auxiliary transformer 厂用辅助变压器，厂用工作变压器，机组厂用变压器，机组副变压器
unit bay 机组间隔，机组间距，机组跨
unit bin system 单元仓储系统
unit boiler and turbine plant 单元机组
unit breaker 机组断路器
unit bunkers 仓群
unit cable 组合电缆【通信用电缆】
unit capacity construction cost 单位容量造价
unit capacity factor 利用系数，利用率
unit capacity 单位容量，单机容量，机组功率，机组容量，单位输出功率
unit cell 单元干电池，晶格单位，单位栅元
unit centralized control 单元集中控制
unit charge 单位电荷
unit circle 单位圆
unit cleaning 单元机组清洗
unit commissioning on load 机组带负荷调试
unit commitment optimization 机组组合优化
unit commitment 机组经济组合，机组组合，机组投入，开停机计划
unit compressive stress 单位压应力
unit-connected turbine-generator transformer 汽轮机、发电机、变压器组合单元
unit connection 连组，单元连接，单元结线
unit constitution 单元组合
unit construction computer 组件式计算机
unit construction cost 单位造价【电】
unit construction 单位组合结构，独立装置，组件，部件结构，构件组合，单位工程
unit contact pressure 单位接触压力
unit control error 单机控制误差
unit control file 部件控制文件
unit control level 机组控制级
unit control room 单元控制室，机组集中控制室
unit control word 部件控制字
unit control 单元控制，部件控制
unit coordinated control 机组协调控制
unit cost 单价，机组价格，单位造价，单位成本，机组成本，机组费用
unit crystal 单晶
unit deformation 单位变形
unit delay time 单位延迟时间
unit derated factor 机组降低出力系数
unit derated hours 降低出力小时
unit derated state 降低出力状态
unit derating capacity 机组降低出力量
unit derating factor 机组降低出力系数
unit derating generation 机组因降低出力少发电量
unit derating hours 机组降低出力小时
unit design 单元设计
unit desk 机组控制台
unit dial 单位盘
unit discharge 单宽流量，单位排放量，比流量
unit discounted cost 单位贴现费用
unit displacement method 单位位移法
unit-distance code 单位间距码
unit dose system 单位剂量系统
unit drainage pit 机组排水槽
United Engineers & Constructors Incorporated 联合工程师和营造者有限公司【美国】
United Kingdom Atomic Energy Authority （英国）原子能管理局

United Nations Industrial Development Organization 联合国工业发展组织
united punching and shearing machine 联合冲剪机
United States Atomic Energy Commission 美国原子能委员会
United States Dollars 美元
United States Energy Commission 美国原子能委员会
United States Environmental Protection Agency 美国环境保护局
United States House Committee of Ways and Means 美国众议院筹款委员会
United States Nuclear Regulatory Commission 美国核管理委员会
united states patent 美国专利
United States Standards 美国（工业）标准
unit efficiency 机组效率
unit electric flux 单位电通量
unit electrostatic charge 单位静电荷
unit elongation 单位伸长，延伸率
uniterruptible power system 不中断电源系统，不停电系统
unite 联合
unit feedback 单位反馈
unit feed system 组合式给水系统
unit graph 单位过程线，单位线
unit heater with axial fan 轴流式暖风机
unit heater with centrifugal fan 离心式暖风机
unit heater 单元加热器，成套加热机组，供暖机组
unit height 机组高度
unit hydrograph duration 单位过程线历时
unit hydrograph method 单位过程线法
unit hydrograph 单位过程线，单位流量曲线图
unit identification system 部件识别系统
unit impulse function 单位脉冲函数
unit impulse input 单位脉冲输入，单位脉冲进料
unit impulse response 单位脉冲响应
unit impulse signal 单位脉冲信号
unit impulse 单位脉冲，单位冲击
unit in charge of construction 施工单位
unit injector system 电子控制式泵喷嘴系统，泵喷嘴系统
unit installation 装机
unit installed capacity 单位装机容量
unit interfacial surface 单位界面
unit interval 单位间隔
unitization 单元化，联合，一元化
unitized construction 组合结构
unitized design 组合设计
unitized machine 统一化电机，成套电机
unitized 统一的，成组的，成套的，通用化的
unit line capacitance （线路）单位电容
unit line reactance per kilometer 每公里线路的单位电抗
unit load command 机组负荷指令
unit load device 集装器，集装化设备
unit load method 单位荷载法
unit load system 单元货载系统，单位装运制
unit load 单位荷载，机组负荷
unit magnetic mass 单位磁质量
unit magnetic pole 单位磁极
unit manhours 单位工时
unit matrix 单位矩阵
unit network interface card 单元网络的接口板，机组网络的接口板
unit network 单元网络，机组网络
unit of capacity 容量单位
unit of error 误差单元
unit of exit width 单位出口宽度
unit of measurement 测量单元，度量单位
unit of structure 结构单元
unit of taxation 纳税单位
unit of water measurement 量水单位
unit operation mode 机组运行方式
unit operation 单机运行，单元运行，单元操作
unitor 连接器，插座连接装置
unit outage 单元机组停运，停机
unit output 机组出力，单位出力，单元出力
unit overhaul 机组大修
unit peaking operating mode 机组调峰运行方式
unit performance test 机组运行性能试验
unit permeance 单位磁导率，磁导系数
unit plant 单元机组，单元设备，单元布置电站，成套动力装置
unit point source 单位点源
unit pole 单极
unit potential 单位电势，单位电位
unit power construction cost 单位功率造价
unit power output 单机出力
unit power plant 成套动力装置
unit power reactivity 单位功率反应性
unit power 单机容量，单机功率，单位功率
unit pressure 比压，单位压力
unit price contract 单价合同
unit price contract 单价合同
unit price 单价，单位造价
unit process 单元作业，单元过程
unit project 单位工程
unit pulse signal 单位脉冲信号
unit pulse 单位脉冲
unit pulverized-coal system 直吹式制粉系统
unit pump system 单体泵系统
unit pump 整体式电动泵
unit radiant panel 块状辐射板
unit rainfall duration 单位降雨历时
unit-ramp signal 单位斜坡信号
unit record device 单元记录装置
unit relay 单元继电器
unit run time 设备运行时间
unit silo supply system coaling sequence 原煤仓供煤系统上煤程序
unit silos 筒仓组
unit site area 单位功率占地面积
unit state 机组状态
unit speed 单位速率
unit start up and commissioning 机组启动试运
unit start-up control system 机组启动控制系统
unit step function 单位阶跃函数
unit step input 单位阶跃输入信号
unit-step response 单位阶跃响应

unit-step signal 单位阶跃信号
unit-step 单位阶跃
unit strain 单位应变
unit strength 单位强度
unit stress 单位应力
unit string 单字符串
unit substation 成套变电所，单元变电所，机组变电站
unit supervisor 机组监督员
unit switchboard 单元机组配电盘
unit switch 组合开关
unit system diagram 单元系统图
unit system of excitation 单元励磁系统
unit system 单元系统，单位制，单元制，单元机组方式
unit task 单元作业
unit testing 单元测试
unit test operation 机组试运转
unit thermal conductance 单位面积热导
unit thermodynamic system 机组热力系统
unit train transportation 专列运输
unit train 专列货车，单元列车，组合列车
unit transformer 厂用变压器，单元变压器，机组变压器
unit trial operation 综合试运行
unit trip 单元机组解扣，机组解列
unit tube of flux 单位通量管
unit-type protective-system 单元保护系统
unit under test 被测部件，被实验部件，在试设备
unituning 单钮调谐，同轴调谐
unit usage factor 单位利用系数【疲劳】
unit value 单位值
unit vector field 单位矢量场
unit vector 单位矢量
unit voltage 单位电压
unit volume change 单位容积变化
unit volume expansion 单位体积膨胀
unit volume 单位容积［体积］
unit water absorption 单位吸水量
unit water content 单位含水量
unit weight of dry soil 干土的容重
unit weight of saturated soil 饱和土容重
unit weight of soil particles 土粒容重
unit weight of soil 土的容重
unit weight of water 水的容重
unit weight 单位权，单位重，单位重量，容重
unity coupler 单一耦合器
unit years 机组使用年，机组运行年数
unity-gain bandwidth 均一增益带宽
unity power-factor test 单位功率因数试验，功率因数等于一的电机试验
unity power factor 单位功率因数
unit 单位，单元，装置，部件，部，机组，单元机组，设备
univalence 单价
univariant function generator 单变量函数发生器
univariant 单变量的
universal arc suppressing reactor 通用消弧电抗器
universal asynchronous receiver 通用异步接收器
universal automatic control and test equipment 通用自动控制及测试设备
universal bender 万能弯管机
universal bevel protractor 万能量角规，万能斜角规
universal bridge 通用电桥，万能电桥
universal buffer-controller 通用缓冲控制器
universal burner 通用喷燃器，通用燃烧器
universal character set 通用符号集
universal chuck 万能卡盘
universal classification system 通用分类系统
universal coil 通用线圈
universal computer 通用计算机
universal connection 万向连接
universal constant 通用常数，普适常数
universal correlation function 通用关系式
universal coupling 通用联轴节，万向联轴节，万向联轴器
universal crane 万能吊车，万能起重机，万向起重机
universal decimal classification 通用十进制分类法
universal decision element 通用计算元件
universal digital computer 通用数字计算机
universal dipole 通用偶极子
universal drafting machine 全能绘图机
universal electric motor 交直流两用电动机
universal equipment 通用设备
universal excavator 万能挖土机
universal fatigue testing machine 万能疲劳试验机
universal fire truck 联用消防车
universal function generator 通用函数发生器
universal function unit 通用函数发生器
universal galvanometer 通用检流计，通用电流计
universal gas constant 通用气体常数
universal gravitation 万有引力
universal grinding machine 万能磨床
universal horizontal milling machine 万能卧铣床
universal hydraulic testing machine 液压万能试验机
universal impact testing machine 通用冲击试验机
universal indicator 通用指示剂
universal instrument 全能仪器，通用量测仪表
universal joint cross 万向节十字头
universal joint extension 通用延长杆
universal joint 万向接头，通用接头，通用［万能］联轴节
universal law of the wall 壁面通用规律
universal leg extension 通用的塔腿延长段
universal level 通用水准器
universal logic block 通用逻辑块
universal manipulator 通用机械手
universal material testing machine （电子）万能材料试验机
universal measuring instrument 通用测量仪表
universal meter 万用表，通用电表，多用电表
universal miller 万能铣床
universal milling machine 万能铣床
universal mill plate 万能板材
universal motor 通用电动机，交直流两用电动机

universal pH test paper 万能 pH 试纸
universal plate 万能板材
universal-pressure boiler UP 锅炉，通用压力锅炉
universal product code 条形码，通用产品码
universal ramp generator 通用斜坡发生器
universal resource identifier 通用资源标识
universal setting gauge 万能调准仪表
universal shunt 通用分流器
universal socket wrench 万能套筒扳手
universal strength testing machine 万能强度试验机
universal surface gauge 划线卡尺，划针盘，万能平面规
universal switch 通用开关
universal synchronous receiver/transmitter 通用同步接收/发送器
universal synchroscope 通用同步示波器
universal table 万能工作台
universal tape-to-tape converter 通用带到带转换器
universal temperature profile 通用温度型
universal tensile testing machine 万能拉力［拉伸］试验机
universal testing instrument 通用测试仪
universal testing machine 万能（材料）试验机
universal time 世界时【UT】
universal tractor 通用拖拉机
universal trailer 万能拖车
Universal Training Reactor 通用训练反应堆【美国】
universal turbulence constant 通用湍流常数
universal type expansion joint 万向膨胀节
universal velocity distribution 通用速度分布
universal xenon curve 氙毒普适曲线，氙毒通用曲线
universal 普遍的，万能的，通用的
universe 宇宙，世界，领域，范围，天地万物
University of Teheran Research Reactor 德黑兰大学研究堆【伊朗】
univertor 变频器
univibrator 单稳态多谐振荡器，单稳态触发器
univocal resolution of force 力的单一解
univoltage 单电压的，单电位的
uniwave signaling 单频信号法
unjammable 抗干扰的，防干扰的
unjoint 脱开
unkilled steel 沸腾钢
unknown loss 未知损失
unknown number 未知数
unknown sample 未知样品
unlabeled 无标号的
unlabelled basic statement 无标号基本语句
unlatched shaft 未闩住的轴
unlatching 脱扣
unlawful diversion 非法转用
unlawful 非法的
unleachable 不能浸出的
unless elsewhere specified 除非另有规定
unless otherwise agreed 除另有协议外
unless otherwise indicated 除非另有规定
unless otherwise noted 除另有注明外
unless otherwise specified 除另有说明外
unless otherwise stated 除另有说明外
unless the context otherwise requires 除上下文另有要求外
unlevel luffing 非水平变幅
unlimited company 无限公司
unlimited liability 无限责任
unlimited power 无限的权利
unlimited stream 无限流，无约束流
unlimited 不受限制的，无限的，不定的，无穷的，无限量的
unlined canal 无衬砌渠道
unlined tunnel 不衬砌隧洞
unlined 未衬砌的，无限量的，无炉衬的，未包覆的
unloaded circuit 无载电路，卸载电路
unloaded sag 无荷弧垂，无载垂度，无荷弛度
unloader 减压器，卸货机，卸料机，卸车机，卸料装置，卸载机
unloading area 卸载场，卸车地点，作业区，卸料区
unloading auger 螺旋卸载机
unloading bay 安装间，卸料间
unloading curve 卸荷曲线
unloading device 卸载装置
unloading equipment 卸载设备
unloading facility 卸料设备
unloading gear 卸荷装置
unloading in motion 行进中卸载
unloading jetty 卸煤码头
unloading joint 除荷节理，去荷节理
unloading machine 卸料机
unloading modulus 卸荷模量
unloading of goods 卸货
unloading point 卸煤点
unloading port 卸货港
unloading quay 卸货码头，卸货站台
unloading schedule 卸料计划，卸料时间表
unloading service yard 卸货作业场地
unloading shed 卸煤棚
unloading station 卸煤装置
unloading tipper hopper 卸载翻斗
unloading track 卸车线
unloading trench 卸煤沟
unloading tunnel 卸煤沟
unloading valve 释荷阀，卸荷阀，卸载阀
unload 卸料【燃料】，减负荷，去负载，放空，去载，甩负荷，卸负荷，卸货
unload time chain 转存时间链
unlocked position 不闭锁的位置
unlocking dual control 分开的双重控制
unlocking rod 开启棒，开锁棒
unlock 未锁的，开锁，除去联锁（保护），松开，解开，释放
unmagnetized 未磁化的，非磁化的
unmanned aerial vehicle 无人机
unmanned plane 无人机
unmanned power station 无人管理电站，自动化电站
unmanned substation 无人变电所，无人值班变电站

unmanned 无人操纵的，无人控制的，无人值班的，无人值守
unmatchable 不可匹配的，不能比拟的
unmatched seal 非匹配封接
unmeasured loss 未计入损失，其他损失
unmendable 不可修理的，不可修正的
unmoderated fission neutron 未经慢化的裂变中子
unmoderated reactor 无慢化剂反应堆
unmoderated 未慢化的，未减速的
unmodified instruction 非修饰指令，基本指令
unmodulated 未调制的
unmonitored control system 无监视的控制系统
unnavigable river 不通航河流
unnavigable 不通航的
unnecessary duplication of effort 事倍功半
unnecessary 不必要
unnotched specimen 没有切口的试件
unnumbered command 无编号指令
unobstructed access 畅通无阻的通道
unobstructed airflow 自由气流
unobstructed 没有障碍的，畅通无阻的
unoccupied 空着的
unofficial visit 非正式访问
unoperated position 非运行位置，停车位置，静止位置
unorganized air supply 无组织进风
unorganized exhaust 无组织排风
unorganized natural ventilation 无组织自然通风
unorthodox winding 不规则绕组
unpackaged technology 不成套技术，分项技术
unpacked cargo 未包装的货物
unpacking 开箱，拆包
unpack numerical part “分出数字部分”的指令
unpack 分出，卸下……，解除……的负担，打开包裹，拆开包装
unparalleled 不平行的，非并联的
unpassable duct 不通行管沟
unpaved 无铺砌的
unperturbed flux 未受扰动通量
unplanned derating hours 非计划降低出力小时
unplanned derating 非计划降低出力，计划外停动时间，计划外停运时间
unplanned maintenance 非计划检查，非计划性检修
unplanned outage factor 非计划停运系数
unplanned outage hours 非计划停运小时
unplanned outage rate 非计划停运率
unplanned outage state 非计划停运状态
unplanned outage 非计划停堆，非计划停机，非计划停运，计划外停电
unplanned shutdown 非计划停堆，非计划停机，非计划停运
unpleasant smell 不悦气味
unpleasant wind condition 不悦风况
unplugging 排除淤积物
unpoisoned fuel 未加毒物燃料，未中毒燃料
unpoisoned reactor 未中毒反应堆
unpolarized burning 自由空间燃烧
unpolarized 未极化的，无偏振的
unpolished surface 未抛光的表面
unpopulated 无人居住的
unpowered ascent 无动力上升
unpowered bogie 无动力转向架
unpowered control 无助力操纵
unpowered flight 无动力飞行
unpowered rotor speed 自转旋翼的转速
unpowered 无功率的，无动力的
unpractical 不现实的
unpredictable work 不可预见工作
unprocessed data 未处理数据
unprocessed material 未加工材料
unproductive 非生产性的
unprofessional operation 违章操作
unprofessional 非专业的
unprotected gangway 无防护的过道
unprotected river bed 无护岸的河床
unprotected 未保护的，无防护设备的
unproven 未经证实的
unqualified 不合格的，不具备条件的
unquantized law 非量子定律
unquote 结束引语，引用结束
unrammed concrete 未捣实混凝土，未夯实混凝土
unreacted core model 未反应核模型
unreacted core 未反应核
unreactive aggregate 惰性骨料，无反应骨料
unreasonable delay 无理的拖延
unreasonable price 不合理价格
unreasonable requirement 不合理要求
unreasonable 不合理的
unrecovered strain 不可恢复应变，非弹性部分的应变
unrectified 未整流的，未调整的
unreeling 退绕放线
unreflected peak 未分辨出的峰，不能分解的峰
unreflected reactor 无反射层反应堆
unreflected system 无反射系统
unreflected 无反射层的，无反射的
unregistered company 未注册公司
unregulated variable 未调变量
unregulated voltage 未调节电压
unregulated 未调的
unreinforced concrete 没有钢筋的混凝土，素混凝土
unreinforced section 不配筋断面
unreinforced 不加固的，无钢筋的
unreliability 不可靠性，不安全性
unremittance of control 控制的不间断性
unremittance 不间断性
unrepresentative 无代表性的
unresolved echo 不清晰回声［回波］
unresolved safety issue 未解决的安全问题
unrestricted dose equivalent index 非限定剂量当量指数
unrestricted flow 无约束水流
unrestricted random sampling 无约束随机抽样，无限制随机采样
unrestricted 无约束的，不限制的
unrest 骚乱，不安，不稳定
unrodded core 未装控制棒的堆芯
unroll 解开，展开

unsafe act　不安全动作，不安全行为
unsafe code　非安全代码
unsafe condition　不安全情况，不安全条件
unsafe error　偏（于）不安全误差
unsafe fuel　不安全燃料
unsafe indication　应急显示
unsafe indicator　应急状态显示器
unsafe range　不安全范围
unsafe region　不安全区
unsafe reset　不安全复位
unsafe temperature　不安全温度
unsafety failure　非安全故障
unsatisfactory　不满意，不能令人满意的
UNSAT＝unsaturated　不饱和的，未饱和的
unsaturated air　未饱和的空气
unsaturated bond　非饱和键
unsaturated cell　不饱和电池
unsaturated current transformer　不饱和电流互感器
unsaturated flow　非饱和土壤水分流动
unsaturated hysteresis　未饱和滞后
unsaturated logic circuit　非饱和逻辑电路
unsaturated permeability　未饱和透水性
unsaturated reactance　不饱和电抗
unsaturated soil　不饱和土，非饱和土壤
unsaturated standard cell　不饱和标准电池
unsaturated steam　不饱和蒸汽，未饱和蒸汽
unsaturated transformer　非饱和变压器
unsaturated zone　非饱和区，含气层，通气区，未饱和区
unsaturated　非饱和的，未饱和的
unsawn timber　粗材
unscheduled downtime　计划外的停工期，设备外部因素停机时间，非计划停用时间
unscheduled maintenance　非计划（性）维护，非计划（性）维修，计划外维修，临时检修
unscheduled outage　非计划停用，非计划停机[停堆]
unscheduled shutdown　非计划停堆
unscheduled withdrawal　非计划提棒
unscheduled　偶然的，计划外的，误动作的
unscreened coal　未筛煤，原煤，未筛选煤
unscreened gravel　未过筛的砾石
unscreened　未分级的，未过筛的
unscrewed　扭松螺丝的，旋出的，扭开的
unscrew　拧松
unsealed source　非密封源
unsealing　未密封，开封，拆封
unseasoned timber　未干燥木材
unseasoned wood　未干燥木材
unseat　打开，罢免，使离职，剥夺……的席位，使失去资格
unsecured loan　无担保贷款，无抵押贷款
unseparated flow　未分离流动
unserviceability　不适用性，不能使用性，不耐用性，运转不安全性，使用不可靠性
unserviceable　不能使用的，不合使用条件的
unsettled weather　不稳定天气
unset　复位，置零，未安装的，未调整的，未整定的
unshared composite beam　无支撑叠合梁
unsharp image　不清晰图片
unshielded　未防护的，未屏蔽的，无防护的，不设屏蔽的
unshocked　未受冲击的，无激波的，无振动的
unshorting　清除短路
unshrouded free standing blade　自由叶片
unshrouded impeller　开式叶轮
unshrouded rotor　无罩风轮
unshrouded wheel　无围带的叶轮
unshunted　未并联的，无分路的，未旁路的
unsigned copy　未签字本，未签字副本
unsintered density　未烧结的密度
unsized coal　非分级煤
unsized　未分级的，未过筛的
unskilled worker　不熟练工人
unskilled　不熟练的
unslaked lime　生石灰
unslotted generator　无槽发电机
unslotted motor　无槽电动机
unslotted rotor turbogenerator　无槽转子汽轮发电机
unslotted rotor　无槽转子
unsmooth　不光滑的，不平滑的，不顺当的
unsoaked　未浸湿的
unsodded　未铺草皮的
unsoldering　脱焊
unsolicited data or unsolicited message　主动数据或主动报文
unsolicited offer　主动报价
unsolved question　未解决的问题
unsound cement　变质水泥，不安定水泥
unsound knot　腐朽木节
unsound　不坚实的，不健全，不安全，不可靠的
unspent　不需花费的
unsplit casing　整体汽缸，整体气缸
unsplit frame　整体机座，整体机组
unsplit　不可分开的，不可拆卸的，整体的，无裂口的
unstability of speed　转速不稳定度
unstability　不稳定性，不安定性，不稳固性
unstable atmosphere　不稳定大气
unstable channel　不稳定河槽
unstable characteristic curve　不稳定的性能曲线
unstable characteristic　不稳定（振动）特性
unstable component　不稳定组分，不稳定成分，非稳定环节
unstable condition　不稳定状态，非稳定状态
unstable crack propagation　不稳定裂纹传播，失稳裂纹扩展
unstable equilibrium point　不稳定平衡点
unstable equilibrium　不稳定平衡，非稳定平衡
unstable fission product　不稳定裂变产物
unstable fluidized bed　不稳定流化床
unstable galloping　不稳定驰振
unstable-head curve　压头不稳定曲线
unstable intermediate　非稳定中间体
unstable isotope　不稳定同位素
unstable nucleate boiling　不稳定泡核沸腾
unstable operation　不稳定运行
unstable oscillation　不稳定振荡
unstable oversteer　不稳定转向

unstable period 不稳定期，不稳定工况
unstable regime 不稳定状态
unstable shock 不稳定激波
unstable stability 不稳定的稳定性，负稳定法
unstable state 非稳态，不稳定状态
unstable stratification 不稳定层结，不稳定层理
unstable system 不稳定系统
unstable wave 不稳定波
unstable 不稳定，不稳定的
unstarred nonterminal 未加星非终结符
unstationary flow 不稳定流
unstationary state 非固定状态
unstayed cover 没有拉撑的平盖板
unstayed 未固定的
unsteadiness 不稳定性，非定常性，不稳定
unsteady airfoil theory 非定常翼型理论
unsteady bernoulli's equation 非定常伯努利方程
unsteady boundary layer 非定常边界层
unsteady combustion 不稳定燃烧
unsteady diffusion 不稳态扩散
unsteady dispersion 不稳定分散体系
unsteady drag 不稳定阻力
unsteady flat head 没有拉撑的平封头
unsteady flow pattern 不稳定流型
unsteady flow 变量流，不稳定流（动），非稳定流，非常定流，非定常流，非恒定流
unsteady fluid flow 非稳定流，不稳定的液体流
unsteady fluidized bed 不稳定流化床
unsteady force 非稳定力
unsteady heat conduction 不稳定热传导
unsteady heat regenerator 不稳定回热器
unsteady heat transfer 不稳定换热
unsteady load 不稳定负荷
unsteady motion 不稳定运动，非定常运动
unsteady non-uniform flow 不稳定变速流
unsteady operating conditions 不稳定运行工况
unsteady running 不稳定运转
unsteady state condition 不稳定工况，非稳条件
unsteady state heat transfer 不稳定传热，非稳态传热
unsteady state 不稳定状态，非定态，变动工况，非稳态
unsteady uniform flow 等速不稳定流
unsteady wave 不稳定波
unsteady 不稳定
unstepped piston 无级活塞
unstratified soil 非层状土，非成层土
unstratified 不成层的，非层状的，无层理的
UNST＝unstable 不稳定的
unsubmerged weir 不淹没堰
unsubmerged 不淹没的
unsuccessful bidder 未得标者，未中标的投标人
unsuccessful forced line energization 线路强送电不成功
unsuccessful party 败诉方
unsuccessful re-closure 重合闸不成功
unsuccessful tenderer 未得标者，未中标的投标人
unsufficient power supply 缺电
unsupercharged engine 无增压发动机
unsupervised learning 非监督学习
UNS(upper neutron shielding) 上部中子屏蔽
unsupported height 自由高度
unsupported length 不支撑的长度【弹簧】，无支撑长度，自由长度
unsurfaced road 土路
unswitched 无开关的【不通过触点，即整体中性】
unsymmetrical balance 不对称平衡
unsymmetrical beam 不对称梁
unsymmetrical circuit 不对称电路
unsymmetrical curve 不对称曲线
unsymmetrical induction machine 不对称感应电机
unsymmetrical load operation 不对称负荷运行
unsymmetrical load 不对称负载，不对称荷载
unsymmetrical matrix 非对称矩阵
unsymmetrical overload 不对称过负荷
unsymmetrical winding 不对称绕组，不对称绕法
unsymmetrical 不对称的
unsymmetry of voltage and current 电压和电流不平衡度
unsymmetry 不对称，不对称性，不匀称
unsystematic risk 可避免的风险
unsystematic 无系统的
untainted 纯净的，未污染的
untamped backfill 未夯实回填土
untanking mass 器身自重，（变压器）不带油箱的重量
untapered key 滑键，平键
untapped 未使用的，未开发的
unterminated line 无端接线
untight 未封闭的，未密闭的，不紧密的，泄漏的
untimbered tunnel 无木撑隧道，无支撑隧洞
untorquing 不扭转
untransposed transmission line 不换位输电线路
untransposed winding 不换位绕组
untransposition 未换位的，不换位的
untreated coil 未处理的线圈
untreated effluent 未处理的废水，未处理污水
untreated sewage 未处理污水，未经处理的废水
untreated water 生水，原水，未处理水
untreated 未经处理的，未清理的，未浸渍过的
untrue 不真实的
untunable magnetron 不可调磁控管
untunable 不可调谐的
untuned blade 不调频叶片，非调频叶片，非调谐叶片
untuned 未调谐的，非调谐的
unturned blade 不调频叶片
untwist angle （叶片的）反扭角
untwisted blade 非扭曲叶片
untwist the cables 电缆解绕
untwist 长叶片反扭曲，反向扭曲，非扭的，解缆
unused balance of letter of credit 信用证未用余额
unused balance 未用余额
unused code 禁用代码
unused portion of letter of credit 信用证未用部分
unusual event recording system 异常事件记录

系统
unusual exposure　异常照射
unusual service condition　不正常工作条件，异常工况
unusual storm condition　异常暴风雨状况
unvegetated　无植被的
unveiled the most comprehensive blueprint　披露[公布]了最全面的蓝图
unvented　没有排气口的，未放气的，无通风的
unverifiable　不能证实的，无法核实的
unvulcanized rubber　非硫化橡胶，生胶
unwanted operation　保护装置误动作，误动作，异常运行
unwanted signal　无用信号
unwanted tripping　误跳闸
unwashed steam　未经清洗的蒸汽，未洗蒸汽
unwatched　不用监视的，自动的
unwatering pit　排水坑
unwatering sump　排水坑
unwatering system　放空水系统
unwater　脱水，排水，使干燥
unweathered　未风化的
unweighted mean　未加权平均数
unwinder　退绕机，拆线机，开卷机
unwinding of loop　循环展开【在程序中】
unwind　（程序）展开
unworkable concrete　不易浇筑的混凝土
unwound core　无绕组铁芯，未嵌绕组铁芯，光铁芯
unwound　未绕的，未嵌线圈的
unyielding support　不沉陷支座
unyielding　不屈服的，不能弯曲的，稳定的，稳固的，坚固的
UO_2 crystal lattice　二氧化铀晶格
uoderprediction　过低预计，过低预报
Upalon　聚氯乙烯的商品名
up and down interface　上和下的接口
up and down method　升降法
up and down plug insertion system　塞子上下移动系统
up and down　起伏的，上下的
upbuilding of delta　三角洲扩建
upcast fault　上投断层
up-coiler　卷绕机
upcoming shipment　下一载货
up-converter　向上变换器
UPC(universal product code)　商品通用代码，条形码
up current　上升气流
upcut milling　向上铣切
update characteristic　更新特性
updated information　更新信息
updated time　更新时间
updated version　更新方案，更新版本，升级版
updated　最新的，更新的，现代化的
update　更新资料，修改，更新，校正，修正，刷新，升级，使现代化
updating maintenance　更新维护
updating of enterprise　企业技术改造
updating time　更新时间
up-down-counter　加减计数器
up-down guidance　上下制导
up-down　上下
updraft boiler　塔式锅炉，直焰锅炉
updraft　上升气流，上曳气流
updrift　逆向推移
upender　翻转装置，翻转机
upending-device pit　倒换装置竖井
upending device　倾斜装置，倾卸装置
upending fixture　翻转固定装置
upending frame　倾倒架，（燃料）起吊架
upend　倾卸，倾翻，翻转，在垂直方向失去平衡
up extreme position　上限位置【夹车器】
upfaulted block　断层上块
UPFC(unified power flow controller)　统一潮流控制器
upfeed distribution　扬水式配水系统
upfeed system　上分式系统
upflow conversion　向上流动转变
upflow filtration process　升流式过滤法
upflow filtration　升流过滤
upflow fluidized bed　上流流化床
upflow regeneration　逆流再生
up flow riser tube panel　垂直上升管屏
upflow tube　溢流管
upflow　上升流，向上流，向上流动
upfold　背斜褶皱，隆皱
upgrade water　扬升水
upgrade welding　上坡焊
upgrade　上坡，升坡，升级，提高，提高等级，加强，上升，提升，改良，不大的改形，淤高，（重水）提浓，再浓集
upgrading column　蒸馏柱，提浓柱【重水】
upgrading studies of existing facilities　现有设施升级改造研究
upgrading of and updating of product　产品升级换代
upgrading of boiler weld　锅炉焊缝加强
upgrading of product　产品升级
upgrading plant　提浓装置，浓缩装置
upgrading system　（重水）提浓系统
uphand welding　仰焊
upheaval　隆起
uphill water diversion　引水上山
uphill welding　仰焊，上坡焊
uphole steam generator　井上注汽锅炉，油田注汽锅炉
upkeep cost　保养费用，维护费用，维修费用
upkeep　保养，维护，维修，检修，维持
upland catchment　高地流域
upland erosion　高地侵蚀
upland moor　高沼地
upland plain　高平原，高原，台地
upland swamp　高地沼泽
upland　高地，高原，山地
uplift cell　上托力计，扬压力计
uplift coefficient　上升系数，上提系数
uplift condition　上拔工况
uplifted bench　上升阶地
uplifted block　上升地块

uplift force 上升力，上托力
uplift intensity factor 上托力强度系数
uplift load 上拔荷载，上拔力
uplift pressure 浮力，向上压力，浮托力，扬压力
uplift reduction coefficient 扬压力折减系数
uplift test （塔的）上拔试验
uplift wind load 上吸风载
uplift 举起，抬起，隆起，上拔，上拔力，上举，上托力，升起，扬压力
uplighting 向上照明
up line 上行线路
up link 上行链路，地与通信卫星的联络
upload 向上（作用的）负荷
upper adjusting ring 上部调节环
upper air chart 高空图
upper air circulation 高空环流
upper air current 高空气流
upper air front 高空锋
upper air sounding 高空探测
upper air synoptic station 高空气象站
upper air 高层大气，高层空气，高空
upper and lower joint hard rolling 上部和下部接头机械滚压
upper and lower limits of exchange rate fluctuation 汇价波动上下限
upper and lower limit 上下限
upper and lower value 高/低值
upper annular ledge 上部环形支座
upper anticyclone 高空反气旋
upper apex of fold 褶皱顶
upper approach channel 上游引航道
upper arm downward movement 上臂向下运动
upper arm rotation 上部旋臂
upper atmosphere pollution 高层大气污染
upper atmosphere 高层大气，高空大气
upper bainite 上贝氏体
upper bank 上游堤岸
upper bar 上层线棒，（鼠笼电机的）上鼠笼条
upper beam 大板梁
upper bend section 上弯段
upper bend 上弯管，向上拱，向上弓起，背斜，鞍廓
upper block 上部装置
upper bound 上界，上限
upper case letter 大写字母
upper case 大写字体，大写字母盘，上盘
upper casing 上汽缸
upper chord member 上弦杆
upper cofferdam 上游围堰
upper coil 上层线圈
upper cold core low 高空冷心低压
upper cold front 高空冷锋
upper computer 上位计算机
upper containment pool 安全壳上部水池
upper control limit 控制上限
upper core barrel 堆芯上围板
upper core internals 堆芯上部构件
upper core plate guide pin bearing pad 上部堆芯板导向销支撑垫
upper core plate 堆芯上隔板，上部堆芯板
upper core structure lifting rig 堆芯上部构件提升设备
upper core structure support skirt 堆芯上部结构支承裙筒
upper core structure 堆芯上部构件
upper core support barrel 堆芯上部结构支承筒体
upper core support plate 堆芯上部支承板
upper core support structure 堆芯上部支撑结构，堆芯上部支承构件
upper course 上游，上游段
upper critical cooling rate 上临界冷却速度
upper critical magnetic flux density 上临界磁通密度
upper critical temperature 上部临界温度
upper culmination 上中天
upper cut-off frequency 上限截止频率
upper cycle 上部级，上级【二元循环】
upper cylinder half 上汽缸，上气缸
upper deck 顶盖
upper destruct limit 破坏极限上限值
upper edge 上缘
upper elastic limit 弹性上限
upper end-shield 上半端罩，（汽轮发电机）上半端盖
upper face of the foundation 基础顶面
upper flange of girder 大梁的上翼缘
upper flexible and lower rigid complex multi-storey building 上柔下刚多层房屋
upper foundation raft 上部底板，上部基础筏基
upper frequency limit 频率上限
upper front 高空锋
upper grid plate 上栅格板，上格板
upper gudgeon 顶枢轴，上耳轴
upper guide bearing 上部导向轴承
upper guide tube assembly 上部导向管组件
upper guide-vane ring 上导叶环
upper half casing 上汽缸，上气缸
upper half plane 上半面
upper harmonic 高次谐波
upper head injection 上封头注水，直接注水
upper head 上封头
upper housing tube 上部外套管
upper hub 上轮毂
upper internals assembly 上部堆内构件组合
upper internals lifting rig 上部堆内构件提升机具
upper internals（outlet）plenum 上部堆内构件（出口）充满介质
upper internals storage stand 上部堆内构件存放台
upper internals 上部堆内构件
upper inversion 高空逆温
upper lateral bracing 上弦横向水平支撑
upper layer of soil 上层土壤
upper layer 上层
upper level cyclone 高空气旋
upper level problem 上级问题
upper level system 上级系统，上层系统
upper level wind data 高空风数据
upper level 高位
upper limb 上翼

upper limit of plasticity　塑性上限
upper limit price　最高限价
upper limit switch　上限开关
upper limit　上限
upper littoral zone　沿岸带上部
upper lock chamber　上游闸室
upper longitudinal bracing　上弦纵向水平支撑
upper/lower core plate　上下堆芯板
upper/lower internals storage ring　上下堆内构件贮存环
upper/lower support grid plate　上下支承栅格板
upper/lower tie plate　上下固定板
upper mantle　上地幔
upper miter wall　上游人字墙
uppermost part　顶部
uppermost reservoir　最上游水库
upper nappe profile　水舌上缘线
upper neutron shielding　上部中子屏蔽
upper normal depth　上游正常水深
upper oil header　上集油箱
upper outer casing　外上缸
upper plastic limit　上塑限，塑性上限
upper plate　顶板，上盘
upper plenum　上贮气腔，上腔室，上联箱
upper pond　上游水池
upper pool level　上池水位
upper pool　上水池，上游进水池
upper position detector　上部位置探测器【螺孔检查】
upper power level　较高功率水平
upper raft　上部底板，上部筏基
upper range-limit　范围上限
upper range value　（作用）范围上限值
upper reach　上游段，上游
upper reflector　上部反射层
upper reservoir　上水库
upper right　右上方
upper ring beam　上部环梁
upper river reach　上游河段
upper roll　上托辊
upper shaft　上轴
upper shelf　上部坪台【冲击试验转变温度】，上部架
upper shell　压力壳芯部上环段
upper shield　上半端罩
upper sideband　上边频带，高旁频带
upper side　上侧，上面，上部，（水平管的）上半部
upper skirt　上部裙筒
upper speed ring　上调速环
upper standard　上级标准
upper station　上游电站
upper steam extraction pipe　上部抽汽管
upper storage reservoir　上池
upper supplementary shielding　上部附加屏
upper support column　上部支撑柱
upper support plate　上支撑板【堆芯】
upper surface brake　上翼面减速板
upper surge tank　上调压井
upper switching value　上切换值
upper trip level　事故保护停堆上限
upper wall　断层上盘，顶壁
upper water level for flood control　防洪高水位
upper water　上层水，上游水
upper wind chart　高空风图
upper winding-bar　上层绕组铜条，上层线棒
upper wind　高空风
upper wing wall　上游翼墙
upper yield limit　屈服上限
upper yield point　上部屈服点【冶金】
upper yield strength　上屈服强度
upper yoke　上铁轭，上磁轭【变压器】
upper　较高的，上面的，上部的，上级的，向上的
up position　竖直放置【焊接】，向上的位置，上升位置
uprated engine　提高出力的发动机，大功率发动机
uprated　提高参数的，提高出力的，大功率的，提高功率的
uprate　提高……的等级，改进……的性能，提高……的价值，提高定额，提高出力，增大额定值
uprating capacity　提高出力的潜力
uprating project　提高出力的设计
upright breakwater　直立式防波堤
upright bucket type steam trap　浮桶式疏水器
upright drill press　立式钻床
upright drill　立式钻床
upright evaporator　竖式蒸发器
upright position　垂直位置，垂直状态
upright post pile　立柱桩
upright resistance　抗拔力
upright shaft　竖轴
upright tension set　正挂耐张串
upright　竖直的，直立的，立式的，立柱，支杆，立杆
uprise　涌高，直立管，向上
uprooting foundation　上拔基础
UPR＝upper　上部的
uprush limit　上爬界限
uprush　冲岸浪，垂直急流，涌浪
upscale protection　上限保护
upscattering　增能散射
upset allowance　顶锻留量
upset bolt　膨径螺栓
upset butt welding　电阻对接焊，电阻对焊
upset condition　扰动工况，失常工况，运行事故
upset current　顶锻电流
upset force　顶锻力
upset speed　顶锻速度
upset test　扰动试验
upset thread　膨径螺纹
upset time　顶锻时间
upsetting arm　倾覆力臂
upsetting couple　倾覆力偶
upsetting level　倾覆水平
upsetting moment　倾覆力矩
upsetting test　顶锻试验，镦粗试验
upsetting thread　膨径螺纹
upset welding　电阻对接焊
upset　干扰，扰动，顶锻，翻转，扰乱，失调

upside down mounting　倒装法
upside down　颠倒
upslope fog　上坡雾
upslope overlay welding　上坡焊
upslope wind　上坡风
upslope　上坡
up-spray nozzle　上喷式喷嘴
upstand beam　上翻梁，直立梁
up state　可用状态
upstream apron　上游防冲铺砌，上游护坦，闸前护坦
upstream batter　上游坡
upstream bay　上游河湾
upstream benefit　上游受益
upstream berme　迎水面马道
upstream blanket　上游铺盖
upstream bottom current　向源底层流
upstream catchment　上游流域
upstream chamber　进口室
upstream core　上游心墙
upstream economizer　逆流式省煤器
upstream end　进汽端，上游端
upstream face　上游面，入水面
upstream facing of weir　堰的上游护面
upstream failure indication　上游故障指示
upstream fill　上游坝体，上游填土体
upstream gate　上游闸门
upstream inclined filter　上游斜反滤层
upstream inflow　上游来水
upstream influence　上游影响
upstream maximum water level　上游最高水位
upstream normal water depth　上游正常水深
upstream nosing　上游鼻端
upstream pier nose　墩的上游端
upstream plant　上游电站
upstream pressure controller　阀前压力调节阀
upstream pressure limit　上游压力限制
upstream pressure　上游压力，进口压力，阀前压力
upstream radius of crest　坝顶上游面曲率半径【水电】
upstream river　上游河流
upstream section　该截面之上游区段，入口区段
upstream shell　上游坝壳
upstream side　上游侧
upstream slope of dam　坝的上游面坡度，坝的迎水坡
upstream slope protection　上游护坡
upstream slope　上游坡，上游坡度，迎水面坡度，迎水坡
upstream spray pattern　逆喷（式）
upstream surge tank　上游调压井
upstream toe of dam　上游坝趾
upstream toe　上游坝脚
upstream total head　上游总水头
upstream tributary　上游支流
upstream water level　上游水位
upstream water line　上游水面线，上游水位线
upstream water pressure　上游水压力
upstream water　上游河水
upstream wind seed　上游风速
upstream zone　上流区
upstream　上游流，迎面流，阀前的，逆流，上游，上段，（坝的）迎水面，上游的，逆流的，进口的
UPS（uninterrupted power supply）　不间断供电电源，不间断电源，不停电电源
upsurge　上涌浪，涌浪
upswell　隆起
uptake by an organ　人体器官吸收
uptake header　垂直联箱
uptake ratio　摄取率
uptake　上升烟道，上升井，上风口，吸风管，进汽道，摄取，吸收
upthrow fault　上投断层
upthrown block　上升盘
upthrust fault　上冲断层
upthrust　浮力，向上推力，仰冲
up time　可用时间，正常运行时间，工作时间
up-to-date map　最新地图
up-to-date　最新的，最现代化的
up to present　到现在为止
up to specification　符合规定
up to standard　合格，达标
up to the time of　截至……时为止
uptown　住宅区，住宅区的，在住宅区，在城镇非商业区
uptrend　向上（发展）的趋势
upturned bucket　上挑鼻坎
upturned　上翻的，翻转的
upturn　情况好转，使……转朝上，翻转
U-pump　U形泵
up valley wind　进谷风
upward air current　向上气流
upward compatibility　向上兼容性
upward component　向上分量
upward current of air　上升气流
upward current　上升（气）流
upward flow　上行流，上升流，向上流动
upward load　向上荷载
upward-moving agglomerate　向上运动的颗粒团
upward-moving dispersed solids　向上运动的分散固体颗粒
upward pressure　向上压力
upward side　上游侧
upward tendency　上升趋势
upward valve　反作用阀门
upward velocity　上升速度
upward vertical position welding　垂直往上焊接【工作姿势】
upward welding in the inclined position　上坡焊
upwarp　向上挠曲，向上翘曲，翘起，上翘
upwash　气流上洗，气流向上偏斜，上洗流，上升流，上洗
upwave　逆波
upwelling area　地下水上升区，上升流区
upwelling current　上升气流，涌升流
upwelling phenomena　上涌现象
upwelling region　上升流区
upwelling sedimentation　上升流沉积
upwelling　喷出，上喷，上涌，上涌水流，涌升流，上升流

upwind direction 上风向
upwind effect 上吹效应
upwind fetch 上风吹程
upwind method 逆风法
upwind rotor 上风式风轮
upwind type of WECS 上风式风能转换系统
upwind wind turbine 迎风式风电机组
upwind WTGS 上风向式风电机组
upwind 迎风，顶风，上升气流，上风向
uquivalence statement 等价语句
uranami welding electrode 封底焊条
uranium aluminide fuel 铝铀合金燃料
uranium americium mixed oxide 铀镅混合氧化物
uranium and heavy-water reactor 铀重水反应堆
uranium assay 铀含量，铀成分
uranium carbide 碳化铀
uranium concentrate 黄饼，铀浓缩物，铀精矿
uranium content meter 含铀量测定器
uranium conversion plant 铀转化厂
uranium cycle 铀循环
uranium deuterium lattice 铀重水栅格
uranium dioxide pellet 二氧化铀芯块
uranium dioxide 铀氧化物，二氧化铀
uranium enriched fuel 浓缩铀燃料，富集铀燃料
uranium fissium fuel 铀和裂变产物合金燃料
uranium free from its daughters 无子体铀
uranium from seawater plant 海水提铀工厂
uranium fuel cycle 铀燃料循环
uranium fuel element 铀燃料元件
uranium-fuelled reactor 铀燃料反应堆
uranium graphite channel type reactor 铀石墨管道式堆
uranium graphite lattice 铀石墨栅格
uranium graphite reactor 铀石墨反应堆
uranium hexafluoride 六氟化铀
uranium lattice 铀栅格
uranium loading 铀萃取量
uranium mixed-oxide sol 钍铀混合氧化物溶胶
uranium molybdenum alloy 铀钼合金
uranium oxide pellet 氧化铀芯块
uranium plutonium alloy 铀钚合金
uranium plutonium carbide 碳化铀钚，铀钚碳化物
uranium plutonium cycle 铀钚循环
uranium plutonium fuel cycle 铀钚燃料循环
uranium plutonium oxide 氧化铀钚，铀钚氧化物
uranium poisoning 铀中毒
uranium reactor 铀堆，铀反应堆
uranium recovery 铀的回收，铀回收
uranium rod lattice 铀棒栅格
uranium rod 铀棒
uranium salvage operation 废铀回收操作
uranium scrap 铀切屑
uranium separative work 铀分离功
uranium series 铀族
uranium silicon alloy 铀硅合金
uranium slug 铀块
uranium tetrafluoride 四氟化铀
uranium thorium cycle 铀钍循环
uranium thorium reactor 铀钍反应堆
uranium trioxide 三氧化铀
uranium water lattice 铀水栅格
uranium zirconium alloy 铀锆合金
uranium zirconium hydride reactor 铀氢化锆反应堆
uranium 铀
uranous-uranium oxide 八氧化三铀
Uranus 尤拉纽斯镍铬合金钢
uranyl acetate 醋酸双氧铀
uranyl compound 铀酰化合物
uranyl fluoride 氟化铀酰，铀酰氟化物
uranyl nitrate dihydrate 二水合硝酸铀酰
uranyl nitrate hexahydrate 六水合硝酸铀铣
uranyl nitrate monohydrate 一水合硝酸铀酰
uranyl nitrate trihydrate 三水合硝酸铀酰
uranyl nitrate 硝酸铀，硝酸铀酰
uranyl sulfate 硫酸铀酰
uranyl 双氧铀，铀酰
urban air pollution source 城市空气污染源
urban air pollution 城市空气污染
urban architecture 城市建筑
urban area 城市地区，市区
urban atmosphere 城市大气
urban boundary layer 城市边界层
urban canyon 城市街谷，都市峡谷
urban center 市中心
urban climate 城市气候
urban complex 城市建筑群
urban construction design institute 城建设计院
urban construction 城市建设
urban development planning 城市发展规划
urban development 城市发展
urban diffusion parameter 城市扩散参数
urban distribution network 城市电网，城市配电网
urban domestic heating 城市居民区增温
urban ecology 城市生态学
urban effect 城市效应
urban environment 城市环境
urban expansion 城区扩展
urban facilities 城市设施
urban freeway 城市高速道路，都会区高速公路
urban heating rate 城市供热率
urban heat island 城市热岛
urban heat plume 城市热羽流
urban hydrology 城市水文学
urban industry heating 城市工业增温
urban infrastructure 城市基础建设
urban investigation 城市调查
urbanization 城市化
urban mixing layer 城市混合层
urban planning 城市规划，城镇规划
urban power distribution 城市配电
urban power network planning 城市电力网规划
urban power network 城市电力网
urban power supply 城市供电
urban redevelopment 城市改造
urban renewal planning 城市更新规划
urban renewal 城市改造，城市更新，城市重建
urban resident 市民
urban road system 城市道路系统
urban road 城市型道路

urban runoff 城市径流
urban sewage 城市污水
urban sewerage 城市径流，城市排水
urban sociology 城市社会学
urban survey 城市调查
urban terrain 城市地形
urban traffic 城市交通
urban transportation 城市运输
urban utilities 城市公用事业
urban waste 城市废物
urban water management 城市水管理
urban water supply 城市供水
urban wind 市区风
urbine gear motor 盘车装置电机
URDG(uniform rules for demand guarantees) 见索即付保函统一规则
urea-formaldehyde plastics 脲醛塑料
urea-formaldehyde resin 脲甲醛树脂
urgency signal 紧急信号
urgent alarm 紧急报警
urgent cable 加急电报，加急电缆
urgent call 紧急呼叫，紧急通话
urgent document 急件
urgent need equipment 急需设备
urgent session 紧急会议
urgent signal 紧急信号
urgent task 紧急任务
urgent telegram 急电
urgent 紧急的，急迫的
urge the fire 把火吹旺，提高炉膛负荷
urge （炉膛）加负荷，催促，推进，驱策，力劝，规劝，极力主张
uric acid 尿酸
urinalysis 尿分析（法），尿检
urinal 小便池
U-rod U形棒【管干甩击限位器】
urotropine 乌洛托品，六次甲基四胺
urushiol resin paint 漆酚树脂漆
usability of electrode 焊条工艺性
usability （焊条）使用性能
usable area 使用面积
usable capacity 可用的容积，有效容积，可用库容
usable enthalpy drop 可用焓降
usable floor area 使用面积
usable storage of reservoir 水库有效库容
usable storage 可用库容，有效库容
US advanced light water reactor 美国改进型轻水反应堆
usage and maintenance 使用（与）维护
usage aperture 可用孔径
usage factor 利用系数【疲劳】，利用率，利用因子
usage ratio 利用率
usage 使用，用法，运用，使用率
usance bill 远期汇票
USCD(unified specific creepage distance) 统一爬电比距
US Committee-for Energy Awareness 美国能源了解委员会
USCS(unified soil classification system) 土的统一分类系统
USD(United States Dollars) 美元
use charge 燃料使用费
used energy end point 汽轮机实际终态焓
used-fuel reprocessing plant 废燃料后处理厂
used oil 废油
use factor （锅炉）利用系数，利用率，利用因子，利用因数，使用因素
useful area 有效面积，可用面积
useful capacity 可用容量，有效容量
useful compression ratio 有效压缩比
useful cross section 有效截面，有效面积
useful depth 有用深度
useful discharge 有效流量
useful energy 有效能
useful flow 有效流量
useful flux 有效磁通，有用磁通
useful heat drop 可用热降，有效热降
useful height 有效高度
useful life period 使用寿命期限，使用寿命，有效期限
useful life 使用寿命，使用期（限），有效期（限）
usefulness of project 工程实用性
usefulness 效益，有效性，有用，有益
useful neutron 有用中子，能产生裂变的中子
useful output at terminal 发电机出线端净出力
useful output of coupling 联轴器端净出力
useful output power 有效输出功率
useful output 有效出力，有效功率，有用出力，有效输出
useful power 有效功率，有用功率
useful range 有效范围
useful ratio 利用率
useful refrigerating effect 有效制冷量
useful storage 兴利库容
useful thermal power 有用热功率
useful time 有效寿命
useful voltage 有效电压，有用电压
useful volume 有效容积
useful working exchange capacity 有效工作交换容量
useful work 有效功，有用功
useful 有用的
U. S. Environment Protection Agency 美国环境保护局
use of waste heat 废热利用
USEPA(United States Environmental Protection Agency) 美国环境保护局
user adjustment 用户调整
user attribute data set 用户属性数据集
user certificate 用户证明
user-friendly interface 用户友好界面
user identification 用户鉴定
user I/O device 用户输入输出设备
user link 用户线
user mode 使用方式，使用状态
user-oriented 面向用户的
user-programmable 用户可编程的
user program 用户程序
user's manuals 用户手册

user work area　用户工作区
user　用户，消费者，使用者
uses of waste heat　废热利用
use's quality cost　使用质量成本
use　利用，使用
U-shape bolt　U形螺栓
U-shaped coil　U形线圈
U-shaped expansion joint　U形膨胀节
U-shaped firing　U形火焰燃烧
U-shaped frame　U形框架
U-shaped holding fixture　U形夹具
U-shaped magnet　马蹄形磁铁
U-shaped valley　U形谷，平底谷
U-shaped　马蹄形的
U-shape hanger chains　U形曲链片吊挂装置
USMVC(unicast sampled measured value control)　单播采样测量值控制
U. S. S(ultimate shear stress)　极限剪应力
ustert　干变性土，干新成土
U-stirrup　U形箍筋，开口箍筋
ustoll　干软土
ustox　干氧化土
usual packing　惯用包装
usual practical unit　常用实用单位
usual practice　惯例
usual route　惯驶航线
usual service conditions　正常工作条件
usual terms　普通条件
usualuse dynamic brake　通用发电制动器
USVCB(unicast sampled value control block)　单播采样值控制块
USVC(unicast sampled value control)　单播采样值控制
utensil　器具，用具，器皿
U-ThO_2 kernel　铀二氧化钍芯核
utilance　空间利用系数
utilidor　保温管道
utiliscope　工业电视装置
utilities company　公共事业公司
utilities cost　公用事业费用
utilities equipment　动力设备
utilities　公用事业【气，水，压缩空气，电力等】
utility analysis　效用分析
utility area　生活区
utility bill　物业账单，水电气费账单，公共设施账单
utility boiler　电站锅炉，公用事业锅炉
utility certificate　实用证书
utility communication architecture　公用事业通信体系结构
utility connection　公用工程接头
utility control center　应用管理中心
utility control program　应用控制程序
utility cost analysis　效用成本分析
utility criterion　效用标准
utility distribution company　配电公司
utility factor　利用率，(设备)利用系数
utility function　效用函数
utility gas plant　公用煤气厂
utility grid　公用电网
utility line　公用事业管线，电力线路
utility load dispatcher　电力公司负荷调度程序
utility model　实用模型
Utility Nuclear Power Oversight Committee　电力公司核动力监督委员会【美国】
utility plant　公用设施工厂，公用事业电厂
utility pole　电杆
utility power plant　公用事业电站
utility power　公用电力，公用电源，公用动力
utility program　实用程序，辅助程序，应用程序
utility room　杂物间，杂用间，多功能间
utility routine　实用程序
utility snips　通用剪，实用剪
utility software　实用软件
utility station　公用电站
utility tape processor　应用磁带处理机
utility theory　效用理论
utility-type unit　电站型机组，公用型机组
utility　公用事业，公用电站，中心电站，效用，实用，电力部门，电力公司
utilizability method　可用能分析法，分析法
utilizable discharge　可用流量
utilization coefficient　利用率，利用系数
utilization cost　利用费用
utilization efficiency　利用效率
utilization equipment　利用的设备，电用户，消费者，用电设备
utilization factor of construction ground　施工现场利用系数
utilization factor of leaving velocity　余速利用系数
utilization factor on the strength of prestressed anchors　预应力锚栓强度利用系数
utilization factor　利用系数，利用率，利用因子，利用因数
utilization fraction　利用系数
utilization heat rate　利用热耗
utilization hours　利用小时数
utilization of electrical energy　用电
utilization of energy　能量利用
utilization of resources　资源利用
utilization of water resources　水利资源利用
utilization period　利用时间，运行时数
utilization rate of wind energy　风能利用率
utilization rate　利用率
utilization ratio　利用率
utilization stage　利用阶段【运行和维修】
utilization　利用，使用，应用
utilizing foreign capital　利用外资
utmost　最大的，极度的，极端的，极度，极端
UTM(universal testing machine)　万能（材料）试验机
U-trap　存水弯
utterly baseless　毫无依据的
uttermost lowest air temperature　极端最低气温
U-tube draft gauge　U形管风压表
U-tube evaporator　U形管蒸发器
U-tube gauge　U形管压力计
U-tube heat exchanger　U形管热交换器
U-tube manometer　U形管压力计
U-tube saturated steam generator　U形管饱和蒸汽发生器
U-tube steam generator　U形管蒸汽发生器

U-tube type feed water heater U形管式给水加热器
U-tube type heater U形管式加热器
U-tube U形管，U型管
UT(ultrasonic test) 超声探伤
UT(univeral time) 世界时间
U-type evaporator U形蒸发器
U type flow meter U型流动仪
UU-test(unconsolidated-undrained test) 不固结不排水剪切试验
uvala 灰岩盆，岩溶谷
UV erasable PROM 紫外可擦可编只读存储器
uviol glass 透紫外线玻璃
uviol 透紫（外线）玻璃
U-vortex 马蹄形涡流
UW＝upwind 迎风，顶风，上升气流，上风向
U-Zr alloy 铀-锆合金

V

vacamatic 真空自动式
vacancy diffusion 空位扩散
vacancy rate 闲置率
vacancy 空位，空穴，空缺，闲置
vacant-number signal 空号信号
vacant operation 空转，空载运行
vacant run 无载运行
vacant 空的
vacate 腾空，留出
vacation and holiday pay 假期和节日工资
vacation pay 假日津贴［工资］
vacation 假期，假日，休假
vaccum breaking 破坏真空
vaccum deaerator 真空除气器
vaccum down 真空度降低
vaccum extraction 抽真空
vaccum leakage test 真空严密性试验
vaccum pump 真空泵
vaccum reduced 真空度降低
vaccum system 真空系统
vaccum 真空
VA characteristics curve 伏安特性曲线
Vac-metal 镍铬电热线合金
vac-sorb 真空吸附
vacuate 抽成真空
vacu-forming 真空造型
vacuity 真空，真空度
vacuo-junction 真空热电偶
vacuole 液泡，空泡，空隙
vacuometer 低压计，真空计
vacuseal 真空密封
vacustat 真空计，真空稳定度
vacuum air pump 真空泵
vacuum air-removed system 抽气真空系统
vacuum and pressure relief valve 真空压力释放阀
vacuum anemometer 真空风速表
vacuum annealing 真空退火
vacuum apparatus 真空设备
vacuum arc extinction 真空灭弧
vacuum arc 真空电弧
vacuum area 真空区（域）
vacuum arrester 真空避雷器
vacuumatic measuring 真空计量
vacuum augmentor 真空增强器
vacuum bag 真空袋
vacuum belt filter 皮带真空吸滤器
vacuum blower house 负压风机房
vacuum blower 负压风机
vacuum box 真空箱【焊缝检验】
vacuum brake 真空制动器
vacuum brazing 真空硬钎焊
vacuum breaker valve 真空破坏阀
vacuum breaker 真空破坏器【凝汽器】，真空断路器，真空解除设施，真空泄放阀，破真空阀
vacuum breaking valve 真空破坏阀，破真空阀
vacuum breaking 真空破坏，真空断裂
vacuum break shutdown 破坏真空停机
vacuum capacitance 真空电容
vacuum capacitor 真空电容器
vacuum casting 真空铸造
vacuum centrifugal pump 离心式真空泵
vacuum chamber 真空室，真空箱，压力室
vacuum circuit breaker 真空断路器
vacuum cleaner 真空吸尘器，真空吸尘装置，吸尘器
vacuum cleaning installation 真空吸尘装置
vacuum cleaning mobile 真空清扫车
vacuum cleaning system 真空清洗系统
vacuum cleaning 吸尘，真空清除法，真空清扫，真空吸尘法
vacuum concrete 真空混凝土
vacuum connector 真空接头，真空接触器
vacuum control check valve 真空逆止阀
vacuum controller 真空控制器
vacuum control panel 真空控制盘
vacuum-cooling 真空冷却
vacuum corer 真空岩芯提取器
vacuum creation 真空形成
vacuum curve 真空曲线
vacuum deaeration 真空除氧，真空排气法
vacuum deaerator 真空除氧器
vacuum decreasing rate 真空下降率
vacuum degasifier 真空（式）除气器
vacuum degassing tank 真空除气箱
vacuum degas 真空除气
vacuum degree 真空度
vacuum de-loading （低）真空减负荷
vacuum deposition 真空沉积
vacuum desiccator 真空干燥器
vacuum diffusion pump 真空扩散泵
vacuum dilatometer 真空膨胀计
vacuum discharge 真空放电
vacuum distillation column 真空蒸馏塔
vacuum down （凝汽器的）真空度降低
vacuum drainage 真空排水
vacuum drier 真空干燥器
vacuum-drive gyroscope 真空陀螺仪
vacuum dust collection system 真空除尘系统
vacuum dust extraction 真空吸尘

vacuum electrical insulation 真空电绝缘
vacuum envelope 真空密封外壳
vacuum establishing 建立真空
vacuum evaporation 真空蒸发
vacuum evaporator 真空蒸发器
vacuum extraction 抽真空
vacuum extractor 抽真空设备
vacuum fan 抽风机，真空风扇
vacuum filter 真空过滤器，真空滤池
vacuum filtration 真空过滤
vacuum floatation 真空浮选法
vacuum fly ash handling system 负压飞灰处理系统
vacuum forepump 前置真空泵
vacuum form concrete 真空模板混凝土
vacuum form 真空模板
vacuum furnace 真空炉
vacuum gauge pressure 真空表压力
vacuum gauge 真空计，真空度仪，真空规，真空表
vacuum grouting 真空灌浆
vacuum head 吸入侧真空
vacuum heater 真空加热器
vacuum heating system 真空采暖系统
vacuum helium leak detector 真空氦气探漏仪
vacuum hood 真空罩
vacuum hose 真空软管
vacuum-impregnated 真空浸渍的
vacuum impregnation 真空浸渍
vacuum indicator 真空度指示器，真空试验器，真空表
vacuum induction melting 真空感应熔炼
vacuum insulation 真空绝缘
vacuum interferometer 真空干涉仪
vacuum interrupter 真空断路器，真空隔离开关
vacuum ionization gauge 电离真空计
vacuumizer （抽）真空机
vacuumize 抽真空，真空处理
vacuum junction 真空热电偶
vacuum lamination 真空层合
vacuum lamp 真空白炽灯
vacuum leakage test 真空泄漏试验
vacuum-leak detector 真空漏气指示器
vacuum lifting 真空起吊
vacuum lightning arrester 真空避雷器
vacuum lightning protector 真空避雷器
vacuum lining 真空衬护
vacuum load of condenser 凝汽器真空吸力
vacuum lock 真空栓，真空闸，真空锁定
vacuum manometer 真空压力表，真空压力计
vacuum measurement 真空测量
vacuum meter 真空表
vacuum multistage evaporator 多级真空蒸发器
vacuum oil filter 真空滤油机
vacuum-operated load reducer 真空下降负荷限制器
vacuum-operated unloading device 凝汽器真空监视器
vacuum packaging 真空包装
vacuum panel 真空模板
vacuum pay-off and trip gear 凝汽器真空监视器
vacuum photo tube 真空光电元件，真空光电管
vacuum preloading 真空预压法
vacuum pressure gauge 真空压力计，联成计
vacuum pressure impregnation 真空压力浸渍
vacuum pressure pump 真空压力泵
vacuum pressure 真空压力
vacuum-processed concrete 真空作业混凝土
vacuum process 真空作业
vacuum profile 真空式剖面
vacuum pump removal system 真空泵抽气装置
vacuum pump 真空泵
vacuum refrigeration 真空制冷
vacuum relay 真空继电器
vacuum relief valve 真空减压阀，真空释放阀，真空泄放阀
vacuum return-line heating system 真空采暖系统
vacuum returnline system 真空回水系统
vacuum sampling tube 真空取样器，真空取样管
vacuum sand blasting machine 真空喷砂机
vacuum sealing component 真空密封元件
vacuum seal 真空密封，真空封接
vacuum seasoning 真空干燥
vacuum sleeve 真空套管
vacuum spray evaporator 真空喷雾蒸发器
vacuum spraying 真空喷涂
vacuum surge tank 真空缓冲箱
vacuum sweeper 真空吸尘器
vacuum switchgear 真空开关装置
vacuum switch 真空开关
vacuum system 真空系统，真空设备
vacuum tank 抽气排水筒，真空罐，真空箱，真空柜
vacuum test 真空试验，真空严密性试验
vacuum thermocouple 真空热电偶
vacuum tight test 真空严密性试验
vacuum-tight 真空密封的
vacuum trap 真空阱
vacuum-treated concrete 真空作业混凝土
vacuum trip device 低真空保护装置，低真空脱口装置
vacuum tube accelerometer 电子管加速计
vacuum tube adapter 电子管适配器
vacuum tube arrester 真空管避雷器
vacuum tube bridge 电子管电桥
vacuum tube characteristic 真空管特性
vacuum tube circuit 电子管电路
vacuum tube detector 真空管检波器
vacuum tube generator 真空管振荡器
vacuum tube microammeter 电子管微安计
vacuum tube millivoltammeter 电子管毫伏表
vacuum tube modulator 真空管调制器
vacuum tube operator 电子管算子
vacuum tube register 电子管寄存器
vacuum tube relay 电子管继电器，真空管继电器
vacuum tube switch 真空管开关
vacuum tube transmitter 真空管发送器
vacuum tube voltmeter 真空管伏特计，真空管电压表，电子管伏特计，电子管电压表
vacuum tube 真空管，电子管
vacuum-type ash conveyer 负压气力输灰机

vacuum-type tensiometer 真空式张力计
vacuum under nappe 水舌下的真空
vacuum unloading and trip gear 真空下降卸载及脱扣装置
vacuum up 真空度升高
vacuum valve 真空管，电子管，真空阀
vacuum variable capacitor 真空可变电容器
vacuum 真空，真空的，真空度，真空吸尘器
VAC＝vacuum 真空
vadose region 渗流区
vadose-water discharge 非饱和区土壤水流量，渗漏水排出量
vadose water 含气带水，渗流水
vadose zone 含气层，上层滞水带，渗流区
V-aerial V形天线
vagabond current 地电流，（地中）杂散电流
vail 有利于
vain 无效的
valence band 价带
valence 原子价，化学价，化合价，效价
vale 峪
validated test coupon 有效试样，确认的试样
validate 证实，确证，确认（有效性），使生效，使合法化，批准，使有法律效力
validation check 有效性检查
validation of contract 契约效力
validation of signature agent 有效签字代理人
validation period of loan 贷款有效期
validation period 有效期限
validation phase 审批阶段
validation report 批准报告，确认定报告，证实报告
validation 确认，生效，使生效，有效
valid certificate 有效证明，有效证书
valid contract 有效合同
valid for negotiation 议付有效
valid graphics editor 有效图形编辑器
valid grid assets 电网有效资产
validity check 有效性检查
validity of an award 裁决有效性
validity of arbitration award 仲裁裁决的效力
validity of bids 投标有效性
validity of claims 索赔有效性
validity of contract 合同的有效性
validity of credit 信用证有效性
validity of tenders 投标有效性
validity period of a bid 投标书有效期
validity stamp 有效印章
validity test 有效测试，有效性检验
validity 有效，有效性，真实性，合法性，正当，效力，有效期限
valid period 有效期限
valid program 有效程序
valid SAIC design and layout tools 有效的专用集成电路设计和布线工具
valid time of forecast 预报时效
vallance 壁式隐灯
valley angles in hoppers 料斗谷角【壁交线与水平夹角】
valley angle 谷角
valley ash yard 山谷灰场
valley axis 河谷轴线
valley bog 低位沼泽
valley bottom 谷底
valley breeze 谷风
valley consumptive use 流域耗水量
valley crossing 跨谷测量
valley cross section 河谷（横）断面
valley current 最小电流，谷值电流
valley development project 流域开发工程
valley flashing 天沟泛水
valley flat 河漫滩
valley floor plain 谷底平原
valley floor 谷底
valley flow 山谷气流
valley fog 谷雾
valley gutter 谷沟，天沟，檐沟
valley in valley structure 迭谷构造
valley in valley 迭谷
valley load 低谷负荷，负荷谷值
valley meander 峡谷曲流
valley plain 谷底
valley point current 谷值电流
valley rafter 沟椽
valley slope wind 谷坡风
valley slope 山谷坡，河谷坡降
valley surface defect 凹陷【表面缺陷】
valley terrace 河谷阶地
valley tile 槽形瓦
valley tract 河谷地区，谷区，中游地段
valley trenching 河谷下切
valley voltage 谷点电压，谷值电压，最小电压
valley wind circulation 谷风环流
valley wind 谷风
valley 谷，山谷，流域，（曲线的）凹部，谷值，最低值，天沟，峡谷
Valloy 瓦洛锆合金
Valray 瓦尔莱合金
valuable material store-room 贵重材料库
valuation of variations of project 工程变更估价
valuation 评估，评价
valubreeder 钍铀增殖反应堆
value added tax 增值税
value analysis framework 价值分析框架
value analysis method 价值分析法，价格分析法
value analysis 价值分析，值分析
value assignment 赋值
value bill 估价单
value call 值调用，调值
value control amplifier 值控制放大器
value date 起息日
valued at 估价为……
value engineering 工程经济学，价值工程（学）
value of export 出口额
value of import 进口额
value of industrial output 工业产值
value of quantity 量值
value of scale division 标度分格值，分格值，格值，刻度值
value preference 价值偏好，价值取向
valuer 鉴定人，评价人
value securities 有价证券

values of thermophysical property 热物理性质的数据
values of viscosity 黏性系数的值
value to risk 风险值
valuevite 绿脆云母
value 价值，估价，数值，值
VAL＝value 数值
valve action 整流作用，阀作用，活门作用
valve actuating gear 阀门传动装置
valve actuator 阀门传动装置，阀门执行机构
valve adapter 电子管适配器，电子管转接器
valve amplifier 电子管放大器
valve area 阀门流通截面积
valve arrester 电子管避雷器，阀式避雷器
valve assembly 阀门总成
valve at turbine inlet 水轮机进口阀
valve authority 阀权度
valve backseated 有后座的阀门
valve barrel 阀缸，阀筒
valve base 电子管管座，阀座
valve block 阀组，给水阀组，给油阀组
valve body assembly 阀体组件
valve body casting 阀体铸件
valve body crotch 阀门岔口
valve body 阀身，阀体
valve bonnet bush 阀盖衬套
valve bonnet gasket 阀盖垫片
valve bonnet 阀盖，阀帽，阀罩
valve bowl 阀箱
valve box 阀门箱
valve breakdown 阀关断，阀闭锁，阀堵塞
valve bush 阀衬
valve cam 阀凸轮
valve cap 阀盖
valve carrier 阀座
valve casing 阀壳
valve chamber 阀室
valve checker 电子管测试器
valve chest 阀室，阀箱，阀门室，蒸汽室
valve clack 阀瓣，阀舌
valve clearance adjustment 阀间隙调整
valve clearance 阀门间隙
valve cone 阀座，阀锥
valve control motor 阀控电动机
valve control pillar 阀门远距离操纵杆
valve control pulse 阀控制脉冲
valve cover 阀套，阀盖
valve cup 阀座
valve damping circuit 阀阻尼回路
valve designation system 电子管型号命名制度，电子管标志法
valve detector 电子管检波器
valved in 阀门装在……
valve disc(/disk) 阀叶【蝶阀】，阀盘，阀板，阀碟
valved-off 阀门隔断的
valved 用阀调节的
valve elbow 带阀弯头
valve electrode 阀电极
valve element 阀元件【避雷器】，电子管元件，阀片，阀芯
valve exercisability bench 阀门操作性能台架
valve filament circuit 电子管灯丝电路
valve fitting 阀门研磨
valve flap 阀瓣
valve flow characteristic 阀门流动特性
valve gallery 阀廊
valve gear 阀动装置
valve guard 阀挡，阀片升程限制器，阀柱护套，阀门限程器，阀挡板
valve guide 阀导，阀导轨
valve handle 阀杆，阀门手柄
valve head 阀头，阀盖，阀顶
valve hood 阀盖，阀帽
valve house 阀室
valve housing 阀壳，阀室，阀套
valve injection 阀注射
valve input impedance 电子管输入阻抗
valve inside 阀芯，气阀芯，气门芯
valve internals 阀门内部件
valve key 阀门扳手，阀门键
valve lag 阀滞后，电子管脚
valveless filter 无阀滤池，无阀滤器
valveless 无活门的
valve level leakage steam system 门杆漏汽系统
valve lever steam leakage piping 门杆漏汽管道
valve lever 阀杆，阀门操纵杆
valve lift 阀门开度，阀门升程
valve liner 阀衬
valve loop correction 调节阀组合特性曲线校正
valve loop curve 调节阀组合特性曲线
valve loop performance curve 调节阀组合特性曲线
valve management program 阀门操纵程序
valve management 阀门管理
valve model 阀门型号
valve monitoring 阀监视，阀控制
valve motor 阀门电动机
valve needle 阀针
valve-off 关闭阀门
valve oil shield 阀防油罩
valve opening 阀门开度，阀开度，阀通口
valve-operated water closet 冲洗阀马桶
valve operating mechanism 阀门操作机构
valve operation platform 阀门操作平台
valve operation test 阀动作试验
valve oscillator 电子管振荡器
valve-out 关闭阀门
valve packing 阀门盘根
valve pad 阀垫
valve phase indicator 阀（相）位指示器，阀位指示器
valve piston 阀活塞，阀塞
valve plug 阀芯，阀塞
valve pocket 阀箱，阀套
valve point curve 阀点曲线
valve point internal efficiency 阀点内效率
valve point performance curve 阀点性能曲线
valve point 阀点
valve port size 阀口尺寸，气门孔口尺寸
valve port 阀口，气门口，阀孔，汽门孔
valve position control 阀门开度控制，阀位控制

valve positioner 阀门位置指示器，阀门定位器，阀位控制器
valve position-flow characteristic 阀位-流量曲线
valve position indicator 阀位置指示器
valve positioning servo loop 阀门定位伺服回路
valve position signal 阀位信号
valve position supervision device 阀位监视装置
valve position tracing 阀位跟踪
valve position transducer 阀位变送器
valve position 阀位置
valve rating 阀门等级
valve reactor 阀电抗器
valve rectifier 电子管整流器
valve refacing machine 磨阀面机，阀面磨光机
valve relay 电子管继电器，阀继电器
valve reseater 阀门修整工具
valve resistance arrester 阀电阻式避雷器
valve rib guide 阀门导向筋
valve rod 阀杆
valve room 阀门间
valve seal 阀止水（密封）
valve seat reamer 汽门座铰刀
valve seat ring 阀座环
valve seat 阀座
valve setting 阀位给定，阀门调整
valve shaft 阀杆
valve size 阀门尺寸，阀门口径
valve sleeve 滑阀套
valves of size greater than 50mm 尺寸大于50毫米的阀门
valves of size less than 50mm 尺寸小于50毫米的阀门
valve spindle 阀轴，阀杆
valve spool 阀芯
valve spring holder 阀的弹簧支架
valve spring housing 阀门弹簧套
valve spring 阀簧
valve station 阀门站
valve steam capacity 阀门的额定蒸汽容量
valve stem ejection 阀杆弹出
valve stem guide hole 导管孔
valve stem guide O ring 阀杆导套O形圈
valve stem guide 阀导杆，阀杆导件，气门导管
valve stem leakoff 阀杆漏汽，阀杆引漏
valve-stem nut 阀杆螺母
valve-stem packing leakage 阀杆密封泄漏
valve stem packing 阀杆填料
valve stem penetration 阀杆贯空件
valve stem ring 阀杆环
valve stem 阀杆
valve stroke 阀行程，阀冲程
valve tandem 阀门串接【卸压阀和隔离阀】
valve tee 带阀三通
valve test 阀门试验
valve travel 阀行程，阀开度
valve trim 阀门内件【与传输液体接触的内件】，阀芯
valve type lightning arrester 阀型避雷器
valve type piston 阀式活塞
valve type surge arrester 阀式避雷器
valve unit control 阀组控制
valve unit protection 阀组保护
valve voltage damper 阀电源阻尼器
valve voltage divider 阀分压器，阀均压器，电子管分压器
valve washer 阀门垫圈，阀座垫圈，滑阀垫板
valve well 阀门井
valve wide open capability 调节（汽）阀全开功率，阀门全开能力
valve wide open 阀门全开
valve winding 阀侧绕组【换流变压器】
valve with ball seat 球座阀
valve with inclined stem 斜杆阀，斜轴阀
valve with straight stem 直轴阀
valve yoke 阀轭
valve 阀，阀门，活门，闸门，真空管
valving-down 用阀调低
valving 阀门关闭，设置阀门
VAMA(variable address multiple access) 可变地址多路存取
vanadium pentoxide 五氧化二钒
vanadium 钒【V】
van container 大型集装箱
Van de Graaff generator 范德格拉夫加速器，静电加速器
van der Waals force 范德华力
vane anemometer 翼式风速计，叶轮风速计，叶轮式风速仪
vane apparatus 导叶装置
vane bend height 冀弯高度
vane camber 叶片弯度
vane cascade 叶栅
vane channel 叶片间流道
vane compressor 叶片式压气机
vane control fan 可调导向叶片式风机
vane control 叶片调节，轴向导向装置调节
vane damper 叶片式风门
vaned diffuser 有叶扩压器
vaned 有叶的，装叶片的
vane efficiency 叶片效率
vane entrance angle 叶片进口角
vane guide （风机）静叶片，叶片导轨
vane-in-body pump 壳体内叶泵
vane-in-rotor pump 转子内叶泵
vaneless diffuser 无叶扩压器
vaneless scroll 无叶蜗壳
vaneless space 无叶空间，无叶扩压室
vaneless-vaned diffuser 无叶有叶混合扩压器
vane motor 叶片式油马达
vane pitch 叶栅栅距
vane pump 叶轮泵【旋转式泵】，叶片泵
vane ratio 导风率
vane setting angle 叶片安装角
vane shear apparatus 十字板剪力仪
vane shear strength 十字板抗剪强度
vane shear test apparatus 十字板剪力试验仪
vane shear test 十字板剪力试验
vane spacing 叶片栅距
vane strength 十字板抗剪强度
vane thickness 叶片厚度
vane tip clearance 叶梢间隙
vane tip 叶片尖端

vane twist 叶片的扭曲度
vane type anemometer 翼式风速计
vane type compressor 叶片式压缩机，叶轮式压缩机
vane type magnetron 翼式磁控管
vane type nozzle block 叶片式喷嘴组
vane type relay 扇形继电器
vane variable capacitor 翼片型可变电容器
vane wheel 叶轮
vane 瓣，叶，叶片，叶轮，风杯，(风洞)导流片，轮叶，桨叶，风叶，导向叶片，翼板，风向标，变化不定的，反复无常的
vanier caliper 游标卡尺
vanishing angle 消没角
vanishing flux 消没通量
vanishing line 投影线
vanishing point 灭点
vanishing tide 缺潮，失潮
vanishing 消没
Van-Stone flange 凡司徒法兰
van 大型运货汽车，拖车
vapor absorption 蒸汽吸收
vapor air heater 暖风机
vapor-air mixture ratio 汽气混合比
vaporization 汽化，汽化作用，蒸发，气化，蒸发作用，气化作用
vapor barrier 防潮层，隔汽层
vapor binding 汽阻塞
vapor blanket 蒸汽覆盖层
vapor bubble 汽泡
vapor burst 蒸汽猝发
vapor circulation 蒸汽循环
vapor cloud 蒸汽云
vapor column 蒸汽柱
vapor composition 蒸汽组成
vapor compression cycle 蒸汽压缩循环
vapor compression evaporator 蒸汽压缩型蒸发器
vapor compression process 蒸汽压缩过程
vapor compression refrigerator 蒸汽压缩式制冷机
vapor concentration 蒸汽浓度，水蒸汽密度
vapor condenser 蒸汽凝结器，蒸汽凝汽器
vapor conditions 蒸汽参数
vapor container spray system 蒸汽容器喷淋系统
vapor content 含汽量
vapor-cooled generator 蒸发冷却发电机
vapor-cooled transformer 蒸发冷却变压器
vapor cooler 蒸汽冷却器
vapor curve 蒸汽压力曲线
vapor cycle condenser 蒸汽循环冷凝器
vapor cycle 蒸汽循环
vapor degreasing 蒸汽除油
vapor density 蒸汽密度
vapor deposition 汽相沉积
vapor dispersion 蒸汽弥散
vapor-dominated hydrothermal resource 以蒸汽为主的热液资源
vapor drying plant 蒸汽干燥装置
vapor eliminator 蒸汽分离器
vapor exhaust fan 油雾排风扇
vapor expansion chamber 蒸汽膨胀室
vapor extraction cowl 抽汽罩
vapor extraction device 抽汽装置
vapor extractor 抽汽器
vapor film 蒸汽膜，蒸汽层
vapor filter 蒸汽过滤器
vapor generator 蒸汽发生器
vaporimeter 蒸汽计，蒸汽压力计，挥发度计
vaporising fuel 气化燃料
vaporization adjustment 汽化调节
vaporization-condensation cycle 气化冷凝循环
vaporization cooling 蒸发冷却
vaporization efficiency 汽化效率，蒸发效率
vaporization heat 汽化热，蒸发热
vaporization losses 蒸发损失
vaporization nucleus 汽化核，汽化中心
vaporization rate 汽化速度，汽化率
vaporization temperature 蒸发温度，汽化温度
vaporization 汽化(作用)，蒸发
vaporizer burner 汽化燃烧器
vaporizer tube 燃油蒸发管
vaporizer 汽化器，蒸发器，喷雾器，雾化器
vaporizing burner 汽化式燃烧器
vaporizing combustor 汽化式燃烧室
vaporizing-oil engine 汽化油发动机
vaporizing point 沸点，蒸发温度
vaporizing property 汽化特性
vaporizing 蒸发
vapor jacket 蒸汽夹套
vapor jump 蒸汽跃迁
vapor liquid equilibrium 汽液平衡
vapor liquid ratio 汽液比
vapor liquid separator 汽液分离器，汽水分离器
vapor lock 汽塞，汽阻
vapor loss 蒸汽损失
vapor nozzle 蒸汽喷嘴
vaporous carryover 气相携带，(盐分的)溶解携带，选择性携带
vaporous water 汽化水
vaporous 蒸汽的，汽态的
vapor phase chromatography 汽相色谱法
vapor phase cooling 蒸发冷却，汽态冷却，汽化冷却
vapor phase corrosion 汽相腐蚀
vapor phase epitaxy 汽相外延
vapor phase inhibitor 汽相缓蚀剂
vapor phase 汽相
vapor pipe 蒸汽管道
vapor plating 气相沉积
vapor plume 蒸汽羽流，蒸汽羽烟
vapor preheater 蒸汽预热器
vapor pressure curve 蒸汽等压线
vapor pressure thermal system 蒸汽压力式感温系统
vapor pressure thermometer 蒸汽压力温度计
vapor pressure thermostat 蒸汽压力温度自动调节器
vapor pressure 蒸汽压，蒸汽压力
vapor proof course 隔汽层
vapor proof engine 防气式电机
vapor proof machine 汽密式电机，防汽电机

vapor proof 抗蒸汽的，隔蒸汽的
vapor recirculation 蒸汽再循环
vapor recovery 蒸汽回收
vapor retarder 蒸汽缓凝剂
vapor riser 导汽管，气体上升管
vapor rise 蒸汽抬升
vapor scrubber 汽化清洗装置，蒸汽清洗装置
vapor seal 汽封
vapors-free 无蒸汽凝结的
vapor space 汽空间，蒸汽腔
vapors-set 蒸汽凝结的
vapor state 汽态，汽化状态
vapor sump 集气池
vapor suppression system 蒸汽弛压系统
vapor table 蒸汽表
vapor tension thermometer 蒸汽压力温度计
vapor tension 蒸汽压力，蒸汽张力
vapor tight 不漏汽的
vapor trail 雾化尾迹
vapor trap 蒸汽阱
vapor vessel 蒸汽容器
vapor 蒸汽，烟雾
vapour barrier 隔汽层
vapour blanket 水面水汽覆盖层
vapour bubble 汽泡
vapour concentration 水汽浓度
vapour density 水汽密度
vapour diffusion 水汽扩散
vapour flow 水汽流
vapour flux 水汽通量
vapour method 汽化法
vapour migration 水汽迁移
vapour pocket 汽囊
vapour pressure difference 饱和差，水汽压差
vapour pressure 水汽压力，蒸汽压力，汽压
vapour quality 蒸汽品质
vapour state 汽态
vapour strainer 蒸汽过滤器
vapour tension 蒸汽张力
vapour trail 水汽尾迹
vapour 水汽，蒸汽，汽，水蒸气，
VAP＝vapor 蒸汽
varactor diode 参量二极管，变容二极管，可变电抗器
varactor frequency modulation 变容二极管调频
varactor frequency multiplier 变容管倍频器，可变电抗倍频器
varactor tuned oscillator 变容二极管调谐振荡器
varactor 可变电抗器，变容二极管
var-hour meter 乏小时计，无功电度表
varhour 乏时，无功伏安小时
variability coefficient 变差系数
variability law 变异定律
variability of runoff 径流变率
variability of stream flow 流量变率
variability 变化性，变异性，波动性，多变性，变率
variable accelerated motion 变加速运动
variable acceleration 变加速度，可变加速度
variable action 可变作用
variable address multiple access 可变地址多路存取
variable address 可变地址
variable air volume air conditioning system 变风量空气调节系统
variable air volume system 变风量系统
variable air volume terminal device 变风量末端装置
variable air volume 变风量
variable-angle nozzle 可调导叶，可转导叶，可调喷嘴
variable-angle 可变角度的
variable-aperture flow meter 变孔口流量计
variable-area flow measuring device 变面积式流量测量仪表
variable-area flowmeter 可变面积流量计，转子流量计，变截面流量计
variable-area nozzle 变截面喷管
variable-area turbine 变截面透平，变截面涡轮机
variable-area type 可变通面积式
variable-area 可变区域，可变面积，变截面的
variable attenuator 可变衰减器
variable backpressure 变化的背压
variable binary scaler 可变二进制计数器
variable blade 可变安装角的叶片
variable budget 可变预算，临时预算
variable camber flap 变弯度襟翼
variable camber 可变弯度
variable capacitance diode 变容二极管，可变电容二极管
variable capacitance parametron 可变电容参变元件
variable capacitance transducer 可变电容传感器
variable capacitance 可变电容，可变容量
variable capacitor with equal capacitance 等容可变电容器
variable capacitor 可变电容，可变电容器
variable-capacity pump 可调流量泵
variable carrier 变幅载波，可控载波
variable-cell method 可变栅元法
variable channel 变化河槽，不稳定水道
variable chord blade 变截面叶片
variable coefficient 可变系数
variable condenser block 可变电容器组
variable condenser 可变电容器
variable-conductance heat pipe 可变热导热管
variable coning 可变锥角
variable connector 可变连接器
variable contact 可变触点
variable cost 可变费用，非固定费用，可变成本
variable cross section wind tunnel 可变截面风洞
variable cross-section 可变截面
variable crystal filter 可变晶体滤波器
variable crystal oscillator 可变晶体振荡器
variable cycle operation 可变周期运行，可变循环运转，可变周期操作
variable cycle 可变循环，可变周期
variable delay line 可变延迟线
variable delivery pump 可变容积泵，可变容量泵，变量输送泵

variable demand 可变需用量
variable density wind tunnel 变密度风洞
variable density 变密度，可变密度
variable digit layout 可变的数字配置
variable-discharge turbine 可变流量透平
variable displacement motor 可变排量马达
variable displacement pump 变量泵，可变排量泵
variable dose rate irradiation facility 可变剂量率辐照设施
variable duration impulse system 脉宽调制系统
variable element 可变参数
variable error 变量误差，可变误差
variable excitation 可变励磁，可变激发
variable expansion 变膨胀
variable factor 可变因子
variable field 可变字段
variable filtration 变速过滤
variable flow control 变流量控制，变流量调节
variable flow pump 变流量泵，可调流量泵
variable-flux reactor 可变中子通量反应堆
variable-focus pyrometer 变焦点高温计
variable-focus 可变焦点
variable frequency AC motor speed control 变频调速
variable frequency control electric motor 变频调速电动机
variable frequency electric drive 变频电气传动
variable frequency generating set 变频发电机组
variable frequency generator 变频发电机
variable frequency induced polarization 变频激发极化
variable frequency oscillator 变频振荡器
variable frequency power source 变频电源
variable frequency starting 变频启动
variable frequency transformer 变频变压器
variable frequency 可变频率
variable function generator 可变函数发生器
variable function 变量函数，可变函数
variable gag 冷却气体可调节流塞
variable gain amplifier 可变增益放大器
variable gain device 可变增益装置
variable gain method 可变增益法
variable gain 可变增益
variable gearing 变速传动装置
variable generator speed fixed-pitch wind turbine 变速定桨距风电机组，VGS-FP风电机组
variable generator speed fixed-pitch 变速定桨距
variable geometry blade 可调形状叶片
variable geometry nozzle 可调导叶，可变叶角喷嘴，可调喷嘴
variable geometry type wind turbine 可变几何翼型风力机
variable geometry 可变几何
variable head permeability test 变水头渗透试验
variable-head permeameter 变水头渗透率仪
variable identifier 变量标识符，变量识别器
variable incidence 可变冲角
variable inductance transducer 可变电感传感器，可变电感换能器
variable inductance 可变电感
variable inductor 可调电感器，可变电感器，可变电感线圈
variable information processing package 可变信息处理包
variable information processing 可变信息处理
variable inlet guide vane 可转进口导叶，可调进口导叶
variable intermittent duty 变载断续工作方式
variable ionization 可变电离
variable-length code 可变字长码
variable length word symbolic assembly system 可变字长符号汇编系统
variable length 可变字长
variable level scram 可变功率紧急停堆
variable load operation 变负荷运行
variable load plant 调峰电厂
variable load power plant 可变负荷发电厂
variable load 变载，可变负荷，可变荷载，变动负荷，变动荷载
variable logic 可变逻辑
variable loss 可变损耗
variable motion 变速运动
variable moving blade axial flow fan 动叶可调轴流式风机
variable moving blade fan 动叶可调风机
variable-mu hexode 变μ六极管，可变放大系数六极管
variable-mutual conductance tube 变跨导管
variable-mu 可变放大系数μ
variable-note buzzer 可调蜂鸣器
variable nozzle 变截面喷嘴，可调喷嘴
variable of state 状态变量
variable operating conditions 可变动运行工况
variable operating cost ranking 可变运营成本排名
variable operating cost 可变运营成本，可变营运费用
variable order 可变指令
variable orifice desuperheater 可调喷孔减温器
variable output circuit 可变输出电路
variable phase output 可变相位输出
variable phase 可变相位
variable pitch blade 变距叶片，变距桨叶
variable pitch impeller 可变节距叶轮
variable pitch propeller 变距螺旋桨
variable pitch rotor 可变桨距风轮
variable pitch turbine 变节距水轮机
variable pitch 可变节距，可变间距
variable power controller 可变功率控制器
variable-precision coding compaction 可变精确编码的数据精简法
variable-pressure deaerator 可变压力除氧器
variable-pressure drop 可变压降
variable-pressure operation 变压运行，滑压运行
variable-pressure wind tunnel 变压式风洞
variable-pressure 滑压，变压，可变压力，变压力，不定压
variable production cost 可变生产成本
variable proportion 可变部分
variable pump 变量泵

variable quantity　变量
variable-radius arch dam　变半径拱坝
variable radix　可变基数
variable range　可变范围，变量范围，调节范围，变动范围
variable rate driver　可变速率激励器
variable rate feeder　可变出力给料机
variable rate　可变利率
variable ratio power divider　可变比功率分配器
variable ratio transformer　变比可调式变压器，调压变压器
variable ratio　可变比
variable reactance　可变电抗
variable reactive compensator　可变无功补偿器
variable reactor　可变电抗器
variable reluctance step motor　反应式步进电动机，变磁阻式步进电动机
variable reluctance type rotor　变磁阻转子，凸极转子
variable reluctance　可变磁阻
variable resistance box　可交电阻箱
variable resistance　可变电阻
variable,resistive load control　可变电阻性负载控制
variable resistor　可变电阻，可变电阻器，变阻器
variable restriction　可变限制
variable rigidity　可变刚度
variable rotor resistance　转子变阻器
variables analysis　变量分析
variable section bar　变断面杆件
variable selectivity　可变选择度，可变选择性
variable setting circulator inlet guide vane　可调整循环风机进口导叶
variable-slope conical silo　变坡锥体筒仓
variable specific heat　可变比热，变比热
variable speech wind turbine　变速风电机组
variable speed constant frequency generator　变速恒频发电机
variable speed control　变速调节
variable speed device　变速装置
variable speed doubly fed induction generator　变速双馈异步发电机
variable speed drill　变速钻
variable speed driver　变速马达，变速传动，变速拖动
variable speed electrical generator　变速发电机
variable speed gear box　变速齿轮箱
variable speed gear drive　变速齿轮传动
variable speed gear　变速齿轮，变速（传动）装置
variable speed input　变速输入
variable speed motor　可变速电动机，变速电动机
variable speed pump　变速泵，调速水泵
variable speed rotor　变速风轮
variable speed scanning　变速扫描
variable speed single-stage centrifugal pump　单级变速离心泵
variable speed tape drive　变速拉带装置
variable speed turbine　变转速透平
variable speed wound rotor induction generator　变速绕线式转子感应发电机，绕线式异步发电机
variable speed　变速，可变速，调速，无级变速
variable static blade fan　静叶可调风机
variable stator blade　可调静叶，可转静叶
variable stator vane　可调静叶，可转静叶
variable stroke engine　可变冲程发动机
variable stroke plunger pump　可变冲程柱塞泵，柱塞变量泵
variable structure control system　变结构控制系统
variable system　变量系统，可变系统
variables　变量
variable temporary duty　变载短时工作方式
variable thermal conductance heat pipe　变热导热管
variable thermal impedance heat pipe　变热阻热管
variable-thickness arch　变厚度拱
variable-threshold decoding　变阈解码，可变阈值解码
variable threshold logic　可变阈值逻辑
variable time-setting relay　可调时间继电器
variable topology random access memory　可变结构随机存取存储器
variable-torque motor　可变转矩电动机
variable transfer address　可变转移地址
variable transformer transducer　可调变压器换能器
variable transformer　可调变压器
variable transmission　变速传动
variable-value control　变值控制，跟踪控制
variable-voltage controller　可变电压控制器
variable-voltage control　变压控制，变压调节，变压调速
variable-voltage regulator　可变电压调节器，调压器
variable voltage speed control　调压调速
variable-voltage transformer　可变电压变压器，变比可调式变压器
variable-voltage　可变电压
variable volume　变容，可变容量
variable water flow system　变水量系统
variable waveguide attenuator　可变波导衰减器
variable wide pulse　可变宽度脉冲
variable wind　多变风
variable word length computer　可变字长计算机
variable word length　可变字长
variable　可变的，变化的，变量，变数，可变因素
variably set inlet guide vane　可调整（风机）进口导叶
variac　自耦调压变压器，自耦变压器
variance analysis　方差分析
variance of wind speed　风速离散值
variance ratio　方差比
variance report　差异报告
variance with the contract　与合同有出入
variance　不同，变化，变异，偏差，方差，偏离值，分歧，差异，离散
variant design　变型设计，变异设计

variant part 变体部分
variant scalar 变量标量
variant 变量，变体
variate-difference method 变量差分法
variational approach 变分法
variational coefficient of capacitance 电容量可变系数
variational form 变分形式
variational invariant 变分不变量
variational method 变分法
variational principle 变分原理，交分原理
variational 变异的，变化的，变分的
variation and dissolution of contract 合同的变更与解除
variation calculus 变分法，变分学
variation coefficient 变差系数，变化系数
variation curve 温度变化曲线
variation equation 变分方程
variation factor 变化系数，变化因数
variation in cross section 沿断面变化
variation in depth 随深度变化
variation in discharge 流量变化
variation in load 负荷变化，负载变化
variation in voltage 电压变动
variation in water level 水位变化
variation margin 价格变动保证金
variation of contract 合同的变更，更改合约的条款
variation of field intensity 场强变化
variation of function 函数变差
variation of parameters 参数变值法
variation of permeance sensitivity 磁导灵敏度变化
variation of project quantity 工程数量变更
variation of sign 符号变更，正负号的替代
variation of stresses 应力变化
variation of the ground topography 地形变化
variation of voltage 电压变化
variation of water quality 水质变化
variation order 变更命令
variation procedure 变更程序
variation range 变动范围
variation rate 变化率
variation with pressure 随压力变化
variation with temperature 随温度变化
variation 变差，变动，变分，变更，变化，变量，变异，偏差
variator 变换器，变速机，变速器，温度变化补偿器
varicap 压控变容器，变容二极管，可变变容器
varied flow function 变流函数
varied flow 变速流
varied quality coal 变质煤
variety speed regulation of WTG 风电机组变速调节
variety 多样性，种类
vari-mu pentode 变 μ 五极管
varindor 可变电感器，变感器
variocoupler 可变耦合器
variodencer 可变电容器
variode 变容二极管
variograph 变压计，变量计
variohm 可变电阻器
variolosser 可变损耗器，可控损耗设备
variometer 可变电感器，变感器
varioplex 可变多路传输器，变工制，变路转换器
various interests 各种权益
various piping system 各种管线系统
various systems 各种系统
variplotter 自动曲线绘制器，自动作图仪
varister 变阻器，压敏电阻器，非线性电阻，可变电阻器，可调电阻器
varistor 压敏电阻器，变阻器，非线性电阻，可变电阻器，可调电阻器
varitran 自耦调压变压器，接触调压器
varitrol 自动调节系统
varivoid filter 变孔隙过滤器
varmeter 无功伏安计，无功功率表，乏计
varnished bias tape 黄蜡带
varnished cambric covered cable 漆布绝缘电缆
varnished cambric 漆布
varnished cloth insulation 漆布绝缘
varnished cloth tape 漆布带
varnished cloth 漆布
varnished fabric 漆布，浸漆织物
varnished glass cloth 玻璃漆布
varnished insulation 浸漆绝缘
varnished paper 漆纸，浸渍绝缘纸
varnished pressboard 漆纸板，浸渍纸板
varnished tape 漆布带
varnished tube 浸渍绝缘管，黄蜡套管
varnished wire 漆包线
varnish film 漆膜
varnish 涂清漆，使有光泽，亮光漆，清漆
varnish 清漆，上清漆，透明漆，涂料，涂漆，凡立水，罩光漆，光泽面，假漆，虚饰，在（某物）上涂清漆，使……有光泽，粉饰
varved clay 层状黏土，成层黏土，带状黏土，明显分层的黏土，纹泥
varved slate 纹板岩
varve 纹泥，成层黏土，每年一度的冲积层
var（volt-ampere reactive） 乏，无功伏安
vary directly as 和……成正比
varying acceleration 变加速度
varying duty 变工况，变负荷，在变动负载下运行，变动负载
varying-field commutator motor 可变磁场式换向器电动机
varying frequency starting 变频启动
varying head 变水头
varying lift 变升力
varying load condition 变负荷工况，变载状态
varying load 变动负载，变动负荷，不定荷载
varying parameter 可变参数
varying-speed motor 变速电动机
varying stress 变应力
varying underflow 变底流
varying-voltage control 变压控制
varying-voltage generator 变压发电机
vary inversely as 和……成反比例
vaseline 矿脂，凡士林

VAT on purchases 增值税进项税额
VAT on sales 增值税销项税额
VAT refund for exported goods 增值税的出口退税
VAT(value added tax) 增值税
vat 桶，槽，箱
vault cooling pipe 堆室冷却管道
vaulted shell 穹壳，双曲壳体
vault head 圆拱顶盖
vault light 地下室照明
vault storage 地下贮藏室
vault-type transformer 地下室型变压器
vault wall 拱顶墙，堆室壁
vault 拱顶，穹窿，地下室，窖，拱，拱形圆顶，夹层，天穹
VA =voltammeter 伏安表
VAV(variable air volume) 变风量
VAWT 垂直轴风轮机
V-ball valve V球阀
V-band V频带，V波段
V-beam sheeting 波纹板
V-belt drive 三角皮带传动
V-belt V形皮带，三角皮带
V-block V形块
VB=vibration 振动
V-clamp V形夹
V-cone blender V锥形混料机
V-connection V形结线，V形接法
V-crimp asbestos-cement sheet V形石棉水泥瓦
VCR(video cassette recorder) 电视盒式录像机
V curve V形曲线，负载特性曲线
VC(value contribution) 价值贡献
V/C(voyage charter) 航次租船，简称程租
VDE 德国电气工程师协会
VDF(voice distribution frame) 音频配线架
VDI 德国工程师协会
VDU mimic 显示器上的生产流程模拟图
VDU(visual display unit) 显像屏，显示器，直视显示装置
vectogram method 向量图法
vectogram 矢量图，向量图
vectograph 矢量图，向量图，向量描记器
vectopluviometer 风向雨量计
vector addition 矢量加法
vector admittance 矢量导纳，复数导纳
vector analysis 矢量分析
vector angle 矢量角
vector area 矢面积
vector balancing 矢量平衡
vector component 分矢量，矢量的分量
vector computer 矢量计算机
vector control 矢量控制
vector current 矢量电流
vector diagram 矢量图
vectored interrupt 向量中断
vector equation 矢量方程（式）
vector field 向量场，矢量场
vector flux 矢通量
vector group of a transformer 变压器的矢量组
vectorial combination 矢量组合
vectorial field 矢量场
vectorial lift 矢量升力
vectorial resultant 矢量合成
vectorial 矢的，矢量的
vector impedance bridge 矢量阻抗电桥
vector impedance 矢量阻抗，复数阻抗
vector Liapunov function 向量李雅普诺夫函数
vectorlyser 矢量分析器
vector phase control 矢量相位控制
vector phase 矢量相位
vector point function 矢量点函数
vector point 矢量点
vector position 矢量位置
vector potential field 矢势场
vector potential 矢位，矢势
vector power factor 矢量功率因数
vector power 矢量功率，复数功率
vector quantity 矢量，向量
vector representation 矢量表示法
vector ring 矢量环
vector rotation 矢量旋转
vectors composition 矢量的合成
vector servomechanism 矢量伺服机构
vector space 矢量空间
vector sum 矢量和
vector triangle 矢量三角形
vector volt-ampere 矢量伏安，复数伏安
vector wind field 向量风场
vector 矢量，向量，引导，引向目标
vedio camera 摄像机
vee connection V形连接
vee connection winding V形绕组，开口三角形绕组
vee corrugated absorber V形波纹吸热板
vee corrugated plate V形波纹板
vee-corrugated V形波纹的
vee-grooved V形槽的
vee-notched ball V形切口的球
vee notch V形槽，V形缺口
veering tendency 自动转弯趋势
veering wind 顺时针旋风
veering 调风，对风，顺时针转向
veer 转向
vee V字形，V字形物，V字形的，开口三角形的
vegetable oil 植物油
vegetable sideboard 存菜柜
vegetable topsoil 腐槽表土
vegetable washing sink 洗菜池
vegetal cover 植被
vegetal discharge 叶面蒸发，植物蒸腾耗水量
vegetational cover 植被
vegetation 植被，自然植被，植物
vegetative cover 植被
vehicle turbine 运输式透平
vehicle with a low flat base 低矮的平板车
vehicle 车辆，机动车，交通工具，运输车，运载工具，传播媒介
vehicular gas turbine 车辆燃气轮机
vehicular load 车辆荷载
vehicular tunnel 行车隧道
vehicular 机动车辆的

veil 面纱，帷幕，帐，遮盖用物
veined structure 脉状结构
vein fissure 脉裂隙
veinlet 细岩脉
vein rock 脉岩
vein stone 脉石
veinule 细岩脉
vein water 脉状水，岩脉水
vein 岩脉
velocimeter 流速仪，速度计
velocities composition 速度的合成
velocities polygon 速度多边形
velocity amplitude 速度幅值
velocity analysis 速度分析
velocity-area gauging station 流速面积测流站
velocity-area method 流速面积法
velocity at stack exit 烟囱出口的流速
velocity-azimuth-depth assembly 流速流向水深仪
velocity change 变速
velocity coefficient 速度系数，流速系数
velocity component 分速度，速度分量
velocity-compounded impulse stage 复速冲动级
velocity-compounded stage 复速级
velocity-compounded turbine 复速级式汽轮机
velocity constraint 速度限值
velocity contour 速度分布图，等流速线
velocity-controlled system 转速控制系统，转速调节系统，速度调节系统，调速系统
velocity curve 水位平均流速（关系）曲线，速度曲线
velocity defect decay 速度亏损衰减
velocity defect law （边界层）速度亏损律
velocity defect profile 速度亏损廓线
velocity defect 速度亏损
velocity diagram 流速分布曲线，流速分布图，速度三角形，速度图
velocity discontinuity 速度不连续
velocity dispersion 速度弥散
velocity distribution curve 流速分布曲线
velocity distribution 流速分布，速度分布
velocity duration curve 速度持续曲线
velocity energy 动能，速度能
velocity error coefficient 速度误差系数
velocity error constant 速度误差常数
velocity factor 流速系数，速度系数
velocity feedback 速度反馈
velocity field 速度场
velocity flowmeter 速度流量计【涡轮，涡流】
velocity fluctuation 速度脉动
velocity function 速度函数
velocity gauging 测流速
velocity gradient 流速梯度，速度梯度，速度陡度
velocity head coefficient 流速水头系数
velocity head correction factor 流速水头校正系数
velocity head difference 流速水头差
velocity head loss 速度头损失
velocity head 流速水头，速度落差，速度水头，速度压头，速度头，动压头，速位差
velocity interval 速度间隔，速度范围
velocity isoline 流速等值线
velocity jump 速度跳跃，速度突跃
velocity lag 速率延迟
velocity layer 涡流层，速度层
velocity limiter 速度限制器
velocity limiting control 速度限值控制
velocity limit 速度极限，速限
velocity logging 速度测井
velocity loss 速度损失
velocity measurement 测流速，速度测定
velocity measuring dynamo 测速用发电机
velocity measuring system 速度测量系统
velocity meter 测速计，测速表，速度计，速度表，流速仪
velocity-modulated tube 速度调制管
velocity modulation 调速，速度调制
velocity of approach 趋近速度
velocity of convergence 收敛速度
velocity of discharge 排出速度，排汽速度，排泄速度，放电速度
velocity of electromagnetic wave 电磁波速度
velocity of exhaust 排汽速度，排气速度
velocity of filtration 过滤速度
velocity of flow 流速
velocity of flue gas 烟气速度，烟道气流速度
velocity of partial sedimentation 部分沉降速度
velocity-of-propagation meter 传播速度测定仪
velocity of water through condenser tubes 通过凝汽器管的水速
velocity of wave propagation 波的传播速度
velocity potential function 速势函数
velocity potential 流速势
velocity pressure 动压，速压，速度头，速度压力，流速压力
velocity profile curve 流速廓线，速度分布曲线
velocity profile 速度分布曲线，流速分布图，流速剖面，速度廓线
velocity pulsation 流速脉动，流速震动
velocity ratio 速比，速度比
velocity reduction 减速
velocity regulation pump 调速泵
velocity relative to the blades 相对叶片的汽流速度
velocity resolution 速度分解
velocity resonance 速度共振，速度谐振
velocity rod 测速杆
velocity seismogragh 速度地震计
velocity sensor 速度传感器
velocity serve 速度伺服机构
velocity shear boundary 速度切变界限
velocity shock 速度冲击
velocity-slip 速度滑移
velocity space 速度空间
velocity spectrum 速度谱
velocity stage impulse turbine 速度级冲动式汽轮机
velocity stage 调速级，速度级
velocity stratification 速度层结
velocity-time diagram 时速图，速度时间图
velocity to prevent erosion 防冲流速
velocity to prevent plant growth 不长草流速

velocity transfer function 速度转移函数
velocity triangle 速度三角形
velocity type transducer 速度传感器
velocity type water meter 速度式水表
velocity variation tube 变速管，速度调制管
velocity variation 速度变化
velocity vector component 速度矢量分量
velocity vector 速度矢量
velocity vertical 测速垂线
velocity 速度，速率
velodyne 测速发电机，转数计传感器
velometer 测速仪，速度表，速度计
Velox boiler 韦洛克斯增压锅炉，增压锅炉
vena contracta tap 缩脉分接头
vena contract pressure tapping 缩流取压孔
vena contract 收缩断面，缩颈流束
vendee 买方，受货人，买主
vender 供货方，卖方
vendor diagrams 供货厂商图，卖主图
vendor drawing review 卖方图纸审查
vendor drawing status report 供货厂商图纸状态报告
vendor drawing 供货厂商图纸，供货图，结构图
vendor inspection 供货厂商检验
vendor quotation 厂商报价
vendor 供货厂商，卖方，供货方，卖主
veneer board 镶板，镶嵌板
veneered brick 镶面砖，釉面砖
veneered wall 镶面墙
veneering 薄片胶合，木板胶合，贴面
veneer reeling machine 卷板机
veneer 胶合板，薄包板，饰面，贴面板，镶面，镶面砖
venetian blind collector 软百叶帘集热器
venetian blind 活动百叶窗，软百叶帘，软百叶窗
venetian shutters 固定百叶窗，遮阳窗
ventage 孔隙，气孔，出口，通风管
vent air 排风，排气
vent and off-gas header 排气总管
vent burner 乏气燃烧器
vent cock 放气旋塞，空气门
vent condenser 乏气凝汽器，余汽凝汽器
vent duct 排汽道，通风道，溢流道
vented air 乏气，排气
vented form 通气模板
vented fuel assembly 透气式燃料组件
vented fuel element 透气式燃料元件
vented fuel rod 透气式燃料棒
vented fuel 透气式燃料
vented irradiation capsule 透气式辐照盒
vented isolation damper 排放隔离调节风门
vented-rod type fuel element 透气棒型燃料元件
vented structure 排风结构
vented system 开口系统
vent fan 排风机，（制粉系统的）乏气风机，排气机
vent filter 排气过滤器
vent finger 通风道隔离片，风道压指
vent gas plume 排出气羽流
vent gas scrubber 排出气体洗涤器
vent gas tower 排气塔
vent grating 通风格子
vent gutter 通风道，通气道
vent hole 放气孔，通风口，通风孔，通气孔
ventiduct 地下通风道，通风道，通风管
ventifact 风棱石，风磨石
ventilated anti-contamination tenting 通风的防污染帐篷
ventilated box 通风箱
ventilated ceiling 通风顶棚
ventilated chamber 通风间
ventilated commutator 风冷式换向器
ventilated frame machine 机壳表面冷却式电机
ventilated machine 风冷电机
ventilated motor 风冷电机，通风式电动机
ventilated protective suit 通风的防护服
ventilated rediator machine 风冷散热器式电机
ventilated rib 通风筋，散热筋
ventilated roof 通风屋顶
ventilated suit 防护衣，气衣
ventilated winding 风冷绕组，空冷绕组
ventilated 风冷的，通风的
ventilate hood 通风柜
ventilate 通风，风冷，风冷的，通气，换气
ventilating arrangement 通风布置，风路布置
ventilating blower 通风机，风扇
ventilating circuit 风路
ventilating cowl 通风帽，通风罩
ventilating device 通风装置
ventilating dryer 通风干燥机
ventilating duct spacer 通风沟隔片，风道隔片
ventilating duct 通风管（道）
ventilating equipment 通风设备
ventilating fan 排气风扇，通风机，风扇，通风扇
ventilating flap 通风瓣
ventilating flue 通风管
ventilating grill 通风格栅
ventilating hole 通风孔
ventilating hood 通风罩，通风柜
ventilating machinery 通风机械
ventilating machine 通风机
ventilating pipe 通风管
ventilating plant room 通风机房
ventilating rate 通风换气次数
ventilating room 通风小室
ventilating shaft 排风井，通风井，通风竖井，通气井
ventilating slot 通风槽
ventilating stack 通风竖管
ventilating system 通风系统
ventilating window 通风窗
ventilation and air-conditioning installation 通风、空调装置
ventilation by extraction 抽气通风
ventilation capacity 通风量
ventilation casement 通风窗
ventilation cell 通风小室
ventilation circuit 风路
ventilation coefficient 通风系数
ventilation column 通风管
ventilation diagram 风路图

ventilation duct spacer　风道隔片
ventilation duct　风道，风沟，通风管道，通风管
ventilation equipment　通风设备
ventilation facilities　通风设备
ventilation fan　通风风机
ventilation from sub-slot　（汽轮发电机）副槽通风
ventilation function　渗透作用
ventilation heating load data per unit building volume　通风体积热指标
ventilation heat loss　通风耗热量
ventilation hole　通风孔
ventilation hood　风机通风罩，通风顶罩
ventilation installation　通风设备，通风装置
ventilation loss　通风损耗
ventilation louver　气窗，通风百叶窗
ventilation nozzle　通风接管
ventilation passage　风道，风巷
ventilation penetration　通风贯穿件
ventilation rate　通风量，换气次数
ventilation resistance　通风阻力，风阻
ventilation shield　通风罩，挡风板
ventilation shroud　通风包壳，通风罩【控制棒驱动机构】
ventilation slot　通风槽
ventilation stack　通风管，通风烟囱，排气烟囱
ventilation station　通风站
ventilation system　通风系统
ventilation test　通风试验
ventilation　通风，通风装置，换气
ventilative　通风的
ventilator　通风机
ventilator fan　通风风扇
ventilator hood　通风罩
ventilator room　通风机室
ventilator　通风机，通风器，送风机，排气风扇，风机
venting chimney of stainless steel　不锈钢排气筒
venting header　排气总管
venting nozzle　排气接管
venting rate　通风率
venting system　排气系统
venting tool　通风装置【控制棒驱动机构】
venting valve　放汽阀，空气门
venting　通风，排，放，消除
ventlight　通风窗
vent line　放气管
vent nozzle　排气喷管，排气连接管，通风孔
vent opening　通风口，通气孔
vent patch　排气孔补片
vent pipe system　排气管系统，放气管系统
vent pipe　放气管，排气管，通风管，通气管（道）
vent plug　排气塞，通风孔塞
vent-port　放气孔
vent roof　通风顶
vent scrap　吹残渣【浇铸】
vent screw　排气螺钉
vent spacer　风道隔片，（各段铁芯间的）通风槽片
vent stack　排气烟囱，通风井道，排泄烟道，通气立管
vent system　通风系统，通气系统，排气系统
vent to atmosphere　向大气中排气
vent tower　排气筒，通风塔
Venturi collector　文丘里除尘器
venture business　风险企业
venture capital investment　风险投资
venture enterprise　风险企业
venture expansion fund　企业发展基金
venture in enterprise management　企业经营风险
venture investment enterprise　风险投资企业
venture port valve　缩孔阀门
Venture scrubber　文丘里除尘器
venture　风险，投机
Venturi classifier　文丘里分粒器
Venturi effect　文丘里效应
Venturi flowmeter　文丘里流量计，文丘里管流量计
Venturi flume　文丘里试槽，文丘里水槽
Venturi measuring　文丘里管测量
Venturi meter coefficient　文丘里流量计系数
Venturi meter　文丘里测量管，文丘里流量计，文丘里管，文丘里速度计
Venturi nozzle　文丘里喷嘴
Venturi pneumatic pyrometer　文丘里气力式高温计
Venturi scrubber　文丘里除尘器，文丘里洗涤器，文丘里洗气器
Venturi scrubbing　文丘里洗气法
Venturi tube　文氏管，文丘里管，文丘里流量计，文丘里流速计
Venturi type wind machine　文丘里型风力机
Venturi type wind turbine　文丘里管式风力机
Venturi　文丘里，文丘里管
vent valve connections　放气阀连接装置
vent valve　放空阀，放气阀，泄气阀，通风阀，通气阀，泄水阀
vent　排气口，出汽口，放气，排气，排气管，放气阀孔，出风口，通风口，通气孔
venue　（会议）地点，会场，审理地，管辖地
verbal agreement　口头协定
verbal assessment　口述评价
verbal contact　口头联系
verbal contract　口头契约
verbal order　口头命令，言语命令
verbal understanding　口头谅解
verbatim record　逐字记录
Verber　维伯【电荷单位】
verdan　通用计数式计算机，万用微分分析器
verdict　裁决，判断
verge board　山墙挡风板，山墙封檐板
verge cross fall　路肩横向坡度
vergence　透镜焦度
verge-perforated　边缘穿孔的
verge　山墙檐口
verification and checkout equipment　验证和检查设备
verification and validation　验证与确认
verification by consent　征得同意进行核查
verification by tests and scrutiny　用抽查方法核对
verification method　验证方法

verification of current meter 流速仪检验
verification procedure 验证程序
verification relay 校核继电器，监控继电器
verification resources 验证手段
verification test 检验试验，验证试验
verification 校验，核对，证实，检验，核实，校准，校正，验证，检定
verified copy 经核实的副本
verified correct 核对无误
verifier 核对员，校验员，检验器，校验器，核对器
verifying attachment 检验用附件，检验装置
verifying the setting 定值校验
verifying 校验，检验
veritcal amplifier 垂直信号放大器
veritcal deaerator 立式除氧器
veritcal displacement 垂直位移
veritcal limit 垂直极限
veritcal linearity 垂直线形，幅度线形
veritcal scan 垂直扫描
vermicular graphite iron 蠕墨铸铁
vermicular 蠕虫状的，蠕动的
vermiculite asphalt slab 沥青蛭石板
vermiculite cement slab 水泥蛭石板
vermiculite concrete 蛭石混凝土
vermiculite plaster finish 蛭石砂浆抹面
vermiculite 蛭石【绝热材料】
vernacular architecture 乡土建筑
vernacular 本国的，本国语的
vernal breeze 春风
vernal circulation 春季环流
vernal equinox 春分，春分点
vernal point 春分点
vernier adjustment 微变调整
vernier calliper 游标卡尺
vernier circle 游标盘
vernier compass 游标罗盘仪
vernier condenser 微调电容器
vernier control 细调节，微调，游标调节
vernier depth gauge 游标深度尺
vernier dial 游标刻度盘
vernier division 游标分划
vernier gauge 游标尺，精密压力表
vernier graduation 游标刻度
vernier labyrinth gland 游标式双面齿迷宫汽封
vernier micrometer 游标千分尺
vernier microscope 游标显微镜
vernier motor 微调电机，(低功率因数的) 微型电动机
vernier reading 游标读数
vernier scale 游标，游标尺，游标刻度
vernier theodolite 游标经纬仪
vernier type gland 游标式汽封，迷宫汽封
vernier valve 微调阀【小阀】
vernier 游标，游尺，游标尺，微调，微动，微调的，微动的
vernitel 高精度数据传送装置
versant 边坡
versatile automatic data exchange 通用自动数据交换器，多用途自动数据交换机
versatile automatic test equipment 多用途自动测试设备
versatile digital computer 通用数字计算机
versatile experimental reactor assembly 通用试验性反应堆装置，多用途试验堆
versatile intermediate pulse experimental reactor 多用途中等脉冲试验反应堆
versatile man 多面手
versatile pulse shaper 通用脉冲形成器
versatile worker 多面手
versatile 万用的，易变的，通用的，多方面的
versatility 多功能性，多面性，通用性，适应性
versine 正矢
versions of the contract 合同文本
version 版本，形式，种类，译文，说明，见解，版次
versus time 随时间
versus 对（比），与……相对，对抗
vertebrate animal 脊椎动物
vertex angle 顶角
vertex band 涡流带
vertex cavity 旋涡空穴
vertex chamber 涡流室
vertex focal distance 顶焦距
vertex form factor 顶点形状因子
vertex lens power 顶焦度
vertex operator 顶点算符
vertex vergence 顶焦度
vertex 顶点，顶，峰
vertical accelerometer 垂直加速计
vertical advection 垂直移流
vertical air tank 立式集气罐
vertical alignment 竖向定线，铅直准线
vertical anchor 锚柱，竖锚
vertical and horizontal communication 纵向和横向沟通【电厂指挥系统上下左右沟通】
vertical angle bench mark 三角高程标志
vertical angle 垂直角，竖直角，高度角，铅直角，直角，对顶角
vertical arrangement drawing of power plant 电厂竖向布置图
vertical arrangement of construction site 施工场地竖向布置
vertical arrangement 竖向布置，垂直排列
vertical-arrayed double-bundled lines 垂直双分裂子导线【输电线路】
vertical axis current meter 竖轴流速仪
vertical axis Darrieus turbine 达里厄垂直轴风力机
vertical axis machine 立式轴向风机
vertical axis rotor-type wind turbine 竖轴式叶轮型风力涡轮机
vertical axis rotor WECS 垂直轴风能转换系统
vertical axis rotor 垂直轴风轮
vertical axis wind machine 垂直轴风力机
vertical axis wind turbine 垂直轴风力机，垂直轴风轮机
vertical axis 立轴，纵轴，垂直轴，铅直轴，竖轴
vertical baffle 竖直挡板
vertical bar of truss 桁架竖杆
vertical bar 竖筋
vertical bending mode 垂直弯曲模态

vertical blanking signal 垂直回描熄灭信号，垂直回描消隐信号
vertical blanking 垂直消隐
vertical board combination 垂直仪表板组合
vertical board 垂直板
vertical boiler 立式锅炉
vertical-boom trenching machine 立臂挖沟机
vertical bottle depth-integrating sampler 竖瓶积深式采样器
vertical bracing 垂直支撑，垂直剪刀撑，竖向支撑
vertical break switch 立式断路开关
vertical breakwater 直立式防波堤
vertical bus 垂直母线
vertical calibration goniometer 垂直校准测向器
vertical can pump 垂直筒式泵
vertical cantilever element 垂直悬臂部分
vertical can type 立式容器型
vertical center line 垂直中心线
vertical channel 垂直管道
vertical circles 垂直圈，（地测）垂直度盘，地平经圈
vertical circulation cell 垂直环流圈
vertical clamp 竖盘制动
vertical clearance 垂直间隙，管子间距，管子间隔，竖向净空
vertical cleat 垂直割理
vertical cliff 垂直崖壁
vertical climatic belt 垂直气候带
vertical closure 立堵
vertical collimation 垂直视准误差
vertical collimator 垂准器
vertical combination crusher 立式复合破碎机
vertical communication 纵向交流，纵向沟通【电厂指挥系统上下沟通】，纵向连通
vertical component seismograph 垂直向地震仪
vertical component 垂直分量，垂直部分
vertical-compound turbine 叠置式汽轮机
vertical concrete spiral 竖轴混凝土蜗壳
vertical configuration 垂直地形，垂直排列
vertical control point 高程控制点
vertical control survey network 高程控制网
vertical control survey 高程控制测量
vertical control 垂直控制，高程控制
vertical convection 垂直对流
vertical conveyer 立式输送机
vertical cooler 立式冷却器，直立式冷却器
vertical coordinate 纵坐标
vertical core rockfill dam 心墙式堆石坝
vertical cross hair 竖十字丝
vertical cross operating activities 竖向交叉作业
vertical cross section 垂直断面
vertical cut 垂直开挖
vertical data processing 纵向数据处理，垂向数据处理
vertical decomposition 纵向分解
vertical deep-well pump 立式深井泵
vertical deflection 垂直挠度
vertical deformation 垂直变形，竖向变形
vertical deviation 垂线偏差
vertical diffusion 垂直扩散
vertical direction 垂直方向
vertical discharge pipe 垂直卸料管
vertical dislocation 垂直错位
vertical displacement 竖向位移，垂直位移，竖直位移
vertical distance 垂直距离
vertical distribution curve 垂直分布曲线
vertical distribution 垂直分布
vertical double-column type machining center 立式双柱加工中心
vertical down position welding 向下立焊
vertical down welding 垂直往下焊接【工作姿势】，向下立焊
vertical drainage 垂直排水，竖井式排水，竖向排水
vertical drain 垂直排水管
vertical drill press 垂直钻床
vertical-drop spillway 直落式溢洪道
vertical eddy 竖直旋涡
vertical element handing pickup hole 垂直元件装卸抓头孔
vertical engine 立式发动机
vertical erosion 垂直侵蚀，下切侵蚀，向下侵蚀
vertical error 垂直误差，高程误差
vertical exaggeration ratio 竖向放大比
vertical expansion rate 竖向膨胀率
vertical experimental hole 垂直试验孔道
vertical face breakwater 直立式防波堤
vertical fault 垂直断层
vertical filter 立式过滤器
vertical fin 垂直尾翼，垂尾
vertical-fired boiler W火焰锅炉
vertical-fired furnace 喷燃器垂直布置式炉膛
vertical firing burner 炉顶布置的燃烧器，垂直布置燃烧器
vertical flange 垂直法兰
vertical flow 垂直流
vertical flux profile 垂直通量分布
vertical fold 垂直褶皱，直立褶皱
vertical format unit 纵向（走纸）格式控制器，垂直（走纸）格式控制器
vertical framed gate 直框式闸门
vertical frequency 帧频，场频
vertical gauge 直立水尺
vertical gradient 垂直梯度
vertical gravity take-up device 垂直拉紧装置
vertical gravity take-up 垂直重力拉紧（装置）
vertical ground acceleration （地震的）垂直地面加速度
vertical gyroscope 垂直陀螺仪
vertical gyro 垂直陀螺
vertical hair 竖丝
vertical handing 垂直搬运
vertical heater 立式加热器
vertical hold control 垂直同步调整，帧同步调整
vertical hold 帧同步，垂直同步
vertical hydraulic turbine 立式水轮机
vertical hydrogen cooler 立式氢冷却器
vertical hydro-generator 立式水轮发电机
vertical idler 立辊
vertical induction motor 立式感应电动机

vertical injection logic 垂直注入逻辑
vertical instability 垂直不稳定
vertical interval 垂直间距，等高距
verticality 垂直，垂直度
vertical joint 垂直缝
vertical key 立销
vertical ladder 直梯
vertical lamination 木材垂直叠合法
vertical-lateral expansion ratio 纵横膨胀比
vertical-lateral shrink age ratio 纵横收缩比
vertical lathe 立车，立式车床
vertical layout drawing 竖向布置图
vertical layout 竖向布置
vertical lift bridge 直升桥
vertical-lift fixed-wheel gate 直升式定轮闸门
vertical-lift gate 上下式闸门，直升平板闸门，直升闸门
vertical line 垂（直）线
vertical load test of pile 桩的下压荷载试验
vertical load 垂直荷载，竖向荷载
vertically-distorted model 竖向变态模型
vertically downward firing 下射燃烧
vertically integrated monopoly 垂直垄断模式【电力企业垄断发电、输电、配电和售电】
vertically mixed estuary 垂直混合河口
vertically mounted 垂直轴
vertically-polarized wave 垂直极化波
vertically sliding door 直升拉门
vertically sliding window 垂直推拉窗，上下推拉窗
vertically-split casing pump 垂直拼合壳体泵
vertically-split casing 垂直拼合泵壳，径向拼合泵壳
vertically-split 垂直拼合，垂直中分的
vertical machine 立式电机，立式机
vertical machining center 立式加工中心
vertical magnet （选择器的）上升磁铁
vertical mass curve 竖向累积曲线
vertical member 垂直杆件
vertical metal spiral setting 竖轴式金属蜗壳装置
vertical mile 立式磨煤机
vertical mixed flow pump 立式混流泵
vertical mixing 垂直混合
vertical motion seismograph 垂直向地震仪
vertical motion 垂直运动，铅直运动，上下运动，竖直运动
vertical movement 垂直位移，垂直移动，铅直运动，竖向位移
vertical mulching 竖向防护层，竖向覆盖层
vertical multijunctions solar cell 垂直多结太阳电池
vertical multi junction 垂直多结
vertical multi-speed motor 立式多速电动机
vertical natural-circulation type shell and tube evaporator 立式自燃循环壳管式蒸发器
vertical normal stress 垂直法向应力
vertical off-normal contact 上升离位触点
vertical one-pipe heating system 垂直单管采暖系统
vertical oscillating transformer 垂直扫描振荡器用变压器，帧扫描振荡器用变压器
vertical output transformer 垂直输出变压器，帧输出变压器
vertical overhead contact system 垂直吊线接触网
vertical panel 立式仪表板，竖直面板
vertical partition plate 垂直隔板【水室】
vertical photography 竖直摄影
vertical pin 立销
vertical pipe hanger 立管吊架
vertical pipe type suspension load sampler 立管式悬移质采样器
vertical pipe 立管
vertical plane projection of coal bed 煤层立面投影图
vertical planer 立式刨床
vertical planing 垂直刨法
vertical plan 垂直面图，立面图，竖向平面布置图
vertical plume growth 羽流垂直增长
vertical plume 垂直羽流
vertical plunger pump 立式柱塞泵
vertical point 天底点
vertical polarization 垂直极化，垂直偏振
vertical position welding 垂直焊接【工作姿势】，立焊
vertical position 立焊位置
vertical press bender 立式弯管机，立式压弯机
vertical pressure 垂直压力
vertical print 竖直摄影照片
vertical projection 垂直投影，竖直投影
vertical pump 垂直泵，立式泵
vertical raceway 垂直电缆管道
vertical redundancy check 垂向冗余码校验
vertical reflector saving 轴向反射层节省
vertical retrace 垂直回描，帧回描
vertical return 竖直回管
vertical-rise cable 竖立电缆，垂直电缆
vertical riser 垂直立管，90°立管，垂直上升
vertical rod 垂直杆，立杆
vertical roller 直轴旋辊
vertical rotary planer 立式旋转刨床
vertical run 垂直槽，垂直敷设【电缆】，垂直管段
vertical sand drain 垂直排水砂井
vertical scale-up 垂直放大
vertical scale 垂直比尺，垂直尺度
vertical scanning frequency 垂直扫描频率
vertical scanning 垂直扫描，纵扫描
vertical screw pump 立式螺旋泵
vertical section 垂直断面，垂直剖面
vertical seismic force 垂直地震力
vertical separation 垂直线间距离，垂直距离
vertical separator 垂直隔块，（线棒的）排间绝缘垫条
vertical shaft arrangement 竖轴式布置
vertical shaft axial-flow full admission turbine 竖轴轴流整周进水式水轮机
vertical shaft current meter 竖轴流速仪
vertical shaft end-suction double volute casing pump 立式端吸双蜗壳泵
vertical shaft hammer crusher 立轴锤式破碎机

vertical shaft impact hammer type crusher 立轴反击锤式复合破碎机
vertical shaft machine 立轴式电机
vertical shaft mixer 立轴混凝土拌和机
vertical shaft motor 立轴式电动机
vertical shaft water turbine 立轴式水轮机
vertical shaft 竖井，竖管，立轴，垂直轴
vertical shearing stress 垂直剪应力
vertical sheave 立式滑轮
vertical shell and U-tube steam generator 立式（壳及）U形管蒸汽发生器
vertical shell boiler 立式锅壳锅炉
vertical ship lift 垂直升船机
vertical single-pipe heating system 垂直单管采暖系统
vertical skew 竖向歪斜
vertical sliding door 竖向推拉门
vertical slipform 垂直滑动模板
vertical spacing 垂直间隔，垂直线间距离，垂直间距
vertical speed brake 垂直速度制动器
vertical speed indicator 垂直速度指示器
vertical speed 垂直方向速度，垂直速度
vertical spindle coal pulveriser 立式磨煤机
vertical spindle disc crusher 立轴圆盘碎矿机
vertical spindle dry dock drainage pump 竖轴干船坞排水泵
vertical spindle mill 立轴磨煤机
vertical spindle motor 立轴式电动机
vertical spindle surface grinder 立轴平面磨床
vertical split 垂直中分面
vertical stability 垂直稳定性，竖向稳定
vertical steam generator 立式蒸汽发生器
vertical stiffener 竖直加劲杆
vertical stirrup 垂直蹬筋，垂直钢箍，箍筋
vertical straight-tube type heat exchanger 立式直管型热交换器
vertical stratification 垂直层理，垂直分层
vertical stress component 垂直分力
vertical sump pump 立式水涡泵
vertical surface 旋回面
vertical survey 垂直测量，高程测量，竖向测量
vertical symmetry 垂直对称
vertical synchronizing signal 垂直同步信号，帧同步信号
vertical system-water heater 立式热网水加热器
vertical tapered draught pipe 正锥形尾水管
vertical temperature gradient 垂直方向温度梯度，垂直温度梯度
vertical temperature variation 垂直方向的温度变化
vertical thread 竖丝，纵丝
vertical transport 垂直输送
vertical tube boiler 立式锅炉
vertical tube coil 垂直管圈
vertical tube evaporator 竖管蒸发器
vertical turbine pump 竖式叶轮泵
vertical turbine 竖轴式水轮机
vertical turbulence 垂直紊动
vertical turbulent diffusion 垂直湍流扩散
vertical type deaerator 立式除氧器
vertical type evaporator 直立管式蒸发器，立式蒸发器
vertical type feed water heater 立式给水加热器
vertical type generator 立式发电机
vertical type highspeed pump 立式高速泵
vertical type hydrogenerator 立式水轮发电机
vertical type motor 立式电动机
vertical type wind turbine 立轴风力机
vertical up position welding 向上立焊
vertical upstream face 垂直上游面
vertical upwelding 垂直往上焊接【工作姿势】
vertical U-tube steam generator 立式U形管蒸汽发生器
vertical velocity component 垂直速度分量
vertical velocity curve 流速垂直分布曲线
vertical velocity 垂直速度
vertical vibration amplitude 垂直振动值
vertical vibration 竖向振动
vertical view 俯瞰图，俯视图，立式图
vertical visibility 垂直能见度
vertical vorticity 垂直涡度
vertical water tube boiler 立式水管锅炉
vertical wave 立波，竖波
vertical weir 垂直堰，直立堰
vertical weld 立焊
vertical whirl 立轴旋涡
vertical wind shear 垂直风切变
vertical wind tunnel 立式风洞
vertical wire 竖直丝
vertical wiring trough 垂直布线槽
vertical zoning of water system 水系统竖向分区
vertical 垂直的【焊接电源的输出特性】，立式的，直立的，垂直面
vertiginous current 旋流
vertimeter 上升速度表，升降速度表
vertistat 空间定向装置
vertoro 变压整流器
VERT＝vertical 垂直的
very abrasive 磨蚀性强
very free flowing 流动性很好
very general algorithm 通用算法
very high active waste 特高放射性废物
very high flux reactor 特高通量反应堆
very high frequency 甚高频
very high level 极高水位，最高水位
very high performance 甚高性能
very high sea 怒涛，狂涛，巨浪，七级风浪
very high temperature gas cooled reactor 超高温气冷反应堆
very high temperature reactor 超高温反应堆
very high voltage cable 甚高压电缆
very high voltage 甚高（电）压
very large scale integration 超大规模集成【电路】
very low frequency communication 甚低频通信
very low frequency highpotential test 甚低频高压试验
very low frequency high voltage test 超低频耐压试验
very low frequency 甚低频
very low level 极低水位，最低水位

very rough sea 巨浪，狂涛，五级风浪
very short wave 甚短波
vesicularity 多孔性
vesicular lava 多孔状溶岩
vesicular structure 多孔结构
vesicular 多孔状，多泡状
vessel bottom head 压力壳下部封头，压力容器底封头
vessel closure head flange 压力壳顶盖法兰
vessel closure head standpipe 压力容器封头竖管
vessel closure head support skirt 压力容器封头支承裙筒
vessel closure head 压力容器封头
vessel core support ledge 堆芯支承凸缘
vessel flange seal surface cleaning equipment 压力壳法兰密封表面清洁机
vessel flange 堆壳法兰，容器法兰
vessel head adapter 压力壳顶盖适配段
vessel head cooling flow nozzle 压力壳顶盖冷却水流管嘴
vessel head installation guide sleeve 压力容器顶盖上的安装导向套
vessel head installation guide 压力容器封头安装导杆
vessel head lifting beam 压力容器封头提升梁
vessel head lifting rig 压力容器封头吊具
vessel head nozzle protection cover 压力容器封头接管保护罩
vessel head plenum 压力壳顶盖腔
vessel head shipping skid 压力容器封头运输托架
vessel head storage ring 压力容器封头贮存环
vessel head storage stand 压力容器封头存放架
vessel head ventilation duct disconnection 压力壳顶盖通风管拆开
vessel head 压力容器封头，压力壳顶盖，容器顶盖
vessel internal pressure 压力容器内压
vessel lower plenum 压力壳下部腔
vessel material surveillance capsule 压力壳材料监督盒
vessel mating surface 压力壳配合面
vessel penetration tube 堆壳仪用表贯穿管【即压力壳仪表贯穿管】
vessel penetration 压力容器贯穿件
vessel reactor 压力容器式反应堆
vessel shell course 压力壳环段
vessel shell ring 压力壳环段
vessel shell 压力容器壳体
vessel shipping cover 压力容器运输罩壳
vessel support lug 压力容器支耳
vessel-to-cavity seal ring 压力壳对换料水池的密封环
vessel-type boiling-water reactor 容器式沸水反应堆
vessel （盛液体或气体）容器，压力壳，压力容器
vestibule main elevators 前厅主要电梯
vestibule 门厅，前厅
vestigial sideband modulation 残留边带调制
vestigial sideband 残留边带
vest-pocket calculator 袖珍计算器，袖珍计算机
vest-pocket 袖珍的
veteran 熟练，经验丰富的人，经验丰富的
veto in detail 部分否决，部分条款否决权
veto power 否决权
veto pressure unloading gear 去压减负载装置
v-f(voltage-frequency) converter 电压频率转换器
VFC(valid for construction) 已验证可施工
VFP(verified for prefabrication) 已验证可预制
VFT(verified for test) 已验证可试验
VFW(variable flux voltage variation) 变磁通调压
V-groove V形槽，V形坡口
VGS-FP(variable generator speed & fixed pitch) 变速定桨距
VHESC 超高效太阳能电池
VHN(Vickers hardness number) 威氏硬度
viability 生存性
viable corona discharge 可见电晕放电
viable 可实施的，切实可行的，有望实现的
via centre 长途电话中心局
viaduct bridge 高架桥
viaduct entrance 高架桥入口，高架桥进口
viaduct parapet wall 高架桥护墙
viaduct pier 高架桥桥墩
viaduct 高架公路，高架桥，旱桥，跨线桥，高架铁路，引桥，栈道
via hole 通路孔
vial 管形瓶，细颈瓶
viameter 测震仪，路程计，车程计，计距器
via pin 通路引线
via resistance 通路电阻
vibramotive force 起振力
vibrated concrete 振捣混凝土
vibrated fluidized bed 振动流化床
vibrate 振动，摆动
vibrating bag filter 振动式布袋除尘器
vibrating board 平板振捣器
vibrating body 振动体
vibrating circuit 振荡电路
vibrating coal feeder 振动给煤机
vibrating coil method 振动线圈法
vibrating compactor 振动压实机，震动压实机
vibrating concrete float 混凝土振平器，混凝土表面振捣装置
vibrating condenser 振荡电容器
vibrating-contact voltage regulator 振动触点式电压调整器
vibrating contact 振动接点
vibrating conveyor 振动输送机
vibrating conveyer 振动输送机
vibrating detector 振动检波器
vibrating disposal 振动式卸料
vibrating drawdown hopper 振动给料斗
vibrating drier 振动干燥器
vibrating drum-type roller 振动滚筒式压路机，振动碾
vibrating feeder 振动给料机，振动加料器
vibrating feed hopper 振动给料斗
vibrating filter 振动过滤器
vibrating galvanometer 振动电流计
vibrating grate 振动炉排

vibrating grinder 摆动研磨机
vibrating grizzly 振动格筛，振动条筛
vibrating isolation 隔振
vibrating load 振动荷载
vibrating machinery 振动器
vibrating marking tool 测振工具
vibrating mill 振动粉碎机
vibrating motor 振动电动机
vibrating needle 振捣棒，振动棒【混凝土】
vibrating pan 振动筛
vibrating plate compactor 平台式振动压实器
vibrating platform 振动台
vibrating rammer 振动夯
vibrating rectifier 振动整流器
vibrating-reed frequency meter 振簧频率计
vibrating-reed instrument 振簧式仪表
vibrating-reed relay 振簧式继电器
vibrating-reed 振簧，振动片
vibrating relay 振动继电器
vibrating rod level switch 振动棒料位开关
vibrating roller 振动式压路机
vibrating screed 板式振动器
vibrating screen 振动筛
vibrating sieve 振动筛
vibrating stoker 振动炉排
vibrating storage pile discharger 贮料堆振动排料点
vibrating string extensometer 振荡弦式引伸仪，振动弦式应变计
vibrating-string strain meter 弦式应变计
vibrating string 振动钢弦
vibrating system 振动系统，振荡系统
vibrating table 振动台，震动台
vibrating tamper 振捣器，振动夯
vibrating tube 振动溜管
vibrating type automatic power factor regulator 振动式自动功率因数调整器
vibrating type automatic voltage regulator 振动式自动电压调整器，振动式自动稳压器
vibrating type circuit breaker 振动式断路器
vibrating wire force transducer 振弦式力传感器
vibrating-wire pressure cell 钢弦式压力盒，振弦式压力盒
vibrating wire strain gauge 钢弦式应变计，振弦式应变计，线震式应变计，弦线式应变计
vibrating wire tele-hygrometer 振弦式遥控湿度计
vibrating wire tensiometer 振弦式张力计
vibrating wire 振动弦线
vibrating 振捣，振动，振动的
vibration absorbed base 振动吸收基础，减振基础
vibration absorbent 消振
vibration absorber 减振器，吸振器，阻尼器，振动阻尼器，振动吸收器，消振器
vibration-absorbing base 减振基础，振动吸收基础
vibration absorption 吸振
vibration acceleration 振动加速度
vibrational behavior 振动性能
vibrational compaction 振动密实，振动压实
vibrational degrees of freedom 振动自由度
vibrational energy level 振动能级
vibrational energy 振动能
vibrational fluidized bed 振动流化床
vibrational force 振动力
vibrational severity concept 振动烈度概念
vibrational specific heat 振动比热
vibrational spectrum 振动能谱
vibrational transition 振荡过渡
vibrational 振动的，振荡的
vibration amplitude 振幅，振动幅度
vibration analysis 振动分析
vibration analyzer 振动分析器
vibration and acoustic insulation material 防振降噪材料
vibration and acoustic monitoring system 振动和噪声监测系统
vibration apparatus 振动试验设备
vibration at fundamental frequency 基频振动
vibration attenuation 振动衰减量
vibration behavior 振动特性
vibration-bin discharger 料仓振动排料器
vibration board 振动板，平板振捣器
vibration characteristics 振动特性
vibration cleaning 振动清灰
vibration coal feeder 振动给煤机
vibration compaction method 振动压实法
vibration compaction 振动压实
vibration damper 振动阻尼器，减振器，缓振器，防振锤，吸振器
vibration damping 振动阻尼，振动衰减
vibration deflection mode 偏移振型
vibration detector 振动指示器，振动探测器，振动测定器，振动检测器
vibration energy 振动能
vibration exciter 激振器
vibration fatigue machine 振动疲劳试验机
vibration fatigue test 振动疲劳试验
vibration field 振动区
vibration filter 震动过滤器
vibration force 振动力
vibration frequency meter 振动频率计
vibration frequency 振动频率
vibration galvanometer 振动式电流计
vibration gauge 振动仪
vibration generator 振子，振动器，激振器，振动发生器
vibration indicator 振动表，振动仪，测振仪，振动指示器
vibration insulating foundation 隔振基础
vibration insulator 隔振体
vibration isolating trench 隔振槽［沟］
vibration isolating 隔振
vibration isolation joint 防振接头，隔振接头
vibration isolation module 隔振模块
vibration isolation mounting 隔振装置
vibration isolation rubber 防振橡皮，隔振橡胶
vibration isolation 隔振装置
vibration isolator 隔振体，隔振器，减震器
vibration level 振动水平
vibration limit 振动限度

vibration liquefaction 振动液化
vibration machine 振动机
vibration magnetometer 振动磁力仪
vibration measurement 振动测量，振动测定
vibration-measuring device 振动测量器
vibration measuring instrument 振动测量仪
vibration mechanics 振动力学
vibration meter 测振计，振动器，振动计
vibration modal analysis 振型分析
vibration mode 振动方式，振动型，振型
vibration monitoring system 振动监视系统
vibration mounting slab 振动装配板
vibration multiplier 振动乘法器
vibration node 振动节点，振动波节
vibration number 振动数
vibration of turbine-generator set 汽轮发电机组振动
vibration open test assembly 振动开口检验组件
vibration pad 防振垫
vibration performance 振动特性
vibration period 振动周期
vibration pickup 振动传感器，测振器，拾振器
vibration pile 振动灌注桩
vibration-proof foundation 防振基础
vibration-proof material 防振材料
vibration-proof motor 防振电动机
vibration-proof structure 防振结构，耐振结构
vibration-proof switch 耐振开关
vibration-proof 防振，抗振
vibration-protected frame 防振机座，抗振机座
vibration protection 振动保护，防振保护
vibration recorder 振动记录仪
vibration relay （防止）振动继电器
vibration resonance 共振
vibration-rotation band 振动转动谱带
vibration sensors 振动传感器
vibration source 振动源，振源
vibration spring isolator 弹簧减振器
vibration stopper 减振器
vibration strength 振动强度
vibration stress 振动应力
vibration suppressed 振动衰减
vibration suppression 减振
vibration switch 振动开关
vibration testing device 振动试验装置
vibration test 振动试验
vibration theory 振动理论，振荡理论
vibration transducer 振动传感器
vibration transfer ratio 振动传递比
vibration transmissibility 振动传递率
vibration triaxial apparatus 振动三轴仪
vibration velocity 振动速度
vibration wave 振动波
vibration 颤振，振动
vibratom 振动球磨机
vibrator inverter 振动式逆变器
vibrator-motor 振动电机，振动器电动机
vibrator power pack 振簧式变压装置
vibrator rectifier 振动整流器
vibrator sunk pile 振入桩，振动沉管桩
vibratory compacted fuel 振动密实的燃料
vibratory compacted 振动密实的
vibratory compaction 振动压实，振动密实
vibratory compactor 振动压实机
vibratory converter 振动换流器
vibratory feeder 振动给料机
vibratory gyroscope 振动陀螺仪
vibratory hammer 振动锤，振动打桩机
vibratory mill 振动球磨机
vibratory pile driver 振动打桩机
vibratory pile driving 振动打桩
vibratory pile extractor 振动拨桩机
vibratory pile hammer 振动桩锤
vibratory regime 振荡状态
vibratory response 振动性能
vibratory roller 振动碾，振动碾压机
vibratory strength 振动强度
vibratory stress 振动应力
vibratory tamper 振捣板
vibratory test 振动试验
vibratory wear 振动磨损
vibratory 振动的
vibrator 振动器，振荡器，振捣器，振子
vibrex pile 振动灌注桩
vibroacoustic monitoring 声振动监测
vibrocast concrete 振动浇置的混凝土
vibro column pile 振冲桩
vibro-compacted fuel 振动密实燃料
vibro-compacted 振动密实的，振动压实的
vibro-compaction 振动密实，振动压实
vibrocompact （核燃料）振动密实
vibroconcrete pile 振实式混凝土桩
vibrocorer 振动取芯器
vibro-densification 振动压密法
vibro-finisher 振动整面机
vibrofloation 振冲洗，振冲，振冲法，振动水冲法，振浮压实法
vibroflotation method 振冲压实法，振冲法
vibroflotation pile 振冲桩
vibroflot 振冲压实器
vibro-fluidized bed 振动流化床
vibrograph 示振仪，示振器
vibrometer 测振计，测振仪，振动计
vibronic spectra 电子振动频谱
vibronic 电子振动的
vibropack 振动子换流器，振动子整流器，振动变流器
vibro-pile 振动灌注桩
vibro-plate 振动板，平板振捣器
vibro-rammer 振捣板
vibro-replacement method 振冲置换法，振捣置换法
vibro-replacement process 振冲置换法，振捣置换法
vibro replacement stone column 振冲碎石桩
vibro-replacement 振冲置换（法），振捣置换（法）
vibroscope 示振仪，振动计
vibroshock 减振器，缓冲器
vibrosieve 振动筛
vibrosinking 振动沉桩
vibrosmoothing trowel 振动抹平镘板

vibrotron　压敏换能器，振敏管，电磁共振器
VIB＝vibrator　振动器
vicarious responsibility　专承责任，替代责任
vice chairman　副董事长
vice chief-engineer in charge of maintenance　检修副总工程师
vice chief-engineer in charge of operation　运行副总工程师
vice director in charge of production　生产副厂长
vice director　副主任
vicenary　二十进制的
vice president of construction　施工副总裁
vice president of engineering　设计副总裁
vice president of plant operation and betterment　电厂运行与改进副总裁
vice president of procurement　采购副总裁
vice section chief　副科长
vice versa　反过来也一样，反之亦然【常指某一问题或观点的另一面】
vice　老虎钳，台钳
vicinity　邻近，附近，附近地区
Vickers diamond hardness　维氏硬度
Vickers hardness meter　里氏硬度计
Vickers hardnessn umber　维氏硬度值
Vickers hardness test　维氏硬度试验
Vickers hardness　维氏硬度
Vickers microhardness test　显微维氏硬度试验
Vickers　维氏硬度计
victim　受害者
Victor　通信中代替字母 V 的词
vide ante　见前
vide infra　见下，见后，参见下文
videlicet　换言之，即，就是说，就是，亦即
video amplifier chain　视频放大器链
video amplifier　视频放大器
video carrier　视频载波
video cassette recorder　电视盒式录像机
video channel　视频通道
video converter　视频变换器，视频变频器
video copier　视频拷贝机
video correlator　视频相关器
video data interrogator　显示数据询问器
video data terminal　可视数据终端
video display unit　显示器
video file　可见文件，视频文件
video frequency　视频
video hard copier　视像硬拷贝机
video image storage system　视频录像储存系统
video limiter　视频限制器，视频限幅器
video phone　电视电话
video photograph　传真照片
video pulse　视频脉冲
video RAM　视频随机存取存储器
video recorder　电视盒式录像机，录像器，视频信号记录器
video recording　视频录象
videoscope　视频示波器
videosignal　视频信号
video system connected to borescope　连到光学孔径仪的电视系统
video tape recorder　磁带录像机，视频带记录器
video tape recording system　磁带录像系统
video tape recording　磁带录像
video telephone set　电视电话机
video telephone　电视电话，可视电话，显像电话
videotex　可视图文
video track　图像跟踪
videotransmitter　视频发射机
video-unit　视频装置，电视摄像器
video　视频的，可视的，视频，电视
vide post　见下，见后，参见后面，参见下文
vide P15　参见第 15 页
vide quod　参见该条
vide supra　见上，见前，参见上文
vidicon　视像管，光导摄像管
vidio presentation　视频显示，检波显示
vierendeel truss　连框桁架
view and size　外形和尺寸
view angle　视场角，视角
viewdata　数据电视检阅，资料电视检阅，可视数据
viewer　观测员，指示器，观测仪，观察灯，观察器，望远镜
view factor　视角系数
view finder　取景器【照相机】
viewing device　观察装置，观察仪器
viewing distance　观察距离，视距
viewing head　视力高度
viewing mechanism　观测机构
viewing port　观察窗，屏蔽窗，观察孔
viewing screen　观察屏
viewing section　观察段
viewing system　观察系统
viewing test　图像观察试验，外观检查
viewing window　观察窗，屏蔽窗
viewing　观察，视力
viewpoint　观点，看法
view port　窥视窗，观察孔，视孔
view window　观察窗
view　观察，检查，观点，视图，景色，意见
vigorous fluidization　活动良好的流化床【沸腾炉】
vigorous vertical mixing　强垂直混合
VI improver　黏度指数改进剂
village automatic exchange　农村自动电话交换机，农村自动电话局
villages and towns　乡镇
village　乡村
Villari effect　维拉里（磁化）效应，磁致伸缩逆效应
villa　别墅
vincent　模锻摩擦压力机
vinculum　线括号，大括号，联系，纽带
vindicate　辩护
vindictive damages　惩罚性损失赔偿费
V-insulating material　V 绝缘材料
vinyl bag　乙烯塑料袋
vinyl chloride　氯乙烯
vinyl ester　乙烯基脂
vinyl gloves　乙烯手套
vinyl lacquer　乙烯漆

Vinylon 维尼纶，聚乙烯醇缩醛纤维的商名
vinyl plastic 乙烯基塑料
vinyl resin 乙烯基树脂，乙烯树脂漆
vinyl tape 聚氯乙烯绝缘带
vinyl tile with nosing 带突缘的聚乙烯塑料面砖
vinyl 乙烯基，乙烯树脂
vinyon 聚乙烯塑料
violate a law 违犯法律
violate a contract 违反合同
violate 侵犯，违犯
violation of agreement 违约
violation of contract 违反合同，违约
violation of law 违法
violation 违反，违犯，违背
violent earthquake 大地震
violent shock 烈震
violent storm 十一级风，暴风，暴风骤雨，急风暴雨
violent tornado 强龙卷风
violent vortices of air 强空气旋涡系
violet cell 紫光电池
violet photocell 紫光电池
violet solar cell 紫光太阳电池
vipac fuel 振动密实核燃料
vipacted fuel 振动密实的燃料
vipact （核燃料）振动密实
VIP(very important person) 贵宾，重要人物
virginal overburden pressure 原始上覆压力
virgin coil 未穿孔的纸带卷，空白纸带卷
virgin compression curve 初压曲线，原始压缩曲线
virgin forest 原生林，原始林，原始森林
virgin fuel 新燃料
virgin isotropic consolidation line 原始各向等压固结曲线，原始等压固结线
virginium 钫
virgin land 处女地，生荒地
virgin medium 未用媒体，空白媒体
virgin neutron flux 原中子通量，未经碰撞的中子通量
virgin neutron 原中子
virgin reactor 未运转反应堆，新堆
virgin rock 原生岩石
virgin soil 生荒地，生荒地土壤
virgin wood land 原始林地
Virial coefficient 维里系数
Virial correction 维里修正
virtral reactor 虚拟反应堆
virtual access method 虚拟存取法
virtual address 虚拟地址，零级地址
virtual airfoil 有效翼型
virtual ampere-turn 虚安匝，有效安匝，工作安匝，工作安匝数
virtual angle of attack 虚迎角
virtual camber effect 虚弯度效应
virtual camber 虚弯度
virtual center of balance （塔式机械）天平虚中心
virtual circuit control program 虚拟线路控制程序
virtual computer 虚拟计算机
virtual deformation 潜变形，假变形，虚（拟）变形
virtual diffusion coefficient 虚拟扩散系数
virtual direct-access storage 虚拟直接存储器
virtual disk 虚拟盘
virtual displacement 假位移，虚位移
virtual duration （冲击波）假想持续时间
virtual earth buffer 虚接地隔离器，虚接地退耦器
virtual earth 虚地
virtual friction-angle 视在摩擦角
virtual get function 虚拟读功能
virtual ground fissure 虚拟地裂缝
virtual groundwater surface 虚拟地下水面
virtual ground 假接地，有效接地，假接地的，有效接地的
virtual height 有效高度，视在高度
virtual image 虚像
virtual impedance 有效阻抗
virtual incidence 虚迎角
virtual kinetic energy 虚动能
virtual leak 虚漏泄，假漏泄
virtual level 虚能级，假想电平
virtual load 虚荷载
virtual local area network 虚拟局域网
virtual machine environment 虚拟机环境
virtual machine 虚拟机，虚拟计算机
virtual manufacturing device 虚拟制造装置
virtual mass 虚质量
virtual memory technique 虚拟存储技术
virtual memory 虚拟存储器
virtual moment 虚力矩
virtual page 虚拟页
virtual preconsolidation pressure 视预固结压力
virtual put function 虚拟写功能
virtual rating 虚额定值，有效额定值
virtual resistance 虚电阻，有效电阻
virtual sequential access method 虚拟顺序存取法
virtual source 虚源
virtual storage 虚拟存储器
virtual support subsystem 虚拟支援子系统
virtual temperature 虚温，虚温度
virtual time to half-value （冲击波）假想到半值时间
virtual total duration （冲击波）假想总持续时间
virtual value 有效值，作用值，实际值
virtual velocity 虚速度
virtual voltage 有效电压，实际电压
virtual work 虚功
virtual zero point 虚零点
virtual zero time 假想零时【即冲击波起点】
virtual 虚拟的
virtue 声誉，优点
virus 病毒，电脑病毒
visa application 签证申请
visa-granting office 签证机关
vis-a-vis the bidding processes 关于投标过程
visa 签证
viscidity 黏性，黏着性，黏度
viscid theory 黏性流体理论
viscid 黏质的，黏的，黏滞的
viscoelastic analysis 黏弹性分析

viscoelastic behaviour 黏弹特性
viscoelastic damper 黏弹性阻尼器
visco-elastic deformation 黏弹性变形
viscoelastic fluid 黏弹性流体
viscoelasticity 黏弹性
visco-elastic material 黏弹性材料
viscoelastic medium 黏弹性介质
visco-elastic plastic body 黏弹塑性体
viscoelastic property 黏弹性能
viscoelastic 黏弹性的
visco-inelasticity 黏非弹性
viscolloid 黏性胶体
viscometer 黏滞计，黏度计
viscometry 黏度测定法
viscoscope 黏度计
viscose glue 胶水
viscosimeter 黏滞计，黏度计
viscosimetry 黏度测定法
viscosity analyzer 黏度分析器，黏度分析仪
viscosity coefficient 黏性系数，黏度系数，黏滞系数
viscosity constant 黏度常数
viscosity control 黏度控制
viscosity effect 黏滞性效应，黏性效应
viscosity energy barrier 黏性能垒
viscosity factor 黏滞因子，黏滞系数，黏滞因数
viscosity fluid 黏性流体
viscosity force 黏性力
viscosity friction 黏性摩擦
viscosity index 黏度指数，黏性指数
viscosity meter 黏度计
viscosity resistance 黏滞阻力
viscosity test 黏滞性试验，黏度试验
viscosity 黏度，黏性，黏滞度
viscous body 黏性体
viscous boundary layer 黏滞边界层
viscous core 黏性核
viscous damper 黏性阻尼器
viscous damping coefficient 黏性阻尼系数
viscous damping structure 黏性阻尼结构
viscous damping 黏性阻尼，黏滞阻尼
viscous dissipation 黏性耗散，黏滞弥散
viscous drag 黏性阻力，黏性摩擦，黏滞曳力，黏滞阻力
viscous factor 黏滞因数
viscous flow equation 黏滞流方程
viscous flow 黏流，滞流，层流，黏性流（动），黏滞流（动）
viscous fluid 黏性流体，黏滞流体
viscous force 黏滞力，黏性力
viscous fracture 黏性破坏
viscous friction coefficient 黏滞摩擦系数
viscous friction force 黏性摩擦力
viscous friction 黏性摩擦，黏滞摩擦
viscous layer 黏性层，黏滞层
viscous loss 黏滞损耗，黏滞损失，摩擦损失
viscous motion 黏滞运动
viscous pressure resistance 黏性压力阻力
viscous pressure 黏滞压力
viscous resistance 黏性阻力，黏滞阻力
viscous shear 黏性剪应力
viscous similarity 黏滞相似性
viscous slag 黏性渣
viscous strain 黏滞应变
viscous stress 黏性应力，黏滞应力
viscous sublayer 黏性底层
viscous surface layer 黏性表面层
viscous trailing vortex 黏滞后缘涡流
viscous vortex 黏性旋涡
viscous wake 黏性尾流
viscous 黏的，黏性，黏性的
vise grip clamp 台钳夹板
vise grippliers 台钳，老虎钳
vise jaw 钳口，台钳口，虎钳口，虎钳钳口
vise tool （机械手）钳夹抓手
vise 老虎钳，虎钳，夹钳，台钳
visibility curve 明视度曲线，可见度曲线
visibility distance 可见距离
visibility factor 能见度因数，可见度因数
visibility limit 能见范围
visibility meter 视度计，能见度计，能见度测定计
visibility range 能见距离
visibility reduction 可见度降低
visibility scale 可见等级
visibility 能见度，可见度，可见性，视野，可视度，清晰度
visible distance 能见距离
visible error 可见误差
visible flame 可见火焰
visible groove 可见沟槽
visible LED 可见光发光二极管
visible light photoresistor 可见光敏电阻器
visible light 可见光
visible penetrant 显露渗透剂
visible plume 可见烟，可见羽流
visible radiation 可见光，可见辐射
visible range monitor 可见光监测器
visible range spectrophotometer 可见光分光光度计
visible ray 可见射线
visible record computer 可见记录计算机
visible refuse 可见矸
visible region 可见区
visible signal 可见信号
visible spectrometry 可见光分光法
visible spectrum 可见光谱
visible tracer 可见示踪物
visible UV laser 可见紫外激光器
visible 可见的
visicode （遥控用）可见符号
visiometer 能见度测定仪
vision electronic recording apparatus 视频电子记录装置，电子视觉记录仪，可视电子记录器
vision field 视域
vision plane 视平面
vision range 视野
vision system 观测系统
vision 版本，版，视度
visit abroad 出国考察
visiting card 名片，参观证

visitors'gallery 参观者走廊
visit user for design quality 设计回访
vis major clause 不可抗力条款
vis major 不可抗力
visoelastic model 黏弹性模型
visor 护面，护目镜，(用脸盔，护目镜等) 遮护，挡板，视度
vista 展望，远景，回顾，前景
visual acuity 视觉灵敏度
visual alarm 灯光报警，可见报警，可视报警，灯光报警信号，可见报警信号
visual analysis subsystem 直观分析子系统
visual and audible alarm 声光报警
visual angle 视角
visual-arc spectrograph 目视弧光摄谱仪
visual assessment 目测评价
visual busy-lamp 占线指示灯，占线灯信号
visual check 目视检查，外观检查
visual colorimetric analysis 目视比色分析
visual colorimetry 目视比色法
visual communication 视觉通信，可见通信
visual comparison 目视比较
visual control 目测检查，外观检查
visual corona 可见电晕
visual critical voltage 可见电晕电压，起晕电压
visual data acquisition 可视数据采集，直观数据采集
visual degree 视度
visual detection method 视觉检测法
visual display console 可见显示控制台
visual display monitor 直观显示监视器
visual display terminal 直观显示终端
visual display unit 图像显示装置，直观显示装置，视觉显示单元，视频显示单元，显像屏
visual display 视觉显示，可见显示，直观显示，可视显示，光学显示，视频显示器
visual estimation 目测，目视估计
visual examination report 目视检验报告
visual examination 表观检验，目视检查，肉眼检视，表面检验，外观检查
visual field 视场，视界
visual frequency 视频
visual identification 肉眼鉴别
visual image processor 可见图像处理机
visual impact 视觉影响，视觉冲击，视觉震撼，视觉效果
visual indication 可见信号，可见指示，直观指示
visual indicator 目测指示器，目视指示器，视觉指示器
visual information display and control 直觉信息显示及控制 (器)
visual inspection of internals 堆内构件的目视检查
visual inspection 目检，外观检查，目视检查，肉眼检查，肉眼鉴定，目视检验
visual instrumentation subsystem 可见测试仪器子系统
visualization 目测，目测法，可视化，形象化，目视观察
visual measurement 目测
visual meteorological conditions 目测气象条件，能见气象条件
visual method 目测法
visual monitor 目视监视器，视频监控器
visual observation 目视观察，直观研究，目视检查，肉眼观测，目视观测，外部观察
visual of sag 弧垂观测
visual optical pyrometer 目视光测高温计
visual persistence 视觉暂留
visual pollution 视觉污染
visual range 可见范围，可见距离，视距，视程
visual readout 目视读出
visual representation 直观表示
visual sensor 视觉传感器
visual signal transmitter 图像信号发射机
visual signal 目视信号，视觉信号
visual soil classification 土的肉眼分类
visual standard sample 目测标准样品
visual standard 目测标准
visual telephone 可视电话
visual test 肉眼检验，外观检验，宏观检测，目视检查
visual titration 目测滴定法
visual transmitter power 电视发射器功率，电视发射机输出功率，图像信号输出功率
visual transmitter 电视发射机
visual tuning indicator 目测调谐指示器
visual 可见的，视觉的
vita glass 维他玻璃
vital bus 主母线
vital circuit 影响安全的电路
vital component 主要部件
vital part 关键部件
vital pieces of equipment 设备的关键零部件
vital power supply 重要电源
vital power system 重要的供电系统
vital system 关键系统
vitals 要害部件，重要器官，重要部件
vitiated air 污浊空气，污染空气
Viton 氟橡胶
vitreous carbon 玻璃碳，玻璃态石墨
vitreous ceramic fuel 玻璃化陶瓷体核燃料
vitreous ceramic tile 上釉瓷砖
vitreous china 上釉瓷，玻璃瓷，玻化瓷器，瓷质陶瓷
vitreous clinker 玻璃状渣
vitreous electricity 玻璃电，正电
vitreous fracture 玻璃状断口
vitreous insulating material 玻璃绝缘材料
vitreous silica 石英玻璃，透明石英
vitreous slag 玻璃状渣
vitreous texture 玻璃状结构
vitreous 玻璃质的，透明的，玻璃状的
vitrification treatment (放射性) 废物玻璃化
vitrification 玻璃固化【核废物】，玻璃化，透明化
vitrified brick 玻璃砖，缸砖，陶砖
vitrified clay pipe 缸瓦管，陶土管
vitrified clay 陶土
vitrified earthenware 玻璃化陶器
vitrified resistor 涂釉电阻器
vitrified tile pipe 陶土瓦管

vitrified tile 玻化瓷砖，瓷化瓦
vitrified 玻璃化的，陶瓷的，是变成玻璃状的
vitrifying point 玻璃化点，玻化点
vitrify 玻璃化
vitrobasalt 玻璃玄武岩
vitroclastic 玻璃状
vitrolite 瓷砖，瓷板
vivianite 蓝铁矿
vivosphere 生物圈
viz. 即，就是，亦即【拉丁语】
V-joint V形缝，V形连接
VLAN identifier 虚拟局域网标识
VLAN(virtual local area network) 虚拟局域网
VLV＝valve 阀门
VLV 阀【DCS画面】
VMJ(vertical multi junction) 垂直多结
VND 越南盾
V-notch MOS V形槽金属氧化物半导体
V-notch test V形缺口试验
V-notch weir 三角形堰
vocabulary 术语
vocal command 声指令
vocal input 声输入
vocal 声的
vocational skill 职业技能
vocational training 职业培训
vocoder 声码器，自动语言合成器
voder 语音合成器
voice access arrangement 声音存取设备
voice amplifier 声音放大器
voice answer back 声音应答装置
voice coder 声码器，语音信号编码器
voice coil motor 音圈电动机
voice communication 电话通信，音频通信
voice data communication 音频数据通信
voice data processing system 声音数据处理系统
voice digitization 声音数字化
voice distribution frame 语音配线盘，电话回线配线架
voice frequency interface 音频接口
voice frequency signaling 音频信令
voice frequency 音频
voice-grade channel 话频级通道，音频级信道
voice interruption priority system 声音中断优先系统
voice modulation 话频调制
voice-operated control 语音操作控制
voice-operated device 声控装置
voice-operated regulator 语音操作调节器
voice-operated relay 话频控制继电器，语控继电器
voice-operated switch 声控开关，语音操作开关
voice-operated 音控制的，语控的
voice operation demonstrator 声控演示器，语音合成器，人工声产生器
voice power 通话功率，语音功率
voiceprint 声印，声波纹
voidable contract 可以取消的合同
voidable 可使无效的
voidage distribution 空隙度分布
voidage disturbance 空隙度扰动
voidage fluctuation 空隙度脉动
voidage fraction 空隙分数，空隙度，空隙率
voidage function 空隙率函数
voidage instability 空泡不稳定性
voidage wave 空隙度波
voidage 空隙率，空隙度，空洞率，空泡量，空穴，空位
voidance 放弃，排泄，排出
void area fraction 孔面积份额，空泡面积份额
void cement ratio 孔隙与水泥的容积比，水泥孔隙比
void coefficient of reactivity 反应性空泡系数
void coefficient 空泡系数，空隙率
void content 空泡含量，空隙量，孔隙度，空隙度，孔隙量，汽泡含量
void contract 无效合同
void detection system 空泡探测系统
void distribution 空隙分布，空泡分布
void drift 空泡漂移
void effect （反应性）空泡效应，空穴效应
void factor 空泡因数，孔隙比
void-filling aggregate 填充性骨料
void formation in the core 堆芯内空泡形成
void formation 汽泡形成，空泡形成
void fraction 空隙分数，空隙度，空隙率，截面含气率，真实容积含汽量，空泡份额，空隙组分，空穴率
void-free coil insulation 无空隙线圈绝缘
void-free reactor 无空穴反应堆
void-free 无孔隙的
void-gas stripping 脱空隙气体
void location method 气孔定位法
void migration 空隙迁移
void of paste 跑浆
void plugging 孔隙堵塞
void rate 空隙率
void ratio in densest state 最小孔隙比
void ratio-pressure curve 孔隙比压力曲线
void ratio-strength curve 孔隙比强度曲线
void ratio test 孔隙比试验
void ratio-time curve 孔隙比时间曲线
void ratio 孔隙比，空隙比，孔隙率，空隙率
void reactivity coefficient 反应性空泡系数
voids and hungry spots 蜂窝麻面
voids and pits 蜂窝麻面
void space 空泡空间，孔隙空间，空隙空间
void transposition 空换位，虚换位
void volume 空泡体积，孔隙容积，孔隙体积
void water 孔隙水
void 空穴，空泡，空隙，气孔，空位，空洞，汽泡，缺乏的，无效的，不具备法律效力的，没人住的，空的，无人的
volatile centent 挥发分
volatile coal 高挥发分煤
volatile corrosion inhibitor 挥发性缓蚀剂，气相缓蚀剂
volatile file 易变文件
volatile fission product 挥发性裂变产物
volatile fuel 挥发性燃料
volatile gas 挥发性气体
volatile inflammable liquid 挥发性可燃液体

volatile inhibitor 挥发性缓蚀剂
volatile matter content 挥发分含量
volatile matter 挥发性物质，易挥发物，挥发物，挥发分
volatile memory 易消存储器，易失性存储器，挥发性存储器，短时间存储器
volatile nuclide 挥发性核素
volatile oil 挥发油
volatile organic material 挥发性有机物
volatile residue 挥发性残渣
volatile solid 挥发性固体
volatile solvent 挥发性溶剂
volatile storage 易失存储器，短时间存储器，瞬时存储器
volatile sulfur 挥发硫
volatile treatment （全）挥发性处理水【水处理方法之一，用于压水堆二回路】
volatile 挥发性的，易发散的，短时间的，挥发物，挥发分
volatilisation 挥发
volatility 挥发性，挥发度，易失性
volatilization temperature 挥发温度
volatilization 挥发
volcanic action 火山作用
volcanic activity 火山活动，火山活动性
volcanic agglomerate 火山块集岩
volcanic ash bed 火山灰层
volcanic ash soil 火山灰土壤
volcanic ash 火山灰
volcanic belt 火山带
volcanic breccia 火山角砾岩
volcanic cinder 火山岩屑，火山渣
volcanic coast 火山岸
volcanic conglomerate 火山砾岩
volcanic debris 火山碎屑
volcanic deposit 火山堆积
volcanic detritus 火山岩屑
volcanic dust 火山尘
volcanic earthquake 火山地震
volcanic ejecta 火山抛出物，火山喷出物
volcanic eruption 火山爆发，火山喷发
volcanic fragment 火山喷出碎片
volcanic glass 火山玻璃
volcanic harbour 火山港
volcanicity 火山活动
volcanic mud 火山泥
volcanic neck 火山颈
volcanic plume 火山羽流
volcanic rock 火山岩
volcanic sand 火山砂
volcanic seaquake 火山海震
volcanic shoreline 火山岸线
volcanic slag 火山熔渣
volcanic smoke 火山烟
volcanic tuff 火山凝灰岩
volcanic water 火山水
volcanism 火山活动，火山现象，火山作用
volcanist 火山学家
volcanologist 火山学家
volcanology 火山学
volcano 火山
volley 齐发爆破
volometer 伏安计，伏安表，万能电表
voloxidation 氧化挥发，氧化挥发法
voltacap 压控变容器
volta cell 伏打电池
volta effect 伏打效应
voltage across poles 极间电压
voltage across the terminals 端电压
voltage adjuster 电压调节器，电压调整器
voltage adjusting range 电压调节范围
voltage adjustment 电压调整
voltage alarm 电压报警
voltage-amplification factor 电压放大系数
voltage amplification 电压放大，电压放大率
voltage amplifier 电压放大器
voltage amplitude 电压振幅
voltage attenuation 电压衰减
voltage balance relay 电压平衡继电器
voltage balancer 均压器
voltage behind transient reactance 暂态电抗内部电压
voltage belt 电压带，电位区
voltage between layers 层间电压
voltage between lines 线间电压
voltage between phases 相间电压
voltage between segments （换向）片间电压
voltage booster 升压变压器，升压器，升压机
voltage boost 电压升高
voltage breakdown 电压崩溃，电压急降
voltage buffer 电压缓冲装置，电压中间转换器
voltage built-up rate 电压上升率，电压建起速度，建压速度
voltage calibrator 电压校准器
voltage change factor 电压变化系数
voltage changer 变压器，电压变换器
voltage characteristic （系统）电压特性
voltage charging circuit 电压充电电路
voltage circuit 电压回路，电压电路
voltage class 电压等级
voltage closed loop 电压闭环
voltage coder 电压编码器
voltage coefficient 电势系数，电压系数
voltage coil 电压线圈
voltage collapse 电压崩溃
voltage commutation 电压换向，电压整流
voltage comparator 电压比较器
voltage comparison encoding 电压比较编码
voltage compensation with series capacitor 串联电容器电压补偿
voltage compensation 电压补偿
voltage compensator 电压补偿器
voltage component 电压分量
voltage contrast 电压对比度
voltage controlled bus 电压控制母线
voltage controlled current source 压控电流源，电压控制电流源
voltage controlled oscillator 控制电压振荡器，电压控制振荡器
voltage controlled overcurrent relay 电压控制过电流继电器
voltage controlled power source 调压电源

voltage controlled voltage source 压控电压源，电压控制电压源
voltage controller 电压调节
voltage control loop 电压控制回路
voltage control system 电压控制系统，调压系统
voltage control 电压控制
voltage converter 电压转换器
voltage corrector 电压校准器，电压校正器
voltage criterion 电压标准，电压判据
voltage current characteristic 伏安特性，电压电流特性
voltage current transformer 电压电流互感器，变压变流器
voltage cut-out 电压开关，电压控制开关
voltage dependent resistivity 压敏电阻率，随电压变化的电阻率
voltage-dependent resistor 随电压变化的电阻器，压敏电阻器
voltage detector 电压检测器，测压器
voltage deviation 电压偏差，电压偏移
voltage difference 电势差，电压差
voltage differential relay 差压继电器，电压差动继电器
voltage digit 电压数字转换器
voltage dip 电压骤降，电压突跌，电压下降，低电压
voltage distortion 电压畸变
voltage distribution computation 电压分布计算
voltage distribution of insulators 绝缘子电压分布
voltage distribution 电压分布
voltage divider 分压器
voltage division 分压
voltage doubler tube 倍压整流管
voltage doubler 倍压器
voltage-doubling circuit 二倍增压电路
voltage-doubling 二倍增压
voltage drift 电压漂移
voltage drive 电压激励，电压驱动
voltage drop along the line 线路电压降
voltage drop compensation directional relay 压降补偿方向继电器
voltage drop compensation 电压降补偿
voltage drop compensator 电压降补偿器
voltage drop in arc 电弧压降
voltage dropping resistor 降压电阻器
voltage drop 电压降落，电压下降，电压降
voltage effect 电压效应
voltage efficiency 电压效率
voltage encoder 电压编码器
voltage endurance capability 耐压能力
voltage endurance test 耐压寿命试验，耐压试验
voltage endurance 耐压【如电机绝缘】，耐电压，耐压特性
voltage response mode 电压响应模式
voltage error 电压误差
voltage factor 电压因数
voltage feedback amplifier 电压反馈放大器
voltage feedback 电压反馈
voltage feed 电压馈电
voltage field 电压场
voltage flicker 电压闪变，电压闪烁
voltage fluctuation and flicker 电压波动和闪变
voltage fluctuation 电压波动
voltage follower 电压跟随器，电压输出器
voltage gain 电压增益，电压放大
voltage generator 测速发电机，电压发生器
voltage grade 电压等级
voltage gradient 电压梯度，电压陡度
voltage grading electrode 分压电极
voltage hunting 电压摆动
voltage imbalance 电压不平衡
voltage impulse 电压脉冲
voltage indicator 电压指示器
voltage integrating meter 电压积分器
voltage inverter switch 电压反相开关
voltage lead 电压引线
voltage leap 电压突变
voltage level test 耐压等级试验，电压电平试验
voltage level 电压电平，电压等级
voltage-life curve 电压寿命曲线
voltage limiter 电压限制器，限压器
voltage limit sensor 电压极限传感器
voltage linearity 电压直线性
voltage loop compensating network 电压环补偿网络
voltage loop 电压波腹，电压环
voltage loss 失（电）压
voltage magnification 电压放大
voltage measurement 电压测量
voltage memory circuit 电压存储电路
voltage meter 电压表
voltage-mode logic circuit 电压型逻辑电路
voltage monitoring node 电压监控点
voltage motor 可调压电动机
voltage multiplier 倍压器，电压倍增器
voltage multiplying circuit 电压倍增电路
voltage node 电压结点，电压波节
voltage of microphonic effect 颤噪电压
voltage operated device 压控装置
voltage-operated GFCI breaker 电压操作接地故障断路器
voltage-operated variable capacitor 压控可变电容器
voltage oscillogram 电压波形图
voltage output converter 电压输出转换器
voltage peak 电压峰值
voltage-phase-angle method 电压相位角度法
voltage phase-balance protection 相电压平衡保护
voltage phasor-diagram 电压相量图
voltage phasor 电压相量
voltage pilot node 电压中枢点
voltage polarity 电压极性
voltage profile 电压分布
voltage proof 耐电压
voltage protection 电压保护
voltage pulsation 电压脉动，电压波动
voltage pulse sequence 电压脉冲次序
voltage pulse 电压脉冲
voltage quadrupler 电压四倍器
voltage quality 电压质量

voltage range multiplier 电压量程倍增器
voltage range 电压范围，电压量程，电压等级
voltage rating 额定电压
voltage ratio type relay 电压比继电器
voltage ratio 变压比，电压比
voltage recovery rate 电压恢复速度
voltage recovery 电压恢复
voltage rectification 电压整流
voltage reference element 基准电压元件，参考电压元件
voltage reference tube 基准电压管
voltage reference 电压标准，基准电压，参考电压
voltage reflection coefficient 电压反射系数
voltage-regulated inverter 可调电压反向换流器
voltage-regulated 电压调节的，可调电压的
voltage-regulating diagram 电压调整图
voltage-regulating factor 电压调整因数，电压调整率
voltage-regulating relay 调压逆变器，电压控制逆变器
voltage-regulating system 电压调节系统，调压系统
voltage-regulating transformer 调压变压器，电压调节变压器
voltage-regulating tube 调压管，稳压管
voltage-regulating 电压调节
voltage regulation curve 电压调整曲线
voltage regulation error 电压调节误差
voltage regulation factor 电压调整率
voltage regulation 电压调整率，电压变动率，电压调节，电压调整
voltage regulator panel 电压调整器面板
voltage regulator response 电压调节器响应
voltage regulator tube 稳压管
voltage regulator 电压调节器，稳压器，调压器
voltage relay 电压继电器【过电压，低电压】
voltage resonance 电压谐振，串联谐振
voltage response time 电压反应时间
voltage response 电压特性曲线，电压响应
voltage responsive relay 电压响应继电器
voltage responsive starter 电压响应式启动器
voltage restoration 电压恢复
voltage sag 电压暂降，电压骤降，电压凹陷
voltage saturation 电压饱和
voltage-second characteristic 伏秒特性
voltage-selected switch 电压选择开关
voltage selection relay 电压选择继电器
voltage sensibility 电压灵敏度
voltage sensing relay 压敏继电器
voltage-sensitive bridge 压敏电桥
voltage-sensitive gate 电压敏感门
voltage-sensitive relay 压敏继电器
voltage-sensitive resistor 压敏电阻器，非线性电阻器
voltage sensor 电压传感器
voltage setpoint 电压整定点
voltage set （多相电路）电压组
voltage-sharing capacitor 均压电容器
voltage-sharing 均压，电压分担
voltage slider 滑线式调压器
voltage source conversion 电压源变换
voltage source inverter 电压源型逆变器
voltage source 电压源
voltage specification 电压规格
voltage spread 电压变化范围，最高和最低电压差
voltage stability calculation 电压稳定计算
voltage stability critical node 电压稳定临界点
voltage stability margin 电压稳定裕度
voltage stability 电压稳定，电压稳定度，电压稳定性
voltage stabilization 电压稳定
voltage stabilized power source 稳压电源
voltage stabilizer 稳压器，电压稳定器，电压稳定装置
voltage stabilizing diode 稳压二极管
voltage stabilizing transformer 稳压变压器
voltage stabilizing tube 稳压管
voltage-standard 电压标准
voltage standing-wave ratio 电压驻波比
voltage step-down circuit 降压电路，电压递降电路
voltage step 电压分级
voltage strength 电压强度
voltage stress 电压应力
voltage supply line （电源）电压输送线
voltage surge 电压浪涌，电压冲击
voltage swell 电压骤升，电压突升
voltage swing 电压摆动
voltage switch 电压开关
voltage synchronizing 电压同步
voltage telemetering 电压遥测，电压远距离测量
voltage temperature coefficient 电压温度系数
voltage terminal 电压端子
voltage-testing screwdriver 电压测试旋凿
voltage-time converter 电压时间变换器
voltage-time curve 伏秒特性曲线
voltage-time response 电压时间响应
voltage-time-to-breakdown curve 击穿电压对击穿时间的关系曲线
voltage time to breakdown 击穿电压时间
voltage-to-current converter 电压电流转换器
voltage to earth 对地电压
voltage to frequency conversion 电压频率转换
voltage to ground 对地电压
voltage to neutral 对中性点电压，相线对中性点电压
voltage-to-number converter 电压数字变换器
voltage-to-number 电压数字
voltage tracking error 电压跟踪误差
voltage transducer 电压互感器［变送器］
voltage transformation 电压变换，变压
voltage transformer 变压器，电压变量器，电压互感器
voltage transient 电压瞬变
voltage triangle 电压三角形
voltage tripler 电压三倍器
voltage-tripling rectifier 三倍电压整流器
voltage tunable magnetron 电压调谐磁控管
voltage tunable oscillator 电压调谐振荡器

voltage-turns ratio 伏匝比
voltage unbalance 电压不平衡
voltage uprating 电压升级
voltage-variable capacitance 随电压而变的电容
voltage-variable capacitor 变电压电容器
voltage variation 电压变化
voltage vector-diagram 电压矢量图
voltage vector 电压矢量
voltage waveform 电压波形
voltage winding 电压绕组
voltage window 电压窗口
voltage with respect to earth 对地电压
voltage withstand test 电气强度试验，耐压试验
voltage 电压，伏特数，电位差
voltaic acid 流电的酸
voltaic arc 电弧，伏打电弧
voltaic battery 伏打电池组，原电池，伏打电池
voltaic cell 伏打电池，伏特电池，原电池，一次电池
voltaic corpuscle 伏打电池
voltaic pile 伏打电堆
voltaic transformer 电流互感器
voltaic 伏打的，电流的，流电的
voltaism 伏打电学
voltalac 油沥青绝缘漆
voltameter 电量表，（电解）电量计，电压计，伏特计
voltammeter 电压电流两用表
voltampere characteristic curve 电压电流特性曲线，伏安特性曲线
voltampere characteristics 伏安特性，伏安特性曲线
voltampere curve 伏安特性曲线
voltampere-hour meter 伏安小时计
voltampere-hour 伏安小时
voltampere meter 伏特安培计，伏安计
voltampere output characteristic 伏安输出特性
voltampere reactive 乏，无功伏安
voltampere 伏安
voltascope 千分伏特计，伏特示波器
Volta's law 伏打定律
volt box 分压器，分压箱
volt efficiency 电压效率
volt gauge 电压表
volticap 变容二极管
voltite 电线被覆绝缘物
volt-line 伏特线，伏秒
voltmeter-ammeter method 伏特-安培计法，电压-电流表法
voltmeter analog-to-digital converter 伏特计模拟数字转换器
voltmeter change-over switch 电压表切换开关
voltmeter switch 伏特计开关
voltmeter-wattmeter method 伏特瓦特计法
voltmeter 电压表，伏特计，伏特表
volt-milliampere meter 伏特毫安计
voltohmist 伏欧计，电压电阻表
volt-ohmmeter 伏欧计，电压电阻表
volt-ohm-milliamperemeter 伏欧毫安计，万用表
voltohmyst 电压电阻表，伏欧计
voltolization process 高压电处理过程，电聚过程
voltolization 无声放电处理，高电压处理，电聚
voltol oil 电聚合油，高电压油
voltol process 高电压处理过程，电聚过程
volts DC 直流电压
volt-second 伏秒
volts per hertz relay 每赫兹伏继电器
volts per meter 伏特/米，伏/米
volts-to-digit converter 电压数字变换器
volts-to-digit 电压数字
volt-time curve 电压时间曲线
volt 伏，伏特
volume absorption 容积吸收，体积吸收
volume activity generator 批量活动发生器
volume activity 体积活度，体（放射性）活度
volume batching 按体积比配料
volume buret 量管
volume capacity 体积交换容量，体积容量
volume charge density 积电荷密度
volume charge 体积电荷
volume coefficient 容积系数
volume concentration 体积浓度
volume conductivity 体积电导率
volume continuity 容积连续
volume contraction 体积收缩
volume control surge tank water level control loop 容积控制波动箱水位控制回路
volume control surge tank 容积控制波动箱
volume control system 容积控制系统
volume control tank 容积控制箱
volume control valve 容积控制阀，体积控制阀
volume control 音量控制，容量控制，容量调节，容积控制
volume-cost-profit analysis 产量成本利润分析，量本利分析
volume coverage （机械手的）作用范围
volume curve of plastic layer 胶质层体积曲线
volume curve 容量曲线，容积曲线
volume decrease 容积减少
volume defect 体积缺陷
volume density of charge 电荷体密度
volume density 体积密度，堆积密度
volume-depth distribution curve 容积深度分布曲线
volume differentiation 体积微分
volume dose 容积中吸收剂量，体积剂量
volume efficiency due to leakage 泄漏引起的容积效率
volume efficiency 容积效率，音量效率
volume electromagnetic energy 体积电磁能
volume element 体积单元，微元体
volume emission 容积发射
volume energy 体积能
volume expansion 体积膨胀
volume factor 体积系数
volume filtration 容积过滤
volume flow meter 容积流量计【如水表】
volume flow-rate 体积流量，容积流量【流径管道横截面的】
volume flow 体积流量，体积流率，容积流量
volume fluctuation （空泡）体积脉动

volume fraction of B　B的体积分数
volume fraction　容积份额，体积分数，容积率
volume gravity　体积重
volume history　（空泡）体积变化图形
volume indicator　容积指示器，容器指示器，音量指示器
volume (insulation) leakage　体积漏泄，绝缘漏泄
volume integral　体积分
% volume in volume　体积百分数，容积百分比，容积比
volume level　强度级，响度级
volume lifetime　体内寿命，体积寿命
volume line　等容线
volume loss　容积损失
volume mass　体积质量
volume method of concrete mix design　混凝土体积比设计法
volume mix　体积配合法
volume modulus of elasticity　体积弹性模数
volume modulus　体积模量
volumenometer　体积计
volume number of molecules　体积分子数
volume number of particles　体积粒子数
volume of bubble　气泡体积
volume of building　建筑体积
volume of current　电流容积，（电枢）安培，电流量
volume of dam　坝体积
volume of duct elements　（磁流体）通道元件体积
volume of earth and stone in building site　土石方工程量
volume of earthworks　土方工程量
volume of excavation　开方量，挖方量
volume of flood tide　涨潮总量
volume of freight traffic　吞吐量
volume of fresh air　新鲜空气量
volume of main power building per kW　每千瓦主厂房容积
volume of production　产量
volume of reservoir　库容
volume of runoff　径流量，径流总量
volume of vaporization　蒸发量
volume-parts per million　体积的百万分之一，容积的百万分之一
volume percentage　容积百分数
volume pipet　刻度吸液管
volume porosity　体积空隙率
volume production　容积产量，成批生产，批量生产
volume range　音量范围
volume rate of flow　容积流速，容积流量，体积流量
volume ratio　体积比
volume reduction factor　减容因子
volume reduction　减容
volume resistance　体积电阻
volume resistivity　体积电阻率，体积电阻系数
volumescope　容积测定仪
volume sharing technique　体积分布测量法
volume shrinkage　体积收缩
volume source　体源
volume specific heat　容积比热
volume strain　体应变
volume-surface mean diameter　体积表面平均直径
volumeter　容积计，容量计，体积计
volume test　容量试验，大量数据检验，容量测定
volumetricalratio　容积比，体积比
volumetrical　容积的，体积的
volumetric analysis　内部缺陷检验，体积检验，滴定分析，容量分析（法）
volumetric apparatus　测容量仪器，容量分析器
volumetric capacity　容量，容积
volumetric change　体积变化
volumetric coefficient　容积系数，体积系数【颗粒材料】
volumetric concentration　体积浓度
volumetric cylinder　量筒
volumetric density　容积密度
volumetric determination　容量测定
volumetric efficiency　容积效率，容量效率
volumetric examination　容积检验，全厚度检查，内部缺陷监测
volumetric exchange rate　体积交换速率
volumetric expansion coefficient　容积膨胀系数，体积膨胀系数
volumetric expansion　容积膨胀，体积膨胀
volumetric factor　容量因数
volumetric feeder　测容给料机，容量送料器
volumetric flask　量瓶，容量瓶
volumetric flow meter　容积式流量计，容积流量计，体积流量计
volumetric flowrate　体积流率，容积流量，体积流量
volumetric flow　容积流量，体积流量
volumetric flux　容积通量
volumetric gas measuring apparatus　气体容量测定仪
volumetric glass　量杯
volumetric heat capacity　容积热容量
volumetric heat exchanger　容积式换热器
volumetric heat release rate　容积热强度，容积热负荷，燃烧室热容强度
volumetric heat transfer coefficient　体积传热系数，容量传质系数
volumetric measurement of discharge　容积法测流
volumetric measurement　容量测定，体积量测法
volumetric meter　测容量仪器
volumetric precipitation method　沉淀滴定法
volumetric quality　体积含汽率，容积含汽量，容积干度
volumetric rate of flow　容积流速，容积流量
volumetric ratio　体积比
volumetric shrinkage limit　体积收缩极限
volumetric shrinkage　体积收缩
volumetric solution　滴定液
volumetric specific heat capacity　体积比热容
volumetric specific heat　容积比热，体积比热
volumetric steam quality　体积含汽率
volumetric strain　体积应变

volumetric tank 容积计量箱
volumetric 容积的，体积的
volumetrix examination 容积测定
volumetry 容积分析，容量分析，容量测定，容积分析法
volume unit 音量单位
volume variation 体积变化，容积变化
volume velocity 体积流速
volume voltameter 容积电量计
volume weight of soil 土壤容重
volume weight 容积重量，容重
volume 量，流量，容量，容积，体积，响度，卷册，一件存储媒体
volumic acceptor number 受主数密度
volumic cross-section 宏观截面
volumic donor number 施主数密度
volumic electron number 本征载流子数密度，电子数密度
volumic hole number 空穴数密度
volumic sound energy 声能密度
volumic total cross-section 宏观总截面
voluminal compressibility 体积压缩性
voluminal volatile matter 体积挥发分［挥发物］
voluminal 体积的
volumometer 容积计
voluntarily 自愿地，自动地
voluntary bankruptcy 自动宣告破产
voluntary cooperation 自愿合作
voluntary indemnity 自动赔偿
voluntary insurance 自愿保险
voluntary participation 自愿参与
volunteer 自愿者，自愿做
volute casing 蜗壳，蜗形外壳，螺旋形外壳
volute chamber 蜗壳腔
volute housing 蜗壳
volute liner 蜗壳衬护套
volute pump 蜗壳式离心泵，蜗壳泵，螺旋泵
volute throat 蜗形体喉部
volute tongue 蜗壳喷口，蜗舌
volute 螺旋形，涡形，涡螺，蜗壳，涡螺壳，蜗室，螺旋形的
VOL＝volume 体积，容积，卷
vomax 电子管电压表
VOM(volt-ohm milliammeter) 伏特、欧姆毫安计
vortex action 涡流作用
vortex anemometer 旋涡风速计
vortex angle 顶角
vortex augmentor 旋涡增强装置
vortex axis 旋涡轴
vortex band 涡流层，涡流带，涡区
vortex-bed 涡流床
vortex breakdown 旋涡破碎
vortex burner 旋涡式喷燃器，旋流式喷燃器，旋流式燃烧器
vortex bursting 旋涡猝发
vortex cavitation 涡流空蚀
vortex cavity 涡流区，旋涡区，涡空泡
vortex chamber structure 涡流室结构
vortex chamber 涡流室，旋涡室
vortex circulation 旋涡环量
vortex cluster 旋涡群
vortex combustion chamber 旋涡式燃烧室，旋涡式燃烧室
vortex concentrator device 旋涡集聚式风能装置
vortex cone 涡流锥，旋锥体
vortex core 旋涡中心，涡核
vortex current 涡流
vortex decay 涡旋衰减
vortex density 涡流密度
vortex doublet 旋涡偶极子
vortex drag 涡阻
vortex eddy 涡流，旋涡
vortex element 涡旋微元
vortex eliminator 防涡流装置【大直径下降管入口处】
vortex equation 涡流方程
vortex excitation 涡激励
vortex excited oscillation 涡激振荡
vortex excited response 涡激响应
vortex field 涡流场
vortex filament 涡流流线，涡旋线，涡线，涡（旋）丝
vortex flow combustion chamber 涡流式燃烧室
vortex flow impeller 涡流叶轮
vortex flowmeter 涡流流量计，旋涡流量计
vortex flow 涡流，有旋流
vortex flux 涡通量
vortex formation region 旋涡形成区
vortex-free discharge 无旋涡水流
vortex-free motion 无涡运动，无旋运动，无旋涡流动
vortex-free 无旋涡，无涡旋的
vortex frequency 涡旋频率
vortex furnace 旋流炉膛
vortex gaseous reactor 涡旋气态反应堆
vortex gasket 涡流垫片
vortex gate 整流栅
vortex-generator 涡流产生器，旋涡发生器，扰流器
vortex grit chamber 沉砂池
vortex induced acoustic vibration 涡致声振
vortex induced displacement 涡致位移
vortex induced noise 涡致噪声
vortex induced oscillator 涡致振荡
vortex induced response 涡致响应
vortex induced vibration 涡致振动
vortex instability 涡旋不稳定性
vortex intensity 旋涡强度，涡流强度
vortex interference 涡流干扰，旋涡干扰
vortex invariant 涡旋不变量
vortex lattice method 涡格法
vortex lattice 涡流栅
vortex layer 旋涡层
vortex lift 涡升力
vortex line 涡线，涡旋线，旋涡线
vortex location 涡流位置
vortex lock-in 锁涡
vortex lock-on 锁涡
vortex motion 旋涡流动，涡旋运动，涡动
vortex pair 双旋涡，涡对，涡偶
vortex path 涡迹
vortex pattern 涡谱，涡流分布

vortex period 旋涡周期
vortex plane rolling-up 涡面卷起
vortex plane 涡面，涡层
vortex pot 涡斗
vortex precession flowmeter 旋进流量计
vortex pump 涡动泵，涡流泵
vortex rail 涡旋尾迹
vortex response 涡响应
vortex rhythm 旋涡节拍
vortex-ring cascade 涡流环叶，涡环级联
vortex ring 涡环
vortex roll-up 涡翻转
vortex row 涡列
vortex scrubber 冲激式除尘器
vortex sensor 涡流检测器
vortex separation 涡流分离，旋涡分离
vortex shedding anemometer 旋涡脱落风速计
vortex shedding device 旋涡流量测量仪表
vortex shedding flowmeter 涡流流量计，涡街流量计，旋涡流量计
vortex shedding frequency 旋涡脱落频率
vortex shedding period 旋涡脱落周期
vortex shedding rhythm 旋涡脱落节拍
vortex shedding 紊流分离，旋涡脱落
vortex sheet rolling-up 涡面卷起
vortex sheet 涡流层，涡面，涡片
vortex street flowmeter 涡街流量计
vortex strength 旋涡强度，涡流强度
vortex strip 涡带
vortex structure 旋涡结构
vortex suppressing basin 消涡池
vortex suppressor 旋涡抑制器
vortex tail 涡流尾迹
vortex theory 旋涡理论
vortex trail 涡旋尾流
vortex train 涡列
vortex trajectory 旋涡轨迹
vortex trunk 涡线，旋涡螺丝
vortex tube 涡管，涡流管，涡旋管
vortex-type flow 涡流，有旋流
vortex-type fuel assembly 旋涡型燃料组件
vortex vector 旋涡矢量
vortex velocity flowmeter 涡流速度流量计
vortex velocity 涡旋速度
vortex wake 尾涡流，旋涡尾流
vortex 旋涡，涡动，涡流，涡旋，涡体，涡面
vortical flow 涡流
vortical fluidized bed 旋转流态化床
vortical singularity 漩涡奇点
vortical 旋涡的，旋转的
vortices 涡流，涡系，旋涡【vortex 的复数】
vorticity effect 涡流影响
vorticity equation 涡流方程，涡度方程
vorticity field 涡旋场
vorticity in isotropic turbulence 各向同性紊流的涡量
vorticity initiation 旋涡发生
vorticity source 涡源
vorticity trajectory 旋度轨迹
vorticity transfer 旋涡转移
vorticity 涡流强度，涡旋强度，涡量，涡度，涡旋，旋量，旋度
vote down 否决
voter comparator switch 表决比较器开关
vote 表决，投票，选票
voting logic 表决逻辑
voting right 表决权
voting shares 有投票权股份
voucher audit 传票审核
voucher 凭证，凭单，证明，证件，证人，收据，传票，担保人
voussoir arch 楔（形）块拱，分块拱
voussoir （楔形）拱石，拱楔块
VO(variation order) 变更指令（合同用语）
vow 誓言
voyage 航程
VPI(vapor phase inhibitor) 汽相缓蚀剂
VPN(Vickers pyramid hardness) 威氏角锥硬度
V-port plug V形插塞
V-port valve 带有V形柱塞的阀
V/P(vacuum pressure) 真空压力
V-ring commutator V形压圈式换向器
V-ring V形环，V形圈
VSAM(virtual sequential access method) 虚拟顺序存取法
V-shaped clamping ring V形夹圈，V形压圈
V-shaped flashtube V形闪光管，V形闪光灯
V-shaped seal V形密封
V-shaped valley V形谷
V-shape folded plate V形折板
VSI(voltage source inverter) 电压源逆变器
V slot V形槽
VSP(vertical speed) 垂直（方向）速度
V-stay V形拉线
V strings of insulators V形绝缘子串
VS(vertical symmetry) 垂直对称
vs 对（比），与……相对
VT(viscocity) chart 黏度温度图
VT(visual examination) 外观检验，目检
VT(voltage transformer) 电压互感器，变压器
V-type ash hopper V形灰斗
V type contact V形触点
V-type furnace bottom （炉膛）冷灰斗
V-type stoker 机械化炉排，机械加煤机
vuggy rock 多孔岩
vulcacite 硫化促进剂
Vulcan clutch 富尔康型离合器，液力离合器
vulcanism 火山现象，火山作用，火山活动
vulcanist 火山学家
vulcanite 硬橡胶，硬橡皮，胶木
vulcanizate 硫化橡胶
vulcanization 硫化作用，硫化，加硫
vulcanized joint 硫化接头
vulcanized rubber 硫化橡胶
vulcanized splice 硫化接头
vulcanizer 硫化机
vulcanize 加硫，硬化，硫化
vulcanizing agent 硫化剂
vulnerability 易损性
vulnerable 易（受）损的，易侵袭的，易受攻击的，脆弱的，薄弱的
V=velocity 速度

VVER type reactor （苏联）水水型动力反应堆
V=voltage 电压
V=volt 伏特
V=volume 容积
VVP 主蒸汽系统【核电站系统代码】
VVR type reactor （前苏联）池式研究反应堆
VV(velocity vector) 速度矢量
VWO(valve wide open) 阀门全开
vycor 耐热玻璃，石英玻璃
V 主蒸汽【核电站系统代码】

W

wabbler 偏心轮，凸轮
wabble 摇摆，不稳定运动，振动，跳动
WACC(weighted average cost of capital) 资本的加权平均成本
wade 涉水，跋涉
wading bird 涉水鸟
wading measurement of discharge 涉水测流
wading method 涉水测流法
wading pool 浅水池
wading rod 测流杆，测深杆
wadi-type depression 干谷低地
wad 潮成平地，潮坪，填块
wafer butterfly valve 无凸缘的蝶阀，薄板蝶阀
wafer check valve 薄板单向阀
wafer diameter 晶片直径
wafer interface 薄片接口
wafer socket 饼形管座
wafer switch 晶片开关，晶质开关
wafer transformer 金属箔变压器
wafer type butterfly valve 对夹式蝶阀，薄片型蝶阀
wafer 圆片，薄片，晶质的，无凸缘的，半导体薄片
waffle raft 蜂窝状的底板
wage contract 工资契约
wage exponent 工资指数
wages and salary administration 工资管理
wages fund 工资基金
wages index 工资指数
wage 薪金
waggle dance （架空线）8字形舞动
Wagner code 瓦格纳码，单差校正码
Wagner earth device 瓦格纳接地装置
wagon balance 车辆秤，地磅，过车称
wagon conveyer 货车运输机
wagon defrosting station 车辆解冻装置
wagon drill 车钻，汽车式钻机，汽车钻机，车装钻机，移动式钻机
wagon dumper 翻车机
wagon jack 车辆起重机
wagon mounted air drill 车载风钻
wagon-mounted drill 车钻
wagon shaker 振车器
wagon tipper house 翻车机室
wagon tipper 翻车机
wagon tippler house 翻车机室
wagon tippler （翻斗车的）翻斗，翻车机，翻车工
wagon unloader 卸车机
wagon unloading facility 卸车设备
wagon 货车，车皮，矿车，铁路货车，小斗车
WAGPCO(West Africa Gas Pipeline Company) 西非天然气管道公司
WAGP(West Africa Gas Project) 西非天然气项目
wainscotting 壁板材料
wainscot 壁板
WAIS(wide area information server) 广域信息服务器（系统）
wait condition 等候条件
waiting charge 待时费
waiting for a job within an enterprise 待岗
waiting-line theory 排队论
waiting loop 等待循环
waiting state 等待状态
waiting time 存储时间，（放射性）冷却时间
waiting track for repair 停留待修线
waiting 等待
waive a claim 放弃索赔（要求）
waiver and estoppels 弃权和禁止翻供，弃权与禁止反言条款
waiver clause 弃权条款，放弃条款
waiver of benefits 津贴放弃
waiver of conditions 放弃条件
waiver of lien 放弃留置权，留置权取消，赦免抵押条款
waiver of premium clause 免付保费条款
waiver of subrogation rights clause 放弃代位权条款
waiver 超差回用，放弃，例外，弃权（声明），拒绝，否认
wake blockage 尾流阻塞
wake boundary 尾流边界
wake bubble 尾流气泡
wake buffeting 尾流抖振
wake capture 尾流捕获
wake cavity region 尾流空穴区
wake circulation 尾流环量
wake closure 尾流封闭区
wake current 尾流
wake decay 尾流衰减
wake drag 尾流阻力
wake effect loss 尾流效应损失
wake energy 尾流能量
wake entrainment 尾流卷挟
wake excitation 尾流激励
wake excited crosswind response 尾流横风响应
wake expansion 尾流扩展
wake flow theory 尾流理论
wake flutter 尾流颤振
wake fraction 尾迹所占体积百分数，伴流率，伴流系数，伴流分数
wake front 尾流前沿
wake galloping 尾流驰振
wake growth 尾流增长
wake induced vibration 尾流致振
wake-in turbine enclosure 可进人的透平罩壳

wakeless bubble 无迹气泡
wake loss 尾流损失
wake mixing 尾流混合
wake momentum thickness 尾流动量厚度
wake of the bubble phase 气泡尾迹相【沸腾炉】
wake of WTG 风电机组尾流
wake oscillator model 尾流振子模型
wake recirculation region 尾流回流区
wake region 尾流区
wake relaxation 尾流松弛
wake resistance 尾流阻力
wake shadow 尾流阴影
wake shedding 尾迹剥落
wake spread 尾流扩展
wake stream 尾流
wake strength 尾流强度
wake Strouhal number 尾流斯特劳哈尔数
wake suction 尾流吸力
wake-survey method 尾迹测量法，尾迹移测法
wake survey 尾流测量
wake turbulence 尾流湍流（度）
wake-up switch 唤醒开关【计算机】
wake vortex resonance 尾流旋涡共振
wake vortex system 尾流涡系
wake vortex 尾流旋涡，尾涡流
wake vorticity 尾流涡量
wake width 尾迹宽度
wake 尾迹，（气流的）尾流，气泡尾涡【沸腾炉】，涡区，扰流，尾波，伴流
wale piece 横撑板
wale 横撑
waling strip 围令木
walk down 信息丢失
Walker phase advancer 沃克相位超前补偿器
walker river 不定床河流
walkie-talkie 对讲机，手提无线电话机
walking beam 摆动梁，平衡梁，摇梁
walking crane 活动吊车，步行式吊车，移动式吊车，执行起重机
walking dragline 移动式索铲
walking excavator 移动式挖土机
walking scoop dredger 移动式铲斗挖泥机
walk-through test 预检，流检
walk-up multistory building 无电梯的多层建筑
walkway clearance 过道宽度
walkway duct 通行管沟
walkway 桥，平台，人行道，步道，过道，通道，走廊，走道
wall absorption 壁吸收
wall action 边界作用，边界影响，壁面作用
wall adhesion 墙附着力
wall arcade 实心连拱廊
wall attachment jet 贴附射流
wall back 墙背
wall base 墙基
wall beam 墙式梁，墙梁
wall bearing construction 承重墙结构，墙承重的结构
wall-bearing structure 承重墙结构
wall between posts 柱间墙
wall between windows 窗间墙
wall blower 水冷壁吹灰器
wall board switch 墙上开关
wall board 墙板
wall boundary layer （风洞）洞壁边界层，壁面边界层
wall box 墙上的盒子，装在墙上的杂用箱柜，炉墙箱壳，暗线箱
wall bracket jib crane 壁装转动伸臂起重机
wall bracket 墙支架，墙上托架
wall bushing 穿墙套管
wall chart 挂图
wall cladding panel 墙面覆盖板
wall cladding 壁覆面，覆层，墙面覆盖层，外墙围护结构
wall constraint （风洞）洞壁约束
wall-cooling surface 水冷受热面，水冷壁面
wall-cooling type 冷壁型的
wall correction factor 壁效应修正系数
wall correction 洞壁修正
wall covering 墙面涂料
wall crane 壁装起重机，墙装起重机，臂上起重机
wall deslagger 水冷壁除渣器
wall drag 器壁曳力
wall duct 墙内风道，墙式电缆槽
walled-in policy 壁垒政策
walled-in 筑墙围的【设备】，有围墙的
wall effect 器壁效应，墙效应
wall enclosure superheater 包墙过热器
wall enclosure tube 包墙管，包覆管
wall entrance insulator 穿墙绝缘导管
wall firing system 墙式燃烧系统
wall firing 侧墙燃烧，墙式燃烧
wall footing 墙基
wall framing 墙构架
wall friction angle 仓壁摩擦角
wall friction coefficient 壁摩擦系数
wall friction 壁面摩擦（力），墙摩擦力
wall heat flux 壁面热通量
wall-heating type 热壁型的
wall-hook 壁钩
wall-hung fixture 挂在墙上的卫生器具
wall insulator 穿墙绝缘子
wall interference 器壁干扰
wall jet 贴壁射流
wall lantern 壁灯
wall liner tube 穿墙管
wall lining 炉壁内衬
Wallman amplifier 渥尔曼放大器
Wallman circuit 渥尔曼电路
wall-mounted box 装在墙上的箱
wall-mounted switchboard 壁装开关板，壁型开关盘
wall-mounted 装在墙上的，吸壁的，墙挂式
wall mounting 墙壁安装（的），壁挂式
wall opening ratio 墙面开洞比
wall panel heating 墙壁辐射采暖
wall panel 护墙板，墙板，镶板
wall paper 贴壁纸，壁纸，墙纸，糊墙纸
wall partition 隔墙
wall-pattern switchboard 墙装交换机

wall penetration　穿墙管，穿墙，墙壁贯件，墙壁贯穿件
wall pier　墙墩
wall plate　承梁板，墙板，承椽板
wall plug　墙装插头
wall pressure hole　壁面测压孔，洞壁测压孔
wall pressure　壁压
wall radiant heating　墙面辐射采暖
wall radiant tube　墙壁辐射管
wall radiator　壁挂式散热器
wall rail　墙梁
wall reaction　器壁反应
wall receptacle　壁插座，墙插座
wall register　墙面风口，壁式空调器
wall roughness　壁面粗糙度
wall sampler　墙壁取样器
wall setting　炉墙
wall shearing stress　器壁剪应力，壁剪应力
wall shear stress　壁面剪应力
wall sleeve　墙套管
wall slug　器壁腾涌
wall socket　墙装插座，壁插座，电器插座
wall stress controller　汽缸壁应力控制器
wall stress　壁应力
wall superheater　墙式过热器
wall support　墙支承，墙支座
wall switch　墙壁开关
wall temperature of critical point　临界点壁温
wall temperature of dry-out point　干涸点壁温
wall temperature probe　壁温测针
wall temperature　壁温
wall thickness transition　（压力容器）壁厚变化
wall thickness　管壁厚度，壁厚
wall thinning　管壁变薄
wall-through bushing insulator　穿墙绝缘套管
wall-through bushing　穿墙套管
wall tile　墙面贴砖，面砖
wall toe　墙脚
wall tube insulator　穿墙瓷管，穿墙套管
wall tube　水冷壁管，墙上进线管，墙管
wall turbulence　壁面湍流，壁面紊流
wall type switchboard　墙装式交换机，墙装式配电盘
wall type　墙挂式，挂式
wall up　把……封住，把……关住
wall with refractory lining　卫燃带
wall　墙，墙壁，墙体，壁，墙坝，势垒【电场】，水冷壁，筑墙
walnut　胡桃木
wamoscope　行波示波管
wamp　浪涌，急变
wandering dune　移动沙丘，游移沙丘
wander　漂移，偏移，游移，漂动，淤满
wand reader　条形码读入器
wand　棒，权杖，嫩枝，识别笔【等于 wand reader】，用扫描笔在……上扫描条形码
wane　没落
want of proof　缺乏证据
want　缺少，需要，希望
WAN(wide area network)　广域网
WAPDA(Water and Power Development Authority)　巴基斯坦水电开发署
warbler　频率摆动器，电抗管调制器，颤音器
warbling alarm　颤音报警
ward commission　受保护人委员会
wardrobe　衣橱，衣室
warehouse and space area　仓库及堆场面积
warehouse and workshop　仓库和机修厂
warehouse crane　仓库用起重机
warehouse in/out charge　进出库费
warehouseman's gloves　仓库工手套
warehouseman　仓库管理员
warehouse　仓库，储存室，货栈
warehousing in/out fee　进出仓费
warehousing　送入仓库【贮存、拣选、补充等】，仓储业务，货物仓储，库存
Warent motor　华伦电动机
ware　制品，成品，器具
warf　码头
warm air curtain　热风幕
warm air heating system　热风采暖系统
warm air mass　暖气团
warm air　热风，热空气
warm anticyclone　暖性反气旋
warm box　保温箱【柜】
warm braw　暖风
warm bubble　热泡
warm climate　温暖气候
warm cloud　暖云
warm-cold air tunnel　冷热空气风洞
warm-compacted　中温密实，中温压块
warm current　暖流
warm cyclone　暖性气旋
warm discharge water　温排水
warm drain　中放射性排放，温排放
warm effluent　中放废水，中放流出液
warm fog　暖雾
warm forging　降温锻造法，中心压实法
warm front　暖锋
warming apparatus　暖机装置
warming coil　加热盘管
warming pipe　暖管，加热管，暖机管
warming up condensate drain-off connections　启动疏水装置
warming up process　暖机过程
warming up the turbine　汽轮机暖机，暖机
warming up　暖机，暖炉，暖管，预热
warming　加温【焊接前】，暖管，暖机，预热，增温
warm laboratory　中等放射性实验室
warm model　中温模化
warm rod　暖棒
warm rolling　中温轧制
warm sector　暖区
warm spot　暖点
warm standby　温储备
warm start-up　暖态启动，半热态启动，温态起动
warm start　暖态启动，半热态启动，温态启动
warm temperate rain forest　暖带雨林
warmth　热情
warm-up bypass valve　暖管旁通阀

warm-up drift 预热式频率漂移，输电功率恢复时频率漂移
warm-up facility for FBC boiler 流化床锅炉点火装置
warm-up light oil burner 启动用轻油喷燃器
warm-up line 加温管线
warm-up of the unit 暖机
warm-up oil fuel trip 加热燃料油跳闸
warm-up period 暖机阶段，暖机期，暖机时间，预热时间，升温时间，予动作时间
warm-up speed 暖机转速
warm-up time 暖机时间，预热时间，升温时间，闭路后至输出正常的时间
warm-up valve 暖机阀
warm-up water 加热水
warm-up 暖机，暖管，预热，升温，慢车启动，加热，加温
warm water intake pipe 温水进水管【海洋温差电站】
warm water irrigation 温水灌溉
warm water meter 暖水表，暖水流量计
warm water pollution 水热污染
warm water port 不冻港
warm working 温加工，半热加
warm work （在临界温度下）中温加工
warm zone 暖温带，温带
warm 温暖的
warner 报警器
warn-free 防抖动，不报警
warning agent 警戒剂
warning alarm 警报
warning apparatus 报警信号器
warning board 警告牌，危险标示牌
warning circuit 警告电路
warning color 警戒色
warning concrete 报警混凝土
warning device 报警装置
warning horn 警告喇叭
warning indicator 警告指示器，报警器
warning label 安全标识
warning lamp 警告灯，指示灯，航空障碍灯
warning light 警告信号灯光，报警信号灯，警告灯，报警灯
warning limit 警戒范围
warning mark 包装储运指示标志，警告标志
warning plate 警告牌
warning poster 危险或警告告示，安全标志
warning screen 警告屏
warning service 报警站
warning signal 警告信号，危险信号，报警信号，警告标志
warning sign 报警信号【放射性保护】，报警信号，警告信号，警告标志，警示牌
warning system 警告系统，报警系统
warning tape 警告信号带
warning water stage 警戒水位
warning 警告
warn 迸发喷射【安全阀】，抖动，警告，警报
warpage 翘曲，扭曲，弯翘，弯曲，热变形，变形
warp and weft direction 经纬向
warped blade 扭叶片
warped clay 淤积黏土
warped surface 翅曲面
warping drum 卷缆筒
warping moment 翘曲力矩
warping stress 翘曲应力
warping winch 牵曳式绞车
warping works 放淤工程
warping 翘面，翘曲，扭曲，变形，歪扭，拖曳，弯翘，放淤
warp land 放淤地，淤泥滩地
warp soil 放淤土
warp 扭曲，变形，翘曲，卷绕，变歪，弯曲处，歪曲，偏差，翘起
warranted efficiency 保证效率
warranted output 保证出力，保证输出功率【电厂】
warrantee 被担保人
warrant fuel consumption 保证燃料耗量
warrant of attorney 授权委托书
warrantor 担保人
warranty clause 保证条款
warranty cost 抵押成本，保修成本
warranty period 保用期，保证期，保修期，抵押期
warranty run 保证期运行
warranty test 保证试验【电厂】，验收试验，认可试验，特性试验
warranty 担保，保证，（商品等）保单，（正当）理由，担保书
warrant 保险期
Warren truss 华伦式屋架
war risk 战争险
wart-type spacer 疣型定位件
washability curve 可洗性曲线
washability 可洗性
wash away 冲走，清洗，清除
wash basin 洗脸槽，洗手池，洗脸池，洗脸盆
wash boring method 水冲钻孔法
wash boring 回水钻探，清水钻进
wash column 冲洗塔
wash down water 冲洗水
wash down 下洗，冲洗，洗涤，水冲洗
wash drilling 冲水钻井法【软基中的】
wash drill 冲水钻孔装置
washed coal 洗选煤，洗煤
washed gas 已清洗烟气，已除尘烟气
washed middling coal 洗中煤
washed ore 洗矿
washed slack 洗煤，洗净的碎煤
washed steam 已清洗蒸汽，已洗蒸汽
washer head screw 圈头螺钉
washer of nut 螺母垫圈
washer screw 衬垫螺丝
washer thermistor 垫圈状热敏电阻器
washery rejects 洗矸
washery slurry 洗煤泥浆
washery waste 洗矸石
washery 洗选厂
washer 垫圈，衬垫，垫片，洗涤器，洗净器，蒸汽清洗器

washes 洗涤物，洗刷剂，洗涤废水
wash heat 渣洗，清洗熔炼
wash house 洗衣间
washing agent 洗涤剂
washing and decontamination system 洗涤和去污系统
washing basin and sewage sink 洗水池和污水槽
washing cell 净化室
washing column 洗涤柱，洗涤塔
washing erosion 冲洗侵蚀
washing filter 洗滤器
washing length 冲洗历时，冲洗时间
washing liquid 洗液
washing liquor 洗液
washing medium 清洗介质
washing of coal 洗煤
washing-off 洗出
washing out 冲坍
washing plant 洗涤装置
washing process （煤的）洗选法
washing screen 冲洗筛
washing section 洗涤段
washing strength regulator 冲洗强度调节器
washing strength 冲洗强度
washing tank 洗涤箱，洗涤槽
washing water 冲洗水
washing 冲刷，冲洗，清洗，洗涤
wash load 冲泻质
wash method 冲洗法
wash mill 洗涤装置
washmittel 洗剂
wash off 清除掉，清洗掉，冲毁，冲掉
washout coefficient 冲洗系数，冲蚀系数
washout deposition rate 冲洗沉积率，冲蚀沉积率
washout factor 冲洗，因子
washout hole 清洗孔
washout parameter 冲洗参数
washout period 洗脱期，清洗期
washout pipette 洗出移液管
washout rate 冲洗率，清除速率
washout ratio 冲洗比，冲刷率
wash out test 冲刷试验
washout valve 冲沙阀
washout 冲坏，冲毁处，冲掉，冲刷，冲蚀，清堵，淘刷，破产，淘汰，失败，惨败
washover 冲益三角洲，越浪堆积，冲溢沉积，溢流
wash pipe 冲洗管
wash plant 洗涤站
wash point penetrometer 水冲式贯入仪
wash port 排水口
wash-prime coat 防冲层
wash primer 活化涂料
wash room 盥洗间，洗手间，厕所
wash sink 洗涤池
wash siphon 冲沙虹吸管
wash stand 盥洗盆
WASH＝washer/washing 冲洗
wash water rate 冲洗强度
wash water tank 冲洗水箱
wash water 钻探用冲水，冲洗水
wash zone 冲刷带
wash 涡流，扰动，刷浆，浅水湾，涂浆，涂刷，洗，冲洗
wastage allowance 材料许可损耗率
wastage area 污水区
wastage of steel 钢材损耗
wastage 废物，废水，磨损量，耗损，消耗，（管壁）耗蚀，管壁变薄现象，减少，磨削，损耗量，损耗，侵蚀，腐蚀，消融
waste activated sludge 废活性污泥
waste air 废气
waste alkali 废碱液
waste assimilation capacity 废物同化能力
waste auxiliary building 核废物辅助厂房，废物辅助厂房
waste back-cycling 废液返回循环
waste bank 废物堆，废土堆，废物库，废石场，矸石堆
waste boiler 废物锅炉
waste burial ground 废物埋葬地
waste burial 废物埋葬
waste calciner 废物焚烧炉，废物焚烧器
waste calcining facility 废物焚烧装置
waste canal 排水沟，退水渠
waste canister 废物罐，废物容器
waste carbon 废（活性）碳
waste category 废物等级
waste caustics 废碱液
waste chemical reagent 废化学试剂
waste chrom liquor 废铬液
waste coal 废煤
waste cock 泄放旋塞
waste collector system 废物收集系统
waste compaction 废物压实
waste component 废物成分
waste condensate pump 废液冷凝水泵，废凝结水泵
waste condensator 废汽凝汽器，回流凝汽器
waste conditioning 废物形态调整
waste confinement （放射性）废物封隔
waste container 废物容器
waste containment （放射性）废物封存
waste control 废物管理
waste convertion techniques 废物转换技术
waste crude oil 废原油
waste cyanide 废氰化物
wasted acid 废酸
waste decay tank 废物衰变箱
waste demineralizer 液态废物除盐器
waste demineralizing plant 液态废物除盐器
wasted energy 损失的能量
waste discharge 排废
waste disposal basin 废水池
waste disposal building （放射性）废物处理厂房
waste disposal deep underground 废物深埋处置
waste disposal site 废物处置场，废物放置场
waste disposal system control panel 废物处置系统控制盘
waste disposal system 废物处置系统
waste disposal work 废物处理厂

waste disposal 处理废弃物，废物处理，废弃物处理，废物弃置
wasted power 耗散功率，耗损功率
waste drains pump 废液排放泵，废液泵
waste drumming 废物装桶
waste drum 废物桶
waste dump 废物堆，废石堆，矸石堆
wasted work 耗功
waste emplacement （放射性）废物安置
waste encapsulation 废物包装
waste evaporator feed pump 废液蒸发器给水泵
waste evaporator package 废液蒸发装置
waste evaporator reflux condenser 废液蒸发回流凝汽器
waste fixation program 废水固化计划
waste fixation system 废水固化系统
waste fluid burning plant 废液燃烧装置
waste fluid 废液
waste form 废物形态
waste from coal gas plant 煤气厂废水
waste from coal mine 煤矿废水
waste fuel 废燃料
waste gas analysis 废气分析
waste gas burning 废气燃烧
waste gas cleaning plant 废气净化设备
waste gas cleaning system 废气净化系统
waste gas cleaning 废气净化
waste gas compressor 废气压缩机
waste gas desulfurization 废气脱硫
waste gas disposal system 废气处置系统
waste gas emission standard 废气排放标准
waste gas header 废气联箱，废气总管
waste gas pollution control 废气排放治理
waste gas purifying equipment 废气净化设备，煤烟处理设施
waste gas supercharger 废气增压器
waste gas superheater 废气过热器
waste gas treatment equipment 废气处理设备
waste gas treatment 废气处理
waste gas vapor trap 废气蒸汽捕集器
waste gas 废气，气态废物
waste gate 弃水闸门
waste heat boiler 废热锅炉，余热锅炉
waste heat capacity 废物散热率
waste heat flue 废热烟道
waste heat generation 废热发电，余热发电
waste heating 废热供暖
waste heat loss 排烟热损失，废热损失
waste heat recovery power plant 废热发电厂，余热发电设备
waste heat recovery unit 废热回收装置
waste heat recovery 废热回收，余热回收
waste heat removal system 废热排出系统
waste heat superheater 废热过热器
waste heat utilization 废热利用
waste heat 废热，余热
waste hold-up tank recirculation pump 废水暂存箱再循环泵，废水滞留箱再循环泵
waste immobilization plant 废物固化工厂
waste incineration plant leachate 垃圾焚烧厂渗沥液
waste incineration power generation project 垃圾焚烧发电项目
waste incineration power plant 垃圾焚烧发电厂
wasteland 荒地
waste liquid 废液，液体废物
waste liquor recovery 废液回收
waste load 废物负载
waste lye 废碱液
waste management advisory programme （放射性）废物管理咨询计划，废物管理咨询计划
waste management program 废物管理规化
waste management 废物管理
waste material boiler 废料锅炉
waste material 废物，废料
waste matrix 废物基质，废物基体
waste matter 废物
waste mine 废矿井
waste monitoring 废物监测
waste neutralizer tank 废物中和箱
waste oil clot 废油凝块
waste oil disposal facility 废油处置设施
waste oil disposal 废油处置
waste oil 废油
waste oxidation basin 废水氧化池
waste package 废物包装
waste pellet 废芯块
waste pile 弃土堆
waste pipe 废水管，排泄管，污水管
waste plant 废水处理厂，废水处理装置，废物处理站
waste processing control room 废物处理控制室
waste processing 废物处理
waste product recovery power plant 利废发电厂
waste product 废品
waste quantity limit 废物量极限
waster calcination demonstration plant 废物煅烧示范工厂
waste receiver tank pump 废液贮箱泵，废液滞留箱泵
waste recipient 接受废水的河流，废水接收器
waste recirculation pump （液体）废物再循环泵
waste recovery power plant 燃垃圾发电厂，燃烧废料发电厂
waste recovery 废物回收，废水回收
waste recycle system 废水再循环系统
waste repository 废物存放场，废物处置库
waste retention system 废物贮存系统
waste retrieval system 废物可回取系统，废物回收系统
waste sampling cabinet 废物取样箱
waste sand and gravel 废砂石
wastes control 废料管理
waste shredder 废物粉碎机
waste silk 废丝
waste sludge 废污泥
waste sluice 弃水闸
waste slurry 废浆液
waste solidification （放射性）废物固化，（放射性）废物固化处理
waste solid 废渣，固体废物
waste stabilization pond 废水稳定塘

waste steam turbine 废汽汽轮机
waste steam 废汽，乏汽，汽轮机排汽
waste stock 废料
waste storage area 废物贮存地区，储存放射性废料场地
waste storage drum 废物贮存桶
waste storage farm 废物贮存场
waste storage 废物贮存
waste tank 废物贮存槽，废物桶
waste treatment equipment 废物处理设备
waste treatment station 废物处理站
waste treatment system 废物处理系统，废水处理系统
waste treatment technique 废物处理技术
waste treatment 废料处理，废物处理
waste uranium 废铀
waste valve 溢流阀，放泄阀，排污阀，废料排出阀
waste vault 废物窖，废物库
wastewater collection system 废水收集系统
wastewater collection 废水收集
wastewater composition 废水成分，污水组成
wastewater dilution 废水稀释
wastewater disinfectant 废水消毒剂
wastewater disposal facility 废水处理设施
wastewater disposal pump 污水泵，污水处理泵
wastewater disposal 污水处理，污水处置
wastewater effluent 废水出水
wastewater engineer 废水治理工程师
wastewater facility 废水处理设备
wastewater flow 废水流量
wastewater from ashpit 灰坑废水
wastewater from car washer 洗车废水
wastewater lift pumping station 废水输送泵站
wastewater neutralization tank 污水中和箱
wastewater neutralizer 废水中和器，废水中和池
wastewater pipe 废水管
wastewater processing 废水处理
wastewater purification plant 废水净化站，污水净化站
wastewater purifying 废水净化
wastewater rate 废水（排放）率
wastewater reclamation plant 废水回收厂
wastewater reclamation 废水改良，废水回收
waste water renovation system 废水净化系统
wastewater renovation 废水净化回收，废水再净，废水重复使用
wastewater reuse 废水再用
wastewater sludge 废水污泥
wastewater solid 废水固渣
waste water storage basin 废水贮存池
wastewater sulfide 硫化物废水
waste water tank 废水箱
wastewater treatment plant 废水处理厂
wastewater treatment process 废水处理过程
waste water treatment project 污水治理工程
wastewater treatment system 废水处理系统
wastewater treatment 废水处理，污水处理
wastewater use 污水利用
waste water 废水，污水
waste weir 弃水堰，退水堰
waste well 污水渗井
waste work 耗功
waste 排放物，尾料，废料损耗，废物，耗损，垃圾，贫料，弃土
WAS(withdrawal authorization sheet) 材料（设备）领用单
watch box 瞭望亭
watchdog timer 看守计时器，程序控制定时器，看门狗定时器［计时器］
watchdog 监察人员，监察机构，监督人员，看门狗，监视器，监控设备
watch glass 表面皿
watch house 值班室
watch tower 瞭望台
water abrasion 水（冲）蚀
water absorbed in the core structures 堆芯结构吸收的水
water-absorbing capacity 吸水能力
water-absorbing power 吸水力
water absorption test 水吸收试验
water absorption 吸水性，吸水率，吸水量，吸水
water absorptivity 吸水率
water accumulation 积水，蓄水
water activity meter 水放射性测量计
water admixing installation 混水装置
water-air-cooled machine 间接水冷式空冷电机
water-air ratio 水气比
water allocation 水量分配
water analysis 水分析
water and land transportation 水陆运输
water and soil conservation 水土保持
water annealing treatment 水韧处理，水冷退火
water application efficiency 水的有效利用系数，供水效率
water application rate 供水率，用水率
water approach flume 来水槽
water appropriation 水的专用权
water area 水域
water aspirator 水抽子
water as refrigerant 冷剂水
water atomizer 水雾化器
water attack （汽轮机）进水，水击，水冲击
water back 贮水器，加热水池，加热管道（炉排）水冷渣挡
water baffle 挡水板
water balance chart 水量平衡图
water balance diagram 水平衡图
water balance equation 水量平衡方程，水平衡方程式
water balance method 水量平衡法，水平衡法
water balance 水量平衡，水平衡
water ballast 水压舱，水压载
water barrel 提水桶
water barrier 拦水建筑物，拦水栅
water bar 挡水条，止水带
water basin thermowell 热电偶套管
water basin 流域，水池
water bath evaportor 水槽蒸发器
water bath 水浴，水槽，水浴锅

water battery 水电池
water-bearing bed 含水层
water-bearing concreae 含水混凝土，湿混凝土
water-bearing deposit 含水沉积层
water-bearing formation 水层，含水层，含水地层，含水岩组，含水建造
water-bearing layer 含水层，蓄水层
water-bearing media 含水介质
water-bearing medium 含水介质
water-bearing rock 含水岩
water-bearing stone 含水石
water bearing strata formation 含水层建造
water-bearing strata 含水层
water-bearing zone 含水层
water bed 水褥
water block 水堵，水封，水箱，水冷头
water bloom 水发绿，水华，水藻大量繁殖
water blower 水枪
water board 水闸板
water body pollution 水体污染
water body 水体，水域
water boiler neutron source 水锅炉中子源，水均匀堆中子源
water boiler reactor 沸水反应堆，水锅炉反应堆，沸水堆
water boiler 水锅炉【即水均匀反应堆】，热水锅炉
water-boiling thermal efficiency of solar cooker 太阳灶煮水热效率
water-borne carrier 水上运输工具
water-borne contaminant 水载污染物，水中的污染物
water-borne disease 水传染病
water-borne material 水成物质，水生物质
water-borne noise 水中噪声
water-borne radioactivity 水载放射性，水中放射性
water-borne sediments 水成沉积物
water-borne waste 水载废物，液体废物，水中的废物，废水
water-borne 带水的，水生的，水中的，水传播的，水运的
water bosh 水封
water bottle 取水样器，水瓶，水筒
water-bound macadam 水结碎石路
water bowl 饮水器
water bowser 供水车
water box cover 水室盖
water box sacrificial anode 水室牺牲阳极（保护法）
water box 水室【凝汽器】，防焦箱【炉排】，水槽，水箱
water brake dynamometer 水力制动马力计，水力测功器
water brake 水力制动，水力制动器
water break 断水，防波堤，减冲设施，破浪堤
water bubble 水泡
water bucket 水斗，水桶，排水吊桶，
water budget 水量平衡，水量收支预算，水平衡
water burst （基坑）涌水量
water capacity 水容积，水容量，水容
water carriage system 水运系统
water carriage 水运
water carrier 含水层，蓄水层
water-carrying section 过水断面
water carryover （蒸汽中）夹带水分
watercart 洒水车，运水车
water cascade 小瀑布
water catcher （去湿用）集水槽，水收集器，水分离器
water catchment area 流域面积，集水地区
water cellar 地下水池，水窖
water cement ratio 水灰比
water cement 水硬性水泥，水灰，水性胶
water centralized treatment facility 废水集中处理设施
water chamber （凝汽器）水室，冷却水套，水夹套，水箱，水室
water channel 水槽，水洞，水渠，水道
water charger 充水装置【水泵】
water charge 水费
water-chemical regime 水化学工况【冷却剂系统】
water chemistry 水化学
water chilled 冷冻水
water chiller unit 水冷冻装置
water chiller 冷冻水装置，水冷却器，水冷冻装置，水冷机组
water chilling unit 冷水机组
water chlorination 加氯水处理，水的加氯处理，水的氯处理
water circuit breaker 水断路器
water circulating pump 循环水泵
water circulation coefficient 水循环系数
water circulation detector 水循环检测器
water circulation pipe 水循环管线
water circulation pump 水循环泵，循环水泵
water circulation system 水循环系统
water circulation 水的循环，水循环
water circulator 水环流器
water cistern 水池，水箱，水塘
water clarification 水澄清作用
water clarifier 澄清池
water classification 水的分类
water closet 厕所
water closing type arrestor 水封式回火防止器
water closing 水封
water-coal slurry 水煤浆
water cock 水龙头
water code 水利法规，水法
water coil 盘形水管
water-collapsible loess-soil 湿陷性大孔土
water collecting area 积水面积，汇水面积
water collection drum 集水箱
water collection pit 集水坑
water collector 集水箱
water column arrester 水柱避雷器
water column correction 水柱高度校正
water column pressure 水柱压力
water column 水柱，水塔
water conditioning 水质处理，水处理
water conductivity meter 水导电率计

water conduit 输水道，输水阀
water conservancy protection 水工保护
water conservancy 水利
water conservation bureau 水利局
water conservation 节约用水，水源保护，水利
water constituent 水成分，水质
water consumption curve 耗水量曲线
water consumption graph 耗水量图
water consumption per capita per day 每人每日用水量
water consumption per GW 百万千瓦耗水指标
water consumption per 1000MW 每一千兆瓦机组耗水量
water consumption 耗水量，水耗，水量消耗，用水量
water contamination 水污染，水质污染，水质污浊
water content coefficient 含水率
water content of steam 蒸汽湿度，蒸汽含湿量
water content profile 含水量纵剖面图
water content ratio 含水比
water content 水分，含水量，水含量，湿度
water control device 水控装置
water control valve 水控制阀
water conveyance facility 输水设施
water conveyance structure 输水构筑物，输水建筑物，输水结构物
water conveyance system 输水系统，水输系统，
water conveyance tunnel 过水隧洞
water conveyance 输水
water conveying capacity 输水能力
water coolant 水冷却剂，冷却水
water-cooled absorptive shielding 隔热水箱
water-cooled air conditioner 水冷式空调器
water-cooled and water-moderated reactor 水冷却慢化（反应）堆，水水（反应）堆
water-cooled baffle 水冷挡墙，水冷折焰板
water-cooled bottom ash hopper 冷灰斗
water-cooled chilling unit 水冷冷水机组
water-cooled condenser 水冷式冷凝器，水冷式凝汽器
water-cooled distributor 水冷布风板【沸腾炉】
water-cooled electro-magnet 水冷电磁铁
water-cooled forced-oil transformer 水冷强制油循环式变压器
water-cooled furnace 水冷炉膛
water-cooled generator 水冷发电机，水内冷发电机
water-cooled grate 水冷炉排
water-cooled hopper bottom 冷灰斗
water-cooled lattice 水冷却栅格
water-cooled low-voltage electrical apparatus 水冷低压电器
water-cooled machine 水冷式机器，水冷电机
water-cooled motor 水冷电动机
water-cooled oil-immersed transformer 水冷油浸变压器
water-cooled oil-insulated transformer 水冷油绝缘变压器
water-cooled packaged air conditioner 水冷式整体空调器
water-cooled reactor 水冷堆，水冷反应堆，水冷电阻器
water-cooled rotor 水冷转子
water-cooled screw conveyor 水冷式螺旋运输机
water-cooled spacer 水冷分隔器，水冷炉排
water-cooled transformer 水冷变压器
water-cooled turbine 水冷式透平
water-cooled turbogenerator 水冷汽轮发电机
water-cooled wall 水冷壁
water-cooled 水冷的，水冷却的，水冷式的
water cooler 水冷却器，水冷器
water cooling system 水冷系统，水冷却系统
water cooling tower 水冷却塔，凉水塔
water cooling 水冷，水冷却
water core 防渗心墙
water correction 海水校正，含水率校正，水质改善
water corrosion 水腐蚀
watercourse bed 常水河床，水道基床
watercourse 航道，水道，渠道，水路
water crop 出水量
water cured concrete 水养混凝土
water curing room 湿养护室
water curing 水处理，水养护
water curtain 水幕
water cushion 水垫，消力池
water cycle 水循环
water-cycling system 水循环系统
water damage 水蚀损伤
water deficiency 水量不足
water-deficient area 缺水地区
water deficit 水量不足，缺水
water delivery gate 输水阀门
water delivery rate 供水流量
water delivery tunnel 输水隧洞
water delivery 输水量，供水，输水
water demand rate 水消耗定额
water demand 用水需求，需水量
water demineralization 水质软化，水脱盐，水软化，除去水中矿物质
water deprivation 脱水，脱水作用
water depth line 水深线
water depth measurement 水深测量
water depth rod 水深测杆
water depth 水深
water desalination 水脱盐
water development 水利开发
water discharge piping 排水管道
water discharge pump 排水泵
water discharge tank 排水池
water discharge tube 排水管
water discharge tunnel 排水隧道
water discharge valve 泄水阀，放水门
water discharge 排水，排水量，泄流量，泄水量
water disinfection 水的消毒
water displacement 排水量
water displacer rod 挤水棒
water disposal system 水处理系统，治水系统
water disposal 水处理
water distribution fill system 配水的充水系统【冷却塔】

water distribution groove 配水槽
water distribution pipe 配水管
water distribution system 配水系统
water distribution 配水，水分布，水分配
water distributor 配水器
water diversion channel 分水槽
water diversion ditch 分水沟
water diversion structure 分水构筑物，引水构筑物
water diversion 引水
water-diverting structure 分水构筑物，引水构筑物
water divider 配水器
water divide 分水界，分水岭
water division box 分水箱
water drainage 排水
water drain pipe 排水管
water drain valve 疏水阀，放水阀
water drain 排水，排水沟，放水，疏水，下水
water draw-off rate 取水流量
water drip cooler 水淋冷却器，喷淋式冷却器
water drip 滴水盘，披水屋檐
water-driven bell 水力警铃
water droplet 水滴，水点
water dropper 水滴集电器【测定空中电位陡度】
water drop 水滴
water drum 水包，水鼓，下汽包，下锅筒，水锅筒
water and dust-proof lamp 防水防尘灯
water ebullition 汽化，水沸腾
water economics 水利经济学
water eddy force 旋涡力
water edge 水边线，水抹线
water ejection pump 射水泵
water ejection tank 射水箱
water ejector 水喷射器，水喷射泵
water-electricity nuclear power plant 海水淡化发电用核电站
water elevation 水平面高度
water elevator 扬水器
water emptying system 放空水系统
water emptying 放空水
water engine 水力发动机
water entrainment 夹带水
water equivalent 水当量
water erosion 水蚀，水力侵蚀
water escape 弃水道
water evaporation cooling 水蒸发冷却
water examination 水质检验
water expansion anchor bar 水胀式锚杆
water expense for power 发电用水费
water-extracting medium 水萃取介质
water extraction structure 取水结构
water extraction 抽水
water extractor 水分离器，干燥器
water face （坝的）迎水面
waterfall erosion 瀑布侵蚀，跌水侵蚀
waterfall line 瀑布线
waterfall 瀑布
water feedback effect 水反馈效应
water feeder 给水器
water-filled fracture 充水裂缝
water-filled 充水的
water filling system 充水系统
water filling test 充水试验，注水试验
water filling 充水
water film coefficient 水膜系数
water film cyclone 水膜除尘器
water film separator 水膜除尘器
water film test 水膜试验
water film 水膜
water filter 滤水器
water filtration 水的过滤
water flow alarm system 水流量警报系统
water flow circuit 水循环回路
water flowmeter 水流量计，水流计，水表
water flow regulator 水流量调节器
water flow relay 水流继电器
water flow retardation 拦滞水流
water flow retarding structure 滞洪建筑物，缓流建筑物
water flow 水流
water flush boring 水冲钻探
water flush drilling 水冲钻探
water flush face 水冲面
water flushing and waste water treatment 水冲洗和废水处理
water flushing system 水冲洗系统
water flushing 水冲洗
water flush 水冲，水泛溢
water flux 水通量
water-free 无水的，不含水的
water front structure 驳岸构筑物，驳岸工程，临水构筑物，海岸构筑物，沿岸构筑物
water front 前沿水域，水边线，岸线
water funnel 水斗
water gain 泛水，泌水
water gap between fuel assemblies 燃料组件间水隙
water gap 水隙【中子物理学】，水口
water gas pipe 水煤气管
water gas shift 水煤气转化
water gas steel pipe 水煤气钢管
water gas welding 水煤气焊接
water gas 水煤气
water gate 水门，水闸
water gauge glass 水表玻璃
water gauge level 水尺高程
water gauge station 水位观测站
water gauge 水标尺，水表，水尺，水量计，水位标尺，水位表，水柱压力计
water gel 水胶炸药
water glass 水玻璃，硅酸钠，水面计玻璃
water-granulated slag 水淬粒状矿渣
water grass 水草
water guide blade 导水叶
water guide mechanism 导水机构
water gun 水枪
water gushing （基坑）涌水量
water hammer action 水锤作用
water hammer analysis 水冲击作用分析，水锤

分析
water hammer arrester 水锤防护装置，水锤消除器
water hammer of heat-supply system 供热系统水击
water hammer pressure 水锤压力
water hammer pump 水锤泵
water hammer test 水锤试验
water hammer wave 水锤波
water hammer 水锤，水击，水冲，水力冲击
water hardening 水淬，用水淬火
water hardness 水的硬度，水质硬度
water head accumulator 液压蓄能器
water head at weir crest 堰顶水头
water header 水流集管，集水池，水联箱
water head 水头，水柱压头，水位差，落差
water heater 水加热器
water heating bath 水加热槽
water heating 热水供暖，水暖
water height 水深
water hemisphere 水半球
water holding capacity 保水能力，持水量，持水能力，含水容量
water holding material 含水材料
water holding power 持水力
water holding 保水，挡水，持水
water hold-up tank 水滞留箱
water hole peaking 水隙中的峰值
water hole 水坑，小池塘
water hose 水龙带
water-hydrogen-hydrogen cooled generator 水氢氢冷发电机
water hydrological cycle 水循环
water hygiene control 水质卫生控制
water immersion test 水浸试验
water immersion wire 潜水电线
water impoundment 蓄水
water in deep layer 深层水
water induction prevention control 防进水控制
water induction 进水，汽机疏水倒流
water inflow （基坑）涌水量，流入水量，来水量
watering can 喷水壶
water ingress 水浸入【废物库或堆芯】
water injection augmentation 喷水增压【压缩机】，喷水加力【压气机】
water injection test hole 注水试验孔
water injection 喷水，注水
water injector 水射器
water inlet 进水口
water inrush 来水，入流，水流涌入
water intake facility 取水设施
water intake on scaffold bridge 栈桥式进水口
water intake pipe 取水管
water intake pumping house 取水泵房
water intake pumping station 取水泵站
water intake structure 取水构筑物，取水结构，进水口结构
water intake tunnel 引水隧道
water intake with filter gallery 渗滤廊道式进水口
water intake 进水口，进水枪，取水口
water interception 水流截断，断流
water inventory 水量【堆冷却剂系统】，水总量
water jacket cooling 水套冷却
water-jacketed condenser 水套凝汽器
water-jacketed motor 水套电动机
water jacket safety valve 水套安全阀
water jacket 水套，水箱
water jet air-ejector 射水抽气器
water jet arrester 水柱避雷器
water jet blower 水喷射器，喷水器
water jet condenser 喷水凝汽器
water jet cutting 水射流切割
water jet driver 射水打桩机
water jet dust absorber 喷雾吸尘器
water jet ejector 喷水器
water jet exhauster 水力引射器
water jet nozzle 水喷射管，水喷嘴
water jet pile-driving 射水打桩法，冲水打桩法
water jet pond 射水池
water jet pump 水力喷射泵，喷水泵，射水泵
water jet suction apparatus 射水抽气器
water jetting at high pressure 高压射水
water jetting process 喷水法
water jetting 水冲洗，射水法
water jet 射水，喷水口，喷水枪，喷射水流
water knockout drum 分水器
water lance 喷水枪
water lancing cleaning 水枪清洗
water law 水法
water leakage in condenser 漏入凝汽器的水
water leakage 漏水
water leg 水夹套【小型锅炉下部】
water-level alarm 水位报警
water level-capacity curve 水位容量（关系）曲线
water level controller 水位控制器
water level control valve 水位控制阀
water level control 水位控制，水位调节
water level duration curve 水位持续曲线，水位历时曲线
water level float 水位指示浮子
water level fluctuation 水面波动，水位波动，水位变动，水位涨落
water level frequency curve 水位频率曲线
water level frequency 水位频率
water level gauge 水位表，水位计
water level hydrograph 水位过程线
water level indicator 水位指示器，水位计，水位表
water level instrumentation 水位监测仪表
water level measuring post 水位尺，水位站
water level observation 水位观测
water level reactivity coefficient 水位反应性系数
water level receiver 水位传感器
water level recorder 水位记录器
water level regulation 水位调整
water level sensor 水位传感器
water level staff 水准尺，水尺
water level 水平面，水平面高度，水位
water-lifting device 提水设备

water-like organic chemicals 亲水有机药品
water lime 熟石灰
water line attack 水线腐蚀
water line corrosion 水线腐蚀
water line 吃水线，水管，水位线，水线
water load power meter 水负载功率计
water load 水负载，水荷载，水力荷载
water lock 存水弯，水封
waterlogged farmland 水涝地
waterlogged ground 水涝地
waterlogged land 水浸地
waterlogged location 水浸塔位
waterlogged soil 渍水土壤
waterlogged 水涝的，渍涝的
waterlogging control 除涝，防涝
waterlogging disaster 涝灾
waterlogging effect 进水效应
waterlogging 积涝，内涝，水涝，水渗，水浸，积水，进水效应
water loss and soil erosion 水土流失
water loss 水量损失
water-loving 亲水的
water lowering 降低水位
water-lubricated bearing 水润滑轴承
water lubrication 水润滑
water main 总水管，给水总管，水母管，供水总管
water makeup 补给水，补水
water management 水利管理，水务管理
water mangle 脱水机
water mark 水迹，水量标，水位标，水位标志，水渍
water mass continuity 水体的连续性
water meadow 漫洪草原，漫水草地，淹水草甸
water-measuring device 量水设备
water-membrane type thermal deaerator 水膜式热力除氧器
water meter chamber 水表井
water metering law 水计量法
water metering 用水计量
water meter load factor 水表记录负载系数
water meter 水量计，水表
water mill 水力磨坊
water-mixing direct connection 混水连接
water moderated reactor 水衰减反应堆，水慢化堆，水慢化反应堆
water-moderated water-cooled reactor 水冷水慢化反应堆
water moderator 水慢化剂
water monitoring 水质监测
water monitor 水位监视器，水位控制器
water motor 水力发动机
water movement 水流运动
water nozzle 喷水嘴
water of constitution 化合水，结构水，组成水
water of crystallization 结晶水
water of hydration 水合水，水化水，化合水
water outcrop 水流出露口
water outlet valve 出水阀，出水口阀门
water outlet 出口水，出水口
water particle 水质点
water parting line 分水线
water parting 流域分界线，分水岭
water penetration 渗水深度，渗水
water-percolating capacity 渗水力
water percolation 渗水，水渗透
water permeability test 透水试验
water permeability 透水性
water phantom 充水体模
water pipe grounding system 水管接地制
water pipe line 水管线路
water pipe 水管
water pit 水池，水坑
water plane 水平面
water planning 水利规划
water plant 自来水厂
water plug 水塞
water pocket （锅炉蒸汽侧管线里的）积水【会导致水击】，袋形水团，（过热器中）水囊，水塞
water policy 水利政策
water pollution control plant 水质污染处理厂
water pollution control 水污染控制，水质污染控制
water pollution index 水污染指数
water pollution monitor 水污染监测仪
water pollution 水污染，水质污染，水质污浊
water pond 储水箱，贮水池
water pool 水池
water potential 水势
water power development 水力开发，水电开发，水力发电开发
water power machinery 水力机械
water power map 水力资源分布图
water power plant 水力发电厂，水电站，水电厂
water power project 水电工程，水力发电工程
water power resources 水力资源，水电资源，水能资源
water power right 水能权
water power station 水力发电厂，水电站，水力发电站
water power survey 水力调查，水力勘察
water power utilization 水能利用
water power 水力发电，水电，水能
water pressure gauge 水压计
water pressure relay 水压继电器
water pressure test hole 压水试验孔
water pressure test in hole 钻孔压水试验
water pressure test 水压试验，压水试验
water pressure 水压，水压力
water pretreatment 水预处理
water processing 水处理
water project 水利工程
waterproof agent 防水剂
waterproof belt 耐水皮带
waterproof carpet 防水毡
waterproof case 防水外壳
waterproof cement 防水水泥
waterproof coating 防水涂层
water-proof compound 防水粉
water-proof concrete 防水混凝土

water-proof course 防水层
water-proof electrical equipment 防水型电气设备
water proofer 防水材料，防水层，防水布
waterproof expansion cement 防水膨胀水泥
waterproof fluorescent lighting fixture 防水荧光灯
waterproofing admixture 防水剂
waterproofing agent 防水剂
waterproofing compound 防水化合物，防水剂【混凝土】
waterproofing layer 防水层
waterproofing material 防水材料，防水剂
waterproofing membrance 防水层【基础】，防水膜，防水薄膜
waterproofing powder 防水粉
waterproofing upstream face 上游防水面
waterproofing works 防水工程
waterproof instrument 防水型仪表
waterproof insulation 防水绝缘
waterproof machine 防水型电机
waterproof membrane 防水隔膜，防水隔墙
waterproof mortar 防水砂浆
waterproof motor 防水型电动机
waterproof of roofs 屋面防水
waterproof packing 防水包装
waterproof paint 防水漆
waterproof paper packing 防潮纸包装
waterproof paper 防水纸
waterproof socket 防水插座
waterproof structure 防水结构
waterproof test 防水试验，防水性试验
waterproof type induction motor 防水型感应电动机
waterproof wax 防水蜡
waterproof wire 防水线，耐水线，防潮线
waterproof 防水的，不透水的，耐水的，水密封的，防湿的，防潮的，防水衣物【复数】，防水布，油布，防水物，防水材料，使防水，使不透水
water property 水的性质
water prospect map 水源勘探图，水源图
water protection 水的保护
water pump bearing 水泵轴承
water pumped storage motor 水泵蓄能电机
water pump house 水泵房
water pumping windmill 抽水风车
water pump shaft 水泵轴
water pump 抽水机，抽水泵，水泵
water purification plant 净水厂，水净化厂
water purification structure 净水构筑物
water purification system 水净化系统
water purification works 净水工厂
water purification 水的净化，水净化
water purifier 净水器，水软化装置
water purifying apparatus 净水设备
water purifying tank 净水槽
water quality analysis 水质分析
water quality change 水质变化
water quality conservation 水质保持
water quality control 水质控制
water quality criteria 水质标准，水质基准，水质判据
water quality evaluation 水质评价
water quality examination 水质检查
water quality forecasting 水质预报
water quality goal 水质目标
water quality management 水质管理
water quality model 水质模型
water quality monitoring apparatus 水质监视计
water quality monitoring system 水质监测系统
water quality monitoring 水质监测
water quality observation 水质观测
water quality pollutant 水质污染物
water quality standard 水质标准
water quality surveillance 水质监测
water quality test 水质试验
water quality 水质
water quenching 水淬火
water radioactivity meter 水放射性测量计
water radiolysis 水辐照分解
water raiser 提水机，扬水机
water raising capacity 扬水能力，扬水高程
water raising engine 提水机，提水机具
water raising machinery 提水机械
water raising machine 扬水机
water ram 水锤扬水机，水力夯锤
water rate 水费，水价，用水率
water reactor 水反应堆，水冷反应堆
water recirculation 水再循环
water recovery apparatus 水回收设备
water recovery condenser 回水凝汽器
water recovery pond 回收水池
water recovery 水回收
water recycle system 水再循环系统
water recycle 水再循环
water recycling 水的重复利用，水的循环
water-reducing admixture 减水外加剂，减水添加剂
water-reducing agent 减水剂
water-reducing retarder 减水缓凝剂
water-reflected system 水反射系统
water-reflected 水反射的，有水反射层的
water regime （土壤的）水分状况，水文情势，水情
water regulator 水量调节器，控水器
water release from reservoir 水库放水，水库泄水
water release pipe 放水管
water release structure 放水建筑，泄水构筑物
water release system 排水系统
water release tunnel 放水隧洞
water release works 放水建筑物
water release 放水，泄水
water removal 除水，去水，脱水
water renovation 水再生，水质改进
water-repellant 防水的，憎水的，疏水的，防潮的，防水剂，憎水剂
water-repellent additive 憎水性外加剂
water-repellent admixture 憎水剂
water-repellent cement 防水水泥，憎水水泥，憎水性水泥
water-repellent concrete 憎水性混凝土

water-repellent envelope 防水外壳
water-repellent film 防水膜，抗水膜
water-repellent varnish 抗水漆，防水漆，防潮漆
water-repelling agent 憎水剂
water requirement 需水量，水分要求
water reservoir 蓄水池，贮水槽
water resistance load tank 水阻负载箱
water resistance 水电阻，水阻力，水阻
water-resistant magnet wire 耐水电磁线
water-resistant paint 防水漆
water-resistant solution 防水浆
water-resisting layer 隔水层
water resistor 水电阻器
water resources assessment 水资源评估
water resources conservation 水资源保护
water resources development 水利资源开发
water resources evaluation system 水资源评价系统
water resources management 水资源管理
water resources plan 水资源规划
water resources policy 水资源政策
water resources research 水资源研究
water resources system planning 水资源系统规划
water resources system 水利资源系统，水利工程系统
water resources 水利资源，水资源
water retaining agent 保水剂【混凝土】
water retaining capacity 保水量，保水能力，水分保持量
water retaining dam 挡水坝，拦水坝
water retaining dike 挡水堤
water retaining facing 挡水面板
water retaining power house 河床式厂房
water retaining property 保水性
water retaining structure 挡水构筑物，拦水构筑物，挡水结构
water retaining type hydroelectric station 河床式水电站
water retaining wall 挡水墙
water retaining 挡水，拦水，保水，固水，挡水的，含水的
water-retentive cement 保水水泥
water retentiveness 保水性，持水能力
water retentivity 保水能力，水分保持量
water return pipe 回水管
water return system 回水系统
water return valve 回水阀
water return 回归水，回水
water reuse 水的反复利用，水的重复利用，水再生
water rheostat 水电阻器，水变阻器
water-rich stage 丰水期
water ring pump 水环泵
water ring vacuum pump 水环式真空泵，水环真空泵
water riser 给水立管
water roll 水辊，水滚
water route 水路
water runoff treatment facility 洗选设施
water sampler 水取样器，采水样器
water sample 水样
water sampling hole 取水样孔
water sampling 采水样，取水样，水样采集
water saturation 水饱和，含水饱和度
water saving pressure regulator 节水式压力调节器，回复式减压阀，省水式压力调节器
water saving 节水
water scraper 刮雨水装置
water screen header 水冷壁联箱
water screen 水滤网，水幕，水帘管，水帘
water screw 推水螺旋桨
water scrubber 水洗涤器
water seal cage 水封环
water-seal chamber 水封室
water-sealed gland 水封【汽轮机】水封套，水封装置
water-sealed joint 存水缝，存水接头，止水接头
water-sealed packing 水封装置
water-sealed rotary gas meter 防水旋转流量计，潮湿气体流量计
water-sealed valve 水封阀
water-sealed 水封的
water seal tank 水封箱
water seal 水封【防止空气漏入炉膛】，安全水封
water seepage 水渗透，渗水
water segregating slurry pump house 水隔离灰浆泵房
water segregating slurry pump 水隔离灰浆泵
water separator 汽水分离器，脱水箱，汽水分离装置
water service building 水处理厂房
water service 供水
watershed area 分水岭地区，分水界域，集水面积，流域面积
watershed boundary line 流域界线
watershed boundary 流域边界，流域界
watershed dam 源头坝
watershed divide 分水岭
watershed elevation 分水岭高程
watershed factor 流域因素
watershed form 流域形状
watershed lag 流域集水滞后时间，流域滞时
watershed leakage 流域渗漏
watershed line 分水界，分水岭，分水线
watershed management 流域管理
watershed map 流域图
watershed model 流域模型
watershed planning 流域规划
watershed protection and flood prevention project 流域保护及防洪工程
watershed sanitation 流域环境卫生
watershed slope 流域坡降
watershed variable 流域变数
watershed 分水界，分水岭，集水区，流域
water shield 水屏蔽
water shock 水流冲击，水激波
water shortage 水量不足，缺水
water-short land 旱地，缺水土壤
water side corrosion 水侧腐蚀

water side deposition 水侧沉积，水侧结垢
water side land 堤外土地，水边地
water side pump house 岸边泵房
water side vacuum pump 水侧真空泵
waterside 岸，海滨，水侧
water sink 渗水坑
water slaked lime 水解石灰
water slope 水坡式升船机
water slug （新蒸汽管道中）水堵塞，水冲击，水击，水塞，蒸汽推动水弹水锤
water sluice project 水闸工程
water sluice 水力除渣，水闸
water-sluicing system （低压头）水力除灰系统
water slurry 水悬浮物，水淤泥，渣水混合物
water smoke 蒸汽雾
water-soaked method 水浸法
water-soaked 被水浸透的
water soaking 水浸，水力除灰，浸水现象
water softener 软水器，软水剂，水质软化剂，水质软化器
water softening agent 硬水软化剂
water softening plant 软水装置
water softening 水的软化，水软化，水软化处理，硬水软化
water-solid ratio 水固比，水灰比
water-solubility 水溶性
water-soluble halogen 水溶性卤素
water-soluble inhibitor 水溶性缓蚀剂
water-soluble preservative 水溶性防腐剂
water-soluble resin 水溶性树脂
water-soluble salt 水溶性盐类
water-soluble solid 水溶固形物
water-soluble 可溶于水的，溶于水的
water-solution enriched reactor 浓缩铀水溶液反应堆，浓缩铀水溶液堆
water soot blower 水吹灰器
water sources survey 水源调查
water source 水源
water space 储水空间，水喷雾器，排水装置，水域
water specification 水质指标
water splitter 分水器【闸墩】
water spouted bed 喷水床
waterspout 水龙卷
water spray cone 水雾锥
water spray cooling 喷水冷却
water spray dust suppression system 水雾除尘系统
water sprayer 淋水装置，水喷雾器，洒水器
water spray fire extinguisher 喷雾式灭火器
water spray fixed system 固定水喷雾灭火系统
water spraying 洒水
water spray plunger-type slurry pump 喷水柱塞式灰浆泵
water spray projector system 喷水装置，水喷雾喷射器系统
water spray scrubber 喷水冲洗器
water spray separator 喷雾式除尘器
water spray system 喷雾洒水系统，救火栓系统，水喷雾系统
water spray tank 喷雾水箱，贮水箱【消防系统】
water spray 洒水，喷水，水喷雾
water spreading system 喷水系统
water spreading 漫流，洒水
water sprinkling tray 淋水盘
water staff 水位标尺，水位尺
water stage discharge curve 水位流量关系曲线
water stage recorder 水位记录器
water stage record 水位记录
water stage transmitter 水位自动记录仪，水位变送器
water stain 水渍
water stake 量水尺
water standard 水质标准
water stand 停潮
water-starved 缺水的
water-steam cycle 汽水循环
water-steam mixture 汽水混合物，水-蒸汽混合物
water-stilling device 消能设备
water stiring system 搅动水系统【灰渣】
water stop flange 止水法兰
waterstop joint 防水密封接头
water stop key 止水键
waterstop strip 止水片，止水条
water stop 止水带，止水，水密装置
water storage basin 蓄水池，贮水池
water-storage capacity 储水量，水侧容积，贮水量
water storage dam 蓄水坝
water storage pond 储水池，蓄水池
water storage reservoir 水库，蓄水库
water storage tank of deaerator 除氧器贮水箱
water storage tank 储水罐，蓄水池，贮水池，贮水箱
water storage works 蓄水工程
water storage 储水量，储水箱，水量储存，蓄水，蓄水量
water storativity 储水率
water strainer 水过滤器，滤水器
water stratification 水的分层
water structures 水构筑物
water subfill 水垫层
water-submerged motor 潜水型电动机
water-submersible electric equipment 潜水电气设备
water supply and drainage system 给排水系统
water supply and sewage 给排水
water supply and sewerage drawing 给排水图纸
water supply and sewerage works 给排水工程，上下水道工程
water supply and waste water systems 供水与废水系统
water supply-drainage discipline 给排水专业
water supply engineering 给水工程
water supply facility island 水工岛
water supply for fire-fighting 消防给水
water supply guarantee rate 供水保证率
water supply header 供水联箱
water supply line 上水道
water supply pipe 给水管，供水管
water supply pump 给水泵，供水泵

water supply rate 供水速率
water supply source 供水水源
water supply system diagram 供水系统图
water supply system for ash handling 除灰供水系统
water supply system for fire prevention 消防给水系统
water supply system for living facilities 生活用水系统
water supply system for process purpose 生产用水系统
water supply system 供水系统，给水系统
water supply tunnel 供水隧洞
water supply valve 补水门
water supply works 供水工程
water supply 供水
water-supporting deck 挡水面板
water surface area 水面面积
water surface elevation 水面高程
water surface fluctuation 水面升降
water surface gradient 水面坡降
water surface profile 水面曲线，水面线
water surface slope 水面比降
water surface undulation 水面起伏
water surface variation 水面变化
water surface 水面，水平面
water surplus 剩余水量，水剩余物
water surveillance network 水监测网
water survey 水的调查，水利勘测
water swelling （直流炉）汽水膨胀
water switch 水压开关
water system survey 水系调查
water system 水系
water table contour plan 地下水面等高线图
water table contour 地下水等高线，地下水等水位线，地下水位等高线
water table depression cone 地下水降落锥
water table gradient 地下水面坡度，地下水位坡降
water table isobath 地下水位等深线
water table outcrop 地下水露头处
water table rise 地下水位上升
water table slope 地下水面比降，地下水面斜率
water table 地下水位，地下水面，潜水面，泄水台，马路边沟，承雨线角
water tagging 示踪测流
water-tamped 水反射的，有水反射层的
water tanker 水罐【消防车】
water tank for heat insulation 隔热水箱
water tank for thermal insulation 隔热水箱
water tank 水箱，水槽，水池，水柜
water technology 水处理技术，水工艺学
water temperature gauge 水温计
water temperature 水温
water tender sink 水池
water tension 水张力
water terminal 港埠，码头
water test 水压试验，液压试验，水检验，水质试验
water thermometer 水温计
water thermostat 恒温水箱
water thrower 喷水器
water-tighness 阻水性
watertight barrier 防水堤，防水墙
watertight box 防渗水箱
watertight building 防水建筑物
watertight bulkhead 水密舱板
watertight cargo space 不漏水的货物空间
watertight compartment 水封室，不透水舱
watertight concrete 不透水混凝土，防水混凝土
watertight connector 水密接头
watertight construction 防水结构
watertight core 不透水心墙
watertight diaphragm 防渗墙，防水层，隔水层
watertight door 气密门
watertight facing arch 防渗护面拱
watertight facing 防渗面层，阻水面层
watertight joint 防水缝，阻水接缝，防水接缝
watertight layer 不透水层，隔水层
watertight machine 不透水型电机，防水型电机
water tightness closure 不透水隔板，水密性封闭
water tightness 不透水性，防水性，阻水性，水封，水密性
watertight screen 阻水帷幕
watertight seal 不透水密封，止水
watertight sleeve 防水套管
watertight stratum 不透水地层
watertight test 水密试验
watertight wall 防水隔膜［隔墙］，不透水墙
watertight 不漏水的，水密的，不透水的，防水，防水的，水密封，水封
water to air cooler 水空气冷却器，水冷空气冷却器
water to air heat pump 水空气热泵
water to air system 水空气系统
water-to-earth ratio 蓄水量与挖方量之比
water tolerance 耐水性，耐水度，水分容许量
water-tolerant solid acid 抗水性固体酸
water-to-steam ratio 水汽比，冷却倍率
water to uranium atomic ratio 水铀原子比
water to uranium mass ratio 水铀质量比
water to uranium volume ratio 水铀体积比
water to water heat exchanger 水水热交换器
water to water system 水水系统
water tower fire truck 举高喷射消防车
water tower 水塔
water trailer 水罐拖车
water transmissibility coefficient 导水系数
water transmissibility 导水性
water transportation 水运
water transport 水路运输
water trap 存水湾，阻水排汽罐，水封，水闸
water-treating chemicals 水处理药品
water-treating equipment 水处理设备
water-treating plant 水处理装置
water-treating system 水处理系统
water treatment building 水处理厂房
water treatment facility 水处理设施
water treatment plant waste 水处理厂废水
water treatment plant 水处理厂，水处理装置，水处理车间

water treatment room 水处理室
water treatment system diagram 锅炉补给水处理系统图
water treatment system 水处理系统，水处理工艺流程
water treatment 软水处理，水处理
water tree 水树枝【绝缘劣化】，水树，水树图
water tube boiler 水管式锅炉
water tube closed heater 水表面式加热器【水在管内流动】
water tube condenser 水管冷凝器
water tube grate 用水管组成的炉排，水冷炉排
water tunnel 排水隧洞，排水通道，水洞，水工隧洞，引水隧道
water turbidity 水的混浊度
water turbine engine 水轮机
water turbine floor 水轮机层
water turbine generator set 水轮发电机组
water turbine generator 水轮发电机
water turbine governor 水轮机调速器
water turbine pump 水轮泵
water turbine regulation system 水轮机调节系统，水轮机调速系统
water turbine 水轮机，水力涡轮机
water turbo-generator 水轮发电机
water uptake 水的吸收，吸水，吸水量
water up to the flange seal surface 水上升到法兰密封表面
water up to vessel midplane 水上升到压力壳中平面【环带线】
water-uranium volume ratio 水铀体积比
water use efficiency 用水效率
water user 用水户，水用户
water use 水的利用，用水
water utilization 水的利用，用水
water value of Engler viscosimeter 恩氏黏度计水值
water valve gland 水阀压盖
water vapor absorption 水蒸气吸收
water vapor arc welding 水蒸气保护电弧焊
water vapor concentration 水蒸气提浓法
water vapor content 水汽含量
water vapor permeability 水蒸气透过性
water vapor tension 水汽张力
water vapor transport 水汽输送
water vapour capacity 水汽含量
water vapour pressure 水汽压力
water vapour saturation 水汽饱和
water vapour tension 水汽张力
water vapour transfer 水汽输送
water vapour transport 水汽输送
water vapour 水汽，水蒸气
water velocity 水流速度，水速
water void ratio 水隙比
water voltameter 水解电量计
water volume 水量，水容量
water wagon 运水车
waterwall bed plate 水管冷却布风板【沸腾炉】
water wall circuit 水冷壁回路
water wall cooling 水冷壁冷却
water wall furnace 水冷壁炉膛
water wall header 水冷壁联箱
water wall panel 水冷壁（管）屏
water wall surface 水冷壁面
water wall tube 水冷壁管
water wall 水冷壁
water-washable penetrant 可水洗渗透剂
water washing 用水清洗，水力冲洗
water-wash test 水洗试验
water-wasting pressure regulator 敞开式减压阀，弃水式压力调节器
water-water cooler 水水冷却器
water-water jet 混水器
water-water reactor 水水反应堆，水水堆
water-water type heat exchanger 水水（式）换热器，水水（式）热交换器
water wave analogy 水波模拟
water wave 水波
water way coal transportation 水路输煤
waterway engineering 水道工程
waterway stabilisation structure 水道加固圬筑物
waterway transportation 水路运输
waterway transport 水运
waterway tunnel 水道隧洞，过水隧洞
waterway 航道，水道，排水沟，航路，水路，排水渠，出水口
water-wet sand 亲水砂
water wheel generator 水轮发电机
water wheel 水车，水轮
water wing 挡水翼墙
water wiper 刮水器
water works 自来水厂，自来水工程
waterworn 被水冲击的
water yearbook 水文年鉴
water year 水文年
water yield improvement 出水量的增加
water-yielding stratum 出水层，供水地层
water yield of river basin 流域总出水量
water yield 产水量，出水量
water 水体，洒水，水，给……浇水，供水
wattage dissipation 功率耗散，损耗瓦数，功率损耗
wattage output 输出瓦数，输出功率
wattage rating 额定功率，额定瓦（特）数
wattage transformer 小功率变压器，小型变压器
wattage 瓦数
watt component 有功分量，有效分量，有功部分，有效部分
watt consumption 功率消耗
watt current 有功电流，有效电流
watt distribution 瓦特分布
wattenschlick 潮泥
wattful component 有功部分，有功分量
wattful current 有功电流，有效电流
wattful loss 有功损耗，欧姆损失，电阻损耗
wattful power 有功功率
wattful 有功的
watt governor 电力调节器，瓦特调速器，飞球调速器，机械弹簧和飞锤式调速器
watt-hour capacity 瓦时容量【蓄电池】
watt-hour constant 瓦时常数
watt-hour consumption 瓦时消耗量

watt-hour demand meter 最大需用瓦时计
watt hour efficiency 瓦时效率
watt-hour meter 瓦时计，电度表，电能表，电表
watt-hour 瓦时，瓦特小时【电功率或电能的一种单位】
watt in 吸收功率，输入功率
wattless component meter 无功分量表，无功电度表，乏时计
wattless component watt-hour meter 无功电度表
wattless component 无功分量，无功部分，虚数部分
wattless current 无功电流
wattless power meter 无功功率计
wattless power 无功功率
wattless 无功的
wattle work 河岸防冲拦淤栅
wattle 编条
wattling 枝条编扎
watt loss 功率损耗
wattmeter 瓦特计，瓦特表，功率表，电（力）表
wattmetric relay 瓦特计式继电器
watt out 输出功率
watt per candle 瓦/烛光
watts active power 瓦茨有功功率
watt-second 瓦秒
watt's fission spectrum 瓦特裂变谱
watt's governor 瓦特调速器，飞球调速器
watts peak 峰瓦
watt transducer 电功率传感器，电功率转换器
watt 瓦，瓦特
wave abrasion 波蚀
wave absorber 电波吸收体，吸浪器，削波器
wave absorbing revetment 消波护岸
wave-absorbing structure 消波结构
wave acoustics 波动声学
wave action 波浪作用，激活作用
wave-activated generator 波浪发电机
wave advance 波浪推进
wave age 波龄
wave agitation 波浪激动
wave amplitude 波浪振幅，波幅
wave analyzer 波形分析器，谐波分析器，波浪分析器
wave angle 电波仰角，波程角
wave antenna 行波天线
wave attack 浪蚀
wave attenuation 波的衰减，波浪衰减
wave band 波段，频带，波底，浪蚀基面
wave base 波浪酌的限界深度，浪基面，波底，波浪基部
wave basin 波浪试验港池
wave bottom 波谷
wave breaker block 防波块体，消浪块体，消波块体
wave breaker 方波堤，防波堤，破浪器，消波器
wave breaking condition 波浪破碎条件
wave breaking point 波浪破碎带，波浪破碎点
wave breaking works 消波工程
wave breaking 波浪破碎
wave-built platform 浪成平台，浪积平台
wave-built structure 波成构造
wave-built terrace 波成台地，浪成阶地，浪积阶地
wave-built 浪成的
wave carrier communication 载波通信
wave carrier 载波
wave changing switch 波段转换开关，波长转换开关
wave channel 波道
wave characteristics 波浪特征
wave chart 波浪图
wave climate 波候
wave clutter 波浪回波干扰
wave coil 波绕线圈
wave component 波组成
wave compression 波的压缩
wave constant 波长常数
wave converter 波形变换器
wave crest length 波峰长度
wave crest 波峰，波顶
wave current 波动电流，波流，波浪流
wave curtain 防波帷幕
wave cut bench 波蚀台地，浪蚀台，浪蚀台地
wave cut cliff 浪蚀崖
wave cut notch 浪蚀洞，海蚀龛，波蚀凹壁
wave cut plain 浪蚀平原
wave cut platform 浪蚀平台，浪蚀台
wave cut terrace 浪蚀阶地
wave cut 波蚀，浪蚀
wave cyclone 波动性气旋
wave damping 波浪衰减
wave data 波浪资料
wave decay 波浪衰减
wave delta 浪成三角洲
wave depression 锋面低压
wave detector 检波器
wave differential equation 波动微分方程
wave diffraction 波浪绕射，波浪衍射
wave directional spectrum 波向谱
wave direction meter 波向计，波向仪
wave direction 波向
wave director 导波体，波导
wave dispersion 波的弥散，波速分散
wave dissipating concrete block 消波混凝土块体
wave distortion 波形畸变
wave disturbance 波扰动，波状扰动，波浪辐散
wave drag 波阻，波阻力，波浪曳力
waved tube 波形管
wave duration 波动历时
waved wire 波形银丝
wave dynamometer 波力计
wave energy air turbine 波能空气涡轮机
wave energy coefficient 波能系数
wave energy extractor 波能发电站
wave energy generator 波能发电机
wave energy power plant 波能发电厂
wave energy 波浪能，波能
wave envelope 波包络
wave equation 波动方程
wave erosion base level 浪蚀基面
wave erosion 波蚀，波浪侵蚀，浪蚀
wave-etched shoreline 浪蚀海岸

wave fence 挡浪栅
wave fetch 波浪行程
wave field 波场
wave filter device 滤水设备
wave filter 滤波器
wave flow 波形流
wave force on pile 桩上波压力
wave force 波浪力，波力
wave forecasting 波浪预报
waveform converter 波形变换器
waveform distortion 波形畸变，波形失真
waveform pulse 波形脉冲
waveform quality 波形质址
waveform recorder 波形记录器
waveform separation 波形分离，波形区分
waveform space 信号空间
waveform synthesizer 波形综合器
waveform test 波形试验
wave frequency 波频，波导频率
wave front angle 波前倾角，波阵面角
wave front matching 波前匹配，波阵面匹配
wave front method 波锋法
wave front reconstruction 波前再现
wave front steepness 脉冲前沿斜率
wave front technique 波阵面技术
wave front 波前，波阵面，冲击波头，波锋，波状面
wave function 波动函数，波函数
wave furrow 波蚀沟
wave gauge 波浪计，测波仪
wave generated current 波生流
wave generating apparatus 波浪发生器
wave generating area 波浪发生区
wave generation 波浪生成
wave generator 波形信号发生器，波形发生器
wave group 波群，波组
waveguide branching filter 波导管分支滤波器
waveguide communication 波导通信
waveguide corner 波导弯头，波导弯角，折波导
waveguide filter 波导管滤波器，波导滤波器
waveguide gasket 波导垫片
waveguide impedance 波导阻抗
waveguide junction 波导管结
waveguide kit 波导卡子
waveguide klystron 波导型调速管
waveguide-magnetron 波导磁控管
waveguide modulator 波导式调制器
waveguide post 波导棒
waveguide reflector 波导反射器
waveguide shim 波导垫圈
waveguide storage 波导存储器，波导管存储器
waveguide switch 波导转换开关，波导转换器，波导开关
waveguide window 波导窗
waveguide 波导，波导管
wave heater 电磁波加热器
wave height coefficient 波高系数
wave height forecast 波高预报
wave height meter 波高仪
wave height record 波高记录
wave height spectrum 波高谱
wave height 波高，浪高
wave hindcasting 波浪后报，波浪追算
wave hollow 波谷
wave impedance 波阻抗
wave-induced current 波生流
wave interference 波干扰，波干涉，波浪干涉
wave launcher 射波器，电波发射器
wavelength shifter 波长转换器
wavelength spectrometer 波长分光仪
wavelength 波长
waveless 平静的，无波的
wavelet 小波，子波，成分波，弱冲波，扰动波，小浪，微波
wave level 波动平面
wave lift 波浪浮托力
wave-like stratification 波状层理
wave line 电波传播方向，波形线
wave loop 波腹
wave maker 波浪发生器
wave manometer 波压计
wave mark 波痕
wave mass 波质量
wave measuring buoy 测波浮标
wave measuring instrument 波浪测量仪器
wave mechanics 波动力学
wavemeter 波长计，波频计，测波仪
wave mode 波型
wave motion 波动，波浪运动
wave motor 波力发动机
wave node 波节
wave normal 波法线，波面法线
wave number 波的数目，波数
wave observation radar 测波雷达
wave observation tower 波浪观测塔
wave observation 波浪观测
wave of infinitely small amplitude 无限小振幅波
wave of oscillation 摆动波，振荡波
wave of translation 平移波
wave origin 波源
wave overtopping 越浪
wave packet 波包，波束，波群
wave path 波径，电波传播路径
wave pattern 波型，波谱，波形图
wave peak 波峰
wave period 波动周期，波周期，波浪周期
wave phase 波相
wave pickup 检波器
wave planation 波浪均夷作用
wave pole 测波杆
wave potential energy 波浪势能
wave power breakwater 拦浪发电站
wave power device 波浪发电装置
wave-powered buoy 波浪发电浮筒
wave power generation 波浪发电，波浪能发电，波力发电，海浪发电
wave power generator 波浪发电设备［装置］
wave power plant 波力发电厂
wave power station 波浪能电站，波力电站，波力发电站，海浪电站
wave power 波浪发电，海浪发电，波浪能，波力

wave pressure diagram 波压图
wave pressure equation 波压公式
wave pressure formula 波压公式
wave pressure meter 波压计
wave pressure 波压力，波压，浪压力
wave process 波动过程
wave profile 波剖面，波形
wave propagation 波传播，波的传播
wave protection works 防浪工程
wave radiation 电波辐射，波辐射
wave range switch 波段开关
wave range 波段
wave ray 波射线，波向线
wave rear 波后
wave reflection 波反射，波浪反射
wave refraction diagram 波浪折射图
wave refraction 波浪折射
wave resistance 波浪阻力，兴波阻力
wave resonator 波浪共振器
wave ridge 波峰，波脊，波颠
wave ripple mark 波涟痕，波痕
wave rose 波浪（玫瑰）图
wave run-up 波浪爬高
waver 波段开关，波形转换器
wave scale 浪级，波级，波浪等级
wave screen 防波栅
wave selector 波型选择器，波段开关
wave separator 电波分离器，分波器
wave shape distortion 波形畸变
wave shaper 波形形成器
wave shape 波形
waveshaping circuit 整形电路
wave simulation 波浪模拟
wave slap 浪拍
wave spectrum 波谱，浪谱
wave splitter 防波堤
waves superposition 波的叠加
wave staff 波高标尺，测波标杆
wave steepness 波的陡度，波陡，波浪陡度
wave-straightened coast 浪成平直海岸
wave stress 波应力
wave surface 波面
wave system 波系，波族
wave tail （冲击）波尾
wave telemetry 电波遥测，电波遥测术
wave theory 波动理论，波动说，波浪理论
wave tilt 波前倾斜
wave time history 波浪时间历程，波浪时间历程曲线
wave train frequency 波列频率
wave train 波列，进行波，波序，同向波族
wave trapper 阻波器，陷波器，陷波电路，防波阱
wave trough 波槽，波谷
wave type compensator 波纹管补偿器
wave type vibration 波形振动
wave variability 波的变化性
wave vector 波矢量，波向量，波矢
wave velocity 波速
wave voltage 波电压
wave wall 坝顶防浪墙，防浪墙【水电】
wave washer 波纹垫圈
wave wash lever 防波堤
wave wash 浪蚀
wave width 波宽
wave winding 波绕组，波形绕法，波形绕组
wave with horizontal crest 平顶波
wave-wound coil 波绕线圈
wave-wound 波绕
wave zone 波带区，波浪区
wave 波，波动，波浪
waviness 波纹，波形，波纹度
wavy arch 波形拱
wavy boundary 波形边界
wavy fin 波纹状肋片
wavy flow 波状流
wavy grain 波形纹理
wavy line 波形线
wavy spacer 波状隔板，波形垫片
wavy texture 波状结构
wavy water surface 波状水面
wavy 波形的，波状的
wax cloth 蜡布
waxed cotton-covered wire 涂蜡沙包线
waxing 打蜡
wax-sealed sample 蜡封土样【测定土的容重试验方法】，蜡封样
wax 蜡，石蜡，涂蜡
way base management system 方法库管理系统
way base 方法库
way bill 托运单，货运单
wayleave agreement 通行权协议
wayleave charge 免费使用电话权
wayleave corridor 通行权地带
wayleave 通行道，通行费，道路通行权，土地通行权，通行权
way of payment 支付方式
way-side inductor 侧向感应线圈
wayside rail 道旁铁轨
way station 中途站，（铁路或公共汽车线路上的）小站
5-way valve manifold 五阀组
3-way valve 三通阀
Way-Wigner correlation 韦维格纳关系式
way 道路，距离，方式
WBMS(way base management system) 方法库管理系统
W-bridge W形电桥
WBSA(World Business Strategist Association) 世界商务策划师联合会
WBS(work breakdown structure) 工作结构分解，（项目）工作分解结构
WB(way base) 方法库
WB(World Bank) 世界银行，世行
WC(water closet) 厕所，盥洗间
WDP(welding data package) 焊接数据包
WD(wind direction) 风向
weak absorber approximation 弱吸收体近似
weak acid cation exchange resin 弱酸阳离子交换树脂
weak acid kation exchanger 弱酸性阳离子交换剂

weak acid solution 弱酸溶液
weak acid 弱酸
weak add cation exchanger 弱酸阳离子交换器
weak base anion exchange resin 弱碱阴离子交换树脂
weak base anion exchanger 弱碱性阴离子交换剂
weak base 弱碱
weak bed 弱岩层
weak-caking coal 弱黏煤
weak caustic anion exchanger 弱碱阴离子交换器
weak concrete 低标号混凝土
weak coupling 弱耦合，疏耦合
weak current control 弱电控制
weak current engineering 弱电工程，弱电工程学
weak current insulator 弱电流绝缘子
weak current line 弱电流线路
weak current switch 弱电流开关
weak current 弱电流，弱电
weak earthquake 弱震
weak echo region 弱回波区
weak echo 弱回波
weak eddies 弱旋涡
weakening ratio （磁场）削弱比
weakening 削弱，减弱，衰减，阻尼，消振
weakest link model 最弱链环模型
weak foundation 软弱地基
weak front 弱锋
weak function 弱函数
weak grid 弱电网
weak indication 微弱显示
weak intensity 弱强度
weak interaction regime 弱相互作用状态
weak jump 弱水跃
weak link 弱环
weak liquor 淡溶液，淡碱液
weak low pressure 弱低压
weakly active components store 低放射性部件贮存库
weakly basic anion exchange 弱碱阴离子交换
weakly caking coal 弱黏煤
weakly coupled system 弱耦合系统
weakly ionized gas 弱电离气体
weakly magnetic ball 弱磁性小球
weak magnetic substance 弱磁性物质
weak market 疲软市场
weak mixture 稀薄混合物，贫混合物，贫燃分混合气
weaknesses 薄弱环节
weakness failure 弱质失效
weakness 缺点，弱点，劣势
weak point 弱点
weak salty soil 弱盐渍土
weak shock wave 弱冲波
weak shock 弱震
weak signal detection 弱信号探测
weak soil 软弱土
weak solution 稀溶液
weak-weathered rock 弱风化岩
weak wind boundary layer 弱风边界层
wealth 财富
wearability 磨损性，耐磨性，磨蚀性
wearable submerged pump 耐磨液下式排污泵
wearable 耐用的，耐磨的
wear abrasion 磨损，磨耗，磨耗量
wear allowance 磨损留量，磨损余量，容许磨耗
wear and tear gauge 磨耗规
wear and tear 磨损，损耗，自然磨损，年常损耗，磨损和毁坏，磨耗和撕裂
wear and waster test 损耗试验
wear back 耐磨衬垫
wear coefficient 磨损系数
wear depth 磨蚀深度
wear-down 磨损，损耗
wear factor 磨损因数
wear free 无磨损
wear hardness 抗磨能力，可抗磨能力，磨损硬度，耐磨性
wear index 磨损指数
wearing bush 防磨套
wearing capacity 耐磨性，磨损量，抗磨性
wearing coat 磨耗层，耐磨层
wearing course 磨损层，耐磨层
wearing depth 磨损深度
wearing detail 损耗件
wearing index 磨损指数
wearing layer 磨损层，耐磨层
wearing part 磨损部件，损耗部件，易损件
wearing plate 防磨板，护板
wearing quality 耐磨性
wearing resistance property 耐磨性
wearing ring clearance 磨损环公差
wearing ring 防磨环，磨损环
wearing sleeve 耐磨套筒
wearing surface 磨损面
wearing test 磨损试验
wear mechanism 磨蚀机理
wear notch 磨蚀缺口
wear off limit 报废界限
wear of volume 体积磨损
wear of water turbine 水轮机磨损
wear of weight 质量磨损
wear out defect 磨损缺陷
wear out failure period 耗损失效期
wear out failure 磨损故障，耗损失效，磨损事故，疲劳故障
wear out period （设备）故障恶化期
wear out 磨损
wear part 磨损部分，消耗品，易损件
wear plate 防磨板
wear-proof 防磨的，耐磨的
wear pump 耐磨泵
wear rate 磨损率
wear-resistance 抗磨，耐磨性
wear resistant coating 耐磨涂层
wear resistant elbow 耐磨弯头
wear resistant material 耐磨材料
wear resistant overlaying welding 耐磨堆焊
wear resistant refractory castings 耐磨耐火浇注料
wear resistant steel 耐磨钢

wear resistant welding rod 耐磨焊条
wear resistant 耐磨的
wear-resisting alloy cast iron pipe 耐磨合金铸铁管
wear-resisting alloy cast steel pipe 耐磨合金铸钢管
wear-resisting alloy steel pipe 耐磨合金钢管
wear-resisting alloy steel 耐磨合金钢
wear-resisting layer 耐磨层
wear-resisting 耐磨
wear ring 承磨环，耐磨环
wear scar 磨蚀疤痕
wear template 防磨板
wear tester 磨损检验器
wear test 耐磨试验，磨损试验，磨耗试验
wearthrough 磨穿
wear trace 磨痕
wear well 经久耐用
wear without current 无电流磨损
wear 磨损【机械】，磨耗，损耗，耐用，磨，耐久性
weathed rock 风化岩石
weather analysis 天气分析
weather anchor 抗风锚
weather anomaly 天气反常，天气异常
weather base 气象站
weather-beaten 风雨侵蚀的
weatherboard 防风雨板，挡风舷，檐板，护墙板，装以护墙板，给……装挡雨板
weather bureau 气象局
weather cap 保护端帽，天气帽，伞形风帽
weather chart 气候图
weather cock 风标，风向标
weather code 天气电码
weather condition 气候条件，气象条件
weather constituent 气象要素
weather data 气候数据，气象参数，天气资料
weathered coal 风化煤
weathered escarpment 风化崖
weathered granite 风化花岗岩
weathered layer 风化层
weathered rock formation 风化岩层
weathered rocky soil 风化岩性土
weathered rock 风化岩，风化岩石
weathered state 风化状态
weathered zone 风化带
weathered 被风化的
weather element 气象要素
weather exposure test 气候暴露实验，大气腐蚀试验
weather flag 气象预报信号，天气信号图，天气图标
weather forecast 天气预报
weather gauge 气压计
weather hazard 天气灾害
weather information 天气情报，天气资料
weathering belt 风化带
weathering capacity 风化能力
weathering crack 风化裂缝
weathering crust 风化壳
weathering degree 风化（程）度，风化度
weathering deposit 风化矿床
weathering disintegration 风化崩解
weathering fissure 风化裂隙
weathering index （煤的）风化指数
weathering layer 风化层
weathering process 风化过程，风化作用
weathering quality 耐风蚀性，耐气候性
weathering shot 风化层爆炸
weathering steel 耐候钢
weathering zone 风化层，风化带，风化区
weathering （煤的）风化，自然时效，自然老化，风化作用，风雨侵蚀，雨蚀
weather instrument 气象仪表
weather map scale 天气图尺度，天气地图比例尺
weather map 天气图
weather modification 气候改造
weather parameter 气象参数，天气参数
weather pattern 气象图
weather phenomena 天气现象
weather phenomenon 天气现象
weather-prognostics 天气形势预报
weather-proof cable 耐风雨线，耐风雨电缆
weather-proof covering 防风雨套
weather-proofing motor 气候防护电动机
weather-proofing 气候防护，耐风雨
weather-proof machine 风雨防护型电机，气候防护型电机
weather-proof structural steels 耐风雨结构钢
weather-proof switch 耐风雨开关
weather-proof wire 耐风雨线
weather-proof 不受天气影响的，密封的，防风雨的，耐风雨的，抗风化的，全天候的
weather prospects 天气展望
weather-protected motor 气候防护型电动机
weather-protected 气候防护的，不受气候影响的
weather protection 防风雨
weather regime 天气状况
weather report 气象报告，天气报告
weather resistance 耐风雨侵蚀能力，耐气候性，耐风化性
weather resistant insulation 耐风雨绝缘
weather resistant 防风雨的
weather satellite 气象卫星，天气卫星
weather sensitive load 天气敏感荷载
weather service station 气象服务站
weather shed 雨棚
weather side 迎风面
weather situation 天气情况，天气形势
weather slating 墙面铺石板
weather sphere 天气层
weather station 测候站，气象站
weather strip 挡风条
weather-tight joint 不透风雨的接合面
weather-tight 不透风雨的，防风雨的，气候防护的
weather trend 天气趋势
weather type 天气型
weather vane 风向标，风标
weathervaning stability 调向稳定性

weathervaning 随风向改变方位，调向，对风
weather warning 气象警报
weather wind tunnel 气象风洞
weather-worn 风化剥蚀的
weather 天气，气候作用，气象，吹干
weave attenuation 波浪衰减
weave bead technique 摆动焊缝技术
weave bead welding 横向摆动焊接，摆动焊接，摆动焊
weave bead 摆动焊道，织动焊珠，交织缝
weave 编织，迂回进行，波状失真
web bar 腹筋，腹杆，抗剪钢筋
web connection of beams to columns 梁柱腹板联结
Weber law 韦伯定理
Weber number 韦伯数【水工模型试验用】
weber 韦伯【磁通单位】
web girder 腹梁
web member 腹杆
web of beam 梁腹
web of girder 大梁腹板
web plate 腹板
web reinforcement 腹筋
website 网络地址
web splice 梁腹拼接
web stiffener 梁腹加劲板
web 腹板，梁腹，网，网状物，万维网，金属薄片，散热片，连接板，联结板
WECS(wind energy conversion system) 风能转换系统，风力发电系统
WEC(Westinghouse Consortium) 西屋联队
WEC(wind energy conversion) 风能转换
wedge analysis 土楔分析法
wedge anchorage 楔形锚固，楔形锚具
wedge anchor 楔形锚具
wedge and shims 插楔开石工具
wedge-and-slot-tube anchor bar 楔管锚杆
wedge assembly 槽楔装配
wedge block 垫铁，楔块
wedge boarding 楔镶板
wedge brick 楔形砖
wedge clamp 楔形夹
wedge contact 楔形触点
wedge cut 楔形掏槽
wedged leading edge 楔形进气边
wedged-shaped oil film 楔型油膜
wedged trailing edge 楔形出气边
wedge effect 楔入效应【即曲率效应】
wedge electrode 楔形电极
wedge flow 楔形流
wedge gate valve 楔形闸阀
wedge gate 楔形闸门
wedge gauge 楔规，楔形块规
wedge groove 楔形槽
wedge hammer 楔锤
wedge HV line 楔形高压线
wedge insert 楔形嵌入件
wedge joint 楔接
wedge key 楔形键
wedge line 楔线
wedge pile 楔形桩
wedge plug lock 插头锁楔
wedge plug 锥塞，楔形塞
wedge roller gate 楔形滚动闸门
wedger 制槽楔机，打槽楔机
wedge-shaped brush 楔形电刷
wedge-shaped curb 楔形路缘石
wedge-shaped deflector 楔形偏导器
wedge-shaped fracture 楔形断裂
wedge-shaped hopper 楔形加料斗
wedge-shaped lightning arrester 楔形隙避雷器
wedge-shaped pile 楔形煤堆
wedge sheet pile 楔形板桩
wedge shoe 楔形底板【夹紧装置】
wedge slide 槽楔滑块
wedge sluice valve 楔形闸门，楔形闸门阀
wedge storage 楔形槽蓄，楔形库容，楔形蓄水体
wedge style 楔形造型
wedge surface 楔面
wedge tenon 楔榫
wedge theory 土楔理论【即土压力理论】，楔体理论
wedge timber-sheet-pile 楔形木板桩，人字缝木板桩
wedge-type anchor clamp 楔形耐张线夹
wedge-type flow 楔型流
wedge-type gate valve 楔形闸阀
wedge-type seal 楔形密封，楔入密封
wedge valve 楔形阀
wedge-wire screen 条缝筛
wedge 楔，楔形，楔形物，槽楔，斜楔，楔块，楔体，楔子，探头【超声波检验用】，打楔子，楔固，楔裂法，楔入
wedging action 楔入作用
wedging bond 楔式握裹
weed cutter 割草机
weed-free 无杂草的
weeds control 杂草防除，野草控制，除草
weedy 多杂草的
weed 藻
weekend mode operation 周末停机运行方式
weekend shutdown 周末停运，周末停役
week infeed 弱馈
weekly average concentration 周平均浓度
weekly load forecasting 周负荷预测
weekly pondage 周调节池容量
weekly regulation 周调节
weekly report 周报
weekly storage plant 周调节电站
weekly storage reservoir 周调节水库
weekly 周报，每周的，周刊的，一周一次的，周刊，每周一次，逐周
weepage 渗漏，滴出
weep drain 泄水管，排水沟，渗水管
weep hole 排气孔，排水孔，流水眼，泄水孔，渗水孔
weeping rock 渗水岩石
weep 渗湿，渗液，渗漏，滴落，滴水，漏水
Weibull distribution 威布尔分布
Weibull form and scale parameter 威布尔形状和尺度参数

Weibull function 威布尔函数
Weibull mode 威布尔模态
Weibull parameter 威布尔参数
Weibull scale parameter 威布尔尺度参数
Weibull shape parameter 威布尔形状参数
Weibull slope 威布尔斜率
weigh anchor 起锚
weigh-batcher 重量配料器
weigh beam 秤杆
weigh belt 称量皮带，皮带秤
weigh-bin hopper 计量煤斗
weigh box 称量器，计量箱
weighbridge platform 轨道衡平台，地磅平台
weighbridge track 轨道衡线
weighbridge 秤桥，地秤，地磅，平台磅
weigh bucket 称料斗
weigh by volume 按体积计重
weigher 平衡器，计量机，磅秤，衡器，台秤
weigh feeder 称重给料机
weigh frame 秤框，秤架
weigh hopper 称料漏斗
weigh-house 过磅处，计量所
weigh idler 计量托辊
weighing accuracy 计量精度
weighing batchbox 重量配料器
weighing belt feeder 带秤带式给料机
weighing bottle 称量瓶
weighing cell 称重传感器
weighing check bin 称量校验仓
weighing crane 称重机，称量输送机
weighing dial 称料计，磅秤计盘
weighing equipment 称量设备，计量设备
weighing function 权函数
weighing hopper 称量料斗，称料漏斗，称重料斗
weighing instruments 衡器
weighing machine 磅秤，称量装置，计量机，称量器，地秤，秤桥
weighing platform 称重平台
weighing scale 秤
weigh 计量，称（量重），权衡，权重，考虑，称……重量，重量为……，具有重要性
weigh-in motion track scale 动态轨道衡
weigh lock 称船闸
weigh platform 称料台
weight aggregate 重骨料
weight agreement 计量协议
weight arrestance test 重量分析试验【过滤器】
weight average 加权平均
weight batcher 按重量配料器
weight batching 重量配合比，按重量配料
weight capacity 重量交换容量
weight cell 衡器
weight coefficient 重量系数，比重量，权重系数，加权系数
weight density 重力密度
weight discharge 重量流量
weighted accumulator 重力式蓄能器
weighted approximation 加权逼近
weighted arithmetic average price 加权算术平均价格
weighted arithmetic mean 加权算术平均
weighted average cost of capital 资本的加权平均成本，加权平均资金成本
weighted average efficiency 加权平均效率，平均运行效率
weighted average head 加权平均水头
weighted average 加权平均值，加权平均，加权平均数
weighted code 加权码
weighted current value 衡重电流值
weighted deviation 加权偏差
weighted dummy upper head 载重的假上部顶盖
weighted factor 衡重系数，加权系数，加权因子
weighted filter 盖重反滤层
weighted function 加权函数
weighted heat rate 加权热耗
weighted integral 加权积分
weighted lever safety valve 杠杆式安全阀
weighted mean temperature 加权平均温度
weighted mean value 加权平均值
weighted mean velocity 加权平均速度
weighted mean 加权平均，加权平均值
weighted monthly mean precipitation 加权月平均降水量
weighted residual approach 加权余量法
weighted safety valve 重量安全阀
weighted sound pressure level 加权声压级
weighted steam rate 加权汽耗
weighted sum 加权和
weighted value of voltage 衡重电压值
weighted value 加权值
weighted 加权的，加重的，计权的，赋予加权的，权重的，载重的，已称重的
weight efficiency 加权效率，重量效率
weight empty 空重，空载，坡重
weight equalizer 配重，平衡器
weight equation 权方程
weight factor 加权系数，加权因子，权重系数
weight flow 重量流量
weight fraction 重量分数，重量分率，重量百分率
weight function 权函数
weight gain 重量增益
weight house 磅房，过秤处，过磅处
weightiest 重大的，重要的，严重的【weighty 的最高级】
weight indicator 负载［荷重］指示器
weighting coefficient 加权系数
weighting factor 加权系数，重量系数，加权因子，权重因子，重量因子
weighting foundation 重力式基础
weighting function 加权函数，权重函数，重量函数，权函数
weighting matrix 加权矩阵
weighting method 加权法
weighting network 加权网络，计权网络
weighting range 加权量程，权重
weighting scale 秤
weighting sequence 称序
weighting 加权，称（量），衡量，加重，衡量
weight in volume 重量容积比
weightless 失重的，无重力的

weight list 重量单，磅码单
weight-loaded governor 载重调节器
weight-loaded pressure relief valve 重锤卸压阀
weight-loaded 载重的
weight loss method 失重法
weight loss 重量损失，失重
weight of measurement 测量的权
weight of pile hammer 打桩锤重量
weight of UO_2 二氧化铀重量
weightometer 自动称重仪，自动秤
weight package handling charge 大件处理费
weight percent abundance （同位素）重量丰度，重量百分比丰度
weight percentage 重量百分比［率，数］
weight potential 重力势
weightrate 重量流量
weights and measures 度量衡，计量，权度
weight span 垂直挡距，重力挡距
weight ton 重量吨
weight to power ratio of solar array 方阵的重量比功率
weight unit 加权单位，重量单位
weight voltameter 重量库仑表，重量库仑计
weight 砝码，重量，重锤，载重，权重，权【统计】，吨位，重体，生活补贴
weir abutment 堰的岸墩
weir crest curve 堰顶曲线
weir crest profile 堰顶剖面
weir crest 堰顶
weir diaphragm valve 堰式隔膜阀
weir end contraction 堰壁收缩
weir formula 堰流公式
weir gauge 堰顶水位计
weir head 堰顶水头
weir lip 堰唇，堰前缘
weir meter 溢流水位计，堰顶水位计
weir method 溢流测定法
weir notch 堰孔
weir pier 堰墩
weir section 堰段
weir sill 堰下槛，堰槛
weir station 堰址测站
weir type flow measuring device 堰式流量测量仪表
weir with inclined crest 斜顶堰
weir with overhead bridge 桥跨堰
weir 坝，堰，溢流用挡墙，溢流道，溢水堰，鱼梁
welcome ceremony 欢迎仪式
welcome gate 迎门
welcome meeting 欢迎会
welcome message 欢迎辞
welcome reception 欢迎宴会
welcome remarks 欢迎辞
welcome speech 欢迎辞
weldability test 焊接性试验，可焊性试验
weldability 焊接性，焊接性能，可焊性
weldable seal membrane 可焊密封薄膜，可焊密封膜片
weldable steel 焊接钢
weldable 可焊的
weld affected zone 焊接热影响区，焊接影响区
weld all around 围焊
weld axis 焊接中心线
weld backing 焊接衬垫，焊缝背衬垫，焊带，焊缝下的垫圈
weld backup ring 焊带，焊缝下的垫圈
weld bead height 焊缝高度
weld bead length 焊缝长度
weld bead stability 焊道稳定性
weld bead 熔敷焊道，线焊，堆焊焊缝，（焊接的）焊道，焊缝，焊珠
weld bonding 胶接点焊，焊接接头
weld bond open 焊缝裂开
weld bond 焊接接头，熔合线
weld commissure 焊接处
weld crack 焊缝开裂，焊接裂纹，焊接裂缝
weld crater 焊口，熔池
weld crosswise 交叉焊接
weld decay 焊缝晶间腐蚀
weld defect 焊接缺陷
weld-deposited cladding 堆焊层，硬质合金堆焊层
welded assembly 焊接装置，焊接组装件
welded-bar fabric 焊接钢筋网
welded bar-mesh reinforcement 焊接钢筋网
welded blade root 焊接叶根
welded blind 焊接堵头
welded bonnet 焊接阀盖，焊接阀帽
welded bridge 焊接桥梁
welded connection 焊接连接
welded construction 焊接结构
welded-contact rectifier 熔接整流器
welded disc rotor 焊接转子
welded disc turbine rotor 焊接转子
welded elbow 焊接弯头，焊接机械肘
welded end plug 锥形焊枪嘴塞【指型套管】
welded fissure 焊接裂缝，焊合的裂缝，焊接裂纹
welded flange 焊接法兰
welded frame 焊接的框架
weld edge 焊缘
welded-in diaphragm 焊接隔板
welded-in stub 焊制管接头
welded joint 焊缝，焊接装配，焊接接头
welded junction 焊口，焊接接头
welded main frame 焊接主框架
welded mesh 焊接网
welded metal plate 焊接钢板
welded metal sheet 焊接钢板
welded-on flange 焊接法兰
welded-on head 焊制端头，焊制封头
welded-pipe construction 焊管结构
welded pipe 焊接管，焊制管，焊接钢管
welded plate girder 焊接板梁
welded pressed-in seat ring 焊接阀座环
welded reinforcement 焊接钢筋
welded rotor 焊接转子
welded seal 焊接密封装置
welded seam 焊缝
welded seat ring 焊接阀座环

welded sheet steel tank （变压器的）钢板焊接油箱
welded splice 焊接接头
welded steel fabric 焊接钢筋网
welded steel pipe 焊接钢管
welded steel plate construction 钢板焊接结构
welded steel reinforcement carcass 焊接钢筋骨架
welded steel structure 焊接钢结构
welded steel tube 焊接钢管
welded steel 焊接钢
welded structure 焊接结构
welded stub connection 焊接管接头
welded stub-in 焊接支管
welded tee 焊制三通
welded tube 焊接管
welded tuff 熔结凝灰岩
welded unit 焊接构件
welded wall 焊制水冷壁
welded wire fabric reinforcement 焊接网加固
welded wire fabric 焊接钢丝网，焊接金属网
welded wire mesh 焊接金属网
welded zone 焊接区
welded 焊接的
weld end 焊头，焊接端
welder helment （焊工）面罩，焊工帽
welder leggings 焊工护腿
welder performance qualification test 焊工技术鉴定试验，焊工技能考试
welder qualification file 电焊工合格证书文件
welder qualification record 电焊工合格证书记录
welder qualification 焊工合格证书，焊工资格（证书）
welder's apron 焊工围裙
welder's booth 焊工的遮棚
welder's foot shield 焊工脚盖
welder's gauntlets 电焊工的长手套
welder's gloves 电焊手套，焊工手套
welder's goggles 焊工护目镜
welder's hand shield 焊工手持护目镜，焊工手盾
welder's helmet 焊工面罩，焊工头盔
welder's hood 焊工防护罩
welder's jacket 焊工短上衣
welder's lifting platform 焊工升降台
welder 焊工，焊机，电焊机，焊接烟尘
weld flaw 焊接缺陷，焊缝缺陷
weld flush 焊缝隆起
weld flux 焊药，焊剂
weld fume 焊接烟尘
weld gauge 焊缝量测仪，焊缝量规
weld groove 焊缝坡口
weld harmful gas 焊接有害气体
welding agent 焊药
welding and erection of pressure parts 承压件的焊接与安装
welding arc voltage 焊弧电压
welding arc 焊接电弧
welding base metal 焊条金属
welding bead 焊道，焊接泡，焊珠
welding bench 焊接工作台
welding blowpipe 焊炬，气焊枪，气焊吹管
welding by both sides 双面焊
welding by one side 单面焊
welding cable 焊接电缆，电焊电缆
welding carriage 焊接滑架，焊接拖板
welding characteristics 焊接特性
welding circuit 焊接电路，焊接回路，焊接电流导电回路
welding cladding 堆焊层
welding code 焊接规程
welding coefficient 焊接系数
welding condition 焊接条件，焊接工艺参数
welding connector （焊接）电缆夹头，焊接电缆接线柱
welding consumables 焊接消耗品，焊接用耗材【主要指焊条、焊剂等】
welding contactor 电焊接触器
welding core 焊条的焊芯
welding crack 焊接裂纹，焊缝裂纹，焊接裂缝
welding current 焊接电流
welding cycle 焊接循环
welding data package 焊接数据包
welding defect 焊接缺陷
welding deformation 焊接变形
welding dynamo 焊接发电机
welding electrode 电焊条，焊接电极，焊条
welding end 焊接端
welding engineer 焊接工程师
welding equipment and accessories 电焊机及附件
welding equipment 焊接设备
welding fabrication 焊接装配
welding fitter 焊接装配工
welding fixture 焊接夹具
welding flame 焊焰
welding flux 焊剂，焊接溶剂，焊药，埋弧焊剂
welding foreman 焊接工长
welding generator 焊接发电机，电焊发电机
welding glove 焊工手套
welding goggles 焊工护目镜
welding groove 焊接坡口
welding ground 电焊地线
welding gun 焊枪
welding head 焊机头，焊接机头，焊头，烧焊枪，烙铁头
welding hood 焊接防护罩
welding impulse 电焊脉冲
welding inspection 焊接工作检查，焊接检查
welding inspector 焊接检查员
welding jig 焊接夹具，焊接支架
welding joint design 焊口设计
welding joint 焊接接头
welding machine stock room 电焊机房
welding machine 电焊机，焊机，焊接装置，焊接机
weld machine 焊机
welding material 焊接材料
welding neck flange 对焊法兰，焊颈法兰，颈法兰
welding-on collar 平焊环
welding operation 焊接操作
welding operator qualification 焊工资格审查

welding operator　焊工
welding parameter　焊接参数
welding piece　焊件
welding pistol　焊枪【热塑性塑料用】
welding plate flange　焊接板式法兰
welding pool　焊接熔池，焊接熔穴
welding positioner　焊接变位机，焊接翻转台，焊接台具
welding position　焊接位置
welding power supply　焊接电源
welding procedure assessment　焊接工艺评定
welding procedure data sheet　焊接程序数据表
welding procedure qualification　焊接程序审定
welding procedure　焊接工艺，焊接程序
welding process specification　焊接工艺规范［规程］
welding process　焊接过程，焊接方法，焊接工艺
welding quality　焊接质量
welding rectifier　焊接整流器，电焊整流器
welding regulator　焊接电流调节器
welding residual deformation　焊接残余变形
welding return cable　焊接返回电缆
welding rod extrusion press　焊条压涂机，焊条挤压机，焊条涂药机
welding rod holder　焊条钳
welding rod　焊条，填充焊丝，铜镍焊条
welding roller　焊接滚轮架
welding scale　焊皮
welding seam　焊缝
welding sequence　焊接顺序，焊接工序
welding set　焊接设备
welding shop　电焊车间
welding side　（接头的）焊接边
welding slag　焊渣
welding spats　（焊接用）护脚
welding speed　焊接速度
welding spot　焊点
welding stick dryer　焊条烘箱
welding stop valve　焊接式截止阀
welding strength　焊接强度
welding stress　焊接应力
welding symbols　焊接符号，焊缝代号，焊缝符号
welding technician　焊接技工
welding technique　焊接技能
welding technology　焊接工艺
welding temperature field　焊接温度场
welding tilter　焊接翻转机
welding torch pipe　焊枪管
welding torch　焊炬，焊枪，焊把
welding transformer　焊接变压器，电焊变压器
welding voltage　焊接电压
welding wire　焊条钢丝，焊丝
welding with backing　衬垫焊
welding with flux backing　焊剂垫焊
welding　焊接，焊缝，施焊
weld initiation　焊接引弧
weld inlay　堆焊层，焊缝镶嵌
weld-in nozzle　焊入式喷嘴
weld interface　熔合线【熔化焊】，熔化线
weld intersection　交叉焊
weld joint intersection　交叉焊缝
weld joint　焊缝，焊接接头
weld junction　熔合线
weld leakage　焊缝泄漏
weld length　焊缝长度，焊缝沿轴线方向的长度
weldless　无焊缝的，无焊的
weld line　焊缝，焊线
weld machined flush　削平补强的焊缝
weldment　焊件，焊接部件，焊接结构，焊接，焊接装配，焊接件
weld metal area　焊缝金属区，焊缝区，焊接金属面积，熔接金属面积
weld metal buildup　焊接金属熔敷【支管焊缝】
weld metal cracking　焊缝裂纹
weld metal overlay　焊敷金属堆层
weld metal　焊缝金属，焊接金属
weld neck flange　带颈对焊法兰，焊接颈状法兰
weld overlap　焊瘤
weld overlay　堆焊层
weld passivation　焊缝钝化
weld pass　焊道
weld penetration　（焊接）熔深，焊穿，焊缝焊透，焊透深度，熔深
weld pitch　焊缝距
weld plugging　焊接管塞
weld preparation　坡口，焊接准备
weld prepping　焊接准备
weld puddle　熔池
weld reinforcement　焊缝补强，焊接补强，余高
weld repair　补焊，修补焊缝
weld ripple　固化波纹【焊缝】，焊缝波纹
weld root　焊接根，焊件剖口，焊根
weld rotation　焊缝转角
weld sealed　焊接密封的
weld seal lip　焊接密封突缘
weld seal　焊接密封
weld seam crack　焊口裂纹
weld seam inspection　焊缝检查
weld seam　焊缝
weld spatter　焊接溅出物
weld starting　焊接引弧
weld steel bend　焊接弯头
weld steel elbow　焊接弯头
weld test coupon　焊接试样
weld testing　焊接试验
weld thermal cycle　焊接热循环
weld thickness　焊缝厚度
weld throat　焊缝喉部【角焊】
weld toe　焊趾
weld width　焊缝宽度
weld with shop beveled ends　工厂加工坡口焊接
weld zone　焊缝区
weld　焊接，焊件，焊接装配，焊接连接，焊缝，焊道
welfare facilities　福利设施
welfare fund　福利基金
welfare　福利
well auger　凿井钻
well borehole　井眼
well borer　凿井工，凿井机

well brandreth 井栏
well capacity 井的出水率
well curb 井栏
well defined goal 明确的目标
well defined right and responsibility 责权明确
well defined water 边界明确的水域
welldrained soil 排水良好的土
welldrained stream basin 排水良好的流域
well drilling technology 成井工艺
well experienced 经验丰富的
well-field 井群区
well flooding 渗井泄洪
well-founded 有依据的
well-graded aggregate 级配良好的骨料
well-graded soil 级配良好的土
well graded 级配良好的
well hydraulics 管井水力学
wellington boot 防水长筒靴
well-known mark 驰名商标
well-known person 知名人士
well log 钻井记录，钻井剖面
well organized vortex 规则旋涡
well point method 井点法
well point 井点
well-pumping test 井的抽水试验
well-rig 打井机具
well separated flow 完全分离流动，无扰分流，无旋分流
well sinker 凿井工
well sinking 凿井，沉井
well-sump type pumping plant 深井式抽水站
well-tempered enterprise 经过考验的企业
well trained staff 训练有素人员
well-tried methods of selection 结果较好的选择方法
well water cut 井中的出水量
well water 井水
well yield 井的出水量
well 地坑，矿井，升降机井道，井
West Africa Gas Pipeline Company 西非天然气管道公司
west africa gas project 西非天然气项目
West-east Gas Transmission Pipeline Project 西气东输工程
westerlies rain belt 西风多雨带
westerlies 西风带
westerly 西风，西的，自西的，向西方的，向西，往西
western cuisine 西餐
western hemisphere 西半球
wester 西风
Westinghouse Electric Corporation （美国）西屋电气公司，威斯汀豪斯公司
west longitude 西经
west northwest 西西北
west southwest 西西南
west 西，西部，西方
wetability 表面可湿润性
wet adiabatic change 湿绝热变化
wet adiabatic curve 湿绝热曲线
wet adiabatic lapse rate 湿绝热递减率
wet adiabat 湿绝热，湿绝热线
wet agitator 湿式搅拌器
wet air 湿空气
wet analysis 湿分析法
wet armature directional valve 湿电枢换向阀
wet ash loading chute 湿灰加料槽
wet ash 湿灰
wet avalanche 湿崩坍
wet ball mill 湿式磨球机
wet basin 封闭式港池
wet-bottom boiler 液态排渣炉，液态排渣锅炉
wet-bottom furnace 液态排渣炉膛
wet-bulb depression 干湿球温差，湿球温差
wet-bulb temperature 湿球温度
wet bulb 湿球
wet chemical route 湿化学方法
wet cleaner 湿式除尘器
wet climate 潮湿气候
wet coal 湿煤
wet-collapse 湿陷
wet-collapsibility 湿陷性
wet collection 湿式除尘器
wet compaction 湿法碾压
wet connection 湿连接
wet consistency of concrete 混凝土塑性稠度
wet cooling system 湿式冷却系统
wet cooling tower 湿式（通风）冷却塔，温冷塔
wet cooling 湿冷却
wet criticality 湿临界
wet cured 湿养护的
wet curing 喷水养护（法）
wet cut 湿开挖，湿挖方
wet cyclone 湿式旋风分离器
wet density 湿密度，湿容重
wet deposition 湿沉降
wet desulphurization 湿法脱硫
wet digestion 湿消化法
wet dock 泊船坞
wet-drill method 湿钻法
wet dust collection 湿式除尘，湿式除尘器
wet dust suppression system 湿式除尘系统
wet excavation 湿开挖
wet feed method 湿式进料法
wet filter 湿式过滤器
wet flashover test of insulator 绝缘子湿闪络试验
wet fraction 水分
wet galvanizing 湿法镀锌
wet gas-holder 湿式气柜，水封储气器
wet laid deposit 水力充填
wetland 湿地
wet lay-up 湿法成型，湿法敷涂层
wet limestone throwaway type 湿式石灰石抛弃型［抛弃法］
wet line correction 湿绳校正
wet mechanical analysis 湿法颗粒分析
wet mechanical draft cooling tower 湿式机械通风冷却塔
wet method 湿法
wet mix shotcrete 湿拌喷射混凝土
wet mix 湿拌

wet monsoon 潮湿季风
wet natural draft cooling tower 湿式自然通风冷却塔
wetness index 年降水量指数，湿润指数
wetness 潮湿，湿润
wet pit pump 湿井坑泵，湿坑泵
wet pit 湿坑，湿井坑【泵体浸没水中】
wet plume 湿羽流
wet power-frequency withstand voltage test 工频湿耐压试验
wet preservation 湿式保护
wet preserve 湿保养
wet pressure loss 湿压损失
wet process 湿处理法【燃料循环】
wet rotor pump 湿转子泵
wet route 湿处理法【燃料循环】
wet sample 湿样
wet sand blasting process 喷湿砂法
wet sand curing 湿沙养护
wet saturated steam 湿饱和蒸汽
wet-screened 湿筛的
wet screening 带水筛分，湿筛分
wet scrubber type dust extraction system 水浴式除尘系统
wet scrubber 湿式洗涤器，湿式烟气脱硫，湿式除尘器，喷水除尘装置
wet season 湿季，雨季
wet separator 湿式分离器
wet slide 水滑道
wet snow zone 湿雪地带
wet snow 湿雪
wet sprinkler system 湿式喷水灭火系统
wet-stator motor 湿定子电动机
wet-stator 湿定子
wet steam turbine 湿蒸汽汽轮机
wet steam 湿蒸汽
wet stirring device 湿式搅拌器
wet sump pump 湿坑泵
wettability 可湿性，吸湿度，润湿性
wetted area 浸湿面积，受潮面积，过水面积，受潮面积
wetted contact 湿触电，降阻触电，湿状态不起弧距离
wetted cross-sectional area 过水断面面积，流体截面积
wetted cross-section 过水断面，湿润断面
wetted perimeter 湿润周界，湿周，润周
wetted surface 润湿面，湿润面
wetted 加湿
wet test on the surface of the insulation 绝缘表面湿试验
wettest stable consistency 最湿稳定稠度
wet test 淋雨试验，湿试验
wetting agent 润湿剂
wetting and drying cycle 干湿循环
wetting and drying test 干湿试验
wetting angle 润湿角
wetting-drying test 干湿循环试验
wetting mechanism 湿润装置
wetting 加湿，润湿，润湿的
wet type cooling tower 湿式冷却塔
wet type watermeter 湿式量水计
wet unit weight 潮湿状态容重，湿容重
wet vent 排湿气孔
wet weight 湿重
wetwell drywell cooling system 湿井干井冷却系
wet well 湿井
wet year 多水年，多雨年，丰水年
wet zone 湿润带
wet 潮湿的，湿的，湿气
WE(westinghouse electric company) 西屋电气公司
WE(with enclosure) 带附件
WFEO(World Federation of Engineering Organizations) 世界工程组织联合会
W-flame boiler W 形火焰锅炉
W-flame firing furnace W 形火焰炉膛
WFOE(wholly foreign-owned enterprise) 外资企业，外商独资企业
WFQC(welding follow up and quality control) 焊接质量跟踪管理
WF(work file) 工作文件
WG(wire gauge) 线规
whack 捣实
whale back dune 鲸背型沙丘
wharfage charge 码头费
wharfage 码头费，码头附加费
wharf apron space 码头前沿区
wharf apron track 码头前沿区轨道
wharf apron 码头前沿
wharf crane 码头起重机
wharf equipment 码头设备
wharf structure 码头构筑物
wharf surface 码头地面
wharf 码头
Wheatstone bridge 惠斯顿电桥
wheel axle 轮轴
wheel barrow 手推车，独轮车，独轮手推车
wheel base 轮距，轴距
wheel boss 轮毂
wheel box 轮箱，齿轮箱
wheel brake 车轮制动器，轮闸
wheel chock 车轮止挡块
wheel clamp(/gripper) control station 夹轮器控制站
wheel clamps(/gripper) 夹轮器
wheel cover 轮罩
wheel discharge capacity 水轮泄流量
wheel ditcher 轮式挖沟机
wheel dozer 轮胎式推土机
wheel dredger 轮式挖泥机
wheeled dozer 轮胎式推煤机
wheel efficiency 轮周效率，水轮效率
wheel flange 轮缘
wheel house 轮箱
wheel hub 轮毂
wheel loader 轮胎式装载机，轮式装载机
wheel load 轮压，轮负荷，车轮载重
wheel mill 轮碾机，碾磨机，中速轮式磨煤机【MPS 型辊盘式中速磨煤机】
wheel-mounted dragline 轮胎式拉铲挖土机
wheel-mounted power shovel 轮胎式动力铲

wheelout breaker 滚出式断路器
wheelout 用车子运出，滚出式
wheel pit 水轮坑
wheel pressure 轮压
wheel ratio 转轮直径比，水轮比值
wheel rib 轮辐
wheel scraper 轮胎式铲运机
wheel set 轮副，轮组，轮轴
wheel span 轮距
wheel stop 轮式止挡器
wheel-track planting 轮迹播种法
wheel track 轮距，轮轨
wheel-tractor-mounted bulldozer 轮胎式拖拉推土机
wheel-tractor-mounted shovel 轮胎式拖拉机铲
wheel tractor 轮胎式拖拉机
wheel tread 轮距
wheel type loader 轮式装载机
wheel-type scraper 轮式铲运机
wheel-type trenching machine 轮胎式挖沟机
wheel work 轮周功
wheel wrench 轮扳手
wheel 车轮，轮，叶轮【汽轮机】
whenever possible 只要有可能，每逢有可能
where applicable 适用时，如适用
where appropriate 在适当情况下，酌情
whereas clause 鉴于条款【合同】
whereas 鉴于……
whereby 凭此（协议，条款），凭借，通过……，借以，与……一致
where possible 在有条件的地方，在可能的情况下
whereto 对于那个，去何处，为什么
whereupon 在那上面，因此，于是
Whetatone bridge 惠斯通电桥
whim 绞车
whip crane 动臂起重机
whip hoist 动臂起重机
whipping 甩击，甩鞭
whip restraint 防甩击装置，防冲击装置
whipstitch 锁缝，包缝
whip-typed tower crane 动臂式塔式起重机
whip 甩击，甩鞭
whirl core 旋动核，旋风核心
whirler crane 旋臂吊车，回转式起重机，旋转起重机
whirler 离心净气器，旋转绞龙
whirl flutter 旋动颤振
whirling current 旋涡流
whirling impeller 旋转的叶轮
whirling motion 涡动，旋转运动
whirling reaction chamber 涡流式反应室
whirling speed 旋转速度
whirling vortices 旋转涡流系
whirling 旋转的，涡流的
whirl plate 折流板
whirlpool 涡流，旋涡，混乱
whirl tube 风洞，风道
whirl velocity 涡流速度
whirlwind air separator 旋风式空气分离器
whirlwind precipitator 旋风除尘器
whirlwind 旋风，旋风般的快速行动，破坏性的力量或事物
whirly 小旋风
whirl 旋转，涡流，旋涡，涡旋
whisper trim 微音结构
whistling signal 啸声信号
white ant 白蚁
white body 白体
white boundary 白边界
white box testing approach 白箱测试法
white brass 白铜
whitecap 白浪，白浪花，浪端的白泡沫
white cast iron 白口铁
white cement 白水泥
white coal 白煤
white coat 白灰罩面
white collar worker 白领工人
white finish coat 白灰罩面
white interface 白交界面
white lead 白铅粉
white metal 白合金
white mica 白云母
whitening 涂白灰
white noise excitation 白噪声激励
white noise field 白噪声速度场，白噪声场
white noise spectrum 白噪声谱
white noise 白噪声
whiteout 乳白天空
white reactivity noise 反应性白噪声
white reflection 白反射
white rust 白锈
whitewash 白色涂料，大白粉，（粉刷用的）石灰水，掩盖，粉刷，刷石灰水，彻底击败
white water 白水【即水中搀气】
white wave 白波
white wax 白蜡
white wood 白木
white 白色，无色透明的锌白，白色颜料，白色的，刷白，漂白
Whittaker-Shannon sampling theorem 惠特克香农采样定理
Whitworth thread 惠氏螺纹
whizzer 离心分离机，离心机
whole-body autoradiograpy 全身放射自显影法
whole-body counter 全身计数器
whole-body dose 全身剂量
whole-body exposure 全身辐照，全身照射量
whole-body gamma measurement 全身伽马测量
whole-body irradiation 全身辐照
whole-body monitor 全身监测仪
whole-body radiation meter 全身射线测量计，全身辐射计，全身放射计
whole-body X-irradiation X射线全身照射
whole circle packed shroud 整圈连接围带
whole core accident 全堆芯事故
whole current method 全流方法
whole deal 整块松板，全松木板，整笔生意
whole energy competition 全电量竞争模式
whole gale 狂风，十级风【风速 245～284 米/秒】
whole life cycle 全生命周期

whole process automation 全过程自动化
whole process bidding 全过程招（投）标
wholesale competition 批发竞争模式
wholesale price 批发价
wholesale rate 趸售电价
wholesaler 批发商
whole scene 全景
whole section member 整体式构件
whole set of equipment 成套设备
whole set of product 整套产品
whole set of thermo-power generating equipment 成套火力发电设备
wholesome water 卫生可饮用水
whole-tide cofferdam 完全挡潮围堰
whole timber 整材
whole 全部，全体
wholly foreign-owned enterprise 外资企业，外商独资企业
wholly-owned subsidiaries 全资子公司
WHO(World Health Organization) 世界卫生组织
WH＝watthour 瓦小时
WIBON(whether in berth or not) 不论靠泊与否
WICCON(whether in custom clearance or not) 不论海关手续办妥与否
wick drain 灯芯式排水井，纸板排水井，排水芯管，排水带，排水板
wicker mat 柳条垫
wicker 柳条
wicket-gate-actuating rod 导叶操纵杆
wicket-gate-adjusting servomotor 导叶调节接力器
wicket gate stem 导叶转动轴，导叶叶柄
wicket gate 导水门，导叶，旋闸，小门，边门，便门
wicket-type gate 旋转式闸门
wicket 放水门，水闸的选装闸门，小门，边门
wide-angle camera 广角摄影机
wide-angle lens 广角透镜
wide-angle multiplex projector 广角多倍投影测图仪
wide-angle nozzle 广角喷嘴
wide-angle photography 广角摄影
wide area data service 广域数据服务
wide area information server 广域信息服务器，广域信息服务系统
wideband data link 宽频带数据链路
wideband data 宽频带数据
wideband excitation 宽带激励
wide-bucket excavator 宽斗挖土机
wide-crested measuring weir 宽顶量水堰
wide-crested weir 宽顶堰
wide-face wheel 宽面轮
wide flanged I-beam 宽缘工字梁，宽翼缘梁
wide flanged 宽翼缘的
wide flange I-section 宽翼工字钢
widely spaced lattice 宽间隔格栅【堆芯】
widening angle 扩大角，扩张角
widening of curve 曲线加宽
widening 加宽，扩展，膨胀，拓宽
wide range speed control 宽调速
wide range strain 大幅度应变
wide range 宽波段的，宽量程的，广范围，宽范围
width at dam crest 坝顶宽度
width average 平均宽度
width meter 宽度计
width of crevice 裂缝宽度
width of step 踏步宽
width of water surface 水面宽度
width 宽度
Wiener filtering 维纳滤波
WIFPON(whether in free pratique or not) 不论通过检疫与否
wildcard character 通配符
wilderness 荒野
wild grass ground 荒草地
wild life animals reserve area 野生动物保护区
wild wood 原始林，自然林
willowbox groin 柳梢捆丁坝
willowbox groyne 柳梢捆丁坝
willow brush 柳梢
willow groin 柳梢丁坝
willow groyne 柳梢丁坝
willow revetment 柳枝护岸
wilting coefficient 枯萎系数
wilting percentage 枯萎百分数
wilting point 凋萎点，凋蔫点，萎蔫点
wilting 枯萎
wimble 螺旋钻，锥，钻孔器，钻
winch barrel 绞车卷筒
winch cable 绞车钢丝绳
winch chain and sprocket 绞车链及链轮
winch drum 绞车卷筒
Winchester disk drive 温彻斯特磁盘机，硬盘机
winch for mine 矿井卷扬机
winch frame 绞车架
winch truck 绞盘车
winch 绞车，卷扬机，绞盘
wind abeam 侧风，横风
wind abrasion 风力侵蚀，风磨蚀，风蚀
windage effect 风阻影响
windage loss 风吹损失，风损，风阻损失，鼓风损失，气流损失
windage resistance 风阻
windage 风力影响，风致偏差，风力修正量，风偏修正，风偏，空气阻力，游隙，余隙
wind allowance 风力裕度
wind aloft observation 高空风观测
wind aloft report 高空风报告
wind aloft 高空风
wind and snow load 风雪荷载，风雪载荷
wind angle 风迎角，风角
wind approach angle 来流风迎角，风接近角
wind area 迎风面积
wind arrow 风矢，风向及风力指示符号，风向指针
wind assessment 风场评价
wind axes system 风轴系，气流坐标系
wind axes 风轴
wind axis wind machine 风轴风力机
wind azimuth 风方位角
wind backwater 风壅水

wind barometer table 风速-气压换算表
wind beam 抗风梁
wind belt 风带
wind blown debris 风吹碎片
wind blown dust 风吹沙尘，风尘
wind blown sand 风积沙，风沙
wind blown snow 风吹雪
wind-blown soil 飞土，风积土
wind-blown 风吹的，风蚀的
wind-borne dust 风载尘
wind-borne load 风携质
wind-borne material 风载物质
wind-borne pollution 风载污染物
wind-borne sediment 风成沉积物
wind-borne snow 风载雪
wind-borne 风传送的
wind box 风箱
wind bracing 风撑，抗风支撑，风拉条
wind break banker 防风墙
wind break fence 防风篱
wind break forest 防风林
wind break system 防风林系统
windbreak 防风林，防风物，防风墙，防风阵，风障
wind calm 静风
wind cap 防风帽
wind carried sand 风积沙
wind catching surface 获风表面
wind caused catastrophic 风灾
wind cave 风穴
wind channel 风洞，风道
wind characteristic 风特性
wind chart 风图
wind chill factor 风冷却因素
wind chill index 风寒指数，风冷指数
wind chill 风寒，寒风
wind circulation 风环流，气流循环
wind class 风级
wind climate 风气候
wind cock 风向标
wind comfort level 风舒适度
wind concentrator 集风器
wind cone 风标，风向袋，风向锥，圆锥风标
wind convection coefficient 风对流系数
wind conversion efficiency 风能转换效率
wind cork 风标
wind correction 风力修正
wind corrosion 风蚀
wind cowl （烟囱）风帽
wind crack （木材）风裂
wind current 风流，气流，吹流
wind damage 风害
wind data logger 风数据记录器
wind data 风数据，风资料
wind deflection 风向偏转
wind deflector 挡风板，风向导流板，导风板，导风装置
wind denivellation 风浪倾斜度
wind deposited soil 风积土
wind deposition 风积
wind deposit 风力堆积，风成堆积物，风沉降
wind description 风描述
wind desiccation 风干作用
wind diagram 风力图，风玫瑰图，风向频率图
wind dial 风速表度盘
wind diffuser 扩风器
wind direction collect 风向采集
wind direction diagram 风向图
wind direction fluctuation 风向波动
wind direction frequency 风向频率
wind direction meter 风向测定器
wind direction recorder 风向自记器
wind direction rose chart 风向玫瑰图
wind direction sector 风向区域
wind direction sensor 风向感受器
wind direction 风向
wind discontinuity 风不连续性
wind dispersion 风弥散作用
wind divide 风向界线
wind down 使（某活动）逐步结束
wind drag load 风阻荷载
wind drag 风阻
wind drainage 暗沟排水
wind drain 阴沟
wind drift sand 流沙，风沙
wind drift 飞砂，漂流，风偏流，吹流，飞沙
wind driven brine pump 风力盐井泵
wind-driven circulation 风生环流，风引起的环流
wind-driven dynamo 风动发电机
wind-driven generating unit 风能发电机组
wind-driven generator 风力发电机
wind driven pump 风力驱动泵
wind driven sawmill 风力据木厂
wind driven snow 风吹雪
wind-driven water pump 风动水泵
wind driven wave 风浪
wind-driven 风动的
wind duration 风期，风时
wind dynamic load 动态风载
wind eddy 风涡
wind effect on agriculture 风对农业影响
wind effect on air pollution 风对空气污染影响
wind effect on building 风对建筑物的影响
wind effect on pedestrian 风对行人的影响
wind effect on structure 风对结构物影响
wind effect phenomena 风力效应现象
wind effect 风效应
wind energy atlas 风能区划图，风能地图集
wind energy collector 风能收集装置
wind energy content 风能蕴藏能
wind energy conversion system 风能转换系统
wind energy conversion 风能转换
wind energy converter 风能转换装置
wind energy density 风能密度
wind energy engineering 风能工程
wind energy extraction limit 风能获取极限值
wind energy farm 风电场
wind energy flux 风能通量
wind energy frequency 风能频率
wind energy pattern factor 风能型因子
wind energy potentiality 风能潜力

wind energy region 风能区划
wind energy resource 风能资源
wind energy rose chart 风能玫瑰图
wind energy rose 风能玫瑰图
wind energy spectrum 风能谱
wind energy system 风能转换系统
wind energy 风力能源，风能
wind engineering 风工程
wind engineer 风能工程师
wind environment 风环境
wind-eroded soil 风蚀土壤
wind erosion basin 风蚀盆地
wind erosion 风力侵蚀，风蚀
winder 绞车，卷扬机，拉丝机
wind excitation 风激励
wind excited acceleration 风激加速度
wind excited oscillation 风激振荡
wind exposure area 受风面积
wind eye 风眼
wind fairing 风嘴【公路科技】
wind farm 风电场，风力发电厂，风能发电机集群
wind fence 风篱，风栅栏，防风栅栏，防风栅
wind fetch 风程
wind field full-scale measurement 风场实测
wind field 风场，风区
wind flag 风旗
wind flow 风流，气流
wind flume 风槽
wind fluttering factor 风振系数
wind flutter noise 风颤振噪声
wind following 风向跟踪，对风
wind force coefficient 风荷载系数
wind force diagram 风力图
wind force scale 风力级，风级
wind force 风力
wind frame 抗风构架
wind frequency curve 风频数曲线
wind furnace 自然通风炉，风炉
wind gap 风口，风谷，风害造成的空隙，风隙
wind gauge 风速表，风速计，风速器，风力计，风压计
wind generated gravity wave 风生重力波
wind generated noise 风致噪声
wind generated sea wave 风致海浪
wind generated water wave 风生水波
wind generation 风力发电
wind generator set 风力发电机组
wind generator 风力发电机
wind girder 抗风桁架
wind gorge 风峡
wind gradient exponent 风速梯度指数
wind gradient 风速梯度
wind gust 阵风
wind hole 风穴
wind indicator 风力指示物
wind induced acceleration 风致加速度
wind induced circulation 风引起的环流
wind induced current 吹流，风生海流，风驱动流
wind induced discomfort 风致不舒适
wind induced heat loss 风致热损失
wind induced internal pressure 风致内压
wind induced load 风致荷载
wind induced motion 风生运动
wind induced oscillation 风致振荡
wind induced pressure 风压
wind induced turbulence 风生湍流
wind induced vibration 风致振动
wind induced wave 风生波
wind industry 风电行业
windiness 多风，招风，有风
wind information 风资料
winding corridor 回廊
winding displacement 绕组位移，排线
winding distribution 绕组分布
winding drum 卷筒
winding engine 卷扬机
winding factor 绕组系数
winding insulation 绕组绝缘
winding loss 绕组铜损耗
winding machine 缠绕机，绕线机
winding pitch 绕组节距，线圈节距
winding resistance 绕组电阻
winding staircase 螺旋楼梯，盘梯，旋转楼梯
winding stream flow 曲折水流
winding technique 绕线工艺
winding temperature 绕组温度，线圈温度
winding type internal rotor 绕线式内部转子
winding up or liquidation of companies 公司的清理或清算
winding up 清理
winding wire 绕包线
winding with non-uniform insulation 分级绝缘线圈［绕组］
winding with uniform insulation 全绝缘线圈
winding 装设绕组，绕组
wind intensity 风强度
wind-laid deposit 风成沉积，风沉积，风积物
wind laid soil 风积土
windlass 起锚机，卷扬机，绞盘，辘轳
wind layer 风层
windless region 无风区
windless 无风
wind loading chain 风载分析链
wind loading code 风载规范
wind loading design criterion 风载设计准则
wind loading standard 风载标准
wind load spectrum 风载谱
wind load 风荷载，风载
windlop 短涌浪，(风吹起的) 短浪，短涌，三角浪
wind loss 风害损失
wind lull 风速暂减，风暂息
wind machine 风力机
wind map 风图
wind measurement mast 测风塔
wind meter 风速计
wind microzonation 风力微区划
windmill anemometer 风车式风速表
windmill curve 风车线，风力出力特性曲线
windmill generator 风力发电机

windmill pump 风车泵，风车驱动泵，风力泵
windmill sail 风力发动机工作叶轮，风车轮叶
windmill 风车，风力机，风力站
wind mixing 风混合
wind moment 风力矩，风载力矩
wind natural ventilation 自然通风
wind noise 风噪声，气流噪声
wind nose 风嘴
wind odd 风壅水，风增水
wind of Beaufort force 2 二级风，蒲福风级
wind of Beaufort force 3 三级风
wind of Beaufort force 1 一级风
wind off test 无风试验
wind onset 风壅水，风增水
wind on test 有风试验
window air conditioner 窗式空调
window alarm 报警光字牌
window apron 窗台板
window bar 窗闩
window blind 窗帘，遮光帘，百叶窗
window board 窗台板
window bolt 窗插销
window butt 窗铰链
window counter tube 窗形计数管
window curtain 窗帘
window fitting 窗配件
window frame 窗框
window groove fastener 窗户挡风毡条座
window hangings 窗帘，窗幔
window hinge 窗铰链
window lintel 窗过梁
window opener 窗开关
window opening 窗洞［口］
window pane 窗格玻璃
window post 窗间柱
window pull 窗拉手
window sash 窗扇
window shutter 百叶窗
window sill wall 窗槛墙
window sill 窗台
window state word 窗口状态字
window-type heat collector 窗式集热器
window-type splicing connector 阴连接管，套接器
window ventilator 通气窗
window 窗户
wind path 风迹
wind porch 风廊
wind power climatology 风能气候学
wind power density 风功率密度
wind power duration curve 风能持续时间曲线
wind powered aeration system 风力曝气系统
wind powered aerator 风力曝气机
wind powered heat pump 风力热泵
wind powered irrigation system 风力提灌系统
wind powered machinery 风力机械
wind power generation 风力发电，风能发电
wind power station utilizing solar energy 太阳能风力发电站
wind power generator 风力发电机
wind power installation 风能装置
wind power penetration limit 风电穿透功率极限
wind power penetration 风电穿透
wind power plant 风电场，风力发电厂，风力发电场
wind power potential 风能潜力
wind power profile 风能廓线
wind power project 风力发电工程项目，风电项目
wind power station 风力发电厂，风电场，风力发电站
wind power system 风能发电系统
wind power technology 风力发电技术
wind power 风电，风功率，风力
wind pressure coefficient 风力系数
wind pressure diagram 风压图
wind pressure factor 风压系数
wind pressure measuring hole 风压测量孔
wind pressure 风压，风压力
wind profile wind shear law 风廓线风切变律
wind profile 风廓线
wind proof capacity 抗风力
wind proof construction 防风结构
wind proof design 耐风设计
wind proof performance 耐风性能
wind propulsion system 风力助航系统
wind protection plantation 防风林
wind protection screen 防风板
wind pump 风力泵
wind reference pressure 基本风压
wind regime 气流状况，气流分布，风况，风情
wind-related loss 风损
wind resistance loss 风阻损耗
wind resistance 风阻力
wind resistant column 抗风柱
wind resistant design 抗风设计
wind resistant feature 抗风特性
wind resisting column 抗风柱
wind resource assessment 风力资源评价
wind resource map 风资源地图
wind resource 风力资源
wind rib 抗风肋
wind riddle 风选筛
wind right 风权
wind ripple 风成波痕
wind roadway 风道
wind rose map 风玫瑰图，风向玫瑰图，风向频率图，风向风速图，风向图
wind rotation 风向转变
wind rotor 风轮
windrowing 废物堆积风干处理
windrow-layered pile 条形分层料堆
windrow stacking 条形堆料法
wind run 风程
wind rush noise 风急速流动噪声
wind scale 风力级，风级，风力等级
wind screen 挡风玻璃，挡风板，防风罩，风挡
wind sector 风区
wind sensitive building 风敏感建筑
wind sensitive structure 风敏感结构
wind sensor assembly 风况传感组件
wind sensor 风感受器

wind setup 风增水，风增壅水，风增大，风潮
wind shadow 风影，背风区
wind shaft 风矢杆
wind shear exponent 风切变指数
wind shear law 风切变律
wind shear 风的切变，风切变，风的切力
wind shielding 挡风（墙）
wind shield wiper 挡风玻璃刮水器
wind shield 挡风板，挡风玻璃，防风罩
wind shift line 风变线，风向突变线
wind shift 风向转变
wind site selection 风场选址
wind site 风场
wind slash 风害迹地，风倒落物
wind sleeve 风向袋
windsock 风向袋
wind span 水平档距，风载档距
wind spectrum 风谱
wind speed alarm 风速警报器
wind speed and direction setter 风速风向给定器
wind speed and direction 风速与风向
wind speed counter 风速计数器
wind speed distribution 风速分布
wind speed duration curve 风速持续时间曲线
wind speed frequency curve 风速频率曲线
wind speed frequency 风速频率
wind speed over water surface 水面风速
wind speed profile coefficient 风（速）廓线系数
wind speed profile 风速廓线
wind speed Reynolds number 风速雷诺数
wind speed scale 风速等级，风速刻度盘，风级
wind speed spectrum 风速谱
wind speed 风速
wind spilling type of governor 风能溢出式限速装置，风能溢出式调速器
wind spilling 风能溢出
wind spout 龙卷风
wind spread sterility 风沙瘠化
wind spun vortex 风动涧旋，风动涡旋
wind statistics 风统计学
windstick 螺旋桨
wind stop 挡风条
windstorm 风暴
wind streak 风带
windstream direction 气流方向
windstream power density 风功率密度
windstream turbulence 风流湍流（度）
windstream 风洞气流，迎面气流
wind strength 风力
wind stress 风应力
wind stripping 防风带
wind structure 风的构造，风结构
wind suction 风吸力
wind survey 测风
wind swept area 受风区，风扫略面积
wind system 风系
wind thrust 风推力
wind tide 风增水
wind-tight 不透风的
wind to weight span ratio 水平档距与垂直档距之比
wind trajectory 风迹
wind transducer 风向传感器
wind truss 抗风桁架
wind tunnel air supply 风洞空气供给
wind tunnel axis system 风洞轴系
wind tunnel bottom 风洞底壁
wind tunnel ceiling 风洞顶壁
wind tunnel centerline 风洞轴线，风洞中心线
wind tunnel choking 风洞壅塞
wind tunnel circuit 风洞回流道，风洞回路
wind tunnel contraction 风洞收缩段，风洞收缩比
wind tunnel control 风洞控制
wind tunnel data 风洞数据
wind tunnel determination 风洞实验测定
wind tunnel diffuser end 风洞扩散段端
wind tunnel diffuser 风洞扩散段
wind tunnel drive 风洞驱动
wind tunnel efficiency 风洞效率
wind tunnel energy ratio 风洞能量比
wind tunnel fan 风洞风扇
wind tunnel fetch 风洞吹程
wind tunnel floor 风洞底壁
wind tunnel geometry 风洞几何形状
wind tunnel heat regenerator 风洞回热器
wind tunnel intake 风洞进气口
wind tunnel laboratory 风洞实验室
wind tunnel measuring technique 风洞（实验）测量方法
wind tunnel modeling technique 风洞建模技术
wind tunnel model test 风洞模型试验
wind tunnel model 风洞模型
wind tunnel noise 风洞噪声
wind tunnel nozzle 风洞喷管
wind tunnel observation 风洞观测
wind tunnel performance 风洞性能
wind tunnel plant 风洞设备
wind tunnel radius 风洞半径
wind tunnel Reynolds number 风洞雷诺数
wind tunnel roof 风洞顶壁
wind tunnel shape 风洞形状
wind tunnel simulation 风洞模拟
wind tunnel solid boundary 风洞实壁边界
wind tunnel speed 风洞风速
wind tunnel study 风洞研究
wind tunnel suspension system 风洞模型悬挂系统
wind tunnel technique 风洞实验技术
wind tunnel test section 风洞试验段
wind tunnel test 风洞试验
wind tunnel time 风洞吹风时间，风洞占用时间
wind tunnel turbulence factor 风洞紊流系数
wind tunnel turbulence 风洞湍流［紊流］
wind tunnel turning vane 风洞导流片
wind tunnel turntable 风洞转盘
wind tunnel wall effect 风洞洞壁效应
wind tunnel wall interference 风洞洞壁干扰
wind tunnel wall 风洞洞壁
wind tunnel window 风洞试验段观察窗
wind tunnel working section 风洞试验段
wind tunnel 风洞

wind turbine certification 风电机组认证
wind turbine cost effectiveness 风电机组的成本效益
wind turbine generating 风力发电
wind turbine generator set 风力发电机组
wind turbine generator system 风力发电机组，风电机组
wind turbine generator unit 风力发电机组
wind turbine generator 风力发电机，风力涡轮发电机
wind turbine international standard 风电机组国际标准
wind turbine penetration 风力机渗透度
wind turbine power performance testing 风电机组功率特性试验
wind turbine safety and design revision 风电机组安全和设计
wind turbine safety and design 风电机组安全和设计
wind turbine simulator 风力机模拟器
wind turbine terminal 风力机端口
wind turbine 风力机，风电机组，风动涡轮，风力涡轮机，风轮机
wind turbulence 风湍流（度）
wind unpredictability 风不可预计性
wind up an account 结算，清盘账户
wind up business 停业
wind uplift load 风吸力
wind uplift resistance 抗风揭性能
wind uplift 风浮力
wind up of a company 公司歇业
wind up 以……告终，结束讲话［会议］，关闭公司［企业］，停止营业，上紧发条，使结束，终结，结束，卷紧，卷起
wind valley 风谷，风口
wind vane 风向标，对风尾舵，风标，风向仪
wind variability 风变性
wind vector 风矢量
wind veering 风向转变
wind velocity gradient 风速梯度
wind velocity profile 风速廓线
wind velocity 风速
wind wake 风尾流
windward drift 向风漂流，迎风漂流
windward face 向风面
windward rudder 迎风机构，迎风装置，上风舵
windward side 迎风面，迎风侧，向风面，上风面
windward slope 迎风坡
windward wall 迎风壁
windward 上风向的，迎风向的，迎风侧，迎风面
wind water channel 风水槽
wind water interface 风水交界面
wind water-lifting set 风力提水机组
wind wave hydraulics 风浪水力学
wind wave tank 风浪水槽，风浪模拟水池
wind wave 风浪
wind wheel anemometer 风轮风速仪
wind wheel 风轮
wind yaw angle 风偏航角
windy district 多风区
windy site 多风地带
wind zone 风带
wind 风，给钟表上发条，蜿蜒，缠绕
wing abutment 翼墩，翼墙，翼式桥台，翼座
wing angle 翼角【托辊侧辊倾角】
wing bending 机翼弯曲
wing butt 翼根
wing cellule 翼组
wing chord 翼弦
wing contour 机翼外形
wing covering 机翼蒙皮
wing dam 翼坝
wing drop 机翼下坠
winged idler 快速托辊
winged pulley 翼形滚筒
winglet 小翼
wing levee 翼堤
wing nut 蝶形螺母
wing pile 翼桩
wing pump 叶片泵，叶轮泵，叶式唧筒
wing rib 翼肋
wing rock 机翼摇摆
wing room 厢房
wing root interference effect 翼根干扰影响
wing root 翼根
wing screen 翼式拦污栅
wing screw 翼形螺钉
wing section 翼剖面
wing shaped core 翼状型心
wing spar 翼梁
wing tip effect 翼尖影响
wing tip vortex 翼梢涡流
wing trench 侧翼截水槽
wing twisting 翼扭转
wing-type pulley 翼形滚筒
wing wall 翼墙
wing 翼缘，机翼，翼
win initial success 初见成效
winner 优胜者
winning bidder 得标人，中标人，投标得标人
winning machine 采煤机
wins fabric 钢丝网
winter concreting 冬季混凝土浇筑，冬季浇注混凝土
winter construction 冬季施工
winter dike 冬堤
winter discharge 冬季流量
winter dormancy 冬季休眠
winter flood 冬汛
winter hardiness 耐寒性，抗寒心态，越冬性
winterization 安装防寒装置，准备过冬
winterized concrete plant 防寒混凝土拌和厂
winterized equipment 防寒设备
winter oil 冬季用油，耐冻油
winter proofing 防冻
winter resistance 耐寒性
winter runoff 冬季径流
winter solstice 冬至，冬至点
winter work cost 冬季施工费
win the bid 中标

win 获胜，赢，成功，收益
wiper arm 刮水器臂
wiper blade 刮水器刮片
wiper 擦拭器，刮板，刮水器，配汽滑块
wipe test 擦拭法检查
wipe 擦拭法检查
wire and conductor 导线
wire armor 钢丝铠装
wire back-tie 模板的反拉钢丝
wirebar 线材
wire basker strainer 铁丝下水口箅子
wire basket 钢丝兜，钢丝笼，铁丝笼，铁石笼
wire box 铁丝笼
wire bridge 钢索吊桥
wire brush 除锈刷，钢丝刷，金属丝刷
wire-by-wire test 逐根导线试验
wire cableway 钢缆索道
wire cable 钢丝绳
wire clamp 线夹
wire clip 钢丝夹，线夹
wire cloth 铁纱
wire coil 线盘，线卷
wire connector 接线器
wire crimp 线缆压接
wire cut brick 丝切砖
wire cutter 钢丝截断器，钢丝（剪）钳，线剪
wired glass 夹丝玻璃
wired program computer 插线程序计算机
wire drawing tongs 紧线钳
wire drawing 拉丝
wired 布线的
wire entanglement 铁丝网障碍物，有刺铁丝网，障碍物
wire extension 焊丝伸出长度
wire fabric 钢丝编织品
wire feeder 送丝机构
wire feed rate 送丝速度
wire feed speed 焊丝供应速度
wire-feed unit 供线装置
wire filter cloth 金属丝滤布
wire-flux combination 金属丝焊药结合体
wire fuse 保险丝
wire gabion dam 铅丝网石龙坝
wire gauge 线径测量，线规
wire gauze 铁丝网
wire glass 钢丝网玻璃
wire grating 铁丝格栅
wire grip 紧线器
wire guy 钢丝牵索
wireless communication system 无线通信系统
wireless communication 无线电通信
wireless direction finding 无线电探向
wireless SCADA system 无线数据采集与监控系统
wireless set 无线电设备
wireless station 无线电台
wireless telegram 无线电报
wireline communication system 有线通信系统
wire locating terminal block 接线板上的导线
wireman 接线工
wire mesh cement 钢丝网水泥
wire mesh gabion construction 铁丝石笼结构
wire mesh hurdle 铁丝网棚
wire mesh reinforced asbestos cement corrugated sheet 钢丝网石棉水泥波瓦
wire mesh reinforcement 钢丝网配筋
wire mesh shielding 金属丝网屏蔽
wire mesh 金属丝网目，金属丝网，钢丝网，铁丝网
wire nail 圆钉
wire netting structure 铁丝网结构
wire-nut connector 导线螺母接头
wire pipe 电线钢管
wire-program control 布线程序控制
wire reel 焊丝盘
wire-reinforced rubber pipe 钢丝橡皮管
wire restraining device 导线约束装置【封套，线夹等】
wire rod 线材，盘条，盘圆，钢丝筋条
wire roll 线材轧辊，钢筋盘条
wire rope scratcher 钢丝绳式刮泥器
wire rope way 缆索道
wire rope 电缆，钢丝绳
wire saw 钢丝锯
wire scratcher 钢丝刮泥器，钢丝刷
wire screen 金属丝网，铁纱
wire shield 金属丝屏蔽
wire size 线径尺寸
wire sounding 钢索测深法
wire spacing 导线间距
wire strain gauge 金属丝应变计
wire stretcher 紧线机
wire stripping tool 导线剥离工具，剥线工具
wire tack 铁丝钉
wire tension draw-tong 张力紧线器
wire terminal 电线接头
wire tie 导线匝带
wire-to-water efficiency 水泵机械效率
wire-to-wire test 线对线试验
wire transfer 电汇
wire transposition 导线换位
wire tripping device 丝脱扣装置
wire twister 导线绞扭器
wire-type bird discourager 针形防鸟装置
wire under voltage 火线
wireway 导线通路
wire-weight gauge 悬锤式水尺
wire-wool pad used to assist arc strike 用来协助引弧的金属棉垫
wirewound connector 线绕接头
wire-wound resistor 绕线电阻器
wirewound 线绕的
wirewrapped connector 绕线接头
wirewrapped dam 钢丝网防护的堆石坝【防洪用】
wirewrapped 缠线的
wire 导体，导线，电线，钢丝，金属线，天线，线，线材
wiring comb 配线梳
wiring compartment 配线隔室
wiring deck 电线层
wiring diagram 布线图，接线图，配线图，线

路图
wiring grommet 布线箍，布线环
wiring gutter 配线隔室
wiring lug 接线片
wiring reinforcement 绑扎钢筋
wiring scheme 接线方案
wiring trough 配线隔室
wiring using helicopter 直升机放线
wiring 布线，接线
with an accuracy of 精确度为……的
with a view to doing sth. 着眼于，考虑到，为了，目的在于
with care 小心
with Chinese features 具有中国特色的
withdraw a bid 撤销投标
withdrawable section 可抽取段
withdrawable 可拔出的，取出的
withdraw a claim 撤销索赔
withdrawal application 拨款申请
withdrawal by notice 通知退出，通知退约
withdrawal force 抽出力
withdrawal of bids 撤标，标书的撤回
withdrawal of control rods 提升控制棒，提出控制棒
withdrawal rate 提棒速度，提棒速率【控制棒】
withdrawal resistance 抗拔力
withdrawal speed 提棒速度【控制棒】
withdraw(al) 拆卸空间，撤回，收回，提款，提取，撤退，取消，退股，缩回，提出（控制棒）
withdraw an offer 取消报价
withdrawn 拔出的【印刷电路板】
withhold at source 从（收入）来源［源头］预扣（缴）
withhold income tax 预扣（缴）所得税，预提所得税
withholding agent 预扣税款代理人
withholding tax 预提税，代扣代缴的预提所得税，代扣所得税
withholding 扣缴
withhold payment 拒绝支付，止付
withhold 拒绝给予，不给，扣留，阻挡，扣除，扣款，扣押，抑制
with immediate effect 即刻生效
with inclusion of 同时包括，同时需要考虑
with in-line replaceability 用串列替换性
within the extent of 在……范围内
within the vicinity of 在……附近
with on-line replaceability 用在线替换性
without appropriation or cash backing 没有拨款或现金支持
without notice 不预先通知，不另行通知
without prejudice to 在不损害……情况下，在不影响……的情况下
without recourse to 不依靠
without recourse 无权追索，无追索权的
with particular average 水渍险
with power-assisted blowdown 带动力辅助的排污［泄放］【可控的排放管】
with power-assisted lift 带动力辅助的提升
with power network in parallel 并行
with recourse 有追索权的
with reference to 关于
with regard to 关于
with resilient coupling 采用弹性联轴节
with respect to 就……而论，关于
with spreader 带扩散器
with spring 带弹簧的【平行座闸阀】
withstand capacity （母线）通流容量
withstand pollution test 污秽耐压试验
withstand test 耐压试验
withstand voltage test 耐受电压试验
withstand voltage 耐受电压，耐压
withstand 承受得住，抵抗，经得起，经受，耐受
with the addition of 外加
with the concurrence of 经……同意
with the exception of 除了……以外，（其余都……）
with the participation of 有……参加［合作］
with the result that 结果是
with the tide 顺潮
with the view of doing sth. 着眼于，考虑到，为了，目的在于
with tight shut-off 严密关断
with wedge 用楔块
witness an operation 见证某项操作
witness clause 结尾条款
witness inspection item 见证检验项目
witness inspection 亲自检查，见证检验
witness point 见证点，通知检验【质保】
witness report 见证报告
witness test 见证试验
witness 见证，目击者，证明，证人
WMO(World Meteorological Organization) 世界气象组织
WMS(work management system) 工作管理系统
WMUP＝warmup 暖机
WNF(welding neck flange) 焊颈法兰
WN-RF(welding neck-raised flange) 焊颈突面法兰
WOA(without attachment) 不带（无）附件
wobble coefficient 摆动系数
wobble pump 手摇泵
wobbler action 偏心作用
wobble shaft 滚转轴
wobble-wheel roller 摆轮式轮胎碾压机
WOE(without enclosure) 不带附件
wok order 工作单
wok site group 工地代表组
wollastonite 硅灰石
wood arch 木拱
wood block 木枕，木块
wood buffer 木缓衡器
wood-burning 烧木柴的
wood chipboard 碎木板
wood chip separator 木屑分离器
wood consumption of power generation project per kW 发电工程每千瓦木材消耗量
wood depot 贮木场
wood drilling machine 木工钻床
wooden-block pavement 木块路面
wooden crib 木笼

wooden cushion pad 木垫块
wooden dam 木坝
wooden envelope 木质蒙皮
wooden fence post 木护栏桩
wooden fender 护舷木
wooden form 木模
wooden hammer 木锤
wooden key 木键
wooden louver 木制百叶窗
wooden maul 木锤
wooden model 木模
wooden nail 木钉
wooden parts 木块
wooden plate girder 木板梁
wooden raft 木筏，木排
wooden rammer 木夯
wooden sheet piling weir 木板桩堰
wooden stave flume 木板渡槽，木板水槽
wooden truss 木桁架
wooden ware 木器
wooden working drill 木工钻头
wood fiber board 木纤维板
wood fibre 木纤维
wood form 木模
wood grabber 抓木工具
wood joint 木接合
woodland 森林
wood lath 木板条
wood oil 桐油
wood pile 木桩
wood planning machine 木工压刨床，刨木机
wood pole 木杆
wood preservatives 木材防腐剂
wood preserving oil 木材防腐油
wood rasp 木锉
wood roof truss 木屋架
wood scraps separator 木屑分离器
wood screw 木螺钉，木螺丝
wood sheet pile 木板桩
wood splinter separator 木片分离器，木块
wood stave flume 板条渡槽
wood steaming 木材蒸干
wood structure 木材构造，木结构
wood veneer 饰面板
wood ware 木器
wood with crooked fibre 纹理扭曲的木材
wood wool slab 木丝板
wood-wool 木刨花，木丝
wood working disk saw 木工圆盘锯
wood working factory 木材加工厂
wood working machinery 木工机械
wood working 木材加工
wood work 木工
wood yard 堆木场
woody coal 木质煤
woody plant 木本植物
wood 树林，木材
wool fabric industry 毛纺织工业
wool tuft technique 贴线法
wool 纤维
word address format 字地址格式
word address register 字地址寄存器
word address 字地址
word control register 字控制寄存器
word description 字描述
word for word addressable 逐字编址
word message 书信电报
word-print message 书信电报
word 字
wor-gear box 蜗轮减速器
workability of concrete 混凝土的和易性
workability 可加工性【材料】，施工性能，和易性，可使用性
workable coal sample 可采煤样
workable concrete 和易性好的混凝土
workable seam-sample of coal 可采煤样
workable system 实用系统
workable 易加工的，易浇筑的
work accident 工伤事故，操作事故
work allocation 工作分配
work analysis program 工作分析程序
work angle 工作角
work approver 工作许可人
work area 施工区，工作区
work bench 工作架，工作台，作业台
workblank 毛坯
work box 工具箱
work breakdown structure 作业分解结构
work conditions 工作环境，工作条件
work connection 加工联接【电弧焊】，工作联系
work cycle 工作周期
work day 工作日，劳动日
work detail 工作细则
workdone for isentropic compression 理论绝热压缩功
work duration 工作期限
worked out section 采空区
worker for the auxilliaries 辅机工
worker for the proper （锅炉）本体工
workers and staff bonus found 职工奖励基金
workers'compensation insurance 工人恤养（补偿）保险，工伤保险，劳工保险
worker 工人，劳力，劳动者
work experience 工作经历
work flow chart 工作流程图
workflow 工作流程
workforce 劳动力
work hardened zone 加工硬化区【金工】
work hardening 加工硬化，冷锻，冷作硬化
work hard 努力工作
workholder 工件夹具
work holding device 工件夹具
work holding jaw 工件夹爪，钳口
work hours control system 工时管理系统
work hour 工作时间
working age 工龄
working area 施工区，工作区，作业区
working assets 运营资产，流动资本
working authorization 授权实施
working barrel 工作筒【一种金属管或泵筒】
working capital costs 工作投资费用
working capital for going into operation 投产所需

的流动资金
working capital turnover rate 流动资金周转率
working capital 流动资金，周转资金，运营资本
working chain 工作链
working chamber of caisson 沉箱工作室
working chamber of pneumatic caisson 气压沉箱作业室
working clearance 作业净距，工作间隙
working clothes 工作服
working conditions chart 工况图
working condition 工况，工作条件，使用条件
working day 工作日
working depth 有效齿高
working diagram 工作图，施工图，作业图
working document 工作文件
working drawing design 施工图设计
working drawing 施工图，加工图，施工详图
working earthing 工作接地
working environment 工作环境
working expenditure 工作费
working expense 运行费用
working face 工作面
working flank 工作齿面
working fluid characteristics 工作流体特性，工质特性
working fluid heat efficiency 工作流体加热效率
working fluid heater 工作介质加热器
working fluid 工作流体
working habit 工作习惯
working head 工作水头
working hour 工时，工作时间
working index 工作指标
working instruction 工作细则
working intensity 施工强度
working joint 工作缝，施工缝
working lamp 安全灯
working language 工作语言
working length 有效长度
working life 工作年限，工作寿命，使用寿命
working load 工作荷载，资用荷载
working map 工作地图，底图
working medium 工作介质，工质
working meeting 碰头会
working order 操作顺序
working pad(/segment) 工作瓦块
working parameter 工作参数，运行参数
working plan 工作规划，工作计划
working platform 工作平台，运转层，操作平台，作息平台
working point 工作点
working position 工作位置
working pressure head 作用压力水头
working pressure 作用压力，使用压力，工作压力，工坠力，试验段压力
working procedure 工作程序
working radius 工作半径，幅度
working range 工作范围
working reference solar cell 工作基准太阳能电池，工作标准太阳电池
working regime 工作制度
working session 工作会议，试验段
working site 施工现场
working space 作业空间
working specification 操作规程
working standard solar cell 工作标准太阳电池
working standard 工作标准
working station 工作站
working storage of reservoir 水库工作库容
working storage 高速暂存存储器，工作库容
working strain 资用应变
working strength 工作强度，资用强度
working stress 工作应力【材料在通常应用中可安全承受的应力】，作用应力
working surface 工作面
working temperature 工作温度
working voltage 运行电压，工作电压
working weight 工作重量
working wind speed 工作风速
working years 工龄
work in hand 现有工作
work input 输入功
work instructions 操作规程
work interruption 中断工作，工作暂停
work interval 工作间隙
work life 工作期限，使用期限，工作寿命
work load analysis 工作负荷分析
work load summary analysis 工作负荷重量分析
work load 工作计划，工作荷载，工作量
work management system 工作管理系统
workmanship 工作质量，工质，工作技巧，手（技）艺，制造工艺
workman 工人，劳动力
work norm 劳动定额
work order 工作订单，工作单，任务单，工作指令
work-out area 采空区
work out 编制出，想出，解决，算出，实现，制定出，消耗完，弄懂
work-out 市场疲软的，疲软的（市场）
work overtime 加班，超时工作
work performance 工作性能
workpiece holder 工件夹具
workpiece 工件
work plan 工作规划
work process 工艺过程
work program 工作大纲，施工大纲，施工组织
work quota standard 劳动定额标准
work ratio of vehicle 车辆工作率
work ratio 工作效率，有效功率比，工作率，功比
work-related injury 因公负伤
work reliability 工作可靠性
workroom sink 工作室水池
works abstract 工程摘要表
works acceptance 工程验收
work schedule 工作计划表，工作进度表
works cost 工程造价
work session control record 工作对话控制记录
work session 碰头会，工作对话，工作会话
work sheet signer 工作票签发人
work sheet 操作单，操作票，工作单
workshop approach 车间引道

workshop assembly　工场装配，车间装配
workshop drawing　加工图
workshop technician　车间技术员，车间技师
workshop　专题讨论会，车间
works inspection certificate　工厂检验证明书
work site delegation　工地代表组，工地代表团，工地代表
work site　建设工地
works progress　工程进程
works scope　工程范围
works suspension　工程停顿，工程暂停
work station for computer aided design　计算机辅助设计工作站
work station　工作站
work stone　料石
work style　工作作风
work suspension　停工
work table　工作台
work zone　施工区，工作区
work　工厂，工程，工作，功，经营，操作
world bank loan project　世行贷款项目
World Bank　世界银行
World Business Strategist Association　世界商务策划师联合会
world climate　世界气候
World Energy Conference　世界能源会议
World Environment Institute　世界环境研究所
world environment　世界环境
World Express Logistics Ltd.　世界快递物流有限公司
World Federation of Engineering Organizations　世界工程组织联合会
World Health Organization　世界卫生组织
world market price　世界市场价格
World Meteorological Organization　世界气象组织
World Radiometric Reference　世界辐射测量基准
world tendency　世界趋势
world time　世界时间
World Trade Organization　世界贸易组织
Worldwide Association of Nuclear Operators　世界核电厂营运者联合会
worldwide fallout　全球性沉降
worldwide federation　世界性联合会
worldwide web　万维网，全球浏览系统
world wind energy association　世界风能协会
WorleyParsons　（澳大利亚）沃利帕森斯集团【跨国工程承包商】
worm and worm gear　涡轮蜗杆副，蜗轮蜗杆，蜗杆与蜗轮
worm drive　蜗杆转动
worm gear drive　蜗轮传动
worm-geared hoist　蜗杆涡轮式起重机，蜗轮卷扬机
worm gear reducer　蜗轮减速器
worm gear speed reducer　蜗轮减速器
worm gear　蜗轮传动装置
wormhole　蛀孔【焊接缺陷】，（焊缝）条虫状气孔
worm motor　蜗轮传动马达
worm-operated bar bender　螺旋式弯钢筋机
worm pipe　蜗形管
worm screw　无头螺丝
worm stem　蜗杆
worm type oil pump　螺杆式油泵
worm wheel　蜗轮
worm　蜗杆，无头螺丝，螺纹，使蠕动，使缓慢前进
worn-out part　磨损件，损坏件
worsen　恶化
worst case design　（按）最不利情况设计
worst-case impact　最不利影响
worth curve　价值曲线【控制棒】
Worthite　沃赛特镍铬耐蚀合金，镍铬钼耐热不绣钢
worth　价值【如控制棒，硼】，值……的
wound core　卷铁芯
wound down　废止
wound gasket　缠绕垫
wound internal rotor　绕线式内部转子
wound rotor asynchronous motor　绕线转子式异步电动机
wound rotor induction generator　绕线转子式感应发电机，绕线式异步发电机
wound rotor motor　绕线式电动机，线绕转子式马达
wound rotor synchronous generator　绕线转子式同步发电机
wound rotor　绕线式转子，绕线型转子，绕线端子
wound stator　绕线式定子
wound thermal insulation　缠绕式保温
wound　绕的，缠绕的，绕缠的，伤口，伤害
woven belt　帆布带，帆布皮带
woven mattress　编梢沉排
woven tape　布卷尺
woven wire fabric　钢丝网
woven wire gate　铁栅栏门
woven wire reinforcement　钢丝网配筋
woven wire window guard　铁丝网窗护栏
woven　织物的，编织的，机织织物，铁丝网，编织，织物
WPA(with particular average)　水渍险
w-plane　W 平面
W point(witness point)　见证点
WPP(wind power plant)　风力发电厂
WPQR(welding procedure qualification report)　焊接工艺评定报告
WPQT(welding procedure qualification test)　焊接工艺评定试验
WPQ(welding procedure qualification)　焊接工艺评定【ASME 标准称之为“焊接程序评定”】
WPS(welding process specification)　焊接工艺规程
WP(working point)　工作点
WP(work plan)　工作计划
WPDS(welding procedure data sheet)　焊接工艺数据表
WQR(welder qualification report)　焊工评定报告
WQT(welder qualification test)　焊工评定试验
WQ(welder qualification)　焊工评定，焊工资格考评
wrap angle of belt　包角【输送带】

wrap-around fluorescent lighting fixture　全密封荧光灯
wrap-around test　环绕测试
wrap-around type solar cell　卷包式太阳电池
wrap-around　环绕式（处理），包覆式
wrapped cable　缠绕缆索
wrapped insulation　捆扎式保温
wrapper　助卷机，包装材料，环段包壳
wrapping machine　打包机，绕包机
wrapping　绕接式
wrap resistance between belt and pulley　胶带和滚筒间的包绕阻力【僵性阻力】
wrap　包裹，包衬，包装，包，卷，捆
wreathed column　螺旋形柱
wreath　圈状物，烟圈
wreckage　残骸，碎片，残余物
wrecking bar　板钉撬棍，救险棒，抢险车
wrecking crew　打捞队，抢险队
wrecking train　抢险列车
wrecking truck　抢修工程车
wreck marking　沉船标志
wrench fault　扭断层
wrench operated valve　扳手操作阀门
wrench spanner　开脚扳手
wrench　扳手，扳钳，扳紧器
wrest　抢夺，用力拧，歪曲，曲解，扭，拧
WRH　墙式再热器【DCS 画面】
WRIG(wound rotor induction generator)　绕线式异步发电机，变速绕线式转子感应发电机
wrinkle　（焊缝的）褶皱
wrist dosimeter　手腕剂量计
wrist plate　肘板
wrist　耳轴，肘节
write accumulator　写累加器
write address counter　写地址计数器
write and compute　写和计算
write buffer　写缓冲器
write-down　削减，减值，调低账面价值，账面价值的故意降低
write drum　写磁鼓
write-enable　允许写入
write end of file　写文件结束
write off　注销
writing　地图注记，书写，文字，笔迹
written application　书面申请
written approval　书面同意
written assessment　书面评定
written authorization　书面批准
written clarification　书面澄清
written confirmation　书面确认
written consent　书面同意
written contract　书面合同
written discharge　书面结清单，书面结清证明
written document　书面单据，书面文件
written error　书写错误
written evidence　书面证据，字据
written form　书面形式
written invitation　书面邀请
written note of default　书面的违约通知
written note　书面通知
written order　书面订货，书面指令
written permission　书面同意
written presentation　书面说明
written report　书面报告
written statement　书面陈述，书面声明
wrong indication　错误显示
wrong　错误的，不正确的，差错，错误，过失
wrought-iron pipe　锻铁管
wrought iron　锻铁，熟钢，熟铁
wrought pipe　锻制钢管，锻制管，熟铁管
wrought-steel pipe　锻制钢管，轧制钢管
wrought-steel　锻钢
WRR(World Radiometric Reference)　世界辐射测量基准
WRSG(wound rotor synchronous generator)　绕线式转子同步发电机
WR(work request)　工作申请
WS＆D(wind speed and direction)　风速与风向
W-shaped valley　W 形谷
WSL(work section leader)　施工队长
WS(work sheet)　订货联系单，工艺卡
WS(work supervisor)　工作负责人
WTD(wind tunnel data)　风洞数据
WTGS(wind turbine generator system)　风力发电机组，风轮发电机系统，风力涡轮发电系统
WTG(wind turbine generator)　风力发电机，风力涡轮发电机
WTO(World Trade Organization)　世界贸易组织
WTR＝water　水
WTS(working time saved)　节省的工作时间
WT%　重量百分比
WVAC　交流工作电压
WVDC　直流工作电压
W＝watt　瓦特
WWDSATPMSHEX(weather working days, Saturday PM, Sundays, holidays excepted)　晴天工作日，星期六下午、星期天和节假日除外
WWDSHEXEIU(weather working days, Sunday, holidays excepted, even if used)　晴天工作日，星期天和节假日除外，即使已使用也除外
WWDSHEXUU(weather working days, Sunday, holidays excepted, unless used)　晴天工作日，星期天和节假日除外，除非已使用
WWDSSHEX(weather working days, Saturday, Sundays, holidays excepted)　晴天工作日，星期六、星期天和节假日除外
WWEA(World Wind Energy Association)　世界风能协会
WWS(wastewater system)　污水系统
WW(water wall)　水冷壁
WW(weather warning)　气象警报
WWW(world wide web)　万维网
wye-body globe valve　Y 阀体球阀，Y 阀
wye-connected　星形连接的
wye connection　Y 连接，星形接法
wye damper　三通挡板
wye level　Y 型水准器，回转水准仪，活镜水准仪，华氏水准仪
wye pipe　叉管
wye tube　叉形管
wye　Y 形三通，斜三通
W　警告【DCS 画面】

X

xalsonte 粗粒砂
xanthic 黄色的，带黄色的
x-axis X 轴，X 坐标轴，横坐标轴
X-bracing 剪刀支撑，交叉支撑
XCA 辅助蒸汽生产系统（辅助锅炉）【核电站系统代码】
XCE 运行前试验用蒸汽生产系统【核电站系统代码】
x-coordinate X 坐标，横坐标
Xe and Sm poisoning 氙和钐中毒
xenolith 捕虏体，捕虏岩
xenon accumulation 氙积累
xenon buildup 氙增长【停堆后】
xenon concentration 氙浓度
xenon decay 氙衰变
xenon effect 氙中毒影响，氙效应
xenon equilibrium concentration 氙平衡浓度
xenon equilibrium poisoning 氙平衡中毒
xenon equilibrium 氙平衡
xenon feedback 氙反馈
xenon instability 氙不稳定性
xenon oscillation 氙振荡
xenon override 氙中毒补偿，克服氙毒，克氙反应性
xenon poisoning computer 氙中毒计算机
xenon poisoning predictor 氙中毒预测器
xenon poisoning 氙中毒
xenon poison 氙毒
xenon reactivity override 克氙反应性
xenon reactivity 氙中毒反应性
xenon simulator 氙中毒模拟机［仿真机］
xenon transient 氙瞬变，氙暂态
xenon variation 氙中毒变化
xenon 氙
xeralf 干热淋溶土
xeransis 除湿，干燥
xeric 干旱的
xerochore 无水沙漠区
xerocline 旱坡
xeroll 干热软土
xerophyte 旱生植物，耐旱植物
xerophytic 耐旱植物的
xerox copy 复印件
XLPE cable 交联聚乙烯电缆
XLPE(cross linked polyethylene) 交联聚乙烯
XLPE insulated and PVC power sheathed cable 交联聚乙烯绝缘并聚氯乙烯护套电力电缆
XLPE insulated cable 交联聚乙烯绝缘电缆
XLPE power cable 交联聚乙烯电力电缆
Xomox 卓玛斯阀门公司
XPA 辅助锅炉燃油系统【核电站系统代码】
X-radiation X 辐射
X-ray apparatus X 光机
X-ray defect detector X 射线探伤仪
X-ray detection X 射线探伤法
X-ray detector X 射线探测仪
X-ray diffraction technique X 射线衍射技术
X-ray diffractometer X 射线衍射仪
X-ray flaw detector X 射线探伤机［器］，X 光探伤仪
X-ray fluorescence spectrometer X 射线荧光光谱仪
X-ray fluorescence X 射线荧光
X-ray generator X 射线机
X-ray goniometer X 射线测角仪
X-ray inspection X 射线检查
X-ray laser X 射线激光器
X-ray radiography X 射线照相
X-ray test X 射线试验
X-ray 伦琴射线，X 射线
X-Y dual plane XY 双平面
xylanthrax 木炭
xylogen 木纤维，木质素
xylograph 木版画，木纹图案
xylometer 木材比重计
xylonite 假象牙，赛璐珞
xylophyta 木本植物
X-Y plotter XY 绘图机，XY 绘图仪
XY plot XY 坐标图
X-Y recorder XY 记录仪
xyster 刮刀
X 辅助蒸汽【核电站系统代码】

yacht basin 快艇碇泊池
y-A connected step-up transformer Y-A 连接升压变压器
yake plate 联板
Yangtze River 长江【即扬子江】
Yankee bond 美国债务
Yankee screwdriver 扬基旋凿【螺丝刀，改锥】
yardage 土方数
yardang 风蚀土脊
yard conveyer 堆场用输送机
yard conveyor system 煤场皮带机系统
yard crane 场内移动起重机，移动吊车
yard lumber 场存木材
yardman 场地工作人员
yard reclaim hopper 煤场取煤斗
yard station 土方站
yard stick 码尺
yard storage 料场贮存
yard trailer 堆场用拖车
yard transfer unit 堆场搬运设备
yard trap 下水道进口截污设备
yard utilization factor 场地利用系数
yard wand 码尺
yard 场地，堆物，（铁路）车场，码，庭院
yarning apparatus 报警装置
yaw acceleration 偏航加速度，侧滑加速度，横摆加速度
yaw action rotor 主动调向风轮
yaw and pitch 偏航与俯仰
yaw angle of rotor shaft 风轮偏角
yaw angle sensor 偏航角传感器

yaw angle 偏航角，侧滑角，横偏角
yaw base 偏航基座，偏航盘
yaw bearing 偏航轴承
yaw brake 偏航制动器
yaw control mechanism 偏航控制机构
yaw control 偏航控制
yaw coupling parameter 偏航耦合参数
yaw coupling 偏航耦合
yaw drive dead band 调向死区
yaw driven upwind wind turbine 偏航驱动迎风式风电机组
yaw drive pinion ratio 偏航驱动小齿轮传动比
yaw drive 偏航驱动
yawed wind 斜向风
yawed 偏航的
yaw error amplifier 偏航误差放大器
yaw error 偏航误差
yawer 偏航控制器
yaw fixed rotor 风定向风轮
yaw gear 偏航齿轮
yaw guy 系塔索
yaw inertia 对风惯性
yawing angle of rotor shaft 风轮偏航角
yawing device 调向装置，对风装置
yawing driven 偏航驱动
yawing mechanism 偏航机构
yawing moment of inertia 横摆惯性矩，偏航惯性矩
yawing moment 偏转力矩，偏航力矩
yawing movement 偏航运动
yawing orientation 偏航调向，偏航对风
yawing speed 偏航速度
yawing 偏航
yaw instability 偏航不稳定性
yawmeter 偏航计，偏航指示器
yaw misalignment 偏航角误差，偏航失调
yaw motor 偏航马达
yaw passive rotor 被动调向风轮，被动对风风轮
yaw probe 方向探头
yaw rate 偏航速率
yaw ring 偏航齿圈
yaw sensor 偏航传感器
yaw system 偏航系统
yaw turntable 偏航转盘
yaw vane 调向尾舵，对风尾舵，偏航翼
yaw 偏航角，偏航
Y-axis Y轴
y-azimuth 基准方向角
Y-bend 分叉弯头
Y-body globe valve Y形阀体截止阀，Y阀体球阀
Y-branch 叉管
Y bus matrix Y母线矩阵
Y-connection 叉形接头
Y-coordinate 纵坐标，Y坐标
Y-Darrieus rotor Y型达里厄式风轮
year-book 年鉴
year-climate 年气候
year-end appraisal of work 年终评比
year-end audit 年终审计
year-end bonus 年终奖
year-end dividend 年终股息
year-end report 年终报告
year-end results 年终业绩
yearly administration cost 年度管理费
yearly amortization cost 年度折旧费
yearly average wind speed 年平均风速
yearly budget 年度预算
yearly coal consumption 年耗煤量
yearly correlation 年相关
yearly energy cost 年度动力费，年度电力费
yearly gas leakage rate 年漏气率
yearly load curve 年负荷曲线
yearly load factor 年负荷因数
yearly maintenance 年度保养，年度维修
yearly maximum load 年最大负荷
yearly maximum utilization hours of unit 年机组最大利用小时
yearly maximum 年最大值，年最大的
yearly mean efficiency 年平均效率
yearly minimum 年最小的，年最小值
yearly output 年产量
yearly progress report 进展情况年度报告
yearly regulation 年调节
yearly runoff 年径流
yearly storage 年蓄水量
yearly variation 年变化，逐年变化
yearly wind speed 年平均风速
yearly 每年，每年的
year of average rainfall 平均雨量年份
year of completion 竣工年
year of grace 宽限年限，缓检年限，公元年，耶稣纪元年
year operating expense 年运行费
year-round air conditioner 全年空调器
year-round air conditioning 全年空调
year-round heating load 常年热负荷
year-round 全年的
year-to-year growth 逐年增长
year-to-year storage 多年调节库容
year-to-year temperature difference 年际温差
year-to-year variation 逐年变化
year with abundance of water 多水年，丰水年
yeasts treatment 酵母处理（法）
yellow cake 黄饼【浓化铀粉块】
yellow deal 黄松板
yellow mud 黄泥
yellow sand 黄沙
yellow sea datum 黄海高程基准面
yellow soil 黄色土壤
yellow 黄色的
Y globe valve Y形球阀
yield criterion 屈服标准，屈服准则
yield curve 产额曲线
yield deflection 屈服偏转
yield distribution 产额分布
yielding ground 松软土地
yielding limit 屈服极限
yielding seat 下沉支座
yielding soil 流动土
yielding 塑流，塑性变形，易变形的

yield locus 屈服轨迹
yield per ion pair 离子对产额
yield point strain 屈服点应变
yield point stress 屈服点应力
yield point 屈服点，降伏点
yield possession 让与所有权，让出所有权
yield ratio 屈强比
yield stock 增益股
yield strain 屈服应变
yield strength at temperature 温度屈服强度
yield strength 屈服强度，软化强度
yield stress ratio 屈服应力比
yield stress 屈服应力
yield surface 屈服面
yield temperature 屈服温度
yield value 塑变值，屈服值
yield 屈服，产量，产率，有效系数，沉陷，生产，获利
yiele point 屈服点，降伏点，击穿点，软化点，屈服值，屈服强度
Y joint Y 接头
Y-level Y 形水准仪，回转水准尺，回转水准仪，活镜水准仪
yoke arm 轭架臂【阻力塞阻件】
yoke assembly 轭架组件【阻力塞阻件】
yoke bonnet 弓形顶盖【阀门】
yoke bush 操作衬套
yoke guide 轭架导向
yoke nut 操作螺母
yoke plate 联板，轭板
yoke ring 轭架环套【阻力塞组件】，轭环
yoke sleeve 阀轭套
yoke type support 叉形支杆，轭式支架
yoke 担梁【指起重机吊梁】，窗头板，磁轭，轭，轭铁，轭架，架，联板，套箍，座
Yokogawa 横河仪表公司【日本】
younger leaf 新生叶
young sea 初生波
Young's modulus 杨氏模量【弹性】
young wave 初生波
Y-pattern check valve Y 形单向阀
Y-piece Y 形件
Y-pipe Y 形管，人字形管
Y-section 三通接头，Y 形接头
Y-splicing sleeve Y 拼合联轴套
Y-track 三叉形轨道，分叉线
Y-tube 叉形管
Y-type pipe 叉管
Y-type strainer Y 形粗滤器，Y 形过滤器
Y-valve Y 阀门，Y 形阀
Y-Y connection 星形星形接法

Z

Z-angle Z 形角铁
Z-axis Z 轴
Z bus matrix Z 母线矩阵
Z-domain Z 域
zeal 热情
zebra crossing 斑马线，人行横道
zebra time 世界时间
zebra wood 条纹木材
Zener diode 稳压二极管，齐纳二极管
zenithal projection 方位投影，天顶投影
zenith angle 天顶角
zenith distance 天顶距
zenith instrument 天顶仪
zenith point 天顶点
zenith telescope 天顶仪
zenith 顶点，极点，天顶
zeolite process 沸石处理法
zeolite softening 沸石软化
zeolite 沸石
zeor sequence impedance 零序阻抗
zeor sequence method 零序方法
zero adjuster 零位调整器，零位调整装置
zero adjustment 对零，对准零点，零点调整，零位调整
zero air void curve 饱和曲线，零气隙曲线
zero air void ratio 饱和孔隙比，饱和气隙比
zero allowance 无公差，零容差
zero and span calibration 零称量距校准
zero balance reading 天平读零
zero balance 零位调整
zero-based budget 零基预算
zero-based conformity 零基一致性
zero-based linearity 零基线性度
zero bias 零偏压
zero bit 零位
zero blowdown 零排污
zero calibration gas 零点校验气
zero capacity head （水泵的）关死扬程
zero circle 零圆
zero code suppression 零代码抑制
zero complement 补码
zero defect management 无缺陷管理
zero defect 无差错，无缺陷
zero deflection 零点偏移
zero degree 零次，零度
zero discharge 零排放，零排污
zero-dispersion slope 零色散斜率
zero-dispersion wavelength 零色散波长
zero displacement height 零风面位移高度
zero drift 零点漂移，零漂
zero elevation 零点迁升，零高程
zero error 零点误差，始点误差
zero field splitting 零磁场分裂
zero frequency component 直流分量，零频分量
zero frequency current 零频电流
zero frequency gain 零频率增益，零频增益
zero frequency 零频率
zero-gas 零气【仪用气】
zero graduation 零刻度
zero-growth 零增长
zero-head range 零水头区
zero heel stress 坝踵零应力
zero-hour mixing 无间歇拌和
zeroing 零位调整，调零
zero-input response 零输入响应
zero-irradiation time 零照射时间

zero isochrone 零等时线
zeroize 补零，填零
zero lateral strain test 无侧向应变试验
zero leakage 零泄漏
zero level release 零级释放
zero level 零级，起点级
zero lift angle of attack 零升力攻角
zero lift chord 零升力弦
zero lift drag rise 零升力时的阻力增量
zero lift drag 零升阻力
zero lift line 零升力线
zero lift moment 零升力矩
zero line 零位线
zero load method 零载法
zero load test 无载试验
zero load 空载
zero location 零位置，基准位置
zero-loss collector efficiency 零损失集热器效率
zero mark 零位刻度，零点标志
zero meridian 零子午线
zero moisture index 零值湿润指数
zero of a measuring instrument 测量仪器仪表的零位
zero of coordinate system 坐标原点
zero of elevation 高程零点
zero-offset 零偏移距，零点偏移，零点漂移，零点误差，零静差
zero of gauge 测站基点，水尺零点
zero of tidal pole 验潮杆零点
zero order release system 零级释放系统
zero order release 零级释放
zero phase sequence current 零相序电流
zero phase sequence reactance at rated current 在额定电流下的零相序电抗
zero phase sequence relay 零相序继电器
zero phase sequence 零相序
zero plane displacement 零风面位移，零面位移
zero point adjustment 零点调整，调零位，对零
zero point error 零点误差
zero point setting 零点设置
zero point suppression 零点抑制
zero point 零点，起始点
zero position 起始位置，零位
zero potential 零电势，零电位
zero power experiment 零功率实验
zero power factor 零值功率因数
zero power kinetics 零功率动力学
zero power noise 零功率噪声
zero power operation 零功率运行
zero power point reactor model 零功率点反应堆模型
zero power reactor 零功率堆
zero power test 零功率试验
zero power 零功率
zero rate of duty 免税
zero rate of penetration 零值贯入度
zero reading correction 零读数修正
zero reading 零点读数，起点读数
zero release fuel reprocessing plant 不排放（放射性废物）的核燃料后处理厂
zero release plant 无放射性排放电站，不排放放射性废物的工厂
zero release reprocessing 不排放废物的后处理，不排放放射性废物的后处理
zero release 零排放
zero scale mark 零标记，零标度标记，零标度线
zero sequence CT 零序电流互感器
zero sequence current method 零序电流法
zero sequence current protection 零序电流保护
zero sequence current relay 零序电流继电器
zero sequence current 零序电流
zero sequence field impedance 零序磁场阻抗
zero sequence impedance 零序阻抗
zero sequence reactance 零序电抗
zero sequence voltage relay 零序电压继电器
zero sequence 零序
zero setting accuracy 置零准确度
zero setting device 置零装置
zero setting scope 置零范围
zero setting 调零，置零，对准零位，零点调整，零位调整
zero shift technique 零点位移法
zero shift 零点迁移，零点漂移
zero signal 零信号
zero slip point 无滑动点
zero slope 平坡
zero-slump concrete 无坍落度混凝土
zero solids treatment 全挥发水处理，无杂质水处理
zero speed switch 零速开关
zero speed 零转速
zero state response 零状态响应
zero state “0”状态，零状态
zero strain velocity 零应变速度
zero-sum game model 零和对策模型
zero-sum game 零和博弈
zero suppression character 零抑制字符，消零字符
zero suppression signal 零点抑制信号
zero suppression 消零【去掉无用的零】，零位抑制，零点压缩，零抑制，零点正迁移
zero-surface 零面
zero-temperature level 零温度层
zero thrust pitch 无推力螺距
zero value suppression 零值抑制
zero voltage 零电压
zero water level 水位零点
zero water pollution 无水质污染
zero wind drag coefficient 无风阻力系数
zero-zone current 零区电流
zero 切断，零，零点，零度
zigzag arch 曲折拱
zigzag connection 曲折形连接
zigzag crack 不规则裂缝
zigzag fault 曲折断层
zigzag fold 曲折褶皱
zigzag line 曲折线，折线
zigzag ruler 曲尺，折尺
zigzag sill 曲折槛
zigzag 之字形
zinc bath 镀锌槽

zinc brass　白铜，锌黄铜
zinc-coated steel wire　镀锌钢丝
zinc-coated steel　镀锌钢，镀锌层钢
zinc-coated　镀锌的，镀锌层的
zinc coating　镀锌，锌面层，涂锌层
zinc-collar sleeve　防腐锌套
zinc dust　锌粉
zinc flashing　锌皮泛水
zinc foil　锌箔
zincification　镀锌，镀锌法
zinc-manganese dioxide dry cell　锌锰干电池
zinc-mercury oxide cell　锌汞电池
zinc-oxide arrester tester　氧化锌避雷器测试仪
zinc-oxide lightning arrester　氧化锌避雷器
zinc-oxide surge arrester　氧化锌避雷器
zinc paint　锌漆
zinc pipe　锌管
zinc plated steel sheet　镀锌铁皮
zinc plated steel　镀锌钢
zinc plated　镀锌的
zinc-plating　镀锌
zinc powder　锌粉
zinc roofing　锌皮屋顶
zinc sheet roofing　锌皮屋顶
zinc sheet　锌皮，锌板
zinc solder　锌焊料
zinc　锌
zircaloy　锆合金
zirconia oxygen indicator　氧化锆氧量计
zirconium alloy　锆合金
zirconium hydride　氢化锆
zirconium melting　锆熔化
zirconium niobium alloy　锆铌合金
zirconium　锆
zirc sponge　海绵锆
ZLC(zero lift chord)　零升力弦
ZL(zero lift)　零升力
ZL(zero line)　零线，基准线
ZL　波兰兹罗提
ZnO lightning arrester　氧化锌避雷器
zonal circulation　纬向环流
zonal embankment　分区填筑
zonal flow　纬向气流
zonal index　纬向指数
zonality　地带性，地区性，区域性
zonal loading　分区装载
zonal mineral　分带矿物
zonal pricing　区域电价法
zonal soil　地带性土壤，区域土，区域性土
zonal structure　环带构造
zonal texture　带状组织
zonal　分区的
zonation　地带性，地理学的地带分布，分带，分区，区划
zone air conditioner　区域空调器
zone-area method　分区面积法
zone coefficient　地区系数
zone control　分段控制，区域控制，分区控制
zoned air　分区送风
zoned earth dam　非均质土坝，分区式土坝，分区土质坝
zone diagram　分区图
zone limit　区域范围
zone loading procedure　分区装料程序
zone load　区域负荷
zone map　分区图
zone of bank storage　河岸调蓄带
zone of capillarity　毛管带
zone of capillary saturation　毛管水饱和层
zone of comfort　舒适区
zone of constant temperature　等温带，恒温带
zone of discontinuity　不连续带
zone of flow establishment　流动形成区
zone of fold　褶皱带
zone of influence　影响区
zone of least resistance　最小阻力区
zone of littoral drift　海岸漂流带，海岸漂沙区
zone of low population density　低人口密度区
zone of metamorphism　变质带
zone of migration　变迁范围，回游区
zone of negative pressure　负压区
zone of occasional distribution　偶然分布带
zone of positive pressure　正压区
zone of reserve flow　逆流区
zone of reservoir inundation　水库淹没区
zone of rock fracture　岩石破碎带
zone of safety　安全区域
zone of saturation　饱水带
zone of soil plastic flow　土的塑性流动区
zone of stagnation　驻区，滞流区
zone of suspended water　悬着水带
zone of weathering　风化带
zone panel　区域盘
zone sampler　区层取样器
zone system　区域制，分区式系统
zone thermostat　区域恒温器
zone time　区时
zone unloading procedure　分区卸料程序
zone　区域，带，区间，区段，地区，地带，区
zoning control　分区管理
zoning ordinance　分区规则
zoning plan　分区规划，分区规划图，区域划分图
zoning regulation　分区规则
zoning　分带，分区，区划
zoogenic rock　动物岩
zooglea　菌胶团
Z-plane　Z 平面
Z-section　Z 形钢
Z-transfer function　Z 传递函数
Z-transform　Z 变换
Z-type elevator　Z 形提升机
zyglo inspection　荧光探伤法

附录 化学元素表

Periodic table of the elements

English Name 英文名称	Chinese Name 中文名称	Symbol 元素符号	Atomic Number 原子序数
hydrogen	氢	H	1
helium	氦	He	2
lithium	锂	Li	3
beryllium	铍	Be	4
boron	硼	B	5
carbon	碳	C	6
nitrogen	氮	N	7
oxygen	氧	O	8
fluorine	氟	F	9
neon	氖	Ne	10
sodium,natrium	钠	Na	11
magnesium	镁	Mg	12
aluminium	铝	Al	13
silicon	硅	Si	14
phosphorus	磷	P	15
sulfur	硫	S	16
chlorine	氯	Cl	17
argon	氩	Ar	18
potassium,kalium	钾	K	19
calcium	钙	Ca	20
scandium	钪	Sc	21
titanium	钛	Ti	22
vanadium	钒	V	23
chromium	铬	Cr	24
manganese	锰	Mn	25
iron,ferrum	铁	Fe	26
cobalt	钴	Co	27
nickel	镍	Ni	28
copper,cuprum	铜	Cu	29
zinc	锌	Zn	30
gallium	镓	Ga	31
germanium	锗	Ge	32
arsenic	砷	As	33

English Name 英文名称	Chinese Name 中文名称	Symbol 元素符号	Atomic Number 原子序数
selenium	硒	Se	34
bromine	溴	Br	35
krypton	氪	Kr	36
rubidium	铷	Rb	37
strontium	锶	Sr	38
yttrium	钇	Y	39
zirconium	锆	Zr	40
niobium	铌	Nb	41
molybdenum	钼	Mo	42
technetium	锝	Tc	43
ruthenium	钌	Ru	44
rhodium	铑	Rh	45
palladium	钯	Pd	46
silver，argentum	银	Ag	47
cadmium	镉	Cd	48
indium	铟	In	49
tin，stannum	锡	Sn	50
antimony，stibium	锑	Sb	51
tellurium	碲	Te	52
iodine	碘	I	53
xenon	氙	Xe	54
caesium	铯	Cs	55
barium	钡	Ba	56
lanthanum	镧	La	57
cerium	铈	Ce	58
praseodymium	镨	Pr	59
neodymium	钕	Nd	60
promethium	钷	Pm	61
samarium	钐	Sm	62
europium	铕	Eu	63
gadolinium	钆	Gd	64
terbium	铽	Tb	65
dysprosium	镝	Dy	66
holmium	钬	Ho	67
erbium	铒	Er	68
thulium	铥	Tm	69
ytterbium	镱	Yb	70
lutetium	镥	Lu	71

English Name 英文名称	Chinese Name 中文名称	Symbol 元素符号	Atomic Number 原子序数
hafnium	铪	Hf	72
tantalum	钽	Ta	73
tungsten，wolfram	钨	W	74
rhenium	铼	Re	75
osmium	锇	Os	76
iridium	铱	Ir	77
platinum	铂	Pt	78
gold，aurum	金	Au	79
mercury，hydrargyrum	汞	Hg	80
thallium	铊	T1	81
lead，plumbum	铅	Pb	82
bismuth	铋	Bi	83
polonium	钋	Po	84
astatine	砹	At	85
radon	氡	Rn	86
francium	钫	Fr	87
radium	镭	Ra	88
actinium	锕	Ac	89
thorium	钍	Th	90
protactinium	镤	Pa	91
uranium	铀	U	92
neptunium	镎	Np	93
plutonium	钚	Pu	94
americium	镅	Am	95
curium	锔	Cm	96
berkelium	锫	Bk	97
californium	锎	Cf	98
einsteinium	锿	Es	99
fermium	镄	Fm	100
mendelevium	钔	Md	101
nobelium	锘	No	102
lawrencium	铹	Lr	103